Berlin [bɜːˈlɪn] **1.** *pr. n.* Berlin *(das).* **2.** *attrib. adj.* Berliner; *(Ling.)* berlinisch

- subject label
  Bereichsangabe
- information on syntax
  syntaktische Angabe

bird [bɜːd] *n.* **a)** Vogel, *der;* ~**s of a feather flock together** *(prov.)* gleich und gleich gesellt sich gern *(Spr.)*

- proverb
  Sprichwort

boot [buːt] **1.** *n.* **a)** Stiefel, *der;* **get the** ~ *(fig. coll.)* rausgeschmissen werden *(ugs.);* **give sb. the** ~ *(fig. coll.)* jmdn. rausschmeißen *(ugs.);* **the** ~ **is on the other foot** *(fig.)* es ist genau umgekehrt; *see also* **big 1 g; [1]die 1 a; heart 1 c**

- idiomatic phrase
  feste Wendung
- swung dash representing the headword
  die Tilde vertritt das Stichwort
- cross-references for additional information
  Verweise auf zusätzliche Informationen

[1]break [breɪk] **1.** *v. t.,* **broke** [brəʊk], **broken** [ˈbrəʊkn] **a)** brechen

- irregular past tenses
  unregelmäßige Verbformen

~ **aˈway 1.** *v. t.* ~ **sth. away [from sth.]** etw. [von etw.] losbrechen *od.* abbrechen. **2.** *v. i.* **a)** ~ **away [from sth.]** [von etw.] losbrechen *od.* abbrechen; *(separate itself/oneself)* sich [von etw.] lösen; *(escape)* [aus etw.] entkommen; **b)** *(Footb.)* sich freilaufen

- grammatical categories
  Gliederung nach grammatischen Gesichtspunkten
- phrasal verbs listed under main verb
  Verben in festen Verbindungen mit Präpositionen oder Adverbien (Phrasal verbs) *im Anschluß an das jeweilige einfache Verb*

~ **ˈdown 1.** *v. i.* **a)** *(fail)* zusammenbrechen; ⟨*Verhandlungen:*⟩ scheitern; **b)** *(cease to function)* ⟨*Auto:*⟩ eine Panne haben; ⟨*Telefonnetz:*⟩ zusammenbrechen; **the machine has broken down** die Maschine funktioniert nicht mehr; **c)** *(be overcome by emotion)* zusammenbrechen; **d)** *(Chem.)* aufspalten

- semantic categories
  Gliederung nach Bedeutungsunterschieden
- sense indicator
  Indikator

The Concise Oxford Duden German Dictionary

THE CONCISE OXFORD DUDEN GERMAN DICTIONARY

Revised Edition

ENGLISH – GERMAN
GERMAN – ENGLISH

Edited by the Dudenredaktion
and the German Section of the
Oxford University Press
Dictionary Department

Chief Editors
M. CLARK
O. THYEN

CLARENDON PRESS • OXFORD
1997

Oxford University Press, Great Clarendon Street, Oxford OX2 6DP

Oxford New York
Athens Auckland Bangkok Bogota Bombay
Buenos Aires Calcutta Cape Town Dar es Salaam
Delhi Florence Hong Kong Istanbul Karachi
Kuala Lumpur Madras Madrid Melbourne
Mexico City Nairobi Paris Singapore
Taipei Tokyo Toronto
and associated companies in
Berlin Ibadan

Oxford is a trade mark of Oxford University Press

Published in the United States by
Oxford University Press Inc., New York

First published 1991
Revised edition 1997

British Library Cataloguing in Publication Data
Data available

Library of Congress Cataloging in Publication Data
Data available

ISBN 0–19–860133–6

Printed in Great Britain by
Clays Ltd.,
Bungay, Suffolk

FOREWORD

The *Concise Oxford–Duden German Dictionary* has been designed to meet the needs of both English-speakers and German-speakers, whether they wish to understand and translate from the foreign language or to communicate effectively and express themselves idiomatically in speech and writing.

Based on the much-acclaimed *Oxford–Duden German Dictionary*, edited by J. B. Sykes and W. Scholze-Stubenrecht, this concise reference book carries all the authority of two of the world's foremost dictionary publishers, Oxford University Press and the Dudenverlag, making use of the unparalleled databases maintained and continually expanded by the two publishers for their celebrated native-speaker dictionaries such as the *Concise Oxford Dictionary* and *Das Große Wörterbuch der Deutschen Sprache*.

For learners of German and English, the *Concise Oxford–Duden German Dictionary* is an ideal study aid, providing numerous examples, clear differentiation of senses, precise guidance on selecting the most appropriate translation, and detailed information on style, usage, and word order. Its up-to-date coverage and wealth of accurate translations make it an authoritative reference tool for school, college, and university students, business people, and all those who require the fullest possible information on German and English in a compact, easy-to-use dictionary for the 1990s.

MICHAEL CLARK
Oxford University Press

EDITORS AND CONTRIBUTORS

in Oxford

Michael Clark
Bernadette Mohan
Maurice Waite
Colin Hope
Peter Lewis
Ursula Lang
Tim Connell
Trish Stableford
Fred McDonald

in Mannheim

Olaf Thyen
Werner Scholze-Stubenrecht
Brigitte Alsleben
Lothar Görke
Cosima Heise
Ulrike Röhrenbeck
Magdalena Seubel
Eva Vennebusch

CONTENTS

PROPRIETARY NAMES

Guide to the use of the Dictionary

Hinweise für die Benutzung des Wörterbuchs

1. Order of entries

Headwords (including compounds, but with the exception of phrasal verbs - see below) are entered in strict alphabetical order, ignoring hyphens, apostrophes, and spaces.

Each English phrasal verb is entered on a new line immediately following the entry for its first element, which is indicated by a swung dash (~).

1. Anordnung der Artikel

Die Stichwörter sind (mit Ausnahme der *Phrasal verbs* im englisch-deutschen Teil; s.u.) alphabetisch angeordnet.

Im englisch-deutschen Teil werden die *Phrasal verbs* auf einer neuen Zeile unmittelbar an das Grundwort angeschlossen, wobei dieses durch eine Tilde repräsentiert wird.

Example/Beispiel: **track**
~ 'down
tracker

Headwords spelt the same but with different meanings are entered separately with a raised number before each.

Where two or more compounds with the same first element occur consecutively, they are given in paragraph-like blocks. The first element is given only once at the beginning of the block and is thereafter represented by a swung dash.

Verschiedene Stichwörter mit gleichem Schriftbild (Homographe) erhalten zur Unterscheidung jeweils eine hochgestellte Ziffer vor dem Anfangsbuchstaben.

Mehrere aufeinanderfolgende als Stichwörter aufgeführte Komposita mit gemeinsamem ersten Element sind im Wörterbuch zu Absätzen zusammengefaßt. Dabei steht das erste Element nur am Anfang des Absatzes; es wird innerhalb des Absatzes durch eine Tilde repräsentiert.

2. The headword

a) Form of the headword

The headword appears in bold type at the beginning of the entry.

German adjectives which strictly speaking have no undeclined form are given with an ellipsis instead of any ending.

2. Das Stichwort

a) Form des Stichworts

Das Stichwort erscheint fett gedruckt am Anfang des Artikels.

Deutsche Adjektive, die eigentlich keine endungslose Form haben, erscheinen ohne Endung, mit Auslassungspunkten.

Examples/Beispiele: **äußer...**
best...
link...

b) Symbols used with headwords

With English headwords:

ˈ shows stress on the following syllable (for more information see 3).

With German headwords:

_ indicates a long vowel or a diphthong, stressed in words of more than one syllable.

b) Zeichen am Stichwort

Am englischen Stichwort kann das folgende Zeichen auftreten:

ˈ Betonungszeichen vor der zu betonenden Silbe. (Näheres s. unter 3.)

Am deutschen Stichwort können die folgenden Zeichen auftreten:

_ waagerechter Strich unter betonten langen Vokalen oder Diphthongen

Examples/Beispiele: **Hieb, Blau, Hörer, amtieren**

. indicates a short vowel, stressed in words of more than one syllable.

. Punkt unter betonten kurzen Vokalen

Examples/Beispiele: **Recht, bitter**

· shows the juncture of elements forming a word.

· Punkt auf der Zeilenmitte, der bei zusammengesetzten Wörtern die Kompositionsfuge markiert

Examples/Beispiele: **Kern·kraft, um·branden**

| shows the juncture of elements forming a compound verb and indicates that the verb is separable.

| senkrechter Strich, der bei zusammengesetzten Verben anzeigt, daß das Verb unfest zusammengesetzt ist.

Examples/Beispiele: **vor|haben, um|werfen**

3. Pronunciation

The symbols used are those of the International Phonetic Alphabet (IPA), which is explained on pp. 1398–1399.

The pronunciation of a headword is normally given in square brackets immediately after it.

Where a *simple headword* has no pronunciation given, it is the same as or can easily be deduced from that of a preceding headword.

The pronunciation of a *German derivative* with none given can be deduced from that

3. Angaben zur Aussprache

Die bei den Angaben verwendeten Zeichen der internationalen Lautschrift der *International Phonetic Association (IPA)* sind auf S. 1398–1399 verzeichnet und erklärt.

In der Regel ist die Aussprache des Stichworts in eckigen Klammern in Lautschrift unmittelbar nach dem Stichwort angegeben. Im übrigen gelten die folgenden Hinweise:

Für mehrere aufeinanderfolgende Stichwörter mit gleicher Aussprache ist diese jeweils nur beim ersten angegeben.

Bei deutschen *Ableitungen* ohne Ausspracheangabe kann die Aussprache vom of its root word. The stress, however, is always shown, by _ or . (see 2b).

Grundwort abgeleitet werden. Bei ihnen wird daher nur die Betonung angegeben, und zwar durch Zeichen am Stichwort selbst (s. 2b).

The pronunciation of a *compound* with none given can be derived from the pronunciations of its elements.

Bei *Komposita* ohne Ausspracheangabe ergibt sich die Aussprache aus der der einzelnen Bestandteile. Bei ihnen ist daher ebenfalls nur die Betonung angegeben (s. 2b).

If stress alone needs to be shown in square brackets, each syllable is represented by a hyphen.

Wenn in einer Ausspracheklammer nur die Betonung angegeben werden soll, steht für jede Silbe ein waagerechter Strich.

Example/Beispiel: **come to 1.** ['--] *v.t.* ... **2.** [-'-] *v.i.*

In blocks of compounds, stress is given as follows:

In Kompositablöcken ist die Betonung folgendermaßen angegeben:

In the English-German section:

Im englisch-deutschen Teil:

If no stress is shown (by the IPA stress mark), it falls on the first element.

Hier wird die Betonung nicht eigens angegeben, wenn sie, wie im Normalfalle, auf dem ersten Element liegt. Anderenfalls jedoch steht vor der betonten Silbe das *IPA*-Betonungszeichen.

Example/Beispiel: **counter: ~measure** ...
~-pro'ductive

In the German-English section:

Im deutsch-englischen Teil:

When the first element at the beginning of the block has a stress mark, this stress applies to all compounds in the block.

Wenn das erste Element zu Beginn des Blocks durch einen untergesetzten Punkt oder Strich als betont markiert ist, gilt dies für alle Komposita im Block.

Example/Beispiel: **Vanille-: ~eis** ...
~geschmack

Exceptions are given in square brackets, with a hyphen standing for each syllable.

Ausnahmen davon werden in eckigen Klammern angegeben, wobei für jede Silbe ein waagerechter Strich steht.

Example/Beispiel: **drei-, Drei-:** ... **~käsehoch** [-'---]

When no stress is shown for the first element at the beginning of a block, the stress of each compound is given individually.

Wenn das erste Element zu Beginn des Blocks nicht als betont markiert ist, ist die Betonung bei jedem Kompositum im Block angegeben.

Example/Beispiel: **nord-, Nord-: ~seite** ['---] ...,
~südlich

4. Grammatical information

a) Nouns

In the German-English section, nouns are denoted by the inclusion of a definite article.

Example/Beispiel: **Tante** ... **die**

If this article is in parentheses, the word is a proper noun and the article is used only in certain circumstances.

Example/Beispiel: **Belgien** ...**(das)**

The definite article is followed by the genitive and plural forms of the noun with the headword represented by a swung dash.

Example/Beispiel: **Tante** ... **die; ~n, ~n**

If only one form is given, it is the genitive, and the word has no plural.

Example/Beispiel: **Schlaf** ... **der; ~[e]s**

The label *n. pl.* or *Pl.* indicates that the noun exists only in the plural.

Examples/Beispiele: **pants** ... *n. pl.*
Ferien ... *Pl.*

b) Verbs

In the English-German section the entries for irregular verbs give their past tense and past participle. Identical forms are given only once.

Examples/Beispiele: **[1]hide** ... **hid** ... **hidden**
make ... **made**

The doubling of a final consonant before **-ed** or **-ing** is also shown.

Example/Beispiel: **[3]bat** *v. t.*, **-tt-**

4. Grammatische Angaben

a) Bei Substantiven

Im deutsch-englischen Teil werden Substantive durch die Angabe des bestimmten Artikels gekennzeichnet.

Steht der Artikel in runden Klammern, handelt es sich um einen Eigennamen, der nur in bestimmten Fällen mit dem bestimmten Artikel gebraucht wird.

Auf den Artikel folgen die Genitiv- und die Pluralform des Substantivs. Dabei steht für das Stichwort die Tilde.

Wird nur eine Form angegeben, so handelt es sich um den Genitiv, das Stichwort hat in diesem Fall keinen Plural.

Die Angabe *n. pl.* bzw. *Pl.* weist darauf hin, daß das Wort nur im Plural vorkommt.

b) Bei Verben

Bei unregelmäßigen Verben werden im englisch-deutschen Text die Stammformen (Präteritum und 2. Partizip) angegeben, wobei gleichlautende Formen nur einmal genannt werden.

Wenn der Endkonsonant eines Verbs bei der Bildung der Formen auf **-ing** und gegebenenfalls **-ed** verdoppelt wird, wird das ebenfalls angegeben.

In the German-English section irregular verbs are labelled *unr.*, and their parts are given on pp. 1402ff.

Unregelmäßige Verben werden im deutsch-englischen Teil mit *unr.* bezeichnet, ihre Stammformen können auf S. 1402ff. nachgeschlagen werden.

Examples/Beispiele: **klingen** *unr. itr. V.*
leihen ... *unr. tr. V.*

Verbs which are conjugated with *sein* rather than *haben* are labelled accordingly.

Verben, die das Perfekt mit dem Hilfsverb *sein* bilden, sind mit einem entsprechenden Hinweis versehen.

Example/Beispiel: **sterben** ... *mit sein*

Separable compound verbs are indicated by a vertical line at the point where the word is split.

Bei unfest zusammengesetzten Verben zeigt ein ins Wort hineingesetzter senkrechter Strich, wo das Verb gegebenenfalls getrennt wird.

Example/Beispiel: **auf|stehen**

c) Adjectives and adverbs

c) Bei Adjektiven und Adverbien

Irregular and, in the German-English section, umlauted comparative and superlative forms are given.

Zu Adjektiven und Adverbien werden unregelmäßige und – im deutsch-englischen Teil – umlautende Steigerungsformen angegeben.

Examples/Beispiele: **bad** ... **worse** ... **worst**
gut ... **besser** ... **best**...
kalt ... **kälter** ... **kältest**...

An adjective marked *attrib.* or *nicht präd.* is used attributively and not predicatively.

Die Angabe *attrib.* bzw. *nicht präd.* bei einem Adjektiv weist darauf hin, daß das Adjektiv nur als Attribut und nicht als Prädikatsteil verwendet wird.

Examples/Beispiele: **giant** ... *attrib. adj.*
abendlich *Adj.; nicht präd.*

An adjective marked *pred.* or *nicht attrib.* is used predicatively and not attributively.

Die Angabe *pred.* bzw. *nicht attrib.* weist darauf hin, daß das Adjektiv nur als Prädikatsteil und nicht als Attribut verwendet wird.

Examples/Beispiele: **afraid** ... *pred. adj.*
hops ... *Adj.; nicht attr.*

d) Compounds

d) Bei Komposita

Compounds are always labelled with their part of speech or gender, but any further

Bei Komposita wird stets die Wortart angegeben. Wenn keine weiteren grammati-

grammatical information is given at the entry for the second element.

schen Angaben gemacht werden, können diese dem Eintrag für das zweite Element des Kompositums entnommen werden.

Examples/Beispiele: **half-life** *n.*
Radau·bruder der

5. Translations and collocators

a) Translations

Normally, one general translation is given for each word or sense of a word. If two or more are given, separated by semicolons, they are synonymous.

Words which are untranslatable because they have no equivalent in the other language (mainly the names of institutions, customs, foods, etc.) are given a short explanation in italic type.

5. Übersetzungen und Kollokatoren

a) Die Übersetzung

Im Normalfall wird für jedes Stichwort bzw. jede Bedeutung eines Stichworts zuerst eine allgemeine Übersetzung gegeben; gelegentlich auch mehrere gleichwertige Übersetzungen, die mit Semikolons aneinandergereiht sind.

Stichwörter, die nicht übersetzt werden können, weil sie in der Zielsprache kein Äquivalent haben (meist Bezeichnungen für Institutionen, Bräuche, Eßwaren u.ä.), sind mit einer kurzen Erklärung in Kursivschrift versehen.

Examples/Beispiele: **mild** ... *n. schwach gehopfte englische Biersorte*
Schützenfest ... *shooting competition with fair*

The symbol ≈ indicates that the translation given is to be taken only as an approximate equivalent.

Das Symbol ≈ zeigt an, daß die vorgeschlagene Übersetzung nur als ungefähres Äquivalent des Stichworts zu verstehen ist.

Examples/Beispiele: **A level** ... ≈ Abitur, *das*
Finanzamt ... ≈ Inland Revenue

b) Collocators

As the choice of the correct translation often depends on the context in which it is to be used, collocators (words with which a translation typically occurs) are frequently supplied for translations of verbs, adjectives, adverbs, and combining forms. They are given in italics in angle brackets.

b) Kollokatoren

Oft hängt die Wahl der richtigen Übersetzung davon ab, mit welchen anderen Wörtern die Übersetzung im Satz verbunden werden soll. Zu vielen Übersetzungen von Verben, Adjektiven, Adverbien und Wortbildungselementen sind deshalb einige typischerweise mit der Übersetzung verbundene Wörter, sogenannte Kollokatoren, angegeben. Sie stehen in Kursivschrift in Winkelklammern.

Examples/Beispiele: **acquire** ... erwerben ⟨*Land, Besitz, ...*⟩; ... annehmen ⟨*Tonfall, Farbe, Gewohnheit*⟩

flink ... nimble ⟨*fingers*⟩; sharp ⟨*eyes*⟩; quick ⟨*hands*⟩

If more than one translation can be used with a given collocator, the translations are separated by commas instead of semicolons.

Wenn Kollokatoren sich auf mehrere gleichwertige Übersetzungen beziehen, sind diese Übersetzungen mit Kommas statt mit Semikolons aneinandergereiht.

Examples/Beispiele: **achieve** ... herstellen, herbeiführen ⟨*Frieden, Harmonie*⟩

kürzen ... shorten, take up ⟨*garment*⟩

With a translation that is used in compounds, other elements with which it typically combines are given as collocators.

Bei Übersetzungen, die ein Wortbildungselement darstellen, werden Wörter angeführt, mit denen die Übersetzung typischerweise kombiniert wird.

Examples/Beispiele: **marine** ... See⟨*versicherung, -recht usw.*⟩

-süchtig ... ⟨*drug-, heroin-, morphine-, etc.*⟩addicted

Thus, **marine law** is translated as **Seerecht,** and **drogensüchtig** as **drug-addicted.**

Die Verbindung **marine law** wird also mit **Seerecht** übersetzt, **drogensüchtig** mit **drug-addicted.**

c) Further information given with translations

c) Zusätzliche Angaben bei Übersetzungen

The prepositions typically following verbs are given and translated.

Bei Verben wird der präpositionale Anschluß des Stichworts angegeben und übersetzt.

Examples/Beispiele: **conceal** ... verbergen (**from** vor + *Dat.*)

sinnieren ... ponder (**über** + *Akk.* over)

Where a German verb takes a case other than the accusative, this is shown, together with any English preposition used to 'translate' it.

Ebenso wird bei deutschen Verben der zum Anschluß an das Stichwort verwendete Kasus und die entsprechende englische Präposition angegeben, sofern es sich nicht um den bei transitiven Verben stets erforderlichen Akkusativ handelt.

Example/Beispiel: **verdächtigen** ... suspect (*Gen.* of)

The indication *attr.* or *attrib.* means that a translation can be used attributively and not predicatively.

Die Angabe *attr.* bzw. *attrib.* weist darauf hin, daß die angegebene Übersetzung nur als Attribut und nicht als Prädikatsteil verwendet werden darf.

Examples/Beispiele: **preferable** ... vorzuziehend *attr.*

achtseitig ... eight-page *attrib.*

The indication *präd.* or *pred.* means that a translation can be used predicatively and not attributively.

Die Angabe *präd.* bzw. *pred.* weist umgekehrt darauf hin, daß die angegebene Übersetzung nur als Prädikatsteil und nicht als Attribut verwendet werden darf.

Examples/Beispiele: **preferable** ... vorzuziehen *präd.*

irreparabel ... beyond repair *pred.*

Attributive use of an English noun is indicated by *attrib.* when it needs a separate translation.

Für den attributiven Gebrauch von englischen Substantiven wird oft eine eigene Übersetzung angegeben. Vor dieser Übersetzung steht dann der Hinweis *attrib.*

Example/Beispiel: **mountain** ... *attrib.* Gebirgs-

6. Phrases

General translations are followed by phrases in which the general translation(s) cannot be used. These include typical uses, fixed phrases, idioms, and proverbs. All are printed in medium bold type and are translated in their entirety. A swung dash is used to represent the headword.

6. Anwendungsbeispiele

Im Anschluß an die allgemeine[n] Übersetzung[en] werden Anwendungsbeispiele für Fälle gegeben, in denen die allgemeine Übersetzung nicht verwendbar ist. Außerdem werden typische Verwendungen des Stichworts, feste Wendungen, Redensarten und Sprichwörter gezeigt. Die Anwendungsbeispiele sind halbfett gedruckt und werden immer als Ganzes übersetzt. Innerhalb der Beispiele repräsentiert die Tilde das Stichwort.

Examples/Beispiele: **giggle** ... **with a ~:** ... **[a fit of] the ~s**

knistern ... **mit etw. ~:** ... **eine ~de Atmosphäre**

To save space, phrases may be combined.

Zur Platzersparnis werden oft mehrere Beispiele zusammengefaßt.

- Two complete phrases separated by a comma are synonymous and share a translation.

- Wenn zwei vollständige Beispiele mit Komma aneinandergereiht sind, sind sie synonym und haben eine gemeinsame Übersetzung.

Examples/Beispiele: **cash** ... **pay [in] ~, pay ~ down** bar zahlen

ausrasten ... **er rastete aus, es rastete bei ihm aus** ... something snapped in him

- Where portions of a phrase or translation are separated by *or* or *od.*, they are synonymous and interchangeable.

- Wenn Teile eines Beispiels oder einer Übersetzung mit *or* bzw. *od.* aneinandergereiht sind, sind diese Teile beliebig gegeneinander austauschbar.

Examples/Beispiele: **decision** ... **come to** *or* **reach a ~:** zu einer Entscheidung kommen

Bankrott ... **seinen ~ anmelden** *od.* **ansagen** *od.* **erklären** declare oneself bankrupt

- Where portions of a phrase or translation are separated by an oblique, they are syntactically interchangeable but have different meanings.

- Wenn Teile eines Beispiels bzw. seiner Übersetzung mit Schrägstrich aneinandergereiht sind, sind sie zwar syntaktisch austauschbar, haben aber nicht die gleiche Bedeutung.

Examples/Beispiele: **beginning** ... **at the ~ of February/the month** Anfang Februar/des Monats

durchschaubar ... **leicht/schwer ~ sein** be easy/difficult to see through

In German phrases and translations, the reflexive pronoun *sich* is accusative unless it is marked *(Dat.)* (= dative) or could only be dative, e.g. **etw. von sich geben; sich/jmdm. Luft zufächeln.**

Das Wort *sich* ist ein Akkusativ, wenn es nicht mit *(Dat.)* gekennzeichnet oder auf Grund des Kontextes eindeutig Dativ ist (wie etwa in **etw. von sich geben; sich/jmdm. Luft zufächeln**).

7. Cross-references

7. Verweise

Cross-references beginning with *see* or *s.* which take the place of a translation refer to a headword at which the translation is to be found.

Verweise mit *see* bzw. *s.* anstelle einer Übersetzung weisen auf ein anderes Stichwort, unter dem die Übersetzung zu finden ist.

Examples/Beispiele: **Primanerin** ... *s.* **Primaner**
fro ... *see* **to 2b**

Cross-references beginning with *see* or *s.* which are followed by a colon and a list of translations occasionally occur at derivatives, such as nouns and adverbs derived from adjectives. They refer the user to the entry containing the indicators, collocators, etc. necessary for distinguishing the translations.

Verweise mit *see* bzw. *s.*, die vor einer Reihe von Übersetzungen stehen, treten gelegentlich bei Ableitungen auf, z. B. bei Substantiven oder Adverbien, die von einem Adjektiv abgeleitet sind. Sie zeigen, wo die zur Unterscheidung der Übersetzungen nötigen Informationen (Indikatoren, Kollokatoren usw.) zu finden sind.

Examples/Beispiele: **decorate** ... *v.t.* **a)** schmücken ⟨*Raum, Straße, Baum*⟩; verzieren ⟨*Kuchen, Kleid*⟩; dekorieren ⟨*Schaufenster*⟩ ...
decoration ... *n.* **a)** *see* **decorate a:** Schmükken, *das;* Verzieren, *das;* Dekoration, *die* ...

verbreiten 1. *tr. V.* **a)** *(bekannt machen)* spread ⟨*rumour, lies, etc.*⟩; ... **b)** *(weitertragen)* spread ⟨*disease, illness, etc.*⟩; disperse ⟨*seeds, spores, etc.*⟩; **c)** *(erwecken)* radiate ⟨*optimism, happiness, calm, etc.*⟩
Verbreitung die; ~, ~en a) *s.* **verbreiten 1a, b, c:** spreading; ... dispersal; radiation

Cross-references beginning with *see also* or *s. auch* refer to headwords at which further information may be found. They either help the user to find a phrase or idiom or refer to an entry which serves as a model for a set of words because it is treated more comprehensively.

Verweise mit *see also* bzw. *s. auch* weisen auf ein Stichwort hin, unter dem zusätzliche Informationen gefunden werden können. Diese Art von Verweis dient entweder zum Auffinden von festen Wendungen usw. oder führt zu einem Stichwort, das als Muster für einen bestimmten Typ besonders ausführlich behandelt wurde.

Examples/Beispiele: **French** ... **2.** *n.* **a)** Französisch, *das; see also* **English 2a**

Französisch ... *s. auch* **Deutsch**

English abbreviations used in the Dictionary/ Im Wörterverzeichnis verwendete englische Abkürzungen

abbr(s).	abbreviation(s)	Commerc.	Commerce, Commercial
abs.	absolute	Communication Res.	Communication Research
adj(s).	adjective(s)	compar.	comparative
Admin.	Administration, Administrative	condit.	conditional
adv.	adverb	conj.	conjunction
Aeronaut.	Aeronautics	Constr.	Construction
Agric.	Agriculture	constr.	construed
Alch.	Alchemy	contr.	contracted form
Amer.	American, America	def.	definite
Anat.	Anatomy	Dent.	Dentistry
Anglican Ch.	Anglican Church	derog.	derogatory
Anglo-Ind.	Anglo-Indian	dial.	dialect
Ant.	Antiquity	Diplom.	Diplomacy
Anthrop.	Anthropology	Dressm.	Dressmaking
arch.	archaic	Eccl.	Ecclesiastical
Archaeol.	Archaeology	Ecol.	Ecology
Archit.	Architecture	Econ.	Economics
art.	article	Educ.	Education
Astrol.	Astrology	Electr.	Electricity
Astron.	Astronomy	ellipt.	elliptical
Astronaut.	Astronautics	emphat.	emphatic
attrib.	attributive	esp.	especially
Austral.	Australian, Australia	Ethnol.	Ethnology
Bacteriol.	Bacteriology	Ethol.	Ethology
Bibl.	Biblical	euphem.	euphemistic
Bibliog.	Bibliography	excl.	exclamation, exclamatory
Biochem.	Biochemistry	expr.	expressing
Biol.	Biology	fem.	feminine
Bookk.	Bookkeeping	fig.	figurative
Bot.	Botany	Footb.	Football
Brit.	British, Britain	Gastr.	Gastronomy
Can.	Canadian, Canada	Geneal.	Genealogy
Chem.	Chemistry	Geog.	Geography
Cinemat.	Cinematography	Geol.	Geology
coll.	colloquial	Geom.	Geometry
collect.	collective	Graph. Arts	Graphic Arts
comb.	combination		

Her.	Heraldry
Hist.	History, Historical
Horol.	Horology
Hort.	Horticulture
Hydraulic Engin.	Hydraulic Engineering
imper.	imperative
impers.	impersonal
incl.	including
Ind.	Indian, India
indef.	indefinite
Information Sci.	Information Science
int.	interjection
interrog.	interrogative
Int. Law	International Law
Ir.	Irish, Ireland
iron.	ironical
joc.	jocular
Journ.	Journalism
lang.	language
Ling.	Linguistics
Lit.	Literature
lit.	literal
Magn.	Magnetism
Managem.	Management
masc.	masculine
Math.	Mathematics
Mech.	Mechanics
Mech. Engin.	Mechanical Engineering
Med.	Medicine
Metalw.	Metalwork
Metaph.	Metaphysics
Meteorol.	Meteorology
Mil.	Military
Min.	Mineralogy
Motor Veh.	Motor Vehicles
Mount.	Mountaineering
Mus.	Music
Mythol.	Mythology
n.	noun
Nat. Sci.	Natural Science
Naut.	Nautical
neg.	negative
N. Engl.	Northern English
ns.	nouns
Nucl. Engin.	Nuclear Engineering
Nucl. Phys.	Nuclear Physics
Num.	Numismatics
N.Z.	New Zealand
obj.	object
Oceanog.	Oceanography
Ornith.	Ornithology
P	Proprietary name
Palaeont.	Palaeontology
Parapsych.	Parapsychology
Parl.	Parliament
pass.	passive
Pharm.	Pharmacy
Philat.	Philately
Philos.	Philosophy
Phonet.	Phonetics
Photog.	Photography
phr(s).	phrase(s)
Phys.	Physics
Physiol.	Physiology
pl.	plural
poet.	poetical
Polit.	Politics
poss.	possessive
postpos.	postpositive
p.p.	past participle
pred.	predicative
pref.	prefix
Prehist.	Prehistory
prep.	preposition
pres.	present
pres. p.	present participle
pr. n.	proper noun
pron.	pronoun
Pros.	Prosody
prov.	proverbial
Psych.	Psychology
p.t.	past tense
Railw.	Railways
RC Ch.	Roman Catholic Church
refl.	reflexive
rel.	relative
Relig.	Religion
Res.	Research
Rhet.	Rhetoric
rhet.	rhetorical
S. Afr.	South African, South Africa
sb.	somebody
Sch.	School
Sci.	Science
Scot.	Scottish, Scotland
Shipb.	Shipbuilding
sing.	singular
sl.	slang
Sociol.	Sociology
Soc. Serv.	Social Services
Soil Sci.	Soil Science
St. Exch.	Stock Exchange
sth.	something
subord.	subordinate
suf.	suffix
superl.	superlative
Surv.	Surveying
symb.	symbol
tech.	technical
Teleph.	Telephony
Telev.	Television
Theol.	Theology
Univ.	University
usu.	usually

v. aux.	auxiliary verb	v.t. & i.	transitive and intransitive verb
Vet. Med.	Veterinary Medicine	W. Ind.	West Indian, West Indies
v.i.	intransitive verb	Woodw.	Woodwork
voc.	vocative	Zool.	Zoology
v. refl.	reflexive verb		
v.t.	transitive verb		

German abbreviations used in the Dictionary/ Im Wörterverzeichnis verwendete deutsche Abkürzungen

a.	anderes; andere
ä.	ähnliches; ähnliche
Abk.	Abkürzung
adj.	adjektivisch
Adj.	Adjektiv
adv.	adverbial
Adv.	Adverb
Akk.	Akkusativ
amerik.	amerikanisch
Amtsspr.	Amtssprache
Anat.	Anatomie
Anthrop.	Anthropologie
Archäol.	Archäologie
Archit.	Architektur
Art.	Artikel
Astrol.	Astrologie
Astron.	Astronomie
A.T.	Altes Testament
attr.	attributiv
Ausspr.	Aussprache
Bauw.	Bauwesen
Bergmannsspr.	Bergmannssprache
berlin.	berlinisch
bes.	besonders
Bez.	Bezeichnung
bibl.	biblisch
bild. Kunst	bildende Kunst
Biol.	Biologie
Bodenk.	Bodenkunde
Börsenw.	Börsenwesen
Bot.	Botanik
BRD	Bundesrepublik Deutschland
brit.	britisch
Bruchz.	Bruchzahl
Buchf.	Buchführung
Buchw.	Buchwesen
Bürow.	Bürowesen
chem.	chemisch
christl.	christlich
Dat.	Dativ
DDR	Deutsche Demokratische Republik
Dekl.	Deklination
Demonstrativpron.	Demonstrativpronomen
d.h.	das heißt
dichter.	dichterisch
Druckerspr.	Druckersprache
Druckw.	Druckwesen
dt.	deutsch
DV	Datenverarbeitung
ehem.	ehemals, ehemalig
Eisenb.	Eisenbahn
elektr.	elektrisch
Elektrot.	Elektrotechnik
Energievers.	Energieversorgung
Energiewirtsch.	Energiewirtschaft
engl.	englisch
etw.	etwas
ev.	evangelisch
fachspr.	fachsprachlich
fam.	familiär
Fem.	Femininum
Ferns.	Fernsehen
Fernspr.	Fernsprechwesen
fig.	figurativ
Finanzw.	Finanzwesen
Fischereiw.	Fischereiwesen
Fliegerspr.	Fliegersprache
Flugw.	Flugwesen
Forstw.	Forstwesen
Fot.	Fotografie
Frachtw.	Frachtwesen
Funkw.	Funkwesen
Gastr.	Gastronomie
Gattungsz.	Gattungszahl
Gaunerspr.	Gaunersprache
geh.	gehoben
Gen.	Genitiv
Geneal.	Genealogie
Geogr.	Geographie
Geol.	Geologie
Geom.	Geometrie
Handarb.	Handarbeit
Handw.	Handwerk
Hausw.	Hauswirtschaft
Her.	Heraldik
hess.	hessisch
Hilfsv.	Hilfsverb
hist.	historisch

Hochschulw.	Hochschulwesen
Holzverarb.	Holzverarbeitung
Indefinitpron.	Indefinitpronomen
indekl.	indeklinabel
Indik.	Indikativ
Inf.	Infinitiv
Informationst.	Informationstechnik
Interj.	Interjektion
iron.	ironisch
intr.	intransitiv
Jagdw.	Jagdwesen
Jägerspr.	Jägersprache
jmd.	jemand
jmdm.	jemandem
jmdn.	jemanden
jmds.	jemandes
Jugendspr.	Jugendsprache
jur.	juristisch
Kardinalz.	Kardinalzahl
kath.	katholisch
Kaufmannsspr.	Kaufmannssprache
Kfz.-W.	Kraftfahrzeugwesen
Kinderspr.	Kindersprache
Kochk.	Kochkunst
Konj.	Konjunktion
Kosew.	Kosewort
Kunstwiss.	Kunstwissenschaft
Kurzf.	Kurzform
Kurzw.	Kurzwort
landsch.	landschaftlich
Landw.	Landwirtschaft
Literaturw.	Literaturwissenschaft
Luftf.	Luftfahrt
ma.	mittelalterlich
MA.	Mittelalter
marx.	marxistisch
Mask.	Maskulinum
Math.	Mathematik
Mech.	Mechanik
Med.	Medizin
Meeresk.	Meereskunde
Met.	Meteorologie
Metall.	Metallurgie
Metallbearb.	Metallbearbeitung
Milit.	Militär
Mineral.	Mineralogie
mod.	modifizierend
Modalv.	Modalverb
Münzk.	Münzkunde
Mus.	Musik
Mythol.	Mythologie
Naturw.	Naturwissenschaft
Neutr.	Neutrum
niederdt.	niederdeutsch
Nom.	Nominativ
nordamerik.	nordamerikanisch
nordd.	norddeutsch
nordostd.	nordostdeutsch
nordwestd.	nordwestdeutsch
ns.	nationalsozialistisch
N.T.	Neues Testament
o.	ohne; oben
o.ä.	oder ähnliches; oder ähnliche
od.	oder
Ordinalz.	Ordinalzahl
orth.	orthodox
ostd.	ostdeutsch
österr.	österreichisch
Päd.	Pädagogik
Paläont.	Paläontologie
Papierdt.	Papierdeutsch
Parapsych.	Parapsychologie
Parl.	Parlament
Part.	Partizip
Perf.	Perfekt
Pers.	Person
pfälz.	pfälzisch
Pharm.	Pharmazie
Philat.	Philatelie
Philos.	Philosophie
Physiol.	Physiologie
Pl.	Plural
Plusq.	Plusquamperfekt
Polizeiw.	Polizeiwesen
Postw.	Postwesen
präd.	prädikativ
Prähist.	Prähistorie
Präp.	Präposition
Präs.	Präsens
Prät.	Präteritum
Pron.	Pronomen
Psych.	Psychologie
Raumf.	Raumfahrt
Rechtsspr.	Rechtssprache
Rechtsw.	Rechtswesen
refl.	reflexiv
regelm.	regelmäßig
Rel.	Religion
Relativpron.	Relativpronomen
rhein.	rheinisch
Rhet.	Rhetorik
röm.	römisch
röm.-kath.	römisch-katholisch
Rundf.	Rundfunk
s.	siehe
S.	Seite
scherzh.	scherzhaft
schles.	schlesisch
schott.	schottisch
Schülerspr.	Schülersprache
Schulw.	Schulwesen
schwäb.	schwäbisch
schweiz.	schweizerisch
Seemannsspr.	Seemannssprache
Seew.	Seewesen

Sexualk.	Sexualkunde	unr.	unregelmäßig
Sg.	Singular	usw.	und so weiter
s.o.	siehe oben	v.	von
Soldatenspr.	Soldatensprache	V.	Verb
Sozialpsych.	Sozialpsychologie	verächtl.	verächtlich
Sozialvers.	Sozialversicherung	veralt.	veraltet; veraltend
Soziol.	Soziologie	Verhaltensf.	Verhaltensforschung
spött.	spöttisch	verhüll.	verhüllend
Spr.	Sprichwort	Verkehrsw.	Verkehrswesen
Sprachw.	Sprachwissenschaft	Vermessungsw.	Vermessungswesen
Steuerw.	Steuerwesen	Versicherungsw.	Versicherungswesen
Stilk.	Stilkunde	vgl.	vergleiche
Studentenspr.	Studentensprache	Vkl.	Verkleinerungsform
s.u.	siehe unten	Völkerk.	Völkerkunde
Subj.	Subjekt	Völkerr.	Völkerrecht
subst.	substantivisch; substantiviert	Volksk.	Volkskunde
		volkst.	volkstümlich
Subst.	Substantiv	vulg.	vulgär
südd.	süddeutsch	Werbespr.	Werbesprache
südwestd.	südwestdeutsch	westd.	westdeutsch
Suff.	Suffix	westfäl.	westfälisch
Sup.	Superlativ	Wiederholungsz.	Wiederholungszahlwort
Textilw.	Textilwesen		
Theol.	Theologie	wiener.	wienerisch
thüring.	thüringisch	Winzerspr.	Winzersprache
Tiermed.	Tiermedizin	Wirtsch.	Wirtschaft
tirol.	tirolisch	Wissensch.	Wissenschaft
tr.	transitiv	Wz.	Warenzeichen
Trenn.	Trennung	Zahnmed.	Zahnmedizin
u.	und	z.B.	zum Beispiel
u.a.	und andere[s]	Zeitungsw.	Zeitungswesen
u.ä.	und ähnliches	Zollw.	Zollwesen
ugs.	umgangssprachlich	Zool.	Zoologie
unbest.	unbestimmt	Zus.	Zusammensetzung
unpers.	unpersönlich	Zusschr.	Zusammenschreibung

A

A, ¹a [eɪ] *n., pl.* **As** *or* **A's a)** *(letter)* A, a, *das;* **from A to Z** von A bis Z; **A road** Straße 1. Ordnung; ≈ Bundesstraße, *die;* **b)** *(Mus.)* A, a, *das;* **A sharp** ais, Ais, *das;* **A flat** as, As, *das;* **c) A 1** *(coll.: first-rate)* eins a *(ugs.)*

²a [ə, *stressed* eɪ] *indef. art.* **a)** ein/eine/ein; **he is a gardener/a Frenchman** er ist Gärtner/Franzose; **she did not say a word** sie sagte kein Wort; **b)** *(per)* pro; **£40 a year** 40 Pfund pro Jahr; **it's 20p a pound** es kostet 20 Pence das Pfund; **two a penny** zwei Stück [für] einen Penny

AA *abbr. (Brit.)* **Automobile Association** *britischer Automobilklub*

aback [ə'bæk] *adv.* **be taken ~:** erstaunt sein

abacus ['æbəkəs] *n., pl.* **~es** *or* **abaci** ['æbəsaɪ] Abakus, *der*

abandon [ə'bændən] **1.** *v. t.* **a)** *(forsake)* verlassen ‹*Ort*›; verlassen, im Stich lassen ‹*Person*›; aussetzen ‹*Kind, Tier*›; aufgeben ‹*Prinzip*›; stehenlassen ‹*Auto*›; aufgeben, fallenlassen ‹*Gedanken, Plan*›; **~ hope** die Hoffnung aufgeben; **~ ship** das Schiff verlassen; **~ ship!** alle Mann von Bord!; **b)** *(surrender)* **~ sth. to the enemy** etw. dem Feind übergeben *od.* überlassen; **c)** *(yield)* **~ oneself to sth.** sich einer Sache *(Dat.)* hingeben. **2.** *n., no pl.* **with ~:** ungezwungen

abandonment [ə'bændənmənt] *n., no pl. (of right, claim)* Preisgabe, *die; (of plan, property)* Aufgabe, *die*

abase [ə'beɪs] *v. t.* demütigen, erniedrigen ‹*Person*›; **~ oneself** sich erniedrigen

abashed [ə'bæʃt] *adj.* beschämt; verlegen

abate [ə'beɪt] *v. i.* [an Stärke *od.* Intensität] abnehmen; nachlassen; ‹*Zorn, Eifer, Sturm:*› abflauen, nachlassen

abattoir ['æbətwɑ:(r)] *n.* Schlachthof, *der; ([part of] building)* Schlachthaus, *das*

abbess ['æbɪs] *n.* Äbtissin, *die*

abbey ['æbɪ] *n.* Abtei, *die*

abbot ['æbət] *n.* Abt, *der*

abbreviate [ə'bri:vɪeɪt] *v. t.* abkürzen ‹*Wort usw.*› **(to** mit)

abbreviation [əbri:vɪ'eɪʃn] *n.* Abkürzung, *die*

ABC [eɪbi:'si:] *n.* **a)** *(alphabet)* ABC, *das;* **b)** *(fig.: rudiments)* ABC, *das;* Einmaleins, *das*

abdicate ['æbdɪkeɪt] *v. t.* abdanken; **~ [the throne]** auf den Thron verzichten

abdication [æbdɪ'keɪʃn] *n.* Abdankung, *die;* Thronverzicht, *der*

abdomen ['æbdəmɪn] *n. (Anat.)* Bauch, *der;* Unterleib, *der;* Abdomen, *das (fachspr.)*

abdominal [æb'dɒmɪnl] *adj.* Bauch-; Abdominal- *(fachspr.)*

abduct [əb'dʌkt] *v. t.* entführen

abduction [əb'dʌkʃn] *n.* Entführung, *die*

aberration [æbə'reɪʃn] *n.* Abweichung, *die;* **mental ~[s]** geistige Verirrung

abet [ə'bet] *v. t.*, **-tt-** helfen (+ *Dat.*); unterstützen; **aid and ~:** Beihilfe leisten (+ *Dat.*)

abeyance [ə'beɪəns] *n.* **be in/fall into ~:** zeitweilig außer Kraft sein/treten

abhor [əb'hɔ:(r)] *v. t.*, **-rr-** hassen; *(loathe)* verabscheuen

abhorrence [əb'hɒrəns] *n., no pl. (loathing)* Abneigung, *die* **(of** gegen)

abhorrent [əb'hɒrənt] *adj.* abscheulich; **be ~ to sb.** jmdm. zuwider sein

abide [ə'baɪd] **1.** *v. i.*, **abode** [ə'bəʊd] *or* **~d a)** *usu.* **~d: ~ by** befolgen ‹*Gesetz, Regel, Vorschrift*›; [ein]halten ‹*Versprechen*›; **b)** *(remain)* bleiben; verweilen *(geh.)*. **2.** *v. t.* ertragen; **I can't ~ dogs** ich kann Hunde nicht ausstehen

ability [ə'bɪlɪtɪ] *n.* **a)** *(capacity)* Können, *das;* Fähigkeit, *die;* **have the ~ to do sth.** etw. können *od. (geh.)* vermögen; **make use of one's ~** *or* **abilities** seine Fähigkeiten einsetzen; **to the best of my ~:** soweit es in meinen Kräften steht; **b)** *no pl. (cleverness)* Intelligenz, *die;* **she is a girl of great ~:** sie ist ein sehr intelligentes Mädchen; **c)** *(talent)* Begabung, *die;* Talent, *das;* Anlagen *Pl.;* **he shows** *or* **has great musical ~:** er ist musikalisch sehr begabt

abject ['æbdʒekt] *adj.* elend; erbärmlich; bitter ‹*Armut*›; demütig ‹*Entschuldigung*›

abjure [əb'dʒʊə(r)] *v. t.* abschwören (+ *Dat.*)

ablaze [ə'bleɪz] *pred. adj.* **be ~:** in Flammen stehen; **be ~ with light** hell erleuchtet sein

able ['eɪbl] *adj.* **a) be ~ to do sth.** etw. können; **I'd love to come but I don't know if I'll be ~ [to]** ich würde sehr gern kommen, aber ich weiß nicht, ob es mir möglich sein wird; **b)** *(competent, talented)* fähig

able: ~-bodied ['eɪblbɒdɪd] *adj.* kräftig; stark; tauglich ‹*Soldat, Matrose*›; **~ 'seaman** *n.* Vollmatrose, *der*

ably ['eɪblɪ] *adv.* geschickt; gekonnt

abnormal [æb'nɔ:ml] *adj.* **a)** abnorm ‹*Gestalt, Größe*›; a[b]normal ‹*Interesse, Verhalten*›; **mentally/physically ~:** geistig/physisch anomal *od.* krank; **b)** *(unusual)* ungewöhnlich; a[b]normal

abnormality [æbnɔ:'mælɪtɪ] *n.* Abnormität, *die;* Anomalie, *die*

abnormally [æb'nɔ:məlɪ] *adv.* ungewöhnlich

aboard [ə'bɔ:d] **1.** *adv.* an Bord; **all ~!** alle Mann an Bord!; *(bus, train)* alle[s] einsteigen! **2.** *prep.* an Bord (+ *Gen.*); **~ the bus/train** im Bus/Zug; **~ ship** an Bord

¹abode [ə'bəʊd] *n. (formal/joc.: dwelling-place)* Wohnstätte, *die;*

Bleibe, *die;* **of no fixed ~:** ohne festen Wohnsitz

[2]**abode** *see* **abide**

abolish [ə'bɒlɪʃ] *v. t.* abschaffen

abolition [æbə'lɪʃn] *n.* Abschaffung, *die*

abominable [ə'bɒmɪnəbl] *adj.* abscheulich; scheußlich; **the A~ Snowman** der Schneemensch; der Yeti

abominably [ə'bɒmɪnəblɪ] *adv.* abscheulich; scheußlich

abomination [əbɒmɪ'neɪʃn] *n.* **a)** *no pl. (abhorrence)* Abscheu, *der* (**of** vor + *Dat.*); **b)** *(object of disgust)* Abscheulichkeit, *die*

aborigine [æbə'rɪdʒɪnɪ] *n.* Ureinwohner, *der;* Urbewohner, *der; (in Australia)* **A~:** [australischer] Ureinwohner

abort [ə'bɔ:t] **1.** *v. i.* eine Fehlgeburt haben; abortieren *(Med.).* **2.** *v. t.* **a)** *(Med.)* **~ a baby** eine Schwangerschaftsunterbrechung durchführen; [ein Baby] abtreiben; **b)** *(fig.: end)* vorzeitig beenden; abbrechen ‹*Projekt, Unternehmen*›

abortion [ə'bɔ:ʃn] *n.* **a)** Schwangerschaftsunterbrechung, *die;* Abtreibung, *die;* **have/get an ~:** die Schwangerschaft unterbrechen lassen; **back-street ~:** illegale Abtreibung *(durch Engelmacherin);* **b)** *(involuntary)* Früh- *od.* Fehlgeburt, *die;* Abort, *der (Med.)*

abortive [ə'bɔ:tɪv] *adj.* mißlungen ‹*Plan*›*;* fehlgeschlagen ‹*Versuch*›

abound [ə'baʊnd] *v. i.* **a)** *(be plentiful)* reichlich *od.* in Hülle und Fülle vorhanden sein *od.* dasein; **b) ~ in sth.** an etw. *(Dat.)* reich sein; **~ with** voll sein von

about [ə'baʊt] **1.** *adv.* **a)** *(all around)* rings[her]um; *(here and there)* überall; **all ~** ringsumher; **strewn/littered ~ all over the room** überall im Zimmer verstreut; **b)** *(near)* **be ~:** dasein; hiersein; **is John ~?** ist John da?; **there was nobody ~:** es war niemand da; **c) be ~ to do sth.** gerade etw. tun wollen; **d)** *(active)* **be out and ~:** aktiv sein; **be up and ~:** aufsein *(ugs.);* **e)** *(approximately)* ungefähr; **[at] ~ 5 p.m.** ungefähr um *od.* gegen 17 Uhr; **f)** *(round)* herum; rum *(ugs.);* **~ turn!,** *(Amer.)* **~ face!** *(Mil.)* kehrt!; **g) [turn and] turn ~** *(in rotation)* abwechselnd. **2.** *prep.* **a)** *(all around)* um [... herum]; **there was litter lying ~ the park/streets** überall im Park/auf den Straßen lag der Abfall herum; **b)** *(with)* **have sth. ~ one** etw. [bei sich] haben; **c)** *(concerning)* über (+ *Akk.*); **an argument/a question ~ sth.** Streit wegen etw./eine Frage zu etw.; **talk/laugh ~ sth.** über etw. *(Akk.)* sprechen/lachen; **cry ~ sth.** wegen etw. weinen; **know ~ sth.** von etw. wissen; **what was it ~?** worum ging es?; **d)** *(occupied with)* **be quick/brief ~ it** beeil dich!; *(in speaking)* fasse dich kurz!; **while you're ~ it** da Sie gerade dabei sind

above [ə'bʌv] **1.** *adv.* **a)** *(position)* oben; oberhalb; *(higher up)* darüber; **up ~:** oben; **from ~:** von oben [herab]; **~ right** rechts oben; oben rechts; **the flat/floor ~:** die Wohnung/das Stockwerk darüber; **b)** *(direction)* nach oben; hinauf; *(upstream)* stromauf[wärts]; **c)** *(earlier in text)* weiter oben; **see ~, p. 123** siehe oben, S. 123. **2.** *prep.* **a)** *(position)* über (+ *Dat.*); *(upstream from)* oberhalb (+ *Gen.*); **~ oneself** *(conceited)* größenwahnsinnig *(ugs.);* **b)** *(direction)* über (+ *Akk.*); **c)** *(more than)* über (+ *Akk.*); **will anyone go ~ £2,000?** bietet jemand mehr als 2000 Pfund?; **be ~ criticism/suspicion** über jede Kritik/jeden Verdacht erhaben sein; **~ all [else]** vor allem; insbesondere; **d)** *(ranking higher than)* über (+ *Dat.*); **she's in the class ~ me** sie ist eine Klasse über mir. **3.** *adj.* obig ‹*Erklärung, Aufzählung, Ziffern*›*; (~-mentioned)* obengenannt. **4.** *n.* **the ~:** das Obige; *(person[s])* der/die Obengenannte/die Obengenannten

a'bove-mentioned *adj.* obengenannt; obenerwähnt

abrasion [ə'breɪʒn] *n.* *(graze)* Hautabschürfung, *die*

abrasive [ə'breɪsɪv] **1.** *adj.* **a)** scheuernd; Scheuer-; **b)** *(fig.: harsh)* aggressiv; herausfordernd ‹*Ton*›. **2.** *n.* Scheuermittel, *das*

abreast [ə'brest] *adv.* **a)** nebeneinander; Seite an Seite; **b)** *(fig.)* **keep ~ of sth.** sich über etw. *(Akk.)* auf dem laufenden halten

abridge [ə'brɪdʒ] *v. t.* kürzen

abroad [ə'brɔ:d] *adv.* im Ausland; *(direction)* ins Ausland; **from ~:** aus dem Ausland

abrupt [ə'brʌpt] *adj.* **a)** *(sudden)* abrupt, plötzlich ‹*Ende, Abreise, Wechsel*›*;* **come to an ~ halt** ‹*Fahrzeug:*› plötzlich *od.* abrupt anhalten; **b)** *(brusque)* schroff, barsch ‹*Art, Ton*›*;* **c)** *(steep)* jäh; steil

abruptly [ə'brʌptlɪ] *adv.* **a)** *(suddenly)* abrupt; plötzlich; **b)** *(brusquely)* schroff; barsch; **c)** *(steeply)* jäh; steil

abscess ['æbsɪs] *n.* *(Med.)* Abszeß, *der*

abscond [əb'skɒnd] *v. i.* sich entfernen

absence ['æbsəns] *n.* **a)** Abwesenheit, *die; (from work)* Fernbleiben, *das;* **his ~s from school** sein Fehlen in der Schule; **b)** *(lack)* **the ~ of sth.** das Fehlen von etw.; **c) ~ [of mind]** Geistesabwesenheit, *die*

absent 1. ['æbsənt] *adj.* **a)** abwesend; **be ~:** nicht dasein; **be ~ from school/work** in der Schule/am Arbeitsplatz fehlen; **be ~ without leave** sich unerlaubt entfernt haben; **b)** *(lacking)* **be ~:** fehlen; **c)** *(abstracted)* geistesabwesend. **2.** [əb'sent] *v. refl.* **~ oneself [from sth.]** [einer Sache *(Dat.)*] fernbleiben

absentee [æbsən'ti:] *n.* Abwesende, *der/die;* **there were a few ~s** ein paar fehlten

absent-minded [æbsənt'maɪndɪd] *adj.* geistesabwesend; *(habitually)* zerstreut

absent-mindedly [æbsənt'maɪndɪdlɪ] *adv.* geistesabwesend

absent-mindedness [æbsənt'maɪndɪdnɪs] *n., no pl.* Geistesabwesenheit, *die*

absolute ['æbsəlu:t, 'æbsəlju:t] *adj.* absolut; unumstößlich ‹*Beweis, Tatsache*›*;* ausgemacht ‹*Lüge, Skandal*›*; (unconditional)* fest ‹*Versprechen*›*;* streng ‹*Verpflichtung*›*;* uneingeschränkt ‹*Macht*›*;* **~ majority** absolute Mehrheit

absolutely ['æbsəlu:tlɪ, 'æbsəlju:tlɪ] *adv.* **a)** absolut; strikt ‹*ablehnen*›*;* völlig ‹*verrückt*›*;* ausgesprochen ‹*kriminell, ekelhaft, schlimm*›*;* **you're ~ right!** du hast völlig recht; **b)** *(positively)* regelrecht; **~ not!** auf keinen Fall!; **c)** [æbsə'lu:tlɪ] *(coll.: yes indeed)* hundertprozentig *(ugs.)*

absolution [æbsə'lu:ʃn, æbsə'lju:ʃn] *n.* *(Relig.) (forgiveness)* Vergebung, *die; (release)* Erlaß, *der* (**from** *Gen.*); **pronounce ~:** [die] Absolution erteilen

absolve [əb'zɒlv] *v. t.* **~ from** entbinden von ‹*Pflichten*›*;* vergeben ‹*Sünde, Verbrechen*›*;* lossprechen von ‹*Schuld*›*; (Relig.)* Absolution erteilen (+ *Dat.*)

absorb [əb'sɔ:b, əb'zɔ:b] *v. t.* **a)** aufsaugen ‹*Flüssigkeit*›*;* aufnehmen ‹*Flüssigkeit, Nährstoff, Wärme*›*; (fig.)* in sich *(Akk.)* aufnehmen ‹*Wissen*›*;* **b)** *(reduce in strength)* absorbieren; abfangen ‹*Schlag, Stoß*›*;* **c)** *(incorporate)* eingliedern, integrieren ‹*Abtei-*

lung, Gemeinde〉; **d)** *(consume)* aufzehren 〈*Kraft, Zeit, Vermögen*〉; **e)** *(engross)* ausfüllen 〈*Person, Interesse, Gedanken*〉

absorbed [əb'sɔ:bd, əb'zɔ:bd] *adj.* versunken; **be/get ~ in sth.** in etw. *(Akk.)* vertieft sein/sich in etw. *(Akk.)* vertiefen

absorbency [əb'sɔ:bənsɪ, əb'zɔ:bənsɪ] *n.* Saugfähigkeit, *die*

absorbent [əb'sɔ:bənt, əb'zɔ:bənt] *adj.* saugfähig

absorbing [əb'sɔ:bɪŋ, əb'zɔ:bɪŋ] *adj.* faszinierend

absorption [əb'sɔ:pʃn, əb'zɔ:pʃn] *n.* **a)** *(incorporation, physical process)* Absorption, *die (fachspr.);* **b)** *(of department, community)* Integration, *die;* **c)** *(engrossment)* Versunkenheit, *die*

abstain [əb'steɪn] *v.i.* **a)** enthaltsam sein; **~ from sth.** sich einer Sache *(Gen.)* enthalten; **b) ~ [from voting]** sich der Stimme enthalten

abstemious [əb'sti:mɪəs] *adj.* enthaltsam

abstention [əb'stenʃn] *n.* **~ from the vote/from voting** Stimmenthaltung, *die;* **how many ~s were there?** wie viele Personen enthielten sich der Stimme?

abstinence ['æbstɪnəns] *n.* Abstinenz, *die*

abstinent ['æbstɪnənt] *adj.* abstinent

abstract 1. ['æbstrækt] *adj.* abstrakt; **~ noun** *(Ling.)* Abstraktum, *das;* **in the ~:** abstrakt. **2.** ['æbstrækt] *n.* **a)** *(summary)* Zusammenfassung, *die;* Abstract, *das (fachspr.); (of book)* Inhaltsangabe, *die;* **b)** *(idea)* Abstraktum, *das.* **3.** [æb'strækt] *v.t. (remove)* wegnehmen

abstruse [æb'stru:s] *adj.* abstrus

absurd [əb'sɜ:d] *adj.* absurd; *(ridiculous)* lächerlich

absurdity [əb'sɜ:dɪtɪ] *n.* Absurdität, *die*

absurdly [əb'sɜ:dlɪ] *adv.* lächerlich

abundance [ə'bʌndəns] *n.* **[an] ~ of sth.** eine Fülle von etw.; **an ~ of love/energy** ein Übermaß an Liebe/Energie *(Dat.);* **in ~:** in Hülle und Fülle

abundant [ə'bʌndənt] *adj.* reich **(in** an + *Dat.*); **an ~ supply of fish/fruit** Fisch/Obst im Überfluß

abundantly [ə'bʌndəntlɪ] *adv.* überreichlich; **I made it ~ clear that ...:** ich habe es überdeutlich zum Ausdruck gebracht, daß ...

abuse 1. [ə'bju:z] *v.t.* **a)** *(misuse)* mißbrauchen 〈*Macht, Recht, Autorität, Vertrauen*〉; *(maltreat)* peinigen, quälen 〈*Tier*〉; **sexually ~:** sexuell mißbrauchen; **b)** *(insult)* beschimpfen. **2.** [ə'bju:s] *n.* **a)** *(misuse)* Mißbrauch, *der;* **b)** *(unjust or corrupt practice)* Mißstand, *der;* **c)** *(insults)* Beschimpfungen *Pl.;* **a term of ~:** ein Schimpfwort

abusive [ə'bju:sɪv] *adj.* beleidigend; **~ language** Beleidigungen; Beschimpfungen; **become** *or* **get ~:** ausfallend werden

abut [ə'bʌt] **1.** *v.i.*, **-tt-: a)** *(border)* **~ on** grenzen an (+ *Akk.*); **b)** *(end)* **~ on/against** stoßen *od.* angrenzen an (+ *Akk.*). **2.** *v.t.* **a)** *(border on)* angrenzen an (+ *Akk.*); **b)** *(end on)* anstoßen an (+ *Akk.*)

abysmal [ə'bɪzml] *adj.* **a)** grenzenlos 〈*Unwissenheit*〉; **b)** *(coll.: bad)* katastrophal *(ugs.)*

abyss [ə'bɪs] *n. (lit. or fig.)* Abgrund, *der*

AC *abbr. (Electr.)* **alternating current** Ws

a/c *abbr.* **account**

acacia [ə'keɪʃə] *n. (Bot.)* Akazie, *die*

academic [ækə'demɪk] **1.** *adj.* akademisch; wissenschaftlich 〈*Fach, Studium*〉. **2.** *n.* Wissenschaftler, *der*/Wissenschaftlerin, *die; (scholar)* Gelehrte, *der/die*

academically [ækə'demɪkəlɪ] *adv.* wissenschaftlich; **be ~ very able** große intellektuelle Fähigkeiten haben

academy [ə'kædəmɪ] *n.* Akademie, *die*

accede [æk'si:d] *v.i.* **a)** *(assent)* zustimmen **(to** *Dat.*); **b)** beitreten **(to** *Dat.*) 〈*Abkommen, Bündnis*〉; antreten 〈*Amt*〉; **~ [to the throne]** den Thron besteigen

accelerate [ək'seləreɪt] **1.** *v.t.* beschleunigen. **2.** *v.i.* sich beschleunigen; 〈*Auto[fahrer], Läufer:*〉 beschleunigen

acceleration [əkselə'reɪʃn] *n.* Beschleunigung, *die*

accelerator [ək'seləreɪtə(r)] *n.* **~ [pedal]** Gas[pedal], *das*

accent 1. ['æksənt] *n.* Akzent, *der; (mark)* Akzent, *der;* Akzentzeichen, *das;* **the ~ is on ...** *(fig.)* der Akzent liegt auf (+ *Dat.*) ... **2.** [ək'sent] *v.t.* betonen

accentuate [ək'sentjʊeɪt] *v.t.* betonen; verstärken 〈*Schmerz, Kummer*〉

accept [ək'sept] *v.t.* **a)** *(be willing to receive)* annehmen; aufnehmen 〈*Mitglied*〉; *(take formally)* entgegennehmen 〈*Dank, Spende, Auszeichnung*〉; übernehmen 〈*Verantwortung, Aufgabe*〉; *(agree to)* annehmen 〈*Vorschlag, Plan, Heiratsantrag, Einladung*〉; **~ sb. for a job/school** jmdm. eine Einstellungszusage geben/jmdn. in eine Schule aufnehmen; **~ sb. for a course** jmdn. in einen Lehrgang aufnehmen; **b)** *(approve)* akzeptieren; **~ sb. as a member of the group** jmdn. als Mitglied der Gruppe anerkennen; **c)** *(acknowledge)* akzeptieren; **it is ~ed that ...:** es ist unbestritten, daß ...; **an ~ed fact** eine anerkannte Tatsache; **d)** *(believe)* **~ sth. [from sb.]** [jmdm.] etw. glauben; **e)** *(tolerate)* hinnehmen

acceptability [əkseptə'bɪlɪtɪ] *n., no pl.* Annehmbarkeit, *die; (of salary, price)* Angemessenheit, *die*

acceptable [ək'septəbl] *adj.* akzeptabel; annehmbar 〈*Preis, Gehalt*〉

acceptance [ək'septəns] *n.* **a)** *(willing receipt)* Annahme, *die; (of gift, offer)* Annahme, *die;* Entgegennahme, *die; (of duty, responsibility)* Übernahme, *die; (in answer)* Zusage, *die; (agreement)* Annahme, *die;* Zustimmung, *die* **(of** zu); **b)** *no pl. (approval)* Billigung, *die;* **c)** *no pl. (acknowledgement)* Anerkennung, *die; (toleration)* Hinnahme, *die*

access ['ækses] *n.* **a)** *no pl., no art. (entering)* Zutritt, *der* **(to** zu); *(by vehicles)* Einfahren, *das* **(into** in + *Akk.*); **this doorway is the only means of ~:** diese Tür ist der einzige Zugang; **b)** *(admission)* **gain** *or* **obtain** *or* **get ~:** Einlaß finden; **c)** *no pl. (opportunity to use or approach)* Zugang, *der* **(to** zu); **d) easy/difficult of ~:** leicht/schwer zugänglich; **e)** *(way [in])* Zugang, *der; (road)* Zufahrt, *die; (door)* Eingang, *der*

accessible [ək'sesɪbl] *adj.* **a)** *(reachable)* **[more] ~ [to sb.]** [besser] erreichbar [für jmdn.]; **b)** *(available, open, understandable)* zugänglich **(to** für)

accession [ək'seʃn] *n.* Amtsantritt, *der;* **~ to the throne** Thronbesteigung, *die*

accessory [ək'sesərɪ] **1.** *adj.* **~ [to sth.]** zusätzlich [zu etw.]. **2.** *n.* **a)** *(accompaniment)* Extra, *das;* **b)** *in pl. (attachments)* Zubehör, *das;* **one of the accessories** eines der Zubehörteile; **c)** *(dress article)* Accessoire, *das*

'access road *n.* Zufahrtsstraße, *die*

accident ['æksɪdənt] *n.* **a)** *(unlucky event)* Unfall, *der;* **road ~:**

Verkehrsunfall, *der;* **have an ~:** einen Unfall haben; **b)** *(chance)* Zufall, *der; (unfortunate chance)* Unglücksfall, *der;* **by ~:** zufällig; **c)** *(mistake)* Versehen, *das;* **by ~:** versehentlich; **d)** *(mishap)* Mißgeschick, *das*

accidental [æksɪ'dentl] *adj. (happening by chance)* zufällig; *(unintended)* unbeabsichtigt

accidentally [æksɪ'dentəlɪ] *adv. (by chance)* zufällig; *(by mistake)* versehentlich

acclaim [ə'kleɪm] *v.t. (welcome)* feiern; *(hail as)* **~ sb. king** jmdn. zum König ausrufen

acclamation [æklə'meɪʃn] *n., no pl.* Beifall, *der*

acclimatization [əklaɪmətaɪ'zeɪʃn] *n. (lit. or fig.)* Akklimatisation, *die*

acclimatize [ə'klaɪmətaɪz] *v.t. (lit. or fig.)* akklimatisieren; **~ sth./sb. to sth.** etw./jmdn. an etw. *(Akk.)* gewöhnen; **get** *or* **become ~d** sich akklimatisieren

accolade ['ækəleɪd, ækə'leɪd] *n. (praise)* **~[s]** Lob, *das; (acknowledgement)* Anerkennung, *die*

accommodate [ə'kɒmədeɪt] *v.t.* **a)** unterbringen; *(hold, have room for)* Platz bieten (+ *Dat.*); **b)** *(oblige)* gefällig sein (+ *Dat.*)

accommodating [ə'kɒmədeɪtɪŋ] *adj.* zuvorkommend

accommodation [əkɒmə'deɪʃn] *n., no pl.* Unterkunft, *die;* **can you provide us with ~ for the night?** können Sie uns ein Nachtquartier besorgen?; **~ is very expensive** Wohnungen/Zimmer sind sehr teuer

accompaniment [ə'kʌmpənɪmənt] *n. (lit. or fig.; also Mus.)* Begleitung, *die*

accompanist [ə'kʌmpənɪst] *n. (Mus.)* Begleiter, *der*/Begleiterin, *die*

accompany [ə'kʌmpənɪ] *v.t. (also Mus.)* begleiten

accomplice [ə'kʌmplɪs] *n.* Komplize, *der*/Komplizin, *die*

accomplish [ə'kʌmplɪʃ] *v.t. (perform)* vollbringen ‹*Tat*›; erfüllen ‹*Aufgabe*›

accomplished [ə'kʌmplɪʃt] *adj.* fähig; **he is an ~ speaker/dancer** er ist ein erfahrener Redner/vollendeter Tänzer

accomplishment [ə'kʌmplɪʃmənt] *n.* **a)** *no pl. (completion)* Vollendung, *die; (of task)* Erfüllung, *die;* **b)** *(achievement)* Leistung, *die; (skill)* Fähigkeit, *die*

accord [ə'kɔ:d] **1.** *v.i.* **~ [with sth.]** [mit etw.] übereinstimmen. **2.** *v.t.* **~ sb. sth.** jmdm. etw. gewähren. **3.** *n.* **a) of one's own ~:** aus eigenem Antrieb; **of its own ~:** von selbst; **b)** *(harmonious agreement)* Übereinstimmung, *die;* **with one ~:** geschlossen

accordance [ə'kɔ:dəns] *n.* **in ~ with** in Übereinstimmung mit; gemäß (+ *Dat.*)

according [ə'kɔ:dɪŋ] *adv.* **a) ~ as** *(depending on how)* je nachdem wie; *(depending on whether)* je nachdem ob; **b) ~ to** nach; **~ to him** *(opinion)* seiner Meinung nach; *(account)* nach seiner Aussage; **~ to circumstances/the season** den Umständen/der Jahreszeit entsprechend

accordingly [ə'kɔ:dɪŋlɪ] *adv. (as appropriate)* entsprechend; *(therefore)* folglich

accordion [ə'kɔ:dɪən] *n.* Akkordeon, *das*

accost [ə'kɒst] *v.t.* ansprechen

account [ə'kaʊnt] *n.* **a)** *(Finance)* Rechnung, *die;* **keep ~s/the ~s** Buch/die Bücher führen; **settle** *or* **square ~s with sb.** *(lit. or fig.)* mit jmdm. abrechnen; **on ~:** auf Rechnung; a conto; **on one's [own] ~:** auf eigene Rechnung; *(fig.)* von sich aus; **b)** *(at bank, shop)* Konto, *das;* **pay sth. into one's ~:** etw. auf sein Konto einzahlen; **draw sth. out of one's ~:** etw. von seinem Konto abheben; **on ~:** auf Rechnung; **c)** *(statement of facts)* Rechenschaft, *die;* **give** *or* **render an ~ for sth.** über etw. *(Akk.)* Rechenschaft ablegen; **call sb. to ~:** jmdn. zur Rechenschaft ziehen; **give a good ~ of oneself** seinen Mann stehen; **d)** *(consideration)* **take ~ of sth., take sth. into ~:** etw. berücksichtigen; **take no ~ of sth./sb., leave sth./sb. out of ~:** etw./jmdn. unberücksichtigt lassen *od.* nicht berücksichtigen; **don't change your plans on my ~:** ändert nicht meinetwegen eure Pläne; **on ~ of** wegen; **on no ~, not on any ~:** auf [gar] keinen Fall; **e)** *(importance)* **of little/no ~:** von geringer/ohne Bedeutung; **f)** *(report)* **an ~ [of sth.]** ein Bericht [über etw. *(Akk.)*]; **give a full ~ of sth.** ausführlich über etw. *(Akk.)* berichten; **by** *or* **from all ~s** nach allem, was man hört

~ for *v.t.* **a)** *(give reckoning)* Rechenschaft *od.* Rechnung ablegen über (+ *Akk.*); **b)** *(explain)* erklären; **c)** *(represent in amount)* ausmachen; ergeben

accountability [əkaʊntə'bɪlɪtɪ] *n., no pl.* Verantwortlichkeit, *die* **(to** gegenüber)

accountable [ə'kaʊntəbl] *adj.* verantwortlich **(for** für); **be ~ to sb.** jmdm. Rechenschaft schuldig sein

accountancy [ə'kaʊntənsɪ] *n., no pl.* Buchhaltung, *die*

accountant [ə'kaʊntənt] *n.* [Bilanz]buchhalter, *der*/[Bilanz]buchhalterin, *die*

accounting [ə'kaʊntɪŋ] *n.* **a)** *no pl. (Finance)* Buchführung, *die;* **b) there's no ~ for taste[s]** über Geschmack läßt sich [nicht] streiten

ac'count number *n.* Kontonummer, *die*

accrue [ə'kru:] *v.i.* ‹*Zinsen:*› auflaufen; **~ to sb.** ‹*Macht:*› jmdm. zuwachsen; ‹*Reichtümer, Einnahmen:*› jmdm. zufließen

accumulate [ə'kju:mjʊleɪt] **1.** *v.t. (gather)* sammeln; machen, *(fachspr.)* akkumulieren ‹*Vermögen*›; *(along the way)* einsammeln; *(produce)* einbringen ‹*Zinsen, Gewinne*› **(for sb.** jmdm.). **2.** *v.i.* ‹*Menge, Staub:*› sich ansammeln; ‹*Schnee, Geld:*› sich anhäufen

accumulation [əkju:mjʊ'leɪʃn] *n.* [An]sammeln, *das; (being accumulated)* Anhäufung, *die; (growth)* Zuwachs, *der* **(of** an + *Dat.*); *(mass)* Menge, *die*

accumulator [ə'kju:mjʊleɪtə(r)] *n. (Electr.)* Akkumulator, *der;* Akku, *der (ugs.);* Sammler, *der*

accuracy ['ækjʊrəsɪ] *n.* Genauigkeit, *die*

accurate ['ækjʊrət] *adj. (precise)* genau; akkurat *(geh.); (correct)* richtig

accurately ['ækjʊrətlɪ] *adv. (precisely)* genau; *(correctly)* richtig

accursed [ə'kɜ:st] *adj. (ill-fated)* verflucht; verwünscht

accusation [ækju:'zeɪʃn] *n.* Anschuldigung, *die* **(of** gegen); Anklage, *die (Rechtsw.)*

accusative [ə'kju:zətɪv] *(Ling.)* **1.** *adj.* Akkusativ-; akkusativisch; **~ case** Akkusativ, *der.* **2.** *n.* Akkusativ, *der*

accuse [ə'kju:z] *v.t. (charge)* beschuldigen; bezichtigen *(geh.); (indict)* anklagen; **~ sb. of cowardice** jmdm. Feigheit vorwerfen; **~ sb. of doing sth.** *or* **of having done sth.** jmdn. beschuldigen, etw. getan zu haben; **~ sb. of theft/murder** jmdn. wegen Diebstahl[s]/Mord[es] anklagen; **the ~d** der/die Angeklagte/die Angeklagten; **point an accusing finger at sb.** *(lit. or fig.)* anklagend mit dem Finger auf jmdn. zeigen

accustom [ə'kʌstəm] *v.t.* **~ sb./**

sth. to sth. jmdn./etw. an etw. *(Akk.)* gewöhnen; **grow/be ~ed to sth.** sich an etw. *(Akk.)* gewöhnen/an etw. *(Akk.)* gewöhnt sein

accustomed [ə'kʌstəmd] *attrib. adj.* gewohnt; üblich

ace [eɪs] *n. (Cards, Tennis, person)* As, *das;* ~ **of trumps/diamonds** Trumpf-/Karoas, *das;* **play one's** ~ *(fig.)* seinen Trumpf ausspielen; **he was within an ~ of doing it** *(hair's breadth)* er hätte es um ein Haar getan

acetylene [ə'setɪli:n] *n.* Acetylen, *das*

ache [eɪk] **1.** *v.i.* **a)** schmerzen; weh tun; **whereabouts does your leg ~?** wo tut [dir] das Bein weh?; **b)** *(fig.: long)* ~ **to do sth.** darauf brennen, etw. zu tun. **2.** *n.* Schmerz, *der;* **~s and pains** Wehwehchen, *die*

achieve [ə'tʃi:v] *v.t.* zustande bringen; ausführen ‹*Aufgabe, Plan*›; erreichen ‹*Ziel, Standard, Absicht*›; herstellen, herbeiführen ‹*Frieden, Harmonie*›; erzielen ‹*Rekord, Leistung, Erfolg*›; erfüllen ‹*Zweck*›

achievement [ə'tʃi:vmənt] *n.* **a)** *no pl. see* **achieve:** Zustandebringen, *das;* Ausführung, *die;* Erreichen, *das;* Herstellung, *die;* Herbeiführung, *die;* Erzielen, *das;* Erfüllung, *die;* **b)** *(thing accomplished)* Leistung, *die;* Errungenschaft, *die*

acid ['æsɪd] **1.** *adj.* sauer. **2.** *n.* Säure, *die*

'acid drop *n. (Brit.)* saurer *od.* saures Drops

acidic [ə'sɪdɪk] *adj.* säuerlich

acidity [ə'sɪdɪtɪ] *n.* Säure, *die;* Acidität, *die (fachspr.);* Säuregrad, *der; (excessive)* Übersäuerung, *die*

acid: ~ **'rain** *n.* saurer Regen; ~ **test** *n. (fig.)* Feuerprobe, *die*

acknowledge [ək'nɒlɪdʒ] *v.t.* **a)** *(admit)* zugeben, eingestehen ‹*Tatsache, Notwendigkeit, Fehler, Schuld*›; *(accept)* sich bekennen zu ‹*einer Verantwortung, Pflicht, Schuld*›; *(take notice of)* grüßen ‹*Person*›; *(recognize)* anerkennen ‹*Autorität, Recht, Forderung, Notwendigkeit*›; **an ~d expert** ein anerkannter Fachmann; ~ **sb./sth. [as** *or* **to be] sth.** jmdn./etw. als etw. anerkennen; **b)** *(express thanks for)* sich erkenntlich zeigen für ‹*Dienste, Bemühungen, Gastfreundschaft*›; erwidern ‹*Gruß*›; **c)** *(confirm receipt of)* bestätigen ‹*Empfang, Bewerbung*›; ~ **a letter** den Empfang eines Briefes bestätigen

acknowledg[e]ment [ək'nɒlɪdʒmənt] *n.* **a)** *(admission of fact, necessity, error, guilt)* Eingeständnis, *das; (acceptance of a responsibility, duty, debt)* Bekenntnis, *das* **(of** zu); *(recognition of authority, right, claim)* Anerkennung, *die;* **b)** *(thanks, appreciation) (of services, friendship)* Dank, *der* **(of** für); *(of greetings)* Erwiderung, *die;* **c)** *(confirmation of receipt)* Bestätigung [des Empfangs/einer Bewerbung]; **letter of ~:** Bestätigungsschreiben, *das;* **'~s'** „Dank"

acne ['æknɪ] *n. (Med.)* Akne, *die*

acorn ['eɪkɔ:n] *n.* Eichel, *die*

acoustic [ə'ku:stɪk] *adj.* akustisch; ~ **guitar** Konzertgitarre, *die*

acoustics [ə'ku:stɪks] *n. pl.* **a)** *(properties)* Akustik, *die;* akustische Verhältnisse; **b)** *constr. as sing. (science)* Akustik, *die*

acquaint [ə'kweɪnt] *v.t.* ~ **sb./oneself with sth.** jmdn./sich mit etw. vertraut machen; **be ~ed with sb.** mit jmdm. bekannt sein

acquaintance [ə'kweɪntəns] *n.* **a)** *no pl.* Vertrautheit, *die;* ~ **with sb.** Bekanntschaft mit jmdm.; **a passing ~:** eine flüchtige Bekanntschaft; **make the ~ of sb.** jmds. Bekanntschaft machen; **b)** *(person)* Bekannte, *der/die*

acquiesce [ækwɪ'es] *v.i.* einwilligen **(in** in + *Akk.*); *(under pressure)* sich fügen

acquiescence [ækwɪ'esəns] *n., no pl. (acquiescing)* Einwilligung, *die* **(in** in + *Akk.*); *(state)* Ergebenheit, *die; (assent)* Zustimmung, *die*

acquiescent [ækwɪ'esənt] *adj.* fügsam; ergeben

acquire [ə'kwaɪə(r)] *v.t.* **a)** sich *(Dat.)* anschaffen ‹*Gegenstände*›; erwerben ‹*Land, Besitz, Wohlstand, Kenntnisse*›; **b)** *(take on)* annehmen ‹*Tonfall, Farbe, Gewohnheit*›; ~ **a taste for sth.** Geschmack an etw. *(Dat.)* gewinnen; **this wine is an ~d taste** an diesen Wein muß man sich erst gewöhnen

acquisition [ækwɪ'zɪʃn] *n.* **a)** *(of goods, wealth, land)* Erwerb, *der; (of knowledge)* Aneignung, *die;* Erwerb, *der; (of habit)* Annahme, *die;* **b)** *(thing)* Anschaffung, *die*

acquisitive [ə'kwɪzɪtɪv] *adj.* raffsüchtig

acquit [ə'kwɪt] *v.t.,* **-tt-: a)** *(Law)* freisprechen; ~ **sb. of sth.** jmdn. von etw. freisprechen; **b)** ~ **oneself well** seine Sache gut machen

acquittal [ə'kwɪtl] *n. (Law)* Freispruch, *der*

acre ['eɪkə(r)] *n.* Acre, *der;* ≈ Morgen, *der*

acrid ['ækrɪd] *adj.* beißend ‹*Geruch, Dämpfe, Rauch*›; bitter ‹*Geschmack*›

acrimonious [ækrɪ'məʊnɪəs] *adj.* bitter; erbittert ‹*Streit, Diskussion*›

acrimony ['ækrɪmənɪ] *n., no pl.* Bitterkeit, *die; (of argument, discussion)* Erbitterung, *die*

acrobat ['ækrəbæt] *n. (lit. or fig.)* Akrobat, *der*/Akrobatin, *die*

acrobatic [ækrə'bætɪk] *adj. (lit. or fig.)* akrobatisch

acrobatics [ækrə'bætɪks] *n., no pl.* Akrobatik, *die*

acronym ['ækrənɪm] *n.* Akronym, *das;* Initialwort, *das*

across [ə'krɒs] **1.** *adv.* **a)** darüber; *(in crossword puzzle)* waagerecht; *(from here to there)* hinüber; **measure** *or* **be 9 miles ~:** 9 Meilen breit sein; **b)** *(on the other side)* drüben; ~ **there/here** [da] drüben/hier drüben; ~ **from** gegenüber von. **2.** *prep.* **a)** über; **protests ~ Canada** Proteste in ganz Kanada; **b)** *(on the other side of)* auf der anderen Seite (+ *Gen.*); ~ **the ocean/river** jenseits des Meeres/Flusses

act [ækt] **1.** *n.* **a)** *(deed)* Tat, *die; (official action)* Akt, *der;* **an ~ of God** höhere Gewalt; **Acts [of the Apostles]** *(Bibl.) constr. as sing.* Apostelgeschichte, *die;* **b)** *(process)* **be in the ~ of doing sth.** gerade dabei sein, etw. zu tun; **he was caught in the ~ [of stealing]** er wurde [beim Stehlen] auf frischer Tat ertappt; **c)** *(in a play)* Akt, *der;* Aufzug, *der (geh.);* **a one-~ play** ein Einakter; **d)** *(theatre performance)* Akt, *der;* Nummer, *die;* **e)** *(pretence)* Theater, *das;* Schau, *die (ugs.);* **it's all an ~ with her** sie tut nur so; **put on an ~** *(coll.)* eine Schau abziehen *(ugs.);* Theater spielen; **f)** *(decree)* Gesetz, *das;* **Act of Parliament** Parlamentsakte, *die.* **2.** *v.t.* spielen ‹*Stück, Rolle*›. **3.** *v.i.* **a)** *(perform actions)* handeln; reagieren; ~ **[up]on** folgen (+ *Dat.*) ‹*Anweisung, Ratschlag*›; **b)** *(behave)* sich verhalten; *(function)* ~ **as sb.** als jmd. fungieren *od.* tätig sein; ~ **as sth.** als etw. dienen; **c)** *(perform special function)* ‹*Person:*› handeln; ‹*Gerät, Ding:*› funktionieren; ‹*Substanz, Mittel:*› wirken **(on** auf + *Akk.*); **d)** *(perform play etc., lit. or fig.)* spielen; schauspielern *(ugs.)*

~ **'up** *v.i. (coll.)* Theater machen *(ugs.); ‹Auto, Magen:›* Zicken machen *(ugs.)*
acting ['æktɪŋ] **1.** *adj. (temporary)* stellvertretend. **2.** *n., no pl. (Theatre etc.)* die Schauspielerei; **an ~ career** eine Karriere als Schauspieler
action ['ækʃn] *n.* **a)** *(doing sth.)* Handeln, *das;* **a man of ~:** ein Mann der Tat; **take ~:** Schritte *od.* etwas unternehmen; **put a plan into ~:** einen Plan in die Tat umsetzen; **put sth. out of ~:** etw. außer Betrieb setzen; **be/be put out of ~:** außer Betrieb sein/gesetzt werden; **a film full of ~:** ein Film mit viel Handlung; **b)** *(effect)* **the ~ of salt on ice** die Wirkung von Salz auf Eis *(Akk.);* **c)** *(act)* Tat, *die;* **d)** *(Theatre)* Handlung, *die;* Geschehen, *das;* **where the ~ is** *(sl.)* wo was los ist *(ugs.);* **e)** *(legal process)* [Gerichts]verfahren, *das;* **bring an ~ against sb.** eine Klage *od.* ein Verfahren gegen jmdn. anstrengen; **f)** *(fighting)* Gefecht, *das;* Kampf, *der;* **he died in ~:** er ist [im Kampf] gefallen; **g)** *(movement)* Bewegung, *die*
actionable ['ækʃənəbl] *adj.* [gerichtlich] verfolgbar *od.* strafbar
action: ~ **'replay** *n.* Wiederholung [in Zeitlupe]; ~ **stations** *n. pl. (Mil.; also fig.)* Stellung, *die;* ~ **stations!** in die Stellungen!
activate ['æktɪveɪt] *v.t.* **a)** in Gang setzen *‹Vorrichtung, Mechanismus›;* auslösen *‹Mechanismus›;* **b)** *(Chem., Phys.)* aktivieren; **~d carbon** *or* **charcoal** Aktivkohle, *die*
active ['æktɪv] *adj.* aktiv; wirksam *‹Kraft, Mittel›;* praktisch *‹Gebrauch, Versuch, Kenntnisse›;* tätig *‹Vulkan›;* **a very ~ child** ein sehr lebhaftes Kind; **take an ~ interest in sth.** regen Anteil an etw. *(Dat.)* nehmen; **take an ~ part in sth.** sich aktiv an etw. *(Dat.)* beteiligen; **on ~ service** *or (Amer.)* **duty** *(Mil.)* im aktiven Dienst
actively ['æktɪvlɪ] *adv.* aktiv
activist ['æktɪvɪst] *n.* Aktivist, *der*/Aktivistin, *die*
activity [æk'tɪvɪtɪ] *n.* **a)** *no pl.* Aktivität, *die;* **military ~:** militärischer Einsatz; **b)** *(efforts)* aktive Tätigkeit; rege [Mit]arbeit; **c)** *usu. in pl. (action)* Aktivität, *die; (occupation)* Betätigung, *die;* **classroom activities** schulische Tätigkeiten; **outdoor activities** Betätigung an der frischen Luft
actor ['æktə(r)] *n.* Schauspieler, *der*
actress ['æktrɪs] *n.* Schauspielerin, *die*
actual ['æktʃʊəl] *adj.* eigentlich, tatsächlich *‹Lage, Gegebenheiten›;* wirklich *‹Name, Gegenstand›;* konkret *‹Beispiel›;* **in ~ fact** tatsächlich; **no ~ crime was committed** es wurde kein eigentliches Verbrechen begangen
actuality [æktʃʊ'ælɪtɪ] *n.* Wirklichkeit, *die;* Realität, *die;* **in ~:** in Wirklichkeit
actually ['æktʃʊəlɪ] *adv. (in fact)* eigentlich; *(by the way)* übrigens; *(believe it or not)* sogar; **~, I must be going** ich muß jetzt wirklich gehen; **he ~ had the cheek to suggest ...:** er hatte tatsächlich die Unverfrorenheit, vorzuschlagen ...
actuary ['æktʃʊərɪ] *n.* Versicherungsmathematiker, *der;* Aktuar, *der*
actuate ['æktʃʊeɪt] *v.t. (activate)* antreiben *‹Maschine›;* auslösen *‹Mechanismus, Reaktion›*
acumen ['ækjʊmen] *n.* Scharfsinn, *der;* **business ~:** Geschäftssinn, *der*
acupuncture ['ækjʊpʌŋktʃə(r)] *n. (Med.)* Akupunktur, *die*
acute [ə'kju:t] *adj.*, **~r** [ə'kju:tə(r)], **~st** [ə'kju:tɪst] **a)** *(Geom.)* **~ angle** spitzer Winkel; **b)** *(Med.)* akut *‹Krankheit, Stadium›;* **c)** *(critical)* akut *‹Gefahr, Situation, Mangel›;* **d)** *(keen)* fein *‹Geruchssinn›;* heftig *‹Schmerz›*
AD *abbr.* **Anno Domini** n. Chr.
ad [æd] *n. (coll.)* Annonce, *die;* Inserat, *das;* **small ad** Kleinanzeige, *die*
adage ['ædɪdʒ] *n.* Sprichwort, *das*
adamant ['ædəmənt] *adj.* unnachgiebig; **be ~ that ...:** darauf bestehen, daß ...
Adam's apple [ædəmz 'æpl] *n.* Adamsapfel, *der*
adapt [ə'dæpt] **1.** *v.t.* **a)** *(adjust)* anpassen **(to** *Dat.*); variieren *‹Kleidung›;* umstellen *‹Maschine›* **(to** auf + *Akk.*); **be ~ed for doing sth.** darauf eingestellt sein, etw. zu tun; **b)** *(modify)* adaptieren, bearbeiten *‹Text, Theaterstück›.* **2.** *v.i.* **a)** *‹Tier, Auge:›* sich anpassen **(to** an + *Akk.*); **b)** *(to surroundings, circumstances)* sich gewöhnen **(to** an + *Akk.*)
adaptable [ə'dæptəbl] *adj.* anpassungsfähig; vielseitig *‹Maschine›;* **be ~ to** *or* **for sth.** an etw. *(Akk.)* angepaßt werden können
adaptation [ædəp'teɪʃn] *n.* **a)** *no pl.* Anpassung, *die* **(to** an + *Akk.*); *(of machine)* Umstellung, *die* **(to** auf + *Akk.*); **b)** *(version)* Adap[ta]tion, *die; (of play, text)* Bearbeitung, *die*
adapter, adaptor [ə'dæptə(r)] *n.* Adapter, *der*
add [æd] **1.** *v.t.* hinzufügen **(to** *Dat.*); hinzufügen, anfügen *‹weitere Worte›;* beisteuern *‹Ideen, Vorschläge›* **(to** zu); dazusetzen *‹Namen, Zahlen›;* **~ two and two** zwei und zwei zusammenzählen; **~ two numbers together** zwei Zahlen addieren; **~ the flour to the liquid** geben Sie das Mehl in die Flüssigkeit. **2.** *v.i.* **~ to** vergrößern *‹Schwierigkeiten, Einkommen›;* verbessern *‹Ruf›*
~ **'up 1.** *v.i.* **a) these figures ~ up to** *or* **make 30 altogether** diese Zahlen ergeben zusammen[gezählt] 30; **these things ~/it ~s up** *(fig. coll.)* all diese Dinge summieren sich/das summiert sich alles; **~ up to sth.** *(fig.)* auf etw. *(Akk.)* hinauslaufen; **b)** *(make sense)* einen Sinn ergeben. **2.** *v.t.* zusammenzählen
addenda [ə'dendə] *n. pl. (in book etc.)* Addenda *Pl.*
adder ['ædə(r)] *n. (Zool.)* Viper, *die*
addict 1. [ə'dɪkt] *v.t.* **be ~ed** süchtig sein; **become ~ed [to sth.]** [nach etw.] süchtig werden; **be ~ed to alcohol/drugs** alkohol-/drogensüchtig sein. **2.** ['ædɪkt] *n.* Süchtige, *der/die; (fig. coll.)* [begeisterte] Anhänger, *der*/Anhängerin, *die;* **become an ~** *(lit.)* süchtig werden; **a TV ~** *(fig. coll.)* ein Fernsehnarr
addiction [ə'dɪkʃn] *n.* Sucht, *die; (fig. coll.)* Fimmel, *der (ugs.);* **an ~ to sth.** die Sucht nach etw.
addictive [ə'dɪktɪv] *adj.* **be ~:** süchtig machen; *(fig. coll.)* zu einer Sucht werden
'adding machine *n.* Rechenmaschine, *die*
addition [ə'dɪʃn] *n.* **a)** *no pl.* Hinzufügen, *das; (of ingredient)* Dazugeben, *das; (adding up)* Addieren, *das; (process)* Addition, *die;* **in ~:** außerdem; **in ~ to** zusätzlich zu; **b)** *(thing added)* Ergänzung, *die* **(to** zu)
additional [ə'dɪʃənl] *adj.* zusätzlich; **~ details** weitere Einzelheiten
additive ['ædɪtɪv] *n.* Zusatz, *der*
address [ə'dres] **1.** *v.t.* **a) ~ sth. to sb./sth.** etw. an jmdn./etw. richten; **b)** *(mark with ~)* adressieren **(to** an + *Akk.*); mit Anschrift versehen; **c)** *(speak to)* anreden *‹Person›;* sprechen zu *‹Zu-*

hörern〉; ~ **sb. as sth.** jmdn. mit etw. *od.* als etw. anreden; **d)** *(give attention to)* angehen 〈*Problem*〉. **2.** *n.* **a)** *(on letter or envelope)* Adresse, *die;* Anschrift, *die; (place of residence)* Wohnsitz, *der;* **of no fixed ~:** ohne festen Wohnsitz; **b)** *(discourse)* Ansprache, *die;* Rede, *die*

ad'dress-book *n.* Adressenbüchlein, *das*

addressee [ædre'si:] *n.* Adressat, *der*/Adressatin, *die;* Empfänger, *der*/Empfängerin, *die*

ad'dress label *n.* Adressenaufkleber, *der*

adept ['ædept, 'ədept] *adj.* geschickt (**in, at** in + *Dat.*)

adequacy ['ædıkwəsı] *n., no pl.* **a)** Angemessenheit, *die;* Adäquatheit, *die;* **b)** *(sufficiency)* **doubt/confirm the ~ of sth.** bezweifeln/bestätigen, daß etw. ausreichend ist *od.* ausreicht; **c)** *(acceptability)* Annehmbarkeit, *die*

adequate ['ædıkwət] *adj.* **a)** angemessen, adäquat (**to** *Dat.*); *(suitable)* passend; **b)** *(sufficient)* ausreichend; **c)** *(acceptable)* annehmbar

adequately ['ædıkwətlı] *adv.* **a)** *(sufficiently)* ausreichend; **b)** *(suitably)* angemessen 〈*gekleidet, qualifiziert usw.*〉

adhere [əd'hıə(r)] *v.i.* **a)** *(stick)* haften, *(by glue)* kleben (**to** an + *Dat.*); ~ **[to each other]** 〈*zwei Dinge:*〉 zusammenkleben; **b)** *(give support)* ~ **to sth./sb.** an jmdm./einer Sache festhalten; **c)** ~ **to** sich halten an (+ *Akk.*) 〈*Abmachung, Versprechen, Regel*〉

adherence [əd'hıərəns] *n., no pl.* *(to programme, agreement, promise, schedule)* Einhalten, *das* (**to** *Gen.*); *(to decision, tradition, principle)* Festhalten, *das* (**to** an + *Dat.*); *(to rule)* Befolgen, *das* (**to** *Gen.*)

adherent [əd'hıərənt] *n.* Anhänger, *der*/Anhängerin, *die*

adhesion [əd'hi:ʒn] *n., no pl.* *(sticking)* Haften, *das, (by glue)* Kleben, *das* (**to** an + *Dat.*)

adhesive [əd'hi:sıv] **1.** *adj.* klebrig; gummiert 〈*Briefmarke, Umschlag*〉; Klebe〈*band, -schicht*〉; **be ~:** kleben/gummiert sein; ~ **plaster** Heftpflaster, *das.* **2.** *n.* Klebstoff, *der;* Klebemittel, *das*

adjacent [ə'dʒeısənt] *adj.* angrenzend; Neben-; ~ **to** *(position)* neben (+ *Dat.*); *(direction)* neben (+ *Akk.*)

adjective ['ædʒıktıv] *n. (Ling.)* Adjektiv, *das;* Eigenschaftswort, *das*

adjoin [ə'dʒɔın] **1.** *v.t.* grenzen an (+ *Akk.*); **the room ~ing ours** das Zimmer neben unserem. **2.** *v.i.* aneinandergrenzen; nebeneinanderliegen; **in the ~ing room** im Zimmer daneben *od.* nebenan

adjourn [ə'dʒɜ:n] **1.** *v.t.* *(break off)* unterbrechen; *(put off)* aufschieben. **2.** *v.i.* *(suspend proceedings)* sich vertagen; ~ **for lunch/for half an hour** eine Mittagspause/eine halbstündige Pause einlegen

adjournment [ə'dʒɜ:nmənt] *n.* *(suspending) (of court)* Vertagung, *die; (of meeting)* Unterbrechung, *die*

adjudge [ə'dʒʌdʒ] *v.t.* *(pronounce)* ~ **sb./sth. [to be] sth.** jmdn./etw. für etw. erklären

adjudicate [ə'dʒu:dıkeıt] *v.i.* *(in court, tribunal)* als Richter tätig sein; *(in contest)* Preisrichter sein (**at** bei, in + *Dat.*)

adjudication [ədʒu:dı'keıʃn] *n.* **a)** *(judging)* Beurteilung, *die;* **b)** *(decision)* Entscheidung, *die*

adjudicator [ə'dʒu:dıkeıtə(r)] *n.* Schiedsrichter, *der*/Schiedsrichterin, *die; (in contest)* Preisrichter, *der*/Preisrichterin, *die*

adjunct ['ædʒʌŋkt] *n.* Anhängsel, *das*

adjust [ə'dʒʌst] **1.** *v.t.* richtig [an]ordnen 〈*Gegenstände, Gliederung*〉; zurechtrücken 〈*Hut, Krawatte*〉; *(regulate)* regulieren, regeln 〈*Geschwindigkeit, Höhe usw.*〉; [richtig] einstellen 〈*Gerät, Motor, Maschine usw.*〉; *(adapt)* entsprechend ändern 〈*Plan, Bedingungen*〉; angleichen 〈*Gehalt, Lohn, Zinsen*〉; ~ **sth. [to sth.]** etw. [an etw. *(Akk.)*] anpassen *od.* [auf etw. *(Akk.)*] einstellen; **'do not ~ your set'** „Störung". **2.** *v.i.* ~ **[to sth.]** sich [an etw. *(Akk.)*] gewöhnen *od.* anpassen; 〈*Gerät:*〉 sich [auf etw. *(Akk.)*] einstellen lassen

adjustable [ə'dʒʌstəbl] *adj.* einstellbar (**to** auf + *Akk.*); verstellbar, justierbar 〈*Gerät*〉; regulierbar 〈*Temperatur*〉

adjustment [ə'dʒʌstmənt] *n.* *(of layout, plan)* Ordnung, *die; (of things)* Anordnung, *die; (of device, engine, machine)* Einstellung, *die; (to situation, life-style)* Anpassung, *die* (**to** an + *Akk.*); *(of eye)* Adaption, *die;* Gewöhnung, *die*

ad-lib ['ædlıb] **1.** *adj.* Stegreif-, improvisiert 〈*Rede, Vortrag*〉. **2.** *v.i.,* **-bb-** *(coll.)* improvisieren

Adm. *abbr.* **Admiral** Adm.

'adman *n.* Werbe-, Reklamefachmann, *der*

admin ['ædmın] *n. (coll.)* Verwaltung, *die;* **an ~ problem** ein Verwaltungsproblem

administer [æd'mınıstə(r)] *v.t.* **a)** *(manage)* verwalten; führen 〈*Geschäfte, Regierung*〉; **b)** *(give, apply)* spenden 〈*Trost*〉; leisten, gewähren 〈*Hilfe, Unterstützung*〉; austeilen, verabreichen 〈*Schläge, Prügel*〉; verabreichen, geben 〈*Medikamente*〉; spenden, geben 〈*Sakramente*〉; ~ **justice** Recht sprechen; ~ **an oath to sb.** jmdn. vereidigen

administration [ədmını'streıʃn] *n.* **a)** Verwaltung, *die;* **b)** *(of sacraments)* Spenden, *das;* Geben, *das; (of medicine)* Verabreichung, *die; (of aid, relief)* Gewährung, *die;* ~ **of justice** Rechtspflege, *die;* ~ **of an oath** Eidesabnahme, *die;* **c)** *(ministry, government)* Regierung, *die; (Amer.: President's period of office)* Amtszeit, *die*

administrative [əd'mınıstrətıv] *adj.* Verwaltungs-; administrativ 〈*Angelegenheit, Geschick*〉; ~ **work** Verwaltungsarbeit, *die;* **an ~ job** ein Verwaltungsposten

administrator [əd'mınıstreıtə(r)] *n.* Administrator, *der;* Verwalter, *der*

admirable ['ædmərəbl] *adj.* bewundernswert; erstaunlich; *(excellent)* vortrefflich

admirably ['ædmərəblı] *adv.* bewundernswert; erstaunlich; *(excellently)* vortrefflich

admiral ['ædmərəl] *n.* **a)** Admiral, *der;* **b)** *(butterfly)* **red ~:** Admiral, *der*

Admiralty ['ædmərəltı] *n.: britisches Marineministerium*

admiration [ædmə'reıʃn] *n., no pl.* Bewunderung, *die* (**of, for** für)

admire [əd'maıə(r)] *v.t.* bewundern (**for** wegen)

admirer [əd'maıərə(r)] *n.* Bewunderer, *der*/Bewunderin, *die; (suitor)* Verehrer, *der*/Verehrerin, *die*

admissible [əd'mısıbl] *adj.* **a)** akzeptabel 〈*Plan, Vorschlag*〉; erlaubt, zulässig 〈*Abweichung, Schreibung*〉; **b)** *(Law)* zulässig

admission [əd'mıʃn] *n.* **a)** *(entry)* Zutritt, *der;* ~ **to university** Zulassung [zum Studium] an einer Universität; ~ **costs** *or* **is 50p** der Eintritt kostet 50 Pence; **b)** *(charge)* Eintritt, *der;* **c)** *(confession)* Eingeständnis, *das* (**of, to** *Gen.*); **by** *or* **on one's own ~:** nach eigenem Eingeständnis

admission: ~ **fee,** ~ **price** *ns.* Eintrittspreis, *der*

admit [əd'mıt] **1.** *v.t.,* **-tt-: a)** *(let*

in) hinein-/hereinlassen; **persons under the age of 16 not ~ted** kein Zutritt für Jugendliche unter 16 Jahren; **~ sb. to a club** jmdn. in einen Klub aufnehmen; **be ~ted to hospital** ins Krankenhaus eingeliefert werden; **b)** *(accept as valid)* **if we ~ that argument/evidence** wenn wir davon ausgehen, daß dieses Argument zutrifft/daß diese Beweise erlaubt sind; **c)** *(acknowledge)* zugeben; eingestehen; **~ to being drunk** zugeben, betrunken zu sein. **2.** *v. i.*, **-tt-: ~ of sth.** etw. zulassen *od.* erlauben

admittance [əd'mɪtəns] *n.* Zutritt, *der;* **no ~ [except on business]** Zutritt [für Unbefugte] verboten

admittedly [əd'mɪtɪdlɪ] *adv.* zugegeben[ermaßen]

admonish [əd'mɒnɪʃ] *v. t.* ermahnen

admonition [ædmə'nɪʃn] *n.* Ermahnung, *die*

ad nauseam [æd 'nɔ:sɪæm, æd 'nɔ:zɪæm] *adv.* bis zum Überdruß

ado [ə'du:] *n.* **without more** *or* **with no further ~:** ohne weiteres Aufhebens

adolescence [ædə'lesns] *n., no art.* die Zeit des Erwachsenwerdens

adolescent [ædə'lesnt] **1.** *n.* Heranwachsende, *der/die.* **2.** *adj.* heranwachsend ‹*Person*›; pubertär ‹*Benehmen*›

adopt [ə'dɒpt] *v. t.* **a)** adoptieren; aufnehmen ‹*Tier*›; **b)** *(take over)* annehmen, übernehmen ‹*Kultur, Sitte*›; annehmen ‹*Glaube, Religion*›; **c)** *(take up)* übernehmen, sich aneignen ‹*Methode*›; einnehmen ‹*Standpunkt, Haltung*›

adoption [ə'dɒpʃn] *n.* **a)** Adoption, *die;* **b)** *(of culture, custom)* Annahme, *die;* Übernahme, *die; (of belief)* Annahme, *die;* **c)** *(taking up) (of method)* Aneignung, *die;* Übernahme, *die; (of point of view)* Einnahme, *die*

adorable [ə'dɔ:rəbl] *adj.* bezaubernd; hinreißend

adoration [ædə'reɪʃn] *n.* **a)** Verehrung, *die;* **b)** *(worship of gods etc.)* Anbetung, *die*

adore [ə'dɔ:(r)] *v. t.* **a)** innig lieben; **his adoring girlfriend** seine schmachtende Freundin; **b)** *(coll.: like greatly)* **~ sth.** für etwas schwärmen; **~ doing sth.** etw. für sein Leben gern tun *(ugs.)*

adorn [ə'dɔ:n] *v. t.* schmücken; **~ oneself** sich schön machen

adornment [ə'dɔ:nmənt] *n.* **a)** *no pl.* Verschönerung, *die;* **b)** *(ornament)* Verzierung, *die;* **~s** Schmuck, *der*

adrenalin, *(Amer.* **P)** [ə'drenəlɪn] *n. (Physiol., Med.)* Adrenalin, *das*

Adriatic [eɪdrɪ'ætɪk] *pr. n.* **~ [Sea]** Adriatisches Meer; Adria, *die*

adrift [ə'drɪft] *pred. adj.* **be ~:** treiben

adroit [ə'drɔɪt] *adj.* geschickt, gewandt (**at** in + *Dat.*)

adulation [ædjʊ'leɪʃn] *n., no pl. (praise)* Beweihräucherung, *die; (admiration of person)* Vergötterung, *die*

adult ['ædʌlt, ə'dʌlt] **1.** *adj.* erwachsen ‹*Person*›; reif ‹*Verhalten*›; ausgewachsen ‹*Tier, Pflanze*›. **2.** *n.* Erwachsene, *der/die;* **'~s only'** „Nur für Erwachsene"; **~ education** Erwachsenenbildung, *die*

adulterate [ə'dʌltəreɪt] *v. t.* verunreinigen; panschen ‹*Wein, Milch*›

adulterous [ə'dʌltərəs] *adj.* ehebrecherisch

adultery [ə'dʌltərɪ] *n., no pl.* Ehebruch, *der*

adulthood ['ædʌlthʊd, ə'dʌlthʊd] *n., no pl.* Erwachsenenalter, *das*

advance [əd'vɑ:ns] **1.** *v. t.* **a)** *(move forward)* vorrücken lassen; **b)** *(put forward)* vorbringen ‹*Plan, Meinung, These*›; **c)** *(bring forward)* vorverlegen ‹*Termin*›; **d)** *(further)* fördern; *(pay before due date)* vorschießen; **~ sb. a week's pay** jmdm. einen Wochenlohn [als] Vorschuß geben; *(loan)* **the bank ~d me two thousand pounds** die Bank lieh mir zweitausend Pfund. **2.** *v. i.* **a)** *(move forward; also Mil.)* vorrücken; ‹*Prozession:*› sich vorwärts bewegen; **~ towards sb./sth.** ‹*Person:*› auf jmdn./etw. zugehen; **b)** *(fig.: make progress)* Fortschritte machen; vorankommen. **3.** *n.* **a)** *(forward movement)* Vorrücken, *das; (fig.: progress)* Fortschritt, *der;* **b)** *usu. in pl. (personal overture)* Annäherungsversuch, *der;* **c)** *(payment beforehand)* Vorauszahlung, *die; (on salary)* Vorschuß, *der* (**on** auf + *Akk.*); *(loan)* Darlehen, *das;* **d) in ~:** im voraus; **send sb./sth. in ~:** jmdn./etw. vorausschicken

advanced [əd'vɑ:nst] *adj.* fortgeschritten; **be ~ in years** in fortgeschrittenem Alter sein; **~ level** *see* **A level**

advance 'guard *n. (lit. or fig.)* Vorhut, *die*

advancement [əd'vɑ:nsmənt] *n., no pl. (furtherance)* Förderung, *die*

advance: ~ 'notice *n.* **a week's ~ notice** Benachrichtigung eine Woche [im] voraus; **give sb. ~ notice of sth.** jmdn. im voraus von etw. in Kenntnis setzen; **~ 'payment** *n.* Vorauszahlung, *die*

advantage [əd'vɑ:ntɪdʒ] *n.* **a)** *(better position)* Vorteil, *der;* **give sb. an ~ over sb.** für jmdn. einen Vorteil gegenüber jmdm. bedeuten; **gain an ~ over sb.** sich *(Dat.)* einen Vorteil gegenüber jmdm. verschaffen; **have an ~ over sb.** jmdm. gegenüber im Vorteil sein; **take [full/unfair] ~ of sth.** etw. [voll/unfairerweise] ausnutzen; **b)** *(benefit)* Vorteil, *der;* **be to one's ~:** für jmdn. von Vorteil sein; **turn sth. to [one's] ~:** etw. ausnutzen

advantageous [ædvən'teɪdʒəs] *adj.* vorteilhaft ‹*Verfahren, Übereinkunft*›; günstig ‹*Lage*›

advent ['ædvənt] *n., no pl.* **a)** *(of thing)* Beginn, *der;* Anfang, *der;* **b)** *no art.* **A~** *(season)* Advent, *der*

adventure [əd'ventʃə(r)] *n.* Abenteuer, *das*

ad'venture playground *n.* Abenteuerspielplatz, *der*

adventurer [əd'ventʃərə(r)] *n.* Abenteurer, *der*

adventurous [əd'ventʃərəs] *adj.* **a)** *(eager for adventure)* abenteuerlustig; **b)** *(filled with adventures)* abenteuerlich

adverb ['ædvɜ:b] *n. (Ling.)* Adverb, *das;* Umstandswort, *das*

adverbial [əd'vɜ:bɪəl] *adj. (Ling.)* adverbial

adversary ['ædvəsərɪ] *n. (enemy)* Widersacher, *der*/Widersacherin, *die; (opponent)* Kontrahent, *der*/Kontrahentin, *die*

adverse ['ædvɜ:s] *adj.* **a)** *(hostile)* ablehnend (**to** gegenüber); **an ~ response** eine abschlägige Antwort; **b)** *(unfavourable)* ungünstig ‹*Bedingung, Entwicklung*›; nachteilig ‹*Auswirkung*›; **c)** *(contrary)* widrig ‹*Wind, Umstände*›

adversity [əd'vɜ:sɪtɪ] *n.* **a)** *no pl.* Not, *die;* **in ~:** in der Not; in Notzeiten; **b)** *usu. in pl. (misfortune)* Widrigkeit, *die*

advert ['ædvɜ:t] *(Brit. coll.) see* **advertisement**

advertise ['ædvətaɪz] **1.** *v. t.* werben für; *(by small ad)* inserieren; ausschreiben ‹*Stelle*›. **2.** *v. i.* werben; *(in newspaper)* inserieren; annoncieren; **~ for sb./sth.** jmdn./etw. [per Inserat] suchen

advertisement [əd'vɜ:tɪsmənt] *n.* Anzeige, *die;* **TV ~:** Fernsehspot, *der;* **classified ~:** Kleinanzeige, *die*

advertiser ['ædvətaɪzə(r)] *n. (in*

newspaper) Inserent, *der*/Inserentin, *die; (on radio, TV)* Auftraggeber/Auftraggeberin [der Werbesendung]

advertising ['ædvətaɪzɪŋ] *n., no pl., no indef. art.* Werbung, *die;* ~ **agency/campaign** Werbeagentur, *die*/-kampagne, *die*

advice [əd'vaɪs] *n., no pl., no indef. art. (counsel)* Rat, *der;* **on sb.'s ~:** auf jmds. Rat *(Akk.)* hin; **a piece of ~:** ein Rat[schlag]; **if you ask** *or* **want my ~:** wenn du meinen Rat hören willst; **ask sb.'s ~ [on sth.]** jmdn. [wegen etw.] um Rat bitten; **take sb.'s ~:** jmds. Rat *(Dat.)* folgen

advisable [əd'vaɪzəbl] *adj.* ratsam

advise [əd'vaɪz] *v. t.* **a)** *(offer advice to)* beraten; **please ~ me** bitte geben Sie mir einen Rat; **~ sb. to do sth.** jmdm. raten, etw. zu tun; **~ sb. not to do** *or* **against doing sth.** jmdm. abraten, etw. zu tun; **b)** *(recommend)* **~ sth.** zu etw. raten; **c)** *(inform)* unterrichten, informieren **(of** über + *Akk.*)

advised [əd'vaɪzd] *adj.* **[well-]~:** wohl überlegt; **be well/better ~:** ‹*Person:*› wohlberaten/besser beraten sein

advisedly [əd'vaɪzɪdlɪ] *adv.* bewußt

adviser, advisor [əd'vaɪzə(r)] *n.* Berater, *der*/Beraterin, *die*

advisory [əd'vaɪzərɪ] *adj.* beratend; **~ committee** Beratungsausschuß, *der;* **in an ~ capacity** in beratender Funktion

advocate 1. ['ædvəkət] *n. (of a cause)* Befürworter, *der*/Befürworterin, *die;* Fürsprecher, *der*/Fürsprecherin, *die; (for a person)* Fürsprecher, *der*/Fürsprecherin, *die; (Law)* [Rechts]anwalt, *der*/[Rechts]anwältin, *die.* **2.** ['ædvəkeɪt] *v. t.* befürworten

advt. *abbr.* **advertisement**

Aegean [iː'dʒiːən] *pr. n.* ~ **[Sea]** Ägäisches Meer

aegis ['iːdʒɪs] *n.* **under the ~ of sb./sth.** unter der Ägide *(geh.) od.* Schirmherrschaft von jmdm./etw.

aerial ['eərɪəl] **1.** *adj.* **a)** *(in the air)* Luft-; **b)** *(atmospheric)* atmosphärisch; **c)** *(Aeronaut.)* Luft-. **2.** *n.* Antenne, *die*

aero- ['eərə] *in comb.* Aero-

aerobatics [eərə'bætɪks] *n.* **a)** *no pl.* Kunstflug, *der;* Aerobatik, *die;* **b)** *pl. (feats of flying skill)* fliegerische Kunststücke

aerobic [eə'rəʊbɪk] *adj.* aerob

aerobics [eə'rəʊbɪks] *n., no pl.* Aerobic, *das*

aerodrome ['eərədrəʊm] *n. (Brit. dated)* Aerodrom, *das (veralt.);* Flugplatz, *der*

aerody'namic [eərəʊdaɪ'næmɪk] *adj.* aerodynamisch

'aerofoil *n.* Tragfläche, *die;* Tragflügel, *der (fachspr.)*

aero'nautical *adj.* aeronautisch

aeronautics [eərə'nɔːtɪks] *n., no pl.* Aeronautik, *die*

'aeroplane *n. (Brit.)* Flugzeug, *das*

aerosol ['eərəsɒl] *n. (spray)* Spray, *der od. das; (container)* ~ **[spray]** Spraydose, *die*

'aerospace *n., no pl., no art.* Erdatmosphäre und Weltraum; *(technology)* Luft- und Raumfahrt, *die*

aesthetic [iːs'θetɪk] *adj.* ästhetisch; schöngeistig ‹*Person*›

aesthetics [iːs'θetɪks] *n., no pl.* Ästhetik, *die*

AF *abbr.* **audio frequency**

afar [ə'fɑː] *adv.* ~ weit fort; in weiter Ferne; **from ~:** aus der Ferne

affable ['æfəbl] *adj.* freundlich

affair [ə'feə(r)] *n.* **a)** *(concern, matter)* Angelegenheit, *die;* **it's not my ~:** es geht mich nichts an; **that's 'his ~:** das ist seine Sache; **b)** *in pl. (everyday business)* Geschäfte *Pl.;* [tägliche] Arbeit; *(business dealings)* Geschäfte *Pl.;* **state of ~s** Lage, *die;* **c)** *(love ~)* Affäre, *die;* **d)** *(occurrence)* Geschichte, *die (ugs.);* Angelegenheit, *die;* **e)** *(coll.: thing)* Ding, *das*

[1]affect [ə'fekt] *v. t. (pretend to have)* nachahmen; imitieren; *(pretend to feel or do)* vortäuschen; spielen; **the boy ~ed indifference** der Junge tat so, als sei es ihm gleichgültig

[2]affect *v. t.* **a)** *(produce effect on)* sich auswirken auf (+ *Akk.*); **b)** *(emotionally)* betroffen machen; **be ~ed by sth.** von etw. betroffen sein; **c)** ‹*Vorschrift:*› betreffen; ‹*Krankheit:*› infizieren ‹*Person*›, befallen ‹*Pflanze*›

affectation [æfek'teɪʃn] *n.* **a)** *(studied display)* Verstellung, *die; (artificiality)* Affektiertheit, *die;* **b)** *no pl. (pretence)* **~ of sth.** Vortäuschung von etw.

affected [ə'fektɪd] *adj.* affektiert; gekünstelt ‹*Sprache, Stil*›

affection [ə'fekʃn] *n.* Zuneigung, *die;* **have** *or* **feel ~ for sb./sth.** für jmdn. Zuneigung empfinden/an etw. *(Dat.)* hängen

affectionate [ə'fekʃənət] *adj.* anhänglich ‹*Person, Kind, [Haus]tier*›; liebevoll ‹*Umarmung*›; zärtlich ‹*Lächeln, Erinnerung*›

affectionately [ə'fekʃənətlɪ] *adv.* liebevoll; **yours ~:** viele Grüße und Küsse

affidavit [æfɪ'deɪvɪt] *n. (Law)* **[sworn] ~:** eidesstattliche Versicherung; **swear an ~:** eine eidesstattliche Versicherung abgeben

affiliate [ə'fɪlɪeɪt] **1.** *v. t. (attach)* **be ~d to** *or* **with sth.** an etw. *(Akk.)* angegliedert sein. **2.** *n. (person)* assoziiertes Mitglied; *(organization)* Zweigorganisation, *die*

affiliation [əfɪlɪ'eɪʃn] *n.* Angliederung, *die* **(to, with** an + *Akk.*)

affinity [ə'fɪnɪtɪ] *n.* **a)** *(relationship, resemblance)* Verwandtschaft, *die* **(to** mit); **b)** *(liking)* Neigung, *die* **(for** zu); **feel an ~ to** *or* **for sb./sth.** sich zu jmdm./etw. hingezogen fühlen

affirm [ə'fɜːm] *v. t. (assert)* bekräftigen ‹*Absicht*›; beteuern ‹*Unschuld*›; *(state as a fact)* bestätigen; **~ sth. to sb.** jmdm. etw. versichern

affirmation [æfə'meɪʃn] *n. (of intention)* Bekräftigung, *die; (of fact)* Bestätigung, *die*

affirmative [ə'fɜːmətɪv] **1.** *adj.* affirmativ; bestätigend ‹*Erklärung*›; bejahend, zustimmend ‹*Antwort*›. **2.** *n.* **answer in the ~:** bejahend antworten

affix [ə'fɪks] *v. t.* **~ sth. to sth.** etw. an etw. *(Dat.)* befestigen; **~ one's signature [to sth.]** seine Unterschrift [unter etw. *(Akk.)*] setzen

afflict [ə'flɪkt] *v. t. (physically)* plagen; *(mentally)* quälen; peinigen; **be ~ed with sth.** von etw. befallen sein

affliction [ə'flɪkʃn] *n.* **a)** *no pl. (distress)* Bedrängnis, *die;* **endure sorrow and ~:** Kummer und Leid ertragen; **b)** *(cause of distress)* Leiden, *das;* **bodily ~s** körperliche Gebrechen

affluence ['æflʊəns] *n., no pl.* **a)** *(wealth)* Reichtum, *der;* **b)** *(plenty)* Überfluß, *der*

affluent ['æflʊənt] *adj.* reich; **the ~ society** die Überflußgesellschaft

afford [ə'fɔːd] *v. t.* **a)** sich *(Dat.)* leisten; **be able to ~ sth.** sich *(Dat.)* etw. leisten können; **be able to ~:** aufbringen können ‹*Geld*›; erübrigen können ‹*Zeit*›; **b)** *(provide)* bieten; gewähren ‹*Schutz*›; bereiten ‹*Vergnügen*›

affray [ə'freɪ] *n.* Schlägerei, *die*

affront [ə'frʌnt] **1.** *v. t. (insult)* beleidigen; *(offend)* kränken. **2.** *n. (insult)* Affront, *der (geh.)* **(to** gegen); Beleidigung, *die* (**to** *Gen.*); *(offence)* Kränkung, *die* (**to** *Gen.*)

Afghan ['æfgæn] **1.** *adj.* afghanisch. **2.** *n.* **a)** *(person)* Afghane, *der*/Afghanin, *die;* **b)** *(language)* Afghanisch, *das; see also* **English 2a**

Afghan 'hound *n.* Afghane, *der*

Afghanistan [æf'gænıstɑ:n] *pr. n.* Afghanistan *(das)*

afield [ə'fi:ld] *adv.* **far ~** *(direction)* weit hinaus; *(place)* weit draußen; **from as far ~ as** von so weit her wie

afloat [ə'fləʊt] *pred. adj.* **a)** *(floating)* über Wasser; flott ‹*Schiff*›; **get a boat ~:** ein Boot flott machen; **b)** *(at sea)* auf See; **be ~:** auf dem Meer treiben

afoot [ə'fʊt] *pred. adj. (under way)* im Gange; **set ~:** in Gang setzen; aufstellen ‹*Plan*›; **plans were ~ to ...:** es gab Pläne, zu ...

aforementioned [ə'fɔ:menʃnd], **aforesaid** [ə'fɔ:sed] *adjs.* obenerwähnt *od.* -genannt

aforethought [ə'fɔ:θɔ:t] *adj.* **with malice ~:** mit Vorbedacht

afraid [ə'freıd] *pred. adj.* **[not] be ~ [of sb./sth.]** [vor jmdm./etw.] [keine] Angst haben; **be ~ to do sth.** Angst davor haben, etw. zu tun; **be ~ of doing sth.** Angst haben, etw. zu tun; **I'm ~ [that] we must assume that ...:** leider müssen wir annehmen, daß ...; **I'm ~ so/not** ich fürchte ja/nein

afresh [ə'freʃ] *adv.* von neuem

Africa ['æfrıkə] *pr. n.* Afrika *(das)*

African ['æfrıkən] **1.** *adj.* afrikanisch; **sb. is ~:** jmd. ist Afrikaner/Afrikanerin. **2.** *n.* **a)** Afrikaner, *der*/Afrikanerin, *die;* **b)** *(Amer.: Negro)* Neger, *der*/Negerin, *die*

Afrikaans [æfrı'kɑ:ns] *n.* Afrikaans, *das; see also* **English 2a**

Afro ['æfrəʊ] *adj.* Afro-; **~ look** Afro-Look, *der*

aft [ɑ:ft] *adv. (Naut., Aeronaut.)* achtern; **go ~:** nach achtern gehen

after ['ɑ:ftə(r)] **1.** *adv.* **a)** *(later)* danach; **two days ~:** zwei Tage danach *od.* später; **b)** *(behind)* hinterher. **2.** *prep.* **a)** *(following in time, as result of)* nach; **~ six months** nach sechs Monaten; **~ you** nach Ihnen; **time ~ time** wieder und wieder; **day ~ day** Tag für Tag; **b)** *(behind)* hinter (+ *Dat.*); **what are you ~?** was suchst du denn?; *(to questioner)* was willst du wirklich wissen?; **she's only ~ his money** sie ist nur hinter seinem Geld her; **c)** *(about)* **ask ~ sb./sth.** nach jmdm./etw. fragen; **d)** *(next in importance to)* nach; **e)** *(in spite of)* nach; **~ all** schließlich; **so you've come ~ all!** du bist also doch gekommen!

after: ~-care *n., no pl. (after hospital stay)* Nachbehandlung, *die; (after prison sentence)* Resozialisierung, *die;* **~-'dinner speech** *n.* Tischrede, *die;* **~-effect** *n., usu. in pl.* Nachwirkung, *die;* **~life** *n.* Leben nach dem Tod

aftermath ['ɑ:ftəmæθ, 'ɑ:ftəmɑ:θ] *n., no pl.* Nachwirkungen *Pl.*; **the ~ of the war** die Folgen *od.* Auswirkungen des Krieges

afternoon [ɑ:ftə'nu:n] *n.* Nachmittag, *der; attrib.* Nachmittags-; **this/tomorrow ~:** heute/morgen nachmittag; **[early/late] in the ~:** am [frühen/späten] Nachmittag; *(regularly)* [früh/spät] nachmittags; **at three in the ~:** um drei Uhr nachmittags; **on Monday ~s/~:** Montag nachmittags/[am] Montag nachmittag; **one ~:** eines Nachmittags; **~s, of an ~:** nachmittags

afters ['ɑ:ftəz] *n. pl. (Brit. coll.)* Nachtisch, *der*

after: ~-sales service *n.* Kundendienst, *der;* **~shave** *n.* Aftershave, *das;* **~-taste** *n.* Nachgeschmack, *der;* **~thought** *n.* nachträglicher Einfall; **be added as an ~thought** erst später hinzukommen

afterwards ['ɑ:ftəwədz] (*Amer.:* **afterward** ['ɑ:ftəwəd]) *adv.* danach

again [ə'gen, ə'geın] *adv.* **a)** *(another time)* wieder; **see a film ~:** einen Film noch einmal sehen; **not ~!** nicht schon wieder!; **~ and ~, time and [time] ~:** immer wieder; **back ~:** wieder zurück; **go back there ~:** wieder dorthin gehen; **half as much/many ~:** noch einmal halb soviel/so viele; **b)** *(besides)* **[there] ~:** außerdem; **c)** *(on the other hand)* **[then/there] ~:** andrerseits

against [ə'genst, ə'geınst] *prep.* **a)** gegen; **as ~:** gegenüber; **protect sth. ~ frost** etw. vor Frost schützen; **be warned ~ doing sth.** davor gewarnt werden, etw. zu tun; **b)** *(in return for)* gegen; **rate of exchange ~ the dollar** Wechselkurs des Dollar

age [eıdʒ] **1.** *n.* **a)** Alter, *das;* **the boys' ~s are 7, 6, and 3** die Jungen sind 7, 6 und 3 Jahre alt; **what ~ are you?, what is your ~?** wie alt bist du?; **at the ~ of** im Alter von; **at what ~:** in welchem Alter; **be six years of ~:** sechs Jahre alt sein; **when I was your ~:** als ich so alt war wie du; **come of ~:** mündig *od.* volljährig werden; *(fig.)* den Kinderschuhen entwachsen; **be/look under ~:** zu jung sein/aussehen; **be** *or* **act your ~** *(coll.)* sei nicht so kindisch; **b)** *(advanced ~)* Alter, *das;* **c)** *(generation)* Generation, *die;* **d)** *(great period)* Zeitalter, *das;* **wait [for] ~s** *or* **an ~ for sb./sth.** *(coll.)* eine Ewigkeit auf jmdn./etw. warten. **2.** *v.i.* altern

aged 1. *adj.* **a)** [eıdʒd] **be ~ five** fünf Jahre alt sein; **a boy ~ five** ein fünfjähriger Junge; **b)** ['eıdʒıd] *(elderly)* bejahrt. **2.** ['eıdʒıd] *n. pl.* **the ~:** die alten Menschen

age: ~-group *n.* Altersgruppe, *die;* **~ limit** *n.* Altersgrenze, *die*

agency ['eıdʒənsı] *n.* **a)** *(action)* Handeln, *das;* **through/by the ~ of sth.** durch [die Einwirkung von] etw.; **through/by the ~ of sb.** durch jmds. Vermittlung; **b)** *(business establishment)* Geschäftsstelle, *die; (news/advertising ~)* Agentur, *die*

agenda [ə'dʒendə] *n. (lit. or fig.)* Tagesordnung, *die*

agent ['eıdʒənt] *n.* **a)** *(substance)* Mittel, *das;* **an oxidizing ~:** ein Oxidationsmittel; **b)** *(one who acts for another)* Vertreter, *der*/Vertreterin, *die;* **c)** *(spy)* Agent, *der*/Agentin, *die*

'age-old *adj.* uralt

aggravate ['ægrəveıt] *v. t.* **a)** verschlimmern ‹*Krankheit, Zustand, Situation*›; verschärfen ‹*Streit*›; **b)** *(coll.: annoy)* aufregen; ärgern; **be ~d by sth.** sich über etw. *(Akk.)* ärgern *od.* aufregen

aggravating ['ægrəveıtıŋ] *adj. (coll.)* ärgerlich; lästig ‹*Kind, Lärm*›

aggravation [ægrə'veıʃn] *n., no pl.* **a)** Verschlimmerung, *die; (of dispute)* Verschärfung, *die;* **b)** *(coll.: annoyance)* Ärger, *der*

aggregate 1. ['ægrıgət] *n. (sum total)* Gesamtmenge, *die; (assemblage)* Ansammlung, *die.* **2.** ['ægrıgət] *adj. (collected into one)* zusammengefügt; *(collective)* gesamt; **the ~ amount** der Gesamtbetrag. **3.** ['ægrıgeıt] *v. t.* **a)** verbinden ‹*Material, Stoff*› (**into** zu); ansammeln ‹*Reichtum*›; **b)** *(unite)* vereinigen

aggregation [ægrı'geıʃn] *n.* Ansammlung, *die;* Aggregation, *die (bes. fachspr.)*

aggression [ə'greʃn] *n.* **a)** *no pl.* Aggression, *die;* **b)** *(unprovoked attack)* Angriff, *der*

aggressive [ə'gresıv] *adj.* aggressiv; angriffslustig ‹*Kämpfer*›; heftig ‹*Angriff*›

aggressively [ə'gresɪvlɪ] *adv.* aggressiv

aggressiveness [ə'gresɪvnɪs] *n., no pl.* Aggressivität, *die*

aggressor [ə'gresə(r)] *n.* Aggressor, *der*

aggrieved [ə'gri:vd] *(resentful)* verärgert; *(offended)* gekränkt

aggro ['ægrəʊ] *n., no pl. (Brit. sl.)* Zoff, *der (ugs.);* Krawall, *der;* **they are looking for ~:** sie suchen Streit

aghast [ə'gɑ:st] *pred. adj.* bestürzt; erschüttert

agile ['ædʒaɪl] *adj.* beweglich; flink, behend[e] ‹*Bewegung*›

agility [ə'dʒɪlɪtɪ] *n., no pl. see* **agile:** Beweglichkeit, *die;* Flinkheit, *die;* Behendigkeit, *die*

agitate ['ædʒɪteɪt] **1.** *v.t.* **a)** *(shake)* schütteln; *(stir up)* aufrühren; **b)** *(disturb)* beunruhigen; erregen. **2.** *v.i.* **~ for/against sth.** für/gegen etw. agitieren

agitation [ædʒɪ'teɪʃn] *n.* **a)** *(shaking)* Schütteln, *das; (stirring up)* Aufrühren, *das;* **b)** *(emotional disturbance)* Erregung, *die;* **c)** *(campaign)* Agitation, *die*

agitator ['ædʒɪteɪtə(r)] *n. (person)* Agitator, *der*

AGM *abbr.* **Annual General Meeting** JHV

agnostic [æg'nɒstɪk] **1.** *adj.* agnostizistisch. **2.** *n.* Agnostiker, *der*/Agnostikerin, *die*

agnosticism [æg'nɒstɪsɪzm] *n., no pl.* Agnostizismus, *der*

ago [ə'gəʊ] *adv.* **ten years ~:** vor zehn Jahren; **[not] long ~:** vor [nicht] langer Zeit; **how long ~ is it that ...?** wie lange ist es her, daß ...?; **no longer ~ than last Sunday** *(only last Sunday)* erst letzten Sonntag

agog [ə'gɒg] *pred. adj.* gespannt **(for** auf + *Akk.*)

agonize ['ægənaɪz] **1.** *v.i.* *(struggle)* ringen; **~ over sth.** sich *(Dat.)* den Kopf über etw. *(Akk.)* zermartern. **2.** *v.t.* **an ~d scream** ein qualerfüllter Schrei; **an agonizing wait** *(fig.)* eine qualvolle Wartezeit

agony ['ægənɪ] *n.* Todesqualen *Pl.;* **suffer ~/agonies** Todesqualen erleiden; **die in ~:** qualvoll sterben; **in an ~ of indecision** *(fig.)* in qualvoller Unentschlossenheit

agony: ~ aunt *n. (coll.)* Briefkastentante, *die (ugs. scherzh.);* **~ column** *n. (Brit. coll.: advice column) Spalte für die „Briefkastentante"*

agree [ə'gri:] **1.** *v.i.* **a)** *(consent)* einverstanden sein; **~ to** *or* **with sth./to do sth.** mit etw. einverstanden sein/damit einverstanden sein, etw. zu tun; **b)** *(hold similar opinion)* einer Meinung sein; **they ~d [with me]** sie waren derselben Meinung [wie ich]; **~ with sb. about** *or* **on sth./that ...:** jmdm. in etw. *(Dat.)* zustimmen/jmdm. darin zustimmen, daß ...; **I ~:** stimmt; **c)** *(reach similar opinion)* **~ on sth.** sich über etw. *(Akk.)* einigen; **d)** *(harmonize; also Ling.)* übereinstimmen **(mit** with); **e) ~ with sb.** *(suit)* jmdm. bekommen. **2.** *v.t. (reach agreement about)* vereinbaren

agreeable [ə'gri:əbl] *adj.* **a)** *(pleasing)* angenehm; erfreulich ‹*Anblick*›; **b)** *(coll.: willing to agree)* **be ~ [to sth.]** [mit etw.] einverstanden sein

agreeably [ə'gri:əblɪ] *adv.* angenehm

agreed [ə'gri:d] *adj.* einig; vereinbart ‹*Summe, Zeit*›; **be ~ that .../about sth.** sich *(Dat.)* darüber einig sein, daß .../ sich *(Dat.)* über etw. *(Akk.)* einig sein; **~!** einverstanden!

agreement [ə'gri:mənt] *n.* **a)** Übereinstimmung, *die; (mutual understanding)* Übereinkunft, *die;* **be in ~ [about sth.]** sich *(Dat.)* [über etw. *(Akk.)*] einig sein; **enter into an ~:** eine Übereinkunft treffen; **come to** *or* **reach an ~ with sb. [about sth.]** mit jmdm. eine Einigung [über etw. *(Akk.)*] erzielen; **b)** *(treaty)* Abkommen, *das;* **c)** *(Law)* Abkommen, *das;* Vertrag, *der;* **legal ~:** rechtliche Vereinbarung; **d)** *(Ling.)* Übereinstimmung, *die*

agricultural [ægrɪ'kʌltʃərl] *adj.* landwirtschaftlich; **~ worker** Landarbeiter, *der*

agriculture ['ægrɪkʌltʃə(r)] *n.* Landwirtschaft, *die*

aground [ə'graʊnd] *pred. adj.* auf Grund gelaufen; **go** *or* **run ~:** auf Grund laufen

ah [ɑ:] *int.* ach; *(of pleasure)* ah

aha [ɑ:'hɑ:] *int.* aha

ahead [ə'hed] *adv.* **a)** *(further forward in space)* voraus; **~ of sb./sth.** vor jmdm./etw.; **keep going straight ~:** gehen Sie immer geradeaus; **b)** *(fig.)* **be ~ of the others** den anderen voraus sein; **be ~ on points** nach Punkten führen; **get ~:** vorwärts kommen; **c)** *(further forward in time)* **~ of us lay three days of intensive training** vor uns lagen drei Tage intensives Training; **finish ~ of schedule** *or* **time** früher als geplant fertig werden

ahoy [ə'hɔɪ] *int. (Naut.)* ahoi

aid [eɪd] **1.** *v.t.* **a) ~ sb. [to do sth.]** jmdm. helfen[, etw. zu tun]; **~ed by** unterstützt von; *see also* **abet; b)** *(promote)* fördern. **2.** *n.* **a)** *no pl. (help)* Hilfe, *die;* **come/go to the ~ of sb.** jmdm. zu Hilfe kommen; **with the ~ of sth./sb.** mit Hilfe einer Sache *(Gen.)*/mit jmds. Hilfe; mit Hilfe von etw./jmdm.; **in ~ of sb./sth.** zugunsten von jmdm./etw.; **b)** *(source of help)* Hilfsmittel, *das* **(to** für)

aide [eɪd] *n.* **a)** *see* **aide-de-camp; b)** *(assistant)* Berater, *der*/Beraterin, *die*

aide-de-camp [eɪddə'kɑ̃:] *n., pl.* **aides-de-camp** [eɪddə'kɑ̃:] *(Mil.)* Adjutant, *der*

Aids [eɪdz] *n., no pl., no art.* Aids *(das)*

ail [eɪl] *v.t. (trouble)* plagen

ailing ['eɪlɪŋ] *adj. (sickly)* kränkelnd; kränklich

ailment ['eɪlmənt] *n.* Gebrechen, *das;* **minor ~:** leichte Erkrankung

aim [eɪm] **1.** *v.t.* ausrichten ‹*Schußwaffe, Rakete*›; **~ sth. at sb./sth.** etw. auf jmdn./ etw. richten; **that remark was not ~ed at you** *(fig.)* diese Bemerkung war nicht gegen Sie gerichtet; **~ a blow/shot at sb.** nach jmdm. schlagen/auf jmdn. schießen. **2.** *v.i.* **a)** zielen **(at** auf + *Akk.*); **~ high/wide** [zu] hoch/[zu] weit zielen; **~ high** *(fig.)* sich *(Dat.)* ein hohes Ziel stecken *od.* setzen; **b)** *(intend)* **~ to do sth.** *or* **at doing sth.** beabsichtigen, etw. zu tun; **~ at** *or* **for sth.** *(fig.)* etw. anstreben. **3.** *n.* Ziel, *das;* **his ~ was true** er hatte genau gezielt; **take ~ [at sth./sb.]** [auf etw./jmdn.] zielen; **take ~ at the target** das Ziel anvisieren

aimless ['eɪmlɪs] *adj.* ziellos ‹*Leben, Aktivität*›; sinnlos ‹*Vorhaben, Beschäftigung*›

aimlessly ['eɪmlɪslɪ] *adv.* ziellos

ain't [eɪnt] *(sl.)* **a) = am not, is not, are not;** *see* **be; b) = has not, have not;** *see* **have 2**

air [eə(r)] **1.** *n.* **a)** Luft, *die;* **be/go on the ~:** senden; ‹*Programm, Sendung:*› gesendet werden; **be/go off the ~:** nicht/nicht mehr senden; ‹*Programm:*› beendet sein/werden; **be in the ~** *(fig.) (be spreading)* ‹*Gerücht, Idee:*› in der Luft liegen; *(be uncertain)* ‹*Plan, Projekt:*› in der Luft hängen; **by ~:** mit dem Flugzeug; **travel by ~:** fliegen; **send a letter by ~:** einen Brief mit *od.* per Luftpost schicken; **from the ~**: aus der Vogelperspektive; **b)** *(appearance)* **there was an ~ of absurdity about the whole exercise** die ganze Übung hatte etwas Absurdes; **c)** *(bearing)* Auftreten, *das; (facial expression)* Miene, *die;* **~s and graces** Allüren *Pl. (abwertend);* **give oneself** *or* **put on ~s** sich aufspielen; **d)** *(Mus.)* Melodie, *die.* **2.** *v. t.* **a)** *(ventilate)* lüften ⟨*Zimmer, Matratze, Kleidung*⟩; **b)** *(finish drying)* nachtrocknen ⟨*Wäsche*⟩; **c)** *(parade)* zur Schau tragen; **d)** *(make public)* [öffentlich] darlegen. **3.** *v. i. (be ventilated)* lüften

air: ~ base *n. (Air Force)* Luftwaffenstützpunkt, *der;* **~-bed** *n.* Luftmatratze, *die;* **~borne** *adj.* **be ~borne** sich in der Luft befinden; **become ~borne** sich in die Luft erheben; **~ brake** *n.* Druckluftbremse, *die; (flap)* Luftbremse, *die;* **~-bubble** *n.* Luftblase, *die;* **~ bus** *n.* Airbus, *der;* **~-conditioned** *adj.* klimatisiert; **~-conditioner** *n.* Klimaanlage, *die;* **~-conditioning** *n., no pl.* Klimatisierung, *die; (system)* Klimaanlage, *die;* **~-cooled** *adj.* luftgekühlt; **~ corridor** *n. (Aeronaut.)* Luftkorridor, *der;* **~ cover** *n.* Deckung aus der Luft

aircraft ['eəkrɑ:ft] *n., pl. same* Luftfahrzeug, *das; (aeroplane)* Flugzeug, *das*

'aircraft-carrier *n. (Navy)* Flugzeugträger, *der*

air: ~ crew *n.* Besatzung, *die;* Flugpersonal, *das;* **~-cushion** *n.* **a)** Luftkissen, *das;* **b)** **~-cushion vehicle** Luftkissenfahrzeug, *das*

airer ['eərə(r)] *n.* Wäscheständer, *der*

air: ~ fare *n.* Flugpreis, *der;* **~field** *n.* Flugplatz, *der* **~ force** *n.* Luftstreitkräfte *Pl.;* Luftwaffe, *die;* **~gun** *n.* Luftgewehr, *das;* **~ hostess** *n.* Stewardeß, *die*

airily ['eərɪlɪ] *adv.* leichthin

airing ['eərɪŋ] *n.* Auslüften, *das;* **these clothes need a good ~**: diese Kleider müssen gründlich gelüftet werden; **~ cupboard** Trockenschrank, *der*

airless ['eəlɪs] *adj.* stickig ⟨*Zimmer, Büro*⟩; windstill ⟨*Nacht*⟩

air: ~ letter *n.* Luftpostleichtbrief, *der;* Aerogramm, *das;* **~lift 1.** *n.* Luftbrücke, *die* **(of** für); **2.** *v. t.* auf dem Luftweg *od.* über eine Luftbrücke transportieren; **~line** *n.* Fluggesellschaft, *die;* Fluglinie, *die;* **~line pilot** [für eine Fluggesellschaft fliegender] Pilot; **~liner** *n.* Verkehrsflugzeug, *das;* **~lock** *n.* **a)** *(of spacecraft etc.)* Luftschleuse, *die;* **b)** *(stoppage)* Luftblase, *die;* **~ mail** *n.* Luftpost, *die;* **by ~ mail** mit *od.* per Luftpost; **~-mail** *v. t.* mit *od.* per Luftpost befördern; **~man** ['eəmən] *n., pl.* **~men** ['eəmən] Flieger, *der;* **~plane** *(Amer.) see* **aeroplane**; **~ pocket** *n. (Aeronaut.)* Luftloch, *das;* **~ pollution** *n.* Luftverschmutzung, *die;* **~port** *n.* Flughafen, *der;* **~ power** *n.* Schlagkraft der Luftwaffe; **~ pressure** *n.* Luftdruck, *der;* **~ pump** *n.* Luftpumpe, *die;* **~ raid** *n.* Luftangriff, *der;* **~-raid precautions** Luftschutz, *der;* **~-raid shelter** Luftschutzraum, *der;* **~-raid siren** Luftschutzsirene, *die;* **~ rifle** *n.* Luftgewehr, *das;* **~-sea 'rescue** *n.* Seenotrettungseinsatz aus der Luft; **~ship** *n.* Luftschiff, *das;* **~sick** *adj.* luftkrank; **~sickness** *n.* Luftkrankheit, *die;* **~space** *n.* Luftraum, *der;* **~ speed** *n. (Aeronaut.)* Eigengeschwindigkeit, *die;* **~strip** *n.* Start-und-Lande-Bahn, *die;* **~ terminal** *n.* [Air-]Terminal, *der od. das;* **~tight** *adj.* luftdicht; **~-to-~** *adj.* Luft-Luft-; **~-to-~ refuelling** Betanken in der Luft; **~ traffic** *n. (Aeronaut.)* Flugverkehr, *der;* **~-traffic control** Flugsicherung, *die;* **~-traffic controller** Fluglotse, *der;* **~waves** *n. pl.* Äther, *der;* **~way** *n.* **a)** *(Aeronaut.)* Luftstraße, *die;* **b)** *(Anat.)* Luftröhre, *die;* **~worthy** *adj. (Aeronaut.)* lufttüchtig

airy ['eərɪ] *adj.* **a)** luftig ⟨*Büro, Zimmer*⟩; **b)** *(superficial)* vage; *(flippant)* leichtfertig

airy-fairy [eərɪ'feərɪ] *adj. (coll. derog.)* aus der Luft gegriffen ⟨*Plan*⟩; versponnen ⟨*Idee, Vorstellung*⟩

aisle [aɪl] *n.* Gang, *der; (of church)* Seitenschiff, *das;* **walk down the ~ with sb.** *(fig.)* mit jmdm. vor den Traualtar treten

aitch [eɪtʃ] *n.* H, h, *das;* **drop one's ~es** das h [im Anlaut] nicht aussprechen

ajar [ə'dʒɑ:(r)] *pred. adj.* **be** *or* **stand ~**: einen Spaltbreit offenstehen; **leave ~** offenlassen

akin [ə'kɪn] *pred. adj.* **a)** verwandt; **b)** *(fig.)* ähnlich; **be ~ to sth.** einer Sache *(Dat.)* ähnlich sein

alabaster ['æləbɑ:stə] *n.* Alabaster, *der*

alacrity [ə'lækrɪtɪ] *n., no pl.* Eilfertigkeit, *die;* **accept with ~**: mit [großer] Bereitwilligkeit annehmen

alarm [ə'lɑ:m] **1.** *n.* **a)** Alarm, *der;* **give** *or* **raise the ~**: Alarm schlagen; **b)** *(fear)* Angst, *die; (uneasiness)* Besorgnis, *die;* **jump up in ~**: erschreckt aufspringen; **c)** *(mechanism)* Alarmanlage, *die; (of ~ clock)* Weckmechanismus, *der; (signal)* Warnsignal, *das;* **sound the ~**: die Alarmanlage betätigen; **d)** *see* **alarm clock. 2.** *v. t.* **a)** *(make aware of danger)* aufschrecken; **b)** *(cause anxiety to)* beunruhigen

a'larm clock *n.* Wecker, *der*

alarming [ə'lɑ:mɪŋ] *adj.* alarmierend

alarmist [ə'lɑ:mɪst] **1.** *n.* Panikmacher, *der.* **2.** *adj.* ⟨*Reden, Behauptungen*⟩ von Panikmachern

alas [ə'læs, ə'lɑ:s] *int.* ach

Albania [æl'beɪnɪə] *pr. n.* Albanien *(das)*

Albanian [æl'beɪnɪən] **1.** *adj.* albanisch; **sb. is ~**: jmd. ist Albaner/Albanerin. **2.** *n.* **a)** *(person)* Albaner, *der*/Albanerin, *die;* **b)** *(language)* Albanisch, *das*

albatross ['ælbətrɒs] *n.* Albatros, *der*

albeit [ɔ:l'bi:ɪt] *conj. (literary)* wenn auch; obgleich *(geh.)*

albino [æl'bi:nəʊ] *n., pl.* **~s** Albino, *der*

album ['ælbəm] *n.* Album, *das*

alchemist ['ælkəmɪst] *n.* Alchimist, *der;* Alchemist, *der*

alchemy ['ælkəmɪ] *n., no pl. (lit. or fig.)* Alchimie, *die;* Alchemie, *die*

alcohol ['ælkəhɒl] *n.* Alkohol, *der*

alcoholic [ælkə'hɒlɪk] **1.** *adj.* alkoholisch; **~ stupor** Vollrausch, *der.* **2.** *n.* Alkoholiker, *der*/Alkoholikerin, *die*

alcoholism ['ælkəhɒlɪzm] *n., no pl.* Alkoholismus, *der;* Trunksucht, *die*

alcove ['ælkəʊv] *n.* Alkoven, *der*

alder ['ɔ:ldə(r)] *n. (Bot.)* Erle, *die*

alderman ['ɔ:ldəmən] *n., pl.* **aldermen** ['ɔ:ldəmən] Stadtrat, *der;* Alderman, *der*

ale [eɪl] *n.* **a)** Ale, *das;* **b)** *(Hist.)* Bier, *das*

'alehouse *n. (Hist.)* [Bier]schenke, *die*

alert [ə'lɜ:t] **1.** *adj.* **a)** *(watchful)* wachsam; **be ~ for trouble** auf der Hut sein; **be ~ to sth.** mit etw. rechnen; **b)** *(mentally lively)* aufgeweckt. **2.** *n. (state of preparedness)* Alarmbereitschaft, *die;* **air-raid ~**: Fliegeralarm, *der;* **be on**

the ~ **[for/against sth.]** [vor etw. *(Dat.)*] auf der Hut sein. **3.** *v.t.* alarmieren; ~ **sb. [to sth.]** jmdn. [vor etw. *(Dat.)*] warnen

A level ['eɪ levl] *n. (Brit. Sch.)* ≈ Abitur, *das;* Abschluß der Sekundarstufe II; **take one's ~s** ≈ das Abitur machen

alga ['ælgə] *n., pl.* **~e** ['ældʒi:, 'ælgi:] *(Bot.)* Alge, *die*

algebra ['ældʒɪbrə] *n. (Math.)* Algebra, *die*

Algeria [æl'dʒɪərɪə] *pr. n.* Algerien *(das)*

Algerian [æl'dʒɪərɪən] **1.** *adj.* algerisch; **sb. is ~:** jmd. ist Algerier/Algerierin. **2.** *n.* Algerier, *der*/Algerierin, *die*

algorithm ['ælgərɪðm] *n. (Math., Computing)* Algorithmus, *der*

alias ['eɪlɪəs] **1.** *adv.* alias. **2.** *n.* angenommener Name; *(of criminal)* falscher Name

alibi ['ælɪbaɪ] *n.* Alibi, *das*

alien ['eɪlɪən] **1.** *adj.* **a)** *(strange)* fremd; **be ~ to sb.** jmdm. fremd sein; **b)** *(foreign)* ausländisch; *(from another world)* außerirdisch. **2.** *n.* **a)** *(Admin.: foreigner)* Ausländer, *der*/Ausländerin, *die;* **b)** *(a being from another world)* Außerirdische, *der/die*

alienate ['eɪlɪəneɪt] *v.t.* befremden ‹*Person*›; **feel ~d from society** sich der Gesellschaft entfremdet fühlen

alienation [eɪlɪə'neɪʃn] *n., no pl.* Entfremdung, *die*

[1]**alight** [ə'laɪt] *v.i.* **a)** aussteigen; **~ from a vehicle** aus einem Fahrzeug aussteigen; **~ from a horse** von einem Pferd absitzen; **b)** ‹*Vogel:*› sich niedersetzen

[2]**alight** *pred. adj. (on fire)* **be/catch ~:** brennen; **set sth. ~:** etw. in Brand setzen

align [ə'laɪn] *v.t.* **a)** *(place in a line)* ausrichten; **the posts must be ~ed** die Pfosten müssen in einer Linie ausgerichtet werden; **b)** *(bring into line)* in eine Linie bringen; **~ the wheels** *(Motor Veh.)* die Spur einstellen

alignment [ə'laɪnmənt] *n.* Ausrichtung, *die;* **in/out of ~:** [genau] ausgerichtet/nicht richtig ausgerichtet

alike [ə'laɪk] **1.** *pred. adj.* ähnlich; *(indistinguishable)* [völlig] gleich. **2.** *adv.* gleich; in gleicher Weise; **this concerns us all ~:** es geht uns alle gleichermaßen an

alimentary [ælɪ'mentərɪ] *adj.* Nahrungs-; **~ canal/organ** Verdauungskanal, *der*/-organ, *das*

alimony ['ælɪmənɪ] *n.* Unterhaltszahlung, *die*

alive [ə'laɪv] *pred. adj.* **a)** lebendig; lebend; **stay ~:** am Leben bleiben; **keep one's hopes ~:** nicht die Hoffnung verlieren; **keep sb.'s hopes ~:** jmdn. noch hoffen lassen; **come ~:** wieder aufleben; **b)** *(aware)* **be ~ to sth.** sich *(Dat.)* einer Sache *(Gen.)* bewußt sein; **c)** *(brisk)* rege; munter; **be ~ and kicking** gesund und munter sein; **look ~!** ein bißchen munter!; **d)** *(swarming)* **be ~ with sth.** von etw. wimmeln

alkali ['ælkəlaɪ] *n., pl.* **~s** *or* **~es** *(Chem.)* Alkali, *das*

alkaline ['ælkəlaɪn] *adj. (Chem.)* alkalisch

all [ɔ:l] **1.** *attrib. adj.* **a)** *(entire extent or quantity of)* ganz; **~ day** den ganzen Tag; **for ~ that** trotz allem; **~ his life** sein ganzes Leben; **~ my money** all mein Geld; mein ganzes Geld; **stop ~ this noise/shouting!** hör mit dem Krach/Geschrei auf!; **thank you for ~ your hard work** danke für all deine Anstrengungen; **get away from it ~:** einmal von allem abschalten; **that says it ~:** das sagt alles; **b)** *(entire number of)* alle; **~ the books** alle Bücher; **~ my books** all[e] meine Bücher; **where are ~ the glasses?** wo sind all die Gläser?; **~ ten men** alle zehn Männer; **we ~ went to bed** wir gingen alle schlafen; **~ the others** alle anderen; **~ Goethe's works** sämtliche Werke Goethes; **why he of ~ people?** warum ausgerechnet er?; **people of ~ ages** Menschen jeden Alters; **All Fools' Day** der 1. April; **c)** *(any whatever)* jeglicher/jegliche/jegliches; **d)** *(greatest possible)* **in ~ innocence** in aller Unschuld; **with ~ speed** so schnell wie möglich. **2.** *n.* **a)** *(~ persons)* alle; **~ present** alle Anwesenden; **one and ~:** [alle] ohne Ausnahme; **~ and sundry** Krethi und Plethi; **~ of us** wir alle; **the happiest/most beautiful of ~:** der/die Glücklichste/die Schönste unter allen; **most of ~:** am meisten; **he ran fastest of ~:** er lief am schnellsten; **b)** *(every bit)* **~ of it/the money** alles/das ganze *od.* alles Geld; **c)** **~ of** *(coll.: as much as)* **be ~ of seven feet tall** gut sieben Fuß groß sein; **d)** *(~ things)* alles; **~ I need is the money** ich brauche nur das Geld; **~ is not lost** es ist nicht alles verloren; **it's ~ or nothing** es geht ums Ganze; **most of ~:** am meisten; **give one's ~:** sein Letztes geben; **it was ~ but impossible** es war fast unmöglich; **~ in ~:** alles in allem; **it's ~ the same** *or* **~ one to me** es ist mir ganz egal *od.* völlig gleichgültig; **that's ~ very well** das ist alles schön und gut; **can I help you at ~?** kann ich Ihnen irgendwie behilflich sein?; **I do not care at ~:** es ist mir völlig gleich; **you are not disturbing me at ~:** du störst mich nicht im geringsten; **were you surprised at ~?** warst du denn überrascht?; **nothing at ~:** gar nichts; **not at ~ happy/well** überhaupt nicht glücklich/gesund; **not at ~!** überhaupt nicht!; *(acknowledging thanks)* gern geschehen!; nichts zu danken!; **if at ~:** wenn überhaupt; **in ~:** insgesamt; **e)** *(Sport)* **two [goals] ~:** zwei zu zwei; *(Tennis)* **thirty ~:** dreißig beide. **3.** *adv.* ganz; **~ but** fast; **he ~ but fell down** er wäre fast heruntergefallen; **~ the better/worse [for that]** um so besser/schlimmer; **I feel ~ the better for it** das hat mir wirklich gutgetan; **~ at once** *(suddenly)* plötzlich; *(simultaneously)* alle[s] zugleich; **~ too soon** allzu schnell; **be ~ 'for sth.** *(coll.)* sehr für etw. sein; **be ~ 'in** *(exhausted)* total *od.* völlig erledigt sein *(ugs.);* **go ~ out [to do sth.]** alles daransetzen[, etw. zu tun]; **be ~ ready [to go]** *(coll.)* fertig [zum Weggehen] sein *(ugs.);* **sth. is ~ right** etw. ist in Ordnung; *(tolerable)* etw. ist ganz gut; **did you get home ~ right?** sind Sie gut nach Hause gekommen?; **I'm ~ right** mir geht es ganz gut; **work out ~ right** gutgehen; klappen *(ugs.);* **that's her, ~ right** das ist sie, ganz recht; **yes, ~ right** ja, gut; **is it ~ right if I go in?** kann ich reingehen?; **it's ~ right by** *or* **with me** das ist mir recht; **lie ~ round the room** überall im Zimmer herumliegen; **I don't think he's ~ there** *(coll.)* ich glaube, er ist nicht ganz da *(ugs.);* **~ the same** trotzdem; **it's ~ the same to me** es ist mir einerlei

Allah ['ælə] *pr. n.* Allah *(der)*

allay [ə'leɪ] *v.t.* **a)** vermindern; zerstreuen ‹*Besorgnis, Befürchtungen*›; **b)** *(alleviate)* stillen ‹*Hunger, Durst*›; lindern ‹*Schmerz*›

all: ~-'clear *n.* Entwarnung, *die;* **sound the ~-clear** entwarnen; **~-day** *adj.* ganztägig

allegation [ælɪ'geɪʃn] *n.* Behauptung, *die;* **make ~s against sb.** Beschuldigungen gegen jmdn. erheben; **reject all ~s of corruption** jeglichen Vorwurf der Korruption zurückweisen

allege [ə'ledʒ] *v.t.* **~ that ...:** be-

haupten, daß ...; ~ **criminal negligence** den Vorwurf grober Fahrlässigkeit erheben
alleged [ə'ledʒd] *adj.*, **allegedly** [ə'ledʒɪdlɪ] *adv.* angeblich
allegiance [ə'li:dʒəns] *n.* Loyalität, *die* (**to** gegenüber)
allegorical [ælɪ'gɒrɪkl] *adj.* allegorisch
allegory ['ælɪgərɪ] *n.* Allegorie, *die*
'all-embracing *adj.* alles umfassend
allergic [ə'lɜ:dʒɪk] *adj.* *(Med.)* allergisch (**to** gegen)
allergy ['ælədʒɪ] *n.* *(Med.)* Allergie, *die* (**to** gegen)
alleviate [ə'li:vɪeɪt] *v. t.* abschwächen
alley ['ælɪ] *n.* [schmale] Gasse; **be up sb.'s** ~ *(coll.)* jmds. Fall sein *(ugs.)*
alliance [ə'laɪəns] *n.* Bündnis, *das; (league)* Allianz, *die;* **in** ~ **with sb./sth.** im Verein mit jmdm./etw.
allied ['ælaɪd] *adj.* **be** ~ **to** *or* **with sb./sth.** mit jmdm./etw. verbündet sein; **the A~ Powers** die Alliierten
alligator ['ælɪgeɪtə(r)] *n.* Alligator, *der*
all: ~-important *adj.* entscheidend; **~-in** *adj.* Pauschal-; **it costs £350 ~-in** es kostet 350 Pfund alles inklusive; **~-in wrestling** Freistilringen, *das*
alliteration [əlɪtə'reɪʃn] *n.* Stabreim, *der;* Alliteration, *die*
'all-night *adj.* die ganze Nacht dauernd ‹*Sitzung*›; nachts durchgehend geöffnet ‹*Gaststätte*›
allocate ['æləkeɪt] *v. t.* ~ **sth. to sb./sth.** jmdm./einer Sache etw. zuweisen *od.* zuteilen
allocation [ælə'keɪʃn] *n.* Verteilung, *die; (ration)* Zuteilung, *die*
allot [ə'lɒt] *v. t.*, **-tt-**: ~ **sth. to sb.** jmdm. etw. zuteilen; **we ~ted two hours to the task** wir haben zwei Stunden für diese Arbeit vorgesehen
allotment [ə'lɒtmənt] *n.* **a)** Zuteilung, *die;* **b)** *(Brit.: plot of land)* ≈ Schrebergarten, *der*
'all-out *attrib. adj.* mit allen [verfügbaren] Mitteln *nachgestellt*
allow [ə'laʊ] **1.** *v. t.* **a)** *(permit)* ~ **sth.** etw. erlauben *od.* zulassen *od.* gestatten; ~ **sb. to do sth.** jmdm. erlauben, etw. zu tun; **be ~ed to do sth.** etw. tun dürfen; **sb. is ~ed sth.** jmdm. ist etw. erlaubt; ~ **sb. in/out/past/through** jmdn. hinein- / hinaus- / vorbei- / durchlassen; ~ **sth. to happen** zulassen, daß etw. geschieht; ~ **sb. a discount** jmdm. Rabatt geben; **b)** *(Law)* bestätigen ‹*Anspruch*›; ~ **the appeal** der Berufung *(Dat.)* stattgeben; **c)** *(Sport)* **the referee ~ed the goal** der Schiedsrichter gab das Tor. **2.** *v. i.* ~ **of sth.** etw. zulassen *od.* erlauben; ~ **for sth.** etw. berücksichtigen
allowable [ə'laʊəbl] *adj.* zulässig
allowance [ə'laʊəns] *n.* **a)** Zuteilung, *die; (money for special expenses)* Zuschuß, *der;* **your luggage** ~ **is 44 kg.** Sie haben 44 kg Freigepäck; **tax ~:** Steuerfreibetrag, *der;* **b) make ~s for sth./sb.** etw./jmdn. berücksichtigen
alloy ['ælɔɪ] *n.* Legierung, *die*
all: ~-'powerful *adj.* allmächtig; **~-purpose** *adj.* Universal-; Allzweck-; **~-round** *adj.* Allround-; **~-'rounder** *n.* Allroundtalent, *das; (Sport)* Allroundspieler, *der*/-spielerin, *die;* **~-time** *adj.* **~-time record** absoluter Rekord; **~-time favourites** *or* **greats** unvergessene Publikumslieblinge; **hit** *or* **reach an ~-time high/low** eine Rekordhöhe/Rekordtiefe erreichen
allude [ə'lju:d, ə'lu:d] *v. i.* ~ **to** sich beziehen auf (+ *Akk.*); *(covertly)* anspielen auf (+ *Akk.*)
allure [ə'ljʊə(r)] **1.** *v. t.* locken; *(fascinate)* faszinieren. **2.** *n., no pl.* Verlockung, *die; (personal charm)* Charme, *der*
alluring [ə'ljʊərɪŋ] *adj.* verlockend; **an** ~ **appeal** eine Verlockung
allusion [ə'lju:ʒn, ə'lu:ʒn] *n.* **a)** Hinweis, *der;* **in an** ~ **to** unter Bezugnahme auf (+ *Akk.*); **b)** *(covert reference)* Anspielung, *die* (**to** auf + *Akk.*)
alluvial [ə'lu:vɪəl] *adj.* *(Geol.)* angeschwemmt
'all-weather *attrib. adj.* Allwetter-
ally 1. [ə'laɪ, 'ælaɪ] *v. t.* ~ **oneself with sb./sth.** sich mit jmdm./etw. verbünden; *see also* **allied. 2.** ['ælaɪ] *n.* Verbündete, *der/die;* **the Allies** die Alliierten
almanac ['ɔ:lmənæk, 'ɒlmənæk] *n.* Almanach, *der*
almighty [ɔ:l'maɪtɪ] **1.** *adj.* **a)** allmächtig; **the A~:** der Allmächtige; **b)** *(sl.: very great, hard, etc.)* mächtig. **2.** *adv.* *(sl.)* mächtig
almond ['ɑ:mənd] *n.* Mandel, *die*
almost ['ɔ:lməʊst] *adv.* fast; beinahe; **she** ~ **fell** sie wäre fast gefallen
alms [ɑ:mz] *n., no pl.* Almosen, *das*
alone [ə'ləʊn] **1.** *pred. adj.* allein; alleine *(ugs.);* **he was not** ~ **in the belief that ...:** er stand nicht allein mit der Überzeugung, daß ... **2.** *adv.* allein; **this fact ~:** schon allein dies
along [ə'lɒŋ] **1.** *prep.* **a)** *(position)* entlang (+ *Dat.*); ~ **one side of the street** auf der einen Straßenseite; **all** ~ **the wall** die ganze *od.* an der ganzen Mauer entlang; **b)** *(direction)* entlang (+ *Akk.*); **walk** ~ **the river-bank/street** am Ufer *od.* das Ufer/die Straße entlanglaufen. **2.** *adv.* **a)** *(onward)* weiter; **he came running ~:** er kam herbei- *od.* angelaufen; **b)** *(with one)* **bring/take sb./sth. ~:** jmdn./etw. mitbringen/mitnehmen; **c)** *(there)* **I'll be** ~ **shortly** ich komme gleich; **d) all ~:** die ganze Zeit [über]
alongside [əlɒŋ'saɪd] **1.** *adv.* daneben; ~ **of** *see* **2. 2.** *prep.* *(position)* neben (+ *Dat.*); *(direction)* neben (+ *Akk.*); *(fig.)* neben (+ *Dat.*); **work** ~ **sb.** mit jmdm. zusammen arbeiten/*(fig.)* zusammenarbeiten
aloof [ə'lu:f] **1.** *adv.* abseits; **hold** ~ **from sb.** sich von jmdm. fernhalten; **keep ~:** Distanz wahren. **2.** *adj.* distanziert; reserviert
aloud [ə'laʊd] *adv.* laut; **read [sth.] ~:** [etw.] vorlesen
alpha ['ælfə] *n.* *(letter)* Alpha, *das*
alphabet ['ælfəbet] *n.* Alphabet, *das;* Abc, *das*
alphabetical [ælfə'betɪkl] *adj.* alphabetisch; **in** ~ **order** in alphabetischer Reihenfolge
alpine ['ælpaɪn] *adj.* **a)** alpin; Hochgebirgs-; ~ **climate/vegetation** Hochgebirgsklima, *das*/alpine Vegetation; ~ **flowers** Alpen-, Gebirgsblumen
Alps [ælps] *pr. n. pl.* **the ~:** die Alpen
already [ɔ:l'redɪ] *adv.* schon; **it's** ~ **8 o'clock** *or* **8 o'clock ~:** es ist schon 8 Uhr
Alsace [æl'sæs] *pr. n.* Elsaß, *das*
Alsatian [æl'seɪʃn] *n.* [deutscher] Schäferhund
also ['ɔ:lsəʊ] *adv.* auch; *(moreover)* außerdem
altar ['ɔ:ltə(r), 'ɒltə(r)] *n.* **a)** *(Communion table)* Altar, *der;* **lead sb. to the** ~ *(fig.)* jmdn. zum Traualtar führen; **b)** *(for sacrifice)* Opferstätte, *die;* Opfertisch, *der*
alter ['ɔ:ltə(r), 'ɒltə(r)] **1.** *v. t.* ändern; verändern ‹*Stadt, Wohnung*›. **2.** *v. i.* sich verändern; **he has ~ed a lot** *(in appearance)* er hat sich stark verändert; *(in character)* er hat sich sehr geändert
alteration [ɔ:ltə'reɪʃn, ɒltə'reɪʃn]

n. Änderung, *die; (of text)* Abänderung, *die; (of house)* Umbau, *der*

altercation [ɔ:ltə'keɪʃn, ɒltə'keɪʃn] *n.* Auseinandersetzung, *die;* Streiterei, *die*

alternate 1. [ɔ:l'tɜ:nət, ɒl'tɜ:nət] *adj.* **a)** *(in turn)* sich abwechselnd; **John and Mary come on ~ days** John und Mary kommen abwechselnd einen um den anderen Tag; *(together)* John und Mary kommen jeden zweiten Tag; **b)** *see* **alternative 1. 2.** ['ɔ:ltəneɪt, 'ɒltəneɪt] *v. t.* abwechseln lassen; **she has only two summer dresses, so she ~s them** sie hat nur zwei Sommerkleider, deshalb trägt sie sie abwechselnd; **he ~s his days off and** *or* **with his working days** er hat abwechselnd einen Tag frei und geht einen Tag zur Arbeit. **3.** *v. i.* sich abwechseln; alternieren *(fachspr.)*

alternately [ɔ:l'tɜ:nətlɪ, ɒl'tɜ:nətlɪ] *adv.* abwechselnd

'alternating current *n. (Electr.)* Wechselstrom, *der*

alternative [ɔ:l'tɜ:nətɪv, ɒl'tɜ:nətɪv] **1.** *adj.* alternativ; Alternativ-; **~ possibility** Ausweich- *od.* Alternativmöglichkeit, *die;* **~ suggestion** Alternativ- *od.* Gegenvorschlag, *der;* **~ route** Alternativstrecke, *die; (to avoid obstruction etc.)* Ausweichstrecke, *die;* **the ~ society** die alternative Gesellschaft; **~ medicine** Alternativmedizin, *die.* **2.** *n.* **a)** *(choice)* Alternative, *die;* Wahl, *die;* **if I had the ~:** wenn ich vor die Wahl *od.* Alternative gestellt würde; **we have no ~ [but to ...]** wir haben keine andere Wahl[, als zu ...]; **b)** *(possibility)* Möglichkeit, *die;* **there is no [other] ~:** es gibt keine Alternative *od.* andere Möglichkeit; **what are the ~s?** welche Alternativen gibt es?

alternatively [ɔ:l'tɜ:nətɪvlɪ, ɒl'tɜ:nətɪvlɪ] *adv.* oder aber; **or ~:** oder aber auch

alternator ['ɔ:ltəneɪtə(r), 'ɒltəneɪtə(r)] *n. (Electr.)* Wechselstromgenerator, *der*

although [ɔ:l'ðəʊ] *conj.* obwohl

altitude ['æltɪtju:d] *n.* Höhe, *die;* **what is our ~?** wie hoch sind wir?; **from this ~:** aus dieser Höhe; **at an ~ of 2,000 ft.** ≈ in einer Höhe von 600 Metern; **at high ~:** in großer Höhe

alto ['æltəʊ] *n., pl.* **~s** *(Mus.) (voice, part)* Alt, *der; (male singer)* Alt, *der;* Altist, *der;* Altsänger, *der; (female singer)* Alt, *der;* Altistin, *die;* Altsängerin, *die*

altogether [ɔ:ltə'geðə(r)] **1.** *adv.* völlig; *(on the whole)* im großen und ganzen; *(in total)* insgesamt; **not ~ [true/convincing]** nicht ganz [wahr/überzeugend]. **2.** *n.* **in the ~** *(coll.)* im Evas-/Adamskostüm

altruism ['æltrʊɪzm] *n., no pl.* Altruismus, *der;* Uneigennützigkeit, *die*

altruistic [æltrʊ'ɪstɪk] *adj.* altruistisch; uneigennützig

aluminium [æljʊ'mɪnɪəm] *(Brit.),* **aluminum** [ə'lu:mɪnəm] *(Amer.) ns.* Aluminium, *das*

always ['ɔ:lweɪz, 'ɔ:lwɪz] *adv. (at all times)* immer; *(repeatedly)* ständig; [an]dauernd *(ugs.); (whatever the circumstances)* jederzeit; **you can ~ come by train if you prefer** ihr könnt ja auch mit der Bahn kommen, wenn euch das lieber ist

AM *abbr.* **amplitude modulation** AM

am *see* **be**

a.m. [eɪ'em] *adv.* vormittags; **[at] one/four ~:** [um] ein/vier Uhr nachts *od.* morgens *od.* früh; **[at] five/eight ~:** [um] fünf/acht Uhr morgens *od.* früh; **[at] nine ~:** [um] neun Uhr morgens *od.* früh *od.* vormittags; **[at] ten/eleven ~:** [um] zehn/elf Uhr vormittags

amalgam [ə'mælgəm] *n.* **a)** *(lit. or fig.: mixture)* Mischung, *die;* **b)** *(alloy)* Amalgam, *das*

amalgamate [ə'mælgəmeɪt] **1.** *v. t.* vereinigen; fusionieren ‹*Firmen*›. **2.** *v. i.* sich vereinigen; ‹*Firmen:*› fusionieren

amalgamation [əmælgə'meɪʃn] *n.* **a)** *(action)* Vereinigung, *die; (of firms)* Fusion, *die;* **b)** *(result)* Vereinigung, *die*

amass [ə'mæs] *v. t.* [ein]sammeln; **~ a [large] fortune** ein [großes] Vermögen anhäufen

amateur ['æmətə(r)] *n.* **a)** *(nonprofessional)* Amateur, *der;* **b)** *(derog.: trifler)* Amateur, *der;* Dilettant, *der;* **c)** *attrib.* Amateur-; Laien-

amateurish ['æmətərɪʃ] *adj. (derog.)* laienhaft; amateurhaft

amaze [ə'meɪz] *v. t.* verblüffen; verwundern; **be ~d [by sth.]** [über etw. *(Akk.)*] verblüfft *od.* verwundert sein

amazement [ə'meɪzmənt] *n., no pl.* Verblüffung, *die;* Verwunderung, *die*

amazing [ə'meɪzɪŋ] *adj. (remarkable)* erstaunlich; *(astonishing)* verblüffend

amazingly [ə'meɪzɪŋlɪ] *adv.* **a)** *as sentence-modifier (remarkably)* erstaunlicherweise; *(astonishingly)* verblüffenderweise; **b)** erstaunlich

¹Amazon ['æməzən] *pr. n.* **the ~:** der Amazonas

²Amazon *n.* **a)** *(Mythol.: female warrior)* Amazone, *die;* **b)** *(fig.)* Mannweib, *das (abwertend);* Amazone, *die (veralt.)*

ambassador [æm'bæsədə(r)] *n.* Botschafter, *der*/Botschafterin, *die;* **~ to a country/court** Botschafter in einem Land/an einem Hof

amber ['æmbə(r)] **1.** *n.* **a)** Bernstein, *der;* **b)** *(traffic light)* Gelb, *das.* **2.** *adj.* Bernstein-; aus Bernstein *nachgestellt; (colour)* bernsteinfarben; gelb ‹*Verkehrslicht*›

ambidextrous [æmbɪ'dekstrəs] *adj.* beidhändig; ambidexter *(fachspr.)*

ambience ['æmbɪəns] *n.* Ambiente, *das (geh.);* Milieu, *das*

ambient ['æmbɪənt] *adj.* Umgebungs-

ambiguity [æmbɪ'gju:ɪtɪ] *n.* Zweideutigkeit, *die; (having several meanings)* Mehrdeutigkeit, *die*

ambiguous [æm'bɪgjʊəs] *adj.*, **ambiguously** [æm'bɪgjʊəslɪ] *adv.* zweideutig; *(with several meanings)* mehrdeutig

ambiguousness [æm'bɪgjʊəsnɪs] *n., no pl.* Zweideutigkeit, *die; (having several meanings)* Mehrdeutigkeit, *die*

ambition [æm'bɪʃn] *n.* Ehrgeiz, *der; (aspiration)* Ambition, *die*

ambitious [æm'bɪʃəs] *adj.* ehrgeizig; ambitioniert *(geh.)* ‹*Person*›

ambitiously [æm'bɪʃəslɪ] *adv.* voller Ehrgeiz; von Ehrgeiz erfüllt

ambivalent [æm'bɪvələnt] *adj.* ambivalent

amble ['æmbl] **1.** *v. i.* schlendern; gemütlich gehen. **2.** *n.* Schlendern, *das*

ambulance ['æmbjʊləns] *n.* Krankenwagen, *der*

ambulance: ~-man *n.* Sanitäter, *der;* **~ service** *n.* Rettungsdienst, *der;* **~ worker** *n.* Sanitäter, *der*/Sanitäterin, *die*

ambush ['æmbʊʃ] **1.** *n. (concealment)* Hinterhalt, *der; (troops concealed)* im Hinterhalt liegende Truppe; **lie in ~** *(lit. or fig.)* im Hinterhalt liegen. **2.** *v. t.* [aus dem Hinterhalt] überfallen

ameliorate [ə'mi:lɪəreɪt] *v. t.* verbessern

amelioration [əmi:lɪə'reɪʃn] *n.* [Ver]besserung, *die*

amen [ɑ:'men, eɪ'men] **1.** *int.* amen. **2.** *n.* Amen, *das*

amenable [ə'mi:nəbl] *adj.* **a)** *(responsive)* zugänglich, aufgeschlossen ‹*Person*› **(to** *Dat.*); **b)** *(subject)* unterworfen ‹*Sache*› **(to** *Dat.*)

amend [ə'mend] *v.t. (correct)* berichtigen; *(improve)* abändern, ergänzen ‹*Gesetzentwurf, Antrag*›; ändern ‹*Verfassung*›

amendment [ə'mendmənt] *n. (to motion)* Abänderungsantrag, *der; (to bill)* Änderungsantrag, *der; (to Constitution)* Änderung, *die* **(to** *Gen.*); Amendement, *das (Dipl.)*

amends [ə'mendz] *n. pl.* **make ~ [to sb.]** es [bei jmdm.] wiedergutmachen; **make ~ for sth.** etw. wiedergutmachen

amenity [ə'mi:nɪtɪ] *n. (pleasant feature) (of residence)* Attraktivität, *die;* Wohnqualität, *die; (of locality)* Attraktivität, *die;* Reiz, *der;* **the amenities of a town** die kulturellen und Freizeiteinrichtungen einer Stadt; **a hotel with every ~:** ein Hotel mit allem Komfort

a'menity centre *n.* Freizeitzentrum, *das*

America [ə'merɪkə] *pr. n.* **a)** Amerika *(das);* **b) the ~s** Nord-, Süd- und Mittelamerika

American [ə'merɪkən] **1.** *adj.* amerikanisch; **sb. is ~:** jmd. ist Amerikaner/Amerikanerin; **~ studies** Amerikanistik, *die.* **2.** *n. (person)* Amerikaner, *der*/Amerikanerin, *die*

American: ~ 'football *n.* Football, *der;* **~ 'Indian 1.** *n.* Indianer, *der*/Indianerin, *die;* **2.** *adj.* indianisch

Americanise *see* **Americanize**

Americanism [ə'merɪkənɪzm] *n.* Amerikanismus, *der*

Americanize [ə'merɪkənaɪz] *v.t.* **a)** amerikanisieren; **b)** *(naturalize)* [in Amerika] einbürgern

amethyst ['æmɪθɪst] *n.* Amethyst, *der*

amiable ['eɪmɪəbl] *adj.* umgänglich; freundlich ‹*Person*›; entgegenkommend ‹*Haltung*›

amicable ['æmɪkəbl] *adj.* freundschaftlich ‹*Gespräch, Beziehungen*›; gütlich ‹*Einigung*›; friedlich ‹*Lösung*›

amicably ['æmɪkəblɪ] *adv.* in [aller] Freundschaft

amid [ə'mɪd] *prep.* inmitten; *(fig.: during)* bei

amidships [ə'mɪdʃɪps] (*Amer.:* **amidship** [ə'mɪdʃɪp]) *adv. (position)* mittschiffs; Mitte Schiff *(Seemannsspr.); (direction)* [nach] mittschiffs

amidst [ə'mɪdst] *see* **amid**

amino acid [əmi:nəʊ 'æsɪd] *n. (Chem.)* Aminosäure, *die*

amiss [ə'mɪs] **1.** *pred. adj. (wrong)* verkehrt; falsch; **is anything ~?** stimmt irgend etwas nicht? **2.** *adv.* **take sth. ~:** etw. übelnehmen; **come** *or* **go ~:** ungelegen kommen

ammeter ['æmɪtə(r)] *n. (Electr.)* Amperemeter, *das*

ammonia [ə'məʊnɪə] *n.* Ammoniak, *das*

ammunition [æmjʊ'nɪʃn] *n., no pl., no indef. art. (lit. or fig.)* Munition, *die*

amnesia [æm'ni:zɪə] *n. (Med.)* Amnesie, *die*

amnesty ['æmnɪstɪ] *n.* Amnestie, *die;* **grant an ~ to sb.** jmdn. amnestieren

amoeba [ə'mi:bə] *n., pl.* **~s** *or* **~e** [ə'mi:bi:] *(Zool.)* Amöbe, *die*

amok [ə'mɒk] *adv.* **run ~:** Amok laufen

among[st] [ə'mʌŋ(st)] *prep.* **a)** unter (+ *Dat.; seltener:* + *Akk.*); **~ us/you/friends** unter uns/euch/Freunden; **~ other things** unter anderem; **~ others** unter anderen; **b)** *(in/into the middle of, surrounded by)* zwischen (+ *Dat./Akk.*); **c)** *(in the practice or opinion of, in the number of)* unter (+ *Dat.*); **~ men/scientists** unter Männern/Wissenschaftlern; **I count him ~ my friends** ich zähle ihn zu meinen Freunden; **d)** *(between)* unter (+ *Dat.; seltener:* + *Akk.*); **share the sweets ~ yourselves** teilt euch die Bonbons; **e)** *(reciprocally)* **they often quarrel ~ themselves** sie streiten oft miteinander; sie streiten sich oft; **f)** *(jointly)* **~ you/them** *etc.* gemeinsam; zusammen

amoral [eɪ'mɒrəl] *adj.* amoralisch

amorous ['æmərəs] *adj.* verliebt

amorphous [ə'mɔ:fəs] *adj.* formlos; amorph ‹*Masse*›; *(fig.)* chaotisch ‹*Stil*›

amount [ə'maʊnt] **1.** *v.i.* **~ to sth.** sich auf etw. *(Akk.)* belaufen; *(fig.)* etw. bedeuten; **all these arguments/proposals don't ~ to much** diese Argumente/Vorschläge bringen alle nicht viel; **my savings don't ~ to very much** meine Ersparnisse sind nicht gerade groß; **what this all ~s to is that ...:** zusammenfassend kann man sagen, daß ... **2.** *n.* **a)** *(total)* Betrag, *der;* Summe, *die; (full significance)* volle Bedeutung *od.* Tragweite; **b) the ~ of a bill** die Höhe einer Rechnung; **c)** *(quantity)* Menge, *die;* **large ~s of money** beträchtliche Geldsummen; **a tremendous ~ of ...** *(coll.)* wahnsinnig viel ... *(ugs.);* **no ~ of money will make me change my mind** und wenn man mir noch soviel Geld gibt: meine Meinung werde ich nicht ändern; *see also* **any 1 e**

amp [æmp] *n.* **a)** *(Electr.)* Ampere, *das;* **b)** *(coll.: amplifier)* Verstärker, *der*

ampere ['æmpeə(r)] *n. (Electr.)* Ampere, *das*

ampersand ['æmpəsænd] *n.* Et-Zeichen, *das*

amphetamine [æm'fetəmi:n, æm'fetəmɪn] *n.* Amphetamin, *das*

amphibian [æm'fɪbɪən] *(Zool.)* **1.** *adj.* amphibisch. **2.** *n.* Amphibie, *die;* Lurch, *der*

amphibious [æm'fɪbɪəs] *adj.* amphibisch; **toads are ~:** Kröten sind Amphibien; **~ vehicle/tank/aircraft** Amphibienfahrzeug, *das*/-panzer, *der*/-flugzeug, *das*

amphitheatre (*Amer.:* **amphitheater**) ['æmfɪθɪətə(r)] *n.* Amphitheater, *das*

ample ['æmpl] *adj.*, **~r** ['æmplə(r)], **~st** ['æmplɪst] **a)** *(spacious)* weitläufig ‹*Garten, Räume*›; groß ‹*Ausdehnung*›; *(extensive, abundant)* reichhaltig ‹*Mahl, Bibliographie*›; weitreichend, umfassend ‹*Vollmachten, Machtbefugnisse*›; **b)** *(enough)* **~ room/food** reichlich Platz/zu essen; **c)** *(stout)* üppig ‹*Busen*›; stattlich ‹*Erscheinung*›

amplification [æmplɪfɪ'keɪʃn] *n.* **a)** *(Electr., Phys.)* Verstärkung, *die;* **b)** *(further explanation)* weitere *od.* zusätzliche Erläuterungen

amplifier ['æmplɪfaɪə(r)] *n.* Verstärker, *der*

amplify ['æmplɪfaɪ] *v.t.* **a)** *(Electr., Phys.)* verstärken; **b)** *(enlarge on)* weiter ausführen, näher *od.* ausführlicher erläutern ‹*Erklärung, Bericht*›

amplitude ['æmplɪtju:d] *n.* **a)** *(Electr.)* Amplitude, *die;* Schwingungsweite, *die;* **b)** *(Phys.)* Amplitude, *die;* größte Ausschlagweite; **c)** *no pl. (breadth)* Breite, *die;* Weite, *die*

amply ['æmplɪ] *adv.* reichlich ‹*breit, belohnen*›; zur Genüge ‹*zeigen, demonstrieren*›

ampoule ['æmpu:l] *n.* Ampulle, *die*

amputate ['æmpjʊteɪt] *v.t.* amputieren

amputation [æmpjʊ'teɪʃn] *n.* Amputation, *die*

amputee [æmpjʊ'ti:] *n.* Amputierte, *der/die*

amulet ['æmjʊlɪt] *n. (lit. or fig.)* Amulett, *das*

amuse [ə'mju:z] *v. t.* **a)** *(interest)* unterhalten; **keep a child ~d** ein Kind richtig beschäftigen; ~ **oneself with sth.** sich mit etw. beschäftigen; ~ **oneself by doing sth.** sich *(Dat.)* die Zeit damit vertreiben, etw. zu tun; **b)** *(make laugh or smile)* belustigen; amüsieren; **be ~d by** *or* **at sth.** sich über etw. *(Akk.)* amüsieren

amusement [ə'mju:zmənt] *n.* Belustigung, *die; (pastime)* Freizeitbeschäftigung, *die*

a'musement arcade *n.* Spielhalle, *die*

amusing [ə'mju:zɪŋ] *adj.*, **amusingly** [ə'mju:zɪŋlɪ] *adv.* amüsant

an [ən, *stressed* æn] *indef. art. see also* ²**a**: ein/eine/ein

anachronism [ə'nækrənɪzm] *n.* Anachronismus, *der*

anachronistic [ənækrə'nɪstɪk] *adj.* anachronistisch; zeitwidrig

anaemia [ə'ni:mɪə] *n., no pl. (Med.)* Blutarmut, *die;* Anämie, *die*

anaemic [ə'ni:mɪk] *adj. (Med.)* blutarm; anämisch; *(fig.)* blutleer; saft- und kraftlos

anaesthesia [ænɪs'θi:zɪə] *n. (Med.)* Narkose, *die;* **general ~:** [Voll]narkose, *die;* Allgemeinanästhesie, *die (fachspr.);* **local ~:** örtliche Betäubung; Lokalanästhesie, *die (fachspr.)*

anaesthetic [ænɪs'θetɪk] *n.* Anästhetikum, *das;* **give sb. an ~:** jmdm. eine Narkose geben; *(local)* jmdn. betäuben; **be under an ~:** in Narkose liegen; **general ~:** Narkotikum, *das;* Narkosemittel, *das;* **local ~:** Lokalanästhetikum, *das*

anaesthetist [ə'ni:sθətɪst] *n. (Med.)* Anästhesist, *der*/Anästhesistin, *die;* Narkose[fach]arzt, *der*/-ärztin, *die*

anaesthetize [ə'ni:sθətaɪz] *v. t.* narkotisieren; betäuben; *(fig.)* abstumpfen (**to** gegenüber)

anagram ['ænəgræm] *n.* Anagramm, *das*

analgesia [ænæl'dʒi:zɪə] *n. (Med.)* Analgesie, *die*

analgesic [ænæl'dʒi:sɪk] *n. (Med.)* Analgetikum, *das*

analog *(Amer.) see* **analogue**

analogous [ə'næləgəs] *adj.* vergleichbar; analog; **be ~ to sth.** einer Sache *(Dat.)* entsprechen

analogue ['ænəlɒg] *n.* Entsprechung, *die;* ~ **computer** Analogrechner, *der*

analogy [ə'nælədʒɪ] *n. (agreement; also Ling.)* Analogie, *die; (similarity)* Parallele, *die;* Analogie, *die;* **draw an ~ between/with** eine Parallele ziehen zwischen (+ *Dat.*)/zu

analyse ['ænəlaɪz] *v. t.* **a)** analysieren; **b)** *(Chem.)* untersuchen (**for** auf + *Akk.*)

analysis [ə'nælɪsɪs] *n., pl.* **analyses** [ə'nælɪsi:z] **a)** Analyse, *die; (Chem., Med.: of sample)* Untersuchung, *die;* **in the final** *or* **last ~:** letzten Endes; **b)** *(Psych.)* Analyse, *die*

analyst ['ænəlɪst] *n.* **a)** Laboratoriumsingenieur, *der;* **b)** *(Econ., Polit., etc.)* Experte, *der;* Fachmann, *der;* **c)** *(Psych.)* Analytiker, *der*/Analytikerin, *die*

analytic [ænə'lɪtɪk], **analytical** [ænə'lɪtɪkl] *adj.* analytisch

analyze *(Amer.) see* **analyse**

anarchic [ə'nɑ:kɪk], **anarchical** [ə'nɑ:kɪkl] *adj.* anarchisch; *(anarchistic)* anarchistisch

anarchism ['ænəkɪzm] *n., no pl.* Anarchismus, *der*

anarchist ['ænəkɪst] *n.* Anarchist, *der*/Anarchistin, *die*

anarchistic [ænə'kɪstɪk] *adj.* anarchistisch

anarchy ['ænəkɪ] *n., no pl.* Anarchie, *die; (fig.: disorder)* Chaos, *das*

anathema [ə'næθəmə] *n., no pl., no art.* **be ~ to sb.** jmdm. verhaßt *od.* ein Greuel sein

anatomical [ænə'tɒmɪkl] *adj.* anatomisch

anatomist [ə'nætəmɪst] *n.* Anatom, *der*

anatomy [ə'nætəmɪ] *n., no pl.* Anatomie, *die*

ancestor ['ænsestə(r)] *n.* Vorfahr, *der;* Ahn[e], *der; (fig.)* Ahn[e], *der*

ancestral [æn'sestrl] *adj.* angestammt ‹*Grundbesitz, Land*›

ancestry ['ænsestrɪ] *n.* **a)** *(lineage)* Abstammung, *die;* Herkunft, *die;* **b)** *(ancestors)* Vorfahren *Pl.*

anchor ['æŋkə(r)] **1.** *n.* Anker, *der;* **lie at ~:** vor Anker liegen; **come to** *or* **cast** *or* **drop ~:** vor Anker gehen; **weigh ~:** den Anker lichten. **2.** *v. t.* **a)** verankern; vor Anker legen; *(secure)* befestigen (**to** an + *Dat.*); **b)** *(fig.)* **be ~ed to sth.** an etw. *(Akk.)* gefesselt sein. **3.** *v. i.* ankern

anchorage ['æŋkərɪdʒ] *n.* Anker+latz, *der*

'anchorman *n.* **a)** *(Sport) (in tug-of-war)* hinterster *od.* letzter Mann; *(in relay race)* Schlußläufer, *der; (Mountaineering)* Seilletzte, *der;* **b)** *(Telev., Radio)* Moderator, *der;* Redakteur im Studio

anchovy ['æntʃəvɪ] *n.* An[s]chovis, *die;* Sardelle, *die*

ancient ['eɪnʃənt] *adj.* **a)** *(belonging to past)* alt; *(pertaining to antiquity)* antik; **that's ~ history** *(fig.)* das ist längst ein alter Hut *(ugs.);* **the ~ Greeks** die alten Griechen; **b)** *(old)* alt; historisch ‹*Gebäude usw.*›; ~ **monument** *(Brit. Admin.)* [offiziell anerkanntes] historisches Denkmal

ancillary [æn'sɪlərɪ] **1.** *adj.* **a)** *(auxiliary)* **be ~ to sth.** für etw. Hilfsdienste leisten; **b)** *(subordinate)* zweitrangig; ~ **industries** Zulieferindustrien. **2.** *n. (Brit.)* Hilfskraft, *die*

and [ənd, *stressed* ænd] *conj.* **a)** und; **two hundred ~ forty** zweihundert[und]vierzig; **a knife, fork, ~ spoon** Messer, Gabel und Löffel; **two ~ two are four** zwei und zwei ist *od.* sind vier; **b)** *expr. condition* und; **take one more step ~ I'll shoot** noch einen Schritt, und ich schieße; **do that ~ you'll regret it** wenn du das tust, wirst du es noch bedauern; **c)** *expr. continuation* und; **she cried ~ cried** sie weinte und weinte; **for weeks ~ weeks/years ~ years** wochen-/jahrelang; **for miles ~ miles** meilenweit; **better ~ better** immer besser

Andes ['ændi:z] *pr. n. pl.* **the ~:** die Anden

anecdotal ['ænɪkdəʊtl] *adj.* anekdotisch; anekdotenhaft

anecdote ['ænɪkdəʊt] *n.* Anekdote, *die*

anemia, anemic *(Amer.) see* **anaem-**

anemone [ə'nemənɪ] *n.* Anemone, *die*

anesthesia *etc. (Amer.) see* **anaesthesia** *etc.*

angel ['eɪndʒl] *n. (lit. or fig.)* Engel, *der;* **be on the side of the ~s** *(fig.)* auf der Seite der Guten stehen; **be an ~ and ...** *(coll.)* sei so lieb und ...

angelic [æn'dʒelɪk] *adj. (like angel[s])* engelhaft; **she looked ~:** sie sah wie ein Engel aus

anger ['æŋgə(r)] **1.** *n., no pl.* Zorn, *der* (**at** über + *Akk.*); *(fury)* Wut, *die* (**at** über + *Akk.*); **be filled with ~:** erzürnt/wütend sein; **in [a moment of] ~:** im Zorn/in der Wut. **2.** *v. t.* verärgern; *(infuriate)* erzürnen *(geh.)*/wütend machen; **be ~ed by sth.** über etw. *(Akk.)* verärgert/erzürnt/wütend sein

¹angle ['æŋgl] **1.** *n.* **a)** *(Geom.)*

Winkel, *der;* **acute/obtuse/right ~:** spitzer/stumpfer/rechter Winkel; **at an ~ of 60°** im Winkel von 60°; **at an ~:** schief; **at an ~ to the wall** schräg zur Wand; **b)** *(corner)* Ecke, *die; (recess)* Winkel, *der;* **c)** *(direction)* Perspektive, *die;* Blickwinkel, *der; (fig.)* Gesichtspunkt, *der;* Aspekt, *der;* **the committee examined the matter from various ~s** der Ausschuß prüfte die Angelegenheit von verschiedenen Seiten; **looking at it from a commercial ~:** aus kaufmännischer Sicht betrachtet. **2.** *v. t.* **a)** [aus]richten; **b)** *(coll.: bias)* färben ‹*Nachrichten, Formulierung*›. **3.** *v. i.* [im Winkel] abbiegen; **the road ~s sharply to the left** die Straße biegt scharf nach links ab

[2]**angle** *v. i. (fish)* angeln; *(fig.)* **~ for sth.** sich um etw. bemühen; **~ for compliments** nach Komplimenten fischen

angled ['æŋgəld] *adj. (angular)* eckig ‹*Form, Figur*›; *(placed obliquely)* schief; *(fig. coll.)* tendenziös, gefärbt ‹*Bericht, Kommentar*›; **acute-/obtuse-/right-~:** spitz-/stumpf-/rechtwinklig

angler ['æŋglə(r)] *n.* Angler, *der*/Anglerin, *die*

Anglican ['æŋglɪkən] **1.** *adj.* anglikanisch. **2.** *n.* Anglikaner, *der*/Anglikanerin, *die*

Anglicize ['æŋglɪsaɪz] *v. t.* anglisieren

angling ['æŋglɪŋ] *n.* Angeln, *das*

Anglo- ['æŋgləʊ] *in comb.* anglo-/Anglo-

Anglo-A'merican 1. *adj.* angloamerikanisch; **an ~ agreement** ein englisch-/britisch-amerikanischer Vertrag. **2.** *n.* Angloamerikaner, *der*/Angloamerikanerin, *die*

Anglo-'Indian 1. *adj.* angloindisch. **2.** *n.* Anglo-Inder, *der*/Anglo-Inderin, *die*

Anglo-'Saxon 1. *n.* Angelsachse, *der*/Angelsächsin, *die; (language)* Angelsächsisch, *das.* **2.** *adj.* angelsächsisch

angrily ['æŋgrɪlɪ] *adv.* verärgert; *(stronger)* zornig

angry ['æŋgrɪ] *adj.* **a)** böse; verärgert ‹*Person, Stimme, Geste*›; *(stronger)* zornig; wütend; **be ~ at *or* about sth.** wegen etw. böse sein; **be ~ with *or* at sb.** mit jmdm. *od.* auf jmdn. böse sein; sich über jmdn. ärgern; **get ~:** böse werden; **get *or* make sb. ~:** jmdn. verärgern; *(stronger)* jmdn. wütend machen; **b)** *(fig.)* drohend, bedrohlich ‹*Wolke, Himmel*›

anguish ['æŋgwɪʃ] *n., no pl.* Qualen *Pl.*

anguished ['æŋgwɪʃt] *adj.* qualvoll; gequält ‹*Herz, Gewissen*›

angular ['æŋgjʊlə(r)] *adj.* **a)** *(having angles)* eckig ‹*Gebäude, Struktur, Gestalt*›; **b)** *(lacking plumpness, stiff)* knochig ‹*Körperbau*›; kantig ‹*Gesicht*›; **c)** *(measured by angle)* angular; winklig; **~ momentum** *(Phys.)* Drehimpuls, *der*

animal ['ænɪməl] **1.** *n.* **a)** Tier, *das; (quadruped)* Vierbeiner, *der; (any living being)* Lebewesen, *das;* **domestic ~:** Haustier, *das;* **b)** *(fig. coll.)* **there is no such ~ as a 'typical' criminal** so etwas wie den „typischen" Verbrecher gibt es gar nicht; **c)** *(fig.: ~ instinct; brute)* Tier, *das.* **2.** *adj.* **a)** tierisch; **~ behaviour/breeding** Tierverhalten, *das*/Tierzucht, *die;* **b)** *(from ~s)* tierisch ‹*Produkt, Klebstoff, Öl*›; **c)** *(carnal, sexual)* körperlich ‹*Triebe, Wünsche, Bedürfnisse*›; tierisch, animalisch ‹*Veranlagung, Natur*›

'animal-lover *n.* Tierfreund, *der*/-freundin, *die*

animate 1. ['ænɪmeɪt] *v. t.* **a)** *(enliven)* beleben; **b)** *(inspire)* anregen; **~ sb. with enthusiasm** jmdn. mit Begeisterung erfüllen; **c)** *(breathe life into)* mit Leben erfüllen. **2.** ['ænɪmət] *adj.* beseelt ‹*Leben, Körper*›; belebt ‹*Objekt, Welt*›; lebendig ‹*Seele*›

animated ['ænɪmeɪtɪd] *adj.* lebhaft ‹*Diskussion, Unterhaltung, Ausdruck, Gebärde*›; **~ cartoon** Zeichentrickfilm, *der*

animatedly ['ænɪmeɪtɪdlɪ] *adv.* lebhaft

animation [ænɪ'meɪʃn] *n.* **a)** *no pl.* Lebhaftigkeit, *die;* **b)** *(Cinemat.)* Animation, *die*

animosity [ænɪ'mɒsɪtɪ] *n.* Animosität, *die (geh.),* Feindseligkeit, *die* (**against, towards** gegen)

aniseed ['ænɪsi:d] *n.* Anis[samen], *der*

ankle ['æŋkl] *n. (joint)* Fußgelenk, *das; (part of leg)* Knöchelgegend, *die;* Fessel, *die*

ankle: ~-deep *adj.* knöcheltief; **~ sock** *n.* Socke, *die; (esp. for children)* Söckchen, *das*

annals ['ænəlz] *n. pl. (lit. or fig.)* Annalen *Pl.*

annex 1. [ə'neks] *v. t.* **a)** *(add)* angliedern (**to** *Dat.*); *(append)* anfügen ‹*Bemerkungen*› (**to** *Dat.*); **b)** *(incorporate)* annektieren ‹*Land, Territorium*›; *(coll.: take without right)* sich *(Dat.)* unter den Nagel reißen *(ugs.)* ‹*Gegenstände*›. **2.** ['æneks] *n.* *(supplementary building)* Anbau, *der; (built-on extension)* Erweiterungsbau, *der; (appendix) (to document)* Zusatz, *der; (to treaty)* Anhang, *der*

annexation [ænɪk'seɪʃn] *n. (of land)* Annexion, *die;* Annektierung, *die*

annexe *see* **annex 2**

annihilate [ə'naɪɪleɪt] *v. t.* **a)** vernichten ‹*Armee, Bevölkerung, Menschheit*›; zerstören ‹*Stadt, Land*›; **b)** *(fig.)* zunichte machen; am Boden zerstören ‹*Person*›

annihilation [ənaɪɪ'leɪʃn] *n.* **a)** *see* **annihilate a:** Vernichtung, *die;* Zerstörung, *die;* **b)** *(fig.)* Verderben, *das;* Untergang, *der*

anniversary [ænɪ'vɜ:sərɪ] *n.* Jahrestag, *der;* **wedding ~:** Hochzeitstag, *der;* **the university celebrated its 500th ~:** die Universität feierte ihr 500jähriges Jubiläum *od.* Bestehen; **the ~ of Shakespeare's birth** [die Wiederkehr von] Shakespeares Geburtstag; **the ~ of his death** sein Todestag

annotate ['ænəteɪt] *v. t.* kommentieren; mit Anmerkungen versehen

announce [ə'naʊns] *v. t.* **a)** bekanntgeben; ansagen ‹*Programm*›; *(over Tannoy etc.)* durchsagen; *(in newspaper)* anzeigen ‹*Heirat usw.*›; **b)** *(make known the approach of; fig.: signify)* ankündigen

announcement [ə'naʊnsmənt] *n.* Bekanntgabe, *die; (over Tannoy etc.)* Durchsage, *die;* **they made an ~ over the radio that ...:** sie gaben im Radio bekannt, daß ...; **did you read the ~ of his death in the paper?** haben Sie seine Todesanzeige in der Zeitung gelesen?

announcer [ə'naʊnsə(r)] *n.* Ansager, *der*/Ansagerin, *die;* Sprecher, *der*/Sprecherin, *die*

annoy [ə'nɔɪ] *v. t.* **a)** ärgern; **his late arrival ~ed me** ich habe mich über sein spätes Kommen geärgert; **b)** *(harass)* schikanieren

annoyance [ə'nɔɪəns] *n.* Verärgerung, *die; (nuisance)* Plage, *die*

annoyed [ə'nɔɪd] *adj.* **be ~ [at *or* with sb./sth.]** ärgerlich [auf *od.* über jmdn./über etw. *(Akk.)*] sein; **be ~ to find that ...:** sich darüber ärgern, daß ...; **he got very ~:** er hat sich darüber sehr geärgert

annoying [ə'nɔɪɪŋ] *adj.* ärgerlich; lästig ‹*Gewohnheit, Person*›; **the ~ part of it is that ...:** das Ärgerliche daran ist, daß ...

annual ['ænjʊəl] **1.** *adj.* **a)** *(reckoned by the year)* Jahres-; ~ **rainfall** jährliche Regenmenge; **b)** *(recurring yearly)* [all]jährlich ‹*Ereignis, Feier*›; Jahres‹*bericht, -hauptversammlung*›. **2.** *n.* **a)** *(Bot.)* einjährige Pflanze; **b)** *(book)* Jahrbuch, *das; (of comic etc.)* Jahresalbum, *das*

annually ['ænjʊəlɪ] *adv. (per year)* jährlich; *(once a year)* [all]jährlich

annuity [ə'nju:ɪtɪ] *n. (grant, sum payable)* Jahresrente, *die; (investment)* Rentenversicherung, *die*

annul [ə'nʌl] *v. t.*, **-ll-** annullieren, für ungültig erklären ‹*Gesetz, Vertrag, Ehe, Testament*›; auflösen ‹*Vertrag*›

annulment [ə'nʌlmənt] *n. (of law, treaty, marriage, will)* Annullierung, *die; (of treaty also)* Auflösung, *die*

Annunciation [ənʌnsɪ'eɪʃn] *n. (Eccl.)* **the ~:** Mariä Verkündigung

anoint [ə'nɔɪnt] *v. t. (esp. Relig.)* salben

anomalous [ə'nɒmələs] *adj.* anomal, anormal ‹*Lage, Verhältnisse, Zustand*›; ungewöhnlich ‹*Situation, Anblick*›

anomaly [ə'nɒməlɪ] *n.* Anomalie, *die;* Absonderlichkeit, *die; (exception)* Ausnahme, *die*

anon. [ə'nɒn] *abbr.* **anonymous [author]** anon.

anonymity [ænə'nɪmɪtɪ] *n.* Anonymität, *die*

anonymous [ə'nɒnɪməs] *adj.* anonym

anorak ['ænəræk] *n.* Anorak, *der*

anorexia [ænə'reksɪə] *n.* Anorexie, *die (Med.);* Magersucht, *die (volkst.);* ~ **nervosa** [ænəreksɪə nɜ:'vəʊsə] nervöse Anorexie *(Med.)*

another [ə'nʌðə(r)] **1.** *pron.* **a)** *(additional)* noch einer/eine/eins; ein weiterer/eine weitere/ein weiteres; **one thing leads to ~:** eines ergibt sich aus dem anderen; **please have ~:** nimm dir doch noch einen; **b)** *(counterpart)* wieder einer/eine/eins; **c)** *(different)* ein anderer/eine andere/ein anderes; **in one way or ~:** so oder so; irgendwie; **for one reason or ~:** aus irgendeinem Grund; *see also* **one 1f, 3b. 2.** *adj.* **a)** *(additional)* noch einer/eine/eins; ein weiterer/eine weitere/ein weiteres; **give me ~ chance** gib mir noch [einmal] eine Chance; **after ~ six weeks** nach weiteren sechs Wochen; ~ **100 pounds** weitere 100 Pfund; **he didn't say ~ word** er sagte nichts mehr; **b)** *(a person like)* ein neuer/eine neue/ein neues; ein zweiter/eine zweite/ein zweites; ~ **Chaplin** ein neuer *od.* zweiter Chaplin; **c)** *(different)* ein anderer/eine andere/ein anderes; **ask ~ person** fragen Sie jemand anderen *od.* anders; ~ **time, don't be so greedy** sei beim nächsten Mal nicht so gierig; **I'll do it ~ time** ich tu's ein andermal; **[and] [there's] ~ thing** [und] noch etwas

answer ['ɑ:nsə(r)] **1.** *n.* **a)** *(reply)* Antwort, *die* (**to** auf + *Akk.*); *(reaction)* Reaktion, *die;* **I tried to phone him, but there was no ~:** ich habe versucht, ihn anzurufen, aber es hat sich niemand gemeldet; **there is no ~ to that** dem ist nichts mehr hinzuzufügen; **by way of [an] ~:** als Antwort; **in ~ to sth.** als Antwort *od.* Reaktion auf etw. *(Akk.);* **b)** *(to problem)* Lösung, *die* (**to** *Gen.*); *(to calculation)* Ergebnis, *das;* **have** *or* **know all the ~s** *(coll.)* alles wissen. **2.** *v. t.* **a)** beantworten ‹*Brief, Frage*›; antworten auf (+ *Akk.*) ‹*Frage, Hilferuf, Einladung, Inserat*›; *(react to)* erwidern ‹*Geste, Schlag*›; eingehen auf (+ *Akk.*), erfüllen ‹*Bitte*›; sich stellen zu ‹*Beschuldigung*›; ~ **sb.** jmdm. antworten; ~ **me!** antworte [mir]!; **b)** ~ **the door/bell** an die Tür gehen; *see also* **telephone 1. 3.** *v. i.* **a)** *(reply)* antworten; ~ **to sth.** sich zu etw. äußern; **b)** *(be responsible)* ~ **for sth.** für etw. die Verantwortung übernehmen; ~ **to sb.** jmdm. [gegenüber] Rechenschaft ablegen; **he has a lot to ~ for** er hat vieles zu verantworten; **c)** *(correspond)* ~ **to a description** einer Beschreibung *(Dat.)* entsprechen; **d)** ~ **to the name of ...:** auf den Namen ... hören; **e)** ~ **back** *(coll.)* widersprechen; Widerworte haben *(ugs.)*

answerable ['ɑ:nsərəbl] *adj. (responsible)* **be ~ to sb.** jmdm. [gegenüber] verantwortlich sein; **be ~ for sb./sth.** für jmdn./etw. verantwortlich sein

'**answering machine** *n. (Teleph.)* Anrufbeantworter, *der*

ant [ænt] *n.* Ameise, *die*

antagonism [æn'tægənɪzm] *n.* Feindseligkeit, *die* (**towards, against** gegenüber); *(between two)* Antagonismus, *der (geh.)*

antagonist [æn'tægənɪst] *n.* Gegner, *der*/Gegnerin, *die; (in debate etc.)* Kontrahent, *der*/Kontrahentin, *die*

antagonistic [æntægə'nɪstɪk] *adj.* feindlich ‹*Mächte, Prinzipien*›; antagonistisch, gegensätzlich ‹*Interessen*›; **be ~ towards sb.** jmdn. anfeinden; **be ~ towards sth.** gegen etw. eingestellt sein

antagonize [æn'tægənaɪz] *v. t.* **a)** sich *(Dat.)* ‹*Person*› zum Feind machen; **b)** *(counteract)* entgegenwirken (+ *Dat.*)

antarctic [ænt'ɑ:ktɪk] **1.** *adj.* antarktisch; **A~ Circle/Ocean** südlicher Polarkreis/Südpolarmeer, *das.* **2.** *pr. n.* **the A~:** die Antarktis

Antarctica [ænt'ɑ:ktɪkə] *pr. n.* die Antarktis

'**ant-eater** *n. (Zool.)* Ameisenfresser, *der*

antecedent [æntɪ'si:dənt] *n. (preceding event)* früherer Umstand; vorangegangenes Ereignis; *(preceding thing)* Vorläufer, *der*

antedate [æntɪ'deɪt] *v. t. (precede)* voraus-, vorangehen (+ *Dat.*)

antediluvian [æntɪdɪ'lu:vɪən] *adj. (lit. or fig.)* vorsintflutlich

antelope ['æntɪləʊp] *n. (Zool.)* Antilope, *die*

antenatal [æntɪ'neɪtl] *adj.* **a)** *(concerning pregnancy)* Schwangerschafts-; Schwangeren-; **b)** *(before birth)* vorgeburtlich

antenna [æn'tenə] *n.* **a)** *pl.* **~e** [æn'teni:] *(Zool.)* Fühler, *der;* Antenne, *die (fachspr.);* **b)** *pl.* **~s** *(Amer.: aerial)* Antenne, *die*

ante-room ['æntɪru:m, 'æntɪrʊm] *n.* Vorraum, *der; (waiting room)* Warteraum, *der*

anthem ['ænθəm] *n.* **a)** *(Eccl. Mus.)* Chorgesang, *der;* **b)** *(song of praise)* Hymne, *die*

anther ['ænθə(r)] *n. (Bot.)* Staubbeutel, *der*

anthill ['ænthɪl] *n.* Ameisenhügel, *der*

anthology [æn'θɒlədʒɪ] *n. (by different writers)* Anthologie, *die; (by one writer)* Auswahl, *die*

anthracite ['ænθrəsaɪt] *n.* Anthrazit, *der*

anthrax ['ænθræks] *n., no pl., no indef. art. (Med., Vet. Med.)* Milzbrand, *der;* Anthrax, *der (fachspr.)*

anthropoid ['ænθrəpɔɪd] *n.* Anthropoid[e], *der;* Menschenaffe, *der*

anthropological [ænθrəpə'lɒdʒɪkl] *adj.* anthropologisch

anthropologist [ænθrə'pɒlədʒɪst] *n.* Anthropologe, *der*/Anthropologin, *die*

anthropology [ænθrə'pɒlədʒɪ] *n., no pl.* Anthropologie, *die*

anti ['æntɪ] *prep.* gegen

anti- ['æntɪ] *pref.* anti-/Anti-

anti-'aircraft *adj. (Mil.)* Flugabwehr-; ~ **gun** Flak, *die;* ~ **battery** Flakbatterie, *die*

antibiotic [æntɪbaɪ'ɒtɪk] *n.* Antibiotikum, *das*

'antibody *n. (Physiol.)* Antikörper, *der*

antic ['æntɪk] *n. (trick)* Mätzchen, *das (ugs.); (of clown)* Possen, *der*

anticipate [æn'tɪsɪpeɪt] *v. t.* **a)** *(expect)* erwarten; *(foresee)* voraussehen; ~ **rain/trouble** mit Regen/Ärger rechnen; **b)** *(discuss or consider before due time)* vorwegnehmen; antizipieren; **c)** *(forestall)* ~ **sb./sth.** jmdm./einer Sache zuvorkommen

anticipation [æntɪsɪ'peɪʃn] *n., no pl.* Erwartung, *die;* **in ~ of sth.** in Erwartung einer Sache *(Gen.);* **with ~:** erwartungsvoll; **thanking you in ~:** Ihnen im voraus dankend

anti'climax *n.* **a)** *(ineffective end)* Abstieg, *der;* Abfall, *der;* **b)** *(Lit.)* Antiklimax, *die*

anti'clockwise 1. *adv.* gegen den Uhrzeigersinn. **2.** *adj.* **in an ~ direction** gegen den *od.* entgegen dem Uhrzeigersinn

anti'cyclone *n. (Meteorol.)* Hochdruckgebiet, *das;* Antizyklone, *die (Met.)*

antidote ['æntɪdəʊt] *n.* Gegengift, -mittel, *das* (**for, against** gegen); *(fig.)* Gegenmittel, *das* (**to** gegen)

'antifreeze *n.* Gefrierschutzmittel, *das;* Frostschutzmittel, *das*

anti-lock 'braking system *n. (Motor Veh.)* Antiblockiersystem, *das*

anti'nuclear *adj.* Anti-Atom-[kraft]-

antipathy [æn'tɪpəθɪ] *n.* Antipathie, *die;* Abneigung, *die;* ~ **to** *or* **for sb./sth.** Abneigung gegen jmdn./etw.

antipodes [æn'tɪpədi:z] *n. pl.* entgegengesetzte *od.* antipodische Teile der Erde; *(Australasia)* Australien und Ozeanien

antiquarian [æntɪ'kweərɪən] *adj.* ~ **bookshop** Antiquariat, *das;* Antiquariatsbuchhandlung, *die;* ~ **bookseller** Antiquar, *der;* Antiquariatsbuchhändler, *der*

antiquated ['æntɪkweɪtɪd] *adj. (old-fashioned)* antiquiert; veraltet; *(out of date)* überholt

antique [æn'ti:k] **1.** *adj.* antik. **2.** *n.* Antiquität, *die;* ~ **dealer** Antiquitätenhändler, *der/*-händlerin, *die;* ~ **shop** Antiquitätenladen, *der*

antiquity [æn'tɪkwɪtɪ] *n., no pl.* **a)** *(ancientness)* Alter, *das;* **b)** *no art. (old times)* Antike, *die*

anti-Semitic [æntɪsɪ'mɪtɪk] *adj.* antisemitisch

anti-Semitism [æntɪ'semɪtɪzm] *n., no pl.* Antisemitismus, *der*

anti'septic 1. *adj.* antiseptisch. **2.** *n.* Antiseptikum, *das*

anti'social *adj.* **a)** asozial; **b)** *(unsociable)* ungesellig ‹*Person*›

antithesis [æn'tɪθəsɪs] *n., pl.* **antitheses** [æn'tɪθəsi:z] *(thing)* Gegenstück, *das* (**of, to** zu)

antivivisectionist [æntɪvɪvɪ'sekʃənɪst] *n.* Vivisektionsgegner, *der/*-gegnerin, *die*

antler ['æntlə(r)] *n.* Geweihstange, *die;* **[pair of] ~s** Geweih, *das*

antonym ['æntənɪm] *n.* Antonym, *das;* Gegen[satz]wort, *das*

Antwerp ['æntwɜ:p] *pr. n.* Antwerpen *(das)*

anus ['eɪnəs] *n. (Anat.)* After, *der*

anvil ['ænvɪl] *n.* Amboß, *der*

anxiety [æŋ'zaɪətɪ] *n.* **a)** *(state)* Angst, *die; (concern about future)* Sorge, *die* (**about** wegen); **anxieties** Sorgen *Pl.;* **cause sb. ~:** jmdm. angst/Sorgen machen; **b)** *(desire)* **his ~ to do sth.** sein Verlangen danach, etw. zu tun

anxious ['æŋkʃəs] *adj.* **a)** *(troubled)* besorgt; **be ~ about sth./sb.** um etw./jmdn. besorgt sein; **we were all so ~ about you** wir haben uns *(Dat.)* alle solche Sorgen um Sie gemacht; **b)** *(eager)* sehnlich; **be ~ for sth.** ungeduldig auf etw. *(Akk.)* warten; **have an ~ desire to do sth.** ängstlich darauf bedacht sein, etw. zu tun; **he is ~ to please** er ist bemüht zu gefallen; **c)** *(worrying)* **an ~ time** eine Zeit banger Sorge

anxiously ['æŋkʃəslɪ] *adv.* **a)** besorgt; **b)** *(eagerly)* sehnsüchtig

any ['enɪ] **1.** *adj.* **a)** *(some)* **have you ~ wool/~ statement to make?** haben Sie Wolle/[irgend]eine Erklärung abzugeben?; **if you have ~ difficulties** wenn du irgendwelche Schwierigkeiten hast; **not ~:** kein/keine; **without ~:** ohne jeden/jede/jedes; **have you ~ idea of the time?** hast du eine Ahnung, wie spät es ist?; **b)** *(one)* ein/eine; **there isn't ~ hood on this coat** dieser Mantel hat keine Kapuze; **a book without ~ cover** ein Buch ohne Deckel; **c)** *(all)* jeder/jede/jedes; **to avoid ~ delay** um jede Verzögerung zu vermeiden; **d)** *(every)* jeder/jede/jedes; **~ time I went there** jedesmal *od.* immer, wenn ich dort hinging; **[at] ~ time** jederzeit; **[at] ~ time of day** zu jeder Tageszeit; **e)** *(whichever)* jeder/jede/jedes [beliebige]; **choose ~ [one] book/~ books you like** suchen Sie sich *(Dat.)* irgendein Buch/irgendwelche Bücher aus; **choose ~ two numbers** nimm zwei beliebige Zahlen; **do it ~ way you like** machen Sie es, wie immer Sie wollen; **[at] ~ time [now]** jederzeit; **~ day/minute [now]** jeden Tag/jede Minute; **you can count on him ~ time** *(coll.)* du kannst dich jederzeit auf ihn verlassen; **I'd prefer Mozart ~ day** *(coll.)* ich würde Mozart allemal *(ugs.) od.* jederzeit vorziehen; **not [just] ~ house** nicht irgendein beliebiges Haus; **take ~ amount you wish** nehmen Sie, soviel Sie wollen; **~ amount of** jede Menge *(ugs.);* **f)** *(an appreciable)* ein nennenswerter/eine nennenswerte/ein nennenswertes; **she didn't stay ~ length of time** sie ist nicht sehr lange geblieben. **2.** *pron.* **a)** *(some) in condit., interrog., or neg. sentence (replacing sing. n.)* einer/eine/ein[e]s; *(replacing collect. n.)* welcher/welche/welches; *(replacing pl. n.)* welche; **not ~:** keiner/keine/kein[e]s/*Pl.* keine; **without ~:** ohne; **I need to buy some sugar, we haven't got ~ at the moment** ich muß Zucker kaufen, wir haben im Augenblick keinen; **Here are some sweets. Would you like ~?** Hier sind ein paar Bonbons. Möchtest du welche?; **hardly ~:** kaum welche/etwas; **do you have ~ of them in stock?** haben Sie [irgend]welche davon vorrätig?; **he is not having ~ of it** *(fig. coll.)* er will nichts davon wissen; **b)** *(no matter which)* irgendeiner/irgendeine/irgendein[e]s/irgendwelche *Pl.;* **Which numbers? – Any between 1 and 10** Welche Zahlen? – Irgendwelche zwischen 1 und 10. **3.** *adv.* **do you feel ~ better today?** fühlen Sie sich heute [etwas] besser?; **if it gets ~ colder** wenn es noch kälter wird; **he didn't seem ~ [the] wiser after that** danach schien er auch nicht klüger zu sein; **I can't wait ~ longer** ich kann nicht [mehr] länger warten; **I don't feel ~ better** mir ist kein bißchen wohler

'anybody *n. & pron.* **a)** *(whoever)* jeder; **~ and everybody** jeder Beliebige; **b)** *(somebody)* [irgend]jemand; **how could ~ be so cruel?** wie kann man nur so grausam sein?; **there wasn't ~ there** es war niemand da; **I've never seen ~ who ...:** ich habe noch keinen gesehen, der ...; **he is a match for ~:** er kann sich mit jedem *od.* jedermann messen; ~ **but** jeder[mann] außer; **it's ~'s match** das Spiel ist offen; **what will happen is ~'s guess** was geschehen wird, [das] weiß keiner; **he's not [just]** ~: er ist nicht [einfach] irgendwer; **c)** *(an important person)* jemand; wer *(ugs.)*; **everybody who was ~ was there** alles, was Rang und Namen hatte, war da

'**anyhow** *adv.* **a)** *see* **anyway; b)** *(haphazardly)* irgendwie; **the furniture was arranged** ~: die Möbel waren wahllos irgendwo hingestellt

'**anyone** *see* **anybody**

'**anything 1.** *n. & pron.* **a)** *(whatever thing)* was [immer]; alles, was; **you may do ~ you wish** Sie können [alles] tun, was Sie möchten; ~ **and everything** alles mögliche; **b)** *(something)* irgend etwas; **is there ~ wrong with you?** fehlt Ihnen [irgend] etwas?; **have you done ~ silly?** hast du [irgend] etwas Dummes gemacht?; **can we do ~ to help you?** können wir Ihnen irgendwie helfen?; **I don't want ~ [further] to do with him** ich möchte nichts [mehr] mit ihm zu tun haben; **c)** *(a thing of any kind)* alles; ~ **like that** so etwas; **as ... as ~** *(coll.)* wahnsinnig ... *(ugs.)*; **not for ~ [in the world]** um nichts in der Welt; ~ **but ...** *(~ except)* alles außer ...; *(far from)* alles andere als; **we don't want [just]** ~: wir wollen nicht einfach irgend etwas [Beliebiges]. **2.** *adv.* **not ~ like as ... as** keineswegs so ... wie

'**anyway** *adv.* **a)** *(in any case, besides)* sowieso; **we wouldn't accept your help** ~: wir würden von Ihnen sowieso keine Hilfe annehmen; **b)** *(at any rate)* jedenfalls; **~, I must go now** wie dem auch sei, ich muß jetzt gehen

'**anywhere 1.** *adv.* **a)** *(in any place) (wherever)* überall, wo; wo [immer]; *(somewhere)* irgendwo; **not ~ near as ... as** *(coll.)* nicht annähernd so ... wie; ~ **but ...**: überall, außer ...; überall, nur nicht ...; **b)** *(to any place) (wherever)* wohin [auch immer]; *(somewhere)* irgendwohin; ~ **but ...**: überallhin, außer ...; überallhin, nur nicht ...; **[just]** ~: [einfach] irgendwohin. **2.** *pron.* **if there's ~ you'd like to see** wenn es irgend etwas gibt, was du sehen möchtest; **have you found ~ to live yet?** haben Sie schon eine Wohnung gefunden?; **there's never ~ open for milk after 6 p.m.** nach 18 Uhr kann man nirgends mehr Milch bekommen; ~ **but ...**: überall, außer ...; überall, nur nicht ...; **[just]** ~: irgendein x-beliebiger Ort

apart [ə'pɑ:t] *adv.* **a)** *(separately)* getrennt; **with one's legs** ~: mit gespreizten Beinen; ~ **from ...** *(except for)* außer ...; bis auf ... (+ *Akk.*); *(in addition to)* außer ...; **b)** *(into pieces)* auseinander; **he took the engine** ~: er nahm den Motor auseinander; **take** ~ *(fig.) (criticize)* auseinandernehmen *(ugs.)* ‹*Theaterstück, Politiker*›; *(analyse)* zergliedern; **c)** ~ **[from]** *(to a distance)* weg [von]; *(at a distance)* **ten kilometres** ~: zehn Kilometer voneinander entfernt

apartheid [ə'pɑ:theɪt] *n., no pl., no art.* Apartheid, *die*

apartment [ə'pɑ:tmənt] *n.* **a)** *(room)* Apartment, *das;* Appartement, *das;* **~s** *(in a mansion etc.)* Räume; Räumlichkeiten *Pl.;* **b)** *(Amer.) see* ¹**flat**

apathetic [æpə'θetɪk] *adj.* apathisch **(about** gegenüber)

apathy ['æpəθɪ] *n., no pl.* Apathie, *die* **(about** gegenüber)

ape [eɪp] **1.** *n.* [Menschen]affe, *der; (apelike person)* Affe, *der.* **2.** *v. t.* nachahmen; nachäffen *(abwertend)*

aperitif [əperɪ'ti:f] *n.* Aperitif, *der*

aperture ['æpətʃə(r)] *n.* Öffnung, *die*

apex ['eipeks] *n., pl.* **~es** *or* **apices** ['eɪpɪsi:z] *(tip)* Spitze, *die; (fig.)* Gipfel, *der;* Höhepunkt, *der*

aphorism ['æfərɪzm] *n.* Aphorismus, *der*

aphrodisiac [æfrə'dɪzɪæk] *n.* Aphrodisiakum, *das*

apices *pl. of* **apex**

apiece [ə'pi:s] *adv.* je; **we took two bags** ~: wir nahmen je zwei Beutel; **they cost a penny** ~: die kosten einen Penny das Stück

aplomb [ə'plɒm] *n.* Sicherheit [im Auftreten]

apocalypse [ə'pɒkəlɪps] *n.* Apokalypse, *die*

apocalyptic [əpɒkə'lɪptɪk] *adj.* apokalyptisch

apocryphal [ə'pɒkrɪfl] *adj.* apokryph

apolitical [eɪpə'lɪtɪkl] *adj.* apolitisch; unpolitisch

apologetic [əpɒlə'dʒetɪk] *adj.* **a)** entschuldigend; **he was most ~ about ...**: er entschuldigte sich vielmals für ...; **b)** *(diffident)* zaghaft ‹*Lächeln, Ton*›; zurückhaltend, bescheiden ‹*Wesen, Art*›

apologetically [əpɒlə'dʒetɪkəlɪ] *adv.* **a)** entschuldigend; **b)** *(diffidently)* zaghaft; bescheiden

apologize [ə'pɒlədʒaɪz] *v. i.* sich entschuldigen; ~ **to sb. for sth./sb.** sich bei jmdm. für etw./jmdn. entschuldigen

apology [ə'pɒlədʒɪ] *n.* **a)** Entschuldigung, *die;* **make an ~ [to sb.] for sth.** sich für etw. [bei jmdm.] entschuldigen; **you owe him an** ~: Sie müssen sich bei ihm entschuldigen; **please accept our apologies** wir bitten vielmals um Entschuldigung; **b)** *(poor example of)* **an ~ for a ...**: ein erbärmliches Exemplar von ...

apoplectic [æpə'plektɪk] *adj.* apoplektisch; ~ **stroke** *or* **fit** Schlaganfall, *der*

apoplexy ['æpəpleksɪ] *n.* Apoplexie, *die (fachspr.);* Schlaganfall, *der*

apostle [ə'pɒsl] *n. (lit. or fig.)* Apostel, *der*

apostrophe [ə'pɒstrəfɪ] *n. (sign)* Apostroph, *der;* Auslassungszeichen, *das*

appal *(Amer.:* **appall)** [ə'pɔ:l] *v. t.,* **-ll-** *(dismay)* entsetzen; *(terrify)* erschrecken; **obscenity ~s her** sie empört sich über Obszönitäten

appalling [ə'pɔ:lɪŋ] *adj. (dismaying)* entsetzlich; *(terrifying)* schrecklich; *(coll.: unpleasant)* fürchterlich

apparatus [æpə'reɪtəs] *n. (equipment)* Gerät, *das; (gymnastic ~)* Geräte *Pl.; (machinery, lit. or fig.)* Apparat, *der;* **a piece of** ~: ein Apparat

apparel [ə'pærəl] *n.* Kleidung, *die;* Gewänder *Pl. (geh.)*

apparent [ə'pærənt] *adj.* **a)** *(clear)* offensichtlich ‹*Ziel, Zweck, Wirkung, Begeisterung, Interesse*›; offenbar ‹*Bedeutung, Wahrheit*›; **it soon became ~ that ...**: es zeigte sich bald, daß ...; **heir** ~: recht- *od.* gesetzmäßiger Erbe; **b)** *(seeming)* scheinbar

apparently [ə'pærəntlɪ] *adv.* **a)** *(clearly)* offensichtlich; offenbar; **b)** *(seemingly)* scheinbar

apparition [æpə'rɪʃn] *n.* **a)** *(appearance)* [Geister]erscheinung, *die;* **b)** *(ghost)* Gespenst, *das*

appeal [ə'pi:l] **1.** *v. i.* **a)** *(Law etc.)* Einspruch erheben *od.* einlegen **(to** bei); ~ **to a court** bei einem Gericht Berufung einlegen; ~ **against sth.** gegen etw. Einspruch/Berufung einlegen; **b)** *(refer)* ~ **to** verweisen auf ‹*Erkenntnisse, Tatsachen*›; **c)** *(make earnest request)* ~ **to sb. for sth./to do sth.** jmdn. um etw. ersuchen/jmdn. ersuchen, etw. zu tun; **d)** *(address oneself)* ~ **to sb./sth.** an jmdn./etw. appellieren; **e)**

(be attractive) ~ **to sb.** jmdm. zusagen. **2.** *n.* **a)** *(Law etc.)* Einspruch, *der* (**to** bei); *(to higher court)* Berufung, *die* (**to** bei); **lodge an ~ with sb.** bei jmdm. Einspruch/Berufung einlegen; **right of ~:** Einspruchs-/Berufungsrecht, *das;* **Court of A~:** Berufungsgericht, *das;* **b)** *(reference)* Berufung, *die;* Verweisung, *die;* **make an ~ to sth.** sich auf etw. *(Akk.)* berufen; auf etw. *(Akk.)* verweisen; **c)** *(request)* Appell, *der;* Aufruf, *der;* **an ~ to sb. for sth.** eine Bitte an jmdn. um etw.; **make an ~ to sb.** an jmdn. appellieren; **d)** *(addressing oneself)* Appell, *der;* Aufruf, *der;* **make an ~ to sb.** einen Appell an jmdn. richten; **e)** *(attraction)* Reiz, *der*

appealing [ə'pi:lɪŋ] *adj.* **a)** *(imploring)* flehend; **b)** *(attractive)* ansprechend ‹*Farbe, Geschichte, Stil*›; verlockend ‹*Essen, Idee*›; reizvoll ‹*Haus, Beruf, Baustil*›; angenehm ‹*Stimme, Charakter*›

appear [ə'pɪə(r)] *v.i.* **a)** *(become visible, be seen, arrive)* erscheinen; ‹*Licht, Mond:*› auftauchen; ‹*Symptom, Darsteller:*› auftreten; *(present oneself)* auftreten; *(Sport)* spielen; **he was ordered to ~ before the court** er wurde vom Gericht vorgeladen; **he ~ed in court charged with murder** er stand wegen Mordes vor Gericht; **b)** *(occur)* vorkommen; ‹*Irrtum:*› vorkommen, auftreten; ‹*Ereignis:*› vorkommen, eintreten; **c)** *(seem)* ~ **[to be] ...:** scheinen ... [zu sein]; **~ to do sth.** scheinen, etw. zu tun; **try to ~ relaxed** versuch, entspannt zu erscheinen; **he could at least ~ to be interested** er könnte zumindest so tun, als ob er interessiert wäre

appearance [ə'pɪərəns] *n.* **a)** *(becoming visible)* Auftauchen, *das;* *(of symptoms)* Auftreten, *das;* *(arrival)* Erscheinen, *das;* *(of performer, speaker, etc.)* Auftritt, *der;* **make an** *or* **one's ~:** erscheinen; **make a public ~:** in der Öffentlichkeit auftreten; **put in an ~:** sich sehen lassen; **b)** *(look)* Äußere, *das;* **outward ~:** äußere Erscheinung; **~s** Äußerlichkeiten *Pl.;* **to judge by ~s, to all ~s** allem Anschein nach; **for the sake of ~s, to keep up ~s** um den Schein zu wahren; **c)** *(semblance)* Anschein, *der;* **~s to the contrary, ...:** entgegen allem Anschein ...; **~s can be deceptive** der Schein trügt; **d)** *(occurrence)* Auftreten, *das;* Vorkommen, *das*

appease [ə'pi:z] *v.t.* **a)** *(make calm)* besänftigen; *(Polit.)* beschwichtigen; **b)** *(soothe)* lindern ‹*Leid, Schmerz, Not*›; mildern ‹*Beunruhigung, Erregung*›; stillen ‹*Hunger, Durst*›

appeasement [ə'pi:zmənt] *n. see* **appease:** Besänftigung, *die;* Beschwichtigung, *die;* Linderung, *die;* Milderung, *die;* Stillen, *das*

append [ə'pend] *v.t.* ~ **sth. to sth.** etw. an etw. *(Akk.)* anhängen; *(add)* etw. einer Sache *(Dat.)* anfügen

appendage [ə'pendɪdʒ] *n.* **a)** Anhängsel, *das;* *(addition)* Anhang, *der;* **b)** *(accompaniment)* Zu-, Beigabe, *die* (**to** zu)

appendices *pl. of* **appendix**

appendicitis [əpendɪ'saɪtɪs] *n.* Blinddarmentzündung, *die* *(volkst.);* Appendizitis, *die* *(fachspr.)*

appendix [ə'pendɪks] *n., pl.* **appendices** [ə'pendɪsi:z] *or* **~es a)** Anhang, *der* (**to** zu); **b)** *(Anat.)* **[vermiform** [vɜ:mɪfɔ:m]**] ~:** Blinddarm, *der (volkst.);* Wurmfortsatz [des Blinddarms]

appertain [æpə'teɪn] *v.i.* ~ **to sth.** *(relate)* sich auf etw. *(Akk.)* beziehen; *(belong)* zu etw. gehören

appetite ['æpɪtaɪt] *n.* **a)** Appetit, *der* (**for** auf + *Akk.*); ~ **for sex** Lust auf Sex; **b)** *(fig.)* Verlangen, *das* (**for** nach); ~ **for knowledge** Wissensdrang, *der*

appetizer ['æpɪtaɪzə(r)] *n.* Appetitanreger, *der*

appetizing ['æpɪtaɪzɪŋ] *adj.* appetitlich

applaud [ə'plɔ:d] **1.** *v.i.* applaudieren; [Beifall] klatschen. **2.** *v.t.* applaudieren (+ *Dat.*); Beifall spenden (+ *Dat.*); *(approve of, welcome)* billigen; *(praise)* loben; anerkennen

applause [ə'plɔ:z] *n., no pl.* Applaus, *der;* *(praise)* Lob, *das;* Anerkennung, *die;* **give ~:** Applaus *od.* Beifall spenden; **get ~:** Applaus *od.* Beifall ernten

apple ['æpl] *n.* Apfel, *der;* **the ~ of sb.'s eye** *(fig.)* jmds. Liebling

apple: ~-cart *n.* **upset the ~-cart** *(fig.)* die Pferde *od.* Gäule scheu machen *(ugs.);* **~-'pie** *n.* gedeckte Apfeltorte; **~ 'sauce** *n.* Apfelmus, *das;* **~-tree** *n.* Apfelbaum, *der*

appliance [ə'plaɪəns] *n.* *(utensil)* Gerät, *das;* *(aid)* Hilfsmittel, *das*

applicable ['æplɪkəbl, ə'plɪkəbl] *adj.* **a)** anwendbar (**to** auf + *Akk.*); **b)** *(appropriate)* geeignet; angebracht; zutreffend ‹*Fragebogenteil usw.*›

applicant ['æplɪkənt] *n.* Bewerber, *der*/Bewerberin, *die* (**for** um); *(claimant)* Antragsteller, *der*/Antragstellerin, *die*

application [æplɪ'keɪʃn] *n.* **a)** *(request)* Bewerbung, *die* (**for** um); *(for passport, licence, etc.)* Antrag, *der* (**for** auf + *Akk.*); ~ **form** Antragsformular, *das;* **available on ~:** auf Anfrage erhältlich; **b)** *(diligence)* Fleiß, *der* (**to** bei); *(with enthusiasm)* Eifer, *der* (**to** für); **c)** *(putting)* Auftragen, *das* (**to** auf + *Akk.*); *(administering)* Anwendung, *die;* *(of heat, liquids)* Zufuhr, *die;* *(employment; of rule etc.)* Anwendung, *die;* **the ~ of new technology** der Einsatz neuer Technologien

apply [ə'plaɪ] **1.** *v.t.* **a)** anlegen ‹*Verband*›; auftragen ‹*Creme, Paste, Farbe*› (**to** auf + *Akk.*); zuführen ‹*Wärme, Flüssigkeit*› (**to** *Dat.*); ~ **the brakes** bremsen; die Bremse betätigen; **b)** *(make use of)* anwenden; **applied linguistics/mathematics** angewandte Sprachwissenschaft / Mathematik; **c)** *(devote)* richten, lenken ‹*Gedanken, Überlegungen, Geist*› (**to** auf + *Akk.*); verwenden ‹*Zeit, Energie*› (**to** auf + *Akk.*); ~ **oneself [to sth.]** sich *(Dat.)* Mühe geben [mit etw.]; sich [um etw.] bemühen. **2.** *v.i.* **a)** *(have relevance)* zutreffen (**to** auf + *Akk.*); *(be valid)* gelten; **things which don't ~ to us** Dinge, die uns nicht betreffen; **b)** *(address oneself)* ~ **[to sb.] for sth.** [jmdn.] um etw. bitten *od.* *(geh.)* ersuchen; *(for passport, licence, etc.)* [bei jmdm.] etw. beantragen; *(for job)* sich [bei jmdm.] um etw. bewerben

appoint [ə'pɔɪnt] *v.t.* **a)** *(fix)* bestimmen; festlegen ‹*Zeitpunkt, Ort*›; **b)** *(choose for a job)* einstellen; *(assign to office)* ernennen; ~ **sb. [to be** *or* **as] sth./to do sth.** jmdn. zu etw. ernennen/jmdn. dazu berufen, etw. zu tun; **~ sb. to sth.** jmdn. in etw. *(Akk.)* einsetzen

appointed [ə'pɔɪntɪd] *adj.* **a)** *(fixed)* vereinbart; verabredet; **b)** **well/badly ~:** gut/schlecht ausgestattet ‹*Zimmer usw.*›

appointment [ə'pɔɪntmənt] *n.* **a)** *(fixing)* Festlegung, *die;* Festsetzung, *die;* **b)** *(assigning to office)* Ernennung, *die;* Berufung, *die;* *(being assigned to office)* Ernennung, *die* (**as** zum/zur); *(to job)* Einstellung, *die;* **~ to a position** Berufung auf einen Posten; **by ~ to Her Majesty the Queen,**

makers of fine confectionery königlicher Hoflieferant für feines Konfekt; **c)** *(post)* Stelle, *die;* Posten, *der;* **a teaching ~**: eine Stelle als Lehrer/Lehrerin; **d)** *(arrangement)* Termin, *der;* **dental ~**: Termin beim Zahnarzt; **make an ~ with sb.** sich *(Dat.)* von jmdm. einen Termin geben lassen; **by ~**: mit Voranmeldung

apportion [ə'pɔ:ʃn] *v.t.* **a)** *(allot)* **~ sth. to sb.** jmdm. etw. zuteilen; **b)** *(portion out)* [gleichmäßig] verteilen (**among** an + *Akk.*)

apposite ['æpəzɪt] *adj.* *(appropriate)* passend; geeignet; *(well-chosen)* treffend; **~ to sth.** zutreffend auf etw. *(Akk.)*

appraisal [ə'preɪzl] *n.* Bewertung, *die*

appraise [ə'preɪz] *v.t.* bewerten

appreciable [ə'pri:ʃəbl] *adj.* *(perceptible)* nennenswert ‹*Unterschied, Einfluß*›; spürbar ‹*Veränderung, Wirkung, Erfolg*›; merklich ‹*Verringerung, Anstieg*›; *(considerable)* beträchtlich; erheblich

appreciably [ə'pri:ʃəblɪ] *adv.* *(perceptibly)* spürbar ‹*verändern*›; merklich ‹*sich unterscheiden*›; *(considerably)* beträchtlich; erheblich

appreciate [ə'pri:ʃɪeɪt, ə'pri:sɪeɪt] **1.** *v.t.* **a)** *([correctly] estimate value or worth of)* [richtig] einschätzen; *(understand)* verstehen; *(be aware of)* sich *(Dat.)* bewußt sein (+ *Gen.*); *(be receptive to)* Gefallen finden an (+ *Dat.*); **~ that/what ...**: verstehen, daß/was ...; **b)** *(be grateful for)* anerkennen; schätzen; *(enjoy)* genießen; **I'd really ~ that** das wäre sehr nett von dir. **2.** *v.i.* im Wert steigen

appreciation [əpri:ʃɪ'eɪʃn, əpri:sɪ'eɪʃn] *n.* **a)** *([right] estimation)* [richtige] Einschätzung; *(understanding)* Verständnis, *das* (**of** für); *(awareness)* Bewußtsein, *das;* *(sensitivity)* Sinn, *der* (**of** für); **b)** *(gratefulness)* Dankbarkeit, *die;* *(enjoyment)* Gefallen, *das* (**of** an + *Dat.*); **c)** *(rise in value)* Wertsteigerung, *die*

appreciative [ə'pri:ʃətɪv] *adj.* **a)** **be ~ of sth./sb.** *(aware of)* fähig sein, etw./jmdn. [richtig] einzuschätzen; **she is very ~ of music** sie hat viel Sinn für Musik; **b)** *(grateful)* dankbar (**of** für); *(approving)* anerkennend

apprehend [æprɪ'hend] *v.t.* **a)** *(arrest)* festnehmen; fassen; **b)** *(perceive)* wahrnehmen; *(understand)* erfassen; begreifen

apprehension [æprɪ'henʃn] *n.* **a)** *(arrest)* Festnahme, *die;* Verhaftung, *die;* **b)** *(uneasiness)* Besorgnis, *die;* **c)** *(conception)* Auffassung, *die;* Ansicht, *die* (**of** über + *Akk.*); *(understanding)* Verständnis, *das*

apprehensive [æprɪ'hensɪv] *adj.* besorgt (**for** um + *Akk.*); **~ of sth.** besorgt wegen etw.; **be ~ that ...**: befürchten, daß ...

apprehensively [æprɪ'hensɪvlɪ] *adv.* besorgt

apprentice [ə'prentɪs] **1.** *n.* *(learner)* Lehrling, *der* (**to** bei); *(beginner)* Neuling, *der;* Anfänger, *der.* **2.** *v.t.* **be ~d [to sb.]** [bei jmdm.] in der Lehre sein *od.* in die Lehre gehen

apprenticeship [ə'prentɪsʃɪp] *n.* *(training)* Lehre, *die* (**to** bei); *(learning period)* Lehrzeit, *die;* Lehrjahre *Pl.;* **serve an/one's ~**: eine/seine Lehre machen; *(fig.)* ein/sein Volontariat machen

approach [ə'prəʊtʃ] **1.** *v.i.* *(in space)* sich nähern; näher kommen; ‹*Sturm usw.:*› aufziehen; *(in time)* nahen; **the train now ~ing platform 1** der auf Gleis 1 einfahrende Zug; **the time is fast ~ing when you will have to ...**: es wird nicht mehr lange dauern und du mußt ... **2.** *v.t.* **a)** *(come near to)* sich nähern (+ *Dat.*); *(set about)* herangehen an (+ *Akk.*); angehen ‹*Problem, Aufgabe, Thema*›; **b)** *(be similar to)* verwandt sein (+ *Dat.*); **c)** *(approximate to)* nahekommen (+ *Dat.*); **the temperature/weight ~es 100 °C/50 kg** die Temperatur/das Gewicht beträgt nahezu 100 °C/50 kg; **d)** *(appeal to)* sich wenden an (+ *Akk.*). **3.** *n.* **a)** [Heran]nahen, *das;* *(treatment)* Ansatz, *der* (**to** zu); *(attitude)* Einstellung, *die* (**to** gegenüber); **b)** *(appeal)* Herantreten, *das* (**to** an + *Akk.*); **make an ~ to sb. concerning sth.** sich wegen etw. an jmdn. wenden; **c)** *(advance)* Annäherungsversuche; **make ~es to sb.** Annäherungsversuche bei jmdm. machen; **d)** *(access)* Zugang, *der;* *(road)* Zufahrtsstraße, *die;* *(fig.)* Zugang, *der;* **e)** *(Aeronaut.)* Landeanflug, *der;* Approach, *der*

approachable [ə'prəʊtʃəbl] *adj.* **a)** *(friendly)* umgänglich; *(receptive)* empfänglich; **b)** *(accessible)* zugänglich; erreichbar

ap'proach road *n.* Zufahrtsstraße, *die*

approbation [æprə'beɪʃn] *n.* *(sanction)* Genehmigung, *die;* *(approval)* Zustimmung, *die*

appropriate 1. [ə'prəʊprɪət] *adj.* *(suitable)* geeignet (**to, for** für); **I feel it is ~ to say a few words** ich halte es für angebracht, ein paar Worte zu sagen; **the ~ authority** die zuständige Behörde. **2.** [ə'prəʊprɪeɪt] *v.t.* **~ sth. [to oneself]** sich *(Dat.)* etw. aneignen

appropriately [ə'prəʊprɪətlɪ] *adv.* gebührend; passend ‹*dekoriert, gekleidet, genannt*›

approval [ə'pru:vl] *n.* **a)** *(sanctioning) (of plan, project, expenditure)* Genehmigung, *die;* *(of proposal, reform, marriage)* Billigung, *die;* *(agreement)* Zustimmung, *die;* Einwilligung, *die* (**for** in + *Akk.*); **b)** *(esteem)* Lob, *das;* Anerkennung, *die;* **does the plan meet with your ~?** findet der Plan Ihre Zustimmung?; **on ~** *(Commerc.)* zur Probe; *(to view)* zur Ansicht

approve [ə'pru:v] **1.** *v.t.* **a)** *(sanction)* genehmigen ‹*Plan, Projekt, Ausgaben*›; billigen ‹*Vorschlag, Reform, Heirat*›; **~d hotel** empfohlenes Hotel; **~d school** *(Brit. Hist.)* Erziehungsheim, *das;* **b)** *(find good)* gutheißen; für gut halten. **2.** *v.i.* **~ of** billigen; zustimmen (+ *Dat.*) ‹*Plan*›; einverstanden sein mit ‹*Tätigkeiten, Gewohnheiten, Verhalten*›

approving [ə'pru:vɪŋ] *adj.* zustimmend, beipflichtend ‹*Worte*›; anerkennend, bewundernd ‹*Blicke*›

approvingly [ə'pru:vɪŋlɪ] *adv. see* **approving**: zustimmend; anerkennend

approximate 1. [ə'prɒksɪmət] *adj.* *(fairly correct)* ungefähr *attr.;* **the figures given here are only ~**: dies hier sind nur ungefähre Zahlen. **2.** [ə'prɒksɪmeɪt] *v.t.* **a)** *(make similar)* **~ sth. to sth.** etw. einer Sache *(Dat.)* anpassen; **b)** *(come near to)* nahekommen (+ *Dat.*); annähernd erreichen (+ *Akk.*). **3.** [ə'prɒksɪmeɪt] *v.i.* **sth. ~s to sth.** etw. gleicht einer Sache *(Dat.)* annähernd

approximately [ə'prɒksɪmətlɪ] *adv.* *(roughly)* ungefähr; *(almost)* fast; **the answer is ~ correct** die Antwort stimmt ungefähr; **very ~**: ganz grob

approximation [əprɒksɪ'meɪʃn] *n.* **a)** Annäherung, *die* (**to** an + *Dat.*); **b)** *(estimate)* Annäherungswert, *der*

Apr. *abbr.* **April** Apr.

apricot ['eɪprɪkɒt] *n.* Aprikose, *die*

April ['eɪprəl] *n.* April, *der;* **~ fool** April[s]narr, *der;* **~ Fool's Day** der 1. April; *see also* **August**

apron ['eɪprən] *n.* *(garment)* Schürze, *die;* **be tied to sb.'s ~ strings** jmdm. an der Schürze *od.* am Schürzenzipfel hängen

apropos [æprə'pəʊ, 'æprəpəʊ] *adv.* ~ **of** in bezug auf (+ *Akk.*); hinsichtlich (+ *Gen.*)

apse [æps] *n.* Apsis, *die*

apt [æpt] *adj.* **a)** *(suitable)* passend ‹*Ausdruck, Geschenk*›; angemessen ‹*Reaktion*›; treffend ‹*Zitat, Bemerkung*›; **b)** *(tending)* **be ~ to do sth.** dazu neigen, etw. zu tun

aptitude ['æptɪtju:d] *n.* **a)** *(propensity)* Neigung, *die;* *(ability)* Begabung, *die;* **linguistic ~:** Sprachbegabung, *die;* **b)** *(suitability)* Eignung, *die*

aptly ['æptlɪ] *adv.* passend; ~ **chosen words** treffend gewählte Worte

aqualung ['ækwəlʌŋ] *n.* Tauchgerät, *das*

aquarium [ə'kweərɪəm] *n., pl.* ~**s** *or* **aquaria** [ə'kweərɪə] Aquarium, *das*

Aquarius [ə'kweərɪəs] *n.* *(Astrol., Astron.)* der Wassermann

aquatic [ə'kwætɪk] *adj.* aquatisch; Wasser-; ~ **plant/bird** Wasserpflanze, *die*/-vogel, *der*

aqueduct ['ækwɪdʌkt] *n.* Aquädukt, *der od. das*

aquiline ['ækwɪlaɪn] *adj.* adlerartig; Adler-; ~ **eye/nose** Adlerauge, *das*/-nase, *die*

Arab ['ærəb] **1.** *adj.* arabisch; ~ **horse** Araber, *der.* **2.** *n.* Araber, *der*/Araberin, *die*

Arabia [ə'reɪbɪə] *pr. n.* Arabien *(das)*

Arabian [ə'reɪbɪən] **1.** *adj.* arabisch; **the ~ Nights** Tausendundeine Nacht. **2.** *n.* Araber, *der*/Araberin, *die*

Arabic ['ærəbɪk] **1.** *adj.* arabisch. **2.** *n.* Arabisch, *das; see also* **English 2 a**

arable ['ærəbl] *adj.* ~ **land** kultivierbares Land; *(cultivated)* Ackerland, *das*

arbiter ['ɑ:bɪtə(r)] *n.* *(judge)* Richter, *der;* *(arbitrator)* Vermittler, *der*

arbitrarily ['ɑ:bɪtrərɪlɪ] *adv.* *(at random)* willkürlich; *(capriciously)* aus einer Laune heraus

arbitrariness ['ɑ:bɪtrərɪnɪs] *n., no pl.* *(randomness)* Willkür, *die*

arbitrary ['ɑ:bɪtrərɪ] *adj.* **a)** *(random)* willkürlich; arbiträr; *(capricious)* launenhaft; launisch ‹*Idee*›; **b)** *(unrestrained)* rücksichtslos ‹*Vorgehen, Bestrafung, Wesen, Haltung*›

arbitrate ['ɑ:bɪtreɪt] **1.** *v. t.* schlichten, beilegen ‹*Streit*›. **2.** *v. i.* ~ **[upon sth.]** [in einer Sache] vermitteln *od.* als Schiedsrichter fungieren

arbitration [ɑ:bɪ'treɪʃn] *n.* Vermittlung, *die;* *(in industry)* Schlichtung, *die;* **go to ~:** einen Schlichter anrufen *od.* einschalten

arbitrator ['ɑ:bɪtreɪtə(r)] *n.* *(mediator)* Vermittler, *der;* *(in industry)* Schlichter, *der;* *(arbiter)* Schiedsrichter, *der;* *(judge)* Richter, *der*

arc [ɑ:k] *n.* **a)** [Kreis]bogen, *der;* **b)** *(Electr.)* Lichtbogen, *der;* ~ **lamp,** ~ **light** Lichtbogenlampe, *die*

arcade [ɑ:'keɪd] *n.* Arkade, *die;* **shopping ~:** Einkaufspassage, *die*

arcane [ɑ:'keɪn] *adj.* geheimnisvoll

arch [ɑ:tʃ] **1.** *n.* Bogen, *der;* *(curvature; of foot)* Wölbung, *die;* *(of bridge)* Bogen, *der;* Joch, *das;* *(vault)* Gewölbe, *das.* **2.** *v. i.* sich wölben; ‹*Ast, Glied:*› sich biegen. **3.** *v. t.* beugen ‹*Rücken, Arm*›; **the cat ~ed its back** die Katze machte einen Buckel

arch- *pref.* Erz-; ~**-villain** Erzschurke, *der;* Erzgauner, *der*

archaeological [ɑ:kɪə'lɒdʒɪkl] *adj.* archäologisch

archaeologist [ɑ:kɪ'ɒlədʒɪst] *n.* Archäologe, *der*/Archäologin, *die*

archaeology [ɑ:kɪ'ɒlədʒɪ] *n.* Archäologie, *die*

archaic [ɑ:'keɪɪk] *adj.* *(out of use)* veraltet; *(antiquated)* altertümlich; überholt ‹*Methode, Gesetz*›

archaism ['ɑ:keɪɪzm] *n.* Archaismus, *der*

archangel ['ɑ:keɪndʒl] *n.* Erzengel, *der*

arch'bishop *n.* Erzbischof, *der*

arch-'enemy *n.* Erzfeind, *der*

archeology *etc.* *(Amer.) see* **archaeology** *etc.*

archer ['ɑ:tʃə(r)] *n.* Bogenschütze, *der*

archery ['ɑ:tʃərɪ] *n., no pl.* Bogenschießen, *das*

archetypal ['ɑ:kɪtaɪpl] *adj.* *(original)* archetypisch *(geh.);* *(typical)* typisch

archetype ['ɑ:kɪtaɪp] *n.* *(original)* Urfassung, *die;* Archetyp, *der;* *(typical specimen)* Prototyp, *der*

archipelago [ɑ:kɪ'peləgəʊ] *n., pl.* ~**s** *or* ~**es** Archipel, *der;* *(islands)* Inselgruppe, *die;* *(sea)* Inselmeer, *das*

architect ['ɑ:kɪtekt] *n.* Architekt, *der*/Architektin, *die*

architectural [ɑ:kɪ'tektʃərl] *adj.* architektonisch

architecture ['ɑ:kɪtektʃə(r)] *n.* **a)** Architektur, *die;* Baukunst, *die (geh.);* *(style)* Bauweise, *die;* Architektur, *die;* **naval/railway ~:** Schiff[s]-/Eisenbahnbau, *der;* **b)** *(structure, lit. or fig.)* Konstruktion, *die;* **c)** *(Computing)* [System]architektur, *die*

archive ['ɑ:kaɪv] **1.** *n., usu. in pl.* Archiv, *das.* **2.** *v. t.* archivieren

'archway *n.* *(vaulted passage)* Gewölbegang, *der;* Tunnel, *der;* *(arched entrance)* Durchgang, *der;* Torbogen, *der*

arctic ['ɑ:ktɪk] **1.** *adj.* *(lit. or fig.)* arktisch; **A~ Circle/Ocean** nördlicher Polarkreis/Nordpolarmeer, *das.* **2.** *pr. n.* **the A~:** die Arktis

ardent ['ɑ:dənt] *adj.* *(eager)* begeistert; *(fervent)* glühend ‹*Bewunderer, Leidenschaft*›; hitzig ‹*Temperament, Wesen*›; brennend ‹*Wunsch*›; leidenschaftlich ‹*Gedicht, Liebesbrief, Anbetung*›; innigst, *(geh.)* inbrünstig ‹*Hoffnung, Liebe*›

ardour (*Brit.; Amer.:* **ardor**) ['ɑ:də(r)] *n.* *(passionate emotion)* Inbrunst, *die (geh.);* *(fervour)* Eifer, *der;* ~ **for reform** Reformeifer, *der*

arduous ['ɑ:djʊəs] *adj.* schwer, anstrengend ‹*Aufgabe, Arbeit*›; hart ‹*Arbeit, Tag*›; beschwerlich ‹*Reise, Aufstieg, Fahrt*›

arduously ['ɑ:djʊəslɪ] *adv.* *(laboriously)* beschwerlich

are *see* **be**

area ['eərɪə] *n.* **a)** *(surface measure)* Flächenausdehnung, *die;* **what is the ~ of your farm?** wie groß ist Ihr Hof?; **b)** *(region)* Gelände, *das;* *(of wood, marsh, desert)* Gebiet, *das;* *(of city, country)* Gegend, *die;* *(of skin, wall, etc.)* Stelle, *die;* **in the Hamburg ~:** im Hamburger Raum; **c)** *(defined space)* Bereich, *der;* **parking/picnic ~:** Park-/Picknickplatz, *der;* **no-smoking ~:** Nichtraucherzone, *die;* **d)** *(subject field)* Gebiet, *das;* **e)** *(scope)* Raum, *der;* ~ **of choice** Wahlmöglichkeiten *Pl.*

arena [ə'ri:nə] *n.* *(at circus, bullfight)* Arena, *die;* *(fig.: scene of conflict)* Bühne, *die;* *(fig.: sphere of action)* Bereich, *der;* **the political ~:** die politische Arena; **enter the ~** *(fig.)* die Arena betreten

aren't [ɑ:nt] *(coll.)* = **are not;** *see* **be**

Argentina [ɑdʒən'ti:nə] *pr. n.* Argentinien *(das)*

Argentine ['ɑ:dʒəntaɪn] **1.** *pr. n.* **the ~:** Argentinien *(das).* **2.** *n. see* **Argentinian 2. 3.** *adj.* argentinisch

Argentinian [ɑ:dʒən'tɪnɪən] **1.** *adj.* argentinisch; **sb. is ~:** jmd. ist Argentinier/Argentinierin. **2.** *n.* Argentinier, *der*/Argentinierin, *die*

argon ['ɑ:gɒn] *n. (Chem.)* Argon, *das*

arguable ['ɑ:gjʊəbl] *adj.* **a)** fragwürdig ‹*Angelegenheit, Punkt*›; **it's ~ whether ...:** es ist noch die Frage, ob ...; **b) it is ~ that ...** *(can reasonably be argued that)* man kann sich auf den Standpunkt stellen, daß ...

arguably ['ɑ:gjʊəblɪ] *adv.* möglicherweise

argue ['ɑ:gju:] **1.** *v. t.* **a)** *(maintain)* **~ that ...:** die Ansicht vertreten, daß ...; **b)** *(treat by reasoning)* darlegen ‹*Grund, Standpunkt, Fakten*›; **c)** *(persuade)* **~ sb. into doing sth.** jmdn. dazu überreden, etw. zu tun; **~ sb. out of doing sth.** [es] jmdm. ausreden, etw. zu tun. **2.** *v. i.* **~ with sb.** sich mit jmdm. streiten; **~ against sb.** jmdm. widersprechen; **~ for/against sth.** sich für/gegen etw. aussprechen; **~ about sth.** sich über etw. *(Akk.)* streiten

argument ['ɑ:gjʊmənt] *n.* **a)** *(reason)* Begründung, *die;* **~s for/against sth.** Argumente für/gegen etw.; **b)** *no pl. (reasoning process)* Argumentieren, *das;* **assume sth. for ~'s sake** etw. rein theoretisch annehmen; **c)** *(debate)* Auseinandersetzung, *die;* **get into an ~/get into ~s with sb.** mit jmdm. in Streit geraten

argumentation [ɑ:gjʊmen'teɪʃn] *n.* Argumentieren, *das*

argumentative [ɑ:gjʊ'mentətɪv] *adj.* widerspruchsfreudig

argy-bargy [ɑ:dʒɪ'bɑ:dʒɪ] *n. (joc.)* Hickhack, *der od. das (ugs.)*

aria ['ɑ:rɪə] *n. (Mus.)* Arie, *die*

Arian ['eərɪən] *see* **Aryan**

arid ['ærɪd] *adj.* **a)** *(dry)* trocken ‹*Klima, Land*›; *(Geog.)* arid; **~ zone** Trockengürtel, *der;* **b)** *(barren)* karg

aridity [ə'rɪdɪtɪ] *n., no pl. see* **arid:** Trockenheit, *die;* Aridität, *die;* Kargheit, *die*

Aries ['eəri:z] *n. (Astrol., Astron.)* der Widder

aright [ə'raɪt] *adv.* recht

arise [ə'raɪz] *v. i.,* **arose** [ə'rəʊz], **arisen** [ə'rɪzn] **a)** *(originate)* entstehen; **b)** *(present itself)* auftreten; ‹*Gelegenheit:*› sich bieten; **a crisis has ~n in Turkey** in der Türkei ist es zu einer Krise gekommen; **c)** *(result)* **~ from** *or* **out of sth.** auf etw. *(Akk.)* zurückzuführen sein; **d)** ‹*Sonne, Nebel:*› aufsteigen; **e)** ‹*See, Sturm:*› anschwellen; **f)** *(rise from the dead)* auferstehen

aristocracy [ærɪ'stɒkrəsɪ] *n.* Aristokratie, *die*

aristocrat ['ærɪstəkræt] *n.* Aristokrat, *der*/Aristokratin, *die*

aristocratic [ærɪstə'krætɪk] *adj.* **a)** aristokratisch; Aristokraten-; adelig; Adels-; **b)** vornehm ‹*Aussehen, Auftreten*›; *(refined)* kultiviert; fein ‹*Manieren, Sitten*›; edel ‹*Geschmack, Wein*›

aristocratically [ærɪstə'krætɪkəlɪ] *adv.* aristokratisch

Aristotle ['ærɪstɒtl] *pr. n.* Aristoteles *(der)*

[1]**arithmetic** [ə'rɪθmətɪk] *n.* **a)** *(science)* Arithmetik, *die;* **b)** *(computation)* Rechnen, *das;* **mental ~:** Kopfrechnen, *das*

[2]**arithmetic** [ærɪθ'metɪk], **arithmetical** [ærɪθ'metɪkl] *adj.* arithmetisch

ark [ɑ:k] *see* **Noah's ark**

[1]**arm** [ɑ:m] *n.* **a)** *(also of sea etc.)* Arm, *der;* **~ in ~:** Arm in Arm; **remain** *or* **keep at ~'s length from sb.** *(fig.)* eine gewisse Distanz zu jmdm. wahren; **as long as sb.'s ~** *(fig.)* ellenlang; **cost sb. an ~ and a leg** *(fig.)* jmdn. eine Stange Geld kosten *(ugs.);* **on sb.'s ~:** an jmds. Arm *(Dat.);* **under one's ~:** unter dem Arm; **take sb. in one's ~s** jmdn. in die Arme nehmen *od. (geh.)* schließen; **with open ~s** *(lit. or fig.)* mit offenen Armen; **b)** *(sleeve)* Ärmel, *der;* **c)** *(support)* Armlehne, *die*

[2]**arm 1.** *n.* **a)** *usu. in pl. (weapon)* Waffe, *die;* **~s race** Rüstungswettlauf, *der;* **small ~s** Handfeuerwaffen; **lay down one's ~s** die Waffen niederlegen; **take up ~s** zu den Waffen greifen; **be up in ~s about sth.** *(fig.)* wegen etw. aufgebracht sein; **b)** *in pl. (heraldic devices)* Wappen, *das.* **2.** *v. t.* **a)** *(furnish with weapons)* bewaffnen; mit Waffen ausrüsten ‹*Schiff*›; **b) ~ oneself with sth.** sich mit etw. wappnen; **c)** *(make able to explode)* scharf machen ‹*Bombe usw.*›

armada [ɑ:'mɑ:də] *n.* Armada, *die*

armadillo [ɑ:mə'dɪləʊ] *n., pl.* **~s** *(Zool.)* Gürteltier, *das*

armament ['ɑ:məmənt] *n.* **~[s]** Kriegsgerät, *das*

arm: ~band *n.* Armbinde, *die;* **~chair 1.** *n.* Sessel, *der;* **2.** *adj.* **~chair critic** Hobby- *od.* Amateurkritiker, *der;* **~chair travel** Reisen in der Phantasie

armed [ɑ:md] *adj.* bewaffnet; **~ forces** Streitkräfte *Pl.;* **~ neutrality** bewaffnete Neutralität

-armed *adj. in comb.* mit ... Armen; **two-~:** zweiarmig

Armenia [ɑ'mi:nɪə] *pr. n.* Armenien *(das)*

Armenian [ɑ:'mi:nɪən] **1.** *adj.* armenisch; **sb. is ~:** jmd. ist Armenier/Armenierin. **2.** *n.* **a)** *(person)* Armenier, *der*/Armenierin, *die;* **b)** *(language)* Armenisch, *das; see also* **English 2 a**

armful ['ɑ:mfʊl] *n.* **an ~ of fruit** ein Armvoll Obst; **flowers by the ~:** ganze Arme voll Blumen

'**armhole** *n.* Armloch, *das*

armistice ['ɑ:mɪstɪs] *n.* Waffenstillstand, *der;* **A~ Day** *Gedenktag des Endes des 1. Weltkriegs*

armless ['ɑ:mlɪs] *adj.* armlos

armor, armored *(Amer.) see* **armour, armoured**

armorial [ɑ:'mɔ:rɪəl] *adj.* Wappen-; **~ bearings** Wappen, *das*

armour ['ɑ:mə(r)] *n. (Brit.)* **a)** *no pl. (Hist.)* Rüstung, *die;* **suit of ~:** Harnisch, *der;* **b)** *no pl. (steel plates)* **~[-plate]** Panzerung, *die;* **c)** *no pl. (~ed vehicles)* Panzerfahrzeuge

armoured ['ɑ:məd] *adj. (Brit.)* **~ car/train** Panzerwagen/-zug, *der;* **~ cable** armiertes Kabel; **~ division** Panzerdivision, *die;* **~ glass** Panzerglas, *das*

arm: ~pit *n.* Achselhöhle, *die;* **~-rest** *n.* Armlehne, *die*

army ['ɑ:mɪ] *n.* **a)** *(fighting force)* Heer, *das;* **b)** *no pl., no indef. art. (military profession)* Militär, *das;* **be in the ~:** beim Militär sein; **go into** *or* **join the ~:** zum Militär gehen; *(as a career)* die Militärlaufbahn einschlagen; **c)** *(large number)* Heer, *das;* **an ~ of workmen/ants** ein Heer von Arbeitern/Ameisen

aroma [ə'rəʊmə] *n.* Duft, *der*

aromatic [ærə'mætɪk] *adj.* aromatisch *(auch Chem.);* duftend ‹*Blütenblätter, Nelken usw.*›

arose *see* **arise**

around [ə'raʊnd] **1.** *adv.* **a)** *(on every side)* **[all] ~:** überall; **he waved his arms ~:** er ruderte mit den Armen; **b)** *(round)* herum; **come ~ to sb.'s house** bei jmdm. vorbeikommen; **show sb. ~:** jmdn. herumführen; **pass the hat ~:** den Hut herumgehen lassen; **get ~ to doing sth.** [endlich] einmal daran denken, etw. zu tun; **[have a] look ~:** sich [ein bißchen] umsehen; **c)** *(coll.: near)* in der Nähe; **we'll always be ~ when you need us** wir werden immer dasein, wenn du uns brauchst; **d)** *(coll.:*

in existence) vorhanden; **there's not much leather ~ these days** zur Zeit gibt es nur wenig Leder; **e)** *(in various places)* **ask/look ~:** herumfragen/-schauen; **he's been ~** *(fig.)* er ist viel herumgekommen. **2.** *prep.* **a)** um [... herum]; **they had their arms ~ each other** sie hielten sich umschlungen; **darkness closed in ~ us** die Dunkelheit umfing uns *(geh.) od.* schloß uns ein; **~ the back of the house** *(position)* hinter dem Haus; *(direction)* hinter das Haus; **b)** *(here and there in)* **we went ~ the town** wir gingen durch die Stadt; **c)** *(approximately at)* **~ 3 o'clock** gegen 3 Uhr; **I saw him somewhere ~ the station** ich habe ihn irgendwo am Bahnhof gesehen; **d)** *(approximately equal to)* ungefähr

arousal [ə'raʊzl] *n. see* **arouse:** Aufwachen, *das;* Erregung, *die;* Erweckung, *die*

arouse [ə'raʊz] *v.t.* **a)** *(awake)* [auf]wecken; **b)** *(excite, also sexually)* erregen; *(call into existence)* erwecken ‹*Interesse, Begeisterung*›; erregen ‹*Haß, Leidenschaften, Verdacht*›

arr. *abbr.* **a)** *(Mus.)* **arranged by** Arr.; **b) arrives** Ank.

arraign [ə'reɪn] *v.t.* anklagen (**for** wegen)

arrange [ə'reɪndʒ] **1.** *v.t.* **a)** *(order)* anordnen; *(adjust)* in Ordnung bringen; **b)** *(Mus., Radio, etc.: adapt)* bearbeiten; **c)** *(settle beforehand)* ausmachen ‹*Termin*›; **d)** *(plan)* planen ‹*Urlaub*›; aufstellen ‹*Stundenplan*›; **don't ~ anything for tomorrow** nimm dir für morgen nichts vor. **2.** *v.i.* **a)** *(plan)* sorgen (**about, for** für); **~ for sb./sth. to do sth.** veranlassen *od.* dafür sorgen, daß jmd./etw. etw. tut; **can you ~ to be at home?** kannst du es so einrichten, daß du zu Hause bist?; **b)** *(agree)* **they ~d to meet the following day** sie verabredeten sich für den nächsten Tag; **~ with sb. about sth.** sich mit jmdm. über etw. *(Akk.)* einigen

arrangement [ə'reɪndʒmənt] *n.* **a)** *(ordering, order)* Anordnung, *die; (thing ordered)* Arrangement, *das;* **seating-~:** Anordnung der Sitze; **b)** *(Mus., Radio, etc.: adapting, adaptation)* Bearbeitung, *die;* **a guitar ~:** eine Bearbeitung *od.* ein Arrangement für Gitarre; **c)** *(settling beforehand)* Vereinbarung, *die; (of plans)* Aufstellung, *die;* **by ~:** nach Vereinbarung; **d)** *in pl. (plans)* Vorkehrungen; **make ~s** Vorkehrungen treffen; **holiday ~s** Urlaubsvorbereitungen; **e)** *(agreement)* Vereinbarung, *die;* **make an ~ to do sth.** vereinbaren, etw. zu tun; **f)** *(resolution)* Einigung, *die;* **I'm sure we can come to some ~ about ...:** wir können uns sicher irgendwie einigen über (+ *Akk.*) ...

arrant ['ærənt] *adj.* Erz‹*lump, -schurke, -lügner, -feigling*›; **~ nonsense** barer Unsinn

array [ə'reɪ] **1.** *v.t. (formal: dress)* schmücken; **~ sb. in sth.** jmdn. in etw. *(Akk.)* kleiden *od. (geh.)* hüllen. **2.** *n.* Reihe, *die*

arrears [ə'rɪəz] *n. pl. (debts)* Schulden *Pl.;* **be in ~ with sth.** mit etw. im Rückstand sein; **be paid in ~:** rückwirkend bezahlt werden

arrest [ə'rest] **1.** *v.t.* **a)** *(stop)* aufhalten; zum Stillstand bringen ‹*Fluß*›; **b)** *(seize)* verhaften, *(temporarily)* festnehmen ‹*Person*›; **c)** *(catch)* erregen ‹*Aufmerksamkeit, Interesse*›. **2.** *n.* **a)** *(stoppage)* Stillstand, *der;* **cardiac ~:** Herzstillstand, *der;* **b)** *(of person)* Verhaftung, *die; (temporary)* Festnahme, *die;* **under ~:** festgenommen

arrival [ə'raɪvl] *n.* **a)** Ankunft, *die; (fig.: at decision etc.)* Gelangen, *das* (**at** zu); *(of mail etc.)* Eintreffen, *das; (coming)* Kommen, *das;* **b)** *(appearance)* Auftauchen, *das;* **c)** *(person)* Ankömmling, *der; (thing)* Lieferung, *die;* **new ~** *(coll.: new-born baby)* Neugeborene, *das;* **new ~s** Neuankömmlinge

arrive [ə'raɪv] *v.i.* **a)** ankommen; **when do we arrive at Frankfurt?** wann kommen wir in Frankfurt an?; **~ at a conclusion/an agreement** zu einem Schluß/einer Einigung kommen; **b)** *(establish oneself)* es schaffen; **c)** *(be brought)* eintreffen; *(coll.: be born)* ankommen; **d)** *(come)* ‹*Stunde, Tag, Augenblick:*› kommen

arrogance ['ærəgəns] *n., no pl.* Arroganz, *die; (presumptuousness)* Anmaßung, *die*

arrogant ['ærəgənt] *adj.* arrogant; *(presumptuous)* anmaßend

arrogantly ['ærəgəntlɪ] *adv.* arrogant; *(presumptuously)* anmaßend; anmaßenderweise ‹*behaupten, verlangen*›

arrow ['ærəʊ] **1.** *n. (missile)* Pfeil, *der; (pointer)* [Hinweis-, Richtungs]pfeil, *der.* **2.** *v.t.* mit einem Pfeil/mit Pfeilen markieren

'arrow-head *n.* Pfeilspitze, *die*

arse [ɑ:s] *n. (coarse)* Arsch, *der (derb)*

'arse-hole *n. (coarse)* Arschloch, *das (derb)*

arsenal ['ɑ:sənl] *n.* [Waffen]arsenal, *das; (fig.)* Arsenal, *das*

arsenic ['ɑ:sənɪk] *n. (Chem.)* **a)** Arsenik, *das;* **b)** *(element)* Arsen, *das*

arson ['ɑ:sn] *n.* Brandstiftung, *die*

arsonist ['ɑ:sənɪst] *n.* Brandstifter, *der*/Brandstifterin, *die*

art [ɑ:t] *n.* **a)** Kunst, *die;* **the ~s** *see* **fine art c; works of ~:** Kunstwerke *Pl.;* **~s and crafts** Kunsthandwerk, *das;* Kunstgewerbe, *das;* **b)** *in pl. (branch of study)* Geisteswissenschaften; **faculty of ~s** philosophische Fakultät; **c)** *(cunning)* List, *die*

'art college *see* **art school**

art deco [ɑ:t 'dekəʊ] *n., no pl.* Art deco, *die*

artefact, artifact ['ɑ:tɪfækt] *n.* Artefakt, *das*

arterial [ɑ:'tɪərɪəl] *adj.* arteriell; **~ road** Hauptverkehrsstraße, *die*

artery ['ɑ:tərɪ] *n.* **a)** *(Anat.)* Schlagader, *die;* Arterie, *die (bes. fachspr.);* **b)** *(fig.: road etc.)* [Haupt]verkehrsader, *die*

artesian [ɑ:'ti:zɪən, ɑ:'ti:ʒən] *adj.* **~ well** artesischer Brunnen

'art-form *n. (form of composition)* [Kunst]gattung, *die; (medium of expression)* Kunst[form], *die*

artful ['ɑ:tfl] *adj.* schlau; **~ dodger** Schlawiner, *der*

artfully ['ɑ:tfəlɪ] *adv.* schlau

artfulness ['ɑ:tflnɪs] *n., no pl.* Schlauheit, *die*

'art gallery *n.* Kunstgalerie, *die*

arthritic [ɑ:'θrɪtɪk] *(Med.)* **1.** *adj.* arthritisch. **2.** *n.* Arthritiker, *der*/Arthritikerin, *die*

arthritis [ɑ:'θraɪtɪs] *n. (Med.)* Arthritis, *die (fachspr.);* Gelenkentzündung, *die*

artichoke ['ɑ:tɪtʃəʊk] *n.* **[globe] ~:** Artischocke, *die;* **Jerusalem** [dʒə'ru:sələm] **~:** Topinamburwurzel, *die*

article ['ɑ:tɪkl] **1.** *n.* **a)** *(of constitution, treaty)* Artikel, *der; (of agreement)* [Vertrags]punkt, *der; (of the law)* Paragraph, *der;* **~s [of association]** Satzung, *die;* **~s of apprenticeship/employment** Lehr-/Arbeitsvertrag, *der;* **~ of faith** *(fig.)* Glaubensbekenntnis, *das (fig.);* **b)** *(in magazine, newspaper)* Artikel, *der; (in technical journal)* Beitrag, *der;* **c)** *(Ling.)* Artikel, *der;* **definite/indefinite ~:** bestimmter/unbestimmter Artikel; **d)** *(thing)* Artikel, *der;* **an ~ of furniture/clothing** ein Möbel-/Kleidungsstück; **an ~ of value** ein Wertgegenstand. **2.** *v.t.* in die

Lehre geben (**to** bei); **be ~d to sb.** bei jmdm. in der Lehre sein

articled ['ɑ:tɪkld] *adj.* ~ **clerk** *(Law)* Rechtspraktikant, *der*/-praktikantin, *die;* ≈ Rechtsreferendar, *der*/-referendarin, *die*

articulate 1. [ɑ:'tɪkjʊlət] *adj.* **a)** *(clear)* verständlich; **b)** *(eloquent)* redegewandt; **be ~/not very ~:** sich gut/nicht sehr gut ausdrücken [können]. **2.** [ɑ:'tɪkjʊleɪt] *v. t.* **a)** *usu. in pass.* durch Gelenke/ein Gelenk verbinden; **~d lorry** Sattelzug, *der;* **b)** *(pronounce)* [deutlich] aussprechen; *(utter, express)* artikulieren. **3.** *v. i.* **a)** *(speak distinctly)* deutlich sprechen; **b)** *(form a joint)* ~ **with sth.** mit etw. Gelenke/ein Gelenk bilden

articulately [ɑ:'tɪkjʊlətlɪ] *adv.* klar

articulation [ɑ:tɪkjʊ'leɪʃn] *n.* **a)** *(clear speech)* deutliche Aussprache; **b)** *(act of speaking)* Artikulation, *die*

artifice ['ɑ:tɪfɪs] *n.* List, *die*

artificial [ɑ:tɪ'fɪʃl] *adj.* **a)** *(not natural)* künstlich; Kunst-; *(not real)* unecht; imitiert; ~ **sweetener** Süßstoff, *der;* ~ **limb** Prothese, *die;* ~ **eye** Glasauge, *das;* **b)** *(affected)* affektiert; *(insincere)* gekünstelt; **she wore an ~ smile for the cameras** für die Fotografen setzte sie ein einstudiertes Lächeln auf

artificial: ~ **ho'rizon** *n.* Kreiselhorizont, *der;* ~ **insemination** [ɑ:tɪfɪʃl ɪnsemɪ'neɪʃn] *n.* künstliche Besamung; ~ **in'telligence** *n.* künstliche Intelligenz

artificiality [ɑ:tɪfɪʃɪ'ælɪtɪ] *n., no pl. see* **artificial:** Künstlichkeit, *die;* Unechtheit, *die;* Affektiertheit, *die;* Gekünsteltheit, *die*

artificial: ~ **'kidney** *see* **kidney machine;** ~ **'language** *n.* Kunstsprache, *die*

artificially [ɑ:tɪ'fɪʃəlɪ] *adv.* **a)** *(unnaturally)* künstlich; **b)** *(affectedly)* affektiert; *(insincerely)* gekünstelt

artificial respi'ration *n.* künstliche Beatmung

artillery [ɑ:'tɪlərɪ] *n.* Artillerie, *die*

artisan ['ɑ:tɪzn, ɑ:tɪ'zæn] *n.* [Kunst]handwerker, *der*

artist ['ɑ:tɪst] *n.* **a)** *(painter, musician, etc.; also fig.)* Künstler, *der*/Künstlerin, *die;* **b)** *see* **artiste**

artiste [ɑ:'ti:st] *n.* Artist, *der*/Artistin, *die*

artistic [ɑ:'tɪstɪk] *adj.* **a)** *(of art)* Kunst-; künstlerisch; ~ **movement** Kunstrichtung, *die;* **b)** *(of artists)* Künstler-; künstlerisch; ~ **circles** Künstlerkreise; **c)** *(made with art)* kunstvoll; Kunst-; **a truly ~ piece of poetry/writing** ein dichterisches/schriftstellerisches Kunstwerk; **d)** *(naturally skilled in art)* künstlerisch begabt; **e)** *(appreciative of art)* kunstverständig; ~ **sense** Kunstverständnis, *das*

artistically [ɑ:'tɪstɪkəlɪ] *adv.* **a)** *(in art)* künstlerisch; **b)** *(with art)* kunstvoll ‹*geschmückt, gestaltet*›

artless ['ɑ:tlɪs] *adj.* **a)** *(guileless)* arglos; **b)** *(simple)* schlicht; ~ **beauty/grace** natürliche Schönheit/Anmut

art nouveau [ɑ: nu:'vəʊ] *n.* Jugendstil, *der*

art: ~ **school** *n.* Kunsthochschule, *die;* **~work** *n.* Illustrationen *Pl.;* Bildmaterial, *das*

arty ['ɑ:tɪ] *adj. (coll.)* auf Künstler machend; **he's an ~ type** er ist so ein Künstlertyp; **~-[and-]crafty** *(joc.)* auf Kunstgewerbe gemacht

Aryan ['eərɪən] **1.** *adj.* indogermanisch. **2.** *n.* **a)** *(language)* Indogermanisch, *das;* **b)** *(person)* Arier, *der*/Arierin, *die (bes. ns.);* Indogermane, *der*/Indogermanin, *die*

as [əz, *stressed* æz] **1.** *adv. in main sentence (in same degree)* **as ... [as ...]** so ... [wie ...]; **half as much** halb soviel; **they did as much as they could** sie taten, was sie konnten; **as good a player [as he]** ein so guter Spieler [wie er]. **2.** *rel. adv. or conj. in subord. clause* **a)** *expr. degree* **[as** *or* **so] ... as ...:** [so ...] wie ...; **as quickly as possible** so schnell wie möglich; **as ... as you can** so ...[, wie] Sie können; **come as quickly as you can** kommen Sie, so schnell Sie können; **quick as a flash** blitzschnell; **b)** *(though)* **... as he** *etc.* **is/was** obwohl er *usw.* ... ist/war; **intelligent as she is, ...:** obwohl sie ziemlich intelligent ist, ...; **safe as it might be, ...:** obwohl es vielleicht ungefährlich ist, ...; **c)** *(however much)* **try as he might/would, he could not concentrate** sosehr er sich auch bemühte, er konnte sich nicht konzentrieren; **d)** *expr. manner* wie; **as it were** sozusagen; gewissermaßen; **as you were!** Kommando zurück!; **e)** *expr. time* als; während; **as and when** wann immer; **as we climbed the stairs** als wir die Treppe hinaufgingen; **as we were talking** während wir uns unterhielten; **f)** *expr. reason* da; **g)** *expr. result* **so ... as to ...:** so ... zu; **would you be so kind as to help us?** würden Sie so freundlich sein und uns helfen?; **h)** *expr. purpose* **so as to ...:** um ... zu ...; **i)** *expr. illustration* wie [zum Beispiel]; **industrial areas, as the north-east of England for example** Industriegebiete wie zum Beispiel der Nordosten Englands. **3.** *prep.* **a)** *(in the function of)* als; **as an artist** als Künstler; **speaking as a parent, ...:** als Mutter/Vater ...; **b)** *(like)* wie; **they regard him as a fool** sie halten ihn für einen Dummkopf. **4.** *rel. pron. (which)* **they danced, as was the custom there** sie tanzten, wie es dort Sitte war; **he was shocked, as were we all** er war wie wir alle schockiert; **the same as ...:** der-/die-/dasselbe wie ...; **such as** wie zum Beispiel; **they enjoy such foreign foods as ...:** sie essen gern ausländische Lebensmittel wie ... **5. as far** *see* **far 1 d; as for ...:** was ... angeht; **as from ...:** von ... an; **as [it] is** wie die Dinge liegen; wie es aussieht; **the place is untidy enough as it is** es ist schon liederlich genug[, wie es jetzt ist]; **as of ...** *(Amer.)* von ... an; **as to** hinsichtlich (+ *Gen.*); **nothing further was mentioned as to holiday plans** von Urlaubsplänen wurde nichts weiter gesagt; **as was** wie es einmal war; **Miss Tay as was** das frühere Fräulein Tay; **as yet** bis jetzt; **as yet the plan is only under discussion** der Plan wird noch diskutiert

a.s.a.p. *abbr.* **as soon as possible**

asbestos [æz'bestɒs, æs'bestɒs] *n.* **a)** Asbest, *der;* **b)** *(mineral)* Amiant, *der*

ascend [ə'send] **1.** *v. i.* **a)** *(go up)* hinaufgehen *od.* -steigen; *(climb up)* hinaufklettern; *(by vehicle)* hinauffahren; *(come up)* heraufkommen; **Christ ~ed into heaven** Christus fuhr auf gen Himmel *(geh.);* **b)** *(rise)* aufsteigen; ‹*Hubschrauber:*› höhersteigen; **c)** *(slope upwards)* ‹*Hügel, Straße:*› ansteigen; **the stairs ~ very steeply** die Treppe ist sehr steil. **2.** *v. t.* **a)** *(go up)* hinaufsteigen, hinaufgehen ‹*Treppe, Leiter, Berg*›; ~ **a rope** an einem Seil hochklettern; **b)** *(come up)* heraufsteigen; **c)** ~ **the throne** den Thron besteigen

ascendancy [ə'sendənsɪ] *n., no pl.* Vorherrschaft, *die*

ascendant [ə'sendənt] *n.* **a)** *(Astrol.)* Aszendent, *der;* **b) in the ~:** im Aufsteigen begriffen

Ascension [ə'senʃn] *n. (Relig.)* **[the] ~:** [Christi] Himmelfahrt

A'scension Day *n.* Himmelfahrtstag, *der*

ascent [ə'sent] *n. (also fig.)* Aufstieg, *der*
ascertain [æsə'teɪn] *v.t.* feststellen; ermitteln ‹*Daten, Fakten*›
ascertainable [æsə'teɪnəbl] *adj.* feststellbar; zu ermittelnd ‹*Daten, Fakten*›
ascetic [ə'setɪk] **1.** *adj.* asketisch. **2.** *n.* **a)** Asket, *der*/Asketin, *die;* **b)** *(Relig. Hist.)* Eremit, *der*
asceticism [ə'setɪsɪzm] *n., no pl.* Askese, *die*
ascribe [ə'skraɪb] *v.t.* **a)** *(regard as belonging)* zuschreiben (**to** *Dat.*); **b)** *(attribute, impute)* zurückführen (**to** auf + *Akk.*)
aseptic [eɪ'septɪk] *adj.* aseptisch
asexual [eɪ'sekʃʊəl] *adj.* asexuell
¹ash [æʃ] *n.* **a)** *(tree)* Esche, *die;* **b)** *(wood)* Eschenholz, *das*
²ash *n. in sing. or pl. (powder)* Asche, *die*
ashamed [ə'ʃeɪmd] *adj.* **be ~:** beschämt sein; sich schämen; **be ~ of sb./sth.** sich jmds./einer Sache wegen schämen; **be/feel ~ for sb./sth.** sich für jmdn./etw. schämen; **be ~ of oneself for doing sth./be ~ to do sth.** sich schämen, etw. zu tun; **I'm ~ to admit that I told a lie** ich muß leider zugeben, daß ich gelogen habe
ash: ~-bin *n.* Mülleimer, *der;* **~ blonde 1.** *adj.* aschblond; **2.** *n.* Aschblonde, *die;* **~-can** *(Amer.) see* **~-bin**
ashen ['æʃn] *adj. (ash-coloured)* aschfarben; aschfahl ‹*Gesicht*›; **~ grey** aschgrau
ashore [ə'ʃɔ:(r)] *adv. (position)* an Land; am Ufer; *(direction)* an Land; ans Ufer
ash: ~-pan *n.* Aschkasten, *der;* **~-tray** *n.* Aschenbecher, *der;* **~-tree** *see* **¹ash a; Ash 'Wednesday** *n.* Aschermittwoch, *der;* **~-wood** *see* **¹ash b**
Asia ['eɪʃə] *pr. n.* Asien *(das);* **~ 'Minor** Kleinasien *(das)*
Asian ['eɪʃən, 'eɪʒən] **1.** *adj.* asiatisch. **2.** *n.* Asiat, *der*/Asiatin, *die*
aside [ə'saɪd] **1.** *adv.* beiseite; zur Seite; **stand ~!** treten Sie zur Seite!; **I pulled the curtain ~:** ich zog den Vorhang zur Seite; **take sb. ~:** jmdn. beiseite nehmen. **2.** *n.* Apart, *das;* Beiseitesprechen, *das*
asinine ['æsɪnaɪn] *adj.* dämlich
ask [ɑ:sk] **1.** *v.t.* **a)** fragen; **~ [sb.] a question** [jmdm.] eine Frage stellen; **~ sb.'s name** nach jmds. Namen fragen; **~ sb. [sth.]** jmdn. [nach etw.] fragen; **~ sb. about sth.** jmdn. nach etw. fragen; **I '~ you!** *(coll.)* ich muß schon sagen!; **if you ~ 'me** *(coll.)* [also,] wenn du mich fragst; **b)** *(seek to obtain)* **~ sth.** um etw. bitten; **how much are you ~ing for that car?** wieviel verlangen Sie für das Auto?; **~ sb. to do sth.** jmdn. [darum] bitten, etw. zu tun; **~ a lot of sb.** viel von jmdm. verlangen; **~ing price** geforderter Preis; **it's yours for the ~ing** du kannst es gern haben; **c)** *(invite)* einladen; **~ sb. to dinner** jmdn. zum Essen einladen; **~ sb. out** jmdn. einladen. **2.** *v.i.* **you may well ~:** du hast allen Grund zu fragen; **~ after sb./sth.** nach jmdm./etw. fragen; **~ for sth./sb.** etw./jmdn. verlangen; **~ for it** *(sl.: invite trouble)* es herausfordern
askance [ə'skæns, ə'skɑ:ns] *adv.* **look ~ at sb./sth.** jmdn. befremdet ansehen/etw. mit Befremden betrachten
askew [ə'skju:] *adv., pred. adj.* schief
asleep [ə'sli:p] *pred. adj.* **a)** *(lit. or fig.)* schlafend; **be/lie ~:** schlafen; **fall ~** *(also euphem.)* einschlafen; **b)** *(numb)* eingeschlafen ‹*Arm, Bein*›
asparagus [ə'spærəgəs] *n.* Spargel, *der*
aspect ['æspekt] *n.* **a)** Aspekt, *der;* **b)** *(position looking in a given direction)* Lage, *die; (front)* Seite, *die;* **have a southern ~:** nach Süden liegen
aspen ['æspən] *n. (Bot.)* Espe, *die*
aspersion [ə'spɜ:ʃn] *n.* Verunglimpfung, *die;* **cast ~s on sb./sth.** jmdn./etw. in den Schmutz ziehen
asphalt ['æsfælt] **1.** *n.* Asphalt, *der.* **2.** *v.t.* asphaltieren
asphyxia [æ'sfɪksɪə] *n., no pl. (Med.)* Asphyxie, *die (fachspr.);* Erstickung, *die*
asphyxiate [æ'sfɪksɪeɪt] *(Med.)* **1.** *v.t.* ersticken; **be ~d by sth.** an etw. *(Dat.)* ersticken. **2.** *v.i.* ersticken
aspic ['æspɪk] *n. (jelly)* Aspik, *der*
aspidistra [æspɪ'dɪstrə] *n. (Bot.)* Schusterpalme, *die*
aspirant [ə'spaɪərənt, 'æspərənt] *adj.* aufstrebend
aspiration [æspə'reɪʃn] *n.* Streben, *das;* **have ~s to sth.** nach etw. streben
aspire [ə'spaɪə(r)] *v.i.* **~ to** *or* **after sth.** nach etw. streben
aspirin ['æspərɪn] *n. (Med.)* Aspirin ⓦⓩ, *das;* Kopfschmerztablette, *die*
aspiring [ə'spaɪərɪŋ] *adj.* aufstrebend
¹ass [æs] *n. (Zool.; also fig.)* Esel, *der;* **make an ~ of oneself** sich blamieren
²ass *(Amer.) see* **arse**
assail [ə'seɪl] *v.t.* **a)** angreifen; **b)** *(fig.)* **~ sb. with questions** jmdn. mit Fragen überschütten; **I was ~ed with doubts** mich überkamen Zweifel
assailant [ə'seɪlənt] *n.* Angreifer, *der*/Angreiferin, *die*
assassin [ə'sæsɪn] *n.* Mörder, *der*/Mörderin, *die*
assassinate [ə'sæsɪneɪt] *v.t.* ermorden; **be ~d** einem Attentat zum Opfer fallen
assassination [əsæsɪ'neɪʃn] *n.* Mord, *der* (**of** an + *Dat.*); **~ attempt** Attentat, *das* (**on** auf + *Akk.*)
assault [ə'sɔ:lt] **1.** *n.* **a)** Angriff, *der; (fig.)* Anschlag, *der;* **verbal ~s** verbale Angriffe; **b)** *(Mil.)* Sturmangriff, *der;* **~ craft** Sturmboot, *das.* **2.** *v.t.* **a)** *(lit. or fig.)* angreifen; **b)** *(Mil.)* stürmen
assemblage [ə'semblɪdʒ] *n.* **a)** *(of things, persons)* Ansammlung, *die;* **b)** *(bringing together)* Zusammentragen, *das; (fitting together)* Zusammensetzen, *das*
assemble [ə'sembl] **1.** *v.t.* **a)** zusammentragen ‹*Beweise, Material, Sammlung*›; zusammenrufen ‹*Personen*›; **b)** *(fit together)* zusammenbauen. **2.** *v.i.* sich versammeln
assembly [ə'semblɪ] *n.* **a)** *(coming together, meeting, deliberative body)* Versammlung, *die; (in school) (tägliche Versammlung aller Schüler und Lehrer zur)* Morgenandacht; **b)** *(fitting together)* Zusammenbau, *der;* Montage, *die;* **c)** *(assembled unit)* Einheit, *die*
as'sembly line *n.* Fließband, *das*
assent [ə'sent] **1.** *v.i.* zustimmen; **~ to sth.** einer Sache *(Dat.)* zustimmen. **2.** *n.* Zustimmung, *die*
assert [ə'sɜ:t] *v.t.* **a)** geltend machen; **~ oneself** sich durchsetzen; **b)** *(declare)* behaupten; beteuern ‹*Unschuld*›
assertion [ə'sɜ:ʃn] *n.* Geltendmachen, *das; (declaration)* Behauptung, *die*
assertive [ə'sɜ:tɪv] *adj.* energisch ‹*Person*›; bestimmt ‹*Ton, Verhalten*›; fest ‹*Stimme*›
assess [ə'ses] *v.t.* **a)** *(evaluate)* einschätzen; beurteilen; **b)** *(value)* schätzen; taxieren; **c)** *(fix amount of)* festsetzen ‹*Steuer, Bußgeld usw.*› (**at** auf + *Akk.*)
assessment [ə'sesmənt] *n.* **a)** *(evaluation)* Einschätzung, *die;* Beurteilung, *die;* **b)** *(valuation)* Schätzung, *die;* Taxierung, *die;* **c)** *(fixing amount of damages or*

fine) Festsetzung, *die; (of tax)* Veranlagung, *die;* **d)** *(tax to be paid)* Steuerbescheid, *der*

asset ['æset] *n.* **a)** Vermögenswert, *der;* **my [personal] ~s** mein [persönlicher] Besitz; **b)** *(fig.) (useful quality)* Vorzug, *der* **(to** für); *(person)* Stütze, *die; (thing)* Hilfe, *die*

assiduous [ə'sɪdjʊəs] *adj.* **a)** *(diligent)* eifrig; *(conscientious)* gewissenhaft; **b)** *(obsequiously attentive)* beflissen

assign [ə'saɪn] *v. t.* **a)** *(allot)* **~ sth. to sb.** jmdm. etw. zuweisen; *(transfer)* jmdm. etw. übereignen; **b)** *(appoint)* zuteilen; **~ sb. to a job/task** jmdn. mit einer Arbeit/Aufgabe betrauen; **~ sb. to do sth.** jmdn. damit betrauen, etw. zu tun; **c)** *(specify)* festsetzen ‹*Zeit, Datum, Grenzwert*›; **d)** *(ascribe)* angeben; **~ a cause to sth.** einen Grund für etw. angeben

assignment [ə'saɪnmənt] *n.* **a)** *(allotting)* Zuteilung, *die; (of property)* Übereignung, *die;* **b)** *(task)* Aufgabe, *die; (Amer. Sch. and Univ.)* Arbeit, *die;* **c)** *(of reason, cause)* Aufgabe, *die*

assimilate [ə'sɪmɪleɪt] *v. t.* **a)** *(make like)* angleichen; **~ sth. with** *or* **to sth.** etw. an etw. *(Akk.)* angleichen; **b)** *(fig.)* aufnehmen ‹*Informationen, Einflüsse usw.*›

assimilation [əsɪmɪ'leɪʃn] *n.* **a)** *(making or becoming like)* Angleichung, *die* **(to, with** an + *Akk.*); **b)** *(fig.: of information, influences, etc.)* Aufnahme, *die*

assist [ə'sɪst] **1.** *v. t. (help)* helfen (+ *Dat.*); voranbringen ‹*Vorgang, Prozeß*›; **~ sb. to do** *or* **in doing sth.** jmdm. helfen, etw. zu tun; **~ sb. with sth.** jmdm. bei etw. helfen. **2.** *v. i.* **a)** *(help)* helfen; **~ with sth./in doing sth.** bei etw. helfen/helfen, etw. zu tun; **b)** *(take part)* mitarbeiten **(in** an + *Dat.*)

assistance [ə'sɪstəns] *n., no pl.* Hilfe, *die;* **give ~ to sb.** jmdm. behilflich sein; **be of ~ [to sb.]** [jmdm.] behilflich sein

assistant [ə'sɪstənt] **1.** *n. (helper)* Helfer, *der*/Helferin, *die; (subordinate)* Mitarbeiter, *der*/Mitarbeiterin, *die; (of professor, artist)* Assistent, *der*/Assistentin, *die; (in shop)* Verkäufer, *der*/Verkäuferin, *die.* **2.** *attrib. adj.* **~ manager** stellvertretender Geschäftsführer; **~ editor** Redaktionsassistent, *der;* **~ professor** *(Amer.)* ≈ Assistenzprofessor, *der*

associate 1. [ə'səʊʃɪət, ə'səʊsɪət] *n.* **a)** *(partner)* Partner, *der*/Partnerin, *die;* Kompagnon, *der; (colleague)* Kollege, *der*/Kollegin, *die; (of gangster)* Komplize, *der*/Komplizin, *die;* **b)** *(subordinate member)* außerordentliches Mitglied. **2.** [ə'səʊʃɪət, ə'səʊsɪət] *adj.* beigeordnet; außerordentlich ‹*Mitglied usw.*›. **3.** [ə'səʊʃɪeɪt, ə'səʊsɪeɪt] *v. t.* **a)** *(join)* in Verbindung bringen; **be ~d** in Verbindung stehen; **b)** *(connect in the mind)* in Verbindung bringen, *(Psych.)* assoziieren **(mit** with); **c)** **~ oneself with sth.** sich einer Sache *(Dat.)* anschließen. **4.** *v. i.* **~ with sb.** mit jmdm. verkehren *od.* Umgang haben

association [əsəʊsɪ'eɪʃn] *n.* **a)** *(organization)* Verband, *der;* Vereinigung, *die;* **articles** *or* **deeds of ~:** Satzung, *die;* **b)** *(mental connection)* Assoziation, *die;* **~ of ideas** Gedankenassoziation, *die;* **have ~s for sb.** bei jmdm. Assoziationen hervorrufen; **c)** **A~ football** *(Brit.)* Fußball, *der;* **d)** *(connection)* Verbindung, *die;* **e)** *(co-operation)* Zusammenarbeit, *die*

assorted [ə'sɔ:tɪd] *adj.* gemischt ‹*Bonbons, Sortiment*›; **cardigans of ~ kinds** verschiedenerlei Strickjacken

assortment [ə'sɔ:tmənt] *n.* Sortiment, *das;* **a good ~ of hats** eine gute Auswahl an Hüten

Asst. *abbr.* **Assistant** Ass.

assuage [ə'sweɪdʒ] *v. t.* stillen; *(soothe)* besänftigen ‹*Person, Ärger*›; lindern ‹*Schmerz, Sorge*›

assume [ə'sju:m] *v. t.* **a)** voraussetzen; ausgehen von; **assuming that ...:** vorausgesetzt, daß ...; **he's not so stupid as we ~d him to be** er ist nicht so dumm, wie wir angenommen haben; **b)** *(undertake)* übernehmen ‹*Amt, Pflichten*›; **c)** *(take on)* annehmen ‹*Namen, Rolle*›; gewinnen ‹*Aspekt, Bedeutung*›; **under an ~d name** unter einem Decknamen

assumption [ə'sʌmpʃn] *n.* **a)** Annahme, *die;* **going on the ~ that ...:** vorausgesetzt, daß ...; **b)** *(undertaking)* Übernahme, *die;* **c)** **the A~** *(Relig.)* Mariä Himmelfahrt

assurance [ə'ʃʊərəns] *n.* **a)** Zusicherung, *die;* **I give you my ~ that ...:** ich versichere Ihnen, daß ...; **I can give you no ~ that ...:** ich kann Ihnen nicht versprechen, daß ...; **b)** *no pl. (self-confidence)* Selbstsicherheit, *die;* **c)** *no pl. (Brit.: insurance)* Versicherung, *die*

assure [ə'ʃʊə(r)] *v. t.* **a)** versichern (+ *Dat.*); **~ sb. of sth.** jmdn. einer Sache *(Gen.)* versichern *(geh.)*; **b)** *(convince)* **~ sb./oneself** jmdn./sich überzeugen; **c)** *(make certain or safe)* gewährleisten; **d)** *(Brit.: insure)* versichern

assured [ə'ʃʊəd] *adj.* gesichert ‹*Tatsache*›; gewährleistet ‹*Erfolg*›; **be ~ of sth.** sich *(Dat.)* einer Sache *(Gen.)* sicher sein

assuredly [ə'ʃʊərɪdlɪ] *adv.* gewiß

aster ['æstə(r)] *n.* Aster, *die*

asterisk ['æstərɪsk] **1.** *n.* Sternchen, *das.* **2.** *v. t.* mit einem Sternchen versehen

astern [ə'stɜ:n] *adv. (Naut., Aeronaut.)* achtern; *(towards the rear)* achteraus; **~ of sth.** hinter etw. *(Dat.)*; **full speed ~!** volle Kraft zurück!

asteroid ['æstərɔɪd] *n. (Astron.)* Asteroid, *der*

asthma ['æsmə] *n. (Med.)* Asthma, *das*

asthmatic [æs'mætɪk] *(Med.)* **1.** *adj.* asthmatisch. **2.** *n.* Asthmatiker, *der*/Asthmatikerin, *die*

astir [ə'stɜ:(r)] *pred. adj.* in Bewegung; *(out of bed)* auf den Beinen

astonish [ə'stɒnɪʃ] *v. t.* erstaunen

astonishing [ə'stɒnɪʃɪŋ] *adj.*, **astonishingly** [ə'stɒnɪʃɪŋlɪ] *adv.* erstaunlich

astonishment [ə'stɒnɪʃmənt] *n., no pl.* Erstaunen, *das;* **in utter ~:** äußerst erstaunt

astound [ə'staʊnd] *v. t.* verblüffen; [sehr] überraschen

astounding [ə'staʊndɪŋ] *adj.* erstaunlich

astrakhan [æstrə'kæn] *n.* Astrachan, *der*

astray [ə'streɪ] **1.** *adv.* in die Irre; **sb. goes ~:** jmd. verirrt sich; **sth. goes ~** *(is mislaid)* etw. wird verlegt; *(is lost)* etw. geht verloren; **lead ~:** irreführen; **go/lead ~** *(fig.)* in die Irre gehen/führen; *(into sin)* vom rechten Weg abkommen/abbringen. **2.** *pred. adj.* **be ~:** sich verirrt haben; *(fig.: be in error)* sich irren

astride [ə'straɪd] **1.** *adv.* rittlings ‹*sitzen*›; breitbeinig ‹*stehen*›; **~ of sth.** rittlings auf etw. *(Dat./Akk.)*. **2.** *prep.* **a)** rittlings auf (+ *Dat.*); **b)** *(extending across)* zu beiden Seiten (+ *Gen.*)

astringent [ə'strɪndʒənt] **1.** *adj.* **a)** herb, streng ‹*Geschmack*›; stechend, beißend ‹*Geruch*›; **b)** *(styptic)* adstringierend *(Med.)*; blutstillend; **c)** *(severe)* scharf. **2.** *n.* Adstringens, *das*

astrologer [ə'strɒlədʒə(r)] *n.* Astrologe, *der*/Astrologin, *die*

astrological [æstrə'lɒdʒɪkl] *adj.* astrologisch
astrology [ə'strɒlədʒɪ] *n., no pl.* Astrologie, *die*
astronaut ['æstrənɔ:t] *n.* Astronaut, *der*/Astronautin, *die*
astronautical [æstrə'nɔ:tɪkl] *adj.* astronautisch
astronautics [æstrə'nɔ:tɪks] *n., no pl.* Raumfahrt, *die*
astronomer [ə'strɒnəmə(r)] *n.* Astronom, *der*/Astronomin, *die*
astronomical [æstrə'nɒmɪkl] *adj.*, **astronomically** [æstrə'nɒmɪkəlɪ] *adv. (lit. or fig.)* astronomisch
astronomy [ə'strɒnəmɪ] *n., no pl.* Astronomie, *die*
astrophysics [æstrəʊ'fɪzɪks] *n., no pl.* Astrophysik, *die*
astute [ə'stju:t] *adj.* scharfsinnig
asylum [ə'saɪləm] *n.* Asyl, *das;* **grant sb. ~:** jmdm. Asyl gewähren; **political ~:** politisches Asyl
asymmetric [æsɪ'metrɪk, eɪsɪ'metrɪk], **asymmetrical** [æsɪ'metrɪkl, eɪsɪ'metrɪkl] *adj.* asymmetrisch; unsymmetrisch
asymmetry [æ'sɪmɪtrɪ, eɪ'sɪmɪtrɪ] *n.* Asymmetrie, *die*
at [ət, *stressed* æt] *prep.* **a)** *expr. place* an (+ *Dat.*); **at the station** am Bahnhof; **at the baker's/butcher's/grocer's** beim Bäcker/Fleischer/Kaufmann; **at the chemist's** in der Apotheke/Drogerie; **at the supermarket** im Supermarkt; **at my mother's** bei meiner Mutter; **at the party** auf der Party; **at the office/hotel** im Büro/Hotel; **at Dover** in Dover; **b)** *expr. time* **at Christmas/Whitsun/Easter** [zu *od.* an] Weihnachten/Pfingsten/Ostern; **at six o'clock** um sechs Uhr; **at midnight** um Mitternacht; **at midday** am Mittag; mittags; **at [the age of] 40** mit 40; im Alter von 40; **at this/the moment** in diesem/im Augenblick *od.* Moment; **at the first attempt** beim ersten Versuch; **c)** *expr. price* **at £2.50 [each]** zu *od.* für [je] 2,50 Pfund; **d) she's still 'at it** sie ist immer noch dabei; **at that** *(at that point)* dabei; *(at that provocation)* daraufhin; *(moreover)* noch dazu
ate *see* **eat**
atheism ['eɪθɪɪzm] *n., no pl.* Atheismus, *der*
atheist ['eɪθɪɪst] *n.* Atheist, *der*/Atheistin, *die*
Athenian [ə'θi:nɪən] **1.** *adj.* athenisch. **2.** *n.* Athener, *der*/Athenerin, *die*
Athens ['æθɪnz] *pr. n.* Athen *(das)*
athlete ['æθli:t] *n.* Athlet, *der*/Athletin, *die;* Sportler, *der*/Sportlerin, *die;* *(runner, jumper)* Leichtathlet, *der*/Leichtathletin, *die;* **~'s foot** *(Med.)* Fußpilz, *der*
athletic [æθ'letɪk] *adj.* sportlich
athletics [æθ'letɪks] *n., no pl.* **a)** Leichtathletik, *die;* **b)** *(Amer.: physical sports)* Sport, *der*
Atlantic [ət'læntɪk] **1.** *adj.* atlantisch; **~ Ocean** Atlantischer Ozean. **2.** *pr. n.* Atlantik, *der*
atlas ['ætləs] Atlas, *der*
atmosphere ['ætməsfɪə(r)] *n.* **a)** *(lit. or fig.)* Atmosphäre, *die;* **b)** *(air in a place)* Luft, *die*
atmospheric [ætmə'sferɪk] *adj.* **a)** atmosphärisch; **b)** *(fig.: evocative)* stimmungsvoll
atmospherics [ætmə'sferɪks] *n. pl. (Radio)* atmosphärische Störungen
atoll ['ætɒl] *n.* Atoll, *das*
atom ['ætəm] *n.* Atom, *das;* **not an ~ of truth** *(fig.)* kein Körnchen Wahrheit
'atom bomb *see* **atomic bomb**
atomic [ə'tɒmɪk] *adj. (Phys.)* Atom-
atomic: ~ 'bomb *n.* Atombombe, *die;* **~ 'energy** *n., no pl.* Atomenergie, *die;* **~ 'power** *n., no pl.* Atomkraft, *die;* **~ 'weight** *n. (Phys., Chem.)* Atomgewicht, *das*
atomize ['ætəmaɪz] *v. t.* atomisieren; zerstäuben ‹*Flüssigkeit*›
atomizer ['ætəmaɪzə(r)] *n.* Zerstäuber, *der*
atone [ə'təʊn] *v. i.* es wiedergutmachen; **~ for sth.** etw. wiedergutmachen
atonement [ə'təʊnmənt] *n.* **a)** *(atoning)* Buße, *die;* *(reparation)* Wiedergutmachung, *die;* **make ~ for sth.** für etw. Buße tun; **b)** *(Relig.)* Versöhnung, *die;* **Day of A~:** Versöhnungsfest, *das*
atrocious [ə'trəʊʃəs] *adj.* grauenhaft; scheußlich ‹*Wetter, Benehmen*›
atrociously [ə'trəʊʃəslɪ] *adv.* grauenhaft; scheußlich ‹*sich benehmen*›
atrocity [ə'trɒsɪtɪ] *n.* **a)** *no pl.* *(extreme wickedness)* Grauenhaftigkeit, *die;* **b)** *(atrocious deed)* Greueltat, *die (geh.);* Grausamkeit, *die*
atrophy ['ætrəfɪ] **1.** *n. (Med.)* Atrophie, *die;* **muscular ~:** Muskelatrophie, *die (Med.);* Muskelschwund, *der.* **2.** *v. i.* atrophieren
attach [ə'tætʃ] **1.** *v. t.* **a)** *(fasten)* befestigen (**to** an + *Dat.*); anhängen ‹*Wagen*› (**to** an + *Dat.*); **please find ~ed ...:** beigeheftet ist ...; **b)** *(assign)* **be ~ed to sth.** einer Sache *(Dat.)* zugeteilt sein; **the research unit is ~ed to the university** die Forschungsabteilung ist der Universität *(Dat.)* angegliedert; **c)** *(fig.: ascribe)* zuschreiben; **~ no blame to sb.** jmdm. keine Schuld geben; **d)** *(attribute)* beimessen; **~ importance to sth.** einer Sache *(Dat.)* Gewicht beimessen. **2.** *v. i.* **no blame ~es to sb.** jmdn. trifft keine Schuld
attaché [ə'tæʃeɪ] *n.* Attaché, *der*
at'taché case *n.* Diplomatenkoffer, *der*
attached [ə'tætʃt] *adj. (emotionally)* **be ~ to sb./sth.** an jmdm./etw. hängen; **become ~ to sb./sth.** jmdn./etw. liebgewinnen
attachment [ə'tætʃmənt] *n.* **a)** *(act or means of fastening)* Befestigung, *die;* **b)** *(accessory)* Zusatzgerät, *das;* **c)** *(ascribing)* Zuordnung, *die;* **d)** *(attribution)* Beimessung, *die;* **e)** *(affection)* Anhänglichkeit, *die* (**to** an + *Akk.*); **have an ~ for sb.** an jmdm. hängen
attack [ə'tæk] **1.** *v. t.* **a)** angreifen; *(ambush, raid)* überfallen; *(fig.: criticize)* attackieren; **b)** ‹*Krankheit:*› befallen; **c)** *(start work on)* in Angriff nehmen; **she ~ed the washing-up** sie machte sich an den Abwasch; **d)** *(act harmfully on)* angreifen ‹*Metall, Oberfläche*›. **2.** *v. i.* angreifen. **3.** *n.* **a)** *(on enemy)* Angriff, *der;* *(on person)* Überfall, *der;* *(fig.: criticism)* Attacke, *die;* Angriff, *der;* **be under ~:** angegriffen werden; **b) make a spirited ~ on sth.** *(start)* etw. beherzt in Angriff nehmen
attacker [ə'tækə(r)] *n. (also Sport)* Angreifer, *der*/Angreiferin, *die*
attacking [ə'tækɪŋ] *adj.* offensiv ‹*Spielweise, Spieler*›
attain [ə'teɪn] **1.** *v. t.* erreichen ‹*Ziel, Wirkung*›; **~ power** an die Macht gelangen; **she ~ed her hope** ihre Hoffnung erfüllte sich. **2.** *v. i.* **~ to sth.** zu etw. gelangen; **~ to success** Erfolg haben
attainable [ə'teɪnəbl] *adj.* erreichbar ‹*Ziel*›; realisierbar ‹*Hoffnung, Ziel*›
attainment [ə'teɪnmənt] *n.* **a)** *no pl.* Verwirklichung, *die;* **b)** *(thing attained)* Leistung, *die*
attempt [ə'tempt] **1.** *v. t.* **a)** versuchen; **~ to do sth.** versuchen, etw. zu tun; **b)** *(try to accomplish)* sich versuchen an (+ *Dat.*); **candidates should ~ 5 out of 10 questions** die Kandidaten sollten 5

von 10 Fragen zu beantworten versuchen. **2.** *n.* Versuch, *der;* **make an ~ to do sth.** den Versuch unternehmen, etw. zu tun; **make an ~ on sb.'s life** ein Attentat *od.* einen Mordanschlag auf jmdn. verüben

attend [ə'tend] **1.** *v. i.* **a)** *(give care and thought)* aufpassen; **~ to sth.** auf etw. *(Akk.)* achten; *(deal with sth.)* sich um etw. kümmern; **b)** *(be present)* anwesend sein; **c)** *(wait)* bedienen; aufwarten *(veralt.);* **~ on sb.** jmdn. bedienen; jmdm. aufwarten *(veralt.).* **2.** *v. t.* **a)** *(be present at)* teilnehmen an (+ *Dat.*); *(go regularly to)* besuchen; **b)** *(follow as a result from)* sich ergeben aus; **be ~ed by sth.** etw. zur Folge haben; **c)** *(wait on)* bedienen; aufwarten *(veralt.)* (+ *Dat.*); **d)** ‹*Arzt:*› behandeln

attendance [ə'tendəns] *n.* **a)** *(being present)* Anwesenheit, *die; (going regularly)* Besuch, *der* **(at** *Gen.*); **regular ~ at school** regelmäßiger Schulbesuch; **b)** *(number of people present)* Teilnehmerzahl, *die;* **c) be in ~:** anwesend sein

at'tendance centre *n. (Brit.)* Jugendarrestanstalt *(in der Freizeitarrest verbüßt wird)*

attendant [ə'tendənt] **1.** *n.* **a) [lavatory] ~:** Toilettenmann, *der*/Toilettenfrau, *die;* **[cloakroom] ~:** Garderobenmann, *der*/Garderobenfrau, *die;* **museum ~:** Museumswärter, *der;* **b)** *(member of entourage)* Begleiter, *der*/Begleiterin, *die.* **2.** *adj.* begleitend; **its ~ risks** die damit verbundenen Risiken

attention [ə'tenʃn] **1.** *n.* **a)** *no pl.* Aufmerksamkeit, *die;* **pay ~ to sb./sth.** jmdn./etw. beachten; **pay ~!** gib acht!; paß auf!; **hold sb.'s ~:** jmds. Interesse wachhalten; **attract [sb.'s] ~:** [jmdn.] auf sich *(Akk.)* aufmerksam machen; **catch sb.'s ~:** jmds. Aufmerksamkeit erregen; **bring sth. to sb.'s ~:** jmdn. auf etw. *(Akk.)* aufmerksam machen; **call** *or* **draw sb.'s ~ to sb./sth.** jmds. Aufmerksamkeit auf jmdn./etw. lenken; **~ Miss Jones** *(on letter)* zu Händen [von] Miss Jones; **b)** *no pl. (consideration)* **give sth. one's personal ~:** sich einer Sache *(Gen.)* persönlich annehmen; **c)** *(Mil.)* **stand to ~:** stillstehen; strammstehen. **2.** *int.* **a)** Achtung; **b)** *(Mil.)* stillgestanden

attentive [ə'tentɪv] *adj.* aufmerksam; **be ~ to sth.** auf etw. *(Akk.)* achten; **be more ~ to one's studies** sich gewissenhafter seinen Studien widmen

attentively [ə'tentɪvlɪ] *adv.* aufmerksam

attentiveness [ə'tentɪvnɪs] *n., no pl.* Aufmerksamkeit, *die*

attest [ə'test] **1.** *v. t.* *(certify validity of)* bestätigen; beglaubigen ‹*Unterschrift, Urkunde*›. **2.** *v. i.* **~ to sth.** etw. bezeugen; *(fig.)* von etw. zeugen

attic ['ætɪk] *n.* **a)** *(storey)* Dachgeschoß, *das;* **b)** *(room)* Dachboden, *der; (habitable)* Dachkammer, *die*

attire [ə'taɪə(r)] *n., no pl.* Kleidung, *die*

attitude ['ætɪtju:d] *n.* **a)** *(posture, way of behaving)* Haltung, *die;* **strike an ~:** eine Haltung einnehmen; **b)** *(mode of thinking)* **~ [of mind]** Einstellung, *die* **(to[wards]** zu); **c)** *(Aeron.)* Fluglage, *die*

attorney [ə'tɜ:nɪ] *n.* **a)** *(legal agent)* Bevollmächtigte, *der/die;* **power of ~:** Vollmacht, *die;* **b)** *(Amer.: lawyer)* [Rechts]anwalt, *der*/[Rechts]anwältin, *die*

Attorney-'General *n., pl.* **Attorneys-General** *oberster Justizbeamter bestimmter Staaten;* ≈ Generalbundesanwalt, *der (Bundesrepublik Deutschland); (in USA)* ≈ Justizminister, *der*

attract [ə'trækt] *v. t.* **a)** *(draw)* anziehen; auf sich *(Akk.)* ziehen ‹*Interesse, Blick, Kritik*›; ‹*Köder, Attraktion:*› anlocken; **b)** *(arouse pleasure in)* anziehend wirken auf (+ *Akk.*); **what ~s me about the girl** was ich an dem Mädchen anziehend finde; **c)** *(arouse interest in)* reizen **(about** an + *Dat.*)

attraction [ə'trækʃn] *n.* **a)** Anziehung, *die; (force, lit. or fig.)* Anziehung[skraft], *die;* **have little ~ for sb.** jmdn. nur wenig reizen; **b)** *(fig.: thing that attracts)* Attraktion, *die; (charm)* Verlockung, *die;* Reiz, *der*

attractive [ə'træktɪv] *adj.* **a)** anziehend; **~ power/force** Anziehungskraft, *die;* **b)** *(fig.)* attraktiv; reizvoll ‹*Vorschlag, Möglichkeit, Idee*›

attractiveness [ə'træktɪvnɪs] *n., no pl.* Attraktivität, *die*

attributable [ə'trɪbjʊtəbl] *adj.* **be ~ to sb./sth.** jmdm./einer Sache zuzuschreiben sein

attribute 1. ['ætrɪbju:t] *n.* **a)** *(quality)* Attribut, *das;* Eigenschaft, *die;* **b)** *(symbolic object)* Attribut, *das.* **2.** [ə'trɪbju:t] *v. t.* *(ascribe, assign)* zuschreiben **(to** *Dat.*); *(refer)* zurückführen **(to** auf + *Akk.*)

attribution [ætrɪ'bju:ʃn] *n.* *(ascribing, assigning)* Zuordnung, *die* **(to** *Dat.*); *(referring)* Zurückführung, *die* **(to** auf + *Akk.*)

attributive [ə'trɪbjʊtɪv] *adj.* *(Ling.)* attributiv

attrition [ə'trɪʃn] *n., no pl.* *(wearing down)* Zermürbung, *die;* **war of ~** *(lit. or fig.)* Zermürbungskrieg, *der*

attune [ə'tju:n] *v. t.* *(make accustomed)* gewöhnen **(to** an + *Akk.*); **be ~d to sth.** auf etw. *(Akk.)* eingestellt sein

aubergine ['əʊbəʒi:n] *n.* Aubergine, *die*

auburn ['ɔ:bən] *adj.* rötlichbraun

auction ['ɔ:kʃn] **1.** *n.* **a)** Auktion, *die;* Versteigerung, *die;* **be put up for ~:** zur Versteigerung kommen; versteigert werden; **Dutch ~:** Abschlag, *der;* **b)** *(Cards)* Bieten, *das.* **2.** *v. t.* versteigern

auctioneer [ɔ:kʃə'nɪə(r)] *n.* Auktionator, *der*/Auktionatorin, *die*

audacious [ɔ:'deɪʃəs] *adj.* *(daring)* kühn; verwegen

audacity [ɔ:'dæsɪtɪ] *n., no pl.* **a)** *(daringness)* Kühnheit, *die;* Verwegenheit, *die;* **b)** *(impudence)* Dreistigkeit, *die*

audibility [ɔ:dɪ'bɪlɪtɪ] *n.* Hörbarkeit, *die*

audible ['ɔ:dɪbl] *adj.* hörbar; **every word was ~:** man konnte jedes Wort hören

audience ['ɔ:dɪəns] *n.* **a)** *(listeners, spectators)* Publikum, *das;* **b)** *(formal interview)* Audienz, *die* **(with** bei); **private ~:** Privataudienz, *die*

audio ['ɔ:dɪəʊ] *adj.* Ton-

audio: ~ typist *n.* Phonotypist, *der*/-typistin, *die;* **~-'visual** *adj.* audiovisuell *(fachspr.)*

audit ['ɔ:dɪt] **1.** *n.* **~ [of the accounts]** Rechnungsprüfung, *die* *(Wirtsch.);* **the ~ of the firm's books** die Revision der Firmengeschäftsbücher. **2.** *v. t.* prüfen

audition [ɔ:'dɪʃn] **1.** *n.* *(singing)* Probesingen, *das; (dancing)* Vortanzen, *das; (acting)* Vorsprechen, *das.* **2.** *v. i.* *(sing)* vorsingen; probesingen; *(dance)* vortanzen; *(act)* vorsprechen; **~ for a part** für eine Rolle vorsprechen. **3.** *v. t.* vorsingen/vortanzen/vorsprechen lassen

auditor ['ɔ:dɪtə(r)] *n.* Buchprüfer, *der*/-prüferin, *die*

auditorium [ɔ:dɪ'tɔ:rɪəm] *n., pl.* **~s** *or* **auditoria** [ɔ:dɪ'tɔ:rɪə] Zuschauerraum, *der*

Aug. *abbr.* **August** Aug.

auger ['ɔ:gə(r)] *n.* Handbohrer, *der;* Stangenbohrer, *der (Technik)*

augment [ɔ:g'ment] *v.t.* verstärken ‹*Armee*›; verbessern ‹*Einkommen*›; aufstocken ‹*Fonds, finanzielle Mittel*›
augur ['ɔ:gə(r)] **1.** *n.* Augur, *der.* **2.** *v.t. (portend)* bedeuten; versprechen ‹*Erfolg*›. **3.** *v.i.* ~ **well/ill for sth./sb.** ein gutes/schlechtes Zeichen für etw./jmdn. sein
augury ['ɔ:gjʊrɪ] *n.* Vorzeichen, *das*
August ['ɔ:gəst] *n.* August, *der;* **in** ~: im August; **last/next** ~: letzten/nächsten August; **the first of/on the first of** ~ *or* **on** ~ **[the] first** der erste/am ersten August; **1st** ~ *(as date on document)* 1. August; **every** ~: jeden August; jedes Jahr im August; **an** ~ **day** ein Augusttag; **from** ~ **to October** von August bis Oktober
august [ɔ:'gʌst] *adj. (venerable)* ehrwürdig; *(noble)* erlaucht
auld [ɔ:ld] *adj. (Scot.) see* **old; for** ~ **lang syne** um der guten, alten Zeiten willen
aunt [ɑ:nt] *n.* Tante, *die*
auntie, aunty ['ɑ:ntɪ] *n. (coll.)* Tantchen, *das; (with name)* Tante, *die*
au pair [əʊ 'peə(r)] **1.** *n.* Au-pair-Mädchen, *das.* **2.** *adj.* ~ **girl** Au-pair-Mädchen, *das*
aura ['ɔ:rə] *n., pl.* ~**e** ['ɔ:ri:] *or* ~**s** Aura, *die*
aural ['ɔ:rl] *adj.* akustisch; aural *(Med.)*
auricle ['ɔ:rɪkl] *n.* Ohrmuschel, *die*
aurora [ɔ:'rɔ:rə] *n., pl.* ~**s** *or* ~**e** [ɔ:'rɔ:ri:] Polarlicht, *das;* ~ **borealis** [bɔ:rɪ'eɪlɪs] Nordlicht, *das*
auspice ['ɔ:spɪs] *n.* **a)** *in pl.* **under the** ~**s of sb./sth.** unter jmds./einer Sache Auspizien *(geh.) od.* Schirmherrschaft; **b)** *(sign)* Auspizium, *das (geh.);* Vorzeichen, *das*
auspicious [ɔ:'spɪʃəs] *adj.* günstig; vielversprechend ‹*Anfang*›
auspiciously [ɔ:'spɪʃəslɪ] *adv.* vielversprechend
Aussie ['ɒzɪ] *(coll.)* **1.** *adj.* australisch. **2.** *n.* Australier, *der*/Australierin, *die*
austere [ɒ'stɪə(r), ɔ:'stɪə(r)] *adj.* **a)** *(morally strict, stern)* streng; unbeugsam ‹*Haltung*›; **b)** *(severely simple)* karg; **c)** *(ascetic)* asketisch ‹*Leben*›
austerely [ɒ'stɪəlɪ, ɔ:'stɪəlɪ] *adv.* **a)** *(strictly, sternly)* streng; **b)** *(severely simply)* karg; ~ **simple** karg und schlicht
austereness [ɒ'stɪənɪs, ɔ:'stɪənɪs] *see* **austerity a, b**
austerity [ɒ'sterɪtɪ, ɔ:'sterɪtɪ] *n.* **a)** *no pl. (moral strictness)* Strenge, *die;* **b)** *no pl. (severe simplicity)* Kargheit, *die;* **c)** *no pl. (lack of luxuries)* wirtschaftliche Einschränkung; **d)** *in pl. (deprivations)* Entbehrungen
Australasia [ɒstrə'leɪʃə, ɔ:strə'leɪʃə] *pr. n.* Australien und der südwestliche Pazifik
Australia [ɒ'streɪlɪə, ɔ:'streɪlɪə] *pr. n.* Australien *(das)*
Australian [ɒ'streɪlɪən, ɔ:'streɪlɪən] **1.** *adj.* australisch; ~ **bear** Beutelbär, *der;* Koala, *der;* **sb. is** ~: jmd. ist Australier/Australierin. **2.** *n.* Australier, *der*/Australierin, *die*
Austria ['ɒstrɪə, 'ɔ:strɪə] *pr. n.* Österreich *(das);* ~**-Hungary** *(Hist.)* Österreich-Ungarn *(das)*
Austrian ['ɒstrɪən, 'ɔ:strɪən] **1.** *adj.* österreichisch. **2.** *n.* Österreicher, *der*/Österreicherin, *die*
authentic [ɔ:'θentɪk] *adj.* authentisch; *(genuine)* authentisch; echt; unverfälscht ‹*Akzent*›
authenticate [ɔ:'θentɪkeɪt] *v.t.* ~ **sth.** die Echtheit einer Sache *(Gen.)* bestätigen; ~ **a report** einen Bericht bestätigen
authentication [ɔ:θentɪ'keɪʃn] *n., no pl.* Bestätigung der Echtheit; *(of report)* Bestätigung, *die*
authenticity [ɔ:θen'tɪsɪtɪ] *n., no pl.* Echtheit, *die;* Authentizität, *die; (of report)* Zuverlässigkeit, *die*
author ['ɔ:θə(r)] **1.** *n.* **a)** *(writer)* Autor, *der*/Autorin, *die; (profession)* Schriftsteller, *der*/Schriftstellerin, *die;* **the** ~ **of the book/article** der Autor *od.* Verfasser des Buches/Artikels; **b)** *(originator)* Vater, *der.* **2.** *v.t. (write)* verfassen
authoritarian [ɔ:θɒrɪ'teərɪən] *adj.* autoritär
authoritative [ɔ:'θɒrɪtətɪv] *adj.* **a)** *(recognized as reliable)* maßgebend; zuverlässig ‹*Bericht, Information*›; *(official)* amtlich; **b)** *(commanding)* respekteinflößend
authoritatively [ɔ:'θɒrɪtətɪvlɪ] *adv.* **a)** *(reliably)* zuverlässig ‹*berichten*›; *(officially)* offiziell; **he talked** ~ **about his specialist field** er sprach als Fachmann über sein Spezialgebiet; **b)** *(commandingly)* mit Bestimmtheit
authority [ɔ:'θɒrɪtɪ] *n.* **a)** *no pl. (power)* Autorität, *die; (delegated power)* Befugnis, *die;* **have the/no** ~ **to do sth.** berechtigt *od.* befugt/nicht befugt sein, etw. zu tun; **have/exercise** ~ **over sb.** Weisungsbefugnis gegenüber jmdm. haben; **on one's own** ~: in eigener Verantwortung; **[be] in** ~: verantwortlich [sein]; **b)** *(body having power)* **the authorities** die Behörde[n]; **the highest legal** ~: die höchste rechtliche Instanz; **c)** *(expert, book, quotation)* Autorität, *die;* **have it on the** ~ **of sb./sth. that ...:** durch jmdn./etw. wissen, daß ...; **have it on good** ~ **that ...:** aus zuverlässiger Quelle wissen, daß ...; **d)** *no pl.* **give** *or* **add** ~ **to sth.** einer Sache *(Dat.)* Gewicht verleihen; **e)** *no pl. (masterfulness)* Souveränität, *die*
authorization [ɔ:θəraɪ'zeɪʃn] *n.* Genehmigung, *die;* **obtain/give** ~: die Genehmigung einholen/erteilen
authorize ['ɔ:θəraɪz] *v.t.* **a)** *(give authority to)* ermächtigen; bevollmächtigen; autorisieren; ~ **sb. to do sth.** jmdn. ermächtigen, etw. zu tun; **b)** *(sanction)* genehmigen
authorship ['ɔ:θəʃɪp] *n., no pl.* Autorschaft, *die;* **of unknown** ~: von einem unbekannten Autor *od.* Verfasser
autistic [ɔ:'tɪstɪk] *adj. (Psych., Med.)* autistisch
auto ['ɔ:təʊ] *n., pl.* ~**s** *(Amer. coll.)* Auto, *das*
auto- ['ɔ:təʊ] *in comb.* auto-/Auto-
autobio'graphic, autobio'graphical *adj.* autobiographisch
autobi'ography *n.* Autobiographie, *die*
autocracy [ɔ:'tɒkrəsɪ] *n.* Autokratie, *die*
autocrat ['ɔ:təkræt] *n.* Autokrat, *der*
autocratic [ɔ:tə'krætɪk] *adj.* autokratisch
autograph ['ɔ:təgrɑ:f] **1.** *n.* **a)** *(signature)* Autogramm, *das;* **b)** *(manuscript)* Autograph, *das.* **2.** *v.t. (sign)* signieren
automat ['ɔ:təmæt] *n. (Amer.)* [Münz]automat, *der*
automate ['ɔ:təmeɪt] *v.t.* automatisieren
automatic [ɔ:tə'mætɪk] **1.** *adj.* automatisch; **his reaction was completely** ~: er reagierte ganz automatisch; ~ **pilot** *see* **autopilot. 2.** *n. (weapon)* automatische Waffe; *(vehicle)* Fahrzeug mit Automatikgetriebe
automatically [ɔ:tə'mætɪkəlɪ] *adv.* automatisch
automation [ɔ:tə'meɪʃn] *n., no pl.* Automation, *die; (automatic control)* Automatisierung, *die*
automaton [ɔ:'tɒmətən] *n., pl.* ~**s** *or* **automata** [ɔ:'tɒmətə] Automat, *der*

automobile ['ɔ:təməbi:l] *n. (Amer.)* Auto, *das*
automotive [ɔ:tə'məʊtɪv] *adj.* Kraftfahrzeug-; ~ **industry** Auto[mobil]industrie, *die*
autonomous [ɔ:'tɒnəməs] *adj. (also Philos.)* autonom
autonomy [ɔ:'tɒnəmɪ] *n., no pl. (also Philos.)* Autonomie, *die*
'**autopilot** *n.* Autopilot, *der;* **[fly] on** ~: mit Autopilot [fliegen]
autopsy ['ɔ:tɒpsɪ, ɔ:'tɒpsɪ] *n.* Autopsie, *die;* Obduktion, *die*
'**autotimer** *n.* [automatische] Schaltuhr
autumn ['ɔ:təm] *n. (lit. or fig.)* Herbst, *der;* **in ~ 1969, in the ~ of 1969** im Herbst 1969; **in early/late** ~: im Frühherbst/Spätherbst; **last/next** ~: letzten/nächsten Herbst
autumnal [ɔ:'tʌmnl] *adj. (lit. or fig.)* herbstlich
auxiliary [ɔ:g'zɪljərɪ] **1.** *adj.* **a)** *(helping)* Hilfs-; **be ~ to sth.** etw. unterstützen *od.* fördern; **b)** *(subsidiary)* zusätzlich; Zusatz-; **c)** *(Ling.)* ~ **verb** Hilfsverb, *das.* **2.** *n.* **a)** Hilfskraft, *die;* **medical** ~: ärztliches Hilfspersonal; **b)** *(Ling.)* Hilfsverb, *das*
avail [ə'veɪl] **1.** *n., no pl., no art.* **be of no** ~: nichts nützen; nutzlos *od.* vergeblich sein; **to no** ~: vergebens. **2.** *v. i.* etwas nützen *od.* fruchten. **3.** *v. t.* nützen; **it will ~ you nothing** es wird dir nichts nützen. **4.** *v. refl.* ~ **oneself of sth.** von etw. Gebrauch machen; ~ **oneself of an opportunity** eine Gelegenheit nutzen
availability [əveɪlə'bɪlɪtɪ] *n., no pl.* Vorhandensein, *das;* **the ~ of sth.** die Möglichkeit, etw. zu bekommen; **the likely ~ of spare parts** die voraussichtliche Lieferbarkeit von Ersatzteilen
available [ə'veɪləbl] *adj.* **a)** *(at one's disposal)* verfügbar; **make sth. ~ to sb.** jmdm. etw. zur Verfügung stellen; **be** ~: zur Verfügung stehen; **b)** *(capable of use)* gültig ⟨*Fahrkarte, Angebot*⟩; **c)** *(obtainable)* erhältlich; lieferbar ⟨*Waren*⟩; verfügbar ⟨*Unterkunft, Daten*⟩; **have sth.** ~: etw. zur Verfügung haben
avalanche ['ævəlɑ:nʃ] *n. (lit. or fig.)* Lawine, *die*
avant-garde [ævɑ̃'gɑ:d] **1.** *adj.* avantgardistisch. **2.** *n.* Avantgarde, *die*
avarice ['ævərɪs] *n., no pl.* Geldgier, *die;* Habsucht, *die*
avaricious [ævə'rɪʃəs] *adj.* geldgierig; habsüchtig
Ave. *abbr.* **Avenue**
Ave [Maria] ['aveɪ (mə'rɪə)] *n.* Ave[-Maria], *das*
avenge [ə'vendʒ] *v. t.* rächen; **be ~d/~ oneself on sb.** sich an jmdm. rächen; **be ~d for sth.** sich für etw. rächen
avenue ['ævənju:] *n. (broad street)* Boulevard, *der; (tree-lined road)* Allee, *die; (fig.)* Weg, *der* **(to** zu); ~ **of approach** Zugang, *der*
aver [ə'vɜ:(r)] *v. t.,* **-rr-** beteuern
average ['ævərɪdʒ] **1.** *n.* **a)** Durchschnitt, *der;* **on [the** *or* **an]** ~: im Durchschnitt; durchschnittlich; im Schnitt *(ugs.);* **above/below** ~: über/unter dem Durchschnitt; **law of ~s** Wahrscheinlichkeitsgesetz, *das;* **b)** *(arithmetic mean)* Mittelwert, *der.* **2.** *adj.* **a)** durchschnittlich; **he is of ~ height** er ist mittelgroß; **b)** *(mediocre)* durchschnittlich; mittelmäßig. **3.** *v. t.* **a)** *(find the ~ of)* den Durchschnitt ermitteln von; **b)** *(amount on ~ to)* durchschnittlich betragen; **the planks ~d three metres in length** die Bretter waren durchschnittlich drei Meter lang; **c)** *(do on ~)* einen Durchschnitt von ... erreichen; **the train ~d 90 m.p.h.** der Zug fuhr im Durchschnitt mit 144 Kilometern pro Stunde. **4.** *v. i.* ~ **out at** im Durchschnitt betragen
averse [ə'vɜ:s] *pred. adj.* **be ~ to sth.** einer Sache *(Dat.)* abgeneigt sein; **be ~ to doing sth.** abgeneigt sein, etw. zu tun
aversion [ə'vɜ:ʃn] *n.* **a)** *no pl. (dislike)* Abneigung, *die;* Aversion, *die;* **have/take an ~ to sth.** eine Abneigung *od.* Aversion gegen etw. haben/bekommen; **b)** *(object)* **my pet ~ is ...**: ein besonderer Greuel ist mir ...
avert [ə'vɜ:t] *v. t.* **a)** *(turn away)* abwenden ⟨*Blick, Gesicht, Aufmerksamkeit*⟩; **b)** *(prevent)* abwenden ⟨*Katastrophe, Schaden, Niederlage*⟩; verhüten ⟨*Unfall*⟩
aviary ['eɪvɪərɪ] *n.* Vogelhaus, *das;* Aviarium, *das*
aviation [eɪvɪ'eɪʃn] *n., no pl., no art.* Luftfahrt, *die;* ~ **fuel** Flugbenzin, *das;* ~ **industry** Flugzeugindustrie, *die*
aviator ['eɪvɪeɪtə(r)] *n.* Flieger, *der*/Fliegerin, *die*
avid ['ævɪd] *adj. (enthusiastic)* begeistert; passioniert; **be ~ for sth.** *(eager, greedy)* begierig auf etw. *(Akk.)* sein
avionics [eɪvɪ'ɒnɪks] *n.* Avionik, *die*
avocado [ævə'kɑ:dəʊ] *n., pl.* **~s**: ~ **[pear]** Avocado[birne], *die*
avoid [ə'vɔɪd] *v. t.* **a)** *(keep away from)* meiden ⟨*Ort*⟩; ~ **an obstacle/a cyclist** einem Hindernis/Radfahrer ausweichen; **b)** *(refrain from)* vermeiden; ~ **doing sth.** vermeiden, etw. zu tun; **c)** *(escape)* vermeiden
avoidable [ə'vɔɪdəbl] *adj.* vermeidbar
avoidance [ə'vɔɪdəns] *n., no pl.* Vermeidung, *die*
avow [ə'vaʊ] *v. t.* bekennen; **an ~ed opponent/supporter** ein erklärter Gegner/Befürworter
avowal [ə'vaʊəl] *n.* Bekenntnis, *das*
avuncular [ə'vʌŋkjʊlə(r)] *adj.* onkelhaft
await [ə'weɪt] *v. t.* erwarten
awake [ə'weɪk] **1.** *v. i.,* **awoke** [ə'wəʊk], **awoken** [ə'wəʊkn] *(lit. or fig.)* erwachen; **~ to sth.** *(fig.)* einer Sache *(Gen.)* gewahr werden. **2.** *v. t.,* **awoke, awoken** *(lit. or fig.)* wecken; **~ sb. to sth.** *(fig.)* jmdm. etw. bewußtmachen. **3.** *pred. adj. (lit. or fig.)* wach; **wide** ~: hellwach; **lie** ~: wach liegen; **be ~ to sth.** *(fig.)* sich *(Dat.)* einer Sache *(Gen.)* bewußt sein
awaken [ə'weɪkn] **1.** *v. t. (esp. fig.) see* **awake 2. 2.** *v. i. (esp. fig.) see* **awake 1**
awakening [ə'weɪknɪŋ] *n.* **a rude** ~ *(fig.)* ein böses Erwachen
award [ə'wɔ:d] **1.** *v. t. (grant)* verleihen, zuerkennen ⟨*Preis, Auszeichnung*⟩; zusprechen ⟨*Sorgerecht, Entschädigung*⟩; gewähren ⟨*Zahlung, Gehaltserhöhung*⟩; ~ **sb. sth.** jmdm. etw. verleihen/zusprechen/gewähren; **he was ~ed the prize** der Preis wurde ihm zuerkannt. **2.** *n.* **a)** *(judicial decision)* Schiedsspruch, *der;* **b)** *(payment)* Entschädigung[ssumme], *die;* *(grant)* Stipendium, *das;* **c)** *(prize)* Auszeichnung, *die;* Preis, *der*
a'ward-winning *adj.* preisgekrönt
aware [ə'weə(r)] *adj.* **a)** *pred. (conscious)* **be ~ of sth.** sich *(Dat.)* einer Sache *(Gen.)* bewußt sein; **be ~ that ...**: sich *(Dat.)* [dessen] bewußt sein, daß ...; **as far as I am** ~: soweit ich weiß; **not that I am ~ of** nicht, daß ich wüßte; **b)** *(well-informed)* informiert
awareness [ə'weənɪs] *n., no pl. (consciousness)* Bewußtsein, *das*
awash [ə'wɒʃ] *pred. adj.* auf gleicher Höhe mit dem Wasserspiegel; **be** ~ *(flooded)* unter Wasser stehen
away [ə'weɪ] **1.** *adv.* **a)** *(at a distance)* entfernt; ~ **in the distance**

weit in der Ferne; **play ~** *(Sport)* auswärts spielen; **Christmas is still months ~**: bis Weihnachten dauert es noch Monate; **b)** *(to a distance)* weg; fort; **~ with you/him!** weg *od.* fort mit dir/ihm!; **throw sth. ~**: etw. wegwerfen *od.* fortwerfen; **~ we go!** los geht's!; **c)** *(absent)* nicht da; **be ~ on business** geschäftlich außer Haus sein; **be ~ [from school] with a cold** wegen einer Erkältung [in der Schule] fehlen; **d) die/fade ~**: verhallen; **the water has all boiled ~**: das ganze Wasser ist verkocht; **e)** *(constantly)* unablässig; **work ~ on sth.** ohne Unterbrechung an etw. *(Dat.)* arbeiten; **laugh ~ at sth.** unablässig über etw. *(Akk.)* lachen; **f)** *(without delay)* gleich ‹*fragen usw.*›; **fire ~** *(lit. or fig.)* losschießen *(ugs.)*. **2.** *adj. (Sport)* auswärts *präd.*; Auswärts-; **~ team** Gastmannschaft, *die*

awe [ɔ:] **1.** *n.* Ehrfurcht, *die* (of vor + *Dat.*); **be *or* stand in ~ of sb.** jmdn. fürchten; **hold sb. in ~**: jmdn. ehrfürchtig respektieren. **2.** *v.t.* Ehrfurcht einflößen (+ *Dat.*); **be ~d by sth.** sich von etw. beeindrucken *od.* einschüchtern lassen

'awe-inspiring *adj.* ehrfurchtgebietend; beeindruckend

awesome ['ɔ:səm] *adj.* überwältigend; eindrucksvoll ‹*Schweigen*›; übergroß ‹*Verantwortung*›

awe: ~-stricken, ~-struck *adjs.* [von Ehrfurcht] ergriffen; ehrfurchtsvoll ‹*Ausdruck, Staunen*›

awful ['ɔ:fl] *adj. (coll.)* furchtbar; fürchterlich; **too ~ for words** unbeschreiblich schlecht; **an ~ lot of money/people** ein Haufen Geld/Leute *(ugs.)*; **an ~ long time/way** eine furchtbar lange Zeit/ein furchtbar weiter Weg

awfully ['ɔ:fəlɪ, 'ɔ:flɪ] *adv. (coll.)* furchtbar; **not ~**: nicht besonders; **thanks ~**: tausend Dank

awkward ['ɔ:kwəd] *adj.* **a)** *(ill-adapted for use)* ungünstig; **be ~ to use** unhandlich sein; **the parcel is ~ to carry** das Paket ist schlecht zu tragen; **b)** *(clumsy)* unbeholfen; **be at an ~ age** in einem schwierigen Alter sein; **c)** *(embarrassing, embarrassed)* peinlich; **feel ~**: sich unbehaglich fühlen; **d)** *(difficult)* schwierig, unangenehm ‹*Person*›; ungünstig ‹*Zeitpunkt*›; schwierig, peinlich ‹*Lage, Dilemma*›

awkwardly ['ɔ:kwədlɪ] *adv.* **a)** *(badly)* ungünstig ‹*geformt, angebracht*›; **b)** *(clumsily)* ungeschickt, unbeholfen ‹*gehen, sich ausdrücken*›; ungeschickt, unglücklich ‹*fallen, sich ausdrücken*›; **c)** *(embarrassingly)* peinlicherweise; **d)** *(unfavourably)* ungünstig ‹*gelegen*›

awkwardness ['ɔ:kwədnɪs] *n., no pl. see* **awkward: a)** Unhandlichkeit, *die;* **b)** Unbeholfenheit, *die;* **c)** Peinlichkeit, *die;* **d)** *(of person)* unangenehmes Wesen; *(of situation, position)* Schwierigkeit, *die*

awl [ɔ:l] *n.* Ahle, *die;* Pfriem, *der*

awning ['ɔ:nɪŋ] *n. (on wagon)* Plane, *die; (on house)* Markise, *die; (of tent)* Vordach, *das*

awoke, awoken *see* **awake 1, 2**

awry [ə'raɪ] *adv.* schief; **go ~** *(fig.)* schiefgehen *(ugs.)*; ‹*Plan:*› fehlschlagen

axe (*Amer.:* **ax**) [æks] **1.** *n.* Axt, *die;* Beil, *das;* **have an ~ to grind** *(fig.)* sein eigenes Süppchen kochen *(ugs.)*. **2.** *v.t. (reduce)* [radikal] kürzen; *(eliminate)* [radikal] einsparen ‹*Stellen*›; *(dismiss)* entlassen; *(abandon)* aufgeben ‹*Projekt*›

axes *pl. of* **axe, axis**

axiom ['æksɪəm] *n.* Axiom, *das*

axiomatic [æksɪə'mætɪk] *adj.* axiomatisch; **I have taken it as ~ that ...**: ich gehe von dem Grundsatz aus, daß ...

axis ['æksɪs] *n., pl.* **axes** ['æksi:z] Achse, *die;* **~ of rotation** Rotationsachse, *die*

axle ['æksl] *n.* Achse, *die*

ayatollah [aɪə'tɒlə] *n.* Ajatollah, *der*

azalea [ə'zeɪlɪə] *n. (Bot.)* Azalee, *die*

Aztec ['æztek] **1.** *adj.* aztekisch. **2.** *n.* Azteke, *der*/Aztekin, *die*

azure ['æʒjə(r), 'eɪʒjə(r)] **1.** *n.* Azur[blau], *das.* **2.** *adj.* azurblau

B

B, b [bi:] *n., pl.* **Bs** *or* **B's a)** *(letter)* B, b, *das;* **B road** Straße 2. Ordnung; ≈ Landstraße, *die;* **~ film** *or (Amer.)* **movie** Vorfilm, *der;* **b) B** *(Mus.)* H, h, *das;* **B flat** B, b, *das*

BA *abbr. (Univ.)* **Bachelor of Arts;** *see also* **B.Sc.**

baa [bɑ:] **1.** *n.* Blöken, *das.* **2.** *v.i.,* **baaed** *or* **baa'd** [bɑ:d] mähen; blöken

babble ['bæbl] **1.** *v.i.* **a)** *(talk incoherently)* stammeln; **b)** *(talk foolishly)* [dumm] schwatzen; **c)** *(talk excessively)* **~ away** *or* **on** quasseln *(ugs.)*; **d)** ‹*Bach:*› plätschern. **2.** *v.t. (utter incoherently)* stammeln. **3.** *n.* **a)** *(incoherent speech)* Gestammel, *das; (childish or foolish speech)* Gelalle, *das;* **b)** *(murmur of water)* Geplätscher, *das*

baboon [bə'bu:n] *n.* Pavian, *der*

baby ['beɪbɪ] *n.* **a)** Baby, *das;* **have a ~/be going to have a ~**: ein Kind bekommen; **she has a young ~**: sie hat ein kleines Baby; **a ~ boy/girl** ein kleiner Junge/ein kleines Mädchen; **throw out *or* away the ~ with the bathwater** *(fig.)* das Kind mit dem Bade ausschütten; **be left holding *or* carrying the ~** *(fig.)* die Sache ausbaden müssen *(ugs.)*; der Dumme sein *(ugs.)*; **it's your/his** *etc.* **~** *(fig.)* das ist dein/sein *usw.* Bier *(ugs.)*; **b)** *(youngest member)* Jüngste, *der/die; (male also)* Benjamin, *der;* **the ~ of the family** das Küken der Familie; **c)** *(childish person)* **be a ~**: sich wie ein kleines Kind benehmen; **d)** *(young animal)* Junge, *das;* **~ bird/giraffe** Vogeljunge, *das*/Giraffenjunge, *das;* **e)** *(small thing)* **be a ~**: winzig sein; **f)** *(sl.: sweetheart)* Schatz, *der; (in pop song also)* Baby, *das;* **g)** *(sl.: young woman)* Kleine, *die (ugs.)*

baby: ~ clothes *n. pl.* Babykleidung, *die;* **~ food** *n.* Babynahrung, *die*

babyish ['beɪbɪɪʃ] *adj.* kindlich ‹*Aussehen*›; kindisch ‹*Benehmen, Person*›

baby: ~-minder *n.* Tagesmutter, *die;* **~ powder** *n.* Babypuder, *der;* **~-sit** *v.i., forms as* **sit** babysitten *(ugs.)*; auf das Kind/die Kinder aufpassen; **~-sitter** *n.* Babysitter, *der;* **~-sitting** *n.* Babysitting, *das;* **~-snatcher** ['beɪbɪsnætʃə(r)] *n.* Kindesentführer, *der;* **~-snatching** *n.* Kindesentführung, *die;* **~-talk** *n.* Babysprache, *die*

bachelor ['bætʃələ(r)] *n.* **a)** *(unmarried man)* Junggeselle, *der;* **b)** *(Univ.)* **B~ of Arts/Science** Bakkalaureus der philosophischen

Fakultät/der Naturwissenschaften
bachelor: ~ flat *n.* Junggesellenwohnung, *die;* **~ girl** *n.* Junggesellin, *die*
bacillus [bə'sɪləs] *n., pl.* **bacilli** [bə'sɪlaɪ] *(Biol., Med.)* Bazillus, *der*
back [bæk] **1.** *n.* **a)** *(of person, animal)* Rücken, *der;* **stand ~ to ~:** Rücken an Rücken stehen; **as soon as my ~ was turned** *(fig.)* sowie ich den Rücken gedreht hatte; **turn one's ~ on sb.** jmdm. den Rücken zuwenden; *(fig.: abandon sb.)* jmdn. im Stich lassen; **turn one's ~ on sth.** *(fig.)* sich um etw. nicht kümmern; **get** *or* **put sb.'s ~ up** *(fig.)* jmdn. wütend machen; **be glad to see the ~ of sb./sth.** *(fig.)* froh sein, jmdn./etw. nicht mehr sehen zu müssen; **have one's ~ to the wall** *(fig.)* mit dem Rücken zur Wand stehen; **get off my ~** *(fig. coll.)* laß mich zufrieden; **have sb./sth. on one's ~** *(fig.)* jmdn./etw. am Hals haben *(ugs.);* **put one's ~ into sth.** *(fig.)* sich für etw. mit allen Kräften einsetzen; **b)** *(outer or rear surface)* Rücken, *der; (of vehicle)* Heck, *das; (inside car)* Rücksitz, *der;* **the car went into the ~ of me** *(coll.)* das Auto ist mir hinten reingefahren *(ugs.);* **with the ~ of one's hand** mit dem Handrücken; **know sth. like the ~ of one's hand** *(fig.)* etw. wie seine Westentasche kennen; **the ~ of one's/the head** der Hinterkopf; **the ~ of the leg** die Wade; **c)** *(of book) (spine)* [Buch]rücken, *der; (final pages)* Ende, *das;* **at the ~ [of the book]** hinten [im Buch]; **d)** *(of dress)* Rücken, *der; (of knife)* [Messer]rücken, *der;* **e)** *(more remote part)* hinterer Teil; **at the ~ [of sth.]** hinten [in etw. *(Dat.)*]; im hinteren Teil [von etw.]; **f)** *(of chair)* [Rücken]lehne, *die; (of house, cheque)* Rückseite, *die; (~ wall)* Rückseite, *die;* Rückwand, *die;* **~ to front** verkehrt rum; **please get to the ~ of the queue** bitte, stellen Sie sich hinten an; **in ~ of sth.** *(Amer.)* hinter etw. *(Dat.);* **g)** *(Sport: player)* Verteidiger, *der;* **h)** *(of ship)* Kiel, *der.* **2.** *adj., no comp.; superl.* **~most** ['bækməʊst] **a)** *(situated behind)* hinter...; **b)** *(of the past)* früher; **~ issue** alte Ausgabe; **c)** *(overdue)* rückständig ‹*Lohn, Steuern*›. **3.** *adv.* **a)** *(to the rear)* zurück; **step ~:** zurücktreten; **b)** *(behind)* zurück; weiter hinten; **we passed a pub two miles ~:** wir sind vor zwei Meilen an einem Pub vorbeigefahren; **~ and forth** hin und her; **~ of sth.** *(Amer.)* hinter etw. *(Dat.);* **c)** *(at a distance)* **the house stands a long way ~ from the road** das Haus steht weit von der Straße zurück; **d)** *(to original position, home)* [wieder] zurück; **I got my letter ~:** ich habe meinen Brief zurückbekommen; **the journey ~:** die Rückfahrt/der Rückflug; **there and ~:** hin und zurück; **e)** *(to original condition)* wieder; **f)** *(in the past)* zurück; **go a long way ~:** weit zurückgehen; **a week/month ~:** vor einer Woche/vor einem Monat; **g)** *(in return)* zurück; **I got a letter ~:** er/sie hat mir wiedergeschrieben. **4.** *v.t.* **a)** *(assist)* helfen (+ *Dat.*); unterstützen ‹*Person, Sache*›; **b)** *(bet on)* wetten *od.* setzen auf (+ *Akk.*) ‹*Pferd, Gewinner, Favorit*›; **~ the wrong/right horse** *(lit. or fig.)* aufs falsche/richtige Pferd setzen *(ugs.);* **c)** *(cause to move back)* zurücksetzen [mit] ‹*Fahrzeug*›; rückwärts gehen lassen' ‹*Pferd*›; **d)** *(put or act as a ~ to)* [an der Rückseite] verstärken; **e)** *(endorse)* indossieren ‹*Wechsel, Scheck*›; **f)** *(lie at the ~ of)* **~ sth.** hinten an etw. *(Akk.)* grenzen; **g)** *(Mus.)* begleiten. **5.** *v.i.* zurücksetzen; **~ into/out of sth.** rückwärts in etw. *(Akk.)*/aus etw. fahren; **~ on to sth.** hinten an etw. *(Akk.)* grenzen
back: ~ache *n., no pl.* Rückenschmerzen *Pl.;* **~-bencher** [bæk'bentʃə(r)] *n.* *(Brit. Parl.)* [einfacher] Abgeordneter/[einfache] Abgeordnete; *(derog.)* Hinterbänkler, *der (abwertend);* **~bone** *n.* Wirbelsäule, *die;* Rückgrat, *das;* **~-breaking** *adj.* äußerst mühsam; gewaltig ‹*Anstrengung*›; **~-breaking work** Knochenarbeit, *die;* **~ 'burner** *n.* **put sth. on the ~ burner** *(fig. coll.)* etw. zurückstellen; **~chat** *n., no pl. (coll.)* [freche] Widerrede; **~-comb** *v.t.* zurückkämmen; **~date** *v.t.* zurückdatieren (**to** auf + *Akk.*); **~ 'door** *n.* Hintertür, *die (auch fig.)*
backer ['bækə(r)] *n.* Geldgeber, *der; (of horse)* Wetter, *der*
back'fire *v.i.* knallen; *(fig.)* fehlschlagen; **it ~fired on me/him** *etc.* der Schuß ging nach hinten los *(ugs.)*
backgammon ['bækgæmən] *n.* Backgammon, *das*
back: ~ground *n.* **a)** *(lit. or fig.)* Hintergrund, *der; (social status)* Herkunft, *die;* **be in the ~ground** im Hintergrund stehen; **against this ~ground** vor diesem Hintergrund; **~ground music** Hintergrundmusik, *die;* **b) ~ground [information]** Hintergrundinformation, *die;* **~hand** *(Tennis etc.)* **1.** *adj.* Rückhand-; **2.** *n.* Rückhand, *die;* **~handed** [bæk'hændɪd] *adj.* **a) ~-handed stroke** *(Tennis)* Rückhandschlag, *der;* **b)** *(fig.)* indirekt; zweifelhaft ‹*Kompliment*›; **~hander** [bæk'hændə(r)] *n.* **a)** *(stroke)* Rückhandschlag, *der (Tennis usw.); (blow)* Schlag [mit dem Handrücken]; **b)** *(sl.: bribe)* Schmiergeld, *das*
backing ['bækɪŋ] *n.* **a)** *(material)* Rückenverstärkung, *die;* **b)** *(support)* Unterstützung, *die*
'backlash *n.* Rückstoß, *der; (fig.)* Gegenreaktion, *die;* **a right-wing ~:** eine Gegenbewegung nach rechts
backless ['bæklɪs] *adj.* rückenfrei ‹*Kleid*›
back: ~list *n.* Verzeichnis der lieferbaren Titel; **~log** *n.* Rückstand, *der;* **~log of work** Arbeitsrückstand, *der;* **~ 'number** *n. (of periodical, magazine)* alte Nummer; **~pack 1.** *n.* Rucksack, *der;* **2.** *v.i.* mit dem Rucksack [ver]reisen; **~-'pedal** *v.i.* **a)** die Pedale rückwärts treten; *(brake)* mit dem Rücktritt bremsen; **b)** *(fig.)* einen Rückzieher machen *(ugs.);* **~rest** *n.* Rückenlehne, *die;* **~ 'room** *n.* Hinterzimmer, *das;* **~scratching** *n. (fig. coll.)* **[mutual] ~scratching** Klüngelei, *die (abwertend);* **~ 'seat** *n.* Rücksitz, *der; (in bus, coach)* hinterer Sitzplatz; **take a ~ seat** *(fig. coll.)* in den Hintergrund treten; **~-seat driver** [--'-] *besserwisserischer Beifahrer, der immer dazwischenredet;* **~side** *n.* Hinterteil, *das (ugs.);* Hintern, *der (ugs.);* **get [up] off one's ~side** seinen Hintern heben *(ugs.);* **~sight** *n.* Visier, *das;* **~slapping** *adj. (fig.)* plump-vertraulich; **~slider** ['bækslaɪdə(r)] *n.* Abtrünnige, *der/die;* **~space** *v.i.* die Rücktaste betätigen; **~'stage 1.** *adj.* hinter der Bühne; **2.** *adv.* **a) go ~stage** hinter die Bühne gehen; **wait ~stage** hinter der Bühne warten; **b)** *(fig.)* hinter den Kulissen; **~ 'stairs** *n. pl.* Hintertreppe, *die;* **~stitch 1.** *v.t. & i.* steppen; **2.** *n.* Steppstich, *der;* **~ street** *n.* kleine Seitenstraße; **~stroke** *n. (Swimming)* Rückenschwimmen, *das;* **~ talk** *(Amer.) see* **~chat; ~track** *v.i.* wieder zurückgehen; *(fig.)* eine Kehrtwendung machen

backward ['bækwəd] **1.** *adj.* **a)** rückwärts gerichtet; Rückwärts-; ~ **movement** Rückwärtsbewegung, *die;* **b)** *(slow, retarded)* zurückgeblieben ⟨*Kind*⟩; *(underdeveloped)* rückständig, unterentwickelt ⟨*Land, Region*⟩; ~ **in sth.** in etw. *(Dat.)* zurückgeblieben. **2.** *adv. see* **backwards**

backwardness ['bækwədnɪs] *n., no pl.* **a)** *(reluctance, shyness)* Zurückhaltung, *die;* **b)** *(of child)* Zurückgebliebenheit, *die; (of country, region)* Rückständigkeit, *die;* **the child's ~ in school** daß das Kind in der Schule zurückgeblieben ist/war

backwards ['bækwədz] *adv.* **a)** nach hinten; **the child fell [over] ~ into the water** das Kind fiel rückwärts ins Wasser; **bend** *or* **fall** *or* **lean over ~ to do sth.** *(fig. coll.)* sich zerreißen, um etw. zu tun *(ugs.);* **b)** *(oppositely to normal direction)* rückwärts; **~ and forwards** *(to and fro, lit. or fig.)* hin und her; **c)** *(into a worse state)* **go ~:** sich verschlechtern; **d)** *(into past)* **look ~:** an frühere Zeiten denken; **e)** *(reverse way)* rückwärts; von hinten nach vorn; **you're doing everything ~:** du machst ja alles verkehrt herum; **know sth. ~:** etw. in- und auswendig kennen

back: ~wash *n.* Rückstrom, *der* **(from** *Gen.*); *(fig.)* Auswirkungen *Pl.;* **~water** *n.* totes Wasser; *(fig.)* Kaff, *das (ugs. abwertend);* ~ **'yard** *n.* Hinterhof, *der;* **in one's own ~ yard** *(fig.)* vor der eigenen Haustür

bacon ['beɪkn] *n.* [Frühstücks]speck, *der;* **~ and eggs** Eier mit Speck; **bring home the ~** *(fig. coll.)* es schaffen; **save one's ~** *(fig.)* die eigene *od.* seine Haut retten

bacteria *pl. of* **bacterium**

bacterial [bæk'tɪərɪəl] *adj.* bakteriell

bacteriological [bæktɪərɪə'lɒdʒɪkl] *adj.* bakteriologisch

bacteriology [bæktɪərɪ'ɒlədʒɪ] *n.* Bakteriologie, *die*

bacterium [bæk'tɪərɪəm] *n., pl.* **bacteria** [bæk'tɪərɪə] Bakterie, *die*

bad [bæd] **1.** *adj.,* **worse** [wɜ:s], **worst** [wɜ:st] **a)** schlecht; *(worthless)* wertlos, ungedeckt ⟨*Scheck*⟩; *(rotten)* schlecht, verdorben ⟨*Fleisch, Fisch, Essen*⟩; faul ⟨*Ei, Apfel*⟩; *(unpleasant)* schlecht, unangenehm ⟨*Geruch*⟩; **sth. gives sb. a ~ name** etw. trägt jmdm. einen schlechten Ruf ein; **sb. gets a ~ name** jmd. kommt in Verruf; **she is in ~ health** sie hat eine angegriffene Gesundheit; **[some] ~ news** schlechte *od.* schlimme Nachrichten; **~ breath** Mundgeruch, *der;* **he is having a ~ day** er hat einen schwarzen Tag; **~ business** ein schlechtes Geschäft; **it is a ~ business** *(fig.)* das ist eine schlimme Sache; **in the ~ old days** in den schlimmen Jahren; **not ~** *(coll.)* nicht schlecht; nicht übel; **sth. is not a ~ idea** etw. ist keine schlechte Idee; **not half ~** *(coll.)* [gar] nicht schlecht; **sth. is too ~** *(coll.)* etw. ist ein Jammer; **too ~!** *(coll.)* so ein Pech! *(auch iron.);* **go ~:** schlecht werden; **b)** *(noxious)* schlecht; schädlich; **c)** *(wicked)* schlecht; *(immoral)* schlecht; verdorben; *(naughty)* ungezogen, böse ⟨*Kind, Hund*⟩; **d)** *(offensive)* **[use] ~ language** Kraftausdrücke [benutzen]; **e)** *(in ill health)* **she's ~ today** es geht ihr heute schlecht; **I have a ~ pain/finger** ich habe schlimme Schmerzen/*(ugs.)* einen schlimmen Finger; **be in a ~ way** in schlechtem Zustand sein; **f)** *(serious)* schlimm, böse ⟨*Sturz, Krise*⟩; schwer ⟨*Fehler, Krankheit, Unfall, Erschütterung*⟩; hoch ⟨*Fieber*⟩; schrecklich ⟨*Feuer*⟩; **g)** *(coll.: regretful)* **a ~ conscience** ein schlechtes Gewissen; **feel ~ about sth./not having done sth.** etw. bedauern/bedauern, daß man etw. nicht getan hat; **I feel ~ about him/her** ich habe seinetwegen/ihretwegen ein schlechtes Gewissen; **h)** *(Commerc.)* **a ~ debt** eine uneinbringliche Schuld *(Wirtsch.). See also* **worse 1; worst 1. 2.** *n.* **be £100 to the ~:** mit 100 Pfund in der Kreide stehen *(ugs.);* **go to the ~:** auf die schiefe Bahn geraten

baddy ['bædɪ] *n. (coll.)* Schurke, *der;* **the goodies and the baddies** die Guten und die Bösen *(oft iron.)*

bade *see* **bid 1 c**

badge [bædʒ] *n.* **a)** *(as sign of office, membership, support)* Abzeichen, *das; (larger)* Plakette, *die;* **b)** *(symbol)* Symbol, *das*

badger ['bædʒə(r)] **1.** *n.* Dachs, *der.* **2.** *v.t.* **~ sb. [into doing/to do sth.]** jmdm. keine Ruhe lassen[, bis er/sie etw. tut]; **~ sb. with questions** jmdn. mit Fragen löchern *(ugs.)*

'badger baiting *n.* Dachshetze, *die*

badly ['bædlɪ] *adv.,* **worse** [wɜ:s], **worst** [wɜ:st] **a)** schlecht; **b)** *(seriously)* schwer ⟨*verletzt, beschädigt*⟩; sehr ⟨*schief sein, knarren*⟩; **he hurt himself ~:** er hat sich *(Dat.)* schwer verletzt; **be ~ beaten** schwer verprügelt werden *(ugs.); (in game, battle)* vernichtend geschlagen werden; **c)** *(urgently)* dringend; **want sth. ~:** sich *(Dat.)* etw. sehr wünschen; **d)** *(coll.: regretfully)* **feel ~ about sth.** etw. [sehr] bedauern. *See also* **worse 2; worst 2**

badminton ['bædmɪntən] *n.* Federball, *der; (als Sport)* Badminton, *das*

bad-tempered [bæd'tempəd] *adj.* griesgrämig

baffle ['bæfl] **1.** *v.t.* **~ sb.** jmdm. unverständlich sein; jmdn. vor ein Rätsel stellen. **2.** *n.* **~[-plate]** Prallfläche, *die (Technik)*

baffled ['bæfld] *adj.* verwirrt; **be ~:** vor einem Rätsel stehen

bafflement ['bæflmənt] *n.* Verwirrung, *die*

baffling ['bæflɪŋ] *adj.* rätselhaft

bag [bæg] **1.** *n.* **a)** Tasche, *die; (sack)* Sack, *der; (hand~)* [Hand]tasche, *die; (of plastic)* Beutel, *der; (small paper ~)* Tüte, *die;* **a ~ of cement** ein Sack Zement; **be a ~ of bones** *(fig.)* nur Haut und Knochen sein; **[whole] ~ of tricks** *(fig.)* Trickkiste, *die (ugs.);* **his nomination is in the ~** *(fig. coll.)* er hat die Nominierung in der Tasche *(ugs.);* **b)** *(Hunting: amount of game)* Jagdbeute, *die;* Strecke, *die (Jägerspr.);* **c)** *in pl. (sl.: large amount)* **~s of** jede Menge *(ugs.);* **d) have ~s under** *or* **below one's eyes** Tränensäcke haben; **e)** *(sl. derog.: woman)* **[old] ~:** alte Schlampe *(ugs. abwertend).* **2.** *v.t.,* **-gg-: a)** *(put in sacks)* in Säcke füllen; *(put in plastic ~s)* in Beutel füllen; *(put in small paper ~s)* in Tüten füllen; **b)** *(Hunting)* erlegen, erbeuten ⟨*Tier*⟩; **c)** *(claim possession of)* sich *(Dat.)* schnappen *(ugs.)*

bagful ['bægfʊl] *n. see* **bag 1 a: a ~ of** ein Sack [voll]/eine Tasche/ein Beutel/eine Tüte [voll]

baggage ['bægɪdʒ] *n.* Gepäck, *das;* **mental/cultural ~** *(fig.)* geistiges/kulturelles Rüstzeug

baggage: ~ car *n. (Amer.)* Gepäckwagen, *der;* **~ check** *n.* Gepäckkontrolle, *die;* **~ reclaim** *n.* Gepäckausgabe, *die*

bagginess ['bægɪnɪs] *n., no pl.* Schlaffheit, *die*

baggy ['bægɪ] *adj.* weit [geschnitten] ⟨*Kleid*⟩; *(through long use)* ausgebeult ⟨*Hose*⟩

bag: ~pipe[s] *n.* Dudelsack,

der; **~piper** *n.* Dudelsackpfeifer, *der*

bah [bɑ:] *int.* bah

Bahamas [bə'hɑ:məz] *pr. n. pl.* **the ~:** die Bahamainseln

[1]**bail** [beɪl] **1.** *n.* **a)** Kaution, *die; (personal)* Bürgschaft, *die;* **grant sb. ~:** jmdm. die Freilassung gegen Kaution bewilligen; **be [out] on ~:** gegen Kaution auf freiem Fuß sein; **put** *or* **release sb. on ~:** jmdn. gegen Kaution freilassen; **b)** *(person[s] acting as surety)* Bürge, *der;* **go ~ for sb.** für jmdn. Bürge sein. **2.** *v. t.* **a)** *(release)* gegen Kaution freilassen; **b)** *(go ~ for)* bürgen für; **~ sb. out** jmdn. gegen Bürgschaft freibekommen; *(fig.)* jmdm. aus der Klemme helfen *(ugs.)*

[2]**bail** *n. (Cricket)* Querstab, *der*

[3]**bail** *v. t. (scoop)* **~ [out]** ausschöpfen

~ out *v. i. (Aeronaut.)* ‹*Pilot:*› abspringen, *(Fliegerspr.)* aussteigen

bailey ['beɪlɪ] *n. (wall)* Burgmauer, *die; (outer court)* Zwinger, *der; (inner court)* Burghof, *der;* **the Old B~:** das Old Bailey *(oberster Strafgerichtshof für London)*

bailiff ['beɪlɪf] *n.* ≈ Justizbeamte, *der;* Büttel, *der (veralt.); (issuing writs, making arrests)* Gerichtsvollzieher, *der*

bairn [beən] *n. (Scot./N. Engl./literary)* Kind, *das*

bait [beɪt] **1.** *v. t.* mit einem Köder versehen ‹*Falle*›; beködern ‹*Angelhaken*›. **2.** *n. (lit. or fig.)* Köder, *der*

bake [beɪk] **1.** *v. t.* **a)** *(cook)* backen; **~d beans** gebackene Bohnen [in Tomatensoße]; **b)** *(harden)* brennen ‹*Ziegel, Keramik*›. **2.** *v. i.* backen; gebacken werden; **I'm baking!** *(fig.)* mir ist wahnsinnig heiß! *(ugs.)*

'**bakehouse** *n.* Backstube, *die*

baker ['beɪkə(r)] *n.* Bäcker, *der;* **at the ~'s** beim Bäcker; in der Bäckerei; **go to the ~'s** zum Bäcker *od.* zur Bäckerei gehen; **a ~'s dozen** 13 Stück

bakery ['beɪkərɪ] *n.* Bäckerei, *die*

baking ['beɪkɪŋ] *adv.* **it's ~ hot today** eine Hitze wie im Backofen ist das heute

baking: ~-dish *n.* Auflaufform, *die;* **~-powder** *n.* Backpulver, *das;* **~-sheet** *n.* Backblech, *das;* **~-soda** *n.* Natron, *das;* **~-tin** *n.* Backform, *die;* **~-tray** *n.* Kuchenblech, *das*

Balaclava [bælə'klɑ:və] *n.* **~ [helmet]** Balaklavamütze, *die (Wollmütze, die Kopf und Hals bedeckt und nur das Gesicht frei läßt)*

balalaika [bælə'laɪkə] *n.* Balalaika, *die*

balance ['bæləns] **1.** *n.* **a)** *(instrument)* Waage, *die;* **~[-wheel]** Unruh, *die;* **b)** *(fig.)* **be** *or* **hang in the ~:** in der Schwebe sein; **c)** *(even distribution)* Gleichgewicht, *das; (due proportion)* ausgewogenes Verhältnis; **strike a ~ between** den Mittelweg finden zwischen (+ *Dat.*); **d)** *(counterpoise, steady position)* Gleichgewicht, *das;* **keep/lose one's ~** das Gleichgewicht halten/verlieren; *(fig.)* sein Gleichgewicht bewahren/verlieren; **off [one's] ~** *(lit. or fig.)* aus dem Gleichgewicht; **e)** *(preponderating weight or amount)* Bilanz, *die;* **f)** *(Bookk.: difference)* Bilanz, *die; (state of bank account)* Kontostand, *der; (statement)* Auszug, *der;* **on ~** *(fig.)* alles in allem; **~ sheet** Bilanz, *die;* **g)** *(Econ.)* **~ of payments** Zahlungsbilanz, *die;* **~ of trade** Handelsbilanz, *die;* **h)** *(remainder)* Rest, *der.* **2.** *v. t.* **a)** *(weigh up)* abwägen; **~ sth. with** *or* **by** *or* **against sth. else** etw. gegen etw. anderes abwägen; **b)** *(bring into or keep in ~)* balancieren; auswuchten ‹*Rad*›; **~ oneself** balancieren; **c)** *(equal, neutralize)* ausgleichen; **~ each other, be ~d** sich *(Dat.)* die Waage halten; **d)** *(make up for, exclude dominance of)* ausgleichen; **e)** *(Bookk.)* bilanzieren. **3.** *v. i.* **a)** *(be in equilibrium)* balancieren; **balancing act** *(lit. or fig.)* Balanceakt, *der;* **b)** *(Bookk.)* ausgeglichen sein

balanced ['bælənst] *adj.* ausgewogen; ausgeglichen ‹*Person, Team, Gemüt*›

balcony ['bælkənɪ] *n.* Balkon, *der*

bald [bɔ:ld] *adj.* **a)** kahl ‹*Person, Kopf*›; kahlköpfig, glatzköpfig ‹*Person*›; **he is ~:** er ist kahl[köpfig] *od.* hat eine Glatze; **go ~:** eine Glatze bekommen; **b)** *(plain)* einfach, schmucklos ‹*Rede, Prosa*›; knapp, nackt ‹*Behauptung*›; **c)** *(coll.: worn smooth)* abgefahren ‹*Reifen*›

balderdash ['bɔ:ldədæʃ] *n., no pl., no indef. art.* Unsinn, *der*

bald-headed [bɔ:ld'hedɪd] *adj.* glatzköpfig; kahlköpfig

balding ['bɔ:ldɪŋ] *adj.* mit beginnender Glatze *nachgestellt;* **be ~:** kahl werden

baldly ['bɔ:ldlɪ] *adv.* unverhüllt; offen; knapp und klar ‹*zusammenfassen, umreißen*›

baldness ['bɔ:ldnɪs] *n., no pl. see* **bald a, b:** Kahlheit, *die;* Einfachheit, *die;* Knappheit, *die*

bale [beɪl] **1.** *n.* Ballen, *der.* **2.** *v. t. (pack)* in Ballen verpacken; zu Ballen binden ‹*Heu*›

baleful ['beɪlfl] *adj.* unheilvoll; *(malignant)* böse

balk [bɔ:k, bɔ:lk] **1.** *n. (timber beam)* Balken, *der.* **2.** *v. t.* [be]hindern; **they were ~ed in their plan** ihr Plan wurde blockiert. **3.** *v. i.* sich sträuben **(at** gegen); ‹*Pferd:*› scheuen **(at** vor + *Dat.*)

Balkan ['bɔ:lkn] **1.** *adj.* Balkan-. **2.** *n. pl.* **the B~s** der Balkan

[1]**ball** [bɔ:l] **1.** *n.* **a)** Kugel, *die;* **the animal rolled itself into a ~:** das Tier rollte sich [zu einer Kugel] zusammen; **b)** *(Sport, incl. Golf, Polo)* Ball, *der; (Billiards etc., Croquet)* Kugel, *die;* **keep one's eye on the ~** *(fig.)* die Sache im Auge behalten; **start the ~ rolling** *(fig.)* den Anfang machen; **be on the ~** *(fig. coll.)* voll dasein *(salopp); (be alert)* auf Zack sein *(ugs.);* **play ~:** Ball spielen; *(fig. coll.: co-operate)* mitmachen; **c)** *(missile)* Kugel, *die;* **d)** *(round mass)* Kugel, *die; (of wool, string, fluff, etc.)* Knäuel, *das;* **e)** *(Anat.)* Ballen, *der;* **~ of the hand/foot** Hand-/Fußballen, *der;* **f)** *in pl. (coarse: testicles)* Eier *Pl. (derb);* **~s!** *(fig.)* Scheiß! *(salopp abwertend).* **2.** *v. t.* zusammenballen; ballen ‹*Faust*›

[2]**ball** *n. (dance)* Ball, *der;* **have [oneself] a ~** *(fig. sl.)* sich riesig amüsieren *(ugs.)*

ballad ['bæləd] *n.* Ballade, *die*

ballast ['bæləst] *n.* **a)** Ballast, *der;* **b)** *(coarse stone etc.)* Schotter, *der*

ball: ~-'bearing *n. (Mech.)* Kugellager, *das;* **~boy** *n.* Balljunge, *der;* **~cock** *n.* Schwimmer[regel]ventil, *das (Technik);* **~ control** *n.* Ballführung, *die*

ballerina [bælə'ri:nə] *n.* Ballerina, *die;* **prima ~:** Primaballerina, *die*

ballet ['bæleɪ] *n.* Ballett, *das;* **~ dancer** Ballettänzer, *der/*-tänzerin, *die*

ball: ~ game *n.* **a)** Ballspiel, *das;* **b)** *(Amer.)* Baseballspiel, *das;* **a whole new ~ game** *(fig. coll.)* eine ganz neue Geschichte *(ugs.);* **~girl** *n.* Ballmädchen, *das*

ballistic [bə'lɪstɪk] *adj.* ballistisch; **~ missile** ballistische Rakete

ballistics [bə'lɪstɪks] *n., no pl.* Ballistik, *die*

balloon [bə'lu:n] **1.** *n.* **a)** Ballon, *der;* **hot-air ~:** Heißluftballon, *der;* **when the ~ goes up** *(fig.)* wenn es losgeht *(ugs.);* **b)** *(toy)*

Luftballon, *der;* **c)** *(coll.: in strip cartoon etc.)* Sprechblase, *die.* **2.** *v. i.* sich blähen

ballot ['bælət] **1.** *n.* **a)** *(voting)* Abstimmung, *die;* **[secret] ~:** geheime Wahl; **hold** *or* **take a ~:** abstimmen; **b)** *(vote)* Stimme, *die;* **c)** *(ticket, paper)* Stimmzettel, *der.* **2.** *v. i.* abstimmen; **~ for sb./sth.** für jmdn./etw. stimmen

ballot: ~-box *n.* Wahlurne, *die;* **~-paper** *n.* Stimmzettel, *der*

ball: ~-pen, ~-point, ~-point 'pen *ns.* Kugelschreiber, *der;* **~room** *n.* Tanzsaal, *der;* **~room dancing** *n.* Gesellschaftstanz, *der*

balls-up ['bɔ:lzʌp] *n.* *(coarse)* Scheiß, *der (salopp abwertend);* **make a ~ of sth.** bei etw. Scheiße bauen *(derb)*

ballyhoo [bælɪ'hu:] *n.* *(publicity)* [Reklame]rummel, *der (ugs.)*

balm [bɑ:m] *n. (lit. or fig.)* Balsam, *der*

balmy ['bɑ:mɪ] *adj.* **a)** *(soft, mild)* mild; **b)** *see* **barmy**

balsa ['bɔ:lsə, 'bɒlsə] *n.* **~[-wood]** Balsaholz, *das*

balsam ['bɔ:lsəm, 'bɒlsəm] *n. (lit. or fig.)* Balsam, *der*

Balt [bɔ:lt, bɒlt] *n.* Balte, *der*/Baltin, *die*

Baltic ['bɔ:ltɪk, 'bɒltɪk] **1.** *pr. n.* Ostsee, *die.* **2.** *adj.* baltisch; **~ coast** Ostseeküste, *die;* **the ~ Sea** die Ostsee; **the ~ States** das Baltikum

baluster ['bæləstə(r)] *n.* Geländerpfosten, *der*

balustrade [bælə'streɪd] *n.* Balustrade, *die*

bamboo [bæm'bu:] *n.* Bambus, *der*

bamboozle [bæm'bu:zl] *v. t. (sl.)* **a)** *(mystify)* verblüffen; **b)** *(cheat)* reinlegen *(ugs.);* **~ sb. into doing sth.** jmdn. so reinlegen, daß er etw. tut

ban [bæn] **1.** *v. t.*, **-nn-** verbieten; **~ sb. from doing sth.** jmdm. verbieten, etw. zu tun; **he was ~ned from driving/playing** er erhielt Fahr-/Spielverbot; **~ sb. from a place** jmdm. die Einreise/den Zutritt *usw.* verbieten; **~ sb. from a pub/the teaching profession** jmdm. Lokalverbot erteilen/jmdn. vom Lehrberuf ausschließen. **2.** *n.* Verbot, *das;* **place a ~ on sth.** etw. mit einem Verbot belegen; **the ~ placed on these drugs** das Verbot dieser Drogen

banal [bə'nɑ:l, bə'næl] *adj.* banal

banality [bə'nælɪtɪ] *n.* Banalität, *die*

banana [bə'nɑ:nə] *n.* Banane, *die;* **a hand of ~s** eine Hand Bananen *(fachspr.);* **go ~s** *(Brit. sl.)* verrückt werden *(salopp)*

banana: ~ republic *n. (derog.)* Bananenrepublik, *die (abwertend);* **~ skin** *n.* Bananenschale, *die;* **~ 'split** *n.* Bananensplit, *das*

band [bænd] **1.** *n.* **a)** Band, *das;* **a ~ of light/colour** ein Streifen Licht/Farbe; **b)** *(range of values)* Bandbreite, *die (fig.);* **c)** *(Radio)* **long/medium ~:** Langwellen-/Mittelwellenband, *das;* **d)** *(organized group)* Gruppe, *die; (of robbers, outlaws, etc.)* Bande, *die;* **e)** *(of musicians)* [Musik]kapelle, *die; (pop group, jazz ~)* Band, *die;* Gruppe, *die; (dance ~)* [Tanz]kapelle, *die; (military ~)* Militärkapelle, *die.* **2.** *v. t.* **a)** **~ sth.** ein Band um etw. machen; **b)** *(mark with stripes)* bändern. **3.** *v. i.* **~ together [with sb.]** sich [mit jmdm.] zusammenschließen

bandage ['bændɪdʒ] **1.** *n.* *(for wound, fracture)* Verband, *der; (for fracture, as support)* Bandage, *die.* **2.** *v. t.* verbinden ‹*[offene] Wunde usw.*›; bandagieren ‹*[verstauchtes] Gelenk usw.*›

'bandbox *n.* Bandschachtel, *die (veralt.);* ≈ Hutschachtel, *die*

bandit ['bændɪt] *n.* Bandit, *der*

banditry ['bændɪtrɪ] *n.* Banditen[un]wesen, *das*

band: ~master *n.* Kapellmeister, *der;* **~-saw** *n.* Bandsäge, *die*

bandsman ['bændzmən] *n., pl.* **bandsmen** ['bændzmən] Mitglied der/einer Kapelle/Band

band: ~stand *n.* Musikertribüne, *die;* **~wagon** *n.* **climb** *or* **jump on [to] the ~wagon** *(fig.)* auf den fahrenden Zug aufspringen *(fig.)*

¹bandy ['bændɪ] *v. t.* **a)** herumerzählen *(ugs.)* ‹*Geschichte*›; **insults were being bandied about** Beschimpfungen flogen hin und her; **b)** *(exchange)* wechseln; **they were ~ing blows** sie tauschten Schläge aus; **don't ~ words with me** ich wünsche keine Diskussion

²bandy *adj.* krumm; **he has ~ legs** *or* **is ~-legged** er hat O-Beine *(ugs.);* **~-legged person** O-beinige Person *(ugs.)*

bane [beɪn] *n.* Ruin, *der;* **he is the ~ of my life** er ist der Nagel zu meinem Sarg *(ugs.)*

bang [bæŋ] **1.** *v. t.* **a)** knallen *(ugs.);* schlagen; zuknallen *(ugs.),* zuschlagen ‹*Tür, Fenster, Deckel*›; **~ one's head on** *or* **against the ceiling** mit dem Kopf an die Decke knallen *(ugs.) od.* schlagen; **he ~ed the nail in** er haute den Nagel rein *(ugs.);* **b)** *(sl.: copulate with)* bumsen *(salopp).* **2.** *v. i.* **a)** *(strike)* **~ [against sth.]** [gegen etw.] schlagen *od.* *(ugs.)* knallen; **~ at the door** gegen die Tür hämmern; **b)** *(make sound of blow or explosion)* knallen; ‹*Trommeln:*› dröhnen; **~ away at sth.** *(shoot)* auf etw. *(Akk.)* ballern *(ugs.);* **~ shut** zuknallen *(ugs.);* zuschlagen. **3.** *n.* **a)** *(blow)* Schlag, *der;* **b)** *(noise)* Knall, *der;* **the party went off with a ~** *(fig.)* die Party war eine Wucht *(ugs.).* **4.** *adv.* **a)** *(with impact)* mit voller Wucht; **b)** *(explosively)* **go ~** ‹*Gewehr, Feuerwerkskörper:*› krachen; **c)** **~ goes sth.** *(fig.: sth. ends suddenly)* aus ist es mit etw.; **~ went £50** 50 Pfund waren weg; **d)** **~ off** *(coll.: immediately)* sofort; **e)** *(coll.: exactly)* genau; **you are ~ on time** du bist pünktlich auf die Minute *(ugs.)*

banger ['bæŋə(r)] *n. (sl.)* **a)** *(sausage)* Würstchen, *das;* **b)** *(firework)* Kracher, *der (ugs.);* **c)** *(car)* Klapperkiste, *die (ugs.)*

Bangladesh [bæŋglə'deʃ] *pr. n.* Bangladesch *(das)*

bangle ['bæŋgl] *n.* Armreif, *der*

banish ['bænɪʃ] *v. t.* verbannen **(from** aus); bannen ‹*Furcht*›

banishment ['bænɪʃmənt] *n.* Verbannung, *die* **(from** aus)

banister ['bænɪstə(r)] *n.* **a)** *(uprights and rail)* [Treppen]geländer, *das;* **b)** *usu. in pl. (upright)* Geländerpfosten, *der*

banjo ['bændʒəʊ] *n., pl.* **~s** *or* **~es** Banjo, *das*

¹bank [bæŋk] **1.** *n.* **a)** *(slope)* Böschung, *die;* **b)** *(at side of river)* Ufer, *das;* **c)** *(in bed of sea or river)* Bank, *die;* **d)** *(mass)* **a ~ of clouds/fog** eine Wolken-/Nebelbank. **2.** *v. t.* **a)** überhöhen ‹*Straße, Kurve*›; **b)** *(heap)* **~ [up]** aufschichten; **~ [up] the fire with coal** Kohlen auf das Feuer schichten; **c)** in die Kurve legen ‹*Flugzeug*›. **3.** *v. i.* ‹*Flugzeug:*› sich in die Kurve legen

²bank **1.** *n. (Commerc., Finance, Gaming)* Bank, *die;* **central ~:** Zentralbank, *die.* **2.** *v. i.* **a)** **~ at/with ...:** ein Konto haben bei ...; **b)** **~ on sth.** *(fig.)* auf etw. *(Akk.)* zählen. **3.** *v. t.* zur Bank bringen

³bank *n. (row)* Reihe, *die*

bank: ~ account *n.* Bankkonto, *das;* **~ balance** *n.* Kontostand, *der;* **~ card** *n.* Scheckkarte, *die;* **~ charges** *n. pl.* Kontoführungskosten; **~ clerk** *n.* Bankangestellte, *der/die*

banker ['bæŋkə(r)] *n. (Commerc., Finance)* Bankier, *der;* Banker, *der (ugs.)*
banker's: ~ **card** *see* **bank card;** ~ **order** *n.* Bankanweisung, *die*
bank 'holiday *n.* **a)** Bankfeiertag, *der;* **b)** *(Brit.: public holiday)* Feiertag, *der*
banking ['bæŋkɪŋ] *n.* Bankwesen, *das;* **a career in** ~: die Banklaufbahn
bank: ~ **manager** *n.* Zweigstellenleiter [einer/der Bank]; ~**note** *n.* Banknote, *die;* Geldschein, *der;* ~ **raid** *n.* Banküberfall, *der;* ~ **rate** *n.* Diskontsatz, *der;* ~**-robber** *n.* Bankräuber, *der*
bankrupt ['bæŋkrʌpt] **1.** *n.* **a)** *(Law)* Gemeinschuldner, *der;* **b)** *(insolvent debtor)* Bankrotteur, *der.* **2.** *adj. (lit. or fig.)* bankrott; **go** ~: in Konkurs gehen; Bankrott machen. **3.** *v. t.* bankrott machen
bankruptcy ['bæŋkrʌptsɪ] *n.* Konkurs, *der;* Bankrott, *der;* ~ **proceedings** Konkursverfahren, *das*
'bank statement *n.* Kontoauszug, *der*
banner ['bænə(r)] **1.** *n.* **a)** *(flag, ensign; also fig.)* Banner, *das;* **b)** *(on two poles)* Spruchband, *das;* Transparent, *das.* **2.** *adj.* ~ **headline** Balkenüberschrift, *die*
banns [bænz] *n. pl.* Aufgebot, *das;* **publish/put up the** ~: das Aufgebot verkünden/aushängen
banquet ['bæŋkwɪt] **1.** *n.* Bankett, *das (geh.).* **2.** *v. i.* [festlich] tafeln *(geh.);* ~**ing-hall** Bankettsaal, *der*
banshee ['bænʃi:] *n. (Ir., Scot.)* Banshee, *die (Myth.);* ≈ Weiße Frau
bantam ['bæntəm] *n.* Zwerg-, Bantamhuhn, *das*
'bantamweight *n. (Boxing etc.)* Bantamgewicht, *das; (person also)* Bantamgewichtler, *der*
banter ['bæntə(r)] *n.* **a)** heiterer Spott; **b)** *(remarks)* Spöttelei, *die*
bap [bæp] *n.* ≈ Brötchen, *das*
baptise *see* **baptize**
baptism ['bæptɪzm] *n.* Taufe, *die;* ~ **of fire** *(fig.)* Feuertaufe, *die*
baptismal [bæp'tɪzml] *adj.* Tauf-; ‹*Wiedergeburt, Reinigung*› durch die Taufe
Baptist ['bæptɪst] **1.** *n.* **a)** Baptist, *der*/Baptistin, *die;* **b) John the** ~: Johannes der Täufer. **2.** *adj.* **the** ~ **Church/a** ~ **chapel** die Kirche/eine Kapelle der Baptisten
baptize [bæp'taɪz] *v. t.* taufen; **be ~d a Catholic/Protestant** katholisch/protestantisch getauft werden
bar [bɑ:(r)] **1.** *n.* **a)** *(long piece of rigid material)* Stange, *die; (shorter, thinner also)* Stab, *der; (of gold, silver)* Barren, *der;* **a** ~ **of soap** ein Stück Seife; **a** ~ **of chocolate** ein Riegel Schokolade; *(slab)* eine Tafel Schokolade; **b)** *(Sport)* Stab, *der; (cross~)* [Sprung]latte, *die;* **parallel** ~**s** Barren, *der;* **high** *or* **horizontal** ~: Reck, *das;* **c)** *(heating element)* Heizelement, *das (Elektrot.);* **d)** *(band)* Streifen, *der; (on medal)* silberner Querstreifen; **e)** *(rod, pole)* Stange, *die; (of cage, prison)* Gitterstab, *der;* **behind** ~**s** *(in prison)* hinter Gittern; *(into prison)* hinter Gitter; **f)** *(barrier, lit. or fig.)* Barriere, *die* **(to** für); **a** ~ **on recruitment/promotion** ein Einstellungs-/Beförderungsstopp; **g)** *(for refreshment)* Bar, *die; (counter)* Theke, *die;* **h)** *(Law: place at which prisoner stands)* ≈ Anklagebank, *die;* **the prisoner at the** ~: der/die Angeklagte; **i)** *(Law: particular court)* Gerichtshof, *der;* **be called to the** ~: als Anwalt vor höheren Gerichten zugelassen werden; **the Bar** die höhere Anwaltschaft; **j)** *(Mus.)* Takt, *der;* **k)** *(sandbank, shoal)* Barre, *die;* Sandbank, *die.* **2.** *v. t.*, **-rr-: a)** *(fasten)* verriegeln; ~**red window** vergittertes Fenster; **b)** *(obstruct)* sperren ‹*Straße, Weg*› **(to** für); ~ **sb.'s way** jmdm. den Weg versperren; **c)** *(prohibit, hinder)* verbieten; ~ **sb. from doing sth.** jmdn. daran hindern, etw. zu tun. **3.** *prep.* abgesehen von; ~ **none** ohne Einschränkung
barb [bɑ:b] *n.* Widerhaken, *der; (fig.)* Gehässigkeit, *die*
barbarian [bɑ:'beərɪən] **1.** *n. (lit. or fig.)* Barbar, *der.* **2.** *adj. (lit. or fig.)* barbarisch
barbaric [bɑ:'bærɪk] *adj.*, **barbarically** [bɑ:'bærɪkəlɪ] *adv.* barbarisch
barbarism ['bɑ:bərɪzm] *n., no pl.* Barbarei, *die*
barbarity [bɑ:'bærɪtɪ] *n.* **a)** *no pl.* Grausamkeit, *die;* **with** ~: äußerst barbarisch; **b)** *(instance)* Barbarei, *die*
barbarous ['bɑ:bərəs] *adj.* barbarisch
barbecue ['bɑ:bɪkju:] **1.** *n.* **a)** *(party)* Grillparty, *die;* **b)** *(food)* Grillgericht, *das;* **c)** *(frame)* Grill, *der.* **2.** *v. t.* grillen
barbed wire [bɑ:bd 'waɪər] *n.* ~ **[fence]** Stacheldraht[zaun], *der*
barber ['bɑ:bə(r)] *n.* Friseur, *der;* Barbier, *der (veralt.);* **go to the** ~**'s** zum Friseur gehen; ~**'s pole** *spiralig rot und weiß gestreifter Stab als Ladenschild des Friseurs*
barbican ['bɑ:bɪkən] *n. (Hist.)* Barbakane, *die;* Torvorwerk, *das*
barbiturate [bɑ:'bɪtjʊrət] *n. (Chem.)* Barbiturat, *das*
bar: ~ **chart** *n.* Stabdiagramm, *das;* ~ **code** *n.* Strichcode, *der*
bard [bɑ:d] *n.* Barde, *der*
bare [beə(r)] **1.** *adj.* **a)** nackt; **with** *or* **in** ~ **feet** barfuß; **b)** *(hatless)* **with one's head** ~: ohne Hut; **c)** *(leafless)* kahl; **d)** *(unfurnished)* kahl; nackt ‹*Boden*›; **e)** *(unconcealed)* **lay** ~ **sth.** etw. aufdecken; **f)** *(unadorned)* nackt ‹*Wahrheit, Tatsache, Wand*›; **g)** *(empty)* leer; **h)** *(scanty)* knapp ‹*Mehrheit*›; [sehr] gering ‹*Menge, Teil*›; **i)** *(mere)* äußerst ‹*Notwendige*›; karg ‹*Essen, Leben*›; **the** ~ **necessities of life** das zum Leben Notwendigste; **j)** *(without tools)* **with one's** ~ **hands** mit den *od.* seinen bloßen Händen; **k)** *(unprovided with)* ~ **of** ohne. **2.** *v. t.* entblößen ‹*Kopf, Arm, Bein*›; bloßlegen ‹*Draht eines Kabels*›; blecken ‹*Zähne*›
bare: ~**back 1.** *adj.* ‹*Reiter, Reiten*› auf ungesatteltem Pferd; **2.** *adv.* ohne Sattel; ~**faced** ['beəfeɪsd] *adj. (fig.)* schamlos; ~**foot 1.** *adj.* barfüßig; **2.** *adv.* barfuß; ~**handed** [beə'hændɪd] **1.** *adj.* **he was** ~**handed** *(without gloves)* er trug keine Handschuhe; *(without weapon)* er war unbewaffnet; **2.** *adv.* mit bloßen Händen; ~**headed** [beə'hedɪd] **1.** *adj.* **he was** ~**headed** er trug keine Kopfbedeckung; **2.** *adv.* ohne Kopfbedeckung; ~**-legged** *adj.* mit bloßen Beinen
barely ['beəlɪ] *adv.* **a)** *(only just)* kaum; knapp ‹*vermeiden, entkommen*›; **b)** *(scantily)* karg
bargain ['bɑ:gɪn] **1.** *n.* **a)** *(agreement)* Abmachung, *die;* **into the** ~, *(Amer.)* **in the** ~: darüber hinaus; **make** *or* **strike a** ~ **to do sth.** sich darauf einigen, etw. zu tun; **b)** *(thing bought)* Kauf, *der;* **a good/bad** ~: ein guter/schlechter Kauf; **c)** *(thing offered cheap)* günstiges Angebot; *(thing bought cheaply)* guter Kauf. *See also* **best 3 e; hard 1 a. 2.** *v. i.* **a)** *(discuss)* handeln; ~ **for sth.** um etw. handeln; **b)** ~ **for** *or* **on sth.** *(expect sth.)* mit etw. rechnen; **more than one had** ~**ed for** mehr als man erwartet hatte
bargain: ~ **'basement** *n.* Tiefgeschoß mit Sonderangeboten;

~ **counter** *n.* Tisch mit Sonderangeboten

bargaining ['bɑ:gɪnɪŋ] *n.* Handel, *der; (negotiating)* Verhandlungen; ~ **position** Verhandlungsposition, *die*

bargain: ~ '**offer** *n.* Sonderangebot, *das;* ~ '**price** *n.* Sonderpreis, *der*

barge [bɑ:dʒ] **1.** *n.* Kahn, *der.* **2.** *v. i.* **a)** *(lurch)* ~ **into sb.** jmdn. anrempeln; **b)** ~ **in** *(intrude)* hineinplatzen/hereinplatzen *(ugs.)* **(on** bei)

bargee [bɑ:'dʒi:] *n. (Brit.)* Flußschiffer, *der*

'**barge-pole** *n.* **I wouldn't touch that with a ~!** *(fig.)* ich würde das nicht mit der Beißzange anfassen! *(ugs.)*

baritone ['bærɪtəʊn] *(Mus.)* **1.** *n.* Bariton, *der; (voice, part also)* Baritonstimme, *die.* **2.** *adj.* Bariton-

[1]**bark** [bɑ:k] **1.** *n. (of tree)* Rinde, *die.* **2.** *v. t. (graze)* aufschürfen

[2]**bark 1.** *n. (of dog; also fig.)* Bellen, *das;* **his ~ is worse than his bite** *(fig.)* ≈ Hunde, die bellen, beißen nicht. **2.** *v. i.* **a)** *(lit. or fig.)* bellen; ~ **at sb.** jmdn. anbellen; **be ~ing up the wrong tree** auf dem Holzweg sein; **b)** *(bellow)* brüllen. **3.** *v. t.* **a)** bellen; **b)** *(bellow)* **~[out] orders to sb.** jmdm. Befehle zubrüllen

barley ['bɑ:lɪ] *n.* Gerste, *die; see also* **pearl barley**

barley: ~**corn** *n. (grain)* Gerstenkorn, *das;* ~ **sugar** *n.* Gerstenzucker, *der;* ~ **water** *n.* Gerstenwasser, *das (veralt.)*

bar: ~**maid** *n. (Brit.)* Bardame, *die;* ~**man** ['bɑ:mən] *n., pl.* ~**men** ['bɑ:mən] Barmann, *der*

barmy ['bɑ:mɪ] *adj. (sl.: crazy)* bescheuert *(salopp)*

barn [bɑ:n] *n.* Scheune, *die*

barnacle ['bɑ:nəkl] *n. (Zool.)* Rankenfüßer, *der*

barn: ~ **dance** *n.* ≈ Schottische, *der;* ~-**owl** *n.* Schleiereule, *die;* ~**yard** *n.* Wirtschaftshof, *der*

barometer [bə'rɒmɪtə(r)] *n. (lit. or fig.)* Barometer, *das*

barometric [bærə'metrɪk] *adj.* barometrisch; ~ **pressure** Luftdruck, *der*

baron ['bærn] *n.* **a)** *(holder of title)* Baron, *der;* Freiherr, *der;* **b)** *(powerful person)* Papst, *der (fig.);* **press ~:** Pressezar, *der;* **oil ~:** Ölmagnat, *der*

baroness ['bærənɪs] *n.* Baronin, *die;* Freifrau, *die*

baronet ['bærənɪt] *n.* Baronet, *der*

baronial [bə'rəʊnɪəl] *adj.* freiherrlich

baroque [bə'rɒk] **1.** *n.* Barock, *das.* **2.** *adj.* **a)** barock; Barock- ⟨*malerei, -musik usw.*⟩; **b)** *(grotesque)* barock

barque [bɑ:k] *n.* Bark, *die*

[1]**barrack** ['bærək] *n. usu. in pl., often constr. as sing.* Kaserne, *die*

[2]**barrack 1.** *v. i.* buhen *(ugs.).* **2.** *v. t.* ausbuhen *(ugs.)*

barrack-'square *n.* Kasernenhof, *der*

barracuda [bærə'ku:də] *n., pl. same or* ~s Barrakuda, *der*

barrage ['bærɑ:ʒ] *n.* **a)** *(Mil.)* Sperrfeuer, *das; (fig.)* **a ~ of questions** ein Bombardement von Fragen; **a ~ of cheers** stürmischer Jubel; **b)** *(dam)* Talsperre, *die*

'**barrage balloon** *n.* Sperrballon, *der*

barrel ['bærl] *n.* **a)** *(vessel)* Faß, *das; (measure)* Barrel, *das;* **be over a ~** *(fig.)* in der Klemme sitzen *(ugs.);* **have sb. over a ~** *(fig.)* jmdn. in der Zange haben *(ugs.);* **scrape the ~** *(fig.)* das Letzte zusammenkratzen *(ugs.);* **b)** *(of gun)* Lauf, *der; (of cannon etc.)* Rohr, *das*

barrel: ~-**chested** ['bærltʃestɪd] *adj.* mit einem breiten, gewölbten Brustkorb *nachgestellt;* ~-**organ** *n. (Mus.)* Leierkasten, *der;* Drehorgel, *die*

barren ['bærn] *adj.* **a)** *(infertile)* unfruchtbar; **b)** *(meagre, dull)* nutzlos ⟨*Handlung, Arbeit*⟩; mager ⟨*Ergebnis*⟩; unfruchtbar ⟨*Periode, Beziehung*⟩; fruchtlos ⟨*Diskussion*⟩

barrenness ['bærnnɪs] *n., no pl. see* **barren:** Unfruchtbarkeit, *die;* Nutzlosigkeit, *die;* Magerkeit, *die;* Fruchtlosigkeit, *die*

barricade [bærɪ'keɪd] **1.** *n.* Barrikade, *die.* **2.** *v. t.* verbarrikadieren

barrier ['bærɪə(r)] *n.* **a)** *(fence)* Barriere, *die; (at railway, frontier)* Schranke, *die;* **b)** *(gate of railway-station)* Sperre, *die;* **c)** *(obstacle, lit. or fig.)* Barriere, *die;* **a ~ to progress** ein Hindernis für den Fortschritt

'**barrier cream** *n.* Schutzcreme, *die*

barring ['bɑ:rɪŋ] *prep.* außer im Falle (+ *Gen.*); ~ **accidents** falls nichts passiert

barrister ['bærɪstə(r)] *n.* **a)** *(Brit.)* **~[-at-law]** Barrister, *der;* ≈ [Rechts]anwalt/-anwältin vor höheren Gerichten, *der;* **b)** *(Amer.: lawyer)* [Rechts]anwalt, *der*/-anwältin, *die*

barroom ['bɑ:ru:m] *n. (Amer.)* Bar, *die*

[1]**barrow** ['bærəʊ] *n.* **a)** Karre, *die;* Karren, *der;* **b)** *see* **wheelbarrow**

[2]**barrow** *n. (Archaeol.)* Hügelgrab, *das*

'**barrow-boy** *n. (Brit.)* Straßenhändler, *der*

'**bartender** *n.* Barkeeper, *der*

barter ['bɑ:tə(r)] **1.** *v. t.* [ein]tauschen **(for** für *od.* **gegen);** ~ **away sth.** etw. verspielen *(fig.).* **2.** *v. i.* Tauschhandel treiben. **3.** *n.* Tauschhandel, *der*

basalt ['bæsɔ:lt] *n.* Basalt, *der*

[1]**base** [beɪs] **1.** *n.* **a)** *(of lamp, pyramid, wall, mountain, microscope)* Fuß, *der; (of cupboard, statue)* Sockel, *der; (fig.) (support)* Basis, *die; (principle)* Ausgangsbasis, *die; (main ingredient)* Hauptbestandteil, *der; (of make-up)* Grundlage, *die;* **b)** *(Mil.)* Basis, *die;* Stützpunkt, *der; (fig.: for sightseeing)* Ausgangspunkt, *der;* **c)** *(Baseball)* Mal, *das;* **get to first ~** *(fig. coll.)* [wenigstens] etwas erreichen; **d)** *(Archit., Geom., Surv., Math.)* Basis, *die;* **e)** *(Chem.)* Base, *die.* **2.** *v. t.* **a)** gründen **(on** auf + *Akk.*); **be ~d on sth.** sich auf etw. *(Akk.)* gründen; ~ **one's hopes on sth.** seine Hoffnung auf etw. *(Akk.)* gründen; **a book ~d on newly discovered papers** ein Buch, das auf neu entdeckten Dokumenten basiert; **b)** *in pass.* **be ~d in Paris** *(permanently)* in Paris sitzen; *(temporarily)* in Paris sein; **computer-~d accountancy** Buchführung über Computer; **land-~d forces** landgestützte Streitkräfte; **c)** ~ **oneself on** sich stützen auf (+ *Akk.*)

[2]**base** *adj.* **a)** *(morally low)* niederträchtig; niedrig ⟨*Beweggrund*⟩; **b)** *(cowardly)* feige; *(selfish)* selbstsüchtig; *(mean)* niederträchtig

baseball ['beɪsbɔ:l] *n.* Baseball, *der*

'**base camp** *n.* Basislager, *das*

baseless ['beɪslɪs] *adj.* unbegründet

'**baseline** *n.* Grundlinie, *die*

basement ['beɪsmənt] *n.* Kellergeschoß, *das; (esp. in department store)* Untergeschoß, *das;* **a ~ flat** eine Kellerwohnung

base 'metal *n.* unedles Metall

baseness ['beɪsnɪs] *n., no pl. see* [2]**base a:** Niederträchtigkeit, *die;* Niedrigkeit, *die*

'**base rate** *n. (Finance)* Eckzins, *der*

bases *pl. of* [1]**base** *or* **basis**

bash [bæʃ] **1.** *v. t.* [heftig] schlagen; ~ **one's head against sth.** sich *(Dat.)* den Kopf [heftig] an etw.

(Dat.) anschlagen. **2.** *n.* **a)** heftiger Schlag; **b)** *(sl.: attempt)* Versuch, *der;* **have a ~ at sth.** etw. mal versuchen

bashful ['bæʃfl] *adj.* schüchtern

basic ['beɪsɪk] *adj.* **a)** *(fundamental)* grundlegend; Grund‹*struktur, -prinzip, -bestandteil, -wortschatz, -lohn*›; Haupt‹*problem, -grund, -sache*›; **be ~ to sth.** wesentlich für etw. sein; **have a ~ knowledge of sth.** Grundkenntnisse einer Sache *(Gen.)* haben; **a ~ working day** ein normaler Arbeitstag; **b)** *(Chem.)* basisch. *See also* **basics**

basically ['beɪsɪkəlɪ] *adv.* im Grunde; grundsätzlich ‹*übereinstimmen*›; *(mainly)* hauptsächlich

Basic 'English *n.* Basic English, *das; auf einem sehr einfachen Grundwortschatz beruhendes Englisch*

basics ['beɪsɪks] *n. pl.* **stick to the ~:** beim Wesentlichen bleiben; **the ~ of maths/cooking** die Grundlagen der Mathematik/das ABC der Kochkunst

basil ['bæzɪl] *n. (Bot.)* **[sweet] ~:** Basilikum, *das*

basilica [bə'zɪlɪkə] *n. (Archit., Eccl.)* Basilika, *die*

basin ['beɪsn] *n.* Becken, *das; (wash-~)* Waschbecken, *das; (bowl)* Schüssel, *die;* **the Amazon ~:** das Amazonasbecken

basis ['beɪsɪs] *n., pl.* **bases** ['beɪsi:z] **a)** *(ingredient)* Grundbestandteil, *der;* **b)** *(foundation, principle, common ground)* Basis, *die;* **on a purely friendly ~:** auf rein freundschaftlicher Basis; **on a first come first served ~:** nach dem Prinzip „Wer zuerst kommt, mahlt zuerst“; **c)** *(beginning)* Ausgangspunkt, *der*

bask [bɑ:sk] *v. i.* **a)** sich [wohlig] wärmen; sich aalen *(ugs.)*; **~ in the sun** sich sonnen; **b)** *(fig.)* sich sonnen (**in** in + *Dat.*)

basket ['bɑ:skɪt] *n.* **a)** Korb, *der; (smaller, for bread etc.)* Körbchen, *das; (of chip-pan)* Drahteinsatz, *der;* **b)** *(quantity)* **a ~ [full] of apples** ein Korb [voll] Äpfel

'basketball *n.* Basketball, *der*

basketful ['bɑ:skɪtfʊl] *see* **basket b**

basketry ['bɑ:skɪtrɪ] *see* **basketwork**

basket: ~ weave *n.* Panamabindung, *die (Weberei);* **~work** *n. (art)* Korbflechterei, *die; (activity)* Korbflechten, *das; (collectively)* Korbwaren; **a piece of ~work** ein Korbgeflecht

basking 'shark *n.* Riesenhai, *der*

Basle [bɑ:l] *pr. n.* Basel *(das)*

Basque [bæsk, bɑ:sk] **1.** *adj.* baskisch; **the ~ Country** das Baskenland. **2.** *n.* **a)** Baske, *der*/Baskin, *die;* **b)** *(language)* Baskisch, *das*

bas-relief ['bæsrɪli:f] *n. (Art)* Basrelief, *das*

¹bass [bæs] *n., pl. same or* **~es** *(Zool.)* Barsch, *der*

²bass [beɪs] *(Mus.)* **1.** *n.* **a)** Baß, *der; (voice, part also)* Baßstimme, *die;* **b)** *(coll.) (double-~)* [Kontra]baß, *der; (~ guitar)* Baß, *der.* **2.** *adj.* Baß-

bass [beɪs]: **~ clef** *n. (Mus.)* Baßschlüssel, *der;* **~ drum** *n.* große Trommel

basset ['bæsɪt] *n.* **~[-hound]** Basset, *der*

bass guitar [beɪs gɪ'tɑ:(r)] *n.* Baßgitarre, *die*

bassoon [bə'su:n] *n. (Mus.)* Fagott, *das*

bastard ['bɑ:stəd] **1.** *adj.* **a)** unehelich; **b)** *(hybrid)* verfälscht ‹*Sprache, Stil*›; **c)** *(Bot., Zool.)* Bastard-. **2.** *n.* **a)** uneheliches Kind; **b)** *(coll.) (disliked person)* Mistkerl, *der (derb); (disliked thing)* Scheißding, *das (derb);* **the poor ~!** *(unfortunate person)* das arme Schwein! *(ugs.)*

¹baste [beɪst] *v. t. (stitch)* heften

²baste *v. t.* [mit Fett/Bratensaft] begießen ‹*Fleisch*›

bastion ['bæstɪən] *n. (lit. or fig.)* Bastei, *die*

¹bat [bæt] *n. (Zool.)* Fledermaus, *die;* **blind as a ~** *(fig.)* blind wie ein Maulwurf; **have ~s in the belfry** *(coll.)* einen Dachschaden haben *(ugs.)*

²bat 1. *n.* **a)** *(Sport)* Schlagholz, *das; (for table-tennis)* Schläger, *der;* **do sth. off one's own ~** *(fig.)* etw. auf eigene Faust tun; **right off the ~** *(Amer. fig.)* sofort; **b)** *(act of using ~)* Schlag, *der.* **2.** *v. t. & i.* **-tt-** schlagen

³bat *v. t.,* **-tt-: not ~ an eyelid** *(fig.)* nicht mit der Wimper zucken

batch [bætʃ] *n.* **a)** *(of loaves)* Schub, *der;* **b)** *(of people)* Gruppe, *die; (of books, papers, etc.)* Stapel, *der; (of rules, regulations)* Bündel, *das*

batch: ~ 'processing *n. (Computing)* Stapelverarbeitung, *die;* **~ production** *n.* Stapelfertigung, *die*

¹bate [beɪt] *v. t.* **with ~d breath** mit angehaltenem Atem

²bate *n. (Brit. sl.)* Rage, *die (ugs.);* **be in a [terrible] ~:** [schrecklich] in Rage sein; **get/fly into a ~:** in Rage geraten

bath [bɑ:θ] **1.** *n., pl.* **~s** [bɑ:ðz] **a)** Bad, *das;* **have** *or* **take a ~:** ein Bad nehmen; **b)** *(vessel)* **~[-tub]** Badewanne, *die;* **room with ~:** Zimmer mit Bad; **c)** *usu. in pl. (building)* Bad, *das;* **[swimming-]~s** Schwimmbad, *das.* **2.** *v. t. & i.* baden

bath: B~ 'bun *n.* ≈ Rosinenbrötchen mit Zuckerguß; **~-cap** *n.* Duschhaube, *die;* **~ chair** *n.* Rollstuhl, *der;* **~ cube** *n.* Badesalzwürfel, *der*

bathe [beɪð] **1.** *v. t.* **a)** baden; **b)** *(moisten)* baden ‹*Wunde, Körperteil*›; **~d with** *or* **in sweat** schweißüberströmt; **c) ~d in sunlight** von der Sonne beschienen. **2.** *v. i.* baden; **go bathing** baden gehen. **3.** *n.* Bad, *das (im Meer usw.);* **have a ~:** baden

bather ['beɪðə(r)] *n.* Badende, *der/die*

bathing: ~ beach *n.* Badestrand, *der;* **~-cap** *n.* Bademütze, *die;* **~-costume** *n.* Badeanzug, *der;* **~-suit** *n.* Badeanzug, *der;* **~-trunks** *n. pl.* Badehose, *die*

bath: ~-mat Badematte, *die;* **~robe** *n.* Bademantel, *der;* **~room** *n.* Badezimmer, *das;* **~ salts** *n. pl.* Badesalz, *das;* **~-time** *n.* Badezeit, *die;* **~-towel** *n.* Badetuch, *das;* **~-tub** *see* **bath 1 b**; **~-water** *n.* Badewasser, *das; see also* **baby a**

batik [bə'ti:k] *n.* Batik, *der od. die*

batman ['bætmən] *n., pl.* **batmen** ['bætmən] *(Mil.)* [Offiziers]bursche, *der*

baton ['bætn] *n.* **a)** *(staff of office)* Stab, *der;* **b)** *(truncheon)* Schlagstock, *der;* **c)** *(Mus.)* Taktstock, *der;* **d)** *(for relay race)* [Staffel]stab, *der*

batsman ['bætsmən] *n., pl.* **batsmen** ['bætsmən] *(Sport)* Schlagmann, *der*

battalion [bə'tæljən] *n. (lit. or fig.)* Bataillon, *das*

batten ['bætn] **1.** *n. (Constr., Naut.)* Latte, *die.* **2.** *v. t. (Naut.)* **~ down** [ver]schalken ‹*Luke*›

¹batter ['bætə(r)] **1.** *v. t.* **a)** *(strike)* einschlagen auf (+ *Akk.*); **~ down/in** einschlagen; **~ sth. to pieces** etw. zerschmettern; **b)** *(attack with artillery)* beschießen; **c)** *(bruise, damage)* übel zurichten; mißhandeln ‹*Baby, Ehefrau*›; **~ed by the gales** vom Sturm stark beschädigt; **a ~ed car** ein verbeultes Auto. **2.** *v. i.* heftig klopfen; **they ~ed at** *or* **against the door** sie hämmerten gegen die Tür

²batter *n. (Cookery)* [Back]teig, *der; (for pancake)* [Eierkuchen]teig, *der*

'battering ram *n.* Rammbock, *der*

battery ['bætərɪ] *n.* **a)** *(series; also Mil., Electr.)* Batterie, *die;* **a ~ of specialists** *(fig.)* eine ganze Reihe von Spezialisten; **b)** *(Law)* **[assault and] ~:** tätlicher Angriff

battery: ~-charger *n.* Batterieladegerät, *das;* **~ 'farming** *n.* Batteriehaltung, *die;* **~ 'hen** *n.* Batteriehuhn, *das*

battle ['bætl] **1.** *n.* **a)** *(fight)* Schlacht, *die;* **the ~ at Amman** die Schlacht bei Amman; **do** *or* **give ~:** kämpfen; **join ~ with sb.** jmdm. eine Schlacht liefern; **die in ~:** [in der Schlacht] fallen; **b)** *(fig.: contest)* Kampf, *der;* **~ of wits** geistiger Wettstreit; **sth. is half the ~:** mit etw. ist schon viel gewonnen. **2.** *v.i.* kämpfen (**with** *or* **against** mit *od.* gegen, **for** für). **3.** *v.t.* **~ one's way through the crowd** sich durch die Menge kämpfen

battle: ~axe *n.* **a)** Streitaxt, *die;* **b)** *(coll.: woman)* Schreckschraube, *die (ugs. abwertend);* **~dress** *n. (Mil.) (for general service)* Arbeitsanzug, *der; (for field service)* Kampfanzug, *der;* **~field, ~ground** *ns.* Schlachtfeld, *das*

battlement ['bætlmənt] *n., usu. in pl.* Zinne, *die*

'battleship *n.* Schlachtschiff, *das*

batty ['bætɪ] *adj. (sl.)* bekloppt *(salopp);* **go** *or* **become ~:** überschnappen *(ugs.)*

bauble ['bɔ:bl] *n.* Flitter, *der*

Bavaria [bə'veərɪə] *pr. n.* Bayern *(das)*

Bavarian [bə'veərɪən] **1.** *adj.* bay[e]risch; **sb. is ~:** jmd. ist Bayer/Bayerin. **2.** *n.* **a)** *(person)* Bayer, *der*/Bayerin, *die;* **b)** *(dialect)* Bay[e]risch[e], *das*

bawdy ['bɔ:dɪ] *adj.* zweideutig, *(stronger)* obszön ‹*Sprache, Geschichte, Witz*›; obszön ‹*Person*›

bawl [bɔ:l] **1.** *v.t.* brüllen; **~ sth. at sb.** jmdm. etw. zubrüllen; **~ sb. out** *(coll.)* jmdn. zusammenstauchen *(ugs.).* **2.** *v.i.* brüllen; **~ out to sb.** nach jmdm. brüllen; **~ at sb.** jmdn. anbrüllen

¹bay [beɪ] *n. (of sea)* Bucht, *die; (larger also)* Golf, *der*

²bay *n.* **a)** *(space in room)* Erker, *der;* **b) loading-~:** Ladeplatz, *der;* **[parking-]~:** Stellplatz, *der;* **sick-~** *(Navy)* Schiffshospital, *das; (Mil.)* Sanitätsbereich, *der; (in school, college, office)* Krankenzimmer, *das*

³bay 1. *n. (bark)* Gebell, *das;* **hold** *or* **keep sb./sth. at ~** *(fig.)* sich *(Dat.)* jmdn./etw. vom Leib halten. **2.** *v.i.* bellen; **~ at sb./sth.** jmdn./etw. anbellen

⁴bay *n. (Bot.)* Lorbeer[baum], *der*

⁵bay 1. *adj.* braun ‹*Pferd*›. **2.** *n.* Braune, *der*

'bay-leaf *n. (Cookery)* Lorbeerblatt, *das*

bayonet ['beɪənɪt] **1.** *n.* Bajonett, *das;* **with fixed ~s** mit aufgepflanzten Bajonetten. **2.** *v.t.* mit dem Bajonett aufspießen

bay 'window *n.* Erkerfenster, *das*

bazaar [bə'zɑ:(r)] *n. (oriental market)* Basar, *der; (sale)* [Wohltätigkeits]basar, *der*

BBC *abbr.* **British Broadcasting Corporation** BBC, *die*

BC *abbr.* **before Christ** v. Chr.

be [bi:] *v., pres. t.* **I am** [əm, *stressed* æm], *neg. (coll.)* **ain't** [eɪnt], **he is** [ɪz], *neg. (coll.)* **isn't** ['ɪznt]; **we are** [ə(r), *stressed* ɑ:(r)], *neg. (coll.)* **aren't** [ɑ:nt]; *p. t.* **I was** [wəz, *stressed* wɒz], *neg. (coll.)* **wasn't** ['wɒznt], **we were** [wə(r), *stressed* wɜ:(r), weə(r)], *neg. (coll.)* **weren't** [wɜ:nt, weənt]; *pres. p.* **being** ['bi:ɪŋ]; *p.p.* **been** [bɪn, *stressed* bi:n] **1.** *copula* **a)** *indicating quality or attribute* sein; **she'll be ten next week** sie wird nächste Woche zehn; **she is a mother/an Italian** sie ist Mutter/Italienerin; **being a Frenchman, he likes wine** als Franzose trinkt er gern Wein; **he is being nice to them/sarcastic** er ist nett zu ihnen/jetzt ist er sarkastisch; **b)** *in exclamation* **was she pleased!** war sie [vielleicht] froh!; **aren't you a big boy!** was bist du schon für ein großer Junge!; **c) will be** *indicating supposition* **[I dare say] you'll be a big boy by now** du bist jetzt sicher schon ein großer Junge; **you'll be relieved to hear that** du wirst erleichtert sein, das zu hören; **d)** *indicating physical or mental welfare or state* sein; sich fühlen; **I am hot** mir ist heiß; **I am freezing** mich friert es; **how are you/is she?** wie geht's *(ugs.)*/geht es ihr?; **e)** *identifying the subject* **it is the 5th today** heute haben wir den Fünften; **who's that?** wer ist das?; **it is she, it's her** sie ist's; **if I were you** an deiner Stelle; **f)** *indicating profession, pastime, etc.* **be a teacher/a footballer** Lehrer/Fußballer sein; **she wants to be a surgeon** sie möchte Chirurgin werden; **g)** *with possessive* **it is hers** es ist ihrs; es gehört ihr; **h)** *(cost)* kosten; **how much are the eggs?** was kosten die Eier?; **i)** *(equal)* sein; **two times three is six, two threes are six** zweimal drei ist *od.* sind *od.* gibt sechs; **sixteen ounces is a pound** sechzehn Unzen sind *od.* ergeben ein Pfund; **j)** *(constitute)* bilden; **London is not England** London ist nicht [gleich] England; **k)** *(mean)* bedeuten; **he was everything to her** er bedeutete ihr alles. **2.** *v.i.* **a)** *(exist)* [vorhanden] sein; existieren; **can such things be?** kann es so etwas geben?; kann so etwas vorkommen?; **I think, therefore I am** ich denke, also bin ich; **there is/are ...:** es gibt ...; **for the time being** vorläufig; **Miss Jones that was** das frühere Fräulein Jones; **be that as it may** wie dem auch sei; **b)** *(remain)* bleiben; **I shan't be a moment** *or* **second** ich komme gleich; noch eine Minute; **she has been in her room for hours** sie ist schon seit Stunden in ihrem Zimmer; **let it be** laß es sein; **let him/her be** laß ihn/sie in Ruhe; **how long has he been here?** wie lange ist er schon hier?; **c)** *(happen)* stattfinden; sein; **where will the party be?** wo ist die Party?; wo findet die Party statt?; **d)** *(go, come)* **be off with you!** geh/geht!; **I'm off** *or* **for home** ich gehe jetzt nach Hause; **she's from Australia** sie stammt *od.* ist aus Australien; **are you for London?** wollen Sie nach London?; **e)** *(on visit etc.)* sein; **have you [ever] been to London?** bist du schon einmal in London gewesen?; **has anyone been?** ist jemand dagewesen?; **f) she's been and tidied the room** *(coll.)* sie hat doch wirklich das Zimmer aufgeräumt; **the children have been at the biscuits** die Kinder waren an den Keksen *(ugs.);* **I've been into this matter** ich habe mich mit der Sache befaßt. **3.** *v. aux.* **a)** *forming passive* werden; **the child was found** das Kind wurde gefunden; **German is spoken here** hier wird Deutsch gesprochen; **b)** *forming continuous tenses, active* **he is reading** er liest [gerade]; er ist beim Lesen; **I am leaving tomorrow** ich reise morgen [ab]; **the train was departing when I got there** der Zug fuhr gerade ab, als ich ankam; **c)** *forming continuous tenses, passive* **the house is/was being built** das Haus wird/wurde [gerade] gebaut; **d)** *expressing obligation* **be to** sollen; **I am to inform you** ich soll Sie unterrichten; **he is to be**

admired er ist zu bewundern; **e)** *expressing arrangement* **be to** sollen; **the Queen is to arrive at 3 p.m.** die Königin soll um 15 Uhr eintreffen; **f)** *expressing possibility* **it was not to be seen** es war nicht zu sehen; **I was not to be side-tracked** ich ließ mich nicht ablenken; **g)** *expressing destiny* **they were never to meet again** sie sollten sich nie wieder treffen; **h)** *expressing condition* **if I were to tell you that ..., were I to tell you that ...:** wenn ich dir sagen würde, daß ... **4. bride-/husband-to-be** zukünftige Braut/zukünftiger Ehemann; **mother-/father-to-be** werdende Mutter/werdender Vater; **the be-all and end-all** das A und O

beach [bi:tʃ] **1.** *n.* Strand, *der;* **on the ~:** am Strand. **2.** *v. t.* auf [den] Strand setzen ⟨*Schiff usw.*⟩; ans Ufer ziehen ⟨*Boot, Wal*⟩

beach: ~-ball *n.* Wasserball, *der;* **~-comber** ['bi:tʃkəʊmə(r)] *n.* Strandgutsammler, *der;* **~-head** *n. (Mil.)* Brückenkopf, *der;* **~ wear** *n.* Strandkleidung, *die*

beacon ['bi:kn] *n.* **a)** Leucht-, Signalfeuer, *das; (Naut.)* Leuchtbake, *die;* **b)** *(radio station)* Funkfeuer, *das;* **c)** *(signal light)* Signalleuchte, *die; (for aircraft)* Landelicht, *das*

bead [bi:d] *n.* **a)** Perle, *die;* **~s** Perlen *Pl.;* Perlenkette, *die;* **tell one's ~s** den Rosenkranz beten; **~s of dew/perspiration** *or* **sweat** Tau-/Schweißtropfen; **b)** *(gun-sight)* Korn, *das*

beady ['bi:dɪ] *adj.* **~ eyes** Knopfaugen; **those ~ eyes of hers don't miss anything** ihrem wachsamen Blick entgeht nichts; **I've got my ~ eye on you** ich lasse dich nicht aus den Augen

'beady-eyed *adj.* mit Knopfaugen *nachgestellt; (watchful)* mit wachen Augen *nachgestellt*

beagle ['bi:gl] *n.* Beagle, *der*

[1]**beak** [bi:k] *n.* Schnabel, *der; (of turtle, octopus)* Mundwerkzeug, *das; (fig.: hooked nose)* Hakennase, *die;* Zinken, *der (salopp)*

[2]**beak** *n. (Brit. sl.: magistrate, judge)* Kadi, *der (ugs.)*

beaked [bi:kt] *adj.* geschnäbelt

beaker ['bi:kə(r)] *n.* **a)** Becher, *der;* **b)** *(Chem.)* Becherglas, *das*

'be-all *see* **be 4**

beam [bi:m] **1.** *n.* **a)** *(timber etc.)* Balken, *der;* **b)** *(Naut.) (ship's breadth)* [größte] Schiffsbreite; *(side of ship)* [Schiffs]seite, *die;* **on the port ~:** backbords; **broad in the ~** *(fig. coll.)* breithüftig; **c)** *(ray etc.)* [Licht]strahl, *der;* **~ of light** Lichtstrahl, *der;* **the car's headlamps were on full ~:** die Scheinwerfer des Wagens waren aufgeblendet; **d)** *(Aeronaut., Mil., etc.: guide)* Peil- *od.* Leitstrahl, *der;* **be off ~** *(fig. coll.)* danebenliegen *(ugs.);* **e)** *(smile)* Strahlen, *das.* **2.** *v. t.* ausstrahlen. **3.** *v. i.* **a)** *(shine)* strahlen; **the sun ~ed down** die Sonne strahlte vom Himmel; **b)** *(smile)* strahlen; **~ at sb.** jmdn. anstrahlen

'beam-ends *n. pl.* **be on one's ~** *(fig.)* pleite *(ugs.) od.* in großer Geldnot sein

beaming ['bi:mɪŋ] *adj.* strahlend

bean [bi:n] *n.* **a)** Bohne, *die;* **full of ~s** *(fig. coll.)* putzmunter *(ugs.);* **he hasn't [got] a ~** *(fig. sl.)* er hat keinen roten Heller *(ugs.);* **b)** *(Amer. sl.: head)* Birne, *die (fig. salopp)*

bean: ~ bag *n.* **a)** *mit Bohnen gefülltes Säckchen zum Spielen;* **b)** *(cushion)* Knautschsessel, *der;* **~feast** *n. (Brit. coll.)* Gelage, *das;* **~pole** *n. (lit. or fig.)* Bohnenstange, *die;* **~sprout** *n.* Sojabohnenkeim, *der;* **~stalk** *n.* Bohnenstengel, *der*

[1]**bear** [beə(r)] *n.* **a)** Bär, *der;* **b)** *(Astron.)* **Great/Little B~:** Großer/Kleiner Bär

[2]**bear 1.** *v. t.,* **bore** [bɔ:(r)], **borne** [bɔ:n] **a)** *(show)* tragen ⟨*Wappen, Inschrift, Unterschrift*⟩; aufweisen, zeigen ⟨*Merkmal, Spuren, Ähnlichkeit, Verwandtschaft*⟩; **~ a resemblance** *or* **likeness to sb.** Ähnlichkeit mit jmdm. haben; **b)** *(be known by)* tragen, führen ⟨*Namen, Titel*⟩; **c) ~ some/little relation to sth.** einen gewissen/wenig Bezug zu etw. haben; **d)** *(poet./formal: carry)* tragen ⟨*Waffe, Last*⟩; mit sich führen ⟨*Geschenk, Botschaft*⟩; **I was borne along by the fierce current** die starke Strömung trug mich mit [sich]; **e)** *(endure, tolerate)* ertragen ⟨*Schmerz, Kummer*⟩; *with neg.* aushalten ⟨*Schmerz*⟩; ausstehen ⟨*Geruch, Lärm, Speise*⟩; **f)** *(sustain)* tragen, übernehmen ⟨*Verantwortlichkeit, Kosten*⟩; auf sich *(Akk.)* nehmen ⟨*Schuld*⟩; tragen, aushalten ⟨*Gewicht*⟩; **g)** *(be fit for)* vertragen; **it does not ~ repeating** *or* **repetition** das läßt sich unmöglich wiederholen; **it does not ~ thinking about** daran darf man gar nicht denken; **~ comparison with sth.** den *od.* einen Vergleich mit etw. aushalten; **h)** *(give birth to)* gebären ⟨*Kind, Junges*⟩; *see also* **born; i)** *(yield)* tragen ⟨*Blumen, Früchte usw.*⟩; **~ fruit** *(fig.)* Früchte tragen *(geh.).* **2.** *v. i.,* **bore, borne a) ~ left** ⟨*Person:*⟩ sich links halten; **the path ~s to the left** der Weg führt nach links; **b) bring to ~:** aufbieten ⟨*Kraft, Energie*⟩; ausüben ⟨*Druck*⟩; **bring one's influence to ~:** seinen Einfluß geltend machen

~ a'way *v. t.* wegtragen; davontragen ⟨*Preis usw.*⟩; **be borne away** fort- *od.* davongetragen werden

~ 'down *v. i.* **~ down on sb./sth.** auf jmdn./etw. zusteuern; ⟨*Wagen:*⟩ auf jmdn./etw. zufahren *od.* -steuern

~ 'off *see* **~ away**

~ on *see* **~ upon**

~ 'out *v. t. (fig.)* bestätigen ⟨*Bericht, Erklärung*⟩; **~ sb. out** jmdm. recht geben

~ 'up *v. i.* durchhalten; **~ up well under sth.** etw. gut ertragen

~ upon *v. t. (relate to)* sich beziehen auf (+ *Akk.*)

~ with *v. t.* Nachsicht haben mit

bearable ['beərəbl] *adj.* zum Aushalten *nachgestellt;* erträglich ⟨*Situation, Beruf*⟩

'bear cub *n.* Bärenjunge, *das*

beard [bɪəd] **1.** *n.* Bart, *der;* **full ~:** Vollbart, *der.* **2.** *v. t.* **~ the lion in his den** *(fig.)* sich in die Höhle des Löwen wagen

bearded ['bɪədɪd] *adj.* bärtig; **be ~:** einen Bart haben

bearer ['beərə(r)] *n. (carrier)* Träger, *der*/Trägerin, *die; (of letter, message, cheque, banknote)* Überbringer, *der*/Überbringerin, *die*

'bear-hug *n.* kräftige Umarmung

bearing ['beərɪŋ] *n.* **a)** *(behaviour)* Verhalten, *das; (deportment)* [Körper]haltung, *die;* **b)** *(relation)* Zusammenhang, *der;* Bezug, *der;* **have some/no ~ on sth.** relevant/irrelevant *od.* von Belang/belanglos für etw. sein; **c)** *(Mech. Engin.)* Lager, *das;* **d)** *(compass ~)* Position, *die;* **take a compass ~:** den Kompaßkurs feststellen; **get one's ~s** sich orientieren; *(fig.)* sich zurechtfinden; **I have lost my ~s** *(lit. or fig.)* ich habe die Orientierung verloren

bear: ~ market *n. (St. Exch.)* Markt mit fallenden Preisen; **~skin** *n.* **a)** Bärenfell, *das;* **b)** *(Mil.)* Bärenfellmütze, *die*

beast [bi:st] *n.* Tier, *das; (ferocious, wild)* Bestie, *die; (fig.: brutal person)* roher, brutaler Mensch; Bestie, *die (abwertend); (disliked person)* Scheusal, *das* *(abwertend)*; **it was a ~ of a winter** das war ein scheußlicher Winter

beastly ['bi:stlɪ] *adj., adv. (coll.)* scheußlich

beat [bi:t] **1.** *v.t.*, **beat, beaten** ['bi:tn] **a)** *(strike repeatedly)* schlagen ‹*Trommel, Rhythmus, Eier, Teig*›; klopfen ‹*Teppich*›; hämmern ‹*Gold, Silber usw.*›; **~ one's breast** *(lit. or fig.)* sich *(Dat.)* an die Brust schlagen; **b)** *(hit)* schlagen; [ver]prügeln; **c)** *(defeat)* schlagen ‹*Mannschaft, Gegner*›; *(surmount)* in den Griff bekommen ‹*Inflation, Arbeitslosigkeit, Krise*›; **~ the deadline** den Termin noch einhalten; **d)** *(surpass)* brechen ‹*Rekord*›; übertreffen ‹*Leistung*›; **hard to ~**: schwer zu schlagen; **you can't ~** *or* **nothing ~s French cuisine** es geht [doch] nichts über die französische Küche; **~ that!** das soll mal einer nachmachen!; **~ everything** *(coll.)* alles in den Schatten stellen; **~ sb. to it** jmdm. zuvorkommen; **can you ~ it?** ist denn das zu fassen?; **e)** *(circumvent)* umgehen; **~ the system** sich gegen das bestehende System durchsetzen; **f)** *(perplex)* **it ~s me how/why ...**: es ist mir ein Rätsel wie/warum ...; **g) ~ time** den Takt schlagen; **h)** *p.p.* **beat: I'm ~** *(coll.: exhausted)* ich bin erledigt *(ugs.)*. *See also* **beaten 2. 2.** *v.i.*, **beat, beaten a)** *(throb)* ‹*Herz:*› schlagen, klopfen; ‹*Puls:*› schlagen; **my heart seemed to stop ~ing** ich dachte, mir bleibt das Herz stehen; **b)** ‹*Sonne:*› brennen (**on** auf + *Akk.*); ‹*Wind, Wellen:*› schlagen (**on** auf + *Akk.*, **against** gegen); ‹*Regen, Hagel:*› prasseln, trommeln (**against** gegen); **c) ~ about the bush** um den [heißen] Brei herumreden *(ugs.)*; **d)** *(knock)* klopfen (**at** an + *Dat.*); **e)** *(Naut.)* kreuzen. **3.** *n.* **a)** *(stroke, throbbing)* Schlag, *der; (rhythm)* Takt, *der;* **his heart missed a ~**: ihm stockte das Herz; **b)** *(Mus.)* Schlag, *der; (of metronome, baton)* Taktschlag, *der;* **c)** *(of policeman, watchman)* Runde, *die; (area)* Revier, *das;* **be off sb.'s [usual] ~** *(fig.)* nicht in jmds. Fach schlagen

~ 'back *v.t.* zurückschlagen ‹*Feind*›

~ 'down 1. *v.i.* ‹*Sonne:*› herniederbrennen; ‹*Regen:*› niederprasseln. **2.** *v.t.* **a)** einschlagen ‹*Tür*›; **b)** *(in bargaining)* herunterhandeln

~ 'in *v.t.* einschlagen

~ 'off *v.t.* abwehren ‹*Angriff*›

~ 'out *v.t.* heraushämmern ‹*Rhythmus*›; aushämmern ‹*Metall*›; ausschlagen ‹*Feuer*›

~ 'up *v.t.* zusammenschlagen ‹*Person*›; schlagen ‹*Sahne usw.*›

beaten ['bi:tn] **1.** *see* **beat 1, 2. 2.** *adj.* **a) off the ~ track** *(remote)* weit abgelegen; **b)** *(hammered)* gehämmert ‹*Silber, Gold*›

beater ['bi:tə(r)] *n.* **a)** *(Cookery)* Rührbesen, *der;* **b)** *(Hunting)* Treiber, *der*

beatify [bɪ'ætɪfaɪ] *v.t. (Relig.)* seligsprechen

beating ['bi:tɪŋ] *n.* **a)** *(punishment)* **a ~**: Schläge *Pl.*; Prügel *Pl.*; **give sb. a good ~**: jmdm. eine gehörige Tracht Prügel verpassen *(ugs.)*; **b)** *(defeat)* Niederlage, *die;* **c)** *(surpassing)* **take some/a lot of ~**: nicht leicht zu übertreffen sein

'beat-up *adj. (coll.)* ramponiert *(ugs.)*

beau [bəʊ] *n., pl.* **~x** [bəʊz] *or* **~s a)** *(dandy)* Dandy, *der (geh.)*; **b)** *(Amer.: boy-friend)* Verehrer, *der*

beaut [bju:t] *(Austral., NZ, & Amer. sl.) n.* Prachtexemplar, *das*

beautician [bju:'tɪʃn] *n.* Kosmetiker, *der*/Kosmetikerin, *die*

beautiful ['bju:tɪfl] *adj.* **a)** [ausgesprochen] schön; wunderschön ‹*Augen, Aussicht, Blume, Kleid, Morgen, Musik, Schmuck*›; **b)** *(enjoyable, impressive)* großartig

beautifully ['bju:tɪfəlɪ] *adv.* wunderbar; *(coll.: very well)* prima *(ugs.)*; *(coll.: very)* schön ‹*weich, warm*›

beautify ['bju:tɪfaɪ] *v.t.* verschönern; *(adorn)* [aus]schmücken

beauty ['bju:tɪ] *n.* **a)** *no pl.* Schönheit, *die; (of action, response)* Eleganz, *die; (of idea, simplicity, sacrifice)* Größe, *die;* **~ is only skin deep** man kann nicht nach dem Äußeren urteilen; **b)** *(person or thing)* Schönheit, *die; (animal)* wunderschönes Tier; **she is a real ~**: sie ist wirklich eine Schönheit; **c)** *(exceptionally good specimen)* Prachtexemplar, *das;* **that last goal was a ~**: dieses letzte Tor war ein Bilderbuchtor; **d)** *(beautiful feature)* Schöne, *das;* **her eyes are her great ~**: das Schöne an ihr sind ihre Augen; **the ~ of it/of living in California** das Schöne *od.* Gute daran/am Leben in Kalifornien

beauty: ~ competition, ~ contest *ns.* Schönheitswettbewerb, *der;* **~ parlour** *see* **~ salon; ~ queen** *n.* Schönheitskönigin, *die;* **~ salon** *n.* Kosmetiksalon, *der;* **~ spot** *n.* Schönheitsfleck, *der; (place)* schönes Fleckchen [Erde]

beaux *pl. of* **beau**

beaver ['bi:və(r)] **1.** *n.* **a)** *pl. same or* **~s** Biber, *der;* **b)** *(fur)* Biber[pelz], *der.* **2.** *v.i. (Brit.)* **~ away** eifrig arbeiten (**at** an + *Dat.*)

becalmed [bɪ'kɑ:md] *adj.* **be ~** in einer Flaute *od.* Windstille treiben

became *see* **become**

because [bɪ'kɒz] **1.** *conj.* weil; **that is ~ you don't know German** das liegt daran, daß du kein Deutsch kannst. **2.** *adv.* **~ of** wegen (+ *Gen.*); **don't come just ~ of me** nur meinetwegen brauchen Sie nicht zu kommen; **~ of which he ...**: weswegen er ...

beck [bek] *n.* **be at sb.'s ~ and call** jmdm. zur Verfügung stehen

beckon ['bekn] **1.** *v.t.* **a)** winken; **~ sb. in/over** jmdn. herein-/herbei- *od.* herüberwinken; **b)** *(fig.: invite)* locken. **2.** *v.i.* **a) ~ to sb.** jmdm. winken; **b)** *(fig.: be inviting)* locken

become [bɪ'kʌm] **1.** *copula*, **became, become** werden; **~ a politician/dentist** Politiker/Zahnarzt werden; **~ a nuisance/rule** zu einer Plage/zur Regel werden. **2.** *v.i.*, **became, become** werden; **what has ~ of him?** was ist aus ihm geworden?; **what has ~ of that book?** wo ist das Buch geblieben? **3.** *v.t.*, **became, become a)** *see* **befit; b)** *(suit)* **~ sb.** jmdm. stehen; zu jmdm. passen

becoming [bɪ'kʌmɪŋ] *adj.* **a)** *(fitting)* schicklich *(geh.)*; **b)** *(flattering)* vorteilhaft ‹*Hut, Kleid, Frisur*›

bed [bed] **1.** *n.* **a)** Bett, *das; (without bedstead)* Lager, *das;* **in ~**: im Bett; **~ and board** Unterkunft und Verpflegung; **~ and breakfast** Zimmer mit Frühstück; **get into/out of ~**: ins *od.* zu Bett gehen/aufstehen; **go to ~**: ins *od.* zu Bett gehen; **go to ~ with sb.** *(fig.)* mit jmdm. ins Bett gehen *(ugs.)*; **make the ~**: das Bett machen; **put sb. to ~**: jmdn. ins *od.* zu Bett bringen; **life isn't a** *or* **is no ~ of roses** *(fig.)* das Leben ist kein reines Vergnügen; **have got out of ~ on the wrong side** *(fig.)* mit dem linken Fuß zuerst aufgestanden sein; **as you make your ~ so you must lie on it** *(prov.)* wie man sich bettet, so liegt man; **take to one's ~**: sich krank ins Bett legen; **b)** *(flat base)* Unterlage, *die; (of machine)* Bett, *das; (of road, railway, etc.)* Unterbau, *der;* **c)** *(in garden)* Beet, *das;* **d)** *(of sea, lake)* Grund, *der;* Boden, *der; (of river)* Bett, *das;* **e)** *(layer)* Schicht, *die.* **2.** *v. t.*, **-dd-: a)** ins Bett legen; **b)** *(fig. coll.)* beschlafen ‹*Frau*›; **c)** *(plant)* setzen ‹*Pflanze, Sämling*›.

~ 'down 1. *v. t.* **the troops were ~ded down in a barn** die Soldaten wurden über Nacht in einer Scheune einquartiert. **2.** *v. i.* kampieren

~ 'in *v. t.* einlassen

~ 'out *v. t.* auspflanzen ‹*Pflanze*›

bed: ~bug *n.* [Bett]wanze, *die;* **~clothes** *n. pl.* Bettzeug, *das*

bedding ['bedɪŋ] *n., no pl., no indef. art.* Matratze und Bettzeug; *(for animal)* Streu, *das*

'bedding plant *n.* Freilandpflanze, *die*

beddy-byes ['bedɪbaɪz] *n. (child lang.)* Heiabett, *das (Kinderspr.);* **off to ~:** ab in die Heia

bedeck [bɪ'dek] *v. t.* schmücken; **~ed with flags** mit Fahnen geschmückt

bedevil [bɪ'devl] *v. t., (Brit.)* **-ll-: a)** *(spoil)* verderben; durcheinanderbringen ‹*System*›; **b)** *(afflict)* heimsuchen; ‹*Pech:*› verfolgen

bed: ~fellow *n.* Bettgenosse, *der*/-genossin, *die;* **make** *or* **be strange ~fellows** *(fig.)* ‹*Personen:*› ein merkwürdiges Gespann sein; ‹*Staaten, Organisationen:*› eine eigenartige Kombination sein; **~jacket** *n.* Bettjacke, *die*

bedlam ['bedləm] *n.* Chaos, *das;* Durcheinander, *das*

'bed-linen *n.* Bettwäsche, *die*

bedouin ['bedu:ɪn] *n., pl. same* Beduine, *der*/Beduinin, *die*

bed: ~pan *n.* Bettpfanne, *die;* **~post** *n.* Bettpfosten, *der*

bedraggled [bɪ'drægəld] *adj.* [naß und] verschmutzt *od.* schmutzig

bed: ~ridden *adj.* bettlägerig; **~rock** *n.* Felssohle, *die; (fig.)* Basis, *die;* **get** *or* **reach down to ~rock** *(fig.)* zum Kern der Sache kommen; **~room** *n.* Schlafzimmer, *das;* **~side** *n.* Seite des Bettes, *die;* **be at the ~side** am Bett sein; **~side table/lamp** Nachttisch, *der*/Nachttischlampe, *die;* **~side reading** Bettlektüre, *die;* **have a good ~side manner** ‹*Arzt:*› gut mit Kranken umgehen können; **~-sit, ~-'sitter** *ns. (coll.),* **~-'sitting-room** *n. (Brit.)* Wohnschlafzimmer, *das;* **~sore** *n.* wundgelegene Stelle; **~spread** *n.* Tagesdecke, *die;* **~stead** ['bedsted] *n.* Bettgestell, *das;* **~time** *n.* Schlafenszeit, *die;* **it's past the children's ~time** die Kinder müßten schon im Bett sein; **will you have it finished by ~time?** bist du vor dem Schlafengehen damit fertig?; **a ~time story** eine Gutenachtgeschichte

beduin *see* **bedouin**

'bed-wetting *n.* Bettnässen, *das*

bee [bi:] *n.* Biene, *die;* **she's such a busy ~** *(fig.)* sie ist so ein fleißiges Mädchen; **as busy as a ~** *(fig.)* bienenfleißig; **she has a ~ in her bonnet about punctuality** sie hat einen Pünktlichkeitsfimmel *(ugs.)*

beech [bi:tʃ] *n.* **a)** *(tree)* Buche, *die;* **b)** *(wood)* Buche, *die;* Buchenholz, *das; attrib.* buchen

beech: ~-nut *n.* [Buch]ecker, *die;* **~wood** *see* **beech b**

beef [bi:f] **1.** *n.* **a)** *no pl.* Rind[fleisch], *das;* **b)** *no pl. (coll.: muscles)* Muskeln; **have plenty of ~:** sehr muskulös sein; **c)** *usu. in pl.* **beeves** [bi:vz] *or (Amer.)* **~s** *(ox)* Mastrind, *das;* **d)** *pl.* **~s** *(sl.: complaint)* Meckerei, *die (ugs.).* **2.** *v. t. (sl.)* **~ up** stärken. **3.** *v. i. (sl.)* meckern *(ugs.)* **(about** über + *Akk.*)

beef: ~burger *n.* Beefburger, *der;* **~eater** *n. (Brit.)* Beefeater, *der;* **~steak** *n.* Beefsteak, *das*

beefy ['bi:fɪ] *adj.* **a)** *(like beef)* wie Rindfleisch *nachgestellt;* Rindfleisch-; **b)** *(coll.) (muscular)* muskulös; *(fleshy)* massig

bee: ~hive *n.* Bienenstock, *der; (rounded)* Bienenkorb, *der; (fig.: scene of activity)* Taubenschlag, *der;* **~-keeper** *n.* Imker, *der*/Imkerin, *die;* **~-keeping** *n.* Imkerei, *die;* **~-line** *n.* **make a ~-line for sth./sb.** schnurstracks auf etw./jmdn. zustürzen

been *see* **be**

beep [bi:p] **1.** *n.* Piepton, *der; (of car horn)* Tuten, *das.* **2.** *v. i.* piepen; ‹*Signalhorn:*› hupen

beer [bɪə(r)] *n.* Bier, *das;* **order two ~s** zwei Bier bestellen; **brew various ~s** verschiedene Biere *od.* Biersorten brauen; **small ~** *(fig.: trifles)* Kleinigkeiten *Pl.;* **that firm's turnover is only small ~:** der Umsatz dieser Firma ist kaum der Rede wert

beer: ~-barrel *n.* Bierfaß, *das;* **~ belly** *n. (coll.)* Bierbauch, *der (ugs.);* **~-bottle** *n.* Bierflasche, *die;* **~-can** *n.* Bierdose, *die;* **~ cellar** *n.* Bierkeller, *der;* **~-crate** *n.* Bierkasten, *der;* **~-drinker** *n.* Biertrinker, *der;* **~ garden** *n.* Biergarten, *der;* **~-glass** *n.* Bierglas, *das;* **~-mat** *n.* Bierdeckel, *der;* Bieruntersetzer, *der;* **~-mug** *n.* Bierkrug, *der*

beeswax ['bi:zwæks] *n.* Bienenwachs, *das*

beet [bi:t] *n.* Rübe, *die*

beetle ['bi:tl] *n.* Käfer, *der*

'beetroot *n.* rote Beete *od.* Rübe

befall [bɪ'fɔ:l] **1.** *v. i., forms as* **fall 2** sich begeben *(geh.);* geschehen. **2.** *v. t., forms as* **fall 2** widerfahren (+ *Dat.*)

befit [bɪ'fɪt] *v. t.,* **-tt-** sich ziemen *od.* gebühren für *(geh.);* **it ill ~s you to do that** es steht Ihnen schlecht an, das zu tun; **she behaved as ~ted a lady** sie benahm sich, wie es sich für eine Dame gebührte

befitting [bɪ'fɪtɪŋ] *adj.* gebührend *(geh.);* schicklich *(geh.)* ‹*Benehmen*›

before [bɪ'fɔ:(r)] **1.** *adv.* **a)** *(of time)* vorher; zuvor; **the day ~:** am Tag zuvor; **long ~:** lange vorher *od.* zuvor; **not long ~:** kurz vorher; **the noise continued as ~:** der Lärm ging nach wie vor weiter; **you should have told me so ~:** das hättest du mir vorher *od.* früher *od.* eher sagen sollen; **I've seen that film ~:** ich habe den Film schon [einmal] gesehen; **I've heard that ~:** das habe ich schon einmal gehört; **b)** *(ahead in position)* vor[aus]; **c)** *(in front)* voran. **2.** *prep.* **a)** *(of time)* vor (+ *Dat.*); **the day ~ yesterday** vorgestern; **the year ~ last** vorletztes Jahr; **the year ~ that** das Jahr davor; **it was [well] ~ my time** das war [lange] vor meiner Zeit; **since ~ the war** schon vor dem Krieg; **~ now** vorher; früher; **~ Christ** vor Christus; vor Christi Geburt; **he got there ~ me** er war vor mir da; **~ then** vorher; **~ long** bald; **~ leaving, he phoned/I will phone** bevor er wegging, rief er an/bevor ich weggehe, rufe ich an; **~ tax** brutto; vor [Abzug *(Dat.)* der] Steuern; **b)** *(position)* vor (+ *Dat.*); *(direction)* vor (+ *Akk.*); **~ my very eyes** vor meinen Augen; **go ~ a court of law** vor ein Gericht kommen; **appear ~ the judge** vor dem Richter erscheinen; *see also* **carry 1 a; c)** *(awaiting)* **have one's life ~ one** sein Leben noch vor sich *(Dat.)* haben; *(confronting)* **the matter ~ us** das uns *(Dat.)* vorliegende Thema; **the task ~ us** die Aufgabe, die vor uns *(Dat.)* liegt; **d)** *(more important than)* vor (+ *Dat.*); **he puts work ~ everything** die Arbeit ist ihm wichtiger als alles andere. **3.** *conj.* bevor; **it'll be ages ~ I finish this** es wird eine Ewigkeit dauern, bis ich damit fertig bin

beforehand [bɪ'fɔ:hænd] *adv.* vorher; *(in anticipation)* im voraus; **I found out about it** ~: ich habe es schon vorher herausgefunden

befriend [bɪ'frend] *v. t.* **a)** *(act as a friend to)* sich anfreunden mit; **b)** *(help)* sich annehmen (+ *Gen.*)

beg [beg] **1.** *v. t.*, **-gg-**: **a)** betteln um; erbetteln ‹*Lebensunterhalt*›; **b)** *(ask earnestly)* bitten; **she ~ged to come with us** sie bat darum, mit uns kommen zu dürfen; **I ~ to differ** da bin ich [aber] anderer Meinung; **~ sb. for sth.** jmdn. um etw. bitten; **c)** *(ask earnestly for)* **~ sth.** um etw. bitten; **~ sth. of sb.** etw. von jmdm. erbitten; **~ a favour [of sb.]** [jmdn.] um einen Gefallen bitten; **~ forgiveness** um Verzeihung bitten; *see also* **pardon 1 b; d) ~ the question** *(evade difficulty)* der Frage *(Dat.)* ausweichen. **2.** *v. i.*, **-gg-** ‹*Bettler:*› betteln **(for** um); ‹*Hund:*› Männchen machen; betteln; **a ~ging letter** ein Bettelbrief; **go [a-]~ging** keinen Abnehmer finden

began *see* **begin**

beggar ['begə(r)] *n.* **a)** Bettler, *der*/Bettlerin, *die;* **b)** *(coll.: person)* Arme, *der/die;* **poor** ~: armer Teufel

begin [bɪ'gɪn] **1.** *v. t.*, **-nn-, began** [bɪ'gæn], **begun** [bɪ'gʌn] **~ sth.** [mit] etw. beginnen; **~ a new bottle** eine neue Flasche anbrechen; **she began life here** sie verbrachte ihre ersten Lebensjahre hier; **~ school** in die Schule kommen; **~ doing** *or* **to do sth.** anfangen *od.* beginnen, etw. zu tun; **I began to slip** ich kam ins Rutschen; **I am ~ning to get annoyed** so langsam werde ich ärgerlich; **the film does not ~ to compare with the book** der Film läßt sich nicht annähernd mit dem Buch vergleichen. **2.** *v. i.* **-nn-, began, begun** anfangen; beginnen *(oft geh.);* **~ning next month** vom nächsten Monat an; **~ at the beginning** von vorne anfangen; **~ with sth./sb.** bei *od.* mit etw./jmdm. anfangen *od.* beginnen; **to ~ with** zunächst *od.* zuerst einmal; **it is the wrong book, to ~ with** das ist schon einmal das falsche Buch

beginner [bɪ'gɪnə(r)] *n.* Anfänger, *der*/Anfängerin, *die;* **~'s luck** Anfängerglück, *das*

beginning [bɪ'gɪnɪŋ] *n.* Anfang, *der;* Beginn, *der;* **at** *or* **in the** ~: am Anfang; **at the ~ of February/the month** Anfang Februar/des Monats; **at the ~ of the day** zu Beginn des Tages; **from ~ to end** von Anfang bis Ende; von vorn bis hinten; **from the [very]** ~: [ganz] von Anfang an; **have its ~s in sth.** seine Anfänge *od.* seinen Ursprung in etw. *(Dat.)* haben; **[this is] the ~ of the end** [das ist] der Anfang vom Ende; **go back to the** ~: wieder von vorne anfangen

begonia [bɪ'gəʊnɪə] *n. (Bot.)* Begonie, *die;* Schiefblatt, *das*

begrudge [bɪ'grʌdʒ] *v. t.* **a)** *(envy)* **~ sb. sth.** jmdm. etw. mißgönnen; **b)** *(give reluctantly)* **I ~ the time I have to spend** es ist mir leid um die Zeit; **c)** *(be dissatisfied with)* **~ doing sth.** etw. ungern tun

beguile [bɪ'gaɪl] *v. t. (delude)* betören; verführen; **~ sb. into doing sth.** jmdn. dazu verführen, etw. zu tun; **be ~d by sb./sth.** sich von jmdm./etw. täuschen lassen

begun *see* **begin**

behalf [bɪ'hɑ:f] *n., pl.* **behalves** [bɪ'hɑ:vz] **on** *or (Amer.)* **in ~ of sb./sth.** *(as representing sb./sth.)* für jmdn./etw.; *(more formally)* im Namen von jmdm./etw.

behave [bɪ'heɪv] **1.** *v. i.* **a)** sich verhalten; sich benehmen; **he ~s more like a friend to them** er behandelt sie mehr wie Freunde; **~ well/badly** sich gut/schlecht benehmen; **~ well/badly towards sb.** jmdn. gut/schlecht behandeln; **well-/ill-** *or* **badly/nicely ~d** brav/ungezogen/lieb *(ugs.);* **b)** *(~ well)* brav sein; sich benehmen; **~!** benimm dich! **2.** *v. refl.* **~ oneself** sich benehmen; **~ yourself!** benimm dich!

behavior, behaviorism *(Amer.) see* **behaviour, behaviourism**

behaviour [bɪ'heɪvjə(r)] *n.* *(conduct)* Verhalten, *das* **(towards** gegenüber); Benehmen, *das* **(towards** gegenüber); *(of child)* Betragen, *das;* **be on one's best** ~: sein bestes Benehmen an den Tag legen; **put sb. on his/her best** ~: jmdm. raten, sich gut zu benehmen

behaviourism [bɪ'heɪvjərɪzm] *n., no pl. (Psych.)* Behaviorismus, *der*

behead [bɪ'hed] *v. t.* enthaupten, köpfen ‹*Person*›

beheld *see* **behold**

behest [bɪ'hest] *n. (literary)* **at sb.'s** ~: auf jmds. Geheiß *(Akk.)*

behind [bɪ'haɪnd] **1.** *adv.* **a)** *(at rear of sb./sth.)* hinten; **from** ~: von hinten; **he glanced ~ before moving off** er schaute nach hinten, bevor er losfuhr; **we'll follow on** ~: wir kommen hinterher; **b)** *(further back)* **be miles** ~: kilometerweit zurückliegen; **leave sb.** ~: jmdn. hinter sich *(Dat.)* lassen (*see also* **d**); **fall** ~: zurückbleiben; *(fig.)* in Rückstand geraten; **lag** ~: zurückbleiben; *(fig.)* im Rückstand sein; **be** ~: hinten sein; *(be late)* im Verzug sein; **c)** *(in arrears)* **be/get ~ with one's rent** mit der Miete im Verzug sein/in Verzug geraten; **d)** *(remaining after sb.'s departure)* **leave sb./sth.** ~: jmdn./etw. zurücklassen (*see also* **b**); **he left his gloves ~ by mistake** er ließ seine Handschuhe versehentlich liegen; **stay** ~: dableiben; *(as punishment)* nachsitzen. **2.** *prep.* **a)** *(at rear of, on other side of; fig.: hidden by)* hinter (+ *Dat.*); **he stepped out from ~ the wall** er trat hinter der Mauer hervor; **he came from ~ her** er kam von hinten; **one ~ the other** hintereinander; **~ sb.'s back** *(fig.)* hinter jmds. Rücken *(Dat.);* **b)** *(towards rear of)* hinter (+ *Akk.*); *(fig.)* **I don't want to go ~ his back** ich will nicht hinter seinem Rücken handeln; **put ~ one** vergessen; **put the past ~ one** einen Strich unter die Vergangenheit ziehen; **c)** *(further back than)* hinter (+ *Dat.*); **they were miles ~ us** sie lagen meilenweit hinter uns *(Dat.)* zurück; **be ~ the times** nicht auf dem laufenden sein; **fall ~ sb./sth.** hinter jmdn./etw. zurückfallen; **d)** *(past)* hinter (+ *Dat.*); **all that trouble is ~ me** ich habe den ganzen Ärger hinter mir; **e)** *(later than)* **~ schedule/time** im Rückstand; **f)** *(in support of)* hinter (+ *Dat.*); **I'm right ~ you** ich stehe hinter dir; **the man ~ the project** der Mann, der hinter dem Projekt steht; **g)** *(remaining after departure of)* **she left nothing ~ her** sie hinterließ nichts. **3.** *n. (buttocks)* Hintern, *der (ugs.)*

behindhand [bɪ'haɪndhænd] *pred. adj.* **a) be ~ with one's rent** mit der Miete im Verzug sein; **b)** *(backward)* **be ~ in doing sth.** etw. zurückhaltend tun

behold [bɪ'həʊld] *v. t.*, **beheld** [bɪ'held] *(arch./literary)* **a)** erblicken *(geh.);* **b)** *in imper.* siehe/sehet

beholder [bɪ'həʊldə(r)] *n.* **beauty is in the eye of the** ~: schön ist, was gefällt

beige [beɪʒ] **1.** *n.* Beige, *das.* **2.** *adj.* beige

being ['bi:ɪŋ] **1.** *pres. part. of* **be. 2.** *n.* **a)** *no pl., no art. (existence)* Dasein, *das;* Leben, *das;* Existenz, *die;* **bring into** ~: einführen; **call into** ~: ins Leben rufen; **come into** ~: entstehen; **when the new**

system comes into ~: wenn das neue System eingeführt wird; **b)** *(anything, esp. person, that exists)* Wesen, *das;* Geschöpf, *das*

belabour *(Brit.; Amer.:* **belabor)** [bɪ'leɪbə(r)] *v. t.* **a)** *(beat)* einschlagen auf (+ *Akk.*); *(fig.)* überhäufen; **b)** *see* **labour 3b**

belated [bɪ'leɪtɪd] *adj.* verspätet

belatedly [bɪ'leɪtɪdlɪ] *adv.* verspätet; nachträglich

belch [beltʃ] **1.** *v. i.* heftig aufstoßen; rülpsen *(ugs.).* **2.** *v. t.* ausstoßen ‹*Rauch, Flüche usw.*›. **3.** *n.* Rülpser, *der (ugs.)*

beleaguer [bɪ'li:gə(r)] *v. t. (lit. or fig.)* belagern

belfry ['belfrɪ] *n.* Glockenturm, *der*

Belgian ['beldʒən] **1.** *n.* Belgier, *der*/Belgierin, *die.* **2.** *adj.* belgisch; **sb. is ~:** jmd. ist Belgier/Belgierin

Belgium ['beldʒəm] *pr. n.* Belgien *(das)*

Belgrade [bel'greɪd] *pr. n.* Belgrad *(das)*

belie [bɪ'laɪ] *v. t.,* **belying** [bɪ'laɪɪŋ] *(fail to fulfil)* enttäuschen ‹*Versprechen, Vorstellung*›; *(give false notion of)* hinwegtäuschen über (+ *Akk.*) ‹*Tatsachen, wahren Zustand*›

belief [bɪ'li:f] *n.* Glaube[n], *der;* **~ in sth.** Glaube an etw. *(Akk.);* **beyond ~:** unglaublich; **it is my ~ that ...:** ich bin der Überzeugung, daß ...; **in the ~ that ...:** in der Überzeugung, daß ...; **to the best of my ~:** meines Wissens

believable [bɪ'li:vəbl] *adj.* glaubhaft; glaubwürdig

believe [bɪ'li:v] **1.** *v. i.* **a) ~ in sth.** *(put trust in truth of)* an etw. *(Dat.)* glauben; **I ~ in free medical treatment for all** ich bin für die kostenlose ärztliche Behandlung aller; **I don't ~ in going to the dentist** ich halte nicht viel von Zahnärzten; **b)** *(have faith)* glauben **(in** an + *Akk.*) ‹*Gott, Himmel usw.*›; **c)** *(suppose, think)* glauben; denken; **I ~ so/not** ich glaube schon/nicht. **2.** *v. t.* **a) ~ sth.** etw. glauben; **I can well ~ it** das glaub' ich gerne; **if you ~ that, you'll ~ anything** wer's glaubt, wird selig *(ugs. scherzh.);* **~ it or not** ob du es glaubst oder nicht; **would you ~ [it]** *(coll.)* stell dir mal vor *(ugs.);* **~ sb.** jmdm. glauben; **I don't ~ you** das glaube ich dir nicht; **~ [you] me** glaub/glaubt mir!; **I couldn't ~ my eyes/ears** ich traute meinen Augen/Ohren nicht; **b)** *(be of opinion that)* glauben; der Überzeugung sein; **he is ~d to be in the London area** man vermutet ihn im Raum London; **people ~d her to be a witch** die Leute hielten sie für eine Hexe; **make ~ [that ...]** so tun, als ob ...

believer [bɪ'li:və(r)] *n.* **a)** Gläubige, *der/die;* **b) be a great *or* firm ~ in sth.** viel von etw. halten

Belisha beacon [bəli:ʃə 'bi:kn] *n. (Brit.) gelbes Blinklicht an Zebrastreifen*

belittle [bɪ'lɪtl] *v. t.* herabsetzen

bell [bel] *n.* **a)** Glocke, *die; (smaller)* Glöckchen, *das;* **clear as a ~:** glockenklar; *(understandable)* [ganz] klar und deutlich; **b)** *(device to give ~-like sound)* Klingel, *die;* **c)** *(ringing)* Läuten, *das;* **the ~ has gone** es hat geläutet *od.* geklingelt; **d)** *(Boxing)* Gong, *der*

bell-bottomed ['belbɒtəmd] *adj.* ausgestellt

belle [bel] *n.* Schönheit, *die;* Schöne, *die;* **~ of the ball** Ballkönigin, *die*

belles-lettres [bel'letr] *n. pl.* schöngeistige Literatur; Belletristik, *die*

bellicose ['belɪkəʊs] *adj.* kriegerisch ‹*Stimmung, Nation*›; streitsüchtig ‹*Person*›

belligerent [bɪ'lɪdʒərənt] **1.** *adj.* **a)** *(eager to fight)* kriegerisch ‹*Nation*›; streitlustig ‹*Person, Benehmen*›; aggressiv ‹*Rede*›; **b)** *(fighting a war)* kriegführend. **2.** *n.* kriegführende Partei

bellow ['beləʊ] **1.** *v. i.* ‹*Tier, Person:*› brüllen; **~ at sb.** jmdn. anbrüllen. **2.** *v. t.* **~ out** brüllen ‹*Befehl*›

bellows ['beləʊz] *n. pl.* Blasebalg, *der;* **a pair of ~:** ein Blasebalg

bell: ~-pull *n.* Klingelzug, *der;* **~-push** *n.* Klingel, *die;* **~-ringer** *n.* Glöckner, *der;* **~-ringing** *n.* Glockenläuten, *das;* **~-shaped** *adj.* glockenförmig; **~-tent** *n.* Rundzelt, *das;* **~-tower** *n.* Glockenturm, *der*

belly ['belɪ] *n.* Bauch, *der; (stomach)* Magen, *der*

belly: ~ache 1. *n.* Bauchschmerzen *Pl.;* Bauchweh, *das (ugs.);* **2.** *v. i. (sl.)* jammern **(about** über + *Akk.*); **~-button** *n. (coll.)* Bauchnabel, *der;* **~-dance** *n.* Bauchtanz, *der;* **~-dancer** *n.* Bauchtänzerin, *die*

bellyful ['belɪfʊl] *n.* **have had a ~ of sth.** *(fig.)* von etw. die Nase voll haben *(ugs.)*

belong [bɪ'lɒŋ] *v. i.* **a)** *(be rightly assigned)* **~ to sb./sth.** jmdm./zu etw. gehören; **b) ~ to** *(be member of)* **~ to a club** einem Verein angehören; **she ~s to a trade union/the club** sie ist Mitglied einer Gewerkschaft/des Vereins; **c)** *(be rightly placed)* **feel that one doesn't ~:** das Gefühl haben, fehl am Platze zu sein *od.* daß man nicht dazugehört; **the cutlery ~s in this drawer** das Besteck gehört in diese Schublade

belongings [bɪ'lɒŋɪŋz] *n. pl.* Habe, *die;* Sachen *Pl.;* **personal ~:** persönlicher Besitz; persönliches Eigentum; **all our ~:** unser ganzes Hab und Gut

beloved [bɪ'lʌvɪd] **1.** *adj.* geliebt; teuer; **be ~** [bɪ'lʌvd] **by *or* of sb.** von jmdm. geliebt werden; jmdm. lieb und teuer sein; **in ~ memory of sb.** in treuem Angedenken an jmdn. **2.** *n.* Geliebte, *der/die*

below [bɪ'ləʊ] **1.** *adv.* **a)** *(position)* unten; unterhalb; *(lower down)* darunter; *(downstream)* weiter unten; **down ~:** unten; **from ~:** von unten [herauf]; **b)** *(direction)* nach unten; hinunter; hinab *(geh.);* **~ left** links unten; unten links; **c)** *(later in text)* unten; **see [p. 123] ~:** siehe unten[, S. 123]; **please sign ~:** bitte hier unterschreiben; **d)** *(downstairs) (position)* unten; *(direction)* nach unten; *(Naut.)* unter Deck; **go ~** *(Naut.)* unter Deck gehen; **the flat/floor ~:** die Wohnung/das Stockwerk darunter *od.* unter uns/ihnen *usw.* **2.** *prep.* **a)** *(position)* unter (+ *Dat.*); unterhalb (+ *Gen.*); *(downstream from)* unterhalb (+ *Gen.*); **b)** *(direction)* unter (+ *Akk.*); **c)** *(ranking lower than)* unter (+ *Dat.*); **she's in the class ~ me** sie ist eine Klasse unter mir

belt [belt] **1.** *n.* **a)** Gürtel, *der; (for carrying tools, weapons, etc.)* Gurt, *der; (on uniform)* Koppel, *das;* **hit below the ~** *(lit. or fig.)* unter die Gürtellinie schlagen; *see also* **tighten 1a; b)** *(strip)* Gurt, *der; (region)* Gürtel, *der;* **industrial ~:** Industrierevier, *das;* **c)** *(Mech. Engin.: drive-~)* Riemen, *der;* **d)** *(sl.: heavy blow)* Schlag, *der.* **2.** *v. t. (sl.: hit hard)* schlagen; **I'll ~ you [one]** ich hau' dir eine runter *(ugs.).* **3.** *v. i. (sl.)* **~ up/down the motorway** über die Autobahn rasen

~ along *v. i. (sl.)* rasen *(ugs.)*

~ out *v. t. (sl.)* schmettern; voll herausbringen ‹*Rhythmus*›

~ 'up *v. i.* **a)** *(Amer. coll., Brit. coll. joc.: put seat-~ on)* sich anschnallen; **b)** *(Brit. sl.: be quiet)* die Klappe halten *(salopp)*

belying *see* **belie**

bemoan [bɪ'məʊn] *v. t.* beklagen
bemused [bɪ'mju:zd] *adj.* verwirrt
bench [bentʃ] *n.* **a)** Bank, *die;* **b)** *(Law)* **on the ~:** auf dem Richterstuhl; **c)** *(office of judge)* Richteramt, *das;* **d)** *(Brit. Parl.)* Bank, *die;* Reihe, *die;* **e)** *(work-table)* Werkbank, *die*
'**bench-mark** *n. (fig.)* Maßstab, *der;* Fixpunkt, *der*
bend [bend] **1.** *n.* **a)** Kurve, *die;* **there is a ~ in the road** die Straße macht eine Kurve; **a ~ in the river** eine Flußbiegung; **be round the ~** *(fig. coll.)* spinnen *(ugs.);* verrückt sein *(ugs.);* **go round the ~** *(fig. coll.)* überschnappen *(ugs.);* durchdrehen *(ugs.);* **drive sb. round the ~** *(fig. coll.)* jmdn. wahnsinnig *od.* verrückt machen *(ugs.);* **b) the ~s** *(Med. coll.)* Taucherkrankheit, *die.* **2.** *v. t.*, **bent** [bent] **a)** biegen; verbiegen ⟨*Nadel, Messer, Eisenstange, Ast*⟩; spannen ⟨*Bogen*⟩; beugen ⟨*Arm, Knie*⟩; anwinkeln ⟨*Arm, Bein*⟩; krumm machen ⟨*Finger*⟩; **~ sth. back/forward/up/down** etw. nach hinten/vorne/oben/unten biegen; *see also* **rule 1 a; b) be bent on sth.** auf etw. *(Akk.)* erpicht sein; **he bent his mind to the problem** er dachte ernst über das Problem nach. **3.** *v. i.*, **bent** sich biegen; sich krümmen; ⟨*Äste:*⟩ sich neigen; **the road ~s** die Straße macht eine Kurve; **the river ~s/~s in and out** der Fluß macht eine Biegung/schlängelt sich
~ 'down *v. i.* sich bücken; sich hinunterbeugen
~ 'over *v. i.* sich bücken; sich nach vorn beugen; *see also* **backwards a**
bended ['bendɪd] *adj.* **on ~ knee[s]** auf [den] Knien
beneath [bɪ'ni:θ] **1.** *prep.* **a)** *(unworthy of)* **~ sb.** jmds. unwürdig; unter jmds. Würde *(Dat.);* **~ contempt** verachtenswert; **b)** *(arch./literary: under)* unter (+ *Dat.*). **2.** *adv. (arch./literary)* darunter
benediction [benɪ'dɪkʃn] *n. (Relig.)* Segnung, *die*
benefactor ['benɪfæktə(r)] *n.* Stifter, *der;* Gönner, *der*
beneficial [benɪ'fɪʃl] *adj.* nutzbringend; vorteilhaft ⟨*Einfluß*⟩; **be ~ to sth./sb.** zum Nutzen von etw./jmdm. sein
beneficiary [benɪ'fɪʃərɪ] *n.* Nutznießer, *der*/Nutznießerin, *die*
benefit ['benɪfɪt] **1.** *n.* **a)** Vorteil, *der;* **be of ~ to sb./sth.** jmdm./einer Sache von Nutzen sein; **with the ~ of** mit Hilfe (+ *Gen.*); **for sb.'s ~:** in jmds. Interesse *(Dat.);* **give sb. the ~ of the doubt** im Zweifelsfall zu jmds. Gunsten entscheiden; **b)** *(allowance)* Beihilfe, *die;* **social security ~:** Sozialhilfe, *die;* **supplementary ~** *(Brit.)* zusätzliche Hilfe zum Lebensunterhalt; **unemployment ~:** Arbeitslosenunterstützung, *die;* **sickness ~:** Krankengeld, *das;* **child ~** *(Brit.)* Kindergeld, *das;* **c) ~ [performance/match/concert]** Benefizveranstaltung, *die*/-spiel, *das*/-konzert, *das.* **2.** *v. t.* **~ sb./sth.** jmdm./einer Sache nützen *od.* guttun. **3.** *v. i.* **~ by/from sth.** von etw. profitieren
Benelux ['benɪlʌks] *pr. n.* **the ~ countries** die Beneluxländer
benevolence [bɪ'nevələns] *n., no pl. see* **benevolent a:** Güte, *die;* Milde, *die;* Wohlwollen, *das*
benevolent [bɪ'nevələnt] *adj.* **a)** *(desiring to do good)* gütig; mild ⟨*Herrscher*⟩; wohlwollend ⟨*Behörde, Despot*⟩; **b)** *attrib. (charitable)* wohltätig, mildtätig ⟨*Institution, Verein*⟩
benign [bɪ'naɪn] *adj.* **a)** gütig ⟨*Person, Aussehen, Verständnis*⟩; wohlwollend ⟨*Person, Verhalten*⟩; mild, heilsam ⟨*Klima, Sonne*⟩; günstig ⟨*Stern, Einfluß*⟩; **b)** *(Med.)* gutartig, *(fachspr.)* benigne ⟨*Tumor*⟩
benignly [bɪ'naɪnlɪ] *adv.* gütig; wohlwollend
bent [bent] **1.** *see* **bend 2, 3. 2.** *n.* Neigung, *die;* Hang, *der;* **have a ~ for sth.** einen Hang zu etw. *od.* eine Vorliebe für etw. haben; **people with** *or* **of an artistic ~:** Menschen mit einer künstlerischen Ader *od.* Veranlagung. **3.** *adj.* **a)** krumm; gebogen; **b)** *(Brit. sl.: corrupt)* link *(salopp);* nicht ganz sauber *(salopp)* ⟨*Händler usw.*⟩
benzene ['benzi:n] *n. (Chem.)* Benzol, *das*
benzine ['benzi:n] *n.* Leichtbenzin, *das*
bequeath [bɪ'kwi:ð] *v. t.* **a) ~ sth. to sb.** jmdm. etw. vermachen; **b)** *(fig.)* überliefern ⟨*Legende, Zeugnisse*⟩; vererben ⟨*Tradition*⟩
bequest [bɪ'kwest] *n.* Vermächtnis, *das* (**to** an + *Akk.*); **make a ~ to sb. of sth.** jmdm. etw. vermachen
berate [bɪ'reɪt] *v. t.* schelten
bereave [bɪ'ri:v] *v. t.* **be ~d [of sb.]** jmdn. verlieren; **the ~d** der/die Hinterbliebene/die Hinterbliebenen
bereavement [bɪ'ri:vmənt] *n.* Trauerfall, *der*
bereft [bɪ'reft] *pred. adj.* **be ~ of sth.** etw. verloren haben
beret ['bereɪ, 'berɪ] *n.* Baskenmütze, *die; (Mil.)* Barett, *das*
berk [bɜ:k] *n. (Brit. sl.)* Dussel, *der (ugs.);* Blödmann, *der (salopp)*
Berlin [bɜ:'lɪn] **1.** *pr. n.* Berlin *(das).* **2.** *attrib. adj.* Berliner; *(Ling.)* berlinisch
Berliner [bɜ:'lɪnə(r)] *n.* Berliner, *der*/Berlinerin, *die*
Bermuda shorts [bəmju:də 'ʃɔ:ts] *n. pl.* Bermudashorts *Pl.*
Berne [bɜ:n] *pr. n.* Bern *(das)*
berry ['berɪ] *n.* Beere, *die*
berserk [bə'sɜ:k, bə'zɜ:k] *adj.* rasend; **go ~:** durchdrehen *(ugs.)*
berth [bɜ:θ] **1.** *n.* **a) give sb./sth. a wide ~** *(fig.)* einen großen Bogen um jmdn./etw. machen; **b)** *(ship's place at wharf)* Liegeplatz, *der;* **c)** *(sleeping-place) (in ship)* Koje, *die;* Kajütenbett, *das; (in train)* Schlafwagenbett, *das; (in aircraft)* Sleeper, *der.* **2.** *v. t.* festmachen ⟨*Schiff*⟩. **3.** *v. i.* ⟨*Schiff:*⟩ festmachen, anlegen
beseech [bɪ'si:tʃ] *v. t.*, **besought** [bɪ'sɔ:t] *or* **~ed** *(literary)* anflehen ⟨*Person*⟩; **~ sb. to do sth.** jmdn. anflehen *od.* inständig bitten, etw. zu tun
beset [bɪ'set] *v. t.*, **-tt-, beset** heimsuchen; plagen; ⟨*Probleme, Versuchungen:*⟩ bedrängen; **~ by doubts** von Zweifeln geplagt
beside [bɪ'saɪd] *prep.* **a)** *(close to)* neben *(position:* + *Dat.; direction:* + *Akk.);* an (+ *Dat.*); **~ the sea/lake** am Meer/See; **walk ~ the river** am Fluß entlanggehen; **b)** *(compared with)* neben (+ *Dat.*); **c) ~ oneself with joy/grief** außer sich vor Freude/Kummer
besides [bɪ'saɪdz] **1.** *adv.* außerdem; **he was a historian ~:** er war außerdem noch Historiker; **do/say sth. [else] ~:** sonst noch etw. tun/sagen. **2.** *prep.* außer; **~ which, he was late** und obendrein *od.* außerdem kam er zu spät
besiege [bɪ'si:dʒ] *v. t.* belagern
besought *see* **beseech**
bespectacled [bɪ'spektəkld] *adj.* bebrillt
best [best] **1.** *adj. superl. of* **good: a)** best...; **be ~ [of all]** am [allerlbesten sein; **the ~ thing about it** das Beste daran; **the ~ thing to do is to apologize** das beste ist, sich zu entschuldigen; **may the ~ man win!** auf daß der Beste gewinnt!; **b)** *(most advantageous)* best...; günstigst...; **which** *or* **what is the ~ way?** wie ist es am besten *od.* günstigsten?; **think it ~ to do**

sth. es für das beste halten, etw. zu tun; **c)** *(greatest)* **[for] the ~ part of an hour** fast eine ganze Stunde. **2.** *adv. superl. of* ²**well 2** am besten; **like sth. ~ of all** etw. am liebsten mögen; **as ~ we could** so gut wir konnten; **he is the person ~ able to do it** er ist der Fähigste, um das zu tun. **3.** *n.* **a) the ~:** der/die/das Beste; **their latest record is their ~:** ihre letzte Platte ist die beste; **b)** *(clothes)* beste Sachen; Sonntagskleider *Pl.;* **wear one's [Sunday] ~:** seine Sonntagskleider tragen; **c) play the ~ of three [games]** um zwei Gewinnsätze spielen; **get the ~ out of sth./sb.** das Beste aus etw./jmdm. herausholen; **he is not in the ~ of health** es geht ihm nicht sehr gut; **bring out the ~ in sb.** jmds. beste Seiten zum Vorschein bringen; **all the ~!** *(coll.)* alles Gute!; **d) the ~** *pl.* die Besten; **they are the ~ of friends** sie sind die besten Freunde; **with the ~ of intentions** in bester Absicht; **from the ~ of motives** aus den edelsten Motiven [heraus]; **e) at ~:** bestenfalls; **be at one's ~:** in Hochform sein; **[even] at the ~ of times** schon normalerweise; **hope for the ~:** das Beste hoffen; **do one's ~:** sein bestes *od.* möglichstes tun; **do the ~ you can** machen Sie es so gut Sie können; **look one's ~:** möglichst gut aussehen; **make the ~ of it/things** das Beste daraus machen; **make the ~ of a bad job** *or* **bargain** *(coll.)* das Beste daraus machen; **to the ~ of one's ability** nach besten Kräften; **to the ~ of my belief/knowledge** meines Wissens. **4.** *v.t.* *(Sport)* schlagen; *(outwit)* übervorteilen

'best-dressed *attrib. adj.* bestgekleidet

bestial ['bestɪəl] *adj.* *(of or like a beast)* tierisch; *(brutish, barbarous)* barbarisch; *(savage)* brutal; *(depraved)* bestialisch; tierisch

best: ~-kept *attrib. adj.* bestgepflegt; bestgehütet ‹*Geheimnis*›; **the ~-kept village in England** das schönste Dorf Englands; **~-known** *attrib. adj.* bekanntest...; **~-loved** *attrib. adj.* meistgeliebt; **~ 'man** *n.* Trauzeuge, *der (des Bräutigams)*

bestow [bɪ'stəʊ] *v.t.* verleihen ‹*Titel*›, schenken ‹*Wohlwollen, Gunst*›, zuteil werden lassen ‹*Ehre, Segnungen*› (**[up]on** *Dat.*)

best: ~ 'seller *n.* Bestseller, *der;* *(author)* Bestsellerautor, *der;* **~-selling** *attrib. adj.* meistverkauft ‹*Schallplatte*›; **~-selling book/novel** Bestseller, *der;* **a ~-selling author/novelist** ein Bestsellerautor

bet [bet] **1.** *v.t.*, **-tt-, ~** *or* **~ted a)** wetten; **I ~ him £10** ich habe mit ihm um 10 Pfund gewettet; **he ~ £10 on that horse** er hat 10 Pfund auf das Pferd gesetzt; **b)** *(coll.: be confident)* wetten; **I ~ he's late** wetten, daß er zu spät kommt?; **~ [you] I 'can** und ob ich kann; **you '~ [I am/he will** *etc.*] und ob; allerdings. **2.** *v.i.*, **-tt-, ~** *or* **~ted** wetten; **~ on sth.** auf etw. *(Akk.)* setzen; **[do you] want to ~?** [wollen wir] wetten? **3.** *n.* **a)** Wette, *die;* *(sum)* Wetteinsatz, *der;* **make** *or* **have a ~ with sb. on sth.** mit jmdm. über etw. *(Akk.)* wetten; **b)** *(fig. coll.: choice)* Tip, *der;* **be a bad/good/safe ~:** ein schlechter/guter/sicherer Tip sein; **be sb.'s best ~:** das beste sein; **my ~ is that ...:** ich wette, daß ...

beta ['bi:tə] *n.* *(letter)* Beta, *das*

betide [bɪ'taɪd] *v.t.* **woe ~ you if ...:** wehe dir, wenn ...

betoken [bɪ'təʊkn] *v.t.* ankündigen ‹*Frühjahr, Krieg*›

betray [bɪ'treɪ] *v.t.* verraten (**to** an + *Akk.*); mißbrauchen ‹*jmds. Vertrauen*›; **~ the fact that ...:** verraten, daß ...

betrayal [bɪ'treɪəl] *n.* Verrat, *der;* **an act of ~:** ein Verrat

betrothal [bɪ'trəʊðl] *n.* *(arch.)* Verlöbnis, *das*

betrothed [bɪ'trəʊðd] *(arch.)* **1.** *adj.* versprochen *(veralt.)* (**to** *Dat.*). **2.** *n.* Anverlobter, *der*/Anverlobte, *die (veralt.)*

better ['betə(r)] **1.** *adj. compar. of* **good 1** besser; **something ~:** etwas Besseres; **do you know of anything ~?** kennst du etwas Besseres?; **that's ~:** so ist's schon besser; **~ and ~:** immer besser; **~ still, let's phone** oder noch besser: Rufen wir doch an; **be much ~** *(recovered)* sich viel besser fühlen; **he is much ~ today** es geht ihm heute schon viel besser; **get ~** *(recover)* gesund werden; **be getting ~:** auf dem Wege der Besserung sein; **I am/my ankle is getting ~:** mir/meinem Knöchel geht es besser; **so much the ~:** um so besser; **she is none/much the ~ for it** das hat ihr nichts/sehr genützt; **my/his ~ half** *(joc.)* meine/seine bessere Hälfte *(scherzh.);* **[for] the ~ part of an hour** fast eine ganze Stunde; *see also* **all 3. 2.** *adv. compar. of* ²**well 2: a)** *(in a ~ way)* besser; **b)** *(to a greater degree)* mehr; **I like Goethe ~ than Schiller** ich mag Goethe lieber als Schiller; **he is ~ liked than Carter** er ist beliebter als Carter; **c) you ought to know ~ than to ...:** du solltest es besser wissen und nicht ...; **you'd ~ not tell her** Sie erzählen es ihr besser nicht; **I'd ~ be off now** ich gehe jetzt besser; **hadn't you ~ ask first?** sollten Sie nicht besser zuerst fragen?; **you'd ~!** das will ich aber auch hoffen; *see also* **better off. 3.** *n.* **a)** Bessere, *das;* **get the ~ of sb./sth.** jmdn./etw. unterkriegen *(ugs.);* **exhaustion got the ~ of him** Erschöpfung übermannte ihn; **be a change for the ~:** eine vorteilhafte Veränderung sein; **for ~ or for worse** was immer daraus werden wird; **I thought ~ of it** ich habe es mir anders überlegt; **b)** *in pl.* **one's ~s** Leute, die über einem stehen *od.* die einem überlegen sind. **4.** *v.t.* **a)** *(surpass)* übertreffen; **b)** *(improve)* verbessern; **~ oneself** *(rise socially)* sich verbessern

better: ~ 'off *adj.* **a)** *(financially)* [finanziell] besser gestellt; **b) he is ~ off than I am** ihm geht es besser als mir; **be ~ off than sb.** besser als jmd. dran sein *(ugs.);* **be ~ off without sth./sb.** ohne etw./jmdn. besser dran sein; **~-than-average** *attrib. adj.* überdurchschnittlich [gut/viel]

betting ['betɪŋ] **1.** *n.* Wetten, *das;* **there was heavy ~ on that horse** auf das Pferd wurde sehr viel gesetzt; **what's the ~ it rains?** *(fig.)* ob es wohl regnen wird? **2.** *attrib. adj.* Wett-; **I'm not a ~ man** ich wette nicht

betting: ~ office, ~-shop *ns.* Wettbüro, *das*

between [bɪ'twi:n] **1.** *prep.* **a)** zwischen *(position: + Dat., direction: + Akk.);* **~ then and now** zwischen damals und jetzt; **[in] ~:** zwischen; **b)** *(amongst)* unter (+ *Dat.*); **the work was divided ~ the volunteers** die Arbeit wurde zwischen den Freiwilligen aufgeteilt; **~ us we had 40p** wir hatten zusammen 40 Pence; **~ ourselves, ~ you and me** unter uns *(Dat.)* gesagt; **that's [just] ~ ourselves** das bleibt aber unter uns *(Dat.);* **c)** *(by joint action of)* **~ them/the four of them they dislodged the stone** gemeinsam/zu viert lösten sie den Stein. **2.** *adv.* **[in] ~:** dazwischen; *(in time)* zwischendurch; **the space ~:** der Zwischenraum

bevel ['bevl] **1.** *n.* *(slope)* Schräge, *die;* **~ edge** Schrägkante, *die.* **2.** *v.t.*, *(Brit.)* **-ll-** abschrägen

beverage ['bevərɪdʒ] *n. (formal)* Getränk, *das*

bewail [bɪ'weɪl] *v.t.* beklagen; *(lament)* bejammern

beware [bɪ'weə(r)] *v.t. & i.; only in imper. and inf.* ~ **[of] sb./sth.** sich vor jmdm./etw. hüten *od.* in acht nehmen; ~ **of doing sth.** sich davor hüten, etw. zu tun; **'~ of pickpockets'** „vor Taschendieben wird gewarnt": **'~ of the dog'** „Vorsicht, bissiger Hund!"

bewilder [bɪ'wɪldə(r)] *v.t.* verwirren; **be ~ed by sth.** durch *od.* von etw. verwirrt werden/sein

bewildering [bɪ'wɪldərɪŋ] *adj.* verwirrend

bewilderment [bɪ'wɪldəmənt] *n., no pl.* Verwirrung, *die;* **in total ~:** völlig verwirrt

bewitch [bɪ'wɪtʃ] *v.t.* verzaubern; verhexen; *(fig.)* bezaubern

bewitching [bɪ'wɪtʃɪŋ] *adj.* bezaubernd

beyond [bɪ'jɒnd] **1.** *adv.* **a)** *(in space)* jenseits; *(on other side of wall, mountain range, etc.)* dahinter; **the world ~:** das Jenseits; **b)** *(in time)* darüber hinaus. **2.** *prep.* **a)** *(at far side of)* jenseits (+ *Gen.*); **when we get ~ the river, we'll stop** wenn wir den Fluß überquert haben, machen wir halt; **b)** *(in space: after)* nach; **c)** *(later than)* nach; **she never looks** *or* **sees ~ the present** sie sieht *od.* blickt nie über die Gegenwart hinaus; **d)** *(out of reach or comprehension or range)* über ... (+ *Akk.*) hinaus; **it's [far** *or (coll.)* **way] ~ me/him** *etc. (too difficult)* das ist mir/ihm *usw.* [bei weitem] zu schwer; *(incomprehensible)* das ist mir/ihm *usw.* [völlig] unverständlich; ~ **reproach** tadellos; **e)** *(surpassing, exceeding)* mehr als; **they're living ~ their means** sie leben über ihre Verhältnisse; **f)** *(more than)* weiter als; **g)** *(besides)* außer; ~ **this/that** weiter. **3.** *n.* **the B~:** das Jenseits; **at the back of ~:** am Ende der Welt

bias ['baɪəs] **1.** *n.* **a)** *(tendency)* Neigung, *die;* **have a ~ towards** *or* **in favour of sth./sb.** etw./jmdn. bevorzugen; **have a ~ against sth./sb.** gegen etw./jmdn. eingenommen sein; **b)** *(prejudice)* Voreingenommenheit, *die;* **be without ~:** unvoreingenommen sein. **2.** *v.t.,* **-s-** *or* **-ss-** beeinflussen; **be ~ed towards** *or* **in favour of sth./sb.** für etw./jmdn. eingestellt sein; **they are ~ed in favour of women** sie bevorzugen Frauen; **be ~ed against sth./sb.** gegen etw./jmdn. voreingenommen sein; **a ~ed account** eine gefärbte *od.* tendenziöse Darstellung

bib [bɪb] *n.* **a)** *(for baby)* Lätzchen, *das;* **b)** *(of apron etc.)* Latz, *der*

Bible ['baɪbl] *n.* **a)** *(Christian)* Bibel, *die;* **b)** *(of other religion)* heiliges Buch; *(fig.: authoritative book)* Bibel, *die*

biblical ['bɪblɪkl] *adj.* biblisch; Bibel-

bibliography [bɪblɪ'ɒgrəfɪ] *n.* Bibliographie, *die*

bicarbonate [baɪ'kɑ:bəneɪt] *n. (Cookery)* ~ **[of soda]** Natron, *das*

bicentenary [baɪsen'ti:nərɪ, baɪsen'tenərɪ], **bicentennial** [baɪsen'tenɪəl] **1.** *adjs.* Zweihundertjahr-. **2.** *ns.* Zweihundertjahrfeier, *die*

biceps ['baɪseps] *n. (Anat.)* Bizeps, *der*

bicker ['bɪkə(r)] *v.i.* ~ **[with sb. about** *or* **over sth.]** [sich mit jmdm. um etw.] zanken *od.* streiten

bicycle ['baɪsɪkl] **1.** *n.* **a)** Fahrrad, *das;* **ride a ~:** [mit dem] Fahrrad fahren; radfahren; **by ~:** mit dem [Fahr]rad; **b)** *(attrib.)* Fahrrad-; ~ **clip/rack** Hosenklammer, *die*/Fahrradständer, *der.* **2.** *v.i.* radfahren

bid [bɪd] **1.** *v.t.* **a) -dd-, bid** *(at auction)* bieten; **b) -dd-, bid** *(Cards)* reizen; **c) -dd-, bade** [bæd] *or* **bid, bidden** ['bɪdn] *or* **bid: ~ sb. welcome** jmdn. willkommen heißen; ~ **sb. goodbye** sich von jmdm. verabschieden. **2.** *v.i.,* **-dd-, bid a)** werben **(for** um); **the President is ~ding for re-election** der Präsident bewirbt sich um die Wiederwahl; ~ **fair to be sth.** etw. zu werden versprechen; **b)** *(at auction)* bieten; **c)** *(Cards)* bieten; reizen. **3.** *n.* **a)** *(at auction)* Gebot, *das;* **b)** *(attempt)* Bemühung, *die;* **make a ~ for sth.** sich um etw. bemühen; **he made a strong ~ for the Presidency** er griff nach dem Präsidentenamt; **the prisoner made a ~ for freedom** der Gefangene versuchte, die Freiheit zu erlangen; **c)** *(Cards)* Ansage, *die;* **make no ~:** passen; **it's your ~:** Sie bieten!

bidden *see* **bid 1c**

bidder ['bɪdə(r)] *n.* Bieter, *der*/Bieterin, *die;* **the highest ~:** der/die Höchstbietende

bidding ['bɪdɪŋ] *n.* **a)** *(at auction)* Steigern, *das;* Bieten, *das;* **open the ~:** das erste Gebot machen; **b)** *(Cards)* Bieten, *das;* Reizen, *das*

bide [baɪd] *v.t.* ~ **one's time** den rechten Augenblick abwarten

bidet ['bi:deɪ] *n.* Bidet, *das*

biennial [baɪ'enɪəl] **1.** *adj.* **a)** *(lasting two years)* zweijährig; **b)** *(once every two years)* zweijährlich. **2.** *n. (Bot.)* zweijährige Pflanze

bier [bɪə(r)] *n.* Totenbahre, *die*

biff [bɪf] *(sl.)* **1.** *n.* Klaps, *der (ugs.).* **2.** *v.t.* hauen; **he ~ed me on the head with a book** er hat mir ein Buch auf den Kopf geknallt *(ugs.)*

bifocal [baɪ'fəʊkl] **1.** *adj.* Bifokal-. **2.** *n. in pl.* Bifokalgläser *Pl.*

big [bɪg] **1.** *adj.* **a)** *(in size)* groß; schwer, heftig ‹*Explosion, Zusammenstoß*›; schwer ‹*Unfall, Niederlage*›; hart ‹*Konkurrenz*›; reichlich ‹*Mahlzeit*›; **earn ~ money** das große Geld verdienen; **he is a ~ man/she is a ~ woman** *(fat)* er/sie ist wohlbeleibt; ~ **words** geschraubte Ausdrücke (*see also* **g**); **in a ~ way** *(coll.)* im großen Stil; **b)** *(of largest size, larger than usual)* groß ‹*Appetit, Zehe, Buchstabe*›; **c) ~ger** *(worse)* schwerer; **~gest** *(worst)* größt...; **he is the ~gest liar/idiot** er ist der größte Lügner/Idiot; **d)** *(grown up, elder)* groß; **e)** *(important)* groß; wichtig ‹*Nachricht, Entscheidung*›; **f)** *(coll.: outstanding)* groß ‹*Augenblick, Chance*›; **g)** *(boastful)* **get** *or* **grow/be too ~ for one's boots** *(coll.)* größenwahnsinnig werden/sein *(ugs.);* ~ **talk** Großsprecherei, *die;* ~ **words** große Worte (*see also* **a**); **h)** *(coll.: generous)* großzügig; nobel *(oft iron.);* **i)** *(coll.: popular)* **be ~** ‹*Schauspieler, Popstar:*› gut ankommen. *See also* **idea d. 2.** *adv.* **talk ~:** groß daherreden *(ugs.);* **think ~:** im großen Stil planen

bigamist ['bɪgəmɪst] *n.* Bigamist, *der*/Bigamistin, *die*

bigamy ['bɪgəmɪ] *n.* Bigamie, *die*

big: ~ **'bang** *n.* Urknall, *der;* **Big 'Brother** *n.* der Große Bruder; ~ **'business** *n.* das Großkapital; ~ **'deal** *see* [1]**deal 3a;** ~ **'dipper** *n. (Brit.)* Achterbahn, *die;* ~ **game** *n.* Großwild, *das;* **~-game hunting** Großwildjagd, *die;* **~head** *n. (coll.)* Fatzke, *der (ugs. abwertend);* **~-'headed** *adj. (coll.)* eingebildet; **~-'hearted** *adj.* großherzig; ~ **mouth** *n. (fig. coll.)* **a)** ['-'-] **have a ~ mouth** ein Schwätzer/eine Schwätzerin sein *(ugs.);* **b)** ['--] **be a ~ mouth** ein Angeber/eine Angeberin sein *(ugs.);* ~ **'name** *n. (person)* Größe, *die;* ~ **'noise** *n. (sl.)* hohes Tier *(ugs.)*

bigot ['bɪgət] *n.* Eiferer, *der*/Eiferin, *die; (Relig.)* bigotter Mensch

bigoted ['bɪgətɪd] *adj.* eifernd; *(Relig.)* bigott

big: ~ **shot** *see* ~ **noise;** ~ **time** *n.* **be in the** ~ **time** *(coll.)* eine große Nummer sein *(ugs.)*; **make it [in]to** *or* **hit the** ~ **time** *(sl.)* groß herauskommen *(ugs.)*; ~ **'top** *n.* Zirkuszelt, *das;* ~ **'wheel** *n.* **a)** *(at fair)* Riesenrad, *das;* **b)** *(sl.: person)* hohes Tier *(ugs.)*; ~**wig** *n. (coll.)* hohes Tier *(ugs.)*

bike [baɪk] *(coll.)* **1.** *n. (bicycle)* Rad, *das; (motor cycle)* Maschine, *die.* **2.** *v.i. (by bicycle)* radfahren; radeln *(ugs., bes. südd.)*; mit dem Fahrrad fahren; *(by motor cycle)* [mit dem] Motorrad fahren

bikini [bɪ'ki:nɪ] *n.* Bikini, *der;* ~ **briefs** Slip, *der*

bilateral [baɪ'lætərl] *adj.* bilateral

bilberry ['bɪlbərɪ] *n.* Blau-, Heidelbeere, *die*

bile [baɪl] *n. (Physiol.)* Gallenflüssigkeit, *die*

bilingual [baɪ'lɪŋgwəl] *adj.* zweisprachig

bilious ['bɪljəs] *adj. (Med.)* Gallen-; *(fig.: peevish)* verdrießlich; ~ **attack** Gallenanfall, *der*

¹bill [bɪl] **1.** *n. (of bird)* Schnabel, *der.* **2.** *v.i.* ‹*Vögel:*› schnäbeln; ‹*Personen:*› sich liebkosen; ~ **and coo** ‹*Vögel:*› schnäbeln und gurren; ‹*Personen:*› [miteinander] turteln

²bill 1. *n.* **a)** *(Parl.)* Gesetzentwurf, *der;* Gesetzesvorlage, *die;* **b)** *(note of charges)* Rechnung, *die;* **could we have the** ~**, please?** wir möchten zahlen; **a** ~ **for £10** eine Rechnung über 10 Pfund *(Akk.)*; *(amount)* **a large** ~**:** eine hohe Rechnung; **a** ~ **of £10** eine Rechnung von 10 Pfund; **c)** *(poster)* Plakat, *das;* **'[stick] no** ~**s'** „Plakate ankleben verboten"; **d)** ~ **of fare** Speisekarte, *die;* **e)** *(Amer.: banknote)* Banknote, *die;* [Geld]schein, *der;* **f)** *(Commerc.)* ~ **[of exchange]** Wechsel, *der;* Tratte, *die (fachspr.)*; ~ **of lading** Konnossement, *das;* Seefrachtbrief, *der.* **2.** *v.t.* **a)** *(announce)* ankündigen; **b)** *(charge)* eine Rechnung ausstellen (+ *Dat.*); ~ **sb. for sth.** jmdm. etw. in Rechnung stellen *od.* berechnen

'billboard *n.* Reklametafel, *die*

billet ['bɪlɪt] **1.** *n.* Quartier, *das;* Unterkunft, *die; (for soldiers)* Truppenunterkunft, *die.* **2.** *v.t.* unterbringen, einquartieren (**with, on** bei; **in** in + *Dat.*)

'billfold *n. (Amer.)* Brieftasche, *die*

billiard ['bɪljəd]: ~**-ball** *n.* Billardkugel, *die;* ~**-cue** *n.* Queue, *das;* Billardstock, *der;* ~**-player** *n.* Billardspieler, *der;* ~**-room** *n.* Billardzimmer, *das*

billiards ['bɪljədz] *n.* Billard[spiel], *das;* **a game of** ~**:** eine Partie Billard

'billiard-table *n.* Billardtisch, *der*

billion ['bɪljən] *n.* **a)** *(thousand million)* Milliarde, *die;* **b)** *(Brit.: million million)* Billion, *die*

billow ['bɪləʊ] **1.** *n.* ~ **of smoke** Rauchwolke, *die;* ~ **of fog** Nebelschwaden, *der.* **2.** *v.i.* ‹*Ballon, Segel:*› sich [auf]blähen; ‹*See, Meer:*› wogen, sich [auf]türmen; ‹*Rauch:*› in Schwaden aufsteigen; ‹*Kleid, Vorhang:*› sich bauschen

billy-goat ['bɪlɪgəʊt] *n.* Ziegenbock, *der*

bin [bɪn] *n.* **a)** *(for storage)* Behälter, *der; (for bread)* Brotkasten, *der;* **b)** *(for rubbish) (inside house)* Abfalleimer, *der;* Mülleimer, *der; (outside house)* Mülltonne, *die; (in public place)* Abfallkorb, *der*

binary ['baɪnərɪ] *adj.* binär

bind [baɪnd] **1.** *v.t.*, **bound** [baʊnd] **a)** *(tie)* fesseln ‹*Person, Tier*›; *(bandage)* wickeln, binden ‹*Glied*›; verbinden ‹*Wunde*› **(with** mit); **he was bound hand and foot** er war/wurde an Händen und Füßen gefesselt; **b)** *(fasten together)* zusammenbinden; *(fig.: unite)* verbinden; **c)** *(Bookb.)* binden; **d) be bound up with sth.** *(fig.)* eng mit etw. verbunden sein; **e)** *(oblige)* ~ **sb./oneself to sth.** jmdn./sich an etw. *(Akk.)* binden; **be bound to do sth.** *(required)* verpflichtet sein, etw. zu tun; **be bound by law** von Gesetzes wegen verpflichtet sein; **f) be bound to do sth.** *(certain)* etw. ganz bestimmt tun; **it is bound to rain** es wird bestimmt *od.* sicherlich regnen; **g) I'm bound to say that ...** *(feel obliged)* ich muß schon sagen, daß ...; **h)** *(Cookery)* binden; **i)** *(Law)* ~ **sb. over [to keep the peace]** jmdn. verwarnen *od.* rechtlich verpflichten[, die öffentliche Ordnung zu wahren]. **2.** *v.i.*, **bound a)** *(cohere)* binden; ‹*Lehm, Ton:*› fest *od.* hart werden; ‹*Zement:*› abbinden; **b)** *(be restricted)* blockieren; ‹*Kolben:*› sich festfressen. **3.** *n.* **a)** *(coll.: nuisance)* **be a** ~**:** recht lästig sein; **what a** ~**!** wie unangenehm *od.* lästig!; **b) be in a** ~ *(Amer. sl.)* in einer Klemme sitzen *(ugs.)*

binder ['baɪndə(r)] *n.* **a)** *(substance)* Bindemittel, *das;* Binder, *der;* **b)** *(book~)* Buchbinder, *der/*-binderin, *die;* **c)** *(cover) (for papers)* Hefter, *der; (for magazines)* Mappe, *die*

binding ['baɪndɪŋ] **1.** *adj.* bindend, verbindlich ‹*Vertrag, Abkommen*› **(on** für). **2.** *n.* **a)** *(cover of book)* [Buch]einband, *der;* **b)** *(on ski)* Bindung, *die*

bindweed ['baɪndwi:d] *n. (Bot.)* Winde, *die*

binge [bɪndʒ] *n. (sl.: drinking bout)* Sauferei, *die (salopp)*; **go/be out on a** ~**:** auf Sauftour gehen/sein *(salopp)*

bingo ['bɪŋgəʊ] **1.** *n., no pl.* Bingo, *das; attrib.* ~ **hall** Bingohalle, *die.* **2.** *int.* peng; zack

'bin-liner *n.* Müllbeutel, *der*

binoculars [bɪ'nɒkjʊləz] *n. pl.* **[pair of]** ~**:** Fernglas, *das;* Binokular, *das*

bio- ['baɪəʊ] *in comb.* Bio-; Lebens-

bio'chemical *adj.* biochemisch

bio'chemistry *n.* Biochemie, *die*

biodegradable [baɪəʊdɪ'greɪdəbl] *adj.* biologisch abbaubar

biographer [baɪ'ɒgrəfə(r)] *n.* Biograph, *der*/Biographin, *die*

biographic [baɪə'græfɪk], **biographical** [baɪə'græfɪkl] *adj.* biographisch

biography [baɪ'ɒgrəfɪ] *n.* Biographie, *die; (branch of literature)* biographische Literatur

biological [baɪə'lɒdʒɪkl] *adj.* biologisch

biological 'warfare *n.* biologische Kriegführung; Bakterienkrieg, *der*

biologist [baɪ'ɒlədʒɪst] *n.* Biologe, *der*/Biologin, *die*

biology [baɪ'ɒlədʒɪ] *n.* Biologie, *die*

biotech'nology *n.* Biotechnik, *die*

bipartite [baɪ'pɑ:taɪt] *adj. (having two parts)* zweiteilig; *(involving two parties)* zweiseitig ‹*Dokument, Abkommen*›

biplane ['baɪpleɪn] *n.* Doppeldecker, *der*

birch [bɜ:tʃ] *n.* **a)** *(tree)* Birke, *die;* **b)** *(for punishment)* [Birken]rute, *die*

bird [bɜ:d] *n.* **a)** Vogel, *der;* ~**s of a feather flock together** *(prov.)* gleich und gleich gesellt sich gern *(Spr.)*; **it's [strictly] for the** ~**s** *(sl.)* das kannste vergessen *(salopp)*; **kill two** ~**s with one stone** *(fig.)* zwei Fliegen mit einer Klappe schlagen; **a** ~ **in the hand is worth two in the bush** *(prov.)* ein Spatz in der Hand ist besser als eine Taube auf dem Dach *(Spr.)*; **a little** ~ **told me** mein kleiner Finger sagt mir das; **b)** *(sl.: girl)* Mieze, *die (salopp);* **c)** *no art. (sl.: imprisonment)* Knast, *der (ugs.);* **do ~:** Knast schieben *(salopp). See also* **early bird**

bird: ~-bath *n.* Vogelbad, *das;* **~cage** *n.* Vogelkäfig, *der;* Vogelbauer, *das od. der;* **~-call** *n.* Vogelruf, *der*

birdie ['bɜ:dɪ] *n.* Vögelchen, *das*

bird: ~ sanctuary *n.* Vogelschutzgebiet, *das;* **~'s-eye 'view** *n.* Vogelperspektive, *die;* **have/get a ~'s-eye view of sth.** *(lit. or fig.)* etw. aus der Vogelperspektive sehen; **~'s nest** *n.* Vogelnest, *das;* **~-table** *n.* Futterstelle für Vögel; **~-watcher** *n.* Vogelbeobachter, *der/*-beobachterin, *die;* **~-watching** *n., no pl., no indef. art.* das Beobachten von Vögeln

Biro, (P) ['baɪrəʊ] *n., pl.* **~s** Kugelschreiber, *der;* Kuli, *der (ugs.)*

birth [bɜ:θ] *n.* **a)** Geburt, *die;* **at the/at ~:** bei der Geburt; **[deaf] from *or* since ~:** von Geburt an [taub]; **date and place of ~:** Geburtsdatum und -ort; **give ~** ⟨*Frau:*⟩ entbinden; ⟨*Tier:*⟩ jungen; **she gave ~ prematurely** sie hatte eine Frühgeburt; **give ~ to** zur Welt bringen; **b)** *(of movement, fashion, etc.)* Aufkommen, *das; (of party, company)* Gründung, *die; (of nation, idea)* Geburt, *die; (of new era)* Anbruch, *der;* Geburt, *die;* **give ~ to sth.** etw. entstehen lassen; **c)** *(parentage)* Geburt, *die;* Abkunft, *die (geh.);* **of humble ~:** von niedriger Abstammung; **of high ~:** von hoher Geburt; [von] edler Abkunft *(geh.);* **be a German by ~:** [ein] gebürtiger Deutscher/[eine] gebürtige Deutsche sein

birth: ~ certificate *n.* Geburtsurkunde, *die;* **~ control** *n.* Geburtenkontrolle *od.* -regelung, *die;* **~day** *n.* Geburtstag, *der; attrib.* Geburtstags⟨*karte, -feier, -geschenk*⟩; **when is your ~day?** wann haben Sie Geburtstag?; **[be] in his/her ~day suit** im Adams-/Evakostüm [sein]; **~mark** *n.* Muttermal, *das;* **~place** *n.* Geburtsort, *der; (house)* Geburtshaus, *das;* **~ rate** *n.* Geburtenrate *od.* -ziffer, *die;* **~right** *n.* Geburtsrecht, *das*

biscuit ['bɪskɪt] **1.** *n.* **a)** *(Brit.)* Keks, *der;* **coffee and ~s** Kaffee und Gebäck; **~ tin** Keksdose, *die;* **b)** *(Amer.: roll)* [weiches] Brötchen; **c)** *(colour)* Beige, *das. See also* **take 1 c. 2.** *adj.* beige

bisect [baɪ'sekt] *v.t. (into halves)* in zwei Hälften teilen; halbieren; *(into two)* in zwei Teile teilen

bisexual [baɪ'seksjʊəl] **1.** *adj.* **a)** *(Biol.)* zwittrig; doppelgeschlechtig; **b)** *(attracted by both sexes)* bisexuell. **2.** *n.* Bisexuelle, *der/die*

bishop ['bɪʃəp] *n.* **a)** *(Eccl.)* Bischof, *der;* **b)** *(Chess)* Läufer, *der*

bison ['baɪsn] *n. (Zool.)* **a)** *(Amer.: buffalo)* Bison, *der;* **b)** *(European)* Wisent, *der*

¹bit [bɪt] *n.* **a)** *(for horse)* Gebiß, *das;* Gebißstange, *die;* **take the ~ between one's teeth** *(fig.)* aufmüpfig werden *(ugs.);* **b)** *(of drill)* [Bohr]einsatz, *der;* Bohrer, *der*

²bit *n.* **a)** *(piece)* Stück, *das; (smaller)* Stückchen, *das;* **a little ~:** ein kleines Stückchen; **a ~ of cheese/sugar/wood/coal** ein bißchen *od.* etwas Käse/Zucker/ein Stück Holz/etwas Kohle; **a ~ of trouble/luck** ein wenig Ärger/Glück; **the best ~s** die besten Teile; **it cost quite a ~:** es kostete ziemlich viel; **~ by ~:** Stück für Stück; *(gradually)* nach und nach; **smashed to ~s** in tausend Stücke zersprungen; **~s and pieces** Verschiedenes; **do one's ~:** seinen Teil tun; **b) a ~** *(somewhat):* **a ~ tired/too early** ein bißchen müde/zu früh; **a little ~, just a ~:** ein klein bißchen; **quite a ~:** um einiges ⟨*besser, stärker, hoffnungsvoller*⟩; **c) a ~ of** *(rather):* **be a ~ of a coward/bully** ein ziemlicher Feigling sein/den starken Mann markieren *(ugs.);* **a ~ of a disappointment** eine ganz schöne Enttäuschung; **d)** *(short time)* **[for] a ~:** eine Weile; **wait a ~ longer** noch ein Weilchen warten; **e)** *(short distance)* **a ~:** ein Stückchen; **a ~ closer** ein bißchen näher; **f)** *(Amer.)* **two/four/six ~s** 25/50/75 Cent

³bit *n. (Computing)* Bit, *das*

⁴bit *see* **bite 1, 2**

bitch [bɪtʃ] **1.** *n.* **a)** *(dog)* Hündin, *die;* **b)** *(sl. derog.: woman)* Miststück, *das.* **2.** *v.i. (coll.)* meckern *(ugs.)* **(about** über + *Akk.*)

bite [baɪt] **1.** *v.t.,* **bit** [bɪt], **bitten** ['bɪtn] beißen; *(sting)* ⟨*Moskito usw.:*⟩ stechen; **~ one's nails** an den Nägeln kauen; *(fig.)* wie auf Kohlen sitzen; **~ one's lip** *(lit. or fig.)* sich *(Dat.)* auf die Lippen beißen; **he won't ~ you** *(fig. coll.)* er wird dich schon nicht beißen; **~ the hand that feeds one** *(fig.)* sich [seinem Gönner gegenüber] undankbar zeigen; **~ the dust** *(fig.)* daran glauben müssen *(ugs.);* **what's biting *or* bitten you?** *(fig. coll.)* was ist mit dir los?; was hast du denn? **2.** *v.i.,* **bit, bitten a)** beißen; *(sting)* stechen; ⟨*Rad:*⟩ fassen, greifen; ⟨*Schraube:*⟩ fassen; *(take bait, lit. or fig.)* anbeißen; **b)** *(have an effect)* sich auswirken; greifen. **3.** *n.* **a)** *(act)* Biß, *der; (piece)* Bissen, *der; (wound)* Bißwunde, *die; (by mosquito etc.)* Stich, *der;* **he took a ~ of the apple** er biß in den Apfel; **can I have a ~?** darf ich mal [ab]beißen?; **b)** *(taking of bait)* [An]beißen, *das;* **I haven't had a ~ all day** es hat den ganzen Tag noch keiner angebissen; **c)** *(food)* Happen, *der;* Bissen, *der;* **I haven't had a ~ [to eat] since breakfast** ich habe seit dem Frühstück nichts mehr gegessen; **have a ~ to eat** eine Kleinigkeit essen; **d)** *(incisiveness)* Bissigkeit, *die;* Schärfe, *die*

~ 'off *v.t.* abbeißen; **the dog bit off the man's ear** der Hund hat dem Mann ein Ohr abgebissen; **~ sb.'s head off** *(fig.)* jmdm. den Kopf abreißen; **~ off more than one can chew** *(fig.)* sich *(Dat.)* zuviel zumuten; sich übernehmen

biting ['baɪtɪŋ] *adj. (stinging)* beißend; schneidend ⟨*Kälte, Wind*⟩; *(sarcastic)* scharf ⟨*Angriff, Worte*⟩; beißend ⟨*Kritik*⟩; bissig ⟨*Bemerkung, Kommentar*⟩

bitten *see* **bite 1, 2**

bitter ['bɪtə(r)] **1.** *adj.* **a)** bitter; **~ lemon** *(drink)* Bitter lemon, *das;* **b)** *(fig.)* scharf, heftig ⟨*Antwort, Bemerkung, Angriff*⟩; bitter ⟨*Kampf, Kälte, Enttäuschung, Tränen*⟩; verbittert ⟨*Person*⟩; erbittert ⟨*Feind*⟩; scharf, bitterkalt ⟨*Wind, Wetter*⟩; streng ⟨*Winter*⟩; **to the ~ end** bis zum bitteren Ende; **be/feel ~ [about sth.]** [über etw. *(Akk.)*] bitter *od.* verbittert sein. **2.** *n. (Brit.)* bitteres Bier *(halbdunkles, obergäriges Bier)*

bitterly ['bɪtəlɪ] *adv.* bitterlich ⟨*weinen, sich beschweren*⟩; bitter ⟨*erwidern*⟩; erbittert ⟨*kämpfen, sich widersetzen*⟩; scharf ⟨*kritisieren*⟩; **~ cold** bitterkalt; **be ~ opposed to sth.** ein erbitterter Gegner einer Sache *(Gen.)* sein

bitterness ['bɪtənɪs] *n., no pl. see* **bitter 1:** Bitterkeit, *die;* Schärfe, *die;* Heftigkeit, *die;* Verbitterung, *die;* bittere Kälte

bitter-'sweet *adj. (lit. or fig.)* bittersüß

bitty ['bɪtɪ] *adj.* zusammengestoppelt *(abwertend)*

bitumen ['bɪtjʊmən] *n.* Bitumen, *das*

bivouac ['bɪvʊæk] **1.** *n.* Biwak, *das;* Lager, *das.* **2.** *v.i.,* **-ck-** biwakieren; im Freien übernachten

bizarre [bɪˈzɑː(r)] *adj.* bizarr; *(eccentric)* exzentrisch
blab [blæb] *v.i., -bb- (coll.)* quatschen *(abwertend)*
black [blæk] **1.** *adj.* **a)** schwarz; *(very dark)* dunkel; **b) B~** *(dark-skinned)* schwarz; **B~ man/woman/child** Schwarze, *der/* Schwarze, *die/*schwarzes Kind; **B~ people** Schwarze *Pl.*; **B~ Africa** Schwarzafrika *(das)*; **c)** *(looking gloomy)* düster; **things look ~:** es sieht böse *od.* düster aus; **d)** *(wicked)* schwarz ‹*Gedanken*›; **he is not as ~ as he is painted** er ist nicht so schlecht, wie er dargestellt wird; **give sb. a ~ look** jmdn. finster ansehen; **e)** *(dismal)* **a ~ day** ein schwarzer Tag; **f)** *(macabre)* schwarz ‹*Witz, Humor*›. **2.** *n.* **a)** *(colour)* Schwarz, *das;* **b) B~** *(person)* Schwarze, *der/die;* **c)** *(credit)* **[be] in the ~:** in den schwarzen Zahlen [sein]. **3.** *v.t.* **a)** *(blacken)* schwärzen; **~ sb.'s eye** jmdm. ein blaues Auge machen; **b)** *(boycott)* bestreiken ‹*Betrieb*›; boykottieren ‹*Arbeit*›
~ 'out 1. *v.t.* verdunkeln. **2.** *v.i.* das Bewußtsein verlieren
black: ~ and 'blue *pred. adj.* grün und blau; **~ and 'white 1.** *pred. adj. (in writing)* schwarz auf weiß; *(Cinemat., Photog., etc.)* schwarzweiß; *(fig.: comprising only opposite extremes)* Schwarzweiß-; **2.** *n.* **[sth. is there/down] in ~ and white** *(in writing)* [etw. steht] schwarz auf weiß [geschrieben]; **this film is in ~ and white** dieser Film ist in Schwarzweiß; **see/portray** *etc.* **things in ~ and white** *(fig.)* schwarzweiß malen; **~-and-white** *attrib. adj.* Schwarzweiß-; **~berry** [ˈblækbərɪ] *n.* Brombeere, *die;* **go ~berrying** Brombeeren pflücken gehen; **~bird** *n.* Amsel, *die;* **~board** *n.* [Wand]tafel, *die;* **~ 'books** *n. pl.* **be in sb.'s ~ books** bei jmdm. schlecht angeschrieben sein; **~ 'box** *n. (flight recorder)* Flugschreiber, *der;* **~ 'bread** *n.* Schwarzbrot, *das;* **B~ Country** *n. (Brit.) Industriegebiet von Staffordshire und Warwickshire;* **~'currant** *n.* schwarze Johannisbeere; **B~ 'Death** *n.* Schwarzer Tod; **~ e'conomy** *n.* Schattenwirtschaft, *die*
blacken [ˈblækn] *v.t.* **a)** *(make dark[er])* verfinstern ‹*Himmel*›; *(make black[er])* schwärzen; **b)** *(fig.: defame)* verunglimpfen; **~ sb.'s [good] name** jmds. [guten] Namen beschmutzen
black: ~ 'eye *n.* blaues Auge *(fig.);* Veilchen, *das (ugs.);* **~-eyed** *adj.* schwarzäugig; **be ~-eyed** schwarze Augen haben; **B~ 'Forest** *pr. n.* Schwarzwald, *der;* **B~ Forest 'gateau** *n.* Schwarzwälder [Kirschtorte], *die;* **~head** *n.* Mitesser, *der;* **~ 'hole** *n. (Astron.)* schwarzes Loch; **~ 'ice** *n.* Glatteis, *das;* **~jack** *n. (Cards)* Vingt-[et-]un, *das;* **~leg 1.** *n. (Brit.: strikebreaker)* Streikbrecher, *der/*-brecherin, *die;* **2.** *v.i.* Streikbrecher/-brecherin sein; **~ list** *n.* schwarze Liste; **~list** *v.t.* auf die schwarze Liste setzen; **~mail 1.** *v.t.* erpressen; **2.** *n.* Erpressung, *die;* **B~ Maria** [blæk məˈraɪə] *n.* grüne Minna *(ugs.);* **~ 'mark** *n. (fig.)* Makel, *der;* **~ 'market** *n.* schwarzer Markt
blackness [ˈblæknɪs] *n., no pl.* **a)** *(black colour)* Schwärze, *die;* **b)** *(darkness)* Finsternis, *die; (fig.: wickedness)* Abscheulichkeit, *die*
black: ~-out *n.* **a)** Verdunkelung, *die; (Theatre, Radio)* Blackout, *der;* **news ~-out** Nachrichtensperre, *die;* **b)** *(Med.)* **I had a ~-out** ich verlor das Bewußtsein; **~ 'pudding** *n.* Blutwurst, *die;* **B~ 'Sea** *pr. n.* Schwarze Meer, *das;* **~smith** *n.* Schmied, *der;* **~ spot** *n. (fig.)* schwarzer Fleck; *(dangerous)* Gefahrenstelle, *die;* **~ 'tie** *n.* schwarze Fliege *(zur Smokingjacke getragen);* **~ 'widow** *n. (Zool.)* Schwarze Witwe
bladder [ˈblædə(r)] *n.* Blase, *die*
blade [bleɪd] *n.* **a)** *(of sword, knife, dagger, razor, plane)* Klinge, *die; (of chisel, scissors, shears)* Schneide, *die; (of saw, oar, paddle, spade, propeller)* Blatt, *das; (of paddle-wheel, turbine)* Schaufel, *die;* **b)** *(of grass etc.)* Spreite, *die;* **c)** *(sword)* Schwert, *das*
blame [bleɪm] **1.** *v.t.* **a)** *(hold responsible)* **~ sb. [for sth.]** jmdm. die Schuld [an etw. *(Dat.)*] geben; **don't ~ me [if ...]** geben Sie nicht mir die Schuld[, wenn ...]; **~ sth. [for sth.]** etw. [für etw.] verantwortlich machen; **be to ~ [for sth.]** an etw. *(Dat.)* schuld sein; **~ sth. on sb./sth.** *(coll.)* jmdn./etw. für etw. verantwortlich machen; **b)** *(reproach)* **~ sb./oneself** jmdm./sich Vorwürfe machen; **I don't ~ you/him** *(coll.)* ich kann es Ihnen/ihm nicht verdenken; **don't ~ yourself** machen Sie sich *(Dat.)* keine Vorwürfe; **have only oneself to ~:** die Schuld bei sich selbst suchen müssen. **2.** *n. (responsibility)* Schuld, *die;* **lay** *or* **put the ~ [for sth.] on sb.** jmdm. [an etw. *(Dat.)*] die Schuld geben; **get the ~:** die Schuld bekommen; **take the ~ [for sth.]** die Schuld [für etw.] auf sich *(Akk.)* nehmen
blameless [ˈbleɪmlɪs] *adj.* untadelig
blameworthy [ˈbleɪmwɜːðɪ] *adj.* tadelnswert
blanch [blɑːnʃ] **1.** *v.t. (whiten)* bleichen; abziehen ‹*Mandeln*›; *(make pale)* erbleichen lassen. **2.** *v.i. (grow pale)* bleich werden
blancmange [bləˈmɒnʒ] *n.* Flammeri, *der*
bland [blænd] *adj. (gentle, suave)* verbindlich; freundlich ‹*Art, Stimmung*›; *(not irritating, not stimulating)* mild ‹*Medizin, Nahrung*›; *(unexciting)* farblos
blandishment [ˈblændɪʃmənt] *n. (flattery)* Schmeichelei, *die; (cajolery)* Beschwatzen, *das*
blandness [ˈblændnɪs] *n., no pl. see* **bland:** Verbindlichkeit, *die;* Freundlichkeit, *die;* Milde, *die;* Farblosigkeit, *die*
blank [blæŋk] **1.** *adj.* **a)** leer; kahl ‹*Wand, Fläche*›; **b)** *(empty)* frei; **leave a ~ space** Platz frei lassen; **c)** *(fig.)* leer, ausdruckslos ‹*Gesicht, Blick*›; **look ~:** ein verdutztes Gesicht machen; **my mind went ~:** ich hatte ein Brett vor dem Kopf. **2.** *n.* **a)** *(space)* Lücke, *die;* **his memory was a ~:** er hatte keinerlei Erinnerung; **b)** *(document with ~s)* Vordruck, *der;* **c) draw a ~** *(fig.)* kein Glück haben; **d)** *(cartridge)* Platzpatrone, *die*
blank: ~ 'cartridge *n.* Platzpatrone, *die;* **~ 'cheque** *n.* Blankoscheck, *der; (fig.)* Blankovollmacht, *die*
blanket [ˈblæŋkɪt] **1.** *n.* **a)** Decke, *die;* **wet ˈ~** *(fig.)* Trauerkloß, *der;* **b)** *(thick layer)* Decke, *die;* **~ of snow/fog** Schnee-/Nebeldecke, *die.* **2.** *v.t.* zudecken. **3.** *adj.* umfassend; **~ agreement** Pauschalabkommen, *das*
blankly [ˈblæŋklɪ] *adv.* verdutzt
blank 'verse *n. (Pros.)* Blankvers, *der*
blare [bleə(r)] **1.** *v.i.* ‹*Lautsprecher:*› plärren; ‹*Trompete:*› schmettern. **2.** *v.t.* **~ [out]** [hinaus]plärren ‹*Worte*›; [hinaus]schmettern ‹*Melodie*›. **3.** *n. see* **1:** Plärren, *das;* Schmettern, *das*
blasé [ˈblɑːzeɪ] *adj.* blasiert
blaspheme [blæsˈfiːm] *v.i.* lästern
blasphemy [ˈblæsfəmɪ] *n.* Blasphemie, *die*
blast [blɑːst] **1.** *n.* **a)** *(gust)* **a ~ [of**

wind] ein Windstoß; **b)** *(sound)* Tuten, *das;* **give one ~ of the horn** einmal ins Horn stoßen; **c) at full ~** *(fig.)* auf Hochtouren *Pl.;* **d)** *(of explosion)* Druckwelle, *die; (coll.: explosion)* Explosion, *die.* **2.** *v. t.* **a)** *(blow up)* sprengen ‹*Felsen*›; *(coll.: kick)* donnern ‹*Fußball*›; **b)** *(curse)* ~ **you/him!** zum Teufel mit dir/ihm! **3.** *v. i. (coll.: shoot)* **start ~ing away** drauflosschießen (**at** auf + *Akk.*). **4.** *int.* **[oh] ~!** [oh] verdammt!

~ **'off** *v. i.* abheben

blasted ['blɑ:stɪd] *adj. (damned)* verdammt *(salopp)*

blast: ~-furnace *n.* Hochofen, *der;* **~-off** *n.* Abheben, *das*

blatant ['bleɪtənt] *adj.* **a)** *(flagrant)* offensichtlich; **b)** *(unashamed)* unverhohlen; unverfroren ‹*Lüge*›

blatantly ['bleɪtəntlɪ] *adv. see* **blatant:** offensichtlich; unverhohlen; unverfroren

[1]**blaze** [bleɪz] **1.** *n.* **a)** *(fire)* Feuer, *das; (in building)* Feuer, *das;* Brand, *der;* **b)** *(display)* **a ~ of lights** ein Lichtermeer; **a ~ of colour** eine Farbenpracht; ein Farbenmeer; **in a ~ of glory** mit Glanz und Gloria; **c)** *(sl.)* **go to ~s!** scher dich zum Teufel! *(salopp);* **like ~s** wie verrückt *(ugs.)* ‹*arbeiten, rennen usw.*›; **what the ~s [...]?** was zum Teufel [...]? *(salopp);* **how/where/who/why the ~s ...?** wie/wo/wer/warum zum Teufel ...? *(salopp).* **2.** *v. i.* **a)** *(burn)* brennen; **the house was already blazing when the firemen arrived** das Haus stand schon in Flammen als die Feuerwehr ankam; **a blazing fire** ein helloderndes Feuer; **the blazing sun** die glühende Sonne; **b)** *(emit light)* strahlen; **c)** *(fig.: with anger etc.)* ‹*Augen:*› glühen; **a blazing row** ein heftiger Streit

~ **a'way** *v. i.* [drauf]losschießen (**at** auf + *Akk.*)

~ **'up** *v. i.* aufflammen

[2]**blaze** *v. t.* ~ **a** *or* **the trail** *(fig.)* den Weg bahnen

blazer ['bleɪzə(r)] *n.* Blazer, *der*

bleach [bli:tʃ] **1.** *v. t.* bleichen ‹*Wäsche, Haar, Knochen*›. **2.** *v. i.* bleichen. **3.** *n.* Bleichmittel, *das*

bleak [bli:k] *adj.* **a)** *(bare)* öde ‹*Landschaft usw.*›; karg ‹*Zimmer*›; **b)** *(chilly)* rauh; kalt ‹*Wetter, Tag*›; **c)** *(unpromising)* düster; ~ **prospect[s]** trübe Aussichten

bleary ['blɪərɪ] *adj.* trübe ‹*Augen*›; **look ~-eyed** verschlafen aussehen

bleat [bli:t] *v. i.* ‹*Schaf, Kalb:*› blöken; ‹*Ziege:*› meckern; *(fig.)* jammern; *(plaintively)* meckern

bled *see* **bleed**

bleed [bli:d] **1.** *v. i.,* **bled** [bled] bluten. **2.** *v. t.,* **bled** *(draw blood from, lit. or fig.)* zur Ader lassen

bleeding ['bli:dɪŋ] **1.** *n. (loss of blood)* Blutung, *die.* **2.** *adj. (Brit. coarse: damned)* Scheiß- *(derb).* **3.** *adv. (Brit. coarse)* ~ **awful** total beschissen *(derb);* ~ **stupid** saublöd *(salopp)*

bleep [bli:p] **1.** *n.* Piepen, *das;* **two faint ~s** zwei schwache Piepser. **2.** *v. i.* ‹*Geigerzähler, Funksignal:*› piepen. **3.** *v. t.* ~ **sb.** jmdn. über seinen Kleinempfänger *od. (ugs.)* Piepser rufen

bleeper ['bli:pə(r)] *n.* Kleinempfänger, *der;* Piepser, *der (ugs.)*

blemish ['blemɪʃ] **1.** *n.* **a)** *(stain)* Fleck, *der;* **b)** *(defect, lit. or fig.)* Makel, *der; (in character)* Fehler, *der.* **2.** *v. t.* **a)** *(spoil)* verunstalten; **b)** *(fig.)* ~ **sth.** einer Sache *(Dat.)* schaden

blend [blend] **1.** *v. t.* **a)** *(mix)* mischen ‹*Whisky-, Tee-, Tabaksorten*›; **b)** *(make indistinguishable)* vermischen. **2.** *v. i.* **a)** sich mischen lassen; ~ **in with/into sth.** [gut] zu etw. passen/mit etw. verschmelzen; **b)** ‹*Whisky-, Tee-, Tabaksorten:*› sich [harmonisch] verbinden. **3.** *n.* Mischung, *die*

blender ['blendə(r)] *n.* **a)** *(person)* [Ver]mischer, *der;* **b)** *(apparatus)* Mixer, *der;* Mixgerät, *das*

bless [bles] *v. t.,* **blessed** [blest] *or (poet.)* **blest** [blest] *(consecrate, pronounce blessing on)* segnen; **[God] ~ you** Gottes Segen; *(as thanks)* das ist sehr lieb von dir/Ihnen; *(to person sneezing)* Gesundheit!; **goodbye and God ~:** Wiedersehen, [und] mach's/macht's gut!; ~ **me!, well I'm blest!, ~ my soul!** du meine Güte! *(ugs.);* ~ **me if it isn't Sid** ja das ist doch Sid!

blessed *adj.* ['blesɪd, *pred.* blest] **a) be ~ with sth.** *(also iron.)* mit etw. gesegnet sein; **b)** *(revered)* heilig ‹*Gott, Mutter Maria*›; *(in Paradise)* selig; *(RC Ch.: beatified)* selig; *(blissful)* beglückend; **c)** *attrib. (euphem.: cursed)* verdammt *(salopp)*

blessing ['blesɪŋ] *n.* **a)** *(divine favour, grace at table)* Segen, *der;* **do sth. with sb.'s ~** *(fig.)* etw. mit jmds. Segen tun *(ugs.);* **give sb./sth. one's ~** *(fig.)* jmdm./etw. seinen Segen geben *(ugs.);* **b)** *(divine gift)* Segnung, *die;* **count one's ~s** *(fig.)* dankbar sein; **c)** *(fig. coll.: welcome thing)* Segen, *der;* **what a ~!** welch ein Segen!; **be a ~ in disguise** sich schließlich doch noch als Segen erweisen

blest *see* **bless;** *(poet.) see* **blessed b**

blew *see* [1]**blow 1, 2**

blight [blaɪt] **1.** *n.* **a)** *(plant disease)* Brand, *der; (fig.)* Geißel, *die;* **b)** *(fig.: unsightly urban area)* Schandfleck, *der.* **2.** *v. t.* **a)** *(affect with ~)* **be ~ed** von Brand befallen werden/sein; **b)** überschatten ‹*Freude, Leben*›; *(frustrate)* zunichte machen ‹*Hoffnung*›; **a ~ed area** eine heruntergekommene Gegend

blighter ['blaɪtə(r)] *n. (Brit. coll.)* **a) the poor ~:** der arme Kerl; **b)** *(derog.)* Lümmel, *der (abwertend)*

blimey ['blaɪmɪ] *interj. (Brit. sl.)* Mensch *(salopp)*

blind [blaɪnd] **1.** *adj.* **a)** blind ‹*Person, Tier*›; **a ~ man/woman** ein Blinder/eine Blinde; **as ~ as a bat** stockblind *(ugs.);* ~ **in one eye** auf einem Auge blind; **go** *or* **become ~:** blind werden; **turn a ~ eye [to sth.]** *(fig.)* [bei etw.] ein Auge zudrücken; **b)** *(Aeronaut.)* ~ **landing/flying** Blindlandung, *die*/Blindflug, *der;* **c)** *(unreasoning)* blind ‹*Vorurteil, Weigerung, Gehorsam, Vertrauen*›; **d)** *(oblivious)* **be ~ to sth.** blind gegenüber etw. sein; **e)** *(not ruled by purpose)* blind ‹*Wut, Zorn*›; dunkel ‹*Instinkt*›; kopflos ‹*Panik*›. **2.** *adv.* **a)** blindlings; **the pilot had to fly/land ~:** der Pilot mußte blind fliegen/landen; **b)** *(completely)* ~ **drunk** stockbetrunken *(ugs.);* **swear ~:** hoch und heilig versichern. **3.** *n.* **a)** *(screen)* Jalousie, *die; (of cloth)* Rouleau, *das; (of shop)* Markise, *die;* **b)** *(Amer. Hunting: hide)* Jagdschirm, *der;* **c)** *(pretext)* Vorwand, *der; (cover)* Tarnung, *die;* **be a ~ for sth.** als Tarnung für etw. dienen; **d)** *pl.* **the ~:** die Blinden *Pl.;* **it's [a case of] the ~ leading the ~** *(fig.)* das ist, wie wenn ein Blinder einen Lahmen [spazieren]führt. **4.** *v. t. (lit. or fig.)* blenden; **be ~ed** *(accidentally)* das Augenlicht verlieren; ~ **sb. with science** jmdn. mit großen Worten beeindrucken

blind: ~ **'alley** *n. (lit. or fig.)* Sackgasse, *die;* ~ **'corner** *n.* unübersichtliche Ecke; ~ **'date** *n.* Verabredung mit einem/einer Unbekannten; **~fold 1.** *v. t.* die Augen verbinden (+ *Dat.*); **2.** *adj.* mit verbundenen Augen *nachgestellt*

blinding ['blaɪndɪŋ] *adj.* blendend ‹*Licht, Sonnenlicht, Blitz*›;

grell ⟨*Strahl*⟩; **a ~ headache** rasende Kopfschmerzen *Pl.*

blindly ['blaɪndlɪ] *adv.* [wie] blind; *(fig.)* blindlings

blind man's 'buff *n.* Blindekuh *o. Art.*

blindness ['blaɪndnɪs] *n., no pl.* Blindheit, *die*

'blind spot *n. (Anat.)* blinder Fleck; *(Motor Veh.)* toter Winkel; *(fig.: weak spot)* schwacher Punkt

blink [blɪŋk] **1.** *v. i.* **a)** blinzeln; **b)** *(shine intermittently)* blinken; *(shine momentarily)* aufblinken. **2.** *v. t.* **~ one's eyes** mit den Augen zwinkern. **3.** *n.* **a)** Blinzeln, *das;* **b)** *(sl.)* **be on the ~:** kaputt sein *(ugs.)*

blinker ['blɪŋkə(r)] **1.** *n. in pl.* Scheuklappen; **have/put ~s on** *(lit. or fig.)* Scheuklappen tragen/anlegen. **2.** *v. t.* Scheuklappen anlegen (+ *Dat.*); **~ed** *(fig.)* borniert

blinking ['blɪŋkɪŋ] *(Brit. coll. euphem.)* **1.** *adj.* verflixt *(ugs.).* **2.** *adv.* verflixt *(ugs.);* **it's ~ raining** verflixt [und zugenäht], es regnet

blip [blɪp] *n.* **a)** *(sound) (of bursting bubble)* leiser Knall; *(on magnetic tape)* leises Knacken; **b)** *(Radar: image)* Echozeichen, *das*

bliss [blɪs] *n. (joy)* [Glück]seligkeit, *die;* Glück, *das*

blissful ['blɪsfl] *adj.* [glück]selig; **~ ignorance** *(iron.)* selige Unwissenheit

blister ['blɪstə(r)] **1.** *n. (on skin, plant, metal, paintwork)* Blase, *die.* **2.** *v. t.* Blasen hervorrufen auf (+ *Dat.*) ⟨*Haut, Metall, Anstrich*⟩. **3.** *v. i.* ⟨*Haut:*⟩ Blasen bekommen; ⟨*Metall, Anstrich:*⟩ Blasen werfen

blistering ['blɪstərɪŋ] *adj.* ätzend ⟨*Kritik*⟩; **a ~ attack** ein erbitterter Angriff

'blister pack *n.* Sichtpackung, *die*

blithely ['blaɪðlɪ] *adv.* **~ ignore sth.** sich unbekümmert über etw. *(Akk.)* hinwegsetzen

blithering ['blɪðərɪŋ] *adj. (coll.) (utter)* total; völlig; **a ~ idiot** ein alter Idiot *(salopp)*

blitz [blɪts] *(coll.)* **1.** *n.* **a)** *(Hist.)* Luftangriff, *der* (**on** auf + *Akk.*); **during the [London] B~:** während der Luft- *od.* Bombenangriffe [auf London]; **b)** *(fig.: attack)* Großaktion, *die (fig.);* **have a ~ on one's room** in seinem Zimmer gründlich saubermachen. **2.** *v. t.* [schwer] bombardieren

blizzard ['blɪzəd] *n.* Schneesturm, *der*

bloated ['bləʊtɪd] *adj.* **a)** *(having overeaten)* aufgedunsen; **I feel ~:** ich bin voll *(ugs.);* **b) be ~ with pride** aufgeblasen sein

blob [blɒb] *n.* **a)** *(drop)* Tropfen, *der; (small mass)* Klacks, *der (ugs.); (of butter etc.)* Klecks, *der;* **b)** *(spot of colour)* Fleck, *der*

bloc [blɒk] *n. (Polit.)* Block, *der;* **the Eastern ~/Eastern ~ countries** der Ostblock/die Ostblockstaaten; **the Western ~ [countries]** die westlichen Staaten

block [blɒk] **1.** *n.* **a)** *(large piece)* Klotz, *der;* **~ of wood** Holzklotz, *der;* **b)** *(for chopping on)* Hackklotz, *der;* **c)** *(for beheading on)* Richtblock, *der;* **d)** *(large mass of concrete or stone; building-stone)* Block, *der;* **e)** *(sl.: head)* **knock sb.'s ~ off** jmdm. eins überziehen *(salopp);* **f)** *(of buildings)* [Häuser]block, *der;* **~ of flats/offices** Wohnblock, *der*/Bürohaus, *das;* **g)** *(Amer.: area between streets)* Block, *der;* **h)** *(large quantity)* Masse, *die;* **a ~ of shares** ein Aktienpaket; **a ~ of seats** mehrere nebeneinanderliegende Sitze; **i)** *(pad of paper)* Block, *der;* **j)** *(obstruction)* Verstopfung, *die;* **k)** *(mental barrier)* **a mental ~:** eine geistige Sperre; Mattscheibe *o. Art. (salopp);* **a psychological ~:** ein psychologischer Block; **l) ~ and tackle** Flaschenzug, *der.* **2.** *v. t.* **a)** *(obstruct)* blockieren, versperren ⟨*Tür, Straße, Durchgang, Sicht*⟩; verstopfen ⟨*Nase*⟩; blockieren ⟨*Fortschritt*⟩; abblocken ⟨*Ball, Torschuß*⟩; **b)** *(Commerc.)* einfrieren ⟨*Investitionen, Guthaben*⟩

~ 'off *v. t.* [ab]sperren ⟨*Straße*⟩; blockieren ⟨*Rohr, Verkehr*⟩

~ 'out *v. t.* ausschließen ⟨*Licht, Lärm*⟩

~ 'up *v. t.* verstopfen; versperren, blockieren ⟨*Eingang*⟩

blockade [blɒ'keɪd] **1.** *n.* Blockade, *die.* **2.** *v. t.* blockieren

blockage ['blɒkɪdʒ] *n.* Block, *der; (of pipe, gutter)* Verstopfung, *die*

block: ~-buster *n.* Knüller, *der (ugs.);* **~ 'capital** *n.* Blockbuchstabe, *der;* **~head** *n.* Dummkopf, *der (abwertend);* **~ 'letters** *n. pl.* Blockschrift, *die*

bloke [bləʊk] *n. (Brit. coll.)* Typ, *der (ugs.)*

blond [blɒnd] *see* **blonde 1**

blonde [blɒnd] **1.** *adj.* blond ⟨*Haar, Person*⟩; hell ⟨*Teint*⟩. **2.** *n.* Blondine, *die*

blood [blʌd] *n.* **a)** Blut, *das;* **sb.'s ~ boils** *(fig.)* jmd. ist in Rage; **it makes my ~ boil** es bringt mich in Rage; **sb.'s ~ turns** *or* **runs cold** *(fig.)* jmdm. erstarrt das Blut in den Adern; **be after** *or* **out for sb.'s ~** *(fig.)* es auf jmdn. abgesehen haben; **it's like getting ~ out of** *or* **from a stone** das ist fast ein Ding der Unmöglichkeit; **do sth. in cold ~** *(fig.)* etw. kaltblütig tun; **b)** *(relationship)* Blutsverwandtschaft, *die;* **~ is thicker than water** *(prov.)* Blut ist dicker als Wasser

blood: ~ bank *n.* Blutbank, *die;* **~-bath** *n.* Blutbad, *das;* **~ cell** *n.* Blutkörperchen, *das;* **~ clot** *n.* Blutgerinnsel, *das;* **~-curdling** *adj.* grauenerregend; **~ donor** *see* **donor b; ~ group** *n.* Blutgruppe, *die;* **~hound** *n.* Bluthund, *der; (fig.)* Spürhund, *der*

bloodless ['blʌdlɪs] *adj.* **a)** *(without bloodshed)* unblutig; **b)** *(without blood, pale)* blutleer

blood: ~-lust *n.* Blutgier, *die;* **~-money** *n.* Blutgeld, *das;* **~-poisoning** *n.* Blutvergiftung, *die;* **~ pressure** *n.* Blutdruck, *der;* **~-red** *adj.* blutrot; **~ relation** *n.* Blutsverwandte, *der/die;* **~ sample** *n.* Blutprobe, *die;* **~shed** *n.* Blutvergießen, *das;* **~shot** *adj.* blutunterlaufen; **~ sports** *n. pl.* Hetzjagd, *die;* **~ stain** *n.* Blutfleck, *der;* **~-stained** *adj. (lit. or fig.)* blutbefleckt; **~stream** *n.* Blutstrom, *der;* **~thirsty** *adj.* blutdürstig *(geh.);* blutrünstig; **~ transfusion** *n.* Bluttransfusion, *die;* **~-vessel** *n.* Blutgefäß, *das*

bloody ['blʌdɪ] **1.** *adj.* **a)** blutig; *(running with blood)* blutend; **b)** *(sl.: damned)* verdammt *(salopp);* **you ~ fool!** du Vollidiot! *(salopp);* **~ hell!** verdammt noch mal! *(salopp);* **c)** *(Brit.) as intensifier* einzig; **that/he is a ~ nuisance** das ist vielleicht ein Mist *(salopp)*/der geht einem vielleicht *od.* ganz schön auf den Wecker *(ugs.).* **2.** *adv.* **a)** *(sl.: damned)* verdammt *(salopp);* **b)** *(Brit.) as intensifier* verdammt *(salopp);* **not ~ likely!** denkste! *(salopp).* **3.** *v. t. (make ~)* blutig machen; *(stain with blood)* mit Blut beflecken

bloody-'minded *adj.* stur *(ugs.)*

bloom [blu:m] **1.** *n.* **a)** Blüte, *die;* **be in ~:** in Blüte stehen; **b)** *(on fruit)* Flaum, *der; (flush)* rosige Gesichtsfarbe; **c)** *(prime)* **in the ~ of youth** in der Blüte der Jugend. **2.** *v. i.* blühen; *(fig.: flourish)* in Blüte stehen

bloomers ['blu:məz] *n. pl.* [Damen]pumphose, *die*

blossom ['blɒsəm] **1.** *n.* **a)**

(flower) Blüte, *die;* **b)** *no pl., no indef. art. (mass of flowers)* Blüte, *die;* **be in ~:** in [voller] Blüte stehen *od.* sein; **have come into ~:** blühen. **2.** *v. i.* **a)** blühen; **b)** *(fig.)* blühen; ‹*Person:*› aufblühen

blot [blɒt] **1.** *n.* **a)** *(spot of ink)* Tintenklecks, *der; (stain)* Fleck, *der; (blemish)* Makel, *der;* Schandfleck, *der;* **b)** *(fig.)* Makel, *der;* **a ~ on sb.'s character** ein Fleck auf jmds. weißer Weste *(fig.).* **2.** *v. t.,* **-tt-: a)** *(dry)* ablöschen ‹*Tinte, Schrift, Papier*›; **b)** *(spot with ink)* beklecksen; **~ one's copy-book** *(fig. coll.)* sich unmöglich machen

~ 'out *v. t.* **a)** *(obliterate)* einen Klecks machen auf (+ *Akk.*); unleserlich machen ‹*Schrift*›; **b)** *(obscure)* verdecken ‹*Sicht*›; **c)** auslöschen ‹*Leben, Menschheit, Erinnerung*›

blotter ['blɒtə(r)] *n.* Schreibunterlage, *die*

'blotting-paper *n.* Löschpapier, *das*

blouse [blaʊz] *n.* Bluse, *die*

[1]**blow** [bləʊ] **1.** *v. i.,* **blew** [blu:], **blown** [bləʊn] **a)** ‹*Wind:*› wehen; ‹*Sturm:*› blasen; ‹*Luft:*› ziehen; **there is a gale ~ing out there** es stürmt draußen; **b)** *(exhale)* blasen; **~ on one's hands to warm them** in die Hände hauchen, um sie zu wärmen; **~ hot and cold** *(fig.)* einmal hü und einmal hott sagen; **c)** *(puff, pant)* ‹*Person:*› schwer atmen, schnaufen; ‹*Tier:*› schnaufen; **d)** *(be sounded by ~ing)* geblasen werden; ‹*Trompete, Flöte, Horn, Pfeife usw.:*› ertönen; **e)** *(melt)* ‹*Sicherung, Glühfaden:*› durchbrennen. **2.** *v. t.,* **blew, blown** *(see also* **k**): **a)** *(breathe out)* [aus]blasen, ausstoßen ‹*Luft, Rauch*›; **b)** *(send by ~ing)* **~ sb. a kiss** jmdm. eine Kußhand zuwerfen; **c)** blasen ‹*Blätter, Schnee, Staub usw.*›; **d)** *(make by ~ing)* blasen ‹*Glas*›; machen ‹*Seifenblasen*›; **e)** *(sound)* blasen ‹*Trompete, Flöte, Horn, Pfeife usw.*›; **~ one's own trumpet** *(fig.)* sein Eigenlob singen; **f)** *(clear)* **~ one's nose** sich *(Dat.)* die Nase putzen; **g)** *(send flying)* schleudern; **~ sth. to pieces** etw. in die Luft sprengen; **h)** *(cause to melt)* durchbrennen lassen ‹*Sicherung, Glühlampe*›; durchhauen *(ugs.)* ‹*Sicherung*›; **i)** *(break into)* sprengen, aufbrechen ‹*Tresor, Safe*›; **j)** *(sl.: reveal)* verraten ‹*Plan, Komplizen*›; **k)** *p. t., p. p.* **~ed** *(sl.: curse)* **[well,] I'm** *or* **I'll be ~ed** ich werde verrückt! *(salopp);* **~ you, Jack!** du kannst mich mal gern haben! *(salopp);* **~!** [so ein] Mist! *(ugs.);* **~ the expense** es ist doch Wurscht, was es kostet *(ugs.);* **l)** *(sl.: squander)* verpulvern, verplempern *(ugs.)* ‹*Geld, Mittel, Erbschaft*›; **~ it** *(lose opportunity)* es vermasseln *(salopp)*

~ 'off 1. *v. i.* weggeblasen werden. **2.** *v. t.* wegblasen

~ 'out 1. *v. t.* **a)** *(extinguish)* ausblasen ‹*Kerze, Lampe*›; **b)** *(by explosion)* **the explosion blew all the windows out** durch die Explosion flogen alle Fensterscheiben raus; **~ sb.'s/one's brains out** jmdm./sich eine Kugel durch den Kopf jagen *(ugs.).* **2.** *v. i.* ‹*Reifen:*› platzen; ‹*Kerze, Lampe:*› ausgeblasen werden. **3.** *v. refl.* ‹*Sturm:*› sich legen

~ 'over 1. *v. i.* umgeblasen werden; ‹*Streit, Sturm:*› sich legen. **2.** *v. t.* umblasen

~ 'up 1. *v. t.* **a)** *(shatter)* [in die Luft] sprengen; **b)** *(inflate)* aufblasen ‹*Ballon*›; aufpumpen ‹*Reifen*›; **c)** *(coll.: reprove)* in der Luft zerreißen *(ugs.);* **d)** *(coll.: enlarge)* vergrößern ‹*Foto, Seite*›; **e)** *(coll.: exaggerate)* hochspielen, aufbauschen ‹*Ereignis, Bericht*›. **2.** *v. i.* **a)** *(explode)* explodieren; **b)** *(arise suddenly)* ‹*Krieg, Sturm, Konflikt:*› ausbrechen; **c)** *(lose one's temper)* [vor Wut] explodieren *(ugs.)*

[2]**blow** *n.* **a)** *(stroke)* Schlag, *der; (with axe)* Hieb, *der; (jolt, push)* Stoß, *der;* **in** *or* **at one ~** *(lit. or fig.)* mit einem Schlag; **come to ~s** handgreiflich werden; **a ~-by-~ account** ein Bericht in allen Einzelheiten; **b)** *(disaster)* [schwerer] Schlag, *der (fig.)* **(to** für); **come as** *or* **be a ~ to sb.** ein schwerer Schlag für jmdn. sein

blow: ~-dry *v. t.* mit dem Fön frisieren; **~lamp** *n.* Lötlampe, *die*

blown *see* [1]**blow 1, 2**

blow: ~-out *n.* **a)** *(burst tyre)* Reifenpanne, *die;* **b)** *(coll.: meal)* feudales Essen *(ugs.);* **~pipe** *n.* *(weapon)* Blasrohr, *das;* **~torch** *(Amer.) see* **~lamp; ~-up** *n. (coll.: enlargement)* Vergrößerung, *die*

blowy ['bləʊɪ] *adj.* windig

blubber ['blʌbə(r)] **1.** *n. (whalefat)* Walfischspeck, *der.* **2.** *v. i. (coll.: weep)* heulen *(ugs.)*

blue [blu:] **1.** *adj.* **a)** blau; **be ~ with cold/rage** blau gefroren/rot vor Zorn sein; **b)** *(depressed)* **be/feel ~:** niedergeschlagen sein/sich bedrückt *od.* deprimiert fühlen; **c)** *(pornographic)* pornographisch; Porno-; **~ jokes** unanständige Witze. **2.** *n.* **a)** *(colour)* Blau, *das;* **b)** *(sky)* Himmelsblau, *das;* **out of the ~** *(fig.)* aus heiterem Himmel *(ugs.);* **c) the ~s** *(melancholy)* Niedergeschlagenheit; **have the ~s** niedergeschlagen *od.* deprimiert sein; **d) the ~s** *(Mus.)* der Blues; **play/sing the ~s** Blues spielen/singen

blue: ~ 'baby *n. (Med.)* blausüchtiger Säugling; **~bell** *n.* *(campanula)* [blaue Wiesen]glockenblume, *die; (wild hyacinth)* Sternhyazinthe, *die;* **~berry** [blu:bərɪ] *n.* Heidelbeere, *die;* Blaubeere, *die;* **~ 'blood** *n.* blaues Blut; **~bottle** *n.* Schmeißfliege, *die;* **~ 'cheese** *n.* Blauschimmelkäse, *der;* **~-'collar** *adj.* **~-collar worker** Arbeiter, *der*/Arbeiterin, *die;* **~-collar union** Arbeitergewerkschaft, *die;* **~-eyed** *adj.* blauäugig; **be ~-eyed** blaue Augen haben; **~-eyed 'boy** *n. (fig. coll.)* Goldjunge, *der;* **~ 'jeans** *n. pl.* Blue jeans *Pl.;* **~ 'moon** *n.* **once in a ~ moon** alle Jubeljahre *(ugs.);* **~print** *n.* **a)** Blaupause, *die;* **b)** *(fig.)* Plan, *der;* Entwurf, *der;* **~stocking** *n.* Blaustrumpf, *der;* **~ tit** *n. (Ornith.)* Blaumeise, *die*

[1]**bluff** [blʌf] **1.** *n. (act)* Täuschungsmanöver, *das;* Bluff, *der (ugs.); see also* **call 2 c. 2.** *v. i. & t.* bluffen *(ugs.)*

[2]**bluff 1.** *n. (headland)* Kliff, *das;* Steilküste, *die; (inland)* Steilhang, *der.* **2.** *adj. (abrupt, blunt, frank, hearty)* rauhbeinig *(ugs.)*

bluish ['blu:ɪʃ] *adj.* bläulich

blunder ['blʌndə(r)] **1.** *n.* [schwerer] Fehler; **make a ~:** einen [schweren] Fehler machen. **2.** *v. i.* **a)** *(make mistake)* einen [schweren] Fehler machen; **b)** *(move blindly)* tappen

blunt [blʌnt] **1.** *adj.* **a)** stumpf; **a ~ instrument** ein stumpfer Gegenstand; **b)** *(outspoken)* direkt; unverblümt; **c)** *(uncompromising)* glatt *(ugs.)* ‹*Ablehnung*›. **2.** *v. t.* **~ [the edge of]** stumpf machen ‹*Messer, Schwert, Säge*›; dämpfen ‹*Begeisterung, Mut*›; mildern ‹*Trauer, Enttäuschung*›

bluntly ['blʌntlɪ] *adv.* **a)** *(outspokenly)* direkt, unverblümt ‹*sprechen, antworten*›; **b)** *(uncompromisingly)* glatt ‹*ablehnen*›

bluntness ['blʌntnɪs] *n., no pl. see* **blunt 1:** Stumpfheit, *die;* Direktheit, *die;* Unverblümtheit, *die*

blur [blɜ:(r)] **1.** *v. t.,* **-rr-: a)** *(smear)* verwischen, verschmieren ‹*Schrift, Seite*›; **b)** *(make indistinct)* verwischen ‹*Schrift, Farben, Konturen*›; **become ~red** ‹*Farben, Schrift*:› verwischt werden; **c)** *(dim)* trüben ‹*Sicht, Wahrnehmung*›; **my vision is ~red** ich sehe alles verschwommen. **2.** *n.* **a)** *(smear)* [verschmierter] Fleck, *der;* **b)** *(dim image)* verschwommener Fleck

blurb [blɜ:b] *n.* Klappentext, *der;* Waschzettel, *der*

blurt [blɜ:t] *v.t.* hervorstoßen ‹*Worte, Beschimpfung*›; **~ sth. out** mit etw. herausplatzen *(ugs.)*

blush [blʌʃ] **1.** *v.i.* **a)** rot werden; **make sb. ~:** jmdn. rot werden lassen; **b)** *(be ashamed)* sich schämen (**at** bei). **2.** *n.* **a)** *(reddening)* Erröten, *das (geh.);* **spare sb.'s ~es** jmdn. nicht in Verlegenheit bringen; **b)** *(rosy glow)* Röte, *die*

bluster ['blʌstə(r)] *v.i.* **a)** ‹*Wind*:› tosen, brausen; **b)** ‹*Person*:› sich aufplustern *(ugs.)*

blustery ['blʌstərı] *adj.* stürmisch ‹*Wetter, Wind*›

boa constrictor ['bəʊə kənstrıktə(r)] *n.* Boa constrictor, *die*

boar [bɔ:(r)] *n.* **a)** *(male pig)* Eber, *der;* **b)** *(wild)* Keiler, *der*

board [bɔ:d] **1.** *n.* **a)** Brett, *das;* **bare ~s** bloße Dielen; **b)** *(black~)* Tafel, *die;* **c)** *(notice-~)* Schwarzes Brett; **d)** *(in game)* Brett, *das;* **e)** *(spring~)* [Sprung]brett, *das;* **f)** *(meals)* Verpflegung, *die;* **~ and lodging** Unterkunft und Verpflegung; **full ~:** Vollpension, *die;* **g)** *(Admin. etc.)* Amt, *das;* Behörde, *die;* **gas/water/electricity ~:** Gas-/Wasser- / Elektrizitätsversorgungsgesellschaft, *die;* **~ of inquiry** Untersuchungsausschuß, *der;* **h)** *(Commerc., Industry)* **~ [of directors]** Vorstand, *der;* **i)** *(Naut., Aeronaut., Transport)* **on ~:** an Bord; **on ~ the ship/plane** an Bord des Schiffes/Flugzeugs; **j) the ~s** *(Theatre)* die Bühne; **k) go by the ~:** ins Wasser fallen; **above ~:** korrekt; **across the ~:** pauschal. **2.** *v.t. (go on ~)* **~ the ship/plane** an Bord des Schiffes/Flugzeugs gehen; **~ the train/bus** in den Zug/Bus einsteigen. **3.** *v.i.* **a)** *(lodge)* [in Pension] wohnen (**with** bei); **b)** *(~ an aircraft)* an Bord gehen; **'flight L 5701 now ~ing [at] gate 15'** „Passagiere des Fluges L 5701 bitte zum Flugsteig 15" **~ 'up** *v.t.* mit Brettern vernageln

boarder ['bɔ:də(r)] *n.* **a)** *(lodger)* Pensionsgast, *der;* **b)** *(Sch.)* Internatsschüler, *der*/-schülerin, *die*

'board-game *n.* Brettspiel, *das*

boarding: ~-house *n.* Pension, *die;* **~-school** *n.* Internat, *das*

board: ~ meeting *n.* Vorstands- / Aufsichtsrats- / Verwaltungsratssitzung, *die;* **~room** *n.* Sitzungssaal, *der*

boast [bəʊst] **1.** *v.i.* prahlen (**of, about** mit). **2.** *v.t.* prahlen mit; *(possess)* sich rühmen (+ *Gen.*). **3.** *n.* **a)** Prahlerei, *die;* **b)** *(cause of pride)* Stolz, *der*

boastful ['bəʊstfl] *adj.* prahlerisch; großspurig ‹*Erklärung, Behauptung*›

boat [bəʊt] *n.* **1. a)** Boot, *das;* **ship's ~:** Beiboot, *das;* **go by ~** mit dem Schiff fahren; **be in the same ~** *(fig.)* im gleichen Boot sitzen; **b)** *(ship)* Schiff, *das;* **c)** *(for sauce etc.)* Sauciere, *die.* **2.** *v.i.* **go ~ing** eine Bootsfahrt machen

boater ['bəʊtə(r)] *n.* **a)** *(person)* Bootsfahrer, *der*/-fahrerin, *die;* **b)** *(hat)* steifer Strohhut

boat: ~-hook *n.* Bootshaken, *der;* **~-house** *n.* Bootshaus, *das;* **~load** *n.* Bootsladung, *die;* **~ race** *n.* Regatta, *die;* **~swain** ['bəʊsn] *n.* Bootsmann, *der;* **~-train** *n.* Zug mit Schiffsanschluß

[1]**bob** [bɒb] **1.** *v.i.*, **-bb-: a) ~ [up and down]** sich auf und nieder bewegen; *(jerkily)* auf und nieder schnellen; ‹*[Pferde]schwanz*:› [auf und nieder] wippen; **~ up** hochschnellen; **b)** *(curtsy)* knicksen. **2.** *n. (curtsy)* Knicks, *der*

[2]**bob 1.** *n. (hair-style)* Bubikopf, *der.* **2.** *v.t.*, **-bb-** kurz schneiden ‹*Haar*›; **wear one's hair ~bed** einen Bubikopf tragen

[3]**bob** *n., pl. same (Brit. coll.)* **a)** *(Hist.: shilling)* Schilling, *der;* **she's not short of a ~ or two** *(fig.)* sie hat schon ein paar Mark; **b)** *(5p)* Fünfer, *der (ugs.);* **two/ten ~:** 10/50 Pence

[4]**bob** *n. (~-sled)* Bob, *der*

bobbin ['bɒbın] *n.* Spule, *die*

bobble ['bɒbl] *n.* Pompon, *der;* Bommel, *die (bes. nordd.)*

bobby ['bɒbı] *n. (Brit. coll.)* Bobby, *der (ugs.)*

bob: ~-sled, ~-sleigh *ns.* Bob[schlitten], *der*

bode [bəʊd] *v.i.* **~ ill/well** nichts Gutes/einiges erhoffen lassen

bodice ['bɒdıs] *n. (part of dress)* Oberteil, *das; (undergarment, part of dirndl)* Mieder, *das*

bodily ['bɒdılı] **1.** *adj.* körperlich; **~ harm** Körperverletzung, *die;* **~ needs** leibliche Bedürfnisse; **~ organs** Körperorgane. **2.** *adv.* **he lifted her ~:** er hob sie einfach hoch

body ['bɒdı] *n.* **a)** *(of person)* Körper, *der;* Leib, *der (geh.); (of animal)* Körper, *der;* **enough to keep ~ and soul together** genug, um am Leben zu bleiben; **b)** *(corpse)* Leiche, *die;* Leichnam, *der (geh.);* **over my dead ~!** nur über meine Leiche; **c)** *(coll.: person)* Mensch, *der; (woman also)* Person, *die;* **d)** *(group of persons)* Gruppe, *die; (having a particular function)* Organ, *das;* **government ~:** staatliche Einrichtung; **e)** *(mass)* **a huge ~ of water** große Wassermassen; **f)** *(main portion)* Hauptteil, *der;* **g)** *(Motor Veh.)* Karosserie, *die; (Railw.)* Aufbau, *der;* **h)** *(collection)* Sammlung, *die;* **a ~ of knowledge** ein Wissensschatz; **a ~ of facts** Tatsachenmaterial, *das;* **i)** *(of wine)* Körper, *der*

body: ~-building *n.* Bodybuilding, *das;* **~guard** *n. (single)* Leibwächter, *der; (group)* Leibwache, *die;* **~ odour** *n.* Körpergeruch, *der;* **~ search** *n.* Leibesvisitation, *die;* **~work** *n., no pl. (Motor Veh.)* Karosserie, *die*

boffin ['bɒfın] *n. (Brit. sl.)* Eierkopf, *der (salopp)*

bog [bɒg] **1.** *n.* **a)** Moor, *das; (marsh, swamp)* Sumpf, *der;* **b)** *(Brit. sl.: lavatory)* Lokus, *der (salopp).* **2.** *v.t.*, **-gg-: be ~ged down** festsitzen *(fig.);* nicht weiterkommen; **get ~ged down in details** *(fig.)* sich in Details verzetteln

boggle ['bɒgl] *v.i. (be startled)* sprachlos sein; **the mind ~s [at the thought]** bei dem Gedanken wird einem schwindlig

boggy ['bɒgı] *adj.* sumpfig; morastig

bogus ['bəʊgəs] *adj.* falsch; gefälscht ‹*Schmuck, Dokument*›; **~ firm** Schwindelfirma, *die*

Bohemia [bəʊ'hi:mıə] *pr. n.* Böhmen *(das)*

Bohemian [bəʊ'hi:mıən] **1.** *adj.* **a)** *(socially unconventional)* unkonventionell; unbürgerlich; **a ~ person** ein Bohemien; **b)** *(Geog.)* böhmisch. **2.** *n.* **a)** *(socially unconventional person)* Bohemien, *der;* **b)** *(native of Bohemia)* Böhme, *der*/Böhmin, *die*

[1]**boil** [bɔıl] **1.** *v.i.* **a)** kochen; *(Phys.)* sieden; **the kettle's ~ing** das Wasser [im Kessel] kocht; **b)** *(fig.)* ‹*Wasser, Wellen*:› schäumen, brodeln; **c)** *(fig.: be angry)* kochen; schäumen (**with** vor + *Dat.*); **d)** *(fig. coll.: be hot)* **I'm ~ing** mir ist heiß; **be ~ing [hot]** sehr heiß sein. **2.** *v.t.* kochen; **~ sth. dry** etw. verkochen; **it is necessary to ~ the water** man muß

das Wasser abkochen; **~ed potatoes** Salzkartoffeln; **~ the kettle** das Wasser heiß machen; **~ed sweet** *(Brit.)* hartes [Frucht]bonbon. **3.** *n.* Kochen, *das;* **come to/go off the ~:** zu kochen anfangen/aufhören; *(fig.)* sich zuspitzen/sich wieder beruhigen; **bring to the ~:** zum Kochen bringen

~ a'way *v.i.* **a)** *(continue ~ing)* weiterkochen; **b)** *(evaporate completely)* verkochen

~ 'down 1. *v.i.* einkochen; **~ down to sth.** *(fig.)* auf etw. *(Akk.)* hinauslaufen. **2.** *v.t.* einkochen

~ 'over *v.i.* überkochen

~ 'up 1. *v.t.* kochen. **2.** *v.i.* kochen; *(fig.)* sich zuspitzen

²boil *n. (Med.)* Furunkel, *der*

boiler ['bɔɪlə(r)] *n.* **a)** Kessel, *der;* **b)** *(hot-water tank)* Boiler, *der*

boiler: ~-room *n.* Kesselraum, *der;* **~ suit** *n.* Overall, *der*

'boiling-point *n.* Siedepunkt, *der;* **be at/reach ~** *(fig.)* auf dem Siedepunkt sein/den Siedepunkt erreichen

boisterous ['bɔɪstərəs] *adj.* **a)** *(noisily cheerful)* ausgelassen; **b)** *(rough)* wild

boisterously ['bɔɪstərəslɪ] *adv. see* **boisterous:** ausgelassen; wild

bold [bəʊld] *adj.* **a)** *(courageous)* mutig; *(daring)* kühn; **b)** *(forward)* keck; kühn ‹*Worte*›; **make so ~ [as to ...]** so kühn sein[, zu ...]; **c)** *(striking)* auffallend, kühn ‹*Farbe, Muster*›; kräftig ‹*Konturen*›; fett ‹*Schlagzeile*›; **bring out in ~ relief** deutlich hervortreten lassen; **d)** *(vigorous)* kühn; ausdrucksvoll ‹*Stil, Beschreibung*›; **e)** *(Printing)* fett; **in ~ [type]** im Fettdruck

boldly ['bəʊldlɪ] *adv.* **a)** *(courageously)* mutig; *(daringly)* kühn; **b)** *(forwardly)* dreist; **c)** mit kühnem Schwung ‹*malen*›; auffällig ‹*mustern*›

boldness ['bəʊldnɪs] *n., no pl.* **a)** *(courage, daring)* Kühnheit, *die;* **b)** *(forwardness)* Dreistigkeit, *die;* **c)** *(strikingness)* Kühnheit, *die; (of description, style)* Ausdruckskraft, *die*

Bolivia [bə'lɪvɪə] *pr. n.* Bolivien *(das)*

Bolivian [bə'lɪvɪən] **1.** *adj.* bolivianisch. **2.** *n.* Bolivianer, *der*/Bolivianerin, *die*

bollard ['bɒlɑːd] *n. (Brit.)* Poller, *der*

Bolshevik ['bɒlʃɪvɪk] *n.* **a)** *(Hist.)* Bolschewik, *der;* **b)** *(coll.: revolutionary)* Bolschewist, *der*/Bolschewistin, *die (ugs.)*

bolshie, bolshy ['bɒlʃɪ] *adj. (sl.: uncooperative)* aufsässig; rotzig *(salopp)*

bolster ['bəʊlstə(r)] **1.** *n. (pillow)* Nackenrolle, *die.* **2.** *v.t. (fig.)* stärken; **~ sb. up** jmdm. Mut machen; **~ sth. up** etw. stärken

bolt [bəʊlt] **1.** *n.* **a)** *(on door or window)* Riegel, *der; (on gun)* Kammerverschluß, *der;* **b)** *(metal pin)* Schraube, *die; (without thread)* Bolzen, *der;* **c)** *(of crossbow)* Bolzen, *der;* **d) ~ [of lightning]** Blitz[strahl], *der;* **[like] a ~ from the blue** *(fig.)* wie ein Blitz aus heiterem Himmel; **e)** *(sudden dash)* **make a ~ for freedom** einen Fluchtversuch machen. **2.** *v.i.* **a)** davonlaufen; ‹*Pferd:*› durchgehen; **~ out of the shop** aus dem Laden rennen; **b)** *(Hort., Agric.)* vorzeitig Samen bilden; ‹*Salat, Kohl:*› schießen. **3.** *v.t.* **a)** *(fasten with ~)* verriegeln; **~ sb. in/out** jmdn. einsperren/aussperren; **b)** *(fasten with ~s with/without thread)* verschrauben/mit Bolzen verbinden; **~ sth. to sth.** etw. an etw. *(Akk.)* schrauben/mit Bolzen befestigen; **c)** *(gulp down)* **~ [down]** hinunterschlingen ‹*Essen*›. **4.** *adv.* **~ upright** kerzengerade

'bolt-hole *n. (lit. or fig.)* Schlupfloch, *das*

bomb [bɒm] **1.** *n.* **a)** Bombe, *die;* **go like a ~** *(fig. coll.)* ein Bombenerfolg sein; **go down a ~ with** *(fig. sl.)* ein Bombenerfolg sein bei; **b)** *(sl.: large sum of money)* **a ~:** 'ne Masse Geld *(ugs.).* **2.** *v.t.* bombardieren

bombard [bɒm'bɑːd] *v.t.* beschießen; *(fig.)* bombardieren

bombardment [bɒm'bɑːdmənt] *n.* Beschuß, *der; (fig.)* Bombardierung, *die*

bombastic [bɒm'bæstɪk] *adj.* bombastisch; schwülstig

'bomb-disposal *n.* Räumung von Bomben

bomber ['bɒmə(r)] *n.* **a)** *(Air Force)* Bomber, *der (ugs.);* **b)** *(terrorist)* Bombenattentäter, *der*/-attentäterin, *die;* Bombenleger, *der*/-legerin, *die (ugs.)*

'bomber jacket *n.* Bomberjacke, *die*

bombing ['bɒmɪŋ] *n.* Bombardierung, *die*

bomb: ~ scare *n.* Bombendrohung, *die;* **~-shell** *n.* Bombe, *die; (fig.)* Sensation, *die;* **come as a** *or* **be something of a ~-shell** wie eine Bombe einschlagen; **~-site** *n.* Trümmergrundstück, *das*

bona fide [bəʊnə 'faɪdɪ] *adj.* echt

bonanza [bə'nænzə] *n.* **a)** *(unexpected success)* Goldgrube, *die (fig.);* **b)** *(large output)* reiche Ausbeute

bond [bɒnd] **1.** *n.* **a)** Band, *das;* **b)** *in pl. (shackles, lit. or fig.)* Fesseln; **c)** *(uniting force)* Band, *das;* **d)** *(adhesion)* Verbindung, *die;* **e)** *(Commerc.: debenture)* Anleihe, *die;* Schuldverschreibung, *die;* **f)** *(agreement)* Übereinkommen, *das;* **g)** *(Insurance)* ≈ Vertrauensschadenversicherung, *die.* **2.** *v.t.* **a)** kleben (to an + *Akk.*); **b)** *(Commerc.)* unter Zollverschluß nehmen

bondage ['bɒndɪdʒ] *n., no pl. (lit. or fig.)* Sklaverei, *die*

bonded ['bɒndɪd] *adj. (Commerc.)* unter Zollverschluß; **~ goods** Zollagergut, *das;* **~ warehouse** Zollager, *das*

bone [bəʊn] **1.** *n.* **a)** Knochen, *der; (of fish)* Gräte, *die;* **~s** *(fig.: remains)* Gebeine *Pl. (geh.);* **be chilled to the ~** *(fig.)* völlig durchgefroren sein; **work one's fingers to the ~** *(fig.)* bis zum Umfallen arbeiten; **I feel it in my ~s** *(fig.)* ich habe es im Gefühl; **the bare ~s** *(fig.)* die wesentlichen Punkte; **close to the** *or* **near the ~** *(fig.: indecent)* gewagt; **b)** *(material)* Knochen, *der;* **c)** *(stiffener) (in collar)* Kragenstäbchen, *das; (in corset)* Korsettstange, *die;* **d)** *(subject of dispute)* **find a ~ to pick with sb.** mit jmdm. ein Hühnchen zu rupfen haben *(ugs.);* **~ of contention** Zankapfel, *der;* **make no ~s about sth./doing sth.** keinen Hehl aus etw. machen/sich nicht scheuen, etw. zu tun. **2.** *v.t.* den/die Knochen herauslösen aus, ausbeinen ‹*Fleisch, Geflügel*›; entgräten ‹*Fisch*›

bone: ~ 'china *n.* Knochenporzellan, *das;* **~-'dry** *adj.* knochentrocken *(ugs.);* **~-'idle, ~ 'lazy** *adjs.* stinkfaul *(salopp);* **~-meal** *n.* Knochenmehl, *das;* **~-shaker** *n.* Klapperkiste, *die (salopp)*

bonfire ['bɒnfaɪə(r)] *n.* **a)** *(at celebration)* Freudenfeuer, *das;* **B~ Night** *(Brit.)* [Abend des] Guy Fawkes Day *(mit Feuerwerk);* **b)** *(for burning rubbish)* Feuer, *das*

bonkers ['bɒŋkəz] *adj. (sl.)* verrückt *(salopp);* wahnsinnig *(ugs.)*

bonnet ['bɒnɪt] *n.* **a)** *(woman's)* Haube, *die; (child's)* Häubchen, *das;* **b)** *(Brit. Motor Veh.)* Motor-od. Kühlerhaube, *die*

bonny ['bɒnɪ] *adj.* **a)** *(healthy-looking)* prächtig ‹*Baby*›; gesund ‹*Gesicht*›; **b)** *(Scot. and N. Engl.: comely)* hübsch

bonsai ['bɒnsaɪ] *n.* **a)** *(tree)* Bonsai[baum], *der;* **b)** *no pl., no art. (method)* Bonsai, *das*

bonus ['bəʊnəs] *n.* **a)** zusätzliche Leistung; **b)** *(to shareholders, insurance-policy holder)* Bonus, *der; (to employee)* **Christmas ~:** Weihnachtsgratifikation, *die*

bony ['bəʊnɪ] *adj.* **a)** *(of bone)* beinern; knöchern; Knochen-; *(like bone)* knochenartig; **b)** *(big-boned)* grobknochig; **c)** *(skinny)* knochendürr *(ugs.);* spindeldürr; **d)** *(full of bones)* grätig ⟨*Fisch*⟩; ⟨*Fleisch*⟩ mit viel Knochen

boo [bu:] **1.** *int. to surprise sb.* huh; *expr. disapproval, contempt* buh; **he wouldn't say '~' to a goose** er ist sehr schüchtern. **2.** *n.* Buh, *das (ugs.).* **3.** *v.t.* ausbuhen *(ugs.);* **he was ~ed off the stage** er wurde so ausgebuht, daß er die Bühne verließ *(ugs.).* **4.** *v.i.* buhen *(ugs.)*

boob [bu:b] *(Brit. sl.)* **1.** *n.* **a)** *(mistake)* Fehler, *der;* Schnitzer, *der (ugs.);* **b)** *(breast)* Titte, *die (derb).* **2.** *v.i.* einen Schnitzer machen *(ugs.)*

booby ['bu:bɪ]: **~ prize** *n. Preis für den schlechtesten Teilnehmer an einem Wettbewerb;* **~ trap** *n.* **a)** *Falle, mit der man jmdm. einen Streich spielen will;* **b)** *(Mil.)* versteckte Sprengladung; **~-trap** *v.t.* **a)** [für einen Streich] präparieren; **b)** *(Mil.)* **the door had been ~-trapped** an der Tür war eine versteckte Sprengladung angebracht worden

book [bʊk] **1.** *n.* **a)** Buch, *das;* **be a closed ~ [to sb.]** *(fig.)* [jmdm. *od.* für jmdn.] ein Buch mit sieben Siegeln sein; **throw the ~ at sb.** *(fig.)* jmdn. kräftig zusammenstauchen *(ugs.);* **bring to ~** *(fig.)* zur Rechenschaft ziehen; **in my ~** *(fig.)* meiner Ansicht *od.* Meinung nach; **be in sb.'s good/bad ~s** *(fig.)* bei jmdm. gut/schlecht angeschrieben sein; **I can read you like a ~** *(fig.)* ich kann in dir lesen wie in einem Buch; **take a leaf out of sb.'s ~** *(fig.)* sich *(Dat.)* jmdn. zum Vorbild nehmen; **you could take a leaf out of his ~:** du könntest dir von ihm eine Scheibe abschneiden *(ugs.);* **b)** *in pl. (records, accounts)* Bücher; **do the ~s** die Abrechnung machen; **balance the ~s** die Bilanz machen *od.* ziehen; *see also* **keep 1 h; c)** *in pl. (list of members)* **be on the ~s** auf der [Mitglieds]liste *od.* im Mitgliederverzeichnis stehen; **d)** *(record of bets)* Wettbuch, *das;* **make** *or* **keep a ~ on sth.** Wetten auf etw. *(Akk.)* annehmen; **e) ~ of tickets** Fahrscheinheft, *das;* **~ of stamps/matches** Briefmarkenheft/Streichholzbriefchen, *das.* **2.** *v.t.* **a)** buchen ⟨*Reise, Flug, Platz [im Flugzeug]*⟩; [vor]bestellen ⟨*Eintrittskarte, Tisch, Zimmer, Platz [im Theater]*⟩; anmelden ⟨*Telefongespräch*⟩; engagieren, verpflichten ⟨*Künstler, Orchester*⟩; **be fully ~ed** ⟨*Vorstellung:*⟩ ausverkauft sein; ⟨*Flug[zeug]:*⟩ ausgebucht sein; ⟨*Hotel:*⟩ voll belegt *od.* ausgebucht sein; **b)** *(enter in ~)* eintragen; *(for offence)* aufschreiben *(ugs.)* **(for** wegen); **c)** *(issue ticket to)* **we are ~ed on a flight to Athens** man hat für uns einen Flug nach Athen gebucht. **3.** *v.i.* buchen; *(for travel, performance)* vorbestellen

~ 'in 1. *v.i.* sich eintragen; **we ~ed in at the Ritz** wir sind im Ritz abgestiegen. **2.** *v.t.* **a)** *(make reservation for)* Zimmer/ein Zimmer vorbestellen *od.* reservieren für; **b)** *(register)* eintragen

~ 'up 1. *v.i.* buchen. **2.** *v.t.* buchen; **the guest-house is ~ed up** die Pension ist ausgebucht *od.* voll belegt

book: ~case *n.* Bücherschrank, *der;* **~ club** *n.* Buchklub, *der;* **~-ends** *n. pl.* Buchstützen

bookie ['bʊkɪ] *n. (coll.)* Buchmacher, *der*

booking ['bʊkɪŋ] *n.* **a)** Buchung, *die; (of ticket)* Bestellung, *die; (of table, room, seat)* Vorbestellung, *die;* **make/cancel a ~:** buchen/eine Buchung rückgängig machen; *(for tickets)* bestellen/abbestellen; **change one's ~:** umbuchen; *(for tickets)* umbestellen; **b)** *(of performer)* Engagement, *das*

booking: ~-clerk *n.* Schalterbeamte, *der*/-beamtin, *die;* Fahrkartenverkäufer, *der*/-verkäuferin, *die;* **~-office** *n. (in station)* [Fahrkarten]schalter, *der; (in theatre)* [Theater]kasse, *die; (selling tickets in advance)* Vorverkaufsstelle, *die*

book: ~-jacket *n.* Schutzumschlag, *der;* **~keeper** *n.* Buchhalter, *der*/-halterin, *die;* **~keeping** *n.* Buchführung, *die;* Buchhaltung, *die*

booklet ['bʊklɪt] *n.* Broschüre, *die*

book: ~maker *n. (in betting)* Buchmacher, *der;* **~mark, ~marker** *ns.* Lese- *od.* Buchzeichen, *das;* **~ review** *n.* Buchbesprechung, *die;* **~seller** *n.* Buchhändler, *der*/-händlerin, *die;* **~shelf** *n.* Bücherbord, *das;* **on my ~shelves** in meinen Bücherregalen; **~shop** *n.* Buchhandlung, *die;* **~stall** *n.* Bücherstand, *der;* **~store** *(Amer.) see* **~shop; ~ token** *n.* Büchergutschein, *der;* **~worm** *n. (lit. or fig.)* Bücherwurm, *der*

¹boom [bu:m] *n.* **a)** *(for camera or microphone)* Ausleger, *der;* **b)** *(Naut.)* Baum, *der;* **c)** *(floating barrier)* [schwimmende] Absperrung

²boom 1. *v.i.* **a)** dröhnen; ⟨*Kanone, Wellen, Brandung:*⟩ dröhnen, donnern; **b)** ⟨*Geschäft, Verkauf, Stadt, Gebiet:*⟩ sich sprunghaft entwickeln; ⟨*Preise, Aktien:*⟩ rapide steigen. **2.** *n.* **a)** *(of gun, waves)* Dröhnen, *das;* Donnern, *das;* **b)** *(in business)* [sprunghafter] Aufschwung; Boom, *der; (in prices)* [rapider] Anstieg; **a ~ year** ein Boomjahr; **c)** *(period of economic expansion)* Hochkonjunktur, *die;* Boom, *der*

~ 'out 1. *v.i.* ⟨*Stimme:*⟩ dröhnen; ⟨*Kanone:*⟩ donnern, dröhnen. **2.** *v.t.* brüllen ⟨*Kommando, Befehl*⟩

boomerang ['bu:məræŋ] **1.** *n. (lit. or fig.)* Bumerang, *der.* **2.** *v.i. (fig.)* sich als Bumerang erweisen

'boom town *n. Stadt in sprunghaftem Aufschwung*

boon [bu:n] *n. (blessing)* Segen, *der;* Wohltat, *die*

boor [bʊə(r)] *n.* Rüpel, *der (abwertend)*

boorish ['bʊərɪʃ] *adj.*, **boorishly** ['bʊərɪʃlɪ] *adv.* flegelhaft *(abwertend);* rüpelhaft *(abwertend)*

boost [bu:st] **1.** *v.t.* **a)** steigern; ankurbeln ⟨*Wirtschaft*⟩; in die Höhe treiben ⟨*Preis, Wert, Aktienkurs*⟩; stärken, heben ⟨*Selbstvertrauen, Moral*⟩; **b)** *(Electr.)* erhöhen ⟨*Spannung*⟩. **2.** *n.* Auftrieb, *der; (increase)* Zunahme, *die;* **give sb./sth. a ~:** jmdn./einer Sache Auftrieb geben; **be given a ~:** Auftrieb erhalten

booster ['bu:stə(r)] *n. (Med.)* **~ [shot** *or* **injection]** Auffrischimpfung, *die*

boot [bu:t] **1.** *n.* **a)** Stiefel, *der;* **get the ~** *(fig. coll.)* rausgeschmissen werden *(ugs.);* **give sb. the ~** *(fig. coll.)* jmdn. rausschmeißen *(ugs.);* **the ~ is on the other foot** *(fig.)* es ist genau umgekehrt; **b)** *(Brit.: of car)* Kofferraum, *der.* **2.** *v.t.* **a)** *(coll.)* treten; kicken *(ugs.)* ⟨*Ball*⟩; **~ sb. out** *(fig. coll.)* jmdn. rausschmeißen *(ugs.);* **b)** *(Computing)* **~ [up]** laden

booth [bu:ð] *n.* **a)** Bude, *die;* **b)** *(telephone ~)* Telefonzelle, *die;* **c)** *(polling-~)* Wahlkabine, *die*

'bootleg *adj.* geschmuggelt; *(sold/made)* schwarz *(ugs.) od.* illegal verkauft/gebrannt

booty ['bu:tɪ] *n., no pl.* Beute, *die*

booze [bu:z] *(coll.)* **1.** *v. i.* saufen *(derb).* **2.** *n., no pl. (drink)* Alkohol, *der*

'booze-up *n. (coll.)* Besäufnis, *das (salopp);* **have a ~:** einen heben *(ugs.)*

bop [bɒp] *(coll.)* **1.** *v. i. (zur Popmusik)* tanzen. **2.** *n.* Tanz, *der (zur Popmusik);* **have a ~:** tanzen

bordello [bɔ:'deləʊ] *n., pl.* ~s *(Amer.)* Bordell, *das*

border ['bɔ:də(r)] **1.** *n.* **a)** Rand, *der; (of table-cloth, handkerchief, dress)* Bordüre, *die;* **b)** *(of country)* Grenze, *die;* **c)** *(flowerbed)* Rabatte, *die.* **2.** *attrib. adj.* Grenz⟨*stadt, -gebiet, -streit*⟩. **3.** *v. t.* **a)** *(adjoin)* [an]grenzen an (+ *Akk.*); **be ~ed by** [an]grenzen an (+ *Akk.*); **b)** *(put a ~ to, act as ~ to)* umranden; einfassen; **c)** *(resemble closely)* grenzen an (+ *Akk.*). **4.** *v. i.* ~ **on** *see* **3a, c**

'borderline 1. *n.* Grenzlinie, *die; (fig.)* Grenze, *die.* **2.** *adj.* **sb./sth. is ~** *(fig.)* jmd. ist/etw. liegt auf der Grenze; **a ~ case/candidate/type** *(fig.)* ein Grenzfall

¹bore [bɔ:(r)] **1.** *v. t. (make hole in)* bohren. **2.** *v. i. (drill)* bohren **(for** nach). **3.** *n.* **a)** *(of firearm, engine cylinder)* Bohrung, *die; (of tube, pipe)* Innendurchmesser, *der;* **b)** *(calibre)* Kaliber, *das*

²bore 1. *n.* **a)** *(nuisance)* **it's a real ~:** es ist wirklich ärgerlich; **what a ~!** wie ärgerlich!; **b)** *(dull person)* Langweiler, *der (ugs. abwertend).* **2.** *v. t. (weary)* langweilen; **sb. is ~d with sth.** etw. langweilt jmdn.; **I'm ~d** ich langweile mich; ich habe Langeweile; **~ sb. to death** *or* **to tears** *(coll.)* jmdn. zu Tode langweilen

³bore *see* **²bear**

boredom ['bɔ:dəm] *n., no pl.* Langeweile, *die*

'borehole *n.* Bohrloch, *das*

boring ['bɔ:rɪŋ] *adj.* langweilig

born [bɔ:n] **1. be ~:** geboren werden; **I was ~ in England** ich bin *od.* wurde in England geboren; **I wasn't ~ yesterday** *(fig.)* ich bin nicht von gestern *(ugs.);* **be ~ blind/lucky** blind von Geburt sein/ein Glückskind sein; **be ~ a poet** zum Dichter geboren sein. **2.** *adj.* **a)** geboren; ~ **again** *(fig.)* wiedergeboren; **in all my ~ days** *(fig. coll.)* in meinem ganzen Leben; *see also* **breed 1c; b)** *(destined to be)* **be a ~ orator** der geborene Redner sein

borne *see* **²bear**

borough ['bʌrə] *n.* **a)** *(Brit.: town sending members to Parliament)* Stadt[bezirk] mit Vertretung im Parlament; **b)** *(Amer.)* **the ~ of ...** *(town)* die Stadt ...; *(village)* die Gemeinde ...; **c)** *(Amer.: district of New York or Alaska)* Verwaltungsbezirk

borrow ['bɒrəʊ] **1.** *v. t.* **a)** [sich *(Dat.)*] ausleihen; [sich *(Dat.)*] borgen; entleihen, ausleihen ⟨*Buch, Schallplatte usw. aus der Leihbücherei*⟩; [sich *(Dat.)*] leihen ⟨*Geld von der Bank*⟩; [sich *(Dat.)*] leihen, [sich *(Dat.)*] borgen ⟨*Geld*⟩; **b)** *(fig.)* übernehmen ⟨*Idee, Methode, Meinung*⟩; entlehnen ⟨*Wort*⟩; **sb. is living on ~ed time** jmds. Uhr ist abgelaufen. **2.** *v. i.* borgen; *(from bank)* Kredit aufnehmen **(from** bei)

borrower ['bɒrəʊə(r)] *n. (from bank)* Kreditnehmer, *der; (from library)* Entleiher, *der*

borrowing ['bɒrəʊɪŋ] *n. (from bank)* Kreditaufnahme, *die* **(from** bei); *(from library)* Entleihen, *das;* Ausleihen, *das; (fig.)* Übernahme, *die*

bos'n ['bəʊsn] *see* **boatswain**

bosom ['bʊzm] *n.* **a)** *(person's breast)* Brust, *die;* Busen, *der (bes. dichter.);* **b)** *(fig.: enfolding relationship)* Schoß, *der (geh.);* **in the ~ of one's family** im Schoße der Familie; *attrib.* **a ~ friend** ein guter Freund; ein Busenfreund

boss [bɒs] *(coll.)* **1.** *n.* Boß, *der (ugs.);* Chef, *der.* **2.** *v. t.* ~ **sb. [about** *or* **around]** jmdn. herumkommandieren *(ugs.)*

bossy ['bɒsɪ] *adj. (coll.)* herrisch; **don't be so ~:** hör auf herumzukommandieren *(ugs.)*

bosun, bo'sun ['bəʊsn] *see* **boatswain**

botanical [bə'tænɪkl] *adj.* botanisch; ~ **garden[s]** botanischer Garten

botanist ['bɒtənɪst] *n.* Botaniker, *der*/Botanikerin, *die*

botany ['bɒtənɪ] *n., no pl.* Botanik, *die;* Pflanzenkunde, *die*

botch [bɒtʃ] *v. t.* **a)** *(bungle)* pfuschen bei *(ugs. abwertend)* ⟨*Reparatur, Arbeit*⟩; **a ~ed job** eine gepfuschte Arbeit *(ugs. abwertend);* **b)** *(repair badly)* [notdürftig] flicken

~ **'up** *v. t.* **a)** *(bungle)* verpfuschen *(ugs. abwertend);* **b)** *(repair badly)* [notdürftig] flicken

both [bəʊθ] **1.** *adj.* beide; **we ~ like cooking** wir kochen beide gern; **~ [the] brothers** beide Brüder; **~ our brothers** unsere beiden Brüder; **you can't have it ~ ways** beides [zugleich] geht nicht; *see also* **cut 2a. 2.** *pron.* beide; **~ [of them] are dead** beide sind tot; **they are ~ dead** sie sind beide tot; **~ of you/them are ...:** ihr seid/sie sind beide ...; **for them ~:** für sie beide; **go along to bed, ~ of you** ihr geht jetzt ins Bett, alle beide. **3.** *adv.* ~ **A and B** sowohl A als [auch] B; **~ you and I** wir beide

bother ['bɒðə(r)] **1.** *v. t.* **a)** *in pass. (take trouble)* **I can't be ~ed [to do it]** ich habe keine Lust[, es zu machen]; **I can't be ~ed with details like that** ich kann mich nicht mit solchen Kleinigkeiten abgeben *od.* befassen; **b)** *(annoy)* lästig sein *od.* fallen (+ *Dat.*); ⟨*Lärm, Licht:*⟩ stören; ⟨*Schmerz, Wunde, Zahn, Rücken:*⟩ zu schaffen machen (+ *Dat.*); **I'm sorry to ~ you, but ...:** es tut mir leid, daß ich Sie damit belästigen muß, aber ...; **don't ~ me now** laß mich jetzt in Ruhe!; **c)** *(worry)* Sorgen machen (+ *Dat.*); ⟨*Problem, Frage:*⟩ beschäftigen; **I'm not ~ed about him/the money** seinetwegen/wegen des Geldes mache ich mir keine Gedanken; **what's ~ing you/is something ~ing you?** was hast du denn/hast du etwas? **2.** *v. i. (trouble oneself)* **don't ~ to do it** Sie brauchen es nicht zu tun; **you needn't have ~ed to come** Sie hätten wirklich nicht zu kommen brauchen; **you needn't/shouldn't have ~ed** das wäre nicht nötig gewesen; **don't ~!** nicht nötig!; **~ with sth./sb.** sich mit etw./jmdm. aufhalten; **~ about sth./sb.** sich *(Dat.)* über etw./jmdn. Gedanken machen. **3.** *n.* **a)** *(nuisance)* **what a ~!** wie ärgerlich!; **it's a real/such a ~:** es ist wirklich lästig; **b)** *(trouble)* Ärger, *der;* **it's no ~ [for me]** es macht mir gar nichts aus; **the children were no ~ at all** ich hatte/wir hatten mit den Kindern überhaupt keine Schwierigkeiten; **have a spot of ~ with sth.** Schwierigkeiten mit etw. haben; **go to the ~ of doing sth.** sich *(Dat.)* die Mühe machen, etw. zu tun. **4.** *int. (coll.)* wie ärgerlich!

bottle ['bɒtl] **1.** *n.* **a)** Flasche, *die;* **a beer-~:** eine Bierflasche; **a ~ of beer** eine Flasche Bier; **b)** *(fig. coll.: alcoholic drink)* **be too fond of the ~:** dem Alkohol zu sehr zugetan sein; **be on the ~:** trinken; *see also* **hit 1h. 2.** *v. t.* **a)** *(put into ~s)* in Flaschen [ab]füllen; **~d beer** Flaschenbier, *das;* **~d gas** Flaschengas, *das;* **b)** *(preserve in jars)* einmachen

~ **'up** *v.t. (conceal)* in sich *(Dat.)* aufstauen
bottle: ~ **bank** *n.* Altglasbehälter, *der;* ~**-fed** *adj.* mit der Flasche gefüttert; ~**-green** *adj.* flaschengrün; ~**-neck** *n. (fig.)* Flaschenhals, *der (ugs.); (in production process also)* Engpaß, *der;* ~**-opener** *n.* Flaschenöffner, *der;* ~**-party** *n.* Bottle-Party, *die;* ~**-top** *n.* Flaschenverschluß, *der*
bottom ['bɒtəm] **1.** *n.* **a)** *(lowest part)* unteres Ende; *(of cup, glass, box, chest)* Boden, *der; (of valley, canyon, crevasse, well, shaft)* Sohle, *die; (of hill, slope, cliff, stairs)* Fuß, *der;* **the ~ of the valley** die Talsohle; **[be] at the ~ of the page/list** unten auf der Seite/Liste [sein]; ~ **up** auf dem Kopf; verkehrt herum; ~**s up!** *(coll.)* hoch die Tassen!; **the ~ fell** *or* **dropped out of her world/the market** *(fig.)* für sie brach eine Welt zusammen/der Markt brach zusammen; **b)** *(buttocks)* Hinterteil, *das (ugs.);* Po[dex], *der (fam.);* **c)** *(of chair)* Sitz, *der;* Sitzfläche, *die;* **d)** *(of sea, lake)* Grund, *der;* **go to the ~:** [ver]sinken; **touch ~:** Grund haben; *(fig.)* den Tiefpunkt erreichen; **e)** *(farthest point)* **at the ~ of the garden/street** hinten im Garten/am Ende der Straße; **f)** *(underside)* Unterseite, *die;* **g)** *(fig.)* **start at the ~:** ganz unten anfangen; **be ~ of the class/league** der/die Letzte in der Klasse sein/Tabellenletzte[r] sein; **h)** *usu. in pl.* ~**[s]** *(of track suit, pyjamas)* Hose, *die;* **i)** *(fig.: basis, origin)* **be at the ~ of sth.** hinter etw. *(Dat.)* stecken *(ugs.);* **get to the ~ of sth.** einer Sache *(Dat.)* auf den Grund kommen; **at ~:** im Grunde genommen; **j)** *(Naut.)* Schiffsboden, *der;* **k)** *(Brit. Motor Veh.)* **in ~:** im ersten Gang. **2.** *adj.* **a)** *(lowest)* unterst...; *(lower)* unter...; **b)** *(fig.: last)* letzt...; **be ~:** der/die/das Letzte sein
bottom 'drawer *n. (fig.)* Aussteuer, *die;* **put sth. [away] in one's ~ drawer** etw. für die Aussteuer beiseite legen
bottomless ['bɒtəmlɪs] *adj.* bodenlos; unendlich tief ‹*Meer, Ozean*›; *(fig.: inexhaustible)* unerschöpflich
bottom 'line *n. (fig. coll.)* **the ~:** das Fazit
bough [baʊ] *n.* Ast, *der*
bought *see* **buy 1**
boulder ['bəʊldə(r)] *n.* Felsbrocken, *der*
bounce [baʊns] **1.** *v.i.* **a)** springen; ~ **up and down on sth.** auf etw. *(Dat.)* herumspringen; **b)** *(coll.)* ‹*Scheck:*› nicht gedeckt sein. **2.** *v.t.* aufspringen lassen ‹*Ball*›; **he ~d the baby on his knee** er ließ das Kind auf den Knien reiten. **3.** *n.* **a)** *(rebound)* Aufprall, *der;* **on the ~:** beim Aufprall; **b)** *(rebounding power)* ≈ Elastizität, *die; (fig.: energy)* Schwung, *der*
~ **'back** *v.i.* zurückprallen; *(fig.)* ‹*Person:*› [plötzlich] wieder dasein
~ **'off 1.** *v.i.* abprallen. **2.** *v.t.* ~ **sth. off sth.** etw. von etw. abprallen lassen; ~ **off sth.** von etw. abprallen
bouncer ['baʊnsə(r)] *n. (coll.)* Rausschmeißer, *der (ugs.)*
bouncing ['baʊnsɪŋ] *adj.* kräftig, stramm ‹*Baby*›
bouncy ['baʊnsɪ] *adj.* **a)** gut springend ‹*Ball*›; federnd ‹*Matratze, Bett*›; **b)** *(fig.: lively)* munter
[1]**bound** [baʊnd] **1.** *n.* **a)** *usu. in pl. (limit)* Grenze, *die;* **within the ~s of possibility** *or* **the possible** im Bereich des Möglichen; **go beyond the ~s of decency** die Grenzen des Anstands verletzen; **sth. is out of ~s [to sb.]** der Zutritt zu etw. ist [für jmdn.] verboten; **keep within the ~s of reason/propriety** vernünftig/im Rahmen bleiben; **b)** *(of territory)* Grenze, *die.* **2.** *v.t., usu. in pass.* begrenzen
[2]**bound 1.** *v.i. (spring)* hüpfen; springen; ~ **into the room** ins Zimmer stürzen; **the dog came ~ing up** der Hund kam angesprungen. **2.** *n. (spring)* Satz, *der;* **at** *or* **with one ~:** mit einem Satz
[3]**bound** *pred. adj.* **be ~ for home/Frankfurt** auf dem Heimweg/nach Frankfurt unterwegs sein; **homeward ~:** auf dem Weg nach Hause; **where are you ~ for?** wohin geht die Reise?; **all passengers ~ for Zürich** alle Passagiere nach Zürich
[4]**bound** *see* **bind 1, 2**
boundary ['baʊndərɪ] *n.* Grenze, *die*
boundless ['baʊndlɪs] *adj.* grenzenlos
bountiful ['baʊntɪfl] *adj. (generous)* großzügig; gütig ‹*Gott*›; *(plentiful)* reichlich ‹*Ernte, Gaben, Ertrag*›
bounty ['baʊntɪ] *n. (reward)* Kopfgeld, *das; (for capturing animal)* Fangprämie, *die*
bouquet [bʊ'keɪ, bəʊ'keɪ, 'bu:keɪ] *n.* **a)** *(bunch of flowers)* Bukett, *das;* [Blumen]strauß, *der;* **b)** *(perfume of wine)* Bukett, *das;* Blume, *die*
bourbon ['bɜ:bən, 'bʊəbən] *n. (Amer.)* ~ **[whiskey]** Bourbon, *der*
bourgeois ['bʊəʒwɑ:] **1.** *n., pl. same* **a)** *(middle-class person)* Bürger, *der*/Bürgerin, *die;* **b)** *(person with conventional ideas, selfish materialist)* Spießbürger, *der (abwertend);* Spießer, *der*/Spießerin, *die (abwertend).* **2.** *adj.* **a)** *(middle-class)* bürgerlich; **b)** *(conventional, selfishly materialist)* spießbürgerlich *(abwertend)*
bourgeoisie [bʊəʒwɑ:'zi:] *n.* **a)** Bürgertum, *das;* **b)** *(capitalist class)* Bourgeoisie, *die (marx.)*
bout [baʊt] *n.* **a)** *(spell)* Periode, *die;* **b)** *(contest)* Wettkampf, *der;* **c)** *(fit)* Anfall, *der;* **he's out on one of his drinking ~s again** er ist mal wieder auf einer seiner Zechtouren *(ugs.)*
boutique [bu:'ti:k] *n.* Boutique, *die*
[1]**bow** [bəʊ] **a)** *(curve, weapon, Mus.)* Bogen, *der;* **have two strings to one's ~:** eine Alternative haben; **b)** *(tied knot or ribbon)* Schleife, *die*
[2]**bow** [baʊ] **1.** *v.i.* **a)** *(submit)* sich beugen **(to** *Dat.***)**; **b)** ~ **[down to** *or* **before sb./sth.]** *(bend)* sich [vor jmdm./etw.] verbeugen *od.* verneigen; **c)** *(incline head)* ~ **[to sb.]** sich [vor jmdm.] verbeugen. **2.** *v.t. (cause to bend)* beugen; ~**ed down by** *or* **with care/responsibilities/age** *(fig.)* von Sorgen/Verpflichtungen niedergedrückt/vom Alter gebeugt. **3.** *n.* Verbeugung, *die*
[3]**bow** [baʊ] *n. (Naut.) usu. in pl.* Bug, *der;* **in the ~s** im Bug; **on the ~:** am Bug
bowel ['baʊəl] *n.* **a)** *(Anat.)* ~**s** *pl., (Med.)* ~**:** Darm, *der;* **b)** *in pl. (interior)* Innere, *das*
[1]**bowl** [bəʊl] *n.* **a)** *(basin)* Schüssel, *die; (shallower)* Schale, *die;* **mixing/washing-up ~:** Rühr-/Abwaschschüssel, *die;* **soup-~:** Suppentasse, *die;* **sugar-~:** Zuckerdose, *die;* **a ~ of water** eine Schüssel/Schale Wasser; **a ~ of soup** eine Tasse Suppe; **b)** *(of WC)* Schüssel, *die; (of spoon)* Schöpfteil, *der; (of pipe)* [Pfeifen]kopf, *der*
[2]**bowl 1.** *n.* **a)** *(ball)* Kugel, *die; (in skittles)* [Kegel]kugel, *die; (in tenpin bowling)* [Bowling]kugel, *die;* **b)** *in pl. (game)* Bowlsspiel, *das;* Bowls, *das.* **2.** *v.i.* **a)** *(play ~s)* Bowls spielen; *(play skittles)* kegeln; *(play ten-pin bowling)*

bowlen; **b)** *(Cricket)* werfen. **3.** *v. t.* **a)** *(roll)* rollen lassen; ~ **sb. over** *(fig.)* jmdn. überwältigen *od. (ugs.)* umhauen; **b)** *(Cricket etc.)* werfen; ~ **the batsman [out]/side out** den Schlagmann/die Mannschaft ausschlagen

bow [bəʊ]: **~-'legged** *adj.* krummbeinig; O-beinig *(ugs.)*; **be ~-legged** krumme Beine *od. (ugs.)* O-Beine haben; **~-'legs** *n. pl.* krumme Beine; O-Beine *Pl. (ugs.)*

¹bowler ['bəʊlə(r)] *n.* *(Cricket)* Werfer, *der*

²bowler *n.* ~ **[hat]** Bowler, *der*

bowling ['bəʊlɪŋ] *n.* **[ten-pin]** ~: Bowling, *das;* **go** ~: bowlen gehen

bowling: ~-alley *n.* *(for tenpin ~)* Bowlingbahn, *die;* *(for skittles)* Kegelbahn, *die;* **~-green** *n.* *Rasenfläche für Bowls*

bow [bəʊ]: **~-tie** *n.* Fliege, *die;* [Smoking-/Frack]schleife, *die;* **~-window** *n.* Erkerfenster, *das*

¹box [bɒks] *n.* **a)** *(container)* Kasten, *der;* *(bigger)* Kiste, *die;* *(made of cardboard, thin wood, etc.)* Schachtel, *die;* **a ~ of cigars** eine Schachtel Zigarren; **pencil-~:** Federkasten, *der;* **jewellery-~:** Schmuckkasten, *der;* **cardboard ~:** [Papp]karton, *der;* *(smaller)* [Papp]schachtel, *die;* **shoe-~:** Schuhkarton, *der;* ~ **of matches** Streichholzschachtel, *die;* **b) the** ~ *(coll.: television)* der Kasten *(ugs. abwertend);* die Flimmerkiste *(scherzh.);* **c)** *(in theatre etc.)* Loge, *die*

~ **'in** *v. t.* **a)** *(enclose in ~)* in einem Gehäuse unterbringen; **b)** *(enclose tightly)* einklemmen; **feel ~ed in** sich eingeengt fühlen

²box 1. *n.* *(slap, punch)* Schlag, *der;* **he gave him a ~ on the ear[s]** er gab ihm eine Ohrfeige. **2.** *v. t.* **a)** *(slap, punch)* schlagen; **he ~ed his ears** *or* **him round the ears** er ohrfeigte ihn; **get one's ears ~ed** eine Ohrfeige bekommen; **b)** *(fight with fists)* ~ **sb.** gegen jmdn. boxen. **3.** *v. i.* boxen **(with, against** gegen)

box: ~ camera *n.* Box, *die;* **~car** *n.* *(Amer. Railw.)* gedeckter [Güter]wagen

boxer ['bɒksə(r)] *n.* **a)** Boxer, *der;* **b)** *(dog)* Boxer, *der*

'boxer shorts *n. pl.* Boxershorts *Pl.*

boxing ['bɒksɪŋ] *n.* Boxen, *das;* **professional/amateur ~:** Berufs-/Amateurboxen, *das*

boxing: B~ Day *n.* *(Brit.)* zweiter Weihnachtsfeiertag; **~-glove** *n.* Boxhandschuh, *der;* **~-match** *n.* Boxkampf, *der;* **~-ring** *n.* Boxring, *der*

box: ~ number *n.* *(at newspaper office)* Chiffre, *die;* *(at post office)* Postfach, *das;* **~-office** *n.* Kasse, *die;* **~-room** *n.* *(Brit.)* Abstellraum, *der*

boy [bɔɪ] **1.** *n.* **a)** Junge, *der;* **baby ~:** kleiner Junge; **~s' school** Jungenschule, *die;* **a ~'s name** ein Jungenname; **[my]** ~ *(as address)* [mein] Junge; **~s will be ~s** so sind Jungs/Männer nun mal; **jobs for the ~s** Vetternwirtschaft, *die (abwertend);* **b)** *(servant)* Boy, *der.* **2.** *int.* **[oh]** ~! Junge, Junge! *(ugs.)*

boycott ['bɔɪkɒt] **1.** *v. t.* boykottieren. **2.** *n.* Boykott, *der*

'boy-friend *n.* Freund, *der*

boyish ['bɔɪɪʃ] *adj.* jungenhaft

BR *abbr.* **British Rail[ways]** *britische Eisenbahngesellschaft*

bra [brɑː] *n.* BH, *der (ugs.)*

brace [breɪs] **1.** *n.* **a)** *(buckle)* Schnalle, *die;* *(connecting piece)* Klammer, *die;* *(Dent.)* [Zahn]spange, *die;* [Zahn]klammer, *die;* **b)** *in pl.* *(trouser-straps)* Hosenträger; **c)** *pl. same (pair)* **a/two ~ of** zwei/vier; **d)** *(Printing, Mus.)* geschweifte Klammer; Akkolade, *die;* **e)** *(strut)* Strebe, *die.* **2.** *v. t.* **a)** *(fasten)* befestigen; *(stretch)* spannen; *(string up)* anspannen; *(with struts)* stützen; **b)** *(support)* stützen. **3.** *v. refl.* ~ **oneself [up]** *(fig.)* sich zusammennehmen; ~ **oneself [up] for sth.** *(fig.)* sich auf etw. *(Akk.)* [innerlich] vorbereiten

brace and 'bit *n.* Bohrwinde, *die*

bracelet ['breɪslɪt] *n.* *(band)* Armband, *das;* *(chain)* Kettchen, *das;* *(bangle)* Armreif, *der*

bracing ['breɪsɪŋ] *adj.* belebend

bracken ['brækn] *n.* [Adler]farn, *der*

bracket ['brækɪt] **1.** *n.* **a)** *(support, projection)* Konsole, *die;* *(lamp-support)* Lampenhalter, *der;* **b)** *(mark)* Klammer, *die;* **open/close ~s** Klammer auf/zu; **c)** *(group)* Gruppe, *die.* **2.** *v. t.* **a)** *(enclose in ~s)* einklammern; **b)** *(couple with brace)* mit einer Klammer verbinden; *(fig.)* in Verbindung bringen

brackish ['brækɪʃ] *adj.* brackig

brag [bræg] **1.** *v. i.,* **-gg-** prahlen **(about** mit). **2.** *v. t.,* **-gg-** prahlen; **he ~s that he has a Rolls Royce** er prahlt damit, daß er einen Rolls-Royce hat

braggart ['brægət] *n.* Prahler, *der*/Prahlerin, *die*

braid [breɪd] **1.** *n.* **a)** *(plait)* Flechte, *die (geh.);* Zopf, *der;* *(band entwined with hair)* Haarband, *das;* **b)** *(decorative woven band)* Borte, *die;* **c)** *(on uniform)* Litze, *die;* *(with metal threads)* Tresse, die. **2.** *v. t.* **a)** *(plait; arrange in ~s)* flechten; **b)** zusammenbinden ‹*Haare*›; **c)** *(trim with ~)* mit Borten/Litzen/Tressen besetzen

Braille [breɪl] *n.* Blindenschrift, *die*

brain [breɪn] **1.** *n.* **a)** Gehirn, *das;* **have [got] sex/money on the ~:** nur Sex/Geld im Kopf haben; **use your ~[s]** gebrauch deinen Verstand; **he's got a good ~:** er ist ein kluger Kopf; **b)** *in pl.* *(Gastr.)* Hirn, *das;* **c)** *(coll.: clever person)* **she's the ~[s] of the class** sie ist die Intelligenteste in der Klasse. **2.** *v. t.* **I'll ~ you!** *(coll.)* du kriegst gleich eins auf die Rübe! *(ugs.)*

brain: ~-child *n.* *(coll.)* Geistesprodukt, *das;* **~-drain** *n.* *(coll.)* Abwanderung [von Wissenschaftlern]

brainless ['breɪnlɪs] *adj.* *(stupid)* hirnlos

brain: ~storm *n.* **a)** Anfall geistiger Umnachtung; **b)** *(Amer. coll.)* *see* **~wave;** ~ **surgeon** *n.* Gehirnchirurg, *der;* **~-teaser** *n.* Denk[sport]aufgabe, *die;* ~ **tumour** *n.* Gehirntumor, *der;* **~wash** *v. t.* einer Gehirnwäsche unterziehen; **~wash sb. into doing sth.** jmdm. [ständig] einreden, etw. zu tun; **~washing** *n.* Gehirnwäsche, *die;* **~wave** *n.* *(coll.: inspiration)* genialer Einfall

brainy ['breɪnɪ] *adj.* intelligent

braise [breɪz] *v. t.* *(Cookery)* schmoren

brake [breɪk] **1.** *n.* *(apparatus; coll.: pedal etc.)* Bremse, *die;* **sth. acts as a ~ on sth.** etw. bremst etw.; **put on** *or* **apply the ~s** die Bremse betätigen; *(fig.)* zurückstecken; **put the ~[s] on sth.** *(fig.)* etw. bremsen. **2.** *v. t. & i.* bremsen; ~ **hard** scharf bremsen

brake: ~-block *n.* Bremsklotz, *der;* **~-drum** *n.* Bremstrommel, *die;* ~ **fluid** *n.* Bremsflüssigkeit, *die;* **~-light** *n.* Bremslicht, *das;* ~ **lining** *n.* Bremsbelag, *der;* **~-pad** *n.* Bremsbelag, *der;* **~-shoe** *n.* Bremsbacke, *die*

braking ['breɪkɪŋ] *n.* Bremsen, *das;* ~ **distance** Bremsweg, *der*

bramble ['bræmbl] *n.* **a)** *(shrub)* Dornenstrauch, *der;* *(blackberry-bush)* Brombeerstrauch, *der;* **b)** *(fruit)* Brombeere, *die*

bran [bræn] *n.* Kleie, *die*

branch [brɑːnʃ] **1.** *n.* **a)** *(bough)*

Ast, *der; (twig)* Zweig, *der;* **b)** *(of nerve, artery, antlers)* Ast, *der; (of river)* [Neben]arm, *der; (local establishment)* Zweigstelle, *die; (shop)* Filiale, *die.* **2.** *v. i.* **a)** sich verzweigen; **b)** *(diverge)* ~ **into sth.** sich in etw. *(Akk.)* aufspalten

~ '**off** *v. i.* abzweigen; *(fig.)* sich abspalten

~ '**out** *v. i.* **a)** *see* ~ **2 a; b)** *(~ off)* abzweigen; **c)** *(fig.)* ~ **out into sth.** sich auch mit etw. befassen

branch: ~ **line** *n. (Railw.)* Nebenstrecke, *die;* ~ **office** *n.* Zweigstelle, *die*

brand [brænd] **1.** *n.* **a)** *(trade mark)* Markenzeichen, *das; (goods of particular make)* Marke, *die; (fig.: type)* Art, *die;* ~ **of washing-powder/soap** Waschpulvermarke, *die*/Seifenmarke, *die;* **b)** *(permanent mark, stigma)* Brandmal, *das; (on sheep, cattle)* Brandzeichen, *das.* **2.** *v. t.* **a)** mit einem Brandzeichen markieren ‹*Tier*›; **b)** *(stigmatize [as])* ~ **[as]** brandmarken als ‹*Verräter, Verbrecher usw.*›; **c)** *(Brit.: label with trade mark)* mit einem Markenzeichen versehen; ~**ed goods** Markenware, *die*

brandish ['brændɪʃ] *v. t.* schwenken; schwingen ‹*Waffe*›

brand: ~ **name** *n.* Markenname, *der;* ~-'**new** *adj.* nagelneu *(ugs.)*

brandy ['brændɪ] *n.* Weinbrand, *der*

brash [bræʃ] *adj. (self-assertive)* dreist; *(garish)* auffällig ‹*Kleidung*›

brass [brɑːs] **1.** *n.* **a)** Messing, *das;* **do sth. as bold as** ~: die Unverfrorenheit haben, etw. zu tun; **b)** *(inscribed tablet)* Grabplatte aus Messing; **c) [horse-]**~**es** Messinggeschirr, *das;* **d) the** ~ *(Mus.)* das Blech *(fachspr.);* die Blechbläser; **e)** *see* **brassware; f)** *no pl., no indef. art. (Brit. sl.: money)* Kies, *der (salopp);* **g) [top]** ~ *(coll.: officers, leaders of industry etc.)* hohe Tiere *(ugs.).* **2.** *attrib. adj.* Messing-; ~ **player** *(Mus.)* Blechbläser, *der*

brass 'band *n.* Blaskapelle, *die*

brassière ['bræzjə(r)] *n. (formal)* Büstenhalter, *der*

brass: ~ '**plate** *n.* Messingschild, *das;* ~ **rubbing** *n.* **a)** *no pl., no indef. art.* Frottage, *die (von Messingtafeln);* **b)** *(impression)* Frottage, *die (einer Messingtafel);* ~ '**tacks** *n. pl.* **get** *or* **come down to** ~ **tacks** *(sl.)* zur Sache kommen; ~**ware** *n., no pl.* Messingteile; *(utensils, candlesticks, etc.)* Messinggerät, *das*

brat [bræt] *n. (derog.: child)* Balg, *das od. der (ugs., meist abwertend); (young rascal)* Flegel, *der*

bravado [brə'vɑːdəʊ] *n., pl.* ~**es** *or* ~**s** Mut, *der;* **do sth. out of** ~: so waghalsig sein, etw. zu tun; *(as pretence)* den starken Mann markieren wollen und etw. tun *(ugs.)*

brave [breɪv] **1.** *adj.* mutig; *(able to endure sth.)* tapfer; **be** ~! nur Mut/sei tapfer! **2.** *n.* [indianischer] Krieger. **3.** *v. t.* trotzen (+ *Dat.*); mutig gegenübertreten (+ *Dat.*) ‹*Kritiker, Interviewer*›; ~ **it out** sich durch nichts einschüchtern lassen

bravely ['breɪvlɪ] *adv.* mutig; *(showing endurance)* tapfer

bravery ['breɪvərɪ] *n., no pl.* Mut, *der; (endurance)* Tapferkeit, *die*

bravo [brɑː'vəʊ] *int.* bravo

brawl [brɔːl] **1.** *v. i.* sich schlagen. **2.** *n.* Schlägerei, *die*

brawn [brɔːn] *n.* **a)** *(muscle)* Muskel, *der; (muscularity)* Muskeln; **b)** *(Gastr.)* ≈ Preßkopf, *der*

brawny ['brɔːnɪ] *adj.* muskulös

bray [breɪ] **1.** *n. (of ass)* Iah, *das.* **2.** *v. i.* ‹*Esel:*› iahen, schreien

brazen ['breɪzn] **1.** *adj.* dreist; *(shameless)* schamlos. **2.** *v. t.* ~ **it out** *(deny guilt)* es abstreiten; *(not admit guilt)* es nicht zugeben

brazier ['breɪzɪə(r), 'breɪʒə(r)] *n.* Kohlenbecken, *das*

Brazil [brə'zɪl] *n.* **a)** *pr. n.* Brasilien *(das);* **b)** *see* **Brazil nut**

Brazilian [brə'zɪlɪən] **1.** *adj.* brasilianisch; **sb. is** ~: jmd. ist Brasilianer/Brasilianerin. **2.** *n.* Brasilianer, *der*/Brasilianerin, *die*

Bra'zil nut *n.* Paranuß, *die*

breach [briːtʃ] **1.** *n.* **a)** *(violation)* Verstoß, *der* **(of** gegen); ~ **of the peace** Störung von Ruhe und Ordnung; *(by noise only)* ruhestörender Lärm; ~ **of contract** Vertragsbruch, *der;* ~ **of promise** Wortbruch, *der;* **b)** *(of relations)* Bruch, *der;* **c)** *(gap)* Bresche, *die; (fig.)* Riß, *der;* **step into the** ~ *(fig.)* in die Bresche treten *od.* springen. **2.** *v. t.* eine Bresche schlagen in (+ *Akk.*); **the wall/dike was** ~**ed** in die Mauer wurde eine Bresche geschlagen/der Deich wurde durchbrochen

bread [bred] **1.** *n.* **a)** Brot, *das;* **a piece of** ~ **and butter** ein Butterbrot; **[some]** ~ **and butter** [ein paar] Butterbrote; ~ **and butter** *(fig.)* tägliches Brot; ~ **and water** *(lit. or fig.)* Wasser und Brot; **know which side one's** ~ **is buttered** wissen, wo etwas zu holen ist; **b)** *(sl.: money)* Kies, *der (salopp).* **2.** *v. t.* panieren

bread: ~-**bin** *n.* Brotkasten, *der;* ~-**board** *n.* [Brot]brett, *das;* ~**crumb** *n.* Brotkrume, *die;* ~**crumbs** *(for coating e.g. fish)* Paniermehl, *das;* ~-**knife** *n.* Brotmesser, *das;* ~**line** *n.* **live on/below the** ~**line** *(fig.)* gerade noch/nicht einmal mehr das Notwendigste zum Leben haben

breadth [bredθ] *n.* **a)** *(broadness)* Breite, *die;* **b)** *(extent)* Weite, *die;* **with his** ~ **of experience/knowledge** bei seiner großen Erfahrung/bei seiner umfassenden Kenntnis

'**bread-winner** *n.* Ernährer, *der*/Ernährerin, *die*

break [breɪk] **1.** *v. t.,* **broke** [brəʊk], **broken** ['brəʊkn] **a)** brechen; *(so as to damage)* zerbrechen; kaputtmachen *(ugs.);* aufschlagen ‹*Ei zum Kochen*›; zerreißen ‹*Seil*›; *(fig.: interrupt)* unterbrechen; brechen ‹*Bann, Zauber, Schweigen*›; ~ **sth. in two/in pieces** etw. in zwei Teile/in Stücke brechen; **the TV/my watch is broken** der Fernseher/meine Uhr ist kaputt *(ugs.);* **b)** *(fracture)* sich *(Dat.)* brechen; *(pierce)* verletzen ‹*Haut*›; **he broke his leg** er hat sich *(Dat.)* das Bein gebrochen; ~ **one's/sb.'s back** *(fig.)* sich/jmdn. kaputtmachen *(ugs.);* ~ **the back of sth.** *(fig.)* bei etw. das Schwerste hinter sich bringen; ~ **open** aufbrechen; **c)** *(violate)* brechen ‹*Vertrag, Versprechen*›; verletzen, verstoßen gegen ‹*Regel, Tradition*›; nicht einhalten ‹*Verabredung*›; überschreiten ‹*Grenze*›; ~ **the law** gegen das Gesetz verstoßen; **d)** *(destroy)* zerstören, ruinieren ‹*Freundschaft, Ehe*›; **e)** *(surpass)* brechen ‹*Rekord*›; **f)** *(abscond from)* ~ **jail** [aus dem Gefängnis] ausbrechen; **g)** *(weaken)* brechen, beugen ‹*Stolz*›; zusammenbrechen lassen ‹*Streik*›; ~ **sb.'s heart** jmdm. das Herz brechen; ~ **sb.** *(crush)* jmdn. fertigmachen *(ugs.);* ~ **the habit** es sich *(Dat.)* abgewöhnen; *see also* **make 1 o; h)** *(cushion)* auffangen ‹*Schlag, jmds. Fall*›; **i)** *(make bankrupt)* ruinieren; ~ **the bank** die Bank sprengen; **it won't** ~ **the bank** *(fig. coll.)* es kostet kein Vermögen; **j)** *(reveal)* ~ **the news that ...:** melden, daß ...; **k)** *(solve)* entschlüsseln, entziffern ‹*Kode, Geheimschrift*›; **l)** *(Tennis)* ~ **service/sb.'s service** den Aufschlag des Gegners/jmds. Aufschlag durchbrechen. *See also* **broken 2. 2.** *v. i.,* **broke, broken a)** kaputtgehen *(ugs.);* entzweigehen; ‹*Faden, Seil:*› [zer]reißen; ‹*Glas, Tasse, Teller:*› zerbrechen; ‹*Eis:*› brechen; **sb.'s heart is ~ing** jmdm. bricht das Herz; **~ in two/ in pieces** entzweibrechen; **b)** *(crack)* ‹*Fenster-, Glasscheibe:*› zerspringen; **my back was nearly ~ing** ich brach mir fast das Kreuz; **c)** *(sever links)* **~ with sb./ sth.** mit jmdm./etw. brechen; **d) ~ into** einbrechen in (+ *Akk.*) ‹*Haus*›; aufbrechen ‹*Safe*›; **he broke into a sweat** ihm brach der Schweiß aus; **~ into a trot/run** *etc.* zu traben/laufen *usw.* anfangen; **~ into a banknote** eine Banknote anbrechen; **~ out of prison** *etc.* aus dem Gefängnis *usw.* ausbrechen; **e) ~ free** *or* **loose [from sb./sb.'s grip]** sich [von jmdm./ aus jmds. Griff] losreißen; **~ free/loose [from prison]** [aus dem Gefängnis] ausbrechen; **f)** ‹*Welle:*› sich brechen (**on/against** an + *Dat.*); **g)** ‹*Wetter:*› umschlagen; **h)** ‹*Wolkendecke:*› aufreißen; **i)** ‹*Tag:*› anbrechen; **j)** ‹*Sturm:*› losbrechen; **k) sb.'s voice is ~ing** jmd. kommt in den Stimmbruch; *(with emotion)* jmdm. bricht die Stimme; **l)** *(have interval)* **~ for coffee/lunch** [eine] Kaffee-/Mittagspause machen; **m)** *(become public)* bekanntwerden. **3.** *n.* **a)** Bruch, *der; (of rope)* Reißen, *das;* **~ [of service]** *(Tennis)* Break, *der od. das;* **a ~ with sb./sth.** ein Bruch mit jmdm./ etw.; **~ of day** Tagesanbruch, *der;* **b)** *(gap)* Lücke, *die; (Electr.: in circuit)* Unterbrechung, *die;* **c)** *(sudden dash)* **they made a sudden ~ [for it]** sie stürmten plötzlich davon; **d)** *(interruption)* Unterbrechung, *die;* **e)** *(pause, holiday)* Pause, *die;* **during the commercial ~s on TV** während der Werbespots im Fernsehen; **take** *or* **have a ~:** [eine] Pause machen; **f)** *(coll.: fair chance, piece of luck)* Chance, *die;* **lucky ~:** große Chance; **that was a bad ~ for him** das war Pech für ihn

~ a'way 1. *v. t.* **~ sth. away [from sth.]** etw. [von etw.] losbrechen *od.* abbrechen. **2.** *v. i.* **a) ~ away [from sth.]** [von etw.] losbrechen *od.* abbrechen; *(separate itself/ oneself)* sich [von etw.] lösen; *(escape)* [aus etw.] entkommen; **b)** *(Footb.)* sich freilaufen

~ 'down 1. *v. i.* **a)** *(fail)* zusammenbrechen; ‹*Verhandlungen:*› scheitern; **b)** *(cease to function)* ‹*Auto:*› eine Panne haben; ‹*Telefonnetz:*› zusammenbrechen; **the machine has broken down** die Maschine funktioniert nicht mehr; **c)** *(be overcome by emotion)* zusammenbrechen; **d)** *(Chem.)* aufspalten. **2.** *v. t.* **a)** *(demolish)* aufbrechen ‹*Tür*›; **b)** *(suppress)* brechen ‹*Widerstand*›; niederreißen ‹*Barriere, Schranke*›; **c)** *(analyse)* aufgliedern

~ 'in 1. *v. i.* einbrechen. **2.** *v. t.* **a)** *(accustom to habit)* eingewöhnen; *(tame)* zureiten ‹*Pferd*›; **b)** einlaufen ‹*Schuhe*›

~ into *see* **~ 2d**

~ 'off 1. *v. t.* abbrechen; auflösen ‹*Verlobung*›; **~ it off [with sb.]** sich von jmdm. trennen. **2.** *v. i.* **a)** abbrechen; **b)** *(cease)* aufhören

~ 'out *v. i. (escape, appear)* ausbrechen; **~ out in spots/a rash** *etc.* Pickel/einen Ausschlag *usw.* bekommen; **he broke out in a cold sweat** ihm brach der kalte Schweiß aus

~ out of *see* **~ 2d**

~ 'through *v. t. & i.* durchbrechen

~ 'up 1. *v. t.* **a)** *(~ into pieces)* zerkleinern; ausschlachten ‹*Auto*›; abwracken ‹*Schiff*›; aufbrechen ‹*Erde*›; **b)** *(disband)* auflösen; auseinanderreißen ‹*Familie*›; zerstreuen ‹*Menge*›; **~ it up!** *(coll.)* auseinander!; **c)** *(end)* zerstören ‹*Freundschaft, Ehe*›. **2.** *v. i.* **a)** *(~ into pieces, lit. or fig.)* zerbrechen; ‹*Erde, Straßenoberfläche:*› aufbrechen; ‹*Eis:*› brechen; **b)** *(disband)* sich auflösen; ‹*Schule:*› schließen; ‹*Schüler, Lehrer:*› in die Ferien gehen; **c)** *(cease)* abgebrochen werden; *(end relationship)* **~ up [with sb.]** sich [von jmdm.] trennen

breakable ['breɪkəbl] **1.** *adj.* zerbrechlich. **2.** *n. in pl.* zerbrechliche Dinge

breakage ['breɪkɪdʒ] *n.* **a)** *(breaking)* Zerbrechen, *das;* **b)** *(result of breaking)* Bruchschaden, *der;* **~s must be paid for** zerbrochene Ware muß bezahlt werden

break: ~away 1. *n.* Ausbrechen, *das;* **2.** *adj. (Brit.)* abtrünnig; **~away group** Splittergruppe, *die;* **~down** *n.* **a)** *(fig.: collapse)* **a ~down in the system** *(fig.)* ein Zusammenbruch des Systems; **b)** *(mechanical failure)* Panne, *die; (in machine)* Störung, *die; attrib.* **~down service** Pannendienst, *der;* **~down truck/van** Abschleppwagen, *der;* **c)** *(health or mental failure)* Zusammenbruch, *der;* **d)** *(analysis)* Aufschlüsselung, *die;* **e)** *(Chem.)* Aufspaltung, *die*

breaker ['breɪkə(r)] *n. (wave)* Brecher, *der*

breakfast ['brekfəst] **1.** *n.* Frühstück, *das;* **for ~:** zum Frühstück; **eat** *or* **have [one's] ~:** frühstücken. **2.** *v. i.* frühstücken

breakfast: ~ cereal *n.* ≈ Frühstücksflocken *Pl.;* **~ 'television** *n.* Frühstücksfernsehen, *das;* **~-time** *n.* Frühstückszeit, *die*

'break-in *n.* Einbruch, *der;* **there has been a ~ at the bank** in der *od.* die Bank ist eingebrochen worden

'breaking-point *n.* Belastungsgrenze, *die;* **be at ~** *(mentally)* die Grenze der Belastbarkeit erreicht haben

break: ~neck *adj.* halsbrecherisch; **~-out** *n.* Ausbruch, *der;* **~through** *n.* Durchbruch, *der;* **~-up** *n.* **a)** *(disbanding, dispersal)* Auflösung, *die;* **b)** *(of relationship)* Bruch, *der;* **~water** *n.* Wellenbrecher, *der*

breast [brest] *n. (lit. or fig.)* Brust, *die;* **make a clean ~ [of sth.]** *(fig.)* [etw.] offen bekennen

breast: ~bone *n.* Brustbein, *das;* **~ cancer** *n.* Brustkrebs, *der;* **~-feed** *v. t. & i.* stillen; **~-feeding** *n.* das Stillen; **~-stroke** *n. (Swimming)* Brustschwimmen, *das*

breath [breθ] *n.* **a)** Atem, *der;* **have bad ~:** Mundgeruch haben; **say sth. below** *or* **under one's ~:** etw. vor sich *(Akk.)* hin murmeln; **a ~ of fresh air** ein wenig frische Luft; **waste one's ~:** seine Worte verschwenden; **she caught her ~:** ihr stockte der Atem; **hold one's ~:** den Atem anhalten; **get one's ~ back** wieder zu Atem kommen; **be out of/short of ~:** außer Atem *od.* atemlos sein/kurzatmig sein; **take sb.'s ~ away** *(fig.)* jmdm. den Atem verschlagen; **b)** *(one respiration)* Atemzug, *der;* **take** *or* **draw a [deep] ~:** [tief] einatmen; **in the same ~:** im selben Atemzug; **c)** *(air movement, whiff)* Hauch, *der;* **there wasn't a ~ of air** es regte sich kein Lüftchen

breathalyser, (P) (*Amer.:* **breathalyzer**) ['breθəlaɪzə(r)] *n.* Alcotest-Röhrchen Ⓦⓩ, *das;* **~ test** Alcotest Ⓦⓩ, *der*

breathe [bri:ð] **1.** *v. i. (lit. or fig.)* atmen; **~ in** einatmen; **~ out** ausatmen; **~ into sth.** [sanft] in etw. *(Akk.)* [hinein]blasen. **2.** *v. t.* **a) ~ one's last** seinen letzten Atemzug tun; **~ [in/out]** ein-/ausatmen; **b)** *(utter)* hauchen; **don't ~ a word about** *or* **of this to anyone** sag kein Sterbenswörtchen darüber zu irgend jemandem

breather ['bri:ðə(r)] *n. (brief pause)* Verschnaufpause, *die*

breathing ['bri:ðɪŋ] *n.* Atmen, *das*

breathing: ~-apparatus *n.* **a)** *(Med.)* Beatmungsgerät, *das;* **b)** *(of fireman etc.)* Atemschutzgerät, *das;* **~-space** *n. (time to breathe)* Zeit zum Luftholen; *(pause)* Atempause, *die*

breathless ['breθlɪs] *adj.* atemlos (**with** vor + *Dat.*); **leave sb. ~** *(lit. or fig.)* jmdm. den Atem nehmen

breathlessness ['breθlɪsnɪs] *n., no pl.* Atemlosigkeit, *die; (caused by smoking or illness)* Kurzatmigkeit, *die*

breath: ~-taking *adj.* atemberaubend; **~ test** *n.* Alcotest ⓦⓩ, *der*

bred *see* **breed 1, 2**

breech [bri:tʃ] *n.* [Geschütz]verschluß, *der*

breeches ['brɪtʃɪz] *n. pl. (short trousers)* **[pair of] ~**: [Knie]bundhose, *die;* **[riding-] ~**: Reithose, *die*

breed [bri:d] **1.** *v. t.,* **bred** [bred] **a)** *(be the cause of)* erzeugen; hervorrufen; **b)** *(raise)* züchten ⟨*Tiere, Pflanzen*⟩; **c)** *(bring up)* erziehen; **he was born and bred in London** er ist in London geboren und aufgewachsen. **2.** *v. i.,* **bred** sich vermehren; ⟨*Vogel:*⟩ brüten; ⟨*Tier:*⟩ Junge haben; **they ~ like flies** *or* **rabbits** sie vermehren sich wie die Kaninchen. **3.** *n.* Art, *die; (of animals)* Rasse, *die;* **~s of cattle** Rinderrassen; **the Jersey ~ [of cattle]** das Jerseyrind; **what ~ of dog is that?** zu welcher Rasse gehört dieser Hund?

breeder ['bri:də(r)] *n.* Züchter, *der;* **be a ~ of sth.** etw. züchten; **dog-/horse-~**: Hunde-/Pferdezüchter, *der*

breeding ['bri:dɪŋ] *n.* Erziehung, *die;* **[good] ~**: gute Erziehung; **have ~**: eine gute Erziehung genossen haben

breeze [bri:z] **1.** *n. (gentle wind)* Brise, *die;* **there is a ~**: es weht eine Brise. **2.** *v. i. (coll.)* **~ along** dahinrollen; *(on foot)* dahinschlendern; **~ in** hereingeschneit kommen *(ugs.)*

breezy ['bri:zɪ] *adj.* **a)** *(windy)* windig; **b)** *(coll.: brisk and carefree)* [frisch und] unbekümmert

brevity ['brevɪtɪ] *n.* Kürze, *die*

brew [bru:] **1.** *v. t.* **a)** brauen ⟨*Bier*⟩; keltern ⟨*Apfelwein*⟩; **~ [up]** kochen ⟨*Kaffee, Tee, Kakao usw.*⟩; **~ up** *abs.* Tee kochen; **b)** *(fig.: put together)* **~ [up]** [zusammen]brauen *(ugs.)* ⟨*Mischung*⟩; ausbrüten *(ugs.)* ⟨*Plan usw.*⟩. **2.** *v. i.* **a)** ⟨*Bier, Apfelwein:*⟩ gären; ⟨*Kaffee, Tee:*⟩ ziehen; **b)** *(fig.: gather)* ⟨*Unwetter:*⟩ sich zusammenbrauen; ⟨*Rebellion, Krieg:*⟩ drohen. **3.** *n.* Gebräu, *das (abwertend); (~ed beer/tea)* Bier, *das*/Tee, *der*

brewer ['bru:ə(r)] *n.* **a)** *(person)* Brauer, *der;* **b)** *(firm)* Brauerei, *die*

brewery ['bru:ərɪ] *n.* Brauerei, *die*

briar *see* **brier**

bribe [braɪb] **1.** *n.* Bestechung, *die;* **a ~ [of £100]** ein Bestechungsgeld [in Höhe von 100 Pfund]; **take a ~/~s** sich bestechen lassen; **he won't accept ~s** er ist unbestechlich; **offer sb. a ~**: jmdn. bestechen wollen. **2.** *v. t.* bestechen; **~ sb. to do/into doing sth.** jmdn. bestechen, damit er etw. tut

bribery ['braɪbərɪ] *n.* Bestechung, *die*

bric-à-brac ['brɪkəbræk] *n.* Antiquarisches; *(smaller things)* Nippsachen *Pl.*

brick [brɪk] **1.** *n.* **a)** *(block)* Ziegelstein, *der;* Backstein, *der; (clay)* Lehmziegel, *der;* **be** *or* **come down on sb. like a ton of ~s** *(coll.)* jmdn. unheimlich fertigmachen *od.* zusammenstauchen *(ugs.);* **b)** *(toy)* Bauklötzchen, *das.* **2.** *adj.* Ziegelstein-; Backstein-. **3.** *v. t.* **~ up/in** zu-/einmauern

brick: ~layer *n.* Maurer, *der;* **~laying** *n.* Mauern, *das;* **~-red** *adj.* ziegelrot; **~ 'wall** *n.* Backsteinmauer, *die;* **bang one's head against a ~ wall** *(fig.)* mit dem Kopf gegen die Wand rennen *(fig.)*

bridal ['braɪdl] *adj. (of bride)* Braut-; *(of wedding)* Hochzeits-; **~ couple/suite** Brautpaar, *das*/Hochzeitssuite, *die*

bride [braɪd] *n.* Braut, *die*

'bridegroom *n.* Bräutigam, *der*

bridesmaid ['braɪdzmeɪd] *n.* Brautjungfer, *die*

¹bridge [brɪdʒ] **1.** *n.* **a)** *(lit. or fig.)* Brücke, *die;* **cross that ~ when you come to it** *(fig.)* alles zu seiner Zeit; **b)** *(Naut.)* [Kommando]brücke, *die;* **c)** *(of nose)* Nasenbein, *das;* Sattel, *der;* **d)** *(of violin, spectacles)* Steg, *der;* **e)** *(Dent.)* [Zahn]brücke, *die.* **2.** *v. t.* eine Brücke bauen *od.* errichten *od.* schlagen über (+ *Akk.*)

²bridge *n. (Cards)* Bridge, *das*

'bridging loan *n. (Commerc.)* Überbrückungskredit, *der*

bridle ['braɪdl] **1.** *n.* Zaumzeug, *das;* Zaum, *der.* **2.** *v. t.* **a)** aufzäumen ⟨*Pferd*⟩; **b)** *(fig.: restrain)* zügeln ⟨*Zunge*⟩; im Zaum halten ⟨*Leidenschaft*⟩

bridle: ~-path, ~-road *ns.* Saumpfad, *der; (for horses)* Reitweg, *der*

¹brief [bri:f] *adj.* **a)** *(of short duration)* kurz; gering, geringfügig ⟨*Verspätung*⟩; **b)** *(concise)* knapp; **in ~, to be ~**: kurz gesagt; **make** *or* **keep it ~**: es kurz machen; **be ~**: sich kurz fassen

²brief **1.** *n.* **a)** *(Law: summary of facts)* Schriftsatz, *der;* **b)** *(Brit. Law: piece of work)* Mandat, *das;* **c)** *(instructions)* Instruktionen *Pl.;* Anweisungen *Pl.* **2.** *v. t.* **a)** *(Brit. Law)* mit der Vertretung eines Falles betrauen; **b)** *(Mil.: instruct)* Anweisungen *od.* Instruktionen geben (+ *Dat.*); **c)** *(inform, instruct)* unterrichten; informieren

'brief-case *n.* Aktentasche, *die*

briefing ['bri:fɪŋ] *n.* **a)** Briefing, *das; (of reporters or press)* Unterrichtung, *die; (before raid etc.)* Einsatzbesprechung, *die;* **b)** *(instructions)* Instruktionen *Pl.;* Anweisungen *Pl.; (information)* Informationen *Pl.*

briefly ['bri:flɪ] *adv.* **a)** *(for a short time)* kurz; **b)** *(concisely)* knapp; kurz; **[to put it] ~, ...**: kurz gesagt ...

briefs [bri:fs] *n. pl.* **[pair of] ~**: Slip, *der*

brier [braɪə(r)] *n. (Bot.: rose)* Wilde Rose

brig [brɪg] *n. (Naut.)* Brigg, *die*

brigade [brɪ'geɪd] *n. (Mil.)* Brigade, *die;* **the old ~** *(fig.)* die alte Garde

brigadier[-general] [brɪgə'dɪə(r) ('dʒenrl)] *n. (Mil.)* Brigadegeneral, *der*

bright [braɪt] **1.** *adj.* **a)** hell ⟨*Licht, Stern, Fleck*⟩; grell ⟨*Scheinwerfer[licht], Sonnenlicht*⟩; strahlend ⟨*Sonnenschein, Stern, Augen*⟩; glänzend ⟨*Metall, Augen*⟩; leuchtend, lebhaft ⟨*Farbe, Blume*⟩; **~ blue** *etc.* leuchtend blau *usw.*; **a ~ day** ein heiterer Tag; **~ intervals/periods** Aufheiterungen; **the ~ lights of the city** *(fig.)* der Glanz der Großstadt; **look on the ~ side** *(fig.)* die Sache positiv sehen; **b)** *(cheerful)* fröhlich, heiter ⟨*Person, Charakter, Stimmung*⟩; strahlend ⟨*Lächeln*⟩; freundlich ⟨*Zimmer, Farbe*⟩; **c)** *(clever)* intelligent; **he is a ~ boy** er ist ein heller *od.* aufgeweckter Junge; **d)** *(hopeful)* vielversprechend ⟨*Zukunft*⟩; glänzend ⟨*Aussichten*⟩. **2.** *adv.* **a)** hell; **b)** **~ and early** in aller Frühe

brighten ['braɪtn] **1.** *v. t.* **~ [up]** **a)**

aufhellen ‹*Farbe*›; **b)** *(make more cheerful)* aufhellen, aufheitern ‹*Zimmer*›. **2.** *v. i.* ~ **[up] a)** ‹*Himmel:*› sich aufhellen; **the weather** *or* **it is ~ing [up]** es klärt sich auf; **b)** *(become more cheerful)* ‹*Person:*› vergnügter werden; ‹*Gesicht:*› sich aufhellen; ‹*Aussichten:*› sich verbessern

brightly ['braɪtlɪ] *adv.* **a)** hell ‹*scheinen, glänzen*›; glänzend ‹*poliert*›; **b)** *(cheerfully)* gutgelaunt

brightness ['braɪtnɪs] *n., no pl. see* **bright 1: a)** Helligkeit, *die;* Grelle, *die;* Grellheit, *die;* Strahlen, *das;* Glanz, *der;* Leuchtkraft, *die;* **b)** Fröhlichkeit, *die;* Heiterkeit, *die;* **c)** Intelligenz, *die*

brill [brɪl] *adj. (Brit. sl.)* super *(ugs.)*

brilliance ['brɪljəns] *n., no pl. see* **brilliant: a)** Helligkeit, *die;* Funkeln, *das;* Leuchten, *das;* **b)** Genialität, *die;* **c)** Glanz, *der*

brilliant ['brɪljənt] *adj.* **a)** *(bright)* hell ‹*Licht*›; strahlend ‹*Sonne*›; funkelnd ‹*Diamant, Stern*›; leuchtend ‹*Farbe*›; **b)** *(highly talented)* genial ‹*Person, Erfindung, Gedanke, Schachzug, Leistung*›; glänzend ‹*Verstand, Aufführung, Vorstellung, Idee*›; bestechend ‹*Theorie, Argument*›; **c)** *(illustrious)* glänzend ‹*Karriere, Erfolg, Sieg*›; **a ~ achievement** eine Glanzleistung

brilliantly ['brɪljəntlɪ] *adv.* **a)** hell ‹*scheinen, funkeln, schimmern*›; **b)** *(with great talent)* brillant; **c)** *(illustriously)* glänzend ‹*erfolgreich sein, triumphieren*›

brim [brɪm] **1.** *n.* **a)** *(of cup, bowl, hollow)* Rand, *der;* **full to the ~:** randvoll; **b)** *(of hat)* [Hut]krempe, *die.* **2.** *v. i.,* **-mm-: be ~ming with sth.** randvoll mit etw. sein; *(fig.)* strotzen vor etw. *(Dat.)*

~ **'over** *v. i.* übervoll sein

brim-'full *pred. adj.* randvoll **(with** mit); **be ~ of energy/curiosity** *(fig.)* vor Energie *(Dat.)* sprühen/vor Neugierde *(Dat.)* platzen

brindled ['brɪndld] *adj.* gestreift ‹*Katze*›; gestromt ‹*Kuh, Hund*›

brine [braɪn] *n.* Salzwasser, *das; (for preserving)* [Salz]lake, *die*

bring [brɪŋ] *v. t.,* **brought** [brɔːt] **a)** bringen; *(as a present or favour)* mitbringen; **~ sth. with one** etw. mitbringen; **I haven't brought my towel** ich habe mein Handtuch nicht mitgebracht *od.* dabei; **~ sth. [up]on oneself/sb.** sich selbst/jmdm. etw. einbrocken; **b)** *(result in)* [mit sich] bringen; **~ tears to sb.'s eyes** jmdm. Tränen in die Augen treiben; **c)** *(persuade)* **~ sb. to do sth.** jmdn. dazu bringen *od.* bewegen, etw. zu tun; **I could not ~ myself to do it** ich konnte es nicht über mich bringen, es zu tun; **d)** *(initiate, put forward)* **~ a charge/legal action against sb.** gegen jmdn. [An]klage erheben/einen Prozeß anstrengen; **~ a complaint** eine Beschwerde vorbringen; **e)** *(be sold for, earn)* [ein]bringen ‹*Geldsumme*›

~ **a'bout** *v. t.* verursachen; herbeiführen; **~ it about that ...:** es zustande bringen, daß ...

~ **a'long** *v. t.* **a)** mitbringen; **b)** *see* **~ on b**

~ **'back** *v. t.* **a)** *(return)* zurückbringen; *(from a journey)* mitbringen; **b)** *(recall)* in Erinnerung bringen *od.* rufen; **~ sth. back to sb.** ‹*Musik, Foto usw.:*› jmdn. an etw. *(Akk.)* erinnern; **~ back memories** Erinnerungen wachrufen *od.* wecken; **c)** *(restore, reintroduce)* wieder einführen ‹*Sitten, Todesstrafe*›; **~ sb. back to life** jmdn. wiederbeleben

~ **'down** *v. t.* **a)** herunterbringen; **b)** *(shoot down out of the air)* abschießen; herunterholen *(ugs.)*; **c)** *(land)* herunterbringen ‹*Flugzeug, Drachen*›; **d)** *(kill, wound)* zur Strecke bringen ‹*Person, Tier*›; erlegen ‹*Tier*›; **e)** *(reduce)* senken ‹*Preise, Inflationsrate, Fieber*›; **f)** *(cause to fall)* zu Fall bringen ‹*Gegner, Fußballer*›; *(fig.)* stürzen, zu Fall bringen ‹*Regierung*›; *see also* **house 1 e**

~ **'forward** *v. t.* **a)** nach vorne bringen; **b)** *(draw attention to)* vorlegen ‹*Beweise*›; vorbringen ‹*Argument*›; zur Sprache bringen ‹*Fall, Angelegenheit, Frage*›; **c)** *(move to earlier time)* vorverlegen ‹*Termin*› **(to** auf + *Akk.*); **d)** *(Bookk.)* übertragen

~ **'in** *v. t.* **a)** hereinbringen; auftragen ‹*Essen*›; einbringen ‹*Ernte*›; **b)** *(yield)* einbringen ‹*Verdienst, Summe*›; **c)** *(Law)* **~ in a verdict of guilty/not guilty** einen Schuldspruch fällen/auf Freispruch erkennen; **d)** *(call in)* hinzuziehen, einschalten ‹*Experten*›

~ **'off** *v. t.* **a)** *(rescue)* retten; in Sicherheit bringen; **b)** *(conduct successfully)* zustande bringen

~ **'on** *v. t.* **a)** *(cause)* verursachen; **brought on by ...** ‹*Krankheit*› infolge von ...; **b)** *(advance progress of)* wachsen *od.* sprießen lassen ‹*Blumen, Getreide*›; weiterbringen, fördern ‹*Schüler, Sportler*›; **c)** *(Sport)* einsetzen

~ **'out** *v. t.* **a)** herausbringen; **b)** *(show clearly)* hervorheben, betonen ‹*Unterschied*›; verdeutlichen ‹*Bedeutung*›; herausbringen ‹*Farbe*›; **c)** *(cause to appear)* herausbringen ‹*Pflanzen, Blüte*›; **the crisis brought out the best in him** die Krise brachte seine besten Seiten zum Vorschein *od.* ans Licht; **d)** *(begin to sell)* einführen ‹*Produkt*›; herausbringen ‹*Buch, Zeitschrift*›

~ **'round** *v. t.* **a)** mitbringen ‹*Bekannte, Freunde usw.*›; vorbeibringen ‹*Gegenstände*›; **b)** *(restore to consciousness)* wieder zu sich bringen ‹*Ohnmächtigen*›; **c)** *(win over)* überreden; herumkriegen *(ugs.)*; **~ sb. round to one's way of thinking** jmdn. von seiner Meinung überzeugen; **d) ~ a conversation round to sth.** ein Gespräch auf etw. *(Akk.)* lenken

~ **'to** *v. t. (restore to consciousness)* wieder zu sich bringen

~ **'up** *v. t.* **a)** heraufbringen; **b)** *(educate)* erziehen; **c)** *(rear)* aufziehen; großziehen; **d)** *(call attention to)* zur Sprache bringen ‹*Angelegenheit, Thema, Problem*›; **e)** *(vomit)* erbrechen

bring-and-'buy [sale] *n.* [Wohltätigkeits]basar, *der*

brink [brɪŋk] *n. (lit. or fig.)* Rand, *der;* **be on the ~ of doing sth.** nahe daran sein, etw. zu tun; **be on the ~ of ruin/success** am Rand des Ruins sein *od.* stehen/dem Erfolg greifbar nahe sein

brisk [brɪsk] *adj.* flott ‹*Gang, Bedienung*›; forsch ‹*Person, Art*›; frisch ‹*Wind*›; *(fig.)* rege ‹*Handel, Nachfrage*›; lebhaft ‹*Geschäft*›

briskly ['brɪsklɪ] *adv.* flott; **the wind blew ~:** es wehte ein frischer Wind; **sell ~:** sich gut verkaufen

bristle ['brɪsl] **1.** *n.* **a)** Borste, *die;* **be made of ~:** aus Borsten bestehen; **b) ~s** *(of beard)* [Bart]stoppeln. **2.** *v. i.* **a) ~ [up]** ‹*Haare:*› sich sträuben; **b) ~ with** *(fig.: have many)* strotzen vor (+ *Dat.*); **c) ~ [up]** *(fig.: become angry)* ‹*Person:*› ungehalten reagieren

bristly ['brɪslɪ] *adj.* borstig; stopp[e]lig ‹*Kinn*›

Brit [brit] *n. (coll.)* Brite, *der*/Britin, *die;* Engländer, *der*/Engländerin, *die (ugs.)*

Britain ['brɪtn] *pr. n.* Großbritannien *(das)*

British ['brɪtɪʃ] **1.** *adj.* britisch; **he/she is ~:** er ist Brite/sie ist Britin; **sth. is ~:** etw. ist aus Großbritannien. **2.** *n. pl.* **the ~:** die Briten

Britisher ['brɪtɪʃə(r)] *n.* Brite, *der*/Britin, *die*

British 'Isles *pr.n. pl.* Britische Inseln

Briton ['brɪtn] *n.* Brite, *der*/Britin, *die*

Brittany ['brɪtənɪ] *pr.n.* Bretagne, *die*

brittle ['brɪtl] *adj.* spröde; zerbrechlich ‹*Glas*›; schwach ‹*Knochen*›; brüchig ‹*Gestein*›

broach [brəʊtʃ] *v.t.* **a)** anzapfen; anstechen ‹*Faß*›; **b)** *(fig.)* anschneiden ‹*Thema*›

broad [brɔ:d] *adj.* **a)** breit; *(extensive)* weit ‹*Ebene, Meer, Land, Felder*›; ausgedehnt ‹*Fläche*›; **grow ~er** breiter werden; sich verbreitern; **make sth. ~er** etw. verbreitern; **it's as ~ as it is long** *(fig.)* es ist gehupft wie gesprungen *(ugs.)*; **b)** *(explicit)* deutlich, klar ‹*Hinweis*›; **a ~ hint** ein Wink mit dem Zaunpfahl *(scherzh.)*; **c)** *(clear, main)* grob; wesentlich ‹*Fakten*›; **in ~ outline** in groben *od.* großen Zügen; *see also* **daylight a**; **d)** *(generalized)* allgemein; **in the ~est sense** im weitesten Sinne; **as a ~ indication** als Faustregel; **e)** *(strongly regional)* stark ‹*Akzent*›; breit ‹*Aussprache*›

broad 'bean *n.* Saubohne, *die;* dicke Bohne

broadcast ['brɔ:dkɑ:st] **1.** *n.* *(Radio, Telev.)* Sendung, *die;* *(live)* Übertragung, *die.* **2.** *v.t.*, **broadcast a)** *(Radio, Telev.)* senden; übertragen ‹*Livesendung, Sportveranstaltung*›; **b)** *(spread)* verbreiten ‹*Gerücht, Nachricht*›. **3.** *v.i.*, **broadcast** ‹*Rundfunk-, Fernsehstation:*› senden; ‹*Redakteur usw.:*› [im Rundfunk/Fernsehen] sprechen. **4.** *adj.* *(Radio, Telev.)* im Rundfunk/Fernsehen gesendet; Rundfunk-/Fernseh-

broadcaster ['brɔ:dkɑ:stə(r)] *n.* *(Radio, Telev.) jmd., der durch häufige Auftritte im Rundfunk und Fernsehen, besonders als Interviewpartner, Diskussionsteilnehmer od. Kommentator, bekannt ist*

broadcasting ['brɔ:dkɑ:stɪŋ] *n., no pl.* *(Radio, Telev.)* Senden, *das;* *(live)* Übertragen, *das;* **work in ~:** beim Funk arbeiten

broaden ['brɔ:dn] **1.** *v.t.* **a)** verbreitern; **b)** *(fig.)* ausweiten ‹*Diskussion*›; **~ one's mind** seinen Horizont erweitern. **2.** *v.i.* breiter werden; sich verbreitern; *(fig.)* sich erweitern

broadly ['brɔ:dlɪ] *adv.* **a)** deutlich ‹*hinweisen*›; breit ‹*grinsen, lächeln*›; **b)** *(in general)* allgemein ‹*beschreiben*›; **~ speaking** allgemein gesprochen

broad: ~'minded *adj.* tolerant; **~sheet** *n.* Flugblatt, *das;* **~-'shouldered** *adj.* breitschultrig; **~side** *n.* **~side on [to sth.]** mit der Breitseite [nach etw.]; **fire [off] a ~side** *(lit. or fig.)* eine Breitseite abfeuern

brocade [brə'keɪd] *n.* Brokat, *der*

broccoli ['brɒkəlɪ] *n.* Brokkoli, *der*

brochure ['brəʊʃə(r)] *n.* Broschüre, *die;* Prospekt, *der*

[1]**brogue** [brəʊg] *n.* *(shoe)* Budapester, *der*

[2]**brogue** *n.* *(accent)* irischer Akzent

broil [brɔɪl] *v.t.* braten; *(on gridiron)* grillen

broiler ['brɔɪlə(r)] *n.* **a)** *(chicken)* Brathähnchen, *das;* [Gold]broiler, *der (regional);* **b)** *(utensil)* Grill, *der;* Bratrost, *der*

broke [brəʊk] **1.** *see* **break 1, 2.** **2.** *pred. adj.* *(coll.)* pleite *(ugs.)*; **go ~:** pleite gehen; **go for ~** *(sl.)* alles riskieren

broken ['brəʊkn] **1.** *see* **break 1, 2.** **2.** *adj.* **a)** zerbrochen; gebrochen ‹*Bein, Hals usw.*›; verletzt ‹*Haut*›; abgebrochen ‹*Zahn*›; gerissen ‹*Seil*›; kaputt *(ugs.)* ‹*Uhr, Fernsehen, Fenster*›; **~ glass** Glasscherben; **get ~:** zerbrechen / brechen / reißen / kaputtgehen; **he got a ~ arm** er hat sich *(Dat.)* den Arm gebrochen; **b)** *(uneven)* uneben ‹*Fläche*›; **c)** *(imperfect)* gebrochen; **in ~ English** in gebrochenem Englisch; **d)** *(fig.)* ruiniert ‹*Ehe*›; gebrochen ‹*Person, Herz, Stimme*›; **come from a ~ home** aus zerrütteten Familienverhältnissen kommen

broken: ~-down *adj.* baufällig ‹*Gebäude*›; kaputt *(ugs.)* ‹*Wagen, Maschine*›; **~-hearted** [brəʊkn'hɑ:tɪd] *adj.* untröstlich

broker ['brəʊkə(r)] *n.* *(Commerc., Insurance, St. Exch.)* Makler, *der*

brolly ['brɒlɪ] *n.* *(Brit. coll.)* [Regen]schirm, *der*

bromide ['brəʊmaɪd] *n.* *(Chem.)* Bromsalz, *das*

bromine ['brəʊmi:n] *n.* *(Chem.)* Brom, *das*

bronchial ['brɒŋkɪəl] *adj.* *(Anat., Med.)* bronchial; Bronchial-; **~ tubes** Bronchien

bronchitis [brɒŋ'kaɪtɪs] *n., no pl.* *(Med.)* Bronchitis, *die*

bronze [brɒnz] **1.** *n.* **a)** *(metal, work of art, medal)* Bronze, *die;* **the B~ Age** die Bronzezeit; **b)** *(colour)* Bronze[farbe], *die.* **2.** *attrib. adj.* Bronze-; *(coloured like ~)* bronzefarben; bronzen. **3.** *v.t.* bräunen ‹*Gesicht, Haut*›. **4.** *v.i.* braun werden

bronzed [brɒnzd] *adj.* [sonnen]gebräunt; braun[gebrannt]

brooch [brəʊtʃ] *n.* Brosche, *die*

brood [bru:d] **1.** *n.* **a)** Brut, *die;* *(of hen)* Küken *Pl.*; **b)** *(joc.: children)* Kinderschar, *die.* **2.** *v.i.* **a)** *(think)* [vor sich *(Akk.)* hin] brüten; **~ over** *or* **upon sth.** über etw. *(Akk.)* [nach]grübeln; **b)** *(sit)* ‹*Vogel:*› brüten

broody ['bru:dɪ] *adj.* brütig; **~ hen** Glucke, *die*

[1]**brook** [brʊk] *n.* Bach, *der*

[2]**brook** *v.t.* dulden; **~ no nonsense/delay** keinen Unfug/Aufschub dulden

broom [bru:m] *n.* **a)** Besen, *der;* **a new ~** *(fig.)* ein neuer Besen; **b)** *(Bot.)* Ginster, *der*

broom: ~-cupboard *n.* Besenschrank, *der;* **~stick** *n.* Besenstiel, *der*

Bros. *abbr.* **Brothers** Gebr.

broth [brɒθ] *n.* *(thin soup)* Bouillon, *die;* [Fleisch]brühe, *die*

brothel ['brɒθl] *n.* Bordell, *das*

brother ['brʌðə(r)] *n.* **a)** Bruder, *der;* **my/your** *etc.* **~s and sisters** meine/deine *usw.* Geschwister; **have you any ~s or sisters?** haben Sie Geschwister?; **b)** *(fellow member of trade union)* Kollege, *der*

brotherhood ['brʌðəhʊd] *n.* **a)** *no pl.* Brüderschaft, *die;* brüderliches Verhältnis; **b)** *(association)* Bruderschaft, *die*

'brother-in-law *n., pl.* **brothers-in-law** Schwager, *der*

brotherly ['brʌðəlɪ] *adj.* brüderlich

brought *see* **bring**

brow [braʊ] *n.* **a)** *(eye~)* Braue, *die;* **b)** *(forehead)* Stirn, *die;* **c)** *(of hill)* [Berg]kuppe, *die*

'browbeat *v.t., forms as* **beat 1** unter Druck setzen; einschüchtern; **~ sb. into doing sth.** jmdn. so unter Druck setzen, daß er etw. tut

brown [braʊn] **1.** *adj.* braun. **2.** *n.* Braun, *das.* **3.** *v.t.* **a)** bräunen ‹*Haut, Körper*›; **b)** *(Cookery)* [an]bräunen; anbraten ‹*Fleisch*›; **c)** *(Brit. sl.)* **be ~ed off with sth./sb.** etw./jmdn. satt haben *(ugs.)*; **be ~ed off with doing sth.** es satt haben, etw. zu tun *(ugs.)*. **4.** *v.i.* **a)** ‹*Haut:*› bräunen; **b)** *(Cookery)* ‹*Fleisch:*› braun werden

brown: ~ 'ale *n.* dunkles Starkbier; **~ 'bear** *n.* Braunbär, *der;* **~ 'bread** *n.* ≈ Mischbrot, *das;* *(wholemeal)* Vollkornbrot, *das*

brownie ['braʊnɪ] *n.* **a) the B~s**

die Wichtel *(Pfadfinderinnen von 7–11 Jahren);* **b)** *(elf)* Heinzelmännchen, *das*

brownish ['braʊnɪʃ] *adj.* bräunlich

brown: ~ '**paper** *n.* Packpapier, *das;* ~ '**rice** *n.* Naturreis, *der;* ~ '**sugar** *n.* brauner Zucker

browse [braʊz] **1.** *v.i.* **a)** ⟨*Vieh:*⟩ weiden; ⟨*Wild:*⟩ äsen; ~ **on sth.** etw. fressen; **b)** *(fig.)* ~ **through a magazine** in einer Zeitschrift blättern. **2.** *n.* **have a ~:** sich umsehen

Bruges [bru:ʒ] *pr. n.* Brügge *(das)*

bruise [bru:z] **1.** *n.* **a)** *(Med.)* blauer Fleck; **b)** *(on fruit)* Druckstelle, *die.* **2.** *v.t.* quetschen ⟨*Obst, Pflanzen*⟩; ~ **oneself/one's leg** sich stoßen/sich am Bein stoßen; **he was badly ~d** er hat sich *(Dat.)* starke Prellungen zugezogen. **3.** *v.i.* ⟨*Person:*⟩ blaue Flecken bekommen; ⟨*Obst:*⟩ Druckstellen bekommen

brunette [bru:'net] **1.** *n.* Brünette, *die.* **2.** *adj.* brünett

Brunswick ['brʌnzwɪk] **1.** *pr. n.* Braunschweig *(das).* **2.** *attrib. adj.* Braunschweiger

brunt [brʌnt] *n.* **bear the ~ of the attack/financial cuts** von dem Angriff/von den Einsparungen am meisten betroffen sein

brush [brʌʃ] **1.** *n.* **a)** Bürste, *die; (for sweeping)* Hand-, Kehrbesen, *der; (with short handle)* Handfeger, *der; (for scrubbing)* [Scheuer]bürste, *die; (for painting or writing)* Pinsel, *der;* **b)** *(quarrel, skirmish)* Zusammenstoß, *der;* **have a ~ with the law** mit dem Gesetz in Konflikt kommen; **c)** *(light touch)* flüchtige Berührung; **d)** **give your hair/teeth a ~:** bürste dir die Haare/putz dir die Zähne; **give your shoes/clothes a ~:** bürste deine Schuhe/Kleider ab. **2.** *v.t.* **a)** *(sweep)* kehren; fegen; abbürsten ⟨*Kleidung*⟩; ~ **one's teeth/hair** sich *(Dat.)* die Zähne putzen/die Haare bürsten; **b)** *(Cookery)* bepinseln, bestreichen ⟨*Teigwaren, Gebäck*⟩; **c)** *(touch in passing)* flüchtig berühren; streifen. **3.** *v.i.* ~ **by** *or* **against** *or* **past sb./sth.** jmdn./etw. streifen

~ **a'side** *v.t.* beiseite schieben ⟨*Person, Hindernis*⟩; abtun, vom Tisch wischen ⟨*Einwand, Zweifel, Beschwerde*⟩

~ **a'way** *v.t.* abbürsten ⟨*Schmutz, Staub usw.*⟩; *(with hand or cloth)* abwischen; wegwischen

~ **'down** *v.t.* abbürsten ⟨*Kleidungsstück*⟩

~ **'off** *v.t.* **a)** abbürsten ⟨*Schmutz, Staub usw.*⟩; *(with hand or cloth)* abwischen; wegwischen; **b)** *(fig.: rebuff)* abblitzen lassen *(ugs.)*

~ **'up 1.** *v.t.* **a)** zusammenfegen ⟨*Krümel*⟩; **b)** auffrischen ⟨*Sprache, Kenntnisse*⟩. **2.** *v.i.* ~ **up on** auffrischen

brush: **~-off** *n.* Abfuhr, *die;* **give sb. the ~-off** jmdm. einen Korb geben *(ugs.);* **~-stroke** *n.* Pinselstrich, *der;* **~-up** *n.* **have a wash and ~-up** sich frisch machen

brusque [brʊsk, brʌsk] *adj.* schroff

Brussels ['brʌslz] *pr. n.* Brüssel *(das)*

Brussels 'sprouts *n. pl.* Rosenkohl, *der;* Kohlsprossen *(österr.)*

brutal ['bru:tl] *adj.* brutal; *(fig.)* brutal, schonungslos ⟨*Offenheit*⟩

brutality [bru:'tælɪtɪ] *n.* Brutalität, *die*

brutally ['bru:təlɪ] *adv.* brutal

brute [bru:t] **1.** *n.* **a)** *(animal)* Bestie, *die;* **b)** *(brutal person)* Rohling, *der;* brutaler Kerl *(ugs.); (thing)* höllische Sache; **a ~ of a problem** *(fig.)* ein höllisches Problem; **a drunken ~:** ein brutaler Trunkenbold. **2.** *attrib. adj. (without capacity to reason)* vernunftlos; irrational; **by ~ force** mit roher Gewalt

brutish ['bru:tɪʃ] *adj.* brutal ⟨*Flegel*⟩; tierisch ⟨*Leidenschaften, Gelüste*⟩

BS *abbr.* **British Standard** Britische Norm

B. Sc. [bi:es'si:] *abbr.* **Bachelor of Science** Bakkalaureus der Naturwissenschaften; **he is a** *or* **has a ~:** ≈ er hat ein Diplom in Naturwissenschaften

BST *abbr.* **British Summer Time** Britische Sommerzeit

bubble ['bʌbl] **1.** *n.* **a)** Blase, *die; (small)* Perle, *die; (fig.)* Seifenblase, *die;* **blow ~s** [Seifen]blasen machen; **b)** *(domed canopy)* [Glas]kuppel, *die.* **2.** *v.i.* **a)** *(form ~s)* ⟨*Wasser, Schlamm, Lava:*⟩ Blasen bilden; ⟨*Suppe, Flüssigkeiten:*⟩ brodeln; *(make sound of ~s)* ⟨*Bach, Quelle:*⟩ plätschern; **b)** *(fig.)* ~ **with sth.** vor etw. *(Dat.)* übersprudeln

~ **'over** *v.i.* überschäumen; ~ **over with excitement/joy** *(fig.)* vor Aufregung übersprudeln/vor Freude überquellen

~ **'up** *v.i.* ⟨*Gas:*⟩ in Blasen aufsteigen; ⟨*Wasser:*⟩ aufsprudeln

bubble: ~ **bath** *n.* Schaumbad, *das;* ~ **gum** *n.* Bubble-Gum, *der;* Ballonkaugummi, *der*

bubbly ['bʌblɪ] **1.** *adj.* **a)** sprudelnd; schäumend ⟨*Bade-, Spülwasser*⟩; **b)** *(fig. coll.)* quirlig *(ugs.)* ⟨*Person*⟩. **2.** *n. (Brit. coll.)* Schampus, *der (ugs.)*

Bucharest [bju:kə'rest] *pr. n.* Bukarest *(das)*

[1]**buck** [bʌk] **1.** *n. (deer, chamois)* Bock, *der; (rabbit, hare)* Rammler, *der.* **2.** *v.i.* ⟨*Pferd:*⟩ bocken. **3.** *v.t.* ~ **[off]** ⟨*Pferd:*⟩ abwerfen

[2]**buck** *n. (coll.)* **pass the ~ to sb.** *(fig.)* jmdm. die Verantwortung aufhalsen; **the ~ stops here** *(fig.)* die Verantwortung liegt letzten Endes bei mir

[3]**buck** *(coll.)* **1.** *v.i.* ~ **'up a)** *(make haste)* sich ranhalten *(ugs.);* ~ **up!** los, schnell!; auf, los!; **b)** *(cheer up)* ~ **up!** Kopf hoch! **2.** *v.t.* ~ **'up a)** *(cheer up)* aufmuntern; **b)** ~ **one's ideas up** *(coll.)* sich zusammenreißen

[4]**buck** *n. (Amer. and Austral. sl.: dollar)* Dollar, *der;* **make a fast ~:** eine schnelle Mark machen *(ugs.)*

bucket ['bʌkɪt] **1.** *n.* Eimer, *der;* **a ~ of water** ein Eimer [voll] Wasser; **kick the ~** *(fig. sl.)* ins Gras beißen *(salopp).* **2.** *v.i.* **the rain** *or* **it is ~ing down** es gießt wie aus Kübeln *(ugs.)*

bucketful ['bʌkɪtfʊl] *n.* Eimer [voll]

buckle ['bʌkl] **1.** *n.* Schnalle, *die.* **2.** *v.t.* **a)** zuschnallen; ~ **sth. on** etw. anschnallen; ~ **sth. up** etw. zuschnallen; **b)** *(crumple)* verbiegen ⟨*Stoßstange, Rad*⟩. **3.** *v.i.* ⟨*Rad, Metallplatte:*⟩ sich verbiegen

~ **'down** *v.i.* sich dahinterklemmen; ~ **down to a task** sich hinter eine Aufgabe klemmen

buck: **~-tooth** *n.* vorstehender Zahn; Raffzahn, *der (ugs.);* **~wheat** ['bʌkwi:t] *n. (Agric.)* Buchweizen, *der*

bud [bʌd] **1.** *n.* Knospe, *die;* **come into ~/be in ~:** knospen; Knospen treiben; **nip sth. in the ~** *(fig.)* etw. im Keim ersticken. **2.** *v.i.,* **-dd-** knospen; Knospen treiben; ⟨*Baum:*⟩ ausschlagen; **a ~ding painter/actor** *(fig.)* ein angehender Maler/Schauspieler

Buddhism ['bʊdɪzm] *n.* Buddhismus, *der*

Buddhist ['bʊdɪst] **1.** *n.* Buddhist, *der*/Buddhistin, *die.* **2.** *adj.* buddhistisch

buddy ['bʌdɪ] *n. (coll.)* Kumpel, *der (ugs.)*

budge [bʌdʒ] **1.** *v.i.* ⟨*Person, Tier:*⟩ sich [von der Stelle] rühren; ⟨*Gegenstand:*⟩ sich bewegen, nachgeben; *(fig.: change opinion)* nachgeben. **2.** *v.t.* **a)** bewegen; **I**

can't ~ this screw ich kriege diese Schraube nicht los; **b)** *(fig.: change opinion)* abbringen; **he refuses to be ~d** er läßt sich nicht umstimmen

budgerigar ['bʌdʒərɪgɑ:(r)] *n.* Wellensittich, *der*

budget ['bʌdʒɪt] **1.** *n.* Budget, *das;* Etat, *der;* Haushalt[splan], *der;* **keep within ~:** seinen Etat nicht überschreiten; **~ meal/holiday** preisgünstige Mahlzeit/Ferien. **2.** *v. i.* planen; **~ for sth.** etw. [im Etat] einplanen

budgie ['bʌdʒɪ] *n. (coll.)* Wellensittich, *der*

buff [bʌf] **1.** *adj.* gelbbraun. **2.** *n.* **a)** *(coll.: enthusiast)* Fan, *der (ugs.);* **b)** *(colour)* Gelbbraun, *das.* **3.** *v. t. (polish)* polieren, [blank] putzen ‹*Metall, Schuhe usw.*›

buffalo ['bʌfələʊ] *n., pl.* **~es** *or same* Büffel, *der*

buffer ['bʌfə(r)] *n.* Prellbock, *der; (on vehicle; also fig.)* Puffer, *der*

[1]**buffet** ['bʌfɪt] *v. t.* schlagen; **~ed by the wind/waves** vom Wind geschüttelt/von den Wellen hin und her geworfen

[2]**buffet** ['bʊfeɪ] *n. (Brit.)* **a)** *(place)* Büfett, *das;* **~ car** *(Railw.)* Büfettwagen, *der;* **b)** *(meal)* Imbiß, *der;* **~ lunch/supper/meal** Büfettessen, *das;* **a cold ~:** ein kaltes Büfett

bug [bʌg] **1.** *n.* **a)** Wanze, *die;* **b)** *(Amer.: small insect)* Insekt, *das;* Käfer, *der;* **c)** *(coll.: virus)* Bazillus, *der;* **d)** *(coll.: disease)* Infektion, *die;* Krankheit, *die;* **catch a ~:** sich *(Dat.)* was holen *(ugs.);* **e)** *(coll.: concealed microphone)* Wanze, *die (ugs.);* **f)** *(coll.: defect)* Macke, *die (salopp).* **2.** *v. t.,* **-gg-: a)** *(coll.: install microphone in)* verwanzen ‹*Zimmer*› *(ugs.);* abhören ‹*Telefon, Konferenz*›; **b)** *(sl.) (annoy)* nerven *(salopp);* den Nerv töten (+ *Dat.*) *(ugs.); (bother)* beunruhigen; **what's ~ging you?** was ist mit dir?

bugbear ['bʌgbeə(r)] *n.* Problem, *das;* Sorge, *die*

bugger ['bʌgə(r)] **1.** *n.* **a)** *(coarse) (fellow)* Bursche, *der (ugs.);* Macker, *der (salopp); as insult* Scheißkerl, *der (derb);* **b)** *(coarse: thing)* Scheißding, *das (derb).* **2.** *v. t. (coarse: damn)* **~ you/him** *(dismissive)* du kannst/der kann mich mal *(derb);* **~ this car/him!** *(angry)* dieses Scheißauto/dieser Scheißkerl! *(derb);* **~ it!** ach du Scheiße! *(derb); (in surprise)* **well, ~ me** *or* **I'll be ~ed!** ach du Scheiße! *(derb)*

~ a'bout, ~ a'round *(coarse)* **1.** *v. i.* Scheiß machen *(derb);* rumblödeln *(ugs.);* **~ about with sth.** mit etw. rumfummeln *(ugs.).* **2.** *v. t.* verarschen *(derb)*

~ 'off *v. i. (coarse)* abhauen *(ugs.)*

~ 'up *v. t. (coarse)* verkorksen *(ugs.)*

buggy ['bʌgɪ] *n. (pushchair)* Sportwagen, *der*

bugle ['bju:gl] *n.* Bügelhorn, *das*

build [bɪld] **1.** *v. t.,* **built** [bɪlt] **a)** bauen; errichten ‹*Gebäude, Damm*›; mauern ‹*Schornstein, Kamin*›; zusammenbauen *od.* -setzen ‹*Fahrzeug*›; **the house is still being built** das Haus ist noch im Bau; **~ sth. from** *or* **out of sth.** etw. aus etw. machen *od.* bauen; **b)** *(fig.)* aufbauen ‹*System, Gesellschaft, Reich, Zukunft*›; schaffen ‹*bessere Zukunft, Beziehung*›; begründen ‹*Ruf*›. **2.** *v. i.,* **built a)** bauen; **b)** *(fig.)* **~ on one's successes** auf seinen Erfolgen aufbauen. **3.** *n.* Körperbau, *der*

~ 'in *v. t.* einbauen

~ into *v. t.* **~ sth. into sth.** etw. in etw. *(Akk.)* einbauen

~ on *v. t.* **a)** aufbauen auf (+ *Dat.*); bebauen ‹*Gelände*›; **b)** *(attach)* **~ sth. on to sth.** etw. an etw. *(Akk.)* anbauen

~ 'up 1. *v. t.* **a)** bebauen ‹*Land, Gebiet*›; **b)** *(accumulate)* aufhäufen ‹*Reserven, Mittel, Kapital*›; **c)** *(strengthen)* stärken ‹*Gesundheit, Widerstandskraft*›; widerstandsfähig machen, kräftigen ‹*Person, Körper*›; **d)** *(increase)* erhöhen, steigern ‹*Produktion, Kapazität*›; stärken ‹*[Selbst]vertrauen*›; **~ up sb.'s hopes [unduly]** jmdm. [falsche] Hoffnung machen; **e)** *(develop)* aufbauen ‹*Firma, Geschäft*›. **2.** *v. i.* **a)** ‹*Spannung, Druck:*› zunehmen, ansteigen; ‹*Musik:*› anschwellen; ‹*Lärm:*› sich steigern (**to** in + *Akk.*); **~ up to a crescendo** sich zu einem Crescendo steigern; **b)** ‹*Schlange, Rückstau:*› sich bilden; ‹*Verkehr:*› sich verdichten, sich stauen

builder ['bɪldə(r)] *n.* **a)** Erbauer, *der;* **b)** *(contractor)* Bauunternehmer, *der;* **~'s labourer** Bauarbeiter, *der*

building ['bɪldɪŋ] *n.* **a)** *no pl.* Bau, *der;* **b)** *(structure)* Gebäude, *das; (for living in)* Haus, *das*

building: ~ contractor *n.* Bauunternehmer, *der;* **~-site** *n.* Baustelle, *die;* **~ society** *n. (Brit.)* Bausparkasse, *die*

'build-up *n.* **a)** *(publicity)* Reklame[rummel], *der;* **give sb./sth. a good ~:** jmdn./etw. groß ankündigen; **b)** *(approach to climax)* Vorbereitungen *Pl.* (**to** für); **c)** *(increase)* Zunahme, *die; (of forces)* Verstärkung, *die;* **a ~ of traffic** ein [Verkehrs]stau

built *see* **build 1, 2**

built: ~-in *adj.* **a)** eingebaut; **a ~-in cupboard/kitchen** ein Einbauschrank/eine Einbauküche; **b)** *(fig.: instinctive)* angeboren; **~-up** *adj.* bebaut; **~-up area** Wohngebiet, *das; (Motor Veh.)* geschlossene Ortschaft

bulb [bʌlb] *n.* **a)** *(Bot., Hort.)* Zwiebel, *die;* **b)** *(of lamp)* [Glüh]birne, *die;* **c)** *(of thermometer, chemical apparatus)* [Glas]kolben, *der*

Bulgaria [bʌl'geərɪə] *pr. n.* Bulgarien *(das)*

Bulgarian [bʌl'geərɪən] **1.** *adj.* bulgarisch; **he/she is ~:** er ist Bulgare/sie ist Bulgarin. **2.** *n.* **a)** *(person)* Bulgare, *der*/Bulgarin, *die;* **b)** *(language)* Bulgarisch, *das; see also* **English 2a**

bulge [bʌldʒ] **1.** *n.* **a)** Ausbeulung, *die;* ausgebeulte Stelle; *(in line)* Bogen, *der; (in tyre)* Wulst, *der od. die;* **b)** *(coll.: increase)* Anstieg, *der* (**in** *Gen.*). **2.** *v. i.* **a)** *(swell outwards)* sich wölben; **b)** *(be full)* vollgestopft sein

bulging ['bʌldʒɪŋ] *adj.* prall gefüllt ‹*Einkaufstasche usw.*›; vollgestopft ‹*Hosentasche, Kiste*›; rund ‹*Bauch*›

bulk [bʌlk] *n.* **a)** *(large quantity)* **in ~:** in großen Mengen; **b)** *(large shape)* massige Gestalt; **c)** *(size)* Größe, *die;* **d)** *(volume)* Menge, *die;* Umfang, *der;* **e)** *(greater part)* **the ~ of the money** der Groß- *od.* Hauptteil des Geldes; **the ~ of the population** die Mehrheit der Bevölkerung; **f)** *(Commerc.)* **in ~** *(loose)* lose; unabgefüllt ‹*Wein*›; *(wholesale)* en gros

bulky ['bʌlkɪ] *adj.* sperrig ‹*Gegenstand*›; massig, wuchtig ‹*Gestalt, Körper*›; unförmig ‹*Kleidungsstück*›; *(unwieldy)* unhandlich ‹*Gegenstand, Paket*›

bull [bʊl] *n.* **a)** Bulle, *der; (for ~fight)* Stier, *der;* **like a ~ in a china shop** *(fig.)* wie ein Elefant im Porzellanladen; **take the ~ by the horns** *(fig.)* den Stier bei den Hörnern fassen *od.* packen; **b)** *(whale, elephant)* Bulle, *der;* **c)** *see* **bull's eye**

bull: ~dog *n.* Bulldogge, *die;* **~dog clip** Flügelklammer, *die;* **~doze** *v. t.* **a)** planieren ‹*Boden*›; mit der Planierraupe wegräumen ‹*Gebäude*›; **b)** *(fig.: force)* **~doze**

sb. into doing sth. jmdn. dazu zwingen, etw. zu tun; **~dozer** ['bʊldəʊzə(r)] *n.* Planierraupe, *die*

bullet ['bʊlɪt] *n.* [Gewehr-, Pistolen]kugel, *die*

'**bullet-hole** *n.* Einschuß, *der;* Einschußloch, *das*

bulletin ['bʊlɪtɪn] *n.* Bulletin, *das*

'**bulletin-board** *n. (Amer.)* Anschlagtafel, *die; (Sch., Univ.)* Schwarzes Brett

'**bulletproof** *adj.* kugelsicher

bull: ~fight *n.* Stierkampf, *der;* **~-fighter** *n.* Stierkämpfer, *der;* **~-fighting** *n.* Stierkämpfe; **~finch** *n. (Ornith.)* Gimpel, *der;* **~frog** *n.* Ochsenfrosch, *der*

bullion ['bʊljən] *n., no pl., no indef. art.* **gold/silver ~:** ungemünztes Gold/Silber; *(ingots)* Gold-/Silberbarren *Pl.*

'**bull market** *n. (St. Exch.)* Haussemarkt, *der*

bullock ['bʊlək] *n.* Ochse, *der*

bull: ~ring *n.* Stierkampfarena, *die;* **~'s-eye** *n. (of target)* Schwarze, *das;* **score a ~'s-eye** *(lit. or fig.)* ins Schwarze treffen; **~shit** *n. (coarse)* Scheiße, *die (salopp abwertend)*

bully ['bʊlɪ] **1.** *n.: jmd., der gern Schwächere schikaniert bzw. tyrannisiert; (esp. schoolboy etc.)* ≈ Rabauke, *der (abwertend); (boss)* Tyrann, der. **2.** *v. t. (persecute)* schikanieren; *(frighten)* einschüchtern; **~ sb. into/out of doing sth.** jmdn. so sehr einschüchtern, daß er etw. tut/läßt

bullying ['bʊlɪɪŋ] **1.** *n.* Schikanieren, *das.* **2.** *adj.* tyrannisch

bulrush ['bʊlrʌʃ] *n.* **a)** *(Bot.)* Teichsimse, *die;* **b)** *(Bibl.)* Rohr, *das*

bulwark ['bʊlwək] *n. (rampart)* Wall, *der;* Bollwerk, *das (auch fig.)*

[1]**bum** [bʌm] *n. (Brit. sl.)* Hintern, *der (ugs.);* Arsch, *der (derb)*

[2]**bum** *(sl.)* **1.** *n. (Amer.)* **a)** *(tramp)* Penner, *der (salopp abwertend);* Berber, *der (salopp);* **b)** *(lazy dissolute person)* Penner, *der (salopp abwertend);* Gammler, *der (ugs. abwertend).* **2.** *adj.* mies *(ugs.).* **3.** *v.i.,* **-mm-: ~ [about** *or* **around]** rumgammeln *(ugs.).* **4.** *v. t.,* **-mm-** schnorren *(ugs.)* ‹*Zigaretten usw.*› **(off** bei)

bumble-bee ['bʌmblbi:] *n.* Hummel, *die*

bumbling ['bʌmblɪŋ] *adj.* stümperhaft

bumf [bʌmf] *n. (Brit. sl.)* Papierkram, *der (ugs.)*

bump [bʌmp] **1.** *n.* **a)** *(sound)* Bums, *der; (impact)* Stoß, *der;* **this car has had a few ~s** der Wagen hat schon einige Dellen abgekriegt; **b)** *(swelling)* Beule, *die;* **c)** *(hump)* Buckel, *der (ugs.).* **2.** *adv.* bums; rums, bums. **3.** *v. t.* **a)** anstoßen; **I ~ed the chair against the wall** ich stieß mit dem Stuhl an die Wand; **b)** *(hurt)* **~ one's head/knee** sich am Kopf/Knie stoßen. **4.** *v. i.* **a) ~ against sb./sth.** jmdn./an etw. *(Akk.) od.* gegen etw. stoßen; **b)** *(move with jolts)* rumpeln

~ into *v. t.* **a)** stoßen an (+ *Akk.*) *od.* gegen; *(with vehicle)* fahren gegen ‹*Mauer, Baum*›; **~ into sb.** jmdn. anstoßen; *(with vehicle)* jmdn. anfahren; **b)** *(meet by chance)* zufällig [wieder]treffen

~ 'off *v. t. (sl.)* kaltmachen *(salopp)*

~ up *v. t. (coll.)* aufschlagen ‹*Preise*›; aufbessern ‹*Gehalt*›

bumper ['bʌmpə(r)] **1.** *n.* **a)** *(Motor Veh.)* Stoßstange, *die;* **b)** *(Amer. Railw.)* Puffer, *der.* **2.** *adj.* Rekord‹*ernte, -jahr*›

'**bumper car** *n.* [Auto]skooter, *der*

bumpkin ['bʌmpkɪn] *n.* **[country] ~:** [Bauern]tölpel, *der (abwertend)*

bumpy ['bʌmpɪ] *adj.* holp[e]rig ‹*Straße, Fahrt, Fahrzeug*›; uneben ‹*Fläche*›; unruhig ‹*Flug*›

bun [bʌn] *n.* **a)** süßes Brötchen; *(currant ~)* Korinthenbrötchen, *das;* **b)** *(hair)* [Haar]knoten, *der*

bunch [bʌntʃ] *n.* **a)** *(of flowers)* Strauß, *der; (of grapes, bananas)* Traube, *die; (of parsley, radishes)* Bund, *das;* **~ of flowers/grapes** Blumenstrauß, *der*/Traube, *die;* **~ of keys** Schlüsselbund, *der;* **b)** *(lot)* Anzahl, *die;* **a whole ~ of ...:** ein ganzer Haufen ... *(ugs.);* **the best** *or* **pick of the ~:** der/die/das Beste [von allen]; **c)** *(group)* Haufen, *der (ugs.)*

~ 'up 1. *v. i.* ‹*Personen:*› zusammenrücken; ‹*Kleid, Stoff:*› sich zusammenknüllen. **2.** *v. t.* zusammenraffen ‹*Kleid*›

bundle ['bʌndl] **1.** *n.* Bündel, *das; (of papers)* Packen, *der; (of hay)* Bund, *das; (of books)* Stapel, *der; (of fibres, nerves)* Strang, *der;* **she's a ~ of mischief/energy** *(fig.)* sie hat nichts als Unfug im Kopf/ist ein Energiebündel. **2.** *v. t.* **a)** bündeln; **b) ~ sth. into the suitcase/back of the car** etw. in den Koffer stopfen/hinten ins Auto werfen; **~ sb. into the car** jmdn. ins Auto verfrachten

~ 'up *v. t. (put in ~s)* bündeln

bung [bʌŋ] **1.** *n.* Spund[zapfen], *der.* **2.** *v. t. (sl.)* schmeißen *(ugs.)*

~ 'up *v. t.* **be/get ~ed up** verstopft sein/verstopfen

bungalow ['bʌŋgələʊ] *n.* Bungalow, *der*

bungle ['bʌŋgl] *v. t.* stümpern bei

bungler ['bʌŋglə(r)] *n.* Stümper, *der (abwertend)*

bungling ['bʌŋglɪŋ] *adj.* stümperhaft ‹*Versuch*›; **~ person** Stümper, der

[1]**bunk** [bʌŋk] *n. (in ship, aircraft, lorry)* Koje, *die; (in sleeping-car, room)* Bett, *das*

[2]**bunk** *n. (sl.: nonsense)* Quatsch, *der (salopp);* Mist, *der (salopp)*

[3]**bunk** *n. (Brit. sl.)* **do a ~:** türmen *(salopp)*

'**bunk-bed** *n.* Etagenbett, *das*

bunker ['bʌŋkə(r)] *n. (fuel-~, Mil., Golf)* Bunker, *der*

bunny ['bʌnɪ] *n.* Häschen, *das*

bunting *n., no pl. (flags)* [bunte] Fähnchen; Wimpel *Pl.*

buoy [bɔɪ] **1.** *n.* Boje, *die.* **2.** *v. t.* **~ [up]** über Wasser halten; *(fig.: support, sustain)* aufrechterhalten; **I was ~ed [up] by the thought that ...:** der Gedanke, daß ..., ließ mich durchhalten

buoyancy ['bɔɪənsɪ] *n.* **a)** Auftrieb, *der;* **b)** *(fig.)* Schwung, *der;* Elan, *der*

buoyant ['bɔɪənt] *adj.* **a)** Auftrieb habend; schwimmend; **be [more] ~:** [einen größeren] Auftrieb haben; [besser] schwimmen; **b)** *(fig.)* rege ‹*Markt*›; heiter ‹*Person*›; federnd ‹*Schritt*›

burble ['bɜ:bl] *v. i.* **a)** *(speak lengthily)* **~ [on] about sth.** von etw. ständig quasseln *(ugs.);* **b)** *(make a murmuring sound)* brummeln *(ugs.)*

burden ['bɜ:dn] **1.** *n. (lit. or fig.)* Last, *die;* **beast of ~:** Lasttier, *das;* **become a ~:** zur Last werden; **be a ~ to sb.** für jmdn. eine Belastung sein; *(less serious)* jmdm. zur Last fallen. **2.** *v. t.* belasten; *(fig.)* **~ sb./oneself with sth.** jmdn./sich mit etw. belasten

burdensome ['bɜ:dnsəm] *adj. (fig.)* lästig ‹*Person, Pflicht, Verantwortung*›; **become/be ~ to sb.** jmdm. zur Last werden/fallen

bureau ['bjʊərəʊ] *n., pl.* **~x** ['bjʊərəʊz] *or* **~s a)** *(Brit.: writing-desk)* Schreibschrank, *der; (Amer.: chest of drawers)* Kommode, *die;* **b)** *(office)* Büro, *das; (department)* Abteilung, *die; (Amer.: of government)* Amt, *das*

bureaucracy [bjʊə'rɒkrəsɪ] *n.* Bürokratie, *die*

bureaucrat ['bjʊərəkræt] *n.* Bürokrat, *der*/Bürokratin, *die (abwertend)*

bureaucratic [bjʊərə'krætɪk] *adj.*, **bureaucratically** [bjʊərə'krætɪkəlɪ] bürokratisch

burglar ['bɜ:glə(r)] *n.* Einbrecher, *der*

'**burglar alarm** *n.* Alarmanlage, *die*

burglarize ['bɜ:gləraɪz] *(Amer.) see* **burgle**

burglary ['bɜ:glərɪ] *n.* Einbruch, *der; (offence)* [Einbruchs]diebstahl, *der*

burgle ['bɜ:gl] *v. t.* einbrechen in (+ *Akk.*); **the shop/he was ~d** in dem Laden/bei ihm wurde eingebrochen

Burgundy ['bɜ:gəndɪ] *pr. n.* Burgund *(das)*

burgundy *n.* Burgunder[wein], *der*

burial ['berɪəl] *n.* Bestattung, *die;* Begräbnis, *das; (funeral)* Beerdigung, *die;* ~ **at sea** Seebestattung, *die*

'**burial-service** *n.* Trauerfeier, *die*

burlesque [bɜ:'lesk] *n.* **a)** Varieté, *das;* **b)** *(book, play)* Burleske, *die; (parody)* Parodie, *die*

burly ['bɜ:lɪ] *adj.* kräftig; stämmig; stramm ⟨*Soldat*⟩

Burma ['bɜ:mə] *pr. n.* Birma *(das)*

Burmese [bɜ:'mi:z] **1.** *adj.* birmanisch; **sb. is ~:** jmd. ist Birmane/Birmanin. **2.** *n., pl. same* **a)** *(person)* Birmane, *der*/Birmanin, *die;* **b)** *(language)* Birmanisch, *das; see also* **English 2 a**

[1]**burn** [bɜ:n] **1.** *n. (on the skin)* Verbrennung, *die; (on material)* Brandfleck, *der; (hole)* Brandloch, *das.* **2.** *v. t.,* **~t** *or* **~ed a)** verbrennen; ~ **a hole in sth.** ein Loch in etw. *(Akk.)* brennen; ~ **one's boats** *or* **bridges** *(fig.)* alle Brücken hinter sich *(Dat.)* abbrechen; **b)** *(use as fuel)* als Brennstoff verwenden ⟨*Gas, Öl usw.*⟩; heizen mit ⟨*Kohle, Holz, Torf*⟩; verbrauchen ⟨*Strom*⟩; *(use up)* verbrauchen ⟨*Treibstoff*⟩; verfeuern ⟨*Holz, Kohle*⟩; ~ **coal in the stove** den Ofen mit Kohle feuern; **c)** *(injure)* verbrennen; ~ **oneself/one's hand** sich verbrennen/sich *(Dat.)* die Hand verbrennen; ~ **one's fingers, get one's fingers ~t** *(fig.)* sich *(Dat.)* die Finger verbrennen *(fig.);* **d)** *(spoil)* anbrennen lassen ⟨*Fleisch, Kuchen*⟩; **be ~t** angebrannt sein; **e)** *(cause ~ing sensation to)* verbrennen; **f)** *(put to death)* ~ **sb. [at the stake]** jmdn. [auf dem Scheiterhaufen] verbrennen; **g)** *(corrode)* ätzen; verätzen ⟨*Haut*⟩. **3.** *v. i.,* **~t** *or* **~ed a)** brennen; ~ **to death** verbrennen; **b)** *(blaze)* ⟨*Feuer:*⟩ brennen; ⟨*Gebäude:*⟩ in Flammen stehen, brennen; **c)** *(give light)* ⟨*Lampe, Kerze, Licht:*⟩ brennen; **d)** *(be injured)* sich verbrennen; **she/her skin ~s easily** sie bekommt leicht einen Sonnenbrand; **e)** *(be spoiled)* ⟨*Kuchen, Milch, Essen:*⟩ anbrennen; **f)** *(be corrosive)* ätzen; ätzend sein

~ '**down 1.** *v. t.* niederbrennen. **2.** *v. i.* ⟨*Gebäude:*⟩ niederbrennen, abbrennen; *(less brightly)* ⟨*Feuer, Kerze:*⟩ herunterbrennen

~ '**out 1.** *v. t.* **a)** ausbrennen; **b)** *(fig.)* **feel ~ed out** sich erschöpft fühlen; ~ **oneself out** sich völlig verausgaben. **2.** *v. i.* **a)** ⟨*Kerze, Feuer:*⟩ erlöschen, ausgehen; ⟨*Rakete[nstufe]:*⟩ ausbrennen; **b)** *(Electr.)* durchbrennen

~ '**up 1.** *v. t.* verbrennen; verbrauchen ⟨*Energie*⟩. **2.** *v. i.* **a)** *(begin to blaze)* auflodern; **b)** *(be destroyed)* ⟨*Rakete, Meteor, Satellit:*⟩ verglühen

[2]**burn** *n. (Scot.)* Bach, *der*

burner ['bɜ:nə(r)] *n.* Brenner, *der*

burning ['bɜ:nɪŋ] **1.** *adj.* **a)** brennend; **b)** *(fig.)* glühend ⟨*Leidenschaft, Haß, Wunsch*⟩; brennend ⟨*Wunsch, Frage, Problem, Ehrgeiz*⟩. **2.** *n.* Brennen, *das;* **a smell of ~:** ein Brandgeruch

burnish ['bɜ:nɪʃ] *v. t.* polieren

burnt *see* [1]**burn 2, 3**

burnt '**offering** *n.* Brandopfer, *das; (fig. joc.: burnt food)* angebranntes Essen

burp [bɜ:p] *(coll.)* **1.** *n.* Rülpser, *der (ugs.); (of baby)* Bäuerchen, *das (fam.).* **2.** *v. i.* rülpsen *(ugs.);* aufstoßen. **3.** *v. t.* ein Bäuerchen machen lassen *(fam.)* ⟨*Baby*⟩

burrow ['bʌrəʊ] **1.** *n.* Bau, *der.* **2.** *v. t.* graben ⟨*Loch, Höhle, Tunnel*⟩; ~ **one's way under/through sth.** einen Weg *od.* Gang unter etw. *(Dat.)* durch/durch etw. graben. **3.** *v. i.* [sich *(Dat.)*] einen Gang graben; ~ **into sth.** *(fig.)* sich in etw. *(Akk.)* einarbeiten; ~ **through sth.** *(fig.)* sich durch etw. hindurchwühlen

bursar ['bɜ:sə(r)] *n. Verwalter der geschäftlichen Angelegenheiten einer Schule/Universität*

bursary ['bɜ:sərɪ] *n.* Kasse, *die; (scholarship)* Stipendium, *das*

burst [bɜ:st] **1.** *n.* **a)** *(split)* Bruch, *der;* **a ~ in a pipe** ein Rohrbruch; **b)** *(outbreak of firing)* Feuerstoß, *der;* Salve, *die;* **c)** *(fig.)* **a ~ of applause** ein Beifallsausbruch; **there was a ~ of laughter** man brach in Lachen aus. **2.** *v. t.,* **burst** zum Platzen bringen; platzen lassen ⟨*Luftballon*⟩; platzen ⟨*Reifen*⟩; sprengen ⟨*Kessel*⟩; ~ **pipe** Rohrbruch, *der;* **the river ~ its banks** der Fluß trat über die Ufer; **he [almost] ~ a blood-vessel** *(fig.)* ihn traf [fast] der Schlag; ~ **the door open** die Tür aufbrechen *od.* aufsprengen; ~ **one's sides with laughing** *(fig.)* vor Lachen beinahe platzen. **3.** *v. i.,* **burst a)** platzen; ⟨*Granate, Bombe, Kessel:*⟩ explodieren; ⟨*Damm:*⟩ brechen; ⟨*Flußufer:*⟩ überschwemmt werden; ⟨*Furunkel, Geschwür:*⟩ aufgehen, aufplatzen; ⟨*Knospe:*⟩ aufbrechen; ~ **open** ⟨*Tür, Deckel, Kiste, Koffer:*⟩ aufspringen; **b)** *(be full to overflowing)* **be ~ing with sth.** zum Bersten voll sein mit etw.; **be ~ing with pride/impatience** *(fig.)* vor Stolz/Ungeduld platzen; **be ~ing with excitement** *(fig.)* vor Aufregung außer sich sein; **I can't eat any more. I'm ~ing** *(fig.)* Ich kann nichts mehr essen. Ich platze [gleich] *(ugs.);* **be ~ing to say/do sth.** *(fig.)* es kaum abwarten können, etw. zu sagen/tun; **c)** *(appear, come suddenly)* ~ **through sth.** etw. durchbrechen

~ '**in** *v. i.* hereinplatzen; hereinstürzen; ~ **in [up]on sb./sth.** bei jmdm./etw. hereinplatzen

~ **into** *v. t.* **a)** eindringen in (+ *Akk.*); **we ~ into the room** wir stürzten ins Zimmer; **b)** ~ **into tears/laughter** in Tränen/Gelächter ausbrechen; ~ **into flames** in Brand geraten

~ '**out** *v. i.* **a)** herausstürzen; ~ **out of a room** aus einem Raum [hinaus]stürmen *od.* stürzen; **b)** *(exclaim)* losplatzen; **c)** ~ **out laughing/crying** in Lachen/Tränen ausbrechen

burton ['bɜ:tn] *n. (Brit. sl.)* **go for a ~** *(be destroyed)* kaputtgehen *(ugs.);* futsch gehen *(salopp); (be lost)* hopsgehen *(salopp)*

bury ['berɪ] *v. t.* **a)** begraben; beisetzen *(geh.)* ⟨*Toten*⟩; **where is Marx buried?** wo ist *od.* liegt Marx begraben?; **b)** *(hide)* vergraben; *(fig.)* begraben; ~ **the hatchet** *or (Amer.)* **tomahawk** *(fig.)* das Kriegsbeil begraben; ~ **one's face in one's hands** das Gesicht in den Händen vergraben; **c)** *(bring underground)* eingraben; abdecken ⟨*Wurzeln*⟩; **the houses were buried by a landslide** die Häuser wurden durch einen Erdrutsch verschüttet; **d)** *(plunge)* ~ **one's teeth in sth.** seine Zähne in etw. *(Akk.)* graben *od.* schlagen; **e)** ~ **oneself in one's**

studies/books sich in seine Studien vertiefen/in seinen Büchern vergraben

bus [bʌs] **1.** *n., pl.* ~es (*Amer.:* ~ses) [Auto-, Omni]bus, *der;* **go by** ~: mit dem Bus fahren. **2.** *v. i., (Amer.)* -ss- mit dem Bus fahren. **3.** *v. t.,* -ss- *(Amer.)* mit dem Bus befördern

busby ['bʌzbɪ] *n. (Brit.)* Kalpak, *der; (worn by guardsmen)* Bärenfellmütze, *die*

bus: ~ **company** *n.* ≈ Verkehrsbetrieb, *der;* ~-**conductor** *n.* Busschaffner, *der;* ~ **depot** *see* ~ **garage;** ~-**driver** *n.* Busfahrer, *der;* ~ **fare** *n.* [Bus]fahrpreis, *der;* ~ **garage** *n.* Busdepot, *das*

bush [bʊʃ] *n.* **a)** *(shrub)* Strauch, *der;* Busch, *der;* **b)** *(woodland)* Busch, *der*

bushy ['bʊʃɪ] *adj.* buschig

busily ['bɪzɪlɪ] *adv.* eifrig

business ['bɪznɪs] *n.* **a)** *(trading operation)* Geschäft, *das; (company, firm)* Betrieb, *der; (large)* Unternehmen, *das;* **b)** *no pl. (buying and selling)* Geschäfte *Pl.;* **on** ~: geschäftlich; **he's in the wool** ~: er ist in der Wollbranche; ~ **is** ~ *(fig.)* Geschäft ist Geschäft; **set up in** ~: ein Geschäft *od.* eine Firma gründen; **go out of** ~: pleite gehen *(ugs.);* **go into** ~: Geschäftsmann/-frau werden; **do** ~ **[with sb.]** [mit jmdm.] Geschäfte machen; **be in** ~: Geschäftsmann/-frau sein; **c)** *(task, duty, province)* Aufgabe, *die;* Pflicht, *die;* **that is 'my ~/none of 'your** ~: das ist meine Angelegenheit/nicht deine Sache; **what** ~ **is it of yours?** was geht Sie das an?; **mind your own** ~: kümmere dich um deine [eigenen] Angelegenheiten!; **he has no** ~ **to do that** er hat kein Recht, das zu tun; **d)** *(matter to be considered)* Angelegenheit, *die;* **'any other** ~**'** „Sonstiges"; **e)** *(serious work)* **get down to [serious]** ~: [ernsthaft] zur Sache kommen; *(Commerc.)* an die Arbeit gehen; **mean** ~: es ernst meinen; ~ **before pleasure** erst die Arbeit, dann das Vergnügen; **f)** *(derog.: affair)* Sache, *die;* Geschichte, *die (ugs.)*

business: ~ **address** *n.* Geschäftsadresse, *die;* ~ **card** *n.* Geschäftskarte, *die;* ~ **hours** *n. pl.* Geschäftszeit, *die; (in office)* Dienstzeit, *die;* ~ **letter** *n.* Geschäftsbrief, *der;* ~-**like** *adj.* geschäftsmäßig ‹*Art*›; sachlich, nüchtern ‹*Untersuchung*›; geschäftstüchtig ‹*Person*›; ~ **lunch** *n.* Arbeitsessen, *das;* ~**man** *n.* Geschäftsmann, *der;* ~ **school** *n.* kaufmännische Fachschule; ~ **studies** *n. pl.* Wirtschaftslehre, *die;* ~ **trip** *n.* Geschäftsreise, *die;* ~**woman** *n.* Geschäftsfrau, *die*

busker ['bʌskə(r)] *n.* Straßenmusikant, *der*

bus: ~ **lane** *n. (Brit.)* Busspur, *die;* ~-**ride** *n.* Busfahrt, *die;* ~-**route** *n.* Buslinie, *die;* ~ **service** *n.* Omnibusverkehr, *der; (specific service)* Busverbindung, *die;* ~ **shelter** *n.* Wartehäuschen, *das;* ~-**station** *n.* Omnibusbahnhof, *der;* ~-**stop** *n.* Bushaltestelle, *die*

[1]**bust** [bʌst] *n.* **a)** *(sculpture)* Büste, *die;* **b)** *(woman's bosom)* Busen, *der;* ~ **[measurement]** Oberweite, *die*

[2]**bust** *(coll.)* **1.** *adj.* **a)** *(broken)* kaputt *(ugs.);* **b)** *(bankrupt)* bankrott; pleite *(ugs.);* **go** ~: pleite gehen. **2.** *v. t.,* ~**ed** *or* **bust** *(break)* kaputtmachen *(ugs.);* ~ **sth. open** etw. aufbrechen. **3.** *v. i.,* ~**ed** *or* **bust** kaputtgehen *(ugs.)*

'bus-ticket *n.* Busfahrkarte, *die;* Busfahrschein, *der*

bustle ['bʌsl] **1.** *v. i.* ~ **about** geschäftig hin und her eilen; **the town centre was bustling with activity** im Stadtzentrum herrschte ein reges Treiben. **2.** *v. t.* jagen *(ugs.);* treiben *(ugs.).* **3.** *n.* Betrieb, *der; (of fair, streets also)* reges Treiben (**of** auf, in + *Dat.*)

bustling ['bʌslɪŋ] *adj.* belebt ‹*Stadt, Markt usw.*›; geschäftig ‹*Person, Art*›; rege ‹*Tätigkeit*›

'bust-up *n. (coll.)* Krach, *der (ugs.);* **have a** ~: Krach haben *(ugs.);* sich verkrachen *(ugs.)*

busy ['bɪzɪ] **1.** *adj.* **a)** *(occupied)* beschäftigt; **I'm** ~ **now** ich habe jetzt zu tun; **be** ~ **at** *or* **with sth.** mit etw. beschäftigt sein; **he was** ~ **packing** er war mit Packen beschäftigt *od.* war gerade beim Packen; **b)** *(full of activity)* arbeitsreich ‹*Leben*›; ziemlich hektisch ‹*Zeit*›; belebt ‹*Stadt*›; ausgelastet ‹*Person*›; rege ‹*Verkehr*›; **a** ~ **road** eine verkehrsreiche *od.* vielbefahrene Straße; **the office was** ~ **all day** im Büro war den ganzen Tag viel los; **I'm/he's a** ~ **man** ich habe/er hat viel zu tun; **c)** *(Amer. Teleph.)* besetzt. **2.** *v. refl.* ~ **oneself with sth.** sich mit etw. beschäftigen; ~ **oneself [in] doing sth.** sich damit beschäftigen, etw. zu tun

'busybody *n.* G[e]schaftlhuber, *der (südd., österr.);* **don't be such a** ~: misch dich nicht überall ein

but 1. [bət, *stressed* bʌt] *conj.* **a)** *co-ordinating* aber; **Sue wasn't there,** ~ **her sister was** Sue war nicht da, dafür aber ihre Schwester; **we tried to do it** ~ **couldn't** wir haben es versucht, aber nicht gekonnt; ~ **I 'did!** hab' ich doch!; **b)** *correcting after a negative* sondern; **not that book** ~ **this one** nicht das Buch, sondern dieses; **not only ...** ~ **also** nicht nur ..., sondern auch; **c)** *subordinating* ohne daß; **never a week passes** ~ **he phones** keine Woche vergeht, ohne daß er anruft. **2.** [bət] *prep.* außer (+ *Dat.*); **all** ~ **three** alle außer dreien; **the next** ~ **one/two** der/die/das über-/überübernächste; **the last** ~ **one/two** der/die/das vor-/vorvorletzte. **3.** [bət] *adv.* nur; bloß; **if I could** ~ **talk to her ...:** wenn ich [doch] nur mit ihr sprechen könnte ...; **we can** ~ **try** wir können es immerhin versuchen. **4.** [bʌt] *n.* Aber, *das;* **no** ~**s [about it]!** kein Aber!

butane ['bju:teɪn] *n. (Chem.)* Butan, *das*

butch [bʊtʃ] *adj.* betont männlich ‹*Frau, Kleidung, Frisur*›; betont maskulin, *(salopp)* macho ‹*Mann*›

butcher ['bʊtʃə(r)] **1.** *n.* **a)** Fleischer, *der;* Metzger, *der (bes. westd., südd.);* Schlachter, *der (nordd.);* ~**'s [shop]** Fleischerei, *die;* Metzgerei, *die (bes. westd., südd.); see also* **baker; b)** *(fig.: murderer)* [Menschen]schlächter, *der.* **2.** *v. t.* schlachten; *(fig.: murder)* niedermetzeln; abschlachten

butchery ['bʊtʃərɪ] *n.* **a)** ~ **[trade or business]** Fleischerhandwerk, *das;* **b)** *(fig.: needless slaughter)* Metzelei, *die*

butler ['bʌtlə(r)] *n.* Butler, *der*

[1]**butt** [bʌt] *n. (vessel)* Faß, *das; (for rainwater)* Tonne, *die*

[2]**butt** *n.* **a)** *(end)* dickes Ende; *(of rifle)* Kolben, *der;* **b)** *(of cigarette, cigar)* Stummel, *der*

[3]**butt** *n.* **a)** *(object of teasing or ridicule)* Zielscheibe, *die;* Gegenstand, *der;* **b)** *in pl. (shooting-range)* Schießstand, *der;* Waffenjustierstand, *der*

[4]**butt 1.** *n. (push) (by person)* [Kopf]stoß, *der; (by animal)* Stoß [mit den Hörnern]. **2.** *v. i.* ‹*Person:*› [mit dem Kopf] stoßen; ‹*Stier, Ziege:*› [mit den Hörnern] stoßen. **3.** *v. t.* ‹*Person:*› mit dem Kopf stoßen; ‹*Stier, Ziege:*› mit den Hörnern stoßen; ~ **sb. in the stomach** jmdm. mit dem Kopf in den Bauch stoßen

~ **'in** *v.i. (fig. coll.)* dazwischenreden; **may I ~ in?** darf ich mal kurz stören?

butter ['bʌtə(r)] **1.** *n.* Butter, *die;* **he looks as if ~ wouldn't melt in his mouth** *(fig.)* er sieht aus, als ob er kein Wässerchen trüben könnte; **melted ~:** zerlassene Butter. **2.** *v.t.* buttern; mit Butter bestreichen

~ **'up** *v.t.* ~ **sb. up** jmdm. Honig um den Mund *od.* Bart schmieren *(fig.)*

butter: ~**-bean** *n.* Mondbohne, *die;* Limabohne, *die;* ~**cup** *n. (Bot.)* Butterblume, *die;* ~**-dish** *n.* Butterdose, *die;* ~**-fingers** *n. sing.* Tolpatsch, *der (beim Fangen usw.)*

butterfly ['bʌtəflaɪ] *n.* **a)** Schmetterling, *der;* **have butterflies [in one's stomach]** *(fig. coll.)* ein flaues Gefühl im Magen haben; **b)** *see* **butterfly stroke**

'butterfly stroke *n. (Swimming)* Delphinstil, *der*

butter: ~**-knife** *n.* Buttermesser, *das;* ~**milk** *n.* Buttermilch, *die;* ~**scotch** *n.* Buttertoffee, *das*

buttock ['bʌtək] *n. (of person)* Hinterbacke, *die;* ~**s** Gesäß, *das*

button ['bʌtn] **1.** *n. (on clothing, of electric bell, etc.)* Knopf, *der.* **2.** *v.t. (fasten)* zuknöpfen; ~ **one's lip** *(Amer. sl.)* die Klappe halten *(salopp).* **3.** *v.i.* [zu]geknöpft werden

~ **'up 1.** *v.t.* zuknöpfen; *(fig.)* erledigen ‹*Job*›; **have the deal [all] ~ed up** das Geschäft unter Dach und Fach haben *(ugs.).* **2.** *v.i.* [zu]geknöpft werden

button: ~**hole 1.** *n.* **a)** Knopfloch, *das;* **b)** *(Brit.: flowers worn in coat-lapel)* Knopflochsträußchen, *das; (single flower)* Knopflochblume, *die;* Blume im Knopfloch; **2.** *v.t. (detain)* zu fassen kriegen *(ugs.);* **he was ~holed by X** X hat sich *(Dat.)* ihn geschnappt *(ugs.);* ~ **'mushroom** *n.* Champignon, *der;* ~**-through** *adj.* durchgeknöpft ‹*Kleid*›

buttress ['bʌtrɪs] **1.** *n.* **a)** *(Archit.)* Mauerstrebe, *die;* **b)** *(fig.)* Stütze, *die.* **2.** *v.t.* ~ **[up]** *(fig.)* [unter]stützen; untermauern ‹*Argument*›

buxom ['bʌksəm] *adj.* drall

buy [baɪ] **1.** *v.t.,* **bought** [bɔ:t] **a)** kaufen; lösen ‹*Fahrkarte*›; ~ **sb./oneself sth.** jmdm./sich etw. kaufen; ~ **and sell goods** Waren an- und verkaufen; ~ **sb. a pint** jmdm. einen Halben ausgeben; **he bought them a round** er spendierte ihnen eine Runde; **b)** *(fig.)* erkaufen ‹*Sieg, Ruhm, Frieden*›; einsparen, gewinnen ‹*Zeit*›; **c)** *(bribe)* bestechen; kaufen *(ugs.);* erkaufen ‹*Zustimmung*›; **d)** *(sl.) (believe)* schlucken *(ugs.);* glauben; *(accept)* akzeptieren; **I'll ~ that** *(believe)* ich glaube es [mal]. **2.** *n.* [Ein]kauf, *der;* **be a good ~:** preiswert sein

~ **'in** *v.t.* einkaufen ‹*Vorräte, Fleisch usw.*›

~ **'off** *v.t.* auszahlen ‹*Forderung*›; abfinden ‹*Anspruchserhebenden*›

~ **'out** *v.t.* auszahlen ‹*Aktionär, Partner*›; aufkaufen ‹*Firma*›

~ **'up** *v.t.* aufkaufen

buyer ['baɪə(r)] *n.* **a)** Käufer, *der*/Käuferin, *die;* **potential ~:** Kaufinteressent, *der;* **b)** *(Commerc.)* Einkäufer, *der*/Einkäuferin, *die*

buzz [bʌz] **1.** *n.* **a)** *(of insect)* Summen, *das; (of large insect)* Brummen, *das; (of smaller or agitated insect)* Schwirren, *das;* **b)** *(sound of buzzer)* Summen, *das;* **give one's secretary a ~:** über den Summer seine Sekretärin rufen; **c)** *(of conversation, movement)* Gemurmel, *das;* **d)** *(sl.: telephone call)* [Telefon]anruf, *der;* **give sb. a ~:** jmdn. anrufen; **e)** *(sl.: thrill)* Nervenkitzel, *der (ugs.).* **2.** *v.i.* **a)** *see* **1a:** ‹*Insekt:*› summen/brummen/schwirren; **b)** *(signal with buzzer)* [mit dem Summer] rufen; **c)** ~ **with excitement** in heller Aufregung sein; **the rumour set the office ~ing** das Gerücht versetzte das Büro in helle Aufregung; **my ears are ~ing** mir sausen die Ohren. **3.** *v.t. (Aeronaut.)* dicht vorbeifliegen an (+ *Dat.*)

~ **a'bout,** ~ **a'round 1.** *v.i.* herumschwirren; *(fig.)* ‹*Person:*› herumsausen. **2.** *v.t.* ~ **around sth.** um etw. [herum]schwirren

~ **'off** *v.i. (sl.)* abhauen *(salopp);* abzischen *(salopp)*

buzzer ['bʌzə(r)] *n.* Summer, *der*

'buzz-word *n.* Schlagwort, *das*

[1]by [baɪ] **1.** *prep.* **a)** *(near, beside)* an (+ *Dat.*); bei; *(next to)* neben; **by the window/river** am Fenster/Fluß; **she sat by me** sie saß neben mir; **b)** *(to position beside)* zu; **c)** *(about, in the possession of)* bei; **have sth. by one** etw. bei sich haben; **d) north-east by east** Nordost auf Ost; **e) by herself** *etc. see* **herself a; f)** *(along)* entlang; **by the river** am *od.* den Fluß entlang; **g)** *(via)* über (+ *Akk.*); **leave by the door/window** zur Tür hinausgehen/zum Fenster hinaussteigen; **we came by the quickest/shortest route** wir sind die schnellste/kürzeste Strecke gefahren; **h)** *(passing)* vorbei an (+ *Dat.*); **run/drive by sb./sth.** an jmdm./etw. vorbeilaufen/vorbeifahren; **i)** *(during)* bei; **by day/night** bei Tag/Nacht; tagsüber/nachts; **j)** *(through the agency of)* von; **written by ...:** geschrieben von ...; **k)** *(through the means of)* durch; **he was killed by lightning/a falling chimney** er ist vom Blitz/von einem umstürzenden Schornstein erschlagen worden; **heated by gas/oil** mit Gas/Öl geheizt; gas-/ölbeheizt; **by bus/ship** *etc.* mit dem Bus/Schiff *usw.;* **by air/sea** mit dem Flugzeug/Schiff; **have children by sb.** Kinder von jmdm. haben; **l)** *(not later than)* bis; **by now/this time** inzwischen; **by next week she will be in China** nächste Woche ist sie schon in China; **by the time this letter reaches you** bis Dich dieser Brief erreicht; **by the 20th** bis zum 20.; **m)** *indicating unit of time* pro; *indicating unit of length, weight, etc.* -weise; **by the second/minute/hour** pro Sekunde/Minute/Stunde; **rent a house by the year** ein Haus für jeweils ein Jahr mieten; **you can hire a car by the day or by the week** man kann sich *(Dat.)* ein Auto tageweise oder wochenweise mieten; **day by day/month by month, by the day/month** *(as each day/month passes)* Tag für Tag/Monat für Monat; **cloth by the metre** Stoff am Meter; **sell sth. by the packet/ton/dozen** etw. paket-/tonnenweise/im Dutzend verkaufen; **10 ft. by 20 ft.** 10 [Fuß] mal 20 Fuß; **n)** *indicating amount* **by the thousands** zu Tausenden; **one by one** einzeln; **two by two/three by three/four by four** zu zweit/dritt/viert; **little by little** nach und nach; **o)** *indicating factor* durch; **8 divided by 2 is 4** 8 geteilt durch 2 ist 4; **p)** *indicating extent* um; **wider by a foot** um einen Fuß breiter; **win by ten metres** mit zehn Metern Vorsprung gewinnen; **passed by nine votes to two** mit neun zu zwei Stimmen angenommen; **q)** *(according to)* nach; **by my watch** nach meiner Uhr; **r)** *in oaths* bei; **by [Almighty] God** bei Gott[, dem Allmächtigen]. **2.** *adv.* **a)** *(past)* vorbei; **drive/run/flow by** vorbeifahren/-laufen/-fließen; **b)** *(near)* **close/near by** in der Nähe; **c) by and large** im großen und ganzen; **by and by** nach und nach; *(in past)* nach einer Weile

[2]by *see* **[2]bye**

[1]**bye** [baɪ] *int. (coll.)* tschüs *(ugs.)*; ~ **[for] now!** bis später!; tschüs! *(ugs.)*
[2]**bye** *n.* **by the ~ = by the way** *see* **way 1 g**
bye-law *see* **by-law**
'**by-election** *n.* Nachwahl, *die*
'**bygone 1.** *n.* **let ~s be ~s** die Vergangenheit ruhen lassen. **2.** *adj.* **[in] ~ days** [in] vergangene[n] Tage[n]
'**by-law** *n. (esp. Brit.)* Verordnung, *die;* **the park ~s** die Parkordnung
'**bypass 1.** *n. (road)* Umgehungsstraße, *die; (channel; also Electr.)* Nebenleitung, *die; (Med.)* Bypass, *der;* ~ **surgery** *(Med.)* eine Bypassoperation / Bypassoperationen. **2.** *v. t.* **a) the road ~es the town** die Straße führt um die Stadt herum; **b)** *(fig.: ignore)* übergehen
'**by-product** *n.* Nebenprodukt, *das*
'**by-road** *n.* Nebenstraße, *die;* Seitenstraße, *die*
bystander ['baɪstændə(r)] *n.* Zuschauer, *der*/Zuschauerin, *die*
byte [baɪt] *n. (Computing)* Byte, *das*
'**byway** *n.* Seitenweg, *der*
'**byword** *n. (notable example)* Inbegriff, *der* (**for** *Gen.*)
Byzantine [bɪ'zæntaɪn, baɪ'zæntaɪn] *adj.* byzantinisch

C

C, c [si:] *n., pl.* **Cs** *or* **C's a)** *(letter)* C, c, *das;* **b) C** *(Mus.)* C, c, *das;* **C sharp** cis, Cis, *das*
C. *abbr.* **a) Celsius** C; **b) Centigrade** C; **c)** *(Geogr.)* **Cape; d)** *(Pol.)* **Conservative**
c. *abbr.* **a) circa** ca.; **b) cent[s]** c
© *symb.* **copyright** ©
ca. *abbr.* **circa** ca.
cab [kæb] *n.* **a)** *(taxi)* Taxi, *das;* **b)** *(of lorry, truck)* Fahrerhaus, *das; (of train)* Führerstand, *der*
cabaret ['kæbəreɪ] *n.* Varieté, *das; (satirical)* Kabarett, *das*
cabbage ['kæbɪdʒ] *n.* **a)** Kohl, *der;* **red ~:** Rotkohl, *der;* **a [head of] ~:** ein Kopf Kohl; ein Kohlkopf; **b)** *(coll.: incapacitated person)* **become a ~:** dahinvegetieren
cabbage 'white *n. (Zool.)* Kohlweißling, *der*
cabby ['kæbɪ] *(coll.),* '**cab-driver** *ns.* Taxifahrer, *der*
cabin ['kæbɪn] *n.* **a)** *(in ship) (for passengers)* Kabine, *die; (for crew)* Kajüte, *die; (in aircraft)* Kabine, *die;* **b)** *(simple dwelling)* Hütte, *die*
cabin: ~-boy *n. (Naut.)* Kabinensteward, *der;* ~ **cruiser** *n.* Kajütboot, *das*
cabinet ['kæbɪnɪt] *n.* **a)** Schrank, *der; (in bathroom, for medicines)* Schränkchen, *das; (display ~)* Vitrine, *die;* **b)** *(Polit.)* Kabinett, *das*
cabinet: ~-maker *n.* Möbeltischler, *der;* **C~ 'Minister** *n.* Minister, *der*/Ministerin, *die*
cable ['keɪbl] **1.** *n.* **a)** *(rope)* Kabel, *das; (of ~-car etc.)* Seil, *das;* **b)** *(Electr., Teleph.)* Kabel, *das;* **c)** *(telegram)* Kabel, *das (veralt.);* [Übersee]telegramm, *das.* **2.** *v. t. (transmit)* telegraphisch durchgeben, kabeln ‹*Mitteilung, Nachricht*›; *(inform)* ~ **sb.** jmdm. kabeln
cable: ~-car *n.* Drahtseilbahn, *die; (in street)* gezogene Straßenbahn; ~ '**railway** *n.* Standseilbahn, *die;* ~ '**television** *n.* Kabelfernsehen, *das*
caboodle [kə'bu:dl] *n., no pl. (sl.)* **the whole ~ :** der ganze Kram *(ugs.);* das ganze Gelumpe *(ugs.)*
cache [kæʃ] **1.** *n.* geheimes [Waffen-/Proviant-]lager. **2.** *v. t.* verstecken
cackle ['kækl] **1.** *n.* **a)** *(clucking of hen)* Gackern, *das;* **b)** *(laughter)* [meckerndes] Gelächter; *(laugh)* **he gave a loud ~:** er prustete los *(ugs.).* **2.** *v. i.* **a)** ‹*Henne:*› gackern; **b)** *(laugh)* meckernd lachen
cacophony [kə'kɒfənɪ] *n.* Kakophonie, *die (geh.);* Mißklang, *der*
cactus ['kæktəs] *n., pl.* **cacti** ['kæktaɪ] *or* **~es** Kaktus, *der*
caddie ['kædɪ] *(Golf)* **1.** *n.* Caddie, *der.* **2.** *v. i.* ~ **for sb.** jmds. Caddie sein
[1]**caddy** ['kædɪ] *n.* Behälter, *der; (tin)* Büchse, *die;* Dose, *die*
[2]**caddy** *see* **caddie**
cadet [kə'det] *n.* Offiziersschüler, *der;* **naval/police ~:** Marinekadett/Anwärter für den Polizeidienst
cadge [kædʒ] **1.** *v. t.* schnorren *(ugs.);* [sich *(Dat.)*] erbetteln; **could I ~ a lift?** können Sie mich vielleicht [ein Stück] mitnehmen? **2.** *v. i.* schnorren *(ugs.)*
cadger ['kædʒə(r)] *n.* Schnorrer, *der (ugs.)*
Caesar ['si:zə(r)] *n.* Cäsar, Caesar *(der)*
Caesarean, Caesarian [sɪ'zeərɪən] *adj. & n.* ~ **[section]** Kaiserschnitt, *der*
café, cafe ['kæfeɪ] *n.* Café, *das*
cafeteria [kæfɪ'tɪərɪə] *n.* Selbstbedienungsrestaurant, *das*
caftan ['kæftæn] *n.* Kaftan, *der*
cage [keɪdʒ] **1.** *n.* **a)** Käfig, *der; (for small birds)* Bauer, *das;* **b)** *(of lift)* Fahrkabine, *die.* **2.** *v. t.* einsperren; käfigen *(fachspr.)* ‹*Vögel*›
'**cage-bird** *n.* Käfigvogel, *der*
cagey ['keɪdʒɪ] *adj. (coll.) (wary)* vorsichtig (**about** bei); *(secretive, uncommunicative)* zugeknöpft *(ugs.)*
cagily ['keɪdʒɪlɪ] *adv. (coll.)* vorsichtig
caginess ['keɪdʒɪnɪs] *n. (coll.) (caution)* Vorsicht, *die; (secretiveness)* Zugeknöpftheit, *die (ugs.)*
cagoule [kə'gu:l] *n.* [leichter, knielanger] Anorak
cairn [keən] *n.* Steinpyramide, *die*
Cairo ['kaɪərəʊ] *pr. n.* Kairo *(das)*
cajole [kə'dʒəʊl] *v. t.* ~ **sb. into sth./into doing sth.** jmdm. etw. einreden/jmdm. einreden, etw. zu tun
cake [keɪk] **1.** *n.* **a)** Kuchen, *der;* **a piece of ~:** ein Stück Kuchen/Torte; *(fig. coll.)* ein Kinderspiel *(ugs.);* **a slice of ~:** eine Scheibe Kuchen; **go** *or* **sell like hot ~s** weggehen wie warme Semmeln *(ugs.);* **you cannot have your ~ and eat it** *(fig.)* beides auf einmal geht nicht; *see also* **take 1 c; b)** *(block)* **a ~ of soap** ein Riegel *od.* Stück Seife. **2.** *v. t. (cover)* verkrusten; **~d with dirt/blood** schmutz-/blutverkrustet. **3.** *v. i. (form a mass)* verklumpen
cake: ~-shop *n.* Konditorei, *die;* **~-slice** *n.* Tortenheber, *der;* **~-stand** *n.* Etagere, *die*
cal. *abbr.* **calorie[s]** cal.
calamity [kə'læmɪtɪ] *n.* Unheil, *das;* Unglück, *das*
calcium ['kælsɪəm] *n.* Kalzium, *das;* Calcium, *das (fachspr.)*
calculate ['kælkjʊleɪt] **1.** *v. t.* **a)** *(ascertain)* berechnen; *(by estimating)* ausrechnen; **b)** *(plan)* **be ~d to do sth.** darauf abzielen, etw. zu tun; **c)** *(Amer. coll.: suppose)* schätzen *(ugs.).* **2.** *v. i.* **a)** *(Math.)* rechnen; **b) ~ on doing sth.** damit rechnen, etw. zu tun
calculated ['kælkjʊleɪtɪd] *adj.*

vorsätzlich ‹*Handlung, Straftat*›; bewußt ‹*Zurückhaltung, Affront*›; kalkuliert ‹*Risiko*›

calculating ['kælkjʊleɪtɪŋ] *adj.* berechnend

calculation [kælkjʊ'leɪʃn] *n.* **a)** *(result)* Rechnung, *die;* **he is out in his ~s** er hat sich verrechnet; **b)** *(calculating)* Berechnung, *die;* **c)** *(forecast)* Schätzung, *die;* **by my ~s** nach meiner Schätzung

calculator ['kælkjʊleɪtə(r)] *n.* Rechner, *der*

calculus ['kælkjʊləs] *n., pl.* **calculi** ['kælkjʊlaɪ] *or* **~es** *(Math. etc.)* Analysis, *die;* **[the] differential/integral ~:** [die] Differential-/Integralrechnung

Calcutta [kæl'kʌtə] *pr. n.* Kalkutta *(das)*

calendar ['kælɪndə(r)] *n.* **a)** Kalender, *der; attrib.* Kalender‹*woche, -monat, -jahr*›; **[church] ~:** Kirchenkalender, *der;* **b)** *(register, list)* Verzeichnis, *das*

[1]**calf** [kɑ:f] *n., pl.* **calves** [kɑ:vz] **a)** Kalb, *das; (leather)* Kalbsleder, *das;* **b)** *(of deer)* Kalb, *das; (of elephant, whale, rhinoceros)* Junge, *das*

[2]**calf** *n., pl.* **calves** *(Anat.)* Wade, *die*

'**calfskin** *n. (leather)* Kalbsleder, *das*

caliber *(Amer.) see* **calibre**

calibrate ['kælɪbreɪt] *v.t.* kalibrieren

calibration [kælɪ'breɪʃn] *n.* Kalibrierung, *die*

calibre ['kælɪbə(r)] *n. (Brit.)* **a)** *(diameter)* Kaliber, *das;* **b)** *(fig.)* Format, *das;* **a man of your ~:** ein Mann von Ihrem Format

calico ['kælɪkəʊ] **1.** *n., pl.* **~es** Kattun, *der.* **2.** *adj.* Kattun-

California [kælɪ'fɔ:nɪə] *pr. n.* Kalifornien *(das)*

caliper *see* **calliper**

call [kɔ:l] **1.** *v.i.* **a)** *(shout)* rufen; **~ to sb.** jmdm. zurufen; **~ [out] for help** um Hilfe rufen; **~ [out] for sb.** nach jmdm. rufen; **~ after sb.** jmdm. hinterherrufen; **b)** *(pay brief visit)* [kurz] besuchen **(at** *Akk.*); vorbeikommen *(ugs.)* **(at** bei); ‹*Zug:*› halten **(at** in + *Dat.*); **~ at a port/station** einen Hafen anlaufen/an einem Bahnhof halten; **~ on sb.** jmdn. besuchen; bei jmdm. vorbeigehen *(ugs.);* **the postman ~ed to deliver a parcel** der Postbote war da und brachte ein Päckchen; **~ round** vorbeikommen *(ugs.);* **c)** *(telephone)* **who is ~ing, please?** wer spricht da, bitte?; **thank you for ~ing** vielen Dank für Ihren Anruf!; *(broadcast)* **this is London ~ing** hier spricht *od.* ist London. **2.** *v.t.* **a)** *(cry out)* rufen; aufrufen ‹*Namen, Nummer*›; **b)** *(cry to)* rufen ‹*Person*›; **c)** *(summon)* rufen; *(to a duty, to do sth.)* aufrufen; **~ sb.'s bluff** es darauf ankommen lassen *(ugs.);* **that was ~ed in question** das wurde in Frage gestellt *od.* in Zweifel gezogen; **please ~ me a taxi** *or* **~ a taxi for me** bitte rufen Sie mir ein Taxi; **d)** *(radio/telephone)* rufen/anrufen; *(initially)* Kontakt aufnehmen mit; **don't ~ us, we'll ~ you** wir sagen Ihnen Bescheid; **e)** *(rouse)* wecken; **f)** *(announce)* einberufen ‹*Konferenz*›; ausrufen ‹*Streik*›; **~ a halt to sth.** mit etw. Schluß machen; **~ time** *(in pub)* ≈ „Feierabend" rufen; **g)** *(name)* nennen; **he is ~ed Bob** er heißt Bob; **you can ~ him by his first name** ihr könnt ihn mit Vornamen anreden; **what is it ~ed in English?** wie heißt das auf englisch?; **~ sb. names** jmdn. beschimpfen; **h)** *(consider)* nennen; **i)** *(Cards etc.)* ansagen. **3.** *n.* **a)** *(shout, cry)* Ruf, *der;* **a ~ for help** ein Hilferuf; **can you give me a ~ at 6 o'clock?** können Sie mich um 6 Uhr wecken?; **remain/be within ~:** in Rufweite bleiben/sein; **on ~:** dienstbereit; **b)** *(of bugle, whistle)* Signal, *das;* **c)** *(visit)* Besuch, *der;* **make** *or* **pay a ~ on sb., make** *or* **pay sb. a ~:** jmdn. besuchen; **have to pay a ~** *(coll.: need lavatory)* mal [verschwinden] müssen *(ugs.);* **d)** *(telephone ~)* Anruf, *der;* Gespräch, *das;* **give sb. a ~:** jmdn. anrufen; **make a ~:** ein Telefongespräch führen; **receive a ~:** einen Anruf erhalten; **e)** *(invitation, summons)* Aufruf, *der;* **the ~ of the sea/the wild** der Ruf des Meeres/der Wildnis; **~ of nature** natürlicher Drang; **answer the ~ of duty** der Pflicht gehorchen; **f)** *(need, occasion)* Anlaß, *der;* Veranlassung, *die;* **g)** *(esp. Comm.: demand)* Abruf, *der;* **have many ~s on one's purse/time** finanziell/zeitlich sehr in Anspruch genommen sein; **h)** *(Cards etc.)* Ansage, *die;* **it's your ~:** du mußt ansagen

~ a'way *v.t.* wegrufen; abrufen

~ 'back 1. *v.t.* zurückrufen. **2.** *v.i.* *(come back)* zurückkommen; noch einmal vorbeikommen *(ugs.)*

~ 'down *v.t. (invoke)* herabflehen *(geh.)* ‹*Segen*›; herausfordern ‹*Unwillen, Tadel*›; **~ down curses on sb.'s head** jmdn. verfluchen

~ for *v.t.* **a)** *(send for, order)* [sich *(Dat.)*] kommen lassen, bestellen ‹*Taxi, Essen, Person*›; **b)** *(collect)* abholen ‹*Person, Güter*›; **'to be ~ed for'** „wird abgeholt"; **c)** *(require, demand)* erfordern; verlangen; **that remark was not ~ed for** die Bemerkung war unangebracht; **this ~s for a celebration** das muß gefeiert werden

~ 'in 1. *v.i.* vorbeikommen *(ugs.);* **I'll ~ in on you** ich komme bei dir vorbei *(ugs.);* **I'll ~ in at your office** ich komme bei dir im Büro vorbei *(ugs.).* **2.** *v.t.* **a)** aus dem Verkehr ziehen ‹*Waren, Münzen*›; **b) ~ in a specialist** einen Fachmann/Facharzt zu Rate ziehen

~ 'off *v.t. (cancel)* absagen ‹*Treffen, Verabredung*›; rückgängig machen ‹*Geschäft*›; lösen ‹*Verlobung*›; *(stop, end)* abbrechen, *(ugs.)* abblasen ‹*Streik*›; **~ off your dogs!** rufen Sie Ihre Hunde zurück!

~ on *see* **~ [up]on**

~ 'out 1. *v.t.* alarmieren ‹*Truppen*›; rufen ‹*Wache*›; zum Streik aufrufen ‹*Arbeitnehmer*›. **2.** *v.i. see* **~ 1 a**

~ 'up *v.t.* **a)** *(imagine, recollect)* wachrufen ‹*Erinnerungen, Bilder*›; [herauf]beschwören, erwecken ‹*böse Erinnerungen, Phantasien*›; **b)** *(summon)* anrufen, beschwören ‹*Teufel, Geister*›; **c)** *(telephone)* anrufen; **d)** *(Mil.)* einberufen

~ [up]on *v.t.* **~ upon God** Gott anrufen; **~ upon sb.'s generosity/sense of justice** an jmds. Großzügigkeit/Gerechtigkeitssinn *(Akk.)* appellieren; **~ [up]on sb. to do sth.** jmdn. auffordern, etw. zu tun

'**call-box** *n.* Telefonzelle, *die;* Fernsprechzelle, *die (Amtsspr.)*

caller ['kɔ:lə(r)] *n. (visitor)* Besucher, *der*/Besucherin, *die; (on telephone)* Anrufer, *der*/Anruferin, *die*

'**call-girl** *n.* Callgirl, *das*

calligraphy [kə'lɪgrəfɪ] *n.* Kalligraphie, *die;* Schönschreiben, *das*

calling ['kɔ:lɪŋ] *n.* **a)** *(occupation, profession)* Beruf, *der;* **b)** *(divine summons)* Berufung, *die*

calliper ['kælɪpə(r)] *n.* **a)** *in pl.* **[pair of] ~s** Tasterzirkel, *der;* **b) ~ [splint]** *(Med.)* Beinschiene, *die*

callous ['kæləs] *adj.* gefühllos; herzlos ‹*Handlung, Verhalten*›

callow ['kæləʊ] *adj.* unreif ‹*Junge, Student*›; grün *(ugs.)* ‹*Jüngling*›

call: ~-sign, ~-signal *ns.* Rufzeichen, *das;* ~**-up** *n. (Mil.)* Einberufung, *die*

calm [kɑ:m] **1.** *n.* **a)** *(stillness)* Stille, *die; (serenity)* Ruhe, *die;* **b)** *(windless period)* Windstille, *die;* **the ~ before the storm** *(lit. or fig.)* die Ruhe vor dem Sturm. **2.** *adj.* **a)** *(tranquil, quiet, windless)* ruhig; **keep ~:** ruhig bleiben; Ruhe bewahren; **b)** *(coll.: self-confident)* gelassen. **3.** *v.t.* besänftigen ‹*Leidenschaften, Zorn*›; **~ sb. [down]** jmdn. beruhigen. **4.** *v.i.* **~ [down]** sich beruhigen; ‹*Sturm:*› abflauen

calmly ['kɑ:mlɪ] *adv.* ruhig; gelassen

calmness ['kɑ:mnɪs] *n., no pl.* Ruhe, *die; (of water)* Stille, *die*

Calor gas, (P) ['kælə gæs] *n.* Butangas, *das*

calorie ['kælərɪ] *n.* Kalorie, *die*

calve [kɑ:v] *v.i.* kalben

calves *pl. of* [1,2]**calf**

cam [kæm] *n.* Nocken, *der*

camber ['kæmbə(r)] *n.* Wölbung, *die*

Cambodia [kæm'bəʊdɪə] *pr. n. (Hist.)* Kambodscha *(das)*

Cambodian [kæm'bəʊdɪən] *(Hist.)* **1.** *adj.* kambodschanisch. **2.** *n.* Kambodschaner, *der*/Kambodschanerin, *die*

came *see* **come**

camel ['kæml] *n. (Zool.)* Kamel, *das*

cameo ['kæmɪəʊ] *n., pl.* **~s a)** *(carving)* Kamee, *die;* **b)** *(minor role)* [winzige] Nebenrolle

camera ['kæmərə] *n.* Kamera, *die; (for still pictures)* Fotoapparat, *der;* Kamera, *die*

'cameraman *n.* Kameramann, *der*

Cameroon ['kæməru:n] *pr. n.* Kamerun *(das)*

camomile ['kæməmaɪl] *n. (Bot.)* Kamille, *die;* **~ tea** Kamillentee, *der*

camouflage ['kæməflɑ:ʒ] **1.** *n. (lit. or fig.)* Tarnung, *die.* **2.** *v.t. (lit. or fig.)* tarnen

[1]**camp** [kæmp] **1.** *n.* Lager, *das; (Mil.)* Feldlager, *das;* **two opposing ~s** *(fig.)* zwei entgegengesetzte Lager. **2.** *v.i.* **~ [out]** campen; *(in tent)* zelten; **go ~ing** Campen/Zelten fahren/gehen

[2]**camp 1.** *adj.* **a)** *(affected)* affektiert ‹*Person, Art, Benehmen*›; **b)** *(exaggerated)* übertrieben ‹*Gestik, Ausdrucksform*›. **2.** *n.* Manieriertheit, *die.* **3.** *v.t.* **~ it up** zu dick auftragen *(ugs.)*

campaign [kæm'peɪn] **1.** *n.* **a)** *(Mil.)* Feldzug, *der;* **b)** *(organized course of action)* Kampagne, *die;* **c)** *(for election)* Wahlkampf, *der; see also* **presidential. 2.** *v.i.* **~ for sth.** sich für etw. einsetzen; **~ against sth.** gegen etw. etwas unternehmen; **be ~ing** *(for election)* ‹*Politiker:*› im Wahlkampf stehen; **~ hard** einen intensiven Wahlkampf führen

campaigner [kæm'peɪnə(r)] *n.* **a)** Vorkämpfer, *der*/Vorkämpferin, *die;* **b)** *(veteran)* Veteran, *der;* alter Kämpfer; **an old ~:** ein alter Kämpfer *od. (veralt.)* Kämpe; **c) ~ for ...:** Anhänger, *der*/Anhängerin, *die* (+ *Gen.*); **~ against ...:** Gegner, *der*/Gegnerin, *die* (+ *Gen.*)

'camp-bed *n.* Campingliege, *die*

camper ['kæmpə(r)] *n.* **a)** Camper, *der*/Camperin, *die;* **b)** *(vehicle)* Wohnmobil, *das; (adapted minibus)* Campingbus, *der*

camp: ~**-fire** *n.* Lagerfeuer, *das;* ~**-follower** Marketender, *der*/Marketenderin, *die; (fig.: disciple, follower)* Mitläufer, *der*/Mitläuferin, *die*

camping ['kæmpɪŋ] *n.* Camping, *das; (in tent)* Zelten, *das*

camping: ~**-ground** *(Amer.) see* **~ site; ~ holiday** *n.* Campingurlaub, *der;* **~ site** *n.* Campingplatz, *der*

'campsite *n.* Campingplatz, *der*

campus ['kæmpəs] *n.* Campus, *der;* Hochschulgelände, *das*

'camshaft *n.* Nockenwelle, *die*

[1]**can** [kæn] **1.** *n.* **a)** *(milk ~, watering-~)* Kanne, *die; (for oil, petrol)* Kanister, *der; (Amer.: for refuse)* Eimer, *der;* Tonne, *die;* **a ~ of paint** eine Büchse Farbe; *(with handle)* ein Eimer Farbe; **carry the ~** *(fig. sl.)* die Sache ausbaden *(ugs.);* **b)** *(container for preserving)* [Konserven]dose, *die;* [Konserven]büchse, *die;* **a ~ of tomatoes/sausages** eine Dose *od.* Büchse Tomaten/Würstchen; **a ~ of beer** eine Dose Bier; **c)** *(Amer. sl.: lavatory)* Lokus, *der (ugs.).* **2.** *v.t.,* **-nn-** eindosen; einmachen ‹*Obst*›

[2]**can** *v. aux., only in pres.* **can,** *neg.* **cannot** ['kænət], *(coll.)* **can't** [kɑ:nt], *past* **could** [kʊd], *neg. (coll.)* **couldn't** ['kʊdnt] können; *(have right, be permitted)* dürfen; können; **as much as one ~:** so viel man kann; **as ... as '~ be** wirklich sehr ...; **~ do** *(coll.)* kein Problem; **he can't be more than 40** er kann nicht über 40 sein; **you can't smoke in this compartment** in diesem Abteil dürfen Sie nicht rauchen; **I can't hear what you're saying** ich kann Sie nicht verstehen; **how [ever] could you do this to me?** wie konnten Sie mir das bloß antun?; **I could have killed him** ich hätte ihn umbringen können; **[that] could be [so]** das könnte *od.* kann sein

Canada ['kænədə] *pr. n.* Kanada *(das)*

Canadian [kə'neɪdɪən] **1.** *adj.* kanadisch; **sb. is ~:** jmd. ist Kanadier/Kanadierin. **2.** *n.* Kanadier, *der*/Kanadierin, *die;* **the French/English ~s** die Franko-/Anglokanadier

canal [kə'næl] *n.* Kanal, *der;* **the Panama C~:** der Panamakanal

ca'nal boat *n. langes, enges Boot zum Befahren der Kanäle*

Canaries [kə'neərɪz] *pr. n. pl.* Kanarische Inseln *Pl.*

canary [kə'neərɪ] *n.* Kanarienvogel, *der*

canary: ~ **'yellow** *n.* Kanariengelb, *das;* ~**-yellow** *adj.* kanariengelb

cancan ['kænkæn] *n.* Cancan, *der*

cancel ['kænsl] **1.** *v.t., (Brit.)* **-ll-: a)** *(call off)* absagen ‹*Besuch, Urlaub, Reise, Sportveranstaltung*›; ausfallen lassen ‹*Veranstaltung, Vorlesung, Zug, Bus*›; fallenlassen ‹*Pläne*›; *(annul, revoke)* rückgängig machen ‹*Einladung, Vertrag*›; zurücknehmen ‹*Befehl*›; stornieren ‹*Bestellung, Auftrag*›; streichen ‹*Schuld[en]*›; kündigen ‹*Abonnement*›; abbestellen ‹*Zeitung*›; aufheben ‹*Klausel, Gesetz, Recht*›; **the match had to be ~led** das Spiel mußte ausfallen; **the lecture has been ~led** die Vorlesung fällt aus; **b)** *(balance, neutralize)* aufheben; **the arguments ~ each other out** die Argumente heben sich gegenseitig auf; **c)** entwerten ‹*Briefmarke, Fahrkarte*›; ungültig machen ‹*Scheck*›. **2.** *v.i., (Brit.)* **-ll-: ~ [out]** sich [gegenseitig] aufheben

cancellation [kænsə'leɪʃn] *n. see* **cancel 1: a)** Absage, *die;* Ausfall, *der;* Ausfallen, *das;* Fallenlassen, *das;* Rückgängigmachen, *das;* [Zu]rücknahme, *die;* Stornierung, *die;* Streichung, *die;* Kündigung, *die;* Abbestellung, *die;* Aufhebung, *die;* **b)** Aufhebung, *die;* **c)** Entwertung, *die;* Ungültigmachen, *das*

cancer ['kænsə(r)] *n.* **a)** *(Med.)* Krebs, *der;* **~ of the liver** Leberkrebs, *der;* **b) C~** *(Astrol., Astron.)* der Krebs; *see also* **tropic**

cancerous ['kænsərəs] *adj.* Krebs‹*geschwulst, -geschwür*›; krebsartig ‹*Wucherung*›

candelabra [kændɪ'lɑ:brə] *n.* Leuchter, *der; (large)* Kandelaber, *der*
candid ['kændɪd] *adj.* offen; ehrlich ‹*Ansicht, Bericht*›
candidate ['kændɪdət, 'kændɪdeɪt] *n. (Polit., examinee)* Kandidat, *der*/Kandidatin, *die;* ~ **for Mayor** Bürgermeisterkandidat/-kandidatin
candidly ['kændɪdlɪ] *adv.* offen; ehrlich
candle [kændl] *n.* Kerze, *die;* **burn the ~ at both ends** *(fig.)* sich *(Dat.)* zuviel aufladen; **the game is not worth the ~** *(fig.)* die Sache ist nicht der Mühe *(Gen.)* wert
candle: ~-light *n.* Kerzenlicht, *das;* **~stick** *n.* Kerzenhalter, *der; (elaborate)* Leuchter, *der;* **~wick** *n. (material)* Frottierplüsch, *der*
candour *(Brit.; Amer.:* **candor**) ['kændə(r)] *n.* Offenheit, *die;* Ehrlichkeit, *die*
candy ['kændɪ] **1.** *n. (Amer.) (sweets)* Süßigkeiten *Pl.; (sweet)* Bonbon, *das od. der.* **2.** *v.t.* kandieren ‹*Früchte*›; **candied lemon/orange peel** Zitronat/Orangeat, *das*
'**candy-floss** *n.* Zuckerwatte, *die*
cane [keɪn] **1.** *n.* **a)** *(stem of bamboo, rattan, etc.)* Rohr, *das; (of raspberry, blackberry)* Sproß, *der;* **b)** *(material)* Rohr, *das;* **c)** *(stick)* [Rohr]stock, *der;* **get the ~:** eine Tracht Prügel bekommen. **2.** *v.t. (beat)* [mit dem Stock] schlagen
cane: ~ chair *n.* Rohrstuhl, *der;* **~-sugar** *n.* Rohrzucker, *der*
canine ['keɪnaɪn] **1.** *adj.* **a)** *(of dog[s])* Hunde-; **b)** ~ **tooth** Eckzahn, *der;* Augenzahn, *der.* **2.** *n.* Eckzahn, *der;* Augenzahn, *der*
canister ['kænɪstə(r)] *n.* Büchse, *die;* Dose, *die; (for petrol, oil, DDT, etc.)* Kanister, *der*
cannabis ['kænəbɪs] *n. (drug)* Haschisch, *das;* Marihuana, *das*
canned [kænd] *adj.* **a)** Dosen-; in Dosen *nachgestellt;* ~ **fish/meat/fruit** Fisch-/Fleisch-/Obstkonserven *Pl.;* ~ **beer** Dosenbier; ~ **food** [Lebensmittel]konserven *Pl.;* **b)** *(sl.: drunk)* abgefüllt *(ugs.);* **c)** *(recorded)* aufgezeichnet; ~ **music** Musikkonserve, *die*
cannibal ['kænɪbl] *n.* Kannibale, *der*/Kannibalin, *die;* Menschenfresser, *der*/-fresserin, *die (ugs.)*
cannibalism ['kænɪbəlɪzm] *n.* Kannibalismus, *der;* Menschenfresserei, *die (ugs.)*
cannibalize ['kænɪbəlaɪz] *v.t.* ausschlachten ‹*Auto, Flugzeug, Maschine usw.*›
cannily ['kænɪlɪ] *adv. (shrewdly)* schlau; *(cautiously)* vorsichtig
cannon ['kænən] **1.** *n.* Kanone, *die.* **2.** *v.i. (Brit.)* ~ **against sth.** gegen etw. prallen; ~ **into sb./sth.** mit etw./jmdm. zusammenprallen
cannon: ~-ball *n. (Hist.)* Kanonenkugel, *die;* **~-fodder** *n.* Kanonenfutter, *das (salopp abwertend)*
cannot *see* ²**can**
canny ['kænɪ] *adj.* **a)** *(shrewd)* schlau; bauernschlau *(ugs.); (thrifty)* sparsam; **b)** *(cautious, wary)* vorsichtig; umsichtig
canoe [kə'nu:] **1.** *n.* Paddelboot, *das; (Indian ~, Sport)* Kanu, *das.* **2.** *v.i.* paddeln; *(in Indian ~, Sport)* Kanu fahren
canoeing [kə'nu:ɪŋ] *n.* Paddeln, *das; (Sport)* Kanusport, *der*
canoeist [kə'nu:ɪst] *n.* Paddelbootfahrer, *der*/-fahrerin, *die; (Sport)* Kanute, *der*/Kanutin, *die*
canon ['kænən] *n.* **a)** *(rule)* Grundregel, *die;* **b)** *(priest)* Kanoniker, *der;* **c)** *(list of sacred books)* Kanon, *der; (fig.)* **the Shakespearean ~:** das Gesamtwerk Shakespeares
canonisation, canonise *see* **canoniz-**
canonization [kænənaɪ'zeɪʃn] *n.* Kanonisation, *die*
canonize ['kænənaɪz] *v.t.* heiligsprechen
'**can-opener** *n.* Dosen-, Büchsenöffner, *der*
canopy ['kænəpɪ] *n.* Baldachin, *der (auch fig.); (over entrance)* Vordach, *das*
¹**cant** [kænt] **1.** *v.t.* kippen; ankippen, kanten ‹*Faß*›; ~ **over** umkippen. **2.** *v.i.* sich neigen. **3.** *n. (tilted position)* Schräglage, *die*
²**cant** *n.* **a)** *(derog.: language of class, sect, etc.)* Kauderwelsch, *das (abwertend);* **thieves' ~:** Rotwelsch, *das;* **b)** *(insincere talk)* scheinheiliges Gerede
can't [kɑ:nt] *(coll.)* = **cannot;** *see* ²**can**
cantankerous [kæn'tæŋkərəs] *adj.* streitsüchtig; knurrig *(ugs.)*
cantata [kæn'tɑ:tə] *n. (Mus.)* Kantate, *die*
canteen [kæn'ti:n] *n.* **a)** Kantine, *die;* **b)** *(case of cutlery)* Besteckkasten, *der*
canter ['kæntə(r)] **1.** *n.* Handgalopp, *der;* Kanter, Canter, *der (fachspr.).* **2.** *v.i.* leicht galoppieren; kantern *(fachspr.)*
cantilever ['kæntɪli:və] *n.* **a)** *(bracket)* Konsole, *die;* Kragplatte, *die;* **b)** *(beam)* Träger, *der*
'**cantilever bridge** *n.* Auslegerbrücke, *die*
canton ['kæntɒn] *n.* Kanton, *der*
Cantonese [kæntə'ni:z] **1.** *adj.* kantonesisch. **2.** *n., pl. same* **a)** *(person)* Kantonese, *der*/Kantonesin, *die;* **b)** *(language)* Kantonesisch *(das)*
canvas ['kænvəs] *n.* **a)** *(cloth)* Leinwand, *die; (for tents, tarpaulins, etc.)* Segeltuch, *das;* **under ~:** im Zelt; *(Naut.)* unter Segel; **b)** *(Art)* Leinwand, *die; (painting)* Gemälde, *das; (for embroidery)* Kanevas, *der;* Gitterleinwand, *die*
canvass ['kænvəs] **1.** *v.t.* **a)** *(solicit votes in or from)* Wahlwerbung treiben in ‹*einem Wahlkreis, Gebiet*›; Wahlwerbung treiben bei ‹*Wählern*›; **b)** *(Brit.: propose)* vorschlagen ‹*Plan, Idee*›. **2.** *v.i.* werben (**on behalf of** für); ~ **for votes** um Stimmen werben
canvasser ['kænvəsə(r)] *n. (for votes)* Wahlhelfer, *der*/Wahlhelferin, *die*
canvassing ['kænvəsɪŋ] *n. (for votes)* Wahlwerbung, *die*
canyon ['kænjən] *n.* Cañon, *der*
cap [kæp] **1.** *n.* **a)** Mütze, *die; (nurse's, servant's)* Haube, *die; (bathing-~)* Badekappe, *die; (with peak)* Schirmmütze, *die; (skull-~)* Kappe, *die;* Käppchen, *das; (Univ.) viereckige akademische Kopfbedeckung;* ≈ Barett, *das;* **if the ~ fits, [he** *etc.* **should] wear it** *(fig.)* wem die Jacke paßt, der soll sie sich *(Dat.)* anziehen; **with ~ in hand** *(fig.)* demütig; **b)** *(device to seal or close)* [Verschluß]kappe, *die; (petrol ~, radiator-~)* Verschluß, *der; (on milk-bottle)* Deckel, *der; (of shoe)* Kappe, *die;* **c)** *(Brit. Sport) Ziermütze als Zeichen der Aufstellung für die [National]mannschaft; (player)* Nationalspieler, *der*/-spielerin, *die;* **d)** *(contraceptive)* Pessar, *das.* **2.** *v.t.*, **-pp-: a)** verschließen ‹*Flasche*›; mit einer Schutzkappe versehen ‹*Zahn*›; **b)** *(Brit. Sport: award ~ to)* aufstellen; **c)** *(crown with clouds or snow or mist)* bedecken; **d)** *(follow with sth. even more noteworthy)* überbieten ‹*Geschichte, Witz usw.*›; **to ~ it all** obendrein
capability [keɪpə'bɪlɪtɪ] *n.* Fähigkeit, *die;* Vermögen, *das (geh.)*
capable ['keɪpəbl] *adj.* **a) be ~ of sth.** ‹*Person:*› zu etw. imstande sein; **show him what you are ~ of** zeig ihm, wozu du imstande bist *od.* wessen du fähig bist; **b)** *(gifted, able)* fähig

capably ['keɪpəblɪ] *adv.* gekonnt, kompetent ‹*leiten, führen*›
capacitor [kə'pæsɪtə(r)] *n.* *(Electr.)* Kondensator, *der*
capacity [kə'pæsɪtɪ] *n.* **a)** *(power)* Aufnahmefähigkeit, *die; (to do things)* Leistungsfähigkeit, *die;* **b)** *no pl. (maximum amount)* Fassungsvermögen, *das;* **the machine is working to ~:** die Maschine ist voll ausgelastet; **a seating ~ of 300** 300 Sitzplätze; **filled to ~** ‹*Saal, Theater*› bis auf den letzten Platz besetzt; *attrib.* **the film drew ~ audiences/houses for ten weeks** zehn Wochen lang waren alle Vorstellungen dieses Films ausverkauft; **c)** *(measure)* Rauminhalt, *der;* Volumen, *das;* **d)** *(position)* Eigenschaft, *die;* Funktion, *die;* **in his ~ as critic/lawyer** *etc.* in seiner Eigenschaft als Kritiker/Anwalt *usw.*
¹cape [keɪp] *n.* *(garment)* Umhang, *der;* Cape, *das; (part of coat)* Pelerine, *die*
²cape *n.* *(Geog.)* Kap, *das;* **the C~ [of Good Hope]** das Kap der guten Hoffnung; **C~ Horn** Kap Hoorn *(das);* **C~ Town** Kapstadt *(das)*
¹caper ['keɪpə(r)] **1.** *n.* **a)** *(frisky movement)* Luftsprung, *der;* **b)** *(wild behaviour)* Kapriole, *die;* **c)** *(sl.: activity, occupation)* Masche, *die (salopp).* **2.** *v. i.* **~ [about]** [herum]tollen; [umher]tollen
²caper *n.* *(Gastr.)* **~s** Kapern *Pl.*
capful ['kæpfʊl] *n.* **one ~:** der Inhalt einer Verschlußkappe
capillary [kə'pɪlərɪ] **1.** *adj.* Kapillar‹*gefäß*›; **~ tube** Kapillare, *die (fachspr.).* **2.** *n.* Kapillare, *die (fachspr.)*
capital ['kæpɪtl] **1.** *adj.* **a)** Todes‹*strafe, -urteil*›; Kapital‹*verbrechen*›; **b)** *attrib.* Groß-, *(fachspr.)* Versal‹*buchstabe*›; **~ letters** Großbuchstaben; Versalien *(fachspr.);* **with a ~ A** *etc.* mit großem A *usw. od. (fachspr.)* mit Versal-A *usw.*; **c)** *attrib. (principal)* Haupt‹*stadt*›; **d)** *(Commerc.)* **~ sum/expenditure** Kapitalbetrag, *der*/-aufwendungen *Pl.* **2.** *n.* **a)** *(letter)* Großbuchstabe, *der;* **[large] ~s** Großbuchstaben; Versalien *(fachspr.);* **small ~s** Kapitälchen *(fachspr.);* **write one's name in [block] ~s** seinen Namen in Blockbuchstaben schreiben; **b)** *(city, town)* Hauptstadt, *die;* **c)** *(stock, accumulated wealth)* Kapital, *das;* **make ~ out of sth.** *(fig.)* aus etw. Kapital schlagen *(ugs.)*
capital 'gains tax *n.* *(Brit.)* Steuer auf Kapitalgewinn, *die*
capitalise *see* **capitalize**
capitalism ['kæpɪtəlɪzm] *n.* Kapitalismus, *der*
capitalist ['kæpɪtəlɪst] **1.** *n.* Kapitalist, *der*/Kapitalistin, *die.* **2.** *adj.* kapitalistisch
capitalize ['kæpɪtəlaɪz] **1.** *v. t.* groß schreiben ‹*Buchstaben, Wort*›. **2.** *v. i. (fig.)* **~ on sth.** von etw. profitieren; aus etw. Kapital schlagen *(ugs.)*
capitulate [kə'pɪtjʊleɪt] *v. i.* kapitulieren
capitulation [kəpɪtjʊ'leɪʃn] *n.* Kapitulation, *die*
caprice [kə'pri:s] *n.* *(change of mind or conduct)* Laune, *die;* Kaprice, *die (geh.); (inclination)* Willkür, *die*
capricious [kə'prɪʃəs] *adj.* launisch; kapriziös *(geh.)*
Capricorn ['kæprɪkɔ:n] *n.* *(Astrol., Astron.)* der Steinbock; *see also* **tropic**
capsize [kæp'saɪz] **1.** *v. t.* zum Kentern bringen. **2.** *v. i.* kentern
capstan ['kæpstən] *n.* *(Naut.)* Winde, *die;* Spill, *das (Seemannsspr.)*
capsule ['kæpsjʊl] *n.* Kapsel, *die*
captain ['kæptɪn] **1.** *n.* **a)** Kapitän, *der; (in army)* Hauptmann, *der; (in navy)* Kapitän [zur See]; **~ of a ship** Schiffskapitän, *der;* **~ of industry** *(fig.)* Industriekapitän, *der (ugs.);* **b)** *(Sport)* Kapitän, *der;* Spielführer, *der*/-führerin, *die.* **2.** *v. t.* befehligen ‹*Soldaten, Armee*›; **~ a team** Mannschaftskapitän sein
captaincy ['kæptɪnsɪ] *n.* *(Sport)* Führung, *die*
caption ['kæpʃn] *n.* **a)** *(heading)* Überschrift, *die;* **b)** *(wording under photograph/drawing)* Bildunterschrift, *die; (Cinemat., Telev.)* Untertitel, *der*
captivate ['kæptɪveɪt] *v. t.* fesseln *(fig.);* gefangennehmen *(fig.)*
captivating ['kæptɪveɪtɪŋ] *adj.* bezaubernd; einnehmend ‹*Lächeln*›
captive ['kæptɪv] **1.** *adj.* gefangen; **be taken ~:** gefangengenommen werden; **hold sb. ~:** jmdn. gefangenhalten. **2.** *n.* Gefangener, *der*/Gefangene, *die*
captive 'audience *n.* unfreiwilliges Publikum
captivity [kæp'tɪvɪtɪ] *n.* Gefangenschaft, *die;* **in ~:** in [der] Gefangenschaft; **be held in ~:** gefangengehalten werden
captor ['kæptə(r)] *n.* *(of city, country)* Eroberer, *der;* **his ~:** der, der/die, die ihn gefangennahm
capture ['kæptʃə(r)] **1.** *n.* **a)** *(of thief etc.)* Festnahme, *die; (of town)* Einnahme, *die;* **b)** *(thing or person ~d)* Fang, *der.* **2.** *v. t.* **a)** festnehmen ‹*Person*›; [ein]fangen ‹*Tier*›; einnehmen ‹*Stadt*›; ergattern ‹*Preis*›; **~ sb.'s heart** jmds. Herz gewinnen; **they ~d the city from the Romans** sie nahmen den Römern die Stadt ab; **b)** *(Chess etc.)* schlagen ‹*Figur*›; **c)** *(Computing)* erfassen ‹*Daten*›
car [kɑ:(r)] *n.* **a)** *(motor ~)* Auto, *das;* Wagen, *der;* **by ~:** mit dem Auto *od.* Wagen; **b)** *(railway-carriage etc.)* Wagen, *der;* **c)** *(Amer.: lift-cage)* Fahrkabine, *die*
carafe [kə'ræf] *n.* Karaffe, *die*
caramel ['kærəmel] *n.* **a)** *(toffee)* Karamelle, *die;* Karamelbonbon, *das;* **b)** *(burnt sugar or syrup)* Karamel, *der*
carat ['kærət] *n.* Karat, *das;* **a 22-~ gold ring** ein 22karätiger Goldring
caravan ['kærəvæn] *n.* **a)** *(Brit.)* Wohnwagen, *der; (used for camping)* Wohnwagen, *der;* Caravan, *der;* **b)** *(company of merchants, pilgrims, etc.)* Karawane, *die*
caravan: ~ park, ~ site *ns.* Campingplatz für Wohnwagen
caraway ['kærəweɪ] *n.* Kümmel, *der*
'caraway seed *n.* Kümmelkorn, *das; in pl.* Kümmel, *der*
carbohydrate [kɑ:bə'haɪdreɪt] *n.* *(Chem.)* Kohle[n]hydrat, *das*
carbolic [kɑ:'bɒlɪk] *adj.* Karbol‹*säure*›; **~ soap** Karbolseife, *die*
'car bomb *n.* Autobombe, *die*
carbon ['kɑ:bən] *n.* **a)** Kohlenstoff, *der;* **b)** *(copy)* Durchschlag, *der; (paper)* Kohlepapier, *das*
carbonate ['kɑ:bəneɪt] *v. t.* mit Kohlensäure versetzen ‹*Getränke*›
carbon: ~ 'copy *n.* Durchschlag, *der; (fig.) (imitation)* Nachahmung, *die;* Abklatsch, *der (abwertend); (identical counterpart)* Ebenbild, *das;* **~ di'oxide** *n.* *(Chem.)* Kohlendioxid, *das;* **~ paper** *n.* Kohlepapier, *das*
carbuncle ['kɑ:bʌŋkl] *n.* *(abscess)* Karbunkel, *der*
carburettor (*Amer.:* **carburetor**) [kɑ:bə'retə(r)] *n.* Vergaser, *der*
carcass (*Brit. also:* **carcase**) ['kɑ:kəs] *n.* **a)** *(dead body)* Kadaver, *der; (at butcher's)* Rumpf, *der;* **b)** *(of ship, fortification, etc.)* Skelett, *das*
carcinogen [kɑ:'sɪnədʒən] *n.* *(Med.)* Karzinogen, *das (fachspr.);* Krebserreger, *der*

carcinogenic [kɑːsɪnəˈdʒenɪk] *adj. (Med.)* karzinogen *(fachspr.)*; krebserregend

car: ~ coat *n.* Autocoat, *der;* **~ crash** *n.* Autounfall, *der*

card [kɑːd] *n.* **a)** *(playing-~)* Karte, *die;* **read the ~s** Karten lesen; **be on the ~s** *(fig.)* zu erwarten sein; **put [all] one's ~s on the table** *(fig.)* [alle] seine Karten auf den Tisch legen; **have another ~ up one's sleeve** *(fig.)* noch einen Trumpf in der Hand haben; **b)** *in pl. (game)* Karten *Pl.;* **play ~s** Karten spielen; **c)** *(~board, post~, visiting~, greeting ~)* Karte, *die;* **let me give you my ~:** ich gebe Ihnen meine Karte; **d)** *in pl. (coll.: employee's documents)* Papiere *Pl.;* **ask for/get one's ~s** sich *(Dat.)* seine Papiere geben lassen/seine Papiere kriegen *(ugs.)*

card: ~board *n.* Pappe, *die;* Pappkarton, *der; (fig. attrib.)* klischeehaft ‹*Figur*›; **~board box** *see* [1]**box a; ~ file** *n.* Kartei, *die; (large)* Kartothek, *die;* **~-game** *n.* Kartenspiel, *das*

cardiac [ˈkɑːdɪæk] *adj.* Herz-

cardigan [ˈkɑːdɪɡən] *n.* Strickjacke, *die*

cardinal [ˈkɑːdɪnl] **1.** *adj. (fundamental)* grundlegend ‹*Frage, Doktrin, Pflicht*›; Kardinal‹*fehler, -problem*›; *(chief)* hauptsächlich, Haupt‹*argument, -punkt, -merkmal*›. **2.** *n.* **a)** *(Eccl.)* Kardinal, *der;* **b)** *see* **cardinal number**

cardinal: ~ number *n.* Grund-, Kardinalzahl, *die;* **~ sin** *n.* Todsünde, *die*

ˈ**card index** *n.* Kartei, *die*

cardiogram [ˈkɑːdɪəʊɡræm] *n.* Kardiogramm, *das (Med.)*

cardiology [kɑːdɪˈɒlədʒɪ] *n.* Kardiologie, *die*

care [keə(r)] **1.** *n.* **a)** *(anxiety)* Sorge, *die;* **she hasn't got a ~ in the world** sie hat keinerlei Sorgen; **b)** *(pains)* Sorgfalt, *die;* **take ~:** sich bemühen; **he takes great ~ over his work** er gibt sich *(Dat.)* große Mühe mit seiner Arbeit; **c)** *(caution)* Vorsicht, *die;* **take ~:** aufpassen; **take ~ to do sth.** darauf achten, etw. zu tun; **take more ~!** paß [doch] besser auf!; **d)** *(attention)* **medical ~:** ärztliche Betreuung; **old people need special ~:** alte Menschen brauchen besondere Fürsorge; **e)** *(concern)* **~ for sb./sth.** die Sorge um jmdn./etw.; **f)** *(charge)* Obhut, *die (geh.);* **be in ~:** in Pflege sein; **put sb. in ~:** jmdn. in Pflege geben; **take ~ of sb./sth.** *(ensure safety of)* auf jmdn./etw. aufpassen; *(attend to, dispose of)* sich um jmdn./etw. kümmern; **take ~ of oneself** für sich selbst sorgen; *(as to health)* sich schonen; **take ~ [of yourself]!** mach's gut! *(ugs.).* **2.** *v. i.* **a)** **~ for** *or* **about sb./sth.** *(feel interest)* sich für jmdn./etw. interessieren; **b)** **~ for** *or* **about sb./sth.** *(like)* jmdn./etw. mögen; **would you ~ for a drink?** möchten Sie etwas trinken?; **c)** *(feel concern)* **I don't ~ [whether/how/what** *etc.*] es ist mir gleich[, ob/wie/was *usw.*]; **for all I ~** *(coll.)* von mir aus *(ugs.);* **I couldn't ~ less** *(coll.)* es ist mir völlig einerlei *od. (ugs.)* egal; **what do I ~?** *(coll.)* mir ist es egal *(ugs.);* **who ~s?** *(coll.)* was soll's *(ugs.);* **d)** *(wish)* **~ to do sth.** etw. tun mögen; **e)** **~ for sb./sth.** *(look after)* sich um jmdn./etw. kümmern; **well ~d for** gepflegt; gut versorgt ‹*Person*›; gut erhalten ‹*Auto*›

career [kəˈrɪə(r)] **1.** *n.* **a)** *(way of livelihood)* Beruf, *der;* **a teaching ~:** der Beruf des Lehrers; **take up a ~ in journalism** *or* **as a journalist** den Beruf des Journalisten ergreifen; **b)** *(progress in life)* [berufliche] Laufbahn; *(very successful)* Karriere, *die.* **2.** *v. i.* rasen; ‹*Pferd, Reiter:*› galoppieren; **go ~ing down the hill** den Hügel hinunterrasen

career: ~ ˈdiplomat *n.* Berufsdiplomat, *der;* **~ girl** *n.* Karrierefrau, *die;* **~s adviser** *n.* Berufsberater, *der/*-beraterin, *die;* **~s office** *n.* Berufsberatung[sstelle], *die;* **~ woman** *n.* Karrierefrau, *die*

ˈ**carefree** *adj.* sorgenfrei

careful [ˈkeəfl] *adj.* **a)** *(thorough)* sorgfältig; *(watchful, cautious)* vorsichtig; **[be] ~!** Vorsicht!; **be ~ to do sth.** darauf achten, etw. zu tun; **he was ~ not to mention the subject** er war darum bemüht, das Thema nicht zu erwähnen; **be ~ that ...:** darauf achten, daß ...; **be ~ of sb./sth.** *(take care of)* mit jmdm./etw. vorsichtig sein; *(be cautious of)* sich vor jmdm./etw. in acht nehmen; **be ~ [about] how/what/where** *etc.* darauf achten, wie/was/wo *usw.;* **be ~ about sth.** auf etw. *(Akk.)* achten; **be ~ about sb.** auf jmdn. aufpassen *od.* achten; **be ~ with sb./sth.** vorsichtig mit jmdm./etw. umgehen; **b)** *(showing care)* sorgfältig; **a ~ piece of work** ein sorgfältig gearbeitetes Stück; **after ~ consideration** nach reiflicher Überlegung; **pay ~ attention to what he says** achte genau auf das, was er sagt

carefully [ˈkeəfəlɪ] *adv. (thoroughly)* sorgfältig; *(attentively)* aufmerksam; *(cautiously)* vorsichtig; **watch ~:** gut aufpassen

careless [ˈkeəlɪs] *adj.* **a)** *(inattentive)* unaufmerksam; *(thoughtless)* gedankenlos; unvorsichtig, leichtsinnig ‹*Fahrer*›; nachlässig ‹*Arbeiter*›; **be ~ about** *or* **of sb./sth.** wenig auf jmdn./etw. achten; **~ with sb./sth.** unvorsichtig mit jmdm./etw.; **be ~ [about** *or* **of] how/what** *etc.* wenig darauf achten, wie/was *usw.;* **b)** *(showing lack of care)* unordentlich, nachlässig ‹*Arbeit*›; unachtsam ‹*Fahren*›; **a [very] ~ mistake** ein [grober] Flüchtigkeitsfehler

carelessly [ˈkeəlɪslɪ] *adv. (without care)* nachlässig; *(thoughtlessly)* gedankenlos; unvorsichtig, leichtsinnig ‹*fahren*›

carelessness [ˈkeəlɪsnɪs] *n., no pl. (lack of care)* Nachlässigkeit, *die; (thoughtlessness)* Gedankenlosigkeit, *die*

caress [kəˈres] **1.** *n.* Liebkosung, *die.* **2.** *v. t.* liebkosen; **~ [each other]** sich *od.* einander liebkosen

care: ~taker *n.* **a)** Hausmeister, *der/*-meisterin, *die;* **b)** **~taker government** Übergangsregierung, *die;* **~-worn** *adj.* von Sorgen gezeichnet

ˈ**car ferry** *n.* Autofähre, *die*

cargo [ˈkɑːɡəʊ] *n., pl.* **~es** *or (Amer.)* **~s** Fracht, *die;* Ladung, *die*

cargo: ~ boat, ~ ship *ns.* Frachter, *der;* Frachtschiff, *das*

Caribbean [kærɪˈbiːən] **1.** *n.* **the ~:** die Karibik. **2.** *adj.* karibisch

caribou [ˈkærɪbuː] *n., pl. same (Zool.)* Karibu, *der od. das*

caricature [ˈkærɪkətjʊə(r)] **1.** *n.* Karikatur, *die; (in mime)* Parodie, *die;* **do a ~ of sb.** jmdn. karikieren/parodieren. **2.** *v. t.* karikieren

ˈ**carload** *n.* Wagenladung, *die*

carnage [ˈkɑːnɪdʒ] *n.* Gemetzel, *das*

carnal [ˈkɑːnl] *adj.* körperlich; sinnlich; fleischlich *(geh.)*

carnation [kɑːˈneɪʃn] *n. (Bot.)* [Garten]nelke, *die*

carnet [ˈkɑːneɪ] *n. (of motorist)* Triptyk, *das; (of camper)* Ausweis für Camper

carnival [ˈkɑːnɪvl] *n.* **a)** *(festival)* Volksfest, *das;* **b)** *(pre-Lent festivities)* Karneval, *der;* Fastnacht, *die;* Fasching, *der (bes. südd., österr.)*

carnivore ['kɑ:nɪvɔ:(r)] *n.* *(animal)* Fleischfresser, *der;* *(plant)* fleischfressende Pflanze

carnivorous [kɑ:'nɪvərəs] *adj.* fleischfressend

carol ['kærl] *n.* **[Christmas]** ~: Weihnachtslied, *das;* ~ **concert,** ~**-singing** weihnachtliches Liedersingen; ~ **singers** *Leute, die von Haus zu Haus gehen und Weihnachtslieder vortragen;* ≈ Weihnachtssänger

carouse [kə'raʊz] *v. i.* zechen *(veralt., noch scherzh.)*

carousel [kærʊ'sel] *n.* **a)** *(conveyor system)* Ausgabeband, *das;* **b)** *see* **carrousel**

[1]**carp** [kɑ:p] *n., pl. same (Zool.)* Karpfen, *der*

[2]**carp** *v. i.* nörgeln; ~ **at sb./sth.** an jmdm./etw. herumnörgeln *(ugs.)*

'**car-park** *n.* Parkplatz, *der;* *(underground)* Tiefgarage, *die;* *(building)* Parkhaus, *das*

carpenter ['kɑ:pɪntə(r)] *n.* Zimmermann, *der;* *(for furniture)* Tischler, *der*/Tischlerin, *die*

carpentry ['kɑ:pɪntrɪ] *n.* **a)** *(art)* Zimmerhandwerk, *das;* *(in furniture)* Tischlerhandwerk, *das;* **b)** *(woodwork)* **[piece of]** ~: Tischlerarbeit, *die*

carpet ['kɑ:pɪt] **1.** *n.* **a)** Teppich, *der;* **[fitted]** ~: Teppichboden, *der;* **stair-**~: [Treppen]läufer, *der;* **be on the** ~ *(coll.: be reprimanded)* zusammengestaucht werden *(ugs.);* **sweep sth. under the** ~ *(fig.)* etw. unter den Teppich kehren *(ugs.);* **b)** *(expanse)* ~ **of flowers** Blumenteppich, *der.* **2.** *v. t.* **a)** *(cover)* [mit Teppich(boden)] auslegen; *(fig.)* bedecken; **b)** *(coll.: reprimand)* **be** ~**ed for sth.** wegen etw. zusammengestaucht werden *(ugs.)*

carpeting ['kɑ:pɪtɪŋ] *n.* Teppich[boden], *der;* **wall-to-wall** ~: Teppichboden, *der*

carpet: ~**-slipper** *n.* Hausschuh, *der;* ~**-sweeper** *n.* Teppichkehrer, *der*

car: ~ **phone** *n.* Autotelefon, *das;* ~ **pool** *n.* Fahrgemeinschaft, *die;* *(of a firm etc.)* Fahrzeugpark, *der;* ~**-port** *n.* Einstellplatz, *der;* ~ '**radio** *n.* Autoradio, *das*

carriage ['kærɪdʒ] *n.* **a)** *(horse-drawn vehicle)* Kutsche, *die;* ~ **and pair/four/six** *etc.* Zwei-/Vier-/Sechsspänner *usw., der;* **b)** *(Railw.)* [Eisenbahn]wagen, *der;* **c)** *(Mech., of typewriter)* Schlitten, *der;* **d)** *no pl. (conveying, being conveyed)* Transport, *der;* **e)** *(cost of conveying)* Frachtkosten *Pl.;* ~ **paid** frachtfrei; **f)** *(bearing)* Haltung, *die*

carriage: ~ **clock** *n.* Reiseuhr, *die;* ~**way** *n.* Fahrbahn, *die*

carrier ['kærɪə(r)] *n.* **a)** *(bearer)* Träger, *der;* **b)** *(Commerc.) (firm)* Transportunternehmen, *das;* *(person)* Transportunternehmer, *der;* **c)** *(on bicycle etc.)* Gepäckträger, *der;* **d)** *see* **carrier-bag**

carrier: ~**-bag** *n.* Tragetasche, *die;* Tragetüte, *die;* ~ **pigeon** *n.* Brieftaube, *die;* **by** ~ **pigeon** mit der Taubenpost

carrion ['kærɪən] *n.* Aas, *das*

'**carrion crow** *n.* Rabenkrähe, *die*

carrot ['kærət] *n.* **a)** Möhre, *die;* Karotte, *die;* **grated** ~**[s]** geraspelte Möhren *od.* Karotten; **b)** *(fig.)* Köder, *der;* **with** ~ **and stick** mit Zuckerbrot und Peitsche

carrousel [karʊ'sel] *n.* *(Amer.)* Karusell, *das*

carry ['kærɪ] **1.** *v. t.* **a)** *(transport)* tragen; *(with emphasis on destination)* bringen; ⟨*Strom:*⟩ spülen; ⟨*Verkehrsmittel:*⟩ befördern; ~ **all before one** *(fig.)* nicht aufzuhalten sein; **b)** *(conduct)* leiten; ~ **sth. into effect** etw. in die Tat umsetzen; **c)** *(support)* tragen; *(contain)* fassen; ~ **responsibility** Verantwortung tragen; **d)** *(have with one)* ~ **[with one]** bei sich haben *od.* tragen; tragen ⟨*Waffe, Kennzeichen*⟩; **e)** *(possess)* besitzen ⟨*Autorität, Gewicht*⟩; *see also* **conviction b; f)** *(hold)* **she carries herself well** sie hat eine gute Haltung; **g)** *(prolong)* ~ **modesty/altruism** *etc.* **to excess** die Bescheidenheit/den Altruismus *usw.* bis zum Exzeß treiben; ~ **things to extremes** die Dinge auf die Spitze treiben; **h)** *(Math.: transfer)* im Sinn behalten; ~ **one** eins im Sinn; **i)** *(win)* durchbringen ⟨*Antrag, Gesetzentwurf, Vorschlag*⟩; **the motion is carried** der Antrag ist angenommen; ~ **one's audience with one** das Publikum überzeugen; ~ **the day** den Sieg davontragen. **2.** *v. i.* ⟨*Stimme, Laut:*⟩ zu hören sein

~ **a'way** *v. t.* forttragen; *(by force)* fortreißen; *(fig.)* **be** *or* **get carried away** *(be inspired)* hingerissen sein (**by** von); *(lose self-control)* sich hinreißen lassen

~ '**back** *v. t.* **a)** *(return)* zurückbringen; **b)** *see* **take back b**

~ '**forward** *v. t.* *(Bookk.)* vortragen

~ '**off** *v. t.* **a)** *(from place)* davontragen; *(as owner or possessor)* mit sich nehmen; *(cause to die)* dahinraffen *(geh.);* **b)** *(abduct)* entführen ⟨*Person*⟩; **c)** *(win)* gewinnen ⟨*Preis, Medaille*⟩; erringen ⟨*Sieg*⟩; **d)** ~ **it/sth. off [well]** es/etw. [gut] zustande bringen

~ '**on 1.** *v. t.* *(continue)* fortführen ⟨*Tradition, Diskussion, Arbeit*⟩; ~ **on the firm** die Firma übernehmen; ~ **on [doing sth.]** weiterhin etw. tun. **2.** *v. i.* **a)** *(continue)* weitermachen; ~ **on with a plan/project** einen Plan/ein Projekt weiterverfolgen; **b)** *(coll.: behave in unseemly manner)* sich danebenbenehmen *(ugs.);* *(make a fuss)* Theater machen *(ugs.);* **c)** ~ **on with sb.** *(have affair)* mit jmdm. ein Verhältnis haben

~ '**out** *v. t.* durchführen ⟨*Plan, Programm, Versuch*⟩; in die Tat umsetzen ⟨*Vorschlag, Absicht, Vorstellung*⟩; ausführen ⟨*Anweisung, Auftrag*⟩; halten ⟨*Versprechen*⟩; vornehmen ⟨*Verbesserungen*⟩; wahr machen ⟨*Drohung*⟩

~ '**over** *v. t.* **a)** *(postpone)* vertagen (**to** auf + *Akk.*); **b)** *(St. Exch.)* prolongieren; **c)** *see* ~ **forward**

~ '**through** *v. t.* *(complete)* durchführen

carry: ~**-cot** *n.* Babytragetasche, *die;* ~**-on** *n.* *(sl.)* Theater, *das* *(ugs.)*

'**carsick** *adj.* **children are often** ~: Kindern wird beim Autofahren oft schlecht

cart [kɑ:t] **1.** *n.* Karren, *der;* Wagen, *der;* **horse and** ~: Pferdewagen, *der;* **put the** ~ **before the horse** *(fig.)* das Pferd beim Schwanz aufzäumen. **2.** *v. t.* **a)** karren; **b)** *(fig. sl.: carry with effort)* schleppen

~ '**off** *v. t.* *(coll.)* abtransportieren

carte blanche [kɑ:t 'blɑ̃ʃ] *n.* unbeschränkte Vollmacht

cartel [kɑ:'tel] *n.* Kartell, *das*

'**cart-horse** *n.* Arbeitspferd, *das*

cartilage ['kɑ:tɪlɪdʒ] *n.* Knorpel, *der*

cartography [kɑ:'tɒgrəfɪ] *n.* Kartographie, *die*

carton ['kɑ:tn] *n.* [Papp]karton, *der;* **a** ~ **of milk** eine Tüte Milch; **a** ~ **of cigarettes** eine Stange Zigaretten; **a** ~ **of yoghurt** ein Becher Joghurt

cartoon [kɑ:'tu:n] *n.* **a)** *(amusing drawing)* humoristische Zeichnung; Cartoon, *der;* *(satirical illustration)* Karikatur, *die;* *(sequence of drawings)* [humoristische] Bilderserie; Cartoon, *der;* **b)** *(film)* Zeichentrickfilm, *der*

cartoonist [kɑ:'tu:nɪst] *n.* Cartoonist, *der*/Cartoonistin, *die*

cartridge ['kɑ:trɪdʒ] *n.* **a)** *(case for explosive)* Patrone, *die;* **b)** *(spool of film, cassette)* Kassette, *die;* **c)** *(of record player)* Tonabnehmer, *der;* **d)** *(of pen)* Patrone, *die*

'**cartridge paper** *n.* Zeichenpapier, *das*

cart: ~-track *n.* ≈ Feldweg, *der;* **~-wheel** *n.* **a)** Wagenrad, *das;* **b)** *(Gymnastics)* Rad, *das;* **turn** *or* **do ~-wheels** radschlagen

carve [kɑ:v] **1.** *v. t.* **a)** *(cut up)* tranchieren ⟨*Fleisch*⟩; **b)** *(from wood)* schnitzen; *(from stone)* meißeln; **~ sth. out of wood/stone** etw. aus Holz schnitzen/aus Stein meißeln. **2.** *v. i.* **a)** tranchieren; **b)** **~ in wood/stone** in Holz schnitzen/in Stein meißeln

~ out *v. t.* heraushauen

~ up *v. t.* aufschneiden ⟨*Fleisch*⟩; aufteilen ⟨*Erbe, Land*⟩

carver ['kɑ:və(r)] *n.* **a)** *(in wood)* [Holz]schnitzer, *der;* *(in stone)* Bildhauer, *der;* **b)** *in pl. (knife and fork)* Tranchierbesteck, *das*

carving ['kɑ:vɪŋ] *n.* **a)** *(in or from wood)* Schnitzerei, *die;* **a ~ of a madonna in wood** eine holzgeschnitzte Madonna; **b)** *(in or from stone)* Skulptur, *die;* *(on stone)* eingeritztes Bild

'**carving: ~-fork** *n.* Tranchiergabel, *die;* **~-knife** *n.* Tranchiermesser, *das*

cascade [kæs'keɪd] **1.** *(lit. or fig.)* Kaskade, *die.* **2.** *v. i.* [in Kaskaden] herabstürzen

[1]**case** [keɪs] *n.* **a)** *(instance, matter)* Fall, *der;* **if that's the ~:** wenn das so ist; **it is [not] the ~ that ...:** es trifft [nicht] zu *od.* stimmt [nicht], daß ...; **it seems to be the ~ that they have ...:** sie scheinen tatsächlich ... zu haben; **as is generally the ~ with ...:** wie das normalerweise bei ... der Fall ist; **as the ~ may be** je nachdem; **in ~ ...:** falls ...; für den Fall, daß ... *(geh.);* **[just] in ~** *(to allow for all possibilities)* für alle Fälle; **in ~ of fire/danger** bei Feuer/Gefahr; **in ~ of emergency** im Notfall; **in the ~ of** bei; **in any ~** *(regardless of anything else)* jedenfalls; **I don't need it in any ~:** ich brauche es sowieso nicht; **in no ~** *(certainly not)* auf keinen Fall; **in that ~:** in diesem Fall; **b)** *(Med., Police, Soc. Serv., etc., or coll.: person afflicted)* Fall, *der;* **he is a mental/psychiatric ~:** er ist ein Fall für den Psychiater; **c)** *(Law)* Fall, *der;* *(action)* Verfahren, *das;* **the ~ for the prosecution/defence** die Anklage/Verteidigung; **put one's ~:** seinen Fall darlegen; **d)** *(fig.: set of arguments)* Fall, *der;* *(valid set of arguments)* **have a [good] ~ for doing sth./for sth.** gute Gründe haben, etw. zu tun/für etw. haben; **make out a ~ for sth.** Argumente für etw. anführen; **e)** *(Ling.)* Fall, *der;* Kasus, *der (fachspr.);* **f)** *(fig. coll.) (comical person)* ulkiger Typ *(ugs.);* *(comical woman)* ulkige Nudel *(ugs.)*

[2]**case 1.** *n.* **a)** Koffer, *der;* *(small)* Handkoffer, *der;* *(brief-~)* [Akten]tasche, *die;* *(for musical instrument)* Kasten, *der;* **b)** *(sheath)* Hülle, *die;* *(for spectacles, cigarettes)* Etui, *das;* *(for jewellery)* Schmuckkassette, *die;* **c)** *(crate)* Kiste, *die;* **~ of oranges** Kiste [mit] Apfelsinen; **d)** *(glass box)* Vitrine, *die;* **[display-]~:** Schaukasten, *der;* **e)** *(cover)* Gehäuse, *das.* **2.** *v. t.* **a)** *(box)* verpacken; **b)** *(sl.: examine)* **~ the joint** sich *(Dat.)* den Laden mal ansehen *(ugs.)*

case 'history *n.* **a)** *(record)* [Vor]geschichte, *die;* **b)** *(Med.)* Krankengeschichte, *die*

casement ['keɪsmənt] *n.* [Fenster]flügel, *der*

'**casement window** *n.* Flügelfenster, *das*

case: ~-study *n.* Fallstudie, *die;* **~work** *n., no pl., no indef. art.* [auf den Einzelfall bezogene] Sozialarbeit; **~worker** *n.* [Einzelfälle betreuender] Sozialarbeiter

cash [kæʃ] **1.** *n., no pl., no indef. art.* **a)** Bargeld, *das;* **payment in ~ only** nur Barzahlung; **pay [in] ~, pay ~ down** bar zahlen; **we haven't got the ~:** wir haben [dafür] kein Geld; **be short of ~:** knapp bei Kasse sein *(ugs.);* **~ on delivery** per Nachnahme; **b)** *(Banking etc.)* Geld, *das;* **can I get ~ for these cheques?** kann ich diese Schecks einlösen? **2.** *v. t.* einlösen ⟨*Scheck*⟩

~ in 1. ['--] *v. t.* sich *(Dat.)* gutschreiben lassen ⟨*Scheck*⟩. **2.** [-'-] *v. i.* **~ in on sth.** *(lit. or fig.)* von etw. profitieren

cash: ~ and 'carry *n. Verkaufssystem, bei dem der Kunde bar bezahlt und die Ware selbst nach Hause transportiert;* cash and carry; **~-and-carry[store]** Cash-and-carry-Laden, *der;* **~-box** *n.* Geldkassette, *die;* **~card** *n.* Geldautomatenkarte, *die;* **~ desk** *n. (Brit.)* Kasse, *die;* **~ dispenser** *n.* Geldautomat, *der*

cashew ['kæʃu:] *n.* **a)** *(nut) see* **cashew-nut; b)** *(tree)* Nierenbaum, *der (Bot.)*

'**cashew-nut** *n.* Cashewnuß, *die*

'**cash-flow** *n. (Econ.)* Cash-flow, *der*

cashier [kæ'ʃɪə(r)] *n.* Kassierer, *der*/Kassiererin, *die;* **~'s office** Kasse, *die*

cashmere ['kæʃmɪə(r)] *n.* Kaschmir, *der*

cash: ~ payment *n.* Barzahlung, *die;* **~point** *n.* Geldautomat, *der;* **~ register** *n.* [Registrier]kasse, *die*

casing ['keɪsɪŋ] *n.* Gehäuse, *das*

casino [kə'si:nəʊ] *n., pl.* **~s** Kasino, *das;* *(for gambling also)* Spielkasino, *das;* Spielbank, *die*

cask [kɑ:sk] *n.* Faß, *das*

casket ['kɑ:skɪt] *n.* **a)** Schatulle, *die (veralt.);* Kästchen, *das;* **b)** *(Amer.: coffin)* Sarg, *der*

Caspian Sea [kæspɪən 'si:] *pr. n.* Kaspische Meer, *das*

casserole ['kæsərəʊl] *n.* *(food, vessel)* Schmortopf, *der*

cassette [kə'set, kæ'set] *n.* Kassette, *die*

cassette: ~-deck Kassettendeck, *das;* **~ recorder** *n.* Kassettenrecorder, *der*

cassock ['kæsək] *n. (Eccl.)* Soutane, *die*

cast [kɑ:st] **1.** *v. t.,* **cast a)** *(throw)* werfen; **~ sth. adrift** etw. abtreiben lassen; **~ loose** losmachen; **~ an** *or* **one's eye over sth.** einen Blick auf etw. *(Akk.)* werfen; *(fig.)* Licht in etw. *(Akk.)* bringen; **~ the line/net** die Angel[schnur]/das Netz auswerfen; **~ a shadow [on/over sth.]** *(lit. or fig.)* einen Schatten [auf etw. *(Akk.)*] werfen; **~ one's vote** seine Stimme abgeben; **~ one's mind back to sth.** an etw. *(Akk.)* zurückdenken; **b)** *(shed)* verlieren ⟨*Haare, Winterfell*⟩; abwerfen ⟨*Gehörn, Blätter, Hülle*⟩; **the snake ~s its skin** die Schlange häutet sich; **~ aside** *(fig.)* beiseite schieben ⟨*Vorschlag*⟩; ablegen ⟨*Vorurteile, Gewohnheiten*⟩; vergessen ⟨*Sorgen, Vorstellungen*⟩; fallenlassen ⟨*Freunde, Hemmungen*⟩; **c)** *(shape, form)* gießen; **d)** *(calculate)* stellen ⟨*Horoskop*⟩; **e)** *(assign role[s] of)* besetzen; **~ Joe as sb./in the role of sb.** jmdn./jmds. Rolle mit Joe besetzen; **~ a play/film** die Rollen [in einem Stück/Film] besetzen. **2.** *n.* **a)** *(Med.)* Gipsverband, *der;* **b)** *(set of actors)* Besetzung, *die;* **c)** *(model)* Abdruck, *der;* **d)** *(Fishing) (throw of net)* Auswerfen, *das;* *(throw of line)* Wurf, *der*

~ a'bout, ~ around *v. i.* **~ about** *or* **around [to find** *or* **for sth.]** sich [nach etw.] umsehen

~ **a'way** *v. t.* **a)** wegwerfen; **b) be ~ away on an island** auf einer Insel stranden
~ **'off 1.** *v. t.* **a)** ablegen ‹*alte Kleider*›; **b)** *(Naut.)* losmachen. **2.** *v. i. (Naut.)* ablegen
~ **'up** *v. t. (wash up)* an Land spülen

castanet [kæstə'net] *n., usu. in pl. (Mus.)* Kastagnette, *die*

castaway ['kɑ:stəweɪ] *n.* Schiffbrüchige, *der/die*

caste [kɑ:st] *n. (lit. or fig.)* **a)** Kaste, *die;* **b)** *no pl., no art. (class system)* Kastenwesen, *das*

caster ['kɑ:stə(r)] *see* **castor**

castigate ['kæstɪgeɪt] *v. t. (punish)* züchtigen *(geh.); (criticize)* geißeln *(geh.)*

castigation [kæstɪ'geɪʃn] *n. (punishment)* Züchtigung, *die (geh.); (criticism)* Geißelung, *die (geh.)*

casting ['kɑ:stɪŋ] *n.* **a)** *(Metallurgy: product)* Gußstück, *das; (Art)* Abguß, *der;* **b)** *(Theatre, Cinemat.)* Rollenbesetzung, *die*

casting 'vote *n.* ausschlaggebende Stimme *(des Vorsitzenden bei Stimmengleichheit)*

cast: ~ 'iron *n.* Gußeisen, *das;* **~-iron** *adj.* gußeisern; *(fig.)* eisern ‹*Konstitution, Magen*›; handfest, triftig ‹*Grund*›; hieb- und stichfest ‹*Alibi, Beweis*›; hundertprozentig ‹*Garantie*›

castle ['kɑ:sl] *n.* **a)** *(stronghold)* Burg, *die; (mansion)* Schloß, *das;* **Windsor C~:** Schloß Windsor; **~s in the air** *or* **in Spain** Luftschlösser; **b)** *(Chess)* Turm, *der*

castor ['kɑ:stə(r)] *n.* **a)** *(sprinkler)* Streuer, *der;* **b)** *(wheel)* Rolle, *die;* Laufrolle, *die (Technik)*

castor: ~ 'oil *n.* Rizinusöl, *das;* Kastoröl, *das (Kaufmannsspr.);* **~ sugar** *n. (Brit.)* Raffinade, *die*

castrate [kæ'streɪt] *v. t.* kastrieren

castration [kæ'streɪʃn] *n.* Kastration, *die*

casual ['kæʒjʊəl, 'kæzjʊəl] **1.** *adj.* ungezwungen; zwanglos; leger ‹*Kleidung*›; beiläufig ‹*Bemerkung*›; flüchtig ‹*Bekannter, Bekanntschaft, Blick*›; unbekümmert, unbeschwert ‹*Haltung, Einstellung*›; salopp ‹*Ausdrucksweise*›; lässig ‹*Auftreten*›; **be ~ about sth.** etw. auf die leichte Schulter nehmen; **~ sex** Sex ohne feste Bindung. **2.** *n.* **a)** *in pl. (clothes)* Freizeitkleidung, *die;* **b)** *see* **casual labourer; c)** *see* **casual shoe**

casual: ~ 'labour *n., no pl.* Gelegenheitsarbeit, *die;* **~ 'labourer** *n.* Gelegenheitsarbeiter, *der*

casually ['kæʒjʊəlɪ, 'kæzjʊəlɪ] *adv.* ungezwungen; zwanglos; beiläufig ‹*bemerken*›; flüchtig ‹*anschauen*›; salopp ‹*sich ausdrücken*›; leger ‹*sich kleiden*›

casual 'shoe *n.* Freizeitschuh, *der*

casualty ['kæʒjʊəltɪ, 'kæzjʊəltɪ] *n.* **a)** *(injured person)* Verletzte, *der/die; (in battle)* Verwundete, *der/die; (dead person)* Tote, *der/die;* **b)** *(fig.)* Opfer, *das*

cat [kæt] *n.* **a)** Katze, *die;* **she-~** Kätzin, *die;* [weibliche] Katze; **tom-~:** Kater, *der;* **the [great] Cats** die Großkatzen; **the ~ family** die Familie der Katzen; **play ~ and mouse with sb.** Katz und Maus mit jmdm. spielen *(ugs.);* **let the ~ out of the bag** *(fig.)* die Katze aus dem Sack lassen; **be like a ~ on hot bricks** wie auf glühenden Kohlen sitzen; **look like something the ~ brought in** *(fig.)* aussehen wie unter die Räuber gefallen; **curiosity killed the ~** *(fig.)* sei nicht so neugierig; **[fight] like ~ and dog** wie Hund und Katze [sein]; **not a ~ in hell's chance** nicht die geringste Chance; **a ~ may look at a king** *(prov.)* das ist doch auch nur ein Mensch; **put the ~ among the pigeons** *(fig.)* für Aufregung sorgen; **rain ~s and dogs** in Strömen regnen; **no room to swing a ~** *(fig.)* kaum Platz zum Umdrehen; **b)** *see* **cat-o'-nine-tails**

cataclysm ['kætəklɪzm] *n.* [Natur]katastrophe, *die*

cataclysmic [kætə'klɪzmɪk] *adj.* katastrophal; verheerend

catacomb ['kætəku:m, 'kætəkəʊm] *n.* Katakombe, *die*

catalog *(Amer.),* **catalogue** ['kætəlɒg] **1.** *n.* Katalog, *der;* **subject ~:** Sachkatalog, *der (Buchw.).* **2.** *v. t.* katalogisieren

catalyst ['kætəlɪst] *n. (Chem.; also fig.)* Katalysator, *der*

catalytic converter [kætəlɪtɪk kən'vɜ:tə(r)] *n. (Motor Veh.)* Katalysator, *der*

catamaran [kætəmə'ræn] *n. (Naut.)* Katamaran, *der*

catapult ['kætəpʌlt] **1.** *n.* Katapult, *das.* **2.** *v. t.* katapultieren

cataract ['kætərækt] *n.* **a)** *(lit. or fig.)* Katarakt, *der;* **b)** *(Med.)* grauer Star

catarrh [kə'tɑ:(r)] *n.* Schleimabsonderung, *die*

catastrophe [kə'tæstrəfɪ] *n.* Katastrophe, *die;* **end in ~:** in einer Katastrophe enden

catastrophic [kætə'strɒfɪk] *adj.* katastrophal

cat: ~ burglar *n.* Fassadenkletterer, *der*/Fassadenkletterin, *die;* **~call 1.** *n.* ≈ Pfiff, *der;* **2.** *v. i.* ≈ pfeifen

catch [kætʃ] **1.** *v. t.,* **caught** [kɔ:t] **a)** *(capture)* fangen; *(lay hold of)* fassen; packen; **~ sb. by the arm** jmdn. am Arm packen *od.* fassen; **~ hold of sb./sth.** jmdn./etw. festhalten; *(to stop oneself falling)* sich an jmdm./etw. festhalten; **b)** *(intercept motion of)* auffangen; fangen ‹*Ball*›; **~ a thread** einen Faden vernähen; **get sth. caught** *or* **~ sth. on/in sth.** mit etw. an/in etw. *(Dat.)* hängenbleiben; **I got my finger caught** *or* **caught my finger in the door** ich habe mir den Finger in der Tür eingeklemmt; **get caught on/in sth.** an/in etw. *(Dat.)* hängenbleiben; **c)** *(travel by)* nehmen; *(manage to see)* sehen; *(be in time for)* [noch] erreichen; [noch] kriegen *(ugs.)* ‹*Bus, Zug*›; [noch] erwischen *(ugs.)* ‹*Person*›; **did you ~ her in?** hast du sie zu Hause erwischt? *(ugs.);* **d)** *(surprise)* **~ sb. at/doing sth.** jmdn. bei etw. erwischen *(ugs.)*/[dabei] erwischen, wie er etw. tut *(ugs.);* **caught in a thunderstorm** vom Sturm überrascht; **I caught myself thinking how ...:** ich ertappte mich bei dem Gedanken, wie ...; **e)** *(become infected with, receive)* sich *(Dat.)* zuziehen *od. (ugs.)* holen; **~ sth. from sb.** sich bei jmdm. mit etw. anstecken; **~ [a] cold** sich erkälten/sich *(Dat.)* einen Schnupfen holen; *(fig.)* übel dran sein; **~ it** *(fig. coll.)* etwas kriegen *(ugs.);* **you'll ~ it from me** du kannst von mir was erleben *(ugs.);* **f)** *(arrest)* **~ sb.'s attention** jmds. Aufmerksamkeit erregen; **~ sb.'s fancy** jmdm. gefallen; jmdn. ansprechen; **~ the Speaker's eye** *(Parl.)* das Wort erhalten; **~ sb.'s eye** jmdm. auffallen; ‹*Gegenstand:*› jmdm. ins Auge fallen; *(be impossible to overlook)* jmdm. ins Auge springen; **g)** *(hit)* **~ sb. on/in sth.** jmdn. auf/in etw. *(Akk.)* treffen; **~ sb. a blow [on/in sth.]** jmdm. einen Schlag [auf/in etw. *(Akk.)*] versetzen; **h)** *(grasp in thought)* verstehen; mitbekommen; **did you ~ his meaning?** hast du verstanden *od.* mitbekommen, was er meint?; **i)** *see* **~ out a. 2.** *v. i.,* **caught a)** *(begin to burn)* [anfangen zu] brennen; **b)** *(become fixed)* hängenbleiben; ‹*Haar, Faden:*› sich verfangen; **my coat caught on a nail** ich blieb mit meinem Mantel an einem Nagel hängen. **3.** *n.* **a)** *(of ball)*

make [several] good ~es [mehrmals] gut fangen; **b)** *(amount caught, lit. or fig.)* Fang, *der;* **c)** *(trick, difficulty)* Haken, *der* **(in** an + *Dat.*); **the ~ is that ...:** der Haken an der Sache ist, daß ...; **it's ~-22** [kætʃtwentɪ'tu:] *(coll.)* es ist ein Teufelskreis; **d)** *(fastener)* Verschluß, *der; (of door)* Schnapper, *der;* **e)** *(Cricket etc.)* ≈ Fang, *der; Abfangen des Balles, das den Schlagmann aus dem Spiel bringt;* **f)** *(catcher)* **he is a good ~:** er kann gut fangen

~ 'on *v. i. (coll.)* **a)** *(become popular)* [gut] ankommen *(ugs.);* sich durchsetzen; **b)** *(understand)* begreifen; kapieren *(ugs.)*

~ 'out *v. t.* **a)** *(detect in mistake)* [bei einem Fehler] ertappen; **b)** *(take unawares)* erwischen *(ugs.)*

~ 'up 1. *v. t.* **a)** *(reach)* **~ sb. up** jmdn. einholen; *(in quality, skill)* mit jmdm. mitkommen; **b)** *(absorb)* **be caught up in sth.** in etw. *(Dat.)* [völlig] aufgehen; **c)** *(snatch)* packen; **sth. gets caught up in sth.** etw. verfängt sich in etw. *(Dat.).* **2.** *v. i. (get level)* **~ up** einholen; **~ up with sb.** *(in quality, skill)* mit jmdm. mitkommen; **~ up on sth.** etw. nachholen; **I'm longing to ~ up on your news** ich bin gespannt, was für Neuigkeiten du hast

catching ['kætʃɪŋ] *adj.* ansteckend

catchment area ['kætʃmənt eərɪə] *n. (lit. or fig.)* Einzugsgebiet, *das*

catch: ~-phrase *n.* Slogan, *der;* **~word** *n. (slogan)* Schlagwort, *das*

catchy ['kætʃɪ] *adj.* eingängig; **a ~ song** ein Ohrwurm *(ugs.)*

catechism ['kætɪkɪzm] *n. (Relig.)* Katechismus, *der*

categorical [kætɪ'gɒrɪkl] *adj.*, **categorically** [kætɪ'gɒrɪkəlɪ] *adv.* kategorisch

category ['kætɪgərɪ] *n.* Kategorie, *die*

cater ['keɪtə(r)] *v. i.* **a)** *(provide or supply food)* **~ [for sb./sth.]** [für jmdn./etw.] [die] Speisen und Getränke liefern; **~ for weddings** Hochzeiten ausrichten; **b)** *(provide requisites etc.)* **~ for sb./sth.** auf jmdn./etw. eingestellt sein; **~ for all ages** jeder Altersgruppe etwas bieten

caterer ['keɪtərə(r)] *n.* Lieferant von Speisen und Getränken; Caterer, *der (fachspr.)*

catering ['keɪtərɪŋ] *n.* **a)** *(trade)* **~ [business]** Gastronomie, *die;* **b)** *(service)* Lieferung von Speisen und Getränken; Catering, *das (fachspr.);* **do the ~:** für Speisen und Getränke sorgen; **~ firm/service** *see* **caterer**

caterpillar ['kætəpɪlə(r)] *n.* **a)** *(Zool.)* Raupe, *die;* **b) C~ [tractor] (P)** *(Mech.)* Raupenfahrzeug, *das*

caterpillar: ~ 'track, ~ 'tread *ns.* Raupen-, Gleiskette, *die*

caterwaul ['kætəwɔ:l] *v. i.* ⟨*Katze:*⟩ schreien, [laut] miauen; ⟨*Sänger:*⟩ jaulen *(abwertend)*

cathedral [kə'θi:drl] *n.* **~ [church]** Dom, *der;* Kathedrale, *die*

Catherine wheel ['kæθrɪn wi:l] *n. (firework)* Feuerrad, *das*

catheter ['kæθɪtə(r)] *n. (Med.)* Katheter, *der*

catholic ['kæθəlɪk, 'kæθlɪk] **1.** *adj.* **a)** *(all-embracing)* umfassend; vielseitig ⟨*Interessen*⟩; **b) C~** *(Relig.)* katholisch. **2.** *n.* **C~:** Katholik, *der*/Katholikin, *die*

Catholicism [kə'θɒlɪsɪzm] *n. (Relig.)* Katholizismus, *der*

catkin ['kætkɪn] *n. (Bot.)* Kätzchen, *das*

cat: ~nap *n.* Nickerchen, *das (ugs.);* kurzes Schläfchen; **~-o'-'nine-tails** *n.* neunschwänzige Katze; **~'s-eye** *n.* **a)** *(stone)* Katzenauge, *das;* **b)** *(Brit.: reflector)* Bodenrückstrahler, *der (Verkehrsw.)*

cattle ['kætl] *n. pl.* Vieh, *das;* Rinder *Pl.*

cattle: ~-breeding *n.* Rinderzucht, *die;* Viehzucht, *die;* **~-market** *n.* Viehmarkt, *der;* **~-truck** *n.* Viehtransporter, *der; (Railw.)* Viehwagen, *der*

catty ['kætɪ] *adj.* gehässig

'catwalk *n.* Laufsteg, *der*

caucus ['kɔ:kəs] *n. (Brit. derog., Amer.)* **a)** *(committee) den Wahlkampf und die Richtlinien der Politik bestimmendes regionales Gremium einer Partei;* **b)** *(party meeting) den Wahlkampf und die Richtlinien der Politik bestimmende Sitzung der regionalen Parteiführung*

caught *see* **catch 1, 2**

cauldron ['kɔ:ldrən, 'kɒldrən] *n.* Kessel, *der*

cauliflower ['kɒlɪflaʊə(r)] *n.* Blumenkohl, *der*

causal ['kɔ:zl] *adj.* kausal

cause [kɔ:z] **1.** *n.* **a)** *(what produces effect)* Ursache, *die* **(of** für *od. Gen.*); *(person)* Verursacher, *der*/Verursacherin, *die;* **be the ~ of sth.** etw. verursachen; **b)** *(reason)* Grund, *der;* Anlaß, *der;* **~ for/to do sth.** Grund *od.* Anlaß zu etw./, etw. zu tun; **no ~ for concern** kein Grund zur Beunruhigung; **without good ~:** ohne triftigen Grund; **c)** *(aim, object of support)* Sache, *die;* **freedom is our common ~:** Freiheit ist unser gemeinsames Anliegen *od.* Ziel; **be a lost ~:** aussichtslos sein; verlorene Liebesmühe sein *(ugs.);* **[in] a good ~:** [für] eine gute Sache. **2.** *v. t.* **a)** *(produce)* verursachen; erregen ⟨*Aufsehen, Ärgernis*⟩; hervorrufen ⟨*Verstimmung, Unruhe, Verwirrung*⟩; **b)** *(give)* **~ sb. worry/pain** *etc.* jmdm. Sorge/ Schmerzen *usw.* bereiten; **~ sb. trouble/bother** jmdm. Umstände machen; **c)** *(induce)* **~ sb. to do sth.** jmdn. veranlassen, etw. zu tun

causeway ['kɔ:zweɪ] *n.* Damm, *der*

caustic ['kɔ:stɪk] *adj.* beißend ⟨*Spott*⟩; bissig ⟨*Bemerkung, Worte*⟩; spitz, scharf ⟨*Zunge*⟩; **~ soda** Ätznatron, *das*

cauterize (cauterise) ['kɔ:təraɪz] *v. t. (Med.)* kauterisieren

caution ['kɔ:ʃn] **1.** *n.* **a)** Vorsicht, *die;* **use ~:** vorsichtig sein; **b)** *(warning)* Warnung, *die; (warning and reprimand)* Verwarnung, *die;* **just a word of ~:** noch ein guter Rat. **2.** *v. t. (warn)* warnen; *(warn and reprove)* verwarnen **(for** wegen); **~ sb. against sth./doing sth.** jmdn. vor etw. *(Dat.)* warnen/ davor warnen, etw. zu tun; **~ sb. to/not to do sth.** jmdn. ermahnen, etw. zu tun/nicht zu tun

cautious ['kɔ:ʃəs] *adj.* vorsichtig; *(circumspect)* umsichtig

cautiously ['kɔ:ʃəslɪ] *adv.* vorsichtig; *(circumspectly)* umsichtig

cavalier [kævə'lɪə(r)] **1.** *n.* Kavalier, *der.* **2.** *adj. (offhand)* keck; *(arrogant)* anmaßend

cavalry ['kævəlrɪ] *n., constr. as sing. or pl.* Kavallerie, *die*

cave [keɪv] **1.** *n.* Höhle, *die.* **2.** *v. i.* Höhlen erforschen

~ 'in *v. i.* einbrechen; *(fig.) (collapse)* zusammenbrechen; *(submit)* nachgeben

caveat ['kævɪæt] *n. (warning)* Warnung, *die*

cave: ~-dweller, ~-man *ns.* Höhlenbewohner, *der; (fig.)* Wilde, *der*

cavern ['kævən] *n. (cave, lit. or fig.)* Höhle, *die*

cavernous ['kævənəs] *adj. (like a cavern)* höhlenartig

caviare (caviar) ['kævɪɑ:(r), kævɪ'ɑ:(r)] *n.* Kaviar, *der*

cavil ['kævɪl] *v. i., (Brit.)* **-ll-** kritteln *(abwertend);* **~ at/about sth.** etw. bekritteln *(abwertend)*

cavity ['kævɪtɪ] *n.* Hohlraum, *der; (in tooth)* Loch, *das;* **nasal ~:** Nasenhöhle, *die*

'cavity wall *n.* Hohlmauer, *die*

cavort [kə'vɔ:t] *v.i. (sl.)* ~ **[about** *or* **around]** herumtollen *(ugs.)*

caw [kɔ:] **1.** *n.* Krächzen, *das.* **2.** *v.i.* krächzen

cayenne [keɪ'en] *n.* ~ **['pepper]** Cayennepfeffer, *der*

CB *abbr.* **citizens' band** CB

CBI *abbr.* **Confederation of British Industry** *britischer Unternehmerverband*

cc [si:'si:] *abbr.* **cubic centimetre(s)** cm^3

CD *abbr.* **a) civil defence; b) Corps Diplomatique** CD; **c) compact disc** CD; **CD player** CD-Spieler, *der*

cease [si:s] **1.** *v.i.* aufhören; **without ceasing** ununterbrochen. **2.** *v.t.* **a)** *(stop)* aufhören; ~ **doing** *or* **to do sth.** aufhören, etw. zu tun; **sth. has ~d to exist** etw. existiert *od.* besteht nicht mehr; **it never ~s to amaze me** ich kann nur immer darüber staunen; **b)** *(end)* aufhören mit; einstellen ‹*Bemühungen, Versuche*›; ~ **'fire** *(Mil.)* das Feuer einstellen

'cease-fire *n.* Waffenruhe, *die; (signal)* Befehl zur Feuereinstellung

ceaseless ['si:slɪs] *adj.* endlos; unaufhörlich ‹*Anstrengung*›; ständig ‹*Wind, Regen, Lärm*›

cedar ['si:də(r)] *n.* **a)** Zeder, *die;* **b)** *see* **cedar-wood**

'cedar-wood *n.* Zedernholz, *das*

cede [si:d] *v.t. (surrender)* abtreten ‹*Land, Rechte*› **(to** *Dat.*, an + *Akk.*)

ceiling ['si:lɪŋ] *n.* **a)** Decke, *die;* **b)** *(upper limit)* Maximum, *das;* **c)** *(Aeronaut.)* Gipfelhöhe, *die*

celebrate ['selɪbreɪt] **1.** *v.t. (observe)* feiern; *(Eccl.)* zelebrieren, lesen ‹*Messe*›. **2.** *v.i.* feiern

celebrated ['selɪbreɪtɪd] *adj.* gefeiert, berühmt ‹*Person*›; berühmt ‹*Gebäude, Werk usw.*›

celebration [selɪ'breɪʃn] *n.* **a)** *(observing)* Feiern, *das; (party etc.)* Feier, *die;* **in ~ of** aus Anlaß (+ *Gen.*); *(with festivities)* zur Feier (+ *Gen.*); **this calls for a ~!** das muß gefeiert werden!; **b)** *(performing)* **the ~ of the wedding/christening** die Trauung[szeremonie]/Taufe; **the ~ of Communion** die Feier der Kommunion

celebrity [sɪ'lebrɪtɪ] *n.* **a)** *no pl. (fame)* Berühmtheit, *die;* **b)** *(person)* Berühmtheit, *die*

celery ['selərɪ] *n.* [Bleich-, Stangen]sellerie, *der od. die*

celestial [sɪ'lestɪəl] *adj.* **a)** *(heavenly)* himmlisch; **b)** *(of the sky)* Himmels-

celibate ['selɪbət] *adj.* zölibatär *(Rel.);* ehelos; **remain ~:** im Zölibat leben *(Rel.);* ehelos bleiben

cell [sel] *n. (also Biol., Electr.)* Zelle, *die*

cellar ['selə(r)] *n.* Keller, *der*

cellist ['tʃelɪst] *n. (Mus.)* Cellist, *der*/Cellistin, *die*

cello ['tʃeləʊ] *n., pl.* **~s** *(Mus.)* Cello, *das*

Cellophane, cellophane, (P) ['seləfeɪn] *n.* Cellophan ®, *das*

'cell phone *n.* Mobiltelefon, *das*

cellular ['seljʊlə(r)] *adj.* **a)** porös ‹*Mineral, Gestein, Substanz*›; *(Biol.: of cells)* zellular; Zell-; **b)** *(with open texture)* luftdurchlässig; atmungsaktiv *(Werbespr.)*

celluloid ['seljʊlɔɪd] *n.* **a)** Zelluloid, *das;* **b)** *(cinema films)* Kino, *das;* ~ **hero** Leinwandheld, *der*

cellulose ['seljʊləʊs, 'seljʊləʊz] *n. (Chem.)* Zellulose, *die*

Celsius ['selsɪəs] *adj.* Celsius

Celt [kelt, selt] *n.* Kelte, *der*/Keltin, *die*

Celtic ['keltɪk, 'seltɪk] **1.** *adj.* keltisch. **2.** *n.* Keltisch, *das*

cement [sɪ'ment] **1.** *n.* **a)** *(Building)* Zement, *der; (mortar)* [Zement]mörtel, *der;* **b)** *(sticking substance)* Klebstoff, *der; (for mending broken vases etc. also)* Kitt, *der.* **2.** *v.t.* **a)** mit Zement/Mörtel zusammenfügen; *(stick together)* zusammenkleben; *(fig.)* zusammenkitten; zementieren ‹*Freundschaft, Beziehung*›; **b)** *(apply ~ to)* zementieren/mörteln

cemetery ['semɪtərɪ] *n.* Friedhof, *der*

cenotaph ['senətɑ:f, 'senətæf] *n.* Kenotaph, *das;* Zenotaph, *das*

censor ['sensə(r)] **1.** *n.* **a)** Zensor, *der;* **b)** *(judge)* Kritiker, *der.* **2.** *v.t.* zensieren

censorship ['sensəʃɪp] *n.* Zensur, *die*

censure ['senʃə(r)] **1.** *n.* Tadel, *der;* **propose a vote of ~:** einen Tadelsantrag stellen. **2.** *v.t.* tadeln

census ['sensəs] *n.* Zählung, *die;* **[national]** ~: Volkszählung, *die;* Zensus, *der*

cent [sent] *n.* Cent, *der*

centenarian [sentɪ'neərɪən] *n.* Hundertjährige, *der/die*

centenary [sen'ti:nərɪ, sen'tenərɪ] **1.** *adj.* ~ **celebrations/festival** Hundertjahrfeier, *die.* **2.** *n.* Hundertjahrfeier, *die*

centennial [sen'tenɪəl] **1.** *adj.* hundertjährig; *(occurring every 100 years)* Jahrhundert-. **2.** *n. see* **centenary 2**

center *(Amer.) see* **centre**

centi- ['sentɪ] *in comb.* Zenti-

'centigrade *see* **Celsius**

centime ['sɑ̃ti:m] *n.* Centime, *der*

'centimetre (*Brit.; Amer.:* **centimeter**) *n.* Zentimeter, *der*

centipede ['sentɪpi:d] *n.* Hundertfüßer, *der (Zool.);* ≈ Tausendfüßler, *der*

central ['sentrl] *adj.* zentral; **be ~ to sth.** von zentraler Bedeutung für etw. sein; **in ~ London** im Zentrum von London

Central: ~ **A'merica** *pr. n.* Mittelamerika *(das);* ~ **'Europe** *pr. n.* Mitteleuropa *(das);* ~ **Euro'pean** *adj.* mitteleuropäisch; **c~ 'heating** *n.* Zentralheizung, *die*

centralize (centralise) ['sentrəlaɪz] *v.t.* zentralisieren

centrally ['sentrəlɪ] *adv.* **a)** *(in centre)* zentral; **b)** *(in leading place)* an zentraler Stelle

central: ~ **'nervous system** *n. (Anat., Zool.)* Zentralnervensystem, *das;* ~ **'processing unit** *n. (Computing)* Zentraleinheit, *die;* ~ **reser'vation** *n. (Brit.)* Mittelstreifen, *der;* ~ **'station** *n.* Hauptbahnhof, *der*

centre ['sentə(r)] *(Brit.)* **1.** *n.* **a)** Mitte, *die; (of circle, globe)* Mitte, *die;* Zentrum, *das;* Mittelpunkt, *der;* **b)** *(town ~)* Innenstadt, *die;* Zentrum, *das;* **c)** *(filling of chocolate)* Füllung, *die;* **d)** *(Polit.)* Mitte, *die;* **e)** *(Sport: player)* Mittelfeldspieler, *der*/-spielerin, *die; (Basketball)* Center, *der;* **f) she likes to be the ~ of attraction/attention** *(fig.)* sie steht gern im Mittelpunkt [des Interesses]. **2.** *adj.* mittler...; ~ **party** *(Polit.)* Partei der Mitte. **3.** *v.i.* ~ **on sth.** sich auf etw. *(Akk.)* konzentrieren; **the novel ~s on Prague** Prag steht im Mittelpunkt des Romans; ~ **[a]round sth.** um etw. kreisen. **4.** *v.t.* **a)** *(place in ~)* in der Mitte anbringen; in der Mitte aufhängen ‹*Bild, Lampe*›; zentrieren ‹*Überschrift*›; **b)** *(concentrate)* ~ **sth. on sth.** etw. auf etw. *(Akk.)* konzentrieren; **be ~d [a]round sth.** etw. zum Mittelpunkt haben; **c)** *(Football, Hockey)* [nach innen] flanken

centre: ~-**'forward** *n. (Sport)* Mittelstürmer, *der*/-stürmerin, *die;* ~-**'half** *n. (Sport)* Mittelläufer, *der*/-läuferin, *die; (Football also)* Vorstopper, *der*/-stopperin, *die;* ~-**piece** *n. (ornament)* ≈ Tafelschmuck, *der (in der Mitte* *der Tafel); (principal item)* Kernstück, *das;* ~ **'spread** *n.* Doppelseite in der Mitte

centrifugal [sentrɪ'fju:gl] *adj.* zentrifugal; ~ **force** Zentrifugalkraft, *die;* Fliehkraft, *die*

centrifuge ['sentrɪfju:dʒ] *n.* Zentrifuge, *die*

centurion [sen'tjʊərɪən] *n. (Roman Hist.)* Zenturio, *der*

century ['sentʃərɪ] *n.* **a)** *(hundred-year period from a year ..00)* Jahrhundert, *das; (hundred years)* hundert Jahre; **b)** *(Cricket)* hundert Läufe

ceramic [sɪ'ræmɪk] **1.** *adj.* keramisch; Keramik⟨*vase, -kacheln*⟩. **2.** *n.* Keramik, *die*

ceramics [sɪ'ræmɪks] *n., no pl.* Keramik, *die*

cereal ['sɪərɪəl] *n.* **a)** *(kind of grain)* Getreide, *das;* **b)** *(breakfast dish)* Getreideflocken *Pl.*

cerebral ['serɪbrl] *adj.* **a)** *(of the brain)* Gehirn⟨*tumor, -blutung, -schädigung*⟩; zerebral *(Anat.);* **b)** *(intellectual)* intellektuell

ceremonial [serɪ'məʊnɪəl] **1.** *adj.* feierlich; *(prescribed for ceremony)* zeremoniell. **2.** *n.* Zeremoniell, *das*

ceremonious [serɪ'məʊnɪəs] *adj.* formell; förmlich ⟨*Höflichkeit*⟩; *(according to prescribed ceremony)* zeremoniell

ceremony ['serɪmənɪ] *n.* **a)** Feier, *die; (formal act)* Zeremonie, *die;* **b)** *no pl., no art. (formalities)* Zeremoniell, *das;* **stand on ~:** Wert auf Förmlichkeiten legen; **without [great] ~:** ohne große Förmlichkeit

cerise [sə'ri:z, sə'ri:s] **1.** *adj.* kirschrot. **2.** *n.* Kirschrot, *das*

cert [sɜ:t] *n. (Brit. coll.)* **a) that's a ~:** das steht fest; **b)** *(as winner)* todsicherer Tip *(ugs.)*

certain ['sɜ:tn, 'sɜ:tɪn] *adj.* **a)** *(settled)* bestimmt ⟨*Zeitpunkt*⟩; **b)** *(unerring)* sicher; *(sure to happen)* unvermeidlich; sicher ⟨*Tod*⟩; **for ~:** bestimmt; **I [don't] know for ~ when ...:** ich weiß [nicht] genau, wann ...; **I can't say for ~ that ...:** ich kann nicht mit Bestimmtheit sagen, daß ...; **make ~ of sth.** *(ensure)* für etw. sorgen; *(examine and establish)* sich einer Sache *(Gen.)* vergewissern; **we made ~ of a seat on the train** wir sicherten uns einen Sitzplatz im Zug; **c)** *(indisputable)* unbestreitbar; **d)** *(confident)* sicher; **of that I'm quite ~:** dessen bin ich [mir] ganz sicher; **e) be ~ to do sth.** *(inevitably)* etw. bestimmt tun; **f)** *(particular but as yet unspecified)* bestimmt; **g)** *(slight; existing but probably not already known)* gewiß; **to a ~ extent** in gewisser Weise; **a ~ Mr Smith** ein gewisser Herr Smith

certainly ['sɜ:tnlɪ, 'sɜ:tɪnlɪ] **a)** *(admittedly)* sicher[lich]; *(definitely)* bestimmt; *(clearly)* offensichtlich; **b)** *(in answer)* [aber] gewiß; [aber] sicher; **[most] ~ 'not!** auf [gar] keinen Fall!

certainty ['sɜ:tntɪ, 'sɜ:tɪntɪ] *n.* **a) be a ~:** sicher sein; feststehen; **regard sth. as a ~:** etw. für sicher halten; **b)** *(absolute conviction, sure fact, assurance)* Gewißheit, *die;* **~ of** *or* **about sth./sb.** Gewißheit über etw./jmdn.; **~ that ...:** Gewißheit [darüber], daß ...; **with some ~:** mit einiger Sicherheit; **with ~, for a ~:** mit Sicherheit *od.* Bestimmtheit

certifiable ['sɜ:tɪfaɪəbl] *adj.* **a)** nachweislich; überprüfbar ⟨*Ergebnis*⟩; **b)** *(as insane)* unzurechnungsfähig ⟨*Person*⟩

certificate [sə'tɪfɪkət] *n.* Urkunde, *die; (of action performed)* Schein, *der;* **doctor's ~:** ärztliches Attest

certify ['sɜ:tɪfaɪ] *v. t.* **a)** bescheinigen; bestätigen; *(declare by certificate)* berechtigen; **this is to ~ that ...:** hiermit wird bescheinigt *od.* bestätigt, daß ...; **b)** *(declare insane)* für unzurechnungsfähig erklären

certitude ['sɜ:tɪtju:d] *n.* Gewißheit, *die*

cervical [sɜ:'vaɪkl, 'sɜ:vɪkl] *adj. (Anat.: of cervix)* Gebärmutterhals-; zervikal *(Anat.);* **~ smear test** [Gebärmutterhals]abstrich, *der*

cervix ['sɜ:vɪks] *n., pl.* **cervices** ['sɜ:vɪsi:z] *(Anat.: of uterus)* Gebärmutterhals, *der*

Cesarean, Cesarian *(Amer.) see* **Caesarean**

cessation [se'seɪʃn] *n.* Ende, *das; (interval)* Nachlassen, *das*

cesspit ['sespɪt] *n.* **a)** *(refuse pit)* Abfallgrube, *die;* **b)** *see* **cesspool**

cesspool ['sespu:l] *n.* Senk- *od.* Jauchegrube, *die*

cf. *abbr.* **compare** vgl.

chafe [tʃeɪf] **1.** *v. t. (make sore)* aufscheuern; wund scheuern; *(rub)* reiben. **2.** *v. i.* ⟨*Person, Tier:*⟩ sich scheuern; ⟨*Gegenstand:*⟩ scheuern (**[up]on, against** an + *Dat.*)

chaff [tʃɑ:f] *(husks of corn, etc.)* Spreu, *die; (cattle-food)* Häcksel, *das*

chaffinch ['tʃæfɪnʃ] *n. (Ornith.)* Buchfink, *der*

chagrin ['ʃægrɪn] **1.** *n.* Kummer, *der;* Verdruß, *der;* **much to sb.'s ~:** zu jmds. großen Kummer *od.* Verdruß. **2.** *v. t.* bekümmern; **be** *or* **feel ~ed at** *or* **by sth.** wegen etw. niedergeschlagen sein

chain [tʃeɪn] **1.** *n.* **a)** Kette, *die; (fig.)* Fessel, *die; (jewellery)* [Hals]kette, *die;* **door-~:** Tür- *od.* Sicherungskette, *die;* **b)** *(series)* Kette, *die;* Reihe, *die;* **~ of events** Reihe *od.* Kette von Ereignissen; **~ of mountains** Gebirgskette, *die;* **~ of shops/hotels** Laden-/Hotelkette, *die;* **c)** *(measurement)* Chain, *das* (≈ *20 m*). **2.** *v. t. (lit. or fig.)* **~ sb./sth. to sth.** jmdn./etw. an etw. *(Akk.)* [an]ketten

chain: ~ re'action *n. (Chem., Phys.; also fig.)* Kettenreaktion, *die;* **~-saw** *n.* Kettensäge, *die;* **~-smoke** *v. t. & i.* Kette rauchen *(ugs.);* **~-smoker** *n.* Kettenraucher, *der*/-raucherin, *die;* **~ store** *n.* Kettenladen, *der (Wirtsch.)*

chair [tʃeə(r)] **1.** *n.* **a)** Stuhl, *der; (arm~, easy ~)* Sessel, *der;* **b)** *(professorship)* Lehrstuhl, *der;* **c)** *(at meeting)* Vorsitz, *der; (chairman)* Vorsitzende, *der/die;* **be** *or* **preside in/take the ~:** den Vorsitz haben *od.* führen/übernehmen. **2.** *v. t. (preside over)* den Vorsitz haben *od.* führen bei

chair: ~-back *n.* Rückenlehne, *die;* **~-lift** *n.* Sessellift, *der;* **~man** ['tʃeəmən] *n., pl.* **chairmen** ['tʃeəmən] Vorsitzende, *der/die;* Präsident, *der*/Präsidentin, *die*

chairmanship ['tʃeəmənʃɪp] *n.* Vorsitz, *der*

chair: ~person *n.* Vorsitzende, *der/die;* **~woman** *n.* Vorsitzende, *die*

chaise longue [ʃeɪz 'lɒŋ] *n.* Chaiselongue, *die*

chalet ['ʃæleɪ] *n.* Chalet, *das*

chalice ['tʃælɪs] *n. (poet./Eccl.)* Kelch, *der*

chalk [tʃɔ:k] **1.** *n.* Kreide, *die;* **as white as ~:** kreidebleich; **not by a long ~** *(Brit. coll.)* bei weitem nicht; **as different as ~ and cheese** so verschieden wie Tag und Nacht. **2.** *v. t.* mit Kreide schreiben/malen/zeichnen *usw.*

~ 'up *v. t.* **a)** [mit Kreide] an- *od.* aufschreiben; **b)** *(fig.: register)* für sich verbuchen können ⟨*Erfolg*⟩; **c) ~ it up** *(fig.)* es auf die Rechnung setzen

challenge ['tʃælɪndʒ] **1.** *n.* **a)** *(to contest or duel; also Sport)* Herausforderung, *die* (**to** *Gen.*); **issue a ~ to sb.** jmdn. herausfordern; **b)** *(of sentry)* Aufforderung, *die;*

(call for password) Anruf, *der;* **c)** *(person, task)* Herausforderung, *die.* **2.** *v. t.* **a)** *(to contest etc.)* herausfordern (**to** zu); ~ **sb. to a duel** jmdn. zum Duell [heraus]fordern; **b)** *(fig.)* auffordern; ~ **sb.'s authority** jmds. Autorität *od.* Befugnis in Frage stellen; **c)** *(demand password etc. from)* ‹*Wachposten:*› anrufen; **d)** *(question)* in Frage stellen; anzweifeln; ~ **a verdict** ein Urteil kritisieren

challenger ['tʃælɪndʒə(r)] *n.* Herausforderer, *der*/Herausforderin, *die*

challenging ['tʃælɪndʒɪŋ] *adj.* herausfordernd; fesselnd, faszinierend ‹*Problem*›; anspruchsvoll ‹*Arbeit*›

chamber ['tʃeɪmbə(r)] *n.* **a)** *(poet./arch.: room)* Gemach, *das (geh.); (bedroom)* [Schlaf]gemach, *das (geh.);* **b) Upper/Lower C~** *(Parl.)* Ober-/Unterhaus, *das;* **c)** *(Anat.; in machinery, etc.)* Kammer, *die*

chamber: ~maid *n.* Zimmermädchen, *das;* ~ **music** *n.* Kammermusik, *die;* **C~ of 'Commerce** *n.* Industrie- und Handelskammer, *die;* **~-pot** *n.* Nachttopf, *der*

chameleon [kə'mi:lɪən] *n. (Zool.; also fig.)* Chamäleon, *das*

chamois ['ʃæmwɑ:] *n., pl. same* ['ʃæmwɑ:z] **a)** *(Zool.)* Gemse, *die;* **b)** *(leather)* Chamois[leder], *das;* **~-leather** ['ʃæmwɑ:-, 'ʃæmɪ-] Chamoisleder, *das*

champagne [ʃæm'peɪn] *n.* Sekt, *der; (from Champagne)* Champagner, *der*

champion ['tʃæmpɪən] **1.** *n.* **a)** *(defender)* Verfechter, *der*/Verfechterin, *die;* **b)** *(Sport)* Meister, *der*/Meisterin, *die;* Champion, *der;* **world ~:** Weltmeister, *der*/-meisterin, *die;* **c)** *(animal or plant best in contest)* Sieger, *der;* **be a ~:** prämiert *od.* preisgekrönt sein; **d)** *attrib.* ~ **dog** preisgekrönter Hund; ~ **boxer** Champion im Boxen. **2.** *v. t.* verfechten ‹*Sache*›; ~ **a person** sich für eine Person einsetzen

championship ['tʃæmpɪənʃɪp] *n.* **a)** *(Sport)* Meisterschaft, *die;* **defend the ~:** den Titel *od.* die Meisterschaft verteidigen; *attrib.* ~ **title/match** Titel, *der*/Titelkampf, *der;* **b)** *(advocacy)* ~ **of a cause** Engagement für eine Sache

chance [tʃɑ:ns] **1.** *n.* **a)** *no art. (fortune)* Zufall, *der; attrib.* Zufalls-; zufällig; **leave sth. to ~:** es dem Zufall *od.* Schicksal überlassen; **pure ~:** reiner Zufall; **by ~:** zufällig; durch Zufall; **b)** *(trick of fate)* Zufall, *der;* **could you by any ~ give me a lift?** könntest du mich vielleicht mitnehmen?; **c)** *(opportunity)* Chance, *die;* Gelegenheit, *die; (possibility)* Chance, *die;* Möglichkeit, *die;* **give sb. a ~:** jmdm. eine Chance geben; **give sb. half a ~:** jmdm. nur die [geringste] Chance geben; **given the ~:** wenn ich *usw.* die Gelegenheit dazu hätte; **give sth. a ~ to do sth.** einer Sache *(Dat.)* Gelegenheit geben, etw. zu tun; **get a/the ~ to do sth.** eine/die Gelegenheit haben, etw. zu tun; **this is my big ~:** das ist die Chance für mich; **now's your ~!** das ist deine Chance!; **on the [off] ~ of doing sth./that ...:** in der vagen Hoffnung, etw. zu tun/daß ...; **stand a ~ of doing sth.** die Chance haben, etw. zu tun; **d)** *in sing. or pl. (probability)* **have a good/fair ~ of doing sth.** gute Aussichten haben, etw. zu tun; **[is there] any ~ of your attending?** besteht eine Chance, daß Sie kommen können?; **there is every/not the slightest ~ that...:** es ist sehr gut möglich/es besteht keine Möglichkeit, daß ...; **the ~s are that ...:** es ist wahrscheinlich, daß ...; **e)** *(risk)* **take one's ~:** es darauf ankommen lassen; **take a ~/~s** ein Risiko/Risiken eingehen; es riskieren; **take a ~ on sth.** es bei etw. auf einen Versuch ankommen lassen. **2.** *v. t.* riskieren; ~ **it** es riskieren *od.* darauf ankommen lassen; **we'll have to ~ that happening** wir müssen es riskieren; ~ **one's arm** *(Brit. coll.)* es riskieren

chancel ['tʃɑ:nsl] *n. (Eccl.)* Altarraum, *der; (choir)* Chor, *der*

chancellor ['tʃɑ:nsələ(r)] *n. (Polit., Law, Univ.)* Kanzler, *der;* **C~ of the Exchequer** *(Brit.)* Schatzkanzler, *der*

chancery ['tʃɑ:nsərɪ] *n.* **a) C~** *(Brit. Law)* Gerichtshof des Lordkanzlers; **b)** *(Brit. Diplom.)* ≈ Botschaft, *die*/Gesandtschaft, *die*

chancy ['tʃɑ:nsɪ] *adj.* riskant; gewagt

chandelier [ʃændə'lɪə(r)] *n.* Kronleuchter, *der*

change ['tʃeɪndʒ] **1.** *n.* **a)** *(of name, address, life-style, outlook, condition, etc.)* Änderung, *die; (of job, surroundings, government, etc.)* Wechsel, *der;* **there has been a ~ of plan** der Plan ist geändert worden; **a ~ in the weather** ein Witterungs- *od.* Wetterumschlag; **a ~ for the better/worse** eine Verbesserung/Verschlechterung; **a ~ of air would do her good** eine Luftveränderung täte ihr gut; **the ~ [of life]** die Wechseljahre; **a ~ of heart** ein Sinneswandel; **b)** *no pl., no art. (process of changing)* Veränderung, *die;* **be for/against ~:** für/gegen eine Veränderung sein; **c)** *(for the sake of variety)* Abwechslung, *die;* **[just] for a ~:** [nur so] zur Abwechslung; **make a ~** *(be different)* mal etwas anderes sein (**from** als); **a ~ is as good as a rest** *(prov.)* Abwechslung wirkt Wunder; **d)** *no pl., no indef. art. (money)* Wechselgeld, *das;* **[loose** *or* **small]** **~:** Kleingeld, *das;* **give ~,** *(Amer.)* **make ~:** herausgeben; **give sb. 40p in ~:** jmdm. 40p [Wechselgeld] herausgeben; **can you give me ~ for 50p?** können Sie mir 50p wechseln?; **I haven't got ~ for a pound** ich kann auf ein Pfund nicht herausgeben; **[you can] keep the ~:** behalten Sie den Rest; [es] stimmt so; **e) a ~ [of clothes]** *(fresh clothes)* Kleidung zum Wechseln. **2.** *v. t.* **a)** *(switch)* wechseln; auswechseln ‹*Glühbirne, Batterie, Zündkerzen*›; ~ **one's clothes** sich umziehen; ~ **one's address/name** seine Anschrift/seinen Namen ändern; ~ **trains/buses** umsteigen; ~ **schools/one's doctor** die Schule/den Arzt wechseln; **he's always changing jobs** er wechselt ständig den Job; ~ **the bed** das Bett frisch beziehen; ~ **the baby** das Baby [frisch] wickeln *od.* trockenlegen; **b)** *(transform)* verwandeln; *(alter)* ändern; ~ **sth./sb. into sth./sb.** etw./jmdn. in etw./jmdn. verwandeln; ~ **direction** die Richtung ändern; **c)** *(exchange)* eintauschen; ~ **seats** die Plätze tauschen; ~ **seats with sb.** mit jmdm. den Platz tauschen; **take sth. back to the shop and ~ it for sth.** etw. [zum Laden zurückbringen und] gegen etw. umtauschen; **d)** *(in currency or denomination)* wechseln ‹*Geld*›; ~ **one's money into Deutschmarks** sein Geld in DM umtauschen. **3.** *v. i.* **a)** *(alter)* sich ändern; ‹*Person, Land:*› sich verändern; ‹*Wetter:*› umschlagen, sich ändern; **she'll never ~!** sie wird sich nie ändern!; **wait for the lights to ~:** warten, daß es grün/rot wird; ~ **for the better** sich verbessern; **conditions ~d for the worse** die Lage verschlechterte sich; **b)** *(into something else)* sich verwandeln; **the wind ~s from east to west** der Wind dreht von Ost nach West; **c)** *(exchange)* tauschen; ~ **with sb.** mit jmdm. tauschen; **d)** *(put on other clothes)* sich umziehen; ~ **out of/into sth.** etw. ausziehen/anziehen; **e)** *(take different train or bus)* umsteigen

~ **'over** *v.i.* **a)** ~ **over from sth. to sth.** von etw. zu etw. übergehen; **they ~d over from one system to another** sie stellten das System auf ein anderes um; **b)** *(exchange places)* die Plätze wechseln; *(Sport)* [die Seiten] wechseln

~ **'round** *v.t.* umstellen ‹*Möbel, Tagesordnung[spunkte]*›; umräumen ‹*Zimmer*›

changeable ['tʃeɪndʒəbl] *adj.* veränderlich; unbeständig ‹*Charakter, Wetter*›; wankelmütig ‹*Person*›; wechselhaft, veränderlich ‹*Wetter*›; wechselnd ‹*Wind, Stimmung*›

changeless ['tʃeɪndʒlɪs] *adj.* unveränderlich

changing ['tʃeɪndʒɪŋ] *adj.* wechselnd; sich ändernd

'changing-room *n.* *(Brit.)* **a)** *(Sport)* Umkleideraum, *der;* **b)** *(in shop)* Umkleidekabine, *die*

channel ['tʃænl] **1.** *n.* **a)** Kanal, *der; (gutter)* Rinnstein, *der; (navigable part of waterway)* Fahrrinne, *die;* **the C~** *(Brit.)* der [Ärmel]kanal; **b)** *(fig.)* Kanal, *der;* **your application will go through the usual ~s** Ihre Bewerbung wird auf dem üblichen Weg weitergeleitet; **c)** *(Telev., Radio)* Kanal, *der;* **d)** *(on recording tape etc.)* Spur, *die;* **e)** *(groove)* Rille, *die.* **2.** *v.t., (Brit.)* **-ll-** *(fig.: guide, direct)* lenken, richten (**into** auf + *Akk.*)

Channel: ~ **Islands** *pr. n. pl.* Kanalinseln *Pl.;* ~ **'Tunnel** *n.* [Ärmel]kanaltunnel, *der*

chant [tʃɑ:nt] **1.** *v.t.* **a)** *(Eccl.)* singen; **b)** *(utter rhythmically)* skandieren. **2.** *v.i.* **a)** *(Eccl.)* singen; **b)** *(utter slogans etc.)* Sprechchöre anstimmen. **3.** *n.* **a)** *(Eccl., Mus.)* Gesang, *der;* **b)** *(sing-song)* Singsang, *der*

chaos ['keɪɒs] *n., no indef. art.* Chaos, *das*

chaotic [keɪ'ɒtɪk] *adj.* chaotisch

[1]**chap** [tʃæp] *n.* *(Brit. coll.)* Bursche, *der;* Kerl, *der;* **old ~:** alter Knabe *(ugs.)*

[2]**chap** **1.** *v.t.*, **-pp-** aufplatzen lassen. **2.** *n. usu. in pl.* Riß, *der*

chapel ['tʃæpl] *n.* Kapelle, *die*

chaperon ['ʃæpərəʊn] **1.** *n.* Anstandsdame, *die; (joc.)* Anstandswauwau, *der (ugs. scherzh.).* **2.** *v.t.* beaufsichtigen; *(escort)* begleiten

chaplain ['tʃæplɪn] *n.* Kaplan, *der*

chapter ['tʃæptə(r)] *n.* **a)** *(of book)* Kapitel, *das;* **give ~ and verse for sth.** etw. hieb- und stichfest belegen; **b)** *(fig.)* ~ **in** *or* **of sb.'s life** Abschnitt in jmds. Leben; **c)** *(Eccl.)* Kapitel, *das*

[1]**char** [tʃɑ:(r)] *v.t. & i.*, **-rr-** *(burn)* verkohlen

[2]**char** *n.* *(Brit.: cleaner)* Putzfrau, *die*

character ['kærɪktə(r)] *n.* **a)** *(mental or moral qualities, integrity)* Charakter, *der;* **be of good ~:** ein guter Mensch sein; einen guten Charakter haben; **a woman of ~:** eine Frau mit Charakter; **strength of ~:** Charakterstärke, *die;* **b)** *no pl. (individuality, style)* Charakter, *der;* **the town has a ~ all of its own** die Stadt hat einen ganz eigenen Charakter; **have no ~:** charakterlos *od.* ohne Charakter sein; **c)** *(in novel etc.)* Charakter, *der; (part played by sb.)* Rolle, *die;* **be in/out of ~** *(fig.)* typisch/untypisch sein; **his behaviour was quite out of ~** *(fig.)* sein Betragen war ganz und gar untypisch für ihn; **d)** *(coll.: extraordinary person)* Original, *das;* **be [quite] a ~/a real ~:** ein [echtes/richtiges] Original sein; **e)** *(coll.: individual)* Mensch, *der; (derog.)* Individuum, *das*

character: ~ **actor** *n.* Chargenspieler, *der;* ~ **actress** *n.* Chargenspielerin, *die*

characterisation, characterise *see* **characteriz-**

characteristic [kærɪktə'rɪstɪk] **1.** *adj.* charakteristisch (**of** für). **2.** *n.* charakteristisches Merkmal; **one of the main ~s** eines der charakteristischsten Merkmale

characteristically [kærɪktə'rɪstɪkəlɪ] *adv.* in charakteristischer Weise

characterization [kærɪktəraɪ'zeɪʃn] *n.* Charakterisierung, *die*

characterize ['kærɪktəraɪz] *v.t.* charakterisieren

characterless ['kærɪktəlɪs] *adj.* nichtssagend

charade [ʃə'rɑ:d] *n.* Scharade, *die; (fig.)* Farce, *die*

charcoal ['tʃɑ:kəʊl] *n.* **a)** Holzkohle, *die; (for drawing)* Kohle, *die;* **b)** *see* **charcoal grey**

charcoal: ~ **'grey** *n.* [Kohlen]grau, *das;* ~ **pencil** *n.* Kohlestift, *der*

charge [tʃɑ:dʒ] **1.** *n.* **a)** *(price)* Preis, *der; (payable to telephone company, bank, authorities, etc., for services)* Gebühr, *die;* **is there a ~ for it?** kostet das etwas?; **b)** *(care)* Verantwortung, *die; (task)* Auftrag, *der; (person entrusted)* Schützling, *der;* **be in ~ of a child** ein Kind betreuen; **the patients in** *or* **under her ~:** die ihr anvertrauten Patienten; **be under sb.'s ~:** unter jmds. Obhut stehen; **the officer/teacher in ~:** der diensthabende Offizier/der verantwortliche Lehrer; **be in ~:** die Verantwortung haben; **be in ~ of sth.** für etw. die Verantwortung haben; *(be the leader)* etw. leiten; **put sb. in ~ of sth.** jmdn. mit der Verantwortung für etw. betrauen; **take ~:** die Verantwortung übernehmen; **take ~ of sth.** *(become responsible for)* etw. übernehmen; **c)** *(Law: accusation)* Anklage, *die;* **make a ~ against sb.** jmdn. beschuldigen; **bring a ~ of sth. against sb.** jmdn. wegen etw. beschuldigen/verklagen; **press ~s** Anzeige erstatten; **on a ~ of** wegen; **d)** *(allegation)* Beschuldigung, *die;* **e)** *(attack)* Angriff, *der;* Attacke, *die;* **f)** *(of explosives etc.)* Ladung, *die;* **g)** *(of electricity)* Ladung, *die;* **put the battery on ~:** die Batterie an das Ladegerät anschließen. **2.** *v.t.* **a)** *(demand payment of or from)* ~ **sb. sth.**, ~ **sth. to sb.** jmdm. etw. berechnen; ~ **sb. £1 for sth.** jmdm. ein Pfund für etw. berechnen ; ~ **sth. [up] to sb.'s account** jmds. Konto mit etw. belasten; **b)** *(Law: accuse)* anklagen; ~ **sb. with sth.** jmdn. wegen etw. anklagen; **c)** *(formal: entrust)* ~ **sb. with sth.** jmdn. mit etw. betrauen; **d)** *(load)* laden ‹*Gewehr*›; **e)** *(Electr.)* laden; [auf]laden ‹*Batterie*›; **~d with emotion** *(fig.)* voller Gefühl; **f)** *(rush at)* angreifen; **g)** *(formal: command)* befehlen; ~ **sb. to do sth.** jmdm. befehlen, etw. zu tun. **3.** *v.i.* **a)** *(attack)* angreifen; **~!** Angriff!; Attacke!; ~ **at sb./sth.** jmdn./etw. angreifen; **he ~d into a wall** *(fig.)* er krachte gegen eine Mauer; **b)** *(coll.: hurry)* sausen

chargeable ['tʃɑ:dʒəbl] *adj.* **be ~ to sb.** auf jmds. Kosten gehen

'charge card *n.* Kreditkarte, *die*

chargé d'affaires [ʃɑ:ʒeɪ dæ'feə(r)] *n., pl.* **chargés d'affaires** [ʃɑ:ʒeɪ dæ'feə(r)] Chargé d'affaires, *der;* [diplomatischer] Geschäftsträger

charger ['tʃɑ:dʒə(r)] *n.* *(Electr.)* [Batterie]ladegerät, *das*

chariot ['tʃærɪət] *n.* *(Hist.)* [zweirädriger] Streitwagen

charisma [kə'rɪzmə] *n., pl.* **~ta** [kə'rɪzmətə] Charisma, *das*

charismatic [kærɪz'mætɪk] *adj.* charismatisch

charitable ['tʃærɪtəbl] *adj.* **a)** *(generous, lenient)* großzügig; **b)** *(of or for charity)* karitativ

charity ['tʃærɪtɪ] *n.* **a)** *(Christian love)* Nächstenliebe, *die;* **b)** *(kindness)* Güte, *die;* **c)** *(generosity in giving)* Wohltätigkeit, *die;* **live on ~/accept ~:** von Almosen leben/Almosen annehmen; **~ begins at home** *(prov.)* man muß zuerst an die eigenen Leute denken; **give money to ~:** Geld für wohltätige Zwecke spenden; **d)** *(institution)* wohltätige Organisation

charlady ['tʃɑ:leɪdɪ] *(Brit.) see* **²char**

charlatan ['ʃɑ:lətən] *n.* Scharlatan, *der*

charm [tʃɑ:m] **1.** *n.* **a)** *(act)* Zauber, *der; (thing)* Zaubermittel, *das; (words)* Zauberspruch, *der;* Zauberformel, *die;* **lucky ~:** Glücksbringer, *der;* **work like a ~:** Wunder wirken; **b)** *(talisman)* Talisman, *der;* **c)** *(trinket)* Anhänger, *der;* **d)** *(attractiveness)* Reiz, *der; (of person)* Charme, *der;* **turn on the ~** *(coll.)* auf charmant machen *(ugs.).* **2.** *v. t.* **a)** *(captivate)* bezaubern; **b)** *(by magic)* verzaubern; **lead a ~ed life** unter einem Glücksstern geboren sein

charming ['tʃɑ:mɪŋ] *adj.* bezaubernd; **~!** *(iron.)* [wie] charmant! *(iron.)*

chart [tʃɑ:t] **1.** *n.* **a)** *(map)* Karte, *die;* **weather ~:** Wetterkarte, *die;* **b)** *(graph etc.)* Schaubild, *das; (diagram)* Diagramm, *das;* **c)** *(tabulated information)* Tabelle, *die;* **the ~s** *(of pop records)* die Hitliste. **2.** *v. t.* graphisch darstellen; *(map)* kartographisch erfassen; *(fig.: describe)* schildern

charter ['tʃɑ:tə(r)] **1.** *n.* Charta, *die; (of foundation also)* Gründungs- *od.* Stiftungsurkunde, *die; (fig.)* Freibrief, *der.* **2.** *v. t. (Transport)* chartern ‹*Schiff, Flugzeug*›; mieten ‹*Bus*›

chartered ac'countant *n.* *(Brit.)* Wirtschaftsprüfer, *der/-prüferin, die*

'charter flight *n.* Charterflug, *der*

charwoman ['tʃɑ:wʊmən] *see* **²char**

chary ['tʃeərɪ] *adj.* **a)** *(sparing, ungenerous)* zurückhaltend (**of** mit); **b)** *(cautious)* vorsichtig; **be ~ of doing sth.** darauf bedacht sein, etw. nicht zu tun

chase [tʃeɪs] **1.** *n.* Verfolgungsjagd, *die;* **car ~:** Verfolgungsjagd im Auto; **give ~ [to the thief]** [dem Dieb] hinterherjagen. **2.** *v. t. (pursue)* jagen; **~ sth.** *(fig.)* einer Sache *(Dat.)* nachjagen. **3.** *v. i.* **~ after sb./sth.** hinter jmdm./etw. herjagen

~ a'bout, ~ a'round *v. i.* herumrasen *(ugs.)*

~ a'way *v. t.* wegjagen

~ round *see* **~ around**

~ 'up *(coll.) v. t.* ausfindig machen

chasm ['kæzm] *n.* *(lit. or fig.)* Kluft, *die*

chassis ['ʃæsɪ] *n., pl. same* ['ʃæsɪz] *(Motor Veh.)* Chassis, *das;* Fahrgestell, *das*

chaste [tʃeɪst] *adj.* keusch

chasten ['tʃeɪsn] *v. t.* **a)** züchtigen *(geh.);* strafen; **b)** *(fig.)* dämpfen ‹*Stimmung*›; demütigen ‹*Person*›

chastening ['tʃeɪsənɪŋ] *adj.* ernüchternd

chastise [tʃæ'staɪz] *v. t.* **a)** *(punish)* züchtigen *(geh.);* bestrafen; **b)** *(thrash)* züchtigen *(geh.)*

chastisement [tʃæ'staɪzmənt] *n.* Züchtigung, *die (geh.);* Strafe, *die*

chastity ['tʃæstɪtɪ] *n., no pl.* Keuschheit, *die*

chat [tʃæt] **1.** *n.* **a)** Schwätzchen, *das;* **have a ~ about sth.** sich über etw. *(Akk.)* unterhalten; **b)** *no pl., no indef. art. (~ting)* Geplauder, *das.* **2.** *v. i.,* **-tt-** plaudern; **~ with** *or* **to sb. about sth.** sich mit jmdm. über etw. *(Akk.)* unterhalten

~ 'up *v. t. (Brit. coll.)* sich heranmachen an (+ *Akk.*) *(ugs.); (amorously)* anmachen *(ugs.)*

'chat show *n.* Talk-Show, *die*

chattel ['tʃætl] *n., usu. in pl.* **~[s]** bewegliche Habe *(geh.)*

chatter ['tʃætə(r)] **1.** *v. i.* **a)** schwatzen; **b)** *(rattle)* ‹*Zähne:*› klappern; **his teeth ~ed** er klapperte mit den Zähnen. **2.** *n.* **a)** Schwatzen, *das;* **b)** *(of teeth)* Klappern, *das*

chatterbox ['tʃætəbɒks] *n.* Quasselstrippe, *die (ugs.); (child)* Plappermäulchen, *das*

chatty ['tʃætɪ] *adj.* gesprächig

chauffeur ['ʃəʊfə(r), ʃəʊ'fɜ:(r)] **1.** *n.* Fahrer, *der;* Chauffeur, *der;* **~-driven car** Wagen mit Chauffeur. **2.** *v. t.* fahren

chauvinism ['ʃəʊvɪnɪzm] *n., no pl.* Chauvinismus, *der;* **male ~:** männlicher Chauvinismus

chauvinist ['ʃəʊvɪnɪst] *n.* Chauvinist, *der/*Chauvinistin, *die;* **male ~/[male] ~ pig** Chauvinist, *der/*Chauvinistenschwein, *das*

chauvinistic [ʃəʊvɪ'nɪstɪk] *adj.* chauvinistisch

cheap [tʃi:p] **1.** *adj.* **a)** billig; *(at reduced rate)* verbilligt; **be ~ and nasty** billiger Ramsch sein; **be ~ at the price** sehr preiswert sein; *(fig.)* es wert sein; **on the ~** *(coll.)* billig; **b)** *(worthless)* billig ‹*Aussehen*›; gemein ‹*Lügner*›; schäbig ‹*Verhalten, Betragen*›. **2.** *adv.* billig; **be going ~:** besonders günstig sein *(ugs.)*

cheapen ['tʃi:pn] *v. t. (fig.)* herabsetzen; **~ oneself** sich [selbst] herabsetzen

cheaply ['tʃi:plɪ] *adv. see* **cheap 1:** billig; gemein; schäbig

cheat [tʃi:t] **1.** *n.* **a)** *(person)* Schwindler, *der/*Schwindlerin, *die;* **b)** *(act)* Schwindel, *der;* **that's a ~!** das ist Betrug! **2.** *v. t.* betrügen; **~ sb./sth. [out] of sth.** jmdn./etw. um etw. betrügen. **3.** *v. i.* betrügen; *(Sch.)* täuschen; **~ at cards** beim Kartenspielen mogeln

¹check [tʃek] **1.** *n.* **a)** *(stoppage, thing that restrains)* Hindernis, *das; (restraint)* Kontrolle, *die;* **[hold** *or* **keep sth.] in ~:** [etw.] unter Kontrolle [halten]; **act as a ~ upon sth.** etw. unter Kontrolle halten; **b)** *(for accuracy)* Kontrolle, *die;* **make a ~ on sth./sb.** etw./jmdn. überprüfen *od.* kontrollieren; **keep a ~ on** überprüfen; kontrollieren; überwachen ‹*Verdächtigen*›; **c)** *(Amer.: bill)* Rechnung, *die;* **d)** *(Amer.) see* **cheque; e)** *(Chess)* Schach, *das;* **be in ~:** im Schach stehen. **2.** *v. t.* **a)** *(restrain)* unter Kontrolle halten; unterdrücken ‹*Ärger, Lachen*›; **~ oneself** sich beherrschen; **b)** *(examine accuracy of)* nachprüfen; nachsehen ‹*Hausaufgaben*›; kontrollieren ‹*Fahrkarte*›; *(Amer.: mark with tick)* abhaken. **3.** *v. i.* **~ on sth.** etw. überprüfen; **~ with sb.** bei jmdm. nachfragen. **4.** *int. (Chess)* Schach

~ 'in 1. *v. t. (at airport)* **~ in one's luggage** sein Gepäck abfertigen lassen *od.* einchecken. **2.** *v. i. (arrive at hotel)* ankommen; *(report one's arrival)* sich melden; *(at airport)* einchecken

~ 'off *v. t.* abhaken

~ 'out 1. *v. t.* überprüfen. **2.** *v. i.* **~ out [of one's hotel]** abreisen

~ 'over *v. t.* durchsehen

~ 'up *v. i.* überprüfen; **~ up on sb./sth.** jmdn./etw. überprüfen *od.* kontrollieren

²check *n. (pattern)* Karo, *das*

checked [tʃekt] *adj. (patterned)* kariert

checkerboard ['tʃekəbɔ:d] *n.* *(Amer.)* Schachbrett, *das*

checkers ['tʃekəz] *n., no pl. (Amer.) see* **draughts**
'**check-in** *n.* Abfertigung, *die; attrib.* Abfertigungs-
'**checking account** *n. (Amer.)* Girokonto, *das*
check: ~-**list** *n.* Checkliste, *die;* ~**mate 1.** *n.* [Schach]matt, *das;* **2.** *int.* [schach]matt; ~-**out** *n.* Abreise, *die;* ~-**out [desk** *or* **point** *or* **counter]** Kasse, *die;* ~-**point** *n.* Kontrollpunkt, *der;* ~-**up** *n. (Med.)* Untersuchung, *die*
Cheddar ['tʃedə(r)] *n.* Cheddar[käse], *der*
cheek [tʃi:k] *n.* **a)** Backe, *die;* Wange, *die (geh.);* **turn the other** ~ *(fig.)* die andere Wange darbieten; **b)** *(impertinence)* Frechheit, *die;* **have the** ~ **to do sth.** die Frechheit besitzen, etw. zu tun
cheekily ['tʃi:kılı] *adv.*, **cheeky** ['tʃi:kı] *adj.* frech
cheep [tʃi:p] **1.** *v.i.* piep[s]en. **2.** *n.* Piep[s]en, *das*
cheer [tʃıə(r)] **1.** *n.* **a)** *(applause)* Beifallsruf, *der;* **give three** ~**s for sb.** jmdn. [dreimal] hochleben lassen; **b)** *in pl. (Brit. coll.: as a toast)* prost; **c)** *in pl. (Brit. coll.: thank you)* danke; **d)** *in pl. (Brit. coll.: goodbye)* tschüs *(ugs.).* **2.** *v.t.* **a)** *(applaud)* ~ **sth./sb.** etw. bejubeln/jmdm. zujubeln; **b)** *(gladden)* aufmuntern; aufheitern. **3.** *v.i.* jubeln
~ '**on** *v.t.* anfeuern ⟨*Sportler, Wettkämpfer*⟩
~ '**up 1.** *v.t.* aufheitern. **2.** *v.i.* bessere Laune bekommen; ~ **up!** Kopf hoch!
cheerful ['tʃıəfl] *adj. (in good spirits)* fröhlich; gutgelaunt; *(bright, pleasant)* heiter; erfreulich ⟨*Aussichten*⟩; lustig ⟨*Feuer*⟩
cheerfully ['tʃıəfəlı] *adv.* vergnügt; **the fire blazed** ~: das Feuer brannte lustig
cheerily ['tʃıərılı] *adv.* fröhlich
cheering ['tʃıərıŋ] **1.** *adj.* **a)** *(gladdening)* fröhlich stimmend; **b)** *(applauding)* jubelnd. **2.** *n.* Jubeln, *das*
cheerio [tʃıərı'əʊ] *int. (Brit. coll.)* tschüs *(ugs.)*
cheerless ['tʃıəlıs] *adj.* freudlos; düster ⟨*Aussichten*⟩
cheery ['tʃıərı] *adj.* fröhlich
cheese [tʃi:z] *n.* **a)** *(food)* Käse, *der;* ~s Käsesorten; **b)** *(whole)* Käselaib, *der*
cheese: ~**board** *n.* Käseplatte, *die;* ~**cake** *n.* Käsetorte, *die;* ~**cloth** *n.* [indischer] Baumwollstoff
cheesed off [tʃi:zd 'ɒf] *adj. (Brit. sl.)* angeödet *(ugs.)*

cheetah ['tʃi:tə] *n. (Zool.)* Gepard, *der*
chef [ʃef] *n.* Küchenchef, *der; (as profession)* Koch, *der*
chemical ['kemıkl] **1.** *adj.* chemisch. **2.** *n.* Chemikalie, *die*
chemical 'warfare *n.* chemische Krieg[s]führung
chemist ['kemıst] *n.* **a)** *(scientist)* Chemiker, *der*/Chemikerin, *die;* **b)** *(Brit.: pharmacist)* Drogist, *der*/Drogistin, *die;* ~**'s [shop]** Drogerie, *die*
chemistry ['kemıstrı] *n., no pl.* **a)** *no indef. art.* Chemie, *die;* **b)** *(fig.)* unerklärliche Wirkungskraft
chemistry: ~ **laboratory** *n.* Chemiesaal, *der;* ~ **set** *n.* Chemiebaukasten, *der*
cheque [tʃek] *n.* Scheck, *der;* **write a** ~: einen Scheck ausfüllen; **pay by** ~: mit [einem] Scheck bezahlen
cheque: ~-**book** *n.* Scheckbuch, *das;* ~ **card** *n.* Scheckkarte, *die*
chequered ['tʃekəd] *adj.* **a)** kariert; **b)** *(fig.)* bewegt ⟨*Geschichte, Leben, Laufbahn*⟩
cherish ['tʃerıʃ] *v.t.* **a)** *(value and keep)* hegen ⟨*Hoffnung, Gefühl*⟩; in Ehren halten ⟨*[Erinnerungs]gegenstand*⟩; **b)** *(foster)* ~ **sb.** [liebevoll] für jmdn. sorgen
cherry ['tʃerı] **1.** *n.* Kirsche, *die.* **2.** *adj.* kirschrot
cherry: ~ **blossom** *n.* Kirschblüte, *die;* ~ '**brandy** *n.* Cherry Brandy, *der;* ≈ Kirschlikör, *der;* ~-**stone** *n.* Kirschkern, *der*
cherub ['tʃerəb] *n., pl.* ~**s** *(Art)* Putte, *die; (child)* Engelchen, *das*
chess [tʃes] *n., no pl., no indef. art.* das Schach[spiel]; **play** ~: Schach spielen
chess: ~-**board** *n.* Schachbrett, *das;* ~-**man** *n.* Schachfigur, *die;* ~-**player** *n.* Schachspieler, *der*/-spielerin, *die*
chest [tʃest] *n.* **a)** Kiste, *die; (for clothes or money)* Truhe, die; **b)** *(part of body)* Brust, *die;* **get sth. off one's** ~ *(fig. coll.)* sich *(Dat.)* etw. von der Seele reden
chest: ~-**expander** *n.* Expander, *der;* ~-**measurement** *n.* Brustumfang, *der*
chestnut ['tʃesnʌt] **1.** *n.* **a)** Kastanie, *die;* **b)** *(colour)* Kastanienbraun, *das;* **c)** *(stale story or topic)* **[old]** ~: alte *od.* olle Kamelle *(ugs.);* **d)** *(horse)* Fuchs, *der.* **2.** *adj. (colour)* ~**[-brown]** kastanienbraun
'**chestnut-tree** *n.* Kastanienbaum, *der*

chest of 'drawers Kommode, *die*
chew [tʃu:] **1.** *v.t.* kauen; ~ **one's finger-nails** an den [Finger]nägeln kauen; *see also* **bite off; cud. 2.** *v.i.* kauen **(on** auf + *Dat.*); ~ **on** *or* **over sth.** *(fig.)* sich *(Dat.)* etw. durch den Kopf gehen lassen. **3.** *n.* Kauen, *das*
'**chewing-gum** *n.* Kaugummi, *der od. das*
chewy ['tʃu:ı] *adj.* zäh ⟨*Fleisch, Bonbon*⟩
chic [ʃi:k] **1.** *adj.* schick; elegant. **2.** *n.* Schick, *der*
chicane [ʃı'keın] *n. (Sport)* Schikane, *die*
chick [tʃık] *n.* **a)** Küken, *das;* **b)** *(sl.: young woman)* Biene, *die (ugs.)*
chicken ['tʃıkın] **1.** *n.* **a)** Huhn, *das; (grilled, roasted)* Hähnchen, *das;* **don't count your** ~**s [before they are hatched]** *(prov.)* man soll den Pelz nicht verkaufen, ehe man den Bären erlegt hat; **b)** **she's no** ~ *(coll.: is no longer young)* sie ist nicht mehr die Jüngste; **c)** *(coll.: coward)* Angsthase, *der.* **2.** *adj. (coll.)* feig[e]. **3.** *v.i.* ~ **out** *(sl.)* kneifen
chicken: ~-**and-'egg** *adj.* Huhn-Ei-⟨*Frage*⟩; ~-**feed** *n.* **a)** Hühnerfutter, *das;* **b)** *(fig. coll.)* eine lächerliche Summe; ~-**pox** *n. (Med.)* Windpocken *Pl.;* ~ '**soup** *n.* Hühnersuppe, *die*
'**chick-pea** *n.* Kichererbse, *die*
chicory ['tʃıkərı] *n. (plant)* Chicorée, *der od. die; (for coffee)* Zichorie, *die*
chief [tʃi:f] **1.** *n.* **a)** *(of state, town, clan)* Oberhaupt, *das; (of tribe)* Häuptling, *der;* **b)** *(of department)* Leiter, *der; (coll.: one's superior, boss)* Chef, *der;* Boss, *der;* ~ **of police** Polizeipräsident, *der;* ~ **of staff** *(of a service)* Generalstabschef, *der; (commander)* Stabschef, *der.* **2.** *adj., usu. attrib.* **a)** Ober-; ~ **engineer** erster Maschinist *(Seew.);* **[Lord] C**~ **Justice** *(Brit.)* [Lord] Oberrichter, *der;* **b)** *(first in importance, influence, etc.)* Haupt-; ~ **reason/aim** Hauptgrund, *der*/-ziel, *das*
chiefly ['tʃi:flı] *adv.* hauptsächlich; vor allem
chieftain ['tʃi:ftən] *n. (of Highland clan)* Oberhaupt, *das; (of tribe)* Stammesführer, *der*
chiffon ['ʃıfɒn] **1.** *n. (Textiles)* Chiffon, *der.* **2.** *adj.* Chiffon-
chihuahua [tʃı'wɑ:wə] *n.* Chihuahua, *der*
chilblain ['tʃılbleın] *n.* Frostbeule, *die*

child [tʃaɪld] *n., pl.* ~**ren** ['tʃɪldrən] Kind, *das;* **when I was a** ~: als ich klein war; **[be] with** ~ *(dated)* schwanger [sein]

child: ~**birth** *n.* Geburt, *die;* ~**care** *n.* **a)** Betreuung von Kindern; **b)** *(social services department)* Kinderfürsorge, *die*

childhood ['tʃaɪldhʊd] *n.* Kindheit, *die;* **in** ~: als Kind; **from** *or* **since** ~: schon als Kind; **be in one's second** ~: an Altersschwachsinn leiden

childish ['tʃaɪldɪʃ] *adj.,* **childishly** ['tʃaɪldɪʃlɪ] *adv.* kindlich; *(derog.)* kindisch

childishness ['tʃaɪldɪʃnɪs] *n., no pl. (derog.: behaviour)* kindisches Benehmen

childless ['tʃaɪldlɪs] *adj.* kinderlos

childlike ['tʃaɪldlaɪk] *adj.* kindlich

child: ~**-minder** *n. (Brit.)* Tagesmutter, *die;* ~ '**prodigy** *n.* Wunderkind, *das;* ~**-proof** *adj.* kindersicher

children *pl. of* **child**

child: ~**'s play** *n., no pl. (fig.)* **it's** ~**'s play!** es ist ein Kinderspiel; ~ '**welfare** *n.* Kinderfürsorge, *die*

Chile ['tʃɪlɪ] *pr. n.* Chile *(das)*

Chilean ['tʃɪlɪən] **1.** *adj.* chilenisch; **sb. is** ~: jmd. ist Chilene/Chilenin. **2.** *n.* Chilene, *der*/Chilenin, *die*

chili *see* **chilli**

chill [tʃɪl] **1.** *n.* **a)** *(cold sensation)* Frösteln, *das; (feverish shivering)* Schüttelfrost, *der; (illness)* Erkältung, *die;* **catch a** ~: sich verkühlen *od.* erkälten; **b)** *(unpleasant coldness)* Kühle, *die; (fig.)* Abkühlung, *die;* **take the** ~ **off [sth.]** etw. leicht erwärmen; **there's a** ~ **in the air** es ist ziemlich kühl [draußen]. **2.** *v.t.* kühlen. **3.** *adj. (literary; lit. or fig.)* kühl

chilli ['tʃɪlɪ] *n., pl.* ~**es** Chili, *der;* ~ **con carne** [tʃɪlɪ kɒn 'kɑ:nɪ] *(Gastr.)* Chili con carne

chilling ['tʃɪlɪŋ] *adj. (fig.)* ernüchternd; frostig ⟨*Art, Worte, Blick*⟩

chilly ['tʃɪlɪ] *adj. (lit. or fig.)* kühl; **I am rather** ~: mir ist ziemlich kühl

chime [tʃaɪm] **1.** *n.* **a)** Geläute, *das;* **b)** *(set of bells)* Glockenspiel, *das.* **2.** *v.i.* läuten; ⟨*Turmuhr:*⟩ schlagen; **chiming clock** Schlaguhr, *die.* **3.** *v.t.* erklingen lassen ⟨*Melodie*⟩; schlagen ⟨*Stunde, Mitternacht*⟩

~ '**in** *v.i.* **a)** *(Mus.)* einstimmen; *(fig.)* übereinstimmen (**with** mit); **b)** *(interject remark)* sich [in die Unterhaltung] einmischen

chimney ['tʃɪmnɪ] *n. (of house, factory, etc.)* Schornstein, *der; (of house also)* Kamin, *der (bes. südd.);* **the smoke goes up the** ~: der Rauch zieht durch den Kaminschacht ab; **come down the** ~: durch den Schornstein kommen; **smoke like a** ~ *(fig.)* wie ein Schlot rauchen

chimney: ~**-breast** *n.* Kaminmantel, *der;* ~**-pot** *n.* ≈ Schornsteinkopf, *der;* ~**-sweep** *n.* Schornsteinfeger, *der*

chimp [tʃɪmp] *(coll.),* **chimpanzee** [tʃɪmpən'zi:] *ns.* Schimpanse, *der*

chin [tʃɪn] *n.* Kinn, *das;* **keep one's** ~ **up** *(fig.)* den Kopf nicht hängen lassen; ~ **up!** Kopf hoch!; **take it on the** ~ *(endure sth. courageously)* es mit Fassung tragen

China ['tʃaɪnə] *pr. n.* China *(das)*

china *n.* Porzellan, *das; (crockery)* Geschirr, *das*

china 'clay *n.* Porzellanerde, *die*

China: ~**man** ['tʃaɪnəmən] *n., pl.* ~**men** ['tʃaɪnəmən] *(derog.)* Chinese, *der;* ~ '**tea** *n.* Chinatee, *der*

chinchilla [tʃɪn'tʃɪlə] *n. (Zool.)* Chinchilla, *die*

Chinese [tʃaɪ'ni:z] **1.** *adj.* chinesisch; **sb. is** ~: jmd. ist Chinese/Chinesin. **2.** *n.* **a)** *pl. same (person)* Chinese, *der*/Chinesin, *die;* **b)** *(language)* Chinesisch, *das; see also* **English 2a**

Chinese 'lantern *n.* Lampion, *der*

[1]**chink** [tʃɪŋk] *n.* **a)** Spalt, *der;* **a** ~ **in sb.'s armour** *(fig.)* jmds. schwache Stelle; **b)** **a** ~ **of light** ein Lichtspalt

[2]**chink** **1.** *n. (sound) see* [1]**clink 1.** **2.** *v.i. & t. see* [1]**clink 2, 3**

'**chin-strap** *n.* Kinnriemen, *der*

chintz [tʃɪnts] *n.* Chintz, *der*

'**chin-wag** *(coll.)* **1.** *n.* Schwatz, *der.* **2.** *v.i.* schwatzen

chip [tʃɪp] **1.** *n.* **a)** Splitter, *der;* **have a** ~ **on one's shoulder** *(fig.)* einen Komplex haben; **b)** *in pl. (Brit.: fried potatoes)* Pommes frites *Pl.; (Amer.: crisps)* Kartoffelchips *Pl.;* **c)** **there is a** ~ **in this cup/paintwork** diese Tasse ist angeschlagen/etwas Farbe ist abgeplatzt; **d)** *(Gambling)* Chip, *der;* Jeton, *der;* **when the** ~**s are down** *(fig. coll.)* wenn's ernst wird; **e)** *(Electronics)* Chip, *der.* **2.** *v.t.,* **-pp-:** **a)** anschlagen ⟨*Geschirr*⟩; ~ **[off]** abschlagen; **the paint is** ~**ped** die Farbe ist abgesprungen; **b)** ~**ped potatoes** Pommes frites

~ '**in** *(coll.)* **1.** *v.i.* **a)** *(interrupt)* sich einmischen; **b)** *(contribute money)* etwas beisteuern; ~ **in with £5** sich mit 5 Pfund an etw. *(Dat.)* beteiligen. **2.** *v.t. (contribute)* beisteuern

'**chipboard** *n.* Spanplatte, *die*

chipmunk ['tʃɪpmʌŋk] *n. (Zool.)* Chipmunk, *das*

chipolata [tʃɪpə'lɑ:tə] *n.* kleine *Wurst;* Chipolata, *die*

'**chip-pan** *n.* Friteuse, *die*

chippings ['tʃɪpɪŋz] *n. pl. (Road Constr.)* Splitt, *der;* '**loose** ~' „Rollsplitt"

'**chip shop** *n. (Brit. coll.)* Frittenbude, *die (ugs.)*

chiropodist [kɪ'rɒpədɪst] *n.* Fußpfleger, *der*/-pflegerin, *die*

chiropody [kɪ'rɒpədɪ] *n.* Fußpflege, *die*

chirp [tʃɜ:p] **1.** *v.i.* zwitschern; ⟨*Sperling:*⟩ tschilpen; ⟨*Grille:*⟩ zirpen. **2.** *n. see* **1:** Zwitschern, *das;* Tschilpen, *das;* Zirpen, *das*

chirrup ['tʃɪrəp] **1.** *v.i.* zwitschern; ⟨*Sperling:*⟩ tschilpen. **2.** *n.* Zwitschern, *das; (of sparrow)* Tschilpen, *das*

chisel ['tʃɪzl] **1.** *n.* Meißel, *der; (for wood)* Stemmeisen, *das;* Beitel, *der.* **2.** *v.t., (Brit.)* **-ll-** meißeln; *(in wood)* hauen; stemmen

chit [tʃɪt] *n. (note)* Notiz, *die; (certificate)* Zeugnis, *das*

chit-chat ['tʃɪttʃæt] *n.* Plauderei, *die*

chivalrous ['ʃɪvlrəs] *adj.,* **chivalrously** ['ʃɪvlrəslɪ] *adv.* ritterlich

chivalry ['ʃɪvlrɪ] *n., no pl.* **a)** Ritterlichkeit, *die;* **b)** **Age of C**~: Ritterzeit, *die*

chives [tʃaɪvz] *n. pl.* Schnittlauch, *der*

chiv[v]y ['tʃɪvɪ] *v.t.* hetzen; ~ **sb. into doing sth.** jmdn. drängen, etw. zu tun; ~ **sb. about sth.** jmdn. wegen etw. drängen

~ **a'long** *v.t.* antreiben

chloride ['klɔ:raɪd] *n. (Chem.)* Chlorid, *das*

chlorinate ['klɔ:rɪneɪt] *v.t.* chloren

chlorine ['klɔ:ri:n] *n.* Chlor, *das*

chloroform ['klɒrəfɔ:m] **1.** *n.* Chloroform, *das.* **2.** *v.t.* chloroformieren

chlorophyll ['klɒrəfɪl] *n. (Bot.)* Chlorophyll, *das*

choc-ice ['tʃɒkaɪs] *n.* Eis mit Schokoladenüberzug

chock [tʃɒk] **1.** *n.* Bremsklotz, *der.* **2.** *v.t.* blockieren

'**chock-a-block** *pred. adj.* vollgepfropft

'**chock-full** *pred. adj.* gestopft voll *(ugs.);* ~ **with sth.** mit etw. vollgepfropft

chocolate ['tʃɒkələt, 'tʃɒklət] **1.** *n.* Schokolade, *die; (sweet with ~ coating)* Praline, *die;* **drinking ~:** Trinkschokolade, *die.* **2.** *adj.* **a)** *(with flavour of ~)* Schokoladen-; **b)** *(with colour of ~)* **~[-brown]** schokoladenbraun

chocolate: ~ 'biscuit *n.* Schokoladenkeks, *der;* **~-box 1.** *n.* Pralinenschachtel, *die;* **2.** *adj.* *(fig.)* kitschig

choice [tʃɔɪs] **1.** *n.* **a)** Wahl, *die;* **take your ~:** suchen Sie sich *(Dat.)* einen/eine/eins aus; **make a [good] ~:** eine [gute] Wahl treffen; **give sb. the ~:** jmdm. die Wahl lassen; **the ~ is yours** Sie haben die Wahl; **do sth. from ~:** etw. freiwillig tun; **have no ~ but to do sth.** keine andere Wahl haben, als etw. zu tun; **leave sb. no ~:** jmdm. keine [andere] Wahl lassen; **you have several ~s** Sie haben mehrere Möglichkeiten; **b)** *(thing chosen)* **his ~ of wallpaper was ...:** die Tapete, die er sich ausgesucht hatte, war ...; **c)** *(variety)* Auswahl, *die;* **there is a ~ of three** es gibt drei zur Auswahl; **be spoilt for ~:** die Qual der Wahl haben; **have a ~:** die Auswahl haben. **2.** *adj.* ausgewählt; **~ fruit** Obst erster Wahl

choir ['kwaɪə(r)] *n.* Chor, *der*

'choirboy *n.* Chorknabe, *der*

choke [tʃəʊk] **1.** *v. t.* **a)** *(lit. or fig.)* ersticken; **b)** *(strangle)* **~ [to death]** erdrosseln; **c)** *(fill chock-full)* vollstopfen; *(block up)* verstopfen. **2.** *v. i.* *(temporarily)* keine Luft [mehr] bekommen; *(permanently)* ersticken (**on** an + *Dat.*). **3.** *n.* *(Motor Veh.)* Choke, *der*

~ 'back *v. t.* unterdrücken ‹*Wut*›; zurückhalten ‹*Tränen*›; hinunterschlucken *(ugs.)* ‹*Wut, Worte*›

choker ['tʃəʊkə(r)] *n.* *(high collar)* Stehkragen, *der; (necklace)* Halsband, *das*

cholera ['kɒlərə] *n.* *(Med.)* Cholera, *die*

cholesterol [kə'lestərɒl] *n.* *(Med.)* Cholesterin, *das*

choose [tʃu:z] **1.** *v. t.*, **chose** [tʃəʊz], **chosen** ['tʃəʊzn] **a)** *(select)* wählen; *(from a group)* auswählen; **~ sb. as** *or* **to be** *or* **for leader** jmdn. zum Anführer wählen; **b)** *(decide)* **~/~ not to do sth.** sich dafür/dagegen entscheiden, etw. zu tun; **there's nothing/not much/little to ~ between them** sie unterscheiden sich in nichts/nicht sehr/nur wenig voneinander. **2.** *v. i.*, **chose, chosen** wählen; **when I ~:** wenn es mir paßt; **~ from sth.** aus etw./*(from several)* unter etw. *(Dat.)* [aus]wählen

choos[e]y ['tʃu:zɪ] *adj.* *(coll.)* wählerisch

¹chop [tʃɒp] **1.** *n.* **a)** Hieb, *der;* **b)** *(of meat)* Kotelett, *das;* **c)** *(coll.)* **get the ~** *(be dismissed)* rausgeworfen werden *(ugs.);* **sth. gets the ~:** etw. wird abgeschafft; **give sb. the ~:** jmdn. rauswerfen *(ugs.);* **be due for the ~:** die längste Zeit existiert haben *(ugs.).* **2.** *v. t.*, **-pp-: a)** hacken ‹*Holz*›; kleinschneiden ‹*Fleisch, Gemüse*›; **~ped herbs** gehackte Kräuter; **b)** *(Sport)* schneiden ‹*Ball*›. **3.** *v. i.*, **-pp-: ~ [away] at sth.** auf etw. *(Akk.)* einhacken; **~ through the bone** den Knochen durchhacken

~ 'down *v. t.* fällen ‹*Baum*›; umhauen ‹*Busch, Pfosten*›

~ 'off *v. t.* abhacken

~ 'up *v. t.* kleinschneiden ‹*Fleisch, Gemüse*›; zerhacken ‹*Möbel*›

²chop *v. i.*, **-pp-: she's always ~ping and changing** sie überlegt es sich *(Dat.)* dauernd anders

chopper ['tʃɒpə(r)] *n.* **a)** *(axe)* Beil, *das; (cleaver)* Hackbeil, *das;* **b)** *(coll.: helicopter)* Hubschrauber, *der*

chopping-board ['tʃɒpɪŋbɔ:d] *n.* Hackbrett, *das*

choppy ['tʃɒpɪ] *adj.* bewegt; kabbelig *(Seemannsspr.)*

'chopstick *n.* [Eß]stäbchen, *das*

choral ['kɔ:rl] *adj.* Chor-

'choral society *n.* Gesangverein, *der*

¹chord [kɔ:d] *n.* **strike a [familiar/responsive] ~ with sb.** *(fig.)* bei jmdm. eine Saite zum Erklingen bringen/bei jmdm. Echo finden; **touch the right ~** *(fig.)* den richtigen Ton anschlagen *od.* treffen

²chord *n.* *(Mus.)* Akkord, *der*

chore [tʃɔ:(r)] *n.* [lästige] Routinearbeit; **do the household ~s** die üblichen Hausarbeiten erledigen; **writing letters is a ~:** Briefe zu schreiben ist eine lästige Pflicht

choreographer [kɒrɪ'ɒgrəfə(r)] *n.* Choreograph, *der*/Choreographin, *die*

choreography [kɒrɪ'ɒgrəfɪ] *n.* Choreographie, *die*

chorister ['kɒrɪstə(r)] *n.* *(choirboy)* Chorknabe, *der*

chortle ['tʃɔ:tl] **1.** *v. i.* vor Lachen glucksen. **2.** *n.* Glucksen, *das*

chorus ['kɔ:rəs] **1.** *n.* **a)** *(utterance)* Chor, *der;* **say sth. in ~:** etw. im Chor sagen; **b)** *(of singers)* Chor, *der; (of dancers)* Ballett, *das;* **c)** *(of song)* Chorus, *der;* **d)** *(composition)* Chor, *der.* **2.** *v. t.* im Chor singen/sprechen

'chorus-girl *n.* [Revue]girl, *das*

chose, chosen *see* **choose**

choux [ʃu:] *n.* **~ [pastry]** Brandteig, *der*

chow [tʃaʊ] *n.* **a)** *(dog)* Chow-Chow, *der;* **b)** *(Amer. sl.: food)* Futterage, *die (ugs.);* Futter, *das (salopp)*

chowder ['tʃaʊdə(r)] *n.* *(Amer.)* *Suppe od. Eintopf mit Fisch od. Muscheln, Pökelfleisch od. Schinken, Milch, Kartoffeln u. Gemüse*

Christ [kraɪst] **1.** *n.* Christus *(der); see also* **before 2 a. 2.** *int.* *(sl.)* **[oh] ~!, ~ almighty!** Herrgott noch mal! *(ugs.)*

christen ['krɪsn] *v. t.* **a)** taufen; **she was ~ed Martha** sie wurde [auf den Namen] Martha getauft; **b)** *(coll.: use for first time)* einweihen *(ugs. scherzh.)*

christening ['krɪsənɪŋ] *n.* Taufe, *die;* **her ~ will be next Sunday** sie wird nächsten Sonntag getauft

Christian ['krɪstjən] **1.** *adj.* christlich. **2.** *n.* Christ, *der*/Christin, *die*

Christianity [krɪstɪ'ænɪtɪ] *n., no pl., no art.* das Christentum

'Christian name *n.* Vorname, *der*

Christmas ['krɪsməs] *n.* Weihnachten, *das od. Pl.;* **merry** *or* **happy ~:** frohe *od.* fröhliche Weihnachten; **what did you get for ~?** was hast du zu Weihnachten bekommen?; **at ~:** [zu *od.* an] Weihnachten

Christmas: ~ cake *n.* Weihnachtskuchen, *der; mit Marzipan und Zuckerguß verzierter, reichhaltiger Gewürzkuchen;* **~ card** *n.* Weihnachtskarte, *die;* **~ 'carol** *n.* Weihnachtslied, *das;* **~ 'Day** *n.* erster Weihnachtsfeiertag; **~ 'Eve** *n.* Heiligabend, *der;* **~ present** *n.* Weihnachtsgeschenk, *das;* **~ 'pudding** *n.* Plumpudding, *der;* **~ time** *n.* Weihnachtszeit, *die;* **at ~ time** in der *od.* zur Weihnachtszeit; **~ tree** *n.* Weihnachtsbaum, *der*

chromatic [krə'mætɪk] *adj.* chromatisch

chrome [krəʊm] *n.* **a)** *(chromium-plate)* Chrom, *das;* **b)** *(colour)* Chromgelb, *das*

chromium ['krəʊmɪəm] *n.* Chrom, *das*

chromium: ~-plate 1. *n.* Chrom, *das;* **2.** *v. t.* verchromen; **~-plated** *adj.* verchromt

chromosome ['krəʊməsəʊm] *n.* *(Biol.)* Chromosom, *das*

chronic ['krɒnɪk] *adj.* **a)** chronisch; ~ **sufferers from arthritis** Personen, die an chronischer Arthritis leiden; **b)** *(Brit. coll.: bad, intense)* katastrophal *(ugs.)*; **be ~:** eine [einzige] Katastrophe sein *(ugs.)*

chronicle ['krɒnɪkl] **1.** *n.* **a)** Chronik, *die;* **b)** *(account)* Schilderung, *die.* **2.** *v.t.* [chronologisch] aufzeichnen

chronological [krɒnə'lɒdʒɪkl] *adj.* chronologisch

chronology [krə'nɒlədʒɪ] *n.* Chronologie, *die; (table)* Zeittafel, *die*

chrysalis ['krɪsəlɪs] *n., pl.* ~es *(Zool.)* **a)** *(pupa)* Chrysalide, *die (Zool.);* Puppe, *die;* **b)** *(case enclosing pupa)* Puppenhülle, *die*

chrysanth [krɪ'sænθ] *(coll.),* **chrysanthemum** [krɪ'sænθɪməm] *ns. (Bot.)* **a)** *(flower)* Chrysantheme, *die;* **b)** *(plant)* Chrysanthemum, *das;* Wucherblume, *die*

chubby ['tʃʌbɪ] *adj.* pummelig; rundlich ⟨*Gesicht*⟩; ~ **cheeks** Pausbacken *(fam.)*

[1]**chuck** [tʃʌk] *v.t. (coll.)* **a)** *(throw)* schmeißen *(ugs.)*; **b)** *(throw out)* wegschmeißen *(ugs.)*; ~ **the whole thing** alles hinschmeißen

~ **a'way** *v.t. (coll.)* wegschmeißen *(ugs.)*; *(fig.: waste)* zum Fenster rauswerfen *(ugs.)* ⟨*Geld*⟩ **(on** für); vertun ⟨*Chance, Gelegenheit*⟩

~ **'out** *v.t. (coll.)* wegschmeißen *(ugs.)*; *(fig.: eject)* rausschmeißen *(ugs.)*

[2]**chuck** *n. (of drill, lathe)* Futter, *das*

chuckle ['tʃʌkl] **1.** *v.i.* leise [vor sich hin] lachen **(at** über + *Akk.*). **2.** *n.* leises, glucksendes Lachen; **have a ~ [to oneself] about sth.** leise über etw. *(Akk.)* vor sich hin lachen

chuffed [tʃʌft] *pred. adj. (Brit. sl.)* zufrieden **(about, at, with** über + *Akk.*); **be ~:** sich freuen

chug [tʃʌg] **1.** *v.i.,* **-gg-** ⟨*Motor:*⟩ tuckern. **2.** *n.* Tuckern, *das*

chum [tʃʌm] *n. (coll.)* Kumpel, *der (salopp)*

chunk [tʃʌŋk] *n.* dickes Stück; *(broken off)* Brocken, *der*

chunky ['tʃʌŋkɪ] *adj.* **a)** *(containing chunks)* ⟨*Orangenmarmelade, Hundefutter*⟩ mit ganzen Stücken; **b)** *(small and sturdy, short and thick)* stämmig; **c)** *(made of thick material)* dick ⟨*Pullover, Strickjacke*⟩

church [tʃɜ:tʃ] *n.* **a)** Kirche, *die;* **in** *or* **at ~:** in der Kirche; **go to ~:** in die *od.* zur Kirche gehen; **b) C~** *(body)* die Kirche; **the C~ of England** die Kirche von England

church: ~**-goer** *n.* Kirchgänger, *der/*-gängerin, *die;* ~**warden** *n.* Kirchenvorsteher, *der/*-vorsteherin, *die;* ~**yard** *n.* Friedhof, *der (bei einer Kirche)*

churlish ['tʃɜ:lɪʃ] *adj. (ill-bred)* ungehobelt; *(surly)* griesgrämig

churn [tʃɜ:n] **1.** *n. (Brit.)* **a)** *(for making butter)* Butterfaß, *das;* **b)** *(milk-can)* Milchkanne, *die.* **2.** *v.t.* **a)** ~ **butter** buttern; **b)** aufwühlen ⟨*Wasser, Schlamm*⟩. **3.** *v.i.* ⟨*Meer:*⟩ wallen *(geh.)*; ⟨*Schiffsschraube:*⟩ wirbeln; **my stomach was ~ing** mir drehte sich der Magen um

~ **'out** *v.t.* massenweise produzieren *(ugs.)*

~ **'up** *v.t.* aufwühlen

chute [ʃu:t] *n.* Schütte, *die; (for persons)* Rutsche, *die*

chutney ['tʃʌtnɪ] *n.* Chutney, *das*

CI *abbr. (Brit.)* **Channel Islands**

CIA *abbr. (Amer.)* **Central Intelligence Agency** CIA, *der od. die*

cicada [sɪ'kɑ:də] *n. (Zool.)* Zikade, die

CID *abbr. (Brit.)* **Criminal Investigation Department** C. I. D.; **the ~:** die Kripo

cider ['saɪdə(r)] *n.* ≈ Apfelwein, *der*

cider: ~ **apple** *n.* Mostapfel, *der;* ~**-press** *n.* Mostpresse, *die*

cig [sɪg] *n. (coll.)* Glimmstengel, *der (ugs.)*

cigar [sɪ'gɑ:(r)] *n.* Zigarre, *die*

cigarette [sɪgə'ret] *n.* Zigarette, *die*

cigarette: ~ **card** *n.* Zigarettenbild, *das;* ~**-case** *n.* Zigarettenetui, *das;* ~**-end** *n.* Zigarettenstummel, *der;* ~**-lighter** *n.* Feuerzeug, *das; (in car)* Zigarettenanzünder, *der;* ~**-packet** *n.* Zigarettenschachtel, *die;* ~**-paper** *n.* Zigarettenpapier, *das*

cigar: ~ **lighter** *see* **cigarette-lighter;** ~**-shaped** *adj.* zigarrenförmig

C.-in-C. *abbr. (Mil.)* **Commander-in-Chief**

cinch [sɪntʃ] *n. (sl.)* *(easy thing)* Klacks, *der (ugs.)*; Kinderspiel, *das; (Amer.: sure thing)* todsichere Sache *(ugs.)*

cinder ['sɪndə(r)] *n.* ausgeglühtes Stück Holz/Kohle; **~s** Asche, *die;* **burnt to a ~:** völlig verkohlt

Cinderella [sɪndə'relə] *n.* Aschenbrödel, *das;* Aschenputtel, *das*

cine ['sɪnɪ]: ~ **camera** *n.* Filmkamera, *die;* ~ **film** *n.* Schmalfilm, *der*

cinema ['sɪnɪmə] *n.* **a)** *(Brit.: building)* Kino, *das;* **go to the ~:** ins Kino gehen; **what's on at the ~?** was gibt's im Kino?; **b)** *no pl., no art. (cinematography)* Kinematographie, *die;* **c)** *(films, film production)* Film, *der*

'cinema-goer *n. (Brit.)* Kinogänger, *der/*-gängerin, *die*

cinematography [sɪnɪmə'tɒgrəfɪ] *n., no pl.* Kinematographie, *die*

cinnamon ['sɪnəmən] *n.* Zimt, *der*

cipher ['saɪfə(r)] *n.* **a)** *(code, secret writing)* Chiffre, *die;* Geheimschrift, *die;* **in ~:** chiffriert; **b)** *(symbol for zero)* [Ziffer] Null, *die;* **c)** *(fig.: nonentity)* Nummer, *die*

circa ['sɜ:kə] *prep.* zirka

circle ['sɜ:kl] **1.** *n.* **a)** *(also Geom.)* Kreis, *der;* **fly/stand in a ~:** im Kreis fliegen/stehen; **run round in ~s** *(fig. coll.)* hektisch herumlaufen *(ugs.)*; **go round in ~s** im Kreis laufen; *(fig.)* sich im Kreis drehen; ~ **of friends** Freundeskreis, *der;* **come full ~** *(fig.)* zum Ausgangspunkt zurückkehren; **b)** *(seats in theatre or cinema)* Rang, *der.* **2.** *v.i.* kreisen; *(walk in a ~)* im Kreis gehen. **3.** *v.t.* **a)** *(move in a ~ round)* umkreisen; **the aircraft ~d the airport** das Flugzeug kreiste über dem Flughafen; **b)** *(draw ~ round)* einkreisen

~ **'back** *v.i.* auf einem Umweg zurückkehren

~ **'round** *v.i.* kreisen

circuit ['sɜ:kɪt] *n.* **a)** *(Electr.)* Schaltung, *die; (path of current)* Stromkreis, *der;* **b)** *(Motor-racing)* Rundkurs, *der;* **c)** *(journey round)* Runde, *die; (by car etc.)* Rundfahrt, *die*

circuitous [sə'kju:ɪtəs] *adj.* umständlich; **the path followed a ~ route** der Pfad machte einen weiten Bogen

circuitry ['sɜ:kɪtrɪ] *n.* *(Electr.)* Schaltungen *Pl.*

circular ['sɜ:kjʊlə(r)] **1.** *adj.* **a)** *(round)* kreisförmig; **b)** *(moving in circle)* Kreis⟨*bahn, -bewegung*⟩; **c)** *(Logic)* **that argument is ~:** das ist ein Zirkelschluß *od.* -beweis. **2.** *n.* *(letter, notice)* Rundschreiben, *das; (advertisement)* Werbeprospekt, *der*

circular: ~ **'letter** *see* **circular 2;** ~ **'saw** *n.* Kreissäge, *die;* ~ **'tour** *n.* *(Brit.)* Rundfahrt, *die* **(of** durch)

circulate ['sɜ:kjʊleɪt] **1.** *v.i.* ⟨*Blut, Flüssigkeit:*⟩ zirkulieren; ⟨*Geld, Gerüchte:*⟩ kursieren; ⟨*Nachrichten:*⟩ sich herumsprechen; ⟨*Verkehr:*⟩ fließen; ⟨*Personen, Wein usw.:*⟩ die Runde machen *(ugs.)*. **2.** *v.t.* in Umlauf setzen ⟨*Ge-*

rücht⟩; verbreiten ⟨*Nachricht, Information*⟩; zirkulieren lassen ⟨*Aktennotiz, Rundschreiben*⟩; herumgehen lassen ⟨*Buch, Bericht*⟩ (**around** in + *Dat.*)

circulation [sɜ:kjʊ'leɪʃn] *n.* **a)** *(Physiol.)* Kreislauf, *der;* Zirkulation, *die (Med.)*; *(of sap, water, atmosphere)* Zirkulation, *die;* **poor** ~ *(Physiol.)* schlechte Durchblutung; Kreislaufstörungen *Pl.*; **b)** *(of news, rumour, publication)* Verbreitung, *die;* **have a wide** ~: große Verbreitung finden; **c)** *(of notes, coins)* Umlauf, *der;* **withdraw from** ~: aus dem Umlauf ziehen; **put/come into** ~: in Umlauf bringen/kommen; **d)** *(fig.)* **be back in** ~: wieder auf dem Posten sein; **be out of** ~: aus dem Verkehr gezogen sein *(ugs. scherzh.)*; **e)** *(number of copies sold)* verkaufte Auflage

circumcise ['sɜ:kəmsaɪz] *v. t.* beschneiden

circumcision [sɜ:kəm'sɪʒn] *n.* Beschneidung, *die*

circumference [sə'kʌmfərəns] *n.* Umfang, *der; (periphery)* Kreislinie, *die*

circumflex ['sɜ:kəmfleks] *n.* Zirkumflex, *der*

circumnavigate [sɜ:kəm'nævɪgeɪt] *v. t.* umfahren; *(by sail)* umsegeln

circumscribe ['sɜ:kəmskraɪb] *v. t. (lay down limits of)* eingrenzen; einschränken ⟨*Macht, Handlungsfreiheit usw.*⟩

circumspect ['sɜ:kəmspekt] *adj.* umsichtig

circumstance ['sɜ:kəmstəns] *n.* **a)** *usu. in pl.* Umstand, *der;* **in** *or* **under the** ~**s** unter den gegebenen *od.* diesen Umständen; **in certain** ~**s** unter [gewissen] Umständen; **under no** ~**s** unter [gar] keinen Umständen; **b)** *in pl. (financial state)* Verhältnisse

circumstantial [sɜ:kəm'stænʃl] *adj.* ~ **evidence** Indizienbeweise; **be purely** ~ ⟨*Beweis:*⟩ nur auf Indizien gegründet sein

circumvent [sɜ:kəm'vent] *v. t.* umgehen; hinters Licht führen

circus ['sɜ:kəs] *n.* Zirkus, *der*

cirrhosis [sɪ'rəʊsɪs] *n., pl.* **cirrhoses** [sɪ'rəʊsi:z] *(Med.)* Zirrhose, *die;* ~ **of the liver** Leberzirrhose, *die*

cissy ['sɪsɪ] *see* **sissy**

cistern ['sɪstən] *n.* Wasserkasten, *der; (in roof)* Wasserbehälter, *der*

citadel ['sɪtədəl] *n.* Zitadelle, *die*

citation [saɪ'teɪʃn] *n.* **a)** *no pl. (citing)* Zitieren, *das;* **b)** *(quotation)* Zitat, *das*

cite [saɪt] *v. t. (quote)* zitieren; anführen ⟨*Beispiel*⟩

citizen ['sɪtɪzən] *n.* **a)** *(of town, city)* Bürger, *der*/Bürgerin, *die;* **b)** *(of state)* [Staats]bürger, *der*/-bürgerin, *die;* **he is a British** ~: er ist britischer Staatsbürger *od.* Brite; ~**'s arrest** Festnahme durch eine Zivilperson; ~**s' band radio** CB-Funk, *der; (radio set)* CB-Funkgerät, *das*

citizenship ['sɪtɪzənʃɪp] *n.* Staatsbürgerschaft, *die*

citric acid [sɪtrɪk 'æsɪd] *n.* *(Chem.)* Zitronensäure, *die*

citrus ['sɪtrəs] *n.* ~ **[fruit]** Zitrusfrucht, *die*

city ['sɪtɪ] *n.* **a)** [Groß]stadt, *die;* **the C**~: die [Londoner] City; das Londoner Banken- und Börsenviertel; **b)** *(Brit.)* Stadt, *die (Ehrentitel für bestimmte Städte, meist Bischofssitze)*; **c)** *(Amer.)* ≈ Stadtgemeinde, *die;* **d)** *attrib.* [Groß]stadt⟨*leben, -verkehr*⟩; ~ **lights** Lichter der Großstadt

city 'centre *n.* Stadtzentrum, *das;* Innenstadt, *die*

civic ['sɪvɪk] *adj.* **a)** *(of citizens, citizenship)* [Staats]bürger-; [staats]bürgerlich; **my** ~ **responsibility** meine Verantwortung als Staatsbürger; **b)** *(of city)* Stadt-; städtisch; ~ **centre** Verwaltungszentrum der Stadt

civies *see* **civvies**

civil ['sɪvɪl, 'sɪvl] *adj.* **a)** *(not military)* zivil; **in** ~ **life** im Zivilleben; **b)** *(polite, obliging)* höflich; **c)** *(Law)* Zivil⟨*gerichtsbarkeit, -prozeß, -verfahren*⟩; zivilrechtlich; **d)** *(of citizens)* bürgerlich; Bürger⟨*krieg, -recht, -pflicht*⟩

civil: ~ **avi'ation** *n.* Zivilluftfahrt, *die;* ~ **de'fence** *n.* Zivilschutz, *der;* ~ **diso'bedience** *n.* ziviler Ungehorsam; ~ **engi'neer** *n.* Bauingenieur, *der*/-ingenieurin, *die;* ~ **engi'neering** *n.* Hoch- und Tiefbau, *der*

civilian [sɪ'vɪljən] **1.** *n.* Zivilist, *der.* **2.** *adj.* Zivil-; **wear** ~ **clothes** Zivil[kleidung] tragen

civilise *etc. see* **civiliz-**

civility [sɪ'vɪlɪtɪ] *n., no pl.* Höflichkeit, *die*

civilization [sɪvɪlaɪ'zeɪʃn] *n.* Zivilisation, *die*

civilize ['sɪvɪlaɪz] *v. t.* zivilisieren

civilized ['sɪvɪlaɪzd] *adj.* zivilisiert; *(refined)* kultiviert

civil 'law *n.* Zivilrecht, *das*

civilly ['sɪvɪlɪ, 'sɪvəlɪ] *adv.* höflich

civil: ~ **'marriage** *n.* Ziviltrauung, *die;* standesamtliche Trauung; ~ **'rights** *n. pl.* Bürgerrechte; ~ **'servant** *n.* ≈ [Staats]beamte, *der*/-beamtin, *die;* **C**~ **'Service** *n.* öffentlicher Dienst; ~ **'war** *n.* Bürgerkrieg, *der*

civvies ['sɪvɪz] *n. pl. (Brit. sl.)* Zivil, *das;* Zivilklamotten *Pl. (ugs.)*

Civvy Street ['sɪvɪ stri:t] *n., no pl., no art. (Brit. sl.)* das Zivilleben

clad [klæd] *adj. (arch./literary)* gekleidet (**in** in + *Akk.*)

cladding ['klædɪŋ] *n.* Verkleidung, *die*

claim [kleɪm] **1.** *v. t.* **a)** *(demand as one's due property)* Anspruch erheben auf (+ *Akk.*), beanspruchen ⟨*Thron, Gebiete*⟩; fordern ⟨*Lohnerhöhung, Schadenersatz*⟩; beantragen ⟨*Arbeitslosenunterstützung, Sozialhilfe usw.*⟩; abholen ⟨*Fundsache*⟩; ~ **one's luggage** sein Gepäck [ab]holen; **b)** *(represent oneself as having)* für sich beanspruchen, in Anspruch nehmen ⟨*Sieg*⟩; **c)** *(profess, contend)* behaupten; **the new system is** ~**ed to have many advantages** das neue System soll viele Vorteile bieten; **d)** *(result in loss of)* fordern ⟨*Opfer, Menschenleben*⟩. **2.** *v. i.* **a)** *(Insurance)* Ansprüche geltend machen; **b)** *(for costs)* ~ **for damages/expenses** Schadenersatz fordern/sich *(Dat.)* Auslagen rückerstatten lassen. **3.** *n.* **a)** Anspruch, *der* (**to** auf + *Akk.*); **lay** ~ **to sth.** auf etw. *(Akk.)* Anspruch erheben; **make too many** ~**s on sth.** etw. zu sehr in Anspruch nehmen; **b)** *(assertion)* **make** ~**s about sth.** Behauptungen über etw. *(Akk.)* aufstellen; **c)** *(pay* ~*)* Forderung, *die* (**for** nach); **d)** ~ **[for expenses]** Spesenabrechnung, *die* (**for** über + *Akk.*); ~ **for damages** Schadenersatzforderung, *die;* **e) stake a** ~ **to sth.** *(fig.)* ein Anrecht auf etw. *(Akk.)* anmelden

~ **'back** *v. t.* zurückfordern

claimant ['kleɪmənt] *n. (for rent rebate, state benefit)* Antragsteller, *der*/-stellerin, *die; (for inheritance)* Erbberechtigte, *der/die*

clairvoyant [kleə'vɔɪənt] **1.** *n.* Hellseher, *der*/Hellseherin, *die.* **2.** *adj.* hellseherisch

clam [klæm] **1.** *n.* Klaffmuschel, *die;* **shut up like a** ~ *(fig.)* ausgesprochen wortkarg werden. **2.** *v. i.,* **-mm-:** ~ **up** *(coll.)* den Mund nicht [mehr] aufmachen

clamber ['klæmbə(r)] *v. i.* klettern; ⟨*Baby:*⟩ krabbeln; ~ **up a wall** auf eine Mauer klettern; eine Mauer hochklettern

clammy ['klæmɪ] *adj.* feucht; kalt und schweißig ⟨*Hände, Gesicht,*

Haut›; klamm ‹*Kleidung usw.*›; naßkalt ‹*Luft usw.*›
clamour (*Brit.; Amer.:* **clamor**) ['klæmə(r)] **1.** *n.* **a)** *(noise, shouting)* Lärm, *der;* lautes Geschrei; **b)** *(protest)* [lautstarker] Protest; *(demand)* [lautstarke] Forderung (**for** nach). **2.** *v. i.* **a)** *(shout)* schreien; **b)** *(protest, demand)* ~ **against sth.** gegen etw. [lautstark] protestieren; ~ **for sth.** nach etw. schreien; ~ **to be let out** lautstark fordern, herausgelassen zu werden
clamp [klæmp] **1.** *n.* Klammer, *die; (Woodw.)* Schraubzwinge, *die; see also* **wheel-clamp. 2.** *v. t.* **a)** klemmen; einspannen ‹*Werkstück*›; *(Med.)* klammern; ~ **two pieces of wood together** zwei Holzstücke miteinander verklammern; **b)** ~ **a vehicle** an einem Fahrzeug eine Parkkralle anbringen. **3.** *v. i.* ~ **down on sb./sth.** gegen jmdn./etw. rigoros vorgehen
'**clamp-down** *n.* rigoroses Vorgehen (**on** gegen)
clan [klæn] *n.* Sippe, *die; (of Scottish Highlanders)* Clan, *der*
clandestine [klæn'destın] *adj.* heimlich
clang [klæŋ] **1.** *n. (of bell)* Läuten, *das; (of hammer)* Klingen, *das; (of sword)* Klirren, *das.* **2.** *v. i.* ‹*Glocke:*› läuten; ‹*Hammer:*› klingen; ‹*Schwert:*› klirren
clanger ['klæŋə(r)] *n. (Brit. sl.)* Schnitzer, *der (ugs.);* **drop a ~:** sich *(Dat.)* einen Schnitzer leisten *(ugs.)*
clank [klæŋk] **1.** *n.* Klappern, *das; (of sword, chain)* Klirren, *das.* **2.** *v. i.* klappern; ‹*Schwert, Kette:*› klirren; ‹*Kette:*› rasseln. **3.** *v. t.* klirren mit ‹*Schwert, Kette*›
clap [klæp] **1.** *n.* **a)** Klatschen, *das;* **give sb. a ~:** jmdm. applaudieren *od.* Beifall klatschen; **b)** *(slap)* Klaps, *der (ugs.);* **c)** ~ **of thunder** Donnerschlag, *der.* **2.** *v. i.*, **-pp-** klatschen. **3.** *v. t.*, **-pp-:** **a)** ~ **one's hands** in die Hände klatschen; ~ **sth.** etw. beklatschen; ~ **sb.** jmdm. Beifall klatschen; **b)** ~ **sb. in prison** jmdn. ins Gefängnis werfen *od. (ugs.)* stecken; ~ **eyes on sb./sth.** jmdn./etw. zu Gesicht bekommen; **c)** **~ped out** *(sl.)* schrottreif *(ugs.)* ‹*Auto, Flugzeug*›; kaputt *(ugs.)* ‹*Person, Idee*›
clapper ['klæpə(r)] *n. (of bell)* Klöppel, *der;* Schwengel, *der*
clapping ['klæpıŋ] *n., no pl.* Beifall, *der;* Applaus, *der*
claptrap ['klæptræp] *n., no pl.* **a)** *(pretentious assertions)* [leere] Phrasen; **b)** *(coll.: nonsense)* Geschwafel, *das (ugs. abwertend)*
claret ['klærət] **1.** *n.* roter Bordeauxwein. **2.** *adj.* weinrot
clarification [klærıfı'keıʃn] *n.* Klärung, *die; (explanation)* Klarstellung, *die*
clarify ['klærıfaı] *v. t.* klären ‹*Situation, Problem usw.*›; *(by explanation)* klarstellen; erläutern ‹*Aussage, Bemerkung usw.*›
clarinet [klærı'net] *n. (Mus.)* Klarinette, *die*
clarinettist (*Amer.:* **clarinetist**) [klærı'netıst] *n. (Mus.)* Klarinettist, *der*/Klarinettistin, *die*
clarity ['klærıtı] *n., no pl.* Klarheit, *die*
clash [klæʃ] **1.** *v. i.* **a)** scheppern *(ugs.);* ‹*Becken:*› dröhnen; ‹*Schwerter:*› aneinanderschlagen; **b)** *(meet in conflict)* zusammenstoßen (**with** mit); **c)** *(disagree)* sich streiten; ~ **with sb.** mit jmdm. eine Auseinandersetzung haben; **d)** *(be incompatible)* aufeinanderprallen; ‹*Interesse, Ereignis:*› kollidieren (**with** mit); ‹*Persönlichkeit, Stil:*› nicht zusammenpassen (**with** mit); ‹*Farbe:*› sich beißen *(ugs.)* (**with** mit). **2.** *v. t.* gegeneinanderschlagen. **3.** *n.* **a)** *(of cymbals)* Dröhnen, *das; (of swords)* Aneinanderschlagen, *das;* **b)** *(meeting in conflict)* Zusammenstoß, *der;* **c)** *(disagreement)* Auseinandersetzung, *die;* **d)** *(incompatibility)* Unvereinbarkeit, *die; (of personalities, styles, colours)* Unverträglichkeit, *die; (of events)* Überschneidung, *die*
clasp [klɑ:sp] **1.** *n.* **a)** Verschluß, *der; (of belt)* Schnalle, *die;* **b)** *(embrace)* Umarmung, *die;* **c)** *(grasp)* Griff, *der.* **2.** *v. t.* **a)** *(embrace)* drücken (**to** an + *Akk.*); **b)** *(grasp)* umklammern
class [klɑ:s] **1.** *n.* **a)** *(group in society)* Gesellschaftsschicht, *die;* Klasse, *die (Soziol.); (system)* Klassensystem, *das;* **b)** *(Educ.)* Klasse, *die; (Sch.: lesson)* Stunde, *die; (Univ.: seminar etc.)* Übung, *die;* **a French ~:** eine Französischstunde; **c)** *(group [according to quality])* Klasse, *die;* **be in a ~ by itself** *or* **on its own/of one's own** *or* **by oneself** eine Klasse für sich sein; **d)** *(coll.: quality)* Klasse, *die (ugs.).* **2.** *v. t.* einordnen; ~ **sth. as sth.** etw. als etw. einstufen
class: ~-conscious *adj.* klassenbewußt; **~-consciousness** *n.* Klassenbewußtsein, *das;* ~ **distinction** *n.* Klassenunterschied, *der*
classic ['klæsık] **1.** *adj.* klassisch. **2.** *n.* **a)** *in pl. (classical studies)* Altphilologie, *die;* **b)** *(book, play, film)* Klassiker, *der*
classical ['klæsıkl] *adj.* klassisch; ~ **studies** Altphilologie, *die;* **the ~ world** die Antike; ~ **education** humanistische [Schul]bildung
classicist ['klæsısıst] *n.* Altphilologe, *der*/-philologin, *die*
classifiable ['klæsıfaıəbl] *adj.* klassifizierbar
classification [klæsıfı'keıʃn] *n.* Klassifikation, *die*
classified ['klæsıfaıd] *adj.* **a)** *(arranged in classes)* gegliedert; unterteilt; ~ **advertisement** Kleinanzeige, *die;* **b)** *(officially secret)* geheim
classify ['klæsıfaı] *v. t.* klassifizieren; ~ **books by subjects** Bücher nach Fachgebieten [ein]ordnen
classless ['klɑ:slıs] *adj.* klassenlos ‹*Gesellschaft*›
class: ~-mate *n.* Klassenkamerad, *der*/-kameradin, *die;* **~-room** *n. (Sch.)* Klassenzimmer, *das;* Klasse, *die;* ~ **struggle,** ~ **war** *ns.* Klassenkampf, *der*
classy ['klɑ:sı] *adj. (coll.)* klasse *(ugs.);* nobel ‹*Vorort, Hotel*›
clatter ['klætə(r)] **1.** *n.* Klappern, *das;* **the kettle fell with a ~ to the ground** der Kessel fiel scheppernd zu Boden. **2.** *v. i.* **a)** klappern; **b)** *(move or fall with a ~)* poltern. **3.** *v. t.* klappern mit
clause [klɔ:z] *n.* **a)** Klausel, *die;* **b)** *(Ling.)* Teilsatz, *der;* **[subordinate] ~:** Nebensatz, *der*
claustrophobia [klɒstrə'fəʊbıə] *n., no pl. (Psych.)* Klaustrophobie, *die*
claustrophobic [klɒstrə'fəʊbık] *adj.* beengend ‹*Ort, Atmosphäre*›; an Klaustrophobie leidend ‹*Person*›
claw [klɔ:] **1.** *n. (of bird, animal)* Kralle, *die; (of crab, lobster, etc.)* Schere, *die; (foot with ~)* Klaue, *die.* **2.** *v. t.* kratzen. **3.** *v. i.* ~ **at sth.** sich an etw. *(Akk.)* krallen
~ '**back** *v. t.* wiedereintreiben ‹*Geld, Unterstützung*›; wettmachen ‹*Defizit*›
clay [kleı] *n.* Lehm, *der; (for pottery)* Ton, *der*
clay 'pigeon shooting *n.* Tontaubenschießen, *das*
clean [kli:n] **1.** *adj.* **a)** sauber; frisch ‹*Wäsche, Hemd*›; **b)** *(unused, fresh)* sauber; *(free of defects)* einwandfrei; sauber; **make a ~ start** noch einmal neu anfangen; **come ~** *(coll.) (confess)* aus-

packen *(ugs.)*; *(tell the truth)* mit der Wahrheit [he]rausrücken *(ugs.)*; **c)** *(regular, complete)* glatt ‹*Bruch*›; glatt, sauber ‹*Schnitt*›; **make a ~ break [with sth.]** *(fig.)* einen Schlußstrich [unter etw. *(Akk.)*] ziehen; **d)** *(coll.: not obscene or indecent)* sauber; stubenrein *(scherzh.)* ‹*Witz*›; **be good ~ fun** völlig harmlos sein; **e)** *(sportsmanlike, fair)* sauber. **2.** *adv.* glatt; einfach ‹*vergessen*›; **we're ~ out of whisky** wir haben überhaupt keinen Whisky mehr; **the fox got ~ away** der Fuchs ist uns/ihnen *usw.* glatt entwischt. **3.** *v. t.* saubermachen; putzen ‹*Zimmer, Haus, Fenster, Schuh*›; reinigen ‹*Teppich, Möbel, Käfig, Kleidung, Wunde*›; fegen, kehren ‹*Kamin*›; *(with cloth)* aufwischen ‹*Fußboden*›; **~ one's hands/teeth** sich *(Dat.)* die Hände waschen/Zähne putzen. **4.** *v.i.* sich reinigen lassen. **5.** *n.* **this carpet needs a good ~:** dieser Teppich muß gründlich gereinigt werden; **give your shoes a ~:** putz deine Schuhe

~ 'out *v.t.* **a)** *(remove dirt from)* saubermachen; ausmisten ‹*Stall*›; **b)** *(sl.)* **~ sb. out** *(take all sb.'s money)* jmdn. [total] schröpfen *(ugs.)*; **the tobacconist was ~ed out of cigarettes** beim Tabakhändler war alles an Zigaretten aufgekauft worden

~ 'up 1. *v.t.* **a)** aufräumen ‹*Zimmer, Schreibtisch*›; beseitigen ‹*Trümmer, Unordnung*›; **b) ~ oneself up** sich saubermachen; *(get washed)* sich waschen; **c)** *(fig.)* säubern ‹*Stadt*›; aufräumen mit ‹*Korruption usw.*›. **2.** *v.i.* **a)** aufräumen; **b)** *(coll.: make money)* absahnen *(ugs.)*

'clean-cut *adj.* klar [umrissen]; **his ~ features** seine klar geschnittenen Gesichtszüge

cleaner ['kli:nə(r)] *n.* **a)** *(person)* Raumpfleger, *der*/-pflegerin, *die*; *(woman also)* Putzfrau, *die*; **b)** *(vacuum ~)* Staubsauger, *der*; *(substance)* Reinigungsmittel, *das*; **c)** *usu. in pl. (dry-~)* Reinigung, *die*; **take sth. to the ~s** etw. in die Reinigung bringen; **take sb. to the ~s** *(sl.)* jmdn. bis aufs Hemd ausziehen *(ugs.)*

cleanliness ['klenlɪnɪs] *n., no pl.* Reinlichkeit, *die*; Sauberkeit, *die*

cleanly ['klenlɪ] *adj.* sauber

cleanness ['kli:nnɪs] *n., no pl.* **a)** Sauberkeit, *die*; **b) the ~ of the ship's lines** die klare Linienführung des Schiffes; **c)** *(regularity of cut or break)* Glätte, *die*

'clean-out *n.* **give sth. a ~:** etw. saubermachen

cleanse [klenz] *v. t.* **a)** *(spiritually purify)* läutern; **~d of** *or* **from sin** von der Sünde befreit; **b)** *(clean)* [gründlich] reinigen

cleanser ['klenzə(r)] *n.* **a)** Reinigungsmittel, *das*; Reiniger, *der*; **b)** *(for skin)* Reinigungscreme, *die*

'clean-shaven *adj.* glattrasiert

cleansing ['klenzɪŋ]: **~ cream** *n.* Reinigungscreme, *die*; **~ department** *n.* Stadtreinigung, *die*

clear [klɪə(r)] **1.** *adj.* **a)** klar; rein ‹*Haut, Teint*›; **b)** *(distinct)* scharf ‹*Bild, Foto, Umriß*›; deutlich ‹*Abbild*›; klar ‹*Ton*›; klar verständlich ‹*Wort*›; **c)** *(obvious, unambiguous)* klar ‹*Aussage, Vorteil, Vorsprung, Mehrheit, Sieg, Fall*›; **make oneself ~:** sich deutlich *od.* klar [genug] ausdrücken; **make sth. ~:** etw. deutlich zum Ausdruck bringen; **make it ~ [to sb.] that ...:** [jmdm.] klar und deutlich sagen, daß ...; **d)** *(free)* frei; *(Horse-riding)* fehlerfrei ‹*Runde*›; **be ~ of suspicion** nicht unter Verdacht stehen; **we're in the ~** *(free of suspicion)* auf uns fällt kein Verdacht; *(free of trouble)* wir haben es geschafft; **be three points ~:** drei Punkte Vorsprung haben; **e)** *(complete)* **three ~ days/lines** drei volle *od.* volle drei Tage/Zeilen; **f)** *(open, unobstructed)* frei; **keep sth. ~** *(not block)* etw. frei halten; **have a ~ run** freie Fahrt haben; **all ~** *(one will not be detected)* die Luft ist rein *(ugs.)*; *see also* **all-clear; the way is [now] ~ [for sb.] to do sth.** *(fig.)* es steht [jmdm.] nichts [mehr] im Wege, etw. zu tun; **g)** *(discerning)* klar; **keep a ~ head** einen klaren *od.* kühlen Kopf bewahren; **h)** *(certain, confident)* **be ~ [on** *or* **about sth.]** sich *(Dat.)* [über etw. *(Akk.)*] im klaren sein. **2.** *adv.* **keep ~ of sth./sb.** etw./jmdn. meiden; **'keep ~'** *(don't approach)* „Vorsicht [Zug *usw.*]"; **please stand** *or* **keep ~ of the door** bitte von der Tür zurücktreten; **move sth. ~ of sth.** etw. von etw. wegräumen; **the driver was pulled ~ of the wreckage** man zog den Fahrer aus dem Wrack seines Wagens. **3.** *v. t.* **a)** *(make ~)* klären ‹*Flüssigkeit*›; **~ the air** lüften; *(fig.)* die Atmosphäre reinigen; **b)** *(free from obstruction)* räumen ‹*Straße*›; abräumen ‹*Regal, Schreibtisch*›; freimachen ‹*Abfluß, Kanal*›; **~ the streets of snow** den Schnee von den Straßen räumen; **~ a space for sb./sth.** für jmdn./etw. Platz machen; **~ one's throat** sich räuspern; *see also* **deck 1 a; way 1 f; c)** *(make empty)* räumen; leeren ‹*Briefkasten*›; **~ the room** das Zimmer räumen; **~ the table** den Tisch abräumen; **~ one's desk** seinen Schreibtisch ausräumen; **~ one's plate** seinen Teller leer essen; **d)** *(remove)* wegräumen; beheben ‹*Verstopfung*›; **~ sth. out of the way** etw. aus dem Weg räumen; **e)** *(pass over without touching)* nehmen ‹*Hindernis*›; überspringen ‹*Latte*›; **f)** *(show to be innocent)* freisprechen; **~ oneself** seine Unschuld beweisen; **~ sb. of sth.** jmdn. von etw. freisprechen; **~ one's name** seine Unschuld beweisen; **g)** *(declare fit to have secret information)* für unbedenklich erklären; **h)** *(get permission for)* **~ sth. with sb.** etw. von jmdm. genehmigen lassen; *(give permission for)* **~ a plane for take-off/landing** einem Flugzeug Start-/Landeerlaubnis erteilen; **i)** *(at customs)* **~ customs** vom Zoll abgefertigt werden; **j)** *(pay off)* begleichen ‹*Schuld*›. **4.** *v. i.* **a)** *(become ~)* klar werden; sich klären; ‹*Wetter, Himmel:*› sich aufheitern; *(fig.)* ‹*Gesicht:*› sich aufhellen; **b)** *(disperse)* ‹*Nebel:*› sich verziehen

~ a'way 1. *v. t.* wegschaffen; *(from the table)* abräumen ‹*Geschirr, Besteck*›. **2.** *v. i.* **a)** abräumen; **b)** *(disperse)* ‹*Nebel:*› sich verziehen

~ 'off 1. *v. t.* begleichen ‹*Schulden*›; abzahlen ‹*Hypothek*›; aufarbeiten ‹*Rückstand*›. **2.** *v. i.* *(coll.)* abhauen *(salopp)*

~ 'out 1. *v. t.* ausräumen. **2.** *v. i.* *(coll.)* verschwinden

~ 'up 1. *v. t.* **a)** beseitigen ‹*Unordnung*›; wegräumen ‹*Abfall*›; aufräumen ‹*Platz, Sachen*›; **b)** *(explain, solve)* klären. **2.** *v. i.* **a)** aufräumen; Ordnung machen; **b)** *(become ~)* ‹*Wetter:*› sich aufhellen; **c)** *(disappear)* ‹*Symptome, Ausschlag:*› zurückgehen

clearance ['klɪərəns] *n.* **a)** *(of obstruction)* Beseitigung, *die*; *(of forest)* Abholzung, *die*; **b)** *(to land/take off)* Lande-/Startererlaubnis, *die*; **c)** *(security ~) Einstufung als unbedenklich [im Sinne der Sicherheitsbestimmungen]*; *(document)* ≈ Sonderausweis, *der*; **d)** *(clear space)* Spielraum, *der*; *(headroom)* lichte Höhe; **e)** *(Sport)* Abwehr, *die*; **make a poor ~:** schlecht abwehren

clearance: ~ **order** *n.* Räumungsbefehl, *der;* ~ **sale** *n.* Räumungsverkauf, *der*

'**clear-cut** *adj.* klar umrissen; klar ‹*Abgrenzung, Ergebnis, Entscheidung*›; [gestochen] scharf ‹*Umriß, Raster*›

clearing ['klɪərɪŋ] *n.* *(land)* Lichtung, *die*

'**clearing bank** *n.* *(Commerc.)* Clearingbank, *die*

clearly ['klɪəlɪ] *adv.* **a)** *(distinctly)* klar; deutlich ‹*sprechen*›; **b)** *(obviously, unambiguously)* eindeutig; klar ‹*denken*›

clearness ['klɪənɪs] *n., no pl. see* **clear 1 a–c**: Klarheit, *die;* Reinheit, *die;* Schärfe, *die;* Deutlichkeit, *die*

clear: ~**-out** *n.* Entrümpelung, *die;* **have a ~-out** eine Aufräum- *od.* Entrümpelungsaktion starten; ~**-up** *n.* Aufräumen, *das;* **have a [good] ~-up** [gründlich] aufräumen; ~**way** *n.* *(Brit.)* Straße mit Halteverbot

cleat [kli:t] *n.* *(to prevent rope from slipping)* Klampe, *die* *(Seemannsspr.)*

cleavage ['kli:vɪdʒ] *n.* **a)** *(act of splitting)* Spaltung, *die;* **b)** *(coll.: between breasts)* Dekolleté, *das*

cleave [kli:v] *v.t.,* **~d** *or* **clove** [kləʊv] *or* **cleft** [kleft], **~d** *or* **cloven** ['kləʊvn] *or* **cleft** *(literary)* **a)** *(split)* spalten; **b)** *(make way through)* durchpflügen ‹*Wellen, Wasser*›. *See also* **²cleft 2; cloven 2**

cleaver ['kli:və(r)] *n.* Hackbeil, *das*

clef [klef] *n.* *(Mus.)* Notenschlüssel, *der*

¹cleft [kleft] *n.* Spalte, *die*

²cleft 1. *see* **cleave. 2.** *adj.* gespalten; ~ **palate** Gaumenspalte, *die;* **be [caught] in a ~ stick** *(fig.)* in der Klemme sitzen *(ugs.)*

clematis ['klemətɪs, klə'meɪtɪs] *n.* *(Bot.)* Klematis, *die*

clemency ['klemənsɪ] *n., no pl.* Nachsicht, *die;* **show ~ to sb.** jmdm. gegenüber Nachsicht walten lassen

clench [klentʃ] *v.t.* **a)** *(close tightly)* zusammenpressen; ~ **one's fist** *or* **fingers** die Faust ballen; **with [one's] ~ed fist** mit geballter Faust; ~ **one's teeth** die Zähne zusammenbeißen; **b)** *(grasp firmly)* umklammern

clergy ['klɜ:dʒɪ] *n. pl.* Geistlichkeit, *die;* Klerus, *der*

clergyman ['klɜ:dʒɪmən] *n., pl.* **~men** ['klɜ:dʒɪmən] Geistliche, *der*

cleric ['klerɪk] *n.* Kleriker, *der*

clerical ['klerɪkl] *adj.* **a)** *(of clergy)* klerikal; geistlich; **b)** *(of or by clerk)* Büro‹*arbeit, -personal*›; ~ **error** Schreibfehler, *der;* ~ **worker** Büroangestellte, *der/die;* Bürokraft, *die*

clerk [klɑ:k] *n.* **a)** Angestellte, *der/die;* *(in bank)* Bankangestellte, *der/die;* *(in office)* Büroangestellte, *der/die;* **b)** *(in charge of records)* Schriftführer, *der/* Schriftführerin, *die*

clever ['klevə(r)] *adj.,* **~er** ['klevərə(r)], **~est** ['klevərɪst] **a)** gescheit; klug; **be ~ at mathematics/thinking up excuses** gut in Mathematik/findig im Ausdenken von Entschuldigungen sein; **b)** *(skilful)* geschickt; **be ~ with one's hands** geschickte Hände haben; **c)** *(ingenious)* brillant, geistreich ‹*Idee, Argument, Rede, Roman, Gedicht*›; geschickt ‹*Täuschung, Vorgehen*›; glänzend *(ugs.)* ‹*Idee, Erfindung, Mittel*›; **d)** *(smart, cunning)* clever; raffiniert ‹*Schritt, Taktik, Täuschung*›; schlau, raffiniert ‹*Person*›

'**clever Dick** *n.* *(coll. derog.)* Schlaumeier, *der (ugs.)*

cleverly ['klevəlɪ] *adv.* **a)** klug; **b)** *(skilfully)* geschickt; **c)** *(cunningly)* trickreich

cleverness ['klevənɪs] *n., no pl.* **a)** Klugheit, *die;* *(talent)* Begabung, *die* **(at** für); **b)** *(skill)* Geschicklichkeit, *die;* **c)** *(ingenuity)* Brillanz, *die;* **d)** *(smartness)* Cleverneß, *die;* Raffiniertheit, *die;* *(of person also)* Schläue, *die*

cliché ['kli:ʃeɪ] *n.* Klischee, *das*

click [klɪk] **1.** *n.* Klicken, *das.* **2.** *v.t.* zuschnappen lassen ‹*Schloß, Tür*›; ~ **the shutter of a camera** den Verschluß einer Kamera auslösen; ~ **one's tongue** mit der Zunge schnalzen. **3.** *v.i.* **a)** klikken; ‹*Absätze, Stricknadeln:*› klappern; **b)** *(sl.: fall into context)* **it's just ~ed** ich hab's *(ugs.);* ~ **with sb.** *(sl.)* mit jmdm. gleich prima auskommen *(ugs.)*

client ['klaɪənt] *n.* **a)** *(of lawyer, social worker)* Klient, *der/*Klientin, *die;* *(of architect)* Auftraggeber, *der/*-geberin, *die;* **b)** *(customer)* Kunde, *der/*Kundin, *die*

clientele [kli:ɒn'tel] *n.* *(of shop)* Kundschaft, *die*

cliff [klɪf] *n.* *(on coast)* Kliff, *das;* *(inland)* Felswand, *die*

'**cliff-hanger** *n.* Thriller, *der*

climate ['klaɪmət] *n.* Klima, *das;* **the ~ of opinion** *(fig.)* die allgemeine Meinung

climax ['klaɪmæks] **1.** *n.* Höhepunkt, *der.* **2.** *v.i.* seinen Höhepunkt erreichen

climb [klaɪm] **1.** *v.t.* hinaufsteigen ‹*Treppe, Leiter, Hügel, Berg*›; hinaufklettern ‹*Mauer, Seil, Mast*›; klettern auf ‹*Baum*›; ‹*Auto:*› hinaufkommen ‹*Hügel*›; **this mountain had never been ~ed before** dieser Berg war noch nie zuvor bestiegen worden; **the prisoners escaped by ~ing the wall** die Gefangenen entkamen, indem sie über die Mauer kletterten. **2.** *v.i.* **a)** klettern **(up** auf + *Akk.*); ~ **into/out of** steigen in (+ *Akk.*)/aus ‹*Auto, Bett*›; ~ **aboard** einsteigen; **b)** ‹*Flugzeug, Sonne:*› aufsteigen; **c)** *(slope upwards)* ansteigen. **3.** *n.* *(ascent)* Aufstieg, *der;* *(of aeroplane)* Steigflug, *der*

~ '**down** *v.i.* **a)** hinunterklettern; *(from horse)* absteigen; **b)** *(fig.: retreat, give in)* nachgeben; einlenken; ~ **down over an issue** in einer Frage nachgeben

'**climb-down** *n.* Rückzieher, *der (ugs.)*

climber ['klaɪmə(r)] *n.* **a)** *(mountaineer)* Bergsteiger, *der;* *(of cliff, rock-face)* Kletterer, *der;* **b)** *(plant)* Kletterpflanze, *die*

climbing: ~**-boot** *n.* Kletterschuh, *der;* ~**-frame** *n.* Klettergerüst, *das*

clinch [klɪntʃ] **1.** *v.t.* zum Abschluß bringen ‹*Angelegenheit*›; perfekt machen *(ugs.)* ‹*Geschäft, Handel*›; **that ~es it** damit ist der Fall klar. **2.** *n.* *(Boxing)* Clinch, *der*

cling [klɪŋ] *v.i.,* **clung** [klʌŋ] **a)** ~ **to sth./sb.** sich an etw./jmdn. klammern; ‹*Schmutz:*› einer Sache/jmdm. anhaften; ‹*Staub:*› sich auf etw./jmdn. setzen; **his sweat-soaked shirt clung to his back** das durchgeschwitzte Hemd klebte ihm am Rücken; ~ **together** aneinanderhaften; **b)** *(remain stubbornly faithful)* ~ **to sb./sth.** sich an jmdn./etw. klammern

'**cling film** *n.* Klarsichtfolie, *die*

clinic ['klɪnɪk] *n.* [Abteilung einer] Klinik; *(occasion)* Sprechstunde, *die;* *(private hospital)* Privatklinik, *die;* **dental ~:** Zahnklinik, *die*

clinical ['klɪnɪkl] *adj.* **a)** *(Med.)* klinisch ‹*Medizin, Tod*›; **b)** *(dispassionate)* nüchtern; *(coldly detached)* kühl

¹clink [klɪŋk] **1.** *n.* *(of glasses, bottles)* Klirren, *das;* *(of coins, keys)* Klimpern, *das.* **2.** *v.i.* ‹*Flaschen, Gläser:*› klirren; ‹*Münzen, Schlüssel:*› klimpern. **3.** *v.t.* klirren mit ‹*Glas*›

²clink *n. (sl.: prison)* Knast, *der (salopp)*

¹clip [klɪp] **1.** *n.* **a)** Klammer, *die; (for paper)* Büroklammer, *die; (of pen)* Klipp, *der;* **b)** *(piece of jewellery)* Klipp, *der;* Clip, *der.* **2.** *v. t.*, **-pp-: ~ sth. [on] to sth.** etw. an etw. *(Akk.)* klammern; **~ papers together** Schriftstücke zusammenklammern

~ 'on *v. t.* anlegen ‹*Ohrring*›; anstecken ‹*Brosche, Mikrophon*›

²clip 1. *v. t.*, **-pp-: a)** *(cut)* schneiden ‹*Fingernägel, Haar, Hecke*›; scheren ‹*Wolle*›; stutzen ‹*Flügel*›; **b)** scheren ‹*Schaf*›; trimmen ‹*Hund*›; **c)** lochen, entwerten ‹*Fahrkarte*›. **2.** *n.* **a)** *(of fingernails, hedge)* Schneiden, *das; (of sheep)* Schur, *die; (of dog)* Trimmen, *das;* **give the hedge a ~:** die Hecke schneiden; **b)** *(extract from film)* [Film]ausschnitt, *der;* **c)** *(blow with hand)* Schlag, *der;* **~ round** *or* **on** *or* **over the ear** Ohrfeige, *die*

clip: ~board *n.* Klemmbrett, *das;* **~-joint** *n. (sl. derog.)* Nepplokal, *das (ugs. abwertend);* **~-on** *adj.* **~-on sun-glasses** eine Sonnenbrille zum Aufstecken

clipped [klɪpt] *adj.* abgehackt ‹*Wörter*›

clipper ['klɪpə(r)] *n. (Naut.)* Klipper, *der*

clipping ['klɪpɪŋ] *n.* **a)** *(piece clipped off)* Schnipsel, *der od. das;* **b)** *(newspaper cutting)* Ausschnitt, *der*

clique [kli:k] *n.* Clique, *die*

cloak [kləʊk] **1.** *n.* Umhang, *der;* Mantel, *der (hist.);* **under the ~ of darkness** im Schutz der Dunkelheit; **use sth. as a ~ for sth.** etw. als Deckmantel für etw. benutzen; **a ~ of secrecy** ein Mantel des Schweigens. **2.** *v. t.* **a)** [ein]hüllen; **b)** *(fig.)* **~ed in mist/darkness** in Nebel/Dunkel gehüllt; **sth. is ~ed in secrecy** über etw. *(Akk.)* wird der Mantel des Schweigens gebreitet

cloak: ~-and-'dagger *adj.* mysteriös; Spionage‹*stück, -tätigkeit*›; **~room** *n.* Garderobe, *die; (Brit. euphem.: lavatory)* Toilette, *die;* **~room attendant** Garderobier, *der*/Garderobiere, *die*/Toilettenmann, *der*/-frau, *die*

clock [klɒk] **1.** *n.* **a)** Uhr, *die;* **[work] against the ~:** gegen die Zeit [arbeiten]; **beat the ~ [by ten minutes]** [10 Minuten] früher fertig werden; **put** *or* **turn the ~ back** *(fig.)* die Zeit zurückdrehen; **round the ~:** rund um die Uhr; **watch the ~** *(fig.)* [dauernd] auf die Uhr sehen *(weil man ungeduldig auf den Arbeitsschluß wartet);* **b)** *(coll.) (speedometer)* Tacho, *der (ugs.); (milometer)* ≈ Kilometerzähler, *der; (taximeter)* Taxameter, *das.* **2.** *v. t.* **~ [up]** zu verzeichnen haben ‹*Sieg, Zeit, Erfolg*›; erreichen ‹*Geschwindigkeit*›; zurücklegen ‹*Entfernung*›

~ 'in, ~ 'on *v. i.* [bei Arbeitsantritt] stechen *od.* stempeln

~ 'off, ~ 'out *v. i.* [bei Arbeitsschluß] stechen *od.* stempeln

clock: ~-face *n.* Zifferblatt, *das;* **~ tower** *n.* Uhr[en]turm, *der*

'clockwise 1. *adv.* im Uhrzeigersinn. **2.** *adj.* im Uhrzeigersinn nachgestellt

'clockwork *n.* Uhrwerk, *das;* **as regular as ~** *(fig.)* absolut regelmäßig; **go like ~** *(fig.)* klappen wie am Schnürchen *(ugs.)*

clod [klɒd] *n.* Klumpen, *der; (of earth)* Scholle, *die*

clog [klɒg] **1.** *n.* Holzschuh, *der; ([fashionable] wooden-soled shoe)* Clog, *der.* **2.** *v. t.*, **-gg-: ~ [up]** verstopfen ‹*Rohr, Poren*›; blockieren ‹*Rad, Maschinerie*›; **be ~ged [up] with sth.** mit etw. verstopft/durch etw. blockiert sein

cloister ['klɔɪstə(r)] *n.* **a)** *(covered walk)* Kreuzgang, *der;* **b)** *(convent, monastery; monastic life)* Kloster, *das*

cloistered ['klɔɪstəd] *adj. (fig.)* klösterlich ‹*Abgeschiedenheit, Dasein*›

clone [kləʊn] *(Biol.)* **1.** *n.* Klon, *der; (fig.: copy)* [schlechte] Kopie. **2.** *v. t.* klonen

close 1. [kləʊs] *adj.* **a)** *(near in space)* dicht; nahe; **be ~ to sth.** nahe bei *od.* an etw. *(Dat.)* sein; **how ~ is London to the South coast?** wie weit ist London von der Südküste entfernt?; **you're too ~ to the fire** du bist zu dicht *od.* nah am Feuer; **I wish we lived ~r to your parents** ich wünschte, wir würden näher bei deinen Eltern wohnen; **be ~ to tears/breaking-point** den Tränen/einem Zusammenbruch nahe sein; **at ~ quarters, the building looked less impressive** aus der Nähe betrachtet, wirkte das Gebäude weniger imposant; **at ~ range** aus kurzer Entfernung; **b)** *(near in time)* nahe (**to** an + *Dat.*); **c)** eng ‹*Freund, Freundschaft, Beziehung, Zusammenarbeit, Verbindung*›; nahe ‹*Verwandte, Bekanntschaft*›; **be/become ~ to sb.** jmdm. nahestehen/nahekommen; **d)** *(rigorous, painstaking)* eingehend, genau ‹*Untersuchung, Prüfung, Befragung usw.*›; **pay ~ attention** genau aufpassen; **e)** *(stifling)* stickig ‹*Luft, Raum*›; drückend, schwül ‹*Wetter*›; **f)** *(nearly equal)* hart ‹*[Wett]kampf, Spiel*›; knapp ‹*Ergebnis*›; **a ~ race** ein Kopf-an-Kopf-Rennen; **that was a ~ call** *or* **shave** *or* **thing** *(coll.)* das war knapp!; **g)** *(nearly matching)* wortgetreu ‹*Übersetzung*›; getreu, genau ‹*Imitation, Kopie*›; groß ‹*Ähnlichkeit*›; **be the ~st equivalent to sth.** einer Sache *(Dat.)* am ehesten entsprechen; **bear a ~ resemblance to sb.** jmdm. sehr ähnlich sehen; **h)** eng ‹*Schrift*›. **2.** [kləʊs] *adv.* **a)** *(near)* nah[e]; **come ~ to the truth** der Wahrheit nahekommen; **be ~ at hand** in Reichweite sein; **~ by** in der Nähe; **~ by the river** nahe am Fluß; **~ on 60 years** fast 60 Jahre; **~ on 2 o'clock** kurz vor 2 [Uhr]; **~ to sb./sth.** nahe bei jmdm./etw.; **don't stand so ~ to the edge of the cliff** stell dich nicht so nah *od.* dicht an den Rand des Kliffs; **~ together** dicht beieinander; **it brought them ~r together** *(fig.)* es brachte sie einander näher; **~ behind** dicht dahinter; **be/come ~ to tears** den Tränen nahe sein; **b)** fest ‹*schließen*›; genau ‹*hinsehen*›; **on looking ~r** bei genauerem Hinsehen. **3.** [kləʊz] *v. t.* **a)** *(shut)* schließen, *(ugs.)* zumachen ‹*Augen, Tür, Fenster, Geschäft*›; zuziehen ‹*Vorhang*›; *(declare shut)* schließen ‹*Laden, Geschäft, Fabrik, Betrieb, Werk, Zeche*›; stillegen ‹*Betrieb, Werk, Zeche, Bahnlinie*›; sperren ‹*Straße, Brücke*›; **b)** *(conclude)* schließen, beenden ‹*Besprechung, Rede, Diskussion*›; schließen ‹*Versammlung, Sitzung*›; **~ an account** ein Konto auflösen; **c)** *(make smaller)* schließen *(auch fig.)* ‹*Lücke*›. **4.** [kləʊz] *v. i.* **a)** *(shut)* sich schließen; ‹*Tür:*› zugehen *(ugs.)*, sich schließen; **the door/lid doesn't ~ properly** die Tür/der Deckel schließt nicht richtig; **b)** ‹*Laden, Geschäft, Fabrik:*› schließen, *(ugs.)* zumachen; *(permanently)* ‹*Betrieb, Werk, Zeche:*› geschlossen *od.* stillgelegt werden; ‹*Geschäft:*› geschlossen werden, *(ugs.)* zumachen; **c)** *(come to an end)* zu Ende gehen; enden; *(finish speaking)* schließen. **5.** *n.* **a)** [kləʊz] *no pl.* Ende, *das;* Schluß, *der;* **come** *or* **draw to a ~:** zu Ende gehen; **bring** *or* **draw sth. to a ~:** einer Sache *(Dat.)* ein Ende bereiten;

etw. zu Ende bringen; **b)** [kləʊs] *(cul-de-sac)* Sackgasse, *die*

~ [kləʊz] '**down 1.** *v.t.* schließen; *(ugs.)* zumachen; stillegen ‹*Werk, Zeche*›; einstellen ‹*Betrieb, Arbeit*›. **2.** *v.i.* geschlossen werden; zugemacht werden *(ugs.)*; ‹*Werk, Zeche:*› stillgelegt werden; *(Brit.)* ‹*Rundfunkstation:*› Sendeschluß haben

~ '**in** *v.i.* ‹*Nacht, Dunkelheit:*› hereinbrechen; ‹*Tage:*› kürzer werden; ~ **in [up]on sb./sth.** *(draw nearer)* sich jmdm./etw. nähern; *(draw around)* jmdn./etw. umzingeln

~ '**off** *v.t.* [ab]sperren; abriegeln

~ '**up 1.** *v.i.* **a)** aufrücken; **b)** ‹*Blume:*› sich schließen; **c)** *(lock up)* abschließen. **2.** *v.t.* abschließen

closed [kləʊzd] *adj.* **a)** *(no longer open)* geschlossen ‹*Laden, Geschäft, Fabrik*›; **we're ~:** wir haben geschlossen; **'~'** „Geschlossen"; **the subject is ~:** das Thema ist [für mich] erledigt; **b)** *(restricted)* [der Öffentlichkeit] nicht frei zugänglich

'**closed-circuit** *adj.* ~ **television** interne Fernsehanlage; *(for supervision)* Fernsehüberwachungsanlage, *die*

close-down ['kləʊzdaʊn] *n.* **a)** *(closing)* Schließung, *die*; *(of works, railway, mine)* Stillegung, *die*; *(of project, operation)* Einstellung, *die*; **b)** *(Radio, Telev.)* Sendeschluß, *der*

closed '**shop** *n.* Closed Shop, *der*; **we have** *or* **operate a ~ in this factory** in unserer Fabrik besteht Gewerkschaftszwang

close-fitting ['kləʊsfɪtɪŋ] *adj.* enganliegend; knapp sitzend ‹*Anzug*›

closely ['kləʊslɪ] *adv.* **a)** dicht; **follow me ~:** bleib *od.* geh dicht hinter mir!; **look ~ at** genau betrachten; **look ~ into** *(fig.)* näher untersuchen; **b)** *(intimately)* eng; **we're not ~ related** wir sind nicht nah miteinander verwandt; **c)** *(rigorously, painstakingly)* genau; genau, eingehend ‹*befragen, prüfen*›; streng, scharf ‹*bewachen*›; **a ~ guarded secret** ein streng *od.* sorgsam gehütetes Geheimnis; **d)** *(nearly equally)* ~ **fought/contested** hart umkämpft; **e)** *(exactly)* genau; ~ **resemble sb.** jmdm. sehr ähneln; **f)** ~ **printed/written** eng bedruckt/beschrieben; ~ **reasoned** *(fig.)* schlüssig

closeness ['kləʊsnɪs] *n., no pl.* **a)** *(nearness in space or time)* Nähe, *die*; **b)** *(intimacy)* Enge, *die*; **c)** *(rigorousness)* Genauigkeit, *die*; *(of questioning)* Nachdrücklichkeit, *die*; **d)** *(of atmosphere, air)* Schwüle, *die*; **e)** *(exactness)* **the ~ of the fit** der genaue Sitz; **the ~ of a translation** die Worttreue einer Übersetzung

close season ['kləʊs si:zn] *n.* Schonzeit, *die*

closet ['klɒzɪt] *n.* **a)** *(Amer.: cupboard)* Schrank, *der*; **come out of the ~** *(fig.)* sich nicht länger verstecken; **b)** *(water-~)* Klosett, *das*

closeted ['klɒzɪtɪd] *adj.* **be ~ together/with sb.** eine Besprechung/mit jmdm. eine Besprechung hinter verschlossenen Türen haben

close-up ['kləʊsʌp] *n.* *(Cinemat., Telev.)* ~ **[picture/shot]** Nahaufnahme *die*; *(of face etc.)* Großaufnahme, *die*; **in ~:** in Nahaufnahme/Großaufnahme

closing ['kləʊzɪŋ]: ~ **date** *n.* *(for competition)* Einsendeschluß, *der*; *(to take part)* Meldefrist, *die*; **the ~ date for applications for the job is ...:** Bewerbungen bitte bis zum ... einreichen; **~-time** *n.* *(of public house)* Polizeistunde, *die*; *(of shop)* Ladenschlußzeit, *die*

closure ['kləʊʒə(r)] *n.* **a)** *(closing)* Schließung, *die*; *(of factory, pit also)* Stillegung, *die*; *(of road, bridge)* Sperrung, *die*; **b)** *(cap, stopper)* [Flaschen]verschluß, *der*

clot [klɒt] **1.** *n.* **a)** Klumpen, *der*; **b)** *(Brit. sl.: stupid person)* Trottel, *der (ugs. abwertend)*. **2.** *v.i.*, **-tt-** ‹*Blut:*› gerinnen; ‹*Sahne:*› klumpen

cloth [klɒθ] *n., pl.* **~s** [klɒθs] **a)** Stoff, *der*; Tuch, *das*; **cut one's coat according to one's ~** *(fig.)* sich nach der Decke strecken *(ugs.)*; **b)** *(piece of ~)* Tuch, *das*; *(dish-~)* Spültuch, *das*; *(table-~)* Tischtuch, *das*; [Tisch]decke, *die*; *(duster)* Staubtuch, *das*

clothe [kləʊð] *v.t.* *(lit. or fig.)* kleiden

clothes [kləʊðz] *n. pl.* Kleider *Pl.*; *(collectively)* Kleidung, *die*; **with one's ~ on** angezogen; **put one's ~ on** sich anziehen; **take one's ~ off** sich ausziehen

clothes: ~-brush *n.* Kleiderbürste, *die*; **~-hanger** *n.* Kleiderbügel, *der*; **~-horse** *n.* Wäscheständer, *der*; **~-line** *n.* Wäscheleine, *die*; **~-peg** *(Brit.)*, **~-pin** *(Amer.)* *ns.* Wäscheklammer, *die*

clothing ['kləʊðɪŋ] *n., no pl.* Kleidung, *die*; **article of ~:** Kleidungsstück, *das*

clotted cream [klɒtɪd 'kri:m] *n.* *sehr fetter Rahm*

cloud [klaʊd] **1.** *n.* **a)** Wolke, *die*; *(collective)* Bewölkung, *die*; **go round with** *or* **have one's head in the ~s** *(fig.)* *(be unrealistic)* in den Wolken schweben; *(be absent-minded)* mit seinen Gedanken ganz woanders sein; **every ~ has a silver lining** *(prov.)* es hat alles sein Gutes; **b)** ~ **of dust/smoke** Staub-/Rauchwolke, *die*; **c)** *(fig.: cause of gloom or suspicion)* dunkle Wolke; **he left under a ~:** unter zweifelhaften Umständen schied er aus dem Dienst. **2.** *v.t.* **a)** verdunkeln ‹*Himmel*›; blind machen ‹*Fenster[scheibe], Spiegel*›; **b)** *(fig.: cast gloom or trouble on)* trüben ‹*Glück, Freude, Aussicht*›; überschatten ‹*Zukunft*›; *(make unclear)* trüben ‹*Urteilsvermögen, Verstand, Bewußtsein*›

~ '**over** *v.i.* sich bewölken; ‹*Spiegel:*› beschlagen

cloud: ~burst *n.* Wolkenbruch, *der*; **C~-'cuckoo-land** *n.* Wolkenkuckucksheim, *das (geh.)*

cloudless ['klaʊdlɪs] *adj.* wolkenlos

cloudy ['klaʊdɪ] *adj.* bewölkt, bedeckt, wolkig ‹*Himmel*›; trübe ‹*Wetter, Flüssigkeit, Glas*›

clout [klaʊt] **1.** *n.* **a)** *(coll.: hit)* Schlag, *der*; **b)** *(coll.: power, influence)* Schlagkraft, *die*. **2.** *v.t.* *(coll.)* hauen *(ugs.)*; ~ **sb. round the ear** jmdm. eins hinter die Ohren geben *(ugs.)*; ~ **sb. [one]** jmdm. eine runterhauen *(salopp)*

[1]**clove** [kləʊv] *n.* Brutzwiebel, *die*; *(of garlic)* [Knoblauch]zehe, *die*

[2]**clove** *n.* *(spice)* [Gewürz]nelke, *die*

[3]**clove** *see* **cleave**

cloven ['kləʊvn] **1.** *see* **cleave. 2.** *adj.* ~ **foot/hoof** Spaltfuß, *der* *(veralt.)*/Spalthuf, *der (veralt.)*; *(of devil)* Pferdefuß, *der*

clover ['kləʊvə(r)] *n.* Klee, *der*

'**clover-leaf** *n.* *(also Road Constr.)* Kleeblatt, *das*

clown [klaʊn] **1.** *n.* **a)** Clown, *der*; **act** *or* **play the ~:** den Clown spielen; **b)** *(ignorant person)* Dummkopf, *der (ugs.)*; *(ill-bred person)* ungehobelter Klotz. **2.** *v.i.* ~ **[about** *or* **around]** den Clown spielen *(abwertend)*

cloy [klɔɪ] *v.t.* übersättigen; überfüttern

cloying ['klɔɪɪŋ] *adj.* *(lit. or fig.)* süßlich

club [klʌb] **1.** *n.* **a)** Keule, *die*; *(golf-~)* Golfschläger, *der*; **b)** *(association)* Klub, *der*; Club, *der*; Verein, *der*; **join the ~** *(fig.)* mitmachen; **join the** *or* **welcome to the ~!** *(fig.)* du also auch!; **c)**

(premises) Klub, *der; (buildings/ grounds)* Klubhaus/-gelände, *das;* **d)** *(Cards)* Kreuz, *das;* **the ace/seven of ~s** das Kreuzas/die Kreuzsieben. **2.** *v.t.*, **-bb-** *(beat)* prügeln; *(with ~)* knüppeln. **3.** *v.i.*, **-bb-: ~ together** sich zusammentun; *(in order to buy something)* zusammenlegen

'clubhouse *n.* Klubhaus, *das*

cluck [klʌk] *v.i.* gackern; *(to call chicks)* glucken

clue [klu:] *n.* **a)** *(fact, principle)* Anhaltspunkt, *der; (in criminal investigation)* Spur, *die;* **b)** *(fig. coll.)* **give sb. a ~:** jmdm. einen Tip geben; **not have a ~:** keine Ahnung haben *(ugs.);* **c)** *(in crossword)* Frage, *die*

clueless ['klu:lɪs] *adj. (coll. derog.)* unbedarft *(ugs.)* ‹*Person*›

clump [klʌmp] **1.** *n. (of trees, bushes, flowers)* Gruppe, *die; (of grass)* Büschel, *das.* **2.** *v.i. (tread)* stapfen. **3.** *v.t.* zusammengruppieren; in Gruppen anordnen

clumsiness ['klʌmzɪnɪs] *n., no pl. see* **clumsy:** Schwerfälligkeit, *die;* Plumpheit, *die*

clumsy ['klʌmzɪ] *adj.* **a)** *(awkward)* schwerfällig, unbeholfen ‹*Person, Bewegungen*›; plump ‹*Form, Figur*›; tolpatschig ‹*Heranwachsender*›; **b)** *(ill-contrived)* plump ‹*Verse, Nachahmung*›; unbeholfen ‹*Worte*›; primitiv ‹*Vorrichtung, Maschine*›; **c)** *(tactless)* plump

clung *see* **cling**

cluster ['klʌstə(r)] **1.** *n.* **a)** *(of grapes, berries)* Traube, *die; (of fruit, flowers, curls)* Büschel, *das; (of trees, shrubs)* Gruppe, *die;* **b)** *(of stars, cells)* Haufen, *der.* **2.** *v.i.* **~ [a]round sb./sth.** sich um jmdn./etw. scharen *od.* drängen

clutch [klʌtʃ] **1.** *v.t.* umklammern. **2.** *v.i.* **~ at sth.** nach etw. greifen; *(fig.)* sich an etw. *(Akk.)* klammern. **3.** *n.* **a)** *in pl. (fig.: control)* **fall into sb.'s ~es** jmdm. in die Klauen fallen; **b)** *(Motor Veh., Mech.)* Kupplung, *die;* **let in the ~, put the ~ in** einkuppeln; **disengage the ~, let the ~ out** auskuppeln

clutter ['klʌtə(r)] **1.** *n.* Durcheinander, *das.* **2.** *v.t.* **~ [up] the table/room** überall auf dem Tisch/im Zimmer herumliegen; **be ~ed [up] with sth.** ‹*Zimmer:*› mit etw. vollgestopft sein; ‹*Tisch:*› mit etw. übersät sein

cm. *abbr.* **centimetre[s]** cm

CND *abbr. (Brit.)* **Campaign for Nuclear Disarmament** Kampagne für atomare Abrüstung

Co. *abbr.* **a) company** Co.; **and Co.** [ənd kəʊ] *(coll.)* und Co. *(ugs.);* **b) county**

c/o *abbr.* **care of** bei; c/o

coach [kəʊtʃ] **1.** *n.* **a)** *(road vehicle)* Kutsche, *die; (state ~)* [Staats]karosse, *die;* **b)** *(railway carriage)* Wagen, *der;* **c)** *(bus)* [Reise]bus, *der;* **by ~:** mit dem Bus; **d)** *(tutor)* Privat- *od.* Nachhilfelehrer, *der/*-lehrerin, *die; (sport instructor)* Trainer, *der/* Trainerin, *die.* **2.** *v.t.* trainieren; **~ a pupil for an examination** einen Schüler auf eine Prüfung vorbereiten

coach: ~-station *n.* Busbahnhof, *der;* **~ tour** *n.* Rundreise [im Omnibus]; Omnibusreise, *die*

coagulate [kəʊ'ægjʊleɪt] **1.** *v.t.* gerinnen lassen; koagulieren *(fachspr.).* **2.** *v.i.* gerinnen; koagulieren *(fachspr.)*

coal [kəʊl] *n.* **a)** Kohle, *die; (hard ~)* Steinkohle, *die;* **b)** *(piece of ~)* Stück Kohle; **live ~s** Glut, *die;* **haul sb. over the ~s** *(fig.)* jmdm. die Leviten lesen *(ugs.);* **carry ~s to Newcastle** *(fig.)* Eulen nach Athen tragen *(fig.)*

coal: ~-cellar *n.* Kohlenkeller, *der;* **~-dust** *n.* Kohlenstaub, *der;* **~-face** *n.* Streb, *der;* **at the ~-face** im Streb *od.* vor Ort; **~field** *n.* Kohlenrevier, *das;* **~ fire** *n.* Kohlenfeuer, *das*

coalition [kəʊə'lɪʃn] *n. (Polit.)* Koalition, *die*

coal: ~-merchant *n.* Kohlenhändler, *der;* **~-mine** *n.* [Kohlen]bergwerk, *das;* **~-miner** *n.* [im Kohlenbergbau tätiger] Grubenarbeiter; **~-mining** *n.* Kohlenbergbau, *der;* **~-oil** *n. (Amer.)* Paraffin, *das*

coarse [kɔ:s] *adj.* **a)** *(in texture)* grob; rauh, grob ‹*Haut, Teint*›; **b)** *(unrefined, rude, obscene)* derb; roh ‹*Geschmack, Kraft*›; primitiv ‹*Person, Geist*›; ungehobelt ‹*Manieren, Person*›; gemein ‹*Lachen, Witz, Geräusch*›

coast [kəʊst] **1.** *n.* Küste, *die.* **2.** *v.i.* **a)** *(ride)* im Freilauf fahren; **b)** *(fig.: progress)* **they are just ~ing along in their work** sie tun bei der Arbeit nur das Nötigste; **he ~s through every examination** er schafft jede Prüfung spielend

coastal ['kəʊstl] *adj.* Küsten-

coaster ['kəʊstə(r)] *n.* **a)** *(mat)* Untersetzer, *der;* **b)** *(ship)* Küstenmotorschiff, *das;* Kümo, *das*

coast: ~guard *n.* **a)** *(person)* Angehörige[r] der Küstenwacht; **b)** *(organization)* Küstenwache, -wacht, *die;* **~line** *n.* Küste, *die*

coat [kəʊt] **1.** *n.* **a)** Mantel, *der;* **b)** *(layer)* Schicht, *die;* **c)** *(animal's hair, fur, etc.)* Fell, *das;* **d)** *see* **coating. 2.** *v.t.* überziehen; *(with paint)* streichen

coated ['kəʊtɪd] *adj.* belegt ‹*Zunge*›; **~ with dust/sugar** staubbedeckt/mit Zucker überzogen

coat: ~-hanger *n.* Kleiderbügel, *der;* **~-hook** *n.* Kleiderhaken, *der*

coating ['kəʊtɪŋ] *n. (of paint)* Anstrich, *der; (of dust, snow, wax, polish, varnish)* Schicht, *die*

coat of 'arms *n.* Wappen, *das*

co-author [kəʊ'ɔ:θə(r)] *n.* Mitautor, *der/*-autorin, *die*

coax [kəʊks] *v.t.* überreden; **~ sb. into doing sth.** jmdn. herumkriegen *(ugs.),* etw. zu tun; **~ a smile/ some money out of sb.** jmdm. ein Lächeln/etw. Geld entlocken

cob [kɒb] *n.* **a)** *(nut)* Haselnuß, *die;* **b)** *(swan)* männlicher Schwan; **c)** *see* **corn-cob**

cobalt ['kəʊbɔ:lt, 'kəʊbɒlt] *n. (element)* Kobalt, *das*

cobber ['kɒbə(r)] *n. (Austral. and NZ coll.)* Kumpel, *der (ugs.)*

¹cobble ['kɒbl] **1.** *n.* Pflaster-, Kopfstein, *der;* Katzenkopf, *der.* **2.** *v.t.* pflastern ‹*Straße*›; **~d streets** Straßen mit Kopfsteinpflaster

²cobble *v.t. (put together, mend)* flicken

cobbler ['kɒblə(r)] *n.* Schuster, *der*

'cobble-stone *see* **¹cobble 1**

'cob-nut *see* **cob a**

cobra ['kɒbrə] *n.* Kobra, *die*

cobweb ['kɒbweb] *n.* Spinnengewebe, *das;* Spinnennetz, *das*

cocaine [kə'keɪn] *n.* Kokain, *das*

cock [kɒk] **1.** *n.* **a)** *(bird, lobster, crab, salmon)* Männchen, *das; (domestic fowl)* Hahn, *der;* **b)** *(spout, tap, etc.)* Hahn, *der;* **c)** *(coarse: penis)* Schwanz, *der (salopp).* **2.** *v.t.* **a)** aufstellen, *(fig.)* spitzen ‹*Ohren*›; **b) a ~ed hat** ein Hut mit hoher Krempe; *(triangular hat)* ein Dreispitz; **knock sb./ sth. into a ~ed hat** *(fig.: surpass)* jmdn./etw. weit übertreffen; jmdn./etw. in den Sack stecken *(fig. ugs.);* **c) ~ a/the gun** den Hahn spannen

~ 'up *v.t. (Brit. sl.)* versauen *(salopp)*

cock-a-doodle-doo [kɒkədu:dl'du:] *n.* Kikeriki, *das*

cock-a-hoop [kɒkə'hu:p] *adj.* überschwenglich; *(boastful)* triumphierend

cock-and-'bull story *n.* Lügengeschichte, *die*

cockatoo [kɒkə'tu:] *n.* Kakadu, *der*
cockchafer ['kɒktʃeɪfə(r)] *n.* Maikäfer, *der*
cockerel ['kɒkərəl] *n.* junger Hahn
cock-eyed ['kɒkaɪd] *adj.* **a)** *(crooked)* schief; **b)** *(absurd)* verrückt
cockle [kɒkl] *n.* Herzmuschel, *die*
cockney ['kɒknɪ] **1.** *adj.* Cockney-. **2.** *n.* **a)** *(person)* waschechter Londoner/waschechte Londonerin; Cockney, *der;* **b)** *(dialect)* Cockney, *das*
'**cockpit** *n.* Cockpit, *das*
cockroach ['kɒkrəʊtʃ] *n.* [Küchen-, Haus-]schabe, *die*
cocksure [kɒk'ʃʊə(r)] *adj.* **a)** *(convinced)* todsicher; **b)** *(self-confident)* selbstsicher
cocktail ['kɒkteɪl] *n.* Cocktail, *der*
cocktail: ~ cabinet *n.* Hausbar, *die;* **~ dress** *n.* Cocktailkleid, *das;* **~ party** *n.* Cocktailparty, *die;* **~-shaker** *n.* Mixbecher, *der*
'**cock-up** *n. (Brit. sl.)* Schlamassel, *der (ugs.);* **make a ~ of sth.** bei etw. Scheiße bauen *(derb)*
cocky ['kɒkɪ] *adj.* anmaßend
cocoa ['kəʊkəʊ] *n.* Kakao, *der*
coconut ['kəʊkənʌt] *n.* Kokosnuß, *die*
coconut: ~ 'matting *n.* Kokosmatten *Pl.;* **~ milk** *n.* Kokosmilch, *die;* **~ palm** *n.* Kokospalme, *die;* **~ shy** *n.* Wurfbude, *die*
cocoon [kə'ku:n] **1.** *n.* **a)** *(Zool.)* Kokon, *der;* **b)** *(covering)* Hülle, *die.* **2.** *v.t.* einmummen
cod [kɒd] *n., pl. same* Kabeljau, *der; (in Baltic)* Dorsch, *der*
COD *abbr.* **cash on delivery; collect on delivery** *(Amer.)* p. Nachn.
coddle ['kɒdl] *v.t.* [ver]hätscheln ⟨*Kind*⟩; verwöhnen ⟨*Kranken*⟩
code [kəʊd] **1.** *n.* **a)** *(collection of statutes etc.)* Kodex, *der;* Gesetzbuch, *das;* **~ of honour** Ehrenkodex, *der;* **~s of behaviour** Verhaltensnormen; **b)** *(system of signals)* Kode, Code, *der; (coded word, etc.)* Chiffre, *die;* **be in ~:** verschlüsselt sein; **put sth. into ~:** etw. verschlüsseln. **2.** *v.t.* chiffrieren; verschlüsseln
code: ~-name *n.* Deckname, *der;* **~-number** *n.* Kenn-, Tarnzahl, *die;* **~-word** *n.* Kennwort, *das*
codger ['kɒdʒə(r)] *n. (coll.)* Knacker, *der (salopp)*
cod-liver 'oil *n.* Lebertran, *der*
co-driver ['kəʊdraɪvə(r)] *n.* Beifahrer, *der*/-fahrerin, *die*
coed ['kəʊed] *(esp. Amer. coll.)* **1.** *n.* Studentin, *die.* **2.** *adj.* koedukativ; Koedukations-; **~ school** gemischte Schule
coeducation [kəʊedjʊ'keɪʃn] *n.* Koedukation, *die*
coeducational [kəʊedjʊ'keɪʃənl] *adj.* koedukativ; Koedukations-
coefficient [kəʊɪ'fɪʃənt] *n. (Math., Phys.)* Koeffizient, *der*
coerce [kəʊ'ɜ:s] *v.t.* zwingen; **~ sb. into sth.** jmdn. zu etw. zwingen; **~ sb. into doing sth.** jmdn. dazu zwingen, etw. zu tun
coercion [kəʊ'ɜ:ʃn] *n.* Zwang, *der*
coexist [kəʊɪg'zɪst] *v.i.* ⟨*Ideen, Überzeugungen:*⟩ nebeneinander bestehen, koexistieren; **~ [together] with sb./sth.** neben jmdm./etw. bestehen; mit jmdm./etw. koexistieren
coexistence [kəʊɪg'zɪstəns] *n.* Koexistenz, *die*
C. of E. [si:əv'i:] *abbr.* **Church of England**
coffee ['kɒfɪ] *n.* Kaffee, *der;* **drink** *or* **have a cup of ~:** eine Tasse Kaffee trinken; **three black/white ~s** drei [Tassen] Kaffee ohne/mit Milch
coffee: ~ bar *n.* Café, *das;* **~-bean** *n.* Kaffeebohne, *die;* **~-break** *n.* Kaffeepause, *die;* **~-cup** *n.* Kaffeetasse, *die;* **~-filter** *n.* Kaffeefilter, *der;* **~-grinder** *n.* Kaffeemühle, *die;* **~-grounds** *n. pl.* Kaffeesatz, *der;* **~ morning** *n.* Morgenkaffee, *der;* **~-pot** *n.* Kaffeekanne, *die;* **~ shop** *n.* Kaffeestube, *die;* Café, *das; (selling ~-beans etc.)* Kaffeegeschäft, *das;* **~-table** *n.* Couchtisch, *der*
coffin ['kɒfɪn] *n.* Sarg, *der*
cog [kɒg] *n. (Mech.)* Zahn, *der;* **be just a ~ [in the wheel/machine]** *(fig.)* bloß ein Rädchen im Getriebe sein
cogent ['kəʊdʒənt] *adj. (convincing)* überzeugend ⟨*Argument*⟩; zwingend ⟨*Grund*⟩; *(valid)* stichhaltig ⟨*Kritik, Analyse*⟩
cogitate ['kɒdʒɪteɪt] *v.i. (formal/joc.)* nachsinnen, nachdenken **(on** über + *Akk.*)
cognac ['kɒnjæk] *n.* Cognac, *der* Ⓦ
cognate ['kɒgneɪt] *adj. (Ling.)* verwandt
cog: ~-railway *n. (esp. Amer.)* Zahnradbahn, *die;* **~-wheel** *n.* Zahnrad, *das*
cohabit [kəʊ'hæbɪt] *v.i.* zusammenleben
cohere [kəʊ'hɪə(r)] *v.i.* ⟨*Teile, Ganzes, Gruppe:*⟩ zusammenhalten
coherence [kəʊ'hɪərəns] *n.* Zusammenhang, *der;* Kohärenz, *die (geh.); (in work, system, form)* Geschlossenheit, *die*
coherent [kəʊ'hɪərənt] *adj.* **a)** *(cohering)* zusammenhängend; **b)** *(fig.)* zusammenhängend; kohärent *(geh.);* in sich *(Dat.)* geschlossen ⟨*System, Ganzes, Werk, Aufsatz, Form*⟩
coherently [kəʊ'hɪərəntlɪ] *adv.* zusammenhängend; im Zusammenhang
cohesive [kəʊ'hi:sɪv] *adj.* geschlossen, in sich *(Dat.)* ruhend ⟨*Ganzes, Einheit, Form*⟩; stimmig ⟨*Stil, Argument*⟩; kohäsiv ⟨*Masse, Mischung*⟩
coil [kɔɪl] **1.** *v.t.* **a)** *(arrange)* aufwickeln; **the snake ~ed itself round a branch** die Schlange wand sich um einen Ast; **b)** *(twist)* aufdrehen; **the snake ~ed itself up** die Schlange rollte sich auf. **2.** *v.i.* **a)** *(twist)* **~ round sth.** etw. umschlingen; **b)** *(move sinuously)* sich winden; ⟨*Rauch:*⟩ sich ringeln. **3.** *n.* **a) ~s of rope/wire/piping** aufgerollte Seile *Pl.*/aufgerollter Draht/aufgerollte Leitungen *Pl.;* **b)** *(single turn of ~ed thing)* Windung, *die;* **c)** *(length of ~ed rope etc.)* Stück, *das;* **d)** *(contraceptive device)* Spirale, *die;* **e)** *(Electr.)* Spule, *die*
coin [kɔɪn] **1.** *n.* Münze, *die; (metal money)* Münzen *Pl.;* **the other side of the ~** *(fig.)* die Kehrseite der Medaille. **2.** *v.t.* **a)** *(invent)* prägen ⟨*Wort, Redewendung*⟩; **..., to ~ a phrase** *(iron.)* ..., um mich ganz originell auszudrücken; **b)** *(make)* prägen ⟨*Geld*⟩
'**coin-box telephone** *n.* Münzfernsprecher, *der*
coincide [kəʊɪn'saɪd] *v.i.* **a)** *(in space)* sich decken; **~ with one another** sich decken; **b)** *(in time)* ⟨*Ereignisse, Veranstaltungen:*⟩ zusammenfallen; **c)** *(agree together)* übereinstimmen **(with** mit)
coincidence [kəʊ'ɪnsɪdəns] *n.* Zufall, *der;* **by pure** *or* **sheer ~:** rein zufällig; **by a curious ~:** durch einen merkwürdigen Zufall
coincidental [kəʊɪnsɪ'dentl] *adj.* zufällig
coincidentally [kəʊɪnsɪ'dentəlɪ] *adv.* gleichzeitig; *(by coincidence)* zufälligerweise
Coke, (P) [kəʊk] *n. (drink)* Coke, *das* Ⓦ
coke [kəʊk] *n.* Koks, *der*
colander ['kʌləndə(r)] *n.* Sieb, *das;* Durchschlag, *der*
cold [kəʊld] **1.** *adj.* **a)** kalt; **I feel**

~: ich friere; mir ist kalt; **her hands/feet were** ~: sie hatte kalte Hände/Füße; **b)** kalt ⟨*Intellekt, Herz*⟩; [betont] kühl ⟨*Person, Ansprache, Aufnahme, Begrüßung*⟩; eiskalt ⟨*Handlung*⟩; **leave sb.** ~: jmdn. kaltlassen *(ugs.)*; **c)** *(sl.: unconscious)* bewußtlos; k.o. *(ugs.)*; **he laid him out** ~: er schlug ihn k.o.; **d)** *(sexually frigid)* [gefühls]kalt; **e)** *(chilling, depressing)* kalt ⟨*Farbe*⟩; nackt ⟨*Tatsache, Statistik*⟩. **2.** *adv.* kalt. **3.** *n.* **a)** Kälte, *die;* **be left out in the** ~ *(fig.)* links liegengelassen werden; **b)** *(illness)* Erkältung, *die;* ~ **[in the head]** Schnupfen, *der*

cold: ~**-blooded** ['kəʊldblʌdɪd] *adj.* **a)** wechselwarm ⟨*Tier*⟩; kaltblütig *(selten)*; ~**-blooded animals** Kaltblüter *Pl.;* wechselwarme Tiere; **b)** *(callous)* kaltblütig ⟨*Person, Mord*⟩; ~ **cream** *n.* Cold Cream, *die od. das*

coldly ['kəʊldlɪ] *adv.* [betont] kühl; [eis]kalt ⟨*handeln*⟩

cold: ~ **'storage** *n.* Kühllagerung, *die;* ~ **store** *n.* Kühlhaus, *das;* ~ **'sweat** *n.* kalter Schweiß; **break out in a** ~ **sweat** in kalten Schweiß ausbrechen; ~ **'turkey** *n. (Amer. sl.)* Totalentzug, *der;* Cold turkey, *der (Drogenjargon);* ~ **'war** *n.* kalter Krieg

coleslaw ['kəʊlslɔ:] *n.* Kohl-, Krautsalat, *der*

colic ['kɒlɪk] *n.* Kolik, *die*

collaborate [kə'læbəreɪt] *v.i.* **a)** *(work jointly)* zusammenarbeiten; ~ **[with sb.] on sth.** zusammen [mit jmdm.] an etw. *(Dat.)* arbeiten; ~ **[with sb.] on** *or* **in doing sth.** mit jmdm. bei etw. zusammenarbeiten; **b)** *(co-operate with enemy)* kollaborieren

collaboration [kəlæbə'reɪʃn] *n.* Zusammenarbeit, *die;* *(with enemy)* Kollaboration, *die*

collaborator [kə'læbəreɪtə(r)] *n.* Mitarbeiter, *der*/-arbeiterin, *die;* *(with enemy)* Kollaborateur, *der*/Kollaborateurin, *die*

collage ['kɒlɑ:ʒ] *n.* Collage, *die*

collapse [kə'læps] **1.** *n.* **a)** *(of person) (physical or mental breakdown)* Zusammenbruch, *der;* *(heart-attack; of lung, blood-vessel, circulation)* Kollaps, *der;* **b)** *(of tower, bridge, structure, wall, roof)* Einsturz, *der;* **c)** *(fig.: failure)* Zusammenbruch, *der;* *(of negotiations, plans, hopes)* Scheitern, *das.* **2.** *v.i.* **a)** ⟨*Person:*⟩ zusammenbrechen; ⟨*Lunge, Gefäß, Kreislauf:*⟩ kollabieren; ~ **with laughter** *(fig.)* sich vor Lachen kugeln; **b)** ⟨*Zelt:*⟩ in sich zusammenfallen; ⟨*Tisch, Stuhl:*⟩ zusammenbrechen; ⟨*Turm, Brücke, Gebäude, Mauer, Dach:*⟩ einstürzen; **c)** *(fig.: fail)* ⟨*Verhandlungen, Pläne, Hoffnungen:*⟩ scheitern; ⟨*Geschäft, Unternehmen usw.:*⟩ zusammenbrechen; **d)** *(fold down)* ⟨*Regenschirm, Fahrrad, Tisch:*⟩ sich zusammenklappen lassen

collapsible [kə'læpsɪbl] *adj.* Klapp-, zusammenklappbar ⟨*Stuhl, Tisch, Fahrrad*⟩; Falt-, faltbar ⟨*Boot*⟩

collar ['kɒlə(r)] **1.** *n.* **a)** Kragen, *der;* **with** ~ **and tie** mit Krawatte; **hot under the** ~ *(fig.)* *(embarrassed)* verlegen; *(angry)* wütend; **b)** *(for dog)* [Hunde]halsband, *das.* **2.** *v.t. (seize)* am Kragen kriegen *(ugs.)*; schnappen *(ugs.)*

'collar-bone *n. (Anat.)* Schlüsselbein, *das*

collate [kə'leɪt] *v.t.* **a)** *(Bibliog.: compare)* kollationieren *(Buchw.)* ⟨*Manuskripte, Druckbögen*⟩; ~ **a copy with the original** eine Abschrift mit dem Original vergleichen; **b)** *(put together)* zusammenstellen ⟨*Daten, Beweismaterial*⟩

collateral [kə'lætərl] *n. (Finance)* ~ **[security]** Sicherheiten *Pl.*

colleague ['kɒli:g] *n.* Kollege, *der*/Kollegin, *die*

collect [kə'lekt] **1.** *v.i.* **a)** *(assemble)* sich versammeln; **b)** *(accumulate)* ⟨*Staub, Müll usw.:*⟩ sich ansammeln. **2.** *v.t.* **a)** *(assemble)* sammeln; aufsammeln ⟨*Müll, leere Flaschen usw.*⟩; ~ **[up] one's belongings** seine Siebensachen *(ugs.)* zusammensuchen; ~ **dust** Staub anziehen; **b)** *(coll.: fetch, pick up)* abholen ⟨*Person, Dinge*⟩; ~ **a parcel from the post office** ein Paket bei *od.* auf der Post abholen; ~ **sb. from the station** jmdn. am Bahnhof *od.* von der Bahn abholen; **c)** eintreiben ⟨*Steuern, Zinsen, Schulden*⟩; kassieren ⟨*Miete, Fahrgeld*⟩; beziehen ⟨*Zahlungen, Sozialhilfe*⟩; **d)** *(as hobby)* sammeln ⟨*Münzen, Bücher, Briefmarken, Gemälde usw.*⟩; **e)** ~ **one's wits/thoughts** seine Gedanken sammeln

collected [kə'lektɪd] *adj.* **a)** *(gathered)* gesammelt; **b)** *(calm)* gesammelt; gelassen

collection [kə'lekʃn] *n.* **a)** *(collecting)* Sammeln, *das;* *(of rent, fares)* Kassieren, *das;* *(of taxes, interest, debts)* Eintreiben, *das;* *(coll.: of goods, persons)* Abholen, *das;* **b)** *(amount of money collected)* Sammlung, *die;* *(in church)* Kollekte, *die;* **c)** *(of mail)* Abholung, *die;* *(from post-box)* Leerung, *die;* **d)** *(group collected)* *(of coins, books, stamps, paintings, etc.)* Sammlung, *die;* *(of fashionable clothes)* Kollektion, *die;* *(of people)* Ansammlung, *die;* **e)** *(accumulated quantity)* Ansammlung, *die*

collective [kə'lektɪv] **1.** *adj.* kollektiv *nicht präd.;* gesamt *nicht präd.* **2.** *n.* Genossenschaftsbetrieb, *der*

collective: ~ **'bargaining** *n.* Tarifverhandlungen *Pl.;* ~ **'noun** *n. (Ling.)* Kollektivum, *das*

collector [kə'lektə(r)] *n.* *(of stamps, coins, etc.)* Sammler, *der*/Sammlerin, *die;* *(of taxes)* Einnehmer, *der*/Einnehmerin, *die;* *(of rent, cash)* Kassierer, *der*/Kassiererin, *die*

college ['kɒlɪdʒ] *n.* **a)** *(esp. Brit.: independent corporation in university)* College, *das;* **b)** *(place of further education)* Fach[hoch]schule, *die;* **go to** ~ *(esp. Amer.)* studieren; **start** ~ *(esp. Amer.)* sein Studium aufnehmen; **c)** *(esp. Brit.: school)* Internatsschule, *die;* Kolleg, *das*

College of Edu'cation Pädagogische Hochschule

collide [kə'laɪd] *v.i.* **a)** *(come into collision)* zusammenstoßen **(with** mit); ⟨*Schiff:*⟩ kollidieren; **b)** *(be in conflict)* zusammenprallen; kollidieren

collie ['kɒlɪ] *n.* Collie, *der*

colliery ['kɒljərɪ] *n.* Kohlengrube, *die*

collision [kə'lɪʒn] *n.* **a)** *(colliding)* Zusammenstoß, *der;* *(between ships)* Kollision, *die;* **come into** ~: zusammenstoßen; ⟨*Schiffe:*⟩ in Kollision geraten, kollidieren; **a head-on** ~ **between a car and a bus** ein Frontalzusammenstoß eines PKW mit einem Bus; **b)** *(fig.)* Konflikt, *der;* Kollision, *die*

col'lision course *n. (lit. or fig.)* Kollisionskurs, *der;* **on a** ~: auf Kollisionskurs

collocator ['kɒləkeɪtə(r)] *n.* *(Ling.)* Kollokator, *der*

colloquial [kə'ləʊkwɪəl] *adj.* umgangssprachlich; ~ **language** Umgangssprache, *die*

collusion [kə'lju:ʒn, kə'lu:ʒn] *n.* geheime Absprache

Cologne [kə'ləʊn] **1.** *pr. n.* Köln *(das).* **2.** *attrib. adj.* Kölner

cologne *see* **eau-de-Cologne**

Colombia [kə'lɒmbɪə] *pr. n.* Kolumbien *(das)*

¹colon ['kəʊlən] *n.* Doppelpunkt, *der*

²colon ['kəʊlən, 'kəʊlɒn] *n.* *(Anat.)* Grimmdarm, *der*

colonel [kɜ:nl] *n.* Oberst, *der*

colonial [kə'ləʊnɪəl] *adj.* Kolonial-; kolonial

colonialism [kə'ləʊnɪəlɪzm] *n.* Kolonialismus, *der*

colonisation, colonise *see* **colonizcolonist** ['kɒlənɪst] *n.* Siedler, *der*/Siedlerin, *die;* Kolonist, *der*/ Kolonistin, *die*

colonization [kɒlənaɪ'zeɪʃn] *n.* Kolonisation, *die;* Kolonisierung, *die*

colonize ['kɒlənaɪz] *v.t.* kolonisieren; besiedeln ⟨*unbewohntes Gebiet*⟩

colonnade [kɒlə'neɪd] *n.* *(Archit.)* Säulengang, *der;* Kolonnade, *die*

colony ['kɒlənɪ] *n.* Kolonie, *die*

color *etc.* *(Amer.) see* **colour** *etc.*

coloration [kʌlə'reɪʃn] *n.* **a)** *(act of colouring)* Kolorierung, *die;* **b)** *(colour)* Färbung, *die*

colossal [kə'lɒsl] *adj.* ungeheuer; gewaltig ⟨*Bauwerk*⟩; riesenhaft, kolossal ⟨*Mann, Statue*⟩

colour ['kʌlə(r)] *(Brit.)* **1.** *n.* **a)** Farbe, *die;* **primary ~s** Grundfarben *Pl.;* **secondary ~s** Mischfarben *Pl.;* **what ~ is it?** welche Farbe hat es?; **b)** *(complexion)* [Gesichts]farbe, *die;* **change ~:** die Farbe ändern; *(go red/pale)* rot/ blaß werden; **he is/looks a bit off ~ today** ihm ist heute nicht besonders gut/er sieht heute nicht besonders gut aus; **c)** *(racial)* Hautfarbe, *die;* **d)** *(character, tone, quality, etc.)* Charakter, *der;* **add ~ to a story** einer Erzählung Farbe *od.* Kolorit geben; **local ~:** Lokalkolorit, *das;* **e)** *in pl.* *(ribbon, dress, etc., worn as symbol of party, club, etc.)* Farben *Pl.;* **show one's [true] ~s** sein wahres Gesicht zeigen; **f)** *in pl.* *(national flag)* Farben *Pl.;* **g)** *(flag)* Fahne, *die;* *(of ship)* Flagge, *die;* **pass with flying ~s** *(fig.)* glänzend abschneiden; **nail one's ~s to the mast** *(fig.)* Farbe bekennen. **2.** *v.t.* **a)** *(give ~ to)* Farbe geben (+ *Dat.*); **b)** *(paint)* malen; **~ in** ausmalen ⟨*Bild, Figur*⟩; **~ a wall red** eine Wand rot anmalen; **c)** *(stain, dye)* färben ⟨*Material, Stoff*⟩; **d)** *(misrepresent)* [schön]färben ⟨*Nachrichten, Bericht*⟩; **e)** *(fig.: influence)* beeinflussen. **3.** *v.i.* *(blush)* **~ [up]** erröten; rot werden

colouration *(Brit.) see* **coloration**

colour: ~ bar *n.* Rassenschranke, *die;* **~-blind** *adj.* farbenblind; **a ~-blind person** ein Farbenblinder/eine Farbenblinde

coloured ['kʌləd] *(Brit.)* **1.** *adj.* **a)** farbig; **~ pencil** Farbstift, *der;* **b)** *(of non-white descent)* farbig; **~ people** Farbige *Pl.* **2.** *n.* Farbige, *der/die*

colour: ~-fast *adj.* farbecht; **~ film** *n.* Farbfilm, *der*

colourful ['kʌləfl] *adj.* *(Brit.)* bunt; farbenfroh, bunt ⟨*Bild, Schauspiel*⟩; farbig, anschaulich ⟨*Sprache, Stil, Bericht*⟩; buntbewegt ⟨*Zeitepoche, Leben*⟩

colouring ['kʌlərɪŋ] *n.* *(Brit.)* **a)** *(colours)* Farben *Pl.;* **b)** *(facial complexion)* Teint, *der;* **c) ~ [matter]** *(in food etc.)* Farbstoff, *der*

colourless ['kʌləlɪs] *adj.* *(Brit.)* **a)** *(without colour)* farblos ⟨*Flüssigkeit, Gas*⟩; *(pale)* blaß ⟨*Teint*⟩; *(dull-hued)* grau, düster ⟨*Bild, Stoff, Himmel*⟩; **b)** *(fig.)* farblos, langweilig ⟨*Geschichte, Schilderung*⟩; unauffällig ⟨*Person*⟩

colour: ~ photograph *n.* Farbfotografie, -aufnahme, *die;* **~ photography** *n.* Farbfotografie, *die;* **~ scheme** *n.* Farb[en]zusammenstellung, *die;* **~ supplement** *n.* Farbbeilage, *die;* **~ television** *n.* **a)** Farbfernsehen, *das;* **b)** *(set)* Farbfernsehgerät, *das;* **~ transparency** *n.* Farbdia, *das*

colt [kəʊlt] *n.* [Hengst]fohlen, *das*

column ['kɒləm] *n.* **a)** *(Archit., of smoke)* Säule, *die;* **b)** *(division of page, table, etc.)* Spalte, *die;* Kolumne, *die;* **c)** *(in newspaper)* Spalte, *die;* Kolumne, *die;* **the sports ~:** der Sportteil; **the gossip ~:** die Klatschspalte *(ugs. abwertend);* **d)** *(of troops, vehicles, ships)* Kolonne, *die*

columnist ['kɒləmɪst] *n.* Kolumnist, *der*/Kolumnistin, *die*

coma ['kəʊmə] *n.* *(Med.)* Koma, *das;* **be in a ~:** im Koma liegen; **go into a ~:** ins Koma fallen

comb [kəʊm] **1.** *n.* Kamm, *der;* **give one's hair a ~:** sich *(Dat.)* die Haare kämmen. **2.** *v.t.* **a)** kämmen ⟨*Haare, Flachs, Wolle*⟩; **~ sb.'s/one's hair** jmdm./sich die Haare kämmen; jmdn./sich kämmen; **b)** *(search)* durchkämmen ⟨*Gelände, Wald*⟩

combat ['kɒmbæt] **1.** *n.* Kampf, *der.* **2.** *v.t.* *(fig.: strive against)* bekämpfen

combatant ['kɒmbətənt] *n.* *(in war)* Kombattant, *der;* *(in duel)* Kämpfer, *der*

combative ['kɒmbətɪv] *adj.* streitlustig

combed [kəʊmd] *adj.* gekämmt

combination [kɒmbɪ'neɪʃn] *n.* Kombination, *die;* **in ~:** zusammen

combi'nation lock *n.* Kombinationsschloß, *das*

combine 1. [kəm'baɪn] *v.t.* **a)** *(join together)* kombinieren; zusammenfügen **(into** zu); **b)** *(possess together)* vereinigen; in sich *(Dat.)* vereinigen ⟨*Eigenschaften*⟩. **2.** [kəm'baɪn] *v.i.* **a)** *(join together)* ⟨*Stoffe:*⟩ sich verbinden; **b)** *(co-operate)* zusammenwirken; ⟨*Parteien:*⟩ sich zusammentun. **3.** ['kɒmbaɪn] *n.* **a)** *(Commerc.)* Konzern, *der;* **b)** *(machine)* **~ [harvester]** Mähdrescher, *der;* Kombine, *die*

combined [kəm'baɪnd] *adj.* vereint; **a ~ operation** eine gemeinsame Operation

combustible [kəm'bʌstɪbl] *adj.* brennbar

combustion [kəm'bʌstʃn] *n.* Verbrennung, *die;* **~ chamber** *(of jet engine)* Brennkammer, *die;* *(of internal-~ engine)* Verbrennungsraum, *der*

come [kʌm] *v.i.*, **came** [keɪm], **come** [kʌm] **a)** kommen; **~ here!** komm [mal] her!; **[I'm] coming!** [ich] komme schon!; **~ running** angelaufen kommen; **~ running into the room** ins Zimmer gerannt kommen; **not know whether** *or* **if one is coming or going** nicht wissen, wo einem der Kopf steht; **they came to a house/town** sie kamen zu einem Haus/in eine Stadt; **Christmas/Easter is coming** bald ist Weihnachten/Ostern; **he has ~ a long way** er kommt von weit her; **~ to sb.'s notice** *or* **attention/knowledge** jmdm. auffallen/zu Ohren kommen; **the train came into the station** der Zug fuhr in den Bahnhof ein; **b)** *(occur)* kommen; *(in list etc.)* stehen; **c)** *(become, be)* **the shoe-laces have ~ undone** die Schnürsenkel sind aufgegangen; **the handle has ~ loose** der Griff ist lose; **it all came right in the end** es ging alles gut aus; **have ~ to believe/realize that ...:** zu der Überzeugung/Einsicht gelangt sein, daß ...; **d)** *(become present)* kommen; **in the coming week/month** kommende Woche/kommenden Monat; **to ~** *(future)* künftig; **in years to ~:** in künftigen Jahren; **for some time to ~:** [noch] für einige Zeit; **e)** *(be result)* kommen; **nothing came of it** es ist nichts daraus geworden; **the suggestion came from him** der Vorschlag war *od.* stammte von ihm; **f)** *(happen)* **how ~s it that you ...?** wie kommt es, daß du ...?; **how ~?** *(coll.)* wieso?; weshalb?; **~ what may** komme, was wolle *(geh.)*; ganz gleich, was kommt; **g)** *(be available)* ‹*Waren*:› erhältlich sein; **this dress ~s in three sizes** dies Kleid gibt es in drei Größen *od.* ist in drei Größen erhältlich; **h)** *(sl.: play a part)* **~ the bully with sb.** bei jmdm. den starken Mann markieren *(salopp)*; **don't ~ the innocent with me** spiel mir nicht den Unschuldsengel vor! *(ugs.)*; **don't ~ that game with me!** komm mir bloß nicht mit dieser Tour *od.* Masche! *(salopp)*

~ a'bout *v. i.* passieren; **how did it ~ about that ...?** wie kam es, daß ...?

~ across 1. [--'-] *v. i.* **a)** *(be understood)* ‹*Bedeutung*:› verstanden werden; ‹*Mitteilung, Rede*:› ankommen *(ugs.)*; **b)** *(coll.: make an impression)* wirken (**as** wie). **2.** ['---] *v. t.* **~ across sb./sth.** jmdm./einer Sache begegnen; **have you ~ across my watch?** ist dir meine Uhr begegnet? *(ugs.)*

~ a'long *v. i. (coll.)* **a)** *(hurry up)* **~ along!** komm/kommt!; nun mach/macht schon! *(ugs.)*; **b)** *(make progress)* **~ along nicely** gute Fortschritte machen; **c)** *(arrive, present oneself/itself)* ‹*Person*:› ankommen; ‹*Gelegenheit, Stelle*:› sich bieten; **he'll take any job that ~s along** er nimmt jeden Job, der sich ihm bietet; **d)** *(to place)* mitkommen (**with** mit)

~ a'way *v. i.* **a)** *(leave)* weggehen; **b)** *(become detached)* sich lösen (**from** von); **c)** *(be left)* **~ away with the impression/feeling that ...:** mit dem Eindruck/Gefühl gehen, daß ...

~ 'back *v. i.* **a)** *(return)* zurückkommen; ‹*Gedächtnis, Vergangenes*:› wiederkehren; **b)** *(return to memory)* **it will ~ back [to me]** es wird mir wieder einfallen; **c) ~ back [into fashion]** wiederkommen; wieder in Mode kommen; **d)** *(retort)* **~ back at sb. with sth.** jmdm. etw. entgegnen

~ between *v. t.* treten zwischen (+ *Akk.*)

~ by 1. ['--] *v. t. (obtain, receive)* kriegen *(ugs.)*; bekommen **2.** [-'-] *v. i.* vorbeikommen

~ 'down *v. i.* **a)** *(collapse)* herunterfallen; runterfallen *(ugs.)*; *(fall)* ‹*Schnee, Regen, Preis*:› fallen; **b)** *(~ to place regarded as lower)* herunterkommen; runterkommen *(ugs.)*; *(~ southwards)* runterkommen *(ugs.)*; **c)** *(land)* [not]landen; *(crash)* abstürzen; **~ down in a field** auf einem Acker [not]landen/auf einen Acker stürzen; **d)** *(be passed on)* ‹*Sage, Brauch*:› überliefert werden; **e) ~ down to** *(reach)* reichen bis; **f) ~ down to** *(be reduced to)* hinauslaufen auf (+ *Akk.*); **g) ~ down to** *(be a question of)* ankommen (**to** auf + *Akk.*); **h)** *(suffer change for the worse)* **she has ~ down in the world** sie hat einen Abstieg erlebt; **~ down to sth.** *(be forced to resort to sth.)* auf etw. *(Akk.)* angewiesen sein; **i) ~ down with** bekommen ‹*Krankheit*›

~ 'in *v. i.* **a)** *(enter)* hereinkommen; reinkommen *(ugs.)*; **~ in!** herein!; **b)** ‹*Flut*:› kommen; **c)** *(be received)* ‹*Nachrichten, Bericht*:› hereinkommen; **d)** *(in radio communication)* melden; **C~ in, Tom, ~ in, Tom. Over** Tom melden, Tom melden. Ende; **e)** *(contribut to discussion etc.)* sich einschalten; **f)** *(become fashionable)* in Mode kommen; aufkommen; **g)** *(in race)* einlaufen als *od.* durchs Ziel gehen als ‹*erster usw.*›; **h)** *(play a part)* **where do I ~ in?** welche Rolle soll ich spielen?; **~ in on sth.** sich an etw. *(Dat.)* beteiligen; **i) ~ in for** erregen ‹*Bewunderung, Aufmerksamkeit*›; auf sich *(Akk.)* ziehen, hervorrufen ‹*Kritik*›

~ into *v. t.* **a)** *(enter)* hereinkommen in (+ *Akk.*); ‹*Zug*:› einfahren in ‹*Bahnhof*›; ‹*Schiff*:› einlaufen in ‹*Hafen*›; **b)** *(inherit)* erben ‹*Vermögen*›; **c)** *(play a part)* **wealth does not ~ into it** Reichtum spielt dabei keine Rolle; **where do I ~ into it?** welche Rolle soll ich [dabei] spielen?

~ near *v. t.* **~ near [to] doing sth.** drauf und dran sein, etw. zu tun *(ugs.)*

~ off 1. [-'-] *v. i.* **a)** *(become detached)* ‹*Griff, Knopf*:› abgehen; *(be removable)* sich abnehmen lassen; ‹*Fleck*:› weg-, rausgehen *(ugs.)*; **b)** *(fall from sth.)* runterfallen; **c)** *(emerge from contest etc.)* abschneiden; **d)** *(succeed)* ‹*Pläne, Versuche*:› Erfolg haben, *(ugs.)* klappen; **e)** *(take place)* stattfinden; **their wedding/holiday did not ~ off** aus ihrer Hochzeit/ihrem Urlaub wurde nichts. **2.** ['--] *v. t.* **~ off a horse/bike** vom Pferd/Fahrrad fallen; **~ 'off it!** *(coll.)* nun mach mal halblang! *(ugs.)*

~ on 1. [-'-] *v. i.* **a)** *(continue coming, follow)* kommen; **~ on!** komm, komm/kommt, kommt!; *(encouraging)* na, komm; *(impatient)* na, komm schon; *(incredulous)* ach komm!; **I'll ~ on later** ich komme später nach; **b)** *(make progress)* **my work is coming on very well** meine Arbeit macht gute Fortschritte; **c)** *(begin to arrive)* ‹*Nacht, Dunkelheit, Winter*:› anbrechen; **d)** *(appear on stage or scene)* auftreten. **2.** ['--] *v. t. see* **~ upon**

~ 'out *v. i.* **a)** herauskommen; **~ out [on strike]** in den Streik treten; **b)** *(appear, become visible)* ‹*Sonne, Knospen, Blumen*:› herauskommen, *(ugs.)* rauskommen; ‹*Sterne*:› zu sehen sein; **c)** *(be revealed)* ‹*Wahrheit, Nachrichten*:› herauskommen, *(ugs.)* rauskommen; **d)** *(be published, declared, etc.)* herauskommen; rauskommen *(ugs.)*; ‹*Ergebnisse, Zensuren*:› bekanntgegeben werden; **e)** *(declare oneself)* **~ out for** *or* **in favour of sth.** sich für etw. aussprechen; **~ out against sth.** sich gegen etw. aussprechen; **f)** *(be covered)* **she came out in a rash** sie bekam einen Ausschlag; **g)** *(be removed)* ‹*Fleck, Schmutz*:› rausgehen *(ugs.)*; **h) ~ out with** herausrücken mit *(ugs.)* ‹*Wahrheit, Fakten*›; loslassen *(ugs.)* ‹*Flüche, Bemerkungen*›

~ 'over 1. *v. i.* **a)** *(~ from some distance)* herüberkommen; **b)** *(change sides or opinions)* **~ over to sb./sth.** sich jmdm./einer Sache anschließen; **c)** *see* **~ across 1 b; d) she came over funny/dizzy** ihr wurde auf einmal ganz komisch/schwindlig *(ugs.)*. **2.** *v. t. (coll.)* kommen über (+ *Akk.*); **what has ~ over him?** was ist über ihn gekommen?

~ 'round *v. i.* **a)** *(make informal visit)* vorbeischauen; **b)** *(recover)* wieder zu sich kommen; **c)** *(be converted)* es sich [anders] *(Dat.)* überlegen; **he came round to my way of thinking** er hat sich meiner Auffassung *(Dat.)* angeschlossen; **d)** *(recur)* **Christmas ~s round again** wir haben wieder Weihnachten

~ 'through 1. *v. i.* durchkommen. **2.** *v. t. (survive)* überleben

~ to 1. ['--] *v. t.* **a)** *(amount to)* ‹*Rechnung, Gehalt, Kosten*:› sich belaufen auf (+ *Akk.*); **his plans came to nothing** aus seinen Plänen wurde nichts; **he/it will never ~ to much** aus ihm wird nichts Besonderes werden/daraus wird nicht viel; **b)** *(inherit)* erben ‹*Vermögen*›; **c)** *(arrive at)* **what is the world coming to?** wohin ist es mit der Welt gekommen?; **this is what he has ~ to** so weit ist es also mit ihm gekommen. **2.** [-'-] *v.i.* wieder zu sich kommen

~ to'gether *v.i.* ‹*Personen:*› zusammenkommen; ‹*Ereignisse:*› zusammenfallen

~ under *v.t.* **a)** *(be classed as or among)* kommen unter (+ *Akk.*); **b)** *(be subject to)* geraten *od.* kommen unter (+ *Akk.*)

~ 'up *v.i.* **a)** *(~ to place regarded as higher)* hochkommen; heraufkommen; *(~ northwards)* raufkommen *(ugs.)*; **he ~s up to London every other weekend** er kommt jedes zweite Wochenende nach London; **b) ~ up to sb.** auf jmdn. zukommen; **c)** *(arise out of ground)* herauskommen; rauskommen *(ugs.)*; **d)** *(be discussed)* ‹*Frage, Thema:*› angeschnitten werden, aufkommen; ‹*Name:*› genannt werden; ‹*Fall:*› verhandelt werden; **e)** *(present itself)* sich ergeben; **~ up for sale/renewal** zum Kauf angeboten werden/erneuert werden müssen; **f) ~ up to** *(reach)* reichen bis an (+ *Akk.*); *(be equal to)* entsprechen (+ *Dat.*) ‹*Erwartungen, Anforderungen*›; **g) ~ up against sth.** *(fig.)* auf etw. *(Akk.)* stoßen; **h) ~ up with** vorbringen ‹*Vorschlag*›; wissen ‹*Lösung, Antwort*›; haben ‹*Erklärung, Idee*›

~ upon *v.t.* *(meet by chance)* begegnen (+ *Dat.*)

~ with *v.t.* *(be supplied together with)* **this model ~s with ...:** zu diesem Modell gehört ...

'come-back *n.* **a)** *(return to profession etc.)* Comeback, *das;* **b)** *(sl.: retort)* Reaktion, *die*

comedian [kə'mi:dɪən] *n.* Komiker, *der*

comedienne [kəmi:dɪ'en, kəmedɪ'en] *n.* Komikerin, *die*

'come-down *n.* *(loss of prestige etc.)* Abstieg, *der*

comedy ['kɒmɪdɪ] *n.* **a)** Lustspiel, *das;* Komödie, *die;* **b)** *(humour)* Witz, *der;* Witzigkeit, *die*

comely ['kʌmlɪ] *adj.* gutaussehend; ansehnlich

comer ['kʌmə(r)] *n.* **the competition is open to all ~s** an dem Wettbewerb kann sich jeder beteiligen; **the first ~:** derjenige, der zuerst kommt

comet ['kɒmɪt] *n.* *(Astron.)* Komet, *der*

comeuppance [kʌm'ʌpəns] *n.* **get one's ~:** die Quittung kriegen *(fig.)*

comfort ['kʌmfət] **1.** *n.* **a)** *(consolation)* Trost, *der;* **it is a ~/no ~ to know that ...:** es ist tröstlich/alles andere als tröstlich zu wissen, daß ...; **he takes ~ from the fact that ...:** er tröstet sich mit der Tatsache, daß ...; **b)** *(physical well-being)* Behaglichkeit, *die;* **live in great ~:** sehr behaglich *od.* bequem leben; **c)** *(person)* Trost, *der;* **d)** *in pl.* *(things that make life easy)* Komfort, *der o. Pl.* **2.** *v.t.* trösten; *(give help to)* sich annehmen (+ *Gen.*)

comfortable ['kʌmfətəbl] *adj.* **a)** bequem ‹*Bett, Sessel, Schuhe, Leben*›; komfortabel ‹*Haus, Hotel, Zimmer*›; *(fig.)* ausreichend ‹*Einkommen, Rente*›; **a ~ victory** ein leichter Sieg; **a ~ majority** eine gute Mehrheit; **b)** *(at ease)* **be/feel ~:** sich wohl fühlen; **make yourself ~:** machen Sie es sich *(Dat.)* bequem

comfortably ['kʌmfətəblɪ] *adv.* bequem; komfortabel ‹*eingerichtet*›; gut, leicht ‹*gewinnen*›; **they are ~ off** es geht ihnen gut

comforting ['kʌmfətɪŋ] *adj.* beruhigend ‹*Gedanke*›; tröstend ‹*Worte*›; wohlig ‹*Wärme*›

'comfort station *n.* *(Amer.)* öffentliche Toilette

comfy ['kʌmfɪ] *adj.* *(coll.)* bequem; gemütlich ‹*Haus, Zimmer*›

comic ['kɒmɪk] **1.** *adj.* komisch; humoristisch ‹*Dichtung, Dichter*›. **2.** *n.* **a)** *(comedian)* Komiker, *der*/Komikerin, *die;* **b)** *(periodical)* Comic-Heft, *das;* **c)** *(amusing person)* Witzbold, *der;* ulkiger Vogel *(ugs.)*

comical ['kɒmɪkl] *adj.* ulkig; komisch

coming ['kʌmɪŋ] **1.** *see* **come. 2.** *n.* *(of person)* Ankunft, *die; (of time)* Beginn, *der; (of institution)* Einführung, *die;* **~s and goings** das Kommen und Gehen

comma ['kɒmə] *n.* Komma, *das*

command [kə'mɑ:nd] **1.** *v.t.* **a)** *(order, bid)* befehlen (**sb.** jmdm.); **b)** *(be in ~ of)* befehligen ‹*Schiff, Armee, Streitkräfte*›; *(have authority over or control of)* gebieten über (+ *Akk.*) *(geh.)*; beherrschen; **c)** *(have at one's disposal)* verfügen über (+ *Akk.*) ‹*Gelder, Ressourcen, Wortschatz*›; **d)** *(deserve and get)* verdient haben ‹*Achtung, Respekt*›; **e) the hill ~s a fine view of ...:** der Berg bietet eine schöne Aussicht auf ... (+ *Akk.*). **2.** *n.* **a)** Kommando, *das; (in writing)* Befehl, *der;* **at** *or* **by sb.'s ~:** auf jmds. Befehl *(Akk.)* [hin]; **b)** *(exercise or tenure)* Kommando, *das;* Befehlsgewalt, *die;* **be in ~ of an army/ship** eine Armee/ein Schiff befehligen; **have/take ~ of ...:** das Kommando über (+ *Akk.*) ... haben/übernehmen; **officer in ~:** befehlshabender Offizier; **c)** *(control, mastery, possession)* Beherrschung, *die;* **have a good ~ of French** das Französische gut beherrschen

commandant [kɒmən'dænt] *n.* Kommandant, *der*

commandeer [kɒmən'dɪə(r)] *v.t.* **a)** *(take arbitrary possession of)* sich *(Dat.)* aneignen; requirieren *(scherzh.)*; **b)** *(seize for military service)* einziehen ‹*Männer*›; beschlagnahmen, requirieren ‹*Pferde, Vorräte, Gebäude*›

commander [kə'mɑ:ndə(r)] *n.* **a)** Führer, *der;* Leiter, *der;* **b)** *(naval officer below captain)* Fregattenkapitän, *der;* **c) C~-in-Chief** Oberbefehlshaber, *der*

commanding [kə'mɑ:ndɪŋ] *adj.* **a)** gebieterisch ‹*Persönlichkeit, Erscheinung, Stimme*›; imposant, eindrucksvoll ‹*Statur, Gestalt*›; **b)** beherrschend ‹*Ausblick, Lage*›

commanding 'officer *n.* Befehlshaber, *der*/Befehlshaberin, *die*

commandment [kə'mɑ:ndmənt] *n.* Gebot, *das;* **the Ten C~s** die Zehn Gebote

commando [kə'mɑ:ndəʊ] *n., pl.* **~s a)** *(unit)* Kommando, *das;* Kommandotrupp, *der;* **b)** *(member of ~)* Angehöriger eines Kommando[trupp]s

com'mand performance *n.* königliche Galavorstellung

commemorate [kə'meməreɪt] *v.t.* gedenken (+ *Gen.*)

commemoration [kəmemə'reɪʃn] *n.* *(act)* Gedenken, *das;* **in ~ of** zum Gedenken an (+ *Akk.*); **the ~ of sb.'s death** das Gedenken an jmds. Tod *(Akk.)*

commemorative [kə'memərətɪv] *adj.* Gedenk-

commence [kə'mens] *v.t. & i.* beginnen; **building ~d** mit dem Bau wurde begonnen; **~ to do** *or* **~ doing sth.** beginnen, etw. zu tun

commencement [kə'mensmənt] *n.* Beginn, *der*

commend [kə'mend] *v.t.* **a)** *(praise)* loben; **~ sb. [up]on sth.** jmdn. wegen etw. loben; **~ sb./sth. to sb.** jmdm. jmdn./etw. empfehlen; **b)** *(entrust or commit to person's care)* anvertrauen

commendable [kə'mendəbl] *adj.* lobenswert; löblich

commendation [kɒmen'deɪʃn] *n.* *(praise)* Lob, *das; (official)* Be-

lobigung, *die; (award)* Auszeichnung, *die*
commensurate [kə'menʃərət, kə'mensjərət] *adj.* ~ **to** *or* **with** entsprechend (+ *Dat.*); **be ~ to** *or* **with sth.** einer Sache *(Dat.)* entsprechen
comment ['kɒment] **1.** *n.* Bemerkung, *die* (**on** über + *Akk.*); *(marginal note)* Anmerkung, *die* (**on** über + *Akk.*); **no ~!** *(coll.)* kein Kommentar! **2.** *v. i.* ~ **on sth.** über etw. *(Akk.)* Bemerkungen machen; **he ~ed that ...:** er bemerkte, daß ...; ~ **on a text/manuscript** einen Text/ein Manuskript kommentieren
commentary ['kɒməntərɪ] *n.* **a)** *(series of comments, treatise)* Kommentar, *der* (**on** zu); *(comment)* Erläuterung, *die* (**on** zu); **b)** *(Radio, Telev.)* [**live** *or* **running**] **~:** Live-Reportage, *die*
commentate ['kɒmənteɪt] *v. i.* ~ **on sth.** etw. kommentieren
commentator ['kɒmənteɪtə(r)] *n.* Kommentator, *der*/Kommentatorin, *die; (Sport)* Reporter, *der*/Reporterin, *die*
commerce ['kɒmɜːs] *n.* Handel, *der; (between countries)* Handel[sverkehr], *der*
commercial [kə'mɜːʃl] **1.** *adj.* **a)** Handels-; kaufmännisch ‹*Ausbildung*›; **b)** *(interested in financial return)* kommerziell. **2.** *n.* Werbespot, *der*
commercial 'art *n.* Gebrauchs-, Werbegraphik, *die*
commercialise *see* **commercialize**
commercialism [kə'mɜːʃəlɪzm] *n.* Kommerzialismus, *der*
commercialize [kə'mɜːʃəlaɪz] *v. t.* kommerzialisieren
commercial: ~ **'television** *n.* kommerzielles Fernsehen; Werbefernsehen, *das;* ~ **'traveller** *n.* Handelsvertreter, *der*/-vertreterin, *die;* ~ **'vehicle** *n.* Nutzfahrzeug, *das*
commiserate [kə'mɪzəreɪt] *v. i.* ~ **with sb.** mit jmdm. mitfühlen; *(express one's commiseration)* jmdm. sein Mitgefühl aussprechen (**on** zu)
commiseration [kəmɪzə'reɪʃn] *n.* **a)** Mitgefühl, *das;* **b)** *in sing. or pl. (condolence)* Teilnahme, *die;* Beileid, *das*
commission [kə'mɪʃn] **1.** *n.* **a)** *(official body)* Kommission, *die;* **b)** *(instruction, piece of work)* Auftrag, *der;* **c)** *(in armed services)* Ernennungsurkunde, *die;* **get one's ~:** zum Offizier ernannt werden; **resign one's ~:** aus dem Offiziersdienst ausscheiden; **d)** *(pay of agent)* Provision, *die;* **sell goods on ~:** Waren auf Provisionsbasis verkaufen; **e) in/out of** ~ ‹*Kriegsschiff*› in/außer Dienst; ‹*Auto, Maschine, Lift usw.*› in/außer Betrieb. **2.** *v. t.* **a)** beauftragen ‹*Künstler*›; in Auftrag geben ‹*Gemälde usw.*›; **b)** *(empower)* bevollmächtigen; **~ed officer** Offizier, *der;* **c)** *(give command of ship to)* zum Kapitän ernennen; **d)** *(prepare for service)* in Dienst stellen ‹*Schiff*›; **e)** *(bring into operation)* in Betrieb setzen ‹*Kraftwerk, Fabrik*›
commissionaire [kəmɪʃə'neə(r)] *n. (esp. Brit.)* Portier, *der*
commissioner [kə'mɪʃənə(r)] *n.* **a)** *(person appointed by commission)* Beauftragte, *der/die; (of police)* Präsident, *der;* **b)** *(member of commission)* Kommissions-, Ausschußmitglied, *das;* **c)** *(representative of supreme authority)* Kommissar, *der;* **d) C~ for Oaths** Notar, *der*/Notarin, *die*
commit [kə'mɪt] *v. t.*, **-tt-: a)** *(perpetrate)* begehen, verüben ‹*Mord, Selbstmord, Verbrechen, Raub*›; begehen ‹*Dummheit, Bigamie, Fehler, Ehebruch*›; **b)** *(pledge, bind)* ~ **oneself/sb. to doing sth.** sich/jmdn. verpflichten, etw. zu tun; ~ **oneself to a course of action** sich auf eine Vorgehensweise festlegen; **c)** *(entrust)* anvertrauen (**to** *Dat.*); ~ **sth. to a person/a person's care** jmdm. etw. anvertrauen/etw. jmds. Obhut *(Dat.)* anvertrauen; **d)** ~ **sb. for trial** jmdn. dem Gericht überstellen
commitment [kə'mɪtmənt] *n. (to course of action or opinion)* Verpflichtung (**to** gegenüber); *(by dedication)* Engagement, *das* (**to** für)
committed [kə'mɪtɪd] *adj.* **a)** verpflichtet (**to** zu); festgelegt (**to** auf + *Akk.*); **b)** *(morally dedicated)* engagiert
committee [kə'mɪtɪ] *n.* Ausschuß, *der (auch Parl.);* Komitee, *das*
commodity [kə'mɒdɪtɪ] *n.* **a)** *(utility item)* **household ~:** Haushaltsartikel, *der;* **a rare/precious ~** *(fig.)* etwas Seltenes/Kostbares; **b)** *(St. Exch.)* [vertretbare] Ware; *(raw material)* Rohstoff, *der*
common ['kɒmən] **1.** *adj.*, **~er** ['kɒmənə(r)], **~est** ['kɒmənɪst] **a)** *(belonging equally to all)* gemeinsam ‹*Ziel, Interesse, Sache, Unternehmung, Vorteil, Merkmal, Sprache*›; ~ **to all birds** allen Vögeln gemeinsam; **b)** *(belonging to the public)* öffentlich; **the ~ good** das Gemeinwohl; **a ~ belief** [ein] allgemeiner Glaube; **have the ~ touch** volkstümlich sein; **c)** *(usual)* gewöhnlich; normal; *(frequent)* häufig ‹*Vorgang, Erscheinung, Ereignis, Erlebnis*›; allgemein verbreitet ‹*Sitte, Wort, Redensart*›; ~ **honesty/courtesy** [ganz] normale Ehrlichkeit/Höflichkeit; **d)** *(without rank or position)* einfach; **e)** *(vulgar)* gemein; gewöhnlich *(abwertend),* ordinär *(ugs. abwertend)* ‹*Ausdrucksweise, Mundart, Aussehen, Benehmen*›. **2.** *n.* **a)** *(land)* Gemeindeland, *das;* Allmende, *die;* **b) have sth./nothing/a lot in ~ [with sb.]** etw./nichts/viel [mit jmdm.] gemein[sam] haben
common: ~ **'cold** *n.* Erkältung, *die;* ~ **de'nominator** *n. (Math.)* gemeinsame Nenner, *der*
commoner ['kɒmənə(r)] *n.* Bürgerliche, *der/die*
common: ~ **'knowledge** *n.* **it's [a matter of] ~ knowledge that ...:** es ist allgemein bekannt, daß ...; **~-law** *adj.* **she's his ~-law wife** sie lebt mit ihm in eheähnlicher Gemeinschaft
commonly ['kɒmənlɪ] *adv.* **a)** *(generally)* im allgemeinen; **b)** *(vulgarly)* gewöhnlich *(abwertend)*
Common 'Market *n.* Gemeinsamer Markt
commonplace ['kɒmənpleɪs] **1.** *n. (platitude)* Gemeinplatz, *der; (anything usual or trite)* Alltäglichkeit, *die.* **2.** *adj.* nichtssagend, banal ‹*Bemerkung, Buch*›; alltäglich ‹*Angelegenheit, Ereignis*›
common: C~ 'Prayer *n.* Liturgie, *die (der Kirche von England);* **~-room** *n. (Brit.)* Gemeinschaftsraum, *der; (for lecturers)* Dozentenzimmer, *das*
Commons ['kɒmənz] *n. pl.* **the [House of] ~:** das Unterhaus
common: ~ **'sense** *n.* gesunder Menschenverstand; **~-sense** *adj.* vernünftig; gesund ‹*Ansicht, Standpunkt*›
Commonwealth ['kɒmənwelθ] *n.* **the [British] ~ [of Nations]** das Commonwealth
commotion [kə'məʊʃn] *n.* Tumult, *der*
communal ['kɒmjʊnl] *adj.* **a)** *(of or for the community)* gemeindlich; ~ **living/life** Gemeinschaftsleben, *das;* **b)** *(for the common use)* gemeinsam; Gemeinschafts‹*-küche, -schüssel, -grab*›

[1]**commune** ['kɒmju:n] *n.* Kommune, *die*

[2]**commune** [kə'mju:n] *v.i.* ~ **with sb./sth.** mit jmdm./etw. Zwiesprache halten *(geh.)*

communicate [kə'mju:nɪkeɪt] **1.** *v.t.* übertragen ‹*Wärme, Bewegung, Krankheit*›; übermitteln ‹*Nachrichten, Informationen*›; vermitteln ‹*Gefühle, Ideen*›. **2.** *v.i.* **a)** ~ **with sb.** mit jmdm. kommunizieren; **b)** *(have common door)* verbunden sein

communication [kəmju:nɪ'keɪʃn] *n.* **a)** *(of disease, motion, heat, etc.)* Übertragung, *die; (of news, information)* Übermittlung, *die; (of ideas)* Vermittlung, *die;* **b)** *(information given)* Mitteilung, *die* (**to** an + *Akk.*); **c)** *(interaction with sb.)* Verbindung, *die;* **be in ~ with sb.** mit jmdm. in Verbindung stehen; **d)** *in pl. (conveying information)* Kommunikation, *die; (science, practice)* Kommunikationswesen, *das*

communication: ~-cord *n.* Notbremse, *die;* **~s satellite** *n.* Nachrichten- *od.* Kommunikationssatellit, *der*

communicative [kə'mju:nɪkətɪv] *adj.* gesprächig

communion [kə'mju:nɪən] *n.* **a)** **[Holy] C~** *(Protestant Ch.)* das [heilige] Abendmahl; *(RC Ch.)* die [heilige] Kommunion; **receive** *or* **take [Holy] C~:** das [heilige] Abendmahl/die [heilige] Kommunion empfangen; **b)** *(fellowship)* Gemeinschaft, *die*

communiqué [kə'mju:nɪkeɪ] *n.* Kommuniqué, *das*

communism ['kɒmjʊnɪzm] *n.* Kommunismus, *der;* **C~:** der Kommunismus

Communist, communist ['kɒmjʊnɪst] **1.** *n.* Kommunist, *der*/Kommunistin, *die.* **2.** *adj.* kommunistisch

community [kə'mju:nɪtɪ] *n.* **a)** *(organized body)* Gemeinwesen, *das;* **the Jewish ~:** die jüdische Gemeinde; **a ~ of monks** eine Mönchsgemeinde; **b)** *no pl. (public)* Öffentlichkeit, *die*

community: ~ centre *n.* Gemeindezentrum, *das;* ~ **'charge** *n. (Brit.)* Gemeindesteuer, *die;* ~ **re'lations** *n. pl.* Verhältnis zwischen den Bevölkerungsgruppen; ~ **'service** *n. [freiwilliger od. als Strafe auferlegter] sozialer Dienst*

commute [kə'mju:t] **1.** *v.t.* **a)** umwandeln ‹*Strafe*› (**to** in + *Akk.*); **b)** *(change to sth. different)* umwandeln. **2.** *v.i.* pendeln

commuter [kə'mju:tə(r)] *n.* Pendler, *der*/Pendlerin, *die*

[1]**compact** [kəm'pækt] **1.** *adj.* kompakt; komprimiert ‹*Stil*›. **2.** *v.t.* zusammenpressen

[2]**compact** ['kɒmpækt] *n.* Puderdose [mit Puder(stein)]

[3]**compact** ['kɒmpækt] *n. (agreement)* Vertrag, *der*

compact disc ['kɒmpækt 'dɪsk] *n.* Compact Disc, *die;* Kompaktschallplatte, *die*

companion [kəm'pænjən] *n.* **a)** *(one accompanying)* Begleiter, *der*/Begleiterin, *die;* **b)** *(associate)* Kamerad, *der*/Kameradin, *die;* **c)** *(matching thing)* Gegenstück, *das;* Pendant, *das*

companionable [kəm'pænjənəbl] *adj.* freundlich

companionship [kəm'pænjənʃɪp] *n.* Gesellschaft, *die; (fellowship)* Kameradschaft, *die*

company ['kʌmpənɪ] *n.* **a)** *(persons assembled, companionship)* Gesellschaft, *die;* **expect/receive ~:** Besuch *od.* Gäste *Pl.* erwarten/empfangen; **for ~:** zur Gesellschaft; **two is ~, three is a crowd** zu zweit ist es gemütlich, ein Dritter stört; **keep sb. ~:** jmdm. Gesellschaft leisten; **part ~ with sb./sth.** sich von jmdm./etw. trennen; **b)** *(firm)* Gesellschaft, *die;* Firma, *die;* ~ **car** Firmenwagen, *der;* **c)** *(of actors)* Truppe, *die;* Ensemble, *das;* **d)** *(Mil.)* Kompanie, *die;* **e)** *(Navy)* **ship's ~:** Besatzung, *die*

comparable ['kɒmpərəbl] *adj.* vergleichbar (**to, with** mit)

comparative [kəm'pærətɪv] **1.** *adj.* **a)** vergleichend ‹*Anatomie, Sprachwissenschaft usw.*›; **b)** *(estimated by comparison)* **the ~ merits/advantages of the proposals** die Vorzüge/Vorteile der Vorschläge im Vergleich; **c)** *(relative)* relativ; **in ~ comfort** relativ *od.* verhältnismäßig komfortabel; **d)** *(Ling.)* komparativ *(fachspr.);* **a ~ adjective/adverb** ein Adjektiv/Adverb im Komparativ. **2.** *n. (Ling.)* Komparativ, *der*

comparatively [kəm'pærətɪvlɪ] *adv.* **a)** *(by means of comparison)* vergleichend; **b)** *(relatively)* relativ; verhältnismäßig

compare [kəm'peə(r)] **1.** *v.t.* vergleichen (**to, with** mit); ~ **two/three** *etc.* **things** zwei/drei *usw.* Dinge [miteinander] vergleichen; **~d with** *or* **to sb./sth.** verglichen mit *od.* im Vergleich zu jmdm./etw. **2.** *v.i.* sich vergleichen lassen. **3.** *n. (literary)* **beyond** *or* **without ~:** unvergleichlich

comparison [kəm'pærɪsn] *n. (act of comparing, simile)* Vergleich, *der;* **in** *or* **by ~ [with sb./sth.]** im Vergleich [zu jmdm./etw.]; **there's no ~ between them** man kann sie einfach nicht vergleichen

compartment [kəm'pɑ:tmənt] *n. (in drawer, desk, etc.)* Fach, *das; (of railway carriage)* Abteil, *das*

compass ['kʌmpəs] *n.* **a)** *in pl.* **[a pair of] ~es** ein Zirkel; **b)** *(for navigating)* Kompaß, *der;* **c)** *(extent)* Gebiet, *das; (fig.: scope)* Rahmen, *der*

compassion [kəm'pæʃn] *n., no pl.* Mitgefühl, *das* (**for, on** mit)

compassionate [kəm'pæʃənət] *adj.* mitfühlend; **on ~ grounds** aus persönlichen Gründen; *(for family reasons)* aus familiären Gründen

compatibility [kəmpætɪ'bɪlɪtɪ] *n., no pl. (consistency, mutual tolerance)* Vereinbarkeit, *die; (of people)* Zueinanderpassen, *das*

compatible [kəm'pætɪbl] *adj. (consistent, mutually tolerant)* vereinbar; zueinander passend ‹*Personen*›; aufeinander abgestimmt, zueinander passend, *(Computing)* kompatibel ‹*Geräte, Maschinen*›

compatriot [kəm'pætrɪət, kəm'peɪtrɪət] *n.* Landsmann, *der*/-männin, *die*

compel [kəm'pel] *v.t.,* **-ll-** zwingen; ~ **sb. to do sth.** jmdn. [dazu] zwingen, etw. zu tun

compelling [kəm'pelɪŋ] *adj.* bezwingend

compendium [kəm'pendɪəm] *n., pl.* **~s** *or* **compendia** [kəm'pendɪə] Abriß, *der; (summary)* Kompendium, *das;* ~ **of games** Spielemagazin, *das*

compensate ['kɒmpenseɪt] **1.** *v.i.* ~ **for sth.** etw. ersetzen; ~ **for injury** *etc.* für Verletzung *usw.* Schaden[s]ersatz leisten. **2.** *v.t.* ~ **sb. for sth.** jmdn. für etw. entschädigen

compensation [kɒmpen'seɪʃn] *n.* Ersatz, *der; (for damages, injuries, etc.)* Schaden[s]ersatz, *der; (for requisitioned property)* Entschädigung, *die*

compère ['kɒmpeə(r)] *(Brit.)* **1.** *n.* Conférencier, *der.* **2.** *v.t.* konferieren ‹*Show*›

compete [kəm'pi:t] *v.i.* konkurrieren (**for** um); *(Sport)* kämpfen; ~ **with sb./sth.** mit jmdm./etw. konkurrieren; ~ **with one another** miteinander wetteifern

competence ['kɒmpɪtəns], **competency** ['kɒmpɪtənsɪ] *n.* **a)**

(ability) Fähigkeiten *Pl.*; **b)** *(Law)* Zuständigkeit, *die*
competent ['kɒmpɪtənt] *adj.* fähig; befähigt; **not ~ to do sth.** nicht kompetent, etw. zu tun
competently ['kɒmpɪtəntlɪ] *adv.* sachkundig; kompetent
competition [kɒmpɪ'tɪʃn] *n.* **a)** *(contest)* Wettbewerb, *der; (in magazine)* Preisausschreiben, *das;* **b)** *(those competing)* Konkurrenz, *die; (Sport)* Gegner *Pl.*
competitive [kəm'petɪtɪv] *adj.* **a)** Leistungs-; **~ sports** Wettkampf- *od.* Leistungssport, *der;* **~ spirit** Konkurrenz- *od.* Wettbewerbsdenken, *das;* **b)** *(comparable with rivals)* leistungs-, wettbewerbsfähig ‹*Preis, Unternehmen*›
competitor [kəm'petɪtə(r)] *n.* Konkurrent, *der*/Konkurrentin, *die; (in contest, race)* Teilnehmer, *der*/-nehmerin, *die; (for job)* Mitbewerber, *der*/-bewerberin, *die;* **our ~s** unsere Konkurrenz
compilation [kɒmpɪ'leɪʃn] *n.* Zusammenstellung, *die*
compile [kəm'paɪl] *v.t.* zusammenstellen; verfassen ‹*Wörterbuch, Reiseführer*›
compiler [kəm'paɪlə(r)] *n.* Verfasser, *der*/Verfasserin, *die*
complacency [kəm'pleɪsənsɪ] *n., no pl.* Selbstzufriedenheit, *die*
complacent [kəm'pleɪsənt] *adj.* selbstzufrieden; selbstgefällig
complain [kəm'pleɪn] *v.i.* sich beklagen *od.* beschweren **(about, at** über + *Akk.*); **~ of sth.** über etw. *(Akk.)* klagen
complaint [kəm'pleɪnt] *n.* **a)** Beanstandung, *die;* Beschwerde, *die;* Klage, *die; (formal accusation, expression of grief)* Klage, *die;* **b)** *(ailment)* Leiden, *das*
complement 1. ['kɒmplɪmənt] *n.* **a)** *(what completes)* Vervollständigung, *die;* **b)** *(full number)* **a [full] ~:** die volle Zahl; *(of people)* die volle Stärke; **the ship's ~:** die volle Schiffsbesatzung; **c)** *(Ling.)* Ergänzung, *die.* **2.** ['kɒmplɪment] *v.t.* ergänzen
complementary [kɒmplɪ'mentərɪ] *adj.* **a)** *(completing)* ergänzend; **b)** *(completing each other)* einander ergänzend
complete [kəm'pli:t] **1.** *adj.* **a)** vollständig; *(in number)* vollzählig; komplett; **a ~ edition** eine Gesamtausgabe; **b)** *(finished)* fertig; abgeschlossen ‹*Arbeit*›; **c)** *(absolute)* völlig; total, komplett ‹*Idiot, Reinfall, Ignoranz*›; absolut ‹*Chaos, Katastrophe*›; vollkommen ‹*Ruhe*›; total, *(ugs.)* blutig ‹*Anfänger, Amateur*›; **a ~ stranger** ein völlig Fremder. **2.** *v.t.* **a)** *(finish)* beenden; fertigstellen ‹*Gebäude, Arbeit*›; abschließen ‹*Vertrag*›; **b)** *(make whole)* vervollkommnen, vollkommen machen ‹*Glück*›; vervollständigen ‹*Sammlung*›; **c)** *(make whole amount of)* vollzählig machen; **d)** ausfüllen ‹*Fragebogen, Formular*›
completely [kəm'pli:tlɪ] *adv.* völlig; absolut ‹*erfolgreich*›
completion [kəm'pli:ʃn] *n.* Beendigung, *die; (of building, work)* Fertigstellung, *die; (of contract)* Abschluß, *der; (of questionnaire, form)* Ausfüllen, *das;* **on ~ of the course** nach Abschluß des Kurses
complex ['kɒmpleks] **1.** *adj.* **a)** *(complicated)* kompliziert; **b)** *(composite)* komplex. **2.** *n. (also Psych.)* Komplex, *der;* **a [building] ~:** ein Gebäudekomplex
complexion [kəm'plekʃn] *n.* Gesichtsfarbe, *die;* Teint, *der; (fig.)* Gesicht, *das;* **that puts a different ~ on the matter** dadurch sieht die Sache schon anders aus
complexity [kəm'pleksɪtɪ] *n. see* **complex 1:** Kompliziertheit, *die;* Komplexität, *die*
compliance [kəm'plaɪəns] *n.* **a)** *(action)* Zustimmung, *die* **(with** zu); **b)** *(submission)* Unterwürfigkeit, *die*
compliant [kəm'plaɪənt] *adj.* unterwürfig
complicate ['kɒmplɪkeɪt] *v.t.* komplizieren
complicated ['kɒmplɪkeɪtɪd] *adj.* kompliziert
complication [kɒmplɪ'keɪʃn] *n.* **a)** Kompliziertheit, *die;* **b)** *(circumstance; also Med.)* Komplikation, *die*
complicity [kəm'plɪsɪtɪ] *n.* Mittäterschaft, *die* **(in** bei)
compliment 1. ['kɒmplɪmənt] *n.* **a)** *(polite words)* Kompliment, *das;* **pay sb. a ~ [on sth.]** jmdm. [wegen etw.] ein Kompliment machen; **return the ~:** das Kompliment erwidern; *(fig.)* zurückschlagen; **b)** *in pl. (formal greetings)* Grüße *Pl.*; Empfehlung, *die.* **2.** ['kɒmplɪment] *v.t.* **~ sb. on sth.** jmdm. Komplimente wegen etw. machen
complimentary [kɒmplɪ'mentərɪ] *adj.* **a)** *(expressing compliment)* schmeichelhaft; **b)** *(given free)* Frei-
comply [kəm'plaɪ] *v.i.* **~ with sth.** sich nach etw. richten; **he refused to ~:** er wollte sich nicht danach richten
component [kəm'pəʊnənt] **1.** *n.* Bestandteil, *der; (of machine)* [Einzel]teil, *das.* **2.** *adj.* **a ~ part** ein Bestandteil
compose [kəm'pəʊz] *v.t.* **a)** *(make up)* bilden; **be ~d of** sich zusammensetzen aus; **b)** *(construct)* verfassen ‹*Rede, Gedicht, Liedertext, Libretto*›; abfassen, aufsetzen ‹*Brief*›; **c)** *(Mus.)* komponieren; **d)** *(calm)* **~ oneself** sich zusammennehmen
composed [kəm'pəʊzd] *adj. (calm)* gefaßt
composer [kəm'pəʊzə(r)] *n.* **a)** *(of music)* Komponist, *der*/Komponistin, *die;* **b)** *(of poem etc.)* Verfasser, *der*/Verfasserin, *die*
composition [kɒmpə'zɪʃn] *n.* **a)** *(act)* Zusammenstellung, *die; (construction)* Herstellung, *die;* **b)** *(constitution) (of soil, etc.)* Zusammensetzung, *die; (of picture)* Aufbau, *der;* **c)** *(piece of writing)* Darstellung, *die; (essay)* Aufsatz, *der; (piece of music)* Komposition, *die;* **d)** *(construction in writing) (of sentences)* Konstruktion, *die; (of prose, verse)* Verfassen, *das; (Mus.)* Komposition, *die*
compost ['kɒmpɒst] **1.** *n.* Kompost, *der.* **2.** *v.t.* kompostieren
compost: ~ heap, ~ pile *ns.* Komposthaufen, *der*
composure [kəm'pəʊʒə(r)] *n.* Gleichmut, *der;* **lose/regain one's ~:** die Fassung verlieren/wiederfinden; **upset sb.'s ~:** jmdn. aus der Fassung bringen
compote ['kɒmpəʊt] *n.* Kompott, *das*
[1]**compound 1.** ['kɒmpaʊnd] *adj.* **a)** *(of several ingredients)* zusammengesetzt; **b)** *(of several parts)* kombiniert; **a ~ word** ein zusammengesetztes Wort; eine Zusammensetzung; **c)** *(Zool.)* **~ eye** Facettenauge, *das;* **d)** *(Med.)* **~ fracture** komplizierter Bruch. **2.** ['kɒmpaʊnd] *n.* **a)** *(Ling.)* Kompositum, *das;* Zusammensetzung, *die;* **b)** *(Chem.)* Verbindung, *die.* **3.** [kəm'paʊnd] *v.t. (increase, complicate)* verschlimmern ‹*Schwierigkeiten, Verletzung usw.*›
[2]**compound** ['kɒmpaʊnd] *n. (enclosed space)* umzäuntes Gebiet *od.* Gelände; **prison ~:** Gefängnishof, *der*
compound 'interest *n. (Finance)* Zinseszinsen *Pl.*
comprehend [kɒmprɪ'hend] *v.t.* begreifen; verstehen
comprehensible [kɒmprɪ'hensɪbl] *adj.* verständlich **(to** *Dat.*)
comprehension [kɒmprɪ'henʃn]

n. **a)** *(understanding)* Verständnis, *das;* **b)** ~ **[exercise/test]** Übung zum Textverständnis
comprehensive [kɒmprɪ'hensɪv] **1.** *adj.* **a)** *(inclusive)* umfassend; universal ‹*Verstand*›; **b)** ~ **school** Gesamtschule, *die;* **go** ~ ‹*Schule:*› zur Gesamtschule [gemacht] werden; **c)** *(Insurance)* Vollkasko-; ~ **policy** Vollkaskoversicherung, *die.* **2.** *n. (Sch.)* Gesamtschule, *die*
comprehensively [kɒmprɪ'hensɪvlɪ] *adv.* umfassend; ~ **beaten** deutlich geschlagen
compress 1. [kəm'pres] *v.t.* **a)** *(squeeze)* zusammenpressen **(into** zu); **b)** komprimieren ‹*Luft, Gas, Bericht*›. **2.** ['kɒmpres] *n. (Med.)* Kompresse, *die*
compressed 'air *adj.* Druck-, Preßluft, *die*
compression [kəm'preʃn] *n.* Kompression, *die*
compressor [kəm'presə(r)] *n.* Kompressor, *der*
comprise [kəm'praɪz] *v.t. (include)* umfassen; *(not exclude)* einschließen; *(consist of)* bestehen aus; *(make up)* bilden
compromise ['kɒmprəmaɪz] **1.** *n.* Kompromiß, *der.* **2.** *v.i.* Kompromisse/einen Kompromiß schließen; ~ **with sb. over sth.** mit jmdm. einen Kompromiß in etw. *(Dat.)* schließen. **3.** *v.t. (bring under suspicion)* kompromittieren; *(bring into danger)* schaden (+ *Dat.*); ~ **oneself** sich kompromittieren
compulsion [kəm'pʌlʃn] *n. (also Psych.)* Zwang, *der;* **be under no** ~ **to do sth.** keineswegs etw. tun müssen
compulsive [kəm'pʌlsɪv] *adj.* **a)** zwanghaft; **he is a** ~ **eater/gambler** er leidet unter Eßzwang/er ist dem Spiel verfallen; **b)** *(irresistible)* **this book is** ~ **reading** von diesem Buch kann man sich nicht losreißen
compulsorily [kəm'pʌlsərɪlɪ] *adv.* zwangsweise
compulsory [kəm'pʌlsərɪ] *adj.* obligatorisch; **be** ~: obligatorisch *od.* Pflicht sein
compunction [kəm'pʌŋkʃn] *n.* Schuldgefühle *Pl.*
computation [kɒmpjʊ'teɪʃn] *n.* Berechnung, *die*
compute [kəm'pju:t] *v.t.* berechnen **(at** auf + *Akk.*)
computer [kəm'pju:tə(r)] *n.* Computer, *der*
computerize (computerise) [kəm'pju:təraɪz] *v.t.* computerisieren
computer: ~-'operated *adj.* computergesteuert; ~ **program** *n.* Programm, *das;* ~ **programmer** *n.* Programmierer, *der*/Programmiererin, *die;* ~ **terminal** *n.* Terminal, *das*
comrade ['kɒmreɪd, 'kɒmrɪd] *n.* Kamerad, *der*/Kameradin, *die;* **~-in-arms** Kampfgefährte, *der*
comradeship ['kɒmreɪdʃɪp] *n., no pl.* Kameradschaft, *die*
[1]**con** [kɒn] *(coll.)* **1.** *n. (trick)* Schwindel, *der.* **2.** *v.t.,* **-nn- a)** *(swindle)* reinlegen *(ugs.);* ~ **sb. out of sth.** jmdm. etw. abschwindeln *od. (ugs.)* abgaunern; **b)** *(persuade)* beschwatzen *(ugs.);* ~ **sb. into sth.** jmdm. etw. aufschwatzen *(ugs.)*
[2]**con** *see* [1]**pro**
concave ['kɒnkeɪv] *adj.* konkav; ~ **mirror/lens** Konkav- *od.* Hohlspiegel, *der*/Konkavlinse, *die*
conceal [kən'si:l] *v.t.* verbergen **(from** vor + *Dat.*); ~ **the true state of affairs from sb.** jmdm. den wirklichen Sachverhalt verheimlichen
concealed [kən'si:ld] *adj.* verdeckt; ~ **lighting** indirekte Beleuchtung
concealment [kən'si:lmənt] *n. see* **conceal:** Verbergen, *das;* Verheimlichung, *die*
concede [kən'si:d] *v.t. (admit, allow)* zugeben; *(grant)* zugestehen, einräumen ‹*Recht, Privileg*›
conceit [kən'si:t] *n., no pl. (vanity)* Einbildung, *die*
conceited [kən'si:tɪd] *adj.* eingebildet
conceivable [kən'si:vəbl] *adj.* vorstellbar; **it's scarcely ~ that ...:** man kann sich *(Dat.)* kaum vorstellen, daß ...
conceivably [kən'si:vəblɪ] *adv.* möglicherweise; **he cannot ~ have done it** er kann es unmöglich getan haben
conceive [kən'si:v] **1.** *v.t.* **a)** empfangen ‹*Kind*›; **b)** *(form in mind)* sich *(Dat.)* vorstellen *od.* denken; haben, kommen auf (+ *Akk.*) ‹*Idee, Plan*›; ~ **a dislike for sb./sth.** eine Abneigung gegen jmdn./etw. entwickeln; **c)** *(think)* meinen; glauben. **2.** *v.i.* **a)** *(become pregnant)* empfangen; **b)** ~ **of sth.** sich *(Dat.)* etw. vorstellen
concentrate ['kɒnsəntreɪt] **1.** *v.t.* ~ **one's efforts [up]on sth.** seine Bemühungen auf etw. *(Akk.)* konzentrieren; ~ **one's mind on sth.** sich auf etw. *(Akk.)* konzentrieren. **2.** *v.i.* sich konzentrieren **(on** auf + *Akk.*); ~ **on doing sth.** sich darauf konzentrieren, etw. zu tun. **3.** *n.* Konzentrat, *das*
concentrated ['kɒnsəntreɪtɪd] *adj.* konzentriert
concentration [kɒnsən'treɪʃn] *n. (also Chem.)* Konzentration, *die;* **power[s] of ~:** Konzentrationsfähigkeit, *die;* **lose one's ~:** sich nicht mehr konzentrieren können
concen'tration camp *n.* Konzentrationslager, *das;* KZ, *das*
concept ['kɒnsept] *n.* Begriff, *der; (idea)* Vorstellung, *die*
conception [kən'sepʃn] **a)** *(idea)* Vorstellung, *die* **(of** von); **b)** *(conceiving)* **great powers of ~:** ein großes Vorstellungsvermögen; **c)** *(of child)* Empfängnis, *die*
conceptual [kən'septjʊəl] *adj.* begrifflich
concern [kən'sɜ:n] **1.** *v.t.* **a)** *(affect)* betreffen; **so far as ... is ~ed** was ... betrifft; **'to whom it may ~'** ≈ „Bestätigung"; *(on certificate, testimonial)* ≈ „Zeugnis"; **b)** *(interest)* ~ **oneself with** *or* **about sth.** sich mit etw. befassen; **c)** *(trouble)* **the news/her health greatly ~s me** ich bin über diese Nachricht tief beunruhigt/ihre Gesundheit bereitet mir große Sorgen. **2.** *n.* **a)** *(relation)* **have no ~ with sth.** mit etw. nichts zu tun haben; **b)** *(anxiety)* Besorgnis, *die; (interest)* Interesse, *das;* **express ~:** Sorge ausdrücken; **c)** *(matter)* Angelegenheit, *die;* **that's no ~ of mine** das geht mich nichts an; **d)** *(firm)* Unternehmen, *das*
concerned [kən'sɜ:nd] *adj.* **a)** *(involved)* betroffen; *(interested)* interessiert; **the people ~:** die Betroffenen; **where work/health is ~:** wenn es um die Arbeit/die Gesundheit geht; **as** *or* **so far as I'm ~:** was mich betrifft *od.* anbelangt; **b)** *(implicated)* verwickelt **(in** in + *Akk.*); **c)** *(troubled)* besorgt; **I am ~ to hear that ...:** ich höre mit Sorge, daß ...; **I was ~ at the news** die Nachricht beunruhigte mich
concerning [kən'sɜ:nɪŋ] *prep.* bezüglich
concert ['kɒnsət] *n.* **a)** *(of music)* Konzert, *das;* **b) work in ~ with sb.** mit jmdm. zusammenarbeiten
concerted [kən'sɜ:tɪd] *adj.* vereint; gemeinsam
concert: ~-goer *n.* Konzertbesucher, *der*/-besucherin, *die;* **~-hall** *n.* Konzertsaal, *der; (building)* Konzerthalle, *die*
concertina [kɒnsə'ti:nə] *n. (Mus.)* Konzertina, *die*
concerto [kən'tʃeətəʊ] *n., pl.* **~s**

or **concerti** [kən'tʃeəti:] *(Mus.)* Konzert, *das*

concert: ~ **pianist** *n.* Konzertpianist, *der*/-pianistin, *die;* ~ **pitch** *n.* Kammerton, *der*

concession [kən'seʃn] *n.* Konzession, *die*

concessionary [kən'seʃənərɪ] *adj.* Konzessions-; ~ **rate/fare** ermäßigter Tarif

conciliate [kən'sɪlɪeɪt] *v. t.* **a)** *(reconcile)* in Einklang bringen ‹*Gegensätze, Theorien*›; **b)** *(pacify)* besänftigen

conciliation [kənsɪlɪ'eɪʃn] *n.* **a)** *(reconcilement)* Versöhnung, *die;* **b)** *(pacification)* Besänftigung, *die;* **c)** *(in industrial relations)* Schlichtung, *die*

conciliatory [kən'sɪljətərɪ] *adj.* versöhnlich; *(pacifying)* beschwichtigend

concise [kən'saɪs] *adj.* kurz und prägnant; knapp, konzis ‹*Stil*›; **be** ~ ‹*Person:*› sich knapp fassen; **a** ~ **dictionary** ein Handwörterbuch

concisely [kən'saɪslɪ] *adv.* kurz und prägnant; knapp, konzis ‹*schreiben*›

conclude [kən'klu:d] **1.** *v. t.* **a)** *(end)* beschließen; beenden; **b)** *(infer)* schließen; folgern; **c)** *(reach decision)* beschließen; **d)** *(agree on)* schließen ‹*Bündnis, Vertrag*›. **2.** *v. i.* *(end)* schließen

concluding [kən'klu:dɪŋ] *attrib. adj.* abschließend

conclusion [kən'klu:ʒn] *n.* **a)** *(end)* Abschluß, *der;* **in** ~**:** zum Abschluß; **b)** *(result)* Ausgang, *der;* **c)** *(decision reached)* Beschluß, *der;* **d)** *(inference)* Schluß, *der;* *(Logic)* [Schluß]folgerung, *die*

conclusive [kən'klu:sɪv] *adj.* schlüssig

conclusively [kən'klu:sɪvlɪ] *adv.* schlüssig ‹*beweisen, belegen*›

concoct [kən'kɒkt] *v. t.* zubereiten; zusammenbrauen ‹*Trank*›; *(fig.)* sich *(Dat.)* ausdenken ‹*Geschichte*›; sich *(Dat.)* zurechtlegen ‹*Ausrede, Alibi*›

concoction [kən'kɒkʃn] *n.* *(drink)* Gebräu, *das*

concord ['kɒŋkɔ:d, 'kɒnkɔ:d] *n.* *(agreement)* Eintracht, *die*

concourse ['kɒŋkɔ:s, 'kɒnkɔ:s] *n.* *(of public building)* Halle, *die;* **station** ~**:** Bahnhofshalle, *die*

concrete ['kɒnkri:t] **1.** *adj.* *(specific)* konkret; ~ **noun** *(Ling.)* Konkretum, *das.* **2.** *n.* Beton, *der; attrib.* Beton-; aus Beton *präd.* **3.** *v. t.* betonieren; *(embed in* ~*)* ~ **[in]** einbetonieren

'concrete-mixer *n.* Betonmischer, *der*

concur [kən'kɜ:(r)] *v. i.,* **-rr-** *(agree)* ~ **[with sb.] [in sth.]** [jmdm.] [in etw. *(Dat.)*] zustimmen *od.* beipflichten

concurrent [kən'kʌrənt] *adj.* gleichzeitig; **be** ~ **with sth.** gleichzeitig mit etw. stattfinden; ~ **sentences** zu einer Gesamtstrafe zusammengefaßte Einzelstrafen

concurrently [kən'kʌrəntlɪ] *adv.* gleichzeitig; **run** ~ ‹*Gefängnisstrafen:*› zu einer Gesamtstrafe zusammengefaßt sein/werden

concuss [kən'kʌs] *v. t.* **be** ~**ed** eine Gehirnerschütterung haben

concussion [kən'kʌʃn] *n.* *(Med.)* Gehirnerschütterung, *die*

condemn [kən'dem] *v. t.* **a)** *(censure)* verdammen; **b)** *(Law: sentence)* verurteilen; *(fig.)* verdammen; ~ **sb. to death** jmdn. zum Tode verurteilen; **a** ~**ed man** ein zum Tode Verurteilter; ~**ed cell** Todeszelle, *die;* **c)** *(declare unfit)* für unbewohnbar erklären ‹*Gebäude*›; für ungenießbar erklären ‹*Fleisch*›

condemnation [kɒndem'neɪʃn] *n.* **a)** *(censure)* Verdammung, *die;* **b)** *(Law: conviction)* Verurteilung, *die*

condensation [kɒnden'seɪʃn] *n.* **a)** *no pl.* *(condensing)* Kondensation, *die;* **b)** *(what is condensed)* Kondensat, *das;* *(water)* Kondenswasser, *das;* **c)** *(abridgement)* [Ver]kürzung, *die;* *(abridged form)* Kurzfassung, *die*

condense [kən'dens] **1.** *v. t.* **a)** komprimieren; ~**d milk** Kondensmilch, *die;* **b)** *(Phys., Chem.)* kondensieren; **c)** *(make concise)* zusammenfassen; **in a** ~**d form** in verkürzter Form. **2.** *v. i.* kondensieren

condenser [kən'densə(r)] *n.* *(of steam-engine; Electr.)* Kondensator, *der*

condescend [kɒndɪ'send] *v. i.* ~ **to do sth.** sich dazu herablassen, etw. zu tun; ~ **to sb.** jmdn. von oben herab behandeln

condescending [kɒndɪ'sendɪŋ] *adj.* herablassend

condescension [kɒndɪ'senʃn] *n.* Herablassung, *die*

condiment ['kɒndɪmənt] *n.* Gewürz, *das*

condition [kən'dɪʃn] **1.** *n.* **a)** *(stipulation)* [Vor]bedingung, *die;* Voraussetzung, *die;* **make it a** ~ **that ...:** es zur Bedingung machen, daß ...; **on [the]** ~ **that ...:** unter der Voraussetzung *od.* Bedingung, daß ...; **b)** *(in pl.: circumstances)* Umstände *Pl.;* **weather/light** ~**s** Witterungsverhältnisse/Lichtverhältnisse; **under** *or* **in present** ~**s** unter den gegenwärtigen Umständen *od.* Bedingungen; **living/working** ~**s** Unterkunfts-/Arbeitsbedingungen; **c)** *(of athlete, etc.)* Kondition, *die;* Form, *die;* *(of thing)* Zustand, *der;* *(of invalid, patient, etc.)* Verfassung, *die;* **keep sth. in good** ~**:** etw. in gutem Zustand erhalten; **be out of** ~**/in [good]** ~ ‹*Person:*› schlecht/gut in Form sein; **d)** *(Med.)* Leiden, *das;* **have a heart/lung** *etc.* ~**:** ein Herz-/Lungenleiden *usw.* haben. **2.** *v. t.* bestimmen

conditional [kən'dɪʃənl] *adj.* **a)** bedingt; **be** ~ **[up]on sth.** von etw. abhängen; **b)** *(Ling.)* konditional; ~ **clause** Konditionalsatz, *der*

conditionally [kən'dɪʃənəlɪ] *adv.* mit *od.* unter Vorbehalt

condolence [kən'dəʊləns] *n.* Anteilnahme, *die;* Mitgefühl, *das;* *(on death)* Beileid, *das;* **letter of** ~**:** Beileidsbrief, *der;* Kondolenzbrief, *der*

condom ['kɒndɒm] *n.* Kondom, *das od. der;* Präservativ, *das*

condominium [kɒndə'mɪnɪəm] *n.* *(Amer.)* Appartementhaus [mit Eigentumswohnungen]; *(single dwelling)* Eigentumswohnung, *die*

condone [kən'dəʊn] *v. t.* **a)** hinwegsehen über (+ *Akk.*); *(approve)* billigen; **b)** *(Law)* in Kauf nehmen; stillschweigend billigen

conducive [kən'dju:sɪv] *adj.* **be** ~ **to sth.** einer Sache *(Dat.)* förderlich sein; zu etw. beitragen

conduct 1. ['kɒndʌkt] *n.* **a)** *(behaviour)* Verhalten, *das;* **good** ~**:** gute Führung; **b)** *(way of* ~*ing)* Führung, *die;* *(of inquiry, operation)* Durchführung, *die.* **2.** [kən'dʌkt] *v. t.* **a)** *(Mus.)* dirigieren; **b)** führen ‹*Geschäfte, Krieg, Gespräch*›; durchführen ‹*Operation, Untersuchung*›; **c)** *(Phys.)* leiten ‹*Wärme, Elektrizität*›; **d)** ~ **oneself** sich verhalten; **e)** *(guide)* führen; **a** ~**ed tour [of a museum/factory]** eine [Museums-/Werks]führung

conduction [kən'dʌkʃn] *n.* *(Phys.)* Leitung, *die*

conductor [kən'dʌktə(r)] *n.* **a)** *(Mus.)* Dirigent, *der*/Dirigentin, *die;* **b)** *(of bus, tram)* Schaffner, *der;* *(Amer.: of train)* Zugführer, *der;* Schaffner, *der* *(ugs.);* **c)** *(Phys.)* Leiter, *der*

conductress [kən'dʌktrɪs] *n.* Schaffnerin, *die*

conduit ['kɒndɪt, 'kɒndjʊɪt] *n.* Leitung, *die;* Kanal, *der (auch fig.)*

cone [kəʊn] *n.* **a)** Kegel, *der;* Konus, *der (fachspr.); (traffic ~)* Leitkegel, *der;* **b)** *(Bot.)* Zapfen, *der;* **c) ice-cream ~:** Eistüte, *die*

confection [kən'fekʃn] *n.* Konfekt, *das*

confectioner [kən'fekʃənə(r)] *n. (maker)* Hersteller von Süßigkeiten; *(retailer)* Süßwarenhändler, *der;* **~'s [shop]** Süßwarengeschäft, *das;* **~s' sugar** *(Amer.)* Puderzucker, *der*

confectionery [kən'fekʃənərɪ] *n.* Süßwaren *Pl.*

confederate [kən'fedərət] **1.** *adj.* verbündet. **2.** *n.* Verbündete, *der/die; (accomplice)* Komplize, *der*/Komplizin, *die*

confederation [kənfedə'reɪʃn] *n.* **a)** *(Polit.)* [Staaten]bund, *der;* **b)** *(alliance)* Bund, *der;* **C~ of British Industry** *britischer Unternehmerverband*

confer [kən'fɜ(r)] **1.** *v. t.,* **-rr-: ~ a title/degree/knighthood [up]on sb.** jmdm. einen Titel/Grad verleihen/jmdn. zum Ritter schlagen. **2.** *v. i.,* **-rr-: ~ with sb.** sich mit jmdm. beraten

conference ['kɒnfərəns] *n.* **a)** *(meeting)* Konferenz, *die;* Tagung, *die;* **b)** *(consultation)* Beratung, *die; (business discussion)* Besprechung, *die;* **be in ~:** in einer Besprechung sein

conference: ~-room *n.* Konferenzraum, *der; (smaller)* Besprechungszimmer, *das;* **~-table** *n.* Konferenztisch, *der*

conferment [kən'fɜ:mənt] *n.* Verleihung, *die*

confess [kən'fes] **1.** *v. t.* **a)** zugeben; gestehen; **b)** *(Eccl.)* beichten. **2.** *v. i.* **a) ~ to sth.** etw. gestehen; **b)** *(Eccl.)* beichten **(to sb.** jmdm.)

confession [kən'feʃn] *n.* **a)** *(of offence etc.; thing confessed)* Geständnis, *das;* **on** *or* **by one's own ~:** nach eigenem Geständnis; **b)** *(Eccl.: of sins etc.)* Beichte, *die;* **c)** *(Relig.: denomination)* Konfession, *die;* **d)** *(Eccl.: confessing)* Bekenntnis, *das*

confessional [kən'feʃənl] *(Eccl.) n. (stall)* Beichtstuhl, *der*

confessor [kən'fesə(r)] *n. (Eccl.)* Beichtvater, *der*

confetti [kən'fetɪ] *n.* Konfetti, *das*

confidant ['kɒnfɪdænt, kɒnfɪ'dænt] *n.* Vertraute, *der*

confidante ['kɒnfɪdænt, kɒnfɪ'dænt] *n.* Vertraute, *die*

confide [kən'faɪd] **1.** *v. i.* **~ in sb.** sich jmdm. anvertrauen; **~ to sb. about sth.** jmdm. etw. anvertrauen. **2.** *v. t.* **~ sth. to sb.** jmdm. etw. anvertrauen; **he ~d that he ...:** er gestand, daß er ...

confidence ['kɒnfɪdəns] *n.* **a)** *(firm trust)* Vertrauen, *das;* **have [complete** *or* **every/no] ~ in sb./sth.** [volles/kein] Vertrauen zu jmdm./etw. haben; **have [absolute] ~ that ...:** [absolut] sicher sein, daß ...; **b)** *(assured expectation)* Gewißheit, *die;* Sicherheit, *die;* **c)** *(self-reliance)* Selbstvertrauen, *das;* **d) in ~:** im Vertrauen; **this is in [strict] ~:** das ist [streng] vertraulich; **take sb. into one's ~:** jmdn. ins Vertrauen ziehen; **e)** *(thing told in ~)* Vertraulichkeit, *die*

confidence: ~ game *(Amer.) see* **~ trick; ~ man** *n.* Trickbetrüger, *der;* Bauernfänger, *der (ugs.);* **~ trick** *n. (Brit.)* Trickbetrug, *der;* Bauernfängerei, *die (ugs.);* **~ trickster** *(Brit.) see* **~ man**

confident ['kɒnfɪdənt] *adj.* **a)** *(trusting, fully assured)* zuversichtlich **(about** in bezug auf + *Akk.*); **be ~ that ...:** sicher sein, daß ...; **be ~ of sth.** auf etw. *(Akk.)* vertrauen; **b)** *(self-assured)* selbstbewußt

confidential [kɒnfɪ'denʃl] *adj.* vertraulich

confidentiality [kɒnfɪdenʃɪ'ælɪtɪ] *n., no pl.* Vertraulichkeit, *die*

confidentially [kɒnfɪ'denʃəlɪ] *adv.* vertraulich

confidently ['kɒnfɪdəntlɪ] *adv.* zuversichtlich

configuration [kənfɪgjʊ'reɪʃn] *n.* **a)** *(arrangement, outline)* Gestaltung, *die;* **b)** *(Computing)* Konfiguration, *die*

configure [kən'fɪgə(r)] *v. t. (Computing)* konfigurieren

confine [kən'faɪn] *v. t.* **a)** einsperren; **be ~d to bed/the house** ans Bett/Haus gefesselt sein; **be ~d to barracks** keinen Ausgang bekommen; **b)** *(fig.)* **~ sb./sth. to sth.** jmdn./etw. auf etw. *(Akk.)* beschränken; **~ oneself to sth./doing sth.** sich auf etw. *(Akk.)* beschränken/sich darauf beschränken, etw. zu tun

confined [kən'faɪnd] *adj.* begrenzt

confinement [kən'faɪnmənt] *n. (imprisonment)* Einsperrung, *die;* **put/keep sb. in ~:** jmdn. in Haft nehmen/halten

confines ['kɒnfaɪnz] *n. pl.* Grenzen

confirm [kən'fɜ:m] *v. t.* **a)** bestätigen; **b)** *(Protestant Ch.)* konfirmieren; *(RC Ch.)* firmen

confirmation [kɒnfə'meɪʃn] *n.* **a)** Bestätigung, *die;* **b)** *(Protestant Ch.)* Konfirmation, *die;* Einsegnung, *die; (RC Ch.)* Firmung, *die*

confirmed [kən'fɜ:md] *adj. (unlikely to change)* eingefleischt ⟨*Junggeselle*⟩; überzeugt ⟨*Atheist, Vegetarier*⟩

confiscate ['kɒnfɪskeɪt] *v. t.* beschlagnahmen; konfiszieren; **~ sth. from sb.** jmdm. etw. wegnehmen

confiscation [kɒnfɪs'keɪʃn] *n.* Beschlagnahme, *die*

conflict 1. ['kɒnflɪkt] *n.* **a)** *(fight)* Kampf, *der; (prolonged)* Krieg, *der;* **come into ~ with sb./sth.** mit jmdm./etw. in Konflikt geraten; **be in ~ with sb./sth.** *(fig.)* mit jmdm./etw. im Kampf liegen; **b)** *(clashing)* Konflikt, *der.* **2.** [kən'flɪkt] *v. i. (be incompatible)* sich *(Dat.)* widersprechen; **~ with sth.** einer Sache *(Dat.)* widersprechen

conflicting [kən'flɪktɪŋ] *adj.* widersprüchlich

conform [kən'fɔ:m] *v. i.* **a)** entsprechen **(to** *Dat.*); **b)** *(comply)* **~ to** *or* **with sth./with sb.** sich nach etw./jmdm. richten

conformism [kən'fɔ:mɪzm] *n.* Konformismus, *der*

conformist [kən'fɔ:mɪst] *n.* Konformist, *der*/Konformistin, *die*

conformity [kən'fɔ:mɪtɪ] *n.* Übereinstimmung, *die* **(with, to** mit)

confound [kən'faʊnd] *v. t.* **a) ~ it!** verflixt noch mal! *(ugs.);* **b)** *(confuse)* verwirren; **c)** *(discomfit)* ins Unrecht setzen

confounded [kən'faʊndɪd] *adj. (coll. derog.)* verdammt

confront [kən'frʌnt] *v. t.* **a)** gegenüberstellen; konfrontieren; **~ sb. with sth./sb.** jmdn. mit etw./[mit] jmdm. konfrontieren; **b)** *(stand facing)* gegenüberstehen (+ *Dat.*); **c)** *(face in defiance)* ins Auge sehen (+ *Dat.*)

confrontation [kɒnfrən'teɪʃn] *n.* Konfrontation, *die*

confuse [kən'fju:z] *v. t.* **a)** *(disorder)* durcheinanderbringen; verwirren; *(blur)* verwischen; **~ the issue** den Sachverhalt unklar machen; **it simply ~s matters** das verwirrt die Sache nur; **b)** *(mix up mentally)* verwechseln; **c)** *(perplex)* konfus machen; verwirren

confused [kən'fju:zd] *adj.* konfus; wirr ⟨*Gedanken, Gerüchte*⟩;

verworren ‹*Lage, Situation*›; *(embarrassed)* verlegen
confusing [kən'fju:zɪŋ] *adj.* verwirrend
confusion [kən'fju:ʒn] *n.* **a)** *(disordering)* Verwirrung, *die; (mixing up)* Verwechslung, *die;* **b)** *(state)* Verwirrung, *die; (embarrassment)* Verlegenheit, *die;* **throw sb./sth. into ~:** jmdn./etw. [völlig] durcheinanderbringen
conga ['kɒŋgə] *n.* Conga, *die*
congeal [kən'dʒi:l] **1.** *v.i.* gerinnen. **2.** *v.t.* gerinnen lassen
congenial [kən'dʒi:nɪəl] *adj. (agreeable)* angenehm
congenital [kən'dʒenɪtl] *adj.* angeboren; **a ~ idiot** ein von Geburt an Schwachsinniger
conger ['kɒŋgə(r)] *n. (Zool.)* ~ **[eel]** Meer- *od.* Seeaal, *der*
congest [kən'dʒest] *v.t.* verstopfen
congested [kən'dʒestɪd] *adj.* überfüllt, verstopft ‹*Straße*›; **my nose is ~:** ich habe eine verstopfte Nase
congestion [kən'dʒestʃn] *n. (of traffic etc.)* Stauung, *die;* **nasal ~:** verstopfte Nase
conglomerate 1. [kən'glɒməreɪt] *v.i.* sich zusammenballen; *(fig.)* sich versammeln. **2.** [kən'glɒmərət] *n. (Commerc.)* Großkonzern, *der*
conglomeration [kənglɒmə'reɪʃn] *n.* Konglomerat, *das; (collection)* Ansammlung, *die*
congratulate [kən'grætjʊleɪt] *v.t.* gratulieren (+ *Dat.*); **~ sb./oneself [up]on sth.** jmdm./sich zu etw. gratulieren
congratulation [kəngrætjʊ'leɪʃn] **1.** *int.* **~s!** herzlichen Glückwunsch! **(on** zu). **2.** *n.* **a)** *in pl.* Glückwünsche *Pl.;* **b)** *(action)* Gratulation, *die*
congregate ['kɒŋgrɪgeɪt] *v.i.* sich versammeln
congregation [kɒŋgrɪ'geɪʃn] *n. (Eccl.)* Gemeinde, *die*
congress ['kɒŋgres] *n.* **a)** *(meeting of heads of state etc.)* Kongreß, *der;* **a party ~:** ein Parteitag; **b) C~** *(Amer.: legislature)* der Kongreß
congressional [kən'greʃənl] *adj.* Kongreß-
Congressman ['kɒŋgresmən] *n., pl.* **Congressmen** ['kɒŋgresmən] *(Amer.)* Kongreßabgeordnete, *der*
congruent ['kɒŋgrʊənt] *adj. (Geom.)* kongruent
conic ['kɒnɪk] *adj.* Kegel-
conical ['kɒnɪkl] *adj.* konisch; kegelförmig
conifer ['kɒnɪfə(r)] *n.* Nadelbaum, *der*
conjecture [kən'dʒektʃə(r)] **1.** *n.* Mutmaßung, *die (geh.);* Vermutung, *die.* **2.** *v.t.* mutmaßen *(geh.);* vermuten. **3.** *v.i. (guess)* Mutmaßungen *(geh.) od.* Vermutungen anstellen
conjugal ['kɒndʒʊgl] *adj.* ehelich; **~ bliss/worries** Eheglück, *das*/Ehesorgen
conjugate ['kɒndʒʊgeɪt] *v.t. (Ling.)* konjugieren
conjugation ['kɒndʒʊ'geɪʃn] *n. (Ling.)* Konjugation, *die*
conjunction [kən'dʒʌŋkʃn] *n.* **a)** Verbindung, *die;* **in ~ with sb./sth.** in Verbindung mit jmdm./etw.; **b)** *(Ling.)* Konjunktion, *die;* Bindewort, *das*
conjure ['kʌndʒə(r)] *v.i.* zaubern; **conjuring trick** Zaubertrick, *der*
~ **'up** *v.t.* beschwören ‹*Geister, Teufel*›; *(fig.)* heraufbeschwören
conjurer, conjuror ['kʌndʒərə(r)] *n.* Zauberkünstler, *der*/-künstlerin, *die;* Zauberer, *der*/Zauberin, *die*
conk [kɒŋk] *v.i.* ~ **'out** *(coll.)* ‹*Maschine, Auto usw.:*› den Geist aufgeben *(scherzh.),* kaputtgehen *(ugs.)*
conker ['kɒŋkə(r)] *n. (horse-chestnut)* [Roß]kastanie, *die;* **play ~s** *ein Wettspiel mit Kastanien machen*
'con-man *(coll.) see* **confidence man**
connect [kə'nekt] **1.** *v.t.* **a)** verbinden **(to, with** mit); *(Electr.)* anschließen **(to, with** an + *Akk.*); **b)** *(associate)* verbinden; **~ sth. with sth.** etw. mit etw. verbinden *od.* in Verbindung bringen; **be ~ed with sb./sth.** mit jmdm./etw. in Verbindung stehen. **2.** *v.i.* **~ with sth.** mit etw. zusammenhängen *od.* verbunden sein; ‹*Zug, Schiff usw.:*› Anschluß haben an etw. *(Akk.)*
~ **'up** *v.t.* anschließen
connected [kə'nektɪd] *adj. (logically joined)* zusammenhängend; *(related)* verwandt
connecting: ~ door *n.* Verbindungstür, *die;* **~ rod** *n. (Mech. Engin.)* Pleuelstange, *die*
connection [kə'nekʃn] *n.* **a)** *(act, state)* Verbindung, *die; (Electr.; of telephone)* Anschluß, *der;* **cut the ~:** die Verbindung abbrechen; **b)** *(fig.: of ideas)* Zusammenhang, *der;* **in ~ with sth.** im Zusammenhang mit etw.; **c)** *(part)* Verbindung, *die;* Verbindungsstück, *das;* **d)** *(train, boat, etc.)* Anschluß, *der;* **miss/catch or make a ~:** einen Anschluß verpassen/erreichen *od. (ugs.)* kriegen
connexion *(Brit.) see* **connection**
conning-tower ['kɒnɪŋtaʊə(r)] *n. (Naut.)* Kommandoturm, *der*
connivance [kə'naɪvəns] *n.* stillschweigende Duldung
connive [kə'naɪv] *v.i.* **~ at sth.** über etw. *(Akk.)* hinwegsehen; etw. stillschweigend dulden; **~ with sb.** mit jmdm. gemeinsame Sache machen **(in** bei)
connoisseur [kɒnə'sɜ:(r)] *n.* Kenner, *der*
connotation [kɒnə'teɪʃn] *n.* Assoziation, *die;* Konnotation, *die (Sprachw.)*
conquer ['kɒŋkə(r)] *v.t.* besiegen ‹*Gegner, Leidenschaft, Gewohnheit*›; erobern ‹*Land*›; bezwingen ‹*Berg, Gegner*›
conqueror ['kɒŋkərə(r)] *n.* Sieger, *der*/Siegerin, *die* **(of** über + *Akk.*); *(of a country)* Eroberer, *der*
conquest ['kɒŋkwest] *n.* Eroberung, *die*
conscience ['kɒnʃəns] *n.* Gewissen, *das;* **have a good** *or* **clear/bad** *or* **guilty ~:** ein gutes/schlechtes Gewissen haben; **with a clear** *or* **easy ~:** mit gutem Gewissen; **have sth. on one's ~:** wegen etw. ein schlechtes Gewissen haben
conscientious [kɒnʃɪ'enʃəs] *adj.* pflichtbewußt; *(meticulous)* gewissenhaft; **~ objector** Wehrdienstverweigerer [aus Gewissensgründen]
conscientiously [kɒnʃɪ'enʃəslɪ] *adv.* pflichtbewußt; *(meticulously)* gewissenhaft
conscious ['kɒnʃəs] *adj.* **a) I was ~ that ...:** mir war bewußt, daß ...; **but he is not ~ of it** aber es ist ihm nicht bewußt; **b)** *pred. (awake)* bei Bewußtsein *präd.;* **c)** *(realized by doer)* bewußt ‹*Handeln, Versuch, Bemühung*›
consciously ['kɒnʃəslɪ] *adv.* bewußt
consciousness ['kɒnʃəsnɪs] *n., no pl.* **a)** Bewußtsein, *das;* **lose/recover** *or* **regain ~:** das Bewußtsein verlieren/wiedererlangen; **b)** *(totality of thought; perception)* Bewußtsein, *das*
conscript 1. [kən'skrɪpt] *v.t.* einberufen ‹*Soldaten*›; ausheben ‹*Armee*›. **2.** ['kɒnskrɪpt] *n.* Einberufene, *der/die*
conscription [kən'skrɪpʃn] *n.* Einberufung, *die; (compulsory military service)* Wehrpflicht, *die*
consecrate ['kɒnsɪkreɪt] *v.t. (Eccl.; also fig.)* weihen

consecration [kɒnsɪ'kreɪʃn] *n.* *(Eccl.; also fig.)* Weihe, *die*
consecutive [kən'sekjʊtɪv] *adj.* aufeinanderfolgend ‹*Monate, Jahre*›; fortlaufend ‹*Zahlen*›; **this is the fifth ~ day that ...**: heute ist schon der fünfte Tag, an dem ...
consecutively [kən'sekjʊtɪvlɪ] *adv.* hintereinander
consensus [kən'sensəs] *n.* Einigkeit, *die;* **the general ~ is that ...**: es besteht allgemeine Einigkeit darüber, daß ...
consent [kən'sent] **1.** *v. i.* zustimmen; **~ to do sth.** einwilligen, etw. zu tun. **2.** *n.* **a)** *(agreement)* Zustimmung, *die* **(to** zu); Einwilligung, *die* **(to** in + *Akk.*); **by common** *or* **general ~**: nach allgemeiner Auffassung; *(as wished by all)* auf allgemeinen Wunsch; **age of ~**: *Alter, in dem man hinsichtlich Heirat und Geschlechtsleben nicht mehr als minderjährig gilt;* ≈ Ehemündigkeitsalter, *das;* **b)** *(permission)* Zustimmung, *die*
consequence ['kɒnsɪkwəns] *n.* **a)** *(result)* Folge, *die;* **in ~**: folglich; **in ~ of** als Folge (+ *Gen.*); **as a ~**: infolgedessen; **b)** *(importance)* Bedeutung, *die;* **be of no ~**: unerheblich *od.* ohne Bedeutung sein
consequent ['kɒnsɪkwənt] *adj.* *(resultant)* daraus folgend; *(following in time)* darauffolgend
consequently ['kɒnsɪkwəntlɪ] *adv.* infolgedessen; folglich
conservation [kɒnsə'veɪʃn] *n.* **a)** *(preservation)* Schutz, *der;* Erhaltung, *die; (wise utilization)* sparsamer Umgang **(of** mit); **wildlife ~**: Schutz wildlebender Tierarten; **b)** *(Phys.)* **~ of energy/momentum** Erhaltung der Energie/des Impulses
conser'vation area *n.* *(Brit.)* *(rural)* Landschaftsschutzgebiet, *das; (urban)* unter Denkmalschutz stehendes Gebiet
conservationist [kɒnsə'veɪʃənɪst] *n.* Naturschützer, *der/*-schützerin, *die*
conservatism [kən'sɜ:vətɪzm] *n.* Konservati[vi]smus, *der*
conservative [kən'sɜ:vətɪv] **1.** *adj.* **a)** *(averse to change)* konservativ; **b)** *(not too high)* vorsichtig, eher zu niedrig ‹*Zahlen, Schätzung*›; **c)** *(avoiding extremes)* konservativ ‹*Geschmack, Ansichten, Baustil*›; **d) C~** *(Brit. Polit.)* konservativ; **the C~ Party** die Konservative Partei. **2.** *n.* **C~** *(Brit. Polit.)* Konservative, *der/die*
conservatively [kən'sɜ:vətɪvlɪ] *adv.* vorsichtig, eher zu niedrig ‹*geschätzt*›
conservatory [kən'sɜ:vətərɪ] *n.* Wintergarten, *der*
conserve [kən'sɜ:v] **1.** *v. t.* erhalten ‹*Gebäude, Kunstwerk, Wälder*›; schonen ‹*Gesundheit, Kräfte*›. **2.** *n. often in pl.* Eingemachte, *das*
consider [kən'sɪdə(r)] *v. t.* **a)** *(look at)* betrachten; *(think about)* **~ sth.** an etw. *(Akk.)* denken; **b)** *(weigh merits of)* denken an (+ *Akk.*); **he's ~ing emigrating** er denkt daran, auszuwandern; **c)** *(reflect)* sich *(Dat.)* überlegen; **d)** *(regard as)* halten für; **I ~ him [to be** *or* **as] a swindler** ich halte ihn für einen Betrüger; **e)** *(allow for)* berücksichtigen; **~ other people's feelings** auf die Gefühle anderer Rücksicht nehmen; **all things ~ed** alles in allem
considerable [kən'sɪdərəbl] *adj.* beträchtlich; erheblich ‹*Schwierigkeiten, Ärger*›; groß ‹*Freude, Charakterstärke*›; eingehend ‹*Überlegung*›; *(Amer.: large)* ansehnlich ‹*Gebäude, Edelstein*›
considerably [kən'sɪdərəblɪ] *adv.* erheblich; *(in amount)* beträchtlich
considerate [kən'sɪdərət] *adj.* rücksichtsvoll **(towards** gegenüber); *(thoughtfully kind)* entgegenkommend
consideration [kənsɪdə'reɪʃn] *n.* **a)** Überlegung, *die; (meditation)* Betrachtung, *die;* **take sth. into ~**: etw. berücksichtigen *od.* bedenken; **give sth. one's ~**: etw. in Erwägung ziehen; **the matter is under ~**: die Angelegenheit wird geprüft; **leave sth. out of ~**: etw. unberücksichtigt lassen; **b)** *(thoughtfulness)* Rücksichtnahme, *die* **(for** auf + *Akk.*); **show ~ for sb.** Rücksicht auf jmdn. nehmen; **c)** *(sth. as reason)* Umstand, *der;* **d)** *(payment)* **for a ~**: gegen Entgelt
considered [kən'sɪdəd] *adj.* **a) ~ opinion** ernsthafte Überzeugung; **b) be highly ~ [by others]** [bei anderen] in hohem Ansehen stehen
considering [kən'sɪdərɪŋ] *prep.* **~ sth.** wenn man etw. bedenkt
consign [kən'saɪn] *v. t.* **a)** anvertrauen **(to** *Dat.*); **~ sth. to the scrap-heap** *(lit. or fig.)* etw. auf den Schrotthaufen werfen; **b)** *(Commerc.)* übersenden, *(fachspr.)* konsignieren ‹*Güter*› **(to** an + *Akk.*)
consignment [kən'saɪnmənt] *n.* *(Commerc.)* **a)** *(consigning)* Übersendung, *die* **(to** an + *Akk.*); **b)** *(goods)* Sendung, *die; (large)* Ladung, *die*
consist [kən'sɪst] *v. i.* **a) ~ of** bestehen aus; **b) ~ in** bestehen in (+ *Dat.*)
consistency [kən'sɪstənsɪ] *n.* **a)** *(density)* Konsistenz, *die;* **b)** *(being consistent)* Konsequenz, *die*
consistent [kən'sɪstənt] *adj.* **a)** *(compatible)* [miteinander] vereinbar; **be ~ with sth.** mit etw. übereinstimmen; mit etw. vereinbar sein; **b)** *(uniform)* beständig; gleichbleibend ‹*Qualität*›; einheitlich ‹*Vorgehen, Darstellung*›
consistently [kən'sɪstəntlɪ] *adv.* in Übereinstimmung ‹*handeln*›; einheitlich ‹*gestalten*›; konsistent ‹*denken*›; konsequent ‹*behaupten, verfolgen, handeln*›
consolation [kɒnsə'leɪʃn] *n.* **a)** *(act)* Tröstung, *die;* Trost, *der;* **words of ~**: Worte des Trostes; **b)** *(consoling circumstance)* Trost, *der*
conso'lation prize *n.* Trostpreis, *der*
[1]**console** [kən'səʊl] *v. t.* trösten
[2]**console** ['kɒnsəʊl] *n.* **a)** *(Mus.)* Spieltisch, *der;* **b)** *(panel)* [Schalt]pult, *das*
consolidate [kən'sɒlɪdeɪt] *v. t.* **a)** konsolidieren ‹*Stellung, Einfluß, Macht*›; **b)** *(combine)* zusammenlegen ‹*Territorien, Grundstücke, Firmen*›; konsolidieren ‹*Anleihen, Schulden*›
consolidation [kənsɒlɪ'deɪʃn] *n., no pl.* **a)** Konsolidierung, *die;* **b)** *(combining)* Zusammenlegung, *die*
consoling [kən'səʊlɪŋ] *adj.* tröstlich
consommé [kən'sɒmeɪ] *n.* *(Gastr.)* Kraftbrühe, *die*
consonant ['kɒnsənənt] *n.* Konsonant, *der;* Mitlaut, *der*
[1]**consort** ['kɒnsɔ:t] *n.* Gemahl, *der/*Gemahlin, *die*
[2]**consort** [kən'sɔ:t] *v. i.* *(keep company)* verkehren **(with** mit)
consortium [kən'sɔ:tɪəm] *n., pl.* **consortia** [kən'sɔ:tɪə] Konsortium, *das*
conspicuous [kən'spɪkjʊəs] *adj.* **a)** *(clearly visible)* unübersehbar; **b)** *(obvious, noticeable)* auffallend
conspicuously [kən'spɪkjʊəslɪ] *adv.* **a)** *(very visibly)* unübersehbar; **b)** *(obviously)* auffallend
conspiracy [kən'spɪrəsɪ] *n.* *(conspiring)* Verschwörung, *die; (plot)* Komplott, *das;* **~ of silence** verabredetes Stillschweigen
conspirator [kən'spɪrətə(r)] *n.* Verschwörer, *der/*Verschwörerin, *die*

conspiratorial [kənspɪrə'tɔ:rɪəl] *adj.* verschwörerisch

conspire [kən'spaɪə(r)] *v. i. (lit. or fig.)* sich verschwören

constable ['kʌnstəbl, 'kɒnstəbl] *n.* **a)** *(Brit.) see* **police constable; b)** *(Brit.)* **Chief C~:** ≈ Polizeipräsident, *der*/-präsidentin, *die*

constabulary [kən'stæbjʊlərɪ] **1.** *n.* Polizei, *die; (unit)* Polizeieinheit, *die.* **2.** *adj.* Polizei-

constancy ['kɒnstənsɪ] *n.* **a)** *(steadfastness)* Standhaftigkeit, *die;* **b)** *(faithfulness)* Treue, *die;* **c)** *(unchangingness)* Beständigkeit, *die*

constant ['kɒnstənt] **1.** *adj.* **a)** *(unceasing)* ständig; anhaltend ‹*Regen*›; **there was a ~ stream of traffic** der Verkehr floß ununterbrochen; **b)** *(unchanging)* gleichbleibend; konstant; **c)** *(steadfast)* standhaft; **d)** *(faithful)* treu. **2.** *n. (Phys., Math.)* Konstante, *die*

constantly ['kɒnstəntlɪ] *adv.* **a)** *(unceasingly)* ständig; **b)** *(unchangingly)* konstant; **c)** *(steadfastly)* standhaft

constellation [kɒnstə'leɪʃn] *n.* Sternbild, *das*

consternation [kɒnstə'neɪʃn] *n.* Bestürzung, *die; (confusion)* Aufregung, *die;* **in ~:** bestürzt/aufgeregt; **be filled with ~:** sehr bestürzt/aufgeregt sein

constipation [kɒnstɪ'peɪʃn] *n.* Verstopfung, *die*

constituency [kən'stɪtjʊənsɪ] *n.* *(voters)* Wählerschaft, *die (eines Wahlkreises); (area)* Wahlkreis, *der*

constituent [kən'stɪtjʊənt] **1.** *adj.* **~ part** Bestandteil, *der.* **2.** *n.* **a)** *(component part)* Bestandteil, *der;* **b)** *(member of constituency)* Wähler, *der*/Wählerin, *die (eines Wahlkreises)*

constitute ['kɒnstɪtju:t] *v. t.* **a)** *(form, be)* sein; **~ a threat to** eine Gefahr sein für; **b)** *(make up)* bilden; begründen ‹*Anspruch*›; **c)** *(establish)* gründen ‹*Partei, Organisation*›

constitution [kɒnstɪ'tju:ʃn] *n.* **a)** *(of person)* Konstitution, *die;* **b)** *(mode of State organization)* Staatsform, *die;* **c)** *(body of laws and principles)* Verfassung, *die*

constitutional [kɒnstɪ'tju:ʃənl] **1.** *adj.* **a)** *(of bodily constitution)* konstitutionell; **b)** *(Polit.) (of constitution)* der Verfassung *nachgestellt; (authorized by or in harmony with constitution)* verfassungsmäßig; konstitutionell ‹*Monarchie*›; **~ law** Verfassungsrecht, *das.* **2.** *n.* Spaziergang, *der*

constrain [kən'streɪn] *v. t.* zwingen

constraint [kən'streɪnt] *n.* **a)** Zwang, *der;* **b)** *(limitation)* Einschränkung, *die*

constrict [kən'strɪkt] *v. t.* verengen

constriction [kən'strɪkʃn] *n.* Verengung, *die*

construct 1. [kən'strʌkt] *v. t.* **a)** *(build)* bauen; *(fig.)* aufbauen; erstellen ‹*Plan*›; **b)** *(Ling.; Geom.: draw)* konstruieren. **2.** ['kɒnstrəkt] *n.* Konstrukt, *das*

construction [kən'strʌkʃn] *n.* **a)** *(constructing)* Bau, *der; (of sentence)* Konstruktion, *die; (fig.: of plan, syllabus)* Erstellung, *die;* **~ work** Bauarbeiten *Pl.;* **be under ~:** im Bau sein; **b)** *(thing constructed)* Bauwerk, *das; (fig.)* Gebilde, *das;* **c)** *(Ling.; Geom.: drawing)* Konstruktion, *die;* **d)** *(interpretation)* Deutung, *die*

constructive [kən'strʌktɪv] *adj.* konstruktiv

construe [kən'stru:] *v. t.* auslegen; auffassen; **I ~d his words as meaning that ...:** ich habe ihn so verstanden, daß ...

consul ['kɒnsl] *n.* Konsul, *der*

consular ['kɒnsjʊlə(r)] *adj.* konsularisch; **~ rank** Rang eines Konsuls

consulate ['kɒnsjʊlət] *n.* Konsulat, *das*

consult [kən'sʌlt] **1.** *v. i.* sich beraten **(with** mit); **~ together** sich miteinander beraten. **2.** *v. t. (seek information from)* konsultieren; befragen ‹*Orakel*›; fragen, konsultieren, zu Rate ziehen ‹*Arzt, Fachmann*›; **~ a list/book** in einer Liste/einem Buch nachsehen; **~ one's watch** auf die Uhr sehen; **~ a dictionary** in einem Wörterbuch nachschlagen

consultant [kən'sʌltənt] **1.** *n.* **a)** *(adviser)* Berater, *der*/Beraterin, *die;* **b)** *(physician)* ≈ Chefarzt, *der*/-ärztin, *die.* **2.** *attrib. adj. see* **consulting**

consultation [kɒnsʌl'teɪʃn] *n.* Beratung, *die* **(on** über + *Akk.*); **have a ~ with sb.** sich mit jmdm. beraten; **by ~ of a dictionary/of an expert** durch Konsultation eines Wörterbuchs/Experten; **act in ~ with sb.** in Absprache mit jmdm. handeln

consulting [kən'sʌltɪŋ] *attrib. adj.* beratend ‹*Architekt, Ingenieur*›

consumable [kən'sju:məbl] *adj.* **a)** kurzlebig ‹*Konsumgüter*›; **b)** *(edible, drinkable)* genießbar

consume [kən'sju:m] *v. t.* **a)** *(use up)* verbrauchen; ‹*Person:*› aufwenden, ‹*Sache:*› kosten ‹*Zeit, Energie*›; **b)** *(destroy)* vernichten; *(eat, drink)* konsumieren; verkonsumieren *(ugs.);* **c)** *(fig.)* **be ~d with love/passion** sich in Liebe/Leidenschaft verzehren; **be ~d with jealousy/envy** sich vor Eifersucht/Neid verzehren *(geh.)*

consumer [kən'sju:mə(r)] *n.* *(Econ.)* Verbraucher, *der*/Verbraucherin, *die;* Konsument, *der*/Konsumentin, *die*

consumer: ~ goods *n. pl.* Konsumgüter; **~ pro'tection** *n.* Verbraucherschutz, *der*

consummate 1. [kən'sʌmət] *adj.* **a)** *(perfect)* vollkommen; **with ~ ease** mühelos; **b)** *(accomplished)* perfekt; **a ~ artist** ein vollendeter Künstler. **2.** ['kɒnsəmeɪt] *v. t.* vollenden, zum Abschluß bringen ‹*Diskussion, Geschäftsverhandlungen*›; vollziehen ‹*Ehe*›

consummation [kɒnsə'meɪʃn] *n.* *(of marriage)* Vollzug, *der*

consumption [kən'sʌmpʃn] *n.* **a)** *(using up, eating, drinking)* Verbrauch, *der* **(of** an + *Dat.*); *(act of eating or drinking)* Verzehr, *der* **(of** von); **~ of fuel/sugar** Kraftstoff-/Zuckerverbrauch, *der;* **~ of alcohol** Alkoholkonsum, *der;* **b)** *(Econ.)* Verbrauch, *der;* Konsum, *der;* **c)** *(Med. dated)* Schwindsucht, *die (veralt.)*

cont. *abbr.* **continued** Forts.

contact 1. ['kɒntækt] *n.* **a)** *(state of touching)* Berührung, *die;* Kontakt, *der; (fig.)* Verbindung, *die;* Kontakt, *der;* **point of ~:** Berührungspunkt, *der;* **be in ~ with sth.** etw. berühren; **be in ~ with sb.** *(fig.)* mit jmdm. in Verbindung stehen *od.* Kontakt haben; **come in** *or* **into ~ [with sth.]** [mit etw.] in Berührung kommen; **come into ~ with sb./sth.** *(fig.)* mit jmdm./etw. etwas zu tun haben; **make ~ with sb.** *(fig.)* mit jmdm. Kontakt aufnehmen; **lose ~ with sb.** *(fig.)* den Kontakt mit jmdm. verlieren; **b)** *(Electr.: connection)* Kontakt, *der;* **make/break a ~:** einen Kontakt herstellen/unterbrechen. **2.** ['kɒntækt, kən'tækt] *v. t.* **a)** *(get into touch with)* sich in Verbindung setzen mit; **can I ~ you by telephone?** sind Sie telefonisch zu erreichen?; **b)** *(begin dealings with)* Kontakt aufnehmen mit

contact: ~ lens *n.* Kontaktlinse, *die;* **~ man** *n.* Kontaktmann, *der;* Mittelsmann, *der*

contagious [kən'teɪdʒəs] *adj. (lit. or fig.)* ansteckend

contain [kən'teɪn] *v. t.* **a)** *(hold as contents, include)* enthalten; *(comprise)* umfassen; **b)** *(prevent from spreading; also Mil.)* aufhalten; *(restrain)* unterdrücken; **he could hardly ~ himself for joy** er konnte vor Freude kaum an sich *(Akk.)* halten

container [kən'teɪnə(r)] *n.* Behälter, *der; (cargo ~)* Container, *der;* **cardboard/wooden ~:** Pappkarton, *der*/Holzkiste, *die*

containerize [kən'teɪnəraɪz] *v. t.* in Container verpacken

con'tainer ship *n.* Containerschiff, *das*

contaminate [kən'tæmɪneɪt] *v. t.* verunreinigen; *(with radioactivity)* verseuchen

contamination [kəntæmɪ'neɪʃn] *n.* Verunreinigung, *die; (with radioactivity)* Verseuchung, *die*

contemplate ['kɒntəmpleɪt] *v. t.* **a)** betrachten; *(mentally)* nachdenken über (+ *Akk.*); **b)** *(expect)* rechnen mit; *(consider)* in Betracht ziehen; **~ sth./doing sth.** an etw. *(Akk.)* denken/daran denken, etw. zu tun

contemplation [kɒntəm'pleɪʃn] *n.* **a)** Betrachtung, *die; (mental)* Nachdenken, *das* **(of** über + *Akk.*); **b)** *(expectation)* Erwartung, *die; (consideration)* Erwägung, *die*

contemplative [kən'templətɪv, 'kɒntəmpleɪtɪv] *adj.* besinnlich; kontemplativ *(geh.)*

contemporary [kən'tempərərɪ] **1.** *adj.* zeitgenössisch; *(present-day)* heutig; zeitgenössisch; **A is ~ with B** A und B finden zur gleichen Zeit statt. **2.** *n.* **a)** *(person belonging to same time)* Zeitgenosse, *der*/-genossin, *die* **(to** von); **we were contemporaries** *or* **he was a ~ of mine at university/school** er war ein Studienkollege *od.* Kommilitone/Schulkamerad von mir; **b)** *(person of same age)* Altersgenosse, *der*/-genossin, *die;* **they are contemporaries** sie sind gleichaltrig *od.* Altersgenossen

contempt [kən'tempt] *n.* **a)** Verachtung, *die* **(of, for** für); **b)** *(disregard)* Mißachtung, *die;* **c) have** *or* **hold sb. in ~:** jmdn. verachten; *see also* **beneath 1 a; d)** *(Law)* **~ of court** ≈ Ungebühr vor Gericht

contemptible [kən'temptɪbl] *adj.* verachtenswert

contemptuous [kən'temptjʊəs] *adj.* verächtlich; überheblich ‹*Person*›; **be ~ of sth./sb.** etw./jmdn. verachten

contend [kən'tend] **1.** *v. i.* **a)** *(strive)* **~ [with sb. for sth.]** [mit jmdm. um etw.] kämpfen; **b)** *(struggle)* **be able/have to ~ with** fertig werden können/müssen mit; **I've got enough to ~ with at the moment** ich habe schon so genug um die Ohren *(ugs.).* **2.** *v. t.* **~ that ...:** behaupten, daß ...

contender [kən'tendə(r)] *n.* Bewerber, *der*/Bewerberin, *die*

[1]**content** ['kɒntent] *n.* **a)** *in pl.* Inhalt, *der; (of medicine)* Zusammensetzung, *die;* **the ~s of the room had all been damaged** alles im Zimmer war beschädigt worden; **[table of] ~s** Inhaltsverzeichnis, *das;* **b)** *(amount contained)* Gehalt, *der* **(of** an + *Dat.*); **c)** *(constituent elements, substance)* Gehalt, *der*

[2]**content** [kən'tent] **1.** *pred. adj.* zufrieden **(with** mit); **be ~ to do sth.** bereit sein, etw. zu tun. **2.** *n.* **to one's heart's ~:** nach Herzenslust. **3.** *v. t.* zufriedenstellen; befriedigen; **~ oneself with sth./sb.** sich mit etw./jmdm. zufriedengeben

contented [kən'tentɪd] *adj.* zufrieden **(with** mit); glücklich ‹*Kindheit, Ehe, Leben*›

contentedly [kən'tentɪdlɪ] *adv.* zufrieden

contention [kən'tenʃn] *n.* **a)** *(dispute)* Streit, *der;* **sth. is the subject of much ~:** etw. wird heftig diskutiert; **b)** *(point asserted)* Behauptung, *die*

contentious [kən'tenʃəs] *adj.* strittig ‹*Punkt, Frage, Thema*›; umstritten ‹*Verhalten, Argument*›

contentment [kən'tentmənt] *n.* Zufriedenheit, *die*

contest 1. ['kɒntest] *n. (competition)* Wettbewerb, *der; (Sport)* Wettkampf, *der.* **2.** [kən'test] *v. t.* **a)** *(dispute)* bestreiten; anfechten ‹*Anspruch, Recht*›; in Frage stellen ‹*Behauptung, These*›; **b)** *(fight for)* kämpfen um; **c)** *(Brit.) (compete in)* kandidieren bei; *(compete for)* kandidieren für

contestant [kən'testənt] *n. (competitor)* Teilnehmer, *der*/Teilnehmerin, *die* **(in** an + *Dat.*, bei); *(in fight)* Gegner, *der*/Gegnerin, *die*

context ['kɒntekst] *n.* Kontext, *der;* **in/out of ~:** im/ohne Kontext; **in this ~:** in diesem Zusammenhang

continent ['kɒntɪnənt] *n.* Kontinent, *der;* Erdteil, *der;* **the ~s of Europe, Asia, Africa** die Erdteile Europa, Asien, Afrika; **the C~:** das europäische Festland; der Kontinent

continental [kɒntɪ'nentl] **1.** *adj.* **a)** kontinental; **~ Europe** Kontinentaleuropa *(das);* **b) C~** *(mainland European)* kontinental[europäisch]. **2.** *n.* **C~:** Kontinentaleuropäer, *der*/-europäerin, *die*

continental: ~ 'breakfast *n.* kontinentales Frühstück *(im Unterschied zum englischen Frühstück);* **~ quilt** *(Brit.) n.* [Stepp]federbett, *das*

contingency [kən'tɪndʒənsɪ] *n.* *(chance event)* Eventualität, *die; (possible event)* Eventualfall, *der;* **~ plan** Alternativplan, *der*

contingent [kən'tɪndʒənt] **1.** *adj.* **a)** *(fortuitous)* zufällig; **b)** *(conditional)* abhängig **([up]on** von). **2.** *n. (Mil.; also fig.)* Kontingent, *das*

continual [kən'tɪnjʊəl] *adj. (frequently happening)* ständig; *(without cessation)* unaufhörlich; **there have been ~ quarrels** es gab ständig *od.* dauernd Streit

continually [kən'tɪnjʊəlɪ] *adv.* *(frequently)* ständig; immer wieder; *(without cessation)* unaufhörlich; **~ tired** immer müde

continuance [kən'tɪnjʊəns] *n.* Fortbestand, *der; (of happiness, noise, rain)* Fortdauer, *die*

continuation [kəntɪnjʊ'eɪʃn] *n.* Fortsetzung, *die;* **a ~ of these good relations** eine Fortdauer dieser guten Beziehungen

continue [kən'tɪnju:] **1.** *v. t.* fortsetzen; **'to be ~d'** „Fortsetzung folgt"; **'~d on page 2'** „Fortsetzung auf S. 2"; **~ doing** *or* **to do sth.** etw. weiter tun; **it ~d to rain** es regnete weiter; **it ~s to be a problem** es ist weiterhin ein Problem; **'...', he ~d** „...", fuhr er fort. **2.** *v. i.* **a)** *(persist)* ‹*Wetter, Zustand, Krise usw.:*› andauern; *(persist in doing etc. sth.)* weitermachen *(ugs.);* nicht aufhören; *(last)* dauern; **if the rain ~s** wenn der Regen anhält; **if you ~ like this** wenn Sie so weitermachen *(ugs.);* **~ with sth.** mit etw. fortfahren; **~ with a plan** einen Plan weiterverfolgen; **~ on one's way** seinen Weg fortsetzen; **b)** *(stay)* bleiben; **~ in power** an der Macht bleiben

continued [kən'tɪnju:d] *adj.* fortgesetzt ‹*Bemühungen*›; **~ existence** Weiterbestehen, *das*

continuity [kɒntɪ'nju:ɪtɪ] *n., no pl.* Kontinuität, *die*

conti'nuity girl *n.* Skriptgirl, *das*

continuous [kən'tɪnjʊəs] *adj.* **a)** ununterbrochen; anhaltend ‹*Regen, Sonnenschein, Anstieg*›; ständig ‹*Kritik, Streit, Änderung*›; fortlaufend ‹*Mauer*›; durchgezogen ‹*Linie*›; **b)** *(Ling.)*

~ **[form]** Verlaufsform, *die;* **present** ~ *or* ~ **present/past** ~ *or* ~ **past** Verlaufsform des Präsens/Präteritums

continuously [kən'tɪnjʊəslɪ] *adv.* *(in space)* durchgehend; *(in time or sequence)* ununterbrochen; ständig ⟨*sich ändern*⟩

continuum [kən'tɪnjʊəm] *n., pl.* **continua** [kən'tɪnjʊə] Kontinuum, *das*

contort [kən'tɔ:t] *v.t.* verdrehen *(auch fig.);* verzerren ⟨*Gesicht, Gesichtszüge*⟩; verrenken, verdrehen ⟨*Körper*⟩

contortion [kən'tɔ:ʃn] *n.* Verzerrung, *die; (of body)* Verdrehung, *die;* Verrenkung, *die*

contour ['kɒntʊə(r)] *n.* Kontur, *die;* ~ **map/line** Höhenlinienkarte, *die*/Höhenschichtlinie, *die*

contraband ['kɒntrəbænd] **1.** *n.* Schmuggelware, *die.* **2.** *adj.* geschmuggelt; ~ **goods** Schmuggelware, *die*

contraception [kɒntrə'sepʃn] *n.* Empfängnisverhütung, *die*

contraceptive [kɒntrə'septɪv] **1.** *adj.* empfängnisverhütend; ~ **device/method** Verhütungsmittel, *das*/-methode, *die.* **2.** *n.* Verhütungsmittel, *das*

contract 1. ['kɒntrækt] *n.* Vertrag, *der;* ~ **of employment** Arbeitsvertrag, *der;* **be under ~ to do sth.** vertraglich verpflichtet sein, etw. zu tun; **exchange ~s** *(Law)* die Vertragsurkunden austauschen. **2.** [kən'trækt] *v.t.* **a)** *(cause to shrink, make smaller)* schrumpfen lassen; *(draw together)* zusammenziehen; **b)** *(become infected with)* sich *(Dat.)* zuziehen; ~ **sth. from sb.** sich mit etw. bei jmdm. anstecken; ~ **sth. from ...:** an etw. *(Dat.)* durch ... erkranken; **c)** *(incur)* machen ⟨*Schulden*⟩. **3.** [kən'trækt] *v.i.* **a)** *(enter into agreement)* Verträge/einen Vertrag schließen; ~ **for sth.** etw. vertraglich zusichern; ~ **to do sth.** sich vertraglich verpflichten, etw. zu tun; **b)** *(shrink, become smaller, be drawn together)* sich zusammenziehen

~ **'out 1.** *v.i.* ~ **out [of sth.]** sich [an etw. *(Dat.)*] nicht beteiligen; *(withdraw)* [aus etw.] aussteigen *(ugs.).* **2.** *v.t.* ~ **work out [to another firm]** Arbeit [an eine andere Firma] vergeben

contract bridge ['kɒntrækt 'brɪdʒ] *n.* Kontraktbridge, *das*

contraction [kən'trækʃn] *n.* **a)** *(shrinking)* Kontraktion, *die (Physik);* **b)** *(Physiol.: of muscle)* Zusammenziehung, *die;* Kontraktion, *die (Med.);* **c)** *(Ling.)* Kontraktion, *die;* **d)** *(catching)* Ansteckung, *die* (**of** mit)

contractor [kən'træktə(r)] *n.* Auftragnehmer, *der*/-nehmerin, *die*

contractual [kən'træktjʊəl] *adj.* vertraglich

contradict [kɒntrə'dɪkt] *v.t.* widersprechen (+ *Dat.*)

contradiction [kɒntrə'dɪkʃn] *n.* Widerspruch, *der* (**of** gegen); **in ~ to sth./sb.** im Widerspruch *od.* Gegensatz zu etw./jmdm.; **be a ~ to** *or* **of sth.** im Widerspruch zu etw. stehen; **a ~ in terms** ein Widerspruch in sich selbst

contradictory [kɒntrə'dɪktərɪ] *adj.* widersprechend; *(mutually opposed)* widersprüchlich

contra-flow ['kɒntrəfləʊ] *n.* Gegenverkehr auf einem Fahrstreifen

contralto [kən'træltəʊ] *n., pl.* ~**s** *(Mus.)* **a)** *(voice)* Alt, *der; (very low)* Kontraalt, *der;* **b)** *(singer)* Altistin, *die;* Alt, *der (selten); (with very low voice)* Kontraalt, *der*

contraption [kən'træpʃn] *n.* *(coll.) (machine)* Apparat, *der (ugs.); (device)* [komisches] Gerät

contrary ['kɒntrərɪ] **1.** *adj.* **a)** entgegengesetzt; **be ~ to sth.** im Gegensatz zu etw. stehen; **the result was ~ to expectation** das Ergebnis entsprach nicht den Erwartungen; **b)** *(opposite)* entgegengesetzt; **c)** [kən'treərɪ] *(coll.: perverse)* widerspenstig; widerborstig. **2.** *n.* **the ~:** das Gegenteil; **be/do completely the ~:** das genaue Gegenteil sein/tun; **on the ~:** im Gegenteil. **3.** *adv.* ~ **to sth.** entgegen einer Sache; ~ **to expectation** wider Erwarten

contrast 1. [kən'trɑ:st] *v.t.* gegenüberstellen; ~ **sth. with sth.** etw. von etw. [deutlich] abheben. **2.** [kən'trɑ:st] *v.i.* ~ **with sth.** mit etw. kontrastieren; sich von etw. abheben. **3.** ['kɒntrɑ:st] *n.* **a)** Kontrast, *der* (**with** zu); **what a ~!** welch ein Gegensatz!; **in ~, ...:** im Gegensatz dazu, ...; **[be] in ~ with sth.** im Gegensatz *od.* Kontrast zu etw. [stehen]; **b)** *(thing)* **a ~ to sth.** ein Gegensatz zu etw.; *(person)* **be a ~ to sb.** [ganz] anders sein als jmd.

contrasting [kən'trɑ:stɪŋ] *adj.* gegensätzlich; kontrastierend ⟨*Farbe*⟩; *(very different)* sehr unterschiedlich

contravene [kɒntrə'vi:n] *v.t.* verstoßen gegen ⟨*Recht, Gesetz*⟩

contravention [kɒntrə'venʃn] *n.* Verstoß, *der* (**of** gegen); **be in ~ of sth.** im Widerspruch zu etw. stehen

contretemps ['kɔ̃trətɑ̃] *n., pl. same* ['kɔ̃trətɑ̃z] Mißgeschick, *das;* Malheur, *das (ugs.)*

contribute [kən'trɪbju:t] **1.** *v.t.* ~ **sth. [to** *or* **towards sth.]** etw. [zu etw.] beitragen/*(co-operatively)* beisteuern; ~ **money towards sth.** für etw. Geld beisteuern/*(for charity)* spenden; **he regularly ~s articles to the 'Guardian'** er schreibt regelmäßig für den „Guardian". **2.** *v.i.* **everyone ~d towards the production** jeder trug etwas zur Aufführung bei; ~ **to charity** für karitative Zwecke spenden; ~ **to sb.'s misery/disappointment** jmds. Kummer/Enttäuschung vergrößern; ~ **to a newspaper** für eine Zeitung schreiben; ~ **to the success of sth.** zum Erfolg einer Sache *(Gen.)* beitragen

contribution [kɒntrɪ'bju:ʃn] *n.* **a)** **make a ~ to a fund** etw. für einen Fonds spenden; **the ~ of clothing and money to sth.** das Spenden von Kleidern und Geld für etw.; **b)** *(thing contributed)* Beitrag, *der; (for charity)* Spende, *die* (**to** für); **~s of clothing and money** Kleider- und Geldspenden; **make a ~ to sth.** einen Beitrag zu etw. leisten

contributor [kən'trɪbjʊtə(r)] *n.* **a)** *(giver)* Spender, *der*/Spenderin, *die;* **b)** *(to encyclopaedia, dictionary, etc.)* Mitarbeiter, *der*/Mitarbeiterin, *die* (**to** *Gen.*); **be a regular ~ to the 'Guardian'** regelmäßig für den 'Guardian' schreiben

'con trick *(coll.) see* **confidence trick**

contrite ['kɒntraɪt] *adj.* zerknirscht

contrition [kən'trɪʃn] *n.* Reue, *die*

contrivance [kən'traɪvəns] *n.* **a)** *(contriving)* Plan, *der;* **b)** *(inventing)* Ersinnen, *das;* **c)** *(device)* Gerät, *das*

contrive [kən'traɪv] *v.t.* **a)** *(manage)* ~ **to do sth.** es fertigbringen *od.* zuwege bringen, etw. zu tun; **they ~d to meet** es gelang ihnen, sich zu treffen; **b)** *(devise)* sich *(Dat.)* ausdenken; ersinnen *(geh.)*

contrived [kən'traɪvd] *adj.* künstlich

control [kən'trəʊl] **1.** *n.* **a)** *(power of directing, restraint)* Kontrolle, *die* (**of** über + *Akk.*); *(management)* Leitung, *die;* **governmental ~:** Regierungsgewalt, *die;* **have ~ of sth.** die Kontrolle über etw. *(Akk.)* haben; *(take decisions)* für etw. zuständig sein; **take ~ of** die Kontrolle übernehmen über (+ *Akk.*); **keep ~ of sth.** etw. unter Kontrolle halten; **be in ~ [of sth.]** die Kontrolle [über etw. *(Akk.)*] haben; **be in ~ of the situation** die Situation unter Kontrolle haben; **[go** *or* **get] out of ~:** außer Kontrolle [geraten]; **[get sth.] under ~:** [etw.] unter Kontrolle [bringen]; **gain ~ of sth.** etw. unter Kontrolle bekommen; **lose/regain ~ of oneself** die Beherrschung verlieren/wiedergewinnen; **have some/complete/no ~ over sth.** eine gewisse/die absolute/keine Kontrolle über etw. *(Akk.)* haben; **b)** *(device)* Regler, *der;* **~s** *(as a group)* Schalttafel, *die; (of TV, stereo system)* Bedienungstafel, *die;* **be at the ~s** ‹*Fahrer, Pilot:*› am Steuer sitzen. **2.** *v. t.*, **-ll-: a)** *(have ~ of)* kontrollieren; steuern, lenken ‹*Auto*›; **he ~s the financial side of things** er ist für die Finanzen zuständig; **~ling interest** *(Commerc.)* Mehrheitsbeteiligung, *die;* **b)** *(hold in check)* beherrschen; zügeln ‹*Zorn, Ungeduld, Temperament*›; *(regulate)* kontrollieren; regulieren ‹*Geschwindigkeit, Temperatur*›; einschränken ‹*Export, Ausgaben*›; regeln ‹*Verkehr*›

control: ~ centre *n.* Kontrollzentrum, *das;* **~ desk** *n.* Schaltpult, *das;* **~ panel** *n.* Schalttafel, *die;* **~ room** *n.* Kontrollraum, *der; (Radio, Telev.)* Regieraum, *der; (in power station)* Schaltwarte, *die;* **~ tower** *n.* Kontrollturm, *der*

controversial [kɒntrə'vɜ:ʃl] *adj.* umstritten ‹*Mode, Kunstwerk, Gesetz, Idee*›; strittig ‹*Frage, Punkt, Angelegenheit*›; *(given to controversy)* streitsüchtig

controversy ['kɒntrəvɜ:sɪ, kən'trɒvəsɪ] *n.* Kontroverse, *die;* Auseinandersetzung, *die;* **much ~:** eine längere Kontroverse *od.* Auseinandersetzung

contusion [kən'tju:ʒn] *n.* Prellung, *die*

conundrum [kə'nʌndrəm] *n. (auf einem Wortspiel beruhendes)* Rätsel

conurbation [kɒnɜ:'beɪʃn] *n.* Konurbation, *die (Soziol.);* ≈ Stadtregion, *die*

convalesce [kɒnvə'les] *v. i.* genesen; rekonvaleszieren *(Med.)*

convalescence [kɒnvə'lesəns] *n.* Genesung, *die;* Rekonvaleszens, *die (Med.)*

convalescent [kɒnvə'lesənt] **1.** *adj.* rekonvaleszent *(Med.).* **2.** *n.* Rekonvaleszent, *der/*Rekonvaleszentin, *die (Med.);* Genesende, *der/die*

convection [kən'vekʃn] *n. (Phys., Meteorol.)* Konvektion, *die;* **~ current** Konvektionsstrom, *der*

convector [kən'vektə(r)] *n.* Konvektor, *der*

convene [kən'vi:n] **1.** *v. t.* einberufen. **2.** *v. i.* zusammenkommen; ‹*Gericht, gewählte Vertreter:*› zusammentreten; ‹*Konferenz, Versammlung:*› beginnen

convener [kən'vi:nə(r)] *n. (Brit.) jmd., der eine Versammlung einberuft/leitet*

convenience [kən'vi:nɪəns] *n.* **a)** *no pl. (suitableness)* Annehmlichkeit, *die;* **its ~ to** *or* **for the city centre** seine günstige Lage zum Stadtzentrum; **b)** *(personal satisfaction)* Bequemlichkeit, *die;* **for sb.'s ~, for ~'s sake** zu jmds. Bequemlichkeit; **at your ~:** wann es Ihnen paßt; **c)** *(advantage)* **be a ~ to sb.** angenehm *od.* praktisch für jmdn. sein; **d)** *(advantageous thing)* Annehmlichkeit, *die;* **e)** *(esp. Admin.: toilet)* Toilette, *die;* **public ~:** öffentliche Toilette *od. (Amtsspr.)* Bedürfnisanstalt

con'venience food *n.* Fertignahrung, *die*

convenient [kən'vi:nɪənt] *adj.* **a)** *(suitable, not troublesome)* günstig; *(useful)* praktisch; angenehm; **be ~ to** *or* **for sb.** günstig für jmdn. sein; **would it be ~ to you?** würde es Ihnen passen?; **it's not very ~ at the moment** es paßt im Augenblick nicht gut; **b)** *(of easy access)* **be ~ to** *or* **for sth.** günstig zu etw. liegen; **a ~ taxi** ein Taxi, das gerade dasteht/angefahren kommt

conveniently [kən'vi:nɪəntlɪ] *adv.* **a)** günstig ‹*gelegen, angebracht*›; leicht ‹*gesehen werden*›; **we're ~ situated for the shops** wir haben es nicht weit zu den Geschäften; **b)** *(opportunely)* angenehmerweise

convenor *see* **convener**

convent ['kɒnvənt] *n.* Kloster, *das*

convention [kən'venʃn] *n.* **a)** *(a practice)* Brauch, *der;* **it is the ~ to do sth.** es ist Brauch, etw. zu tun; **b)** *no art. (established customs)* Konvention, *die;* **break with ~:** sich über die Konventionen hinwegsetzen; **c)** *(formal assembly)* Konferenz, *die;* **d)** *(agreement between States)* Konvention, *die (bes. Völkerrecht)*

conventional [kən'venʃənl] *adj.* konventionell; *(not spontaneous)* formell

conventionally [kən'venʃənəlɪ] *adv.* konventionell

converge [kən'vɜ:dʒ] *v. i.* **~ [on each other]** aufeinander zulaufen; ‹*Gedanken, Meinungen, Ansichten:*› sich [einander] annähern; **~ on sb.** auf jmdn. zulaufen

convergence [kən'vɜ:dʒəns] *n.* Annäherung, *die;* Konvergenz, *die (geh.); (of roads, rivers)* Zusammentreffen, *das*

convergent [kən'vɜ:dʒənt] *adj.* aufeinander zulaufend

conversant [kən'vɜ:sənt] *pred. adj.* vertraut (**with** mit)

conversation [kɒnvə'seɪʃn] *n.* Unterhaltung, *die;* Gespräch, *das; (in language-teaching)* Konversation, *die;* **be in ~ [with sb.]** sich [mit jmdm.] unterhalten; **be deep in ~:** in ein Gespräch vertieft sein; **make [polite] ~ with sb.** mit jmdm. Konversation machen; **come up in ~:** gesprächsweise erwähnt werden; **have a ~ with sb.** mit jmdm. ein Gespräch führen

conversational [kɒnvə'seɪʃənl] *adj.* gesprächig ‹*Person*›; ungezwungen ‹*Art*›; **~ English** gesprochenes Englisch

[1]**converse** [kən'vɜ:s] *v. i. (formal)* **~ [with sb.] [about** *or* **on sth.]** sich [mit jmdm.] [über etw. *(Akk.)*] unterhalten

[2]**converse** ['kɒnvɜ:s] **1.** *adj.* entgegengesetzt; umgekehrt ‹*Fall, Situation*›. **2.** *n.* Gegenteil, *das*

conversely [kən'vɜ:slɪ] *adv.* umgekehrt

conversion [kən'vɜ:ʃn] *n.* **a)** *(transforming)* Umwandlung, *die* (**into** in + *Akk.*); **b)** *(adaptation, adapted building)* Umbau, *der;* **do a ~ on sth.** etw. umbauen; **c)** *(of person)* Bekehrung, *die* (**to** zu); Konversion, *die (Rel.);* **d)** *(to different units or expression)* Übertragung, *die* (**into** in + *Akk.*); **e)** *(Theol., Psych., Phys.)* Konversion, *die; (calculation)* Umrechnung, *die;* **f)** *(Rugby, Amer. Footb.)* Erhöhung, *die*

convert 1. [kən'vɜ:t] *v. t.* **a)** *(transform, change in function)* umwandeln (**into** in + *Akk.*); **b)** *(adapt)* **~ sth. [into sth.]** etw. [zu etw.] umbauen; **c)** *(bring over)* **~ sb. [to sth.]** *(lit. or fig.)* jmdn. [zu etw.] bekehren; **d)** *(to different units or expressions)* übertragen (**into** in + *Akk.*); **e)** *(calculate)* umrechnen (**into** in + *Akk.*); **f)** *(Rugby, Amer. Footb.)* erhöhen. **2.** [kən-

'vɜ:t] *v. i.* **a)** ~ **into sth.** sich in etw. *(Akk.)* umwandeln lassen; **b)** *(be adaptable)* sich umbauen lassen; **c)** *(to new method etc.)* umstellen (**to** auf + *Akk.*). **3.** ['kɒnvɜ:t] *n. (Relig.)* Konvertit, *der*/Konvertitin, *die*

convertible [kən'vɜ:tɪbl] **1.** *adj.* **a) be** ~ **into sth.** *(transformable)* sich in etw. *(Akk.)* umwandeln lassen; **b)** *(able to be altered)* **be** ~ **[into sth.]** sich zu etw. umbauen lassen. **2.** *n.* Kabrio[lett], *das; (with four or more seats)* Kabriolimousine, *die*

convex ['kɒnveks] *adj.* konvex; *attrib.* Konvex‹*linse, -spiegel*›

convey [kən'veɪ] *v. t.* **a)** *(transport)* befördern; *(transmit)* übermitteln ‹*Nachricht, Grüße*›; **b)** *(impart)* vermitteln; **words cannot** ~ **it** Worte können es nicht wiedergeben; **the message** ~**ed nothing whatever to me** die Nachricht sagte mir überhaupt nichts

conveyance [kən'veɪəns] *n.* **a)** *(transportation)* Beförderung, *die;* **b)** *(formal: vehicle)* Beförderungsmittel, *das;* **c)** *(Law)* Übertragung, *die;* Überschreibung, *die*

conveyancing [kən'veɪənsɪŋ] *n. (Law)* ~ **[of property]** [Eigentums]übertragung, *die*

conveyer, conveyor [kən'veɪə(r)] *n.* Förderer, *der (Technik);* ~ **[belt]** *(Industry)* Förderband, *das; (in manufacture also)* Fließband, *das*

convict 1. ['kɒnvɪkt] *n.* Strafgefangene, *der/die.* **2.** [kən'vɪkt] *v. t.* **a)** *(declare guilty)* für schuldig befinden; verurteilen; **be** ~**ed** verurteilt werden; **b)** *(prove guilty)* ~ **sb. of sth.** jmdn. einer Sache *(Gen.)* überführen

conviction [kən'vɪkʃn] *n.* **a)** *(Law)* Verurteilung, *die* (**for** wegen); **have you [had] any previous** ~**s?** sind Sie vorbestraft?; **b)** *(settled belief)* Überzeugung, *die;* **it is their** ~ **that ...:** sie sind der Überzeugung, daß ...; **carry** ~: überzeugend sein

convince [kən'vɪns] *v. t.* überzeugen; ~ **sb. that ...:** jmdn. davon überzeugen, daß ...; **be** ~**d that ...:** davon überzeugt sein, daß ...

convincing [kən'vɪnsɪŋ] *adj.*, **convincingly** [kən'vɪnsɪŋlɪ] *adv.* überzeugend

convivial [kən'vɪvɪəl] *n.* fröhlich

convoluted ['kɒnvəlu:tɪd] *adj.* **a)** *(twisted)* verschlungen; **b)** *(complex)* kompliziert

convoy ['kɒnvɔɪ] *n.* Konvoi, *der;* **in** ~: im Konvoi

convulse [kən'vʌls] *v. t.* **a) be** ~**d** von Krämpfen geschüttelt werden; **b)** *(shake, lit. or fig.)* erschüttern

convulsion [kən'vʌlʃn] *n.* **a)** *in pl.* Schüttelkrampf, *der (Med.);* Krämpfe; **b)** *(shaking, lit. or fig.)* Erschütterung, *die*

coo [ku:] **1.** *int. (of person)* oh; *(of dove)* ruckedigu. **2.** *n. (of dove)* **the** ~**[s]** das Gurren. **3.** *v. i.* gurren; ‹*Baby:*› gurren *(fig.)*

cook [kʊk] **1.** *n.* Koch, *der*/Köchin, *die.* **2.** *v. t.* **a)** garen; zubereiten, kochen ‹*Mahlzeit*›; *(fry, roast)* braten; *(boil)* kochen; **how would you** ~ **this piece of meat?** wie würden Sie dieses Stück Fleisch zubereiten?; ~**ed in the oven** im Backofen zubereitet *od. (Kochk.)* gegart; ~**ed meal** warme Mahlzeit; *abs.* **do you** ~ **with gas or electricity?** kochen Sie mit Gas oder mit Strom?; **she knows how to** ~: sie kann gut kochen *od.* kocht gut; ~ **sb.'s goose [for him]** *(fig.)* jmdm. alles verderben; **b)** *(fig. coll.: falsify)* frisieren *(ugs.).* **3.** *v. i.* kochen; garen *(Kochk.);* **the meat was** ~**ing slowly** das Fleisch garte langsam; **what's** ~**ing?** *(fig. coll.)* was liegt an? *(ugs.)*

~ **'up** *v. t.* sich *(Dat.)* ausbrüten, *(ugs.)* aushecken ‹*Plan*›; erfinden ‹*Geschichte*›

cooker ['kʊkə(r)] *n.* **a)** *(Brit.: stove)* Herd, *der;* **electric/gas** ~: Elektroherd/Gasherd, *der;* **b)** *(fruit)* **are those apples eaters or** ~**s?** sind diese Äpfel zum Essen oder zum Kochen?

cookery ['kʊkərɪ] *n.* Kochen, *das*

'cookery book *n. (Brit.)* Kochbuch, *das*

cookhouse ['kʊkhaʊs] *n. (Mil.)* Feldküche, *die*

cookie ['kʊkɪ] *n.* **a)** *(Scot.: plain bun)* Plätzchen, *das;* **b)** *(Amer.: biscuit)* Keks, *der*

cooking ['kʊkɪŋ] *n.* Kochen, *das;* **German** ~: die deutsche Küche; **do one's own** ~: für sich selbst kochen; **do the** ~: kochen

cooking: ~ **apple** *n.* Kochapfel, *der;* ~ **fat** *n.* Bratfett, *das;* ~ **salt** *n.* Speisesalz, *das;* ~ **utensil** *n.* Küchengerät, *das*

'cook-out *n. (Amer.)* ≈ Grillparty, *die*

cool [ku:l] **1.** *adj.* **a)** kühl; luftig ‹*Kleidung*›; **'store in a** ~ **place'** „kühl aufbewahren"; **b)** *(calm)* **he kept** *or* **stayed** ~: er blieb ruhig *od.* bewahrte die Ruhe; **play it** ~ *(coll.)* ruhig bleiben; cool vorgehen *(salopp);* **he was** ~, **calm, and collected** er war ruhig und gelassen; **keep a** ~ **head** einen kühlen Kopf bewahren; **c)** *(unemotional, unfriendly)* kühl; *(calmly audacious)* kaltblütig. **2.** *n.* Kühle, *die.* **3.** *v. i.* abkühlen; **the weather has** ~**ed** es ist kühler geworden; *(fig.)* **our relationship has** ~**ed** unsere Beziehung ist kühler geworden; ~ **towards sb./sth.** an jmdm./etw. das Interesse verlieren. **4.** *v. t.* kühlen; *(from high temperature)* abkühlen; *(fig.)* abkühlen ‹*Leidenschaft, Raserei*›; ~ **one's heels** *(fig.)* lange warten

~ **'down 1.** *v. i.* **a)** ‹*Tee:*› abkühlen; ‹*Luft:*› sich abkühlen; **b)** *(fig.)* sich beruhigen. **2.** *v. t.* abkühlen

~ **'off 1. a)** *v. i.* abkühlen; **the weather has** ~**ed off** es ist kühler geworden; **we need a few minutes to** ~ **off** wir brauchen ein paar Minuten, um uns abzukühlen; **b)** *(fig.)* sich beruhigen; ‹*Zorn, Begeisterung, Interesse:*› sich legen, nachlassen. **2.** *v. t.* abkühlen; *(fig.)* beruhigen

'cool box *n.* Kühlbox, *die*

cooler ['ku:lə(r)] *n.* Kühler, *der*

coolie ['ku:lɪ] *n.* Kuli, *der*

coolly ['ku:llɪ] *adv.* **a)** kühl; **b)** *(fig.) (calmly)* ruhig; *(unemotionally, in unfriendly manner)* kühl; *(audaciously)* kaltblütig; unverfroren ‹*verlangen, fordern*›

coolness ['ku:lnɪs] *n., no pl.* Kühle, *die; (fig.) (calmness)* Ruhe, *die; (unemotional nature, unfriendliness)* Kühle, *die; (audacity)* Kaltblütigkeit, *die*

coop [ku:p] *n. (cage)* Geflügelkäfig, *der; (for poultry)* Hühnerstall, *der; (fowl-run)* Auslauf, *der*

co-operate [kəʊ'ɒpəreɪt] *v. i.* mitarbeiten (**in** bei); *(with each other)* zusammenarbeiten (**in** bei); *(not obstruct)* mitmachen *(ugs.);* ~ **with sb.** mit jmdm. zusammenarbeiten

co-operation [kəʊɒpə'reɪʃn] *n.* Mitarbeit, *die;* Zusammenarbeit, *die;* Kooperation, *die;* **with the** ~ **of** unter Mitarbeit von; **in** ~ **with** in Zusammenarbeit mit

co-operative [kəʊ'ɒpərətɪv] **1.** *adj.* kooperativ; *(helpful)* hilfsbereit. **2.** *n.* Genossenschaft, *die;* Kooperative, *die (bes. in der ehemaligen DDR); (shop)* Genossenschaftsladen, *der;* **workers'** ~: Produktivgenossenschaft, *die*

co-opt [kəʊ'ɒpt] *v. t.* kooptieren; hinzuwählen; **be** ~**ed [on] to a committee** von einem Komitee kooptiert werden

co-ordinate 1. [kəʊ'ɔ:dɪnət] *n.* **a)** *(Math.)* Koordinate, *die;* **b)** *in pl. (clothes)* Kombination, *die.* **2.** [kəʊ'ɔ:dɪneɪt] *v. t.* koordinieren; **co-ordinating conjunction** koordinierende Konjunktion

co-ordination [kəʊɔ:dɪ'neɪʃn] *n.* Koordination, *die*

¹cop [kɒp] *n. (sl.: police officer)* Bulle, *der (salopp)*

²cop *(sl.)* **1.** *v. t.*, **-pp-: a) when ..., you'll ~ it** *(be punished)* wenn ..., dann kannst du was erleben; **b) they ~ped it** *(were killed)* sie mußten dran glauben *(salopp).* **2.** *n.* **it's a fair ~!** guter Fang!; **no ~, not much ~:** nichts Besonderes

~ **'out** *v. i. (sl.)* **a)** *(escape)* abhauen *(salopp);* ~ **out of society** [aus der Gesellschaft] aussteigen *(ugs.);* **b)** *(give up)* alles hinwerfen *(ugs.)*

cope [kəʊp] *v. i.* ~ **with sb./sth.** mit jmdm./etw. fertig werden; ~ **with a handicapped child** mit einem behinderten Kind zurechtkommen

Copenhagen [kəʊpn'heɪgn] *pr. n.* Kopenhagen *(das)*

copier ['kɒpɪə(r)] *n. (machine)* Kopiergerät, *das;* Kopierer, *der (ugs.)*

co-pilot ['kəʊpaɪlət] *n.* Kopilot, *der*/Kopilotin, *die*

copious ['kəʊpɪəs] *adj. (plentiful)* reichhaltig; *(informative)* umfassend

'cop-out *n. (sl.)* Drückebergerei, *die (ugs. abwertend);* **that's a ~:** das ist Drückebergerei

¹copper ['kɒpə(r)] **1.** *n.* **a)** Kupfer, *das;* **b)** *(coin)* Kupfermünze, *die;* **a few ~s** etwas Kupfergeld; **c)** *(for laundry)* Waschkessel, *der.* **2.** *attrib. adj.* **a)** *(made of ~)* kupfern; Kupfer‹*münze, -kessel, -rohr*›; **b)** *(coloured like ~)* kupferfarben; kupfern

²copper *(Brit. sl.) see* **¹cop**

copper: ~ **'beech** *n.* Blutbuche, *die;* **~-coloured** *adj.* kupferfarben; **~plate 1.** *n.* **a)** *(metal plate)* Kupferplatte, *die;* **b)** *(print)* Kupferstich, *der;* **2.** *adj.* **~plate writing** ≈ Schönschrift, *die*

coppice ['kɒpɪs], **copse** [kɒps] *ns.* Wäldchen, *das;* Niederwald, *der (Forstw.)*

copula ['kɒpjʊlə] *n. (Ling.)* Kopula, *die*

copulate ['kɒpjʊleɪt] *v. i.* kopulieren

copulation [kɒpjʊ'leɪʃn] *n.* Kopulation, *die*

copy ['kɒpɪ] **1.** *n.* **a)** *(reproduction)* Kopie, *die; (imitation)* Nachahmung, *die; (with carbon paper etc.) (typed)* Durchschlag, *der; (written)* Durchschrift, *die;* **b)** *(specimen)* Exemplar, *das;* **have you a ~ of today's 'Times'?** haben Sie die „Times" von heute?; **send three copies of the application** die Bewerbung in dreifacher Ausfertigung schicken; **top ~:** Original, *das.* **2.** *v. t.* **a)** *(make ~ of)* kopieren; *(by photocopier)* [foto]kopieren; *(transcribe)* abschreiben; **b)** *(imitate)* nachahmen. **3.** *v. i.* **a)** kopieren; ~ **from sb./sth.** jmdn./etw. kopieren; **b)** *(in exam etc.)* abschreiben; ~ **from sb./sth.** bei jmdm./aus etw. abschreiben

~ **'out** *v. t.* abschreiben

'copy-book *attrib. adj.* wie im Bilderbuch *nachgestellt; see also* **blot 2 b**

copyright ['kɒpɪraɪt] **1.** *n.* Copyright, *das;* Urheberrecht, *das;* **be out of ~:** gemeinfrei [geworden] sein; **protected by ~:** urheberrechtlich geschützt. **2.** *adj.* urheberrechtlich geschützt

'copy typist *n.* Schreibkraft *(die nur nach schriftlichen Vorlagen arbeitet)*

coral ['kɒrl] **1.** *n.* Koralle, *die.* **2.** *attrib. adj.* korallen; Korallen‹*insel, -riff, -rot*›

cord [kɔ:d] *n.* **a)** Kordel, *die;* **b)** *(cloth)* Cord, *der*

cordial ['kɔ:dɪəl] **1.** *adj.* herzlich; **a ~ dislike for sb.** eine tiefempfundene Abneigung gegenüber jmdm. **2.** *n. (drink)* Sirup, *der*

cordially ['kɔ:dɪəlɪ] *adv.* herzlich; ~ **dislike sb.** eine tiefempfundene Abneigung gegenüber jmdm. haben

cordon ['kɔ:dn] **1.** *n.* Kordon, *der; see also throw around b.* **2.** *v. t.* ~ **[off]** absperren; abriegeln

corduroy ['kɔ:dərɔɪ, 'kɔ:djʊrɔɪ] *n.* Cordsamt, *der*

core [kɔ:(r)] **1.** *n.* **a)** *(of fruit)* Kerngehäuse, *das;* **b)** *(Geol.) (rock sample)* [Bohr]kern, *der; (of earth)* [Erd]kern, *der;* **c)** *(fig.: innermost part)* **rotten to the ~:** verdorben bis ins Mark; **English to the ~:** durch und durch englisch; **shake sb. to the ~:** jmdn. zutiefst erschüttern. **2.** *v. t.* entkernen ‹*Apfel, Birne*›

co-respondent [kəʊrɪ'spɒndənt] *n.* Mitbeklagte, *der/die (im Scheidungsprozeß)*

corgi ['kɔ:gɪ] *n.* **[Welsh] ~:** Welsh Corgi, *der*

coriander [kɒrɪ'ændə(r)] *n.* Koriander, *der;* ~ **seed** Koriander, *der*

cork [kɔ:k] **1.** *n.* **a)** *(bark)* Kork, *der;* **b)** *(bottle-stopper)* Korken, *der.* **2.** *v. t.* zukorken; verkorken

~ **'up** *v. t.* zukorken; verkorken

'corkscrew *n.* Korkenzieher, *der*

cormorant ['kɔ:mərənt] *n. (Ornith.)* Kormoran, *der*

¹corn [kɔ:n] *n.* **a)** *(cereal)* Getreide, *das; (esp. rye, wheat also)* Korn, *das;* **[sweet] ~** *(maize)* Mais, *der;* ~ **on the cob** [gekochter/gerösteter] Maiskolben; **b)** *(seed)* Korn, *das*

²corn *n. (on foot)* Hühnerauge, *das*

corn: **~-cob** *n.* Maiskolben, *der;* ~ **dolly** *n.* Strohpuppe, *die*

cornea ['kɔ:nɪə] *n. (Anat.)* Hornhaut, *die;* Cornea, *die (fachspr.)*

corned beef [kɔ:nd 'bi:f] *n.* Corned beef, *das*

corner ['kɔ:nə(r)] **1.** *n.* **a)** Ecke, *die; (curve)* Kurve, *die;* **on the ~:** an der Ecke/in der Kurve; **at the ~:** an der Ecke; ~ **of the street** Straßenecke, *die;* **cut [off] a/the ~:** eine/die Kurve schneiden; **cut ~s** *(fig.)* auf die schnelle arbeiten *(ugs.);* **[sth. is] just [a]round the ~:** [etw. ist] gleich um die Ecke; **Christmas is just round the ~** *(fig. coll.)* Weihnachten steht vor der Tür; **turn the ~:** um die Ecke biegen; **he has turned the ~ now** *(fig.)* er ist jetzt über den Berg *(ugs.);* **b)** *(hollow angle between walls)* Ecke, *die; (of mouth, eye)* Winkel, *der;* **c)** *(Boxing, Wrestling)* Ecke, *die;* **d)** *(secluded place)* Eckchen, *das;* Plätzchen, *das; (remote region)* Winkel, *der;* **from the four ~s of the earth** aus aller Welt; **e)** *(Hockey, Footb.)* Ecke, *die;* **f)** *(Commerc.)* Corner, *der;* Schwänze, *die.* **2.** *v. t.* **a)** *(drive into ~)* in eine Ecke treiben; *(fig.)* in die Enge treiben; **have [got] sb. ~ed** jmdn. in der Falle haben; **b)** *(Commerc.)* ~ **the market in coffee** die Kaffeevorräte aufkaufen; den Kaffeemarkt aufschwänzen *(fachspr.).* **3.** *v. i.* die Kurve nehmen; ~ **well/badly** ‹*Fahrzeug:*› eine gute/schlechte Kurvenlage haben

corner: ~ **flag** *n. (Sport)* Eckfahne, *die;* **~-kick** *n. (Footb.)* Eckball, *der;* Eckstoß, *der;* ~ **seat** *n.* Ecksitz, *der;* ~ **shop** *n.* Tante-Emma-Laden, *der (ugs.);* **~-stone** *n.* Eckstein, *der; (fig.)* Eckpfeiler, *der*

cornet ['kɔ:nɪt] *n.* **a)** *(Brit.: wafer)* [Eis]tüte, *die;* Eishörnchen, *das;* **b)** *(Mus.)* Kornett, *das*

corn: **~field** *n.* Kornfeld, *das; (Amer.)* Maisfeld, *das;* **~flakes** *n. pl.* Corn-flakes *Pl.;* **~flour** *n.* **a)** *(Brit.: ground maize)* Maismehl, *das;* **b)** *(flour of rice etc.)*

Stärkemehl, *das;* **~flower** *n.* Kornblume, *die*

cornice ['kɔ:nɪs] *n. (Archit.)* Kranzgesims, *das*

Cornish ['kɔ:nɪʃ] **1.** *adj.* kornisch. **2.** *n.* Kornisch, *das*

'**cornstarch** *(Amer.) see* **cornflour a**

corny ['kɔ:nɪ] *adj. (coll.) (old-fashioned)* altmodisch ‹*Witz usw.*›; *(trite)* abgedroschen *(ugs.)*

corollary [kə'rɒlərɪ] *n. (proposition)* Korollar[ium], *das (Logik); (consequence)* [logische *od.* natürliche] Folge

coronary ['kɒrənərɪ] **1.** *adj. (Anat.)* koronar. **2.** *n. (Med.) see* **coronary thrombosis**

coronary throm'bosis *n. (Med.)* Koronarthrombose, *die*

coronation [kɒrə'neɪʃn] *n.* Krönung, *die*

coroner ['kɒrənə(r)] *n.* Coroner, *der; Beamter, der gewaltsame od. unnatürliche Todesfälle untersucht*

coronet ['kɒrənet] *n.* Krone, *die*

corpora *pl. of* **corpus**

[1]**corporal** ['kɔ:pərl] *adj.* körperlich

[2]**corporal** *n.* Korporal, *der (hist.; österr.);* ≈ Hauptgefreite, *der*

corporate ['kɔ:pərət] *adj.* körperschaftlich; **~ body, body ~:** Körperschaft, *die*

corporation [kɔ:pə'reɪʃn] *n.* **a)** *(civic authority)* **[municipal] ~:** Gemeindeverwaltung, *die; (of borough, city)* Stadtverwaltung, *die;* **b)** *(united body)* Körperschaft, *die;* Korporation, *die*

corporeal [kɔ:'pɔ:rɪəl] *adj.* **a)** *(bodily)* körperlich; **b)** *(material)* materiell; stofflich

corps [kɔ:(r)] *n., pl. same* [kɔ:z] Korps, *das*

corpse [kɔ:ps] *n.* Leiche, *die;* Leichnam, *der (geh.)*

corpulent ['kɔ:pjʊlənt] *adj.* korpulent

corpus ['kɔ:pəs] *n., pl.* **corpora** ['kɔ:pərə] Sammlung, *die;* Korpus, *das*

Corpus Christi [kɔ:pəs 'krɪstɪ] *n. (Eccl.)* Fronleichnam *(der);* Fronleichnamsfest, *das*

corpuscle ['kɔ:pəsl] *n. (Anat.)* **[blood] ~:** Blutkörperchen, *das*

corral [kə'rɑ:l] *n. (Amer.)* Pferch, *der*

correct [kə'rekt] **1.** *v.t.* **a)** *(amend)* korrigieren; verbessern, korrigieren ‹*Fehler, Formulierung, jmds. Englisch/Deutsch*›; **~ me if I'm wrong** ich könnte mich natürlich irren; **b)** *(counteract)* ausgleichen ‹*etw. Schädliches*›; **c)** *(admonish)* zurechtweisen **(for** wegen). **2.** *adj.* richtig; korrekt; *(precise)* korrekt; akkurat; **that is ~:** das stimmt; **have you the ~ time?** haben Sie die genaue Uhrzeit?; **am I ~ in assuming that ...?** gehe ich recht in der Annahme, daß ...?

correction [kə'rekʃn] *n.* Korrektur, *die;* **the pupils had to write out** *or* **do their ~s** die Schüler mußten die Verbesserung *od.* Berichtigung schreiben

corrective [kə'rektɪv] *adj.* korrigierend; **take ~ action** korrigierend eingreifen

correctly [kə'rektlɪ] *adv.* richtig; korrekt; *(precisely)* korrekt; akkurat; **behave very ~:** sich sehr korrekt benehmen

correlate ['kɒrɪleɪt] **1.** *v.i.* einander entsprechen; **~ with** *or* **to sth.** einer Sache *(Dat.)* entsprechen. **2.** *v.t.* **~ sth. with sth.** etw. zu etw. in Beziehung setzen

correlation [kɒrɪ'leɪʃn] *n.* [Wechsel]beziehung, *die;* Korrelation, *die (bes. Math., Naturw.); (connection)* Zusammenhang, *der*

correspond [kɒrɪ'spɒnd] *v.i.* **a)** *(be analogous, agree in amount)* **~ [to each other]** einander entsprechen; **~ to sth.** einer Sache *(Dat.)* entsprechen; **b)** *(agree in position)* **~ [to sth.]** [mit etw.] übereinstimmen; *(be in harmony)* **~ [with** *or* **to sth.]** [mit etw.] zusammenpassen; **c)** *(communicate)* **~ with sb.** mit jmdm. korrespondieren

correspondence [kɒrɪ'spɒndəns] *n.* **a)** Übereinstimmung **(with, to** mit); **b)** *(communication, letters)* Briefwechsel, *der;* Korrespondenz, *die*

correspondence: ~ college *n.* Fernschule, *die;* **~ column** *n.* Rubrik „Leserbriefe"; **~ course** *n.* Fernkurs, *der*

correspondent [kɒrɪ'spɒndənt] *n.* **a)** Briefschreiber, *der/*-schreiberin, *die; (pen-friend)* Brieffreund, *der/*-freundin, *die; (to newspaper)* Leserbriefschreiber, *der/*-schreiberin, *die;* **b)** *(Radio, Telev., Journ., etc.)* Berichterstatter, *der/*-erstatterin, *die;* Korrespondent, *der/*Korrespondentin, *die*

corresponding [kɒrɪ'spɒndɪŋ] *adj.* entsprechend **(to** *Dat.***)**

correspondingly [kɒrɪ'spɒndɪŋlɪ] *adv.* entsprechend

corridor ['kɒrɪdɔ:(r)] *n.* **a)** *(inside passage)* Flur, *der;* Gang, *der;* Korridor, *der; (outside passage)* Galerie, *die;* **in the ~s of power** *(fig.)* in den politischen Schaltstellen; **b)** *(Railw.)* [Seiten]gang, *der*

corroborate [kə'rɒbəreɪt] *v.t.* bestätigen

corroboration [kərɒbə'reɪʃn] *n.* Bestätigung, *die;* **in ~ of sth.** als *od.* zur Bestätigung einer Sache *(Gen.)*

corrode [kə'rəʊd] **1.** *v.t.* zerfressen; korrodieren, zerfressen ‹*Metall, Gestein*›. **2.** *v.i.* zerfressen werden; ‹*Gestein, Metall:*› korrodieren, zerfressen werden

corrosion [kə'rəʊʒn] *n.* Zerfall, *der; (of metal, stone)* Korrosion, *die*

corrosive [kə'rəʊsɪv] **1.** *adj.* zerstörend; korrosiv *(bes. Chemie, Geol.);* ätzend ‹*Chemikalien*›; *(fig.)* zerstörerisch. **2.** *n.* Korrosion verursachender Stoff

corrugate ['kɒrʊgeɪt] *v.t.* zerfurchen; **~d cardboard/paper** Wellpappe, *die;* **~d iron** Wellblech, *das*

corrugation [kɒrʊ'geɪʃn] *n.* **a)** Zerfurchung, *die;* **b)** *(wrinkle, ridge mark)* Furche, *die; (ridge made by bending)* Rille, *die*

corrupt [kə'rʌpt] **1.** *adj. (depraved)* verkommen; verdorben *(geh.); (influenced by bribery)* korrupt. **2.** *v.t. (deprave)* korrumpieren; *(bribe)* bestechen

corruption [kə'rʌpʃn] *n.* **a)** *(moral deterioration)* Verdorbenheit, *die (geh.);* **b)** *(use of corrupt practices)* Korruption, *die;* **c)** *(perversion)* Korrumpierung, *die*

corset ['kɔ:sɪt] *n., in sing. or pl.* Korsett, *das*

Corsica ['kɔ:sɪkə] *pr. n.* Korsika *(das)*

Corsican ['kɔ:sɪkən] **1.** *adj.* korsisch; **sb. is ~:** jmd. ist Korse/Korsin. **2.** *n. (person)* Korse, *der/*Korsin, *die*

cortège [kɔ:'teɪʒ] *n.* Trauerzug, *der*

cortisone ['kɔ:tɪzəʊn] *n.* Kortison, *das (Med.);* Cortison, *das (fachspr.)*

corvette [kɔ:'vet] *n. (Naut.)* Korvette, *die*

[1]**cos** [kɒs] *n.* Römischer Salat; Sommerendivie, *die*

[2]**cos, 'cos** [kɒz] *(coll.) see* **because**

cosh [kɒʃ] *(Brit. coll.)* **1.** *n.* Totschläger, *der;* Knüppel, *der.* **2.** *v.t.* niederknüppeln

cosily ['kəʊzɪlɪ] *adv.* bequem; gemütlich, behaglich ‹*plaudern, wohnen*›

cosine ['kəʊsaɪn] *n. (Math.)* Kosinus, *der*

cosmetic [kɒz'metɪk] **1.** *adj. (lit. or fig.)* kosmetisch; **~ surgery**

Schönheitschirurgie, *die*. **2.** *n.* Kosmetikum, *das*
cosmic ['kɒzmɪk] *adj. (lit. or fig.)* kosmisch; ~ **radiation** *or* **rays** kosmische Strahlung
cosmonaut ['kɒzmənɔ:t] *n.* Kosmonaut, *der*/Kosmonautin, *die*
cosmopolitan [kɒzmə'pɒlɪtən] *adj.* kosmopolitisch
cosmos ['kɒzmɒs] *n.* Kosmos, *der*
Cossack ['kɒsæk] *n.* Kosak, *der*
cosset ['kɒsɪt] *v.t.* [ver]hätscheln
cost [kɒst] **1.** *n.* **a)** Kosten *Pl.*; **the ~ of bread/gas/oil** der Brot-/Gas-/Ölpreis; **the ~ of heating a house** die Heizkosten für ein Haus; **regardless of ~, whatever the ~:** ganz gleich, was es kostet; **b)** *(fig.)* Preis, *der;* **at all ~s, at any ~:** um jeden Preis; **at the ~ of sth.** auf Kosten einer Sache *(Gen.)*; **whatever the ~:** koste es, was es wolle; **to my/his** *etc.* **~:** zu meinem/seinem *usw.* Nachteil; **as I know to my ~:** wie ich aus bitterer Erfahrung weiß; *see also* [1]**count 2a; c)** *in pl. (Law)* [Gerichts]kosten *Pl.* **2.** *v.t.* **a)** *p.t., p.p.* **cost** *(lit. or fig.)* kosten; **how much does it ~?** was kostet es?; **whatever it may ~:** koste es, was es wolle; **~ sb. dear[ly]** jmdm. *od.* jmdn. teuer zu stehen kommen; **b)** *p.t., p.p.* **~ed** *(Commerc.: fix price of)* **~ sth.** den Preis für etw. kalkulieren
co-star ['kəʊstɑ:(r)] *(Cinemat., Theatre)* **1.** *n.* **be a/the ~:** eine der Hauptrollen/die zweite Hauptrolle spielen. **2.** *v.i.*, **-rr-** eine der Hauptrollen spielen
'**cost-effective** *adj.* rentabel
coster[monger] ['kɒstə(mʌŋgə(r))] *n. (Brit.)* Straßenhändler, *der*/-händlerin, *die*
costing ['kɒstɪŋ] *n.* **a)** *(estimation of costs)* Kostenberechnung, *die;* **b)** *(costs)* Kosten *Pl.*
costly ['kɒstlɪ] *adj.* **a)** teuer; kostspielig; **b)** *(fig.)* **a ~ victory** ein teuer erkaufter Sieg; **a ~ error** ein folgenschwerer Irrtum
cost: ~ **of** '**living** *n.* Lebenshaltungskosten *Pl;* ~ **price** *n.* Selbstkostenpreis, *der*
costume ['kɒstju:m] *n.* Kleidermode, *die; (theatrical ~)* Kostüm, *das;* **historical ~s** historische Kostüme
cosy ['kəʊzɪ] **1.** *adj.* gemütlich; behaglich ⟨*Atmosphäre*⟩; bequem ⟨*Sessel*⟩; **feel ~:** sich wohl *od.* behaglich fühlen; **be ~:** es gemütlich haben. **2.** *n. see* **tea-cosy**
cot [kɒt] *n. (Brit.: child's bed)* Kinderbett, *das*
cottage ['kɒtɪdʒ] *n.* Cottage, *das;* Häuschen, *das*
cottage: ~ '**cheese** *n.* Hüttenkäse, *der;* ~ **industry** *n.* Heimarbeit, *die;* ~ '**pie** *n. mit Kartoffelbrei überbackenes Hackfleisch*
cotter ['kɒtə(r)] *n.* ~**[-pin]** Splint, *der*
cotton ['kɒtn] **1.** *n.* Baumwolle, *die; (thread)* Baumwollgarn, *das; (cloth)* Baumwollstoff, *der.* **2.** *attrib. adj.* Baumwoll-. **3.** *v.i.* ~ '**on** *(coll.)* kapieren *(ugs.)*
cotton: **~-mill** *n.* Baumwollspinnerei, *die;* ~ **plant** *n.* Baumwollpflanze, *die;* **~-reel** *n.* [Näh]garnrolle, *die;* ~ '**waste** *n.* Putzwolle, *die;* ~ '**wool** *n.* Watte, *die;* **~-wool ball** Wattebausch, *der;* **wrap sb. up** *or* **keep sb. in ~ wool** *(fig.)* jmdn. in Watte packen
couch [kaʊtʃ] **1.** *n. (sofa)* Couch, *die.* **2.** *v.t.* formulieren
couchette [ku:'ʃet] *n. (Railw.)* Liegewagen, *der; (berth)* Liegesitz, *der*
cough [kɒf] **1.** *n. (act of ~ing, condition)* Husten, *der;* **give a ~:** husten; **have a [bad] ~:** [einen schlimmen] Husten haben. **2.** *v.i.* **a)** husten; **b)** ⟨*Motor:*⟩ stottern. **3.** *v.t.* ~ **out** [her]aushusten; ~ **up** [her]aushusten; *(sl.: pay)* ausspucken *(ugs.)*
coughing ['kɒfɪŋ] *n.* Husten, *das;* Gehuste, *das*
cough: ~ **medicine** *n.* Hustenmittel, *das;* ~ **mixture** *n.* Hustensaft, *der*
could *see* [2]**can**
couldn't ['kʊdnt] *(coll.)* = **could not;** *see* [2]**can**
council ['kaʊnsl] *n.* **a)** Ratsversammlung, *die;* **b)** *(administrative/advisory body)* Rat, *der;* **local ~:** Gemeinderat, *der;* **city/town ~:** Stadtrat, *der*
council: ~ **estate** *n.* Wohnviertel mit Sozialwohnungen; ~ **flat** *n.* Sozialwohnung, *die;* ~ **house** *n.* Haus des sozialen Wohnungsbaus; ~ **housing** *n.* sozialer Wohnungsbau
councillor ['kaʊnsələ(r)] *n.* Ratsmitglied, *das;* **town ~:** Stadtrat, *der*/-rätin, *die*
council: ~ **of** '**war** *n. (lit. or fig.)* Kriegsrat, *der;* ~ **tax** *n. (Brit.)* Gemeindesteuer, *die*
counsel ['kaʊnsl] **1.** *n.* **a)** *(consultation)* Beratung, *die;* **take/hold ~ with sb. [about sth.]** sich mit jmdm. [über etw. *(Akk.)*] beraten; **b)** Rat[schlag], *der;* **keep one's own ~:** seine Meinung für sich behalten; **c)** *pl. same (Law)* Rechtsanwalt, *der*/-anwältin, *die;* **~ for the defence** Verteidiger, *der*/Verteidigerin, *die;* **~ for the prosecution** Anklagevertreter, *der*/-vertreterin, *die;* Staatsanwalt, *der*/-anwältin, *die;* **Queen's/King's C~:** Anwalt/Anwältin der Krone; Kronanwalt, *der*/-anwältin, *die.* **2.** *v.t.*, *(Brit.)* **-ll-** *(advise)* beraten; **~ sb. to do sth.** jmdm. raten *od.* den Rat geben, etw. zu tun
counselling (*Amer.:* **counseling**) ['kaʊnsəlɪŋ] *n.* Beratung, *die*
counsellor, *(Amer.)* **counselor** ['kaʊnsələ(r)] *n.* Berater, *der*/Beraterin, *die;* **marriage-guidance ~:** Eheberater, *der*/-beraterin, *die*
[1]**count** [kaʊnt] **1.** *n.* **a)** Zählen, *das;* Zählung, *die;* **keep ~ [of sth.]** [etw.] zählen; **lose ~:** beim Zählen durcheinandergeraten; **lose ~ of sth.** etw. gar nicht mehr zählen können; **have/take/make a ~:** zählen; **on the ~ of three** bei „drei"; **b)** *(Law)* Anklagepunkt, *der;* **on that ~** *(fig.)* in diesem Punkt; **c)** *(Boxing)* Auszählen, *das;* **be out for the ~:** ausgezählt werden; *(fig.)* hinüber sein *(ugs.)*. **2.** *v.t.* **a)** zählen; **~ ten** bis zehn zählen; **~ the votes** die Stimmen [aus]zählen; **~ again** nachzählen; **~ the pennies** *(fig.)* jeden Pfennig umdrehen; **~ the cost** *(fig.)* unter den Folgen zu leiden haben; **b)** *(include)* mitzählen; **be ~ed against sb.** gegen jmdn. sprechen; **not ~ing** abgesehen von; *see also* **nothing 1a; c)** *(consider)* halten für; **~ oneself lucky** sich glücklich schätzen können. **3.** *v.i.* **a)** zählen; **~ [up] to ten** bis zehn zählen; **~ing from now** von jetzt an [gerechnet]; ab jetzt; **b)** *(be included)* zählen; **every moment ~s** jede Sekunde zählt; **~ against sb.** gegen jmdn. sprechen; **~ for much/little** viel/wenig zählen
~ '**in** *v.t.* mitrechnen; **you can ~ me in** ich bin dabei
~ **on** *v.t.* **~ on sb./sth.** sich auf jmdn./etw. verlassen
~ '**out** *v.t.* **a)** *(one by one)* abzählen; **b)** *(exclude)* **[you can] ~ me out** ich komme/mache nicht mit; **c)** *(Boxing)* auszählen
~ '**up** *v.t.* zusammenzählen; zusammenrechnen
~ **upon** *see* ~ **on**
[2]**count** *n. (nobleman)* Graf, *der*
'**countdown** *n.* Countdown, *der od. das*
countenance ['kaʊntɪnəns] **1.** *n.* **a)** *(literary: face)* Antlitz, *das (dichter.)*; **b)** *(formal: expression)* Gesichtsausdruck, *der.* **2.** *v.t.* *(formal: approve)* gutheißen

[1]**counter** ['kaʊntə(r)] *n.* **a)** *(in shop)* Ladentisch, *der; (in cafeteria, restaurant, train)* Büfett, *das; (in post office, bank)* Schalter, *der;* ~ **clerk** Schalterbeamte, *der*/-beamtin, *die;* **[buy/sell sth.] under the** ~ *(fig.)* [etwas] unter dem Ladentisch [kaufen/verkaufen]; **b)** *(disc for games)* Spielmarke, *die;* **c)** *(apparatus for counting)* Zähler, *der*

[2]**counter 1.** *adj.* entgegengesetzt; Gegen-/gegen-. **2.** *v. t.* **a)** *(oppose, contradict)* begegnen (+ *Dat.*); **b)** *(take action against)* kontern. **3.** *v. i. (take opposing action)* antworten. **4.** *adv.* **act** ~ **to** zuwiderhandeln (+ *Dat.*); **go** ~ **to** zuwiderlaufen (+ *Dat.*)

counter: ~'**act** *v. t.* entgegenwirken (+ *Dat.*); ~**-attack** *(lit. or fig.)* **1.** *n.* Gegenangriff, *der;* **2.** *v. t.* ~**-attack sb.** gegen jmdn. einen Gegenangriff richten; **3.** *v. i.* zurückschlagen; ~**-attraction** *n.* **a)** *(rival)* Konkurrenz, *die;* **b)** *(of contrary tendency)* entgegengesetzte Anziehungskraft; ~**balance** *v. t.* ein Gegengewicht bilden zu; *(fig.: neutralize)* ausgleichen; ~-'**clockwise** *see* **anticlockwise;** ~-'**espionage** *n.* Spionageabwehr, *die*

counterfeit ['kaʊntəfɪt, 'kaʊntəfi:t] **1.** *adj.* falsch, unecht ‹*Schmuck*›; falsch, gefälscht ‹*Unterschrift, Münze, Banknote*›; ~ **money** Falschgeld, *das.* **2.** *v. t.* fälschen

counterfeiter ['kaʊntəfɪtə(r)] *n.* Fälscher, *der*/Fälscherin, *die*

counter: ~**foil** *n.* Kontrollabschnitt, *der;* ~**-intelligence** *see* ~**-espionage**

countermand [kaʊntə'mɑ:nd] *v. t. (revoke)* widerrufen

counter: ~**measure** *n.* Gegenmaßnahme, *die;* ~**-offensive** *n. (Mil.)* Gegenoffensive, *die;* ~**part** *n.* Gegenstück, *das* (of zu); ~**-pro'ductive** *adj.* das Gegenteil des Gewünschten bewirkend; **sth. is** ~**-productive** etw. bewirkt das Gegenteil des Gewünschten; ~**sign** *v. t.* gegenzeichnen; ~**sink** *v. t.*, ~**sunk** ['kaʊntəsʌŋk] *(Woodw., Metalw.)* senken ‹*Loch*›; versenken ‹*Schraube*›; ~**weight** *n.* Gegengewicht, *das*

countess ['kaʊntɪs] *n.* Gräfin, *die*

countless ['kaʊntlɪs] *adj.* zahllos; ~ **numbers of** eine zahllose Menge von

country ['kʌntrɪ] *n.* **a)** Land, *das;* **sb.'s [home]** ~: jmds. Heimat; **fight/die for one's** ~: für sein [Vater]land kämpfen/sterben; **farming** ~: Ackerland, *das;* **b)** *(rural district)* Land, *das; (countryside)* Landschaft, *die;* **[be/live** *etc.*] **in the** ~: auf dem Land [sein/leben *usw.]*; **to the** ~: aufs Land; **c)** *(Brit.: population)* Volk, *das;* **appeal** *or* **go to the** ~: den Wähler entscheiden lassen

country: ~ '**dancing** *n.* Kontertanz, *der;* ~ '**house** *n.* Landhaus, *das;* ~**man** ['kʌntrɪmən] *n., pl.* ~**men** ['kʌntrɪmən] **a)** *(national)* Landsmann, *der;* **[my/her** *etc.*] **fellow** ~**man** [mein/ihr *usw.*] Landsmann; **b)** *(rural)* Landbewohner, *der;* ~**side** *n.* **a)** *(rural areas)* Land, *das;* **b)** *(rural scenery)* Landschaft, *die;* ~**-wide** *adj.* landesweit; ~**woman** *n.* **a)** *(national)* Landsmännin, *die;* **b)** *(rural)* Landbewohnerin, *die*

county ['kaʊntɪ] *n. (Brit.)* Grafschaft, *die*

county: ~ '**council** *n.* Grafschaftsrat, *der;* ~ '**town** *n. (Brit.) Verwaltungssitz einer Grafschaft*

coup [ku:] *n.* **a)** Coup, *der;* **b)** *see* **coup d'état**

coup d'état [ku: deɪ'tɑ:] *n.* Staatsstreich, *der*

coupé ['ku:peɪ] (*Amer.:* **coupe** [ku:p]) *n.* Coupé, *das*

couple [kʌpl] **1.** *n.* **a)** *(pair)* Paar, *das; (married)* [Ehe]paar, *das;* **b) a** ~ **[of]** *(a few)* ein paar; *(two)* zwei; **a** ~ **of people/things/days/weeks** *etc.* ein paar/zwei Leute/Dinge/Tage/Wochen *usw.* **2.** *v. t.* **a)** *(associate)* verbinden; **b)** *(fasten together)* koppeln

couplet ['kʌplɪt] *n. (Pros.)* Verspaar, *das; (rhyming)* Reimpaar, *das*

coupling ['kʌplɪŋ] *n. (Railw., Mech. Engin.)* Kupplung, *die*

coupon ['ku:pɒn] *n. (for rationed goods)* Marke, *die; (in advertisement)* Gutschein, *der;* Coupon, *der; (entry-form for football pool etc.)* Tippschein, *der*

courage ['kʌrɪdʒ] *n.* Mut, *der;* **have/lack the** ~ **to do sth.** den Mut haben/nicht den Mut haben, etw. zu tun; **take one's** ~ **in both hands** sein Herz in beide Hände nehmen

courageous [kə'reɪdʒəs] *adj.*, **courageously** [kə'reɪdʒəslɪ] *adv.* mutig

courgette [kʊə'ʒet] *n. (Brit.)* Zucchino, *der*

courier ['kʊrɪə(r)] *n.* **a)** *(Tourism)* Reiseleiter, *der*/-leiterin, *die;* **b)** *(messenger)* Kurier, *der*

course [kɔ:s] *n.* **a)** *(of ship, plane)* Kurs, *der;* **change [one's]** ~ *(lit. or fig.)* den Kurs wechseln; ~ **[of action]** Vorgehensweise, *die;* **what are our possible** ~**s of action?** welche Möglichkeiten haben wir?; **the most sensible** ~ **would be to ...:** das Vernünftigste wäre, zu ...; **the** ~ **of nature/history** der Lauf der Dinge/Geschichte; **run** *or* **take its** ~: seinen/ihren Lauf nehmen; **let things take their** ~: den Dingen ihren Lauf lassen; **off/on** ~: vom Kurs abgekommen/auf Kurs; **b) of** ~: natürlich; **[do sth.] as a matter of** ~: [etw.] selbstverständlich [tun]; **c)** *(progression)* Lauf, *der;* **in due** ~: zu gegebener Zeit; **in the** ~ **of the lesson/the day/his life** im Lauf[e] der Stunde/des Tages/seines Lebens; **d)** *(of river etc.)* Lauf, *der;* **e)** *(of meal)* Gang, *der;* **f)** *(Sport)* Kurs, *der; (for race)* Rennstrecke, *die;* **[golf-]**~**:** [Golf]platz, *der;* **g)** *(Educ.)* Kurs[us], *der; (for employee also)* Lehrgang, *der; (book)* Lehrbuch, *das;* **go to** *or* **attend/do a** ~ **in sth.** einen Kurs in etw. *(Dat.)* besuchen/machen; **h)** *(Med.)* **a** ~ **of treatment** eine Kur

court [kɔ:t] **1.** *n.* **a)** *(yard)* Hof, *der;* **b)** *(Sport)* Spielfeld, *das; (Tennis, Squash also)* Platz, *der;* **c)** *(of sovereign)* Hof, *der;* **hold** ~ *(fig.)* hofhalten *(scherzh.);* **d)** *(Law)* Gericht, *das;* ~ **of law** *or* **justice** Gerichtshof, *der;* **take sb. to** ~: jmdn. vor Gericht bringen *od.* verklagen; **appear in** ~: vor Gericht erscheinen. **2.** *v. t.* **a)** *(woo)* ~ **sb.** jmdn. umwerben; ~**ing couple** Liebespärchen, *das;* **b)** *(fig.)* suchen ‹*Gunst, Ruhm, Gefahr*›; **he is** ~**ing disaster/danger** er wandelt am Rande des Abgrunds *(fig. geh.)*

courteous ['kɜ:tɪəs] *adj.* höflich

courtesy ['kɜ:təsɪ] *n.* Höflichkeit, *die;* **by** ~ **of the museum** mit freundlicher Genehmigung des Museums

'**courtesy light** *n. (Motor Veh.)* Innenbeleuchtung, *die*

'**court-house** *n. (Law)* Gerichtsgebäude, *das*

courtier ['kɔ:tɪə(r)] *n.* Höfling, *der*

court: ~ '**martial** *n., pl.* ~**s martial** *(Mil.)* Kriegsgericht, *das;* **be tried by** ~ **martial** vor das/ein Kriegsgericht kommen; ~-'**martial** *v. t.*, *(Brit.)* **-ll-** vor das/ein Kriegsgericht stellen; ~**room** *n. (Law)* Gerichtssaal, *der*

courtship ['kɔ:tʃɪp] *n.* Werben, *das*

court: ~ shoe *n.* Pumps, *der;* **~yard** *n.* Hof, *der*

cousin ['kʌzn] *n.* **[first] ~:** Cousin, *der*/Cousine, *die;* Vetter, *der*/*(veralt.)* Base, *die;* **[second] ~:** Cousin/Cousine zweiten Grades

cove [kəʊv] *n.* *(Geog.)* [kleine] Bucht

covenant ['kʌvənənt] **1.** *n.* formelle Übereinkunft. **2.** *v. t.* *(also Law)* [vertraglich] vereinbaren

Coventry ['kɒvəntrɪ] *n.* **send sb. to ~** *(fig.)* jmdn. [demonstrativ] schneiden

cover ['kʌvə(r)] **1.** *n.* **a)** *(piece of cloth)* Decke, *die;* *(of cushion, bed)* Bezug, *der;* *(lid)* Deckel, *der;* *(of hole, engine, typewriter, etc.)* Abdeckung, *die;* **put a ~ on** *or* **over** zudecken; abdecken ‹*Loch, Fußboden, Grab, Fahrzeug, Maschine*›; beziehen ‹*Kissen, Bett*›; **b)** *(of book)* Einband, *der;* *(of magazine)* Umschlag, *der;* *(of record)* [Platten]hülle, *die;* **read sth. from ~ to ~:** etw. von vorn bis hinten lesen; **on the [front/back] ~:** auf dem [vorderen/hinteren] Buchdeckel; *(of magazine)* auf der Titelseite/hinteren Umschlagseite; **c)** *(Post: envelope)* [Brief]umschlag, *der;* **under plain ~:** in neutralem Umschlag; **[send sth.] under separate ~:** [etw.] mit getrennter Post [schicken]; **d)** *in pl.* *(bedclothes)* Bettzeug, *das;* **e)** *(hiding-place, shelter)* Schutz, *der;* **take ~ [from sth.]** Schutz [vor etw. *(Dat.)*] suchen; **[be/go] under ~** *(from bullets etc.)* in Deckung [sein/gehen]; **under ~** *(from rain)* überdacht ‹*Sitzplatz*›; regengeschützt; **keep sth. under ~:** etw. abgedeckt halten; **under ~ of darkness** im Schutz der Dunkelheit; **f)** *(Mil.: supporting force)* Deckung, *die;* **g)** *(protection)* Deckung, *die;* **give sb./sth. ~:** jmdm. Deckung geben; **h)** *(pretence)* Vorwand, *der;* *(false identity, screen)* Tarnung, *die;* **i)** *(Insurance)* **[insurance] ~:** Versicherung, *die;* **get ~ against sth.** sich gegen etw. versichern; **have adequate ~:** ausreichend versichert sein. **2.** *v. t.* **a)** bedecken; **~ a book with leather** ein Buch in Leder binden; **~ a chair with chintz** einen Stuhl mit Chintz beziehen; **~ a pan with a lid** eine Pfanne mit einem Deckel zudecken; **she ~ed her face with her hands** sie verbarg das Gesicht in den Händen; **the roses are ~ed with greenfly** die Rosen sind voller Blattläuse; **sb. is ~ed in** *or* **with confusion/shame** *(fig.)* jmd. ist ganz verlegen/sehr beschämt; **b)** *(conceal, lit. or fig.)* verbergen; *(for protection)* abdecken; **c)** *(travel)* zurücklegen; **d)** *in p. p.* *(having roof)* überdacht; **e)** *(deal with)* behandeln; *(include)* abdecken; **f)** *(Journ.)* berichten über (+ *Akk.*); **g) ~ expenses** die Kosten decken; **£10 will ~ my needs for the journey** 10 Pfund werden für die Reisekosten reichen; **h)** *(shield)* decken; **I'll keep you ~ed** ich gebe dir Deckung; **i) ~ oneself** *(fig.)* sich absichern; *(Insurance)* **~ oneself against sth.** sich gegen etw. versichern; **j)** *(aim gun at)* in Schach halten *(ugs.)*; **I've got you ~ed** ich habe meine Waffe auf dich gerichtet

~ for *v. t.* einspringen für

~ 'in *v. t.* überdachen; *(fill in)* zuschütten

~ 'up 1. *v. t.* *(conceal)* zudecken; *(fig.)* vertuschen. **2.** *v. i.* *(fig.: conceal)* es vertuschen; **~ up for sb.** jmdn. decken

coverage ['kʌvərɪdʒ] *n., no pl.* **a)** *(Journ., Radio, Telev.: treatment)* Berichterstattung, *die* **(of** über + *Akk.*); **newspaper/broadcast ~:** Berichterstattung in der Presse/in Funk und Fernsehen; **give sth. [full/limited] ~:** [ausführlich/kurz] über etw. *(Akk.)* berichten; **b)** *(Advertising)* Abdeckung des Marktes

coverall ['kʌvərɔ:l] *n., usu. in pl.* *(esp. Amer.)* Overall, *der;* *(for baby)* Strampelanzug, *der*

cover: ~ charge *n.* [Preis für das] Gedeck; **~ girl** *n.* Covergirl, *das*

covering ['kʌvərɪŋ] *n.* *(material)* Decke, *die;* *(of chair, bed)* Bezug, *der*

covering: ~ letter *n.* Begleitbrief, *der;* **~ note** *n.* [kurzes] Begleitschreiben

'cover note *n.* *(Insurance)* Deckungskarte, *die*

covert ['kʌvət] *adj.* versteckt

'cover-up *n.* Verschleierung, *die*

covet ['kʌvɪt] *v. t.* begehren *(geh.)*

covetous ['kʌvɪtəs] *adj.* begehrlich *(geh.)*

[1]cow [kaʊ] *n.* **a)** Kuh, *die;* **till the ~s come home** *(fig. coll.)* bis in alle Ewigkeit *(ugs.)*; **b)** *(sl. derog.: woman)* Kuh, *die (salopp abwertend)*

[2]cow *v. t.* einschüchtern; **~ sb. into submission** jmdn. so einschüchtern, daß er sich unterordnet

coward ['kaʊəd] *n.* Feigling, *der*

cowardice ['kaʊədɪs] *n.* Feigheit, *die*

cowardly ['kaʊədlɪ] *adj.* feig[e]

'cowboy *n.* Cowboy, *der;* *(Brit. coll.: unscrupulous businessman, tradesman, etc.)* Betrüger, *der*

cower ['kaʊə(r)] *v. i.* sich ducken; *(squat)* kauern

'cowherd *n.* Kuhhirte, *der*

cowl [kaʊl] *n.* **a)** *(of monk)* Kutte, *die;* *(hood)* Kapuze, *die;* **b)** *(of chimney)* Schornsteinaufsatz, *der*

co-worker ['kəʊwɜ:kə(r)] *n.* Kollege, *der*/Kollegin, *die*

cow: ~-shed *n.* Kuhstall, *der;* **~slip** *n.* Schlüsselblume, *die*

cox [kɒks] **1.** *n.* Steuermann, *der.* **2.** *v. t. & i.* *(esp. Rowing)* steuern

coxswain ['kɒksweɪn, 'kɒksn] *see* **cox 1**

coy [kɔɪ] *adj.* gespielt schüchtern; geziert ‹*Benehmen, Ausdruck*›

coyote [kə'jəʊtɪ, 'kɔɪəʊt] *n.* *(Zool.)* Kojote, *der*

cozily, cozy *(Amer.) see* **cos-**

crab [kræb] *n.* Krabbe, *die*

'crab-apple *n.* Holzapfel, *der*

crack [kræk] **1.** *n.* **a)** *(noise)* Krachen, *das;* **give sb./have a fair ~ of the whip** *(fig.)* jmdm. eine Chance geben/eine Chance haben; **b)** *(in china, glass, eggshell, ice, etc.)* Sprung, *der;* *(in rock)* Spalte, *die;* *(chink)* Spalt, *der;* **there's a ~ in the ceiling** die Decke hat einen Riß; **c)** *(blow)* Schlag, *der;* **d)** *(coll.: try)* Versuch, *der;* **have a ~ at sth./at doing sth.** etw. in Angriff nehmen/versuchen, etw. zu tun; **e) the/at the ~ of dawn** *(coll.)* der/bei Tagesanbruch; **f)** *(coll.: wisecrack)* [geistreicher] Witz; **g)** *(sl.: drug)* Crack, *der.* **2.** *adj.* *(coll.)* erstklassig. **3.** *v. t.* **a)** *(break, lit. or fig.)* knacken ‹*Nuß, Problem*›; knacken *(salopp)* ‹*Safe, Kode*›; **b)** *(make a ~ in)* anschlagen ‹*Porzellan, Glas*›; **c) ~ a whip** mit einer Peitsche knallen; **~ the whip** *(fig.)* Druck machen *(ugs.)*; **d) ~ a joke** einen Witz machen. **4.** *v. i.* **a)** ‹*Porzellan, Glas:*› einen Sprung/Sprünge bekommen; ‹*Haut:*› aufspringen, rissig werden; ‹*Eis:*› Risse bekommen; **b)** *(make sound)* ‹*Peitsche:*› knallen; ‹*Gelenk:*› knacken; ‹*Gewehr:*› krachen; **c)** *(coll.)* **get ~ing!** mach los! *(ugs.)*; **let's get ~ing** fangen wir endlich an; **get ~ing [with sth.]** [mit etw.] loslegen *(ugs.)*

~ 'down *v. i.* *(coll.)* **~ down [on sb./sth.]** [gegen jmdn./etw.] [hart] vorgehen

~ 'up *(coll.)* **1.** *v. i.* ‹*Flugzeug usw.:*› auseinanderbrechen; ‹*Gesellschaft, Person:*› zusammenbrechen. **2.** *v. t.* **she/it is not all she/it**

is ~ed up to be so toll ist sie/es nun auch wieder nicht[, wie sie/es dargestellt wird]

cracked [krækt] *adj.* **a)** gesprungen ‹*Porzellan, Ziegel, Glas*›; rissig, aufgesprungen ‹*Haut, Erdboden*›; rissig ‹*Verputz*›; **b)** *(coll.: crazy)* übergeschnappt *(ugs.)*

cracker ['krækə(r)] *n.* **a)** **[Christmas] ~** ≈ Knallbonbon, *der od. das;* **b)** *(firework)* Knallkörper, *der;* **c)** *(biscuit)* Cracker, *der*

crackers ['krækəz] *pred. adj. (Brit. coll.)* übergeschnappt *(ugs.)*

crackle [krækl] **1.** *v. i.* knistern; ‹*Feuer:*› prasseln. **2.** *n.* Knistern, *das; (of fire)* Prasseln, *das*

crackling ['kræklɪŋ] *n., no pl., no indef. art. (Cookery)* Kruste, *die*

'crackpot *n. (coll.)* Spinner, *der*/Spinnerin, *die (ugs.); attrib.* **~ ideas/schemes** hirnrissige Ideen/Pläne *(abwertend)*

cradle [kreɪdl] **1.** *n. (cot, lit. or fig.)* Wiege, *die;* **from the ~ to the grave** von der Wiege bis zur Bahre. **2.** *v. t.* wiegen; **~ sb./sth. in one's arms** jmdn. in den Armen/etw. im Arm halten

craft [krɑ:ft] *n.* **a)** *(trade)* Handwerk, *das; (art)* Kunsthandwerk, *das;* **b)** *no pl. (skill)* Kunstfertigkeit, *die;* **c)** *no pl. (cunning)* List, *die;* **d)** *pl. same (boat)* Boot, *das*

craftsman ['krɑ:ftsmən] *n., pl.* **craftsmen** ['krɑ:ftsmən] Handwerker, *der*

craftsmanship ['krɑ:ftsmənʃɪp] *n., no pl. (skilled workmanship)* handwerkliches Können

crafty ['krɑ:ftɪ] *adj.* listig

crag [kræg] *n.* Felsspitze, *die*

craggy ['krægɪ] *adj. (rugged)* zerklüftet; zerfurcht ‹*Gesicht*›; *(rocky)* felsig

cram [kræm] **1.** *v. t.,* **-mm-: a)** *(overfill)* vollstopfen *(ugs.); (force)* stopfen; **the bus was ~med** der Bus war gerammelt voll *(ugs.) od.* war überfüllt; **b)** *(for examination)* **~ pupils** mit Schülern pauken *(ugs.);* **c)** *(feed to excess)* mästen. **2.** *v. i.,* **-mm-** *(for examination)* büffeln *(ugs.);* pauken *(ugs.)*

cramp [kræmp] **1.** *n. (Med.)* Krampf, *der;* **suffer an attack of ~:** einen Krampf bekommen; **have ~ [in one's leg/arm]** einen Krampf [im Bein/Arm] haben. **2.** *v. t. (confine)* einengen; **~ [up]** zusammenpferchen; **~ sb.'s style** jmdn. einengen

cramped [kræmpt] *adj.* eng ‹*Raum*›; gedrängt ‹*Handschrift*›

cranberry ['krænbərɪ] *n.* Preiselbeere, *die*

crane [kreɪn] **1.** *n.* **a)** *(machine)* Kran, *der;* **b)** *(Ornith.)* Kranich, *der.* **2.** *v. t.* **~ one's neck** den Hals recken. **3.** *v. i.* den Hals recken; **~ forward** den Hals [nach vorn] recken

'crane-fly *n.* Schnake, *die*

¹crank [kræŋk] **1.** *n. (Mech. Engin.)* [Hand]kurbel, *die.* **2.** *v. t.* **~ [up]** ankurbeln

²crank *n.* Irre, *der/die (salopp);* **health ~:** Gesundheitsfanatiker, *der*/-fanatikerin, *die (ugs.)*

'crankshaft *n. (Mech. Engin.)* Kurbelwelle, *die*

cranky ['kræŋkɪ] *adj.* **a)** *(eccentric)* schrullig; verschroben; **b)** *(Amer.: ill-tempered)* griesgrämig

cranny ['krænɪ] *n.* Ritze, *die; see also* **nook**

crap [kræp] *(coarse)* **1.** *n.* **a)** *(faeces)* Scheiße, *die (derb);* **have a ~:** scheißen *(derb);* **b)** *(nonsense)* Scheiß, *der (salopp abwertend).* **2.** *v. i.,* **-pp-** scheißen *(derb)*

craps [kræps] *n. pl. (Amer.: dice game)* Craps, *das*

crash [kræʃ] **1.** *n.* **a)** *(noise)* Krachen, *das;* **fall with a ~:** mit einem lauten Krach fallen; **a sudden ~ of thunder** ein plötzlicher Donnerschlag; **b)** *(collision)* Zusammenstoß, *der;* **plane/train ~:** Flugzeug- / Eisenbahnunglück, *das;* **have a ~:** einen Unfall haben; **in a [car] ~:** bei einem [Auto]unfall; **c)** *(Finance etc.)* Zusammenbruch, *der.* **2.** *v. i.* **a)** *(make a noise, go noisily)* krachen; **b)** *(have a collision)* einen Unfall haben; ‹*Flugzeug, Flieger:*› abstürzen; **~ into sth.** gegen etw. krachen; **c)** *(Finance etc., Computing)* zusammenbrechen. **3.** *v. t.* **a)** *(smash)* schmettern; **b)** *(cause to have collision)* einen Unfall haben mit

crash: ~ barrier *n.* Leitplanke, *die;* **~ course** *n.* Intensivkurs, *der;* **~-helmet** *n.* Sturzhelm, *der;* **~-land 1.** *v. t.* **~-land a plane** mit einem Flugzeug bruchlanden; **2.** *v. i.* bruchlanden

crass [kræs] *adj.* kraß; grob ‹*Benehmen*›; haarsträubend ‹*Dummheit, Unwissenheit*›; *(very stupid)* strohdumm

crate [kreɪt] *n.* Kiste, *die;* **a ~ of beer/lemonade** ein Kasten Bier/Limonade

crater ['kreɪtə(r)] *n.* Krater, *der*

cravat [krə'væt] *n. (scarf)* Halstuch, *das; (necktie)* Krawatte, *die*

crave [kreɪv] **1.** *v. t.* **a)** *(beg)* erbitten; erflehen ‹*Gnade*›; **b)** *(long for)* sich sehnen nach. **2.** *v. i.* **~ for** *or* **after** *see* **1**

craving ['kreɪvɪŋ] *n.* Verlangen, *das;* **have a ~ for sth.** ein [dringendes] Verlangen nach etw. haben

crawl [krɔ:l] **1.** *v. i.* **a)** kriechen; **the baby/insect ~s along the ground** das Baby/Insekt krabbelt über den Boden; **b)** *(coll.: behave abjectly)* kriechen *(abwertend);* **~ to sb.** vor jmdm. buckeln *od.* kriechen; **c)** **be ~ing** *(be covered or filled)* wimmeln (**with** von); **d)** *see* **creep 1 b. 2.** *n.* **a)** Kriechen, *das; (of insect, baby also)* Krabbeln, *das; (slow speed)* Schneckentempo, *das;* **move/go at a ~:** sich im Schneckentempo bewegen/im Schneckentempo fahren; **b)** *(swimming-stroke)* Kraulen, *das*

crawler lane ['krɔ:lə leɪn] *n.* Kriechspur, *die*

crayfish ['kreɪfɪʃ] *n., pl. same* Flußkrebs, *der*

crayon ['kreɪən] *n. (pencil)* **[coloured] ~:** Buntstift, *der; (of wax)* Wachsmalstift, *der; (of chalk)* Kreidestift, *der*

craze [kreɪz] **1.** *n.* Begeisterung, *die;* Fimmel, *der (ugs. abwertend);* **there's a ~ for doing sth.** es ist gerade große Mode, etw. zu tun. **2.** *v. t.* **be [half] ~d with pain/grief** *etc.* [halb] wahnsinnig vor Schmerz/Kummer *usw.* sein; **a ~d look/expression [on sb.'s face]** ein vom Wahnsinn verzerrtes Gesicht

crazy ['kreɪzɪ] *adj.* **a)** *(mad)* verrückt; wahnsinnig; **go ~:** verrückt *od.* wahnsinnig werden; **drive** *or* **send sb. ~:** jmdn. verrückt *od.* wahnsinnig machen *(ugs.);* **b)** *(coll.: enthusiastic)* **be ~ about sb./sth.** nach jmdm./etw. verrückt sein *(ugs.);* **c)** **~ paving** gestückeltes Pflaster

creak [kri:k] **1.** *n. (of gate, door)* Quietschen, *das; (of floor-board, door, chair)* Knarren, *das.* **2.** *v. i.* ‹*Tor, Tür:*› quietschen; ‹*Diele, Tür, Stuhl:*› knarren

cream [kri:m] **1.** *n.* **a)** Sahne, *die;* **b)** *(Cookery) (sauce)* Sahnesoße, *die; (dessert)* Creme, *die;* **~ of mushroom soup** Champignoncremesuppe, *die;* **c)** *(cosmetic preparation)* Creme, *die;* **d)** *(fig.: best)* Beste, *das;* **the ~ of society** die Creme der Gesellschaft; **e)** *(colour)* Creme, *das.* **2.** *adj.* **~[-coloured]** creme[farben]. **3.** *v. t.* cremig rühren *od.* schlagen; schaumig rühren ‹*Butter*›; **~ed potatoes** Kartoffelpüree, *das*

~ 'off *v. t.* **~ off the best players** die besten Spieler wegschnappen *(ugs.)*

cream: ~ **cake** *n.* Cremetorte, *die; (with whipped ~)* Sahnetorte, *die;* ~ **'cheese** *n.* ≈ Frischkäse, *der;* ~ **'tea** *n.* Tee mit Marmeladetörtchen und Sahne

creamy ['kri:mɪ] *adj. (with cream)* sahnig; *(like cream)* cremig

crease [kri:s] **1.** *n. (pressed)* Bügelfalte, *die; (accidental; in skin)* Falte, *die; (in fabric)* Falte, *die;* Knitter, *der;* **put a ~ in trousers** Bügelfalten in Hosen bügeln. **2.** *v.t. (press)* eine Falte/Falten bügeln in (+ *Akk.*); *(accidentally)* knittern; *(extensively)* zerknittern. **3.** *v.i.* Falten bekommen; knittern

'crease-resistant *adj.* knitterfrei

create [krɪ'eɪt] **1.** *v.t.* **a)** schaffen; erschaffen *(geh.);* verursachen ‹*Verwirrung*›; machen ‹*Eindruck*›; ‹*Sache:*› mit sich bringen, ‹*Person:*› machen ‹*Schwierigkeiten*›; ~ **a scene** eine Szene machen; ~ **a sensation** für eine Sensation sorgen; **b)** *(design)* schaffen; kreieren ‹*Mode, Stil*›; **c)** *(invest with rank)* ernennen; ~ **sb. a peer** jmdn. zum Peer erheben *od.* ernennen. **2.** *v.i. (Brit. coll.: make a fuss)* Theater machen *(ugs.)*

creation [krɪ'eɪʃn] *n.* **a)** *no pl. (act of creating)* Schaffung, *die; (of the world)* Erschaffung, *die;* Schöpfung, *die (geh.);* **b)** *no pl. (all created things)* Schöpfung, *die;* **c)** *(Fashion)* Kreation, *die*

creative [krɪ'eɪtɪv] *adj.* schöpferisch; kreativ

creator [krɪ'eɪtə(r)] *n.* Schöpfer, *der*/Schöpferin, *die;* **the C~:** der Schöpfer

creature ['kri:tʃə(r)] *n.* **a)** *(created being)* Geschöpf, *das;* **all living ~s** alle Lebewesen; **b)** *(human being)* Geschöpf, *das; (derog.)* Kerl, *der (abwertend); (woman)* **the ~ with the red hair** die mit den roten Haaren *(ugs.);* ~ **of habit** Gewohnheitsmensch, *der*

crèche [kreʃ] *n.* [Kinder]krippe, *die*

credential [krɪ'denʃl] *n., usu. in pl. (testimonial)* Zeugnis, *das*

credibility [kredɪ'bɪlɪtɪ] *n.* Glaubwürdigkeit, *die*

credible ['kredɪbl] *adj.* glaubwürdig ‹*Person, Aussage*›

credibly ['kredɪblɪ] *adv.* glaubwürdig; glaubhaft

credit ['kredɪt] **1.** *n.* **a)** *no pl. (commendation)* Anerkennung, *die; (honour)* Ehre, *die;* **give sb. [the] ~ for sth.** jmdm. für etw. Anerkennung zollen *(geh.);* **take the ~ for sth.** die Anerkennung für etw. einstecken; **[we must give] ~ where ~ is due** Ehre, wem Ehre gebührt; **it is [much *or* greatly/little] to sb.'s/sth.'s ~ that ...:** es macht jmdm./einer Sache [große/wenig] Ehre, daß ...; **it is to his ~ that ...:** es ehrt ihn, daß ...; **be a ~ to sb./sth.** jmdm./einer Sache Ehre machen; **b) ~s, ~ titles** *(at beginning of film)* Vorspann, *der; (at end)* Nachspann, *der;* **c)** *no pl., no art. (belief)* Glaube, *der;* **gain ~:** an Glaubwürdigkeit gewinnen; **d)** *no pl. (Commerc.)* Kredit, *der;* **give [sb.] ~:** [jmdm.] Kredit geben; **their ~ is excellent** sie sind unbedingt kreditwürdig; **e)** *no pl. (Finance, Bookk.)* Guthaben, *das;* **be in ~** ‹*Konto:*› im Haben sein; ‹*Person:*› mit seinem Konto im Haben sein; **f)** *(fig.)* **have sth. to one's ~:** etw. vorzuweisen haben; **he's cleverer than I gave him ~ for** er ist klüger, als ich dachte. **2.** *v.t.* **a)** *(believe)* glauben; **b)** *(accredit)* ~ **sb. with sth.** jmdm. etw. zutrauen; ~ **sth. with sth.** einer Sache *(Dat.)* etw. zuschreiben; **c)** *(Finance, Bookk.)* gutschreiben; ~ **£10 to sb./sb.'s account** jmdm./jmds. Konto 10 Pfund gutschreiben

creditable ['kredɪtəbl] *adj.* anerkennenswert

credit: ~ **card** *n.* Kreditkarte, *die;* ~ **note** *n.* Gutschein, *der*

creditor ['kredɪtə(r)] *n.* Gläubiger, *der*/Gläubigerin, *die*

credit: ~ **sale** *n.* Kreditkauf, *der;* ~ **side** *n. (Finance)* Habenseite, *die; (fig.)* **on the ~ side she has experience** für sie spricht ihre Erfahrung; ~ **squeeze** *n.* Kreditrestriktion, *die;* **~-worthy** *adj.* kreditwürdig

credulity [krɪ'dju:lɪtɪ] *n., no pl.* Leichtgläubigkeit, *die*

credulous ['kredjʊləs] *adj.* leichtgläubig

creed [kri:d] *n. (lit. or fig.)* Glaubensbekenntnis, *das*

creek [kri:k] *n.* **a)** *(Brit.: inlet on sea-coast)* [kleine] Bucht; **b)** *(short arm of river)* [kurzer] Flußarm; **c) be up the ~** *(coll.: be in difficulties or trouble)* in der Klemme *od.* Tinte sitzen *(ugs.)*

creep [kri:p] **1.** *v.i.,* **crept** [krept] **a)** kriechen; *(move timidly, slowly, stealthily)* schleichen; ~ **and crawl** *(fig.)* kriechen; **b) make sb.'s flesh ~:** jmdm. eine Gänsehaut über den Rücken jagen. **2.** *n., in pl. (coll.)* **give sb. the ~s** jmdm. nicht [ganz] geheuer sein

~ **'in** *v.i.* [sich] hinein-/hereinschleichen; *(fig.)* ‹*Irrtum, Enttäuschung usw.:*› sich einschleichen

~ **'on** *v.i.* **time is ~ing on** die Zeit verrinnt [unaufhaltsam]

~ **'up** *v.i. (approach)* sich anschleichen; ~ **up on sb.** sich an jmdn. anschleichen

creeper ['kri:pə(r)] *n. (Bot.) (growing along ground)* Kriechpflanze, *die; (growing up wall etc.)* Kletterpflanze, *die*

creepy ['kri:pɪ] *adj.* unheimlich; gruselig, schaurig ‹*Geschichte, Film*›

cremate [krɪ'meɪt] *v.t.* einäschern

cremation [krɪ'meɪʃn] *n.* Einäscherung, *die*

crematorium [kremə'tɔ:rɪəm] *n., pl.* **crematoria** [kremə'tɔ:rɪə] *or* **~s** Krematorium, *das*

creosote ['kri:əsəʊt] **1.** *n.* Kreosot, *das.* **2.** *v.t.* mit Kreosot behandeln

crêpe [kreɪp] *n.* Krepp, *der*

crept *see* **creep 1**

crescendo [krɪ'ʃendəʊ] *n., pl.* **~s** *(Mus.)* Crescendo, *das; (fig.)* Zunahme, *die;* **reach a ~** *(fig. coll.)* einen Höhepunkt erreichen

crescent ['kresənt] **1.** *n.* **a)** Mondsichel, *die; (as emblem)* Halbmond, *der;* **~-shaped** halbmondförmig; **b)** *(Brit.: street)* [kleinere] halbkreisförmige Straße. **2.** *adj.* **the ~ moon** die Mondsichel

cress [kres] *n.* Kresse, *die*

crest [krest] *n.* **a)** *(on bird's or animal's head)* Kamm, *der;* **b)** *(top of mountain or wave)* Kamm, *der;* **[be/ride] on the ~ of a *or* the wave** *(fig.)* ganz oben [sein/schwimmen]; **c)** *(Her.)* Helmzier, *die; (emblem)* Emblem, *das*

'crestfallen *adj. (fig.)* niedergeschlagen

Crete [kri:t] *pr. n.* Kreta *(das)*

cretin ['kretɪn] *n.* **a)** *(Med.)* Kretin, *der;* **b)** *(coll.: fool)* Trottel, *der (ugs. abwertend)*

crevasse [krɪ'væs] *n.* Gletscherspalte, *die*

crevice ['krevɪs] *n.* Spalt, *der*

crew [kru:] **1.** *n.* **a)** *(of ship, aircraft, etc.)* Besatzung, *die;* Crew, *die; (excluding officers)* Mannschaft, *die;* Crew, *die; (Sport)* Mannschaft, *die;* Crew, *die;* **b)** *(associated body)* Gruppe, *die; (set, often derog.)* Haufen, *der;* **a motley ~:** ein bunt zusammengewürfelter Haufen. **2.** *v.i.* die Mannschaft/Mitglied der Mannschaft sein. **3.** *v.t.* ~ **a boat** Mitglied der Mannschaft/die Mannschaft eines Bootes sein

ˈ**crew cut** *n.* Bürstenschnitt, *der*
crib [krɪb] **1.** *n.* **a)** *(cot)* Gitterbett, *das;* **b)** *(model of manger-scene; manger)* Krippe, *die;* **c)** *(coll.: translation)* Klatsche, *die (Schülerspr.).* **2.** *v.t.,* **-bb-** *(coll.: plagiarize)* abkupfern *(salopp)*
crick [krɪk] **1.** *n.* **a ~ [in one's neck/back]** ein steifer Hals/Rücken. **2.** *v.t.* **~ one's neck/back** einen steifen Hals/Rücken bekommen
[1]**cricket** [ˈkrɪkɪt] *n. (Sport)* Kricket, *das;* **it's/that's not ~** *(Brit. dated coll.)* das ist nicht die feine Art *(ugs.)*
[2]**cricket** *n. (Zool.)* Grille, *die*
cricket: ~ ball *n.* Kricketball, *der;* **~ bat** Schlagholz, *das*
cricketer [ˈkrɪkɪtə(r)] *n.* Kricketspieler, *der/*-spielerin, *die*
cricket: ~ match *n.* Kricketspiel, *das;* **~ pitch** *n.* Kricketfeld, *das (zwischen den Toren)*
crime [kraɪm] *n.* **a)** Verbrechen, *das;* **b)** *collect., no pl.* **a wave of ~:** eine Welle von Straftaten; **~ doesn't pay** Verbrechen lohnen sich nicht; **c)** *(fig. coll.: shameful action)* Sünde, *die*
ˈ**crime rate** *n.* Kriminalitätsrate, *die*
criminal [ˈkrɪmɪnl] **1.** *adj.* **a)** *(illegal)* kriminell; strafbar; *(concerned with criminals and crime)* Straf-; **~ act** *or* **deed/offence** Straftat, *die;* **take ~ proceedings against sb.** strafrechtlich gegen jmdn. vorgehen; **b)** *(fig. coll.)* kriminell *(ugs.);* **it's a ~ waste** es ist eine sträfliche Verschwendung. **2.** *n.* Kriminelle, *der/die*
criminal ˈ**law** *n.* Strafrecht, *das*
criminally [ˈkrɪmɪnəlɪ] *adv.* kriminell; *(according to criminal law)* strafrechtlich
criminal ˈ**record** *n.* Strafregister, *das;* **have a ~:** vorbestraft sein
criminology [krɪmɪˈnɒlədʒɪ] *n.* Kriminologie, *die*
crimson [ˈkrɪmzn] **1.** *adj.* purpurrot; **turn ~** ‹*Himmel:*› sich blutrot färben; *(with anger)* ‹*Person:*› rot anlaufen; *(blush)* puterrot werden. **2.** *n.* Purpurrot, *das*
cringe [krɪndʒ] *v.i.* zusammenzucken; ‹*Hund:*› sich ducken, kuschen; **~ at sth.** bei etw. zusammenzucken; **~ away** *or* **back [from sb./sth.]** [vor jmdm./etw.] zurückschrecken; **it makes me ~** *(in disgust)* da wird mir schlecht
cringing [ˈkrɪndʒɪŋ] *adj.* kriecherisch *(abwertend)*
crinkle [krɪŋkl] **1.** *n.* Knick, *der; (in fabric)* Knitterfalte, *die; (in hair)* Kräusel, *die.* **2.** *v.t.* knicken; zerknittern ‹*Stoff, Papier*›; kräuseln ‹*Haar*›. **3.** *v.i.* ‹*Stoff, Papier:*› knittern; ‹*Haar:*› sich kräuseln
crinoline [ˈkrɪnəlɪn, ˈkrɪnəliːn] *n. (Hist.)* Krinoline, *die*
cripple [krɪpl] **1.** *n. (lit. or fig.)* Krüppel, *der.* **2.** *v.t.* zum Krüppel machen; *(fig.)* lähmen
crippled [krɪpld] *adj.* verkrüppelt ‹*Arm, Baum, Bettler*›; **be ~ with rheumatism** durch Rheuma gelähmt sein; **industry was ~ by the strikes** die Streiks haben die ganze Industrie lahmgelegt
crisis [ˈkraɪsɪs] *n., pl.* **crises** [ˈkraɪsiːz] Krise, *die*
crisp [krɪsp] **1.** *adj.* knusprig ‹*Brot, Keks, Speck*›; knackig ‹*Apfel, Gemüse*›; trocken ‹*Herbstblätter, Zweige*›; frisch [gebügelt/gestärkt] ‹*Wäsche*›; [druck]frisch ‹*Banknote*›; verharscht ‹*Schnee*›; scharf ‹*Umrisse, Kanten*›; knapp [und klar] ‹*Stil*›. **2.** *n.* **a)** *usu. in pl. (Brit.: potato ~)* [Kartoffel]chip, *der;* **b) be burned to a ~:** verbrannt sein
ˈ**crispbread** *n.* Knäckebrot, *das*
crispy [ˈkrɪspɪ] *adj.* knusprig ‹*Brot, Keks, Speck*›; knackig ‹*Apfel, Gemüse*›
criss-cross [ˈkrɪskrɒs] **1.** *adj.* **~ pattern** Muster aus gekreuzten Linien. **2.** *adv.* kreuz und quer. **3.** *v.t. (intersect repeatedly)* wiederholt schneiden
criterion [kraɪˈtɪərɪən] *n., pl.* **criteria** [kraɪˈtɪərɪə] Kriterium, *das*
critic [ˈkrɪtɪk] *n.* Kritiker, *der/*Kritikerin, *die;* **literary ~:** Literaturkritiker, *der/*-kritikerin, *die*
critical [ˈkrɪtɪkl] *adj.* **a)** kritisch; **be ~ of sb./sth.** jmdn./etw. kritisieren; **cast a ~ eye over sth.** etw. mit kritischen Augen betrachten; **the play received ~ acclaim** das Stück fand die Anerkennung der Kritik; **b)** *(involving risk, crucial)* kritisch ‹*Zustand, Punkt, Phase*›; entscheidend ‹*Faktor, Test*›
critically [ˈkrɪtɪkəlɪ] *adv.* kritisch; **be ~ ill** ernstlich krank sein
criticise *see* **criticize**
criticism [ˈkrɪtɪsɪzm] *n.* Kritik, *die* **(of** an + *Dat.***); come in for a lot of ~:** heftig kritisiert werden; **be open to ~:** der Kritik ausgesetzt sein; **literary ~:** Literaturkritik, *die*
criticize [ˈkrɪtɪsaɪz] *v.t.* kritisieren **(for** wegen**); ~ sb. for sth.** jmdn. wegen etw. kritisieren
critique [krɪˈtiːk] *n.* Kritik, *die*
croak [krəʊk] **1.** *n. (of frog)* Quaken, *das; (of raven, person)* Krächzen, *das.* **2.** *v.i.* ‹*Frosch:*› quaken; ‹*Rabe, Person:*› krächzen. **3.** *v.t.* krächzen
crochet [ˈkrəʊʃeɪ, ˈkrəʊʃɪ] **1.** *n.* Häkelarbeit, *die;* **~ hook** Häkelhaken, *der.* **2.** *v.t., p.t. and p.p.* **~ed** [ˈkrəʊʃeɪd, ˈkrəʊʃɪd] häkeln
[1]**crock** [krɒk] *n. (pot)* Topf, *der (aus Ton)*
[2]**crock** *(coll.) n.* **a)** *(person)* Wrack, *das (fig.);* **b)** *(vehicle)* [Klapper]kiste, *die (ugs.)*
crockery [ˈkrɒkərɪ] *n.* Geschirr, *das*
crocodile [ˈkrɒkədaɪl] *n.* Krokodil, *das; (skin)* Krokodilleder, *das*
ˈ**crocodile tears** *n. pl.* Krokodilstränen *Pl. (ugs.)*
crocus [ˈkrəʊkəs] *n.* Krokus, *der*
croft [krɒft] *n. (Brit.)* **a)** [kleines] Stück Acker-/Weideland; **b)** *(smallholding)* [kleines] Pachtgut
croissant [ˈkrwɑːsɑ̃] *n.* Hörnchen, *das*
crony [ˈkrəʊnɪ] *n.* Kumpel, *der; (female)* Freundin, *die;* **they were old cronies** sie waren gute, alte Freunde
crook [krʊk] **1.** *n.* **a)** *(coll.: rogue)* Gauner, *der;* **b)** *(staff)* Hirtenstab, *der; (of bishop)* [Krumm]stab, *der;* **c)** *(hook)* Haken, *der;* **d)** *(of arm)* [Arm]beuge, *die.* **2.** *v.t.* biegen; **~ one's finger** seinen Finger krümmen
crooked [ˈkrʊkɪd] *adj.* krumm; schief ‹*Lächeln*›; *(fig.: dishonest)* betrügerisch; **the picture is ~:** das Bild hängt schief; **you've got your hat on ~:** dein Hut sitzt schief; **a ~ person** *(fig.)* ein Gauner; **~ dealings** krumme Geschäfte
croon [kruːn] *v.t. & i.* [leise] singen; ‹*Popsänger:*› schmachtend singen, schnulzen *(ugs. abwertend)*
crooner [ˈkruːnə(r)] *n.* Schnulzensänger, *der (ugs. abwertend)*
crop [krɒp] **1.** *n.* **a)** *(Agric.)* [Feld]frucht, *die; (season's total yield)* Ernte, *die; (fig.)* [An]zahl, *die;* **~ of apples** Apfelernte, *die;* **b)** *(of bird)* Kropf, *der;* **c)** *(of whip)* [Peitschen]stiel, *der;* **d)** *(of hair)* kurzer Haarschnitt; *(style)* Kurzhaarfrisur, *die.* **2.** *v.t.,* **-pp-** *(cut off)* abschneiden; *(cut short)* stutzen ‹*Bart, Haare, Hecken, Flügel*›; ‹*Tier:*› abweiden ‹*Gras*›
~ ˈ**up** *v.i.* **a)** *(occur)* auftauchen; *(be mentioned)* erwähnt werden
cropper [ˈkrɒpə(r)] *n. (coll.)* **come a ~:** einen Sturz bauen *(ugs.); (fig.)* auf die Nase fallen *(ugs.)*
croquet [ˈkrəʊkeɪ, ˈkrəʊkɪ] *n.* Krocket[spiel], *das*
croquette [krəˈket] *n. (Cookery)* Krokette, *die*

cross [krɒs] **1.** *n.* **a)** Kreuz, *das; (monument)* [Gedenk]kreuz, *das; (sign)* Kreuzzeichen, *das;* **the C~:** das Kreuz [Christi]; **b)** *(~-shaped thing or mark)* Kreuz[zeichen], *das;* **c)** *(mixture, compromise)* Mittelding, *das* (**between** zwischen + *Dat.*); Mischung, *die* (**between** aus); **d)** *(affliction, cause of trouble)* Kreuz, *das;* **e)** *(intermixture of breeds)* Kreuzung, *die;* **f)** *(Footb.)* Querpaß, *der; (Boxing)* Cross, *der.* **2.** *v. t.* **a)** [über]kreuzen; **~ one's arms/legs** die Arme verschränken/die Beine übereinanderschlagen; **~ one's fingers** *or* **keep one's fingers ~ed [for sb.]** *(fig.)* [jmdm.] die *od.* den Daumen drücken/halten; **I got a ~ed line** *(Teleph.)* es war jemand in der Leitung; **b)** *(go across)* kreuzen; überqueren ‹*Straße, Gewässer, Gebirge*›; durchqueren ‹*Land, Wüste, Zimmer*›; **~ the road** über die Straße gehen; **we can ~** *abs.* die Straße ist frei; **the bridge ~es the river** die Brücke führt über den Fluß; **~ sb.'s mind** *(fig.)* jmdm. einfallen; **~ sb.'s path** *(fig.)* jmdm. über den Weg laufen *(ugs.)*; **c)** *(Brit.)* **~ a cheque** einen Scheck zur Verrechnung ausstellen; **a ~ed cheque** ein Verrechnungsscheck; **d)** *(make sign of ~ on)* **~ oneself** sich bekreuzigen; **e)** *(cause to interbreed)* kreuzen; *(~-fertilize)* kreuzbefruchten. **3.** *v. i.* *(meet and pass)* aneinander vorbeigehen; **~ [in the post]** ‹*Briefe:*› sich kreuzen. **4.** *adj.* **a)** *(transverse)* Quer-; **~ traffic** kreuzender Verkehr; **b)** *(coll.: peevish)* verärgert; ärgerlich ‹*Worte*›; **sb. will be ~:** jmd. wird ärgerlich *od.* böse werden; **be ~ with sb.** böse auf jmdn. *od.* mit jmdm. sein

~ 'off *v. t.* streichen; **~ a name off a list** einen Namen von einer Liste streichen

~ 'out *v. t.* ausstreichen

~ 'over *v. t.* überqueren; *abs.* hinübergehen

cross: ~bar *n.* **a)** [Fahrrad]stange, *die;* **b)** *(Sport)* Querlatte, *die;* **~-bencher** ['krɒsbentʃə(r)] *n.* Abgeordnete, *der/die weder der Regierungspartei noch der Opposition angehört;* **~bow** ['krɒsbəʊ] *n.* Armbrust, *die;* **~-breed 1.** *n.* Hybride, *die; (animal)* Bastard, *der;* **2.** *v. t.* kreuzen; **~-Channel** *adj.* **~-Channel traffic/ferry** Verkehr/Fähre über den Kanal; **~-check 1.** *n.* Gegenprobe, *die;* **2.** *v. t.* [nochmals] nachprüfen; nachkontrollieren; **~-country 1.** *adj.* Querfeldein-; **~-country running** Crosslauf, *der;* **2.** *adv.* querfeldein; **~-examination** *n.* Kreuzverhör, *das;* **~-examine** *v. t.* ins Kreuzverhör nehmen; **~-eyed** ['krɒsaɪd] *adj.* [nach innen] schielend; **be ~-eyed** schielen; **~-'fertilize** *v. t.* fremdbestäuben; **~-fire** *n.* *(lit. or fig.)* Kreuzfeuer, *das*

crossing ['krɒsɪŋ] *n.* **a)** *(act of going across)* Überquerung, *die;* **b)** *(road or rail intersection)* Kreuzung, *die;* **c)** *(pedestrian ~)* Überweg, *der*

cross-legged ['krɒslegd] *adv.* mit gekreuzten Beinen; *(with feet across thighs)* im Schneidersitz

crossly ['krɒslɪ] *adv. (coll.)* verärgert

cross: ~-patch *n.* Griesgram, *der;* Miesepeter, *der;* **~ 'purposes** *n. pl.* **talk at ~ purposes** aneinander vorbeireden; **~-'reference 1.** *n.* Querverweis, *der;* **2.** *v. t.* verweisen ‹*Person, Stichwort*› (**to** auf + *Akk.*); mit Querverweisen versehen ‹*Eintrag, Werk*›; **~roads** *n. sing.* Kreuzung, *die; (fig.)* Wendepunkt, *der;* **be at a/the ~roads** *(fig.)* am Scheideweg stehen; **~-section** *n.* Querschnitt, *der; (fig.)* repräsentative Auswahl; **a ~-section of the population** ein Querschnitt durch die Bevölkerung; **~-wind** *n.* Seitenwind, *der;* **~word** *n.* **~word [puzzle]** Kreuzworträtsel, *das*

crotchet ['krɒtʃɪt] *n. (Brit. Mus.)* Viertelnote, *die*

crouch [kraʊtʃ] *v. i.* [sich zusammen]kauern; **~ down** sich niederkauern; ‹*Person:*› sich hinhocken

croupier ['kru:pɪə(r), 'kru:pɪeɪ] *n.* Croupier, *der*

crow [krəʊ] **1.** *n.* **a)** *(bird)* Krähe, *die;* **as the ~ flies** Luftlinie; **b)** *(cry of cock)* Krähen, *das.* **2.** *v. i.* **a)** ‹*Hahn:*› krähen; **b)** *(exult)* **~ over** [hämisch] frohlocken über (+ *Akk.*)

'crowbar *n.* Brechstange, *die*

crowd [kraʊd] **1.** *n.* **a)** *(large number of persons)* Menschenmenge, *die;* **~[s] of people** Menschenmassen *Pl.*; **stand out from the ~:** aus der Menge herausragen; **b)** *(mass of spectators, audience)* Zuschauermenge, *die;* **c)** *(multitude)* breite Masse; **follow the ~** *(fig.)* mit der Herde laufen; **d)** *(coll.: company, set)* Clique, *die;* **e)** *(large number of things)* Menge, *die.* **2.** *v. t.* **a)** *(fill, occupy, cram)* füllen; **~ sth. with sth.** etw. mit etw. vollstopfen; **the streets were ~ed with people** die Straßen waren voll mit Leuten; **b)** *(fig.: fill)* ausfüllen. **3.** *v. i.* *(collect)* sich sammeln; **~ around sb./sth.** sich um jmdn./etw. drängen *od.* scharen

~ 'out *v. t.* herausdrängen

crowded ['kraʊdɪd] *adj.* überfüllt; voll ‹*Programm*›; ereignisreich ‹*Tag, Leben, Karriere*›

crown [kraʊn] **1.** *n.* **a)** Krone, *die;* **the C~:** die Krone; **b)** *(of head)* Scheitel, *der; (of tree, tooth)* Krone, *die; (of hat)* Kopfteil, *das; (thing that forms the summit)* Gipfel, *der.* **2.** *v. t.* **a)** krönen; **~ sb. king/queen** jmdn. zum König/zur Königin krönen; **b)** *(put finishing touch to)* krönen; **to ~ [it] all** zur Krönung des Ganzen; *(to make things even worse)* um das Maß vollzumachen; **c)** *(Dent.)* überkronen

Crown 'Court *n. (Brit. Law)* Krongericht, *das*

crowning ['kraʊnɪŋ] **1.** *n.* Krönung, *die.* **2.** *adj.* krönend

crown: ~ 'jewels *n. pl.* Kronjuwelen; **C~ 'prince** *n. (lit. or fig.)* Kronprinz, *der*

'crow's-nest *n. (Naut.)* Krähennest, *das;* Mastkorb, *der*

crucial ['kru:ʃl] *adj.* entscheidend (**to** für)

crucifix ['kru:sɪfɪks] *n.* Kruzifix, *das*

crucifixion [kru:sɪ'fɪkʃn] *n.* Kreuzigung, *die*

crucify ['kru:sɪfaɪ] *v. t.* kreuzigen

crude [kru:d] *adj.* **a)** *(in natural or raw state)* roh; Roh-; **~ oil/ore** Rohöl, *das*/Roherz, *das;* **b)** *(fig.: rough, unpolished)* primitiv; simpel; grob ‹*Entwurf, Skizze*›; **c)** *(rude, blunt)* ungehobelt, ungeschliffen ‹*Person, Benehmen*›; grob, derb ‹*Worte*›; ordinär ‹*Witz*›

crudeness ['kru:dnɪs] *n., no pl.* **a)** *(roughness)* Primitivität, *die; (of theory, design, plan)* Skizzenhaftigkeit, *die;* **b)** *(rudeness, bluntness) (of person, behaviour, manners)* Ungeschliffenheit, *die; (of words)* Derbheit, *die; (of joke)* Geschmacklosigkeit, *die*

crudity ['kru:dɪtɪ] *n.* **a)** *no pl. see* **crudeness; b)** *(crude remark)* Grobheit, *die*

cruel ['kru:əl] *adj., (Brit.)* **-ll-** grausam; **be ~ to animals** ein Tierquäler sein; **be ~ to be kind** in jmds. Interesse unbarmherzig sein müssen

cruelty ['kru:əltɪ] *n.* Grausamkeit, *die;* **~ to animals** Tierquälerei, *die;* **~ to children** Kindesmißhandlung, *die*

cruise [kru:z] **1.** *v.i.* **a)** *(sail for pleasure)* eine Kreuzfahrt machen; **b)** *(at random)* ‹*Fahrzeug, Fahrer:*› herumfahren; **c)** *(at economical speed)* ‹*Fahrzeug:*› mit Dauergeschwindigkeit fahren; ‹*Flugzeug:*› mit Reisegeschwindigkeit fliegen; **cruising speed** Reisegeschwindigkeit, *die*. **2.** *n.* Kreuzfahrt, *die;* **go on** *or* **for a ~:** eine Kreuzfahrt machen

ˈ**cruise missile** *n.* Marschflugkörper, *der*

cruiser [ˈkru:zə(r)] *n.* Kreuzer, *der*

crumb [krʌm] *n.* Krümel, *der;* Brösel, *der; (fig.)* Brocken, *der;* **~[s] of comfort** kleiner Trost

crumble [ˈkrʌmbl] **1.** *v.t.* zerbröckeln ‹*Brot*›; zerkrümeln ‹*Keks, Kuchen*›. **2.** *v.i.* ‹*Brot, Kuchen:*› krümeln; ‹*Gestein:*› [zer]bröckeln; ‹*Mauer:*› zusammenfallen. **3.** *n. (Cookery) mit Streuseln bestreutes und überbackenes [Apfel-, Rhabarber- usw.]dessert*

crumbly [ˈkrʌmblɪ] *adj.* krümelig ‹*Keks, Kuchen, Brot*›; bröckelig ‹*Gestein, Erde*›

crumpet [ˈkrʌmpɪt] *n. weiches Hefeküchlein zum Toasten*

crumple [ˈkrʌmpl] **1.** *v.t.* **a)** *(crush)* zerdrücken; zerquetschen; **b)** *(ruffle, wrinkle)* zerknittern ‹*Kleider, Papier, Stoff*›; **~ [up] a piece of paper** ein Stück Papier zerknüllen. **2.** *v.i.* ‹*Kleider, Stoff, Papier:*› knittern

crunch [krʌntʃ] **1.** *v.t.* [geräuschvoll] knabbern ‹*Keks, Zwieback*›. **2.** *v.i.* ‹*Schnee, Kies:*› knirschen; ‹*Eis:*› [zer]splittern; **the wheels ~ed on the gravel** der Kies knirschte unter den Rädern. **3.** *n. (~ing noise)* Knirschen, *das;* **when it comes to the ~:** wenn es hart auf hart geht

crunchy [ˈkrʌntʃɪ] *adj.* knusprig ‹*Gebäck, Nüsse*›; knackig ‹*Apfel*›

crusade [kru:ˈseɪd] *n. (Hist.; also fig.)* Kreuzzug, *der*

crusader [kru:ˈseɪdə(r)] *n. (Hist.)* Kreuzfahrer, *der*

crush [krʌʃ] **1.** *v.t.* **a)** *(compress with violence)* quetschen; auspressen ‹*Trauben, Obst*›; *(kill, destroy)* zerquetschen; zermalmen; **b)** *(reduce to powder)* zerstampfen; zermahlen; zerstoßen ‹*Gewürze, Tabletten*›; **c)** *(fig.: subdue, overwhelm)* niederwerfen, niederschlagen ‹*Aufstand*›; vernichten ‹*Feind*›; zunichte machen ‹*Hoffnungen*›; **d)** *(crumple, crease)* zerknittern ‹*Kleid, Stoff*›; zerdrücken, verbeulen ‹*Hut*›. **2.** *n.* **a)** *(crowded mass)* Gedränge, *das;* **b)** *(sl.: infatuation)* Schwärmerei, *die;* **have a ~ on sb.** in jmdn. verknallt sein *(ugs.)*

crust [krʌst] *n.* **a)** *(of bread)* Kruste, *die;* **b)** *(hard surface)* Kruste, *die;* **the earth's ~:** die Erdkruste; **c)** *(of pie)* Teigdeckel, *der*

crustacean [krʌˈsteɪʃn] *n.* Krusten- *od.* Krebstier, *das*

crusty [ˈkrʌstɪ] *adj.* **a)** *(crisp)* knusprig; **b)** *(irritable)* barsch

crutch [krʌtʃ] *n. (lit. or fig.)* Krücke, *die*

crux [krʌks] *n., pl.* **~es** *or* **cruces** [ˈkru:si:z] *(decisive point)* Kern[punkt], *der;* **the ~ of the matter** der springende Punkt bei der Sache

cry [kraɪ] **1.** *n.* **a)** *(of grief)* Schrei, *der; (of words)* Schreien, *das;* Geschrei, *das; (of hounds or wolves)* Heulen, *das;* **a ~ of pain/rage** ein Schmerzens-/Wutschrei; **a far ~ from ...** *(fig.)* etwas ganz anderes als ...; **b)** *(appeal, entreaty)* Appell, *der;* **a ~ for help** ein Hilferuf; **c)** *(fit or spell of weeping)* **have a good ~:** sich ausweinen. **2.** *v.t.* **a)** rufen; *(loudly)* schreien; **b)** *(weep)* weinen; **~ one's eyes out** sich *(Dat.)* die Augen ausweinen; **~ oneself to sleep** sich in den Schlaf weinen. **3.** *v.i.* **a)** rufen; *(loudly)* schreien; **~ [out] for sth./sb.** nach etw./jmdm. rufen *od.* schreien; **~ with pain** vor Schmerz[en] schreien; **b)** *(weep)* weinen **(over** wegen**)**; **~ for sth.** nach etw. weinen; *(fig.)* einer Sache *(Dat.)* nachweinen; **c)** ‹*Möwe:*› schreien

~ ˈoff *v.i.* absagen; einen Rückzieher machen *(ugs.)*

~ ˈout *v.i.* aufschreien; *see also* **~ 3 a**

crying [ˈkraɪɪŋ] *attrib. adj.* weinend ‹*Kind*›; schreiend ‹*Unrecht*›; dringend ‹*Bedürfnis, Notwendigkeit*›; **it is a ~ shame** es ist eine wahre Schande

crypt [krɪpt] *n.* Krypta, *die*

cryptic [ˈkrɪptɪk] *adj.* **a)** *(secret, mystical)* geheimnisvoll; **b)** *(obscure in meaning)* undurchschaubar; kryptisch

crystal [ˈkrɪstl] **1.** *n.* **a)** *(Chem., Min., etc.)* Kristall, *der;* **b)** *see* **crystal glass. 2.** *adj. (made of ~ glass)* kristallen

crystal: **~ ˈball** *n.* Kristallkugel, *die;* **~-clear** *adj.* kristallklar; kristallen *(geh.); (fig.)* glasklar; **~-gazing** *n.* Hellseherei, *die;* **~ ˈglass** *n.* Bleikristall, *das;* Kristallglas, *das*

crystallisation, crystallise *see* **crystalliz-**

crystallization [krɪstəlaɪˈzeɪʃn] *n.* Kristallbildung, *die;* Kristallisation, *die*

crystallize [ˈkrɪstəlaɪz] **1.** *v.t.* auskristallisieren ‹*Salze*›; kandieren ‹*Früchte*›. **2.** *v.i.* kristallisieren; *(fig.)* feste Form annehmen

cub [kʌb] *n.* **a)** Junge, *das; (of wolf, fox, dog)* Welpe, *der;* Junge, *das;* **b) Cub** *see* **Cub Scout**

Cuba [ˈkju:bə] *n.* Kuba *(das)*

Cuban [ˈkju:bn] **1.** *adj.* kubanisch; **sb. is ~:** jmd. ist Kubaner/Kubanerin. **2.** *n.* Kubaner, *der*/Kubanerin, *die*

cubby[-hole] [ˈkʌbɪ(-həʊl)] *n.* Kämmerchen, *das; (snug place)* Kuschelecke, *die*

cube [kju:b] *n.* **a)** Würfel, *der;* Kubus, *der (fachspr.);* **b)** *(Math.)* dritte Potenz

cube ˈroot *n.* Kubikwurzel, *die*

cubic [ˈkju:bɪk] *adj.* **a)** würfelförmig; **b)** *(of three dimensions)* Kubik-; **~ metre/centimetre/foot/yard** Kubikmeter / -zentimeter/ -fuß/-yard, *der*

cubicle [ˈkju:bɪkl] *n.* **a)** *(sleeping-compartment)* Bettnische, *die;* **b)** *(for dressing, private discussion, etc.)* Kabine, *die*

ˈ**Cub Scout** *n.* Wölfling, *der*

cuckoo [ˈkʊku:] **1.** *n.* Kuckuck, *der.* **2.** *adj. (sl.)* meschugge *nicht attr. (salopp)*

ˈ**cuckoo clock** *n.* Kuckucksuhr, *die*

cucumber [ˈkju:kʌmbə(r)] *n.* [Salat]gurke, *die;* **be as cool as a ~** *(fig.)* einen kühlen Kopf behalten

cud [kʌd] *n.* wiedergekäutes Futter; **chew the ~:** wiederkäuen

cuddle [ˈkʌdl] **1.** *n.* Liebkosung, *die;* **give sb. a ~:** jmdn. drücken *od.* in den Arm nehmen; **have a ~:** schmusen. **2.** *v.t.* schmusen mit; hätscheln ‹*kleines Kind*›. **3.** *v.i.* schmusen

cuddly [ˈkʌdlɪ] *adj. (given to cuddling)* verschmust

cuddly ˈtoy *n.* Kuscheltier, *das*

cudgel [ˈkʌdʒl] **1.** *n.* Knüppel, *der;* **take up the ~s for sb./sth.** *(fig.)* [energisch] für jmdn./etw. eintreten. **2.** *v.t., (Brit.)* **-ll-** knüppeln

[1]**cue** [kju:] *n. (Billiards etc.)* Queue, *das;* Billardstock, *der*

[2]**cue** *n.* **a)** *(Theatre)* Stichwort, *das; (Music)* Stichnoten *Pl.; (Cinemat., Broadcasting)* Zeichen zum Aufnahmebeginn; **be/speak/play on ~:** rechtzeitig einsetzen; **b)** *(sign when or how to act)* Wink, *der;* **take one's ~ from sb.** *(lit. or fig.)* sich nach jmdm. richten

[1]**cuff** [kʌf] *n.* **a)** Manschette, *die;* **off the ~** *(fig.)* aus dem Stegreif; **b)** *(Amer.: trouser turn-up)* [Hosen]aufschlag, *der*

[2]**cuff** *v.t.* **~ sb.'s ears, ~ sb. over the ears** jmdm. eins hinter die Ohren geben *(ugs.);* **~ sb.** jmdm. einen Klaps geben

'**cuff-link** *n.* Manschettenknopf, *der*

cuisine [kwɪ'zi:n] *n.* Küche, *die*

cul-de-sac ['kʌldəsæk] *n., pl.* **culs-de-sac** ['kʌldəsæk] Sackgasse, *die*

culinary ['kʌlɪnərɪ] *adj.* kulinarisch

cull [kʌl] *v.t.* erlegen; *(shoot)* abschießen

culminate ['kʌlmɪneɪt] *v.i.* gipfeln; kulminieren; **~ in sth.** in etw. *(Dat.)* seinen Höchststand erreichen

culmination [kʌlmɪ'neɪʃn] *n.* Höhepunkt, *der*

culottes [kju:'lɒt] *n. pl.* Hosenrock, *der*

culpable ['kʌlpəbl] *adj.* schuldig ‹*Person*›; strafbar ‹*Handlung*›; **~ negligence** grobe Fahrlässigkeit

culprit ['kʌlprɪt] *n.* *(guilty of crime)* Schuldige, *der/die;* Täter, *der*/Täterin, *die; (guilty of wrong)* Übeltäter, *der*/-täterin, *die*

cult [kʌlt] *n.* Kult, *der; attrib.* Kult‹*film, -figur usw.*›

cultivate ['kʌltɪveɪt] *v.t.* **a)** *(for crops)* kultivieren; bestellen, bebauen ‹*Feld, Land*›; **b)** anbauen, züchten ‹*Pflanzen*›; **c)** *(fig.)* kultivieren, entwickeln ‹*Geschmack*›; kultivieren ‹*Freundschaft, Gefühl, Gewohnheit*›; entwickeln ‹*Kunst, Fertigkeit*›

cultivation [kʌltɪ'veɪʃn] *n.* *(lit. or fig.)* Kultivierung, *die; (of a skill)* Entwicklung, *die;* **~ of land** Landbau, *der;* Pflanzenbau, *der*

cultural ['kʌltʃərl] *adj.* kulturell ‹*Entwicklung, Ereignis, Interessen, Beziehungen*›; **~ revolution/anthropology** Kulturrevolution/-anthropologie, *die*

culture ['kʌltʃə(r)] *n.* **a)** Kultur, *die;* **b)** *(Agric., of bacteria)* Kultur, *die; (tillage of the soil)* Landbau, *der; (rearing, production)* Zucht, *die*

cultured ['kʌltʃəd] *adj.* **a)** *(cultivated, refined)* kultiviert; gebildet; **b)** **~ pearl** Zuchtperle, *die*

cumbersome ['kʌmbəsəm] *adj.* lästig, hinderlich ‹*Kleider*›; sperrig ‹*Gepäck, Pakete*›; schwerfällig ‹*Bewegung, Stil, Arbeitsweise*›

cumulate ['kju:mjʊleɪt], **cumulation** [kju:mjʊ'leɪʃn] *see* **accumul-**

cumulative ['kju:mjʊlətɪv] *adj.* **a)** *(increased by successive additions)* kumulativ *(geh.);* **~ strength/effect** Gesamtstärke/-wirkung, *die;* **~ evidence** Häufung von Beweismaterial; **b)** *(formed by successive additions)* zusätzlich; Zusatz-

cunning ['kʌnɪŋ] **1.** *n.* Schläue, *die;* Gerissenheit, *die.* **2.** *adj.* schlau; gerissen

cup [kʌp] **1.** *n.* **a)** Tasse, *die;* **b)** *(prize, competition)* Pokal, *der;* **c)** *(~ful)* Tasse, *die;* **a ~ of coffee/tea** eine Tasse Kaffee/Tee; **it's [not] my ~ of tea** *(fig. coll.)* das ist [nicht] mein Fall *(ugs.);* **d)** *(of brassière)* Körbchen, *das.* **2.** *v.t.*, **-pp-: a) ~ one's chin in one's hand** das Kinn in die Hand stützen; **b)** *(make ~-shaped)* hohl machen; **~ one's hand to one's ear** die Hand ans Ohr halten

cupboard ['kʌbəd] *n.* Schrank, *der*

'**cupboard love** *n.* geheuchelte Zuneigung

Cup 'Final *n. (Footb.)* Pokalendspiel, *das*

cupful ['kʌpfʊl] *n.* Tasse, *die;* **a ~ of water** eine Tasse Wasser

cupola ['kju:pələ] *n.* Kuppel, *die*

curable ['kjʊərəbl] *adj.* heilbar

curate ['kjʊərət] *n. (Eccl.)* Kurat, *der;* Hilfsgeistliche, *der*

curator [kjʊə'reɪtə(r)] *n.* *(of museum)* Direktor, *der*/Direktorin, *die*

curb [kɜ:b] **1.** *v.t. (lit. or fig.)* zügeln. **2.** *n.* **a)** *(chain or strap for horse)* Kandare, *die;* **b)** *see* **kerb**

curdle ['kɜ:dl] **1.** *v.t. (lit. or fig.)* gerinnen lassen. **2.** *v.i. (lit. or fig.)* gerinnen

curds [kɜ:dz] *n. pl.* ≈ Quark, *der*

cure [kjʊə(r)] **1.** *n.* **a)** *(thing that ~s)* [Heil]mittel, *das* **(for** gegen); *(fig.)* Mittel, *das;* **b)** *(restoration to health)* Heilung, *die;* **c)** *(treatment)* Behandlung, *die;* **take a ~ at a spa** in *od.* zur Kur gehen. **2.** *v.t.* **a)** heilen; kurieren; **~ sb. of a disease** jmdn. von einer Krankheit heilen; **b)** *(fig.)* kurieren; **c)** *(preserve)* [ein]pökeln ‹*Fleisch*›; räuchern ‹*Fisch*›; trocknen ‹*Häute, Tabak*›

curfew ['kɜ:fju:] *n.* Ausgangssperre, *die*

curio ['kjʊərɪəʊ] *n., pl.* **~s** Kuriosität, *die*

curiosity [kjʊərɪ'ɒsɪtɪ] *n.* **a)** *(desire to know)* Neugier[de], *die* **(about** in bezug auf + *Akk.*); **~ killed the cat** *(fig.)* die Neugier ist schon manchem zum Verhängnis geworden; **b)** *(strange or rare object)* Wunderding, *das;* Rarität, *die; (strange matter)* Kuriosität, *die*

curious ['kjʊərɪəs] *adj.* **a)** *(inquisitive)* neugierig; *(eager to learn)* wißbegierig; **be ~ about sth.** *(eagerly awaiting)* auf etw. *(Akk.)* neugierig sein; **be ~ about sb.** in bezug auf jmdn. neugierig sein; **be ~ to know sth.** etw. gern wissen wollen; **b)** *(strange, odd)* merkwürdig; seltsam; **how [very] ~!** [sehr] seltsam!

curiously ['kjʊərɪəslɪ] *adv.* neugierig ‹*fragen, gucken*›; seltsam, merkwürdig ‹*sprechen, sich verhalten*›; **~ [enough]** *as sentence-modifier* merkwürdigerweise; seltsamerweise

curl [kɜ:l] **1.** *n.* **a)** *(of hair)* Locke, *die;* **b)** *(sth. spiral or curved inwards)* **the ~ of a leaf/wave** ein gekräuseltes Blatt/eine gekräuselte Welle; **a ~ of smoke** ein Rauchkringel. **2.** *v.t.* **a)** *(cause to form coils)* locken; *(tightly)* kräuseln; **she ~ed her hair** sie legte ihr Haar in Locken *(Akk.);* **b)** *(bend, twist)* kräuseln ‹*Blätter, Lippen*›. **3.** *v.i.* **a)** *(grow in coils)* sich locken; *(tightly)* sich kräuseln; **b)** ‹*Straße, Fluß:*› sich winden, sich schlängeln

~ 'up 1. *v.t.* hochbiegen; **~ oneself up** sich zusammenrollen. **2.** *v.i.* sich zusammenrollen; **~ up with a book** es sich *(Dat.)* mit einem Buch gemütlich machen

curler ['kɜ:lə(r)] *n.* Lockenwickler, *der;* **in ~s** mit Lockenwicklern

curlew ['kɜ:lju:] *n. (Ornith.)* Brachvogel, *der*

curling ['kɜ:lɪŋ] *n.* Curling, *das;* ≈ Eisschießen, *das*

curly ['kɜ:lɪ] *adj.* lockig, *(tightly)* kraus ‹*Haar*›

'**curly-haired** ['kɜ:lɪheəd] *adj.* lockenköpfig

currant ['kʌrənt] *n.* **a)** *(dried fruit)* Korinthe, *die;* **b)** *(fruit)* Johannisbeere, *die*

currency ['kʌrənsɪ] *n.* **a)** *(money)* Währung, *die;* **foreign currencies** Devisen *Pl.;* **b)** *(other commodity)* Zahlungsmittel, *das;* **c)** *(prevalence) (of word, idea, story, rumour)* Verbreitung, *die; (of expression)* Gebräuchlichkeit, *die*

current ['kʌrənt] **1.** *adj.* **a)** *(in general circulation or use)* kursierend, umlaufend ‹*Geld, Geschichte, Gerücht*›; verbreitet ‹*Meinung*›; gebräuchlich ‹*Wort*›; gängig ‹*Redensart*›; **b)** laufend ‹*Jahr, Monat*›; **in the ~ year** in diesem Jahr; **c)** *(belonging to the present time)* aktuell ‹*Ereignis,*

Mode›; Tages‹*politik, -preis*›; gegenwärtig ‹*Krise, Aufregung*›; ~ **issue/edition** letzte Ausgabe/neueste Auflage; ~ **affairs** Tagespolitik, *die;* aktuelle Fragen. **2.** *n.* **a)** *(of water, air)* Strömung, *die;* **air/ocean** ~: Luft-/Meeresströmung, *die;* **swim against/with the** ~: gegen den/mit dem Strom schwimmen; **b)** *(Electr.)* Strom, *der; (intensity)* Stromstärke, *die;* **c)** *(running stream)* Strömung, *die;* **d)** *(tendency of events, opinions, etc.)* Tendenz, *die;* Trend, *der*

'**current account** *n.* Girokonto, *das*

currently ['kʌrəntlɪ] *adv.* gegenwärtig; momentan; **he is ~ writing a book** er schreibt gerade *od.* zur Zeit an einem Buch

curriculum [kə'rɪkjʊləm] *n., pl.* **curricula** [kə'rɪkjʊlə] Lehrplan, *der*

curriculum vitae [kərɪkjʊləm 'vi:taɪ] *n.* Lebenslauf, *der*

[1]**curry** ['kʌrɪ] *(Cookery) n.* Curry[gericht], *das*

[2]**curry** *v.t.* ~ **favour [with sb.]** sich [bei jmdm.] einschmeicheln

'**curry-powder** *n.* Currypulver, *das*

curse [kɜ:s] **1.** *n.* **a)** Fluch, *der;* **be under a** ~: unter einem Fluch stehen; **b)** *(great evil)* Geißel, *die;* Plage, *die.* **2.** *v.t.* **a)** *(utter ~ against)* verfluchen; **b)** *(as oath)* ~ **it/you!** verflucht!; verdammt!; **c)** *(afflict)* strafen. **3.** *v.i.* fluchen (**at** über + *Akk.*)

cursed ['kɜ:sɪd] *adj.* verflucht

cursor ['kɜ:sə(r)] *n.* Läufer, *der; (on screen)* Cursor, *der*

cursory ['kɜ:sərɪ] *adj.* flüchtig ‹*Blick*›; oberflächlich ‹*Untersuchung, Bericht, Studium*›

curt [kɜ:t] *adj.* kurz und schroff ‹*Brief, Mitteilung*›; kurz angebunden ‹*Person, Art*›

curtail [kɜ:'teɪl] *v.t.* kürzen; abkürzen ‹*Urlaub*›; beschneiden ‹*Macht*›

curtain ['kɜ:tən] *n.* **a)** Vorhang, *der; (with net ~s)* Übergardine, *die;* **draw** *or* **pull the ~s** *(open)* die Vorhänge aufziehen; *(close)* die Vorhänge zuziehen; **draw** *or* **pull back the ~s** die Vorhänge aufziehen; **b)** *(fig.)* **a ~ of fog/mist** ein Nebelschleier; **a ~ of smoke/flames/rain** eine Rauch-/Flammen-/Regenwand; **c)** *(Theatre)* Vorhang, *der; (end of play)* Schlußszene, *die; (rise of ~ at start of play)* Aufgehen des Vorhanges; Aktbeginn, *der; (fall of ~ at end of scene)* Fallen des Vorhanges; Aktschluß, *der;* **the ~ rises/falls** der Vorhang hebt sich/fällt

~ '**off** *v.t.* mit einem Vorhang abteilen

curtain: ~-**call** *n.* Vorhang, *der;* ~ **rail** *n.* Gardinenstange, *die;* ~-**raiser** ['kɜ:tənreɪzə(r)] *n.* [kurzes] Vorspiel; *(fig.)* Auftakt, *der;* ~-**rod** *n.* Gardinenstange, *die;* ~-**track** *n.* Gardinenleiste, *die*

curtsy (curtsey) ['kɜ:tsɪ] **1.** *n.* Knicks, *der;* **make** *or* **drop a ~ to sb.** vor jmdm. einen Knicks machen. **2.** *v.i.* ~ **to sb.** vor jmdm. knicksen

curvaceous [kɜ:'veɪʃəs] *adj.* *(coll.)* kurvenreich *(ugs.);* **a ~ figure** eine üppige Figur

curvature ['kɜ:vətʃə(r)] *n.* Krümmung, *die*

curve [kɜ:v] **1.** *v.t.* krümmen. **2.** *v.i.* ‹*Straße, Fluß:*› *(once)* eine Biegung machen, *(repeatedly)* sich winden; ‹*Horizont:*› sich krümmen; ‹*Linie:*› einen Bogen machen; **the road ~s round the town** die Straße macht einen Bogen um die Stadt. **3.** *n.* **a)** Kurve, *die;* **b)** *(surface; curved form or thing)* Rundung, *die*

curved [kɜ:vd] *adj.* krumm; gebogen; gekrümmt ‹*Raum, Linie*›

cushion ['kʊʃn] **1.** *n.* **a)** Kissen, *das;* **b)** *(for protection)* Kissen, *das;* Polster, *das.* **2.** *v.t.* **a)** [aus]polstern ‹*Stuhl*›; **b)** *(absorb)* dämpfen ‹*Aufprall, Stoß*›

cushy ['kʊʃɪ] *adj.* *(coll.)* bequem

cuss [kʌs] *(coll.)* **1.** *n.* Fluch, *der;* Beschimpfung, *die;* **sb. does not give** *or* **care a ~:** jmdm. ist es vollkommen schnuppe *(ugs.);* **he/it is not worth a tinker's ~:** er/es ist keinen Pfifferling *od.* roten Heller wert *(ugs.).* **2.** *v.i.* fluchen; schimpfen. **3.** *v.t.* verfluchen; beschimpfen

cussed ['kʌsɪd] *adj.* *(coll.: perverse, obstinate)* stur *(ugs.)*

cussedness ['kʌsɪdnɪs] *n., no pl.* Sturheit, *die;* **from sheer ~:** aus reiner Sturheit

custard ['kʌstəd] *n.* **a)** ~ **[pudding]** ≈ Vanillepudding, *der;* **b)** *(sauce)* ≈ Vanillesoße, *die*

custard: ~-'**pie** *n.* *(pie) Kuchen mit einer Füllung aus Vanillepudding; (in comedy)* Sahnetorte, *die;* ~ **powder** *n.* Vanillesoßenpulver, *das*

custodian [kʌs'təʊdɪən] *n.* *(of public building)* Wärter, *der*/Wärterin, *die; (of park, museum)* Wächter, *der*/Wächterin, *die; (of valuables, traditions, culture)* Hüter, *der*/Hüterin, *die*

custody ['kʌstədɪ] *n.* **a)** *(guardianship, care)* Obhut, *die;* **be in the ~ of sb.** unter jmds. Obhut *(Dat.)* stehen; **the mother was given [the] ~ of the children** die Kinder wurden der Mutter zugesprochen; **b)** *(imprisonment)* **[be] in ~:** in Haft [sein]; **take sb. into ~:** jmdn. verhaften *od.* festnehmen

custom ['kʌstəm] *n.* **a)** Brauch, *der;* Sitte, *die;* **it was his ~ to smoke a cigar after dinner** er pflegte nach dem Essen eine Zigarre zu rauchen; **b)** *in pl.* *(duty on imports)* Zoll, *der;* **[the] C~s** *(government department)* der Zoll; **c)** *(Law)* Gewohnheitsrecht, *das;* **d)** *(business patronage)* Kundschaft, *die (veralt.);* **we should like to have your ~:** wir hätten Sie gern zum/zur *od.* als Kunden/Kundin

customary ['kʌstəmərɪ] *adj.* üblich

'**custom-built** *adj.* spezial[an]gefertigt; ~ **clothes** *(Amer.)* maßgeschneiderte Kleidung

customer ['kʌstəmə(r)] *n.* **a)** Kunde, *der*/Kundin, *die; (of restaurant)* Gast, *der; (of theatre)* Besucher, *der*/Besucherin, *die;* **b)** *(coll.: person)* Kerl, *der (ugs.);* **a queer/an awkward ~:** ein schwieriger Kunde *(ugs.)*

'**custom-made** *adj.* spezial[an]gefertigt; maßgeschneidert ‹*Kleidung*›

customs: ~ **clearance** *n.* Zollabfertigung, *die;* ~ **declaration** *n.* Zollerklärung, *die;* ~ **duty** *n.* Zoll, *der;* ~ **inspection** *n.* Zollkontrolle, *die;* ~ **officer** *n.* Zollbeamter, *der*/-beamtin, *die*

cut [kʌt] **1.** *v.t.,* **-tt-, cut a)** *(penetrate, wound)* schneiden; ~ **one's finger/leg** sich *(Dat. od. Akk.)* in den Finger/ins Bein schneiden; **he ~ himself on broken glass** er hat sich an einer Glasscherbe geschnitten; **the remark ~ him to the quick** *(fig.)* die Bemerkung traf ihn ins Mark; **b)** *(divide) (with knife)* schneiden; durchschneiden ‹*Seil*›; *(with axe)* durchhacken; ~ **sth. in half/two/three** etw. halbieren/zweiteilen/dreiteilen; ~ **one's ties** *or* **links** alle Verbindungen abbrechen; ~ **no ice with sb.** *(fig. sl.)* keinen Eindruck auf jmdn. machen; **c)** *(detach, reduce)* abschneiden; schneiden, stutzen ‹*Hecke*›; mähen ‹*Getreide, Gras*›; ~ *(p.p.)* **flowers** Schnittblumen; ~ **one's nails** sich *(Dat.)* die Nägel schneiden; **d)** *(shape, fashion)* schleifen ‹*Glas, Edelstein, Kristall*›; hauen, schlagen ⟨*Stufen*⟩; ~ **a key** einen Schlüssel feilen *od.* anfertigen; ~ **figures in wood/stone** Figuren aus Holz schnitzen/aus Stein hauen; **e)** *(meet and cross)* ⟨*Straße, Linie, Kreis*:⟩ schneiden; **f)** *(fig.: renounce, refuse to recognize)* schneiden; **g)** *(carve)* [auf]schneiden ⟨*Fleisch, Geflügel*⟩; abschneiden ⟨*Scheibe*⟩; **h)** *(reduce)* senken ⟨*Preise*⟩; verringern, einschränken ⟨*Menge, Produktion*⟩; mindern ⟨*Qualität*⟩; kürzen ⟨*Ausgaben, Lohn*⟩; verkürzen ⟨*Arbeitszeit, Urlaub*⟩; *(cease, stop)* einstellen ⟨*Dienstleistungen, Lieferungen*⟩; abstellen ⟨*Strom*⟩; **i)** *(absent oneself from)* schwänzen ⟨*Schule, Unterricht*⟩; **j)** ~ **one's losses** höherem Verlust vorbeugen; **k)** ~ **sth. short** *(lit. or fig.: interrupt, terminate)* etwas abbrechen; ~ **sb. short** jmdn. unterbrechen; *(impatiently)* jmdm. ins Wort fallen; **to ~ a long story short** der langen Rede kurzer Sinn; **l)** *(Cards)* abheben; **m)** ~ **a tooth** einen Zahn bekommen; **n) be ~ and dried** genau festgelegt *od.* abgesprochen sein. **2.** *v.i.*, **-tt-, cut a)** ⟨*Messer, Schwert usw.*:⟩ schneiden; ⟨*Papier, Tuch, Käse*:⟩ sich schneiden lassen; ~ **both ways** *(fig.)* ein zweischneidiges Schwert sein *(fig.)*; **b)** *(cross, intersect)* sich schneiden; **c)** *(pass)* ~ **through** *or* **across the field/park** [quer] über das Feld/durch den Park gehen; **d)** *(Cinemat.) (stop the cameras)* abbrechen; *(go quickly to another shot)* überblenden (**to** zu). **3.** *n.* **a)** *(act of cutting)* Schnitt, *der;* **b)** *(stroke, blow) (with knife)* Schnitt, *der; (with sword, whip)* Hieb, *der; (injury)* Schnittwunde, *die;* **c)** *(reduction) (in wages, expenditure, budget)* Kürzung, *die; (in prices)* Senkung, *die; (in working hours, holiday, etc.)* Verkürzung, *die; (in services)* Verringerung, *die; (in production, output, etc.)* Einschränkung, *die;* **d)** *(of meat)* Stück, *das;* **e)** *(coll.: commission, share)* Anteil, *der;* **f)** *(of hair: style)* [Haar]schnitt, *der; (of clothes)* Schnitt, *der;* **g)** *(in play, book, etc.)* Streichung, *die; (in film)* Schnitt, *der;* **make ~s** Streichungen/Schnitte vornehmen

~ **a'way** *see* ~ **off a**

~ **'back 1.** *v.t.* **a)** *(reduce)* einschränken ⟨*Produktion*⟩; verringern ⟨*Investitionen*⟩; **b)** *(prune)* stutzen. **2.** *v.i. (reduce)* ~ **back on sth.** etw. einschränken

~ **'down 1.** *v.t.* **a)** *(fell)* fällen; **b)** *(kill)* töten; **c)** *(reduce)* einschränken; ~ **sb. down to size** *(fig.)* jmdn. auf seinen Platz verweisen. **2.** *v.i. (reduce)* ~ **down on sth.** etw. einschränken

~ **'in** *v.i.* **a)** *(come in abruptly, interpose)* sich einschalten; ~ **in on sb./sth.** jmdn./etw. unterbrechen; **b)** *(after overtaking)* schneiden; **c)** ⟨*Motor usw.*:⟩ sich einschalten

~ **'off** *v.t.* **a)** *(remove by ~ting)* abschneiden; abtrennen; *(with axe etc.)* abschlagen; **b)** *(interrupt, make unavailable)* abschneiden ⟨*Zufuhr*⟩; abstellen ⟨*Strom, Gas, Wasser*⟩; unterbrechen ⟨*Telefongespräch, Sprecher am Telefon*⟩; **c)** *(isolate)* abschneiden; **be ~ off by the snow/tide** durch den Schnee/die Flut abgeschnitten sein; **d)** *(prevent, block)* abschneiden; **their retreat was ~ off** ihnen wurde der Rückzug abgeschnitten; **e)** *(exclude from contact)* ~ **sb. off from the outside world** jmdn. von der Außenwelt abschneiden; ~ **oneself off** sich absondern

~ **'out 1.** *v.t.* **a)** *(remove by ~ting)* ausschneiden (**out of** aus); **b)** *(stop doing or using)* aufhören mit; ~ **out cigarettes/alcohol** aufhören, Zigaretten zu rauchen/Alkohol zu trinken; ~ **it** *or* **that out!** *(coll.)* hör/hört auf damit!; **c) be ~ out for sth.** für etw. geeignet sein; **he was not ~ out to be a teacher** er war nicht zum Lehrer gemacht. **2.** *v.i.* ⟨*Motor*:⟩ aussetzen; ⟨*Gerät*:⟩ sich abschalten

~ **'up** *v.t.* zerschneiden; in Stücke schneiden ⟨*Fleisch, Gemüse*⟩; **be ~ up about sth.** *(fig.)* zutiefst betroffen über etw. *(Akk.)* sein

'cut-back *n. (reduction)* Kürzung, *die*

cute [kju:t] *adj. (coll., esp. Amer.)* süß, niedlich ⟨*Kind, Mädchen*⟩; entzückend ⟨*Stadt, Haus*⟩

cut 'glass *n.* Kristall[glas], *das*

cuticle ['kju:tɪkl] *n.* Epidermis, *die (fachspr.)*; Oberhaut, *die; (of nail)* Nagelhaut, *die*

cutlery ['kʌtlərɪ] *n.* Besteck, *das*

cutlet ['kʌtlɪt] *n.* **a)** *(of mutton or lamb)* Kotelett, *das;* **b) veal ~:** Frikandeau, *das;* **c)** *(minced meat etc. in shape of ~)* Hacksteak, *das*

cut: ~-off *n.* Trennung, *die; attrib.* **~-off point** Trennungslinie, *die;* **~-price** *adj.* herabgesetzt; **~-price goods** Waren zu herabgesetzten Preisen; **~-rate** *adj.* verbilligt; herabgesetzt

cutter ['kʌtə(r)] *n.* **a)** *(person) (of cloth)* Zuschneider, *der/*-schneiderin, *die; (of films)* Cutter, *der/*Cutterin, *die;* **b)** *(machine)* Schneidmaschine, *die;* **c)** *(Naut.)* Kutter, *der*

'cutthroat 1. *n.* Strolch, *der; (murderer)* Killer, *der (ugs.).* **2.** *adj.* **a)** mörderisch, gnadenlos ⟨*Wettbewerb*⟩; **b)** ~ **razor** Rasiermesser, *das*

cutting ['kʌtɪŋ] **1.** *adj.* beißend ⟨*Bemerkung, Antwort*⟩; ~ **edge** Schneide, *die.* **2.** *n.* **a)** *(esp. Brit.: from newspaper)* Ausschnitt, *der;* **b)** *(esp. Brit.: excavation for railway, road etc.)* Einschnitt, *der;* **c)** *(of plant)* Ableger, *der*

cuttle[fish] ['kʌtl(fɪʃ)] *n.* Tintenfisch, *der;* Sepia, *die (fachspr.)*

c.v. *abbr.* **curriculum vitae**

cwt. *abbr.* **hundredweight** ≈ Ztr.

cyanide ['saɪənaɪd] *n.* Cyanid, *das*

cyclamen ['sɪkləmən] *n. (Bot.)* Alpenveilchen, *das*

cycle ['saɪkl] **1.** *n.* **a)** Zyklus, *der; (period of completion)* Turnus, *der;* ~ **per second** *(Phys., Electr.)* Schwingung pro Sekunde; **b)** *(bicycle)* Rad, *das.* **2.** *v.i.* radfahren; mit dem [Fahr]rad fahren; **go cycling** radfahren

'cycle-track *n.* Rad[fahr]weg, *der; (for racing)* Radrennbahn, *die*

cyclic ['saɪklɪk], **cyclical** ['saɪklɪkl] *adj.* zyklisch

cyclist ['saɪklɪst] *n.* Radfahrer, *der/*-fahrerin, *die*

cyclone ['saɪkləʊn] *n. (system of winds)* Tiefdruckgebiet, *das;* Zyklon, *die (fachspr.); (violent hurricane)* Zyklon, *der*

cygnet ['sɪgnɪt] *n.* junger Schwan

cylinder ['sɪlɪndə(r)] *n. (also Geom., Motor Veh.)* Zylinder, *der; (for compressed or liquefied gas)* Gasflasche, *die; (of diving apparatus)* [Sauerstoff]flasche, *die; (of typewriter, mower)* Walze, *die*

cylinder: ~ block *n.* Motorblock, *der;* **~ head** *n.* Zylinderkopf, *der*

cylindrical [sɪ'lɪndrɪkl] *adj.* zylindrisch

cymbal ['sɪmbl] *n. (Mus.)* Beckenteller, *der;* **~s** Becken *Pl.*

cynic ['sɪnɪk] *n.* Zyniker, *der*

cynical ['sɪnɪkl] *adj.* zynisch; bissig ⟨*Artikel, Bemerkung, Worte*⟩

cynicism ['sɪnɪsɪzm] *n.* Zynismus, *der*

cypher *see* **cipher**

cypress ['saɪprɪs] *n.* Zypresse, *die*

Cypriot ['sɪprɪət] **1.** *adj.* zyprisch; zypriotisch. **2.** *n.* Zypriot, *der/*Zypriotin, *die*

Cyprus ['saıprəs] *pr. n.* Zypern *(das)*
Czech [tʃek] **1.** *adj.* tschechisch. **2.** *n.* **a)** *(language)* Tschechisch, *das;* **b)** *(person)* Tscheche, *der*/ Tschechin, *die*
Czechoslovakia [tʃekəʊslə'vækıə] *n.* die Tschechoslowakei
Czechoslovakian [tʃekəʊslə'vækıən] **1.** *adj.* tschechoslowakisch. **2.** *pr. n.* Tschechoslowake, *der*/Tschechoslowakin, *die*

D

D, d [di:] *n., pl.* **Ds** *or* **D's a)** *(letter)* D, d, *das;* **b) D** *(Mus.)* D, d, *das;* **D sharp** dis, Dis, *das;* **D flat** des, Des, *das*
d. *abbr.* **a) died** gest.; **b)** *(Brit. Hist.)* **penny/pence** d.
DA *abbr. (Amer.)* **District Attorney**
[1]**dab** [dæb] **1.** *n.* Tupfer, *der.* **2.** *v.t.,* **-bb-** *(press with sponge etc.)* abtupfen; ~ **sth. on** *or* **against sth.** etw. auf etw. *(Akk.)* tupfen. **3.** *v.i.,* **-bb-:** ~ **at sth.** etw. ab- *od.* betupfen
[2]**dab** *(Brit. coll.: expert)* **1.** *n.* Könner, *der.* **2.** *adj.* geschickt; **be a ~ hand at cricket/making omelettes** ein As im Kricket/Eierkuchenbacken sein *(ugs.)*
dabble ['dæbl] **1.** *v.t. (wet slightly)* befeuchten; ~ **one's feet in the water** mit den Füßen im Wasser planschen. **2.** *v.i.* ~ **in/at sth.** sich in etw. *(Dat.)* versuchen
dachshund ['dækshʊnd] *n.* Dackel, *der*
dad [dæd] *n. (coll.)* Vater, *der*
daddy ['dædı] *n. (coll.)* Vati, *der (fam.);* Papa, *der (fam.)*
daddy-'long-legs *n. sing. (Zool.)* **a)** *(crane-fly)* Schnake, *die;* **b)** *(Amer.: harvestman)* Weberknecht, *der;* Kanker, *der*
daffodil ['dæfədıl] *n.* Gelbe Narzisse; Osterglocke, *die*
daft [dɑ:ft] *adj.* doof *(ugs.);* blöd[e] *(ugs.)*
dagger ['dægə(r)] *n.* Dolch, *der;* **be at ~s drawn with sb.** *(fig.)* mit jmdm. auf Kriegsfuß stehen; **look ~s at sb.** jmdm. finstere Blicke zuwerfen
dago ['deıgəʊ] *n., pl.* **~s** *or* **~es** *(sl. derog.: Spaniard, Portuguese, Italian)* Welsche, *der (veralt. abwertend);* Kanake, *der (derb abwertend)*
dahlia ['deılıə] *n.* Dahlie, *die*
daily ['deılı] **1.** *adj.* täglich; ~ **[news]paper** Tageszeitung, *die.* **2.** *adv.* täglich; jeden Tag; *(constantly)* Tag für Tag. **3.** *n.* **a)** *(newspaper)* Tageszeitung, *die;* **b)** *(Brit. coll.: charwoman)* Reinemachefrau, *die*
dainty ['deıntı] **1.** *adj.* zierlich; anmutig ‹*Bewegung, Person*›; zart, fein ‹*Gesichtszüge*›. **2.** *n. (lit. or fig.)* Delikatesse, *die;* Leckerbissen, *der*
dairy ['deərı] *n.* **a)** Molkerei, *die;* **b)** *(shop)* Milchladen, *der*
dairy: ~ **cattle** *n.* Milchvieh, *das;* **~man** ['deərımən] *n., pl.* **~men** ['deərımən] Milchmann, *der;* ~ **produce** *n.,* ~ **products** *n. pl.* Molkereiprodukte
dais ['deııs, 'deıs] *n.* Podium, *das*
daisy ['deızı] *n.* Gänseblümchen, *das; (ox-eye)* Margerite, *die*
dale [deıl] *n. (literary/N. Engl.)* Tal, *das; see also* **up 2 a**
dally ['dælı] *v.i.* **a)** ~ **with sb.** mit jmdm. spielen *od.* leichtfertig umgehen; *(flirt)* mit jmdm. schäkern *(ugs.) od.* flirten; **b)** *(idle, loiter)* [herum]trödeln *(ugs.);* ~ **[over sth.]** mit etw. trödeln *(ugs.)*
Dalmatian [dæl'meıʃn] *n.* Dalmatiner, *der*
[1]**dam** [dæm] **1.** *n.* [Stau]damm, *der; (made by beavers)* Damm, *der.* **2.** *v.t.,* **-mm-: a)** *(lit. or fig.)* ~ **[up/back] sth.** etw. abblocken; **b)** *(furnish or confine with ~)* aufstauen
[2]**dam** *n. (Zool.)* Muttertier, *das*
damage ['dæmıdʒ] **1.** *n.* **a)** *no pl.* Schaden, *der;* **do a lot of ~ to sb./sth.** jmdm./einer Sache großen Schaden zufügen; **b)** *in pl. (Law)* Schaden[s]ersatz, *der.* **2.** *v.t.* **a)** beschädigen; **smoking can ~ one's health** Rauchen gefährdet die Gesundheit; **b)** *(detract from)* schädigen
damaging ['dæmıdʒıŋ] *adj.* schädlich (**to** für)
dame [deım] *n.* **a) D~** *(Brit.)* Dame *(Titel der weiblichen Träger verschiedener Orden im Ritterstand);* **b) D~** *(literary/poet.: title of woman of rank)* Dame, *die;* **c)** *(arch./poet./joc./Amer. sl.)* Weib, *das*
damfool ['dæmfu:l] *(coll.)* **1.** *adj.* idiotisch *(ugs.);* blöd *(ugs.).* **2.** *n.* Idiot, *der (ugs.)*
dammit ['dæmıt] *int. (coll.)* verdammt noch mal *(ugs.);* **as near as ~:** jedenfalls so gut wie *(ugs.)*
damn [dæm] **1.** *v.t.* **a)** *(condemn, censure)* verreißen ‹*Buch, Film, Theaterstück*›; **b)** *(doom to hell, curse)* verdammen; **c)** *(coll.)* ~ **[it]!** verflucht [noch mal]! *(ugs.);* ~ **you/him!** hol' dich/ihn der Teufel! *(salopp);* **[well,] I'll be** *or* **I'm ~ed** ich werd' verrückt *(ugs.);* **[I'll be** *or* **I'm] ~ed if I know** ich habe nicht die leiseste Ahnung. **2.** *n.* **a)** *(curse)* Fluch, *der;* **b) he didn't give** *or* **care a ~ [about it]** ihm war es völlig Wurscht *(ugs.).* **3.** *adj.* verdammt *(ugs.).* **4.** *adv.* verdammt
damnation [dæm'neıʃn] **1.** *n.* Verdammnis, *die.* **2.** *int.* verdammt [noch mal] *(ugs.)*
damned [dæmd] *(coll.)* **1.** *adj.* **a)** *(infernal, unwelcome)* verdammt *(ugs.);* **what a ~ nuisance!** verdammter Mist! *(ugs.);* **c) do/try one's ~est** sein möglichstes tun. **2.** *adv.* verdammt *(ugs.);* **I should ~ well hope so** das will ich aber [auch] schwer hoffen *(ugs.)*
damning ['dæmıŋ] *adj.* vernichtend ‹*Urteil, Kritik, Worte*›; belastend ‹*Beweise*›
damp [dæmp] **1.** *adj.* feucht; **a ~ squib** *(fig.)* ein Reinfall. **2.** *v.t.* **a)** befeuchten; **b)** ~ **[down] a fire** ein Feuer ersticken; **c)** *(Mus., Phys.)* dämpfen; **d)** dämpfen ‹*Eifer, Begeisterung*›; ~ **sb.'s spirits** jmdm. den Mut nehmen. **3.** *n.* Feuchtigkeit, *die*
'**damp course** *see* **damp-proof**
dampen ['dæmpn] *see* **damp 2 a, d**
damper ['dæmpə(r)] *n.* **a) put a ~ on sth.** einer Sache *(Dat.)* einen Dämpfer aufsetzen; **b)** *(Mus.)* Dämpfer, *der;* **c)** *(in flue)* Luftklappe, *die*
'**damp-proof** *adj.* feuchtigkeitsbeständig; ~ **course** Sperrschicht, *die (gegen aufsteigende Bodenfeuchtigkeit)*
damsel ['dæmzl] *n. (arch./literary)* Maid, *die (veralt.);* **a ~ in distress** *(joc.)* eine hilflose junge Dame
damson ['dæmzn] *n.* Haferpflaume, *die*
dance [dɑ:ns] **1.** *v.i.* tanzen; *(jump about, skip)* herumtanzen. **2.** *v.t.* **a)** tanzen; **b)** *(move up and down)* schaukeln. **3.** *n.* **a)** Tanz, *der;* **lead sb. a [merry] ~** *(fig.)* jmdn. [schön] an der Nase herumführen; **b)** *(party)* Tanzveran-

staltung, *die; (private)* Tanzparty, *die*

dance: ~-band *n.* Tanzkapelle, *die;* **~-hall** *n.* Tanzsaal, *der*

dancer ['dɑ:nsə(r)] *n.* Tänzer, *der*/Tänzerin, *die*

dancing ['dɑ:nsɪŋ]: **~-girl** *n.* Tänzerin, *die;* **~-partner** *n.* Tanzpartner, *der*/-partnerin, *die*

dandelion ['dændɪlaɪən] *n.* Löwenzahn, *der*

dandruff ['dændrʌf] *n.* [Kopf]schuppen *Pl.*

dandy ['dændɪ] *n.* Dandy, *der (geh.);* Geck, *der (abwertend)*

Dane [deɪn] *n.* Däne, *der*/Dänin, *die*

danger ['deɪndʒə(r)] *n.* Gefahr, *die;* **a ~ to sb./sth.** eine Gefahr für jmdn./etw.; **'~!'** „Vorsicht!"; **there is [a] ~ of war** es besteht Kriegsgefahr; **in ~:** in Gefahr; **be in ~ of doing sth.** ‹*Person:*› Gefahr laufen, etw. zu tun; ‹*Sache:*› drohen, etw. zu tun; **out of ~:** außer Gefahr

danger: ~ list *n.* **be on/off the ~ list** in/außer Lebensgefahr sein; **~ money** *n.* Gefahrenzulage, *die*

dangerous ['deɪndʒərəs] *adj.*, **dangerously** ['deɪndʒərəslɪ] *adv.* gefährlich

'danger signal *n.* Warnzeichen, *das*

dangle ['dæŋgl] **1.** *v.i.* baumeln **(from** an + *Dat.*). **2.** *v.t.* baumeln lassen; **~ sth. in front of sb.** *(fig.)* jmdm. etw. in Aussicht stellen

Danish ['deɪnɪʃ] **1.** *adj.* dänisch; **sb. is ~:** jmd. ist Däne/Dänin. **2.** *n.* Dänisch, *das; see also* **English 2 a**

Danish: ~ 'blue *n.* dänischer Blauschimmelkäse; **~ 'pastry** *n.* Plunderstück, *das*

dank [dæŋk] *adj.* feucht

Danube ['dænju:b] *pr. n.* Donau, *die*

dapper ['dæpə(r)] *adj.* adrett; schmuck *(veralt.)*

dappled ['dæpld] *adj.* gesprenkelt; gefleckt ‹*Pferd, Kuh*›

dare [deə(r)] **1.** *v.t.* , *pres.* **he ~** *or* **~s**, *neg.* **~ not,** *(coll.)* **~n't** [deənt] **a)** *(venture)* [es] wagen; sich *(Akk.)* trauen; **if you ~ [to] give away the secret** wenn du es wagst, das Geheimnis zu verraten; **we ~ not/~d not** *or (coll.)* **didn't ~ tell him the truth** wir wagen/wagten [es] nicht *od.* trauen/trauten uns nicht, ihm die Wahrheit zu sagen; **you wouldn't ~:** das wagst du nicht; du traust dich nicht; **just you/don't you ~!** untersteh dich!; **how ~ you!** was fällt dir ein!; *(formal)* was erlauben Sie sich!; **I ~ say** *(supposing)* ich nehme an; *(confirming)* das glaube ich gern; **b)** *(challenge)* **~ sb. to do sth.** jmdn. dazu aufstacheln, etw. zu tun; **I ~ you!** trau dich! **2.** *n.* **do sth. for/as a ~:** etw. als Mutprobe tun

'daredevil *n.* Draufgänger, *der*/-gängerin, *die*

daring ['deərɪŋ] **1.** *adj.* kühn; waghalsig ‹*Kunststück, Tat*›. **2.** *n., no pl.* Kühnheit, *die*

dark [dɑ:k] **1.** *adj.* **a)** dunkel; dunkel, finster ‹*Nacht, Haus, Straße*›; *(gloomy)* düster; **b)** dunkel ‹*Farbe*›; *(brown-complexioned)* dunkelhäutig; *(dark-haired)* dunkelhaarig; **~-blue/-brown** *etc.* dunkelblau/-braun *usw.*; **c)** *(evil)* finster; **d)** *(cheerless)* finster; düster ‹*Bild*›. **2.** *n.* **a)** Dunkel, *das;* **in the ~:** im Dunkeln; **b) keep sb. in the ~ about/as to sth.** jmdn. über etw. *(Akk.)* im dunkeln lassen; **it was a shot in the ~:** es war aufs Geratewohl geraten/versucht; **a leap in the ~:** ein Sprung ins Ungewisse

'Dark Ages *n. pl.* [frühes] Mittelalter

darken ['dɑ:kn] **1.** *v.t.* **a)** verdunkeln; **b)** *(fig.)* verdüstern; **never ~ my door again!** du betrittst mir meine Schwelle nicht mehr! **2.** *v.i.* ‹*Zimmer:*› dunkel werden; ‹*Wolken, Himmel:*› sich verfinstern

dark: ~ 'glasses *n. pl.* dunkle Brille; **~-haired** ['dɑ:kheəd] *adj.* dunkelhaarig; **~ 'horse** *n. (fig.: secretive person)* **be a ~ horse** ein stilles Wasser sein

darkness ['dɑ:knɪs] *n., no pl.* Dunkelheit, *die*

'dark-room *n.* Dunkelkammer, *die*

darling ['dɑ:lɪŋ] **1.** *n.* Liebling, *der;* **she was his ~:** sie war seine Liebste *od.* sein Schatz. **2.** *adj.* geliebt

¹darn [dɑ:n] **1.** *v.t.* stopfen. **2.** *n.* gestopfte Stelle

²darn *(sl.: damn)* **1.** *v.t.* **~ you** *etc.*! zum Kuckuck mit dir *usw.*! *(salopp);* **~ [it]!** verflixt [und zugenäht]! *(ugs.).* **2.** *adj.* verflixt *(ugs.)*

darned [dɑ:nd] *(sl.)* **1.** *adj.* verflixt *(ugs.).* **2.** *adv.* verflixt *(ugs.)*

darning ['dɑ:nɪŋ] *n.* Stopfen, *das*

'darning-needle *n.* Stopfnadel, *die*

dart [dɑ:t] **1.** *n.* **a)** *(missile)* Pfeil, *der;* **b)** *(Sport)* Wurfpfeil, *der;* **~s** *sing. (game)* Darts, *das.* **2.** *v.i.* sausen

'dartboard *n.* Dartscheibe, *die*

dash [dæʃ] **1.** *v.i. (move quickly)* sausen; *(coll.: hurry)* sich eilen; **~ down/up [the stairs]** [die Treppe] hinunter-/hinaufstürzen. **2.** *v.t.* **a)** *(shatter)* **~ sth. [to pieces]** etw. [in tausend Stücke] zerschlagen *od.* zerschmettern; **b)** *(fling)* schleudern; schmettern; **c)** *(frustrate)* **sb.'s hopes are ~ed** jmds. Hoffnungen haben sich zerschlagen. **3.** *n.* **a) make a ~ for sth.** zu etw. rasen *(ugs.);* **make a ~ for shelter** rasch Schutz suchen; **make a ~ for freedom** plötzlich versuchen, wegzulaufen; **b)** *(horizontal stroke)* Gedankenstrich, *der;* **c)** *(Morse signal)* Strich, *der;* **d)** *(small amount)* Schuß, *der;* **a ~ of salt** eine Prise Salz

~ a'way *v.i. (rush)* davonjagen; *(coll: hurry)* **they had to ~ away** sie mußten schnell weg

~ 'off 1. *v.i. see* **~ away. 2.** *v.t.* rasch schreiben

'dashboard *n. (Motor Veh.)* Armaturenbrett, *das*

dashing ['dæʃɪŋ] *adj.* schneidig

data ['deɪtə, 'dɑ:tə] *n. pl., constr. as pl. or sing.* Daten *Pl.*

data: ~ bank *n.* Datenbank, *die;* **~-handling, ~ 'processing** *ns.* Datenverarbeitung, *die;* **~ 'processor** *n.* Datenverarbeitungsanlage, *die;* **~ pro'tection** *n.* Datenschutz, *der*

¹date [deɪt] *n. (Bot.)* Dattel, *die*

²date 1. *n.* **a)** Datum, *das; (on coin etc.)* Jahreszahl, *die;* **~ of birth** Geburtsdatum, *das;* **b)** *(coll.: appointment)* Verabredung, *die;* **have/make a ~ with sb.** mit jmdm. verabredet sein/sich mit jmdm. verabreden; **go [out] on a ~ with sb.** mit jmdm. ausgehen; **c)** *(Amer. coll.: person)* Freund, *der*/Freundin, *die;* **d) be out of ~:** altmodisch sein; *(expired)* nicht mehr gültig sein; **to ~:** bis heute. *See also* **up to date. 2.** *v.t.* **a)** datieren; **b)** *(coll: make seem old)* alt machen. **3.** *v.i.* **a) ~ back to/~ from a certain time** aus einer bestimmten Zeit stammen; **b)** *(coll.: become out of ~)* aus der Mode kommen

dated ['deɪtɪd] *adj. (coll.)* altmodisch

date: ~-line *n. (Geog.)* Datumsgrenze, *die;* **~-palm** *n.* Dattelpalme, *die;* **~-stamp 1.** *n.* Datumsstempel, *der;* **2.** *v.t.* abstempeln

dative ['deɪtɪv] *(Ling.)* **1.** *adj.* Dativ; dativisch; **~ case** Dativ, *der.* **2.** *n.* Dativ, *der*

daub [dɔ:b] *v.t.* **a)** *(coat)* bewer-

fen; *(smear, soil)* beschmieren; **b)** *(lay crudely)* schmieren
daughter ['dɔ:tə(r)] *n. (lit. or fig.)* Tochter, *die*
'**daughter-in-law** *n., pl.* **daughters-in-law** Schwiegertochter, *die*
daunt [dɔ:nt] *v.t.* entmutigen; schrecken *(geh.)*; **nothing ~ed** unverzagt
dawdle ['dɔ:dl] *v.i.* bummeln *(ugs.)*
dawn [dɔ:n] **1.** *v.i.* **a)** dämmern; **day[light] ~ed** der Morgen dämmerte; **b)** *(fig.)* ‹*Zeitalter:*› anbrechen; ‹*Idee:*› aufkommen; **sth. ~s on** *or* **upon sb.** etw. dämmert jmdm.; **hasn't it ~ed on you that ...?** ist dir nicht langsam klargeworden, daß ...? **2.** *n.* [Morgen]dämmerung, *die;* **from ~ to dusk** von früh bis spät; **at ~:** im Morgengrauen
dawn '**chorus** *n.* morgendlicher Gesang der Vögel
day [deɪ] *n.* **a)** Tag, *der;* **all ~ [long]** den ganzen Tag [lang]; **take all ~** *(fig.)* eine Ewigkeit brauchen; **all ~ and every ~:** tagaus, tagein; **to this ~, from that ~ to this** bis zum heutigen Tag; **for two ~s** zwei Tage [lang]; **what's the ~** *or* **what ~ is it today?** welcher Tag ist heute?; **twice a ~:** zweimal täglich *od.* am Tag; **in a ~/two ~s** *(within)* in *od.* an einem Tag/in zwei Tagen; **[on] the ~ after/before** am Tag danach/davor; **[the] next/[on] the following/[on] the previous ~:** am nächsten/folgenden/vorhergehenden Tag; **the ~ before yesterday/after tomorrow** vorgestern/übermorgen; **the other ~:** neulich; **from this/that ~ [on]** von heute an/von diesem Tag an; **one of these [fine] ~s** eines [schönen] Tages; **some ~:** eines Tages; irgendwann einmal; **for the ~:** für einen Tag; **~ after ~:** Tag für Tag; **~ by ~, from ~ to ~:** von Tag zu Tag; **~ in ~ out** tagaus, tagein; **call it a ~** *(end work)* Feierabend machen; *(more generally)* Schluß machen; **at the end of the ~** *(fig.)* letzten Endes; **it's not my ~:** ich habe [heute] einen schlechten Tag; **b)** *in sing. or pl. (period)* **in the ~s when ...:** zu der Zeit, als ...; **these ~s** heutzutage; **in those ~s** damals; zu jener Zeit; **in this ~ and age** heutzutage; **have seen/known better ~s** bessere Tage gesehen/gekannt haben; **those were the ~s** das waren noch Zeiten; **in one's ~:** zu seiner Zeit; *(during lifetime)* in seinem Leben; **every dog has its ~:** jeder hat einmal seine Chance; **it has had its ~:** es hat ausgedient *(ugs.)*; **c)** *(victory)* **win** *or* **carry the ~:** den Sieg davontragen
-day *adj. in comb.* -tägig; **three-~[s]-old** drei Tage alt; **five-~ week** Fünftagewoche, *die*
day: ~-boy *n. (Brit.)* externer Schüler; **~break** *n.* Tagesanbruch, *der;* **at ~break** bei Tagesanbruch; **~-dream 1.** *n.* Tagtraum, *der;* **2.** *v.i.* träumen; **~-dreamer** *n.* Tagträumer, *der*/-träumerin, *die;* **~-girl** *n. (Brit.)* externe Schülerin; **~light** *n.* **a)** *(light of ~)* Tageslicht, *das;* **go on working while it's still ~light** weiterarbeiten, solange es noch hell ist; **in broad ~light** am hellichten Tag[e]; **~light saving [time]** Sommerzeit, *die;* **b)** *(dawn)* **at** *or* **by/before ~light** bei/vor Tagesanbruch; **c)** *(fig.)* **I see ~light** ich denke, die Situation lichtet sich; **it's ~light robbery** es ist der reine Wucher; **~ re'turn** *n.* Tagesrückfahrkarte, *die;* **~ shift** *n.* Tagschicht, *die;* **be on [the] ~ shift** Tagschicht haben; **~time** *n.* Tag, *der;* **in** *or* **during the ~time** während des Tages; **~-to-~** *adj.* [tag]täglich; **~-to-~ life** Alltagsleben, *das;* **~ trip** *n.* Tagesausflug, *der;* **~ tripper** *n.* Tagesausflügler, *der*/-ausflüglerin, *die*
daze [deɪz] *v.t.* benommen machen; **be ~d** benommen sein (**at** von)
dazzle ['dæzl] *v.t. (lit., or fig.: delude)* blenden; *(fig.: confuse, impress)* überwältigen
DC *abbr. (Electr.)* **direct current** GS
D-Day ['di:deɪ] *n. Tag der Landung der Alliierten in der Normandie*
DDT *abbr.* DDT, *das*
deacon ['di:kn] *n.* Diakon, *der*
dead [ded] **1.** *adj.* **a)** tot; **[as] ~ as a doornail/as mutton** mausetot *(ugs.)*; **I wouldn't be seen ~ in a place like that** *(coll.)* keine zehn Pferde würden mich an solch einen Ort bringen *(ugs.)*; **b)** tot ‹*Materie*›; erloschen ‹*Vulkan, Gefühl, Interesse*›; verbraucht, leer ‹*Batterie*›; tot ‹*Telefon, Leitung, Saison, Kapital, Ball, Sprache*›; **the phone has gone ~:** die Leitung ist tot; **the motor is ~:** der Motor läuft nicht; **c)** *expr. completeness* plötzlich ‹*Halt*›; völlig ‹*Stillstand*›; genau ‹*Mitte*›; **~ silence** *or* **quiet** Totenstille, *die;* **~ calm** Flaute, *die;* **~ faint** [totenähnliche] Ohnmacht; **d)** *(benumbed)* taub; **e)** *(exhausted)* erschöpft; kaputt *(ugs.)*. **2.** *adv.* **a)** *(completely)* völlig; **~ straight** schnurgerade; **~ tired** todmüde; **~ easy** *or* **simple/slow** kinderleicht/ganz langsam; '**~ slow**' „besonders langsam fahren"; **~ drunk** stockbetrunken *(ugs.)*; **be ~ against sth.** absolut gegen etw. sein; **b)** *(exactly)* **~ on target** genau im Ziel; **~ on time** auf die Minute; **~ on two [o'clock]** Punkt zwei [Uhr]. **3.** *n.* **a) in the ~ of winter/night** mitten im Winter/in der Nacht; **b)** *pl.* **the ~:** die Toten *Pl.*
dead-'beat *adj. (exhausted)* völlig zerschlagen
deaden ['dedn] *v.t.* dämpfen; abstumpfen ‹*Gefühl*›; betäuben ‹*Nerv, Körperteil, Schmerz*›
dead: ~ 'end *n. (closed end)* Absperrung, *die; (street; also fig.)* Sackgasse, *die;* **~-end** *attrib. adj.* **a) ~-end street/road** Sackgasse, *die;* **b)** *(fig.)* aussichtslos; **she's in a ~-end job** in ihrem Job hat sie keine Aufstiegschancen; **~ 'heat** *n.* totes Rennen; **~ 'letter** *n.* **a)** *(law)* Gesetz, das nicht angewendet wird; **be a ~ letter** nur noch auf dem Papier bestehen; **b)** *(letter)* unzustellbarer Brief; **~line** *n.* [letzter] Termin; **meet the ~line** den Termin einhalten; **set a ~line for sth.** eine Frist für etw. setzen; **~lock** *n.* völliger Stillstand; **come to a** *or* **reach [a] ~lock/be at ~lock** an einem toten Punkt anlangen/angelangt sein; **the negotiations had reached ~lock** die Verhandlungen waren festgefahren; **~ 'loss** *n. (coll.) (worthless thing)* totaler Reinfall *(ugs.)*; *(person)* hoffnungsloser Fall *(ugs.)*
deadly ['dedlɪ] **1.** *adj.* tödlich; *(fig. coll.: awful)* fürchterlich; *(very boring)* todlangweilig; *(very dangerous)* lebensgefährlich; **~ enemy** Todfeind, *der;* **I'm in ~ earnest about this** es ist mir todernst damit. **2.** *adv.* tod; *(extremely)* äußerst; **~ pale** totenblaß; **~ dull** todlangweilig
deadly 'nightshade *n. (Bot.)* Tollkirsche, *die*
dead: ~-'on *adj., adv.* [ganz] genau; **~pan** *adj.* unbewegt; **he had a ~pan expression** er verzog keine Miene; **D~ 'Sea** *pr. n.* Tote Meer, *das;* **~ weight** *n. (inert mass)* Eigengewicht, *das; (fig.)* schwere Bürde
deaf [def] *adj.* **a)** taub; **~ and dumb** taubstumm; **~ in one ear** auf einem Ohr taub; **b)** *(insensitive)* **be ~ to sth.** kein Ohr für etw. haben; *(fig.)* taub gegenüber etw.

sein; **turn a ~ ear [to sth./sb.]** sich [gegenüber etw./jmdm.] taub stellen; **fall on ~ ears** kein Gehör finden

'deaf-aid *n.* Hörgerät, *das*

deafen ['defn] *v.t.* **~ sb.** bei jmdm. zur Taubheit führen; **I was ~ed by the noise** *(fig.)* ich war von dem Lärm wie betäubt

deafening ['defənıŋ] ohrenbetäubend ⟨*Lärm, Musik, Geschrei*⟩

deaf 'mute *n.* Taubstumme, *der/die*

deafness ['defnıs] *n., no pl.* Taubheit, *die*

¹deal [di:l] **1.** *v.t.,* **dealt** [delt] **a)** *(Cards)* austeilen; **who ~t the cards?** wer hat gegeben?; **b) ~ sb. a blow** *(lit. or fig.)* jmdm. einen Schlag versetzen. **2.** *v.i.,* **dealt a)** *(do business)* **~ with sb.** mit jmdm. Geschäfte machen; **~ in sth.** mit etw. handeln; **b)** *(occupy oneself)* **~ with sth.** sich mit etw. befassen; *(manage)* mit etw. fertig werden; **c)** *(take measures)* **~ with sb.** mit jmdm. fertig werden. **3.** *n.* **a)** *(coll.: arrangement, bargain)* Geschäft, *das;* **make a ~ with sb.** mit jmdm. ein Geschäft abschließen; **it's a ~!** abgemacht!; **big ~!** *(iron.)* na und?; **fair ~** *(treatment)* faire *od.* gerechte Behandlung; **raw** *or* **rough ~:** ungerechte Behandlung; **b)** *(coll.: agreement)* **make** *or* **do a ~ with sb.** mit jmdm. eine Vereinbarung treffen; **c)** *(Cards)* **it's your ~:** du gibst **~ 'out** *v.t.* verteilen

²deal *n.* **a great** *or* **good ~,** *(coll.)* **a ~:** viel; *(often)* ziemlich viel; **a great** *or* **good ~ of,** *(coll.)* **a ~ of** eine [ganze] Menge

dealer ['di:lə(r)] *n.* **a)** *(trader)* Händler, *der;* **he's a ~ in antiques** er ist Antiquitätenhändler *od.* handelt mit Antiquitäten; **b)** *(Cards)* Geber, *der;* **he's the ~:** er gibt

dealing ['di:lıŋ] *n.* **have ~s with sb.** mit jmdm. zu tun haben

dealt *see* **¹deal 1, 2**

dean [di:n] *n.* **a)** *(Eccl.)* Dechant, *der;* Dekan, *der;* **b)** *(in college, university, etc.)* Dekan, *der*

dear [dıə(r)] **1.** *adj.* **a)** *(beloved; also iron.)* lieb; geliebt; *(sweet; also iron.)* entzückend; **my ~ sir/madam** [mein] lieber Herr/[meine] liebe Dame; **my ~ man/woman** guter Mann/gute Frau; **my ~ child/girl** [mein] liebes Kind/liebes Mädchen; **sb./sth. is [very] ~ to sb.['s heart]** jmd. liebt jmdn./etw. [über alles]; **sb. holds sb./sth. ~:** jmd./etw. liegt jmdm. [sehr] am Herzen; **run for ~ life** um sein Leben rennen; **b)** *(beginning letter)* **D~ Sir/Madam** Sehr geehrter Herr/Sehr verehrte gnädige Frau; **D~ Mr Jones/Mrs Jones** Sehr geehrter Herr Jones/Sehr verehrte Frau Jones; **D~ Malcolm/Emily** Lieber Malcolm/Liebe Emily; **c)** *(expensive)* teuer. **2.** *int.* **~, ~!, ~ me!, oh ~!** [ach] du liebe *od.* meine Güte! **3.** *n.* **a) she is a ~:** sie ist ein Schatz; **b) [my] ~** *(to wife, husband, younger relative)* [mein] Liebling; [mein] Schatz; *(to little girl/boy)* [meine] Kleine/[mein] Kleiner; **~est** Liebling *(der)*. **4.** *adv.* teuer

dearly ['dıəlı] *adv.* **a)** von ganzem Herzen; **I'd ~ love to do that** ich würde das liebend gern tun; **b)** *(at high price)* teuer

dearth [dɜ:θ] *n.* Mangel, *der* (**of** an + *Dat.*); **there is no ~ of sth.** es fehlt nicht an etw. *(Dat.)*

death [deθ] *n.* **a)** Tod, *der;* **after ~:** nach dem Tod; **meet one's death** den Tod finden *(geh.);* **catch one's ~ [of cold]** *(coll.)* sich *(Dat.)* den Tod holen *(ugs.);* **... to ~:** zu Tode ...; **bleed to ~:** verbluten; **freeze to ~:** erfrieren; **beat sb. to ~:** jmdn. totschlagen; **I'm scared to ~** *(fig.)* mir ist angst und bange (**about** vor + *Dat.*); **be sick to ~ of sth.** *(fig.)* etw. gründlich satt haben; **[fight] to the ~:** auf Leben und Tod [kämpfen]; **be at ~'s door** an der Schwelle des Todes stehen; **b)** *(instance)* Todesfall, *der*

death: ~bed *n.* **on one's ~bed** auf dem Sterbebett; **~ certificate** *n.* Totenschein, *der*

deathly ['deθlı] **1.** *adj.* tödlich; **~ stillness/hush** Totenstille, *die.* **2.** *adv.* tödlich; **~ pale** totenblaß; **~ still/quiet** totenstill

death: ~ penalty *n.* Todesstrafe, *die;* **~ rate** *n.* Sterblichkeitsziffer, *die;* **~ sentence** *n.* Todesurteil *das;* **~'s head** *n.* Totenkopf, *der;* **~-toll** *n.* Zahl der Todesopfer; **~-trap** *n.* lebensgefährliche Sache; **~-warrant** *n.* Exekutionsbefehl, *der; (fig.)* Todesurteil, *das;* **~-watch [beetle]** *n. (Zool.)* Totenuhr, *die*

debar [dı'ba:(r)] *v.t.,* **-rr-** ausschließen; **~ sb. from doing sth.** jmdn. davon ausschließen, etw. zu tun

debase [dı'beıs] *v.t.* **a)** verschlechtern; herabsetzen, entwürdigen ⟨*Person*⟩; **~ oneself** sich erniedrigen; **b) ~ the coinage** den Wert der Währung mindern

debatable [dı'beıtəbl] *adj. (questionable)* fraglich

debate [dı'beıt] **1.** *v.t.* debattieren über (+ *Akk.*); **be ~d** diskutiert *od.* debattiert werden. **2.** *n.* Debatte, *die;* **there was much ~ about whether ...:** es wurde viel darüber debattiert, ob ...

debauchery [dı'bɔ:tʃərı] *n. (literary)* Ausschweifung, *die*

debenture [dı'bentʃə(r)] *n. (Finance)* Schuldverschreibung, *die*

debility [dı'bılıtı] *n.* Schwäche, *die*

debit ['debıt] **1.** *n. (Bookk.)* Soll, *das;* **~ balance** Lastschrift, *die;* **~ side** *(Finance)* Sollseite, *die.* **2.** *v.t.* belasten; **~ sb./sb.'s account with a sum** jmdn./jmds. Konto mit einer Summe belasten

debonair [debə'neə(r)] *adj.* frohgemut

debrief [di:'bri:f] *v.t. (coll.)* befragen *(bei Rückkehr von einem Einsatz usw.)*

debris ['debri:, 'deıbri:] *n., no pl.* Trümmer *Pl.*

debt [det] *n.* Schuld, *die;* **National D~:** Staatsverschuldung, *die;* **be in ~:** verschuldet sein; **get** *or* **run into ~:** in Schulden geraten; sich verschulden; **get out of ~:** aus den Schulden herauskommen; **be in sb.'s ~:** in jmds. Schuld stehen

'debt collector *n.* Inkassobevollmächtigte, *der/die*

debtor ['detə(r)] *n.* Schuldner, *der*/Schuldnerin, *die*

debug [di:'bʌg] *v.t.,* **-gg-** *(coll.) (remove microphones from)* von Wanzen befreien; *(remove defects from)* von Fehlern befreien

debunk [di:'bʌŋk] *v.t. (coll.) (remove false reputation from)* entlarven; *(expose falseness of)* bloßstellen

début (*Amer.:* **debut**) ['deıbu:, 'deıbju:] *n.* Debüt, *das;* **make one's ~:** debütieren

débutante (*Amer.:* **debutante**) ['debju:tɑ:nt, 'deıbju:tɑ:nt] *n.* Debütantin, *die*

Dec. *abbr.* **December** Dez.

decade ['dekeıd] *n.* Jahrzehnt, *das;* Dekade, *die*

decadence ['dekədəns] *n.* Dekadenz, *die*

decadent ['dekədənt] *adj.* dekadent

decamp [dı'kæmp] *v.i.* verschwinden *(ugs.)*

decant [dı'kænt] *v.t.* abgießen; dekantieren ⟨*Wein*⟩

decanter [dı'kæntə(r)] *n.* Karaffe, *die*

decapitate [dı'kæpıteıt] *v.t.* köpfen

decathlon [dı'kæθlən] *n. (Sport)* Zehnkampf, *der*

decay [dɪ'keɪ] **1.** *v.i.* **a)** *(become rotten)* verrotten; [ver]faulen; ‹*Zahn:*› faul *od. (fachspr.)* kariös werden; ‹*Gebäude:*› zerfallen; **b)** *(decline)* verfallen. **2.** *n.* **a)** *(rotting)* Verrotten, *das; (of tooth)* Fäule, *die; (of building)* Zerfall, *der;* **b)** *(decline)* Verfall, *der*

decease [dɪ'si:s] *n. (Law/formal)* Ableben, *das (geh.)*

deceased [dɪ'si:st] *(Law/formal)* **1.** *adj.* verstorben. **2.** *n.* Verstorbene, *der/die*

deceit [dɪ'si:t] *n.* Täuschung, *die;* Betrug, *der; (being deceitful)* Falschheit, *die*

deceitful [dɪ'si:tfl] *adj.* falsch ‹*Person, Art, Charakter*›; hinterlistig ‹*Trick*›

deceitfulness [dɪ'si:tflnɪs] *n., no pl. see* **deceitful:** Falschheit, *die;* Hinterlistigkeit, *die*

deceive [dɪ'si:v] *v.t.* täuschen; *(be unfaithful to)* betrügen; ~ **sb. into doing sth.** jmdn. [durch Täuschung] dazu bringen, etw. zu tun; ~ **oneself** sich täuschen; *(delude oneself)* sich *(Dat.)* etwas vormachen *(ugs.)*

December [dɪ'sembə(r)] *n.* Dezember, *der; see also* **August**

decency ['di:sənsɪ] *n. (propriety)* Anstand, *der; (of manners, literature, language)* Schicklichkeit, *die (geh.); (fairness, respectability)* Anständigkeit, *die;* **it is [a matter of] common ~:** es ist eine Frage des Anstands

decent ['di:sənt] *adj.* **a)** *(seemly)* schicklich *(geh.);* anständig ‹*Person*›; **b)** *(passable, respectable)* annehmbar; anständig ‹*Person, ugs. auch Preis, Gehalt*›

decentralize (decentralise) [di:'sentrəlaɪz] *v.t.* dezentralisieren

deception [dɪ'sepʃn] *n.* **a)** *(deceiving, trickery)* Betrug, *der; (being deceived)* Täuschung, *die;* **use ~:** betrügen; **b)** *(trick)* Betrügerei, *die*

deceptive [dɪ'septɪv] *adj.* trügerisch

decibel ['desɪbel] *n.* Dezibel, *das*

decide [dɪ'saɪd] **1.** *v.t.* **a)** *(settle, judge)* entscheiden über (+ *Akk.*); ~ **that ...:** entscheiden, daß ...; **b)** *(resolve)* ~ **that ...:** beschließen, daß ...; ~ **to do sth.** sich entschließen, etw. zu tun. **2.** *v.i.* sich entscheiden **(in favour of** zugunsten von, **on** für); ~ **against doing sth.** sich dagegen entscheiden, etw. zu tun

decided [dɪ'saɪdɪd] *adj.* **a)** *(unquestionable)* entschieden; eindeutig; **b)** *(not hesitant)* bestimmt

decidedly [dɪ'saɪdɪdlɪ] *adv.* **a)** *(unquestionably)* entschieden; deutlich; **b)** *(firmly)* bestimmt

decider [dɪ'saɪdə(r)] *n. (game)* Entscheidungsspiel, *das*

deciduous [dɪ'sɪdjʊəs] *adj. (Bot.)* ~ **leaves** Blätter, die abgeworfen werden; ~ **tree** laubwerfender Baum; ≈ Laubbaum, *der*

decimal ['desɪml] **1.** *adj.* Dezimal-; **go ~:** sich auf das Dezimalsystem umstellen. **2.** *n.* Dezimalbruch, *der*

decimal: ~ **'coinage,** ~ **'currency** *ns.* Dezimalwährung, *die;* ~ **'fraction** *n.* Dezimalbruch, *der*

decimalize (decimalise) ['desɪməlaɪz] *v.t. (express as decimal)* als Dezimalzahl schreiben; *(convert to decimal system)* dezimalisieren

decimal: ~ **'place** *n.* Dezimale, *die;* **calculate sth. to five ~ places** etw. auf fünf Stellen nach dem Komma ausrechnen; ~ **'point** *n.* Komma, *das;* ~ **system** *n.* Dezimalsystem, *das*

decimate ['desɪmeɪt] *v.t.* dezimieren

decipher [dɪ'saɪfə(r)] *v.t.* entziffern

decision [dɪ'sɪʒn] *n.* Entscheidung, *die* **(on** über + *Akk.*); **it's 'your ~:** die Entscheidung liegt ganz bei dir; **come to** *or* **reach a ~:** zu einer Entscheidung kommen; **make** *or* **take a ~:** eine Entscheidung treffen

decisive [dɪ'saɪsɪv] *adj.* **a)** *(conclusive)* entscheidend; **b)** *(decided)* entschlußfreudig ‹*Person*›; bestimmt ‹*Charakter, Art*›

deck [dek] **1.** *n.* **a)** *(of ship)* Deck, *das;* **above ~:** auf Deck; **below ~[s]** unter Deck; **clear the ~s [for action** *etc.*] das Schiff klarmachen [zum Gefecht *usw.*]; **on ~:** an Deck; **all hands on ~!** alle Mann an Deck!; **b)** *(of bus etc.)* Deck, *das;* **the upper ~:** das Oberdeck; **c)** *(tape ~)* Tape-deck, *das; (record ~)* Plattenspieler, *der.* **2.** *v.t.* ~ **sth. [with sth.]** etw. [mit etw.] schmücken

~ **'out** *v.t.* herausputzen ‹*Person*›; [aus]schmücken ‹*Raum*›

'deck-chair *n.* Liegestuhl, *der; (on ship)* Liege- *od.* Deckstuhl, *der*

declaim [dɪ'kleɪm] *v.i.* eifern; deklamieren *(veralt.)*

declaration [deklə'reɪʃn] *n.* Erklärung, *die; (at customs)* Deklaration, *die;* ~ **of war** Kriegserklärung, *die;* **make a ~:** eine Erklärung abgeben

declare [dɪ'kleə(r)] *v.t.* **a)** *(announce)* erklären; *(state explicitly)* kundtun *(geh.)* ‹*Wunsch, Absicht*›; Ausdruck verleihen (+ *Dat.*) *(geh.)* ‹*Hoffnung*›; **b)** *(pronounce)* ~ **sth./sb. [to be] sth.** etw./jmdn. für etw. erklären

declassify [di:'klæsɪfaɪ] *v.t.* freigeben

declension [dɪ'klenʃn] *n. (Ling.)* Deklination, *die*

decline [dɪ'klaɪn] **1.** *v.i.* **a)** *(fall off)* nachlassen; ‹*Moral:*› sinken, nachlassen; ‹*Preis, Anzahl:*› sinken, zurückgehen; ‹*Gesundheitszustand:*› sich verschlechtern; **b)** *(refuse)* ~ **with thanks** *(also iron.)* dankend ablehnen. **2.** *v.t.* **a)** *(refuse)* ablehnen; ~ **to do sth.** [es] ablehnen, etw. zu tun; **b)** *(Ling.)* deklinieren. **3.** *n.* Nachlassen, *das* **(in** *Gen.*); **a ~ in prices/numbers** ein Sinken der Preise/Anzahl; **be on the ~:** nachlassen

declutch [di:'klʌtʃ] *v.i. (Motor Veh.)* auskuppeln; **double-~:** Zwischengas geben

decode [di:'kəʊd] *v.t.* dekodieren, dechiffrieren ‹*Mitteilung, Signal*›; entschlüsseln ‹*Schrift, Hieroglyphen*›

decompose [di:kəm'pəʊz] *v.i.* sich zersetzen

decomposition [di:kɒmpə'zɪʃn] *n.* Zersetzung, *die*

decompression [di:kəm'preʃn] *n.* Dekompression, *die*

decontaminate [di:kən'tæmɪneɪt] *v.t.* dekontaminieren *(fachspr.);* entseuchen

decontamination [di:kəntæmɪ'neɪʃn] *n.* Dekontamination, *die (fachspr.);* Entseuchung, *die*

décor ['deɪkɔ:(r)] *n.* Ausstattung, *die*

decorate ['dekəreɪt] *v.t.* **a)** schmücken ‹*Raum, Straße, Baum*›; verzieren ‹*Kuchen, Kleid*›; dekorieren ‹*Schaufenster*›; *(with wallpaper)* tapezieren; *(with paint)* streichen; **b)** *(invest with order etc.)* auszeichnen

decoration [dekə'reɪʃn] *n.* **a)** *see* **decorate a:** Schmücken, *das;* Verzieren, *das;* Dekoration, *die;* Tapezieren, *das;* Streichen, *das;* **b)** *(adornment) (thing)* Schmuck, *der; (in shop window)* Dekoration, *die;* **c)** *(medal etc.)* Auszeichnung, *die;* **d)** *in pl.* **Christmas ~s** Weihnachtsschmuck, *der*

decorative ['dekərətɪv] *adj.* dekorativ

decorator ['dekəreɪtə(r)] *n.* Maler, *der*/Malerin, *die; (paperhanger)* Tapezierer, *der*/Tapeziererin, *die*

decorum [dɪ'kɔ:rəm] *n.* Schicklichkeit, *die (geh.);* **behave with ~:** sich schicklich benehmen

decoy [dɪ'kɔɪ, 'di:kɔɪ] *n. (Hunting; also person)* Lockvogel, *der*

decrease 1. [dɪ'kri:s] *v. i.* abnehmen; ‹*Anzahl, Einfuhr, Produktivität:*› abnehmen, zurückgehen; ‹*Stärke, Gesundheit:*› nachlassen; **~ in value/size/weight** an Wert/Größe/Gewicht verlieren; **~ in price** im Preis fallen. **2.** [dɪ'kri:s] *v. t.* reduzieren; [ver]mindern ‹*Wert, Lärm, Körperkraft*›; schmälern ‹*Popularität, Macht*›. **3.** ['di:kri:s] *n.* Rückgang, *der; (in weight, stocks)* Abnahme, *die; (in strength, power, energy)* Nachlassen, *das; (in value, noise)* Minderung, *die;* **a ~ in speed** eine Minderung der Geschwindigkeit; **be on the ~** *see* **1**

decree [dɪ'kri:] **1.** *n.* **a)** *(ordinance)* Dekret, *das;* Erlaß, *der;* **b)** *(Law)* Urteil, *das;* **~ nisi/absolute** vorläufiges/endgültiges Scheidungsurteil. **2.** *v. t. (ordain)* verfügen

decrepit [dɪ'krepɪt] *adj.* altersschwach; *(dilapidated)* heruntergekommen ‹*Haus, Stadt*›

decry [dɪ'kraɪ] *v. t.* verwerfen

dedicate ['dedɪkeɪt] *v. t.* **a) ~ sth. to sb.** jmdm. etw. widmen; **b)** *(give up)* **~ one's life to sth.** sein Leben einer Sache *(Dat.)* weihen; **c)** *(devote solemnly)* weihen

dedicated ['dedɪkeɪtɪd] *adj.* **a)** *(devoted)* **be ~ to sth./sb.** nur für etw./jmdn. leben; **b)** *(devoted to vocation)* hingebungsvoll; **a ~ teacher** ein Lehrer mit Leib und Seele

dedication [dedɪ'keɪʃn] *n.* **a)** Widmung, *die* (**to** *Dat.*); **b)** *(devotion)* Hingabe, *die*

deduce [dɪ'dju:s] *v. t.* ableiten, schließen auf (**from** aus); **~ from sth. that ...:** aus etw. schließen, daß ...

deduct [dɪ'dʌkt] *v. t.* abziehen (**from** von)

deductible [dɪ'dʌktɪbl] *adj.* **be ~:** einbehalten werden [können]

deduction [dɪ'dʌkʃn] *n.* **a)** *(deducting)* Abzug, *der;* **b)** *(deducing, thing deduced)* Ableitung, *die;* **c)** *(amount)* Abzüge *Pl.*

deductive [dɪ'dʌktɪv] *adj.* deduktiv

deed [di:d] *n.* **a)** Tat, *die;* **b)** *(Law)* [gesiegelte] Urkunde

deem [di:m] *v. t.* erachten für; **[as] I ~ed** wie mir schien

deep [di:p] **1.** *adj.* **a)** *(lit. or fig.)* tief; **water ten feet ~:** drei Meter tiefes Wasser; **take a ~ breath** tief Atem holen; **ten feet ~ in water** drei Meter tief unter Wasser; **be ~ in thought/prayer** in Gedanken/im Gebet versunken sein; **be ~ in debt** hoch verschuldet sein; **be standing three ~:** drei hintereinander stehen; **b)** *(profound)* tief ‹*Grund*›; gründlich ‹*Studium, Forschung*›; tiefgründig ‹*Bemerkung*›; **give sth. ~ thought** über etw. *(Akk.)* gründlich nachdenken; **he's a ~ one** *(coll.)* er ist ein stilles Wasser *(ugs.)*; **c)** *(heartfelt)* tief; aufrichtig ‹*Interesse, Dank*›. **2.** *adv.* tief; **still waters run ~** *(prov.)* stille Wasser sind tief *(Spr.)*; **~ down** *(fig.)* im Innersten

deepen ['di:pn] **1.** *v. t.* **a)** tiefer machen; vertiefen; **b)** *(increase, intensify)* vertiefen; intensivieren ‹*Farbe*›. **2.** *v. i.* **a)** tiefer werden; **b)** *(intensify)* sich vertiefen

deep: ~-'freeze 1. *n.* (*Amer.:* **P**) Tiefkühltruhe, *die;* **2.** *v. t.* tiefgefrieren; **~-fried** *adj.* fritiert

deeply ['di:plɪ] *adv. (lit. or fig.)* tief; äußerst ‹*interessiert, dankbar, selbstbewußt*›; **be ~ in love** sehr verliebt sein; **be ~ indebted to sb.** jmdm. sehr zu Dank verpflichtet sein

deep: ~-rooted *adj.* tief ‹*Abneigung*›; tiefverwurzelt ‹*Tradition*›; **~-sea** *adj.* Tiefsee; **~-'seated** *adj.* tief sitzend

deer [dɪə(r)] *n., pl. same* Hirsch, *der; (roe ~)* Reh, *das*

deer: ~-park *n.* Wildpark, *der;* **~skin** *n.* Rehleder, *das;* **~-stalker** ['dɪəstɔ:kə(r)] *n. (hat)* ≈ Sherlock-Holmes-Mütze, *die*

deface [dɪ'feɪs] *v. t.* verunstalten; verschandeln ‹*Gebäude*›

defamation [defə'meɪʃn, di:fə'meɪʃn] *n.* Diffamierung, *die*

defamatory [dɪ'fæmətərɪ] *adj.* diffamierend

defame [dɪ'feɪm] *v. t.* diffamieren; beschmutzen ‹*Name, Ansehen*›

default [dɪ'fɔ:lt, dɪ'fɒlt] **1.** *n.* **in ~ of** mangels (+ *Gen.*); in Ermangelung *(geh.)* (+ *Gen.*); **lose/go by ~:** durch Abwesenheit verlieren/nicht zur Geltung kommen; **win by ~:** durch Nichterscheinen des Gegners gewinnen. **2.** *v. i.* versagen; **~ on one's payments/debts** seinen Zahlungsverpflichtungen nicht nachkommen

defeat [dɪ'fi:t] **1.** *v. t.* **a)** *(overcome)* besiegen; zu Fall bringen ‹*Antrag, Vorschlag*›; **b)** *(baffle)* **sth. ~s me** ich kann etw. nicht begreifen; *(frustrate)* **the task has ~ed us** diese Aufgabe hat uns überfordert; **~ the object/purpose of sth.** etw. völlig sinnlos machen. **2.** *n. (being ~ed)* Niederlage, *die; (~ing)* Sieg, *der* (**of** über + *Akk.*)

defeatism [dɪ'fi:tɪzm] *n.* Defätismus, *der*

defeatist [dɪ'fi:tɪst] **1.** *n.* Defätist, *der.* **2.** *adj.* defätistisch

defecate ['defəkeɪt] *v. i.* Kot ausscheiden; defäkieren *(Med.)*

defect 1. ['di:fekt, dɪ'fekt] *n.* **a)** *(lack)* Mangel, *der;* **b)** *(shortcoming)* Fehler, *der; (in construction, body, mind, etc. also)* Defekt, *der.* **2.** [dɪ'fekt] *v. i.* überlaufen (**to** zu)

defection [dɪ'fekʃn] *n.* Abfall, *der; (desertion)* Flucht, *die*

defective [dɪ'fektɪv] *adj.* **a)** *(faulty)* defekt ‹*Maschine*›; fehlerhaft ‹*Material, Arbeiten, Methode, Plan*›; **sb./sth. is ~ in sth.** es mangelt jmdm./einer Sache an etw. *(Dat.)*; **b)** *(mentally deficient)* geistig gestört

defector [dɪ'fektə(r)] *n.* Überläufer, *der*/-läuferin, *die; (from a cause or party)* Abtrünnige, *der/die*

defence [dɪ'fens] *n. (Brit.)* **a)** *(defending)* Verteidigung, *die; (of body against disease)* Schutz, *der;* **in ~ of** zur Verteidigung (+ *Gen.*); **b)** *(thing that protects, means of resisting attack)* Schutz, *der;* **c)** *(justification)* Rechtfertigung, *die;* **in sb.'s ~:** zu jmds. Verteidigung; **d)** *(military resources)* Verteidigung, *die;* **e)** *in pl. (fortification)* Befestigungsanlagen *Pl.*; **f)** *(Sport, Law)* Verteidigung, *die;* **the case for the ~:** die Verteidigung; **~ witness** Zeuge/Zeugin der Verteidigung

defenceless [dɪ'fenslɪs] *adj. (Brit.)* wehrlos

defend [dɪ'fend] **1.** *v. t.* **a)** *(protect)* schützen (**from** vor + *Dat.*); *(by fighting)* verteidigen; **b)** *(uphold by argument, speak or write in favour of)* verteidigen; verteidigen, rechtfertigen ‹*Politik, Handeln*›; **c)** *(Sport, Law)* verteidigen. **2.** *v. i. (Sport)* verteidigen

defendant [dɪ'fendənt] *n. (Law) (accused)* Angeklagte, *der/die; (sued)* Beklagte, *der/die*

defender [dɪ'fendə(r)] *n. (also Sport)* Verteidiger, *der*/Verteidigerin, *die*

defense, defenseless *(Amer.) see* **defence, defenceless**

defensive [dɪ'fensɪv] **1.** *adj.* **a)** *(protective)* defensiv ‹*Strategie, Handlung*›; **~ player** Defensivspieler, *der;* **~ wall** Schutzwall, *der;* **b)** *(excessively self-justifying)* **he's always so ~ when he's**

criticized er will sich immer um jeden Preis rechtfertigen, wenn er kritisiert wird. **2.** *n.* Defensive, *die;* **be on the ~:** in der Defensive sein

[1]**defer** [dɪ'fɜ:(r)] *v. t.*, **-rr-** aufschieben

[2]**defer** *v. i.*, **-rr-:** ~ **[to sb.]** sich [jmdm.] beugen; ~ **to sb.'s wishes** sich jmds. Wünschen fügen

deference ['defərəns] *n.* Respekt, *der;* Ehrerbietung, *die (geh.);* **in ~ to sb./sth.** aus Achtung vor jmdm./etw.

deferential [defə'renʃl] *adj.* respektvoll; groß ⟨*Respekt*⟩; **be ~ to sb./sth.** jmdm./einer Sache mit Respekt begegnen

deferment [dɪ'fɜ:mənt] *n.* Aufschub, *der*

defiance [dɪ'faɪəns] *n.* Aufsässigkeit, *die; (open disobedience)* Mißachtung, *die;* **in ~ of sb./sth.** jmdm./einer Sache zum Trotz

defiant [dɪ'faɪənt] *adj.*, **defiantly** [dɪ'faɪəntlɪ] *adv.* aufsässig

deficiency [dɪ'fɪʃənsɪ] *n.* **a)** *(lack)* Mangel, *der* **(of, in** an + *Dat.*); **nutritional ~:** Ernährungsmangel, *der;* **b)** *(inadequacy)* Unzulänglichkeit, *die*

deficient [dɪ'fɪʃənt] *adj.* **a)** *(not having enough)* **sb./sth. is ~ in sth.** jmdm./einer Sache mangelt es an etw. *(Dat.);* **be [mentally] ~:** geistig behindert sein; **b)** *(not being enough)* nicht ausreichend; *(in quality also)* unzulänglich

deficit ['defɪsɪt] *n.* Defizit, *das* **(of** an + *Dat.*)

[1]**defile** ['di:faɪl] *n. (gorge)* Hohlweg, *der*

[2]**defile** [dɪ'faɪl] *v. t.* **a)** verschandeln; verpesten ⟨*Luft*⟩; **b)** *(desecrate)* beflecken ⟨*Unschuld, Reinheit*⟩

define [dɪ'faɪn] *v. t.* definieren; **be ~d [against sth.]** sich [gegen etw.] abzeichnen; **~ one's position** *(fig.)* Stellung beziehen **(on** zu)

definite ['defɪnɪt] *adj. (having exact limits)* bestimmt; *(precise)* eindeutig, definitiv ⟨*Antwort, Entscheidung*⟩; eindeutig ⟨*Beschluß, Verbesserung, Standpunkt*⟩; eindeutig, klar ⟨*Vorteil*⟩; klar umrissen ⟨*Ziel, Plan, Thema*⟩; klar ⟨*Konzept, Linie, Vorstellung*⟩; deutlich ⟨*Konturen, Umrisse*⟩; genau ⟨*Zeitpunkt*⟩; **you don't seem to be very ~:** Sie scheinen sich nicht ganz sicher zu sein; **but that is not yet ~:** aber das ist noch nicht endgültig

definitely ['defɪnɪtlɪ] **1.** *adv.* eindeutig ⟨*festlegen, größer sein, verbessert, erklären*⟩; endgültig ⟨*entscheiden, annehmen*⟩; fest ⟨*vereinbaren*⟩; **she's ~ going to America** sie fährt auf jeden Fall nach Amerika. **2.** *int. (coll.)* na, klar *(ugs.)*

definition [defɪ'nɪʃn] *n.* **a)** Definition, *die;* **by ~:** per definitionem *(geh.);* **b)** *(making or being distinct, degree of distinctness)* Schärfe, *die;* **improve the ~ on the TV** den Fernseher schärfer einstellen

definitive [dɪ'fɪnɪtɪv] *adj.* **a)** *(decisive)* endgültig, definitiv ⟨*Beschluß, Antwort, Urteil*⟩; **b)** *(most authoritative)* maßgeblich

deflate [dɪ'fleɪt] **1.** *v. t.* **a)** ~ **a tyre/balloon** die Luft aus einem Reifen/Ballon ablassen; **b)** *(cause to lose conceitedness)* ernüchtern; **c)** *(Econ.)* deflationieren. **2.** *v. i. (Econ.)* deflationieren

deflation [dɪ'fleɪʃn] *n. (Econ.)* Deflation, *die*

deflationary [dɪ'fleɪʃənərɪ] *adj. (Econ.)* deflationär

deflect [dɪ'flekt] *v. t* beugen ⟨*Licht*⟩; **~ sb./sth. [from sb./sth.]** jmdn./etw. [von jmdm./einer Sache] ablenken

deflection, *(Brit.)* **deflexion** [dɪ'flekʃn] *n. (deviation)* Ablenkung, *die*

deform [dɪ'fɔ:m] *v. t.* **a)** *(deface)* deformieren; verunstalten; **b)** *(misshape)* verformen

deformed [dɪ'fɔ:md] *adj.* entstellt ⟨*Gesicht*⟩; verunstaltet ⟨*Person, Körperteil*⟩

deformity [dɪ'fɔ:mɪtɪ] *n. (being deformed)* Mißgestalt, *die; (malformation)* Verunstaltung, *die*

defraud [dɪ'frɔ:d] *v. t.* **~ sb. [of sth.]** jmdn. [um etw.] betrügen

defray [dɪ'freɪ] *v. t.* bestreiten ⟨*Kosten*⟩

defrost [di:'frɒst] *v. t.* auftauen ⟨*Speisen*⟩; abtauen ⟨*Kühlschrank*⟩; enteisen ⟨*Windschutzscheibe, Fenster*⟩

deft [deft] *adj.*, **deftly** ['deftlɪ] *adv.* sicher und geschickt

defunct [dɪ'fʌŋkt] *adj.* defekt ⟨*Maschine*⟩; veraltet ⟨*Gesetz*⟩; eingegangen ⟨*Zeitung*⟩; überholt, vergessen ⟨*Brauch, Idee, Mode*⟩

defuse [di:'fju:z] *v. t. (lit. or fig.)* entschärfen

defy [dɪ'faɪ] *v. t.* **a)** *(resist openly)* **~ sb.** jmdm. trotzen *od.* Trotz bieten; *(refuse to obey)* **~ sb./sth.** sich jmdm./einer Sache widersetzen; **b)** *(present insuperable obstacles to)* widerstehen; **it defies explanation** das spottet jeder Erklärung

degenerate 1. [dɪ'dʒenəreɪt] *v. i.* ~ **[into sth.]** [zu etw.] verkommen *od.* degenerieren. **2.** [dɪ'dʒenərət] *adj.* degeneriert

degeneration [dɪdʒenə'reɪʃn] *n.* Degeneration, *die*

degradation [degrə'deɪʃn] *n. (abasement)* Erniedrigung, *die*

degrade [dɪ'greɪd] *v. t. (abase)* erniedrigen; herabsetzen ⟨*Ansehen, Maßstab*⟩

degrading [dɪ'greɪdɪŋ] *adj.* entwürdigend; erniedrigend

degree [dɪ'gri:] *n.* **a)** *(Math., Phys.)* Grad, *der;* **an angle/a temperature of 45 ~s** ein Winkel/eine Temperatur von 45 Grad; **b)** *(stage in scale or extent)* Grad, *der;* **by ~s** allmählich; **a certain ~ of imagination** ein gewisses Maß an Phantasie; **to some** *or* **a certain ~:** [bis] zu einem gewissen Grad; **c)** *(academic rank)* [akademischer] Grad; **take/receive a ~ in sth.** einen akademischen Grad in etw. *(Dat.)* erwerben/verliehen bekommen; **have a ~ in physics/maths** einen Hochschulabschluß in Physik/Mathematik haben

dehydrate [di:'haɪdreɪt] *v. t.* das Wasser entziehen (+ *Dat.*), austrocknen ⟨*Körper*⟩; **~d** dehydratisiert *(fachspr.);* getrocknet

de-ice [di:'aɪs] *v. t.* enteisen

de-icer [di:'aɪsə(r)] *n.* Defroster, *der*

deign [deɪn] *v. t.* **~ to do sth.** sich [dazu] herablassen, etw. zu tun

deity ['di:ɪtɪ] *n.* Gottheit, *die*

dejected [dɪ'dʒektɪd] *adj.* niedergeschlagen

dejection [dɪ'dʒekʃn] *n.* Niedergeschlagenheit, *die*

delay [dɪ'leɪ] **1.** *v. t. (postpone)* verschieben; *(make late)* aufhalten; verzögern ⟨*Ankunft, Abfahrt*⟩; *(hinder)* aufhalten; **be ~ed** ⟨*Veranstaltung:*⟩ verspätet *od.* später erfolgen. **2.** *v. i. (wait)* warten; *(loiter)* trödeln *(ugs.);* **don't ~:** warte nicht damit; **~ in doing sth.** zögern, etw. zu tun. **3.** *n.* **a)** Verzögerung, *die* **(to** bei); **what's the ~ now?** weshalb geht es jetzt nicht weiter?; **without ~:** unverzüglich; **b)** *(Transport)* Verspätung, *die;* **trains are subject to ~:** es ist mit Zugverspätungen zu rechnen

delayed-action [dɪleɪd'ækʃn] *adj.* **~ bomb** Bombe mit Zeitzünder; **~ mechanism** *(Photog.)* Selbstauslöser, *der*

delectable [dɪ'lektəbl] *adj.* köstlich

delegate 1. ['delɪgət] *n.* Delegierte, *der/die.* **2.** ['delɪgeɪt] *v. t.* **a)** *(depute)* delegieren; **b)** *(commit)*

~ **sth. [to sb.]** etw. [an jmdn.] delegieren; *abs.* **he does not know how to ~:** er will alles selbst erledigen

delegation [delɪ'geɪʃn] *n.* Delegation, *die* (**to** an + *Akk.*)

delete [dɪ'li:t] *v.t.* streichen (**from** in + *Dat.*); *(Computing)* löschen; ~ **where inapplicable** Nichtzutreffendes streichen

deletion [dɪ'li:ʃn] *n.* Streichung, *die; (Computing)* Löschung, *die*

deliberate 1. [dɪ'lɪbərət] *adj.* **a)** *(intentional)* absichtlich; bewußt ‹*Lüge, Irreführung*›; vorsätzlich ‹*Verbrechen*›; **b)** *(unhurried and considered)* bedächtig. **2.** [dɪ'lɪbəreɪt] *v.i.* **a)** *(think carefully)* ~ **on sth.** über etw. *(Akk.)* [sorgfältig] nachdenken; **b)** *(debate)* ~ **over** *or* **on** *or* **about sth.** über etw. *(Akk.)* beraten

deliberately [dɪ'lɪbərətlɪ] *adv.* **a)** *(intentionally)* absichtlich; mit Absicht; vorsätzlich ‹*ein Verbrechen begehen*›; **b)** *(with full consideration)* **[very]** ~**:** [ganz] bewußt; **c)** *(in unhurried manner)* bedächtig

deliberation [dɪlɪbə'reɪʃn] *n.* **a)** *no pl. (unhurried nature)* Bedächtigkeit, *die;* **b)** *no pl. (careful consideration)* Überlegung, *die;* **c)** *(discussion)* Beratung, *die*

delicacy ['delɪkəsɪ] *n.* **a)** *(tactfulness and care)* Feingefühl, *das;* Delikatesse, *die (geh.);* **b)** *(fineness)* Zartheit, *die;* **c)** *(weakliness)* Zartheit, *die;* **d)** *(need of discretion etc.)* Delikatheit, *die;* **e)** *(food)* Delikatesse, *die*

delicate ['delɪkət] *adj.* **a)** *(easily injured)* empfindlich ‹*Organ*›; zart ‹*Gesundheit, Konstitution*›; *(sensitive)* sensibel, empfindlich ‹*Person, Natur*›; empfindlich ‹*Waage, Instrument*›; **b)** *(requiring careful handling)* empfindlich; *(fig.)* delikat, heikel ‹*Frage, Angelegenheit, Problem*›; **c)** *(fine, of exquisite quality)* zart; delikat; *(dainty)* delikat; **d)** *(subtle)* fein; **e)** *(deft, light)* geschickt; zart; **f)** *(tactful)* taktvoll; behutsam

delicatessen [delɪkə'tesən] *n.* Feinkostgeschäft, *das;* Delikatessengeschäft, *das*

delicious [dɪ'lɪʃəs] *adj.* köstlich, lecker ‹*Speise, Geschmack*›

delight [dɪ'laɪt] **1.** *v.t.* erfreuen. **2.** *v.i.* **sb. ~s in doing sth.** es macht jmdm. Freude, etw. zu tun. **3.** *n.* **a)** *(great pleasure)* Freude, *die* (**at** über + *Akk.*); ~ **in sth./in doing sth.** Freude an etw. *(Dat.)* /daran, etw. zu tun; **to my ~:** zu meiner Freude; **sb. takes ~ in doing sth.** es macht jmdm. Freude, etw. zu tun; **b)** *(cause of pleasure)* Vergnügen, *das*

delighted [dɪ'laɪtɪd] *adj.* freudig ‹*Schrei*›; **be ~** ‹*Person:*› hocherfreut sein; **be ~ by** *or* **with sth.** sich über etw. *(Akk.)* freuen; **be ~ to do sth.** sich freuen, etw. zu tun

delightful [dɪ'laɪtfl] *adj.* wunderbar; köstlich ‹*Geschmack, Klang*›; reizend ‹*Person, Landschaft*›

delightfully [dɪ'laɪtfəlɪ] *adv.* wunderbar; bezaubernd ‹*singen, tanzen, hübsch*›

delimit [dɪ'lɪmɪt] *v.t.* begrenzen ‹*Gebiet, Region*›; *(fig.)* eingrenzen

delineate [dɪ'lɪnɪeɪt] *v.t.* *(draw)* zeichnen; *(describe)* darstellen

delinquency [dɪ'lɪŋkwənsɪ] *n., no pl.* Kriminalität, *die*

delinquent [dɪ'lɪŋkwənt] **1.** *n.* *(bes. jugendlicher)* Randalierer, *der.* **2.** *adj.* kriminell

delirious [dɪ'lɪrɪəs] *adj.* **a)** delirant *(Med.);* **be ~:** im Delirium sein; **b)** *(wildly excited)* **be ~ [with sth.]** außer sich *(Dat.)* [vor etw. *(Dat.)*] sein

delirium [dɪ'lɪrɪəm] *n.* Delirium, *das*

deliver [dɪ'lɪvə(r)] *v.t.* **a)** *(utter)* halten ‹*Rede, Vorlesung, Predigt*›; vorbringen ‹*Worte*›; vortragen ‹*Verse*›; *(pronounce)* verkünden ‹*Urteil, Meinung, Botschaft*›; **b)** werfen ‹*Ball*›; versetzen ‹*Stoß, Schlag, Tritt*›; vortragen ‹*Angriff*›; **c)** *(hand over)* bringen; liefern ‹*Ware*›; zustellen ‹*Post, Telegramm*›; überbringen ‹*Botschaft*›; ~ **sth. to the door** etw. ins Haus liefern; ~ **[the goods]** *(fig.)* es schaffen *(ugs.); (fulfil promise)* halten, was man versprochen hat; **d)** *(give up)* aushändigen; **e)** *(render)* geben, liefern ‹*Bericht*›; stellen ‹*Ultimatum*›; **f)** *(assist in giving birth, aid in being born)* entbinden; **g)** *(save)* ~ **sb./sth. from sb./sth.** jmdn./etw. von jmdm./etw. erlösen

deliverance [dɪ'lɪvərəns] *n.* Erlösung, *die* (**from** von)

delivery [dɪ'lɪvərɪ] *n.* **a)** *(handing over)* Lieferung, *die; (of letters, parcels)* Zustellung, *die;* **take ~ of sth.** etw. annehmen; **pay on ~:** bei Lieferung bezahlen; *(Post)* per Nachnahme bezahlen; **b)** *(manner of uttering)* Vortragsweise, *die;* Vortrag, *der;* **c)** *(childbirth)* Entbindung, *die*

delivery: ~ date *n.* Liefertermin, *der;* ~ **van** *n.* Lieferwagen, *der*

dell [del] *n.* [bewaldetes] Tal

delphinium [del'fɪnɪəm] *n. (Bot.)* Rittersporn, *der*

delta ['deltə] *n.* Delta, *das*

delude [dɪ'lju:d, dɪ'lu:d] *v.t.* täuschen; ~ **sb. into believing that ...:** jmdm. weismachen, daß ...

deluge ['delju:dʒ] **1.** *n.* **a)** *(rain)* sintflutartiger Regen; **b)** *(Bibl.)* **the D~:** die Sintflut. **2.** *v.t. (lit. or fig.)* überschwemmen

delusion [dɪ'lju:ʒn, dɪ'lu:ʒn] *n.* Illusion, *die; (as symptom of madness)* Wahnvorstellung, *die;* **be under a ~:** einer Täuschung unterliegen; **be under the ~ that ...:** sich *(Dat.)* der Täuschung hingeben, daß ...

de luxe [də'lʌks, də'lu:ks] *adj.* Luxus-

delve [delv] *v.i.* ~ **into sth. [for sth.]** tief in etw. *(Akk.)* greifen[, um etw. herauszuholen]

demagogue (*Amer.:* **demagog**) ['deməgɒg] *n.* Demagoge, *der*/Demagogin, *die*

demand [dɪ'mɑ:nd] **1.** *n.* **a)** *(request)* Forderung, *die* (**for** nach); **final ~:** letzte Mahnung; **b)** *(desire for commodity)* Nachfrage, *die* (**for** nach); **by popular ~:** auf vielfachen Wunsch; **sth./sb. is in [great] ~:** etw. ist [sehr] gefragt/jmd. ist [sehr] begehrt; **c)** *(claim)* **make ~s on sb.** jmdn. beanspruchen. **2.** *v.t.* **a)** *(ask for, require, need)* verlangen (**of, from** von); fordern ‹*Recht, Genugtuung*›; ~ **to know/see sth.** etw. zu wissen/zu sehen verlangen; **b)** *(insist on being told)* unbedingt wissen wollen; **he ~ed my business** er fragte mich nachdrücklich, was ich wünschte

demanding [dɪ'mɑ:ndɪŋ] *adj.* anspruchsvoll

demarcate ['di:mɑ:keɪt] *v.t.* festlegen ‹*Grenze*›; demarkieren *(geh.)*

demarcation [di:mɑ:'keɪʃn] *n.* *(of frontier)* Demarkation, *die (geh.)*

demar'cation dispute *n.* Streit um die Abgrenzung der Zuständigkeitsbereiche

demeaning [dɪ'mi:nɪŋ] *adj.* erniedrigend

demeanour (*Brit.; Amer.:* **demeanor**) [dɪ'mi:nə(r)] *n.* Benehmen, *das*

demented [dɪ'mentɪd] *adj.* wahnsinnig

demerara [demə'reərə] *n.* ~ **[sugar]** brauner Zucker

demi- ['demɪ] *pref.* Halb-

'demigod *n.* Halbgott, *der*

demilitarize **(demilitarise)**

[di:'mılıtəraız] *v.t.* entmilitarisieren

demise [dı'maız] *n. (death)* Ableben, *das (geh.); (fig.)* Verschwinden, *das; (of firm, party, creed, etc.)* Untergang, *der*

demist [di:'mıst] *v.t. (Brit.)* trockenblasen; *(with cloth etc.)* trockenreiben

demister [di:'mıstə(r)] *n. (Brit.)* Defroster, *der;* Gebläse, *das*

demo ['deməʊ] *n., pl.* **~s** *(coll.)* Demo, *die (ugs.)*

demob [di:'mɒb] *(Brit. coll.) v.t.,* **-bb-** aus dem Kriegsdienst entlassen

demobilize (demobilise) [di:'məʊbılaız] *v.t.* demobilisieren ‹*Armee, Kriegsschiff*›; aus dem Kriegsdienst entlassen ‹*Soldat*›

democracy [dı'mɒkrəsı] *n.* Demokratie, *die*

democrat ['deməkræt] *n.* Demokrat, *der*/Demokratin, *die;* **D~** *(Amer. Polit.)* Demokrat, *der*/Demokratin, *die*

democratic [demə'krætık] *adj.* demokratisch; **D~ Party** *(Amer. Polit.)* Demokratische Partei

democratically [demə'krætıkəlı] *adv.* demokratisch

demolish [dı'mɒlıʃ] *v.t.* **a)** *(pull down)* abreißen; *(break to pieces)* zerstören; demolieren; **b)** abschaffen ‹*System, Privilegien*›; widerlegen, umstoßen ‹*Theorie*›; entkräften ‹*Einwand*›; zerstören ‹*Legende, Mythos*›

demolition [demə'lıʃn, di:mə'lıʃn] *n. see* **demolish a:** Abriß, *der;* Zerstörung, *die;* Demolierung, *die; attrib.* **~ contractors** Abbruchunternehmen, *das;* **~ work** Abbruchsarbeit, *die*

demon ['di:mən] *n.* **a)** Dämon, *der;* **b)** *(person)* Teufel, *der*

demonstrable ['demənstrəbl, dı'mɒnstrəbl] *adj.* beweisbar

demonstrably ['demənstrəblı, dı'mɒnstrəblı] *adv.* nachweislich

demonstrate ['demənstreıt] **1.** *v.t.* **a)** *(by examples, experiments, etc.)* zeigen; demonstrieren; *(show, explain)* vorführen ‹*Vorrichtung, Gerät*›; **b)** *(be, provide, proof of)* beweisen; **c)** zeigen ‹*Gefühl, Bedürfnis, Gutwilligkeit*›. **2.** *v.i.* **a)** *(protest etc.)* demonstrieren; **b) ~ on sth./sb.** etw./jmdn. als Demonstrationsobjekt benutzen

demonstration [demən'streıʃn] *n.* **a)** *(also meeting, procession)* Demonstration, *die;* **b)** *(showing of appliances etc.)* Vorführung, *die;* **give sb. a ~ of sth.** jmdm. etw. vorführen; **c)** *(proof)* Beweis, *der*

demonstrative [dı'mɒnstrətıv] *adj.* **a)** offen ‹*Person*›; **b)** *(Ling.)* Demonstrativ-; hinweisend

demonstrator ['demənstreıtə(r)] *n. (protestor etc.)* Demonstrant, *der*/Demonstrantin, *die*

demoralisation, demoralise *see* **demoraliz-**

demoralization [dımɒrəlaı'zeıʃn] *n.* Demoralisierung, *die*

demoralize [dı'mɒrəlaız] *v.t.* demoralisieren

demote [di:'məʊt] *v.t.* degradieren **(to** zu)

demotion [di:'məʊʃn] *n.* Degradierung, *die* **(to** zu)

demur [dı'mɜ:(r)] *v.i.,* **-rr-** Einwände erheben **(to** gegen)

demure [dı'mjʊə(r)] *adj.* **a)** *(affectedly quiet and serious)* betont zurückhaltend; **b)** *(grave, composed)* ernst; gesetzt ‹*Benehmen*›

den [den] *n.* **a)** *(of wild beast)* Höhle, *die;* **fox's ~:** Fuchsbau, *der;* **b) ~ of thieves, thieves' ~:** Diebeshöhle, *die;* Diebesnest, *das;* **c)** *(coll.: small room)* Bude, *die (ugs.)*

denationalize (denationalise) [di:'næʃənəlaız] *v.t.* privatisieren

denial [dı'naıəl] *n. (refusal)* Verweigerung, *die; (of request, wish)* Ablehnung, *die*

denigrate ['denıgreıt] *v.t.* verunglimpfen

denim ['denım] *n.* **a)** *(fabric)* Denim ®, *der;* Jeansstoff, *der;* **~ jacket** Jeansjacke, *die;* **b)** *in pl. (garment)* Bluejeans *Pl.*

Denmark ['denmɑ:k] *pr. n.* Dänemark *(das)*

denomination [dınɒmı'neıʃn] *n.* **a)** *(class of units)* Einheit, *die;* **coins/paper money of the smallest ~:** Münzen/Papiergeld mit dem geringsten Nennwert; **b)** *(Relig.)* Glaubensgemeinschaft, *die;* Konfession, *die*

denominator [dı'nɒmıneıtə(r)] *n. (Math.)* Nenner, *der; see also* **common denominator**

denote [dı'nəʊt] *v.t.* **a)** *(indicate)* hindeuten auf (+ *Akk.*); **~ that ...:** darauf hindeuten, daß ...; **b)** *(designate)* bedeuten

dénouement [deı'nu:mɑ̃] *n.* Ausgang, *der;* Auflösung, *die*

denounce [dı'naʊns] *v.t. (inform against)* denunzieren **(to** bei); *(accuse publicly)* beschuldigen; **~ sb. as a spy** jmdn. beschuldigen, ein Spion zu sein

dense [dens] *adj.* **a)** dicht; massiv ‹*Körper*›; **b)** *(crowded together)* dichtgedrängt; eng ‹*Schrift*›; **c)** *(stupid)* dumm; **he's pretty ~:** er ist ziemlich schwer von Begriff

densely ['denslı] *adv.* dicht; **~ packed** dichtgedrängt

denseness ['densnıs] *n., no pl.* **a)** Dichte, *die;* **b)** *(stupidity)* Begriffsstutzigkeit, *die*

density ['densıtı] *n. (also Phys.)* Dichte, *die;* **population ~:** Bevölkerungsdichte, *die*

dent [dent] **1.** *n.* Beule, *die; (fig. coll.)* Loch, *das.* **2.** *v.t.* einbeulen; eindellen *(ugs.)* ‹*Holz, Tisch*›; *(fig.)* anknacksen *(ugs.)*

dental ['dentl] *adj.* Zahn-; **~ care** Zahnpflege, *die*

'dental surgeon *n.* Zahnarzt, *der*/-ärztin, *die*

dentist ['dentıst] *n.* Zahnarzt, *der*/-ärztin, *die;* **at the ~['s]** beim Zahnarzt

dentistry ['dentıstrı] *n., no pl.* Zahnheilkunde, *die*

denture ['dentʃə(r)] *n.* **~[s]** Zahnprothese, *die;* [künstliches] Gebiß; **partial ~:** Teilprothese, *die*

denunciation [dınʌnsı'eıʃn] *n.* Denunziation, *die; (public accusation)* Beschuldigung, *die*

deny [dı'naı] *v.t.* **a)** *(declare untrue)* bestreiten; zurückweisen ‹*Beschuldigung*›; **there is no ~ing the fact that ...:** es läßt sich nicht bestreiten *od.* leugnen, daß ...; **~ all knowledge of sth.** bestreiten, irgendetwas von etw. zu wissen; **b)** *(refuse)* verweigern; **~ sb. sth.** jmdm. etw. verweigern; **c)** *(disavow, repudiate; refuse access to)* verleugnen; ablehnen ‹*Verantwortung*›

deodorant [di:'əʊdərənt] **1.** *adj.* **~ spray** Deo[dorant]spray, *der od. das.* **2.** *n.* Deodorant, *das*

dep. *abbr.* **departs** *(Railw.)* Abf.; *(Aeronaut.)* Abfl.

depart [dı'pɑ:t] **1.** *v.i.* **a)** *(go away, take one's leave)* weggehen; fortgehen; **b)** *(set out, start, leave)* abfahren; ‹*Flugzeug:*› abfliegen; *(on one's journey)* abreisen; **c)** *(fig.: deviate)* **~ from sth.** von etw. abweichen. **2.** *v.t. (literary)* **~ this life/world** aus dem Leben/aus dieser Welt scheiden *(geh.)*

departed [dı'pɑ:tıd] **1.** *adj. (deceased)* dahingeschieden *(geh. verhüll.).* **2.** *n.* **the ~:** der/die Dahingeschiedene/die Dahingeschiedenen *(geh. verhüll.)*

department [dı'pɑ:tmənt] *n.* **a)** *(of municipal administration)* Amt, *das; (of State administration)* Ministerium, *das; (of university)* Seminar, *das; (of shop)* Abteilung, *die;* **D~ of Employment/Education** Arbeits-/Erziehungsministerium, *das;* **the personnel**

~: die Personalabteilung; **b)** *(fig.: area of activity)* Ressort, *das;* **it's not my ~** *(not my responsibility)* dafür bin ich nicht zuständig

departmental [di:pɑ:t'mentl] *adj. see* **department a**: Amts-; Ministerial-; Seminar-; Abteilungs-

de'partment store *n.* Kaufhaus, *das*

departure [dɪ'pɑ:tʃə(r)] *n.* **a)** *(going away)* Abreise, *die;* **b)** *(deviation)* **~ from sth.** Abweichen von etw.; **c)** *(of train, bus, ship)* Abfahrt, *die; (of aircraft)* Abflug, *der;* **d) point of ~**: Ansatzpunkt, *der;* **this product is a new ~ for us** mit diesem Produkt schlagen wir einen neuen Weg ein

departure: ~ lounge *n.* Abflughalle, *die;* **~ time** *n. (of train, bus)* Abfahrtzeit, *die; (of aircraft)* Abflugzeit, *die*

depend [dɪ'pend] *v.i.* **a) ~ [up]on** abhängen von; **it [all] ~s on whether/what/how ...**: das hängt [ganz] davon ab *od.* kommt ganz darauf an, ob/was/wie ...; **that ~s** es kommt darauf an; **~ing on how ...**: je nachdem, wie ...; **b)** *(rely, trust)* **~ [up]on** sich verlassen auf (+ *Akk.*); *(have to rely on)* angewiesen sein auf (+ *Akk.*)

dependable [dɪ'pendəbl] *adj.* verläßlich; zuverlässig

dependant [dɪ'pendənt] *n.* Abhängige, *der/die*

dependence [dɪ'pendəns] *n.* **a)** Abhängigkeit, *die* (**[up]on** von); **b)** *(reliance)* **put** *or* **place ~ [up]on sb.** sich auf jmdn. verlassen

dependency [dɪ'pendənsɪ] *n.* *(country)* Territorium, *das*

dependent [dɪ'pendənt] **1.** *n. see* **dependant. 2.** *adj.* **a)** *(also Ling.)* abhängig; **be ~ on sth.** von etw. abhängen *od.* abhängig sein; **b) be ~ on** *(be unable to do without)* angewiesen sein auf (+ *Akk.*); abhängig sein von ‹*Droge*›

depict [dɪ'pɪkt] *v.t.* darstellen

depilatory [dɪ'pɪlətərɪ] *n.* Enthaarungsmittel, *das*

deplete [dɪ'pli:t] *v.t.* erheblich verringern; **our stores are ~d** unser Vorrat ist zusammengeschrumpft

depletion [dɪ'pli:ʃn] *n.* Verringerung, *die*

deplorable [dɪ'plɔ:rəbl] *adj.* beklagenswert

deplore [dɪ'plɔ:(r)] *v.t.* **a)** *(disapprove of)* verurteilen; **b)** *(bewail, regret)* beklagen; **sth. is to be ~d** etw. ist beklagenswert

deploy [dɪ'plɔɪ] **1.** *v.t. (also Mil.)* einsetzen. **2.** *v.i. (Mil.)* eingesetzt werden

deployment [dɪ'plɔɪmənt] *n.* Einsatz, *der*

depopulate [di:'pɒpjʊleɪt] *v.t.* entvölkern

deport [dɪ'pɔ:t] *v.t.* deportieren; *(from country)* ausweisen

deportation [di:pɔ:'teɪʃn] *n.* Deportation, *die; (from country)* Ausweisung, *die*

depose [dɪ'pəʊz] *v.t.* absetzen

deposit [dɪ'pɒzɪt] **1.** *n.* **a)** *(in bank)* Depot, *das; (credit)* Guthaben, *das; (Brit.: at interest)* Sparguthaben, *das;* **make a ~**: etwas einzahlen; **b)** *(payment as pledge)* Kaution, *die; (first instalment)* Anzahlung, *die;* **pay a ~**: eine Kaution zahlen; eine Anzahlung leisten; **there is a five pence ~ on the bottle** auf der Flasche sind fünf Pence Pfand; **c)** *(of sand, mud, lime, etc.)* Ablagerung, *die; (of ore, coal, oil)* Lagerstätte, *die; (in glass, bottle)* Bodensatz, *der.* **2.** *v.t.* **a)** *(put down in a place)* ablegen; abstellen ‹*etw. Senkrechtes, auch Tablett, Teller usw.*›; absetzen ‹*Mitfahrer*›; **b)** *(leave lying)* ‹*Wasser usw.*:› ablagern; **c)** *(in bank)* deponieren, [auf ein Konto] einzahlen ‹*Geld*›; *(Brit.: at interest)* [auf ein Sparkonto] einzahlen

de'posit account *n. (Brit.)* Sparkonto, *das*

depositor [dɪ'pɒzɪtə(r)] *n. (Banking)* Einleger, *der*/Einlegerin, *die*

depository [dɪ'pɒzɪtərɪ] *n. (storehouse)* Lagerhaus, *das; (place for safe keeping)* Aufbewahrungsort, *der; (fig.)* Fundgrube, *die*

depot ['depəʊ] *n.* **a)** Depot, *das;* **b)** *(storehouse)* Lager, *das;* **c) [bus] ~** *(Brit.)* Depot, *das;* Omnibusgarage, *die; (Amer.: bus station)* Omnibusbahnhof, *der; (Amer.: railway station)* Bahnhof, *der*

depraved [dɪ'preɪvd] *adj.* verdorben; lasterhaft ‹*Gewohnheit*›

deprecate ['deprɪkeɪt] *v.t. (disapprove of)* mißbilligen

depreciate [dɪ'pri:ʃɪeɪt, dɪ'pri:sɪeɪt] **1.** *v.t.* abwerten. **2.** *v.i.* an Wert verlieren

depreciation [dɪpri:ʃɪ'eɪʃn, dɪpri:sɪ'eɪʃn] *n. (of money, currency, property)* Wertverlust, *der*

depress [dɪ'pres] *v.t.* **a)** *(deject)* deprimieren; **b)** *(push or pull down)* herunterdrücken; **c)** *(reduce activity of)* unterdrücken; sich nicht entfalten lassen ‹*Handel, Wirtschaftswachstum*›

depressant [dɪ'presənt] *(Med.)* **1.** *adj.* beruhigend; sedativ *(fachspr.).* **2.** *n.* Beruhigungsmittel, *das;* Sedativ[um], *das (fachspr.)*

depressed [dɪ'prest] *adj.* deprimiert ‹*Person, Stimmung*›; geschwächt ‹*Industrie*›; **~ area** unter [wirtschaftlicher] Depression leidendes Gebiet

depressing [dɪ'presɪŋ] *adj.*, **depressingly** [dɪ'presɪŋlɪ] *adv.* deprimierend

depression [dɪ'preʃn] *n.* **a)** Depression, *die;* **b)** *(sunk place)* Vertiefung, *die;* **c)** *(Meteorol.)* Tief[druckgebiet], *das;* **d)** *(Econ.)* **the D~**: die Weltwirtschaftskrise; **economic ~**: Wirtschaftskrise, *die;* Depression, *die*

depressive [dɪ'presɪv] *adj.* bedrückend; deprimierend

deprival [dɪ'praɪvl], **deprivation** [deprɪ'veɪʃn] *ns.* Entzug, *der; (of one's rights, liberties, or title)* Aberkennung, *die*

deprive [dɪ'praɪv] *v.t.* **a) ~ sb. of sth.** jmdm. etw. nehmen; *(debar from having)* jmdm. etw. vorenthalten; **~ sb. of citizenship** jmdm. die Staatsbürgerschaft aberkennen; **be ~d of light** nicht genug Licht haben; **b)** *(prevent from having normal life)* benachteiligen

deprived [dɪ'praɪvd] *adj.* benachteiligt ‹*Kind, Familie usw.*›

Dept. *abbr.* **Department** Amt/Min./Seminar/Abt.

depth [depθ] *n.* **a)** *(lit. or fig.)* Tiefe, *die;* **at a ~ of 3 metres** in einer Tiefe von 3 Metern; **3 feet in ~**: 3 Fuß tief; **what is the ~ of the pond?** wie tief ist der Teich?; **from/in the ~s of the forest/ocean** aus/in der Tiefe des Waldes/des Ozeans; **in the ~s of winter** im tiefen Winter; **b) in ~**: gründlich, intensiv ‹*studieren*›; **an in-~ study/analysis** *etc.* eine gründliche Untersuchung/Analyse *usw.;* **c) be out of one's ~**: nicht mehr stehen können; keinen Grund mehr unter den Füßen haben; *(fig.)* ins Schwimmen kommen *(ugs.);* **get out of one's ~** *(lit. or fig.)* den Grund unter den Füßen verlieren

'depth-charge *n.* Wasserbombe, *die*

deputation [depjʊ'teɪʃn] *n.* Abordnung, *die;* Delegation, *die*

depute [dɪ'pju:t] *v.t.* **a)** *(commit task or authority to)* **~ sb. to do sth.** jmdn. beauftragen, etw. zu tun; **b)** *(appoint as deputy)* **~ sb. to do sth.** jmdn. [als Stellvertreter] damit betrauen, etw. zu tun

deputize (deputise) ['depjʊtaɪz] *v.i.* als Stellvertreter einspringen; **~ for sb.** jmdn. vertreten

deputy ['depjʊtɪ] *n.* **a)** [Stell]vertreter, *der*/-vertreterin, *die; at-*

trib. stellvertretend; **act as ~ for sb.** jmdn. vertreten; **b)** *(parliamentary representative)* Abgeordnete, *der/die*

derail [dɪ'reɪl, di:'reɪl] *v.t.* **be ~ed** entgleisen

derailment [dɪ'reɪlmənt, di:'reɪlmənt] *n.* Entgleisung, *die*

deranged [dɪ'reɪndʒd] *adj.* **[mentally] ~:** geistesgestört

derelict ['derɪlɪkt] **1.** *adj.* verlassen und verfallen. **2.** *n. (person)* Ausgestoßene, *der/die*

dereliction [derɪ'lɪkʃn] *n.* **a)** Vernachlässigung, *die; (state)* verkommener Zustand; **b) ~ of duty** Pflichtverletzung, *die*

deride [dɪ'raɪd] *v.t. (treat with scorn)* sich lustig machen über (+ *Akk.*); *(laugh scornfully at)* verlachen

derision [dɪ'rɪʒn] *n.* Spott, *der;* **be an object of ~:** Zielscheibe des Spottes sein

derisive [dɪ'raɪsɪv] *adj. (ironical)* spöttisch; *(scoffing)* verächtlich

derisory [dɪ'raɪsərɪ, dɪ'raɪzərɪ] *adj.* **a)** *(ridiculously inadequate)* lächerlich; **b)** *(scoffing)* verächtlich; *(ironical)* spöttisch

derivation [derɪ'veɪʃn] *n.* **a)** *(obtaining from a source)* Herleitung, *die;* **b)** *(extraction, origin)* Herkunft, *die;* **c)** *(Ling.)* Ableitung, *die;* Derivation, *die (fachspr.)*

derivative [dɪ'rɪvətɪv] **1.** *adj.* abgeleitet; *(lacking originality)* nachahmend; epigonal. **2.** *n. (word)* Ableitung, *die*

derive [dɪ'raɪv] **1.** *v.t.* **~ sth. from sth.** etw. aus etw. gewinnen; **the river ~s its name from a Greek god** der Name des Flusses geht auf eine griechische Gottheit zurück; **~ pleasure from sth.** Freude an etw. *(Dat.)* haben. **2.** *v.i.* **~ from** beruhen auf (+ *Dat.*); **the word ~s from Latin** das Wort stammt *od.* kommt aus dem Lateinischen

dermatitis [dɜ:mə'taɪtɪs] *n. (Med.)* Hautentzündung, *die*

derogatory [dɪ'rɒgətərɪ] *adj.* abfällig; abschätzig; **~ sense [of a word]** abwertende Bedeutung [eines Wortes]

derrick ['derɪk] *n.* [Derrick]kran, *der; (over oil-well)* Bohrturm, *der*

derv [dɜ:v] *n. (Brit. Motor Veh.)* Diesel[kraftstoff], *der*

dervish ['dɜ:vɪʃ] *n.* Derwisch, *der*

descant ['deskænt] *n. (Mus.)* Diskant, *der*

'descant recorder *n. (Mus.)* Sopranflöte, *die*

descend [dɪ'send] **1.** *v.i.* **a)** *(go down)* hinuntergehen/-steigen/-klettern/-fahren; *(come down)* herunterkommen; *(sink)* niedergehen **(on** auf + *Dat.*); **the lift ~ed** der Aufzug fuhr nach unten; **~ in the lift** mit dem Aufzug nach unten fahren; **b)** *(slope downwards)* abfallen; **the hill ~s into/towards the sea** der Hügel fällt zum Meer hin ab; **c)** *(in quality, thought, etc.)* herabsinken; **d)** *(in pitch)* fallen; sinken; **e)** *(make sudden attack)* **~ on sth.** über etw. *(Akk.)* herfallen; **~ on sb.** *(lit., or fig.: arrive unexpectedly)* jmdn. überfallen; **f)** *(fig.: lower oneself)* **~ to sth.** sich zu etw. erniedrigen; **g)** *(derive)* abstammen **(from** von); *(have origin)* zurückgehen **(from** auf + *Akk.*). **2.** *v.t. (go/come down)* hinunter- / heruntergehen / -steigen / -klettern/-fahren; hinab-/herabsteigen *(geh.)*

descendant [dɪ'sendənt] *n.* Nachkomme, *der;* **be ~s/a ~ of** abstammen von

descended [dɪ'sendɪd] *adj.* **be ~ from sb.** von jmdm. abstammen

descent [dɪ'sent] *n.* **a)** *(of person)* Abstieg, *der; (of parachute, plane, bird, avalanche)* Niedergehen, *das;* **b)** *(way)* Abstieg, *der;* **c)** *(slope)* Abfall, *der;* **the ~ was very steep** das Gefälle war sehr stark; **d)** *(lineage)* Abstammung, *die;* Herkunft, *die;* **be of Russian ~:** russischer Abstammung sein

describe [dɪ'skraɪb] *v.t.* **a)** beschreiben; schildern; **~ [oneself] as ...:** [sich] als ... bezeichnen; **b)** *(move in, draw)* beschreiben ⟨*Kreis, Bogen, Kurve*⟩

description [dɪ'skrɪpʃn] *n.* **a)** Beschreibung, *die;* Schilderung, *die;* **he answers [to]** *or* **fits the ~:** er entspricht der Beschreibung *(Dat.);* **b)** *(sort, class)* **cars of every ~:** Autos aller Art; **c)** *(designation)* Bezeichnung, *die*

descriptive [dɪ'skrɪptɪv] *adj.* **a)** anschaulich; beschreibend ⟨*Lyrik*⟩; deskriptiv ⟨*Analyse*⟩; **b)** *(not expressing feelings or judgements)* deskriptiv

desecrate ['desɪkreɪt] *v.t.* entweihen; schänden

desegregate [di:'segrɪgeɪt] *v.t.* die Rassentrennung aufheben an (+ *Dat.*)

¹desert [dɪ'zɜ:t] *n. in pl. (what is deserved)* Verdienste *Pl.;* **get one's [just] ~s** das bekommen, was man verdient hat

²desert ['dezət] **1.** *n.* Wüste, *die; (fig.)* Einöde, *die;* **the Sahara ~:** die Wüste Sahara. **2.** *adj.* öde; Wüsten⟨*klima, -stamm*⟩

³desert [dɪ'zɜ:t] **1.** *v.t.* verlassen; im Stich lassen ⟨*Frau, Familie usw.*⟩. **2.** *v.i.* ⟨*Soldat:*⟩ desertieren

deserted [dɪ'zɜ:tɪd] *adj.* verlassen; **the streets were ~:** die Straßen waren wie ausgestorben

deserter [dɪ'zɜ:tə(r)] *n.* Deserteur, *der;* Fahnenflüchtige, *der/die*

desertion [dɪ'zɜ:ʃn] *n.* Verlassen, *das; (Mil.)* Desertion, *die;* Fahnenflucht, *die;* **~ to the enemy** Überlaufen zum Feind

desert island [dezət 'aɪlənd] *n.* einsame Insel

deserve [dɪ'zɜ:v] *v.t.* verdienen; **he ~s to be punished** er verdient [es], bestraft zu werden; **what have I done to ~ this?** womit habe ich das verdient?; **he got what he ~d** er hat es nicht besser verdient

deservedly [dɪ'zɜ:vɪdlɪ] *adv.* verdientermaßen

deserving [dɪ'zɜ:vɪŋ] *adj.* **a)** *(worthy)* verdienstvoll; **donate money to a ~ cause** Geld für einen guten Zweck geben; **b) be ~ of sth.** etw. verdienen

desiccated ['desɪkeɪtɪd] *adj.* getrocknet; *(fig.)* vertrocknet ⟨*Person*⟩

design [dɪ'zaɪn] **1.** *n.* **a)** *(preliminary sketch)* Entwurf, *der;* **b)** *(pattern)* Muster, *das;* **c)** *no art. (art)* Design, *das;* Gestaltung, *die (geh.);* **d)** *(established form of a product)* Entwurf, *der; (of machine, engine, etc.)* Bauweise, *die;* **e)** *(general idea, construction from parts)* Konstruktion, *die;* **f)** *in pl.* **have ~s on sb./sth.** es auf jmdn./etw. abgesehen haben; **g)** *(purpose)* Absicht, *die;* **by ~:** mit Absicht; absichtlich; **h)** *(end in view)* Ziel, *das.* **2.** *v.t.* **a)** *(draw plan of, sketch)* entwerfen; konstruieren, entwerfen ⟨*Maschine, Fahrzeug, Flugzeug*⟩; **b) be ~ed to do sth.** ⟨*Maschine, Werkzeug, Gerät:*⟩ etw. tun sollen; **c)** *(set apart)* vorsehen; **be ~ed for sb./sth.** für jmdn./etw. gedacht *od.* vorgesehen sein

designate 1. ['dezɪgnət] *postpos. adj.* designiert. **2.** ['dezɪgneɪt] *v.t.* **a)** *(serve as name of, describe)* bezeichnen; *(serve as distinctive mark of)* kennzeichnen; **b)** *(appoint to office)* designieren *(geh.)*

designation [dezɪg'neɪʃn] *n.* **a)** Bezeichnung, *die;* **b)** *(appointing to office)* Designation, *die*

designer [dɪ'zaɪnə(r)] *n.* Designer, *der*/Designerin, *die; (of machines, buildings)* Konstrukteur, *der*/Konstrukteurin, *die; attrib.* Modell⟨*kleidung, -jeans*⟩

desirable [dɪˈzaɪərəbl] *adj.* **a)** *(worth having or wishing for)* wünschenswert; **'knowledge of French ~'** „Französischkenntnisse erwünscht"; **b)** *(causing desire)* attraktiv; begehrenswert ‹*Frau*›

desire [dɪˈzaɪə(r)] **1.** *n.* **a)** *(wish, request)* Wunsch, *der* (**for** nach); *(longing)* Sehnsucht, *die* (**for** nach); **~ to do sth.** Wunsch, etw. zu tun; **I have no ~ to see him** ich habe nicht den Wunsch, ihn zu sehen; **b)** *(thing ~d)* **she is my heart's ~:** sie ist die Frau meines Herzens; **c)** *(lust)* Verlangen, *das;* **fleshly ~s** fleischliche Begierden. **2.** *v. t.* **a)** *(wish)* sich *(Dat.)* wünschen; *(long for)* sich sehnen nach; **he only ~d her happiness** er wollte nur ihr Glück; **leave much to be ~d** viel zu wünschen übriglassen; **b)** *(request)* wünschen; **c)** *(sexually)* begehren ‹*Mann, Frau*›

desirous [dɪˈzaɪərəs] *pred. adj.* *(formal)* **be ~ of sth.** etw. wünschen

desist [dɪˈzɪst, dɪˈsɪst] *v. i. (literary)* einhalten *(geh.);* **~ from sth.** von etw. ablassen *(geh.);* **~ in one's efforts to do sth.** von seinen Bemühungen ablassen, etw. zu tun

desk [desk] *n.* **a)** Schreibtisch, *der; (in school)* Tisch, *der; (teacher's raised ~)* Pult, *das;* **~ copy** Arbeitsexemplar, *das;* **b)** *(for cashier)* Kasse, *die; (for receptionist)* Rezeption, *die;* **information ~:** Auskunft, *die;* **sales ~:** Verkauf, *der;* **c)** *(section of newspaper office)* Ressort, *das*

desk: ~ calendar, ~ diary *ns.* Tischkalender, *der;* **~ lamp** *n.* Schreibtischlampe, *die;* **~-top** *adj.* **~-top publishing** Desktop publishing, *das;* **~-top computer** Tischcomputer, *der*

desolate 1. [ˈdesələt] *adj.* **a)** *(ruinous, neglected, barren)* trostlos ‹*Haus, Ort*›; desolat ‹*Zustand*›; **b)** *(uninhabited)* öde; verlassen; **c)** *(forlorn, wretched)* trostlos ‹*Leben*›; verzweifelt ‹*Schrei*›. **2.** [ˈdesəleɪt] *v. t.* **a)** *(devastate)* verwüsten ‹*Land*›; **b)** *(make wretched)* in Verzweiflung stürzen

desolation [desəˈleɪʃn] *n.* **a)** *(desolating)* Verwüstung, *die;* **b)** *(neglected or barren state)* Öde, *die; (state of ruin)* Verwüstung, *die;* **c)** *(loneliness, being forsaken)* Verlassenheit, *die;* **d)** *(wretchedness)* Verzweiflung, *die*

despair [dɪˈspeə(r)] **1.** *n.* **a)** Verzweiflung, *die;* **b)** *(cause)* **be the ~ of sb.** jmdn. zur Verzweiflung bringen. **2.** *v. i.* verzweifeln; **~ of doing sth.** die Hoffnung aufgeben, etw. zu tun; **~ of sth.** die Hoffnung auf etw. *(Akk.)* aufgeben

despatch *(Brit.) see* **dispatch**

desperate [ˈdespərət] *adj.* **a)** verzweifelt; *(coll.: urgent)* dringend; **get** *or* **become ~:** verzweifeln; **feel ~:** verzweifelt sein; **be ~ for sth.** etw. dringend brauchen; **b)** extrem ‹*Maßnahmen, Lösung*›; **c)** verzweifelt ‹*Lage, Situation*›

desperately [ˈdespərətlɪ] *adv.* **a)** verzweifelt; **be ~ ill** *or* **sick** todkrank sein; **b)** *(appallingly, shockingly, extremely)* schrecklich *(ugs.)*

desperation [despəˈreɪʃn] *n.* Verzweiflung, *die;* **out of** *or* **in [sheer] ~:** aus [lauter] Verzweiflung

despicable [ˈdespɪkəbl] *adj.* verabscheuungswürdig

despise [dɪˈspaɪz] *v. t.* verachten

despite [dɪˈspaɪt] *prep.* trotz; **~ what she said** ungeachtet dessen, was sie sagte

despondency [dɪˈspɒndənsɪ] *n., no pl.* Niedergeschlagenheit, *die*

despondent [dɪˈspɒndənt] *adj.* niedergeschlagen; bedrückt; **be ~ about sth.** wegen etw. *od.* über etw. *(Akk.)* bedrückt sein; **feel ~:** niedergeschlagen sein; **grow** *or* **get ~:** mutlos werden

despot [ˈdespət] *n.* Despot, *der*

despotic [dɪˈspɒtɪk] *adj.* despotisch

despotism [ˈdespətɪzm] *n.* Despotie, *die; (political system)* Despotismus, *der*

dessert [dɪˈzɜ:t] *n.* **a)** süße Nachspeise; **b)** *(Brit.: after dinner)* Dessert, *das;* Nachtisch, *der*

dessert: ~spoon *n.* Dessertlöffel, *der;* **~spoonful** *n.* Eßlöffel, *der;* **a ~spoonful** ein Eßlöffel

destination [destɪˈneɪʃn] *n.* *(of person)* Reiseziel, *das; (of goods)* Bestimmungsort, *der; (of train, bus)* Zielort, *der;* **arrive at one's ~:** am Ziel ankommen

destine [ˈdestɪn] *v. t.* bestimmen; **~ sb. for sth.** jmdn. für etw. bestimmen; ‹*Schicksal:*› jmdn. für etw. vorbestimmen; **be ~d to do sth.** dazu ausersehen *od.* bestimmt sein, etw. zu tun; **we were ~d [never] to meet again** wir sollten uns [nie] wiedersehen

destiny [ˈdestɪnɪ] *n.* **a)** Schicksal, *das;* Los, *das;* **b)** *no art. (power)* das Schicksal

destitute [ˈdestɪtju:t] *adj.* mittellos; **the ~:** die Mittellosen

destitution [destɪˈtju:ʃn] *n., no pl.* Armut, *die;* Not, *die*

destroy [dɪˈstrɔɪ] *v. t.* **a)** zerstören; kaputtmachen *(ugs.)* ‹*Tisch, Stuhl, Uhr, Schachtel*›; vernichten ‹*Ernte, Papiere, Dokumente*›; **b)** *(kill, annihilate)* vernichten ‹*Feind, Insekten*›; **the dog will have to be ~ed** der Hund muß eingeschläfert werden; **c)** *(fig.)* zunichte machen ‹*Hoffnungen, Chancen*›; ruinieren ‹*Zukunft*›; zerstören ‹*Glück, Freundschaft*›

destroyer [dɪˈstrɔɪə(r)] *n.* *(also Naut.)* Zerstörer, *der*

destruction [dɪˈstrʌkʃn] *n.* **a)** Zerstörung, *die;* **b)** *(cause of ruin)* Untergang, *der*

destructive [dɪˈstrʌktɪv] *adj.* zerstörerisch; verheerend ‹*Sturm, Feuer, Krieg*›; zersetzend ‹*Einfluß, Haltung, Tendenz*›; destruktiv ‹*Person, Kritik, Vorstellung, Einfluß, Ziel*›

desultory [ˈdesəltərɪ] *adj.* **a)** sprunghaft; zwanglos, ungezwungen ‹*Gespräch*›; **b)** *(unmethodical)* planlos

detach [dɪˈtætʃ] *v. t.* **a)** entfernen; ablösen ‹*Aufgeklebtes*›; abbrechen ‹*Angewachsenes*›; abtrennen ‹*zu Entfernendes*›; abnehmen ‹*wieder zu Befestigendes*›; abhängen ‹*Angekuppeltes*›; herausnehmen ‹*innen Befindliches*›; **b)** *(Mil., Navy)* abkommandieren (**from** aus)

detachable [dɪˈtætʃəbl] *adj.* abnehmbar; herausnehmbar ‹*Futter*›

detached [dɪˈtætʃt] *adj.* **a)** *(impartial)* unvoreingenommen; *(unemotional)* unbeteiligt; **b)** **~ house** Einzelhaus, *das*

detachment [dɪˈtætʃmənt] *n.* **a)** *(detaching) see* **detach a:** Entfernen, *das;* Ablösen, *das;* Abbrechen, *das;* Abtrennen, *das;* Abnehmen, *das;* Abhängen, *das;* Herausnehmen, *das;* **b)** *(Mil., Navy)* Abteilung, *die;* **c)** *(independence of judgement)* Unvoreingenommenheit, *die*

detail [ˈdi:teɪl] **1.** *n.* **a)** *(item)* Einzelheit, *die;* Detail, *das;* **enter** *or* **go into ~s** ins Detail gehen; auf Einzelheiten eingehen; **b)** *(dealing with things item by item)* **in ~:** Punkt für Punkt; **in great** *or* **much ~:** in allen Einzelheiten; **go into ~:** ins Detail gehen; auf Einzelheiten eingehen; **c)** *(in building, picture, etc.)* Detail, *das.* **2.** *v. t.* **a)** *(list)* einzeln aufführen; **b)** *(Mil.)* abkommandieren

detailed [ˈdi:teɪld] *adj.* detailliert; eingehend ‹*Studie*›

detain [dɪˈteɪn] *v. t.* **a)** *(keep in confinement)* festhalten; *(take*

into confinement) verhaften; **b)** *(delay)* aufhalten
detainee [dɪteɪ'ni:] *n.* Verhaftete, *der/die*
detect [dɪ'tekt] *v. t.* entdecken; bemerken ‹*Trauer, Verärgerung*›; wahrnehmen ‹*Bewegung*›; aufdecken ‹*Irrtum, Verbrechen*›; durchschauen ‹*Beweggrund*›; feststellen ‹*Strahlung*›
detectable [dɪ'tektəbl] *adj.* feststellbar; wahrnehmbar ‹*Bewegung*›
detection [dɪ'tekʃn] *n.* **a)** *see* **detect:** Entdeckung, *die;* Bemerken, *das;* Wahrnehmung, *die;* Aufdeckung, *die;* Durchschauen, *das;* Feststellung, *die;* **try to escape ~:** versuchen, unentdeckt zu bleiben; **b)** *(work of detective)* Ermittlungsarbeit, *die*
detective [dɪ'tektɪv] **1.** *n.* Detektiv, *der; (policeman)* Kriminalbeamte, *der*/Kriminalbeamtin, *die;* **private ~:** Privatdetektiv, *der.* **2.** *attrib. adj.* Kriminal-; **~ work** Ermittlungsarbeit, *die;* **~ story** Detektivgeschichte, *die*
detector [dɪ'tektə(r)] *n.* Detektor, *der*
détente [deɪ'tɑ̃t] *n. (Polit.)* Entspannung, *die*
detention [dɪ'tenʃn] *n.* **a)** Festnahme, *die; (confinement)* Haft, *die;* **b)** *(Sch.)* Nachsitzen, *das*
deter [dɪ'tɜ:(r)] *v. t.,* **-rr-** abschrecken; **~ sb. from sth.** jmdn. von etw. abhalten; **~ sb. from doing sth.** jmdn. davon abhalten, etw. zu tun; **be ~red by sth.** sich durch etw. abschrecken lassen
detergent [dɪ'tɜ:dʒənt] *n.* Reinigungsmittel, *das; (for washing)* Waschmittel, *das*
deteriorate [dɪ'tɪərɪəreɪt] *v. i.* sich verschlechtern; ‹*Haus:*› verfallen, verkommen; ‹*Holz, Leder:*› verrotten; **his work has ~d** seine Arbeit hat nachgelassen
deterioration [dɪtɪərɪə'reɪʃn] *n. see* **deteriorate:** Verschlechterung, *die;* Verfall, *der;* Verrottung, *die*
determinate [dɪ'tɜ:mɪnət] *adj.* **a)** *(limited, finite)* begrenzt; **b)** *(distinct)* bestimmt
determination [dɪtɜ:mɪ'neɪʃn] *n.* **a)** *(ascertaining, defining)* Bestimmung, *die;* **b)** *(resoluteness)* Entschlossenheit, *die;* **with [sudden] ~:** [kurz] entschlossen; **c)** *(intention)* [feste] Absicht
determine [dɪ'tɜ:mɪn] **1.** *v. t.* **a)** *(decide)* beschließen; **b)** *(make decide)* veranlassen; **~ sb. to do sth.** jmdn. dazu veranlassen, etw. zu tun; **c)** *(be a decisive factor for)* bestimmen; **d)** *(ascertain, define)* feststellen; bestimmen. **2.** *v. i. (decide)* **~ on doing sth.** beschließen, etw. zu tun
determined [dɪ'tɜ:mɪnd] *adj.* **a)** *(resolved)* **be ~ to do** *or* **on doing sth.** fest entschlossen sein, etw. zu tun; **sb. is ~ that ...:** es ist für jmdn. beschlossene Sache, daß ...; **b)** *(resolute)* entschlossen; resolut ‹*Person*›
deterrence [dɪ'terəns] *n.* Abschreckung, *die*
deterrent [dɪ'terənt] **1.** *adj.* abschreckend. **2.** *n.* Abschreckungsmittel, *das* (**to** für)
detest [dɪ'test] *v. t.* verabscheuen; **~ doing sth.** es verabscheuen, etw. zu tun
detestable [dɪ'testəbl] *adj.* verabscheuenswert
detestation [di:te'steɪʃn] *n., no pl.* Abscheu, *der* (**of** vor + *Dat.*)
detonate ['detəneɪt] **1.** *v. i.* detonieren. **2.** *v. t.* zur Explosion bringen; zünden
detonation [detə'neɪʃn] *n.* Detonation, *die*
detonator ['detəneɪtə(r)] *n.* Sprengkapsel, *die;* Detonator, *der*
detour ['di:tʊə(r)] *n.* Umweg, *der; (in a road)* Bogen, *der;* Schleife, *die; (diversion)* Umleitung, *die;* **make a ~:** einen Umweg machen
detract [dɪ'trækt] *v. i.* **~ from sth.** etw. beeinträchtigen
detraction [dɪ'trækʃn] *n.* Beeinträchtigung, *die* (**from** *Gen.*)
detriment ['detrɪmənt] *n.* **to the ~ of sth.** zum Nachteil *od.* Schaden einer Sache *(Gen.);* **without ~ to** ohne Schaden für
detrimental [detrɪ'mentl] *adj.* schädlich; **be ~ to sth.** einer Sache *(Dat.)* schaden *od. (geh.)* abträglich sein
detritus [dɪ'traɪtəs] *n., no pl.* Überbleibsel, *das*
[1]**deuce** [dju:s] *n. (Tennis)* Einstand, *der*
[2]**deuce** *n. (coll.)* **who/where/what** *etc.* **the ~:** wer/wo/was *usw.* zum Teufel? *(salopp);* **there will be the ~ to pay** da ist der Teufel los *(ugs.)*
Deutschmark ['dɔɪtʃmɑ:k] *n.* Deutsche Mark
devaluation [di:vælju:'eɪʃn] *n. (also Econ.)* Abwertung, *die*
devalue [di:'vælju:] *v. t. (also Econ.)* abwerten
devastate ['devəsteɪt] *v. t.* verwüsten; verheeren; *(fig.)* niederschmettern
devastating ['devəsteɪtɪŋ] *adj.* verheerend; *(fig.)* niederschmetternd ‹*Nachricht, Analyse*›; vernichtend ‹*Spielweise, Kritik*›
devastation [devə'steɪʃn] *n., no pl.* Verwüstung, *die;* Verheerung, *die*
develop [dɪ'veləp] **1.** *v. t.* **a)** *(also Photog.)* entwickeln; aufbauen ‹*Handel, Handelszentrum*›; entfalten ‹*Persönlichkeit, Individualität*›; erschließen ‹*natürliche Ressourcen*›; **~ a business from scratch** ein Geschäft neu aufziehen; **b)** *(expand; make more sophisticated)* weiterentwickeln; ausbauen ‹*Verkehrsnetz, System, Handel, Verkehr, Position*›; **~ sth. further** etw. weiterentwickeln; **c)** *(begin to exhibit, begin to suffer from)* annehmen ‹*Gewohnheit*›; bei sich entdecken ‹*Vorliebe*›; bekommen ‹*Krankheit, Fieber, Lust*›; entwickeln ‹*Talent, Stärke*›; erkranken an (+ *Dat.*) ‹*Krebs, Tumor*›; **~ a taste for sth.** Geschmack an etw. *(Akk.)* finden; **the car ~ed a fault** an dem Wagen ist ein Defekt aufgetreten; **d)** *(construct buildings etc. on, convert to new use)* erschließen; sanieren ‹*Altstadt*›. **2.** *v. i.* **a)** sich entwickeln (**from** aus; **into** zu); ‹*Defekt, Symptome, Erkrankungen:*› auftreten; **b)** *(become fuller)* sich [weiter]entwickeln (**into** zu)
developer [dɪ'veləpə(r)] *n.* **a)** *(Photog.)* Entwickler, *der;* **b)** *(person who develops real estate)* ≈ Bauunternehmer, *der;* **c) late** *or* **slow ~** *(person)* Spätentwickler, *der*
de'veloping country *n.* Entwicklungsland, *das*
development [dɪ'veləpmənt] *n.* **a)** *(also Photog.)* Entwicklung, *die* (**from** aus, **into** zu); *(of individuality, talent)* Entfaltung, *die; (of natural resources etc.)* Erschließung, *die;* **b)** *(expansion)* Ausbau, *der;* Weiterentwicklung, *die;* **c)** *(of land etc.)* Erschließung, *die;* **d)** *(full-grown state)* Vollendung, *die;* **e)** *(developed product or form)* **a ~ of sth.** eine Fortentwicklung *od.* Weiterentwicklung einer Sache
de'velopment area *n. (Brit.)* Entwicklungsgebiet, *das*
deviant ['di:vɪənt] *adj.* abweichend
deviate ['di:vɪeɪt] *v. i. (lit. or fig.)* abweichen
deviation [di:vɪ'eɪʃn] *n.* Abweichung, *die*
device [dɪ'vaɪs] *n.* **a)** Gerät, *das; (as part of sth.)* Vorrichtung, *die;* **nuclear ~:** atomarer Sprengkörper; **b)** *(plan, scheme)* List, *die;* **c)**

leave sb. to his own ~s jmdn. sich *(Dat.)* selbst überlassen

devil ['devl] *n.* **a)** *(Satan)* **the D~:** der Teufel; **b)** *or* **D~** *(coll.)* **who/where/what** *etc.* **the ~?** wer/wo/was *usw.* zum Teufel? *(salopp);* **the ~ take him!** hol' ihn der Teufel! *(salopp);* **the ~!** Teufel auch! *(salopp);* **there will be the ~ to pay** da ist der Teufel los *(ugs.);* **[you can] go to the ~!** scher dich zum Teufel! *(salopp);* **work like the ~:** wie ein Besessener arbeiten; **run like the ~:** wie der Teufel rennen *(ugs.);* **between the ~ and the deep [blue] sea** in einer Zwickmühle *(ugs.);* **better the ~ one knows** lieber das bekannte Übel; **speak** *or* **talk of the ~ [and he will appear]** wenn man vom Teufel spricht[, kommt er]; **c) a** *or* **the ~ of a mess** ein verteufelter Schlamassel *(ugs.);* **have the ~ of a time** es verteufelt schwer haben; **d)** *(able, clever person)* As, *das (ugs.);* **he's a clever ~:** er ist ein schlauer Hund *(ugs.);* **you ~!** *(ugs.)* du Schlingel!; **a poor ~:** ein armer Teufel; **lucky ~:** Glückspilz, *der (ugs.);* **cheeky ~:** Frechdachs, *der (fam., meist scherzh.)*

devilish ['devəlɪʃ] *adj. (lit. or fig.)* teuflisch ‹*Künste, Zauberei*›

'devil-may-care *adj.* sorglos-unbekümmert

devilment ['devlmənt] *n. (mischief)* Unfug, *der; (wild spirits)* Übermut, *der*

devil's 'advocate *n.* Advocatus Diaboli, *der*

devious ['di:vɪəs] *adj.* **a)** *(winding)* verschlungen; **take a ~ route** einen Umweg fahren; **b)** *(unscrupulous, insincere)* verschlagen ‹*Person*›; hinterhältig ‹*Person, Methode, Tat*›

devise [dɪ'vaɪz] *v.t.* entwerfen; schmieden ‹*Pläne*›; kreieren ‹*Mode, Stil*›; ausarbeiten ‹*Programm*›

devoid [dɪ'vɔɪd] *adj.* **~ of sth.** *(lacking)* ohne etw.; bar einer Sache *(Gen.) (geh.); (free from)* frei von etw.

devolution [di:və'lu:ʃn] *n. (deputing, delegation)* Übertragung, *die; (Polit.)* Dezentralisierung, *die*

devote [dɪ'vəʊt] *v.t.* widmen; bestimmen ‹*Geld*› **(to** für); **~ one's thoughts/energy to sth.** sein Denken/seine Energie auf etw. *(Akk.)* verwenden

devoted [dɪ'vəʊtɪd] *adj.* treu; ergeben ‹*Diener*›; aufrichtig ‹*Freundschaft, Liebe, Verehrung*›; **he is very ~ to his work/his wife** er geht in seiner Arbeit völlig auf/liebt seine Frau innig

devotee [devə'ti:] *n.* Anhänger, *der*/Anhängerin, *die; (of music, art)* Liebhaber, *der*/Liebhaberin, *die*

devotion [dɪ'vəʊʃn] *n.* **a)** *(addiction, loyalty, devoutness)* **~ to sb./sth.** Hingabe an jmdn./etw.; **~ to music/the arts** Liebe zur Musik/Kunst; **~ to duty** Pflichteifer, *der;* **b)** *(devoting)* Weihung, *die*

devour [dɪ'vaʊə(r)] *v.t.* verschlingen

devout [dɪ'vaʊt] *adj.* fromm; sehnlich ‹*Wunsch*›; inständig ‹*Hoffnung*›

dew [dju:] *n.* Tau, *der*

dewy ['dju:ɪ] *adj.* taufeucht

'dewy-eyed *adj.* naiv; **go all ~** ganz feuchte Augen bekommen

dexterity [dek'sterɪtɪ] *n., no pl. (skill)* Geschicklichkeit, *die*

dextrous ['dekstrəs] *adj.* geschickt

diabetes [daɪə'bi:ti:z] *n., pl. same (Med.)* Zuckerkrankheit, *die;* Diabetes, *der (fachspr.)*

diabetic [daɪə'betɪk, daɪə'bi:tɪk] *(Med.)* **1.** *adj.* **a)** *(of diabetes)* diabetisch; **b)** *(having diabetes)* diabetisch *(Med.);* zuckerkrank; **c)** *(for diabetics)* Diabetiker‹*nahrung, -schokolade usw.*›. **2.** *n.* Diabetiker, *der*/Diabetikerin, *die*

diabolic [daɪə'bɒlɪk], **diabolical** [daɪə'bɒlikl] *adj.* teuflisch; diabolisch; *(coll.: extremely bad)* mörderisch *(ugs.)* ‹*Hitze*›; teuflisch *(ugs.)* ‹*Kälte, Wetter*›

diagnose [daɪəg'nəʊz] *v.t.* diagnostizieren ‹*Krankheit*›; feststellen ‹*Fehler*›

diagnosis [daɪəg'nəʊsɪs] *n., pl.* **diagnoses** [daɪəg'nəʊsi:z] **a)** *(of disease)* Diagnose, *die;* **make a ~:** eine Diagnose stellen; **b)** *(of difficulty, fault)* Feststellung, *die*

diagnostic [daɪəg'nɒstɪk] *adj.* diagnostisch

diagonal [daɪ'ægənl] **1.** *adj.* diagonal. **2.** *n.* Diagonale, *die*

diagonally [daɪ'ægənəlɪ] *adv.* diagonal

diagram ['daɪəgræm] *n.* **a)** *(sketch)* schematische Darstellung; **I'll make a ~ to show you how to get there** ich zeichne Ihnen auf, wie Sie dorthin kommen; **b)** *(graphic or symbolic representation; Geom.)* Diagramm, *das*

dial ['daɪəl] **1.** *n.* **a)** *(of clock or watch)* Zifferblatt, *das;* **b)** *(of gauge, meter, etc.; on radio or television)* Skala, *die;* **c)** *(Teleph.)* Wählscheibe, *die.* **2.** *v.t., (Brit.)* **-ll-** *(Teleph.)* wählen; **~ [London] direct** [nach London] durchwählen. **3.** *v.i., (Brit.)* **-ll-** *(Teleph.)* wählen

dialect ['daɪəlekt] *n.* Dialekt, *der;* Mundart, *die*

dialling *(Amer.:* **dialing)** ['daɪəlɪŋ]: **~ code** *n.* Vorwahl, *die;* Ortsnetzkennzahl, *die (Amtsspr.);* **~ tone** *n.* Freizeichen, *das;* Wählton, *der (fachspr.)*

dialogue ['daɪəlɒg] *n.* Dialog, *der*

'dial tone *(Amer.) see* **dialling tone**

dialysis [daɪ'ælɪsɪs] *n., pl.* **dialyses** [daɪ'ælɪsi:z] **a)** *(Chem.)* Dialyse, *die;* **b)** *(Med.)* [Hämo]dialyse, *die (fachspr.);* Blutwäsche, *die*

diameter [daɪ'æmɪtə(r)] *n.* Durchmesser, *der*

diametrical [daɪə'metrɪkl] *adj.,* **diametrically** [daɪə'metrɪkəlɪ] *adv.* diametral

diamond ['daɪəmənd] **1.** *n.* **a)** Diamant, *der;* **b)** *(figure)* Raute, *die;* Rhombus, *der;* **c)** *(Cards)* Karo, *das; see also* **club 1d. 2.** *adj. (made of ~[s])* diamanten; *(set with ~[s])* diamantenbesetzt; Diamant‹*ring, -staub, -schmuck*›

diamond: ~ 'jubilee *n.* 60jähriges/75jähriges Jubiläum; **~ 'wedding** *n.* diamantene Hochzeit

diaper ['daɪəpə(r)] *n. (Amer.)* Windel, *die*

diaphragm ['daɪəfræm] *n.* Diaphragma, *das (fachspr.); (Anat. also)* Zwerchfell, *das; (contraceptive also)* Pessar, *das*

diarrhoea *(Amer.:* **diarrhea)** [daɪə'ri:ə] *n.* Durchfall, *der;* Diarrhö[e], *die (Med.)*

diary ['daɪərɪ] *n.* **a)** Tagebuch, *das;* **keep a ~:** [ein] Tagebuch führen; **b)** *(for appointments)* Terminkalender, *der;* **pocket/desk ~:** Taschen-/Tischkalender, *der*

dice [daɪs] **1.** *n., pl. same* **a)** *(cube)* Würfel, *der;* **throw ~:** würfeln; **throw ~ for sth.** etw. auswürfeln; **no ~!** *(fig. coll.)* kommt nicht in Frage!; **b)** *in sing. (game)* Würfelspiel, *das;* **play ~:** würfeln. **2.** *v.i.* **~ with death** mit seinem Leben spielen. **3.** *v.t. (Cookery)* würfeln

dicey ['daɪsɪ] *adj. (sl.)* riskant

dichotomy [daɪ'kɒtəmɪ, dɪ'kɒtəmɪ] *n.* Dichotomie, *die*

dick [dɪk] *n. (sl.: detective)* Schnüffler, *der (ugs. abwertend)*

dicky ['dɪkɪ] *adj. (Brit. sl.)* mies *(ugs. abwertend);* klapprig *(ugs.)* ‹*Herz*›

'dicky-bird *n. (child lang./coll.)* Piepvogel, *der (Kinderspr.)*

dictate 1. [dɪk'teɪt] *v.t. & i.* diktieren; *(prescribe)* vorschreiben;

~ to Vorschriften machen (+ *Dat.*); **I will not be ~d to** ich lasse mir keine Vorschriften machen. **2.** ['dɪkteɪt] *n., usu. in pl.* Diktat, *das*

dic'tating-machine *n.* Diktiergerät, *das*

dictation [dɪk'teɪʃn] *n.* Diktat, *das;* **take a ~:** ein Diktat aufnehmen

dictator [dɪk'teɪtə(r)] *n. (lit. or fig.)* Diktator, *der;* **be a ~** *(fig.)* diktatorisch sein

dictatorial [dɪktə'tɔ:rɪəl] *adj.* diktatorisch

dictatorship [dɪk'teɪtəʃɪp] *n. (lit. or fig.)* Diktatur, *die*

diction ['dɪkʃn] *n.* Diktion, *die (geh.)*

dictionary ['dɪkʃənərɪ] *n.* Wörterbuch, *das*

did *see* [1]**do**

didactic [dɪ'dæktɪk, daɪ'dæktɪk] *adj.* **a)** didaktisch; **b)** *(authoritarian)* schulmeisterlich

diddle ['dɪdl] *v. t. (sl.)* übers Ohr hauen *(ugs.);* **~ sb. out of sth.** jmdm. etw. abluchsen *(salopp)*

didn't ['dɪdnt] *(coll.)* = **did not;** *see* [1]**do**

[1]**die** [daɪ] **1.** *v. i.,* **dying** ['daɪɪŋ] **a)** sterben; ‹*Tier, Pflanze:*› eingehen, *(geh.)* sterben; ‹*Körperteil:*› absterben; **be dying** sterben; **~ from** *or* **of sth.** an etw. *(Dat.)* sterben; **~ of a heart attack/a brain tumour** einem Herzanfall/ Hirntumor erliegen; **~ from one's injuries** seinen Verletzungen erliegen; **sb. would ~ rather than do sth.** um nichts in der Welt würde jmd. etw. tun; **never say ~** *(fig.)* nur nicht den Mut verlieren; **b)** *(fig.)* **be dying for sth.** etw. unbedingt brauchen; **be dying for a cup of tea** nach einer Tasse Tee lechzen; **be dying to do sth.** darauf brennen, etw. zu tun; **be dying of boredom** vor Langeweile sterben; **~ with** *or* **of shame** sich zu Tode schämen; **c)** *(disappear)* in Vergessenheit geraten; ‹*Gefühl, Liebe, Ruhm:*› vergehen; ‹*Ton:*› verklingen; ‹*Flamme:*› verlöschen. **2.** *v. t.,* **dying: ~ a natural/ violent death** eines natürlichen/ gewaltsamen Todes sterben

~ 'down *v. i.* ‹*Sturm, Wind, Protest, Aufruhr:*› sich legen; ‹*Flammen:*› kleiner werden; ‹*Feuer:*› herunterbrennen; ‹*Lärm:*› leiser werden; ‹*Kämpfe:*› nachlassen

~ 'off *v. i.* ‹*Pflanzen, Tiere:*› [nacheinander] eingehen; ‹*Blätter:*› [nacheinander] absterben; ‹*Personen:*› [nacheinander] sterben

~ 'out *v. i.* aussterben

[2]**die** *n., pl.* **dice** [daɪs] *(formal)* Würfel, *der;* **the ~ is cast** die Würfel sind gefallen; **as straight** *or* **true as a ~:** schnurgerade ‹*Weg, Linie*›

'die-hard 1. *n.* hartnäckiger Typ; *(reactionary)* Ewiggestrige, *der/ die.* **2.** *adj.* hartnäckig; *(dyed-in-the-wool)* eingefleischt; *(reactionary)* ewiggestrig

diesel ['di:zl] *n.* **~ [engine]** Diesel[motor], *der;* **~ [lorry/car]** Diesel, *der;* **~ [train]** *(Railw.)* [Zug mit] Dieseltriebwagen; **~ [fuel]** Diesel[kraftstoff], *der*

'diesel oil *n.* Dieseltreibstoff, *der*

diet ['daɪət] **1.** *n.* **a)** *(for slimming)* Diät, *die;* Schlankheitskur, *die;* **be/go on a ~:** eine Schlankheitskur *od.* Diät machen; **b)** *(Med.)* Diät, *die;* Schonkost, *die;* **c)** *(habitual food)* Kost, *die.* **2.** *v. i.* eine Schlankheitskur *od.* Diät machen

dietitian (dietician) [daɪə'tɪʃn] *n.* Diätassistent, *der/*-assistentin, *die*

differ ['dɪfə(r)] *v. i.* **a)** *(vary, be different)* sich unterscheiden; **opinions/ideas ~:** die Meinungen/ Vorstellungen gehen auseinander; **tastes/temperaments ~:** die Geschmäcker *(ugs.)* / Temperamente sind verschieden; **~ from sb./sth. in that ...:** sich von jmdm./etw. dadurch *od.* darin unterscheiden, daß ...; **b)** *(disagree)* anderer Meinung sein

difference ['dɪfərəns] *n.* **a)** Unterschied, *der;* **~ in age** Altersunterschied, *der;* **have a ~ of opinion [with sb.]** eine Meinungsverschiedenheit [mit jmdm.] haben; **it makes a ~:** es ist ein *od. (ugs.)* macht einen Unterschied; **what ~ would it make if ...?** was würde es schon ausmachen, wenn ...?; **make all the ~ [in the world]** ungeheuer viel ausmachen; **make no ~ [to sb.]** [jmdm.] nichts ausmachen; **b)** *(between amounts)* Differenz, *die;* **pay the ~:** den Rest[betrag] bezahlen; **split the ~:** sich *(Dat.)* den Rest[betrag] teilen; **c)** *(dispute)* **have a ~ with sb.** mit jmdm. eine Auseinandersetzung haben; **settle one's ~s** seine Differenzen beilegen

different ['dɪfərənt] *adj.* verschieden; *(pred. also)* anders; *(attrib. also)* ander...; **be ~ from** *or (esp. Brit.)* **to** *or (Amer.)* **than ...:** anders sein als ...; **~ viewpoints/cultures** unterschiedliche Standpunkte/Kulturen; **how are they ~?** worin *od.* wodurch unterscheiden sie sich?

differential [dɪfə'renʃl] *n.* **a)** *(Commerc.)* **[wage] ~:** [Einkommens]unterschied, *der;* **price ~s** Preisunterschiede; **b)** *(Motor Veh.)* Differential[getriebe], *das*

differentiate [dɪfə'renʃɪeɪt] **1.** *v. t.* unterscheiden. **2.** *v. i.* **a)** *(recognize the difference)* unterscheiden; differenzieren; **b)** *(treat sth. differently)* einen Unterschied machen; differenzieren

differently ['dɪfərəntlɪ] *adv.* anders (**from,** *esp. Brit.* **to** als); **~ [to** *or* **from each other]** verschieden; *(with different result, at various times)* unterschiedlich

differing ['dɪfərɪŋ] *adj.* unterschiedlich

difficult ['dɪfɪkəlt] *adj.* **a)** schwer; schwierig; **he finds it ~ to do sth.** ihm fällt es schwer, etw. zu tun; **make things ~ for sb.** es jmdm. nicht leicht machen; **b)** *(unaccommodating)* schwierig; **he is being ~:** er macht Schwierigkeiten; **he is ~ to get on with** es ist schwer, mit ihm auszukommen

difficulty ['dɪfɪkəltɪ] *n.* **a)** Schwierigkeit, *die;* **with [great] ~:** [sehr] mühsam; **with the greatest ~:** unter größten Schwierigkeiten; **without [great] ~:** ohne große Probleme; mühelos; **have ~ [in] doing sth.** Schwierigkeiten haben, etw. zu tun; **b)** *usu. in pl. (trouble)* **be in ~** *or* **difficulties** in Schwierigkeiten sein; **fall** *or* **get into difficulties** in Schwierigkeiten kommen *od.* geraten

diffident ['dɪfɪdənt] *adj.* zaghaft; *(modest)* zurückhaltend

diffuse 1. [dɪ'fju:z] *v. t.* verbreiten; diffundieren *(fachspr.).* **2.** [dɪ'fju:z] *v. i.* sich ausbreiten (**through** in + *Dat.*); diffundieren *(fachspr.).* **3.** [dɪ'fju:s] *adj.* diffus

diffusion [dɪ'fju:ʒn] *n.* Verbreitung, *die*

dig [dɪg] **1.** *v. i.,* **-gg-, dug** [dʌg] **a)** graben (**for** nach); **b)** *(Archaeol.: excavate)* Ausgrabungen machen; graben. **2.** *v. t.,* **-gg-, dug a)** graben; **~ a hole [in sth.]** ein Loch [in etw. *(Akk.)*] graben; **b)** *(turn up with spade etc.)* umgraben; **c)** *(Archaeol.)* ausgraben; **d)** *(sl.: appreciate)* stark finden *(Jugendspr.); (understand)* schnallen *(salopp).* **3.** *n.* **a)** Grabung, *die;* **b)** *(Archaeol. coll.)* Ausgrabung, *die; (site)* Ausgrabungsort, *der;* **c)** *(fig.)* Anspielung, *die* (**at** auf + *Akk.*); **have** *or* **make a ~ at sb./ sth.** eine [spitze] Bemerkung über jmdn./etw. machen

~ 'in 1. *v. i. (Mil.)* sich eingraben;

(fig.) sich festsetzen. **2.** *v. t.* **a)** *(Mil.)* eingraben; ~ **oneself in** sich eingraben; *(fig.)* sich etablieren; **b)** *(thrust)* **the cat dug its claws in** die Katze krallte sich fest; ~ **one's heels** *or* **toes in** *(fig. coll.)* sich auf die Hinterbeine stellen *(ugs.)*; **c)** *(mix with soil)* eingraben

~ **'out** *v. t. (lit. or fig.)* ausgraben

~ **'up** *v. t.* umgraben ⟨*Garten, Rasen, Erde*⟩; ausgraben ⟨*Pflanzen, Knochen, Leiche, Schatz*⟩

digest 1. [dɪ'dʒest, daɪ'dʒest] *v. t.* **a)** *(assimilate, lit or fig.)* verdauen; **b)** *(consider)* durchdenken. **2.** ['daɪdʒest] *n. (periodical)* Digest, *der od. das*

digestible [dɪ'dʒestɪbl, daɪ'dʒestɪbl] *adj.* verdaulich

digestion [dɪ'dʒestʃn, daɪ'dʒestʃn] *n.* Verdauung, *die*

digestive [dɪ'dʒestɪv, daɪ'dʒestɪv] **1.** *adj.* Verdauungs-; ~ **biscuit** *(Brit.) see* **2. 2.** *n. (Brit.: biscuit)* Keks, *der (aus Vollkornmehl)*

digger ['dɪgə(r)] *n. (Mech.)* Bagger, *der*

digit ['dɪdʒɪt] *n.* **a)** *(numeral)* Ziffer, *die;* **a six-~ number** eine sechsstellige Zahl; **b)** *(Zool., Anat.) (finger)* Finger, *der; (toe)* Zehe, *die*

digital ['dɪdʒɪtl] *adj.* digital; ~ **clock/watch** Digitaluhr, *die;* ~ **computer** Digitalrechner, *der;* ~ **recording** Digitalaufnahme, *die;* ~ **audio tape** Digitaltonband, *das*

dignified ['dɪgnɪfaɪd] *adj.* würdig; *(self-respecting)* würdevoll

dignify ['dɪgnɪfaɪ] *v. t.* **a)** *(make stately)* Würde verleihen (+ *Dat.*); **b)** *(give distinction to)* Glanz verleihen (+ *Dat.*); auszeichnen ⟨*Person*⟩; **c)** *(give grand title to)* aufwerten *(fig.)*

dignitary ['dɪgnɪtərɪ] *n.* Würdenträger, *der;* **dignitaries** *(prominent people)* Honoratioren

dignity ['dɪgnɪtɪ] *n.* Würde, *die;* **be beneath one's ~:** unter seiner Würde sein

digress [daɪ'gres] *v. i.* abschweifen (**from** von, **on** zu)

digression [daɪ'greʃn] *n.* Abschweifung, *die; (passage)* Exkurs, *der*

digs [dɪgz] *n. pl. (Brit. coll.)* Bude, *die (ugs.)*

dike [daɪk] *n.* **a)** *(flood-wall)* Deich, *der;* **b)** *(ditch)* Graben, *der;* **c)** *(causeway)* Damm, *der*

dilapidated [dɪ'læpɪdeɪtɪd] *adj.* verfallen ⟨*Gebäude*⟩; verwahrlost ⟨*Äußeres, Erscheinung*⟩

dilapidation [dɪlæpɪ'deɪʃn] *n., no pl.* Verfall, *der*

dilate [daɪ'leɪt] **1.** *v. i.* sich weiten. **2.** *v. t.* ausdehnen

dilation [daɪ'leɪʃn] *n.* Dilatation, *die; (Phys. also)* Ausdehnung, *die; (Med. also)* Erweiterung, *die*

dilatory ['dɪlətərɪ] *adj.* langsam; zögernd ⟨*Antwort, Reaktion*⟩; *(causing delay)* **be ~** in sich *(Dat.)* [viel] Zeit lassen bei

dilemma [dɪ'lemə, daɪ'lemə] *n.* Dilemma, *das;* **be on the horns of** *or* **faced with a ~:** vor einem Dilemma stehen

dilettante [dɪlɪ'tæntɪ] *n., pl.* **dilettanti** [dɪlɪ'tænti:] *or* **~s** Dilettant, *der*/Dilettantin, *die;* Laie, *der*

diligence ['dɪlɪdʒəns] *n.* Fleiß, *der; (purposefulness)* Eifer, *der*

diligent ['dɪlɪdʒənt] *adj.* fleißig; *(purposeful)* eifrig; sorgfältig, gewissenhaft ⟨*Arbeit, Suche*⟩

diligently ['dɪlɪdʒəntlɪ] *adv.* fleißig; *(purposefully)* eifrig

dill [dɪl] *n. (Bot.)* Dill, *der*

dilly-dally ['dɪlɪdælɪ] *v. i. (coll.)* trödeln

dilute 1. [daɪ'lju:t, 'daɪlju:t] *adj.* verdünnt. **2.** [daɪ'lju:t] *v. t.* **a)** verdünnen; **b)** *(fig.)* abschwächen; entschärfen

dim [dɪm] **1.** *adj.* **a)** schwach, trüb ⟨*Licht, Flackern*⟩; matt, gedeckt ⟨*Farbe*⟩; dämmrig, dunkel ⟨*Zimmer*⟩; undeutlich, verschwommen ⟨*Gestalt*⟩; **grow ~:** schwächer werden; **b)** *(fig.)* blaß; verschwommen; **in the ~ and distant past** in ferner Vergangenheit; **c)** *(indistinct)* schwach, getrübt ⟨*Seh-, Hörvermögen*⟩; **d)** *(coll.: stupid)* beschränkt; **e) take a ~ view of sth.** *(coll.)* von etw. nicht erbaut sein. **2.** *v. i.,* **-mm-** *(lit. or fig.)* schwächer werden. **3.** *v. t.,* **-mm-** verdunkeln; *(fig.)* trüben; dämpfen; ~ **the lights** *(Theatre, Cinemat.)* die Lichter langsam verlöschen lassen

dime [daɪm] *n. (Amer.)* Zehncentstück, *das;* ≈ Groschen, *der (ugs.)*

dimension [dɪ'menʃn, daɪ'menʃn] *n. (lit. or fig.)* Dimension, *die; (measurement)* Abmessung, *die*

diminish [dɪ'mɪnɪʃ] **1.** *v. i.* nachlassen; ⟨*Zahl:*⟩ sich verringern; ⟨*Vorräte, Autorität, Einfluß:*⟩ abnehmen; ⟨*Wert, Bedeutung, Ansehen:*⟩ geringer werden; ~ **in value/number** an Wert verlieren/an Zahl od. zahlenmäßig abnehmen. **2.** *v. t.* vermindern; verringern; *(fig.)* herabwürdigen ⟨*Person*⟩; schmälern ⟨*Ansehen, Ruf*⟩

diminished [dɪ'mɪnɪʃt] *adj.* geringer ⟨*Wert, Anzahl, Einfluß, Popularität*⟩; vermindert ⟨*Stärke, Fähigkeit*⟩; ~ **responsibility** *(Law)* verminderte Zurechnungsfähigkeit

diminishing [dɪ'mɪnɪʃɪŋ] *adj.* sinkend; abnehmend ⟨*Vorräte*⟩; schwindend ⟨*Kraft, Einfluß, Macht*⟩

diminutive [dɪ'mɪnjʊtɪv] **1.** *adj.* **a)** winzig; **b)** *(Ling.)* diminutiv. **2.** *n. (Ling.)* Diminutiv[um], *das*

dimly ['dɪmlɪ] *adv.* schwach; undeutlich ⟨*sehen*⟩; ungefähr ⟨*begreifen*⟩; **I ~ remember it** ich erinnere mich noch dunkel daran

dimple ['dɪmpl] *n.* Grübchen, *das; (on golf-ball etc.)* kleine Vertiefung

dim: ~-wit *n. (coll.)* Dummkopf, *der (ugs.)*; **~-witted** ['dɪmwɪtɪd] *adj. (coll.)* dusselig *(salopp)*

din [dɪn] **1.** *n.* Lärm, *der.* **2.** *v. t.,* **-nn-: ~ sth. into sb.** jmdm. etw. einhämmern *od.* einbleuen

dine [daɪn] *v. i. (at midday/in the evening)* [zu Mittag/zu Abend] essen *od. (geh.)* speisen; ~ **off/on sth.** *(eat)* etw. [zum Mittag-/Abendessen] verzehren; ~ **off sth.** *(eat from)* von etw. speisen

~ **'out** *v. i.* **a)** auswärts [zu Mittag/Abend] essen; **b) ~ out on sth.** wegen etw. zum Essen eingeladen werden

diner ['daɪnə(r)] *n.* Gast, *der (zum Abendessen)*

ding-dong ['dɪŋdɒŋ] *n.* Bimbam, *das*

dinghy ['dɪŋgɪ, 'dɪŋɪ] *n.* Ding[h]i, *das; (inflatable)* Schlauchboot, *das*

dingo ['dɪŋgəʊ] *n., pl.* **~es** Dingo, *der*

dingy ['dɪndʒɪ] *adj.* schmuddelig

dining ['daɪnɪŋ]: ~ **area** *n.* ≈ Eßecke, *die;* **~-car** *n. (Railw.)* Speisewagen, *der;* **~-chair** *n.* Eßzimmerstuhl, *der;* **~-hall** *n.* Speisesaal, *der;* **~-room** *n. (in private house)* Eßzimmer, *das; (in hotel etc.)* Speisesaal, *der;* **~-table** *n.* Eßtisch, *der*

dinkum ['dɪŋkəm] *(Austral. and NZ coll.) adj.* astrein *(ugs.)*

dinner ['dɪnə(r)] *n.* Essen, *das; (at midday also)* Mittagessen, *das; (in the evening also)* Abendessen, *das; (formal event)* Diner, *das;* **have** *or* **eat [one's] ~:** zu Mittag/Abend essen; **go out to ~:** [abends] essen gehen; **be having** *or* **eating [one's] ~:** gerade beim Essen sein; **have people to** *or* **for ~:** Gäste zum Essen haben

dinner: ~-dance *n. Abendessen mit anschließendem Tanz;* **~-jacket** *n. (Brit.)* Dinnerjacket, *das;* ~ **lady** *n. (Brit.) Serviererin beim Mittagessen in der Schule;* ~**-table** *n.* Eßtisch, *der;* ~**-time** *n.* Essenszeit, *die;* **at** ~**-time** zur Essenszeit; *(12–2 p.m.)* mittags

dinosaur ['daɪnəsɔ:(r)] *n.* Dinosaurier, *der*

dint [dɪnt] *n.* **by** ~ **of** durch; **by** ~ **of doing sth.** indem jmd. etw. tut

diocesan [daɪ'ɒsɪsən] *adj. (Eccl.)* diözesan

diocese ['daɪəsɪs] *n. (Eccl.)* Diözese, *die*

dioxin [daɪ'ɒksɪn] *n. (Chem.)* Dioxin, *das*

dip [dɪp] **1.** *v.t.*, **-pp-: a)** [ein]tauchen (**in** in + *Akk.*); **she** ~**ped her hand into the sack** sie griff in den Sack; **b)** *(Agric.)* dippen ‹*Schaf*›; **c)** *(Brit. Motor Veh.)* ~ **one's [head]lights** abblenden; **[drive with** *or* **on]** ~**ped headlights** [mit] Abblendlicht [fahren]. **2.** *v.i.*, **-pp-: a)** *(go down)* sinken; **b)** *(incline downwards, lit. or fig.)* abfallen. **3.** *n.* **a)** *(~ping)* [kurzes] Eintauchen; **b)** *(coll.: bathe)* [kurzes] Bad; **c)** *(in road)* Senke, *die;* **d)** *(Gastr.)* Dip, *der*

~ **into** *v.t.* **a)** greifen in (+ *Akk.*); *(fig.)* ~ **into one's pocket** *or* **purse** tief in die Tasche greifen; ~ **into one's savings** seine Ersparnisse angreifen; **b)** *(look cursorily at)* einen flüchtigen Blick werfen in (+ *Akk.*)

diphtheria [dɪf'θɪərɪə] *n. (Med.)* Diphtherie, *die*

diphthong ['dɪfθɒŋ] *n.* Diphthong, *der (fachspr.);* Doppellaut, *der*

diploma [dɪ'pləʊmə] *n. (Educ.)* Diplom, *das*

diplomacy [dɪ'pləʊməsɪ] *n. (Polit.; also fig.)* Diplomatie, *die*

diplomat ['dɪpləmæt] *n. (Polit.; also fig.)* Diplomat, *der*/Diplomatin, *die*

diplomatic [dɪplə'mætɪk] *adj.*, **diplomatically** [dɪplə'mætɪkəlɪ] *adv. (Polit.; also fig.)* diplomatisch

diplomatic: ~ **'bags** *n. pl.* Kuriergepäck, *das;* ~ **corps** *n.* diplomatisches Korps

dip: ~**-stick** *n.* [Öl-/Benzin]meßstab, *der;* ~**-switch** *n. (Brit. Motor Veh.)* Abblendschalter, *der*

dire ['daɪə(r)] *adj.* **a)** *(dreadful)* entsetzlich; furchtbar; **b)** *(extreme)* ~ **necessity** dringende Notwendigkeit; **be in** ~ **need of sth.** etw. dringend benötigen; **be in** ~ **[financial] straits** in einer ernsten [finanziellen] Notlage sein

direct [dɪ'rekt, daɪ'rekt] **1.** *v.t.* **a)** *(turn)* richten **(to[wards]** auf + *Akk.*); ~ **sb.'s attention to sth.** jmds. Aufmerksamkeit auf etw. *(Akk.)* lenken; **the remark was** ~**ed at you** die Bemerkung galt dir; **the bomb/missile was** ~**ed at** die Bombe/das Geschoß galt (+ *Dat.*); ~ **sb. to a place** jmdm. den Weg zu einem Ort weisen *od.* sagen; **b)** *(control)* leiten; beaufsichtigen ‹*Arbeitskräfte, Arbeitsablauf*›; regeln, dirigieren ‹*Verkehr*›; **c)** *(order)* anweisen; ~ **sb. to do sth.** jmdn. anweisen, etw. zu tun; **as** ~**ed [by the doctor]** wie [vom Arzt] verordnet; **d)** *(Theatre, Cinemat., Telev., Radio)* Regie führen bei. **2.** *adj.* **a)** direkt; durchgehend ‹*Zug*›; unmittelbar ‹*Ursache, Gefahr, Auswirkung*›; *(immediate)* unmittelbar, persönlich ‹*Erfahrung, Verantwortung, Beteiligung*›; **b)** *(diametrical)* genau ‹*Gegenteil*›; direkt ‹*Widerspruch*›; diametral ‹*Gegensatz*›; **c)** *(frank)* direkt; offen; glatt ‹*Absage*›. **3.** *adv.* direkt

direct: ~ **'current** *n. (Electr.)* Gleichstrom, *der;* ~ **dialling** *n.* Durchwahl, *die;* **we will soon have** ~ **dialling** wir werden bald ein Durchwahlsystem haben; ~ **'hit** *n.* Volltreffer, *der*

direction [dɪ'rekʃn, daɪ'rekʃn] *n.* **a)** *(guidance)* Führung, *die; (of firm, orchestra)* Leitung, *die; (of play, film, TV or radio programme)* Regie, *die;* **b)** *usu. in pl. (order)* Anordnung, *die;* ~**s [for use]** Gebrauchsanweisung, *die;* **on** *or* **by sb.'s** ~**:** auf jmds. Anordnung *(Akk.)* [hin]; **give sb.** ~**s to the museum/to York** jmdm. den Weg zum Museum/nach York beschreiben; **c)** *(point moved towards or from, lit. or fig.)* Richtung, *die;* **from which** ~**?** aus welcher Richtung?; **travel in a southerly** ~**/in the** ~ **of London** in südliche[r] Richtung/in Richtung London reisen; **sense of** ~**:** Orientierungssinn, *der; (fig.)* Orientierung, *die;* **lose all sense of** ~ *(lit. or fig.)* jede Orientierung verlieren

di'rection-indicator *n. (Motor Veh.)* [Fahrt]richtungsanzeiger, *der*

directive [dɪ'rektɪv, daɪ'rektɪv] *n.* Weisung, *die;* Direktive, *die*

directly [dɪ'rektlɪ, daɪ'rektlɪ] **1.** *adv.* **a)** direkt; unmittelbar ‹*folgen, verantwortlich sein*›; **b)** *(exactly)* direkt; wörtlich ‹*zitieren, abschreiben*›; **c)** *(at once)* direkt; umgehend; *(shortly)* gleich; sofort. **2.** *conj. (Brit. coll.)* sowie

directness [dɪ'rektnɪs, daɪ'rektnɪs] *n., no pl.* **a)** *(of route, course)* Geradheit, *die;* **b)** *(fig.)* Direktheit, *die*

director [dɪ'rektə(r), daɪ'rektə(r)] *n.* **a)** *(Commerc.)* Direktor, *der*/Direktorin, *die; (of project)* Leiter, *der*/Leiterin, *die;* **board of** ~**s** Aufsichtsrat, *der;* **b)** *(Theatre, Cinemat., Telev., Radio)* Regisseur, *der*/Regisseurin, *die*

directorship [dɪ'rektəʃɪp, daɪ'rektəʃɪp] *n. (Commerc.)* Leitung, *die;* **hold two** ~**s** in zwei Aufsichtsräten sein

directory [dɪ'rektərɪ, daɪ'rektərɪ] *n. (telephone* ~*)* Telefonbuch, *das; (of tradesmen etc.)* Branchenverzeichnis, *das; attrib.* ~ **enquiries** *(Brit.),* ~ **information** *(Amer.)* [Fernsprech]auskunft, *die*

direct 'speech *n. (Ling.)* direkte Rede

dirge [dɜ:dʒ] *n.* **a)** *(for the dead)* Grabgesang, *der;* **b)** *(mournful song)* Klagegesang, *der*

dirt [dɜ:t] *n., no pl.* **a)** Schmutz, *der;* Dreck, *der (ugs.);* **be covered in** ~**:** ganz schmutzig sein; *(stronger)* vor Schmutz starren; ~ **cheap** spottbillig; **treat sb. like** ~**:** jmdn. wie [den letzten] Dreck behandeln *(salopp);* **b)** *(soil)* Erde, *die*

dirty ['dɜ:tɪ] **1.** *adj.* **a)** schmutzig; dreckig *(ugs.);* **get one's shoes/hands** ~**:** sich *(Dat.)* die Schuhe/Hände schmutzig machen; **get sth.** ~**:** etw. schmutzig machen; **b)** ~ **weather** stürmisches Wetter; Dreckwetter, *das (ugs. abwertend);* **c)** ~ **look** *(coll.)* giftiger Blick; **d)** *(fig.: obscene)* schmutzig; schlüpfrig; *(sexually illicit)* **spend a** ~ **weekend together** ein Liebeswochenende zusammen verbringen; ~ **old man** alter Lustmolch *(ugs. abwertend);* geiler alter Bock *(salopp abwertend);* **e)** *(despicable, sordid)* schmutzig ‹*Lüge, Gerücht, Geschäft*›; dreckig *(salopp abwertend),* gemein ‹*Lügner, Betrüger*›; *(unsportsmanlike)* unfair; **do the** ~ **on sb.** *(coll.)* jmdn. [he]reinlegen *(ugs.);* ~ **trick** gemeiner Trick; ~ **work** *(coll.)* schmutziges Geschäft; **do sb.'s/the** ~ **work** sich *(Dat.)* für jmdn./sich *(Dat.)* die Finger schmutzig machen. **2.** *v.t.* schmutzig machen; beschmutzen

dirty 'word *n.* unanständiges Wort

disability [dɪsə'bɪlɪtɪ] *n.* Behinderung, *die;* **suffer from** *or* **have a** ~**:** behindert sein

disable [dɪ'seɪbl] *v.t.* **a)** ~ **sb. [physically]** jmdn. zum Invaliden machen; **be ~d by sth.** durch etw. behindert sein; **b)** *(make unable to fight)* kampfunfähig machen ‹*Feind, Schiff, Panzer, Flugzeug*›
disabled [dɪ'seɪbld] **1.** *adj.* **a)** behindert; **physically/mentally** ~: körperbehindert/geistig behindert; **b)** *(unable to fight)* kampfunfähig ‹*Schiff, Panzer, Flugzeug*›. **2.** *n. pl.* **the [physically/ mentally]** ~: die [Körper]behinderten/[geistig] Behinderten
disablement [dɪ'seɪblmənt] *n., no pl.* Behinderung, *die*
disadvantage [dɪsəd'vɑ:ntɪdʒ] **1.** *n.* Nachteil, *der;* **be at a** ~: im Nachteil sein; benachteiligt sein; **be to sb.'s/sth.'s** ~: sich zu jmds. Nachteil/zum Nachteil einer Sache auswirken. **2.** *v.t.* benachteiligen
disadvantaged [dɪsəd'vɑ:ntɪdʒd] *adj.* benachteiligt
disadvantageous [dɪsædvən'teɪdʒəs] *adj.* nachteilig
disagree [dɪsə'gri:] *v.i.* **a)** anderer Meinung sein; ~ **with sb./sth.** mit jmdm./etw. nicht übereinstimmen; ~ **[with sb.] about** *or* **over sth.** sich [mit jmdm.] über etw. *(Akk.)* nicht einig sein; **b)** *(quarrel)* eine Auseinandersetzung haben; **c)** *(be mutually inconsistent)* nicht übereinstimmen; **d)** ~ **with sb.** *(have bad effects on)* jmdm. nicht bekommen
disagreeable [dɪsə'gri:əbəl] *adj.* unangenehm
disagreement [dɪsə'gri:mənt] *n.* **a)** *(difference of opinion)* Uneinigkeit, *die;* **be in** ~: geteilter Meinung sein; **be in** ~ **with sb./sth.** mit jmdm./etw. nicht übereinstimmen; **b)** *(strife, quarrel)* Meinungsverschiedenheit, *die;* **c)** *(discrepancy)* Diskrepanz, *die*
disallow [dɪsə'laʊ] *v.t.* nicht gestatten; abweisen ‹*Antrag, Anspruch, Klage*›; *(refuse to admit)* nicht anerkennen; nicht gelten lassen; *(Sport)* nicht geben ‹*Tor*›
disappear [dɪsə'pɪə(r)] *v.i.* verschwinden; ‹*Brauch, Kunst, Tierart:*› aussterben
disappearance [dɪsə'pɪərəns] *n.* Verschwinden, *das; (of customs; extinction)* Aussterben, *das*
disappoint [dɪsə'pɔɪnt] *v.t.* enttäuschen; **be ~ed in** *or* **by** *or* **with sb./sth.** von jmdm./etw. enttäuscht sein
disappointing [dɪsə'pɔɪntɪŋ] *adj.* enttäuschend
disappointment [dɪsə'pɔɪntmənt] *n.* Enttäuschung, *die*
disapproval [dɪsə'pru:vl] *n.* Mißbilligung, *die*
disapprove [dɪsə'pru:v] **1.** *v.i.* dagegen sein; ~ **of sb./sth.** jmdn. ablehnen/etw. mißbilligen; ~ **of sb. doing sth.** es mißbilligen, wenn jmd. etw. tut. **2.** *v.t.* mißbilligen
disapproving [dɪsə'pru:vɪŋ] *adj.* mißbilligend
disarm [dɪ'sɑ:m] *v.t. (lit. or fig.)* entwaffnen
disarmament [dɪ'sɑ:məmənt] *n.* Abrüstung, *die; attrib.* ~ **talks** Abrüstungsgespräche
disarming [dɪ'sɑ:mɪŋ] *adj.* entwaffnend
disarray [dɪsə'reɪ] *n.* Unordnung, *die; (confusion)* Wirrwarr, *der;* **be in** ~: in Unordnung sein; *see also* **throw 1 b**
disaster [dɪ'zɑ:stə(r)] *n.* **a)** Katastrophe, *die;* **air** ~: Flugzeugunglück, *das;* **a railway/mining** ~: ein Eisenbahn-/Grubenunglück; **end in** ~: in einer Katastrophe enden; **b)** *(complete failure)* Fiasko, *das;* Katastrophe, *die*
di'saster area *n.* Katastrophengebiet, *das*
disastrous [dɪ'zɑ:strəs] *adj.* katastrophal; verhängnisvoll ‹*Irrtum, Entscheidung, Politik*›; verheerend, katastrophal ‹*Überschwemmung, Wirbelsturm*›
disband [dɪs'bænd] **1.** *v.t.* auflösen. **2.** *v.i.* sich auflösen
disbelief [dɪsbɪ'li:f] *n.* Unglaube, *der;* **in** ~: ungläubig
disbelieve [dɪsbɪ'li:v] *v.t.* ~ **sb./ sth.** jmdm. etw. nicht glauben
disc [dɪsk] *n.* **a)** Scheibe, *die;* **b)** *(gramophone record)* [Schall]platte, *die;* **c)** *(Computing)* **[magnetic]** ~: Magnetplatte, *die;* **floppy** ~: Floppy disk, *die;* Diskette, *die;* **hard** ~ *(fixed)* Festplatte, *die*
discard 1. [dɪ'skɑ:d] *v.t.* **a)** wegwerfen; ablegen ‹*Kleidung*›; fallenlassen ‹*Vorschlag, Idee, Person*›; **b)** *(Cards)* abwerfen. **2.** ['dɪskɑ:d] *n.* Ausschuß, *der*
disc: ~ **brake** *n.* Scheibenbremse, *die;* ~ **drive** *n. (Computing)* Diskettenlaufwerk, *das*
discern [dɪ'sɜ:n] *v.t.* wahrnehmen; **sth. can be ~ed** etw. ist zu erkennen; ~ **from sth. whether ...:** an etw. *(Dat.)* erkennen, ob ...
discernible [dɪ'sɜ:nɪbl] *adj.* erkennbar
discerning [dɪ'sɜ:nɪŋ] *adj.* fein ‹*Gaumen, Ohr, Geschmack*›; scharf ‹*Auge*›; urteilsfähig ‹*Richter, Kritiker*›; kritisch ‹*Leser, Kunde, Zuschauer, Kommentar*›
discharge 1. [dɪs'tʃɑ:dʒ] *v.t.* **a)** *(dismiss, allow to leave)* entlassen **(from** aus); freisprechen ‹*Angeklagte*›; *(exempt from liabilities)* befreien **(from** von); **b)** abschießen ‹*Pfeil, Torpedo*›; ablassen ‹*Flüssigkeit, Gas*›; absondern ‹*Eiter*›; **c)** *(fire)* abfeuern ‹*Gewehr, Kanone*›; **d)** erfüllen ‹*Pflicht, Verbindlichkeiten, Versprechen*›; bezahlen ‹*Schulden*›. **2.** [dɪs'tʃɑ:dʒ] *v.i.* entladen werden; ‹*Schiff auch:*› gelöscht werden; ‹*Batterie:*› sich entladen. **3.** [dɪs'tʃɑ:dʒ, 'dɪstʃɑ:dʒ] *n.* **a)** *(dismissal)* Entlassung, *die* **(from** aus); *(of defendant)* Freispruch, *der; (exemption from liabilities)* Befreiung, *die;* **b)** *(emission)* Ausfluß, *der; (of gas)* Austritt, *der; (of pus)* Absonderung, *die; (Electr.)* Entladung, *die; (of gun)* Abfeuern, *das;* **c)** *(of debt)* Begleichung, *die; (of duty)* Erfüllung, *die*
disciple [dɪ'saɪpl] *n.* **a)** *(Relig.)* Jünger, *der;* **b)** *(follower)* Anhänger, *der*/Anhängerin, *die*
disciplinarian [dɪsɪplɪ'neərɪən] *n.* [strenger] Erzieher
disciplinary ['dɪsɪplɪnərɪ, dɪsɪ'plɪnərɪ] *adj.* Disziplinar-; disziplinarisch
discipline ['dɪsɪplɪn] **1.** *n.* Disziplin, *die;* **maintain** ~: die Disziplin aufrechterhalten. **2.** *v.t.* **a)** disziplinieren; **b)** *(punish)* bestrafen; *(physically also)* züchtigen *(geh.)*
disciplined ['dɪsɪplɪnd] *adj.* diszipliniert
'disc jockey *n.* Diskjockey, *der*
disclaim [dɪs'kleɪm] *v.t.* abstreiten
disclaimer [dɪs'kleɪmə(r)] *n.* Gegenerklärung, *die; (Law)* Verzichterklärung, *die*
disclose [dɪs'kləʊz] *v.t.* enthüllen; bekanntgeben ‹*Information, Nachricht*›; **he didn't** ~ **why he'd come** er verriet nicht, warum er gekommen war
disclosure [dɪs'kləʊʒə(r)] *n.* Enthüllung, *die; (of information, news)* Bekanntgabe, *die*
disco ['dɪskəʊ] *n., pl.* **~s** *(coll.: discothèque, party)* Disko, *die*
discolor *(Amer.) see* **discolour**
discoloration [dɪskʌlə'reɪʃn] *n.* Verfärbung, *die*
discolour [dɪs'kʌlə(r)] *(Brit.)* **1.** *v.t.* verfärben; *(fade)* ausbleichen. **2.** *v.i.* sich verfärben
discolouration *(Brit.) see* **discoloration**
discomfit [dɪs'kʌmfɪt] *v.t.* verunsichern
discomfiture [dɪs'kʌmfɪtʃə(r)] *n.* Verunsicherung, *die*

discomfort [dɪs'kʌmfət] *n.* **a)** *no pl. (uneasiness of body)* Beschwerden *Pl.;* **b)** *no pl. (uneasiness of mind)* Unbehagen, *das;* **c)** *(hardship)* Unannehmlichkeit, *die*

disconcert [dɪskən'sɜ:t] *v. t.* irritieren

disconnect [dɪskə'nekt] *v. t.* **a)** abtrennen; **b)** *(Electr., Teleph.)* ~ **the electricity from a house** ein Haus von der Stromversorgung abtrennen; ~ **the TV** den Stecker des Fernsehers herausziehen; **if you don't pay your telephone bill you will be ~ed** wenn Sie Ihre Telefonrechnung nicht bezahlen, wird Ihr Telefon abgestellt

disconnected [dɪskə'nektɪd] *adj.* **a)** abgetrennt; abgestellt ‹*Telefon*›; **is the cooker/TV ~?** ist der Stecker beim Herd/Fernseher herausgezogen?; **b)** *(incoherent)* unzusammenhängend

disconsolate [dɪs'kɒnsələt] *adj.* untröstlich

discontent [dɪskən'tent] *n.* Unzufriedenheit, *die*

discontented [dɪskən'tentɪd] *adj.* unzufrieden (**with, about** mit)

discontentment [dɪskən'tentmənt] *n., no pl.* Unzufriedenheit, *die*

discontinue [dɪskən'tɪnju:] *v. t.* einstellen; abbestellen ‹*Abonnement*›; abbrechen ‹*Behandlung*›

discord ['dɪskɔ:d] *n.* **a)** Zwietracht, *die; (quarrelling)* Streit, *der;* **b)** *(Mus.)* Dissonanz, *die*

discordant [dɪ'skɔ:dənt] *adj.* **a)** *(conflicting)* gegensätzlich; **b)** *(dissonant)* mißtönend

discothèque ['dɪskətek] *n.* Diskothek, *die*

discount 1. ['dɪskaʊnt] *n. (Commerc.)* Rabatt, *der;* **give** *or* **offer [sb.] a ~ on sth.** [jmdm.] Rabatt auf etw. *(Akk.)* geben *od.* gewähren; ~ **for cash** Skonto, *der od. das;* **at a ~:** mit Rabatt; *(fig.)* nicht gefragt. 2. [dɪ'skaʊnt] *v. t. (disbelieve)* unberücksichtigt lassen; *(discredit)* widerlegen ‹*Beweis, Theorie*›; *(underrate)* zu gering einschätzen

discourage [dɪ'skʌrɪdʒ] *v. t.* **a)** *(dispirit)* entmutigen; **b)** *(advise against)* abraten; ~ **sb. from sth.** jmdm. von etw. abraten; **c)** *(stop)* abhalten ‹*Person*›; verhindern ‹*Handlung*›; ~ **sb. from doing sth.** jmdn. davon abhalten, etw. zu tun

discouragement [dɪ'skʌrɪdʒmənt] *n.* **a)** Entmutigung, *die;* **b)** *(deterrent)* Abschreckung, *die;* **c)** *(depression)* Mutlosigkeit, *die*

discouraging [dɪ'skʌrɪdʒɪŋ] *adj.* entmutigend

discourteous [dɪs'kɜ:tɪəs] *adj.* unhöflich

discourtesy [dɪs'kɜ:təsɪ] *n.* Unhöflichkeit, *die*

discover [dɪ'skʌvə(r)] *v. t.* entdecken; *(by search)* herausfinden

discoverer [dɪ'skʌvərə(r)] *n.* Entdecker, *der*/Entdeckerin, *die*

discovery [dɪ'skʌvərɪ] *n.* Entdeckung, *die;* **voyage of ~:** Entdeckungsreise, *die*

discredit [dɪs'kredɪt] 1. *n.* **a)** *no pl.* Mißkredit, *der;* **bring ~ on sb./sth., bring sb./sth. into ~:** jmdn./etw. in Mißkredit *(Akk.)* bringen; **b)** *(sb. or sth. that ~s)* **be a ~ to sb./sth.** jmdm./einer Sache keine Ehre machen. 2. *v. t.* **a)** *(disbelieve)* keinen Glauben schenken (+ *Dat.*); *(cause to be disbelieved)* unglaubwürdig machen; **b)** *(disgrace)* diskreditieren *(geh.);* in Verruf bringen

discreet [dɪ'skri:t] *adj.*, **~er** [dɪ'skri:tə(r)], **~est** [dɪ'skri:tɪst] diskret; taktvoll; *(unobtrusive)* diskret; dezent ‹*Parfüm, Kleidung*›

discreetly [dɪ'skri:tlɪ] *adv.* diskret; dezent ‹*gekleidet*›

discrepancy [dɪ'skrepənsɪ] *n.* Diskrepanz, *die*

discrete [dɪ'skri:t] *adj.* eigenständig; *(Math., Phys.)* diskret

discretion [dɪ'skreʃn] *n.* **a)** *(prudence)* Umsicht, *die; (reservedness)* Diskretion, *die;* **use ~:** diskret sein; ~ **is the better part of valour** *(prov.)* Vorsicht ist besser als Nachsicht *(ugs. scherzh.);* **b)** *(liberty to decide)* Ermessen, *das;* **leave sth. to sb.'s ~:** etw. in jmds. Ermessen *(Akk.)* stellen; **at sb.'s ~:** nach jmds. Ermessen; **use one's ~:** nach eigenem Ermessen *od.* Gutdünken handeln

discriminate [dɪ'skrɪmɪneɪt] *v. i.* **a)** *(distinguish)* unterscheiden; ~ **between [two things]** unterscheiden zwischen [zwei Dingen]; **b)** ~ **against/in favour of sb.** jmdn. diskriminieren/bevorzugen

discriminating [dɪ'skrɪmɪneɪtɪŋ] *adj.* kritisch ‹*Urteil, Auge, Kunde, Kunstsammler*›; fein ‹*Geschmack, Gaumen, Ohr*›

discrimination [dɪskrɪmɪ'neɪʃn] *n.* **a)** *(discernment)* [kritisches] Urteilsvermögen; **b)** *(differential treatment)* Diskriminierung, *die* (**against** *Gen.*); ~ **against Blacks/women** Diskriminierung von Schwarzen/Frauen; ~ **in favour of** Bevorzugung (+ *Gen.*); **racial ~:** Rassendiskriminierung, *die*

discus ['dɪskəs] *n. (Sport)* Diskus, *der*

discuss [dɪ'skʌs] *v. t.* **a)** *(talk about)* besprechen; **I'm not willing to ~ this matter at present** ich möchte jetzt nicht darüber sprechen; **b)** *(debate)* diskutieren über (+ *Akk.*); *(examine)* erörtern; diskutieren

discussion [dɪ'skʌʃn] *n.* **a)** *(conversation)* Gespräch, *das; (more formal)* Unterredung, *die;* **b)** *(debate)* Diskussion, *die; (examination)* Erörterung, *die;* **be under ~:** zur Diskussion stehen

disdain [dɪs'deɪn] 1. *n.* Verachtung, *die;* **with ~:** verächtlich; **a look of ~:** ein verächtlicher *od.* geringschätziger Blick. 2. *v. t.* verachten; verächtlich ablehnen ‹*Rat, Hilfe*›; ~ **to do sth.** zu stolz sein, etw. zu tun

disdainful [dɪs'deɪnfl] *adj.* verächtlich, geringschätzig ‹*Lachen, Ton, Blick, Kommentar*›; **look ~:** verächtlich dreinblicken

disease [dɪ'zi:z] *n. (lit. or fig.)* Krankheit, *die*

diseased [dɪ'zi:zd] *adj. (lit. or fig.)* krank

disembark [dɪsɪm'bɑ:k] 1. *v. t.* ausschiffen. 2. *v. i.* von Bord gehen

disembodied [dɪsɪm'bɒdɪd] *adj.* körperlos ‹*Seele, Geist*›; geisterhaft ‹*Stimme*›

disenchant [dɪsɪn'tʃɑ:nt] *v. t.* **a)** entzaubern *(geh.);* **b)** *(disillusion)* ernüchtern; **he became ~ed with her/it** sie/es hat ihn desillusioniert

disengage [dɪsɪn'geɪdʒ] *v. t.* lösen (**from** aus, von); ~ **the clutch** auskuppeln

disentangle [dɪsɪn'tæŋgl] *v. t.* **a)** *(extricate)* befreien (**from** aus); *(fig.)* herauslösen (**from** aus); **b)** *(unravel)* entwirren

disfavour *(Brit.; Amer.:* **disfavor)** [dɪs'feɪvə(r)] *n. (displeasure, disapproval)* Mißfallen, *das; (being out of favour)* Ungnade, *die;* **incur sb.'s ~:** jmds. Unwillen erregen

disfigure [dɪs'fɪgə(r)] *v. t.* entstellen

disfigurement [dɪs'fɪgəmənt] *n.* Entstellung, *die*

disgorge [dɪs'gɔ:dʒ] *v. t.* ausspucken; ausspeien *(geh.); (fig.)* ausspeien *(geh.)*

disgrace [dɪs'greɪs] 1. *n., no pl.* **a)** *(ignominy)* Schande, *die;* Schmach, *die (geh.); (deep disfavour)* Ungnade, *die;* **bring ~ on sb./sth.** Schande über jmdn./etw. bringen; **b) be a ~ [to sb./sth.]** [für jmdn./etw.] eine Schande sein. **2.** *v.t.* ⟨*Person:*⟩ Schande machen (+ *Dat.*); ⟨*Person, Handlung:*⟩ Schande bringen über (+ *Akk.*); ~ **oneself** sich blamieren

disgraceful [dɪs'greɪsfl] *adj.* erbärmlich; miserabel ⟨*Handschrift*⟩; skandalös ⟨*Benehmen, Enthüllung, Bedingungen, Verstoß, Behandlung, Tat*⟩; **it's [absolutely** *or* **really** *or* **quite]** ~: es ist [wirklich] ein Skandal

disgracefully [dɪs'greɪsfəlɪ] *adv.* erbärmlich; schändlich ⟨*verraten, betrügen, behandeln*⟩; **behave** ~: sich schändlich *od.* *(geh.)* schimpflich benehmen

disgruntled [dɪs'grʌntld] *adj.* verstimmt

disguise [dɪs'gaɪz] **1.** *v.t.* **a)** verkleiden ⟨*Person*⟩; verstellen ⟨*Stimme*⟩; tarnen ⟨*Gegenstand*⟩; ~ **oneself** sich verkleiden; **b)** *(misrepresent)* verschleiern; **there is no disguising the fact that ...:** es läßt sich nicht verheimlichen, daß ...; **c)** *(conceal)* verbergen. **2.** *n.* Verkleidung, *die; (fig.)* Maske, *die;* **wear a** ~: verkleidet sein; **in the** ~ **of** verkleidet als; **in** ~: verkleidet

disgust [dɪs'gʌst] **1.** *n.* *(nausea)* Ekel, *der* (**at** vor + *Dat.*); *(revulsion)* Abscheu, *der* (**at** vor + *Dat.*); *(indignation)* Empörung, *die* (**at** über + *Akk.*); **in/with** ~: angewidert; *(with indignation)* empört. **2.** *v.t.* anwidern; *(fill with nausea)* anwidern; ekeln; *(fill with indignation)* empören

disgusted [dɪs'gʌstɪd] *adj.* angewidert; *(nauseated)* angewidert; angeekelt; *(indignant)* empört

disgusting [dɪs'gʌstɪŋ] *adj.* widerlich; widerwärtig; *(nauseating also)* ekelhaft

dish [dɪʃ] *n.* **a)** *(for food)* Schale, *die; (flatter)* Platte, *die; (deeper)* Schüssel, *die;* **b)** *in pl. (crockery)* Geschirr, *das;* **wash** *or (coll.)* **do the** ~**es** Geschirr spülen; abwaschen; **c)** *(type of food)* Gericht, *das;* **d)** *(coll.: woman, girl)* klasse Frau *(ugs.);* **e)** *(receptacle)* Schale, *die*

~ **'out** *v.t.* **a)** austeilen ⟨*Essen*⟩; **b)** *(coll.: distribute)* verteilen

~ **'up** *v.t.* auftragen, servieren ⟨*Essen*⟩

'dishcloth *n.* **a)** *(for washing)* Abwaschlappen, *der;* Spültuch, *das;* **b)** *(Brit.: for drying)* Geschirrtuch, *das*

dishearten [dɪs'hɑ:tn] *v.t.* entmutigen; **be** ~**ed** den Mut verlieren/verloren haben

disheartening [dɪs'hɑ:tənɪŋ] *adj.* entmutigend

dishevelled (*Amer.:* **disheveled**) [dɪ'ʃevld] *adj.* unordentlich ⟨*Kleidung*⟩; zerzaust ⟨*Haar, Bart*⟩; ungepflegt ⟨*Erscheinung*⟩

dishonest [dɪ'sɒnɪst] *adj.* unehrlich ⟨*Person*⟩; unaufrichtig ⟨*Person, Antwort*⟩; unlauter *(geh.)* ⟨*Geschäftsgebaren, Vorhaben*⟩; unredlich ⟨*Geschäftsmann*⟩; unreell ⟨*Geschäft, Gewinn*⟩; **be** ~ **with sb.** unehrlich *od.* unaufrichtig gegen jmdn. sein

dishonestly [dɪ'sɒnɪstlɪ] *adv.* unehrlich; unaufrichtig; unlauter *(geh.)* ⟨*handeln*⟩; unredlich ⟨*sich verhalten*⟩

dishonesty [dɪ'sɒnɪstɪ] *n.* Unehrlichkeit, *die;* Unaufrichtigkeit, *die; (of methods)* Unlauterkeit, *die (geh.)*

dishonor *etc. (Amer.) see* **dishonour** *etc.*

dishonour [dɪ'sɒnə(r)] **1.** *n.* Schande, *die.* **2.** *v.t.* beleidigen

dishonourable [dɪ'sɒnərəbl] *adj.* unehrenhaft

dish: ~**-rack** *n.* Abtropfgestell, *das; (in dishwasher)* Geschirrwagen, *der;* ~**-towel** *n.* Geschirrtuch, *das;* ~**washer** *n.* Geschirrspülmaschine, *die;* Geschirrspüler, *der (ugs.);* ~**-water** *n.* Abwaschwasser, *das;* Spülwasser, *das*

disillusion [dɪsɪ'lju:ʒn, dɪsɪ'lu:ʒn] **1.** *n., no pl.* Desillusion, *die* (**with** über + *Akk.*). **2.** *v.t.* ernüchtern

disillusioned [dɪsɪ'lju:ʒnd, dɪsɪ'lu:ʒnd] *adj.* desillusioniert; **become** ~ **with sth.** seine Illusionen über etw. *(Akk.)* verlieren

disillusionment [dɪsɪ'lju:ʒnmənt, dɪsɪ'lu:ʒnmənt] *n.* Desillusionierung, *die*

disincentive [dɪsɪn'sentɪv] *n.* Hemmnis, *das;* **act as** *or* **be a** ~ **to sb. to do sth.** jmdn. davon abhalten, etw. zu tun

disinclination [dɪsɪnklɪ'neɪʃn] *n.* Abneigung, *die* (**for, to** gegen)

disincline [dɪsɪn'klaɪn] *v.t.* abgeneigt machen (**for, to** gegen)

disinclined [dɪsɪn'klaɪnd] *adj.* abgeneigt

disinfect [dɪsɪn'fekt] *v.t.* desinfizieren

disinfectant [dɪsɪn'fektənt] **1.** *adj.* desinfizierend. **2.** *n.* Desinfektionsmittel, *das*

disingenuous [dɪsɪn'dʒenjʊəs] *adj.* unaufrichtig

disintegrate [dɪs'ɪntɪgreɪt] **1.** *v.i.* zerfallen; *(shatter suddenly)* zerbersten; *(fig.)* sich auflösen. **2.** *v.t.* zerstören

disintegration [dɪsɪntɪ'greɪʃn] *n.* Zerfall, *der; (fig.)* Auflösung, *die*

disinter [dɪsɪn'tɜ:(r)] *v.t.,* **-rr-** ausgraben

disinterested [dɪs'ɪntrəstɪd, dɪs'ɪntrɪstɪd] *adj.* **a)** *(impartial)* unvoreingenommen; unparteiisch; *(free from selfish motive)* selbstlos; uneigennützig; **b)** *(coll.: uninterested)* desinteressiert

disjointed [dɪs'dʒɔɪntɪd] *adj.* unzusammenhängend; zusammenhanglos

disk *see* **disc**

diskette [dɪ'sket] *n. (Computing)* Diskette, *die*

dislike [dɪs'laɪk] **1.** *v.t.* nicht mögen; ~ **sb./sth. greatly** *or* **intensely** jmdn./etw. ganz und gar nicht leiden können; **I don't** ~ **it** ich finde es nicht schlecht; ~ **doing sth.** es nicht mögen, etw. zu tun; etw. ungern tun. **2.** *n.* **a)** *no pl.* Abneigung, *die* (**of, for** gegen); **she took an instant** ~ **to him/the house** sie empfand sofort eine Abneigung gegen ihn/das Haus; **b)** *(object)* **one of my greatest** ~**s is ...:** zu den Dingen, die ich am wenigsten leiden kann, gehört ...

dislocate ['dɪsləkeɪt] *v.t. (Med.)* ausrenken; auskugeln ⟨*Schulter, Hüfte*⟩

dislodge [dɪs'lɒdʒ] *v.t.* entfernen (**from** aus); *(detach)* lösen (**from** von)

disloyal [dɪs'lɔɪəl] *adj.* illoyal (**to** gegenüber); treulos ⟨*Freund, Ehepartner*⟩; **be** ~: nicht loyal sein

disloyalty [dɪs'lɔɪəltɪ] *n.* Illoyalität, *die* (**to** gegenüber); *(to spouse, friend)* Treulosigkeit, *die*

dismal ['dɪzməl] *adj.* trist; düster; trostlos ⟨*Landschaft, Ort*⟩; *(coll.: feeble)* kläglich ⟨*Zustand, Leistung, Versuch*⟩; **a** ~ **failure** ein völliger Reinfall *(ugs.)*

dismantle [dɪs'mæntl] *v.t.* zerlegen; demontieren; *(fig.)* demontieren; abbauen ⟨*Schuppen, Gerüst*⟩

dismay [dɪs'meɪ] **1.** *v.t.* bestürzen; **he was** ~**ed to hear that ...:** mit Bestürzung hörte er, daß ... **2.** *n.* Bestürzung, *die* (**at** über + *Akk.*); **watch in** *or* **with** ~: bestürzt zusehen

dismiss [dɪs'mɪs] *v.t.* **a)** entlassen; auflösen, aufheben ⟨*Versammlung*⟩; **b)** *(from the mind)* verwerfen; *(reject)* ablehnen; *(treat very briefly)* abtun

dismissal [dɪs'mɪsl] *n.* **a)** Entlassung, *die; (of gathering etc.)* Auflösung, *die;* Aufhebung, *die;* **b)** *(from the mind)* Aufgabe, *die;*

(rejection) Ablehnung, *die; (very brief treatment)* Abtun, *das*
dismissive [dɪs'mɪsɪv] *adj.* abweisend; *(disdainful)* abschätzig; **be ~ about sth.** etw. abtun
dismount [dɪs'maʊnt] **1.** *v.i.* absteigen. **2.** *v.t.* abwerfen ‹*Reiter*›
disobedience [dɪsə'bi:dɪəns] *n.* Ungehorsam, *der;* **act of ~:** ungehorsames Verhalten
disobedient [dɪsə'bi:dɪənt] *adj.* ungehorsam; **be ~ to sb.** jmdm. nicht gehorchen
disobey [dɪsə'beɪ] *v.t.* nicht gehorchen (+ *Dat.*); nicht befolgen, mißachten ‹*Befehl, Vorschrift usw.*›; übertreten ‹*Gesetz*›
disobliging [dɪsə'blaɪdʒɪŋ] *adj.* ungefällig
disorder [dɪs'ɔ:də(r)] *n.* **a)** Unordnung, *die;* Durcheinander, *das;* **everything was in [complete] ~:** alles war ein einziges[, heilloses] Durcheinander; **the meeting broke up in ~:** die Versammlung endete in einem heillosen Durcheinander; **b)** *(rioting, disturbance)* Unruhen *Pl.;* **c)** *(Med.)* [Funktions]störung, *die;* **suffer from a mental ~:** geisteskrank sein; **a stomach/liver ~:** ein Magen-/Leberleiden
disordered [dɪs'ɔ:dəd] *adj.* unordentlich; ungeordnet ‹*Wortschwall, Gedanken[gang]*›
disorderly [dɪs'ɔ:dəlɪ] *adj.* **a)** *(untidy)* unordentlich; ungeordnet ‹*Ansammlung*›; **b)** *(unruly)* undiszipliniert; aufrührerisch ‹*Mob*›; **~ conduct** ungebührliches Benehmen
disorganization [dɪsɔ:gənaɪ'zeɪʃn] *n., no pl.* Desorganisation, *die; (muddle)* Durcheinander, *das*
disorganize [dɪs'ɔ:gənaɪz] *v.t.* durcheinanderbringen
disorganized [dɪs'ɔ:gənaɪzd] *adj.* chaotisch
disorient [dɪs'ɔ:rɪənt], **disorientate** [dɪs'ɔ:rɪənteɪt] *v.t.* die Orientierung nehmen (+ *Dat.*); *(fig.)* verwirren
disorientated [dɪs'ɔ:rɪənteɪtɪd], **disoriented** [dɪs'ɔ:rɪəntɪd] *adj.* verwirrt, desorientiert
disown [dɪs'əʊn] *v.t.* verleugnen
disparage [dɪ'spærɪdʒ] *v.t.* herabsetzen
disparagement [dɪ'spærɪdʒmənt] *n.* Herabsetzung, *die*
disparaging [dɪ'spærɪdʒɪŋ] *adj.* abschätzig
disparate ['dɪspərət] *adj.* [völlig] verschieden; disparat *(geh.)*
disparity [dɪ'spærɪtɪ] *n.* Unterschied, *der; (lack of parity)* Ungleichheit, *die*
dispassionate [dɪ'spæʃənət] *adj.* *(impartial)* unvoreingenommen
dispatch [dɪ'spætʃ] **1.** *v.t.* **a)** *(send off)* schicken; **~ sb. [to do sth.]** jmdn. entsenden *(geh.)* [um etw. zu tun]; **b)** *(deal with)* erledigen; **c)** *(kill)* töten. **2.** *n.* **a)** *(official report, Journ.)* Bericht, *der;* **b)** *(sending off)* Absenden, *das; (of troops, messenger, delegation)* Entsendung, *die (geh.)*
dispel [dɪ'spel] *v.t.,* **-ll-** vertreiben; zerstreuen ‹*Besorgnis, Befürchtung*›; verdrängen, unterdrücken ‹*Gefühl, Erinnerung*›
dispensable [dɪ'spensəbl] *adj.* entbehrlich
dispensary [dɪ'spensərɪ] *n.* *(Pharm.)* Apotheke, *die*
dispensation [dɪspen'seɪʃn] *n.* **a)** *(distribution)* Verteilung, *die* (**to** an + *Akk.*); *(of favours)* Gewährung, *die;* **b)** *(exemption)* Sonderregelung, *die*
dispense [dɪ'spens] **1.** *v.i.* **~ with** verzichten auf (+ *Akk.*); *(do away with)* überflüssig machen. **2.** *v.t.* **a)** *(distribute, administer)* verteilen (**to** an + *Akk.*); gewähren ‹*Gastfreundschaft*›; zuteil werden lassen ‹*Gnade*›; **the machine ~s hot drinks** der Automat gibt heiße Getränke aus; **b)** *(Pharm.)* dispensieren *(fachspr.)*
dispenser [dɪ'spensə(r)] *n.* *(vending machine)* Automat, *der; (container)* Spender, *der*
dispensing 'chemist *n.* Apotheker, *der*/Apothekerin, *die*
dispersal [dɪ'spɜ:sl] *n.* Zerstreuung, *die; (diffusion)* Ausbreitung, *die; (of mist, oil-slick)* Auflösung, *die*
disperse [dɪ'spɜ:s] **1.** *v.t.* zerstreuen; *(dispel)* auflösen ‹*Dunst, Öl*›; vertreiben ‹*Wolken, Gase*›. **2.** *v.i.* sich zerstreuen
dispersion [dɪ'spɜ:ʃn] *n.* *(scattering)* Zerstreuung, *die; (diffusion)* Ausbreitung, *die*
dispirited [dɪ'spɪrɪtɪd] *adj.* entmutigt
dispiriting [dɪ'spɪrɪtɪŋ] *adj.* entmutigend
displace [dɪs'pleɪs] *v.t.* **a)** *(move from place)* verschieben; **b)** *(supplant)* ersetzen; *(crowd out)* verdrängen
displaced 'person *n.* Vertriebene, *der/die*
displacement [dɪs'pleɪsmənt] *n.* **a)** *(moving)* Verschiebung, *die;* **b)** *(supplanting)* Ersetzung, *die;* **c)** *(Naut.: weight displaced)* [Wasser]verdrängung, *die*
display [dɪ'spleɪ] **1.** *v.t.* **a)** zeigen, tragen ‹*Abzeichen*›; aufstellen ‹*Trophäe*›; *(to public view)* ausstellen; *(on notice-board)* aushängen; *(standing)* aufstellen ‹*Schild*›; *(attached)* aufhängen ‹*Schild, Fahne*›; **b)** *(flaunt)* zur Schau stellen; **c)** *(Commerc.)* ausstellen. **2.** *n.* **a)** Aufstellung, *die; (to public view)* Ausstellung, *die;* **a ~ of ill-will/courage** eine Demonstration von jmds. Übelwollen/Mut; **b)** *(exhibition)* Ausstellung, *die; (Commerc.)* Auslage, *die;* **a fashion ~:** eine Modenschau; **an air ~:** eine Flugschau; **be on ~:** ausgestellt werden; **c)** *(ostentatious show)* Zurschaustellung, *die;* **make a ~ of one's affection** seine Gefühle zur Schau stellen; **d)** *(Computing etc.)* Display, *das;* Anzeige, *die*
display: ~ cabinet, ~ case *see* [2]**case 1 d**
displease [dɪs'pli:z] *v.t.* **a)** *(earn disapproval of)* **~ sb.** jmds. Mißfallen erregen; **b)** *(annoy)* verärgern; **be ~d [with sb./at sth.]** [über jmdn./etw.] verärgert sein
displeasure [dɪs'pleʒə(r)] *n., no pl.* Mißfallen, *das* (**at** über + *Akk.*)
disposable [dɪ'spəʊzəbl] *adj.* **a)** *(to be thrown away after use)* Wegwerf-; **~ bottle/container/syringe** Einwegflasche/-behälter/-spritze; **be ~:** nach Gebrauch weggeworfen werden; **b)** *(available)* verfügbar; **~ income** verfügbares Einkommen
disposal [dɪ'spəʊzl] *n.* **a)** *(getting rid of, killing)* Beseitigung, *die; (of waste)* Entsorgung, *die;* **b)** *(putting away)* Forträumen, *das;* **c)** *(settling)* Erledigung, *die;* **d)** *(treating)* Abhandlung, *die;* **e)** *(control)* Verfügung, *die;* **place** *or* **put sth./sb. at sb.'s [complete] ~:** jmdm. etw./jmdn. [ganz] zur Verfügung stellen; **have sth./sb. at one's ~:** etw./jmdn. zur Verfügung haben; **be at sb.'s ~:** jmdm. zur Verfügung stehen
dispose [dɪ'spəʊz] *v.t.* **a)** *(make inclined)* **~ sb. to do sth.** jmdn. dazu veranlassen, etw. zu tun; **b)** *(arrange)* anordnen; *(Mil.)* aufstellen ‹*Truppen*›
~ of *v.t.* **a)** *(do as one wishes with)* **~ of sth./sb.** über etw./jmdn. frei verfügen; **b)** *(kill, get rid of)* beseitigen ‹*Rivalen, Leiche, Abfall*›; erlegen, töten ‹*Gegner, Drachen*›; **c)** *(put away)* wegräumen; **d)** *(eat up)* aufessen; verputzen *(ugs.)*; **e)** *(settle, finish)* erledigen; **f)** *(disprove)* widerlegen
disposed [dɪ'spəʊzd] *adj.* **be ~ to sth.** zu etw. neigen; **be ~ to do sth.**

dazu neigen, etw. zu tun; **be well/ill ~ towards sb.** jmdm. wohl/übel gesinnt sein; **be well/ill ~ towards sth.** einer Sache *(Dat.)* positiv/ablehnend gegenüberstehen

disposition [dɪspə'zɪʃn] *n.* **a)** *(arrangement)* Aufstellung, *die; (of seating, figures)* Anordnung, *die;* **b)** *(temperament)* Veranlagung, *die;* Disposition, *die;* **she has a/is of a rather irritable ~:** sie ist ziemlich reizbar; **c)** *(inclination)* Hang, *der;* Neigung, *die* **(towards** zu); **have a ~ to do sth./to[wards] sth.** dazu neigen, etw. zu tun/zu etw. neigen

dispossess [dɪspə'zes] *v. t.* **~ sb. of sth.** jmdm. etw. entziehen; *(fig.)* jmdm. etw. rauben

disproportion [dɪsprə'pɔ:ʃn] *n.* Mißverhältnis, *das*

disproportionate [dɪsprə'pɔ:ʃənət] *adj.* vom Normalen abweichend; unangemessen; **be [totally] ~ to sth.** in einem [völligen] Mißverhältnis *od.* in [gar] keinem Verhältnis zu etw. stehen

disprove [dɪs'pru:v] *v. t.* widerlegen

disputable [dɪ'spju:təbl] *adj.* strittig

dispute [dɪ'spju:t] **1.** *n.* **a)** *no pl. (controversy)* Streit, *der;* **a matter of much ~:** eine sehr umstrittene Frage; **that is [not] in ~:** darüber wird [nicht] gestritten; **be beyond ~:** außer Frage stehen; **b)** *(argument)* Streit, *der* **(over** um); *see also* **industrial dispute. 2.** *v. t.* **a)** sich streiten über (+ *Akk.*); **~ whether .../how ...:** sich darüber streiten, ob .../wie ...; **b)** *(oppose)* bestreiten; anfechten ‹*Rechtsanspruch*›; angreifen ‹*Entscheidung*›; **c)** *(contend for)* streiten um

disqualification [dɪskwɒlɪfɪ'keɪʃn] *n.* Ausschluß, *der* **(from** von); *(Sport)* Disqualifikation, *die*

disqualify [dɪs'kwɒlɪfaɪ] *v. t.* **a)** *(debar)* ausschließen **(from** von); *(Sport)* disqualifizieren; **b)** *(make unfit)* ungeeignet machen; **~ sb./sth. for sth.** jmdn./etw. für etw. ungeeignet machen

disquiet [dɪs'kwaɪət] *n.* Unruhe, *die*

disregard [dɪsrɪ'gɑ:d] **1.** *v. t.* ignorieren; nicht berücksichtigen ‹*Tatsache*›; **~ a request** einer Bitte *(Dat.)* nicht nachkommen. **2.** *n.* Mißachtung, *die* **(of, for** *Gen.*); *(of wishes, feelings)* Gleichgültigkeit, *die* **(for, of** gegenüber)

disrepair [dɪsrɪ'peə(r)] *n. (of building)* schlechter [baulicher] Zustand; *(of furniture etc.)* schlechter Zustand

disreputable [dɪs'repjʊtəbl] *adj.* zwielichtig; übelbeleumdet ‹*Person*›; verrufen ‹*Etablissement, Gegend*›; schäbig ‹*Aussehen*›

disrepute [dɪsrɪ'pju:t] *n.* Verruf, *der; (of area)* Verrufenheit, *die;* **bring sb./sth. into ~:** jmdn./etw. in Verruf bringen

disrespect [dɪsrɪ'spekt] *n.* Mißachtung, *die;* **show ~ for sb./sth.** keine Achtung vor jmdm./etw. haben

disrespectful [dɪsrɪ'spektfl] *adj.* respektlos

disrupt [dɪs'rʌpt] *v. t.* unterbrechen; stören ‹*Klasse, Sitzung*›

disruption [dɪs'rʌpʃn] *n.* Unterbrechung, *die; (of class, meeting)* Störung, *die*

disruptive [dɪs'rʌptɪv] *adj.* störend

dissatisfaction [dɪsætɪs'fækʃn] *n., no pl.* Unzufriedenheit, *die* **(with** mit)

dissatisfied [dɪ'sætɪsfaɪd] *adj.* unzufrieden **(with** mit)

dissect [dɪ'sekt] *v. t.* **a)** *(cut into pieces)* zerschneiden, zerlegen **(into** in + *Akk.*); **b)** *(Med., Biol.)* präparieren

dissection [dɪ'sekʃn] *n.* **a)** *(cutting into pieces)* Zerlegung, *die;* **b)** *(Med., Biol.)* Präparation, *die*

disseminate [dɪ'semɪneɪt] *v. t. (lit. or fig.)* verbreiten

dissension [dɪ'senʃn] *n.* Dissens, *der;* Streit, *der* **(on** über + *Akk.*); **~s** Streitigkeiten

dissent [dɪ'sent] **1.** *v. i.* **a)** *(refuse to assent)* nicht zustimmen; **~ from sth.** mit etw. nicht übereinstimmen; **b)** *(disagree)* **~ from sth.** von etw. abweichen. **2.** *n.* Ablehnung, *die; (from majority)* Abweichung, *die*

dissenter [dɪ'sentə(r)] *n.* Andersdenkende, *der/die*

dissertation [dɪsə'teɪʃn] *n. (spoken)* Vortrag, *der; (written)* Abhandlung, *die; (for Ph.D.)* Dissertation, *die*

disservice [dɪ'sɜ:vɪs] *n.* **do sb. a ~:** jmdm. einen schlechten Dienst erweisen

dissident ['dɪsɪdənt] **1.** *adj.* andersdenkend; **hold a ~ view** *or* **opinion** eine abweichende Meinung vertreten. **2.** *n.* Dissident, *der/*Dissidentin, *die*

dissimilar [dɪ'sɪmɪlə(r)] *adj.* unähnlich; unterschiedlich, verschieden ‹*Ideen, Ansichten, Geschmack*›; **be ~ to sth./sb.** anders als etw./jmd. sein

dissimilarity [dɪsɪmɪ'lærɪtɪ] *n.* Unähnlichkeit, *die*

dissipate ['dɪsɪpeɪt] *v. t.* **a)** *(dispel)* auflösen ‹*Nebel, Dunst*›; zerstreuen ‹*Befürchtungen, Zweifel*›; **b)** *(fritter away)* vergeuden; durchbringen ‹*Vermögen, Erbschaft*›

dissipated ['dɪsɪpeɪtɪd] *adj.* ausschweifend; zügellos

dissipation [dɪsɪ'peɪʃn] *n. (intemperate living)* Ausschweifung, *die*

dissociate [dɪ'səʊʃɪeɪt, dɪ'səʊsɪeɪt] *v. t.* trennen; **~ oneself from sth./sb.** sich von etw./jmdm. distanzieren

dissolute ['dɪsəlu:t, 'dɪsəlju:t] *adj. (licentious)* ausschweifend; zügellos ‹*Benehmen*›

dissolution [dɪsə'lu:ʃn, dɪsə'lju:ʃn] *n.* **a)** *(disintegration)* Zersetzung, *die;* **b)** *(undoing, dispersal)* Auflösung, *die*

dissolve [dɪ'zɒlv] **1.** *v. t.* auflösen. **2.** *v. i.* sich auflösen; **~ into tears/laughter** in Tränen/Gelächter ausbrechen

dissonance ['dɪsənəns] *n. (Mus.)* Dissonanz, *die*

dissonant ['dɪsənənt] *adj. (Mus.)* dissonant

dissuade [dɪ'sweɪd] *v. t.* **~ sb. from sth.** jmdn. von etw. abbringen; **~ sb. from doing sth.** jmdn. davon abbringen, etw. zu tun

distance ['dɪstəns] **1.** *n.* **a)** Entfernung, *die* **(from** zu); **their ~ from each other** die räumliche Entfernung zwischen ihnen; **keep [at] a [safe] ~ [from sb./sth.]** jmdm./einer Sache nicht zu nahe kommen; **b)** *(fig.: aloofness)* Abstand, *der;* **keep one's ~ [from sb./sth.]** Abstand [zu jmdm./etw.] wahren; **c)** *(way to cover)* Strecke, *die;* Weg, *der; (gap)* Abstand, *der;* **from this ~:** aus dieser Entfernung; **at a ~ of ... [from sb./sth.]** in einer Entfernung von ... [von jmdm./etw.]; **a short ~ away** ganz in der Nähe; **d)** *(remoter field of vision)* Ferne, *die;* **in/into the ~:** in der/die Ferne; **e)** *(distant point)* Entfernung, *die;* **at a ~/[viewed] from a ~:** von weitem; **f)** *(space of time)* Abstand, *der;* **at a ~ of 20 years** aus einem Abstand von 20 Jahren. **2.** *v. t.* **~ oneself from sb./sth.** sich von jmdm./etwas distanzieren

distant ['dɪstənt] *adj.* **a)** *(far)* fern; **be ~ [from sb.]** weit [von jmdm.] weg sein; **b)** *(fig.: remote)* entfernt ‹*Ähnlichkeit, Verwandtschaft, Verwandte, Beziehung*›; **it's a ~ prospect/possibility** das ist Zukunftsmusik; **c)** *(in time)* fern;

in the ~ past/future in ferner Vergangenheit/Zukunft; **d)** *(cool)* reserviert, distanziert ‹*Person, Haltung*›

distaste [dɪs'teɪst] *n.* Abneigung, *die;* **[have] a ~ for sb./sth.** eine Abneigung gegen jmdn./etw. [haben]; **in ~:** aus Abneigung

distasteful [dɪs'teɪstfl] *adj.* unangenehm; **be ~ to sb.** jmdm. zuwider sein

[1]**distemper** [dɪ'stempə(r)] **1.** *n.* *(paint)* Temperafarbe, *die.* **2.** *v.t.* mit Temperafarbe bemalen

[2]**distemper** *n.* *(animal disease)* Staupe, *die*

distend [dɪ'stend] *v.t.* aufblähen, auftreiben ‹*Leib, Bauch*›; blähen ‹*Nüstern*›; erweitern ‹*Gefäße, Darm, Ader*›

distil, *(Amer.)* **distill** [dɪ'stɪl] *v.t.,* **-ll-** *(lit. or fig.)* destillieren; brennen ‹*Branntwein*›; **~ sth. from sth.** *(fig.)* etw. aus etw. [heraus]destillieren

distillation [dɪstɪ'leɪʃn] *n.* Destillation, *die; (fig.)* Herausdestillieren, *das; (result)* Destillat, *das*

distiller [dɪ'stɪlə(r)] *n.* Destillateur, *der;* Branntweinbrenner, *der*

distillery [dɪ'stɪlərɪ] *n.* [Branntwein]brennerei, *die*

distinct [dɪ'stɪŋkt] *adj.* **a)** *(different)* verschieden; **keep two things ~:** zwei Dinge auseinanderhalten; **as ~ from** im Unterschied zu; **b)** *(clearly perceptible, decided)* deutlich; klar ‹*Stimme, Sicht*›; **c)** *(separate)* unterschiedlich

distinction [dɪ'stɪŋkʃn] *n.* **a)** *(making a difference)* Unterscheidung, *die;* **by way of ~, for ~:** zur Unterscheidung; **b)** *(difference)* Unterschied, *der;* **make** *or* **draw a ~ between A and B** einen Unterschied zwischen A und B machen; **c) have the ~ of being ...** ‹*Person:*› sich dadurch auszeichnen, daß man ... ist; **d) gain** *or* **get a ~ in one's examination** das Examen mit Auszeichnung bestehen; **a scientist of ~:** ein Wissenschaftler von Rang [und Namen]

distinctive [dɪ'stɪŋktɪv] *adj.* unverwechselbar; **be ~ of sth.** für etw. charakteristisch sein

distinctly [dɪ'stɪŋktlɪ] *adv.* **a)** *(clearly)* deutlich; **b)** *(decidedly)* merklich

distinguish [dɪ'stɪŋgwɪʃ] **1.** *v.t.* **a)** *(make out)* erkennen; **b)** *(differentiate)* unterscheiden; *(characterize)* kennzeichnen; **c)** *(make prominent)* **~ oneself [by sth.]** sich [durch etw.] hervortun; **~ oneself by doing sth.** sich dadurch hervortun, daß man etw. tut. **2.** *v.i.* unterscheiden; **~ between persons/things** Personen/Dinge auseinanderhalten

distinguishable [dɪ'stɪŋgwɪʃəbl] *adj.* erkennbar

distinguished [dɪ'stɪŋgwɪʃt] *adj.* **a)** *(eminent)* namhaft, angesehen ‹*Persönlichkeit, Schule, Firma*›; glänzend ‹*Laufbahn*›; **a ~ politician** ein Politiker von Rang [und Namen]; **b)** *(looking eminent)* vornehm, *(geh.)* distinguiert ‹*Aussehen, Mensch*›

distort [dɪ'stɔ:t] *v.t.* **a)** verzerren; ‹*Schmerz, Krankheit:*› entstellen; **b)** *(misrepresent)* entstellt *od.* verzerrt wiedergeben; verdrehen ‹*Worte, Wahrheit*›

distortion [dɪ'stɔ:ʃn] *n.* Verzerrung, *die*

distract [dɪ'strækt] *v.t.* ablenken; **~ sb.['s attention from sth.]** jmdn. [von etw.] ablenken

distracted [dɪ'stræktɪd] *adj.* **a)** *(mad)* von Sinnen *nachgestellt;* außer sich *nachgestellt; (worried)* besorgt; beunruhigt; **b)** *(mentally far away)* abwesend

distraction [dɪ'strækʃn] *n.* **a)** *(frenzy)* Wahnsinn, *der;* **drive sb. to ~:** jmdn. wahnsinnig machen *od.* zum Wahnsinn treiben; **b)** *(diversion)* Ablenkung, *die;* **c)** *(interruption)* Störung, *die;* **be a ~:** ein Störfaktor sein; **d)** *(amusement)* Zerstreuung, *die*

distraught [dɪ'strɔ:t] *adj.* aufgelöst (**with** vor + *Dat.*); verstört ‹*Blick, Gesichtsausdruck*›

distress [dɪ'stres] **1.** *n.* **a)** *(anguish)* Kummer, *der* (**at** über + *Akk.*); **b)** *(suffering caused by want)* Not, *die;* Elend, *das;* **c)** *(danger)* **an aircraft/a ship in ~:** ein Flugzeug in Not/ein Schiff in Seenot; **d)** *(exhaustion)* Erschöpfung, *die; (severe pain)* Qualen *Pl.* **2.** *v.t.* **a)** *(worry)* bedrücken; bekümmern; *(cause anguish to)* ängstigen; *(upset)* nahegehen (+ *Dat.*); mitnehmen; **we were most ~ed** wir waren zutiefst betroffen; **b)** *(exhaust)* erschöpfen

distressed [dɪ'strest] *adj.* **a)** *(anguished)* leidvoll; betrübt; **b)** *(impoverished)* notleidend ‹*Volkswirtschaft, Dritte Welt*›; verarmt ‹*Adel*›; armselig ‹*Verhältnisse*›

distressing [dɪ'stresɪŋ] *adj.* **a)** *(upsetting)* erschütternd; **be ~ to sb.** jmdn. sehr belasten; **b)** *(regrettable)* beklagenswert

di'stress signal *n.* Notsignal, *das*

distribute [dɪ'strɪbju:t] *v.t.* verteilen (**to** an + *Akk.*, **among** unter + *Akk.*)

distribution [dɪstrɪ'bju:ʃn] *n.* Verteilung, *die* (**to** an + *Akk.*, **among** unter + *Akk.*); *(Econ.: of goods)* Distribution, *die (fachspr.);* Vertrieb, *der; (of films)* Verleih, *der;* **the ~ of wealth** die Vermögensverteilung

distributor [dɪ'strɪbjʊtə(r)] *n.* **a)** Verteiler, *der*/Verteilerin, *die; (Econ.)* Vertreiber, *der; (firm)* Vertrieb, *der; (of films)* Verleih[er], *der;* **b)** *(Motor Veh.)* [Zünd]verteiler, *der*

district ['dɪstrɪkt] *n.* **a)** *(administrative area)* Bezirk, *der;* **b)** *(Brit.: part of county)* Distrikt, *der;* **c)** *(Amer.: political division)* Wahlkreis, *der;* **d)** *(tract of country, area)* Gegend, *die;* **country ~s** ländliche Gegenden

district: ~ at'torney *n.* *(Amer. Law)* [Bezirks]staatsanwalt, *der*/-anwältin, *die;* **~ 'nurse** *n.* *(Brit.)* Gemeindeschwester, *die*

distrust [dɪs'trʌst] **1.** *n.* Mißtrauen, *das* (**of** gegen). **2.** *v.t.* mißtrauen (+ *Dat.*); *(because of bad experiences)* mit Argwohn *od.* Mißtrauen begegnen (+ *Dat.*)

distrustful [dɪs'trʌstfl] *adj.* mißtrauisch; **be ~ of sb./sth.** jmdm./einer Sache nicht trauen

disturb [dɪ'stɜ:b] *v.t.* **a)** *(break calm of)* stören; aufscheuchen ‹*Vögel*›; aufhalten, behindern ‹*Fortschritt*›; **'do not ~!'** „bitte nicht stören!"; **~ing the peace** Ruhestörung, *die;* **b)** *(move from settled position)* durcheinanderbringen; **c)** *(worry)* beunruhigen; *(agitate)* nervös machen; **don't be ~ed** beunruhigen Sie sich nicht

disturbance [dɪ'stɜ:bəns] *n.* **a)** *(interruption)* Störung, *die; (nuisance)* Belästigung, *die;* **b)** *(agitation, tumult)* Unruhe, *die;* **political ~s** politische Unruhen

disturbed [dɪ'stɜ:bd] *adj.* besorgt ‹*Eindruck, Ausdruck*›; unruhig ‹*Nacht*›; **be [mentally] ~:** geistig gestört sein

disuse [dɪs'ju:s] *n.* *(discontinuance)* Außer-Gebrauch-Kommen, *das; (disappearance)* Verschwinden, *das; (abolition)* Abschaffung, *die;* **fall into ~:** außer Gebrauch kommen

disused [dɪs'ju:zd] *adj.* stillgelegt ‹*Bergwerk, Eisenbahnlinie*›; leerstehend ‹*Gebäude*›; ausrangiert *(ugs.)* ‹*Fahrzeug, Möbel*›

ditch [dɪtʃ] **1.** *n.* Graben, *der; (at side of road)* Straßengraben, *der.* **2.** *v.t.* *(sl.: abandon)* sitzenlassen

⟨*Familie, Freunde*⟩*;* sausenlassen *(ugs.)* ⟨*Plan*⟩

'ditchwater *n.* stehendes, fauliges Wasser; **[as] dull as** ~: sterbenslangweilig

dither ['dɪðə(r)] **1.** *v.i.* schwanken. **2.** *n. (coll.)* **be all of a** ~ *or* **in a** ~: am Rotieren sein *(ugs.)*

ditto ['dɪtəʊ] *n., pl.* ~**s: p. 5 is missing, p. 19** ~: S. 5 fehlt, ebenso S. 19; ~ **marks** Unterführungszeichen, *das;* **I'm hungry. – D**~: Ich habe Hunger. – Ich auch

ditty ['dɪtɪ] *n.* Weise, *die*

divan [dɪ'væn] *n.* **a)** *(couch, bed)* [Polster]liege, *die;* **b)** *(long seat)* Chaiselongue, *die*

di'van bed *see* **divan a**

dive [daɪv] **1.** *v.i.,* ~**d** *or (Amer.)* **dove** [dəʊv] **a)** einen Kopfsprung machen; springen; *(when already in water)* [unter]tauchen; **b)** *(plunge downwards)* ⟨*Vogel, Flugzeug usw.:*⟩ einen Sturzflug machen; ⟨*Unterseeboot usw.:*⟩ abtauchen *(Seemannsspr.),* tauchen; **c)** *(dart down)* sich hinwerfen; **d)** *(dart)* ~ **[out of sight]** sich schnell verstecken. **2.** *n.* **a)** *(plunge)* Kopfsprung, *der; (of bird, aircraft, etc.)* Sturzflug, *der* (**towards** auf + *Akk.*); *(of submarine etc.)* [Unter]tauchen; **b)** *(sudden darting movement)* Sprung, *der;* **c)** *(coll.: disreputable place)* Spelunke, *die (abwertend)*

~ **'in** *v.i.* [mit dem Kopf voraus] hineinspringen

dive: ~**-bomb** *v.t. (Mil.)* im Sturzflug bombardieren; ~**-bomber** *n. (Mil.)* Sturzkampfflugzeug, *das*

diver ['daɪvə(r)] *n.* **a)** *(Sport)* Kunstspringer, *der*/-springerin, *die;* **b)** *(as profession)* Taucher, *der*/Taucherin, *die*

diverge [daɪ'vɜ:dʒ] *v.i.* **a)** auseinandergehen; **here the road** ~**s from the river** hier entfernt die Straße sich vom Fluß; **b)** *(fig.)* ⟨*Berufswege, Pfade:*⟩ sich trennen; *(from norm etc.)* abweichen; **c)** ⟨*Meinungen, Aussichten:*⟩ voneinander abweichen

divergence [daɪ'vɜ:dʒəns] *n.* **a)** Divergenz, *die (fachspr.);* Auseinandergehen, *das;* **b)** *(fig.)* ~ **of opinions/views** Meinungsverschiedenheit, *die*

divergent [daɪ'vɜ:dʒənt] *adj.* **a)** divergent *(fachspr.);* auseinandergehend, -laufend ⟨*Routen, Wege*⟩; **b)** *(differing)* unterschiedlich, voneinander abweichend ⟨*Ansichten, Methoden*⟩

diverse [daɪ'vɜ:s] *adj.* **a)** *(unlike)* verschieden[artig]; unterschiedlich; **b)** *(varied)* vielseitig, breit gefächert ⟨*[Aus]bildung, Interessen, Kenntnisse*⟩*;* bunt [gewürfelt] ⟨*Mischung*⟩

diversify [daɪ'vɜ:sɪfaɪ] **1.** *v.t.* abwechslungsreich[er] gestalten. **2.** *v.i. (Commerc.)* diversifieren

diversion [daɪ'vɜ:ʃn] *n.* **a)** *(diverting of attention)* Ablenkung, *die;* **b)** *(feint)* Ablenkungsmanöver, *das;* **create a** ~: ein Ablenkungsmanöver durchführen; **c)** *no pl. (recreation)* Unterhaltung, *die; (distraction)* Zerstreuung, *die;* Abwechslung, *die;* **d)** *(amusement)* [Möglichkeit der] Freizeitbeschäftigung; **e)** *(of river, traffic)* Ableitung, *die;* **f)** *(Brit.: alternative route)* Umleitung, *die*

diversionary [daɪ'vɜ:ʃənərɪ] *adj.* Ablenkungs⟨*angriff, -bombardement, -manöver*⟩

diversity [daɪ'vɜ:sɪtɪ] *n.* Vielfalt, *die;* ~ **of opinion** Meinungsvielfalt, *die*

divert [daɪ'vɜ:t] *v.t.* **a)** umleiten ⟨*Verkehr, Fluß, Fahrzeug*⟩*;* ablenken ⟨*Aufmerksamkeit*⟩*;* lenken ⟨*Energien, Aggressionen*⟩*;* **b)** *(distract)* ablenken; **c)** *(entertain)* unterhalten

diverting [daɪ'vɜ:tɪŋ] *adj. (entertaining)* unterhaltsam

divest [daɪ'vest, dɪ'vest] *v.t.* ~ **sb./sth. of sth.** *(deprive)* jmdn./etw. einer Sache *(Gen.)* berauben

divide [dɪ'vaɪd] **1.** *v.t.* **a)** teilen; *(subdivide)* aufteilen; *(with precision)* einteilen; *(into separated pieces)* zerteilen; ~ **sth. in[to] parts** *(separate)* etw. [in Stücke *(Akk.)*] aufteilen; ~ **sth. into halves/quarters** etw. halbieren/vierteln; ~ **sth. in two** etw. [in zwei Teile] zerteilen; **b)** *(by marking out)* ~ **sth. into sth.** etw. in etw. *(Akk.)* unterteilen; **c)** *(part by marking)* trennen; ~ **sth./sb. from** *or* **and sth./sb.** etw./jmdn. von etw./jmdm. trennen; **d)** *(mark off)* ~ **sth. from sth. else** etw. von etw. anderem abgrenzen; **dividing line** Trennungslinie, *die;* **e)** *(distinguish)* unterscheiden; **f)** *(cause to disagree)* entzweien; **be** ~**d over an issue** in einer Angelegenheit nicht einig sein; **be** ~**d against itself** zerstritten sein; **g)** *(distribute)* aufteilen (**among** unter + *Akk. od. Dat.*); **h)** *(Math.)* dividieren *(fachspr.),* teilen (**by** durch); ~ **three into nine** neun durch drei dividieren *od.* teilen. **2.** *v.i.* **a)** *(separate)* ~ **[in** *or* **into parts]** sich [in Teile] teilen; ⟨*Buch, Urkunde usw.:*⟩ sich [in Teile] gliedern, [in Teile] gegliedert sein; ~ **into two** sich in zwei Teile teilen; **b)** ~ **[from sth.]** von etw. abzweigen; **c)** *(Math.)* ~ **[by a number]** sich [durch eine Zahl] dividieren *(fachspr.) od.* teilen lassen

~ **'off 1.** *v.t.* trennen; ~ **off an area** einen Bereich abtrennen *od.* abteilen. **2.** *v.i.* ~ **off from sth.** sich von etw. trennen

~ **'out** *v.t.* ~ **sth. out [among/between persons]** etw. unter Personen *(Akk. od. Dat.)* aufteilen

~ **'up 1.** *v.t.* aufteilen. **2.** *v.i.* ~ **up into sth.** sich in etw. *(Akk.)* aufteilen lassen

divided 'skirt *n.* Hosenrock, *der*

dividend ['dɪvɪdend] *n.* **a)** *(Commerc., Finance)* Dividende, *die;* **b)** *in pl. (fig.: benefit)* Vorteil, *der;* **pay** ~**s** sich auszahlen *od.* rentieren; **reap the** ~**s** die Früchte ernten

divider [dɪ'vaɪdə(r)] *n.* **a)** *(screen)* Trennwand, *die;* **b)** ~**s** *pl.* Stechzirkel, *der*

divine [dɪ'vaɪn] **1.** *adj.,* ~**r** [dɪ'vaɪnə(r)], ~**st** [dɪ'vaɪnɪst] **a)** göttlich; *(devoted to God)* gottgeweiht; **b)** *(coll.: delightful)* traumhaft. **2.** *v.t.* deuten

diving ['daɪvɪŋ] *n. (Sport)* Kunstspringen, *das*

diving: ~**-board** *n.* Sprungbrett, *das;* ~**-suit** *n.* Taucheranzug, *der*

divinity [dɪ'vɪnɪtɪ] *n.* **a)** *(god)* Gottheit, *die;* **b)** *no pl. (being a god)* Göttlichkeit, *die;* **c)** *no pl. (theology)* Theologie, *die*

divisible [dɪ'vɪzɪbl] *adj.* **a)** *(separable)* aufteilbar; **be** ~ **into ...:** sich in ... aufteilen lassen; **b)** *(Math.)* **be** ~ **[by a number]** [durch eine Zahl] teilbar sein

division [dɪ'vɪʒn] *n.* **a)** *see* **divide 1 a:** Teilung/Auf-/Ein-/Zerteilung, *die;* **b)** *(parting) (of things)* Abtrennung, *die; (of persons)* Trennung, *die; (marking off)* Abgrenzung, *die;* **c)** *(distinguishing)* Unterscheidung, *die;* Abgrenzung, *die* (**from** gegenüber); **d)** *(distributing)* Verteilung, *die* (**between/among** an + *Akk.*); *(sharing)* Teilen, *das;* **e)** *(disagreement)* Unstimmigkeit, *die;* **f)** *(Math.)* Teilen, *das;* Dividieren, *das;* Division, *die (fachspr.);* **do** ~: dividieren; **long** ~: ausführliche Division *(mit Aufschreiben der Zwischenprodukte);* **short** ~: verkürzte Division *(ohne Aufschreiben der Zwischenprodukte);* **g)** *(separation in voting)* Abstimmung [durch Hammelsprung]; **h)** *(part)* Unterteilung, *die;* Abschnitt, *der;* **i)** *(section)* Abteilung, *die; (group)* Gruppe, *die; (Mil. etc.)* Division, *die; (of police)* Einheit, *die;* **j)** *(Footb. etc.)* Liga, *die;* Spielklasse, *die; (in British football)* Division, *die*

di'vision sign *n.* Divisionszeichen, *das*

divisive [dɪ'vaɪsɪv] *adj.* spalterisch

divisor [dɪ'vaɪzə(r)] *n. (Math.)* Divisor, *der;* Teiler, *der*

divorce [dɪ'vɔ:s] **1.** *n.* **a)** [Ehe]scheidung, *die;* **want a ~:** sich scheiden lassen wollen; **get** *or* **obtain a ~:** sich scheiden lassen; *attrib.* **~ court** Scheidungsgericht, *das;* **~ proceedings** [Ehe]scheidungsverfahren, *das;* **b)** *(fig.)* Trennung, *die.* **2.** *v. t.* **a)** *(dissolve marriage of)* scheiden ‹*Ehepartner*›; **b) ~ one's husband/wife** sich von seinem Mann/seiner Frau scheiden lassen

divot ['dɪvət] *n. (Golf)* ausgehacktes Rasenstück

divulge [daɪ'vʌldʒ] *v. t.* preisgeben; enthüllen ‹*Identität*›; bekanntgeben ‹*Nachrichten*›

Dixie ['dɪksɪ] *n.* **a)** die Südstaaten [der USA]; **b)** *(Mus.)* Dixie, *der*

'Dixieland *n.* **a)** *(Mus.)* Dixie[land], *der;* **b)** *see* **Dixie a**

DIY *abbr.* **do-it-yourself**

dizzy ['dɪzɪ] *adj.* **a)** *(giddy)* schwind[e]lig; **I feel ~:** mir ist schwindlig; **he felt ~:** ihm wurde schwindlig; **b)** *(making giddy)* schwindelerregend

DJ [di:'dʒeɪ] *abbr.* **disc jockey** Diskjockey, *der*

¹do [də, *stressed* du:] **1.** *v. t., neg. coll.* **don't** [dəʊnt], *pres. t.* **he does** [dʌz], *neg. (coll.)* **doesn't** ['dʌznt], *p. t.* **did** [dɪd], *neg. (coll.)* **didn't** ['dɪdnt], *pres. p.* **doing** ['du:ɪŋ], *p.p.* **done** [dʌn] **a)** *(perform)* machen ‹*Hausaufgaben, Hausarbeit, Examen, Handstand*›; vollbringen ‹*Tat*›; tun, erfüllen ‹*Pflicht*›; tun, verrichten ‹*Arbeit*›; ausführen ‹*Malerarbeiten*›; vorführen ‹*Trick, Striptease, Nummer, Tanz*›; durchführen ‹*Test*›; aufführen ‹*Stück*›; singen ‹*Lied*›; mitmachen ‹*Rennen, Wettbewerb*›; spielen ‹*Musikstück, Rolle*›; tun ‹*Buße*›; **do the shopping/washing up/cleaning** einkaufen [gehen]/abwaschen/saubermachen; **do a lot of reading/walking** *etc.* viel lesen/spazierengehen *usw.;* **do a dance/the foxtrot** tanzen/Foxtrott tanzen; **have nothing to do** nichts zu tun haben; **do sth. to sth./sb.** etw. mit etw./jmdm. machen; **what can I do for you?** was kann ich für Sie tun?; *(in shop)* was darf's sein?; **do sth. about sth./sb.** etw. gegen etw./jmdn. unternehmen; **not know what to do with oneself** nicht wissen, was man machen soll; **that does it** jetzt reicht's *(ugs.);* **that's done it** *(caused a change for the worse)* das hat das Faß zum Überlaufen gebracht; *(caused a change for the better)* das hätten wir; **that will/should do it** so müßte es gehen; *(is enough)* das müßte genügen; **do a Garbo** *(coll.)* es der Garbo *(Dat.)* gleichtun; **how many miles has this car done?** wie viele Kilometer hat der Wagen gefahren?; **the car does/was doing about 100 m.p.h./does 45 miles to the gallon** das Auto schafft/fuhr mit ungefähr 160 Stundenkilometer/frißt *(ugs.) od.* braucht sechs Liter pro 100 Kilometer; **b)** *(spend)* **do a spell in the armed forces** eine Zeitlang bei der Armee sein; **how much longer have you to do at college?** wie lange mußt du noch aufs College gehen?; **c)** *(produce)* machen ‹*Übersetzung, Kopie*›; anfertigen ‹*Bild, Skulptur*›; herstellen ‹*Artikel, Produkte*›; schaffen ‹*Pensum*›; **d)** *(provide)* haben ‹*Vollpension, Mittagstisch*›; *(coll.: offer for sale)* führen; **e)** *(prepare)* machen ‹*Bett, Frühstück*›; *(work on)* machen *(ugs.),* fertig machen ‹*Garten, Hecke*›; *(clean)* saubermachen; putzen ‹*Schuhe, Fenster*›; machen *(ugs.)* ‹*Treppe*›; *(arrange)* [zurecht]machen ‹*Haare*›; fertig machen ‹*Korrespondenz, Zimmer*›; *(make up)* schminken ‹*Lippen, Augen, Gesicht*›; machen *(ugs.)* ‹*Nägel*›; *(cut)* schneiden ‹*Nägel*›; schneiden ‹*Gras, Hecke*›; *(paint)* machen *(ugs.)* ‹*Zimmer*›; streichen ‹*Haus, Möbel*›; *(attend to)* sich kümmern um ‹*Bücher, Rechnungen, Korrespondenz*›; *(repair)* in Ordnung bringen; **f)** *(cook)* braten; **well done** durch[gebraten]; **g)** *(solve)* lösen ‹*Problem, Rätsel*›; machen ‹*Puzzle, Kreuzworträtsel*›; **h)** *(study, work at)* machen; haben ‹*Abiturfach*›; **i)** *(sl.: swindle)* reinlegen *(ugs.);* **do sb. out of sth.** jmdn. um etw. bringen; **j)** *(sl.: defeat, kill)* fertigmachen *(ugs.);* **k)** *(traverse)* schaffen ‹*Entfernung*›; **l)** *(sl.: undergo)* absitzen, *(salopp)* abreißen ‹*Strafe*›; **m)** *(coll.: visit)* besuchen; **do Europe in three weeks** Europa in drei Wochen absolvieren *od.* abhaken *(ugs.);* **n)** *(satisfy)* zusagen (+ *Dat.*); *(suffice for, last)* reichen (+ *Dat.*). **2.** *v. i., forms as* **1: a)** *(act)* tun; *(perform)* spielen; **you can do just as you like** du kannst machen, was du willst; **do as they do** mach es wie sie; **do or die** kämpfen oder untergehen; **b)** *(fare)* **how are you doing?** wie geht's dir?; **c)** *(get on)* vorankommen; *(in exams)* abschneiden; **how are you doing at school?** wie geht es in der Schule?; **do well/badly at school** gut/schlecht in der Schule sein; **d) how do you do?** *(formal)* guten Tag/Morgen/Abend!; **e)** *(coll.: manage)* **how are we doing for time?** wie steht es mit der Zeit *od. (ugs.)* sieht es mit der Zeit aus?; **f)** *(serve purpose)* es tun; *(suffice)* [aus]reichen; *(be suitable)* gehen; **that won't do** das geht nicht; **that will do!** jetzt aber genug!; **g)** *(be usable)* **do for** *or* **as sth.** als etw. benutzt werden können; **h)** *(happen)* **what's doing?** was ist los?; **there's nothing doing on the job market** es tut sich nichts auf dem Arbeitsmarkt *(ugs.);* **Nothing doing. He's not interested** Nichts zu machen *(ugs.).* Er ist nicht interessiert. *See also* **doing; done. 3.** *v. substitute, forms as* **1: a)** *replacing v.: usually not translated;* **you mustn't act as he does** du darfst nicht so wie er handeln; **b)** *replacing v. and obj. etc.* **he read the Bible every day as his father did before him** er las täglich in der Bibel, wie es schon sein Vater vor ihm getan hatte *od.* wie schon vor ihm sein Vater; **as they did in the Middle Ages** wie sie es im Mittelalter taten; **c)** *as ellipt. aux.* **You went to Paris, didn't you? – Yes, I did** Du warst doch in Paris, oder *od.* nicht wahr? – Ja[, stimmt *od.* war ich]; **d)** *with 'so', 'it', etc.* **I knew John Lennon. – So did I** Ich kannte John Lennon. – Ich auch; **go ahead and do it** nur zu; **e)** *in tag questions* **I know you from somewhere, don't I?** wir kennen uns doch irgendwoher, nicht? **4.** *v. aux.* + *inf. as pres. or past, forms as* **1: a)** *for special emphasis* **I do love Greece** Griechenland gefällt mir wirklich gut; **I do apologize** es tut mit wirklich leid; **you do look glum** du siehst ja so bedrückt aus; **but I tell you, I did see him** aber ich sage dir doch, daß ich ihn gesehen habe; **b)** *for inversion* **little did he know that ...:** er hatte keine Ahnung, daß ...; **c)** *in questions* **do you know him?** kennst du ihn?; **what does he want?** was will er?; **didn't they look wonderful?** haben sie nicht wunderhübsch ausgesehen?; **d)** *in negation* **I don't** *or* **do not wish to take part** ich möchte nicht teilnehmen; **e)** *in neg. commands* **don't** *or* **do not expect to find him in a good mood** erwarten Sie nicht, daß Sie ihn in guter Stimmung antreffen; **children, do not forget ...:** Kinder, vergeßt [ja] nicht ...; **don't be so noisy!** seid [doch] nicht so laut!; **don't!** tu's/tut's/tun Sie's nicht!; **f)** + *inf. as imper. for emphasis etc.* **do sit down, won't you?** bitte setzen Sie sich doch!; **do be quiet, Paul!** Paul, sei doch mal ruhig!; **do hurry up!** beeil dich doch!

do a'way with *v.t.* abschaffen

'do by *v.t.* **do well by sb.** jmdn. gut behandeln; **he felt hard done by** er fühlte sich zurückgesetzt *od.* schlecht behandelt

do 'down *v.t. (coll.)* schlechtmachen; heruntermachen *(ugs.)*

'do for *v.t.* **a)** *see* [1]**do 2 g**; **b)** *(coll.: destroy)* **do for sb.** jmdn. fertigmachen *od.* schaffen *(ugs.)*; **do for sth.** etw. kaputtmachen *(ugs.)*; **if we don't do better next time we're done for** wenn wir das nächste Mal nicht besser sind, sind wir erledigt; **c)** *(Brit. coll.: keep house for)* **do for sb.** für jmdn. sorgen; ‹*Putzfrau:*› für *od.* bei jmdm. putzen

do 'in *v.t. (sl.)* kaltmachen *(salopp)*

do 'out *v.t. (clean)* saubermachen; *(redecorate)* streichen; *(in wallpaper)* tapezieren; *(decorate, furnish)* herrichten

do 'up 1. *v.t.* **a)** *(fasten)* zumachen; binden ‹*Schnürsenkel, Fliege*›; **b)** *(wrap)* einpacken; verpacken; *(arrange)* zurechtmachen; **c)** *(adorn)* zurechtmachen ‹*Menschen*›; dekorieren ‹*Haus*›. **2.** *v.i.* ‹*Kleid, Reißverschluß, Knopf usw.:*› zugehen

'do with *v.t.* **a)** *(get by with)* auskommen mit; *(get benefit from)* **I could do with a glass of orange-juice** ich könnte ein Glas Orangensaft vertragen *(ugs.)*; **he could do with a good hiding** eine Tracht Prügel würde ihm nicht schaden; **b) have to do with** zu tun haben mit; **have something/nothing to do with sth./sb.** etwas/nichts mit etw./jmdm. zu tun haben

'do without *v.t.* **do without sth.** ohne etw. auskommen; auf etw. *(Akk.)* verzichten

[2]**do** [du:] *n., pl.* **dos** *or* **do's** [du:z] **a)** *(sl.: swindle)* Schwindel, *der;* krumme Sache *(ugs.)*; **b)** *(Brit. coll.: festivity)* Feier, *die;* Fete, *die (ugs.)*; **c)** *in pl.* **the dos and don'ts** die Ge- und Verbote (**of** *Gen.*)

doc [dɒk] *n. (coll.)* Doktor, *der (ugs.)*

docile ['dəʊsaɪl] *adj.* sanft; *(submissive)* unterwürfig

[1]**dock** [dɒk] **1.** *n.* **a)** Dock, *das;* **the ship came into ~:** das Schiff ging in[s] Dock; **be in ~:** im Dock liegen; **b)** *usu. in pl. (area)* Hafen, *der;* **down by the ~[s]** unten im Hafen. **2.** *v.t. (Naut.)* [ein]docken; *(Astronaut.)* docken. **3.** *v.i. (Naut.)* anlegen; *(Astronaut.)* docken

[2]**dock** *n. (in lawcourt)* Anklagebank, *die;* **stand/be in the ~** *(lit. or fig.)* ≈ auf der Anklagebank sitzen

[3]**dock** *v.t.* **a)** *(cut short)* kupieren ‹*Hund, Pferd, Schwanz*›; **b)** kürzen ‹*Lohn, Stipendium usw.*›; **he had his pay ~ed by £14, he had £14 ~ed from his pay** sein Lohn wurde um 14 Pfund gekürzt

docker ['dɒkə(r)] *n.* Hafenarbeiter, *der*

'dockyard *n.* Schiffswerft, *die*

doctor ['dɒktə(r)] **1.** *n.* **a)** Arzt, *der*/Ärztin, *die;* Doktor, *der (ugs.)*; *as title* Doktor, *der; as address* Herr/Frau Doktor; **~'s orders** ärztliche Anweisung; **just what the ~ ordered** [ganz] genau das richtige!; **b)** *(Amer.: dentist)* Zahnarzt, *der*/-ärztin, *die;* **c)** *(Amer.: veterinary surgeon)* Tierarzt, *der*/-ärztin, *die;* **d)** *(holder of degree)* Doktor, *der;* **D~ of Medicine** Doktor der Medizin. **2.** *v.t. (coll.) (falsify)* verfälschen ‹*Dokumente, Tonbänder*›; frisieren *(ugs.)* ‹*Bilanzen, Bücher*›; *(adulterate)* panschen *(ugs.)* ‹*Wein*›; verwürzen ‹*Gericht*›

doctorate ['dɒktərət] *n.* Doktorwürde, *die;* **do a ~:** seinen Doktor machen *(ugs.)*; promovieren

doctrinaire [dɒktrɪ'neə(r)] *adj.* doktrinär

doctrine ['dɒktrɪn] *n.* **a)** *(principle)* Lehre, *die;* **the ~ of free speech** der Grundsatz der Redefreiheit; **b)** *(body of instruction)* Doktrin, *die;* Lehrmeinung, *die*

document 1. ['dɒkjʊmənt] *n.* Dokument, *das;* Urkunde, *die.* **2.** ['dɒkjʊment] *v.t.* **a)** *(prove by ~[s])* dokumentieren; **b) be well ~ed** ‹*Leben, Zeit usw.:*› gut belegt sein

documentary [dɒkjʊ'mentərɪ] **1.** *adj.* dokumentarisch, urkundlich ‹*Beweis*›; *(factual)* dokumentarisch; **~ film** Dokumentarfilm, *der.* **2.** *n. (film)* Dokumentarfilm, *der*

documentation [dɒkjʊmen'teɪʃn] *n.* **a)** *(documenting)* Dokumentation, *die;* **b)** *(material)* beweiskräftige Dokumente

doddery ['dɒdərɪ] *adj.* tatterig *(ugs.)* ‹*alter Mann*›; zittrig ‹*Beine, Bewegungen*›

doddle ['dɒdl] *n. (Brit. coll.)* Kinderspiel, *das (fig.)*

dodge [dɒdʒ] **1.** *v.i.* **a)** *(move quickly)* ausweichen; **~ behind the hedge** hinter die Hecke springen; **~ out of the way** zur Seite springen; **b)** *(move to and fro)* ständig in Bewegung sein; **~ through the traffic** sich durch den Verkehr schlängeln. **2.** *v.t.* ausweichen (+ *Dat.*) ‹*Schlag, Hindernis usw.*›; entkommen (+ *Dat.*) ‹*Polizei, Verfolger*›; *(avoid)* sich drücken vor (+ *Dat.*) ‹*Wehrdienst*›; umgehen ‹*Steuer*›; aus dem Weg gehen (+ *Dat.*) ‹*Frage, Problem*›; **~ doing sth.** es umgehen, etw. zu tun. **3.** *n.* **a)** *(move)* Sprung zur Seite; **b)** *(trick)* Trick, *der;* **he's up to all the ~s** er ist mit allen Wassern gewaschen

dodgem ['dɒdʒəm] *n.* [Auto]skooter, *der; in pl.* [Auto]skooterbahn, *die;* **have a ride/go on the ~s** Autoskooter fahren

dodgy ['dɒdʒɪ] *adj. (Brit. coll.) (unreliable)* unsicher; schwach ‹*Knie, Herz usw.*›; *(awkward)* verzwickt; vertrackt; *(tricky)* kniffig; *(risky)* gewagt; heikel

dodo ['dəʊdəʊ] *n., pl.* **~s** *or* **~es** Dodo, *der;* Dronte, *die;* **[as] dead as the** *or* **a ~:** völlig ausgestorben

doe [dəʊ] *n.* **a)** *(deer)* Damtier, *das;* Damgeiß, *die;* **b)** *(hare)* Häsin, *die;* **c)** *(rabbit)* [Kaninchen]weibchen, *das*

DOE *abbr. (Brit.)* **Department of the Environment** Umweltministerium, *das*

does [dʌz] *see* [1]**do**

doesn't ['dʌznt] *(coll.)* = **does not;** *see* [1]**do**

doff [dɒf] *v.t.* lüften, ziehen ‹*Hut*›

dog [dɒg] **1.** *n.* **a)** Hund, *der;* **not [stand** *or* **have] a ~'s chance** nicht die geringste Chance [haben]; **dressed up/done up like a ~'s dinner** *(coll.)* aufgeputzt wie ein Pfau *(ugs.)*; ‹*Frau:*› aufgetakelt wie eine Fregatte *(ugs.)*; **give a ~ a bad name** einmal in Verruf gekommen, bleibt man immer verdächtig; **go to the ~s** vor die Hunde gehen *(ugs.)*; **a ~ in the manger** ein Biest, das keinem was gönnt; **~-in-the-manger** mißgünstig ‹*Benehmen*›; **be like a ~ with two tails** sich freuen wie ein Schneekönig *(ugs.)*; **the ~s** *(Brit. coll.: greyhound-racing)* das Windhundrennen; **b)** *(male ~)* Rüde, *der;* **c)** *(despicable person; coll.: fellow)* Hund, *der (derb);* **wise old ~/cunning [old] ~:** schlauer Fuchs *(ugs.)*. **2.** *v. t.*, **-gg-** verfolgen; *(fig.)* heimsuchen; verfolgen

dog: ~-biscuit *n.* Hundekuchen, *der;* **~-breeder** *see* **breeder; ~-collar** *n.* **a)** [Hunde]halsband, *das;* **b)** *(joc.: clerical collar)* Kollar, *das;* **~-eared** ['dɒgɪəd] *adj.* **a ~-eared book** ein Buch mit Eselsohren; **~fight** *n.* **a)** Hundekampf, *der; (fig.)* Handgemenge, *das;* **b)** *(between aircraft)* Luftkampf, *der*

dogged ['dɒgɪd] *adj.* hartnäckig ‹*Weigerung, Verurteilung*›; zäh ‹*Durchhaltevermögen, Ausdauer*›; beharrlich ‹*Haltung, Kritik*›

doggerel ['dɒgərəl] *n.* Knittelvers, *der*

doggie *see* **doggy**

doggo ['dɒgəʊ] *adv. (sl.)* **lie ~:** sich nicht mucksen *(ugs.)*

doggy ['dɒgɪ] *n. (coll.)* Hündchen, *das*

'doggy-bag *n. (coll.) Tüte, in der man Essensreste [bes. von einer Mahlzeit im Restaurant] mit nach Hause nimmt*

dog: ~house *n.* **a)** *(Amer.)* Hundehütte, *die;* **b) be in the ~house** *(sl.: in disgrace)* in Ungnade sein; **~-leg** *n.* Knick, *der;* **~ licence** *n.* Hundesteuerbescheinigung, *die*

dogma ['dɒgmə] *n.* Dogma, *das*

dogmatic [dɒg'mætɪk] *adj.* dogmatisch; **be ~ about sth.** in etw. *(Dat.)* dogmatisch sein

dogmatism ['dɒgmətɪzm] *n.* Dogmatismus, *der*

do-gooder [du:'gʊdə(r)] *n.* Wohltäter, *der (iron.)*; *(reformer)* Weltverbesserer, *der (iron.)*

dog: ~rose *n. (Bot.)* Hundsrose, *die;* **~sbody** *n. (Brit. coll.)* Mädchen für alles; **~'s life** *n.* **a ~'s life** ein Hundeleben; **~-'tired** *adj.* hundemüde; **~-trot** *n.* gemächlicher Trott; **~-watch** *n. (Naut.) (from 4 p.m. to 6 p.m./from 6 p.m. to 8 p.m.)* 1./2. Plattfuß, *der (Seemannsspr.)*; **~wood** *n. (Bot.)* Hartriegel, *der;* Hornstrauch, *der*

doily ['dɔɪlɪ] *n.* [Spitzen-, Zier]deckchen, *das*

doing ['du:ɪŋ] **1.** *pres. p. of* **¹do. 2.** *n.* **a)** *vbl. n. of* **¹do; b)** *no pl.* Tun, *das;* **be [of] sb.'s ~:** jmds. Werk sein; **it was not [of]** *or* **none of his ~:** er hatte nichts damit zu tun; **that takes a lot of/some ~:** da gehört sehr viel/schon etwas dazu; **c)** *in pl.* **sb.'s ~s** *(actions)* jmds. Tun und Treiben; **the ~s** *(sl.)* die Dinger *(ugs.)*; *(thing with unknown name)* das Dings *(ugs.)*

do-it-yourself [du:ɪtjə'self] **1.** *adj.* Do-it-yourself-. **2.** *n.* Heimwerken, *das*

doldrums ['dɒldrəmz] *n. pl.* **a) in the ~** *(in low spirits)* niedergeschlagen; **b)** *(Naut.)* **in the ~:** ohne Wind; *(fig.)* in einer Flaute

dole [dəʊl] **1.** *n. (Brit. coll.)* **the ~:** Stempelgeld, *das (ugs.)*; Stütze, *die (ugs.)*; **be/go on the ~:** stempeln gehen *(ugs.)*. **2.** *v. t.* **~ out** verteilen; austeilen

doleful ['dəʊlfl] *adj.* traurig ‹*Augen, Blick, Gesichtsausdruck*›

doll [dɒl] **1.** *n.* **a)** Puppe, *die;* **b)** *(sl.: young woman)* Mieze, *die (ugs.)*. **2.** *v. t.* **~ up** herausputzen

dollar ['dɒlə(r)] *n.* Dollar, *der;* **feel/look like a million ~s** *(coll.)* sich pudelwohl fühlen *(ugs.)*/tipptopp aussehen *(ugs.)*; **sixty-four [thousand] ~ question** *(lit. or fig.)* Preisfrage, *die*

dollar: ~ 'bill *n.* Dollarnote, *die;* Dollarschein, *der;* **~ sign** *n.* Dollarzeichen, *das*

'dollhouse *(Amer.) see* **doll's house**

dollop ['dɒləp] *n. (coll.)* Klacks, *der (ugs.)*

'doll's house *n.* Puppenhaus, *das*

dolly ['dɒlɪ] *n.* Puppe, *die;* Püppchen, *das; (child language)* Püppi, *die (Kinderspr.)*

'dolly-bird *n. (Brit. coll.)* Mieze, *die (ugs.)*

Dolomites ['dɒləmaɪts] *pr. n. pl.* **the ~:** die Dolomiten

dolphin ['dɒlfɪn] *n.* Delphin, *der*

dolt [dəʊlt] *n.* Tölpel, *der*

domain [də'meɪn] *n.* **a)** *(estate)* Gut, *das;* Ländereien *Pl.*; **b)** *(field)* Domäne, *die (geh.)*; [Arbeits-/Wissens-/Aufgaben]gebiet, *das*

dome [dəʊm] *n.* Kuppel, *die*

domestic [də'mestɪk] *adj.* **a)** *(household)* häuslich ‹*Verhältnisse, Umstände*›; *(family)* familiär ‹*Atmosphäre, Angelegenheit, Reibereien*›; ‹*Wasserversorgung, Ölverbrauch*› der privaten Haushalte; **~ servant** Hausgehilfe, *der*/-gehilfin, *die;* **~ help** Haushaltshilfe, *die;* **b)** *(of one's own country)* inländisch; einheimisch ‹*Produkt*›; innenpolitisch ‹*Problem, Auseinandersetzungen*›; **c)** *(kept by man)* **~ animal** Haustier, *das;* **~ rabbit/cat** Hauskaninchen, *das*/Hauskatze, *die*

domesticated [də'mestɪkeɪtɪd] *adj.* **a)** domestiziert *(fachspr.)*, gezähmt ‹*Tier*›; **b)** häuslich ‹*Person*›

domestic 'science *n., no pl.* Hauswirtschaftslehre, *die*

domicile ['dɒmɪsaɪl] **1.** *n.* Heimat, *die.* **2.** *v. t.* ansiedeln

dominance ['dɒmɪnəns] *n., no pl.* Dominanz, *die;* Vorherrschaft, *die* **(over** über + *Akk.*); *(of colours etc.)* Vorherrschen, *das*

dominant ['dɒmɪnənt] *adj.* dominierend *(geh.)*; beherrschend; hervorstechend, herausragend ‹*[Wesens]merkmal, Eigenschaft*›; vorherrschend ‹*Kultur, Farbe, Geschmack*›; **have a ~ position** eine beherrschende Stellung einnehmen; **be ~ over** dominieren über (+ *Akk.*)

dominate ['dɒmɪneɪt] **1.** *v. t.* beherrschen. **2.** *v. i.* **a) ~ over sb./sth.** jmdn./etw. beherrschen; **b)** *(be the most influential)* dominieren

domination [dɒmɪ'neɪʃn] *n., no pl.* [Vor]herrschaft, *die* **(over** über + *Akk.*); **under Roman ~:** unter römischer Herrschaft

domineering [dɒmɪ'nɪərɪŋ] *adj.* herrisch, herrschsüchtig ‹*Person*›

dominion [də'mɪnjən] *n.* **a)** *(control)* Herrschaft, *die* **(over** über + *Akk.*); **[be] under Roman ~:** unter römischer Herrschaft [stehen]; **b)** *usu. in pl. (territory of sovereign or government)* Reich, *das;* **c)** *(Commonwealth Hist.)* Dominion, *das*

domino ['dɒmɪnəʊ] *n., pl.* **~es** Domino[stein], *der;* **~es** *sing. (game)* Domino[spiel], *das;* **play ~es** Domino spielen

¹don [dɒn] *n. (Univ.)* [Universitäts]dozent, *der (bes. in Oxford und Cambridge)*

²don *v. t.*, **-nn-** anlegen *(geh.)*; anziehen ‹*Mantel usw.*›; aufsetzen ‹*Hut*›

donate [də'neɪt] *v. t.* spenden ‹*Organe*›; stiften, spenden ‹*Geld, Kleidung*›

donation [də'neɪʃn] *n.* Spende, *die* **(to[wards]** für); *(large-scale)* Stiftung, *die;* **a ~ of money/clothes** eine Geld-/Kleiderspende; **make a ~ of £1,000 [to charity]** 1 000 Pfund [für wohltätige Zwecke] spenden *od.* stiften

done [dʌn] *adj.* **a)** *(coll.: acceptable)* **it's not ~ [in this country]** das macht man [hierzulande] nicht; **it's [not] the ~ thing** es ist [nicht] üblich; **b)** *as int. (accepted)* abgemacht!; **c)** *(finished)* **be ~:** vorbei sein; **be ~ with sth.** mit etw. fertig sein; **is your plate ~**

with? brauchen Sie Ihren Teller noch?; **d) have ~ [doing sth.]** *(have stopped)* aufgehört haben, etw. zu tun; **have ~ with sth./doing sth.** *(stop)* mit etw. aufhören/aufhören, etw. zu tun

donkey ['dɒŋkɪ] *n. (lit. or fig.)* Esel, *der;* **she could talk the hind leg[s] off a ~!** *(fig.)* die kann einem die Ohren abreden! *(ugs.)*

'donkey-work *n.* Schwerarbeit, *die*

donor ['dəʊnə(r)] *n.* **a)** *(of gift)* Schenker, *der*/Schenkerin, *die; (to institution etc.)* Stifter, *der*/Stifterin, *die;* **b)** *(of blood, organ, etc.)* Spender, *der*/Spenderin, *die;* **blood ~:** Blutspender, *der*/-spenderin, *die*

don't [dəʊnt] **1.** *v.i. (coll.)* = **do not;** *see* [1]**do. 2.** *n.* Nein, *das;* Verbot, *das;* **dos and ~s** *see* [2]**do c**

doodle ['du:dl] **1.** *v.i.* ≈ Männchen malen; [herum]kritzeln. **2.** *n.* Kritzelei, *die*

doom [du:m] **1.** *n. (fate)* Schicksal, *das; (ruin)* Verhängnis, *das;* **meet one's ~:** vom Schicksal heimgesucht *od. (geh.)* ereilt werden. **2.** *v.t.* verurteilen; verdammen; **~ sb./sth. to sth.** jmdn./eine Sache zu etw. verdammen *od.* verurteilen; **be ~ed to fail** *or* **failure** zum Scheitern verurteilt sein; **be ~ed** verloren sein

doomsday ['du:mzdeɪ] *n.* der Jüngste Tag; **till ~** *(fig.)* bis zum Jüngsten Tag

door [dɔ:(r)] *n.* **a)** Tür, *die; (of castle, barn)* Tor, *das;* **'~s open at 7'** „Einlaß ab 7 Uhr"; **he put his head round the ~:** er streckte den Kopf durch die Tür; **lay sth. at sb.'s ~** *(fig.)* jmdm. etw. anlasten *od.* zur Last legen; **next ~:** nebenan; **live next ~ to sb.** neben jmdm. wohnen; **from ~ to ~:** von Haus zu Haus; von Tür zu Tür; **b)** *(fig.: entrance)* Zugang, *der* **(to** zu); **all ~s are open/closed to him** ihm stehen alle Türen offen/sind alle Türen verschlossen; **close the ~ to sth.** etw. unmöglich machen; **have/get one's foot in the ~:** mit einem Fuß *od.* Bein drin sein/hineinkommen; **leave the ~ open for sth.** die Tür für *od.* zu etw. offenhalten; **open the ~ to** *or* **for sth.** etw. möglich machen; **show sb. the ~:** jmdm. die Tür weisen; **c)** *(~way)* [Tür]eingang, *der;* **walk through the ~:** zur Tür hineingehen/hereinkommen; **shop ~:** Geschäftseingang, *der;* **d) out of ~s** im Freien; draußen; **go out of ~s** nach draußen gehen

door: ~bell *n.* Türklingel, *die;* **~-handle** *n.* Türklinke, *die;* **~-keeper** *n.* Pförtner, *der;* Portier, *der;* **~knob** *n.* Türknopf, -knauf, *der;* **~-knocker** *see* **knocker a; ~man** *n.* Portier, *der;* **~mat** *n.* Fußmatte, *die; (fig.)* Fußabtreter, *der;* **~post** *n.* Türpfosten, *der;* **~step** *n.* Eingangsstufe, *die;* Türstufe, *die;* **on one's/the ~step** *(fig.)* vor jmds./der Tür; **~stop** *n.* Türanschlag, *der; (stone, wedge, etc.)* Türstopper, *der;* **~-to-~** *adj.* **~-to-~ collection** Haussammlung, *die;* **~way** *n.* Eingang, *der*

dope [dəʊp] **1.** *n.* **a)** *(stimulant)* Aufputschmittel, *das; (sl.: narcotic)* Stoff, *der (salopp);* **~ test** Dopingkontrolle, *die;* **b)** *(sl.: information)* Informationen *Pl.;* **c)** *(coll.: fool)* Dussel, *der (ugs.).* **2.** *v.t.* dopen ‹*Pferd, Athleten*›; *(administer narcotic to)* Rauschgift verabreichen (+ *Dat.*); *(stupefy)* betäuben

dopey ['dəʊpɪ] *adj. (sl.)* benebelt *(ugs.)*

dormant ['dɔ:mənt] *adj.* untätig ‹*Vulkan*›; ruhend ‹*Tier, Pflanze*›; verborgen, schlummernd ‹*Talent, Fähigkeiten*›; **lie ~** ‹*Tier:*› schlafen; ‹*Pflanze, Ei:*› ruhen; ‹*Talent, Fähigkeiten:*› schlummern

dormer ['dɔ:mə(r)] *n.* **~ [window]** Mansardenfenster, *das*

dormitory ['dɔ:mɪtərɪ] *n.* **a)** Schlafsaal, *der;* **b)** *attrib.* **~ suburb** *or* **town** Schlafstadt, *die*

dormouse ['dɔ:maʊs] *n., pl.* **dormice** ['dɔ:maɪs] Haselmaus, *die*

dos *pl. of* [2]**do**

dosage ['dəʊsɪdʒ] *n.* **a)** *(giving of medicine)* Dosierung, *die;* **b)** *(size of dose)* Dosis, *die*

dose [dəʊs] **1.** *n.* **a)** *(lit. or fig.)* Dosis, *die;* **take a ~ of medicine** Medizin [ein]nehmen; **in small ~s** *(fig.)* in kleinen Mengen; **b)** *(amount of radiation)* Strahlen-, Bestrahlungsdosis, *die.* **2.** *v.t.* **~ sb. with sth.** jmdm. etw. geben *od.* verabreichen

doss-house ['dɒshaʊs] *n. (Brit. sl.)* Nachtasyl, *das*

dossier ['dɒsɪə(r), 'dɒsjeɪ] *n.* Akte, *die; (bundle of papers)* Dossier, *das*

dot [dɒt] **1.** *n.* **a)** Punkt, *der; (smaller)* Pünktchen, *das;* **b) on the ~:** auf den Punkt genau. **2.** *v.t.,* **-tt-: a)** *(mark with ~)* mit Punkten/einem Punkt markieren; **b) ~ one's i's/j's** i-/j-Punkte machen; **c)** *(mark as with ~s)* [be]sprenkeln; **the sky was ~ted with stars** der Himmel war von Sternen übersät; **d)** *(scatter)* verteilen

dotage ['dəʊtɪdʒ] *n.* **be in one's ~:** senil sein

dote [dəʊt] *v.i.* **[absolutely] ~ on sb./sth.** jmdn./etw. abgöttisch lieben

doting ['dəʊtɪŋ] *adj.* vernarrt

dotted ['dɒtɪd] *adj.* gepunktet ‹*Kleid, Linie*›; **sign on the ~ line** *(fig.)* unterschreiben

dotty ['dɒtɪ] *adj. (coll.)* **a)** *(silly)* dümmlich; **be ~ over** *or* **about sb./sth.** in jmdn./etw. vernarrt sein; **b)** *(feeble-minded)* schrullig *(ugs. abwertend);* vertrottelt *(ugs. abwertend);* **go ~:** vertrotteln; **c)** *(absurd)* blödsinnig *(ugs.);* verrückt ‹*Idee*›

double ['dʌbl] **1.** *adj.* **a)** *(consisting of two parts etc.)* doppelt ‹*Anstrich, Stofflage, Sohle*›; **b)** *(twofold)* doppelt ‹*Sandwich, Futter, Fenster, Boden*›; **c)** *(with pl.: two)* zwei ‹*Punkte, Klingen*›; **d)** *(for two persons)* Doppel-; **~ seat** Doppelsitz, *der;* **~ bed/room** Doppelbett, *das*/-zimmer, *das;* **e) folded ~:** einmal *od.* einfach gefaltet; **be bent ~ with pain** sich vor Schmerzen *(Dat.)* krümmen; **f)** *(having some part ~)* Doppel‹*adler, -heft, -stecker*›; **g)** *(dual)* doppelt ‹*Sinn, [Verwendungs]zweck*›; **h)** *(twice as much)* doppelt ‹*Anzahl*›; **a room ~ the size of this** ein doppelt so großes Zimmer wie dieses; **be ~ the height/width/time** doppelt so hoch/breit/lang sein; **be ~ the cost** doppelt so teuer sein; **at ~ the cost** zum doppelten Preis; **i)** *(twice as many)* doppelt so viele wie; **j)** *(of twofold size etc.)* doppelt ‹*Portion, Lautstärke, Kognak, Whisky*›; **k)** *(of extra size etc.)* doppelt so groß ‹*Anstrengung, Mühe, Schwierigkeit, Problem, Anreiz*›; **l)** *(deceitful)* falsch ‹*Spiel*›. **2.** *adv.* doppelt. **3.** *n.* **a)** *(~ quantity)* Doppelte, *das;* **b)** *(~ measure of whisky etc.)* Doppelte, *der; (~ room)* Doppelzimmer, *das;* **c)** *(twice as much)* das Doppelte; doppelt soviel; *(twice as many)* doppelt so viele; **~ or quits** doppelt oder nichts; **d)** *(duplicate person)* Doppelgänger, *der*/-gängerin, *die;* **e) at the ~:** unverzüglich; *(Mil.)* aufs schnellste; **f)** *(pair of victories)* Doppelerfolg, *der; (pair of championships)* Double, *das;* Doppel, *das;* **g)** *in pl. (Tennis etc.)* Doppel, *das;* **women's** *or* **ladies'/men's/mixed ~s** Damen-/Herrendoppel, *das*/gemischtes Doppel. **4.** *v.t.* verdoppeln; *(make ~)* doppelt nehmen ‹*Decke*›. **5.** *v.i.* **a)** sich verdoppeln; **b)** *(have two functions)* doppelt verwendbar sein; **the sofa ~s as a bed** man kann das Sofa auch als Bett benutzen

~ **'back** *v.i.* kehrtmachen *(ugs.)*

~ **'up 1.** *v.i.* **a)** sich krümmen; ~ **up with pain** sich vor Schmerzen *(Dat.)* krümmen; **b)** *(fig.)* ~ **up with laughter** sich vor Lachen krümmen; **c)** *(share quarters)* sich *(Dat.)* eine Unterkunft teilen; *(in hotel etc.)* sich *(Dat.)* ein Zimmer teilen. **2.** *v.t.* *(fold)* einmal falten

double: ~ **'agent** *n.* Doppelagent, *der*/-agentin, *die;* **~-barrelled** *(Amer.:* **~-barreled)** ['dʌblbærəld] *adj.* doppelläufig; **~-barrelled surname** *(Brit.)* Doppelname, *der;* **~-bass** [dʌbl'beɪs] *n.* *(Mus.)* Kontrabaß, *der;* **~-'check** *v.t.* **a)** *(verify twice)* zweimal kontrollieren; **b)** *(verify in two ways)* zweifach überprüfen; ~ **'chin** *n.* Doppelkinn, *das;* ~ **'cream** *n.* Sahne mit hohem Fettgehalt; **~-'cross 1.** *n.* Doppelspiel, *das;* **2.** *v.t.* ein Doppelspiel treiben mit; reinlegen *(ugs.);* **~-dealing 1.** [--'--] *n.* Betrügerei, *die;* **2.** ['----] *adj.* betrügerisch; **~-decker** [dʌbldekə(r)] **1.** ['----] *adj.* Doppeldecker-; Doppelstock- *(Amtsspr.);* **~-decker bus** Doppeldeckerbus, *der;* **a ~-decker sandwich** ein doppelter Sandwich; ein Doppeldecker *(ugs.);* **2.** [--'--] *n.* Doppeldecker, *der;* **~-de'clutch** *see* **declutch;** ~ **'door** *n.* *(door with two parts)* Flügeltür, *die;* *(twofold door)* Doppeltür, *die;* ~ **'Dutch** *see* **Dutch 2c; ~-edged** *adj.* *(lit. or fig.)* zweischneidig

double entendre [du:bl ɑ̃'tɑ̃dr] *n.* Zweideutigkeit, *die*

double: ~ **'feature** *n.* Doppelprogramm, *das;* **~-'glazed** *adj.* Doppel⟨*fenster*⟩; ~ **'glazing** *n.* Doppelverglasung, *die;* **~-'jointed** *adj.* sehr gelenkig; **~-'parking** *n.* Parken in der zweiten Reihe; **~-quick 1.** ['---] *adj.* **a) in ~-quick time/at a ~-quick pace** im Laufschritt; **b)** *(fig.)* ganz schnell; **2.** [--'-] *adv.* *(Mil.)* im Laufschritt; *(fig.)* ganz schnell; ~ **'room** *n.* Doppelzimmer, *das;* **~-'spaced** *adj.* mit doppeltem Zeilenabstand *nachgestellt;* ~ **'take** *n.* **he did a ~ take a moment after he saw her walk by** nachdem sie vorbeigegangen war, stutzte er und sah ihr nach; ~ **'time** *n.* doppelter Stundenlohn; **be on ~ time** 100% Zuschlag bekommen; ~ **'vision** *n.* *(Med.)* Doppeltsehen, *das*

doubly ['dʌblɪ] *adv.* doppelt; **make ~ sure that ...:** [ganz] besonders darauf achten, daß ...

doubt [daʊt] **1.** *n.* **a)** Zweifel, *der;* **~[s] [about** *or* **as to sth./as to whether ...]** *(as to future)* Ungewißheit, *(as to fact)* Unsicherheit [über etw. *(Akk.)*/darüber, ob ...]; **~[s] about** *or* **as to sth., ~ of sth.** *(inclination to disbelieve)* Zweifel an etw. *(Dat.);* **there's no ~ that ...:** es besteht kein Zweifel daran, daß ...; **~[s]** *(hesitations)* Bedenken *Pl.;* **have [one's] ~s about doing sth.** [seine] Bedenken haben, ob man etw. tun soll [oder nicht]; **when** *or* **if in ~:** im Zweifelsfall; **no ~** *(certainly)* gewiß; *(probably)* sicherlich; *(admittedly)* wohl; **cast ~ on sth.** etw. in Zweifel ziehen; **b)** *no pl.* *(uncertain state of things)* Ungewißheit, *die;* **be in ~:** ungewiß sein; **beyond [all] ~, without [a] ~:** ohne [jeden] Zweifel. **2.** *v.t.* anzweifeln; zweifeln an (+ *Dat.*); **she ~ed him** sie zweifelte an ihm; **I don't ~ that** *or* **it** ich zweifle nicht daran; **I ~ whether** *or* **if** *or* **that ...:** ich bezweifle, daß ...

doubter ['daʊtə(r)] *n.* Zweifler, *der*/Zweiflerin, *die*

doubtful ['daʊtfl] *adj.* **a)** *(sceptical)* skeptisch ⟨*Mensch, Wesen*⟩; **b)** *(showing doubt)* ungläubig ⟨*Gesicht, Blick, Stirnrunzeln*⟩; **c)** *(uncertain)* zweifelnd; **be ~ as to** *or* **about sth.** an etw. *(Dat.)* zweifeln; **d)** *(causing doubt)* fraglich; **e)** *(uncertain in meaning etc.)* ungewiß ⟨*Ergebnis, Ausgang, Herkunft, Aussicht*⟩; *(questionable)* zweifelhaft ⟨*Ruf, Charakter, Wert, Autorität*⟩; *(ambiguous)* unklar ⟨*Bedeutung*⟩; *(unsettled)* unsicher ⟨*Lage*⟩; **f)** *(unreliable)* zweifelhaft ⟨*Maßstab, Stütze*⟩; **g)** *(giving reason to suspect evil)* bedenklich ⟨*Gewohnheit, Spiel, Botschaft*⟩

doubtfully ['daʊtfəlɪ] *adv.* skeptisch

doubtless ['daʊtlɪs] *adv.* **a)** *(certainly)* gewiß; **b)** *(probably)* sicherlich; **c)** *(admittedly)* wohl

dough [dəʊ] *n.* **a)** Teig, *der;* **b)** *(sl.: money)* Knete, *die (salopp)*

'doughnut *n.* [Berliner] Pfannkuchen, *der*

dour [dʊə(r)] *adj.* hartnäckig; düster ⟨*Blick, Gesicht*⟩

douse [daʊs] *v.t.* **a)** *(extinguish)* ausmachen ⟨*Licht, Kerze, Feuer*⟩; **b)** *(throw water on)* übergießen ⟨*Feuer, Flamme, Menschen*⟩

¹dove [dʌv] *n.* *(also Polit.)* Taube, *die*

²dove *see* **dive 1**

dove: ~cot, ~-cote ['dʌvkɒt] *n.* Taubenschlag, *der;* **~tail 1.** *n.* *(Carpentry)* Schwalbenschwanzverbindung, *die;* **2.** *v.i.* *(fig.: fit together)* ⟨*Vorbereitungen, Zeitpläne:*⟩ aufeinander abgestimmt sein

dowager ['daʊədʒə(r)] *n.* Witwe von Stand

dowdy ['daʊdɪ] *adj.* unansehnlich; *(shabby)* schäbig

dowel ['daʊəl] *(Carpentry)* *n.* [Holz]dübel, *der*

¹down [daʊn] *n.* *(Geog.)* [baumloser] Höhenzug; *in pl.* Downs *Pl.* *(an der Süd- und Südostküste Englands)*

²down *n.* **a)** *(of bird)* Daunen *Pl.;* Flaum, *der;* **b)** *(hair)* Flaum, *der*

³down 1. *adv.* **a)** *(to lower place, to ~stairs, southwards)* runter *(bes. ugs.);* herunter/hinunter *(bes. schriftsprachlich);* *(in lift)* abwärts; *(in crossword puzzle)* senkrecht; **[right] ~ to sth.** [ganz] bis zu etw. her-/hinunter; **go ~ to the shops/the end of the road** zu den Läden/zum Ende der Straße hinuntergehen; **b)** *(Brit.: from capital)* raus *(bes. ugs.);* heraus/hinaus *(bes. schriftsprachlich);* **get ~ to Reading from London** von London nach Reading raus-/hinausfahren; **come ~ from Edinburgh to London** von Edinburgh nach London [he]runterkommen; **c)** *(of money: at once)* sofort; **pay for sth. cash ~:** etw. [in] bar bezahlen; **d)** *(into prostration)* nieder⟨*fallen, -geschlagen werden*⟩; **shout the place/house ~** *(fig.)* schreien, daß die Wände zittern; **e)** *(on to paper)* **copy sth. ~ from the board** etw. von der Tafel abschreiben; **f)** *(on programme)* **put a meeting ~ for 2 p.m.** ein Treffen für *od.* auf 14 Uhr ansetzen; **g)** *as int.* runter! *(bes. ugs.);* *(to dog)* leg dich!; nieder!; *(Mil.)* hinlegen!; **~ with imperialism/the president!** nieder mit dem Imperialismus/dem Präsidenten!; **h)** *(in lower place, ~stairs, in fallen position, in south)* unten; **~ on the floor** auf dem Fußboden; **low/lower ~:** tief/tiefer unten; **~ under the table** unter dem Tisch; **wear one's hair ~:** sein Haar offen tragen; **~ there/here** da/hier unten; **X metres ~:** X Meter tief; **his flat is on the next floor ~:** seine Wohnung ist ein Stockwerk tiefer; **~ in Wales/in the country** weit weg in Wales/draußen auf dem Lande; **~ south** unten im Süden *(ugs.);* **~ south/east** *(Amer.)* in den Südstaaten/im Osten; ~ **[on the floor]** *(Boxing)* am Boden; auf den Brettern; ~ **and out** *(Boxing)* k.o.; *(fig.)* fertig *(ugs.);* **i)** *(prostrate)* auf dem Fußboden/der Erde; **be ~ with an illness** eine Krankheit haben; **j)** *(on paper)* **be ~ in writing/on paper/in print** niedergeschrieben/zu Papier gebracht/gedruckt sein; **k)** *(on programme)* angesetzt ‹*Termin, Treffen*›; **l)** *(facing ~wards, bowed)* zu Boden; **keep one's eyes ~:** zu Boden sehen; **be ~** *(brought to the ground)* am Boden liegen; **m)** *(in depression)* ~ **[in the mouth]** niedergeschlagen; **n)** *(now cheaper)* [jetzt] billiger; **o) be ~ to ...** *(have only ... left)* nichts mehr haben außer ...; **we're ~ to our last £100** wir haben nur noch 100 Pfund; **now it's ~ to him to do something** nun liegt es bei *od.* an ihm, etwas zu tun; **p)** *(to reduced consistency or size)* **thin gravy ~:** Soße verdünnen; **the water had boiled right ~:** das Wasser war fast verdampft; **wear the soles ~:** die Sohlen ablaufen; **q)** *(including lower limit)* **from ... ~ to ...:** von ... bis zu ... hinunter; **r)** *(in position of lagging or loss)* weniger; **be three points/games ~:** mit drei Punkten/Spielen zurückliegen; **be ~ on one's earnings of the previous year** weniger verdienen als im Vorjahr; **be ~ on one's luck** eine Pechsträhne haben. *See also* **up 1. 2.** *prep.* **a)** *(~wards along, from top to bottom of)* runter *(bes. ugs.);* herunter/hinunter *(bes. schriftsprachlich);* **lower ~ the river** weiter unten am Fluß; **fall ~ the stairs/steps** die Treppe/Stufen herunterstürzen; **his eye travelled ~ the list** sein Auge wanderte über die Liste; **walk ~ the hill/road** den Hügel/die Straße heruntergehen; **b)** *(~wards through)* durch; **c)** *(~wards into)* rein in (+ *Akk.*) *(bes. ugs.);* hinein in (+ *Akk.*) *(bes. schriftsprachlich);* **fall ~ a hole/ditch** in ein Loch/einen Graben fallen; **d)** *(~wards over)* über (+ *Akk.*); **spill water all ~ one's skirt** sich *(Dat.)* Wasser über den Rock gießen; **e)** *(~wards in time)* **the tradition has continued ~ the ages** die Tradition ist von Generation zu Generation weitergegeben worden; **f)** *(along)* **come ~ the street** die Straße herunter- *od.* entlangkommen; **go ~ the pub/disco** *(Brit. coll.)* in die Kneipe/Disko gehen; **g)** *(at or in a lower position in or on)* [weiter] unten; **further ~ the ladder/coast** weiter unten auf der Leiter/an der Küste; **live just ~ the road** ein Stück weiter unten in der Straße wohnen; **h)** *(from top to bottom along)* an (+ *Dat.*); **~ the side of a house** an der Seite eines Hauses; **i)** *(all over)* überall auf (+ *Dat.*); **I've got coffee [all] ~ my skirt** mein ganzer Rock ist voll Kaffee; **j)** *(Brit. coll.: in, at)* **~ the pub/café/town** in der Kneipe/im Café/in der Stadt. **3.** *adj.* *(directed ~wards)* nach unten führend ‹*Rohr, Kabel*›; ‹*Rolltreppe*› nach unten; nach unten gerichtet ‹*Kolbenhub, Sog*›; aus der Hauptstadt herausführend ‹*Bahnlinie*›. **4.** *v.t.* *(coll.)* **a)** *(knock ~)* auf die Bretter schicken ‹*Boxer*›; **b)** *(drink ~)* leer machen *(ugs.)* ‹*Flasche, Glas*›; schlucken *(ugs.)* ‹*Getränk*›; **c) ~ tools** *(cease work)* zu arbeiten aufhören; *(take a break)* die Arbeit unterbrechen; *(go on strike)* die Arbeit niederlegen; **d)** *(shoot ~)* abschießen, *(ugs.)* runterholen ‹*Flugzeug*›. **5.** *n.* *(coll.)* **have a ~ on sb./sth.** jmdn./etw. auf dem Kieker haben *(ugs.); see also* **up 4**

down: ~-and-out *n.* Stadtstreicher, *der*/Stadtstreicherin, *die;* Penner, *der*/Pennerin, *die (ugs.);* **~beat** *n.* *(Mus.)* erster/betonter Taktteil; **~cast** *adj.* niedergeschlagen ‹*Blick, Gesicht*›; **~fall** *n.* Untergang, *der;* **~grade** *v.t.* niedriger einstufen; **~-hearted** *adj.* niedergeschlagen; **~hill 1.** ['--] *adj.* bergab führend ‹*Fahrt*›; ‹*Strecke, Weg*› bergab; **he's on the ~hill path** *(fig.)* es geht bergab mit ihm; **be ~hill all the way** *(fig.)* ganz einfach sein; **2.** [-'-] *adv.* bergab; **come ~hill** den Berg herunterkommen; **3.** ['--] *n.* **a)** Gefällstrecke, *die;* **b)** *(Skiing)* Abfahrtslauf, *der;* **~land** *n.* [baumloses] Hügelland; **~-market** *adj.* weniger anspruchsvoll; **go ~-market** weniger anspruchsvoll werden; **~ payment** *n.* Anzahlung, *die;* **~pipe** *n.* [Regenab]fallrohr, *das;* **~pour** *n.* Regenguß, *der;* **~right 1.** *adj.* ausgemacht ‹*Frechheit, Dummheit, Idiot, Lügner*›; glatt ‹*Lüge*›; **2.** *adv.* geradezu; ausgesprochen; **it would be ~right stupid to do that** es wäre eine ausgemachte Dummheit, das zu versuchen; **~stage** *adv.* *(Theatre)* im Vordergrund der Bühne; **move ~stage** sich zum Vordergrund der Bühne bewegen; **~stairs 1.** [-'-] *adv.* die Treppe hinunter ‹*gehen, fallen, kommen*›; unten ‹*wohnen, sein*›; **2.** ['--] *adj.* im Parterre *od.* Erdgeschoß *nachgestellt;* Parterre‹*wohnung*›; **3.** [-'-] *n.* Untergeschoß, *das;* **~stream 1.** [-'-] *adv.* flußabwärts; **2.** ['--] *adj.* flußabwärts gelegen ‹*Ort*›; **~ time** *n.* *(Computing)* Ausfallzeit, *die;* **~-to-earth** *adj.* praktisch, nüchtern ‹*Person*›; realistisch ‹*Plan, Vorschlag*›; **~town** *(Amer.)* **1.** *adj.* im Stadtzentrum *nachgestellt;* **~town Manhattan** das Stadtzentrum Manhattan; **2.** *adv.* ins Stadtzentrum ‹*gehen, fahren*›; im Stadtzentrum ‹*leben, liegen, sein*›; **~trodden** *adj.* geknechtet; unterdrückt; **~ 'under** *(coll.)* **1.** *adv.* in/*(to)* nach Australien/Neuseeland; **2.** *n.* *(Australia)* Australien *(das); (New Zealand)* Neuseeland *(das)*

downward ['daʊnwəd] **1.** *adj.* nach unten *nachgestellt;* nach unten gerichtet; **~ movement/trend** *(lit. or fig.)* Abwärtsbewegung, *die*/-trend, *der;* **~ gradient** *or* **slope** Gefälle, *das.* **2.** *adv.* abwärts ‹*sich bewegen*›; nach unten ‹*sehen, gehen*›; *see also* **face down[ward]**

downwards ['daʊnwədz] *see* **downward 2**

'downwind *adv.* mit dem Wind; vor dem Wind ‹*segeln*›; **be ~ of sth.** im Windschatten einer Sache *(Gen.)* sein

dowry ['daʊrɪ] *n.* Mitgift, *die (veralt.);* Aussteuer, *die*

doyley *see* **doily**

doz. *abbr.* **dozen** Dtzd.

doze [dəʊz] **1.** *v.i.* dösen *(ugs.);* [nicht tief] schlafen; **lie dozing** im Halbschlaf liegen. **2.** *n.* Nickerchen, *das (ugs.)*

~ **'off** *v.i.* eindösen *(ugs.)*

dozen ['dʌzn] *n.* **a)** *pl. same* *(twelve)* Dutzend, *das;* **six ~ bottles of wine** sechsmal zwölf Flaschen Wein; **a ~ times/reasons** *(fig. coll.: many)* dutzendmal/Dutzende von Gründen; **half a ~:** sechs; **b)** *pl.* **~s** *(set of twelve)* Dutzend, *das;* **by the ~** *(in twelves)* im Dutzend; *(fig. coll.: in great numbers)* in großen Mengen; **c)** *in pl.* *(coll.: many)* Dutzende *Pl.;* **~s of times** dutzendmal

Dr *abbr.* **a) Doctor** *(as prefix to name)* Dr.; **b) debtor** Sch.

drab [dræb] *adj.* **a)** *(dull brown)* gelblich braun; *(dull-coloured)* matt; **b)** *(dull, monotonous)* langweilig ‹*Ort, Gebäude*›

draft [drɑ:ft] **1.** *n.* **a)** *(rough copy)* *(of speech)* Konzept, *das;* *(of treaty, parliamentary bill)* Entwurf, *der; attrib.* ~ **copy/version** Konzept, *das;* **b)** *(plan of work)* Skizze, *die;* [Bau-, Riß-]zeichnung, *die;* **c)** *(Mil.: detaching for special duty)* Sonderkommando, *das; (Brit.: those detached)* Abkommandierte *Pl.;* **d)** *(Amer. Mil.: conscription)* Einberufung, *die; (those conscripted)* Wehrpflichtige *Pl.;* Einberufene *Pl.;* **e)** *(Commerc.: cheque drawn)* Wechsel, *der;* Tratte, *die;* **f)** *(Amer.) see* **draught. 2.** *v. t.* **a)** *(make rough copy of)* entwerfen; **b)** *(Mil.)* abkommandieren; **c)** *(Amer. Mil.: conscript)* einberufen

drafty *(Amer.) see* **draughty**

drag [dræg] **1.** *n.* **a)** *(difficult progress)* **it was a long ~ up the hill** der Aufstieg auf den Hügel war ein ganz schöner Schlauch *(ugs.);* **b)** *(obstruction)* Hindernis, *das* (**on** für); Hemmnis, *das* (**on** für); **c)** *no pl. (sl.: women's dress worn by men)* Frauenkleider *Pl.* **2.** *v. t.,* **-gg-: a)** [herum]schleppen; ~ **one's feet** *or* **heels** *(fig.)* sich *(Dat.)* Zeit lassen (**over, in** mit); **b)** *(move with effort)* ~ **oneself** sich schleppen; ~ **one's feet** [mit den Füßen] schlurfen; **c)** *(fig. coll.: take despite resistance)* **he ~ged me to a dance** er schleifte mich *(ugs.)* zu einer Tanzveranstaltung; ~ **sb. into sth.** jmdn. in etw. *(Akk.)* hineinziehen; **d)** *(search)* [mit einem Schleppnetz] absuchen ‹*Fluß-, Seegrund*›. **3.** *v. i.,* **-gg-: a)** schleifen; ~ **on** *or* **at a cigarette** *(coll.)* an einer Zigarette ziehen; **b)** *(fig.: pass slowly)* sich [hin]schleppen

~ **'in** *v. t.* hineinziehen

~ **'on** *v. i. (continue)* sich [da]hinschleppen; **time ~ged on** die Zeit verstrich; ~ **on for months** sich über Monate hinziehen

~ **'out** *v. t. (protract unduly)* hinausziehen

'drag-net *n. (lit. or fig.)* Schleppnetz, *das; (fig.)* Netz, *das*

dragon ['drægn] *n.* Drache, *der; (fig.: person)* Drachen, *der*

'dragon-fly *n. (Zool.)* Libelle, *die*

drain [dreɪn] **1.** *n.* **a)** Abflußrohr, *das; (underground)* Kanalisationsrohr, *das; (grating at roadside)* Gully, *der;* **down the ~** *(fig. coll.)* für die Katz *(ugs.);* **go down the ~** *(fig. coll.)* für die Katz sein *(ugs.);* **that was money [thrown] down the ~** *(fig. coll.)* das Geld war zum Fenster hinausgeworfen *(ugs.);* **b)** *(fig.: constant demand)* Belastung, *die* (**on** *Gen.*). **2.** *v. t.* **a)** trockenlegen ‹*Teich*›; entwässern ‹*Land*›; ableiten ‹*Wasser*›; **b)** *(Cookery)* abgießen ‹*Wasser, Kartoffeln, Gemüse*›; **c)** *(drink all contents of)* austrinken; **d)** *(fig.: deprive)* ~ **a country of its wealth** *or* **resources/~ sb. of his energy** ein Land/jmdn. auslaugen. **3.** *v. i.* **a)** ‹*Geschirr, Gemüse:*› abtropfen; ‹*Flüssigkeit:*› ablaufen; **b) the colour ~ed from her face** *(fig.)* die Farbe wich aus ihrem Gesicht

drainage ['dreɪnɪdʒ] *n.* Entwässerung, *die; (system)* Entwässerungssystem, *das; (of city, house, etc.)* Kanalisation, *die*

'drain-pipe *n.* **a)** *(to carry off rain-water)* Regen[abfall]rohr, *das;* **b)** *(to carry off sewage)* Abwasserleitung, *die; (underground)* Kanalisationsleitung, *die*

drake [dreɪk] *n.* Enterich, *der*

dram [dræm] *n.* **a)** *(Pharm.) (weight)* Drachme, *die;* **b)** *(small drink)* Schlückchen, *das (ugs.)*

drama ['drɑːmə] *n. (lit. or fig.)* Drama, *das; (dramatic art)* Schauspielkunst, *die; attrib.* ~ **critic** Theaterkritiker, *der*

dramatic [drə'mætɪk] *adj. (lit. or fig.)* dramatisch; ~ **art** Dramatik, *die*

dramatise *see* **dramatize**

dramatist ['dræmətɪst] *n.* Dramatiker, *der*/Dramatikerin, *die*

dramatize ['dræmətaɪz] *v. t. (lit. or fig.)* dramatisieren

drank *see* **drink 2, 3**

drape [dreɪp] **1.** *v. t.* **a)** *(cover, adorn)* ~ **oneself/sb. in sth.** sich/jmdn. in etw. *(Akk.)* hüllen; **b)** *(put loosely)* legen; drapieren. **2.** *n.* **a)** *(cloth)* Tuch, *das;* **b)** *usu. in pl. (Amer.: curtain)* Vorhang, *der*

draper ['dreɪpə(r)] *n. (Brit.)* Textilkaufmann, *der;* **the ~'s [shop]** das Textilgeschäft

drapery ['dreɪpərɪ] *n.* **a)** *(Brit.: cloth)* Stoffe; **b)** *(Brit.: trade)* Textilgewerbe, *das;* ~ **shop** Textilgeschäft, *das;* **c)** *(arrangement of cloth)* Draperie, *die*

drastic ['dræstɪk] *adj.* drastisch; erheblich ‹*Wandel, Verbesserung*›; durchgreifend, rigoros ‹*Mittel*›; dringend ‹*Bedarf*›; einschneidend ‹*Veränderung*›; erschreckend ‹*Mangel*›; **something ~ will have to be done** drastische Maßnahmen müssen ergriffen werden

drat [dræt] *v. t. (coll.)* ~ **[it]/him!** verflucht!/verfluchter Kerl! *(salopp)*

draught [drɑːft] *n.* **a)** *(of air)* [Luft]zug, *der;* **be [sitting] in a ~:** im Zug sitzen; **there's a ~ [in here]** es zieht [hier]; **feel the ~** *(fig. sl.)* [finanziell] in der Klemme sitzen *(ugs.);* **b) [beer] on ~:** [Bier] vom Faß; **c)** *(swallow)* Zug, *der; (amount)* Schluck, *der;* **d)** *(Naut.)* Tiefgang, *der*

draught: ~ **'beer** *n.* Faßbier, *das;* **~-board** *n. (Brit.)* Damebrett, *das;* **~-proof 1.** *adj.* winddicht; **2.** *v. t.* winddicht machen

draughts [drɑːfts] *n., no pl. (Brit.)* Damespiel, *das*

draughtsman ['drɑːftsmən] *n., pl.* **draughtsmen** ['drɑːftsmən] *(Brit.)* Zeichner, *der*/Zeichnerin, *die*

draughty ['drɑːftɪ] *adj.* zugig

draw [drɔː] **1.** *v. t.,* **drew** [druː], **drawn** [drɔːn] **a)** *(pull)* ziehen; ~ **the curtains/blinds** *(open)* die Vorhänge aufziehen/die Jalousien hochziehen; *(close)* die Vorhänge zuziehen/die Jalousien herunterlassen; ~ **the bolt** *(unfasten)* den Riegel zurückschieben; **b)** *(attract, take in)* anlocken ‹*Publikum, Menge, Kunden*›; **be ~n to sb.** von jmdm. angezogen werden; **he refused to be ~n** er ließ sich nichts entlocken; **c)** *(take out)* herausziehen; ziehen (**from** aus); ~ **money from the bank/one's account** Geld bei der Bank holen/von seinem Konto abheben; ~ **water from a well** Wasser an einem Brunnen holen *od.* schöpfen; **d)** *(derive, elicit)* finden; ~ **comfort from sth.** Trost in etw. *(Dat.)* finden; ~ **reassurance/encouragement from sth.** Zuversicht/Mut aus etw. schöpfen; **e)** *(get as one's due)* erhalten; bekommen; beziehen ‹*Gehalt, Rente, Arbeitslosenunterstützung*›; **f)** *(select at random)* ~ **straws** Lose ziehen; ~ **cards from a pack** Karten von einem Haufen abheben; ~ **a winner** ein Gewinnlos ziehen; **g)** *(trace)* ziehen ‹*Strich*›; zeichnen ‹*geometrische Figur, Bild*›; ~ **the line at sth.** *(fig.)* bei etw. nicht mehr mitmachen; **h)** *(formulate)* ziehen ‹*Parallele, Vergleich*›; herstellen ‹*Analogie*›; herausstellen ‹*Unterschied*›; **i)** *(end with neither side winner)* unentschieden beenden ‹*Spiel*›; **the match was ~n** das Spiel ging unentschieden aus. **2.** *v. i.,* **drew, drawn a)** *(make one's way, move)* ‹*Person:*› gehen; ‹*Fahrzeug:*› fahren; ~ **into sth.** ‹*Zug:*› in etw. *(Akk.)* einfahren; ‹*Schiff:*› in etw. *(Akk.)* einlaufen; ~ **towards sth.** sich einer Sache *(Dat.)* nähern; ~ **to an end** zu Ende gehen; **b)** *(~ lots)* ziehen; losen; ~ **[for partners]** [die Partner] auslosen. **3.** *n.* **a)** *(raffle)* Tombola, *die; (for matches, contests)* Auslosung, *die;* **b)** *([result of] drawn game)* Unentschieden, *das;* **end in a ~:** mit einem Unentschieden enden; **c)** Attraktion, *die; (film, play)* Publikumserfolg, *der;* **d) be quick/slow on the ~:** den Finger schnell/zu langsam am Abzug haben

~ a'side *v.t.* zur Seite ziehen; **~ sb. aside** jmdn. beiseite nehmen

~ a'way *v.i.* **a)** *(move ahead)* **~ away from sth./sb.** sich von etw. entfernen/jmdm. davonziehen; **b)** *(set off)* losfahren

~ 'back 1. *v.t.* zurückziehen; aufziehen ⟨*Vorhang*⟩. **2.** *v.i.* zurückweichen; *(fig.)* sich zurückziehen

~ 'in 1. *v.i.* **a)** *(move in and stop)* einfahren; **the car drew in to the side of the road** das Auto fuhr an den Straßenrand heran; **b)** ⟨*Tage:*⟩ kürzer werden; ⟨*Abende:*⟩ länger werden. **2.** *v.t. (fig.)* hineinziehen ⟨*Person*⟩; zum Mitmachen überreden

~ on 1. [-'-] *v.i.* ⟨*Zeit:*⟩ vergehen; *(approach)* ⟨*Winter, Nacht:*⟩ nahen. **2.** ['--] *v.t.* anziehen ⟨*Kleidung*⟩; **b)** zurückgreifen auf ⟨*Ersparnisse, Vorräte*⟩; schöpfen aus ⟨*Wissen, Erfahrungen*⟩

~ 'out 1. *v.t. (extend)* ausdehnen; in die Länge ziehen; **long ~n out** ausgedehnt. **2.** *v.i.* **a)** abfahren; **the train drew out of the station** der Zug fuhr aus dem Bahnhof aus; **b)** ⟨*Tage:*⟩ länger werden; ⟨*Abende:*⟩ kürzer werden

~ 'up 1. *v.t.* **a)** *(formulate)* abfassen; aufsetzen ⟨*Vertrag*⟩; aufstellen ⟨*Liste*⟩; entwerfen ⟨*Plan, Budget*⟩; **b)** *(pull closer)* heranziehen; **c) ~ oneself up [to one's full height]** sich [zu seiner vollen Größe] aufrichten. **2.** *v.i.* [an]halten

~ upon *see* **~ on 2 b**

draw: ~back *n.* Nachteil, *der;* **~bridge** *n.* Zugbrücke, *die*

drawer *n.* **a)** [drɔ:(r), 'drɔ:ə(r)] *(in furniture)* Schublade, *die;* **b)** *in pl.* [drɔ:z] *(dated/joc.: woman's garment)* Schlüpfer *Pl.*

drawing ['drɔ:ɪŋ] *n. (sketch)* Zeichnung, *die*

drawing: ~-board *n.* Zeichenbrett, *das;* **so it's back to the ~-board** dann müssen wir wohl wieder von vorne beginnen; **~-office** *n.* Konstruktionsbüro, *das;* **~-pin** *n. (Brit.)* Reißzwecke, *die;* **~-room** *n.* Salon, *der*

drawl [drɔ:l] **1.** *v.i.* gedehnt sprechen. **2.** *v.t.* dehnen; gedehnt aussprechen. **3.** *n.* gedehntes Sprechen; **speak with a ~:** gedehnt sprechen

drawn [drɔ:n] **1.** *see* **draw 1, 2. 2.** *adj.* verzogen ⟨*Gesicht*⟩; **look ~** *(from tiredness)* abgespannt aussehen; *(from worries)* abgehärmt aussehen

'draw-string *n.* Durchziehband, *das*

dread [dred] **1.** *v.t.* sich sehr fürchten vor (+ *Dat.*); **the ~ed day/moment** der gefürchtete Tag/ Augenblick; **I ~ to think [what may have happened]** ich mag gar nicht daran denken[, was passiert sein könnte]. **2.** *n., no pl. (terror)* Angst, *die;* **be** *or* **live in ~ of sth./ sb.** in [ständiger] Furcht vor etw./ jmdm. leben

dreadful ['dredfl] *adj.* schrecklich; furchtbar; *(coll.: very bad)* fürchterlich; **I feel ~** *(unwell)* ich fühle mich scheußlich *(ugs.); (embarrassed)* es ist mir furchtbar peinlich

dreadfully ['dredfəlɪ] *adv.* **a)** schrecklich; entsetzlich, furchtbar ⟨*leiden*⟩; *(coll.: very badly)* grauenhaft; fürchterlich; **b)** *(coll.: extremely)* schrecklich; furchtbar

dream [dri:m] **1.** *n.* **a)** Traum, *der;* **have a ~ about sb./sth.** von jmdm./etw. träumen; **it was all a bad ~:** das ganze war wie ein böser Traum; **in a ~:** im Traum; **go/ work like a ~** *(coll.)* wie eine Eins fahren/funktionieren *(ugs.);* **b)** *(ambition, vision)* Traum, *der;* **never in one's wildest ~s** nicht in seinen kühnsten Träumen; **c)** *attrib.* traumhaft; Traum⟨*haus, -auto, -urlaub*⟩. **2.** *v.i.,* **~t** [dremt] *or* **~ed** träumen (**about, of** von); *(while awake)* vor sich *(Akk.)* hin träumen; **he wouldn't ~ of doing it** *(fig.)* er würde nicht im Traum daran denken, das zu tun. **3.** *v.t.,* **~t** *or* **~ed** träumen; **she never ~t that she'd win** sie hätte sich *(Dat.)* nie träumen lassen, daß sie gewinnen würde

~ 'up *v.t.* sich *(Dat.)* ausdenken

dreamer ['dri:mə(r)] *n. (day-~)* Träumer, *der*/Träumerin, *die*

dreamless ['dri:mlɪs] *adj.* traumlos

dreamt *see* **dream 2, 3**

dreary ['drɪərɪ] *adj.* trostlos; monoton ⟨*Musik*⟩; langweilig ⟨*Unterricht, Lehrbuch*⟩

dredge [dredʒ] *v.t.* ausbaggern; **~ [up]** *(fig.)* ausgraben

dredger ['dredʒə(r)] *n.* Bagger, *der; (boat)* Schwimmbagger, *der*

dregs [dregz] *n. pl.* **a)** [Boden]satz, *der;* **drain one's glass to the ~:** sein Glas bis zur Neige leeren; **b)** *(fig.)* Abschaum, *der*

drench [drentʃ] *v.t.* durchnässen; **get completely ~ed, get ~ed to the skin** naß bis auf die Haut werden

dress [dres] **1.** *n.* **a)** *(woman's or girl's frock)* Kleid, *das;* **b)** *no pl. (clothing)* Kleidung, *die;* **articles of ~:** Kleidungsstücke; **c)** *no pl. (manner of dressing)* Kleidung, *die.* **2.** *v.t.* **a)** *(clothe)* anziehen; **be ~ed** angezogen sein; **be well ~ed** gut gekleidet sein; **get ~ed** sich anziehen; **b)** *(provide clothes for)* einkleiden ⟨*Familie*⟩; **c)** *(deck, adorn)* schmücken; beflaggen ⟨*Schiff*⟩; dekorieren ⟨*Schaufenster*⟩; **d)** *(Med.)* verbinden, versorgen ⟨*Wunde*⟩; **e)** *(Cookery)* zubereiten; **f)** *(treat, prepare)* gerben ⟨*tierische Häute, Felle*⟩; *(put finish on)* appretieren ⟨*Gewebe, Holz, Leder*⟩. **3.** *v.i.* *(wear clothes)* sich anziehen; sich kleiden; *(get ~ed)* sich anziehen; **~ for dinner** sich zum Abendessen umziehen

~ 'down *v.t. (fig.)* zurechtweisen

~ 'up 1. *v.t.* **a)** *(in formal clothes)* feinmachen; **b)** *(disguise)* verkleiden; **c)** *(smarten)* verschönern. **2.** *v.i.* **a)** *(wear formal clothes)* sich feinmachen; **b)** *(disguise oneself)* sich verkleiden

dressage ['dresɑ:ʒ] *n.* Dressurreiten, *das*

'dress circle *n. (Theatre)* erster Rang

¹dresser ['dresə(r)] *n.* Anrichte, *die;* Büfett, *das*

²dresser *n.* **a) he's a careless/elegant ~:** er kleidet sich nachlässig/elegant; **b)** *(Theatre)* Garderobier, *der*/Garderobiere, *die*

dressing ['dresɪŋ] *n.* **a)** *no pl.* Anziehen, *das;* **b)** *(Cookery)* Dressing, *das;* **c)** *(Med.)* Verband, *der;* **d)** *(Agric.)* Dünger, *der*

dressing: ~ 'down *n.* **give sb. a ~ down** jmdm. eine Standpauke halten *(ugs.);* **~-gown** *n.* Bademantel, *der;* **~-room** *n.* **a)** *(of actor or actress)* [Schauspieler]garderobe, *die;* [Künstler]garderobe, *die;* **b)** *(for games-players)* Umkleideraum, *der;* **~-table** *n.* Frisierkommode, *die*

dress: ~maker *n.* Damenschneider, *der*/-schneiderin, *die;* **~making** *n.* Damenschneiderei, *die;* **~ rehearsal** *n. (lit. or fig.)* Generalprobe, *die;* **~ shirt** *n.* Smokinghemd, *das;* **~ suit** *n.* Abendanzug, *der;* **~-uniform** *n.* *(Mil.)* Paradeuniform, *die*

drew *see* **draw 1, 2**

dribble ['drɪbl] **1.** *v.i.* **a)** *(trickle)* tropfen; **b)** *(slobber)* ⟨*Baby:*⟩ sabbern; **c)** *(Sport)* dribbeln. **2.** *v.t.* **a)** ‹*Baby:*› kleckern; **b)** *(Sport)* dribbeln mit ‹*Ball*›

dribs *n. pl.* ~ **and drabs** [drɪbz n 'dræbz] kleine Mengen; **in ~ and drabs** kleckerweise *(ugs.)*

dried [draɪd] *adj.* getrocknet; ~ **fruit[s]** Dörr- *od.* Backobst, *das;* ~ **milk/meat** Trockenmilch, *die*/ -fleisch, *das*

¹drier *see* **dry 1**

²drier ['draɪə(r)] *n.* *(for hair)* Trockenhaube, *die; (hand-held)* Fön ®, *der;* Haartrockner, *der; (for laundry)* [Wäsche]trockner, *der*

driest *see* **dry 1**

drift [drɪft] **1.** *n.* **a)** *(flow, steady movement)* Wanderung, *die;* **b)** *(fig.: trend, shift, tendency)* Tendenz, *die;* **c)** *(flow of air or water)* Strömung, *die;* **d)** *(Naut., Aeronaut.: deviation from course)* Abdrift, *die (fachspr.);* **e)** *(of snow or sand)* Verwehung, *die;* **f)** *(fig.: gist, import)* das Wesentliche; **get** *or* **catch the ~ of sth.** etw. im wesentlichen verstehen. **2.** *v.i.* **a)** *(be borne by current; fig.: move passively or aimlessly)* treiben; ‹*Wolke:*› ziehen; ~ **out to sea** aufs Meer hinaustreiben; ~ **off course** abtreiben; ~ **into crime** in die Kriminalität [ab]driften; ~ **into unconsciousness** in Bewußtlosigkeit versinken; **months ~ed by** die Monate vergingen; **b)** *(coll.: come or go casually)* ~ **in** hereinschneien *(ugs.);* ~ **out** abziehen *(ugs.);* **c)** *(form ~s)* zusammengeweht werden; **~ing sand** Treibsand, *der*

drifter ['drɪftə(r)] *n.* **a)** *(Naut.)* Drifter, *der;* **b)** *(person)* **be a ~:** sich treiben lassen

drift: ~-net *n.* Treibnetz, *das;* **~wood** *n.* Treibholz, *das*

drill [drɪl] **1.** *n.* **a)** *(tool)* Bohrer, *der; (Dent.)* Bohrinstrument, *das; (Carpentry, Building)* Bohrmaschine, *die;* **b)** *(Mil.: training)* Drill, *der;* **c)** *(Educ.; also fig.)* Übung, *die;* **d)** *(Brit. coll.: agreed procedure)* Prozedur, *die;* **know the ~:** wissen, wie es gemacht wird. **2.** *v.t.* **a)** *(bore)* bohren ‹*Loch, Brunnen*›; an-, ausbohren ‹*Zahn*›; ~ **sth.** *(right through)* etw. durchbohren; **b)** *(Mil.: instruct)* drillen; **c)** *(Educ.; also fig.)* ~ **sb. in sth., ~ sth. into sb.** mit jmdm. etw. systematisch einüben; jmdm. etw. eindrillen *(ugs.).* **3.** *v.i.* bohren (**for** nach)

drill: ~-bit *n.* Bohrer, *der;* **~-chuck** *n.* Bohrfutter, *das*

drink [drɪŋk] **1.** *n.* **a)** Getränk; *das;* **have a ~:** [etwas] trinken; **would you like a ~ of milk?** möchten Sie etwas Milch [trinken]?; **give sb. a ~ [of fruit-juice]** jmdm. etwas [Fruchtsaft] zu trinken geben; **b)** *(glass of alcoholic liquor)* Glas, *das; (not with food)* Drink, *der;* Glas, *das;* **have a ~:** ein Glas trinken; **let's have a ~!** trinken wir einen!; **he has had a few ~s** er hat einige getrunken *(ugs.);* **c)** *no pl., no art. (intoxicating liquor)* Alkohol, *der;* **[strong] ~:** scharfe *od.* hochprozentige Getränke; **the worse for ~:** betrunken; **drive sb. to ~:** jmdn. zum Trinker werden lassen. **2.** *v.t.*, **drank** [dræŋk], **drunk** [drʌŋk] trinken ‹*Kaffee, Glas Milch, Flasche Whisky*›; ~ **down** *or* **off [in one gulp]** [in einem *od.* auf einen Zug] austrinken; ~ **oneself to death** sich zu Tode trinken. **3.** *v.i.*, **drank, drunk** trinken; ~ **from a bottle** aus einer Flasche trinken; **~ing and driving, ~-driving** Alkohol am Steuer; ~ **to sb./sth.** auf jmdn./etw. trinken

~ **'in** *v.t.* einsaugen ‹*Luft; fig.: Schönheit*›; begierig aufnehmen ‹*Worte, Geschichten*›

~ **'up** *v.t. & i.* austrinken

drinkable ['drɪŋkəbl] *adj.* trinkbar

drinker ['drɪŋkə(r)] *n.* Trinker, *der*/Trinkerin, *die*

drinking ['drɪŋkɪŋ]: ~ **fountain** *n.* Trinkbrunnen, *der;* **~-'up time** *n. (Brit.) Zeit zwischen dem Ende des Ausschanks und der Schließung der Gaststätte (meist 10 Minuten);* **~-water** *n.* Trinkwasser, *das*

drip [drɪp] **1.** *v.i.*, **-pp-: a)** tropfen; *(overflow in drops)* triefen; **be ~ping with moisture** triefend naß sein; **b)** *(fig.)* **be ~ping with** überladen sein mit ‹*Schmuck*›; triefen von *od.* vor ‹*Ironie, Sentimentalität usw.*›. **2.** *v.t.*, **-pp-** tropfen lassen. **3.** *n.* **a)** Tropfen, *das;* **b)** *(Med.)* Tropfinfusion, *die;* **the patient was on a ~:** der Patient hing am Tropf; **c)** *(coll.: feeble person)* Schlappschwanz, *der (salopp abwertend)*

drip-dry *(Textiles)* **1.** [-'-] *v.i.* knitterfrei trocknen. **2.** ['--] *adj.* bügelfrei; schnelltrocknend

dripping ['drɪpɪŋ] **1.** *adv.* ~ **wet** tropf- *od. (ugs.)* patsch- *od. (ugs.)* klitschnaß. **2.** *n. (Cookery)* Schmalz, *das;* **bread and ~:** Schmalzbrot, *das*

drive [draɪv] **1.** *n.* **a)** Fahrt, *die;* **a nine-hour ~, a ~ of nine hours** eine neunstündige Autofahrt; **have a long ~ to work** eine lange Anfahrt zur Arbeit haben; **b)** *(street)* Straße, *die;* **c)** *(private road)* Zufahrt, *die; (entrance to large building)* Auffahrt, *die;* **d)** *(energy to achieve)* Tatkraft, *die;* **e)** *(Commerc., Polit., etc.: vigorous campaign)* Aktion, *die;* Kampagne, *die;* **export/sales/recruiting ~:** Export- / Verkaufs- / Anwerbekampagne, *die;* **f)** *(Psych.)* Trieb, *der;* **g)** *(Motor Veh.: position of steering-wheel)* **left-hand/right-hand ~:** Links-/Rechtssteuerung *od.* -lenkung, *die;* **be left-hand ~:** Linkssteuerung haben; **h)** *(Motor Veh., Mech. Engin.: transmission of power)* Antrieb, *der;* **front-wheel/rear-wheel ~:** Front-/Heckantrieb, *der.* **2.** *v.t.*, **drove** [drəʊv], **driven** ['drɪvn] **a)** fahren ‹*Auto, Lkw, Route, Strecke, Fahrgast*›; lenken ‹*Kutsche, Streitwagen*›; treiben ‹*Tier*›; **b)** *(as job)* ~ **a lorry/train** Lkw-Fahrer/Lokomotivführer sein; **c)** *(compel to move)* vertreiben; ~ **sb. out of** *or* **from a place/country** jmdn. von einem Ort/aus einem Land vertreiben; **d)** *(chase, urge on)* treiben ‹*Vieh, Wild*›; **e)** *(fig.)* ~ **sb. to sth.** jmdn. zu etw. treiben; ~ **sb. out of his mind** *or* **wits** jmdn. in den Wahnsinn treiben; **f)** ‹*Wind, Wasser:*› treiben; **be ~n off course** abgetrieben werden; **g)** *(cause to penetrate)* ~ **sth. into sth.** etw. in etw. *(Akk.)* treiben; **h)** *(power)* antreiben ‹*Mühle, Maschine*›; **be steam-~n** *or* **~n by steam** dampfgetrieben sein; **i)** *(incite to action)* antreiben; ~ **oneself [too] hard** sich [zu sehr] schinden. **3.** *v.i.*, **drove, driven a)** fahren; **in Great Britain we ~ on the left** bei uns in Großbritannien ist Linksverkehr; ~ **at 30 m.p.h.** mit 50 km/h fahren; **learn to ~:** [Auto]fahren lernen; den Führerschein machen *(ugs.);* **can you ~?** kannst du Auto fahren?; **b)** *(go by car)* mit dem [eigenen] Auto fahren; **c)** ‹*Hagelkörner, Wellen:*› schlagen; **clouds were driving across the sky** Wolken jagten über den Himmel

~ **at** *v.t.* hinauswollen auf (+ *Akk.*); **what are you driving at?** worauf wollen Sie hinaus?

~ **a'way 1.** *v.i.* wegfahren. **2.** *v.t.* **a)** wegfahren, wegbringen ‹*Ladung, Fahrzeug*›; *(chase away)* wegjagen; **b)** *(fig.)* zerstreuen ‹*Bedenken, Befürchtungen*›

~ **'back** *v.t. (force to retreat)* zurückschlagen ‹*Eindringlinge*›

~ **'off 1.** *v.i.* **a)** wegfahren; **b)** *(Golf)* abschlagen. **2.** *v.t. (repel)* zurückschlagen ‹*Angreifer*›

~ '**on 1.** *v.i.* weiterfahren. **2.** *v.t.* *(impel)* treiben (**to** zu)
~ '**out** *v.t.* hinauswerfen ‹*Person*›; hinausjagen ‹*Hund*›
~ '**up 1.** *v.i.* vorfahren (**to** vor + *Dat.*). **2.** *v.t.* hochtreiben ‹*Kosten*›
'**drive-in** *adj.* Drive-in-; ~ **bank** Bank mit Autoschalter; ~ **cinema** *or (Amer.)* **movie [theater]** Autokino, *das*
drivel ['drɪvl] *n.* Gefasel, *das (ugs. abwertend)*; **talk** ~: faseln *(ugs. abwertend)*
driven *see* **drive 2, 3**
'**drive-on** *adj.* ~ **car ferry** Autofährschiff, *das*
driver ['draɪvə(r)] *n.* Fahrer, *der*/Fahrerin, *die; (of locomotive)* Führer, *der*/Führerin, *die;* **be in the ~'s seat** *(fig.)* das Steuer in der Hand haben *(fig.)*
drive: ~-shaft *n.* Antriebswelle, *die;* **~way** *see* ~ **1d**
driving ['draɪvɪŋ] **1.** *n.* Fahren, *das;* **his ~ is awful** er fährt furchtbar. **2.** *adj.* **a)** ~ **rain** peitschender Regen; **b)** *(fig.)* treibend
driving: ~-instructor *n.* Fahrlehrer, *der*/-lehrerin, *die;* **~-lesson** *n.* Fahrstunde, *die;* **~-licence** *n.* Führerschein, *der;* **~-mirror** *n.* Rückspiegel, *der;* **~-school** *n.* Fahrschule, *die;* **~-seat** *n.* Fahrersitz, *der;* **be in the ~-seat** das Steuer in der Hand haben *(fig.)*; **~-test** *n.* Fahrprüfung, *die;* **take/pass/fail one's ~-test** die Fahrprüfung ablegen/bestehen/nicht bestehen
drizzle ['drɪzl] **1.** *n.* Sprühregen, *der;* Nieseln, *das.* **2.** *v.i.* **it's drizzling** es nieselt; **drizzling rain** Nieselregen, *der*
dromedary ['drɒmɪdərɪ] *n.* *(Zool.)* Dromedar, *das*
drone [drəʊn] **1.** *n.* **a)** *(of bees, flies)* Summen, *das; (of machine)* Brummen, *das;* **b)** *(derog.: monotonous tone of speech)* Geleier, *das;* **c)** *(Zool.: bee; Aeronaut.)* Drohne, *die.* **2.** *v.i.* **a)** *(buzz, hum)* ‹*Biene:*› summen; ‹*Maschine:*› brummen; **b)** *(derog.)* ‹*Rezitator:*› leiern. **3.** *v.t.* leiern
drool [dru:l] *v.i.* ~ **over sb./sth.** über jmdn./etw. in Verzückung geraten
droop [dru:p] *v.i.* **a)** herunterhängen; ‹*Blume:*› den Kopf hängen lassen; **her head ~ed forwards** ihr Kopf sank nach vorn; **his eyelids were ~ing** ihm fielen die Augen zu; **b)** ‹*Person:*› ermatten
drop [drɒp] **1.** *n.* **a)** Tropfen, *der;* **~s of rain/dew/blood/sweat** Regen-/Tau-/Bluts-/Schweißtropfen; ~ **by** ~, **in ~s** tropfenweise; **be a ~ in the ocean** *or* **in the** *or* **a bucket** *(fig.)* ein Tropfen auf einen heißen Stein sein; *(fig.: small amount)* **[just] a** ~: [nur] ein kleiner Tropfen; **b)** *(fig. coll.: of alcohol)* Gläschen, *das;* **have had a ~ too much** ein Glas über den Durst getrunken haben *(ugs.)*; **c)** *in pl. (Med.)* Tropfen *Pl.*; **d)** *(vertical distance)* **there was a ~ of 50 metres from the roof to the ground below** vom Dach bis zum Boden waren es 50 Meter; **e)** *(abrupt descent of land)* plötzlicher Abfall; Absturz, *der;* **f)** *(fig.: decrease)* Rückgang, *der;* ~ **in temperature/prices** Temperatur-/Preisrückgang, *der;* **a ~ in the cost of living** ein Sinken der Lebenshaltungskosten; **a ~ in salary/wages/income** eine Gehalts-/Lohn-/Einkommensminderung. **2.** *v.i.*, **-pp-: a)** *(fall) (accidentally)* [herunter]fallen; *(deliberately)* sich [hinunter]fallen lassen; ~ **out of** *or* **from sb.'s hand** jmdm. aus der Hand fallen; **b)** *(sink to ground)* ‹*Person:*› fallen; ~ **to the ground** umfallen; zu Boden fallen; ~ **[down] dead** tot umfallen; ~ **dead!** *(sl.)* scher dich zum Teufel!; ~ **into bed/an armchair** ins Bett/in einen Sessel sinken; **be fit** *or* **ready to** ~ *(coll.)* zum Umfallen müde sein; **c)** *(in amount etc.)* sinken; ‹*Wind:*› abflauen, sich legen; ‹*Stimme:*› sich senken; ‹*Kinnlade:*› herunterfallen; **d)** *(move, go)* ~ **back** *(Sport)* zurückfallen; **e)** *(fall in ~s)* ‹*Flüssigkeit:*› tropfen (**from** aus); **f)** ~ **[back] into one's old routine** in den alten Trott verfallen; ~ **into the habit** *or* **way of doing sth.** die Gewohnheit annehmen, etw. zu tun; **g)** *(cease)* **the affair was allowed to** ~: man ließ die Angelegenheit auf sich *(Dat.)* beruhen; **h) let** ~: beiläufig erwähnen ‹*Tatsache, Absicht*›; fallenlassen ‹*Bemerkung*›; **let [it] ~ that/when ...:** beiläufig erwähnen, daß/wann ... **3.** *v.t.*, **-pp-: a)** *(let fall)* fallen lassen; abwerfen ‹*Bomben, Flugblätter, Nachschub*›; absetzen ‹*Fallschirmjäger, Truppen*›; ~ **a letter in the letter-box** einen Brief einwerfen; ~ **the latch on the door** den Türriegel vorlegen; **b)** *(by mistake)* fallen lassen; **she ~ped crumbs on the floor/ juice on the table** ihr fielen Krümel auf den Boden/tropfte Saft auf den Tisch; **he ~ped the glass** ihm fiel das Glas herunter; **c)** *(let fall in ~s)* tropfen; **d)** *(utter casually)* fallenlassen ‹*Namen*›; ~ **a hint** eine Anspielung machen; **e)** *(send casually)* ~ **sb. a note** *or* **line** jmdm. [ein paar Zeilen] schreiben; **f)** *(set down, unload from car)* absetzen ‹*Mitfahrer, Fahrgast*›; **g)** *(omit) (in writing)* auslassen; *(in speech)* nicht aussprechen; ~ **a name from a list** einen Namen von einer Liste streichen; **h)** *(discontinue, abandon)* fallenlassen ‹*Plan, Thema, Anklage*›; einstellen ‹*Untersuchung, Ermittlungen*›; beiseite lassen ‹*Formalitäten*›; aufgeben, Schluß machen mit ‹*Verstellung, Heuchelei*›; ~ **it!** laß das!; **shall we ~ the subject?** lassen Sie uns [lieber] das Thema wechseln; **i)** ~ **sb. from a team** jmdn. aus einer Mannschaft nehmen; **j)** ~ **one's voice** die Stimme senken; **k) ~ped handlebars** Rennlenker, *der*
~ '**by** *v.i.* vorbeikommen
~ '**in 1.** *v.t.* *(deliver)* vorbeibringen. **2.** *v.i.* **a)** hineinfallen; **b)** *(visit)* hereinschauen; vorbeikommen; ~ **in on sb.** *or* **at sb.'s house** bei jmdm. hereinschauen
~ '**off 1.** *v.i.* **a)** *(fall off)* abfallen; *(become detached)* abgehen; **b)** *(fall asleep)* einnicken; **c)** *(decrease)* ‹*Teilnahme, Geschäft:*› zurückgehen; ‹*Unterstützung, Interesse:*› nachlassen. **2.** *v.t.* **a)** *(fall off)* abfallen von; **b)** *(set down)* absetzen ‹*Fahrgast*›
~ '**out** *v.i.* **a)** *(fall out)* herausfallen (**of** aus); **b)** *(withdraw beforehand)* seine Teilnahme absagen; *(withdraw while in progress)* aussteigen *(ugs.)* (**of** aus); **c)** *(cease to take part)* aussteigen *(ugs.)* (**of** aus); ‹*Student:*› das Studium abbrechen; ~ **out [of society]** aussteigen *(ugs.)*
~ '**round** *v.i.* vorbeikommen
droplet ['drɒplɪt] *n.* Tröpfchen, *das*
'**drop-out** *n.* *(coll.) (from college etc.)* Abbrecher, *der*/Abbrecherin, *die; (from society)* Aussteiger, *der*/Aussteigerin, *die (ugs.)*
droppings ['drɒpɪŋz] *n. pl.* Mist, *der; (of horse)* Pferdeäpfel *Pl.*
'**drop-shot** *n.* *(Tennis etc.)* Stoppball, *der*
drought [draʊt], *(Amer., Scot., Ir./poet.)* **drouth** [draʊθ] *n.* Dürre, *die;* **a period of** ~: eine Dürreperiode
drove *see* **drive 2, 3**
drown [draʊn] **1.** *v.i.* ertrinken. **2.** *v.t.* **a)** ertränken; **be ~ed** ertrinken; **b)** *(fig.)* ~ **one's sorrows** seine Sorgen ertränken; **c)** übertönen ‹*Geräusch, Musik*›

drowse [draʊz] *v. i.* [vor sich hin]dösen

drowsy ['draʊzɪ] *adj.* **a)** *(half asleep)* schläfrig; *(on just waking)* verschlafen; **b)** *(soporific)* einschläfernd

drudge [drʌdʒ] *n.* Schwerarbeiter, *der (fig.);* Kuli, *der (ugs.)*

drudgery ['drʌdʒərɪ] *n.* Schufterei, *die;* Plackerei, *die*

drug [drʌg] **1.** *n.* **a)** *(Med., Pharm.)* Medikament, *das;* [Arznei]mittel, *das;* **b)** *(narcotic, opiate, etc.)* Droge, *die;* Rauschgift, *das.* **2.** *v. t.,* **-gg-: he was ~ged and kidnapped** er wurde betäubt und entführt; **~ sb.'s food/drink** jmds. Essen/Getränk *(Dat.)* ein Betäubungsmittel beimischen

drug: ~ addict *n.* Drogen- *od.* Rauschgiftsüchtige, *der/die;* **~ addiction** *n.* Drogen- *od.* Rauschgiftsucht, *die*

druggist ['drʌgɪst] *n.* Drogist, *der*/Drogistin, *die*

'drugstore *n. (Amer.)* Drugstore, *der*

Druid ['dru:ɪd] *n.* Druide, *der*

drum [drʌm] **1.** *n.* **a)** Trommel, *die;* **b)** *in pl. (in jazz or pop)* Schlagzeug, *das; (section of band etc.)* Trommeln *Pl.;* **c)** *(container for oil etc.)* Faß, *das.* **2.** *v. i.,* **-mm-** trommeln. **3.** *v. t.,* **-mm-: ~ one's fingers on the desk** mit den Fingern auf den Tisch trommeln

~ into *v. t.* **~ sth. into sb.** jmdm. etw. einhämmern *(ugs.)*

~ 'up *v. t.* auftreiben ‹*Kunden, Unterstützung*›; zusammentrommeln *(ugs.)* ‹*Helfer, Anhänger*›; anbahnen ‹*Geschäfte*›

drum: ~ 'major *n. (Mil.)* Tambourmajor, *der;* **~ majorette** [drʌm meɪdʒə'ret] *n.* Tambourmajorette, *die*

drummer ['drʌmə(r)] *n.* Schlagzeuger, *der*

'drumstick *n.* **a)** *(Mus.)* Trommelschlegel, *der;* **b)** *(of fowl)* Keule, *die*

drunk [drʌŋk] **1.** *adj.* **be ~:** betrunken sein; **get ~ [on gin]** [von Gin] betrunken werden; *(intentionally)* sich [mit Gin] betrinken; **be ~ as a lord** *(coll.)* voll wie eine Haubitze sein *(ugs.);* **~ in charge [of a vehicle]** betrunken am Steuer. **2.** *n.* Betrunkene, *der/die*

drunkard ['drʌŋkəd] *n.* Trinker, *der*/Trinkerin, *die*

drunken ['drʌŋkn] *attrib. adj.* **a)** betrunken; *(habitually drunk)* versoffen *(derb);* **b) a ~ brawl** *or* **fight** eine Schlägerei zwischen Betrunkenen; **~ driving** Trunkenheit am Steuer

drunkenness ['drʌŋknnɪs] *n., no pl.* **a)** *(temporary)* Betrunkenheit, *die;* **b)** *(habitual)* Trunksucht, *die*

dry [draɪ] **1.** *adj.,* **drier** ['draɪə(r)], **driest** ['draɪɪst] **a)** trocken; trocken, *(very ~)* herb ‹*Wein*›; ausgetrocknet ‹*Fluß, Flußbett*›; **go ~:** austrocknen; **as ~ as a bone** völlig trocken; **~ shave/shampoo** Trockenrasur, *die*/-shampoo, *das;* **b)** *(not rainy)* trocken ‹*Wetter, Klima*›; **c)** *(coll.: thirsty)* durstig; **I'm a bit ~:** ich habe eine trockene Kehle; **d)** ausgetrocknet, versiegt ‹*Brunnen*›; **e)** *(fig.)* trocken ‹*Humor*›; *(impassive, cold)* kühl ‹*Art, Bemerkung usw.*›; **f)** *(dull)* trocken ‹*Stoff, Bericht, Vorlesung*›. **2.** *v. t.* **a)** trocknen ‹*Haare, Wäsche*›; abtrocknen ‹*Geschirr, Baby*›; **~ oneself** sich abtrocknen; **~ one's eyes** *or* **tears/hands** sich *(Dat.)* die Tränen abwischen/die Hände abtrocknen; **b)** *(preserve)* trocknen ‹*Kräuter, Holz, Blumen*›; dörren ‹*Obst, Fleisch*›. **3.** *v. i.* trocknen; trocken werden

~ 'out 1. *v. t.* **a)** trocknen; **b)** einer Entziehungskur unterziehen ‹*Alkoholiker, Drogenabhängigen*›. **2.** *v. i.* trocknen

~ 'up 1. *v. t.* abtrocknen. **2.** *v. i.* **a)** *(~ the dishes)* abtrocknen; **b)** ‹*Brunnen, Quelle:*› versiegen; ‹*Fluß, Teich:*› austrocknen; **c)** *(fig.)* ‹*Ideen, Erfindergeist:*› versiegen

dry: ~-'clean *v. t.* chemisch reinigen; **have sth. ~-cleaned** etw. in die Reinigung geben; **~-'cleaners** *n. pl.* chemische Reinigung; **~-'cleaning** *n.* chemische Reinigung; **~ 'dock** *n.* Trockendock, *das*

dryer *see* **²drier**

drying-'up *n.* Abtrocknen, *das;* **do the ~:** abtrocknen

dry 'land *n.* Festland, *das;* **be back on ~:** wieder festen Boden unter den Füßen haben

dryness ['draɪnɪs] *n., no pl. (lit. or fig.)* Trockenheit, *die*

dry: ~ 'rot *n.* Trockenfäule, *die;* **~ 'run** *n. (coll.)* Probelauf, *der*

dual [dju:əl] *adj.* doppelt; Doppel-; **~ role/function** Doppelrolle, *die*/-funktion, *die*

dual: ~ 'carriageway *n. (Brit.)* zweispurige Straße; **~-'purpose** *adj.* zweifach verwendbar

¹dub [dʌb] *v. t.,* **-bb-** *(Cinemat.)* synchronisieren

²dub *v. t.,* **-bb-: a) ~ sb. [a] knight** jmdn. zum Ritter schlagen; **b)** *(call, nickname)* titulieren

dubious ['dju:bɪəs] *adj.* **a)** *(doubting)* unschlüssig; **I'm ~ about accepting the invitation** ich weiß nicht recht, ob ich die Einladung annehmen soll; **b)** *(suspicious, questionable)* zweifelhaft

dubiously ['dju:bɪəslɪ] *adv.* **a)** *(doubtingly)* unschlüssig; **b)** *(suspiciously)* dubios

duchess ['dʌtʃɪs] *n.* Herzogin, *die*

duchy ['dʌtʃɪ] *n.* Herzogtum, *das*

duck [dʌk] **1.** *n.* **a)** *pl.* **~s** *or (collect.) same* Ente, *die;* **wild ~:** Wildente, *die;* **it was [like] water off a ~'s back** *(fig.)* das lief alles an ihm/ihr *usw.* ab; **take to sth. like a ~ to water** bei etw. gleich in seinem Element sein; **b)** *(Brit. coll.: dear)* **[my] ~:** Schätzchen; **c)** *(Cricket)* **be out for a ~:** ohne einen Punkt zu machen aus sein. **2.** *v. i.* **a)** *(bend down)* sich [schnell] ducken; **b)** *(coll.: move hastily)* türmen *(ugs.).* **3.** *v. t.* **a) ~ sb. [in water]** jmdn. untertauchen; **b) ~ one's head** den Kopf einziehen

duck: ~-boards *n. pl.* Lattenrost, *der;* **~-egg** *n.* Entenei, *das*

duckie ['dʌkɪ] *see* **duck 1 b**

ducking ['dʌkɪŋ] *n.* [Ein-, Unter]tauchen, *das;* **give sb. a ~:** jmdn. untertauchen

duckling ['dʌklɪŋ] *n.* Entenküken, *das; (as food)* junge Ente

'duck-pond *n.* Ententeich, *der*

duct [dʌkt] *n. (for fluid, gas, cable)* [Rohr]leitung, *die;* Rohr, *das; (for air)* Ventil, *das*

dud [dʌd] *(coll.)* **1.** *n.* **a)** *(useless thing)* Niete, *die (ugs.); (counterfeit)* Fälschung, *die; (banknote)* Blüte, *die (ugs.);* **this battery/ball-point is a ~:** diese Batterie/dieser Kugelschreiber taugt nichts; **b)** *(bomb etc.)* Blindgänger, *der.* **2.** *adj.* **a)** mies *(ugs.);* schlecht; *(fake)* gefälscht; **a ~ banknote** eine Blüte *(ugs.);* **b) a ~ bullet/shell/bomb** ein Blindgänger

dude [du:d] *n. (Amer. sl.)* feiner Pinkel aus der Stadt *(ugs.)*

dudgeon ['dʌdʒn] *n.* **in high ~:** äußerst empört

due [dju:] **1.** *adj.* **a)** *(owed)* geschuldet; zustehend ‹*Eigentum, Recht usw.*›; **the share/reward ~ to him** der Anteil, der/die Belohnung, die ihm zusteht; **the amount ~:** der zu zahlende Betrag; **there's sth. ~ to me, I've got sth. ~, I'm ~ for sth.** mir steht etw. zu; **b)** *(immediately payable, lit. or fig.)* fällig; **be more than ~** *(fig.)* überfällig sein; **c)** *(that it is proper to give, use)* gebührend; geziemend *(geh.);* angemessen ‹*Belohnung*›; reiflich ‹*Überle-*

gung⟩; **be ~ to sb.** jmdm. gebühren; **recognition ~ to sb.** Anerkennung, die jmdm. gebührt; **with all ~ respect, madam** bei allem gebotenen Respekt, meine Dame; **with ~ allowance** *or* **regard** unter gebührender Berücksichtigung (**for** *Gen.*); **with ~ caution/care** mit der nötigen Vorsicht/Sorgfalt; **they were given ~ warning** sie wurden hinreichend gewarnt; **in ~ time** rechtzeitig; **d)** *(attributable)* **~ to negligence** auf Grund von Nachlässigkeit; **the mistake was ~ to negligence** der Fehler war durch Nachlässigkeit verursacht; **it's ~ to her that we missed the train** ihretwegen verpaßten wir den Zug; **be ~ to the fact that ...:** darauf zurückzuführen sein, daß ...; **e)** *(scheduled, expected, under instructions)* **be ~ to do sth.** etw. tun sollen; **I'm ~** *(my plan is)* **to leave tomorrow** ich werde morgen abfahren; **be ~ [to arrive]** ankommen sollen; **the train is now ~:** der Zug müßte jetzt planmäßig ankommen; **when are we ~ to land?** wann landen wir?; **the baby is ~ in two weeks' time** das Baby kommt in zwei Wochen; **f)** *(likely to get, deserving)* **be ~ for sth.** etw. verdienen; **he is ~ for promotion** seine Beförderung ist fällig. **2.** *adv.* **a) ~ north** genau nach Norden; **b) ~ to** auf Grund (+ *Gen.*); aufgrund (+ *Gen.*). **3.** *n.* **a)** *in pl. (debt)* Schulden *Pl.*; **pay one's ~s** seine Schulden bezahlen; **b)** *no pl. (fig.: just deserts, reward)* **sb.'s ~:** das, was jmdm. zusteht; **that was no more than his ~:** das hatte er auch verdient; **give sb. his ~:** jmdm. Gerechtigkeit widerfahren lassen; **c)** *usu. in pl. (fee)* Gebühr, *die;* **membership ~s** Mitgliedsbeiträge *Pl.*

duel ['dju:əl] **1.** *n.* **a)** Duell, *das; (Univ.)* Mensur, *die;* **fight a ~:** ein Duell/eine Mensur austragen; **b)** *(fig.: contest)* Kampf, *der;* **~ of wits** geistiger Wettstreit. **2.** *v.i., (Brit.)* **-ll-** sich duellieren; *(Univ.)* eine Mensur austragen *od.* schlagen

duet [dju:'et] *n. (Mus.) (for voices)* Duett, *das; (instrumental)* Duo, *das*

duffle ['dʌfl]: **~ bag** *n.* Matchbeutel, *der; (waterproof, also)* Seesack, *der;* **~ coat** *n.* Dufflecoat, *der*

dug *see* **dig 1, 2**

'**dug-out** *n.* **a)** *(canoe)* Einbaum, *der;* **b)** *(Mil.)* Unterstand, *der*

duke [dju:k] *n.* Herzog, *der*

dukedom ['dju:kdəm] *n.* **a)** *(territory)* Herzogtum, *das;* **b)** *(rank)* Herzogwürde, *die*

dulcimer ['dʌlsɪmə(r)] *n. (Mus.)* Hackbrett, *das*

dull [dʌl] **1.** *adj.* **a)** *(stupid)* beschränkt; *(slow to understand)* begriffsstutzig *(abwertend);* **b)** *(boring)* langweilig; stumpfsinnig ⟨*Arbeit, Routine*⟩; **c)** *(gloomy)* trübe ⟨*Wetter, Tag*⟩; **d)** *(not bright)* matt, stumpf ⟨*Farbe, Glanz, Licht, Metall*⟩; trübe ⟨*Augen*⟩; blind ⟨*Spiegel*⟩; *(not sharp)* dumpf ⟨*Geräusch, Aufprall, Schmerz, Gefühl*⟩; **e)** *(listless)* lustlos; **f)** *(blunt)* stumpf. **2.** *v.t.* **a)** *(make less acute)* schwächen; trüben; betäuben ⟨*Schmerz*⟩; **b)** *(make less bright or sharp)* stumpf werden lassen; verblassen lassen ⟨*Farbe*⟩; **c)** *(blunt)* stumpf machen; **d)** *(fig.)* dämpfen ⟨*Freude, Enthusiasmus*⟩; abstumpfen ⟨*Geist, Sinne, Verstand, Vorstellungskraft*⟩

dullness ['dʌlnɪs] *n., no pl.* **a)** *(stupidity)* Beschränktheit, *die; (slowwittedness)* Begriffsstutzigkeit, *die (abwertend);* [geistige] Trägheit; **b)** *(boringness)* Langweiligkeit, *die; (of work, life, routine)* Stumpfsinn, *der;* **c)** *(of colour, light, metal)* Stumpfheit, *die;* Mattheit, *die*

dull-witted [dʌl'wɪtɪd] *see* **dull 1a**

duly ['dju:lɪ] *adv.* **a)** *(rightly, properly)* ordnungsgemäß; **b)** *(sufficiently)* ausreichend; hinreichend

dumb [dʌm] **1.** *adj.*, **~er** ['dʌmə(r)], **~est** ['dʌmɪst] **a)** stumm; **a ~ person** ein Stummer/eine Stumme; **~ animals** *or* **creatures** die Tiere; die stumme Kreatur *(dichter.)*; **he was [struck] ~ with amazement** vor Staunen verschlug es ihm die Sprache; **b)** *(coll.: stupid)* doof *(ugs.)*; **act ~:** sich dumm stellen *(ugs.)*; **a ~ blonde** eine dümmliche Blondine *(ugs.)*. **2.** *n. pl.* **the ~:** die Stummen; **the deaf and ~:** die Taubstummen

'**dumb-bell** *n.* Hantel, *die*

dumbfound [dʌm'faʊnd] *v.t.* sprachlos machen; verblüffen

dumbfounded [dʌm'faʊndɪd] *adj.* sprachlos; verblüfft

dumb: ~ show *n.* **in ~ show** durch Mimik; **~ 'waiter** *n.* **a)** *(trolley)* stummer Diener; **b)** *(lift)* Speisenaufzug, *der*

dummy ['dʌmɪ] **1.** *n.* **a)** *(of tailor)* Schneiderpuppe, *die; (in shop)* Modepuppe, *die;* Schaufensterpuppe, *die; (of ventriloquist)* Puppe, *die; (stupid person)* Dummkopf, *der (ugs.)*; Doofi, *der (ugs.)*; **like a stuffed ~:** wie ein Ölgötze *(ugs.)*; **b)** *(imitation)* Attrappe, *die;* Dummy, *der; (Commerc.)* Schaupackung, *die;* **c)** *(esp. Brit.: for baby)* Schnuller, *der.* **2.** *attrib. adj.* unecht; blind ⟨*Tür, Fenster*⟩; Übungs- *(Mil.)*; **~ gun** Gewehrattrappe, *die;* **~ run** Probelauf, *der*

dump [dʌmp] **1.** *n.* **a)** *(place)* Müllkippe, *die; (heap)* Müllhaufen, *der; (permanent)* Müllhalde, *die;* **b)** *(Mil.)* Depot, *das;* Lager, *das;* **c)** *(coll. derog.: unpleasant place)* Dreckloch, *das (salopp abwertend); (boring town)* Kaff, *das (ugs. abwertend).* **2.** *v.t.* **a)** *(dispose of)* werfen; *(deposit)* abladen, kippen ⟨*Sand, Müll usw.*⟩; *(leave)* lassen; *(place)* abstellen; **b)** *(Commerc.: send abroad)* zu Dumpingpreisen verkaufen; **c)** *(fig. coll.: abandon)* abladen *(ugs.)*

dumpling ['dʌmplɪŋ] *n.* **a)** *(Gastr.)* Kloß, *der;* **apple ~:** Apfel im Schlafrock; **b)** *(coll.: short, plump person)* Tönnchen, *das (ugs.)*

dumps [dʌmps] *n. pl. (coll.)* **be** *or* **feel [down] in the ~:** ganz down sein *(ugs.)*

'**dump truck** *n.* Kipper, *der*

dumpy ['dʌmpɪ] *adj.* pummelig *(ugs.)*

dun [dʌn] **1.** *adj.* graubraun. **2.** *n.* Graubraun, *das*

dunce [dʌns] *n.* Niete, *die (ugs. abwertend);* **the ~ of the class** das Schlußlicht der Klasse *(ugs.)*; **~'s cap** *(Hist.)* Spotthut, *der (für schlechte Schüler)*

dune [dju:n] *n.* Düne, *die*

dung [dʌŋ] *n.* Dung, *der;* Mist, *der*

dungarees [dʌŋgə'ri:z] *n. pl.* Latzhose, *die;* **a pair of ~:** eine Latzhose

dungeon ['dʌndʒn] *n.* Kerker, *der;* Verlies, *das*

'**dunghill** *n.* Misthaufen, *der*

dunk [dʌŋk] *v.t.* tunken; stippen *(bes. nordd.)*

duo ['dju:əʊ] *n., pl.* **~s** Paar, *das;* **comedy ~:** Komikerpaar, *das*

duodenal [dju:ə'di:nl] *adj. (Anat.)* duodenal *(fachspr.)*; Zwölffingerdarm-

duodenum [dju:ə'di:nəm] *n. (Anat.)* Duodenum, *das (fachspr.)*; Zwölffingerdarm, *der*

dupe [dju:p] **1.** *v.t.* düpieren *(geh.)*; übertölpeln; **be ~d [into doing sth.]** sich übertölpeln lassen [und etw. tun]. **2.** *n.* Dumme, *der/*

die; Gelackmeierte, *der/die (salopp scherzh.)*

duplex ['dju:pleks] *adj. (esp. Amer.) (two-storey)* zweistöckig ‹*Wohnung*›; *(two-family)* Zweifamilien‹*haus*›

duplicate 1. ['dju:plıkət] *adj.* **a)** *(identical)* Zweit; ~ **key** Nach- *od.* Zweitschlüssel, *der;* ~ **copy** Zweit- *od.* Abschrift, *die;* Doppel, *das;* **b)** *(twofold)* doppelt. **2.** ['dju:plıkət] *n.* **a)** Kopie, *die; (second copy of letter/document/key)* Duplikat, *das;* **b) prepare/complete sth. in** ~: etw. in doppelter Ausfertigung machen/ausfüllen. **3.** ['dju:plıkeıt] *v.t.* **a)** *(make a copy of, make in* ~*)* ~ **sth.** eine zweite Anfertigung von etw. machen; etw. nachmachen *(ugs.);* **b)** *(be exact copy of)* genau gleichen (+ *Dat.*); **c)** *(on machine)* vervielfältigen; **d)** *(unnecessarily)* [unnötigerweise] noch einmal tun

duplication [dju:plı'keıʃn] *n.* Wiederholung, *die*

duplicator ['dju:plıkeıtə(r)] *n.* Vervielfältigungsgerät, *das*

duplicity [dju:'plısıtı] *n.* Falschheit, *die*

durability [djʊərə'bılıtı] *n., no pl.* **a)** *(of friendship, peace, etc.)* Dauerhaftigkeit, *die; (of person)* Unverwüstlichkeit, *die;* **b)** *(of garment, material)* Haltbarkeit, *die;* Strapazierfähigkeit, *die*

durable ['djʊərəbl] **1.** *adj.* **a)** dauerhaft ‹*Friede, Freundschaft usw.*›; **b)** *(resisting wear)* solide; strapazierfähig, haltbar ‹*Kleidung, Stoff*›; widerstandsfähig ‹*Metall, Bauelement*›; ~ **goods** *see* **2. 2.** *n. in pl.* **consumer** ~**s** langlebige Konsumgüter

duration [djʊə'reıʃn] *n.* Dauer, *die;* **be of short/long** ~: von kurzer/langer Dauer sein

duress [djʊə'res, 'djʊəres] *n., no pl.* Zwang, *der*

during ['djʊərıŋ] *prep.* während; *(at a point in)* in (+ *Dat.*); ~ **the night** während *od.* in der Nacht

dusk [dʌsk] *n.* [Abend]dämmerung, *die;* Einbruch der Dunkelheit; **at** ~: bei Einbruch der Dunkelheit

dusky ['dʌskı] *adj.* dunkelhäutig ‹*Person, Schönheit*›

dust [dʌst] **1.** *n., no pl.* Staub, *der. See also* **bite 1; raise 1 b. 2.** *v.t.* **a)** abstauben ‹*Möbel*›; ~ **a room/house** in einem Zimmer/Haus Staub wischen; **b)** *(sprinkle; also Cookery)* ~ **sth. with sth.** etw. mit etw. bestäuben; *(with talc etc.)* etw. mit etw. pudern. **3.** *v.i.* Staub wischen

dust: ~**bin** *n. (Brit.)* Mülltonne, *die;* Abfalltonne, *die;* ~**-cart** *n. (Brit.)* Müllwagen, *der;* ~**-cover** *n. (on record-player)* Abdeckhaube, *die; (on book) see* **dust-jacket**

duster ['dʌstə(r)] *n.* Staubtuch, *das*

dusting ['dʌstıŋ] *n. see* **dust 2 a:** Abstauben, *das;* Staubwischen, *das;* **give a room a** ~: in einem Zimmer Staub wischen

dust: ~**-jacket** *n.* Schutzumschlag, *der;* ~**man** ['dʌstmən] *n., pl.* ~**men** ['dʌstmən] *(Brit.)* Müllwerker, *der;* Müllmann, *der;* ~**pan** *n.* Kehrschaufel, *die;* ~**-sheet** *n.* Staubdecke, *die;* ~**-trap** *n.* Staubfänger, *der (abwertend);* ~**-up** *n. (coll.)* Krach, *der (ugs.)*

dusty ['dʌstı] *adj.* staubig ‹*Straße, Stadt, Zimmer*›; verstaubt ‹*Bücher, Möbel*›

Dutch [dʌtʃ] **1.** *adj.* **a)** holländisch; niederländisch; **sb. is** ~: jmd. ist Holländer/Holländerin; **b) go** ~ **[with sb.] [on sth.]** *(coll.)* getrennte Kasse [mit jmdm.] [bei etw.] machen. **2.** *n.* **a)** *constr. as pl.* **the** ~: die Holländer *od.* Niederländer; **b)** *(language)* Holländisch, *das;* Niederländisch, *das;* **c) it was all double** ~ **to him** das waren alles böhmische Dörfer für ihn. *See also* **English 2 a**

Dutch: ~ **'auction** *see* **auction 1 a;** ~ **'barn** *n.* offene Scheune; ~ **'courage** *n.* angetrunkener Mut; **give oneself** *or* **get** ~ **courage** sich *(Dat.)* Mut antrinken; ~ **'elm disease** *n. (Bot.)* Ulmensterben, *das;* ~**man** ['dʌtʃmən] *n., pl.* ~**men** ['dʌtʃmən] **a)** Holländer, *der;* Niederländer, *der;* **b)** *(fig. coll.)* **or I'm a** ~**man** oder ich will Emil heißen *(ugs.);* ~**woman** *n.* Holländerin, *die;* Niederländerin, *die*

dutiable ['dju:tıəbl] *adj. (Customs)* zollpflichtig; abgabenpflichtig

dutiful ['dju:tıfl] *adj.* pflichtbewußt ‹*Ehefrau, Arbeiter, Bürger*›

duty ['dju:tı] *n.* **a)** *no pl. (obligation)* Pflicht, *die;* Verpflichtung, *die;* ~ **calls** die Pflicht ruft; **have a** ~ **to do sth.** die Pflicht haben, etw. zu tun; **do one's** ~ **[by sb.]** [jmdm. gegenüber] seine Pflicht [und Schuldigkeit] tun; **b)** *(specific task, esp. professional)* Aufgabe, *die;* Pflicht, *die;* **take up one's duties** seinen Dienst antreten; **your duties will consist of ...:** zu Ihren Aufgaben gehören ...; **the** ~**-nurse** die diensthabende Schwester; **on** ~: im Dienst; **be on** ~: Dienst haben; **go/come on** ~ **at 7 p.m.** um 19 Uhr seinen Dienst antreten; **off** ~: nicht im Dienst; **be off** ~: keinen Dienst haben; ‹*ab ... Uhr*› dienstfrei sein; **go/come off** ~ **at 8 a.m.** seinen Dienst um 8 Uhr beenden; **c)** *(Econ.: tax)* Zoll, *der;* **pay** ~ **on sth.** Zoll für etw. bezahlen; etw. verzollen

duty: ~**-bound** *adj.* **be/feel [oneself]** ~**-bound to do sth.** verpflichtet sein/sich verpflichtet fühlen, etw. zu tun; ~**-free** *adj.* zollfrei ‹*Ware, Preis*›; ~**-frees** *n. pl. (coll.)* zollfreie Waren; ~**-free 'shop** *n.* Duty-free-Shop, *der;* ~ **officer** *n. (Mil.)* Offizier vom Dienst

duvet ['du:veı] *n.* Federbett, *das*

dwarf [dwɔ:f] **1.** *n., pl.* ~**s** *or* **dwarves** [dwɔ:vz] **a)** *(person)* Liliputaner, *der*/Liliputanerin, *die;* Zwerg, *der*/Zwergin, *die (auch abwertend);* **b)** *(Mythol.)* Zwerg, *der*/Zwergin, *die.* **2.** *adj.* Zwerg‹*baum, -stern*›. **3.** *v.t.* **a)** *(cause to look small)* klein erscheinen lassen; **b)** *(fig.)* in den Schatten stellen

dwell [dwel] *v.i.,* **dwelt** [dwelt] *(literary)* wohnen; weilen *(geh.)*

~ **[up]on** *v.t. (in discussion)* sich länger *od.* ausführlich befassen mit; *(in thought)* in Gedanken verweilen bei

dwelling ['dwelıŋ] *n. (Admin. lang./literary)* Wohnung, *die*

'dwelling-place *n.* Wohnsitz, *der*

dwelt *see* **dwell**

dwindle ['dwındl] *v.i.* ~ **[away]** abnehmen; ‹*Unterstützung, Interesse:*› nachlassen; ‹*Vorräte, Handel, Hoheitsgebiet:*› schrumpfen; ‹*Macht, Einfluß, Tageslicht:*› schwinden *(geh.);* ~ **away to nothing** dahinschwinden

dye [daı] **1.** *n.* **a)** *(substance)* Färbemittel, *das;* **b)** *(colour)* Farbe, *die.* **2.** *v.t.,* ~**ing** ['daııŋ] färben; ~**d-in-the-wool** eingefleischt, *(ugs.)* in der Wolle gefärbt ‹*Konservative, Reaktionär usw.*›

'dyestuff *n.* Färbermittel, *das*

dying ['daııŋ] **1.** *adj.* **a)** sterbend ‹*Person, Tier*›; eingehend ‹*Pflanze*›; absterbend ‹*Baum*›; aussterbend ‹*Kunst, Kultur, Tradition, [Tier]art, Menschenschlag*›; zuendegehend ‹*Jahr*›; **b)** *(related to time of death)* letzt...; **to my** ~ **day** bis an mein Lebensende. **2.** *n. pl.* **the** ~: die Sterbenden. *See also* **'die**

dyke *see* **dike**

dynamic [daı'næmık] *adj.,*

dynamically [daɪ'næmɪkəlɪ] *adv. (lit. or fig.; also Mus.)* dynamisch
dynamism ['daɪnəmɪzm] *n.* Dynamik, *die*
dynamite ['daɪnəmaɪt] **1.** *n.* **a)** Dynamit, *das;* **b)** *(fig.: politically dangerous thing)* Sprengstoff, *der;* **c)** *(fig.: sensational person or thing)* **be ~** ‹*Person:*› eine Wucht sein *(salopp);* ‹*Sache:*› eine Sensation sein. **2.** *v. t.* mit Dynamit sprengen
dynamo ['daɪnəməʊ] *n., pl.* **~s** Dynamomaschine, *die; (of car)* Lichtmaschine, *die; (of bicycle)* Dynamo, *der*
dynasty ['dɪnəstɪ] *n. (lit. or fig.)* Dynastie, *die*
dysentery ['dɪsəntərɪ] *n. (Med.)* Ruhr, *die;* Dysenterie, *die (fachspr.)*
dyslexia [dɪs'leksɪə] *n. (Med., Psych.)* Dyslexie, *die (fachspr.);* Lesestörung, *die*
dyslexic [dɪs'leksɪk] *(Med., Psych.)* **1.** *adj.* dyslektisch *(fachspr.);* **a ~ child** ein Kind mit einer Lesestörung. **2.** *n.* Dyslektiker, *der*/Dyslektikerin, *die (fachspr.);* Mensch mit einer Lesestörung
dyspepsia [dɪs'pepsɪə] *n. (Med.)* Dyspepsie, *die (fachspr.);* Verdauungsstörung, *die*

E

E, e [i:] *n., pl.* **Es** *or* **E's a)** *(letter)* E, e, *das;* **b) E** *(Mus.)* E, e, *das;* **E flat** es, Es, *das*
E. *abbr.* **a) east** O; **b) eastern** ö.
each [i:tʃ] **1.** *adj.* jeder/jede/jedes; **they cost** *or* **are a pound ~:** sie kosten ein Pfund pro Stück *od.* je[weils] ein Pfund; **they ~ have ...:** sie haben jeder ...; jeder von ihnen hat ...; **books at £1 ~:** Bücher zu je einem Pfund *od.* für je ein Pfund; **two teams with 10 players ~:** zwei Mannschaften mit je 10 Spielern; **I gave them a book ~** *or* **~ a book** ich habe jedem von ihnen ein Buch *od.* ihnen je ein Buch gegeben; **~ one of them** jeder/jede/jedes einzelne von ihnen. **2.** *pron.* **a)** jeder/jede/jedes; **have some of ~:** von jedem etwas nehmen/haben *usw.;* **b) ~ other** sich [gegenseitig]; **they are cross with ~ other** sie sind böse aufeinander; **they wore ~ other's hats** jeder trug den Hut des anderen; **be in love with ~ other** ineinander verliebt sein; **live next door to ~ other** Tür an Tür wohnen
eager ['i:gə(r)] *adj.* eifrig; **be ~ to do sth.** etw. unbedingt tun wollen; **be ~ for sth.** etw. unbedingt haben wollen
eagerly ['i:gəlɪ] *adv.* eifrig ‹*ja sagen, zustimmen*›; gespannt, ungeduldig ‹*warten, aufblicken*›; **look forward ~ to sth.** sich sehr auf etw. *(Akk.)* freuen
eagerness ['i:gənɪs] *n., no pl.* Eifer, *der;* **~ to learn** Lerneifer, *der;* Lernbegier[de], *die;* **~ to succeed** Erfolgshunger, *der*
eagle ['i:gl] *n.* Adler, *der*
'eagle-eyed *adj.* adleräugig
[1]**ear** [ɪə(r)] *n.* **a)** Ohr, *das;* **~, nose, and throat hospital/specialist** Hals-Nasen-Ohren-Klinik, *die*/-Arzt, *der*/-Ärztin, *die;* **smile from ~ to ~:** von einem Ohr zum anderen strahlen *(ugs.);* **be out on one's ~** *(fig. coll.)* auf der Straße stehen *(ugs.);* **up to one's ~s in work/debt** bis zum Hals in Arbeit/Schulden; **have a word in sb.'s ~:** jmdm. ein Wort im Vertrauen sagen; **keep one's ~s open** *(fig.)* die Ohren offenhalten; **have/keep an ~ to the ground** sein Ohr ständig am Puls der Masse haben *(ugs. scherzh.);* **be[come] all ~s** [plötzlich] ganz Ohr sein; **go in [at] one ~ and out [at] the other** *(coll.)* zum einen Ohr herein, zum anderen wieder hinausgehen; **b)** *no pl. (sense)* Gehör, *das;* **have an ~** *or* **a good ~/no ~ for music** ein [gutes]/kein Gehör für Musik haben; **play by ~** *(Mus.)* nach dem Gehör spielen; *see also* **play 3a**
[2]**ear** *n. (Bot.)* Ähre, *die;* **~ of corn** Kornähre, *die*
ear: ~ache *n. (Med.)* Ohrenschmerzen *Pl.;* **~-drum** *n. (Anat.)* Trommelfell, *das*
earl [3:l] *n.* Graf, *der*
early ['3:lɪ] **1.** *adj.* früh; **I am a bit ~:** ich bin etwas zu früh gekommen *od. (ugs.)* dran; **the train was 10 minutes ~:** der Zug kam 10 Minuten zu früh; **have an ~ night** früh ins Bett gehen; **~ riser** Frühaufsteher, *der*/-aufsteherin, *die;* **at the earliest** frühestens; **in the ~ afternoon/evening** am frühen Nachmittag/Abend; **into the ~ hours** bis in die frühen Morgenstunden; **at/from an ~ age** in jungen Jahren/von klein auf; **at an ~ stage, in its ~ stages** im Frühstadium; **an ~ work of an author** ein Frühwerk eines Autors. **2.** *adv.* früh; **~ next week** Anfang der nächsten Woche; **~ next Wednesday** nächsten Mittwoch früh; **~ in June** Anfang Juni; **as ~ as tomorrow** schon *od.* bereits morgen; **the earliest I can come is Friday** ich kann frühestens Freitag kommen; **~ on** schon früh; **earlier on this week/year** früher in der Woche/im Jahr
early: ~ bird *n. (joc.) jmd., der etw. frühzeitig tut; (getting up)* Frühaufsteher, *der*/-aufsteherin, *die;* **the ~ bird catches the worm** *(prov.)* Morgenstunde hat Gold im Munde *(Spr.);* **~ 'closing** *n.* **it is ~ closing** die Geschäfte haben nachmittags geschlossen; **~-'closing day** *n. Tag, an dem die Geschäfte nachmittags geschlossen haben;* **~-'warning** *attrib. adj.* Frühwarn-
ear: ~mark *v. t. (fig.)* vorsehen; **~-muffs** *n. pl.* Ohrenschützer *Pl.*
earn [3:n] *v. t.* **a)** ‹*Person, Tat, Benehmen:*› verdienen; **~ed income** Einkommen aus Arbeit; **it ~ed him much respect** es trug ihm viel Respekt ein; **b)** *(bring in as income or interest)* einbringen; **c)** *(incur)* eintragen; einbringen
earnest ['3:nɪst] **1.** *adj.* **a)** *(serious)* ernsthaft; **b)** *(ardent)* innig ‹*Wunsch, Gebet, Hoffnung*›; leidenschaftlich ‹*Appell*›. **2.** *n.* **in ~:** mit vollem Ernst; **this time I'm in ~ [about it]** diesmal ist es mir Ernst *od.* meine ich es ernst [damit]
earnestly ['3:nɪstlɪ] *adv.* ernsthaft
earnings ['3:nɪŋz] *n. pl.* Verdienst, *der; (of business etc.)* Ertrag, *der*
ear: ~phones *n. pl.* Kopfhörer, *der;* **~-plug** *n.* Ohropax, *das*Ⓦᴢ; **~-ring** *n.* Ohrring, *der;* **~shot** *n.* **out of/within ~shot** außer/in Hörweite; **~-splitting** *adj.* ohrenbetäubend
earth [3:θ] **1.** *n.* **a)** *(land, soil)* Erde, *die; (ground)* Boden, *der;* **be brought/come down** *or* **back to ~ [with a bump]** *(fig.)* [schnell] wieder auf den Boden der Tatsachen zurückgeholt werden/zurückkommen; **b)** *or* **E~** *(planet)* Erde, *die;* **c)** *(world)* Erde, *die;* **on ~** *(existing anywhere)* auf der Welt; **nothing on ~ will stop me** keine Macht der Welt kann mich auf-

halten; **how/what** *etc.* **on ~ ...?** wie/was *usw.* in aller Welt ...?; **who on ~ is that?** wer ist das bloß?; **what on ~ do you mean?** was meinst du denn nur?; **where on ~ has she got to?** wo ist sie denn bloß hingegangen?; **look like nothing on ~** *(be unrecognizable)* nicht zu erkennen sein; *(look repellent)* furchtbar aussehen; **d)** *(of animal)* Bau, *der;* **have gone to ~** *(fig.)* untergetaucht sein; **run to ~** *(fig.)* aufspüren; **e)** *(coll.)* **charge/cost/pay the ~:** ein Vermögen *od. (ugs.)* eine ganze Stange Geld verlangen/kosten/bezahlen; **f)** *(Brit. Electr.)* Erde, *die;* Erdung, *die.* **2.** *v. t. (Brit. Electr.)* erden

~ 'up *v. t.* mit Erde bedecken

earthenware ['ɜ:θnweə(r)] **1.** *n., no pl. (pots etc.)* Tonwaren *Pl.* **2.** *adj.* Ton-; tönern

earthly ['ɜ:θlɪ] *adj.* irdisch; **no ~ use** *etc. (coll.)* nicht der geringste Nutzen *usw.*

earth: ~-moving *adj.* **~-moving vehicle** Fahrzeug für Erdarbeiten; **~quake** *n.* Erdbeben, *das;* **~-shaking, ~-shattering** *adjs. (fig.)* weltbewegend; **~worm** *n.* Regenwurm, *der*

earthy ['ɜ:θɪ] *adj.* **a)** erdig; **b)** derb ⟨*Person*⟩

earwig ['ɪəwɪg] *n.* Ohrwurm, *der*

ease [i:z] **1.** *n.* **a)** *(freedom from pain or trouble)* Ruhe, *die;* **b)** *(leisure)* Muße, *die; (idleness)* Müßiggang, *der;* **c)** *(freedom from constraint)* Entspanntheit, *die;* **at [one's] ~:** entspannt; behaglich; **be** *or* **feel at [one's] ~:** sich wohl fühlen; **put** *or* **set sb. at his ~:** jmdm. die Befangenheit nehmen; **d) with ~** *(without difficulty)* mit Leichtigkeit. **2.** *v. t.* **a)** *(relieve)* lindern ⟨*Schmerz, Kummer*⟩; *(make lighter, easier)* erleichtern ⟨*Last*⟩; entspannen ⟨*Lage*⟩; **b)** *(give mental ~ to)* erleichtern; **~ sb.'s mind** jmdn. beruhigen; **c)** *(relax, adjust)* lockern ⟨*Griff, Knoten*⟩; verringern ⟨*Druck, Spannung, Geschwindigkeit*⟩; **d)** *(cause to move)* behutsam bewegen; **~ the clutch in** die Kupplung langsam kommen lassen; **~ the cap off a bottle** eine Flasche vorsichtig öffnen. **3.** *v. i.* **a)** ⟨*Belastung, Druck, Wind, Sturm:*⟩ nachlassen; **b) ~ off** *or* **up** *(begin to take it easy)* sich entspannen; *(drive more slowly)* ein bißchen langsamer fahren

easel ['i:zl] *n.* Staffelei, *die*

easily ['i:zɪlɪ] *adv.* **a)** leicht; **b)** *(without doubt)* zweifelsohne; **it is ~ a hundred metres deep** es ist gut und gerne 100 m tief

easiness ['i:zɪnɪs] *n.* Leichtigkeit, *die*

east [i:st] **1.** *n.* **a)** *(direction)* Osten, *der;* **the ~:** Ost *(Met., Seew.);* **in/to[wards]/from the ~:** im/nach/von Osten; **to the ~ of** östlich von; östlich (+ *Gen.*); **b)** *usu.* **E~** *(also Polit.)* Osten, *der;* **from the E~:** aus dem Osten. **2.** *adj.* östlich; Ost⟨*küste, -wind, -grenze, -tor*⟩. **3.** *adv.* ostwärts; nach Osten; **~ of** östlich von; östlich (+ *Gen.*)

East: ~ 'Africa *pr. n.* Ostafrika *(das);* **~ Ber'lin** *pr. n. (Hist.)* Ost-Berlin *(das);* **e~bound** *adj.* ⟨*Zug, Verkehr usw.*⟩ in Richtung Osten; **~ 'End** *n. (Brit.)* Londoner Osten

Easter ['i:stə(r)] *n.* Ostern, *das od. Pl.;* **at ~:** [zu *od.* an] Ostern; **next/last ~:** nächste/letzte Ostern

Easter: ~ 'Day *n.* Ostersonntag, *der;* **~ egg** *n.* Osterei, *das*

easterly ['i:stəlɪ] **a)** *(in position or direction)* östlich; **in an ~ direction** nach Osten; **b)** *(from the east)* ⟨*Wind*⟩ aus östlichen Richtungen

eastern ['i:stən] *adj.* östlich; Ost⟨*grenze, -hälfte, -seite*⟩; **~ Germany** Ostdeutschland

Eastern 'Europe *pr. n.* Osteuropa *(das)*

easternmost ['i:stənməʊst] *adj.* östlichst...

Easter: ~ 'Sunday *see* **Easter Day**; **~ week** *n., no art.* Osterwoche, *die*

East: ~ 'German *(Hist.)* **1.** *adj.* ostdeutsch; **2.** *n.* Ostdeutsche, *der/die;* **~ 'Germany** *pr. n. (Hist.)* Ostdeutschland *(das)*

eastward ['i:stwəd] **1.** *adj.* nach Osten gerichtet; *(situated towards the east)* östlich; **in an ~ direction** nach Osten; [in] Richtung Osten. **2.** *adv.* ostwärts; **they are ~bound** sie fahren nach *od.* [in] Richtung Osten. **3.** *n.* Osten, *der*

eastwards [i:stwədʒ] *adv.* ostwärts

easy ['i:zɪ] **1.** *adj.* **a)** *(not difficult)* leicht; **~ to clean/see** *etc.* leicht zu reinigen/sehen *usw.*; **it is ~ to see that ...:** es ist offensichtlich, daß ...; man sieht sofort, daß ...; **it's as ~ as anything** *(coll.)* es ist kinderleicht; **it is ~ for him to talk** er hat leicht *od.* gut reden; **on ~ terms** auf Raten ⟨*kaufen*⟩; **b)** *(free from pain, anxiety, etc.)* sorglos, angenehm ⟨*Leben, Zeit*⟩; **make it** *or* **things ~ for sb.** es jmdm. leichtmachen; **c)** *(free from constraint, strictness, etc.)* ungezwungen; unbefangen ⟨*Art*⟩; **he is ~ to get on with/work with** mit ihm kann man gut auskommen / zusammenarbeiten; **I'm ~** *(coll.)* es ist mir egal. **2.** *adv.* leicht; **easier said than done** leichter gesagt als getan; **~ does it** immer langsam *od.* sachte; **go ~:** vorsichtig sein; **go ~ on** *or* **with** sparsam umgehen mit; **go ~ on** *or* **with sb.** mit jmdm. nachsichtig sein; **take it ~!** *(calm down!)* beruhige dich!; **take it** *or* **things** *or* **life ~:** sich nicht übernehmen

easy: ~ 'chair *n.* Sessel, *der;* **~-going** *adj. (calm, placid)* gelassen; *(lax)* nachlässig

eat [i:t] **1.** *v. t.,* **ate** [et, eɪt], **eaten** ['i:tn] **a)** ⟨*Person:*⟩ essen; ⟨*Tier:*⟩ fressen; **he won't ~ you!** *(fig.)* er wird dich schon nicht fressen *(ugs.);* **~ sb. out of house and home** jmdn. arm essen; **what's ~ing you?** *(coll.)* was hast du denn?; **~ one's words** seine Worte zurücknehmen; **b)** *(destroy, consume, make hole in)* fressen; **~ its way into/through sth.** sich in etw. *(Akk.)* hineinfressen/durch etw. hindurchfressen. **2.** *v. i.,* **ate, eaten a)** ⟨*Person:*⟩ essen; ⟨*Tier:*⟩ fressen; **b)** *(make a way by gnawing or corrosion)* **~ into** sich hineinfressen in (+ *Akk.*); **~ through** sich durchfressen durch

~ 'out *v. i.* essen gehen

~ 'up 1. *v. t.* **a)** *(consume)* ⟨*Person:*⟩ aufessen; ⟨*Tier:*⟩ auffressen; **b)** *(traverse rapidly)* **our car ~s up the miles** unser Auto frißt die Meilen nur so *(ugs.).* **2.** *v. i.* aufessen

eatable ['i:təbl] *adj.* genießbar; eßbar

eaten *see* **eat**

eater ['i:tə(r)] *n.* Esser, *der*/Esserin, *die;* **a big ~:** ein guter Esser

eating ['i:tɪŋ] *n.* Essen, *das;* **make good ~:** ein gutes Essen sein; **not for ~:** nicht zum Essen [geeignet]

'eating apple *n.* Eßapfel, *der*

eau-de-Cologne [əʊdəkə'ləʊn] *n.* Eau de Cologne, *das;* Kölnisch Wasser, *das*

eaves [i:vz] *n. pl.* Dachgesims, *das*

eaves: ~drop *v. i.* lauschen; **~drop on sth./sb.** etw./jmdn. belauschen; **~dropper** ['i:vsdrɒpə(r)] *n.* Lauscher, *der*/Lauscherin, *die*

ebb [eb] **1.** *n.* **a)** *(of tide)* Ebbe, *die;* **the tide is on the ~:** es ist Ebbe; **b)** *(decline, decay)* Niedergang, *der;* **their morale was at its lowest ~:** ihre Moral war auf dem Tiefpunkt angelangt; **the ~ and**

flow das Auf und Ab. **2.** *v. i.* **a)** *(flow back)* zurückgehen; **b)** *(recede, decline)* schwinden; ~ **away** dahinschwinden

'ebb-tide *n.* Ebbe, *die*

ebony ['ebənɪ] **1.** *n.* Ebenholz, *das.* **2.** *adj.* Ebenholz⟨*baum*⟩; ebenholzfarben ⟨*Haar, Haut*⟩; ~ **box** *etc.* Kiste *usw.* aus Ebenholz

ebullient [ɪ'bʌlɪənt, ɪ'bʊlɪənt] *adj.* überschwenglich; überschäumend ⟨*Temperament, Laune*⟩

EC *abbr.* **European Community** EG

eccentric [ɪk'sentrɪk] **1.** *adj.* exzentrisch. **2.** *n.* Exzentriker, *der*/Exzentrikerin, *die*

eccentricity [eksən'trɪsɪtɪ] *n.* Exzentrizität, *die*

ecclesiastical [ɪkli:zɪ'æstɪkl] *adj.* kirchlich; Kirchen⟨*gebäude, -amt, -jahr*⟩; ~ **music** geistliche Musik; Kirchenmusik, *die*

echelon ['eʃəlɒn] *see* **upper 1 b**

echo ['ekəʊ] **1.** *n., pl.* ~**es a)** Echo, *das;* **b)** *(fig.)* Anklang, *der* (**of** an + *Akk.*). **2.** *v. i.* **a)** ⟨*Ort:*⟩ hallen (**with** von); **it** ~**es in here** hier gibt es ein Echo; **b)** *(Geräusch:)* widerhallen. **3.** *v. t.* **a)** *(repeat)* zurückwerfen; **b)** *(repeat words of)* echoen; wiederholen; *(imitate words or opinions of)* widerspiegeln

éclair [eɪ'kleə(r)] *n.* Eclair, *das*

eclipse [ɪ'klɪps] **1.** *n. (Astron.)* Eklipse, *die (fachspr.);* Finsternis, *die;* ~ **of the sun, solar** ~: Sonnenfinsternis, *die;* ~ **of the moon, lunar** ~: Mondfinsternis, *die.* **2.** *v. t.* **a)** verfinstern ⟨*Sonne, Mond*⟩; **b)** *(fig.: outshine, surpass)* in den Schatten stellen

ecological [i:kə'lɒdʒɪkl] *adj.* ökologisch

ecology [i:'kɒlədʒɪ] *n.* Ökologie, *die*

economic [i:kə'nɒmɪk, ekə'nɒmɪk] *adj.* **a)** *(of economics)* Wirtschafts⟨*politik, -system, -modell*⟩; ökonomisch, wirtschaftlich ⟨*Entwicklung, Zusammenbruch*⟩; **b)** *(giving adequate return)* wirtschaftlich ⟨*Miete*⟩

economical [i:kə'nɒmɪkl, ekə'nɒmɪkl] *adj.* wirtschaftlich; ökonomisch; sparsam ⟨*Person*⟩; **be** ~ **with sth.** mit etw. haushalten; **the car is** ~ **to run** das Auto ist wirtschaftlich

economically [i:kə'nɒmɪkəlɪ, ekə'nɒmɪkəlɪ] *adv.* **a)** *(with reference to economics)* wirtschaftlich; **b)** *(not wastefully)* sparsam; **be** ~ **minded** wirtschaftlich denken

economics [i:kə'nɒmɪks, ekə'nɒmɪks] *n., no pl.* **a)** Wirtschaftswissenschaft, *die (meist Pl.);* [politische] Ökonomie; **b)** *(economic considerations)* wirtschaftlicher Aspekt

economise *see* **economize**

economist [ɪ'kɒnəmɪst] *n.* Wirtschaftswissenschaftler, *der*/-wissenschaftlerin, *die*

economize [ɪ'kɒnəmaɪz] *v. i.* sparen; ~ **on sth.** etw. sparen

economy [ɪ'kɒnəmɪ] *n.* **a)** *(frugality)* Sparsamkeit, *die; (of effort, motion)* Wirtschaftlichkeit, *die;* **b)** *(instance)* Einsparung, *die;* **make economies** zu Sparmaßnahmen greifen; **c)** *(of country etc.)* Wirtschaft, *die*

economy: ~ **class** *n.* Touristenklasse, *die;* ~ **size** *n.* Haushaltspackung, *die;* Sparpackung, *die;* **an** ~**-size packet of salt** eine Haushaltspackung Salz

ecstasy ['ekstəsɪ] *n.* Ekstase, *die;* Verzückung, *die;* **be in/go into ecstasies [over sth.]** in Ekstase [über etw. *(Akk.)*] sein/geraten

ecstatic [ɪk'stætɪk] *adj.,* **ecstatically** [ɪk'stætɪkəlɪ] *adv.* ekstatisch; verzückt

ECU, ecu ['eɪkju:] *abbr.* **European currency unit** Ecu, *der od. die*

Ecuador [ekwa'dɔ:(r)] *pr. n.* Ekuador *(das)*

ecumenical [i:kjʊ'menɪkl, ekjʊ'menɪkl] *adj. (Relig.)* ökumenisch

eczema ['eksɪmə] *n. (Med.)* Ekzem, *das (fachspr.);* Hautausschlag, *der*

eddy ['edɪ] *n.* **a)** *(whirlpool)* Strudel, *der;* **b)** *(of smoke etc.)* Wirbel, *der*

edge [edʒ] **1.** *n.* **a)** *(of knife, razor, weapon)* Schneide, *die;* **the knife has lost its** ~: das Messer ist stumpf geworden *od.* ist nicht mehr scharf; **take the** ~ **off sth.** etw. stumpf machen; *(fig.)* etw. abschwächen; **that took the** ~ **off our hunger** das nahm uns erst einmal den Hunger; **be on** ~ **[about sth.]** [wegen etw.] nervös *od.* gereizt sein; **set sb.'s teeth on** ~: jmdm. durch Mark und Bein gehen; **have the** ~ **[on sb./sth.]** *(coll.)* jmdm./einer Sache überlegen *od.* *(ugs.)* über sein; **b)** *(of solid, bed, brick, record, piece of cloth)* Kante, *die; (of dress)* Saum, *der;* ~ **of a table** Tischkante, *die;* **roll off the** ~ **of the table** vom Tisch hinunterrollen; **c)** *(boundary) (of sheet of paper, road, forest, desert, cliff)* Rand, *der; (of sea, lake, river)* Ufer, *das; (of estate)* Grenze, *die;* ~ **of the paper/road** Papier-/Straßenrand, *der;* **on the** ~ **of sth.** *(fig.)* am Rande einer Sache *(Gen.).* **2.** *v. i. (move cautiously)* sich schieben; ~ **along sth.** sich an etw. *(Dat.)* entlangschieben; ~ **away from sb./sth.** sich allmählich von jmdm./etw. entfernen; ~ **out of the room** sich aus dem Zimmer stehlen. **3.** *v. t.* **a)** *(furnish with border)* säumen ⟨*Straße, Platz*⟩; besetzen ⟨*Kleid, Hut*⟩; einfassen ⟨*Garten, Straße*⟩; **b)** *(push gradually)* [langsam] schieben; ~ **one's way through a crowd** sich [langsam] durch eine Menschenmenge schieben *od.* drängen

edgeways ['edʒweɪz], **edgewise** ['edʒwaɪz] *adv.* **a)** mit der Schmalseite voran; **stand sth.** ~: etw. hochkant stellen; **b)** *(fig.)* **I can't get a word in** ~! ich komme überhaupt nicht zu Wort!

edging ['edʒɪŋ] *n. (of dress)* Borte, *die; (of garden)* Einfassung, *die*

edgy ['edʒɪ] *adj.* nervös

edible ['edɪbl] *adj.* eßbar; genießbar

edict ['i:dɪkt] *n.* Erlaß, *der;* Edikt, *das (hist.)*

edifice ['edɪfɪs] *n.* Gebäude, *das*

edit ['edɪt] *v. t.* **a)** *(act as editor of)* herausgeben ⟨*Zeitung*⟩; **b)** *(prepare for publication)* redigieren ⟨*Buch, Artikel, Manuskript*⟩; **c)** *(prepare an edition of)* bearbeiten; ~ **the works of Homer** die Werke Homers neu herausgeben; **d)** schneiden, cutten, montieren ⟨*Film, Bandaufnahme*⟩

edition [ɪ'dɪʃn] *n.* **a)** *(form of work, one copy; also fig.)* Ausgabe, *die;* **paperback** ~: Taschenbuchausgabe, *die;* **first** ~: Erstausgabe, *die;* **b)** *(printing)* Auflage, *die;* **morning/evening** ~ **of a newspaper** Morgen-/Abendausgabe einer Zeitung

editor ['edɪtə(r)] *n.* **a)** *(who prepares the work of others)* Redakteur, *der*/Redakteurin, *die; (of particular work)* Bearbeiter, *der*/Bearbeiterin, *die; (scholarly)* Herausgeber, *der*/-geberin, *die;* **b)** *(of newspaper or periodical)* Herausgeber, *der*/-geberin, *die;* **sports/business** ~: Sport-/Wirtschaftsredakteur, *der*

editorial [edɪ'tɔ:rɪəl] **1.** *n.* Leitartikel, *der.* **2.** *adj. (of an editor)* redaktionell; Redaktions⟨*assistent, -angestellte*⟩; ~ **department** Redaktion, *die*

EDP *abbr.* **electronic data processing** EDV

educate ['edjʊkeɪt] *v. t.* **a)** *(bring up)* erziehen; **b)** *(provide school-*

ing for) **he was ~d at Eton and Cambridge** er hat seine Ausbildung in Eton und Cambridge erhalten; **c)** *(give intellectual and moral training to)* bilden; **~ oneself** sich [weiter]bilden; **d)** *(train)* schulen ⟨*Geist, Körper*⟩; [aus]bilden ⟨*Geschmack*⟩; **~ oneself to do sth.** sich dazu erziehen, etw. zu tun

educated ['edjʊkeɪtɪd] *adj.* gebildet; **make an ~ guess** eine wohlbegründete *od.* fundierte Vermutung anstellen

education [edjʊ'keɪʃn] *n.* *(instruction)* Erziehung, *die; (course of instruction)* Ausbildung, *die; (system)* Erziehungs[- und Ausbildungs]wesen, *das; (science)* Erziehungswissenschaften *Pl.;* Pädagogik, *die;* **~ is free** die Schulausbildung ist kostenlos; **receive a good ~:** eine gute Ausbildung genießen

educational [edjʊ'keɪʃənl] *adj.* pädagogisch; erzieherisch; Lehr⟨*film, -spiele, -anstalt*⟩; Erziehungs⟨*methoden, -arbeit*⟩

educationalist [edjʊ'keɪʃənəlɪst] *n.* Pädagoge, *der*/Pädagogin, *die;* Erziehungswissenschaftler, *der*/-wissenschaftlerin, *die*

educator ['edjʊkeɪtə(r)] *n.* Pädagoge, *der*/Pädagogin, *die;* Erzieher, *der*/Erzieherin, *die*

Edwardian [ed'wɔ:dɪən] **1.** *adj.* Edwardianisch. **2.** *n.* Edwardianer, *der*

EEC *abbr.* **European Economic Community** EWG

eel [i:l] *n.* Aal, *der*

eerie ['ɪərɪ] *adj.* unheimlich ⟨*Ort, Gebäude, Form*⟩; schaurig ⟨*Klang*⟩; schauerlich ⟨*Schrei*⟩

efface [ɪ'feɪs] *v. t.* **a)** *(rub out)* beseitigen ⟨*Inschrift*⟩; **b)** *(fig.: obliterate)* auslöschen; tilgen *(geh.)*

effect [ɪ'fekt] **1.** *n.* **a)** *(result)* Wirkung, *die* **(on** auf + *Akk.***)**; **her words had little ~ on him** ihre Worte erzielten bei ihm nur eine geringe Wirkung; **the ~s of sth. on sth.** die Auswirkungen einer Sache *(Gen.)* auf etw. *(Akk.);* die Folgen einer Sache *(Gen.)* für etw.; **with the ~ that ...:** mit der Folge *od.* dem Resultat, daß ...; **take ~:** wirken; die erwünschte Wirkung erzielen; **in ~:** in Wirklichkeit; praktisch; **b)** *no art. (impression)* Wirkung, *die;* Effekt, *der;* **solely** *or* **only for ~:** nur des Effekts wegen; aus reiner Effekthascherei *(abwertend);* **c)** *(meaning)* Inhalt, *der;* Sinn, *der;* **or words to that ~:** oder etwas in diesem Sinne; **we received a letter to the ~ that ...:** wir erhielten ein Schreiben des Inhalts, daß ...; **d)** *(operativeness)* Kraft, *die;* Gültigkeit, *die;* **be in ~:** gültig *od.* in Kraft sein; **come into ~:** gültig *od.* wirksam werden; ⟨*bes. Gesetz:*⟩ in Kraft treten; **put into ~:** in Kraft setzen ⟨*Gesetz*⟩; verwirklichen ⟨*Plan*⟩; **take ~:** in Kraft treten; **with ~ from Monday** mit Wirkung von Montag; **e)** *in pl. (property)* Vermögenswerte *Pl.;* Eigentum, *das;* **personal ~s** persönliches Eigentum; Privateigentum, *das;* **household ~s** Hausrat, *der.* **2.** *v. t.* durchführen; herbeiführen ⟨*Einigung*⟩; erzielen ⟨*Übereinstimmung, Übereinkommen*⟩; tätigen ⟨*Umsatz, Kauf*⟩; abschließen ⟨*Versicherung*⟩; leisten ⟨*Zahlung*⟩

effective [ɪ'fektɪv] *adj.* **a)** *(having an effect)* wirksam ⟨*Mittel*⟩; effektiv ⟨*Maßnahmen*⟩; **be ~** ⟨*Arzneimittel:*⟩ wirken; **b)** *(in operation)* gültig; **~ from/as of** mit Wirkung von; **the law is ~ as from 1 September** das Gesetz tritt ab 1. September in Kraft *od.* wird ab 1. September wirksam; **c)** *(powerful in effect)* überzeugend ⟨*Rede, Redner, Worte*⟩; **d)** *(striking)* wirkungsvoll; effektvoll; **e)** *(existing)* wirklich, tatsächlich ⟨*Hilfe*⟩; effektiv ⟨*Gewinn, Umsatz*⟩

effectively [ɪ'fektɪvlɪ] *adv.* *(in fact)* effektiv; *(with effect)* wirkungsvoll; effektvoll

effectiveness [ɪ'fektɪvnɪs] *n., no pl.* Wirksamkeit, *die;* Effektivität, *die*

effeminate [ɪ'femɪnət] *adj.* unmännlich; *(geh.)* effeminiert

effervesce [efə'ves] *v. i.* sprudeln

effervescence [efə'vesəns] *n., no pl.* Sprudeln, *das; (fig.)* Übersprudeln, *das;* Überschäumen, *das*

effervescent [efə'vesənt] *adj.* sprudelnd; *(fig.)* übersprudelnd, überschäumend ⟨*Freude, Verhalten*⟩; **~ tablets** Brausetabletten

effete [e'fi:t] *adj.* *(exhausted, worn out)* verbraucht; saft- und kraftlos ⟨*Person*⟩; überlebt ⟨*System*⟩

efficacious [efɪ'keɪʃəs] *adj.* wirksam ⟨*Methode, Mittel, Medizin*⟩

efficiency [ɪ'fɪʃənsɪ] *n.* **a)** *(of person)* Fähigkeit, *die;* Tüchtigkeit, *die; (of machine, factory, engine)* Leistungsfähigkeit, *die; (of organization, method)* Rationalität, *die;* Effizienz, *die (geh.);* **b)** *(Mech., Phys.)* Wirkungsgrad, *der*

efficient [ɪ'fɪʃənt] *adj.* effizient *(geh.);* fähig ⟨*Person*⟩; tüchtig ⟨*Arbeiter, Sekretärin*⟩; leistungsfähig ⟨*Maschine, Abteilung, Fabrik*⟩; rationell ⟨*Methode, Organisation*⟩

efficiently [ɪ'fɪʃəntlɪ] *adv.* einwandfrei; gut; effizient *(geh.)*

effigy ['efɪdʒɪ] *n.* Bildnis, *das;* **hang/burn sb. in ~:** jmdn. in effigie hängen/verbrennen *(geh.)*

effluent ['eflʊənt] *n.* Abwässer *Pl.*

effort ['efət] *n.* **a)** *(exertion)* Anstrengung, *die;* Mühe, *die;* **make an/every ~** *(physically)* sich anstrengen; *(mentally)* sich bemühen; **without [any] ~:** ohne Anstrengung; mühelos; **[a] waste of time and ~:** vergebliche Liebesmüh; **make every possible ~ to do sth.** jede nur mögliche Anstrengung machen, etw. zu tun; **he makes no ~ at all** er gibt sich überhaupt keine Mühe; **b)** *(attempt)* Versuch, *der;* **in an ~ to do sth.** beim Versuch, etw. zu tun; **make no ~ to be polite** sich *(Dat.)* nicht die Mühe machen, höflich zu sein; **c)** *(coll.: result)* Leistung, *die;* **that was a pretty poor ~:** das war ein ziemlich schwaches Bild *(ugs.);* **whose is this rather poor ~?** welcher Stümper hat das denn verbrochen? *(ugs.);* **the book was one of his first ~s** das Buch war einer seiner ersten Versuche

effortless ['efətlɪs] *adj.* mühelos; leicht; flüssig, leicht ⟨*Stil*⟩

effrontery [ɪ'frʌntərɪ] *n.* Dreistigkeit, *die;* **have the ~ to do sth.** die Stirn besitzen, etw. zu tun *(geh.)*

effusive [ɪ'fju:sɪv] *adj.* überschwenglich; exaltiert *(geh.)* ⟨*Person, Stil, Charakter*⟩

e.g. [i:'dʒi:] *abbr.* **for example** z. B.

[1]egg [eg] *n.* Ei, *das;* **a bad ~** *(fig. coll.: person)* eine üble Person; **have** *or* **put all one's ~s in one basket** *(fig. coll.)* alles auf eine Karte setzen; **as sure as ~s is** *or* **are ~s** *(coll.)* so sicher wie das Amen in der Kirche *(ugs.)*

[2]egg *v. t.* **~ sb. on [to do sth.]** jmdn. anstacheln[, etw. zu tun]

egg: ~-cup *n.* Eierbecher, *der;* **~ 'custard** *n.* Eierkrem, *die;* **~head** *n. (coll.)* Eierkopf, *der (abwertend);* **~-plant** *n.* Aubergine, *die;* **~-shaped** *adj.* eiförmig; **~shell** *n.* Eierschale, *die;* **~-spoon** *n.* Eierlöffel, *der;* **~-timer** *n.* Eieruhr, *die;* **~-whisk** *n.* Schneebesen, *der;* **~-white** *n.* Eiweiß, *das;* **~ yolk** *n.* Eigelb, *das;* Eidotter, *der od. das*

ego ['i:gəʊ] *n., pl.* **~s a)** *(Psych.)* Ego, *das; (Metaphys.)* Ich, *das;* **b)**

(self-esteem) Selbstbewußtsein, *das;* **inflated ~:** übersteigertes Selbstbewußtsein; **boost sb.'s ~:** jmds. Selbstbewußtsein stärken; jmdm. Auftrieb geben

egocentric [i:gəʊ'sentrɪk] *adj.* egozentrisch; ichbezogen

egoism ['i:gəʊɪzm] *n., no pl.* **a)** *(systematic selfishness)* Egoismus, *der;* Selbstsucht, *die (abwertend);* **b)** *(arrogance)* Selbstherrlichkeit, *die*

egoist ['i:gəʊɪst] *n.* Egoist, *der/* Egoistin, *die*

egotism ['i:gətɪzm] *n., no pl.* **a)** Egotismus, *der (fachspr.);* Ichbezogenheit, *die;* **b)** *(self-conceit)* Egotismus, *der;* Selbstgefälligkeit, *die*

egotist ['i:gətɪst] *n.* Egotist, *der/* Egotistin, *die (fachspr.); (self-centred person)* Egozentriker, *der/*Egozentrikerin, *die*

egotistic [i:gə'tɪstɪk], **egotistical** [i:gə'tɪstɪkl] *adj.* **a)** ichbezogen ⟨*Rede*⟩; **b)** selbstsüchtig, selbstgefällig *(abwertend)* ⟨*Person*⟩

Egypt ['i:dʒɪpt] *pr. n.* Ägypten *(das)*

Egyptian [ɪ'dʒɪpʃn] **1.** *adj.* ägyptisch; **sb. is ~:** jmd. ist Ägypter/ Ägypterin. **2.** *n. (person)* Ägypter, *der/*Ägypterin, *die*

eh [eɪ] *int. (coll.) expr. inquiry or surprise* wie?; wie bitte?; *inviting assent* nicht [wahr]?; *asking for sth. to be repeated or explained* was?; hä? *(salopp);* **wasn't that good, eh?** war das nicht gut?; **let's not have any more fuss, eh?** Schluß mit dem Theater, ja? *(ugs.)*

eider ['aɪdə(r)]: **~down** *n.* Daunenbett, *das;* Federbett, *das;* **~ duck** *n.* Eiderente, *die*

eight [eɪt] **1.** *adj.* acht; **at ~:** um acht; **it's ~ [o'clock]** es ist acht [Uhr]; **half past ~:** halb neun; **~ thirty** acht Uhr dreißig; **~ ten/ fifty** zehn nach acht/vor neun; *(esp. in timetable)* acht Uhr zehn/ fünfzig; **around ~, at about ~:** gegen acht [Uhr]; **half ~** *(coll.)* halb neun; **girl of ~:** Mädchen von acht Jahren; **~-year-old boy** achtjähriger Junge; **~-year-old** Achtjähriger, *der/*Achtjährige, *die;* **be ~ [years old]** acht [Jahre alt] sein; **at [the age of] ~, aged ~:** mit acht Jahren; im Alter von acht Jahren; **he won ~-six** er hat acht zu sechs gewonnen; **Book/Volume/ Part/Chapter E~:** Buch/Band/ Teil/Kapitel acht; achtes Buch/ achter Band/achter Teil/achtes Kapitel; **~-figure number** achtstellige Zahl; **~-page** achtseitig; **~-storey[ed] building** achtstöckiges *od.* achtgeschossiges Gebäude; **~-sided** achtseitig; **bet at ~ to one** acht zu eins wetten; **~ times** achtmal. **2.** *n.* **a)** *(number, symbol)* Acht, *die;* **the first/last ~:** die ersten/letzten acht; **there were ~ of us** wir waren [zu] acht; **come ~ at a time/in ~s** acht auf einmal/ zu je acht kommen; **stack the boxes in ~s** die Kisten zu achten stapeln; **the [number] ~ [bus]** die Buslinie Nr. 8; der Achter *(ugs.);* **b)** *(8-shaped figure)* **[figure of] ~:** Achter, *der (ugs.);* Acht, *die;* **c)** *(Cards)* **~ [of hearts/trumps]** [Herz-/Trumpf]acht, *die;* **d)** *(size)* **a size ~ dress** ein Kleid [in] Größe 8; **wear size ~ shoes** [Schuh]größe 8 haben *od.* tragen; **wear an ~, be size ~:** Größe 8 tragen *od.* haben; **e)** *(Rowing: crew)* Achtermannschaft, *die*

eighteen [eɪ'ti:n] **1.** *adj.* achtzehn; *see also* **eight 1. 2.** *n.* Achtzehn, *die;* **~ seventy** achtzehnhundertsiebzig; **in the ~ seventies** in den siebziger Jahren des neunzehnten Jahrhunderts; *see also* **eight 2 a, d**

eighteenth [eɪ'ti:nθ] **1.** *adj.* achtzehnt...; *see also* **eighth 1. 2.** *n. (fraction)* Achtzehntel, *das; see also* **eighth 2**

eighth [eɪtθ] **1.** *adj.* acht...; **be/ come ~:** achter sein/als achter ankommen; **~ largest** achtgrößt... **2.** *n. (in sequence)* achte, *der/die/das; (in rank)* Achte, *der/ die/das; (fraction)* Achtel, *das;* **be the ~ to do sth.** der/die/das achte sein, der/die/das etw. tut; **the ~ of May** der achte Mai; **the ~ [of the month]** der Achte [des Monats]

'eighth-note *n. (Amer. Mus.)* Achtelnote, die

eightieth ['eɪtɪɪθ] *adj.* achtzigst...; *see also* **eighth 1**

eighty ['eɪtɪ] **1.** *adj.* achtzig; *see also* **eight 1. 2.** *n.* Achtzig, *die;* **be in one's eighties** in den Achtzigern sein; **be in one's early/late eighties** Anfang/Ende Achtzig sein; **the eighties** *(years)* die achtziger Jahre; **the temperature will be rising [well] into the eighties** die Temperatur steigt auf [gut] über 80 Grad Fahrenheit; *see also* **eight 2 a**

eighty: ~-'first *etc. adj.* einundachtzigst... *usw.; see also* **eighth 1; ~-'one** *etc.* **1.** *adj.* einundachtzig *usw.; see also* **eight 1; 2.** *n.* Einundachtzig *usw., die; see also* **eight 2 a**

Eire ['eərə] *pr. n.* Irland, *das;* Eire, *das*

either ['aɪðə(r), 'i:ðə(r)] **1.** *adj.* **a)** *(each)* **at ~ end of the table** an beiden Enden des Tisches; **on ~ side of the road** auf beiden Seiten der Straße; **~ way** so oder so; **b)** *(one or other)* [irgend]ein ... [von beiden]; **take ~ one** nimm einen/eine/eins von [den] beiden. **2.** *pron.* **a)** *(each)* beide *Pl.;* **~ is possible** beides ist möglich; **I can't cope with ~:** ich kann mit keinem von beiden fertig werden; **I don't like ~ [of them]** ich mag beide nicht; **b)** *(one or other)* einer/eine/ ein[e]s [von beiden]; **~ of the buses** jeder der beiden Busse; beide Busse. **3.** *adv.* **a)** *(any more than the other)* auch [nicht]; **'I don't like that ~:** ich mag es auch nicht; **I don't like 'that ~:** auch das mag ich nicht; **b)** *(moreover, furthermore)* noch nicht einmal; **there was a time, and not so long ago ~:** früher, noch gar nicht einmal so lange her. **4.** *conj.* **~ ... or ...:** entweder ... oder ...; *(after negation)* weder ... noch ...

ejaculate [ɪ'dʒækjʊleɪt] **1.** *v.t.* ausstoßen ⟨*Fluch, Gebet*⟩. **2.** *v.i.* *(Physiol.)* ejakulieren

ejaculation [ɪdʒækjʊ'leɪʃn] *n.* **a)** *(cry)* Ausruf, *der;* **b)** *(Physiol.)* Ejakulation, *die;* Samenerguß, *der*

eject [ɪ'dʒekt] **1.** *v.t. (from hall, meeting)* hinauswerfen **(from** aus); *(from machine-gun)* auswerfen. **2.** *v.i.* sich hinauskatapultieren

ejection [ɪ'dʒekʃn] *n. (of intruder etc.)* Vertreibung, *die; (of heckler, drunk)* Hinauswurf, *der; (of empty cartridge)* Auswerfen, *das*

ejector seat [ɪ'dʒektə si:t] *n.* Schleudersitz, *der*

eke [i:k] *v.t.* **~ out** strecken ⟨*Vorräte, Essen, Einkommen*⟩; **~ out a living** *or* **an existence** sich *(Dat.)* seinen Lebensunterhalt [notdürftig *od.* mühsam] verdienen

elaborate 1. [ɪ'læbərət] *adj.* kompliziert; ausgefeilt ⟨*Stil*⟩; durchorganisiert ⟨*Studium, Forschung*⟩; kunstvoll [gearbeitet] ⟨*Arrangement, Verzierung, Kleidungsstück*⟩; üppig ⟨*Menü*⟩. **2.** [ɪ'læbəreɪt] *v.t.* weiter ausarbeiten; weiter ausführen ⟨*Arbeit, Plan, Thema*⟩. **3.** [ɪ'læbəreɪt] *v.i.* mehr ins Detail gehen; **could you ~ [on that]?** könnten Sie das näher ausführen?

elapse [ɪ'læps] *v.i.* ⟨*Zeit:*⟩ vergehen

elastic [ɪ'læstɪk] **1.** *adj.* **a)** ela-

stisch; **b)** *(fig.: flexible)* flexibel. **2.** *n.* *(~ band)* Gummiband, *das;* *(fabric)* elastisches Material
elasticated [ɪ'læstɪkeɪtɪd] *adj.* elastisch
elastic 'band *n.* Gummiband, *das*
elasticity [ɪlæs'tɪsɪtɪ] *n., no pl.* **a)** Elastizität, *die;* **b)** *(fig.: flexibility)* Flexibilität, *die*
elated [ɪ'leɪtɪd] *adj.* freudig erregt; **~ mood, ~ state of mind** Hochstimmung, *die;* **be** *or* **feel ~:** in Hochstimmung sein
elation [ɪ'leɪʃn] *n., no pl.* freudige Erregung
elbow ['elbəʊ] **1.** *n.* *(also of garment)* Ell[en]bogen, *der;* **at one's ~:** bei sich; in Reichweite. **2.** *v.t.* **~ one's way** sich mit den Ellenbogen einen Weg bahnen; sich drängeln *(ugs.);* **~ sb. aside** jmdn. mit den Ellenbogen zur Seite stoßen; **~ sb. out** *(fig.)* jmdn. hinausdrängeln
elbow: ~-grease *n., no pl.* *(joc.)* Muskelkraft, *die;* **~-room** *n.* *(lit. or fig.)* Ell[en]bogenfreiheit, *die;* *(fig.)* Spielraum, *der*
[1]**elder** ['eldə(r)] **1.** *attrib. adj.* älter... **2.** *n.* **a) our ~s and betters** die Älteren mit mehr Lebenserfahrung; **the village ~s** die Dorfältesten; **b)** *(official in Church)* [Kirchen]älteste, *der/die*
[2]**elder** *n.* *(Bot.)* Holunder, *der*
'elderberry *n.* Holunderbeere, *die*
elderly ['eldəlɪ] **1.** *adj.* älter; **my parents are both quite ~ now** meine Eltern sind beide inzwischen ziemlich alt geworden. **2.** *n. pl.* **the ~:** ältere Menschen
elder 'statesman *n.* Elder Statesman, *der (Politik)*
eldest ['eldɪst] *adj.* ältest...
elect [ɪ'lekt] **1.** *postpos. adj.* gewählt; **the President ~:** der gewählte *od.* designierte Präsident. **2.** *v.t.* **a)** wählen; **~ sb. chairman/MP** *etc.* jmdn. zum Vorsitzenden/Abgeordneten *usw.* wählen; **~ sb. to the Senate** jmdn. in den Senat wählen; **b)** *(choose)* **~ to do sth.** sich dafür entscheiden, etw. zu tun
election [ɪ'lekʃn] *n.* Wahl, *die;* **presidential ~s** *(Amer.)* Präsidentschaftswahlen *Pl.;* **general/local ~:** allgemeine/kommunale Wahlen; **~ as chairman** Wahl zum Vorsitzenden; **~ results** Wahlergebnisse *Pl.*
e'lection campaign *n.* Wahlkampf, *der*
electioneering [ɪlekʃə'nɪərɪŋ] *n., no pl.* Agitation, *die* (**for** für)
elector [ɪ'lektə(r)] *n.* Wähler, *der*/Wählerin, *die;* Wahlberechtigte, *der/die*
electoral [ɪ'lektərl] *adj.* Wahl‹*liste, -system, -bezirk, -berechtigung*›; Wähler‹*liste, -wille*›
electorate [ɪ'lektərət] *n.* Wähler *Pl.;* Wählerschaft, *die*
electric [ɪ'lektrɪk] *adj.* elektrisch ‹*Strom, Feld, Licht, Orgel usw.*›; Elektro‹*kabel, -motor, -herd, -kessel*›; *(fig.)* spannungsgeladen ‹*Atmosphäre*›; elektrisierend ‹*Wirkung*›
electrical [ɪ'lektrɪkl] *adj.* elektrisch ‹*Defekt, Kontakt*›; Elektro‹*abteilung, -handel, -geräte*›
electrical: ~ engi'neer *n.* Elektroingenieur, *der*/-ingenieurin, *die;* **~ engi'neering** *n.* Elektrotechnik, *die*
electrically [ɪ'lektrɪkəlɪ] *adv.* elektrisch; *(fig.)* [wie] elektrisiert
electric: ~ 'blanket *n.* Heizdecke, *die;* **~ 'chair** *n.* elektrischer Stuhl; **~ 'fire** *n.* [elektrischer] Heizofen; Heizstrahler, *der*
electrician [ɪlek'trɪʃn] *n.* Elektriker, *der*/Elektrikerin, *die*
electricity [ɪlek'trɪsɪtɪ] *n., no pl.* **a)** Elektrizität, *die;* **b)** *(supply)* Strom, *der;* **c)** *(fig.)* Spannung, *die*
electricity: ~ bill *n.* Stromrechnung, *die;* **~ meter** *n.* Stromzähler, *der*
electric 'shock *n.* Stromschlag, *der;* [elektrischer] Schlag
electrify [ɪ'lektrɪfaɪ] *v.t.* **a)** *(convert)* elektrifizieren ‹*Eisenbahnstrecke*›; **b)** *(fig.)* elektrisieren
electrocute [ɪ'lektrəkju:t] *v.t.* durch Stromschlag töten
electrocution [ɪlektrə'kju:ʃn] *n.* Tod durch Stromschlag
electrode [ɪ'lektrəʊd] *n.* Elektrode, *die*
electromag'netic *adj.* elektromagnetisch
electron [ɪ'lektrɒn] *n.* Elektron, *das*
electronic [ɪlek'trɒnɪk, elek'trɒnɪk] *adj.* elektronisch; Elektronen‹*uhr, -orgel, -blitz*›; **~ newsgathering** elektronische Berichterstattung
electronics [ɪlek'trɒnɪks, elek'trɒnɪks] *n., no pl.* Elektronik, *die*
electron 'microscope *n.* Elektronenmikroskop, *das*
e'lectroplate *v.t.* galvanisieren
elegance ['elɪgəns] *n., no pl.* Eleganz, *die*
elegant ['elɪgənt] *adj.,* **elegantly** ['elɪgəntlɪ] *adv.* elegant
element ['elɪmənt] *n.* **a)** *(component part)* Element, *das;* **an ~ of truth** ein Körnchen Wahrheit; **an ~ of chance/danger in sth.** eine gewisse Zufälligkeit/Gefahr bei etw.; **b)** *(Chem.)* Element, *das;* Grundstoff, *der;* **c)** *in pl.* *(weather)* Elemente *Pl.;* **d) be in one's ~** *(fig.)* in seinem Element sein; **e)** *(Electr.)* Heizelement, *das;* **f)** *in pl.* *(rudiments of learning)* Grundlagen *Pl.;* Elemente *Pl.*
elementary [elɪ'mentərɪ] *adj.* elementar; grundlegend ‹*Fakten, Wissen*›; schlicht ‹*Fabel, Stil*›; Grundschul‹*lehrer, -bildung*›; Grund‹*stufe, -kurs, -ausbildung, -rechnen, -kenntnisse*›; Ausgangs‹*text, -thema*›; Anfangs‹*stadium*›; **course in ~ German** Grundkurs in Deutsch
ele'mentary school *n.* Grundschule, *die*
elephant ['elɪfənt] *n.* Elefant, *der*
elevate ['elɪveɪt] *v.t.* [empor]heben ‹*Gerät, Gegenstand*›; aufrichten ‹*Blick, Geschützrohr*›
elevated ['elɪveɪtɪd] *adj.* **a)** *(raised)* gehoben ‹*Stellung*›; erhöht ‹*Lage, Plazierung*›; aufgeschüttet ‹*Damm, Straße*›; **b)** *(above ground level)* Hoch‹*bahn, -straße*›; **c)** *(formal, dignified)* gehoben ‹*Stil, Rede, Wortwahl*›
elevation [elɪ'veɪʃn] *n.* **a)** *(of mind, thought)* Erhebung, *die;* *(state)* Erhabenheit, *die;* **b)** *(height)* Höhe, *die;* **~ of the ground** Bodenerhebung, *die;* Anhöhe *die;* **c)** *(drawing, diagram)* Aufriß, *der*
elevator ['elɪveɪtə(r)] *n.* **a)** *(machine)* Förderwerk, *das;* Elevator, *der;* **b)** *(Amer.) see* **lift 3b**
eleven [ɪ'levn] **1.** *adj.* elf; *see also* **eight 1. 2.** *n.* Elf, *die; see also* **eight 2a, d**
elevenses [ɪ'levnzɪz] *n. sing. or pl.* *(Brit. coll.)* ≈ zweites Frühstück [gegen elf Uhr]
eleventh [ɪ'levnθ] **1.** *adj.* elft...; **at the ~ hour** im letzten Augenblick; in letzter Minute; *see also* **eighth 1. 2.** *n.* *(fraction)* Elftel, *das; see also* **eighth 2**
elf [elf] *n., pl.* **elves** [elvz] **a)** *(Mythol.)* Elf, *der*/Elfe, *die;* **b)** *(mischievous creature)* [boshafter] Schelm; Kobold, *der*
elicit [ɪ'lɪsɪt] *v.t.* entlocken ‹*Antwort, Auskunft, Wahrheit, Geheimnis*› (**from** *Dat.*); gewinnen ‹*Unterstützung*› (**amongst** bei)
eligibility [elɪdʒɪ'bɪlɪtɪ] *n., no pl.* *(fitness)* Qualifikation, *die;* *(for a job)* Eignung, *die;* *(entitlement)* Berechtigung, *die* (**for** zu)

eligible ['elɪdʒɪbl] *adj.* **be ~ for sth.** *(fit)* für etw. qualifiziert *od.* geeignet sein; *(entitled)* zu etw. berechtigt sein; **be ~ to do sth.** etw. tun dürfen; **an ~ bachelor** ein begehrter Junggeselle

eliminate [ɪ'lɪmɪneɪt] *v.t.* **a)** *(remove)* beseitigen ‹*Zweifel, Fehler, Gegner*›; ausschließen ‹*Möglichkeit*›; **b)** *(exclude)* ausschließen; **the team was ~d in the third round** die Mannschaft schied in der dritten Runde aus

elimination [ɪlɪmɪ'neɪʃn] *n.* **a)** *(removal)* Beseitigung, *die;* **process of ~:** Ausleseverfahren, *das;* **by a process of ~:** durch Eliminierung; **b)** *(exclusion)* Ausschluß, *der; (Sport)* Ausscheiden, *das*

élite [eɪ'li:t] *n.* Elite, *die*

élitism [eɪ'li:tɪzm] *n.* Elitedenken, *das*

élitist [eɪ'li:tɪst] **1.** *adj.* Elite‹*denken*›. **2.** *n.* elitär Denkender/Denkende

elixir [ɪ'lɪksə(r)] *n.* Heilmittel, *das;* **~ [of life]** [Lebens]elixier, *das*

Elizabethan [ɪlɪzə'bi:θn] *adj.* elisabethanisch

elk [elk] *n., pl.* **~s** *or same (moose)* Riesenelch, *der*

ellipse [ɪ'lɪps] *n.* Ellipse, *die*

elliptical [ɪ'lɪptɪkl] *adj.* elliptisch; Ellipsen‹*bogen, -bahn*›

elm [elm] *n.* Ulme, *die*

elocution [elə'kju:ʃn] *n., no pl.* Sprechkunst, *die;* **give lessons in ~:** Sprechunterricht geben

elongate ['i:lɒŋgeɪt] *v.t.* länger werden lassen ‹*Schatten*›; strecken ‹*Körper*›; recken ‹*Hals*›

elongated ['i:lɒŋgeɪtɪd] *adj.* langgestreckt ‹*Gestalt, Gliedmaße*›; langgereckt ‹*Hals*›

elongation [i:lɒŋ'geɪʃn] *n.* Verlängerung, *die; (of limbs, neck)* [Aus]recken, *das; (of forms, shapes)* Strecken, *das*

elope [ɪ'ləʊp] *v.i.* durchbrennen *(ugs.)*

elopement [ɪ'ləʊpmənt] *n.* Durchbrennen, *das (ugs.)*

eloquence ['eləkwəns] *n.* Beredtheit, *die;* **a man of great ~:** ein sehr beredter Mann

eloquent ['eləkwənt] *adj.* **a)** gewandt ‹*Stil, Redner*›; beredt ‹*Person*›; **b)** *(fig.)* beredt ‹*Blick, Schweigen*›

else [els] *adv.* **a)** *(besides, in addition)* sonst [noch]; **anybody/anything ~?** sonst noch jemand/etwas?; **don't mention it to anybody ~:** erwähnen Sie es gegenüber niemandem sonst; **somebody/something ~:** [noch] jemand anders/noch etwas; **everybody/everything ~:** alle anderen/alles andere; **nobody ~:** niemand sonst; sonst niemand; **nothing ~:** sonst *od.* weiter nichts; **that is something ~ again** das ist wieder etwas anderes; **anywhere ~?** anderswo? *(ugs.);* woanders?; **not anywhere ~:** sonst nirgendwo; **somewhere ~:** anderswo *(ugs.);* woanders; **go somewhere ~:** anderswohin *(ugs.) od.* woandershin gehen; **everywhere ~:** auch sonst überall; **nowhere ~:** sonst nirgendwo; **little ~:** kaum noch etwas; nur noch wenig; **much ~:** [noch] vieles andere *od.* mehr; **not much ~:** nicht mehr viel; nur noch wenig; **who/what/when/how ~?** wer/was/wann/wie sonst noch?; **where ~?** wo/wohin sonst noch?; **b)** *(instead)* ander...; **sb. ~'s hat** der Hut von jmd. anders *od.* jmd. anderem *(ugs.);* **anybody/anything ~?** [irgend] jemand anders/etwas anderes?; **anyone ~ but Joe would have realized that** jeder [andere] außer Joe hätte das bemerkt; **somebody/something ~:** jemand anders/etwas anderes; **everybody/everything ~:** alle anderen/alles andere; **nobody/nothing ~:** niemand anders/nichts anderes; **there's nothing ~ for it** es hilft nichts; **anywhere ~?** anderswo? *(ugs.);* woanders?; **somewhere ~:** anderswo *(ugs.);* woanders; **go somewhere ~:** woandershin gehen; **his mind was/his thoughts were somewhere ~:** im Geist/mit seinen Gedanken war er woanders; **everywhere ~:** überall anders; überall sonst; **nowhere ~:** nirgendwo sonst; **there's not much ~ we can do but ...:** wir können kaum etwas anderes tun, als ...; **who ~ [but]?** wer anders [als]?; **what ~ can I do?** was kann ich anderes machen?; **why ~ would I have done it?** warum hätte ich es sonst getan?; **where ~ could we go?** wohin könnten wir statt dessen gehen?; **how ~ would you do it?** wie würden Sie es anders *od.* sonst machen?; **c)** *(otherwise)* sonst; anderenfalls; **or ~:** oder aber; **do it or ~ ...!** tun Sie es, sonst ...!; **do it or ~!** *(coll.)* tu es gefälligst!

'elsewhere *adv.* woanders; **go ~:** woandershin gehen; **his mind was/his thoughts were ~:** im Geist/mit seinen Gedanken war er woanders

elucidate [ɪ'lu:sɪdeɪt] *v.t.* erläutern; aufklären ‹*Geheimnis*›

elude [ɪ'lu:d] *v.t. (avoid)* ausweichen (+ *Dat.*) ‹*Person, Angriff, Blick, Frage*›; *(escape from)* entkommen (+ *Dat.*); **~ the police** sich dem Zugriff der Polizei entziehen; **the name ~s me at the moment** der Name fällt mir im Moment nicht ein

elusive [ɪ'lu:sɪv] *adj.* **a)** *(avoiding grasp or pursuit)* schwer zu erreichen ‹*Person*›; schwer zu fassen ‹*Straftäter*›; scheu ‹*Fuchs, Waldbewohner*›; **b)** *(short-lived)* flüchtig ‹*Freude, Glück*›; **c)** *(hard to define)* schwer definierbar

elves *pl. of* **elf**

emaciated [ɪ'meɪsɪeɪtɪd, ɪ'meɪʃɪeɪtɪd] *adj.* ausgemergelt; abgezehrt

emanate ['eməneɪt] *v.i.* **a)** *(originate)* ausgehen **(from** von); **b)** *(proceed, issue)* ausgestrahlt werden **(from** von); **c)** *(formal: be sent out)* ‹*Befehle:*› erteilt *od.* erlassen werden; ‹*Briefe, Urkunden:*› ausgestellt *od.* ausgefertigt werden

emancipate [ɪ'mænsɪpeɪt] *v.t.* emanzipieren

emancipated [ɪ'mænsɪpeɪtɪd] *adj.* emanzipiert ‹*Frau, Vorstellung, Einstellung*›; **become ~:** sich emanzipieren; **~ slave** freigelassener Sklave

emancipation [ɪmænsɪ'peɪʃn] *n.* Emanzipation, *die; (of slave)* Freilassung, *die*

embalm [ɪm'bɑ:m] *v.t.* einbalsamieren

embankment [ɪm'bæŋkmənt] *n.* Damm, *der;* **[railway] ~:** Bahndamm, *der;* **the Thames E~:** die Themse-Uferstraße *(in London)*

embargo [ɪm'bɑ:gəʊ] **1.** *n., pl.* **~es** Embargo, *das.* **2.** *v.t.* mit einem Embargo belegen

embark [ɪm'bɑ:k] **1.** *v.t.* einschiffen ‹*Passagiere, Waren*›. **2.** *v.i.* **a)** sich einschiffen **(for** nach); **b)** *(engage)* **~ [up]on sth.** etw. in Angriff nehmen

embarkation [embɑ:'keɪʃn] *n.* Einschiffung, *die*

embarrass [ɪm'bærəs] *v.t.* in Verlegenheit bringen; **be ~ed by lack of money** in Geldverlegenheit sein

embarrassed [ɪm'bærəst] *adj.* verlegen ‹*Person, Blick, Lächeln, Benehmen, Schweigen*›; **feel ~:** verlegen sein; **now don't be ~!** geniere dich nicht!; **make sb. feel ~:** jmdn. verlegen machen

embarrassing [ɪm'bærəsɪŋ] *adj.* peinlich ‹*Benehmen, Schweigen, Situation, Augenblick, Frage, Thema*›; beschämend ‹*Großzügigkeit*›; verwirrend ‹*Auswahl*›

embarrassment [ɪm'bærəsmənt] *n.* Verlegenheit, *die; (instance)* Peinlichkeit, *die;* **much to his ~:** zu seiner großen Verlegenheit

embassy ['embəsɪ] *n.* Botschaft, *die*

embed [ɪm'bed] *v.t.*, **-dd-: a)** *(fix)* einlassen; ~ **sth. in concrete** etw. einbetonieren; **b)** *(fig.)* **be firmly ~ded in sth.** fest in etw. *(Dat.)* verankert sein

embellish [ɪm'belɪʃ] *v.t.* **a)** *(beautify)* schmücken; beschönigen ‹*Wahrheit*›; **b)** ausschmücken ‹*Geschichte, Bericht*›

embellishment [ɪm'belɪʃmənt] *n.* **a)** *(arrangement)* Verzierung, *die;* **b)** *no pl. (ornamentation of story)* Ausschmückung, *die*

ember ['embə(r)] *n., usu. in pl. (lit. or fig.)* Glut, *die*

embezzle [ɪm'bezl] *v.t.* unterschlagen

embezzlement [ɪm'bezlmənt] *n.* Unterschlagung, *die*

embitter [ɪm'bɪtə(r)] *v.t.* vergiften ‹*Beziehungen*›; verschärfen ‹*Auseinandersetzung*›; verbittern ‹*Person*›

emblem ['embləm] *n. (symbol)* Emblem, *das;* Wahrzeichen, *das*

embodiment [ɪm'bɒdɪmənt] *n.* Verkörperung, *die*

embody [ɪm'bɒdɪ] *v.t.* verkörpern

emboss [ɪm'bɒs] *v.t.* prägen ‹*Metall, Papier, Leder usw.*›; **an ~ed design** ein erhabenes Muster

embrace [ɪm'breɪs] **1.** *v.t.* **a)** umarmen; **b)** *(fig.: surround)* umgeben; **c)** *(accept)* wahrnehmen ‹*Gelegenheit*›; annehmen ‹*Angebot*›; **d)** *(adopt)* annehmen; ~ **a cause** eine Sache zu seiner eigenen machen; ~ **Catholicism** sich zum Katholizismus bekennen; **e)** *(include)* umfassen. **2.** *v.i.* sich umarmen. **3.** *n.* Umarmung, *die*

embroider [ɪm'brɔɪdə(r)] *v.t.* sticken ‹*Muster*›; besticken ‹*Tuch, Kleid*›; *(fig.)* ausschmücken ‹*Erzählung, Wahrheit*›

embroidery [ɪm'brɔɪdərɪ] *n.* **a)** Stickerei, *die;* **b)** *no pl. (fig.: ornament)* Ausschmückungen *Pl.*

embroil [ɪm'brɔɪl] *v.t.* **become/be ~ed in a war/dispute** in einen Krieg/Streit verwickelt werden/sein

embryo ['embrɪəʊ] *n., pl.* **~s** Embryo, *der;* **in ~** *(fig.)* im Keim

embryonic [embrɪ'ɒnɪk] *adj. (Biol., fig.)* Embryonal-; unausgereift ‹*Vorstellung*›

emend [ɪ'mend] *v.t. (Lit.)* emendieren *(fachspr.);* berichtigen

emerald ['emərəld] **1.** *n.* **a)** Smaragd, *der;* **b)** ~ **[green]** Smaragdgrün, *das.* **2.** *adj.* **a)** smaragdgrün; **b) the E~ Isle** *(Ireland)* die Grüne Insel

emerge [ɪ'mɜːdʒ] *v.i.* **a)** *(come out)* auftauchen (**from** aus, **from behind** hinter + *Dat.*, **from beneath** *or* **under** unter + *Dat.* hervor); **the sun ~d from behind the clouds** die Sonne trat hinter den Wolken hervor; **the caterpillar ~d from the egg** die Raupe schlüpfte aus dem Ei; **b)** *(arise)* hervorgehen (**from** aus); ‹*Leben:*› entstammen (**from** + *Dat.*); **difficulties may ~:** es können Schwierigkeiten auftreten; **c)** *(become known)* ‹*Wahrheit:*› an den Tag kommen; **it ~s that ...:** es zeigt sich *od.* stellt sich heraus, daß ...

emergence [ɪ'mɜːdʒəns] *n.* **a)** *(rising out of liquid)* Auftauchen, *das;* **b)** *(coming forth)* Hervortreten, *das; (of school of thought, new ideas)* Aufkommen, *das*

emergency [ɪ'mɜːdʒənsɪ] **1.** *n.* **a)** Notfall, *der;* **in an** *or* **in case of ~:** im Notfall; ~ **[case]** *(Med.)* Notfall, *der;* **b)** *(Polit.)* Ausnahmezustand, *der;* **declare a state of ~:** den Ausnahmezustand erklären. **2.** *adj.* Not‹*bremse, -ruf, -ausgang, -landung*›; ~ **ward** Unfallstation, *die*

emergent [ɪ'mɜːdʒənt] *adj.* jung, aufstrebend ‹*Volk*›

emery ['emərɪ]: **~-board** *n.* Schleifbrett, *das; (strip for fingernails)* Sandblattfeile, *die;* **~-paper** *n.* Schmirgelpapier, *das*

emetic [ɪ'metɪk] *(Med.) n.* Emetikum, *das (fachspr.);* Brechmittel, *das*

emigrant ['emɪgrənt] *n.* Auswanderer, *der*/Auswanderin, *die;* Emigrant, *der*/Emigrantin, *die*

emigrate ['emɪgreɪt] *v.i.* auswandern, emigrieren (**to** nach, **from** aus)

emigration [emɪ'greɪʃn] *n.* Auswanderung, *die,* Emigration, *die* (**to** nach, **from** aus)

émigré ['emɪgreɪ] *n.* Emigrant, *der*/Emigrantin, *die*

eminence ['emɪnəns] *n. no pl. (distinguished superiority)* hohes Ansehen; **person of great ~:** bedeutender *od.* hochangesehener Mensch

eminent ['emɪnənt] *adj.* **a)** *(distinguished)* bedeutend, hochangesehen ‹*Redner, Gelehrter, Künstler*›; ~ **guest** hoher Gast; **b)** *(remarkable)* ausnehmend ‹*Eigenschaft*›

eminently ['emɪnəntlɪ] *adv.* ausnehmend; vorzüglich ‹*geeignet*›; überaus ‹*erfolgreich*›; ~ **respectable** hochangesehen

emissary ['emɪsərɪ] *n.* Abgesandte, *der/die*

emission [ɪ'mɪʃn] *n.* **a)** *(giving off or out)* Aussendung, *die; (of vapour)* Ablassen, *das; (of liquid)* Ausscheidung, *die;* ~ **of light/heat** Licht-/Wärmeausstrahlung, *die;* **b)** *(thing given off)* Abstrahlung, *die*

emit [ɪ'mɪt] *v.t.*, **-tt-** aussenden ‹*Strahlen*›; ausstrahlen ‹*Wärme, Licht*›; ausstoßen ‹*Rauch, Schrei*›; ausscheiden ‹*Flüssigkeit*›; abgeben ‹*Geräusch*›

emotion [ɪ'məʊʃn] *n.* **a)** *(state)* Ergriffenheit, *die;* Bewegtheit, *die;* **be overcome with ~:** von Gefühl übermannt sein; **show no ~:** keine Gefühlsregung zeigen; **b)** *(feeling)* Gefühl, *das;* Emotion, *die*

emotional [ɪ'məʊʃənl] *adj.* **a)** *(of emotions)* emotional; Gefühls‹*ausdruck, -leben, -erlebnis, -reaktion*›; Gemüts‹*zustand, -störung*›; gefühlsgeladen ‹*Worte, Musik, Geschichte, Film*›; gefühlvoll ‹*Stimme, Ton*›; **b)** *(liable to excessive emotion)* leicht erregbar ‹*Person*›

emotionally [ɪ'məʊʃənəlɪ] *adv.* emotional; ~ **exhausted/disturbed** seelisch erschöpft/gestört; **get ~ involved with sb.** eine gefühlsmäßige Bindung mit jmdm. eingehen

emotive [ɪ'məʊtɪv] *adj.* emotional; gefühlsbetont; emotiv *(Psych., Sprachw.)*

empathy ['empəθɪ] *n.* Empathie, *die (Psych.);* Einfühlung, *die*

emperor ['empərə(r)] *n.* Kaiser, *der*

emphasis ['emfəsɪs] *n., pl.* **emphases** ['emfəsiːz] **a)** *(in speech etc.)* Betonung, *die;* **the ~ is on sth.** die Betonung liegt auf etw. *(Dat.);* **lay** *or* **put ~ on sth.** etw. betonen; **b)** *(intensity)* Nachdruck, *der;* **with ~:** nachdrücklich; **c)** *(importance attached)* Gewicht, *das;* **lay** *or* **put [considerable] ~ on sth.** [großes] Gewicht auf etw. *(Akk.)* legen; **the ~ has shifted** der Akzent hat sich verlagert

emphasize (emphasise) ['emfəsaɪz] *v.t. (lit. or fig.)* betonen; *(attach importance to)* Gewicht auf etw. *(Akk.)* legen

emphatic [ɪm'fætɪk] *adj.* nachdrücklich; *(forcible)* demonstrativ ‹*Rückzug, Ablehnung*›; eindringlich ‹*Demonstration*›; **be quite ~ that ...**: durchaus darauf bestehen, daß ...

emphatically [ɪm'fætɪkəlɪ] *adv.* nachdrücklich; eindringlich ‹*sprechen*›; *(decisively)* entschieden ‹*bestreiten usw.*›

empire ['empaɪə(r)] *n.* **a)** Reich, *das;* **b)** *(commercial organization)* Imperium, *das (fig.)*

empirical [ɪm'pɪrɪkl] *adj.* empirisch; empirisch begründet ‹*Argument, Wissen, Schlußfolgerung*›

employ [ɪm'plɔɪ] **1.** *v.t.* **a)** *(take on)* einstellen; *(have working for one)* beschäftigen; **be ~ed by a company** bei einer Firma beschäftigt sein; **b)** *(use services of)* **~ sb. on sth.** jmdn. für etw. einsetzen; **~ sb. to do sth.** jmdn. dafür einsetzen, etw. zu tun; **c)** *(use)* einsetzen **(for, in, on** für); anwenden ‹*Methode, List*› **(for, in, on** bei). **2.** *n., no pl., no indef. art.* **be in the ~ of sb.** bei jmdm. beschäftigt sein; in jmds. Diensten stehen *(veralt.)*

employee (*Amer.:* **employe**) [emplɔɪ'i:, em'plɔɪi:] *n.* Angestellte, *der/die; (in contrast to employer)* Arbeitnehmer, *der/*-nehmerin, *die;* **the firm's ~s** die Belegschaft der Firma

employer [ɪm'plɔɪə(r)] *n.* Arbeitgeber, *der/*-geberin, *die*

employment [ɪm'plɔɪmənt] *n., no pl.* **a)** *(work)* Arbeit, *die;* **be in/without regular ~**: eine/keine feste Anstellung haben; **b)** *(regular trade or profession)* Beschäftigung, *die*

em'ployment agency *n.* Stellenvermittlung, *die*

empower [ɪm'paʊə(r)] *v.t.* *(authorize)* ermächtigen; *(enable)* befähigen

empress ['emprɪs] *n.* Kaiserin, *die*

emptiness ['emptɪnɪs] *n., no pl.* *(lit. or fig.)* Leere, *die*

empty ['emptɪ] **1.** *adj.* **a)** leer; frei ‹*Sitz, Parkplatz*›; **~ of sth.** ohne etw.; **b)** *(coll.: hungry)* **I feel a bit ~**: ich bin ein bißchen hungrig; **c)** *(fig.) (foolish)* dumm; hohl ‹*Kopf*›; *(meaningless)* leer. **2.** *n.* *(bottle)* leere Flasche; *(container)* leerer Behälter. **3.** *v.t.* **a)** *(remove contents of)* leeren; *(finish using contents of)* aufbrauchen; *(eat/drink whole contents of)* leer essen ‹*Teller*›/leeren ‹*Glas*›; **b)** *(transfer)* umfüllen **(into** in + *Akk.*); *(pour)* schütten **(over** über + *Akk.*). **4.** *v.i.* **a)** *(become ~)* sich leeren; **b)** *(discharge)* **~ into** ‹*Fluß, Abwasserkanal:*› münden in (+ *Akk.*)

empty: ~-handed [emptɪ'hændɪd] *pred. adj.* mit leeren Händen; **~-headed** ['emptɪhedɪd] *adj.* hohlköpfig *(abwertend)*

EMS *abbr.* **European Monetary System** EWS

emu ['i:mju:] *n. (Ornith.)* Emu, *der*

emulate ['emjʊleɪt] *v.t.* nacheifern (+ *Dat.*)

emulsion [ɪ'mʌlʃn] *n.* **a)** Emulsion, *die;* **b)** *see* **emulsion paint**

e'mulsion paint *n.* Dispersionsfarbe, *die*

enable [ɪ'neɪbl] *v.t.* **~ sb. to do sth.** es jmdm. ermöglichen, etw. zu tun; **~ sth. [to be done]** etw. ermöglichen

enact [ɪ'nækt] *v.t.* **a)** *(make law)* erlassen; **b)** *(act out)* aufführen ‹*Theaterstück*›; spielen ‹*Rolle*›

enamel [ɪ'næml] **1.** *n.* **a)** Emaille, *die;* Email, *das;* *(paint)* Lack, *der;* **b)** *(Anat.)* [Zahn]schmelz, *der.* **2.** *attrib. adj.* emailliert. **3.** *v.t., (Brit.)* **-ll-** emaillieren

encase [ɪn'keɪs] *v.t.* einschließen

enchant [ɪn'tʃɑ:nt] *v.t.* **a)** *(bewitch)* verzaubern; **b)** *(delight)* entzücken

enchanting [ɪn'tʃɑ:ntɪŋ] *adj.* *(delightful)* entzückend; bezaubernd

enchantment [ɪn'tʃɑ:ntmənt] *n.* *(delight)* Entzücken, *das* **(with** über + *Akk.*)

encircle [ɪn'sɜ:kl] *v.t.* **a)** einkreisen; ‹*Bäume, Zaun usw.:*› umgeben; **b)** *(mark with circle)* einkreisen ‹*Buchstabe, Antwort*›

encl. *abbr.* **enclosed, enclosure[s]** Anl.

enclave ['enkleɪv] *n. (lit. or fig.)* Enklave, *die*

enclose [ɪn'kləʊz] *v.t.* **a)** *(surround)* umgeben; *(shut up or in)* einschließen; **~ land with barbed wire** Land mit Stacheldraht einzäunen; **b)** *(put in envelope with letter)* beilegen **(with, in** *Dat.*); **please find ~d, ~d please find** als Anlage übersenden wir Ihnen; anbei erhalten Sie; **the ~d brochure** der beiliegende Prospekt; **a cheque is ~d** beiliegend finden Sie einen Scheck

enclosure [ɪn'kləʊʒə(r)] *n.* **a)** *(act)* Einzäunung, *die;* **b)** *(place) (in zoo)* Gehege, *das;* *(paddock)* Koppel, *die;* **c)** *(fence)* Umzäunung, *die;* **d)** *(with letter)* Anlage, *die*

encode [ɪn'kəʊd] *v.t.* verschlüsseln; chiffrieren

encore ['ɒŋkɔ:(r)] **1.** *int.* Zugabe. **2.** *n.* Zugabe, *die;* **give an ~**: eine Zugabe spielen. **3.** *v.t.* als Zugabe verlangen ‹*Lied, Tanz usw.*›

encounter [ɪn'kaʊntə(r)] **1.** *v.t.* **a)** *(as adversary)* treffen auf (+ *Akk.*); **b)** *(by chance)* begegnen (+ *Dat.*); **c)** *(meet with)* stoßen auf (+ *Akk.*) ‹*Problem, Schwierigkeit, Kritik, Widerstand usw.*›. **2.** *n.* **a)** *(in combat)* Zusammenstoß, *der;* **b)** *(chance meeting, introduction)* Begegnung, *die*

encourage [ɪn'kʌrɪdʒ] *v.t.* **a)** *(stimulate, incite)* ermutigen; **bread ~s rats** Brot lockt Ratten an; **b)** *(promote)* fördern; **~ a smile/a response from sb.** jmdm. ein Lächeln/eine Reaktion entlocken; **~ bad habits** schlechte Angewohnheiten unterstützen; **we do not ~ smoking** wir unterstützen es nicht, daß geraucht wird; **c)** *(urge)* **~ sb. to do sth.** jmdn. dazu ermuntern, etw. zu tun; **d)** *(cheer)* **be [much] ~d by sth.** durch etw. neuen Mut schöpfen; **we were ~d to hear ...**: wir schöpften neuen Mut, als wir hörten ...

encouragement [ɪn'kʌrɪdʒmənt] *n.* **a)** *(support, incitement)* Ermutigung, *die* **(from** durch); **give sb. ~**: jmdn. ermutigen; **get** *or* **receive ~ from sth.** durch etw. ermutigt werden; **b)** *(urging)* Ermunterung, *die;* **c)** *(stimulus)* Ansporn, *der*

encouraging [ɪn'kʌrɪdʒɪŋ] *adj.* ermutigend

encroach [ɪn'krəʊtʃ] *v.i.* *(lit. or fig.)* **~ [on sth.]** [in etw. (*Akk.*)] eindringen; **the sea is ~ing [on the land]** das Meer dringt vor; **~ on sb.'s time** jmds. Zeit immer mehr in Anspruch nehmen

encrust [ɪn'krʌst] *v.t.* **~ed with diamonds** über und über mit Diamanten besetzt

encumber [ɪn'kʌmbə(r)] *v.t.* **a)** *(hamper)* behindern; **~ oneself/sb. with sth.** sich/jmdn. mit etw. belasten; **b)** *(burden)* **~ sb. with debt** jmdn. mit Schulden belasten

encumbrance [ɪn'kʌmbrəns] *n.* **a)** *(impediment)* Hindernis, *das* **(to** für); **b)** *(burden)* Last, *die*

encyclopaedia [ɪnsaɪklə'pi:dɪə] *n.* Lexikon, *das;* Enzyklopädie, *die*

encyclopaedic [ensaɪklə'pi:dɪk, ɪnsaɪklə'pi:dɪk] *adj.* enzyklopädisch

end [end] **1.** *n.* **a)** *(farthest point)* Ende, *das;* *(of nose, hair, tail, branch, finger)* Spitze, *die;* **that was the ~** *(coll.) (no longer tolerable)* da war Schluß *(ugs.); (very bad)* das war das Letzte *(ugs.);* **at an ~**: zu Ende; **come to an ~**: enden (*see also* **1 g**); **my patience has come to** *or* **is now at an ~**: meine

Geduld ist jetzt am Ende; **look at a building/a pencil ~ on** ein Gebäude von der Schmalseite/einen Bleistift von der Spitze her betrachten; **from ~ to ~**: von einem Ende zum anderen; **~ to ~**: längs hintereinander; **lay ~ to ~**: aneinanderreihen; **keep one's ~ up** *(fig.)* seinen Mann stehen; **make [both] ~s meet** *(fig.)* [mit seinem Geld] zurechtkommen; **no ~** *(coll.)* unendlich viel; **there is no ~ to sth.** *(coll.)* etw. nimmt kein Ende; **put an ~ to sth.** einer Sache *(Dat.)* ein Ende machen; **b)** *(of box, packet, tube, etc.)* Schmalseite, *die; (top/bottom surface)* Ober-/Unterseite, *die;* **on ~** *(upright)* hochkant; **sb.'s hair stands on ~** *(fig.)* jmdm. stehen die Haare zu Berge *(ugs.);* **c)** *(remnant)* Rest, *der; (of cigarette, candle)* Stummel, *der;* **d)** *(side)* Seite, *die;* **be on the receiving ~ of sth.** etw. abbekommen *od.* einstecken müssen; **how are things at your ~?** wie sieht es bei dir aus?; **e)** *(half of sports pitch or court)* Spielfeldhälfte, *die;* **f)** *(of swimming-pool)* **deep/shallow ~ [of the pool]** tiefer/flacher Teil [des Schwimmbeckens]; **g)** *(conclusion, lit. or fig.)* Ende, *das; (of lesson, speech, story, discussion, meeting, argument, play, film, book, sentence)* Schluß, *der;* Ende, *das;* **by the ~ of the week/meeting** als die Woche herum war/als die Versammlung zu Ende war; **at the ~ of 1987/March** Ende 1987/März; **that's the ~ of 'that** *(fig.)* damit ist die Sache erledigt; **be at an ~**: zu Ende sein; **bring a meeting** *etc.* **to an ~**: eine Versammlung *usw.* beenden; **come to an ~**: ein Ende nehmen *(see also* **1 a**); **have come to the ~ of sth.** mit etw. fertig sein; **in the ~**: schließlich; **on ~**: ununterbrochen *(see also* **b**); **h)** *(downfall, destruction)* Ende, *das; (death)* Ende, *das (geh. verhüll.);* **meet one's ~**: den Tod finden *(geh.);* **sb. comes to a bad ~**: es nimmt ein böses *od.* schlimmes Ende mit jmdm.; **i)** *(purpose, object)* Ziel, *das;* Zweck, *der;* **be an ~ in itself** *(the only purpose)* das eigentliche Ziel sein; **the ~ justifies the means** der Zweck heiligt die Mittel; **with this ~ in view** mit diesem Ziel vor Augen; **to this/what ~**: zu diesem/welchem Zweck. **2.** *v.t.* **a)** *(bring to an ~)* beenden; kündigen ‹*Abonnement*›; **~ one's life/days** *(spend last part of life)* sein Leben/seine Tage beschließen; **b)** *(put an ~ to, destroy)* ein Ende setzen (+ *Dat.*); **~ it [all]** *(coll.: kill oneself)* [mit dem Leben] Schluß machen *(ugs.);* **c)** *(stand as supreme example of)* **a feast/race** *etc.* **to ~ all feasts/races** *etc.* ein Fest/Rennen *usw.*, das alles [bisher Dagewesene] in den Schatten stellt. **3.** *v.i.* enden; **where will it all ~?** wo soll das noch hinführen?; **the match ~ed in a draw** das Spiel ging unentschieden aus

~ 'up *v.i.* enden; **~ up [as] a teacher/an alcoholic** *(coll.)* schließlich Lehrer/zum Alkoholiker werden; **I always ~ up doing all the work** *(coll.)* am Ende bleibt die ganze Arbeit immer an mir hängen

'end-all *see* **be 4**

endanger [ɪn'deɪndʒə(r)] *v.t.* gefährden; **an ~ed species** eine vom Aussterben bedrohte Art

endear [ɪn'dɪə(r)] *v.t.* **~ sb./sth./oneself to sb.** jmdn./etw./sich bei jmdm. beliebt machen

endearing [ɪn'dɪərɪŋ] *adj.* reizend; gewinnend ‹*Lächeln, Art*›

endearment [ɪn'dɪəmənt] *n.* Zärtlichkeit, *die;* **term of ~**: Kosename, *der*

endeavour *(Brit.; Amer.:* **endeavor)** [ɪn'devə(r)] **1.** *v.i.* **~ to do sth.** sich bemühen, etw. zu tun. **2.** *n.* Bemühung, *die; (attempt)* Versuch, *der;* **make every ~ to do sth.** alle Anstrengungen unternehmen, um etw. zu tun; **despite his best ~s** obwohl er sich nach Kräften bemühte

endemic [en'demɪk] *adj.* verbreitet

ending ['endɪŋ] *n.* Schluß, *der; (of word)* Endung, *die*

endive ['endaɪv] *n.* Endivie, *die*

endless ['endlɪs] *adj.* endlos; *(coll.: innumerable)* unzählig; *(eternal, infinite)* unendlich; unendlich lang ‹*Liste*›; **have an ~ wait, wait an ~ time** endlos lange warten

endlessly ['endlɪslɪ] *adv.* unaufhörlich ‹*streiten, schwatzen*›

endorse [ɪn'dɔːs] *v.t.* **a)** *(sign one's name on back of)* indossieren ‹*Scheck, Wechsel*›; **b)** *(support)* beipflichten (+ *Dat.*) ‹*Meinung, Aussage*›; billigen, gutheißen ‹*Entscheidung, Handlung, Einstellung*›; unterstützen ‹*Vorschlag, Kandidaten*›; **c)** *(Brit.: make entry regarding offence on)* einen Strafvermerk machen auf (+ *Akk. od. Dat.*)

endorsement [ɪn'dɔːsmənt] *n.* **a)** *(of cheque)* Indossament, *das;* **b)** *(support)* Billigung, *die; (of proposal, move, candidate)* Unterstützung, *die;* **c)** *(Brit.: entry regarding offence)* Strafvermerk, *der*

endow [ɪn'daʊ] *v.t.* **a)** *(give permanent income to)* [über Stiftungen/eine Stiftung] finanzieren ‹*Einrichtung, Krankenhaus usw.*›; stiften ‹*Preis, Lehrstuhl*›; **b)** *(fig.)* **be ~ed with charm/a talent for music** *etc.* Charme/musikalisches Talent *usw.* besitzen

endowment [ɪn'daʊmənt] *n.* **a)** *(endowing, fund, etc.)* Stiftung, *die;* **b)** *(talent)* Begabung, *die*

en'dowment policy *n.* abgekürzte *od.* gemischte Lebensversicherung

'end-product *n. (lit. or fig.)* Endprodukt, *das; (fig.)* Resultat, *das*

endurable [ɪn'djʊərəbl] *adj.* erträglich

endurance [ɪn'djʊərəns] *n.* **a)** Widerstandskraft, *die; (ability to withstand strain)* Ausdauer, *die; (patience)* Geduld, *die;* **past or beyond ~**: unerträglich; **b)** *(lastingness)* Dauerhaftigkeit, *die*

en'durance test *n.* Belastungsprobe, *die*

endure [ɪn'djʊə(r)] **1.** *v.t.* *(undergo, tolerate)* ertragen; *(submit to)* über sich ergehen lassen. **2.** *v.i.* fortdauern

enduring [ɪn'djʊərɪŋ] *adj.* dauerhaft; beständig ‹*Glaube, Tradition*›

enema ['enɪmə] *n. (Med.)* Einlauf, *der;* Klistier, *das (Med.)*

enemy ['enəmɪ] **1.** *n. (lit. or fig.)* Feind, *der* **(of, to** *Gen.***)**; **make an ~ of sb.** sich *(Dat.)* jmdn. zum Feind machen; **be one's own worst ~**: sich *(Dat.)* selbst im Wege stehen. **2.** *adj.* feindlich; **destroyed by ~ action** durch Feindeinwirkung zerstört

energetic [enə'dʒetɪk] *adj.* **a)** *(very active)* energiegeladen; tatkräftig ‹*Mitarbeiter*›; lebhaft ‹*Kind*›; **I don't feel ~ enough** ich habe nicht genug Energie; **b)** *(vigorous)* schwungvoll; entschieden, energisch ‹*Zustimmung, Ablehnung*›; kräftig ‹*Rühren*›

energetically [enə'dʒetɪkəlɪ] *adv.* schwungvoll; entschieden ‹*sich äußern*›

energy ['enədʒɪ] *n.* **a)** *(vigour)* Energie, *die; (active operation)* Kraft, *die;* **save your ~!** schone deine Kräfte!; **I've no ~ left** ich habe keine Energie mehr; **b)** *in pl. (individual's powers)* Kraft, *die;* **c)** *(Phys.)* Energie, *die;* **sources of ~**: Energiequellen

energy: ~ **crisis** *n.* Energiekrise, *die;* **~-giving** *adj.* energiespendend; **~-saving** adj. energiesparend

enervate ['enəveɪt] *v.t.* schwächen

enfeeble [ɪn'fi:bl] *v.t.* schwächen

enforce [ɪn'fɔ:s] *v.t.* **a)** durchsetzen; sorgen für ‹*Disziplin*›; ~ **the law** dem Gesetz Geltung verschaffen; **~d** erzwungen ‹*Schweigen*›; unfreiwillig ‹*Untätigkeit*›; **b)** *(give more force to)* Nachdruck verleihen (+ *Dat.*)

enforceable [ɪn'fɔ:səbl] *adj.* durchsetzbar

enforcement [ɪn'fɔ:smənt] *n.* Durchsetzung, *die*

ENG *abbr.* **electronic news-gathering** EB

engage [ɪn'geɪdʒ] **1.** *v.t.* **a)** *(hire)* einstellen ‹*Arbeiter*›; engagieren ‹*Sänger*›; **b)** *(employ busily)* beschäftigen (**in** mit); *(involve)* verwickeln (**in** in + *Akk.*); **c)** *(attract and hold fast)* wecken [und wachhalten] ‹*Interesse*›; auf sich *(Akk.)* ziehen ‹*Aufmerksamkeit*›; fesseln ‹*Person*›; in Anspruch nehmen ‹*Konzentration*›; gewinnen ‹*Sympathie, Unterstützung*›; **d)** *(enter into conflict with)* angreifen; **e)** *(Mech.)* ~ **the clutch/gears** einkuppeln/einen Gang einlegen. **2.** *v.i.* **a)** ~ **in sth.** sich an etw. *(Dat.)* beteiligen; ~ **in politics** sich politisch engagieren; ~ **in a sport** eine Sportart betreiben; **b)** *(Mech.)* ineinandergreifen

engaged [ɪn'geɪdʒd] *adj.* **a)** *(to be married)* verlobt; **be ~ [to be married] [to sb.]** [mit jmdm.] verlobt sein; **become** *or* **get ~ [to be married] [to sb.]** sich [mit jmdm.] verloben; **b)** *(bound by promise)* verabredet; **be otherwise ~:** etwas anderes vorhaben; **c)** *(occupied with business)* beschäftigt; **d)** *(occupied or used by person)* besetzt ‹*Toilette, Taxi*›; **e)** *(Teleph.)* besetzt; **you're always ~:** bei dir ist immer besetzt; ~ **signal** *or* **tone** *(Brit.)* Besetztzeichen, *das*

engagement [ɪn'geɪdʒmənt] *n.* **a)** *(to be married)* Verlobung, *die* (**to** mit); **b)** *(appointment made with another)* Verabredung, *die;* **have a previous** *or* **prior ~:** schon anderweitig festgelegt sein; **c)** *(booked appearance)* Engagement, *das;* **d)** *(Mil.)* Kampfhandlung, *die*

en'gagement ring *n.* Verlobungsring, *der*

engaging [ɪn'geɪdʒɪŋ] *adj.* bezaubernd; gewinnend ‹*Lächeln*›; einnehmend ‹*Persönlichkeit, Art*›

engender [ɪn'dʒendə(r)] *v.t.* zur Folge haben; erzeugen

engine ['endʒɪn] *n.* **a)** Motor, *der;* *(rocket/jet ~)* Triebwerk, *das;* **b)** *(locomotive)* Lok[omotive], *die*

'engine-driver *n.* *(Brit.)* Lok[omotiv]führer, *der*

engineer [endʒɪ'nɪə(r)] **1.** *n.* **a)** Ingenieur, *der*/Ingenieurin, *die;* *(service ~, installation ~)* Techniker, *der*/Technikerin, *die;* **b)** *(maker or designer of engines)* Maschinenbauingenieur, *der;* **c)** **[ship's] ~:** Maschinist, *der.* **2.** *v.t.* **a)** *(coll.: contrive)* arrangieren; entwickeln ‹*Plan*›; **b)** *(manage construction of)* konstruieren

engineering [endʒɪ'nɪərɪŋ] *n., no pl.* **a)** Technik, *die;* **b)** *attrib.* technisch ‹*Arbeiten, Fähigkeiten*›; ~ **science** Ingenieurwesen, *das;* ~ **company** *or* **firm** Maschinenbaufirma, *die*

'engine-room *n.* Maschinenhaus, *das;* Maschinenraum, *der*

England ['ɪŋglənd] *pr. n.* England *(das)*

English ['ɪŋglɪʃ] **1.** *adj.* englisch; **he/she is ~:** er ist Engländer/sie ist Engländerin. **2.** *n.* **a)** *(language)* Englisch, *das;* **say sth. in ~:** etw. auf englisch sagen; **speak ~:** Englisch sprechen; **be speaking ~:** englisch sprechen; **I [can] speak/read ~:** ich spreche Englisch/kann Englisch lesen; **I cannot** *or* **do not speak/read ~:** ich spreche kein Englisch/kann Englisch nicht lesen; **translate into/from [the] ~:** ins Englische/aus dem Englischen übersetzen; **write sth. in ~:** etw. [auf *od.* in] englisch schreiben; **her ~ is very good** sie schreibt/spricht ein sehr gutes Englisch; **the King's/Queen's ~:** die englische Hochsprache; **Old ~:** Altenglisch, *das;* **in plain ~:** in einfachen Worten; **b)** *pl.* **the ~:** die Engländer. *See also* **pidgin English**

English: ~ **'Channel** *pr. n.* **the ~ Channel** der [Ärmel]kanal; **~man** ['ɪŋglɪʃmən] *n., pl.* **~men** ['ɪŋglɪʃmən] Engländer, *der;* **~woman** *n.* Engländerin, *die*

engrave [ɪn'greɪv] *v.t.* gravieren; ~ **sth. with a name** *etc.* einen Namen *usw.* in etw. *(Akk.)* [ein]gravieren

engraving [ɪn'greɪvɪŋ] *n.* **a)** *(design, marks)* Gravur, *die;* **b)** *(Art: print)* Stich, *der;* *(from wood)* Holzschnitt, *der*

engross [ɪn'grəʊs] *v.t.* **be ~ed in sth.** in etw. *(Akk.)* vertieft sein; **become** *or* **get ~ed in sth.** sich in etw. *(Akk.)* vertiefen

engrossing [ɪn'grəʊsɪŋ] *adj.* fesselnd

engulf [ɪn'gʌlf] *v.t.* *(lit. or fig.)* verschlingen; **the house was ~ed in flames** das Haus stand in hellen Flammen

enhance [ɪn'hɑ:ns] *v.t.* verbessern ‹*Aussichten, Stellung*›; erhöhen ‹*Wert, [An]reiz, Macht, Schönheit*›; steigern ‹*Qualität, Wirkung*›; heben ‹*Aussehen*›

enhancement [ɪn'hɑ:nsmənt] *n.* *see* **enhance:** Verbesserung, *die;* Erhöhung, *die;* Steigerung, *die;* Hebung, *die*

enigma [ɪ'nɪgmə] *n.* Rätsel, *das*

enigmatic [enɪg'mætɪk] *adj.* rätselhaft

enjoin [ɪn'dʒɔɪn] *v.t.* ~ **sb. [not] to do sth.** jmdn. eindringlich ermahnen, etw. [nicht] zu tun

enjoy [ɪn'dʒɔɪ] **1.** *v.t.* **a) I ~ed the film** der Film hat mir gefallen; **are you ~ing your meal?** schmeckt dir das Essen?; **he ~s reading** er liest gern; **he ~s music** er mag Musik; **we really ~ed seeing you again** wir haben uns wirklich gefreut, euch wiederzusehen; **b)** *(have use of)* genießen ‹*Recht, Privileg, Vorteil*›; sich erfreuen (+ *Gen.*) ‹*hohen Einkommens*›. **2.** *v. refl.* sich amüsieren; **we thoroughly ~ed ourselves in Spain** wir hatten viel Spaß in Spanien; ~ **yourself at the theatre!** viel Spaß im Theater!

enjoyable [ɪn'dʒɔɪəbl] *adj.* schön; angenehm ‹*Empfindung, Unterhaltung, Arbeit*›; unterhaltsam ‹*Buch, Film, Stück*›

enjoyment [ɪn'dʒɔɪmənt] *n.* *(delight)* Vergnügen, *das* (**of** an + *Dat.*)

enlarge [ɪn'lɑ:dʒ] **1.** *v.t.* vergrößern; *(widen)* erweitern ‹*Wissen*›. **2.** *v.i.* **a)** sich vergrößern; größer werden; *(widen)* sich verbreitern; **b)** ~ **[up]on sth.** etw. weiter ausführen

enlargement [ɪn'lɑ:dʒmənt] *n.* **a)** Vergrößerung, *die;* **b)** *(further explanation)* weitere Ausführung

enlighten [ɪn'laɪtn] *v.t.* aufklären (**on, as to** über + *Akk.*); **let me ~ you** ich will es dir erklären

enlightened [ɪn'laɪtnd] *adj.* aufgeklärt

enlightenment [ɪn'laɪtnmənt] *n., no pl.* Aufklärung, *die*

enlist [ɪn'lɪst] **1.** *v.t.* **a)** *(Mil.)* anwerben; **b)** *(obtain)* gewinnen. **2.** *v.i.* in die Armee/Marine eintreten

enlistment [ɪn'lɪstmənt] *n.* *(Mil.)* Anwerbung, *die*

enliven [ɪn'laɪvn] *v.t.* beleben; in

Schwung bringen *(ugs.)* ‹*Person, Schulklasse usw.*›; lebhafter gestalten ‹*Tanz, Unterricht*›

enmity ['enmɪtɪ] *n.* Feindschaft, *die*

enormity [ɪ'nɔ:mɪtɪ] *n.* **a)** *(atrocity)* Ungeheuerlichkeit, *die (abwertend);* **b)** *see* **enormousness**

enormous [ɪ'nɔ:məs] *adj.* **a)** enorm; riesig, gewaltig ‹*Figur, Tier, Fluß, Wüste, Menge*›; gewaltig, enorm ‹*Veränderung, Unterschied, Liebe, Haß, Widerspruch, Größe, Ausgabe, Kraft*›; ungeheuer ‹*Mut, Charme, Problem*›; **b)** *(fat)* ungeheuer dick

enormously [ɪ'nɔ:məslɪ] *adv.* ungeheuer; enorm

enormousness [ɪ'nɔ:məsnɪs] *n., no pl.* ungeheure Größe; Riesenhaftigkeit, *die; (of size, length, height)* ungeheures Ausmaß

enough [ɪ'nʌf] **1.** *adj.* genug; genügend; **there's ~ room** *or* **room ~:** es ist Platz genug *od.* genügend Platz; **more than ~:** mehr als genug. **2.** *n., no pl., no art.* genug; **be ~ to do sth.** genügen, etw. zu tun; **are there ~ of us?** sind wir genug [Leute]?; **four people are quite ~:** vier Leute genügen völlig; **that [amount] will be ~ to go round** das reicht für alle; **~ of ...:** genug von ...; **are there ~ of these books to go round?** reichen diese Bücher für alle?; **[that's] ~ [of that]!** [jetzt ist es] genug!; **~ of your nonsense!** Schluß mit dem Unsinn!; **have had ~ [of sb./sth.]** genug [von jmdm./etw.] haben; **I've had ~!** jetzt reicht's mir aber!; jetzt habe ich aber genug!; **more than ~:** mehr als genug; **[that's] ~ about ...:** genug über ... *(Akk.)* geredet; **~ said** mehr braucht man dazu nicht zu sagen; **~ is ~:** mal muß es auch genug sein *(ugs.)*; **it's ~ to make you weep** es ist zum Weinen; **as if that were not ~:** als ob das noch nicht genügte. **3.** *adv.* genug; **the meat is not cooked ~:** das Fleisch ist nicht genügend durch; **he is not trying hard ~:** er gibt sich nicht genug *od.* genügend Mühe; **they were friendly ~ towards us** sie waren soweit recht nett zu uns; **oddly/funnily ~:** merkwürdiger-/ *(ugs.)* komischerweise; **sure ~:** natürlich; **be good/kind ~ to do sth.** so gut sein, etw. zu tun

enquire *etc. see* **inquir-**

enrage [ɪn'reɪdʒ] *v.t.* wütend machen; reizen ‹*wildes Tier*›; **be ~d by sth.** über etw. *(Akk.)* wütend werden; **be ~d at sb./sth.** auf jmdn./etw. wütend sein

enrich [ɪn'rɪtʃ] *v.t.* **a)** *(make wealthy)* reich machen; **b)** *(fig.)* bereichern; anreichern ‹*Nahrungsmittel, Boden, Uran*›; verbessern ‹*Haut*›

enrichment [ɪn'rɪtʃmənt] *n. (lit. or fig.)* Bereicherung, *die; (of soil, food, uranium)* Anreicherung, *die*

enrol (*Amer.:* **enroll**) [ɪn'rəʊl] **1.** *v.i.*, **-ll-** sich anmelden; sich einschreiben [lassen]; *(Univ.)* sich einschreiben; sich immatrikulieren; **~ for a course** sich zu einem Kurs anmelden. **2.** *v.t.*, **-ll-** einschreiben ‹*Studenten, Kursteilnehmer*›; anwerben ‹*Rekruten*›; aufnehmen ‹*Schüler, Mitglied, Rekrut*›; **~ sb. for a course/the army** jmdn. für einen Kurs annehmen/in die Armee aufnehmen

enrolment (*Amer.:* **enrollment**) [ɪn'rəʊlmənt] *n.* Anmeldung, *die; (Univ.)* Immatrikulation, *die;* Einschreibung, *die; (in army)* Eintritt, *der*

en route [ɑ̃ 'ru:t] *adv.* unterwegs; auf dem Weg; **~ to Scotland/for Perth** unterwegs *od.* auf dem Weg nach Schottland/Perth

ensemble [ɑ̃'sɑ̃bl] *n.* Ensemble, *das*

ensign ['ensaɪn, 'ensn] *n.* **a)** *(banner)* Hoheitszeichen, *das;* **b)** *(Brit.)* **blue/red/white ~:** Flagge der britischen Marinereserve/ Handelsflotte/Marine

enslave [ɪn'sleɪv] *v.t.* versklaven

ensnare [ɪn'sneə(r)] *v.t. (lit. or fig.)* fangen

ensue [ɪn'sju:] *v.i.* **a)** *(follow)* sich daran anschließen; **b)** *(result)* sich daraus ergeben; **~ from sth.** sich aus etw. ergeben

ensure [ɪn'ʃʊə(r)] *v.t.* **a) ~ that ...** *(satisfy oneself that)* sich vergewissern, daß ...; *(see to it that)* gewährleisten, daß ...; **b)** *(secure)* **~ sth.** etw. gewährleisten; **this will ~ victory for the Party** dies wird der Partei den Sieg sichern

entail [ɪn'teɪl] *v.t.* mit sich bringen; **what exactly does your job ~?** worin besteht Ihre Arbeit ganz genau?; **sth. ~s doing sth.** etw. bedeutet, daß man etw. tun muß

entangle [ɪn'tæŋgl] *v.t.* **a)** *(catch)* sich verfangen lassen; **get [oneself] or become ~d in or with sth.** sich in etw. *(Dat.)* verfangen; **be ~d in sth.** sich in etw. *(Dat.)* verfangen haben; **b)** *(fig.: involve)* verwickeln; **be/become ~d in sth.** in etw. *(Akk.)* verwickelt sein/ werden; **c)** *(make tangled)* völlig durcheinanderbringen; **get sth. ~d [with sth.]** etw. [mit etw.] durcheinanderbringen

entanglement [ɪn'tæŋglmənt] *n.* **a)** Verwicklung, *die;* **b)** *(fig.: involvement)* **his ~ in a divorce case** seine Verwicklung in eine Scheidungsaffäre; **c)** *(entangled things)* Durcheinander, *das; (Mil.)* [Draht]verhau, *der*

enter ['entə(r)] **1.** *v.i.* **a)** *(go in)* hineingehen; ‹*Fahrzeug:*› hineinfahren; *(come in)* hereinkommen; *(walk into room)* eintreten; *(come on stage)* auftreten; **~ Macbeth** *(Theatre)* Auftritt Macbeth; **~ into a building/another world** ein Gebäude/eine andere Welt betreten; **'E~!'** „Herein!"; **b)** *(announce oneself as competitor in race etc.)* sich zur Teilnahme anmelden (**for** an + *Dat.*). **2.** *v.t.* **a)** *(go into)* [hinein]gehen in (+ *Akk.*); ‹*Fahrzeug:*› [hinein]fahren in (+ *Akk.*); ‹*Flugzeug:*› [hinein]fliegen in (+ *Akk.*); betreten ‹*Gebäude, Zimmer*›; eintreten in (+ *Akk.*) ‹*Zimmer*›; einlaufen in (+ *Akk.*) ‹*Hafen*›; einreisen in (+ *Akk.*) ‹*Land*›; *(drive into)* hineinfahren in (+ *Akk.*); *(come into)* [herein]kommen in (+ *Akk.*); **has it ever ~ed your mind that ...?** ist dir nie der Gedanke gekommen, daß ...?; **b)** *(become a member of)* beitreten (+ *Dat.*) ‹*Verein, Organisation, Partei*›; eintreten in (+ *Akk.*) ‹*Kirche, Kloster*›; ergreifen ‹*Beruf*›; **~ the army/[the] university** zum Militär/auf die *od.* zur Universität gehen; **~ teaching/medicine** den Lehr-/ Arztberuf ergreifen; **~ the law** die juristische Laufbahn einschlagen; **c)** *(participate in)* sich beteiligen an (+ *Dat.*) ‹*Diskussion, Unterhaltung*›; teilnehmen an (+ *Dat.*) ‹*Rennen, Wettbewerb*›; **d)** *(write)* eintragen (**in** in + *Akk.*); **~ sth. in a dictionary/an index** etw. in ein Wörterbuch/ein Register aufnehmen; **e) ~ sb./ sth./one's name for** jmdn./etw./ sich anmelden für ‹*Rennen, Wettbewerb, Prüfung*›

~ into *v.t.* **a)** *(engage in)* anknüpfen ‹*Gespräch*›; sich beteiligen an (+ *Dat.*) ‹*Diskussion, Debatte, Wettbewerb*›; aufnehmen ‹*Beziehung, Verhandlungen*›; *(bind oneself by)* eingehen ‹*Verpflichtung, Ehe, Beziehung*›; schließen ‹*Vertrag*›; **b)** *(form part of)* Bestandteil sein von; **that doesn't ~ into it at all** das hat damit gar nichts zu tun

~ on *v.t.* beginnen ‹*Karriere,*

Laufbahn, Amtsperiode›; in Angriff nehmen ‹*Aufgabe, Projekt*›
~ **'up** *v. t.* eintragen
~ **upon** *see* ~ **on**
enterprise ['entəpraɪz] *n.* **a)** *(undertaking)* Unternehmen, *das;* **commercial** ~: Handelsunternehmen, *das;* **free/private** ~: freies/privates Unternehmertum; **b)** *no indef. art. (readiness to undertake new ventures)* Unternehmungsgeist, *der*
enterprising ['entəpraɪzɪŋ] *adj.* unternehmungslustig; rührig ‹*Geschäftsmann*›; kühn ‹*Reise, Gedanke, Idee*›
entertain [entə'teɪn] *v. t.* **a)** *(amuse)* unterhalten; **we were greatly ~ed by ...:** wir haben uns köstlich über ... *(Akk.)* amüsiert; **b)** *(receive as guest)* bewirten; ~ **sb. to lunch/dinner** *(Brit.)* jmdn. zum Mittag-/Abendessen einladen; **c)** *(have in the mind)* haben ‹*Meinung, Vorstellung*›; hegen *(geh.)* ‹*Gefühl, Vorurteil, Verdacht, Zweifel, Groll*›; *(consider)* in Erwägung ziehen; **he would never ~ the idea of doing that** er würde es nie ernstlich erwägen, das zu tun
entertainer [entə'teɪnə(r)] *n.* Entertainer, *der*/Entertainerin, *die*
entertaining [entə'teɪnɪŋ] **1.** *adj.* unterhaltsam. **2.** *n., no pl., no indef. art.* **they enjoy** ~: sie haben gern Gäste; **do some** *or* **a bit of/a lot of** ~: manchmal/sehr oft Gäste einladen; **she's not very good at** ~: sie ist keine sehr gute Gastgeberin
entertainment [entə'teɪnmənt] *n.* **a)** *(amusement)* Unterhaltung, *die;* **the world of** ~: die Welt des Showbusineß; **b)** *(public performance, show)* Veranstaltung, *die*
enthral (*Amer.:* **enthrall**) [ɪn'θrɔ:l] *v. t.*, **-ll-: a)** *(captivate)* gefangennehmen *(fig.);* **b)** *(delight)* begeistern; entzücken
enthrone [ɪn'θrəʊn] *v. t.* inthronisieren
enthuse [ɪn'θju:z, ɪn'θu:z] *(coll.)* **1.** *v. i.* in Begeisterung ausbrechen (**about, over** über + *Akk.*). **2.** *v. t.* begeistern
enthusiasm [ɪn'θju:zɪæzəm, ɪn'θu:zɪæzəm] *n.* **a)** *no pl.* Enthusiasmus, *der;* Begeisterung, *die* (**for, about** für); **b)** *(thing about which sb. is enthusiastic)* Leidenschaft, *die*
enthusiast [ɪn'θju:zɪæst, ɪn'θu:zɪæst] *n.* Enthusiast, *der; (for sport, pop music)* Fan, *der;* **a DIY** ~: ein begeisterter Heimwerker
enthusiastic [ɪnθju:zɪ'æstɪk, ɪnθu:zɪ'æstɪk] *adj.* begeistert (**about** von); **not be very ~ about doing sth.** keine große Lust haben, etw. zu tun
enthusiastically [ɪnθju:zɪ'æstɪkəlɪ, ɪnθu:zɪ'æstɪkəlɪ] *adv.* begeistert
entice [ɪn'taɪs] *v. t.* locken (**into** in + *Akk.*); ~ **sb./sth. [away] from sb./sth.** jmdn./etw. von jmdm./etw. fortlocken; ~ **sb. into doing** *or* **to do sth.** jmdn. dazu verleiten, etw. zu tun
enticement [ɪn'taɪsmənt] *n.* *(thing)* Lockmittel, *das*
enticing [ɪn'taɪsɪŋ] *adj.* verlockend
entire [ɪn'taɪə(r)] *adj.* **a)** *(whole)* ganz; **b)** *(intact)* vollständig ‹*Ausgabe, Buch, Manuskript, Service*›; **remain** ~: unversehrt bleiben
entirely [ɪn'taɪəlɪ] *adv.* **a)** *(wholly)* völlig; **not ~ suitable for the occasion** dem Anlaß nicht ganz angemessen; **b)** *(solely)* ganz ‹*für sich behalten*›; allein, voll ‹*verantwortlich sein*›; **it's up to you** ~: es liegt ganz bei dir
entirety [ɪn'taɪərətɪ] *n., no pl.* **in its** ~: in seiner/ihrer Gesamtheit
entitle [ɪn'taɪtl] *v. t.* **a)** *(give title of)* ~ **a book/film ...:** einem Buch/Film den Titel ... geben; **b)** *(give rightful claim)* berechtigen (**to** zu); ~ **sb. to do sth.** jmdn. berechtigen *od.* jmdm. das Recht geben, etw. zu tun; **be ~d to [claim] sth.** Anspruch auf etw. *(Akk.)* haben; **be ~d to do sth.** das Recht haben, etw. zu tun
entitlement [ɪn'taɪtlmənt] *n.* *(rightful claim)* Anspruch, *der* (**to** auf + *Akk.*)
entity ['entɪtɪ] *n. (thing that exists)* **[separate]** ~: eigenständiges Gebilde
entomologist [entə'mɒlədʒɪst] *n.* Entomologe, *der*/Entomologin, *die*
entomology [entə'mɒlədʒɪ] *n.* Entomologie, *die;* Insektenkunde, *die*
entourage [ɒntʊ'rɑ:ʒ] *n.* Gefolge, *das*
entrails ['entreɪlz] *n. pl.* Eingeweide; Gedärm, *das*
[1]**entrance** ['entrəns] *n.* **a)** *(entering)* Eintritt, *der* (**into** in + *Akk.*); *(of troops)* Einzug, *der; (of vehicle)* Einfahrt, *die;* **b)** *(on to stage, lit. or fig.)* Auftritt, *der;* **make an** *or* **one's** ~: seinen Auftritt haben; **c)** *(way in)* Eingang, *der* (**to** *Gen. od.* zu); *(for vehicle)* Einfahrt, *die;* **d)** *no pl., no art. (right of admission)* Aufnahme, *die* (**to** in + *Akk.*); ~ **to the concert is by ticket only** man kommt nur mit einer Eintrittskarte in das Konzert; **e)** *(fee)* Eintritt, *der*
[2]**entrance** [ɪn'trɑ:ns] *v. t.* hinreißen; bezaubern; **be ~d by** *or* **with sth.** von etw. hingerissen *od.* bezaubert sein
entrance ['entrəns]: ~ **examination** *n.* Aufnahmeprüfung, *die;* ~ **fee** *n.* Eintrittsgeld, *das; (for competition)* Teilnahmegebühr, *die; (on joining club)* Aufnahmegebühr, *die;* ~ **hall** *n.* Eingangshalle, *die*
entrancing [ɪn'trɑ:nsɪŋ] *adj.* bezaubernd; hinreißend
entrant ['entrənt] *n.* **a)** *(into a profession etc.)* Anfänger, *der*/Anfängerin, *die;* **b)** *(for competition, race, etc.)* Teilnehmer, *der*/Teilnehmerin, *die* (**for** *Gen.*, an + *Dat.*)
entrap [ɪn'træp] *v. t.*, **-pp-** *(trick)* ~ **sb. into doing sth.** jmdn. verleiten, etw. zu tun
entreat [ɪn'tri:t] *v. t.* anflehen
entreaty [ɪn'tri:tɪ] *n.* flehentliche Bitte
entrecôte ['ɒntrəkəʊt] *n. (Gastr.)* ~ **[steak]** Entrecote, *das*
entrench [ɪn'trentʃ] *v. t.* **become ~ed** *(fig.)* ‹*Vorurteil, Gedanke:*› sich festsetzen; ‹*Tradition:*› sich verwurzeln
entrepreneur [ɒntrəprə'nɜ:(r)] *n.* Unternehmer, *der*/Unternehmerin, *die*
entrust [ɪn'trʌst] *v. t.* ~ **sb. with sth.** jmdm. etw. anvertrauen; ~ **sb./sth. to sb./sth.** jmdn./etw. jmdm./einer Sache anvertrauen; ~ **a task to sb.,** ~ **sb. with a task** jmdn. mit einer Aufgabe betrauen
entry ['entrɪ] *n.* **a)** Eintritt, *der* (**into** in + *Akk.*); *(of troops)* Einzug, *der; (into organization)* Beitritt, *der* (**into** zu); *(into country)* Einreise, *die; (ceremonial entrance)* [feierlicher] Einzug; **gain ~ to the house** ins Haus gelangen; **'no ~'** *(for people)* „Zutritt verboten"; *(for vehicle)* „Einfahrt verboten"; **a 'no ~' sign** ein Schild mit der Aufschrift „Zutritt/Einfahrt verboten"; **b)** *(on to stage)* Auftritt, *der;* **c)** *(way in)* Eingang, *der; (for vehicle)* Einfahrt, *die;* **d)** *(registration, item registered)* Eintragung, *die* (**in, into** in + *Akk. od. Dat.*); *(in dictionary, encyclopaedia, year-book, index)* Eintrag, *der;* **make an** ~: eine Eintragung vornehmen; **e)** *(person or thing in competition)* Nennung, *die; (set of answers etc.)* Lösung, *die*

entry: ~ **fee** *see* **entrance fee;** ~ **form** *n.* Anmeldeformular, *das; (for competition)* Teilnahmeschein, *der;* ~ **permit** *n.* Einreiseerlaubnis, *die;* ~ **visa** *n.* Einreisevisum, *das*

entwine [ɪn'twaɪn] *v.t.* ~ **sth. round sb./sth.** etw. um jmdn./etw. schlingen *od. (geh.)* winden; ~ **sth. with sth.** etw. mit etw. umschlingen *od. (geh.)* umwinden

enumerate [ɪ'nju:məreɪt] *v.t.* [einzeln] aufzählen

enumeration [ɪnju:mə'reɪʃn] *n.* Aufzählung, *die*

enunciate [ɪ'nʌnsɪeɪt] *v.t.* artikulieren

enunciation [ɪnʌnsɪ'eɪʃn] *n.* Artikulation, *die;* [deutliche] Aussprache

envelop [ɪn'veləp] *v.t.* [ein]hüllen (**in** in + *Akk.*); **be ~ed in flames** ganz von Flammen umgeben sein

envelope ['envələʊp, 'ɒnvələʊp] *n.* [Brief]umschlag, *der*

enviable ['envɪəbl] *adj.* beneidenswert

envious ['envɪəs] *adj.* neidisch (**of** auf + *Akk.*)

environment [ɪn'vaɪərənmənt] *n.* **a)** *(natural surroundings)* **the** ~: die Umwelt; **the Department of the E~** *(Brit.)* das Umweltministerium; **b)** *(surrounding objects, region)* Umgebung, *die; (social surroundings)* Milieu, *das;* **physical/working** ~: Umwelt, *die*/Arbeitswelt, *die;* **home/family** ~: häusliches Milieu/Familienverhältnisse *Pl.*

environmental [ɪnvaɪərən'mentl] *adj.* Umwelt-

environmentalist [ɪnvaɪərən'mentəlɪst] *n.* Umweltschützer, *der*/-schützerin, *die*

envisage [ɪn'vɪzɪdʒ] *v.t.* sich *(Dat.)* vorstellen; **what do you ~ doing [about it]?** was gedenkst du [in der Sache] zu tun?

envoy ['envɔɪ] *n. (messenger)* Bote, *der*/Botin, *die; (Diplom. etc.)* Gesandte, *der*/Gesandtin, *die*

envy ['envɪ] **1.** *n.* **a)** Neid, *der;* **feelings of** ~: Neidgefühle; **b)** *(object)* **his new sports car was the ~ of all his friends** alle seine Freunde beneideten ihn um seinen neuen Sportwagen. **2.** *v.t.* beneiden; ~ **sb. sth.** jmdn. um etw. beneiden; **I don't ~ you** dich kann ich nicht beneiden

enzyme ['enzaɪm] *n. (Chem.)* Enzym, *das*

ephemeral [ɪ'femərl] *adj.* ephemer[isch] *(geh.);* kurzlebig

epic ['epɪk] **1.** *adj.* **a)** episch; **b)** *(of heroic type or scale, lit. or fig.)* monumental; ~ **film** Filmepos, *das.* **2.** *n.* Epos, *das*

epicentre (*Brit.; Amer.:* **epicenter**) ['epɪsentə(r)] *n.* Epizentrum, *das*

epidemic [epɪ'demɪk] *(Med.; also fig.)* **1.** *adj.* epidemisch. **2.** *n.* Epidemie, *die*

epigram ['epɪgræm] *n. (Lit.)* Epigramm, *das;* Sinngedicht, *das*

epilepsy ['epɪlepsɪ] *n. (Med.)* Epilepsie, *die*

epileptic [epɪ'leptɪk] *(Med.)* **1.** *adj.* epileptisch; *see also* **[1]fit a. 2.** *n.* Epileptiker, *der*/Epileptikerin, *die*

epilogue (*Amer.:* **epilog**) ['epɪlɒg] *n. (Lit.)* Epilog, *der*

Epiphany [ɪ'pɪfənɪ] *n.* **[Feast of the]** ~: Epiphanias, *das;* Dreikönigsfest, *das*

episcopal [ɪ'pɪskəpl] *adj.* episkopal; bischöflich

episode ['epɪsəʊd] *n.* **a)** Episode, *die;* **b)** *(instalment of serial)* Folge, *die*

epistle [ɪ'pɪsl] *n. (Bibl., Lit., or usu. joc.: letter)* Epistel, *die*

epitaph ['epɪtɑ:f] *n.* Epitaph, *das;* Grab[in]schrift, *die*

epithet ['epɪθet] *n.* **a)** Beiname, *der;* **b)** *(Ling.)* Epitheton, *das (fachspr.);* Beiwort, *das*

epitome [ɪ'pɪtəmɪ] *n.* Inbegriff, *der*

epitomize [ɪ'pɪtəmaɪz] *v.t.* ~ **sth.** der Inbegriff einer Sache *(Gen.)* sein

epoch ['i:pɒk, 'epɒk] *n.* Epoche, *die*

'epoch-making *adj.* epochal ‹*Bedeutung*›; epochemachend ‹*Entdeckung*›

equable ['ekwəbl] *adj.* ausgeglichen ‹*Wesen, Person, Klima*›; *(equally proportioned)* ausgewogen ‹*Maße, System, Proportionen*›

equal ['i:kwl] **1.** *adj.* **a)** gleich; ~ **in** *or* **of ~ height/weight/size/importance** *etc.* gleich hoch/schwer/groß/wichtig *usw.;* **not ~ in length** verschieden lang; **divide a cake into ~ parts/portions** einen Kuchen in gleich große Stücke/Portionen aufteilen; ~ **amounts of milk and water** gleich viel Milch und Wasser; **be ~ in size to sth.** ebenso groß wie etw. sein; **Michael came ~ third** *or* **third ~ with Richard in the class exams** bei den Klassenprüfungen kam Michael zusammen mit Richard auf den dritten Platz; **be on ~ terms [with sb.]** [mit jmdm.] gleichgestellt sein; **all/other things being** ~: wenn nichts dazwischen kommt; **b) be ~ to sth./sb.** *(strong, clever, etc. enough)* einer Sache/jmdm. gewachsen sein; **be ~ to doing sth.** imstande sein, etw. zu tun; **c) they were all given ~ treatment** sie wurden alle gleich behandelt; **d)** *(evenly balanced)* ausgeglichen. **2.** *n.* Gleichgestellte, *der/die;* **be sb.'s/sth.'s** ~: jmdm. ebenbürtig sein/einer Sache *(Dat.)* gleichkommen; **he/she/it has no** *or* **is without** ~: er/sie/es hat nicht seines-/ihresgleichen. **3.** *v.t., (Brit.)* **-ll-: a)** *(be equal to)* ~ **sb./sth. [in sth.]** jmdm./einer Sache [in etw. *(Dat.)*] entsprechen; **three times four ~s twelve** drei mal vier ist [gleich] zwölf; **b)** *(do sth. equal to)* ~ **sb.** es jmdm. gleichtun

equalise, equaliser *see* **equaliz-**

equality [ɪ'kwɒlɪtɪ] *n.* Gleichheit, *die; (equal rights)* Gleichberechtigung, *die;* **racial** ~: Gleichberechtigung der Rassen; ~ **between the sexes** Gleichheit von Mann und Frau

equalize ['i:kwəlaɪz] **1.** *v.t.* ausgleichen ‹*Druck, Temperatur*›. **2.** *v.i. (Sport)* den Ausgleich[streffer] erzielen

equalizer ['i:kwəlaɪzə(r)] *n. (Sport)* Ausgleich[streffer], *der*

equally ['i:kwəlɪ] *adv.* **a)** ebenso; **be ~ close to a and b** von a und b gleich weit entfernt sein; **the two are ~ gifted** die beiden sind gleich begabt; **b)** *(in equal shares)* in gleiche Teile ‹*aufteilen*›; gleichmäßig ‹*verteilen*›; **c)** *(according to the same rule and measurement)* in gleicher Weise; gleich ‹*behandeln*›

equal oppor'tunity *n.* Chancengleichheit, *die*

'equals sign *n. (Math.)* Gleichheitszeichen, *das*

equanimity [ekwə'nɪmɪtɪ] *n., no pl.* Gleichmut, *der*

equate [ɪ'kweɪt] *v.t.* ~ **sth. [to** *or* **with sth.]** etw. [einer Sache *(Dat.)* *od.* mit etw.] gleichsetzen

equation [ɪ'kweɪʒn] *n. (Math., Chem.)* Gleichung, *die*

equator [ɪ'kweɪtə(r)] *n. (Geog., Astron.)* Äquator, *der*

equestrian [ɪ'kwestrɪən] *adj.* reiterlich; Reit‹*turnier, -talent*›

equidistant [i:kwɪ'dɪstənt] *adj.* gleich weit entfernt (**from** von)

equilateral [i:kwɪ'lætərl] *adj. (Math.)* gleichseitig

equilibrium [i:kwɪ'lɪbrɪəm] *n., pl.* **equilibria** [i:kwɪ'lɪbrɪə] *or* **~s** Gleichgewicht, *das;* **mental/emotional** ~: geistige/emotionale Ausgeglichenheit; **in** ~: im Gleichgewicht

equinox ['i:kwɪnɒks, 'ekwɪnɒks] *n.* Tagundnachtgleiche, *die*
equip [ɪ'kwɪp] *v.t.*, **-pp-** ausrüsten ⟨*Fahrzeug, Armee, Person*⟩; ausstatten ⟨*Zimmer, Küche*⟩; **fully ~ped** komplett ausgerüstet/ausgestattet
equipment [ɪ'kwɪpmənt] *n.* Ausrüstung, *die; (of kitchen, laboratory, etc.)* Ausstattung, *die; (sth. needed for activity)* Geräte; **breathing/recording ~:** Sauerstoffgerät, *das*/Aufnahmegeräte; **climbing/diving ~:** Bergsteiger-/Taucherausrüstung, *die*
equitable ['ekwɪtəbl] *adj.* gerecht; **in an ~ manner** gerecht
equity ['ekwɪtɪ] *n.* **a)** *(fairness)* Gerechtigkeit, *die;* **b)** *in pl. (stocks and shares without fixed interest)* [Stamm]aktien
equivalence [ɪ'kwɪvələns] *n.* **a)** *(being equivalent)* Gleichwertigkeit, *die; (of two amounts)* Wertgleichheit, *die;* **b)** *(having equivalent meaning)* ~ **[in meaning]** Bedeutungsgleichheit, *die*
equivalent [ɪ'kwɪvələnt] **1.** *adj.* **a)** *(equal, having same result)* gleichwertig; *(corresponding)* entsprechend; **be ~ to sth.** einer Sache *(Dat.)* entsprechen; **be ~ to doing sth.** dasselbe sein, wie wenn man etw. tut; **b)** *(meaning the same)* äquivalent *(Sprachw.);* entsprechend; **these two words are [not] ~ in meaning** diese beiden Wörter sind [nicht] bedeutungsgleich. **2.** *n.* **a)** *(~ or corresponding thing or person)* Pendant, *das,* Gegenstück, *das* (**of** zu); **b)** *(word etc. having same meaning)* Entsprechung, *die* (**of** zu); Äquivalent, *das* (**of** für); **c)** *(thing having same result)* **be the ~ of sth.** einer Sache *(Dat.)* entsprechen
equivocal [ɪ'kwɪvəkl] *adj.* **a)** *(ambiguous)* zweideutig; **b)** *(questionable)* zweifelhaft
equivocate [ɪ'kwɪvəkeɪt] *v.i.* ausweichen
er [3:(r)] *int.* äh
era ['ɪərə] *n.* Ära, *die*
eradicate [ɪ'rædɪkeɪt] *v.t.* ausrotten
erase [ɪ'reɪz] *v.t.* **a)** *(rub out)* auslöschen; *(with rubber, knife)* ausradieren; **b)** *(obliterate)* tilgen *(geh.)* (**from** aus); **c)** *(from recording tape; also Computing)* löschen
eraser [ɪ'reɪzə(r)] *n.* **[pencil] ~:** Radiergummi, *der;* **[blackboard] ~:** *Block mit Filzbelag o.ä. zum Löschen von Kreideschrift*
erect [ɪ'rekt] **1.** *adj.* **a)** *(upright, vertical; also fig.)* aufrecht; gerade ⟨*Rücken, Wuchs*⟩; **b)** *(Physiol.)* erigiert. **2.** *v.t.* errichten; aufbauen ⟨*Gerüst*⟩; aufstellen ⟨*Standbild, Verkehrsschild*⟩; aufschlagen, aufstellen ⟨*Zelt*⟩
erection [ɪ'rekʃn] *n.* **a)** *see* **erect 2:** Errichtung, *die;* Aufbau, *der;* Aufstellen, *das;* Aufschlagen, *das;* **b)** *(structure)* Bauwerk, *das; (other than a building)* Konstruktion, *die;* **c)** *(Physiol.)* Erektion, *die*
ergonomics [3:gə'nɒmɪks] *n., no pl.* Ergonomie, *die*
ermine ['3:mɪn] *n.* **a)** *(fur; also Her.)* Hermelin, *der;* **b)** *(Zool.)* Hermelin, *das*
erode [ɪ'rəʊd] *v.t.* **a)** ⟨*Säure, Rost:*⟩ angreifen; ⟨*Wasser, Regen, Meer:*⟩ auswaschen; ⟨*Wasser, Regen, Meer, Wind:*⟩ erodieren *(Geol.);* **b)** *(fig.)* unterminieren
erosion [ɪ'rəʊʒn] *n.* **a)** *see* **erode a:** Angreifen, *das;* Auswaschung, *die;* Erosion, *die (Geol.);* **b)** *(fig.)* Unterminierung, *die*
erotic [ɪ'rɒtɪk] *adj.*, **erotically** [ɪ'rɒtɪkəlɪ] *adv.* erotisch
err [3:(r)] *v.i.* sich irren; **to ~ is human** *(prov.)* Irren ist menschlich; **let's ~ on the safe side and ...:** um sicher zu gehen, wollen wir ...
errand ['erənd] *n.* Botengang, *der; (shopping)* Besorgung, *die;* **go on** *or* **run an ~:** einen Botengang/eine Besorgung machen
errand: ~-boy *n.* Laufbursche, *der;* Bote[njunge], *der;* **~-girl** *n.* Laufmädchen, *das;* Botin, *die*
erratic [ɪ'rætɪk] *adj.* unregelmäßig; sprunghaft ⟨*Wesen, Person, Art*⟩; unbeständig ⟨*Charakter, Leistung*⟩; launenhaft ⟨*Verhalten*⟩; ungleichmäßig ⟨*Bewegung, Verlauf*⟩
erroneous [ɪ'rəʊnɪəs] *adj.* falsch; irrig ⟨*Schlußfolgerung, Eindruck, Ansicht, Auffassung, Annahme*⟩
erroneously [ɪ'rəʊnɪəslɪ] *adv.* fälschlich; irrigerweise
error ['erə(r)] *n.* **a)** *(mistake)* Fehler, *der;* **gross ~ of judgement** grobe Fehleinschätzung; **b)** *(wrong opinion)* Irrtum, *der;* **realize the ~ of one's ways** seine Fehler einsehen; **in ~:** irrtümlich[erweise]
erudite ['erʊdaɪt] *adj.* gelehrt ⟨*Abhandlung, Vortrag*⟩; gebildet, gelehrt ⟨*Person*⟩
erudition [erʊ'dɪʃn] *n., no pl.* Gelehrsamkeit, *die (geh.)*
erupt [ɪ'rʌpt] *v.i.* **a)** ⟨*Vulkan, Geysir:*⟩ ausbrechen; **~ with anger/into a fit of rage** *(fig.)* einen Wutanfall bekommen; **b)** ⟨*Hautausschlag:*⟩ ausbrechen
eruption [ɪ'rʌpʃn] *n.* *(of volcano, geyser)* Ausbruch, *der;* Eruption, *die (Geol.)*
escalate ['eskəleɪt] **1.** *v.i.* sich ausweiten (**into** zu); eskalieren *(geh.)* (**into** zu); ⟨*Preise, Kosten:*⟩ [ständig] steigen. **2.** *v.t.* ausweiten (**into** zu); eskalieren *(geh.)* (**into** zu)
escalator ['eskəleɪtə(r)] *n.* Rolltreppe, *die*
escalope ['eskələʊp] *n.* *(Gastr.)* Schnitzel, *das*
escapade ['eskəpeɪd] *n.* Eskapade, *die (geh.)*
escape [ɪ'skeɪp] **1.** *n.* **a)** *(lit. or fig.)* Flucht, *die* (**from** aus); *(from prison)* Ausbruch, *der* (**from** aus); **there is no ~** *(lit. or fig.)* es gibt kein Entkommen; **~ vehicle** Fluchtfahrzeug, *das;* **make one's ~ [from sth.]** [aus etw.] entkommen; **have a narrow ~:** gerade noch einmal davonkommen; **have a lucky ~:** glücklich davonkommen; **b)** *(leakage of gas etc.)* Austritt, *der;* Entweichen, *das.* **2.** *v.i.* **a)** *(lit. or fig.)* fliehen (**from** aus); entfliehen *(geh.)* (**from** *Dat.*); *(successfully)* entkommen (**from** *Dat.*); *(from prison)* ausbrechen (**from** aus); ⟨*Großtier:*⟩ ausbrechen; ⟨*Kleintier:*⟩ entlaufen (**from** *Dat.*); ⟨*Vogel:*⟩ entfliegen (**from** *Dat.*); **while trying to ~:** auf der Flucht; **~d prisoner/convict** entflohener Gefangener/Sträfling; **b)** *(leak)* ⟨*Gas:*⟩ ausströmen; ⟨*Flüssigkeit:*⟩ auslaufen; **c)** *(avoid harm)* davonkommen; **~ alive** mit dem Leben davonkommen. **3.** *v.t.* **a)** entkommen (+ *Dat.*) ⟨*Verfolger, Angreifer, Feind*⟩; entgehen (+ *Dat.*) ⟨*Bestrafung, Gefangennahme, Tod, Entdeckung*⟩; verschont bleiben von ⟨*Katastrophe, Krankheit, Zerstörung, Auswirkungen*⟩; **she narrowly ~d being killed** sie wäre fast getötet worden; **b)** *(not be remembered by)* entfallen sein (+ *Dat.*); **c) ~ sb.['s notice]** *(not be seen)* jmdm. entgehen; **~ notice** nicht bemerkt werden; **~ sb.'s attention** jmds. Aufmerksamkeit *(Dat.)* entgehen
escape: ~ attempt *n.* Fluchtversuch, *der; (from prison)* Ausbruchsversuch, *der;* **~ route** *n.* Fluchtweg, *der;* **~ valve** *n.* Sicherheitsventil, *das*
escapism [ɪ'skeɪpɪzm] *n.* Realitätsflucht, *die*
escort 1. ['eskɔ:t] *n.* **a)** *(armed guard)* Eskorte, *die;* Geleitschutz, *der (Milit.);* **police ~:** Polizeieskorte, *die;* **with an ~, under ~:** mit einer Eskorte; **b)** *(person[s] protecting or guiding)* Begleitung,

die; **be sb.'s ~:** jmdn. begleiten; **c)** *(hired companion)* Begleiter, *der*/Begleiterin, *die; (woman also)* ≈ Hostess, *die.* **2.** [ɪ'skɔ:t] *v. t.* **a)** begleiten; *(Mil.)* eskortieren; **b)** *(take forcibly)* bringen

escort ['eskɔ:t]: ~ **agency** *n. Agentur für Begleiter/Begleiterinnen;* ~ **vessel** *n. (Navy)* Geleitschiff, *das*

Eskimo ['eskɪməʊ] **1.** *adj.* Eskimo-. **2.** *n.* **a)** *no pl. (language)* Eskimoisch, *das; see also* **English 2a; b)** *pl.* ~s *or same* Eskimo, *der*/Eskimofrau, *die;* **the ~[s]** die Eskimos

esoteric [esə'terɪk, i:sə'terɪk] *adj.* esoterisch *(geh.)*

ESP *abbr. (Psych.)* **extra-sensory perception** ASW

especial [ɪ'speʃl] *attrib. adj.* [ganz] besonder...

especially [ɪ'speʃəlɪ] *adv.* besonders; **what ~ do you want to see?** was möchten Sie insbesondere sehen?; ~ **as** zumal; **more ~:** ganz besonders

Esperanto [espə'ræntəʊ] *n., no pl.* Esperanto, *das; see also* **English 2a**

espionage ['espɪənɑ:ʒ] *n.* Spionage, *die*

esplanade [esplə'neɪd, esplə'nɑ:d] *n.* Esplanade, *die (geh.)*

espouse [ɪ'spaʊz] *v. t.* eintreten für

espresso [e'spresəʊ] *n., pl.* ~s *(coffee)* Espresso, *der*

e'spresso bar *n.* Espressobar, *die;* Espresso, *das*

Esq. *abbr.* **Esquire** ≈ Hr.; *(on letter)* ≈ Hrn.; **Jim Smith, ~:** Hr./Hrn. Jim Smith

essay ['eseɪ] *n.* Essay, *der;* Aufsatz, *der (bes. Schulw.)*

essence ['esəns] *n.* **a)** Wesen, *das; (gist)* Wesentliche, *das; (of problem, teaching)* Kern, *der;* **in ~:** im wesentlichen; **be of the ~:** von entscheidender Bedeutung sein; **b)** *(Cookery)* Essenz, *die*

essential [ɪ'senʃl] **1.** *adj.* **a)** *(fundamental)* wesentlich ‹*Unterschied, Merkmal, Aspekt*›; entscheidend ‹*Frage*›; **b)** *(indispensable)* unentbehrlich; lebenswichtig ‹*Nahrungsmittel, Güter*›; unabdingbar ‹*Erfordernis, Qualifikation, Voraussetzung*›; unbedingt notwendig ‹*Bestandteile, Maßnahmen, Ausrüstung*›; wesentlich, entscheidend ‹*Rolle*›; ~ **to life** lebensnotwendig *od.* -wichtig; **it is [absolutely** *or* **most] ~ that ...:** es ist unbedingt notwendig, daß ... **2.** *n., esp. in pl.* **a)** *(indispensable element)* Notwendigste, *das;* **the bare ~s** das Allernotwendigste; **b)** *(fundamental element)* Wesentliche, *das;* **the ~s of French grammar** die Grundzüge der französischen Grammatik

essentially [ɪ'senʃəlɪ] *adv.* im Grunde

establish [ɪ'stæblɪʃ] **1.** *v. t.* **a)** *(set up, create, found)* schaffen ‹*Einrichtung, Präzedenzfall, Ministerposten*›; gründen ‹*Organisation, Institut*›; errichten ‹*Geschäft, Lehrstuhl, System*›; einsetzen, bilden ‹*Regierung, Ausschuß*›; herstellen ‹*Kontakt, Beziehungen*› **(with** zu); aufstellen ‹*Rekord*›; ins Leben rufen, begründen ‹*Bewegung*›; ~ **one's authority** sich *(Dat.)* Autorität verschaffen; ~ **law and order** Recht und Ordnung herstellen; **b)** *(secure acceptance for)* etablieren; **become ~ed** sich einbürgern; ~ **one's reputation** sich *(Dat.)* einen Namen machen; **c)** *(prove)* beweisen ‹*Schuld, Unschuld, Tatsache*›; unter Beweis stellen ‹*Können*›; nachweisen ‹*Anspruch*›; **d)** *(discover)* feststellen; ermitteln ‹*Umstände, Aufenthaltsort*›. **2.** *v. refl.* ~ **oneself [at** *or* **in a place]** sich [an einem Ort] niederlassen

established [ɪ'stæblɪʃt] *adj.* **a)** eingeführt ‹*Geschäft usw.*›; bestehend ‹*Ordnung*›; etabliert ‹*Schriftsteller*›; **b)** *(accepted)* üblich; etabliert ‹*Gesellschaftsordnung*›; geltend ‹*Norm*›; fest ‹*Brauch*›; feststehend ‹*Tatsache*›; **become ~:** sich durchsetzen; **c)** *(Eccl.)* ~ **church/religion** Staatskirche/-religion, *die*

establishment [ɪ'stæblɪʃmənt] *n.* **a)** *(setting up, creation, foundation)* Gründung, *die; (of government, committee)* Einsetzung, *die; (of movement)* Begründung, *die; (of relations)* Schaffung, *die;* **b)** *(institution)* **[business] ~:** Unternehmen, *das;* **commercial/industrial ~:** Handels-/Industrieunternehmen, *das;* **c)** *(Brit.)* **the E~:** das Establishment

estate [ɪ'steɪt] *n.* **a)** *(landed property)* Gut, *das;* **b)** *(Brit.) (housing ~)* [Wohn]siedlung, *die; (industrial ~)* Industriegebiet, *das; (trading ~)* Gewerbegebiet, *das;* **c)** *(total assets) (of deceased person)* Erbmasse, *die (Rechtsspr.);* Nachlaß, *der; (of bankrupt)* Konkursmasse, *die (Wirtsch., Rechtsspr.)*

estate: ~ **agent** *n. (Brit.)* Grundstücksmakler, *der;* Immobilienmakler, *der;* ~ **car** *n. (Brit.)* Kombiwagen, *der*

esteem [ɪ'sti:m] **1.** *n., no pl.* Wertschätzung, *die (geh.)* **(for** *Gen.,* für); **hold sb./sth. in [high** *or* **great] ~:** [hohe *od.* große] Achtung vor jmdm./etw. haben. **2.** *v. t.* **a)** *(think favourably of)* schätzen; **highly** *or* **much** *or* **greatly ~ed** hochgeschätzt *(geh.);* sehr geschätzt; **b)** *(consider)* ~ **[as]** erachten für *(geh.);* ansehen als

estimate 1. ['estɪmət] *n.* **a)** *(of number, amount, etc.)* Schätzung, *die;* **at a rough ~:** grob geschätzt; **b)** *(of character, qualities, etc.)* Einschätzung, *die;* **c)** *(Commerc.)* Kostenvoranschlag, *der.* **2.** ['estɪmeɪt] *v. t.* schätzen ‹*Größe, Entfernung, Zahl, Umsatz*› **(at** auf + *Akk.*); einschätzen ‹*Fähigkeiten, Durchführbarkeit, Aussichten*›

estimation [estɪ'meɪʃn] *n.* Schätzung, *die; (of situation etc.)* Einschätzung, *die;* Beurteilung, *die;* **in sb.'s ~:** nach jmds. Schätzung; **go up/down in sb.'s ~:** in jmds. Achtung steigen/sinken

Estonia [e'stəʊnɪə] *pr. n.* Estland *(das)*

estrange [ɪ'streɪndʒ] *v. t.* entfremden **(from** *Dat.*); **be/become ~d from sb.** jmdm. entfremdet sein/sich jmdm. entfremden

estuary ['estjʊərɪ] *n. (Geog.)* Mündung, *die*

etc. *abbr.* **et cetera** usw.

etcetera [et'setərə, ɪt'setərə] und so weiter; et cetera

etch [etʃ] *v. t.* **a)** ätzen **(on** auf *od.* in + *Akk.*); *(on metal also)* ‹*bes. Künstler:*› radieren; **b)** *(fig.)* einprägen **(in, on** *Dat.*)

etching ['etʃɪŋ] *n.* Ätzung, *die; (piece of art)* Radierung, *die*

eternal [ɪ'tɜ:nl, i:'tɜ:nl] *adj.* **a)** ewig; **life ~:** das ewige Leben; ~ **triangle** Dreiecksverhältnis, *das;* **b)** *(coll.: unceasing)* ewig *(ugs.)*

eternity [ɪ'tɜ:nɪtɪ, i:'tɜ:nɪtɪ] *n.* **a)** Ewigkeit, *die;* **for all ~:** [bis] in alle Ewigkeit; **b)** *(coll.: long time)* Ewigkeit, *die (ugs.)*

ether ['i:θə(r)] *n.* Äther, *der*

ethic ['eθɪk] *n.* Ethik, *die (geh.)*

ethical ['eθɪkl] *adj.* **a)** *(relating to morals)* ethisch; ~ **philosophy** Ethik, *die;* **b)** *(morally correct)* moralisch einwandfrei

ethics ['eθɪks] *n., no pl.* **a)** Moral, *die; (moral philosophy)* Ethik, *die;* **b)** *usu. constr. as pl. (moral code)* Ethik, *die (geh.);* **professional ~:** Berufsethos, *das*

Ethiopia [i:θɪ'əʊpɪə] *pr. n.* Äthiopien *(das)*

Ethiopian [i:θɪ'əʊpɪən] **1.** *adj.* äthiopisch. **2.** *n.* Äthiopier, *der*/Äthiopierin, *die*

ethnic ['eθnɪk] *adj.* **a)** ethnisch; Volks‹*gruppe, -musik, -tanz*›; **b)** *(from specified group)* Volks‹*chinesen, -deutsche*›

ethos ['i:θɒs] *n.* *(guiding beliefs)* Gesinnung, *die; (fundamental values)* Ethos, *das (geh.)*

etiquette ['etɪket] *n.* Etikette, *die;* **breach of ~:** Verstoß gegen die Etikette

etymological [etɪmə'lɒdʒɪkl] *adj. (Ling.)* etymologisch

etymology [etɪ'mɒlədʒɪ] *n. (Ling.)* Etymologie, *die*

eucalyptus [ju:kə'lɪptəs] *n.* **a)** ~ **[oil]** *(Pharm.)* Eukalyptusöl, *das;* **b)** *(Bot.)* Eukalyptus[baum], *der*

Eucharist ['ju:kərɪst] *n. (Eccl.)* Eucharistie, *die*

eulogy ['ju:lədʒɪ] *n.* Lobrede, *die; (Amer.: funeral oration)* Grabrede, *die*

eunuch ['ju:nək] *n.* Eunuch, *der*

euphemism ['ju:fəmɪzm] *n.* Euphemismus, *der (bes. Sprachw.);* verhüllende Umschreibung

euphemistic [ju:fə'mɪstɪk] *adj.* euphemistisch *(bes. Sprachw.);* verhüllend

euphoria [ju:'fɔ:rɪə] *n., no pl.* Euphorie, *die (geh.)*

euphoric [ju:'fɔ:rɪk] *adj.* euphorisch *(geh.)*

eureka [jʊə'ri:kə] *int.* heureka *(geh.);* ich hab's *(ugs.)*

Euro- ['jʊərəʊ] *in comb.* euro-/Euro-

'Eurodollar *n. (Econ.)* Eurodollar, *der*

Europe ['jʊərəp] *pr. n.* **a)** Europa *(das);* **b)** *(Brit. coll.: EC)* EG, *die;* **go into ~:** der EG beitreten

European [jʊərə'pi:ən] **1.** *adj.* europäisch; **sb. is ~:** jmd. ist Europäer/Europäerin. **2.** *n.* Europäer, *der*/Europäerin, *die*

European: ~ Eco'nomic Community *n.* Europäische Wirtschaftsgemeinschaft; **~ 'Monetary System** *n.* Europäisches Währungssystem; **~ 'Parliament** *n.* Europäisches Parlament

euthanasia [ju:θə'neɪzɪə] *n.* Euthanasie, *die*

evacuate [ɪ'vækjʊeɪt] *v.t.* **a)** *(remove from danger, clear of occupants)* evakuieren (**from** aus); **b)** *(esp. Mil.: cease to occupy)* räumen

evacuation [ɪvækjʊ'eɪʃn] *n.* **a)** *(removal of people or things, clearance of place)* Evakuierung, *die* (**from** aus); **b)** *(esp. Mil.)* **the ~ of a territory** die Räumung eines Gebietes

evade [ɪ'veɪd] *v.t.* ausweichen (+ *Dat.*) ‹*Angriff, Angreifer, Blick, Problem, Schwierigkeit, Tatsache, Frage, Thema*›; sich entziehen (+ *Dat.*) ‹*Verhaftung, Ergreifung, Wehrdienst, Gerechtigkeit, Pflicht, Verantwortung*›; entkommen (+ *Dat.*) ‹*Polizei, Verfolger, Verfolgung*›; hinterziehen ‹*Steuern, Zölle*›; umgehen ‹*Gesetz, Vorschrift*›; **~ doing sth.** vermeiden, etw. zu tun

evaluate [ɪ'væljʊeɪt] *v.t.* **a)** *(value)* schätzen ‹*Wert, Preis, Schaden, Kosten*›; **b)** *(appraise)* einschätzen; auswerten ‹*Daten*›

evaluation [ɪvæljʊ'eɪʃn] *n.* **a)** Schätzung, *die;* **b)** *(appraisal)* Einschätzung, *die; (of data)* Auswertung, *die*

evangelical [i:væn'dʒelɪkl] *adj.* **a)** *(Protestant)* evangelikal; **b)** *(evangelizing)* missionarisch *(fig.)*

evangelise *see* **evangelize**

evangelism [ɪ'vændʒəlɪzm] *n., no pl.* Evangelisation, *die*

evangelist [ɪ'vændʒəlɪst] *n.* Evangelist, *der*

evangelize [ɪ'vændʒəlaɪz] *v.t.* evangelisieren

evaporate [ɪ'væpəreɪt] *v.i.* **a)** verdunsten; **b)** *(fig.)* sich in Luft auflösen; ‹*Furcht, Begeisterung:*› verfliegen

evaporated 'milk *n.* Kondensmilch, *die*

evaporation [ɪvæpə'reɪʃn] *n.* Verdunstung, *die*

evasion [ɪ'veɪʒn] *n.* **a)** *(avoidance)* Umgehung, *die; (of duty)* Vernachlässigung, *die; (of responsibility, question)* Ausweichen, *das* (**of** vor + *Dat.*); **b)** *(evasive statement)* Ausrede, *die;* **~s** Ausflüchte *Pl.*

evasive [ɪ'veɪsɪv] *adj.* ausweichend ‹*Antwort*›; **be/become [very] ~:** [ständig] ausweichen; **be ~ about sth.** um etw. herumreden; **take ~ action** ein Ausweichmanöver machen

Eve [i:v] *pr. n. (Bibl.)* Eva *(die)*

eve *n.* Vorabend, *der; (day)* Vortag, *der;* **the ~ of** der Abend/Tag vor (+ *Dat.*); der Vorabend/Vortag (+ *Gen.*)

even ['i:vn] **1.** *adj.,* **~er** ['i:vənə(r)], **~est** ['i:vənɪst] **a)** *(smooth, flat)* eben ‹*Boden, Fläche*›; **make sth. ~:** etw. ebnen; **b)** *(level)* gleich hoch ‹*Stapel, Stuhl-, Tischbein*›; gleich lang ‹*Vorhang, Stuhl-, Tischbein usw.*›; **be of ~ height/length** gleich hoch/lang sein; **~ with** genauso hoch/lang wie; **on an ~ keel** *(fig.)* ausgeglichen; **c)** *(straight)* gerade ‹*Saum, Kante*›; **d)** *(parallel)* parallel (**with** zu); **e)** *(regular)* regelmäßig ‹*Zähne*›; *(steady)* gleichmäßig ‹*Schrift, Rhythmus, Atmen, Schlagen*›; stetig ‹*Fortschritt*›; **f)** *(equal)* gleich [groß] ‹*Menge, Abstand*›; gleichmäßig ‹*Verteilung, Aufteilung*›; **the odds are ~, it's an ~ bet** die Chancen stehen fünfzig zu fünfzig *od. (ugs.)* fifty-fifty; **break ~:** die Kosten decken; **g)** *(balanced)* im Gleichgewicht; **h)** *(quits, fully revenged)* **be** *or* **get ~ with sb.** es jmdm. heimzahlen; **i)** *(divisible by two, so numbered)* gerade ‹*Zahl, Seite, Hausnummer*›. **2.** *adv.* **a)** sogar; selbst; **hard, unbearable ~:** hart, ja unerträglich; **do sth. ~ without being told** etw. auch ohne Aufforderung tun; **b)** *with negative* **not** *or* **never ~ ...:** [noch] nicht einmal ...; **without ~ saying goodbye** ohne wenigstens auf Wiedersehen zu sagen; **c)** *with compar. adj. or adv.* sogar noch ‹*komplizierter, weniger, schlimmer usw.*›; **d) ~ if Arsenal win** selbst wenn Arsenal gewinnt; **~ if Arsenal won** selbst wenn Arsenal gewinnen würde; *(fact)* obgleich Arsenal gewann; **~ so** [aber] trotzdem *od.* dennoch; **~ now/then** selbst *od.* sogar jetzt/dann

~ 'out *v.t.* **a)** *(make smooth)* glätten; **b)** ausgleichen ‹*Unterschiede*›

~ 'up *v.t.* ausgleichen; **so as to ~ things up** zum Ausgleich

evening ['i:vnɪŋ] *n.* **a)** Abend, *der; attrib.* Abend‹*vorstellung, -ausgabe, -messe*›; **this/tomorrow ~:** heute/morgen abend; **during the ~:** am Abend; **[early/late] in the ~:** am [frühen/späten] Abend; *(regularly)* [früh/spät] abends; **at eight in the ~:** um acht Uhr abends; **on Wednesday ~s/~:** Mittwoch abends/am Mittwoch abend; **one ~:** eines Abends; **~s, of an ~:** abends; **b)** *(coll: greeting)* 'n Abend! *(ugs.)*

evening: ~ class *n.* Abendkurs, *der;* **take** *or* **do ~ classes in pottery** *etc.* Abendkurse im Töpfern *usw.* besuchen; **~ dress** *n.* **a)** *no pl.* Abendkleidung, *die;* **in [full] ~ dress** in Abendkleidung; **b)** Abendkleid, *das;* **~ gown** *n.* Abendkleid, *das;* **~ 'paper** *n.* Abendzeitung, *die*

evenly ['i:vnlɪ] *adv.* gleichmäßig; **be ~ spaced** den gleichen Abstand voneinander haben; **the runners are ~ matched** die Läufer sind einander ebenbürtig

'even-numbered *adj.* gerade

'evensong *n. (Eccl.)* Abendandacht, *die*

event [ɪ'vent] *n.* **a) in the ~ of his dying** *or* **death** im Falle seines Todes; **in the ~ of sickness/war** im Falle einer Krankheit/im Kriegsfalle; **in that ~:** in dem Falle; **b)** *(outcome)* **in any/either ~ = in any case** *see* **¹case a; at all ~s** auf jeden Fall; **in the ~:** letzten Endes; **c)** *(occurrence)* Ereignis, *das;* **d)** *(Sport)* Wettkampf, *der*

even-'tempered *adj.* ausgeglichen

eventful [ɪ'ventfl] *adj.* ereignisreich ‹*Tag, Zeiten*›; bewegt ‹*Leben, Jugend, Zeiten*›

eventual [ɪ'ventjʊəl] *adj.* **predict sb.'s ~ downfall** vorhersagen, daß jmd. schließlich zu Fall kommen wird; **the rise of Napoleon and his ~ defeat** der Aufstieg Napoleons und schließlich seine Niederlage

eventuality [ɪventjʊ'ælɪtɪ] *n.* Eventualität, *die;* **in certain eventualities** in bestimmten [möglichen] Fällen; **be ready for all eventualities** auf alle Eventualitäten gefaßt sein

eventually [ɪ'ventjʊəlɪ] *adv.* schließlich

ever ['evə(r)] *adv.* **a)** *(always, at all times)* immer; stets; **for ~:** für immer ‹*weggehen, gelten*›; ewig ‹*lieben, dasein, leben*›; **for ~ and ~:** immer und ewig; **for ~ and a day** eine Ewigkeit; **~ since [then]** seit [dieser Zeit]; **~ since I've known her** solange ich sie kenne; **~ since I can remember** soweit ich zurückdenken kann; **b)** *in comb. with compar. adj. or adv.* noch; immer; **get ~ deeper into debt** sich noch *od.* immer mehr verschulden; **~ further** noch immer weiter; **c)** *in comb. with participles etc.* **~-increasing** ständig zunehmend; **~-present** allgegenwärtig; **d)** *(at any time)* je[mals]; **not ~:** noch nie; **~ before** je zuvor; **never ~:** nie im Leben; **nothing ~ happens** es passiert nie etwas; **his best performance ~:** seine beste Vorstellung überhaupt; **it hardly ~ rains** es regnet so gut wie nie; **don't you ~ do that again!** mach das bloß nicht noch mal!; **better than ~:** besser denn je; **as ~:** wie gewöhnlich; *(iron.)* wie gehabt; **if I ~ catch you doing that again** wenn ich dich dabei noch einmal erwische; **the greatest tennis-player ~:** der größte Tennisspieler, den es je gegeben hat; **e)** *(coll.) emphasizing question* **what ~ does he want?** was will er nur?; **how ~ did I drop it?/could I have dropped it?** wie konnte ich es nur fallen lassen?; **why ~ not?** warum denn nicht?; **f)** *intensifier before* **~ he opened his mouth** noch bevor er seinen Mund aufmachte; **as soon as ~ I can** so bald wie irgend möglich; **I'm ~ so sorry** *(coll.)* mir tut es ja so leid; **thanks ~ so [much]** *(coll.)* vielen herzlichen Dank; **it was ~ such a shame** *(coll.)* es war so schade

'evergreen 1. *adj.* **a)** immergrün ‹*Baum, Strauch*›; **b)** *(fig.)* immer wieder aktuell ‹*Problem, Thema*›; immer wieder gern gehört ‹*Lied, Schlager, Sänger*›; **~ song** Evergreen, *der.* **2.** *n.* immergrüne Pflanze/immergrüner Baum

ever'lasting *adj.* **a)** *(eternal)* immerwährend; ewig ‹*Leben, Höllenqualen, Gott, Gedenken*›; unvergänglich ‹*Ruhm, Ehre*›; **b)** *(incessant)* ewig *(ugs.);* endlos

everlastingly [evə'lɑ:stɪŋlɪ] *adv.* **a)** *(eternally)* ewig ‹*leben, leiden*›; **b)** *(incessantly)* ewig *(ugs.);* ständig

ever'more *adv.* auf ewig; **for ~:** in [alle] Ewigkeit

every ['evrɪ] *adj.* **a)** *(each single)* jeder/jede/jedes; **have ~ reason** allen Grund haben; **~ [single] time/on ~ [single] occasion** [aber auch] jedesmal; **he ate ~ last** *or* **single biscuit** *(coll.)* er hat die ganzen Kekse aufgegessen *(ugs.);* **~ one** jeder/jede/jedes [einzelne]; **b)** *after possessive adj.* **your ~ wish** all[e] deine Wünsche; **his ~ thought** all[e] seine Gedanken; **c)** *(indicating recurrence)* **she comes [once] ~ day** sie kommt jeden Tag [einmal]; **~ three/few days** alle drei/paar Tage; **~ other** *(~ second, or fig.: almost ~)* jeder/jede/jedes zweite; **~ now and then** *or* **again, ~ so often, ~ once in a while** hin und wieder; **d)** *(the greatest possible)* unbedingt, uneingeschränkt ‹*Vertrauen*›; voll ‹*Beachtung*›; all ‹*Respekt, Aussicht*›; **I wish you ~ happiness/success** ich wünsche dir alles Gute/viel Erfolg

'everybody *n. & pron.* jeder; **~ else** alle anderen; **~ knows ~ else round here** hier kennt jeder jeden; **he asked ~ to be quiet** er bat alle um Ruhe; **opera isn't [to] ~'s taste** Oper ist nicht jedermanns Sache

'everyday *attrib. adj.* alltäglich; Alltags‹*kleidung, -sprache*›; **in ~ life** im Alltag; im täglichen Leben

everyone ['evrɪwʌn, 'evrɪwən] *see* **everybody**

'everyplace *(Amer.) see* **everywhere**

'everything *n. & pron.* **a)** alles; **~ else** alles andere; **~ interesting/valuable** alles Interessante/Wertvolle; **there's a [right] time for ~:** alles zu seiner Zeit; **b)** *(coll.: all that matters)* alles; **looks aren't ~:** das Aussehen [allein] ist nicht alles

'everywhere *adv.* **a)** *(in every place)* überall; **b)** *(to every place)* **go ~:** überall hingehen/-fahren; **~ you go/look** wohin man auch geht/sieht

evict [ɪ'vɪkt] *v.t.* **~ sb. [from his/her home]** jmdn. zur Räumung [seiner Wohnung] zwingen

eviction [ɪ'vɪkʃn] *n.* Zwangsräumung, *die;* **the ~ of the tenant** die zwangsweise Vertreibung des Mieters [aus seiner Wohnung]

evidence ['evɪdəns] *n.* **a)** Beweis, *der;* **be ~ of sth.** etw. beweisen; **provide ~ of sth.** den Beweis *od.* Beweise für etw. liefern; **there was no ~ of a fight** nichts deutete auf einen Kampf hin; **b)** *(Law)* Beweismaterial, *das; (testimony)* [Zeugen]aussage, *die;* **give ~:** [als Zeuge] aussagen; **piece of ~:** Beweisstück, *das; (statement)* Beweis, *der;* **c) be [much] in ~:** [stark] in Erscheinung treten; **he was nowhere in ~:** er war nirgends zu sehen; **sth. is very much in ~:** überall sieht man etw.

evident ['evɪdənt] *adj.* offensichtlich; deutlich ‹*Verbesserung*›; **be ~ to sb.** jmdm. klar sein; **it soon became ~ that ...:** es stellte sich bald heraus, daß ...

evidently ['evɪdəntlɪ] *adv.* offensichtlich

evil ['i:vl, 'i:vɪl] **1.** *adj.* **a)** böse; schlecht ‹*Charakter, Beispiel, Einfluß, System*›; übel, verwerflich ‹*Praktiken*›; **b)** *(unlucky)* verhängnisvoll, unglückselig ‹*Tag, Stunde*›; **~ days** *or* **times** schlechte *od.* schlimme Zeiten; **c)** *(disagreeable)* übel ‹*Geruch, Geschmack*›. **2.** *n.* **a)** *no pl. (literary)* Böse, *das;* **the root of all ~:** die Wurzel allen Übels; **b)** *(bad thing)* Übel, *das;* **necessary ~:** notwendiges Übel; **the lesser ~:** das kleinere Übel

evil: ~-doer ['i:vldu:ə(r)] *n.* Übeltäter, *der*/Übeltäterin, *die;* **~-'minded** *adj.* bösartig; **~-smelling** *adj.* übelriechend

evince [ɪ'vɪns] *v.t.* ‹*Person:*› an den Tag legen; ‹*Äußerung, Handlung:*› zeugen von

evocation [evə'keɪʃn] *n.* Heraufbeschwören, *das*

evocative [ɪ'vɒkətɪv] *adj.* *(thought-provoking)* aufrüttelnd *(fig.)*; **be ~ of sth.** an etw. *(Akk.)* erinnern; etw. heraufbeschwören; **an ~ scent** ein Duft, der Erinnerungen weckt

evoke [ɪ'vəʊk] *v. t.* heraufbeschwören; hervorrufen ‹*Bewunderung, Überraschung, Wirkung*›; erregen ‹*Interesse*›

evolution [i:və'lu:ʃn, evə'lu:ʃn] *n.* **a)** *(development)* Entwicklung, *die;* **b)** *(Biol.: of species etc.)* Evolution, *die;* **theory of ~:** Evolutionstheorie, *die*

evolutionary [i:və'lu:ʃənərɪ, evə'lu:ʃənərɪ] *adj.* evolutionär

evolve [ɪ'vɒlv] **1.** *v. i.* sich entwickeln (**out of, from** aus; **into** zu). **2.** *v. t.* entwickeln (**from** aus)

ewe [ju:] *n.* Mutterschaf, *das*

ex- *pref.* Ex-‹*Freundin, Präsident, Champion*›; Alt‹*[bundes]kanzler, -bundespräsident*›; ehemalig

exacerbate [ek'sæsəbeɪt] *v. t.* verschlimmern ‹*Schmerz, Krankheit, Wut*›; verschlechtern ‹*Zustand*›; verschärfen ‹*Lage*›

exact [ɪg'zækt] **1.** *adj.* genau; exakt, genau ‹*Daten, Berechnung*›; **those were his ~ words** das waren genau seine Worte; **on the ~ spot where ...:** genau an der Stelle, wo ...; **could you give me the ~ money?** könnten Sie mir das Geld passend geben? **2.** *v. t.* fordern, verlangen; erheben ‹*Gebühr*›

exacting [ɪg'zæktɪŋ] *adj.* anspruchsvoll; streng ‹*Lehrer, Maßstab*›; hoch ‹*Anforderung, Maßstab*›

exactitude [ɪg'zæktɪtju:d] *n., no pl.* Genauigkeit, *die*

exactly [ɪg'zæktlɪ] *adv.* **a)** genau; **when ~** *or* **~ when did he leave?** wann genau ging er?; **at ~ the right moment** genau im richtigen Moment; **~!** genau!; **at four o'clock ~:** Punkt vier Uhr; **not ~** *(coll. iron.)* nicht gerade; **b)** *(with perfect accuracy)* [ganz] genau

exactness [ɪg'zæktnɪs] *n., no pl.* Genauigkeit, *die*

exaggerate [ɪg'zædʒəreɪt] *v. t.* übertreiben; **you are exaggerating his importance** du machst ihn wichtiger, als er ist

exaggerated [ɪg'zædʒəreɪtɪd] *adj.* übertrieben

exaggeration [igzædʒə'reɪʃn] *n.* Übertreibung, *die;* **it is a wild/is no ~ to say that ...:** es ist stark/nicht übertrieben, wenn man sagt, daß ...

exalt [ɪg'zɔ:lt] *v. t.* [lob]preisen

exalted [ɪg'zɔ:ltɪd] *adj.* **a)** *(high-ranking)* hoch; **b)** *(lofty, sublime)* hoch ‹*Ideal*›; erhaben ‹*Thema, Stil, Stimmung, Gedanke*›

exam [ɪg'zæm] *(coll.) see* **examination c**

examination [ɪgzæmɪ'neɪʃn] *n.* **a)** *(inspection)* Untersuchung, *die; (of accounts)* [Über]prüfung, *die;* **be under ~:** untersucht *od.* überprüft werden; **b)** *(Med.)* Untersuchung, *die;* **undergo an ~:** sich untersuchen lassen; **c)** *(test of knowledge or ability)* Prüfung, *die; (final ~ at university)* Examen, *das;* **d)** *(Law) (of witness, accused)* Verhör, *das;* Vernehmung, *die; (of case)* Untersuchung, *die*

exami'nation-paper *n.* **a)** ~[s] schriftliche Prüfungsaufgaben; **b)** *(with candidate's answers)* ≈ Klausurarbeit, *die*

examine [ɪg'zæmɪn] *v. t.* **a)** *(inspect)* untersuchen (**for** auf + *Akk.*); prüfen ‹*Dokument, Gewissen, Geschäftsbücher*›; **b)** *(Med.)* untersuchen; **c)** *(test knowledge or ability of)* prüfen (**in** in + *Dat.*); **~ sb. on his knowledge of French** jmds. Französischkenntnisse prüfen; **d)** *(Law)* verhören; vernehmen

examinee [ɪgzæmɪ'ni:] *n.* Prüfungskandidat, *der*/-kandidatin, *die;* Prüfling, *der; (Univ. also)* Examenskandidat, *der*/-kandidatin, *die*

examiner [ɪg'zæmɪnə(r)] *n.* Prüfer, *der*/Prüferin, *die;* **board of ~s** Prüfungsausschuß, *der*

example [ɪg'zɑ:mpl] *n.* Beispiel, *das;* **by way of [an] ~:** als Beispiel; **take sth. as an ~:** etw. zum Beispiel nehmen; **for ~:** zum Beispiel; **set an ~** *or* **a good ~ to sb.** jmdm. ein Beispiel geben; **make an ~ of sb.** ein Exempel an jmdm. statuieren

exasperate [ɪg'zæspəreɪt, ɪg'zɑ:spəreɪt] *v. t.* *(irritate)* verärgern; *(infuriate)* zur Verzweiflung bringen; **be ~d at** *or* **by sb./sth.** über jmdn./etw. verärgert/verzweifelt sein; **become** *or* **get ~d [with sb.]** sich [über jmdn.] ärgern

exasperating [ɪg'zæspəreɪtɪŋ, ɪg'zɑ:spəreɪtɪŋ] *adj.* ärgerlich; ‹*Aufgabe*› die einen zur Verzweiflung bringt; **be ~:** einen zur Verzweiflung bringen

exasperation [ɪgzæspə'reɪʃn, ɪgzɑ:spə'reɪʃn] *n. see* **exasperate:** Ärger, *der*/Verzweiflung, *die* (**with** über + *Akk.*); **in ~:** verärgert/verzweifelt

excavate ['ekskəveɪt] *v. t.* **a)** ausschachten; *(with machine)* ausbaggern; fördern, abbauen ‹*Erz, Metall*›; **b)** *(Archaeol.)* ausgraben

excavation [ekskə'veɪʃn] *n.* **a)** Ausschachtung, *die; (with machine)* Ausbaggerung, *die; (of ore, metals)* Förderung, *die;* Abbau, *der;* **b)** *(Archaeol.)* Ausgrabung, *die; (place)* Ausgrabungsstätte, *die*

excavator ['ekskəveɪtə(r)] *n.* *(machine)* Bagger, *der*

exceed [ɪk'si:d] *v. t.* **a)** *(be greater than)* übertreffen (**in** an + *Dat.*); ‹*Kosten, Summe, Anzahl:*› übersteigen (**by** um); **not ~ing** bis zu; **b)** *(go beyond)* überschreiten; hinausgehen über (+ *Akk.*) ‹*Auftrag, Befehl*›

exceedingly [ɪk'si:dɪŋlɪ] *adv.* äußerst; ausgesprochen ‹*häßlich, dumm*›

excel [ɪk'sel] **1.** *v. t.,* **-ll-** übertreffen; **~ oneself** *(lit. or iron.)* sich selbst übertreffen. **2.** *v. i.,* **-ll-** sich hervortun (**at, in** in + *Dat.*)

excellence ['eksələns] *n.* hervorragende Qualität

excellency ['eksələnsɪ] *n.* Exzellenz, *die*

excellent ['eksələnt] *adj.* ausgezeichnet; hervorragend; exzellent *(geh.)*; vorzüglich ‹*Wein, Koch, Speise*›

except [ɪk'sept] **1.** *prep.* **~ [***(coll.)* **for]** außer (+ *Dat.*); **~ for** *(in all respects other than)* bis auf (+ *Akk.*); abgesehen von; **~ [for the fact] that ...,** *(coll.)* **~ ...:** abgesehen davon, daß ...; **there was nothing to be done ~ [to] stay there** man konnte nichts anderes tun als dableiben. **2.** *v. t.* ausnehmen (**from** bei); **~ed** ausgenommen

excepting [ɪk'septɪŋ] *prep.* außer (+ *Dat.*); **not ~ Peter** Peter nicht ausgenommen; **~ that ...,** *(coll.)* **~ ...:** abgesehen davon, daß ...

exception [ɪk'sepʃn] *n.* **a)** Ausnahme, *die;* **with the ~ of** mit Ausnahme (+ *Gen.*); **with the ~ of her/myself** mit Ausnahme von ihr/mir; **the ~ proves the rule** *(prov.)* Ausnahmen bestätigen die Regel; **make an ~ [of/for sb.]** [bei jmdm.] eine Ausnahme machen; **b) take ~ to sth.** an etw. *(Dat.)* Anstoß nehmen

exceptional [ɪk'sepʃənl] *adj.* außergewöhnlich; **in ~ cases** in Ausnahmefällen

exceptionally [ɪk'sepʃənəlɪ] *adv.* **a)** *(as an exception)* ausnahmsweise; **b)** *(remarkably)* ungewöhnlich; außergewöhnlich

excerpt ['eksɜ:pt] *n.* Auszug, *der* (**from, of** aus)

excess [ɪk'ses] *n.* **a)** *(inordinate degree or amount)* Übermaß, *das* (**of** an + *Dat.*); **eat/drink to ~:** übermäßig essen/trinken; **in ~:** im Übermaß; **b)** *esp. in pl. (im-*

moderate act) Exzeß, *der; (savage also)* Ausschreitung, *die;* **c)** **be in ~ of sth.** etw. übersteigen; **in ~ of a million** über eine Million; **d)** *(surplus)* Überschuß, *der* (**of** an + *Dat.*); **~ weight** Übergewicht, *das;* **e)** *(esp. Brit. Insurance)* Selbstbeteiligung, *die*

excess ['ekses]: **~ 'baggage** *n.* Mehrgepäck, *das;* **~ 'fare** *n.* Mehrpreis, *der;* **pay the ~ fare** nachlösen

excessive [ɪk'sesɪv] *adj.* übermäßig; übertrieben ‹*Forderung*›; zu stark ‹*Schmerz, Belastung*›; unmäßig ‹*Esser, Trinker*›

excessively [ɪk'sesɪvlɪ] *adv.* übertrieben; unmäßig ‹*essen, trinken*›

exchange [ɪks'tʃeɪndʒ] **1.** *v.t.* **a)** tauschen ‹*Plätze, Zimmer, Ringe, Küsse*›; umtauschen, wechseln ‹*Geld*›; austauschen ‹*Adressen, [Kriegs]gefangene, Erinnerungen, Gedanken, Erfahrungen*›; wechseln ‹*Blicke, Worte, Ringe*›; **~ letters** einen Briefwechsel führen; **~ blows/insults** sich schlagen/sich gegenseitig beleidigen; **b)** *(give in place of another)* eintauschen (**for** für, gegen); umtauschen ‹*[gekaufte] Ware*› (**for** gegen); austauschen ‹*Spion*› (**for** gegen). **2.** *v.i.* tauschen. **3.** *n.* **a)** Tausch, *der; (of prisoners, spies, compliments, greetings, insults)* Austausch, *der;* **an ~ of ideas/blows** ein Meinungsaustausch/Handgreiflichkeiten *Pl.;* **in ~:** dafür; **in ~ for sth.** für etw.; **b)** *(Educ.)* Austausch, *der; attrib.* Austausch-; **c)** *(of money)* Umtausch, *der;* **~ [rate], rate of ~:** Wechselkurs, *der;* **~ rate mechanism** Wechselkursmechanismus, *der;* **d)** *see* **telephone exchange**

exchequer [ɪks'tʃekə(r)] *n. (Brit.)* Schatzamt, *das;* Finanzministerium, *das*

[1]**excise** ['eksaɪz] *n.* Verbrauchsteuer, *die;* **Customs and E~ [Department]** *(Brit.) Amt für Zölle und Verbrauchsteuer*

[2]**excise** [ɪk'saɪz] *v.t.* **a)** *(from book, article)* entfernen (**from** aus); *(from film also)* herausschneiden (**from** aus); **b)** *(Med.)* entfernen; exzidieren *(fachspr.)*

excitable [ɪk'saɪtəbl] *adj.* leicht erregbar

excite [ɪk'saɪt] *v.t.* **a)** *(thrill)* begeistern; **she was/became ~d by the idea** die Idee begeisterte sie; **b)** *(agitate)* aufregen; **be/become ~d by sth.** sich über etw. *(Akk.)* aufregen *od.* erregen; **c)** *(stimulate sexually)* erregen

excited [ɪk'saɪtɪd] *adj.* **a)** *(thrilled)* aufgeregt (**at** über + *Akk.*); **you don't seem very ~ [about it]** du scheinst [davon] nicht sehr begeistert zu sein; **it's nothing to get ~ about** es ist nichts Besonderes; **don't get ~, it's only Tom** keine Aufregung, es ist nur Tom; **b)** *(agitated)* erregt; aufgeregt; **it's nothing to get ~ about** es besteht kein Grund zur Aufregung; **don't get ~, it's only Tom** keine Panik, es ist nur Tom; **don't get so ~:** reg dich nicht so auf; **c)** *(sexually)* erregt

excitement [ɪk'saɪtmənt] *n., no pl.* Aufregung, *die; (enthusiasm)* Begeisterung, *die*

exciting [ɪk'saɪtɪŋ] *adj.* aufregend; *(full of suspense)* spannend

exclaim [ɪk'skleɪm] **1.** *v.t.* ausrufen; **~ that ...:** rufen, daß ... **2.** *v.i.* aufschreien

exclamation [eksklə'meɪʃn] *n.* Ausruf, *der*

excla'mation mark *n.* Ausrufezeichen, *das*

exclude [ɪk'sklu:d] *v.t.* **a)** *(keep out)* ausschließen (**from** von); **sb. is ~d from a profession/the Church** jmdm. ist die Ausübung eines Berufes/die Zugehörigkeit zur Kirche verwehrt; **b)** *(leave out of account)* nicht berücksichtigen (**from** bei)

excluding [ɪk'sklu:dɪŋ] *prep.* **~ drinks/VAT** Getränke ausgenommen/ohne Mehrwertsteuer

exclusion [ɪk'sklu:ʒn] *n.* Ausschluß, *der;* **[talk about sth.] to the ~ of everything else** ausschließlich [über etw. *(Akk.)* sprechen]

exclusive [ɪk'sklu:sɪv] *adj.* **a)** *(not shared)* alleinig ‹*Besitzer, Kontrolle*›; Allein‹*eigentum*›; *(Journ.)* Exklusiv‹*bericht, -interview*›; **~ right** Alleinrecht, *das;* **have ~ rights** die Alleinrechte/Exklusivrechte haben; **b)** *(select)* exklusiv; **c)** *(excluding)* ausschließlich; **~ of** ohne; **~ of drinks** Getränke ausgenommen; **the price is ~ of postage** Versandkosten sind im Preis nicht inbegriffen; **be mutually ~:** sich gegenseitig ausschließen

exclusively [ɪk'sklu:sɪvlɪ] *adv.* ausschließlich; *(Journ.)* exklusiv

excommunicate [ekskə'mju:nɪkeɪt] *v.t. (Eccl.)* exkommunizieren

excommunication [ekskəmju:nɪ'keɪʃn] *n. (Eccl.)* Exkommunikation, *die*

excrement ['ekskrɪmənt] *n. in sing. or pl.* Exkremente *Pl. (bes. Med.);* Kot, *der (geh.)*

excruciating [ɪk'skru:ʃɪeɪtɪŋ] *adj.* unerträglich; qualvoll ‹*Tod*›; **an ~ pun** ≈ ein schlimmer Kalauer

excursion [ɪk'skɜ:ʃn] *n.* Ausflug, *der;* **day ~:** Tagesausflug, *der*

excusable [ɪk'skju:zəbl] *adj.* entschuldbar; verzeihlich

excuse 1. [ɪk'skju:z] *v.t.* **a)** *(forgive, exonerate)* entschuldigen; **~ oneself** *(apologize)* sich entschuldigen; **~ me** Entschuldigung; Verzeihung; **please ~ me** bitte entschuldigen Sie; **~ me[, what did you say]?** *(Amer.)* Verzeihung[, was haben Sie gesagt]?; **~ me if I don't get up** entschuldigen Sie, wenn ich nicht aufstehe; **~ sb. sth.** etw. bei jmdm. entschuldigen; **I can be ~d for confusing them** es ist verzeihlich, daß ich sie verwechselt habe; **b)** *(release)* befreien; **~ sb. [from] sth.** jmdn. von etw. befreien; **c)** *(allow to leave)* entschuldigen; **~ oneself** sich entschuldigen; **if you will ~ me** wenn Sie mich bitte entschuldigen wollen; **you are ~d** ihr könnt gehen; **may I be ~d?** *(euphem.: to go to the toilet)* darf ich mal austreten? **2.** [ɪk'skju:s] *n.* Entschuldigung, *die;* **give** *or* **offer an ~ for sth.** sich für etw. entschuldigen; **there is no ~ for what I did** was ich getan habe, ist nicht zu entschuldigen; **I'm not trying to make ~s, but ...:** das soll keine Entschuldigung sein, aber ...; **any ~ for a drink!** zum Trinken gibt es immer einen Grund!

ex-di'rectory *adj. (Brit. Teleph.)* Geheim‹*nummer, -anschluß*›

execute ['eksɪkju:t] *v.t.* **a)** *(kill)* hinrichten; exekutieren *(Milit.);* **b)** *(put into effect)* ausführen; durchführen ‹*Vorschrift, Gesetz*›; **c)** *(Law)* vollstrecken ‹*Testament*›; unterzeichnen ‹*Urkunde*›

execution [eksɪ'kju:ʃn] *n.* **a)** *(killing)* Hinrichtung, *die;* Exekution, *die (Milit.);* **b)** *(putting into effect)* Ausführung, *die; (of instruction, law)* Durchführung, *die;* **in the ~ of one's duty** in Erfüllung seiner Pflicht

executioner [eksɪ'kju:ʃənə(r)] *n.* Scharfrichter, *der*

executive [ɪg'zekjʊtɪv] **1.** *n.* **a)** *(person)* leitender Angestellter/leitende Angestellte; **b)** *(administrative body)* **the ~** *(of government)* die Exekutive; *(of political organization, trade union)* der Vorstand. **2.** *adj.* **a)** *(Commerc.)* leitend ‹*Stellung, Funktion*›; **b)** *(relating to government)* exekutiv

executive com'mittee *n.* [geschäftsführender] Vorstand

executor [ɪg'zekjʊtə(r)] *n. (Law)* Testamentsvollstrecker, *der*

exemplary [ɪg'zemplərɪ] *adj.* **a)** *(model)* vorbildlich; **b)** *(deterrent)* exemplarisch

exemplify [ɪg'zemplɪfaɪ] *v. t.* veranschaulichen

exempt [ɪg'zempt] **1.** *adj.* befreit (**from** von). **2.** *v. t.* befreien (**from** von)

exemption [ɪg'zempʃn] *n.* Befreiung, *die*

exercise ['eksəsaɪz] **1.** *n.* **a)** *no pl., no indef. art. (physical exertion)* Bewegung, *die;* **take ~:** sich *(Dat.)* Bewegung verschaffen; **b)** *(task set, activity; also Mus., Sch.)* Übung, *die;* **the object of the ~:** der Sinn der Übung; **c)** *(to improve fitness)* [Gymnastik]übung, *die;* **d)** *no pl. (employment, application)* Ausübung, *die;* **e)** *usu. in pl. (Mil.)* Übung, *die.* **2.** *v. t.* **a)** ausüben ‹*Recht, Macht, Einfluß*›; walten lassen ‹*Vorsicht*›; sich üben in (+ *Dat.*) ‹*Zurückhaltung, Diskretion*›; **b)** **~ the mind** die geistigen Fähigkeiten herausfordern; **c)** *(physically)* trainieren ‹*Körper, Muskeln*›; bewegen ‹*Pferd*›. **3.** *v. i.* sich *(Dat.)* Bewegung verschaffen

'exercise book *n.* [Schul]heft, *das*

exert [ɪg'zɜ:t] **1.** *v. t.* aufbieten ‹*Kraft, Beredsamkeit*›; ausüben ‹*Einfluß, Druck, Macht*›. **2.** *v. refl.* sich anstrengen

exertion [ɪg'zɜ:ʃn] *n.* **a)** *no pl. (of strength, force)* Aufwendung, *die; (of influence, pressure, force)* Ausübung, *die;* **b)** *(effort)* Anstrengung, *die*

exhale [eks'heɪl] **1.** *v. t.* ausatmen. **2.** *v. i.* ausatmen; exhalieren *(Med.)*

exhaust [ɪg'zɔ:st] **1.** *v. t.* **a)** *(use up)* erschöpfen; erschöpfend behandeln ‹*Thema*›; **b)** *(tire)* erschöpfen; **have ~ed oneself** sich völlig verausgabt haben. **2.** *n. (Motor Veh.)* **a)** **~ [system]** Auspuff, *der;* **b)** *(what is expelled)* Auspuffgase *Pl.*

exhausted [ɪg'zɔ:stɪd] *adj.* erschöpft

exhausting [ɪg'zɔ:stɪŋ] *adj.* anstrengend

exhaustion [ɪg'zɔ:stʃn] *n., no pl.* Erschöpfung, *die*

exhaustive [ɪg'zɔ:stɪv] *adj.* umfassend

ex'haust-pipe *n. (Motor Veh.)* Auspuffrohr, *das*

exhibit [ɪg'zɪbɪt] **1.** *v. t.* **a)** *(display)* vorzeigen; *(show publicly)* ausstellen; **b)** *(manifest)* zeigen ‹*Mut, Verachtung, Symptome, Neigung, Angst*›. **2.** *n.* **a)** Ausstellungsstück, *das;* **b)** *(Law)* Beweisstück, *das*

exhibition [eksɪ'bɪʃn] *n.* **a)** *(public display)* Ausstellung, *die;* **b)** *(derog.)* **make an ~ of oneself** sich unmöglich aufführen

exhibitionist [eksɪ'bɪʃənɪst] *n.* Exhibitionist, *der*/Exhibitionistin, *die*

exhibitor [ɪg'zɪbɪtə(r)] *n.* Aussteller, *der*/Ausstellerin, *die*

exhilarated [ɪg'zɪləreɪtɪd] *adj.* belebt; *(gladdened)* fröhlich gestimmt; *(stimulated)* angeregt

exhilarating [ɪg'zɪləreɪtɪŋ] *adj.* belebend; fröhlich stimmend ‹*Nachricht, Musik, Anblick*›

exhilaration [ɪgzɪlə'reɪʃn] *n.* **[feeling of] ~:** Hochgefühl, *das*

exhort [ɪg'zɔ:t] *v. t.* [ernsthaft] ermahnen

exhortation [eksɔ:'teɪʃn] *n.* Ermahnung, *die*

exile ['eksaɪl, 'egzaɪl] **1.** *n.* **a)** Exil, *das; (forcible also)* Verbannung, *die* (**from** aus); **in ~:** im Exil; **into ~:** ins Exil; **b)** *(person, lit. or fig.)* Verbannte, *der/die.* **2.** *v. t.* verbannen

exist [ɪg'zɪst] *v. i.* **a)** *(be in existence)* existieren; ‹*Zweifel, Gefahr, Problem, Brauch, Einrichtung:*› bestehen; **fairies do ~:** es gibt Feen; **the biggest book that has ever ~ed** das größte Buch aller Zeiten; **b)** *(survive)* existieren; überleben; **~ on sth.** von etw. leben; **c)** *(be found)* **sth. ~s only in Europe** es gibt etw. nur in Europa

existence [ɪg'zɪstəns] *n.* **a)** *(existing)* Existenz, *die;* **doubt sb.'s ~/the ~ of sth.** bezweifeln, daß es jmdn./etw. gibt; **be in ~:** existieren; **the only such plant in ~:** die einzige Pflanze dieser Art, die es gibt; **come into ~:** entstehen; **go out of ~:** verschwinden; **b)** *(mode of living)* Dasein, *das; (survival)* Existenz, *die*

existential [egzɪ'stenʃl] *adj. (Philos.)* existentiell

existentialism [egzɪ'stenʃəlɪzm] *n., no pl. (Philos.)* Existentialismus, *der*

existing [ɪg'zɪstɪŋ] *adj.* bestehend ‹*Ordnung, Schwierigkeiten*›; gegenwärtig ‹*Lage, Führung, Stand der Dinge*›

exit ['eksɪt] **1.** *n.* **a)** *(way out)* Ausgang, *der* (**from** aus); *(for vehicle)* Ausfahrt, *die;* **b)** *(from stage)* Abgang, *der;* **make one's ~:** abgehen; **c)** *(from room)* Hinausgehen, *das.* **2.** *v. i.* **a)** hinausgehen (**from** aus); *(from stage)* abgehen (**from** von); **b)** *(Theatre: as stage direction)* ab

exit: ~ permit *n.* Ausreiseerlaubnis, *die;* **~ visa** *n.* Ausreisevisum, *das*

exodus ['eksədəs] *n.* Auszug, *der;* Exodus, *der (geh.);* **general ~:** allgemeiner Aufbruch

exonerate [ɪg'zɒnəreɪt] *v. t.* entlasten

exorbitant [ɪg'zɔ:bɪtənt] *adj.* [maßlos] überhöht ‹*Preis, Miete, Gewinn, Anforderung, Rechnung*›; maßlos ‹*Ehrgeiz, Forderung*›; **£10 – that's ~!** 10 Pfund – das ist unverschämt viel! *(ugs.)*

exorcise *see* **exorcize**

exorcism ['eksɔ:sɪzm] *n.* Exorzismus, *der;* Teufelsaustreibung, *die*

exorcist ['eksɔ:sɪst] *n.* Exorzist, *der*

exorcize ['eksɔ:saɪz] *v. t.* austreiben; exorzieren

exotic [ɪg'zɒtɪk] *adj.* exotisch

expand [ɪk'spænd] **1.** *v. i.* **a)** *(get bigger)* sich ausdehnen; ‹*Unternehmen, Stadt, Staat:*› expandieren; ‹*Institution:*› erweitert werden; **~ into sth.** zu etw. anwachsen; **b)** **~ on a subject** ein Thema weiter ausführen. **2.** *v. t.* **a)** *(enlarge)* ausdehnen; erweitern ‹*Horizont, Wissen*›; dehnen ‹*Körper*›; **~ sth. into sth.** etw. zu etw. erweitern; **b)** *(Commerc.: develop)* erweitern; **~ the economy** das Wirtschaftswachstum fördern; **c)** *(amplify)* weiter ausführen ‹*Gedanken, Notiz, Idee*›

expanse [ɪk'spæns] *n.* [weite] Fläche; **~ of water** Wasserfläche, *die*

expansion [ɪk'spænʃn] *n.* **a)** Ausdehnung, *die; (of territorial rule also)* Expansion, *die; (of knowledge, building)* Erweiterung, *die;* **b)** *(Commerc.)* Expansion, *die*

expansive [ɪk'spænsɪv] *adj.* offen; *(responsive)* zugänglich

expatriate [eks'pætrɪət, eks'peɪtrɪət] **1.** *attrib. adj.* im Ausland lebend. **2.** *n. (exile)* Exilant, *der*/Exilantin, *die; (foreigner)* Ausländer, *der*/Ausländerin, *die*

expect [ɪk'spekt] *v. t.* **a)** erwarten; **~ to do sth.** damit rechnen, etw. zu tun; **~ sb. to do sth.** damit rechnen, daß jmd. etw. tut; **I ~ you'd like something to eat** ich nehme an, daß du gern etwas essen möchtest; **don't ~ me to help you out** von mir hast du keine Hilfe zu erwarten; **it is ~ed that ...:** man erwartet, daß ...; **that was [not] to be ~ed** das war [auch nicht] zu erwarten; **be ~ing a baby/child** ein Baby/Kind erwarten; **~ sb. to do sth.** von jmdm. erwarten, daß er etw. tut; ~ **sth. from** *or* **of sb.** etw. von jmdm. erwarten; **b)** *(coll.: think, suppose)* glauben; **I ~ so** ich glaube schon; **I don't ~ so** ich glaube nicht; **I ~ it was/he did** *etc.* das glaube ich gern

expectancy [ɪk'spektənsɪ] *n., no pl.* Erwartung, *die*

expectant [ɪk'spektənt] *adj.* **a)** erwartungsvoll; **b)** ~ **mother** werdende Mutter

expectantly [ɪk'spektəntlɪ] *adv.* erwartungsvoll; gespannt ⟨*warten*⟩

expectation [ekspek'teɪʃn] *n.* **a)** *no pl. (expecting)* Erwartung, *die;* **in the ~ of sth.** in Erwartung einer Sache *(Gen.);* **b)** *usu. in pl. (thing expected)* Erwartung, *die;* **come up to ~[s]/sb.'s ~s** den/jmds. Erwartungen entsprechen; **contrary to ~** *or* **to all ~s** wider Erwarten

expediency [ɪk'spi:dɪənsɪ] *n.* Zweckmäßigkeit, *die*

expedient [ɪk'spi:dɪənt] **1.** *adj.* **a)** *(appropriate, advantageous)* angebracht; **b)** *(politic)* zweckmäßig. **2.** *n.* Mittel, *das*

expedite ['ekspɪdaɪt] *v. t. (hasten)* beschleunigen; vorantreiben

expedition [ekspɪ'dɪʃn] *n.* **a)** Expedition, *die;* **b)** *(Mil.)* Feldzug, *der;* **c)** *(excursion)* Ausflug, *der*

expel [ɪk'spel] *v. t.,* **-ll-: a)** ausweisen; ~ **from school** von der Schule verweisen; ~ **sb. from a country** jmdn. aus einem Land ausweisen; ~ **from a club** aus einem Verein ausschließen; **b)** *(with force)* vertreiben **(from** aus)

expend [ɪk'spend] *v. t.* **a)** aufwenden **([up]on** für); **b)** *(use up)* aufbrauchen **([up]on** für)

expendable [ɪk'spendəbl] *adj.* **a)** *(inessential)* entbehrlich; **b)** *(used up in service)* zum Verbrauch bestimmt

expenditure [ɪk'spendɪtʃə(r)] *n.* **a)** *(amount spent)* Ausgaben *Pl.* **(on** für); **b)** *(using up of fuel or effort)* Aufwand, *der* **(of** an + *Dat.*)

expense [ɪk'spens] *n.* **a)** Kosten *Pl.;* **at one's own ~:** auf eigene Kosten; **go to the ~ of travelling first-class** sogar noch das Geld für die erste Klasse ausgeben; **go to some/great ~:** sich in Unkosten/große Unkosten stürzen; **b)** *(expensive item)* teure Angelegenheit; **be** *or* **prove a great** *or* **big ~:** mit großen Ausgaben verbunden sein; **c)** *usu. in pl. (Commerc. etc.)* Spesen *Pl.;* **with [all] ~s paid** auf Spesen; **d)** *(fig.)* Preis, *der;* **[be] at the ~ of sth.** auf Kosten von etw. [gehen]; **at sb.'s ~:** auf jmds. Kosten *(Akk.)*

ex'pense account *n.* Spesenabrechnung, *die*

expensive [ɪk'spensɪv] *adj.* teuer

experience [ɪk'spɪərɪəns] **1.** *n.* **a)** *no pl., no indef. art.* Erfahrung, *die;* **have ~ of sth./sb.** Erfahrung in etw. *(Dat.)* /mit jmdm. haben; **have ~ of doing sth.** Erfahrung darin haben, etw. zu tun; **learn from ~:** durch eigene *od.* aus eigener Erfahrung lernen; **in/from my [own] [previous] ~:** nach meiner/aus eigener Erfahrung; **b)** *(event)* Erfahrung, *die;* Erlebnis, *das.* **2.** *v. t.* erleben; stoßen auf (+ *Akk.*) ⟨*Schwierigkeiten*⟩; kennenlernen ⟨*Lebensweise*⟩; empfinden ⟨*Hunger, Kälte, Schmerz*⟩

experienced [ɪk'spɪərɪənst] *adj.* erfahren **(in** in + *Dat.*); **an ~ eye** ein geschulter Blick

experiment 1. [ɪk'sperɪmənt] *n.* **a)** Experiment, *das,* Versuch, *der* **(on** an + *Dat.*); **do an ~:** ein Experiment machen; **b)** *(fig.)* Experiment, *das;* **as an ~:** versuchsweise. **2.** [ɪk'sperɪment] *v. i.* experimentieren **(on** an + *Dat.,* **with** mit)

experimental [ɪksperɪ'mentl] *adj.* **a)** experimentell; Experimental⟨*physik, -psychologie*⟩; Experimentier⟨*theater*⟩; Versuchs⟨*labor, -bedingungen*⟩; Versuchs⟨*tier*⟩; **at the/an ~ stage** im Versuchsstadium; **b)** *(fig.: tentative)* vorläufig

experimentation [ɪksperɪmen'teɪʃn] *n.* Experimentieren, *das*

expert ['ekspɜ:t] **1.** *adj.* **a)** ausgezeichnet; **be ~ in** *or* **at sth.** Fachmann *od.* Experte in etw. *(Dat.)* sein; **be ~ in** *or* **at doing sth.** etw. ausgezeichnet können; **b)** *(of an ~)* fachmännisch; ~ **witness** sachverständiger Zeuge; **an ~ opinion** die Meinung eines Fachmanns; ~ **knowledge** Fachkenntnis, *die.* **2.** *n.* Fachmann, *der;* Experte, *der*/Expertin, *die; (Law)* Sachverständige, *der/die;* **be an ~ in** *or* **at/on sth.** Fachmann *od.* Experte in etw. *(Dat.)*/für etw. sein

expertise [ekspɜ:'ti:z] *n.* Fachkenntnisse; *(skill)* Können, *das*

expertly ['ekspɜ:tlɪ] *adv.* meisterhaft; fachmännisch ⟨*reparieren, beraten, beurteilen*⟩

expire [ɪk'spaɪə(r)] *v. i.* **a)** *(become invalid)* ablaufen; ⟨*Vertrag, Amtszeit:*⟩ auslaufen; **b)** *(literary: die)* versterben *(geh.)*

expiry [ɪk'spaɪərɪ] *n.* Ablauf, *der*

explain [ɪk'spleɪn] **1.** *v. t., also abs.* erklären; erläutern ⟨*Grund, Motiv, Gedanken*⟩; darlegen ⟨*Absicht, Beweggrund*⟩; **how do you ~ that?** wie erklären Sie sich *(Dat.)* das? **2.** *v. refl.* **a)** *often abs. (justify one's conduct)* **please ~ [yourself]** bitte erklären Sie mir das; **he refused to ~:** er wollte mir keine Erklärung dafür geben; **b)** *(make one's meaning clear)* **please ~ yourself** bitte erklären Sie das [näher]

~ **a'way** *v. t.* eine [plausible] Erklärung finden für

explanation [eksplə'neɪʃn] *n.* Erklärung, *die;* **need ~:** einer Erklärung *(Gen.)* bedürfen

explanatory [ɪk'splænətərɪ] *adj.* erklärend; erläuternd ⟨*Bemerkung*⟩

expletive [ɪk'spli:tɪv, ek'spli:tɪv] *n.* Kraftausdruck, *der*

explicable [ɪk'splɪkəbl] *adj.* erklärbar

explicit [ɪk'splɪsɪt] *adj. (stated in detail)* ausführlich; *(openly expressed)* offen; unverhüllt; *(definite)* klar; ausdrücklich ⟨*Zustimmung, Erwähnung*⟩

explicitly [ɪk'splɪsɪtlɪ] *adv.* ausdrücklich

explode [ɪk'spləʊd] **1.** *v. i. (lit. or fig.)* explodieren; ⟨*Bevölkerung:*⟩ rapide zunehmen. **2.** *v. t.* **a)** zur Explosion bringen; **b)** *(fig.)* widerlegen ⟨*Vorstellung, Doktrin, Theorie*⟩

exploit 1. ['eksplɔɪt] *n. (feat; also joc.: deed)* Heldentat, *die.* **2.** [ɪk'splɔɪt] *v. t.* **a)** *(derog.)* ausbeuten ⟨*Arbeiter, Kolonie usw.*⟩; ausnutzen ⟨*Gutmütigkeit, Freund, Unwissenheit*⟩; **b)** *(utilize)* nutzen; nützen; ausnutzen ⟨*Gelegenheit, Situation*⟩; ausbeuten ⟨*Grube*⟩

exploitation [eksplɔɪ'teɪʃn] *n. see* **exploit 2:** Ausbeutung, *die;* Ausnutzung, *die;* Nutzung, *die*

exploration [eksplə'reɪʃn] *n.* **a)** Erforschung, *die; (of town, house)* Erkundung, *die;* **voyage of ~:** Entdeckungsreise, *die;* **b)** *(fig.)* Untersuchung, *die*

exploratory [ɪk'splɒrətərɪ] *adj.* Forschungs-; ~ **talks** Sondierungsgespräche; ~ **operation** *(Med.)* explorative Operation

explore [ɪk'splɔ:(r)] *v. t.* **a)** erforschen; erkunden ⟨*Stadt, Haus*⟩; **b)** *(fig.)* untersuchen

explorer [ɪk'splɔ:rə(r)] *n.* Entdeckungsreisende, *der/die;* **Arctic ~:** Arktisforscher, *der*/-forscherin, *die*

explosion [ɪk'spləʊʒn] *n.* **a)** *(lit.; fig.: rapid increase)* Explosion, *die; (noise)* [Explosions]knall, *der;* **b)** *(fig.: of anger etc.)* Ausbruch, *der*

explosive [ɪk'spləʊsɪv] **1.** *adj.* **a)** explosiv; ~ **device** Sprengkörper, *der;* **b)** *(fig.)* explosiv; brisant ‹*Thema*›. **2.** *n.* Sprengstoff, *der;* **high** ~: hochexplosiver Stoff

exponent [ɪk'spəʊnənt] *n.* Vertreter, *der*/Vertreterin, *die*

exponential [ekspə'nenʃl] *adj.* exponentiell; Exponential-

exponentially [ekspə'nenʃəlɪ] *adv.* exponentiell

export 1. [ɪk'spɔ:t] *v. t.* exportieren; ausführen; ~**ing country** Ausfuhrland, *das;* **oil-**~**ing countries** [erd]ölexportierende Länder. **2.** ['ekspɔ:t] *n.* **a)** *(process, amount* ~*ed)* Export, *der;* Ausfuhr, *die;* *(*~*ed articles)* Exportgut, *das;* Ausfuhrgut, *das;* ~**s of sugar** Zuckerexporte *od.* -ausfuhren; **b)** *attrib.* Export‹*leiter, -handel, -markt, -kaufmann*›

exporter [ɪk'spɔ:tə(r)] *n.* Exporteur, *der*

expose [ɪk'spəʊz] **1.** *v. t.* **a)** *(uncover)* freilegen; entblößen ‹*Haut, Körper, Knie*›; **b)** *(make known)* offenbaren ‹*Schwäche, Tatsache, Geheimnis, Plan*›; aufdecken ‹*Irrtum, Mißstände, Verbrechen, Verrat*›; entlarven ‹*Täter, Verräter, Spion*›; **c)** *(subject)* ~ **to sth.** einer Sache *(Dat.)* aussetzen; **d)** *(Photog.)* belichten. **2.** *v. refl.* sich [unsittlich] entblößen

exposed [ɪk'spəʊzd] *adj.* *(unprotected)* ungeschützt

exposition [ekspə'zɪʃn] *n.* **a)** *(statement)* Darstellung, *die;* **b)** *(exhibition)* Ausstellung, *die*

expostulate [ɪk'spɒstjʊleɪt] *v. i.* protestieren

expostulation [ɪkspɒstjʊ'leɪʃn] *n.* Protest, *der*

exposure [ɪk'spəʊʒə(r)] *n.* **a)** *(to air, cold, etc.)* *(being exposed)* Ausgesetztsein, *das;* *(exposing)* Aussetzen, *das;* **die of/suffer from** ~ **[to cold]** an Unterkühlung *(Dat.)* sterben/leiden; **indecent** ~: Entblößung in schamverletzender Weise; **media** ~: Publicity, *die;* **b)** *(of fraud etc.)* Enthüllung, *die;* *(of criminal)* Entlarvung, *die;* **c)** *(Photog.)* *(exposing time)* Belichtung, *die;* *(picture)* Aufnahme, *die*

ex'posure meter *n.* *(Photog.)* Belichtungsmesser, *der*

expound [ɪk'spaʊnd] *v. t.* darlegen ‹*Theorie, Doktrin*› **(to** *Dat.***)**

express [ɪk'spres] **1.** *v. t.* **a)** *(indicate)* ausdrücken; **b)** *(put into words)* äußern ‹*Meinung, Wunsch*›; zum Ausdruck bringen ‹*Dank, Bedauern, Liebe*›; ~ **sth. in another language** etw. in einer anderen Sprache ausdrücken; ~ **oneself** sich ausdrücken; **c)** *(represent by symbols)* ausdrücken ‹*Zahl, Wert*›. **2.** *attrib. adj.* **a)** Eil‹*brief, -bote usw.*›; Schnell‹*paket, -sendung*›; *see also* **express train; b)** *(particular)* besonder...; bestimmt; ausdrücklich ‹*Absicht*›; **c)** *(stated)* ausdrücklich ‹*Wunsch, Befehl usw.*›. **3.** *adv.* als Eilsache ‹*senden*›. **4.** *n.* *(train)* Schnellzug, *der;* D-Zug, *der*

expression [ɪk'spreʃn] *n.* Ausdruck, *der;* **the** ~ **on his face, his facial** ~: sein Gesichtsausdruck; **full of/without** ~: ausdrucksvoll/-los

expressionism [ɪk'spreʃənɪzm] *n., no pl.* Expressionismus, *der*

expressive [ɪk'spresɪv] *adj.* ausdrucksvoll; vielsagend ‹*Schweigen*›

expressively [ɪk'spresɪvlɪ] *adv.* ausdrucksvoll

expressly [ɪk'spreslɪ] *adv.* ausdrücklich

express: ~ **'train** *n.* Schnellzug, *der;* D-Zug, *der;* ~**way** *n.* *(Amer.)* Schnell[verkehrs]straße, *die*

expulsion [ɪk'spʌlʃn] *n.* *(from school, college)* Verweisung, *die* **(from** von); *(from country)* Ausweisung, *die* **(from** aus); *(from club)* Ausschluß, *der* **(from** aus)

expurgate ['ekspəgeɪt] *v. t.* ~**d version/edition** zensierte Fassung/Ausgabe

exquisite ['ekskwɪzɪt, ɪk'skwɪzɪt] *adj.* **a)** erlesen; exquisit, bezaubernd ‹*Aussicht, Muster, Melodie, Anmut*›; **b)** heftig ‹*Schmerz, Freude*›; unerträglich ‹*Leiden, Schmerzen*›

exquisitely ['ekskwɪzɪtlɪ, ɪk'skwɪzɪtlɪ] *adv.* **a)** vorzüglich; kunstvoll ‹*verziert, geschnitzt*›; **b)** *(acutely)* äußerst; außerordentlich

extend [ɪk'stend] **1.** *v. t.* **a)** *(stretch out)* ausstrecken ‹*Arm, Bein, Hand*›; ausziehen ‹*Leiter, Teleskop*›; ausbreiten ‹*Flügel*›; ~ **one's hand to sb.** jmdm. die Hand reichen; **the table can be** ~**ed** der Tisch ist ausziehbar; **b)** *(make longer) (in space)* verlängern; ausdehnen ‹*Grenze*›; ausbauen ‹*Bahnlinie, Straße*›; *(in time)* verlängern; verlängern lassen ‹*Leihbuch, Visum*›; ~ **the time limit** den Termin hinausschieben; **c)** *(enlarge)* ausdehnen ‹*Einfluß, Macht*›; erweitern ‹*Wissen, Wortschatz, Bedeutung, Freundeskreis, Besitz, Geschäft*›; ausbauen, vergrößern ‹*Haus, Geschäft*›; **d)** *(offer)* gewähren, zuteil werden lassen ‹*[Gast]freundschaft, Schutz, Hilfe, Kredit*› **(to** *Dat.***)**; *(accord)* aussprechen ‹*Dank, Einladung, Glückwunsch*› **(to** *Dat.***)**; ausrichten ‹*Gruß*› **(to** *Dat.***)**; ~ **a welcome to sb.** jmdn. willkommen heißen. **2.** *v. i.* sich erstrecken; **the wall** ~**s for miles** die Mauer zieht sich meilenweit hin; **the season** ~**s from November to March** die Saison geht von November bis März

extension [ɪk'stenʃn] *n.* **a)** *(stretching out) (of arm, leg, hand)* [Aus]strecken, *das;* *(of wings)* Ausbreiten, *das;* **b)** *(prolonging)* Verlängerung, *die;* *(of road, railway)* Ausbau, *der;* **ask for an** ~: um Verlängerung bitten; **be granted** *or* **get an** ~: Verlängerung bekommen; **c)** *(enlargement) (of power, influence, research, frontier)* Ausdehnung, *die;* *(of enterprise, trade, knowledge)* Erweiterung, *die;* **d)** *(additional part) (of house)* Anbau, *der;* *(of office, university, hospital, etc.)* Erweiterungsbau, *der;* **e)** *(telephone)* Nebenanschluß, *der;* *(number)* Apparat, *der*

ex'tension lead *n.* *(Brit.)* Verlängerungsschnur, *die*

extensive [ɪk'stensɪv] *adj.* ausgedehnt ‹*Ländereien, Reisen, Stadt, Wald, Besitz[tümer], Handel, Forschungen*›; umfangreich ‹*Reparatur, Investitionen, Wissen, Nachforschungen, Studien, Auswahl, Sammlung*›; beträchtlich ‹*Schäden, Geldmittel, Anstrengungen*›; weitreichend ‹*Änderungen, Reformen, Einfluß*›; ausführlich ‹*Bericht, Einleitung*›

extensively [ɪk'stensɪvlɪ] *adv.* beträchtlich ‹*ändern, beschädigen*›; ausführlich ‹*berichten, schreiben*›

extent [ɪk'stent] *n.* **a)** *(space)* Ausdehnung, *die;* *(of wings)* Spannweite, *die;* **b)** *(scope) (of knowledge, power, authority)* Umfang, *der;* *(of damage, disaster)* Ausmaß, *das;* *(of debt, loss)* Höhe, *die;* **to what** ~**?** inwieweit?; **to a great** *or* **large** ~: in hohem Maße; **to some** *or* **a certain** ~: in gewissem Maße; **to a greater or lesser** ~: mehr oder weniger; **to such an** ~ **that ...:** in solchem Maße, daß ...

extenuating [ɪk'stenjʊeɪtɪŋ] *adj.* ~ **circumstances** mildernde Umstände

exterior [ɪk'stɪərɪə(r)] **1.** *adj.* **a)** äußer...; Außen‹*fläche, -wand,*

-anstrich⟩; **b)** *([coming from] outside)* äußer...; außerhalb gelegen. **2.** *n.* **a)** Äußere, *das; (of house)* Außenwände *Pl.*; **b)** *(appearance)* Äußere, *das*

exterminate [ɪk'stɜ:mɪneɪt] *v.t.* ausrotten; vertilgen ⟨*Ungeziefer*⟩

extermination [ɪkstɜ:mɪ'neɪʃn] *n.* Ausrottung, *die; (of pests)* Vertilgung, *die*

external [ɪk'stɜ:nl] *adj.* **a)** äußer...; Außen⟨*fläche, -druck, -winkel, -abmessungen*⟩; **purely ~:** nur *od.* rein äußerlich; **b)** *(applied to outside)* äußerlich ⟨*Heilmittel*⟩; **for ~ use only** nur äußerlich anzuwenden; **c)** *(of foreign affairs)* Außen⟨*minister, -handel, -politik*⟩

externally [ɪk'stɜ:nəlɪ] *adv.* äußerlich

extinct [ɪk'stɪŋkt] *adj.* erloschen ⟨*Vulkan, Leidenschaft, Liebe, Hoffnung*⟩; ausgestorben ⟨*Art, Rasse, Volk, Gattung*⟩; **become ~:** aussterben

extinction [ɪk'stɪŋkʃn] *n., no pl.* Aussterben, *das;* **threatened with ~:** vom Aussterben bedroht

extinguish [ɪk'stɪŋgwɪʃ] *v.t.* löschen; erlöschen lassen ⟨*Liebe, Hoffnung*⟩; auslöschen ⟨*Leben*⟩

extinguisher [ɪk'stɪŋgwɪʃə(r)] *n.* *(for fire)* Feuerlöscher, *der*

extol [ɪk'stəʊl, ɪk'stɒl] *v.t.*, **-ll-** rühmen; preisen

extort [ɪk'stɔ:t] *v.t.* erpressen (**out of, from** von); **~ a confession from sb.** ein Geständnis aus jmdm. herauspressen

extortion [ɪk'stɔ:ʃn] *n.* *(of money, taxes)* Erpressung, *die;* **£50? That's sheer ~!** 50 Pfund? Das ist ja Wucher!

extortionate [ɪk'stɔ:ʃənət] *adj.* Wucher⟨*preis, -zinsen usw.*⟩; horrend ⟨*Gebühr, Steuer*⟩; maßlos überzogen ⟨*Forderung*⟩

extra ['ekstrə] **1.** *adj.* **a)** *(additional)* zusätzlich; Mehr⟨*arbeit, -kosten, -ausgaben, -aufwendungen*⟩; Sonder⟨*bus, -zug*⟩; **~ charge** Aufpreis, *der;* **all we need is an ~ hour/three pounds** wir brauchen nur noch eine Stunde/drei Pfund [zusätzlich]; **b)** *(more than is necessary)* überzählig ⟨*Exemplar, Portion*⟩; **an ~ pair of gloves** noch ein *od.* ein zweites Paar Handschuhe. **2.** *adv.* **a)** *(more than usually)* besonders; extra ⟨*lang, stark, fein*⟩; überaus ⟨*froh*⟩; **an ~ large blouse** eine Bluse in Übergröße; **an ~ special occasion** eine ganz besondere Gelegenheit; **b)** *(additionally)* extra; **packing and postage ~:** zuzüglich Verpackung und Porto. **3.** *n.* **a)** *(added to services, salary, etc.)* zusätzliche Leistung; *(on car etc. offered for sale)* Extra, *das;* **b)** *(sth. with ~ charge)* **be an ~:** zusätzlich berechnet werden; **c)** *(in play, film, etc.)* Statist, *der*/Statistin, *die*

extract 1. ['ekstrækt] *n.* **a)** *(substance)* Extrakt, *der (fachspr. auch: das);* **b)** *(from book, music, etc.)* Auszug, *der;* Extrakt, *der (geh.).* **2.** [ɪk'strækt] *v.t.* **a)** ziehen, *(fachspr.)* extrahieren ⟨*Zahn*⟩; **~ sth. from sb.** *(fig.)* etw. aus jmdm. herausholen; **~ a promise/confession from sb.** jmdm. ein Versprechen/Geständnis abpressen; **~ papers from a folder** einem Aktenordner Unterlagen entnehmen; **b)** *(obtain)* extrahieren; **~ the juice from apples** Äpfel entsaften; **~ metal from ore** Metall aus Erz gewinnen; **c)** *(derive)* erfassen ⟨*Bedeutung, Hauptpunkte*⟩

extraction [ɪk'strækʃn] *n.* **a)** *(of tooth)* Extraktion, *die; (of juice, honey, metal)* Gewinnung, *die;* **b) be of German ~:** deutscher Abstammung *od.* Herkunft sein

extractor fan [ɪk'stræktə fæn] *n.* Entlüfter, *der*

extradite ['ekstrədaɪt] *v.t.* **a)** ausliefern ⟨*Verbrecher*⟩; **b)** *(obtain extradition of)* **~ sb.** jmds. Auslieferung erwirken

extradition [ekstrə'dɪʃn] *n.* Auslieferung, *die;* **~ treaty** Auslieferungsvertrag, *der*

extra-'marital *adj.* außerehelich

extra'mural *adj.* *(Univ.)* außerhalb der Universität *nachgestellt*

extraneous [ɪk'streɪnɪəs] *adj.* **a)** *(from outside)* von außen; **b)** *(irrelevant)* belanglos; **be ~ to sth.** für etw. ohne Belang sein

extraordinarily [ɪk'strɔ:dɪnərɪlɪ, ekstrə'ɔ:dɪnərɪlɪ] *adv.* außergewöhnlich; überaus ⟨*merkwürdig*⟩

extraordinary [ɪk'strɔ:dɪnərɪ, ekstrə'ɔ:dɪnərɪ] *adj.* *(exceptional)* außergewöhnlich; *(unusual, peculiar)* ungewöhnlich ⟨*Gabe*⟩; merkwürdig ⟨*Zeichen, Benehmen, Angewohnheit*⟩; außerordentlich ⟨*Verdienste, Einfluß*⟩; *(additional)* außerordentlich ⟨*Versammlung*⟩; **how ~!** wie seltsam!

extra-'sensory *adj.* **~ perception** außersinnliche Wahrnehmung

extra 'time *n.* *(Sport)* **after ~:** nach einer Verlängerung; **play ~:** in die Verlängerung gehen

extravagance [ɪk'strævəgəns] *n.* **a)** *no pl. (being extravagant)* Extravaganz, *die; (of claim, wish, order, demand)* Übertriebenheit, *die; (of words, thoughts, ideas)* Verstiegenheit, *die; (with money)* Verschwendungssucht, *die;* **b)** *(extravagant thing)* Luxus, *der*

extravagant [ɪk'strævəgənt] *adj.* **a)** *(wasteful)* verschwenderisch; aufwendig ⟨*Lebensstil*⟩; teuer ⟨*Geschmack*⟩; **b)** *(immoderate)* übertrieben ⟨*Benehmen, Lob, Eifer, Begeisterung usw.*⟩; **c)** *(beyond bounds of reason)* abwegig ⟨*Theorie, Frage, Einfall*⟩

extravagantly [ɪk'strævəgəntlɪ] *adv.* extravagant ⟨*ausstatten, sich kleiden*⟩; verschwenderisch ⟨*benutzen, verbrauchen*⟩; luxuriös, aufwendig ⟨*leben*⟩; überschwenglich ⟨*loben*⟩

extreme [ɪk'stri:m] **1.** *adj.* **a)** *(outermost, utmost)* äußerst... ⟨*Spitze, Rand, Ende*⟩; extrem, kraß ⟨*Gegensätze*⟩; **at the ~ edge/left** ganz am Rand/ganz links; **in the ~ North** im äußersten Norden; **b)** *(reaching high degree)* extrem; gewaltig ⟨*Entfernung, Unterschied*⟩; höchst... ⟨*Gefahr*⟩; äußerst... ⟨*Notfall, Höflichkeit, Bescheidenheit*⟩; stärkst... ⟨*Schmerzen*⟩; heftigst... ⟨*Zorn*⟩; tiefst... ⟨*Haß, Dankbarkeit*⟩; größt... ⟨*Wichtigkeit*⟩; **c)** *(not moderate)* extrem ⟨*Person, Ideen, Kritik*⟩; **~ right-wing views** rechtsextreme Ansichten; **d)** *(severe)* drastisch ⟨*Maßnahme*⟩. **2.** *n.* Extrem, *das;* [krasser] Gegensatz; **the ~s of wealth and poverty** größter Reichtum und äußerste Armut; **~s of temperature** extreme Temperaturunterschiede; **go to ~s** vor nichts zurückschrecken; **go to the other ~:** ins andere Extrem verfallen; **go from one ~ to another** von *od.* aus einem Extrem ins andere fallen; **... in the ~:** äußerst ...; *see also* **carry 1 g**

extremely [ɪk'stri:mlɪ] *adv.* äußerst; **Did you enjoy the party? – Yes, ~:** Hat dir die Party gefallen? – Ja, sehr sogar!

extremist [ɪk'stri:mɪst] *n.* **a)** Extremist, *der*/Extremistin, *die;* **right-wing ~:** Rechtsextremist, *der*/-extremistin, *die;* **b)** *attrib.* extremistisch

extremity [ɪk'stremɪtɪ] *n.* **a)** *(of branch, road)* äußerstes Ende; *(of region)* Rand, *der;* **b)** *in pl. (hands and feet)* Extremitäten *Pl.*

extricate ['ekstrɪkeɪt] *v.t.* **~ sth. from sth.** etw. aus etw. herausziehen; **~ oneself/sb. from sth.** sich/jmdn. aus etw. befreien

extrovert ['ekstrəvɜ:t] **1.** *n.* extrovertierter Mensch; **be an ~:** extrovertiert sein. **2.** *adj.* extrovertiert

extroverted ['ekstrəvɜ:tɪd] *adj.* extrovertiert

exuberance [ɪg'zju:bərəns] *n.* **a)** *(vigour)* Überschwang, *der;* ~ **of youth** jugendlicher Überschwang; **b)** *(of language, style)* Lebendigkeit, *die*

exuberant [ɪg'zju:bərənt] *adj.* **a)** *(overflowing)* überschäumend ‹*Kraft, Freude, Eifer*›; **b)** *(effusive)* überschwenglich

exude [ɪg'zju:d] *v. t.* absondern ‹*Flüssigkeit, Harz*›; ausströmen ‹*Geruch*›; *(fig.)* ausstrahlen ‹*Charme, Zuversicht*›

exult [ɪg'zʌlt] *v. i. (literary)* frohlocken *(geh.)* **(in, at, over** über + *Akk.*)

exultant [ɪg'zʌltənt] *adj. (literary)* jubelnd; **be ~:** jubeln

exultation [egzʌl'teɪʃn] *n.* Jubel, *der*

eye [aɪ] **1.** *n.* **a)** Auge, *das;* **~s** *(look, glance, gaze)* Blick, *der;* **the sun/light is [shining] in my ~s** die Sonne/das Licht blendet mich; **out of the corner of one's ~:** aus den Augenwinkeln; **with one's own** *or* **very ~s** mit eigenen Augen; **before sb.'s very ~s** vor jmds. Augen *(Dat.);* **measure a distance by ~:** einen Abstand nach Augenmaß schätzen; **paint/draw sth. by ~:** etw. nach der Natur malen/zeichnen; **look sb. in the ~:** jmdm. gerade in die Augen sehen; **be unable to take one's ~s off sb./sth.** die Augen *od.* den Blick nicht von jmdm./etw. abwenden können; **keep an ~ on sb./sth.** auf jmdn./etw. aufpassen; **have [got] an ~** *or* **one's ~[s] on sb./sth.** ein Auge auf jmdn./etw. geworfen haben; **I've got my ~ on you!** ich lasse dich nicht aus den Augen!; **keep an ~ open** *or* **out [for sb./sth.]** [nach jmdm./etw.] Ausschau halten; **keep one's ~s open** die Augen offenhalten; **keep one's ~s open** *or* *(coll.)* **peeled** *or* *(coll.)* **skinned for sth.** nach etw. Ausschau halten; **with one's ~s open** *(fig.)* mit offenen Augen; **with one's ~s shut** *(fig.) (without full awareness)* blind; *(with great ease)* im Schlaf; **be all ~s** gespannt zusehen; **[an] ~ for [an] ~:** Auge um Auge; **have an ~ to sth./doing sth.** auf etw. *(Akk.)* bedacht sein/darauf bedacht sein, etw. zu tun; **that was one in the ~ for him** *(coll.)* das war ein Schlag ins Kontor *(ugs.)* für ihn; **see ~ to ~ [on sth. with sb.]** [mit jmdm.] einer Meinung [über etw. *(Akk.)*] sein; **be up to one's ~s** *(fig.)* bis über beide Ohren drinstecken *(ugs.);* **be up to one's ~s in work/debt** bis über beide Ohren in der Arbeit/in Schulden stecken *(ugs.);* **have a keen/good ~ for sth.** einen geschärften/einen sicheren *od.* den richtigen Blick für etw. haben; **b)** *(of needle, fish-hook)* Öhr, *das; (metal loop)* Öse, *die.* **2.** *v. t.*, **~ing** *or* **eying** ['aɪɪŋ] beäugen; **~ sb. up and down** jmdn. von oben bis unten mustern

eye: ~ball *n.* Augapfel, *der;* **~brow** *n.* Augenbraue, *die;* **raise an ~brow** *or* **one's ~brows [at sth.]** *(fig.: in surprise)* die Stirn runzeln (**at** über + *Akk.*); **it will raise a few ~brows** das wird einiges Stirnrunzeln hervorrufen; **~-catching** *adj.* ins Auge springend *od.* fallend ‹*Inserat, Plakat, Buchhülle usw.*›; **~-drops** *n. pl. (Med.)* Augentropfen *Pl.*

eyeful ['aɪfʊl] *n. (coll.)* **get an ~ [of sth.]** einiges [von etw.] zu sehen bekommen

'**eyelash** *n.* Augenwimper, *die*

eyelet ['aɪlɪt] *n.* Öse, *die*

eye: ~-level *n.* Augenhöhe, *die; attrib.* in Augenhöhe *nachgestellt;* **~lid** *n.* Augenlid, *das;* **~-opener** *n. (surprise, revelation)* Überraschung, *die;* **the book was an ~-opener to the public** das Buch hat der Öffentlichkeit *(Dat.)* die Augen geöffnet; **~piece** *n. (Optics)* Okular, *das;* **~-shade** *n.* Augenschirm, *der;* **~-shadow** *n.* Lidschatten, *der;* **~sight** *n.* Sehkraft, *die;* **have good ~sight** gute Augen haben; **his ~sight is poor** er hat schlechte Augen; **~sore** *n.* Schandfleck, *der (abwertend);* **the building is an ~sore** das Gebäude beleidigt das Auge; **~-strain** *n.* Überanstrengung der Augen; **~-tooth** *n.* Eckzahn, *der;* **~wash** *n.* **a)** *(Med.: lotion)* Augenwasser, *das;* **b)** *(coll.) (nonsense)* Gewäsch, *das (ugs. abwertend); (concealment)* Augen[aus]wischerei, *die (ugs.);* **~witness** *n.* Augenzeuge, *der*/-zeugin, *die; attrib.* **~witness account** *or* **report** Augenzeugenbericht, *der*

eyrie ['ɪərɪ] *n. (nest)* Horst, *der*

F

F, f [ef] *n., pl.* **Fs** *or* **F's a)** *(letter)* F, f, *das;* **b) F** *(Mus.)* F, f, *das;* **F sharp** fis, Fis, *das*

F. *abbr.* **a) Fahrenheit** F; **b) franc** F

f. *abbr.* **a) female** weibl.; **b) feminine** f.; **c) focal length** f; **f/8** *(Photog.)* Blende 8; **d) following [page]** f.; **e) forte** f

fable ['feɪbl] *n.* **a)** *(myth, lie)* Märchen, *das;* **b)** *(thing that does not really exist, brief story)* Fabel, *die*

fabled ['feɪbld] *adj.* **a) it is ~ that ...:** es heißt, daß ...; **b)** *(mythical)* Fabel‹*land, -wesen, -tier*›; **c)** *(celebrated)* berühmt (**for** für)

fabric ['fæbrɪk] *n.* **a)** *(material, construction, texture)* Gewebe, *das;* **woven/knitted ~:** Web-/Strickware, *die;* **b)** *(of building)* bauliche Substanz; **c)** *(fig.)* **the ~ of society** die Struktur der Gesellschaft

fabricate ['fæbrɪkeɪt] *v. t.* **a)** *(invent)* erfinden; *(forge)* fälschen; **b)** *(manufacture)* herstellen

fabrication [fæbrɪ'keɪʃn] *n.* **a)** *(of story etc.)* Erfindung, *die;* **the story is [a] pure ~:** die Geschichte ist frei erfunden; **b)** *(manufacture)* Herstellung, *die*

fabulous ['fæbjʊləs] *adj.* **a)** sagenhaft; Fabel‹*tier, -wesen*›; **b)** *(coll.: marvellous)* fabelhaft *(ugs.)*

façade [fə'sɑ:d] *n. (lit. or fig.)* Fassade, *die*

face [feɪs] **1.** *n.* **a)** Gesicht, *das;* **wash one's ~:** sich *(Dat.)* das Gesicht waschen; **go blue in the ~** *(with cold)* blau im Gesicht werden; **go red** *or* **purple in the ~** *(with exertion or passion or shame)* rot im Gesicht werden; **the stone struck me in the ~:** der Stein traf mich ins Gesicht; **bring A and B ~ to ~:** A und B einander *(Dat.)* gegenüberstellen; **meet sb. ~ to ~:** jmdn. persönlich kennenlernen; **come** *or* **be brought ~ to ~ with sb.** mit jmdm. konfrontiert werden; **come ~ to ~ with the fact that ...:** vor der Tatsache stehen, daß ...; **in [the] ~ of sth.** *(despite)* trotz; **slam the door in sb.'s ~:** jmdm. die Tür vor der Nase zuknallen *(ugs.)*; **fall [flat] on one's ~** *(lit. or fig.)* auf die Nase fallen *(ugs.)*; **look sb./sth. in the ~:** jmdm./einer Sache ins Gesicht sehen; **show one's ~:** sich sehen *od.* blicken lassen; **tell sb. to his ~ what ...:** jmdm. [offen] ins Gesicht sagen, was ...; **till one is blue in the ~:** bis man verrückt wird *(ugs.)*; **save one's ~:** das Gesicht wahren *od.* retten; **lose ~ [with sb.] [over sth.]** das Gesicht [vor jmdm.] [wegen etw.] verlieren; **make** *or* **pull a ~/~s [at sb.]** *(to show dislike)* ein Gesicht/Gesichter machen *od.* ziehen; *(to amuse or frighten)* eine Grimasse/Grimassen schneiden; **don't make a ~!** mach nicht so ein Gesicht!; **on the ~ of it** dem Anschein nach; **put a brave ~ on it** gute Miene zum bösen Spiel machen; **b)** *(front) (of mountain, cliff)* Wand, *die; (of building)* Stirnseite, *die; (of clock, watch)* Zifferblatt, *das; (of coin, medal, banknote, playing-card)* Vorderseite, *die; (of golf-club, cricket-bat, hockey-stick, tennis-racket)* Schlagfläche, *die;* **c)** *(surface)* **the ~ of the earth** die Erde; **disappear off** *or* **from the ~ of the earth** spurlos verschwinden; **d)** *(Geom.; also of crystal, gem)* Fläche, *die;* **e)** *see* **type-face.** *See also* **face down[ward]; face up[ward]. 2.** *v. t.* **a)** *(look towards)* sich wenden zu; **sb. ~s the front** jmd. sieht nach vorne; **[stand] facing one another** sich *(Dat.) od. (meist geh.)* einander gegenüber [stehen]; **the window ~s the garden/front** das Fenster geht zum Garten/zur Straße hinaus; **sit facing the engine** *(in a train)* in Fahrtrichtung sitzen; **b)** *(fig.: have to deal with)* ins Auge sehen (+ *Dat.*) ‹*Tod, Vorstellung*›; gegenübertreten (+ *Dat.*) ‹*Kläger*›; sich stellen (+ *Dat.*) ‹*Anschuldigung, Kritik*›; stehen vor (+ *Dat.*) ‹*Ruin, Entscheidung*›; **~ trial for murder, ~ a charge of murder** sich wegen Mordes vor Gericht verantworten müssen; **c)** *(not shrink from)* ins Auge sehen (+ *Dat.*) ‹*Tatsache, Wahrheit*›; mit Fassung gegenübertreten (+ *Dat.*) ‹*Kläger*›; **~ the music** *(fig.)* die Suppe auslöffeln *(ugs.)*; **let's ~ it** *(coll.)* machen wir uns *(Dat.)* doch nichts vor *(ugs.)*; **d) be ~d with sth.** sich einer Sache *(Dat.)* gegenübersehen; **~d with these facts** mit diesen Sachen konfrontiert; **e)** *(coll.: bear)* verkraften. **3.** *v. i.* **~ forwards/backwards** ‹*Person, Bank, Sitz:*› in/entgegen Fahrtrichtung sitzen/aufgestellt sein; **in which direction was he facing?** in welche Richtung blickte er?; **stand facing away from sb.** mit dem Rücken zu jmdm. stehen; **~ away from the road/on to the road/east[wards]** *or* **to[wards] the east** ‹*Fenster, Zimmer:*› nach hinten/vorn/Osten liegen; **the side of the house ~s to[wards] the sea** die Seite des Hauses liegt zum Meer

~ 'up to *v. t.* ins Auge sehen (+ *Dat.*); sich abfinden mit ‹*Möglichkeit*›; auf sich nehmen ‹*Verantwortung*›

face: ~-cloth *n.* Waschlappen [für das Gesicht]; **~-cream** *n.* Gesichtscreme, *die;* **~ 'down[ward]** *adv.* mit der Vorderseite nach unten; **lie ~ down[ward]** ‹*Person/Buch:*› auf dem Bauch/Gesicht liegen; **~-flannel** *(Brit.) see* **~-cloth**

faceless ['feɪslɪs] *adj. (anonymous)* anonym *(fig.)*

face: ~-lift *n.* **a)** Facelifting, *das;* **have** *or* **get a ~-lift** sich liften lassen; **b)** *(fig.)* Verschönerung, *die;* **~-saving** *adj.* zur Wahrung des Gesichts *nachgestellt;* **as a ~-saving gesture** um das Gesicht zu wahren

facet ['fæsɪt] *n.* **a)** *(of cut stone etc.)* Facette, *die;* **b)** *(aspect)* Seite, *die;* **every ~:** alle Seiten

facetious [fə'si:ʃəs] *adj.* [gewollt] witzig; **[not] be ~ [about sth.]** [keine] Witze [über etw. *(Akk.)*] machen *(ugs.)*

face: ~-to-~ *adj.* unmittelbar ‹*Gegenüberstellung*›; persönlich ‹*Gespräch, Treffen*›; **~ 'up[ward]** *adv.* mit der Vorderseite nach oben; **lie ~ up[ward]** ‹*Person:*› auf dem Rücken liegen; *(open)* ‹*Buch:*› aufgeschlagen liegen; **~ value** *n. (Finance)* Nennwert, *der;* **accept sth. at [its] ~ value** *(fig.)* etw. für bare Münze nehmen; **take sb. at [his/her] ~ value** *(fig.)* jmdn. nach seinem Äußeren beurteilen

facial ['feɪʃl] **1.** *adj.* Gesichts-. **2.** *n.* Gesichtsmassage, *die;* **have a ~:** sich *(Dat.)* das Gesicht massieren lassen

facile ['fæsaɪl] *adj. (often derog.)* leicht ‹*Sieg, Aufgabe*›; nichtssagend, banal ‹*Bemerkung*›

facilitate [fə'sɪlɪteɪt] *v. t.* erleichtern

facility [fə'sɪlɪtɪ] *n.* **a)** *esp. in pl.* Einrichtung, *die;* **cooking/washing facilities** Koch-/Waschgelegenheit, *die;* **sports facilities** Sportanlagen; **b)** *(opportunity)* Möglichkeit, *die;* **c)** *(ease)* Leichtigkeit, *die; (dexterity)* Gewandtheit, *die*

facing ['feɪsɪŋ] *n.* **a)** *(on garment)* Aufschlag, *der;* Besatz, *der;* **b)** *(covering)* Verkleidung, *die*

facsimile [fæk'sɪmɪlɪ] *n.* **a)** Faksimile, *das;* **b)** *(Telecommunications) see* **fax 1**

fact [fækt] *n.* **a)** *(true thing)* Tatsache, *die;* **~s and figures** Fakten und Zahlen; **the ~ remains that ...:** Tatsache bleibt: ...; **the true ~s of the case** *or* **matter** der wahre Sachverhalt; **know for a ~ that ...:** genau *od.* sicher wissen, daß ...; **is that a ~?** *(coll.)* Tatsache? *(ugs.)*; **and that's a ~:** und daran gibt's nichts zu zweifeln *(ugs.)*; **the reason lies in the ~ that ...:** der Grund besteht darin, daß ...; **face [the] ~s** den Tatsachen ins Gesicht sehen; **it is a proven ~ that ...:** es ist erwiesen, daß ...; **the ~ [of the matter] is that ...:** die Sache ist die, daß ...; **[it is a] ~ of life** [das ist die] harte *od.* rauhe Wirklichkeit; **tell** *or* **teach sb. the ~s of life** *(coll. euphem.)* jmdn. [sexuell] aufklären; **b)** *(reality)* Wahrheit, *die;* Tatsachen *Pl.;* **distinguish ~ from fiction** Fakten und Fiktion *(geh.) od.* Dichtung und Wahrheit unterscheiden; **in ~:** tatsächlich; **I don't think he'll come back; in ~ I know he won't** ich glaube nicht, daß er zurückkommt, ich weiß es sogar; *see also* **matter 1d; c)** *(thing assumed to be ~)* Faktum, *das;* **deny the ~ that ...:** [die Tatsache] abstreiten, daß ...

'fact-finding *attrib. adj.* Erkundungs‹*fahrt, -trupp*›; **~ committee/trip** Untersuchungsausschuß, *der*/Informationsreise, *die*

faction ['fækʃn] *n.* Splittergruppe, *die*

factor ['fæktə(r)] *n. (also Math.)* Faktor, *der*

factory ['fæktərɪ] *n.* Fabrik, *die;* Werk, *das*

factory: ~ farm *n.* [voll]automatisierter landwirtschaftlicher Betrieb; Agrarfabrik, *die (abwertend)*; **~ ship** *n.* Fabrikschiff, *das;* **~ worker** *n.* Fabrikarbeiter, *der*/-arbeiterin, *die*

factual ['fæktʃʊəl] *adj.* sachlich ‹*Bericht, Darlegung*›; auf Tatsachen beruhend ‹*Punkt, Beweis*›; **~ error** Sachfehler, *der*

faculty ['fækəltɪ] *n.* **a)** *(physical capability)* Fähigkeit, *die;* Vermögen, *das;* ~ **of sight/speech/hearing/thought** Seh-/Sprach-/Hör-/Denkvermögen, *das;* **b)** *(mental power)* **in [full] possession of [all] one's [mental] faculties** im [Voll]besitz [all] seiner [geistigen] Kräfte; **c)** *(Univ.: department)* Fakultät, *die;* Fachbereich, *der;* **d)** *(Amer. Sch., Univ.: staff)* Lehrkörper, *der*

fad [fæd] *n.* Marotte, *die;* Spleen, *der (ugs.)*

fade [feɪd] **1.** *v. i.* **a)** *(droop, wither)* ‹*Blätter, Blumen:*› [ver]welken, welk werden; **b)** *(lose freshness, vigour)* verblassen; [v]erlöschen; ‹*Läufer:*› langsamer werden; ‹*Schönheit:*› verblühen; **c)** *(lose colour)* bleichen; ~ **[in colour]** [ver]bleichen; **d)** *(grow pale, dim)* **the light ~d [into darkness]** es dunkelte; **e)** *(fig.: lose strength)* ‹*Erinnerung:*› verblassen; ‹*Eingebung, Kreativität, Optimismus:*› nachlassen; ‹*Freude, Lust, Liebe:*› erlöschen; ‹*Ruhm:*› verblassen; ‹*Traum, Hoffnung:*› zerrinnen; **f)** *(grow faint)* ‹*Laut:*› verklingen; ~ **into the distance** in der Ferne entschwinden; ‹*Laut, Stimme:*› in der Ferne verklingen; **g)** *(blend)* übergehen **(into** in + *Akk.*). **2.** *v. t.* ausbleichen ‹*Vorhang, Teppich, Farbe*›

~ **a'way** *v. i.* schwinden; ‹*Laut:*› verklingen **(into** in + *Dat.*); ‹*Erinnerung, Augenlicht, Kraft:*› nachlassen; ‹*Interesse, Hoffnung:*› erlöschen; *(joc.)* ‹*dünne Person:*› immer weniger werden *(scherzh.)*

~ **'in** *(Radio, Telev., Cinemat.) v. t.* einblenden

~ **'out** *(Radio, Telev., Cinemat.)* **1.** *v. i.* ausgeblendet werden. **2.** *v. t.* *(Radio, Telev., Cinemat.)* ausblenden

faded ['feɪdɪd] *adj.* welk ‹*Blume, Blatt, Laub*›; verblichen ‹*Stoff, Jeans, Farbe, Gemälde, Ruhm, Teppich*›; verblüht ‹*Schönheit*›

'fade-in *n. (Radio, Telev., Cinemat.)* Einblendung, *die*

'fade-out *n. (Radio, Telev., Cinemat.)* Ausblendung, *die*

faeces ['fi:si:z] *n. pl.* Fäkalien *Pl.*

fag [fæg] **1.** *v. i.*, **-gg-** *(toil)* sich [ab]schinden *(ugs.)* **([away] at** mit). **2.** *v. t.*, **-gg-:** ~ **sb. [out]** jmdn. schlauchen *(ugs.);* ~ **oneself out** sich [ab]schinden *(ugs.);* **be ~ged out** geschlaucht sein *(ugs.).* **3.** *n.* **a)** *(Brit. coll.)* Schinderei, *die (ugs. abwertend);* **b)** *(sl.: cigarette)* Glimmstengel, *der (ugs. scherzh.)*

'fag-end *n.* **a)** *(remnant)* Schluß, *der;* Ende, *das;* **b)** *(sl.: cigarette-end)* Kippe, *die (ugs.)*

faggot (*Amer.:* **fagot**) ['fægət] *n.* **a)** Reisigbündel, *das;* **b)** *usu. in pl. (Gastr.)* Leberknödel, *der*

Fahrenheit ['færənhaɪt] *adj.* Fahrenheit

fail [feɪl] **1.** *v. i.* **a)** *(not succeed)* scheitern **(in** mit); ~ **in one's duty** seine Pflicht versäumen; ~ **as a human being/a doctor** als Mensch/Arzt versagen; **b)** *(miscarry, come to nothing)* scheitern; fehlschlagen; **if all else ~s** wenn alle Stricke *od.* Stränge reißen *(ugs.);* **c)** *(become bankrupt)* Bankrott machen; **d)** *(in examination)* nicht bestehen **(in** *Akk.*); **e)** *(become weaker)* ‹*Augenlicht, Gehör, Gedächtnis, Stärke:*› nachlassen; ‹*Mut:*› sinken; **his health is ~ing** sein Gesundheitszustand verschlechtert sich; **f)** *(break down, stop)* ‹*Versorgung:*› zusammenbrechen; ‹*Motor, Radio:*› aussetzen; ‹*Generator, Batterie, Pumpe:*› ausfallen; ‹*Bremse, Herz:*› versagen; **g)** ‹*Ernte:*› schlecht ausfallen. **2.** *v. t.* **a)** ~ **to do sth.** *(not succeed in doing)* etw. nicht tun [können]; ~ **to reach a decision** zu keinem Entschluß kommen; ~ **to achieve one's purpose/aim** seine Absicht/sein Ziel verfehlen; **b)** *(be unsuccessful in)* nicht bestehen ‹*Prüfung*›; **c)** *(reject)* durchfallen lassen *(ugs.)* ‹*Prüfling*›; **d)** ~ **to do sth.** *(not do)* etw. nicht tun; *(neglect to do)* [es] versäumen, etw. zu tun; **not ~ to do sth.** etw. tun; **I ~ to see why ...:** ich sehe nicht ein, warum ...; **e)** *(not suffice for)* im Stich lassen; **words ~ sb.** jmdm. fehlen die Worte. **3.** *n.* **without ~:** auf jeden Fall; garantiert

failed [feɪld] *attrib. adj.* nicht bestanden ‹*Prüfung*›; durchgefallen *(ugs.)* ‹*Prüfling*›; gescheitert ‹*Geschäft, Ehe, Versuch*›

failing ['feɪlɪŋ] **1.** *n.* Schwäche, *die.* **2.** *prep.* ~ **that** *or* **this** andernfalls; wenn nicht. **3.** *adj.* sich verschlechternd ‹*Gesundheitszustand*›; nachlassend ‹*Kraft*›; sinkend ‹*Mut*›; dämmrig ‹*Licht*›

'fail-safe *adj.* ausfallsicher; abgesichert ‹*Methode*›; Failsafe-‹*Vorkehrung, Prinzip*› *(fachspr.)*

failure ['feɪljə(r)] *n.* **a)** *(omission, neglect)* Versäumnis, *das;* ~ **to do sth.** das Versäumnis, etw. zu tun; ~ **to observe the rule** Nichtbeachtung der Regel; ~ **to pass an exam** Nichtbestehen einer Prüfung; **b)** *(lack of success)* Scheitern, *das;* **end in ~:** scheitern; **c)** *(unsuccessful person or thing)* Versager, *der;* **the party/play was a ~:** das Fest/Stück war ein Mißerfolg; **our plan/attempt was a ~:** unser Plan/Versuch war fehlgeschlagen; **d)** *(of supply)* Zusammenbruch, *der; (of engine, generator)* Ausfall, *der;* **signal/engine ~:** Ausfall des Signals/des Motors; **power ~:** Stromausfall, *der;* **crop ~:** Mißernte, *die;* **e)** *(bankruptcy)* Zusammenbruch, *der*

faint [feɪnt] **1.** *adj.* **a)** matt ‹*Licht, Farbe, Stimme, Lächeln*›; schwach ‹*Geruch, Duft*›; leise ‹*Geräusch, Stimme, Ton*›; entfernt ‹*Ähnlichkeit*›; undeutlich ‹*Umriß, Linie, Gestalt, Spur, Fotokopie*›; leise ‹*Hoffnung, Verdacht, Ahnung*›; gering ‹*Chance*›; **not have the ~est idea** nicht die geringste *od.* blasseste Ahnung haben; **b)** *(giddy, weak)* matt; schwach; **she felt ~:** ihr war schwindelig; **c)** *(feeble)* schwach ‹*Lob, Widerstand*›; zaghaft ‹*Versuch, Bemühung*›. **2.** *v. i.* ohnmächtig werden, in Ohnmacht fallen **(from** vor + *Dat.*). **3.** *n.* Ohnmacht, *die*

faint-hearted ['feɪnthɑ:tɪd] *adj.* hasenherzig *(abwertend);* zaghaft ‹*Versuch*›

faintly ['feɪntlɪ] *adv.* undeutlich ‹*markieren, hören*›; kaum ‹*sichtbar*›; schwach ‹*riechen, scheinen*›; entfernt ‹*sich ähneln*›; wenig ‹*interessieren*›; leicht ‹*enttäuschen*›; zaghaft ‹*lächeln*›

faintness ['feɪntnɪs] *n., no pl.* **a)** *(of marking, outline)* Undeutlichkeit, *die; (of resemblance)* Entferntheit, *die; (of colour)* Mattheit, *die;* **the ~ of the light** das schwache Licht; **b)** *(dizziness)* Schwäche, *die*

[1]fair [feə(r)] *n.* **a)** *(gathering)* Markt, *der; (with shows, merry-go-rounds)* Jahrmarkt, *der;* **b)** *see* **fun-fair; c)** *(exhibition)* Messe, *die;* **antiques/book/trade ~:** Antiquitäten- / Buch- / Handelsmesse, *die*

[2]fair **1.** *adj.* **a)** *(just)* gerecht; begründet ‹*Beschwerde, Annahme*›; berechtigt ‹*Frage*›; fair ‹*Spiel, Kampf, Prozeß, Preis, Handel*›; *(representative)* typisch, markant ‹*Beispiel, Kostprobe*›; **be ~ with** *or* **to sb.** gerecht gegen jmdn. *od.* zu jmdm. sein; **it's only ~ to do sth./for sb. to do sth.** es ist nur recht und billig, etw. zu tun/daß jmd. etw. tut; **that's not ~:** das ist ungerecht *od.* unfair; ~ **enough!** *(coll.)* dagegen ist nichts einzuwenden; *(OK)* na gut; **all's ~ in**

love and war in der Liebe und im Krieg ist alles erlaubt; ~ **play** Fairneß, *die;* **b)** *(not bad, pretty good)* ganz gut ‹*Bilanz, Vorstellung, Anzahl, Kenntnisse, Chance*›; ziemlich ‹*Maß, Geschwindigkeit*›; **a ~ amount of work** ein schönes Stück Arbeit; **c)** *(favourable)* schön ‹*Wetter, Tag, Abend*›; günstig ‹*Wetterlage, Wind*›; heiter ‹*Wetter, Tag*›; **d)** *(blond)* blond ‹*Haar, Person*›; *(not dark)* hell ‹*Teint, Haut*›; hellhäutig ‹*Person*›; **e)** *(poet. or literary: beautiful)* hold *(dichter. veralt.)* ‹*Maid, Prinz, Gesicht*›; **the ~ sex** das schöne Geschlecht. **2.** *adv.* **a)** fair ‹*kämpfen, spielen*›; gerecht ‹*behandeln*›; **b)** *(coll.: completely)* völlig; **the sight ~ took my breath away** der Anblick hat mir glatt *(ugs.)* den Atem verschlagen; **c) ~ and square** *(honestly)* offen und ehrlich; *(accurately)* voll, genau ‹*schlagen, treffen*›. **3.** *n.* **~'s ~** *(coll.)* Gerechtigkeit muß sein

fair: ~ground *n.* Festplatz, *der;* **~-haired** ['feəheəd] *adj.* blond

fairly ['feəlɪ] *adv.* **a)** fair ‹*kämpfen, spielen*›; gerecht ‹*bestrafen, beurteilen, behandeln*›; **b)** *(tolerably, rather)* ziemlich; **c)** *(completely)* völlig; **it ~ took my breath away** es hat mir glatt *(ugs.)* den Atem verschlagen; **d)** *(actually)* richtig; **I ~ jumped for joy** ich habe einen regelrechten Freudensprung gemacht; **e) ~ and squarely** *(honestly)* offen und ehrlich; **beat sb. ~ and squarely** jmdn. nach allen Regeln der Kunst *(ugs.)* besiegen

fair-'minded *adj.* unvoreingenommen

fairness ['feənɪs] *n., no pl.* Gerechtigkeit, *die;* **in all ~ [to sb.]** um fair [gegen jmdn.] zu sein

fair: ~-sized ['feəsaɪzd] *adj.* recht ansehnlich; **~way** *n.* **a)** *(channel)* Fahrrinne, *die;* **b)** *(Golf)* Fairway, *das*

fairy ['feərɪ] *n. (Mythol.)* Fee, *die; (in a household)* Kobold, *der*

fairy: ~ 'godmother *n. (lit. or fig.)* gute Fee; **F~land** *n. (land of fairies)* Feenland, *das; (enchanted region)* Märchenland, *das;* **~ lights** *n. pl.* kleine farbige Lichter; **~ 'ring** *n. (Bot.)* Hexenring, *der;* **~ story** *see* **~-tale 1; ~-tale 1.** *n. (lit. or fig.)* Märchen, *das;* **2.** *adj.* Märchen‹*landschaft*›; märchenhaft schön ‹*Szene, Wirkung, Kleid*›

fait accompli [feɪt æ'kɔ̃pli:, feɪt ə'kɒmplɪ] *n.* vollendete Tatsache

faith [feɪθ] *n.* **a)** *(reliance, trust)* Vertrauen, *das;* **have ~ in sb./sth.** Vertrauen zu jmdm./etw. haben; auf jmdn./etw. vertrauen; **lose ~ in sb./sth.** das Vertrauen zu jmdm./etw. verlieren; **b)** *([religious] belief)* Glaube, *der;* **different Christian ~s** verschiedene christliche Glaubensrichtungen; **c) keep ~ with sb.** jmdm. treu bleiben *od.* die Treue halten; **d) in good ~:** ohne Hintergedanken; *(unsuspectingly)* in gutem Glauben; **in bad ~:** in böser Absicht

faithful ['feɪθfl] **1.** *adj.* **a)** *(loyal)* treu **(to** *Dat.***)**; **b)** *(conscientious)* pflichttreu; [ge]treu ‹*Diener*›; **c)** *(accurate)* [wahrheits]getreu; originalgetreu ‹*Wiedergabe, Kopie*›. **2.** *n. pl.* **the ~:** die Gläubigen; **the party ~:** treue Anhänger der Partei

faithfully ['feɪθfəlɪ] *adv.* **a)** *(loyally)* treu ‹*dienen*›; hoch und heilig, fest ‹*versprechen*›; **b)** originalgetreu ‹*wiedergeben*›; genau ‹*befolgen*›; **c) yours ~** *(in letter)* mit freundlichen Grüßen; *(more formally)* hochachtungsvoll

faith: ~-healer *n.* Gesundbeter, *der*/-beterin, *die;* **~-healing** *n.* Gesundbeten, *das*

fake [feɪk] **1.** *adj.* unecht; gefälscht ‹*Dokument, Banknote, Münze*›. **2.** *n.* **a)** *(imitation)* Imitation, *die; (painting)* Fälschung, *die;* **b)** *(person)* Schwindler, *der*/Schwindlerin, *die.* **3.** *v. t.* **a)** fälschen ‹*Unterschrift, Gemälde*›; vortäuschen ‹*Krankheit, Unfall*›; erfinden ‹*Geschichte*›; **b)** *(alter so as to deceive)* verfälschen

fakir ['feɪkɪə(r)] *n.* Fakir, *der*

falcon ['fɔ:lkn, 'fɔ:kn] *n. (Ornith.)* Falke, *der*

fall [fɔ:l] **1.** *n.* **a)** *(act or manner of ~ing)* Fallen, *das; (of person)* Sturz, *der;* **~ of snow/rain** Schnee-/Regenfall, *der;* **in a ~:** bei einem Sturz; **have a ~:** stürzen; **b)** *(collapse, defeat)* Fall, *der; (of dynasty, empire)* Untergang, *der; (of government)* Sturz, *der;* **c)** *(slope)* Abfall, *der* **(to** zu, nach**)**; **d)** *(Amer.: autumn)* Herbst, *der.* **2.** *v. i.,* **fell** [fel], **~en** ['fɔ:ln] **a)** fallen; ‹*Person:*› [hin]fallen, stürzen; ‹*Pferd:*› stürzen; **~ off sth., ~ down from sth.** von etw. [herunter]fallen; **~ down [into] sth.** in etw. *(Akk.)* [hinein]fallen; **~ to the ground** zu Boden fallen; **~ down dead** tot umfallen; **~ down the stairs** die Treppe herunter-/hinunterfallen; **~ [flat] on one's face** *(lit. or fig.)* auf die Nase fallen *(ugs.);* **~ into the trap** in die Falle gehen; **~ from a great height** aus großer Höhe abstürzen; **rain/snow is ~ing** es regnet/schneit; **~ from power** entmachtet werden; **b)** *(fig.)* ‹*Nacht, Dunkelheit:*› hereinbrechen; ‹*Abend:*› anbrechen; ‹*Stille:*› eintreten; **c)** *(fig.: be uttered)* fallen; **~ from sb.'s lips** über jmds. Lippen *(Akk.)* kommen; **let ~ a remark** eine Bemerkung fallenlassen; **d)** *(become detached)* ‹*Blätter:*› [ab]fallen; **~ out** ‹*Haare, Federn:*› ausfallen; **e)** *(sink to lower level)* sinken; ‹*Barometer:*› fallen; ‹*Absatz, Verkauf:*› zurückgehen; **~ into sin/temptation** eine Sünde begehen/der Versuchung er- *od.* unterliegen; **f)** *(subside)* ‹*Wasserspiegel, Gezeitenhöhe:*› fallen; ‹*Wind:*› sich legen; **g)** *(show dismay)* **his/her face fell** er/sie machte ein langes Gesicht *(ugs.);* **h)** *(be defeated)* ‹*Festung, Stadt:*› fallen; ‹*Monarchie, Regierung:*› gestürzt werden; ‹*Reich:*› untergehen; **the fortress fell to the enemy** die Festung fiel dem Feind in die Hände; **i)** *(perish)* ‹*Soldat:*› fallen; **the ~en** die Gefallenen; **j)** *(collapse, break)* einstürzen; **~ to pieces, ~ apart** ‹*Buch, Wagen:*› auseinanderfallen; **~ apart at the seams** an den Nähten aufplatzen; **k)** *(come by chance, duty, etc.)* fallen **(to** an + *Akk.***)**; **it fell to me** *or* **to my lot to do it** das Los, es tun zu müssen, hat mich getroffen; **~ into decay** ‹*Gebäude:*› verfallen; **~ ill** krank werden; **~ into a swoon** *or* **faint** in Ohnmacht fallen; **they fell to fighting among themselves** es kam zu einer Schlägerei zwischen ihnen; **l)** ‹*Auge, Strahl, Licht, Schatten:*› fallen **(upon** auf + *Akk.***)**; **m)** *(have specified place)* liegen **(on, to** auf + *Dat.,* **within** in + *Dat.***)**; **~ into** *or* **under a category** in *od.* unter eine Kategorie fallen; **n)** *(occur)* fallen **(on** auf + *Akk.***)**

~ a'bout *v. i.* **~ about [laughing** *or* **with laughter]** sich [vor Lachen] kringeln *(ugs.)*

~ a'way *v. i. (have slope)* abfallen **(to** zu**)**

~ 'back *v. i.* zurückweichen; ‹*Armee:*› sich zurückziehen; *(lag)* zurückbleiben

~ 'back on *v. t.* zurückgreifen auf (+ *Akk.*)

~ behind 1. ['---] *v. t.* zurückfallen hinter (+ *Akk.*). **2.** [--'-] *v. i.* zurückbleiben; **~ behind with sth.** mit etw. in Rückstand geraten

~ 'down *v. i.* **a)** *see* **~ 2 a; b)** *(collapse)* ‹*Brücke, Gebäude:*› einstürzen; ‹*Person:*› hinfallen; **~ down**

[on sth.] *(fig. coll.)* [bei etw.] versagen

~ **for** *v. t. (coll.)* **a)** *(~ in love with)* sich verknallen in *(ugs.)*; **b)** *(be persuaded by)* hereinfallen auf (+ *Akk.*) *(ugs.)*

~ **'in** *v. i.* **a)** hineinfallen; **b)** *(Mil.)* antreten (**for** zu); ~ **in!** angetreten!; **c)** *(collapse)* ‹*Gebäude, Wand usw.:*› einstürzen

~ **'in with** *v. t.* **a)** *(meet and join)* stoßen zu; **b)** *(agree)* beipflichten (+ *Dat.*) ‹*Person, Meinung, Vorschlag usw.*›; eingehen auf (+ *Akk.*) ‹*Plan, Person, Bitte*›

~ **'off** *v. i.* **a)** *see* ~ **2a**; **b)** ‹*Nachfrage, Produktion, Anzahl:*› zurückgehen; ‹*Dienstleistungen, Gesundheit, Geschäft:*› sich verschlechtern; ‹*Interesse:*› nachlassen

~ **on** *v. t. (be borne by)* ~ **on sb.** jmdm. zufallen; ‹*Verdacht, Schuld, Los:*› auf jmdn. fallen

~ **'out** *v. i.* **a)** herausfallen; **b)** *see* ~ **2d**; **c)** *(quarrel)* ~ **out [with sb. over sth.]** sich [mit jmdm. über etw. *(Akk.)*] [zer]streiten; **d)** *(come to happen)* vonstatten gehen; **see how things** ~ **out** abwarten, wie sich die Dinge entwickeln

~ **over 1.** *v. t.* **a)** ['---] *(stumble over)* fallen über (+ *Akk.*); ~ **over one's own feet** über seine eigenen Füße stolpern; **b)** [-'--] ~ **over oneself to do sth.** *(fig. coll.)* sich vor Eifer überschlagen, um etw. zu tun *(ugs.)*. **2.** [-'--] *v. i.* umfallen; ‹*Person:*› [hin]fallen

~ **'through** *v. i. (fig.)* ins Wasser fallen *(ugs.)*

fallacious [fə'leɪʃəs] *adj.* **a)** *(containing a fallacy)* irrig; ~ **conclusion** Fehlschluß, *der;* **b)** *(deceptive, delusive)* irreführend

fallacy ['fæləsɪ] *n.* **a)** *(delusion, error)* Irrtum, *der;* **b)** *(unsoundness, delusiveness)* Irrigkeit, *die*

fallen *see* **fall 2**

'fall guy *n. (sl.: scapegoat)* Prügelknabe, *der (ugs.)*

fallible ['fælɪbl] *adj.* **a)** *(liable to err)* fehlbar ‹*Person*›; **b)** *(liable to be erroneous)* nicht unfehlbar

'fall-off *n. (in quality)* [Ver]minderung, *die* (**in** *Gen.*); *(in quantity)* Rückgang, *der* (**in** *Gen.*); **there has been a** ~ **in support for the government** die Regierung hat an Rückhalt verloren

Fallopian tube [fə'ləʊpɪən tju:b] *n. (Anat.)* Eileiter, *der*

'fall-out *n.* radioaktiver Niederschlag; *(fig.: side-effects)* Abfallprodukte; *attrib.* ~ **shelter** Atombunker, *der*

fallow ['fæləʊ] *adj. (lit. or fig.)* brachliegend; ~ **ground/land** Brache, *die*/Brachland, *das;* **lie** ~ *(lit. or fig.)* brachliegen

'fallow deer *n.* Damhirsch, *der*

false [fɔ:ls, fɒls] *adj.* falsch; Fehl‹*deutung, -urteil*›; Falsch‹*meldung, -eid, -aussage*›; treulos ‹*Geliebte[r]*›; gefälscht ‹*Urkunde, Dokument*›; künstlich ‹*Wimpern, Auge*›; geheuchelt ‹*Bescheidenheit*›; **under a** ~ **name** unter falschem Namen

false: ~ **a'larm** *n.* blinder Alarm; ~ **'bottom** *n.* doppelter Boden

falsehood ['fɔ:lshʊd, 'fɒlshʊd] *n.* **a)** *no pl. (falseness)* Unrichtigkeit, *die;* **b)** *(untrue thing)* Unwahrheit, *die*

falsely ['fɔ:lslɪ, 'fɒlslɪ] *adv.* **a)** *(dishonestly)* unaufrichtig ‹*sprechen*›; falsch ‹*schwören*›; **b)** *(incorrectly, unjustly)* falsch ‹*auslegen, verstehen*›; fälschlich[erweise] ‹*annehmen, glauben, behaupten, beschuldigen*›

false 'move *see* **false step**

falseness ['fɔ:lsnɪs, 'fɒlsnɪs] *n., no pl.* **a)** *(incorrectness)* Unrichtigkeit, *die;* Falschheit, *die;* **b)** *(faithlessness)* Treulosigkeit, *die* (**to** gegenüber)

false: ~ **pre'tences** *n. pl.* Vorspiegelung falscher Tatsachen; ~ **'start** *n. (Sport; also fig.)* Fehlstart, *der;* ~ **'step** *n. (lit. or fig.)* falscher Schritt; ~ **'teeth** *n. pl.* [künstliches] Gebiß; Prothese, *die*

falsetto [fɔ:l'setəʊ, fɒl'setəʊ] *n., pl.* ~**s** *(voice)* Kopfstimme, *die; (Mus.: of man)* Falsett, *das*

falsify ['fɔ:lsɪfaɪ, 'fɒlsɪfaɪ] *v. t. (alter)* fälschen; *(misrepresent)* verfälschen ‹*Tatsache, Geschichte, Ereignis, Wahrheit*›

falsity ['fɔ:lsɪtɪ, 'fɒlsɪtɪ] *n., no pl.* **a)** *(incorrectness)* Falschheit, *die;* **b)** *(falsehood)* Unwahrheit, *die*

falter ['fɔ:ltə(r), 'fɒltə(r)] **1.** *v. i.* **a)** *(waver)* stocken; ‹*Mut:*› sinken; ~ **in one's determination** in seiner Entschlossenheit schwankend werden; **b)** *(stumble)* wanken; **with** ~**ing steps** mit [sch]wankenden Schritten. **2.** *v. t.* ~ **sth.** etw. stammeln

fame [feɪm] *n., no pl.* Ruhm, *der;* **rise to** ~**:** zu Ruhm kommen *od.* gelangen; **ill** ~**:** schlechter Ruf

famed [feɪmd] *adj.* berühmt (**for** für, wegen)

familiar [fə'mɪljə(r)] *adj.* **a)** *(well acquainted)* bekannt; **be** ~ **with sb.** jmdn. näher kennen; **b)** *(having knowledge)* vertraut (**with** mit); **c)** *(well known)* vertraut; bekannt ‹*Gesicht, Name, Lied*›; *(common, usual)* geläufig ‹*Ausdruck*›; gängig ‹*Vorstellung*›; **he looks** ~**:** er kommt mir bekannt vor; **d)** *(informal)* familiär ‹*Ton, Begrüßung*›; ungezwungen ‹*Art, Sprache, Stil*›; **e)** *(presumptuous)* plump-vertraulich *(abwertend)*

familiarise *see* **familiarize**

familiarity [fəmɪlɪ'ærɪtɪ] *n.* **a)** *no pl. (acquaintance)* Vertrautheit, *die;* **b)** *no pl. (relationship)* familiäres Verhältnis; **c)** *(of action, behaviour)* Vertraulichkeit, *die;* ~ **breeds contempt** *(prov.)* zu große Vertraulichkeit erzeugt Verachtung

familiarize [fə'mɪljəraɪz] *v. t.* vertraut machen (**with** mit)

family ['fæməlɪ] *n.* **a)** Familie, *die; attrib.* Familien-; familiär ‹*Hintergrund*›; **be one of the** ~**:** zur Familie gehören; **start a** ~**:** eine Familie gründen; **run in the** ~**:** in der Familie liegen; **b)** *(group, race)* Geschlecht, *das*

family: ~ **al'lowance** *n.* Kindergeld, *das;* ~ **'doctor** *n.* Hausarzt, *der;* ~ **'income supplement** *n. (Brit.)* ≈ Familienzulage, *die;* ~ **man** *n.* Familienvater, *der; (home-loving man)* häuslich veranlagter Mann; ~ **name** *n.* Familienname, *der;* Nachname, *der;* ~ **'planning** *n.* Familienplanung, *die;* ~ **'planning clinic** *n.* ≈ Familienberatung[sstelle], *die;* ~ **'tree** *n.* Stammbaum, *der*

famine ['fæmɪn] *n.* Hungersnot, *die*

famished ['fæmɪʃt] *adj.* ausgehungert; **I'm absolutely** ~ *(coll.)* ich sterbe vor Hunger *(ugs.)*

famous ['feɪməs] *adj.* berühmt; **a** ~ **victory** ein rühmlicher Sieg

[1]**fan** [fæn] **1.** *n.* **a)** *(held in hand)* Fächer, *der;* **b)** *(apparatus)* Ventilator, *der.* **2.** *v. t.*, **-nn-** fächeln ‹*Gesicht*›; anfachen ‹*Feuer*›; ~ **oneself/sb.** sich/jmdm. Luft zufächeln; ~ **the flame[s]** *(fig.)* das Feuer schüren

~ **'out 1.** *v. t.* fächern; auffächern ‹*Spielkarten*›. **2.** *v. i.* fächern; ‹*Soldaten:*› ausfächern

[2]**fan** *n. (devotee)* Fan, *der*

fanatic [fə'nætɪk] **1.** *adj.* fanatisch. **2.** *n.* Fanatiker, *der*/Fanatikerin, *die*

fanatical [fə'nætɪkl] *see* **fanatic 1**

fanaticism [fə'nætɪsɪzm] *n.* Fanatismus, *der*

'fan belt *n. (Motor Veh.)* Keilriemen, *der*

fancier ['fænsɪə(r)] *n.* Liebhaber, *der*/Liebhaberin, *die*

fanciful ['fænsɪfl] *adj.* versponnen ‹*Person*›; abstrus, über-

spannt ⟨*Vorstellung, Gedanke*⟩; phantastisch ⟨*Gemälde, Design*⟩
'fan club *n.* Fanklub, *der*
fancy ['fænsɪ] **1.** *n.* **a)** *(taste, inclination)* **he has taken a ~ to a new car/her** ein neues Auto/sie hat es ihm angetan; **take** *or* **catch sb.'s ~:** jmdm. gefallen; jmdn. ansprechen; **b)** *(whim)* Laune, *die;* **just as the ~ takes me** ganz nach Lust und Laune; **tickle sb.'s ~:** jmdn. reizen; **c)** *(notion)* merkwürdiges Gefühl; *(delusion, belief)* Vorstellung, *die;* **d)** *(faculty of imagining)* Phantasie, *die;* **e)** *(mental image)* Phantasievorstellung, *die;* **just a ~:** nur Einbildung. **2.** *attrib. adj.* **a)** *(ornamental)* kunstvoll ⟨*Arbeit, Muster*⟩; **~ jewellery** Modeschmuck, *der;* **nothing ~:** etwas ganz Schlichtes; **b)** *(extravagant)* stolz *(ugs.);* **~ prices** gepfefferte Preise *(ugs.).* **3.** *v. t.* **a)** *(imagine)* sich *(Dat.)* einbilden; **b)** *(coll.) in imper. as excl. of surprise* **~ meeting you here!** na, so etwas, Sie hier zu treffen!; **~ that!** sieh mal einer an!; also so etwas!; **c)** *(suppose)* glauben; denken; **..., I ~:** ..., möchte ich meinen; **d)** *(wish to have)* mögen; **what do you ~ for dinner?** was hättest du gern zum Abendessen?; **he fancies [the idea of] doing sth.** er würde etw. gern tun; **do you think she fancies him?** glaubst du, sie mag ihn?; **e)** *(coll.: have high opinion of)* **~ oneself** von sich eingenommen sein; **~ oneself as a singer** sich für einen [großen] Sänger halten; **~ one's/sb.'s chances** seine/jmds. Chancen hoch einschätzen
fancy: ~ 'dress *n.* [Masken]kostüm, *das;* **in ~ dress** kostümiert; *attrib.* **~-dress party** Kostümfest, *das;* **~-'free** *adj.* frei und ungebunden; **~ goods** *n. pl.* Geschenkartikel
fanfare ['fænfeə(r)] *n.* Fanfare, *die*
fang [fæŋ] *n.* **a)** *(canine tooth)* Reißzahn, *der;* Fang[zahn], *der;* *(of boar; joc.: of person)* Hauer, *der;* **b)** *(of snake)* Giftzahn, *der*
fan: ~ heater *n.* Heizlüfter, *der;* **~light** *n.* Oberlicht, *das;* *(~-shaped)* Fächerfenster, *das* *(Archit.);* **~ mail** *n.* Fanpost, *die;* **~-shaped** *adj.* fächerförmig
fantasia [fæn'teɪzɪə] *n.* *(Mus.)* Fantasie, *die*
fantastic [fæn'tæstɪk] *adj.* **a)** *(grotesque, quaint)* bizarr; *(fanciful)* phantastisch; **b)** *(coll.: excellent, extraordinary)* phantastisch *(ugs.)*
fantastically [fæn'tæstɪkəlɪ] *adv.* **a)** phantastisch; **b)** *(coll.: excellently, extraordinarily)* phantastisch *(ugs.)*
fantasy ['fæntəzɪ] *n.* Phantasie, *die;* *(mental image)* Phantasiegebilde, *das*
far [fɑ:(r)] **1.** *adv.*, **farther, further; farthest, furthest a)** *(in space)* weit; **~ away** weit entfernt; **~ [away] from** weit entfernt von; **see sth. from ~ away** etw. aus der Ferne sehen; **I won't be ~ away** ich werde ganz in der Nähe sein; **~ above/below** hoch über/tief unter (+ *Dat.*); *adv.* hoch oben/tief unten; **fly as ~ as Munich** bis [nach] München fliegen; **~ and wide** weit und breit; **from ~ and near** *or* **wide** von fern und nah; **b)** *(in time)* weit; **~ into the night** bis spät *od.* tief in die Nacht; **as ~ back as I can remember** soweit ich zurückdenken kann; **c)** *(by much)* weit; **~ too** viel zu; **~ longer/better** weit[aus] länger/besser; **d)** *(fig.)* **as ~ as** *(to whatever extent, to the extent of)* so weit [wie]; **I haven't got as ~ as phoning her** ich bin noch nicht dazu gekommen, sie anzurufen; **not as ~ as I know** nicht, daß ich wüßte; **as ~ as I remember/know** soweit ich mich erinnere/weiß; **go so ~ as to do sth.** so weit gehen und etw. tun; **in so ~ as** insofern *od.* insoweit als; **so ~** *(until now)* bisher; **so ~ so good** so weit, so gut; **by ~:** bei weitem; **better by ~:** weitaus besser; **~ from easy/good** alles andere als leicht/gut; **~ from it!** ganz im Gegenteil!; **go too ~:** zu weit gehen; **carry** *or* **take sth. too ~:** etw. zu weit treiben. **2.** *adj.*, **farther, further; farthest, furthest a)** *(remote)* weit entfernt; *(remote in time)* fern; **in the ~ distance** in weiter Ferne; **b)** *(more remote)* weiter entfernt; **the ~ bank of the river/side of the road** das andere Flußufer/die andere Straßenseite
'far-away *attrib. adj.* **a)** *(remote in space)* entlegen; abgelegen; *(remote in time)* fern; **b)** *(dreamy)* verträumt ⟨*Stimme, Blick, Augen*⟩
farce [fɑ:s] *n.* **a)** Farce, *die;* **b)** *(Theatre)* Posse, *die;* Farce, *die*
farcical ['fɑ:sɪkl] *adj.* **a)** *(absurd)* farcenhaft; absurd; **b)** *(Theatre)* possenhaft ⟨*Stück, Element*⟩
fare [feə(r)] **1.** *n.* **a)** *(price)* Fahrpreis, *der;* *(money)* Fahrgeld, *das;* **what** *or* **how much is the ~?** was kostet die Fahrt/*(by air)* der Flug/*(by boat)* die Überfahrt?; **any more ~s?** noch jemand ohne [Fahrschein]?; **b)** *(passenger)* Fahrgast, *der;* **c)** *(food)* Kost, *die.* **2.** *v. i.* *(get on)* **I don't know how he is faring/how he ~d on his travels** ich weiß nicht, wie es ihm geht/wie es ihm auf seinen Reisen ergangen ist
Far: ~ 'East *n.* **the ~ East** der Ferne Osten; Fernost *o. Art.;* **~ 'Eastern** *adj.* fernöstlich; des Fernen Ostens *nachgestellt*
'fare-stage *n.* Teilstrecke, *die;* *(end of section)* Zahlgrenze, *die*
farewell [feə'wel] **1.** *int.* leb[e] wohl *(veralt.).* **2.** *n.* **a) make one's ~s** sich verabschieden; **b)** *attrib.* **~ speech/gift** Abschiedsrede, *die/*-geschenk, *das*
far: ~-fetched ['fɑ:fetʃt] *adj.* weit hergeholt; an *od.* bei den Haaren herbeigezogen *(ugs.);* **~-flung** *adj.* *(widely spread)* weit ausgedehnt; *(distant)* weit entfernt
farm [fɑ:m] **1.** *n.* [Bauern]hof, *der;* *(larger)* Gut, *das;* Gutshof, *der;* *(in English-speaking countries outside Europe)* Farm, *die;* **poultry/chicken ~:** Geflügel-/Hühnerfarm, *die;* **~ bread/eggs** Landbrot, *das*/Landeier *Pl.;* **~ animals** Vieh, *das.* **2.** *v. t.* bebauen, bewirtschaften ⟨*Land*⟩; züchten ⟨*Lachs, Forellen*⟩. **3.** *v. i.* Landwirtschaft betreiben
~ 'out *v. t.* **a)** verpachten ⟨*Land*⟩; **b)** vergeben ⟨*Arbeit*⟩ **(to** an + *Akk.*)
farmer ['fɑ:mə(r)] *n.* Landwirt, *der/*-wirtin, *die;* Bauer, *der/* Bäuerin, *die;* **poultry ~:** Geflügelzüchter, *der/*-züchterin, *die*
farm: ~-hand *n.* Landarbeiter, *der/*-arbeiterin, *die;* **~house** *n.* Bauernhaus, *das;* *(larger)* Gutshaus, *das*
farming ['fɑ:mɪŋ] *n, no pl., no indef. art.* Landwirtschaft, *die;* **~ of crops** Ackerbau, *der;* **~ of animals** Viehzucht, *die*
farm: ~stead ['fɑ:msted] *n.* Bauernhof, *der;* Gehöft, *das;* **~-worker** *n.* Landarbeiter, *der/* -arbeiterin, *die;* **~yard** *n.* Hof, *der*
Faroes ['feərəʊz] *pr. n. pl.* Färöer *Pl.*
far: ~-off *adj.* *(in space)* [weit] entfernt; *(in time)* fern; **~-out** *adj.* **a)** *(distant)* [weit] entfernt; **b)** *(fig. coll.: excellent)* toll *(ugs.);* super *(ugs.);* **~-reaching** *adj.* weitreichend ⟨*Konsequenzen, Bedeutung, Wirkung*⟩; **~-seeing** *adj.* weitblickend; **~-sighted** *adj.* **a)** *(able to see a great distance)* scharfsichtig; **b)** *(having foresight)* weitblickend

fart [fɑːt] *(coarse)* **1.** *v.i.* furzen *(derb).* **2.** *n.* Furz, *der (derb)*

farther ['fɑːðə(r)] *see* **further 1 a, 2 a**

farthest ['fɑːðɪst] *see* **furthest**

farthing ['fɑːðɪŋ] *n. (Brit. Hist.)* Farthing, *der*

Far 'West *n. (Amer.)* **the ~:** der Westen der USA

fascinate ['fæsɪneɪt] *v.t.* fesseln; faszinieren *(geh.);* **it ~s me how ...:** ich finde es faszinierend, wie ...

fascinated ['fæsɪneɪtɪd] *adj.* fasziniert

fascinating ['fæsɪneɪtɪŋ] *adj.* faszinierend *(geh.);* bezaubernd; hochinteressant ‹*Thema, Faktum*›; spannend, fesselnd ‹*Buch*›

fascination [fæsɪ'neɪʃn] *n., no pl.* Faszination, *die (geh.); (quality of fascinating)* Zauber, *der;* Reiz, *der;* **have a ~ for sb.** einen besonderen Reiz auf jmdn. ausüben

Fascism ['fæʃɪzm] *n.* Faschismus, *der*

Fascist ['fæʃɪst] **1.** *n.* Faschist, *der*/Faschistin, *die.* **2.** *adj.* faschistisch

fashion ['fæʃn] **1.** *n.* **a)** Art [und Weise]; **talk/behave in a peculiar ~:** merkwürdig sprechen/sich merkwürdig verhalten; **walk crab-~/in a zigzag ~:** im Krebsgang/Zickzack gehen; **after** *or* **in the ~ of** im Stil *od.* nach Art von; **after** *or* **in a ~:** schlecht und recht; einigermaßen; **b)** *(custom, esp. in dress)* Mode, *die;* **the latest summer/autumn ~s** die neusten Sommer-/Wintermodelle; **it is the ~:** es ist Mode *od.* modern; **be all the ~:** große Mode *od.* groß in Mode sein; **in ~:** in Mode; modern; **be out of ~:** aus der Mode *od.* nicht mehr modern sein; **come into/go out of ~:** in Mode/aus der Mode kommen; **it was the ~ in those days** das war damals Sitte *od.* Brauch. **2.** *v.t.* formen, gestalten (**out of, from** aus; **[in]to** zu)

fashionable ['fæʃənəbl] *adj.* modisch ‹*Kleider, Person, Design*›; vornehm ‹*Gegend, Hotel, Restaurant*›; Mode‹*farbe, -krankheit, -wort, -autor*›; **it isn't ~ any more** es ist nicht mehr modern *od.* in Mode

fashionably ['fæʃənəblɪ] *adv.* modisch ‹*sich kleiden*›

fashion: ~-conscious *adj.* modebewußt; **~ magazine** *n.* Modezeitschrift, *die;* **~ show** *n.* Mode[n]schau, *die*

¹fast [fɑːst] **1.** *v.i.* fasten; **a day of ~ing** ein Fast[en]tag. **2.** *n. (going without food)* Fasten, *das; (hunger-strike)* Hungerstreik, *der;* **a 40-day ~:** eine Fastenzeit von 40 Tagen

²fast 1. *adj.* **a)** *(fixed, attached)* fest; **the rope is ~:** das Tau ist fest[gemacht]; **make [the boat] ~:** das Boot festmachen; **hard and ~:** fest; bindend, verbindlich ‹*Regeln*›; **b)** *(not fading)* farbecht ‹*Stoff*›; echt, beständig ‹*Farbe*›; **c)** *(rapid)* schnell; **~ train** Schnellzug, *der;* D-Zug, *der;* **~ speed** hohe Geschwindigkeit; **he is a ~ worker** *(lit. or fig.)* er arbeitet schnell; *(in amorous activities)* er geht mächtig ran *(ugs.);* **pull a ~ one [on sb.]** *(sl.)* jmdn. übers Ohr hauen *od.* reinlegen *(ugs.);* **d) be ~ [by ten minutes], be [ten minutes] ~** ‹*Uhr:*› [zehn Minuten] vorgehen. **2.** *adv.* **a)** *(lit. or fig.)* fest; **hold ~ to sth.** sich an etw. *(Dat.)* festhalten; *(fig.)* an etw. *(Dat.)* festhalten; **b)** *(soundly)* **be ~ asleep** fest schlafen; *(when one should be awake)* fest eingeschlafen sein; **c)** *(quickly)* schnell; **not so ~!** nicht so hastig!; **d) play ~ and loose with sb.** mit jmdm. ein falsches *od.* doppeltes Spiel treiben

fasten ['fɑːsn] **1.** *v.t.* **a)** festmachen, befestigen (**on, to** an + *Dat.*); festziehen, anziehen ‹*Schraube*›; zumachen ‹*Kleid, Spange, Knöpfe, Jacke*›; schließen ‹*Tür, Fenster*›; anstecken ‹*Brosche*› (**to** an + *Akk.*); **~ sth. together with a clip** etw. zusammenheften; **~ one's safety-belt** sich anschnallen; **~ up one's shoes** seine Schuhe binden *od.* schnüren; **b)** heften ‹*Blick*› (**[up]on** auf + *Akk.*); **~ one's attention on sb.** jmdm. seine Aufmerksamkeit zuwenden. **2.** *v.i.* **a)** sich schließen lassen; **the skirt ~s at the back** der Rock wird hinten zugemacht; **b) ~ [up]on sth.** *(single out)* etw. herausgreifen; *(seize upon)* etw. aufs Korn nehmen *(ugs.)*

fastener ['fɑːsnə(r)], **fastening** ['fɑːsnɪŋ] *ns.* Verschluß, *der*

'fast food *n. im Schnellrestaurant angebotenes Essen;* Fast food, *das;* **~ restaurant** Schnellrestaurant, *das*

fastidious [fæ'stɪdɪəs] *adj. (hard to please)* heikel, *(ugs.)* pingelig (**about** in bezug auf + *Akk.*); *(carefully selective)* wählerisch (**about** in bezug auf + *Akk.*)

'fast lane *n.* Überholspur, *die;* **life in the ~** *(fig.)* Leben auf vollen Touren *(ugs.)*

fat [fæt] **1.** *adj.* **a)** dick; fett *(abwertend);* rund ‹*Wangen, Gesicht*›; fett ‹*Schwein*›; **grow** *or* **get ~:** dick werden; **b)** fett ‹*Essen, Fleisch, Brühe*›; **c)** *(fig.)* dick ‹*Bündel, Buch, Zigarre*›; üppig, fett ‹*Gewinn, Gehalt, Scheck*›; **d)** *(sl. iron.)* **~ lot of good 'you are** du bist mir 'ne schöne Hilfe *(iron.);* **a ~ lot [of good it would do me]** [das würde mir] herzlich wenig [helfen]. **2.** *n.* Fett, *das;* **low in ~:** fettarm ‹*Nahrungsmittel*›; **put on ~:** Fett ansetzen; **run to ~:** [zu] dick werden; **the ~ is in the fire** *(fig.)* der Teufel ist los *(ugs.);* **live off** *or* **on the ~ of the land** *(fig.)* wie die Made im Speck leben *(ugs.)*

fatal ['feɪtl] *adj.* **a)** *(ruinous, disastrous)* verheerend (**to** für); fatal; schicksalsschwer ‹*Tag, Moment*›; **it would be ~:** das wäre das Ende; **b)** *(deadly)* tödlich ‹*Unfall, Verletzung*›; **deal sb. a ~ blow** jmdm. einen vernichtenden Schlag versetzen

fatalism ['feɪtəlɪzm] *n., no pl.* Fatalismus, *der (geh.);* Schicksalsergebenheit, *die*

fatalist ['feɪtəlɪst] *n.* Fatalist, *der*/Fatalistin, *die*

fatalistic [feɪtə'lɪstɪk] *adj.* fatalistisch; schicksalsergeben ‹*Person*›

fatality [fə'tælɪtɪ] *n.* Todesfall, *der; (in car crash, war, etc.)* [Todes]opfer, *das*

fatally ['feɪtəlɪ] *adv.* tödlich ‹*verwunden*›; *(disastrously)* verhängnisvoll; unwiderstehlich ‹*attraktiv*›; **be ~ wrong** *or* **mistaken** einem verhängnisvollen Irrtum unterliegen; **be ~ ill** todkrank sein

fate [feɪt] *n.* Schicksal, *das;* **an accident** *or* **stroke of ~:** eine Fügung des Schicksals

fated ['feɪtɪd] *adj. (doomed)* zum Scheitern verurteilt ‹*Plan, Projekt*›; **be ~ to fail** *or* **to be unsuccessful** zum Scheitern verurteilt sein; **be ~:** unter einem ungünstigen Stern stehen

fateful ['feɪtfl] *adj.* **a)** *(important, decisive)* schicksalsschwer ‹*Tag, Stunde, Entscheidung*›; entscheidend ‹*Worte*›; **b)** *(controlled by fate)* schicksalhaft ‹*Begegnung, Treffen, Ereignis*›; **c)** *(prophetic)* schicksalverkündend; *(of misfortune)* unheilverkündend

fat: ~head *n.* Dummkopf, *der (ugs.);* **~-headed** *adj.* dumm; blöd *(ugs.)*

father ['fɑːðə(r)] **1.** *n.* **a)** *(lit. or fig.)* Vater, *der;* **become a ~:** Vater werden; **he is a** *or* **the ~ of six**

er hat sechs Kinder; **like ~ like son** der Apfel fällt nicht weit vom Stamm *(ugs. scherzh., Spr.)*; **[our heavenly] F~**: [unser himmlischer] Vater; **b)** *(priest)* Pfarrer, *der; (monk)* Pater, *der;* **F~** *(as title: priest)* Herr Pfarrer; *(as title: monk)* Pater. **2.** *v. t.* zeugen

father: F~ 'Christmas *n.* der Weihnachtsmann; **~-figure** *n.* Vaterfigur, *die*

fatherhood ['fɑ:ðəhʊd] *n., no pl.* Vaterschaft, *die*

father: ~-in-law *n., pl.* **~s-in-law** Schwiegervater, *der;* **~land** *n.* Vaterland, *das*

fatherly ['fɑ:ðəlɪ] *adj.* väterlich

fathom ['fæðəm] **1.** *n. (Naut.)* Fathom, *das (geh.);* Faden, *der.* **2.** *v. t.* **a)** *(measure)* mit dem Lot messen; **b)** *(fig.: comprehend)* verstehen; **~ sb./sth. out** jmdn./etw. ergründen

fatigue [fə'ti:g] **1.** *n.* **a)** Ermüdung, *die;* Erschöpfung, *die;* **extreme ~**: Übermüdung, *die;* **b)** *(of metal etc.)* Ermüdung, *die.* **2.** *v. t.* ermüden; **be/look ~d** erschöpft sein/aussehen

fatten ['fætn] *v. t.* herausfüttern ‹*Person*›; mästen ‹*Tier*›

fattening ['fætnɪŋ] *adj.* dick machend ‹*Nahrungsmittel*›; **~ foods** Dickmacher *Pl. (ugs.);* **be ~**: dick machen

fatty ['fætɪ] **1.** *adj.* **a)** fett ‹*Fleisch*›; fetthaltig ‹*Nahrung, Speise*›; fettig ‹*Substanz*›; **b)** *(consisting of fat)* Fett-. **2.** *n. (coll.)* Dickerchen, *das (scherzh.)*

fatuous ['fætjʊəs] *adj.* albern; töricht; einfältig ‹*Grinsen*›

faucet ['fɔ:sɪt] *n. (Amer.)* Wasserhahn, *der*

fault [fɔ:lt, fɒlt] **1.** *n.* **a)** Fehler, *der;* **to a ~**: allzu übertrieben; übermäßig; **find ~ [with sb./sth.]** etwas [an jmdm./etw.] auszusetzen haben; **b)** *(responsibility)* Schuld, *die;* Verschulden, *das;* **whose ~ was it?** wer war schuld [daran]?; **it's all your own ~!** das ist deine eigene Schuld!; **it isn't my ~**: es ist nicht meine Schuld; **be at ~**: im Unrecht sein; **c)** *(Tennis etc.)* Fehler, *der;* **double ~**: Doppelfehler, *der;* **d)** *(in gas or water supply; Electr.)* Defekt, *der;* **e)** *(Geol.)* Verwerfung, *die.* **2.** *v. t.* Fehler finden an (+ *Dat.*); etwas auszusetzen haben an (+ *Dat.*)

fault: ~finder *n.* Krittler, *der/* Krittlerin, *die;* **~finding 1.** *n.* Krittelei, *die;* **2.** *adj.* krittelig

faultless ['fɔ:ltlɪs, 'fɒltlɪs] *adj.* einwandfrei; fehlerlos, fehlerfrei ‹*Übersetzung, Englisch*›

faulty ['fɔ:ltɪ, 'fɒltɪ] *adj.* fehlerhaft; unzutreffend ‹*Argument*›; defekt ‹*Gerät usw.*›; **~ design** Fehlkonstruktion, *die*

fauna ['fɔ:nə] *n., pl.* **~e** ['fɔ:ni:] *or* **~s** *(Zool.)* Fauna, *die*

faux pas [fəʊ 'pɑ:] *n., pl. same* [fəʊ 'pɑ:z] Fauxpas, *der*

favor *etc. (Amer.) see* **favour** *etc.*

favour ['feɪvə(r)] *(Brit.)* **1.** *n.* **a)** Gunst, *die;* Wohlwollen, *das;* **find/lose ~ with sb.** ‹*Sache:*› bei jmdm. Anklang finden/jmdm. nicht mehr gefallen; ‹*Person:*› jmds. Wohlwollen gewinnen/verlieren; **be in ~ [with sb.]** [bei jmdm.] beliebt sein; ‹*Idee, Kleidung usw.:*› [bei jmdm.] in Mode sein; **be out of ~ [with sb.]** [bei jmdm.] unbeliebt sein; ‹*Idee, Kleidung usw.:*› [bei jmdm.] nicht mehr in Mode sein; **b)** *(kindness)* Gefallen, *der;* Gefälligkeit, *die;* **ask a ~ of sb., ask sb. a ~**: jmdn. um einen Gefallen bitten; **do sb. a ~, do a ~ for sb.** jmdm. einen Gefallen tun; **as a ~**: aus Gefälligkeit; **as a ~ to sb.** jmdm. zuliebe; **c)** *(support)* **be in ~ of sth.** für etw. sein; **in ~ of** zugunsten (+ *Gen.*); **all those in ~**: alle, die dafür sind; **in sb.'s ~**: zu jmds. Gunsten; **d)** *(partiality)* Begünstigung, *die;* **show ~ to[wards] sb.** jmdn. begünstigen. **2.** *v. t.* **a)** *(approve)* für gut halten, gutheißen ‹*Plan, Idee, Vorschlag*›; *(think preferable)* bevorzugen; **I ~ the first proposal** ich bin für den ersten Vorschlag; **b)** *(oblige)* beehren **(with** mit) *(geh.);* **c)** *(treat with partiality)* bevorzugen; **d)** *(prove advantageous to)* begünstigen

favourable ['feɪvərəbl] *adj. (Brit.)* **a)** günstig ‹*Eindruck, Licht*›; gewogen ‹*Haltung, Einstellung*›; wohlmeinend ‹*Urteil*›; **be ~ to[wards] sth.** ‹*Person:*› einer Sache *(Dat.)* positiv gegenüberstehen; **b)** *(praising)* freundlich ‹*Erwähnung*›; positiv, günstig ‹*Bericht[erstattung], Bemerkung*›; **c)** *(promising)* vielversprechend; gut ‹*Omen, Zeichen*›; **d)** *(helpful)* günstig **(to** für) ‹*Wetter, Wind, Umstand*›; **e) give sb. a ~ answer** jmdm. eine Zusage geben

favourably ['feɪvərəblɪ] *adv. (Brit.)* **a)** wohlwollend ‹*ansehen, anhören, denken, urteilen*›; **be ~ impressed with sb./sth.** von jmdm./etw. sehr angetan sein; **be ~ disposed towards sb./sth.** jmdm./einer Sache positiv gegenüberstehen; **b)** lobend ‹*erwähnen*›; positiv ‹*vermerken*›

favoured ['feɪvəd] *adj. (Brit.)* *(privileged)* bevorzugt; *(well-liked)* Lieblings‹*platz, -buch, -gericht*›

favourite ['feɪvərɪt] *(Brit.)* **1.** *adj.* Lieblings-. **2.** *n.* **a)** *(film/food/pupil etc.)* Lieblingsfilm, *der*/-essen, *das*/-schüler, *der usw.; (person)* Liebling, *der;* **this/he is my ~**: das/ihn mag ich am liebsten; **b)** *(Sport)* Favorit, *der*/Favoritin, *die*

favouritism ['feɪvərɪtɪzm] *n., no pl. (Brit.)* Begünstigung, *die; (when selecting sb. for a post etc.)* Günstlingswirtschaft, *die*

¹fawn [fɔ:n] **1.** *n.* **a)** *(fallow deer)* [Dam]kitz, *das; (buck)* Bockkitz, *das; (doe)* Geißkitz, *das;* **b)** *(colour)* Rehbraun, *das.* **2.** *adj.* rehfarben; **~ colour** Rehbraun, *das*

²fawn *v. i.* **a)** ‹*Hund:*› [bellen und] mit dem Schwanz wedeln; **b)** *(behave servilely)* **~ [on *or* upon sb.]** [vor jmdm.] katzbuckeln *(abwertend)*

fax [fæks] **1.** *n.* [Tele]fax, *das;* Fernkopie, *die.* **2.** *v. t.* faxen; fernkopieren

'fax machine *n.* Faxgerät, *das;* Fernkopierer, *der*

FBI *abbr. (Amer.)* **Federal Bureau of Investigation** FBI, *das*

fear [fɪər] **1.** *n.* **a)** Furcht, *die,* Angst, *die* **(of** vor + *Dat.*); **~ of death *or* dying/heights** Todes-/Höhenangst, *die;* **~ of doing sth.** Angst *od.* Furcht davor, etw. zu tun; **in ~ of being caught** in der Angst, gefaßt zu werden; **strike ~ into sb.** jmdn. in Angst versetzen; **b)** *(object of ~)* Furcht, *die; in pl.* Befürchtungen *Pl.;* **c)** *(anxiety for sb.'s/sth.'s safety)* Sorge, *die* **(for** um); **go *or* be in ~ of one's life** Angst um sein Leben haben; **d)** *(coll.: risk)* Gefahr, *die;* **no ~!** *(coll.)* keine Bange! *(ugs.).* **2.** *v. t.* **a)** *(be afraid of)* **~ sb./sth.** vor jmdm./etw. Angst haben; jmdn./etw. fürchten; **~ to do *or* doing sth.** Angst haben *od.* sich fürchten, etw. zu tun; **you have nothing to ~**: Sie haben nichts zu befürchten; **~ the worst** das Schlimmste befürchten; **b)** *(be worried about)* befürchten; **~ [that ...]** fürchten[, daß ...]. **3.** *v. i.* sich fürchten; **~ for sb./sth.** um jmdn./etw. bangen *(geh.) od.* fürchten; **never ~** *(also joc. iron.)* keine Bange *(ugs.)*

fearful ['fɪəfl] *adj.* **a)** *(terrible)* furchtbar; **b)** *(frightened)* ängstlich; **be ~ of sth./sb.** vor etw./jmdm. Angst haben; **be ~ of doing sth.** Angst [davor] haben, etw. zu tun

fearfully ['fɪəfəlɪ] *adv.* ängstlich
fearless ['fɪəlɪs] *adj.* furchtlos; **be ~ [of sth./sb.]** keine Angst [vor etw./jmdm.] haben *od.* kennen
fearlessly ['fɪəlɪslɪ] *adv.* furchtlos; ohne Angst
fearsome ['fɪəsəm] *adj.* furchteinflößend; furchterregend
feasibility [fi:zɪ'bɪlɪtɪ] *n., no pl.* Durchführbarkeit, *die; (of method)* Anwendbarkeit, *die; (possibility)* Möglichkeit, *die*
feasible ['fi:zɪbl] *adj.* durchführbar ‹*Plan, Vorschlag*›; anwendbar ‹*Methode*›; *(possible)* möglich
feast [fi:st] **1.** *n.* **a)** *(Relig.)* Fest, *das;* **movable/immovable ~:** beweglicher/unbeweglicher Feiertag; **b)** *(banquet)* Festessen, *das;* **a ~ for the eyes/ears** eine Augenweide/ein Ohrenschmaus. **2.** *v. i.* schlemmen; schwelgen; **~ on sth.** sich an etw. *(Dat.)* gütlich tun. **3.** *v. t.* festlich bewirten; *(fig.)* **he ~ed his eyes on her beauty** er labte sich an ihrer Schönheit *(geh.)*
feat [fi:t] *n. (action)* Meisterleistung, *die; (thing)* Meisterwerk, *das;* **a ~ of intellect/strength** eine intellektuelle Meisterleistung/ein Kraftakt
feather ['feðə(r)] **1.** *n.* **a)** Feder, *die; (on arrow)* [Pfeil]feder, *die;* **as light as a ~:** federleicht; **a ~ in sb.'s cap** *(fig. coll.)* ein Grund für jmdn., stolz zu sein; **you could have knocked me down with a ~:** ich war völlig von den Socken *(ugs.);* **b)** *collect. (plumage)* Gefieder, *das. See also* **bird a. 2.** *v. t.* **~ one's nest** *(fig.)* auf seinen finanziellen Vorteil bedacht sein
feather: ~ **'bed** *n.* mit Federn gefüllte Matratze; **~-bed** *v. t.* [ver]hätscheln; **~-brained** ['feðəbreɪnd] *adj.* schwachköpfig *(ugs.);* ~ **'duster** *n.* Federwisch, *der;* **~weight** *n.* **a)** *(very light thing/person)* Fliegengewicht, *das;* **b)** *(Boxing etc.)* Federgewicht, *das; (person also)* Federgewichtler, *der*
feathery ['feðərɪ] *adj.* **a)** *(covered with feathers)* befiedert; gefiedert; **b)** *(feather-like) (in quality)* federnartig; *(in weight)* federleicht; locker ‹*Kuchenteig*›
feature ['fi:tʃə(r)] **1.** *n.* **a)** *usu. in pl. (part of face)* Gesichtszug, *der;* **b)** *(distinctive characteristic)* [charakteristisches] Merkmal; **be a ~ of sth.** charakteristisch für etw. sein; **make a ~ of sth.** etw. [sehr] betonen *od.* herausstellen; **c)** *(Journ. etc.)* Reportage, *die;* Feature, *das;* **d)** *(Cinemat.)* **~ [film]** Hauptfilm, *der;* Spielfilm *der;* **e)** *(Radio, Telev.)* **~ [programme]** Feature, *das.* **2.** *v. t. (make attraction of)* vorrangig vorstellen; *(give special prominence to) (in film)* in der Hauptrolle zeigen; *(in show)* als Stargast präsentieren. **3.** *v. i.* **a)** *(be ~)* vorkommen; **b)** *(be [important] participant)* **~ in sth.** eine [bedeutende] Rolle bei etw. spielen
featureless ['fi:tʃəlɪs] *adj.* eintönig
Feb. *abbr.* **February** Febr.
February ['febrʊərɪ] *n.* Februar, *der; see also* **August**
feces *(Amer.) see* **faeces**
feckless ['feklɪs] *adj. (feeble)* schwächlich ‹*Person*›; nutzlos, vertan ‹*Leben*›; *(inefficient)* untauglich; *(aimless)* ziellos
fed [fed] **1.** *see* **feed 1, 2. 2.** *pred. adj. (sl.)* **be/get ~ up with sb./sth.** jmdn./etw. satt haben *(ugs.);* **be/get ~ up with doing sth.** es satt haben, etw. zu tun *(ugs.)*
federal ['fedərl] *adj.* Bundes-; föderativ ‹*System*›; föderalistisch ‹*Partei usw.*›
federation [fedə'reɪʃn] *n. (group of states)* Bündnis, *das;* Föderation, *die; (society)* Bund, *der*
fee [fi:] *n.* **a)** Gebühr, *die; (of doctor, lawyer, etc.)* Honorar, *das; (of performer)* Gage, *die;* **registration ~:** Aufnahmegebühr, *die;* **school ~s** Schulgeld, *das;* **b)** *(administrative charge)* Bearbeitungsgebühr, *die*
feeble ['fi:bl] *adj.* **a)** *(weak)* schwach; **b)** *(deficient)* schwächlich; *(in resolve, argument)* halbherzig; **c)** *(lacking energy)* schwach ‹*Leistung, Kampf, Applaus*›; wenig überzeugend ‹*Argument, Entschuldigung, Erklärung*›; zaghaft, kläglich ‹*Versuch, Bemühung*›; lahm *(ugs.)* ‹*Witz*›; **d)** *(indistinct)* schwach ‹*[Licht]schein, Herzschlag*›
feeble-'minded *adj.* **a)** töricht; **b)** *(Psych.)* geistesschwach
feed [fi:d] **1.** *v. t.,* **fed** [fed] **a)** *(give food to)* füttern; **~ sb./an animal with sth.** jmdm. etw. zu essen/einem Tier [etw.] zu fressen geben; **~ a baby/an animal on *or* with sth.** ein Baby/Tier mit etw. füttern; **~ [at the breast]** stillen; **~ oneself** allein *od.* ohne Hilfe essen; **b)** *(provide food for)* ernähren; **~ sb./an animal on *or* with sth.** jmdn./ein Tier mit etw. ernähren; **c)** *(give out)* verfüttern ‹*Viehfutter*› **(to** an + *Akk.*); **d)** *(keep supplied)* speisen ‹*Wasserreservoir*›; *(supply with material)* versorgen; **~ a film into the projector** einen Film in das Vorführgerät einlegen; **~ data into the computer** Daten in den Computer eingeben. **2.** *v. i.,* **fed** ‹*Tier:*› fressen **(from** aus); ‹*Person:*› essen **(off** von); **~ on sth.** ‹*Tier:*› etw. fressen; ‹*Person:*› sich von etw. [er]nähren. **3.** *n.* **a)** *(instance of eating) (of animals)* Fressen, *das; (of baby)* Mahlzeit, *die;* **have [quite] a ~:** [ordentlich] futtern *(ugs.);* [kräftig] zulangen; **b)** *(fodder)* **[cattle/pig] ~:** [Vieh-/Schweine]futter, *das*
~ **'back** *v. t.* zurückleiten; weiterleiten, -geben ‹*Informationen*›; **be fed back** zurückfließen
'feedback *n.* **a)** *(information about result, response)* Reaktion, *die;* Feedback, *das (fachspr.);* **b)** *(Electr.)* Rückkopplung, *die*
feeding ['fi:dɪŋ]: **~-bottle** *n.* [Saug]flasche, *die;* **~-time** *n.* Fütterungszeit, *die*
feel [fi:l] **1.** *v. t.,* **felt** [felt] **a)** *(explore by touch)* befühlen; **~ sb.'s pulse** jmdm. den Puls fühlen; **~ one's way** sich *(Dat.)* seinen Weg ertasten; *(fig.: try sth. out)* sich vorsichtig vor[an]tasten; **b)** *(perceive by touch)* fühlen; *(become aware of)* bemerken; *(be aware of)* merken; *(have sensation of)* spüren; **c)** empfinden ‹*Mitleid, Dank, Eifersucht*›; verspüren ‹*Drang, Wunsch*›; **~ the cold/heat** unter der Kälte/Hitze leiden; **~ one's age** sein Alter spüren; **make itself felt** zu spüren sein; *(have effect)* sich bemerkbar machen; **d)** *(experience)* empfinden; *(be affected by)* zu spüren bekommen; **e)** *(have vague or emotional conviction)* **~ [that] ...:** das Gefühl haben, daß ...; **f)** *(think)* **~ [that] ...:** glauben, daß ...; **if that's what you ~ about the matter** wenn du so darüber denkst. **2.** *v. i.,* **felt a) ~ [about] in sth. [for sth.]** in etw. *(Dat.)* [nach etw.] [herum]suchen; **~ [about] [after *or* for sth.] with sth.** mit etw. [nach etw.] [umher]tasten; **b)** *(have sense of touch)* fühlen; **c)** *(be conscious that one is)* sich ... fühlen; **~ angry/delighted/disappointed** böse/froh/enttäuscht sein; **I felt such a fool** ich kam mir wie ein Idiot vor; **~ inclined to do sth.** dazu neigen, etw. zu tun; **the child did not ~ loved/wanted** das Kind hatte das Gefühl, ungeliebt/unerwünscht zu sein; **~ quite hopeful** guter Hoffnung sein; **I felt sorry for him** er tat mir leid; **how do you ~ today?** wie fühlst du dich *od.* wie geht es dir heute?; **~ like sth./doing sth.**

(coll.: wish to have/do) auf etw. *(Akk.)* Lust haben/Lust haben, etw. zu tun; **do you ~ like a cup of tea?** möchtest du eine Tasse Tee?; **we ~ as if** *or* **as though ...:** es kommt uns vor, als ob ...; *(have the impression that)* wir haben das Gefühl, daß ...; **how do you ~ about the idea?** was hältst du von der Idee?; **if that's how** *or* **the way you ~ about it** wenn du so darüber denkst; **d)** *(be emotionally affected)* **~ passionately/bitterly about sth.** sich für etw. begeistern/über etw. *(Akk.)* verbittert sein; **e)** *(be consciously perceived as)* sich ... anfühlen; **~ like sth.** sich wie etw. anfühlen; **it ~s nice/uncomfortable** es ist ein angenehmes/unangenehmes Gefühl. **3.** *n.* **have a silky ~:** sich seidig anfühlen; **let me have a ~:** laß mich mal fühlen; **get/have a ~ for sth.** *(fig.)* ein Gespür für etw. bekommen/haben

~ for *v.t.* **~ for sb.** mit jmdm. Mitleid haben

~ 'out *v.t. (sound out)* **~ sb. out** jmds. Ansichten feststellen

~ with *v.t.* Mitgefühl haben mit

feeler ['fi:lə(r)] *n.* Fühler, *der;* **put out ~s** *(fig.)* seine Fühler ausstrecken

feeling ['fi:lɪŋ] *n.* **a)** *(sense of touch)* **[sense of] ~:** Tastsinn, *der;* **have no ~ in one's legs** kein Gefühl in den Beinen haben; **b)** *(physical sensation, emotion)* Gefühl, *das;* **what are your ~s for each other?** was empfindet ihr füreinander?; **say sth. with ~:** etw. mit Nachdruck sagen; **~s were running high** Emotionen wurden geweckt; **bad ~** *(jealousy)* Neid, *der; (annoyance)* Verstimmung, *die;* **c)** *in pl. (sensibilities)* Gefühle; **hurt sb.'s ~s** jmdn. verletzen; **d)** *(belief)* Gefühl, *das;* **have a/the ~ [that] ...:** das Gefühl haben, daß ...; **e)** *(sentiment)* Ansicht, *die;* **the general ~ was that ...:** man war allgemein der Ansicht, daß ...

feet *pl. of* **foot**

feign [feɪn] *v.t.* vorspiegeln; vortäuschen; **~ ignorance** sich dumm stellen; **~ to do sth.** vorgeben, etw. zu tun

feint [feɪnt] *n. (Boxing, Fencing)* Finte, *die;* **make a ~:** eine Finte ausführen; fintieren

feline ['fi:laɪn] **1.** *adj. (of cat[s])* Katzen-; *(catlike)* katzenhaft. **2.** *n.* Katze, *die;* **the ~s** die Katzen *od. (fachspr.)* Feliden

[1]**fell** *see* **fall 2**

[2]**fell** [fel] *v.t.* **a)** *(cut down)* fällen ‹*Baum*›; **b)** *(strike down)* niederstrecken ‹*Gegner*›

[3]**fell** *n. (Brit.)* **a)** *(in names: hill)* Berg, *der;* **b)** *(stretch of high moorland)* Hochmoor, *das*

[4]**fell** *adj.* **at** *or* **in one ~ swoop** auf einen Schlag

fellow ['feləʊ] **1.** *n.* **a)** *usu. in pl. (comrade)* Kamerad, *der;* **a good ~:** ein guter Kumpel *(ugs.);* **b)** *usu. in pl. (equal)* Gleichgestellte, *der/die;* **c)** *(Brit. Univ.)* Fellow, *der; (member of academy or society)* Fellow, *der;* Mitglied, *das;* **d)** *(coll.: man, boy)* Bursche, *der (ugs.);* Kerl, *der (ugs.);* **well, young ~:** nun, junger Mann; **old** *or* **dear ~:** alter Junge *od.* Knabe *(ugs.).* **2.** *attrib. adj.* Mit-; **~ worker** Kollege, *der*/Kollegin, *die;* **~ man** *or* **human being** Mitmensch, *der;* **~ sufferer** Leidensgenosse, *der*/-genossin, *die;* **my ~ teachers/workers** *etc.* meine Lehrer-/Arbeitskollegen *usw.;* **~ student** Kommilitone, *der*/Kommilitonin, *die*

fellow: ~ 'countryman *see* **countryman a; ~-'feeling** *n.* **a)** *(sympathy)* Mitgefühl, *das;* **have a ~-feeling for sb.** mit jmdm. fühlen; **b)** *(mutual understanding)* Zusammengehörigkeitsgefühl, *das*

fellowship ['feləʊʃɪp] *n.* **a)** *no pl. (companionship)* Gesellschaft, *die;* **b)** *no pl. (community of interest)* Zusammengehörigkeit, *die;* **c)** *(Univ. etc.)* Status eines Fellows; Fellowship, *die*

fellow-'traveller *n.* Mitreisende, *der/die*

felony ['felənɪ] *n.* Kapitalverbrechen, *das*

[1]**felt** [felt] *n. (cloth)* Filz, *der;* **~ hat** Filzhut, *der*

[2]**felt** *see* **feel 1, 2**

felt[-tipped] pen [felt(tɪpt) 'pen] *n.* Filzstift, *der*

female ['fi:meɪl] **1.** *adj.* weiblich; Frauen‹*stimme, -station, -chor, -verein*›; **~ animal/bird/fish/insect** Weibchen, *das;* **~ child/doctor** Mädchen, *das*/Ärztin, *die.* **2.** *n.* **a)** *(person)* Frau, *die; (foetus, child)* Mädchen, *das; (animal)* Weibchen, *das;* **b)** *(derog.: woman)* Weib[sbild], *das (ugs. abwertend)*

feminine ['femɪnɪn] *adj.* **a)** *(of women)* weiblich; Frauen‹*angelegenheit, -problem, -leiden*›; **b)** *(womanly)* fraulich; feminin; **c)** *(Ling.)* weiblich; feminin *(fachspr.)*

feminism ['femɪnɪzm] *n., no pl.* Feminismus, *der*

feminist ['femɪnɪst] **1.** *n.* Feministin, *die*/Feminist, *der;* Frauenrechtlerin, *die*/-rechtler, *der.* **2.** *adj.* feministisch; Feministen‹*bewegung, -blatt, -gruppe*›

femur ['fi:mə(r)] *n., pl.* **~s** *or* **femora** ['femərə] *(Anat.)* Oberschenkelknochen, *der;* Femur, *der (fachspr.)*

fen [fen] *n.* Sumpfland, *das;* Fenn, *das;* **the Fens** die Fens

fence [fens] **1.** *n.* **a)** Zaun, *der;* **sit on the ~** *(fig.)* sich nicht einmischen; sich neutral verhalten; **b)** *(for horses to jump)* Hindernis, *das;* **c)** *(sl.: receiver)* Hehler, *der*/Hehlerin, *die.* **2.** *v.i. (Sport)* fechten. **3.** *v.t. (surround with fence)* einzäunen; *(fig.)* absichern (**with** durch)

~ 'in *v.t.* einzäunen; *(fig.)* einengen (**with** durch)

~ 'off *v.t.* abzäunen

fencer ['fensə(r)] *n.* Fechter, *der*/Fechterin, *die*

fencing ['fensɪŋ] *n., no pl.* **a)** Einzäunen, *das;* **b)** *(Sport/Hist.)* Fechten, *das; attrib.* Fecht-; **c)** *(fences)* Zäune *Pl.*

fend [fend] *v.i.* **~ for oneself** für sich selbst sorgen; *(in hostile surroundings)* sich allein durchschlagen

~ 'off *v.t.* abwehren; von sich fernhalten

fender ['fendə(r)] *n.* **a)** *(for fire)* Kaminschutz, *der;* **b)** *(Amer.) (car mudguard or wing)* Kotflügel, *der; (bicycle mudguard)* Schutzblech, *das*

fennel ['fenl] *n. (Bot.)* Fenchel, *der*

ferment 1. [fə'ment] *v.i. (lit. or fig.)* gären. **2.** *v.t.* zur Gärung bringen; *(fig.)* heraufbeschwören ‹*Unzufriedenheit, Unruhe*›. **3.** ['fɜ:ment] *n.* **a)** *(fermentation)* Gärung, *die;* Fermentation, *die (fachspr.);* **b)** *(agitation)* Unruhe, *die;* Aufruhr, *der;* **in ~:** in Unruhe *od.* Aufruhr

fermentation [fɜ:men'teɪʃn] *n.* Gärung, *die;* Fermentation, *die (fachspr.)*

fern [fɜ:n] *n.* Farnkraut, *das*

ferocious [fə'rəʊʃəs] *adj.* wild ‹*Tier, Person, Aussehen, Blick, Lachen*›; grimmig ‹*Stimme*›; heftig ‹*Schlag, Kampf, Stoß*›; *(fig.)* scharf ‹*Kritik, Angriff*›; heftig ‹*Streit, Auseinandersetzung*›

ferocity [fə'rɒsɪtɪ] *n., no pl. see* **ferocious:** Wildheit, *die;* Grimmigkeit, *die;* Heftigkeit, *die;* Schärfe, *die*

ferret ['ferɪt] **1.** *n.* Frettchen, *das.* **2.** *v.i.* **~ [about** *or* **around]** herum-

stöbern *(ugs.)*; herumschnüffeln *(abwertend)*
~ **'out** *v.t.* aufspüren; aufstöbern *(ugs.)*
ferrule ['feru:l, 'ferl] *n.* Zwinge, *die*
ferry ['ferı] **1.** *n.* **a)** Fähre, *die;* **b)** *(service)* Fährverbindung, *die;* Fähre, *die (ugs.).* **2.** *v.t.* **a)** *(convey in boat)* ~ **[across** *or* **over]** übersetzen; **b)** *(transport)* befördern, bringen ‹*Güter, Personen*›
ferry: ~**-boat** *n.* Fährboot, *das;* ~**man** ['ferımən] *n., pl.* ~**men** ['ferımən] Fährmann, *der*
fertile ['fɜ:taıl] *adj.* **a)** *(fruitful)* fruchtbar (**in** an + *Dat.*); **have a ~ imagination** viel Phantasie haben; **b)** *(capable of developing)* befruchtet; **c)** *(able to become parent)* fortpflanzungsfähig
fertilisation, fertilise, fertiliser *see* **fertiliz-**
fertility [fɜ:'tılıtı] *n., no pl.* **a)** *(lit. or fig.)* Fruchtbarkeit, *die;* **b)** *(ability to become parent)* Fortpflanzungsfähigkeit, *die*
fer'tility drug *n. (Med.)* Hormonpräparat, *das (zur Steigerung der Fruchtbarkeit)*
fertilization [fɜ:tılaı'zeıʃn] *n. (Biol.)* Befruchtung, *die*
fertilize ['fɜ:tılaız] *v.t.* **a)** *(Biol.)* befruchten; **b)** *(Agric.)* düngen
fertilizer ['fɜ:tılaızə(r)] *n.* Dünger, *der*
fervent ['fɜ:vənt] *adj.* leidenschaftlich; inbrünstig ‹*Gebet, Wunsch, Hoffnung*›; glühend ‹*Verehrer, Liebe, Haß*›
fervour *(Brit.; Amer.:* **fervor***)* ['fɜ:və(r)] *n.* Leidenschaftlichkeit, *die; (of love, belief)* Inbrunst, *die*
fester ['festə(r)] *v.i. (lit or fig.)* eitern
festival ['festıvl] *n.* **a)** *(feast day)* Fest, *das;* **b)** *(performances, plays, etc.)* Festival, *das;* Festspiele *Pl.; (rock ~, jazz ~, single event)* Festival, *das*
festive ['festıv] *adj.* festlich; **the ~ season** die Weihnachtszeit
festivity [fe'stıvıtı] *n.* **a)** *no pl. (gaiety)* Feststimmung, *die;* **b)** *(festive celebration)* Feier, *die;* **festivities** Feierlichkeiten *Pl.*
festoon [fe'stu:n] **1.** *n.* Girlande, *die.* **2.** *v.t.* schmücken (**with** mit)
fetal *(Amer.) see* **foetal**
fetch [fetʃ] **1.** *v.t.* **a)** holen; *(collect)* abholen (**from** von); ~ **sb. sth.,** ~ **sth. for sb.** jmdm. etw. holen; **b)** *(be sold for)* erzielen ‹*Preis*›; **my car ~ed £500** ich habe für den Wagen 500 Pfund bekommen; **c)** *(deal)* ~ **sb. a blow/punch** jmdm. einen Schlag versetzen. **2.** *v.i.* ~ **and carry [for sb.]** [bei jmdm.] Mädchen für alles sein *(ugs.)*
~ **'up** *v.i. (coll.)* landen *(ugs.)*
fetching ['fetʃıŋ] *adj.* einnehmend, gewinnend ‹*Lächeln, Stimme, Wesen, Benehmen*›
fête [feıt] **1.** *n.* **a)** [Wohltätigkeits]basar, *der;* **b)** *(festival)* Fest, *das;* Feier, *die.* **2.** *v.t.* feiern
fetid ['fetıd] *adj.* stinkend; übelriechend; ~ **smell/odour/stench** Gestank, *der*
fetish ['fetıʃ] *n.* Fetisch, *der;* **she has a ~ about tidiness** Sauberkeit ist bei ihr zur Manie geworden
fetishism ['fetıʃızm] *n.* Fetischismus, *der*
fetishist ['fetıʃıst] *n.* Fetischist, *der*/Fetischistin, *die*
fetlock ['fetlɒk] *n.* Köte, *die*
fetter ['fetə(r)] **1.** *n.* **a)** *(shackle)* Fußfessel, *die;* **b)** *in pl. (bonds; fig.: captivity)* Fesseln *Pl.* **2.** *v.t.* fesseln; *(fig.)* hemmen ‹*Fortschritt, Entwicklung*›
fettle ['fetl] *n.* **be in good** *or* **fine/poor ~:** sich in guter/schlechter Verfassung befinden
fetus *(Amer.) see* **foetus**
feud [fju:d] **1.** *n.* Zwist, *der;* Zwistigkeiten *Pl.; (Hist./fig.)* Fehde, *die.* **2.** *v.i.* ~ **[with sb./each other]** [mit jmdm./miteinander] im Streit liegen
feudal ['fju:dl] *adj.* Feudal-; feudalistisch; **in ~ Britain** im feudalistischen England
'feudal system *n. (Hist.)* Feudalsystem, *das*
fever ['fi:və(r)] *n.* **a)** *no pl. (Med.: high temperature)* Fieber, *das;* **have a [high] ~:** [hohes] Fieber haben; **a ~ of 105 °F** 40,5 °C Fieber; **b)** *(Med.: disease)* Fieberkrankheit, *die;* **c)** *(nervous excitement)* Erregung, *die;* Aufregung, *die;* **in a ~ of anticipation** im Fieber der Erwartung
feverish ['fi:vərıʃ] *adj.* **a)** *(Med.)* fiebrig; Fieber‹*zustand, -traum*›; **be ~:** fiebern; Fieber haben; **b)** fiebrig ‹*Erwartung*›; fieberhaft ‹*Aufregung, Eifer, Kampf, Eile*›
'fever pitch *n.* Siedepunkt, *der (fig.)*; **reach ~:** auf dem Siedepunkt angelangt sein; **at ~:** auf dem Siedepunkt
few [fju:] **1.** *adj.* **a)** *(not many)* wenige; ~ **people** [nur] wenige [Leute]; **very ~ housewives know that** das wissen die wenigsten Hausfrauen; **his ~ belongings** seine paar Habseligkeiten; **[all] too ~ people** [viel] zu wenig Leute; **~ and far between** rar; **they were ~ in number** sie waren nur sehr wenige; **a ~ ...:** wenige ...; **not a ~ ...:** eine ganze Reihe ...; **[just** *or* **only] a ~ trouble-makers** einige [wenige] Störenfriede; **just a ~ words from you** nur ein paar Worte von dir; **b)** *(some)* wenige; **a ~ ...:** einige *od.* ein paar ...; **every ~ minutes** alle paar Minuten; **a good ~ [...]/quite a ~ [...]** *(coll.)* eine ganze Menge [...]/ziemlich viele [...]. **2.** *n.* **a)** *(not many)* wenige; **a ~:** wenige; **a very ~:** nur wenige; **the ~:** die wenigen; **~ of us/them** nur wenige von uns/nur wenige [von ihnen]; **~ of the people** nur wenige [Leute]; **just a ~ of you/her friends** nur ein paar von euch/ihrer Freunde; **not a ~ of them** eine ganze Reihe von ihnen; **not a ~:** nicht wenige; **b)** *(some)* **the/these/those ~ who** diejenigen, die; **there were a ~ of us who ...:** es gab einige unter uns, die ...; **with a ~ of our friends** mit einigen *od.* ein paar unserer Freunde; **a ~ [more] of these biscuits** [noch] ein paar von diesen Keksen; **a good ~/quite a ~** *(coll.)* eine ganze Menge/ziemlich viele [Leute]
fewer ['fju:ə(r)] *adj.* weniger; **become ~ and ~:** immer weniger werden
fewest ['fju:ıst] **1.** *adj.* **[the] ~ [...]** die wenigsten [...]. **2.** *n.* **the ~ [of us/them]** die wenigsten [von uns/ihnen]; **at the ~:** mindestens
fiancé [fı'ɑ̃seı] *n.* Verlobte, *der*
fiancée [fı'ɑ̃seı] *n.* Verlobte, *die*
fiasco [fı'æskəʊ] *n., pl.* ~**s** Fiasko, *das*
fib [fıb] **1.** *n.* Flunkerei, *die (ugs.);* **tell ~s** flunkern *(ugs.);* schwindeln; **that was a ~:** das war geschwindelt. **2.** *v.i.,* **-bb-** schwindeln; flunkern *(ugs.)*
fibber ['fıbə(r)] *n.* Flunkerer, *der (ugs.);* Schwindler, *der*/Schwindlerin, *die*
fibre *(Brit.; Amer.:* **fiber***)* ['faıbə(r)] *n.* **a)** Faser, *die;* **b)** *(substance consisting of ~s)* [Faser]gewebe, *das;* **c)** *(roughage)* Ballaststoffe *Pl.;* **d) moral ~:** Charakterstärke, *die*
fibre: ~**glass** *(Amer.:* **fiber glass***)* *n. (fibrous glass)* Glasfaser, *die; (plastic)* glasfaserverstärkter Kunststoff; ~ **'optics** *n.* Faseroptik, *die*
fibrous ['faıbrəs] *adj.* faserig; Faser‹*gewebe, -holz, -stoff*›
fiche [fi:ʃ] *n., pl. same or* ~**s** *see* **microfiche**
fickle ['fıkl] *adj.* unberechenbar; launisch
fiction ['fıkʃn] *n.* **a)** *(literature)* er-

zählende Literatur; **b)** *(thing feigned or imagined)* **a ~/~s** eine Erfindung

fictional ['fɪkʃənl] *adj.* belletristisch; erfunden ‹*Geschichte*›; **~ literature** erzählende Literatur; **~ characters** fiktive Figuren

'**fiction-writer** *n.* Belletrist, *der*/Belletristin, *die*

fictitious [fɪk'tɪʃəs] *adj.* **a)** *(counterfeit)* fingiert; unwahr ‹*Behauptung, Darstellung*›; **b)** *(assumed)* falsch ‹*Name, Identität*›; **c)** *(imaginary)* [frei] erfunden ‹*Person, Figur, Geschichte*›

fiddle ['fɪdl] **1.** *n.* **a)** *(Mus.) (coll./derog.)* Fidel, *die; (violin for traditional or folk music)* Geige, *die;* Fidel, *die;* **[as] fit as a ~:** kerngesund; **play first/second ~** *(fig.)* die erste/zweite Geige spielen *(ugs.);* **play second ~ to sb.** in jmds. Schatten *(Dat.)* stehen; **b)** *(sl.: swindle)* Gaunerei, *die;* **it's all a ~:** das ist alles Schiebung *(ugs.);* **be on the ~:** krumme Dinger machen *(ugs.).* **2.** *v.i.* **a) ~ about** *(coll.: waste time)* herumtrödeln *(ugs.);* **~ about with sth.** *(work on to adjust etc.)* an etw. *(Dat.)* herumfummeln *(ugs.); (tinker with)* an etw. *(Dat.)* herumbasteln *(ugs.);* **~ with sth.** *(play with)* mit etw. herumspielen; **b)** *(sl.: deceive)* krumme Dinger drehen *(ugs.).* **3.** *v.t. (sl.) (falsify)* frisieren *(ugs.)* ‹*Bücher, Rechnungen*›; *(get by cheating)* [sich *(Dat.)*] ergaunern *(ugs.)*

fiddler ['fɪdlə(r)] *n.* **a)** *(player)* Geiger, *der*/Geigerin, *die;* **b)** *(sl.: swindler etc.)* Gauner, *der*/Gaunerin, *die (abwertend)*

'**fiddlesticks** *int. (coll.)* dummes Zeug *(ugs.);* Schnickschnack *(ugs.)*

fiddling ['fɪdlɪŋ] *adj.* **a)** *(petty)* belanglos; **b)** *see* **fiddly**

fiddly ['fɪdlɪ] *adj. (coll.)* **a)** *(awkward to do)* knifflig; **b)** *(awkward to use)* umständlich

fidelity [fɪ'delɪtɪ] *n.* **a)** *(faithfulness)* Treue, *die* (**to** zu); **b)** *(Radio, Telev., etc.)* Wiedergabetreue, *die; (of sound)* Klangtreue, *die; (of picture)* Bildtreue, *die*

fidget ['fɪdʒɪt] **1.** *n.* **a) have/get the ~s** zappelig sein/werden *(ugs.);* **b)** *(person)* Zappelphilipp, *der (ugs.).* **2.** *v.i.* **~ [about]** [herum]zappeln *(ugs.);* herumrutschen

fidgety ['fɪdʒɪtɪ] *adj.* unruhig ‹*Person, Pferd, Stimmung*›; zappelig ‹*Kind*›; nervös ‹*Bewegungen, Zuckungen*›

field [fi:ld] **1.** *n.* **a)** *(cultivated)* Feld, *das;* Acker, *der; (for grazing)* Weide, *die; (meadow)* Wiese, *die;* **work in the ~s** auf dem Feld arbeiten; **b)** *(area rich in minerals etc.)* Lagerstätte, *die;* **gas-~:** Gasfeld, *das;* **c)** *(battlefield)* Schlachtfeld, *das; (fig.)* Feld, *das;* **leave sb. a clear** *or* **the ~** *(fig.)* jmdm. das Feld überlassen; **d)** *(playing ~)* Sportplatz, *der; (ground marked out for game)* Platz, *der;* [Spiel]feld, *das;* **send sb. off the ~:** jmdn. vom Platz schicken; **take the ~:** das Spielfeld betreten; **e)** *(competitors in sports event)* Feld, *das; (fig.)* Teilnehmerkreis, *der;* **f)** *(area of operation, subject area, etc.)* Fach, *das;* [Fach]gebiet, *das;* **in the ~ of medicine** auf dem Gebiet der Medizin; **~ of vision** *or* **view** Blickfeld, *das;* **g)** *(Phys.)* **magnetic/gravitational ~:** Magnet-/Gravitationsfeld, *das.* **2.** *v.i. (Cricket, Baseball, etc.)* als Fänger spielen. **3.** *v.t.* **a)** *(Cricket, Baseball, etc.) (stop)* fangen ‹*Ball*›; *(stop and return)* auffangen und zurückwerfen; **b)** *(put into ~)* aufstellen, aufs Feld schicken ‹*Mannschaft, Spieler*›; **c)** *(fig.: deal with)* fertig werden mit; parieren ‹*Fragen*›

'**field-day** *n.* **have a ~:** seinen großen Tag haben

fielder ['fi:ldə(r)] *n.* *(Cricket, Baseball, etc.)* Feldspieler, *der*

field: ~ events *n. pl.* *(Sport)* technische Disziplinen; **~-glasses** *n. pl.* Feldstecher, *der;* **F~ 'Marshal** *n. (Brit. Mil.)* Feldmarschall, *der;* **~ mouse** *n.* Brandmaus, *die;* **~ sports** *n. pl.* Sport im Freien *(bes. Jagen und Fischen);* **~-test 1.** *n. see* **~-trial; 2.** *v.t.* in der Praxis erproben; **~-trial** *n.* Feldversuch, *der;* **~-trip** *n.* Exkursion, *die;* **~-work** *n. (of surveyor etc.)* Arbeit im Gelände; *(of sociologist, collector of scientific data, etc.)* Feldforschung, *die;* **~-worker** *n.* Feldforscher, *der*/-forscherin, *die*

fiend [fi:nd] *n.* **a)** *(very wicked person)* Scheusal, *das;* Unmensch, *der;* **b)** *(evil spirit)* böser Geist; **c)** *(coll.: mischievous or tiresome person)* Plagegeist, *der;* **d)** *(devotee)* Fan, *der;* **fresh-air ~:** Frischluftfanatiker, *der*/-fanatikerin, *die*

fiendish ['fi:ndɪʃ] *adj.* **a)** *(fiendlike)* teuflisch; **b)** *(extremely awkward)* höllisch

fiendishly ['fi:ndɪʃlɪ] *adv.* **a)** teuflisch; **b) ~ clever** *(coll.)* gerissen und schlau; **c)** *(extremely awkwardly)* höllisch

fierce ['fɪəs] *adj.* **a)** *(violently hostile)* wild; erbittert ‹*Widerstand, Kampf*›; wuchtig ‹*Schlag*›; heftig ‹*Angriff*›; **b)** *(raging)* wütend; grimmig ‹*Haß, Wut*›; scharf ‹*Kritik*›; wild ‹*Tier*›; **c)** heftig ‹*Andrang, Streit*›; heiß ‹*Wettbewerb*›; leidenschaftlich ‹*Stolz, Wille*›; **d)** *(unpleasantly strong or intense)* unerträglich; **e)** *(violent in action)* hart ‹*Bremsen, Ruck*›

fiercely ['fɪəslɪ] *adv.* **a)** heftig ‹*angreifen, Widerstand leisten*›; wütend, grimmig ‹*brüllen*›; **b)** wütend ‹*toben*›; aufs heftigste ‹*kritisieren, bekämpfen*›; **c)** äußerst ‹*stolz, unabhängig sein*›; wild ‹*entschlossen, kämpfen*›

fiery ['faɪərɪ] *adj.* **a)** *(consisting of or flaming with fire)* glühend; feurig ‹*Atem*›; *(looking like fire)* feurig; **b)** *(producing burning sensation)* feurig ‹*Geschmack, Gewürz*›; scharf ‹*Getränk*›; **c)** *(irascible, impassioned)* hitzig ‹*Temperament*›; feurig ‹*Rede, Redner*›; **have a ~ temper** ein Hitzkopf sein

fiesta [fi:'estə] *n.* Fest, *das*

fife [faɪf] *n.* Pfeife, *die*

fifteen [fɪf'ti:n] **1.** *adj.* fünfzehn; *see also* **eight 1. 2.** *n.* Fünfzehn, *die; see also* **eight 2 a, d; eighteen 2**

fifteenth [fɪf'ti:nθ] **1.** *adj.* fünfzehnt...; *see also* **eighth 1. 2.** *n.* *(fraction)* Fünfzehntel, *das; see also* **eighth 2**

fifth [fɪfθ] **1.** *adj.* fünft...; *see also* **eighth 1. 2.** *n. (in sequence)* fünfte, *der/die/das; (in rank)* Fünfte, *der/die/das; (fraction)* Fünftel, *das; see also* **eighth 2**

fiftieth ['fɪftɪɪθ] **1.** *adj.* fünfzigst...; *see also* **eighth 1. 2.** *n.* *(fraction)* Fünfzigstel, *das; see also* **eighth 2**

fifty ['fɪftɪ] **1.** *adj.* fünfzig; *see also* **eight 1. 2.** *n.* Fünfzig, *die; see also* **eight 2 a; eighty 2**

fifty: ~-'~ *adv., adj.* fifty-fifty *(ugs.);* halbe-halbe *(ugs.);* **go ~-~:** fifty-fifty *od.* halbpart machen; **~-'first** *etc. adj.* einundfünfzigst... *usw.; see also* **eighth 1; ~-'one** *etc.* **1.** *adj.* einundfünfzig *usw.; see also* **eight 1; 2.** *n.* Einundfünfzig *usw., die; see also* **eight 2 a**

fig [fɪg] *n.* Feige, *die;* **not care** *or* **give a ~ about sth.** sich keinen Deut für etw. interessieren

fig. *abbr.* **figure** Abb.

fight [faɪt] **1.** *v.i.,* **fought** [fɔ:t] **a)** *(lit. or fig.)* kämpfen; *(with fists)* sich schlagen; **~ shy of sb./sth.** jmdm./einer Sache aus dem Weg gehen; **b)** *(squabble)* [sich] streiten, [sich] zanken (**about** wegen). **2.** *v. t.*, **fought a)** *(in battle)* ~ **sb./sth.** gegen jmdn./etw. kämpfen; *(using fists)* ~ **sb.** sich mit jmdm. schlagen; ‹*Boxer:*› gegen jmdn. boxen; **b)** *(seek to overcome)* bekämpfen; *(resist)* ~ **sb./sth.** gegen jmdn./etw. ankämpfen; **c)** ~ **a battle** einen Kampf austragen; **be ~ing a losing battle** *(fig.)* auf verlorenem Posten stehen *od.* kämpfen; **d)** führen ‹*Kampagne*›; kandidieren bei ‹*Wahl*›; **e)** ~ **one's way** sich *(Dat.)* den Weg freikämpfen; *(fig.)* sich *(Dat.)* seinen Weg bahnen; ~ **one's way to the top** *(fig.)* sich an die Spitze kämpfen. **3.** *n.* **a)** Kampf, *der* (**for** um); *(brawl)* Schlägerei, *die;* **make a ~ of it, put up a ~:** sich wehren; *(fig.)* sich zur Wehr setzen; **give in without a ~** *(fig.)* klein beigeben; **b)** *(squabble)* Streit, *der;* **they are always having ~s** zwischen ihnen gibt es dauernd Streit; **c)** *(ability to ~)* Kampffähigkeit, *die; (appetite for ~ing)* Kampfgeist, *der;* **all the ~ had gone out of him** *(fig.)* sein Kampfgeist war erloschen

~ **against** *v. t. (lit. or fig.)* kämpfen gegen; ankämpfen gegen ‹*Wellen, Wind, Gefühle*›

~ **'back 1.** *v. i.* zurückschlagen; sich zur Wehr setzen. **2.** *v. t. (suppress)* zurückhalten

~ **'down** *v. t.* zurückhalten

~ **for** *v. t. (lit. or fig.)* kämpfen für; ~ **for one's life** um sein Leben kämpfen

~ **'off** *v. t. (lit. or fig.)* abwehren; abwimmeln *(ugs.)* ‹*Reporter, Fans, Bewunderer*›; bekämpfen ‹*Erkältung*›; ~ **off the desire** dem Wunsch widerstehen

~ **'out** *v. t. (lit. or fig.)* ausfechten

~ **over** *v. t.* **a)** *(~ with regard to)* [sich] streiten über (+ *Akk.*); **b)** *(~ to gain possession of)* kämpfen um; *(squabble to gain possession of)* [sich] streiten um

~ **with** *v. t.* **a)** kämpfen mit; **b)** *(squabble with)* [sich] streiten mit

fighter ['faɪtə(r)] *n.* **a)** Kämpfer, *der*/Kämpferin, *die; (warrior)* Krieger, *der; (boxer)* Fighter, *der;* **b)** *(aircraft)* Kampfflugzeug, *das;* ~ **pilot** Jagdflieger, *der*

fighting ['faɪtɪŋ] **1.** *adj.* Kampf‹*truppen, -schiff*›. **2.** *n.* Kämpfe

fighting: ~ **'chance** *n.* **have a ~ chance of succeeding/of doing sth.** Aussicht auf Erfolg haben/gute Chancen haben, etw. zu tun; ~ **'fit** *adj.* topfit *(ugs.);* ~ **'words** *n. pl. (coll.)* Kampfparolen

'fig-leaf *n. (lit. or fig.)* Feigenblatt, *das*

figment ['fɪgmənt] *n.* Hirngespinst, *das;* **a ~ of one's** *or* **the imagination** pure Einbildung

'fig-tree *n.* Feigenbaum, *der*

figurative ['fɪgjʊrətɪv, 'fɪgərətɪv] *adj.* übertragen; figurativ *(Sprachw.)*

figure ['fɪgə(r)] **1.** *n.* **a)** *(shape)* Form, *die;* **b)** *(Geom.)* Figur, *die;* **c)** *(one's bodily shape)* Figur, *die;* **keep one's ~:** sich *(Dat.)* seine Figur bewahren; **lose one's ~:** dick werden; **d)** *(person as seen)* Gestalt, *die; (literary ~)* Figur, *die; (historical etc. ~)* Persönlichkeit, *die;* **a fine ~ of a man/woman** eine stattliche Erscheinung; **e)** *(simile etc.)* ~ **[of speech]** Redewendung, *die; (Rhet.)* Redefigur, *die;* **f)** *(illustration)* Abbildung, *die;* **g)** *(Dancing, Skating)* Figur, *die;* **h)** *(numerical symbol)* Ziffer, *die; (number so expressed)* Zahl, *die; (amount of money)* Betrag, *der;* **double ~s** zweistellige Zahlen; **go** *or* **run into three ~s** sich auf dreistellige Zahlen belaufen; **three-/four-~:** drei-/vierstellig; **i)** *in pl. (accounts, result of calculations)* Zahlen *Pl.;* **can you check my ~s?** kannst du mal nachrechnen? **2.** *v. t.* **a)** *(picture mentally)* sich *(Dat.)* vorstellen; **b)** *(calculate)* schätzen. **3.** *v. i.* **a)** vorkommen; erscheinen; *(in play)* auftreten; **children don't ~ in her plans for the future** Kinder spielen in ihren Zukunftsplänen keine Rolle; **b)** *(coll.: be likely, understandable)* **that ~s** das kann gut sein

~ **'out** *v. t.* **a)** *(work out by arithmetic)* ausrechnen; **b)** *(Amer.: estimate)* ~ **out that ...:** damit rechnen, daß ...; **c)** *(understand)* verstehen; **I can't ~ him out** ich werde nicht schlau aus ihm; **d)** *(ascertain)* herausfinden

figure: **~-head** *n. (lit. or fig.)* Galionsfigur, *die;* **~-skating** *n.* Eiskunstlauf, *der*

Fiji ['fi:dʒi:] *pr. n.* Fidschi *(das)*

filament ['fɪləmənt] *n.* **a)** Faden, *der;* **b)** *(conducting wire or thread)* Glühfaden, *der*

filch [fɪltʃ] *v. t.* stibitzen *(ugs.)*

[1]file [faɪl] **1.** *n.* Feile, *die; (nail-~)* [Nagel]feile, *die.* **2.** *v. t.* feilen ‹*Fingernägel*›; mit der Feile bearbeiten ‹*Holz, Eisen*›

~ **a'way** *v. t.* abfeilen

~ **'down** *v. t.* abfeilen

[2]file 1. *n.* **a)** *(holder)* Ordner, *der; (box)* Kassette, *die;* [Dokumenten]schachtel, *die;* **on ~:** in der Kartei/in *od.* bei den Akten; **put sth. on ~:** etw. in die Akten/Kartei aufnehmen; **b)** *(set of papers)* Ablage, *die; (as cards)* Kartei, *die;* **open/keep a ~ on sb./sth.** eine Akte über jmdn./etw. anlegen/führen. **2.** *v. t.* **a)** *(place in a ~)* [in die Kartei] einordnen/[in die Akten] aufnehmen; ablegen *(Bürow.);* **b)** *(submit)* einreichen ‹*Antrag*›; **c)** ‹*Journalist:*› einsenden ‹*Bericht*›

~ **a'way** *v. t.* ablegen *(Bürow.)*

[3]file 1. *n.* Reihe, *die;* **[in] single** *or* **Indian ~:** [im] Gänsemarsch. **2.** *v. i.* in einer Reihe hintereinander hergehen

~ **a'way** *v. i.* [einer nach dem anderen] weggehen

~ **'off** *see* ~ **away**

'file card *n.* Karteikarte, *die*

filibuster ['fɪlɪbʌstə(r)] **1.** *n. (obstruction)* Verschleppungstaktik, *die;* Filibuster, *das;* **2.** *v. i.* obstruieren; Dauerreden halten

filigree ['fɪlɪgri:] *n. (lit. or fig.)* Filigran, *das*

filing ['faɪlɪŋ] *n.* **~s** *(particles)* Späne

filing: ~ **cabinet** *n.* Aktenschrank, *der;* ~ **clerk** *n.* Archivkraft, *die*

Filipino [fɪlɪ'pi:nəʊ] **1.** *adj.* philippinisch. **2.** *n., pl.* **~s** Filipino, *der*/Filipina, *die*

fill [fɪl] **1.** *v. t.* **a)** *(make full)* ~ **sth. [with sth.]** etw. [mit etw.] füllen; **~ed with** voller ‹*Reue, Bewunderung, Neid, Verzweiflung*› (**at** über + *Akk.*); **be ~ed with people/flowers/fish** *etc.* voller Menschen/Blumen/Fische *usw.* sein; **b)** *(occupy whole capacity of, spread over)* füllen; besetzen ‹*Sitzplätze*›; *(fig.)* ausfüllen ‹*Gedanken, Zeit*›; **the room was ~ed to capacity** der Raum war voll besetzt; ~ **the bill** *(fig.)* den Erwartungen entsprechen; *(be appropriate)* angemessen sein; **c)** *(pervade)* erfüllen; **light ~ed the room** Licht strömte in das Zimmer; **d)** *(block up)* füllen ‹*Lücke*›; füllen, *(veralt.)* plombieren ‹*Zahn*›; **e)** *(Cookery) (stuff)* füllen; *(put layer of sth. solid in)* belegen; *(put layer of sth. spreadable in)* bestreichen; **f)** *(hold)* innehaben ‹*Posten*›; versehen ‹*Amt*›; *(take up)* ausfüllen ‹*Position*›; *(appoint sb. to)* besetzen ‹*Posten, Lehrstuhl*›. **2.** *v. i.* ~ **[with sth.]** sich [mit etw.] füllen; *(fig.)* sich [mit etw.] erfüllen. **3.** *n.* **eat/drink one's ~:** sich satt essen/trinken; **have had one's ~** seinen Hunger und Durst gestillt haben; **have had one's ~ of sth./doing sth.** genug von etw. haben/etw. zur Genüge getan haben

~ **'in 1.** *v. t.* **a)** füllen; zuschütten,

auffüllen ⟨*Erdloch*⟩; **b)** *(complete)* ausfüllen; ergänzen ⟨*Auslassungen*⟩; **c)** *(insert)* einsetzen; **d)** überbrücken ⟨*Zeit*⟩; **e)** *(coll.: inform)* ~ **sb. in [on sth.]** jmdn. [über etw. *(Akk.)*] ins Bild setzen. **2.** *v. i.* ~ **in for sb.** für jmdn. einspringen

~ **'out 1.** *v. t.* **a)** *(enlarge to proper size or extent)* ausfüllen; **b)** *(Amer.: complete)* ausfüllen ⟨*Formular usw.*⟩. **2.** *v. i.* **a)** *(become enlarged)* sich ausdehnen; **b)** *(become plumper)* voller werden

~ **'up 1.** *v. t.* **a)** *(make full)* ~ **sth. up [with sth.]** etw. [mit etw.] füllen; ~ **up sb.'s glass** jmdm. noch einmal einschenken; **b)** *(put petrol into)* ~ **up [the tank]** tanken; ~ **her up!** *(coll.)* voll[tanken]!; **c)** auffüllen ⟨*Loch*⟩; **d)** *(complete)* ausfüllen ⟨*Formular usw.*⟩. **2.** *v. i.* ⟨*Theater, Zimmer, Zug usw.:*⟩ sich füllen; ⟨*Becken, Spülkasten:*⟩ vollaufen

filler ['fɪlə(r)] *n. (to fill cavity)* Füllmasse, *die*

'filler cap *n.* Tankverschluß, *der*

fillet ['fɪlɪt] **1.** *n. (Gastr.)* Filet, *das;* ~ **[steak]** *(slice)* Filetsteak, *das; (cut)* Filet, *das;* ~ **of pork/cod** Schweine-/Kabeljaufilet, *das.* **2.** *v. t.* filetieren; *(remove bones from)* entgräten ⟨*Fisch*⟩

filling ['fɪlɪŋ] **1.** *n.* **a)** *(for teeth)* Füllung, *die;* Plombe, *die (veralt.);* **have a** ~: sich *(Dat.)* einen Zahn füllen lassen; **b)** *(for pancakes etc.)* Füllung, *die; (for sandwiches etc.)* Belag, *der; (for spreading)* Aufstrich, *der.* **2.** *adj.* sättigend

'filling station *n.* Tankstelle, *die*

fillip ['fɪlɪp] *n. (stimulus)* Anreiz, *der;* Ansporn, *der;* **give sb. a** ~: jmdn. anspornen

filly ['fɪlɪ] *n.* junge Stute; Stutfohlen, *das*

film [fɪlm] **1.** *n.* **a)** *(thin layer)* Schicht, *die;* ~ **[of oil/slime]** [Öl-/Schmier]film, *der;* **b)** *(Photog.; Cinemat.: story etc.)* Film, *der;* **c)** *in pl. (cinema industry)* Kino, *das;* Film, *der;* **go into** ~**s** zum Kino *od.* Film gehen; **d)** *no pl. (as artform)* der Film. **2.** *v. t.* filmen; drehen ⟨*Kinofilm, Szene*⟩; verfilmen ⟨*Buch usw.*⟩

film: ~ **clip** *see* [2]**clip 2 b**; ~ **crew** *n.* Kamerateam, *das;* ~ **director** *n.* Filmregisseur, *der*/-regisseurin, *die;* ~ **projector** *n.* Projektor, *der;* ~ **script** *n.* Drehbuch, *das;* ~ **set** *n.* Dekoration, *die;* ~ **show** *n.* Filmvorführung, *die;* ~ **star** *n.* Filmstar, *der;* ~**-strip** *n.* Filmstreifen, *der*

Filofax, (P) ['faɪləʊfæks] *n.* ≈ Terminplaner, *der*

filter ['fɪltə(r)] **1.** *n.* **a)** Filter, *der;* **b)** *(Brit.) (route)* Abbiegespur, *die; (light)* grünes Licht für Abbieger. **2.** *v. t.* filtern. **3.** *v. i.* **a)** ⟨*Flüssigkeiten:*⟩ sickern; **b)** *(make way gradually)* ~ **through/into sth.** durch etw. hindurch-/in etw. *(Akk.)* hineinsickern; **c)** *(at road junction)* sich einfädeln

~ **'out 1.** *v. t. (lit. or fig.)* herausfiltern. **2.** *v. i.* durchsickern

filter: ~ **lane** *n. (Brit.)* Abbiegespur, *die;* ~**-tip** *n.* **a)** Filter, *der;* **b)** ~**-tip [cigarette]** Filterzigarette, *die*

filth [fɪlθ] *n., no pl.* **a)** *(disgusting dirt)* Dreck, *der;* **b)** *(obscenity)* Schmutz [und Schund]

filthy ['fɪlθɪ] **1.** *adj.* **a)** *(disgustingly dirty)* dreckig *(ugs.);* schmutzig; *(fig.)* widerlich ⟨*Angewohnheit*⟩; **b)** *(vile)* gemein ⟨*Lügner, Trick*⟩; ~ **lucre** schnöder Mammon *(abwertend, auch scherzh.);* **c)** *(obscene)* schweinisch *(ugs.);* obszön, unflätig ⟨*Sprache*⟩; **a** ~ **devil** ein Schweinigel *(ugs.).* **2.** *adv.* ~ **dirty** völlig verdreckt *(ugs.);* ~ **rich** *(coll.)* stinkreich *(ugs.)*

fin [fɪn] *n. (Zool.; on boat)* Flosse, *die; (flipper)* [Schwimm]flosse, *die*

final ['faɪnl] **1.** *adj.* **a)** *(ultimate)* letzt...; End⟨*spiel, -stadium, -stufe, -ergebnis*⟩; Schluß⟨*bericht, -szene, -etappe, -phase*⟩; ~ **examination** Abschlußprüfung, *die;* **give a** ~ **wave** ein letztes Mal winken; **b)** *(conclusive)* endgültig ⟨*Urteil, Entscheidung*⟩; **is this your** ~ **decision/word?** ist das Ihr letztes Wort?; **I'm not coming with you, and that's** ~! ich komme nicht mit, und damit basta! *(ugs.).* **2.** *n.* **a)** *(Sport etc.)* Finale, *das; (of quiz game)* Endrunde, *die;* **b)** *in pl. (examination)* Abschlußprüfung, *die; (at university)* Examen, *das*

finale [fɪ'nɑːlɪ] *n.* Finale, *das*

finalise *see* **finalize**

finalist ['faɪnəlɪst] *n.* Teilnehmer/Teilnehmerin in der Endausscheidung; *(Sport)* Finalist, *der*/Finalistin, *die*

finality [faɪ'nælɪtɪ] *n., no pl.* Endgültigkeit, *die; (of tone of voice)* Entschiedenheit, *die*

finalize ['faɪnəlaɪz] *v. t.* [endgültig] beschließen; unter Dach und Fach bringen ⟨*Geschäft, Vertrag*⟩; *(complete)* zum Abschluß bringen

finally ['faɪnəlɪ] *adv.* **a)** *(in the end)* schließlich; *(expressing impatience etc.)* endlich; **b)** *(in conclusion)* zum Schluß; *(once for all)* ein für allemal

finance [faɪ'næns, fɪ'næns, 'faɪnæns] **1.** *n.* **a)** *in pl. (resources)* Finanzen *Pl.;* **b)** *(management of money)* Geldwesen, *das;* **c)** *(support)* Gelder *Pl. (ugs.);* Geldmittel *Pl.* **2.** *v. t.* finanzieren; finanziell unterstützen ⟨*Person*⟩

financial [faɪ'nænʃl, fɪ'nænʃl] *adj.* finanziell; Finanz⟨*mittel, -quelle, -experte, -lage*⟩; Geld⟨*mittel, -geber, -sorgen*⟩; Wirtschafts⟨*nachrichten, -bericht*⟩

financially [faɪ'nænʃəlɪ, fɪ'nænʃəlɪ] *adv.* finanziell

financier [faɪ'nænsɪə(r), fɪ'nænsɪə(r)] *n.* Finanzier, *der*

finch [fɪnʃ] *n. (Ornith.)* Fink[envogel], *der*

find [faɪnd] **1.** *v. t.,* **found** [faʊnd] **a)** *(get possession of by chance)* finden; *(come across unexpectedly)* entdecken; ~ **that ...:** herausfinden *od.* entdecken, daß ...; **he was found dead/injured** er wurde tot/verletzt aufgefunden; **b)** *(obtain)* finden ⟨*Zustimmung, Erleichterung, Trost, Gegenliebe*⟩; **have found one's feet** *(be able to walk)* laufen können; *(be able to act by oneself)* auf eigenen Füßen stehen; **c)** *(recognize as present)* sehen ⟨*Veranlassung, Schwierigkeit*⟩; *(acknowledge or discover to be)* finden; ~ **no difficulty in doing sth.** etw. nicht schwierig finden; ~ **sb. in/out** jmdn. antreffen/nicht antreffen; ~ **sb./sth. to be ...:** feststellen, daß jmd./etw. ... ist/war; **d)** *(discover by trial or experience to be or do)* für ... halten; **do you** ~ **him easy to get on with?** finden Sie, daß sich gut mit ihm auskommen läßt?; **she** ~**s it hard to come to terms with his death** es fällt ihr schwer, sich mit seinem Tod abzufinden; ~ **sth. necessary** etw. für nötig befinden *od.* erachten; ~ **sth./sb. to be ...:** herausfinden, daß etw./jmd. ... ist/war; **you will** ~ **[that] ...:** Sie werden sehen *od.* feststellen, daß ...; **e)** *(discover by search)* finden; **want to** ~ **sth.** etw. suchen; ~ **[again]** wiederfinden; **f)** *(succeed in obtaining)* finden ⟨*Zeit, Mittel und Wege, Worte*⟩; auftreiben ⟨*Geld, Gegenstand*⟩; aufbringen ⟨*Kraft, Energie*⟩; ~ **it in oneself** *or* **one's heart to do sth.** es über sich *od.* übers Herz bringen, etw. zu tun; **g)** *(ascertain by study or calculation or inquiry)* finden; ~ **what time the train leaves** herausfinden, wann der Zug [ab]fährt; ~ **one's way home** nach Hause zurückfinden; **h)** *(supply)* besorgen; ~ **sb. sth.** *or* **sth. for sb.** jmdn. mit etw. versorgen; **all found** bei freier Kost und Logis. **2.** *n.* **a)** Fund, *der;* **make a ~/two ~s** fündig/zweimal fündig werden; **b)** *(person)* Entdeckung, *die*

~ for *v. t. (Law)* **~ for the defendant/plaintiff** zugunsten der Verteidigung/des Klägers entscheiden; **~ for the accused** auf Freispruch erkennen

~ 'out *v. t.* **a)** *(discover, devise)* herausfinden; bekommen ‹*Informationen*›; **manage to ~ out how ...:** herausbekommen, wie ...; **~ out about** *(get information on)* sich informieren über (+ *Akk.*); *(learn of)* erfahren von; **b)** *(detect in offence, act of deceit, etc.)* erwischen, ertappen ‹*Dieb usw.*›

findable ['faɪndəbl] *pred. adj.* **be [easily] ~:** [leicht] zu finden sein

finder ['faɪndə(r)] *n. (of sth. lost)* Finder, *der*/Finderin, *die; (of sth. unknown)* Entdecker, *der*/Entdeckerin, *die;* **~s keepers** *(coll.)* wer's findet, dem gehört's *(ugs.)*

finding ['faɪndɪŋ] *n. usu. in pl. (conclusion)* Ergebnis, *das; (verdict)* Urteil, *das*

[1]fine [faɪn] **1.** *n.* Geldstrafe, *die; (for minor offence)* Bußgeld, *das.* **2.** *v. t.* mit einer Geldstrafe belegen; **we were ~d £10** wir mußten ein Bußgeld von 10 Pfund bezahlen; **be ~d for speeding** ein Bußgeld wegen überhöhter Geschwindigkeit zahlen müssen

[2]fine 1. *adj.* **a)** *(of high quality)* gut; hochwertig ‹*Qualität, Lebensmittel*›; fein ‹*Besteck, Gewebe, Spitze*›; edel ‹*Holz, Wein*›; **b)** *(delicately beautiful)* zart ‹*Porzellan, Spitze*›; fein ‹*Muster, Kristall, Stickerei, Gesichtszüge*›; **c)** *(refined)* edel ‹*Empfindungen*›; fein ‹*Taktgefühl, Geschmack*›; **sb.'s ~r feelings** das Gute in jmdm.; **d)** *(delicate in structure or texture)* fein; **e)** *(thin)* fein; hauchdünn; **cut it ~:** knapp kalkulieren; **we'd be cutting it ~ if ...** es wird etwas knapp werden, wenn ...; **f)** *(in small particles)* [hauch]fein ‹*Sand, Staub*›; **~ rain** Nieselregen, *der;* **g)** *(sharp, narrow-pointed)* scharf ‹*Spitze, Klinge*›; spitz ‹*Nadel, Schreibfeder*›; **h) ~ print** *see* **small print; i)** *(capable of delicate discrimination)* fein ‹*Gehör*›; scharf ‹*Auge*›; genau ‹*Werkzeug*›; empfindlich ‹*Meßgerät*›; **j)** *(perceptible only with difficulty)* fein ‹*Unterschied, Nuancen*›; *(precise)* klein ‹*Detail*›; **the ~r points** die Feinheiten; **k)** *(excellent)* schön; ausgezeichnet ‹*Sänger, Schauspieler*›; **a ~ time to do sth.** *(iron.)* ein passender Zeitpunkt, etw. zu tun *(iron.)*; **you 'are a ~ one!** *(iron.)* du bist mir vielleicht einer! *(ugs.)*; **l)** *(satisfactory)* schön; gut; **that's ~ with** *or* **by me** ja, ist mir recht; **everything is ~:** es ist alles in Ordnung; **m)** *(well conceived or expressed)* schön ‹*Worte, Ausdruck usw.*›; gelungen ‹*Rede, Übersetzung usw.*›; **n)** *(of handsome appearance or size)* schön; stattlich ‹*Mann, Baum, Tier*›; **o)** *(in good health or state)* gut; **feel ~:** sich wohl fühlen; **How are you? – F~, thanks** Wie geht es Ihnen? – Gut, danke; **p)** *(bright and clear)* schön ‹*Wetter, Sommerabend*›; **~ and sunny** heiter und sonnig; **q)** *(ornate)* prächtig ‹*Kleidung*›; **r)** *(affectedly ornate)* geziert; schönklingend ‹*Worte*›. **2.** *adv.* **a)** *(into small particles)* fein ‹*mahlen, raspeln, hacken*›; **b)** *(coll.: well)* gut

fine 'art *n.* **a)** *(subject)* bildende Kunst; **b) get sth. [down] to a ~:** etw. zu einer richtigen Kunst entwickeln; **c) the ~s** die Schönen Künste

finely ['faɪnlɪ] *adv.* **a)** *(exquisitely, delicately)* fein; genau ‹*ausbalanciert*›; **b) a ~-sharpened blade** eine sorgfältig geschärfte Klinge; **a ~-drawn line** eine fein *od.* dünn [aus]gezogene Linie; **c)** *(into small particles)* fein ‹*mahlen*›

finery ['faɪnərɪ] *n., no pl.* Pracht, *die; (garments etc.)* Staat, *der*

finesse [fɪ'nes] *n. (refinement)* Feinheit, *die; (of diplomat)* Gewandtheit, *die; (delicate manipulation)* Finesse, *die*

fine-tooth 'comb *n.* **go through a manuscript** *etc.***/house** *etc.* **with a ~** *(fig.)* ein Manuskript *usw.* Punkt für Punkt durchgehen/ein Haus *usw.* durchkämmen

finger ['fɪŋgə(r)] **1.** *n.* Finger, *der;* **lay a ~ on sb.** *(fig.)* jmdm. ein Härchen krümmen *(ugs.)*; **they never lift** *or* **raise a ~ to help her** *(fig.)* sie rühren keinen Finger, um ihr zu helfen; **pull** *or* **take one's ~ out** *(fig. sl.)* Dampf dahinter machen *(ugs.)*; **point a** *or* **one's ~ at sb./sth.** mit dem Finger/*(fig. ugs.)* mit Fingern auf jmdn./etw. zeigen; **put the ~ on sb.** *(fig. sl.)* jmdn. verpfeifen *(ugs. abwertend)*; **put** *or* **lay one's ~ on sth.** *(fig.)* etw. genau ausmachen; **sth. slips through sb.'s ~s** etw. gleitet jmdm. durch die Finger; **his ~s are [all] thumbs, he is all ~s and thumbs** er hat zwei linke Hände *(ugs.)*; **a ~ of toast** ein Streifen Toast. **2.** *v. t.* berühren ‹*Ware*›; greifen ‹*Akkord*›; *(toy or meddle with)* befingern; herumfingern an (+ *Dat.*)

finger: ~-board *n.* Griffbrett, *das;* **~-bowl** *n.* Fingerschale, *die;* **~-end** Fingerspitze, *die;* **~-mark** *n.* Fingerabdruck, *der;* **~-nail** *n.* Fingernagel, *der;* **~print 1.** *n.* Fingerabdruck, *der;* **2.** *v. t.* **~print sb.** jmdm. die Fingerabdrücke abnehmen; **~tip** *n.* Fingerspitze, *die;* **have sth. at one's ~tips** *(fig.)* etw. aus dem Effeff können *od.* im kleinen Finger haben *(ugs.)*

finicky ['fɪnɪkɪ] *adj.* heikel ‹*Person*›; kniff[e]lig ‹*Arbeit, Stickerei*›

finish ['fɪnɪʃ] **1.** *v. t.* **a)** *(bring to an end)* beenden ‹*Unterhaltung*›; erledigen ‹*Arbeit*›; abschließen ‹*Kurs, Ausbildung*›; **have ~ed sth.** etw. fertig haben; mit etw. fertig sein; **have you ~ed the letter/book?** hast du den Brief/das Buch fertig?; **~ writing/reading sth.** etw. zu Ende schreiben/lesen; **have you quite ~ed?** sind Sie fertig?; **b)** *(get through)* aufessen ‹*Mahlzeit*›; auslesen ‹*Buch, Zeitung*›; austrinken ‹*Flasche, Glas*›; **c)** *(kill)* umbringen; *(coll.: overcome)* schaffen *(ugs.)*; *(overcome completely)* bezwingen ‹*Feind*›; *(ruin)* zugrunde richten; **any more stress would ~ him** noch mehr Streß würde ihn kaputtmachen *(ugs.)*; **it almost ~ed me!** das hat mich fast geschafft! *(ugs.)*; **d)** *(perfect)* vervollkommnen; den letzten Schliff geben (+ *Dat.*); **~ a seam** einen Saum vernähen; **e)** *(complete manufacture of by surface treatment)* eine schöne Oberfläche geben (+ *Dat.*); glätten ‹*Papier, Holz*›; appretieren ‹*Gewebe, Leder*›; **the ~ed article** *or* **product** das fertige Produkt. **2.** *v. i.* **a)** *(reach the end)* aufhören; ‹*Geschichte, Episode:*› enden; **when does the concert ~?** wann ist das Konzert aus?; **b)** *(come to end of race)* das Ziel erreichen; **~ first** als erster durchs Ziel gehen; erster werden; **~ badly/well** nicht durchhalten/einen guten Endspurt haben; **c) ~ by doing sth.** zum Schluß etw. tun. **3.** *n.* **a)** *(termination, cause of ruin)* Ende, *das;* **it would be the ~ of him as a politician** das würde das Ende seiner Karriere als Politiker bedeuten; **b)** *(point at which race etc. ends)* Ziel, *das;* **arrive at the ~:** das Ziel erreichen; durchs Ziel gehen; **c)** *(what serves to give completeness)* letzter Schliff; **a ~ to sth.** die Vervollkommnung *od.*

Vollendung einer Sache; **d)** *(mode of finishing)* [technische] Ausführung; Finish, *das;* **paintwork with a matt/gloss ~:** Matt-/Hochglanzlack, *der*

~ 'off *v.t.* **a)** *see* **~ 1c, d; b)** *(provide with ending)* abschließen; beenden; **c)** *(finish or trim neatly)* sauber verarbeiten

~ 'up *v.i.* **a)** *see* **~ 2c; b) = end up**

~ with *v.t.* **a)** *(complete one's use of)* **have you ~ed with the sugar?** brauchen Sie den Zucker noch?; **have ~ed with a book** ein Buch aus- *od.* fertiggelesen *od.* zu Ende gelesen haben; **b)** *(end association with)* brechen mit; **she ~ed with her boy-friend** sie hat mit ihrem Freund Schluß gemacht

finishing: ~-post *n.* Zielpfosten, *der;* **~ 'touch** *n.* **as a ~ touch to sth.** zur Vollendung *od.* Vervollkommnung einer Sache; um eine Sache abzurunden; **put the ~ touches to sth.** einer Sache *(Dat.)* den letzten Schliff geben

finite ['faınaıt] **a)** *(bounded)* begrenzt; **~ number** *(Math.)* endliche Zahl; **b)** *(Ling.)* finit

Finland ['fınlənd] *pr. n.* Finnland *(das)*

Finn [fın] *n.* Finne, *der*/Finnin, *die*

Finnish ['fınıʃ] **1.** *adj.* finnisch; **sb. is ~:** jmd. ist Finne/Finnin; **the ~ language** das Finnische. **2.** *n.* Finnisch, *das; see also* **English 2a**

fiord [fjɔ:d] *n.* Fjord, *der*

fir [fɜ:(r)] *n.* **a)** *(tree)* Tanne, *die;* **b)** *(wood)* Tanne, *die;* Tannenholz, *das*

'fir-cone *n.* Tannenzapfen, *der*

fire ['faıə(r)] **1.** *n.* **a)** Feuer, *das;* **set ~ to sth.** ‹*Person:*› etw. anzünden; **be on ~:** brennen *(auch fig.);* in Flammen stehen; **catch ~:** Feuer fangen; ‹*Wald, Gebäude:*› in Brand geraten; **set sth. on ~:** etw. anzünden; *(in order to destroy)* etw. in Brand stecken; *(deliberately)* Feuer an etw. *(Akk.)* legen; **b)** *(in grate)* [offenes] Feuer; *(electric or gas ~)* Heizofen, *der; (in the open air)* Lagerfeuer, *das;* **open ~:** Kaminfeuer, *das;* **turn up the ~** *(electric)* die Heizung/*(gas)* das Gas höher drehen *od.* aufdrehen; **play with ~** *(lit. or fig.)* mit dem Feuer spielen; **light the ~:** den Ofen anstecken; *(in grate)* das [Kamin]feuer anmachen; **c)** *(destructive burning)* Brand, *der;* **in case of ~:** bei Feuer; **where's the ~?** *(coll. iron.)* wo brennt's denn? **d)** *(fervour)* Feuer, *das;* **the ~ with which he speaks** die Leidenschaft, mit der er spricht; **e)** *(firing of guns)* Schießen, *das;* Schießerei, *die;* **pistol ~:** [Pistolen]schüsse; **cannon ~:** Kanonenfeuer, *das;* **line of ~** *(lit. or fig.)* Schußlinie, *die;* **be/come under ~:** beschossen werden/unter Beschuß geraten. **2.** *v.t.* **a)** *(fill with enthusiasm)* begeistern, in Begeisterung versetzen ‹*Person*›; **b)** *(supply with fuel)* befeuern ‹*Ofen*›; [be]heizen ‹*Lokomotive*›; **c)** *(discharge)* abschießen ‹*Gewehr*›; abfeuern ‹*Kanone*›; **~ one's gun/pistol/rifle at sb.** auf jmdn. schießen; **d)** *(propel from gun etc.)* abgeben, abfeuern ‹*Schuß*›; **~ questions at sb.** jmdn. mit Fragen bombardieren; Fragen auf jmdn. abfeuern; **e)** *(coll.: dismiss)* feuern *(ugs.)* ‹*Angestellten*›; **f)** brennen ‹*Tonwaren, Ziegel*›. **3.** *v.i.* **a)** *(shoot)* schießen; feuern; **~!** [gebt] Feuer!; **be the first to ~:** das Feuer eröffnen; **~ at/on sth./sb.** auf etw./jmdn. schießen; **b)** ‹*Motor:*› zünden

~ a'way *v.i.* *(fig. coll.)* losschießen *(fig. ugs.);* **~ away!** schieß los!; fang an!

fire: ~-alarm *n.* Feuermelder, *der;* **~arm** *n.* Schußwaffe, *die;* **~-ball** *n.* *(ball of flame)* Feuerball, *der;* **~ brigade** *(Brit.),* **~ department** *(Amer.)* *ns.* Feuerwehr, *die;* **~-drill** *n.* Probe[feuer]alarm, *der; (for ~men)* Feuerwehrübung, *die;* **~-eater** *n.* Feuerschlucker, *der;* **~-engine** *n.* Löschfahrzeug, *das;* **~-escape** *n.* *(staircase)* Feuertreppe, *die; (ladder)* Feuerleiter, *die;* **~ extinguisher** *n.* Feuerlöscher, *der;* **~fly** *n.* Leuchtkäfer, *der;* **~-guard** *n.* Kamingitter, *das;* **~ hazard** *n.* Brandrisiko, *das;* **~ insurance** *n.* Feuer- *od.* Brandversicherung, *die;* **~-lighter** *n.* *(Brit.)* Feueranzünder, *der;* **~man** ['faıəmən] *n., pl.* **~men** ['faıəmən] **a)** *(member of ~ brigade)* Feuerwehrmann, *der;* **b)** *(Railw.)* Heizer, *der;* **~place** *n.* Kamin, *der;* **~proof 1.** *adj.* feuerfest; **2.** *v.t.* feuerfest machen; **~-resistant** *adj.* feuerbeständig; **F~ Service** *n.* Feuerwehr, *die;* **~side** *n.* Kaminecke, *die;* **at** *or* **by the ~side** am Kamin; **~ station** *n.* Feuerwache, *die;* **~wood** *n.* Brennholz, *das;* **~work** *n.* **a)** Feuerwerkskörper, *der;* **~work display** Feuerwerk, *das;* **b)** *in pl.* *(display)* Feuerwerk, *das;* **there were** *or* **it caused ~works** *(fig.)* da war was los *od.* flogen die Funken *(ugs.)*

firing ['faıərıŋ] *n.* **a)** *(of pottery)* Brennen, *das;* **b)** *no pl.* *(of guns)* Abfeuern, *das;* **we could hear ~ in the distance** in der Ferne konnten wir Schüsse hören

firing: ~-line *n.* *(lit. or fig.)* Feuerlinie, *die;* **~-squad** *n.* *(at military execution)* Exekutionskommando, *das*

[1]firm [fɜ:m] *n.* Firma, *die;* **~ of architects/decorators** Architektenbüro, *das*/Malerbetrieb, *der*

[2]firm 1. *adj.* **a)** fest; stabil ‹*Verhältnis, Konstruktion, Stuhl*›; straff ‹*Busen*›; verbindlich ‹*Angebot*›; **be on ~ ground again** *(lit. or fig.)* wieder festen Boden unter den Füßen haben; **they are ~ friends** sie sind gut befreundet; **the chair is not ~:** der Stuhl ist wacklig *od.* wackelt; **b)** *(resolute)* entschlossen ‹*Blick*›; bestimmt, entschieden ‹*Ton*›; **be a ~ believer in sth.** fest an etw. *(Akk.)* glauben; **c)** *(insisting on obedience etc.)* bestimmt; **be ~ with sb.** jmdm. gegenüber bestimmt auftreten. **2.** *adv.* **stand ~!** *(fig.)* sei standhaft!; **hold ~ to sth.** *(fig.)* an einer Sache festhalten. **3.** *v.t.* fest werden lassen; festigen, straffen ‹*Muskulatur, Körper*›

firmly ['fɜ:mlı] *adv.* **a)** fest; **a ~-built structure** eine stabile Konstruktion; **b)** *(resolutely)* beharrlich ‹*unterstützen, sich widersetzen*›; bestimmt, energisch ‹*reden*›

firmness ['fɜ:mnıs] *n., no pl.* **a)** *(solidity)* Festigkeit, *die; (of foundations, building)* Stabilität, *die;* **b)** *(resoluteness)* Entschlossenheit, *die; (of voice)* Bestimmtheit, *die;* **c)** *(insistence on obedience etc.)* Bestimmtheit, *die*

first [fɜ:st] **1.** *adj.* erst...; *(for the ~ time ever)* Erst‹*aufführung, -besteigung*›; *(of an artist's ~ achievement)* Erstlings‹*film, -roman, -stück, -werk*›; **he was ~ to arrive** er kam als erster an; **for the [very] ~ time** zum [aller]ersten Mal; **the ~ two** die ersten beiden *od.* zwei; **come in ~** *(win race)* [das Rennen] gewinnen; **head/feet ~:** mit dem Kopf/den Füßen zuerst *od.* voran; **~ thing in the morning** gleich frühmorgens; *(coll.: tomorrow)* gleich morgen früh; **~ things ~** *(coll.)* eins nach dem anderen; **he's always [the] ~ to help** er ist immer als erster zur Stelle, wenn Hilfe benötigt wird; **not know the ~ thing about sth.** von einer Sache nicht das geringste verstehen. **2.** *adv.* **a)** *(before anyone else)* zuerst; als erster/erste ‹*sprechen, ankommen*›; *(before* *anything else)* an erster Stelle ⟨*stehen, kommen*⟩; *(when listing: firstly)* zuerst; als erstes; **ladies ~!** Ladies first!; den Damen der Vortritt!; **you [go] ~** *(as invitation)* Sie haben den Vortritt; bitte nach Ihnen; **~ come ~ served** wer zuerst kommt, mahlt zuerst *(Spr.)*; **say ~ one thing and then another** erst so̱ und dann wieder so̱ sagen *(ugs.)*; **b)** *(beforehand)* vorher; **... but ~ we must ...:** ... aber zuerst *od.* erst müssen wir ...; **c)** *(for the ~ time)* zum ersten Mal; das erste Mal; erstmals ⟨*bekanntgeben, sich durchsetzen*⟩; **d)** *(in preference)* eher; lieber; **e) ~ of all** zuerst; *(in importance)* vor allem; **~ and foremost** *(basically)* zunächst einmal. **3.** *n.* **a) the ~** *(in sequence)* der/die/das erste; *(in rank)* der/die/das Erste; **be the ~ to arrive** als erster/erste ankommen; **she is the ~ in the class** sie ist Klassenbeste; **this is the ~ I've heard of it** das höre ich zum ersten Mal; **b) at ~:** zuerst; anfangs; **from the ~:** von Anfang an; **from ~ to last** von Anfang bis Ende; **c)** *(day)* **the ~ of May** der erste Mai; **the ~ [of the month]** der Erste [des Monats]

first: ~ 'aid *n.* Erste Hilfe; **give [sb.] ~ aid** [jmdm.] Erste Hilfe leisten; **~-born 1.** *adj.* erstgeboren; **2.** *n.* Erstgeborene, *der/die;* **~ 'class** *n.* **a)** erste Kategorie; **b)** *(Transport, Post)* erste Klasse; **c)** *(Brit. Univ.)* ≈ Eins, *die;* **~-class 1.** ['--] *adj.* **a)** *(of the ~ class)* erster Klasse *nachgestellt;* Erster-Klasse-⟨*Fahrkarte, Abteil, Passagier, Post, Brief usw.*⟩; **~-class stamp** *Briefmarke für einen Erster-Klasse-Brief;* **b)** *(excellent)* erstklassig; **2.** [-'-] *adv.* erster Klasse ⟨*fahren*⟩; **send a letter ~-class** *einen Brief mit Erster-Klasse-Post schicken;* **~ 'cousin** *see* **cousin; ~ e'dition** *n.* Erstausgabe, *die;* **~ 'gear** *n., no pl. (Motor Veh.)* erster Gang; *see also* **gear 1 a; ~hand** *adj.* aus erster Hand *nachgestellt;* **from ~hand experience** aus eigener Erfahrung; **~ 'light** *n.* **at ~ light** im *od.* beim Morgengrauen

firstly ['fɜ:stlɪ] *adv.* zunächst [einmal]; *(followed by 'secondly')* erstens

first: ~ name *n.* Vorname, *der;* **~ 'night** *n. (Theatre)* Premiere, *die;* **~ 'officer** *n. (Naut.)* Erster Offizier; **~-rate** *adj.* erstklassig; **~ school** *n. (Brit.)* ≈ Grundschule, *die*

'fir tree *see* **fir a**

fiscal ['fɪskl] *adj.* fiskalisch; finanzpolitisch

fish [fɪʃ] **1.** *n., pl. same* **a)** Fisch, *der;* **~ and chips** Fisch mit Pommes frites; **[be] like a ~ out of water** [sich] wie ein Fisch auf dem Trockenen [fühlen]; **there are plenty more ~ in the sea** *(fig. coll.)* es gibt noch andere auf der Welt; **b)** *(coll.: person)* **queer ~:** komischer Kauz; **big ~:** großes Tier *(ugs., scherzh.)*. **2.** *v. i.* **a)** fischen; *(with rod)* angeln; **go ~ing** fischen/angeln gehen; **b)** *(fig. coll.) (try to get information)* auf Informationen aussein; *(delve)* **~ around in one's bag** in der Tasche herumsuchen. **3.** *v. t.* fischen; *(with rod)* angeln

~ for *v. t.* **a)** fischen/angeln; fischen/angeln auf (+ *Akk.*) *(Anglerjargon)*; **b)** *(fig. coll.)* suchen nach

~ 'out *v. t. (fig. coll.)* herausfischen *(ugs.)*

fish: ~-bone *n.* [Fisch]gräte, *die;* **~ cake** *n. (Cookery)* Fischfrikadelle, *die*

fisherman ['fɪʃəmən] *n., pl.* **fishermen** ['fɪʃəmən] Fischer, *der;* *(angler)* Angler, *der*

fishery ['fɪʃərɪ] *n.* *(part of sea)* Fischfanggebiet, *das;* Fischereigewässer, *das*

fish: ~ 'finger *n.* Fischstäbchen, *das;* **~-hook** *n.* Angelhaken, *der*

fishing: ~ boat *n.* Fischerboot, *das;* **~-net** *n.* Fischernetz, *das;* **~-rod** *n.* Angelrute, *die;* **~-tackle** *n.* Angelgeräte

fish: ~-knife *n.* Fischmesser, *das;* **~monger** ['fɪʃmʌŋgə(r)] *n.* *(Brit.)* Fischhändler, *der/*-händlerin, *die;* **~-net** *n.* Fischnetz, *das;* **~-net stockings** Netzstrümpfe; **~ shop** *n.* Fischgeschäft, *das;* **~-slice** *n.* Wender, *der*

fishy ['fɪʃɪ] *adj.* **a)** fischartig; **~ taste/smell** Fischgeschmack/-geruch, *der;* **b)** *(coll.: suspect)* verdächtig

fission ['fɪʃn] *n.* *(Nucl. Phys.)* [Kern]spaltung, *die;* Fission, *die (fachspr.)*

fissure ['fɪʃə(r)] *n.* Riß, *der*

fist [fɪst] *n.* Faust, *die*

fistful ['fɪstfʊl] *n.* Handvoll, *die*

¹fit [fɪt] *n.* **a)** Anfall, *der;* **~ of coughing** Hustenanfall, *der;* **epileptic ~:** epileptischer Anfall; **b)** *(fig.)* [plötzliche] Anwandlung; **have** *or* **throw a ~:** einen Anfall bekommen; **[almost] have** *or* **throw a ~** *(fig.)* [fast] Zustände kriegen *(ugs.)*; **be in ~s of laughter** sich vor Lachen biegen; **sb./sth. has sb. in ~s [of laughter]** jmd. ruft dröhnendes Gelächter bei jmdm. hervor

²fit 1. *adj.* **a)** *(suitable)* geeignet; **~ to eat** *or* **to be eaten/for human consumption** eßbar/zum Verzehr geeignet; **b)** *(worthy)* würdig; wert; **c)** *(right and proper)* richtig; **see** *or* **think ~ [to do sth.]** es für richtig *od.* angebracht halten[, etw. zu tun]; **d)** *(ready)* **be ~ to drop** zum Umfallen müde sein; **e)** *(healthy)* gesund; fit *(ugs.)*; in Form *(ugs.)*; **~ for duty** *or* **service** dienstfähig *od.* -tauglich; *see also* **fiddle 1 a. 2.** *n.* Paßform, *die;* **it is a good/bad ~:** es sitzt *od.* paßt gut/nicht gut; **I can just get it in the suitcase, but it's a tight ~** *(fig.)* ich kriege es noch in den Koffer, aber nur gerade so *(ugs.)*. **3.** *v. t.*, **-tt-: a)** ⟨*Kleider:*⟩ passen (+ *Dat.*); ⟨*Schlüssel:*⟩ passen in (+ *Akk.*); ⟨*Deckel, Bezug:*⟩ passen auf (+ *Akk.*); **b)** anpassen ⟨*Kleidungsstück, Brille*⟩; **c)** *(correspond to, suit)* entsprechen (+ *Dat.*); *(make correspond)* abstimmen (**to** auf + *Akk.*); anpassen (**to** an + *Akk.*); **d)** *(put into place)* anbringen (**to** an + *Dat. od. Akk.*); einbauen ⟨*Motor, Ersatzteil*⟩; einsetzen ⟨*Scheibe, Tür, Schloß*⟩; *(equip)* ausstatten. **4.** *v. i.*, **-tt-** passen; *(agree)* zusammenpassen; übereinstimmen; **~ well** ⟨*Kleidungsstück:*⟩ gut sitzen

~ 'in 1. *v. t.* **a)** unterbringen; **b)** *(to a schedule)* einen Termin geben (+ *Dat.*); unterbringen, einschieben ⟨*Treffen, Besuch, Sitzung*⟩. **2.** *v. i.* **a)** hineinpassen; **b)** *(be in accordance)* **~ in with sth.** mit etw. übereinstimmen; **~ in with sb.'s plan/ideas** in jmds. Plan/Konzept *(Akk.)* passen; **c)** *(settle harmoniously)* ⟨*Person:*⟩ sich anpassen (**with** an + *Akk.*); **~ in easily with a group** sich leicht in eine Gruppe einfügen

~ 'out *v. t.* ausstatten; *(for expedition etc.)* ausrüsten

fitful ['fɪtfl] *adj.* unbeständig; unruhig ⟨*Schlaf*⟩; launisch ⟨*Brise*⟩

fitment ['fɪtmənt] *n.* *(piece of furniture)* Einrichtungsgegenstand, *der;* *(piece of equipment)* Zubehörteil, *das*

fitness ['fɪtnɪs] *n., no pl.* **a)** *(physical)* Fitneß, *die;* **b)** *(suitability)* Eignung, *die;* *(appropriateness)* Angemessenheit, *die*

fitted ['fɪtɪd] *adj.* **a)** *(suited)* geeignet (**for** für, zu); **b)** tailliert, auf Taille gearbeitet ⟨*Kleider*⟩; **~ carpet** Teppichboden, *der;* **~ kitchen/cupboards** Einbauküche, *die*/Einbauschränke

fitter ['fɪtə(r)] *n.* Monteur, *der; (of pipes)* Installateur, *der; (of machines)* Maschinenschlosser, *der;* **electrical ~:** Elektriker, *der*

fitting ['fɪtɪŋ] **1.** *adj. (appropriate)* passend; angemessen; geeignet ‹*Moment, Zeitpunkt*›; günstig, passend ‹*Gelegenheit*›; *(becoming)* schicklich *(geh.)* ‹*Benehmen*›. **2.** *n.* **a)** *usu. in pl. (fixture)* Anschluß, *der;* **~s** *(furniture)* Ausstattung, *die;* **b)** *(of clothes)* Anprobe, *die*

five [faɪv] **1.** *adj.* **a)** fünf; *see also* **eight 1; b)** ~ **o'clock shadow** [nachmittäglicher] Stoppelbart *(ugs.).* **2.** *n.* Fünf, *die; see also* **eight 2 a, c, d**

fiver ['faɪvə(r)] *n. (coll.) (Brit.)* Fünfpfundschein, *der; (Amer.)* Fünfdollarschein, *der*

fix [fɪks] **1.** *v. t.* **a)** *(place firmly, attach, prevent from moving)* befestigen; festmachen; ~ **sth. to/on sth.** etw. an/auf etw. *(Dat.)* befestigen *od.* festmachen; ~ **shelves to the wall/a handle on the door** Regale an der Wand/eine Klinke an der Tür anbringen; ~ **sth. in one's mind** sich *(Dat.)* etw. fest einprägen; **b)** *(direct steadily)* richten ‹*Blick, Gedanken, Augen*› (**[up]on** auf + *Akk.*); **c)** *(decide, specify)* festsetzen, festlegen ‹*Termin, Preis, Strafe, Grenze*›; *(settle, agree on)* ausmachen; **it was ~ed that ...:** es wurde beschlossen *od.* vereinbart, daß ...; **d)** *(repair)* in Ordnung bringen; reparieren; **e)** *(arrange)* arrangieren; ~ **a rehearsal for Friday** eine Probe für *od.* auf Freitag *(Akk.)* ansetzen; **nothing definite has been ~ed yet** es ist noch nichts Endgültiges vereinbart *od.* ausgemacht; **f)** *(manipulate fraudulently)* manipulieren ‹*Rennen, Kampf*›; **the whole thing was ~ed** das war eine abgekartete Sache *(ugs.);* **g)** *(Amer. coll.: prepare)* machen ‹*Essen, Kaffee, Drink*›; **h)** *(sl.: deal with)* in Ordnung bringen; regeln; ~ **sb.** *(get even with)* es jmdm. heimzahlen; *(kill)* jmdn. kaltmachen *(salopp).* **2.** *n.* **a)** *(coll.: predicament)* Klemme, *die (ugs.);* **be in a ~:** in der Klemme sein *(ugs.);* **b)** *(sl.: of drugs)* Fix, *der (Drogenjargon)*

~ **on** *v. t.* **a)** [-'-] anbringen; **b)** ['--] *(decide on)* sich entscheiden für

~ **'up** *v. t.* **a)** *(arrange)* arrangieren; festsetzen, ausmachen ‹*Termin, Treffpunkt*›; **we've nothing ~ed up for tonight** wir haben noch nichts vor [für] heute abend; **b)** *(provide)* versorgen; *(provide with accommodation)* unterbringen; ~ **sb. up with sth.** jmdm. etw. verschaffen *od.* besorgen

fixation [fɪk'seɪʃn] *n.* Fixierung, *die*

fixed [fɪkst] *adj.* **a)** *pred. (coll.: placed)* **how are you/is he** *etc.* ~ **for cash/fuel?** wie sieht's bei dir/ihm *usw.* mit dem Geld/Treibstoff aus? *(ugs.);* **b)** *(not variable)* fest; starr ‹*Lächeln, Gesichtsausdruck*›; ~ **assets** Anlagevermögen, *das*

fixture ['fɪkstʃə(r)] *n.* **a)** *(furnishing)* eingebautes Teil; *(accessory)* festes Zubehörteil; **~s and fittings** Ausstattung und Installationen; **b)** *(Sport)* Veranstaltung, *die*

fizz [fɪz] **1.** *v. i.* [zischend] sprudeln. **2.** *n. (effervescence)* Sprudeln, *das*

fizzle ['fɪzl] *v. i.* zischen

~ **'out** *v. i.* ‹*Feuerwerk:*› zischend verlöschen; ‹*Begeisterung:*› sich legen; ‹*Kampagne:*› im Sande verlaufen

fizzy ['fɪzɪ] *adj.* sprudelnd; ~ **lemonade** Brause[limonade], *die;* ~ **drinks** kohlensäurehaltige Getränke

flab [flæb] *n. (coll.)* Fett, *das;* Speck, *der (ugs.)*

flabbergast ['flæbəgɑ:st] *v. t.* verblüffen; umhauen *(ugs.)*

flabby ['flæbɪ] *adj.* schlaff ‹*Muskeln, Bauch, Fleisch, Hände, Wangen, Brüste*›; wabbelig *(ugs.),* schwammig ‹*Bauch, Fleisch*›

¹flag [flæg] *n.* Fahne, *die; (small paper etc. device)* Fähnchen, *das; (national ~, on ship)* Flagge, *die;* **keep the ~ flying** *(fig.)* die Fahne hochhalten

~ **'down** *v. t.* [durch Winken] anhalten

²flag *v. i.,* **-gg-** ‹*Person:*› abbauen; ‹*Kraft, Interesse, Begeisterung:*› nachlassen; **business is ~ging** die Geschäfte lassen nach

'flag-day *n. (Brit.) Tag der Straßensammlung für wohltätige Zwecke*

flagon ['flægən] *n.* Kanne, *die*

'flag-pole *n.* Flaggenmast, *der*

flagrant ['fleɪgrənt] *adj.* eklatant; flagrant ‹*Verstoß*›; *(scandalous)* ungeheuerlich; himmelschreiend ‹*Unrecht*›

flag: ~ship *n. (Navy)* Flaggschiff, *das; (fig. attrib.)* führend...; **~stone** *n.* Steinplatte, *die; (for floor)* Fliese, *die; in pl. (pavement)* Straßenpflaster, *das*

flail [fleɪl] *v. i.* [wild] um sich schlagen; **with arms ~ing he tried to keep his balance** mit den Armen fuchtelnd, versuchte er, das Gleichgewicht zu halten

flair [fleə(r)] *n.* Gespür, *das; (special ability)* Talent, *das;* [natürliche] Begabung; **have a ~ for sth.** *(talent)* ein Talent *od.* eine Begabung für etw. haben

flak [flæk] *n.* Flakfeuer, *das (Milit.);* **get a lot of ~ for sth.** *(fig.)* wegen etw. [schwer] unter Beschuß geraten

flake [fleɪk] **1.** *n. (of snow, soap, cereals)* Flocke, *die; (of dry skin)* Schuppe, *die; (of enamel, paint)* ≈ Splitter, *der.* **2.** *v. i.* ‹*Stuck, Verputz, Stein:*› abbröckeln; ‹*Farbe, Rost, Emaille:*› abblättern; ‹*Haut:*› sich schuppen

flaky ['fleɪkɪ] *adj.* bröcklig ‹*Farbe, Gips, Rost*›; blättrig ‹*Kruste*›; schuppig ‹*Haut*›; ~ **pastry** Blätterteig, *der*

flamboyance [flæm'bɔɪəns] *n.* Extravaganz, *die; (of clothes, lifestyle)* Pracht, *die*

flamboyant [flæm'bɔɪənt] *adj.* extravagant; prächtig ‹*Farben, Federkleid*›

flame [fleɪm] *n.* **a)** Flamme, *die;* **be in ~s** in Flammen stehen; **burst into ~:** in Brand geraten; **b)** *(joc.: boy-/girl-friend)* Flamme, *die (ugs.);* **old ~:** alte Flamme *(ugs. veralt.)*

flamenco [flə'meŋkəʊ] *n., pl.* **~s** Flamenco, *der*

flaming ['fleɪmɪŋ] **1.** *adj.* **a)** *(bright-coloured)* feuerrot; flammend ‹*Rot, Abendhimmel*›; **b)** *(very hot)* glühend heiß; *(coll.: passionate)* heftig, leidenschaftlich ‹*Auseinandersetzung*›; **c)** *(coll.: damned)* verdammt. **2.** *adv. (coll.: damned)* **he is too ~ idle** *or* **lazy** er ist, verdammt noch mal, einfach zu faul *(ugs.)*

flamingo [flə'mɪŋgəʊ] *n., pl.* **~s** *or* **~es** *(Ornith.)* Flamingo, *der*

flan [flæn] *n.* **[fruit] ~:** [Obst]torte, *die*

Flanders ['flɑ:ndəz] *pr. n.* Flandern *(das)*

flange [flændʒ] *n.* Flansch, *der*

flank [flæŋk] *n. (of person)* Seite, *die; (of animal; also Mil.)* Flanke, *die*

flannel ['flænl] *n.* **a)** *(fabric)* Flanell, *der;* **b)** *(Brit.: for washing oneself)* Waschlappen, *der;* **c)** *(Brit. sl.: verbose nonsense)* Geschwafel, *das (ugs. abwertend)*

flap [flæp] **1.** *v. t.,* **-pp-** schlagen; ~ **its wings** mit den Flügeln schlagen; *(at short intervals)* [mit den Flügeln] flattern. **2.** *v. i.,* **-pp-: a)** ‹*Flügel:*› schlagen; ‹*Segel, Fahne,*

Vorhang:› flattern; **b) sb.'s ears were ~ping** *(was very interested)* jmd. spitzte die Ohren; **c)** *(fig. coll.: panic)* die Nerven verlieren. **3.** *n.* **a)** Klappe, *die; (seal on envelope, tongue of shoe)* Lasche, *die;* **b)** *(fig. coll.: panic)* **be in a ~:** furchtbar aufgeregt sein

flare [fleə(r)] **1.** *v. i.* **a)** *(blaze)* flackern; *(fig.)* ausbrechen; **tempers ~d** die Gemüter erhitzten sich; **b)** *(widen)* sich erweitern. **2.** *n.* **a)** *(as signal; also Naut.)* Leuchtsignal, *das; (from pistol)* Leuchtkugel, *die;* **b)** *(blaze of light)* Lichtschein, *der;* **c)** *(widening)* **skirt/trousers with ~s** ausgestellter Rock/ausgestellte Hose

~ 'up *v. i.* **a)** *(burn more fiercely)* aufflackern; auflodern; **b)** *(break out)* [wieder] ausbrechen; **c)** *(become angry)* aufbrausen; aus der Haut fahren *(ugs.)*

flash [flæʃ] **1.** *n.* **a)** *(of light)* Aufleuchten, *das;* Aufblinken, *das; (as signal)* Lichtsignal, *das;* **~ of lightning** Blitz, *der;* **~ in the pan** *(fig. coll.)* Zufallstreffer, *der;* **b)** *(Photog.)* Blitzlicht, *das;* **c)** *(fig.)* **~ of genius** *or* **inspiration** *or* **brilliance** Geistesblitz, *der;* **~ of insight** *or* **intuition** Eingebung, *die;* **d)** *(instant)* **be over in a ~:** gleich *od.* im Nu vorbei sein. **2.** *v. t.* **a)** aufleuchten lassen; **~ the/one's headlights** aufblenden; die Lichthupe betätigen; **b)** *(fig.)* **her eyes ~ed fire** ihre Augen sprühten Feuer *od.* funkelten böse; **~ sb. a smile/glance** jmdm. ein Lächeln/einen Blick zuwerfen; **c)** *(display briefly)* kurz zeigen; **~ one's money about** *or* **around** mit [dem] Geld um sich werfen *(ugs.);* **d)** *(Communications)* durchgeben. **3.** *v. i.* **a)** aufleuchten; **the lightning ~ed** es blitzte; **~ at sb. with one's headlamps** jmdn. anblinken *od.* mit der Lichthupe anblenden; **b)** *(fig.)* **her eyes ~ed in anger** ihre Augen blitzten vor Zorn; **c)** *(move swiftly)* **~ by** *or* **past** vorbeiflitzen *(ugs.); (fig.)* ‹*Zeit, Ferien:*› wie im Fluge vergehen; **d)** *(burst suddenly into perception)* **sth. ~ed through my mind** etw. schoß mir durch den Kopf; **e)** *(Brit. sl.: expose oneself)* sich [unsittlich] entblößen

flash: ~back *n. (Cinemat. etc.)* Rückblende, *die* (**to** auf + *Akk.*); **~ bulb** *n. (Photog.)* Blitzbirnchen, *das;* **~-gun** *n. (Photog.)* Blitzlichtgerät, *das;* **~light** *n.* **a)** *(for signals)* Blinklicht, *das;* **b)** *(Photog.)* Blitzlicht, *das;* **c)** *(Amer.)* Taschenlampe, *die;* **~-point** *n.* Flammpunkt, *der; (fig.)* Siedepunkt, *der*

flashy ['flæʃɪ] *adj.* auffällig; protzig *(ugs. abwertend)*

flask [flɑ:sk] *n.* **a)** *see* **Thermos; vacuum flask; b)** *(for wine, oil)* [bauchige] Flasche; *(Chem.)* Kolben, *der*

¹flat [flæt] *n. (Brit.: dwelling)* Wohnung, *die*

²flat 1. *adj.* **a)** flach; eben ‹*Fläche*›; platt ‹*Nase, Reifen*›; **spread the blanket ~ on the ground** die Decke glatt auf dem Boden ausbreiten; **b)** *(fig.) (monotonous)* eintönig; *(dull)* lahm *(ugs.);* fade; *(stale)* schal, abgestanden ‹*Bier, Sekt*›; *(Electr.)* leer ‹*Batterie*›; **fall ~:** nicht ankommen *(ugs.);* seine Wirkung verfehlen; **c)** *(downright)* glatt *(ugs.);* **[and] that's ~:** und damit basta *(ugs.);* **d)** *(Mus.)* [um einen Halbton] erniedrigt ‹*Note*›. **2.** *adv.* **a)** *(coll.: completely)* **~ broke** total pleite; **b)** *(coll.: exactly)* **in two hours ~:** in genau zwei Stunden. **3.** *n.* **a)** flache Seite; **~ of the hand** Handfläche, *die;* **b)** *(level ground)* Ebene, *die;* **c)** *(Mus.)* erniedrigter Ton

flat: ~-chested [flæt'tʃestɪd] *adj.* flachbrüstig; flachbusig; **~ 'feet** *n. pl.* Plattfüße; **~-fish** *n.* Plattfisch, *der;* **~-footed** [flæt'fʊtɪd] *adj.* plattfüßig

flatly ['flætlɪ] *adv.* rundweg; glatt *(ugs.)*

flatness ['flætnɪs] *n., no pl.* Flachheit, *die; (of nose)* Plattheit, *die*

flat: ~ 'out *adv.* **he ran/worked ~ out** er rannte/arbeitete, so schnell er konnte; **~ race** *n.* Flachrennen, *das*

flatten ['flætn] **1.** *v. t.* flach *od.* platt drücken ‹*Schachtel*›; dem Erdboden gleichmachen ‹*Stadt, Gebäude*›. **2.** *v. refl.* **~ oneself against sth.** sich flach *od.* platt gegen etw. drücken

flatter ['flætə(r)] **1.** *v. t.* schmeicheln (+ *Dat.*). **2.** *v. refl.* **~ oneself [on being/having sth.]** sich *(Dat.)* einbilden[, etw. zu sein/haben]

flatterer ['flætərə(r)] *n.* Schmeichler, *der*/Schmeichlerin, *die*

flattering ['flætərɪŋ] *adj.* schmeichelhaft; schmeichlerisch ‹*Person*›; vorteilhaft ‹*Kleid, Licht, Frisur*›

flattery ['flætərɪ] *n.* Schmeichelei, *die*

flat 'tyre *n.* Reifenpanne, *die; (the tyre itself)* platter Reifen

flatulence ['flætjʊləns] *n.* Blähungen; Flatulenz, *die (Med.)*

flaunt [flɔ:nt] *v. t.* zur Schau stellen

flautist ['flɔ:tɪst] *n.* Flötist, *der*/Flötistin, *die*

flavor *etc. (Amer.) see* **flavour** *etc.*

flavour ['fleɪvə(r)] *(Brit.)* **1.** *n.* **a)** Aroma, *das; (taste)* Geschmack, *der;* **the dish lacks ~:** das Gericht schmeckt fade; **b)** *(fig.)* Touch, *der (ugs.);* Anflug, *der.* **2.** *v. t.* **a)** abschmecken; würzen; **b)** *(fig.)* Würze verleihen (+ *Dat.*)

flavouring ['fleɪvərɪŋ] *n. (Brit.)* Aroma, *das*

flaw [flɔ:] *n. (imperfection)* Makel, *der; (in plan, argument)* Fehler, *der; (in goods)* Mangel, *der*

flawless ['flɔ:lɪs] *adj.* **a)** makellos ‹*Schönheit*›; einwandfrei, fehlerlos ‹*Aussprache, Verarbeitung*›; **b)** *(masterly)* vollendet ‹*Aufführung, Wiedergabe*›; **c)** lupenrein ‹*Edelstein*›

flax [flæks] *n.* **a)** *(Bot.)* Flachs, *der;* **b)** *(Textiles: fibre)* Flachsfaser, *die;* Flachs, *der*

'flaxen-haired *adj.* flachsblond

flay [fleɪ] *v. t.* **a)** häuten; **b)** *(fig.: criticize)* heruntermachen *(ugs.)*

flea [fli:] *n.* Floh, *der;* **send sb. away** *or* **off with a ~ in his/her ear** *(fig. coll.)* jmdn. abblitzen lassen *(ugs.)*

'flea-bite *n.* Flohbiß, *der;* **it's just a ~** *(fig.)* es ist nur eine Kleinigkeit *od. (ugs.)* ein Klacks

fleck [flek] **1.** *n.* **a)** Tupfen, *der; (small)* Punkt, *der; (blemish on skin)* Fleck, *der;* **b)** *(speck)* Flocke, *die.* **2.** *v. t.* sprenkeln

fled *see* **flee**

fledg[e]ling ['fledʒlɪŋ] *n.* Jungvogel, *der*

flee [fli:] **1.** *v. i.*, **fled** [fled] fliehen; **~ from sth./sb.** aus etw./vor jmdm. flüchten *od.* fliehen. **2.** *v. t.*, **fled** fliehen aus; **~ the country** aus dem Land fliehen *od.* flüchten

fleece [fli:s] **1.** *n.* Vlies, *das;* [Schaf]fell, *das.* **2.** *v. t. (fig.)* ausplündern; *(charge excessively)* neppen *(ugs. abwertend)*

fleecy ['fli:sɪ] *adj.* flauschig; **~ cloud** Schäfchenwolke, *die*

fleet [fli:t] *n.* **a)** *(Navy)* Flotte, *die;* **b)** *(in operation together) (vessels)* Flotte, *die; (aircraft)* Geschwader, *das;* **c)** *(under same ownership)* Flotte, *die (fig.);* **he owns a ~ of cars** ihm gehört ein ganzer Wagenpark

fleeting ['fli:tɪŋ] *adj.* flüchtig; vergänglich ‹*Natur, Schönheit*›; **~ visit** Stippvisite, *die (ugs.)*

Flemish ['flemɪʃ] **1.** *adj.* flämisch. **2.** *n.* Flämisch, *das; see also* **English 2 a**

flesh [fleʃ] *n., no pl., no indef. art.* **a)** Fleisch, *das;* ~ **and blood** Fleisch und Blut; **b)** *(of fruit, plant)* [Frucht]fleisch, *das;* **c)** *(fig.: body)* Fleisch, *das (geh.);* **go the way of all** ~: den Weg allen Fleisches gehen *(geh.)*

flesh-wound ['fleʃwu:nd] *n.* Fleischwunde, *die*

fleshy ['fleʃɪ] *adj. (fat, boneless)* fett; fleischig ‹*Hände*›

flew *see* **²fly 1, 2**

¹flex [fleks] *n. (Brit. Electr.)* Kabel, *das*

²flex *v. t.* **a)** *(Anat.)* beugen ‹*Arm, Knie*›; **b)** ~ **one's muscles** *(lit. or fig.)* seine Muskeln spielen lassen

flexibility [fleksɪ'bɪlɪtɪ] *n., no pl.* **a)** Biegsamkeit, *die;* Elastizität, *die;* **b)** *(fig.)* Flexibilität, *die*

flexible ['fleksɪbl] *adj.* **a)** biegsam; elastisch; **b)** *(fig.)* flexibel; dehnbar ‹*Vorschriften*›; schwach ‹*Wille*›; ~ **working hours** *or* **time** gleitende Arbeitszeit

flexitime ['fleksɪtaɪm] *(Brit.),* **flextime** ['flekstaɪm] *(Amer.) ns.* Gleitzeit, *die*

flick [flɪk] **1.** *n.* ~ **of the wrist** kurze, schnelle Drehung des Handgelenks; **a** ~ **of the switch** ein einfaches Klicken des Schalters. **2.** *v. t.* schnippen; anknipsen ‹*Schalter*›; verspritzen ‹*Tinte*›; ~ **one's fingers** mit den Fingern schnipsen

~ **through** *v. t.* durchblättern

flicker ['flɪkə(r)] **1.** *v. i.* flackern; ‹*Fernsehapparat:*› flimmern; **a smile ~ed round her lips** ein Lächeln spielte um ihre Lippen. **2.** *n.* Flackern, *das; (of TV)* Flimmern, *das; (fig.)* Aufflackern, *das; (of smile)* Anflug, *der*

'flick-knife *n. (Brit.)* Schnappmesser, *das*

¹flight [flaɪt] *n.* **a)** *(flying)* Flug, *der;* **in** ~: im Flug; **b)** *(journey)* Flug, *der; (migration of birds)* Zug, *der;* **c)** ~ **[of stairs** *or* **steps]** Treppe, *die;* **d)** *(flock of birds)* Schwarm, *der;* **e)** *(Air Force)* ≈ Staffel, *die;* **in the first** *or* **top** ~ *(fig.)* in der Spitzengruppe

²flight *n. (fleeing)* Flucht, *die;* **take [to]** ~: die Flucht ergreifen

flight: ~ **attendant** *n.* Flugbegleiter, *der/-begleiterin, die;* ~**-deck** *n.* **a)** *(of aircraft-carrier)* Flugdeck, *das;* **b)** *(of aircraft)* Cockpit, *das;* ~ **path** *n. (Aeronaut.)* Flugweg, *der;* ~**-recorder** *n.* Flugschreiber, *der*

flimsy ['flɪmzɪ] *adj.* **a)** dünn; fadenscheinig ‹*Kleidung, Vorhang*›; nicht [sehr] haltbar ‹*Verpackung*›; **b)** *(fig.)* fadenscheinig *(abwertend)* ‹*Entschuldigung, Argument*›

flinch [flɪntʃ] *v. i.* **a)** zurückschrecken; ~ **from sth.** vor einer Sache zurückschrecken; ~ **from one's responsibilities** sich seinen Pflichten entziehen; **b)** *(wince)* zusammenzucken

fling [flɪŋ] **1.** *n.* **a)** *(fig.: attempt)* **have a** ~ **at sth., give sth. a** ~: es mit etw. versuchen; **b)** *(fig.: indulgence)* **have one's** ~: sich ausleben. **2.** *v. t.,* **flung** [flʌŋ] **a)** werfen; ~ **back one's head** den Kopf zurückwerfen; ~ **sth. away** *(lit. or fig.)* etw. fortwerfen; ~ **down the money** das Geld hinschmeißen *(ugs.);* ~ **on one's jacket** [sich *(Dat.)*] die Jacke überwerfen; **b)** *(fig.)* ~ **sb. into jail** jmdn. ins Gefängnis werfen; ~ **caution to the winds/**~ **aside one's scruples** alle Vorsicht/seine Skrupel über Bord werfen. **3.** *v. refl.,* **flung a)** ~ **oneself at sb.** sich auf jmdn. stürzen; ~ **oneself in front of/upon** *or* **on to sth.** sich vor/auf etw. *(Akk.)* werfen; **b)** *(fig.)* ~ **oneself into sth.** sich in etw. *(Akk.)* stürzen

flint [flɪnt] *n.* Feuerstein, *der;* Flint, *der (veralt.)*

flip [flɪp] **1.** *n.* Schnipsen, *das;* **give sth. a** ~: etw. hochschnipsen. **2.** *v. t.,* **-pp-** schnipsen; ~ **[over]** *(turn over)* umdrehen. **3.** *v. i. (sl.)* ausflippen *(ugs.)*

~ **through** *v. t.* durchblättern

flippant ['flɪpənt] *adj.* unernst; leichtfertig

flipper ['flɪpə(r)] *n.* Flosse, *die*

flipping ['flɪpɪŋ] *(Brit. sl.) adj., adv.* verdammt *(salopp)*

'flip side *n.* B-Seite, *die*

flirt [flɜ:t] **1.** *n.* **he/she is just a** ~: er/sie will nur flirten. **2.** *v. i.* **a)** ~ **[with sb.]** [mit jmdm.] flirten; **b)** *(fig.)* ~ **with sth.** mit etw. liebäugeln; ~ **with danger/death** die Gefahr [leichtfertig] herausfordern/mit dem Leben spielen

flirtation [flɜ:'teɪʃn] *n.* Flirt, *der*

flirtatious [flɜ:'teɪʃəs] *adj.* kokett ‹*Blick, Art*›

flit [flɪt] *v. i.,* **-tt-** huschen; **recollections/thoughts ~ted through his mind** Erinnerungen/Gedanken schossen ihm durch den Kopf

float [fləʊt] **1.** *v. i.* **a)** *(on water)* treiben; ~ **away** wegtreiben; **b)** *(through air)* schweben; ~ **across sth.** ‹*Wolke, Nebel:*› über etw. *(Akk.)* ziehen; **c)** *(fig.)* ~ **about** *or* **[a]round** umgehen; im Umlauf sein; **d)** *(sl.: move casually)* ~ **[around** *or* **about]** herumziehen *(ugs.);* **e)** *(Finance)* floaten. **2.** *v. t.* **a)** *(convey by water, on rafts)* flößen; *(set afloat)* flott machen ‹*Schiff*›; **b)** *(fig.: circulate)* in Umlauf bringen; **c)** *(Finance)* floaten lassen; freigeben; **d)** *(Commerc.)* ausgeben, auf den Markt bringen ‹*Aktien*›; gründen ‹*Unternehmen*›; lancieren ‹*Plan, Idee*›. **3.** *n.* **a)** *(for carnival)* Festwagen, *der; (Brit.: delivery cart)* Wagen, *der;* **b)** *(petty cash)* Bargeld, *das;* **c)** *(Angling)* Floß, *das (fachspr.);* Schwimmer, *der*

floating ['fləʊtɪŋ] *adj.* treibend; schwimmend ‹*Hotel*›

floating 'voter *n.* Wechselwähler, *der/-wählerin, die*

flock [flɒk] **1.** *n.* **a)** *(of sheep, goats; also Eccl.)* Herde, *die; (of birds)* Schwarm, *der;* **b)** *(of people)* Schar, *die.* **2.** *v. i.* strömen; ~ **round sb.** sich um jmdn. scharen; ~ **in/out/together** [in Scharen] hinein-/heraus-/zusammenströmen

floe [fləʊ] *n.* Eisscholle, *die*

flog [flɒg] *v. t.,* **-gg-: a)** auspeitschen; ~ **a dead horse** *(fig.)* seine Kraft und Zeit verschwenden; ~ **sth. to death** *(fig.)* etw. zu Tode reiten; **b)** *(Brit. sl.: sell)* verscheuern *(salopp)*

flood [flʌd] **1.** *n.* **a)** Überschwemmung, *die;* **the F~** *(Bibl.)* die Sintflut; *attrib.* ~ **area** Überschwemmungsgebiet, *das;* **b)** *(of tide)* Flut, *die.* **2.** *v. i.* **a)** ‹*Fluß:*› über die Ufer treten; **there's danger of ~ing** es besteht Überschwemmungsgefahr; **b)** *(fig.)* strömen. **3.** *v. t.* **a)** überschwemmen; *(deluge)* unter Wasser setzen; **the cellar was ~ed** der Keller stand unter Wasser; **b)** *(fig.)* überschwemmen; **~ed with light** lichtdurchflutet

flood: ~**gate** *n. (Hydraulic Engin.)* Schütze, *die;* **open the ~gates to sth.** *(fig.)* einer Sache *(Dat.)* Tür und Tor öffnen; ~**light 1.** *n.* Scheinwerfer, *der; (illumination in a broad beam)* Flutlicht, *das;* **2.** *v. t.,* **~lit** ['flʌdlɪt] anstrahlen ‹*Bauwerk*›; beleuchten ‹*Weg, Straße*›; ~ **water** *n.* Hochwasser, *das; (in motion)* anflutendes Wasser

floor [flɔ:(r)] **1.** *n.* **a)** Boden, *der; (of room)* [Fuß]boden, *der;* **take the** ~ *(dance)* sich aufs Parkett begeben (*see also* **c**); **b)** *(storey)* Stockwerk, *das;* **first** ~ *(Amer.)* Erdgeschoß, *das;* **first** ~ *(Brit.),*

second ~ *(Amer.)* erster Stock; **ground** ~: Erdgeschoß, *das;* Parterre, *das;* **c)** *(in debate, meeting)* Sitzungssaal, *der; (Parl.)* Plenarsaal, *der;* **be given** *or* **have the** ~: das Wort haben; **take the** ~ *(Amer.: speak)* das Wort ergreifen (*see also* **a**). **2.** *v. t.* **a)** *(confound)* überfordern; *(overcome, defeat)* besiegen; **b)** *(knock down)* zu Boden schlagen *od.* strecken

floor: ~**board** *n.* Dielenbrett, *das;* ~**-cloth** *n. (Brit.)* Scheuertuch, *das*

flooring ['flɔ:rɪŋ] *n.* Fußboden[belag], *der*

floor: ~**-polish** *n.* Bohnerwachs, *das;* ~ **show** *n.* ≈ Unterhaltungsprogramm, *das*

floozie (floosie) ['flu:zɪ] *n. (coll.)* Flittchen, *das (ugs. abwertend)*

flop [flɒp] **1.** *v. i.*, **-pp-: a)** plumpsen; **she** ~**ped into a chair** sie ließ sich in einen Sessel plumpsen; **b)** *(coll.: fail)* fehlschlagen; ein Reinfall sein *(ugs.);* ‹*Theaterstück, Show:*› durchfallen. **2.** *n. (coll.: failure)* Reinfall, *der (ugs.);* Flop, *der (ugs.)*

floppy ['flɒpɪ] *adj.* weich und biegsam; ~ **disc** *see* **disc c;** ~ **ears/hat** Schlappohren/Schlapphut, *der*

flora ['flɔ:rə] *n., pl.* ~**e** ['flɔ:ri:] *or* ~**s** Flora, *die*

floral ['flɔ:rl, 'flɒrl] *adj.* geblümt ‹*Kleid, Stoff, Tapete*›; Blumen‹*gesteck, -arrangement, -muster*›

Florence ['flɒrəns] *pr. n.* Florenz *(das)*

florid ['flɒrɪd] *adj.* **a)** *(over-ornate)* schwülstig *(abwertend);* blumig ‹*Stil, Redeweise*›; **b)** *(high-coloured)* gerötet ‹*Teint*›

florist ['flɒrɪst] *n.* Florist, *der*/ Floristin, *die;* ~**'s [shop]** Blumenladen, *der*

flotilla [flə'tɪlə] *n.* Flottille, *die*

flotsam ['flɒtsəm] *n.* ~ **[and jetsam]** Treibgut, *das*

flounce [flaʊns] *v. i.* stolzieren

[1]**flounder** ['flaʊndə(r)] *v. i.* taumeln; *(stumble, lit. or fig.)* stolpern

[2]**flounder** *n. (Zool.)* Flunder, *die*

flour ['flaʊə(r)] *n.* Mehl, *das*

flourish ['flʌrɪʃ] **1.** *v. i.* **a)** gedeihen; ‹*Handel, Geschäft:*› florieren, gutgehen; **b)** *(be active)* seine Blütezeit erleben *od.* haben. **2.** *v. t.* schwingen. **3.** *n.* **a) do sth. with a** ~: etw. schwungvoll *od.* mit einer schwungvollen Bewegung tun; **b)** *(in writing)* Schnörkel, *der;* **c)** *(Mus.: fanfare)* Fanfare, *die*

flout [flaʊt] *v. t.* mißachten; sich hinwegsetzen über (+ *Akk.*) ‹*Ratschlag, Wunsch, öffentliche Meinung*›

flow [fləʊ] **1.** *v. i.* **a)** fließen; ‹*Körner, Sand:*› rinnen, rieseln; ‹*Gas:*› strömen; **the river** ~**ed over its banks** der Fluß trat über die Ufer; **b)** *(fig.)* fließen; ‹*Personen:*› strömen; **keep the traffic** ~**ing smoothly** den Verkehr fließend halten; **c)** *(abound)* ~ **freely** *or* **like water** reichlich *od.* in Strömen fließen; **d)** ~ **from** *(be derived from)* sich ergeben aus. **2.** *n.* **a)** Fließen, *das; (progress)* Fluß, *der; (volume)* Durchflußmenge, *die;* ~ **of water/people** Wasser-/Menschenstrom, *der;* ~ **of electricity/information/conversation** Strom-/Informations-/ Gesprächsfluß, *der;* **b)** *(of tide, river)* Flut, *die*

~ **a'way** *v. i.* abfließen

'**flow chart** *n.* Flußdiagramm, *das*

flower ['flaʊə(r)] **1.** *n.* **a)** *(blossom)* Blüte, *die; (plant)* Blume, *die;* **come into** ~: zu blühen beginnen; **b)** *no pl. (fig.: best part)* Zierde, *die;* **in the** ~ **of youth** in der Blüte der Jugend. **2.** *v. i.* blühen; *(fig.)* erblühen (**into** zu)

flower: ~ **arrangement** *n.* Blumenarrangement, *das; (smaller also)* Gesteck, *das;* ~**-bed** *n.* Blumenbeet, *das*

flowered [flaʊəd] *adj.* geblümt ‹*Stoff, Teppich, Tapete*›; **purple-**~: purpurblühend ‹*Pflanze*›

flower: ~**-garden** *n.* Blumengarten, *der;* ~**pot** *n.* Blumentopf, *der;* ~**-shop** *n.* Blumenladen, *der;* ~**-show** *n.* Blumenschau, *die*

flowery ['flaʊərɪ] *adj.* geblümt ‹*Stoff, Muster*›; blumig ‹*Duft, Wein*›; *(fig.)* blumig ‹*Sprache, Ausdruck*›

flowing ['fləʊɪŋ] *adj.* fließend; wallend ‹*Haar, Bart, Gewand*›

flown *see* [2]**fly 1, 2**

flu [flu:] *n. (coll.)* Grippe, *die;* **get** *or* **catch [the]** ~: Grippe bekommen

fluctuate ['flʌktjʊeɪt] *v. i.* schwanken

fluctuation [flʌktjʊ'eɪʃn] *n.* Schwankung, *die;* Fluktuation, *die (bes. Wirtsch., Soziol.)*

flue [flu:] *n.* **a)** *(in chimney)* Rauchabzug, *der;* **b)** *(for passage of hot air)* Luftkanal, *der*

fluency ['flu:ənsɪ] *n.* Gewandtheit, *die; (in speaking)* Redegewandtheit, *die*

fluent ['flu:ənt] *adj.* gewandt ‹*Stil, Redeweise, Redner, Schreiber, Erzähler*›; **be** ~ **in Russian, speak** ~ **Russian, be a** ~ **speaker of Russian** fließend Russisch sprechen

fluff [flʌf] **1.** *n.* Flusen; Fusseln; *(on birds, rabbits, etc.)* Flaum, *der.* **2.** *v. t.* **a) the bird** ~**ed itself/ its feathers [up]** der Vogel plusterte sich/seine Federn auf; **b)** *(sl.: bungle)* verpatzen *(ugs.)*

fluffy ['flʌfɪ] *adj.* [flaum]weich ‹*Kissen, Küken, Haar*›; flauschig ‹*Spielzeug, Stoff, Decke*›; locker ‹*Omelett*›; schaumig ‹*Eiweiß*›

fluid ['flu:ɪd] **1.** *n.* **a)** *(liquid)* Flüssigkeit, *die;* **b)** *(liquid or gas)* Fluid, *das (Technik, Chemie).* **2.** *adj.* **a)** *(liquid)* flüssig; **b)** *(liquid or gaseous)* fluid *(Technik, Chemie);* **c)** *(fig.)* ungewiß, unklar ‹*Lage*›

fluke [flu:k] *n.* *(piece of luck)* Glücksfall, *der;* **by a** *or* **some [pure]** ~: [nur] durch einen glücklichen Zufall

fluky ['flu:kɪ] *adj.* glücklich ‹*Zufall, Zusammentreffen, Sieg*›; zufällig ‹*Ergebnis, Relikt*›; Zufalls‹*treffer, -ergebnis*›

flung *see* **fling 2, 3**

flunkey, flunky ['flʌŋkɪ] *n. (usu. derog.)* Lakai, *der (abwertend)*

fluorescent [flu:ə'resənt] *adj.* fluoreszierend; ~ **material** Leuchtstoff, *der; (fabric)* fluoreszierendes Material

fluorescent: ~ **'lamp,** ~ **'light** *ns.* Leuchtstofflampe, *die (Elektrot.);* ≈ Neonlampe, *die*

fluoride ['flu:əraɪd] *n.* Fluorid, *das;* ~ **toothpaste** fluorhaltige Zahnpasta

flurry ['flʌrɪ] **1.** *n.* **a)** Aufregung, *die;* **there was a sudden** ~ **of activity** es herrschte plötzlich rege Betriebsamkeit; **b)** *(of rain/snow)* [Regen-/Schnee]schauer, *der.* **2.** *v. t.* durcheinanderbringen; **don't let yourself be flurried** laß dich nicht nervös *od. (ugs.)* verrückt machen

[1]**flush** [flʌʃ] **1.** *v. i.* rot werden; erröten (**with** vor + *Dat.*). **2.** *v. t.* ausspülen ‹*Becken*›; durch-, ausspülen ‹*Rohr*›; ~ **the toilet** *or* **lavatory** spülen. **3.** *n.* **a)** *(blush)* Erröten, *das;* **hot** ~**es** Hitzewallungen; **b)** *(elation)* **in the [first]** ~ **of victory** *or* **conquest** im [ersten] Siegestaumel

[2]**flush** *adj.* **a)** *(level)* bündig; **be** ~ **with sth.** mit etw. bündig abschließen; **b)** *usu. pred. (plentiful)* reichlich vorhanden *od.* im Umlauf ‹*Geld*›; **be** ~ **[with money]** gut bei Kasse sein *(ugs.)*

flushed [flʌʃt] *adj.* gerötet ‹*Wan-*

gen, Gesicht⟩; ~ **with pride** vor Stolz glühend
flush 'out *v. t.* aufscheuchen *(fig.)* ⟨*Spion, Verbrecher*⟩
fluster ['flʌstə(r)] *v. t.* aus der Fassung bringen
flustered ['flʌstəd] *adj.* **be/become** ~: nervös sein/werden
flute [flu:t] *n. (Mus.)* Flöte, *die*
flutter ['flʌtə(r)] **1.** *v. i.* **a)** ⟨*Vogel, Motte, Papier, Vorhang, Fahne, Segel, Drachen, Flügel:*⟩ flattern; ⟨*Blumen, Gräser usw.:*⟩ schaukeln; **b)** *(beat abnormally)* ⟨*Herz:*⟩ schneller *od.* höher schlagen. **2.** *v. t.* flattern mit ⟨*Flügel*⟩; ~ **one's eyelashes** mit den Wimpern klimpern; ~ **one's eyelashes at sb.** jmdm. mit den Wimpern zuklimpern. **3.** *n.* **a)** Flattern, *das;* **b)** *(fig.) (stir)* [leichte] Unruhe; *(nervous state)* Aufregung, *die;* **c)** *(Brit. sl.: bet)* Wette, *die;* **have a** ~: ein paar Scheinchen riskieren *(ugs.)*
flux [flʌks] *n. (change)* **be in a state of** ~: im Fluß sein; sich verändern
[1]**fly** [flaɪ] *n. (Zool.)* Fliege, *die;* **the only** ~ **in the ointment** *(fig.)* der einzige Haken [bei der Sache] *(ugs.);* **he wouldn't hurt a** ~ *(fig.)* er kann keiner Fliege etwas zuleide tun; **[there are] no flies on him** *(fig. sl.)* ihm kann man nichts vormachen *(ugs.)*
[2]**fly 1.** *v. i.*, **flew** [flu:], **flown** [fləʊn] **a)** fliegen; ~ **about/away** *or* **off** umher-/weg- *od.* davonfliegen; **b)** *(float, flutter)* fliegen; **rumours are ~ing about** *(fig.)* es gehen Gerüchte um; **c)** *(move quickly)* fliegen; **come ~ing towards sb.** jmdm. entgegengeflogen kommen; ~ **open** auffliegen; **knock** *or* **send sb./sth. ~ing** jmdn./etw. umstoßen; ~ **into a temper** *or* **rage** einen Wutanfall bekommen; **d)** *(fig.)* ~ **[by** *or* **past]** wie im Fluge vergehen; **how time flies!, doesn't time ~!** wie die Zeit vergeht!; **e)** ⟨*Fahne:*⟩ gehißt sein; **f)** *(attack angrily)* ~ **at sb.** *(lit. or fig.)* über jmdn. herfallen; **let** ~: zuschlagen; *(fig.: use strong language)* losschimpfen; **let** ~ **with** abschießen ⟨*Pfeil, Rakete, Gewehr*⟩; werfen ⟨*Stein*⟩; **g)** *(flee)* fliehen; *(coll.: depart hastily)* eilig aufbrechen; **I really must** ~ *(coll.)* jetzt muß ich aber schnell los. **2.** *v. t.*, **flew, flown a)** fliegen ⟨*Flugzeug, Fracht, Einsatz*⟩; fliegen über (+ *Akk.*) ⟨*Strecke*⟩; *(travel over)* überfliegen; überqueren; ~ **Concorde/Lufthansa** mit der Concorde/mit Lufthansa fliegen; **b)** führen ⟨*Flagge*⟩; ~ **a kite** einen Drachen steigen lassen. **3.** *n. in sing. or pl. (on trousers)* Hosenschlitz, *der*
~ **'in 1.** *v. i. (arrive in aircraft)* [mit dem Flugzeug] eintreffen **(from** aus); *(come in to land)* landen. **2.** *v. t.* landen ⟨*Flugzeug*⟩; *(bring by aircraft)* einfliegen
~ **'off** *v. i.* **a)** abfliegen; **b)** *(become detached)* abgehen; ⟨*Hut:*⟩ wegfliegen
~ **'out 1.** *v. i.* abfliegen **(of** von). **2.** *v. t.* ausfliegen
'fly-by-night 1. *adj.* zwielichtig. **2.** *n. jmd., der sich nachts heimlich aus dem Staub macht*
flyer ['flaɪə(r)] *n.* **a)** *(pilot)* Flieger, *der*/Fliegerin, *die;* **b)** *(handbill)* Handzettel, *der*
flying ['flaɪɪŋ] **1.** *adj.* Kurz-; ~ **visit** Stippvisite, *die (ugs.).* **2.** *n.* Fliegen, *das; attrib.* Flug⟨*wetter, -zeit, -geschwindigkeit, -erfahrung*⟩
flying: ~ **bomb** *n.* V-Waffe, *die;* ~ **fish** *n.* fliegender Fisch; ~ **'jump,** ~ **'leap** *ns.* Sprung mit Anlauf; großer Satz *(ugs.);* ~ **machine** *n.* Luftfahrzeug, *das;* Flugmaschine, *die (veralt.);* ~ **'saucer** *n.* fliegende Untertasse; ~ **'start** *n. (Sport)* fliegender Start; **get off to** *or* **have a** ~ **start** *(fig.)* einen glänzenden Start haben
fly: ~**leaf** *n.* Vorsatzblatt, *das;* ~**over** *n. (Brit.)* [Straßen]überführung, *die;* Fly-over, *der;* ~**-past** *n.* Luftparade, *die;* ~**-spray** *n.* Insektenspray, *der od. das;* ~**-swatter** *n.* Fliegenklatsche, *die;* ~**weight** *n. (Boxing etc.)* Fliegengewicht, *das; (person also)* Fliegengewichtler, *der;* ~**wheel** *n.* Schwungrad, *das*
FM *abbr.* **frequency modulation** FM
foal [fəʊl] *n.* Fohlen, *das*
foam [fəʊm] **1.** *n.* **a)** Schaum, *der;* **b)** *see* **foam rubber. 2.** *v. i. (lit. or fig.)* schäumen **(with** vor + *Dat.*); ~ **at the mouth** Schaum vorm Mund haben; *(fig. coll.)* [vor Wut] schäumen
foam: ~ **bath** *n.* Schaumbad, *das;* ~ **'rubber** *n.* Schaumgummi, *der*
fob [fɒb] *v. t.*, **-bb-:** ~ **sb. off with sth.** jmdn. mit etw. abspeisen *(ugs.)*
focal ['fəʊkl]: ~ **'distance,** ~ **'length** *ns.* Brennweite, *die*
foc's'le ['fəʊksl] *see* **forecastle**
focus ['fəʊkəs] **1.** *n., pl.* ~**es** *or* **foci** ['fəʊsaɪ] **a)** *(Optics, Photog.)* Brennpunkt, *der; (focal length)* Brennweite, *die; (adjustment of eye or lens)* Scharfeinstellung, *die;* **out of/in** ~: unscharf/scharf eingestellt ⟨*Kamera, Teleskop*⟩; unscharf/scharf ⟨*Foto, Film, Vordergrund usw.*⟩; **get sth. in** ~ *(fig.)* etw. klarer erkennen; **b)** *(fig.: centre, central object)* Mittelpunkt, *der;* **be the** ~ **of attention** im Brennpunkt des Interesses stehen. **2.** *v. t.*, **-s-** *or* **-ss-: a)** *(Optics, Photog.)* einstellen **(on** auf + *Akk.*); ~ **one's eyes on sth./sb.** die Augen auf etw./jmdn. richten; **b)** *(concentrate)* bündeln ⟨*Licht, Strahlen*⟩; *(fig.)* konzentrieren **(on** auf + *Akk.*). **3.** *v. i.*, **-s-** *or* **-ss-: a) the camera ~es automatically** die Kamera hat automatische Scharfeinstellung; **b)** ⟨*Licht, Strahlen:*⟩ sich bündeln; *(fig.)* sich konzentrieren **(on** auf + *Akk.*)
fodder ['fɒdə(r)] *n.* [Vieh]futter, *das*
foe [fəʊ] *n. (poet./rhet.)* Feind, *der*
foetal ['fi:tl] *adj.* fötal; fetal
foetid ['fi:tɪd] *see* **fetid**
foetus ['fi:təs] *n.* Fötus, *der;* Fetus, *der*
fog [fɒg] *n.* Nebel, *der;* **be in a [complete]** ~ *(fig.)* [völlig] verunsichert sein
'fog-bound *adj.* **a)** *(surrounded)* in Nebel gehüllt; **b)** *(immobilized)* durch Nebel festgehalten
foggy ['fɒgɪ] *adj.* **a)** neblig; **b)** *(fig.)* nebelhaft ⟨*Vorstellung, Sprache, Bewußtsein*⟩; **[I] haven't the foggiest [idea** *or* **notion]** *(coll.)* [ich] hab' keinen blassen Schimmer *(ugs.)*
fog: ~**-horn** *n. (Naut.)* Nebelhorn, *das;* ~**-lamp,** ~**-light** *ns. (Motor Veh.)* Nebelscheinwerfer, *der*
fogy ['fəʊgɪ] *n.* **[old]** ~: [alter *od.* rückständiger] Opa *(salopp)*/[alte *od.* rückständige] Oma *(salopp)*
foible ['fɔɪbl] *n.* Eigenheit, *die*
[1]**foil** [fɔɪl] *n.* **a)** *(metal as thin sheet)* Folie, *die;* **b)** *(to wrap or cover food etc.)* Folie, *die;* **c)** *(sb./sth. contrasting)* ≈ Kontrast, *der*
[2]**foil** *v. t.* vereiteln ⟨*Versuch, Plan, Flucht*⟩; durchkreuzen ⟨*Vorhaben, Plan*⟩
[3]**foil** *n. (sword)* Florett, *das*
foist [fɔɪst] *v. t.* ~ **[off] on to** *or* **[up]on sb.** jmdm. andrehen *(ugs.)* ⟨*schlechte Waren*⟩; jmdm. zuschieben ⟨*Schuld, Verantwortung*⟩; auf jmdn. abwälzen ⟨*Probleme, Verantwortung*⟩; ~ **oneself on sb.** sich jmdm. aufdrängen
fold [fəʊld] **1.** *v. t.* **a)** *(double over on itself)* [zusammen]falten; zu-

sammenlegen ‹*Laken, Wäsche*›; **b)** *(embrace)* ~ **sb. in one's arms** jmdn. in die Arme schließen; **c)** ~ **one's arms** die Arme verschränken; **d)** *(envelop)* ~ **sth./sb. in sth.** etw./jmdn. in etw. *(Akk.)* einhüllen. **2.** *v. i.* **a)** *(become ~ed)* sich zusammenlegen; sich zusammenfalten; **b)** *(collapse)* zusammenklappen; *(go bankrupt)* Konkurs *od.* Bankrott machen; **c)** *(be able to be ~ed)* sich falten lassen; **it ~s easily** es ist leicht zu falten. **3.** *n.* **a)** *(doubling)* Falte, *die;* **b)** *(act of ~ing)* Faltung, *die;* **c)** *(line made by ~ing)* Kniff, *der*

~ **a'way 1.** *v. t.* zusammenklappen. **2.** *v. i.* zusammenklappbar sein; sich zusammenklappen lassen

~ **'back 1.** *v. t.* zurückschlagen, aufschlagen ‹*Laken*›; zurückklappen ‹*Rücksitz*›; umknicken ‹*Papier*›. **2.** *v. i.* sich zurückschlagen lassen

~ **'down 1.** *v. t.* zusammenklappen; *(~ back)* zurückschlagen. **2.** *v. i.* sich zusammenklappen lassen; *(~ back)* sich zurückschlagen lassen

~ **'out** *v. i.* ‹*Landkarte:*› sich auseinanderfalten lassen; ‹*Tisch:*› sich hochklappen lassen

~ **'up 1.** *v. t.* **a)** zusammenfalten; zusammenlegen ‹*Laken, Wäsche*›; **b)** zusammenklappen ‹*Stuhl, Tisch usw.*›. **2.** *v. i.* **a)** sich zusammenfalten lassen; **b)** ‹*Stuhl, Tisch usw.:*› sich zusammenklappen lassen

folder ['fəʊldə(r)] *n.* Mappe, *die*

folding: ~ 'door *n.* Falttür, *die;* ~ **'doors** *n. pl.* Falttür, *die; (of hangar, barn, etc.)* Falttor, *das*

foliage ['fəʊlɪɪdʒ] *n., no pl. (leaves)* Blätter *Pl.; (of tree also)* Laub, *das*

folk [fəʊk] *n., pl. same or ~s* **a)** *(a people)* Volk, *das;* **b)** *in pl.* **~[s]** *(people)* Leute *Pl.; (people in general)* die Leute; **[the] rich/poor ~:** die Reichen/Armen; **old ~[s]** alte Leute; **c)** *in pl.* **~s** *(coll., as address: people, friends)* Leute *Pl. (ugs.);* **d)** *in pl.* **~s** *(coll.: one's relatives)* Verwandte *Pl.;* Leute *Pl. (ugs.);* **e)** *attrib. (of the people, traditional)* Volks-

folk: ~-dance *n.* Volkstanz, *der;* **~lore** *n.* **a)** *(traditional beliefs)* [volkstümliche] Überlieferung; Folklore, *die;* **b)** *(study)* Volkskunde, *die;* Folklore, *die;* **~-music** *n.* Volksmusik, *die;* **~-song** *n.* Volkslied, *das; (modern)* Folksong, *der;* **~-tale** *n.* Volksmärchen, *das*

follow ['fɒləʊ] **1.** *v. t.* **a)** folgen (+ *Dat.*); **you're being ~ed** Sie werden verfolgt; **b)** *(go along)* folgen (+ *Dat.*); entlanggehen/-fahren ‹*Straße usw.*›; **c)** *(come after in order or time)* folgen (+ *Dat.*); folgen auf (+ *Akk.*); **A is ~ed by B** auf A folgt B; **d)** *(accompany)* [nach]folgen (+ *Dat.*); **e)** *(provide with sequel)* ~ **sth. with sth.** einer Sache *(Dat.)* etw. folgen lassen; **f)** *(result from)* die Folge sein von; hervorgehen aus; **g)** *(treat or take as guide or leader)* folgen (+ *Dat.*); sich orientieren an (+ *Dat.*); *(adhere to)* anhängen (+ *Dat.*); **h)** *(act according to)* folgen (+ *Dat.*) ‹*Prinzip, Instinkt, Trend*›; verfolgen ‹*Politik*›; befolgen ‹*Vorschrift, Regel, Anweisung, Rat, Warnung*›; handeln nach ‹*Gefühl, Wunsch*›; sich halten an (+ *Akk.*) ‹*Konventionen, Diät, Maßstab*›; **i)** *(keep up with mentally, grasp meaning of)* folgen (+ *Dat.*); **do you ~ me?, are you ~ing me?** verstehst du, was ich meine?; **j)** *(be aware of the present state or progress of)* verfolgen ‹*Ereignisse, Nachrichten, Prozeß*›. **2.** *v. i.* **a)** *(go, come)* ~ **after sb./sth.** jmdm./einer Sache folgen; **b)** *(go or come after person or thing)* folgen; ~ **in the wake of sth.** etw. ablösen; auf etw. *(Akk.)* folgen; **c)** *(come next in order or time)* folgen; **as ~s** wie folgt; **d)** ~ **from sth.** *(result)* die Folge von etw. sein; *(be deducible)* aus etw. folgen

~ **'on** *v. i. (continue)* ~ **on from sth.** die Fortsetzung von etw. sein

~ **'through 1.** *v. t.* zu Ende verfolgen; durchziehen *(ugs.)*. **2.** *v. i. (Sport)* durchschwingen

~ **'up** *v. t.* **a)** *(add further action etc. to)* ausbauen ‹*Erfolg, Sieg*›; **b)** *(investigate further)* nachgehen (+ *Dat.*) ‹*Hinweis*›; **c)** *(consider further)* berücksichtigen ‹*Bitte, Angebot*›

follower ['fɒləʊə(r)] *n.* Anhänger, *der*/Anhängerin, *die*

following ['fɒləʊɪŋ] **1.** *adj.* **a)** *(now to be mentioned)* folgend; **in the ~ way** folgendermaßen; **the ~:** folgendes; *(persons)* folgende; **b)** ~ **wind** Rückenwind, *der.* **2.** *prep.* nach. **3.** *n.* Anhängerschaft, *die*

'follow-up *n.* Fortsetzung, *die;* **as a ~:** im Anschluß (**to** an + *Akk.*); *attrib.* ~ **letter/visit** Nachfaßbrief, *der*/-besuch, *der (Werbespr.)*

folly ['fɒlɪ] *n.* **a)** Torheit, *die (geh.);* **it would be [sheer] ~:** es wäre [äußerst] töricht *(geh.);* **b)** *(costly structure considered useless)* nutzloser Prunkbau

foment [fə'ment, fəʊ'ment] *v. t.* schüren

fond [fɒnd] *adj.* **a)** *(tender)* zärtlich; *(affectionate)* liebevoll ‹*Blick*›; lieb ‹*Erinnerung*›; **be ~ of sb.** jmdn. mögen *od.* gern haben; **be ~ of doing sth.** etw. gern tun; **I'm not very ~ of sweets** ich mache mir nicht viel aus Süßigkeiten; **b)** *(foolishly credulous or hopeful)* kühn ‹*Hoffnung, Traum*›; gutgläubig ‹*Person*›; allzu zuversichtlich ‹*Glaube*›

fondant ['fɒndənt] *n.* Fondant, *der od. das*

fondle ['fɒndl] *v. t.* streicheln

fondness ['fɒndnɪs] *n., no pl. (tenderness)* Zärtlichkeit, *die; (affection)* Liebe, *die;* ~ **for sth./doing sth.** *(special liking)* Vorliebe für etw./dafür, etw. zu tun

fondue ['fɒndju:, 'fɒndu:] *n. (Gastr.)* Fondue, *das od. die*

font [fɒnt] *n.* Taufstein, *der*

food [fu:d] *n.* **a)** *no pl., no art.* Nahrung, *die; (for animals)* Futter, *das;* **b)** *no pl., no art. (as commodity)* Lebensmittel *Pl.;* **c)** *no pl. (in solid form)* Essen, *das;* **some ~:** etwas zu essen; **he's very keen on Italian ~:** er mag die italienische Küche; er ißt gern italienisch; **d)** *(particular kind)* Nahrungsmittel, *das;* Kost, *die; (for animals)* Futter, *das;* **canned ~s** Konserven *Pl.;* **e)** *(fig.)* ~ **for thought** Stoff zum Nachdenken

food: ~ poisoning *n.* Lebensmittelvergiftung, *die;* **~-processor** *n.* Küchenmaschine, *die;* **~ shop, ~ store** *ns.* Lebensmittelgeschäft, *das;* **~stuff** *n.* Nahrungsmittel, *das*

fool [fu:l] **1.** *n.* **a)** Dummkopf, *der (ugs.);* **what a ~ I am!** wie dumm von mir!; **be no** *or* **nobody's ~:** nicht dumm *od. (ugs.)* nicht auf den Kopf gefallen sein; **make a ~ of oneself** sich lächerlich machen; **b)** *(Hist.: jester, clown)* Narr, *der;* **c)** *(dupe)* **make a ~ of sb.** jmdn. zum Narren halten. **2.** *v. i.* herumalbern *(ugs.)*. **3.** *v. t.* **a)** *(cheat)* ~ **sb. into doing sth.** jmdn. [durch Tricks] dazu bringen, etw. zu tun; **b)** *(dupe)* täuschen; hereinlegen *(ugs.);* **you could have ~ed me** *(iron.)* ach, was du nicht sagst!

~ **a'bout, ~ a'round** *v. i. (play the ~)* herumalbern *(ugs.); (idle)* herumtrödeln *(ugs.);* ~ **about** *or* **around with sth./sb.** mit etw./jmdm. herumspielen

~ **with** *v. t.* [herum]spielen mit

foolhardy ['fu:lhɑ:dɪ] *adj.* tollkühn ⟨*Handlung, Behauptung, Person*⟩; draufgängerisch ⟨*Person*⟩

foolish ['fu:lɪʃ] *adj.* **a)** töricht; verrückt *(ugs.)* ⟨*Idee, Vorschlag*⟩; **don't do anything ~:** mach keinen Unsinn; **what a ~ thing to do/say** wie kann man nur so etwas Dummes tun/sagen; **b)** *(ridiculous)* albern *(ugs.)* ⟨*Verhalten*⟩; blöd, dumm *(ugs.)* ⟨*Grinsen, Bemerkung*⟩; lächerlich ⟨*Aussehen*⟩

foolishly ['fu:lɪʃlɪ] *adv., as sentence-modifier* törichterweise

fool: ~proof *adj. (not open to misuse)* wasserdicht *(fig.)*; *(infallible)* absolut sicher; *(that cannot break down)* narrensicher *(ugs.)*; **~scap** ['fu:lskæp, 'fu:lzkæp] *n.* **a)** *(size of paper)* Kanzleiformat, *das;* **b)** *(paper of this size)* Kanzleipapier, *das;* **~'s 'paradise** *n.* Traumwelt, *die*

foot [fʊt] **1.** *n., pl.* **feet** [fi:t] **a)** Fuß, *der;* **at sb.'s feet** zu jmds. Füßen; **put one's best ~ forward** *(fig.) (hurry)* sich beeilen; *(do one's best)* sein Bestes tun; **feet first** mit den Füßen zuerst *od.* voran; **go into sth. feet first** *(fig.)* sich Hals über Kopf *(ugs.)* in etw. hineinstürzen; **have one ~ in the grave** *(fig.)* mit einem Fuß im Grabe stehen; **have both [one's] feet on the ground** *(fig.)* mit beiden Beinen [fest] auf der Erde stehen; **on ~:** zu Fuß; **on one's/its feet** *(lit. or fig.)* auf den Beinen; **put one's ~ down** *(fig.) (be firmly insistent or repressive)* energisch werden; *(accelerate motor vehicle)* [Voll]gas geben; **put one's ~ in it** *(fig. coll.)* ins Fettnäpfchen treten *(ugs.)*; **put one's feet up** die Beine hochlegen; **start [off]** *or* **get off on the right/wrong ~** *(fig.)* einen guten/schlechten Start haben; **set ~ in/on sth.** etw. betreten; **be rushed off one's feet** *(fig.)* in Trab gehalten werden *(ugs.)*; **stand on one's own [two] feet** *(fig.)* auf eigenen Füßen stehen; **rise** *or* **get to one's feet** sich erheben; aufstehen; **never put a ~ wrong** *(fig.)* nie etwas falsch machen; **get/have cold feet** kalte Füße kriegen/gekriegt haben *(ugs.)*; **b)** *(far end)* unteres Ende; *(of bed)* Fußende, *das;* *(lowest part)* Fuß, *der;* **at the ~ of the list/page** unten auf der Liste/Seite; **c)** *(of stocking etc.)* Fuß, *der;* Füßling, *der;* **d)** *(Pros.: metrical unit)* [Vers]fuß, *der;* **e)** *pl.* **feet** *or same (linear measure)* Fuß, *der (30,48 cm)*; **7 ~** *or* **feet** 7 Fuß; **f)** *(base)* Fuß, *der;* *(of statue, pillar)* Sockel, *der.* **2.** *v. t. (pay)* **~ the bill** die Rechnung bezahlen

foot-and-'mouth [disease] *n.* Maul- und Klauenseuche, *die*

football ['fʊtbɔ:l] *n. (game, ball)* Fußball, *der*

'**football boot** *n.* Fußballschuh, *der*

footballer ['fʊtbɔ:lə(r)] *n.* Fußballspieler, *der*/-spielerin, *die*

football: ~ pitch *n.* Fußballplatz, *der;* **~ pools** *n. pl.* **the ~ pools** das Fußballtoto

foot: ~-brake *n.* Fußbremse, *die;* **~-bridge** *n.* Steg, *der;* *(across road, railway, etc.)* Fußgängerbrücke, *die;* **~hill** *n., usu. pl.* [Gebirgs]ausläufer, *der;* **~hold** *n.* Halt, *der;* *(fig.)* Stützpunkt, *der;* **get a ~hold** *(fig.)* Fuß fassen

footing ['fʊtɪŋ] *n.* **a)** *(fig.: status)* Stellung, *die;* **be on an equal ~ [with sb.]** [jmdm.] gleichgestellt sein; **place sth. on a firm ~:** etw. auf eine feste Basis stellen; **be on a war ~:** sich im Kriegszustand befinden; **b)** *(foothold)* Halt, *der*

foot: ~lights *n. pl. (Theatre)* Rampenlicht, *das;* **~loose** *adj.* ungebunden; **~loose and fancy-free** frei und ungebunden; **~man** ['fʊtmən] *n., pl.* **~men** ['fʊtmən] Lakai, *der;* Diener, *der;* **~note** *n.* Fußnote, *die;* **~path** *n. (path)* Fußweg, *der;* **~print** *n.* Fußabdruck, *der;* **~prints in the snow** Fußspuren im Schnee; **~-rest** *n.* Fußstütze, *die;* **~step** *n.* Schritt, *der;* **follow** *or* **tread in sb.'s ~steps** *(fig.)* in jmds. Fußstapfen treten; **~stool** *n.* Fußbank, *die;* Fußschemel, *der;* **~wear** *n., no pl., no indef. art.* Schuhe *Pl.*; Schuhwerk, *das;* Fußbekleidung, *die (Kaufmannsspr.)*; **~work** *n., no pl. (Sport, Dancing)* Beinarbeit, *die*

for [fə(r), *stressed* fɔ:(r)] **1.** *prep.* **a)** *(representing, on behalf of, in exchange against)* für; *(in place of)* für; anstelle von; **what is the German ~ 'buzz'?** wie heißt „bzzz" auf Deutsch?; **b)** *(in defence, support, or favour of)* für; **be ~ doing sth.** dafür sein, etw. zu tun; **it's each [man]** *or* **every man ~ himself** jeder ist auf sich selbst gestellt; **c)** *(to the benefit of)* für; **do sth. ~ sb.** für jmdn. etw. tun; **d)** *(with a view to)* für; *(conducive[ly] to)* zu; **they invited me ~ Christmas/Monday/supper** sie haben mich zu Weihnachten/für Montag/zum Abendessen eingeladen; **what is it ~?** wofür/wozu ist das? **be saving up ~ sth.** auf etw. *(Akk.)* sparen; **e)** *(being the motive of)* für; *(having as purpose)* zu; **reason ~ living** Grund zu leben; **a dish ~ holding nuts** eine Schale für Nüsse; **f)** *(to obtain, win, save)* **a request ~ help** eine Bitte um Hilfe; **study ~ a university degree** auf einen Hochschulabschluß hin-studieren; **phone ~ a doctor** nach einem Arzt telefonieren; **take sb. ~ a ride in the car/a walk** jmdn. im Auto spazierenfahren/mit jmdm. einen Spaziergang machen; **work ~ a living** für den Lebensunterhalt arbeiten; **run/jump** *etc.* **~ it** loslaufen/-springen *usw.*; **g)** *(to reach)* nach; **set out ~ England/the north/an island** nach England/Norden/zu einer Insel aufbrechen; **h)** *(to be received by)* für; **that's Jim ~ you** das sieht Jim mal wieder ähnlich; **i)** *(as regards)* **checked ~ accuracy** auf Richtigkeit geprüft; **be dressed/ready ~ dinner** zum Dinner angezogen/fertig sein; **open ~ business** eröffnet; **have sth. ~ breakfast/pudding** etw. zum Frühstück/Nachtisch haben; **enough ... ~:** genug ... für; **that's quite enough ~ me** das reicht mir völlig; **too ... ~:** zu ... für; **there is nothing ~ it but to do sth.** es gibt keine andere Möglichkeit, als etw. zu tun; **j)** *(to the amount of)* **cheque/ bill ~ £5** Scheck/Rechnung über *od.* in Höhe von 5 Pfund; **k)** *(to affect, as if affecting)* für; **things don't look very promising ~ the business** was die Geschäfte angeht, sieht das alles nicht sehr vielversprechend aus; **it is wise/advisable ~ sb. to do sth.** es ist vernünftig/ratsam, daß jmd. etw. tut; **it's hopeless ~ me to try and explain the system** es ist sinnlos, dir das System erklären zu wollen; **l)** *(as being)* für; **what do you take me ~?** wofür hältst du mich?; **I/you** *etc.* **~ one** ich/ du *usw.* für mein[en]/dein[en] *usw.* Teil; **m)** *(on account of, as penalty of)* wegen; **famous/well-known ~ sth.** berühmt/ bekannt wegen *od.* für etw.; **jump/ shout ~ joy** vor Freude in die Luft springen/schreien; **were it not ~ you/ your help, I should not be able to do it** ohne dich/deine Hilfe wäre ich nicht dazu in der Lage; **n)** *(on the occasion of)* **~ the first time** zum ersten Mal; **why can't you help ~ once?** warum kannst du nicht einmal helfen?; **what shall I give him ~ his birthday?** was soll ich ihm zum Geburtstag schenken?;

o) *(in spite of)* ~ **all ...:** trotz ...; ~ **all that, ...:** trotzdem ...; **p)** *(on account of the hindrance of)* vor (+ *Dat.*); ~ **fear of ...:** aus Angst vor (+ *Dat.*); **but** ~ **..., except** ~ **...:** wenn nicht ... gewesen wäre, [dann] ...; **q)** *(so far as concerns)* ~ **all I know/care ...:** möglicherweise/was mich betrifft, ...; ~ **one thing, ...:** zunächst einmal ...; **r)** *(considering the usual nature of)* für; **not bad** ~ **a first attempt** nicht schlecht für den ersten Versuch; **s)** *(during)* seit; **we've/we haven't been here** ~ **three years** wir sind seit drei Jahren hier/nicht mehr hier gewesen; **we waited** ~ **hours/three hours** wir warteten stundenlang/drei Stunden lang; **how long are you here** ~? *(coll.)* wie lange bleiben Sie hier?; **sit here** ~ **now** *or* ~ **the moment** bleiben Sie im Augenblick hier sitzen; **t)** *(to the extent of)* **walk** ~ **20 miles/**~ **another 20 miles** 20 Meilen [weit] gehen/weiter gehen. **2.** *conj. (since, as proof)* denn

forage ['fɒrɪdʒ] **1.** *n. (food for horses or cattle)* Futter, *das.* **2.** *v.i.* auf Nahrungssuche sein; ~ **for sth.** auf der Suche nach etw. sein

foray ['fɒreɪ] *n.* Streifzug, *der; (Mil.)* Ausfall, *der*

forbad, forbade *see* **forbid**

[1]**forbear** ['fɔ:beə(r)] *n., usu. in pl.* Vorfahr, *der*

[2]**forbear** [fɔ:'beə(r)] *v.i.*, **forbore** [fɔ:'bɔ:(r)], **forborne** [fɔ:'bɔ:n] **a)** *(refrain)* ~ **from doing sth.** davon Abstand nehmen, etw. zu tun; **b)** *(be patient)* sich gedulden

forbearance [fɔ:'beərəns] *n., no pl.* Nachsicht, *die*

forbid [fə'bɪd] *v.t.*, **-dd-, forbade** [fə'bæd, fə'beɪd] *or* **forbad** [fə'bæd], **forbidden** [fə'bɪdn] **a)** ~ **sb. to do sth.** jmdm. verbieten, etw. zu tun; ~ **[sb.] sth.** [jmdm.] etw. verbieten; **it is ~den [to do sth.]** es ist verboten *od.* nicht gestattet[, etw. zu tun]; **b)** *(make impossible)* nicht zulassen; nicht erlauben; **God/Heaven** ~ **[that ...]!** Gott/der Himmel bewahre[, daß ...]!

forbidding [fə'bɪdɪŋ] *adj.* furchteinflößend ‹*Aussehen, Stimme*›; unwirtlich ‹*Landschaft*›; *(fig.)* düster ‹*Aussicht*›

forbore, forborne *see* [2]**forbear**

force [fɔ:s] **1.** *n.* **a)** *no pl. (strength, power)* Stärke, *die; (of bomb, explosion, attack, storm)* Wucht, *die; (physical strength)* Kraft, *die;* **achieve sth. by brute** ~**:** etw. mit roher Gewalt erreichen; **b)** *no pl. (fig.: power, validity)* Kraft, *die;* **by** ~ **of** auf Grund (+ *Gen.*); **argue with much** ~**:** sehr überzeugend argumentieren; **in** ~ *(in effect)* in Kraft; **come into** ~ ‹*Gesetz usw.*› in Kraft treten; **put in[to]** ~**:** in Kraft setzen; **c)** *(coercion, violence)* Gewalt, *die;* **use** *or* **employ** ~ **[against sb.]** Gewalt [gegen jmdn.] anwenden; **by** ~**:** gewaltsam; mit Gewalt; **d)** *(organized group) (of workers)* Kolonne, *die;* Trupp, *der; (of police)* Einheit, *die; (Mil.)* Armee, *die;* **the ~s** die Armee; **be in the ~s** beim Militär sein; **e)** *(forceful agency or person)* Kraft, *die;* Macht, *die;* **there are ~s in action/at work here ...:** hier walten Kräfte/sind Kräfte am Werk ...; **he is a** ~ **in the land** *(fig.)*/**a** ~ **to be reckoned with** er ist ein einflußreicher Mann im Land/eine Macht, die nicht zu unterschätzen ist; **f)** *(meaning)* Bedeutung, *die;* **g)** *(Phys.)* Kraft, *die.* **2.** *v.t.* **a)** zwingen; ~ **sb./oneself [to do sth.]** jmdn./sich zwingen[, etw. zu tun]; **be ~d to do sth.** gezwungen sein *od.* sich gezwungen sehen, etw. zu tun; **I was ~d to accept/into accepting the offer** *(felt obliged)* ich fühlte mich verpflichtet, das Angebot anzunehmen; ~ **sb.'s hand** *(fig.)* jmdn. zwingen zu handeln; **b)** *(take by ~)* ~ **sth. from sb.** jmdm. etw. entreißen; **he ~d it out of her hands** er riß es ihr aus der Hand; ~ **a confession from sb.** *(fig.)* jmdn. zu einem Geständnis zwingen; **c)** *(push)* ~ **sth. into sth.** etw. in etw. *(Akk.)* [hinein]zwängen; **d)** *(impose, inflict)* ~ **sth. [up]on sb.** jmdm. etw. aufzwingen *od.* aufnötigen; **he ~d his attentions on her** er drängte sich ihr mit seinen Aufmerksamkeiten auf; **e)** *(break open)* ~ **[open]** aufbrechen; **f)** *(effect by violent means)* sich *(Dat.)* erzwingen ‹*Zutritt*›; ~ **one's way in[to a building]** sich *(Dat.)* mit Gewalt Zutritt [zu einem Gebäude] verschaffen; **g)** *(produce with effort)* sich zwingen zu; ~ **a smile** sich zu einem Lächeln zwingen

~ **'down** *v.t.* **a)** drücken ‹*Preis*›; **b)** zur Landung zwingen ‹*Flugzeug*›; **c)** *(make oneself eat)* herunterwürgen *(ugs.)* ‹*Nahrung*›

~ **'up** *v.t.* hochtreiben ‹*Preis*›

forced [fɔ:st] *adj.* **a)** *(contrived, unnatural)* gezwungen; gewollt ‹*Geste, Vergleich, Metapher*›; gekünstelt ‹*Benehmen*›; **b)** *(compelled by force)* erzwungen; Zwangs‹*arbeit, -anleihe*›

forced: ~ **'landing** *n.* Notlandung, *die;* ~ **'march** *n. (Mil.)* Gewaltmarsch, *der*

'force-feed *v.t.* zwangsernähren

forceful ['fɔ:sfl] *adj.* stark ‹*Persönlichkeit, Charakter*›; energisch ‹*Person, Art, Maßnahme*›; schwungvoll ‹*Rede-, Schreibweise*›; eindrucksvoll ‹*Sprache*›; eindringlich ‹*Worte*›

forceps ['fɔ:seps] *n., pl. same* **[pair of]** ~**:** Zange, *die*

forcible ['fɔ:sɪbl] *adj.* gewaltsam

forcibly ['fɔ:sɪblɪ] *adv.* gewaltsam; mit Gewalt

ford [fɔ:d] **1.** *n.* Furt, *die.* **2.** *v.t.* durchqueren; *(wade through)* durchwaten

fore [fɔ:(r)] **1.** *adj., esp. in comb.* vorder...; Vorder‹*teil, -front usw.*›. **2.** *n.* **[be/come] to the** ~**:** im Vordergrund [stehen]/in den Vordergrund [rücken]. **3.** *adv. (Naut.)* vorn; ~ **and aft** längs[schiffs]

'forearm *n.* Unterarm, *der*

forebear *see* [1]**forbear**

foreboding [fɔ:'bəʊdɪŋ] *n.* Vorahnung, *die; (unease caused by premonition)* ungutes Gefühl

'forecast 1. *v.t.*, ~ *or* **~ed** vorhersagen. **2.** *n.* Voraussage, *die; (Meteorol.)* [Wetter]vorhersage, *die*

forecastle ['fəʊksl] *n. (Naut.)* Back, *die*

'forecourt *n.* Vorhof, *der;* ~ **attendant** ≈ Tankwart, *der*

'forefather *n., usu. in pl.* Vorfahr, *der;* **our ~s** unsere Vorväter

'forefinger *n.* Zeigefinger, *der*

'forefront *n.* **[be] in the** ~ **of** in vorderster Linie (+ *Gen.*) [stehen]

forego *see* **forgo**

foregoing ['fɔ:gəʊɪŋ, fɔ:'gəʊɪŋ] *adj.* vorhergehend

'foregone *adj.* **be a** ~ **conclusion** *(be predetermined)* von vornherein feststehen; *(be certain)* so gut wie sicher sein

'foreground *n.* Vordergrund, *der*

'forehand *(Tennis etc.)* **1.** *adj.* Vorhand-. **2.** *n.* Vorhand, *die*

forehead ['fɒrɪd, 'fɔ:hed] *n.* Stirn, *die*

foreign ['fɒrɪn] *adj.* **a)** *(from abroad)* ausländisch; Fremd‹*herrschaft, -kapital, -sprache*›; fremdartig ‹*Gebräuche*›; ~ **worker** Gastarbeiter, *der/*-arbeiterin, *die;* **he is** ~**:** er ist Ausländer; **b)** *(abroad)* fremd; Auslands‹*reise, -niederlassung, -markt*›; ~ **country** Ausland, *das;* ~ **travel** Reisen ins Ausland; **c)** *(related to countries abroad)* außenpolitisch; Außen‹*politik, -handel*›; ~ **affairs** auswärtige

Angelegenheiten; **d)** *(from outside)* fremd; ~ **body** *or* **substance** Fremdkörper, *der;* **e)** *(alien)* fremd; **be ~ to sb./sb.'s nature** jmdm. fremd sein/nicht jmds. Art sein

foreign: ~ **'aid** *n.* Entwicklungshilfe, *die;* **F~ and 'Commonwealth Office** *n. (Brit.)* Außenministerium, *das;* ~ **corre'spondent** *n. (Journ.)* Auslandskorrespondent, *der*/-korrespondentin, *die*

foreigner ['fɒrɪnə(r)] *n.* Ausländer, *der*/Ausländerin, *die*

foreign: ~ **ex'change** *n. (dealings)* Devisenhandel, *der; (currency)* Devisen *Pl.;* ~ **'language** *n.* Fremdsprache, *die;* ~ **'legion** *n.* Fremdenlegion, *die;* **F~ 'Minister** *n.* Außenminister, *der;* **F~ Office** *n. (Brit. Hist./coll.)* Außenministerium, *das;* **F~ 'Secretary** *n. (Brit.)* Außenminister, *der;* ~ **service** *n.* diplomatischer Dienst

'foreleg *n.* Vorderbein, *das*

foreman ['fɔ:mən] *n., pl.* **foremen** ['fɔ:mən] **a)** Vorarbeiter, *der;* **b)** *(Law)* Sprecher [der Geschworenen/*(in Germany)* der Schöffen]

foremost ['fɔ:məʊst, 'fɔ:məst] *adj.* **a)** vorderst...; **b)** *(fig.)* führend

'forename *n.* Vorname, *der*

forensic [fə'rensɪk] *adj.* gerichtlich; forensisch *(fachspr.);* ~ **medicine** Gerichtsmedizin, *die*

'foreplay *n.* Vorspiel, *das*

'forerunner *n.* Vorläufer, *der*/Vorläuferin, *die*

foresaw *see* **foresee**

foresee [fɔ:'si:] *v. t., forms as* **see** voraussehen

foreseeable [fɔ:'si:əbl] *adj.* vorhersehbar; **in the ~ future** in nächster Zukunft

foreseen *see* **foresee**

fore'shadow *v. t.* vorausahnen lassen; vorausdeuten auf (+ *Akk.*)

fore'shorten *v. t.* **a)** *(Art, Photog.)* [perspektivisch] verkürzen; **b)** *(shorten, condense)* verkürzen

'foresight *n., no pl.* Weitblick, *der;* Voraussicht, *die*

'foreskin *n. (Anat.)* Vorhaut, *die*

forest ['fɒrɪst] *n.* Wald, *der; (commercially exploited)* Forst, *der*

fore'stall *v. t.* zuvorkommen (+ *Dat.*); *(prevent by prior action)* vermeiden

forester ['fɒrɪstə(r)] *n.* Förster, *der*

forestry ['fɒrɪstrɪ] *n.* Forstwirtschaft, *die; (science)* Forstwissenschaft, *die*

'foretaste *n.* Vorgeschmack, *der*

fore'tell *v. t.,* **foretold** vorhersagen; voraussagen

'forethought *n. (prior deliberation)* [vorherige] Überlegung; *(care for the future)* Vorausdenken, *das*

foretold *see* **foretell**

forever [fə'revə(r)] *adv.* **a)** *(constantly, persistently)* ständig; **b)** *(Amer.)* = **for ever;** *see* **ever a**

fore'warn *v. t.* vorwarnen; **~ed is fore'armed** *(prov.)* wer gewarnt ist, ist gewappnet

fore'warning *n.* Vorwarnung, *die*

'foreword *n.* Vorwort, *das*

forfeit ['fɔ:fɪt] **1.** *v. t.* verlieren *(auch fig.);* einbüßen *(geh., auch fig.);* verwirken *(geh.)* ‹*Recht, jmds. Gunst*›. **2.** *n.* Strafe, *die*

forgave *see* **forgive**

¹forge [fɔ:dʒ] **1.** *n.* **a)** *(workshop)* Schmiede, *die;* **b)** *(blacksmith's hearth)* Esse, *die; (furnace for melting or refining metal)* Schmiedeofen, *der.* **2.** *v. t.* **a)** schmieden **(into** zu); **b)** *(fig.)* schmieden ‹*Plan, Verbindung*›; schließen ‹*Vereinbarung, Freundschaft*›; **c)** *(counterfeit)* fälschen

²forge *v. i.* ~ **ahead** [das Tempo] beschleunigen; ‹*Wettläufer:*› vorstoßen; *(fig.)* vorankommen; Fortschritte machen

forger ['fɔ:dʒə(r)] *n.* Fälscher, *der*/Fälscherin, *die*

forgery ['fɔ:dʒərɪ] *n.* Fälschung, *die*

forget [fə'get] **1.** *v. t.,* **-tt-, forgot** [fə'gɒt], **forgotten** [fə'gɒtn] vergessen; *(~ learned ability)* verlernen; vergessen; **gone but not forgotten** in bleibender Erinnerung; **I ~ his name** *(have forgotten)* ich habe seinen Namen vergessen; ~ **doing sth./having done sth.** vergessen, daß man etw. getan hat; **don't ~ that ...:** vergiß nicht *od.* denk[e] daran, daß ...; ~ **how to dance** das Tanzen verlernen; **and don't you ~ it** *(coll.)* vergiß das ja nicht; ~ **it!** *(coll.)* schon gut!; vergiß es! **2.** *v. i.,* **-tt-, forgot, forgotten** es vergessen; ~ **about sth.** etw. vergessen; ~ **about it!** *(coll.)* schon gut!; **I forgot about Joe** ich habe gar nicht an Joe gedacht. **3.** *v. refl.,* **-tt-, forgot, forgotten a)** *(act unbecomingly)* sich vergessen; **b)** *(neglect one's own interests)* sich selbst vergessen

forgetful [fə'getfl] *adj.* **a)** *(absent-minded)* vergeßlich; **b)** ~ **of sth.** ohne an etw. *(Akk.)* zu denken; **be ~ of one's duty** seine Pflicht vernachlässigen

forgetfulness [fə'getflnɪs] *n., no pl.* Vergeßlichkeit, *die*

for'get-me-not *n. (Bot.)* Vergißmeinnicht, *das*

forgettable [fə'getəbl] *adj.* **easily ~:** leicht zu vergessen

forgive [fə'gɪv] *v. t.,* **forgave** [fə'geɪv], **forgiven** [fə'gɪvn] vergeben ‹*Sünden*›; verzeihen ‹*Unrecht*›; entschuldigen, verzeihen ‹*Unterbrechung, Neugier, Ausdrucksweise*›; ~ **sb.** [**sth.** *or* **for sth.**] jmdm. [etw.] verzeihen *od. (geh.)* vergeben; **God ~ me** möge Gott mir vergeben; **am I ~n?** verzeihst du mir?; **you are ~n** ich verzeihe dir; ~ **me for saying so, but ...:** entschuldigen *od.* verzeihen Sie[, daß ich es sage], [aber] ...

forgiveness [fə'gɪvnɪs] *n., no pl.* Verzeihung, *die; (esp. of sins)* Vergebung, *die (geh.);* **ask/beg** [**sb.'s**] **~:** [jmdn.] um Verzeihung/ *(geh.)* Vergebung bitten

forgiving [fə'gɪvɪŋ] *adj.* versöhnlich

forgo [fɔ:'gəʊ] *v. t., forms as* **go** verzichten auf (+ *Akk.*)

forgone *see* **forgo**

forgot, forgotten *see* **forget**

fork [fɔ:k] **1.** *n.* **a)** Gabel, *die;* **the knives and ~s** das Besteck; **b)** *([point of] division into branches)* Gabelung, *die; (one branch)* Abzweigung, *die; (of tree)* Astgabel, *die.* **2.** *v. i.* **a)** *(divide)* sich gabeln; **b)** *(turn)* abbiegen; ~ [**to the**] **left** [**for**] [nach] links abbiegen [nach]

~ **'out,** ~ **'up** *(sl.)* **1.** *v. t.* lockermachen *(ugs.).* **2.** *v. i.* ~ **out** *or* **up** [**for sth.**] [für etw.] blechen *(ugs.)*

forked [fɔ:kt] *adj.* gegabelt

forked 'lightning *n., no pl., no indef. art.* Linienblitz, *der*

'fork-lift truck *n.* Gabelstapler, *der*

forlorn [fə'lɔ:n] *adj.* **a)** *(desperate)* verzweifelt; ~ **hope** *(faint hope)* verzweifelte Hoffnung; *(desperate enterprise)* aussichtsloses Unterfangen; **b)** *(forsaken)* [einsam und] verlassen

form [fɔ:m] **1.** *n.* **a)** *(type, style)* Form, *die;* ~ **of address** [Form der] Anrede; **in human ~:** in menschlicher Gestalt; in Menschengestalt; **in the ~ of** in Form von *od.* + *Gen.;* **in book ~:** in Buchform; als Buch; **b)** *no pl. (shape, visible aspect)* Form, *die;* Gestalt, *die;* **take ~:** Gestalt annehmen *od.* gewinnen; **c)** *(printed sheet)* Formular, *das;* **d)** *(Brit. Sch.)* Klasse, *die;* **e)** *(bench)* Bank, *die;* **f)** *no pl., no indef. art. (Sport: physical condition)* Form, *die;* **peak ~:** Bestform, *die;* **out of**

~: außer Form; nicht in Form; **in [good]** ~ *(lit. or fig.)* [gut] in Form; **she was in great ~ at the party** *(fig.)* bei der Party war sie groß in Form; **on/off** ~ *(lit. or fig.)* in/nicht in Form; **g)** *(Sport: previous record)* bisherige Leistungen; **on/judging by [past/present]** ~ *(fig.)* nach der Papierform; **true to** ~ *(fig.)* wie üblich *od.* zu erwarten; **h)** *(etiquette)* **for the sake of** ~: der Form halber; **good/bad** ~: gutes/schlechtes Benehmen; **i)** *(figure)* Gestalt, *die;* **j)** *(Ling.)* Form, *die.* **2.** *v.t.* **a)** *(make; also Ling.)* bilden; **be ~ed from sth.** aus etw. entstehen; **b)** *(shape, mould)* formen, gestalten **(into** zu); *(fig.)* formen ‹*Charakter usw.*›; **c)** sich *(Dat.)* bilden ‹*Meinung, Urteil*›; gewinnen ‹*Eindruck*›; fassen ‹*Entschluß, Plan*›; kommen zu ‹*Schluß*›; *(acquire, develop)* entwickeln ‹*Vorliebe, Gewohnheit, Wunsch*›; schließen ‹*Freundschaft*›; **d)** *(constitute, compose, be, become)* bilden; **Schleswig once ~ed [a] part of Denmark** Schleswig war einmal ein Teil von Dänemark; **e)** *(establish, set up)* bilden ‹*Regierung*›; gründen ‹*Bund, Verein, Firma, Partei, Gruppe*›. **3.** *v.i. (come into being)* sich bilden; ‹*Idee:*› sich formen, Gestalt annehmen

formal ['fɔ:ml] *adj.* **a)** formell; förmlich ‹*Person, Art, Einladung, Begrüßung*›; steif ‹*Person, Begrüßung*›; *(official)* offiziell; **wear ~ dress** *or* **clothes** Gesellschaftskleidung tragen; **b)** *(explicit)* formell; ~ **education/knowledge** ordentliche Schulbildung/reales Wissen

formality [fɔ:'mælɪtɪ] *n.* **a)** *(requirement)* Formalität, *die;* **b)** *no pl. (being formal, ceremony)* Förmlichkeit, *die*

formalize ['fɔ:məlaɪz] *v.t.* **a)** *(specify)* formalisieren; **b)** *(make official)* formell bekräftigen

format ['fɔ:mæt] *n.* **a)** *(of book) (layout)* Aufmachung, *die; (shape and size)* Format, *das;* **b)** *(Telev., Radio)* Aufbau, *der*

formation [fɔ:'meɪʃn] *n.* **a)** *no pl. (forming) (of substance, object)* Bildung, *die; (of character)* Formung, *die; (of plan)* Entstehung, *die; (establishing)* Gründung, *die;* **b)** *(Mil., Aeronaut., Dancing)* Formation, *die*

formative ['fɔ:mətɪv] *adj.* formend, prägend ‹*Einfluß*›; **the ~ years of life** die entscheidenden Lebensjahre

former ['fɔ:mə(r)] *attrib. adj.* **a)** *(earlier)* früher; *(ex-)* ehemalig; Ex-; **in ~ times** früher; **b)** *(first-mentioned)* **in the ~ case** im ersteren Fall; **the ~:** der/die/das erstere; *pl.* die ersteren

formerly ['fɔ:məlɪ] *adv.* früher

Formica, (P) [fɔ:'maɪkə] *n.* ≈ Resopal, *das* Ⓦⓩ

formidable ['fɔ:mɪdəbl] *adj.* gewaltig; ungeheuer; bedrohlich, gefährlich ‹*Gegner, Herausforderung*›; *(awe-inspiring)* formidabel; beeindruckend

formula ['fɔ:mjʊlə] *n., pl.* ~s *or (esp. as tech. term)* ~e ['fɔ:mjʊli:] **a)** *(also Math., Chem., Phys.)* Formel, *die;* **b)** *(set form)* Schema, *das; (prescription, recipe)* Rezeptur, *die; (fig.)* Rezept, *das*

formulate ['fɔ:mjʊleɪt] *v.t.* formulieren; *(devise)* entwickeln

fornicate ['fɔ:nɪkeɪt] *v.i.* Unzucht treiben; huren *(abwertend)*

forsake [fə'seɪk] *v.t.,* **forsook** [fə'sʊk], **~n** [fə'seɪkn] **a)** *(give up)* entsagen *(geh.)* (+ *Dat.*); verzichten auf (+ *Akk.*); **b)** *(desert)* verlassen

forsaken [fə'seɪkn] *adj.* verlassen

forsook *see* **forsake**

fort [fɔ:t] *n. (Mil.)* Fort, *das;* **hold the ~** *(fig.)* die Stellung halten

forte ['fɔ:teɪ, fɔ:t] *n.* Stärke, *die;* starke Seite *(ugs.)*

forth [fɔ:θ] *adv.* **a) and so ~:** und so weiter; **b) from this/that day** *etc.* ~: von diesem/jenem Tag *usw.* an; von Stund an *(geh.)*

forthcoming ['fɔ:θkʌmɪŋ, fɔ:θ'kʌmɪŋ] *adj.* **a)** *(approaching)* bevorstehend; *(about to appear)* in Kürze zu erwartend ...; in Kürze anlaufend ‹*Film*›; in Kürze erscheinend ‹*Ausgabe, Buch usw.*›; **be ~:** bevorstehen; *(about to appear)* in Kürze zu erwarten sein/anlaufen/erscheinen; **b)** *pred. (made available)* **be ~** ‹*Geld, Antwort:*› kommen; ‹*Hilfe:*› geleistet werden; **not be ~:** ausbleiben; **c)** *(responsive)* mitteilsam ‹*Person*›

forthright ['fɔ:θraɪt] *adj.* direkt; offen ‹*Blick*›

forthwith [fɔ:θ'wɪθ, fɔ:θ'wɪð] *adv.* unverzüglich

fortieth ['fɔ:tɪɪθ] **1.** *adj.* vierzigst...; *see also* **eighth 1. 2.** *n. (fraction)* Vierzigstel, *das; see also* **eighth 2**

fortification [fɔ:tɪfɪ'keɪʃn] *n. (Mil.)* **a)** *no pl. (fortifying)* Befestigung, *die;* **b)** *usu. in pl. (defensive works)* Befestigung, *die;* Festungsanlage, *die*

fortify ['fɔ:tɪfaɪ] *v.t.* **a)** *(Mil.)* befestigen; **b)** *(strengthen, lit. or fig.)* stärken; **c)** aufspriten ‹*Wein*›

fortitude ['fɔ:tɪtju:d] *n., no pl.* innere Stärke

fortnight ['fɔ:tnaɪt] *n.* vierzehn Tage; zwei Wochen; **a ~ [from] today** heute in vierzehn Tagen

fortnightly ['fɔ:tnaɪtlɪ] **1.** *adj.* vierzehntäglich; zweiwöchentlich. **2.** *adv.* alle vierzehn Tage; alle zwei Wochen

fortress ['fɔ:trɪs] *n.* Festung, *die*

fortuitous [fɔ:'tju:ɪtəs] *adj.,* **fortuitously** [fɔ:'tju:ɪtəslɪ] *adv.* zufällig

fortunate ['fɔ:tʃʊnət, 'fɔ:tʃənət] *adj.* glücklich; **it is ~ for sb. [that ...]** es ist jmds. Glück[, daß ...]; **sb. is ~ to be alive** jmd. kann von Glück sagen *od.* reden, daß er noch lebt; **it was very ~ that ...:** es war ein Glück, daß ...

fortunately ['fɔ:tʃʊnətlɪ, 'fɔ:tʃənətlɪ] *adv. (luckily)* glücklicherweise; zum Glück

fortune ['fɔ:tʃən, 'fɔ:tʃu:n] *n.* **a)** *(private wealth)* Vermögen, *das;* **make a ~:** ein Vermögen machen; **b)** *(prosperous condition)* Glück, *das; (of country)* Wohl, *das;* **c)** *(luck, destiny)* Schicksal, *das;* **bad/good ~:** Pech, *das*/Glück, *das;* **by sheer good ~ there was ...:** es war reines Glück, daß ... war; **thank one's good ~ that ...:** dem Glück dafür danken, daß ...; **tell sb.'s ~:** jmdm. wahrsagen *od.* sein Schicksal vorhersagen; **tell ~s** wahrsagen

fortune: ~-teller *n.* Wahrsager, *der*/Wahrsagerin, *die;* **~-telling** *n., no pl.* Wahrsagerei, *die*

forty ['fɔ:tɪ] **1.** *adj.* vierzig; **have ~ 'winks** ein Nickerchen *(fam.)* machen *od.* halten; *see also* **eight 1. 2.** *n.* Vierzig, *die; see also* **eight 2 a; eighty 2**

forty: ~-'first *etc. adj.* einundvierzigst... *usw.; see also* **eighth 1; ~-'one** *etc.* **1.** *adj.* einundvierzig *usw.; see also* **eight 1; 2.** *n.* Einundvierzig *usw., die; see also* **eight 2 a**

forum ['fɔ:rəm] *n. (also Roman Hist.)* Forum, *das*

forward ['fɔ:wəd] **1.** *adv.* **a)** *(in direction faced)* vorwärts; **take three steps ~:** drei Schritte vortreten; **b)** *(towards end of room etc. faced)* nach vorn; vor‹*laufen, -rücken, -schieben*›; **c)** *(closer)* heran; **he came ~ to greet me** er kam auf mich zu, um mich zu begrüßen; **d)** *(ahead, in advance)* voraus‹*schicken, -gehen*›; **e)** *(into future)* voraus‹*schauen, -denken*›; **f) come ~** *(present oneself)* ‹*Zeuge, Helfer:*› sich melden. **2.** *adj.* **a)** *(directed ahead)* vorwärts gerichtet; nach vorn *nachgestellt;* **b)** *(at or to the front)* Vorder-; vorder...; **c)** *(advanced)* frühreif ‹*Kind, Pflanze, Getreide*›; fortschrittlich ‹*Vorstellung, Ansicht, Maßnahme*›; **d)** *(bold)* dreist; **e)** *(Commerc.)* Termin‹*geschäft, -verkauf*›; Zukunfts‹*planung*›. **3.** *n. (Sport)* Stürmer, *der*/Stürmerin, *die*. **4.** *v. t.* **a)** *(send on)* nachschicken ‹*Brief, Paket, Post*› (**to** an + *Akk.*); *(dispatch)* abschicken ‹*Waren*› (**to** an + *Akk.*); **'please ~'** „bitte nachsenden"; **~ing address** Nachsendeanschrift, *die;* **b)** *(pass on)* weiterreichen, weiterleiten ‹*Vorschlag, Plan*› (**to** an + *Akk.*); **c)** *(promote)* voranbringen ‹*Karriere, Vorbereitung*›

forwards ['fɔ:wədz] *see* **forward 1 a, b**

forwent *see* **forgo**

fossil ['fɒsɪl] *n.* Fossil, *das;* ~ **fuel** fossiler Brennstoff

fossilize (fossilise) ['fɒsɪlaɪz] *v. t.* fossilisieren lassen *(Paläont.);* versteinern lassen *(auch fig.);* **~d** fossil *(Paläont.)*

foster ['fɒstə(r)] **1.** *v. t.* **a)** *(encourage)* fördern; pflegen ‹*Freundschaft*›; *(harbour)* hegen *(geh.);* **b)** in Pflege haben ‹*Kind*›. **2.** *adj.* **~-:** Pflege‹*kind, -mutter, -eltern, -sohn usw.*›

fought *see* **fight 1, 2**

foul [faʊl] **1.** *adj.* **a)** *(offensive to the senses)* abscheulich; übel ‹*Geruch, Geschmack*›; **b)** *(polluted)* verschmutzt ‹*Wasser, Luft*›; **c)** *(sl.: awful)* scheußlich *(ugs.);* mies *(ugs. abwertend);* **d)** *(morally vile)* anstößig, unanständig ‹*Sprache, Gerede*›; niederträchtig ‹*Verleumdung, Tat*›; **e)** *(unfair)* unerlaubt, unredlich ‹*Mittel*›; ~ **play** *(Sport)* Foulspiel, *das;* **the police do not suspect ~ play** die Polizei vermutet kein Verbrechen; **f) fall** *or* **run ~ of** *(fig.)* kollidieren *od.* in Konflikt geraten mit ‹*Vorschrift, Gesetz, Polizei*›. **2.** *n. (Sport)* Foul, *das;* **commit a ~:** foulen; ein Foul begehen. **3.** *v. t.* **a)** *(make ~)* beschmutzen *(auch fig.);* verunreinigen *(abwertend);* verpesten ‹*Luft*›; **b)** *(be entangled with)* sich verfangen in (+ *Dat.*); **c)** *(Sport)* foulen

~ **'up** *v. t. (coll.: spoil)* vermasseln *(salopp)*

foul: ~-mouthed ['faʊlmaʊðd] *adj.* unanständig; unflätig; **~-smelling** *adj.* übelriechend

¹found [faʊnd] *v. t.* **a)** *(establish)* gründen; stiften ‹*Krankenhaus, Kloster*›; begründen ‹*Wissenschaft, Religion, Glauben, Kirche*›; **b)** *(fig.: base)* begründen; ~ **sth. [up]on sth.** etw. auf etw. *(Akk.)* gründen; **be ~ed [up]on sth.** [sich] auf etw. *(Akk.)* gründen

²found *see* **find 1**

foundation [faʊn'deɪʃn] *n.* **a)** *(establishing)* Gründung, *die; (of hospital, monastery)* Stiftung, *die; (of school of painting, of religion)* Begründung, *die;* **b)** *(institution)* Stiftung, *die;* **c)** *usu. in pl.* **~[s]** *(underlying part of building; also fig.)* Fundament, *das;* **be without** *or* **have no ~** *(fig.)* unbegründet sein; **lay the ~ of/for sth.** *(fig.)* das Fundament *od.* die Grundlage zu etw. legen; **d)** *(cosmetic)* Grundierung, *die*

foun'dation-stone *n. (lit. or fig.)* Grundstein, *der*

¹founder ['faʊndə(r)] *n.* Gründer, *der*/Gründerin, *die; (of hospital, or with an endowment)* Stifter, *der*/Stifterin, *die*

²founder *v. i.* **a)** ‹*Schiff:*› sinken, untergehen; **b)** *(fig.: fail)* sich zerschlagen

foundry ['faʊndrɪ] *n. (Metallurgy)* Gießerei, *die*

fount [faʊnt, fɒnt] *n. (Printing)* Schrift, *die*

fountain ['faʊntɪn] *n.* **a)** *(jet[s] of water)* Fontäne, *die; (structure)* Springbrunnen, *der;* **b)** *(fig.: source)* Quelle, *die*

'fountain-pen *n.* Füllfederhalter, *der;* Füller, *der (ugs.)*

four [fɔ:(r)] **1.** *adj.* vier; *see also* **eight 1. 2.** *n.* Vier, *die;* **on all ~s** auf allen vieren *(ugs.); see also* **eight 2 a, c, d**

four: ~-door *attrib. adj.* viertürig ‹*Auto*›; **~-footed** ['fɔ:fʊtɪd] *adj.* vierfüßig; **~-leaf clover, ~-leaved clover** *n.* vierblättriges Kleeblatt; **~-letter 'word** *n.* vulgärer Ausdruck; **~-poster** *n.* Himmelbett, *das*

foursome ['fɔ:səm] *n.* **a)** Quartett, *das;* **go in** *or* **as a ~:** zu viert gehen; **b)** *(Golf)* Vierer, *der*

'four-stroke *adj. (Mech. Engin.)* Viertakt‹*motor, -verfahren*›

fourteen [fɔ:'ti:n] **1.** *adj.* vierzehn; *see also* **eight 1. 2.** *n.* Vierzehn, *die; see also* **eight 2 a, d; eighteen 2**

fourteenth [fɔ:'ti:nθ] **1.** *adj.* vierzehnt...; *see also* **eighth 1. 2.** *n. (fraction)* Vierzehntel, *das; see also* **eighth 2**

fourth [fɔ:θ] **1.** *adj.* **a)** viert...; **the ~ finger** der kleine Finger; *see also* **eighth 1; b) ~ dimension** vierte Dimension. **2.** *n. (in sequence)* vierte, *der/die/das; (in rank)* Vierte, *der/die/das; (fraction)* Viertel, *das; see also* **eighth 2**

fourth 'gear *n., no pl. (Motor Veh.)* vierter Gang; *see also* **gear 1 a**

fourthly ['fɔ:θlɪ] *adv.* viertens

four-wheel 'drive *n. (Motor Veh.)* Vier- *od.* Allradantrieb, *der*

fowl [faʊl] *n., pl.* **~s** *or same* Haushuhn, *das; (collectively)* Geflügel, *das*

fox [fɒks] **1.** *n.* Fuchs, *der.* **2.** *v. t.* verwirren

fox: ~ cub *n.* Fuchswelpe, *der;* **~glove** *n. (Bot.)* Fingerhut, *der;* **~-hunt** *n.* Fuchsjagd, *die;* **~-'terrier** *n.* Foxterrier, *der;* **~trot** *n.* Foxtrott, *der*

foyer ['fɔɪeɪ, 'fwɑjeɪ] *n.* Foyer, *das*

fraction ['frækʃn] *n.* **a)** *(Math.)* Bruch, *der;* **b)** *(small part)* Bruchteil, *der;* **the car missed me by a ~ of an inch** das Auto hätte mich um Haaresbreite überfahren

fractional ['frækʃənl] *adj.*, **fractionally** ['frækʃənəlɪ] *adv. (fig.)* geringfügig

fractious ['frækʃəs] *adj. (unruly)* aufsässig; *(peevish)* quengelig ‹*Kind*›

fracture ['fræktʃə(r)] **1.** *n. (also Med.)* Bruch, *der.* **2.** *v. t. (also Med.)* brechen; ~ **one's jaw** *etc.* sich *(Dat.)* den Kiefer *usw.* brechen; ~ **one's skull** sich *(Dat.)* einen Schädelbruch zuziehen. **3.** *v. i. (Med.)* brechen

fragile ['frædʒaɪl] *adj.* **a)** zerbrechlich; zart ‹*Teint, Hand*›; **'~ – handle with care'** „Vorsicht, zerbrechlich!"; **b)** *(fig.)* unsicher ‹*Frieden*›; zart ‹*Gesundheit, Konstitution*›; zerbrechlich ‹*alte Frau*›

fragment 1. ['frægmənt] *n.* Bruchstück, *das; (of document, conversation)* Fetzen, *der; (of china)* Scherbe, *die; (Lit., Mus.)* Fragment, *das.* **2.** [fræg'ment] *v. t. & i.* zersplittern

fragmentary ['frægməntərɪ] *adj.* bruchstückhaft; fragmentarisch

fragmented [fræg'mentɪd] *adj.* bruchstückhaft

fragrance ['freɪgrəns] *n.* Duft, *der*

fragrant ['freɪgrənt] *adj.* duftend

frail [freɪl] *adj.* zerbrechlich; zart ‹*Gesundheit*›; gebrechlich ‹*Greis, Greisin*›; schwach ‹*Stimme*›

frailty ['freɪltɪ] *n.* **a)** *no pl.* Zerbrechlichkeit, *die; (of health)* Zartheit, *die;* **b)** *esp. in pl. (fault)* Schwäche, *die*

frame [freɪm] **1.** *n.* **a)** *(of vehicle, bicycle)* Rahmen, *der; (of easel, rucksack, bed, umbrella)* Gestell, *das; (of ship, aircraft)* Gerüst, *das;* **b)** *(border)* Rahmen, *der;* **[spectacle] ~s** [Brillen]gestell, *das;* **c)** *(of person, animal)* Körper, *der;* **d)** *(Photog., Cinemat.)* [Einzel]bild, *das.* **2.** *v. t.* **a)** rahmen ‹*Bild, Spiegel*›; **b)** *(compose)* formulieren ‹*Frage, Antwort, Satz*›; aufbauen ‹*Rede, Aufsatz*›; *(devise)* entwerfen ‹*Gesetz, Politik, Plan*›; ausarbeiten ‹*Plan, Methode, Denksystem*›; **c)** *(sl.: incriminate unjustly)* ~ **sb.** jmdm. etwas anhängen *(ugs.)*

frame: **~-up** *n. (coll.)* abgekartetes Spiel *(ugs.)*; **~work** *n. (of ship etc., fig.: of project)* Gerüst, *das;* **[with]in the ~work of** *(as part of)* im Rahmen (+ *Gen.*); *(in relation to)* im Zusammenhang mit

franc [fræŋk] *n.* Franc, *der; (Swiss)* Franken, *der*

France [frɑ:ns] *pr. n.* Frankreich *(das)*

franchise ['fræntʃaɪz] *n.* **a)** Stimmrecht, *das; (esp. for Parliament)* Wahlrecht, *das;* **b)** *(Commerc.)* Lizenz, *die*

[1]**frank** *adj. (candid)* offen ‹*Bekenntnis, Aussprache, Blick, Gesicht, Person*›; freimütig ‹*Geständnis, Äußerung*›; **be ~ with sb.** zu jmdm. offen sein; **to be [quite] ~** *(as sentence-modifier)* offen gesagt

[2]**frank** *v. t. (Post)* **a)** *(in lieu of postage stamp)* freistempeln; **b)** *(put postage stamp on)* frankieren

frankfurter ['fræŋkfɜ:tə(r)] (*Amer.:* **frankfurt** ['fræŋkfɜ:t]) *n.* Frankfurter [Würstchen]

frankincense ['fræŋkɪnsens] *n.* Weihrauch, *der*

franking-machine ['fræŋkɪŋməʃi:n] *n. (Brit. Post)* Frankiermaschine, *die;* Freistempler, *der*

frankly ['fræŋklɪ] *adv. (candidly)* offen; frank und frei; *(honestly)* offen *od.* ehrlich gesagt; *(openly, undisguisedly)* unverhohlen ‹*kritisch, materialistisch usw.*›

frankness ['fræŋknɪs] *n., no pl.* Offenheit, *die;* Freimütigkeit, *die*

frantic ['fræntɪk] *adj.* **a)** verzweifelt ‹*Hilferufe, Gestikulieren*›; **be ~ with fear/rage** *etc.* außer sich *(Dat.)* sein vor Angst/Wut *usw.*; **drive sb. ~:** jmdn. in den Wahnsinn treiben; **b)** *(very anxious, noisy, uncontrolled)* hektisch ‹*Aktivität, Suche, Getriebe*›

fraternal [frə'tɜ:nl] *adj.* brüderlich

fraternise *see* **fraternize**

fraternity [frə'tɜ:nɪtɪ] *n.* **a) the teaching/medical/legal ~:** die Lehrer-/Ärzte-/Juristenzunft; die Zunft der Lehrer/Ärzte/Juristen; **b)** *(Amer. Univ.)* [studentische] Verbindung; **c)** *no pl. (brotherliness)* Brüderlichkeit, *die*

fraternize ['frætənaɪz] *v. i.* ~ **[with sb.]** sich verbrüdern [mit jmdm.]

fraud [frɔ:d] *n.* **a)** *no pl. (cheating, deceit)* Betrug, *der;* Täuschung, *die; (Law)* [arglistige] Täuschung; **b)** *(trick, false thing)* Schwindel, *der;* **c)** *(person)* Betrüger, *der*/Betrügerin, *die;* Schwindler, *der*/Schwindlerin, *die*

fraudulent ['frɔ:djʊlənt] *adj.* betrügerisch

fraught [frɔ:t] *adj.* **a) be ~ with danger** voller Gefahren sein; **~ with obstacles/difficulties** voller Hindernisse/Schwierigkeiten; **b)** *(coll.: distressingly tense)* stressig *(ugs.)* ‹*Atmosphäre, Situation, Diskussion*›; gestreßt *(ugs.)* ‹*Person*›

[1]**fray** [freɪ] *n. (fight)* [Kampf]getümmel, *das; (noisy quarrel)* Streit, *der;* **be eager/ready for the ~** *(lit. or fig.)* kampflustig/kampfbereit sein; **enter** *or* **join the ~** *(lit. or fig.)* sich in den Kampf *od.* ins Getümmel stürzen

[2]**fray 1.** *v. i.* [sich] durchscheuern; ‹*Hosenbein, Teppich, Seilende:*› ausfransen; **our nerves/tempers began to ~** *(fig.)* wir verloren langsam die Nerven/unsere Gemüter erhitzten sich. **2.** *v. t.* durchscheuern; ausfransen ‹*Hosenbein, Teppich, Seilende*›

freak [fri:k] *n.* **a)** *(monstrosity)* Mißgeburt, *die; (plant)* mißgebildete Pflanze; **~ of nature** Laune der Natur; *attrib.* ungewöhnlich ‹*Wetter, Ereignis*›; völlig überraschend ‹*Sieg, Ergebnis*›; **b)** *(sl.: fanatic)* Freak, *der;* **health ~:** Gesundheitsfanatiker, *der*

freckle ['frekl] *n.* Sommersprosse, *die*

freckled ['frekld] *adj.* sommersprossig

free [fri:] **1.** *adj.*, **freer** ['fri:ə(r)], **freest** ['fri:ɪst] **a)** frei; **get ~:** freikommen; sich befreien; **go ~** *(escape unpunished)* straffrei ausgehen; **let sb. go ~** *(leave captivity)* jmdn. freilassen; *(unpunished)* jmdn. freisprechen; **set ~:** freilassen; *(fig.)* erlösen; **~ of sth.** *(without)* frei von etw.; **~ of charge/cost** gebührenfrei/kostenlos; **~ and easy** ungezwungen; locker *(ugs.)*; **give ~ rein to sth.** einer Sache *(Dat.)* freien Lauf lassen; **b)** *(having liberty)* **sb. is ~ to do sth.** es steht jmdm. frei, etw. zu tun; **you're ~ to choose** du kannst frei [aus]wählen; **leave sb. ~ to do sth.** es jmdm. ermöglichen, etw. zu tun; **feel ~!** nur zu! *(ugs.)*; **feel ~ to correct me** du darfst mich gerne korrigieren; **it's a ~ country** *(coll.)* wir leben in einem freien Land; **~ from sth.** frei von etw.; **~ from pain/troubles** schmerz-/sorgenfrei; **c)** *(provided without payment)* kostenlos; frei ‹*Überfahrt, Unterkunft, Versand, Verpflegung*›; Frei‹*karte, -exemplar, -fahrt*›; Gratis‹*probe, -vorstellung*›; **'admission ~'** „Eintritt frei"; **have a ~ ride on the train** umsonst mit der Bahn fahren; **for ~** *(coll.)* umsonst; **d)** *(not occupied, not reserved, not being used)* frei; **~ time** Freizeit, *die;* **when would you be ~ to start work?** wann könnten Sie mit der Arbeit anfangen?; **he's ~ in the mornings** er hat morgens Zeit; **e)** *(generous)* **be ~ with sth.** mit etw. großzügig umgehen; **f)** *(frank, open)* offen; freimütig; **g)** *(not strict)* frei ‹*Übersetzung, Interpretation, Bearbeitung usw.*›. **2.** *adv. (without cost or payment)* gratis; umsonst. **3.** *v. t. (set at liberty)* freilassen; *(disentangle)* befreien **(of, from** von); **~ sb./oneself from** jmdn./sich befreien von ‹*Tyrannei, Unterdrückung, Tradition*›; jmdn./sich befreien aus ‹*Gefängnis, Sklaverei, Umklammerung*›; **~ sb./oneself of** jmdn./sich befreien *od.* freimachen von

freebie ['fri:bɪ] *(Amer. coll.) n.* Gratisgeschenk, *das*

freedom ['fri:dəm] *n.* **a)** Freiheit, *die;* **give sb. his ~:** jmdn. freigeben; *(from prison, slavery)* jmdn. freilassen; **~ of the press** Pressefreiheit, *die;* **~ of action/speech/movement** Handlungs-/Rede-/Bewegungsfreiheit, *die;* **b)** *(privilege)* **[give sb.** *or* **present sb. with] the ~ of the city** [jmdm.] die Ehrenbürgerrechte [verleihen]

'freedom fighter *n.* Freiheitskämpfer, *der*/-kämpferin, *die*

free: **~ 'enterprise** *n.* freies Unternehmertum; **~ 'fall** *n.* freier Fall; **~-for-all** *n.* [allgemeine] Schlägerei; *(less violent)* [allgemeines] Gerangel; **~ 'gift** *n.* Gratisgabe, *die;* **~hold 1.** *n.* Besitzrecht, *das;* **2.** *adj.* Eigentums-; **~hold land** freier Grundbesitz; **~ house** *n. (Brit.) brauereiunabhängiges Wirtshaus;* **~lance 1.** *n.* freier Mitarbeiter/freie Mitarbeiterin; **2.** *adj.* freiberuflich; **3.** *v. i.* freiberuflich arbeiten; **~'loader** *n. (sl.)* Nassauer, *der (ugs.)*

freely ['fri:lɪ] *adv.* **a)** *(willingly)* großzügig; freimütig ‹*eingestehen*›; **b)** *(without restriction, loosely)* frei; **c)** *(frankly)* offen

free: ~ '**market** *n.* *(Econ.)* freier Markt; **F~mason** *n.* Freimaurer, *der*

freer *see* **free 1**

'**free-range** *adj.* freilaufend ‹*Huhn*›; ~ **eggs** Eier von freilaufenden Hühnern

freesia ['fri:zɪə] *n.* *(Bot.)* Freesie, *die*

free '**speech** *n.* Redefreiheit, *die*

freest *see* **free 1**

free: ~ '**trade** *n.* Freihandel, *der;* ~**way** *n.* *(Amer.)* Autobahn, *die;* ~-**wheel** *v. i.* im Freilauf fahren; *(fig.: drift)* sich treiben lassen

freeze [fri:z] **1.** *v. i.*, **froze** [frəʊz], **frozen** ['frəʊzn] **a)** frieren; **it will** ~ *(Meteorol.)* es wird Frost geben; **b)** *(become covered with ice)* ‹*See, Fluß, Teich:*› zufrieren; ‹*Straße:*› vereisen; **c)** *(solidify)* ‹*Flüssigkeit:*› gefrieren; ‹*Rohr, Schloß:*› einfrieren; **d)** *(become rigid)* steif frieren; *(fig.)* ‹*Lächeln:*› gefrieren *(geh.)*; **e)** *(be or feel cold)* sehr frieren; *(fig.)* erstarren ‹**with** vor + *Dat.*›; ‹*Blut:*› gefrieren *(geh.)*; **my hands are freezing** meine Hände sind eiskalt; ~ **to death** erfrieren; *(fig.)* bitterlich frieren; **f)** *(make oneself motionless)* erstarren. **2.** *v. t.*, **froze, frozen a)** zufrieren lassen ‹*Teich, Fluß*›; gefrieren lassen ‹*Rohr*›; *(fig.)* erstarren lassen; **we were frozen stiff** *(fig.)* wir waren steif gefroren; **b)** *(preserve)* tiefkühlen, tiefgefrieren ‹*Lebensmittel*›; **c)** einfrieren ‹*Kredit, Guthaben, Gelder, Löhne, Preise usw.*›; **d)** *(fig.)* erstarren lassen. **3.** *n.* *(fixing)* Einfrieren, *das* (**on** *Gen.*); **price/wage** ~: Preis-/Lohnstopp, *der*

~ '**up 1.** *v. i.* ‹*Fluß, Teich:*› zufrieren; ‹*Schloß, Rohr:*› einfrieren. **2.** *v. t. see* **1:** zufrieren/einfrieren lassen

'**freeze-dry** *v. t.* gefriertrocknen

freezer ['fri:zə(r)] *n.* *(deep-freeze)* Tiefkühltruhe, *die;* Gefriertruhe, *die;* **[upright]** ~: Tiefkühlschrank, *der;* Gefrierschrank, *der; attrib.* ~ **compartment** Tiefkühlfach, *das;* Gefrierfach, *das*

freezing ['fri:zɪŋ] **1.** *adj.* *(lit. or fig.)* frostig; ~ **temperatures** Temperaturen unter null Grad; **it is** ~ **in here** es ist eiskalt hier drinnen. **2.** *n., no pl.* *(~-point)* **above/below** ~: über/unter dem/den Gefrierpunkt

'**freezing-point** *n.* Gefrierpunkt, *der*

freight [freɪt] **1.** *n.* Fracht, *die.* **2.** *v. t.* befrachten

'**freight car** *n.* *(Amer. Railw.)* Güterwagen, *der*

freighter ['freɪtə(r)] *n.* *(ship)* Frachter, *der;* Frachtschiff, *das;* *(aircraft)* Frachtflugzeug, *das*

French [frentʃ] **1.** *adj.* französisch; **he/she is** ~: er ist Franzose/sie ist Französin. **2.** *n.* **a)** Französisch, *das; see also* **English 2 a; b)** *constr. as pl.* **the** ~: die Franzosen

French: ~ '**bean** *n.* *(Brit.)* Gartenbohne, *die;* [grüne] Bohne; ~ **Ca'nadian** *n.* Frankokanadier, *der*/-kanadierin, *die;* ~-**Ca'nadian** *adj.* frankokanadisch; ~ '**dressing** *n.* Vinaigrette, *die;* ~ '**fries** *n. pl.* Pommes frites *Pl.*; ~ '**horn** *n.* *(Mus.)* [Wald]horn, *das;* ~ '**letter** *n.* *(Brit. coll.)* Pariser, *der (salopp);* ~**man** ['frentʃmən] *n., pl.* ~**men** ['frentʃmən] Franzose, *der;* ~ '**polish** *n.* Schellackpolitur, *die;* ~ '**window** *n., in sing. or pl.* französisches Fenster; ~**woman** *n.* Französin, *die*

frenetic [frɪ'netɪk] *adj.* verzweifelt ‹*Hilferuf, Versuch*›

frenzied ['frenzɪd] *adj.* rasend; wahnsinnig ‹*Tat*›

frenzy ['frenzɪ] *n.* **a)** *(derangement)* Wahnsinn, *der;* **b)** *(fury, agitation)* Raserei, *die;* **in a** ~ **of despair/passion** in einem Anfall von Verzweiflung/von wilder Leidenschaft übermannt

frequency ['fri:kwənsɪ] *n.* **a)** Häufigkeit, *die;* **b)** *(Phys., Statistics)* Frequenz, *die*

frequent 1. ['fri:kwənt] *adj.* **a)** häufig; **it's a** ~ **occurrence** es kommt häufig vor; **b)** *(habitual, constant)* eifrig ‹*[Kino-, Theater]besucher, Briefschreiber*›. **2.** [frɪ'kwent] *v. t.* frequentieren *(geh.)*; häufig besuchen ‹*Café, Klub usw.*›; **much** ~**ed** stark frequentiert *(geh.)*

frequently ['fri:kwəntlɪ] *adv.* häufig

fresco ['freskəʊ] *n., pl.* ~**es** *or* ~**s a)** *no pl., no art.* *(method)* Freskomalerei, *die;* **b)** *(a painting)* Fresko, *das*

fresh [freʃ] **1.** *adj.* **a)** frisch; neu ‹*Beweise, Anstrich, Ideen*›; frisch, neu ‹*Energie, Mut, Papierbogen*›; **a** ~ **approach** ein neuer Ansatz; ~ **supplies** Nachschub, *der* (**of** an + *Dat.*); **make a** ~ **start** noch einmal von vorn anfangen; *(fig.)* neu beginnen; ~ **from** *or* **off the press** druckfrisch; frisch aus der Presse; **get some** ~ **air** frische Luft schnappen *(ugs.)*; **as** ~ **as a daisy/as paint** ganz frisch; *(in appearance)* frisch wie der junge Morgen *(meist scherzh.)*; **b)** *(cheeky)* keck; **get** ~ **with sb.** jmdm. frech kommen *(ugs.)*. **2.** *adv.* frisch; **we're** ~ **out of eggs** *(coll.)* uns sind gerade die Eier ausgegangen

freshen ['freʃn] *v. i.* *(increase)* ‹*Wind:*› auffrischen

~ '**up** *v. i.* sich frisch machen

freshly ['freʃlɪ] *adv.* frisch

freshman ['freʃmən] *n., pl.* **freshmen** ['freʃmən] Erstsemester, *das*

freshness ['freʃnɪs] *n., no pl.* Frische, *die;* *(of idea, approach)* Neuartigkeit, *die*

fresh: ~ '**water** *n.* Süßwasser, *das;* ~**water** *adj.* Süßwasser-

[1]**fret** [fret] *v. i.*, **-tt-** beunruhigt sein; besorgt sein; **don't** ~**!** sei unbesorgt!; ~ **at** *or* **about** *or* **over sth.** sich über etw. *(Akk.) od.* wegen etw. aufregen

[2]**fret** *n.* *(Mus.)* Bund, *der*

fretful ['fretfl] *adj.* *(peevish)* verdrießlich; quengelig *(ugs.)* ‹*Kleinkind*›; *(restless)* unruhig

Freudian ['frɔɪdɪən] *adj.* freudianisch; ~ **slip** Freudsche Fehlleistung

Fri. *abbr.* **Friday** Fr.

friar ['fraɪə(r)] *n.* Ordensbruder, *der*

fricassee ['frɪkəsi:, frɪkə'si:] *(Cookery) n.* Frikassee, *das*

friction ['frɪkʃn] *n.* Reibung, *die*

Friday ['fraɪdeɪ, 'fraɪdɪ] **1.** *n.* Freitag, *der;* **on** ~: [am] Freitag; **on a** ~, **on** ~**s** freitags; **we got married on a** ~: wir haben an einem Freitag geheiratet; ~ **13 August** Freitag, der 13. August; *(at top of letter etc.)* Freitag, den 13. August; **on** ~ **13 August** am Freitag, den *od.* dem 13. August; **next/last** ~: [am] nächsten/letzten *od.* vergangenen Freitag; **we were married a year [ago] last/next** ~: vergangenen/kommenden Freitag vor einem Jahr haben wir geheiratet; **[last]** ~**'s newspaper** die Zeitung vom [letzten] Freitag; **Good** ~: Karfreitag, *der;* **man/girl** ~: Mädchen für alles *(ugs.)*. **2.** *adv.* *(coll.)* **a)** ~ **[week]** Freitag [in einer Woche]; **b)** ~**s** freitags; Freitag *(ugs.)*; **she comes** ~**s** sie kommt freitags

fridge [frɪdʒ] *n.* *(Brit. coll.)* Kühlschrank, *der*

fried *see* [1]**fry**

friend [frend] *n.* **a)** Freund, *der*/Freundin, *die;* **be** ~**s with sb.**

mit jmdm. befreundet sein; **make ~s [with sb.]** [mit jmdm.] Freundschaft schließen; **a ~ in need is a ~ indeed** *(prov.)* Freunde in der Not gehn hundert *od.* tausend auf ein Lot *(Spr.)*; **between ~s** unter Freunden; **~s in high places** einflußreiche Freunde; **b) the Society of F~s** die Quäker

friendless ['frendlɪs] *adj.* ohne Freund[e] *nachgestellt*

friendliness ['frendlɪnɪs] *n., no pl.* Freundlichkeit, *die*

friendly ['frendlɪ] **1.** *adj.* **a)** freundlich **(to** zu); freundschaftlich ‹*Rat, Beziehungen, Wettkampf, Gespräch*›; **be on ~ terms** *or* **be ~ with sb.** mit jmdm. auf freundschaftlichem Fuße stehen; **b)** *(not hostile)* freundlich [gesinnt] ‹*Bewohner*›; befreundet ‹*Staat*›; zutraulich ‹*Tier*›; **c)** *(well-wishing)* wohlwollend ‹*Erwähnung*›. **2.** *n. (Sport)* Freundschaftsspiel, *das*

friendship ['frendʃɪp] *n.* Freundschaft, *die;* **strike up a ~ with sb.** sich mit jmdm. anfreunden

Friesian ['fri:zɪən, 'fri:ʒən] *(Agric.)* **1.** *adj.* schwarzbunt. **2.** *n.* Schwarzbunte, *die*

frigate ['frɪgət] *n. (Naut.)* Fregatte, *die*

fright [fraɪt] *n.* Schreck, *der;* Schrecken, *der;* **take ~:** erschrecken; **give sb. a ~:** jmdm. einen Schreck[en] einjagen; **get** *or* **have a ~:** einen Schreck[en] bekommen; **be** *or* **look a ~:** zum Fürchten aussehen *(ugs.)*

frighten ['fraɪtn] *v.t.* ‹*Explosion, Schuß:*› erschrecken; ‹*Gedanke, Drohung:*› angst machen (+ *Dat.*); **be ~ed at** *or* **by sth.** vor etw. *(Dat.)* erschrecken; **~ sb. out of his wits** jmdn. furchtbar erschrecken; **be ~ed to death** *(fig.)* zu Tode erschrocken sein

~ a'way, ~ 'off *v.t.* vertreiben; *(put off)* abschrecken

frightened ['fraɪtnd] *adj.* verängstigt; angsterfüllt ‹*Stimme*›; **be ~ [of sth.]** [vor etw. *(Dat.)*] Angst haben

frightening ['fraɪtnɪŋ] *adj.* furchterregend

frightful ['fraɪtfl] *adj.* furchtbar; schrecklich; *(coll.: terrible)* furchtbar *(ugs.)*

frightfully ['fraɪtfəlɪ] *adv.* furchtbar; schrecklich; *(coll.: extremely)* furchtbar *(ugs.)*

frigid ['frɪdʒɪd] *adj. (formal, unfriendly)* frostig; *(sexually unresponsive)* frigid[e] ‹*Frau*›

frill [frɪl] *n.* **a)** *(ruffled edge)* Rüsche, *die;* **b)** *in pl. (embellishments)* Beiwerk, *das;* Ausschmückungen *(fig.)*; **with no ~s** ‹*Ferienhaus, Auto*› ohne besondere Ausstattung

frilly ['frɪlɪ] *adj.* mit Rüschen besetzt; Rüschen‹*kleid, -bluse*›

fringe [frɪndʒ] *n.* **a)** *(bordering)* Fransen; Fransenkante, *die* **(on** an + *Dat.*); **b)** *(hair)* [Pony]fransen *(ugs.)*; **c)** *(edge)* Rand, *der; attrib.* Rand‹*geschehen, -gruppe, -gebiet*›; **live on the ~[s] of the city** in den Randgebieten der Stadt wohnen; **lunatic ~:** Extremisten; *attrib.* **~ benefits** zusätzliche Leistungen

frisk [frɪsk] **1.** *v.i.* **~ [about]** [herum]springen. **2.** *v.t. (coll.)* filzen *(ugs.)*

frisky ['frɪskɪ] *adj.* munter

[1]fritter ['frɪtə(r)] *n. (Cookery)* **apple/sausage ~s** Apfelstücke/Würstchen in Pfannkuchenteig

[2]fritter *v.t.* **~ away** vergeuden; verplempern *(ugs.)*

frivolity [frɪ'vɒlɪtɪ] *n., no pl.* Oberflächlichkeit, *die;* Leichtfertigkeit, *die*

frivolous ['frɪvələs] *adj.* **a)** *(not serious)* frivol; extravagant ‹*Kleidung*›; **b)** *(trifling, futile)* belanglos

fro [frəʊ] *see* **to 2 b**

frock [frɒk] *n.* Kleid, *das*

frog [frɒg] *n.* Frosch, *der;* **have a ~ in the** *or* **one's throat** *(coll.)* einen Frosch im Hals haben *(ugs.)*

frog: ~man ['frɒgmən] *n., pl.* **~men** ['frɒgmən] Froschmann, *der;* **~-march** *v.t. (carry)* zu viert an Händen und Füßen tragen; *(hustle)* ≈ im Polizeigriff abführen; **~-spawn** *n.* Froschlaich, *der*

frolic ['frɒlɪk] *v.i.*, **-ck-:** **~ [about** *or* **around]** [herum]springen

from [frəm, *stressed* frɒm] *prep.* **a)** *expr. starting-point* von; *(~ within)* aus; **[come] ~ Paris/Munich** aus Paris/München [kommen]; **~ Paris to Munich** von Paris nach München; **where have you come ~?** woher kommen Sie?; **b)** *expr. beginning* von; **~ the year 1972 we never saw him again** seit 1972 haben wir ihn nie mehr [wieder]gesehen; **~ tomorrow [until ...]** von morgen an [bis ...]; **start work ~ 2 August** am 2. August anfangen zu arbeiten; **~ now on** von jetzt an; **~ then on** seitdem; **c)** *expr. lower limit* von; **blouses [ranging] ~ £2 to £5** Blusen [im Preis] zwischen 2 und 5 Pfund; **dresses ~ £20 [upwards]** Kleider von 20 Pfund aufwärts *od.* ab 20 Pfund; **~ 4 to 6 eggs** 4 bis 6 Eier; **~ the age of 18 [upwards]** ab 18 Jahre *od.* Jahren; **~ a child** *(since childhood)* schon als Kind; **d)** *expr. distance* von; **be a mile ~ sth.** eine Meile von etw. entfernt sein; **away ~ home** von zu Hause weg; **e)** *expr. removal, avoidance* von; *expr. escape* vor (+ *Dat.*); **f)** *expr. change* von; **~ ... to ...:** von ... zu ...; *(relating to price)* von ... auf ...; **~ crisis to crisis, ~ one crisis to another** von einer Krise zur anderen; **g)** *expr. source, origin* aus; **pick apples ~ a tree** Äpfel vom Baum pflücken; **buy everything ~ the same shop** alles im selben Laden kaufen; **where do you come ~?, where are you ~?** woher kommen Sie?; **~ the country** vom Land; **h)** *expr. viewpoint* von [... aus]; **i)** *expr. giver, sender* von; **take it ~ me that ...:** laß dir gesagt sein, daß ...; **j)** *(after the model of)* **painted ~ life/nature** nach dem Leben/nach der Natur gemalt; **k)** *expr. reason, cause* **she was weak ~ hunger/tired ~ so much work** sie war schwach vor Hunger/müde von der vielen Arbeit; **~ what I can see/have heard ...:** wie ich das sehe/wie ich gehört habe, ...; **l)** *with adv.* von ‹*unten, oben, innen, außen*›; **m)** *with prep.* **~ behind/under[neath] sth.** hinter/unter etw. *(Dat.)* hervor

front [frʌnt] **1.** *n.* **a)** Vorderseite, *die; (of door)* Außenseite, *die; (of house)* Vorderfront, *die; (of queue)* vorderes Ende; *(of procession)* Spitze, *die; (of book)* vorderer Deckel; **in** *or* **at the ~ [of sth.]** vorn [in etw. *position: Dat., movement: Akk.*]; **sit in the ~ of the car** vorne sitzen; **the index is at the ~:** das Register ist vorn; **to the ~:** nach vorn; **in ~:** vorn[e]; **be in ~ of sth./sb.** vor etw./jmdm. sein; **walk in ~ of sb.** *(preceding)* vor jmdm. gehen; *(to position)* vor jmdn. gehen; **he was murdered in ~ of his wife** er wurde vor den Augen seiner Frau ermordet; **b)** *(Mil.; also fig.)* Front, *die;* **on the Western ~:** an der Westfront; **be attacked on all ~s** an allen Fronten/*(fig.)* von allen Seiten angegriffen werden; **c)** *(at seaside)* Strandpromenade, *die;* **d)** *(Meteorol.)* Front, *die;* **cold/warm ~:** Kalt-/Warmluftfront, *die;* **e)** *(outward appearance)* Aussehen, *das; (bluff)* Fassade, *die (oft abwertend); (pretext, façade)* Tarnung, *die;* **put on a brave ~:** nach außen hin gefaßt bleiben; **it's all a ~:** das ist alles nur Fassade *(abwer-*

tend). **2.** *adj.* vorder...; Vorder-⟨*rad, -zimmer, -zahn*⟩; ~ **garden** Vorgarten, *der;* ~ **row** erste Reihe
frontage ['frʌntɪdʒ] *n.* **a)** *(extent)* Frontbreite, *die;* **b)** *(façade)* Fassade, *die*
frontal ['frʌntl] *adj.* **a)** Frontal-; **b)** *(Art)* frontal ⟨*Darstellung*⟩; **[full]** ~: frontal dargestellt ⟨*Akt*⟩
front: ~ **'bench** *n. (Brit. Parl.)* vorderste Bank; ~ **'door** *n. (of flat)* Wohnungstür, *die; (of house; also fig.)* Haustür, *die*
frontier ['frʌntɪə(r)] *n. (lit. or fig.)* Grenze, *die*
frontispiece ['frʌntɪspi:s] *n.* Frontispiz, *das;* Titelbild, *das*
front: ~ **'line** *n.* Front[linie], *die;* ~ **'page** *n.* Titelseite, *die;* **make the ~ page** auf die Titelseite kommen; ~ **runner** *n. (fig.)* Spitzenkandidat, *der;* ~ **'seat** *n. (in theatre)* Platz in den ersten Reihen; *(in car)* Vordersitz, *der; (in bus, coach)* vorderer Sitzplatz
frost [frɒst] **1.** *n.* Frost, *der; (frozen dew or vapour)* Reif, *der;* **ten degrees of ~** *(Brit.)* zehn Grad minus. **2.** *v. t.* **a)** *(esp. Amer. Cookery)* mit Zucker bestreuen; *(ice)* glasieren; **b) ~ed glass** Mattglas, *das*
~ **'over 1.** *v. t.* **be ~ed over** vereist sein. **2.** *v. i.* vereisen
frost: **~-bite** *n.* Erfrierung, *die;* **~-bitten** *adj.* durch Frost geschädigt; **sb. is ~-bitten** jmd. hat Erfrierungen; **his toes are ~-bitten** er hat Frost *od.* Erfrierungen in den Zehen
frosting ['frɒstɪŋ] *n. (esp. Amer. Cookery)* Zucker, *der; (icing)* Glasur, *die*
frosty ['frɒstɪ] *adj. (lit. or fig.)* frostig; *(with hoar-frost)* bereift
froth [frɒθ] **1.** *n. (foam)* Schaum, *der.* **2.** *v. i.* schäumen; ~ **at the mouth** Schaum vor dem Mund haben
frothy ['frɒθɪ] *adj.* schaumig; schäumend ⟨*Bier, Brandung, Maul*⟩
frown [fraʊn] **1.** *v. i.* die Stirn runzeln (**at, [up]on** über + *Akk.*); ~ **at sth./sb.** etw./jmdn. stirnrunzelnd ansehen. **2.** *n.* Stirnrunzeln, *das;* **with a [deep/worried/puzzled] ~:** mit [stark/sorgenvoll/verwirrt] gerunzelter Stirn; **a ~ of disapproval** ein mißbilligender Blick
froze *see* **freeze 1, 2**
frozen ['frəʊzn] **1.** *see* **freeze 1, 2. 2.** *adj.* **a)** gefroren, zugefroren ⟨*Fluß, See*⟩; erfroren ⟨*Tier, Person, Pflanze*⟩; eingefroren ⟨*Wasserleitung*⟩; **I am ~ stiff** *(fig.)* ich bin ganz steif gefroren; **my hands are ~** *(fig.)* meine Hände sind eiskalt; **b)** *(to preserve)* tiefgekühlt; ~ **food** Tiefkühlkost, *die*
frugal ['fru:gl] *adj.* sparsam ⟨*Hausfrau*⟩; genügsam ⟨*Lebensweise, Person*⟩; frugal ⟨*Mahl*⟩
fruit [fru:t] *n.* Frucht, *die; (collectively)* Obst, *das;* Früchte; **bear ~** *(lit. or fig.)* Früchte tragen
fruiterer ['fru:tərə(r)] *n.* Obsthändler, *der/*-händlerin, *die*
fruitful ['fru:tfl] *adj. (lit. or fig.)* fruchtbar
fruition [fru:'ɪʃn] *n.* **bring to ~:** verwirklichen ⟨*Plan, Ziel*⟩; **come to ~** ⟨*Plan:*⟩ Wirklichkeit werden
'fruit juice *n.* Fruchtsaft, *der*
fruitless ['fru:tlɪs] *adj.* nutzlos ⟨*Versuch*⟩; fruchtlos ⟨*Verhandlung, Bemühung, Suche*⟩
fruit: ~ **machine** *n. (Brit.)* Spielautomat, *der;* ~ **'salad** *n.* Obstsalat, *der;* **~-tree** *n.* Obstbaum, *der*
fruity ['fru:tɪ] *adj.* **a)** fruchtig ⟨*Geschmack, Wein*⟩; **b)** *(coll.: rich in tone)* volltönend ⟨*Stimme*⟩; herzhaft ⟨*Lachen*⟩
frump [frʌmp] *n. (derog.)* Vogelscheuche, *die (ugs.)*
frustrate [frʌ'streɪt, 'frʌstreɪt] *v. t.* vereiteln, durchkreuzen ⟨*Plan, Vorhaben, Versuch*⟩; zunichte machen ⟨*Hoffnung, Bemühungen*⟩; enttäuschen ⟨*Erwartung*⟩
frustrated [frʌ'streɪtɪd, 'frʌstreɪtɪd] *adj.* frustriert
frustrating [frʌ'streɪtɪŋ, 'frʌstreɪtɪŋ] *adj.* frustrierend; ärgerlich ⟨*Angewohnheit*⟩
frustration [frʌ'streɪʃn] *n.* Frustration, *die; (of plans, efforts)* Scheitern, *das*
[1]**fry** [fraɪ] *v. t.* braten; **fried eggs/potatoes** Spiegeleier/Bratkartoffeln
~ **'up** *v. t.* aufbraten ⟨*Reste*⟩
[2]**fry** *n. (young fishes etc.)* Brut, *die;* **'small ~** *(fig.)* unbedeutende Leute
frying-pan ['fraɪɪŋpæn] *n.* Bratpfanne, *die;* **[fall/jump] out of the ~ into the fire** vom Regen in die Traufe [kommen] *(ugs.)*
fry: **~pan** *(Amer.) see* **frying-pan; ~-up** *n.* Pfannengericht, *das*
ft. *abbr.* **feet, foot** ft.
fuchsia ['fju:ʃə] *n. (Bot.)* Fuchsie, *die*
fuck [fʌk] *(coarse)* **1.** *v. t. & i.* ficken *(vulg.);* **[oh,] ~!, [oh,] ~ it!** [au,] Scheiße! *(derb);* ~ **you!** leck mich am Arsch! *(derb).* **2.** *n. (act)* Fick, *der (vulg.)*
fuddle ['fʌdl] *v. t.* **a)** *(intoxicate)* **slightly ~d** [leicht] beschwipst *(ugs.);* **b)** *(confuse)* verwirren
fuddy-duddy ['fʌdɪdʌdɪ] *(sl.)* **1.** *adj.* verkalkt *(ugs.).* **2.** *n.* Fossil, *das (fig.)*
[1]**fudge** [fʌdʒ] *n. (sweet)* Karamelbonbon, *der od. das*
[2]**fudge 1.** *v. t.* frisieren *(ugs.)* ⟨*Geschäftsbücher*⟩; sich *(Dat.)* aus den Fingern saugen ⟨*Ausrede, Geschichte, Entschuldigung*⟩. **2.** *n.* Schwindel, *der*
fuel ['fju:əl] **1.** *n.* Brennstoff, *der; (for vehicle)* Kraftstoff, *der; (for ship, aircraft, spacecraft)* Treibstoff, *der;* **add ~ to the flames** *or* **fire** *(fig.)* Öl ins Feuer gießen. **2.** *v. t., (Brit.)* **-ll-** auftanken ⟨*Schiff, Flugzeug*⟩; *(fig.: stimulate)* Nahrung geben (+ *Dat.*) ⟨*Verdacht, Spekulationen*⟩; anheizen ⟨*Inflation*⟩
fuel: ~ **consumption** *n. (of vehicle)* Kraftstoffverbrauch, *der;* ~ **pump** *n.* Kraftstoffpumpe, *die;* ~ **tank** *n. (of vehicle)* Kraftstofftank, *der; (for storage)* Kraftstoffbehälter, *der*
fug [fʌg] *n. (coll.)* Mief, *der (salopp)*
fugitive ['fju:dʒɪtɪv] **1.** *adj. (lit. or fig.)* flüchtig. **2.** *n.* **a)** Flüchtige, *der/die;* **be a ~ from justice/from the law** auf der Flucht vor der Justiz/dem Gesetz sein; **b)** *(exile)* Flüchtling, *der*
fugue [fju:g] *n. (Mus.)* Fuge, *die*
fulfil (*Amer.:* **fulfill**) [fʊl'fɪl] *v. t.,* **-ll-** erfüllen; stillen ⟨*Verlangen, Bedürfnisse*⟩; entsprechen (+ *Dat.*) ⟨*Erwartungen*⟩; ausführen ⟨*Befehl*⟩; halten ⟨*Versprechen*⟩; **be fulfilled** ⟨*Wunsch, Hoffnung, Prophezeiung:*⟩ sich erfüllen; **be** *or* **feel fulfilled [in one's job]** [in seinem Beruf] Erfüllung finden
fulfilment (*Amer.:* **fulfillment**) [fʊl'fɪlmənt] *n.* Erfüllung, *die; (of an order)* Ausführung, *die;* **bring sth. to ~:** etw. erfüllen
full [fʊl] **1.** *adj.* **a)** voll; **the jug is ~ of water** der Krug ist voll Wasser; **the bus was completely ~:** der Bus war voll besetzt; ~ **of hatred/holes** voller Haß/Löcher; **be ~ up** *(coll.)* voll [besetzt] sein; ⟨*Behälter:*⟩ randvoll sein; ⟨*Liste:*⟩ voll sein; ⟨*Flug:*⟩ völlig ausgebucht sein (*see also* **c**); **b)** ~ **of** *(engrossed with):* **be ~ of oneself/one's own importance** sehr von sich eingenommen sein/sich sehr wichtig nehmen; **she's been ~ of it ever since** seitdem spricht sie von nichts anderem [mehr]; **the newspapers are ~ of the crisis** die Zeitungen sind voll von Berichten über die Krise; **c)** *(replete with*

food) voll ‹*Magen*›; satt ‹*Person*›; **I'm ~ [up]** *(coll.)* ich bin voll [bis obenhin] *(ugs.)* (*see also* **a**); **d)** *(comprehensive)* ausführlich, umfassend ‹*Bericht, Beschreibung*›; *(satisfying)* vollwertig ‹*Mahlzeit*›; erfüllt ‹*Leben*›; *(complete)* ganz ‹*Stunde, Tag, Jahr, Monat, Semester, Seite*›; voll ‹*Name, Fahrpreis, Gehalt, Bezahlung, Unterstützung, Mitgefühl, Verständnis*›; **[the] ~ details** alle Einzelheiten; **in ~ daylight** am hellichten Tag; **the moon is ~:** es ist Vollmond; **in ~ bloom** in voller Blüte; **~ member** Vollmitglied, *das;* **in ~ view of sb.** [direkt] vor jmds. Augen; **at ~ speed** mit Höchstgeschwindigkeit; **be at ~ strength** ‹*Mannschaft, Ausschuß, Kabinett:*› vollzählig sein; **e)** *(intense in quality)* hell, voll ‹*Licht*›; voll ‹*Klang, Stimme, Aroma*›; **f)** *(rounded, plump)* voll ‹*Gesicht, Busen, Lippen, Mund, Segel*›; füllig ‹*Figur*›; weit geschnitten ‹*Rock*›. **2.** *n.* **a) in ~:** vollständig; **write your name [out] in ~:** schreiben Sie Ihren Namen aus; **b) enjoy sth. to the ~:** etw. in vollen Zügen genießen. **3.** *adv.* **a)** *(very)* **know ~ well that ...:** ganz genau *od.* sehr wohl wissen, daß ...; **b)** *(exactly, directly)* genau; **~ in the face** direkt ins Gesicht ‹*schlagen, scheinen*›; **look sb. ~ in the face** jmdn. voll ansehen

full: ~-blooded ['fʊlblʌdɪd] *adj. (vigorous)* vollblütig; **~-blown** *adj.* ausgewachsen ‹*Skandal*›; ausgereift ‹*Theorie, Plan, Gedanke*›; umfassend ‹*Bericht*›; **~ 'board** *n.* Vollpension, *die;* **~-bodied** ['fʊlbɒdɪd] *adj.* vollmundig, *(fachspr.)* körperreich ‹*Wein*›; voll ‹*Ton, Klang*›; **~-cream milk** *n.* Vollmilch, *die;* **~ em'ployment** *n.* Vollbeschäftigung, *die;* **~-grown** *adj.* ausgewachsen ‹*Person, Tier*›; **~ 'house** *n. (Theatre)* ausverkauftes *od.* volles Haus; **~ 'length** *adv.* der Länge nach ‹*hinfallen, liegen*›; **~-length** *adj.* abendfüllend ‹*Film, Theaterstück*›; **~-length portrait** Ganzporträt, *das;* **~-length dress** langes Kleid; **~ 'marks** *n. pl., no art.* die höchste Bewertung; *(Sch., Univ.)* die beste Note; **~ marks!** *(fig. coll.)* ausgezeichnet!; **give sb. ~ marks** *(fig.)* jmdm. höchstes Lob zollen; **~ 'moon** *n.* Vollmond, *der*

fullness ['fʊlnɪs] *n., no pl. (of skirt)* weiter Schnitt; *(of figure)* Fülligkeit, *die; (of face)* Rundheit, *die;* **in the ~ of time** *(literary)* wenn die Zeit dafür gekommen ist/als die Zeit dafür gekommen war

full: ~-page *adj.* ganzseitig; **~ 'point** Punkt, *der;* **~-scale** *adj.* **a)** in Originalgröße *nachgestellt;* **b)** großangelegt ‹*Werbekampagne, Untersuchung, Suchaktion*›; umfassend ‹*Umarbeitung, Revision*›; **~ 'stop** *n.* **a)** Punkt, *der;* **b)** *(fig. coll.)* **come to a ~ stop** zum Stillstand kommen; **I'm not going, ~ stop!** ich gehe nicht, [und damit] basta! *(ugs.);* **~ 'time** *adv.* ganztags ‹*arbeiten*›; **~-time** *adj.* ganztägig; Ganztags‹*arbeit, -beschäftigung*›

fully ['fʊlɪ] *adv.* **a)** voll [und ganz]; fest ‹*entschlossen*›; ausführlich ‹*erklären usw.*›; restlos ‹*überzeugt*›; **b)** *(at least)* **~ two hours** volle zwei Stunden; **~ three weeks ago** vor gut drei Wochen

fully: ~-fledged ['fʊlɪfledʒd] *attrib. adj.* flügge ‹*Vogel*›; *(fig.)* [ganz] selbständig; **~-qualified** *attrib. adj.* vollqualifiziert

fulsome ['fʊlsəm] *adj.* übertrieben ‹*Lob, Kompliment*›

fumble ['fʌmbl] *v.i.* **~ at** *or* **with** [herum]fingern an (+ *Dat.*); **~ with one's papers** in seinen Papieren kramen *(ugs.);* **~ in one's pockets for sth.** in seinen Taschen nach etw. fingern *od. (ugs.)* kramen; **~ for the light-switch** nach dem Lichtschalter tasten; **~ [about** *or* **around] in the dark** im Dunkeln herumtasten

fume [fju:m] **1.** *n. in pl.* **petrol/ammonia ~s** Benzin-/Ammoniakdämpfe; **~s of wine/whisky** Alkohol-/Whiskydunst, *der.* **2.** *v.i.* vor Wut schäumen; **~ at sb.** auf *od.* über jmdn. wütend sein; **~ at** *or* **about sth.** wegen etw. wütend sein

fumigate ['fju:mɪgeɪt] *v.t.* ausräuchern

fun [fʌn] **1.** *n.* Spaß, *der;* **have ~ doing sth.** Spaß daran haben, etw. zu tun; **have ~!** viel Spaß!; **make ~ of** *or* **poke ~ at sb./sth.** sich über jmdn./etw. lustig machen; **in ~:** im Spaß; **for ~, for the ~ of it** zum Spaß; **spoil the** *or* **sb.'s ~:** jmdm. den Spaß verderben; **sth. is [good** *or* **great/no] ~:** etw. macht [großen/keinen] Spaß; **it's no ~ being unemployed** es ist kein Vergnügen, arbeitslos zu sein; **we had the usual ~ and games with him** *(iron.: trouble)* wir hatten wieder das übliche Theater mit ihm *(ugs.).* **2.** *adj. (coll.)* lustig; amüsant

function ['fʌŋkʃn] **1.** *n.* **a)** *(role)* Aufgabe, *die;* **in his ~ as surgeon** in seiner Funktion *od.* Eigenschaft als Chirurg; **b)** *(mode of action)* Funktion, *die;* **c)** *(reception)* Empfang, *der; (official ceremony)* Feierlichkeit, *die;* **d)** *(Math.)* Funktion, *die.* **2.** *v.i.* ‹*Maschine, System, Organisation:*› funktionieren; ‹*Organ:*› arbeiten; **~ as** *(have the ~ of)* fungieren als; *(serve as)* dienen als

functional ['fʌŋkʃənl] *adj.* **a)** *(useful, practical)* funktionell; **b)** *(working)* funktionsfähig; **be ~ again** wieder funktionieren

fund [fʌnd] **1.** *n.* **a)** *(of money)* Fonds, *der;* **b)** *(fig.: stock, store)* Fundus, *der* (**of** von, an + *Dat.*); **c)** *in pl. (resources)* Mittel *Pl.;* Gelder *Pl.;* **be short of ~s** knapp *od.* schlecht bei Kasse sein *(ugs.).* **2.** *v.t.* finanzieren

fundamental [fʌndə'mentl] *adj.* grundlegend (**to** für); elementar ‹*Bedürfnisse*›; *(primary, original)* Grund‹*struktur, -form, -typus*›

fundamentally [fʌndə'mentəlɪ] *adv.* grundlegend; von Grund auf ‹*verschieden, ehrlich*›; **~ opposed to sth.** grundsätzlich gegen etw.; **man is ~ good** der Mensch ist von Natur aus gut

funeral ['fju:nərl] *n.* **a)** Beerdigung, *die;* **b)** *attrib.* **~ director** Bestattungsunternehmer, *der;* **~ procession** Leichenzug, *der (geh.);* **~ service** Trauerfeier, *die;* **c)** *(sl.: concern)* **that's my ~:** das ist mein Problem

funereal [fju:'nɪərɪəl] *adj.* düster; **~ expression** Trauermiene, *die (ugs.);* trauervolle Miene

'fun-fair *n. (Brit.)* Jahrmarkt, *der*

fungus ['fʌŋgəs] *n., pl.* **fungi** ['fʌŋgaɪ, 'fʌndʒaɪ] *or* **~es** Pilz, *der*

funicular [fju:'nɪkjʊlə(r)] *adj.* **~ [railway]** [Stand]seilbahn, *die*

funk [fʌŋk] *(sl.)* **1.** *n.* Bammel, *der (salopp);* Schiß, *der (salopp);* **be in a [blue] ~:** [mächtig] Bammel *od.* Schiß haben *(salopp).* **2.** *v.t.* kneifen vor (+ *Dat.*) *(ugs.);* **he ~ed it** er hat gekniffen *(ugs.)*

funnel ['fʌnl] **1.** *n.* **a)** *(cone)* Trichter, *der;* **b)** *(of ship etc.)* Schornstein, *der.* **2.** *v.i., (Brit.)* **-ll-** strömen

funnily ['fʌnɪlɪ] *adv.* komisch; **~ enough** komischerweise *(ugs.)*

funny ['fʌnɪ] *adj.* **a)** *(comical)* komisch; lustig; witzig ‹*Person, Einfall, Bemerkung*›; **are you being** *or* **trying to be ~?** das soll wohl ein Witz sein?; **b)** *(strange)* komisch; seltsam; **the ~ thing 'is that ...:** das Komische [daran] ist, daß ...; **have a ~ feeling that ...:**

das komische Gefühl haben, daß ...; **there's something ~ going on here** hier ist doch was faul *(ugs.)*
'funny-bone *n. (Anat.)* Musikantenknochen, *der*
fur [fɜ:(r)] **1.** *n.* **a)** *(coat of animal)* Fell, *das; (for or as garment)* Pelz, *der; attrib.* **~ coat/hat** Pelzmantel, *der*/-mütze, *die;* **trimmed/lined with ~:** mit Pelz besetzt *od.* verbrämt/gefüttert; **b)** *(coating formed by hard water)* Wasserstein, *der; (in kettle)* Kesselstein, *der.* **2.** *v. i.,* **-rr-: the kettle has/pipes have ~red [up]** im Kessel hat sich Kesselstein/in den Rohren hat sich Wasserstein gebildet
furious ['fjʊərɪəs] *adj.* wütend; heftig ‹*Streit, Kampf, Sturm, Lärm*›; wild ‹*Tanz, Sturm, Tempo, Kampf*›; **be ~ with sb./at sth.** wütend auf jmdn./über etw. *(Akk.)* sein
furiously ['fjʊərɪəslɪ] *adv.* wütend; wild ‹*kämpfen, tanzen*›; wie wild *(ugs.)* ‹*arbeiten, in die Pedale treten*›
furlong ['fɜ:lɒŋ] *n.* Achtelmeile, *die*
furnace ['fɜ:nɪs] *n.* Ofen, *der; (blast-~)* Hochofen, *der; (smelting-~)* Schmelzofen, *der*
furnish ['fɜ:nɪʃ] *v. t.* **a)** möblieren; **live in ~ed accommodation** möbliert wohnen; **~ing fabrics** Möbel- und Vorhangstoffe; **b)** *(provide, supply)* liefern ‹*Vorräte*›; **~ sb. with sth.** jmdm. etw. liefern
furnishings ['fɜ:nɪʃɪŋz] *n. pl.* Einrichtungsgegenstände
furniture ['fɜ:nɪtʃə(r)] *n., no pl.* Möbel *Pl.;* **piece of ~:** Möbel[stück], *das*
furniture: ~ polish *n.* Möbelpolitur, *die;* **~ van** *n.* Möbelwagen, *der*
furore [fjʊə'rɔ:rɪ] (*Amer.:* **furor** ['fjʊərɔ:(r)]) *n.* **create** *or* **cause a ~:** Furore machen; *(cause a scandal)* einen Skandal verursachen
furrier ['fʌrɪə(r)] *n. (one who prepares fur)* Kürschner, *der*/Kürschnerin, *die; (dealer)* Pelzhändler, *der*/-händlerin, *die*
furrow ['fʌrəʊ] **1.** *n. (lit. or fig.)* Furche, *die.* **2.** *v. t. (mark with wrinkles)* **~ed face** zerfurchtes Gesicht
furry ['fɜ:rɪ] *adj.* haarig; flauschig ‹*Stoff*›; belegt ‹*Zunge*›
further ['fɜ:ðə(r)] **1.** *adj. compar. of* **far: a)** *(of two)* ander...; *(in space)* weiter entfernt; **on the ~ bank of the river/side of town** am anderen Ufer/Ende der Stadt; **b)** *(additional)* weiter...; **till ~ notice/orders** bis auf weiteres; **will there be anything ~?** darf es noch etwas sein?; haben Sie sonst noch einen Wunsch?; **~ details** *or* **particulars** weitere *od.* nähere Einzelheiten. **2.** *adv. compar. of* **far: a)** weiter; **not let it go any ~** *(keep it secret)* es nicht weitersagen; **until you hear ~ from us** bis Sie wieder von uns hören; **nothing was ~ from his thoughts** nichts lag ihm ferner; **b)** *(moreover)* außerdem. **3.** *v. t.* fördern; **~ one's career** beruflich vorankommen
furtherance ['fɜ:ðərəns] *n., no pl.* Förderung, *die;* Unterstützung, *die;* **in ~ of sth.** zur Förderung *od.* Unterstützung einer Sache *(Gen.)*
further edu'cation *n.* Weiterbildung, *die; (for adults also)* Erwachsenenbildung, *die*
furthermore [fɜ:ðə'mɔ:(r)] *adv.* außerdem
furthermost ['fɜ:ðəməʊst] *adj.* äußerst...; entlegenst...; **to the ~ ends of the earth** bis ans Ende der Welt
furthest ['fɜ:ðɪst] **1.** *adj. superl. of* **far** am weitesten entfernt; **ten miles at the ~:** höchstens zehn Meilen. **2.** *adv. superl. of* **far** am weitesten ‹*springen, laufen*›; am weitesten entfernt ‹*sein, wohnen*›
furtive ['fɜ:tɪv] *adj.* verstohlen; **his ~ behaviour** seine offenkundige Bemühtheit, nicht aufzufallen
furtively ['fɜ:tɪvlɪ] *adv.* verstohlen
fury ['fjʊərɪ] *n.* **a)** Wut, *die; (of storm, sea, battle, war)* Wüten, *das;* **in a ~:** wütend; **fly into a/be in a ~:** einen Wutanfall bekommen/haben; **b) like ~** *(coll.)* wie wild *(ugs.)*
¹fuse [fju:z] **1.** *v. t. (blend)* verschmelzen **(into** zu). **2.** *v. i. (blend)* **~ together** miteinander verschmelzen; **~ with sth.** *(fig.)* sich mit etw. verbinden
²fuse *n.* **[time-]~:** [Zeit]zünder, *der; (cord)* Zündschnur, *die*
³fuse *(Electr.)* **1.** *n.* Sicherung, *die.* **2.** *v. t.* **~ the lights** die Sicherung [für die Lampen] durchbrennen lassen. **3.** *v. i.* **the lights have ~d** die Sicherung [für die Lampen] ist durchgebrannt
'fuse-box *n. (Electr.)* Sicherungskasten, *der*
fuselage ['fju:zəlɑ:ʒ] *n. (Aeronaut.)* [Flugzeug]rumpf, *der*
fusion ['fju:ʒn] *n.* **a)** Verschmelzung, *die; (fig.)* Verbindung, *die;* **b)** *(Phys.)* Fusion, *die*
fuss [fʌs] **1.** *n.* Theater, *das (ugs.);* **without any ~:** ohne großes Theater *(ugs.);* **kick up a ~:** ein großes Theater machen; **make a ~ [about sth.]** Aufhebens [von etw.] *od.* einen Wirbel [um etw.] machen; **make a ~ of** [einen] Wirbel machen um ‹*Person, Tier*›. **2.** *v. i.* Wirbel machen; *(get agitated)* sich [unnötig] aufregen; **she is always ~ing over sb./sth.** sie macht immer ein Theater mit jmdm./etw. *(ugs.)*
fussy ['fʌsɪ] *adj.* **a)** *(fastidious)* eigen; penibel; **be ~ about one's food** *or* **what one eats** mäklig im Essen sein *(ugs.);* **I'm not ~** *(I don't mind)* ich bin nicht wählerisch; **b)** *(full of unnecessary decoration)* verspielt
futile ['fju:taɪl] *adj.* vergeblich ‹*Versuch, Bemühungen, Vorschlag usw.*›; zum Scheitern verurteilt ‹*Plan, Vorgehen usw.*›
futility [fju:'tɪlɪtɪ] *n., no pl. (of effort, attempt, etc.)* Vergeblichkeit, *die; (of plan)* Zwecklosigkeit, *die; (of war)* Sinnlosigkeit, *die*
future ['fju:tʃə(r)] **1.** *adj.* **a)** [zu]künftig; **at some ~ date** zu einem späteren Zeitpunkt; **b)** *(Ling.)* futurisch; **~ tense** Futur, *das;* Zukunft, *die;* **~ perfect** Futur II, *das.* **2.** *n.* **a)** Zukunft, *die;* **a man with a ~:** ein Mann mit Zukunft; **in ~:** in Zukunft; künftig; **see sb. in the near ~:** jmdn. demnächst sehen; **there's no ~ in it** das hat keine Zukunft; **b)** *(Ling.)* Futur, *das;* Zukunft, *die*
futuristic [fju:tʃə'rɪstɪk] *adj.* futuristisch
fuze *(Amer.) see* **²fuse**
fuzz [fʌz] *n.* **a)** *(fluff)* Flaum, *der;* **b)** *(frizzy hair)* Kraushaar, *das;* **c)** *no pl. (sl.: police)* Polente, *die (salopp)*
fuzzy ['fʌzɪ] *adj.* **a)** *(like fluff)* flaumig; **b)** *(frizzy)* kraus; **c)** *(blurred)* verschwommen; unscharf

G

G, g [dʒi:] *n., pl.* **Gs** *or* **G's a)** *(letter)* G, g, *das;* **b) G** *(Mus.)* G, g, *das;* **G sharp** gis, Gis, *das;* **G flat** ges, Ges, *das*

g. *abbr.* **a) gram[s]** g; **b) gravity** g

gab [gæb] *n. (coll.)* **have the gift of the ~:** reden können

gabble ['gæbl] **1.** *v.i. (inarticulately)* brabbeln *(ugs.); (volubly)* schnattern *(fig.)*. **2.** *v.t.* herunterschnurren *(salopp)* ‹*Gebet, Gedicht*›. **3.** *n.* Gebrabbel, *das (ugs.)*

gable ['geɪbl] *n.* **a)** Giebel, *der;* **b)** *see* **gable-end**

gabled ['geɪbld] *adj.* gegiebelt; Giebel‹*dach, -haus*›

'**gable-end** *n.* Giebelseite, *die*

gad [gæd] *v.i.,* **-dd-** *(coll.)* ~ **about** *or* **around** herumziehen; sich herumtreiben *(ugs. abwertend)*

gadget ['gædʒɪt] *n.* Gerät, *das; (larger)* Apparat, *der;* **~s** *(derog.)* [technischer] Krimskrams *(ugs.)*

gadgetry ['gædʒɪtrɪ] *n., no pl.* [hochtechnisierte] Ausstattung

Gaelic ['geɪlɪk, 'gælɪk] **1.** *adj.* gälisch. **2.** *n.* Gälisch, *das; see also* **English 2 a**

gaff [gæf] *n. (sl.)* **blow the ~:** plaudern (**on** über + *Akk.*)

gaffe [gæf] *n.* Fauxpas, *der;* Fehler, *der;* **make** *or* **commit a ~:** einen Fauxpas begehen

gaffer ['gæfə(r)] *n. (coll.)* **a)** *(old fellow)* Alte, *der;* **b)** *(Brit.: boss)* Boß, *der (ugs.)*

gag [gæg] **1.** *n.* **a)** Knebel, *der;* **b)** *(joke)* Gag, *der.* **2.** *v.t.,* **-gg-:** ~ **sb.** jmdn. knebeln; *(fig.)* jmdn. zum Schweigen bringen

gaga ['gɑːgɑː] *adj. (sl.: senile)* verkalkt *(ugs.);* **go ~:** verkalken *(ugs.)*

gage *(Amer.) see* **gauge**

gaggle ['gægl] *n.* ~ **[of geese]** Schar [Gänse], *die*

gaiety ['geɪətɪ] *n., no pl.* Fröhlichkeit, *die*

gaily ['geɪlɪ] *adv.* **a)** fröhlich; **b)** *(brightly, showily)* in leuchtenden Farben ‹*bemalt, geschmückt*›; ~ **coloured** farbenfroh

gain [geɪn] **1.** *n.* **a)** Gewinn, *der;* **be to sb.'s ~:** für jmdn. von Vorteil sein; **ill-gotten ~s** unrechtmäßig erworbener Besitz; **b)** *(increase)* Zunahme, *die* (**in** an + *Dat.*). **2.** *v.t.* **a)** *(obtain)* gewinnen; finden ‹*Zugang, Zutritt*›; erwerben ‹*Wissen, Ruf*›; erlangen ‹*Freiheit, Ruhm*›; erzielen ‹*Vorteil, Punkte*›; verdienen ‹*Lebensunterhalt, Geldsumme*›; ~ **possession of sth.** in den Besitz einer Sache *(Gen.)* kommen; **b)** *(win)* gewinnen ‹*Preis, Schlacht*›; erringen ‹*Sieg*›; **c)** *(obtain as increase)* ~ **weight/five pounds [in weight]** zunehmen/fünf Pfund zunehmen; ~ **speed** schneller werden; **d)** *(reach)* gewinnen *(geh.),* erreichen ‹*Gipfel, Ufer*›; **e)** *(become fast by)* **my watch ~s two minutes a day** meine Uhr geht pro Tag zwei Minuten vor. **3.** *v.i.* **a)** *(make a profit)* ~ **by sth.** von etw. profitieren; **b)** *(obtain increase)* ~ **in influence/prestige** an Einfluß/ Prestige gewinnen; ~ **in wisdom** weiser werden; ~ **in knowledge** sein Wissen vergrößern; ~ **in weight** zunehmen; **c)** *(become fast)* ‹*Uhr:*› vorgehen; **d)** ~ **on sb.** *(come closer)* jmdm. [immer] näher kommen; *(increase lead)* den Vorsprung zu jmdm. vergrößern

gainful ['geɪnfl] *adj.* bezahlt; ~ **employment** Erwerbstätigkeit, *die*

gait [geɪt] *n.* Gang, *der;* **with a slow ~:** mit langsamen Schritten

gal. *abbr.* **gallon[s]** gal.; gall.

gala ['gɑːlə, 'geɪlə] *n.* **a)** *(fête)* Festveranstaltung, *die; attrib.* Gala‹*abend, -diner, -vorstellung*›; **b)** *(Brit. Sport)* Sportfest, *das;* **swimming ~:** Schwimmfest, *das*

galaxy ['gæləksɪ] *n.* **a)** *(star system)* Galaxie, *die;* **b)** *(Milky Way)* **the G~:** die Milchstraße

gale [geɪl] *n.* Sturm, *der;* ~ **force** Sturmstärke, *die*

[1]**gall** [gɔːl] *n.* **a)** *(Physiol.)* Galle, *die;* **b)** *(sl.: impudence)* Unverschämtheit, *die;* Frechheit, *die*

[2]**gall** *v.t. (fig.) (annoy)* ärgern; *(vex)* schmerzen; **be ~ed by sth.** unter etw. *(Dat.)* leiden

gallant *adj.* **a)** ['gælənt] *(brave)* tapfer; *(chivalrous)* ritterlich; **b)** ['gælənt, gə'lænt] *(attentive to women)* galant

gallantly *adv.* **a)** ['gæləntlɪ] *(bravely)* tapfer; **b)** ['gæləntlɪ, gə'læntlɪ] *(with courtesy)* galant

gallantry ['gæləntrɪ] *n.* **a)** *(bravery)* Tapferkeit, *die;* **b)** *(courtesy)* Galanterie, *die (geh.)*

'**gall-bladder** *n. (Anat.)* Gallenblase, *die*

galleon ['gælɪən] *n. (Hist.)* Galeone, *die*

gallery ['gælərɪ] *n.* **a)** *(Archit.)* Galerie, *die;* **b)** *(Theatre)* dritter Rang; **play to the ~** *(fig. coll.)* für die Galerie spielen; **c)** *(art ~) (building)* Galerie, *die; (room)* Ausstellungsraum, *der*

galley ['gælɪ] *n.* **a)** *(Hist.)* Galeere, *die;* **b)** *(kitchen) (of ship)* Kombüse, *die; (of aircraft)* Bordküche, *die;* **c)** *(Printing)* Satzschiff, *das;* ~ **[proof]** [Druck]fahne, *die*

'**galley-slave** *n.* Galeerensklave, *der*

Gallic ['gælɪk] *adj.* gallisch

Gallicism ['gælɪsɪzm] *n.* **a)** Gallizismus, *der;* **b)** *(characteristic)* französische Eigenart

galling ['gɔːlɪŋ] *adj.* **a)** *(irritating)* ärgerlich; **b)** *(humiliating)* erniedrigend

gallivant [gælɪ'vænt] *v.i. (coll.)* herumziehen *(ugs.)* **(about, around** in + *Dat.*)

gallon ['gælən] *n.* Gallone, *die;* **drink ~s of water** *etc. (fig. coll.)* literweise Wasser *usw.* trinken

gallop ['gæləp] **1.** *n.* Galopp, *der;* **at a ~:** im Galopp. **2.** *v.i.* **a)** ‹*Pferd, Reiter:*› galoppieren; **b)** *(fig.)* ~ **through** im Galopp *(ugs.)* durchlesen ‹*Buch*›; rasch herunterspielen ‹*Musikstück*›; im Galopp *(ugs.)* erledigen ‹*Arbeit*›; **~ing inflation** *(fig.)* galoppierende Inflation

gallows ['gæləʊz] *n. sing.* Galgen, *der*

'**gallstone** *n. (Med.)* Gallenstein, *der*

galore [gə'lɔː(r)] *adv.* im Überfluß; in Hülle und Fülle

galvanize (galvanise) ['gælvənaɪz] *v.t.* **a)** *(fig.: rouse)* wachrütteln ‹*Volk, Partei usw.*›; ~ **sb. into action** jmdn. veranlassen, sofort aktiv zu werden; **b)** *(coat with zinc)* verzinken

gambit ['gæmbɪt] *n. (Chess)* Gambit, *das; (fig.: trick, device)* Schachzug, *der;* **[opening]** ~ *(fig.)* einleitender Schachzug; *(in conversation)* einleitende Bemerkung

gamble ['gæmbl] **1.** *v.i.* **a)** [um Geld] spielen; ~ **at cards/on horses** mit Karten um Geld spielen/auf Pferde wetten; **b)** *(fig.)* spekulieren; ~ **on the Stock Exchange/in oil shares** an der Börse/ in Öl[aktien] spekulieren; ~ **on sth.** sich auf etw. *(Akk.)* verlassen. **2.** *v.t.* **a)** verspielen; **b)** *(fig.)* riskieren, aufs Spiel setzen ‹*Vermögen*›. **3.** *n. (lit. or fig.)* Glücksspiel, *das;* **take a ~:** ein Wagnis auf sich *(Akk.)* nehmen

~ **a'way** *v.t.* verspielen; *(on the Stock Exchange)* verspekulieren

gambler ['gæmblə(r)] *n.* Glücksspieler, *der*

gambling ['gæmblɪŋ] *n.* Spiel[en], *das;* Glücksspiel, *das; (on horses, dogs)* Wetten, *das*

'**gambling debts** *n. pl.* Spielschulden

gambol ['gæmbl] *v.i., (Brit.)* **-ll-** ‹*Kind, Lamm:*› herumspringen

[1]**game** [geɪm] *n.* **a)** *(form of contest)* Spiel, *das; (a contest) (with ball)* Spiel, *das; (at [table-]tennis, chess, cards, billiards, cricket)* Partie, *die;* **have** *or* **play a ~ of tennis/chess** *etc.* **[with sb.]** eine Partie Tennis/Schach *usw.* [mit

jmdm.] spielen; **have** *or* **play a ~ of football [with sb.]** Fußball [mit jmdm.] spielen; **be on/off one's ~**: gut in Form/nicht in Form sein; **beat sb. at his own ~** *(fig.)* jmdn. mit seinen eigenen Waffen schlagen *(geh.)*; **play the ~** *(fig.)* sich an die Spielregeln halten *(fig.)*; **[I'll show her that] two can play at that ~** *(fig.)* was sie kann, kann ich auch; **b)** *(fig.: scheme, undertaking)* Vorhaben, *das;* **play sb.'s ~**: jmdm. in die Hände arbeiten; *(for one's own benefit)* jmds. Spiel mitspielen; **the ~ is up** *(coll.)* das Spiel ist aus; **give the ~ away** alles verraten; **what's his ~?** *(coll.)* was hat er vor?; **what's the ~?** *(coll.)* was soll das?; **c)** *(business, activity)* Gewerbe, *das;* Branche, *die;* **be new to the ~** *(fig.)* neu im Geschäft sein *(auch fig. ugs.)*; **be/go on the ~** ‹*Prostituierte:*› anschaffen gehen *(salopp)*; **d)** *(diversion)* Spiel, *das; (piece of fun)* Scherz, *der;* Spaß, *der;* **don't play ~s with me** versuch nicht, mich auf den Arm zu nehmen *(ugs.)*; **e)** *in pl. (athletic contests)* Spiele; *(in school) (sports)* Schulsport, *der; (athletics)* Leichtathletik, *die;* **good at ~s** gut im Sport; **f)** *(portion of contest)* Spiel, *das;* **two ~s all** zwei beide; zwei zu zwei; **~ to Graf** *(Tennis)* Spiel Graf; **~, set, and match** *(Tennis)* Spiel, Satz und Sieg; **g)** *no pl. (Hunting, Cookery)* Wild, *das;* **fair ~** *(fig.)* Freiwild, *das;* **easy ~** *(fig. coll.)* leichte Beute; **big ~**: Großwild, *das*

[2]**game** *adj.* mutig; **be ~ to do sth.** *(be willing)* bereit sein, etw. zu tun; **be ~ for sth./anything** zu etw./allem bereit sein

'**gamekeeper** *n.* Wildheger, *der*

gamely ['geɪmlɪ] *adv.* mutig

game: ~ park, ~ reserve *ns.* Wildreservat, *das*

gamesmanship ['geɪmzmənʃɪp] *n., no pl.* Gerissenheit *od.* Gewieftheit *(ugs.)* beim Spiel

'**game-warden** *n.* Wildhüter, *der*

gaming ['geɪmɪŋ]: **~-machine** *n.* Münzspielgerät, *das;* **~-table** *n.* Spieltisch, *der*

gamma ['gæmə] *n. (letter)* Gamma, *das*

'**gamma rays** *n. pl. (Phys.)* Gammastrahlen *Pl.*

gammon ['gæmən] *n. (ham cured like bacon)* Räucherschinken, *der*

gamut ['gæmət] *n. (fig.: range)* Skala, *die;* **run the whole ~ of ...**: die ganze Skala von ... durchgehen

gander ['gændə(r)] *n.* **a)** *(Ornith.)* Gänserich, *der;* **b)** *(sl.: look, glance)* **have a ~ at/round sth.** sich *(Dat.)* etw. ansehen

gang [gæŋ] **1.** *n.* **a)** *(of workmen, slaves, prisoners)* Trupp, *der;* **b)** *(of criminals)* Bande, *die;* Gang, *die;* **~ of thieves/criminals/terrorists** Diebes-/Verbrecher-/Terroristenbande, *die;* **c)** *(coll.: group of friends etc.)* Haufen, *der;* Bande, *die (scherzh.)*. **2.** *v. i.* **a)** **~ up [with sb.]** *(join)* sich [mit jmdm.] zusammentun *(ugs.)*; **b)** **~ up against** *or* **on** *(coll.: combine against)* sich zusammenschließen gegen

gangling ['gæŋglɪŋ] *adj.* schlaksig *(ugs.)* ‹*Person, Gang, Gestalt*›

'**gangplank** *n. (Naut.)* Laufplanke, *die*

gangrene ['gæŋgri:n] *n. (Med.)* Gangrän, *die od. das;* Brand, *der*

gangrenous ['gæŋgrɪnəs] *adj. (Med.)* brandig

gangster ['gæŋstə(r)] *n.* Gangster, *der*

'**gangway** *n.* **a)** *(for boarding ship or plane)* Gangway, *die;* **b)** *(Brit.: between rows of seats)* Gang, *der;* **leave a ~** *(fig.)* einen Durchgang freilassen

gantry ['gæntrɪ] *n. (crane)* Portal, *das; (on road)* Schilderbrücke, *die; (Railw.)* Signalbrücke, *die; (Astronaut.)* Startrampe, *die*

gaol [dʒeɪl] *(Brit. in official use) see* **jail**

gaoler ['dʒeɪlə(r)] *(Brit. in official use) see* **jailer**

gap [gæp] *n.* **a)** Lücke, *die;* **a ~ in the curtains** ein Spalt im Vorhang; **b)** *(in time)* Pause, *die;* **c)** *(fig.: contrast, divergence in views etc.)* Kluft, *die;* **fill a ~**: eine Lücke füllen *od.* schließen; **stop** *or* **close** *or* **bridge a ~**: eine Kluft überbrücken *od.* überwinden

gape [geɪp] *v. i.* **a)** *(open mouth)* den Mund aufsperren; *(be open wide)* ‹*Schnabel, Mund:*› aufgesperrt sein; ‹*Loch, Abgrund, Wunde:*› klaffen; **b)** *(stare)* Mund und Nase aufsperren *(ugs.)*; **~ at sb./sth.** jmdn./etw. mit offenem Mund anstarren

gaping ['geɪpɪŋ] *adj.* **a)** *(open wide)* gähnend ‹*Loch*›; klaffend ‹*Wunde*›; **b)** *(staring)* erstaunt starrend

garage ['gærɑ:ʒ, 'gærɪdʒ] *n.* **a)** *(for parking)* Garage, *die;* **bus ~**: Busdepot, *das;* **b)** *(for selling petrol)* Tankstelle, *die; (for repairing cars)* [Kfz-]Werkstatt, *die; (for selling cars)* Autohandlung, *die*

garb [gɑ:b] *n.* Tracht, *die;* **strange ~**: seltsame Kleidung

garbage ['gɑ:bɪdʒ] *n.* **a)** Abfall, *der;* Müll, *der;* **b)** *(fig.: rubbishy literature)* Schund, *der;* **c)** *(coll.: nonsense)* Quatsch, *der (salopp)*

garbage: ~ can *(Amer.) see* **dustbin; ~ dis'posal unit, ~ disposer** ['gɑ:bɪdʒ dɪspəʊzə(r)] *ns.* Abfallvernichter, *der;* Müllwolf, *der*

garble ['gɑ:bl] *v. t.* **a)** verstümmeln, entstellen ‹*Bericht, Korrespondenz, Tatsache*›; **b)** *(confuse)* durcheinanderbringen

garden [gɑ:dn] *n.* Garten, *der;* **lead sb. up the ~ path** *(fig. coll.)* jmdn. an der Nase herumführen *(ugs.)*

garden: ~ centre *n.* Gartencenter, *das;* **~ 'city** *n.* Gartenstadt, *die*

gardener ['gɑ:dnə(r)] *n.* Gärtner, *der*/Gärtnerin, *die*

gardening ['gɑ:dnɪŋ] *n.* Gartenarbeit, *die; attrib.* Garten‹*gerät, -buch, -handschuh*›; **he likes ~**: er gärtnert gern

garden: ~ party *n.* Gartenfest, *das;* **~ 'shed** *n.* Geräteschuppen, *der*

gargle ['gɑ:gl] **1.** *v. i.* gurgeln. **2.** *n.* **a)** *(liquid)* Gurgelmittel, *das;* **b)** *(act)* **have a ~**: gurgeln

gargoyle ['gɑ:gɔɪl] *n. (Archit.)* Wasserspeier, *der*

garish ['geərɪʃ] *adj.* grell ‹*Farbe, Licht, Beleuchtung*›; knallbunt ‹*Kleidung, Verzierung, Muster*›

garishly ['geərɪʃlɪ] *adv.* grell ‹*beleuchten*›; knallbunt ‹*kleiden, tapezieren*›

garland ['gɑ:lənd] *n.* Girlande, *die;* **~ of flowers/laurel** Blumen-/Lorbeerkranz, *der*

garlic ['gɑ:lɪk] *n.* Knoblauch, *der*

garment ['gɑ:mənt] *n.* Kleidungsstück, *das; in pl. (clothes)* Kleidung, *die;* Kleider

garnish ['gɑ:nɪʃ] **1.** *v. t. (lit. or fig.)* garnieren. **2.** *n. (Cookery)* Garnierung, *die*

garret ['gærɪt] *n. (room on top floor)* Dachkammer, *die*

garrison ['gærɪsn] **1.** *n.* Garnison, *die.* **2.** *v. t.* **a)** *(furnish with ~)* mit einer Garnison belegen ‹*Stadt*›; **b)** *(place as ~)* in Garnison legen ‹*Truppen*›

'**garrison town** *n.* Garnison[s]stadt, *die*

garrulous ['gærʊləs] *adj. (talkative)* gesprächig; geschwätzig

garter ['gɑ:tə(r)] *n.* Strumpfband, *das*

gas [gæs] **1.** *n., pl.* **~es** ['gæsɪz] **a)** Gas, *das;* **natural ~**: Erdgas, *das;* **cook by** *or* **with ~**: mit Gas kochen; **b)** *(Amer. coll.: petrol)* Benzin, *das;* **c)** *(anaesthetic)* Narkotikum, *das;* Lachgas, *das;* **d)** *(for lighting)* Leuchtgas, *das.* **2.** *v. t.,* **-ss-** mit Gas vergiften. **3.** *v. i.,* **-ss-** *(coll.: talk idly)* schwafeln *(ugs. abwertend)* **(about** von)

gas: ~bag *n. (sl. derog.: talker)* Schwafler, *der*/Schwaflerin, *die (ugs. abwertend);* **~ cylinder** *n.* Gasflasche, *die;* **~ 'fire** *n.* Gasofen, *der;* **~-fired** ['gæsfaɪəd] *adj.* mit Gas betrieben; Gas‹*boiler, -ofen usw.*›

gash [gæʃ] **1.** *n. (wound)* Schnittwunde, *die; (cleft)* [klaffende] Spalte; *(in sack etc.)* Schlitz, *der.* **2.** *v. t.* aufritzen ‹*Haut*›; aufschlitzen ‹*Sack*›; **~ one's finger/knee** sich *(Dat. od. Akk.)* in den Finger schneiden/sich *(Dat.)* das Knie aufschlagen

'gasholder *n.* Gasbehälter, *der*

gasket ['gæskɪt] *n.* Dichtung, *die*

gas: ~ lamp *n.* Gaslampe, *die; (in street etc.)* Gaslaterne, *die;* **~light** *n.* **a)** *see* **~ lamp; b)** *no pl. (illumination)* Gaslicht, *das;* **~ lighter** *n.* **a)** [Gas]anzünder, *der;* **b)** *(cigarette-lighter)* Gasfeuerzeug, *das;* **~ main** *n.* Hauptgasleitung, *die;* **~man** *n. (fitter)* Gasinstallateur, *der; (meter-reader, collector)* Gasableser, *der;* Gasmann, *der (ugs.);* **~ mask** *n.* Gasmaske, *die;* **~ meter** *n.* Gaszähler, *der*

gasoline (gasolene) ['gæsəli:n] *n. (Amer.)* Benzin, *das*

gasometer [gæ'sɒmɪtə(r)] *n.* Gasometer, *der*

'gas oven *n.* Gasherd, *der*

gasp [gɑ:sp] **1.** *v. i.* nach Luft schnappen **(with** vor); **make sb. ~** *(fig.)* jmdm. den Atem nehmen; **leave sb. ~ing [with sth.]** jmdm. [vor etw.] den Atem verschlagen *od.* rauben; **he was ~ing for air** *or* **breath** er rang nach Luft. **2.** *v. t.* **~ out** hervorstoßen ‹*Bitte, Worte*›. **3.** *n.* Keuchen, *das;* **give a ~ of fear/surprise** vor Furcht/Überraschung die Luft einziehen; **be at one's last ~:** in den letzten Zügen liegen *(ugs.)*

gas: ~-pipe *n.* Gasleitung, *die;* **~ ring** *n.* Gasbrenner, *der;* **~ station** *n. (Amer.)* Tankstelle, *die;* **~ stove** *n.* Gasherd, *der*

gassy ['gæsɪ] *adj. (fizzy)* sprudelnd; schäumend ‹*Bier*›

gas: ~ tank *n.* **a)** Gastank, *der;* **b)** *(Amer.: petrol tank)* Benzintank, *der;* **~ tap** *n.* Gashahn, *der*

gastric ['gæstrɪk]: **~ 'flu** *(coll.),* **~ influ'enza** *ns.* Darmgrippe, *die;* **~ 'ulcer** *n.* Magengeschwür, *das*

gastronomic [gæstrə'nɒmɪk] *adj.* gastronomisch; kulinarisch ‹*Genüsse*›

gastronomy [gæ'strɒnəmɪ] *n.* Gastronomie, *die;* **French ~:** französische Küche

'gasworks *n. sing., pl. same* Gaswerk, *das*

gate [geɪt] *n.* **a)** *(lit. or fig.)* Tor, *das; (barrier)* Sperre, *die; (to field etc.)* Gatter, *das; (in garden fence)* [Garten]pforte, *die; (Railw.: of level crossing)* [Bahn]schranke, *die; (in airport)* Flugsteig, *der;* **b)** *(Sport: number to see match)* Besucher[zahl], *die*

gateau ['gætəʊ] *n., pl.* **~s** *or* **~x** ['gætəʊz] Torte, *die*

gate: ~crash *v. t.* ohne Einladung einfach hingehen zu; **~crasher** ['geɪtkræʃə(r)] *n.* ungeladener Gast; **~-money** *n.* Eintrittsgelder *Pl.;* Einnahmen *Pl.;* **~post** *n.* Torpfosten, *der;* **between you and me and the ~post** *(coll.)* unter uns *(Dat.)* gesagt; **~way** *n.* **a)** *(~, lit. or fig.)* Tor, *das* **(to** zu); **b)** *(Archit.) (structure)* Torbau, *der; (frame)* Torbogen, *der*

gather ['gæðə(r)] **1.** *v. t.* **a)** sammeln; zusammentragen ‹*Informationen*›; pflücken ‹*Obst, Blumen*›; **~ sth. [together]** etw. zusammensuchen *od.* -sammeln; **~ [in] the harvest** die Ernte einbringen; **b)** *(infer, deduce)* schließen **(from** aus); **~ from sb. that ...:** von jmdm. erfahren, daß ...; **as far as I can ~:** soweit ich weiß; **as you will have ~ed** wie Sie sicherlich vermutet haben; **c) ~ speed/force** schneller/stärker werden; **d)** *(summon up)* **~ [together]** zusammennehmen ‹*Kräfte, Mut*›; **~ oneself [together]** sich zusammennehmen; **~ one's thoughts** seine Gedanken ordnen; **~ one's breath/strength** [wieder] zu Atem kommen/Kräfte sammeln; **e) she ~ed her shawl round her neck** sie schlang den Schal um den Hals; **f)** *(Sewing)* ankrausen. **2.** *v. i.* **a)** sich versammeln; ‹*Wolken:*› sich zusammenziehen; ‹*Staub:*› sich ansammeln; ‹*Schweißperlen:*› sich sammeln; **be ~ed [together]** versammelt sein; **~ round** zusammenkommen; **~ round sb./sth.** sich um jmdn./etw. versammeln; **b)** *(increase)* zunehmen; **darkness was ~ing** es wurde dunkler. **3.** *n. in pl. (Sewing)* Kräusel[falten]

~ 'up *v. t.* **a)** *(bring together and pick up)* aufsammeln; zusammenpacken ‹*Habseligkeiten, Werkzeug*›; **b)** *(draw)* hochraffen ‹*Rock*›; *(summon)* sammeln ‹*Kräfte, Gedanken usw.*›

gathering ['gæðərɪŋ] *n. (assembly, meeting)* Versammlung, *die*

gauche [gəʊʃ] *adj.* linkisch

gaudy ['gɔ:dɪ] *adj.* protzig *(abwertend);* grell ‹*Farben*›

gauge [geɪdʒ] **1.** *n.* **a)** *(standard measure)* [Normal]maß, *das; (of rail)* Spurweite, *die;* **narrow ~:** Schmalspur, *die;* **b)** *(instrument)* Meßgerät, *das; (for dimensions of tools or wire)* Lehre, *die;* **c)** *(fig.: criterion, test)* Kriterium, *das;* Maßstab, *der.* **2.** *v. t.* **a)** *(measure)* messen; **b)** *(fig.)* beurteilen **(by** nach)

gaunt [gɔ:nt] *adj.* hager; *(from suffering)* verhärmt; karg ‹*Landschaft*›

[1]gauntlet ['gɔ:ntlɪt] *n.* Stulpenhandschuh, *der;* **fling** *or* **throw down the ~** *(fig.)* jmdm. den Fehdehandschuh hinwerfen

[2]gauntlet *n.* **run the ~:** Spießruten laufen

gauze [gɔ:z] *n. (fabric, wire)* Gaze, *die*

gave *see* **give 1, 2**

gawky ['gɔ:kɪ] *adj.* linkisch; unbeholfen; *(lanky)* schlaksig *(ugs.)*

gay [geɪ] **1.** *adj.* **a)** fröhlich; fidel *(ugs.)* ‹*Person, Gesellschaft*›; **b)** *(showy, bright-coloured)* farbenfroh ‹*Stoff, Ausstattung*›; fröhlich, lebhaft ‹*Farbe*›; **c)** *(coll.: homosexual)* schwul *(ugs.);* Schwulen‹*lokal, -blatt*›. **2.** *n. (coll.)* Schwule, *der (ugs.)*

gaze [geɪz] *v. i.* blicken; *(more fixedly)* starren; **~ at sb./sth.** jmdn./etw. anstarren *od.* ansehen

gazelle [gə'zel] *n.* Gazelle, *die*

gazette [gə'zet] *n.* **a)** *(Brit.: official journal)* Amtsblatt, *das;* **b)** *(newspaper)* Zeitung, *die*

gazetteer [gæzɪ'tɪə(r)] *n.* alphabetisches [Orts]verzeichnis

gazump [gə'zʌmp] *v. t. (sl.) durch nachträgliches Überbieten um die Chance bringen, ein Haus zu kaufen*

GB *abbr.* **Great Britain** GB

GCE *abbr. (Brit. Hist.)* **General Certificate of Education**

GCSE *abbr. (Brit.)* **General Certificate of Secondary Education**

GDR *abbr. (Hist.)* **German Democratic Republic** DDR, *die*

gear [gɪə(r)] **1.** *n.* **a)** *(Motor Veh.)* Gang, *der;* **first** *or* **bottom/top ~** *(Brit.)* der erste/höchste Gang; **high/low ~:** hoher/niedriger Gang; **change ~:** schalten; **change into second/a higher/lower ~:** in den zweiten Gang/in einen höheren/niedrigeren Gang schalten; **a bicycle with ten-speed ~s** ein Fahrrad mit Zehngangschaltung; **put the car into ~:** einen Gang einlegen; **leave the car in ~:** den Gang drin lassen; **b)** *(combination of wheels, levers, etc.)* Getriebe, *das;* **c)** *(clothes)* Aufmachung, *die;* **travelling ~:** Reisekleidung, *die;* **d)** *(equipment, tools)* Gerät, *das;* Ausrüstung, *die.* **2.** *v.t. (adjust, adapt)* ausrichten (**to** auf + *Akk.*)

gear: ~box *n.* Getriebekasten, *der;* **five-speed ~box** Fünfganggetriebe, *das;* **~-lever,** *(Amer.)* **~-shift, ~-stick** *ns.* Schalthebel, *der;* **~wheel** *n.* Zahnrad, *das*

gee [dʒi:] *int. (coll.)* Mann *(salopp);* Mensch [Meier] *(salopp)*

geese *pl. of* **goose**

gee 'whiz *see* **gee**

geezer ['gi:zə(r)] *n. (sl.: old man)* Opa, *der (ugs. scherzh. od. abwertend)*

Geiger counter ['gaɪgə kaʊntə(r)] *n. (Phys.)* Geigerzähler, *der*

gel [dʒel] **1.** *n.* Gel, *das.* **2.** *v.i.,* **-ll-: a)** gelatinieren; gelieren; **b)** *(fig.)* Gestalt annehmen

gelatin ['dʒelətɪn], *(esp. Brit.)* **gelatine** ['dʒeləti:n] *n.* Gelatine, *die*

gelding ['geldɪŋ] *n.* kastriertes Tier; *(male horse)* Wallach, *der*

gelignite ['dʒelɪgnaɪt] *n.* Gelatinedynamit, *das*

gem [dʒem] *n.* **a)** Edelstein, *der; (cut also)* Juwel, *das od. der;* **b)** *(fig.)* Juwel, *das;* Perle, *die; (choicest part)* Glanzstück, *das*

Gemini ['dʒemɪnaɪ, 'dʒemɪnɪ] *n. (Astrol., Astron.)* Zwillinge *Pl.*

gen [dʒen] *n. (Brit. sl.)* notwendige Angaben

Gen. *abbr.* **General** Gen.

gender ['dʒendə(r)] *n.* **a)** *(Ling.)* [grammatisches] Geschlecht; Genus, *das;* **b)** *(coll.: one's sex)* Geschlecht, *das*

gene [dʒi:n] *n. (Biol.)* Gen, *das*

genealogy [dʒi:nɪ'ælədʒɪ] *n.* Genealogie, *die (fachspr.); (pedigree)* Ahnentafel, *die (geh.)*

genera *pl. of* **genus**

general ['dʒenrl] **1.** *adj.* **a)** allgemein; **the ~ public** weite Kreise der Öffentlichkeit *od.* Bevölkerung; **in ~ use** allgemein verbreitet; **his ~ health/manner** sein Allgemeinbefinden/sein Benehmen im allgemeinen; **he has had a good ~ education** er hat eine gute Allgemeinbildung; **b)** *(prevalent, widespread, usual)* allgemein; weitverbreitet ‹*Übel, Vorurteil, Aberglaube, Ansicht*›; **it is the ~ custom** *or* **rule** es ist allgemein üblich *od.* ist Sitte *od.* Brauch; **c)** *(not limited in application)* allgemein; *(true of [nearly] all cases)* allgemeingültig; generell; **as a ~ rule, in ~:** im allgemeinen; **d)** *(not detailed, vague)* allgemein; ungefähr, vage ‹*Vorstellung, Beschreibung, Ähnlichkeit usw.*›; **the ~ idea** *or* **plan is that we ...:** wir haben uns das so vorgestellt, daß wir ... **2.** *n. (Mil.)* General, *der*

general: ~ anaes'thetic *see* **anaesthetic; G~ Certificate of Edu'cation** *n. (Brit. Hist.) (ordinary level)* ≈ mittlere Reife; *(advanced level)* ≈ Abitur, *das;* **G~ Certificate of Secondary Edu'cation** *n. (Brit.) Abschluß der Sekundarstufe;* **~ e'lection** *see* **election**

generality [dʒenə'rælɪtɪ] *n.* **a)** **talk in generalities** verallgemeinern; **b)** *(majority) (of mankind, electorate, etc.)* Großteil, *der; (of voters, individuals, etc.)* Mehrheit, *die*

generalisation, generalise *see* **generaliz-**

generalization [dʒenrəlaɪ'zeɪʃn] *n.* Verallgemeinerung, *die*

generalize ['dʒenrəlaɪz] **1.** *v.t.* verallgemeinern. **2.** *v.i.* **~ [about sth.]** [etw.] verallgemeinern; **~ about the French** die Franzosen alle über einen Kamm scheren

general 'knowledge *n.* Allgemeinwissen, *das*

generally ['dʒenrəlɪ] *adv.* **a)** *(extensively)* allgemein; **~ available** überall erhältlich; **b)** **~ speaking** im allgemeinen; **c)** *(usually)* im allgemeinen; normalerweise; **d)** *(summarizing the situation)* ganz allgemein

general: ~ 'manager *n.* [leitender] Direktor/[leitende] Direktorin; **~ 'practice** *n. (Med.)* Allgemeinmedizin, *die;* **~ prac'titioner** *n. (Med.)* Arzt/Ärztin für Allgemeinmedizin; **~ 'staff** *n.* Generalstab, *der;* **~ 'strike** *n.* Generalstreik, *der*

generate ['dʒenəreɪt] *v.t. (produce)* erzeugen (**from** aus); *(result in)* führen zu

generating station ['dʒenəreɪtɪŋ steɪʃn] *n.* Elektrizitätswerk, *das*

generation [dʒenə'reɪʃn] *n.* **a)** Generation, *die;* **the present/rising ~:** die heutige/heranwachsende *od.* junge Generation; **first-/second-~ computers** *etc.* Computer *usw.* der ersten/zweiten Generation; *attrib.* **~ gap** Generationsunterschied, *der;* **b)** *(production)* Erzeugung, *die;* **~ of electricity** Stromerzeugung, *die*

generator ['dʒenəreɪtə(r)] *n.* Generator, *der; (in motor car also)* Lichtmaschine, *die*

generic [dʒɪ'nerɪk] *adj.* **a)** **~ term** *or* **name** Ober- *od.* Gattungsbegriff, *der;* **b)** *(Biol.)* Gattungs‹*name, -bezeichnung*›

generosity [dʒenə'rɒsɪtɪ] *n.* Großzügigkeit, *die; (magnanimity)* Großmut, *die*

generous ['dʒenərəs] *adj.* **a)** großzügig; *(noble-minded)* großmütig; **b)** *(ample, abundant)* großzügig; reichhaltig ‹*Mahl*›; reichlich ‹*Nachschub, Vorrat, Portion*›; üppig ‹*Figur, Formen, Mahl*›

generously ['dʒenərəslɪ] *adv.* großzügig; *(magnanimously)* großmütig

genesis ['dʒenɪsɪs] *n., pl.* **geneses** ['dʒenɪsi:z] **a)** **G~** *no pl.* Schöpfungsgeschichte, *die;* **b)** *(origin)* Herkunft, *die; (development into being)* Entstehung, *die*

genetic [dʒɪ'netɪk] *adj.,* **genetically** [dʒɪ'netɪkəlɪ] *adv.* genetisch

genetics [dʒɪ'netɪks] *n., no pl.* Genetik, *die*

Geneva [dʒɪ'ni:və] **1.** *pr. n.* Genf *(das);* **Lake ~:** der Genfer See. **2.** *attrib. adj.* Genfer

genial ['dʒi:nɪəl] *adj. (jovial, kindly)* freundlich; *(sociable)* jovial, leutselig ‹*Person, Art*›

geniality [dʒi:nɪ'ælɪtɪ] *n., no pl.* Freundlichkeit, *die*

genital ['dʒenɪtl] **1.** *n. in pl.* Geschlechtsorgane; Genitalien. **2.** *adj.* Geschlechts‹*teile, -organe*›

genitive ['dʒenɪtɪv] *(Ling.)* **1.** *adj.* Genitiv-; genitivisch; **~ case** Genitiv, *der.* **2.** *n.* Genitiv, *der*

genius ['dʒi:nɪəs] *n., pl.* **~es** *or* **genii** ['dʒi:nɪaɪ] **a)** *pl.* **~es** *(person)* Genie, *das;* **b)** *(natural ability; also iron.)* Talent, *das;* Begabung, *die; (extremely great)* Genie, *das;* **a man of ~:** ein genialer Mensch; ein Genie

Genoa ['dʒenəʊə] *pr. n.* Genua *(das)*

genocide ['dʒenəsaɪd] *n.* Völkermord, *der*

genre [ʒɑ̃rə] *n.* Genre, *das;* Gattung, *die*

gent [dʒent] *n.* **a)** *(coll./joc.)* Gent, der *(iron.);* **b)** **~s'** Herren‹*friseur, -ausstatter*›; **c)** **the G~s** *(Brit. coll.)* die Herrentoilette

genteel [dʒen'ti:l] *adj.* vornehm; fein

Gentile ['dʒentaɪl] **1.** *n.* Nichtjude, *der/*-jüdin, *die.* **2.** *adj.* nichtjüdisch

gentility [dʒen'tɪlɪtɪ] *n., no pl.* Vornehmheit, *die*

gentle ['dʒentl] *adj.*, ~**r** ['dʒentlə(r)], ~**st** ['dʒentlɪst] sanft; sanftmütig ‹*Wesen*›; liebenswürdig, freundlich ‹*Person, Verhalten, Ausdrucksweise*›; leicht, schwach ‹*Brise*›; ruhig ‹*Fluß, Wesen*›; leise ‹*Geräusch*›; gemäßigt ‹*Tempo*›; mäßig ‹*Hitze*›; gemächlich ‹*Tempo, Schritte, Spaziergang*›; sanft ‹*Abhang usw.*›; mild ‹*Reinigungsmittel, Shampoo usw.*›; wohlig ‹*Wärme*›; zahm, lammfromm ‹*Tier*›; **be ~ with sb./sth.** sanft mit jmdm./etw. umgehen; **a ~ reminder/hint** ein zarter Wink/eine zarte Andeutung; **the ~ sex** das zarte Geschlecht *(ugs. scherzh.)*

gentleman ['dʒentlmən] *n., pl.* **gentlemen** ['dʒentlmən] **a)** *(man of good manners and breeding)* Gentleman, *der;* **b)** *(man)* Herr, *der;* **[Ladies and] Gentlemen!** meine [Damen und] Herren!; **Gentlemen, ...** *(in formal, business letter)* Sehr geehrte Herren!; **~'s agreement** Gentleman's Agreement, *das;* **gentlemen's** Herren‹*friseur, -schneider*›

gentlemanly ['dʒentlmənlɪ] *adj.* gentlemanlike *nicht attrib.*; eines Gentlemans *nachgestellt*

gentleness ['dʒentlnɪs] *n., no pl.* Sanftheit, *die; (of nature)* Sanftmütigkeit, *die; (of shampoo, cleanser, etc.)* Milde, *die*

gently ['dʒentlɪ] *adv. (tenderly)* zart; zärtlich; *(mildly)* sanft; *(carefully)* vorsichtig; behutsam; *(quietly, softly)* leise; *(moderately)* sanft; *(slowly)* langsam; **she broke the news to him ~:** sie brachte ihm die Nachricht schonend bei; **~ does it!** immer sachte! *(ugs.)*; **~!** [sachte] sachte!

gentry ['dʒentrɪ] *n. pl.* niederer Adel; Gentry, *die*

genuine ['dʒenjʊɪn] *adj.* **a)** *(real)* echt; authentisch ‹*Text*›; **the ~ article** die echte Ausgabe *(fig.)*; **b)** *(true)* aufrichtig; wahr ‹*Grund, Not*›; echt ‹*Tränen*›; ernsthaft, ernstgemeint ‹*Angebot*›

genuinely ['dʒenjʊɪnlɪ] *adv.* wirklich

genus ['dʒi:nəs, 'dʒenəs] *n., pl.* **genera** ['dʒenərə] *(Biol.)* Gattung, *die*

geographer [dʒɪ'ɒgrəfə(r)] *n.* Geograph, *der*/Geographin, *die*

geographical [dʒi:ə'græfɪkl] *adj.*, **geographically** [dʒi:ə'græfɪkəlɪ] *adv.* geographisch

geography [dʒɪ'ɒgrəfɪ] *n.* Geographie, *die;* Erdkunde, *die (Schulw.)*

geological [dʒi:ə'lɒdʒɪkl] *adj.*, **geologically** [dʒi:ə'lɒdʒɪkəlɪ] *adv.* geologisch

geologist [dʒɪ'ɒlədʒɪst] *n.* Geologe, *der*/Geologin, *die*

geology [dʒɪ'ɒlədʒɪ] *n.* Geologie, *die*

geometric [dʒi:ə'metrɪk], **geometrical** [dʒi:ə'metrɪkl] *adj.*, **geometrically** [dʒi:ə'metrɪkəlɪ] *adv.* geometrisch

geometry [dʒɪ'ɒmɪtrɪ] *n.* Geometrie, *die*

Georgian ['dʒɔ:dʒɪən] *adj. (Brit. Hist.)* georgianisch

geranium [dʒə'reɪnɪəm] *n.* Geranie, *die;* Pelargonie, *die*

gerbil ['dʒɜ:bɪl] *n. (Zool.)* Wüstenmaus, *die;* Rennmaus, *die*

geriatric [dʒerɪ'ætrɪk] *adj.* geriatrisch

germ [dʒɜ:m] *n. (lit. or fig.)* Keim, *der;* **I don't want to catch your ~s** ich möchte mich nicht bei dir anstecken; **wheat ~:** Weizenkeim, *der*

German ['dʒɜ:mən] **1.** *adj.* deutsch; **a ~ person** ein Deutscher/eine Deutsche; **the ~ people** die Deutschen; **he/she is ~:** er ist Deutscher/sie ist Deutsche; **he is a native ~ speaker** seine Muttersprache ist Deutsch. *See also* **East German 1; West German 1. 2.** *n.* **a)** *(person)* Deutsche, *der/die;* **he/she is a ~:** er ist Deutscher/sie ist Deutsche; **b)** *(language)* Deutsch, *das;* **High ~:** Hochdeutsch, *das;* **Low ~:** Niederdeutsch, *das. See also* **East German 2; English 2 a; West German 2**

German Democratic Re'public *pr. n. (Hist.)* Deutsche Demokratische Republik

Germanic [dʒɜ:'mænɪk] *adj.* germanisch

German: ~ **'measles** *n. sing.* Röteln *Pl.*; ~ **'shepherd [dog]** *n.* [deutscher] Schäferhund

Germany ['dʒɜ:mənɪ] *pr. n.* Deutschland *(das);* **Federal Republic of ~:** Bundesrepublik Deutschland, *die; see also* **East Germany; West Germany**

germinate ['dʒɜ:mɪneɪt] **1.** *v. i.* keimen; *(fig.)* entstehen. **2.** *v. t.* zum Keimen bringen

germination [dʒɜ:mɪ'neɪʃn] *n.* Keimung, *die;* Keimen, *das*

germ 'warfare *n.* Bakterienkrieg, *der;* biologische Kriegführung

gerund ['dʒerənd] *n. (Ling.)* Gerundium, *das*

gestation [dʒe'steɪʃn] *n. (of animal)* Trächtigkeit, *die; (of woman)* Schwangerschaft, *die*

gesticulate [dʒe'stɪkjʊleɪt] *v. i.* gestikulieren

gesticulation [dʒestɪkjʊ'leɪʃn] *n.* ~[**s** *pl.*] Gesten

gesture ['dʒestʃə(r)] **1.** *n.* Geste, *die (auch fig.);* Gebärde, *die (geh.);* **a ~ of resignation** eine resignierte Geste. **2.** *v. i.* gestikulieren; **~ to sb. to do sth.** jmdm. zu verstehen geben *od. (geh.)* jmdm. bedeuten, etw. zu tun. **3.** *v. t.* **~ sb. to do sth.** jmdm. bedeuten, etw. zu tun *(geh.)*

get [get] **1.** *v. t.*, **-tt-**, *p. t.* **got** [gɒt], *p.p.* **got** *or* (*in comb./arch./ Amer. except in sense* **m) gotten** ['gɒtn] **(got** *also coll. abbr. of* **has got** *or* **have got) a)** *(obtain)* bekommen; kriegen *(ugs.); (by buying)* kaufen; sich *(Dat.)* anschaffen ‹*Auto usw.*›; *(by one's own effort for special purpose)* sich *(Dat.)* besorgen ‹*Visum, Genehmigung, Arbeitskräfte*›; sich *(Dat.)* beschaffen ‹*Geld*›; einholen ‹*Gutachten*›; *(by contrivance)* kommen zu; *(find)* finden ‹*Zeit*›; **where did you ~ that?** wo hast du das her?; **he got him by the leg/arm** er kriegte ihn am Bein/Arm zu fassen; **~ sb. a job/taxi, ~ a job/taxi for sb.** jmdm. einen Job verschaffen/ein Taxi besorgen *od.* rufen; **~ oneself sth./a job** sich *(Dat.)* etw. zulegen/einen Job finden; **you can't ~ this kind of fruit in the winter months** dieses Obst gibt es im Winter nicht zu kaufen; **b)** *(fetch)* holen; **what can I ~ you?** was kann ich Ihnen anbieten?; **is there anything I can ~ you in town?** soll ich dir etwas aus der Stadt mitbringen?; **c) ~ the bus** *etc. (be in time for, catch)* den Bus *usw.* erreichen *od. (ugs.)* kriegen; *(travel by)* den Bus nehmen; **d)** *(prepare)* machen *(ugs.)*, zubereiten ‹*Essen*›; **e)** *(coll.: eat)* essen; **~ something to eat** etwas zu essen holen; *(be given)* etwas zu essen bekommen; **f)** *(gain)* erreichen; **what do I ~ out of it?** was habe ich davon?; **g)** *(by calculation)* herausbekommen; **h)** *(receive)* bekommen; erhalten, *(ugs.)* kriegen ‹*Geldsumme*›; **the country ~s very little sun/rain** die Sonne scheint/es regnet nur sehr wenig in dem Land; **he got his jaw broken in a fight** bei einer Schlägerei wurde ihm der Kiefer gebrochen; **i)** *(receive as penalty)* bekommen, *(ugs.)* kriegen ‹*6 Monate Gefängnis, Geldstrafe, Tracht Prügel*›; **you'll ~ it** *(coll.)* du kriegst Prügel *(ugs.);* es setzt was *(ugs.); (be scolded)* du kriegst was zu hören *(ugs.)*; **j)** *(kill)* töten; erlegen ‹*Wild*›; *(hit, injure)* treffen; **k)** *(win)* bekommen; finden ‹*Anerkennung*›; sich *(Dat.)* verschaffen ‹*Ansehen*›; erzielen ‹*Tor, Punkt, Treffer*›; gewinnen ‹*Preis, Belohnung*›; belegen ‹*ersten usw. Platz*›; ~ **permission** die Erlaubnis erhalten; **l)** *(come to have)* finden ‹*Schlaf, Ruhe*›; bekommen ‹*Einfall, Vorstellung, Gefühl*›; gewinnen ‹*Eindruck*›; *(contract)* bekommen ‹*Kopfschmerzen, Grippe, Malaria)*; ~ **some rest** sich ausruhen; ~ **an idea/a habit from sb.** von jmdm. eine Idee/Angewohnheit übernehmen; **m) have got** *(coll.: have)* haben; **give it all you've got** gib dein Bestes; **have got a toothache/a cold** Zahnschmerzen/eine Erkältung haben *od.* erkältet sein; **have got to do sth.** etw. tun müssen; **something has got to be done [about it]** dagegen muß etwas unternommen werden; **n)** *(succeed in bringing, placing, etc.)* bringen; kriegen *(ugs.)*; **I must ~ a message to her** ich muß ihr eine Nachricht zukommen lassen; **o)** *(bring into some state)* ~ **a machine going** eine Maschine in Gang setzen *od.* bringen; ~ **things going** *or* **started** die Dinge in Gang bringen; ~ **everything packed/prepared** alles [ein]packen/vorbereiten; ~ **sth. ready/done** etw. fertig machen; ~ **one's feet wet** nasse Füße kriegen; ~ **one's hands dirty** sich *(Dat.)* die Hände schmutzig machen; **I didn't ~ much done today** ich habe heute nicht viel geschafft; **you'll ~ yourself thrown out/arrested** du schaffst es noch, daß du rausgeworfen/verhaftet wirst; ~ **sb. talking/drunk/interested** jmdn. zum Reden bringen/betrunken machen/jmds. Interesse wecken; ~ **one's hair cut** sich *(Dat.)* die Haare schneiden lassen; **p)** *(induce)* ~ **sb. to do sth.** jmdn. dazu bringen, etw. zu tun; ~ **sth. to do sth.** es schaffen, daß etw. etw. tut; **I can't ~ the car to start/the door to shut** ich kriege das Auto nicht in Gang/die Tür nicht zu; **q)** *(Radio, Telev.: pick up)* empfangen ‹*Sender*›; **r)** *(contact by telephone)* ~ **sb. [on the phone]** jmdn. [telefonisch] erreichen; **s)** *(answer)* **I'll ~ it!** ich geh' schon!; *(answer doorbell)* ich mach' auf!; *(answer the phone)* ich gehe ran *(ugs.) od.* nehme ab!; **t)** *(coll.: perplex)* in Verwirrung bringen; **you've got me there; I don't know** da bin ich überfragt – ich weiß es nicht; **u)** *(coll.) (understand)* kapieren *(ugs.)*; verstehen ‹*Personen*›; *(hear)* mitkriegen *(ugs.)*; ~ **it?** alles klar? *(ugs.)*; **v)** *(coll.: annoy)* aufregen *(ugs.)*. **2.** *v. i.*, **-tt-, got, gotten a)** *(succeed in coming or going)* kommen; ~ **to London before dark** London vor Einbruch der Dunkelheit erreichen; **when did you get here/to school?** wann bist du gekommen?/wann warst du in der Schule?; **we got as far as Oxford** wir kamen bis Oxford; **how did that ~ here?** wie ist das hierher gekommen?; **b)** *(come to be)* ~ **talking [to sb.]** [mit jmdm.] ins Gespräch kommen; ~ **going** *or* **started** *(leave)* losgehen; aufbrechen; *(start talking)* loslegen *(ugs.)*; *(become lively or operative)* in Schwung kommen; ~ **going on** *or* **with sth.** mit etw. anfangen; **c)** ~ **to know sb.** jmdn. kennenlernen; **he got to like/hate her** mit der Zeit mochte er sie/begann er, sie zu hassen; ~ **to hear of sth.** von etw. erfahren; ~ **to do sth.** *(succeed in doing)* etw. tun können; **d)** *(become)* werden; ~ **ready/washed** sich fertigmachen/waschen; ~ **frightened/hungry** Angst/Hunger kriegen; ~ **excited about sth.** sich auf etw. *(Akk.)* freuen; ~ **interested in sth.** sich für etw. interessieren; ~ **caught in the rain** vom Regen überrascht werden; ~ **well soon!** gute Besserung!

~ **a'bout** *v. i.* **a)** *(move)* sich bewegen; *(travel)* herumkommen; **b)** ‹*Gerücht*:› sich verbreiten

~ **across 1.** [--'-] *v. i.* **a)** *(to/from other side)* rüberkommen *(ugs.)*; **b)** *(coll.: be communicated)* rüberkommen *(ugs.)*; ~ **across [to sb.]** ‹*Person*:› sich [jmdm.] verständlich machen; ‹*Witz, Idee*:› [bei jmdm.] ankommen. **2.** *[stress varies] v. t.* **a)** *(cross)* überqueren; ~ **sb./sth. across [sth.]** *(transport to/from other side)* jmdn./etw. [über etw. *(Akk.)*] hin-/herüberbringen; **b)** *(coll.: communicate)* vermitteln, klarmachen (**to** *Dat.*)

~ **a'long** *v. i.* **a)** *(advance, progress)* ~ **along well** [gute] Fortschritte machen; **how is he ~ting along with his work?** wie kommt er mit seiner Arbeit voran?; **b)** *(manage)* zurechtkommen; **c)** *(agree or live sociably)* auskommen; ~ **along with each other** *or* **together** miteinander auskommen; **d)** *(leave)* sich auf den Weg machen

~ **at** *v. t.* **a)** herankommen an (+ *Akk.*); **b)** *(find out)* [he]rausfinden ‹*Wahrheit, Ursache usw.*›; **c)** *(coll.)* **what are you/is he getting at?** worauf wollen Sie/will er hinaus?; *(referring to)* worauf spielen Sie/spielt er jetzt an?; **d)** *(sl.: attack, taunt)* anmachen *(salopp)*

~ **a'way 1.** *v. i.* **a)** wegkommen; **I can't ~ away from work** ich kann nicht von der Arbeit weg; **there is no ~ting away from the fact that ...:** man kommt nicht um die Tatsache *od.* darum herum, daß ...; ~ **away from it all** *see* **all 1 a**; **b)** *(escape)* entkommen; entwischen *(ugs.)*; **c)** *in imper. (coll.)* ~ **away [with you]!** ach, geh *od.* komm! *(ugs.)*; ach, erzähl mir doch nichts! *(ugs.)*. **2.** *v. t.* wegnehmen; ~ **sth. away from sb.** jmdm. etw. wegnehmen

~ **a'way with** *v. t.* **a)** *(steal and escape with)* entkommen mit; **b)** *(coll.: go unpunished for)* ungestraft davonkommen mit; **the things he ~s away with!** was der sich *(Dat.)* alles erlauben kann!; ~ **away with it** es sich *(Dat.)* erlauben können; *(succeed)* damit durchkommen

~ **'back 1.** *v. i.* **a)** *(return)* zurückkommen; ~ **back home** nach Hause kommen; **b)** *(stand away)* zurücktreten. **2.** *v. t.* **a)** *(recover)* wieder- *od.* zurückbekommen; ~ **one's strength back** wieder zu Kräften kommen; **b)** *(return)* zurücktun; **I can't ~ the lid back on [it]** ich kriege den Deckel nicht wieder drauf *(ugs.)*; **c)** ~ **one's 'own back [on sb.]** *(sl.)* sich [an jmdm.] rächen

~ **'back to** *v. t.* ~ **back to sb.** auf jmdn. zurückkommen; **I'll ~ back to you on that** ich komme darauf noch zurück; ~ **back to work** wieder an die Arbeit gehen

~ **be'hind** *v. i.* zurückbleiben; ins Hintertreffen geraten *(ugs.)*; *(with payments)* in Rückstand geraten

~ **'by 1.** *v. i.* **a)** *(move past)* passieren; vorbeikommen; **let sb. ~ by** jmdn. vorbeilassen; **b)** *(coll.: be acceptable, adequate)* **she should [just about] ~ by in the exam** sie müßte die Prüfung [gerade so] schaffen; **c)** *(coll.: manage)* über die Runden kommen *(ugs.)* (**on** mit). **2.** *v. t.* **a)** *(move past)* ~ **by sb./sth.** an jmdm./etw. vorbeikommen; **b)** *(pass unnoticed)* entgehen (+ *Dat.*)

~ **down 1.** [-'-] *v. i.* **a)** *(come down)* heruntersteigen; *(go down)* hinuntersteigen; **b)** *(leave table)* aufstehen; **c)** *(bend down)* sich bücken; ~ **down on one's knees** niederknien. **2.** *[stress varies] v. t.* **a)** *(come down)* heruntersteigen; herunterkommen; *(go down)* hinuntersteigen; hinuntergehen; **b)** ~ **sb./sth. down** *(manage to bring down)* jmdn./etw. hin-/herunterbringen; *(with difficulty)* jmdn./ etw. hin-/herunterbekommen; *(take down from above)* jmdn./ etw. hin-/herunterholen; **c)** *(swallow)* hinunterschlucken; **d)** *(record, write)* ~ **sth. down [on paper]** etw. schriftlich festhalten *od.* zu Papier bringen; **e)** *(coll.: depress)* fertigmachen *(ugs.)*; **f)** senken ‹*Fieber, Preis*›; *(by bargaining)* herunterdrücken ‹*Preis*›

~ **'down to** *v. t.* ~ **down to sth.** sich an etw. *(Akk.)* machen; ~ **down to writing a letter** sich hinsetzen und einen Brief schreiben

~ **'in 1.** *v. i.* **a)** *(into bus etc.)* einsteigen; *(into bath)* hineinsteigen; *(into bed)* sich hinlegen; *(into room, house, etc.)* eintreten; *(intrude)* eindringen; **b)** *(arrive)* ankommen; *(get home)* heimkommen; **c)** *(be elected)* gewählt werden; **d)** *(obtain place) (at institution etc.)* angenommen werden; *(at university)* einen Studienplatz bekommen. **2.** *v. t.* **a)** *(bring in)* einbringen ‹*Ernte*›; hineinbringen, ins Haus bringen ‹*Einkäufe, Kind*›; einlagern ‹*Kohlen, Kartoffeln*›; reinholen ‹*Wäsche*›; *(Brit.: fetch and pay for)* holen ‹*Getränke*›; **b)** *(coll.: enter)* einsteigen in (+ *Akk.*) ‹*Auto, Zug*›; **c)** *(submit)* abgeben ‹*Artikel, Hausarbeit*›; einreichen ‹*Bewerbung, Bericht*›; **d)** *(receive)* erhalten; reinkriegen *(ugs.)*; **e)** *(send for)* holen; rufen ‹*Arzt, Polizei*›; hinzuziehen ‹*Spezialist*›; **f)** *(fit in)* reinkriegen *(ugs.)*; einschieben ‹*Unterrichtsstunde*›; **try to ~ in a word about sth.** sich zu etw. äußern wollen

~ **'in on** *v. t. (coll.)* sich beteiligen an (+ *Dat.*); ~ **in on the act** mitmischen *(ugs.)*

~ **into** *v. t.* **a)** *(bring into)* fahren ‹*Auto usw.*› in (+ *Akk.*) ‹*Garage*›; bringen in (+ *Akk.*) ‹*Haus, Bett, Hafen*›; **b)** *(enter)* gehen/*(as intruder)* eindringen in (+ *Akk.*) ‹*Haus*›; [ein]steigen in (+ *Akk.*) ‹*Auto usw.*›; [ein]treten in (+ *Akk.*) ‹*Zimmer*›; steigen in (+ *Akk.*) ‹*Wasser*›; **the coach ~s into the station at 9 p.m.** der Bus kommt um 21.00 Uhr am Busbahnhof an; **c)** *(gain admission to)* eingelassen werden in (+ *Akk.*); einen Studienplatz erhalten an (+ *Dat.*) ‹*Universität*›; genommen werden von ‹*Firma*›; ~ **into Parliament** ins Parlament einziehen; **d)** *(coll.)* ~ **into one's clothes** sich anziehen; **I can't ~ into these trousers** ich komme in diese Hose nicht mehr rein *(ugs.)*; **e)** *(penetrate)* [ein]dringen in (+ *Akk.*); **f)** *(begin to undergo)* geraten in (+ *Akk.*); kommen in (+ *Akk.*) ‹*Schwierigkeiten*›; *(cause to undergo)* stürzen in (+ *Akk.*) ‹*Schulden, Unglück*›; bringen in (+ *Akk.*) ‹*Schwierigkeiten*›; **g)** *(accustom to, become accustomed to)* annehmen ‹*Gewohnheit*›; ~ **into the job/work** sich einarbeiten; *see also* **habit a; h)** geraten in (+ *Akk.*) ‹*Wut, Panik*›; **i) what's got into him?** was ist nur in ihn gefahren?

~ **'in with** *v. t. (coll.)* ~ **in [well] with sb.** sich mit jmdm. gut stellen; **he got in with a bad crowd** er geriet in schlechte Gesellschaft

~ **off 1.** [-'-] *v. i.* **a)** *(alight)* aussteigen; *(dismount)* absteigen; **tell sb. where he ~s off** *or* **where to ~ off** *(fig. coll.)* jmdn. in seine Grenzen verweisen; **b)** *(not remain on sb./sth.)* runtergehen; *(from chair)* aufstehen; *(from ladder, table, carpet)* herunterkommen; *(let go)* loslassen; **c)** *(start)* aufbrechen; ~ **off to school/to work** zur Schule/Arbeit losgehen/-fahren; ~ **off to a good** *etc.* **start** einen guten *usw.* Start haben; **d)** *(escape punishment or injury)* davonkommen; ~ **off lightly** glimpflich davonkommen; **e)** *(fall asleep)* einschlafen; **f)** *(leave)* [weg]gehen; ~ **off early** [schon] früh [weg]gehen. **2.** *[stress varies] v. t.* **a)** *(dismount from)* [ab]steigen von ‹*Fahrrad*›; steigen von ‹*Pferd*›; *(alight from)* aussteigen aus ‹*Bus, Zug usw.*›; steigen aus ‹*Boot*›; **b)** *(not remain on)* herunterkommen von ‹*Teppich, Mauer, Leiter, Tisch*›; aufstehen von ‹*Stuhl*›; verschwinden von, verlassen ‹*Gelände*›; ~ **off the subject** vom Thema abkommen; **c)** *(cause to start)* [los]schicken; **it takes ages to ~ the children off to school** es dauert eine Ewigkeit, die Kinder für die Schule fertigzumachen; **d)** *(remove)* ausziehen ‹*Kleidung usw.*›; entfernen ‹*Fleck, Farbe usw.*›; abbekommen ‹*Deckel, Ring*›; ~ **sth. off sth.** etw. von etw. entfernen/abbekommen; ~ **sb. off a subject** jmdn. von einem Thema abbringen; **e)** *(send, dispatch)* abschicken; aufgeben ‹*Telegramm, Paket*›; **f)** *(cause to escape punishment)* davonkommen lassen; **g)** *(not have to do, go to, etc.)* frei haben; ~ **time/a day off [work]** frei/einen Tag frei bekommen; ~ **off work [early]** [früher] Feierabend machen; **I have got the afternoon off** ich habe den Nachmittag frei

~ **on 1.** [-'-] *v. i.* **a)** *(on bicycle)* aufsteigen; *(on horse)* aufsitzen; *(enter vehicle)* einsteigen; **b)** *(make progress)* vorankommen; ~ **on in life/the world** es zu etwas [im Leben] bringen; **c)** *(fare)* **how did you ~ on there?** wie ist es dir dort ergangen?; **he's ~ting on well** es geht ihm gut; **I didn't ~ on too well in my exams** meine Prüfungen sind nicht besonders gut gelaufen *(ugs.)*; **d)** *(become late)* vorrücken; **it's ~ting on for five** es geht auf fünf zu; **it's ~ting on for six months since ...:** es sind bald sechs Monate, seit ...; **time is ~ting on** es wird langsam spät; **e)** *(advance in age)* älter werden; **be ~ting on in years/for seventy** langsam älter werden/auf die Siebzig zugehen; **f) there were ~ting on for fifty people** es waren an die fünfzig Leute da; **g)** *(manage)* zurechtkommen; **h)** *see* ~ **along c. 2.** *[stress varies] v. t.* **a)** *(climb on)* steigen auf (+ *Akk.*) ‹*Fahrrad, Pferd*›; *(enter, board)* einsteigen in (+ *Akk.*) ‹*Zug, Bus, Flugzeug*›; gehen auf (+ *Akk.*) ‹*Schiff*›; **b)** *(put on)* anziehen ‹*Kleider, Schuhe*›; aufsetzen ‹*Hut, Kessel*›; *(load)* [auf]laden auf (+ *Akk.*); ~ **the cover [back] on** den Deckel [wieder] draufbekommen; **c)** *(coll.)* ~ **something on sb.** *(discover sth. incriminating)* etwas gegen jmdn. in der Hand haben

~ **'on to** *v. t.* **a)** *see* ~ **on 2a; b)** *(contact)* sich in Verbindung setzen mit; *(by telephone)* anrufen; **c)** *(realize)* ~ **on to sth.** hinter etw. *(Akk.)* kommen; ~ **on to the fact that ...:** dahinterkommen, daß ...; **d)** *(move on to discuss, study, etc.)* übergehen zu

~ **'on with** *v. t.* **a)** weitermachen mit; **let sb. ~ on with it** *(coll.)* jmdn. [allein weiter]machen lassen; **enough to be ~ting on with** genug für den Anfang *od.* fürs erste; **b)** = ~ **along with** *see* ~ **along a, c**

~ **'out 1.** *v. i.* **a)** *(walk out)* rausgehen **(of** aus); *(drive out)* rausfahren **(of** aus); *(alight)* aussteigen **(of** aus); *(climb out)* rausklettern **(of** aus); ~ **out [of my room]!** raus

[aus meinem Zimmer]!; **b)** *(leak)* austreten **(of** aus); *(escape from cage, jail)* ausbrechen, entkommen **(of** aus); *(fig.)* ‹*Geheimnis:*› herauskommen; ‹*Nachrichten:*› durchsickern. **2.** *v. t.* **a)** *(cause to leave)* rausbringen **(of** aus); *(send out)* rausschicken **(of** aus); *(throw out)* rauswerfen **(of** aus); **~ a stain out/out of sth.** einen Fleck wegbekommen/aus etw. herausbekommen; **b)** *(bring or take out)* herausholen **(of** aus); herausziehen ‹*Korken*›; *(drive out)* herausfahren **(of** aus); **c)** *(withdraw)* abheben ‹*Geld*› **(of** von)

~ 'out of *v. t.* **a)** *(leave)* verlassen ‹*Zimmer, Haus, Stadt, Land*›; *(cause to leave)* entfernen aus; *(extract from)* herausziehen aus; *(bring or take out of)* herausholen; *(leak from)* austreten aus; *(withdraw from)* abheben ‹*Geld*› von; **~ a book out of the library** ein Buch aus der Bibliothek ausleihen; **~ him out of my sight!** schaff ihn mir aus den Augen!; **~ sth. out of one's head** *or* **mind** sich *(Dat.)* etw. aus dem Kopf schlagen; **b)** *(escape)* herauskommen aus; *(avoid)* herumkommen um *(ugs.)*; sich drücken vor (+ *Dat.*) *(ugs.)*‹*Arbeit*›; **c)** *(gain from)* herausholen ‹*Geld*› aus; machen *(ugs.) od.* erzielen ‹*Gewinn*› bei; **~ a word/the truth/a confession out of sb.** aus jmdm. ein Wort/die Wahrheit/ein Geständnis herausbringen; **~ the best/most out of sb./sth.** das Beste/Meiste aus jmdm./etw. herausholen

~ 'over 1. *v. i.* **a)** *(cross)* **~ over to the other side** auf die andere Seite gehen; **b)** *see* **~ across 1 b. 2.** *v. t.* **a)** *(cross)* gehen über (+ *Akk.*); setzen über (+ *Akk.*) ‹*Fluß*›; *(climb)* klettern über (+ *Akk.*); *(cause to cross)* [hinüber]bringen über (+ *Akk.*); **b)** *see* **~ across 2 b; c)** *(overcome, recover from)* überwinden; hinwegkommen über (+ *Akk.*); verwinden *(geh.)* ‹*Verlust*›; sich erholen von ‹*Krankheit*›; **d)** *(fully believe)* **I can't ~ over his cheek/the fact that ...:** solche Frechheit kann ich nicht begreifen/ich kann gar nicht fassen, daß ...

~ 'over with *v. t. (coll.)* **~ sth. over with** etw. hinter sich *(Akk.)* bringen

~ 'past 1. *v. i. see* **~ by 1 a. 2.** *v. t. see* **~ by 2**

~ 'round 1. *v. i.* **a)** *see* **~ about; b) ~ round to doing sth.** dazu kommen, etw. zu tun. **2.** *v. t.* **a)** *(avoid)* umgehen ‹*Gesetz, Bestimmungen*›; **b) ~ round sb.** *(get one's way with)* jmdn. herumkriegen *(ugs.)*; *(persuade)* jmdn. überzeugen **(to** von); **c)** *(overcome)* lösen ‹*Problem usw.*›; überwinden ‹*Hindernis usw.*›; umgehen ‹*Schwierigkeit usw.*›

~ through 1. [-'-] *v. i.* **a)** *(pass obstacle)* durchkommen; *(make contact)* durchkommen *(ugs.)*; Verbindung bekommen **(to** mit); **b)** *(be transmitted)* durchkommen *(ugs.)*; durchdringen **(to** bis zu *od.* nach); **c)** *(win heat or round)* gewinnen; **~ through to the finals** in die Endrunde kommen; **d) ~ through [to sb.]** *(make sb. understand)* sich [jmdm.] verständlich machen; **e)** *(pass)* bestehen; durchkommen *(ugs.)*; **f)** *(be approved)* angenommen werden; durchkommen *(ugs.)*. **2.** *[stress varies] v. t.* **a)** *(pass through)* [durch]kommen durch; **b)** *(help to make contact)* **~ sb. through to** jmdn. verbinden mit; **c)** *(bring)* [durch]bringen; übermitteln ‹*Nachricht*› **(to** *Dat.*); **~ a message through to sb.** jmdm. eine Nachricht zukommen lassen; **d)** *(communicate)* **~ sth. through to sb.** jmdm. etw. klarmachen; **e)** *(pass)* durchkommen bei *(ugs.)*, bestehen ‹*Prüfung*›; **f)** *(consume, use up)* verbrauchen; verqualmen *(ugs. abwertend)* ‹*Zigaretten*›; aufessen ‹*Essen*›; *(spend)* durchbringen ‹*Geld, Vermögen*›; **g)** *(survive)* durchstehen; überstehen; kommen durch; **h)** fertig werden mit, erledigen ‹*Arbeit*›; durchkriegen ‹*Buch*›

~ to *v. t.* **a)** *(reach)* kommen zu ‹*Gebäude*›; erreichen ‹*Person, Ort*›; **he is ~ting to the age when ...:** er wird bald das Alter erreicht haben, wo ...; **I haven't got to the end [of the novel] yet** ich habe [den Roman] noch nicht zu Ende gelesen; **where has the child/the book got to?** wo ist das Kind hin/das Buch hingekommen?; **b)** *(begin)* **~ to doing sth.** anfangen, etw. zu tun; **c) ~ to sb.** *(coll.: annoy)* jmdm. auf die Nerven gehen *(ugs.)*

~ to'gether 1. *v. i.* zusammenkommen; **why not ~ together after work?** wollen wir uns nach Feierabend treffen? **2.** *v. t.* **a)** *(collect)* zusammenbringen; **~ one's things together** seine Sachen zusammenpacken; **b)** *(sl.: organize)* **~ it** *or* **things together** die Dinge auf die Reihe kriegen *(ugs.)*

~ up 1. [-'-] *v. i.* **a)** *(rise from bed, chair, floor; leave table)* aufstehen; **please don't ~ up!** bitte bleiben Sie sitzen!; **b)** *(climb)* [auf]steigen, aufsitzen **(on** auf + *Dat. od. Akk.*); **c)** *(rise, increase in force)* zunehmen; **the sea is ~ting up** die See wird immer wilder. **2.** *[stress varies] v. t.* **a)** *(call, awaken)* wecken; *(cause to leave bed)* aus dem Bett holen; **b)** *(cause to stand up)* aufhelfen (+ *Dat.*); **c)** *(climb)* hinaufsteigen; **your car will not ~ up that hill** dein Auto kommt den Berg nicht hinauf; **d)** *(carry up)* **~ sb./sth. up [sth.]** jmdn./etw. [etw.] her-/hinaufbringen; *(with difficulty)* jmdn. etw. [etw.] her-/hinaufbekommen; **e)** *(organize)* organisieren; auf die Beine stellen; auf die Beine bringen ‹*Personen*›; **f)** *(arrange appearance of, dress up)* zurechtmachen; herrichten ‹*Zimmer*›; **~ sb./oneself up as sb.** jmdn./sich als jmdn. ausstaffieren

~ 'up to *v. t.* **a)** *(reach)* erreichen ‹*Leistungsniveau*›; *(cause to reach)* bringen auf (+ *Akk.*); **b)** *(indulge in)* aussein auf (+ *Akk.*); **~ up to mischief** etwas anstellen; **what have you been ~ting up to?** was hast du getrieben *od.* angestellt?

get: ~-away *n.* Flucht, *die; attrib.* Flucht‹*plan, -wagen*›; **make one's ~-away** entkommen; **~-together** *n. (coll.)* Zusammenkunft, *die; (informal social gathering)* gemütliches Beisammensein; **have a ~-together** sich treffen; zusammenkommen; **~-up** *n. (coll.)* Aufmachung, *die;* **~-up-and-'go** *n. (coll.)* Elan, *der;* Schwung, *der*

geyser *n.* **a)** ['gi:zə(r), 'gaɪzə(r)] *(hot spring)* Geysir, *der;* **b)** ['gi:zə(r)] *(Brit.: water-heater)* Durchlauferhitzer, *der*

Ghana ['gɑ:nə] *pr. n.* Ghana *(das)*

Ghanaian [gɑ:'neɪən] **1.** *adj.* ghanaisch. **2.** *n.* Ghanaer, *der*/Ghanaerin, *die*

ghastly ['gɑ:stlɪ] *adj.* **a)** grauenvoll; gräßlich; entsetzlich ‹*Verletzungen*›; schrecklich ‹*Geschichte, Fehler, Irrtum*›; **b)** *(coll.: objectionable, unpleasant)* scheußlich *(ugs.)*; gräßlich *(ugs.)*; **I feel ~:** ich fühle mich scheußlich

gherkin ['gɜ:kɪn] *n.* Essiggurke, *die*

ghetto ['getəʊ] *n., pl.* **~s** Getto, *das*

ghost [gəʊst] **1.** *n.* Geist, *der;* Gespenst, *das;* **give up the ~:** den *od.* seinen Geist aufgeben *(ver-*

alt., scherzh.); (fig.: give up hope) die Hoffnung aufgeben; **not have the** *or* **a ~ of a chance** nicht die geringste Chance haben. **2.** *v.t.* **~ sb.'s speech** *etc.* für jmdn. eine Rede *usw.* [als Ghostwriter] schreiben

ghostly ['gəʊstlɪ] *adj.* gespenstisch; geisterhaft

ghost: ~ story *n.* Gespenstergeschichte, *die;* **~ town** *n.* Geisterstadt, *die;* **~-writer** *n.* Ghostwriter, *der*

ghoulish ['gu:lɪʃ] *adj.* teuflisch ‹*Freude*›; schaurig ‹*Gelächter*›; makaber ‹*Geschichte*›

GI ['dʒi:aɪ, dʒi:'aɪ] **1.** *adj.* GI-‹*Uniform, Haarschnitt*›. **2.** *n.* GI, *der*

giant ['dʒaɪənt] **1.** *n.* Riese, *der.* **2.** *attrib. adj.* riesig; Riesen- *(ugs.)*

giant 'panda *n.* Bambusbär, *der;* Riesenpanda, *der*

gibber ['dʒɪbə(r)] *v.i.* plappern; ‹*Affe:*› schnattern

gibberish ['dʒɪbərɪʃ] *n.* Kauderwelsch, *das*

gibbon ['gɪbən] *n. (Zool.)* Gibbon, *der*

gibe [dʒaɪb] **1.** *n.* Spöttelei, *die (ugs.);* Stichelei, *die (ugs.).* **2.** *v.i.* **~ at sb./sth.** über jmdn./etw. spötteln

giblets ['dʒɪblɪts] *n. pl.* [Geflügel]klein, *das*

Gibraltar [dʒɪ'brɔ:ltə(r)] *pr. n.* Gibraltar *(das)*

giddiness ['gɪdɪnɪs] *n., no pl.* Schwindel, *der*

giddy ['gɪdɪ] *adj.* schwind[e]lig; schwindelerregend ‹*Höhe, Abgrund*›; **I feel ~:** mir ist schwindlig

gift [gɪft] *n.* **a)** *(present)* Geschenk, *das;* Gabe, *die (geh.);* **it was given to me as a ~:** ich habe es geschenkt bekommen; **a ~ box/pack** eine Geschenkpackung; **b)** *(talent etc.)* Begabung, *die;* **have a ~ for languages/mathematics** sprachbegabt/mathematisch begabt sein; **c)** *(easy task etc.)* **be a ~:** geschenkt sein *(ugs.)*

gifted ['gɪftɪd] *adj.* begabt (**in, at** für); **be ~ in** *or* **at languages** sprachbegabt sein

gift: ~-horse *n.* **never** *or* **don't look a ~-horse in the mouth** *(prov.)* einem geschenkten Gaul schaut man nicht ins Maul *(Spr.);* **~ shop** *n.* Geschenkboutique, *die;* **~ token, ~ voucher** *ns.* Geschenkgutschein, *der;* **~-wrap** *v.t.* als Geschenk einpacken

gigantic [dʒaɪ'gæntɪk] *adj.* gigantisch; riesig; enorm, gewaltig ‹*Verbesserung, Appetit, Portion*›

giggle ['gɪgl] **1.** *n.* **a)** Kichern, *das;* Gekicher, *das;* **with a ~:** kichernd; **[a fit of] the ~s** ein Kicheranfall; **b)** *(coll.) (amusing person)* Witzbold, *der; (amusing thing, joke)* Spaß, *der;* **we did it for a ~:** wir wollten unseren Spaß haben. **2.** *v.i.* kichern

gild [gɪld] *v.t.* vergolden; **~ the lily** des Guten zuviel tun

gill [gɪl] *n., usu. in pl.* Kieme, *die*

gilt [gɪlt] **1.** *n. (gilding)* Goldauflage, *die; (paint)* Goldfarbe, *die.* **2.** *adj.* vergoldet

gilt-edged ['gɪltedʒd] *adj. (Commerc.)* **~ securities/stocks** mündelsichere Wertpapiere

gimmick ['gɪmɪk] *n. (coll.)* Gag, *der;* **a publicity ~:** ein Werbegag

gimmickry ['gɪmɪkrɪ] *n. (coll.)* Firlefanz, *der (ugs.);* Pipifax, *der (ugs.);* **advertising ~:** Werbetricks *od.* -gags

gimmicky ['gɪmɪkɪ] *adj. (coll.)* vergagt

gin [dʒɪn] *n. (drink)* Gin, *der;* **~ and tonic** Gin [und] Tonic, *der*

ginger ['dʒɪndʒə(r)] **1.** *n.* **a)** Ingwer, *der;* **b)** *(colour)* Rötlichgelb, *das.* **2.** *adj.* **a)** *(flavour)* Ingwer‹*gebäck, -geschmack*›; **b)** *(colour)* rötlichgelb; rotblond ‹*Bart, Haare*›

ginger: ~-'ale *n.* Ginger-ale, *das;* **~-'beer** *n.* Ingwerbier, *das;* Ginger-beer, *das;* **~bread** *n.* Pfefferkuchen, *der*

gingerly ['dʒɪndʒəlɪ] *adv.* behutsam; [übertrieben] vorsichtig

gipsy *see* **gypsy**

giraffe [dʒɪ'rɑ:f, dʒɪ'ræf] *n.* Giraffe, *die*

girder ['gɜ:də(r)] *n.* [Eisen-/Stahl]träger, *der*

girdle ['gɜ:dl] *n. (corset)* Hüfthalter, *der;* Hüftgürtel, *der*

girl [gɜ:l] *n.* **a)** Mädchen, *das; (teenager)* junges Mädchen; *([young] woman)* Frau, *die; (daughter)* Mädchen, *das (ugs.);* Tochter, *die;* **baby ~:** kleines Mädchen; **~s' school** Mädchenschule, *die;* **a ~'s name** ein Mädchenname; **[my] ~** *(as address)* [mein] Mädchen; **the ~s** *(female friends)* meine/ihre *usw.* Freundinnen; **the ~ at the cash-desk/switchboard** die Kassiererin/Telefonistin; **b)** *(sweetheart)* Mädchen, *das;* Freundin, *die*

girl: ~ 'Friday *see* **Friday 1; ~-friend** *n.* Freundin, *die;* **~ 'guide** *see* **guide 1c**

girlish ['gɜ:lɪʃ] *adj. (coll.)* mädchenhaft

giro ['dʒaɪrəʊ] *n., pl.* **~s** Giro, *das; attrib.* Giro-; **post office/bank ~:** Postgiro- *od. (veralt.)* Postscheck-/Giroverkehr, *der*

girth [gɜ:θ] *n.* **a)** *(circumference)* Umfang, *der; (at waist)* Taillenumfang, *der;* **b)** *(for horse)* Bauchgurt, *der*

gismo ['gɪzməʊ] *n. (sl.)* Ding, *das (ugs.)*

gist [dʒɪst] *n.* Wesentliche, *das; (of tale, argument, question, etc.)* Kern, *der;* **this is the ~ of what he said** das hat er im wesentlichen gesagt; **get the ~ of sth.** das Wesentliche einer Sache mitbekommen; **could you give me the ~ of it?** könntest du mir sagen, worum es hier geht?

give [gɪv] **1.** *v.t.,* **gave** [geɪv], **given** ['gɪvn] **a)** *(hand over, pass)* geben; *(transfer from one's authority, custody, or responsibility)* überbringen; übergeben (**to** an + *Akk.*); **she gave him her bag to carry** sie gab ihm ihre Tasche zum Tragen; **G~ it to me! I'll do it** Gib her! Ich mache das; **~ me ...** *(on telephone)* geben Sie mir ...; verbinden Sie mich mit ...; **b)** *(as gift)* schenken; *(donate)* spenden; geben; *(bequeath)* vermachen; **~ sb. sth., ~ sth. to sb.** jmdm. etw. schenken; **the book was ~n [to] me by my son** das Buch hat mir mein Sohn geschenkt; **I wouldn't have it if it was ~n [to] me** ich würde es nicht mal geschenkt nehmen; *abs.* **~ towards sth.** zu etw. beisteuern; **~ blood** Blut spenden; **~ [a donation] to charity** für wohltätige Zwecke spenden; **~ and take** *(fig.)* Kompromisse eingehen; *(in marriage etc.)* geben und nehmen; **c)** *(sell)* verkaufen; geben; *(pay)* zahlen; geben *(ugs.); (sacrifice)* geben; opfern; **~ sb. sth. [in exchange] for sth.** jmdm. etw. für etw. [im Tausch] geben; **I would ~ anything** *or* **my right arm/a lot to be there** ich würde alles/viel darum geben, wenn ich dort sein könnte; **d)** *(assign)* aufgeben ‹*Hausaufgaben, Strafarbeit usw.*›; *(sentence to)* geben ‹*10 Jahre Gefängnis usw.*›; **e)** *(grant, award)* geben ‹*Erlaubnis, Arbeitsplatz, Interview, Rabatt, Fähigkeit, Kraft*›; verleihen ‹*Preis, Titel, Orden usw.*›; **be ~n sth.** etw. bekommen; **he was ~n the privilege/honour of doing it** ihm wurde das Vorrecht/die Ehre zuteil, es zu tun; **~ sb. to understand** *or* **believe that ...:** jmdn. glauben lassen, daß ...; **f)** *(entrust sb. with)* übertragen (**to** *Dat.*); **~ sb. the power to do sth.** jmdn. ermächtigen, etw. zu tun; **g)** *(allow sb. to* have) geben ‹Recht, Zeit, Arbeit›; überlassen ‹seinen Sitzplatz›; lassen ‹Wahl, Zeit›; **they gave me [the use of] their car for the weekend** sie überließen mir ihr Auto übers Wochenende; **I will ~ you a day to think it over** ich lasse dir einen Tag Bedenkzeit; **~ yourself time to think about it** laß dir Zeit, und denk darüber nach; **~ me London any day** *or* **time** *or* **every time** *(fig. coll.)* London ist mir zehnmal lieber; **I['ll] ~ you/him** *etc.* **that** *(fig. coll.: grant)* das gebe ich zu; zugegeben; **you've got to ~ it to him** *(fig. coll.)* das muß man ihm lassen; **it cost £5, ~ or take a few pence** es hat so um die fünf Pfund gekostet *(ugs.)*; **~n that** *(because)* da; *(if)* wenn; **~n the right tools** mit dem richtigen Werkzeug; **~n time, I'll do it** wenn ich Zeit habe, mache ich es; **h)** *(offer to sb.)* geben, reichen ‹Arm, Hand usw.›; **please ~ me your attention** ich bitte um Ihre Aufmerksamkeit; **~ sb. in marriage** jmdn. verheiraten; **i)** *(cause sb./sth. to have)* geben; verleihen ‹Charme, Reiz, Gewicht, Nachdruck›; bereiten, machen ‹Freude, Mühe, Kummer›; bereiten, verursachen ‹Schmerz›; bieten ‹Abwechslung, Schutz›; leisten ‹Hilfe›; gewähren ‹Unterstützung›; **I was ~n the guest-room** man gab mir das Gästezimmer; **~ a clear picture** *(Telev.)* ein gutes Bild haben; **~ hope to sb.** jmdm. Hoffnung machen; **~ sb. what for** *(sl.)* es jmdm. geben *(ugs.)*; **j)** *(convey in words, tell, communicate)* angeben ‹Namen, Anschrift, Alter, Grund, Zahl›; nennen ‹Grund, Einzelheiten, Losungswort›; geben ‹Rat, Beispiel, Befehl, Anweisung, Antwort›; fällen ‹Urteil, Entscheidung›; sagen ‹Meinung›; bekanntgeben ‹Nachricht, Ergebnis›; machen ‹Andeutung›; erteilen ‹Verweis, Rüge›; *(present, set forth)* ‹Wörterbuch, Brief:› enthalten; ‹Zeitung:› bringen ‹Bericht›; **~ details of sth.** Einzelheiten einer Sache *(Gen.)* darlegen; **~ sth. a mention** etw. erwähnen; **~ sb. the facts** jmdn. mit den Fakten vertraut *od.* bekannt machen; **she gave us the news** sie teilte es uns mit; **~ sb. a decision** jmdm. eine Entscheidung mitteilen; **~ him my best wishes** richte ihm meine besten Wünsche aus; **don't ~ me 'that!** *(coll.)* erzähl mir [doch] nichts! *(ugs.)*; **k) ~n** *(specified)* gegeben; **l)** *(perform, read, sing, etc.)* geben ‹Vorstellung, Konzert›; halten ‹Vortrag, Seminar›; vorlesen ‹Gedicht, Erzählung›; singen ‹Lied›; spielen ‹Schauspiel, Oper, Musikstück›; **~ us a song** sing mal was; **m)** ausbringen ‹Toast, Trinkspruch›; *(as toast)* **ladies and gentlemen, I ~ you the Queen** meine Damen, meine Herren, auf die Königin *od.* das Wohl der Königin; **n)** *(produce)* geben ‹Licht, Milch›; tragen ‹Früchte›; ergeben ‹Zahlen, Resultat›; erbringen ‹Ernte›; **o)** *(cause to develop)* machen; **sth. ~s me a headache** von etw. bekomme ich Kopfschmerzen; **running ~s me an appetite** Laufen macht mich hungrig; **p)** *(make sb. undergo)* geben; versetzen ‹Schlag, Stoß›; verabreichen *(geh.)*, geben ‹Arznei›; **~ sb. a [friendly] look** jmdm. einen [freundlichen] Blick zuwerfen; **he gave her hand a squeeze** er drückte ihr die Hand; **~ as good as one gets** *(coll.)* es jmdm. mit gleicher Münze heimzahlen; **q)** *(execute, make, show)* geben ‹Zeichen, Stoß, Tritt›; machen ‹Satz, Ruck›; ausstoßen ‹Schrei, Seufzer, Pfiff›; **~ a [little] smile** [schwach] lächeln; **~ sth./sb. a look** sich *(Dat.)* etw./jmdn. ansehen; **r)** *(devote, dedicate)* widmen; **be ~n to sth./doing sth.** zu etw. neigen/etw. gern tun; **~ [it] all one's got** *(coll.)* sein möglichstes tun; **s)** *(be host at)* geben ‹Party, Empfang, Essen usw.›; **t) ~ sb./sth. two months/a year** jmdm./einer Sache zwei Monate/ein Jahr geben. **2.** *v. i.*, **gave, given a)** *(yield, bend)* nachgeben *(auch fig.)*; ‹Knie:› weich werden; ‹Bett:› federn; *(break down)* zusammenbrechen; ‹Brücke:› einstürzen; *(fig.)* nachlassen; **b)** *(lead)* **~ on to the street/garden** ‹Tür usw.:› auf die Straße hinausführen/in den Garten führen. **3.** *n.* **a)** Nachgiebigkeit, *die*; *(elasticity)* Elastizität, *die*; **have [no] ~:** [nicht] nachgeben; **b) ~ and take** *(compromise)* Kompromiß, *der*; *(exchange of concessions)* Geben und Nehmen, *das*

~ a'way *v. t.* **a)** *(without charge, as gift)* verschenken; **b)** *(in marriage)* dem Bräutigam zuführen; **c)** *(distribute)* verteilen, vergeben ‹Preise›; **d)** *(fig.: betray)* verraten

~ 'back *v. t. (lit. or fig.)* zurückgeben; wiedergeben

~ in 1. ['--] *v. t.* abgeben. **2.** [-'-] *v. i.* nachgeben **(to** *Dat.*)

~ 'off *v. t.* ausströmen ‹Rauch, Geruch›; aussenden ‹Strahlen›

~ out 1. ['--] *v. t.* **a)** *(distribute)* verteilen ‹Prospekte, Karten, Preise›; austeilen ‹Stifte, Hefte usw.›; vergeben ‹Arbeit›; **b)** *(declare)* bekanntgeben ‹Nachricht›. **2.** [-'-] *v. i.* ‹Vorräte:› ausgehen; ‹Maschine:› versagen; ‹Kraft:› nachlassen

~ 'over *v. t.* **a) be ~n over to sth.** für etw. beansprucht werden; **b)** *(abandon)* **~ sth./sb. over to sb.** etw. jmdm. überlassen/jmdn. jmdm. ausliefern; **c)** *(coll.: stop)* **~ over [doing sth.]** aufhören[, etw. zu tun]

~ 'up 1. *v. i.* aufgeben. **2.** *v. t.* **a)** *(renounce)* aufgeben; ablegen ‹Gewohnheit›; widmen ‹Zeit›; *(relinquish)* verzichten auf (+ *Akk.*) ‹Territorium, Süßigkeiten›; **~ sth. up** *(abandon habit)* sich *(Dat.)* etw. abgewöhnen; **~ sb./sth. up as a bad job** *(coll.)* jmdn./etw. abschreiben *(ugs.)*; **b) ~ sb. up** *(as not coming)* jmdn. nicht mehr erwarten; *(as beyond help)* jmdn. aufgeben; **c)** *(hand over to police etc.)* übergeben **(to** *Dat.*); **~ oneself up [to sb.]** sich [jmdm.] stellen

~ 'way *v. i.* **a)** *(yield, lit. or fig.)* nachgeben; *(collapse)* ‹Brücke, Balkon:› einstürzen; **his legs gave way under him** er knickte [in den Knien] ein; **~ way to anger** seinem Ärger Luft machen; **~ way to fear** der Angst erliegen; **b)** *(in traffic)* **~ way [to traffic from the right]** [dem Rechtsverkehr] die Vorfahrt lassen; **'G~ Way'** „Vorfahrt beachten"; **c)** *(be succeeded by)* **~ way to sth.** einer Sache *(Dat.)* weichen

'give-away *n. (coll.)* **a)** *(what betrays)* **the tremble in her voice was the ~:** mit ihrer zitternden Stimme hat sie sich verraten; **it was a dead ~:** es verriet alles; **b)** *attrib. (Commerc.)* **~ prices** Schleuderpreise

given *see* **give 1, 2**

giver ['gɪvə(r)] *n.* Geber, *der*/Geberin, *die*; *(donor)* Spender, *der*/Spenderin, *die*

gizmo *see* **gismo**

glacé ['glæseɪ] *adj.* glasiert

glacial ['gleɪsɪəl, 'gleɪʃl] *adj.* **a)** *(icy)* eisig; *(fig.)* eiskalt; **b)** *(Geol.)* Gletscher-

glacier ['glæsɪə(r)] *n.* Gletscher, *der*

glad [glæd] *adj. pred.* froh; **be ~ about sth.** sich über etw. *(Akk.)* freuen; **be ~ that ...:** sich freuen, daß ...; *(be relieved)* froh sein [darüber], daß ...; **[I'm] ~ to meet you** es freut mich *od.* ich freue mich, Sie kennenzulernen; **be ~ to hear sth.** sich freuen, etw. zu hören; *(relieved)* froh sein, etw. zu hören; **he's ~ to be alive** er ist froh, daß er lebt; **..., you'll be ~ to know/hear:** ..., das freut Sie sicherlich; **I'd be ~ to [help you]** aber gern [helfe ich Ihnen]; **Take your gloves. You'll be ~ of them** Nimm deine Handschuhe mit. Du wirst sie gebrauchen können

gladden ['glædn] *v.t.* erfreuen

glade [gleɪd] *n.* Lichtung, *die*

gladiator ['glædɪeɪtə(r)] *n.* Gladiator, *der*

gladiolus [glædɪ'əʊləs] *n., pl.* **gladioli** [glædɪ'əʊlaɪ] *or* **~es** *(Bot.)* Gladiole, *die*

gladly ['glædlɪ] *adv.* **a)** *(willingly)* gern; **b)** *(with joy)* freudig

glamor *(Amer.) see* **glamour**

glamorize (glamorise) ['glæməraɪz] *v.t.* *(add glamour to)* [mehr] Glanz verleihen (+ *Dat.*); *(idealize)* verherrlichen (**into** zu); glorifizieren

glamorous ['glæmərəs] *adj.* glanzvoll; glamourös ‹*Filmstar, Lebenswandel*›; mondän ‹*Kleidung*›; **a ~ job** ein Traumberuf

glamour ['glæmə(r)] *n.* Glanz, *der; (of person)* Ausstrahlung, *die*

glance [glɑ:ns] **1.** *n.* Blick, *der;* **cast** *or* **take** *or* **have a [quick] ~ at sth./sb.** einen [kurzen] Blick auf etw./jmdn. werfen; **at a ~:** auf einen Blick. **2.** *v.i.* **a)** blicken; schauen; **~ at sb./sth.** jmdn./etw. anblicken; **~ at one's watch** auf seine Uhr blicken; **she ~d at herself in the mirror** sie warf einen Blick in den Spiegel; **~ down/up [at sth.]** [auf etw. *(Akk.)*] hinunter-/[zu etw.] aufblicken; **~ through the newspaper** *etc.* die Zeitung *usw.* durchblättern; **~ at the newspaper** *etc.* einen Blick in die Zeitung *usw.* werfen; **~ round [the room]** sich [im Zimmer] umsehen; **b) ~ [off sth.]** abprallen [an etw. *(Dat.)*]; ‹*Messer, Schwert:*› abgleiten [an etw. *(Dat.)*]; **strike sb. a glancing blow** jmdn. nur streifen

gland [glænd] *n.* Drüse, *die*

glandular ['glændjʊlə(r)] *adj.* Drüsen-

glare [gleə(r)] **1.** *n.* **a)** *(dazzle)* grelles Licht; **the ~ of the sun** die grelle Sonne; **amidst the ~/in the full ~ of publicity** *(fig.)* im Rampenlicht der Öffentlichkeit; **b)** *(hostile look)* feindseliger Blick; **with a ~:** feindselig. **2.** *v.i.* **a)** *(glower)* [finster] starren; **~ at sb./sth.** jmdn./etw. anstarren; **b)** ‹*Licht:*› grell scheinen

glaring ['gleərɪŋ] *adj.* *(dazzling)* grell [strahlend/scheinend *usw.*]; gleißend hell ‹*Licht*›; *(fig.: conspicuous)* schreiend; eklatant; grob ‹*Fehler*›; kraß ‹*Gegensatz*›

glasnost ['glæsnɒst] *n.* Glasnost, *die*

glass [glɑ:s] **1.** *n.* **a)** *no pl. (substance)* Glas, *das;* **pieces of/broken ~:** Glasscherben *Pl.; (smaller)* Glassplitter *Pl.;* **b)** *(drinking ~)* Glas, *das;* **a ~ of milk** ein Glas Milch; **wine by the ~:** offener Wein; **c)** *(of spectacles, watch)* Glas, *das; (pane, covering picture)* [Glas]scheibe, *die;* **d)** *in pl. (spectacles)* **[a pair of] ~es** eine Brille. **2.** *attrib. adj.* Glas-

glass: ~-blower *n.* Glasbläser, *der*/Glasbläserin, *die;* **~-blowing** *n.* Glasblasen, *das;* **~ 'fibre** *n.* Glasfaser, *die*

glassful ['glɑ:sfʊl] *n.* Glas, *das* (**of** von); **a ~ of milk** ein Glas Milch

glass: ~house *n.* **a)** *(Brit.: greenhouse)* Gewächshaus, *das;* Glashaus, *das;* **b)** *(Brit. sl.: military prison)* Bunker, *der (Soldatenspr. salopp);* **~ware** *n.* Glas, *das*

glassy ['glɑ:sɪ] *adj.* gläsern; *(fig.)* glasig ‹*Blick*›

glaucoma [glɔ:'kəʊmə] *n.* *(Med.)* Glaukom, *das (fachspr.);* grüner Star

glaze [gleɪz] **1.** *n.* *(on food or pottery)* Glasur, *die; (of paint)* Lasur, *die; (on paper, fabric)* Appretur, *die.* **2.** *v.t.* **a)** *(cover with ~)* glasieren ‹*Eßwaren, Töpferwaren*›; satinieren ‹*Papier, Kunststoff*›; lasieren ‹*Farbe, bemalte Fläche*›; **~d tile** Kachel, *die;* **b)** *(fit with glass)* **~ [in]** verglasen ‹*Fenster, Haus usw.*›. **3.** *v.i.* **~ [over]** ‹*Augen:*› glasig werden

glazier ['gleɪzɪə(r), 'gleɪʒə(r)] *n.* Glaser, *der*

gleam [gli:m] **1.** *n.* **a)** Schein, *der; (fainter)* Schimmer, *der;* **~ of light** Lichtschein, *der;* **b)** *(fig.: faint trace)* Anflug, *der* (**of** von); **~ of hope/truth** Hoffnungsschimmer, *der*/Funke Wahrheit. **2.** *v.i.* ‹*Sonne, Licht:*› scheinen; ‹*Fußboden, Fahrzeug, Stiefel:*› glänzen; ‹*Zähne:*› blitzen; ‹*Augen:*› leuchten

gleaming ['gli:mɪŋ] *adj.* glänzend ‹*Wasser, Metall, Fahrzeug*›

glean [gli:n] *v.t.* **a)** zusammentragen ‹*Informationen, Nachrichten usw.*›; herausfinden ‹*Inhalt eines Briefes usw.*›; **~ sth. from sth.** einer Sache *(Dat.)* etw. entnehmen; **b)** *(Agric.)* nachlesen ‹*Getreide*›

glee [gli:] *n.* Freude, *die; (gloating joy)* Schadenfreude, *die*

gleeful ['gli:fl] *adj.* freudig; vergnügt; *(gloatingly joyful)* schadenfroh; hämisch

glen [glen] *n.* [schmales] Tal

glib [glɪb] *adj.* *(derog.)* aalglatt ‹*Person*›; *(impromptu, offhand)* leicht dahingesagt ‹*Antwort*›; *(facile in the use of words)* zungenfertig ‹*Person*›; flink ‹*Zunge*›; flinkzüngig ‹*Antwort*›

glide [glaɪd] *v.i.* **a)** gleiten; *(through the air)* schweben; **b)** ‹*Segelflugzeug:*› gleiten, schweben; ‹*Flugzeug:*› im Gleitflug fliegen

glider ['glaɪdə(r)] *n.* Segelflugzeug, *das*

gliding ['glaɪdɪŋ] *n.* *(Sport)* Segelfliegen, *das; attrib.* Segelflug-

glimmer ['glɪmə(r)] **1.** *n.* *(of light)* [schwacher] Schein; Schimmer, *der* (**of** von) *(auch fig.); (of fire)* Glimmen, *das.* **2.** *v.i.* glimmen

glimpse [glɪmps] **1.** *n.* [kurzer] Blick; **catch** *or* **have** *or* **get a ~ of sb./sth.** jmdn./etw. [kurz] zu sehen *od.* zu Gesicht bekommen. **2.** *v.t.* flüchtig sehen

glint [glɪnt] **1.** *n.* Schimmer, *der; (reflected flash)* Glitzern, *das; (of eyes)* Funkeln, *das; (of knife, dagger)* Blitzen, *das.* **2.** *v.i.* blinken; glitzern

glisten ['glɪsn] *v.i.* glitzern; *see also* **glitter 1**

glitter ['glɪtə(r)] **1.** *v.i.* glitzern; ‹*Augen, Juwelen, Sterne:*› funkeln; **all that ~s** *or* **glistens is not gold** *(prov.)* es ist nicht alles Gold, was glänzt *(Spr.).* **2.** *n.* Glitzern, *das; (of diamonds)* Funkeln, *das*

gloat [gləʊt] *v.i.* **~ over sth.** *(look at with selfish delight)* sich an etw. *(Dat.)* weiden *od.* ergötzen; *(derive sadistic pleasure from)* sich hämisch über etw. *(Akk.)* freuen

global ['gləʊbl] *adj.* global; weltweit; weltumspannend ‹*Kommunikationssystem*›; **~ warming** globaler Temperaturanstieg

globe [gləʊb] *n.* **a)** *(sphere)* Kugel, *die;* **b)** *(sphere with map)* Globus, *der;* **c)** *(world)* **the ~:** der Erdball

'globe-trotter *n.* Globetrotter, *der;* Weltenbummler, *der*

globular ['glɒbjʊlə(r)] *adj.* kugelförmig

globule ['glɒbju:l] *n.* Kügelchen, *das; (of liquid)* Tröpfchen, *das*

gloom [glu:m] *n.* **a)** *(darkness)* Dunkel, *das (geh.);* **b)** *(despondency)* düstere Stimmung

gloomy ['glu:mɪ] *adj.* **a)** *(dark)* düster; finster; dämmrig ‹*Tag,*

Nachmittag usw.〉; **b)** *(depressing)* düster, finster [stimmend]; bedrückend; *(depressed)* trübsinnig 〈*Person*〉; bedrückt 〈*Gesicht*〉; **he always tends to see the ~ side of things** er sieht immer gleich schwarz; **feel ~ about the future** der Zukunft pessimistisch entgegensehen

glorification [glɔːrɪfɪˈkeɪʃn] *n.* Verherrlichung, *die*

glorify [ˈglɔːrɪfaɪ] *v. t.* *(extol)* verherrlichen; **he's just a glorified messenger-boy** er ist nichts weiter als ein besserer Botenjunge

glorious [ˈglɔːrɪəs] *adj.* **a)** *(illustrious)* ruhmreich 〈*Held, Sieg, Geschichte*〉; **b)** *(delightful)* wunderschön; herrlich

glory [ˈglɔːrɪ] **1.** *n.* **a)** *(splendour)* Schönheit, *die; (majesty)* Herrlichkeit, *die;* **b)** *(fame)* Ruhm, *der;* **c)** **~ [be] to God in the highest** Ehre sei Gott in der Höhe. **2.** *v. i.* **~ in sth./doing sth.** *(be pleased by)* etw. genießen/es genießen, etw. zu tun; *(be proud of)* sich einer Sache *(Gen.)* rühmen/sich rühmen, etw. zu tun; **~ in the name of ...**: den stolzen Namen ... besitzen *od.* führen

[1]**gloss** [glɒs] *n.* **a)** *(sheen)* Glanz, *der;* **~ paint** Lackfarbe, *die;* **b)** *(fig.)* Anstrich, *der*

~ over *v. t.* bemänteln; beschönigen 〈*Fehler*〉; *(conceal)* unter den Teppich kehren *(ugs.)*

[2]**gloss 1.** *n.* [Wort]erklärung, *die.* **2.** *v. t.* glossieren

glossary [ˈglɒsərɪ] *n.* Glossar, *das*

glossy [ˈglɒsɪ] *adj.* glänzend; **~ print** Glanzabzug, *der*

glottal stop [glɒtl ˈstɒp] *n.* *(Phonet.)* Glottisschlag, *der;* Knacklaut, *der*

glove [glʌv] *n.* Handschuh, *der;* **sth. fits sb. like a ~**: etw. paßt jmdm. wie angegossen *(ugs.)*

ˈ**glove compartment** *n.* Handschuhfach, *das*

glow [gləʊ] **1.** *v. i.* **a)** glühen; 〈*Lampe, Leuchtfarbe:*〉 schimmern, leuchten; **b)** *(fig.)* *(with warmth or pride)* 〈*Gesicht, Wangen:*〉 glühen (**with** vor + *Dat.*); *(with health or vigour)* strotzen (**with** vor + *Dat.*); **c)** *(be suffused with warm colour)* [warm] leuchten. **2.** *n.* **a)** Glühen, *das; (of candle, lamp)* Schein, *der; (of embers, sunset)* Glut, *die;* **b)** *(fig.)* Glühen, *das;* **his cheeks had a healthy ~**: seine Wangen hatten eine blühende Farbe

glower [ˈglaʊə(r)] *v. i.* finster dreinblicken; **~ at sb.** jmdn. finster anstarren

glowing [ˈgləʊɪŋ] *adj.* glühend *(auch fig.); (fig.: enthusiastic)* begeistert 〈*Bericht, Beschreibung*〉; **describe sth. in ~ colours** etw. in glühenden Farben beschreiben

ˈ**glow-worm** *n.* Glühwürmchen, *das*

glucose [ˈgluːkəʊs, ˈgluːkəʊz] *n.* Glucose, *die*

glue [gluː] **1.** *n.* Klebstoff, *der.* **2.** *v. t.* **a)** kleben; **~ sth. together/on** etw. zusammen-/ankleben; **~ sth. to sth.** etw. an etw. *(Dat.)* an- *od.* festkleben; **b)** *(fig.)* **be ~d to sth./sb.** an etw./jmdm. kleben *(ugs.);* **their eyes** *or* **they were ~d to the TV screen** sie starrten auf den Bildschirm

glum [glʌm] *adj.* trübsinnig 〈*Person*〉; bedrückt 〈*Gesicht*〉

glut [glʌt] *(Commerc.)* **1.** *n.* Überangebot, *das* (**of** an, von + *Dat.*); **a ~ of apples** eine Apfelschwemme. **2.** *v. t.*, **-tt-** überschwemmen

glutinous [ˈgluːtɪnəs] *adj.* klebrig

glutton [ˈglʌtən] *n.* Vielfraß, *der (ugs.);* **a ~ for punishment** *(iron.)*/**work** *(fig.)* ein Masochist *(fig.)*/ein Arbeitstier *(fig.)*

gluttonous [ˈglʌtənəs] *adj.* gefräßig

gluttony [ˈglʌtənɪ] *n.* Gefräßigkeit, *die*

glycerine [ˈglɪsəriːn] (*Amer.:* **glycerin** [ˈglɪsərɪn]) *n.* Glyzerin, *das*

gm. *abbr.* **gram[s]** g

GMT *abbr.* **Greenwich mean time** GMT; WEZ

gnarled [nɑːld] *adj.* knorrig; knotig 〈*Finger, Hand*〉

gnash [næʃ] *v. t.* **~ one's teeth [in anger]** [vor Zorn] mit den Zähnen knirschen; **~ing of teeth** Zähneknirschen, *das*

gnat [næt] *n.* [Stech]mücke, *die*

gnaw [nɔː] **1.** *v. i.* **~ [away] at sth.** *(lit. or fig.)* an etw. *(Dat.)* nagen; **~ through a rope/sack** ein Seil/einen Sack durchnagen. **2.** *v. t.* nagen an (+ *Dat.*); abnagen 〈*Knochen*〉; kauen an *od.* auf (+ *Dat.*) 〈*Fingernägeln*〉; **~ a hole in sth.** ein Loch in etw. *(Akk.)* nagen

gnawing [ˈnɔːɪŋ] *adj.* nagend 〈*Hunger, Schmerz, Zweifel, Kummer usw.*〉; quälend 〈*Zahnschmerzen, Angst*〉

gnome [nəʊm] *n.* Gnom, *der; (in garden)* Gartenzwerg, *der*

GNP *abbr.* **gross national product** BSP

gnu [nuː, njuː] *n.* *(Zool.)* Gnu, *das*

go [gəʊ] **1.** *v. i., pres.* **he goes** [gəʊz], *p. t.* **went** [went], *pres. p.* **going** [ˈgəʊɪŋ], *p. p.* **gone** [gɒn] **a)** gehen; 〈*Fahrzeug:*〉 fahren; 〈*Flugzeug:*〉 fliegen; 〈*Vierfüßer:*〉 laufen; 〈*Reptil:*〉 kriechen; *(on horseback etc.)* reiten; *(on skis, roller-skates)* laufen; *(in wheelchair, pram, lift)* fahren; **go by bicycle/car/bus/train** *or* **rail/boat** *or* **sea** *or* **ship** mit dem [Fahr]rad/Auto/Bus/Zug/Schiff fahren; **go by plane** *or* **air** fliegen; **go on foot** zu Fuß gehen; laufen *(ugs.);* **as one goes [along]** *(fig.)* nach und nach; **do sth. as one goes [along]** *(lit.)* etw. beim Gehen *od.* unterwegs tun; **go on a journey** eine Reise machen; verreisen; **go first-class/at 50 m.p.h.** erster Klasse reisen *od.* fahren/80 Stundenkilometer fahren; **have far to go** weit zu gehen *od.* zu fahren haben; es weit haben; **the doll/dog goes everywhere with her** sie hat immer ihre Puppe/ihren Hund dabei; **who goes there?** *(sentry's challenge)* wer da?; **there you go** *(coll., giving sth.)* bitte!; da! *(ugs.);* **b)** *(proceed as regards purpose, activity, destination, or route)* 〈*Bus, Zug, Lift, Schiff:*〉 fahren; *(use means of transportation)* fahren; *(fly)* fliegen; *(proceed on outward journey)* weg-, abfahren; *(travel regularly)* 〈*Verkehrsmittel:*〉 verkehren (**from ... to** zwischen + *Dat.* ... und); **his hand went to his pocket** er griff nach seiner Tasche; **go to the toilet/cinema/moon/a museum/a funeral** auf die Toilette/ins Kino gehen/zum Mond fliegen/ins Museum/zu einer Beerdigung gehen; **go to a dance** tanzen gehen; **go to the doctor['s]** *etc.* zum Arzt *usw.* gehen; **go [out] to China** nach China gehen; **go [over] to America** nach Amerika [hinüber]fliegen/-fahren; **go [off] to London** nach London [ab]fahren/[ab]fliegen; **last year we went to Italy** letztes Jahr waren wir in Italien; **go this/that way** hier/da entlanggehen/-fahren; **go out of one's way** einen Umweg machen; *(fig.)* keine Mühe scheuen; **go towards sth./sb.** auf etw./jmdn. zugehen; **don't go on the grass** geh nicht auf den Rasen; **go by sth./sb.** 〈*Festzug usw.:*〉 an etw./jmdm. vorbeiziehen; 〈*Bus usw.:*〉 an etw./jmdm. vorbeifahren; **go in and out [of sth.]** [in etw. *(Dat.)*] ein- und ausgehen; **go into sth.** in etw. *(Akk.)* [hinein]gehen; **go looking for sb.** jmdn. suchen gehen; **go chasing after sth./sb.** hinter etw./jmdm. herrennen *(ugs.);*

go to live in Berlin nach Berlin ziehen; **go to see sb.** jmdn. aufsuchen; **I went to water the garden** ich ging den Garten sprengen; **go and do sth.** [gehen und] etw. tun; **I'll go and get my coat** ich hole jetzt meinen Mantel; **go and see whether ...:** nachsehen [gehen], ob ...; **go on a pilgrimage** *etc.* eine Pilgerfahrt *usw.* machen; **go on TV/the radio** im Fernsehen/Radio auftreten; **I'll go!** ich geh schon!; *(answer phone)* ich geh ran *od.* nehme ab; *(answer door)* ich mache auf; **'you go!** *(to the phone)* geh du mal ran!; **c)** *(start)* losgehen; *(in vehicle)* losfahren; **let's go!** *(coll.)*, fangen wir an!; **here goes!** *(coll.)* dann mal los!; **whose turn is it to go?** *(in game)* wer ist an der Reihe?; **go first** *(in game)* anfangen; **from the word go** *(fig. coll.)* [schon] von Anfang an; **d)** *(pass, circulate, be transmitted)* gehen; **a shiver went up** *or* **down my spine** ein Schauer lief mir über den Rücken *od.* den Rücken hinunter; **go to** *(be given to)* ‹*Preis, Sieg, Gelder, Job:*› gehen an (+ *Akk.*); ‹*Titel, Krone, Besitz:*› übergehen auf (+ *Akk.*); ‹*Ehre, Verdienst:*› zuteil werden *(Dat.)*; **go towards** *(be of benefit to)* zugute kommen (+ *Dat.*); **go according to** *(be determined by)* sich richten nach; **e)** *(make specific motion, do something specific)* **go round** ‹*Rad:*› sich drehen; **there he** *etc.* **goes again** *(coll.)* da, schon wieder!; **here we go again** *(coll.)* jetzt geht das wieder los!; **f)** *(act, work, function effectively)* gehen; ‹*Mechanismus, Maschine:*› laufen; **get the car to go** das Auto ankriegen *(ugs.) od.* starten; **at midnight we were still going** um Mitternacht waren wir immer noch dabei *od.* im Gange; **go by electricity** mit Strom betrieben werden; **keep going** *(in movement)* weitergehen/-fahren; *(in activity)* weitermachen; *(not fail)* sich aufrecht halten; **keep sb. going** *(enable to continue)* jmdn. aufrecht halten; **that'll keep me going** damit komme ich aus; **keep sth. going** etw. in Gang halten; **make sth. go, get/set sth. going** etw. in Gang bringen; **g) go to** *(attend):* **go to work** zur Arbeit gehen; **go to church/school** in die Kirche/die Schule gehen; **go to a comprehensive school** eine Gesamtschule besuchen; auf eine Gesamtschule gehen; **h)** *(have recourse)* **go to the originals** auf die Quellen zurückgreifen; **go to the relevant authority/UN** sich an die zuständige Behörde/UN wenden; **where do we go from here?** *(fig.)* und was nun? *(ugs.)*; **i)** *(depart)* gehen; ‹*Bus, Zug:*› [ab]fahren; ‹*Post:*› rausgehen *(ugs.)*; **I must be going now** ich muß allmählich gehen; **time to go!** wir müssen/ihr müßt *usw.* gehen!; **to go** *(Amer.)* ‹*Speisen, Getränke:*› zum Mitnehmen; **j)** *(euphem.: die)* sterben; **be dead and gone** tot sein; **after I go** wenn ich einmal nicht mehr bin; **k)** *(fail)* ‹*Gedächtnis, Kräfte:*› nachlassen; *(cease to function)* kaputtgehen; ‹*Maschine, Computer usw.:*› ausfallen; ‹*Sicherung:*› durchbrennen; *(break)* brechen; ‹*Seil usw.:*› reißen; *(collapse)* einstürzen; *(fray badly)* ausfransen; **the jacket has gone at the elbows** die Jacke ist an den Ellbogen durchgescheuert; **l)** *(disappear)* verschwinden; ‹*Geruch, Rauch:*› sich verziehen; ‹*Geld, Zeit:*› draufgehen *(ugs.)* **(in, on** für); *(be relinquished)* aufgegeben werden; ‹*Tradition:*› abgeschafft werden; *(be dismissed)* ‹*Arbeitskräfte:*› entlassen werden; **be gone from sight** außer Sicht geraten sein; **my coat/the stain has gone** mein Mantel/der Fleck ist weg; **where has my hat gone?** wo ist mein Hut [geblieben]?; **all his money goes on women** er gibt sein ganzes Geld für Frauen aus; **m)** *(elapse)* ‹*Zeit:*› vergehen; ‹*Interview usw.:*› vorüber-, vorbeigehen; **that has all gone by** das ist jetzt alles vorbei; **in days gone by** in längst vergangenen Zeiten; **n) to go** *(still remaining):* **have sth. [still] to go** [noch] etw. übrig haben; **there's hours to go** es dauert noch Stunden; **one week** *etc.* **to go to ...:** noch eine Woche *usw.* bis ...; **there's only another mile to go** [es ist] nur noch eine Meile; **still have a mile to go** noch eine Meile vor sich *(Dat.)* haben; **one down, two to go** einer ist bereits erledigt, bleiben noch zwei übrig *(salopp)*; **o)** *(be sold)* weggehen *(ugs.)*; verkauft werden; **it went for £1** es ging für 1 Pfund weg; **going! going! gone!** zum ersten! zum zweiten! zum dritten!; **p)** *(run)* ‹*Grenze, Straße usw.:*› verlaufen, gehen; *(afford access, lead)* gehen; führen; *(extend)* reichen; *(fig.)* gehen; **as** *or* **so far as he/it goes** soweit; **q)** *(turn out, progress)* ‹*Ereignis, Projekt, Interview, Abend:*› verlaufen; **go against sb./sth.** ‹*Wahl, Kampf:*› zu jmds./einer Sache Ungunsten ausgehen; ‹*Entscheidung, Urteil:*› zu jmds./einer Sache Ungunsten ausfallen; **how did your holiday/party go?** wie war Ihr Urlaub/Ihre Party?; **how is the book going?** was macht [denn] das Buch?; **things have been going well/badly/smoothly** *etc.* in der letzten Zeit läuft alles gut/schief/glatt *usw.*; **how are things going?, how is it going?** wie steht's *od.* *(ugs.)* läuft's?; **r)** *(be, have form or nature, be in temporary state)* sein; ‹*Sprichwort, Gedicht, Titel:*› lauten; **this is how things go, that's the way it goes** so ist es nun mal; **go against sth.** mit etw. nicht übereinstimmen; **go against one's principles** gegen seine Prinzipien gehen; **go hungry** hungern; hungrig bleiben; **go without food/water** es ohne Essen/Wasser aushalten; **go in fear of one's life** in beständiger Angst um sein Leben leben; *see also* **go against; s)** *(become)* werden; **the tyre has gone flat** der Reifen ist platt; **the phone has gone dead** die Leitung ist tot; **the constituency/York went Tory** der Wahlkreis/York ging an die Tories; **t)** *(have usual place)* kommen; *(belong)* gehören; **where does the box go?** wo kommt *od.* gehört die Kiste hin?; **where do you want this chair to go?** wo soll *od.* kommt der Stuhl hin?; **u)** *(fit)* passen; **go in[to] sth.** in etw. *(Akk.)* gehen *od.* [hinein]passen; **go through sth.** durch etw. [hindurch]gehen *od.* [hindurch]passen; **six into twelve goes twice** sechs geht zweimal in zwölf; **five goes into forty exactly** vierzig durch fünf geht auf; **v)** *(harmonize, match)* passen **(with** zu); **the two colours don't go** die beiden Farben passen nicht zusammen *od.* beißen sich; **w)** *(serve, contribute)* dienen; **the qualities that go to make a leader** die Eigenschaften, die einen Führer ausmachen; **it just goes to show that ...:** daran zeigt sich, daß ...; **x)** *(make sound of specified kind)* machen; *(emit sound)* ‹*Turmuhr, Gong:*› schlagen; ‹*Glocke:*› läuten; **There goes the bell. School is over** Es klingelt. Die Schule ist aus; **the fire alarm went at 3 a.m.** der Feueralarm ging um 3 Uhr morgens los; **a police car with its siren going** ein Polizeiwagen mit eingeschalteter Sirene; **y)** *as intensifier (coll.)* **don't go making** *or* **go and make him angry** verärgere ihn bloß nicht; **don't go looking**

for trouble such keinen Streit; **I gave him a £10 note and, of course, he had to go and lose it** *(iron.)* ich gab ihm einen 10-Pfund-Schein, und er mußte ihn natürlich prompt verlieren; **now you've been and gone and done it!** *(sl.)* du hast ja was Schönes angerichtet! *(ugs. iron.)*; **go tell him I'm ready** *(coll./Amer.)* geh und sag ihm, daß ich fertig bin; **z)** *(coll.: be acceptable or permitted)* erlaubt sein; gehen *(ugs.)*; **everything/anything goes** es ist alles erlaubt; **it/that goes without saying** es/das ist doch selbstverständlich; **what he** *etc.* **says, goes** was er *usw.* sagt, gilt. *See also* **going; gone. 2.** *v.t., forms as* **1: a)** *(Cards)* spielen; **b)** *(sl.)* **go it** es toll treiben; *(work hard)* rangehen; **he has been going it a bit too hard** er hat es etwas zu weit getrieben; **go it!** los!; weiter! **3.** *n., pl.* **goes** [gəʊz] *(coll.)* **a)** *(attempt, try)* Versuch, *der; (chance)* Gelegenheit, *die;* **have a go** es versuchen *od.* probieren; **have a go at doing sth.** versuchen, etw. zu tun; **have a go at sth.** sich an etw. *(Dat.)* versuchen; **let me have/can I have a go?** laß mich [auch ein]mal/kann ich [auch ein]mal? *(ugs.)*; **it's my go** ich bin an der Reihe *od.* dran; **in two/three goes** bei zwei/drei Versuchen; **at the first go** auf Anhieb; **b) have a go at sb.** *(scold)* sich *(Dat.)* jmdn. vornehmen *od.* vorknöpfen *(ugs.)*; *(attack)* über jmdn. herfallen; **c)** *(period of activity)* **in one go** auf einmal; **he downed his beer in one go** er trank sein Bier in einem Zug aus; **d)** *(energy)* Schwung, *der;* **be full of go** voller Schwung *od.* Elan sein; **have plenty of go** einen enormen Schwung *od.* Elan haben; **e)** *(vigorous activity)* **it's all go** es ist alles eine einzige Hetzerei *(ugs.)*; **it's all go at work** es ist ganz schön was los bei der Arbeit; **be on the go** auf Trab sein *(ugs.)*; **keep sb. on the go** jmdn. auf Trab halten *(ugs.)*; **f)** *(success)* **make a go of sth.** mit etw. Erfolg haben; **it's no go** da ist nichts zu machen. **4.** *adj. (coll.)* **all systems go** alles klar

go about 1. [--'-] *v.i.* **a)** *(move from place to place)* herumgehen/-fahren; **go about in groups** in Gruppen herumziehen; **go about in leather gear/dressed like a tramp** in Lederkleidung/wie ein Landstreicher herumlaufen; **go about doing sth.** *(be in the habit of)* etw. immer tun; **b)** *(circulate)* ⟨*Gerücht, Geschichte, Grippe:*⟩ umgehen. **2.** ['---] *v.t.* **a)** *(set about)* erledigen ⟨*Arbeit*⟩; angehen ⟨*Problem*⟩; **how does one go about it?** wie geht man da vor?; **b)** *(busy oneself with)* nachgehen (+ *Dat.*) ⟨*Arbeit usw.*⟩

'go after *v.t. (hunt)* jagen; zu stellen versuchen; *(fig.)* anstreben; sich bemühen um ⟨*Job*⟩

'go against *v.t.* zuwiderhandeln (+ *Dat.*); handeln gegen ⟨*Prinzip, Gesetz*⟩; **go against sb.** sich jmdm. in den Weg stellen *od.* widersetzen; *see also* **go 1 q, r**

go a'head *v.i.* **a)** *(in advance)* vorausgehen (**of** *Dat.*); **You go ahead. I'll meet you there** Geh mal schon vor. Wir treffen uns dann dort; **b)** *(proceed)* weitermachen; *(make progress)* ⟨*Arbeit:*⟩ fortschreiten, vorangehen; **go ahead with a plan** einen Plan durchführen; **go ahead and do it** es einfach machen; **go ahead!** nur zu!

go a'long 1. *v.i.* dahingehen/-fahren; *(attend)* hingehen. **2.** *v.t.* entlanggehen/-fahren

go a'long with *v.t.* **go along with sth.** *(share sb.'s opinion)* einer Sache *(Dat.)* zustimmen; *(agree to)* sich einer Sache *(Dat.)* anschließen; **go along with sb.** mit jmdm. übereinstimmen

go a'round *see* **go about 1 a, b; go round**

'go at *v.t.* **go at sb.** *(attack)* auf jmdn. losgehen; **go at sth./it** *(work at)* sich an etw. *(Akk.)* machen/sich dranmachen

go a'way *v.i.* weggehen; *(on holiday or business)* wegfahren; verreisen; **the problem won't go away** das Problem kann man nicht einfach ignorieren

go 'back *v.i.* **a)** *(return)* zurückgehen/-fahren; *(fig.)* zurückgehen; **go back to a subject** auf ein Thema zurückkommen; **go back to the beginning** noch mal von vorne anfangen; **b)** *(be returned)* zurückgegeben werden; ⟨*Waren:*⟩ zurückgehen (**to** an + *Akk.*); **c)** *(be put back)* ⟨*Uhren:*⟩ zurückgestellt werden

'go by *v.t.* **go by sth.** sich nach etw. richten; *(adhere to)* sich an etw. *(Akk.)* halten; **if the report is anything to go by** wenn man nach dem Bericht gehen kann; **go by appearances** nach dem Äußeren gehen; *see also* **go 1 a, b, m**

go 'down *v.i.* **a)** hinuntergehen/-fahren; ⟨*Taucher:*⟩ [hinunter]tauchen; *(set)* ⟨*Sonne:*⟩ untergehen; *(sink)* ⟨*Schiff:*⟩ sinken, untergehen; *(fall to ground)* ⟨*Flugzeug usw.:*⟩ abstürzen; **go down to the bottom of the garden/to the beach** zum hinteren Ende des Gartens gehen/an den Strand gehen; **b)** *(be swallowed)* hinuntergeschluckt werden; **go down the wrong way** in die falsche Kehle geraten; **c)** *(become less)* sinken; ⟨*Umsatz, Schwellung:*⟩ zurückgehen; ⟨*Vorräte usw.:*⟩ abnehmen; ⟨*Währung:*⟩ fallen; *(become lower)* fallen; *(subside)* ⟨*Wind usw.:*⟩ nachlassen; **go down in sb.'s estimation/in the world** in jmds. Achtung *(Dat.)* sinken/sich verschlechtern; **d) go down well/all right** *etc.* **[with sb.]** [mit jmdm.] gut *usw.* klarkommen *(ugs.)*; ⟨*Film, Schauspieler, Vorschlag:*⟩ [bei jmdm.] gut *usw.* ankommen *(ugs.)*; **that didn't go down [at all] well with his wife** das hat ihm seine Frau nicht abgenommen; **e)** *(be defeated)* unterliegen; **go down to sb.** gegen jmdn. verlieren

go 'down with *v.t.* bekommen ⟨*Krankheit*⟩; *see also* **go down d**

'go for *v.t.* **a)** *(go to fetch)* **go for sb./sth.** jmdn./etw. holen; **b)** *(apply to)* **go for sb./sth.** für jmdn./etw. gelten; **that goes for me too** das gilt auch für mich; ich auch; **c)** *(like)* **go for sb./sth.** jmdn./etw. gut finden. *See also* **go 1 o; going 2 e**

go 'forward *v.i.* **a)** weitergehen/-fahren; *(fig.)* voranschreiten; **b)** *(be put forward)* ⟨*Uhren:*⟩ vorgestellt werden

go 'in *v.i.* **a)** *(go indoors)* hineingehen; reingehen *(ugs.)*; **b)** *(be covered by cloud)* verschwinden; weggehen *(ugs.)*; **c)** *(be learnt)* [in den Kopf] reingehen *(ugs.) See also* **go 1 b, u**

go 'in for *v.t.* **go in for sth.** *(choose as career)* etw. [er]lernen wollen; *(enter)* an etw. *(Dat.)* teilnehmen; *(indulge in, like)* für etw. zu haben sein; *(have as one's hobby, pastime, etc.)* sich auf etw. *(Akk.)* verlegen; **go in for teaching** Lehrer/Lehrerin werden; **go in for wearing loud colours** gern knallige Farben tragen

'go into *v.t.* **a)** *(join)* eintreten in (+ *Akk.*) ⟨*Orden, Geschäft usw.*⟩; gehen in (+ *Akk.*) ⟨*Industrie, Politik*⟩; gehen zu ⟨*Film, Fernsehen, Armee*⟩; beitreten (+ *Dat.*) ⟨*Bündnis*⟩; **go into law/the church** Jurist/Geistlicher werden; **go into nursing** Krankenschwester/-pfleger werden; **go into publishing** ins Verlagswesen gehen; **go into general practice** *(Med.)* sich als allgemeiner Mediziner niederlassen; **b)** *(go and live in)* gehen in (+ *Akk.*) ‹*Krankenhaus, Heim usw.*›; ziehen in (+ *Akk.*) ‹*Wohnung, Heim*›; **c)** [-'-] *(consider)* eingehen auf *(Akk.)*; *(investigate, examine)* sich befassen mit; *(explain)* darlegen; **d)** *(crash into)* [hinein]fahren in (+ *Akk.*); fahren gegen ‹*Baum usw.*›. *See also* **go 1b, u**

go 'in with *v. t.* **go in with sb.** [mit jmdm.] mitmachen

go off 1. [-'-] *v. i.* **a)** *(Theatre)* abgehen; **b) go off with sb./sth.** sich mit jmdm./etw. auf- und davonmachen *(ugs.)*; **his wife has gone off with the milkman** seine Frau ist mit dem Milchmann durchgebrannt *(ugs.)*; **c)** ‹*Alarm, Klingel, Schußwaffe:*› losgehen; ‹*Wecker:*› klingeln; ‹*Bombe:*› hochgehen; **d)** *(turn bad)* schlecht werden; *(turn sour)* sauer werden; *(fig.)* sich verschlechtern; **e)** ‹*Strom, Gas, Wasser:*› ausfallen; **f) go off [to sleep]** einschlafen; **g)** *(be sent)* abgehen **(to** an + *Akk.*); **h) go off well** *etc.* gut *usw.* verlaufen. *See also* **go 1b. 2.** ['--, -'-] *v. t.* *(begin to dislike)* **go off sth.** von etw. abkommen; **go/have gone off sb.** jmdn. nicht mehr mögen; **I have gone off the cinema** ich mache mir nichts mehr aus Kino

go on 1. [-'-] *v. i.* **a)** weitergehen/-fahren; *(by vehicle)* die Reise/Fahrt *usw.* fortsetzen; *(go ahead)* vorausgehen/-fahren; **b)** *(continue)* weitergehen; ‹*Kämpfe:*› anhalten; ‹*Verhandlungen, Arbeiten:*› [an]dauern; *(continue to act)* weitermachen; *(continue to live)* weiterleben; **I can't go on** ich kann nicht mehr; **go on for weeks** *etc.* Wochen *usw.* dauern; **this has been going on for months** das geht schon seit Monaten so; **go on to say** *etc.* fortfahren und sagen *usw.*; **go on and on** kein Ende nehmen wollen; **go on [and on]** *(coll.: chatter)* reden und reden; **go on about sb./sth.** *(coll.)* *(talk)* stundenlang von jmdm./etw. erzählen; *(complain)* sich ständig über jmdn./etw. beklagen; **go on at sb.** *(coll.)* auf jmdm. herumhacken *(ugs.)*; **c)** *(elapse)* ‹*Zeit:*› vergehen; **as time/the years went on** im Laufe der Zeit/Jahre; **d)** *(happen)* passieren; vor sich gehen; **there's more going on in the big cities** in den großen Städten ist mehr los; **what's going on?** was geht vor?; was ist los?; **e) be going on [for] ...** *(be nearly)* fast ... sein; **he is going on [for] ninety** er geht auf die Neunzig zu; **it is going on [for] ten o'clock** es geht auf 10 Uhr zu; **f)** *(behave)* sich benehmen; sich aufführen; **g)** ‹*Kleidung:*› passen; **my dress wouldn't go on** ich kam nicht in mein Kleid rein *(ugs.)*; **h)** *(Theatre)* auftreten; **i)** ‹*Licht:*› angehen; ‹*Strom, Wasser:*› kommen; **go on again** ‹*Strom, Gas, Wasser:*› wiederkommen; **j) go on!** *(proceed)* los, mach schon! *(ugs.)*; *(resume)* fahren Sie fort!; *(coll.: stop talking nonsense)* ach, geh *od.* komm! *(ugs.)*. **2.** ['--] *v. t.* **a)** *(ride on)* fahren mit; **go on the Big Dipper** Achterbahn fahren; **b)** *(continue)* **go on working/talking** *etc.* weiterarbeiten/-reden *usw.*; **go on trying** es weiter[hin] versuchen; **c)** *(coll.: be guided by)* sich stützen auf (+ *Akk.*); **d)** *(begin to receive)* bekommen, erhalten ‹*Arbeitslosengeld, Sozialfürsorge*›; *see also* **dole 1; e)** *(start to take)* nehmen ‹*Medikament, Drogen*›; **go on a diet** eine Abmagerungs- *od.* Schlankheitskur machen; **f)** *(coll.: like)* *see* **much 3d.** *See also* **go 1a, b, l**

go 'on for *see* **go on 1e**

go 'on to *v. t.* übergehen zu; **he went on to become ...:** er wurde schließlich ...

go 'on with *v. t.* **go on with sth.** mit etw. weitermachen; **something/enough to go on with** *or* **be going on with** etwas/genug für den Anfang *od.* fürs erste; **here's £10 to be going on with** hier sind erst [ein]mal 10 Pfund [für den Anfang] *(ugs.)*

go 'out *v. i.* **a)** *(from home)* ausgehen; **go out to work/for a meal** arbeiten/essen gehen; **out you go!** hinaus *od.* *(ugs.)* raus mit dir!; **go out with sb.** *(regularly)* mit jmdm. gehen *(ugs.)*; **b)** *(be extinguished)* ‹*Feuer, Licht, Zigarre usw.:*› ausgehen; **go out like a light** *(fig. coll.: fall asleep)* sofort weg sein *(ugs.)* *od.* einschlafen; **c)** *(ebb)* ‹*Ebbe, Wasser:*› ablaufen, zurückgehen; **the tide has gone out** es ist Ebbe; **d)** *(be issued)* verteilt werden; *(Radio, Telev.: be transmitted)* ausgestrahlt werden

go over 1. [-'--] *v. i.* **a) he went over to the fireplace** er ging zum Kamin hinüber; **we're going over to our friends'** wir fahren zu unseren Freunden; **b)** *(be received)* ‹*Rede, Ankündigung, Plan:*› ankommen **(with** bei); **c)** *(Radio, Telev.)* **go over to sb./sth./Belfast** zu jmdm./in etw. *(Akk.)*/nach Belfast umschalten. *See also* **go 1b; go over to. 2.** ['---, -'--] *v. t.* **a)** *(re-examine, think over, rehearse)* durchgehen; **go over sth./the facts in one's head** *or* **mind** etw. im Geiste durchgehen/die Fakten überdenken; **b)** *(clean)* saubermachen; *(inspect and repair)* durchsehen ‹*Maschine, Auto usw.*›

go 'over to *v. t.* hinübergehen zu; übertreten zu ‹*Glauben, Partei*›; überwechseln zu ‹*Revolutionären*›; ‹*Verräter:*› überlaufen zu ‹*Feind*›; *see also* **go 1b; go over 1a, c**

go round 1. [-'-] *v. i.* **a)** *(call)* **go round and** *or* **to see sb.** jmdn. besuchen; bei jmdm. vorbeigehen *(ugs.)*; **go round to sb.'s house** *(call at)* jmdn. aufsuchen; **b)** *(look round)* sich umschauen; **c)** *(suffice)* reichen; langen *(ugs.)*; **enough coffee to go round** genug Kaffee für alle; **d)** *(spin)* sich drehen; **my head is going round** mir dreht sich alles; **e)** *(circulate)* **the word went round that ...:** es ging die Parole um, daß ... **2.** ['--] *v. t.* **a)** *(inspect)* besichtigen; **b)** ‹*Gürtel:*› herumreichen um ‹*Taille*›

go through 1. [-'-] *v. i.* ‹*Ernennung, Gesetzesvorlage:*› durchkommen; ‹*Geschäft:*› [erfolgreich] abgeschlossen werden; ‹*Antrag, Bewerbung:*› durchgehen; **go through to the final** in die Endrunde kommen. **2.** ['--] *v. t.* **a)** *(execute, undergo)* erledigen ‹*Formalität, Anforderung*›; **b)** *(rehearse)* durchgehen; **c)** *(examine)* durchsehen ‹*Post, Unterlagen*›; *(search)* durchsuchen ‹*Taschen*›; **d)** *(endure)* durchmachen ‹*schwere Zeiten*›; *(suffer)* erleiden ‹*Schmerzen*›; **e)** *(use up)* verbrauchen; durchbringen ‹*Erbschaft*›; aufbrauchen ‹*Vorräte*›. *See also* **go 1u**

go 'through with *v. t.* zu Ende führen; ausführen ‹*Hinrichtung*›

go to'gether *v. i.* **a)** *(coincide)* zusammengehen; **b)** *(match)* zusammenpassen

go 'under *v. i.* *(sink below surface)* untergehen; *(fig.: fail)* ‹*Unternehmen:*› eingehen

go 'up *v. i.* **a)** hinaufgehen/-fahren; ‹*Ballon:*› aufsteigen; *(Theatre)* ‹*Vorhang:*› aufgehen, hochgehen; ‹*Lichter:*› angehen; **b)** *(increase)* ‹*Bevölkerung, Zahl:*› wachsen; ‹*Preis, Wert, Zahl, Niveau:*› steigen; *(in price)* ‹*Ware:*› teurer werden; **c)** *(be constructed)* ‹*Gebäude, Barrikade:*› errichtet werden; **d)** *(be destroyed)* in die

Luft fliegen *(ugs.)*; hochgehen *(ugs.)*
'go with *v.t.* **a)** *(be commonly found together with)* einhergehen mit; **b)** *(be included with)* gehören zu. *See also* **go 1 a, v**
go without 1. ['---] *v.t.* verzichten auf (+ *Akk.*); **have to go without sth.** ohne etw. auskommen müssen. **2.** [--'-] *v.i.* *(willingly)* verzichten; **[have to] go without** *(not from choice)* leer ausgehen
goad [gəʊd] *v.t.* ~ **sb. into sth./doing sth.** jmdn. zu etw. anstacheln/dazu anstacheln, etw. zu tun
~ **'on** *v.t.* ~ **sb. on** jmdn. anstiften
'go-ahead 1. *adj.* *(enterprising)* unternehmungslustig; *(progressive)* fortschrittlich. **2.** *n.* **give sb./sth. the** ~: jmdm./einer Sache grünes Licht geben
goal [gəʊl] *n.* **a)** *(aim)* Ziel, *das;* **attain one's** ~: sein Ziel erreichen; **b)** *(Footb., Hockey)* Tor, *das; (Rugby)* Mal, *das;* **[play] in** ~: im Tor [stehen]; **score/kick a** ~: einen Treffer erzielen
'goalkeeper *n.* Torwart, *der*
goat [gəʊt] *n.* Ziege, *die;* **get sb.'s** ~ *(fig. sl.)* jmdn. aufregen *(ugs.)*
gob [gɒb] *n.* *(sl.)* Schnauze, *die* *(derb abwertend)*
gobble ['gɒbl] **1.** *v.t.* ~ **[down** *or* **up]** hinunterschlingen. **2.** *v.i.* schlingen
~ **up** *v.t.* *(fig. coll.)* verschlingen
gobbledegook, gobbledygook ['gɒbldɪgu:k] *n.* Kauderwelsch, *das*
'go-between *n.* Vermittler, *der*/Vermittlerin, *die*
goblet ['gɒblɪt] *n.* Kelchglas, *das*
goblin ['gɒblɪn] *n.* Kobold, *der*
god [gɒd] *n.* **a)** Gott, *der;* **be** *or* **lie in the lap of the** ~**s** im Schoß der Götter liegen; **b) God** *no pl.* *(Theol.)* Gott; **God knows** *(as God is witness)* weiß Gott *(ugs.)*; **God [only] knows** *(nobody knows)* weiß der Himmel *(ugs.)*; **an act of God** höhere Gewalt; **[oh/my/dear] God!** [ach *od.* o/mein/lieber] Gott!; **good God!** großer *od.* allmächtiger *od.* guter Gott!; **for God's sake!** um Himmels *od.* Gottes willen!; **thank God!** Gott sei Dank!; **God damn it!** zum Teufel noch mal! *(ugs.)*; **God help you/him** *etc.* Gott steh dir/ihm *usw.* bei; **c)** *(fig.)* Gott, *der;* Götze, *der* *(geh., abwertend)*; **d)** *(Theatre)* **the** ~**s** der Olymp *(ugs. scherzh.)*
'God-awful *adj.* *(sl.)* fürchterlich
god: ~child *n.* Patenkind, *das;* **~-dam, ~-damn, ~-damned** *(sl.)* **1.** *adj.* gottverdammt *(derb)*; **[it is] none of your ~-dam business** das geht dich einen Dreck an *(salopp)*; **2.** *adv.* gottverdammt *(derb)*; **you're ~-dam right!** du hast, verdammt noch mal, recht! *(derb)*; **~-daughter** *n.* Patentochter, *die*
goddess ['gɒdɪs] *n.* Göttin, *die*
'godfather *n.* Pate, *der*
'God-forsaken *adj.* gottverlassen
godly ['gɒdlɪ] *adj.* gottgefällig; gottergeben
god: ~mother *n.* Patin, *die;* **~parent** *n.* *(male)* Pate, *der;* *(female)* Patin, *die;* **~parents** Paten; **~send** *n.* Gottesgabe, *die;* **be a ~send to sb.** für jmdn. ein Geschenk des Himmels sein; **~son** *n.* Patensohn, *der*
goes *see* **go**
go-getter ['gəʊgetə(r)] *n.* *(coll.)* Draufgänger, *der*
goggle ['gɒgl] **1.** *n.* *in pl.* **[a pair of] ~s** eine Schutzbrille. **2.** *v.i.* glotzen *(ugs.)*; ~ **at sb./sth.** jmdn./etw. anglotzen *(ugs.)*
'goggle-eyed *adj.* glotzäugig *(ugs.)*
going ['gəʊɪŋ] **1.** *n.* **a) 150 miles in two hours, that is good** ~: 150 Meilen in zwei Stunden, das ist wirklich gut; **the journey was slow** ~: die Reise zog sich [in die Länge]; **this book is heavy** ~: dieses Buch liest sich schwer; **while the** ~ **is good** solange noch Zeit dazu ist *od.* es noch geht; **b)** *(Horse-racing, Hunting, etc.)* Geläuf, *das.* **2.** *adj.* **a)** *(available)* erhältlich; **there is sth.** ~: es gibt etw.; **take any job** ~: jede Arbeit annehmen, die es nur gibt; **b) be** ~ **to do sth.** etw. tun [werden/wollen]; **he's** ~ **to be a ballet-dancer when he grows up** wenn er groß ist, wird er Balletttänzer; **I was** ~ **to say** ich wollte sagen; **I was not** ~ *(did not intend)* **to do sth.** ich hatte nicht die Absicht, etw. zu tun; **c)** *(current)* [derzeit/damals/dann] geltend; **d) a** ~ **concern** eine gesunde Firma; **e) have a lot/nothing** *etc.* ~ **for one** *(coll.)* viel/nichts *usw.* haben, was für einen spricht; **f) to be** ~ **on with** *see* **go on with**
going-'over *n.* **a)** *(coll.: overhaul of engine etc.)* Überholung, *die;* **give sth. a [good etc.]** ~: eine Sache [gründlich *usw.*] durchgehen *od.* durchsehen; **b)** *(sl.: thrashing)* **give sb. a [good]** ~: jmdn. ordentlich verprügeln *(ugs.)*
goings-on [gəʊɪŋz'ɒn] *n. pl.* Ereignisse; Vorgänge; **there have been some strange** ~: es sind seltsame Dinge passiert
go-kart ['gəʊkɑ:t] *n.* Go-Kart, *der*
gold [gəʊld] **1.** *n.* **a)** *no pl., no indef. art.* Gold, *das;* **be worth one's weight in** ~: nicht mit Gold aufzuwiegen sein; **a heart of** ~: ein goldenes Herz; **b)** *(colour, medal)* Gold, *das.* **2.** *attrib. adj.* golden; Gold‹*münze, -stück, -kette, -krone usw.*›
'gold-coloured *adj.* goldfarben
golden ['gəʊldn] *adj.* **a)** golden; ~ **brown** goldbraun; **b)** *(fig.)* golden; einmalig ‹*Gelegenheit*›
golden: ~ **'age** *n.* goldenes Zeitalter; ~ **'eagle** *n.* Steinadler, *der;* ~ **hamster** *n.* Goldhamster, *der;* ~ **'handshake** *n.* Abfindung[ssumme], *die;* ~ **re'triever** *n.* Golden Retriever, *der;* ~ **'rule** *n.* goldene Regel; ~ **syrup** *n.* *(Brit.)* Sirup, *der;* ~ **'wedding** *n.* goldene Hochzeit
gold: ~finch *n.* Stieglitz, *der;* Distelfink, *der;* **~fish** *n.* Goldfisch, *der;* **~fish bowl** *n.* Goldfischglas, *das;* ~ **'leaf** *n.* Blattgold, *das;* ~ **'medal** *n.* Goldmedaille, *die;* ~ **'medallist** *n.* Goldmedaillengewinner, *der*/-gewinnerin, *die;* **~-mine** *n.* Goldmine, *die;* *(fig.)* Goldgrube, *die;* ~ **'plate** *n., no pl., no indef. art.* vergoldete Ware; *(coating)* Goldauflage, *die;* **~smith** *n.* Goldschmied, *der*/-schmiedin, *die*
golf [gɒlf] *n., no pl.* Golf, *das; attrib.* Golf‹*platz, -schlag usw.*›
golf: ~ **ball** *n.* **a)** Golfball, *der;* **b)** *(coll.: in typewriter)* Kugelkopf, *der;* **~-club** *n.* **a)** *(implement)* Golfschläger, *der;* **b)** *(association)* Golfclub, *der;* **~-course** *n.* Golfplatz, *der*
golfer ['gɒlfə(r)] *n.* Golfer, *der*/Golferin, *die;* Golfspieler, *der*/Golfspielerin, *die*
gondola ['gɒndələ] *n.* Gondel, *die*
gondolier [gɒndə'lɪə(r)] *n.* Gondoliere, *der*
gone [gɒn] *pred. adj.* **a)** *(away)* weg; **it's time you were** ~: es ist *od.* wird Zeit, daß du gehst; **he has been** ~ **ten minutes** *(coll.)* er ist seit zehn Minuten fort *od.* weg; **he will be** ~ **a year** er wird ein Jahr lang weg sein; **b)** *(of time: after)* nach; **it's** ~ **ten o'clock** es ist zehn Uhr vorbei; **c)** *(used up)* **be all** ~: alle sein *(ugs.)*; **d) be** ~ **on sb./sth.** *(sl.)* ganz weg von jmdm./etw. sein *(ugs.)*. *See also* **far 1 d; forget 1; go 1 k**
goner ['gɒnə(r)] *n.* *(sl.)* **he is a** ~:

er hat die längste Zeit gelebt *(ugs.)*

gong [gɒŋ] *n.* Gong, *der*

gonorrhoea (*Amer.:* **gonorrhea**) [gɒnə'rɪə] *n. (Med.)* Tripper, *der;* Gonorrhöe, *die (fachspr.)*

goo [gu:] *n. (coll.)* Schmiere, *die (ugs.)*

good [gʊd] **1.** *adj.*, **better** ['betə(r)], **best** [best] **a)** *(satisfactory)* gut; *(reliable)* gut; zuverlässig; *(sufficient)* gut; ausreichend ‹*Vorrat*›; ausgiebig ‹*Mahl*›; *(competent)* gut; geeignet; **his ~ eye/leg** sein gesundes Auge/Bein; **Late again! It's just not ~ enough!** *(coll.)* Schon wieder zu spät. So geht es einfach nicht!; **in ~ time** frühzeitig; **all in ~ time** alles zu seiner Zeit; **take ~ care of sb.** gut für jmdn. sorgen; **be ~ at sth.** in etw. *(Dat.)* gut sein; **be ~ at doing sth.** etw. gut können; **speak ~ English** gut[es] Englisch sprechen; **be ~ with people** *etc.* mit Menschen *usw.* gut *od.* leicht zurechtkommen; **b)** *(favourable, advantageous)* gut; günstig ‹*Gelegenheit, Augenblick, Angebot*›; **a ~ chance of succeeding** gute Erfolgschancen; **too ~ to be true** zu schön, um wahr zu sein; **the ~ thing about it is that ...:** das Gute daran ist, daß ...; **be too much of a ~ thing** zuviel des Guten sein; **you can have too much of a ~ thing** man kann es auch übertreiben; **be ~ for sb./sth.** gut für jmdn./etw. sein; **apples are ~ for you** Äpfel sind gesund; **eat more than is ~ for one** mehr essen, als einem guttut; **it's a ~ thing you told him** nur gut, daß du es ihm gesagt hast; **c)** *(prosperous)* gut; **~ times** eine schöne Zeit; **d)** *(enjoyable)* schön ‹*Leben, Urlaub, Wochenende*›; **the ~ things in life** Annehmlichkeiten; **the ~ old days** die gute alte Zeit; **the ~ life** das angenehme[, sorglose] Leben; **have a ~ time!** viel Spaß *od.* Vergnügen!; **have a ~ journey!** gute Reise!; **it's ~ to be home again** es ist schön, wieder zu Hause zu sein; **Did you have a ~ day at the office?** Wie war es heute im Büro?; **e)** *(cheerful)* gut; angenehm ‹*Patient*›; **~ humour** *or* **spirits** *or* **mood** gute Laune; **feel ~:** sich wohl fühlen; **I'm not feeling too ~** *(coll.)* mir geht es nicht sehr gut; **f)** *(well-behaved)* gut; brav; **be ~!, be a ~ girl/boy!** sei brav *od.* lieb!; **[as] ~ as gold** ganz artig *od.* brav; **g)** *(virtuous)* rechtschaffen; *(kind)* nett; gut ‹*Absicht, Wünsche, Benehmen, Tat*›; **the ~ guy** der Gute; **be ~ to sb.** gut zu jmdm. sein; **would you be so ~ as to** *or* **~ enough to do that?** wären Sie so freundlich *od.* nett, das zu tun?; **how ~ of you!** wie nett von Ihnen!; **that/it is ~ of you** das/es ist nett *od.* lieb von dir; **h)** *(commendable)* gut; **~ for 'you** *etc. (coll.)* bravo!; **~ old Jim** *etc. (coll.)* der gute alte Jim *usw. (ugs.)*; **my ~ man/friend** *(coll.)* mein lieber Herr/Freund *(ugs.; auch iron.)*; **that's a ~ one** *(sl.)* der ist 'gut! *(ugs.)*; *(iron.)* das ist'n Ding! *(ugs.)*; **i)** *(attractive)* schön; gut ‹*Figur, Haltung*›; gepflegt ‹*Erscheinung, Äußeres*›; wohlgeformt ‹*Beine*›; **look ~:** gut aussehen; **j)** *(thorough)* gut; **take a ~ look round** sich gründlich umsehen; **give sth. a ~ polish** etw. ordentlich polieren; **have a ~ weep/rest/sleep** sich richtig ausweinen/ausruhen/[sich] richtig ausschlafen *(ugs.)*; **k)** *(considerable)* [recht] ansehnlich ‹*Menschenmenge*›; ganz schön, ziemlich *(ugs.)* ‹*Stück Wegs, Entfernung, Zeitraum, Strecke*›; gut, anständig ‹*Preis, Erlös*›; hoch ‹*Alter*›; **l)** *(sound, valid)* gut ‹*Grund, Rat, Gedanke*›; berechtigt ‹*Anspruch*›; *(Commerc.)* solide ‹*Kunde*›; sicher ‹*Anleihe, Kredit*›; **~ sense** Vernünftigkeit, *die;* **have the ~ sense to do sth.** so vernünftig sein, etw. zu tun; **m)** *(in greetings)* **~ afternoon/day** guten Tag!; **~ evening/morning** guten Abend/Morgen!; **~ night** gute Nacht!; **n)** *in exclamation* gut; **very ~, sir** sehr wohl!; **~ God/Lord** *etc. see nouns;* **o)** *(best)* gut ‹*Geschirr, Anzug*›; **p)** *(correct, fitting)* gut; *(appropriate)* angebracht; ratsam; **q) as ~ as** so gut wie; **r) make ~** *(succeed)* erfolgreich sein; *(effect)* in die Tat umsetzen; ausführen ‹*Plan*›; erfüllen ‹*Versprechen*›; *(compensate for)* wiedergutmachen ‹*Fehler*›; *(indemnify)* ersetzen ‹*Schaden, Ausgaben*›. *See also* **best 1; better 1. 2.** *adv. as intensifier (coll.)* **~ and ...:** richtig ...; **hit sb. ~ and proper** jmdn. ordentlich verprügeln. *See also* **best 2; better 2. 3.** *n.* **a)** *(use)* Nutzen, *der;* **be some ~ to sb./sth.** jmdm./einer Sache nützen; **he'll never be any ~:** aus dem wird nichts Gutes werden; **is this book any ~?** taugt dieses Buch etwas?; **be no ~ to sb./sth.** für jmdn./etw. nicht zu gebrauchen sein; **it is no/not much ~ doing sth.** es hat keinen/kaum einen Sinn, etw. zu tun; **what's the ~ of ...?, what ~ is ...?** was nützt ...?; **b)** *(benefit)* **for your/his** *etc.* **own ~:** zu deinem/seinem *usw.* Besten *od.* eigenen Vorteil; **for the ~ of mankind/the country** zum Wohl[e] der Menschheit/des Landes; **do no/little ~:** nichts/wenig helfen *od.* nützen; **do sb./sth. ~:** jmdm./einer Sache nützen; ‹*Ruhe, Erholung:*› jmdm./einer Sache guttun; ‹*Arznei:*› jmdm./einer Sache helfen; **I'll tell him, but what ~ will that do?** ich sag es ihm, aber was nützt *od.* hilft das schon?; **this development was all to the ~:** diese Entwicklung war nur von Vorteil; **come home £10 to the ~:** mit 10 Pfund plus nach Hause kommen; **come to no ~:** kein gutes Ende nehmen; **c)** *(goodness)* Gute, *das;* **there's ~ and bad in everyone** in jedem steckt Gutes und Böses; **the difference between ~ and bad** *or* **evil** der Unterschied zwischen Gut und Böse; **d)** *(kind acts)* Gute, *das;* **be up to no ~:** nichts Gutes im Sinn haben *od.* im Schilde führen; **do ~:** Gutes tun; **e) for ~ [and all]** *(finally)* ein für allemal; *(permanently)* für immer [und ewig]; endgültig; **f)** *constr. as pl. (virtuous people)* **the ~:** die Guten; **g)** *in pl. (wares etc.)* Waren; *(Brit. Railw.)* Fracht, *die; attrib.* Güter‹*bahnhof, -wagen, -zug*›; **~s and chattels** Sachen; **h)** *in pl.* **the ~s** *(coll.: what is wanted)* das Gewünschte; das Verlangte; **deliver the ~s** *(fig.)* halten, was man verspricht

good: ~'bye (*Amer.:* **~'by**) **1.** *int.* auf Wiedersehen!; *(on telephone)* auf Wiederhören!; **2.** *n., pl.* **~byes** (*Amer.:* **~bys**) *(saying '~-bye')* Lebewohl, *das (geh.)*; *(parting)* Abschied, *der;* **say ~bye to sb.** jmdm. auf Wiedersehen sagen; **say ~bye** sich verabschieden; **wave ~bye** zum Abschied winken; **say ~bye to sth., kiss sth. ~bye** *(fig.: accept its loss)* etw. abschreiben *(ugs.)*; **~-for-nothing** *(derog.)* **1.** *adj.* nichtsnutzig; **2.** *n.* Taugenichts, *der;* **G~ 'Friday** *see* **Friday 1; ~-humoured** [gʊd'hju:məd] *adj.* gutmütig

goodies ['gʊdɪz] *n. pl. (coll.) (food)* Naschereien; *(sweets)* Süßigkeiten; *(attractive things)* Attraktionen; tolle Sachen

good: ~-'looking *adj.* gutaussehend; **~-natured** [gʊd'neɪtʃəd] *adj.* gutwillig; gutmütig

goodness ['gʊdnɪs] **1.** *n., no pl.* **a)** *(virtue)* Güte, *die;* **b)** *(of food)*

Nährgehalt, *der;* Güte, *die.* **2.** *int.* **[my]** ~ *expr. surprise* meine Güte! *(ugs.);* **[oh] my** ~ *expr. shock* lieber Himmel!; ~ **gracious** *or* **me!** [ach] du lieber Himmel *od.* liebe Güte! *(ugs.);* **for ~' sake** um Himmels willen; ~ **[only] knows** weiß der Himmel *(ugs.)*

goods *see* **good 3g, h**

good: ~-tempered [gʊd'tempəd] *adj.* gutmütig; verträglich ‹*Person*›; **~'will** *n.* **a)** *(friendly feeling)* guter Wille; *attrib.* Goodwill‹*botschaft, -reise usw.*›; **b)** *(willingness)* Bereitwilligkeit, *die;* **c)** *(Commerc.)* Goodwill, *der*

¹goody ['gʊdɪ] *n. (coll.: hero)* Gute, *der/die; see also* **goodies, baddy**

²goody *int. (coll.)* toll; prima

gooey ['gu:ɪ] *adj.,* **gooier** ['gu:ɪə(r)], **gooiest** ['gu:ɪəst] *(coll.)* klebrig

goof [gu:f] *(sl.)* **1.** *n. (gaffe)* Schnitzer, *der (ugs.).* **2.** *v.i.* Mist machen *od.* bauen *(salopp)*

goose [gu:s] *n., pl.* **geese** [gi:s] Gans, *die; see also* **boo 1; cook 2 a**

gooseberry ['gʊzbərɪ] *n.* **a)** Stachelbeere, *die;* **b) play ~:** das fünfte Rad am Wagen sein *(ugs.)*

goose: ~-flesh *n., no pl.* Gänsehaut, *die;* **~-pimples** *n. pl.* **have ~-pimples** eine Gänsehaut haben; **~-step 1.** *n.* Stechschritt, *der;* **2.** *v.i.* im Stechschritt marschieren

¹gore [gɔ:(r)] *v.t.* **be ~d [to death] by a bull** von den Hörnern eines Stieres durchbohrt [und tödlich verletzt] werden

²gore *n. (blood)* Blut, *das*

gorge [gɔ:dʒ] **1.** *n.* Schlucht, *die.* **2.** *v.i. & refl.* ~ **[oneself]** sich vollstopfen **(on** mit) *(ugs.)*

gorgeous ['gɔ:dʒəs] *adj.* **a)** *(magnificent)* prächtig; hinreißend ‹*Frau, Mann, Lächeln*›; *(richly coloured)* farbenprächtig; **b)** *(coll.: splendid)* sagenhaft *(ugs.)*

gorilla [gə'rɪlə] *n.* Gorilla, *der*

gormless ['gɔ:mlɪs] *adj. (Brit. coll.)* dämlich *(ugs.)*

gorse [gɔ:s] *n.* Stechginster, *der*

gory ['gɔ:rɪ] *adj.* **a)** blutbefleckt ‹*Hände*›; blutig ‹*Schlacht*›; **b)** *(fig.: sensational)* blutrünstig

gosh [gɒʃ] *int. (coll.)* Gott!

gosling ['gɒzlɪŋ] *n.* Gänseküken, *das;* Gössel, *das (nordd.)*

'go-slow *n. (Brit.)* Bummelstreik, *der*

gospel ['gɒspl] *n.* Evangelium, *das;* **take sth. as** ~ *(fig.)* etw. für bare Münze nehmen

gospel 'truth *n.* absolute *od.* reine Wahrheit

gossamer ['gɒsəmə(r)] *n.* Altweibersommer, *der; (fig.)* Spinnfäden *(fig.); attrib.* hauchdünn

gossip ['gɒsɪp] **1.** *n.* **a)** *(person)* Klatschbase, *die (ugs. abwertend);* **b)** *(talk)* Schwatz, *der; (malicious)* Klatsch, *der (ugs. abwertend).* **2.** *v.i.* schwatzen; *(maliciously)* klatschen *(ugs. abwertend)*

got *see* **get**

Gothic ['gɒθɪk] *adj.* **a)** gotisch; **b)** *(Lit.)* ~ **novel** Schauerroman, *der*

gotten *see* **get**

gouge [gaʊdʒ] **1.** *v.t.* aushöhlen; ~ **a channel** ‹*Fluß:*› eine Rinne auswaschen. **2.** *n.* Hohleisen, *das*
~ **out** *v.t.* ausschneiden; ~ **sb.'s eye out** jmdm. ein Auge ausstechen

goulash ['gu:læʃ] *n. (Gastr.)* Gulasch, *das od. der*

gourd [gʊəd] *n. (fruit, plant)* [Flaschen]kürbis, *der*

gourmand ['gʊəmənd] *n. (glutton)* Gourmand, *der*

gourmet ['gʊəmeɪ] *n.* Gourmet, *der; attrib.* ~ **meal/restaurant** Feinschmeckergericht, *das*/-lokal, *das*

gout [gaʊt] *n. (Med.)* Gicht, *die*

govern ['gʌvn] **1.** *v.t.* **a)** *(rule)* regieren ‹*Land, Volk*›; *(administer)* verwalten ‹*Provinz, Kolonie*›; **b)** *(dictate)* bestimmen; **be ~ed by sth.** sich von etw. leiten lassen; **c)** *(regulate)* ‹*Vorschriften:*› regeln; **d)** *(Ling.)* verlangen; regieren ‹*Kasus*›. **2.** *v.i.* regieren

governess ['gʌvənɪs] *n.* Gouvernante, *die (veralt.);* Hauslehrerin, *die*

governing ['gʌvənɪŋ] *adj.* **a)** *(ruling)* regierend; **b)** *(guiding)* dominierend ‹*Einfluß*›; ~ **body** leitendes Gremium

government ['gʌvnmənt] *n.* Regierung, *die; attrib.* Regierungs-; ~ **money** Staatsgelder *Pl.;* ~ **securities** *or* **stocks** Staatspapiere *od.* -anleihen

government de'partment *n.* Regierungsstelle, *die*

governor ['gʌvənə(r)] *n.* **a)** *(ruler)* Herrscher, *der;* **b)** *(of province, town, etc.)* Gouverneur, *der;* **c)** *(of State of US)* Gouverneur, *der;* **d)** *(of institution, prison)* Direktor, *der*/Direktorin, *die;* **[board of] ~s** Vorstand, *der; (of school)* Schulleitung, *die; (of bank, company)* Direktorium, *das;* Direktion, *die;* **e)** *(sl.: employer)* Boß, *der (ugs.);* **f)** *(Mech.)* Regler, *der*

Govt. *abbr.* **Government** Reg.

gown [gaʊn] *n.* **a)** [elegantes] Kleid; **bridal ~:** Brautkleid, *das;* **b)** *(official or uniform robe)* Talar, *der;* Robe, *die;* **c)** *(surgeon's overall)* [Operations]kittel, *der*

GP *abbr.* **general practitioner**

GPO *abbr. (Hist.)* **General Post Office** Post, *die*

grab [græb] **1.** *v.t.,* **-bb-** greifen nach; *(seize)* packen; *(capture, arrest)* schnappen *(ugs.);* ~ **sb. by the arm** *etc.* jmdn. am Arm *usw.* packen; ~ **some food** *or* **a bite to eat** *(coll.)* schnell etwas essen; ~ **hold of sb./sth.** sich *(Dat.)* jmdn./etw. schnappen *(ugs.).* **2.** *v.i.,* **-bb-:** ~ **at sth.** nach etw. greifen. **3.** *n.* **a) make a ~ at** *or* **for sb./sth.** nach jmdm./etw. greifen *od.* *(ugs.)* grapschen; **be up for ~s** *(sl.)* zu erwerben sein; ‹*Posten:*› frei sein; **b)** *(Mech.)* Greifer, *der*

grace [greɪs] **1.** *n.* **a)** *(charm)* Anmut, *die (geh.);* Grazie, *die;* **b)** *(attractive feature)* Charme, *der;* **airs and ~s** vornehmes Getue *(ugs. abwertend);* affektiertes Benehmen; **c)** *(accomplishment)* **social ~s** Umgangsformen *Pl.;* **d)** *(decency)* Anstand, *der;* **have the ~ to do sth.** so anständig sein und etw. tun; *(civility)* **with [a] good/bad ~:** bereitwillig/widerwillig; **he accepted my criticism with good/bad ~:** er trug meine Kritik mit Fassung/nahm meine Kritik mit Verärgerung hin; **e)** *(favour)* Wohlwollen, *das;* Gunst, *die;* **he fell from ~:** er fiel in Ungnade; **f)** *(delay)* Frist, *die; (Commerc.)* Zahlungsfrist, *die;* **give sb. a day's ~:** jmdm. einen Tag Aufschub gewähren; **g)** *(prayers)* Tischgebet, *das;* **say ~:** das Tischgebet sprechen; **h)** *in address* **Your G~:** Euer Gnaden. **2.** *v.t.* **a)** *(adorn)* zieren *(geh.);* schmücken; **b)** *(honour)* auszeichnen; ehren

graceful ['greɪsfl] *adj.* elegant; graziös ‹*Bewegung, Eleganz*›; geschmeidig ‹*Katze, Pferd*›

gracefully ['greɪsfəlɪ] *adv.* elegant; graziös ‹*tanzen, sich bewegen*›; **grow old ~:** mit Würde alt werden

gracious ['greɪʃəs] **1.** *adj.* **a)** liebenswürdig; freundlich; ~ **living** kultivierter Lebensstil; **b)** *(merciful)* gnädig. **2.** *int.* **~!, good[ness] ~!, [goodness] ~ me!** [ach] du meine *od.* liebe Güte!

graciously ['greɪʃəslɪ] *adv.* liebenswürdig; freundlich; *(with condescension)* gnädig

grade [greɪd] **1.** *n.* **a)** Rang, *der; (Mil.)* Dienstgrad, *der; (salary ~)* Gehaltsstufe, *die; (of goods)* [Handels-, Güte]klasse, *die; (of textiles)* Qualität, *die; (position,* *level)* Stufe, *die;* **b)** *(Amer. Sch.: class)* Klasse, *die;* **c)** *(Sch., Univ.: mark)* Note, *die;* Zensur, *die;* **d)** *(Amer.: gradient) (ascent)* Steigung, *die; (descent)* Neigung, *die;* **e) make the ~:** es schaffen. **2.** *v. t.* **a)** einstufen ‹*Arbeit nach Gehalt, Schüler nach Fähigkeiten, Leistungen*›; [nach Größe/Qualität] sortieren ‹*Eier, Kartoffeln*›; **b)** *(mark)* benoten; zensieren

'grade school *n. (Amer.)* Grundschule, *die*

gradient ['greɪdɪənt] *n. (ascent)* Steigung, *die; (descent)* Gefälle, *das; (inclined part of road)* Neigung, *die;* **a ~ of 1 in 10** eine Steigung/ein Gefälle von 10%

gradual ['grædʒʊəl] *adj.* allmählich; sanft ‹*Steigung, Gefälle usw.*›

gradually ['grædʒʊəlɪ] *adv.* allmählich; sanft ‹*ansteigen, abfallen*›

graduate 1. ['grædʒʊət] *n.* Graduierte, *der/die; (who has left university)* Akademiker, *der*/Akademikerin, *die;* **university ~:** Hochschulabsolvent, *der*/-absolventin, *die.* **2.** ['grædʒʊeɪt] *v. i.* **a)** einen akademischen Grad/Titel erwerben; **when did you ~?** wann haben Sie Ihr Studium abgeschlossen?; **b)** *(Amer. Sch.)* die [Schul]abschlußprüfung bestehen (**from** an + *Dat.*). **3.** ['grædʒʊeɪt] *v. t. (mark)* mit Gradeinteilung versehen; graduieren *(bes. Technik)* ‹*Thermometer*›

graduation [grædʒʊ'eɪʃn] *n.* **a)** *(Univ.)* Graduierung, *die;* **b)** *(Amer. Sch.)* Entlassung, *die;* **c)** *attrib.* Abschluß-; **d)** *(mark)* Graduation, *die (bes. Technik)*

graffiti [grə'fi:ti:] *n. sing. or pl.* Graffiti *Pl.*

¹graft [grɑ:ft] **1.** *n.* **a)** *(Bot.)* Pfropfreis, *das;* **b)** *(Med.) (operation)* Transplantation, *die (fachspr.); (thing ~ed)* Transplantat, *das;* **c)** *(Brit. sl.: work)* Plackerei, *die (ugs.).* **2.** *v. t.* **a)** *(Bot.)* pfropfen; **b)** *(Med.)* transplantieren *(fachspr.);* verpflanzen. **3.** *v. i.* **a)** pfropfen; **b)** *(Brit. sl.: work)* schuften *(ugs.)*

²graft *n. (coll.) (dishonesty)* Gaunerei, *die; (profit)* Fischzug, *der*

grain [greɪn] *n.* **a)** Korn, *das; (collect.: [species of] corn)* Getreide, *das;* Korn, *das;* **b)** *(particle)* Korn, *das;* **c)** *(unit of weight)* Gran, *das (veralt.);* **a ~ of truth** *(fig.)* ein Gran *od.* Körnchen Wahrheit; **d)** *(texture)* Korn, *das (fachspr.);* Griff, *der; (in wood)* Maserung, *die; (in paper)* Faser, *die;* Faserverlauf, *der; (in leather)* Narbung, *die;* **go against the ~ [for sb.]** *(fig.)* jmdm. gegen den Strich gehen *(ugs.)*

grainy ['greɪnɪ] *adj.* körnig; gemasert ‹*Holz*›; genarbt ‹*Leder*›

gram [græm] *n.* Gramm, *das*

grammar ['græmə(r)] *n. (also book)* Grammatik, *die;* **sth. is bad ~:** etw. ist grammat[ikal]isch nicht richtig *od.* korrekt

grammar: ~ book *n.* Grammatik, *die;* **~ school** *n.* **a)** *(Brit.)* ≈ Gymnasium, *das;* **b)** *(Amer.)* ≈ Realschule, *die*

grammatical [grə'mætɪkl] *adj.* **a)** grammat[ikal]isch richtig *od.* korrekt; **b)** *(of grammar)* grammatisch

grammatically [grə'mætɪkəlɪ] *adv.* grammat[ikal]isch ‹*richtig, falsch*›; **speak English ~:** grammatisch richtiges *od.* korrektes Englisch sprechen

gramme *see* **gram**

gramophone ['græməfəʊn] *n.* Plattenspieler, *der*

granary ['grænərɪ] *n.* Getreidesilo, *der od. das;* Kornspeicher, *der*

grand [grænd] **1.** *adj.* **a)** *(most or very important)* groß; **~ finale** großes Finale; *see also* **²slam; b)** *(final)* **~ total** Gesamtsumme, *die;* **c)** *(splendid)* grandios; *(conducted with solemnity, splendour, etc.)* glanzvoll; **d)** *(distinguished)* vornehm; **e)** *(dignified, lofty)* erhaben; groß ‹*Versprechungen, Pläne, Worte*›; *(noble, admirable)* ehrwürdig; **f)** *(coll.: excellent)* großartig. **2.** *n. (piano)* Flügel, *der*

grandad ['grændæd] *see* **granddad**

grand: ~child *n.* Enkel, *der*/Enkelin, *die;* Enkelkind, *das;* **~-dad[dy]** ['grændæd(ɪ)] *n. (coll./child lang.)* Großpapa, *der (fam.);* Opa, *der (Kinderspr./ugs.);* **~daughter** *n.* Enkelin, *die*

grandeur ['grændʒə(r), 'grændjə(r)] *n.* **a)** Erhabenheit, *die;* **b)** *(splendour of living, surroundings, etc.)* Großartigkeit, *die;* Glanz, *der;* **c)** *(nobility of character)* Größe, *die;* Erhabenheit, *die*

'grandfather *n.* Großvater, *der;* **~ clock** Standuhr, *die*

grandiose ['grændɪəʊs] *adj.* **a)** *(impressive)* grandios; **b)** *(pompous)* bombastisch *(abwertend)*

grandly ['grændlɪ] *adv.* großartig; aufwendig ‹*sich kleiden*›; in großem Stil ‹*leben*›

grand: ~ma *n. (coll./child lang.)* Großmama, *die (fam.);* Oma, *die (Kinderspr./ugs.);* **~mother** *n.* Großmutter, *die;* **~ 'opera** *n.* große Oper; **~pa** *n. (coll./child lang.)* Großpapa, *der (fam.);* Opa, *der (Kinderspr./ugs.);* **~parent** *n. (male)* Großvater, *der; (female)* Großmutter, *die;* **~parents** Großeltern *Pl.;* **~ pi'ano** *n.* [Konzert]flügel, *der;* **G~ Prix** [grɑ̃ 'pri:] *n.* Grand Prix, *der;* **~son** *n.* Enkel, *der;* **~stand** *n.* [Haupt]tribüne, *die*

granite ['grænɪt] *n.* Granit, *der*

granny (grannie) ['grænɪ] *n. (coll./child lang.)* Großmama, *die (fam.);* Oma, *die (Kinderspr./ugs.)*

grant [grɑ:nt] **1.** *v. t.* **a)** *(consent to fulfil)* erfüllen ‹*Wunsch*›; stattgeben (+ *Dat.*) ‹*Gesuch*›; **b)** *(concede, give)* gewähren; bewilligen ‹*Geldmittel*›; zugestehen ‹*Recht*›; erteilen ‹*Erlaubnis*›; **c)** *(in argument)* zugeben; einräumen *(geh.);* **~ed that ...:** zugegeben, daß ...; **take sb./sth. [too much] for ~ed** sich *(Dat.)* jmds. [allzu] sicher sein/etw. für [allzu] selbstverständlich halten. **2.** *n.* Zuschuß, *der; (financial aid [to student])* [Studien]beihilfe, *die; (scholarship)* Stipendium, *das*

granular ['grænjʊlə(r)] *adj.* körnig; granulös *(Med.)*

granulated sugar ['grænjʊleɪtɪd 'ʃʊgə(r)] *n.* [Zucker]raffinade, *die;* Kristallzucker, *der*

granule ['grænju:l] *n.* Körnchen, *das*

grape [greɪp] *n.* Weintraube, *die;* Weinbeere, *die;* **a bunch of ~s** eine Traube; **[it's] sour ~s** *(fig.)* die Trauben hängen zu hoch

grape: ~fruit *n., pl. same* Grapefruit, *die;* **~-juice** *n.* Traubensaft, *der;* **~vine** *n.* **a)** Wein, *der;* **b)** *(fig.)* **I heard [it] on the ~vine that ...:** es wird geflüstert, daß ...

graph [græf, grɑ:f] *n.* graphische Darstellung

graphic ['græfɪk] **1.** *adj.* **a)** graphisch; **~ art[s]** Graphik, *die;* **b)** *(clear, vivid)* plastisch; anschaulich; **in ~ detail** in allen Einzelheiten. **2.** *n.* **a)** *(product)* Graphik, *die;* **b)** *in pl. see* **graphics**

graphically ['græfɪkəlɪ] *adv.* **a)** *(clearly, vividly)* plastisch; anschaulich; **b)** *(by use of graphic methods)* graphisch

graphics ['græfɪks] *n. (design and decoration)* graphische Gestaltung; *(use of diagrams)* graphische Darstellung; **computer ~:** Computergraphik, *die*

graphite ['græfaɪt] *n.* Graphit, *der*

'graph paper *n.* Diagrammpapier, *das*

grapple ['græpl] *v.i.* handgemein werden; ~ **with** *(fig.)* sich auseinandersetzen *od.* *(ugs.)* herumschlagen mit

grasp [grɑ:sp] **1.** *v.i.* ~ **at** *(lit. or fig.)* ergreifen; sich stürzen auf (+ *Akk.*) ‹*Angebot*›. **2.** *v.t.* **a)** *(clutch at, seize)* ergreifen *(auch fig.)*; **manage to** ~: zu fassen bekommen; **b)** *(hold firmly)* festhalten; ~ **sb. in one's arms** jmdn. [fest] in den Armen halten; ~ **the nettle** *(fig.)* das Problem beherzt anpacken; **c)** *(understand)* verstehen; erfassen ‹*Bedeutung*›. **3.** *n.* **a)** *(firm hold)* Griff, *der;* **he had my hand in a firm** ~: er hielt meine Hand mit festem Griff; **sth. is within/beyond sb.'s** ~: etwas ist in/außer jmds. Reichweite *(Dat.)*; **b)** *(mental hold)* **have a good** ~ **of sth.** etw. gut beherrschen; **sth. is beyond/within sb.'s** ~: etw. überfordert jmds. [intellektuelle] Fähigkeiten/kann von jmdm. verstanden werden

grasping ['grɑ:spɪŋ] *adj.* *(greedy)* habgierig

grass [grɑ:s] **1.** *n.* **a)** Gras, *das;* **b)** *no pl.* *(lawn)* Rasen, *der;* **c)** *no pl.* *(grazing, pasture)* Weide, *die;* *(pasture land)* Weideland, *das;* **put** *or* **turn out to** ~: auf die Weide treiben *od.* führen; *(fig.)* in den Ruhestand versetzen; **d)** *(sl.: marijuana)* Grass, *das (ugs.)*; **e)** *(Brit. sl.: police informer)* Spitzel, *der.* **2.** *v.t.* *(cover with turf)* mit Rasen bedecken. **3.** *v.i.* *(Brit. sl.: inform police)* singen *(salopp)*; ~ **on sb.** jmdn. verpfeifen *(ugs.)*

grass: ~hopper *n.* Grashüpfer, *der;* **~land** *n.* Grasland, *das;* *(for grazing)* Weideland, *das;* **~-root[s]** *attrib. adj.* *(Polit.)* Basis-; ~ **roots** *n. pl.* *(fig.)* *(source)* Wurzeln; *(Polit.)* Basis, *die;* ~ **'skirt** *n.* Baströckchen, *das;* ~ **snake** *n.* **a)** *(Brit.: ringed snake)* Ringelnatter, *die;* **b)** *(Amer.: greensnake)* Grasnatter, *die;* ~ **widow** *n.* Strohwitwe, *die (ugs. scherzh.)*; ~ **widower** *n.* Strohwitwer, *der (ugs. scherzh.)*

grassy ['grɑ:sɪ] *adj.* mit Gras bewachsen

[1]**grate** [greɪt] *n.* Rost, *der;* *(fireplace)* Kamin, *der*

[2]**grate** **1.** *v.t.* **a)** *(reduce to particles)* reiben; *(less finely)* raspeln; **b)** *(grind)* ~ **one's teeth in anger** vor Wut mit den Zähnen knirschen; **c)** *(utter in harsh tone)* [durch die Zähne] knirschen. **2.** *v.i.* **a)** *(rub, sound harshly)* knirschen; **b)** ~ **[up]on sb./sb.'s nerves** jmdm. auf die Nerven gehen

grateful ['greɪtfl] *adj.* dankbar (**to** *Dat.*)

gratefully ['greɪtfəlɪ] *adv.* dankbar

grater ['greɪtə(r)] *n.* Reibe, *die;* *(less fine)* Raspel, *die*

gratify ['grætɪfaɪ] *v.t.* **a)** *(please)* freuen; **be gratified by** *or* **with** *or* **at sth.** über etw. *(Akk.)* erfreut sein; **b)** *(satisfy)* befriedigen ‹*Neugier, Bedürfnis, Eitelkeit*›; stillen ‹*Sehnsucht, Verlangen*›

gratifying ['grætɪfaɪɪŋ] *adj.* erfreulich

grating ['greɪtɪŋ] *n.* *(framework)* Gitter, *das*

gratis ['grɑ:tɪs] **1.** *adv.* gratis ‹*bekommen, abgeben*›; umsonst ‹*tun*›. **2.** *adj.* gratis *nicht attr.;* Gratis‹*mahlzeit, -vorstellung usw.*›

gratitude ['grætɪtju:d] *n., no pl.* Dankbarkeit, *die* (**to** gegenüber); **show one's** ~ **to sb.** sich jmdm. gegenüber dankbar zeigen

gratuitous [grə'tju:ɪtəs] *adj.* *(uncalled-far, motiveless)* grundlos; unnötig; *(without logical reason)* unbegründet

gratuity [grə'tju:ɪtɪ] *n.* *(formal: tip)* Trinkgeld, *das*

[1]**grave** [greɪv] *n.* Grab, *das;* **it was as quiet** *or* **silent as the** ~: es herrschte Grabesstille; **dig one's own** ~ *(fig.)* sich *(Dat.)* selbst sein Grab graben *(fig.)*; **he would turn in his** ~ *(fig.)* er würde sich im Grabe herumdrehen

[2]**grave** *adj.* **a)** *(important, dignified, solemn)* ernst; **b)** *(formidable, serious)* schwer, gravierend ‹*Fehler, Verfehlung*›; ernst ‹*Lage, Schwierigkeit*›; groß ‹*Gefahr, Risiko, Verantwortung*›; schlimm ‹*Nachricht, Zeichen*›

'grave-digger *n.* Totengräber, *der*

gravel ['grævl] *n.* Kies, *der;* *attrib.* ~ **path/pit** Kiesweg, *der*/-grube, *die*

gravelly ['grævəlɪ] *adj.* rauh, heiser ‹*Stimme*›

gravely ['greɪvlɪ] *adv.* **a)** *(solemnly)* ernst; **b)** *(seriously)* ernstlich

grave: ~stone *n.* Grabstein, *der;* **~yard** *n.* Friedhof, *der*

gravitate ['grævɪteɪt] *v.i.* **sb. ~s towards sb./sth.** es zieht jmdn. zu jmdn./etw.

gravity ['grævɪtɪ] *n.* **a)** *(importance)* *(of mistake, offence)* Schwere, *die;* *(of situation)* Ernst, *der;* **b)** *(Phys., Astron.)* Gravitation, *die;* Schwerkraft, *die;* **the law/force of** ~: das Gravitationsgesetz/die Schwerkraft; **centre of** ~ *(lit. or fig.)* Schwerpunkt, *der*

gravy ['greɪvɪ] *n.* **a)** *(juices)* Bratensaft, *der;* **b)** *(dressing)* [Braten]soße, *die;* ~ **boat** Sauciere, *die*

gray *etc.* *(Amer.)* *see* **grey** *etc.*

[1]**graze** [greɪz] *v.i.* grasen; weiden

[2]**graze** **1.** *n.* Schürfwunde, *die.* **2.** *v.t.* **a)** *(touch lightly)* streifen; **b)** *(scrape)* abschürfen ‹*Haut*›; zerkratzen ‹*Oberfläche*›; ~ **one's knee/elbow** sich *(Dat.)* das Knie/den Ellbogen aufschürfen

grease [gri:s] **1.** *n.* Fett, *das;* *(lubricant)* Schmierfett, *das.* **2.** *v.t.* einfetten; *(lubricate)* schmieren; **like ~d lightning** *(coll.)* wie ein geölter Blitz *(ugs.)*

grease: ~-gun *n.* Fettpresse, *die (Technik)*; **~-paint** *n.* [Fett]schminke, *die;* **~-proof 'paper** *n.* Pergamentpapier, *das*

greasy ['gri:sɪ] *adj.* **a)** fettig; fett ‹*Essen*›; speckig ‹*Kleidung*›; *(lubricated)* geschmiert; *(slippery, dirty with lubricant)* schmierig; **b)** *(fig.)* schmierig *(abwertend)*

great [greɪt] **1.** *adj.* **a)** *(large)* groß; ~ **big** *(coll.)* riesengroß *(ugs.)*; **a** ~ **many** sehr viele; **b)** *(beyond the ordinary)* groß; sehr gut ‹*Freund*›; **a** ~ **age** ein hohes Alter; **take** ~ **care of/a** ~ **interest in** sich sehr kümmern um/interessieren für; **c)** *(important)* groß ‹*Tag, Ereignis, Attraktion, Hilfe*›; *(powerful, able)* groß ‹*Person, Komponist, Schriftsteller*›; *(impressive)* großartig; **the** ~ **thing is ...:** die Hauptsache ist ...; **Peter the G~:** Peter der Große; **be** ~ **at sth.** *(skilful)* in etw. *(Dat.)* ganz groß sein *(ugs.)*; **be a** ~ **one for sth.** etw. sehr gern tun; **d)** *(coll.: splendid)* großartig; **e)** *(in relationship)* Groß‹*onkel, -tante, -neffe, -nichte*›; Ur‹*großmutter, -großvater, -enkel, -enkelin*›. **2.** *n.* *(person)* Größe, *die;* *as pl.* **the** ~: die Großen [der Geschichte/Literatur *usw.*]; **the ~est** *(sl.)* der/die Größte/die Größten *(ugs.)*

Great: ~ **'Britain** *pr. n.* Großbritannien *(das)*; ~ **'Dane** *n.* deutsche Dogge

Greater 'London *pr. n.* Groß-London

greatly ['greɪtlɪ] *adv.* sehr; höchst ‹*verärgert*›; stark ‹*beeinflußt, beunruhigt*›; bedeutend ‹*verbessert*›; **it doesn't** ~ **matter** es ist nicht so wichtig

greatness ['greɪtnɪs] *n., no pl.* Größe, *die*

Great 'War *n.* erster Weltkrieg

Grecian ['gri:ʃn] *adj.* griechisch

Greece [gri:s] *pr. n.* Griechenland *(das)*

greed [gri:d] *n.* Gier, *die* (**for** nach); *(gluttony)* Gefräßigkeit, *die (abwertend); (of animal)* Freßgier, *die;* ~ **for money/power** Geld-/Machtgier, *die*

greedily ['gri:dılı] *adv.* gierig

greedy ['gri:dı] *adj.* gierig; *(gluttonous)* gefräßig *(abwertend);* **be ~ for sth.** nach etw. gieren; ~ **for money/power** geldgierig/machthungrig

Greek [gri:k] **1.** *adj.* griechisch; **sb. is ~:** jmd. ist Grieche/Griechin. **2.** *n.* **a)** *(person)* Grieche, *der*/Griechin, *die;* **b)** *(language)* Griechisch, *das;* **it's all ~ to me** *(fig.)* das sind mir *od.* für mich böhmische Dörfer; *see also* **English 2a**

green [gri:n] **1.** *adj.* **a)** grün; **have ~ fingers** *(fig.)* eine grüne Hand haben *(ugs.);* ~ **vegetables** Grüngemüse, *das;* **b)** *(Polit.)* **G~:** grün; **he/she is G~:** er ist ein Grüner/sie ist eine Grüne; **the G~s** die Grünen; **c)** *(environmentally safe)* ökologisch; **d)** *(unripe, young)* grün ‹*Obst, Zweig*›; **e) be/turn ~ with envy** vor Neid grün sein/werden; **f)** *(gullible)* naiv; einfältig; *(inexperienced)* grün. **2.** *n.* **a)** *(colour, traffic light)* Grün, *das;* **b)** *(piece of land)* Grünfläche, *die;* **village ~:** Dorfanger, *der;* **c)** *in pl.* *(~ vegetables)* Grüngemüse, *das*

green: ~ **'belt** *n.* Grüngürtel, *der;* ~ **'card** *n. (Insurance)* grüne Karte *(Verkehrsw.)*

greenery ['gri:nərı] *n., no pl.* Grün, *das*

green: ~**fly** *n. (Brit.)* grüne Blattlaus; ~**gage** ['gri:ngeıdʒ] *n.* Reineclaude, *die;* ~**grocer** *n. (Brit.)* Obst- und Gemüsehändler, *der*/-händlerin, *die; see also* **baker;** ~**house** *n.* Gewächshaus, *das;* ~**house effect/gas** *(Ecol.)* Treibhauseffekt, *der/den Treibhauseffekt bewirkendes Gas*

greenish ['gri:nıʃ] *adj.* grünlich

Greenland ['gri:nlənd] *pr. n.* Grönland *(das)*

green: ~ **'light** *n.* **a)** grünes Licht; *(as signal)* Grün, *das;* **b)** *(fig. coll.)* **give sb./get the ~ light** jmdm. grünes Licht geben/grünes Licht erhalten; **G~ 'Paper** *n. (Brit.) öffentliches Diskussionspapier über die Regierungspolitik;* **G~ Party** *n. (Polit.)* die Grünen; ~ **'pepper** *see* **pepper 1b**

Greenwich ['grenıdʒ, 'grenıtʃ] *n.* ~ **[mean] time** Greenwicher Zeit

greet [gri:t] *v.t.* **a)** begrüßen; *(in passing)* grüßen; *(receive)* empfangen; ~ **sb. with sth.** jmdn. mit etw. begrüßen/grüßen/empfangen; **b)** *(meet)* empfangen; ~ **sb.'s eyes/ears** sich jmds. Augen *(Dat.)* darbieten/an jmds. Ohr *(Akk.)* dringen

greeting ['gri:tıŋ] *n.* Begrüßung, *die; (in passing)* Gruß, *der; (words)* Grußformel, *die; (reception)* Empfang, *der;* **please give my ~s to your parents** grüßen Sie bitte Ihre Eltern von mir; **my husband also sends his ~s** mein Mann läßt auch grüßen

'greeting[s] card *n.* Grußkarte, *die; (for anniversary, birthday)* Glückwunschkarte, *die*

gregarious [grı'geərıəs] *adj.* **a)** *(Zool.)* gesellig; *attrib.* Herden-; **b)** *(fond of company)* gesellig

gremlin ['gremlın] *n. (coll. joc.)* ≈ Kobold, *der*

grenade [grı'neıd] *n.* Granate, *die*

grew *see* **grow**

grey [greı] **1.** *adj. (lit. or fig.)* grau; **he or his hair went or turned ~:** er wurde grau *od.* ergraute; ~ **area** *(fig.)* Grauzone, *die.* **2.** *n.* Grau, *das*

'greyhound *n.* Windhund, *der*

greyish ['greııʃ] *adj.* gräulich

grey: ~ **matter** *n. (fig.: intelligence)* graue Zellen; ~ **'squirrel** *n.* Grauhörnchen, *das*

grid [grıd] *n.* **a)** *(grating)* Rost, *der;* **b)** *(of lines)* Gitter[netz], *das;* **c)** *(for supply)* [Versorgungs]netz, *das;* **d)** *(Motor-racing)* Startmarkierung, *die*

grief [gri:f] *n.* **a)** Kummer, *der* (**over, at** über + *Akk.*, um); *(at loss of sb.)* Trauer, *die* (**for** um); **come to ~** *(fail)* scheitern; **b) good ~!** großer Gott!

grievance ['gri:vəns] *n. (complaint)* Beschwerde, *die; (grudge)* Groll, *der;* **air one's ~s** seine Beschwerden vorbringen

grieve [gri:v] **1.** *v.t.* betrüben; bekümmern. **2.** *v.i.* trauern (**for** um); ~ **over sb./sth.** jmdm./einer Sache nachtrauern

grievous ['gri:vəs] *adj.* schwer ‹*Verwundung, Krankheit*›; groß ‹*Schmerz*›; ~ **bodily harm** *(Law)* schwere Körperverletzung

¹grill [grıl] **1.** *v.t.* **a)** *(cook)* grillen; **b)** *(fig.: question)* in die Mangel nehmen *(ugs.).* **2.** *n.* **a)** *(Gastr.)* Grillgericht, *das;* **mixed ~:** Mixed grill, *der;* gemischte Grillplatte; **b)** *(on cooker)* Grill, *der*

grille (²grill) *n.* **a)** *(grating)* Gitter, *das;* **b)** *(Motor Veh.)* [Kühler]grill, *der*

grim [grım] *adj. (stern)* streng; grimmig ‹*Lächeln, Gesicht, Blick, Schweigen, Humor, Entschlossenheit*›; *(unrelenting, merciless, severe)* erbittert ‹*Widerstand, Kampf, Schlacht*›; *(sinister, ghastly)* grauenvoll ‹*Aufgabe, Anblick, Nachricht*›; trostlos ‹*Winter, Tag, Landschaft, Aussichten*›; **hold or hang or cling on [to sth.] like ~ death** sich mit aller Kraft [an etw. *(Dat.)*] festklammern

grimace [grı'meıs] **1.** *n.* Grimasse, *die;* **make a ~:** eine Grimasse machen *od.* schneiden. **2.** *v.i.* Grimassen machen *od.* schneiden; ~ **with pain** vor Schmerz das Gesicht verziehen

grime [graım] *n.* Schmutz, *der; (soot)* Ruß, *der*

grimly ['grımlı] *adv.* grimmig; eisern ‹*entschlossen sein, sich festhalten*›; erbittert ‹*kämpfen*›

grimy ['graımı] *adj.* schmutzig; rußgeschwärzt ‹*Gebäude*›

grin [grın] **1.** *n.* Grinsen, *das.* **2.** *v.i.*, **-nn-** grinsen; ~ **at sb.** jmdn. angrinsen; ~ **and bear it** gute Miene zum bösen Spiel machen

grind [graınd] **1.** *v.t.*, **ground** [graʊnd] **a)** *(reduce to small particles)* ~ **[up]** zermahlen; pulverisieren ‹*Metall*›; mahlen ‹*Kaffee, Pfeffer, Getreide*›; **b)** *(sharpen)* schleifen ‹*Schere, Messer*›; schärfen ‹*Klinge*›; *(smooth, shape)* schleifen ‹*Linse, Edelstein*›; **c)** *(rub harshly)* zerquetschen; ~ **one's teeth** mit den Zähnen knirschen; **d)** *(produce by ~ing)* mahlen ‹*Mehl*›; **e)** *(fig.: oppress, harass)* auspressen *(fig.);* ~**ing poverty** erdrückende Armut. **2.** *v.i.*, **ground:** ~ **to a halt, come to a ~ing halt** ‹*Fahrzeug:*› quietschend zum Stehen kommen; *(fig.)* ‹*Verkehr:*› zum Erliegen kommen; ‹*Maschine:*› stehenbleiben; ‹*Projekt:*› sich festfahren. **3.** *n.* Plackerei, *die (ugs.);* **the daily ~** *(coll.)* der alltägliche Trott

~ **a'way** *v.t.* abschleifen

~ **'down** *v.t. (fig.)* ‹*Tyrann, Regierung:*› unterdrücken; ‹*Armut, Verantwortung:*› erdrücken

grinder ['graındə(r)] *n.* Schleifmaschine, *die; (coffee-~ etc.)* Mühle, *die*

'grindstone *n.* Schleifstein, *der;* **keep one's/sb.'s nose to the ~** *(fig.)* sich dahinterklemmen *(ugs.)*/dafür sorgen, daß jmd. sich dahinterklemmt *(ugs.);* **get back to the ~:** sich wieder an die Arbeit machen

grip [grıp] **1.** *n.* **a)** *(firm hold)* Halt, *der; (fig.: power)* Umklammerung, *die;* **have a ~ on sth.** etw.

festhalten; **loosen one's ~:** loslassen; **get** *or* **take a ~ on oneself** *(fig.)* sich zusammenreißen *(ugs.)*; **have/get a ~ on sth.** *(fig.)* etw. im Griff haben/in den Griff bekommen; **come** *or* **get to ~s with sth./sb.** *(fig.)* mit etw. fertigwerden/sich *(Dat.)* jmdn. vorknöpfen *od.* vornehmen *(ugs.)*; **be in the ~ of** *(fig.)* beherrscht werden von ‹*Angst, Leidenschaft, Furcht*›; heimgesucht werden von ‹*Naturkatastrophe, Armut, Krieg*›; **lose one's ~** *(fig.)* nachlassen; **b)** *(strength or way of ~ping, part which is held)* Griff, *der;* **c)** *(bag)* Reisetasche, *die.* **2.** *v.t.*, **-pp-** greifen nach; ‹*Reifen:*› greifen; *(fig.)* ergreifen; fesseln ‹*Publikum, Aufmerksamkeit*›. **3.** *v.i.*, **-pp-** ‹*Räder, Bremsen usw.:*› greifen

gripe [graɪp] **1.** *n.* **a)** *(sl.: complaint)* Meckern, *das (ugs. abwertend)*; **have a good ~ about sth./at sb.** sich über etw. *(Akk.)* ausschimpfen/jmdn. tüchtig ausschimpfen; **b)** *in pl. (colic)* **the ~s** Bauchschmerzen; Bauchweh *(ugs.)*. **2.** *v.i. (sl.)* meckern *(ugs. abwertend)* (**about** über + *Akk.*)

gripping ['grɪpɪŋ] *adj.* *(fig.)* packend

grisly ['grɪzlɪ] *adj.* grausig

grist [grɪst] *n.* **it's all ~ to the/sb.'s mill** man kann aus allem etwas machen/jmd. versteht es, aus allem etwas zu machen

gristle ['grɪsl] *n.* Knorpel, *der*

grit [grɪt] **1.** *n.* **a)** Sand, *der;* **b)** *(coll.: courage)* Schneid, *der (ugs.)*. **2.** *v.t.*, **-tt-: a)** streuen ‹*vereiste Straßen*›; **b) ~ one's teeth** die Zähne zusammenbeißen *(ugs.)*

gritty ['grɪtɪ] *adj.* sandig

grizzly ['grɪzlɪ] *n.* **~ [bear]** Grislybär, *der*

groan [grəʊn] **1.** *n. (of person)* Stöhnen, *das; (of thing)* Ächzen, *das (fig.)*. **2.** *v.i.* ‹*Person:*› [auf]stöhnen (**at** bei); ‹*Tisch, Planken:*› ächzen *(fig.)*

grocer ['grəʊsə(r)] *n.* Lebensmittelhändler, *der*/-händlerin, *die; see also* **baker**

grocery ['grəʊsərɪ] *n.* **a)** *in pl. (goods)* Lebensmittel *Pl.*; **b) ~ [store]** Lebensmittelgeschäft, *das*

grog [grɒg] *n.* Grog, *der*

groggy ['grɒgɪ] *adj.* groggy *(ugs.) präd.*

groin [grɔɪn] *n.* Leistengegend, *die*

groom [gru:m] **1.** *n.* **a)** *(stableboy)* Stallbursche, *der;* **b)** *(bride~)* Bräutigam, *der.* **2.** *v.t.* **a)** striegeln ‹*Pferd*›; **~ oneself** sich zurechtmachen; **b)** *(fig.: prepare)* **~ sb. for a career** jmdn. auf *od.* für eine Laufbahn vorbereiten

groove [gru:v] *n.* **a)** *(channel)* Nut, *die (bes. Technik); (of gramophone record)* Rille, *die;* **b)** *(fig.: routine)* **be stuck in a ~:** aus dem Trott nicht mehr herauskommen

grope [grəʊp] **1.** *v.i.* tasten (**for** nach); **~ for the right word/truth** nach dem richtigen Wort/der Wahrheit suchen. **2.** *v.t.* **~ one's way [along]** sich [entlang]tasten; *(fig.)* [sich durch]lavieren *(ugs. abwertend)*

[1]**gross** [grəʊs] **1.** *adj.* **a)** *(flagrant)* grob ‹*Fahrlässigkeit, Fehler, Irrtum*›; übel ‹*Laster, Beleidigung*›; schreiend ‹*Ungerechtigkeit*›; **b)** *(obese)* fett *(abwertend)*; **c)** *(coarse, rude)* ordinär *(abwertend)*; **d)** *(total)* Brutto-; **earn £15,000 ~:** 15 000 Pfund brutto verdienen; **~ national product** Bruttosozialprodukt, *das;* **e)** *(dull, not delicate)* grob ‹*Person, Geschmack*›. **2.** *v.t.* [insgesamt] einbringen ‹*Geld*›

[2]**gross** *n., pl. same* Gros, *das;* **by the ~:** en gros

grossly ['grəʊslɪ] *adv.* **a)** *(flagrantly)* äußerst; grob ‹*übertreiben*›; schwer ‹*beleidigen*›; **b)** *(coarsely, rudely)* ordinär ‹*sich benehmen, sprechen*›

grotesque [grəʊ'tesk] *adj.*, **grotesquely** [grəʊ'tesklɪ] *adv.* grotesk

grotto ['grɒtəʊ] *n., pl.* **~es** *or* **~s** Grotte, *die*

grotty ['grɒtɪ] *adj. (Brit. sl.)* mies *(ugs.)*

grouch [graʊtʃ] *(coll.)* **1.** *v.i.* schimpfen; mosern *(ugs.)*. **2.** *n.* **a)** *(person)* Miesepeter, *der (ugs. abwertend)*; **b)** *(cause)* Ärger, *der;* **have a ~ against sb.** auf jmdn. sauer sein *(salopp)*

grouchy ['graʊtʃɪ] *adj.* *(coll.)* griesgrämig

[1]**ground** [graʊnd] **1.** *n.* **a)** Boden, *der;* **work above/below ~:** über/unter der Erde arbeiten; **deep under the ~:** tief unter der Erde; **uneven, hilly ~:** unebenes, hügeliges Gelände; **on high ~:** in höheren Lagen; **b)** *(fig.)* **cut the ~ from under sb.'s feet** jmdm. den Wind aus den Segeln nehmen *(ugs.)*; **be** *or* **suit sb. down to the ~** *(coll.)* genau das richtige für jmdn. sein; **get off the ~** *(coll.)* konkrete Gestalt annehmen; **get sth. off the ~** *(coll.)* etw. in die Tat umsetzen; **go to ~** ‹*Fuchs usw.:*› im Bau verschwinden; ‹*Person:*› untertauchen; **run sb./oneself into the ~** *(coll.)* jmdn./sich kaputtmachen *(ugs.)*; **run a car into the ~** *(coll.)* ein Auto solange fahren, bis es schrottreif ist; **on the ~** *(in practice)* an Ort und Stelle; **thin/thick on the ~:** dünn/dicht gesät; **cover much** *or* **a lot of ~:** weit vorankommen; **give** *or* **lose ~:** an Boden verlieren; **hold** *or* **keep** *or* **stand one's ~:** nicht nachgeben; **c)** *(special area)* Gelände, *das;* **[sports] ~:** Sportplatz, *der;* **[cricket] ~:** Cricketfeld, *das;* **d)** *in pl. (attached to house)* Anlage, *die;* **e)** *(motive, reason)* Grund, *der;* **on the ~[s] of, on ~s of** auf Grund (+ *Gen.*); *(giving as one's reason)* unter Berufung auf (+ *Akk.*); **on the ~s that ...:** unter Berufung auf die Tatsache, daß ...; **on health/religious** *etc.* **~s** aus gesundheitlichen/religiösen *usw.* Gründen; **the ~s for divorce are ...:** als Scheidungsgrund gilt ...; **have no ~s for sth./to do sth.** keinen Grund für etw. haben/keinen Grund haben, etw. zu tun; **have no ~s for complaint** keinen Grund zur Klage haben; **f)** *in pl. (sediment)* Satz, *der; (of coffee)* Kaffeesatz, *der;* **g)** *(Electr.)* Erde, *die.* **2.** *v.t.* **a)** *(cause to run ashore)* auf Grund setzen; **be ~ed** auf Grund gelaufen sein; **b)** *(base, establish)* gründen (**on** auf + *Akk.*); **be ~ed on** gründen auf (+ *Dat.*); **c)** *(Aeronaut.)* am Boden festhalten; *(prevent from flying)* nicht fliegen lassen ‹*Piloten*›. **3.** *v.i.* *(run ashore)* ‹*Schiff:*› auf Grund laufen

[2]**ground 1.** *see* **grind 1, 2. 2.** *adj.* gemahlen ‹*Kaffee, Getreide*›; **~ meat** *(Amer.)* Hackfleisch, *das;* **~ coffee** Kaffeepulver, *das*

ground: ~ control *n.* *(Aeronaut.)* Flugsicherungskontrolldienst, *der;* **~ 'floor** *see* **floor 1 b; ~ frost** *n.* Bodenfrost, *der*

grounding ['graʊndɪŋ] *n.* *(basic knowledge)* Grundkenntnisse *Pl.*; Grundwissen, *das;* **give sb./receive a ~ in sth.** jmdm. die Grundlagen einer Sache *(Gen.)* vermitteln/die Grundlagen einer Sache *(Gen.)* vermittelt bekommen

groundless ['graʊndlɪs] *adj.* unbegründet; **these reports** *etc.* **are ~:** diese Berichte *usw.* entbehren jeder Grundlage

ground: ~ level *n.* **above/below ~ level** oberhalb/unterhalb der ebenen Erde; **on** *or* **at ~ level** ebenerdig; **~-plan** *n.* Grundriß, *der;* ~ '**rice** *n.* Reismehl, *das;* ~ **rule** *n.* **a)** *(Sport)* Platzregel, *die;* **b)** *(basic principle)* Grundregel, *die*

ground: ~**sheet** *n.* Bodenplane, *die;* ~**sman** ['graʊndzmən] *n., pl.* ~**smen** ['graʊndzmən] *(Sport)* Platzwart, *der;* ~ **staff** *n. (Aeronaut.)* Bodenpersonal, *das;* ~**work** *n.* Vorarbeiten *Pl.*

group [gru:p] **1.** *n.* **a)** Gruppe, *die; attrib.* Gruppen⟨*verhalten, -dynamik, -therapie, -diskussion*⟩; ~ **of houses/islands/trees** Häuser-/Insel-/Baumgruppe, *die;* **b)** *(Commerc.)* [Unternehmens]gruppe, *die;* **c)** *see* **pop group. 2.** *v. t.* gruppieren; ~ **books according to their subjects** Bücher nach ihrer Thematik ordnen

group: ~ **captain** *n. (Air Force)* Oberst der Luftwaffe; ~ **practice** *n.* Gemeinschaftspraxis, *die*

¹grouse [graʊs] *n.* **a)** *pl. same* Rauhfußhuhn, *das;* **[red]** ~ *(Brit.)* Schottisches Moorschneehuhn; **b)** *no pl. (as food)* Waldhuhn, *das;* schottisches Moorhuhn

²grouse *(coll.)* **1.** *v.i.* meckern *(ugs.)* (**about** über + *Akk.*). **2.** *n.* Meckerei, *die (ugs.)*

grout [graʊt] *n.* Mörtelschlamm, *der*

grove [grəʊv] *n.* Wäldchen, *das;* Hain, *der (dichter. veralt.)*

grovel ['grɒvl] *v. i., (Brit.)* **-ll-: a)** sich auf die Knie werfen; **be ~ling on the floor** auf dem Fußboden kriechen; **b)** *(fig.: be subservient)* katzbuckeln *(abwertend)*

grow [grəʊ] **1.** *v. i.,* **grew** [gru:], **grown** [grəʊn] **a)** wachsen; ⟨*Bevölkerung:*⟩ zunehmen, wachsen; ~ **out of** *or* **from sth.** *(develop)* sich aus etw. entwickeln; *(from sth. abstract)* von etw. herrühren; ⟨*Situation, Krieg usw.:*⟩ die Folge von etw. sein; ⟨*Plan:*⟩ aus etw. erwachsen; ~ **in** gewinnen an (+ *Dat.*) ⟨*Größe, Bedeutung, Autorität, Popularität, Weisheit*⟩; **b)** *(become)* werden; ~ **used to sth./sb.** sich an etw./jmdn. gewöhnen; ~ **apart** *(fig.)* sich auseinanderleben; ~ **to be sth.** allmählich etw. werden; **he grew to be a man** er wuchs zum Manne heran *(geh.);* ~ **to love/hate** *etc.* **sb./sth.** jmdn./etw. liebenlernen/hassenlernen *usw.;* ~ **to like sb./sth.** nach und nach Gefallen an jmdm./etw. finden. *See also* **growing; grown 2. 2.** *v. t.,* **grew, grown a)** *(cultivate) (on a small scale)* ziehen; *(on a large scale)* anpflanzen; züchten ⟨*Blumen*⟩; **b)** ~ **one's hair [long]** sich *(Dat.)* die Haare [lang] wachsen lassen; ~ **a beard** sich *(Dat.)* einen Bart wachsen lassen

~ **into** *v. t.* **a)** *(become)* werden zu; **b)** *(become big enough for)* hineinwachsen in (+ *Akk.*) ⟨*Kleidung*⟩

~ **on** *v. t.* **it ~s on you** man findet mit der Zeit Gefallen daran

~ '**out of** *v. t.* **a)** *(become too big for)* herauswachsen aus ⟨*Kleidung*⟩; **b)** *(lose eventually)* ablegen ⟨*Angewohnheit*⟩; entwachsen (+ *Dat.*) ⟨*Kindereien*⟩; überwinden ⟨*Zustand*⟩; *see also* ~ **1 a**

~ '**up** *v. i.* **a)** *(spend early years)* aufwachsen; *(become adult)* erwachsen werden; **what do you want to be** *or* **do when you ~ up?** was willst du denn mal werden, wenn du groß bist?; **b)** *(fig.: behave [more] maturely)* erwachsen werden; ~ **up!** werde endlich erwachsen!; **c)** *(develop)* ⟨*Freundschaft, Feindschaft:*⟩ sich entwickeln; ⟨*Legende:*⟩ entstehen; ⟨*Tradition, Brauch:*⟩ sich herausbilden

~ '**up into** *v. t.* werden *od.* sich entwickeln zu

grower ['grəʊə(r)] *n. usu. in comb. (person)* Produzent, *der*/Produzentin, *die;* **fruit-/vegetable-~:** Obst-/Gemüsebauer, *der*

growing ['grəʊɪŋ] *adj.* wachsend; immer umfangreicher werdend ⟨*Sachgebiet*⟩; sich immer mehr verbreitend ⟨*Praktik*⟩

'**growing pains** *n. pl.* Wachstumsschmerzen *Pl.; (fig.)* Anfangsschwierigkeiten *Pl.*

growl [graʊl] **1.** *n.* Knurren, *das; (of bear)* Brummen, *das.* **2.** *v. i.* knurren; ⟨*Bär:*⟩ [böse] brummen; ~ **at sb.** jmdn. anknurren/anbrummen

grown [grəʊn] **1.** *see* **grow. 2.** *adj.* erwachsen; **fully ~:** ausgewachsen

'**grown-up 1.** *n.* Erwachsene, *der/die.* **2.** *adj.* erwachsen; ~ **books/clothes** Bücher/Kleider für Erwachsene

growth [grəʊθ] *n.* **a)** *(of industry, economy, population)* Wachstum, *das* (**of, in** *Gen.*); *(of interest, illiteracy)* Zunahme, *die* (**of, in** *Gen.*); *attrib.* Wachstums⟨*hormon, -rate*⟩; **b)** *(of organisms, amount grown)* Wachstum, *das;* **c)** *(thing grown)* Vegetation, *die;* Pflanzenwuchs, *der;* **d)** *(Med.)* Geschwulst, *die;* Gewächs, *das*

'**growth industry** *n.* Wachstumsindustrie, *die*

grub [grʌb] **1.** *n.* **a)** Larve, *die; (maggot)* Made, *die;* **b)** *(sl.: food)* Fressen, *das (salopp); (provisions)* Fressalien *Pl. (ugs.);* ~**['s] up!** ran an die Futterkrippe! *(ugs.);* **lovely ~!** ein Spitzenfraß! *(salopp).* **2.** *v. i.,* **-bb-** wühlen (**for** nach); ~ **about** [herum]wühlen

grubby ['grʌbɪ] *adj.* schmudd[e]lig *(ugs. abwertend)*

grudge [grʌdʒ] **1.** *v. t.* ~ **sb. sth.** jmdm. etw. mißgönnen; **I don't ~ him his success** ich gönne ihm seinen Erfolg; ~ **doing sth.** *(be unwilling to do sth.)* nicht bereit sein, etw. zu tun; *(do sth. reluctantly)* etw. ungern tun; **I ~ paying £20 for this** es geht mir gegen den Strich, dafür 20 Pfund zu zahlen *(ugs.).* **2.** *n.* Groll, *der;* **have** *or* **hold a ~ against sb.** einen Groll *od. (ugs.)* Haß auf jmdn. haben; jmdm. grollen; **bear sb. a ~** *or* **a ~ against sb.** jmdm. gegenüber nachtragend sein

grudging ['grʌdʒɪŋ] *adj.* widerwillig ⟨*Lob, Bewunderung, Unterstützung*⟩; widerwillig gewährt ⟨*Zuschuß*⟩

grudgingly ['grʌdʒɪŋlɪ] *adv.* widerwillig

gruel ['gru:əl] *n.* Schleimsuppe, *die*

gruelling (*Amer.:* **grueling**) ['gru:əlɪŋ] *adj.* aufreibend; zermürbend; [äußerst] strapaziös ⟨*Reise, Marsch*⟩; mörderisch *(ugs.)* ⟨*Tempo, Rennen*⟩

gruesome ['gru:səm] *adj.* grausig; schaurig

gruff [grʌf] *adj.* barsch; schroff; ruppig ⟨*Benehmen, Wesen*⟩; rauh ⟨*Stimme*⟩

grumble ['grʌmbl] **1.** *v. i.* murren; ~ **about** *or* **over sth.** sich über etw. *(Akk.)* beklagen. **2.** *n. (act)* Murren, *das; (complaint)* Klage, *die;* **without a ~:** ohne Murren

grumbler ['grʌmblə(r)] *n.* Querulant, *der*/Querulantin, *die*

grumpily ['grʌmpɪlɪ] *adv.*, **grumpy** ['grʌmpɪ] *adj.* unleidlich; grantig *(ugs.)*

grunt [grʌnt] **1.** *n.* Grunzen, *das;* **give a ~:** grunzen. **2.** *v. i.* grunzen

'**G-string** *n. (garment)* ≈ Cache-sex, *das;* G-String, *die od. der*

guarantee [gærən'ti:] **1.** *v. t.* **a)** garantieren für; [eine] Garantie geben auf (+ *Akk.*); **the clock is ~d for a year** die Uhr hat ein Jahr Garantie; **~d wage** Garantielohn, *der;* **~d genuine** *etc.* garantiert echt *usw.;* **b)** *(promise)* garantieren *(ugs.); (ensure)* bürgen für ⟨*Qualität*⟩; garantieren ⟨*Erfolg*⟩; **be ~d to do sth.** etw. garantiert tun. **2.** *n.* **a)** *(Commerc. etc.)* Garantie, *die; (document)* Garantieschein, *der;* **there's a year's ~ on**

this radio, this radio has a year's ~: auf dieses Radio gibt es *od.* dieses Radio hat ein Jahr Garantie; **is it still under ~?** ist noch Garantie darauf?; **b)** *(coll.: promise)* Garantie, *die (ugs.)*; **give sb. a ~ that ...**: jmdm. garantieren, daß ...; **be a ~ of sth.** *(ensure)* eine Garantie für etw. sein

guard [gɑ:d] **1.** *n.* **a)** *(Mil.: guardsman)* Wachtposten, *der;* **b)** *no pl. (Mil.: group of soldiers)* Wache, *die;* Wachmannschaft, *die;* **~ of honour** Ehrenwache, *die;* Ehrengarde, *die;* **c) G~s** *(Brit. Mil.: household troops)* Garderegiment, *das;* Garde, *die;* **d)** *(watch; also Mil.)* Wache, *die;* **be on ~**: Wache haben; **keep** *or* **stand ~**: Wache halten *od.* stehen; **keep** *or* **stand ~ over** bewachen; **be on [one's] ~ [against sb./sth.]** *(lit. or fig.)* sich [vor jmdm./etw.] hüten; **be off [one's] ~** *(fig.)* nicht auf der Hut sein; **be caught** *or* **taken off ~** *or* **off one's ~ [by sth.]** *(fig.)* [von etw.] überrascht werden; **put sb. on [his/her] ~**: jmdn. mißtrauisch machen; **under ~**: unter Bewachung; **be [kept/held] under ~**: unter Bewachung stehen; **keep** *or* **hold/put under ~**: bewachen/unter Bewachung stellen; **e)** *(Brit. Railw.)* [Zug]schaffner, *der*/-schaffnerin, *die;* **f)** *(Amer.: prison warder)* [Gefängnis]wärter, *der*/-wärterin, *die;* **g)** *(safety device)* Schutz, *der;* Schutzvorrichtung, *die; (worn on body)* Schutz, *der;* **h)** *(posture) (Boxing, Fencing)* Deckung, *die;* **drop** *or* **lower one's ~**: die Deckung fallen lassen; *(fig.)* seine Reserve aufgeben. **2.** *v. t. (watch over)* bewachen; *(keep safe)* hüten ‹*Geheimnis, Schatz*›; schützen ‹*Leben*›; beschützen ‹*Prominenten*›; **~ sb. against sth.** jmdn. vor etw. *(Dat.)* beschützen

~ against *v. t.* sich hüten vor (+ *Dat.*); verhüten ‹*Unfall*›; vorbeugen (+ *Dat.*) ‹*Krankheit, Gefahr, Irrtum*›; **~ against doing sth.** sich [davor] hüten, etw. zu tun

guard: ~ dog *n.* Wachhund, *der;* **~ duty** *n.* Wachdienst, *der;* **be on** *or* **do ~ duty** Wachdienst haben

guarded ['gɑ:dɪd] *adj.* zurückhaltend; vorsichtig

guardian ['gɑ:dɪən] *n.* **a)** Hüter, *der;* Wächter, *der;* **b)** *(Law)* Vormund, *der*

guard: ~-rail *n.* Geländer, *das;* **~room** *n. (Mil.)* Wachstube, *die;* Wachlokal, *das*

guardsman ['gɑ:dzmən] *n., pl.* **guardsmen** ['gɑ:dzmən] Wachtposten, *der; (in Guards)* Gardist, *der*

guerrilla [gə'rɪlə] *n.* Guerillakämpfer, *der*/-kämpferin, *die; attrib.* Guerilla-

guess [ges] **1.** *v. t.* **a)** *(estimate)* schätzen; *(surmise)* raten; *(surmise correctly)* erraten; raten ‹*Rätsel*›; **can you ~ his weight?** schätz mal, wieviel er wiegt; **~ what!** *(coll.)* stell dir vor!; **you'd never ~ that ...**: man würde nie vermuten, daß ...; **I ~ed as much** das habe ich mir schon gedacht; **b)** *(esp. Amer.: suppose)* **I ~**: ich glaube; ich schätze *(ugs.)*; **I ~ we'll have to** wir müssen wohl; **I ~ so/not** ich glaube schon *od.* ja/nicht *od.* kaum. **2.** *v. i. (estimate)* schätzen; *(make assumption)* vermuten; *(surmise correctly)* es erraten; **~ at sth.** etw. schätzen; *(surmise)* über etw. *(Akk.)* Vermutungen anstellen; **I'm just ~ing** das ist nur eine Schätzung/eine Vermutung; **you've ~ed right/wrong** deine Vermutung ist richtig/falsch; **keep sb. ~ing** *(coll.)* jmdn. im unklaren *od.* ungewissen lassen; **you'll never ~!** darauf kommst du nie! **3.** *n.* Schätzung, *die;* **at a ~**: schätzungsweise; **make** *or* **have a ~**: schätzen; **have a ~!** rate *od.* schätz mal!; **my ~ is [that] ...**: ich schätze, daß ...; **I'll give you three ~es** *(coll.)* dreimal darfst du raten *(ugs.)*

'guesswork *n., no pl., no indef. art.* **be ~**: eine Vermutung sein

guest [gest] *n.* Gast, *der;* **be my ~** *(fig. coll.)* tun Sie sich/tu dir keinen Zwang an; **~ of honour** Ehrengast, *der*

guest: ~-house *n.* Pension, *die;* **~-room** *n.* Gästezimmer, *das*

guffaw [gʌ'fɔ:] **1.** *n.* brüllendes Gelächter; **give a [great] ~**: in brüllendes Gelächter ausbrechen. **2.** *v. i.* brüllend lachen

guidance ['gaɪdəns] *n., no pl., no indef. art.* **a)** *(leadership, direction)* Führung, *die; (by teacher, tutor, etc.)* [An]leitung, *die;* **b)** *(advice)* Rat, *der;* **give sb. ~ on sth.** jmdn. in etw. *(Dat.)* beraten

guide [gaɪd] **1.** *n.* **a)** Führer, *der*/Führerin, *die; (Tourism)* [Fremden]führer, *der*/-führerin, *die; (professional mountain-climber)* [Berg]führer, *der*/-führerin, *die;* **b)** *(indicator)* **be a [good] ~ to sth.** ein [guter] Anhaltspunkt für etw. sein; **be no ~ to sth.** keine Rückschlüsse auf etw. *(Akk.)* zulassen; **c)** *(Brit.)* **[Girl] G~**: Pfadfinderin, *die;* **the G~s** die Pfadfinderinnen; **d)** *(handbook)* Handbuch, *das;* **a ~ to healthier living** ein Ratgeber für ein gesünderes Leben; **e)** *(book for tourists)* [Reise]führer, *der;* **a ~ to York** ein Führer für *od.* durch York. **2.** *v. t.* **a)** führen ‹*Personen, Pflug, Maschinenteil usw.*›; **b)** *(fig.)* bestimmen ‹*Handeln, Urteil*›; anleiten ‹*Schüler, Lehrling*›; **be ~d by sth./sb.** sich von etw./jmdm. leiten lassen

'guidebook *see* **guide 1e**

guided missile [gaɪdɪd 'mɪsaɪl] *n.* Lenkflugkörper, *der*

'guide-dog *n.* **~ [for the blind]** Blinden[führ]hund, *der*

guided tour [gaɪdɪd 'tʊə(r)] *n.* Führung, *die* **(of** durch)

'guideline *n. (fig.)* Richtlinie, *die*

guild [gɪld] *n.* **a)** Verein, *der;* **b)** *(Hist.) (of merchants)* Gilde, *die; (of artisans)* Zunft, *die*

guile [gaɪl] *n., no pl.* Hinterlist, *die*

guillotine ['gɪləti:n] **1.** *n.* **a)** Guillotine, *die;* Fallbeil, *das;* **b)** *(for paper)* Papierschneidemaschine, *die.* **2.** *v. t.* **a)** *(behead)* mit der Guillotine *od.* dem Fallbeil hinrichten; **b)** *(cut)* schneiden

guilt [gɪlt] *n., no pl.* **a)** Schuld, *die* **(of** an + *Dat.*); **b)** *(guilty feeling)* Schuldgefühle *Pl.*

guiltily ['gɪltɪlɪ] *adv.* schuldbewußt

guiltless ['gɪltlɪs] *adj.* unschuldig **(of** an + *Dat.*)

guilty ['gɪltɪ] *adj.* **a)** schuldig; **the ~ person** der/die Schuldige; **be ~ of murder** des Mordes schuldig sein; **find sb. ~/not ~ [of sth.]** jmdn. [an etw. *(Dat.)*] schuldig sprechen/[von etw.] freisprechen; **[return** *or* **find a verdict of] ~/not ~**: [auf] „schuldig"/„nicht schuldig" [erkennen]; **feel ~ about sth./having done sth.** *(coll.)* ein schlechtes Gewissen haben wegen etw./, weil man etw. getan hat; **everyone is/we're all ~ of that** *(coll.)* das tut jeder/das tun wir alle; **b)** *(prompted by guilt)* schuldbewußt ‹*Miene, Blick, Verhalten*›; schlecht ‹*Gewissen*›

guinea ['gɪnɪ] *n. (Hist.)* Guinee, *die*

'guinea-pig *n.* **a)** *(animal)* Meerschweinchen, *das;* **b)** *(fig.: subject of experiment)* Versuchskaninchen, *das (ugs. abwertend)*; **act as ~**: Versuchskaninchen spielen

guise [gaɪz] *n.* Gestalt, *die;* **in the ~ of** in Gestalt (+ *Gen.*)

guitar [gɪ'tɑ:(r)] *n.* Gitarre, *die; attrib.* Gitarren‹*musik, -spieler*›

guitarist [gɪ'tɑ:rɪst] *n.* Gitarrist, *der*/Gitarristin, *die*

gulch [gʌltʃ] *n. (Amer.)* Schlucht, *die;* Klamm, *die*

gulf [gʌlf] *n.* **a)** *(portion of sea)* Golf, *der;* Meerbusen, *der;* **the [Arabian** *or* **Persian] G~:** der [Persische] Golf; **the G~ of Mexico** der Golf von Mexiko; **b)** *(wide difference)* Kluft, *die;* **c)** *(chasm)* Abgrund, *der*

Gulf: ~ **States** *pr. n. pl.* Golfstaaten *Pl.;* ~ **Stream** *pr. n.* Golfstrom, *der;* ~ **War** *n.* Golfkrieg, *der*

gull [gʌl] *n.* Möwe, *die*

gullet ['gʌlɪt] *n.* **a)** *(food-passage)* Speiseröhre, *die;* **b)** *(throat)* Kehle, *die;* Gurgel, *die*

gullible ['gʌlɪbl] *adj.* leichtgläubig; *(trusting)* gutgläubig

gully ['gʌlɪ] *n.* **a)** *(artificial channel)* Abzugskanal, *der;* **b)** *(drain)* Gully, *der;* **c)** *(water-worn ravine)* [Erosions]rinne, *die*

gulp [gʌlp] **1.** *v. t.* hinunterschlingen; hinuntergießen ‹*Getränk*›. **2.** *n.* **a)** *(act of ~ing, effort to swallow)* Schlucken, *das;* **swallow in** *or* **at one ~:** mit einem Schluck herunterstürzen ‹*Getränk*›; in einem Bissen herunterschlingen ‹*Speise*›; **b)** *(large mouthful of drink)* kräftiger Schluck

~ **'down** *see* ~ **1**

[1]**gum** [gʌm] *n., usu. in pl. (Anat.)* ~**[s]** Zahnfleisch, *das*

[2]**gum 1.** *n.* **a)** *(natural substance)* Gummi, *das; (glue)* Klebstoff, *der;* **b)** *(sweet)* Gummibonbon, *der od. das;* **c)** *(Amer.) see* **chewing-gum. 2.** *v. t.,* **-mm-: a)** *(smear with ~)* gummieren ‹*Briefmarken, Etiketten usw.*›; **b)** *(fasten with ~)* kleben

~ **'up** *v. t.* aufkleben

gum: ~**boil** *n.* Zahnfleischabszeß, *der;* ~**boot** *n.* Gummistiefel, *der*

gumption ['gʌmpʃn] *n., no pl., no indef. art. (coll.) (resourcefulness)* Grips, *der (ugs.); (enterprising spirit)* Unternehmungsgeist, *der*

'gum-tree *n.* **be up a** ~ *(fig.)* in der Klemme sitzen *(ugs.)*

gun [gʌn] *n.* **a)** Schußwaffe, *die; (piece of artillery)* Geschütz, *das; (rifle)* Gewehr, *das; (pistol)* Pistole, *die; (revolver)* Revolver, *der;* **big** ~ *(sl.: important person)* hohes *od.* großes Tier *(ugs.);* **be going great ~s** laufen wie geschmiert *(ugs.);* ‹*Person:*› toll in Schwung sein *(ugs.);* **stick to one's ~s** *(fig.)* auf seinem Standpunkt beharren; **b)** *(starting-pistol)* Startpistole, *die;* **jump the ~:** einen Fehlstart verursachen; *(fig.)* vorpreschen; *(by saying sth.)* vorzeitig etwas bekanntwerden lassen

~ **'down** *v. t.* niederschießen

~ **for** *v. t. (fig.)* auf dem Kieker haben *(ugs.)*

gun: ~**-battle** *n.* Schießerei, *die;* ~**boat** *n.* Kanonenboot, *das;* ~**-carriage** *n.* [fahrbare] Geschützlafette; ~**-fight** *n. (Amer. coll.)* Schießerei, *die;* ~**fighter** *n.* Revolverheld, *der;* ~**-fire** *n.* Geschützfeuer, *das; (of small arms)* Schießerei, *die*

gunge [gʌndʒ] *n. (Brit. coll.)* Schmiere, *die*

gunman ['gʌnmən] *n., pl.* ~**men** ['gʌnmən] [mit einer Schußwaffe] bewaffneter Mann

gunner ['gʌnə(r)] *n.* Artillerist, *der; (private soldier)* Kanonier, *der*

gun: ~**point** *see* **point 1b;** ~**powder** *n.* Schießpulver, *das;* **Gunpowder Plot** *(Hist.)* Pulververschwörung, *die;* ~**shot** *n.* **a)** *(shot)* Schuß, *der;* **b) within/out of ~shot** in/außer Schußweite; ~**smith** *n.* Büchsenmacher, *der*

gunwale ['gʌnl] *n. (Naut.)* Schandeck, *das;* Schandeckel, *der; (of rowing-boat)* Dollbord, *der*

gurgle ['gɜ:gl] **1.** *n.* Gluckern, *das; (of brook)* Plätschern, *das.* **2.** *v. i.* gluckern; ‹*Bach:*› plätschern; ‹*Baby:*› lallen; *(with delight)* glucksen

guru ['gʊru:] *n.* **a)** Guru, *der;* **b)** *(mentor)* Mentor, *der*

gush [gʌʃ] **1.** *n.* **a)** *(sudden stream)* Schwall, *der;* **b)** *(effusiveness)* Überschwenglichkeit, *die;* **c)** *(excessive enthusiasm)* Schwärmerei, *die.* **2.** *v. i.* **a)** strömen; schießen; ~ **out** herausströmen; herausschießen; **b)** *(fig.: speak or act effusively)* überschwenglich sein; **c)** *(fig.: speak with excessive enthusiasm)* schwärmen. **3.** *v. t.* **sth. ~es water/oil/blood** Wasser/Öl/Blut schießt aus etw. hervor

gushing ['gʌʃɪŋ] *adj.* **a)** reißend ‹*Strom*›; **b)** *(effusive)* exaltiert

gusset ['gʌsɪt] *n.* Zwickel, *der;* Keil, *der*

gust [gʌst] **1.** *n.* ~ **[of wind]** Windstoß, *der;* Bö[e], *die.* **2.** *v. i.* böig wehen

gusto ['gʌstəʊ] *n., no pl. (enjoyment)* Genuß, *der; (vitality)* Schwung, *der*

gusty ['gʌstɪ] *adj.* böig

gut [gʌt] **1.** *n.* **a)** *(material)* Darm, *der;* **b)** *in pl. (bowels)* Eingeweide *Pl.;* Gedärme *Pl.;* **hate sb.'s ~s** *(coll.)* jmdn. auf den Tod nicht ausstehen können; **sweat** *or* **work one's ~s out** *(coll.)* sich dumm und dämlich schuften *(ugs.);* **c)** *in pl. (fig.: contents)* Innereien *Pl. (scherzh.);* **d)** *in pl. (coll.: courage)* Schneid, *der (ugs.);* Mumm, *der (ugs.);* **e)** *(intestine)* Darm, *der.* **2.** *v. t.,* **-tt-: a)** *(take out ~s of)* ausnehmen; **b)** *(remove or destroy fittings in)* ausräumen; **the house was ~ted [by the fire]** das Haus brannte aus. **3.** *attrib. adj. (instinctive)* gefühlsmäßig ‹*Reaktion*›

gutter ['gʌtə(r)] **1.** *n. (below edge of roof)* Dach- *od.* Regenrinne, *die; (at side of street)* Rinnstein, *der;* Gosse, *die;* **the** ~ *(fig.)* die Gosse. **2.** *v. i.* ‹*Kerze:*› tropfen; ‹*Flamme:*› [immer schwächer] flackern

gutter: ~ **press** *n.* Sensationspresse, *die (abwertend);* ~**snipe** *n.* Gassenjunge, *der (abwertend)*

guttural ['gʌtərl] *adj. (from the throat)* guttural; kehlig

[1]**guy** [gaɪ] *n. (rope)* Halteseil, *das*

[2]**guy** *n.* **a)** *(sl.: man)* Typ, *der (ugs.);* **b)** *in pl. (Amer.: everyone)* **[listen,] you ~s!** [hört mal,] Kinder! *(ugs.);* **c)** *(Brit.: effigy)* Guy-Fawkes-Puppe, *die;* **Guy Fawkes Day** *Festtag (5. November) zum Gedenken an die Pulververschwörung*

'guy-rope *n.* Zelt[spann]leine, *die*

guzzle ['gʌzl] **1.** *v. t. (eat)* hinunterschlingen; *(drink)* hinuntergießen. **2.** *v. i.* schlingen

gym [dʒɪm] *n. (coll.)* **a)** *(gymnasium)* Turnhalle, *die;* **b)** *no pl., no indef. art. (gymnastics)* Turnen, *das*

gymkhana [dʒɪm'kɑ:nə] *n.* Gymkhana, *das*

gymnasium [dʒɪm'neɪzɪəm] *n., pl.* ~**s** *or* **gymnasia** [dʒɪm'neɪzɪə] Turnhalle, *die*

gymnast ['dʒɪmnæst] *n.* Turner, *der*/Turnerin, *die*

gymnastic [dʒɪm'næstɪk] *adj.* turnerisch ‹*Können*›; ~ **equipment** Turngeräte

gymnastics [dʒɪm'næstɪks] *n., no pl.* Gymnastik, *die; (esp. with apparatus)* Turnen, *das; attrib.* Gymnastik-/Turn‹*stunde, -lehrer*›

gym: ~**-shoe** *n.* Turnschuh, *der;* ~**-slip,** ~**-tunic** *ns.* Trägerrock, *der (für die Schule)*

gynaecological [gaɪnɪkə'lɒdʒɪkl] *adj. (Med.)* gynäkologisch

gynaecologist [gaɪnɪ'kɒlədʒɪst] *n. (Med.)* Gynäkologe, *der*/Gynäkologin, *die;* Frauenarzt, *der*/Frauenärztin, *die*

gynaecology [gaɪnɪ'kɒlədʒɪ] *n. (Med.)* Gynäkologie, *die;* Frauenheilkunde, *die*

gynecological *etc. (Amer.) see* **gynaec-**

gypsy (Gypsy) ['dʒɪpsɪ] *n.* Zigeuner, *der*/Zigeunerin, *die*
gyrate [dʒaɪə'reɪt] *v. i.* sich drehen
gyration [dʒaɪə'reɪʃn] *n.* Drehung, *die;* kreiselnde Bewegung
gyroscope ['dʒaɪərəskəʊp] *n.* *(Phys., Naut., Aeronaut.)* Kreisel, *der; (for scientific purposes)* Gyroskop, *das*

H

[1]H, h [eɪtʃ] *n., pl.* **Hs** *or* **H's** ['eɪtʃɪz] *(letter)* H, h, *das*
[2]H *abbr. (on pencil)* **hard** H
habeas corpus [heɪbɪəs 'kɔ:pəs] *n., no pl. (Law)* Anordnung eines Haftprüfungstermins
haberdashery ['hæbədæʃərɪ] *n.* **a)** *(goods) (Brit.)* Kurzwaren *Pl.; (Amer: menswear)* Herrenmoden *Pl.;* **b)** *(shop) (Brit.)* Kurzwarengeschäft, *das; (Amer.)* Herrenmodengeschäft, *das*
habit ['hæbɪt] *n.* **a)** *(set practice)* Gewohnheit, *die;* **good/bad ~:** gute/schlechte [An]gewohnheit; **the ~ of smoking** das [gewohnheitsmäßige] Rauchen; **have a** *or* **the ~ of doing sth.** die Angewohnheit haben, etw. zu tun; **out of ~, from [force of] ~:** aus Gewohnheit; **old ~s die hard** der Mensch ist ein Gewohnheitstier *(ugs.);* **be in the ~ of doing sth.** die Gewohnheit haben, etw. zu tun; **not be in the ~ of doing sth.** es nicht gewohnt sein, etw. zu tun; **get** *or* **fall into a** *or* **the ~ of doing sth.** [es] sich *(Dat.)* angewöhnen, etw. zu tun; **get** *or* **fall into** *or* *(coll.)* **pick up bad ~s** schlechte [An]gewohnheiten annehmen; **get out of the ~ of doing sth.** [es] sich *(Dat.)* abgewöhnen, etw. zu tun; **b)** *(coll.) (addiction)* Süchtigkeit, *die;* [Drogen]abhängigkeit, *die;* **c)** *(dress)* Habit, *der od. das*
habitable ['hæbɪtəbl] *adj.* bewohnbar
habitat ['hæbɪtæt] *n. (of animals, plants)* Habitat, *das (Zool., Bot.);* Lebensraum, *der;* Standort, *der* *(Bot.)*
habitation [hæbɪ'teɪʃn] *n.* **fit/unfit** *or* **not fit for human ~:** bewohnbar/unbewohnbar
habitual [hə'bɪtjʊəl] *adj.* **a)** *(usual)* gewohnt; **b)** *(continual)* ständig; **c)** *(given to habit)* gewohnheitsmäßig ‹*Lügner*›; Gewohnheits‹*trinker*›
habitually [hə'bɪtjʊəlɪ] *adv.* **a)** *(regularly, recurrently)* regelmäßig; **b)** *(incessantly)* ständig
habitué [hə'bɪtjʊeɪ] *n.* regelmäßiger Besucher; *(of hotel, casino, etc.)* Stammgast, *der*
[1]hack [hæk] **1.** *v. t.* hacken ‹*Holz*›; **~ sb./sth. to bits** *or* **pieces** jmdn. zerstückeln/etw. in Stücke hacken; **~ one's way [through/along/out of sth.]** sich *(Dat.)* einen Weg [durch etw./etw. entlang/aus etw. heraus] [frei]schlagen. **2.** *v. i.* **a) ~ at** herumhacken auf (+ *Dat.*); **~ through the undergrowth** sich *(Dat.)* einen Weg durchs Unterholz schlagen; **b) ~ing cough** trockener Husten; Reizhusten, *der*
~ 'off *v. t.* abhacken; abschlagen
[2]hack 1. *n.* **a)** *(writer)* Schreiberling, *der (abwertend);* **newspaper ~:** Zeitungsschreiber, *der;* **publisher's ~:** Lohnschreiber, *der;* **b)** *(hired horse)* Mietpferd, *das.* **2.** *adj.* **a) ~ writer** Lohnschreiber, *der;* **b)** *(mediocre)* Nullachtfünfzehn- *(ugs. abwertend)*
hacker ['hækə(r)] *n. (Computing)* Hacker, *der*
hacking jacket ['hækɪŋ dʒækɪt] *n.* Reitjackett, *das; (sports jacket)* Sportjacke, *die*
hackle ['hækl] *n.* **sb.'s ~s rise** *(fig.)* jmd. gerät in Harnisch; **get sb.'s ~s up, make sb.'s ~s rise** *(fig.)* jmdn. wütend machen
hackney ['hæknɪ]: **~ 'cab, ~ 'carriage** *ns.* Droschke, *die (veralt.);* Taxe, *die*
hackneyed ['hæknɪd] *adj.* abgegriffen; abgedroschen *(ugs.)*
'hack-saw *n.* [Metall]bügelsäge, *die*
had *see* **have 1, 2**
haddock ['hædək] *n., pl. same* Schellfisch, *der;* **smoked ~:** Haddock, *der*
hadn't ['hædnt] *(coll.)* = **had not;** *see* **have 1, 2**
haemoglobin [hi:mə'gləʊbɪn] *n.* *(Anat., Zool.)* Hämoglobin, *das*
haemophilia [hi:mə'fɪlɪə] *n.* *(Med.)* Hämophilie, *die (fachspr.);* Bluterkrankheit, *die*
haemorrhage ['hemərɪdʒ] *(Med.)* **1.** *n.* Hämorrhagie, *die (fachspr.);* Blutung, *die.* **2.** *v. i.* starke Blutungen haben
haemorrhoid ['hemərɔɪd] *n.* Hämorrhoide, *die*
hag [hæg] *n.* **a)** *(old woman)* [alte] Hexe; **b)** *(witch)* Hexe, *die*
haggard ['hægəd] *adj. (worn)* ausgezehrt; *(with worry)* abgehärmt; *(tired)* abgespannt
haggis ['hægɪs] *n. (Gastr.)* Haggis, *der; gefüllter Schafsmagen*
haggle ['hægl] *v. i.* sich zanken (**over, about** wegen); *(over price)* feilschen *(abwertend)* (**over, about** um)
Hague [heɪg] *pr. n.* **The ~:** Den Haag *(das)*
[1]hail [heɪl] **1.** *n.* **a)** *no pl., no indef. art. (Meteorol.)* Hagel, *der;* **b)** *(fig.: shower)* Hagel, *der; (of insults, questions, etc.)* Schwall, *der;* Flut, *die;* **a ~ of bullets/arrows** ein Kugel-/Pfeilhagel *od.* -regen. **2.** *v. i.* **a)** *impers. (Meteorol.)* **it ~s** *or* **is ~ing** es hagelt; **b)** *(fig.)* **~ down** niederprasseln (**on** auf + *Akk.*); **~ down on sb.** ‹*Beschimpfungen, Vorwürfe usw.:*› auf jmdn. einprasseln. **3.** *v. t.* niederhageln *od.* niederprasseln lassen
[2]hail 1. *v. t.* **a)** *(call out to)* anrufen, *(fachspr.)* anpreien ‹*Schiff*›; *(signal to)* heranwinken, anhalten ‹*Taxi*›; **b)** *(acclaim)* zujubeln (+ *Dat.*); bejubeln (**as** als); **~ sb. king** jmdm. als König zujubeln. **2.** *int. (arch.)* sei gegrüßt *(geh.);* **~ Macbeth/to thee, O Caesar** Heil Macbeth/dir, o Cäsar; **H~ Mary** *see* **Ave Maria; ~-fellow-well-met** kumpelhaft
hail: ~stone *n. (Meteorol.)* Hagelkorn, *das;* **~storm** *n.* *(Meteorol.)* Hagelschauer, *der*
hair [heə(r)] *n.* **a)** *(one strand)* Haar, *das;* **without turning a ~** *(fig.)* ohne eine Miene zu verziehen; **not harm a ~ of sb.'s head** *(fig.)* jmdm. kein Haar krümmen; **b)** *collect., no pl. (many strands, mass)* Haar, *das;* Haare *Pl.; attrib.* Haar-; **do one's/sb.'s ~:** sich/jmdm. das Haar machen *(ugs.);* **have** *or* **get one's ~ done** sich *(Dat.)* das Haar *od.* die Haare machen *(ugs.)* lassen; **pull sb.'s ~:** jmdn. an den Haaren ziehen; **he's losing his ~:** ihm gehen die Haare aus; **keep your ~ on!** *(sl.)* geh [mal] nicht gleich an die Decke! *(ugs.);* **let one's ~ down** *(give free expression to one's feelings etc.)* aus sich herausgehen; *(have a good time)* auf den Putz hauen *(ugs.);* **sb.'s ~ stands on end** *(fig.)* jmdm. stehen die Haare zu Berge *(ugs.);* **get in sb.'s ~** *(fig. coll.)* jmdm. auf die Nerven *od.*

den Wecker gehen *od.* fallen *(ugs.)*

hair: ~**brush** *n.* Haarbürste, *die;* ~**cut** *n.* **a)** *(act)* Haareschneiden, *das;* **go for/need a ~cut** zum Friseur gehen/müssen; **give sb. a ~cut** jmdm. die Haare schneiden; **get/have a ~cut** sich *(Dat.)* die Haare schneiden lassen; **b)** *(style)* Haarschnitt, *der;* ~-**do** *n.* *(coll.)* **a) give sb. a ~-do** jmdm. das Haar machen *(ugs.);* **b)** *(style)* Frisur, *die;* ~**dresser** *n.* Friseur, *der*/Friseuse, *die;* **men's ~dresser** Herrenfriseur, *der*/-friseuse, *die;* **ladies' ~dresser** Damenfriseur, *der*/-friseuse, *die;* **go to the ~dresser['s]** zum Friseur gehen; ~-**drier** *n.* Haartrockner, *der;* Fön ⓦ, *der; (with a hood)* Trockenhaube, *die*

-haired [heəd] *adj. in comb.* **black-/frizzy-~:** schwarz-/krausshaarig

hair: ~-**grip** *n. (Brit.)* Haarklammer, *die;* ~ **lacquer** *see* ~**spray;** ~-**line** *n.* **a)** *(edge of hair)* Haaransatz, *der;* **his ~-line is receding, he has a receding ~-line** er bekommt eine Stirnglatze; **b)** *(narrow line)* haarfeine Linie; haarfeiner Strich; **c) ~-line [crack]** haarfeiner Riß; ~-**net** *n.* Haarnetz, *das;* ~-**piece** *n.* Haarteil, *das;* ~**pin** *n.* Haarnadel, *die;* ~**pin 'bend** *n.* Haarnadelkurve, *die;* ~-**raising** ['heəreɪzɪŋ] *adj.* furchterregend; *(very bad)* haarsträubend; mörderisch ‹*Rennstrecke; Abstieg vom Berg usw.*›; ~**'s breadth** *n.* **by [no more than] a ~'s breadth** [nur] um Haaresbreite ‹*verfehlen*›; nur knapp ‹*gewinnen*›; ~-**splitting** *(derog.)* **1.** *adj.* haarspalterisch *(abwertend);* **2.** *n.* Haarspalterei, *die (abwertend);* ~**spray** *n.* Haarspray, *das;* ~-**style** *n.* Frisur, *die*

hairy ['heərɪ] *adj.* **a)** *(having hair)* behaart; flauschig ‹*Schal, Pullover, Teppich*›; **b)** *(sl.: difficult, dangerous)* haarig; **c)** *(sl.: unpleasant, frightening)* eklig *(ugs.)*

hake [heɪk] *n., pl. same (Zool.)* Seehecht, *der*

'halcyon days ['hælsɪən deɪz] *n. pl.* glückliche Zeiten *Pl.*

hale [heɪl] *adj.* kräftig ‹*Körper, Konstitution*›; rege ‹*Geist*›; ~ **and hearty** gesund und munter

half [hɑ:f] **1.** *n., pl.* **halves** [hɑ:vz] **a)** *(part)* Hälfte, *die;* ~ **[of sth.]** die Hälfte [von etw.]; ~ **of Europe** halb Europa; **I've only ~ left** ich habe nur noch die Hälfte; ~ **[of] that** die Hälfte [davon]; **cut sth. in ~** *or* **into [two] halves** etw. in zwei Hälften schneiden; **divide sth. in ~** *or* **into halves** etw. halbieren; **one/two and a ~ hours, one hour/two hours and a ~:** anderthalb *od.* eineinhalb/zweieinhalb Stunden; **she is three and a ~:** sie ist dreieinhalb; **not/never do anything/things by halves** keine halben Sachen machen; **be too cheeky/big by ~:** entschieden zu frech/groß sein; **go halves** *or* **go ~ and ~ [with sb.]** halbe-halbe [mit jmdm.] machen *(ugs.);* **how the other ~ lives** wie andere Leute leben; **that's only** *or* **just** *or* **not the ~ of it** das ist noch nicht alles; **b)** *(coll.: ~-pint)* kleines Glas; *(of beer)* kleines Bier; Kleine, *das (ugs.);* **a ~ of bitter** *etc.* ein kleines Bitter *usw.;* **c)** *(Footb. etc.: period)* Halbzeit, *die.* **2.** *adj.* halb; ~ **the house/books/staff/time** die Hälfte des Hauses/der Bücher/des Personals/der Zeit; **he is drunk ~ the time** *(very often)* er ist fast immer betrunken; ~ **an hour** eine halbe Stunde. **3.** *adv.* **a)** *(to the extent of ~)* zur Hälfte; halb ‹*öffnen, schließen, aufessen, fertig, voll, geöffnet*›; *(almost)* fast ‹*fallen, ersticken, tot sein*›; ~ **as much/many/big/heavy** halb so viel/viele/groß/schwer; ~ **run [and] ~ walk** teils laufen, teils gehen; **I ~ wished/hoped that ...:** ich wünschte mir/hoffte fast, daß ...; **only ~ hear what ...:** nur zum Teil hören, was ...; ~ **listen for/to** mit halbem Ohr horchen auf (+ *Akk.*)/zuhören (+ *Dat.*); ~ **cook sth.** etw. halb gar werden lassen; **b)** *(by the amount of a ~-hour)* halb; ~ **past** *or (coll.)* ~ **one/two/three** *etc.* halb zwei/drei/vier *usw.;* ~ **past twelve** halbeins

half- *in comb.* halb ‹*gar, verbrannt, betrunken, voll, leer*›; ~-**starved** halb verhungert; **a ~-dozen** ein halbes Dutzend; ~-**pound bag/~-litre glass** Halbpfundtüte, *die*/-literglas, *das;* ~-**year** Halbjahr, *das;* halbes Jahr

half: ~-**and-'~ 1.** *n.* **Does it contain a or b? – H~-and-~:** Enthält es a oder b? – Halb und halb; **2.** *adj.* ~-**and-~ mixture of a and b** Mischung, die je zur Hälfte aus a und b besteht; **3.** *adv.* zu gleichen Teilen; **they divide their earnings ~-and-~:** sie teilen ihre Einkünfte gleichmäßig untereinander auf; ~-**baked** [hɑ:f'beɪkt] *adj.* unausgegoren *(abwertend),* unausgereift ‹*Plan, Aufsatz*›; ~-**breed** *n.* **a)** Mischling, *der;* Halbblut, *das;* **b)** *see* **cross-breed 1;** ~-**brother** *n.* Halbbruder, *der;* ~-**caste 1.** *n.* Mischling, *der;* Halbblut, *das;* **2.** *adj.* Mischlings-; ~-**'crown** *n. (Brit. Hist.)* Half-crown, *die;* ~-**hearted** [hɑ:f'hɑ:tɪd] *adj.,* ~-**heartedly** [hɑ:f'hɑ:tɪdlɪ] *adv.* halbherzig; ~-**'hour** *n.* halbe Stunde; ~-**'hourly 1.** *adj.* halbstündlich; halbstündlich verkehrend ‹*Bus usw.*›; **the bus service is ~-hourly** der Bus verkehrt halbstündlich; **2.** *adv.* jede halbe Stunde; halbstündlich; ~-**life** *n. (Phys.)* Halbwertszeit, *die;* ~-**light** *n.* Halblicht, *das;* ~-**'mast** *n.* **be [flown] at ~-mast** ‹*Flagge:*› auf halbmast gehißt sein *od.* stehen; ~ **measure** *n.* **a) a ~ measure of whisky** ein halber Whisky; **b)** *in pl.* halbe Maßnahme; Halbheit, *die (abwertend);* ~ **'moon** *n.* Halbmond, *der;* ~-**note** *(Amer. Mus.) see* **minim;** ~ **'pay** *n.* Ruhegehalt, *das;* Pension, *die;* **be on ~ pay** Ruhegehalt *od.* Pension beziehen; ~**penny** ['heɪpnɪ], *pl. usu.* ~**pennies** ['heɪpnɪz] *for separate coins,* ~**pence** ['heɪpəns] *for sum of money (Brit. Hist.) (coin)* Halfpenny, *der; (sum)* halber Penny; ~-**'pint** *n.* halbes Pint; ~-**'price 1.** *n.* halber Preis; **reduce sth. to ~-price** etw. um die Hälfte heruntersetzen; **2.** *adj.* zum halben Preis *nachgestellt;* **3.** *adv.* zum halben Preis; ~-**sister** *n.* Halbschwester, *die;* ~-**'term** *n. (Brit.)* **a) it is nearly ~-term** das Trimester ist fast zur Hälfte vorüber; **by/at ~-term** bis zur/in der Mitte des Trimesters; **b)** *(holiday)* ~-**term [holiday/break]** Ferien in der Mitte des Trimesters; **before ~-term** in der ersten Trimesterhälfte; ~-**timbered** [hɑ:f'tɪmbəd] *adj.* Fachwerk‹*haus, -bauweise*›; **be ~-timbered** ein Fachwerkbau sein; ~-**'time** *n. (Sport)* Halbzeit, *die; attrib.* ['--] Halbzeit‹*pfiff, -stand*›; **at ~-time** bei *od.* bis zur Halbzeit; *(during interval)* in der Halbzeitpause; ~-**tone** *n. (Amer. Mus.) see* **semitone;** ~-**truth** *n.* Halbwahrheit, *die;* ~-**volley** *n.* Halfvolley, *der;* ~-**'way 1.** *adj.* ~-**way point** Mitte, *die;* ~-**way house** *(compromise)* Kompromiß, *der;* Mittelweg, *der;* **2.** *adv.* die Hälfte des Weges ‹*begleiten, fahren*›; ~**wit** *n.* Schwachkopf, *der;* ~-**witted** ['hɑ:fwɪtɪd] *adj.* dumm; *(mentally deficient)* debil; schwachsinnig; ~-**'yearly 1.** *adj.* halbjährlich; **2.** *adv.* halbjährlich; jedes halbe Jahr

halibut ['hælɪbət] *n., pl. same (Zool.)* Heilbutt, *der*

halitosis [hælɪ'təʊsɪs] *n., pl.* **halitoses** [hælɪ'təʊsi:z] *(Med.)* Halitose, *die (fachspr.)*; schlechter Atem

hall [hɔ:l] *n.* **a)** *(large [public] room)* Saal, *der; (public building)* Halle, *die; (for receptions, banquets)* Festsaal, *der; (in medieval house: principal living-room)* Wohnsaal, *der;* **school/church ~:** Aula, *die*/Gemeindehaus, *das;* **b)** *(Univ.) (residential building)* **~ [of residence]** Studentenwohnheim, *das;* **live in ~:** im [Studenten]wohnheim wohnen; **c)** *(entrance-passage)* Diele, *die;* Flur, *der*

'**hallmark 1.** *n.* [Feingehalts]stempel, *der;* Repunze, *die; (fig.: distinctive mark)* Kennzeichen, *das;* **be the ~ of quality/perfection** *(fig.)* für Qualität/Vollkommenheit bürgen *od.* stehen. **2.** *v. t.* stempeln; repunzieren

hallo [hə'ləʊ] **1.** *int.* **a)** *(to call attention)* hallo; **b)** *(Brit.) see* **hello 1. 2.** *n., pl.* **~s** Hallo, *das*

hallow ['hæləʊ] *v. t.* heiligen; **~ed** geheiligt *(auch fig.);* heilig ‹*Boden*›

Hallowe'en [hæləʊ'i:n] *n.* Halloween, *das; Abend vor Allerheiligen;* **on** *or* **at ~:** [an] Halloween

hall: ~ 'porter *n. (Brit.)* [Hotel]portier, *der;* **~-stand** *n.* [Flur]garderobe, *die*

hallucinate [hə'lu:sɪneɪt] *v. i.* halluzinieren *(Med., Psych.);* Halluzinationen haben

hallucination [həlu:sɪ'neɪʃn] *n. (act)* Halluzinieren, *das; (instance, imagined object)* Halluzination, *die;* Sinnestäuschung, *die*

hallucinogenic [həlu:sɪnə'dʒenɪk] *adj. (Med.)* halluzinogen

'**hallway** *n.* **a)** *see* **hall c; b)** *(corridor)* Flur, *der;* Korridor, *der*

halo ['heɪləʊ] *n., pl.* **~es a)** *(Meteorol.)* Halo, *der (fachspr.);* Hof, *der;* **b)** *(around head)* Heiligen-, Glorienschein, *der*

halt [hɒlt, hɔ:lt] **1.** *n.* **a)** *(temporary stoppage)* Pause, *die; (on march or journey)* Rast, *die;* Pause, *die; (esp. Mil. also)* Halt, *der;* **make a ~:** Rast/eine Pause machen/haltmachen; **call a ~:** eine Pause machen lassen/haltmachen lassen; **let's call a ~:** machen wir eine Pause!; **b)** *(interruption)* Unterbrechung, *die;* **c)** *(Brit. Railw.)* Haltepunkt, *der.* **2.** *v. i.* **a)** *(stop)* ‹*Fußgänger, Tier:*› stehenbleiben; ‹*Fahrer:*› anhalten; *(for a rest)* eine Pause machen; *(esp. Mil.)* haltmachen; **~, who goes there?** *(Mil.)* halt, wer da?; **b)** *(end)* eingestellt werden. **3.** *v. t.* **a)** *(cause to stop)* anhalten; haltmachen lassen ‹*Marschkolonne usw.*›; **b)** *(cause to end)* stoppen ‹*Diskussion*›; einstellen ‹*Projekt*›

halter ['hɒltə(r), 'hɔ:ltə(r)] *n.* **a)** *(for horse)* Halfter, *das;* **b)** *(Dressmaking) (strap)* Nackenträger, *der;* **~ dress/top** Kleid/Oberteil *od.* Top mit Nackenträger

halting ['hɒltɪŋ, hɔ:ltɪŋ] *adj.* schleppend ‹*Stimme, Redeweise, Fortschritt*›; holprig ‹*Verse*›; zögernd ‹*Antwort*›

halve [hɑ:v] *v. t.* **a)** *(divide)* halbieren; **b)** *(reduce)* halbieren; auf *od.* um die Hälfte verringern

halves *pl. of* **half**

ham [hæm] **1.** *n.* **a)** *([meat from] thigh of pig)* Schinken, *der;* **b)** *(sl.) (amateur)* Amateur, *der; (poor actor)* Schmierenkomödiant, *der (abwertend).* **2.** *v. i.,* **-mm-** *(sl.)* überziehen. **3.** *v. t.,* **-mm-** *(sl.)* überzogen spielen

~ 'up *v. t. (sl.)* überzogen spielen ‹*Stück*›; **~ it up** überziehen

hamburger ['hæmbɜ:gə(r)] *n. (beef cake)* Hacksteak, *das; (filled roll)* Hamburger, *der*

ham: ~-fisted [hæm'fɪstɪd], **~-handed** [hæm'hændɪd] *adjs. (sl.)* tolpatschig *(ugs.)* ‹*Person, Art*›

hamlet ['hæmlɪt] *n.* Weiler, *der*

hammer ['hæmə(r)] **1.** *n.* **a)** Hammer, *der;* **go** *or* **be at sth. ~ and tongs** sich bei etw. schwer ins Zeug legen *(ugs.);* **go** *or* **be at it ~ and tongs** *(quarrel)* sich streiten, daß die Fetzen fliegen; **b)** *(of gun)* Hahn, *der;* **c)** *(Athletics)* [Wurf]hammer, *der;* **[throwing] the ~** *(event)* das Hammerwerfen. **2.** *v. t.* **a)** hämmern; *(fig.)* hämmern auf *(Akk.)* ‹*Tasten, Tisch*›; **~ a nail into sth.** einen Nagel in etw. *(Akk.)* hämmern *od.* schlagen; **~ sth. into sb.['s head]** *(fig.)* jmdm. etw. einhämmern; **b)** *(coll.: inflict heavy defeat on)* abservieren *(ugs.)* ‹*Gegner*›; vernichtend schlagen ‹*Feind*›. **3.** *v. i.* hämmern; klopfen; **~ at sth.** an etw. *(Dat.)* [herum]hämmern

~ a'way *v. i.* hämmern; **~ away at** herumhämmern auf (+ *Dat.*)

~ 'out *v. t.* **a)** *(make smooth)* ausklopfen ‹*Delle, Beule*›; ausbeulen ‹*Kotflügel usw.*›; glatt klopfen ‹*Blech usw.*›; **b)** *(fig.: devise)* ausarbeiten ‹*Plan, Methode, Vereinbarung*›

hammock ['hæmək] *n.* Hängematte, *die*

¹hamper ['hæmpə(r)] *n.* **a)** *(basket)* [Deckel]korb, *der;* **b)** *(consignment of food)* Präsentkorb, *der*

²hamper *v. t.* behindern; hemmen ‹*Entwicklung, Wachstum usw.*›

hamster ['hæmstə(r)] *n.* Hamster, *der; see also* **golden hamster**

'**hamstring 1.** *n. (Anat.)* Kniesehne, *die.* **2.** *v. t.,* **hamstrung** *or* **~ed** *(fig.)* lähmen

hand [hænd] **1.** *n.* **a)** *(Anat., Zool.)* Hand, *die;* **eat from** *or* **out of sb.'s ~** *(lit. or fig.)* jmdm. aus der Hand fressen; **get one's ~s dirty** *(lit. or fig.)* sich *(Dat.)* die Hände schmutzig machen; **give sb. one's ~** *(reach, shake)* jmdm. die Hand geben *od.* reichen; **give** *or* **lend [sb.] a ~ [with** *or* **in sth.]** [jmdm.] [bei etw.] helfen; **pass** *or* **go through sb.'s ~s** *(fig.)* durch jmds. Hand *od.* Hände gehen; **~ in ~:** Hand in Hand; **go ~ in ~ [with sth.]** *(fig.)* [mit etw.] Hand in Hand gehen; **the problem/matter in ~:** das vorliegende Problem/die vorliegende Angelegenheit; **hold ~s** Händchen halten *(ugs. scherzh.);* sich bei den Händen halten; **hold sb.'s ~:** jmds. Hand halten; jmdm. die Hand halten; *(fig.: give sb. close guidance)* jmdn. bei der Hand nehmen; *(fig.: give sb. moral support or backing)* jmdm. das Händchen halten *(iron.);* **~s off!** Hände *od.* Finger weg!; **take/keep one's ~s off sb./sth.** jmdn./etw. loslassen/nicht anfassen; **keep one's ~s off sth.** *(fig.)* die Finger von etw. lassen *(ugs.);* **~s up [all those in favour]** wer dafür ist, hebt die Hand!; **~s up!** *(as sign of surrender)* Hände hoch!; **~s down** *(fig.) (easily)* mit links *(ugs.); (without a doubt, by a large margin)* ganz klar *(ugs.);* **turn one's ~ to sth.** sich einer Sache *(Dat.)* zuwenden; **have sth. at ~:** etw. zur Hand haben; **have sb. at ~:** jmdn. bei sich haben; **be at ~** *(be nearby)* in der Nähe sein; *(be about to happen)* unmittelbar bevorstehen; **out of ~** *(summarily)* kurzerhand; **be to ~** *(be readily available, within reach)* zur Hand sein; *(be received)* ‹*Brief, Notiz, Anweisung:*› vorliegen; **fight ~ to ~:** Mann gegen Mann kämpfen; **go/pass from ~ to ~:** von Hand zu Hand gehen; **~ live from ~ to mouth** von der Hand in den Mund leben; **be ~ in glove [with]** unter einer Decke stecken [mit]; **wait on sb. ~ and foot** *(fig.)* jmdn. vorn und hinten bedienen *(ugs.);*

have one's ~s full die Hände voll haben; *(fig.: be fully occupied)* alle Hände voll zu tun haben *(ugs.)*; ~ **on heart** *(fig.)* Hand aufs Herz; **get one's ~s on sb./sth.** jmdn. erwischen *od. (ugs.)* in die Finger kriegen/etw. auftreiben; **lay** *or* **put one's ~ on sth.** etw. finden; **by** ~ *(manually)* mit der *od.* von Hand; *(in handwriting)* handschriftlich; *(by messenger)* durch Boten; **be made by ~:** Handarbeit sein; **b)** *(fig.: authority)* **with a firm/iron ~:** mit starker Hand/eiserner Faust ‹*regieren*›; **he needs a father's ~:** er braucht die väterliche Hand; **get out of ~:** außer Kontrolle geraten; *see also* **take 1f; upper 1a; c)** *in pl. (custody)* **in sb.'s ~s, in the ~s of sb.** *(in sb.'s possession)* in jmds. Besitz; *(in sb.'s care)* in jmds. Obhut; **may I leave the matter in your ~s?** darf ich die Angelegenheit Ihnen überlassen?; **fall into sb.'s ~s** ‹*Person, Geld:*› jmdm. in die Hände fallen; **have [got] sth./sb. on one's ~s** sich um etw./jmdn. kümmern müssen; **he's got such a lot/enough on his ~s at the moment** er hat augenblicklich so viel/genug um die Ohren *(ugs.)*; **have time on one's ~s** [viel] Zeit haben; *(too much)* mit seiner Zeit nichts anzufangen wissen; **take sb./sth. off sb.'s ~s** jmdm. jmdn./etw. abnehmen; **change ~s** den Besitzer wechseln; **d)** *(disposal)* **have sth. in ~:** etw. zur Verfügung haben; *(not used up)* etw. [übrig] haben; **keep in ~:** in Reserve halten ‹*Geld*›; **have on ~:** dahaben; **be on ~:** dasein; **e)** *(share)* **have a ~ in sth.** bei etw. seine Hände im Spiel haben; **take a ~ [in sth.]** sich [an etw. *(Dat.)*] beteiligen; **f)** *(agency)* Wirken, *das (geh.)*; **the ~ of a craftsman has been at work here** hier war ein Handwerker am Werk; **the ~ of God** die Hand Gottes; **suffer/suffer injustice at the ~s of sb.** unter jmdm./jmds. Ungerechtigkeit zu leiden haben; **g)** *(pledge of marriage)* **ask for** *or* **seek sb.'s ~ [in marriage]** um jmds. Hand bitten *od. (geh.)* anhalten; **h)** *(worker)* Arbeitskraft, *die;* Arbeiter, *der; (Naut.: seaman)* Hand, *die (fachspr.)*; Matrose, *der;* **the ship sank with all ~s** das Schiff sank mit der gesamten Mannschaft; **i)** *(person having ability)* **be a good/poor ~ at tennis** ein guter/schwacher Tennisspieler sein; **I'm no ~ at painting** ich kann nicht malen; **j)** *(source)* Quelle, *die;* **at first/second/third ~:** aus erster/zweiter/dritter Hand; *see also* **firsthand; secondhand; k)** *(skill)* Geschick, *das;* **get one's ~ in** wieder in Übung kommen *od. (ugs.)* reinkommen; **keep one's ~ in** in der Übung bleiben; **l)** *(style of ~writing)* Handschrift, *die; (signature)* Unterschrift, *die;* **m)** *(of clock or watch)* Zeiger, *der;* **n)** *(side)* Seite, *die;* **on the right/left ~:** rechts/links; rechter/linker Hand; **on sb.'s right/left ~:** rechts/links von jmdm.; zu jmds. Rechten/Linken; **on every ~** von allen Seiten ‹*umringt sein*›; ringsum ‹*etw. sehen*›; **on the one ~ ..., [but] on the other [~] ...:** einerseits ..., andererseits ...; auf der einen Seite ..., auf der anderen Seite ...; **o)** *(measurement)* Handbreit, *die;* **p)** *(coll.: applause)* Beifall, *der;* Applaus, *der;* **give him a big ~, let's have a big ~ for him** viel Applaus *od.* Beifall für ihn!; **q)** *(cards)* Karte, *die; (period of play)* Runde, *die; see also* **throw in d; 2.** *v.t.* geben; ‹*Überbringer:*› übergeben ‹*Sendung, Lieferung*›; ~ **sth. from one to another** etw. von einem zum anderen weitergeben; ~ **sth. [a]round** *(pass round, circulate)* etw. herumgeben; *(among group)* etw. herumgehen lassen; **you've got to ~ it to them/her** *etc. (fig. coll.)* das muß man ihnen/ihr *usw.* lassen

~ **'back** *v.t. (return)* zurückgeben

~ **'down** *v.t.* **a)** *(pass on)* überliefern ‹*Geschichte, Tradition*›; weitergeben ‹*Gegenstand*› **(to** an + *Akk.*); [weiter]vererben ‹*Erbstück*› **(to** an + *Akk.*); **b)** *(Law)* verhängen ‹*Strafe*› **(to** über + *Akk.*); fällen ‹*Entscheidung*›; verkünden ‹*Urteil*›

~ **'in** *v.t.* abgeben ‹*Klausur, Arbeit, Aufsatz*› **(to, at** bei); einreichen ‹*Petition, Bewerbung*› **(to, at** bei)

~ **'on** *v.t.* weitergeben **(to** an + *Akk.*)

~ **'out** *v.t.* aus-, verteilen **(to** an + *Akk.*, **among** unter + *Dat.*); geben ‹*Ratschläge, Tips, Winke*› **(to** an + *Akk.*)

~ **'over 1.** *v.t.* **a)** *(deliver)* übergeben **(to** *Dat.*); freilassen ‹*Geisel*›; ~ **over your guns/money!** Waffen/Geld her!; **b)** *(transfer)* übergeben *od.* -reichen **(to** *Dat.*); *(pass)* herüber- *od.* rübergeben *od.* -reichen **(to** *Dat.*); *(allow to have)* abgeben. **2.** *v.i. (to next speaker/one's successor)* das Wort/die Arbeit übergeben **(to** an + *Akk.*)

hand- *in comb.* **a)** *(operated by hand, held in the hand)* Hand-; **b)** *(done by hand)* hand‹*gestickt*›; mit der Hand *od.* von Hand ‹*glasiert, verziert, gebacken*›

hand: ~bag *n.* Handtasche, *die;* **~-baggage** *n.* Handgepäck, *das;* **~bill** *n.* Handzettel, *der;* **~book** *n.* Handbuch, *das; (guidebook)* Führer, *der;* **~-brake** *n.* Handbremse, *die;* **~cart** *n.* Handwagen, *der;* **~clap** *n.* **a)** *(single clap)* In-die-Hände-Klatschen, *das;* **give three ~claps** dreimal in die Hände klatschen; **b)** *(applause)* [Hände]klatschen, *das;* ~ **cream** *n.* Handcreme, *die;* **~cuff 1.** *n., usu. in pl.* Handschelle, *die;* **2.** *v.t.* in Handschellen *(Akk.)* legen ‹*Hände*›; **~cuff sb.** jmdm. Handschellen anlegen

handful ['hændfʊl] *n.* **a)** Handvoll, *die;* **a few ~s of nuts** ein paar Handvoll Nüsse; **b)** *(fig. coll.: troublesome person[s] or thing[s])* **these children are/this dog is a real ~:** die Kinder halten/der Hund hält einen ständig auf Trab *(ugs.)*

hand: ~grenade *n.* Handgranate, *die;* **~-gun** *n.* Faustfeuerwaffe, *die;* **~hold** *n.* Halt, *der;* **provide ~holds/a ~hold for sb.** jmdm. Halt bieten

handicap ['hændɪkæp] **1.** *n.* **a)** *(Sport: advantage)* Handikap, *das (fachspr.)*; Vorgabe, *die;* **b)** *(race, competition)* Handikaprennen, *das;* Ausgleichsrennen, *das;* **c)** *(fig.: hindrance)* Handikap, *das;* **have a mental/physical ~:** geistig behindert/körperbehindert sein. **2.** *v.t.,* **-pp-: a)** *(Sport: impose a ~ on)* ein Handikap festlegen für; **b)** *(fig.: put at a disadvantage)* benachteiligen; handikapen *(ugs.)*

handicapped ['hændɪkæpt] **1.** *adj.* behindert; **mentally/physically ~:** geistig behindert/körperbehindert. **2.** *n. pl.* **the [mentally/physically] ~:** die [geistig/körperlich] Behinderten

handicraft ['hændɪkrɑːft] *n.* [Kunst]handwerk, *das; (knitting, weaving, needlework)* Handarbeit, *die*

handily ['hændɪlɪ] *adv.* praktisch; günstig ‹*gelegen*›

handiwork ['hændɪwɜːk] *n., no pl., no indef. art.* **a)** *(working)* handwerkliche Arbeit; **b) this ring is all my own ~:** diesen Ring habe ich selbst gemacht; **whose ~ is this?** *(derog.)* wer hat das [denn] verbrochen *(ugs.)*

handkerchief ['hæŋkətʃɪf, 'hæŋkətʃiːf] *n., pl.* **~s** *or* **hand-**

kerchieves ['hæŋkətʃi:vz] Taschentuch, *das*

handle ['hændl] **1.** *n.* **a)** *(part held)* [Hand]griff, *der; (of bag etc.)* [Trag]griff, *der; (of knife, chisel)* Heft, *das;* Griff, *der; (of axe, brush, comb, broom, saucepan)* Stiel, *der; (of handbag)* Bügel, *der; (of door)* Klinke, *die; (of bucket, watering-can, cup, jug)* Henkel, *der; (of pump)* Schwengel, *der;* **fly off the ~** *(fig. coll.)* an die Decke gehen *(ugs.);* **b)** *(coll.: title)* Titel, *der.* **2.** *v. t.* **a)** *(touch, feel)* anfassen; **'Fragile! H~ with care!'** „Vorsicht! Zerbrechlich!"; **b)** *(deal with)* umgehen mit ‹*Person, Tier, Situation*›; führen ‹*Verhandlung*›; erledigen ‹*Korrespondenz, Telefonat usw.*›; *(cope with)* fertigwerden *od.* zurechtkommen mit ‹*Person, Tier, Situation*›; **c)** *(control)* handhaben ‹*Fahrzeug, Flugzeug*›; **d)** *(process, transport)* umschlagen ‹*Fracht*›; **Heathrow ~s x passengers per year** in Heathrow werden pro Jahr x Passagiere abgefertigt

handlebar ['hændlbɑ:(r)] *n.* Lenkstange, *die;* Lenker, *der;* **~ moustache** Schnauzbart, *der*

handler ['hændlə(r)] *n.* **a)** *(of police-dog)* Hundeführer, *der*/-führerin, *die;* **b) a ~ of stolen goods** ein Hehler

handling ['hændlɪŋ] *n., no pl.* **a)** *(management)* Handhabung, *die; (of troops, workforce, bargaining, discussion)* Führung, *die; (of situation, class, crowd)* Umgang, *der* **(of** mit**)**; **b)** *(use)* Handhabung, *die; (Motor Veh.)* Fahrverhalten, *das;* Handling, *das;* **c)** *(treatment)* Behandlung, *die;* **the child needs firm ~:** das Kind braucht eine feste Hand; **d)** *(processing)* Beförderung, *die; (of passengers)* Abfertigung, *die*

hand: ~luggage *n.* Handgepäck, *das;* **~made** *adj.* handgearbeitet; handgeschöpft ‹*Papier*›; **~-me-down** *n.* abgelegtes *od.* gebrauchtes Kleidungsstück **(from** *Gen.***)**; **~-out** *n.* **a)** *(alms)* Almosen, *das;* Gabe, *die;* **b)** *(information)* Handout, *das; (press release)* Presseerklärung, *die;* **~-painted** *adj.* handbemalt ‹*Gegenstand*›; handgemalt ‹*Muster, Bild*›; **~-'picked** *adj.* sorgfältig ausgewählt; handverlesen *(ugs. scherzh.);* **~rail** *n.* Geländer, *das;* Handlauf, *der (Bauw.); (on ship)* Handläufer, *der;* **~set** *n. (Teleph.)* Handapparat, *der;* **~shake** *n.* Händedruck, *der;* Handschlag, *der*

handsome ['hænsəm] *adj.*, **~r** ['hænsəmə(r)], **~st** ['hænsəmɪst] **a)** *(good-looking)* gutaussehend ‹*Mann, Frau*›; schön, edel ‹*Tier, Möbel*›; **b)** *(generous)* großzügig ‹*Geschenk, Belohnung, Mitgift*›; nobel ‹*Behandlung, Verhalten, Empfang*›; *(considerable)* stattlich, ansehnlich ‹*Vermögen, Summe, Preis*›

handsomely ['hænsəmlɪ] *adv.* großzügig; mit großem Vorsprung ‹*gewinnen*›

hand: ~spring *n.* Handstandüberschlag, *der;* **~stand** *n.* Handstand, *der;* **~-to-~** *adj.* **~-to-~ combat** ein Kampf Mann gegen Mann; **~-to-mouth** *adj.* kärglich, kümmerlich ‹*Leben, Dasein*›; **eke out/lead a ~-to-mouth life/existence** von der Hand in den Mund leben; **~-towel** *n.* [Hände]handtuch, *das;* **~writing** *n.* [Hand]schrift, *die;* **~-'written** *adj.* handgeschrieben

handy ['hændɪ] *adj.* **a)** *(ready to hand)* griffbereit; **keep/have sth. ~:** etw. griffbereit haben; **the house is very ~ for the town centre** von dem Haus aus ist man sehr schnell in der Stadt; **b)** *(useful)* praktisch; nützlich; **come in ~:** sich als nützlich erweisen; **that'll come in ~!** das kann ich gebrauchen!; **c)** *(adroit)* geschickt; **be [quite/very] ~ with sth.** [ganz gut/sehr gut] mit etw. umgehen können

'handyman *n.* Handwerker, *der;* **[home]~:** Heimwerker, *der*

hang [hæŋ] **1.** *v. t.*, **hung** [hʌŋ] *(see also* **e)**: **a)** *(support from above)* hängen; aufhängen ‹*Gardinen*›; **~ sth. from sth.** etw. an etw. *(Dat.)* aufhängen; **b)** *(place on wall)* aufhängen ‹*Bild, Gemälde, Zeichnung*›; **c)** *(paste up)* ankleben ‹*Tapete*›; **d)** *(Cookery)* abhängen lassen ‹*Fleisch, Wild*›; **e)** *p. t., p.p.* **hanged** *(execute)* hängen, *(ugs.)* aufhängen **(for** wegen**)**; **~ oneself** sich erhängen *od. (ugs.)* aufhängen; **I'll be** *or* **I am ~ed if ...** *(fig.)* der Henker soll mich holen, wenn ...; **~ the expense!** die Kosten interessieren mich nicht; **f)** *(let droop)* **~ one's head in shame** beschämt den Kopf senken. **2.** *v. i.*, **hung a)** *(be supported from above)* hängen; ‹*Kleid usw.:*› fallen; **~ from the ceiling** an der Decke hängen; **~ by a rope** an einem Strick hängen; **~ in there!** *(sl.)* halte durch!; **time ~s heavily** *or* **heavy on sb.** die Zeit wird jmdm. lang; **b)** *(be executed)* hängen; **c)** *(droop)* **the dog's ears and tail hung [down]** der Hund ließ die Ohren und den Schwanz hängen; **his head hung** er hielt den Kopf gesenkt. **3.** *n., no pl.* **get the ~ of** *(fig. coll.: get the knack of, understand)* klarkommen mit *(ugs.)* ‹*Gerät, Arbeit*›; **you'll soon get the ~ of it/doing it** du wirst den Bogen bald raushaben *(ugs.)*/wirst bald raushaben, wie man es macht

~ about *(Brit.)*, **~ around 1.** [--'-] *v. i. (loiter about)* herumlungern *(salopp);* **we ~ about** *or* **around there all evening** wir hängen da den ganzen Abend rum *(ugs.).* **2.** ['---] *v. t.* herumlungern an/in/usw. (+ *Dat.*) *(salopp)*

~ 'back *v. i.* **a)** *(be reluctant)* sich zieren; **b)** *(keep to the rear)* zurückbleiben

~ on 1. [-'-] *v. i.* **a)** *(hold fast)* sich festhalten; **~ on to** *(lit.: grasp)* sich festhalten an (+ *Dat.*) ‹*Gegenstand*›; *(fig. coll.: retain)* behalten ‹*Eigentum, Stellung*›; **b)** *(stand firm, survive)* durchhalten; **c)** *(sl.: wait)* warten; **~ on [a minute]!** Moment *od. (ugs.)* Sekunde mal!; **d)** *(coll.: not ring off)* dranbleiben *(ugs.).* **2.** ['--] *v. t.* **~ on sth.** *(fig.)* von etw. abhängen; **~ on sb.'s words** jmdm. gespannt zuhören

~ 'out 1. *v. t.* **a)** aufhängen ‹*Wäsche*›; **b)** heraushängen lassen ‹*Zunge, Tentakel*›. **2.** *v. i.* **a)** *(protrude)* heraushängen; **let it all ~ out** *(fig. sl.)* die Sau rauslassen *(ugs.);* **b)** *(sl.) (reside)* wohnen; seine Bude haben *(ugs.); (be often present)* sich herumtreiben *(ugs.)*

~ to'gether *v. i.* **a)** *(be coherent)* ‹*Teile eines Ganzen:*› sich zusammenfügen; ‹*Aussagen:*› zusammenstimmen; **b)** *(be or remain associated)* zusammenhalten

~ 'up 1. *v. t.* **a)** aufhängen; **~ up sth. on a hook** etw. an einen Haken hängen; **b)** *(sl.)* **be hung up about sth.** ein gestörtes Verhältnis zu etw. haben. **2.** *v. i. (Teleph.)* einhängen; auflegen; **~ up on sb.** einfach einhängen *od.* auflegen

hangar ['hæŋə(r), 'hæŋgə(r)] *n.* Hangar, *der;* Flugzeughalle, *die*

'hangdog *adj.* zerknirscht

hanger ['hæŋə(r)] *n.* **a)** *(for clothes)* Bügel, *der;* **b)** *(loop on clothes etc.)* Aufhänger, *der*

hanger-'on *n.* **there are many hangers-on in every political party** in jeder politischen Partei gibt es viele, denen es nur um den persönlichen Vorteil geht; **the rock group with its usual [crowd of]**

hangers-on die Rockgruppe mit ihrem üblichen Anhang

hang: ~-glider *n.* Hängegleiter, *der;* Drachen, *der;* **~-glider pilot** Drachenflieger, *der*/-fliegerin, *die;* **~-gliding** *n.* Drachenfliegen, *das*

hanging ['hæŋıŋ] **1.** *n.* **a)** *see* **hang 1**: [Auf]hängen, *das;* Ankleben, *das;* Abhängen, *das;* **b)** *(execution)* Hinrichtung [durch den Strang]; **c)** *in pl. (drapery)* Behang, *der.* **2.** *adj.* ~ **basket** Hängekorb, *der*

hang: ~man ['hæŋmən] *n., pl.* **~men** ['hæŋmən] Henker, *der;* **~over** *n.* **a)** *(after-effects)* Kater, *der (ugs.);* **b)** *(remainder)* Relikt, *das;* **~-up** *n. (sl.)* **a)** *(inhibition)* Macke, *die (ugs.);* **have a ~-up about sth.** ein gestörtes Verhältnis zu etw. haben; **b)** *(fixation)* Komplex, *der* (**about** wegen)

hank [hæŋk] *n.* Strang, *der*

hanker ['hæŋkə(r)] *v. i.* ~ **after** *or* **for** ein [heftiges] Verlangen haben nach ‹*Person, etwas Neuem, Zigarette*›; sich *(Dat.)* sehnlichst wünschen ‹*Gelegenheit*›

hankering ['hæŋkərıŋ] *n. (craving)* Verlangen, *das* (**after, for** nach); *(longing)* Sehnsucht, *die* (**after, for** nach)

hanky ['hæŋkı] *n. (coll.)* Taschentuch, *das*

hanky-panky [hæŋkı'pæŋkı] *n., no pl., no indef. art. (sl.)* Mauschelei, *die (abwertend);* **there's been some ~**: es ist gemauschelt worden *(ugs. abwertend)*

Hanover ['hænəʊvə(r)] *pr. n.* Hannover *(das)*

Hansard ['hænsɑ:d] *n.* Hansard, *der; die britischen Parlamentsberichte*

Hanseatic [hænsı'ætık] *adj. (Hist.)* hansisch; ~ **town** Hansestadt, *die;* **the ~ League** der Hansebund

haphazard [hæp'hæzəd] **1.** *adj.* willkürlich ‹*Auswahl*›; unbedacht ‹*Bemerkung*›; **the whole thing was rather ~**: das Ganze geschah ziemlich planlos. **2.** *adv. (at random)* willkürlich; wahllos

haphazardly [hæp'hæzədlı] *adv.* willkürlich; wahllos

happen ['hæpn] *v. i.* **a)** *(occur)* geschehen; ‹*Vorhergesagtes:*› eintreffen; **these things [do] ~**: das kommt vor; **what's ~ing?** was ist los?; **what's ~ing this evening?** was ist für heute abend geplant?; **I can't** *or* **don't see 'that ~ing** das kann ich mir nicht vorstellen; **nothing ever ~s here** hier ist nichts los; **don't let it ~ again!** daß mir das nicht wieder vorkommt!; **that's what ~s!** das kommt davon!; ~ **to sb.** jmdm. passieren; **what has ~ed to him/her arm?** was ist mit ihm/ihrem Arm?; **what can have ~ed to him?** was mag mit ihm los sein?; **it all ~ed so quickly that ...**: es ging alles so schnell, daß ...; **it's all ~ing** *(sl.)* es ist was los *(ugs.);* **b)** *(chance)* ~ **to do sth.** zufällig etw. tun; **it so ~s** *or* **as it ~s I have ...**: zufällig habe ich *od.* ich habe zufällig ...; **how does it ~ that ...?** wie kommt es, daß ...?; **do you ~ to know him?** kennen Sie ihn zufällig?

~ **[up]on** *v. t.* zufällig treffen ‹*Person*›; zufällig finden ‹*Arbeit, Gegenstand*›

happening ['hæpnıŋ] *n.* **a)** *usu. in pl. (event)* Ereignis, *das;* **a regrettable ~**: ein bedauerlicher Vorfall; **b)** *(improvised performance)* Happening, *das*

happily ['hæpılı] *adv.* **a)** glücklich ‹*lächeln*›; fröhlich, vergnügt ‹*spielen, lachen*›; **b)** *(gladly)* mit Vergnügen; **c)** *(aptly)* gut; treffend, passend ‹*ausdrücken, formulieren*›; **d)** *(fortunately)* glücklicherweise; zum Glück; **it ended ~**: es ging gut aus

happiness ['hæpınıs] *n., no pl. see* **happy a**: Glück, *das;* Heiterkeit, *die;* Zufriedenheit, *die*

happy ['hæpı] *adj.* **a)** *(joyful)* glücklich; heiter ‹*Bild, Veranlagung, Ton*›; *(contented)* zufrieden; *(causing joy)* erfreulich ‹*Gedanke, Erinnerung, Szene*›; froh ‹*Ereignis*›; glücklich ‹*Zeiten*›; **I'm not ~ with her work** ich bin mit ihrer Arbeit nicht zufrieden; **not be ~ about sth./doing sth.** nicht froh über etw. *(Akk.)* sein/etw. nicht gern tun; ~ **event** *(euphem.: birth)* freudiges Ereignis *(verhüll.);* **[strike] a ~ medium** den goldenen Mittelweg [wählen]; **b)** *(glad)* **be ~ to do sth.** etw. gern *od.* mit Vergnügen tun; **yes, I'd be ~ to** *(as reply to request)* ja, gern *od.* mit Vergnügen; **c)** *(lucky)* glücklich; **by a ~ chance/coincidence** durch einen glücklichen Zufall

happy: ~ 'ending *n.* Happy-End, *das;* **~-go-'lucky** *adj.* sorglos; unbekümmert

harangue [hə'ræŋ] **1.** *n.* Tirade, *die (abwertend).* **2.** *v. t.* eine Ansprache halten an (+ *Akk.*)

harass ['hærəs] *v. t.* schikanieren; **constantly ~ the enemy** den Feind nicht zur Ruhe kommen lassen; ~ **sb. into doing sth.** jmdm. so sehr zusetzen, daß er etw. tut

harassed ['hærəst] *adj.* geplagt (**with** von); gequält ‹*Blick, Ausdruck*›

harassment ['hærəsmənt] *n.* Schikanierung, *die;* **sexual ~**: [sexuelle] Belästigung

harbour (*Brit.; Amer.:* **harbor**) ['hɑ:bə(r)] **1.** *n.* Hafen, *der;* **in ~**: im Hafen. **2.** *v. t.* Unterschlupf gewähren (+ *Dat.*) ‹*Verbrecher, Flüchtling*›; *(fig.)* hegen *(geh.)* ‹*Groll, Verdacht*›

hard [hɑ:d] **1.** *adj.* **a)** hart; stark, heftig ‹*Regen*›; gesichert ‹*Beweis, Zahlen, Daten, Information*›; **drive a ~ bargain** hart verhandeln; **b)** *(difficult)* schwer; schwierig; **this is ~ to believe** das ist kaum zu glauben; **it is ~ to do sth.** es ist schwer, etw. zu tun; **make it ~ for sb. [to do sth.]** es jmdm. schwermachen[, etw. zu tun]; **[choose to] go about/do sth. the ~ way** es sich *(Dat.)* bei etw. unnötig schwermachen; **learn sth. the ~ way** etw. durch schlechte Erfahrungen lernen; **be ~ of hearing** schwerhörig sein; **be ~ going** ‹*Buch:*› sich schwer lesen; ‹*Arbeit:*› anstrengend sein; **play ~ to get** *(coll.)* so tun, als sei man nicht interessiert; **have a ~ time doing sth.** Schwierigkeiten haben, etw. zu tun; **it's a ~ life** *(joc.)* das Leben ist schwer; **it is [a bit] ~ on him** es ist [schon] schlimm für ihn; **~ luck** *(coll.)* Pech; **c)** *(strenuous)* hart; beschwerlich ‹*Reise*›; leidenschaftlich ‹*Spieler*›; **be a ~ drinker** viel trinken; **try one's ~est to do sth.** sich nach Kräften bemühen, etw. zu tun; **d)** *(vigorous)* heftig ‹*Angriff, Schlag*›; kräftig ‹*Schlag, Stoß, Tritt*›; *(severe)* streng ‹*Winter*›; **e)** *(unfeeling)* hart; **be ~ [up]on sb.** streng mit jmdm. sein; **take a ~ line [with sb. on sth.]** [in bezug auf etw. *(Akk.)*] eine harte Linie [gegenüber jmdm.] vertreten. **2.** *adv.* **a)** *(strenuously)* hart ‹*arbeiten, trainieren*›; fleißig ‹*lernen, studieren, üben*›; genau ‹*überlegen, beobachten*›; scharf ‹*nachdenken*›; gut ‹*aufpassen, zuhören, sich festhalten*›; **concentrate ~/~er** sich sehr/mehr konzentrieren; **try ~**: sich sehr bemühen; **be ~ at work on sth.** an etw. *(Dat.)* intensiv *od.* konzentriert arbeiten; **be ~ 'at it** schwer arbeiten; **b)** *(vigorously)* heftig; fest ‹*schlagen, drücken, klopfen*›; **c)** *(severely, drastically)* hart; streng ‹*zensieren*›; **cut back ~ on sth.** etw. drastisch einschränken; **be ~ up** knapp bei Kasse sein *(ugs.);* **d)**

be ~ put to it [to do sth.] große Schwierigkeiten haben[, etw. zu tun]; **e)** hart ‹*kochen*›; fest ‹*gefrieren [lassen]*›; **set ~:** fest werden

hard: ~ and 'fast *see* ²**fast 1 a; ~back** *(Printing)* **1.** *n.* gebundene Ausgabe; **in ~back** gebunden; mit festem Einband; **2.** *adj.* gebunden; **~bitten** *adj.* hartgesotten; abgebrüht *(ugs.)* ‹*Veteran, Journalist, Karrieremacher*›; **~board** *n.* Hartfaserplatte, *die;* **~-boiled** *adj.* **a)** *(boiled solid)* hartgekocht; **b)** *(fig.)* *(shrewd)* ausgekocht *(ugs.)*; *(tough)* hartgesotten; **~ 'cash** *n.* **a)** *(coins)* Hartgeld, *das;* **b)** *(actual money)* Bargeld, *das;* **in ~ cash** in bar ‹*bezahlen*›; **~ core** *n.* **a)** [-'-] *(nucleus)* harter Kern; *(of a problem)* Kern, *der;* **b)** ['--] *(Brit.: material)* Packlage, *die (Bauw.)*; **~-core** *attrib. adj.* hart ‹*Pornographie*›; zum harten Kern gehörend ‹*Terrorist*›; **~ 'court** *n.* *(Tennis)* Hartplatz, *der;* **~ 'currency** *n.* *(Econ.)* harte Währung; **~-drinking** *attrib. adj.* ‹*Mann/Frau,*› der/ die viel [Alkohol] trinkt; **~ drug** *n.* harte Droge; **~-earned** *adj.* schwer verdient

harden ['hɑ:dn] **1.** *v.t.* **a)** *(make hard)* härten; **b)** *(fig.: reinforce)* **~ sb.'s attitude/conviction** jmdn. in seiner Haltung/Überzeugung bestärken; **c)** *(make robust)* abhärten **(to** gegen); **d)** *(make tough)* unempfindlich machen **(to** gegen); **~ sb./oneself to sth.** jmdn./sich gegenüber etw. hart machen; **he ~ed his heart against her** er verhärtete sich gegen sie. **2.** *v.i.* **a)** *(become hard)* hart werden; **b)** *(become confirmed)* sich verhärten; **c)** *(become severe)* ‹*Gesicht:*› einen harten Ausdruck annehmen; ‹*Gesichtsausdruck:*› hart werden

hardened ['hɑ:dnd] *adj.* **a)** verhärtet ‹*Arterie*›; **b)** *(grown tough)* abgehärtet, unempfindlich **(to, against** gegen); hartgesotten ‹*Verbrecher, Krieger*›; **be/become ~ to sth.** gegen etw. unempfindlich sein/werden

hardening ['hɑ:dnɪŋ] *n.* **a)** *(of steel)* Härten, *das;* **b)** *(of arteries)* Verhärtung, *die*

hard: ~-featured *adj.* ‹*Person*› mit harten Gesichtszügen; **~ 'feelings** *n. pl. (coll.)* **no ~ feelings** schon gut; **~-fought** *adj.* heftig ‹*Kampf*›; hart ‹*Spiel*›; **~-headed** *adj.* sachlich; nüchtern; **~-hearted** [hɑ:d'hɑ:tɪd] *adj.* hartherzig **(towards** gegenüber); **~-'hitting** *adj.* schlagkräftig; *(fig.)* aggressiv ‹*Rede, Politik, Kritik*›; **~ 'labour** *n.* Zwangsarbeit, *die;* **~-line** *adj.* kompromißlos; **~-liner** *n.* Befürworter einer harten Linie **(on** gegenüber); **~-'luck story** *n.* Leidensgeschichte, *die*

hardly ['hɑ:dlɪ] *adv.* kaum; **he can ~ have arrived yet** er kann kaum jetzt schon angekommen sein; **~ anyone** *or* **anybody/anything** kaum jemand/etwas; **~ any wine/beds** kaum Wein/Betten; **~ ever** so gut wie nie; **~ at all** fast überhaupt nicht

hardness ['hɑ:dnɪs] *n., no pl.* Härte, *die;* *(of blow)* Heftigkeit, *die;* *(of person)* Strenge, *die*

hard: ~-nosed ['hɑ:dnəʊzd] *adj.* *(coll.)* abgebrüht; **~ 'pressed** *adj.* hart bedrängt; **be ~ pressed** große Schwierigkeiten haben; **~ sell** *n.* aggressive Verkaufsmethoden; *attrib.* aggressiv ‹*Werbung, [Verkaufs]methode*›

hardship ['hɑ:dʃɪp] *n.* **a)** *no pl., no indef. art.* Not, *die;* Elend, *das;* **b)** *(instance)* Notlage, *die;* **~s** Not, *die;* Entbehrungen; **c)** *(sth. causing suffering)* Unannehmlichkeit, *die*

hard: ~ 'shoulder *n.* Standspur, *die;* **~ware** *n., no pl., no indef. art.* **a)** *(goods)* Eisenwaren *Pl.*; *(for domestic use also)* Haushaltswaren *Pl.*; *attrib.* Eisen-/Haushaltswaren‹*geschäft*›; **b)** *(Computing)* Hardware, *die;* **~-wearing** *adj.* strapazierfähig; **~wood** *n.* Hartholz, *das; attrib.* Hartholz-; **~ 'words** *n. pl.* *(angry)* harte Worte; **~-working** *adj.* fleißig ‹*Person*›

hardy ['hɑ:dɪ] *adj.* **a)** *(robust)* abgehärtet; zäh, robust ‹*Rasse*›; **b)** *(Hort.)* winterhart

hardy: ~ 'annual *n.* *(Hort.)* winterharte einjährige Pflanze; **~ per'ennial** *n.* **a)** *(Hort.)* winterharte mehrjährige Pflanze; **b)** *(fig. joc.)* Dauerbrenner, *der (ugs.)*

hare [heə(r)] **1.** *n.* Hase, *der;* **[as] mad as a March ~** *(fig.)* völlig verrückt *(ugs.)*; **2.** *v.i.* sausen *(ugs.)*; **go haring about** herumsausen *(ugs.)*

hare: ~-brained ['heəbreɪnd] *adj.* unüberlegt; **~'lip** *n.* Hasenscharte, *die*

harem ['hɑ:ri:m] *n.* Harem, *der*

hark [hɑ:k] *v.i.* **a)** *(arch.: listen)* **~!** horch!/horcht!; **b)** *(coll.)* **just ~ at him!** hör ihn dir/hört ihn euch nur an!

~ 'back *v.i.* **~ back to** *(come back to)* zurückkommen auf (+ *Akk.*); zurückgreifen auf (+ *Akk.*) ‹*Tradition*›; wieder anfangen von ‹*alten Zeiten*›; *(go back to)* ‹*Idee, Brauch:*› zurückgehen auf (+ *Akk.*)

harm [hɑ:m] **1.** *n.* Schaden, *der;* **do ~:** Schaden anrichten; **do ~ to sb., do sb. ~:** jmdm. schaden; *(injure)* jmdn. verletzen; **it won't do you any ~** *(iron.)* es würde dir nichts schaden; **do ~ to sth.** einer Sache *(Dat.)* schaden; **sb./sth. comes to no ~:** jmdm./einer Sache passiert nichts; **there is no ~ done** nichts ist passiert; **there's no ~ in doing sth., it will do no ~ to do sth.** *(could be of benefit)* es kann nicht schaden, etw. zu tun; **there's no ~ in asking** Fragen kostet nichts; **it will do more ~ than good** es wird mehr schaden als nützen; **where's** *or* **what's the ~ in it?** was ist denn schon dabei?; **keep out of ~'s way** der Gefahr fernbleiben; von der Gefahr fernhalten ‹*Person*›. **2.** *v.t.* etwas [zuleide] tun (+ *Dat.*); schaden (+ *Dat.*) ‹*Beziehungen, Land, Karriere, Ruf*›

harmful ['hɑ:mfl] *adj.* schädlich **(to** für); schlecht ‹*Angewohnheit*›

harmless ['hɑ:mlɪs] *adj.* harmlos; **make** *or* **render ~:** unschädlich machen; entschärfen ‹*Bombe*›

harmonica [hɑ:'mɒnɪkə] *n.* *(Mus.)* Mundharmonika, *die*

harmonious [hɑ:'məʊnɪəs] *adj., adv.* harmonisch

harmonise *see* **harmonize**

harmonium [hɑ:'məʊnɪəm] *n.* *(Mus.)* Harmonium, *das*

harmonize ['hɑ:mənaɪz] **1.** *v.t.* **a)** *(bring into harmony)* aufeinander abstimmen; **b)** *(Mus.)* harmonisieren. **2.** *v.i.* *(be in harmony)* harmonieren **(with** mit); ‹*Interessen, Ansichten, Wort und Tat:*› miteinander im *od.* in Einklang stehen

harmony ['hɑ:mənɪ] *n.* **a)** Harmonie, *die;* **live in perfect ~:** völlig harmonisch *od.* in vollkommener Harmonie zusammenleben; **be in ~** *see* **harmonize 2; be in ~ with sth.** mit etw. im *od.* in Einklang stehen; **b)** *(Mus.)* Harmonie, *die;* **sing in ~:** mehrstimmig singen

harness ['hɑ:nɪs] **1.** *n.* **a)** Geschirr, *das;* **b)** *(on parachute)* Gurtzeug, *das;* *(for toddler, dog)* Laufgeschirr, *das;* *(for window-cleaner, steeplejack, etc.)* Sicherheitsgürtel, *der;* **die in ~:** in den Sielen sterben. **2.** *v.t.* **a)** *(put ~ on)* anschirren; **~ a horse to a cart** ein Pferd vor einen Wagen spannen; **b)** *(fig.)* nutzen

harp [hɑ:p] **1.** *n.* Harfe, *die.* **2.** *v.i.* **~ on [about] sth.** [immer wieder] von etw. reden; *(critically)* auf etw. *(Dat.)* herumreiten *(salopp);* **don't ~ on about it!** hör auf damit!

harpoon [hɑ:'pu:n] **1.** *n.* Harpune, *die.* **2.** *v.t.* harpunieren

harpsichord ['hɑ:psɪkɔ:d] *n. (Mus.)* Cembalo, *das*

harrow ['hærəʊ] *n.* Egge, *die*

harrowing ['hærəʊɪŋ] *adj.* entsetzlich; *(horrific)* grauenhaft ‹*Anblick, Geschichte*›

harry ['hærɪ] *v.t.* **a) ~ [continuously]** wiederholt angreifen; **b)** *(harass)* bedrängen

harsh [hɑ:ʃ] *adj.* **a)** rauh ‹*Gewebe, Oberfläche, Gegend, Land, Klima*›; schrill ‹*Ton, Stimme*›; grell ‹*Licht, Farbe*›; hart ‹*Bedingungen*›; **b)** *(excessively severe)* [sehr] hart; [äußerst] streng ‹*Richter, Disziplin*›; rücksichtslos ‹*Tyrann, Herrscher, Politik*›; **don't be ~ on him** sei nicht zu streng mit ihm

harvest ['hɑ:vɪst] **1.** *n.* Ernte, *die;* **find/reap a [rich] ~** *(fig.)* einen [tollen] Fang machen. **2.** *v.t.* ernten; lesen ‹*Weintrauben*›

harvester ['hɑ:vɪstə(r)] *n.* **a)** *(machine)* Erntemaschine, *die; see also* **combine 3b; b)** *(person)* Erntearbeiter, *der/*-arbeiterin, *die*

harvest: ~ 'festival *n.* Erntedankfest, *das;* **~ 'home** *n.* Erntefest, *das*

has *see* **have 1, 2**

has-been ['hæzbi:n] *n. (coll.)* **be [a bit of] a ~:** seine besten Jahre hinter sich *(Dat.)* haben

¹hash [hæʃ] *n. (Cookery)* Haschee, *das;* **make a ~ of sth.** *(coll.)* etw. verpfuschen *(ugs.)*

~ 'up *v.t. (coll.)* verpfuschen *(ugs.)*

²hash *n. (coll.: drug)* Hasch, *das (ugs.)*

hashish ['hæʃɪʃ] *n.* Haschisch, *das*

hasn't ['hæznt] = **has not;** *see* **have 1, 2**

hasp [hɑ:sp] *n.* Haspe, *die; (fastener snapping into a lock)* [Schnapp]schloß, *das*

hassle ['hæsl] *(coll.)* **1.** *n.* **~[s]** Krach, *der (ugs.); (trouble, problem)* Ärger, *der;* **it's a real ~:** das ist ein echtes Problem; **it's too much [of a]/such a ~:** das macht zuviel/soviel Umstände. **2.** *v.t.* schikanieren

hassock ['hæsək] *n. (cushion)* Kniekissen, *das*

haste [heɪst] *n., no pl.* Eile, *die; (rush)* Hast, *die;* **in his ~:** in seiner Hast; **more ~, less speed** *(prov.)* eile mit Weile *(Spr.);* **make ~:** sich beeilen

hasten ['heɪsn] **1.** *v.t. (cause to hurry)* drängen; *(accelerate)* beschleunigen. **2.** *v.i.* eilen; **~ to do sth.** sich beeilen, etw. zu tun; **I ~ to add/say** ich muß *od.* möchte gleich hinzufügen/sagen

hastily ['heɪstɪlɪ] *adv. (hurriedly)* eilig; *(precipitately)* hastig; *(rashly)* übereilt; *(quick-temperedly)* hitzig; **judge sb. too ~:** jmdn. vorschnell beurteilen

hasty ['heɪstɪ] *adj. (hurried)* eilig; flüchtig ‹*Skizze, Blick*›; *(precipitate)* hastig; *(rash)* übereilt; *(quick-tempered)* hitzig; **beat a ~ retreat** sich schnellstens zurückziehen *od. (ugs.)* aus dem Staub machen

hat [hæt] *n.* **a)** Hut, *der;* **[sailor's/woollen/knitted] ~:** [Matrosen- / Woll- / Strick]mütze, *die;* **raise one's ~ to sb.** vor jmdm. den Hut ziehen; **take one's ~ off to sb./sth.** *(lit. or fig.)* vor jmdm./etw. den Hut ziehen; **b)** *(fig.)* **at the drop of a ~:** auf der Stelle; **sb. will eat his ~ if ...:** jmd. frißt einen Besen, wenn ... *(salopp);* **be old ~** *(coll.)* ein alter Hut sein *(ugs.);* **produce sth. out of a ~:** etw. aus dem Ärmel schütteln; **pass the ~ round** *(coll.)* den Hut herumgehen lassen; **keep sth. under one's ~:** etw. für sich behalten; **[when he is] wearing his ... ~:** in seiner Rolle als ...

¹hatch [hætʃ] *n.* **a)** *(opening)* Luke, *die;* **down the ~!** *(fig. sl.)* runter damit! *(ugs.);* **b)** *(serving-~)* Durchreiche, *die*

²hatch 1. *v.t. (lit. or fig.)* ausbrüten. **2.** *v.i.* [aus]schlüpfen

~ 'out 1. *v.i.* ausschlüpfen; **the eggs have ~ed out** die Eier sind ausgebrütet. **2.** *v.t.* ausbrüten

'hatchback *n.* **a)** *(door)* Heckklappe, *die;* **b)** *(vehicle)* Schräghecklimousine, *die*

hatchet ['hætʃɪt] *n.* Beil, *das;* **bury the ~** *(fig.)* das Kriegsbeil begraben

hatchet: ~ job *n.* **do a ~ job on sb./sth.** jmdn./etw. in der Luft zerreißen *(salopp);* **~ man a)** *(professional killer)* Killer, *der;* **b)** *(henchman)* Erfüllungsgehilfe, *der (fig. abwertend)*

hate [heɪt] **1.** *n.* **a)** Haß, *der;* **~ for sb.** Haß auf *od.* gegen jmdn.; **b)** *(coll.: object of dislike)* **be sb.'s ~:** jmdm. verhaßt sein; **my pet ~ is ...:** ... hasse ich am meisten. **2.** *v.t.* hassen; **I ~ having to get up at seven** ich hasse es, um sieben Uhr aufstehen zu müssen; **I ~ to say this** *(coll.)* ich sage das nicht gern; **I ~ to think what would have happened if ...** *(coll.)* ich darf gar nicht daran denken, was geschehen wäre, wenn ...

hateful ['heɪtfl] *adj.* abscheulich

hatred ['heɪtrɪd] *n.* Haß, *der;* **feel ~ for** *or* **of sb./sth.** Haß auf *od.* gegen jmdn./etw. empfinden

'hat-stand *n.* Hutständer, *der*

hatter ['hætə(r)] *n.* Hutmacher, *der;* **[as] mad as a ~** *(fig.)* völlig verrückt *(ugs.)*

'hat trick *n.* Hattrick, *der*

haughty ['hɔ:tɪ] *adj.* hochmütig

haul [hɔ:l] **1.** *v.t.* **a)** *(pull)* ziehen; schleppen; **~ down** einholen ‹*Flagge, Segel*›; **b)** *(transport)* transportieren; befördern. **2.** *v.i.* ziehen. **3.** *n.* **a)** Ziehen, *das;* Schleppen, *das;* **b)** *(catch)* Fang, *der; (fig.)* Beute, *die*

haulage ['hɔ:lɪdʒ] *n., no pl.* **a)** *(hauling)* Transport, *der;* **b)** *(charges)* Transportkosten *Pl.*

haunch [hɔ:ntʃ] *n.* **a) sit on one's/its ~es** auf seinem Hinterteil sitzen; **b)** *(Gastr.)* Keule, *die*

haunt [hɔ:nt] **1.** *v.t.* **a) ~ a house/castle** in einem Haus/Schloß spuken *od.* umgehen; **b)** *(fig.: trouble)* ‹*Erinnerung, Gedanke:*› plagen, verfolgen. **2.** *n.* **a favourite ~ of artists** ein beliebter Treffpunkt für Künstler

haunted ['hɔ:ntɪd] *adj.* **a) a ~ house** ein Haus, in dem es spukt; **b)** *(fig.: troubled)* gehetzt ‹*Blick, Eindruck*›

haunting ['hɔ:ntɪŋ] *adj.* sehnsüchtig ‹*Klänge, Musik*›; lastend ‹*Erinnerung*›

have 1. [hæv] *v.t., pres.* **he has** [hæz], *p.t. & p.p.* **had** [hæd] **a)** *(possess)* haben; **I ~ it!** ich hab's[!]; **and what ~ you** *(coll.)* und so weiter; **b)** *(obtain)* bekommen; **let's not ~ any ...:** laß uns ... vermeiden; **come on, let's ~ it!** *(coll.)* rück schon raus damit! *(ugs.);* **c)** *(take)* nehmen; **d)** *(keep)* behalten; haben; **e)** *(eat, drink, etc.)* **~ breakfast/dinner/lunch** frühstücken/zu Abend/zu Mittag essen; **f)** *(experience)* haben ‹*Spaß, Vergnügen*›; **g)** *(suffer)* haben ‹*Krankheit, Schmerz, Enttäuschung, Abenteuer*›; *(show)* haben ‹*Güte, Freundlichkeit, Frechheit*›; **h)** *(engage in)* **~ a game of football** Fußball spielen; **i)** *(accept)* **I won't ~ it** das lasse ich mir nicht bieten; **j)** *(give birth to)* bekommen; **k)** *(coll.: swindle)* **I was had** ich bin [he]reingelegt worden *(ugs.);* **ever been had!** da bist du ganz schön reingefallen *(ugs.)*; **l)** *(know)* **I ~ it on good authority that ...:** ich weiß es aus zuverlässiger Quelle, daß ...; **m)** *(as guest)* **~ sb. to stay** jmdn. zu Besuch haben; **n)** *(summon)* **he had me into his office** er hat mich in sein Büro beordert; **o)** *(in coll. phrases)* **you've had it now** *(coll.)* jetzt ist es aus *(ugs.)*; **this car/dress has had it** *(coll.)* dieser Wagen/dieses Kleid hat ausgedient. **2.** [həv, əv, *stressed* hæv] *v. aux.*, **he has** [həz, əz, *stressed* hæz], **had** [həd, əd, *stressed* hæd] **a)** *forming past tenses* **I ~/I had read** ich habe/hatte gelesen; **I ~/I had gone** ich bin/war gegangen; **having seen him** *(because)* weil ich ihn gesehen habe/hatte; *(after)* wenn ich ihn gesehen habe/nachdem ich ihn gesehen hatte; **if I had known ...:** wenn ich gewußt hätte ...; **b)** *(cause to be)* **~ sth. made/repaired** etw. machen/reparieren lassen; **~ the painters in** die Maler haben; **~ sb. do sth.** jmdn. etw. tun lassen; **~ a tooth extracted** sich *(Dat.)* einen Zahn ziehen lassen; **c) she had her purse stolen** man hat ihr das Portemonnaie gestohlen; **d)** *expr. obligation* **~ to** müssen; **I only ~ to do the washing-up** ich muß nur noch den Abwasch machen; **I ~ only to see him to feel annoyed** ich brauche ihn nur zu sehen, und ich ärgere mich; **he 'has to be guilty** er ist fraglos schuldig. **3.** *n.* **the ~s and the ~-nots** die Besitzenden und die Besitzlosen

~ off *v. t.* **a)** abmachen; **b) ~ it off [with sb.]** *(sl.)* es [mit jmdm.] treiben *(salopp)*

~ 'on *v. t.* **a)** *(wear)* tragen; **b)** *(Brit. coll: deceive)* **~ sb. on** jmdn. auf den Arm nehmen *(ugs.)*

~ 'out *v. t.* **a) ~ a tooth/one's tonsils out** sich *(Dat.)* einen Zahn ziehen lassen/sich *(Dat.)* die Mandeln herausnehmen lassen; **b)** *(discuss and settle)* **~ sth. out** sich über etw. *(Akk.)* offen [mit jmdm.] aussprechen; **~ it out with sb.** mit jmdm. offen sprechen

haven ['heɪvn] *n.* geschützte Anlegestelle; *(fig.)* Zufluchtsort, *der*

have-not *see* **have 3**

haven't ['hævnt] = **have not**; *see* **have 1, 2**

haversack ['hævəsæk] *n.* Brotbeutel, *der*

havoc ['hævək] *n., no pl.* **a)** *(devastation)* Verwüstungen; **cause** *or* **wreak ~:** Verwüstungen anrichten; **b)** *(confusion)* Chaos; **play ~ with sth.** etw. völlig durcheinanderbringen

Hawaii [hə'waɪɪ] *pr. n.* Hawaii *(das)*

Hawaiian [hə'waɪən] **1.** *adj.* hawaiisch. **2.** *n. (person)* Hawaiianer, *der*/Hawaiianerin, *die*

[1]**hawk** [hɔ:k] *n. (also Polit.)* Falke, *der;* **watch sb. like a ~:** jmdn. mit Argusaugen beobachten

[2]**hawk** *v. t. (peddle)* **~ sth.** *(at door)* mit etw. hausieren [gehen]; *(in street)* etw. [auf der Straße] verkaufen

hawker ['hɔ:kə(r)] *n.* Hausierer, *der*/Hausiererin, *die*

hawthorn ['hɔ:θɔ:n] *n. (Bot.) (white)* Weißdorn, *der; (red)* Rotdorn, *der*

hay [heɪ] *n., no pl.* Heu, *das;* **make ~ while the sun shines** *(prov.)* die Zeit nutzen

hay: ~ fever *n., no pl.* Heuschnupfen, *der;* **~making** *n., no pl.* Heuernte, *die;* **~rick, ~stack** *ns.* Heuschober, *der (südd.)*; Heudieme, *die (nordd.)*; *see also* **needle 1**

haywire ['heɪwaɪə(r)] *adj. (coll.)* **go ~** ⟨*Instrument:*⟩ verrückt spielen *(ugs.)*; ⟨*Plan:*⟩ über den Haufen geworfen werden *(ugs.)*

hazard ['hæzəd] **1.** *n.* Gefahr, *die; (on road)* Gefahrenstelle, *die.* **2.** *v. t.* **~ a guess** mit Raten probieren

hazardous ['hæzədəs] *adj. (dangerous)* gefährlich; *(risky)* riskant

haze [heɪz] *n.* Dunst[schleier], *der*

hazel ['heɪzl] **1.** *n. (Bot.)* Haselnußstrauch, *der.* **2.** *adj.* haselnußbraun

'hazel-nut *n.* Haselnuß, *die*

hazy ['heɪzɪ] *adj.* dunstig, diesig ⟨*Wetter, Tag[eszeit]*⟩; verschwommen, unscharf ⟨*Konturen*⟩; *(fig.)* vage

H-bomb ['eɪtʃbɒm] *n.* H-Bombe, *die*

[1]**he** [hɪ, *stressed* hi:] *pron.* er; *referring to personified things or animals which correspond to German feminines/neuters* sie/es; **it was he** *(formal)* er war es; **he who** wer; *see also* **him; himself; his**

[2]**he** [hi:] *int.* haha

head [hed] **1.** *n.* **a)** Kopf, *der;* Haupt, *das (geh.)*; **mind your ~!** Vorsicht, dein Kopf!; *(on sign)* Vorsicht – geringe Durchgangshöhe!; **~ first** mit dem Kopf zuerst/voran; **~ over heels** kopfüber; **~ over heels in love** bis über beide Ohren verliebt *(ugs.)*; **keep one's ~:** einen klaren Kopf behalten; **lose one's ~** *(fig.)* den Kopf verlieren; **be unable to make ~ or tail of sth./sb.** aus etw./jmdm. nicht klug werden; **b)** *(mind)* Kopf, *der;* **in one's ~:** im Kopf; **enter sb.'s ~:** jmdm. in den Sinn kommen; **two ~s are better than one** *(prov.)* zwei Köpfe sind besser als einer; **I've got a good/bad ~ for figures** ich kann gut rechnen/rechnen kann ich überhaupt nicht; **use your ~:** gebrauch deinen Verstand; **not quite right in the ~** *(coll.)* nicht ganz richtig [im Kopf] *(ugs.)*; **get sth. into one's ~:** etw. begreifen; **have got it into one's ~ that ...:** fest [davon] überzeugt sein, daß ...; **the first thing that comes into sb.'s ~:** das erste, was jmdm. einfällt; **c)** *(person)* **a** *or* **per ~:** pro Kopf; **d)** *pl. same (in counting)* Stück [Vieh], *das;* **e)** *in pl. (on coin)* **~s or tails?** Kopf oder Zahl?; **f)** *(working end etc.; also Mus.)* Kopf, *der;* **playback/erasing ~:** Wiedergabe-/Löschkopf, *der;* **g)** *(on beer)* Blume, *die;* **h)** *(highest part)* Kopf, *der; (of stairs)* oberes Ende; *(of list, column)* oberste Reihe; **i)** *(upper or more important end)* Kopf, *der; (of bed)* Kopfende, *das;* **j)** *(leader)* Leiter, *der*/Leiterin, *die;* **~ of government** Regierungschef, *der*/-chefin, *die;* **~ of state** Staatsoberhaupt, *das;* **k)** *see* **headmaster; headmistress. 2.** *attrib. adj.* **~ waiter** Oberkellner, *der;* **~ office** Hauptverwaltung, *die; (Commerc.)* Hauptbüro, *das.* **3.** *v. t.* **a)** *(provide with heading)* überschreiben; betiteln; **~ed notepaper** Briefpapier mit Kopf; **b)** *(stand at top of)* anführen ⟨*Liste*⟩; *(lead)* leiten; führen ⟨*Bewegung*⟩; **c)** *(direct)* **we were ~ed towards Plymouth** wir fuhren mit Kurs auf Plymouth; **d)** *(Footb.)* köpfen; **e)** *(overtake and stop)* **~ sb./sth. [off]** jmdn./etw. abdrängen. **4.** *v. i.* steuern; **~ for London** ⟨*Flugzeug, Schiff:*⟩ Kurs auf London nehmen; ⟨*Auto:*⟩ in Richtung London fahren; **~ towards** *or* **for sb./the buffet** auf jmdn./das Buffet zusteuern; **you're ~ing for trouble** du wirst Ärger bekommen

head: ~ache *n.* Kopfschmerzen *Pl.*; *(fig. coll.)* Problem, *das;* **~board** *n.* Kopfende, *das;* **~-dress** *n.* Kopfschmuck, *der*

-headed ['hedɪd] *adj. in comb.* -köpfig

header ['hedə(r)] *n. (Footb.)* Kopfball, *der*

head: ~gear *n., no pl.* Kopfbedeckung, *die;* **protective ~gear** Kopfschutz, *der;* **~-hunter** *n. (lit. or fig.)* Kopfjäger, *der*

heading ['hedɪŋ] *n.* Überschrift, *die; (in encyclopaedia)* Stichwort, *das; (fig.: category)* Rubrik, *die*

head: ~lamp *n.* Scheinwerfer, *der;* **~lamp flasher** Lichthupe, *die;* **~land** ['hedlənd, 'hedlænd] *n. (Geog.)* Landspitze, *die;* **~light** *n.* Scheinwerfer, *der;* **~line** *n.* Schlagzeile, *die;* **hit the ~lines, be ~line news** Schlagzeilen machen; **the [news] ~lines** *(Radio, Telev.)* die Kurznachrichten; *(within news programme)* der [Nachrichten]überblick

'headlong 1. *adv.* **a)** *(head first)* **fall/plunge ~ into sth.** kopfüber in etw. fallen/springen; **b)** *(uncontrollably)* blindlings. **2.** *attrib. adj.* **~ dive** Kopfsprung, *der*

head: ~man *n.* Häuptling, *der;* **~'master** *n.* Schulleiter, *der;* **~'mistress** *n.* Schulleiterin, *die;* **~-on 1.** ['--] *adj.* frontal; offen ‹*Konfrontation, Konflikt*›; **a ~-on collision** *or* **crash** ein Frontalzusammenstoß; **2.** [-'-] *adv.* frontal; **meet sth./sb. ~-on** *(fig.: resolutely)* einer Sache/jmdm. entschieden entgegentreten; **~-phones** *n. pl.* Kopfhörer, *der;* **~'quarters** *n. sing. or pl.* Hauptquartier, *das; (of firm)* Zentrale, *die;* **~-rest** *n.* Kopfstütze, *die;* **~room** *n., no pl.* [lichte] Höhe, *die; (in car)* Kopffreiheit, *die;* **~scarf** *n.* Kopftuch, *das;* **~'start** *n.* **a ~ start [over sb.]** eine Vorgabe [gegenüber jmdm.]; **~strong** *adj.* eigensinnig; **~'teacher** *see* **headmaster; headmistress; ~way** *n., no pl.* **make ~way** Fortschritte machen; **~wind** *n.* Gegenwind, *der;* **~word** *n.* Stichwort, *das*

heady ['hedɪ] *adj. (intoxicating)* berauschend

heal [hi:l] **1.** *v.t. (lit. or fig.)* heilen; **time ~s all** *(fig.)* die Zeit heilt [alle] Wunden. **2.** *v.i.* **~ [up]** [ver]heilen

healing ['hi:lɪŋ] *n.* Heilung, *die*

health [helθ] *n.* **a)** *no pl. (state)* Gesundheitszustand, *der; (healthiness)* Gesundheit, *die;* **in good ~:** bei guter Gesundheit; **be in poor ~:** in schlechtem gesundheitlichen Zustand sein; **b)** *(toast)* **drink sb.'s ~** auf jmds. Gesundheit trinken; **good** *or* **your ~!** auf deine Gesundheit!

health: ~ centre *n.* medizinisches Versorgungszentrum; Poliklinik, *die;* **~ food** *n.* Reformhauskost, *die;* **~-food shop** Reformhaus, *das:* **~ insurance** *n.* Krankenversicherung, *die;* **~ resort** *n.* Kurort, *der;* **~ service** *n.* Gesundheitsdienst, *der;* **~ visitor** *n.* Krankenschwester/-pfleger im Sozialdienst

healthy ['helθɪ] *adj.* gesund

heap [hi:p] **1.** *n.* **a)** Haufen, *der;* **lying in a ~/in ~s** auf einem/in Haufen liegen; **he was lying in a ~ on the ground** er lag zusammengesackt am Boden; **b)** *(fig. coll.: quantity)* **~s of** jede Menge *(ugs.).* **2.** *v.t.* aufhäufen

hear [hɪə(r)] **1.** *v.t.,* **heard** [hɜ:d] **a)** hören; **they ~d the car drive away** sie hörten den Wagen abfahren; **I can hardly ~ myself think/speak** ich kann keinen klaren Gedanken fassen/kann mein eigenes Wort nicht verstehen; **b)** *(understand)* verstehen; **c)** *(Law)* [an]hören; verhandeln ‹*Fall*›. **2.** *v.i.,* **heard: ~ about sb./sth.** von jmdm./etw. [etwas] hören; **~ from sb.** von jmdm. hören; **he wouldn't ~ of it** er wollte nichts davon hören. **3.** *int.* **H~! H~!** bravo!; richtig!

~'out *v.t.* ausreden lassen

heard *see* **hear 1, 2**

hearing ['hɪərɪŋ] *n.* Gehör, *das;* **have good ~:** gut hören können; **be hard of ~:** schwerhörig sein; **within/out of ~:** in/außer Hörweite

'hearing-aid *n.* Hörgerät, *das*

hearsay ['hɪəseɪ] *n., no pl., no indef. art.* Gerücht, *das;* **it is only ~:** es ist nur ein Gerücht

hearse [hɜ:s] *n.* Leichenwagen, *der*

heart [hɑ:t] *n.* **a)** *(lit. or fig.)* Herz, *das;* **know/learn sth. by ~:** auswendig; **at ~:** im Grunde seines/ihres Herzens; **from the bottom of one's ~:** aus tiefstem Herzen; **my ~ goes out to them** ich verspüre großes Mitleid mit ihnen; **set one's ~ on sth./on doing sth.** sein Herz an etw. *(Akk.)* hängen/daran hängen, etw. zu tun; **take sth. to ~:** sich *(Dat.)* etw. zu Herzen nehmen; *(accept)* beherzigen ‹*Rat*›; **it does my ~ good** es erfreut mein Herz; **somebody after my own ~:** jemand ganz nach meinem Herzen; **not have the ~ to do sth.** nicht das Herz haben, etw. zu tun; **take ~:** Mut schöpfen **(from** bei); **lose ~:** Mut verlieren; **my ~ sank** mein Mut sank; **the ~ of the matter** der wahre Kern der Sache; **b)** *(Cards)* Herz, *das; see also* **club 1 d.** *See also* **[1]break 1 g, 2 a; change 1 a; desire 1 b; gold 1 a**

heart: ~ache *n.* [seelische] Qual; **~ attack** *n.* Herzanfall, *der; (fatal)* Herzschlag, *der;* **~beat** *n.* Herzschlag, *der;* **~-breaking** *adj.* herzzerreißend; **~-broken** *adj.* **she was ~-broken** ihr Herz war gebrochen; **~burn** *n., no pl. (Med.)* Sodbrennen, *das*

hearten ['hɑ:tn] *v.t.* ermutigen

heartening ['hɑ:tənɪŋ] *adj.* ermutigend

heart: ~ failure *n.* Herzversagen, *das;* **~felt** *adj.* tiefempfunden ‹*Beileid*›; aufrichtig ‹*Dankbarkeit*›

hearth [hɑ:θ] *n. Platz vor dem Kamin*

'hearth-rug *n.* Kaminvorleger, *der*

heartily ['hɑ:tɪlɪ] *adv.* von Herzen; **eat ~:** tüchtig essen; **be ~ sick of sth.** etw. herzlich leid sein

heartless ['hɑ:tlɪs] *adj.,* **heartlessly** ['hɑ:tlɪslɪ] *adv.* herzlos; unbarmherzig

heart: ~-rending ['hɑ:trendɪŋ] *adj.* herzzerreißend; **~-searching** *n.* Gewissenserforschung, *die;* **~-shaped** *adj.* herzförmig; **~-throb** *n. (person)* Idol, *das;* **~-to-~** *attrib. adj.* **have a ~-to-~ talk** offen und ehrlich miteinander sprechen; **~ trouble** *n.* Probleme mit dem Herzen; **~-warming** *adj.* herzerfreuend

hearty ['hɑ:tɪ] *adj.* **a)** *(wholehearted)* ungeteilt ‹*Unterstützung, Zustimmung*›; *(enthusiastic, unrestrained)* herzlich; begeistert ‹*Gesang*›; **b)** *(large)* herzhaft ‹*Mahlzeit*›; gesund ‹*Appetit*›; *see also* **hale**

heat [hi:t] **1.** *n.* **a)** *(hotness)* Hitze, *die;* **b)** *(Phys.)* Wärme, *die;* **c)** *(Zool.)* Brunst, *die;* **be in** *or* **on ~:** brünstig sein; **d)** *(Sport)* Vorlauf, *der.* **2.** *v.t.* heizen ‹*Raum*›; erhitzen ‹*Substanz, Lösung*›

~ 'up *v.t.* heiß machen ‹*Essen, Wasser*›

heated ['hi:tɪd] *adj.* hitzig; **a ~ exchange** ein heftiger Schlagabtausch *(fig.)*

heatedly ['hi:tɪdlɪ] *adv.* hitzig

heater ['hi:tə(r)] *n.* Ofen, *der; (for water)* Boiler, *der*

heath [hi:θ] *n.* Heide, *die*

heathen ['hi:ðn] **1.** *adj.* heidnisch. **2.** *n.* Heide, *der*/Heidin, *die*

heather ['heðə(r)] *n.* Heidekraut, *das*

heating ['hi:tɪŋ] *n., no pl.* Heizung, *die*

heat: ~-resistant *adj.* hitzebeständig; **~-stroke** *n.* Hitzschlag, *der;* **~wave** *n.* Hitzewelle, *die*

heave [hi:v] **1.** *v.t.* **a)** *(lift)* heben; wuchten *(ugs.);* **b)** *p.t. & p.p.*

hove [həʊv] *(coll.: throw)* werfen; schmeißen *(ugs.)*; **c)** ~ **a sigh [of relief]** [erleichtert] aufseufzen. **2.** *v.i.* **a)** *(pull)* ziehen; ~ **ho!** hau ruck!; **b)** *(retch)* sich übergeben; **c)** *p.t. & p.p.* **hove** *(move)* ~ **in sight** in Sicht kommen. **3.** *n. (pull)* Zug, *der*

heaven ['hevn] *n.* **a)** Himmel, *der;* **in** ~: im Himmel; **go to** ~: in den Himmel kommen; **it was** ~ **[to her]** *(fig.)* es war der Himmel auf Erden [für sie]; **b)** *in pl., (poet.) in sing. (sky)* Firmament, *das;* **c)** *(God, Providence)* **for H~'s sake** um Gottes *od.* Himmels willen; **thank H~[s]** Gott sei Dank; *see also* **forbid b**

heavenly ['hevnlı] *adj.* **a)** himmlisch; **b)** ~ **body** Himmelskörper, *der*

heavily ['hevılı] *adj.* **a)** schwer; **b)** *(to a great extent)* stark; schwer ‹*bewaffnet*›; tief ‹*schlafen*›; dicht ‹*bevölkert*›; **smoke/drink** ~: ein starker Raucher/Trinker sein; **rely** ~ **on sb./sth.** von jmdm./etw. [vollkommen] abhängig sein; **c)** *(with great force)* **it rained/snowed** ~: es regnete/schneite stark; **fall** ~: hart fallen

heaviness ['hevınıs] *n., no pl.* **a)** *(weight)* Gewicht, *das;* **b)** *(clinging quality)* Schwere, *die*

heavy ['hevı] *adj.* **a)** *(in weight)* schwer; dick ‹*Mantel*›; fest ‹*Schuh*›; ~ **traffic** *(dense)* hohes Verkehrsaufkommen; **b)** *(severe)* schwer ‹*Schaden, Verlust, Strafe, Kampf*›; hoch ‹*Steuern, Schulden, Anforderungen*›; massiv ‹*Druck, Unterstützung*›; **c)** *(excessive)* unmäßig ‹*Trinken, Essen, Rauchen*›; **a** ~ **smoker/drinker** ein starker Raucher/Trinker; **d)** *(violent)* schwer ‹*Schlag, Sturm, Regen, Sturz, Seegang*›; **make** ~ **weather of sth.** *(fig.)* die Dinge unnötig komplizieren; **e)** *(clinging)* schwer ‹*Boden*›; *see also* **going 1a; f)** *(tedious)* schwerfällig; *(serious)* seriös ‹*Zeitung*›; ernst ‹*Musik, Theaterrolle*›

heavy: ~**-duty** *adj.* strapazierfähig ‹*Kleidung, Material*›; schwer ‹*Werkzeug, Maschine*›; ~ **'goods vehicle** *n.* Schwerlastwagen, *der;* ~**-handed** [hevı'hændıd] *adj. (clumsy)* ungeschickt ‹*Person*›; *(oppressive)* unbarmherzig; ~**weight** *n. (Boxing etc.)* Schwergewicht, *das; (person also)* Schwergewichtler, *der; (fig.)* Größe, *die*

Hebrew ['hi:bru:] **1.** *adj.* hebräisch. **2.** *n.* **a)** *(Israelite)* Hebräer, *der*/Hebräerin, *die;* **b)** *no pl. (language)* Hebräisch, *das; see also* **English 2a**

heckle ['hekl] *v.t.* ~ **sb./a speech** jmdn./eine Rede durch Zwischenrufe unterbrechen

heckler ['heklə(r)] *n.* Zwischenrufer, *der*

hectare ['hektɑ:(r), 'hekteə(r)] *n.* Hektar, *das od. der*

hectic ['hektık] *adj.* hektisch

he'd [hıd, *stressed* hi:d] **a)** = **he had; b)** = **he would**

hedge [hedʒ] **1.** *n.* Hecke, *die; (fig.: barrier)* Mauer, *die.* **2.** *v.t.* **a)** mit einer Hecke umgeben; **b)** *(protect)* ~ **one's bets** mit verteiltem Risiko wetten; *(fig.)* nicht alles auf eine Karte setzen. **3.** *v.i. (avoid commitment)* sich nicht festlegen

hedge: ~**hog** ['hedʒhɒg] *n.* Igel, *der;* ~**-hop** *v.i.* im Tiefflug fliegen; ~**row** ['hedʒrəʊ] *n.* Hecke, *die* [als Feldbegrenzung]

heed [hi:d] **1.** *v.t.* beachten; beherzigen ‹*Rat, Lektion*›; ~ **the danger/risk** sich *(Dat.)* der Gefahr/des Risikos bewußt sein. **2.** *n., no art., no pl.* **give** *or* **pay** ~ **to, take** ~ **of** Beachtung schenken (+ *Dat.*); **give** *or* **pay no** ~ **to, take no** ~ **of** nicht beachten

heedless ['hi:dlıs] *adj.* unachtsam; **be** ~ **of sth.** auf etw. *(Akk.)* nicht achten

heel [hi:l] **1.** *n.* **a)** Ferse, *die;* ~ **of the hand** Handballen, *der;* **Achilles'** ~ *(fig.)* Achillesferse, *die;* **bring a dog to** ~: einen Hund bei Fuß rufen; **bring sb. to** ~ *(fig.)* jmdn. auf Vordermann bringen *(ugs.)*; **take to one's** ~**s** *(fig.)* Fersengeld geben *(ugs.)*; *see also* **dig in 2b; b)** *(of shoe)* Absatz, *der; (of stocking)* Ferse, *die;* **down at** ~: abgetreten; *(fig.)* heruntergekommen *(ugs.)*. **2.** *v.t.* ~ **a shoe** einen Schuh mit einem [neuen] Absatz versehen

hefty ['heftı] *adj.* kräftig; *(heavy)* schwer; *(fig.: large)* hoch ‹*Rechnung, Summe, Strafe, Anteil*›; deutlich ‹*Mehrheit*›; stark ‹*Erhöhung*›

heifer ['hefə(r)] *n.* Färse, *die*

height [haıt] *n.* **a)** Höhe, *die; (of person, animal, building)* Größe, *die;* **be three metres in** ~: drei Meter hoch sein; **be six feet in** ~ ‹*Person:*› 1,80 m groß sein; **at a** ~ **of three metres** in einer Höhe von drei Metern; **b)** *(fig.: highest point)* Höhepunkt, *der;* **the** ~ **of folly** der Gipfel der Dummheit

heighten ['haıtn] *v.t.* aufstocken; *(fig.: intensify)* verstärken

heinous ['heınəs] *adj.* schändlich

heir [eə(r)] *n. (lit. or fig.)* Erbe, *der*/Erbin, *die*

heiress ['eərıs] *n.* Erbin, *die*

heirloom ['eəlu:m] *n.* Erbstück, *das; (fig.)* Erbe, *das*

heist [haıst] *(Amer. sl.)* **1.** *n.* Raubüberfall, *der.* **2.** *v.t. (steal)* rauben; *(rob)* ausrauben

held *see* [2]**hold 1, 2**

helical ['helıkl] *adj.* spiralförmig; spiralig

helices *pl. of* **helix**

helicopter ['helıkɒptə(r)] *n.* Hubschrauber, *der*

Heligoland ['helıgəlænd] *pr. n.* Helgoland *(das)*

heliport ['helıpɔ:t] *n.* Heliport, *der*

helium ['hi:lıəm] *n.* Helium, *das*

helix ['hi:lıks] *n., pl.* **helices** ['hi:lısi:z] Spirale, *die*

hell [hel] *n.* **a)** Hölle, *die;* **all** ~ **was let loose** *(fig.)* es war die Hölle los; *see also* **raise 1g; b)** *(coll.)* **[oh]** ~! verdammter Mist! *(ugs.)*; **what the** ~! ach, zum Teufel! *(ugs.)*; **to** *or* **the** ~ **with it!** ich hab's satt *(ugs.)*; **a** *or* **one** ~ **of a [good] party** eine unheimlich gute Party *(ugs.)*; **work/run like** ~: wie der Teufel arbeiten/rennen *(ugs.)*; **it hurt like** ~: es tat höllisch weh *(ugs.)*

he'll [hıl, *stressed* hi:l] = **he will**

hell: ~**-'bent** *adj.* **be** ~ **on doing sth.** *(coll.)* wild entschlossen sein, etw. zu tun *(ugs.)*; ~**fire** *n.* Höllenfeuer, *das*

hellish ['helıʃ] *adj.* höllisch ‹*Qual, Schmerz*›; scheußlich ‹*Arbeit, Zeit*›

hello [hə'ləʊ, he'ləʊ] **1.** *int. (greeting)* hallo; *(surprise)* holla. **2.** *n.* Hallo, *das*

hell's 'angel *n.* Rocker, *der*

helm [helm] *n. (Naut.)* Ruder, *das*

helmet ['helmıt] *n.* Helm, *der*

helmsman ['helmzmən] *n., pl.* **helmsmen** ['helmzmən] *(Naut.)* Rudergänger, *der*

help [help] **1.** *v.t.* **a)** ~ **sb. [to do sth.]** jmdm. helfen [, etw. zu tun]; ~ **oneself** sich *(Dat.)* selbst helfen; **can I** ~ **you?** was kann ich für Sie tun?; *(in shop also)* was möchten Sie bitte?; **b)** *(serve)* ~ **oneself** sich *(Dat.)* nehmen; sich bedienen; ~ **oneself to sth.** sich *(Dat.)* etw. nehmen; *(coll.: steal)* etw. mitgehen lassen *(ugs.)*; **c)** *(avoid)* **if I/you can** ~ **it** wenn es irgend zu vermeiden ist; **not if I can** ~ **it** nicht wenn ich es verhindern kann; **it can't be** ~**ed** es läßt sich nicht ändern; **I can't** ~ **it** *(remedy)* ich kann nichts dafür *(ugs.)*; **d)** *(refrain from)* **I can't** ~

thinking *or* **can't ~ but think that ...:** ich kann mir nicht helfen, ich glaube, ...; **I can't ~ laughing** ich muß einfach lachen. **2.** *n.* Hilfe, *die;* **be of [some]/no/much ~ to sb.** jmdm. eine gewisse/keine/eine große Hilfe sein; **there's no ~ for it** daran läßt sich nichts ändern

~ 'out 1. *v.i.* aushelfen. **2.** *v.t.* **~ sb. out** jmdm. helfen

helper ['helpə(r)] *n.* Helfer, *der*/Helferin, *die; (paid assistant)* Aushilfskraft, *die*

helpful ['helpfl] *adj. (willing)* hilfsbereit; *(useful)* hilfreich; nützlich

helping ['helpɪŋ] **1.** *attrib. adj.* **lend [sb.] a ~ hand [with sth.]** *(fig.)* [jmdm.] [bei etw.] helfen. **2.** *n.* Portion, *die*

helpless ['helplɪs] *adj.* hilflos; *(powerless)* machtlos

helter-skelter ['heltəskeltə(r)] **1.** *adv.* in wildem Durcheinander. **2.** *n. (in fun-fair)* [spiralförmige] Rutschbahn

hem [hem] **1.** *n.* Saum, *der.* **2.** *v.t.,* **-mm-: a)** säumen; **b)** *(surround)* **~ sb./sth. in** *or* **about** jmdn./etw. einschließen; **feel ~med in** *(fig.)* sich eingeengt fühlen

he-man ['hi:mæn] *n.* **a real ~:** ein richtiger Mann

'hemisphere *n.* Halbkugel, *die;* Hemisphäre, *die*

'hem-line *n.* Saum, *der;* **~s are up/down** die Röcke sind kurz/lang

hemo- *(Amer.) see* **haemo-**

hemp [hemp] *n.* **a)** *(Bot., Textiles)* Hanf, *der;* **b)** *(drug)* Haschisch, *das od. der*

hen [hen] *n.* Huhn, *das;* Henne, *die (bes. im Gegensatz zu „Hahn")*

hence [hens] *adv.* **a)** *(therefore)* daher; **b)** *(from this time)* **a week/ten years ~:** in einer Woche/zehn Jahren

hence'forth, hence'forward *advs.* von nun an

henchman ['hentʃmən] *n., pl.* **henchmen** ['hentʃmən] *(derog.)* Handlanger, *der*

henna ['henə] *n. (dye)* Henna, *das*

hen: ~-party *n. (coll.)* [Damen]kränzchen, *das;* **~pecked** ['henpekt] *adj.* **a ~pecked husband** ein Pantoffelheld *(ugs.);* **be ~pecked** unter dem Pantoffel stehen *(ugs.)*

hepatitis [hepə'taɪtɪs] *n. (Med.)* Leberentzündung, *die;* Hepatitis, *die (fachspr.)*

¹her [hə(r), *stressed* hɜ:(r)] *pron.* sie; *as indirect object* ihr; *reflexively* sich; *referring to personified things or animals which correspond to German masculines/neuters* ihn/es; *as indirect object* ihm; **it was ~:** sie war's; **~ and me** *(coll.)* sie und ich; **if I were ~** *(coll.)* wenn ich sie wäre

²her *poss. pron. attrib.* ihr; *referring to personified things or animals which correspond to German masculines/neuters* sein; **she opened ~ eyes/mouth** sie öffnete die Augen/den Mund; **~ father and mother** ihr Vater und ihre Mutter; **she has a room of ~ own** sie hat ein eigenes Zimmer; **he complained about ~ being late** er beklagte sich darüber, daß sie zu spät kam

herald ['herəld] **1.** *n.* **a)** Herold, *der;* **b)** *(messenger)* Bote, *der; (fig.: forerunner)* Vorbote, *der.* **2.** *v.t. (lit. or fig.)* ankündigen

heraldic [he'rældɪk] *adj.* heraldisch

heraldry ['herəldrɪ] *n., no pl.* Wappenkunde, *die;* Heraldik, *die*

herb [hɜ:b] *n.* Kraut, *das; (Cookery)* Gewürzkraut, *das*

herbaceous [hɜ:'beɪʃəs] *adj.* *(Bot.)* krautartig ‹*Pflanze*›; **~ border** Staudenrabatte, *die*

herbal ['hɜ:bl] **1.** *attrib. adj.* Kräuter‹*tee, -arznei*›; ‹*Behandlung*› mit Heilkräutern. **2.** *n.* Pflanzenbuch, *das*

herbivorous [hɜ:'bɪvərəs] *adj.* pflanzenfressend

herd [hɜ:d] **1.** *n.* **a)** Herde, *die; (of wild animals)* Rudel, *das;* **b)** *(fig.)* Masse, *die.* **2.** *v.t.* **a)** *(lit. or fig.)* treiben; **~ [people] together** *(fig.)* [Menschen] zusammenpferchen; **b)** *(tend)* hüten

'herdsman ['hɜ:dzmən] *n., pl.* **~smen** ['hɜ:dzmən] Hirt[e], *der*

here [hɪə(r)] **1.** *adv.* **a)** *(in or at this place)* hier; **Schmidt ~** *(on telephone)* Schmidt; **spring is ~:** der Frühling ist da; **down/in/up ~:** hier unten/drin/oben; **~ goes!** *(coll.)* dann mal los! *(ugs.);* **~, there, and everywhere** überall; **~ you are** *(coll.: giving sth.)* hier; **~ we are** *(on arrival)* da sind *od.* wären wir; **b)** *(to this place)* hierher; **in[to] ~:** hierherein; **come/bring ~:** [hier]herkommen/-bringen; **~ comes the bus** hier *od.* da kommt der Bus. **2. up to ~, as far as ~:** bis hierhin; **from ~ on** von nun an; **where do we go from ~?** *(fig.)* was machen wir jetzt? **3.** *int. (attracting attention)* he

here: ~a'bout[s] *adv.* hier [in dieser Gegend]; **~'after** *adv.* *(formal)* im folgenden; **~'by** *adv.* *(formal)* hiermit

hereditary [hɪ'redɪtərɪ] *adj.* **a)** erblich ‹*Titel, Amt*›; ererbt ‹*Reichtum*›; **~ monarchy/right** Erbmonarchie, *die*/Erbrecht, *das;* **b)** *(Biol.)* angeboren ‹*Instinkt, Verhaltensweise*›

heredity [hɪ'redɪtɪ] *n. (Biol.)* **a)** *(transmission of qualities)* Vererbung, *die;* **b)** *(genetic constitution)* Erbgut, *das*

heresy ['herɪsɪ] *n.* Ketzerei, *die;* Häresie, *die (geh.)*

heretic ['herɪtɪk] *n.* Ketzer, *der*/Ketzerin, *die;* Häretiker, *der*/Häretikerin, *die (geh.)*

heretical [hɪ'retɪkl] *adj.* ketzerisch; häretisch *(geh.)*

here: ~u'pon *adv.* hierauf; **~'with** *adv. (with this)* in der Anlage; **we enclose ~with your cheque** wir legen Ihren Scheck diesem Schreiben bei

heritage ['herɪtɪdʒ] *n. (lit. or fig.)* Erbe, *das*

hermetic [hɜ:'metɪk] *adj.* luftdicht; *(fig.)* hermetisch *(geh.)*

hermetically [hɜ:'metɪkəlɪ] *adv.* hermetisch

hermit ['hɜ:mɪt] *n.* Einsiedler, *der*/Einsiedlerin, *die*

hernia ['hɜ:nɪə] *n., pl.* **~s** *or* **~e** ['hɜ:nɪi:] *(Med.)* Bruch, *der;* Hernie, *die (Med.)*

hero ['hɪərəʊ] *n., pl.* **~es** Held, *der;* *(demigod)* Heros, *der;* **~ of the hour** Held des Tages

heroic [hɪ'rəʊɪk] *adj.* **a)** heldenhaft; heroisch *(geh.);* **b)** *(Lit.)* **~ epic/legend** Heldenepos, *das*/-legende, *die*

heroics [hɪ'rəʊɪks] *n. pl. (language)* Theatralische, *das; (foolhardiness)* Draufgängertum, *das*

heroin ['herəʊɪn] *n., no pl.* Heroin, *das*

heroine ['herəʊɪn] *n.* Heldin, *die;* Heroin, *die (geh.);* Heroine, *die (Theater)*

heroism ['herəʊɪzm] *n., no pl.* Heldentum, *das*

heron ['hern] *n.* Reiher, *der*

'hero-worship 1. *n.* Heldenverehrung, *die.* **2.** *v.t.* vergöttern

herpes ['hɜ:pi:z] *n. (Med.)* Herpes, *der*

herring ['herɪŋ] *n.* Hering, *der*

hers [hɜ:z] *poss. pron. pred.* ihrer/ihre/ihres; der/die/das ihre *od.* ihrige *(geh.);* **the book is ~:** das Buch gehört ihr; **some friends of ~:** ein paar Freunde von ihr; **those children of ~:** ihre Gören *(ugs.);* **~ is a difficult job** sie hat einen schwierigen Job *(ugs.)*

herself [hɜ:'self] *pron.* **a)** *emphat.* selbst; **she ~ said so** sie selbst hat das gesagt; **she saw it ~:** sie hat es selbst gesehen; **she was just being ~:** sie gab sich einfach so wie sie ist; **she is [quite] ~ again** sie ist wieder ganz die alte; *(after an illness)* sie ist wieder auf der Höhe *(ugs.)*; **all right in ~:** im wesentlichen gesund; **[all] by ~** *(on her own, by her own efforts)* [ganz] allein[e]; **b)** *refl.* sich; allein[e] ‹*tun, wählen*›; **she wants to see for ~:** sie will [es] selbst sehen; **younger than/as heavy as ~:** jünger als/so schwer wie sie selbst; **... she thought to ~:** ... dachte sie sich [im stillen]; ... dachte sie bei sich

he's [hɪz, *stressed* hi:z] **a)** = **he is; b)** = **he has**

hesitant ['hezɪtənt] *adj.* zögernd ‹*Reaktion*›; stockend ‹*Rede*›; unsicher ‹*Person, Stimme*›; **be ~ to do sth.** *or* **about doing sth.** Bedenken haben, etw. zu tun

hesitate ['hezɪteɪt] *v.i.* **a)** *(show uncertainty)* zögern; **he who ~s is lost** *(prov.)* man muß die Gelegenheit beim Schopfe fassen; **b)** *(falter)* ins Stocken geraten; **c) ~ to do sth.** Bedenken haben, etw. zu tun

hesitation [hezɪ'teɪʃn] *n.* **a)** *no pl.* *(indecision)* Unentschlossenheit, *die;* **without the slightest ~:** ohne im geringsten zu zögern; **b)** *(instance of faltering)* Unsicherheit, *die;* **c)** *no pl.* *(reluctance)* Bedenken *Pl.*

hessian ['hesɪən] *n.* Sackleinen, *das;* Hessian, *das (fachspr.)*

het [het] *adj.* *(coll.)* **~ up** aufgeregt; **get ~ up over sth.** sich über etw. *(Akk.)* aufregen

heterogeneous [hetərə'dʒi:nɪəs, hetərə'dʒenɪəs] *adj.* ungleichartig; heterogen

heterosexual [hetərəʊ'seksjʊəl] **1.** *adj.* heterosexuell. **2.** *n.* Heterosexuelle, *der/die*

hew [hju:] **1.** *v.t., p.p.* **~n** [hju:n] *or* **~ed** [hju:d] *(cut)* hacken ‹*Holz*›; fällen ‹*Baum*›; losschlagen ‹*Kohle, Gestein*›. **2.** *v.i., p.p.* **~n** *or* **~ed** zuschlagen

hex [heks] *n.* *(Amer.)* **put a ~ on sb./sth.** jmdn./etw. verhexen

hexagon ['heksəgən] *n.* *(Geom.)* Sechseck, *das;* Hexagon, *das (fachspr.)*

hey [heɪ] *int.* he; **~ presto!** simsalabim!

heyday ['heɪdeɪ] *n., no pl.* Blütezeit, *die*

HGV *abbr.* *(Brit.)* **heavy goods vehicle**

hi [haɪ] *int.* hallo *(ugs.)*

hiatus [haɪ'eɪtəs] *n.* *(gap)* Bruch, *der;* *(interruption)* Unterbrechung, *die*

hibernate ['haɪbəneɪt] *v.i.* Winterschlaf halten

hibernation [haɪbə'neɪʃn] *n.* Winterschlaf, *der*

hiccup ['hɪkʌp] **1.** *n.* **a)** Schluckauf, *der;* **have/get [the] ~s** [den] Schluckauf haben/bekommen; **b)** *(fig.: stoppage)* Störung, *die;* **without any ~s** reibungslos. **2.** *v.i.* schlucksen *(ugs.)*; hick machen *(ugs.)*; *(many times)* den Schluckauf haben

hid *see* **[1]hide 1, 2**

hidden *see* **[1]hide 1, 2**

[1]hide [haɪd] **1.** *v.t.,* **hid** [hɪd], **hidden** ['hɪdn] **a)** verstecken ‹*Gegenstand, Person usw.*› **(from** vor + *Dat.*); **~ one's face in one's hands** sein Gesicht in den Händen bergen; **b)** *(keep secret)* verbergen ‹*Gefühle, Sinn, Freude usw.*› **(from** vor + *Dat.*); verheimlichen ‹*Tatsache, Absicht, Grund usw.*› **(from** *Dat.*); **have nothing to ~:** nichts zu verbergen haben; **c)** *(obscure)* verdecken; **~ sth. [from view]** etw. verstecken; *(by covering)* etw. verdecken; ‹*Nebel, Rauch usw.*:› etw. einhüllen. **2.** *v.i.,* **hid, hidden** sich verstecken *od.* verbergen **(from** vor + *Dat.*). **3.** *n.* *(Brit.)* Versteck, *das;* *(hunter's ~)* Ansitz, *der (Jägerspr.)*

~ a'way *v.i.* sich verstecken *od.* verbergen

~ 'out, ~ 'up *v.i.* sich versteckt *od.* verborgen halten

[2]hide *n.* *(animal's skin)* Haut, *die;* *(of furry animal)* Fell, *das;* *(dressed)* Leder, *das;* *(joc.: human skin)* Haut, *die;* Fell, *das;* **tan sb.'s ~:** jmdm. das Fell gerben *od.* versohlen *(salopp)*

hide: ~-and-'seek *n.* Versteckspiel, *das;* **play ~-and-seek** Verstecken spielen; **~bound** *adj.* engstirnig; borniert

hideous ['hɪdɪəs] *adj.* **a)** scheußlich; *(horrific)* entsetzlich; grauenhaft; **b)** *(coll.: unpleasant)* furchtbar *(ugs.)*

'hide-out *n.* Versteck, *das;* *(of bandits, partisans, etc.)* Versteck, *das;* Unterschlupf, *der*

[1]hiding ['haɪdɪŋ] *n.* **go into ~** *(to avoid police, public attention)* untertauchen; **be in ~:** sich versteckt halten; **come out of ~:** wieder auftauchen

[2]hiding *n.* *(coll.: beating)* Tracht Prügel; *(fig.)* Schlappe, *die;* **give sb. a [good] ~:** jmdm. eine [ordentliche] Tracht Prügel verpassen; *(fig.)* jmdm. eine [klare] Abfuhr erteilen; **be on a ~ to nothing** eine undankbare Rolle haben

'hiding-place *n.* Versteck, *das*

hierarchic [haɪə'rɑ:kɪk], **hierarchical** [haɪə'rɑ:kɪkl] *adj.* hierarchisch

hierarchy ['haɪərɑ:kɪ] *n.* Hierarchie, *die*

hieroglyphics ['haɪərə'glɪfɪks] *n. pl.* *(also joc.)* Hieroglyphen

hi-fi ['haɪfaɪ] *(coll.)* **1.** *adj.* Hi-Fi-. **2.** *n.* *(equipment)* Hi-Fi-Anlage, *die*

higgledy-piggledy [hɪgldɪ'pɪgldɪ] **1.** *adv.* wie Kraut und Rüben *(ugs.)*. **2.** *adj.* wirr, kunterbunt ‹*Ansammlung usw.*›

high [haɪ] **1.** *adj.* **a)** hoch ‹*Berg, Gebäude, Mauer*›; **b)** *(above normal level)* hoch ‹*Stiefel*›; **the river/water is ~:** der Fluß/das Wasser steht hoch; **be left ~ and dry** *(fig.)* auf dem trock[e]nen sitzen *(ugs.)*; **c)** *(far above ground or sea level)* hoch ‹*Gipfel, Punkt*›; groß ‹*Höhe*›; **d)** *(to or from far above the ground)* hoch ‹*Aufstieg, Sprung*›; **~ diving** Turmspringen, *das; see also* **bar 1 b; e)** *(of exalted rank)* hoch ‹*Beamter, Amt, Gericht*›; **a ~er court** eine höhere Instanz; **~ and mighty** *(coll.: ~handed)* selbstherrlich; *(coll.: superior)* hochnäsig *(ugs.)*; **be born** *or* **destined for ~er things** zu Höherem geboren *od.* bestimmt sein; **those in ~ places** die Oberen; **f)** *(great in degree)* hoch; groß ‹*Gefallen, Bedeutung*›; stark ‹*Wind*›; **be held in ~ regard/esteem** hohes Ansehen/hohe Wertschätzung genießen; **~ blood pressure** Bluthochdruck, *der;* **have a ~ opinion of sb./sth.** eine hohe Meinung von jmdm./etw. haben *(geh.)*; viel von jmdm./etw. halten; **g)** *(noble, virtuous)* hoch ‹*Ideal, Ziel, Prinzip, Berufung*›; edel ‹*Charakter*›; **of ~ birth** von hoher Geburt *(geh.)*; **h)** *(of time, season)* **it is ~ time you left** es ist *od.* wird höchste Zeit, daß du gehst; **~ noon** Mittag; **~ summer** Hochsommer, *der;* **i)** *(luxurious, extravagant)* üppig ‹*Leben*›; **j)** *(enjoyable)* **have a ~ [old] time** sich bestens amüsieren; **k)** *(coll.: on a drug)* high *nicht attr. (ugs.)* **(on** von); **get ~ on** sich anturnen mit *(ugs.)* ‹*Haschisch, LSD usw.*›; **l)** *(in pitch)* hoch ‹*Ton, Stimme, Lage, Klang usw.*›; **m)** *(slightly decomposed)* angegangen *(landsch.)* ‹*Fleisch*›; **n)** *(Cards)* hoch; **ace is ~:** As ist hoch. **2.** *adv.* **a)** *(in or to a ~ position)* hoch; **~ on our list of priorities** weit oben auf unserer Prioritätenliste; **search** *or* **hunt** *or*

look ~ and low überall suchen; **b)** *(to a ~ level)* hoch; **prices have gone too ~:** die Preise sind zu stark gestiegen; **I'll go as ~ as two thousand pounds** ich gehe bis zweitausend Pfund. **3.** *n.* **a)** *(~est level/figure)* Höchststand, *der; see also* **all-time; b)** *(~ position)* **on ~:** hoch oben *od. (geh., südd., österr.)* droben; *(in heaven)* im Himmel; **c)** *(Meteorol.)* Hoch, *das*

high: ~ 'altar *n. (Eccl.)* Hochaltar, *der;* **~brow** *(coll.) adj.* intellektuell ‹*Person, Gerede usw.*›; hochgestochen *(abwertend)* ‹*Person, Gerede, Musik, Literatur usw.*›; **~ chair** *n. (for baby)* Hochstuhl, *der;* **H~ 'Church** *n.* High Church, *die;* Hochkirche, *die;* **~-class** *adj.* hochwertig ‹*Erzeugnis*›; erstklassig ‹*Unterkunft, Konditor usw.*›; **H~ 'Court [of Justice]** *n. (Brit. Law) oberster Gerichtshof für Zivil- und Strafsachen*

higher ['haɪə(r)]: **~ edu'cation** *n., no pl., no art.* Hochschul[aus]bildung, *die;* **~ mathe'matics** *n.* höhere Mathematik

high: ~ ex'plosive *see* **explosive 2; ~-flown** *adj.* geschwollen *(abwertend)* ‹*Stil, Ausdrucksweise*›; hochfliegend ‹*Ideen, Pläne*›; **~-'flyer** *n. (able person)* Hochbegabte, *der/die;* **~ 'frequency** *n.* hohe Frequenz; *(radio-frequency)* Hochfrequenz, *die;* **~-grade** *adj.* hochwertig; **~-grade steel** Edelstahl, *der;* **~-handed** [haɪ'hændɪd] *adj.* selbstherrlich; **~ 'heel** *n.* **a)** hoher Absatz; **b)** *in pl. (shoes)* hochhackige Schuhe; **~-heeled** [haɪ'hi:ld] *adj.* ‹*Schuhe*› mit hohen Absätzen; **~ jinks** ['haɪ dʒɪŋks] *n. pl.* [übermütige] Ausgelassenheit; **~ jump** *n., no pl.* **a)** *(Sport)* Hochsprung, *der;* **b)** *(fig.: reprimand, punishment)* **he is for the ~ jump** er kann sich auf was gefaßt machen *(ugs.);* **~land** ['haɪlənd] **1.** *n., usu. in pl.* Hochland, *das;* **the H~lands** *(in Scotland)* die Highlands; **2.** *adj.* hochländisch; **~-level** *adj.* ‹*Verhandlungen usw.*› auf hoher Ebene; **~ life** *n., no pl.* **a)** *(life of upper class)* das Leben der Oberschicht; **b)** *(luxurious living)* **the ~ life** das Leben auf großem Fuße; **~light 1.** *n.* **a)** *(outstanding moment)* Höhepunkt, *der;* **b)** *(bright area)* Licht, *das;* **c)** *(in hair) usu. pl.* Strähnchen, *das;* **2.** *v.t.,* **~lighted** ein Schlaglicht werfen auf (+ *Akk.*) ‹*Probleme usw.*›

highly ['haɪlɪ] *adv.* **a)** *(to a high degree)* sehr; äußerst; hoch‹*begabt, -interessant, -angesehen, -bezahlt, -gebildet, -modern, -aktuell*›; leicht ‹*entzündlich*›; stark ‹*gewürzt*›; **feel ~ honoured** sich hoch geehrt fühlen; **I can ~ recommend the restaurant** ich kann dieses Restaurant sehr empfehlen; **b)** *(favourably)* **think ~ of sb./sth., regard sb./sth. ~:** eine hohe Meinung von jmdm./etw. haben

'highly-strung *adj.* übererregbar

high-minded [haɪ'maɪndɪd] *adj.* hochgesinnt ‹*Person*›; hoch, *(geh.)* hehr ‹*Prinzipien, Dienstauffassung usw.*›

Highness ['haɪnɪs] *n.* Hoheit, *die;* **His/Her/Your [Royal] ~:** Seine/Ihre/Eure [Königliche] Hoheit

high: ~-pitched *adj.* **a)** hoch ‹*Ton, Stimme*›; **b)** *(Archit.)* steil ‹*Dach*›; **~ point** *n.* Höhepunkt, *der;* Gipfelpunkt, *der;* **~-powered** ['haɪpaʊəd] *adj.* **a)** *(powerful)* stark ‹*Fahrzeug, Motor, Glühbirne usw.*›; **b)** *(forceful)* dynamisch ‹*Geschäftsmann, Manager usw.*›; **~ 'pressure** *n.* **a)** *(Meteorol.)* Hochdruck, *der;* **an area of ~ pressure** ein Hochdruckgebiet; **b)** *(Mech. Engin.)* Überdruck, *der;* **c)** *(fig.: high degree of activity)* Hochdruck, *der;* **~-pressure** *adj.* Hochdruck-; *(fig.: persuasive)* aggressiv ‹*Verkaufsmethoden*›; **~ 'priest** *n.* Hohepriester, *der;* **~-ranking** *adj.* hochrangig; von hohem Rang *nachgestellt;* **~-rise** *adj.* **~-rise building** Hochhaus, *das;* **~-rise [block of] flats/office block** Wohn-/Bürohochhaus, *das;* **~ road** *n.* Hauptstraße, *die;* **~ school** *n.* ≈ Oberschule, *die;* **~ 'seas** *n. pl.* **the ~ seas** die hohe See; **~ season** *n.* Hochsaison, *die;* **~-speed** *adj.* schnell[fahrend]; **~-speed train** Hochgeschwindigkeitszug, *der;* **~-spirited** *see* **spirited b; ~ 'spirits** *see* **spirit 1g; ~ street** *n.* Hauptstraße, *die;* **~ 'tea** *see* **tea b; ~-tech** ['haɪtek] *adj. (coll.)* High-Tech-; **~ tech** [haɪ'tek] *(coll.),* **~ tech'nology** *ns.* Spitzentechnologie, *die;* Hochtechnologie, *die;* **~-technology** *adj.* hochtechnisiert; High-Tech-; **~ 'tide** *see* **tide 1a; ~ 'treason** *see* **treason; ~-'voltage** *adj. (Electr.)* Hochspannungs-; **~ 'water** *n.* Hochwasser, *das;* **~-'water mark** *n.* Hochwassermarke, *die;* **~way** *n.* **a)** *(public road)* öffentliche Straße; **b)** *(main route)* Verkehrsweg, *der;* **H~way 'Code** *n. (Brit.)* Straßenverkehrsordnung, *die;* **~wayman** ['haɪweɪmən] *n., pl.* **~waymen** ['haɪweɪmən] *(Hist.)* Straßenräuber, *der;* Wegelagerer, *der*

hijack ['haɪdʒæk] **1.** *v.t.* in seine Gewalt bringen; **they ~ed an aircraft to Cuba** sie haben ein Flugzeug nach Kuba entführt. **2.** *n. (of aircraft)* Entführung, *die* (**of** *Gen.*); *(of vehicle)* Überfall, *der* (**of** auf + *Akk.*)

hijacker ['haɪdʒækə(r)] *n.* Entführer, *der; (of aircraft)* Hijacker, *der;* Flugzeugentführer, *der*

hike [haɪk] **1.** *n.* Wanderung, *die;* **go on a ~:** eine Wanderung machen; wandern gehen. **2.** *v.i.* wandern; eine Wanderung machen

hiker ['haɪkə(r)] *n.* Wanderer, *der*/Wanderin, *die*

hilarious [hɪ'leərɪəs] *adj.* urkomisch; rasend komisch *(ugs.)*

hilariously [hɪ'leərɪəslɪ] *adv.* **be ~ funny** rasend komisch sein *(ugs.)*

hilarity [hɪ'lærɪtɪ] *n., no pl.* **a)** *(gaiety)* Fröhlichkeit, *die;* **b)** *(merriment)* übermütige Ausgelassenheit; *(loud laughter)* Heiterkeit, *die*

hill [hɪl] *n.* **a)** Hügel, *der; (higher)* Berg, *der;* **built on a ~:** am Hang gebaut; **be over the ~** *(fig. coll.)* auf dem absteigenden Ast sein *(ugs.); (past the crisis)* über den Berg sein *(ugs.);* **[as] old as the ~s** *(fig.)* uralt; ‹*Person*› [so] alt wie Methusalem; *see also* **up 2a; b)** *(heap)* Hügel, *der; (ant~, dung~, mole~)* Haufen, *der;* **c)** *(sloping road)* Steigung, *die*

hill-billy ['hɪlbɪlɪ] *n. (Amer.)* Hinterwäldler, *der*/Hinterwäldlerin, *die (spött.)*

hillock ['hɪlək] *n.* [kleiner] Hügel

hill: ~side *n.* Hang, *der;* **~top** *n.* [Berg]gipfel, *der*

hilly ['hɪlɪ] *adj.* hüg[e]lig; *(higher)* bergig

hilt [hɪlt] *n.* Griff, *der;* Heft, *das (geh., fachspr.);* **[up] to the ~** *(fig.)* voll und ganz ‹*unterstützen usw.*›

him [ɪm, *stressed* hɪm] *pron.* ihn; *as indirect object* ihm; *reflexively* sich; *referring to personified things or animals which correspond to German feminines/neuters* sie/es; *as indirect object* ihr/ihm; **it was ~:** er war's; **~ and me** *(coll.)* er und ich; **if I were ~:** wenn ich er wäre

Himalayas [hɪmə'leɪəz] *pr. n. pl.* Himalaya, *der*

himself [hɪm'self] *pron.* **a)** *emphat.* selbst; **b)** *refl.* sich. *See also* **herself**

¹hind [haɪnd] *n.* Hirschkuh, *die*

²hind *adj.* hinter...; ~ **legs** Hinterbeine

hinder ['hɪndə(r)] *v. t.* *(impede)* behindern; *(delay)* verzögern ‹*Vollendung einer Arbeit, Vorgang*›; aufhalten ‹*Person*›; ~ **sb. from doing sth.** jmdn. daran hindern, etw. zu tun

Hindi ['hɪndi:] **1.** *adj.* Hindi-. **2.** *n.* Hindi, *das; see also* **English 2 a**

hind: ~**most** *adj.* hinterst...; **it was devil take the ~most** es galt nur noch: Rette sich, wer kann!; ~**quarters** *n. pl.* Hinterteil, *das; (of large quadruped)* Hinterteil, *das;* Hinterhand, *die (fachspr.)*

hindrance ['hɪndrəns] *n.* **a)** *(action)* Behinderung, *die; see also* **²let; b)** *(obstacle)* Hindernis, *das* **(to** für); **he is more of a ~ than a help** er stört mehr, als daß er hilft

'hindsight *n.* **in ~, with [the benefit of] ~:** im nachhinein

Hindu ['hɪndu:, hɪn'du:] **1.** *n.* Hindu, *der.* **2.** *adj.* hinduistisch; Hindu‹*gott, -tempel*›

hinge [hɪndʒ] **1.** *n.* Scharnier, *das; (continuous)* Klavierband, *das;* **off its ~s** ‹*Tür*› aus den Angeln gehoben. **2.** *v. t.* mit Scharnieren/einem Scharnier versehen. **3.** *v. i.* *(fig.)* abhängen (**[up]on** von)

hint [hɪnt] **1.** *n.* **a)** *(suggestion)* Wink, *der;* Hinweis, *der;* **give a ~ that ...:** andeuten, daß ...; *see also* **broad b; drop 3 d; take 1 v; b)** *(slight trace)* Spur, *die* **(of** von); **the ~/no ~ of a smile** der Anflug/nicht die Spur eines Lächelns; **a ~ of aniseed** ein Hauch von Anis; **c)** *(practical information)* Tip, *der* **(on** für). **2.** *v. t.* andeuten; **nothing has yet been ~ed about it** darüber hat man noch nichts herausgelassen *(ugs.).* **3.** *v. i.* ~ **at** andeuten

¹hip [hɪp] *n.* **a)** Hüfte, *die;* **with one's hands on one's ~s** die Arme in die Hüften gestemmt; **b)** *in sing. or pl. (~-measurement)* Hüftumfang, *der;* Hüftweite, *die; (of man, boy)* Gesäßumfang, *der;* Gesäßweite, *die*

²hip *n. (Bot.)* Hagebutte, *die*

hip: ~**-bone** *n. (Anat.)* Hüftbein, *das;* Hüftknochen, *der;* ~**-flask** *n.* Taschenflasche, *die;* ~**-joint** *n. (Anat.)* Hüftgelenk, *das*

hippie ['hɪpɪ] *n. (coll.)* Hippie, *der*

hippo ['hɪpəʊ] *n., pl.* ~**s** *(coll.) see* **hippopotamus**

hip-'pocket *n.* Gesäßtasche, *die*

hippopotamus [hɪpə'pɒtəməs] *n., pl.* ~**es** *or* **hippopotami** [hɪpə'pɒtəmaɪ] *(Zool.)* Nilpferd, *das;* Flußpferd, *das*

hippy ['hɪpɪ] *see* **hippie**

hire [haɪə(r)] **1.** *n.* **a)** *(action)* Mieten, *das; (of servant)* Einstellen, *das;* **b)** *(condition)* **be on ~ [to sb.]** [an jmdn.] vermietet sein; **for** *or* **on ~:** zu vermieten. **2.** *v. t.* **a)** *(employ)* anwerben; engagieren ‹*Anwalt, Berater usw.*›; **b)** *(obtain use of)* mieten; ~ **sth. from sb.** etw. bei jmdm. mieten; **c)** *(grant use of)* vermieten; ~ **sth. to sb.** etw. jmdm. *od.* an jmdn. vermieten

~ **'out** *v. t.* vermieten

hire: ~**-car** *n.* Mietwagen, *der;* Leihwagen, *der;* ~**-'purchase** *n., no pl., no art. (Brit.)* Ratenkauf, *der;* Teilzahlungskauf, *der; attrib.* Raten-; Teilzahlungs-; **pay for/buy sth. on ~-purchase** etw. in Raten bezahlen/auf Raten *od.* Teilzahlung kaufen

his [ɪz, *stressed* hɪz] *poss. pron.* **a)** *attrib.* sein; *referring to personified things or animals which correspond to German feminines/neuters* ihr/sein; *see also* **²her; b)** *pred. (the one[s] belonging to him)* seiner/seine/sein[e]s; der/die/das seine *od.* seinige *(geh.); see also* **hers**

hiss [hɪs] **1.** *n. (of goose, snake, escaping steam, crowd, audience)* Zischen, *das; (of cat, locomotive)* Fauchen, *das.* **2.** *v. i.* ‹*Gans, Schlange, Dampf, Publikum, Menge:*› zischen; ‹*Katze, Lokomotive:*› fauchen. **3.** *v. t.* auszischen ‹*Redner, Schauspieler*›

historian [hɪ'stɔ:rɪən] *n.* **a)** *(writer of history)* Geschichtsschreiber, *der*/-schreiberin, *die;* **b)** *(scholar of history)* Historiker, *der*/Historikerin, *die*

historic [hɪ'stɒrɪk] *adj.* historisch

historical [hɪ'stɒrɪkl] *adj.* **a)** historisch; geschichtlich ‹*Belege, Hintergrund*›; **b)** *(belonging to the past)* in früheren Zeiten üblich ‹*Methode*›

history ['hɪstərɪ] *n.* **a)** *(continuous record)* Geschichte, *die;* **histories** historische Darstellungen; **b)** *no pl., no art.* Geschichte, *die; (study of past events)* Geschichte, *die;* Geschichtswissenschaft, *die;* **make ~:** Geschichte machen; **c)** *(train of events)* Geschichte, *die; (of person)* Werdegang, *der;* **have a ~ of asthma/shop-lifting** schon lange an Asthma leiden/eine Vorgeschichte als Ladendieb haben; **d)** *(eventful past career)* Geschichte, *die*

'history book *n.* Geschichtsbuch, *das*

hit [hɪt] **1.** *v. t.,* **-tt-, hit a)** *(strike with blow)* schlagen; *(strike with missile)* treffen; ‹*Geschoß, Ball usw.:*› treffen; **I've been ~!** *(struck by bullet)* ich bin getroffen!; **I could ~ him** *(fig. coll.)* ich könnte ihm eine runterhauen *(ugs.);* ~ **sb. over the head** jmdm. eins überziehen *(ugs.);* ~ **by lightning** vom Blitz getroffen; **b)** *(come forcibly into contact with)* ‹*Fahrzeug:*› prallen gegen ‹*Mauer usw.*›; ‹*Schiff:*› laufen gegen ‹*Felsen usw.*›; **the aircraft ~ the ground** das Flugzeug schlug auf den Boden auf; ~ **the roof** *or* **ceiling** *(fig. coll.: become angry)* an die Decke *od.* in die Luft gehen *(ugs.);* **c)** *(cause to come into contact)* [an]stoßen; [an]schlagen; ~ **one's head on sth.** mit dem Kopf gegen etw. stoßen; sich *(Dat.)* den Kopf an etw. *(Dat.)* stoßen; **d)** *(fig.: cause to suffer)* ~ **badly** *or* **hard** schwer treffen; **e)** *(fig.: affect)* treffen; **have been ~ by frost/rain** *etc.* durch Frost/Regen *usw.* gelitten haben; **f)** *(fig.: light upon)* finden; stoßen *od.* treffen auf (+ *Akk.*); finden ‹*Bodenschätze*›; **g)** *(fig. coll.: arrive at)* erreichen ‹*Höchstform, bestimmten Ort, bestimmte Höhe, bestimmtes Alter usw.*›; **I think we've ~ a snag** ich glaube, jetzt gibt's Probleme; ~ **town** ankommen; *see also* **all-time; h)** *(fig. coll.: indulge in)* zuschlagen bei (+ *Dat.*) *(salopp);* **[begin to] ~ the bottle** das Trinken anfangen; **i)** *(Cricket)* erzielen ‹*Lauf*›; ~ **the ball for six** *(Brit.)* sechs Läufe auf einmal erzielen; ~ **sb. for six** *(fig.)* jmdn. übertrumpfen. **2.** *v. i.,* **-tt-, hit a)** *(direct a blow)* schlagen; ~ **hard** fest *od.* hart zuschlagen; ~ **at sb./sth.** auf jmdn./etw. einschlagen; ~ **and run** ‹*Autofahrer:*› Fahrer- *od.* Unfallflucht begehen; ‹*Angreifer:*› einen Blitzüberfall machen; **b)** *(come into forcible contact)* ~ **against** *or* **upon sth.** gegen *od.* auf etw. *(Akk.)* stoßen. **3.** *n.* **a)** *(blow)* Schlag, *der;* **b)** *(shot or bomb striking target)* Treffer, *der;* **c)** *(success)* Erfolg, *der;* Knüller, *der (ugs.); (success in entertainment)* Schlager, *der;* Hit, *der (ugs.);* **make a ~:** gut ankommen

~ **'back 1.** *v. t.* zurückschlagen. **2.** *v. i.* zurückschlagen; *(verbally)* kontern; sich wehren; ~ **back at sb.** *(fig.)* jmdm. Kontra geben

~ **'off** *v. t.* **a)** *(characterize)* genau treffen; treffend charakterisieren; **b)** ~ **it off [with each other]** gut miteinander auskommen; ~ **it off with sb.** gut mit jmdm. auskommen

~ '**out** *v.i.* drauflosschlagen; ~ **out at** *or* **against sb./sth.** *(fig.)* jmdn./etw. scharf angreifen
~ **upon** *v.t.* stoßen auf (+ *Akk.*); finden ‹*richtige Antwort, Methode*›; kommen auf (+ *Akk.*) ‹*Idee*›
hit: ~**-and-'miss** *see* ~**-or-miss**; ~**-and-'run** *adj.* unfallflüchtig ‹*Fahrer*›; ~**-and-run accident** Unfall mit Fahrerflucht
hitch [hɪtʃ] **1.** *v.t.* **a)** *(move by a jerk)* rücken; **b)** *(fasten)* [fest]binden ‹*Tier*› (**to** an + *Akk.*); binden ‹*Seil*› (**round** um + *Akk.*); [an]koppeln ‹*Anhänger usw.*› (**to** an + *Akk.*); spannen ‹*Zugtier, -maschine usw.*› (**to** vor + *Akk.*). **2.** *v.i. see* **hitch-hike 1. 3.** *n.* **a)** *(stoppage)* Unterbrechung, *die;* **b)** *(impediment)* Problem, *das;* Schwierigkeit, *die;* **have one** ~: einen Haken haben *(ugs.)*
~ '**up** *v.t.* hochheben ‹*Rock*›
hitch: ~**-hike 1.** *v.i.* per Anhalter fahren; trampen; **2.** *n.* Tramptour, *die;* ~**-hiker** *n.* Anhalter, *der*/Anhalterin, *die;* Tramper, *der*/Tramperin, *die;* ~**-hiking** *n.* Trampen, *das*
hither ['hɪðə(r)] *adv.* *(literary)* hierher; ~ **and thither** *or* **yon** hierhin und dorthin
hitherto ['hɪðətʊ, hɪðə'tu:] *adv.* *(literary)* bisher; bislang
hit: ~ **man** *n.* *(sl.)* Killer, *der (salopp);* ~**-or-'miss** *adj.* *(coll.)* *(random)* unsicher, unzuverlässig ‹*Methode*›; *(careless)* schlampig, schluderig *(ugs. abwertend)* ‹*Arbeit*›; **it was a very** ~**-or-miss affair** das ging alles aufs Geratewohl *(ugs.);* ~ **parade** *n.* Hitparade, *die;* ~ '**record** *n.* Hit, *der* *(ugs.)*
HIV *abbr.* *(Med.)* **human immunodeficiency virus** HIV; **HIV-positive/-negative** HIV-positiv/-negativ
hive [haɪv] *n.* [Bienen]stock, *der;* *(of straw)* Bienenkorb, *der;* **what a** ~ **of industry!** der reinste Bienenstock! *(ugs.)*
~ '**off** *(Brit.)* *v.t.* *(separate and make independent)* verselbständigen; **the firm was** ~**d off from the parent company** die Firma wurde aus der Muttergesellschaft ausgegliedert
HM *abbr.* *(Brit.)* **a)** **Her/His Majesty** I.M./S.M.; **b)** **Her/His Majesty's**
HMS *abbr.* *(Brit.)* **Her/His Majesty's Ship** H.M.S.
ho [həʊ] *int. expr. surprise* oh; nanu; *expr. admiration* oh; *expr. triumph* ha; *drawing attention* he; heda; *expr. derision* haha
hoard [hɔ:d] **1.** *n.* **a)** *(store laid by)* Vorrat, *der;* **make/collect a** ~ **of sth.** etw. horten; **b)** *(fig.: amassed stock)* Sammlung, *die.* **2.** *v.t.* ~ **[up]** horten ‹*Geld, Brennmaterial, Lebensmittel usw.*›; hamstern ‹*Lebensmittel*›
hoarder ['hɔ:də(r)] *n.* Hamsterer, *der*/Hamsterin, *die*
hoarding ['hɔ:dɪŋ] *n.* **a)** *(fence)* Bretterzaun, *der;* Bretterwand, *die;* *(round building-site)* Bauzaun, *der;* **b)** *(Brit.: for advertisements)* Reklamewand, *die;* Plakatwand, *die*
hoar-frost ['hɔ:frɒst] *n.* [Rauh]reif, *der*
hoarse [hɔ:s] *adj.* **a)** *(rough, husky)* heiser, rauh ‹*Stimme*›; *(croaking)* krächzend ‹*Laut*›; *(with emotion)* belegt ‹*Stimme*›; **b)** *(having a dry, husky voice)* heiser
hoary ['hɔ:rɪ] *adj.* **a)** *(grey)* grau; ergraut *(geh.);* *(white)* [schloh]weiß; **b)** *(very old)* altehrwürdig ‹*Gebäude*›; ~ **old joke** uralter Witz
hoax [həʊks] **1.** *v.t.* anführen *(ugs.);* foppen; zum besten haben *od.* halten; ~ **sb. into believing sth.** jmdm. etw. weismachen. **2.** *n.* *(deception)* Schwindel, *der;* *(false report)* Falschmeldung, *die;* Ente, *die (ugs.);* *(practical joke)* Streich, *der;* *(false alarm)* blinder Alarm
hob [hɒb] *n.* *(of cooker)* Kochmulde, *die (Fachspr.);* [Koch]platte, *die;* Kochstelle, *die*
hobble ['hɒbl] **1.** *v.i.* ~ **[about]** [herum]humpeln *od.* -hinken. **2.** *n.* Humpeln, *das;* Hinken, *das*
hobby ['hɒbɪ] *n.* Hobby, *das;* Steckenpferd, *das*
'**hobby-horse** *n.* *(child's toy)* Steckenpferd, *das*
hob: ~**nail** *n.* [starker] Schuh- *od.* Stiefelnagel; ~**nailed** ['hɒbneɪld] *adj.* Nagel‹*schuh, -stiefel*›
hobo ['həʊbəʊ] *n., pl.* ~**es** *(Amer.)* Landstreicher, *der*/-streicherin, *die*
[1]**hock** [hɒk] *n.* *(Brit.: wine)* Rheinwein, *der*
[2]**hock** *(esp. Amer.)* **1.** *v.t.* versetzen. **2.** *n.* **be in** ~: versetzt sein
hockey ['hɒkɪ] *n.* Hockey, *das*
'**hockey-stick** *n.* Hockeystock, *der;* Hockeyschläger, *der*
hocus-pocus [həʊkəs'pəʊkəs] *n.* Zauberei, *die*
hod [hɒd] *n.* Tragmulde, *die*
hoe [həʊ] **1.** *n.* Hacke, *die.* **2.** *v.t.* hacken ‹*Beet, Acker*›
hog [hɒg] **1.** *n.* **a)** *(domesticated pig)* [Mast]schwein, *das;* **go the whole** ~ *(coll.)* Nägel mit Köpfen machen *(ugs.);* **b)** *(fig.: person)* Schwein, *das (derb);* Sau, *die (derb);* Ferkel, *das (derb).* **2.** *v.t.*, **-gg-** *(coll.)* mit Beschlag belegen
Hogmanay ['hɒgməneɪ] *n.* *(Scot., N. Engl.)* Silvester, *der od. das*
hoist [hɔɪst] **1.** *v.t.* **a)** *(raise aloft)* hoch-, aufziehen, hissen ‹*Flagge usw.*›; heißen *(Seemannsspr.)* ‹*Flagge usw.*›; **b)** *(raise by tackle etc.)* hieven ‹*Last*›; setzen ‹*Segel*›. **2.** *n.* *(goods lift)* [Lasten]aufzug, *der.* **3.** *adj.* **be** ~ **with one's own petard** sich in seiner eigenen Schlinge fangen
hoity-toity [hɔɪtɪ'tɔɪtɪ] *adj.* *(coll.)* hochnäsig *(abwertend);* eingebildet; *(petulant)* pikiert
[1]**hold** [həʊld] *n.* *(of ship)* Laderaum, *der;* *(of aircraft)* Frachtraum, *der*
[2]**hold 1.** *v.t.*, **held** [held] **a)** *(grasp)* halten; *(carry)* tragen; *(keep fast)* festhalten; ~ **sb. by the arm** jmdn. am Arm festhalten; **b)** *(support)* ‹*tragendes Teil:*› halten, stützen, tragen ‹*Decke, Dach usw.*›; aufnehmen ‹*Gewicht, Kraft*›; **c)** *(keep in position)* halten; ~ **the door open for sb.** jmdm. die Tür aufhalten; **d)** *(grasp to control)* halten ‹*Kind, Hund, Zügel*›; **e)** *(keep in particular attitude)* ~ **oneself still** stillhalten; ~ **oneself ready** *or* **in readiness** sich bereit *od.* in Bereitschaft halten; ~ **one's head high** *(fig.)* *(be confident)* selbstbewußt sein *od.* auftreten; *(be proud)* den Kopf hoch tragen; **f)** *(contain)* enthalten; bergen ‹*Gefahr, Geheimnis*›; *(be able to contain)* fassen ‹*Liter, Personen usw.*›; **the room** ~**s ten people** in dem Raum haben 10 Leute Platz; der Raum bietet 10 Leuten Platz; ~ **water** ‹*Behälter:*› wasserdicht sein; Wasser halten; *(fig.)* ‹*Argument, Theorie:*› stichhaltig sein, hieb- und stichfest sein; **g)** *(not be intoxicated by)* **he can/can't** ~ **his drink** *or* **liquor** er kann etwas/nichts vertragen; **h)** *(possess)* besitzen; haben; **i)** *(have gained)* halten ‹*Rekord*›; haben ‹*Diplom, Doktorgrad*›; **j)** *(keep possession of)* halten ‹*Stützpunkt, Stadt, Stellung*›; *(Mus.: sustain)* [aus]halten ‹*Ton*›; ~ **one's own** *(fig.)* sich behaupten; ~ **one's position** *(fig.)* auf seinem Standpunkt beharren; **k)** *(occupy)* innehaben, *(geh.)* bekleiden ‹*Posten, Amt, Stellung*›; ~ **office** im Amt sein; ~ **the line** *(Teleph.)* am Apparat bleiben; **l)** *(engross)* fesseln, *(geh.)* gefangenhalten ‹*Auf-*

merksamkeit, Publikum⟩; **m)** *(keep in specified condition)* halten; ~ **the ladder steady** die Leiter festhalten; *see also* **[3]bay 1; ransom 1; n)** *(detain) (in custody)* in Haft halten, festhalten; *(imprison)* festsetzen; inhaftieren; *(arrest)* festnehmen; **be held in a prison** in einem Gefängnis einsitzen; **o)** *(oblige to adhere)* ~ **sb. to the terms of the contract/to a promise** darauf bestehen, daß jmd. sich an die Vertragsbestimmungen hält/daß jmd. ein Versprechen hält *od.* einlöst; **p)** *(Sport: restrict)* ~ **one's opponent [to a draw]** ein Unentschieden [gegen den Gegner] halten *od.* verteidigen; **q)** *(cause to take place)* stattfinden lassen; abhalten ⟨*Veranstaltung, Konferenz, Gottesdienst, Sitzung, Prüfung*⟩; veranstalten ⟨*Festival, Auktion*⟩; austragen ⟨*Meisterschaften*⟩; führen ⟨*Unterhaltung, Gespräch, Korrespondenz*⟩; durchführen ⟨*Untersuchung*⟩; geben ⟨*Empfang*⟩; halten ⟨*Vortrag, Rede*⟩; **be held** stattfinden; **r)** *(restrain)* [fest]halten; ~ **one's fire** [noch] nicht schießen; *(fig.: refrain from criticism)* mit seiner Kritik zurückhalten; **s)** *(coll.: withhold)* zurückhalten; ~ **it!** [einen] Moment mal!; *see also* **horse a; t)** *(think, believe)* ~ **a view** *or* **an opinion** eine Ansicht haben (**on** über + *Akk.*); ~ **that ...:** dafürhalten, daß ...; der Ansicht sein, daß ...; ~ **sb./oneself guilty/blameless** jmdn./sich für schuldig/unschuldig halten (**for** an + *Dat.*); ~ **oneself responsible for sth.** sich für etw. verantwortlich fühlen; ~ **sth. against sb.** jmdm. etw. vorwerfen; *see also* **dear 1a; responsible a. 2.** *v. i.*, **held a)** *(not give way)*, ⟨*Seil, Nagel, Anker, Schloß, Angeklebtes:*⟩ halten; ⟨*Damm:*⟩ [stand]halten; **b)** *(remain unchanged)* anhalten; [an]dauern; ⟨*Wetter:*⟩ sich halten, so bleiben; ⟨*Angebot, Versprechen:*⟩ gelten; **his luck held** er hatte auch weiterhin Glück; **c)** *(remain steadfast)* ~ **to sth.** bei etw. bleiben; an etw. *(Dat.)* festhalten; **d)** *(be valid)* ~ **[good** *or* **true]** gelten; Gültigkeit haben. **3.** *n.* **a)** *(grasp)* Griff, *der;* **grab** *or* **seize** ~ **of sth.** etw. ergreifen; **get** *or* **lay** *or* **take** ~ **of sth.** etw. fassen *od.* packen; **keep** ~ **of sth.** etw. festhalten; **lose one's** ~**:** den Halt verlieren; **take** ~ *(fig.)* sich durchsetzen; ⟨*Krankheit:*⟩ fortschreiten; **get** ~ **of sth.** *(fig.)* etw. bekommen *od.* auftreiben; **get** ~ **of sb.** *(fig.)* jmdn. erreichen; **get a** ~ **on oneself** sich fassen; **have a** ~ **over sb.** jmdn. in der Hand halten; *see also* **catch 1a; b)** *(influence)* Einfluß, *der* (**on, over** auf + *Akk.*); **c)** *(Sport)* Griff, *der;* **there are no** ~**s barred** *(fig.)* alles ist erlaubt; **d)** *(thing to hold by)* Griff, *der;* **e) put on** ~**:** auf Eis legen ⟨*Plan, Programm*⟩

~ **'back 1.** *v. t.* **a)** *(restrain)* zurückhalten; ~ **sb. back from doing sth.** jmdn. [daran] hindern, etw. zu tun; **b)** *(impede progress of)* hindern; **c)** *(withhold)* zurückhalten; ~ **sth. back from sb.** jmdm. etw. vorenthalten. **2.** *v. i.* zögern; ~ **back from doing sth.** zögern, etw. zu tun

~ **'down** *v. t.* **a)** festhalten; *(repress)* unterdrücken; niederhalten ⟨*Volk*⟩; *(fig.: keep at low level)* niedrig halten ⟨*Preise, Löhne usw.*⟩; **b)** *(keep)* sich halten in (+ *Dat.*) ⟨*Stellung, Position*⟩

~ **'forth 1.** *v. t. (offer)* anpreisen. **2.** *v. i.* sich in langen Reden ergehen; ~ **forth about** *or* **on sth.** sich über etw. *(Akk.)* auslassen

~ **'off 1.** *v. t. (keep at bay)* von sich fernhalten, *(ugs.)* sich *(Dat.)* vom Leib halten ⟨*Fans, Presse*⟩; abwehren ⟨*Angriff*⟩. **2.** *v. i.* ⟨*Käufer usw.:*⟩ sich zurückhalten; ⟨*Feind:*⟩ sich ruhig verhalten; ⟨*Regen, Monsun, Winter:*⟩ ausbleiben, auf sich *(Akk.)* warten lassen

~ **'on 1.** *v. t. (keep in position)* [fest]halten. **2.** *v. i.* **a)** sich festhalten; ~ **on to sb./sth.** sich an jmdm./etw. festhalten; *(fig.: retain)* jmdn./etw. behalten; **b)** *(stand firm)* durchhalten; aushalten; **c)** *(Teleph.)* am Apparat bleiben; dranbleiben *(ugs.)*; **d)** *(coll.: wait)* warten; ~ **on!** einen Moment!

~ **'out 1.** *v. t.* **a)** ausstrecken ⟨*Hand, Arm usw.*⟩; ausbreiten ⟨*Arme*⟩; hinhalten ⟨*Tasse, Teller*⟩; **b)** *(fig.: offer)* in Aussicht stellen (**to** *Dat.*); **he did not** ~ **out much hope** er hat mir/dir *usw.* nicht viel Hoffnung gemacht. **2.** *v. i.* **a)** *(maintain resistance)* sich halten; **b)** *(last)* ⟨*Vorräte:*⟩ vorhalten; ⟨*Motor:*⟩ halten; **c)** ~ **out for sth.** etw. herauszuschinden versuchen *(ugs.)*

~ **'over** *v. t.* vertagen (**till** auf + *Akk.*)

~ **'up 1.** *v. t.* **a)** *(raise)* hochhalten; hochheben ⟨*Person*⟩; [hoch]heben ⟨*Hand, Kopf*⟩; **b)** *(fig.: offer as an example)* ~ **sb. up as ...:** jmdn. als ... hinstellen; ~ **sb./sth. up to ridicule/scorn** jmdn./etw. dem Spott/Hohn preisgeben; **c)** *(support)* stützen; tragen ⟨*Dach usw.*⟩; ~ **sth. up with sth.** etw. mit etw. abstützen; **d)** *(delay)* aufhalten; behindern ⟨*Verkehr, Versorgung*⟩; verzögern ⟨*Friedensvertrag*⟩; *(halt)* ins Stocken bringen ⟨*Produktion*⟩; **e)** *(rob)* überfallen [und ausrauben]. **2.** *v. i. (under scrutiny)* sich als stichhaltig erweisen

~ **with** *v. t.* ~**/not** ~ **with sth.** mit etw. einverstanden sein/etw. ablehnen

'holdall [həʊldɔ:l] *n.* Reisetasche, *die*

holder ['həʊldə(r)] *n.* **a)** *(of post)* Inhaber, *der*/Inhaberin, *die;* **b)** *(of title)* Träger, *der*/Trägerin, *die; (Sport)* Titelhalter, -inhaber, *der;* **c)** ⟨*Zigaretten*⟩spitze, *die;* ⟨*Papier-, Feder-, Zahnputzglas*⟩halter, *der*

'hold-up *n.* **a)** *(robbery)* [Raub]überfall, *der;* **b)** *(stoppage)* Unterbrechung, *die; (delay)* Verzögerung, *die*

hole [həʊl] **1.** *n.* **a)** Loch, *das;* **make a** ~ **in sth.** *(fig.)* eine ganze Menge von etw. verschlingen; **pick** ~**s in** *(fig.: find fault with)* zerpflücken *(ugs.)*; auseinandernehmen *(ugs.)*; madig machen *(ugs.)* ⟨*Person*⟩; ~ **in the heart** Loch in der Herzscheidewand; **b)** *(burrow) (of fox, badger, rabbit)* Bau, *der; (of mouse)* Loch, *das;* **c)** *(coll.) (dingy abode)* Loch, *das (salopp abwertend); (wretched place)* Kaff, *das (ugs. abwertend);* Nest, *das (ugs. abwertend);* **d)** *(Golf)* Loch, *das; (space between tee and* ~*)* [Spiel]bahn, *die;* ~ **in one** Hole-in-One, *das;* As, *das.* **2.** *v. t.* **a)** Löcher/ein Loch machen in (+ *Akk.*); **b)** *(Naut.)* **be** ~**d** leckschlagen *(Seemannsspr.)*

~ **'up** *v. i. (Amer. coll.)* sich verkriechen *(ugs.)*

holiday ['hɒlɪdeɪ, 'hɒlɪdɪ] **1.** *n.* **a)** *(day of recreation)* [arbeits]freier Tag; *(day of festivity)* Feiertag, *der;* **tomorrow is a** ~**:** morgen ist frei/Feiertag; **b)** *in sing. or pl. (Brit.: vacation)* Urlaub, *der; (Sch.)* [Schul]ferien *Pl.*; **need a** ~**:** urlaubsreif sein; **have a good** ~**!** schönen Urlaub!; **take** *or* **have a/one's** ~**:** Urlaub nehmen *od.* machen/seinen Urlaub nehmen; **on** ~**, on one's** ~**s** im *od.* in seinem Urlaub. **2.** *attrib. adj.* Urlaubs-/Ferien⟨*stimmung, -pläne*⟩. **3.** *v. i.* Urlaub/Ferien machen; urlauben *(ugs.)*

holiday: ~ **camp** *n.* Feriendorf, *das;* Ferienpark, *der;* ~**-maker** *n.* Urlauber, *der*/Urlauberin, *die;* ~ **resort** *n.* Ferienort, *der*

holiness ['həʊlɪnɪs] *n., no pl.* Heiligkeit, *die;* **His H~:** Seine Heiligkeit

Holland ['hɒlənd] *pr. n.* Holland *(das)*

hollow ['hɒləʊ] **1.** *adj.* **a)** *(not solid)* hohl; Hohl‹*ziegel, -mauer, -zylinder, -kugel*›; **b)** *(sunken)* eingefallen ‹*Wangen, Schläfen*›; hohl, tiefliegend ‹*Augen*›; **c)** *(echoing)* hohl ‹*Ton, Klang*›; **d)** *(fig.: empty)* wertlos; **e)** *(fig.: cynical)* verlogen; leer ‹*Versprechen*›; gequält ‹*Lachen*›. **2.** *n.* [Boden]senke, *die;* [Boden]vertiefung, *die;* **hold sth. in the ~ of one's hand** etw. in der hohlen Hand halten. **3.** *adv.* **beat sb. ~** *(coll.)* jmdn. um Längen schlagen *(ugs.)*. **4.** *v. t.* ~ **out** aushöhlen; graben ‹*Höhle*›

holly ['hɒlɪ] *n. (tree)* Stechpalme, *die;* Ilex, *der (fachspr.)*

'**hollyhock** *n. (Bot.)* Stockrose, *die*

holocaust ['hɒləkɔ:st] *n.* Massenvernichtung, *die;* **the H~:** der Holocaust; die Judenvernichtung

hologram ['hɒləgræm] *n.* Hologramm, *das*

holster ['həʊlstə(r)] *n.* [Pistolen]halfter, *die od. das*

holy ['həʊlɪ] *adj.* heilig; fromm ‹*Zweck*›; ~ **saints** Heilige

Holy: ~ '**Bible** *n.* Heilige Schrift; ~ **Com'munion** *see* **communion a;** ~ '**Ghost** *see* ~ **Spirit;** ~ **Grail** *n.* Heiliger Gral; ~ **Land** *n.* **the ~ Land** das Heilige Land; ~ **Roman 'Empire** *n. (Hist.)* Heiliges Römisches Reich [Deutscher Nation]; ~ '**Spirit** *n. (Relig.)* Heiliger Geist; ~ **Week** *n.* Karwoche, *die*

homage ['hɒmɪdʒ] *n. (tribute)* Huldigung, *die* (**to** an + *Akk.*); **pay** *or* **do ~ to sb./sth.** jmdm./einer Sache huldigen

home [həʊm] **1.** *n.* **a)** Heim, *das; (flat)* Wohnung, *die; (house)* Haus, *das; (household)* [Eltern]haus, *das;* **my ~ is in Leeds** ich bin in Leeds zu Hause *od.* wohne in Leeds; **a ~ of one's own** ein eigenes Zuhause; **leave/have left ~:** aus dem Haus gehen/sein; **live at ~:** im Elternhaus wohnen; **they had no ~/~s [of their own]** sie hatten kein Zuhause; **at ~:** zu Hause; *(not abroad)* im Inland; **be/feel at ~** *(fig.)* sich wohl fühlen; **make sb. feel at ~:** es jmdm. behaglich machen; **make yourself at ~:** fühl dich wie zu Hause; **he is quite at ~ in French** er ist im Französischen ganz gut zu Hause; ~ **from ~:** zweites Zuhause; **b)** *(fig.)* **to take an example nearer ~, ...:** um ein Beispiel zu nehmen, das uns näher liegt, ...; **c)** *(native country)* die Heimat; **at ~:** zu Hause; in der Heimat; **d)** *(institution)* Heim, *das; (coll.: mental ~)* Anstalt, *die (salopp)*. **2.** *adj.* **a)** *(connected with ~)* Haus-; Haushalts‹*gerät usw.*›; **b)** *(done at ~)* häuslich; Selbst‹*backen, ~brauen usw.*›; **c)** *(in the neighbourhood of ~)* nahegelegen; **d)** *(Sport)* Heim‹*spiel, -sieg, -mannschaft*›; ‹*Anhänger, Spieler*› der Heimmannschaft; ~ **ground** eigener Platz; **e)** *(not foreign)* [ein]heimisch; inländisch. **3.** *adv.* **a)** *(to ~)* nach Hause; **on one's way ~:** auf dem Weg nach Hause *od.* Nachhauseweg; **he takes ~ £200 a week after tax** er verdient 200 Pfund netto in der Woche; **nothing to write ~ about** *(coll.)* nichts Besonderes *od.* Aufregendes; **b)** *(arrived at ~)* zu Hause; **be ~ and dry** *(fig.)* aus dem Schneider sein *(ugs.)*; **c)** *(as far as possible)* **push ~:** [ganz] hineinschieben ‹*Schublade*›; ausnutzen ‹*Vorteil*›; **press ~:** [ganz] hinunterdrücken ‹*Hebel*›; forcieren ‹*Angriff*›; [voll] ausnutzen ‹*Vorteil*›; **drive ~:** [ganz] einschlagen ‹*Nagel*›; **d) come** *or* **get ~ to sb.** *(become fully realized)* jmdm. in vollem Ausmaß bewußt werden; *see also* **roost 1. 4.** *v. i.* **a)** ‹*Vogel usw.*:› zurückkehren; **b)** *(be guided)* **these missiles ~ [in] on their targets** diese Flugkörper suchen sich *(Dat.)* ihr Ziel; **c)** ~ **in/on sth.** *(fig.)* etw. herausgreifen

home: ~ **address** *n.* Privatanschrift, *die;* ~ '**brew** *n.* selbstgebrautes Bier; ~**-coming** *n.* Heimkehr, *die;* ~ **com'puter** *n.* Heimcomputer, *der;* **H~ Counties** *n. pl. (Brit.)* **the H~ Counties** die Home Counties; *die Grafschaften um London;* ~ **eco'nomics** *n. sing. see* **domestic science;** ~ '**ground** *n.* **on [one's] ~ ground** auf heimischem Boden; *(fig.)* zu Hause *(ugs.)*; ~**-grown** *adj.* selbstgezogen ‹*Gemüse, Obst*›; ~ '**help** *n. (Brit.)* Haushaltshilfe, *die;* ~**land** *n.* **a)** Heimat, *die;* Heimatland, *das;* **b)** *(in South Africa)* Homeland, *das*

homeless ['həʊmlɪs] **1.** *adj.* obdachlos. **2.** *n.* **the ~:** die Obdachlosen

'**home-loving** *adj.* häuslich

homely ['həʊmlɪ] *adj.* einfach, schlicht ‹*Worte, Stil, Sprache usw.*›; warmherzig ‹*Person*›

home: ~**-made** *adj.* selbstgemacht; selbstgebacken ‹*Brot*›; hausgemacht ‹*Lebensmittel*›; ~ '**movie** *n.* Amateurfilm, *der;* **H~ Office** *n. (Brit.)* Innenministerium, *das*

homeopathic *etc. (Amer.) see* **homoeo-**

home: H~ 'Secretary *n. (Brit.)* Innenminister, *der;* ~**sick** *adj.* heimwehkrank; **become/be ~sick** Heimweh bekommen/haben; ~**spun** *adj.* **a)** *(spun [and woven] at ~)* selbstgesponnen [und -gewoben]; *(of ~ manufacture)* in Heimarbeit gesponnen; **b)** *(unsophisticated)* schlicht; einfach; ~ '**town** *n.* Heimatstadt, *die;* Vaterstadt, *die (geh.)*; *(town of residence)* Wohnort, *der;* ~ '**truth** *n.* unangenehme Wahrheit; **tell sb. a few ~ truths** jmdm. [gehörig] die Meinung sagen

homeward ['həʊmwəd] **1.** *adj.* nach Hause *nachgestellt;* Nachhause‹*weg*›; *(return)* Rück‹*fahrt, -reise, -weg*›; *see also* [3]**bound. 2.** *adv.* nach Hause; heimwärts

'**homework** *n. (Sch.)* Hausaufgaben *Pl.*; **piece of ~:** Hausaufgabe, *die;* **do one's ~** *(fig.)* sich mit der Materie vertraut machen; seine Hausaufgaben machen *(scherzh.)*

homicidal [hɒmɪ'saɪdl] *adj.* gemeingefährlich

homicide ['hɒmɪsaɪd] *n.* Tötung, *die; (manslaughter)* Totschlag, *der*

homily ['hɒmɪlɪ] *n.* **a)** *(sermon)* Homilie, *die (Theol.)*; **b)** *(tedious talk)* Moralpredigt, *die*

homing ['həʊmɪŋ] *attrib. adj.* zielsuchend ‹*Flugkörper, Torpedo*›; ~ **instinct** Heimfindevermögen, *das*

'**homing pigeon** *n.* Brieftaube, *die*

homo- ['həʊməʊ, 'hɒməʊ] *in comb.* homo-/Homo-

homoeopathic [həʊmɪə'pæθɪk, hɒmɪə'pæθɪk] *adj.* homöopathisch

homoeopathy [həʊmɪ'ɒpəθɪ, hɒmɪ'ɒpəθɪ] *n.* Homöopathie, *die*

homogeneous [hɒmə'dʒi:nɪəs, həʊmə'dʒi:nɪəs] *adj.* homogen

homogenize (homogenise) [hə'mɒdʒɪnaɪz] *v. t. (lit. or fig.)* homogenisieren

homonym ['hɒmənɪm] *n. (Ling.)* Homonym, *das*

homo'sexual 1. *adj.* homosexuell. **2.** *n.* Homosexuelle, *der/die*

homosexu'ality *n., no pl.* Homosexualität, *die*

Honduras [hɒn'djʊərəs] *pr. n.* Honduras *(das)*

hone [həʊn] *v.t.* wetzen ‹*Messer, Klinge usw.*›

honest ['ɒnɪst] *adj.* **a)** ehrlich; *(showing righteousness)* redlich; ehrenhaft ‹*Absicht, Tat, Plan*›; ehrlich ‹*Arbeit*›; **the ~ truth** die reine Wahrheit; **make an ~ living** sein Leben auf ehrliche Weise verdienen; **b)** *(unsophisticated)* [gut und] einfach; *(unadulterated)* rein

honestly ['ɒnɪstlɪ] *adv.* ehrlich; redlich ‹*handeln*›; **~!** ehrlich!; *(annoyed)* also wirklich!

honesty ['ɒnɪstɪ] *n.* Ehrlichkeit, *die; (upright conduct)* Redlichkeit, *die;* **in all ~:** ganz ehrlich; **~ is the best policy** *(prov.)* ehrlich währt am längsten *(Spr.)*

honey ['hʌnɪ] *n.* **a)** Honig, *der;* **b)** *(Amer., Ir.: darling)* Schatz, *der (ugs.)*

honey: ~-bee *n.* Honigbiene, *die;* **~comb** *n.* Bienenwabe, *die; (filled with ~)* Honigwabe, *die;* **~moon 1.** *n.* **a)** Flitterwochen *Pl.;* Honigmond, *der (scherzh.); (journey)* Hochzeitsreise, *die;* **go on one's ~moon** in die Flitterwochen fahren; **b)** *(fig.: initial period)* anfängliche Begeisterung; **2.** *v.i.* seine Flitterwochen verbringen; **~suckle** *n. (Bot.)* Geißblatt, *das*

honk [hɒŋk] **1.** *n.* **a)** *(of horn)* Hupen, *das;* **b)** *(of goose or seal)* Schrei, *der.* **2.** *v.i.* **a)** ‹*Fahrzeug, Fahrer:*› hupen; **b)** ‹*Gans, Seehund:*› schreien

honor, honorable *(Amer.) see* **honour, honourable**

honorary ['ɒnərərɪ] *adj.* **a)** ehrenamtlich; Ehren‹*mitglied, -präsident, -doktor, -bürger*›; **b)** *(conferred as an honour)* Ehren-; **~ degree** ehrenhalber verliehener akademischer Grad

honour ['ɒnə(r)] *(Brit.)* **1.** *n.* **a)** *no indef. art. (reputation)* Ehre, *die;* **do ~ to sb./sth.** jmdm./einer Sache zur Ehre gereichen *(geh.);* jmdm./einer Sache Ehre machen; **b)** *(respect)* Hochachtung, *die;* **do sb. ~, do ~ to sb.** jmdm. Ehre erweisen; *(show appreciation of)* jmdn. würdigen; **in ~ of sb.** jmdm. zu Ehren; **in ~ of sth.** um etw. gebührend zu feiern; **c)** *(privilege)* Ehre, *die;* **may I have the ~ [of the next dance]?** darf ich [um den nächsten Tanz] bitten?; **d)** *no art. (ethical quality)* Ehre, *die;* **he is a man of ~:** er ist ein Ehrenmann *od.* Mann von Ehre; **feel [in] ~ bound to do sth.** sich moralisch verpflichtet fühlen, etw. zu tun; **promise [up]on one's ~:** sein Ehrenwort geben; **e)** *(distinction)* Auszeichnung, *die; (title)* Ehrentitel, *der; in pl. (Univ.)* **she gained ~s in her exam, she passed [the exam] with ~s** sie hat das Examen mit Auszeichnung bestanden; **f)** *in pl.* **do the ~s** *(coll.) (introduce guests)* die Honneurs machen; *(serve guests)* den Gastgeber spielen; **g)** *in title* **your H~** *(Brit. Law)* hohes Gericht; Euer Ehren; **h)** *(person or thing that brings credit)* **be an ~ to sb./sth.** jmdm./einer Sache Ehre machen. **2.** *v.t.* **a)** ehren; würdigen ‹*Verdienste, besondere Eigenschaften*›; **be ~ed as an artist** als Künstler Anerkennung finden; **~ sb. with one's presence** *(iron.)* jmdn. mit seiner Gegenwart beehren; **b)** *(acknowledge)* beachten ‹*Vorschriften*›; respektieren ‹*Gebräuche, Rechte*›; **c)** *(fulfil)* sich halten an (+ *Akk.*); *(Commerc.)* honorieren; begleichen ‹*Rechnung, Schuld*›

honourable ['ɒnərəbl] *adj. (Brit.)* **a)** *(worthy of respect)* ehrenwert *(geh.);* **b)** *(bringing credit)* achtbar; *(consistent with honour)* ehrenvoll ‹*Frieden, Rückzug, Entlassung*›; **c)** *(ethical)* rechtschaffen; redlich ‹*Geschäftsgebaren*›; **d)** *in title* **the H~ ...:** ≈ der/die ehrenwerte ...; **the ~ gentleman/lady, the ~ member [for X]** *(Brit. Parl.)* der Herr/die Frau Abgeordnete [für den Wahlkreis X]; ≈ der [verehrte] Herr Kollege/die [verehrte] Frau Kollegin

hood [hʊd] *n.* **a)** Kapuze, *die;* **b)** *(of vehicle) (Brit.: waterproof top)* Verdeck, *das; (Amer.: bonnet)* [Motor]haube, *die*

hoodlum ['hu:dləm] *n. (young thug)* Rowdy, *der (abwertend)*

hoodoo ['hu:du:] *n.* **a)** *(bad spell)* Fluch, *der;* **there is a ~ on ...:** es liegt ein Fluch auf ... (+ *Dat.*); **b)** *(bringer of bad luck)* **be a ~:** Unglück bringen

hoodwink ['hʊdwɪŋk] *v.t.* hinters Licht führen; täuschen

hoof [hu:f] **1.** *n., pl.* **~s** *or* **hooves** [hu:vz] Huf, *der;* **buy cattle on the ~** *(for meat)* Lebendvieh kaufen. *See also* **cloven 2. 2.** *v.t. (sl.) (walk)* **~ it** tippeln *(ugs.)*

hook [hʊk] **1.** *n.* **a)** Haken, *der; (Fishing)* [Angel]haken, *der;* **~ and eye** Haken und Öse; **swallow sth. ~, line, and sinker** *(fig.)* etw. blind glauben; **get sb. off the ~** *(fig. sl.)* jmdn. herauspauken *(ugs.);* **that lets me/him off the ~** *(fig. sl.)* da bin ich/ist er noch einmal davongekommen; **by ~ or by crook** mit allen Mitteln; **b)** *(telephone cradle)* Gabel, *die;* **c)** *(Boxing)* Haken, *der.* **2.** *v.t.* **a)** *(grasp)* mit Haken/mit einem Haken greifen; **b)** *(fasten)* mit Haken/mit einem Haken befestigen (**to** an + *Dat.*); festhaken ‹*Tor*› (**to** an + *Akk.*); haken ‹*Bein, Finger*› (**over** über + *Akk.*, **in** in + *Akk.*); **c) be ~ed [on sth./sb.]** *(sl.) (addicted harmfully)* [von etw./jmdm.] abhängig sein; *(addicted harmlessly)* [auf etw./jmdn.] stehen *(ugs., bes. Jugendspr.); (captivated)* [von etw./jmdm.] fasziniert sein; **d)** *(catch)* an die Angel bekommen ‹*Fisch*›; *(fig.)* sich *(Dat.)* angeln

~ 'on 1. *v.t.* anhaken (**to** an + *Akk.*); anhängen ‹*Wagen, Anhänger*› (**to** an + *Akk.*). **2.** *v.i.* angehakt werden (**to** an + *Akk.*)

~ 'up 1. *v.t.* **a)** festhaken (**to** an + *Akk.*); zuhaken ‹*Kleid*›; **b)** *(Radio and Telev. coll.)* zusammenschalten ‹*Sender*›. **2.** *v.i.* ‹*Kleid:*› mit Haken geschlossen werden

hooked [hʊkt] *adj.* **a)** *(hook-shaped)* hakenförmig; **b)** *(having hook[s])* mit Haken/mit einem Haken versehen. *See also* **hook 2c**

hooker ['hʊkə(r)] *n.* **a)** *(Rugby)* Hakler, *der;* **b)** *(Amer. sl.: prostitute)* Nutte, *die (salopp)*

hook: ~-'nose *n.* Hakennase, *die;* **~-up** *n. (Radio and Telev. coll.)* Zusammenschaltung, *die (zu einer Gemeinschaftssendung)*

hooligan ['hu:lɪgən] *n.* Rowdy, *der*

hooliganism ['hu:lɪgənɪzm] *n., no pl.* Rowdytum, *das*

hoop [hu:p] *n.* Reifen, *der; (in circus, show, etc.)* Springreifen, *der;* **put sb. through the ~[s]** *(fig.)* jmdn. durch die Mangel drehen *(salopp)*

hooray *see* **hurray**

hoot [hu:t] **1.** *v.i.* **a)** *(call out)* johlen; **b)** ‹*Eule:*› schreien; **c)** ‹*Fahrzeug, Fahrer:*› hupen, tuten; ‹*Sirene, Nebelhorn usw.:*› heulen, tuten; **~ at sb./sth.** jmdn./etw. anhupen. **2.** *v.t.* heulen *od.* tuten lassen ‹*Sirene, Nebelhorn*›. **3.** *n.* **a)** *(shout)* **~s of derision/scorn** verächtliches Gejohle; **b)** *(owl's cry)* Schrei, *der;* **c)** *(signal) (of vehicle)* Hupen, *das; (of siren, fog-horn)* Heulen, *das;* Tuten, *das;* **d)** *(coll.)* **I don't care** *or* **give a**

~ *or* two ~s **what you do** es ist mir völlig piepegal *od.* schnuppe *(ugs.),* was du tust

hooter ['hu:tə(r)] *n. (Brit.)* **a)** *(siren)* Sirene, *die;* **b)** *(motor horn)* Hupe, *die*

hoover ['hu:və(r)] *(Brit.)* **1.** *n.* **a)** **H~ (P)** [Hoover]staubsauger, *der;* **b)** *(made by any company)* Staubsauger, *der.* **2.** *v. t. (coll.)* staubsaugen; saugen ⟨*Boden, Teppich*⟩; absaugen ⟨*Möbel*⟩. **3.** *v. i. (coll.)* [staub]saugen

¹hop [hɒp] *n. (Bot.) (plant)* Hopfen, *der; in pl. (cones)* Hopfendolden

²hop 1. *v. i.,* **-pp-: a)** hüpfen; ⟨*Hase:*⟩ hoppeln; **be ~ping mad [about** *or* **over sth.]** *(coll.)* [wegen etw.] fuchsteufelswild sein *(ugs.);* **b)** *(fig. coll.)* ~ **out of bed** aus dem Bett springen; ~ **into the car/on [to] the bus/train/bicycle** sich ins Auto/in den Bus/Zug/aufs Fahrrad schwingen *(ugs.);* ~ **off/out** aussteigen. **2.** *v. t.,* **-pp-: a)** *(jump over)* springen über (+ *Akk.*); **b)** *(coll.: jump aboard)* aufspringen auf (+ *Akk.*); **c)** ~ **it** *(Brit. sl.: go away)* sich verziehen *(ugs.).* **3.** *n.* **a)** *(action)* Hüpfer, *der;* Hopser, *der (ugs.);* **b) keep sb. on the ~** *(Brit. coll.: bustling about)* jmdn. in Trab halten *(ugs.);* **c) catch sb. on the ~** *(Brit. coll.: unprepared)* jmdn. überraschen *od.* überrumpeln; **d)** *(distance flown)* Flugstrecke, *die; (stage of journey)* Teilstrecke, *die;* Etappe, *die*

hope [həʊp] **1.** *n.* Hoffnung, *die;* **give up ~:** die Hoffnung aufgeben; **hold out ~ [for sb.]** [jmdm.] Hoffnung machen; **beyond** *or* **past ~:** hoffnungslos; **in the ~/in ~[s] of sth./doing sth.** in der Hoffnung auf etw. *(Akk.)*/, etw. zu tun; **I have some ~[s] of success** *or* **of succeeding** es besteht die Hoffnung, daß ich Erfolg habe; **set** *or* **put** *or* **place one's ~s on** *or* **in sth./sb.** seine Hoffnung auf etw./jmdn. setzen; **raise sb.'s ~s** jmdm. Hoffnung machen; **high ~s** große Hoffnungen; **have high ~s of sth.** sich *(Dat.)* große Hoffnungen auf etw. *(Akk.)* machen; **not have a ~ [in hell] [of sth.]** *(coll.)* sich *(Dat.)* keine[rlei] Hoffnung [auf etw. *(Akk.)*] machen können; **what a ~!** *(coll.),* **some ~[s]!** *(coll. iron.)* schön wär's!; **be hoping against ~ that ...:** trotz allem die Hoffnung nicht aufgeben, daß ... **2.** *v. i.* hoffen (**for** auf + *Akk.*); **I ~ so/not** hoffentlich/hoffentlich nicht; ich hoffe es/ich hoffe nicht; ~ **for the best** das Beste hoffen. **3.** *v. t.* ~ **to do sth./that sth. may be so** hoffen, etw. zu tun/daß etw. so eintrifft; **I ~ to go to Paris** *(am planning)* ich habe vor, nach Paris zu fahren

hopeful ['həʊpfl] **1.** *adj.* **a)** zuversichtlich; **I'm ~/not ~ that ...:** ich hoffe zuversichtlich/bezweifle, daß ...; **be ~ of sth./of doing sth.** auf etw. *(Akk.)* hoffen/voller Hoffnung sein, etw. zu tun; **b)** *(promising)* vielversprechend; aussichtsreich ⟨*Kapitalanlage, Kandidat*⟩. **2.** *n.* **[young] ~:** hoffnungsvoller junger Mensch

hopefully ['həʊpfəlɪ] *adv.* **a)** *(expectantly)* voller Hoffnung; **b)** *(promisingly)* vielversprechend; **c)** *(coll.: it is hoped that)* hoffentlich; **~, all our problems should now be over** wir wollen hoffen, daß unsere ganzen *(ugs.)* Probleme jetzt beseitigt sind

hopeless ['həʊplɪs] *adj.* **a)** hoffnungslos; **b)** *(inadequate, incompetent)* miserabel; **be ~, be a ~ case** ein hoffnungsloser Fall sein *(ugs.)* (**at** in + *Dat.*); **be ~ at doing sth.** etw. überhaupt nicht können

hopelessly ['həʊplɪslɪ] *adv.* **a)** hoffnungslos; **b)** *(inadequately)* miserabel

hopper ['hɒpə(r)] *n. (Mech.)* Trichter, *der*

'hopscotch *n.* Himmel-und-Hölle-Spiel, *das;* **play ~:** „Himmel und Hölle" spielen

horde [hɔ:d] *n.* große Menge; *(derog.)* Horde, *die;* **in [their] ~s** in Scharen

horizon [hə'raɪzn] *n. (lit. or fig.)* Horizont, *der;* **on/over the ~:** am Horizont; **there is trouble on the ~** *(fig.)* am Horizont tauchen Probleme auf; **there's nothing on the ~** *(fig.)* da ist nichts in Sicht *(ugs.)*

horizontal [hɒrɪ'zɒntl] **1.** *adj.* horizontal; waagerecht; *see also* **bar 1 b. 2.** *n.* Horizontale, *die;* Waagerechte, *die*

horizontally [hɒrɪ'zɒntəlɪ] *adv.* horizontal; *(flat)* waagerecht; flach ⟨*liegen*⟩

hormone ['hɔ:məʊn] *n. (Biol., Pharm.)* Hormon, *das*

horn [hɔ:n] *n.* **a)** *(of animal or devil)* Horn, *das; (of deer)* Geweihstange, *die (Jägerspr.);* **~s** Geweih, *das;* **lock ~s [with sb.]** *(fig.)* [mit jmdm.] die Klinge[n] kreuzen *(geh.);* **draw in one's ~s** *(fig.)* sich zurückhalten; *(restrain one's ambition)* zurückstecken; **b)** *(substance)* Horn, *das;* **c)** *(Mus.)* Horn, *das;* **[French] ~:** [Wald]horn, *das;* **d)** *(of vehicle)* Hupe, *die; (of ship)* [Signal]horn, *das; (of factory)* [Fabrik]sirene, *die;* **sound** *or* **blow** *or* **hoot one's ~ [at sb.]** ⟨*Fahrer:*⟩ [jmdn. an]hupen; **e)** *(Geog.)* **the H~:** das Kap Hoorn. *See also* **dilemma**

horned [hɔ:nd] *adj.* gehörnt

hornet ['hɔ:nɪt] *n.* Hornisse, *die;* **stir up** *or* **walk into a ~s' nest** *(fig.)* in ein Wespennest stechen *od.* greifen *(ugs.)*

'hornpipe *n. (Mus.)* Hornpipe, *die*

'horn-rimmed *adj.* ~ **spectacles** *or* **glasses** Hornbrille, *die*

horny ['hɔ:nɪ] *adj.* **a)** *(hard)* hornig ⟨*Fußsohlen, Haut, Hände*⟩; **b)** *(made of horn)* aus Horn *nachgestellt; (like horn)* hornartig; **c)** *(sl.: sexually aroused)* spitz *(ugs.)*

horoscope ['hɒrəskəʊp] *n.* *(Astrol.)* Horoskop, *das;* **draw up** *or* **cast sb.'s ~:** jmdm. das Horoskop stellen

horrendous [hə'rendəs] *adj.* *(coll.)* schrecklich *(ugs.);* entsetzlich *(ugs.)* ⟨*Dummheit*⟩; horrend ⟨*Preis*⟩

horrible ['hɒrɪbl] *adj.* **a)** grauenhaft; grausig ⟨*Monster, Geschichte*⟩; grauenvoll ⟨*Verbrechen, Alptraum*⟩; schauerlich ⟨*Maske*⟩; **I find all insects ~:** mir graust vor jeder Art von Insekten; **b)** *(coll.: unpleasant, excessive)* grauenhaft *(ugs.);* horrend ⟨*Ausgaben, Kosten*⟩; **I have a ~ feeling that ...:** ich habe das ungute Gefühl, daß ...

horribly ['hɒrɪblɪ] *adv.* **a)** entsetzlich ⟨*entstellt*⟩; scheußlich ⟨*grinsen*⟩; **b)** *(coll.: unpleasantly, excessively)* entsetzlich *(ugs.);* fürchterlich *(ugs.)* ⟨*aufregen*⟩; horrend ⟨*teuer*⟩

horrid ['hɒrɪd] *adj.* scheußlich; **don't be so ~ to me** *(coll.)* sei nicht so garstig zu mir

horrific [hə'rɪfɪk] *adj.* schrecklich; *(coll.)* horrend ⟨*Preis*⟩

horrify ['hɒrɪfaɪ] *v. t.* **a)** *(excite horror in)* mit Schrecken erfüllen; **b)** *(shock, scandalize)* **be horrified** entsetzt sein (**at, by** über + *Akk.*)

horrifying ['hɒrɪfaɪɪŋ] *adj.* grauenhaft; grausig ⟨*Film*⟩; **it is ~ to think that ...:** der Gedanke, daß ..., ist schrecklich

horror ['hɒrə(r)] **1.** *n.* **a)** Entsetzen, *das* (**at** über + *Akk.*); *(repugnance)* Grausen, *das;* **have a ~ of sb./sth./doing sth.** einen Horror vor jmdm./etw. haben/einen Horror davor haben, etw. zu tun *(ugs.);* **b)** *(horrifying quality)* Grauenhaftigkeit, *die;*

(horrifying thing) Greuel, *der; (horrifying person)* Scheusal, *das.* **2.** *attrib. adj.* Horror⟨*comic, -film, -geschichte*⟩

horror: ~-stricken, ~-struck *adjs.* von Entsetzen gepackt

hors-d'œuvre [ɔ:'dɜ:vr, ɔ:'dɜ:v] *n. (Gastr.)* Horsd'œuvre, *das;* ≈ Vorspeise, *die*

horse [hɔ:s] *n.* **a)** Pferd, *das; (adult male)* Hengst, *der;* **be/get on one's high ~** *(fig.)* auf dem hohen Roß sitzen/sich aufs hohe Roß setzen *(ugs.);* **hold your ~s!** *(fig.)* immer sachte mit den jungen Pferden! *(ugs.);* **as strong as a ~:** bärenstark *(ugs.);* **eat/work like a ~:** wie ein Scheunendrescher essen *(salopp)*/wie ein Pferd arbeiten; **I could eat a ~** *(coll.)* ich habe einen Bärenhunger *(ugs.);* **[right** *or* **straight] from the ~'s mouth** *(fig.)* aus erster Hand *od.* Quelle; **it's [a question** *or* **matter of] ~s for courses** *(fig.)* jeder sollte die Aufgaben übernehmen, für die er am besten geeignet ist; **b)** *(Gymnastics)* **[vaulting-]~:** [Sprung]pferd, *das;* **c)** *(framework)* Gestell, *das;* **[clothes-]~:** Wäscheständer, *der*

horse: ~back *n.* **on ~back** zu Pferd; **~-box** *n. (trailer)* Pferdeanhänger, *der; (Motor Veh.)* Pferdetransporter, *der;* **~-'chestnut** *n. (Bot.)* Roßkastanie, *die;* **~-drawn** *attrib. adj.* pferdebespannt; von Pferden gezogen; **~-drawn vehicle** Pferdewagen, *der;* **~hair** *n.* **a)** *(single hair)* Pferdehaar, *das;* **b)** *no pl., no indef. art. (mass of hairs)* Roßhaar, *das;* **~man** ['hɔ:smən] *n., pl.* **~men** ['hɔ:smən] *([skilled] rider)* [guter] Reiter

horsemanship ['hɔ:smənʃɪp] *n., no pl.* **[skills of] ~:** reiterliches Können

horse: ~play *n.* Balgerei, *die;* **~power** *n., pl. same (Mech.)* Pferdestärke, *die;* **a 40 ~power car** ein Auto mit 40 PS; **~-racing** *n.* Pferderennsport, *der;* **~radish** *n.* Meerrettich, *der;* **~shoe** *n.* Hufeisen, *das;* **~whip 1.** *n.* Reitpeitsche, *die;* **2.** *v. t.* auspeitschen; **~woman** *n.* Reiterin, *die*

horsy (horsey) ['hɔ:sɪ] *adj.* **a)** *(horselike)* pferdeähnlich; **b)** *(much concerned with horses)* pferdenärrisch

horticultural [hɔ:tɪ'kʌltʃərl] *adj.* gartenbaulich; Gartenbau⟨*zeitschrift, -ausstellung*⟩

horticulture ['hɔ:tɪkʌltʃə(r)] *n.* Gartenbau, *der*

hose [həʊz] **1.** *n.* Schlauch, *der.* **2.** *v. t.* sprengen

~ 'down *v. t.* abspritzen

'hose-pipe *n.* Schlauch, *der*

hosiery ['həʊʒɪərɪ] *n., no pl.* Strumpfwaren *Pl.*

hospice ['hɒspɪs] *n.* **a)** *(Brit.) (for the destitute)* Heim für Mittellose; *(for the terminally ill)* Sterbeklinik, *die;* **b)** *(for travellers or students)* Hospiz, *das*

hospitable ['hɒspɪtəbl] *adj. (welcoming)* gastfreundlich ⟨*Person, Wesensart*⟩; gastlich ⟨*Haus, Hotel, Klima*⟩

hospital ['hɒspɪtl] *n.* Krankenhaus, *das;* **in ~** *(Brit.),* **in the ~** *(Amer.)* im Krankenhaus; **into** *or* **to ~** *(Brit.),* **to the ~** *(Amer.)* ins Krankenhaus ⟨*gehen, bringen*⟩

hospital: ~ bed *n.* Krankenhausbett, *das;* **~ case** *n.* Fall fürs Krankenhaus

hospitalise *see* **hospitalize**

hospitality [hɒspɪ'tælɪtɪ] *n., no pl. (of person)* Gastfreundschaft, *die*

hospitalize ['hɒspɪtəlaɪz] *v. t.* ins Krankenhaus einweisen

hospital: ~ nurse *n.* Krankenschwester, *die*/Krankenpfleger, *der;* **~ porter** *n.* ≈ Krankenpflegehelfer, *der*/-helferin, *die*

¹host [həʊst] *n. (large number)* Menge, *die;* **in [their] ~s** in Scharen

²host 1. *n.* **a)** Gastgeber, *der*/Gastgeberin, *die;* **be** *or* **play ~ to sb.** jmdn. zu Gast haben; **b)** *(compère)* Moderator, *der.* **2.** *v. t.* **a)** *(act as host at)* Gastgeber sein bei; **b)** *(compère)* moderieren

hostage ['hɒstɪdʒ] *n.* Geisel, *die;* **hold/take sb. ~:** jmdn. als Geisel festhalten/nehmen; **a ~ to fortune** etwas, was einem das Schicksal nehmen kann

hostel ['hɒstl] *n. (Brit.)* **a)** Wohnheim, *das;* **b)** *see* **youth hostel**

hostess ['həʊstɪs] *n.* **a)** Gastgeberin, *die;* **b)** *(in night-club)* Animierdame, *die;* **c)** *(in passenger transport)* Hostess, *die;* **d)** *(compère)* Moderatorin, *die*

hostile ['hɒstaɪl] *adj.* **a)** feindlich; **b)** *(unfriendly)* feindselig **(to, towards** gegenüber); **be ~ to sth.** etw. ablehnen; **c)** *(inhospitable)* unwirtlich; feindselig ⟨*Atmosphäre*⟩

hostility [hɒ'stɪlɪtɪ] *n.* **a)** *no pl. (enmity)* Feindschaft, *die;* **b)** *no pl. (antagonism)* Feindseligkeit, *die* **(to[wards]** gegenüber); **feel no ~ towards anybody** niemandem feindlich gesinnt sein; **c)** *(state of war, act of warfare)* Feindseligkeit, *die*

hot [hɒt] *adj.* **a)** heiß; *(cooked)* warm ⟨*Mahlzeit, Essen*⟩; *(fig.: potentially dangerous, difficult)* heiß *(ugs.)* ⟨*Thema, Geschichte*⟩; ungemütlich, gefährlich ⟨*Lage*⟩; **~ and cold running water** fließend warm und kalt Wasser; **be too ~ to handle** *(fig.)* eine zu heiße Angelegenheit sein *(ugs.);* **make it** *or* **things [too] ~ for sb.** *(fig.)* jmdm. die Hölle heiß machen *(ugs.);* **b)** *(feeling heat)* **I am/feel ~:** mir ist heiß; **c)** *(pungent)* scharf ⟨*Gewürz, Senf usw.*⟩; **d)** *(passionate, lustful)* heiß ⟨*Küsse, Tränen, Umarmung*⟩; **be ~ for sth.** heiß auf etw. *(Akk.)* sein *(ugs.);* **he's really ~ on her** *(sexually)* er ist richtig scharf auf sie *(ugs.);* **e)** *(agitated, angry)* hitzig; **get [all] ~ and bothered** sich [fürchterlich *(ugs.)*] aufregen; **f)** *(coll.: good)* toll *(ugs.);* **be ~ at sth.** in etw. *(Dat.)* [ganz] groß sein *(ugs.);* **I'm not too ~ at that** darin bin ich nicht besonders umwerfend *(ugs.);* **be ~ on sth.** *(knowledgeable)* sich in *od.* mit etw. *(Dat.)* gut auskennen; **g)** *(recent)* noch warm ⟨*Nachrichten*⟩; **this is really ~ [news]** das ist wirklich das Neueste vom Neuen; **h)** *(close)* **you are getting ~/are ~** *(in children's games)* es wird schon wärmer/[jetzt ist es] heiß; **follow ~ on sb.'s heels** jmdm. dicht auf den Fersen folgen *(ugs.);* **i)** *(coll.: in demand)* zugkräftig; **a ~ property** *(singer, actress, etc.)* eine ertragreiche Zugnummer; *(company, invention, etc.)* eine ertragreiche Geldanlage; **j)** *(Sport; also fig.)* heiß *(ugs.)* ⟨*Tip, Favorit*⟩; **k)** *(sl.: illegally obtained)* heiß ⟨*Ware, Geld*⟩. *See also* **¹blow 1 b; cake 1 a; collar 1 a; potato**

~ 'up *(Brit. coll.)* **1.** *v. t.* **a)** *(heat)* warm machen; **b)** *(excite)* auf Touren bringen *(ugs.);* **c)** *(make more exciting)* in Schwung bringen; *(make more dangerous)* verschärfen; **d)** *(intensify)* anheizen *(ugs.).* **2.** *v. i.* **a)** *(rise in temperature)* heiß werden; **the weather ~s up** es wird wärmer; **b)** *(become exciting)* in Schwung kommen; *(become dangerous)* sich verschärfen; **c)** *(become more intense)* sich verstärken; ⟨*Wortgefecht:*⟩ zunehmend hitziger werden

hot: ~ 'air *n. (sl.: idle talk)* leeres Gerede *(ugs.);* **~bed** *n. (Hort.)* Mistbeet, *das;* Frühbeet, *das; (fig.)* Nährboden, *der* **(of** für); *(of vice, corruption, etc.)* Brutstätte, *die* **(of** für)

hotchpotch ['hɒtʃpɒtʃ] *n. (mixture)* Mischmasch, *der (ugs.)* **(of** aus)

hot: ~ cross 'bun *n. mit einem Kreuz aus Teig verziertes Rosinenbrötchen, das am Karfreitag gegessen wird;* **~ 'dog** *n. (coll.)* Hot dog, *das od. der*

hotel [hə'tel, həʊ'tel] *n.* Hotel, *das*

hotelier [hə'telɪə(r)] *n.* Hotelier, *der*

hot: ~foot 1. *adv.* stehenden Fußes; **2.** *v.t.* **~foot it** sich hastig davonmachen; **~head** *n.* Hitzkopf, *der;* **~house** *n.* Treibhaus, *das;* **~ line** *n. (Polit.)* heißer Draht

hotly ['hɒtlɪ] *adv.* heftig

hot: ~plate *n.* Kochplatte, *die; (for keeping food ~)* Warmhalteplatte, *die;* **~ rod** *n. (Motor Veh.)* hochfrisiertes Auto *(ugs.);* **~ seat** *n. (sl.) (uneasy situation)* Folterbank, *die (fig.); (involving heavy responsibility)* **be in the ~ seat** den Kopf hinhalten müssen *(ugs.);* **~-shot** *n. (coll.)* As, *das (ugs.);* **~ spot** *n.* **a)** heiße Gegend; **b)** *(difficult situation)* **get into a ~ spot** in die Bredouille kommen *od.* geraten *(ugs.);* **~ 'stuff** *n., no pl., no art. (sl.)* **sb./sth. is ~ stuff** jmd./etw. ist große Klasse *(ugs.);* **~-tempered** *adj.* heißblütig; **~ 'water** *n. (fig. coll.)* **be in ~ water** in der Bredouille sein *(ugs.);* **~-'water bottle** *n.* Wärmflasche, *die*

hound [haʊnd] **1.** *n.* Jagdhund, *der;* **the [pack of] ~s** *(Brit. Hunting)* die Meute *(Jägerspr.).* **2.** *v.t.* jagen; *(fig.)* verfolgen

~ 'out *v.t.* **a)** *(hunt out)* aufspüren; **b)** *(force to leave)* vertreiben **(of** aus); verjagen **(of** aus)

hour ['aʊə(r)] *n.* **a)** Stunde, *die;* **half an ~:** eine halbe Stunde; **an ~ and a half** anderthalb Stunden; **be paid by the ~:** stundenweise bezahlt werden; **a two-~ session** eine zweistündige Sitzung; **the 24-~** *or* **twenty-four-~ clock** die Vierundzwanzigstundenuhr; **b)** *(time o'clock)* Zeit, *die;* **on the ~:** zur vollen Stunde; **at this late ~:** zu so später Stunde *(geh.);* **at all ~s** zu jeder [Tages- oder Nacht]zeit; *(late at night)* spät in der Nacht; **the small ~s [of the morning]** die frühen Morgenstunden; **0100/0200/1700/1800 ~s** *(on 24-~ clock)* 1.00/2.00/17.00/18.00 Uhr; **c)** *in pl.* **doctor's ~s** Sprechstunde, *die;* **post-office ~s** Schalterstunden der Post; **what ~s do you work?** wie ist deine Arbeitszeit?; **work long ~s** einen langen Arbeitstag haben; **during school ~s** während der Schulstunden *od.* des Unterrichts; **out of/after ~s** *(in office, bank, etc.)* außerhalb der Dienstzeit; *(in shop)* außerhalb der Geschäftszeit; *(in pub)* außerhalb der Ausschankzeit; **d)** *(particular time)* Stunde, *die;* **sb.'s finest ~:** jmds. größte Stunde; **the question** *etc.* **of the ~:** das Problem *usw.* der Stunde; **e)** *(distance)* Stunde, *die;* **they are two ~s from us by train** sie wohnen zwei Bahnstunden von uns entfernt

hour: ~glass *n.* Sanduhr, *die;* Stundenglas, *das (veralt.);* **~-hand** *n.* Stundenzeiger, *der;* kleiner Zeiger; **~-long** *attrib. adj.* einstündig

hourly ['aʊəlɪ] **1.** *adj. (happening every hour)* stündlich; **at ~ intervals** jede Stunde; stündlich; **there are ~ trains to London** jede *od.* alle Stunde fährt ein Zug nach London; **he is paid an ~ rate of £6** er hat einen Stundenlohn von 6 Pfund; **two-~:** zweistündlich. **2.** *adv.* stündlich; **be paid ~:** stundenweise bezahlt werden

house 1. [haʊs] *n., pl.* **~s** ['haʊzɪz] **a)** Haus, *das;* **to/at my ~:** zu mir [nach Hause]/bei mir [zu Hause]; **keep ~ [for sb.]** [jmdm.] den Haushalt führen; **put** *or* **set one's ~ in order** *(fig.)* seine Angelegenheiten in Ordnung bringen; **[as] safe as ~s** absolut sicher; **[get on] like a ~ on fire** *(fig.)* prächtig [miteinander auskommen]; **b)** *(Parl.) (building)* Parlamentsgebäude, *das; (assembly)* Haus, *das;* **the H~** *(Brit.)* das Parlament; *see also* **Commons; lord 1c; parliament; representative 1b; c)** *(institution)* Haus, *das;* **fashion ~:** Modehaus, *das;* **d)** *(inn etc.)* Wirtshaus, *das;* **on the ~:** auf Kosten des Hauses; **e)** *(Theatre) (audience)* Publikum, *das; (performance)* Vorstellung, *die;* **an empty ~:** ein leeres Haus; **bring the ~ down** stürmischen Beifall auslösen; *(cause laughter)* Lachstürme entfesseln. **2.** [haʊz] *v.t.* **a)** *(provide with home)* ein Heim geben (+ *Dat.*); **be ~d in sth.** in etw. *(Dat.)* untergebracht sein; **b)** *(keep, store)* unterbringen; einlagern ⟨*Waren*⟩

house [haʊs]: **~ arrest** *n.* Hausarrest, *der;* **~boat** *n.* Hausboot, *das;* **~breaking** *n., no pl. (burglary)* Einbruch, *der;* **~coat** *n.* Hausmantel, *der;* Morgenmantel, *der;* **~ guest** *n.* Logiergast, *der*

household ['haʊshəʊld] *n.* Haushalt, *der; attrib.* Haushalts-; **~ chores** Hausarbeit, *die*

householder ['haʊshəʊldə(r)] *n. (home-owner)* Wohnungsinhaber, *der*/-inhaberin, *die*

household: ~ 'management *n.* Hauswirtschaft, *die;* **~ 'name** *n.* geläufiger Name; **be a ~ name** ein Begriff sein

house [haʊs]: **~-hunting** *n., no indef. art.* Suche nach einem Haus; **go ~-hunting** sich nach einem Haus umsehen; **~keeper** *n. (woman managing household affairs)* Haushälterin, *die;* Wirtschafterin, *die; (person running own home)* Hausfrau, *die*/Hausmann, *der;* **~keeping** *n.* **a)** *(management)* Hauswirtschaft, *die;* Haushaltsführung, *die;* **b)** *(fig.: maintenance, record-keeping, etc.)* Wirtschaften, *das;* **~maid** *n.* Hausgehilfin, *die;* **~-painter** *n.* Maler, *der*/Malerin, *die;* Anstreicher, *der*/Anstreicherin, *die;* **~-plant** *n.* Zimmerpflanze, *die;* **~-proud** *adj.* **he/she is ~-proud** Ordnung und Sauberkeit [im Haushalt] gehen ihm/ihr über alles; **~-room** *n., no pl., no indef. art.* **find ~-room for sth.** einen Platz für etw. [in der Wohnung] finden; **I wouldn't give it ~-room** so etwas wollte ich nicht im Haus haben; **~-to-~:** *adj.* **make ~-to-~ enquiries** von Haus zu Haus gehen und fragen; **~-train** *v.t. (Brit.)* **~-train a cat/child** eine Katze/ein Kleinkind dazu bringen, daß sie/es stubenrein/sauber wird; **~-trained** *adj. (Brit.)* stubenrein ⟨*Hund, Katze*⟩; sauber ⟨*Kleinkind*⟩; **~-warming** ['haʊswɔːmɪŋ] *n.* **~-warming [party]** Einzugsfeier, *die;* **~wife** *n.* Hausfrau, *die;* **~work** *n., no pl.* Hausarbeit, *die*

housing ['haʊzɪŋ] *n.* **a)** *no pl. (dwellings collectively)* Wohnungen; *(provision of dwellings)* Wohnungsbeschaffung, *die; attrib.* Wohnungs-; **~ programme** Wohnungsbauprogramm, *das;* **b)** *no pl. (shelter)* Unterkunft, *die*

housing: ~ association *n. (Brit.)* Gesellschaft für sozialen Wohnungsbau; **~ benefit** *n. (Brit.)* Wohngeld, *das;* **~ estate** *n. (Brit.)* Wohnsiedlung, *die*

hove *see* **heave 1b, 2c**

hovel ['hɒvl] *n.* [armselige] Hütte; *(joc.)* Bruchbude, *die (ugs. abwertend)*

hover ['hɒvə(r)] *v.i.* **a)** schweben; **b)** *(linger)* sich herumdrücken *(ugs.);* **c)** *(waver)* schwanken; ~

between life and death *(fig.)* zwischen Leben und Tod schweben
hover: ~craft *n., pl. same* Hovercraft, *das;* Luftkissenfahrzeug, *das;* **~ mower** *n.* Luftkissenmäher, *der*
how [haʊ] *adv.* wie; **learn ~ to ride a bike/swim** *etc.* radfahren/schwimmen *usw.* lernen; **this is ~ to do it** so macht man das; **~ do you know that?** woher weißt du das?; **~'s that?** *(~ did that happen?)* wie kommt das [denn]?; *(is that as it should be?)* ist es so gut?; *(will you agree to that?)* was hältst du davon?; **~ so?** wieso [das]?; **~ would it be if ...?** wie wäre es, wenn ...?; **~ is she/the car?** *(after accident)* wie geht es ihr?/was ist mit dem Auto?; **~ 'are you?** wie geht es dir?; *(greeting)* guten Morgen/Tag/Abend!; **~ do you 'do?** *(formal)* guten Morgen/Tag/Abend!; **~ much?** wieviel?; **~ many?** wieviel?; wie viele?; **~ many times?** wie oft?; **~ far** *(to what extent)* inwieweit; **~ right/wrong you are!** da hast du völlig recht/da irrst du dich gewaltig!; **~ naughty of him!** das war aber frech von ihm; **and ~!** *(coll.)* und wie! *(ugs.)*; **~ about ...?** wie ist es mit ...?; *(in invitation, proposal, suggestion)* wie wäre es mit ...?; **~ about tomorrow?** wie sieht's morgen aus? *(ugs.)*; **~ about it?** na, wie ist das?; *(is that acceptable?)* was hältst du davon?
however [haʊ'evə(r)] *adv.* **a)** wie ... auch; egal, wie *(ugs.)*; **I shall never win this race, ~ hard I try** ich werde dieses Rennen nie gewinnen, und wenn ich mich noch so anstrenge *od.* wie sehr ich mich auch anstrenge; **b)** *(nevertheless)* jedoch; aber; **I don't like him very much. H~, he has never done me any harm** Ich mag ihn nicht sehr. Er hat mir allerdings noch nie etwas getan; **~, the rain soon stopped, and ...:** es hörte jedoch *od.* aber bald auf zu regnen, und ...
howitzer ['haʊɪtsə(r)] *n.* *(Mil.)* Haubitze, *die*
howl [haʊl] **1.** *n.* *(of animal)* Heulen, *das;* *(of distress)* Schrei, *der;* **a ~ of pain** *or* **agony** ein Schmerzensschrei; **~s of laughter** brüllendes Gelächter; **~s of derision/scorn** verächtliches Gejohle. **2.** *v.i.* ⟨*Tier, Wind:*⟩ heulen; *(with distress)* schreien; **~ in** *or* **with pain/hunger** *etc.* vor Schmerz/Hunger *usw.* schreien. **3.** *v.t.* [hinaus]schreien
~ 'down *v.t.* niederbrüllen
howler ['haʊlə(r)] *n.* *(coll.: blunder)* Schnitzer, *der (ugs.)*; **make a ~:** sich *(Dat.)* einen Schnitzer leisten
HP *abbr.* **a)** [eɪtʃ'pi:] *(Brit.)* **hire-purchase; on ~:** auf Teilzahlungsbasis; **b) horsepower** PS
HQ *abbr.* **headquarters** HQ
hub [hʌb] *n.* **a)** *(of wheel)* [Rad]nabe, *die;* **b)** *(fig.: central point)* Mittelpunkt, *der;* Zentrum, *das;* **the ~ of the universe** *(fig.)* der Nabel der Welt *(geh.)*
hubbub ['hʌbʌb] *n.* Lärm, *der;* **a ~ of conversation/voices** ein Stimmengewirr
'hub-cap *n.* Radkappe, *die*
huddle ['hʌdl] **1.** *v.i.* sich drängen; *(curl up, nestle)* sich kuscheln; **~ against each other/together** sich aneinanderdrängen/sich zusammendrängen. **2.** *v.t.* **a)** *(put on)* **~ one's coat around one** sich in den Mantel hüllen; **b)** *(crowd together)* [eng] zusammendrängen. **3.** *n.* **a)** *(tight group)* dichtgedrängte Menge *od.* Gruppe; **b)** *(coll.: conference)* Besprechung, *die;* **be in a ~/go [off] in[to] a ~:** die Köpfe zusammenstecken *(ugs.)*
~ 'up *v.i.* *(nestle up)* sich zusammenkauern; *(crowd together)* sich [zusammen]drängen
[1]hue [hju:] *n.* **a)** Farbton, *der;* **the sky took on a reddish ~:** der Himmel färbte sich rötlich; **b)** *(fig.: aspect)* Schattierung, *die*
[2]hue *n.* **~ and cry** *(outcry)* lautes Geschrei; *(protest)* Gezeter, *das (abwertend)*; **raise a ~ and cry against sb./sth.** ein lautes Geschrei/Gezeter über jmdn./etw. anstimmen
huff [hʌf] **1.** *v.i.* **~ and puff** schnaufen und keuchen. **2.** *n.* **be in a ~:** beleidigt *od.* *(ugs.)* eingeschnappt sein; **go off in a ~:** beleidigt *od.* eingeschnappt abziehen *(ugs.)*
hug [hʌg] **1.** *n.* Umarmung, *die;* *(of animal)* Umklammerung, *die;* **give sb. a ~:** jmdn. umarmen. **2.** *v.t.*, **-gg-: a)** umarmen; ⟨*Tier:*⟩ umklammern; **~ sb./sth. to oneself** jmdn./etw. an sich *(Akk.)* drücken *od.* pressen; **the bear ~ged him to death** der Bär drückte ihn zu Tode; **b)** *(keep close to)* sich dicht halten an (+ *Dat.*); ⟨*Schiff, Auto usw.:*⟩ dicht entlangfahren an (+ *Dat.*); **c)** *(fit tightly around)* eng anliegen an (+ *Dat.*); **a pullover that ~s the figure** ein Pullover, der die Figur betont
huge [hju:dʒ] *adj.* riesig; gewaltig ⟨*Unterschied, Verbesserung, Interesse*⟩; **the problem is ~:** das Problem ist außerordentlich schwierig
hulk [hʌlk] *n.* **a)** *(body of ship)* [Schiffs]rumpf, *der;* *(as store etc.)* Hulk, *die od. der (Seew.)*; **b)** *(wreck of car, machine, etc.)* Wrack, *das;* **c)** *(fig.)* *(big thing)* Klotz, *der;* *(big person)* Koloß, *der (ugs. scherzh.)*; **a ~ of a man** ein Klotz von [einem] Mann *(fig. ugs.)*
hulking ['hʌlkɪŋ] *adj.* *(coll.)* *(bulky)* wuchtig; *(clumsy)* klotzig *(abwertend)*; **a ~ great person/thing** ein klobiger Mensch/ein klobiges Etwas
hull [hʌl] *n.* *(Naut.)* Schiffskörper, *der;* *(Aeronaut.)* Rumpf, *der*
hullabaloo [hʌləbə'lu:] *n.* **a)** *(noise)* Radau, *der (ugs.)*; Lärm, *der;* *(of show-business life, city)* Trubel, *der;* **b)** *(controversy)* Aufruhr, *der;* **make a ~ about sth.** viel Lärm um etw. machen
hullo [hə'ləʊ] *see* **hallo; hello**
hum [hʌm] **1.** *v.i.*, **-mm-: a)** summen; ⟨*Motor, Maschine, Kreisel:*⟩ brummen; **~ and ha** *or* **haw** *(coll.)* herumdrucksen *(ugs.)*; **b)** *(coll.: be in state of activity)* voller Leben *od.* Aktivität sein. **2.** *v.t.*, **-mm-** summen ⟨*Melodie, Lied*⟩. **3.** *n.* **a)** Summen, *das;* *(of spinning-top, machinery, engine)* Brummen, *das;* **b)** *(inarticulate sound)* Hm, *das;* **c)** *(of voices, conversation)* Gemurmel, *das;* *(of insects and small creatures)* Gesumme, *das;* *(of traffic)* Brausen, *das.* **4.** *int.* hm
human ['hju:mən] **1.** *adj.* menschlich; **the ~ race** die menschliche Rasse; **I'm only ~:** ich bin auch nur ein Mensch; **~ error** menschliches Versagen; **lack the ~ touch** menschliche Wärme vermissen lassen; **be ~!** sei kein Unmensch!; *see also* **nature d. 2.** *n.* Mensch, *der*
human 'being *n.* Mensch, *der*
humane [hju:'meɪn] *adj.* **a)** human; **b)** *(tending to civilize)* humanistisch
humanise *see* **humanize**
humanism ['hju:mənɪzm] *n., no pl.* **a)** Humanität, *die;* **b)** *(literary culture; also Philos.)* Humanismus, *der*
humanist ['hju:mənɪst] *n.* Humanist, *der*/Humanistin, *die*
humanitarian [hju:mænɪ'teərɪən] **1.** *adj.* humanitär. **2.** *n.* *(philanthropist)* Menschenfreund, *der;* *(promoter of human welfare)* Hu-

manitarist, *der*/Humanitaristin, *die*

humanity [hju:'mænɪtɪ] *n.* **a)** *no pl., no art. (mankind)* Menschheit, *die; (people collectively)* Menschen; **b)** *no pl. (being humane)* Humanität, *die;* Menschlichkeit, *die;* **c)** *in pl. (cultural learning)* **[the] humanities** [die] Geisteswissenschaften

humanize ['hju:mənaɪz] *v.t.* **a)** *(make human)* vermenschlichen; **b)** *(adapt to human use)* den menschlichen Bedürfnissen anpassen; humanisieren ‹*Industrie*›; **c)** *(make humane)* humanisieren ‹*Strafvollzug*›

humanly ['hju:mənlɪ] *adv.* menschlich; *(by human means)* mit menschlichen Mitteln; **do everything ~ possible** alles menschenmögliche tun

humble ['hʌmbl] **1.** *adj.* **a)** *(modest)* bescheiden; ergeben ‹*Untertan, Diener, Gefolgsmann*›; unterwürfig *(oft abwertend)* ‹*Haltung, Knechtschaft*›; **please accept my ~ apologies** ich bitte ergebenst um Verzeihung; **eat ~ pie** klein beigeben; **b)** *(low-ranking)* einfach; niedrig ‹*Status, Rang usw.*›; **c)** *(unpretentious)* einfach; bescheiden ‹*Zuhause, Wohnung, Anfang*›. **2.** *v.t.* **a)** *(abase)* demütigen; **~ oneself** sich demütigen *od.* erniedrigen; **b)** *(defeat decisively)* [vernichtend] schlagen

humbly ['hʌmblɪ] *adv. (with humility)* demütig; ergebenst ‹*um Verzeihung bitten*›; *(in formal address)* höflichst ‹*bitten, ersuchen*›

humbug ['hʌmbʌg] *n.* **a)** *no pl., no art. (deception, nonsense)* Humbug, *der (ugs. abwertend);* **b)** *(Brit.: sweet)* [Pfefferminz]bonbon, *der od. das*

humdrum ['hʌmdrʌm] *adj.* **a)** alltäglich; eintönig ‹*Leben*›; **b)** *(monotonous)* stumpfsinnig; **the ~ routine of life/things** das tägliche Einerlei

humid ['hju:mɪd] *adj.* feucht; humid *(Geogr.)*

humidity [hju:'mɪdɪtɪ] *n.* **a)** *no pl.* Feuchtigkeit, *die;* **b)** *(degree of moisture)* **~ [of the atmosphere]** Luftfeuchtigkeit, *die (Met.)*

humiliate [hju:'mɪlɪeɪt] *v.t.* demütigen; **I was** *or* **felt totally ~d** ich war zutiefst beschämt

humiliation [hju:mɪlɪ'eɪʃn] *n.* Demütigung, *die*

humility [hju:'mɪlɪtɪ] *n.* Demut, *die; (of servant)* Ergebenheit, *die; (absence of pride or arrogance)* Bescheidenheit, *die*

'**humming-bird** *n.* Kolibri, *der*

hummock ['hʌmək] *n. (hillock)* [kleiner] Hügel

humor *(Amer.) see* **humour**

humorist ['hju:mərɪst] *n.* **a)** *(facetious person)* Spaßvogel, *der;* Komiker, *der (fig.);* **b)** *(talker, writer)* Humorist, *der*/Humoristin, *die*

humorless *(Amer.) see* **humourless**

humorous ['hju:mərəs] *adj.* lustig, komisch ‹*Geschichte, Name, Situation*›; witzig ‹*Bemerkung*›; humorvoll ‹*Person*›

humour ['hju:mə(r)] *(Brit.)* **1.** *n.* **a)** *no pl., no indef. art. (faculty, comic quality)* Humor, *der; (of situation)* Komische, *das;* **sense of ~:** Sinn für Humor; **b)** *no pl., no indef. art. (facetiousness)* Witzigkeit, *die;* **c)** *(mood)* Laune, *die;* **in good ~:** gutgelaunt; **be out of ~:** schlechte Laune haben. **2.** *v.t.* **~ sb.** jmdm. seinen Willen lassen; **~ sb.'s taste** jmds. Geschmack *od.* Vorliebe *(Dat.)* entsprechen; **do it just to ~ her/him** tu's doch, damit sie ihren/er seinen Willen hat

humourless ['hju:məlɪs] *adj. (Brit.)* humorlos

hump [hʌmp] **1.** *n.* **a)** *(human)* Buckel, *der;* Höcker, *der (ugs.); (of animal)* Höcker, *der;* **he has a ~ on his back** er hat einen Buckel; **b)** *(mound)* Hügel, *der.* **2.** *v.t. (Brit. sl.: carry)* schleppen

hump: ~back '**bridge** *n.* gewölbte Brücke; **~backed** ['hʌmpbækt] *see* **hunchbacked**

humus ['hju:məs] *n.* Humus, *der*

[1]**hunch** [hʌntʃ] *v.t.* hochziehen ‹*Schultern*›; **sit ~ed in a corner** zusammengekauert in einer Ecke sitzen

~ 'up *v.t.* hochziehen; **~ oneself up** einen Buckel machen

[2]**hunch** *n. (intuitive feeling)* Gefühl, *das*

hunch: ~back *n.* **a)** *(back)* Buckel, *der;* **b)** *(person)* Bucklige, *der/die;* **be a ~back** einen Buckel haben; **~backed** ['hʌntʃbækt] *adj.* buck[e]lig

hundred ['hʌndrəd] **1.** *adj.* **a)** hundert; **a** *or* **one ~:** [ein]hundert; **two/several ~:** zweihundert/mehrere hundert; **a** *or* **one ~ and one** [ein]hundert[und]eins; **a** *or* **one ~ and one people** hundert[und]ein Menschen *od.* Mensch; **the ~ metres race** der Hundertmeterlauf; **b) a ~ [and one]** *(fig.: innumerable)* hundert *(ugs.);* **c) a** *or* **one ~ per cent** hundertprozentig; **I'm not a ~ per cent at the moment** *(fig.)* momentan geht es mir nicht sehr gut. *See also* **eight 1. 2.** *n.* **a)** *(number)* hundert; **a** *or* **one/two ~:** [ein]hundert/zweihundert; **not if I live to be a ~:** nie im Leben; **in** *or* **by ~s** hundertweise; **the seventeen-~s** *etc.* das achtzehnte *usw.* Jahrhundert; **a ~ and one** *etc.* [ein]hundert[und]eins *usw.;* **it's a ~ to one that ...:** die Chancen stehen hundert zu eins, daß ...; **b)** *(symbol, written figure)* Hundert, *die; (~-pound etc. note)* Hunderter, *der;* **c)** *(indefinite amount)* **~s** Hunderte *Pl.;* **~s of times** hundertmal. *See also* **eight 2a**

hundredth ['hʌndrədθ] **1.** *adj.* hundertst...; **a ~ part** ein Hundertstel. **2.** *n. (fraction)* Hundertstel, *das; (in sequence)* hundertste, *der/die/das; (in rank)* Hundertste, *der/die/das*

hundredweight ['hʌndrədweɪt] *n., pl. same or* **~s** *(Brit.)* 50,8 kg; ≈ Zentner, *der*

hung *see* **hang 1, 2**

Hungarian [hʌŋ'geərɪən] **1.** *adj.* ungarisch; **sb. is ~:** jmd. ist Ungar/Ungarin. **2.** *n.* **a)** *(person)* Ungar, *der*/Ungarin, *die;* **b)** *(language)* Ungarisch, *das; see also* **English 2a**

Hungary ['hʌŋgərɪ] *pr. n.* Ungarn *(das)*

hunger ['hʌŋgə(r)] **1.** *n. (lit. or fig.)* Hunger, *der;* **pang[s] of ~:** quälender Hunger; **die of ~:** verhungern; *(fig.: be very hungry)* vor Hunger sterben *(ugs.);* **~ for sth.** *(lit. or fig.)* Hunger nach etw. *(geh.).* **2.** *v.i.* **~ after** *or* **for sb./sth.** [heftiges] Verlangen nach jmdm./etw. haben

'**hunger-strike** *n.* Hungerstreik, *der;* **go on ~:** in den Hungerstreik treten

hung: ~-'over *adj. (coll.)* verkatert *(ugs.);* **~ 'parliament** *n. Parlament, in dem keine Partei die absolute Mehrheit hat*

hungrily ['hʌŋgrɪlɪ] *adv.* **a)** hungrig; **b)** *(fig.: longingly)* sehnsüchtig ‹*an etw. denken*›; [be]gierig ‹*etw. verfolgen*›

hungry ['hʌŋgrɪ] *adj.* **a)** *(feeling hunger)* hungrig; *(regularly feeling hunger)* hungernd; *(showing hunger)* hungrig, gierig ‹*Augen, Blick*›; **be ~:** Hunger haben; hungrig sein; **go ~:** hungern; hungrig bleiben; **b)** *(fig.: eager, avaricious)* **be ~ for sth.** nach etw. hungern *(geh.);* **~ for success/power/knowledge/love** erfolgs-/macht-/bildungs-/liebeshungrig

hunk [hʌŋk] *n.* **a)** *(large piece)*

[großes] Stück; *(of bread)* Brocken, *der;* **b)** *(coll.: large person)* stattliche Erscheinung; **he is a gorgeous great ~:** er ist ein blendend aussehender, stattlicher Mann

hunt [hʌnt] **1.** *n.* **a)** *(pursuit of game)* Jagd, *die;* **b)** *(search)* Suche, *die; (strenuous search)* Jagd, *die;* **be on the ~ for sb./sth.** auf der Suche/Jagd nach jmdm./etw. sein; **c)** *(body of fox-hunters)* Jagd[gesellschaft], *die; (association)* Jagdverband, *der.* **2.** *v. t.* **a)** jagen; Jagd machen auf (+ *Akk.*); **b)** *(search for)* Jagd machen auf (+ *Akk.*) ‹*Mörder usw.*›; fahnden nach ‹*vermißter Person*›; **c)** *(drive, lit. or fig.)* jagen; **he was ~ed out of society** er wurde aus der Gesellschaft ausgestoßen. **3.** *v. i.* **a)** jagen; **go ~ing** jagen; auf die Jagd gehen; **b)** *(seek)* **~ after** *or* **for sb./sth.** nach jmdm./etw. suchen; **the police are ~ing for him** die Polizei ist auf der Suche nach ihm

~ a'bout, ~ a'round *v. i.* **~ about** *or* **around for sb./sth.** [überall] nach jmdm./etw. suchen

~ 'down *v. t.* **a)** *(bring to bay)* hetzen und stellen; **b)** *(pursue and overcome)* zur Strecke bringen ‹*Person*›; abschießen ‹*feindliches Flugzeug*›; **c)** *(fig.: track down)* aufstöbern

~ 'out *v. t.* **a)** *(drive from cover)* aufstöbern; **b)** *(seek out)* suchen; **c)** *(fig.: track down)* ausfindig machen ‹*Tatsachen, Antworten*›

~ 'up *v. t.* aufspüren

hunted ['hʌntɪd] *adj.* **a)** *(pursued)* gejagt; **b)** *(expressing fear)* gejagt, gehetzt ‹*Blick, Gesichtsausdruck*›

hunter ['hʌntə(r)] *n.* **a)** Jäger, *der;* **b)** *(fig.: seeker)* **autograph-~:** Autogrammjäger, *der;* **bargain-~s** *Leute, die ständig auf der Suche nach Sonderangeboten, nach einem Gelegenheitskauf sind;* **c)** *(horse)* Jagdpferd, *das*

hunting ['hʌntɪŋ] *n., no pl.* die Jagd (**of** auf + *Akk.*); das Jagen (**of** *Gen.*)

huntsman ['hʌntsmən] *n., pl.* **huntsmen** ['hʌntsmən] *(hunter)* Jäger, *der; (riding to hounds)* Jagdreiter, *der*

hurdle ['hɜ:dl] **1.** *n.* *(Athletics)* Hürde, *die;* **~ race, ~s** Hürdenlauf, *der;* **fall at the last ~** *(fig.)* an der letzten Hürde scheitern. **2.** *v. t.* überspringen ‹*Zaun, Hecke usw.*›

hurdler ['hɜ:dlə(r)] *n.* *(Athletics)* Hürdenläufer, *der*/-läuferin, *die*

hurl [hɜ:l] *v. t.* werfen; *(violently)* schleudern; *(throw down)* stürzen; **she ~ed herself to her death from a 15th-floor window** sie stürzte sich aus einem Fenster im 15. Stock zu Tode; **~ insults at sb.** jmdm. Beleidigungen ins Gesicht schleudern

hurly-burly ['hɜ:lɪbɜ:lɪ] *n.* Tumult, *der*

hurrah [hə'rɑ:, hʊ'rɑ:], **hurray** [hə'reɪ, hʊ'reɪ] *int.* hurra; **~ for sb./sth.!** jmd./etw. lebe hoch!; **hip, hip, ~!** hipp, hipp, hurra!

hurricane ['hʌrɪkən] *n.* *(tropical cyclone)* Hurrikan, *der; (storm, lit. or fig.)* Orkan, *der*

'hurricane-lamp *n.* Sturmlaterne, *die*

hurried ['hʌrɪd] *adj.* eilig; überstürzt ‹*Abreise*›; eilig *od.* hastig geschrieben ‹*Brief, Aufsatz*›; in Eile ausgeführt ‹*Arbeit*›

hurry ['hʌrɪ] **1.** *n.* **a)** *(great haste)* Eile, *die;* **what is** *or* **why the [big] ~?** warum die Eile?; **in a ~:** eilig; **be in a [great** *or* **terrible] ~:** es [furchtbar] eilig haben; **do sth. in a ~:** etw. in Eile tun; **leave in a ~:** davoneilen; **I need it in a ~:** ich brauche es dringend; **I shan't ask again in a ~** *(coll.)* ich frage so schnell nicht wieder; **be in a/not be in a** *or* **be in no ~ to do sth.** es eilig/nicht eilig haben, etw. zu tun; **b)** *(urgent requirement)* **there is a ~ for sth.** etw. ist sehr gefragt; **what's the [big] ~?** wozu die Eile?; **there's no ~:** es eilt nicht; es hat keine Eile. **2.** *v. t.* *(transport fast)* schnell bringen; *(urge to go or act faster)* antreiben; *(consume fast)* hinunterschlingen ‹*Essen*›; **~ one's work** seine Arbeit in zu großer Eile erledigen. **3.** *v. i.* sich beeilen; *(to or from place)* eilen; **~ downstairs/out/in** nach unten/nach draußen/nach drinnen eilen

~ a'long 1. *v. i.* *(coll.)* sich beeilen. **2.** *v. t.* zur Eile antreiben; beschleunigen ‹*Vorgang*›

~ 'on 1. *v. i.* weitereilen; **I must ~ on** ich muß [rasch] weiter. **2.** *v. t.* antreiben

~ through *v. t.* **a)** [--'-] beschleunigen; **b)** ['---] schnell durcheilen; *(fig.)* möglichst schnell durchziehen *(ugs.)*

~ 'up 1. *v. i.* *(coll.)* sich beeilen. **2.** *v. t.* antreiben; vorantreiben ‹*Vorgang*›

hurt [hɜ:t] **1.** *v. t.*, **hurt a)** weh tun (+ *Dat.*); *(injure physically)* verletzen; **~ one's arm/back** sich *(Dat.)* am Arm/Rücken weh tun; *(injure)* sich *(Dat.)* den Arm/am Rücken verletzen; **you are ~ing me/my arm** du tust mir weh/am Arm weh; **my arm is ~ing me** mein Arm tut [mir] weh; mir tut der Arm weh; **he wouldn't ~ a fly** *(fig.)* er tut keiner Fliege etwas zuleide; **sth. won't** *or* **wouldn't ~ sb.** etw. tut nicht weh; *(fig.)* etw. würde jmdm. nichts schaden *(ugs.)*; **~ oneself** sich *(Dat.)* weh tun; *(injure oneself)* sich verletzen; **b)** *(damage, be detrimental to)* schaden (+ *Dat.*); **sth. won't** *or* **wouldn't ~ sth.** etw. würde einer Sache *(Dat.)* nichts schaden; **c)** *(emotionally)* verletzen, kränken ‹*Person*›; verletzen ‹*Ehrgefühl, Stolz*›; **~ sb.'s feelings** jmdn. verletzen. **2.** *v. i.*, **hurt a)** weh tun; schmerzen; **my leg ~s** mein Bein tut [mir] weh; **does your hand ~?** tut dir die Hand weh?; **b)** *(cause damage, be detrimental)* schaden; **c)** *(cause emotional distress)* weh tun; ‹*Worte, Beleidigungen:*› verletzen; ‹*Person:*› verletzend sein. **3.** *adj.* gekränkt ‹*Tonfall, Miene*›. **4.** *n.* *(emotional pain)* Schmerz, *der*

hurtful ['hɜ:tfl] *adj.* *(emotionally wounding)* verletzend; **what a ~ thing to say/do!** wie kann man nur so etwas Verletzendes sagen/tun!

hurtle ['hɜ:tl] *v. i.* rasen *(ugs.)*; **he went hurtling down the street/round the corner** er raste die Straße hinunter/um die Ecke

husband ['hʌzbənd] *n.* Ehemann, *der;* **my/your/her ~:** mein/dein/ihr Mann; **~ and wife** Mann und Frau

husbandry ['hʌzbəndrɪ] *n., no pl.* **animal/dairy ~:** Viehzucht, *die*/Milchviehhaltung, *die*

hush [hʌʃ] **1.** *n.* **a)** *(silence)* Schweigen, *das;* **a sudden ~ fell over them** sie verstummten plötzlich; **b)** *(stillness)* Stille, *die.* **2.** *v. t.* *(silence)* zum Schweigen bringen; *(still)* beruhigen; besänftigen. **3.** *v. i.* still sein; *(become silent)* verstummen; **~!** still!

~ 'up *v. t.* **a)** *(make silent)* zum Schweigen bringen; **b)** *(keep secret)* **~ sth. up** etw. vertuschen

hushed [hʌʃt] *adj.* gedämpft ‹*Flüstern, Stimme*›

'hush-hush *adj.* *(coll.)* geheim; **keep sth. ~:** etw. geheimhalten

husk [hʌsk] *n.* Schale, *die; (of wheat, grain, rice)* Spelze, *die; (fig.: useless remainder)* Hülse, *die*

¹husky ['hʌskɪ] *adj.* *(hoarse)* heiser

²husky *n.* *(dog)* Eskimohund, *der; (sledge-dog)* Schlittenhund, *der*

hussar [hʊˈzɑː(r)] *n.* *(Mil.)* Husar, *der*

hustings [ˈhʌstɪŋz] *n. pl. constr. as sing. or pl. (proceedings)* Wahlveranstaltungen

hustle [ˈhʌsl] **1.** *v. t.* **a)** drängen **(into** zu); **b)** *(jostle)* anrempeln *(salopp); (thrust)* [hastig] drängen; **the guide ~d the tourists along** der Führer scheuchte die Touristen voran. **2.** *v. i.* **a)** *(push roughly)* **~ through the crowds** sich durch die Menge drängeln; **b)** *(hurry)* hasten; **~ and bustle about** geschäftig hin und her eilen *od.* sausen. **3.** *n.* **a)** *(jostling)* Gedränge, *das;* **b)** *(hurry)* Hetze, *die;* **~ and bustle** Geschäftigkeit, *die; (in street)* geschäftiges Treiben

hut [hʌt] *n.* Hütte, *die; (Mil.)* Baracke, *die*

hutch [hʌtʃ] *n.* Stall, *der*

hyacinth [ˈhaɪəsɪnθ] *n.* *(Bot.)* Hyazinthe, *die*

hybrid [ˈhaɪbrɪd] **1.** *n.* **a)** *(Biol.)* Hybride, *die od. der* **(between** aus); Kreuzung, *die;* **b)** *(fig.: mixture)* Mischung, *die.* **2.** *adj.* **a)** *(Biol.)* hybrid ‹*Züchtung*›; **b)** *(fig.: mixed)* gemischt; Misch‹*kultur, -sprache*›

hydrangea [haɪˈdreɪndʒə] *n.* *(Bot.)* Hortensie, *die*

hydrant [ˈhaɪdrənt] *n.* Hydrant, *der*

hydraulic [haɪˈdrɔːlɪk] *adj.* *(Mech. Engin.)* hydraulisch; **~ engineering** Wasserbau, *der*

hydrocarbon [haɪdrəˈkɑːbən] *n.* *(Chem.)* Kohlenwasserstoff, *der*

hydrochloric acid [haɪdrəklɔːrɪk ˈæsɪd] *n.* *(Chem.)* Salzsäure, *die*

hydroelectric [haɪdrəʊɪˈlektrɪk] *adj.* *(Electr.)* hydroelektrisch; **~ power plant** *or* **station** Wasserkraftwerk, *das*

hydrofoil [ˈhaɪdrəfɔɪl] *n.* *(Naut.: vessel)* Tragflächenboot, *das*

hydrogen [ˈhaɪdrədʒən] *n.* Wasserstoff, *der*

ˈ**hydrogen bomb** *n.* Wasserstoffbombe, *die*

hydrometer [haɪˈdrɒmɪtə(r)] *n.* Hydrometer, *das*

hydroponics [haɪdrəˈpɒnɪks] *n., no pl.* *(Hort.)* Hydroponik, *die (fachspr.);* Hydrokultur, *die*

hydroxide [haɪˈdrɒksaɪd] *n.* *(Chem.)* Hydroxid, *das*

hyena [haɪˈiːnə] *n.* *(Zool.)* Hyäne, *die;* **laugh like a ~:** wie eine Hyäne kreischen

hygiene [ˈhaɪdʒiːn] *n., no pl.* Hygiene, *die;* **dental ~:** Zahnhygiene, *die; see also* **personal**

hygienic [haɪˈdʒiːnɪk] *adj.* hygienisch; **not ~:** unhygienisch

hymn [hɪm] *n.* Hymne, *die*

ˈ**hymn-book** *n.* Gesangbuch, *das*

hype [haɪp] *n.* *(sl.: misleading publicity)* Reklameschwindel, *der (ugs.)*

hyperactive [haɪpəˈræktɪv] *adj.* überaktiv

hyperbola [haɪˈpɜːbələ] *n., pl.* **~s** *or* **~e** [haɪˈpɜːbəliː] *(Geom.)* Hyperbel, *die*

hypercritical [haɪpəˈkrɪtɪkl] *adj.* hyperkritisch

hypermarket [ˈhaɪpəmɑːkɪt] *n.* Verbrauchermarkt, *der*

hypersensitive [haɪpəˈsensɪtɪv] *adj.* hypersensibel; überempfindlich; **be ~ to sth.** überempfindlich auf etw. *(Akk.)* reagieren

hyphen [ˈhaɪfn] **1.** *n.* **a)** Bindestrich, *der;* **b)** *(connecting separate syllables)* Trennungsstrich, *der;* Divis, *das (fachspr.).* **2.** *v. t.* mit Bindestrich schreiben

hyphenate [ˈhaɪfəneɪt] *see* **hyphen 2**

hyphenation [haɪfəˈneɪʃn] *n., no pl.* Kopplung, *die*

hypnosis [hɪpˈnəʊsɪs] *n., pl.* **hypnoses** [hɪpˈnəʊsiːz] Hypnose, *die; (act, process)* Hypnotisierung, *die;* **under ~:** in Hypnose *(Dat.)*

hypnotic [hɪpˈnɒtɪk] *adj.* hypnotisch; *(producing hypnotism)* hypnotisch; hypnotisierend ‹*Wirkung, Blick*›

hypnotism [ˈhɪpnətɪzm] *n.* Hypnotik, *die; (act)* Hypnotisieren, *das*

hypnotist [ˈhɪpnətɪst] *n.* Hypnotiseur, *der*/Hypnotiseuse, *die*

hypnotize [ˈhɪpnətaɪz] *v. t.* *(lit. or fig.)* hypnotisieren; *(fig.: fascinate)* faszinieren

hypochondria [haɪpəˈkɒndrɪə] *n.* Hypochondrie, *die*

hypochondriac [haɪpəˈkɒndrɪæk] *n.* Hypochonder, *der*

hypocrisy [hɪˈpɒkrɪsɪ] *n.* **a)** Heuchelei, *die;* **b)** *(simulation of virtue)* Scheinheiligkeit, *die*

hypocrite [ˈhɪpəkrɪt] *n.* **a)** Heuchler, *der*/Heuchlerin, *die;* **b)** *(person feigning virtue)* Scheinheilige, *der/die*

hypocritical [hɪpəˈkrɪtɪkl] *adj.* heuchlerisch; *(feigning virtue)* scheinheilig

hypodermic [haɪpəˈdɜːmɪk] *(Med.)* **1.** *adj.* subkutan ‹*Injektion*›; **~ syringe** Injektionsspritze, *die.* **2.** *n.* *(syringe)* Injektionsspritze, *die*

hypotenuse [haɪˈpɒtənjuːz] *n.* *(Geom.)* Hypotenuse, *die*

hypothermia [haɪpəˈθɜːmɪə] *n.* *(Med.)* Hypothermie, *die (fachspr.);* Unterkühlung, *die*

hypothesis [haɪˈpɒθɪsɪs] *n., pl.* **hypotheses** [haɪˈpɒθɪsiːz] Hypothese, *die*

hypothetical [haɪpəˈθetɪkl] *adj.* hypothetisch

hysterectomy [hɪstəˈrektəmɪ] *n.* *(Med.)* Hysterektomie, *die*

hysteria [hɪˈstɪərɪə] *n.* Hysterie, *die*

hysterical [hɪˈsterɪkl] *adj.* hysterisch

hysterically [hɪˈsterɪkəlɪ] *adv.* hysterisch; **~ funny** urkomisch

hysterics [hɪˈsterɪks] *n. pl.* *(laughter)* hysterischer Lachanfall; *(crying)* hysterischer Weinkrampf; **have ~:** hysterisch lachen/weinen

Hz *abbr.* **hertz** Hz

I

[1]**I, i** [aɪ] *n., pl.* **Is** *or* **I's** I, i, *das; see also* **dot 2 b**

[2]**I** *pron.* ich; **it is I** *(formal)* ich bin es; *see also* [1]**me;** [2]**mine; my; myself**

I. *abbr.* **a) Island[s]** I.; **b) Isle[s]** I.

IBA *abbr.* *(Brit.)* **Independent Broadcasting Authority** *Kontrollgremium für den privaten Rundfunk und das Privatfernsehen*

Iberia [aɪˈbɪərɪə] *pr. n.* *(Hist., Geog.)* Iberische Halbinsel

Iberian Peninsula [aɪˈbɪərɪən pɪˈnɪnsjʊlə(r)] *pr. n.* *(Geog.)* Iberische Halbinsel

ice [aɪs] **1.** *n.* **a)** *no pl.* Eis, *das;* **feel/be like ~** *(be very cold)* eiskalt sein; **there was ~ over the pond** eine Eisschicht bedeckte den Teich; **fall through the ~:** auf dem Eis einbrechen; **be on ~** *(coll.)* ‹*Plan:*› auf Eis *(Dat.)* liegen *(ugs.);* **put on ~** *(coll.)* auf Eis *(Akk.)* legen *(ugs.);* **be on thin ~** *(fig.)* sich auf dünnes Eis begeben haben; **break the ~** *(fig.: break through reserve)* das Eis brechen; **b)** *(Brit.: ~-cream)* [Speise]eis, *das;* Eiscreme, *die;* **an ~/two ~s** ein/zwei Eis. **2.** *v. t.*

a) *(cool with ~)* [mit Eis] kühlen; **~d coffee/tea** Eiskaffee, *der*/Tee mit Eis; **be ~d** eisgekühlt sein; **b)** glasieren ⟨*Kuchen*⟩

~ 'over *v. i.* ⟨*Gewässer:*⟩ zufrieren; ⟨*Straße:*⟩ vereisen

~ 'up *v. i.* **a)** *(freeze)* ⟨*Wasserleitung:*⟩ einfrieren; **b)** *see* **~ over**

'ice age *n.* Eiszeit, *die*

iceberg ['aɪsbɜ:g] *n.* Eisberg, *der;* **the tip of the ~** *(fig.)* die Spitze des Eisbergs

ice: ~-blue 1. [-'-] *n.* Eisblau, *das;* **2.** ['--] *adj.* eisblau; **~-bound** *adj.* eingefroren ⟨*Schiff*⟩; durch Vereisung abgeschnitten ⟨*Hafen, Küste*⟩; **~box** *n. (Amer.)* Kühlschrank, *der;* **~-breaker** *n. (Naut.)* Eisbrecher, *der;* **~-cold** *adj.* eiskalt; **~-'cream** *n.* Eis, *das;* Eiscreme, *die;* **one ~-cream/two ~-creams** ein/zwei Eis; **~-cube** *n.* Eiswürfel, *der;* **~-floe** *n.* Eisscholle, *die;* **~ hockey** *n.* Eishockey, *das*

Iceland ['aɪslənd] *pr. n.* Island *(das)*

Icelander ['aɪsləndə(r)] *n.* Isländer, *der*/Isländerin, *die*

Icelandic [aɪs'lændɪk] **1.** *adj.* isländisch; **sb. is ~:** jmd. ist Isländer/Isländerin. **2.** *n.* Isländisch, *das; see also* **English 2 a**

ice: ~ 'lolly *see* **lolly a; ~-pack** *n.* **a)** *(to relieve pain)* Eispackung, *die;* **b)** *(to keep food cold)* Kühlakku, *der (ugs.);* **~-rink** *n.* Schlittschuh-, Eisbahn, *die;* **~-skate 1.** *n.* Schlittschuh, *der;* **2.** *v. i.* Schlittschuh laufen; eislaufen; **~-skating** *n.* Schlittschuhlaufen, *das*

icicle ['aɪsɪkl] *n.* Eiszapfen, *der*

icily ['aɪsɪlɪ] *adv.* eisig; *(fig.)* kalt ⟨*ablehnend, lächelnd*⟩; eisig, frostig ⟨*begrüßen, anblicken*⟩

icing ['aɪsɪŋ] *n. (Cookery)* Zuckerguß, *der;* Zuckerglasur, *die;* **[the] ~ on the cake** *(fig.)* das Tüpfelchen auf dem i

'icing sugar *n. (Brit.)* Puderzucker, *der*

icon ['aɪkən, 'aɪkɒn] *n. (Orthodox Ch.)* Ikone, *die*

iconoclastic [aɪkɒnə'klæstɪk] *adj. (lit. or fig.)* bilderstürmerisch

icy ['aɪsɪ] *adj.* **a)** vereist ⟨*Berge, Landschaft, Straße, See*⟩; eisreich ⟨*Region, Land*⟩; **in ~ conditions** bei Eis; **b)** *(very cold)* eiskalt; eisig; *(fig.)* frostig ⟨*Ton*⟩

ID [aɪ'di:] *n.* **ID card, ID plate** *see* **identification; have you [got] some** *or* **any ID?** können Sie sich ausweisen?

I'd [aɪd] **a)** = **I had; b)** = **I would**

idea [aɪ'dɪə] *n.* **a)** *(conception)* Idee, *die;* Gedanke, *der;* **arrive at an ~:** auf eine Idee *od.* einen Gedanken kommen; **the ~ of going abroad** der Gedanke *od.* die Vorstellung, ins Ausland zu fahren; **give/get some ~ of sth.** einen Überblick über etw. *(Akk.)* geben/einen Eindruck von etw. bekommen; **get the ~ [of sth.]** verstehen, worum es [bei etw.] geht; **sb.'s ~ of sth.** *(coll.)* jmds. Vorstellung von etw.; **not my ~ of ...** *(coll.)* nicht, was ich mir unter ... *(Dat.)* vorstelle; **he has no ~** *(coll.)* er hat keine Ahnung *(ugs.);* **b)** *(mental picture)* Vorstellung, *die;* **what gave you 'that ~?** wie bist du darauf gekommen?; **get the ~ that ...:** den Eindruck bekommen, daß ...; **get** *or* **have ~s** *(coll.) (be rebellious)* auf dumme Gedanken kommen *(ugs.); (be ambitious)* sich *(Dat.)* Hoffnungen machen; **put ~s into sb.'s head** jmdn. auf dumme Gedanken bringen; **c)** *(vague notion)* Ahnung, *die;* Vorstellung, *die;* **have you any ~ [of] how ...?** weißt du ungefähr, wie ...?; **you can have no ~ [of] how ...:** du kannst dir gar nicht vorstellen, wie ...; **not have the remotest** *or* **slightest** *or* **faintest** *or (coll.)* **foggiest ~:** nicht die entfernteste *od.* mindeste *od.* leiseste Ahnung haben; **I suddenly had the ~ that ...:** mir kam plötzlich der Gedanke, daß ...; **I've an ~ that ...:** ich habe so eine Ahnung, daß ...; **the [very] ~!, what an ~!** *(coll.)* unvorstellbar!; allein die Vorstellung!; **d)** *(plan)* Idee, *die;* **good ~!** [das ist eine] gute Idee!; **'that's an ~** *(coll.)* das ist eine gute Idee; **that gives me an ~:** das hat mich auf eine Idee gebracht; **the ~ was that ...:** der Plan war, daß ...; **have big ~s** große Rosinen im Kopf haben; **what's the big ~?** *(iron.)* was soll das?; was soll der Blödsinn? *(ugs.)*

ideal [aɪ'dɪəl] **1.** *adj.* **a)** ideal; vollendet ⟨*Ehemann, Gastgeber*⟩; vollkommen ⟨*Glück, Welt*⟩; **b)** *(embodying an idea, existing only in idea)* ideell; gedacht. **2.** *n.* Ideal, *das*

idealise *see* **idealize**

idealism [aɪ'dɪəlɪzm] *n., no pl.* Idealismus, *der*

idealist [aɪ'dɪəlɪst] *n.* Idealist, *der*/Idealistin, *die*

idealistic [aɪdɪə'lɪstɪk] *adj.* idealistisch

idealize [aɪ'dɪəlaɪz] *v. t.* idealisieren

ideally [aɪ'dɪəlɪ] *adv.* ideal; **~, the work should be finished in two weeks** im Idealfalle *od.* idealerweise sollte die Arbeit in zwei Wochen abgeschlossen sein

identical [aɪ'dentɪkl] *adj.* **a)** *(same)* identisch; **the ~ species** dieselbe Art; **b)** *(agreeing in every detail)* identisch; sich *(Dat.)* gleichend; **be ~:** sich *(Dat.)* völlig gleichen; **~ twins** eineiige Zwillinge

identifiable [aɪ'dentɪfaɪəbl] *adj.* erkennbar (**by** an + *Dat.*); nachweisbar ⟨*Stoff, Substanz*⟩; bestimmbar ⟨*Pflanzen-, Tierart*⟩

identification [aɪdentɪfɪ'keɪʃn] *n.* Identifizierung, *die; (of plants or animals)* Bestimmung, *die;* **means of ~:** Ausweispapiere *Pl.;* **have you any means of ~?** können Sie sich ausweisen?; **~ card** [Personal]ausweis, *der;* **~ plate** Kennzeichenschild, *das*

identifi'cation parade *n. (Brit.)* Gegenüberstellung [zur Identifizierung]

identify [aɪ'dentɪfaɪ] **1.** *v. t.* **a)** *(treat as identical)* gleichsetzen (**with** mit); **b)** *(associate)* identifizieren (**with** mit); **c)** *(recognize)* identifizieren; bestimmen ⟨*Pflanze, Tier*⟩; **d)** *(establish)* ermitteln. **2.** *v. i.* **~ with sb.** sich mit jmdm. identifizieren

Identikit, (P) [aɪ'dentɪkɪt] *n.* Phantombild, *das*

identity [aɪ'dentɪtɪ] *n.* **a)** *(sameness)* Übereinstimmung, *die;* **b)** *(individuality, being specified person)* Identität, *die;* **proof of ~:** Identitätsnachweis, *der;* **[case of] mistaken ~:** [Personen]verwechslung, *die;* **c) ~ card/plate** *see* **identification**

i'dentity parade *see* **identification parade**

ideological [aɪdɪə'lɒdʒɪkl, ɪdɪə'lɒdʒɪkl] *adj.* ideologisch

ideology [aɪdɪ'ɒlədʒɪ, ɪdɪ'ɒlədʒɪ] *n.* Ideologie, *die;* Weltanschauung, *die*

idiocy ['ɪdɪəsɪ] *n.* Dummheit, *die;* Idiotie, *die*

idiom ['ɪdɪəm] *n.* [Rede]wendung, *die;* idiomatischer Ausdruck

idiomatic [ɪdɪə'mætɪk] *adj.* idiomatisch

idiosyncrasy [ɪdɪə'sɪŋkrəsɪ] *n.* Eigentümlichkeit, *die;* Eigenheit, *die*

idiosyncratic [ɪdɪəsɪŋ'krætɪk] *adj.* eigenwillig

idiot ['ɪdɪət] *n.* Idiot, *der (ugs.);* Trottel, *der (ugs.)*

idiotic [ɪdɪ'ɒtɪk] *adj.* idiotisch *(ugs.);* **what an ~ thing to do/say** was für ein Schwachsinn

idle ['aɪdl] **1.** *adj.* **a)** *(lazy)* faul; träge; **b)** *(not in use)* außer Betrieb *nachgestellt;* **be** *or* **stand** ~ ‹*Maschinen, Fabrik:*› stillstehen; *see also* ²**lie 2 b; c)** *(having no special purpose)* bloß ‹*Neugier*›; nutzlos, leer ‹*Geschwätz*›; **d)** *(groundless)* unbegründet ‹*Annahme, Mutmaßung*›; bloß, rein ‹*Spekulation, Gerücht*›; **no** ~ **boast** *or* **jest** *(iron.)* kein leeres Versprechen; **e)** *(ineffective)* sinnlos, *(geh.)* müßig ‹*Diskussion, Streit*›; leer ‹*Versprechen*›; **f)** *(unoccupied)* frei ‹*Zeit, Stunden, Tag*›; **g)** *(unemployed)* **be made** ~ ‹*Arbeiter:*› arbeitslos werden. **2.** *v. i.* ‹*Motor:*› leer laufen, im Leerlauf laufen

~ **a'way** *v. t.* vertun ‹*Zeit, Leben*›

idleness ['aɪdlnɪs] *n., no pl. (being unoccupied)* Untätigkeit, *die; (avoidance of work)* Müßiggang, *der (geh.)*

idler ['aɪdlə(r)] *n.* Faulenzer, *der*/Faulenzerin, *die;* Faulpelz, *der (fam.)*

idly ['aɪdlɪ] *adv.* **a)** *(carelessly)* leichtsinnig; gedankenlos; **b)** *(inactively)* untätig; **stand** ~ **by while ...** *(fig.)* untätig zusehen, wie ...; **c)** *(indolently)* faul

idol ['aɪdl] *n.* **a)** *(false god)* Götze, *der; (image of deity)* Götzenbild, *das;* **b)** *(person venerated)* Idol, *das; (thing venerated)* Götze, *der*

idolatry [aɪ'dɒlətrɪ] *n.* Götzenverehrung, *die*

idolize (idolise) ['aɪdəlaɪz] *v. t.* **a)** *(make an idol of)* anbeten; verehren; **b)** *(fig.: venerate)* vergöttern; zum Idol erheben

idyll ['ɪdɪl] Idyll, *das*

idyllic [aɪ'dɪlɪk, ɪ'dɪlɪk] *adj.* idyllisch

i.e. [aɪ'i:] *abbr.* **that is** d. h.; i. e.

if [ɪf] **1.** *conj.* **a)** wenn; **if anyone should ask ...:** falls jemand fragt, ...; wenn jemand fragen sollte, ...; **if you would lend me some money ...:** wenn du mir Geld leihen würdest, ...; **if I knew what to do ...:** wenn ich wüßte, was ich tun soll ...; **if I were you** an deiner Stelle; **better, if anything** vielleicht etwas besser; **tell me what I can do to help, if anything** falls ich irgendwie helfen kann, sag es mir; **if so/not** wenn ja/nein *od.* nicht; **if then/that/at all** wenn überhaupt; **if only for today** wenn auch nur für heute; **if only because/to ...:** schon allein, weil/um ... zu ...; **as if** als ob; **as if you didn't know!** als ob du es nicht gewußt hättest!; **it isn't** *or* **it's not as if we were** *or (coll.)* **we're rich** es ist nicht etwa so, daß wir reich wären; **b)** *(whenever)* [immer] wenn; **c)** *(whether)* ob; **d)** *in excl. of wish* **if I only knew, if only I knew!** wenn ich das nur wüßte!; das wüßte ich gern!; **if only you could have seen it!** wenn du es nur hättest sehen können!; **e)** *expr. surprise etc.* **if it isn't Ronnie!** das ist doch Ronnie!; **f)** *in polite request* **if you will wait a moment** wenn Sie einen Augenblick warten wollen; **if you wouldn't mind holding the door open** wenn Sie so freundlich wären und die Tür aufhielten; **g)** *(though)* und wenn; auch *od.* selbst wenn; **even if he did say that, ...:** selbst wenn er das gesagt hat, ...; **h)** *(despite being)* wenn auch; **likeable, if somewhat rough** liebenswürdig, wenn auch etwas derb. **2.** *n.* Wenn, *das;* **ifs and buts** Wenn und Aber, *das*

igloo ['ɪglu:] *n.* Iglu, *der od. das*

ignite [ɪg'naɪt] **1.** *v. t.* anzünden; entzünden *(geh.).* **2.** *v. i.* sich entzünden

ignition [ɪg'nɪʃn] *n.* **a)** *(igniting)* Zünden, *das;* Entzünden, *das (geh.);* **b)** *(Motor Veh.)* Zündung, *die*

ig'nition key *n. (Motor Veh.)* Zündschlüssel, *der*

ignoble [ɪg'nəʊbl] *adj.* niedrig ‹*Geburt, Herkunft*›; niederträchtig ‹*Person*›; schändlich ‹*Tat*›

ignominious [ɪgnə'mɪnɪəs] *adj.* verwerflich *(geh.)* ‹*Tat, Idee, Praktik*›; *(humiliating)* schändlich

ignominy ['ɪgnəmɪnɪ] *n.* Schande, *die*

ignoramus [ɪgnə'reɪməs] *n.* Ignorant, *der;* Nichtswisser, *der*

ignorance ['ɪgnərəns] *n., no pl.* Ignoranz, *die (abwertend);* Unwissenheit, *die;* **keep sb. in** ~ **of sth.** jmdn. in Unkenntnis über etw. *(Akk.)* lassen; ~ **is bliss** was ich nicht weiß, macht mich nicht heiß *(Spr.);* **his** ~ **of physics** seine mangelnden Kenntnisse in Physik

ignorant ['ɪgnərənt] *adj.* **a)** *(lacking knowledge)* unwissend; ungebildet; **b)** *(behaving in uncouth manner)* unkultiviert *(abwertend);* **c)** *(uninformed)* **be** ~ **of sth.** über etw. *(Akk.)* nicht informiert sein; **remain** ~ **of sth.** über etw. *(Akk.)* nie etwas erfahren; **be** ~ **in** *or* **of mathematics** mangelnde Kenntnisse in Mathematik haben

ignore [ɪg'nɔ:(r)] *v. t.* ignorieren; nicht beachten; nicht befolgen ‹*Befehl, Rat*›; übergehen, überhören ‹*Frage, Bemerkung*›; **I shall** ~ **that remark!** ich habe das nicht gehört!

ilk [ɪlk] *n. (coll.)* **Bill and [others of] his** ~: Bill und seinesgleichen; **he's another of the same** ~: er gehört auch zu derselben Sorte; **people of that** ~: solche Leute

ill [ɪl] **1.** *adj.,* **worse** [wɜ:s], **worst** [wɜ:st] **a)** *(sick)* krank; **be** ~ **with flu** an Grippe *(Dat.)* erkrankt sein; **be** ~ **with worry** vor Sorgen [ganz] krank sein; *see also* **fall 2 k; take 1 i; b)** *(harmful)* ~ **effects** schädliche Wirkungen; **c)** *(unfavourable)* widrig ‹*Schicksal, Umstand*›; ~ **fate** *or* **fortune** *or* **luck** Pech, *das;* **it's an** ~ **wind that blows nobody [any] good** *(prov.)* des einen Leid, des andern Freud' *(Spr.);* **as** ~ **luck would have it** wie es das Unglück wollte. **2.** *n.* **a)** *(evil)* Übel, *das;* **for good or** ~: komme, was will; **b)** *(harm)* Schlechte, *das;* Unglück, *das;* **wish sb.** ~: jmdm. nichts Gutes *od.* nur das Schlechteste wünschen; **speak** ~ **of sb./sth.** Schlechtes über jmdn. *od.* von jmdm./von etw. sagen; **c)** *in pl. (misfortunes)* Mißstände *Pl.* **3.** *adv.,* **worse, worst a)** *(badly)* schlecht, unschicklich ‹*sich benehmen*›; **b)** *(imperfectly)* schlecht, unzureichend ‹*versorgt, ausgestattet*›; **he can** ~ **afford it** er kann es sich *(Dat.)* kaum leisten; ~ **at ease** verlegen

I'll [aɪl] **a)** = **I shall; b)** = **I will**

ill: ~**-advised** *adj.* unklug; schlechtberaten ‹*Kunde*›; **be** ~**-advised** ‹*Person:*› schlecht beraten sein; ~**-behaved** *see* **behave 1 a;** ~**-bred** *adj.* schlecht erzogen ‹*Kind, Jugendlicher*›; unkultiviert *(abwertend)* ‹*Leute, Kerl usw.*›; ~**-conceived** *adj.* schlecht durchdacht; ~**-defined** *adj.* ungenau definiert, unklar ‹*Verfahren, Vorgehen*›

illegal [ɪ'li:gl] *adj.* ungesetzlich; illegal; *(Sport)* regelwidrig; unerlaubt; **it is** ~ **to drive a car without a licence** es ist verboten, ohne Führerschein Auto zu fahren

illegality [ɪlɪ'gælɪtɪ] *n. no pl.* Ungesetzlichkeit, *die*

illegally [ɪ'li:gəlɪ] *adv.* illegal; **bring sth. into the country** ~: etw. illegal einführen

illegible [ɪ'ledʒɪbl] *adj.* unleserlich

illegitimate [ɪlɪ'dʒɪtɪmət] *adj.* **a)** unehelich ‹*Kind*›; **b)** *(not authorized by law)* unrechtmäßig ‹*Machtergreifung, Geschäft*›; mit dem Gesetz unvereinbar ‹*Maß-*

nahme, Vorgehen, Beweggrund⟩; **c)** *(wrongly inferred)* unzulässig

ill: ~ **'fated** *adj.* unglückselig; verhängnisvoll ⟨*Entscheidung, Stunde, Tag*⟩; ~ **'feeling** *n.* Verstimmung, *die;* **cause ~ feeling** böses Blut machen *od.* schaffen; **no ~ feeling[s]?** sind Sie jetzt verstimmt *od.* *(fam.)* böse?; **~-founded** ['ɪlfaʊndɪd] *adj.* haltlos ⟨*Theorie, Gerücht*⟩; **be ~-founded** völlig haltlos sein; **~-gotten** *adj.* unrechtmäßig erworben; ~ **'health** *n.* schwache Gesundheit; **~-humoured** [ɪl'hju:məd] *adj.* schlecht gelaunt

illicit [ɪ'lɪsɪt] *adj.* verboten ⟨*Glücksspiel*⟩; unerlaubt ⟨*[Geschlechts]verkehr, Beziehung*⟩; Schwarz⟨*handel, -verkauf, -arbeit*⟩

'ill-informed *adj.* schlecht informiert; auf Unkenntnis beruhend ⟨*Bemerkung, Schätzung, Urteil*⟩

illiteracy [ɪ'lɪtərəsɪ] *n., no pl.* Analphabetentum, *das*

illiterate [ɪ'lɪtərət] **1.** *adj.* **a)** des Lesens und Schreibens unkundig; analphabetisch ⟨*Bevölkerung*⟩; **he is ~:** er ist Analphabet; **b)** *(showing lack of learning)* primitiv *(abwertend);* **musically ~:** musikalisch völlig unbedarft. **2.** *n.* Analphabet, *der*/Analphabetin, *die*

ill: **~-judged** *adj.* unklug; *(rash)* unüberlegt; leichtfertig; **~-mannered** [ɪl'mænəd] *adj.* rüpelhaft *(abwertend);* ungezogen ⟨*Kind*⟩

illness ['ɪlnɪs] *n.* Krankheit, *die;* **children's ~:** Kinderkrankheit, *die;* **because of ~:** wegen [einer] Krankheit

illogical [ɪ'lɒdʒɪkl] *adj.* unlogisch

ill: **~-tempered** [ɪl'tempəd] *adj.* schlecht gelaunt; **~-timed** *adj.* [zeitlich] ungelegen; unpassend ⟨*Bemerkung*⟩; **~-'treat** *v.t.* mißhandeln ⟨*Lebewesen*⟩; nicht schonend behandeln ⟨*Gegenstand*⟩; **~-'treatment** *n., no pl.* *(of living thing)* Mißhandlung, *die;* *(of object)* wenig pflegliche Behandlung

illuminate [ɪ'lju:mɪneɪt, ɪ'lu:mɪneɪt] *v.t.* **a)** ⟨*Lampe usw.:*⟩ beleuchten; ⟨*Mond, Sonne:*⟩ erleuchten; **b)** *(give enlightenment to)* erleuchten; **c)** *(help to explain)* erhellen; [näher] beleuchten; **d)** *(decorate with lights)* festlich beleuchten; illuminieren; **~d advertisements** Leuchtreklamen; **e)** ausmalen, *(fachspr.)* illuminieren ⟨*Handschriften usw.*⟩

illuminating [ɪ'lju:mɪneɪtɪŋ, ɪ'lu:mɪneɪtɪŋ] *adj.* aufschlußreich

illumination [ɪlju:mɪ'neɪʃn, ɪlu:mɪ'neɪʃn] *n.* **a)** *(lighting)* Beleuchtung, *die;* **b)** *(enlightenment)* Erleuchtung, *die;* **c)** *(decorative lights) often in pl.* **~[s]** Festbeleuchtung, *die;* Illumination, *die*

illusion [ɪ'lju:ʒn, ɪ'lu:ʒn] *n.* Illusion, *die;* *(misapprehension)* falsche Vorstellung; Illusion, *die;* **be under an ~:** sich Illusionen *(Dat.)* hingeben; **be under the ~ that ...:** sich *(Dat.)* einbilden, daß ...; **have no ~s about sb./sth.** sich *(Dat.)* über jmdn./etw. keine Illusionen machen

illusory [ɪ'lju:sərɪ, ɪ'lu:sərɪ] *adj.* *(deceptive)* illusorisch

illustrate ['ɪləstreɪt] *v.t.* **a)** *(serve as example of)* veranschaulichen; illustrieren; **b)** *(elucidate by pictures)* [bildlich] darstellen ⟨*Vorgang, Ablauf*⟩; illustrieren ⟨*Buch, Erklärung*⟩; **c)** *(explain)* verdeutlichen; erläutern; *(make clear by examples)* anschaulicher machen; illustrieren

illustration [ɪlə'streɪʃn] *n.* **a)** *(example)* Beispiel, *das* (**of** für); *(drawing)* Abbildung, *die;* **b)** *(picture)* Abbildung, *die;* Illustration, *die;* **c)** *no pl.* *(with example)* Illustration, *die;* **by way of ~:** zur Illustration *od.* Verdeutlichung

illustrative ['ɪləstrətɪv] *adj.* erläuternd; illustrativ; **be ~ of sth.** beispielhaft für etw. sein

illustrator ['ɪləstreɪtə(r)] *n.* Illustrator, *der*/Illustratorin, *die*

illustrious [ɪ'lʌstrɪəs] *adj.* berühmt ⟨*Person*⟩ (**for** wegen); ruhmreich ⟨*Tat, Herrschaft*⟩

ill 'will *n.* Böswilligkeit, *die*

I'm [aɪm] = **I am**

image ['ɪmɪdʒ] *n.* **a)** Bildnis, *das (geh.);* *(statue)* Standbild, *das;* **b)** *(Optics, Math.)* Bild, *das;* **c)** *(semblance)* Bild, *das;* *(counterpart)* Ebenbild, *das (geh.);* **she is the [very] ~ of her mother** sie ist das [getreue] Ebenbild ihrer Mutter; **d)** *(Lit.: simile, metaphor)* Bild, *das;* **e)** *(mental representation)* Bild, *das;* *(conception)* Vorstellung, *die;* **f)** *(perceived character)* Image, *das;* **improve one's ~:** sein Image aufbessern; **public ~:** Image [in der Öffentlichkeit]

imagery ['ɪmɪdʒərɪ, 'ɪmɪdʒrɪ] *n., no pl.* *(Lit.)* Metaphorik, *die*

imaginable [ɪ'mædʒɪnəbl] *adj.* erdenklich; **the biggest lie ~:** die unverschämteste Lüge, die man sich *(Dat.)* vorstellen kann

imaginary [ɪ'mædʒɪnərɪ] *adj.* imaginär *(geh.);* konstruiert ⟨*Bildnis*⟩; eingebildet ⟨*Krankheit*⟩

imagination [ɪmædʒɪ'neɪʃn] *n.* **a)** *no pl., no art.* Phantasie, *die;* **use your ~!** hab doch ein bißchen Phantasie! *(ugs.);* **b)** *no pl., no art.* *(fancy)* Einbildung, *die;* **catch sb.'s ~:** jmdn. begeistern; **it's just your ~:** das bildest du dir nur ein

imaginative [ɪ'mædʒɪnətɪv] *adj.* phantasievoll; *(showing imagination)* einfallsreich

imagine [ɪ'mædʒɪn] *v.t.* **a)** *(picture to oneself, guess, think)* sich *(Dat.)* vorstellen; **can you ~?** stell dir vor!; ~ **things** sich *(Dat.)* Dinge einbilden[, die gar nicht stimmen]; ~ **sth. to be easy/difficult** *etc.* sich *(Dat.)* etw. leicht/schwer *usw.* vorstellen; **do not ~ that ...:** bilden Sie sich *(Dat.)* bloß nicht ein, daß ...; **as you can ~** wie du dir denken *od.* vorstellen kannst; **b)** *(coll.: suppose)* glauben; **c)** *(get the impression)* ~ **[that] ...:** sich *(Dat.)* einbilden[, daß] ...

imbalance [ɪm'bæləns] *n.* Unausgeglichenheit, *die*

imbecile ['ɪmbɪsi:l, 'ɪmbɪsaɪl] **1.** *adj.* schwachsinnig *(ugs. abwertend).* **2.** *n.* Idiot, *der (ugs.)*

imbibe [ɪm'baɪb] *v.t.* **a)** *(drink)* trinken; **b)** *(fig.: assimilate)* in sich *(Akk.)* aufsaugen

imbue [ɪm'bju:] *v.t.* durchdringen

IMF *abbr.* **International Monetary Fund** IWF, *der*

imitate ['ɪmɪteɪt] *v.t.* **a)** *(mimic)* nachahmen; nachmachen *(ugs.);* ~ **sb.** *(follow example of)* es jmdm. gleichtun; **b)** *(produce sth. like)* kopieren; **c)** *(be like)* imitieren

imitation [ɪmɪ'teɪʃn] **1.** *n.* **a)** *(imitating)* Nachahmung, *die;* **a style developed in ~ of classical models** ein nach klassischen Vorbildern entwickelter Stil; **do ~s of sb.** jmdn. imitieren *od.* nachahmen; **he sings, tells jokes, and does ~s** er singt, erzählt Witze und ahmt andere Leute nach; **b)** *(copy)* Kopie, *die;* *(counterfeit)* Imitation, *die.* **2.** *adj.* imitiert; Kunst⟨*leder, -horn*⟩; ~ **marble/fur** *etc.* Marmor-/Pelzimitation *usw., die*

imitative ['ɪmɪtətɪv, 'ɪmɪteɪtɪv] *adj.* **a)** uneigenständig; **be ~ of sb./sth.** jmdn./etw. nachahmen; **b)** *(prone to copy)* imitativ *(geh.)*

imitator ['ɪmɪteɪtə(r)] *n.* Nachahmer, *der*/Nachahmerin, *die;* Imitator, *der*/Imitatorin, *die*

immaculate [ɪ'mækjʊlət] *adj.* **a)** *(spotless)* makellos ⟨*Kleidung, Weiß*⟩; **b)** *(faultless)* tadellos

immaculately [ɪ'mækjʊlətlɪ] *adv.* **a)** *(spotlessly)* makellos; **b)** *(faultlessly)* tadellos

immaterial [ɪmə'tɪərɪəl] *adj.* unerheblich; **it's quite ~ to me** das ist für mich vollkommen uninteressant

immature [ɪmə'tjʊə(r)] *adj.* unreif; noch nicht voll entwickelt ‹*Lebewesen*›; noch nicht voll ausgereift ‹*Begabung, Talent*›

immaturity [ɪmə'tjʊərɪtɪ] *n. no pl.* Unreife, *die*

immeasurable [ɪ'meʒərəbl] *adj.* unermeßlich

immediate [ɪ'mi:djət] *adj.* **a)** unmittelbar; *(nearest)* nächst... ‹*Nachbar[schaft], Umgebung, Zukunft*›; engst... ‹*Familie*›; unmittelbar ‹*Kontakt*›; **your ~ action must be to ...:** als erstes müssen Sie ...; **his ~ plan is to ...:** zunächst einmal will er ...; **b)** *(occurring at once)* prompt; unverzüglich ‹*Handeln, Maßnahmen*›; umgehend ‹*Antwort*›

immediately [ɪ'mi:djətlɪ] **1.** *adv.* **a)** unmittelbar; direkt; **b)** *(without delay)* sofort. **2.** *conj. (coll.)* sobald

immemorial [ɪmɪ'mɔ:rɪəl] *adj.* undenklich; **from time ~:** seit undenklichen Zeiten

immense [ɪ'mens] *adj.* **a)** ungeheuer; immens; **b)** *(coll.: great)* enorm

immensely [ɪ'menslɪ] *adv.* **a)** ungeheuer; **b)** *(coll.: very much)* unheimlich *(ugs.)*

immensity [ɪ'mensɪtɪ] *n., no pl.* Ungeheuerlichkeit, *die*

immerse [ɪ'mɜ:s] *v. t.* **a)** *(dip)* [ein]tauchen; **b)** *(cause to be under water)* versenken; **~d in water** unter Wasser; **c) be ~d in thought/ one's work** *(fig.: involved deeply)* in Gedanken versunken/ in seine Arbeit vertieft sein

immersion [ɪ'mɜ:ʃn] *n.* Eintauchen, *das*

im'mersion heater *n.* Heißwasserbereiter, *der; (small, portable)* Tauchsieder, *der*

immigrant ['ɪmɪgrənt] **1.** *n.* Einwanderer, *der*/Einwanderin, *die;* Immigrant, *der*/Immigrantin, *die.* **2.** *adj.* Einwanderer-; **~ population** Einwanderer *Pl.;* **~ workers** ausländische Arbeitnehmer

immigrate ['ɪmɪgreɪt] *v. i.* einwandern, immigrieren (**into** nach, **from** aus)

immigration [ɪmɪ'greɪʃn] *n.* Einwanderung *die,* Immigration, *die* (**into** nach, **from** aus); *attrib.* Einwanderungs‹*kontrolle, -gesetz*›; **~ officer** Beamter/Beamte der Einwanderungsbehörde; **go through ~:** durch die Paßkontrolle gehen

imminent ['ɪmɪnənt] *adj.* unmittelbar bevorstehend; drohend ‹*Gefahr*›; **be ~:** unmittelbar bevorstehen/drohen

immobile [ɪ'məʊbaɪl] *adj.* **a)** *(immovable)* unbeweglich; **b)** *(motionless)* bewegungslos

immobilise *see* **immobilize**

immobility [ɪmə'bɪlɪtɪ] *n., no pl.* **a)** *(immovableness)* Unbeweglichkeit, *die;* **b)** *(motionlessness)* Bewegungslosigkeit, *die*

immobilize [ɪ'məʊbɪlaɪz] *v. t.* **a)** *(fix immovably)* verankern; *(fig.)* lähmen; **b)** gegen Wegfahren sichern ‹*Fahrzeug*›

immoderate [ɪ'mɒdərət] *adj.* **a)** *(excessive)* unmäßig ‹*Rauchen, Trinken*›; **b)** *(extreme)* extrem ‹*Ansichten, Politiker*›; maßlos ‹*Lebensstil*›

immodest [ɪ'mɒdɪst] *adj.* **a)** *(impudent)* unbescheiden; **b)** *(improper)* unanständig

immoral [ɪ'mɒrəl] *adj.* **a)** *(not conforming to morality)* unmoralisch; unsittlich; sittenwidrig *(Rechtsspr.);* **b)** *(morally evil)* pervers; **c)** *(in sexual matters)* sittenlos

immorality [ɪmə'rælɪtɪ] *n.* **a)** *no pl.* Unsittlichkeit, *die;* Unmoral, *die;* Sittenwidrigkeit, *die (Rechtsspr.);* **b)** *no pl. (wickedness)* Verdorbenheit, *die;* **c)** *(in sexual matters)* Sittenlosigkeit, *die;* **d)** *(morally wrong act)* Unsittlichkeit, *die*

immortal [ɪ'mɔ:tl] *adj.* unsterblich

immortalise *see* **immortalize**

immortality [ɪmɔ:'tælɪtɪ] *n., no pl.* Unsterblichkeit, *die*

immortalize [ɪ'mɔ:təlaɪz] *v. t.* unsterblich machen

immovable [ɪ'mu:vəbl] *adj.* **a)** unbeweglich; **b)** *(motionless)* bewegungslos; **c)** *(steadfast)* unerschütterlich

immovably [ɪ'mu:vəblɪ] *adv.* fest; **be ~ stuck** feststecken

immune [ɪ'mju:n] *adj.* **a)** *(exempt)* sicher (**from** vor + *Dat.*); geschützt (**from, against** vor + *Dat.*); **b)** *(insusceptible)* unempfindlich (**to** gegen); *(to hints, suggestions, etc.)* unempfänglich (**to** für); immun (**to** gegen); **c)** *(Med.)* immun (**to** gegen)

immunisation, immunise *see* **immunize-**

immunity [ɪ'mju:nɪtɪ] *n.* **a) ~ from prosecution** Schutz vor Strafverfolgung; **give sb. ~ from punishment** ‹*Person:*› jmdn. von der Bestrafung ausnehmen; ‹*Umstand:*› jmdn. vor Strafe schützen; **diplomatic ~:** diplomatische Immunität; **b)** *see* **immune b:** Unempfindlichkeit, *die* (**to** gegen); Unempfänglichkeit, *die* (**to** für); Immunität, *die* (**to** gegen); **c)** *(Med.)* Immunität, *die*

immunization [ɪmjʊnaɪ'zeɪʃn] *n.* *(Med.)* Immunisierung, *die*

immunize ['ɪmjʊnaɪz] *v. t. (Med.)* immunisieren

immunology [ɪmjʊ'nɒlədʒɪ] *n.* *(Med.)* Immunologie, *die*

immutable [ɪ'mju:təbl] *adj.* unveränderlich

imp [ɪmp] *n.* **a)** Kobold, *der;* **b)** *(fig.: mischievous child)* Racker, *der (fam.)*

impact 1. ['ɪmpækt] *n.* **a)** Aufprall, *der* (**on, against** auf + *Akk.*); *(of shell or bomb)* Einschlag, *der; (collision)* Zusammenprall, *der;* **b)** *(fig.: effect)* Wirkung, *die;* **have an ~ on sb./sth.** Auswirkungen auf jmdn./etw. haben; **make an ~ on sb./sth.** Eindruck auf jmdn./etw. machen. **2.** [ɪm'pækt] *v. t.* pressen

impacted [ɪm'pæktɪd] *adj.* *(Dent.)* impaktiert ‹*Zahn*›

impair [ɪm'peə(r)] *v. t.* **a)** *(damage)* beeinträchtigen; schaden (+ *Dat.*) ‹*Gesundheit*›; **b)** *(weaken)* beeinträchtigen; **~ed vision** Sehschwäche, *die;* **~ed hearing** Schwerhörigkeit, *die*

impale [ɪm'peɪl] *v. t.* aufspießen; *(Hist.)* pfählen

impart [ɪm'pɑ:t] *v. t.* **a)** *(give)* [ab]geben (**to** an + *Akk.*); vermachen (**to** *Dat.*); **b)** *(communicate)* kundtun *(geh.)* (**to** *Dat.*); vermitteln ‹*Kenntnisse*› (**to** *Dat.*)

impartial [ɪm'pɑ:ʃl] *adj.* unparteiisch; gerecht ‹*Entscheidung, Behandlung, Urteil*›

impartiality [ɪmpɑ:ʃɪ'ælɪtɪ] *n., no pl.* Unparteilichkeit, *die*

impassable [ɪm'pɑ:səbl] *adj.* unpassierbar (**to** für); *(to vehicles)* unbefahrbar (**to** für)

impasse ['æmpɑ:s, 'ɪmpɑ:s] *n. (lit. or fig.)* Sackgasse, *die;* **the negotiations have reached an ~:** die Verhandlungen sind in eine Sackgasse geraten

impassioned [ɪm'pæʃnd] *adj.* leidenschaftlich

impassive [ɪm'pæsɪv] *adj.* **a)** ausdruckslos; **b)** *(incapable of feeling emotion)* leidenschaftslos

impatience [ɪm'peɪʃəns] *n., no pl.* **a)** Ungeduld, *die* (**at** über + *Akk.*); **b)** *(intolerance)* Unduldsamkeit, *die* (**of** gegen); **c)** *(eager desire)* [ungeduldige] Erwartung (**for** *Gen.*)

impatient [ɪm'peɪʃənt] *adj.* **a)** ungeduldig; ~ **at sth./with sb.** ungeduldig über etw. *(Akk.)*/mit jmdm.; **b)** *(intolerant)* unduldsam **(of** gegen); **c)** *(eagerly desirous)* **be ~ for sth.** etw. kaum erwarten können; **be ~ to do sth.** unbedingt etw. tun wollen

impatiently [ɪm'peɪʃəntlɪ] *adv.* ungeduldig

impeach [ɪm'pi:tʃ] *v.t.* **a)** *(call in question)* in Frage stellen; **b)** *(Law)* anklagen **(of** *Gen.*, wegen)

impeachment [ɪm'pi:tʃmənt] *n.* *(Law)* Impeachment, *das*

impeccable [ɪm'pekəbl] *adj.* makellos; tadellos ⟨*Manieren*⟩

impede [ɪm'pi:d] *v.t.* behindern

impediment [ɪm'pedɪmənt] *n.* **a)** Hindernis, *das* **(to** für); **b)** *(speech defect)* Sprachfehler, *der*

impel [ɪm'pel] *v.t.*, **-ll-** treiben, antreiben ⟨*Turbine usw.*⟩; **feel ~led to do sth.** sich genötigt *od.* gezwungen fühlen, etw. zu tun

impend [ɪm'pend] *v.i.* *(be about to happen)* bevorstehen; ⟨*Gefahr:*⟩ drohen

impenetrable [ɪm'penɪtrəbl] *adj.* **a)** undurchdringlich **(by, to** für); unbezwingbar, uneinnehmbar ⟨*Festung*⟩; **b)** *(inscrutable)* unergründlich

imperative [ɪm'perətɪv] **1.** *adj.* **a)** *(commanding)* gebieterisch *(geh.)* ⟨*Stimme, Geste*⟩; **b)** *(urgent)* dringend erforderlich. **2.** *n.* **a)** Befehl, *der;* **b)** *(Ling.)* Imperativ, *der*

imperceptible [ɪmpə'septɪbl] *adj.* **a)** nicht wahrnehmbar **(to** für); unsichtbar ⟨*Schranke (fig.)*⟩; **be ~ to sb./the senses** von jmdm./den Sinnen nicht wahrgenommen werden können; **b)** *(very slight or gradual)* unmerklich; *(subtle)* kaum zu erkennen *nicht attr.;* kaum zu erkennend *nicht präd.;* minimal ⟨*Unterschied*⟩

imperfect [ɪm'pɜ:fɪkt] **1.** *adj.* **a)** *(not fully formed)* unfertig; *(incomplete)* unvollständig; **slightly ~ stockings/pottery** *etc.* Strümpfe/Keramik *usw.* mit kleinen Fehlern; **b)** *(faulty)* mangelhaft. **2.** *n.* *(Ling.)* Imperfekt, *das*

imperfection [ɪmpə'fekʃn] *n.* **a)** *no pl.* *(incompleteness)* Unvollständigkeit, *die;* **b)** *no pl.* *(faultiness)* Mangelhaftigkeit, *die;* **c)** *(fault)* Mangel, *der*

imperfectly [ɪm'pɜ:fɪktlɪ] *adv.* **a)** *(incompletely)* unvollständig; **b)** *(faultily)* fehlerhaft; mangelhaft

imperial [ɪm'pɪərɪəl] *adj.* **a)** kaiserlich; Reichs⟨*adler, -insignien*⟩; **b)** *(of an emperor)* Kaiser-; **c)** *(fixed by statute)* britisch ⟨*Maße, Gewichte*⟩

imperialism [ɪm'pɪərɪəlɪzm] *n.*, *no pl.* *(derog.)* Imperialismus, *der;* **US/Soviet ~:** der US-/Sowjetimperialismus

imperialist [ɪm'pɪərɪəlɪst] *n.* *(derog.)* Imperialist, *der*/Imperialistin, *die*

imperil [ɪm'perl] *v.t.*, *(Brit.)* **-ll-** gefährden

imperious [ɪm'pɪərɪəs] *adj.* *(overbearing)* herrisch; gebieterisch

imperishable [ɪm'perɪʃəbl] *adj.* alterungsbeständig ⟨*Material*⟩; unverderblich ⟨*Lebensmittel*⟩

impermeable [ɪm'pɜ:mɪəbl] *adj.* undurchlässig

impermissible [ɪmpə'mɪsɪbl] *adj.* unzulässig

impersonal [ɪm'pɜ:sənl] *adj.* unpersönlich

impersonate [ɪm'pɜ:səneɪt] *v.t.* *(for entertainment)* imitieren; nachahmen; *(for purpose of fraud)* sich ausgeben als

impersonation [ɪmpɜ:sə'neɪʃn] *n.* **a)** *(personification)* Verkörperung, *die;* **b)** *(imitation)* Imitation, *die;* Nachahmung, *die;* **he does ~s** er ist Imitator; **do an ~ of sb.** jmdn. imitieren *od.* nachahmen

impersonator [ɪm'pɜ:səneɪtə(r)] *n.* *(entertainer)* Imitator, *der*/Imitatorin, *die*

impertinence [ɪm'pɜ:tɪnəns] *n.* Unverschämtheit, *die*

impertinent [ɪm'pɜ:tɪnənt] *adj.* unverschämt

imperturbable [ɪmpə'tɜ:bəbl] *adj.* gelassen; **be completely ~:** durch nichts zu erschüttern sein; die Ruhe weghaben *(ugs.)*

impervious [ɪm'pɜ:vɪəs] *adj.* **a)** undurchlässig; **~ to water/bullets/rain** wasserdicht/kugelsicher/regendicht; **b) be ~ to sth.** *(fig.)* unempfänglich für etw. sein; **be ~ to argument** Argumenten unzugänglich sein

impetuosity [ɪmpetjʊ'ɒsɪtɪ] *n.* **a)** *no pl.* *(quality)* Impulsivität, *die;* **b)** *(act, impulse)* Ausbruch, *der*

impetuous [ɪm'petjʊəs] *adj.* impulsiv ⟨*Person*⟩; unüberlegt ⟨*Handlung, Entscheidung*⟩; *(vehement)* stürmisch; ungestüm ⟨*Person, Angriff*⟩

impetus ['ɪmpɪtəs] *n.* **a)** Kraft, *die;* *(of impact)* Wucht, *die;* **b)** *(fig.: impulse)* Motivation, *die*

impinge [ɪm'pɪndʒ] *v.i.* **a)** *(make impact)* **~ [up]on sth.** auf etw. *(Akk.)* auftreffen; **b)** *(encroach)* **~ [up]on sth.** auf etw. *(Akk.)* Einfluß nehmen

impish ['ɪmpɪʃ] *adj.* lausbübisch; diebisch ⟨*Freude*⟩

implacable [ɪm'plækəbl] *adj.* unversöhnlich; erbittert ⟨*Gegner*⟩; erbarmungslos ⟨*Verfolgung*⟩; unerbittlich ⟨*Schicksal*⟩

implant 1. [ɪm'plɑ:nt] *v.t.* **a)** *(Med.)* implantieren *(fachspr.)*, einpflanzen **(in** *Dat.*); **b)** *(fig.: instil)* einpflanzen **(in** *Dat.*). **2.** ['ɪmplɑ:nt] *n.* *(Med.)* Implantat, *das*

implausible [ɪm'plɔ:zɪbl] *adj.* unglaubwürdig

implement 1. ['ɪmplɪmənt] *n.* Gerät, *das.* **2.** ['ɪmplɪment] *v.t.* erfüllen ⟨*Versprechen, Vertrag*⟩; einhalten ⟨*Termin usw.*⟩; vollziehen ⟨*Erlaß usw.*⟩; [in die Tat] umsetzen ⟨*Politik, Plan usw.*⟩

implementation [ɪmplɪmen'teɪʃn] *n.* *see* **implement 2:** Erfüllung, *die;* Einhaltung, *die;* Vollzug, *der;* Umsetzung [in die Tat], *die*

implicate ['ɪmplɪkeɪt] *v.t.* *(show to be involved)* belasten ⟨*Verdächtigen usw.*⟩; **be ~d in a scandal** in einen Skandal verwickelt sein

implication [ɪmplɪ'keɪʃn] *n.* **a)** *no pl.* *(implying)* Implikation, *die (geh.);* **by ~:** implizit; implizite *(geh.);* **b)** *no pl.* *(being involved)* Verwicklung, *die* **(in** in + *Akk.*); **c)** *no pl.* *(being affected)* Betroffenheit, *die* **(in** von); **d)** *(thing implied)* Implikation, *die*

implicit [ɪm'plɪsɪt] *adj.* **a)** *(implied)* implizit; unausgesprochen ⟨*Drohung, Zweifel*⟩; **b)** *(virtually contained)* **be ~ in sth.** in etw. *(Dat.)* enthalten sein; **c)** *(resting on authority)* unbedingt; blind ⟨*Vertrauen*⟩

implode [ɪm'pləʊd] **1.** *v.i.* implodieren. **2.** *v.t.* implodieren lassen

implore [ɪm'plɔ:(r)] *v.t.* **a)** *(beg for)* erflehen *(geh.);* flehen um; **'please', she ~d** „bitte", flehte sie; **b)** *(entreat)* anflehen **(for** um)

imploring [ɪm'plɔ:rɪŋ] *adj.* flehend

imploringly [ɪm'plɔ:rɪŋlɪ] *adv.* flehentlich *(geh.)*

imply [ɪm'plaɪ] *v.t.* **a)** *(involve the existence of)* implizieren *(geh.);* *(by inference)* schließen lassen auf (+ *Akk.*); **be implied in sth.** in etw. *(Dat.)* enthalten sein; **b)** *(express indirectly)* hindeuten auf (+ *Akk.*); *(insinuate)* unterstellen; **are you ~ing that ...?** willst du damit etwa sagen, daß ...?

impolite [ɪmpə'laɪt] *adj.*, **~r** [ɪmpə'laɪtə(r)], **~st** [ɪmpə'laɪtɪst] unhöflich; ungezogen ⟨*Kind*⟩

impoliteness [ɪmpə'laɪtnɪs] *n.*, *no pl.* Unhöflichkeit, *die;* *(of child)* Ungezogenheit, *die*

imponderable [ɪm'pɒndərəbl] *adj.* unwägbar

import 1. [ɪm'pɔ:t] *v. t.* **a)** importieren, einführen ‹*Waren*› (**from** aus, **into** nach); **b)** *(signify)* bedeuten. **2.** ['ɪmpɔ:t] *n.* **a)** *(process, amount ~ed)* Import, *der;* Einfuhr, *die;* **b)** *(article ~ed)* Importgut, *das;* **c)** *(meaning, importance)* Bedeutung, *die;* **an event of great ~:** ein sehr bedeutungsvolles Ereignis

importance [ɪm'pɔ:təns] *n., no pl.* **a)** Bedeutung, *die;* Wichtigkeit, *die;* **be of great ~ to sb./sth.** für jmdn./etw. äußerst wichtig sein; **b)** *(significance)* Bedeutung, *die; (of decision)* Tragweite, *die;* **be of/without ~:** wichtig/unwichtig sein; **full of one's own ~:** von seiner eigenen Wichtigkeit überzeugt

important [ɪm'pɔ:tənt] *adj.* **a)** bedeutend; *(in a particular matter)* wichtig (**to** für); **the most ~ thing is ...:** die Hauptsache ist ...; **b)** *(momentous)* wichtig ‹*Entscheidung*›; bedeutsam ‹*Tag*›; **c)** *(having high rank)* wichtig ‹*Persönlichkeit*›; **d)** *(considerable)* beträchtlich; erheblich

importantly [ɪm'pɔ:təntlɪ] *adv.* **a) bear ~ [up]on sth.** auf etw. *(Akk.)* bedeutsame Auswirkungen haben; **more/most ~** *as sentence-modifier* was noch wichtiger/am wichtigsten ist; **b)** *(pompously)* wichtigtuerisch

importer [ɪm'pɔ:tə(r)] *n.* Importeur, *der;* **be an ~ of cotton** Baumwollimporteur sein; ‹*Land:*› Baumwolle importieren

impose [ɪm'pəʊz] **1.** *v. t.* **a)** auferlegen *(geh.)* ‹*Bürde, Verpflichtung*› (**[up]on** *Dat.*); erheben ‹*Steuer, Zoll*› (**on** auf + *Akk.*); verhängen ‹*Kriegsrecht*›; anordnen ‹*Rationierung*›; verhängen ‹*Sanktionen*› (**on** gegen); **~ a ban on sth.** etw. mit einem Verbot belegen; **~ a tax on sth.** etw. mit einer Steuer belegen; **b)** *(compel compliance with)* **~ sth. [up]on sb.** jmdm. etw. aufdrängen. **2.** *v. i.* **a)** *(exert influence)* imponieren; Eindruck machen; **b)** *(take advantage)* **I do not want** *or* **wish to ~:** ich will nicht aufdringlich sein. **3.** *v. refl.* **~ oneself on sb.** sich jmdm. aufdrängen

~ on, ~ upon *v. t.* ausnutzen ‹*Gutmütigkeit, Toleranz usw.*›; **~ on sb.** sich jmdm. aufdrängen

imposing [ɪm'pəʊzɪŋ] *adj.* imposant

imposition [ɪmpə'zɪʃn] *n.* **a)** *no pl. (action)* Auferlegung, *die; (of tax)* Erhebung, *die;* **b)** *no pl. (enforcement)* Durchsetzung, *die;* **c)** *(piece of advantage-taking)* Ausnützung, *die;* **I hope it's not too much of an ~:** ich hoffe, es macht nicht zu viele Umstände

impossibility [ɪmpɒsɪ'bɪlɪtɪ] *n.* **a)** *no pl.* Unmöglichkeit, *die;* **b) that's an absolute ~:** das ist völlig unmöglich *od.* ein Ding der Unmöglichkeit *(ugs.)*

impossible [ɪm'pɒsɪbl] **1.** *adj.* **a)** unmöglich; **it is ~ for me to do it** es ist mir nicht möglich, es zu tun; **b)** *(not easy)* schwer; *(not easily believable)* unmöglich *(ugs.);* **c)** *(coll.: intolerable)* unmöglich *(ugs.).* **2.** *n.* **the ~:** das Unmögliche; Unmögliches; **achieve the ~:** das Unmögliche erreichen

impostor [ɪm'pɒstə(r)] *n.* Hochstapler, *der*/-staplerin, *die; (swindler)* Betrüger, *der*/Betrügerin, *die*

impotence ['ɪmpətəns] *n., no pl.* **a)** *(powerlessness)* Machtlosigkeit, *die;* **b)** *(sexual)* Impotenz, *die*

impotent ['ɪmpətənt] *adj.* **a)** *(powerless)* machtlos; kraftlos ‹*Argument*›; **be ~ to do sth.** nicht in der Lage sein, etw. zu tun; **b)** *(sexually)* impotent

impound [ɪm'paʊnd] *v. t.* **a)** *(shut up)* einpferchen ‹*Vieh*›; einsperren ‹*streunende Hunde usw.*›; **b)** *(take possession of)* beschlagnahmen; requirieren *(Milit.)*

impoverish [ɪm'pɒvərɪʃ] *v. t.* **a)** verarmen lassen; **be/become ~ed** verarmt sein/verarmen; **b)** *(exhaust)* auslaugen ‹*Boden*›

impoverishment [ɪm'pɒvərɪʃmənt] *n., no pl.* **a)** *(making poor)* Verarmung, *die; (being poor)* Armut, *die;* **b)** *(of soil) (process)* Auslaugung, *die; (state)* Ausgelaugtheit, *die*

impracticable [ɪm'præktɪkəbl] *adj.* undurchführbar

imprecise [ɪmprɪ'saɪs] *adj.*, **imprecisely** [ɪmprɪ'saɪslɪ] *adv.* ungenau; unpräzise *(geh.)*

imprecision [ɪmprɪ'sɪʒn] *n.* Ungenauigkeit, *die*

impregnable [ɪm'pregnəbl] *adj.* uneinnehmbar ‹*Festung, Bollwerk*›; einbruch[s]sicher ‹*Tresorraum usw.*›; *(fig.)* unanfechtbar ‹*Ruf, Tugend, Stellung*›

impregnate ['ɪmpregneɪt, ɪm'pregneɪt] *v. t.* imprägnieren

impresario [ɪmprɪ'sɑ:rɪəʊ] *n., pl.* **~s** Intendant, *der*/Intendantin, *die;* Impresario, *der (veralt.)*

impress [ɪm'pres] *v. t.* **a)** *(apply)* drücken; **~ a pattern** *etc.* **on/in sth.** ein Muster *usw.* auf etw. *(Akk.)* aufdrücken/in etw. *(Akk.)* eindrücken; **b)** beeindrucken ‹*Person*›; *abs.* Eindruck machen; **~ sb. favourably/unfavourably** auf jmdn. einen günstigen/ungünstigen Eindruck machen; **c) ~ sth. [up]on sb.** jmdm. etw. einprägen

impression [ɪm'preʃn] *n.* **a)** *(mark)* Abdruck, *der;* **b)** *(Printing) (quantity of copies)* Auflage, *die; (unaltered reprint)* Nachdruck, *der;* **c)** *(effect on person)* Eindruck, *der* (**of** von); *(effect on inanimate things)* Wirkung, *die;* **make an ~ on sb.** Eindruck auf jmdn. machen; **d)** *(impersonation)* **do an ~ of sb.** jmdn. imitieren; **do ~s** andere Leute imitieren; **e)** *(notion)* Eindruck, *der;* **it's my ~ that ...:** ich habe den Eindruck, daß ...; **form an ~ of sb.** sich *(Dat.)* ein Bild von jmdm. machen; **give [sb.] the ~ that .../of being bored** [bei jmdm.] den Eindruck erwecken, als ob .../als ob man sich langweile; **be under the ~ that ...:** der Auffassung *od.* Überzeugung sein, daß ...; *(less certain)* den Eindruck haben, daß ...

impressionable [ɪm'preʃənəbl] *adj.* beeinflußbar; **have an ~ mind, be ~:** sich leicht beeinflussen lassen

impressionism [ɪm'preʃənɪzm] *n., no pl.* Impressionismus, *der*

impressionist [ɪm'preʃənɪst] *n.* Impressionist, *der*/Impressionistin, *die; attrib.* impressionistisch ‹*Kunst usw.*›

impressive [ɪm'presɪv] *adj.* beeindruckend; imponierend

imprint 1. ['ɪmprɪnt] *n.* **a)** Abdruck, *der;* **b)** *(fig.)* Stempel, *der;* **leave one's ~ on sb./sth.** jmdm./einer Sache seinen Stempel aufdrücken. **2.** [ɪm'prɪnt] *v. t.* **a)** *(stamp)* aufdrucken; aufdrücken ‹*Poststempel*›; *(on metal)* aufprägen; **b)** *(fix indelibly)* **sth. is ~ed in** *or* **on sb.'s memory** etw. hat sich jmdm. [unauslöschlich] eingeprägt

imprison [ɪm'prɪzn] *v. t.* in Haft nehmen; **be ~ed** sich in Haft befinden

imprisonment [ɪm'prɪznmənt] *n.* Haft, *die;* **a long term** *or* **period of ~:** eine langjährige Haft- *od.* Freiheitsstrafe

improbability [ɪmprɒbə'bɪlɪtɪ] *n.* Unwahrscheinlichkeit, *die*

improbable [ɪm'prɒbəbl] *adj.* *(not likely)* unwahrscheinlich

impromptu [ɪm'prɒmptju:] **1.** *adj.* improvisiert; **an ~ speech** eine Stegreifrede; **an ~ visit** ein Überraschungsbesuch. **2.** *adv.* aus dem Stegreif

improper [ɪm'prɒpə(r)] *adj.* **a)** *(wrong)* unrichtig; ungeeignet ‹*Werkzeug*›; **b)** *(unseemly)* ungehörig; unpassend; *(indecent)* unanständig; **c)** *(not in accordance with rules of conduct)* unangebracht; unzulässig ‹*Gebühren*›

improperly [ɪm'prɒpəlɪ] *adv.* **a)** *(wrongly)* unrichtig; **use sth. ~:** etw. unsachgemäß gebrauchen; **b)** *(in unseemly fashion)* unpassend; *(indecently)* unanständig

impropriety [ɪmprə'praɪətɪ] *n.* **a)** *no pl. (unseemliness)* Unpassende, *das; (indecency)* Unanständigkeit, *die;* **b)** *(instance of improper conduct)* Unanständigkeit, *die;* **moral ~:** moralisches Fehlverhalten

improve [ɪm'pru:v] **1.** *v.i.* sich verbessern; besser werden; ‹*Person, Wetter:*› sich bessern; **he was ill, but he's improving now** er war krank, aber es geht ihm jetzt schon besser; **things are improving** es sieht schon besser aus. **2.** *v.t.* verbessern; erhöhen, steigern ‹*Produktion*›; ausbessern ‹*Haus usw.*›; verschönern ‹*öffentliche Anlage usw.*›; **~ one's situation** sich verbessern. **3.** *v. refl.* **~ oneself** sich weiterbilden

~ [up]on *v.t.* überbieten ‹*Rekord, Angebot*›; verbessern ‹*Leistung*›

improvement [ɪm'pru:vmənt] *n.* **a)** *no pl.* Verbesserung, *die;* Besserung, *die; (in trading)* Steigerung, *die;* **there is need for ~ in your handwriting** deine Handschrift müßte besser werden; **b)** *(addition)* Verbesserung, *die;* **make ~s to sth.** Verbesserungen an etw. *(Dat.)* vornehmen

improvise ['ɪmprəvaɪz] *v.t.* improvisieren; aus dem Stegreif vortragen ‹*Rede*›

imprudent [ɪm'pru:dənt] *adj.* unklug; *(showing rashness)* unbesonnen

impudence ['ɪmpjʊdəns] *n.* Unverschämtheit, *die; (brazenness)* Dreistigkeit, *die*

impudent ['ɪmpjʊdənt] *adj.*, **impudently** ['ɪmpjʊdəntlɪ] *adv.* unverschämt; *(brazen)* dreist

impugn [ɪm'pju:n] *v.t.* in Zweifel ziehen

impulse ['ɪmpʌls] *n.* **a)** *(act of impelling)* Stoß, *der;* Impuls, *der; (fig.: motivation)* Impuls, *der;* **give an ~ to sth.** einer Sache *(Dat.)* neue Impulse geben; **b)** *(mental incitement)* Impuls, *der;* **be seized with an irresistible ~ to do sth.** von einem unwiderstehlichen Drang ergriffen werden, etw. zu tun; **from pure ~:** rein impulsiv; **act/do sth. on [an] ~:** impulsiv handeln/etw. tun

'impulse buying *n.* Spontankäufe *Pl.*

impulsive [ɪm'pʌlsɪv] *adj.* impulsiv

impulsively [ɪm'pʌlsɪvlɪ] *adv.* impulsiv

impulsiveness [ɪm'pʌlsɪvnɪs] *n., no pl.* Impulsivität, *die*

impunity [ɪm'pju:nɪtɪ] *n., no pl.* **be able to do sth. with ~:** etw. gefahrlos tun können; *(without being punished)* etw. ungestraft tun können

impure [ɪm'pjʊə(r)] *adj.* **a)** *(lit. or fig.)* unrein; *(dirty)* unsauber; schmutzig ‹*Wasser*›; **b)** schmutzig ‹*Gedanke*›

impurity [ɪm'pjʊərɪtɪ] *n.* **a)** *no pl. (lit. or fig.)* Unreinheit, *die; (being dirty)* Unsauberkeit, *die; (of water)* Verschmutzung, *die;* **b)** *in pl. (dirt)* Schmutz, *der;* **c)** *(foreign body)* Fremdkörper, *der;* Fremdstoff, *der*

impute [ɪm'pju:t] *v.t.* **~ sth. to sb./sth.** jmdm./einer Sache etw. zuschreiben; **~ bad intentions to sb.** jmdm. schlechte Absichten unterstellen

in [ɪn] **1.** *prep.* **a)** *(position; also fig.)* in (+ *Dat.*); **in the fields** auf den Feldern; **a ride in a motor car** eine Autofahrt; **shot/wounded in the leg** ins Bein geschossen/am Bein verwundet; **in this heat** bei dieser Hitze; **the highest mountain in the world** der höchste Berg der Welt; **b)** *(wearing as dress)* in (+ *Dat.*); *(wearing as headgear)* mit; **in brown shoes** mit braunen Schuhen; **a lady in black** eine Dame in Schwarz; **c)** *(with respect to)* **two feet in diameter** mit einem Durchmesser von zwei Fuß; **a change in attitude** eine Änderung der Einstellung; *see also* **herself a; itself a; d)** *(as a proportionate part of)* **eight dogs in ten** acht von zehn Hunden; *see also* **gradient; e)** *(as a member of)* in (+ *Dat.*); **be in the Scouts** bei den Pfadfindern sein; **be employed in the Civil Service** als Beamter/Beamtin beschäftigt sein; **f)** *(as content of)* **there are three feet in a yard** ein Yard hat drei Fuß; **what is there in this deal for me?** was springt für mich bei dem Geschäft heraus? *(ugs.);* **there is nothing/not much** *or* **little in it** *(difference)* da ist kein/kein großer Unterschied [zwischen ihnen]; **there is something in what you say** an dem, was Sie sagen, ist etwas dran *(ugs.);* **g)** *expr. identity* in (+ *Dat.*); **have a faithful friend in sb.** an jmdm. einen treuen Freund haben; **h)** *(concerned with)* in (+ *Dat.*); **what line of business are you in?** in welcher Branche sind Sie?; **he's in politics** er ist Politiker; **she's in insurance** sie ist in der Versicherungsbranche tätig; **i) be [not] in it** *(as competitor)* [nicht] dabei *od.* im Rennen sein; **j)** *(with the means of; having as material or colour)* **a message in code** eine verschlüsselte Nachricht; **in writing** schriftlich; **in this way** auf diese Weise; so; **in a few words** mit wenigen Worten; **a dress in velvet** ein Kleid aus Samt; **this sofa is also available in leather/blue** dieses Sofa gibt es auch in Leder/Blau; **write sth. in red** etw. in Rot schreiben; **draw in crayon/ink** *etc.* mit Kreide/Tinte *usw.* zeichnen; *see also* **English 2a; k)** *(while, during)* **in fog/rain** *etc.* bei Nebel/Regen *usw.;* **in the eighties/nineties** in den Achtzigern/Neunzigern; **4 o'clock in the morning/afternoon** 4 Uhr morgens/abends; **in 1990** [im Jahre] 1990; **l)** *(after a period of)* in (+ *Dat.*); **in three minutes/years** in drei Minuten/Jahren; **m)** *(within the ability of)* **have it in one [to do sth.]** fähig sein [, etw. zu tun]; **I didn't know you had it in you** das hätte ich dir nicht zugetraut; **there is no malice in him** er hat nichts Bösartiges an sich *(Dat.);* **n) in that** insofern als; *see also* **far 1d; o) in doing this** *(by so doing)* indem jmd. das tut/tat; dadurch. **2.** *adv.* **a)** *(inside)* hinein‹*gehen usw.*›; *(towards speaker)* herein‹*kommen usw.*›; **is everyone in?** sind alle drin? *(ugs.);* **'In'** „Einfahrt"/„Eingang"; **b)** *(at home, work, etc.)* **be in** dasein; **find sb. in** jmdn. antreffen; **ask sb. in** jmdn. hereinbitten; **he's been in and out all day** er war den ganzen Tag über mal da und mal nicht da; **c)** *(included)* darin; drin *(ugs.);* **cost £50 all in** 50 Pfund kosten, alles inbegriffen; **d)** *(inward)* innen; **e)** *(in fashion)* in *(ugs.);* in Mode; **f)** *(elected)* **be in** gewählt sein; **g)** *(having arrived)* **be in** ‹*Zug, Schiff, Ware, Bewerbung:*› dasein; ‹*Ernte:*› eingebracht sein; **h) sb. is in for sth.** *(about to undergo sth.)*

jmdm. steht etw. bevor; *(taking part in sth.)* jmd. nimmt an etw. *(Dat.)* teil; **we're in for it now!** *(coll.)* jetzt blüht uns was! *(ugs.)*; **have it in for sb.** es auf jmdn. abgesehen haben *(ugs.)*; **i)** *(coll.: as participant, accomplice, observer, etc.)* **be in on the secret/discussion** in das Geheimnis eingeweiht sein/bei der Diskussion dabei sein; **be [well] in with sb.** mit jmdm. [gut] auskommen. **3.** *attrib. adj. (fashionable)* Mode-; **the 'in crowd** die Clique, die gerade in ist *(ugs.)*; **'in joke** Insiderwitz, *der.* **4.** *n.* **know the ins and outs of a matter** sich in einer Sache genau auskennen

inability [ɪnə'bɪlɪtɪ] *n., no pl.* Unfähigkeit, *die*

inaccessibility [ɪnəksesɪ'bɪlɪtɪ] *n., no pl.* Unzugänglichkeit, *die*

inaccessible [ɪnək'sesɪbl] *adj.* unzugänglich

inaccuracy [ɪn'ækjʊrəsɪ] *n.* **a)** *(incorrectness)* Unrichtigkeit, *die;* **b)** *(imprecision)* Ungenauigkeit, *die*

inaccurate [ɪn'ækjʊrət] *adj.* **a)** *(incorrect)* unrichtig; **b)** *(imprecise)* ungenau

inaction [ɪn'ækʃn] *n., no pl., no indef. art.* Untätigkeit, *die*

inactive [ɪn'æktɪv] *adj.* **a)** untätig; **b)** *(sluggish)* träge

inactivity [ɪnæk'tɪvɪtɪ] *n., no pl.* **a)** Untätigkeit, *die;* **b)** *(sluggishness)* Trägheit, *die*

inadequacy [ɪn'ædɪkwəsɪ] *n.* **a)** Unzulänglichkeit, *die;* **b)** *(incompetence)* mangelnde Eignung

inadequate [ɪn'ædɪkwət] *adj.* **a)** unzulänglich; **his response was ~ [to the situation]** seine Antwort war [der Situation] nicht angemessen; **the resources are ~ to his needs** die Mittel reichen für seine Bedürfnisse nicht aus; **b)** *(incompetent)* ungeeignet; **feel ~:** sich überfordert fühlen

inadmissible [ɪnəd'mɪsɪbl] *adj.* unzulässig

inadvertent [ɪnəd'vɜːtənt] *adj.* ungewollt; versehentlich

inadvertently [ɪnəd'vɜːtəntlɪ] *adv.* versehentlich

inadvisable [ɪnəd'vaɪzəbl] *adj.* nicht ratsam; unratsam

inane [ɪ'neɪn] *adj.* dümmlich

inanimate [ɪn'ænɪmət] *adj.* unbelebt

inapplicable [ɪn'æplɪkəbl, ɪnə'plɪkəbl] *adj.* nicht anwendbar (**to** auf + *Akk.*); **delete if ~:** Unzutreffendes [bitte] streichen

inappropriate [ɪnə'prəʊprɪət] *adj.* unpassend; **be ~ for sth.** für etw. nicht geeignet sein; **be ~ to the occasion** dem Anlaß nicht angemessen sein

inapt [ɪn'æpt] *adj.* unpassend

inarticulate [ɪnɑː'tɪkjʊlət] *adj.* **a) she's rather/very ~:** sie kann sich ziemlich/sehr schlecht ausdrücken; **a clever but ~ mathematician** ein kluger Mathematiker, der sich aber nur schlecht ausdrücken kann; **b)** *(indistinct)* unverständlich; inartikuliert *(geh.)*

inasmuch [ɪnəz'mʌtʃ] *adv.* *(formal)* **~ as** insofern als; *(because)* da

inattention [ɪnə'tenʃn] *n., no pl.* Unaufmerksamkeit, *die* (**to** gegenüber)

inattentive [ɪnə'tentɪv] *adj.* unaufmerksam (**to** gegenüber)

inaudible [ɪn'ɔːdɪbl] *adj.*, **inaudibly** [ɪn'ɔːdɪblɪ] *adv.* unhörbar

inaugural [ɪ'nɔːgjʊrl] *adj.* **a)** *(first in series)* Eröffnungs-; **b)** *(by person being inaugurated)* **~ lecture** *or* **address** Antrittsrede, *die*

inaugurate [ɪ'nɔːgjʊreɪt] *v.t.* **a)** *(admit to office)* in sein Amt einführen; **b)** *(begin)* einführen; in Angriff nehmen ⟨*Projekt*⟩

inauspicious [ɪnɔː'spɪʃəs] *adj.* *(ominous)* unheilverkündend; unheilvoll

'inborn *adj.* angeboren (**in** *Dat.*)

in'bred *adj.* **a)** angeboren; **b)** *(impaired by inbreeding)* **they are/have become ~:** bei ihnen herrscht Inzucht

in'breeding *n.* Inzucht, *die*

'in-built *adj.* jmdm./einer Sache eigen

incalculable [ɪn'kælkjʊləbl] *adj.* **a)** *(very great)* unermeßlich; **b)** *(unpredictable)* unabsehbar

incantation [ɪnkæn'teɪʃn] *n.* **a)** *(words)* Zauberspruch, *der;* **b)** *(spell)* Beschwörung, *die*

incapable [ɪn'keɪpəbl] *adj.* **a)** *(lacking ability)* **be ~ of doing sth.** außerstande sein, etw. zu tun; **be ~ of sth.** zu etw. unfähig sein; **b)** *(incompetent)* unfähig

incapacitate [ɪnkə'pæsɪteɪt] *v.t.* unfähig machen; **physically ~d/~d by illness** körperlich/durch Krankheit behindert

incapacity [ɪnkə'pæsɪtɪ] *n., no pl.* Unfähigkeit, *die* (**for** zu)

incarcerate [ɪn'kɑːsəreɪt] *v.t.* einkerkern *(geh.)*

incarnate [ɪn'kɑːnət] *adj.* **be the devil ~:** der leibhaftige Satan sein; **be beauty/wisdom** *etc.* **~:** die personifizierte Schönheit/Weisheit *usw.* sein

incarnation [ɪnkɑː'neɪʃn] *n.* Inkarnation, *die*

incendiary [ɪn'sendɪərɪ] *adj.* **~ attack** Brandstiftung, *die;* **~ device** Brandsatz, *der;* **~ bomb** Brandbombe, *die*

[1]**incense** ['ɪnsens] *n.* Weihrauch, *der*

[2]**incense** [ɪn'sens] *v.t.* erzürnen; erbosen; **be ~d at** *or* **by sth./with sb.** über etw./jmdn. erbost *od.* erzürnt sein

incentive [ɪn'sentɪv] *n.* **a)** *(motivation)* Anreiz, *der;* **b)** *(payment)* finanzieller Anreiz

inception [ɪn'sepʃn] *n.* Einführung, *die;* **from** *or* **since/at its ~:** von Beginn an/zu Beginn

incessant [ɪn'sesənt] *adj.*, **incessantly** [ɪn'sesəntlɪ] *adv.* unablässig; unaufhörlich

incest ['ɪnsest] *n.* Inzest, *der;* Blutschande, *die*

incestuous [ɪn'sestjʊəs] *adj.* *(lit. or fig.)* inzestuös

inch [ɪntʃ] **1.** *n.* **a)** Inch, *der;* Zoll, *der (veralt.)*; **b)** *(small amount)* **~ by ~:** ≈ Zentimeter um Zentimeter; **by ~es** ≈ zentimeterweise; **not give** *or* **yield an ~:** keinen Fingerbreit nachgeben. **2.** *v.t.* ≈ zentimeterweise bewegen; **~ one's way forward** sich Zoll für Zoll vorwärtsbewegen. **3.** *v.i.* ≈ sich zentimeterweise bewegen

incidence ['ɪnsɪdəns] *n.* **a)** *(occurrence)* Auftreten, *das;* Vorkommen, *das;* **b)** *(manner or range of occurrence)* Häufigkeit, *die*

incident ['ɪnsɪdənt] *n.* **a)** *(notable event)* Vorfall, *der; (minor occurrence)* Begebenheit, *die;* Vorkommnis, *das;* **b)** *(clash)* Zwischenfall, *der;* **c)** *(in play, novel, etc.)* Episode, *die*

incidental [ɪnsɪ'dentl] **1.** *adj. (casual)* beiläufig ⟨*Art, Bemerkung*⟩; Neben⟨*ausgaben, -einnahmen*⟩. **2.** *n., in pl.* Nebensächlichkeiten; *(expenses)* Nebenausgaben

incidentally [ɪnsɪ'dentəlɪ] *adv.* *(by the way)* nebenbei [bemerkt]

inci'dental music *n.* Begleitmusik, *die*

'incident room *n. [temporäres] lokales Einsatzzentrum der Polizei*

incinerate [ɪn'sɪnəreɪt] *v.t.* verbrennen

incinerator [ɪn'sɪnəreɪtə(r)] *n.* Verbrennungsofen, *der; (in garden)* Abfallverbrenner, *der*

incipient [ɪn'sɪpɪənt] *adj.* anfänglich; einsetzend ⟨*Schmerzen*⟩; aufkommend ⟨*Zweifel, Angst*⟩

incision [ɪn'sɪʒn] *n.* **a)** *(cutting)* Einschneiden, *das;* **b)** *(cut)* Einschnitt, *der*

incisive [ɪn'saɪsɪv] *adj.* schneidend ‹*Ton*›; scharf ‹*Verstand*›; scharfsinnig ‹*Genie, Kritik, Frage, Bemerkung, Argument*›; präzise ‹*Sprache, Stil*›

incisor [ɪn'saɪzə(r)] *n. (Anat., Zool.)* Schneidezahn, *der*

incite [ɪn'saɪt] *v.t.* anstiften; aufstacheln ‹*Massen, Volk*›

incitement [ɪn'saɪtmənt] *n. (act)* Anstiftung, *die; (of masses, crowd)* Aufstachelung, *die*

inclement [ɪn'klemənt] *adj.* unfreundlich ‹*Wetter*›

inclination [ɪnklɪ'neɪʃn] *n.* **a)** *(slope)* [Ab]hang, *der; (of roof)* Neigung, *die;* **b)** *(preference, desire)* Neigung, *die* (**to, for** für); **my ~ is to let the matter rest** ich neige dazu, die Sache auf sich beruhen zu lassen; **c)** *(liking)* **~ for sb.** Zuneigung für jmdn.

incline 1. [ɪn'klaɪn] *v.t.* **a)** *(bend)* neigen; **b)** *(dispose)* veranlassen; **all her instincts ~d her to stay** alles in ihr drängte sie zu bleiben. **2.** *v.i.* **a)** *(be disposed)* neigen (**to[wards]** zu); **~ to believe that ...:** geneigt sein zu glauben, daß ...; **b)** *(lean)* sich neigen. **3.** ['ɪnklaɪn] *n.* Steigung, *die*

inclined [ɪn'klaɪnd] *adj.* geneigt; **be mathematically ~:** sich für Mathematik interessieren; **if you feel [so] ~:** wenn Sie Lust dazu haben; **he is that way ~:** er neigt dazu

include [ɪn'klu:d] *v.t.* einschließen; *(contain)* enthalten; **his team ~s a number of people who ...:** zu seiner Mannschaft gehören einige, die ...; **..., [the] children ~d ...,** [die] Kinder eingeschlossen; **does that ~ 'me?** gilt das auch für mich?; **your name is not ~d in the list** dein Name steht nicht auf der Liste; **have you ~d the full amount?** haben Sie den vollen Betrag einbezogen?; **~d in the price** im Preis inbegriffen

including [ɪn'klu:dɪŋ] *prep.* einschließlich; **I make that ten ~ the captain** mit dem Kapitän sind das nach meiner Rechnung zehn; **up to and ~ the last financial year** bis einschließlich des letzten Geschäftsjahres; **~ VAT** inklusive Mehrwertsteuer

inclusive [ɪn'klu:sɪv] *adj.* **a)** inklusive *(bes. Kaufmannsspr.);* einschließlich; **be ~ of sth.** etw. einschließen; **from 2 to 6 January ~:** vom 2. bis einschließlich 6. Januar; **b)** *(including everything)* Pauschal-; Inklusiv-; **cost £50 ~:** 50 Pfund kosten, alles inbegriffen

incognito [ɪnkɒg'ni:təʊ] **1.** *adj., adv.* inkognito. **2.** *n.* Inkognito, *das*

incoherent [ɪnkəʊ'hɪərənt] *adj.* zusammenhanglos

income ['ɪnkəm] *n.* Einkommen, *das;* **~s** *(receipts)* Einkünfte *Pl.;* **live within/beyond one's ~:** entsprechend seinen Verhältnissen/über seine Verhältnisse leben

income: ~s policy *n.* Einkommenspolitik, *die;* **~ support** *n. (Brit.)* zusätzliche Hilfe zum Lebensunterhalt; **~ tax** *n.* Einkommensteuer, *die; (on wages, salary)* Lohnsteuer, *die;* **~ tax return** Einkommensteuererklärung, *die*/Lohnsteuererklärung, *die*

'incoming *adj.* **a)** *(arriving)* ankommend; einlaufend ‹*Zug, Schiff*›; landend ‹*Flugzeug*›; einfahrend ‹*Zug*›; eingehend ‹*Telefongespräch, Auftrag*›; **the ~ tide** die Flut; **b)** *(succeeding)* neu ‹*Vorsitzender, Präsident, Mieter, Regierung*›

incommunicado [ɪnkəmju:nɪ'kɑ:dəʊ] *pred. adj.* von der Außenwelt abgeschnitten; **hold sb. ~:** jmdn. ohne Verbindung zur Außenwelt halten

incomparable [ɪn'kɒmpərəbl] *adj.,* **incomparably** [ɪn'kɒmpərəblɪ] *adv.* unvergleichlich

incompatibility [ɪnkəmpætɪ'bɪlɪtɪ] *n., no pl.* **a)** *(inability to harmonize)* Unverträglichkeit, *die;* **b)** *(unsuitability for use together)* Nichtübereinstimmung, *die*

incompatible [ɪnkəm'pætɪbl] *adj.* **a)** *(unable to harmonize)* unverträglich; **they were ~ and they separated** sie paßten nicht zueinander und trennten sich; **b)** *(unsuitable for use together)* unvereinbar; inkompatibel *(Technik);* **c)** *(inconsistent)* unvereinbar

incompetence [ɪn'kɒmpɪtəns], **incompetency** [ɪn'kɒmpɪtənsɪ] *n.* Unfähigkeit, *die;* Unvermögen, *das*

incompetent [ɪn'kɒmpɪtənt] **1.** *adj.* unfähig; unzulänglich ‹*Arbeit*›. **2.** *n.* Unfähige, *der/die*

incomplete [ɪnkəm'pli:t] *adj.,* **incompletely** [ɪnkəm'pli:tlɪ] *adv.* unvollständig

incomprehensible [ɪnkɒmprɪ'hensɪbl] *adj.* unbegreiflich; unverständlich ‹*Sprache, Rede, Theorie, Argument*›

inconceivable [ɪnkən'si:vəbl] *adj.,* **inconceivably** [ɪnkən'si:vəblɪ] *adv.* unvorstellbar

inconclusive [ɪnkən'klu:sɪv] *adj.* ergebnislos; nicht schlüssig ‹*Beweis, Argument*›

incongruity [ɪnkɒŋ'gru:ɪtɪ] *n.* **a)** *no pl. (quality)* Deplaziertheit, *die;* **b)** *(instance)* Absurdität, *die*

incongruous [ɪn'kɒŋgrʊəs] *adj.* **a)** *(inappropriate)* unpassend; **b)** *(inharmonious)* unvereinbar; nicht zusammenpassend ‹*Farben, Kleidungsstücke*›

inconsequential [ɪnkɒnsɪ'kwenʃl] *adj.* belanglos

inconsiderable [ɪnkən'sɪdərəbl] *adj.* unbeträchtlich; unerheblich

inconsiderate [ɪnkən'sɪdərət] *adj.* **a)** *(unkind)* rücksichtslos; **b)** *(rash)* unbedacht; unüberlegt

inconsistency [ɪnkən'sɪstənsɪ] *n.* **a)** *(incompatibility, self-contradiction)* Widersprüchlichkeit, *die* (**with** zu); **b)** *(illogicality)* Inkonsequenz, *die;* **c)** *(irregularity)* Unbeständigkeit, *die;* Inkonsistenz, *die (geh.)*

inconsistent [ɪnkən'sɪstənt] *adj.* **a)** *(incompatible, self-contradictory)* widersprüchlich; **be ~ with sth.** zu etw. im Widerspruch stehen; **b)** *(illogical)* inkonsequent; **c)** *(irregular)* unbeständig; inkonsistent *(geh.)*

inconsolable [ɪnkən'səʊləbl] *adj.* untröstlich

inconspicuous [ɪnkən'spɪkjʊəs] *adj.* unauffällig

incontestable [ɪnkən'testəbl] *adj.* unbestreitbar; unwiderlegbar ‹*Beweis*›

incontinence [ɪn'kɒntɪnəns] *n. (Med.)* Inkontinenz, *die*

incontinent [ɪn'kɒntɪnənt] *adj. (Med.)* inkontinent; **be ~:** an Inkontinenz leiden

incontrovertible [ɪnkɒntrə'vɜ:tɪbl] *adj.* unbestreitbar; unwiderlegbar ‹*Beweis*›

inconvenience [ɪnkən'vi:nɪəns] **1.** *n.* **a)** *no pl. (discomfort, disadvantage)* Unannehmlichkeiten (**to** für); **go to a great deal of ~:** große Unannehmlichkeiten auf sich *(Akk.)* nehmen; **b)** *(instance)* **if it's no ~:** wenn es keine Umstände macht. **2.** *v.t.* Unannehmlichkeiten bereiten (+ *Dat.*); *(disturb)* stören

inconvenient [ɪnkən'vi:nɪənt] *adj.* unbequem; ungünstig ‹*Lage, Standort*›; unpraktisch ‹*Design, Konstruktion, Schnitt*›; **come at an ~ time** zu ungelegener Zeit kommen; **if it is not ~ [to you]** wenn es Ihnen recht ist

incorporate [ɪn'kɔ:pəreɪt] *v.t. (include)* aufnehmen (**in[to], with** in + *Akk.*)

incorporated [ɪn'kɔ:pəreɪtɪd] *adj.* eingetragen ‹*[Handels]gesellschaft*›

incorrect [ɪnkə'rekt] *adj.* **a)** unrichtig; inkorrekt; **be ~:** nicht stimmen; **you are ~ in believing that...:** du irrst, wenn du glaubst, daß...; **b)** *(improper)* inkorrekt

incorrectly [ɪnkə'rektlɪ] *adv.* **a)** unrichtigerweise; falsch ⟨*beantworten, aussprechen*⟩; **b)** *(improperly)* inkorrekt

incorrigible [ɪn'kɒrɪdʒɪbl] *adj.* unverbesserlich

increase 1. [ɪn'kri:s] *v.i.* zunehmen; ⟨*Schmerzen:*⟩ stärker werden; ⟨*Lärm:*⟩ größer werden; ⟨*Verkäufe, Preise, Nachfrage:*⟩ steigen; **~ in weight/size/price** schwerer/größer/teurer werden; **~ in maturity/value/popularity** an Reife/Wert/Popularität *(Dat.)* gewinnen. **2.** *v.t.* **a)** *(make greater)* erhöhen; vermehren ⟨*Besitz*⟩; **b)** *(intensify)* verstärken; **~ one's efforts/commitment** sich mehr anstrengen/engagieren. **3.** ['ɪnkri:s] *n.* **a)** *(becoming greater)* Zunahme, *die* (**in** *Gen.*); *(in measurable amount)* Anstieg, *der* (**in** *Gen.*); *(deliberately caused)* Steigerung, *die* (**in** *Gen.*); **~ in weight/size** Gewichtszunahme, *die*/Vergrößerung, *die;* **~ in popularity** Popularitätsgewinn, *der;* **be on the ~:** [ständig] zunehmen; **b)** *(by reproduction)* Zunahme, *die;* Zuwachs, *der;* **c)** *(amount)* Erhöhung, *die; (of growth)* Zuwachs, *der*

increasing [ɪn'kri:sɪŋ] *adj.* steigend; wachsend; **an ~ number of people** mehr und mehr Menschen

increasingly [ɪn'kri:sɪŋlɪ] *adv.* in zunehmendem Maße; **become ~ apparent** immer deutlicher werden

incredible [ɪn'kredɪbl] *adj.* **a)** *(beyond belief)* unglaublich; **b)** *(coll.) (remarkable)* unglaublich *(ugs.); (wonderful)* toll *(ugs.)*

incredibly [ɪn'kredɪblɪ] *adv.* **a)** unglaublich; **b)** *(coll.: remarkably)* unglaublich *(ugs.);* unwahrscheinlich *(ugs.);* **c)** *as sentence-modifier* es ist/war kaum zu glauben, aber ...

incredulity [ɪnkrɪ'dju:lɪtɪ] *n., no pl.* Ungläubigkeit, *die*

incredulous [ɪn'kredjʊləs] *adj.*, **incredulously** [ɪn'kredjʊləslɪ] *adv.* ungläubig

increment ['ɪnkrɪmənt] *n.* Erhöhung, *die; (amount of growth)* Zuwachs, *der*

incriminate [ɪn'krɪmɪneɪt] *v.t.* belasten; **incriminating evidence** belastendes Material

incubate ['ɪnkjʊbeɪt] *v.t.* bebrüten; *(to hatching)* ausbrüten

incubation [ɪnkjʊ'beɪʃn] *n.* Inkubation, *die (Biol.);* Bebrütung, *die*

incubator ['ɪnkjʊbeɪtə(r)] *n.* Inkubator, *der (Biol., Med.); (for babies also)* Brutkasten, *der*

inculcate ['ɪnkʌlkeɪt] *v.t.* **~ sth. in[to] sb., ~ sb. with sth.** jmdm. etw. einpflanzen

incur [ɪn'kɜ:(r)] *v.t.*, **-rr-** sich *(Dat.)* zuziehen ⟨*Unwillen, Ärger*⟩; **~ a loss** einen Verlust erleiden; **~ debts/expenses/risks** Schulden machen/Ausgaben haben/Risiken eingehen

incurable [ɪn'kjʊərəbl] *adj.* **a)** *(Med.)* unheilbar; **b)** *(fig.)* unheilbar *(ugs.);* unstillbar ⟨*Sehnsucht, Verlangen*⟩; unüberwindbar ⟨*Zurückhaltung, Scheu*⟩

incursion [ɪn'kɜ:ʃn] *n. (invasion)* Eindringen, *das; (by sudden attack)* Einfall, *der*

indebted [ɪn'detɪd] *pred. adj.* **be/feel deeply ~ to sb.** tief in jmds. Schuld *(Dat.)* stehen *(geh.);* **he was ~ to a friend for this information** er verdankte einem Freund diese Information; **be [much] ~ to sb. for sth.** jmdm. für etw. [sehr] verbunden sein *(geh.) od.* zu Dank verpflichtet sein

indecency [ɪn'di:sənsɪ] *n.* Unanständigkeit, *die*

indecent [ɪn'di:sənt] *adj.* **a)** *(immodest, obscene)* unanständig; *see also* **exposure a; b)** *(unseemly)* ungehörig; **with ~ haste** mit unziemlicher Hast *(geh.)*

indecipherable [ɪndɪ'saɪfərəbl] *adj.* unentzifferbar

indecision [ɪndɪ'sɪʒn] *n., no pl.* Unentschlossenheit, *die*

indecisive [ɪndɪ'saɪsɪv] *adj.* **a)** *(not conclusive)* ergebnislos ⟨*Streit, Diskussion*⟩; nichts entscheidend ⟨*Krieg, Schlacht*⟩; **b)** *(hesitating)* unentschlossen

indecisiveness [ɪndɪ'saɪsɪvnɪs] *n., no pl.* **a)** *(inconclusiveness)* Ergebnislosigkeit, *die;* **b)** *(hesitation)* Unentschlossenheit, *die*

indeed [ɪn'di:d] *adv.* **a)** *(in truth)* in der Tat; tatsächlich; **b)** *emphat.* **thank you very much ~:** haben Sie vielen herzlichen Dank; **it was very kind of you ~:** es war wirklich sehr freundlich von Ihnen; **~ it is** in der Tat; allerdings; **yes ~, it certainly is/I certainly did** *etc.* ja, das kann man wohl sagen; **no, ~:** nein, ganz bestimmt nicht; **c)** *(in fact)* ja sogar; **if ~ such a thing is possible** wenn so etwas überhaupt möglich ist; **I feel, ~ I know, she will come** ich habe das Gefühl, [ja] ich weiß sogar, daß sie kommen wird; **d)** *(admittedly)* zugegebenermaßen; zwar; **e)** *interrog.* **~?** wirklich?; ist das wahr?; **f)** *expr. irony, surprise, interest, etc.* **He expects to win – Does he ~!** Er glaubt, daß er gewinnt – Tatsächlich?; **I want a fortnight off work – [Do you] ~!** Ich möchte 14 Tage freihaben – Ach wirklich?

indefatigable [ɪndɪ'fætɪgəbl] *adj.* unermüdlich

indefensible [ɪndɪ'fensɪbl] *adj.* **a)** *(insecure)* unhaltbar; **b)** *(untenable)* unvertretbar; unhaltbar; **c)** *(intolerable)* unverzeihlich

indefinable [ɪndɪ'faɪnəbl] *adj.* undefinierbar

indefinite [ɪn'defɪnɪt] *adj.* **a)** *(vague)* unbestimmt; **b)** *(unlimited)* unbegrenzt

indefinitely [ɪn'defɪnɪtlɪ] *adv.* **a)** *(vaguely)* unbestimmt; **b)** *(unlimitedly)* unbegrenzt; **it can't go on ~:** es kann nicht endlos so weitergehen; **postponed ~:** auf unbestimmte Zeit verschoben

indelible [ɪn'delɪbl] *adj.* unauslöschlich *(auch fig.);* nicht zu entfernend ⟨*Fleck*⟩; **~ ink** Wäschetinte, *die;* **~ pencil** Kopierstift, *der;* Tintenstift, *der*

indelicate [ɪn'delɪkət] *adj.* *(coarse)* ungehörig; *(almost indecent)* geschmacklos; *(slightly tactless)* nicht sehr feinfühlig

indemnify [ɪn'demnɪfaɪ] *v.t.* **a)** *(protect)* **~ sb. against sth.** jmdn. gegen etw. absichern; **b)** *(compensate)* entschädigen

indemnity [ɪn'demnɪtɪ] *n.* **a)** *(security)* Absicherung, *die;* **b)** *(compensation)* Entschädigung, *die*

indent [ɪn'dent] *v.t.* **a)** *(make notches in)* einkerben; **b)** *(form recesses in)* einschneiden in (+ *Akk.*); **c)** *(from margin)* einrücken

indentation [ɪnden'teɪʃn] *n.* **a)** *(indenting, notch)* Einkerbung, *die;* **b)** *(recess)* Einschnitt, *der*

independence [ɪndɪ'pendəns] *n.* Unabhängigkeit, *die*

Inde'pendence Day *n. (Amer.)* Unabhängigkeitstag, *der*

independent [ɪndɪ'pendənt] **1.** *adj.* **a)** unabhängig; **~ income/means** eigenes Einkommen; **b)** *(not wanting obligations)* selbständig. **2.** *n. (Polit.)* Unabhängige, *der/die*

independently [ɪndɪ'pendəntlɪ] *adv.* unabhängig (**of** von); **they work ~:** sie arbeiten unabhängig voneinander

indescribable [ɪndɪ'skraɪbəbl] *adj.* unbeschreiblich

indestructible [ɪndɪ'strʌktɪbl] *adj.* unzerstörbar; unerschütterlich ‹*Glaube*›

indeterminate [ɪndɪ'tɜ:mɪnət] *adj.* **a)** *(not fixed, vague)* unbestimmt ‹*Form, Menge*›; unklar ‹*Konzept, Bedeutung*›; **b)** *(left undecided)* ergebnislos; offen ‹*Rechtsfrage*›

index ['ɪndeks] **1.** *n.* Index, *der;* Register, *das;* ~ **of sources** Quellenverzeichnis, *das.* **2.** *v.t.* **a)** *(furnish with ~)* mit einem Register *od.* Index versehen; **b)** *(enter in ~)* ins Register aufnehmen

index: ~ **finger** *n.* Zeigefinger, *der;* ~**-linked** *adj. (Econ.)* indexiert; dynamisch ‹*Rente*›

India ['ɪndɪə] *pr. n.* Indien *(das)*

Indian ['ɪndɪən] **1.** *adj.* **a)** indisch; **b) [American]** ~: indianisch. **2.** *n.* **a)** Inder, *der*/Inderin, *die;* **b) [American]** ~: Indianer, *der*/Indianerin, *die*

Indian: ~ **'ink** *n. (Brit.)* Tusche, *die;* ~ **'Ocean** *pr. n.* Indischer Ozean; ~ **'summer** *n.* Altweibersommer, *der;* Nachsommer, *der (auch fig.)*

'India rubber *see* **¹rubber a, b**

indicate ['ɪndɪkeɪt] **1.** *v.t.* **a)** *(be a sign of)* erkennen lassen; **b)** *(state briefly)* andeuten; ~ **the rough outlines of a project** ein Projekt kurz umreißen; **c)** *(mark, point out)* anzeigen; **d)** *(suggest, make evident)* zum Ausdruck bringen **(to** gegenüber). **2.** *v.i.* blinken *(bes. Verkehrsw.)*

indication [ɪndɪ'keɪʃn] *n.* [An]zeichen, *das* **(of** *Gen.,* für); **there is every/no ~ that ...:** alles/nichts weist darauf hin, daß ...; **first ~s are that ...:** die ersten Anzeichen deuten darauf hin, daß ...

indicator ['ɪndɪkeɪtə(r)] *n.* **a)** *(instrument)* Anzeiger, *der;* **b)** *(board)* Anzeigetafel, *die;* **c)** *(on vehicle)* Blinker, *der;* **d)** *(fig.: pointer)* Indikator, *der (bes. Wirtsch.)*

indifference [ɪn'dɪfərəns] *n., no pl.* **a)** *(unconcern)* Gleichgültigkeit, *die* **(to[wards]** gegenüber); **b)** *(neutrality)* Indifferenz, *die;* **c)** *(unimportance)* **a matter of ~:** eine Belanglosigkeit; **this is a matter of complete ~ to** *or* **for him** das ist für ihn völlig belanglos

indifferent [ɪn'dɪfərənt] *adj.* **a)** *(without concern or interest)* gleichgültig; unbeteiligt ‹*Beobachter*›; **b)** *(not good)* mittelmäßig; *(fairly bad)* mäßig; *(neither good nor bad)* durchschnittlich; **very ~:** schlecht

indigenous [ɪn'dɪdʒɪnəs] *adj.* einheimisch; eingeboren ‹*Bevölkerung*›

indigestible [ɪndɪ'dʒestɪbl] *adj. (lit. or fig.)* unverdaulich

indigestion [ɪndɪ'dʒestʃn] *n., no pl., no indef. art.* Magenverstimmung, *die; (chronic)* Verdauungsstörungen

indignant [ɪn'dɪgnənt] *adj.* entrüstet **(at, over, about** über + *Akk.*); indigniert ‹*Blick, Geste*›; **grow ~:** sich entrüsten; **it makes me ~:** es regt mich auf

indignation [ɪndɪg'neɪʃn] *n., no pl.* Entrüstung, *die* **(about, at, against, over** über + *Akk.*)

indignity [ɪn'dɪgnɪtɪ] *n.* Demütigung, *die;* **the ~ of my position** das Demütigende [an] meiner Situation

indigo ['ɪndɪgəʊ] **1.** *n., pl.* **~s a)** *(dye)* Indigo, *der od. das;* **b)** *(colour)* ~ **[blue]** Indigoblau, *das.* **2.** *adj.* ~ **[blue]** indigoblau

indirect [ɪndɪ'rekt, ɪndaɪ'rekt] *adj.* indirekt; *(long-winded)* umständlich; **follow an ~ route** nicht den direkten Weg nehmen; **by ~ means** auf Umwegen *(fig.)*

indirectly [ɪndɪ'rektlɪ, ɪndaɪ'rektlɪ] *adv.* indirekt; auf Umwegen ‹*hören, herausfinden*›

indirect: ~ **object** *n. (Ling.)* indirektes Objekt; *(in German)* Dativobjekt, *das;* ~ **speech** *n. (Ling.)* indirekte Rede

indiscreet [ɪndɪ'skri:t] *adj.* indiskret; taktlos ‹*Benehmen*›

indiscretion [ɪndɪ'skreʃn] *n.* **a)** *(conduct)* Indiskretion, *die; (tactlessness)* Taktlosigkeit, *die;* **b)** *(imprudence)* Unbedachtheit, *die;* **c)** *(action)* Unbedachtsamkeit, *die; (love affair)* Affäre, *die;* **d)** *(revelation of official secret etc.)* Indiskretion, *die*

indiscriminate [ɪndɪ'skrɪmɪnət] *adj.* **a)** *(undiscriminating)* unkritisch; **b)** *(unrestrained, promiscuous)* wahllos; willkürlich ‹*Anwendung*›

indispensable [ɪndɪ'spensəbl] *adj.* unentbehrlich **(to** für); unabdingbar ‹*Voraussetzung*›

indisposed [ɪndɪ'spəʊzd] *adj.* **a)** *(unwell)* unpäßlich; indisponiert ‹*Sänger, Schauspieler*›; **b)** *(disinclined)* **be ~ to do sth.** abgeneigt sein, etw. zu tun

indisposition [ɪndɪspə'zɪʃn] *n.* **a)** *(ill health)* Unpäßlichkeit, *die; (of singer, actor)* Indisposition, *die;* **b) an ~ to do sth.** eine Abneigung dagegen, etw. zu tun

indisputable [ɪndɪ'spju:təbl] *adj.,* **indisputably** [ɪndɪ'spju:təblɪ] *adv.* unbestreitbar

indistinct [ɪndɪ'stɪŋkt] *adj.* undeutlich; *(blurred)* verschwommen

indistinguishable [ɪndɪ'stɪŋgwɪʃəbl] *adj.* **a)** *(not distinguishable)* nicht unterscheidbar; **b)** *(imperceptible)* nicht erkennbar; nicht wahrnehmbar ‹*Geräusch*›

individual [ɪndɪ'vɪdjʊəl] **1.** *adj.* **a)** *(single)* einzeln; **b)** *(special, personal)* besonder... ‹*Vorteil, Merkmal*›; ~ **case** Einzelfall, *der;* **c)** *(intended for one)* für eine [einzelne] Person bestimmt; **d)** *(distinctive)* eigentümlich; individuell; **e)** *(characteristic)* eigen; individuell. **2.** *n.* **a)** *(one member)* einzelne, *der/die; (animal)* Einzeltier, *das;* **~s** einzelne; **b)** *(one being)* Individuum, *das;* einzelne, *der/die;* **c)** *(coll.: person)* Individuum, *das (abwertend)*

individualist [ɪndɪ'vɪdjʊəlɪst] *n.* Individualist, *der*/Individualistin, *die*

individuality [ɪndɪvɪdjʊ'ælɪtɪ] *n., no pl.* eigene Persönlichkeit; Individualität, *die*

individually [ɪndɪ'vɪdjʊəlɪ] *adv.* **a)** *(singly)* einzeln; **b)** *(distinctively)* individuell; **c)** *(personally)* persönlich

indivisible [ɪndɪ'vɪzɪbl] *adj.* **a)** *(not divisible)* unteilbar; **b)** *(not distributable)* nicht aufteilbar

indoctrinate [ɪn'dɒktrɪneɪt] *v.t.* indoktrinieren *(abwertend)*

indolence ['ɪndələns] *n., no pl.* Trägheit, *die*

indolent ['ɪndələnt] *adj.* träge

Indonesia [ɪndə'ni:zjə] *pr. n.* Indonesien *(das)*

Indonesian [ɪndə'ni:zjən] **1.** *adj.* indonesisch; **sb. is ~:** jmd. ist Indonesier/Indonesierin. **2.** *n. (person)* Indonesier, *der*/Indonesierin, *die*

'indoor *adj.* ~ **swimming-pool/sports** Hallenbad, *das*/-sport, *der;* ~ **plants** Zimmerpflanzen; ~ **games** Spiele im Haus; *(Sport)* Hallenspiele

indoors [ɪn'dɔ:z] *adv.* drinnen; im Haus; **come/go ~:** nach drinnen *od.* ins Haus kommen/gehen

induce [ɪn'dju:s] *v.t.* **a)** *(persuade)* ~ **sb. to do sth.** jmdn. dazu bringen, etw. zu tun; **b)** *(bring about)* hervorrufen; verursachen, führen zu ‹*Krankheit*›; **c)** *(Med.)* einleiten ‹*Wehen, Geburt*›; herbeiführen ‹*Schlaf*›

inducement [ɪn'dju:smənt] *n.* *(incentive)* Anreiz, *der;* **as an added ~:** als besonderer Anreiz *od.* Ansporn

induction [ɪn'dʌkʃn] *n.* **a)** *(formal*

introduction) Amtseinführung, *die;* **b)** *(initiation)* Einführung, *die* (**into** in + *Akk.*); **c)** *(Med.)* Einleitung, *die;* **d)** *(Electr., Phys., Math., Philos.)* Induktion, *die*

indulge [ɪn'dʌldʒ] **1.** *v. t.* **a)** *(yield to)* nachgeben (+ *Dat.*) ‹*Wunsch, Verlangen, Verlockung*›; frönen *(geh.)* (+ *Dat.*) ‹*Leidenschaft, Neigung*›; **b)** *(please)* verwöhnen; ~ **sb. in sth.** jmdm. in etw. *(Dat.)* nachgeben; ~ **oneself in** schwelgen in *(geh.)* (+ *Dat.*). **2.** *v. i.* **a)** *(allow oneself pleasure)* ~ **in** frönen *(geh.)* (+ *Dat.*) ‹*Leidenschaft, Neigung*›; **b)** *(coll.: take alcoholic drink)* sich *(Dat.)* einen genehmigen *(ugs.)*

indulgence [ɪn'dʌldʒəns] *n.* **a)** Nachsicht, *die;* *(humouring)* Nachgiebigkeit, *die* (**with** gegenüber); **b) sb.'s ~ in sth.** jmds. Hang zu etw.; **c)** *(thing indulged in)* Luxus, *der*

indulgent [ɪn'dʌldʒənt] *adj.* nachsichtig (**with, to[wards]** gegenüber)

industrial [ɪn'dʌstrɪəl] *adj.* **a)** industriell; betrieblich ‹*Ausbildung, Forschung*›; Arbeits‹*unfall, -medizin, -psychologie*›; **b)** *(intended for industry)* Industrie‹*alkohol, -diamant usw.*›

industrial: ~ **'action** *n.* Arbeitskampfmaßnahmen; **take ~ action** in den Ausstand treten; ~ **di'spute** *n.* Arbeitskonflikt, *der;* ~ **'espionage** *n.* Industriespionage, *die;* ~ **'injury** *n.* Arbeitsverletzung, *die*

industrialisation, industrialise *see* **industrializ-**

industrialist [ɪn'dʌstrɪəlɪst] *n.* Industrielle, *der/die*

industrialization [ɪndʌstrɪəlaɪ'zeɪʃn] *n.* Industrialisierung, *die*

industrialize [ɪn'dʌstrɪəlaɪz] *v. i. & t.* industrialisieren

industrial: ~ **re'lations** *n. pl.* Industrial relations *Pl.* *(Wirtsch.); Beziehungen zwischen Arbeitgebern und Gewerkschaften;* **I~ Revo'lution** *n.* *(Hist.)* industrielle Revolution; ~ **un'rest** *n.* Unruhe in der Arbeitnehmerschaft

industrious [ɪn'dʌstrɪəs] *adj.* fleißig; *(busy)* emsig

industry ['ɪndəstrɪ] *n.* **a)** Industrie, *die;* **several industries** mehrere Industriezweige; **b)** *see* **industrious:** Fleiß, *der;* Emsigkeit, *die*

inebriated [ɪ'ni:brɪeɪtɪd] *adj.* *(drunk)* betrunken

inedible [ɪn'edɪbl] *adj.* ungenießbar

ineffective [ɪnɪ'fektɪv] *adj.* **a)** unwirksam; ineffektiv; fruchtlos ‹*Anstrengung, Versuch*›; wirkungslos ‹*Argument*›; **b)** *(inefficient)* untauglich

ineffectiveness [ɪnɪ'fektɪvnɪs] *n., no pl. see* **ineffective:** Unwirksamkeit, *die;* Ineffizienz, *die;* Fruchtlosigkeit, *die;* Wirkungslosigkeit, *die;* Untauglichkeit, *die*

ineffectual [ɪnɪ'fektjʊəl] *adj.* unwirksam; ineffektiv; fruchtlos ‹*Versuch, Bemühung*›; ineffizient ‹*Methode, Person*›

inefficiency [ɪnɪ'fɪʃənsɪ] *n.* Ineffizienz, *die;* *(incapability)* Unfähigkeit, *die*

inefficient [ɪnɪ'fɪʃənt] *adj.* ineffizient; *(incapable)* unfähig; **the worker/machine is ~:** der Arbeiter/die Maschine leistet nicht genug

inelegant [ɪn'elɪgənt] *adj.* **a)** unelegant; schwerfällig ‹*Bewegung, Gang*›; **b)** *(unrefined, unpolished)* ungeschliffen *(abwertend)*

ineligible [ɪn'elɪdʒɪbl] *adj.* ungeeignet; **be ~ for** nicht in Frage kommen für ‹*Beförderung, Position, Mannschaft*›; nicht berechtigt sein zu ‹*Leistungen des Staats usw.*›

inept [ɪ'nept] *adj.* **a)** *(unskilful, clumsy)* unbeholfen; **b)** *(inappropriate)* unpassend, unangebracht ‹*Bemerkung, Eingreifen*›; **c)** *(foolish)* albern

ineptitude [ɪ'neptɪtju:d] *n., no pl.* **a)** *(unskilfulness, clumsiness)* Unbeholfenheit, *die;* **b)** *(of remark, intervention)* Unangebrachtheit, *die;* **c)** *(foolishness)* Albernheit, *die*

inequality [ɪnɪ'kwɒlɪtɪ] *n.* Ungleichheit, *die*

inequitable [ɪn'ekwɪtəbl] *adj.* ungerecht

inert [ɪ'nɜ:t] *adj.* **a)** reglos; *(sluggish)* träge; *(passive)* untätig; **b)** *(Chem.: neutral)* inert

inert 'gas *n.* *(Chem.)* Edelgas, *das*

inertia [ɪ'nɜ:ʃə, ɪ'nɜ:ʃɪə] *n.* *(also Phys.)* Trägheit, *die*

i'nertia reel *n.* Aufrollautomatik, *die;* ~ **seat-belt** Automatikgurt, *der*

inescapable [ɪnɪ'skeɪpəbl] *adj.* unausweichlich ‹*Logik, Schlußfolgerung*›

inessential [ɪnɪ'senʃl] **1.** *adj.* *(not necessary)* unwesentlich; *(dispensable)* entbehrlich. **2.** *n.* Nebensächlichkeit, *die*

inevitability [ɪnevɪtə'bɪlɪtɪ] *n., no pl.* Unvermeidlichkeit, *die;* *(of fate, event)* Unabwendbarkeit, *die*

inevitable [ɪn'evɪtəbl] *adj.* unvermeidlich; unabwendbar ‹*Ereignis, Krieg, Schicksal*›; zwangsläufig ‹*Ergebnis, Folge*›; **bow to the ~:** sich in das Unvermeidliche fügen

inevitably [ɪn'evɪtəblɪ] *adv.* zwangsläufig

inexact [ɪnɪg'zækt] *adj.* ungenau

inexcusable [ɪnɪk'skju:zəbl] *adj.* unverzeihlich; unentschuldbar

inexhaustible [ɪnɪg'zɔ:stɪbl] *adj.* unerschöpflich ‹*Reserven, Quelle, Energie*›; unverwüstlich ‹*Person*›

inexorable [ɪn'eksərəbl] *adj.* unerbittlich

inexpensive [ɪnɪk'spensɪv] *adj.* preisgünstig; **the car is ~ to run** der Wagen ist sparsam im Verbrauch

inexperience [ɪnɪk'spɪərɪəns] *n.* Unerfahrenheit, *die;* Mangel an Erfahrung

inexperienced [ɪnɪk'spɪərɪənst] *adj.* unerfahren; ~ **in sth.** wenig vertraut mit etw.

inexpert [ɪn'ekspɜ:t] *adj.* unerfahren; *(unskilled)* ungeschickt

inexplicable [ɪnek'splɪkəbl] *adj.* unerklärlich

inexpressible [ɪnɪk'spresɪbl] *adj.* unbeschreiblich

inextricable [ɪn'ekstrɪkəbl] *adj.* unentwirrbar

infallibility [ɪnfælɪ'bɪlɪtɪ] *n., no pl.* Unfehlbarkeit, *die*

infallible [ɪn'fælɪbl] *adj.*, **infallibly** [ɪn'fælɪblɪ] *adv.* unfehlbar

infamous ['ɪnfəməs] *adj.* **a)** berüchtigt; **b)** *(wicked)* infam; niederträchtig

infancy ['ɪnfənsɪ] *n.* **a)** frühe Kindheit; **b)** *(fig.: early state)* Frühzeit, *die;* **be in its ~:** noch in den Anfängen *od.* Kinderschuhen stecken

infant ['ɪnfənt] **1.** *n.* kleines Kind. **2.** *adj.* **a)** kindlich; **b)** *(fig.: not developed)* in den Anfängen steckend

infantile ['ɪnfəntaɪl] *adj.* **a)** *(relating to infancy)* kindlich; **b)** *(childish)* kindisch *(abwertend);* infantil *(abwertend)*

infantry ['ɪnfəntrɪ] *n. constr. as sing. or pl.* Infanterie, *die*

'infant school *n.* *(Brit.)* ≈ Vorschule, *die;* *Grundschule für die ersten beiden Jahrgänge*

infatuated [ɪn'fætjʊeɪtɪd] *adj.* betört *(geh.);* verzaubert; **be ~ with sb./oneself** in jmdn./sich selbst vernarrt sein

infatuation [ɪnfætjʊ'eɪʃn] *n.* Vernarrtheit, *die* (**with** in + *Akk.*)

infect [ɪn'fekt] *v. t.* **a)** *(contaminate)* verseuchen; **b)** *(affect with* *disease)* infizieren *(Med.)* **(with** mit); **the wound became ~ed** die Wunde entzündete sich; **c)** *(with enthusiasm etc.)* anstecken

infection [ɪnˈfekʃn] *n.* Infektion, *die;* **throat/ear/eye ~:** Hals-/Ohren-/Augenentzündung, *die*

infectious [ɪnˈfekʃəs] *adj.* **a)** infektiös *(Med.),* ansteckend ‹*Krankheit*›; **be ~** ‹*Person:*› eine ansteckende Krankheit haben; ansteckend sein *(ugs.);* **b)** *(fig.)* ansteckend ‹*Heiterkeit, Begeisterung, Lachen*›

infer [ɪnˈfɜː(r)] *v. t.,* **-rr-** schließen **(from** aus); erschließen ‹*Voraussetzung*›; gewinnen ‹*Kenntnisse*›; ziehen ‹*Schlußfolgerung*›

inference [ˈɪnfərəns] *n.* [Schluß]folgerung, *die*

inferior [ɪnˈfɪərɪə(r)] **1.** *adj.* *(of lower quality)* minderwertig ‹*Ware*›; minder... ‹*Qualität*›; gering ‹*Kenntnis*›; unterlegen ‹*Gegner*›; **~ to sth.** schlechter als etw.; **feel ~:** Minderwertigkeitsgefühle haben; **feel ~ to sb.** sich jmdm. gegenüber unterlegen fühlen. **2.** *n.* Untergebene, *der/die;* **his social ~s** die gesellschaftlich unter ihm Stehenden

inferiority [ɪnfɪərɪˈɒrɪtɪ] *n., no pl.* Unterlegenheit, *die* **(to** gegenüber); *(of goods)* schlechtere Qualität

inferiˈority complex *n. (Psych.)* Minderwertigkeitskomplex, *der*

infernal [ɪnˈfɜːnl] *adj.* **a)** *(of hell)* höllisch; ‹*Geister, Götter*› der Unterwelt; **b)** *(hellish)* teuflisch; **c)** *(coll.: detestable)* verdammt *(salopp)*

inferno [ɪnˈfɜːnəʊ] *n., pl.* **~s** Inferno, *das;* **a blazing ~:** ein flammendes Inferno

infertile [ɪnˈfɜːtaɪl] *adj.* unfruchtbar

infertility [ɪnfɜːˈtɪlɪtɪ] *n., no pl.* Unfruchtbarkeit, *die*

infest [ɪnˈfest] *v. t.* ‹*Ungeziefer, Schädlinge:*› befallen; ‹*Unkraut:*› überwuchern; *(fig.)* heimsuchen

infidelity [ɪnfɪˈdelɪtɪ] *n.* Untreue, *die* **(to** gegenüber)

ˈinfighting *n.* interne Machtkämpfe

infiltrate [ˈɪnfɪltreɪt] **1.** *v. t.* **a)** *(penetrate into)* infiltrieren ‹*feindliche Reihen*›; unterwandern ‹*Partei, Organisation*›; **b)** *(cause to enter)* einschleusen ‹*Agenten*›; **c)** *(esp. Biol., Med.: pass into, permeate)* infiltrieren. **2.** *v. i.* **a)** *(penetrate)* einsickern *(fig.);* **~ into** unterwandern ‹*Partei, Organisation*›; infiltrieren ‹*feindliche Reihen*›; **b)** ‹*Flüssigkeit:*› eindringen

infiltrator [ˈɪnfɪltreɪtə(r)] *n.* Eindringling, *der; (of party, organization)* Unterwanderer, *der*

infinite [ˈɪnfɪnɪt] *adj.* **a)** *(endless)* unendlich; **b)** *(very great)* ungeheuer; unendlich groß

infinitesimal [ɪnfɪnɪˈtesɪml] *adj.* **a)** *(Math.)* infinitesimal; **b)** *(very small)* äußerst gering; winzig ‹*Menge*›

infinitive [ɪnˈfɪnɪtɪv] *n. (Ling.)* Infinitiv, *der*

infinity [ɪnˈfɪnɪtɪ] *n.* Unendlichkeit, *die;* **at ~** *(Geom.)* im Unendlichen ‹*sich schneiden*›; **focus on ~** *(Photog.)* auf unendlich stellen

infirm [ɪnˈfɜːm] *adj.* *(weak)* gebrechlich; *(irresolute)* schwach

infirmary [ɪnˈfɜːmərɪ] *n.* Krankenhaus, *das*

inflame [ɪnˈfleɪm] *v. t.* **a)** *(excite)* entflammen *(geh.);* **b)** *(aggravate)* schüren ‹*Feindschaft, Haß*›; **c)** *(Med.)* **become/be ~d** ‹*Auge, Wunde:*› sich entzünden/entzündet sein

inflammable [ɪnˈflæməbl] *adj.* **a)** *(easily set on fire)* feuergefährlich; leicht entzündlich *od.* entflammbar; **'highly ~'** „feuergefährlich"; **b)** explosiv ‹*Situation*›

inflammation [ɪnfləˈmeɪʃn] *n.* *(Med.)* Entzündung, *die*

inflammatory [ɪnˈflæmətərɪ] *adj.* **a)** aufrührerisch; **an ~ speech** eine Hetzrede *(abwertend);* **b)** *(Med.)* entzündlich

inflatable [ɪnˈfleɪtəbl] **1.** *adj.* aufblasbar; **~ dinghy** Schlauchboot, *das.* **2.** *n.* *(boat)* Schlauchboot, *das*

inflate [ɪnˈfleɪt] *v. t.* **a)** *(distend)* aufblasen; *(with pump)* aufpumpen; **b)** *(Econ.)* in die Höhe treiben ‹*Preise, Kosten*›; inflationieren ‹*Währung*›; **~ the economy** Inflationspolitik betreiben

inflated [ɪnˈfleɪtɪd] *adj.* *(lit or fig.)* aufgeblasen; geschwollen ‹*Stil*›; **have an ~ opinion of oneself** aufgeblasen sein *(ugs. abwertend)*

inflation [ɪnˈfleɪʃn] *n.* **a)** Aufblasen, *das; (with pump)* Aufpumpen, *das;* **b)** *(Econ.)* Inflation, *die*

inflationary [ɪnˈfleɪʃənərɪ] *adj.* *(Econ.)* inflationär

inˈflation-proofed *adj.* mit Inflationsausgleich *nachgestellt*

inflect [ɪnˈflekt] *v. t.* flektieren; beugen

inflexible [ɪnˈfleksɪbl] *adj.* **a)** *(stiff)* unbiegsam; **b)** *(obstinate)* [geistig] unbeweglich ‹*Person*›; wenig flexibel ‹*Einstellung, Meinung*›

inflict [ɪnˈflɪkt] *v. t.* zufügen ‹*Leid, Schmerzen*›, beibringen ‹*Wunde*›, versetzen ‹*Schlag*› **(on** *Dat.*); **~ oneself** *or* **one's company on sb.** sich jmdm. aufdrängen

ˈin-flight *adj.* Bord‹*verpflegung, -programm*›

influence [ˈɪnflʊəns] **1.** *n.* *(also thing, person)* Einfluß, *der;* **exercise ~:** Einfluß ausüben **(over** auf + *Akk.*); **a person of ~:** eine einflußreiche Persönlichkeit; **be a bad/major ~ [on sb.]** einen schlechten/bedeutenden Einfluß [auf jmdn.] ausüben; **be under the ~** *(coll.)* betrunken sein. **2.** *v. t.* beeinflussen; **be too easily ~d** sich zu leicht beeinflussen lassen

influential [ɪnflʊˈenʃl] *adj.* einflußreich ‹*Person*›; **be ~ in sb.'s decision/on sb.'s career** jmdn. in seiner Entscheidung beeinflussen/jmds. Karriere beeinflussen

influenza [ɪnflʊˈenzə] *n.* Grippe, *die*

influx [ˈɪnflʌks] *n.* Zustrom, *der*

inform [ɪnˈfɔːm] **1.** *v. t.* **a)** informieren **(of, about** über + *Akk.*); **I am pleased to ~ you that ...:** ich freue mich, Ihnen mitteilen zu können, daß ...; **keep sb./oneself ~ed** jmdn./sich auf dem laufenden halten; **b)** *(animate, inspire)* durchdringen. **2.** *v. i.* **~ against** *or* **on sb.** jmdn. anzeigen *od.* *(abwertend)* denunzieren **(to** bei)

informal [ɪnˈfɔːml] *adj.* **a)** *(without formality)* zwanglos; ungezwungen ‹*Ton, Sprache*›; leger ‹*Kleidungsstück*›; **'dress: ~'** „keine festliche Garderobe"; **b)** *(unofficial)* informell ‹*Gespräch, Treffen*›

informality [ɪnfɔːˈmælɪtɪ] *n. no pl.* Zwanglosigkeit, *die;* Ungezwungenheit, *die*

informant [ɪnˈfɔːmənt] *n.* Informant, *der*/Informantin, *die;* Gewährsmann, *der*

information [ɪnfəˈmeɪʃn] *n., no pl., no indef. art.* Informationen; **give ~ on sth.** Auskunft über etw. *(Akk.)* erteilen; **piece** *or* **bit of ~:** Information, *die;* **where can we get hold of some ~?** wo können wir Auskunft bekommen?; **for your ~:** zu Ihrer Information; *(iron.)* damit du Bescheid weißt!

information: ~ bureau, ~ centre *ns.* Auskunftsbüro, *das;* **~ desk** *n.* Informationsschalter, *der;* **~ technology** *n.* Informationstechnologie, *die;* Informationstechnik, *die*

informative [ɪnˈfɔːmətɪv] *adj.* informativ; **not very ~:** nicht sehr aufschlußreich ‹*Dokument, Schriftstück*›

informed [ɪn'fɔ:md] *adj.* informiert; fundiert ‹*Schätzung*›

informer [ɪn'fɔ:mə(r)] *n.* Denunziant, *der*/Denunziantin, *die (abwertend);* Informant, *der*/Informantin, *die;* **police ~:** Polizeispitzel, *der (abwertend)*

infra-red [ɪnfrə'red] *adj.* **a)** infrarot; **b)** *(using ~ radiation)* Infrarot-

infrastructure ['ɪnfrəstrʌktʃə(r)] *n.* Infrastruktur, *die*

infrequent [ɪn'fri:kwənt] *adj.* **a)** *(uncommon)* selten; **b)** *(sparse)* vereinzelt

infrequently [ɪn'fri:kwəntlɪ] *adv.* selten

infringe [ɪn'frɪndʒ] **1.** *v.t.* verstoßen gegen. **2.** *v.i.* ~ **[up]on** verstoßen gegen ‹*Recht, Gesetz usw.*›

infringement [ɪn'frɪndʒmənt] *n.* **a)** *(violation)* Verstoß, *der* (**of** gegen); ~ **of the contract** Vertragsverletzung, *die;* Vertragsbruch, *der;* **b)** *(encroachment)* Übergriff, *der* (**on** auf + *Akk.*)

infuriate [ɪn'fjʊərɪeɪt] *v.t.* wütend machen; **be ~d** wütend sein (**by** über + *Akk.*)

infuriating [ɪn'fjʊərɪeɪtɪŋ] *adj.* **she is an ~ person** sie kann einen zur Raserei bringen; **it is ~ when/that ...:** es ist wahnsinnig ärgerlich, wenn/daß ... *(ugs.);* ~ **calmness/slowness** aufreizende Gelassenheit/Langsamkeit

infuse [ɪn'fju:z] *v.t.* **a)** *(instil)* ~ **sth. into sb.,** ~ **sb. with sth.** jmdm. etw. einflößen; **b)** *(steep)* aufgießen ‹*Tee usw.*›

infusion [ɪn'fju:ʒn] *n.* **a)** *(Med.)* Infusion, *die;* **b)** *(imparting)* Einflößen, *das;* **c)** *(steeping)* Aufgießen, *das;* **d)** *(liquid)* Aufguß, *der*

ingenious [ɪn'dʒi:nɪəs] *adj.* **a)** *(resourceful)* einfallsreich; *(skilful)* geschickt; **b)** *(cleverly constructed)* genial ‹*Methode, Idee*›; raffiniert ‹*Spielzeug, Werkzeug, Maschine*›

ingenuity [ɪndʒɪ'nju:ɪtɪ] *n., no pl.* **a)** *(resourcefulness)* Einfallsreichtum, *der; (skill)* Geschicklichkeit, *die;* **b)** *(cleverness of construction)* Genialität, *die*

ingot ['ɪŋgət] *n.* Ingot, *der (Metall.)*

ingrained ['ɪngreɪnd, ɪn'greɪnd] *adj.* **a)** *(embedded)* **hands ~ with dirt** stark verschmutzte Hände; **b)** *(fig.)* tief eingewurzelt ‹*Vorurteil usw.*›

ingratiate [ɪn'greɪʃɪeɪt] *v. refl.* ~ **oneself with sb.** sich bei jmdm. einschmeicheln

ingratiating [ɪn'greɪʃɪeɪtɪŋ] *adj.* schmeichlerisch

ingratitude [ɪn'grætɪtju:d] *n., no pl.* Undankbarkeit, *die* (**to[wards]** gegenüber)

ingredient [ɪn'gri:dɪənt] *n.* Zutat, *die;* **the ~s of a successful marriage** *(fig.)* die Voraussetzungen für eine gute Ehe

ingrowing ['ɪngrəʊɪŋ] *adj.* eingewachsen ‹*Zehennagel usw.*›

inhabit [ɪn'hæbɪt] *v.t.* bewohnen; **the region was ~ed by penguins/the Celts** in der Gegend lebten Pinguine/die Kelten

inhabitant [ɪn'hæbɪtənt] *n.* Bewohner, *der*/Bewohnerin, *die; (of village etc. also)* Einwohner, *der*/Einwohnerin, *die*

inhale [ɪn'heɪl] **1.** *v.t. (breathe in)* einatmen; *(take into the lungs)* inhalieren *(ugs.)* ‹*Zigarettenrauch usw.*›; *(Med.)* inhalieren. **2.** *v.i.* einatmen; *(Med.)* inhalieren; ‹*Raucher:*› inhalieren *(ugs.)*

inherent [ɪn'hɪərənt, ɪn'herənt] *adj.* innewohnend *(geh.);* natürlich ‹*Anmut, Eleganz*›

inherently [ɪn'hɪərəntlɪ, ɪn'herəntlɪ] *adv.* von Natur aus

inherit [ɪn'herɪt] *v.t.* erben

inheritance [ɪn'herɪtəns] *n.* **a)** *(what is inherited)* Erbe, *das;* **come into one's ~:** sein Erbe antreten; **b)** *no pl. (inheriting)* Erbschaft, *die*

inhibit [ɪn'hɪbɪt] *v.t.* hemmen; ~ **sb. from doing sth.** jmdn. daran hindern, etw. zu tun

inhibited [ɪn'hɪbɪtɪd] *adj.* gehemmt

inhibition [ɪnhɪ'bɪʃn] *n.* **a)** Unterdrückung, *die;* **b)** *(Psych.)* Hemmung, *die;* **c)** *(coll.: emotional resistance)* Hemmung, *die*

inhospitable [ɪnhɒ'spɪtəbl] *adj.* **a)** ungastlich ‹*Person, Verhalten*›; **b)** unwirtlich ‹*Gegend, Klima*›

'in-house *adj.* hausintern

inhuman [ɪn'hju:mən] *adj.* unmenschlich ‹*Tyrann, Grausamkeit, Strenge*›; inhuman ‹*Arbeitgeber, Verhalten*›

inhumane [ɪnhju:'meɪn] *adj.* unmenschlich; inhuman *(geh.);* menschenunwürdig ‹*Zustände, Behandlung*›

inimical [ɪ'nɪmɪkl] *adj. (harmful)* schädlich (**to** für)

inimitable [ɪ'nɪmɪtəbl] *adj.* unnachahmlich ‹*Gabe, Fähigkeit*›; einzigartig ‹*Persönlichkeit*›

initial [ɪ'nɪʃl] **1.** *adj.* anfänglich; zu Anfang auftretend ‹*Symptome*›; Anfangs‹*stadium, -schwierigkeiten*›. **2.** *n. esp. in pl.* Initiale, *die.* **3.** *v.t., (Brit.)* **-ll-** abzeichnen ‹*Scheck, Quittung*›; paraphieren ‹*Vertrag, Abkommen usw.*›

initial 'letter *n.* Anfangsbuchstabe, *der*

initially [ɪ'nɪʃəlɪ] *adv.* anfangs; am *od.* zu Anfang

initiate 1. [ɪ'nɪʃɪeɪt] *v.t.* **a)** *(admit)* [feierlich] aufnehmen; initiieren *(Soziol., Völkerk.); (introduce)* einführen (**into** in + *Akk.*); ~ **sb. into sth.** *(into club, group, etc.)* jmdn. in etw. *(Akk.)* aufnehmen; *(into knowledge, mystery, etc.)* jmdn. in etw. *(Akk.)* einweihen; **b)** *(begin)* initiieren *(geh.);* in die Wege leiten ‹*Vorhaben*›; einleiten ‹*Verhandlungen, Reformen*›; eröffnen ‹*Diskussion, Verhandlung, Feindseligkeiten*›. **2.** [ɪ'nɪʃɪət] *n.* Eingeweihte, *der/die*

initiation [ɪnɪʃɪ'eɪʃn] *n.* **a)** *(beginning)* Initiierung, *die (geh.); (of hostilities, discussion, negotiation)* Eröffnung, *die; (of reforms, negotiations)* Einleitung, *die;* **b)** *(admission)* Aufnahme, *die* (**into** in + *Akk.*); *(introduction)* Einführung, *die* (**into** in + *Akk.*); ~ **ceremony** Aufnahmezeremonie, *die*

initiative [ɪ'nɪʃətɪv, ɪ'nɪʃɪətɪv] *n.* **a)** *(power)* **the ~ is ours/lies with them** die Initiative liegt bei uns/ihnen; **b)** *no pl., no indef. art. (ability)* Initiative, *die;* **c)** *(first step)* Initiative, *die;* **take the ~:** die Initiative ergreifen; **on one's own ~:** aus eigener Initiative

inject [ɪn'dʒekt] *v.t.* **a)** [ein]spritzen; injizieren *(Med.);* **b)** *(put fluid into)* ~ **a vein with sth.** etw. in eine Vene spritzen *od. (Med.)* injizieren; **c)** *(administer sth. to)* ~ **sb. with sth.** jmdm. etw. spritzen *od. (Med.)* injizieren; ~ **sb. against smallpox** jmdn. gegen Pocken impfen; **d)** *(fig.)* pumpen ‹*Geld*›; ~ **new life into sth.** einer Sache *(Dat.)* neues Leben geben

injection [ɪn'dʒekʃn] *n.* **a)** *(injecting)* Einspritzung, *die;* Injektion, *die (Med.);* **b)** *(liquid injected)* Injektion, *die;* Injektionslösung, *die;* **c)** *(fig.)* ~ **of money/capital** Geldzuschuß, *der*

injudicious [ɪndʒu:'dɪʃəs] *adj.* unklug; ungünstig ‹*Moment*›

injunction [ɪn'dʒʌŋkʃn] *n.* **a)** *(order)* Verfügung, *die;* **b)** *(Law)* [richterliche] Verfügung

injure ['ɪndʒə(r)] *v.t.* **a)** *(hurt)* verletzen; *(fig.)* verletzen ‹*Stolz, Gefühle*›; kränken ‹*Person*›; **b)** *(impair)* schaden (+ *Dat.*); schädigen ‹*Gesundheit*›

injured ['ɪndʒəd] *adj.* **a)** *(hurt, lit. or fig.)* verletzt; verwundet ‹*Soldat*›; **the ~:** die Verletzten/Verwundeten; **b)** *(wronged)* geschädigt; **the ~ party** *(Law)* der/die Geschädigte; **c)** *(offended)* gekränkt ⟨*Stimme, Blick*⟩; verletzt, beleidigt ⟨*Person*⟩

injurious [ɪn'dʒʊərɪəs] *adj.* **a)** *(wrongful)* ungerecht ⟨*Behandlung*⟩; **b)** *(hurtful)* schädlich; **be ~ to sb./sth.** jmdm./einer Sache schaden

injury ['ɪndʒərɪ] *n.* *([instance of] harm)* Verletzung, *die* (**to** *Gen.*); *(fig.)* Kränkung, *die* (**to** *Gen.*); **add insult to ~:** das Ganze noch schlimmer machen; **do sb./oneself an ~:** jmdm./sich weh tun

injustice [ɪn'dʒʌstɪs] *n.* Ungerechtigkeit, *die*

ink [ɪŋk] **1.** *n.* Tinte, *die; (for stamp-pad)* Farbe, *die; (for drawing)* Tusche, *die; (in printing)* Druckfarbe, *die; (in duplicating, newsprint)* Druckerschwärze, *die.* **2.** *v. t.* **a)** **~ in** mit Tinte/Tusche nachziehen; **~ over** mit Tusche übermalen ⟨*Papier, Blatt*⟩; **b)** *(apply ink to)* einfärben ⟨*Druckform*⟩; mit Farbe schwärzen ⟨*Stempel*⟩

inkling ['ɪŋklɪŋ] *n.* Ahnung, *die;* **I haven't an ~:** ich habe nicht die leiseste Ahnung *od. (ugs.)* keinen blassen Schimmer; **have an ~ of sth.** etw. ahnen

inky ['ɪŋkɪ] *adj.* **a)** *(covered with ink)* tintenbeschmiert; tintig; **b)** *(black)* tintenschwarz; tintig

inland 1. ['ɪnlənd, 'ɪnlænd] *adj.* **a)** *(placed ~)* Binnen-; binnenländisch; **b)** *(carried on ~)* inländisch; Binnen⟨*handel, -verkehr*⟩; Inlands⟨*brief, -paket, -gebühren*⟩. **2.** [ɪn'lænd] *adv.* landeinwärts; im Landesinneren ⟨*leben*⟩

inland: ~ navi'gation *n.* Binnenschiffahrt, *die;* **I~ 'Revenue** *n. (Brit.)* ≈ Finanzamt, *das;* **~ 'sea** *n.* Binnenmeer, *das*

'in-law *n., usu. in pl. (coll.)* angeheirateter Verwandter/angeheiratete Verwandte; **~s** *(parents-in-law)* Schwiegereltern

inlet ['ɪnlet, 'ɪnlɪt] *n.* **a)** [schmale] Bucht; **b)** *(opening)* Einlaßöffnung, *die*

'inmate *n. (of hospital, prison, etc.)* Insasse, *der*/Insassin, *die; (of house)* Bewohner, *der*/Bewohnerin, *die*

inn [ɪn] *n.* **a)** *(hotel)* Herberge, *die (veralt.);* Gasthof, *der;* **b)** *(pub)* Wirtshaus, *das;* Gastwirtschaft, *die*

innards ['ɪnədz] *n. pl. (coll.)* Eingeweide *Pl.; (in animals for slaughter)* Innereien *Pl.*

innate [ɪ'neɪt, 'ɪneɪt] *adj. (inborn)* angeboren

inner ['ɪnə(r)] *adj.* **a)** inner...; Innen⟨*hof, -tür, -fläche, -seite usw.*⟩; **~ ear** *(Anat.)* Innenohr, *das;* **b)** *(fig.)* inner... ⟨*Gefühl, Wesen, Zweifel, Ängste*⟩; verborgen ⟨*Bedeutung*⟩

inner 'city *n.* Innenstadt, *die;* City, *die*

innermost ['ɪnəməʊst] *adj.* innerst...; **one's ~ thoughts** seine geheimsten Gedanken; **in the ~ depths of the forest** im tiefsten Wald

'inner tube *n.* Schlauch, *der*

innings ['ɪnɪŋz] *n., pl. same or (coll.)* **~es** *(Cricket)* Durchgang, *der;* Innings, *das (fachspr.)*

'innkeeper *n.* [Gast]wirt, *der*/-wirtin, *die*

innocence ['ɪnəsəns] *n., no pl.* **a)** Unschuld, *die;* **b)** *(freedom from cunning)* Naivität, *die;* **c)** *(lack of knowledge)* Unkenntnis, *die*

innocent ['ɪnəsənt] **1.** *adj.* **a)** unschuldig (**of an** + *Dat.*); **b)** *(harmless)* harmlos; **c)** *(naïve)* unschuldig; **he is ~ about the ways of the world** er ist völlig unerfahren; **d)** *(pretending to be guileless)* arglos, unschuldig ⟨*Blick, Erscheinung*⟩. **2.** *n.* *(innocent person)* Unschuldige, *der/die*

innocuous [ɪ'nɒkjʊəs] *adj.* *(not injurious)* unschädlich ⟨*Tier, Mittel*⟩; *(inoffensive)* harmlos

innovate ['ɪnəveɪt] *v. i.* Innovationen vornehmen

innovation [ɪnə'veɪʃn] *n.* **a)** *(introduction of something new)* Innovation, *die (geh., fachspr.); (thing introduced)* Neuerung, *die;* **b)** *(change)* [Ver]änderung, *die;* Neuerung, *die;* Innovation, *die (geh., fachspr.)*

innuendo [ɪnju:'endəʊ] *n., pl.* **~es** *or* **~s** versteckte Andeutung; Innuendo, *das (geh.)*

innumerable [ɪ'nju:mərəbl] *adj.* unzählig; zahllos; *(uncountable)* unzählbar

innumerate [ɪ'nju:mərət] *adj.* *(Brit.)* **be ~:** nicht rechnen können

inoculate [ɪ'nɒkjʊleɪt] *v. t.* impfen (**against, for** gegen)

inoculation [ɪnɒkjʊ'leɪʃn] *n.* Impfung, *die;* **give sb. an ~:** jmdn. impfen

inoffensive [ɪnə'fensɪv] *adj.* **a)** *(unoffending)* harmlos; **b)** *(not objectionable)* harmlos ⟨*Bemerkung*⟩; unaufdringlich ⟨*Geruch, Art, Person*⟩

inoperative [ɪn'ɒpərətɪv] *adj.* ungültig; außer Kraft *nicht attr.;* **render sth. ~:** etw. außer Betrieb setzen

inopportune [ɪn'ɒpətju:n] *adj.* inopportun *(geh.);* unangebracht ⟨*Bemerkung*⟩; ungelegen, unpassend ⟨*Augenblick, Besuch*⟩

inordinate [ɪ'nɔ:dɪnət] *adj.* *(immoderate)* unmäßig; ungeheuer ⟨*Menge*⟩; überzogen ⟨*Forderung*⟩; **an ~ amount of work/money** ungeheuer viel Arbeit/eine Unmenge Geld

inorganic [ɪnɔ:'gænɪk] *adj.* *(Chem.)* anorganisch

'in-patient *n.* stationär behandelter Patient/behandelte Patientin; **be an ~:** stationär behandelt werden

'input 1. *n.* *(esp. Computing)* Input, *der od. das; (of capital)* Investition, *die; (of electricity)* Energiezufuhr, *die.* **2.** *v. t.,* **-tt-, ~** *or* **~ted** eingeben ⟨*Daten, Programm*⟩; zuführen ⟨*Strom, Energie*⟩

inquest ['ɪnkwest, 'ɪŋkwest] *n.* *(legal inquiry)* **~ [into the causes of death]** gerichtliche Untersuchung der Todesursache

inquire [ɪn'kwaɪə(r), ɪŋ'kwaɪə(r)] **1.** *v. i.* **a)** *(make search)* Untersuchungen anstellen (**into** über + *Akk.*); **~ into a matter** eine Angelegenheit untersuchen; **b)** *(seek information)* sich erkundigen (**about, after** nach, **of** bei). **2.** *v. t.* sich erkundigen nach, fragen nach ⟨*Weg, Namen*⟩

inquiring [ɪn'kwaɪərɪŋ, ɪŋ'kwaɪərɪŋ] *adj.* fragend; forschend ⟨*Geist*⟩

inquiry [ɪn'kwaɪərɪ, ɪŋ'kwaɪərɪ] *n.* **a)** *(asking)* Anfrage, *die;* **give sb. a look of ~:** jmdn. fragend ansehen; **b)** *(question)* Erkundigung, *die* (**into** über + *Akk.*); **make inquiries** Erkundigungen einziehen; **c)** *(investigation)* Untersuchung, *die;* **hold an ~:** eine Untersuchung durchführen (**into** *Gen.*)

inquiry: ~ desk, ~ office *ns.* Auskunft, *die*

inquisition [ɪnkwɪ'zɪʃn, ɪŋkwɪ'zɪʃn] *n.* **a)** *(search)* Nachforschung, *die* (**into** über + *Akk.*); **b)** *(judicial inquiry)* gerichtliche Untersuchung; *(fig. coll.)* Verhör, *das;* **c)** **I~** *(Hist.)* Inquisition, *die*

inquisitive [ɪn'kwɪzɪtɪv, ɪŋ'kwɪzɪtɪv] *adj.* **a)** *(unduly inquiring)* neugierig; **b)** *(inquiring)* wißbegierig

inquisitiveness [ɪn'kwɪzɪtɪvnɪs, ɪŋ'kwɪzɪtɪvnɪs] *n., no pl. see* **inquisitive:** Neugier[de], *die;* Wißbegier[de], *die*

'inroad *n.* **a)** *(intrusion)* Eingriff, *der* (**on, into** in + *Akk.*); **make ~s into sb.'s savings** jmds. Ersparnisse angreifen; **b)** *(hostile incursion)* Einfall, *der* (**into** in + *Akk.*); Überfall, *der* (**[up]on** auf + *Akk.*)

insane [ɪn'seɪn] *adj.* **a)** geisteskrank; **b)** *(extremely foolish)* wahnsinnig *(ugs.)*; irrsinnig *(ugs.)*

insanitary [ɪn'sænɪtərɪ] *adj.* unhygienisch

insanity [ɪn'sænɪtɪ] *n.* **a)** Geisteskrankheit, *die;* Wahnsinn, *der;* **b)** *(extreme folly)* Irrsinn, *der; (instance)* Verrücktheit, *die*

insatiable [ɪn'seɪʃəbl] *adj.* unersättlich; unstillbar ‹*Verlangen, Neugierde*›

inscribe [ɪn'skraɪb] *v.t.* **a)** *(write)* schreiben; *(on ring etc.)* eingravieren; *(on stone, rock)* einmeißeln; **b)** mit einer Inschrift versehen ‹*Denkmal, Grabstein*›

inscription [ɪn'skrɪpʃn] *n.* *(words inscribed)* Inschrift, *die; (on coin)* Aufschrift, *die*

inscrutable [ɪn'skru:təbl] *adj.* *(mysterious)* unergründlich; geheimnisvoll ‹*Lächeln*›; undurchdringlich ‹*Miene*›

insect ['ɪnsekt] *n.* Insekt, *das;* Kerbtier, *das*

'**insect bite** *n.* Insektenstich, *der*

insecticide [ɪn'sektɪsaɪd] *n.* Insektizid, *das*

'**insect repellent** *n.* Insektenschutzmittel, *das*

insecure [ɪnsɪ'kjʊə(r)] *adj.* **a)** *(unsafe)* unsicher; **b)** *(not firm, liable to give way)* nicht sicher; **c)** *(Psych.)* unsicher; **feel ~:** sich nicht sicher fühlen

insecurity [ɪnsɪ'kjʊərɪtɪ] *n., no pl.* *(also Psych.)* Unsicherheit, *die*

insensibility [ɪnsensɪ'bɪlɪtɪ] *n., no pl.* **a)** *(lack of emotional feeling, indifference)* Gefühllosigkeit, *die* (**to** gegenüber); **b)** *(unconsciousness)* Bewußtlosigkeit, *die;* **c)** *(lack of physical feeling)* Unempfindlichkeit, *die* (**to** gegen)

insensible [ɪn'sensɪbl] *adj.* **a)** *(emotionless)* gefühllos ‹*Person, Art*›; unempfindlich (**to** für); **b)** *(deprived of sensation)* unempfindlich (**to** für); **c)** *(unconscious)* bewußtlos; **d)** *(unaware)* **be ~ of** *or* **to sth.** sich *(Dat.)* einer Sache *(Gen.)* nicht bewußt sein

insensitive [ɪn'sensɪtɪv] *adj.* **a)** *(lacking feeling)* gefühllos ‹*Person, Art*›; **b)** *(unappreciative)* unempfänglich (**to** für); **c)** *(not physically sensitive)* unempfindlich (**to** gegen)

insensitiveness [ɪn'sensɪtɪvnɪs], **insensitivity** [ɪnsensɪ'tɪvɪtɪ] *ns., no pl.* **a)** *(lack of feeling)* Gefühllosigkeit, *die* (**to** gegenüber); **b)** *(unappreciativeness)* Unempfindlichkeit, *die* (**to** für); **c)** *(lack of physical sensitiveness)* Unempfindlichkeit, *die* (**to** gegen); **~ to heat** Hitzeunempfindlichkeit, *die*

inseparable [ɪn'sepərəbl] *adj.* untrennbar; *(fig.)* unzertrennlich ‹*Freunde, Zwillinge usw.*›

insert 1. [ɪn'sɜ:t] *v.t.* **a)** einlegen ‹*Film*›; einwerfen ‹*Münze*›; einsetzen ‹*Herzschrittmacher*›; einstechen ‹*Nadel*›; **~ a piece of paper into the typewriter** ein Blatt Papier in die Schreibmaschine einspannen; **b)** *(introduce into)* einfügen ‹*Wort, Satz usw.*› (**in** in + *Akk.*); **~ an advertisement in 'The Times'** eine Anzeige in die „Times" setzen. **2.** ['ɪnsɜ:t] *n.* *(in magazine)* Beilage, *die; (in book)* Einlage, *die; (printed in newspaper)* Inserat, *das*

insertion [ɪn'sɜ:ʃn] *n.* **a)** *see* **insert 1a:** Einlegen, *das;* Einwerfen, *das;* Einsetzen, *das;* Einstechen, *das;* **b)** *(words etc. in a text)* Einfügung, *die;* Beifügung, *die; (in newspaper)* Inserat, *das*

inset ['--] *n.* *(small map)* Nebenkarte, *die; (small photograph, diagram)* Nebenbild, *das*

inshore ['--] *adj.* Küsten‹*fischerei, -gewässer, -schiffahrt*›

inside 1. [-'-, '--] *n.* **a)** *(internal side)* Innenseite, *die;* **on the ~:** innen; **to/from the ~:** nach/von innen; **lock the door from the ~:** die Tür von innen abschließen; **b)** *(inner part)* Innere, *das;* **c)** *in sing. or pl. (coll.: stomach and bowels)* Eingeweide *Pl.;* Innere, *das;* **d) the wind blew her umbrella ~ out** der Wind hat ihren Regenschirm umgestülpt; **turn a jacket ~ out** eine Jacke nach links wenden. **2.** ['--] *adj.* inner...; Innen‹*wand, -ansicht, -durchmesser*›; *(fig.)* intern; **be on an ~ page** im Inneren [der Zeitung] stehen; **give the ~ story of sth.** etw. von innen beleuchten *(fig.)*; **~ information** interne Informationen; **~ pocket** Innentasche, *die;* **~ lane** Innenspur, *die.* **3.** [-'-] *adv.* **a)** *(on or in the ~)* innen; *(to the ~)* nach innen hinein/herein; *(indoors)* drinnen; **come ~:** hereinkommen; **go ~:** [ins Haus] hineingehen; **b)** *(sl.: in prison)* **be ~:** sitzen *(ugs.)*; **put sb. ~:** jmdn. einlochen *(salopp)*. **4.** [-'-] *prep.* **a)** *(on inner side of)* [innen] in (+ *Dat.*); *(with direction)* in (+ *Akk.*) hinein; **sit/get ~ the house** im Haus sitzen/ins Haus hineinkommen; **b)** *(in less than)* **~ an hour** innerhalb [von] einer Stunde

insider [ɪn'saɪdə(r)] *n.* *(within a society)* Mitglied, *das;* Zugehörige, *der/die;* **~ dealing** *or* **trading** *(Stock Exch.)* Insiderhandel, *der*

insidious [ɪn'sɪdɪəs] *adj.* heimtückisch

'**insight** *n.* Verständnis, *das; (instance)* Einblick, *der* (**into** in + *Akk.*); **gain an ~ into sth.** [einen] Einblick in etw. *(Akk.)* gewinnen *od.* bekommen

insignificant [ɪnsɪg'nɪfɪkənt] *adj.* **a)** unbedeutend; geringfügig ‹*Summe*›; unbedeutend, geringfügig ‹*Unterschied*›; **b)** *(contemptible)* unscheinbar ‹*Person*›

insincere [ɪnsɪn'sɪə(r)] *adj.* unaufrichtig; falsch ‹*Lächeln*›

insincerity [ɪnsɪn'serɪtɪ] *n.* Unaufrichtigkeit, *die; (of smile)* Falschheit, *die*

insinuate [ɪn'sɪnjʊeɪt] *v.t.* **a)** *(introduce)* [auf geschickte Art] einflößen ‹*Propaganda*›; **b)** *(convey)* andeuten (**to sb.** jmdm. gegenüber); unterstellen; **insinuating remarks** Andeutungen; Unterstellungen; **c) ~ oneself into sb.'s favour** sich bei jmdm. einschmeicheln

insinuation [ɪnsɪnjʊ'eɪʃn] *n.* Anspielung, *die* (**about** auf + *Akk.*)

insipid [ɪn'sɪpɪd] *adj.* **a)** fad[e] ‹*Essen*›; schal ‹*Getränk*›; **b)** fad[e] *(ugs.)*, geistlos ‹*Person*›; langweilig ‹*Farbe, Musik*›

insist [ɪn'sɪst] **1.** *v.i.* bestehen (**[up]on** auf + *Dat.*); **~ on doing sth.** darauf bestehen, etw. zu tun; **she ~s on her innocence** sie behauptet beharrlich, unschuldig zu sein. **2.** *v.t.* **a) ~ that ...:** darauf bestehen, daß ...; **b)** *(maintain positively)* **they keep ~ing that ...:** sie beharren *od.* bestehen beharrlich darauf, daß ...

insistence [ɪn'sɪstəns] *n., no pl.* Bestehen, *das* (**on** auf + *Dat.*); **I only came here at your ~:** ich kam nur auf dein Drängen hierher

insistent [ɪn'sɪstənt] *adj.* **a)** beharrlich, hartnäckig ‹*Person*›; aufdringlich ‹*Musik*›; nachdrücklich ‹*Forderung*›; **b)** *([annoyingly] persistent)* penetrant *(abwertend)*

insole ['ɪnsəʊl] *n.* Einlegesohle, *die*

insolence ['ɪnsələns] *n., no pl.* Unverschämtheit, *die;* Frechheit, *die*

insolent ['ɪnsələnt] *adj.*, **insolently** ['ɪnsələntlɪ] *adv.* **a)** *(insulting[ly])* unverschämt; frech; **b)** *(contemptuous[ly])* anmaßend; überheblich

insoluble [ɪn'sɒljʊbl] *adj.* **a)** unlösbar ‹*Problem, Rätsel usw.*›; **b)** unlöslich ‹*Substanz*›; insolubel *(Chem.)* ‹*Verbindung*›

insolvency [ɪn'sɒlvənsɪ] *n.* Insolvenz, *die (bes. Wirtsch.)*; Zahlungsunfähigkeit, *die*

insolvent [ɪn'sɒlvənt] *adj. (unable to pay debts)* insolvent *(bes. Wirtsch.)*; zahlungsunfähig

insomnia [ɪn'sɒmnɪə] *n.* Schlaflosigkeit, *die*

inspect [ɪn'spekt] *v.t.* **a)** *(view closely)* prüfend betrachten; **b)** *(examine officially)* überprüfen; inspizieren, kontrollieren ‹*Räumlichkeiten*›

inspection [ɪn'spekʃn] *n.* Überprüfung, *die; (of premises)* Kontrolle, *die;* Inspektion, *die;* **present/submit sth. for ~:** etw. zur Prüfung vorlegen

inspector [ɪn'spektə(r)] *n.* **a)** *(official) (on bus, train, etc.)* Kontrolleur, *der*/Kontrolleurin, *die;* **b)** *(Brit.: police officer)* ≈ Polizeiinspektor, *der*

inspiration [ɪnspə'reɪʃn] *n.* Inspiration, *die (geh.)*; **get one's ~ from sth.** sich von etw. inspirieren lassen; **sth. is an ~ to sb.** etw. inspiriert jmdn.

inspire [ɪn'spaɪə(r)] *v.t.* **a)** inspirieren *(geh.)*; **in an ~d moment** *(coll.)* in einem Augenblick der Erleuchtung; **b)** *(animate)* inspirieren; anregen; *(encourage)* anspornen; **~d idea** genialer Gedanke; **~d guess** intuitiv richtige Vermutung; **c)** *(instil)* einflößen ‹*Mut, Angst, Respekt*› (**in** *Dat.*); [er]wecken ‹*Vertrauen, Gedanke, Hoffnung*› (**in** in + *Dat.*); hervorrufen ‹*Haß, Abneigung*› (**in** bei); *(incite)* anstiften; anzetteln *(abwertend)* ‹*Unruhen usw.*›

inspiring [ɪn'spaiərɪŋ] *adj.* inspirierend *(geh.)*

instability [ɪnstə'bɪlɪtɪ] *n. (mental, physical)* Labilität, *die; (inconstancy)* Instabilität, *die*

install [ɪn'stɔ:l] *v.t.* **a)** *(establish)* **~ oneself** sich installieren; *(in a house etc.)* sich einrichten; **b)** *(set up for use)* installieren ‹*Heizung, Leitung*›; anschließen ‹*Telefon*›; einbauen ‹*Badezimmer*›; aufstellen, anschließen ‹*Herd*›; **c)** *(place ceremonially)* installieren *(geh.)*; **~ sb. in an office/a post** jmdn. in ein Amt einsetzen

installation [ɪnstə'leɪʃn] *n.* **a)** *(in an office or post)* Amtseinsetzung, *die;* **b)** *(setting up for use)* Installation, *die; (of bathroom etc.)* Einbau, *der; (of telephone, cooker)* Anschluß, *der;* **c)** *(apparatus etc. installed)* Anlage, *die*

instalment *(Amer.:* **installment**) [ɪn'stɔ:lmənt] *n.* **a)** *(partpayment)* Rate, *die;* **pay by** *or* **in ~s** in Raten *od.* ratenweise zahlen; **b)** *(of serial, novel)* Fortsetzung, *die; (of film, radio programme)* Folge, *die*

instance ['ɪnstəns] *n.* **a)** *(example)* Beispiel, *das* (**of** für); **as an ~ of ...:** als [ein] Beispiel für ...; **for ~:** zum Beispiel; **b)** *(particular case)* **in your/this ~:** in deinem/diesem Fall[e]; **c) in the first ~:** zuerst *od.* zunächst einmal; *(at the very beginning)* gleich zu Anfang

instant ['ɪnstənt] **1.** *adj.* **a)** *(occurring immediately)* unmittelbar; sofortig ‹*Wirkung, Linderung, Ergebnis*›; **b) ~ coffee/tea** Instant- *od.* Pulverkaffee/Instanttee, *der;* **~ meal** Fertiggericht, *das.* **2.** *n.* Augenblick, *der;* **come here this ~:** komm sofort *od.* auf der Stelle her; **the ~ he walked in at the door ...:** in dem Augenblick, als er hereintrat, ...; **in an ~:** augenblicklich

instantaneous [ɪnstən'teɪnɪəs] *adj.* unmittelbar; **his reaction was ~:** er reagierte sofort

instantly ['ɪnstəntlɪ] *adv.* sofort

instead [ɪn'sted] *adv.* statt dessen; **~ of doing sth.** [an]statt etw. zu tun; **~ of sth.** anstelle einer Sache *(Gen.)*; **I will go ~ of you** ich gehe an deiner Stelle

'**instep** *n.* **a)** *(of foot)* Spann, *der;* Fußrücken, *der;* **b)** *(of shoe)* Blatt, *das*

instigate ['ɪnstɪgeɪt] *v.t.* **a)** *(urge on)* anstiften (**to** zu); **b)** *(bring about)* initiieren *(geh.)* ‹*Reformen, Projekt usw.*›; anzetteln *(abwertend)* ‹*Streik usw.*›

instigation [ɪnstɪ'geɪʃn] *n.* **a)** *(urging)* Anstiftung, *die;* **at sb.'s ~:** auf jmds. Betreiben *(Akk.)*; **b)** *(bringing about)* Anzettelung, *die (abwertend); (of reforms etc.)* Initiierung, *die (geh.)*

instil *(Amer.:* **instill**) [ɪn'stɪl] *v.t.*, **-ll-** einflößen (**in** *Dat.*); einimpfen (**in** *Dat.*); beibringen ‹*gutes Benehmen, Wissen*› (**in** *Dat.*)

instinct ['ɪnstɪnkt] *n.* **a)** Instinkt, *der;* **~ for survival, survival ~:** Überlebenstrieb, *der;* **b)** *(intuition)* Instinkt, *der;* instinktives Gefühl (**for** für); *(unconscious skill)* natürliche Begabung (**for** für); Sinn, *der* (**for** für)

instinctive [ɪn'stɪnktɪv] *adj.*, **instinctively** [ɪn'stɪnktɪvlɪ] *adv.* instinktiv

institute ['ɪnstɪtju:t] **1.** *n.* Institut, *das.* **2.** *v.t.* einführen ‹*Reform, Brauch, Beschränkung*›; einleiten ‹*Suche, Verfahren, Untersuchung*›; gründen ‹*Gesellschaft*›; anstrengen ‹*Prozeß, Klage*›; schaffen ‹*Posten*›

institution [ɪnstɪ'tju:ʃn] *n.* **a)** *(instituting)* Einführung, *die;* **b)** *(law, custom)* Institution, *die;* **c)** *(coll.: familiar object)* Institution, *die;* **become an ~:** zur Institution werden; **d)** *(institute)* Heim, *das;* Anstalt, *die*

institutional [ɪnstɪ'tju:ʃənl] *adj.* **a)** *(of, like, organized through institutions)* institutionell *(geh.)*; **b)** *(suggestive of typical charitable institutions)* Heim-; Anstalts-

instruct [ɪn'strʌkt] *v.t.* **a)** *(teach)* unterrichten ‹*Klasse, Fach*›; **b)** *(direct, command)* anweisen; die Anweisung erteilen (+ *Dat.*); **c)** *(inform)* unterrichten; **d)** *(Law: appoint)* beauftragen ‹*Anwalt*›

instruction [ɪn'strʌkʃn] *n.* **a)** *(teaching)* Unterricht, *der;* **b)** *esp. in pl. (direction, order)* Anweisung, *die;* Instruktion, *die;* **~ manual/~s for use** Gebrauchsanleitung, *die; (for machine etc.)* Betriebsanleitung, *die*

instructive [ɪn'strʌktɪv] *adj.* aufschlußreich; instruktiv; lehrreich ‹*Erfahrung, Buch*›

instructor [ɪn'strʌktə(r)] *n.* **a)** Lehrer, *der*/Lehrerin, *die; (Mil.)* Ausbilder, *der;* **b)** *(Amer. Univ.)* Dozent, *der*/Dozentin, *die*

instrument ['ɪnstrʊmənt] *n. (also Mus.)* Instrument, *das; (person)* Werkzeug, *das*

instrumental [ɪnstrʊ'mentl] *adj.* **a)** *(serving as instrument or means)* dienlich (**to** *Dat.*); förderlich (**to** *Dat.*); **he was ~ in finding me a post** er hat mir zu einer Stelle verholfen; **b)** *(Mus.)* instrumental; Instrumental‹*musik, -version, -nummer*›

instrumentalist [ɪnstrʊ'mentəlɪst] *n.* Instrumentalist, *der*/Instrumentalistin, *die*

'**instrument panel** *n.* Instrumentenbrett, *das*

insubordinate [ɪnsə'bɔ:dɪnət] *adj.* aufsässig; widersetzlich; *(Mil.)* ungehorsam

insubordination [ɪnsəbɔ:dɪ'neɪʃn] *n., no pl.* Aufsässigkeit, *die;* Widersetzlichkeit, *die; (Mil.)* Gehorsamsverweigerung, *die*

insubstantial [ɪnsəb'stænʃl] *adj.* wenig substantiell *(geh.)*; dürftig ‹*Essen, Kleidung*›; gering[fügig] ‹*Menge, Betrag*›

insufferable [ɪn'sʌfərəbl] *adj.* **a)** *(unbearably arrogant)* unausstehlich; **b)** *(intolerable)* unerträglich

insufficient [ɪnsə'fɪʃənt] *adj.* nicht genügend ‹*Arbeit, Gründe, Geld*›; unzulänglich ‹*Beweise*›; unzureichend ‹*Versorgung, Beleuchtung*›; **give sb. ~ notice** jmdm. nicht rechtzeitig Bescheid geben

insular ['ɪnsjʊlə(r)] *adj.* **a)** *(of an island)* Insel-; insular *(fachspr.)*; **b)** *(fig.)* provinziell *(abwertend)*

insularity [ɪnsjʊ'lærɪtɪ] *n.* Provinzialität, *die (abwertend)*

insulate ['ɪnsjʊleɪt] *v.t.* **a)** *(isolate)* isolieren (**against, from** gegen); **b)** *(detach from surroundings)* isolieren (**from** von)

'**insulating tape** *n.* Isolierband, *das*

insulation [ɪnsjʊ'leɪʃn] *n.* Isolierung, *die*

insulator ['ɪnsjʊleɪtə(r)] *n.* Isolator, *der*

insulin ['ɪnsjʊlɪn] *n. (Med.)* Insulin, *das*

insult 1. ['ɪnsʌlt] *n.* Beleidigung, *die* (**to** *Gen.*); *see also* **injury. 2.** [ɪn'sʌlt] *v.t.* beleidigen

insulting [ɪn'sʌltɪŋ] *adj.* beleidigend

insuperable [ɪn'su:pərəbl, ɪn'sju:pərəbl] *adj.* unüberwindlich

insupportable [ɪnsə'pɔ:təbl] *adj.* *(that cannot be endured)* unerträglich

insurance [ɪn'ʃʊərəns] *n.* **a)** *(insuring)* Versicherung, *die; (fig.)* Sicherheit, *die;* Gewähr, *die;* **take out ~ against/on sth.** eine Versicherung gegen etw. abschließen/etw. versichern lassen; **travel ~:** Reisegepäck- und -unfallversicherung, *die;* **b)** *(sum received)* Versicherungssumme, *die; (sum paid)* Versicherungsbetrag, *der*

insurance: ~ agent Versicherungsvertreter, *der*/-vertreterin, *die;* **~ company** *n.* Versicherungsgesellschaft, *die;* **~ policy** *n.* Versicherungspolice, *die; (fig.)* Sicherheit, *die;* Gewähr, *die*

insure [ɪn'ʃʊə(r)] *v.t.* **a)** *(secure payment to)* versichern ‹*Person*› (**against** gegen); **~ [oneself] against sth.** sich gegen etw. versichern; **b)** *(secure payment for)* ‹*Versicherungsgesellschaft:*› versichern; ‹*Versicherungsnehmer:*› versichern lassen ‹*Gepäck, Gemälde usw.*›

insurgent [ɪn'sɜ:dʒənt] **1.** *attrib. adj.* aufständisch. **2.** *n.* Aufständische, *der/die*

insurmountable [ɪnsə'maʊntəbl] *adj.* unüberwindlich

insurrection [ɪnsə'rekʃn] *n. (uprising)* Aufstand, *der*

intact [ɪn'tækt] *adj.* **a)** *(entire)* unbeschädigt; unversehrt; intakt ‹*Uhr, Maschine usw.*›; **b)** *(unimpaired)* unversehrt; **keep one's reputation ~:** sich *(Dat.)* einen guten Ruf bewahren; **c)** *(untouched)* unberührt; unangetastet

'**intake** *n.* **a)** *(action)* Aufnahme, *die;* **~ of breath** Atemholen, *das;* **b)** *(where water enters channel or pipe)* Einströmungsöffnung, *die; (where air or fuel enters engine)* Ansaugöffnung, *die;* **c)** *(persons or things taken in)* Neuzugänge; *(amount taken in)* aufgenommene Menge; **~ of calories** Kalorienzufuhr, *die*

intangible [ɪn'tændʒɪbl] *adj.* **a)** *(that cannot be touched)* nicht greifbar; **b)** *(that cannot be grasped mentally)* unbestimmbar

integer ['ɪntɪdʒə(r)] *n. (Math.)* ganze Zahl

integral ['ɪntɪgrl] *adj.* **a)** *(of a whole)* wesentlich, integral ‹*Bestandteil*›; **b)** *(whole, complete)* vollständig; vollkommen; **c)** *(forming a whole)* ein Ganzes bildend; integrierend

integrate ['ɪntɪgreɪt] *v.t.* **a)** *(combine into a whole; also Math.)* integrieren; **an ~d Europe** ein vereintes Europa; **b)** *(into society)* integrieren (**in** in + *Akk.*); **c)** *(open to all racial groups)* **~ a school/college** eine Schule/ein College für alle Rassen zugänglich machen

integrated '**circuit** *n. (Electronics)* integrierter Schaltkreis

integration [ɪntɪ'greɪʃn] *n.* **a)** *(integrating; also Math.)* Integration, *die;* **b)** *(ending of segregation)* Integration, *die* (**into** in + *Akk.*); **racial ~:** Rassenintegration, *die*

integrity [ɪn'tegrɪtɪ] *n.* **a)** *(uprightness, honesty)* Redlichkeit, *die; (of business, venture)* Seriosität, *die; (of style)* Echtheit, *die;* Unverfälschtheit, *die;* **b)** *(wholeness)* Einheit, *die;* **territorial ~:** territoriale Integrität

intellect ['ɪntəlekt] *n.* **a)** *(faculty)* Verstand, *der;* Intellekt, *der;* **b)** *(understanding)* Intelligenz, *die;* **powers of ~:** Verstandeskräfte

intellectual [ɪntə'lektjʊəl] **1.** *adj.* **a)** *(of intellect)* intellektuell; geistig ‹*Klima, Interessen, Arbeit*›; abstrakt ‹*Mitgefühl, Sympathie*›; **b)** *(possessing good understanding or intelligence)* geistig anspruchsvoll ‹*Person, Publikum*›. **2.** *n.* Intellektuelle, *der/die*

intelligence [ɪn'telɪdʒəns] *n.* **a)** Intelligenz, *die;* **have the ~ to do sth.** so intelligent sein, etw. zu tun; **b)** *(information)* Informationen *Pl.; (news)* Nachrichten *Pl.;* Meldungen *Pl.;* **c)** *([persons employed in] collecting information)* Nachrichtendienst, *der;* **military ~:** militärischer Geheimdienst

intelligence: ~ quotient *n.* Intelligenzquotient, *der;* **~ test** *n.* Intelligenztest, *der*

intelligent [ɪn'telɪdʒənt] *adj.* intelligent; intelligent geschrieben, geistreich ‹*Buch*›

intelligentsia [ɪntelɪ'dʒentsɪə] *n.* Intelligentsia, *die (geh.);* Intelligenz, *die*

intelligible [ɪn'telɪdʒɪbl] *adj.* verständlich (**to** für)

intemperate [ɪn'tempərət] *adj.* **a)** *(immoderate)* maßlos; überzogen, übertrieben ‹*Verhalten, Bemerkung*›; unmäßig, maßlos ‹*Verlangen, Appetit, Konsum*›; **b)** *(addicted to drinking)* trunksüchtig

intend [ɪn'tend] *v.t.* **a)** *(have as one's purpose)* beabsichtigen; **~ doing sth.** *or* **to do sth.** beabsichtigen, etw. zu tun; **it isn't really what we ~ed** es ist eigentlich nicht das, was wir wollten; **b)** *(design, mean)* **we ~ him to go** wir wollen, daß er geht; er soll gehen; **it was ~ed as a joke** das sollte ein Witz sein; **what do you ~ by that remark?** was willst du mit dieser Bemerkung sagen? *See also* **intended**

intended [ɪn'tendɪd] **1.** *adj.* beabsichtigt ‹*Wirkung*›; erklärt ‹*Ziel*›; absichtlich ‹*Beleidigung*›; **be ~ for sb./sth.** für jmdn./etw. bestimmt *od.* gedacht sein. **2.** *n. (coll.)* Zukünftige, *der/die (ugs.)*

intense [ɪn'tens] *adj.*, **~r** [ɪn'tensə(r)], **~st** [ɪn'tensɪst] **a)** intensiv; stark ‹*Hitze, Belastung*›; stark, heftig ‹*Schmerzen*›; kräftig, intensiv ‹*Farbe*›; äußerst groß ‹*Aufregung*›; ungeheuer ‹*Kälte, Helligkeit*›; **b)** *(eager, ardent)* eifrig, lebhaft ‹*Diskussion*›; stark, ausgeprägt ‹*Interesse*›; brennend, glühend ‹*Verlangen*›; äußerst groß ‹*Empörung, Aufregung, Betrübnis*›; tief ‹*Gefühl*›; rasend ‹*Haß, Eifersucht*›; **c)** *(with strong emotion)* stark gefühlsbetont ‹*Person, Brief*›; *(earnest)* ernst

intensely [ɪn'tenslɪ] *adv.* äußerst ‹*schwierig, verärgert, enttäuscht, kalt*›; ernsthaft, intensiv ‹*studieren*›; intensiv ‹*fühlen*›

intensifier [ɪn'tensɪfaɪə(r)] *n. (Ling.)* intensivierendes Wort

intensify [ɪn'tensɪfaɪ] **1.** *v.t.* intensivieren. **2.** *v.i.* zunehmen;

⟨*Hitze, Schmerzen:*⟩ stärker werden; ⟨*Kampf:*⟩ sich verschärfen

intensity [ɪn'tensɪtɪ] *n.* Intensität, *die; (of feeling also)* Heftigkeit, *die*

intensive [ɪn'tensɪv] *adj.* **a)** *(vigorous, thorough)* intensiv; Intensiv⟨*kurs*⟩; **b)** *(Ling.)* verstärkend; intensivierend; **c)** *(concentrated, directed to a single point or area)* intensiv; heftig ⟨*Beschuß*⟩; gezielt ⟨*Entwicklung*⟩; **d)** *(Econ.)* intensiv ⟨*Landwirtschaft*⟩; **e)** *in comb.* **capital-~/labour-~:** kapital-/arbeitsintensiv

intensive 'care *n.* Intensivpflege, *die (Med.)*; **be in ~:** auf der Intensivstation sein; **~ unit** Intensivstation, *die*

intent [ɪn'tent] **1.** *n.* Absicht, *die;* **by ~:** beabsichtigt; **with ~ to do sth.** *(Law)* in der Absicht *od.* mit dem Vorsatz, etw. zu tun; **to all ~s and purposes** im Grunde; praktisch; *see also* **loiter. 2.** *adj.* **a)** *(resolved)* erpicht, versessen (**[up]on** auf + *Akk.*); **be ~ on achieving sth.** etw. unbedingt erreichen wollen; **b)** *(attentively occupied)* eifrig beschäftigt (**on** mit); **be ~ on one's work** auf seine Arbeit konzentriert sein; **c)** *(earnest, eager)* aufmerksam; konzentriert; forschend ⟨*Blick*⟩

intention [ɪn'tenʃn] *n.* **a)** Absicht, *die;* Intention, *die;* **it was my ~ to visit him** ich hatte die Absicht *od.* beabsichtigte, ihn zu besuchen; **with the best of ~s** in der besten Absicht; **b)** *in pl. (coll.: in respect of marriage)* [Heirats]absichten

intentional [ɪn'tenʃənl] *adj.* absichtlich; vorsätzlich *(bes. Rechtsspr.)*; **it wasn't ~:** es war keine Absicht

intentionally [ɪn'tenʃənəlɪ] *adv.* absichtlich; mit Absicht

intently [ɪn'tentlɪ] *adv.* aufmerksam ⟨*zuhören, lesen, beobachten*⟩

interact [ɪntər'ækt] *v.i.* **a)** ⟨*Ideen:*⟩ sich gegenseitig beeinflussen; ⟨*Chemikalien usw.:*⟩ aufeinander einwirken, miteinander reagieren; **b)** *(Sociol., Psych.)* interagieren

interaction [ɪntər'ækʃn] *n.* **a)** gegenseitige Beeinflussung; *(Chem., Phys.)* Wechselwirkung, *die;* **b)** *(Sociol., Psych.)* Interaktion, *die*

interactive [ɪntər'æktɪv] *adj.* **a)** *(Chem.)* miteinander reagierend; **b)** *(Sociol., Psych., Computing)* interaktiv

intercede [ɪntə'si:d] *v.i.* sich einsetzen (**with** bei; **for, on behalf of** für)

intercept [ɪntə'sept] *v.t.* **a)** *(seize)* abfangen; **b)** *(check, stop)* abwehren ⟨*Schlag, Angriff*⟩; **c)** *(listen in to)* abhören ⟨*Gespräch, Funkspruch*⟩

interceptor [ɪntə'septə(r)] *n. (Air Force)* Abfangjäger, *der*

interchange 1. ['ɪntətʃeɪndʒ] *n.* **a)** *(reciprocal exchange)* Austausch, *der;* **b)** *(road junction)* [Autobahn]kreuz, *das.* **2.** [ɪntə'tʃeɪndʒ] *v.t.* **a)** *(exchange with each other)* austauschen; **b)** *(put each in the other's place)* [miteinander] vertauschen; **c)** *(alternate)* wechseln

interchangeable [ɪntə'tʃeɪndʒəbl] *adj.* austauschbar; synonym ⟨*Wörter, Ausdrücke*⟩

inter-city [ɪntə'sɪtɪ] *adj.* Intercity-; **~ train** Intercity[-Zug], *der*

intercom ['ɪntəkɒm] *n. (coll.)* Gegensprechanlage, *die*

interconnect [ɪntəkə'nekt] **1.** *v.t.* miteinander verbinden; zusammenschalten ⟨*Stromkreise, Verstärker, Lautsprecher*⟩. **2.** *v.i.* miteinander in Zusammenhang stehen; **~ing rooms** miteinander verbundene Zimmer

intercontinental [ɪntəkɒntɪ'nentl] *adj.* interkontinental; Interkontinental⟨*rakete, -reise, -flug*⟩

intercourse ['ɪntəkɔ:s] *n., no pl.* **a)** *(social communication)* Umgang, *der;* **social ~:** gesellschaftlicher Verkehr; **b)** *(sexual ~)* [Geschlechts]verkehr, *der*

interdependence [ɪntədɪ'pendəns] *n.* gegenseitige Abhängigkeit; Interdependenz, *die*

interdependent [ɪntədɪ'pendənt] *adj.* voneinander abhängig; interdependent

interest ['ɪntrəst, 'ɪntrɪst] **1.** *n.* **a)** Interesse, *das;* Anliegen, *das;* **take** *or* **have an ~ in sb./sth.** sich für jmdn./etw. interessieren; **[just] for** *or* **out of ~:** [nur] interessehalber; **with ~:** interessiert (*see also* **c**); **lose ~ in sb./sth.** das Interesse an jmdm./etw. verlieren; **~ in life/food** Lust am Leben/Essen; **be of ~:** interessant *od.* von Interesse sein (**to** für); **this is of no ~ to me** das ist belanglos für mich; **act in one's own/sb.'s ~[s]** im eigenen/in jmds. Interesse handeln; **in the ~[s] of humanity** zum Wohle der Menschheit; **b)** *(thing in which one is concerned)* Angelegenheit, *die;* Belange *Pl.*; **c)** *(Finance)* Zinsen *Pl.*; **at ~:** gegen *od.* auf Zinsen; **with ~** *(fig.: with increased force etc.)* überreichlich; doppelt und dreifach *(ugs.)* (*see also* **a**); **d)** *(financial stake)* Beteiligung, *die;* Anteil, *der;* **declare an ~:** seine Interessen darlegen; **e)** *(legal concern)* [Rechts]anspruch, *der.* **2.** *v.t.* interessieren (**in** für); **be ~ed in sb./sth.** sich für jmdn./etw. interessieren; **sb. is ~ed by sb./sth.** jmd./etw. erregt jmds. Interesse; *see also* **interested**

interested ['ɪntrəstɪd, 'ɪntrɪstɪd] *adj.* **a)** *(taking or showing interest)* interessiert; **be ~ in music/football/sb.** sich für Musik/Fußball/jmdn. interessieren; **be ~ in doing sth.** sich dafür interessieren, etw. zu tun; **he is ~ in buying a car** er würde gern ein Auto kaufen; **not ~ in his work** nicht an seiner Arbeit interessiert; **b)** *(not impartial)* voreingenommen

interesting ['ɪntrəstɪŋ, 'ɪntrɪstɪŋ] *adj.* interessant

interestingly ['ɪntrəstɪŋlɪ, 'ɪntrɪstɪŋlɪ] *adv.* interessant; **~ [enough], ...:** interessanterweise ...

interface ['ɪntəfeɪs] *n.* **a)** *(surface)* Grenzfläche, *die;* **b)** *(Computing)* Schnittstelle, *die*

interfere [ɪntə'fɪə(r)] *v.i.* **a)** *(meddle)* sich einmischen (**in** in + *Akk.*); **~ with sth.** sich an etw. *(Dat.)* zu schaffen machen; **b)** *(come into opposition)* in Konflikt geraten (**with** mit); **~ with sth.** etw. beeinträchtigen; **~ with sb.'s plans** jmds. Pläne durchkreuzen; **c)** *(Radio, Telev.)* stören (**with** *Akk.*)

interference [ɪntə'fɪərəns] *n.* **a)** *(interfering)* Einmischung, *die;* **b)** *(Radio, Telev.)* Störung, *die*

interim ['ɪntərɪm] **1.** *n.* **in the ~:** in der Zwischenzeit. **2.** *adj.* **a)** *(intervening)* dazwischenliegend; **b)** *(temporary, provisional)* vorläufig ⟨*Vereinbarung, Bericht, Anordnung, Zustand, Maßnahme*⟩; Zwischen⟨*lösung, -abkommen, -kredit*⟩; Übergangs⟨*regierung, -regelung, -hilfe*⟩

interior [ɪn'tɪərɪə(r)] **1.** *adj.* **a)** inner...; Innen⟨*fläche, -einrichtung, -wand*⟩; **b)** *(inland)* im Landesinneren befindlich; **c)** *(internal, domestic)* Inlands-. **2.** *n.* **a)** *(inland region)* [Landes]innere, *das;* **b)** *(~ part)* Innere, *das;* **c)** *([picture of] inside of building, room, etc.)* Innere, *das; (picture)* Interieur, *das;* **d)** *(Cinemat.)* Innenaufnahme, *die*

interject [ɪntə'dʒekt] *v.t. (interpose)* einwerfen ⟨*Behauptung, Bemerkung, Frage*⟩; **~ remarks** Einwürfe *od.* Zwischenbemerkungen machen

interjection [ɪntəˈdʒekʃn] *n. (exclamation)* Ausruf, *der; (Ling.)* Interjektion, *die*

interlock [ɪntəˈlɒk] **1.** *v. i.* sich ineinanderhaken; ‹*Teile eines Puzzles:*› sich zusammenfügen. **2.** *v. t. (lock together)* zusammenfügen; verflechten ‹*Fasern*›

interloper [ˈɪntələʊpə(r)] *n.* Eindringling, *der*

interlude [ˈɪntəlu:d, ˈɪntəlju:d] *n.* **a)** *(Theatre: break)* Pause, *die;* **b)** *(occurring in break)* Zwischenspiel, *das;* Intermezzo, *das;* **musical ~:** musikalisches Zwischenspiel; **c)** *(intervening time)* kurze Phase *od.* Periode; **d)** *(event interposed)* Intermezzo, *das*

intermediary [ɪntəˈmi:dɪərɪ] *n.* Vermittler, *der*/Vermittlerin, *die*

intermediate [ɪntəˈmi:djət] **1.** *adj.* **a)** Zwischen-; **b)** *(Educ.)* Mittel‹*stufe, -schule*›. **2.** *n.* fortgeschrittener Anfänger

interminable [ɪnˈtɜ:mɪnəbl] *adj. (lit. or fig.)* endlos

intermingle [ɪntəˈmɪŋgl] **1.** *v. i.* sich vermischen; ‹*Personen:*› miteinander in Kontakt treten. **2.** *v. t.* vermischen

intermission [ɪntəˈmɪʃn] *n.* **a)** *(pause)* Unterbrechung, *die;* **b)** *(period of inactivity)* Pause, *die*

intermittent [ɪntəˈmɪtənt] *adj.* in Abständen auftretend ‹*Signal, Fehler, Geräusch*›; **be ~:** in Abständen auftreten; **there was ~ rain all day** es hat den ganzen Tag mit kurzen Unterbrechungen geregnet

intermittently [ɪntəˈmɪtəntlɪ] *adv.* in Abständen

intern [ɪnˈtɜ:n] *v. t.* gefangenhalten; internieren ‹*Kriegsgefangenen usw.*›

internal [ɪnˈtɜ:nl] *adj.* **a)** inner...; Innen‹*winkel, -durchmesser, -fläche, -druck, -gewinde, -abmessungen*›; **b)** *(Physiol.)* inner... ‹*Blutung, Sekretion, Verletzung*›; **c)** *(intrinsic)* inner... ‹*Logik, Stimmigkeit*›; **d)** *(within country)* inner... ‹*Angelegenheiten, Frieden, Probleme*›; Binnen‹*handel, -markt*›; innenpolitisch ‹*Angelegenheiten, Streitigkeiten, Probleme*›; *(within organization)* [betriebs-/partei]intern ‹*Auseinandersetzung, Post, Verfahren[sweise]*›; inner[betrieblich/-kirchlich/-gewerkschaftlich *usw.*] ‹*Streitigkeiten*›; **e)** *(Med.)* innerlich ‹*Anwendung*›; **f)** *(of the mind)* inner... ‹*Monolog, Regung, Widerstände, Groll*›

internal-comˈbustion engine *n.* Verbrennungsmotor, *der*

internally [ɪnˈtɜ:nəlɪ] *adv.* innerlich; *(within organization)* [partei-/betriebs]intern

internal: ~ ˈ**medicine** *n.* innere Medizin; **I~ ˈRevenue Service** *n. (Amer.)* ≈ Finanzamt, *das*

international [ɪntəˈnæʃənl] **1.** *adj.* international; **~ travel** Auslandsreisen *Pl.;* **~ team** *(Sport)* Nationalmannschaft, *die.* **2.** *n.* **a)** *(Sport: contest)* Länderkampf, *der; (in team sports)* Länderspiel, *das;* **b)** *(Sport: participant)* Internationale, *der/die; (in team sports)* Nationalspieler, *der*/-spielerin, *die*

international: ~ **call** *n.* *(Teleph.)* Auslandsgespräch, *das;* ~ **date-line** *see* **date-line;** ~ ˈ**law** *n.* Völkerrecht, *das;* **I~ ˈMonetary Fund** *n.* internationaler Währungsfonds

internment [ɪnˈtɜ:nmənt] *n.* Internierung, *die*

interplay [ˈɪntəpleɪ] *n.* **a)** *(interaction)* Wechselwirkung, *die;* **b)** *(reciprocal action)* Zusammenspiel, *das*

Interpol [ˈɪntəpɒl] *n.* Interpol, *die*

interpose [ɪntəˈpəʊz] **1.** *v. t.* **a)** *(insert)* dazwischenlegen; **~ sth. between sb./sth. and sb./sth.** etw. zwischen jmdn./etw. und jmdn./etw. bringen; **b)** einwerfen ‹*Frage, Bemerkung*›. **2.** *v. i.* [kurz] unterbrechen

interpret [ɪnˈtɜ:prɪt] **1.** *v. t.* **a)** interpretieren; deuten ‹*Traum, Zeichen*›; **b)** *(between languages)* dolmetschen; **c)** *(decipher)* entziffern ‹*Schrift, Inschrift*›. **2.** *v. i.* dolmetschen

interpretation [ɪntɜ:prɪˈteɪʃn] *n.* **a)** Interpretation, *die; (of dream, symptoms)* Deutung, *die;* **b)** *(deciphering)* Entzifferung, *die*

interpreter [ɪnˈtɜ:prɪtə(r)] *n.* **a)** *(between languages)* Dolmetscher, *der*/Dolmetscherin, *die;* **b)** *(of dreams, hieroglyphics)* Deuter, *der;* **c)** *(performer on stage etc.)* Interpret, *der*/Interpretin, *die*

interrelated [ɪntərɪˈleɪtɪd] *adj.* zusammenhängend ‹*Tatsachen, Ereignisse, Themen*›; verwandt ‹*Sprachen, Fachgebiete*›; **be ~:** zusammenhängen/verwandt sein

interrogate [ɪnˈterəgeɪt] *v. t.* vernehmen ‹*Zeugen, Angeklagten*›; verhören ‹*Angeklagten, Verdächtigen, Spion, Gefangenen*›; ausfragen ‹*Freund, Kind usw.*›

interrogation [ɪnterəˈgeɪʃn] *n.* *(interrogating)* Verhör, *das;* **under ~:** beim Verhör; **be under ~:** verhört werden

interrogative [ɪntəˈrɒgətɪv] *adj.* **a)** *(having question form)* Frage-; fragend ‹*Tonfall*›; **b)** *(Ling.)* Interrogativ‹*pronomen, -adverb, -form*›

interrupt [ɪntəˈrʌpt] **1.** *v. t.* unterbrechen; **~ sb.'s sleep** jmds. Schlaf stören; **don't ~ me when I'm busy** stör mich nicht, wenn ich zu tun habe. **2.** *v. i.* stören; unterbrechen

interruption [ɪntəˈrʌpʃn] *n.* *(of work etc.)* Unterbrechung, *die;* Störung, *die; (of peace, sleep)* Störung, *die; (of services)* [zeitweiliger] Ausfall

intersect [ɪntəˈsekt] **1.** *v. t.* **a)** ‹*Kanäle, Schluchten, [Quarz]adern:*› durchziehen ‹*Land, Boden*›; **b)** *(Geom.)* schneiden. **2.** *v. i.* **a)** ‹*Straßen:*› sich kreuzen; **b)** *(Geom.)* sich schneiden

intersection [ɪntəˈsekʃn] *n.* **a)** *(intersecting; road etc. junction)* Kreuzung, *die;* **b)** *(Geom.)* **[point of] ~:** Schnittpunkt, *der*

intersperse [ɪntəˈspɜ:s] *v. t.* **a)** *(scatter)* [hier und da] einfügen; **b) be ~d with** durchsetzt sein mit

intertwine [ɪntəˈtwaɪn] **1.** *v. t.* flechten (**in** in + *Akk.*). **2.** *v. i.* sich [ineinander] verschlingen

interval [ˈɪntəvl] *n.* **a)** *(intervening space)* Zwischenraum, *der; (intervening time)* [Zeit]abstand, *der;* **at ~s** in Abständen; **after an ~ of three years** nach [Ablauf von] drei Jahren; **b)** *(break; also Brit. Theatre etc.)* Pause, *die;* **c)** *(period)* Pause, *die*

intervene [ɪntəˈvi:n] *v. i.* **a)** [vermittelnd] eingreifen (**in** in + *Akk.*); **if nothing ~s** wenn nichts dazwischenkommt; **b) the intervening years** die dazwischenliegenden Jahre

intervention [ɪntəˈvenʃn] *n.* Eingreifen, *das;* Intervention, *die* *(bes. Politik)*

interview [ˈɪntəvju:] **1.** *n.* **a)** *(for job etc.)* Vorstellungsgespräch, *das;* **b)** *(Journ., Radio, Telev.)* Interview, *das.* **2.** *v. t.* Vorstellungsgespräch[e] führen mit ‹*Stellen-, Studienbewerber*›; interviewen ‹*Politiker, Filmstar, Konsumenten usw.*›; vernehmen ‹*Zeugen*›

interviewer [ˈɪntəvju:ə(r)] *n.* *(reporter, pollster, etc.)* Interviewer, *der*/Interviewerin, *die; (for job etc.)* Leiter/Leiterin des Vorstellungsgesprächs

intestate [ɪnˈtestət] *adj.* Intestat‹*erbe, -erbfolge, -nachlaß*›; **die ~:** ohne Hinterlassung eines Testaments sterben

intestinal [ɪnˈtestɪnl] *adj. (Med.)* Darm-; intestinal *(fachspr.)*

intestine [ɪn'testɪn] *n. in sing. or pl.* Darm, *der;* Gedärme *Pl.*
intimacy ['ɪntɪməsɪ] *n.* **a)** *(state)* Vertrautheit, *die; (close personal relationship)* enges [Freundschafts]verhältnis; **b)** *(euphem.: sexual intercourse)* Intimität, *die*
intimate 1. ['ɪntɪmət] *adj.* **a)** *(close, closely acquainted)* eng ‹*Freund, Freundschaft, Beziehung, Verhältnis*›; vertraulich ‹*Ton*›; **be on ~ terms with sb.** zu jmdm. ein enges *od.* vertrautes Verhältnis haben; **b)** *(euphem.: having sexual intercourse)* intim ‹*Beziehungen*›; **be/become ~ with sb.** mit jmdm. intim sein/werden; **c)** *(from close familiarity)* **~ knowledge of sth.** genaue *od.* intime Kenntnis einer Sache; **d)** *(closely personal)* persönlich ‹*Problem*›; privat ‹*Angelegenheit, Gefühl, Dinge*›; geheim ‹*Gedanken*›; *(euphem.)* Intim‹*bereich, -spray*›. **2.** ['ɪntɪmət] *n. (close friend)* Vertraute, *der/die.* **3.** ['ɪntɪmeɪt] *v.t.* **a) ~ sth. [to sb.]** *(make known)* [jmdm.] etw. mitteilen; *(show clearly)* [jmdm.] etw. deutlich machen *od.* zu verstehen geben; **b)** *(imply)* andeuten
intimately ['ɪntɪmətlɪ] *adv.* genau[estens] ‹*kennen*›; bestens ‹*vertraut*›; eng ‹*verbinden*›
intimation [ɪntɪ'meɪʃn] *n. (hint)* Andeutung, *die; (of trouble, anger)* Anzeichen, *das*
intimidate [ɪn'tɪmɪdeɪt] *v.t.* einschüchtern
intimidation [ɪntɪmɪ'deɪʃn] *n.* Einschüchterung, *die*
into [*before vowel* 'ɪntʊ, *before consonant* 'ɪntə] *prep.* in (+ *Akk.*); *(against)* gegen; **I went out ~ the street** ich ging auf die Straße hinaus; **they disappeared ~ the night** sie verschwanden in die Nacht hinein; **4 ~ 20 = 5** 20 durch 4 = 5; **until well ~ this century** bis weit in unser Jahrhundert hinein; **translate sth. ~ English** etw. ins Englische übersetzen
intolerable [ɪn'tɒlərəbl] *adj.* unerträglich; **it's ~:** es ist nicht auszuhalten
intolerance [ɪn'tɒlərəns] *n., no pl.* Intoleranz, *die,* Unduldsamkeit, *die* (**of** gegenüber)
intolerant [ɪn'tɒlərənt] *adj.* intolerant, unduldsam (**of** gegenüber)
intonation [ɪntə'neɪʃn] *n.* Intonation, *die (Sprachw.);* Sprachmelodie, *die*
intoxicate [ɪn'tɒksɪkeɪt] *v.t. (make drunk)* betrunken machen; **be/become ~d** betrunken sein/werden
intoxicating [ɪn'tɒksɪkeɪtɪŋ] *adj.* berauschend ‹*Wirkung, Schönheit*›; mitreißend ‹*Worte, Rhythmus*›; **~ liquors** alkoholische Getränke
intoxication [ɪntɒksɪ'keɪʃn] *n.* Rausch, *der*
intractable [ɪn'træktəbl] *adj.* widerspenstig ‹*Verhalten, Kind, Tier*›; hartnäckig ‹*Krankheit, Schmerzen, Problem*›
intransigence [ɪn'trænsɪdʒəns, ɪn'trænzɪdʒəns] *n., no pl. see* **intransigent:** Kompromißlosigkeit, *die;* Unnachgiebigkeit, *die;* Intransigenz, *die (geh.);* Unerschütterlichkeit, *die*
intransigent [ɪn'trænsɪdʒənt, ɪn'trænzɪdʒənt] *adj.* kompromißlos, unnachgiebig, *(geh.)* intransigent ‹*Haltung, Einstellung*›; unerschütterlich ‹*Wille, Grundsätze, Glaube*›
'in-tray *n.* Ablage für Eingänge
intrepid [ɪn'trepɪd] *adj.* unerschrocken
intricacy ['ɪntrɪkəsɪ] *n.* **a)** *no pl. (quality)* Kompliziertheit, *die;* **b)** *in pl. (things)* Feinheiten *Pl.*
intricate ['ɪntrɪkət] *adj.* verschlungen ‹*Pfad, Windung*›; kompliziert ‹*System, Muster, Werkstück, Aufgabe*›
intrigue [ɪn'tri:g] **1.** *v.t.* faszinieren; **I'm ~d to find out what ...:** ich bin gespannt darauf, zu erfahren, was ... **2.** *v.i.* **~ against sb.** gegen jmdn. intrigieren; **~ with sb.** mit jmdm. Ränke schmieden *od.* Intrigen spinnen. **3.** [ɪn'tri:g, 'ɪntri:g] *n.* Intrige, *die*
intriguing [ɪn'tri:gɪŋ] *adj.,* **intriguingly** [ɪn'tri:gɪŋlɪ] *adv.* faszinierend
intrinsic [ɪn'trɪnsɪk, ɪn'trɪnzɪk] *adj. (inherent)* innewohnend; inner... ‹*Aufbau, Logik*›; *(essential)* wesentlich, *(Philos.)* essentiell ‹*Eigenschaft, Bestandteil, Mangel*›; **~ value** innerer Wert; *(of sth. concrete)* Eigenwert, *der*
introduce [ɪntrə'dju:s] *v.t.* **a)** *(bring in)* [erstmals] einführen ‹*Ware, Tier, Pflanze*› (**into** in + *Akk.;* **from ... into** von ... nach); einleiten ‹*Maßnahmen*›; einschleppen ‹*Krankheit*›; **b)** einführen ‹*Katheter, Schlauch*› (**into** in + *Akk.*); stecken ‹*Schlüssel, Draht, Rohr, Schlauch*› (**into** in + *Akk.*); **c)** *(bring into use)* einführen ‹*Neuerung, Verfahren, Brauch, Nomenklatur*›; aufbringen ‹*Schlagwort*›; **d)** *(make known)* vorstellen; einführen ‹*Vortragenden*›; **~ oneself/sb. [to sb.]** sich/jmdn. [jmdm.] vorstellen; **I ~d them to each other** ich machte sie miteinander bekannt; **I don't think we've been ~d** ich glaube, wir kennen uns noch nicht; **e)** *(usher in, begin, precede)* einleiten ‹*Buch, Thema, Musikstück, Epoche*›; **f)** *(present)* ankündigen ‹*Programm, Darsteller*›; **g)** *(Parl.)* einbringen ‹*Antrag, Entwurf, Gesetz*›; einleiten ‹*Reform*›
introduction [ɪntrə'dʌkʃn] *n.* **a)** *(of methods, measures, process, machinery)* Einführen, *das;* Einführung, *die; (of rules)* Aufstellung, *die;* **b)** *(formal presentation)* Vorstellung, *die; (into society)* Einführung, *die; (of reform)* Einleiten, *das;* **do the ~s** die Anwesenden miteinander bekannt machen; **letter of ~:** Empfehlungsschreiben, *das;* **c)** *(preliminary matter)* Einleitung, *die*
introductory [ɪntrə'dʌktərɪ] *adj.* einleitend; Einführungs‹*kurs, -vortrag*›; Einleitungs‹*kapitel, -rede*›
introspective [ɪntrə'spektɪv] *adj.* in sich *(Akk.)* gerichtet; verinnerlicht; introspektiv *(geh., Psych.)*
introvert 1. ['ɪntrəvɜ:t] *n.* Introvertierte, *der/die;* introvertierter Mensch; **be an ~:** introvertiert sein. **2.** *adj.* introvertiert
introverted [ɪntrə'vɜ:tɪd] *adj.* introvertiert
intrude [ɪn'tru:d] **1.** *v.i.* stören; **~ [up]on sb.'s grief/leisure time/privacy** jmdn. in seiner Trauer stören/jmds. Freizeit beanspruchen/in jmds. Privatsphäre *(Akk.)* eindringen; **~ in[to] sb.'s affairs/conversation** sich in jmds. Angelegenheiten / Unterhaltung *(Akk.)* einmischen. **2.** *v.t.* aufdrängen (**into, [up]on** *Dat.*)
intruder [ɪn'tru:də(r)] *n.* Eindringling, *der; (Mil.)* Intruder, *der*
intrusion [ɪn'tru:ʒn] *n.* **a)** *(intruding)* Störung, *die;* **b)** *(into building, country, etc.)* [gewaltsames] Eindringen; *(Mil.)* Einmarsch, *der* (**into** in + *Akk.*); **c)** *(forcing oneself in)* Einmischung, *die* (**upon** in + *Akk.*)
intrusive [ɪn'tru:sɪv] *adj.* aufdringlich ‹*Person*›
intuition [ɪntju:'ɪʃn] *n.* Intuition, *die;* **have an ~ that ...:** eine Eingebung haben *od.* intuitiv spüren, daß ...
intuitive [ɪn'tju:ɪtɪv] *adj.* intuitiv; gefühlsmäßig ‹*Ablehnung, Beurteilung*›; instinktiv ‹*Annahme, Gefühl*›
inundate ['ɪnəndeɪt] *v.t.* über-

schwemmen; ⟨*Meer*:⟩ überfluten; *(fig.) (with inquiries, letters)* überschwemmen; *(with work, praise)* überhäufen; **~d with tourists** von Touristen überlaufen

inure [ɪ'njʊə(r)] *v. t.* gewöhnen (**to** an + *Akk.*); *(toughen)* abhärten (**to** gegen); **become ~d to sth.** sich an etw. *(Akk.)* gewöhnen

invade [ɪn'veɪd] *v. t.* **a)** einfallen in ⟨*Gebiet, Staat*⟩; **Poland was ~d by the Germans** die Deutschen marschierten in Polen *(Akk.)* ein; **b)** *(swarm into)* ⟨*Touristen, Kinder*:⟩ überschwemmen; **c)** *(fig.)* ⟨*unangenehmes Gefühl, Krankheit, Schwäche*:⟩ befallen; ⟨*Krankheit, Seuche, Unwetter*:⟩ heimsuchen; **d)** *(encroach upon)* stören ⟨*Ruhe, Frieden*⟩; eindringen in (+ *Akk.*) ⟨*Bereich, Privatsphäre*⟩

invader [ɪn'veɪdə(r)] *n.* Angreifer, *der;* Invasor, *der (bes. Milit.)*

¹invalid 1. ['ɪnvəlɪd] *n. (Brit.)* Kranke, *der/die; (disabled person)* Körperbehinderte, *der/die; (from war injuries)* Kriegsinvalide *der/*-invalidin, *die.* **2.** *adj. (Brit.)* körperbehindert. **3.** ['ɪnvəli:d, ɪnvə'li:d] *v. t.* **~ home** *or* **out** als dienstuntauglich entlassen

²invalid [ɪn'vælɪd] *adj.* nicht schlüssig ⟨*Argument, Behauptung, Folgerung, Theorie*⟩; nicht zulässig ⟨*Annahme*⟩; ungerechtfertigt ⟨*Forderung, Vorwurf*⟩; ungültig ⟨*Fahrkarte, Garantie, Vertrag, Testament, Ehe*⟩

invalidate [ɪn'vælɪdeɪt] *v. t.* aufheben; widerlegen ⟨*Theorie, These, Behauptung*⟩

invalid ['ɪnvəlɪd]: **~ carriage** *n.* Kranken[fahr]stuhl, *der;* **~ chair** *n.* Rollstuhl, *der*

invaluable [ɪn'væljʊəbl] *adj.* unbezahlbar; unersetzlich ⟨*Mitarbeiter, Person*⟩; unschätzbar ⟨*Dienst, Verdienst, Hilfe, Bedeutung*⟩; außerordentlich wichtig ⟨*Rolle, Funktion*⟩; außerordentlich wertvoll ⟨*Rat[schlag]*⟩

invariable [ɪn'veərɪəbl] *adj.* **a)** *(fixed)* unveränderlich ⟨*Wert, Einheit*⟩; **b)** *(always the same)* [stets] gleichbleibend ⟨*Druck, Temperatur, Höflichkeit*⟩

invariably [ɪn'veərɪəblɪ] *adv.* immer; ausnahmslos ⟨*falsch, richtig*⟩

invasion [ɪn'veɪʒn] *n.* **a)** *(of troops, virus, locusts)* Invasion, *die; (of weeds etc.)* massenweise Ausbreitung; *(intrusion)* [überfallartiges] Eindringen (**of** in + *Akk.*); **the ~ of Belgium by German troops** der Einmarsch deutscher Truppen in Belgien; **b)** *(encroachment) see* **invade d**: Störung, *die;* Eindringen, *das*

invective [ɪn'vektɪv] *n.* **a)** *(abusive language)* Beschimpfungen *Pl.;* **b)** *(violent attack in words)* Schmähung, *die;* Invektive, *die (geh.)*

inveigh [ɪn'veɪ] *v. i.* **~ against sb./sth.** über jmdn./etw. schimpfen *od.* sich empören

invent [ɪn'vent] *v. t.* erfinden

invention [ɪn'venʃn] *n.* **a)** *(thing invented, inventing)* Erfindung, *die; (concept)* Idee, *die;* **b)** *(inventiveness)* Erfindungsgabe, *die;* **c)** *(fictitious story)* Erfindung, *die;* Lüge, *die*

inventive [ɪn'ventɪv] *adj.* **a)** schöpferisch ⟨*Person, Kraft, Geist, Begabung*⟩; phantasievoll ⟨*Künstler, Kind*⟩; **b)** *(produced with originality)* originell; einfallsreich

inventor [ɪn'ventə(r)] *n.* Erfinder, *der*/Erfinderin, *die*

inventory ['ɪnvəntərɪ] *n.* **a)** *(list)* Bestandsliste, *die;* **make** *or* **take an ~ of sth.** von etw. ein Inventar aufstellen; **b)** *(stock)* Lagerbestand, *der*

inverse [ɪn'vɜ:s, 'ɪnvɜ:s] **1.** *adj.* umgekehrt ⟨*Reihenfolge*⟩. **2.** *n. (opposite)* Gegenteil, *das; (inversion)* Umkehrung, *die*

inversion [ɪn'vɜ:ʃn] *n.* **a)** *(turning upside down)* Umdrehen, *das;* **b)** *(reversal of role, relation)* Umkehrung, *die;* **c)** *(Ling., Meteorol., Mus.)* Inversion, *die*

invert [ɪn'vɜ:t] *v. t.* **a)** *(turn upside down)* umstülpen; **b)** umkehren ⟨*Wortstellung*⟩; vertauschen ⟨*Wörter*⟩

invertebrate [ɪn'vɜ:tɪbrət, ɪn'vɜ:tɪbreɪt] *(Zool.)* **1.** *adj.* wirbellos. **2.** *n.* wirbelloses Tier

inverted: ~ 'commas *n. pl. (Brit.)* Anführungszeichen *Pl.;* Gänsefüßchen *Pl. (ugs.);* **~ 'snob** *n.* Edelproletarier, *der (salopp);* **~ 'snobbery** *n.* Edelproletariertum, *das (salopp)*

invest [ɪn'vest] **1.** *v. t.* **a)** *(Finance)* anlegen (**in** in + *Dat.*); investieren (**in** in + *Dat. od. Akk.*); **~ time and effort in sth.** Zeit und Mühe in etw. *(Akk.)* investieren; **b) ~ sb. with** jmdm. übertragen ⟨*Aufgabe, Amt, Leitung*⟩; jmdm. verleihen ⟨*Orden, Titel*⟩; **c) ~ sth. with sth.** einer Sache *(Dat.)* etw. verleihen. **2.** *v. i.* investieren (**in** in + *Akk.*, **with** bei); **~ in sth.** *(coll.: buy)* sich *(Dat.)* etw. zulegen *(ugs.)*

investigate [ɪn'vestɪgeɪt] **1.** *v. t.* untersuchen; prüfen ⟨*Rechtsfrage, Material, Methode*⟩; ermitteln ⟨*Produktionskosten*⟩. **2.** *v. i.* nachforschen; ⟨*Kripo, Staatsanwaltschaft*:⟩ ermitteln

investigation [ɪnvestɪ'geɪʃn] *n. see* **investigate**: Untersuchung, *die;* Prüfung, *die;* Ermittlung, *die;* **sth. is under ~**: etw. wird überprüft; **sb. is under ~**: gegen jmdn. wird ermittelt

investigative [ɪn'vestɪgətɪv] *adj.* detektivisch; **~ journalism** Enthüllungsjournalismus, *der*

investigator [ɪn'vestɪgeɪtə(r)] *n.* Ermittler, *der*/Ermittlerin, *die;* **[private] ~**: [Privat]detektiv, *der/*-detektivin, *die*

investiture [ɪn'vestɪtʃə(r)] *n.* Investitur, *die*

investment [ɪn'vestmənt] *n.* **a)** *(of money)* Investition, *die (auch fig.);* Anlage, *die; (fig.)* Einsatz, *der;* Aufwand, *der; attrib.* Investitions-; Anlage-; **~ of capital** Kapitalanlage, *die;* **~ trust** Investmenttrust, *der;* Investmentgesellschaft, *die;* **b)** *(money invested)* angelegtes Geld; **c)** *(property)* Kapitalanlage, *die;* **be a good ~** *(fig.)* sich bezahlt machen

investor [ɪn'vestə(r)] *n.* Investor, *der*/Investorin, *die;* [Kapital]anleger, *der/*-anlegerin, *die;* **small ~s** Kleinanleger

inveterate [ɪn'vetərət] *adj.* **a)** *(deep-rooted)* unüberwindbar ⟨*Vorurteil, Mißtrauen*⟩; unversöhnlich ⟨*Haß*⟩; unverbesserlich ⟨*Faulheit usw.*⟩; **b)** *(habitual)* eingefleischt ⟨*Trinker, Raucher*⟩; unverbesserlich ⟨*Lügner*⟩

invidious [ɪn'vɪdɪəs] *adj.* undankbar ⟨*Aufgabe*⟩; unpassend, unfair ⟨*Vergleich, Bemerkung*⟩

invigorate [ɪn'vɪgəreɪt] *v. t. (make vigorous)* stärken; *(physically)* kräftigen

invincible [ɪn'vɪnsɪbl] *adj.* unbesiegbar; unerschütterlich ⟨*Entschlossenheit, Mut*⟩

inviolable [ɪn'vaɪələbl] *adj.* unantastbar

inviolate [ɪn'vaɪələt] *adj.* unversehrt; ungestört ⟨*Friede, Ruhe*⟩; nicht verletzt ⟨*Abkommen*⟩

invisibility [ɪnvɪzɪ'bɪlɪtɪ] *n.* Unsichtbarkeit, *die*

invisible [ɪn'vɪzɪbl] *adj. (also Econ.)* unsichtbar; *(hidden because of fog etc.; too small)* nicht sichtbar; **~ mending** Kunststopfen, *das*

invitation [ɪnvɪ'teɪʃn] *n. (lit. or fig.)* Einladung, *die;* **at sb.'s ~**: auf jmds. Einladung; **an [open] ~**

to thieves eine Aufforderung zum Diebstahl

invite [ɪn'vaɪt] *v. t.* **a)** *(request to come)* einladen; **before an ~d audience** vor geladenen Gästen; **b)** *(request to do sth.)* auffordern; **she ~d him to accompany her** sie forderte ihn auf *od.* lud ihn ein, sie zu begleiten; **c)** *(bring on)* herausfordern ‹*Kritik, Verhängnis*›; **you're inviting ridicule** du machst dich lächerlich *od.* zum Gespött

inviting [ɪn'vaɪtɪŋ] *adj.* einladend; verlockend ‹*Gedanke, Vorstellung, Aussicht*›; freundlich ‹*Klima*›; ansprechend ‹*Anblick*›

invoice ['ɪnvɔɪs] **1.** *n.* *(bill)* Rechnung, *die; (list)* Lieferschein, *der.* **2.** *v. t.* **a)** *(make ~ for)* eine Rechnung ausstellen für; **b)** *(send ~ to)* **~ sb.** jmdm. eine Rechnung schicken; **~ sb. for sth.** jmdm. etw. in Rechnung stellen

invoke [ɪn'vəʊk] *v. t.* **a)** *(call on)* anrufen; **b)** *(appeal to)* sich berufen auf (+ *Akk.*); **~ an example/ sth. as an example** ein Beispiel/ etw. als Beispiel anführen

involuntarily [ɪn'vɒləntərɪlɪ] *adv.*, **involuntary** [ɪn'vɒləntərɪ] *adj.* unwillkürlich

involve [ɪn'vɒlv] *v. t.* **a)** *(implicate)* verwickeln; **b)** *(draw in as a participant)* **~ sb. in a game/fight** jmdn. an einem Spiel beteiligen/ in eine Schlägerei [mit] hineinziehen; **become** *or* **get ~d in a fight** in eine Schlägerei verwickelt werden; **be ~d in a project** *(employed)* an einem Projekt mitarbeiten; **get ~d with sb.** sich mit jmdm. einlassen; *(sexually, emotionally)* eine Beziehung mit jmdm. anfangen; **c)** *(include)* enthalten; *(contain implicitly)* beinhalten; **d)** *(be necessarily accompanied by)* mit sich bringen; *(require as accompaniment)* erfordern; *(cause, mean)* bedeuten

involved [ɪn'vɒlvd] *adj.* verwickelt; *(complicated)* kompliziert; *(complex)* komplex

involvement [ɪn'vɒlvmənt] *n.* **a)** **his ~ in the company** seine Beteiligung an der Firma; **I don't know the extent of his ~ in this affair** ich weiß nicht, inwieweit er mit dieser Sache zu tun hat; **b)** *(implication)* **~ in a conflict** Einmischung in einen Konflikt; **have an ~ with sb.** *(sexually)* eine Affäre mit jmdm. haben

inward ['ɪnwəd] **1.** *adj.* **a)** *(situated within)* inner...; **b)** *(mental, spiritual)* inner... ‹*Impuls, Regung, Friede, Kampf*›; innerlich *(geh.)* ‹*Leben*›; **c)** *(directed inside)* nach innen gehend; nach innen gerichtet. **2.** *adv.* einwärts ‹*gerichtet, gebogen*›; **open ~:** nach innen öffnen; **an ~-looking person** *(fig.)* ein in sich *(Akk.)* gekehrter Mensch

inwards ['ɪnwədz] *see* **inward 2**

iodine ['aɪədi:n, 'aɪədɪn] *n.* Jod, *das*

ion ['aɪən] *n.* *(Phys., Chem.)* Ion, *das*

iota [aɪ'əʊtə] *n.* Jota, *das (geh.);* **not an** *or* **one ~:** nicht ein Jota *(geh.);* kein Jota *(geh.);* **there's not an ~ of truth in that** daran ist nicht ein Fünkchen Wahrheit

IOU [aɪəʊ'ju:] *n.* Schuldschein, *der*

IQ *abbr.* **intelligence quotient** IQ, *der;* **IQ-test** IQ-Test, *der*

IRA *abbr.* **Irish Republican Army** IRA, *die*

Iran [ɪ'rɑ:n] *pr. n.* Iran, *der*

Iranian [ɪ'reɪnɪən] **1.** *adj.* iranisch; **sb. is ~:** jmd. ist Iraner/Iranerin. **2.** *n.* **a)** *(person)* Iraner, *der*/Iranerin, *die;* **b)** *(Ling.)* Iranisch, *das*

Iraq [ɪ'rɑ:k] *pr. n.* Irak, *der*

Iraqi [ɪ'rɑ:kɪ] **1.** *adj.* irakisch; **sb is ~:** jmd. ist Iraker/Irakerin. **2.** *n.* **a)** *(person)* Iraker, *der*/Irakerin, *die;* **b)** *(Ling.)* Irakisch, *das*

irascible [ɪ'ræsɪbl] *adj.* *(hot-tempered)* aufbrausend; *(irritable)* reizbar

irate [aɪ'reɪt] *adj.* wütend ‹*Person, Menge*›; erbost *(geh.)* ‹*Person*›

Ireland ['aɪələnd] *pr. n.* **[Republic of] ~:** Irland *(das)*

iris ['aɪərɪs] *n.* **a)** *(Anat.)* Iris, *die;* Regenbogenhaut, *die;* **b)** *(Bot.)* Iris, *die;* Schwertlilie, *die*

Irish ['aɪrɪʃ] **1.** *adj.* irisch; **sb. is ~:** jmd. ist Ire/Irin; **~ joke** Irenwitz, *der.* **2.** *n.* **a)** *(language)* Irisch, *das; see also* **English 2 a; b)** *constr. as pl.* **the ~:** die Iren

Irish: ~man ['aɪərɪʃmən] *n., pl.* **~men** ['aɪərɪʃmən] Ire, *der;* **~ Re'public** *pr. n.* Irische Republik; **~ 'Sea** *pr. n.* Irische See; **~woman** *n.* Irin, *die*

irk [ɜ:k] *v. t.* ärgern

irksome ['ɜ:ksəm] *adj.* lästig

iron ['aɪən] **1.** *n.* **a)** *(metal)* Eisen, *das;* **as hard as ~:** eisenhart; **strike while the ~ is hot** *(prov.)* das Eisen schmieden, solange es heiß ist *(Spr.);* **b)** *(tool)* Eisen, *das;* **have several ~s in the fire** mehrere Eisen im Feuer haben *(ugs.);* **c)** *(for smoothing)* Bügeleisen, *das.* **2.** *attrib. adj.* **a)** *(of iron)* eisern; Eisen‹*platte usw.*›; **b)** *(very robust)* eisern ‹*Konstitution*›; **c)** *(unyielding)* eisern; ehern *(geh.)* ‹*Stoizismus*›. **3.** *v. t.* bügeln

~ 'out *v. t.* herausbügeln ‹*Falten*›; *(flatten)* glätten ‹*Papier*›; *(fig.)* beseitigen ‹*Kurve, Unregelmäßigkeit*›; aus dem Weg räumen ‹*Schwierigkeit, Problem*›

Iron 'Curtain *n.* *(fig.)* Eiserner Vorhang

ironic [aɪ'rɒnɪk], **ironical** [aɪ'rɒnɪkl] *adj.* ironisch; **it is ~ that ...:** es ist paradox, daß ...

ironing ['aɪənɪŋ] *n.* Bügeln, *das; (things [to be] ironed)* Bügelwäsche, *die;* **do the ~:** bügeln

'ironing-board *n.* Bügelbrett, *das*

ironmonger ['aɪənmʌŋgə(r)] *n.* *(Brit.)* Eisenwarenhändler, *der*/ -händlerin, *die; see also* **baker**

iron 'ore *n.* Eisenerz, *das*

irony ['aɪrənɪ] *n.* Ironie, *die;* **the ~ was that ...:** die Ironie lag darin, daß ...; das Ironische war, daß ...

irradiate [ɪ'reɪdɪeɪt] *v. t.* *(Phys., Med., Gastr.)* bestrahlen

irrational [ɪ'ræʃənl] *adj.* *(unreasonable)* irrational *(geh.);* vernunftwidrig

irreconcilable [ɪ'rekənsaɪləbl] *adj.* **a)** *(implacably hostile)* unversöhnlich; **b)** *(incompatible)* unvereinbar; unversöhnlich ‹*Gegensätze*›

irrecoverable [ɪrɪ'kʌvərəbl] *adj.* unwiederbringlich verloren; endgültig ‹*Verlust*›

irrefutable [ɪ'refjʊtəbl, ɪrɪ'fju:təbl] *adj.* unwiderlegbar

irregular [ɪ'regjʊlə(r)] *adj.* **a)** unkorrekt ‹*Verhalten, Handlung usw.*›; **b)** *(in duration, order, etc.)* unregelmäßig; **c)** *(abnormal)* sonderbar; eigenartig; **d)** *(not symmetrical)* unregelmäßig; uneben ‹*Oberfläche, Gelände*›; **e)** *(Ling.)* unregelmäßig

irregularity [ɪregjʊ'lærɪtɪ] *n.* **a)** *(of behaviour, action)* Unkorrektheit, *die; (instance also)* Unregelmäßigkeit, *die;* **b)** *(in duration, order, etc.)* Unregelmäßigkeit, *die;* **c)** *(abnormality)* Sonderbarkeit, *die;* Eigenartigkeit, *die;* **d)** *(lack of symmetry)* Unregelmäßigkeit, *die; (of surface)* Unebenheit, *die*

irrelevant [ɪ'relɪvənt] *adj.* belanglos; irrelevant *(geh.);* **be ~ to a subject** für ein Thema ohne Belang *od. (geh.)* irrelevant sein

irreparable [ɪ'repərəbl] *adj.* nicht wiedergutzumachend *nicht präd.;* irreparabel *(geh., Med.)*

irreplaceable [ɪrɪ'pleɪsəbl] *adj.* **a)** *(not replaceable)* nicht ersetzbar; **b)** *(of which the loss cannot be made good)* unersetzlich

irrepressible [ɪrɪ'presɪbl] *adj.* nicht zu unterdrückend *nicht* *präd.;* unbezähmbar ⟨*Neugier, Verlangen*⟩; unerschütterlich ⟨*Optimismus*⟩; unbändig ⟨*Freude, Entzücken*⟩; sonnig ⟨*Gemüt*⟩; **he/she is ~:** er/sie ist nicht unterzukriegen *(ugs.)*

irreproachable [ɪrɪ'prəʊtʃəbl] *adj.* untadelig ⟨*Charakter, Lebenswandel, Benehmen*⟩; unanfechtbar ⟨*Ehrlichkeit*⟩; tadellos ⟨*Kleidung, Manieren*⟩

irresistible [ɪrɪ'zɪstɪbl] *adj.* unwiderstehlich; bestechend ⟨*Argument*⟩

irresolute [ɪ'rezəlu:t, ɪ'rezəlju:t] *adj.* unentschlossen

irrespective [ɪrɪ'spektɪv] *adj.* **~ of** ungeachtet (+ *Gen.*); *(independent of)* unabhängig von

irresponsible [ɪrɪ'spɒnsɪbl] *adj.* verantwortungslos ⟨*Person*⟩; unverantwortlich ⟨*Benehmen*⟩

irresponsibly [ɪrɪ'spɒnsɪblɪ] *adv.* verantwortungslos; unverantwortlich

irretrievable [ɪrɪ'tri:vəbl] *adj.* nicht mehr wiederzubekommen *nicht attr.; (irreversible)* endgültig ⟨*Ruin, Verfall, Verlust*⟩; unheilbar ⟨*Zerrüttung einer Ehe*⟩; ausweglos ⟨*Situation*⟩

irreverent [ɪ'revərənt] *adj.* respektlos; *(towards religious values or the dead)* pietätlos *(geh.)*

irreversible [ɪrɪ'vɜ:sɪbl] *adj.* **a)** *(unalterable)* unabänderlich, unumstößlich ⟨*Entscheidung, Entschluß, Tatsache*⟩; unwiderruflich ⟨*Entschluß, Entscheidung, Anordnung, Befehl usw.*⟩; **b)** *(not reversible)* irreversibel *(geh.)*⟨*Vorgang*⟩; *(inexorable)* unaufhaltsam ⟨*Entwicklung, Verfall*⟩

irrevocable [ɪ'revəkəbl] *adj.* unwiderruflich

irrigate ['ɪrɪgeɪt] *v.t.* bewässern

irrigation [ɪrɪ'geɪʃn] *n.* Bewässerung, *die*

irritability [ɪrɪtə'bɪlɪtɪ] *n. see* **irritable:** Reizbarkeit, *die;* Gereiztheit, *die*

irritable ['ɪrɪtəbl] *adj. (quick to anger)* reizbar; *(temporarily)* gereizt

irritant ['ɪrɪtənt] *n.* Reizstoff, *der*

irritate ['ɪrɪteɪt] *v.t.* **a)** ärgern; **get ~d** ärgerlich werden; **be ~d by** *or* **feel ~d at sth.** sich über etw. *(Akk.)* ärgern; **be ~d with sb.** sich über jmdn. aufregen *od.* ärgern; **b)** *(Med.)* reizen

irritating ['ɪrɪteɪtɪŋ] *adj.* lästig; **I find him ~:** er geht mir auf die Nerven *(ugs.)*

irritation [ɪrɪ'teɪʃn] *n.* **a)** Ärger, *der;* **[source** *or* **cause of] ~:** Ärgernis, *das;* **b)** *(Med.)* Reizung, *die*

is *see* **be**

Islam ['ɪzlɑ:m, 'ɪzlæm, ɪz'lɑ:m] *n.* Islam, *der*

Islamic [ɪz'læmɪk] *adj.* islamisch

island ['aɪlənd] *n. (lit. or fig.)* Insel, *die; see also* **traffic island**

islander ['aɪləndə(r)] *n.* Inselbewohner, *der*/-bewohnerin, *die*

isle [aɪl] *n.* Insel, *die;* Eiland, *das (dichter.); see also* **British Isles**

Isle of Man [aɪl əv 'mæn] *pr. n.* Insel Man, *die*

isn't ['ɪznt] *(coll.)* = **is not;** *see* **be**

isolate ['aɪsəleɪt] *v.t.* isolieren; *(Electr.)* vom Stromkreis trennen

isolated ['aɪsəleɪtɪd] *adj.* **a)** *(single)* einzeln; *(occasional)* vereinzelt; **~ instances/cases** Einzelfälle; **b)** *(solitary)* einsam; *(remote)* abgelegen **(from** von); *(cut off)* abgeschnitten **(from** von)

isolation [aɪsə'leɪʃn] *n.* **a)** *(act)* Isolierung, *die;* Absonderung, *die;* **b)** *(state)* Isoliertheit, *die;* Isolation, *die;* Abgeschnittenheit, *die; (remoteness)* Abgeschiedenheit, *die;* **examine/look at/treat sth. in ~:** etw. isoliert *od.* gesondert betrachten; **~ hospital** Infektionskrankenhaus, *das;* **~ ward** Isolierstation, *die*

isosceles [aɪ'sɒsəli:z] *adj.* *(Geom.)* gleichschenklig

isotope ['aɪsətəʊp] *n.* Isotop, *das*

Israel ['ɪzreɪl] *pr. n.* Israel *(das)*

Israeli [ɪz'reɪlɪ] **1.** *adj.* israelisch. **2.** *n.* Israeli, *der/die*

issue ['ɪʃu:, 'ɪsju:] **1.** *n.* **a)** *(point in question)* Frage, *die;* **contemporary ~s** aktuelle Fragen *od.* Themen; **make an ~ of sth.** etw. aufbauschen; **become an ~:** zum Problem werden; **what is at ~ here?** worum geht es [hier] eigentlich?; **evade** *or* **dodge the ~:** ausweichen; **the point at ~:** der strittige Punkt; worum es geht; **take ~ with sb. over sth.** sich mit jmdm. auf eine Diskussion über etw. *(Akk.)* einlassen; **b)** *(giving out)* Ausgabe, *die; (of document)* Ausstellung, *die; (of shares)* Emission, *die;* **date of ~:** Ausgabedatum, *das; (of document)* Ausstellungsdatum, *das; (of stamps)* Ausgabetag, *der;* **c)** *(of magazine, journal, etc.)* Ausgabe, *die;* **d)** *(total number of copies)* Auflage, *die;* **e)** *(quantity of coins)* Emissionszahl, *die; (quantity of stamps)* Auflage, *die;* **f)** *(result, outcome)* Ergebnis, *das;* Ausgang, *der;* **decide the ~:** den Ausschlag geben; **force the ~:** eine Entscheidung erzwingen. **2.** *v.t.* **a)** *(give out)* ausgeben; ausstellen ⟨*Paß, Visum, Lizenz, Zeugnis, Haft-, Durchsuchungsbefehl*⟩; erteilen ⟨*Lizenz, Befehl*⟩; **~ sb. with sth.** etw. an jmdn. austeilen; **b)** *(publish)* herausgeben ⟨*Publikation*⟩; herausbringen ⟨*Publikation, Münze, Briefmarke*⟩; emittieren ⟨*Wertpapiere*⟩; geben ⟨*Warnung*⟩; **c)** *(supply)* ausgeben **(to** an + *Akk.*); **~ sb. with sth.** jmdn. mit etw. ausstatten; **be ~d with sth.** etw. erhalten. **3.** *v.i.* ⟨*Personen:*⟩ herausströmen **(from** aus); ⟨*Gas, Flüssigkeit:*⟩ austreten **(from** aus); ⟨*Rauch:*⟩ heraus-, hervorquellen **(from** aus); ⟨*Ton, Geräusch:*⟩ hervor-, herausdringen **(from** aus)

IT *abbr.* **information technology**

it [ɪt] *pron.* **a)** *(the thing, animal, young child previously mentioned)* er/sie/es; *as direct obj.* ihn/sie/es; *as indirect obj.* ihm/ihr/ihm; **behind/under it** dahinter/darunter; **b)** *(the person in question)* **who is it?** wer ist da?; **it was the children** es waren die Kinder; **is it you, dad?** bist du es, Vater?; **c)** *subj. of impers. v.* es; **it is snowing/warm** es schneit/ist warm; **it is winter/midnight/ten o'clock** es ist Winter/Mitternacht/zehn Uhr; **d)** *anticipating subj. or obj.* es; **it is typical of her to do that** es ist typisch für sie, so etwas zu tun; **it is absurd talking** *or* **to talk like that** es ist absurd, so zu reden; **it was for our sake that he did it** um unseretwillen hat er es getan; **e)** *as indef. obj.* es; **I can't cope with it any more** ich halte das nicht mehr länger aus; **have a hard time of it** eine schwere Zeit haben; **what is it?** was ist los?; was ist denn?; **f)** *(exactly what is needed)* **That's it! That's exactly what I've been looking for** Das ist es! Genau das habe ich gesucht; **he thinks he's really 'it** er denkt, er ist der Größte *(ugs.);* **g) that's 'it** *(coll.) (that's the problem)* das ist es [eben]; *(that's the end)* jetzt ist Schluß; *(my patience is at an end)* jetzt reicht's [mir]; *(that's true)* genau *(ugs.);* **this is 'it** *(coll.) (the time for action)* es ist soweit; *(the real problem)* das ist es [eben]. *See also* **its; itself**

Italian [ɪ'tæljən] **1.** *adj.* italienisch; **sb. is ~:** jmd. ist Italiener/Italienerin. **2.** *n.* **a)** *(person)* Italiener, *der*/Italienerin, *die;* **b)** *(language)* Italienisch, *das; see also* **English 2 a**

italic [ɪ'tælɪk] **1.** *adj.* kursiv. **2.** *n. in pl.* Kursivschrift, *die;* **in ~s** kursiv

Italy ['ɪtəlɪ] *pr. n.* Italien *(das)*
itch [ɪtʃ] **1.** *n.* **a)** Juckreiz, *der;* Jucken, *das;* **I have an ~:** es juckt mich; **b)** *(restless desire)* Drang, *der;* **I have an ~ to do it** es juckt *(ugs.) od.* reizt mich, es zu tun. **2.** *v. i.* **a)** einen Juckreiz haben; **I'm ~ing** es juckt mich; **it ~es** es juckt; **my back ~es** mein Rücken juckt; es juckt mich am Rücken; **b)** *(feel a desire)* ~ *or* **be ~ing to do sth.** darauf brennen, etw. zu tun; **~ for sth.** sich nach etw. sehnen; **he is ~ing for a fight** er ist nur darauf aus, sich zu prügeln
itchy ['ɪtʃɪ] *adj.* kratzig ‹*Socken, Laken*›; **be ~** ‹*Körperteil:*› jucken; **I've got ~ feet** *(fig. coll.)* mich hält es hier nicht länger; *(by temperament)* mich hält es nirgends lange
it'd ['ɪtəd] *(coll.)* **a)** = **it had; b)** = **it would**
item ['aɪtəm] *n.* **a)** Ding, *das;* Sache, *die; (in shop, catalogue)* Artikel, *der; (in variety show, radio, TV)* Nummer, *die;* **b)** ~ **[of news]** Nachricht, *die;* **c)** *(in account or bill)* Posten, *der; (in list, programme, agenda)* Punkt, *der*
itemize (itemise) ['aɪtəmaɪz] *v. t.* einzeln aufführen; spezifizieren ‹*Rechnung*›
itinerant [ɪ'tɪnərənt, aɪ'tɪnərənt] **1.** *adj.* reisend; umherziehend; Wander‹*prediger, -arbeiter*›. **2.** *n.* Landfahrer, *der/*-fahrerin, *die*
itinerary [aɪ'tɪnərərɪ, ɪ'tɪnərərɪ] *n.* [Reise]route, *die;* [Reise]weg, *der*
it'll [ɪtl] *(coll.)* = **it will**
its [ɪts] *poss. pron. attrib.* sein/ihr/ sein; *see also* **²her**
it's [ɪts] **a)** = **it is; b)** = **it has**
itself [ɪt'self] *pron.* **a)** *emphat.* selbst; **by ~** *(automatically)* von selbst; *(alone)* allein; *(taken in isolation)* für sich; **in ~:** für sich genommen; **he is generosity ~:** er ist die Großzügigkeit in Person; **b)** *refl.* sich; **the machine switches ~ off** die Maschine schaltet sich [von] selbst aus
ITV *abbr. (Brit.)* **Independent Television** *kommerzielles britisches Fernsehprogramm*
I've [aɪv] = **I have**
ivory ['aɪvərɪ] *n.* **a)** *(substance)* Elfenbein, *das; attrib.* elfenbeinern; Elfenbein-; **b)** *(colour)* Elfenbein, *das; attrib.* elfenbeinfarbig
ivory: I~ 'Coast *pr. n.* Elfenbeinküste, *die;* **~ 'tower** *n.* Elfenbeinturm, *der*
ivy ['aɪvɪ] *n.* Efeu, *der*
'Ivy League *n. (Amer.) Eliteuniversitäten im Osten der USA*

J

J, j [dʒeɪ] *n., pl.* **Js** *or* **J's** J, j, *das*
jab [dʒæb] **1.** *v. t.,* **-bb-: a)** *(poke roughly, thrust abruptly)* stoßen; **b)** *(stab)* stechen. **2.** *v. i.,* **-bb-: ~ at sb. [with sth.]** auf jmdn. [mit etw.] einhauen; *(stab at)* auf jmdn. [mit etw.] einstechen. **3.** *n.* **a)** *(abrupt blow)* Schlag, *der; (with stick, elbow)* Stoß, *der; (with needle)* Stich, *der; (Boxing)* Jab, *der;* **b)** *(Brit. coll.: hypodermic injection)* Spritze, *die;* **give sb./oneself a ~:** jmdm./sich eine Spritze verpassen *(ugs.)*
jabber ['dʒæbə(r)] **1.** *v. i.* plappern *(ugs.).* **2.** *v. t.* brabbeln *(ugs.)*
jack [dʒæk] **1.** *n.* **a)** *(Cards)* Bube, *der;* **b)** *(for lifting vehicle wheel)* Wagenheber, *der;* **c)** *(Bowls)* Malkugel, *die.* **2.** *v. t.* **a)** **~ in** *or* **up** *(Brit. sl.: abandon)* [auf]stecken *(ugs.);* **b)** **~ up** *(lift)* aufbocken ‹*Fahrzeug*›; *(fig. sl.: increase)* was draufsatteln auf (+ *Akk.*) *(ugs.)*
jackass ['dʒækæs] *n.* **a)** *(male ass)* Eselhengst, *der;* **b)** *(stupid person)* Esel, *der (ugs.)*
jackdaw ['dʒækdɔ:] *n. (Ornith.)* Dohle, *die*
jacket ['dʒækɪt] *n.* **a)** Jacke, *die; (of suit)* Jackett, *das;* **sports ~:** Sakko, *der;* **~ pocket** Jackentasche, *die*/Jackettasche, *die;* **b)** *(round a boiler etc.)* Mantel, *der;* **c)** *(of book)* Schutzumschlag, *der;* **d)** *(of a potato)* Schale, *die;* **~ potatoes** in der Schale gebackene Kartoffeln; **e)** *(Amer.) see* **sleeve b**
jack: J~ 'Frost *n.* Väterchen Frost *(scherzh.);* **~-in-the-box** *n.* Schachtelteufel, *der;* Kastenteufel, *der;* **~-knife** *v. i.* **the lorry ~-knifed** der Anhänger des Lastwagens stellte sich quer; **~ of 'all trades** *n.* Hansdampf [in allen Gassen]; **~pot** *n.* Jackpot, *der;* **hit the ~pot** *(fig.)* das große Los ziehen; **J~ Robinson** [dʒæk 'rɒbɪnsn] *n.* **before you could say J~ Robinson** im Nu *(ugs.)*

jacuzzi, *(Amer.:* **P)** [dʒə'ku:zɪ] *n.* ≈ Whirlpool, *der*
¹jade [dʒeɪd] *v. t., esp. in p. p. (tire)* ermüden; abstumpfen ‹*Geschmacksnerven*›; **look ~d** abgespannt *od.* erschöpft aussehen
²jade *n.* Jade, *der od. die; (carvings)* Jadearbeiten
jagged ['dʒægɪd] *adj.* gezackt; ausgefranst ‹*Loch/Riß in Kleidungsstücken*›; zerklüftet ‹*Küste*›
jaguar ['dʒægjʊə(r)] *n. (Zool.)* Jaguar, *der*
jail [dʒeɪl] **1.** *n. (place)* Gefängnis, *das; (confinement)* Haft, *die;* **in ~:** im Gefängnis; **be sent to ~:** ins Gefängnis kommen; **go to ~:** ins Gefängnis gehen. **2.** *v. t.* ins Gefängnis bringen
jail: ~bird *n.* Knastbruder, *der (ugs.);* **~break** *n.* Gefängnisausbruch, *der*
jailer, jailor ['dʒeɪlə(r)] *n.* Gefängniswärter, *der/*-wärterin, *die*
jalopy [dʒə'lɒpɪ] *n. (coll.)* Klapperkiste, *die (ugs.)*
¹jam [dʒæm] **1.** *v. t.,* **-mm-: a)** *(squeeze and fix between two surfaces)* einklemmen; **~ sth. into sth.** etw. in etw. *(Akk.)* zwängen; **b)** *(make immovable)* blockieren; *(fig.)* lähmen; lahmlegen; **c)** *(squeeze together in compact mass)* stopfen **(into** in + *Akk.*); **~ together** zusammenpferchen ‹*Personen*›; **d)** *(thrust into confined space)* stopfen **(into** in + *Akk.*); stecken ‹*Schlüssel, Münze*› **(into** in + *Akk.*); **e)** *(block by crowding)* blockieren; versperren, blockieren ‹*Eingang*›; verstopfen, blockieren ‹*Rohr*›; **the switchboard was ~med with calls** sämtliche Leitungen waren durch Anrufe blockiert; **f)** *(Radio)* stören. **2.** *v. i.,* **-mm-: a)** *(become tightly wedged)* sich verklemmen; **b)** *(become unworkable)* ‹*Maschine:*› klemmen. **3.** *n.* **a)** *(crush, stoppage)* Blockierung, *die;* Klemmen, *das;* **b)** *(crowded mass)* Stau, *der;* **c)** *(coll.: dilemma)* Klemme, *die (ugs.);* **be in a ~:** in der Klemme stecken *(ugs.)*
~ 'in *v. t.* hineinzwängen; **we were ~med in** wir waren eingepfercht
~ 'on *v. t.* **~ the brakes [full] on** [voll] auf die Bremse steigen *(ugs.);* eine Vollbremsung machen
~ 'up *v. t.* verstopfen ‹*Straße usw.*›; lahmlegen ‹*System*›; verklemmen ‹*Mechanismus*›
²jam *n.* Marmelade, *die;* Konfitüre, *die (bes. Kaufmannsspr.);* **make ~:** Marmelade einmachen;

sb. wants ~ on it *(fig. coll.)* jmdm. genügt etw. noch nicht

Jamaica [dʒə'meɪkə] *pr. n.* Jamaika *(das)*

jamb [dʒæm] *n. (of doorway, window)* Pfosten, *der*

'**jam-jar** *n.* Marmeladenglas, *das*

jammy ['dʒæmɪ] *adj. (Brit. coll.: lucky)* **~ beggar** Glückspilz, *der (ugs.)*; **that was ~:** das war Schwein *(ugs.)*

jam: ~-packed *adj. (coll.)* knallvoll *(ugs.)*, proppenvoll *(ugs.)* **(with** von); **~ session** *n. (Jazz coll.)* Jam session, *die*; **~ 'tart** *n.* Marmeladentörtchen, *das*

Jan. *abbr.* **January** Jan.

jangle ['dʒæŋgl] **1.** *v.i.* klimpern; ‹*Klingel:*› bimmeln. **2.** *v.t.* rasseln mit; klimpern mit ‹*[Klein]geld*›. **3.** *n.* Geklapper, *das; (of bell)* Schrillen, *das*

janitor ['dʒænɪtə(r)] *n.* **a)** *(doorkeeper)* Portier, *der*; **b)** *(caretaker)* Hausmeister, *der*

January ['dʒænjʊərɪ] *n.* Januar, *der; see also* **August**

Japan [dʒə'pæn] *n.* Japan *(das)*

Japanese [dʒæpə'ni:z] **1.** *adj.* japanisch; **sb. is ~:** jmd. ist Japaner/Japanerin. **2.** *n., pl. same* **a)** *(person)* Japaner, *der*/ Japanerin, *die*; **b)** *(language)* Japanisch, *das*

jape [dʒeɪp] *n.* Scherz, *der*; Spaß, *der*; *(practical joke)* Streich, *der*

[1]**jar** [dʒɑ:(r)] **1.** *n.* **a)** *(harsh or grating sound)* Quietschen, *das*; **b)** *(jolt)* Stoß, *der*; *(thrill of nerves, shock)* Schlag, *der*. **2.** *v.i.*, **-rr-: a)** *(sound discordantly)* quietschen; *(rattle)* ‹*Fenster:*› scheppern *(ugs.)*; **~ on** *or* **against sth.** über etw. *(Akk.)* knirschen; **b)** *(have discordant or painful effect)* **~ [up]on sb./sb.'s nerves** jmdm. auf die Nerven gehen; **~ on the ears** durch Mark und Bein gehen *(ugs. scherzh.)*; **a ~ring sound** ein Geräusch, das einem durch und durch geht. **3.** *v.t.*, **-rr-: a)** *(cause to vibrate)* erschüttern; **b)** *(send shock through)* **~ sb.'s nerves** jmdm. auf die Nerven gehen; **~ one's elbow** sich *(Dat.)* den Ellbogen anschlagen

[2]**jar** *n. (vessel)* Topf, *der*; *(of glass)* Glas, *das*; **~ of jam** *etc.* Topf/ Glas Marmelade *usw.*

jargon ['dʒɑ:gən] *n.* Jargon, *der*

jasmin[e] ['dʒæsmɪn, 'dʒæzmɪn] *n.* Jasmin, *der*

jaundice ['dʒɔ:ndɪs] **1.** *n. (Med.)* Gelbsucht, *die*. **2.** *v.t. usu. in p.p. (fig.: affect with bitterness)* verbittern; **~d** verbittert; *(cynical)* zynisch; **with [a] ~d eye** *(enviously)* neidvoll; mit Neid; **have a very ~d view of life** dem Leben voller Verbitterung gegenüberstehen

jaunt [dʒɔ:nt] *n.* Ausflug, *der*; **be off on/go for a ~:** einen Ausflug machen

jaunty ['dʒɔ:ntɪ] *adj.* unbeschwert; keck ‹*Hut*›

Java ['dʒɑ:və] *pr. n.* Java *(das)*

javelin ['dʒævəlɪn, 'dʒævlɪn] *n.* **a)** Speer, *der*; **b)** *(Sport: event)* Speerwerfen, *das*

jaw [dʒɔ:] *n.* **a)** *(Anat.)* Kiefer, *der*; **his ~ dropped** er ließ die Kinnlade herunterfallen; **upper/ lower ~:** Ober-/Unterkiefer, *der*; **b)** *(of machine)* [Klemm]backe, *die*; **c)** *in pl. (large dangerous mouth)* Rachen, *der*; *(fig.: of fate, death, etc.)* Klauen

jawbone ['dʒɔ:bəʊn] *n.* Kieferknochen, *der*

jay [dʒeɪ] *n.* Eichelhäher, *der*

'**jay-walker** *n. verkehrswidrig die Fahrbahn überquerender Fußgänger*

jazz [dʒæz] **1.** *n.* Jazz, *der*; *attrib.* Jazz‹*musik, -musiker*›; **and all that ~** *(coll.)* und der ganze Kram *(ugs.)*. **2.** *v.t.* **~ up** aufpeppen *(ugs.)*; aufmotzen *(ugs.)*

'**jazz band** *n.* Jazzband, *die*

jealous ['dʒeləs] *adj.* eifersüchtig **(of** auf + *Akk.*)

jealously ['dʒeləslɪ] *adv.* eifersüchtig

jealousy ['dʒeləsɪ] *n.* Eifersucht, *die*

jeans [dʒi:nz] *n. pl.* Jeans *Pl.*; Jeans, *die*; **a pair of ~** ein Paar Jeans; eine Jeans

Jeep, (P) [dʒi:p] *n.* Jeep (Wz), *der*

jeer [dʒɪə(r)] **1.** *v.i.* höhnen *(geh.)*; **~ at sb.** jmdn. verhöhnen; **~ing** höhnisch johlend ‹*Menge, Mob*›. **2.** *v.t.* verhöhnen. **3.** *n.* höhnisches Johlen; *(remark)* höhnische Bemerkung

Jehovah's 'Witness [dʒɪhəʊvəz 'wɪtnɪs] *n. (Relig.)* Zeuge Jehovas

jell [dʒel] *v.i. (set as jelly)* fest werden; gelieren

jelly ['dʒelɪ] *n.* Gelee, *das*; *(dessert)* Götterspeise, *die*; **her legs felt like ~:** sie hatte Pudding in den Knien *(ugs.)*

'**jelly fish** *n.* Qualle, *die*

jemmy ['dʒemɪ] *n. (Brit.)* Brecheisen, *das*

jeopardize (jeopardise) ['dʒepədaɪz] *v.t.* gefährden

jeopardy ['dʒepədɪ] *n., no pl.* Gefahr, *die*; **put sth./sb. in ~:** etw. aufs Spiel setzen/jmdn. in Gefahr bringen; etw./jmdn. gefährden; **in ~:** in Gefahr; gefährdet

jerk [dʒɜ:k] **1.** *n.* **a)** *(sharp sudden pull)* Ruck, *der*; **with a series of ~s** ruckartig; ruckend; **give sth. a ~:** einer Sache *(Dat.)* einen Ruck geben; an etw. *(Dat.)* rucken; **b)** *(involuntary movement)* Zuckung, *die*; Zucken, *das*. **2.** *v.t.* reißen an (+ *Dat.*) ‹*Seil usw.*›; **~ sth. away/back** *etc.* etw. weg-/zurückreißen *usw.*; **~ sth. off/out of sth.** *etc.* etw. von etw. [herunter]reißen/aus etw. [heraus]reißen *usw.* **3.** *v.i.* ruckeln; *(move in a spasmodic manner)* zucken

jerky ['dʒɜ:kɪ] *adj.* abgehackt, holprig ‹*Art zu schreiben/sprechen*›; holprig ‹*Busfahrt*›; holpernd ‹*Fahrzeug*›; ruckartig ‹*Bewegung*›

Jerry ['dʒerɪ] *n. (Brit. dated sl.) (soldier)* Deutsche, *der*

'**jerry-built** *adj.* unsolide gebaut

jersey ['dʒɜ:zɪ] *n.* Pullover, *der*; *(Sport)* Trikot, *das*; Jersey, *das*

jest [dʒest] **1.** *n.* **a)** *(joke)* Scherz, *der*; Witz, *der*; **b)** *no pl. (fun)* Spaß, *der*; **in ~:** im Scherz. **2.** *v.i.* scherzen; Witze machen

jester ['dʒestə(r)] *n.* Spaßmacher, *der*; *(at court)* Hofnarr, *der*; *(fool)* Hanswurst, *der*

Jesuit ['dʒezjʊɪt] *n.* Jesuit, *der*

Jesus ['dʒi:zəs] **1.** *pr. n.* Jesus *(der)*. **2.** *interj. (sl.)* **~ [Christ]!** Herrgott noch mal! *(ugs.)*

[1]**jet** [dʒet] **1.** *n.* **a)** *(stream)* Strahl, *der*; **~ of flame/steam/water** Feuer-/Dampf-/Wasserstrahl, *der*; **b)** *(spout, nozzle)* Düse, *die*; **c)** *(aircraft)* Düsenflugzeug, *das*; Jet, *der*; *(engine)* Düsentriebwerk, *das*. **2.** *v.i.*, **-tt-: a)** *(spurt out)* ‹*Wasser:*› herausschießen **(from** aus); ‹*Gas, Dampf:*› ausströmen **(from** aus); **b)** *(coll.: travel by ~ plane)* jetten *(ugs.)*

[2]**jet** *n. (Min.)* Jett, *der od. das*; Gagat, *der*

jet: ~-black *adj.* pechschwarz; kohlrabenschwarz; **~ engine** *n.* Düsen- *od.* Strahltriebwerk, *das*; **~ lag** *n.* ≈ Zeitverschiebung, *die*; Jet-travel-Syndrom, *das (Med.)*; **~-lagged** *adj.* **sb. is ~-lagged** jmdm. macht die Zeitverschiebung zu schaffen; **~ plane** *n.* Düsenflugzeug, *das*; **~-propelled** *adj.* düsen- *od.* strahlgetrieben; mit Düsen- *od.* Strahlantrieb *nachgestellt*

jetsam ['dʒetsəm] *n.* sinkendes Seewurfgut *(Seew.)*; *(on seashore)* Strandgut, *das*; *see also* **flotsam**

'**jet set** *n.* Jet-set, *der*

jettison ['dʒetɪsən] *v.t.* **a)** *(from ship)* über Bord werfen; *(from aircraft)* abwerfen ‹*Ballast, Bombe*›; *(discard)* wegwerfen; **b)**

(fig.: abandon) aufgeben; über Bord werfen ‹*Plan*›

jetty ['dʒetɪ] *n.* **a)** *(protecting harbour or coast)* [Hafen]mole, *die;* **b)** *(landing-pier)* Landungsbrücke, *die*

Jew [dʒu:] *n.* Jude, *der*/Jüdin, *die*

jewel ['dʒu:əl] *n.* **a)** *(ornament)* [kostbares] Schmuckstück; **~s** *collect.* Schmuck, *der;* Juwelen *Pl.;* **b)** *(precious stone)* Juwel, *das od. der;* [wertvoller] Edelstein; *(of watch)* Stein, *der;* **c)** *(fig.) (person)* Goldstück, *das;* Juwel, *das; (thing)* Kleinod, *das*

jeweller *(Amer.:* **jeweler)** ['dʒu:ələ(r)] *n.* Juwelier, *der*

jewellery *(Brit.),* **jewelry** ['dʒu:əlrɪ] *n.* Schmuck, *der*

Jewess ['dʒu:ɪs] *n.* Jüdin, *die*

Jewish ['dʒu:ɪʃ] *adj.* jüdisch; **he/she is ~:** er ist Jude/sie ist Jüdin

[1]**jib** [dʒɪb] *n.* **a)** *(Naut.) (on sailing ship)* Stagsegel, *das (Seew.); (on yacht or dinghy)* Fock, *die (Seew.);* **b)** *(of crane)* Ausleger, *der*

[2]**jib** *v.i.,* **-bb-: a)** ‹*Pferd usw.:*› bocken; *(because of fright)* scheuen; **b)** *(fig.)* sich sträuben; streiken *(ugs.);* **~ at sth./at doing sth.** sich gegen etw. sträuben/sich dagegen sträuben, etw. zu tun

jibe *see* **gibe**

jiff [dʒɪf], **jiffy** ['dʒɪfɪ] *n. (coll.)* Augenblick, *der;* Moment, *der;* **in a ~:** sofort; gleich

jig [dʒɪg] *n.* **a)** *(dance, music)* Jig, *die;* **b)** *(appliance)* Einspannvorrichtung, *die*

jiggle ['dʒɪgl] **1.** *v.t.* rütteln an, wackeln an (+ *Dat.*). **2.** *v.i.* rütteln; wackeln

'**jigsaw** *n.* **~ [puzzle]** Puzzle, *das*

jilt [dʒɪlt] *v.t.* sitzenlassen *(ugs.)*

jingle ['dʒɪŋgl] **1.** *n.* **a)** Klingeln, *das;* Bimmeln, *das (ugs.); (of cutlery, chains, spurs)* Klirren, *das; (of coins, keys)* Geklimper, *das;* **b)** *(trivial verse)* Wortgeklingel, *das (abwertend); (Commerc.)* Werbespruch, *der;* Jingle, *der (Werbespr.).* **2.** *v.i.* ‹*Metallgegenstände:*› klimpern; ‹*Kasse, Schelle:*› klingeln; ‹*Glöckchen:*› bimmeln. **3.** *v.t.* klingeln mit, *(ugs.)* bimmeln mit ‹*Glöckchen*›; klimpern mit ‹*Münzen, Schlüsseln, Armreifen*›

jingoism ['dʒɪŋgəʊɪzm] *n., no pl.* Chauvinismus, *der (abwertend);* Hurrapatriotismus, *der (ugs. abwertend)*

jinx [dʒɪŋks] **1.** *n. (coll.)* Fluch, *der;* **there seemed to be a ~ on him** er schien vom Pech verfolgt zu sein. **2.** *v.t.* verhexen

jitters ['dʒɪtəz] *n. pl. (coll.)* großes Zittern; Bammel, *der (salopp);* **give sb. the ~:** jmdm. Schiß machen *(salopp)*

jittery ['dʒɪtərɪ] *adj. (nervous)* nervös; *(frightened)* verängstigt

job [dʒɒb] *n.* **a)** *(piece of work)* **~ [of work]** Arbeit, *die;* **I have a little ~ for you** ich habe eine kleine Aufgabe *od.* einen kleinen Auftrag für dich; **do a ~ for sb.** für jmdn. etw. erledigen; **you're doing an excellent ~:** Sie machen das ausgezeichnet; **b)** *(position of employment)* Stelle, *die;* Anstellung, *die;* Job, *der (ugs.);* **he is only doing his ~!** er tut schließlich nur seine Pflicht; **he knows his ~:** er versteht sein Handwerk; **~ vacancies** offene Stellen; *(in newspaper)* „Stellenangebote"; **have ~ security** einen sicheren Arbeitsplatz haben; **just the ~** *(fig. sl.)* genau das richtige; die Sache *(ugs.);* **on the ~:** bei der Arbeit; **out of a ~:** arbeitslos; ohne Stellung; **c)** *(sl.: crime)* [krummes] Ding *(ugs.);* **d)** *(result of work)* Ergebnis, *das;* **make a [good] ~ of sth.** bei etw. gute Arbeit leisten; **e)** *(coll.: difficult task)* [schönes] Stück Arbeit; **I had a [hard** *or* **tough] ~ convincing** *or* **to convince him** es war gar nicht so einfach für mich, ihn zu überzeugen; **f)** *(state of affairs)* **a bad ~:** eine schlimme *od.* üble Sache; **give sb./sth. up as a bad ~** *see* **give up 2 a; a good ~:** ein Glück; **we've finished, and a good ~ too!** wir sind fertig, zum Glück; **it's a good ~ he doesn't know about it!** nur gut, daß er nichts davon weiß!

job: ~centre *n. (Brit.)* Arbeitsvermittlungsstelle, *die;* **~ creation** *n.* Schaffung von Arbeitsplätzen; **~-hunting** *n.* Arbeitssuche, *die;* Stellensuche, *die*

jobless ['dʒɒblɪs] *adj.* beschäftigungslos; arbeitslos

job: ~ 'lot *n.* Partieware, *die (Kaufmannsspr.); (fig.)* Sammelsurium, *das (abwertend);* **~ satisfaction** *see* **satisfaction b; ~-sharing** *n.* Job-sharing, *das*

jockey ['dʒɒkɪ] **1.** *n.* Jockei, *der;* Jockey, *der.* **2.** *v.i.* rangeln **(for** um); **~ for position** *(lit. or fig.)* alles daransetzen, eine möglichst gute Position zu erringen

jock-strap ['dʒɒkstræp] *n.* [Sport]suspensorium, *das*

jocular ['dʒɒkjʊlə(r)] *adj.* lustig, witzig ‹*Bemerkung, Antwort*›; spaßig, scherzhaft ‹*Person*›

jodhpurs ['dʒɒdpəz] *n. pl.* Reithose, *die;* Jodhpur[hose], *die*

jog [dʒɒg] **1.** *v.t.,* **-gg-: a)** *(shake with push or jerk)* rütteln; schütteln; **b)** *(nudge)* [an]stoßen; **c)** *(stimulate)* **~ sb.'s memory** jmds. Gedächtnis *(Dat.)* auf die Sprünge helfen. **2.** *v.i.,* **-gg-: a)** *(move up and down)* auf und ab hüpfen; **b)** *(move at ~trot)* ‹*Pferd:*› [dahin]trotten; **c)** *(run at slow pace)* [in mäßigem Tempo] laufen; traben *(Sport); (for physical exercise)* joggen; [einen] Dauerlauf machen. **3.** *n.* **a)** *(shake, nudge)* Stoß, *der;* Schubs, *der (ugs.);* **b)** *(slow walk or trot) (of horse)* Trott, *der; (of person for physical exercise)* Dauerlauf, *der;* **go for a ~:** joggen gehen

jogger ['dʒɒgə(r)] *n.* Jogger, *der*/Joggerin, *die*

jogging ['dʒɒgɪŋ] *n.* Jogging, *das;* Joggen, *das*

'**jogtrot** *n. (lit. or fig.)* Trott, *der*

join [dʒɔɪn] **1.** *v.t.* **a)** *(put together, connect)* verbinden **(to** mit); **~ two things [together]** zwei Dinge miteinander verbinden; zwei Dinge zusammenfügen; **~ hands** sich *(Dat.)* die Hände reichen; **b)** *(come into company of)* sich gesellen zu; sich zugesellen (+ *Dat.*); *(meet)* treffen; *(come with)* mitkommen mit; sich anschließen (+ *Dat.*); **I'll ~ you in a minute** ich komme gleich nach; **may I ~ you** *(at table)* kann ich mich zu euch setzen?; **do ~ us for lunch** iß doch mit uns zu Mittag; **would you like to ~ me in a drink?** hast du Lust, ein Glas mit mir zu trinken?; **c)** *(become member of)* eintreten in (+ *Akk.*) ‹*Armee, Firma, Orden, Verein, Partei*›; beitreten (+ *Dat.*) ‹*Verein, Partei, Orden*›; **d)** *(take one's place in)* sich einreihen in (+ *Akk.*) ‹*Umzug, Demonstrationszug*›; **e)** ‹*Fluß, Straße:*› münden in (+ *Akk.*). **2.** *v.i.* **a)** *(come together)* ‹*Flüsse:*› sich vereinigen, zusammenfließen; ‹*Straßen:*› sich vereinigen, zusammenlaufen; ‹*Grundstücke:*› aneinandergrenzen, aneinanderstoßen; **b)** *(take part)* **~ with sb.** sich jmdm. anschließen. **3.** *n.* Verbindung, *die; (line)* Nahtstelle, *die*

~ in 1. [-'-] *v.i.* mitmachen **(with** bei); *(in conversation)* sich beteiligen **(with** an + *Dat.*); *(in singing)* einstimmen; mitsingen. **2.** ['--] *v.t.* mitmachen bei ‹*Spiel, Spaß*›; sich beteiligen an (+ *Dat.*) ‹*Spiel, Festlichkeiten, Gespräch*›; mitsingen ‹*Refrain*›; sich anschließen (+ *Dat.*) ‹*Demonstrations-, Umzug*›

~ **'up** **1.** *v.i.* **a)** *(Mil.)* einrücken; Soldat werden; **b)** ⟨*Straßen:*⟩ zusammenlaufen. **2.** *v.t.* miteinander verbinden

joiner ['dʒɔɪnə(r)] *n.* Tischler, *der*/Tischlerin, *die*

joinery ['dʒɔɪnərɪ] *n., no pl.* **a)** *no art. (craft)* Tischlerei, *die;* Tischlerhandwerk, *das;* **b)** *no indef. art. (products)* Tischlerarbeiten

joint [dʒɔɪnt] **1.** *n.* **a)** *(place of joining)* Verbindung, *die; (line)* Nahtstelle, *die; (Building)* Fuge, *die;* **b)** *(Anat., Mech. Engin., etc.)* Gelenk, *das;* **c)** **a ~ [of meat]** ein Stück Fleisch; *(for roasting, roast)* ein Braten; **a ~ of roast beef** ein Rinderbraten; **d)** *(sl.) (place)* Laden, *der; (pub)* Kaschemme, *die (abwertend).* **2.** *adj.* **a)** *(of two or more)* gemeinsam ⟨*Anstrengung, Bericht, Besitz, Projekt, Ansicht, Konto*⟩; **~ venture** Gemeinschaftsunternehmen, *das;* Joint-venture, *das (Wirtsch.);* **b)** Mit⟨*autor, -erbe, -besitzer*⟩

jointly ['dʒɔɪntlɪ] *adv.* gemeinsam

joint 'stock *n. (Econ.)* Gesellschafts- *od.* Aktienkapital, *das;* **~ company** Aktiengesellschaft, *die*

joist [dʒɔɪst] *n. (Building)* Deckenbalken, *der; (steel)* [Decken]träger, *der*

joke [dʒəʊk] **1.** *n.* **a)** Witz, *der;* Scherz, *der;* **sb.'s little ~** *(iron.)* jmds. Scherzchen; **make a ~:** einen Scherz machen; **do sth. for a ~:** etw. spaßeshalber *od.* zum Spaß tun; **tell a ~:** einen Witz erzählen; **have a ~ with sb.** mit jmdm. scherzen *od.* spaßen; **play a ~ on sb.** jmdm. einen Streich spielen; **he can/can't take a ~:** er versteht Spaß/keinen Spaß; **the ~ was on him** er war der Narr; **this is getting beyond a ~:** da hört der Spaß auf; **this is no ~:** das ist nicht zum Lachen; **b)** *(ridiculous thing or circumstance)* Witz, *der (ugs.); (ridiculous person)* Witzfigur, *die;* **treat sth. as a ~:** etw. nicht weiter ernst nehmen. **2.** *v.i.* scherzen, Witze machen (**about** über + *Akk.*); **joking apart** Scherz *od.* Spaß beiseite!; **you have [got] to be joking!** *(coll.)* das soll wohl ein Witz sein!; mach keine Witze!

joker ['dʒəʊkə(r)] *n.* **a)** *(person)* Spaßvogel, *der;* Witzbold, *der (ugs.);* **b)** *(Cards)* Joker, *der*

jolly ['dʒɒlɪ] **1.** *adj. (cheerful)* fröhlich; knallig ⟨*Farbe*⟩; *(multicoloured)* bunt. **2.** *adv. (Brit. coll.)* ganz schön *(ugs.);* sehr ⟨*nett*⟩; **~ good!** ausgezeichnet!; **I should ~ well think so!** das möchte ich auch meinen!

~ **a'long** *v.t.* bei Laune halten

jolt [dʒəʊlt] **1.** *v.t.* **a)** *(shake)* ⟨*Fahrzeug:*⟩ durchrütteln, durchschütteln; **~ sb./sth. out of/on to sth.** jmdn./etw. aus etw./auf etw. *(Akk.)* schleudern *od.* werfen; **b)** *(shock)* aufschrecken; **~ sb. into action** jmdn. auf Trab bringen *(ugs.).* **2.** *v.i.* ⟨*Fahrzeug:*⟩ holpern, rütteln, rumpeln *(ugs.).* **3.** *n.* **a)** *(jerk)* Stoß, *der;* Ruck, *der;* **b)** *(fig.) (shock)* Schock, *der;* Schreck, *der; (surprise)* Überraschung, *die*

Jordan ['dʒɔ:dn] *pr. n.* **a)** *(river)* Jordan, *der;* **b)** *(country)* Jordanien *(das)*

joss-stick ['dʒɒsstɪk] *n.* Räucherstäbchen, *das*

jostle ['dʒɒsl] **1.** *v.i.* **~ [against each other]** aneinanderstoßen. **2.** *v.t.* stoßen

jot [dʒɒt] *n.* **[not] a ~:** [k]ein bißchen

~ **'down** *v.t.* [rasch] aufschreiben *od.* notieren

jotter ['dʒɒtə(r)] *n. (pad)* Notizblock, *der; (notebook)* Notizbuch, *das*

jotting ['dʒɒtɪŋ] *n., usu. pl.* Notiz, *die*

journal ['dʒɜ:nl] *n.* **a)** *(newspaper)* Zeitung, *die; (periodical)* Zeitschrift, *die;* **weekly ~:** Wochenzeitung, *die;* **b)** *(daily record of events)* Tagebuch, *das*

journalism ['dʒɜ:nəlɪzm] *n.* Journalismus, *der*

journalist ['dʒɜ:nəlɪst] *n.* Journalist, *der*/Journalistin, *die*

journalistic [dʒɜ:nə'lɪstɪk] *adj.* journalistisch ⟨*Stil*⟩

journey ['dʒɜ:nɪ] **1.** *n.* Reise, *die; (distance)* Weg, *der;* **a three-hour ~:** eine dreistündige Fahrt; **a ~ by car/train/ship** eine Auto-/Bahn-/Schiffsreise; **go on a ~:** verreisen; eine Reise machen; **go on a train/car ~:** eine Reise mit dem Zug *od.* Zugreise/eine Reise mit dem Auto *od.* Autoreise machen; **~ through life** Lebensreise, *die (geh.).* **2.** *v.i. (formal)* fahren; ziehen

jovial ['dʒəʊvɪəl] *adj.* *(hearty)* herzlich ⟨*Gruß*⟩; *(merry)* fröhlich ⟨*Ausdruck, Person*⟩

jowl [dʒaʊl] *n. (jaw)* Unterkiefer, *der; (lower part of face)* Kinnbacken *Pl.; (double chin)* Doppelkinn, *das; (flabby cheek)* Hängebacke, *die;* **cheek by ~:** dicht nebeneinander

joy [dʒɔɪ] *n.* **a)** Freude, *die;* **wish sb. ~:** jmdm. viel Spaß *od.* Vergnügen wünschen; **I wish you ~ of it** *(also iron.)* ich wünsche dir viel Vergnügen damit; **sing for/weep with ~:** vor Freude *(Dat.)* singen/weinen; **be full of the ~s of spring** *(fig. coll.)* vor Freude ganz aus dem Häuschen sein *(ugs.);* **it was a ~ to look at** es war eine Augenweide; **b)** *no pl., no art. (coll.: success, satisfaction)* Erfolg, *der;* **he didn't get much ~ out of it** es hat ihm nicht viel gebracht; **any ~?** Erfolg gehabt?; was erreicht? *(ugs.)*

joyful ['dʒɔɪfl] *adj.* froh[gestimmt] ⟨*Person*⟩; froh ⟨*Gesicht*⟩; freudig ⟨*Blick, Ereignis, Umarmung, Gesang, Beifall*⟩; freudig, froh ⟨*Nachricht, Kunde*⟩; erfreulich ⟨*Nachricht, Ergebnis, Anblick*⟩

joyous ['dʒɔɪəs] *adj.* freudig ⟨*Anlaß, Ereignis*⟩; froh ⟨*Lachen, Herz*⟩; Freuden⟨*tag, -schrei*⟩

joy: ~-ride *n. (coll.)* Spritztour [im gestohlenen Auto]; **~-rider** *n.* Autodieb *(der den Wagen nur für eine Spritztour gestohlen hat);* **~stick** *n.* **a)** *(Aeronaut. coll.)* Knüppel, *der;* **b)** *(on computer etc.)* Hebel, *der;* Joystick, der *(DV)*

JP *abbr.* **Justice of the Peace**

Jr. *abbr.* **Junior** jun.

jubilant ['dʒu:bɪlənt] *adj.* jubelnd; freudestrahlend ⟨*Miene*⟩; **be ~** ⟨*Person:*⟩ frohlocken

jubilation [dʒu:bɪ'leɪʃn] *n.* Jubel, *der*

jubilee ['dʒu:bɪli:] *n. (anniversary)* Jubiläum, *das*

judge [dʒʌdʒ] **1.** *n.* **a)** Richter, *der*/Richterin, *die;* **b)** *(in contest)* Preisrichter, *der*/-richterin, *die; (Sport)* Kampfrichter, *der*/-richterin, *die;* Schiedsrichter, *der*/-richterin, *die; (in dispute)* Schiedsrichter, *der*/-richterin, *die;* **c)** *(fig.: connoisseur, critic)* Kenner, *der*/Kennerin, *die;* **~ of character** Menschenkenner, *der;* **be a good ~ of sth.** etw. gut beurteilen können; **d)** *(person who decides question)* Schiedsrichter, *der;* **be the ~ of sth.** über etw. *(Akk.)* entscheiden. **2.** *v.t.* **a)** *(pronounce sentence on)* richten *(geh.);* **~ sb.** *(Law)* jmds. Fall entscheiden; **b)** *(try)* verhandeln ⟨*Fall*⟩; **c)** *(act as adjudicator of)* Preisrichter/-richterin sein bei; *(Sport)* Schiedsrichter/-richterin sein bei; **d)** *(form opinion about)* urteilen *od.* ein Urteil fällen über (+ *Akk.*); beurteilen; **~ sth. [to be] necessary** etw. für *od.* als notwendig erachten; **be good at judg-**

ing distances gut Entfernungen schätzen können; **e)** *(decide)* entscheiden ‹*Angelegenheit, Frage*›. **3.** *v.i.* *(form a judgement)* urteilen; **to ~ by its size, ...:** der Größe nach zu urteilen, ...; **judging** *or* **to ~ by the look on his face ...:** nach dem Gesicht zu schließen, das er macht/machte, ...; **judging from what you say, ...:** nach dem, was du sagst, ...; **as far as I can ~, ...:** soweit ich es beurteilen kann, ...

judgement, judgment ['dʒʌdʒmənt] *n.* **a)** Urteil, *das;* **~ was given in favour of/against sb.** das Urteil fiel zu jmds. Gunsten/Ungunsten aus; **pass [a] ~:** ein Urteil abgeben (**on** über + *Akk.*); **in** *or* **according to my ~:** meines Erachtens; **form a ~:** sich *(Dat.)* ein Urteil *od.* eine Meinung bilden; **against one's better ~:** entgegen seiner besseren Einsicht; **b)** *(critical faculty)* Urteilsfähigkeit, *die;* Urteilsvermögen, *das;* **error of ~:** Fehlurteil, *das;* Fehleinschätzung, *die;* **I leave it to your ~:** ich stelle das in Ihr Ermessen; **use your own ~:** verfahren Sie nach Ihrem Gutdünken; **c)** *(trial by God)* **day of ~, J~ Day** Tag des Jüngsten Gerichts; **the last ~:** das Jüngste *od.* Letzte Gericht

judicial [dʒu:'dɪʃl] *adj.* **a)** gerichtlich; richterlich ‹*Gewalt*›; **~ murder** Justizmord, *der;* **b)** *(expressing judgement)* kritisch

judiciary [dʒu:'dɪʃərɪ] *n.* *(Law)* Richterschaft, *die*

judicious [dʒu:'dɪʃəs] *adj.* **a)** *(discerning)* klarblickend; **b)** *(sensible)* besonnen

judiciously [dʒu:'dɪʃəslɪ] *adv.* mit Bedacht

judo ['dʒu:dəʊ] *n., pl.* **~s** Judo, *das*

jug [dʒʌg] **1.** *n.* **a)** Krug, *der;* *(with lid, water-~)* Kanne, *die;* *(small milk-~)* Kännchen, *das;* **a ~ of water** ein Krug/eine Kanne Wasser; **b)** *(sl.: prison)* Loch, *das (salopp).* **2.** *v.t.,* **-gg-** *(Cookery)* schmoren; **~ged hare** Hasenpfeffer, *der*

juggernaut ['dʒʌgənɔ:t] *n.* *(Brit.: lorry)* schwerer Brummer *(ugs.)*

juggle ['dʒʌgl] **1.** *v.i.* **a)** jonglieren; *(perform conjuring tricks)* zaubern; **b) ~ with** *(misrepresent)* jonglieren mit ‹*Fakten, Zahlen*›. **2.** *v.t.* *(lit., or fig.: manipulate)* jonglieren [mit]

juggler ['dʒʌglə(r)] *n.* Jongleur, *der*/Jongleuse, *die*

Jugoslav *etc. see* **Yugoslav** *etc.*

jugular ['dʒʌgjʊlə(r)] *adj. & n.* *(Anat.)* **~ [vein]** Jugularvene, *die (fachspr.);* Drosselvene, *die*

juice [dʒu:s] *n.* **a)** Saft, *der;* **b)** *(sl.)* *(electricity)* Saft, *der (salopp);* *(petrol)* Sprit, *der (ugs.)*

juicy ['dʒu:sɪ] *adj.* **a)** saftig; **b)** *(coll.)* *(racy)* saftig *(ugs.)* ‹*Geschichte, Skandal*›; *(suggestive)* schlüpfrig; *(profitable)* fett *(ugs.)* ‹*Vertrag, Geschäft usw.*›

ju-jitsu [dʒu:'dʒɪtsu:] *n.* Jiu-Jitsu, *das*

juke-box ['dʒu:kbɒks] *n.* Jukebox, *die;* Musikbox, *die*

Jul. *abbr.* **July** Jul.

July [dʒʊ'laɪ] *n.* Juli, *der; see also* **August**

jumble ['dʒʌmbl] **1.** *v.t.* **~ up** *or* **together** durcheinanderbringen; durcheinanderwerfen. **2.** *n.* **a)** Wirrwarr, *der;* Gewirr, *das;* *(muddle)* Durcheinander, *das;* **b)** *no pl., no indef. art.* *(Brit.: articles for ~ sale)* alte *od.* gebrauchte Sachen

'jumble sale *n.* *(Brit.)* Trödelmarkt, *der;* *(for charity)* Wohltätigkeitsbasar, *der*

jumbo ['dʒʌmbəʊ] **1.** *n.* *(jet)* Jumbo, *der.* **2.** *adj.* **~[-sized]** riesig; Riesen- *(ugs.)*

jumbo 'jet *n.* Jumbo-Jet, *der*

jump [dʒʌmp] **1.** *n.* **a)** Sprung, *der;* **always be one ~ ahead of sb.** jmdm. immer um eine Nasenlänge voraus sein *(ugs.);* **b)** *(sudden transition)* Sprung, *der;* sprunghafter Wechsel; *(gap)* Lücke, *die;* **c)** *(abrupt rise)* sprunghafter Anstieg; **~ in value/temperature** plötzliche Wertsteigerung/plötzlicher Temperaturanstieg; **d)** *(Parachuting)* Absprung, *der.* **2.** *v.i.* **a)** springen; ‹*Fallschirmspringer:*› abspringen; **~ to one's feet/from one's seat** aufspringen/vom Sitz aufspringen; **b) ~ to** *(reach overhastily)* voreilig gelangen zu ‹*Annahme, Lösung*›; **~ to conclusions** voreilige Schlüsse ziehen; **c)** *(make sudden movement)* springen; *(start)* zusammenzucken; **~ for joy** einen Freudensprung/Freudensprünge machen; **d)** *(rise suddenly)* ‹*Kosten, Preise usw.:*› sprunghaft steigen, in die Höhe schnellen; **e) ~ to it** *(coll.)* zupacken; **~ to it!** *(coll.)* mach/macht schon! **3.** *v.t.* **a)** springen über (+ *Akk.*); überspringen ‹*Mauer, Zaun usw.*›; **b)** *(move to point beyond)* überspringen; **c)** *(not stop at)* überfahren ‹*rote Ampel*›; **~ the lights** bei Rot [durch]fahren; **d) ~ the rails** *or* **track** ‹*Zug:*› entgleisen; **e) ~ ship** ‹*Seemann:*› [unter Bruch des Heuervertrages vorzeitig] den Dienst quittieren; **f) ~ the queue** *(Brit.)* sich vordrängeln; **g)** *(skip over)* überspringen ‹*Seite, Kapitel usw.*›; **h)** *(attack)* herfallen über (+ *Akk.*)

~ a'bout, ~ a'round *v.i.* herumspringen *(ugs.)*

~ at *v.t.* **a)** anspringen; **b)** *(fig.: accept eagerly)* sofort [beim Schopf] ergreifen ‹*Gelegenheit*›; sofort zugreifen *od.* *(ugs.)* zuschlagen bei ‹*Angebot*›; sofort aufgreifen *od.* *(ugs.)* anspringen auf ‹*Vorschlag*›

~ 'in *v.i.* reinspringen *(ugs.)*

~ 'off 1. *v.i.* abspringen. **2.** *v.t.* **~ off sth.** von etw. springen

~ on 1. ['-'-] *v.i.* aufspringen; **~ on to a bus/train** in einen Bus/Zug springen; **~ on to one's bicycle/horse** sich aufs Fahrrad/Pferd schwingen. **2.** ['--] *v.t.* **~ on a bus/train** in einen Bus/Zug springen; **~ on one's bicycle** sich aufs Fahrrad schwingen

~ 'out *v.i.* hinaus-/herausspringen; **~ out of** springen aus

~ 'up *v.i.* aufspringen (**from** von); **~ up on to sth.** auf etw. *(Akk.)* springen

jumped-up ['dʒʌmptʌp] *adj.* *(coll.)* emporgekommen

jumper ['dʒʌmpə(r)] *n.* **a)** Pullover, *der;* Pulli, *der (ugs.);* **b)** *(Amer.: pinafore dress)* Trägerkleid, *das*

jump: ~-jet *n.* *(Aeronaut.)* Senkrechtstarter, *der;* **~ leads** *n. pl.* *(Brit. Motor Veh.)* Starthilfekabel; **~ suit** *n.* Overall, *der*

jumpy ['dʒʌmpɪ] *adj.* nervös; aufgeregt

Jun. *abbr.* **a) June** Jun.; **b) Junior** jun.

junction ['dʒʌŋkʃn] *n.* **a)** Verbindungspunkt, *der;* Verbindungsstelle, *die;* **b)** *(of railway lines, roads)* ≈ Einmündung, *die;* *(of motorway)* Anschlußstelle, *die;* *(crossroads)* Kreuzung, *die*

'junction box *n.* *(Electr.)* Verteilerkasten, *der*

juncture ['dʒʌŋktʃə(r)] *n.* **at this ~:** zu diesem Zeitpunkt

June [dʒu:n] *n.* Juni, *der; see also* **August**

jungle ['dʒʌŋgl] *n.* Dschungel, *der (auch fig.);* Urwald, *der*

junior ['dʒu:nɪə(r)] **1.** *adj.* **a)** *(below a certain age)* jünger; **b)** *(of lower rank)* rangniedriger ‹*Person*›; einfach ‹*Angestellter*›; **c)** *appended to name (the younger)* **Mr Smith J~:** Mr. Smith junior; **d)** *(Amer. Sch., Univ.)* **~ year** vorletztes Jahr vor der Abschlußprüfung. **2.** *n. (younger person)* Jüngere, *der/die; (person of lower rank)* Untergebene, *der/die;* **be [six years] sb.'s ~:** [sechs Jahre] jünger sein als jmd.

junior: ~ 'minister *n. (Brit.)* ≈ Ministerialdirektor, *der*/-direktorin, *die;* ~ **'partner** *n.* Juniorpartner, *der*/-partnerin, *die;* ~ **school** *n. (Brit.)* Grundschule, *die*

juniper ['dʒu:nɪpə(r)] *n. (Bot.)* Wacholder, *der*

¹junk [dʒʌŋk] **1.** *n. (discarded material)* Trödel, *der (ugs.);* Gerümpel, *das; (trash)* Plunder, *der (ugs.);* Ramsch, *der (ugs.).* **2.** *v. t.* wegwerfen; ausmisten *(ugs.); (fig.)* aufgeben

²junk *n. (ship)* Dschunke, *die*

junk: ~ food *n.* minderwertige Kost; ~**-heap** *n.* **a)** *see* **scrap-heap; b)** *(sl.: old car etc.)* Schrotthaufen, *der (ugs.)*

junkie ['dʒʌŋkɪ] *n. (sl.)* Junkie, *der (Drogenjargon)*

'junk-shop *n.* Trödelladen, *der (ugs.)*

junta ['dʒʌntə] *n.* Junta, *die;* **military ~:** Militärjunta, *die*

Jupiter ['dʒu:pɪtə(r)] *pr. n.* **a)** *(Astron.)* Jupiter, *der;* **b)** *(Roman Mythol.)* Jupiter *(der)*

jurisdiction [dʒʊərɪs'dɪkʃn] *n. (authority)* Jurisdiktion, *die;* Gerichtsbarkeit, *die; (extent)* Zuständigkeit, *die;* **fall** *or* **come under** *or* **within the ~ of sth./sb.** in die Zuständigkeit *od.* den Zuständigkeitsbereich von etw./jmdm. fallen; **have ~ over sb./in a matter** für jmdn./in einer Angelegenheit zuständig sein

juror ['dʒʊərə(r)] *n.* Geschworene, *der/die; (in Germany, in some Austrian courts)* Schöffe, *der*/ Schöffin, *die*

jury ['dʒʊərɪ] *n.* **a)** *(in court)* **the ~:** die Geschworenen; *(in Germany, in some Austrian courts)* die Schöffen; **sit on the ~:** auf der Geschworenen- / Schöffenbank sitzen; **do ~ service** das Amt eines Geschworenen/Schöffen ausüben; **b)** *(in competition)* Jury, *die;* Preisgericht, *das*

'jury-box *n.* Geschworenenbank, *die; (in Germany, in some Austrian courts)* Schöffenbank, *die*

just [dʒʌst] **1.** *adj.* **a)** *(morally right, deserved)* gerecht; anständig, korrekt ‹*Verhalten, Benehmen*›; **b)** *(legally right)* rechtmäßig; **c)** *(justified)* berechtigt ‹*Angst, Zorn, Groll*›; **d)** *(right in amount)* recht, richtig ‹*Proportion, Maß, Verhältnis*›. **2.** *adv.* **a)** *(exactly)* genau; **~ then/enough** gerade da/genug; **~ as** *(exactly as, in the same way as)* genauso wie; *(when)* gerade, als; **~ as you like** ganz wie Sie wünschen; **~ as good/tidy** *etc.* genauso gut/ordentlich *usw.;* **come ~ as you are** komm so, wie du bist; **~ as fast as I can** so schnell wie ich nur kann; **it'll ~ about be enough** *(coll.)* es wird in etwa reichen; **that is ~ 'it** das ist es ja gerade; genau das ist es ja; **that's ~ like him** das ist typisch er *od.* für ihn; **~ 'so** *(in an orderly manner)* ordentlich; *expr. agreement* ganz recht; **b)** *(barely)* gerade [eben]; *(with very little time to spare)* gerade [eben] noch; *(no more than)* nur; **~ under £10** nicht ganz zehn Pfund; **it's ~ possible** das ist gerade noch möglich; **it's ~ after the traffic-lights** es ist direkt hinter der Verkehrsampel; **c)** *(exactly or nearly now or then, in immediate past)* gerade [eben]; [so]eben; *(at this moment)* gerade; **I have ~ seen him** *(Brit.),* **I ~ saw him** *(Amer.)* ich habe ihn gerade [eben] *od.* eben gesehen; **~ now** *(at this moment)* [im Moment] gerade; *(a little time ago)* gerade eben; **not ~ now** im Moment nicht; **d)** *(coll.) (simply)* einfach; *(only)* nur; *esp. with imperatives* mal [eben]; **I've come here ~ to see you** ich bin nur gekommen, um dich zu besuchen; **~ anybody** irgend jemand; **~ look at that!** guck dir das mal an!; **could you ~ turn round?** kannst du dich mal [eben] umdrehen?; **~ come here a moment** komm [doch] mal einen Moment her; **~ a moment, please** einen Moment mal; **~ in case** für alle Fälle; **e)** *(coll.: positively)* einfach; echt *(ugs.);* **that's ~ ridiculous/fantastic** das ist einfach lächerlich/phantastisch; **f)** *(quite)* **not ~ yet** noch nicht ganz; **it is ~ as well that ...:** [es ist] nur gut *od.* es ist doch gut, daß ...; **you might ~ as well ...:** du könntest genausogut ...; **g)** *(coll.: really, indeed)* wirklich; echt *(ugs.);* **That's lovely. – Isn't it ~?** Das ist schön. – Ja, und wie; **~ the same** *(nevertheless)* trotzdem; **that's ~ too bad** das ist Pech

justice ['dʒʌstɪs] *n.* **a)** Gerechtigkeit, *die;* **administer ~:** Recht sprechen; **poetic ~:** ausgleichende Gerechtigkeit; **do ~ to sth.** einer Sache *(Dat.)* gerecht werden; **~ was done in the end** der Gerechtigkeit wurde schließlich Genüge getan; **do oneself ~:** sich richtig zur Geltung bringen; **in ~ to sb.** um jmdm. gerecht zu werden; **with ~:** mit Recht; **b)** *(judicial proceedings)* **bring sb. to ~:** jmdn. vor Gericht bringen *od.* stellen; **let ~ take its course** der Gerechtigkeit ihren Lauf lassen; **c)** *(magistrate)* Schiedsrichter, *der*/ -richterin, *die;* **J~ of the Peace** Friedensrichter, *der*/-richterin, *die*

justifiable [dʒʌstɪ'faɪəbl] *adj.* berechtigt; gerechtfertigt ‹*Maßnahme, Handlung*›

justifiably [dʒʌstɪ'faɪəblɪ] *adv.* zu Recht; berechtigterweise *(Papierdt.)*

justification [dʒʌstɪfɪ'keɪʃn] *n.* Rechtfertigung, *die; (condition of being justified)* Berechtigung, *die;* **with some ~:** mit einigem Recht

justify ['dʒʌstɪfaɪ] *v. t.* **a)** *(show justice of, vindicate)* rechtfertigen; *(demonstrate correctness of)* belegen, beweisen ‹*Behauptung, Argument, Darstellung*›; *(offer adequate grounds for)* begründen ‹*Verhalten, Vorstellung, Behauptung*›; **~ oneself/sth. to sb.** sich/ etw. jmdm. gegenüber *od.* vor jmdm. rechtfertigen; **the end justifies the means** der Zweck heiligt die Mittel; **be justified in doing sth.** etw. zu Recht tun; **b)** *(Printing)* ausschließen

jut [dʒʌt] *v. i.,* **-tt-: ~ [out]** [her]vorragen; herausragen

jute [dʒu:t] *n.* Jute, *die*

juvenile ['dʒu:vənaɪl] **1.** *adj.* **a)** jugendlich, *(geh.)* juvenil ‹*Geschmack, Einstellung*›; Jugend‹*literatur, -mode*›; **~ crime** Jugendkriminalität, *die;* **b)** *(immature)* kindisch *(abwertend);* infantil *(abwertend).* **2.** *n.* Jugendliche, *der/die*

juvenile: ~ court *n. (Law)* Jugendgericht, *das;* **~ de'linquency** *n.* Jugendkriminalität, *die;* **~ de'linquent** *n.* jugendlicher Straftäter/jugendliche Straftäterin

juxtapose [dʒʌkstə'pəʊz] *v. t.* nebeneinanderstellen **(with, to** und)

juxtaposition [dʒʌkstəpə'zɪʃn] *n. (action)* Nebeneinanderstellung, *die; (condition)* Nebeneinander, *das*

K

K, k [keɪ] *n., pl.* **Ks** *or* **K's** K, k, *das*

kale [keɪl] *n. (Bot.)* **[curly] ~:** Grünkohl, *der;* Krauskohl, *der*

kaleidoscope [kə'laɪdəskəʊp] *n. (lit. or fig.)* Kaleidoskop, *das*

Kampuchea [kæmpʊ'tʃi:ə] *pr. n.* Kamputschea *(das)*

kangaroo [kæŋgə'ru:] *n.* Känguruh, *das*

kangaroo 'court *n.* Femegericht, *das;* Feme, *die*

kaput [kæ'pʊt] *pred. adj. (sl.)* kaputt *(ugs.)*

karate [kə'rɑ:tɪ] *n., no pl., no indef. art.* Karate, *das*

kayak ['kaɪæk] *n.* Kajak, *der*

kebab [kɪ'bæb] *n. (Cookery)* Kebab, *der*

keel [ki:l] **1.** *n. (Naut.)* Kiel, *der; see also* **even 1 b. 2.** *v. i.* **~ over a)** *(overturn)* umstürzen; ⟨*Schiff:*⟩ kentern; **b)** *(fall)* ⟨*Person:*⟩ umkippen

keen [ki:n] *adj.* **a)** *(sharp)* scharf ⟨*Messer, Klinge, Schneide*⟩; **b)** *(piercingly cold)* scharf, schneidend ⟨*Wind, Kälte*⟩; *(penetrating, strong)* grell ⟨*Licht*⟩; durchdringend, stechend ⟨*Geruch*⟩; **c)** *(eager)* begeistert, leidenschaftlich ⟨*Fußballfan, Sportler*⟩; ausgeprägt, lebhaft ⟨*Interesse*⟩; heftig ⟨*Konkurrenz, Verlangen*⟩; **be ~ to do sth.** darauf erpicht sein, etw. zu tun; **he's really ~ to win** er will unbedingt gewinnen; **be ~ on doing sth.** etw. gern[e] tun; **not be ~ on sth.** nicht gerade begeistert von etw. sein; **be ~ on sb.** scharf auf jmdn. sein *(ugs.)*; **d)** *(highly sensitive)* scharf ⟨*Augen*⟩; fein ⟨*Sinne*⟩; ausgeprägt ⟨*Sinn für etw.*⟩; **e)** *(intellectually sharp)* scharf ⟨*Verstand, Intellekt*⟩; **f)** *(acute)* heftig, stark ⟨*Schmerzen, Qualen*⟩; **g)** *(Brit.)* niedrig, günstig ⟨*Preis*⟩

keenly ['ki:nlɪ] *adv.* **a)** *(sharply)* scharf ⟨*geschliffen*⟩; **b)** *(eagerly)* eifrig ⟨*arbeiten*⟩; brennend ⟨*interessiert sein*⟩; **c)** *(piercingly)* scharf ⟨*ansehen*⟩; **d)** *(acutely)* **be ~ aware of sth.** sich *(Dat.)* einer Sache *(Gen.)* voll bewußt sein; **feel sth. ~:** etw. deutlich fühlen

keenness ['ki:nnɪs] *n., no pl.* **a)** *(sharpness, coldness, acuteness of sense)* Schärfe, *die;* **b)** *(eagerness)* Eifer, *der;* **c)** *(of intellect)* Schärfe, *die*

keep [ki:p] **1.** *v. t.,* **kept** [kept] **a)** *(observe)* halten ⟨*Versprechen, Schwur usw.*⟩; einhalten ⟨*Verabredung, Vereinbarung, Vertrag, Zeitplan*⟩; **b)** *(guard)* behüten, beschützen ⟨*Person*⟩; hüten ⟨*Herde, Schafe*⟩; schützen ⟨*Stadt, Festung*⟩; verwahren ⟨*Wertgegenstände*⟩; **~ sb. safe** jmdn. beschützen; **~ sth. locked away** etw. unter Verschluß halten *od.* aufbewahren; **c)** *(have charge of)* aufbewahren; verwahren; **d)** *(retain possession of)* behalten; *(not lose or destroy)* aufheben ⟨*Quittung, Rechnung*⟩; **you can ~ it** *(coll.: I do not want it)* das kannst du behalten *od.* dir an den Hut stecken *(ugs.)*; **e)** *(maintain)* unterhalten, instandhalten ⟨*Gebäude, Straße usw.*⟩; pflegen ⟨*Garten*⟩; **neatly kept** gut gepflegt; **f)** *(carry on, manage)* unterhalten, führen, betreiben ⟨*Geschäft, Lokal, Bauernhof*⟩; **g)** halten ⟨*Schweine, Bienen, Hund, Katze usw*⟩; sich *(Dat.)* halten ⟨*Diener, Auto*⟩; **h)** führen ⟨*Tagebuch, Liste usw.*⟩; **~ the books** die Bücher führen; **i)** *(provide for)* versorgen, unterhalten ⟨*Familie*⟩; **~ sb./oneself in cigarettes** *etc.* jmdn./sich mit Zigaretten *usw.* versorgen; **j)** sich *(Dat.)* halten ⟨*Geliebte, Mätresse usw.*⟩; **k)** *(have on sale)* führen ⟨*Ware*⟩; **~ a stock of sth.** etw. [am Lager] haben; **l)** *(maintain in quality, state, or position)* halten ⟨*Rhythmus*⟩; **~ sth. in one's head** etw. [im Kopf] behalten; sich *(Dat.)* etw. merken; **~ sb. waiting** jmdn. warten lassen; **~ the office running smoothly** dafür sorgen, daß im Büro weiterhin alles reibungslos [ab]läuft; **~ sb. alive** jmdn. am Leben halten; **~ the traffic moving** den Verkehr in Fluß halten; **~ sth. shut/tidy** etw. geschlossen/in Ordnung halten; **m)** *(detain)* festhalten; **what kept you?** wo bleibst du denn?; **don't let me ~ you** laß dich [von mir] nicht aufhalten; **~ sb. in prison** jmdn. in Haft halten; **n)** *(restrain, prevent)* **~ sb. from doing sth.** jmdn. davon abhalten *od.* daran hindern, etw. zu tun; **to ~ myself from falling** um nicht zu fallen; **o)** *(reserve)* aufheben; aufsparen; **~ a seat for sb.** jmdm. einen Platz freihalten; **~ it for oneself** es für sich behalten; **~ sth. for later** *etc.* sich *(Dat.)* etw. für später *usw.* aufheben *od.* aufsparen; **p)** *(conceal)* **~ sth. to oneself** etw. für sich behalten; **~ sth. from sb.** jmdm. etw. verheimlichen. **2.** *v. i.,* **kept a)** *(remain in specified place, condition)* bleiben; **~ warm/clean** sich warm/sauber halten; **how are you ~ing?** *(coll.)* wie geht's [dir] denn so? *(ugs.)*; **are you ~ing well?** geht's dir gut?; **~ together** zusammenbleiben; **b)** *(continue in course, direction, or action)* **~ [to the] left/[to the] right/straight on** sich links/rechts halten/immer geradeaus fahren/gehen *usw.*; **'~ left'** *(traffic sign)* „links vorbeifahren"; **~ behind me** halte dich *od.* bleib hinter mir; **~ doing sth.** *(not stop)* etw. weiter tun; *(repeatedly)* etw. immer wieder tun; *(constantly)* etw. dauernd *od.* immer tun; **~ talking/working** *etc.* **until ...:** weiterreden/-arbeiten *usw.*, bis ...; **c)** *(remain good)* ⟨*Lebensmittel:*⟩ sich halten; **what I have to say won't ~:** was ich zu sagen habe, ist eilig *od.* eilt. **3.** *n.* **a)** *(maintenance)* Unterhalt, *der;* **I get £100 a month and my ~:** ich bekomme 100 Pfund monatlich und Logis; **sth. doesn't earn its ~:** etw. zahlt sich nicht aus *(ugs.)*; **you don't earn your ~:** du bist nichts als ein unnützer Esser; **b) for ~s** *(coll.)* auf Dauer; *(to be retained)* zum Behalten; **c)** *(Hist.: tower)* Bergfried, *der*

~ 'after *v. t.* verfolgen; jagen; *(fig.: chivvy)* antreiben

~ at *v. t. (work persistently)* weitermachen mit; **~ 'at it!** nicht nachlassen!

~ a'way 1. *v. i.* wegbleiben *(ugs.)* **(from** von); sich fernhalten **(from** von). **2.** *v. t.* fernhalten **(from** von)

~ 'back 1. *v. i.* zurückbleiben; **~ back!** bleib wo du bist!; **~ back from sth.** von etw. wegbleiben *(ugs.)*. **2.** *v. t.* **a)** *(restrain)* zurückhalten ⟨*Menschenmenge, Tränen*⟩; **b)** *(withhold)* verschweigen ⟨*Informationen, Tatsachen*⟩ **(from** *Dat.*); einbehalten ⟨*Geld, Zahlung*⟩

~ 'down 1. *v. i.* unten bleiben. **2.** *v. t.* **a)** *(oppress, suppress)* unterdrücken ⟨*Volk, Person*⟩; **you can't ~ a good man down** *(prov.)* er/sie *usw.* läßt/lassen sich nicht unterkriegen *(ugs.)*; **b)** *(prevent in-*

crease of) niedrig halten ‹*Steuern, Preise, Zinssatz, Ausgaben, usw.*›; eindämmen ‹*Epidemie*›; ~ **one's weight down** nicht zunehmen; ~ **the weeds down** dafür sorgen, daß das Unkraut nicht überhandnimmt; **c)** *(not raise)* unten lassen ‹*Kopf*›; ~ **your voice down** rede nicht so laut; **d)** *(not vomit)* bei sich behalten ‹*Essen*›

~ **from** *v. t.* ~ **from doing sth.** etw. nicht tun; *(avoid doing)* es vermeiden, etw. zu tun; **I couldn't ~ from smiling** ich mußte einfach lächeln; *see also* ~ **1 n**

~ **'in 1.** *v. i. (remain in favour)* ~ **in with sb.** sich mit jmdm. gut stellen; sich *(Dat.)* jmdn. warmhalten *(ugs.)*. **2.** *v. t.* **a)** unterdrücken ‹*Gefühle*›; einziehen ‹*Bauch*›; **b)** *(Sch.)* nachsitzen lassen ‹*Schüler*›; **be kept in [after school]** nachsitzen müssen

~ **'off 1.** *v. i.* ‹*Person:*› wegbleiben; ‹*Regen, Sturm usw.:*› ausbleiben; **'~ off'** *(on building-site etc.)* „Betreten verboten". **2.** *v. t.* **a)** fernhalten ‹*Person, Tier*›; abhalten ‹*Sonne*›; ~ **sb./sth. off sth.** jmdn./etw. von etw. fernhalten/ abhalten; **b)** *(not go on)* nicht betreten; **'~ off the grass'** „Betreten des Rasens verboten"; **c)** *(not touch)* ~ **off my whisky!** Hände *od.* Finger weg von meinem Whisky!; **d)** *(not eat or drink)* ~ **off chocolate/brandy** keine Schokolade essen/keinen Brandy trinken; ~ **off the drink** keinen Alkohol *od. (ugs.)* nichts trinken; **e)** *(not mention)* vermeiden ‹*Thema*›

~ **'on 1.** *v. i.* **a)** *(continue, persist)* weitermachen **(with** *Akk.*); **b)** *(Brit.: talk tiresomely)* ~ **on about sth.** immer wieder von etw. anfangen; ~ **on at sb. about sth.** jmdm. mit etw. ständig in den Ohren liegen *(ugs.)*. **2.** *v. t.* **a)** ~ **on doing sth.** etw. [immer] weiter tun; *(repeatedly)* etw. immer wieder tun; *(constantly)* etw. dauernd *od.* immer tun; **b)** weiterbeschäftigen, behalten ‹*Angestellten*›; behalten ‹*Wohnung, Auto*›; **c)** anbehalten, anlassen ‹*Kleid, Mantel*›; aufbehalten ‹*Hut*›

~ **'out 1.** *v. i.* draußen bleiben; **'~ out'** „Zutritt verboten". **2.** *v. t.* **a)** *(not let enter)* nicht hereinlassen ‹*Person, Tier*›; **b)** abhalten ‹*Kälte*›; abweisen ‹*Nässe*›

~ **'out of** *v. t.* **a)** *(stay outside)* ~ **out of a room/an area/a country** ein Zimmer/eine Gegend nicht betreten/nicht in ein Land reisen; **b)** *(avoid)* ~ **out of danger** Gefahren meiden; sich nicht in Gefahr begeben; ~ **out of trouble** zurechtkommen; ~ **out of the rain/sun** *etc.* nicht in den Regen/ die Sonne *usw.* gehen; ~ **out of sb.'s way** jmdm. aus dem Weg gehen; **c)** *(not let enter)* nicht hereinlassen in (+ *Akk.*); **d)** *(cause to avoid)* ~ **the dog out of my way** halte mir den Hund vom Leibe *(ugs.)*

~ **to** *v. t.* **a)** *(not leave)* bleiben auf (+ *Dat.*) ‹*Straße, Weg*›; ~ **to the left!** halte dich links!; bleib links!; **b)** *(follow, observe)* sich halten an (+ *Akk.*) ‹*Regeln, Muster, Gesetz, Diät, usw.*›; einhalten ‹*Zeitplan*›; halten ‹*Versprechen*›; ~ **to one's word** Wort halten; **c)** ~ **[oneself] to oneself** für sich bleiben; **they ~ themselves to themselves** sie bleiben unter sich. *See also* ~ **2 b**

~ **'up 1.** *v. i.* **a)** *(proceed equally)* ~ **up with sb./sth.** mit jmdm./etw. Schritt halten; ~ **up with the Joneses** mit den andern gleichziehen; **b)** *(maintain contact)* ~ **up with sb.** mit jmdm. Kontakt halten; ~ **up with sth.** sich über etw. *(Akk.)* auf dem laufenden halten. **2.** *v. t.* **a)** *(prevent from falling)* festhalten ‹*Leiter, Zelt usw.*›; **b)** *(prevent from sinking)* aufrechterhalten ‹*Produktion, Standard usw.*›; auf gleichem Niveau halten ‹*Preise, Löhne usw.*›; **c)** *(maintain)* aufrechterhalten ‹*Bräuche, Freundschaft, jmds. Moral*›; *(keep in repair)* instand *od. (ugs.)* in Schuß halten ‹*Haus*›; *(keep in proper condition)* in Ordnung *od. (ugs.)* in Schuß halten ‹*Garten*›; **d)** *(continue)* weiterhin zahlen ‹*Raten*›; ~ **one's courage/spirits up** den Mut nicht sinken lassen; ~ **one's strength up** sich bei Kräften halten; ~ **it up** weitermachen; ~ **it up!** weiter so!; **he'll never be able to ~ it up** er wird es nicht durchhalten [können]; **e)** *(prevent from going to bed)* am Schlafengehen hindern; **they kept me up all night** sie haben mich die ganze Nacht nicht schlafen lassen

keeper ['ki:pə(r)] *n.* **a)** *see* **gamekeeper; b)** *see* **goalkeeper; c)** *(zoo~)* Tierwärter, *der*/-wärterin, *die;* **d)** *(custodian)* Wärter, *der*/Wärterin, *die*

keep-'fit *n.* Fitneßtraining, *das*

keep-'fit class *n.* Fitneßgruppe, *die;* **go to ~es** zu Fitneßübungen gehen

keeping ['ki:pɪŋ] *n., no pl.* **a)** *no art.* **be in ~ with sth.** einer Sache *(Dat.)* entsprechen; *(be suited to sth.)* zu etw. passen; **b)** *(custody)* **give sth. into sb.'s ~:** jmdm. etw. zur Aufbewahrung [über]geben

'keepsake *n.* Andenken, *das*

keg [keg] *n.* **a)** *(barrel)* [kleines] Faß; Fäßchen, *das;* **b)** *attrib.* ~ **beer** *aus luftdichten Metallbehältern gezapftes, mit Kohlensäure versetztes Bier;* ≈ Faßbier, *das*

ken [ken] *n.* **this is beyond my ~:** das geht über meinen Horizont; *(beyond range of knowledge)* das übersteigt mein Wissen

kennel ['kenl] *n.* **a)** Hundehütte, *die;* **b)** *in pl.* **[boarding] ~s** Hundepension, *die;* **[breeding] ~s** Zwinger, *der*

Kenya ['kenjə, 'ki:njə] *pr. n.* Kenia *(das)*

Kenyan ['kenjən, 'ki:njən] **1.** *adj.* kenianisch; **sb. is ~:** jmd. ist Kenianer/Kenianerin. **2.** *n.* Kenianer, *der*/Kenianerin, *die*

kept *see* **keep 1, 2**

kerb [kɜ:b] *n. (Brit.)* Bordstein, *der*

'kerbstone *n. (Brit.)* Bordstein, *der*

kernel ['kɜ:nl] *n. (lit. or fig.)* Kern, *der;* **a ~ of truth** ein Körnchen Wahrheit

kestrel ['kestrl] *n. (Ornith.)* Turmfalke, *der*

ketchup ['ketʃʌp] *n.* Ketchup, *der od. das*

kettle ['ketl] *n.* [Wasser]kessel, *der;* **a pretty** *or* **fine ~ of fish** *(iron.)* eine schöne Bescherung *(ugs. iron.)*; **a different ~ of fish** eine ganz andere Sache

'kettle-drum *n. (Mus.)* [Kessel]pauke, *die*

key [ki:] **1.** *n.* **a)** *(lit. or fig.)* Schlüssel, *der;* **the ~ to success** der Schlüssel zum Erfolg; **the ~ to the mystery** des Rätsels Lösung; **b)** *(set of answers)* [Lösungs]schlüssel, *der; (to map etc.)* Zeichenerklärung, *die; (to cipher)* Schlüssel, *der;* **c)** *(on piano, typewriter, etc.)* Taste, *die; (on wind instrument)* Klappe, *die;* **d)** *(Mus.)* Tonart, *die;* **sing/play in/off ~:** richtig/falsch singen/spielen. **2.** *attrib. adj.* entscheidend; Schlüssel‹*frage, -position, -rolle, -figur, -industrie*›. **3.** *v. t. (Computing)* eintasten

~ **in** *v. t. (Computing)* eintasten

key: ~board 1. *n. (of piano etc.)* Klaviatur, *die; (of typewriter etc.)* Tastatur, *die;* **2.** *v. t.* tasten; **~boarder** *n.* Taster, *der*/Tasterin, *die*

keyed up [ki:d'ʌp] *adj.* **be all ~:** ganz aufgeregt sein

key: ~hole *n.* Schlüsselloch,

das; ~**note** *n.* **a)** *(Mus.)* Grundton, *der;* **b)** *(fig.)* Grundgedanke, *der;* [Grund]tenor, *der;* ~**note speech** programmatische Rede; ~**ring** *n.* Schlüsselring, *der;* ~ **signature** *n. (Mus.)* Tonartvorzeichnung, *die;* ~**word** *n. (key to cipher)* Schlüsselwort, *das*

kg. *abbr.* **kilogram[s]** kg

khaki ['kɑ:kɪ] **1.** *adj.* khakifarben. **2.** *n. (cloth)* Khaki, *der*

kibbutz [kɪ'bʊts] *n., pl.* **kibbutzim** [kɪbʊt'si:m] Kibbuz, *der*

kick [kɪk] **1.** *n.* **a)** [Fuß]tritt, *der; (Footb.)* Schuß, *der;* **give sb. a ~:** jmdm. einen Tritt geben *od.* versetzen; **give sth. a ~:** gegen etw. treten; **give sb. a ~ in the pants** *(fig. coll.)* jmdm. Feuer unterm Hintern machen *(salopp);* **a ~ in the teeth** *(fig.)* ein Schlag ins Gesicht; **b)** *(Sport: burst of speed)* Spurt, *der;* **c)** *(coll.: sharp effect, thrill)* Kitzel, *der; (of wine)* Feuer, *das;* **he gets a ~ out of it** er hat Spaß daran; es macht ihm Spaß; **do sth. for ~s** etw. zum Spaß tun; **d)** *(recoil of gun)* Rückstoß, *der.* **2.** *v. i.* **a)** treten; ‹*Pferd:*› ausschlagen; ‹*Baby:*› strampeln; ‹*Tänzer:*› das Bein hochwerfen; **~ at sth.** gegen etw. treten; **b)** *(show opposition)* sich zur Wehr setzen (**at, against** gegen). **3.** *v. t.* **a)** einen Tritt geben (+ *Dat.*) ‹*Person, Hund*›; treten gegen ‹*Gegenstand*›; kicken *(ugs.),* schlagen, schießen ‹*Ball*›; **~ the door open/shut** die Tür auf-/zutreten; **he ~ed the ball straight at me** er kickte den Ball genau in meine Richtung; **~ sb. in the teeth** *(fig. coll.)* jmdn. vor den Kopf stoßen; **I could ~ myself!** *(coll.)* ich könnte mir *od.* mich in den Hintern beißen *(salopp);* **b)** *(sl.: abandon)* ablegen ‹*schlechte Angewohnheit*›; aufgeben ‹*Rauchen*›; **~ the habit** es aufstecken *(ugs.)*

~ **a'bout, ~ a'round 1.** *v. t.* **a)** [in der Gegend] herumkicken *(ugs.);* **b)** *(treat badly)* herumstoßen; schikanieren. **2.** *v. i.* **be ~ing about** *or* **around** *(coll.) (be present, alive)* rumhängen *(ugs.); (lie scattered)* rumliegen *(ugs.)*

~ **'in** *v. t. (break, damage)* eintreten

~ **'off 1.** *v. t.* von sich schleudern ‹*Kleidungsstück, Schuhe*›. **2.** *v. i. (Footb.)* anstoßen; ‹*Spiel:*› beginnen; *(fig. coll.: start)* anfangen

~ **'out** *v. t. (force to leave)* hinauswerfen; rausschmeißen *(ugs.);* **get ~ed out** rausfliegen *(ugs.);* **get ~ed out of one's job** [aus der Stellung] fliegen *(ugs.)*

~ **'up** *v. t. (coll.: create)* **~ up a fuss/row** Krach schlagen/anfangen *(ugs.)*

kid [kɪd] **1.** *n.* **a)** *(young goat)* Kitz, *das;* Zickel, *das;* **b)** *(leather)* Ziegenleder, *das; attrib.* Ziegenleder-; **c)** *(coll.: child)* Kind, *das; (Amer. coll.: young person)* Jugendliche, *der/die;* Kid, *das (ugs.);* **it's ~[s'] stuff** *(sl.: easy)* das ist ein Kinderspiel; **~ brother/sister** *(sl.)* kleiner Bruder/kleine Schwester; Brüderchen, *das*/Schwesterchen, *das.* **2.** *v. t.,* **-dd-** *(coll.) (hoax)* anführen *(ugs.);* auf den Arm nehmen *(ugs.); (deceive)* was vormachen (+ *Dat.*) *(ugs.); (tease)* aufziehen *(ugs.);* **~ oneself** sich *(Dat.)* was vormachen *(ugs.).* **3.** *v. i.,* **-dd-** *(coll.)* **be ~ding** Spaß machen *(ugs.);* **you've got to be ~ding!** das ist doch nicht dein Ernst!; **no ~ding** [ganz] im Ernst *od.* ohne Scherz

kiddie ['kɪdɪ] *n. (coll.)* Kindchen, *das*

kid 'glove *n.* Glacéhandschuh, *der;* **handle sb. with ~s** *(fig.)* jmdn. mit Samt- *od.* Glacéhandschuhen anfassen *(ugs.)*

kidnap ['kɪdnæp] *v. t., (Brit.)* **-pp-** entführen ‹*Person*›; *(to obtain ransom)* kidnappen; entführen

kidnapper ['kɪdnæpə(r)] *n.* Entführer, *der*/Entführerin, *die;* Kidnapper, *der*/Kidnapperin, *die*

kidney ['kɪdnɪ] *n. (Anat., Gastr.)* Niere, *die*

kidney: ~ **bean** *n.* Gartenbohne, *die; (scarlet runner bean)* Feuerbohne, *die;* ~ **machine** *n.* künstliche Niere

kill [kɪl] **1.** *v. t.* **a)** töten; *(deliberately)* umbringen; ‹*Rauchen usw.:*› tödliche Folgen haben für; **be ~ed in action** im Kampf fallen; **shoot to ~:** gezielt schießen; **be ~ed in a car crash** bei einem Autounfall ums Leben kommen; **the shock almost ~ed her** sie wäre vor Schreck fast gestorben; **it won't ~ you** *(iron.)* es wird dich [schon] nicht *od.* nicht gleich umbringen; **~ oneself** sich umbringen; **~ oneself laughing** *(fig.)* sich totlachen; **b)** *(coll.: cause severe pain to)* **it is ~ing me** das bringt mich noch um; **my feet are ~ing me** meine Füße tun wahnsinnig weh *(ugs.);* **c)** abtöten ‹*Krankheitserreger, Schmerz, Ungeziefer, Hefe*›; absterben lassen ‹*Bäume, Pflanzen*›; totschlagen ‹*Geschmack*›; verderben ‹*Witz*›; [ab]töten ‹*Gefühl*›; zerstören ‹*Glauben*›; **d) ~ time** sich *(Dat.)* die Zeit vertreiben; die Zeit totschlagen *(abwertend);* **e)** *(obtain meat from)* schlachten ‹*Tier*›; **f)** *(overwhelm)* überwältigen; **dress to ~:** sich herausputzen. **2.** *n.* *(~ing of game)* Abschuß, *der; (prey)* Beute, *die;* **move in for the ~** ‹*Raubtier:*› die Beute anschleichen, zum Sprung auf die Beute ansetzen; *(fig.)* zum entscheidenden Schlag ausholen

~ **'off** *v. t.* vernichten ‹*Feinde, Konkurrenz*›; abschlachten ‹*Vieh*›; sterben lassen ‹*Romanfigur usw.*›; vertilgen ‹*Unkraut*›; scheitern lassen ‹*Projekt*›

killer ['kɪlə(r)] *n.* Mörder, *der*/Mörderin, *die; (murderous ruffian)* Killer, *der (salopp);* **be a ~** ‹*Krankheit:*› tödlich sein; *attrib.* **the ~ instinct** der Instinkt zum Töten; der Killerinstinkt *(Sportjargon)*

'killer whale *n.* Mörderwal, *der*

killing ['kɪlɪŋ] **1.** *n.* **a)** Töten, *das;* Tötung, *die;* **the ~ of the three children** der Mord an den drei Kindern; **b)** *(instance)* Mord[fall], *der;* **c)** *(fig. coll.: great success)* Coup, *der (ugs.);* **make a ~** *(make a great profit)* einen [Mords]reibach machen *(ugs.).* **2.** *adj.* **a)** tödlich; **b)** *(coll.: exhausting)* mörderisch *(ugs.);* **c)** *(coll.: attractive, amusing, etc.)* umwerfend

'killjoy *n.* Spielverderber, *der*/-verderberin, *die*

kiln [kɪln] *n. (for burning/drying)* [Brenn-/Trocken]ofen, *der*

kilo ['ki:ləʊ] *n., pl.* **~s** Kilo, *das*

kilo- ['kɪlə] *pref.* kilo-/Kilo-

'kilogram, 'kilogramme *n.* Kilogramm, *das*

kilometre (*Brit.; Amer.:* **kilometer**) ['kɪləmi:tə(r) *(Brit.),* kɪ'lɒmɪtə(r)] *n.* Kilometer, *der*

'kilowatt *n. (Electr., Phys.)* Kilowatt, *das*

kilt [kɪlt] *n.* Schottenrock, *der*

kilter ['kɪltə(r)] *n.* **be out of ~:** nicht in Ordnung sein

kin [kɪn] *n. (ancestral stock)* Geschlecht, *das; (relatives)* Verwandte; *see also* **kith; next 3 b**

[1]**kind** [kaɪnd] *n.* **a)** *(class, sort)* Art, *die;* **several ~s of apples** mehrere Sorten Äpfel; **all ~s of things/excuses** alles mögliche/alle möglichen Ausreden; **no ... of any ~:** keinerlei ...; **books of every ~:** Bücher aller Art; **be [of] the same ~:** von derselben Sorte *od.* Art sein; **I know your ~:** deine Sorte kenne ich; **something/nothing of the ~:** so etwas Ähnliches/nichts dergleichen; **you'll do nothing of the**

~! das kommt gar nicht in Frage!; **two of a ~**: zwei gleiche; **what ~ is it?** was für einer/eine/eins ist es?; **what ~ of [a] tree is this?** was für ein Baum ist das?; **what ~ of people are they?** was für Leute sind sie?; **what ~ of [a] fool do you take me for?** für wie dumm hältst du mich?; **what ~ of [a] person do you think I am?** für wen hältst du mich?; **the ~ of person we need** der Typ, den wir brauchen; **they are the ~ of people who ...**: sie gehören zu der Sorte von Leuten, die ...; das sind solche Leute, die ...; **this ~ of food/atmosphere** diese Art *od.* solches Essen/solch *od.* so eine Stimmung; **these ~ of people/things** *(coll.)* solche Leute/Sachen; **b)** *(implying vagueness)* **a ~ of ...**: [so] eine Art ...; **~ of interesting/cute** *etc. (coll.)* irgendwie interessant/niedlich *usw. (ugs.)*; **c) in ~** *(not in money)* in Sachwerten; **pay in ~**: in Naturalien zahlen/bezahlen; **pay back** *or* **repay sth. in ~** *(fig.)* etw. mit *od.* in gleicher Münze zurückzahlen

[2]**kind** *adj. (of gentle nature)* liebenswürdig; *(showing friendliness)* freundlich; *(affectionate)* lieb; **have a ~ heart** gutherzig sein; **would you be so ~ as to do that?** wären Sie so freundlich, das zu tun?; **be ~ to animals/children** gut zu Tieren/Kindern sein; **oh, you 'are ~!** sehr nett *od.* liebenswürdig von Ihnen; **how ~!** wie nett [von ihm/ihr/Ihnen *usw.*]!

kindergarten ['kɪndəgɑ:tn] *n.* Kindergarten, *der; (forming part of a school)* ≈ Vorklasse, *die*

kind-hearted [kaɪnd'hɑ:tɪd] *adj.* gutherzig; liebenswürdig ‹*Geste, Handlung*›

kindle ['kɪndl] *v. t. (light)* anzünden, *(geh.)* entzünden ‹*Holz, Feuer*›; entfachen *(geh.)* ‹*Flamme*›; wecken ‹*Interesse, Gefühl*›

kindling ['kɪndlɪŋ] *n., no pl., no indef. art.* Anmachholz, *das*

kindly ['kaɪndlɪ] **1.** *adv.* **a)** freundlich; nett; **..., she said ~**: ..., sagte sie freundlich; **b)** *in polite request etc.* freundlicherweise; **c) he didn't take at all ~ to the suggestion** er konnte sich mit dem Vorschlag gar nicht recht anfreunden; **d) thank sb. ~**: jmdm. herzlich danken; **thank you ~**: herzlichen Dank. **2.** *adj.* freundlich; nett; liebenswürdig; *(kind-hearted)* gütig; wohlwollend; gut ‹*Herz, Tat*›

kindness ['kaɪndnɪs] *n.* **a)** *no pl. (kind nature)* Freundlichkeit, *die;* Liebenswürdigkeit, *die;* **do sth. out of ~**: etw. aus Gefälligkeit tun; **out of the ~ of one's heart** aus reiner Freundlichkeit; **b)** *(kind act)* Gefälligkeit, *die;* **do sb. a ~**: jmdm. eine Gefälligkeit erweisen *od.* einen Gefallen tun

kindred ['kɪndrɪd] **1.** *n., no pl.* **a)** *(blood relationship)* Blutsverwandtschaft, *die;* **b)** *(one's relatives)* Verwandtschaft, *die;* Verwandte. **2.** *adj.* **a)** *(related by blood)* blutsverwandt; **b)** *(fig.: connected)* verwandt

kindred 'spirit *n.* Gleichgesinnte, *der/die*

kinetic [kɪ'netɪk, kaɪ'netɪk] *adj.* kinetisch

king [kɪŋ] *n. (also Chess, Cards)* König, *der;* **live like a ~**: leben wie ein Fürst; **a feast fit for a ~**: ein königliches Mahl

kingdom ['kɪŋdəm] *n.* **a)** Königreich, *das;* **b) the ~ of God** das Reich Gottes; **the ~ of heaven** das Himmelreich; **wait till ~ come** *(sl.)* bis in alle Ewigkeit warten *(ugs.)*; **c)** *(province of nature)* Reich, *das;* **animal ~**: Tierreich, *das*

'kingfisher ['kɪŋfɪʃə(r)] *n. (Ornith.)* Eisvogel, *der*

kingly ['kɪŋlɪ] *adj.* königlich

king: ~maker *n.* Königsmacher, *der;* **~pin** *n. (lit., or fig.)* Hauptstütze, *die;* **-size[d]** *adj.* extragroß; King-size-‹*Zigaretten*›

kink [kɪŋk] **1.** *n.* **a)** *(in pipe, wire, etc.)* Knick, *der; (in hair, wool)* Welle, *die;* **b)** *(fig.: mental peculiarity)* Tick, *der (ugs.)*; Spleen, *der.* **2.** *v. i.* Knicke kriegen; ‹*Haar:*› sich wellen. **3.** *v. t.* knicken

kinky ['kɪŋkɪ] *adj. (coll.: bizarre, perverted)* spleenig; *(sexually)* abartig

kinship ['kɪnʃɪp] *n.* **a)** *(blood relationship)* Blutsverwandtschaft, *die;* **b)** *(similarity)* Ähnlichkeit, *die; (spiritual)* Verwandtschaft, *die*

kinsman ['kɪnzmən] *n., pl.* **kinsmen** ['kɪnzmən] Verwandte, *der*

kinswoman ['kɪnzwʊmən] *n.* Verwandte, *die*

kiosk ['ki:ɒsk] *n.* **a)** Kiosk, *der;* **b)** *(public telephone booth)* [Telefon]zelle, *die*

kip [kɪp] *(Brit. sl.)* **1.** *n. (sleep)* Schlaf, *der;* **have a** *or* **get some ~**: eine Runde pennen *(salopp)*. **2.** *v. i.*, **-pp-** pennen *(salopp)*; **~ down** sich hinhauen *(salopp)*

kipper ['kɪpə(r)] *n.* Kipper, *der;* ≈ Bückling, *der*

kiss [kɪs] **1.** *n.* Kuß, *der;* **the ~ of death** der Todesstoß; **give sb. the ~ of life** *(Brit.)* jmdn. von Mund zu Mund beatmen. **2.** *v. t.* küssen; **~ sb. good night/goodbye** jmdm. einen Gutenacht-/Abschiedskuß geben. **3.** *v. i.* sich küssen

kit [kɪt] **1.** *n.* **a)** *(personal equipment)* Sachen *(ugs.)*; **b)** *(Brit.: set of items)* Set, *das;* **construction/self-assembly ~**: Bausatz, *der;* **repair ~**: Reparatursatz, *der;* Reparaturset, *das;* **c)** *(Brit.: clothing etc.)* **sports ~**: Sportzeug, *das;* Sportsachen *Pl.*; **d)** *(Brit. Mil.)* Ausrüstung, *die; (pack)* [Feld]gepäck, *das; (uniform)* Montur, *die.* **2.** *v. t.*, **-tt-** *(Brit.)* **~ out** *or* **up** *(equip)* ausrüsten; *(give clothes or uniforms to)* einkleiden

'kitbag *n.* Tornister, *der*

kitchen ['kɪtʃɪn] *n.* Küche, *die; attrib.* Küchen-

kitchenette [kɪtʃɪ'net] *n.* kleine Küche; *(alcove)* Kochnische, *die*

kitchen: ~ 'garden *n.* Küchengarten, *der;* **~ 'sink** *n.* [Küchen]ausguß, *der;* Spüle, *die;* **everything but the ~ sink** *(fig.)* der halbe Hausrat; **~ unit** *n.* Küchenelement, *das;* **~ units** Küchenmöbel; **~ utensil** *n.* Küchengerät, *das;* **~-ware** *n.* Küchengeräte

kite [kaɪt] *n.* **a)** *(toy)* Drachen, *der;* **b)** *(Ornith.)* Roter Milan

kith [kɪθ] *n.* **~ and kin** Freunde und Verwandte

kitten ['kɪtn] *n.* **a)** [Katzen]junge, *das;* Kätzchen, *das;* **the cat has had ~s** die Katze hat Junge bekommen; **b)** *(coll.)* **have ~s** *(be upset)* Zustände kriegen *(ugs.)*; **be having ~s** *(be nervous)* am Rotieren sein *(ugs.)*

[1]**kitty** ['kɪtɪ] *n. (kitten)* Kätzchen, *das; (child lang.)* Miez[e], *die (fam.)*

[2]**kitty** *n.* **a)** *(Cards)* [Spiel]kasse, *die;* **b)** *(joint fund)* Kasse, *die*

'kiwi fruit ['ki:wi: fru:t] *n.* Kiwi[frucht], *die*

kleptomania [kleptə'meɪnɪə] *n., no pl. (Psych.)* Kleptomanie, *die*

knack [næk] *n.* **a)** *(faculty)* Talent, *das;* **have a ~ of doing sth.** das Talent haben, etw. zu tun; **get the ~ [of doing sth.]** den Bogen rauskriegen[, wie man etw. macht] *(ugs.)*; **there's a [real] ~ in** *or* **to doing sth.** es gehört schon [einiges] Geschick dazu, etw. zu tun; **have lost the ~**: es nicht mehr zustande bringen *od. (ugs.)* hinkriegen; **b)** *(habit)* **have a ~ of doing sth.** es [mit seltenem Talent] verstehen, etw. zu tun *(iron.)*

knacker ['nækə(r)] *n. (Brit.)* Abdecker, *der*

knackered ['nækəd] *adj. (Brit. sl.)* geschlaucht *(ugs.)*

knapsack ['næpsæk] *n.* Rucksack, *der; (Mil.)* Tornister, *der*

knead [ni:d] *v. t.* kneten

knee [ni:] *n.* **a)** Knie, *das;* **on one's ~s/on bended ~[s]** auf Knien; **be on one's ~s** knien; *(fig.: be defeated)* in die Knie gezwungen sein *(geh.);* **force sb. to his ~s** *(fig.)* jmdn. in die Knie zwingen *(geh.);* **go down on one's ~s [to** *or* **before sb.]** [vor jmdm.] auf die Knie sinken *(geh.);* **b)** *(of animal)* Kniegelenk, *das*

knee: ~cap *n. (Anat.)* Kniescheibe, *die;* **~-deep** *adj.* **a)** knietief; **b)** *(fig.: deeply involved)* **be ~-deep in sth.** bis über den Hals in etw. *(Dat.)* stecken *(ugs.);* **~-high** *adj.* kniehoch; **~-jerk** *n.* Kniesehnenreflex, *der; attrib.* **~-jerk reaction** *(fig.)* automatische Reaktion; **~-joint** *n.* Kniegelenk, *das*

kneel [ni:l] *v. i.,* **knelt** [nelt] *or (esp. Amer.)* **~ed** knien; **~ down** niederknien; **~ [down] to do sth.** niederknien *od.* sich [hin]knien, um etw. zu tun; **~ to sb.** vor jmdm. [nieder]knien

'knee-length *adj.* knielang

knell [nel] *n.* Glockengeläut, *das; (at funeral)* Totengeläut, *das*

knelt *see* **kneel**

knew *see* **know 1**

knickerbockers ['nɪkəbɒkəz] *n. pl.* Knickerbocker *Pl.*

knickers ['nikəz] *n. pl. (Brit.: undergarment)* [Damen]schlüpfer, *der*

knick-knack ['nɪknæk] *n.* **~s** Schnickschnack, *der (ugs.)*

knife [naɪf] **1.** *n., pl.* **knives** [naɪvz] Messer, *das;* **put a ~ into sb.** jmdm. ein Messer zwischen die Rippen jagen; **turn** *or* **twist the ~ [in the wound]** *(fig.)* Salz in die Wunde streuen; **the knives are out [for sb.]** *(fig.)* das Messer wird [für jmdn.] gewetzt; *see also* **fork 1 a. 2.** *v. t. (stab)* einstechen auf (+ *Akk.*); *(kill)* erstechen

knife: ~-edge *n.* Schneide, *die;* **be [balanced] on a ~-edge** *(fig.)* auf des Messers Schneide stehen; **~-point** *see* **point 1 b**

knight [naɪt] **1.** *n.* **a)** *(Hist.)* Ritter, *der;* **b)** *(Chess)* Springer, *der.* **2.** *v. t.* adeln; zum Ritter schlagen *(hist.)*

knighthood ['naɪthʊd] *n. (rank)* Ritterwürde, *die;* **receive one's ~:** geadelt werden; in den Ritterstand erhoben werden *(hist.)*

knit [nɪt] **1.** *v. t.,* **-tt-, knitted** *or (esp. fig.)* **knit a)** stricken ⟨*Kleidungsstück usw.*⟩; **b) ~ a stitch** eine [rechte] Masche stricken; **~ 2, purl 2** zwei rechts, zwei links [stricken]; **c) ~ one's brow** die Stirn runzeln; **d) tightly ~** *(fig.)* festgefügt. **2.** *v. i.* ⟨*Knochenbruch:*⟩ verheilen; ⟨*Knochen:*⟩ zusammenwachsen

~ to'gether *v. i.* ⟨*Knochen:*⟩ zusammenwachsen; ⟨*Knochenbruch:*⟩ zusammenheilen

knitting ['nɪtɪŋ] *n., no pl., no indef. art.* Stricken, *das; (work in process of being knitted)* Strickarbeit, *die;* **do one's/some ~:** stricken

knitting: ~ machine *n.* Strickmaschine, *die;* **~-needle** *n.* Stricknadel, *die;* **~-pattern** *n.* Strickmuster, *das*

'knitwear *n., no pl., no indef. art.* Strickwaren *Pl.*

knives *pl. of* **knife 1**

knob [nɒb] *n.* **a)** *(protuberance)* Verdickung, *die; (on club, tree-trunk, etc.)* Knoten, *der;* **b)** *(on door, walking-stick, etc.)* Knauf, *der; (on radio etc.)* Knopf, *der*

knobbly ['nɒblɪ] *adj.* knotig ⟨*Finger, Stock*⟩; knorrig ⟨*Baum*⟩

knock [nɒk] **1.** *v. t.* **a)** *(strike) (lightly)* klopfen gegen *od.* an (+ *Akk.*); *(forcefully)* schlagen gegen *od.* an (+ *Akk.*); **b)** *(make by striking)* schlagen; **~ a hole in sth.** ein Loch in etw. *(Akk.)* schlagen; **c)** *(drive by striking)* schlagen; **~ sb.'s brains out** jmdm. den Schädel einschlagen; **I'd like to ~ their heads together** *(lit.)* ich könnte ihre Köpfe gegeneinanderschlagen; *(fig.: reprove them)* ich möchte ihnen mal gehörig die Leviten lesen; **d) ~ sb. on the head** jmdm. eins über *od.* auf den Schädel geben; **~ sth. on the head** *(fig.: put an end to)* einer Sache *(Dat.)* ein Ende setzen; **e)** *(sl.: criticize)* herziehen über (+ *Akk.*) *(ugs.);* **don't ~ it** halt dich zurück. **2.** *v. i.* **a)** *(strike) (lightly)* klopfen; *(forcefully)* schlagen; **b)** *(seek admittance)* klopfen (**at** an + *Akk.*). **3.** *n.* **a)** *(rap)* Klopfen, *das;* **there was a ~ on** *or* **at the door** es klopfte an der Tür; **b)** *(blow)* Schlag, *der; (gentler)* Stoß, *der;* **c)** *(fig.: blow of misfortune)* [Schicksals]schlag, *der;* **take a ~:** einen Schlag erleiden

~ a'bout 1. *v. t.* schlagen; verprügeln; **be ~ed about** Schläge *od.* Prügel einstecken müssen. **2.** *v. i.* herumhängen *(ugs.);* ⟨*Gegenstand:*⟩ herumfliegen *(ugs.);* **~ about with sb.** sich mit jmdm. herumtreiben *(ugs.)*

~ a'round *see* **~ about**

~ 'back *v. t. (coll.)* **a)** *(eat quickly)* verputzen *(ugs.); (drink quickly)* hinunterkippen *(ugs.);* **b)** *(cost)* **~ sb. back a thousand** jmdn. um einen Tausender ärmer machen

~ 'down *v. t.* **a)** *(strike to the ground)* niederreißen, umstürzen ⟨*Zaun, Hindernis*⟩; *(with fist or weapon)* niederschlagen; ⟨*Fahrer, Fahrzeug:*⟩ umfahren ⟨*Person*⟩; **b)** *(demolish)* abreißen; abbrechen; **c)** *(sell by auction)* zuschlagen; **~ sth. down to sb.** jmdm. etw. zuschlagen

~ 'off 1. *v. t.* **a)** *(coll.: leave off)* aufhören mit; **~ off work** Feierabend machen; **~ it off!** *(coll.)* hör auf [damit]!; **b)** *(coll.) (produce rapidly)* aus dem Ärmel schütteln *(ugs.);* **c)** *(deduct)* **~ five pounds off the price** es fünf Pfund billiger machen; **d)** *(sl.: steal)* mitgehen lassen *(ugs.);* klauen *(salopp);* **e)** *(sl.: copulate with)* bumsen *(salopp).* **2.** *v. i. (coll.)* Feierabend machen; **~ off for lunch** Mittag machen

~ 'out *v. t.* **a)** *(make unconscious)* bewußtlos umfallen lassen; **b)** *(Boxing)* k.o. schlagen; **c)** *(fig.: defeat)* **be ~ed out** ausscheiden *od. (ugs.)* rausfliegen; **they ~ed us out of the Cup** sie warfen uns aus dem Pokal; **d)** *(sl.: astonish)* umhauen *(salopp);* **e)** *(sl.: exhaust)* kaputtmachen *(ugs.)*

~ 'over *v. t.* umstoßen; ⟨*Fahrer, Fahrzeug:*⟩ umfahren ⟨*Person*⟩

~ to'gether 1. *v. t.* zusammenzimmern *(ugs.)* ⟨*Hütte, Tisch, Bühne*⟩; *see also* **~ 1 c. 2.** *v. i.* **my knees were ~ing together** mir schlotterten die Knie

~ 'up *v. t.* **a)** *(make hastily)* [her]zaubern ⟨*Mahlzeit, Imbiß*⟩; grob skizzieren ⟨*Plan*⟩; **b)** *(score)* erzielen; **c)** *(Brit.: awaken)* durch Klopfen wecken; *(unexpectedly)* herausklopfen; **d)** *(exhaust)* fertigmachen *(ugs.);* **be ~ed up** fertig *od.* groggy sein *(ugs.);* **e)** *(sl.: make pregnant)* dick machen *(derb)*

knock-down *adj.* **a)** *(low)* **~-down cost/prices** minimale Kosten/Schleuderpreise; **b)** *(minimum)* Mindest⟨*preis, -gebot*⟩

knocker ['nɒkə(r)] *n.* **a)** *(on door)* [Tür]klopfer, *der;* **b)** *(sl.: critic)* Beckmesser, *der*

knock: ~-kneed ['nɒkni:d] *adj.* X-beinig ⟨*Person*⟩; **~ 'knees** *n. pl.* X-Beine *Pl.;* **~-out 1.** *n.* **a)**

(blow) Knockout[schlag], *der;* K.-o.[-Schlag], *der;* **b)** *(competition)* Ausscheidungs[wett]kampf, *der;* **c)** *(sl.: outstanding person or thing)* **sb./sth. is a [real] ~-out** jmd./etw. ist eine Wucht *(salopp);* **2.** *adj.* **a) ~-out blow** K.-o.-Schlag, *der;* **b)** Ausscheidungs⟨*spiel, -[wett]kampf, -runde*⟩

knoll [nəʊl] *n.* Anhöhe, *die*

knot [nɒt] **1.** *n.* **a)** Knoten, *der;* **tie sb. [up] in ~s** *(fig. coll.)* jmdn. in Widersprüche verwickeln; **b)** *(in wood)* Ast, *der;* **c)** *(speed unit)* Knoten, *der;* **at a rate of ~s** *(coll.)* mit einem Affenzahn *(salopp).* **2.** *v. t.,* **-tt-: a)** *(tie)* knoten ⟨*Seil, Faden usw.*⟩; knoten ⟨*Schnürsenkel*⟩; knoten, binden ⟨*Krawatte*⟩; **~ threads together** Fäden verknoten; **~ a rope** Knoten in ein Seil machen; **b)** *(entangle)* verfilzen; **c) get ~ted!** *(sl.)* rutsch mir den Buckel runter! *(ugs.)*

knotty ['nɒtɪ] *adj. (fig.: puzzling)* verwickelt

know [nəʊ] **1.** *v. t.,* **knew** [nju:], **~n** [nəʊn] **a)** *(recognize)* erkennen **(by** an + *Dat.,* **for** als + *Akk.*); **b)** *(be able to distinguish)* **~ sth. from sth.** etw. von etw. unterscheiden können; **~ the difference between right and wrong** den Unterschied zwischen Gut und Böse kennen; **he wouldn't ~ the difference** er wüßte den Unterschied nicht; **c)** *(be aware of)* wissen; kennen ⟨*Person*⟩; **I ~ who she is** ich weiß, wer sie ist; **I ~ for a fact that ...:** ich weiß ganz bestimmt, daß ...; **it is ~n that ...:** man weiß, daß ...; es ist bekannt, daß ...; **~ sb./sth. to be ...:** wissen, daß jmd./etw. ... ist; **that's/that might be worth ~ing** das ist gut/wäre wichtig zu wissen; **he doesn't want to ~:** er will nichts davon wissen *od.* hören; **I 'knew it** ich hab's ja geahnt; **'I ~ what** ich weiß was *(ugs.);* **you ~** *(coll.: as reminder)* weißt du [noch]; **you ~ something** *or* **what?** weißt du was?; **you never ~:** man kann nie wissen *(ugs.);* **sb. has [never] been ~n to do sth.** jmd. hat bekanntlich [noch nie] etw. getan; **and he ~s it** und er weiß das auch; **don't I ~ it!** *(coll.)* das weiß ich nur zu gut; **before sb. ~s where he is** ehe jmd. sich's versieht; **what do you ~ [about that]?** *(coll.: that is surprising)* was sagst du dazu?; **sb. is not to ~** *(is not to be told)* jmd. soll nichts wissen **(about, of** von); *(has no way of learning)* jmd. kann nicht wissen; **not ~ what hit one** *(fig.)* gar nicht begreifen, was geschehen ist; **that's all 'you ~ [about it]** das glaubst du vielleicht; **if you 'must ~:** wenn du es unbedingt wissen willst; **~ different** *or* **otherwise** es besser wissen; **~ what's what** wissen, wie es in der Welt zugeht; **how should I ~?** woher soll ich das wissen?; **I might have ~n** das hätte ich mir denken können; **do you ~, ...:** stell dir [mal] vor, ...; **d)** *(have understanding of)* können ⟨*ABC, Einmaleins, Deutsch usw.*⟩; beherrschen ⟨*Grundlagen, Regeln*⟩; sich auskennen mit ⟨*Gerät, Verfahren, Gesetz*⟩; **do you ~ any German?** können Sie etwas Deutsch?; **~ how to mend fuses** wissen, wie man Sicherungen repariert; **~ how to drive a car** Auto fahren können; **he doesn't ~ much about computers** er hat nicht viel Ahnung von Computern; **e)** *(be acquainted with)* kennen; **we have ~n each other for years** wir kennen uns [schon] seit Jahren; **you don't really ~ him** du kennst ihn nicht gut genug; **you ~ what he/it is** *(is like)* du kennst ihn ja/du weißt ja, wie es ist; **f)** *(have experience of)* erleben; erfahren; **he ~s no fear** er kennt keine Furcht; **~ what it is to be hungry** wissen, was es heißt, Hunger zu haben. **2.** *n. (coll.)* **be in the ~:** Bescheid wissen

~ about *v. t.* wissen über (+ *Akk.*); **oh, I didn't ~ about it/that** oh, das habe ich nicht gewußt; **did you ~ about your son's behaviour?** haben Sie gehört, wie sich Ihr Sohn benommen hat?; **I don't ~ about 'that** na, ich weiß nicht [so recht]

~ of *v. t.* wissen von ⟨*Plänen, Vorhaben*⟩; kennen, wissen ⟨*Lokal, Geschäft*⟩; **~ of sb.** von jmdm. gehört haben; **not that I ~ of** nicht, daß ich wüßte

know: ~-all *n. (derog.)* Neunmalkluge, *der/die (spöttisch);* **~-how** *n., no pl., no indef. art.* praktisches Wissen; *(technical expertise)* Know-how, *das*

knowing ['nəʊɪŋ] *adj.* **a)** *(shrewd)* verschmitzt ⟨*Blick, Lachen, Lächeln*⟩; *(indicating possession of inside information)* vielsagend, wissend ⟨*Blick, Lächeln*⟩; **b)** *(derog.: cunning)* verschlagen *(abwertend)*

knowingly ['nəʊɪŋlɪ] *adv.* **a)** *(intentionally)* wissentlich ⟨*lügen, verletzen*⟩; bewußt ⟨*planen*⟩; **b)** *(in a shrewd manner)* verschmitzt ⟨*lachen, blicken*⟩; *(indicating possession of inside information)* vielsagend ⟨*lächeln, anblicken*⟩

knowledge ['nɒlɪdʒ] *n., no pl.* **a)** *(familiarity)* Kenntnisse **(of** in + *Dat.*); **a ~ of this field** Kenntnisse auf diesem Gebiet; **~ of human nature** Menschenkenntnis, *die;* **b)** *(awareness)* Wissen, *das;* **have no ~ of sth.** nichts von etw. wissen; keine Kenntnis von etw. haben *(geh.);* **she had no ~ of it** sie wußte nichts davon; sie war völlig ahnungslos; **sth. came to my ~:** etw. ist mir zu Ohren gekommen; **[not] to my** *etc.* **~:** meines *usw.* Wissens [nicht]; **c)** *(understanding)* **[a] ~ of languages/French** Sprach-/Französischkenntnisse *Pl.;* **sb. with [a] ~ of computers** jmd., der sich mit Computern auskennt; **d)** *no art. (what is known)* Wissen, *das*

knowledgeable ['nɒlɪdʒəbl] *adj.* sachkundig; **be ~ about** *or* **on sth.** viel über etw. *(Akk.)* wissen

known [nəʊn] **1.** *see* **know 1. 2.** *adj.* bekannt; *(generally recognized)* anerkannt

knuckle ['nʌkl] *n.* **a)** *(Anat.)* [Finger]knöchel, *der;* **b)** Hachse, *die;* **~ of pork** Eisbein, *das*

~ 'down *v. i.* **~ down to sth.** sich hinter etw. *(Akk.)* klemmen *(ugs.)*

~ 'under *v. i.* klein beigeben **(to** gegenüber)

'knuckleduster *n.* Schlagring, *der*

KO *abbr.* **knock-out** K. o.

koala [kəʊ'ɑ:lə] *n.* **~ [bear]** *(Zool.)* Koala, *der;* Beutelbär, *der*

Koran [kɔ:'rɑ:n, kə'rɑ:n] *n. (Muslim Relig.)* Koran, *der*

Korea [kə'rɪə] *pr. n.* Korea *(das)*

Korean [kə'ri:ən] **1.** *adj.* koreanisch; **sb. is ~:** jmd. ist Koreaner/Koreanerin. **2.** *n.* **a)** *(person)* Koreaner, *der*/Koreanerin, *die;* **b)** *(language)* Koreanisch, *das; see also* **English 2 a**

kosher ['kəʊʃə(r), 'kɒʃə(r)] *adj.* koscher

kowtow [kaʊ'taʊ] *v. i.* **~ [to sb./sth.]** [vor jmdm./etw.] [s]einen Kotau machen

Kraut [kraʊt] *n. & adj. (sl. derog.) angelsächsische abwertende Bez. für „Deutscher" und „deutsch"*

Kremlin ['kremlɪn] *n.* **the K~:** der Kreml

kudos ['kju:dɒs] *n., no pl., no indef. art. (coll.)* Prestige, *das*

kung fu [kʊŋ'fu:, kʌŋ'fu:] *n.* Kung-Fu, *das*

kW *abbr.* **kilowatt[s]** kW

L

L, l [el] *n., pl.* **Ls** *or* **L's** L, l, *das*
L. *abbr.* **Lake**
£ *abbr.* **pound[s]** £; **cost £5** 5 £ *od.* Pfund kosten
l. *abbr.* **a) litre[s]** l; **b) left** l.; **c) line** Z.
lab [læb] *n. (coll.)* Labor, *das*
label ['leɪbl] **1.** *n.* **a)** *(slip)* Schildchen, *das; (on goods, bottles, jars, in clothes)* Etikett, *das; (tied/stuck to an object)* Anhänger/Aufkleber, *der;* **b)** *(on record)* Label, *das; (record company)* Plattenfirma, *die;* **c)** *(fig.: classifying phrase)* Etikett, *das;* **acquire the ~ of ...:** als ... etikettiert werden. **2.** *v. t., (Brit.)* **-ll-: a)** *(attach ~ to)* etikettieren; *(attach price-tag to)* auszeichnen ‹*Waren*›; *(write on)* beschriften; **b)** *(fig.: classify)* **~ sb./sth. [as] sth.** jmdn./etw. als etw. etikettieren
labor *(Amer.) see* **labour**
laboratory [lə'bɒrətərɪ] *n.* Labor[atorium], *das*
labored, laborer *(Amer.) see* **labour-**
laborious [lə'bɔ:rɪəs] *adj.* mühsam; mühevoll ‹*Forschung, Aufgabe usw.*›; schwerfällig, umständlich ‹*Stil*›
laboriously [lə'bɔ:rɪəslɪ] *adv.* *(with difficulty)* mühevoll; **~ slow** mühsam und schleppend
labour ['leɪbə(r)] *(Brit.)* **1.** *n.* **a)** *(task)* Arbeit, *die;* **sth. is/they did it as a ~ of love** etw. geschieht/sie taten es aus Liebe zur Sache; **b)** *(exertion)* Mühe, *die;* **c)** *(work)* Arbeit, *die;* **cost of ~:** Arbeitskosten *Pl.;* **d)** *(body of workers)* Arbeiterschaft, *die;* **immigrant ~:** eingewanderte Arbeitskräfte; **e)** **L~** *(Polit.)* die Labour Party; **f)** *(childbirth)* Wehen *Pl.;* **be in ~:** in den Wehen liegen; **go into ~:** die Wehen bekommen. *See also* **intensive e. 2.** *v. i.* **a)** *(work hard)* hart arbeiten (**at, on** an + *Dat.*); *(slave away)* sich abmühen (**at, over** mit); **b)** *(strive)* sich einsetzen (**for** für); **c) ~ under a delusion** sich einer Täuschung *(Dat.)* hingeben. **3.** *v. t. (elaborate needlessly)* **~ the point** sich lange darüber verbreiten; **there's no need to ~ the point** du brauchst dich nicht lange darüber zu verbreiten
labour: ~ camp *n.* Arbeitslager, *das;* **L~ Day** *n.* Tag der Arbeit *(in Amerika: erster Montag im September)*
laboured ['leɪbəd] *adj. (Brit.)* mühsam; schwerfällig ‹*Stil*›; mühsam zusammengetragen ‹*Argumente*›; **his breathing was ~:** er atmete schwer
labourer ['leɪbərə(r)] *n. (Brit.)* Arbeiter, *der*/Arbeiterin, *die*
labour: L~ Exchange *n. (Brit. Hist./coll.)* Arbeitsamt, *das;* **~ force** *n.* Arbeitskräfte; **~-market** *n.* Arbeitsmarkt, *der;* **~ pains** *n. pl.* Wehenschmerzen; **L~ Party** *n. (Polit.)* Labour Party, *die;* **~ relations** *n. pl.* Beziehungen zwischen Arbeitgebern und Arbeitnehmern; *(within one company)* Betriebsklima, *das;* **~-saving** *adj.* arbeit[s]sparend ‹*Methode, Vorrichtung*›
Labrador ['læbrədɔ:(r)] *n.* **~ [dog *or* retriever]** Labrador[hund], *der*
labyrinth ['læbərɪnθ] *n.* Labyrinth, *das*
lace [leɪs] **1.** *n.* **a)** *(for shoe)* Schuhband, *das (bes. südd.);* Schnürsenkel, *der (bes. nordd.);* **b)** *(fabric)* Spitze, *die; attrib.* Spitzen-. **2.** *v. t.* **a)** *(fasten)* **~ [up]** [zu]schnüren; **b)** *(pass through)* [durch]ziehen; **c) ~ sth. with alcohol** einen Schuß Alkohol in etw. *(Akk.)* geben; **~d with brandy** mit einem Schuß Weinbrand; **~ sb.'s drink** einen Schuß Alkohol/eine Droge in jmds. Getränk *(Akk.)* geben
lacerate ['læsəreɪt] *v. t.* aufreißen; **her arm was badly ~d** sie hatte tiefe Wunden am Arm
laceration [læsə'reɪʃn] *n.* Rißwunde, *die; (from glass)* Schnittwunde, *die*
'lace-up 1. *attrib. adj.* zum Schnüren *nachgestellt;* **~-up boot** Schnürstiefel, *der.* **2.** *n.* Schnürschuh/-stiefel, *der*
lack [læk] **1.** *n.* Mangel, *der* (**of** an + *Dat.*); **~ of self-consciousness** Unbefangenheit, *die;* **~ of work** Arbeitsmangel, *der;* **there is no ~ of it [for them]** es fehlt [ihnen] nicht daran; **for ~ of sth.** aus Mangel an etw. *(Dat.);* **for ~ of time** aus Zeitmangel. **2.** *v. t.* **sb./sth. ~s sth.** jmdm./einer Sache fehlt es an etw. *(Dat.);* **sb. ~s the ability to do sth.** jmdm. fehlt die Fähigkeit, etw. zu tun; **what he ~s is ...:** woran es ihm fehlt, ist ...; **~ content** inhaltsarm sein. **3.** *v. i.* **I ~ for nothing** mir fehlt es an nichts
lackadaisical [lækə'deɪzɪkl] *adj.* *(unenthusiastic)* gleichgültig; desinteressiert; *(listless)* lustlos
lackey ['lækɪ] *n.* **a)** *(footman)* Lakai, *der;* **b)** *(servant)* Diener, *der*
lacking ['lækɪŋ] *adj.* **be ~** ‹*Geld, Ressourcen usw.:*› fehlen; **he was found to be ~** *(incapable)* es erwies sich, daß er den Ansprüchen nicht genügte
'lacklustre *adj.* trüb; glanzlos ‹*Augen*›; matt ‹*Lächeln*›; langweilig ‹*Aufführung, Party*›
laconic [lə'kɒnɪk] *adj.* **a)** *(concise)* lakonisch; **b)** wortkarg ‹*Person, Naturell*›
lacquer ['lækə(r)] **1.** *n.* Lack, *der.* **2.** *v. t.* lackieren; **~ed wood** Lackholz, *das*
lacy ['leɪsɪ] *adj.* Spitzen-; *(of metalwork)* spitzenartig; Filigran-
lad [læd] *n.* **a)** *(boy)* Junge, *der;* **young ~:** kleiner Junge; **when I was a ~:** als ich noch ein Junge war; **these are my ~s** das sind meine Jungen *od. (ugs.)* Jungs; **b)** *(man)* Typ, *der;* **the ~s** die Jungs *(ugs.);* **he always goes out for a drink with the ~s** er geht immer mit seinen Kumpels einen trinken *(ugs.);* **my ~:** mein Junge *(ugs.);* **c)** *(spirited person)* **be a bit of a ~:** kein Kind von Traurigkeit sein *(ugs.); (one for the ladies)* es mit den Mädchen/Frauen haben *(ugs.)*
ladder ['lædə(r)] **1.** *n.* **a)** *(lit. or fig.)* Leiter, *die; (fig.: means of advancement)* Aufstiegsmöglichkeit, *die;* **have a foot on the ~:** die erste Sprosse auf der Leiter des Erfolgs erklommen haben *(geh.);* **b)** *(Brit.: in tights etc.)* Laufmasche, *die.* **2.** *v. i. (Brit.)* Laufmaschen/eine Laufmasche bekommen. **3.** *v. t. (Brit.)* Laufmaschen/eine Laufmasche machen in (+ *Akk.*)
laddie ['lædɪ] *n.* Jungchen, *das (fam.);* Bubi, *der (bes. südd.)*
laden ['leɪdn] beladen (**with** mit)
la-di-da [lɑ:dɪ'dɑ:] *adj.* affektiert
ladies' ['leɪdɪz]: **~ man** *n.* Frauenheld, *der;* **~ room** *n.* Damentoilette, *die*
ladle ['leɪdl] **1.** *n.* Schöpfkelle, *die;* Schöpflöffel, *der.* **2.** *v. t.* schöpfen
~ 'out *v. t. (lit. or fig.)* austeilen
lady ['leɪdɪ] *n.* **a)** Dame, *die; (English, American, etc. also)* Lady,

die; **~-in-waiting** *(Brit.)* Hofdame, *die;* **ladies' hairdresser** Damenfriseur, *der;* **b) 'Ladies'** „Damen"; **c)** *as form of address in sing. (poet.)* Herrin *(veralt.); in pl.* meine Damen; **Ladies and Gentlemen!** meine Damen und Herren!; **d)** *(Brit.) as title* **L~:** Lady; **my ~:** Mylady; **e)** *(ruling woman)* Herrin, *die;* **~ of the house** Dame des Hauses; **Our L~** *(Relig.)* Unsere Liebe Frau; **f)** *attrib. (female)* **~ clerk** Angestellte, *die;* **~ doctor** Ärztin, *die;* **~ friend** Freundin, *die. See also* **first 2a**

lady: ~bird, *(Amer.)* **~bug** *ns. (Zool.)* Marienkäfer, *der;* **~-killer** *n. (coll.)* Herzensbrecher, *der;* **~like** *adj.* damenhaft; **be ~like** sich wie eine Dame benehmen

ladyship ['leɪdɪʃɪp] *n.* **her/your ~/their ~s** Ihre/Eure Ladyschaft/Ihre Ladyschaften

lady: ~'s-maid *n.* [Kammer]zofe, *die;* **~'s man** *see* **ladies' man**

[1]lag [læg] **1.** *v. i.,* **-gg-** *(lit. or fig.)* zurückbleiben; **~ behind sb./sth.** hinter jmdm./etw. bleiben. **2.** *n. (delay)* Verzögerung, *die; (falling behind)* Zurückbleiben, *das*

[2]lag *v. t.,* **-gg-** *(insulate)* isolieren

lager ['lɑ:gə(r)] *n.* Lagerbier, *das;* **a small ~:** ≈ ein kleines Helles

[1]lagging [lægɪŋ] *n.* **no ~!** nicht zurückbleiben!

[2]lagging *n. (insulation)* Isolierung, *die*

lagoon [lə'gu:n] *n.* Lagune, *die*

laid *see* **[2]lay 1**

'laid-back *adj. (coll.)* gelassen

lain *see* **[2]lie 2**

lair [leə(r)] *n. (of wild animal)* Unterschlupf, *der; (fig.) (of bandits)* Schlupfwinkel, *der; (of children etc.)* Versteck, *das*

laird [leəd] *n. (Scot.)* Gutsbesitzer, *der*

laity ['leɪɪtɪ] *n. pl.* Laien

lake [leɪk] *n.* See, *der;* **the Great L~s** die Großen Seen

lake: L~ Constance [leɪk 'kɒnstəns] *pr. n.* der Bodensee; **L~ District, L~land** ['leɪklənd] *pr. ns. (Brit.)* Lake District, *der (Seenlandschaft im Nordwesten Englands);* **L~ Lucerne** *see* **Lucerne; ~side** *n.* Seeufer, *das;* **by the ~side** am See[ufer]

lama ['lɑ:mə] *n.* Lama, *der*

lamb [læm] **1.** *n.* **a)** Lamm, *das;* **as gentle/meek as a ~:** sanft wie ein Lamm; **one may *or* might as well be hanged *or* hung for a sheep as [for] a ~** *(fig.)* darauf kommt es jetzt auch nicht mehr an; **like a ~ [to the slaughter]** wie ein Lamm [zur Schlachtbank *(geh.)*]; **b)** *no pl. (flesh)* Lamm[fleisch], *das.* **2.** *v. i.* lammen; **~ing season** Lammzeit, *die*

lambaste [læm'beɪst] (**lambast** [læm'bæst]) *v. t. (coll.: thrash, lit. or fig.)* fertigmachen *(ugs.)*

lamb: ~ 'chop *n.* Lammkotelett, *das;* **~ 'cutlet** *n.* Kammkotelett vom Lamm; **~skin** *n. (with wool on)* Lammfell, *das; (as leather)* Schafleder, *das;* **~'s-wool** *n.* Lambswool, *die (Textilw.)*

lame [leɪm] **1.** *adj.* **a)** *(disabled)* lahm; **go ~:** lahm werden; **be ~ in one's right leg** ein lahmes rechtes Bein haben; **the horse was ~ in one leg** das Pferd lahmte auf einem Bein; **b)** *(fig.: unconvincing)* lahm *(ugs. abwertend).* **2.** *v. t.* lahm reiten ‹*Pferd usw.*›; *(fig.: hinder)* lähmen ‹*Person, Fähigkeiten, Kraft*›

lame 'duck *n.* **a)** *(incapable person)* Versager, *der*/Versagerin, *die;* **b)** *(firm)* zahlungsunfähige Firma

lameness ['leɪmnɪs] *n., no pl. (lit.; also fig.: unconvincingness)* Lahmheit, *die*

lament [lə'ment] **1.** *n.* **a)** *(expression of grief)* Klage, *die* (**for** um); **b)** *(dirge)* Klagegesang, *der.* **2.** *v. t.* klagen über (+ *Akk.*) *(geh.);* klagen um *(geh.)* ‹*Freund, Heimat, Glück*›; **~ that ...:** beklagen, daß ... **3.** *v. i.* klagen *(geh.);* **~ over *or* for sth.** etw. beklagen *(geh.);* etw. beweinen; **~ over *or* for sb.** jmdn. beweinen

lamentable ['læməntəbl] *adj.* beklagenswert; kläglich ‹*Versuch, Leistung*›

lamentably ['læməntəblɪ] *adv.* beklagenswert; kläglich ‹*scheitern*›

lamentation [læmən'teɪʃn] *n.* **a)** *no pl., no art. (lamenting)* Wehklagen, *das (geh.);* **b)** *(lament)* [Weh]klage, *die (geh.)*

laminated ['læmɪneɪtɪd] *adj.* lamelliert *(Technik);* **~ glass** Verbundglas, *das*

lamp [læmp] *n.* Lampe, *die; (in street)* [Straßen]laterne, *die;* [Straßen]lampe, *die; (of vehicle)* Licht, *das; (car head~)* Scheinwerfer, *der*

'lamplight *n.* Lampenlicht, *das*

lampoon [læm'pu:n] **1.** *n.* Spottschrift, *die;* Pasquill, *das (geh.).* **2.** *v. t.* verhöhnen; verspotten

lamp: ~post *n.* Laternenpfahl, *der; (taller)* Lichtmast, *der;* **~shade** *n.* Lampenschirm, *der;* **~-standard** *n.* Lichtmast, *der*

lance [lɑ:ns] **1.** *n. (weapon)* Lanze, *die.* **2.** *v. t. (Med.)* mit der Lanzette öffnen

lance-'corporal *n. (Mil.)* Obergefreite, *der*

land [lænd] **1.** *n.* **a)** *no pl., no indef. art. (solid part of the earth)* Land, *das;* **by ~:** auf dem Landweg; **on ~:** zu Lande; *(not in air)* auf dem Boden; *(not in or on water)* an Land; **b)** *no indef. art. (expanse of country)* Land, *das;* **see how the ~ lies** *(fig.)* herausfinden, wie die Dinge liegen; *see also* **[2]lie 1a; c)** *no pl., no indef. art. (ground for farming or building, property)* Land, *das;* **work the ~:** das Land bebauen; **live off the ~:** sich von dem ernähren, was das Land hergibt; **d)** *(country)* Land, *das;* **the greatest in the ~:** der/die Größte im ganzen Land. **2.** *v. t.* **a)** *(set ashore)* [an]landen ‹*Truppen, Passagiere, Waren, Fang*›; **b)** *(Aeronaut.)* landen ‹*[Wasser]flugzeug*›; **c)** *(bring into a situation)* **~ oneself in trouble** sich in Schwierigkeiten bringen; sich *(Dat.)* Ärger einhandeln *(ugs.);* **~ sb. in [the thick of] it** jmdn. [ganz schön] reinreiten *(salopp);* **d)** *(deal)* landen ‹*Schlag*›; **~ sb. one** jmdm. einen Schlag verpassen *(ugs.);* **e)** *(burden)* **~ sb. with sth., ~ sth. on sb.** jmdm. etw. aufhalsen *(ugs.);* **be ~ed with sb./sth.** jmdn. auf dem Hals haben *(ugs.)*/etw. aufgehalst bekommen *(ugs.);* **f) ~ a fish** einen Fisch an Land ziehen; **g)** *(fig.: succeed in obtaining)* an Land ziehen *(ugs.).* **3.** *v. i.* **a)** ‹*Boot usw.:*› anlegen, landen; ‹*Passagier:*› aussteigen (**from** aus); **we ~ed at Dieppe** wir gingen in Dieppe an Land; **b)** *(Aeronaut.)* landen; *(on water)* [auf dem Wasser] aufsetzen; **be about to ~:** zur Landung angesetzt haben; gerade landen; **c)** *(alight)* landen; ‹*Ball:*› aufkommen; **~ on one's feet** auf den Füßen landen; *(fig.)* [wieder] auf die Füße fallen

~ 'back *v. i.* wieder landen *(ugs.)*

~ 'up *v. i.* landen *(ugs.)*

'land-agent *n.* Grundstücksmakler, *der*/-maklerin, *die*

'land breeze *n.* Landwind, *der*

landed ['lændɪd] *adj.* **~ gentry/aristocracy** Landadel, *der*

landing ['lændɪŋ] *n.* **a)** *(of ship)* Landung, *die;* **b)** *(of aircraft)* Landung, *die;* **emergency ~:** Notlandung, *die;* **c)** *(place for disembarkation)* Anlegestelle, *die;* **d)** *(between flights of stairs)* Treppenabsatz, *der; (passage)* Treppenflur, *der*

landing: **~-card** *n.* Landekarte, *die;* **~-craft** *n. (Navy)* Landungsboot, *das;* **~-gear** *n.* Fahrwerk, *das;* **~-stage** *n.* Landungssteg, *der;* Landungsbrücke, *die*

land: **~lady** *n.* **a)** *(of rented property)* Vermieterin, *die;* **b)** *(of public house)* [Gast]wirtin, *die;* **c)** *(of lodgings etc.)* [Pensions]wirtin, *die;* **~-locked** *adj.* vom Land eingeschlossen ‹*Bucht, Hafen*›; ‹*Staat*› ohne Zugang zum Meer; **~lord** *n.* **a)** *(of rented property)* Vermieter, *der;* [Haus]wirt, *der;* **b)** *(of public house)* [Gast]wirt, *der;* **c)** *(of lodgings etc.)* [Pensions]wirt, *der;* **~mark** *n.* **a)** weithin sichtbares Erkennungszeichen; *(Naut.)* Landmarke, *die;* **b)** *(fig.)* Markstein, *der;* **stand as a ~mark** einen Meilenstein bedeuten; **~mass** *n. (Geog.)* Landmasse, *die;* **~-mine** *n. (Mil.)* Landmine, *die;* **~owner** *n.* **[large** *or* **big] ~owner** [Groß]grundbesitzer, *der/*-besitzerin, *die*

landscape ['lændskeɪp, 'lænskeɪp] **1.** *n.* **a)** Landschaft, *die;* **b)** *(picture)* Landschaftsbild, *das;* Landschaft, *die.* **2.** *v.t.* landschaftsgärtnerisch gestalten ‹*Garten, Park*›

'**landscape gardener** *n.* Landschaftsgärtner, *der/*-gärtnerin, *die*

land: **~slide** *n.* **a)** Erdrutsch, *der;* **b)** *(fig.: majority)* Erdrutsch[wahl]sieg, *der; attrib.* **a ~slide victory** ein Erdrutsch[wahl]sieg; **~slip** *see* **~slide a**

lane [leɪn] *n.* **a)** *(in the country)* Landsträßchen, *das; (unmetalled)* [Hecken]weg, *der;* **b)** *(in town)* Gasse, *die;* **c)** *(part of road)* [Fahr]spur, *die;* **slow ~** *(in Britain)* linke Spur; *(on the continent)* rechte Spur; **'get in ~'** „bitte einordnen"; *see also* **fast lane; d)** *(for race)* Bahn, *die*

language ['læŋgwɪdʒ] *n.* **a)** Sprache, *die;* **speak the same ~** *(fig.)* die gleiche Sprache sprechen; **b)** *no pl., no art. (words, wording)* Sprache, *die;* **[style of] ~:** [Sprach]stil, *der;* **use of ~:** Sprachgebrauch, *der;* **c)** *(style)* Ausdrucksweise, *die;* Sprache, *die;* see also **bad 1 d; strong language; d)** *(professional vocabulary)* [Fach]sprache, *die;* **e)** *(Computing)* Sprache, *die*

language: **~ laboratory** *n.* Sprachlabor, *das;* **~-teacher** *n.* Sprachlehrer, *der/*-lehrerin, *die*

languid ['læŋgwɪd] *adj.* **a)** *(sluggish)* träge; **b)** *(inert)* matt

languish ['læŋgwɪʃ] *v.i.* **a)** *(lose vitality)* ermatten *(geh.);* **b)** *(live wretchedly)* **~ under sth.** unter etw. *(Dat.)* schmachten *(geh.);* **~ in prison** im Gefängnis schmachten *(geh.);* **c)** *(pine)* **~ for sth.** nach etw. schmachten *(geh.)*

languor [læŋgə(r)] *n. see* **languorous:** Mattigkeit, *die;* Trägheit, *die*

languorous ['læŋgərəs] *adj.* **a)** *(faint)* matt; **b)** *(inert)* träge

lank [læŋk] *adj.* **a)** *(tall)* hager; **b)** *(limp)* glatt herabhängend ‹*Haar*›

lanky ['læŋkɪ] *adj.* schlaksig *(ugs.);* [dürr und] lang ‹*Arm, Bein*›

lantern ['læntən] *n.* Laterne, *die*

[1]**lap** [læp] *n. (part of body)* Schoß, *der;* **live in the ~ of luxury** *(fig.)* im Überfluß leben; **fall into sb.'s ~** *(fig.)* jmdm. in den Schoß fallen; *see also* **god a**

[2]**lap 1.** *n. (Sport)* Runde, *die;* **on the last ~** *(fig. coll.)* auf der Zielgeraden *(fig.).* **2.** *v.t.,* **-pp-: a)** *(Sport)* überrunden; **b)** *(cause to overlap)* überlappen

[3]**lap 1.** *v.i.,* **-pp-** *(drink)* schlappen; schlecken. **2.** *v.t.,* **-pp-: a)** *(drink)* **~ [up]** [auf]schlappen; [auf]schlecken; **b)** *see* **~ up b; c)** *see* **~ up c**

~ 'up *v.t.* **a)** *(drink) see* **~ 2 a; b)** *(consume greedily)* hinunterschütten; **c)** *(fig.: receive eagerly)* schlucken *(ugs.);* begierig aufnehmen ‹*Lob*›

lapel [lə'pel] *n.* Revers, *das od. (österr.) der*

Lapland ['læplænd] *pr. n.* Lappland *(das)*

Lapp [læp] **1.** *n.* Lappe, *der/*Lappin, *die.* **2.** *adj.* **a)** lappisch; lappländisch; **b)** *(of language)* lappisch

lapse [læps] **1.** *n.* **a)** *(interval)* **a/the ~ of ...:** eine/die Zeitspanne von ...; **a ~ in the conversation** eine Gesprächspause; **b)** *(mistake)* Fehler, *der;* Lapsus, *der (geh.);* **~ of memory** Gedächtnislücke, *die;* **c)** *(deviation)* Verstoß, *der* **(from** gegen); **momentary ~ of concentration** momentane Konzentrationsschwäche. **2.** *v.i.* **a)** *(fail)* versagen; **~ from sth.** etw. vermissen lassen; **b)** *(sink)* **~ into** verfallen in (+ *Akk.*); fallen in (+ *Akk.*) ‹*Schlaf, Koma*›; **c)** *(become void)* ‹*Vertrag, Versicherungspolice usw.:*› ungültig werden; ‹*Plan, Projekt:*› hinfällig werden; ‹*Anspruch:*› verfallen

lapsed [læpst] *adj.* **a)** abgefallen ‹*Christ, Katholik usw.*›; **b)** abgelaufen, ungültig ‹*Paß, Führerschein, Versicherungspolice*›

'**lap-top 1.** *adj.* Laptop‹*gerät, -PC*›. **2.** *n.* Laptop, *der*

larceny ['lɑːsənɪ] *n. (Law)* Diebstahl, *der*

larch [lɑːtʃ] *n.* Lärche, *die*

lard [lɑːd] *n.* Schweineschmalz, *das;* Schweinefett, *das*

larder ['lɑːdə(r)] *n. (room)* Speisekammer, *die; (cupboard)* Speiseschrank, *der*

large [lɑːdʒ] **1.** *adj.* **a)** groß; **a ~ lady** eine stattliche Dame; **~ importer/user** Großimporteur, *der/* Großverbraucher, *der; see also* **life d; b)** *(comprehensive, broad)* umfassend. **2.** *n.* **at ~** *(at liberty)* frei; *(not in prison etc.)* auf freiem Fuß; in Freiheit; *(as a body)* insgesamt; **society at ~:** die Gesellschaft in ihrer Gesamtheit. **3.** *adv. see* [1]**by 2 c;** [2]**loom; write 2 d**

largely ['lɑːdʒlɪ] *adv.* weitgehend

large: **~-scale** *attrib. adj.* großangelegt; ‹*Katastrophe*› großen Ausmaßes; ‹*Modell*› in großem Maßstab; **~-scale manufacture** Massenproduktion, *die;* **~-size[d]** *adj.* groß

largish ['lɑːdʒɪʃ] *adj.* ziemlich groß; recht stattlich ‹*Person*›

[1]**lark** [lɑːk] *n. (Ornith.)* Lerche, *die;* **be up with the ~:** beim *od.* mit dem ersten Hahnenschrei aufstehen

[2]**lark** *(coll.)* **1.** *n.* **a)** *(piece of fun)* Jux, *der (ugs.);* **do sth. for a ~:** etw. aus Jux machen *(ugs.);* **what a ~!** das ist/war spitze! *(ugs.);* **b)** *(Brit.) (form of activity)* Blödsinn, *der (ugs.); (affair)* Geschichte, *die (ugs.).* **2.** *v.i.* **~ [about** *or* **around]** herumalbern *(ugs.)*

larva ['lɑːvə] *n., pl.* **~e** ['lɑːviː] Larve, *die*

laryngitis [lærɪn'dʒaɪtɪs] *n. (Med.)* Kehlkopfentzündung, *die*

larynx ['lærɪŋks] *n., pl.* **larynges** [lə'rɪndʒiːz] *(Anat.)* Kehlkopf, *der;* Larynx, *der (fachspr.)*

lascivious [lə'sɪvɪəs] *adj.* **a)** *(lustful)* lüstern *(geh.);* **b)** *(inciting to lust)* lasziv

laser ['leɪzə(r)] *n.* Laser, *der*

laser: **~ beam** *n.* Laserstrahl, *der;* **~ printer** *n.* Laserdrucker, *der*

lash [læʃ] **1.** *n.* **a)** *(stroke)* [Peitschen]hieb, *der;* **b)** *(part of whip)* *biegsamer Teil der Peitsche; (whipcord)* Peitschenschnur, *die; (as punishment)* **the ~:** die Peitsche; **c)** *(on eyelid)* Wimper, *die.* **2.** *v.i.* **a)** *(make violent movement)* schlagen; ‹*Peitsche, Schlange:*› zuschlagen; **b)** *(strike)* ‹*Welle, Regen:*› peitschen **(against** gegen, **on** auf + *Akk.*); ‹*Person:*› [mit der Peitsche] schlagen (**at** nach). **3.** *v. t.* **a)** *(fasten)* festbinden (**to** an + *Dat.*); ~ **together** zusammenbinden; **b)** *(flog)* mit der Peitsche schlagen; *(as punishment)* auspeitschen; **c)** *(move violently)* schlagen mit; **d)** *(beat upon)* peitschen; **the rain ~ed the windows/roof** der Regen peitschte gegen die Fenster/auf das Dach

~ **'down 1.** *v. t.* festbinden; *(Naut.)* festzurren *(bes. Seemannsspr.)*. **2.** *v. i.* ‹*Regen:*› niederprasseln

~ **'out** *v. i.* **a)** *(hit out)* um sich schlagen; ‹*Pferd:*› ausschlagen; ~ **out at sb.** nach jmdm. schlagen; *(fig.)* über jmdn. herziehen *(ugs.)*; **b)** ~ **out on sth.** *(coll.: spend freely)* sich *(Dat.)* etw. leisten *od.* gönnen

lashings ['læʃɪŋz] *n. pl.* *(large amounts)* ~ **of sth.** Unmengen von etw.

lass [læs], **lassie** ['læsɪ] *ns. (Scot., N. Engl.)* Mädchen, *das*

lasso [lə'su:, 'læsəʊ] **1.** *n., pl.* ~**s** *or* ~**es** Lasso, *das.* **2.** *v. t.* mit dem Lasso fangen

¹last [lɑ:st] **1.** *adj.* letzt...; **be ~ to arrive** als letzter/letzte ankommen; **for the [very] ~ time** zum [aller]letzten Mal; **who was ~?** wer war letzter?; **the ~ two** die letzten beiden; **he came ~:** er war letzter; **second ~, ~ but one** vorletzt...; ~ **but not least** last, not least; nicht zuletzt; ~ **evening/night was windy** gestern abend/gestern *od.* heute nacht war es windig; ~ **evening/week we were out** gestern abend/letzte Woche waren wir aus; **that would be the '~ thing to do in this situation** das wäre das Letzte, was man in dieser Situation tun würde. **2.** *adv.* **a)** [ganz] zuletzt; als letzter/letzte ‹*sprechen, ankommen*›; **b)** *(on ~ previous occasion)* das letzte Mal; zuletzt; **when did you ~ see him** *or* **see him ~?** wann hast du ihn zuletzt *od.* das letzte Mal gesehen? **3.** *n.* **a)** *(mention, sight)* **I shall never hear the ~ of it** das werde ich ständig zu hören bekommen; **you haven't heard the ~ of this matter** das letzte Wort in dieser Sache ist noch nicht gesprochen; **that was the ~ we ever saw of him** das war das letzte Mal, daß wir ihn gesehen haben; **b)** *(person or thing)* letzter...; **these ~:** letztere; **I'm always the ~ to be told** ich bin immer der letzte, der etwas erfährt; **she was the ~ to know about it** sie erfuhr es als letzte; **c)** *(day, moment[s])* **to** *or* **till the ~:** bis zuletzt; *see also* **breathe 2 a; d) at [long] ~:** endlich; schließlich [doch noch]

²last *v. i.* **a)** *(continue)* andauern; ‹*Wetter, Ärger:*› anhalten; ~ **all night** die ganze Nacht dauern; ~ **till** dauern bis; ~ **from ... to ...:** von ... bis ... dauern; **built to ~:** dauerhaft gebaut; **it can't/won't ~:** das geht nicht mehr lange so; **it's too good to ~:** es ist zu schön, um von Dauer zu sein; **b)** *(manage to continue)* es aushalten; **c)** *(suffice)* reichen; **while stocks ~:** solange Vorrat reicht; **this knife will ~ [me] a lifetime** dies Messer hält mein ganzes Leben

³last *n.* *(for shoemaker)* Leisten, *der*

'last-ditch *adj.* ~ **attempt** letzter verzweifelter Versuch

lasting ['lɑ:stɪŋ] *adj.* *(permanent)* bleibend; dauerhaft ‹*Beziehung*›; nachhaltig ‹*Eindruck, Wirkung, Bedeutung*›; nicht nachlassend ‹*Interesse*›

lastly ['lɑ:stlɪ] *adv.* schließlich

last: ~-mentioned *attrib. adj.* letztgenannt; ~ **'minute** *n.* **at the ~ minute** in letzter Minute; **up to the ~ minute** bis zum letzten Augenblick; **~-minute** *attrib. adj.* in letzter Minute *nachgestellt;* ~ **name** *n.* Zuname, *der;* Nachname, *der;* **L~ 'Supper** *n., no pl.* *(Relig.)* **the L~ Supper** das Abendmahl; ~ **'thing** *adv. (coll.)* als letztes; ~ **'word** *n., no pl., no indef. art.* letztes Wort; **be the ~ word** *(fig.)* nicht zu überbieten sein (**in** an + *Dat.*); das letzte sein (**in** an + *Dat.*)

lat. *abbr.* **latitude** Br.

latch [lætʃ] *n.* **a)** *(bar)* Riegel, *der;* **b)** *(spring-lock)* Schnappschloß, *das;* **c) on the ~** *(with lock not in use)* nur eingeklinkt

~ **'on to** *v. t. (coll.)* **a)** *(attach oneself to)* ~ **on to sb.** sich an jmdn. hängen *(ugs.)*; **b)** *(understand)* kapieren *(ugs.)*; **c)** *(be enthusiastic about)* abfahren auf (+ *Akk.*) *(salopp)*

'latchkey *n.* Hausschlüssel, *der;* ~ **child** *(fig.)* Schlüsselkind, *das*

late [leɪt] **1.** *adj.* **a)** spät; *(after proper time)* verspätet; **am I ~?** komme ich zu spät?; **be ~ for the train** den Zug verpassen; **the train is [ten minutes] ~:** der Zug hat [zehn Minuten] Verspätung; **spring is ~ this year** dieses Jahr haben wir einen späten Frühling; ~ **riser** Spätaufsteher, *der*/-aufsteherin, *die;* ~ **entry** verspätete Anmeldung; ~ **shift** Spätschicht, *die;* **it is ~:** es ist [schon] spät; **have a ~ dinner** [erst] spät zu Abend essen; ~ **summer** Spätsommer, *der;* **in ~ July** Ende Juli; **b)** *(deceased)* verstorben; **c)** *(former)* ehemalig; vormalig; **d)** *(recent)* letzt...; **in ~ times** in letzter Zeit. *See also* **later; latest. 2.** *adv.* **a)** *(after proper time)* verspätet; **[too] ~:** zu spät; **they got home very ~:** sie kamen [erst] sehr spät nach Hause; **better ~ than never** lieber spät als gar nicht; **b)** *(far on in time)* spät; ~ **in August** Ende August; ~ **last century** [gegen] Ende des letzten Jahrhunderts; ~ **in life** erst im fortgeschrittenen Alter; **c)** *(at or till a ~ hour)* spät; **be up/sit up ~:** bis spät in die Nacht *od.* lange aufbleiben; **work ~ at the office** [abends] lange im Büro arbeiten; **d)** *(formerly)* ~ **of ...:** ehemals wohnhaft in ...; ehemaliger Mitarbeiter ‹*einer Firma*›; **e)** *(at ~ stage)* **she was seen as ~ as yesterday** sie wurde gestern noch gesehen; **[a bit] ~ in the day** *(fig. coll.)* reichlich spät. **3.** *n.* **of ~:** in letzter Zeit

'latecomer *n.* Zuspätkommende, *der/die*

lately ['leɪtlɪ] *adv.* in letzter Zeit; **till ~:** bis vor kurzem

lateness ['leɪtnɪs] *n., no pl.* **a)** *(being after due time)* Verspätung, *die;* **b)** *(being far on in time)* **the ~ of the performance** der späte Beginn der Vorstellung; **the ~ of the hour** die späte *od.* vorgerückte Stunde

latent ['leɪtənt] *adj.* latent [vorhanden]

later ['leɪtə(r)] **1.** *adv.* später; ~ **on** später; ~ **[on] the same day** im weiteren Verlauf des Tages; später am Tag; **see you ~:** bis nachher; bis später. **2.** *adj.* später; *(more recent)* neuer; jünger; **at a ~ date** zu einem späteren Zeitpunkt; später

lateral ['lætərl] *adj.* seitlich (**to** von); Seiten‹*flügel, -ansicht*›; ~ **thinker** Querdenker, *der;* ~ **thinking** Querdenken, *das*

laterally ['lætərəlɪ] *adv.* seitlich

latest ['leɪtɪst] *adj.* **a)** *(modern)* neu[e]st...; **the very ~ thing** das Allerneu[e]ste; **b)** *(most recent)* letzt...; **have you heard the ~?** wissen Sie schon das Neu[e]ste?; **what's the ~?** was gibt's Neues?; **c) at [the] ~/the very ~:** spätestens/allerspätestens

latex ['leɪteks] *n., pl.* ~**es** *or* **latices** ['leɪtɪsi:z] Latex, *der*

lath [lɑ:θ] *n., pl.* ~**s** [lɑ:θs, lɑ:ðz] Latte, *die*

lathe [leɪð] *n.* Drehbank, *die*

lather ['lɑ:ðə(r), 'læðə(r)] **1.** *n.* **a)** *(froth)* [Seifen]schaum, *der;* **b)** *(sweat)* Schweiß, *der;* **get [oneself] into a ~ [about sth.]** *(fig.)* sich [über etw. *(Akk.)*] aufregen. **2.** *v. t.* einschäumen; einseifen

Latin ['lætɪn] **1.** *adj.* **a)** lateinisch; **b)** *(of Southern Europeans)* romanisch; südländisch ‹*Temperament*›. **2.** *n.* Latein, *das; see also* **English 2 a**

Latin: ~ **A'merica** *pr. n.* Lateinamerika *(das);* ~**-A'merican 1.** *adj.* lateinamerikanisch; **2.** *n.* Lateinamerikaner, *der*/Lateinamerikanerin, *die;* ~ **Quarter** *n.* Quartier Latin, *das*

latitude ['lætɪtju:d] *n.* **a)** *(freedom)* Freiheit, *die;* **b)** *(Geog.)* [geographische] Breite; *(of a place)* Breite, *die;* ~**s** *(regions)* Breiten *Pl.;* ~ **40° N.** 40° nördlicher Breite

latrine [lə'tri:n] *n.* Latrine, *die*

latter ['lætə(r)] *attrib. adj.* **a)** letzter...; **the ~:** der/die/das letztere; *pl.* die letzteren; **b)** *(later)* letzt...; **the ~ half of the century** die zweite Hälfte des Jahrhunderts; **the ~ part of the year** die zweite Jahreshälfte

latterly ['lætəlɪ] *adv.* in letzter Zeit

Latvia ['lætvɪə] *pr. n.* Lettland *(das)*

Latvian ['lætvɪən] **1.** *adj.* lettisch. **2.** *n.* **a)** *(person)* Lette, *der*/Lettin, *die;* **b)** *(language)* Lettisch, *das; see also* **English 2 a**

laudable ['lɔ:dəbl] *adj.* lobenswert

laugh [lɑ:f] **1.** *n.* Lachen, *das; (loud and continuous)* Gelächter, *das;* **have a [good] ~ about sth.** [herzlich] über etw. *(Akk.)* lachen; **give a loud ~:** laut auflachen; **this line always gets a ~:** diese Zeile bringt immer einen Lacher; **have the last ~:** derjenige sein, der zuletzt lacht *(fig.);* **he is always good for a ~:** bei ihm gibt es immer etwas zu lachen; **sb./sth. is a ~ a minute** bei jmdm./etw. muß man alle Augenblicke lachen; **for ~s** zum *od.* aus Spaß; **for a ~:** [so] zum Spaß. **2.** *v. i.* lachen; ~ **out loud** laut auflachen; **I ~ed till I cried** ich habe Tränen gelacht; ~ **at sb./sth.** *(in amusement)* über jmdn./etw. lachen; *(jeer)* jmdn. auslachen/etw. verlachen; über jmdn./etw. lachen; ~ **in sb.'s face** jmdm. ins Gesicht lachen; **he who ~s last ~s longest** *(prov.)* wer zuletzt lacht, lacht am besten *(Spr.);* **don't make me ~** *(coll. iron.)* daß ich nicht lache! **3.** *v. t.* lachen; ~ **oneself silly** sich krank- *od.* schieflachen *(ugs.)*

~ **'off** *v. t.* mit einem Lachen abtun

laughable ['lɑ:fəbl] *adj.* lachhaft *(abwertend);* lächerlich

laughing ['lɑ:fɪŋ] *n.* **be no ~ matter** nicht zum Lachen sein

'laughing-gas *n.* Lachgas, *das*

laughingly ['lɑ:fɪŋlɪ] *adv.* lachend; **what is ~ called ...** *(iron.)* was sich ... nennt *(spött.)*

'laughing-stock *n.* **make sb. a ~, make a ~ of sb.** jmdn. zum Gespött machen

laughter ['lɑ:ftə(r)] *n.* Lachen, *das; (loud and continuous)* Gelächter, *das*

[1]**launch** [lɔ:nʃ] **1.** *v. t.* **a)** zu Wasser lassen, aussetzen ‹*Rettungsboot, Segelboot*›; vom Stapel lassen ‹*neues Schiff*›; *(propel)* werfen, abschießen ‹*Harpune*›; schleudern ‹*Speer*›; abschießen ‹*Torpedo*›; ~ **a rocket into space** eine Rakete ins All schießen; **b)** *(fig.)* lancieren *(bes. Wirtsch.);* auf den Markt bringen ‹*Produkt*›; vorstellen ‹*Buch, Schallplatte, Sänger*›; auf die Bühne bringen ‹*Theaterstück*›; gründen ‹*Firma*›; ~ **an attack** einen Angriff durchführen. **2.** *v. i.* ~ **into a song** ein Lied anstimmen; ~ **into a long speech** eine lange Rede vom Stapel lassen *(ugs.)*

~ **'out** *v. i. (fig.)* ~ **out into films/a new career/on one's own** sich beim Film versuchen/beruflich etwas ganz Neues anfangen/sich selbständig machen

[2]**launch** *n. (boat)* Barkasse, *die*

launching: ~ **pad** *n.* [Raketen]abschußrampe, *die;* ~ **site** *n.* [Raketen]abschußbasis, *die*

'launch pad *see* **launching pad**

launder ['lɔ:ndə(r)] *v. t.* **a)** waschen und bügeln; **b)** *(fig.)* waschen ‹*Geld*›

launderette [lɔ:ndə'ret], **laundrette** [lɔ:n'dret], *(Amer.)* **laundromat** ['lɔ:ndrəmæt] *ns.* Waschsalon, *der*

laundry ['lɔ:ndrɪ] *n.* **a)** *(place)* Wäscherei, *die;* **b)** *(clothes etc.)* Wäsche, *die;* **do the ~:** Wäsche waschen

laurel ['lɒrl] *n. (emblem of victory)* Lorbeer[kranz], *der;* **rest on one's ~s** *(fig.)* sich auf seinen Lorbeeren ausruhen *(ugs.)*

lav [læv] *n. (coll.)* Klo, *das (ugs.)*

lava ['lɑ:və] *n.* Lava, *die*

lavatory ['lævətərɪ] *n.* Toilette, *die*

lavatory: ~**-paper** *see* **toilet-paper;** ~**-seat** *see* **toilet-seat**

lavender ['lævɪndə(r)] *n. (Bot.)* Lavendel, *der*

lavish ['lævɪʃ] **1.** *adj. (generous)* großzügig; überschwenglich ‹*Lob, Liebe*›; verschwenderisch ‹*Ausgaben*›; *(abundant)* üppig; **be ~ of** *or* **with sth.** nicht mit etw. geizen; **be too ~ with sth.** mit etw. übertreiben. **2.** *v. t.* ~ **sth. on sb.** jmdn. mit etw. überhäufen *od.* überschütten

lavishly ['lævɪʃlɪ] *adv.* großzügig; verschwenderisch ‹*Geld ausgeben*›; herrschaftlich ‹*eingerichtet*›

law [lɔ:] *n.* **a)** *no pl. (body of established rules)* Gesetz, *das;* Recht, *das;* **the ~ forbids/allows sth. to be done** nach dem Gesetz ist es verboten/erlaubt, etw. zu tun; **according to/under British** *etc.* **~:** nach britischem *usw.* Recht; **break the ~:** gegen das Gesetz verstoßen; **be against the ~:** gegen das Gesetz sein; **under the** *or* **by** *or* **in ~:** nach dem Gesetz; **be/become ~:** vorgeschrieben sein/werden; **lay down the ~:** Vorschriften machen (**to** *Dat.*); **lay down the ~ on/about sth.** sich zum Experten für etw. aufschwingen; ~ **enforcement** Durchführung der Gesetze/des Gesetzes; **b)** *no pl., no indef. art. (control through ~)* Gesetz, *das;* ~ **and order** Ruhe und Ordnung; **be above the ~:** über dem Gesetz stehen; **outside the ~:** außerhalb der Legalität; **c)** *(statute)* Gesetz, *das;* **there ought to be a ~ against it/people like you** so etwas sollte/Leute wie du sollten verboten werden; **be a ~ unto oneself** machen, was man will; **d)** *no pl., no indef. art. (litigation)* Rechtswesen, *das;* Gerichtswesen, *das;* **go to ~ [over sth.]** [wegen etw.] vor Gericht gehen; [wegen etw.] den Rechtsweg beschreiten; **have the ~ on sb.** *(coll.)* jmdm. die Polizei auf den Hals schicken *(ugs.);* jmdn. vor den Kadi schleppen *(ugs.);* **take the ~ into one's own hands** sich *(Dat.)* selbst Recht verschaffen; **e)** *no pl., no indef. art. (profession)* **practise ~:** Jurist/Juristin sein; **f)** *no pl., no art. (Univ.: jurisprudence)* Jura *o. Art.;* Rechtswissenschaft, *die; attrib.* Rechts-; **Faculty of Law** juristische Fakultät; ~ **school** *(Amer.)* juristische Fakultät; **g)** *no indef. art. (branch of ~)* **commercial ~:** Handelsrecht, *das;* **h)** *(Sci., Philos., etc.)* Gesetz, *das;* ~ **of nature, natural ~** Naturgesetz, *das*

law: ~**-abiding** ['lɔ:əbaɪdɪŋ] *adj.*

gesetzestreu; **~-breaker** *n.* Gesetzesbrecher, *der*/-brecherin, *die;* Rechtsbrecher, *der*/-brecherin, *die;* **~court** *n.* Gerichtsgebäude, *das; (room)* Gerichtssaal, *der;* **~ firm** *n. (Amer.)* Anwaltskanzlei, *die*

lawful ['lɔ:fl] *adj.* rechtmäßig, legitim ‹*Besitzer, Erbe*›; legitim, ehelich ‹*Tochter, Sohn, Nachkomme*›; legal, gesetzmäßig ‹*Vorgehen, Maßnahme*›

lawfully ['lɔ:fəlɪ] *adv.* legal; auf legalem Weg[e] ‹*erwerben*›

lawless ['lɔ:lɪs] *adj.* gesetzlos

lawn [lɔ:n] *n. (grass)* Rasen, *der;* **~s** Rasenflächen

lawn: ~-mower *n.* Rasenmäher, *der;* **~-seed** *n.* Grassamen, *der;* **~ 'tennis** *n.* Rasentennis, *das*

'law suit *n.* Prozeß, *der*

lawyer ['lɔ:jə(r), 'lɔɪə(r)] *n.* Rechtsanwalt, *der*/Rechtsanwältin, *die*

lax [læks] *adj.* lax; **be ~ about hygiene/paying the rent** *etc.* es mit der Hygiene/der Zahlung der Miete *usw.* nicht sehr genau nehmen

laxative ['læksətɪv] *(Med.)* **1.** *adj.* abführend; stuhlgangfördernd. **2.** *n.* Abführmittel, *das;* Laxativ[um] *das (fachspr.)*

laxity ['læksɪtɪ], **laxness** ['læksnɪs] *ns.* Laxheit, *die*

¹lay [leɪ] *adj.* **a)** *(Relig.)* laikal; Laien‹*bruder, -schwester, -predigt*›; **b)** *(inexpert)* laienhaft

²lay 1. *v.t.,* **laid** [leɪd] **a)** legen, [ver]legen ‹*Teppichboden, Rohr, Gleis, Steine, Kabel, Leitung*›; legen ‹*Parkett, Fliesen, Fundament*›; anlegen ‹*Straße, Gehsteig*›; *see also* **hand 1 a; b)** *(fig.)* **~ one's case before sb.** jmdm. seinen Fall vortragen; **~ one's plans/ideas before sb.** jmdm. seine Pläne/Vorstellungen unterbreiten; *see also* **blame 2; open 1 d; c)** *(impose)* auferlegen ‹*Verantwortung, Verpflichtung*› **(on** *Dat.*); **~ weight on sth.** Gewicht auf etw. *(Akk.)* legen; **d)** *(wager)* **I'll ~ you five to one that ...:** ich wette mit dir fünf zu eins, daß ... **~ a wager on sth.** eine Wette auf etw. *(Akk.)* abschließen; auf etw. *(Akk.)* wetten; **e)** *(prepare)* **~ the table** den Tisch decken; **~ three places for lunch** drei Gedecke zum Mittagessen auflegen; **~ the breakfast things** den Frühstückstisch decken; **f)** *(Biol.)* legen ‹*Ei*›; **g)** *(devise)* schmieden ‹*Plan*›; bannen ‹*Geist, Gespenst*›; **h)** *(sl.: copulate with)* **~ a woman** eine Frau vernaschen *od.* aufs Kreuz legen *(salopp).* **2.** *n. (sl.: sexual partner)* **she's a good/an easy ~:** sie ist gut im Bett/steigt mit jedem ins Bett *(ugs.)*

~ a'bout *v.t. (coll.)* **~ about sb.** auf jmdn. einschlagen; *(scold)* jmdn. ausschimpfen

~ a'side *v.t.* beiseite *od.* zur Seite legen, weglegen ‹*angefangene Arbeit*›; beiseite *od.* auf die Seite legen ‹*Geld*›

~ 'by *v.t.* beiseite *od.* auf die Seite legen; **have some money laid by** etwas [Geld] auf der hohen Kante haben *(ugs.)*

~ 'down *v.t.* **a)** hinlegen; **~ sth. down on the table** etw. auf den Tisch legen; **b)** *(give up)* niederlegen ‹*Amt, Waffen*›; *(deposit)* hinterlegen ‹*Geld*›; **~ down one's arms** sich ergeben; die Waffen strecken *(geh.)*; **~ down one's life for sth./sb.** sein Leben für etw./jmdn. [hin]geben; **c)** *(formulate)* festlegen ‹*Regeln, Richtlinien, Bedingungen*›; aufstellen ‹*Grundsätze, Regeln, Norm*›; festsetzen ‹*Preis*›; *(in a contract, constitution)* verankern; niederlegen; *see also* **law a**

~ 'in *v.t.* einlagern; sich eindecken mit

~ into *v.t. (coll.)* **~ into sb.** auf jmdn. losgehen; über jmdn. herfallen; *(fig.)* jmdn. zusammenstauchen *(ugs.)*

~ 'off 1. *v.t.* **a)** *(from work)* vorübergehend entlassen; **b)** *(coll.) (stop)* **~ off it!** laß das!; hör auf damit!; *(stop attacking, lit. or fig.)* **~ off him!** laß ihn in Ruhe! **2.** *v.i. (coll.: stop)* aufhören

~ 'on *v.t.* **a)** *(provide)* sorgen für ‹*Getränke, Erfrischungen, Unterhaltung*›; bereitstellen ‹*Transportmittel*›; organisieren ‹*Theaterbesuch, Stadtrundfahrt*›; anschließen ‹*Gas, Strom*›; **b)** *(apply)* auftragen ‹*Farbe usw.*›; *see also* **trowel a**

~ 'out *v.t.* **a)** *(spread out)* ausbreiten; *(ready for use)* zurechtlegen; **~ out sth. for sb. to see** etw. vor jmdm. ausbreiten; **b)** *(for burial)* aufbahren; **c)** *(arrange)* anlegen ‹*Garten, Park, Wege*›; das Layout machen für ‹*Buch*›; **d)** *(coll.: knock unconscious)* **~ sb. out** jmdn. außer Gefecht setzen; **e)** *(spend)* ausgeben

~ 'up *v.t.* **a)** *(store)* lagern; **you're ~ing up trouble/problems for yourself [later on]** *(fig.)* du handelst dir [für später] nur Ärger/Schwierigkeiten ein; **b)** *(put out of service)* [vorübergehend] aus dem Verkehr ziehen ‹*Fahrzeug*›; **I was laid up in bed for a week** ich mußte eine Woche das Bett hüten

³lay *see* **²lie 2**

lay: ~about *n. (Brit.)* Gammler, *der (ugs. abwertend);* Nichtstuer, *der (abwertend);* **~-by** *n., pl.* **~-bys** *(Brit.)* Parkbucht, *die;* Haltebucht, *die*

layer ['leɪə(r)] *n.* Schicht, *die;* **several ~s of paper** mehrere Lagen Papier

'layer cake *n.* Schichttorte, *die*

layette [leɪ'et] *n.* **[baby's] ~:** Babyausstattung, *die*

lay: ~man ['leɪmən] *n., pl.* **~men** ['leɪmən] Laie, *der;* **~-off** *n.* **a)** *(temporary dismissal)* vorübergehende Entlassung; **b)** *(Sport; coll.: break from work)* Pause, *die;* **~out** *n. (of house, office)* Raumaufteilung, *die; (of garden, park)* Gestaltung, *die;* Anlage, *die; (of book, magazine, advertisement)* Gestaltung, *die;* Layout, *das*

laze [leɪz] *v.i.* faulenzen; **~ around** *or* **about** herumfaulenzen *(ugs.)*

lazily ['leɪzɪlɪ] *adv.* faul; *(sluggishly)* träge

laziness ['leɪzɪnɪs] *n., no pl.* Faulheit, *die; (sluggishness)* Trägheit, *die*

lazy ['leɪzɪ] *adj.* faul; träge ‹*Geste, Sprechweise*›; träge fließend ‹*Fluß*›; **have a ~ day on the beach** einen Tag am Strand faulenzen

'lazy-bones *n. sing.* Faulpelz, *der*

lb. *abbr.* **pound[s]** ≈ Pfd.

LCD *abbr.* **liquid crystal display** LCD

L-driver ['eldraɪvə(r)] *(Brit.) see* **learner-driver**

¹lead [led] *n.* **a)** *(metal)* Blei, *das;* **b)** *(in pencil)* [Bleistift]mine, *die*

²lead [li:d] **1.** *v.t.,* **led** [led] **a)** führen; **~ sb. by the hand** jmdn. an der Hand führen; **~ sb. by the nose** *(fig.)* jmdn. nach seiner Pfeife tanzen lassen; **~ sb. into trouble** *(fig.)* jmdm. Ärger einbringen; **this is ~ing us nowhere** *(fig.)* das führt zu nichts; **b)** *(fig.: influence, induce)* **~ sb. to do sth.** jmdn. veranlassen, etw. zu tun; **be easily led** sich leicht beeinflussen lassen; **that ~s me to believe that ...:** das läßt mich glauben, daß ...; **he led me to suppose/believe that ...:** er gab mir Grund zu der Annahme/er machte mich glauben, daß ...; **c)** führen ‹*Leben*›; **~ a life of misery/a miserable existence** ein erbärmliches Dasein führen/eine kümmerliche Existenz fristen; **d)** *(be first in)* anführen; **~ the world in electrical engineering** auf dem

Gebiet der Elektrotechnik in der ganzen Welt führend sein; **Smith led Jones by several yards/seconds** *(Sport)* Smith hatte mehrere Yards/Sekunden Vorsprung vor Jones; **e)** *(direct, be head of)* anführen ‹*Bewegung, Abordnung*›; leiten ‹*Diskussion, Veranstaltung, Ensemble*›; ‹*Dirigent:*› leiten ‹*Orchester, Chor*›; ‹*Konzertmeister:*› führen ‹*Orchester*›; **~ a party** Vorsitzender/Vorsitzende einer Partei sein. **2.** *v. i.*, **led a)** ‹*Straße usw., Tür:*› führen; **~ to the town/to the sea** zur Stadt/ans Meer führen; **~ to confusion** Verwirrung stiften; **one thing led to another** es kam eins zum anderen; **b)** *(be first)* führen; *(go in front)* vorangehen; *(fig.: be leader)* an der Spitze stehen; **~ by 3 metres** mit 3 Metern in Führung liegen; 3 Meter Vorsprung haben; **~ in the race** das Rennen anführen. **3.** *n.* **a)** *(precedent)* Beispiel, *das; (clue)* Anhaltspunkt, *der;* **follow sb.'s ~, take one's ~ from sb.** jmds. Beispiel *(Dat.)* folgen; **b)** *(first place)* Führung, *die;* **be in the ~:** in Führung liegen; an der Spitze liegen; **move** *or* **go into the ~, take the ~:** sich an die Spitze setzen; in Führung gehen; **c)** *(amount, distance)* Vorsprung, *der;* **d)** *(on dog etc.)* Leine, *die;* **on a ~:** an der Leine; **put a dog on the ~:** einen Hund anleinen; **e)** *(Electr.)* Kabel, *das;* Leitung, *die;* **f)** *(Theatre)* Hauptrolle, *die; (player)* Hauptdarsteller, *der/*-darstellerin, *die*

~ a'way *v. t.* abführen ‹*Gefangenen, Verbrecher*›

~ 'off 1. *v. t.* **a)** *(take away)* abführen; **b)** *(begin)* beginnen. **2.** *v. i.* beginnen

~ 'on 1. *v. t.* **a)** *(entice)* **~ sb. on** jmdn. reizen; **he's ~ing you on** er versucht, dich zu reizen; **b)** *(deceive)* auf den Leim führen; **she's just ~ing him on** sie hält ihn nur zum Narren; **c)** *(take further)* **that ~s me on to my next point** das bringt mich zu meinem nächsten Punkt. **2.** *v. i.* **a)** *imper. (go first)* **~ on!** geh vor!; **b) ~ing on from what you have just said, ...:** um fortzuführen, was Sie eben sagten, ...; **~ on to the next topic** *etc.* zum nächsten Thema *usw.* führen

~ 'up to *v. t.* [schließlich] führen zu; *(aim at)* hinauswollen auf (+ *Akk.*)

leaden ['ledn] *adj.* **a)** bleiern; **b)** *(fig.)* bleiern ‹*Schlaf, Augenlider, Glieder*›

leader ['li:də(r)] *n.* **a)** Führer, *der/*Führerin, *die; (of political party)* Vorsitzende, *der/die; (of gang, rebels)* Anführer, *der/*Anführerin, *die; (of expedition, project)* Leiter, *der/*Leiterin, *die; (of deputation)* Sprecher, *der/*Sprecherin, *die; (of tribe)* [Stammes]häuptling, *der;* Stammesführer, *der;* **the Egyptian/Labour ~:** der ägyptische Präsident/der Vorsitzende der Labour Party; **union/the Labour ~s** Gewerkschaftsvorsitzende/die Führenden der Labour Party; **have the qualities of a ~:** Führungsqualitäten haben; **b)** *(one who is first)* **he is a ~ in his field** er ist eine führende Kapazität auf seinem Gebiet; *(in race etc.)* **be the ~:** in Führung liegen; **c)** *(Brit. Journ.)* Leitartikel, *der;* **d)** *(Mus.) (leading performer)* Leader, *der/*Leaderin, *die; (Brit.: principal first violinist)* Konzertmeister, *der/*-meisterin, *die*

leadership ['li:dəʃɪp] *n.* **a)** Führung, *die; (capacity to lead)* Führungseigenschaften *Pl.;* **under the ~ of** unter [der] Führung von; **b)** *(leaders)* Führung[sspitze], *die;* **~ of the party** Parteivorsitz, *der*

lead-free ['ledfri:] *adj.* bleifrei

leading ['li:dɪŋ] *adj.* führend; *(in first position)* ‹*Läufer, Pferd, Auto*› an der Spitze; **~ role** Hauptrolle, *die; (fig.)* führende Rolle

leading: ~ 'article *n. (Brit. Journ.)* Leitartikel, *der;* **~ 'lady** *n.* Hauptdarstellerin, *die;* **~ 'light** *n.* herausragende Persönlichkeit; *(expert)* führende Kapazität; **~ 'man** *n.* Hauptdarsteller, *der;* **~ 'question** *n.* Suggestivfrage, *die*

lead: ~ pencil [led 'pensl] *n.* Bleistift, *der;* **~-poisoning** ['ledpɔɪzənɪŋ] *n.* Bleivergiftung, *die;* **~ story** ['li:d stɔ:rɪ] *n. (Journ.)* Titelgeschichte, *die*

leaf [li:f] **1.** *n., pl.* **leaves** [li:vz] **a)** Blatt, *das;* **shake like a ~:** zittern wie Espenlaub; **be in ~:** grün sein; **come into ~:** grün werden; **b)** *(of paper)* Blatt, *das;* **a ~ of paper** ein Blatt Papier; **turn over a new ~** *(fig.)* einen neuen Anfang machen; sich ändern; *see also* **book 1 a; c)** *(of table) (hinged/sliding flap)* Platte, *die; (for inserting)* Einlegebrett, *das.* **2.** *v. i.* **~ through sth.** etw. durchblättern; in etw. *(Dat.)* blättern

leaflet ['li:flɪt] *n.* [Hand]zettel, *der; (with instructions)* Gebrauchsanweisung, *die; (advertising)* Reklamezettel, *der; (political)* Flugblatt, *das*

leafy ['li:fɪ] *adj.* belaubt; **a ~ country lane** eine baumbestandene Landstraße

league [li:g] *n.* **a)** *(agreement)* Bündnis, *das;* Bund, *der; (in history)* Liga, *die;* **be in ~ with sb.** mit jmdm. im Bunde sein *od.* stehen; **those two are in ~ [together]** die beiden stecken unter einer Decke *(ugs.);* **b)** *(Sport)* Liga, *die;* **I am not in his ~, he is out of my ~** *(fig.)* ich komme nicht an ihn heran; **be in the big ~** *(fig.)* es geschafft haben

league: ~ 'football *n.* Ligafußball, *der;* **~ game, ~ match** *ns.* Ligaspiel, *das;* **L~ of 'Nations** *n. (Hist.)* Völkerbund, *der;* **~ table** *n.* Tabelle, *die (Sport)*

leak [li:k] **1.** *n.* **a)** *(hole)* Leck, *das; (in roof, ceiling, tent)* undichte Stelle; **there's a ~ in the tank** der Tank ist leck; der Tank hat ein Leck; **spring a ~** ‹*Schiff:*› leckschlagen *(Seemannsspr.);* ‹*Gas-, Flüssigkeitsbehälter:*› ein Leck bekommen; **stop the ~:** das Leck abdichten *od.* stopfen; **b)** *(escaping fluid/gas)* durch ein Leck austretende Flüssigkeit/austretendes Gas; **c)** *(instance)* **a gas/oil ~, a ~ of gas/oil** ein Austreten von Gas/Öl; **there has been a gas/oil ~:** es ist Gas/Öl ausgetreten; **d)** *(fig.: of information)* undichte Stelle; **e)** *(Electr.)* Elektrizitätsverlust, *der; (path or point)* Fehlerstelle, *die.* **2.** *v. t.* **a)** austreten lassen; **the pipe is ~ing water/gas** aus dem lecken Rohr tritt Wasser/Gas aus; **b)** *(fig.: disclose)* durchsickern lassen; **~ sth. to sb.** jmdm. etw. zuspielen. **3.** *v. i.* **a)** *(escape)* austreten **(from** aus); *(enter)* eindringen **(in** in + *Akk.*); **b)** ‹*Faß, Tank, Schiff:*› lecken; ‹*Rohr, Leitung, Dach:*› undicht sein; ‹*Gefäß, Füller:*› auslaufen; **the roof ~s** es regnet durch das Dach; **c)** *(fig.)* **~ [out]** durchsickern

leakage ['li:kɪdʒ] *n.* **a)** Auslaufen, *das; (of fluid, gas)* Ausströmen, *das; (fig.: of information)* Durchsickern, *das;* **b)** *(substance, amount)* **the ~ is increasing** das Leck wird größer; **mop up the ~:** das ausgelaufene Wasser *usw.* aufwischen

leaky ['li:kɪ] *adj.* undicht; leck ‹*Schiff, Boot, Tank*›

[1]**lean** [li:n] **1.** *adj.* mager; hager ‹*Person, Gesicht*›; **we had a ~ time [of it]** es ging uns sehr schlecht. **2.** *n. (meat)* Magere, *das*

[2]**lean 1.** *v. i.*, **~ed** [li:nd, lent] *or (Brit.)* **~t** [lent] **a)** sich beugen; **~ against the door** sich gegen die

Tür lehnen; ~ **out of the window** sich aus dem Fenster lehnen; ~ **down/forward** sich herab-/vorbeugen; ~ **back** sich zurücklehnen; **b)** *(support oneself)* ~ **against/on sth.** sich gegen/an etw. *(Akk.)* lehnen; ~ **on sth.** *(from above)* sich auf etw. *(Akk.)* lehnen; ~ **on sb.'s arm** sich auf jmds. Arm *(Akk.)* stützen; **c)** *(be supported)* lehnen (**against** an + *Dat.*); **d)** *(fig.: rely)* ~ **[up]on sb.** auf jmdn. bauen; **e)** *(stand obliquely)* sich neigen; **f)** *(fig.: tend)* ~ **to[wards] sth.** zu etw. neigen. **2.** *v.t.*, ~**ed** *or (Brit.)* ~**t** lehnen (**against** gegen *od.* an + *Akk.*). **3.** *n.* Neigung, *die;* **have a definite ~ to the right** eine deutliche Neigung nach rechts aufweisen

~ **over 1.** ['---] *v.t.* sich neigen über (+ *Akk.*). **2.** [-'--] *v.i.* ‹*Person:*› sich hinüberbeugen; *(forwards)* sich verbeugen; *see also* **backwards a**

leaning ['li:nɪŋ] *n.* Hang, *der;* Neigung, *die*

leanness ['li:nnɪs] *n., no pl. (of person, face)* Hagerkeit, *die*

leant *see* [2]**lean 1, 2**

leap [li:p] **1.** *v.i.*, ~**ed** [li:pt, lept] *or* ~**t** [lept] **a)** springen; ‹*Herz:*› hüpfen; ~ **to one's feet** aufspringen; ~ **out of/up from one's chair** aus seinem Sessel/von seinem Stuhl aufspringen; ~ **back in shock** vor Entsetzen zurückspringen; **b)** *(fig.)* ~ **to sb.'s defence** jmdm. beispringen *(geh.);* ~ **at the chance** die Gelegenheit beim Schopf packen. **2.** *v.t.*, ~**ed** *or* ~**t** *(jump over)* überspringen; springen *od.* setzen über (+ *Akk.*). **3.** *n.* Sprung, *der;* **with** *or* **in one ~:** mit einem Satz; **by ~s and bounds** *(fig.)* mit Riesenschritten ‹*vorangehen*›; sprunghaft ‹*zunehmen*›; *see also* **dark 2 b**

'**leap-frog 1.** *n.* Bockspringen, *das.* **2.** *v.i.*, **-gg-** Bockspringen machen; ~ **over sb.** einen Bocksprung über jmdn. machen. **3.** *v.t.*, **-gg-** *(fig.)* überspringen

leapt *see* **leap 1, 2**

'**leap year** *n.* Schaltjahr, *das*

learn [lɜ:n] **1.** *v.t.*, **learnt** [lɜ:nt] *or* **learned** [lɜ:nd, lɜ:nt] **a)** lernen; *(with emphasis on completeness of result)* erlernen; ~ **sth. by** *or* **from experience** etw. durch [die] *od.* aus der Erfahrung lernen; ~ **sth. from sb./a book/an example** etw. von jmdm./aus einem Buch/am Beispiel lernen; **I am ~ing [how] to play tennis** ich lerne Tennis spielen; *see also* **lesson b; rope e;** **b)** *(find out)* erfahren; lernen; *(by oral information)* hören; *(by observation)* erkennen; merken; *(by thought)* erkennen; *(be informed of)* erfahren; **I ~ed from the newspaper that ...:** ich habe in der Zeitung gelesen *od.* aus der Zeitung erfahren, daß ... **2.** *v.i.*, **learnt** *or* **learned a)** lernen; **be slow to ~:** langsam lernen; **you'll soon ~:** du wirst es bald lernen; **will you never ~?** du lernst es wohl nie!; **some people never ~:** mancher lernt's nie; ~ **by one's mistakes** aus seinen Fehlern lernen; ~ **about sth.** etwas über etw. *(Akk.)* lernen; **b)** *(get to know)* erfahren (**of** von)

learned ['lɜ:nɪd] *adj.* gelehrt; wissenschaftlich ‹*Gesellschaft, Zeitschrift*›

learner ['lɜ:nə(r)] *n.* Lernende, *der/die; (beginner)* Anfänger, *der*/Anfängerin, *die;* **be a slow/quick ~:** langsam/schnell lernen; **the car is driven by a ~:** ein Fahrschüler steuert den Wagen

learner-'driver *n. (Brit.)* Fahrschüler/-schülerin *(der/die unter Aufsicht fährt)*

learning ['lɜ:nɪŋ] *n. (scholarship)* Wissen, *das; (of person)* Gelehrsamkeit, *die*

learnt *see* **learn**

lease [li:s] **1.** *n. (of land, business premises)* Pachtvertrag, *der; (of house, flat, office)* Mietvertrag, *der;* **be on [a] ~:** gepachtet/gemietet sein; **give sb./sth. a new ~ of life** jmdm. Auftrieb geben/etw. wieder in Schuß bringen *(ugs.).* **2.** *v.t.* **a)** *(grant ~ on)* verpachten ‹*Grundstück, Geschäft, Rechte*›; vermieten ‹*Haus, Wohnung, Büro*›; **b)** *(take ~ on)* pachten ‹*Grundstück, Geschäft, Rechte*›; mieten ‹*Haus, Wohnung, Büro*›; leasen ‹*Auto*›

lease: ~**back** *n.* Verpachtung an den Verkäufer; ~**hold** *see* **lease 2: 1.** *n.* **have the ~hold of** *or* **on sth.** etw. gepachtet *od.* in Pacht/gemietet haben; **2.** *adj.* gepachtet/gemietet; ~**holder** *n. see* **lease 2:** Pächter, *der*/Pächterin, *die;* Mieter, *der*/Mieterin, *die*

leash [li:ʃ] *see* [2]**lead 3 d**

least [li:st] **1.** *adj. (smallest)* kleinst...; *(in quantity)* wenigst...; *(in status)* geringst...; **that's the ~ of our problems** das ist unser geringstes Problem; *see also* [1]**last 1.** **2.** *n.* Geringste, *das;* **the ~ I can do** das mindeste, was ich tun kann; **the ~ he could do would be to apologize** er könnte sich wenigstens entschuldigen; **to say the ~ [of it]** gelinde gesagt; **at ~:** mindestens; *(if nothing more; anyway)* wenigstens; **at the [very] ~:** [aller]mindestens; **not [in] the ~:** nicht im geringsten. **3.** *adv.* am wenigsten; **not ~ because ...:** nicht zuletzt deshalb, weil ...; ~ **of all** am allerwenigsten; **the ~ likely answer** die unwahrscheinlichste Lösung

leather ['leðə(r)] **1.** *n.* Leder, *das; (things made of ~)* Lederwaren *Pl.* **2.** *adj.* ledern; Leder‹*jacke, -mantel, -handschuh*›

leathery ['leðərɪ] *adj.* ledern

[1]**leave** [li:v] *n., no pl.* **a)** *(permission)* Erlaubnis, *die; (official approval)* Genehmigung, *die;* **grant** *or* **give sb. ~ to do sth.** jmdm. gestatten, etw. zu tun; **be absent without ~:** sich unerlaubt entfernt haben; **get ~ from sb. to do sth.** von jmdm. die Erlaubnis bekommen, etw. zu tun; **by ~ of sb.** mit jmds. Genehmigung; **by your ~** *(formal)* mit Ihrer Erlaubnis; **b)** *(from duty or work)* Urlaub, *der;* ~ **[of absence]** Beurlaubung, *die;* Urlaub, *der (auch Mil.);* **go on ~:** in Urlaub gehen; **be on ~:** Urlaub haben; in Urlaub sein; **c) take one's ~** *(say farewell)* sich verabschieden; Abschied nehmen *(geh.);* **he must have taken ~ of his senses** er muß von Sinnen sein

[2]**leave** *v.t.*, **left** [left] **a)** *(make or let remain, lit. or fig.)* hinterlassen; **he left a message with me for Mary** er hat bei mir eine Nachricht für Mary hinterlassen; ~ **sb. to do sth.** es jmdm. überlassen, etw. zu tun; **6 from 10 ~s 4** 10 weniger 6 ist 4; *(in will)* ~ **sb. sth.,** ~ **sth. to sb.** jmdm. etw. hinterlassen; **b)** *(by mistake)* vergessen; **I left my gloves in your car** ich habe meine Handschuhe in deinem Auto liegenlassen *od.* vergessen; **c) be left with** nicht loswerden ‹*Gefühl, Verdacht*›; übrigbehalten ‹*Geld*›; zurückbleiben mit ‹*Schulden, Kind*›; **I was left with the job of clearing up** es blieb mir überlassen, aufzuräumen; **d)** *(refrain from doing, using, etc., let remain undisturbed)* stehenlassen ‹*Abwasch, Essen*›; sich *(Dat.)* entgehen lassen ‹*Gelegenheit*›; **e)** *(let remain in given state)* lassen; ~ **the door open/the light on** die Tür offenlassen/das Licht anlassen; ~ **the book lying on the table** das Buch auf dem Tisch liegenlassen; ~ **sb. in the dark** *(fig.)* jmdn. im dunkeln lassen; ~ **one's clothes all over the room** seine Kleider im ganzen Zimmer her-

umliegen lassen; ~ **sb. alone** *(allow to be alone)* jmdn. allein lassen; *(stop bothering)* jmdn. in Ruhe lassen; ~ **sth. alone** etw. in Ruhe lassen; ~ **it at that** *(coll.)* es dabei bewenden lassen; **f)** *(refer, entrust)* ~ **sth. to sb./sth.** etw. jmdm./einer Sache überlassen; **I ~ the matter entirely in your hands** ich lege diese Angelegenheit ganz in Ihre Hand/Hände; ~ **it to me** laß mich nur machen; **g)** *(go away from)* verlassen; ~ **home at 6 a.m.** um 6 Uhr früh von zu Hause weggehen/-fahren; **the plane ~s Bonn at 6 p.m.** das Flugzeug fliegt um 18 Uhr von Bonn ab; ~ **Bonn at 6 p.m.** *(by car, in train)* um 18 Uhr von Bonn abfahren; *(by plane)* um 18 Uhr in Bonn abfliegen; ~ **the road** *(crash)* von der Fahrbahn abkommen; ~ **the rails** *or* **tracks** entgleisen; **the train ~s the station** der Zug rollt aus dem Bahnhof; **I left her at the bus stop** *(parted from)* an der Bushaltestelle haben wir uns getrennt; *(set down)* ich habe sie an der Bushaltestelle abgesetzt; ~ **the table** vom Tisch aufstehen; *abs.* **the train ~s at 8.30 a.m.** der Zug fährt *od.* geht um 8.30 Uhr; ~ **for Paris** nach Paris fahren/fliegen; **it is time to ~:** wir müssen gehen *od.* aufbrechen; ~ **on the 8 a.m. train/flight** mit dem Acht-Uhr-Zug fahren/der Acht-Uhr-Maschine fliegen; **h)** *(quit permanently)* verlassen; ~ **school** die Schule verlassen; *(prematurely)* von der Schule abgehen; ~ **work** aufhören zu arbeiten; **i)** *(desert)* verlassen; ~ **sb. for another man/woman** jmdn. wegen eines anderen Mannes/einer anderen Frau verlassen; **he was left for dead** man ließ ihn zurück, weil man ihn für tot hielt

~ **a'side** *v. t.* beiseite lassen

~ **be'hind** *v. t.* **a)** zurücklassen; **b)** *(by mistake) see* ~ **b**

~ **'off** *v. t.* **a)** *(cease to wear)* auslassen *(ugs.);* nicht anziehen; **b)** *(discontinue)* aufhören mit; *abs.* aufhören; **has it left off raining?** hat es aufgehört zu regnen?

~ **'out** *v. t.* auslassen

~ **'over** *v. t.* **a)** *(Brit.: not deal with till later)* zurückstellen; **b) be left over** übrig[geblieben] sein

-leaved [li:vd] *adj. in comb.* -blätt[e]rig

leaven ['levn] *n.* Treibmittel, *das; (fermenting dough)* Sauerteig, *der*

Lebanon ['lebənən] *pr. n.* **[the]** ~: [der] Libanon

lecherous ['letʃərəs] *adj.* lüstern *(geh.);* geil *(abwertend)*

lechery ['letʃərɪ] *n.* Wollust, *die (geh.)*

lectern ['lektən] *n. (for Bible etc.)* Lektionar[ium], *das; (for singers)* Notenpult, *das*

lecture ['lektʃə(r)] **1.** *n.* **a)** Vortrag, *der; (Univ.)* Vorlesung, *die;* **give [sb.] a ~ on sth.** [vor jmdm.] einen Vortrag/eine Vorlesung über etw. *(Akk.)* halten; **b)** *(reprimand)* Strafpredigt, *die (ugs.);* **give sb. a ~:** jmdm. eine Strafpredigt halten. **2.** *v. i.* ~ **[to sb.] [on sth.]** [vor jmdm.] einen Vortrag/ *(Univ.)* eine Vorlesung [über etw. *(Akk.)*] halten; *(give ~s)* [vor jmdm.] Vorträge/*(Univ.)* Vorlesungen [über etw. *(Akk.)*] halten. **3.** *v. t. (scold)* ~ **sb.** jmdm. eine Strafpredigt halten *(ugs.)*

lecturer ['lektʃərə(r)] *n.* **a)** Vortragende, *der/die;* **b)** *(Univ.)* Lehrbeauftragte, *der/die;* **senior ~:** Dozent, *der*/Dozentin, *die;* **be a ~ in French** Dozent/Dozentin für Französisch sein

lecture: **~-room** *n.* Vortragsraum, *der; (Univ.)* Vorlesungsraum, *der;* **~-theatre** *n.* Hörsaal, *der*

led *see* [2]**lead 1, 2**

ledge [ledʒ] *n.* **a)** Vorsprung, *der;* Sims, *der od. das;* **b)** *(of rock)* [schmaler] Vorsprung

ledger ['ledʒə(r)] *n. (Commerc.)* Hauptbuch, *das*

lee [li:] *n.* **a)** *(shelter)* Schutz, *der;* **in/under the ~ of** im Schutz (+ *Gen.*); **b)** ~ **[side]** *(Naut.)* Leeseite, *die*

leech [li:tʃ] *n.* [Blut]egel, *der*

leek [li:k] *n.* Porree, *der;* Lauch, *der; (as Welsh emblem)* Lauch, *der;* **I like ~s** ich mag Porree *od.* Lauch; **three ~s** drei Stangen Porree/Lauch

leer [lɪə(r)] **1.** *n.* **[suggestive/sneering]** ~: anzüglicher/spöttischer Blick. **2.** *v. i.* [anzüglich/spöttisch/*(lustfully)* begehrlich] blicken; ~ **at sb.** jmdm. einen anzüglichen / spöttischen / begehrlichen [Seiten]blick zuwerfen

lees [li:z] *n. pl.* Bodensatz, *der*

leeward ['li:wəd, *(Naut.)* 'lu:əd] *(esp. Naut.)* **1.** *adj.* **to/on the ~ side of the ship** nach/in Lee. **2.** *n.* Leeseite, *die;* **to ~:** leewärts; nach Lee

'leeway *n.* **a)** *(Naut.)* Leeweg, *der;* Abdrift, *die;* **b)** *(fig.)* Spielraum, *der;* **allow** *or* **give sb. ~:** jmdm. Spielraum lassen

[1]**left** *see* [2]**leave**

[2]**left** [left] **1.** *adj.* **a)** *(opposite of right)* link...; **on the ~ side** auf der linken Seite; links; *see also* **turn 1 c; b) L~** *(Polit.)* link... **2.** *adv.* nach links; ~ **of the road** links von der Straße. **3.** *n.* **a)** *(~-hand side)* linke Seite; **move to the ~:** nach links rücken; **on** *or* **to the ~ [of sb./sth.]** links [von jmdm./etw.]; **on** *or* **to my ~, to the ~ of me** links von mir; zu meiner Linken; **drive on the ~:** links fahren; **b)** *(Polit.)* **the L~:** die Linke; **be on the L~ of the Party** dem linken Flügel der Partei angehören; **c)** *(Boxing)* Linke, *die;* **d)** *(in marching)* **~, right, ~, right, ~, ...** *(Mil.)* links, zwo, drei, vier, links, ...

left: ~ **'hand** *n.* **a)** linke Hand; Linke, *die;* **b)** *(~ side)* **on** *or* **at sb.'s ~ hand** zu jmds. Linken; links von jmdm.; **~-hand** *adj.* link...; **~-hand bend** Linkskurve, *die;* **on your ~-hand side you see ...:** links *od.* zur Linken sehen Sie ...; *see also* **drive 1 g; ~-handed** [left'hændɪd] **1.** *adj.* **a)** linkshändig; ‹*Werkzeug*› für Linkshänder; **be ~-handed** Linkshänder/Linkshänderin sein; **b)** *(turning to ~)* linksgängig, linksdrehend ‹*Schraube, Gewinde*›; **2.** *adv.* linkshändig; mit der linken Hand; ~ **'luggage [office]** *n. (Brit. Railw.)* Gepäckaufbewahrung, *die;* **~-over** *attrib. adj.* übriggeblieben; **~-overs** *n. pl.* Reste; *(fig.)* Relikte; Überbleibsel *(ugs.);* ~ **'wing** *n.* linker Flügel; **~-wing** *adj.* **a)** *(Sport)* Linksaußen‹*spieler, -position*›; **b)** *(Polit.)* link...; linksgerichtet; Links‹*intellektueller, -extremist, -radikalismus*›; **~-'winger** *n.* **a)** *(Sport)* Linksaußen, *der;* **b)** *(Polit.)* Angehöriger/Angehörige des linken Flügels

leg [leg] **1.** *n.* **a)** Bein, *das;* **upper/lower ~:** Ober-/Unterschenkel, *der;* **artificial ~:** Beinprothese, *die;* **wooden ~:** Holzbein, *das;* **as fast as my ~s would carry me** so schnell mich die Füße trugen; **give sb. a ~ up on to a horse/over the gate** jmdm. auf ein Pferd/über das Gatter helfen; **be on one's last ~s** sich kaum noch auf den Beinen halten können; *(be about to die)* mit einem Fuß *od.* Bein im Grabe stehen; **the car is on its last ~s** das Auto macht es nicht mehr lange *(ugs.);* **pull sb.'s ~** *(fig.)* jmdn. auf den Arm nehmen *(ugs.);* **not have a ~ to stand on** *(fig.)* nichts in der Hand haben *(fig.);* **stretch one's ~s** sich *(Dat.)* die Beine vertreten; **b)** *(of table, chair, etc.)* Bein, *das;* **c)**

trouser-~s Hosenbeine; **d)** *(Gastr.)* Keule, *die;* ~ **of lamb/veal** Lamm-/Kalbskeule, *die;* **e)** *(of journey)* Etappe, *die;* Teilstrecke, *die;* **f)** *(Sport coll.)* Durchgang, *der; (of relay race)* Teilstrecke, *die.* **2.** *v. t.,* **-gg-:** ~ **it** *(coll.)* die Beine in die Hand *od.* unter die Arme nehmen *(ugs.)*

legacy ['legəsı] *n.* Vermächtnis, *das (Rechtsspr.);* Erbschaft, *die; (fig.)* Erbe, *das;* **leave sb. sth. as a** ~ *(lit. or fig.)* jmdm. etw. hinterlassen

legal ['li:gl] *adj.* **a)** *(concerning the law)* juristisch; Rechts‹*beratung, -berater, -streit, -schutz*›; gesetzlich ‹*Vertreter*›; rechtlich ‹*Gründe, Stellung*›; *(of the law)* Gerichts‹*kosten*›; **in** ~ **matters/affairs** in Rechtsfragen/-angelegenheiten; **seek** ~ **advice** sich juristisch beraten lassen; **he is a member of the** ~ **profession** er ist Jurist; **b)** *(required by law)* gesetzlich vorgeschrieben ‹*Mindestalter, Zeitraum*›; gesetzlich ‹*Verpflichtung*›; gesetzlich verankert ‹*Recht*›; **c)** *(lawful)* legal; rechtsgültig ‹*Vertrag, Testament*›; gesetzlich zulässig ‹*Grenze, Höchstwert*›; **it is ~/not ~ to do sth.** es ist rechtlich zulässig/gesetzlich verboten, etw. zu tun; **make sth. ~:** etw. legalisieren

legal: ~ **'action** *n.* Gerichtsverfahren, *das;* Prozeß, *der;* **take** ~ **action against sb.** gerichtlich gegen jmdn. vorgehen; ~ **'aid** *n.* ≈ Prozeßkostenhilfe, *die*

legality [lı'gælıtı] *n.* Legalität, *die;* Rechtmäßigkeit, *die*

legalize ['li:gəlaız] *v. t.* legalisieren

legally ['li:gəlı] *adv.* rechtlich ‹*zulässig, verpflichtet, begründet, unhaltbar, möglich*›; gesetzlich ‹*verankert, verpflichtet*›; vor dem Gesetz ‹*verantwortlich*›; legal ‹*durchführen, abwickeln, erwerben*›; ~ **valid/binding** rechtsgültig/-verbindlich; **be** ~ **entitled to sth.** einen Rechtsanspruch auf etw. *(Akk.)* haben

legation [lı'geıʃn] *n.* *(Diplom.)* Gesandtschaft, *die; (residence also)* Gesandtschaftsgebäude, *das*

legend ['ledʒənd] *n.* Sage, *die; (of life of saint etc.; unfounded belief)* Legende, *die;* ~ **has it that ...:** es geht die Sage, daß ...

legendary ['ledʒəndərı] *adj.* **a)** legendenhaft; *(described in legend)* legendär; sagenhaft; **b)** *(coll.: famous)* sagenhaft *(ugs.);* legendär

-legged [legd, legıd] *adj. in comb.* -beinig; **two-~:** zweibeinig

leggy ['legı] *adj.* langbeinig; hochbeinig; ‹*Junge, Fohlen*› mit [staksigen] langen Beinen

legibility [ledʒı'bılıtı] *n., no pl.* Leserlichkeit, *die*

legible ['ledʒıbl] *adj.* leserlich; **easily/scarcely** ~**:** leicht/kaum lesbar

legion ['li:dʒn] *n.* Legion, *die*

legionnaire [li:dʒə'neə(r)] *n.* Legionär, *der*

legislate ['ledʒısleıt] *v. i.* Gesetze verabschieden; ~ **for/against sth.** Gesetze zum Schutz von/gegen etw. einbringen

legislation [ledʒıs'leıʃn] *n.* **a)** *(laws)* Gesetze; **b)** *(legislating)* Gesetzgebung, *die*

legislative ['ledʒıslətıv] *adj.* gesetzgebend; *(created by legislature)* gesetzgeberisch

legislator ['ledʒısleıtə(r)] *n.* Gesetzgeber, *der*

legislature ['ledʒısleıtʃə(r)] *n.* Legislative, *die*

legitimacy [lı'dʒıtıməsı] *n., no pl.* **a)** Rechtmäßigkeit, *die;* Legitimität, *die;* **b)** *(of child)* Ehelichkeit, *die*

legitimate [lı'dʒıtımət] *adj.* **a)** ehelich, legitim ‹*Kind*›; **b)** *(lawful)* legitim; rechtmäßig ‹*Besitzer, Regierung*›; legal ‹*Vorgehen, Weg, Geschäft, Gewinn*›; **c)** *(valid)* berechtigt; stichhaltig; legitim *(geh.)* ‹*Argument*›; ausreichend ‹*Entschuldigung*›; triftig ‹*Grund*›

legitimize (legitimise) [lı'dʒıtımaız] *v. t.* legitimieren; [durch Heirat] ehelich machen ‹*Kind*›

'leg-room *n., no pl., no indef. art.* Beinfreiheit, *die*

leisure ['leʒə(r)] *n.* Freizeit, *die; (for relaxation)* Muße, *die; attrib.* Freizeit‹*kleidung, -beschäftigung, -zentrum, -industrie*›; **a life/day of ~:** ein Leben/Tag der Muße *(geh.);* **do sth. at ~:** etw. in Ruhe tun; **do sth. at one's ~:** sich *(Dat.)* Zeit mit etw. lassen; ~ **time** *or* **hours** Freizeit, *die*

leisurely ['leʒəlı] **1.** *adj.* gemächlich. **2.** *adv.* langsam; ohne Hast

lemming ['lemıŋ] *n.* Lemming, *der*

lemon ['lemən] *n.* **a)** Zitrone, *die;* **b)** *(colour)* Zitronengelb, *das*

lemonade [lemə'neıd] *n.* [Zitronen]limonade, *die*

lemon: ~ **curd** [lemən 'kɜ:d] *n.* Zitronencreme, *die;* **~-juice** *n.* Zitronensaft, *der;* ~ **'sole** *n.* Seezunge, *die;* **~-yellow** *adj.* zitronengelb

lend [lend] **1.** *v. t.,* **lent** [lent] **a)** leihen; ~ **sth. to sb.** jmdm. etw. leihen; **b)** *(give, impart)* geben; zur Verfügung stellen ‹*Dienste*›; verleihen ‹*Würde, Glaubwürdigkeit, Zauber*›; ~ **one's support to sth.** etw. unterstützen; ~ **one's name/authority to sth.** seinen Namen/guten Namen für etw. hergeben. **2.** *v. refl.,* **lent: the book ~s itself/does not ~ itself to use as a learning aid** das Buch eignet sich/eignet sich nicht als Lehrmittel; **the system ~s itself to manipulation** das System bietet sich zur Manipulation an

lender ['lendə(r)] *n.* Verleiher, *der*/Verleiherin, *die*

length [leŋθ, leŋkθ] *n.* **a)** *(also Horse-racing, Rowing, Fashion)* Länge, *die;* **a road four miles in ~:** eine vier Meilen lange Straße; **be six feet** *etc.* **in ~:** sechs Fuß *usw.* lang sein; **the room is twice the** ~ **of yours** das Zimmer ist doppelt so lang wie deins; **travel the** ~ **and breadth of the British Isles** überall auf den Britischen Inseln herumreisen; **a list the** ~ **of my arm** *(fig.)* eine ellenlange Liste; **win by a ~:** mit einer Länge siegen; **b)** *(of time)* Länge, *die;* **a short** ~ **of time** kurze Zeit; **the play was three hours in ~:** das Stück dauerte drei Stunden; **c) at** ~ *(for a long time)* lange; *(eventually)* schließlich; **at [great]** ~ *(in great detail)* lang und breit; sehr ausführlich; **at some ~:** ziemlich ausführlich; **d) go to any/great** *etc.* **~s** alles nur/alles Erdenkliche tun; **e)** *(piece of material)* Länge, *die;* Stück, *das*

lengthen ['leŋθən, 'leŋkθən] **1.** *v. i.* länger werden. **2.** *v. t.* verlängern; länger machen ‹*Kleid*›

lengthways ['leŋθweız, 'leŋkθweız] *adv.* der Länge nach; längs

lengthwise ['leŋθwaız, 'leŋkθwaız] *see* **lengthways**

lengthy ['leŋθı, 'leŋkθı] *adj.* überlang

lenience ['li:nıəns], **leniency** ['li:nıənsı] *ns., no pl.* Nachsicht, *die;* Milde, *die;* **show ~:** Milde walten lassen; Nachsicht zeigen

lenient ['li:nıənt] *adj.* **a)** *(tolerant)* nachsichtig; **b)** *(mild)* mild ‹*Urteil, Strafe*›

lens [lenz] *n.* **a)** *(Optics, Anat.)* Linse, *die; (in spectacles)* Glas, *das;* **b)** *(Photog.)* Objektiv, *das*

Lent [lent] *n.* Fastenzeit, *die*

lent *see* **lend**

lentil ['lentıl] *n.* Linse, *die*

Leo ['li:əʊ] *n., pl.* **~s** *(Astrol., Astron.)* der Löwe

leopard ['lepəd] *n. (Zool.)* Leopard, *der*

leotard ['li:ətɑ:d] *n.* Turnanzug, *der*

leper ['lepə(r)] *n.* Leprakranke, *der/die*

leprosy ['leprəsɪ] *n. (Med.)* Lepra, *die*

lesbian ['lezbɪən] **1.** *n.* Lesbierin, *die.* **2.** *adj.* lesbisch

lesion ['li:ʒn] *n. (Med.)* Läsion, *die (fachspr.);* Verletzung, *die*

less [les] **1.** *adj.* weniger; **of ~ value/importance/account** *or* **note** weniger wertvoll/wichtig/bedeutend; **his chances are ~ than mine** seine Chancen sind geringer als meine; **for ~ time** kürzere Zeit; **the pain is getting ~:** der Schmerz läßt nach; **~ talking, please** etwas mehr Ruhe, bitte. **2.** *adv.* weniger; **I think ~/no ~ of him after what he did** ich halte nicht mehr so viel/nicht weniger von ihm, seit er das getan hat; **~ and ~:** immer weniger; **~ and ~ [often]** immer seltener; **~ so** weniger; **the ~ so because ...:** um so weniger, als *od.* weil ...; **even** *or* **still/far** *or* **much ~:** noch/viel weniger. **3.** *n., no pl., no indef. art.* weniger; **~ and ~:** immer weniger; **the ~ said [about it] the better** je weniger man darüber sagt, um so besser; **in ~ than no time** *(joc.)* in Null Komma nichts *(ugs.);* **~ of that!** *(coll.)* Schluß damit!; **~ of your cheek!** *(coll.)* sei nicht so frech! **4.** *prep. (deducting)* **ten ~ three is seven** zehn weniger drei ist sieben

lessee [le'si:] *n. see* **lease 2:** Pächter, *der*/Pächterin, *die;* Mieter, *der*/Mieterin, *die*

lessen ['lesn] **1.** *v. t. (reduce)* verringern; lindern ‹*Schmerz*›. **2.** *v. i. (become less)* sich verringern; ‹*Fieber:*› sinken, fallen; ‹*Schmerz:*› nachlassen

lesser ['lesə(r)] *attrib. adj.* geringer...; weniger bedeutend... ‹*Schauspieler, Werk*›

lesson ['lesn] *n.* **a)** *(class)* [Unterrichts]stunde, *die; (in textbook)* Lektion, *die;* **I like her ~s** mir gefällt ihr Unterricht; **give ~s** Privatstunden *od.* -unterricht geben; **give Italian ~s** Italienischunterricht *od.* -stunden geben; **[give] ~s in/on** Unterricht [erteilen] in (+ *Dat.*); **take piano ~s with sb.** bei jmdm. Klavierstunden nehmen; **b)** *(fig.: example, warning)* Lektion, *die;* Lehre, *die;* **teach sb. a ~:** jmdm. eine Lektion erteilen; ‹*Vorfall usw.:*› jmdm. eine Lehre sein; **be a ~ to sb.** jmdm. eine Lehre sein; **learn one's** *or* **a ~ from sth.** aus etw. eine Lehre ziehen; **I have learnt my ~:** das soll mir eine Lehre sein

lest [lest] *conj. (literary)* damit ... nicht; **he ran away ~ he [should] be seen** er rannte weg, um nicht gesehen zu werden; **I was afraid ~ he [should] come back before I was ready** ich fürchtete, daß er zurückkommen würde, bevor ich fertig war

[1]**let** [let] **1.** *v. t.,* **-tt-, let a)** *(allow to)* lassen; **~ sb. do sth.** jmdn. etw. tun lassen; **don't ~ things get you down/worry you** laß dich nicht entmutigen/mach dir keine Sorgen; **don't ~ him upset you** reg dich seinetwegen nicht auf; **I'll come if you will ~ me** ich komme, wenn ich darf; **~ sb./sth. alone** jmdn./etw. in Ruhe lassen; **~ alone** *(far less)* geschweige denn; **~ sb. be** jmdn. in Ruhe *od.* Frieden lassen; **~ go [of] sth./sb.** etw./jmdn. loslassen; **~ sb. go** *(from captivity)* jmdn. freilassen; **~ go** *(release hold)* loslassen; *(neglect)* herunterkommen lassen ‹*Haus*›; *(~ pass)* durchgehen lassen ‹*Bemerkung*›; **~ it go [at that]** es dabei belassen *od.* bewenden lassen; **~ oneself go** *(neglect oneself)* sich vernachlässigen; *(abandon self-restraint)* sich gehenlassen; **~ loose** loslassen; **b)** *(cause to)* **~ sb. know** jmdn. wissen lassen; **~ sb. think that ...:** jmdn. in dem Glauben lassen, daß ...; **I will ~ you know as soon as ...:** ich gebe Ihnen Bescheid, sobald ...; **c)** *(release)* ablassen ‹*Wasser*› **(out of, from** aus); lassen ‹*Luft*› **(out of** aus); **d)** *(Brit.: rent out)* vermieten ‹*Haus, Wohnung, Büro*›; verpachten ‹*Gelände, Grundstück*›; **'to ~'** „zu vermieten". **2.** *v. aux.,* **-tt-, let a)** *in exhortations* lassen; **~ us suppose that ...:** nehmen wir [nur] einmal an, daß ...; **Let's go to the cinema. – Yes, ~'s/No, ~'s not** *or* **don't ~'s** Komm/Kommt, wir gehen ins Kino. – Ja, gut/Nein, lieber nicht; **b)** *in command, challenge* lassen; **~ them come in** sie sollen hereinkommen; lassen Sie sie herein; **never ~ it be said that ...:** keiner soll sagen, daß ...; **[just] ~ him try!** das soll er [nur] mal wagen! **3.** *n. (Brit.)* **rent a flat on a short ~:** eine Wohnung für kurze Zeit mieten

~ **'down** *v. t.* **a)** *(lower)* herunter-/hinunterlassen; *see also* **hair b; b)** *(deflate)* die Luft [heraus]lassen aus; **c)** *(Dressm.)* auslassen ‹*Saum, Ärmel, Kleid, Hose*›; **d)** *(disappoint, fail)* im Stich lassen; **I ~ myself down in the exam** ich habe in der Prüfung enttäuschend abgeschnitten

~ **'in** *v. t.* **a)** *(admit)* herein-/hineinlassen; *(fig.)* die Tür öffnen (+ *Dat.*); **~ oneself/sb. in** sich *(Dat.)* [die Tür] aufschließen/jmdm. aufmachen; **my shoes are ~ting in water** meine Schuhe sind undicht; **b)** *(Dressm.)* enger machen; einnähen; **c) ~ oneself in for sth.** sich auf etw. *(Akk.)* einlassen; **d) ~ sb. in on a secret/plan** *etc.* jmdn. in ein Geheimnis/einen Plan *usw.* einweihen

~ **into** *v. t.* **a)** *(admit into)* lassen in (+ *Akk.*); **b)** *(fig.: acquaint with)* **~ sb. into a secret** jmdn. in ein Geheimnis einweihen

~ **'off** *v. t.* **a)** *(excuse)* laufenlassen *(ugs.); (allow to go)* gehen lassen; **~ sb. off lightly/with a fine** jmdn. glimpflich/mit einer Geldstrafe davonkommen lassen; **~ sb. off sth.** jmdm. etw. erlassen; **b)** abbrennen ‹*Feuerwerk*›; **c)** *(allow to escape)* ablassen ‹*Dampf, Flüssigkeit*›; **d)** *(allow to alight)* aussteigen lassen

~ **'on** *(sl.)* **1.** *v. i.* **don't ~ on!** nichts verraten! **2.** *v. t.* **a) sb. ~ on to me that ...:** man hat mir gesteckt, daß ... *(ugs.);* **b)** *(pretend)* **~ on that ...:** so tun, als ob ... *(ugs.)*

~ **'out** *v. t.* **a)** *(open door for)* **~ sb./an animal out** jmdn./ein Tier heraus-/hinauslassen; **Don't get up. I'll ~ myself out** Bleiben Sie sitzen. Ich finde schon allein hinaus; **b)** *(allow out)* rauslassen *(ugs.);* gehen lassen; **c)** *(emit)* ausstoßen ‹*Schrei*›; hören lassen ‹*Lachen, Seufzer*›; **~ out a groan** aufstöhnen; **d)** *(reveal)* verraten, ausplaudern ‹*Geheimnis*›; **~ out that ...:** durchsickern lassen, daß ...; **e)** *(Dressm.)* auslassen; **f)** *(Brit.: rent out) see* [1]**let 1d; g)** *(from duty)* **On Saturday? That ~s me out** Samstag? Da falle ich schon mal aus

~ **'through** *v. t.* durchlassen

~ **'up** *v. i. (coll.)* nachlassen; **don't you ever ~ up?** wirst du überhaupt nicht müde?

[2]**let** *n.* **without ~ [or hindrance]** *(formal/Law)* ohne jede Behinderung

'let-down *n.* Enttäuschung, *die*

lethal ['li:θl] *adj.* tödlich; *(fig.)* vernichtend

lethargic [lɪ'θɑ:dʒɪk] *adj.* träge; *(apathetic)* lethargisch; *(causing lethargy)* träge machend

lethargy ['leθədʒɪ] *n.* Trägheit, *die; (apathy)* Lethargie, *die*

letter [letə(r)] *n.* **a)** *(written communication)* Brief, *der* **(to** an + *Akk.*); *(official communication)* Schreiben, *das;* **a ~ of appointment** eine [briefliche] Anstellungszusage; **b)** *(of alphabet)* Buchstabe, *der;* **write in capital/small ~s** mit Groß-/Kleinbuchstaben schreiben; **have ~s after one's name** Ehrentitel/einen Ehrentitel haben; **c)** *(fig.)* **to the ~:** buchstabengetreu; aufs Wort; **the ~ of the law** der Buchstabe des Gesetzes; **d)** *in pl. (literature)* Literatur, *die;* **man of ~s** Homme de lettres, *der;* Literat, *der*

letter: ~-bomb *n.* Briefbombe, *die;* **~-box** *n.* Briefkasten, *der;* **~-head, ~-heading** *ns.* Briefpapier mit Briefkopf; *(heading)* Briefkopf, *der*

lettering ['letərɪŋ] *n.* Typographie, *die; (on book-cover)* Aufschrift, *die; (carved)* Inschrift, *die*

lettuce ['letɪs] *n.* [Kopf]salat, *der*

leukaemia, *(Amer.)* **leukemia** [lu:'ki:mɪə] *n.* Leukämie, *die*

level ['levl] **1.** *n.* **a)** Höhe, *die; (storey)* Etage, *die; (fig.: steady state)* Niveau, *das; (fig.: basis)* Ebene, *die;* **the water rose to the ~ of the doorstep** das Wasser stieg bis zur Türschwelle; **be on a ~ [with sb./sth.]** sich auf gleicher *od.* einer Höhe [mit jmdm./etw.] befinden; *(fig.)* auf dem gleichen Niveau sein [wie jmd./etw.]; **on the ~** *(fig. coll.)* ehrlich; **find one's ~** *(fig.)* seinen Platz finden; **b)** *(height)* **at waist/roof-top** *etc.* **~:** in Taillen-/Dachhöhe *usw.;* **c)** *(relative amount)* **sugar/alcohol ~:** [Blut]zucker-/Alkoholspiegel, *der;* **noise ~:** Geräuschpegel, *der;* **d)** *(social, moral, or intellectual plane)* Niveau, *das; (degree of achievement etc.)* Grad, *der* **(of** an + *Dat.*); **talks at the highest ~ [of government]** Gespräche auf höchster [Regierungs]ebene; **e)** *(instrument to test horizontal)* Wasserwaage, *die.* **2.** *adj.* **a)** waagerecht; flach ‹*Land*›; eben ‹*Boden, Land*›; **a ~ spoonful of flour** ein gestrichener Löffel Mehl; **the picture is not ~:** das Bild hängt nicht gerade; **b)** *(on a ~)* **be ~ [with sth./sb.]** auf gleicher Höhe [mit etw./jmdm.] sein; *(fig.)* [mit etw./jmdm.] gleichauf liegen; **the two pictures are not ~:** die beiden Bilder hängen nicht gleich hoch; **draw/keep ~ with a rival** mit einem Gegner gleichziehen/auf gleicher Höhe bleiben; **c)** *(fig.: steady, even)* ausgeglichen ‹*Leben, Temperament*›; ausgewogen ‹*Stil*›; **keep a ~ head** einen kühlen Kopf bewahren; **d) do one's ~ best** *(coll.)* sein möglichstes tun. **3.** *v. t., (Brit.)* **-ll-: a)** *(make ~ 2a)* ebnen; **b)** *(aim)* richten ‹*Blick, Gewehr, Rakete*› **(at, against** auf + *Akk.*); *(fig.)* richten ‹*Kritik usw.*› **(at, against** gegen); erheben ‹*Anklage, Vorwurf*› **(at, against** gegen); **c)** *(raze)* dem Erdboden gleichmachen ‹*Stadt, Gebäude*›

~ 'off 1. *v. t.* glatt machen. **2.** *v. i.* *(Aeronaut.)* die Flughöhe beibehalten

~ 'out 1. *v. t.* einebnen. **2.** *v. i.* **a)** *see* **~ off 2; b)** *(fig.)* sich ausgleichen; ‹*Preise, Markt:*› sich beruhigen

level: ~ 'crossing *n. (Brit. Railw.)* [schienengleicher] Bahnübergang; **~-'headed** *adj.* besonnen; **remain ~-headed** einen kühlen Kopf bewahren

lever ['li:və(r)] **1.** *n.* **a)** Hebel, *der; (crowbar)* Brechstange, *die;* **b)** *(fig.)* Druckmittel, *das.* **2.** *v. t.* **~ sth. open** etw. aufhebeln; **~ sth. up** etw. hochhebeln

leverage ['li:vərɪdʒ] *n.* **a)** Hebelwirkung, *die; (action of lever)* Hebelkraft, *die;* **b)** *(fig.: influence)* **give sb. [a lot of] ~:** jmds. Position [sehr] stärken

levity ['levɪtɪ] *n.* **a)** *(frivolity)* Unernst, *der*

levy ['levɪ] **1.** *n.* **a)** [Steuer]erhebung, *die;* **b)** *(tax)* Steuer, *die.* **2.** *v. t. (exact)* erheben ‹*Steuern, Beträge*›; **~ a fine on sb./a tax on sth.** jmdn. mit einer Geldstrafe/etw. mit einer Steuer belegen

lewd [lju:d] *adj.* geil *(oft abwertend),* lüstern *(geh.)* ‹*Person*›; anzüglich ‹*Blick, Geste*›; schlüpfrig, unanständig ‹*Lied, Witz*›

lexicography [leksɪ'kɒgrəfɪ] *n., no pl.* Lexikographie, *die*

lexicon ['leksɪkən] *n.* Wörterbuch, *das;* Lexikon, *das (veralt.)*

liability [laɪə'bɪlɪtɪ] *n.* **a)** *no pl. (legal obligation)* Haftung, *die;* **limited ~** *(Brit.)* beschränkte Haftung; **~ to pay tax[es]** *or* **for taxation** Steuerpflicht, *die;* **b)** *no pl. (proneness to disease etc.)* Anfälligkeit, *die* **(to** für); **c)** *(handicap)* Belastung, *die* **(to** für)

liable ['laɪəbl] *pred. adj.* **a)** *(legally bound)* **be ~ for sth.** für etw. haftbar sein *od.* haften; **be ~ to pay tax[es]** steuerpflichtig sein; **b)** *(prone)* **be ~ to sth.** ‹*Sache:*› leicht etw. haben; ‹*Person:*› zu etw. neigen; **be ~ to do sth.** ‹*Sache:*› leicht etw. tun; ‹*Person:*› dazu neigen, etw. zu tun; **c)** *(likely)* **difficulties are ~ to occur** mit Schwierigkeiten muß man rechnen; **she is ~ to change her mind** es kann durchaus sein, daß sie ihre Meinung ändert; **it is ~ to be cold there** im allgemeinen ist es dort kalt

liaise [lɪ'eɪz] *v. i. (coll.)* eine Verbindung herstellen

liaison [lɪ'eɪzɒn] *n.* **a)** *(cooperation)* Zusammenarbeit, *die; (connection)* Verbindung, *die;* **b)** *(illicit relation)* Verhältnis, *das;* Liaison, *die (geh.)*

liar ['laɪə(r)] *n.* Lügner, *der*/Lügnerin, *die*

Lib [lɪb] *abbr.* **a) Liberal** Lib.; **b)** *(coll.)* **liberation**

libel ['laɪbl] **1.** *n. (schriftliche)* Verleumdung. **2.** *v. t., (Brit.)* **-ll-** *(schriftlich)* verleumden

libellous *(Amer.:* **libelous)** ['laɪbələs] *adj.* verleumderisch

liberal ['lɪbərl] **1.** *adj.* **a)** *(generous, abundant)* großzügig; **a ~ amount of** reichlich; **b)** *(not strict)* liberal; frei ‹*Auslegung*›; **c)** *(open-minded; also Polit.)* liberal; **the L~ Democrats** *(Brit.)* die Liberaldemokraten. **2.** *n.* **L~** *(Polit.)* Liberale, *der/die*

liberality [lɪbə'rælɪtɪ] *n., no pl. (generosity)* Großzügigkeit, *die* **(to** gegenüber)

liberally ['lɪbərəlɪ] *adv. (generously)* großzügig; *(abundantly)* reichlich

liberate ['lɪbəreɪt] *v. t.* befreien **(from** aus)

liberation [lɪbə'reɪʃn] *n.* Befreiung, *die* **(from** aus); *see also* **Women's Liberation**

liberator ['lɪbəreɪtə(r)] *n.* Befreier, *der*/Befreierin, *die*

liberty ['lɪbətɪ] *n.* Freiheit, *die;* **you are at ~ to come and go as you please** es steht Ihnen frei, zu kommen und zu gehen, wie Sie wollen; **be at ~:** auf freiem Fuß sein; **set sb. at ~:** jmdn. auf freien Fuß setzen; **take the ~ to do** *or* **of doing sth.** sich *(Dat.)* die Freiheit nehmen, etw. zu tun; **take liberties with sb.** sich *(Dat.)* Freiheiten gegen jmdn. herausnehmen *(ugs.);* **take liberties with sth.** mit etw. allzu frei umgehen

libido [lɪ'bi:dəʊ] *n. (Psych.)* Libido, *die*

Libra ['li:brə, 'lɪbrə] *n. (Astrol., Astron.)* die Waage

librarian [laɪ'breərɪən] *n.* Bibliothekar, *der*/Bibliothekarin, *die*

library ['laɪbrərɪ] *n.* Bibliothek, *die;* Bücherei, *die;* **reference ~:** Präsenzbibliothek, *die;* **public ~:** öffentliche Bücherei

library: ~ **book** *n.* Buch aus der Bibliothek *od.* Bücherei; ~ **ticket** *n.* Lesekarte, *die*

libretto [lɪ'bretəʊ] *n., pl.* **libretti** [lɪ'breti:] *or* ~**s** Libretto, *das*

Libya ['lɪbɪə] *pr. n.* Libyen *(das)*

Libyan ['lɪbɪən] *adj.* libysch

lice *pl. of* **louse a**

licence ['laɪsəns] **1.** *n.* **a)** *(official permit)* [behördliche] Genehmigung; Lizenz, *die;* Konzession, *die (Amtsspr.); (driving-~)* Führerschein, *der;* **gun** ~: Waffenschein, *der;* **b)** *([excessive] liberty of action)* [uneingeschränkte] Handlungsfreiheit; **c)** *(licentiousness)* Unzüchtigkeit, *die;* Zügellosigkeit, *die;* **d) poetic** ~: dichterische Freiheit. **2.** *v. t. see* **license 1**

license ['laɪsəns] **1.** *v. t.* ermächtigen; ~**d** ‹*Händler, Makler, Buchmacher*› mit [einer] Lizenz; ~**d to sell alcoholic beverages** *(formal)* [für den Ausschank von alkoholischen Getränken] konzessioniert; **the restaurant is ~d to sell drinks** das Restaurant hat eine Schankerlaubnis *od.* -konzession; **licensing hours** *(in public house)* Ausschankzeiten; **licensing laws** Schankgesetze; ≈ Gaststättengesetz, *das;* ~**d premises** Gaststätte mit Schankerlaubnis; **get a car ~d,** ~ **a car** ≈ die Kfz-Steuer für ein Auto bezahlen. **2.** *n. (Amer.) see* **licence 1**

licensee [laɪsən'si:] *n.* Lizenzinhaber, *der;* Konzessionsinhaber, *der; (of bar)* Wirt, *der*/Wirtin, *die*

licentious [laɪ'senʃəs] *adj.* zügellos, ausschweifend ‹*Leben, Person*›; unzüchtig ‹*Benehmen*›; freizügig ‹*Buch, Theaterstück*›

lichen ['laɪkn, 'lɪtʃn] *n.* Flechte, *die*

lick [lɪk] **1.** *v. t.* **a)** lecken; ~ **a stamp** eine Briefmarke anlecken *od.* belecken; ~ **one's lips** *(lit. or fig.)* sich *(Dat.)* die Lippen lecken; ~ **sth./sb. into shape** *(fig.)* etw./jmdn. auf Vordermann bringen *(ugs.);* ~ **one's wounds** *(lit. or fig.)* seine Wunden lecken; **b)** *(play gently over)* ‹*Flammen, Feuer:*› [empor]züngeln an (+ *Dat.*); **c)** *(sl.: beat)* verdreschen *(ugs.); (fig.)* bewältigen, meistern ‹*Problem*›; *(in contest)* eine Abfuhr erteilen (+ *Dat.*). **2.** *n.* **a)** *(act)* Lecken, *das;* **give a door a** ~ **of paint** eine Tür [oberflächlich] überstreichen; **b)** *(sl.: fast pace)* **at a great** *or* **at full** ~: mit einem Affenzahn *(ugs.)*

~ **'off** *v. t.* ablecken

~ **'up** *v. t.* auflecken

lid [lɪd] *n.* **a)** Deckel, *der;* **take the** ~ **off sth.** *(fig.)* etw. aufdecken; **put the [tin]** ~ **on sth.** *(Brit. sl.) (be the final blow)* einer Sache *(Dat.)* die Krone aufsetzen; *(put an end to)* etw. stoppen; **b)** *(eyelid)* Lid, *das*

[1]**lie** [laɪ] **1.** *n.* **a)** *(false statement)* Lüge, *die;* **tell ~s/a** ~: lügen; **no, I tell a** ~, ... *(coll.)* nein, nicht daß ich jetzt lüge, ... *(ugs.);* **white** ~: Notlüge, *die;* **give the** ~ **to sth.** etw. Lügen strafen; **b)** *(thing that deceives)* [einzige] Lüge *(fig.);* Schwindel, *der (abwertend).* **2.** *v. i.,* **lying** ['laɪɪŋ] lügen; ~ **to sb.** jmdn. be- *od.* anlügen

[2]**lie 1.** *n. (direction, position)* Lage, *die;* **the** ~ **of the land** *(Brit. fig.: state of affairs)* die Lage der Dinge; die Sachlage. **2.** *v. i.,* **lying** ['laɪɪŋ], **lay** [leɪ], **lain** [leɪn] **a)** liegen; *(assume horizontal position)* sich legen; **many obstacles** ~ **in the way of my success** *(fig.)* viele Hindernisse verstellen mir den Weg zum Erfolg; **she lay asleep/resting on the sofa** sie lag auf dem Sofa und schlief/ruhte sich aus; ~ **still/dying** still liegen/im Sterben liegen; **b)** ~ **idle** ‹*Feld, Garten:*› brachliegen; ‹*Maschine, Fabrik:*› stillstehen; ‹*Gegenstand:*› [unbenutzt] herumstehen *(ugs.);* **let sth./things** ~: etw./die Dinge ruhen lassen; **how do things** ~**?** wie liegen die Dinge?; **c)** *(be buried)* [begraben] liegen; **d)** *(be situated)* liegen; **Austria ~s to the south of Germany** Österreich liegt südlich von Deutschland; **e)** *(be spread out to view)* **the valley/plain/desert lay before us** vor uns lag das Tal/die Ebene/die Wüste; **a brilliant career lay before him** *(fig.)* eine glänzende Karriere lag vor ihm; **f)** *(Naut.)* ~ **at anchor/in harbour** vor Anker/im Hafen liegen; **g)** *(fig.)* ‹*Gegenstand:*› liegen; **her interest ~s in languages** ihr Interesse liegt auf sprachlichem Gebiet; **I will do everything that ~s in my power to help** ich werde alles tun, was in meiner Macht steht, um zu helfen

~ **a'bout,** ~ **a'round** *v. i.* herumliegen *(ugs.)*

~ **'back** *v. i. (recline against sth.)* sich zurücklegen; *(in sitting position)* sich zurücklehnen

~ **'down** *v. i.* sich hinlegen; **take sth. lying down** *(fig.)* etw. ruhig *od.* tatenlos hinnehmen

~ **'in** *v. i. (Brit. coll.: stay in bed)* liegenbleiben

~ **'up** *v. i. (hide)* sich versteckt halten

'lie-detector *n.* Lügendetektor, *der*

'lie-in *n. (coll.) (Brit.: extra time in bed)* **have a** ~: [sich] ausschlafen

lieu [lju:, lu:] *n.* **in** ~ **of sth.** anstelle einer Sache *(Gen.);* **get money/holiday in** ~: statt dessen Geld/Urlaub bekommen

lieutenant [lef'tenənt, ləf'tenənt] *n.* **a)** *(Army)* Oberleutnant, *der; (Navy)* Kapitänleutnant, *der;* **b)** *(Amer.: policeman)* ≈ Polizeioberkommissar, *der*

life [laɪf] *n., pl.* **lives** [laɪvz] **a)** Leben, *das;* **it is a matter of** ~ **and death** es geht [dabei] um Leben und Tod; *(fig.: it is of vital importance)* es ist äußerst wichtig **(to** für); **come to** ~ ‹*Bild, Statue:*› lebendig werden; **run** *etc.* **for one's** ~: um sein Leben rennen *usw.;* **I cannot for the** ~ **of me** ich kann beim besten Willen nicht; **lose one's** ~: sein Leben verlieren; **many lives were lost** viele Menschen kamen ums Leben; **without loss of** ~: ohne Todesopfer; ~ **is not worth living** das Leben ist nicht lebenswert; **marry early in** ~: früh heiraten; **late in** ~: erst im fortgeschrittenen Alter; **for** ~: lebenslänglich ‹*inhaftiert*›; **he's doing** ~ *(coll.)* er sitzt lebenslänglich *(ugs.);* **get** ~ *(sl.)* lebenslänglich kriegen *(ugs.);* **expectation of** ~: Lebenserwartung, *die;* **get the fright/shock of one's** ~ *(coll.)* zu Tode erschrecken/den Schock seines Lebens bekommen *(ugs.);* **have the time of one's** ~: sich hervorragend amüsieren; **he will do anything for a quiet** ~: für ihn ist die Hauptsache, daß er seine Ruhe hat; **make** ~ **easy for oneself/sb.** es sich *(Dat.)*/jmdm. leichtmachen; **make** ~ **difficult for oneself/sb.** sich *(Dat.)*/jmdm. das Leben schwermachen; **this is the** ~**!** *expr. content* so läßt sich's leben!; **that's** ~, **~'s like that** so ist das Leben [nun mal]; **not on your** ~ *(coll.)* nie im Leben! *(ugs.);* **save one's/sb.'s** ~: sein Leben/jmdm. das Leben retten; **sth. is as much as sb.'s** ~ **is worth** mit etw. setzt jmd. sein Leben aufs Spiel; **take one's [own]** ~: sich *(Dat.)* das Leben nehmen; **take one's** ~ **in one's hands** sein Leben riskieren; **b)** *(energy, animation)* Leben, *das;* **be the** ~ **and soul of the party** der Mittelpunkt der Party sein; **there is still** ~ **in sth.** in etw. *(Dat.)* steckt noch Leben; **c)** *(living things and their activity)* Leben, *das;* **bird/insect** ~: die Vogelwelt/die Insekten; **d)** *(living* *form or model)* **draw sb. from ~:** jmdn. nach dem Leben zeichnen; **as large as ~** *(~-size)* lebensgroß; *(in person)* in voller Schönheit *(ugs. scherzh.)*; **larger than ~:** überzeichnet; **true to ~:** wahrheitsgetreu; **e)** *(specific aspect)* ‹*Privat-, Wirtschafts-, Dorf*›leben, *das;* **in this ~** *(on earth)* in diesem Leben; **the other** *or* **the future** *or* **the next ~** *(in heaven)* das zukünftige Leben [nach dem Tode]; **eternal** *or* **everlasting ~:** ewiges Leben; **daily ~:** das Alltagsleben; **see ~:** etwas von der Welt sehen; **f)** *(of battery, light-bulb, etc.)* Lebensdauer, *die*

life: ~-and-death *adj.* ‹*Kampf*› auf Leben und Tod; *(fig.)* überaus wichtig ‹*Frage, Brief*›; **~ assurance** *n. (Brit.)* Lebensversicherung, *die;* **~belt** *n.* Rettungsring, *der;* **~-blood** *n.* Blut, *das;* *(fig.)* Lebensnerv, *der;* **~boat** *n.* Rettungsboot, *das;* **~buoy** *n.* *(ring-shaped)* Rettungsring, *der;* **~ cycle** *n.* Lebenszyklus, *der;* **~-guard** *n.* **a)** *(soldiers)* Leibwache, *die;* **b)** *(expert swimmer)* Rettungsschwimmer, *der*/-schwimmerin, *die;* **~ 'history** *n.* Lebensgeschichte, *die;* **~ insurance** *n.* Lebensversicherung, *die;* **~-jacket** *n.* Schwimmweste, *die*

lifeless ['laɪflɪs] *adj.* leblos; unbelebt ‹*Gegend, Planet*›; *(fig.)* farblos ‹*Stimme, Rede, Aufführung*›; ‹*Stadt*› ohne Leben

life: ~like *adj.* lebensecht; **~line** *n.* **a)** *(rope)* Rettungsleine, *die;* **b)** *(fig.)* [lebenswichtige] Verbindung; *(support)* Rettungsanker, *der;* **~long** *adj.* lebenslang; **sb.'s ~long friend** *(future)* jmds. Freund fürs Leben; *(past)* jmds. Freund seit der Kindheit; **~ 'member** *n.* Mitglied auf Lebenszeit; **~-raft** *n.* Rettungsfloß, *das;* **~-saving** *n.* Rettungsschwimmen, *das; attrib.* Rettungs‹*gerät, -technik*›; lebensrettend ‹*Medikament*›; **~ sentence** *n.* lebenslängliche Freiheitsstrafe; **get a ~ sentence** lebenslänglich bekommen; **~-size, ~-sized** *adj.* lebensgroß; in Lebensgröße *nachgestellt;* **~-span** *n.* Lebenserwartung, *die;* *(Biol.)* Lebensdauer, *die;* **~ story** *n.* Lebensgeschichte, *die;* **~-style** *n.* Lebensstil, *der;* **~-support** *n.* **~-support system** lebenserhaltende Apparate; **~time** *n.* Lebenszeit, *die;* *(Phys.)* Lebensdauer, *die; attrib.* lebenslang; **once in a ~time** einmal im Leben; **during my ~time** während meines Lebens; **the chance of a ~time** eine einmalige Gelegenheit

lift [lɪft] **1.** *v. t.* **a)** heben; *(slightly)* anheben; *(fig.)* erheben ‹*Seele, Gemüt, Geist*›; **~ sb.'s spirits** jmds. Stimmung heben; **b)** *(sl.: steal)* klauen *(salopp)*; **c)** *(sl.: plagiarize)* abkupfern *(salopp)* **(from** aus); **d)** *(end)* aufheben ‹*Verbot, Beschränkung, Blockade*›. **2.** *v. i.* **a)** *(disperse)* sich auflösen; **b)** *(rise)* ‹*Stimmung:*› sich aufhellen; ‹*Herz:*› höher schlagen. **3.** *n.* **a)** *(ride in vehicle)* Mitfahrgelegenheit, *die;* **get a ~ [with** *or* **from sb.]** [von jmdm.] mitgenommen werden; **give sb. a ~:** jmdn. mitnehmen; **would you like a ~?** möchtest du mitfahren?; **b)** *(Brit.: in building)* Aufzug, *der;* Fahrstuhl, *der;* **c)** *(~ing)* Heben, *das*

~ 'down *v. t.* herunterheben

~ off *v. t. & i.* abheben

~ 'up *(raise)* hochheben; *(turn upwards)* heben ‹*Kopf*›

lift: ~-attendant, *n. (Brit.)* Aufzugführer, *der;* **~-off** *n. (Astronaut.)* Abheben, *das*

ligament ['lɪgəmənt] *n. (Anat.)* Band, *das;* Ligament[um], *das (fachspr.)*

ligature ['lɪgətʃə(r)] *n.* Bandage, *die; (in surgery)* Ligaturfaden, *der*

[1]light [laɪt] **1.** *n.* **a)** Licht, *das;* **in a good ~:** bei gutem Licht; **be in sb.'s ~:** jmdm. im Licht sein; **at first ~:** bei Tagesanbruch; **while the ~ lasts** solange es [noch] hell ist; **~ of day** *(lit. or fig.)* Tageslicht, *das;* **b)** *(electric lamp)* Licht, *das; (fitting)* Lampe, *die;* **go out like a ~** *(fig.)* sofort weg sein *(ugs.)*; **c)** *(signal to ships)* Leuchtfeuer, *das;* **d)** *in sing. or pl. (signal to traffic)* Ampel, *die;* **at the third set of ~s** an der dritten Ampel; **e)** *(to ignite)* Feuer, *das;* **have you got a ~?** haben Sie Feuer?; **put a/set ~ to sth.** etw. anzünden; **f) throw** *or* **shed ~ [up]on sth.** Licht in etw. *(Akk.)* bringen ; **bring sth. to ~:** etw. ans [Tages]licht bringen; *see also* **see 1 a; g)** *in pl. (beliefs, abilities)* **according to one's ~s** nach bestem Wissen [und Gewissen]; **h)** *(aspect)* **in that ~:** aus dieser Sicht; **seen in this ~:** so gesehen; **in the ~ of** *(taking into consideration)* angesichts (+ *Gen.*); **show sb. in a bad ~:** ein schlechtes Licht auf jmdn. werfen; **put sb. in a good/bad ~:** jmdn. in einem guten/schlechten Licht erscheinen lassen. **2.** *adj.* hell; **~-blue/-brown** *etc.* hellblau/-braun *usw.* **3.** *v. t.,* **lit** [lɪt] *or* **~ed a)** *(ignite)* anzünden; **b)** *(illuminate)* erhellen; **~ sb.'s/one's way** jmdm./sich leuchten. **4.** *v. i.,* **lit** *or* **~ed** ‹*Feuer, Zigarette:*› brennen, sich anzünden lassen

~ 'up 1. *v. i.* **a)** *(become lit)* erleuchtet werden; **b)** *(become bright)* aufleuchten **(with** vor); **c)** *(begin to smoke a cigarette etc.)* sich *(Dat.)* eine anstecken *(ugs.)*. **2.** *v. t.* **a)** *(illuminate)* erleuchten; **b)** *(make bright)* erhellen; **c)** anzünden ‹*Zigarette usw.*›

[2]light 1. *adj.* **a)** leicht; **[for] ~ relief** [als] kleine Abwechslung; **be a ~ sleeper** einen leichten Schlaf haben; **b)** *(small in amount)* gering; **traffic is ~ on these roads** auf diesen Straßen herrscht nur wenig Verkehr; **c)** *(not important)* leicht; **make ~ of sth.** etw. bagatellisieren; **d)** *(nimble)* leicht ‹*Schritt, Bewegungen*›; **have ~ fingers** *(steal)* gern lange Finger machen *(ugs.)*; **e)** *(easily borne)* leicht ‹*Krankheit, Strafe*›; gering ‹*Steuern*›; mild ‹*Strafe*›; **f) with a ~ heart** *(carefree)* leichten *od.* frohen Herzens; **g) feel ~ in the head** *(giddy)* leicht benommen sein. **2.** *adv.* **travel ~:** mit wenig *od.* leichtem Gepäck reisen

[3]light *v. i.,* **lit** [lɪt] *or* **~ed** *(come by chance)* **~ [up]on sth.** auf etw. *(Akk.)* kommen *od.* stoßen

light: ~-bulb *n.* Glühbirne, *die;* **~-coloured** *adj.* hell

lighted ['laɪtɪd] *adj.* brennend ‹*Kerze, Zigarette*›; angezündet ‹*Streichholz*›; beleuchtet ‹*Zimmer, Pfad, Schild, Vitrine*›

[1]lighten ['laɪtn] **1.** *v. t.* **a)** *(make less heavy)* leichter machen; **b)** *(make less oppressive)* leichter machen ‹*Arbeit, Aufgabe*›; erleichtern ‹*Gewissen*›; **~ sb.'s burden** jmdn. entlasten. **2.** *v. i.* *(become less heavy)* leichter werden

[2]lighten 1. *v. t.* *(make brighter)* aufhellen; heller machen ‹*Raum*›. **2.** *v. i.* sich aufhellen

lighter ['laɪtə(r)] *n.* *(device)* Feuerzeug, *das;* *(in car)* Zigarettenanzünder, *der*

light: ~-fingered ['laɪtfɪŋgəd] *adj.* langfing[e]rig; **~-headed** [laɪt'hedɪd] *adj.* leicht benommen; **~-hearted** ['laɪthɑ:tɪd] *adj.* **a)** *(gay, humorous)* unbeschwert; heiter; **b)** *(optimistic, casual)* unbekümmert; **~house** *n.* Leuchtturm, *der;* **~house-keeper** *n.* Leuchtturmwärter, *der;* **~ 'industry** *n.* Leichtindustrie, *die*

lighting ['laɪtɪŋ] *n.* Beleuchtung, *die*

lighting-'up time *n.* Zeit zum Einschalten der Beleuchtung

lightly ['laɪtlɪ] *adv.* **a)** *(not heavily)* leicht; **sleep ~:** einen leichten Schlaf haben; **b)** *(in a small degree)* leicht; **c)** *(without serious consideration)* leichtfertig; **d)** *(cheerfully, deprecatingly)* leichthin; **not treat sth. ~:** etw. nicht auf die leichte Schulter nehmen; **take sth. ~:** etw. nicht [so] ernst nehmen; **e)** *(nimbly)* behend; **f) get off ~** *(not receive heavy penalty)* glimpflich davonkommen; *see also* **let off a**

'lightmeter *n.* Lichtmesser, *der; (exposure meter)* Belichtungsmesser, *der*

[1]lightness ['laɪtnɪs] *n., no pl.* **a)** *(having little weight, lit. or fig.)* Leichtigkeit, *die;* **b)** *(of penalty, weather)* Milde, *die;* **c) ~ of heart/spirit** Heiterkeit/Unbekümmertheit, *die;* **d)** *(lack of concern)* Leichtfertigkeit, *die;* **e)** *(agility of movement)* Leichtigkeit, *die*

[2]lightness *n. (brightness, paleness of colour)* Helligkeit, *die*

lightning ['laɪtnɪŋ] **1.** *n., no pl., no indef. art.* Blitz, *der;* **flash of ~:** Blitz, *der;* **like ~** *(coll.)* wie der Blitz *(ugs.);* **[as] quick as ~** *(coll.)* schnell wie der Blitz *(ugs.);* **like greased ~** *(coll.)* wie ein geölter Blitz *(ugs.).* **2.** *adj.* Blitz-; **with ~ speed** blitzschnell

'lightning-conductor *n. (lit. or fig.)* Blitzableiter, *der*

light: ~weight 1. *adj.* **a)** leicht; **b)** *(fig.: of little consequence)* unmaßgeblich; **2.** *n. (Boxing etc.)* Leichtgewicht, *das; (person also)* Leichtgewichtler, *der; (fig.)* Leichtgewicht, *das (fig.);* **~-year** *n.* Lichtjahr, *das*

lignite ['lɪgnaɪt] *n.* Braunkohle, *die*

[1]like [laɪk] **1.** *adj.* **a)** *(resembling)* wie; **your dress is ~ mine** dein Kleid ist so ähnlich wie meins; **your dress is very ~ mine** dein Kleid ist meinem sehr ähnlich; **in a case ~ that** in so einem Fall; **there was nothing ~ it** es gab nichts Vergleichbares; **what is sb./sth. ~?** wie ist jmd./etw.?; **what's he ~ to talk to?** wie redet es sich mit ihm?; **more ~ twelve** eher zwölf; **that's [a bit] more ~ it** *(coll.: better)* das ist schon [etwas] besser; *(coll.: nearer the truth)* das stimmt schon eher; **they are nothing ~ each other** sie sind sich *(Dat.)* nicht im geringsten ähnlich; **nothing ~ as** *or* **so good/bad/many** *etc.* **as ...:** bei weitem nicht so gut/schlecht/viele *usw.* wie ...; **b)** *(characteristic of)* typisch für ⟨*dich, ihn usw.*⟩; **it's just ~ you to be late!** du mußt natürlich wieder zu spät kommen!; **c)** *(similar)* ähnlich; **be as ~ as two peas in a pod** sich *(Dat.)* gleichen wie ein Ei dem andern; **~ father, ~ son** *(prov.)* der Apfel fällt nicht weit vom Stamm *(Spr.).* **2.** *prep. (in the manner of)* wie; **[just] ~ that** [einfach] so. **3.** *conj. (coll.)* **a)** *(in same or similar manner as)* wie; **he is not shy ~ he used to be** er ist nicht mehr so schüchtern wie früher; **b)** *(coll.: for example)* etwa; beispielsweise. **4.** *n.* **a)** *(equal)* **his/her ~:** seines-/ihresgleichen; **the ~s of me/you** *(coll.)* meines-/deinesgleichen; **b)** *(similar things)* **the ~:** so etwas; **and the ~:** und dergleichen

[2]like 1. *v.t. (be fond of, wish for)* mögen; **~ it or not** ob es dir/ihm *usw.* gefällt oder nicht; **~ vegetables** Gemüse mögen; gern Gemüse essen; **~ doing sth.** etw. gern tun; **would you ~ a drink/to borrow the book?** möchtest du etwas trinken/dir das Buch leihen?; **would you ~ me to do it?** möchtest du, daß ich es tue?; **I'd ~ it back soon** ich hätte es gern bald zurück; **I didn't ~ to disturb you** ich wollte dich nicht stören; **I ~ 'that!** *(iron.)* so was hab' ich gern! *(ugs. iron.);* **how do you ~ it?** wie gefällt es dir?; **how does he ~ living in America?** wie gefällt es ihm in Amerika?; **how would you ~ an ice-cream?** was hältst du von einem Eis?; **if you ~** *expr. assent* wenn du willst *od.* möchtest; *expr. limited assent* wenn man so will. **2.** *n., in pl.* **~s and dislikes** Vorlieben und Abneigungen

likeable ['laɪkəbl] *adj.* nett; sympathisch

likelihood ['laɪklɪhʊd] *n.* Wahrscheinlichkeit, *die;* **what is the ~ of this happening?** wie wahrscheinlich ist es, daß dies geschieht?; **in all ~:** aller Wahrscheinlichkeit nach

likely ['laɪklɪ] **1.** *adj.* **a)** *(probable)* wahrscheinlich; glaubhaft ⟨*Geschichte*⟩; voraussichtlich ⟨*Bedarf, Zukunft*⟩; **be the ~ reason/source** wahrscheinlich der Grund/die Ursache sein; **do you think it ~?** hältst du es für wahrscheinlich?; **is it ~ that he'd do that?** ist ihm so etwas zuzutrauen?; **[that's] a ~ story** *(iron.)* wer's glaubt, wird selig *(ugs. scherzh.);* **b)** *(to be expected)* wahrscheinlich; **there are ~ to be [traffic] hold-ups** man muß mit [Verkehrs]staus rechnen; **they are [not] ~ to come** sie werden wohl *od.* wahrscheinlich [nicht] kommen; **is it ~ to rain tomorrow?** wird es morgen wohl regnen?; **this is not ~ to happen** es ist unwahrscheinlich, daß das geschieht; **the candidate most ~ to succeed** der Kandidat mit den größten Erfolgsaussichten; **c)** *(promising, apparently suitable)* geeignet ⟨*Person, Ort, Methode, Weg*⟩; **we've looked in all the ~ places** wir haben an allen in Frage kommenden Stellen gesucht. **2.** *adv. (probably)* wahrscheinlich; **very** *or* **more than** *or* **quite** *or* **most ~:** höchstwahrscheinlich; sehr wahrscheinlich; **as ~ as not** höchstwahrscheinlich; **not ~!** *(coll.)* auf keinen Fall!

'like-minded *adj.* gleichgesinnt

liken ['laɪkn] *v.t.* **~ sth./sb. to sth./sb.** etw./jmdn. mit etw./jmdm. vergleichen

likeness ['laɪknɪs] *n.* **a)** *(resemblance)* Ähnlichkeit, *die* **(to** mit); **b)** *(guise)* Aussehen, *das;* Gestalt, *die;* **c)** *(Porträt)* Bild, *das*

likewise ['laɪkwaɪz] *adv.* ebenso; **do ~:** das gleiche tun

liking ['laɪkɪŋ] *n.* Vorliebe, *die;* **take a ~ to sb./sth.** an jmdm./etw. Gefallen finden; **sth. is [not] to sb.'s ~:** etw. ist [nicht] nach jmds. Geschmack

lilac ['laɪlək] **1.** *n.* **a)** *(Bot.)* Flieder, *der;* **b)** *(colour)* Zartlila, *das.* **2.** *adj.* zartlila; fliederfarben

lilt [lɪlt] *n. (Scot./literary)* schwingender Rhythmus

lily ['lɪlɪ] *n.* Lilie, *die;* **~ of the valley** Maiglöckchen, *das*

limb [lɪm] *n.* **a)** *(Anat.)* Glied, *das;* **~s** Glieder; Gliedmaßen; **a danger to life and ~:** eine Gefahr für Leib und Leben; **b) be out on a ~** *(fig.)* exponiert sein

limber up [lɪmbər'ʌp] *v.i.* sich einlaufen/einspielen *usw.; (loosen up)* die Muskeln lockern

limbo ['lɪmbəʊ] *n., pl.* **~s a)** *(neglect, oblivion)* Vergessenheit, *die;* **vanish into ~:** spurlos verschwinden; **b) be in ~** *(be pending)* in der Schwebe sein; *(be abandoned)* abgeschrieben sein

[1]lime [laɪm] *n.* **[quick]~:** [ungelöschter] Kalk

[2]lime *n. (fruit)* Limone, *die*

[3]lime *see* **lime-tree**

lime: ~-green 1. *adj.* [leuchtend] hellgrün; **2.** *n.* Hellgrün, *das;* **~light** *n. (fig.: attention)* **be in the ~light** im Rampenlicht [der Öffentlichkeit] stehen

limerick ['lɪmərɪk] *n.* Limerick, *der*

'**lime-tree** *n.* Linde, *die*

limit ['lɪmɪt] **1.** *n.* **a)** *usu. in pl. (boundary)* Grenze, *die;* **b)** *(point or line that may not be passed)* Limit, *das;* **set** *or* **put a ~ on sth.** etw. begrenzen *od.* beschränken; **be over the ~** ‹*Autofahrer:*› zu viele Promille haben; ‹*Reisender:*› Übergepäck haben; **there is a ~ to what I can spend/do** ich kann nicht unbegrenzt Geld ausgeben/ meine Möglichkeiten sind auch nur begrenzt; **there is a ~ to my patience** meine Geduld ist begrenzt; **lower/upper ~:** Untergrenze/Höchstgrenze, *die;* **without ~:** unbegrenzt; **within ~s** innerhalb gewisser Grenzen; **c)** *(coll.)* **this is the ~!** das ist [doch] die Höhe!; **he/she is the [very] ~:** er/sie ist [einfach] unmöglich *(ugs.).* **2.** *v. t.* begrenzen (**to** auf + *Akk.*); einschränken ‹*Freiheit*›

limitation [lɪmɪ'teɪʃn] *n.* **a)** *(act)* Beschränkung, *die; (of freedom)* Einschränkung, *die;* **b)** *(condition) (of extent)* Begrenzung, *die; (of amount)* Beschränkung, *die;* **know one's ~s** seine Grenzen kennen

limited ['lɪmɪtɪd] *adj.* **a)** *(restricted)* begrenzt; **~ company** *(Brit.)* Gesellschaft mit beschränkter Haftung; **~ edition** limitierte Auflage; **b)** *(intellectually narrow)* beschränkt *(abwertend)*

limitless ['lɪmɪtlɪs] *adj.* grenzenlos

limousine ['lɪmʊzi:n] *n.* Limousine, *die (mit Trennscheibe)*

[1]**limp** [lɪmp] **1.** *v. i. (lit. or fig.)* hinken; **the ship managed to ~ into port** das Schiff schaffte es mit Müh und Not in den Hafen. **2.** *n.* Hinken, *das;* **walk with a ~:** hinken

[2]**limp** *adj. (not stiff, lit. or fig.)* schlaff; welk ‹*Blumen*›

limpet ['lɪmpɪt] *n. (Zool.)* Napfschnecke, *die*

limpid ['lɪmpɪd] *adj.* klar

linchpin ['lɪntʃpɪn] *n. (fig.: essential element)* Kernstück, *das;* **he is the ~ of the company** mit ihm steht und fällt die Firma

[1]**line** [laɪn] **1.** *n.* **a)** *(string, cord, rope, etc.)* Leine, *die;* **[fishing-]~:** [Angel]schnur, *die;* **b)** *(telephone or telegraph cable)* Leitung, *die;* **our company has 20 ~s** unsere Firma hat 20 Anschlüsse; **get me a ~ to Washington** verbinden Sie mich mit Washington; **bad ~:** schlechte Verbindung; *see also* [2]**hold 1 k; c)** *(long mark; also Math., Phys.)* Linie, *die; (less precise or shorter)* Strich, *der; (Telev.)* Zeile, *die;* **d)** *in pl. (outline of car, ship, etc.)* Linien *Pl.;* **e)** *(boundary)* Linie, *die;* **lay sth. on the ~ [for sb.]** [jmdm.] etw. rundheraus sagen; **put oneself on the ~:** ein Risiko eingehen; **f)** *(row)* Reihe, *die; (Amer.: queue)* Schlange, *die;* **~ of trees** Baumreihe, *die;* **bring sb. into ~:** dafür sorgen, daß jmd. nicht aus der Reihe tanzt *(ugs.);* **come** *or* **fall into ~:** sich in die Reihe stellen; ‹*Gruppe:*› sich in einer Reihe aufstellen; *(fig.)* nicht mehr aus der Reihe tanzen *(ugs.);* **be in ~ [with sth.]** [mit etw.] in einer Linie liegen; **be in ~ for promotion** Aussicht auf Beförderung haben; **be in/out of ~ with sth.** *(fig.)* mit etw. in/nicht in Einklang stehen; **somewhere along the ~:** irgendwann einmal; **g)** *(row of words on a page)* Zeile, *die;* **~s** *(actor's part)* Text, *der;* **drop me a ~:** schreib mir ein paar Zeilen; **he gave the boy 100 ~s** *(Sch.)* er ließ den Jungen 100 Zeilen abschreiben; **h)** *(system of transport)* Linie, *die;* **[shipping] ~:** Schiffahrtslinie, *die;* **i)** *(series of persons or things)* Reihe, *die; (generations of family)* Linie, *die;* **be third in ~ to the throne** dritter in der Thronfolge sein; **j)** *(direction, course)* Richtung, *die;* **on the ~s of** nach Art (+ *Gen.*); **on similar ~s** auf ähnliche Art; **be on the right/wrong ~s** in die richtige/falsche Richtung gehen; **along** *or* **on the same ~s** in der gleichen Richtung; **be on the same ~s** die gleiche Richtung verfolgen; **~ of thought** Gedankengang, *der;* **take a strong ~ with sb.** jmdm. gegenüber bestimmt *od.* energisch auftreten; **~ of action** Vorgehensweise, *die;* **k)** *(Railw.)* Bahnlinie, *die; (track)* Gleis, *das;* **the Waterloo ~, the ~ to Waterloo** die Linie nach Waterloo; **this is the end of the ~ [for you]** *(fig.)* dies ist das Aus [für dich]; **l)** *(field of activity)* Branche, *die; (academic)* Fachrichtung, *die;* **what's your ~?** in welcher Branche sind Sie?/was ist Ihre Fachrichtung?; **he's in the building ~:** er ist in der Baubranche; **be in the ~ of duty/business** zu den Pflichten/zum Geschäft gehören; **m)** *(Commerc.: product)* Artikel, *der;* Linie, *die (fachspr.);* **n)** *(Fashion)* Linie, *die;* **o)** *(Mil.: series of defences)* Linie, *die;* **enemy ~s** feindliche Stellungen *od.* Linien. **2.** *v. t.* **a)** *(mark with ~s)* linieren ‹*Papier*›; **a ~d face** ein faltiges Gesicht; **b)** *(stand at intervals along)* säumen *(geh.)* ‹*Straße, Strecke*›

~ 'up 1. *v. t.* antreten lassen ‹*Gefangene, Soldaten usw.*›; [in einer Reihe] aufstellen ‹*Gegenstände*›; *(fig.)* **I've got a nice little job/a surprise ~d up for you** ich hab da eine nette kleine Beschäftigung/ eine Überraschung für dich *(ugs.).* **2.** *v. i.* ‹*Gefangene, Soldaten:*› antreten; ‹*Läufer:*› Aufstellung nehmen; *(queue up)* sich anstellen

[2]**line** *v. t.* füttern ‹*Kleidungsstück*›; auskleiden ‹*Magen, Nest*›; ausschlagen ‹*Schublade usw.*›; **~ one's pockets** *(fig.)* sich *(Dat.)* die Taschen füllen

lineage ['lɪnɪɪdʒ] *n.* Abstammung, *die*

linear ['lɪnɪə(r)] *adj.* linear

linen ['lɪnɪn] **1.** *n.* **a)** Leinen, *das;* **b)** *(shirts, sheets, clothes, etc.)* Wäsche, *die.* **2.** *adj.* Leinen‹*faden, -bluse, -laken*›; Lein‹*tuch*›

linen: ~-basket *n. (Brit.)* Wäschekorb, *der;* **~ cupboard** *n.* Wäscheschrank, *der*

liner *n. (ship)* Linienschiff, *das;* **ocean ~:** [Ozean-]Liner, *der*

linesman ['laɪnzmən] *n., pl.* **linesmen** ['laɪnzmən] *(Sport)* Linienrichter, *der*

'**line-up** *n.* Aufstellung, *die*

linger ['lɪŋgə(r)] *v. i.* **a)** *(remain, wait)* verweilen *(geh.);* bleiben; *(persist)* fortbestehen; **b)** *(dwell)* **~ over** *or* **up[on] a subject** *etc.* bei einem Thema *usw.* verweilen; **~ over a meal** lange beim Essen sitzen

lingerie ['læ̃ʒəri:] *n.* **[women's] ~:** Damenunterwäsche, *die*

lingering ['lɪŋgərɪŋ] *adj.* anhaltend; verbleibend ‹*Zweifel*›; langwierig ‹*Krankheit*›; langsam ‹*Tod*›; nachklingend ‹*Melodie*›

lingo ['lɪŋgəʊ] *n., pl.* **~es a)** *(derog./joc.: language)* Sprache, *die;* **b)** *(jargon)* Fachjargon, *der*

linguist ['lɪŋgwɪst] *n.* **a)** Sprachkundige, *der/die;* **she's a good ~:** sie kann mehrere Sprachen; **b)** *(philologist)* Linguist, *der*/Linguistin, *die*

linguistic [lɪŋ'gwɪstɪk] *adj. (of ~s)* linguistisch; sprachwissenschaftlich; *(of language)* sprachlich; Sprach-

linguistics [lɪŋ'gwɪstɪks] *n., no pl.* Linguistik, *die;* Sprachwissenschaft, *die*

liniment ['lɪnɪmənt] *n.* Liniment, *das (Med.);* Einreib[e]mittel, *das*

lining ['laɪnɪŋ] *n. (of clothes)* Futter, *das; (of stomach)* Magenschleimhaut, *die; (of objects, containers, machines, etc.)* Auskleidung, *die*

link [lɪŋk] **1.** *n.* **a)** *(of chain)* Glied, *das;* **b)** *(connecting part)* Bindeglied, *das;* Verbindung, *die;* **road/rail ~:** Straßen-/Zugverbindung, *die;* **what is the ~ between these two?** was verbindet diese beiden?; **sever all ~s with sb.** alle Bindungen zu jmdm. lösen; **c)** *see* **linkman a. 2.** *v.t.* **a)** *(connect)* verbinden; **how are these events ~ed?** was haben diese Ereignisse miteinander zu tun?; **~ sb. with sth.** jmdn. mit etw. in Verbindung bringen; **b) ~ hands** sich bei den Händen halten; **~ arms** sich unterhaken. **3.** *v.i.* **~ together** sich zusammenfügen

~ 'up 1. *v.t.* miteinander verbinden; ankoppeln ‹*Wagen, Raumschiff usw.*› **(to** an + *Akk.*); miteinander in Verbindung bringen ‹*Fakten usw.*›. **2.** *v.i.* **~ up with sb.** sich mit jmdm. zusammentun *od.* zusammenschließen; **the spacecraft ~ed up** die Raumschiffe wurden angekoppelt; **this road ~s up with the M3** diese Straße mündet in die M3

linkage ['lɪŋkɪdʒ] *n.* **a)** Verbindung, *die;* **b)** *(system of links or bars)* Gestänge, *das*

linkman *n.* **a)** Verbindungsmann, *der;* **b)** *(Radio, Telev.)* Moderator, *der*/Moderatorin, *die*

links [lɪŋks] *sing. or pl.* **[golf] ~:** Golfplatz, *der*

lino ['laɪnəʊ] *n., pl.* **~s** Linoleum, *das*

linoleum [lɪ'nəʊlɪəm] *n.* Linoleum, *das*

linseed ['lɪnsi:d] *n.* Leinsamen, *der*

linseed 'oil *n.* Leinöl, *das*

lint [lɪnt] *n.* Mull, *der*

lintel ['lɪntl] *n. (Archit.)* Sturz, *der*

lion ['laɪən] *n.* Löwe, *der;* **the ~'s share** der Löwenanteil

lioness ['laɪənɪs] *n.* Löwin, *die*

'lion tamer ['laɪən teɪmə(r)] *n.* Löwenbändiger, *der*

lip [lɪp] *n.* **a)** Lippe, *die;* **lower/upper ~:** Unter-/Oberlippe, *die;* **bite one's ~** *(lit. or fig.)* sich *(Dat.)* auf die Lippen beißen; **escape sb.'s ~s** jmds. Lippen *(Dat.)* entschlüpfen; **lick one's ~s** *(lit. or fig.)* sich *(Dat.)* die Lippen lecken; **not let a word pass one's ~s** kein Wort über seine Lippen kommen lassen; **keep a stiff upper ~** *(fig.)* Haltung bewahren; **b)** *(of saucer, cup, crater)* [Gieß]rand, *der; (of jug)* Schnabel, *der;* Tülle, *die;* **c)** *(sl.: impudence)* **give sb. some ~:** jmdm. gegenüber eine dicke Lippe riskieren *(ugs.);* **none of your ~!** keine frechen Bemerkungen!

lip: ~-read 1. *v.i.* von den Lippen lesen; **2.** *v.t.* **be able to ~-read what sb. says** jmdm. von den Lippen ablesen können, was er/sie sagt; **~-reading** *n.* Lippenlesen, *das;* **~-service** *n.* **pay** *or* **give ~-service to sth.** ein Lippenbekenntnis zu etw. ablegen; **~stick** *n.* Lippenstift, *der*

liquefy ['lɪkwɪfaɪ] **1.** *v.t.* verflüssigen. **2.** *v.i.* sich verflüssigen

liqueur [lɪ'kjʊə(r)] *n.* Likör, *der*

liquid ['lɪkwɪd] **1.** *adj.* **a)** flüssig; **b)** *(Commerc.)* liquid; **~ assets** flüssige Mittel. **2.** *n.* Flüssigkeit, *die*

liquidate ['lɪkwɪdeɪt] *v.t.* **a)** *(Commerc.)* liquidieren; **b)** *(eliminate, kill)* liquidieren; beseitigen

liquidation [lɪkwɪ'deɪʃn] *n. (Commerc.)* Liquidation, *die*

liquid crystal dis'play *n.* Flüssigkeitskristallanzeige, *die*

liquidity [lɪ'kwɪdɪtɪ] *n., no pl.* **a)** flüssiger Zustand; **b)** *(Commerc.)* Liquidität, *die*

liquidize ['lɪkwɪdaɪz] *v.t.* auflösen; *(Cookery)* [im Mixer] pürieren

liquidizer ['lɪkwɪdaɪzə(r)] *n.* Mixer, *der*

liquor ['lɪkə(r)] *n. (drink)* Alkohol, *der;* Spirituosen; **be able to carry** *or* **hold one's ~:** etwas vertragen können; **hard** *or* **strong ~:** hochprozentiger Alkohol

liquorice ['lɪkərɪs] *n. (root)* Süßholz, *das; (preparation)* Lakritze, *die*

'liquor store *n. (Amer.)* Spirituosenladen, *der*

Lisbon ['lɪzbən] *pr. n.* Lissabon *(das)*

lisp [lɪsp] **1.** *v.i. & t.* lispeln. **2.** *n.* Lispeln, *das;* **speak with a ~:** lispeln; **have a bad ~:** stark lispeln

lissom ['lɪsəm] *adj.* geschmeidig

¹list [lɪst] **1.** *n.* Liste, *die;* **shopping ~:** Einkaufszettel, *der.* **2.** *v.t.* aufführen; auflisten

²list *(Naut.)* **1.** *n.* Schlagseite, *die;* **have a pronounced ~:** deutlich Schlagseite haben. **2.** *v.i.* Schlagseite haben

listen ['lɪsn] *v.i.* zuhören; **~ to music/the radio** Musik/Radio hören; **just ~ to the noise they are making!** hör dir bloß mal an, was sie für einen Lärm machen!; **they ~ed to his words** sie hörten ihm zu; **you never ~ to what I say** du hörst mir nie zu; **~ [out] for sth./sb.** auf etw. *(Akk.)* horchen/horchen, ob jmd. kommt; **~ to sth./sb.** *(pay heed)* auf etw./jmdn. hören; **he wouldn't ~** *(heed)* er wollte nicht hören

~ 'in *v.i.* **a)** *(Radio)* hören **(on, to** *Akk.*); **b)** *(eavesdrop)* mithören **(on, to** *Akk.*)

listener ['lɪsnə(r)] *n.* **a)** Zuhörer, *der*/Zuhörerin, *die;* **b)** *(Radio)* Hörer, *der*/Hörerin, *die*

listless ['lɪstlɪs] *adj.* lustlos

'list price *n.* Katalogpreis, *der*

lit [lɪt] *see* **¹light 3, 4; ³light**

litany ['lɪtənɪ] *n. (lit. or fig.)* Litanei, *die*

liter *(Amer.) see* **litre**

literacy ['lɪtərəsɪ] *n., no pl.* Lese- und Schreibfertigkeit, *die;* **adult ~ classes** Kurse für Analphabeten

literal ['lɪtərl] *adj.* **a)** wörtlich; **take sth. in a ~ sense** etw. wörtlich nehmen; **b)** *(not exaggerated)* buchstäblich; **the ~ truth** die reine Wahrheit; **c)** *(coll.: with some exaggeration)* wahr

literally ['lɪtərəlɪ] *adv.* **a)** wörtlich; **take sth./sb. ~:** etw./was jmd. sagt, wörtlich nehmen; **b)** *(actually)* buchstäblich; **c)** *(coll.: with some exaggeration)* geradezu

literary ['lɪtərərɪ] *adj.* literarisch; *(not colloquial)* gewählt

literary: ~ 'agent *n.* Literaturagent, *der*/-agentin, *die;* **~ 'critic** *n.* Literaturkritiker, *der*/-kritikerin, *die*

literate ['lɪtərət] *adj. (able to read and write)* des Lesens und Schreibens kundig; *(educated)* gebildet

literature ['lɪtərətʃə(r), 'lɪtrətʃə(r)] *n.* **a)** Literatur, *die;* **b)** *(writings on a subject)* [Fach]literatur, *die* **(on** zu); **c)** *(coll.: printed matter)* Literatur, *die;* Informationsmaterial, *das*

lithe [laɪð] *adj.* geschmeidig

lithograph ['lɪθəgrɑ:f] **1.** *n.* Lithographie, *die.* **2.** *v.t.* lithographieren

Lithuania [lɪθjʊ'eɪnɪə] *pr. n.* Litauen *(das)*

litigation [lɪtɪ'geɪʃn] *n.* Rechtsstreit, *der;* **in ~:** rechtshängig

litmus ['lɪtməs] *n.* Lackmus, *das od. der*

litre ['li:tə(r)] *n. (Brit.)* Liter, *der od. das*

litter ['lɪtə(r)] **1.** *n.* **a)** *(rubbish)* Abfall, *der;* Abfälle; **b)** *(for animals)* Streu, *die;* **c)** *(young)* Wurf, *der.* **2.** *v.t.* verstreuen; **papers were ~ed about the room** im Zimmer lagen überall Zeitungen herum

litter: ~-basket *n.* Abfallkorb, *der;* **~-bin** *n.* Abfalleimer, *der*

little ['lɪtl] **1.** *adj.*, **~r** ['lɪtlə(r)], **~st** ['lɪtlɪst] (*Note: it is more common to use the compar. and superl. forms* **smaller, smallest**) **a)** *(small)* klein; **~ town/book/dog** kleine Stadt/kleines Buch/kleiner Hund; *(showing affection or amusement)* Städtchen, *das*/Büchlein, *das*/Hündchen, *das;* **~ toe** kleine Zehe; **you poor ~ thing!** du armes kleines Ding!; **I know your ~ ways** ich kenne deine Tricks; **b)** *(young)* klein; **the ~ ones** die Kleinen; **my ~ sister** meine kleine Schwester; **c)** *(short)* klein ‹*Person*›; **a ~ way** ein kleines *od.* kurzes Stück; **after a ~ while** nach kurzer Zeit; **d)** *(not much)* wenig; **you have ~ time left** dir bleibt nicht mehr viel Zeit; **there is very ~ tea left** es ist kaum noch Tee *od.* nur noch ganz wenig Tee da; **make a nice ~ profit** *(coll. iron.)* einen hübschen Gewinn machen *(ugs.)*; **a ~ ...** *(a small quantity of)* etwas ...; ein wenig *od.* bißchen ...; **no ~ ...:** nicht wenig ...; **e)** *(trivial)* klein. **2.** *n.* wenig; **but ~:** nur wenig; **~ or nothing** kaum etwas; so gut wie nichts; **[do] not a ~:** einiges [tun]; **not a ~ angry** *etc.* ziemlich verärgert *usw.;* **there was ~ we could do** wir konnten nur wenig tun; **a ~** *(a small quantity)* etwas; ein wenig *od.* bißchen; *(somewhat)* ein wenig; **think ~ of sb.** gering von jmdm. denken; **a ~ after eight** kurz nach acht; **we see very ~ of one another** wir sehen sehr wenig voneinander; **~ by ~:** nach und nach. **3.** *adv.*, **less** [les], **least** [li:st] **a)** *(not at all)* **she ~ thought that ...:** sie dachte nicht im geringsten daran, daß ...; **he ~ suspected/knew what ...:** er hatte nicht die geringste Ahnung/wußte überhaupt nicht, was ...; **b)** *(to only a small extent)* **~ as he liked it** sowenig es ihm auch gefiel; **he writes ~ now** er schreibt nur noch wenig; **~ more/less than ...:** kaum mehr/weniger als ...; **that is ~ less than ...:** das grenzt schon an (+ *Akk.*) ...

little 'finger *n.* kleiner Finger; **twist sb. round one's ~:** jmdn. um den [kleinen] Finger wickeln *(ugs.)*

liturgy ['lɪtədʒɪ] *n.* Liturgie, *die*

¹live [laɪv] **1.** *adj.* **a)** *attrib. (alive)* lebend; **b)** *(Radio, Telev.)* **~ performance** Live-Aufführung, *die;* **~ broadcast** Live-Sendung, *die;* Direktübertragung, *die;* **c)** *(topical)* aktuell ‹*Thema, Frage*›; **d)** *(Electr.)* stromführend; **e)** *(unexploded)* scharf ‹*Munition usw.*›; **f)** *(glowing)* glühend ‹*Kohle*›; **g)** *(joc.: actual)* **real ~:** richtig. **2.** *adv. (Radio, Telev.)* live ‹*übertragen usw.*›

²live [lɪv] **1.** *v. i.* **a)** leben; **~ and let ~:** leben und leben lassen; **~ by sth.** von etw. leben; **you'll ~** *(iron.)* du wirst's [schon] überleben *(iron.)*; **as long as I ~ I shall never ...:** mein Leben lang werde ich nicht ...; **~ to see** [mit]erleben; **she will ~ to regret it** sie wird es noch bereuen; **you ~ and learn** man lernt nie aus; **~ through sth.** etw. durchmachen *(ugs.)*; *(survive)* etw. überleben; **~ to a ripe old age/to be a hundred** ein hohes Alter erreichen/hundert Jahre alt werden; **long ~ the queen!** lang lebe die Königin!; **b)** *(make permanent home)* wohnen; leben; **~ together** zusammenleben; **~ with sb.** mit jmdm. zusammenleben; **~ with sth.** *(lit. or fig.)* mit etw. leben. **2.** *v. t.* leben; **~ it up** das Leben in vollen Zügen genießen; *(have a good time)* einen draufmachen *(ugs.)*

~ 'down *v. t.* Gras wachsen lassen über (+ *Akk.*); **he will never be able to ~ it down** das wird ihm ewig anhängen

~ 'in *v. i. (Brit.)* ‹*Personal, Koch usw.:*› im Haus wohnen; ‹*Student, Krankenschwester:*› im Wohnheim wohnen

~ on 1. ['--] *v. t.* leben von. **2.** [-'-] *v. i.* weiterleben

~ out 1. [-'-] *v. i. (Brit.)* außerhalb wohnen. **2.** ['--] *v. t.* **a)** *(survive)* überleben; **b)** *(complete, spend)* verbringen; **they had ~d out their lives as fishermen** sie waren ihr Leben lang Fischer gewesen

~ 'up to *v. t.* gerecht werden (+ *Dat.*); **~ up to one's principles/faith** nach seinen Prinzipien/seinem Glauben leben; **~ up to one's reputation** seinem Ruf Ehre machen

livelihood ['laɪvlɪhʊd] *n.* Lebensunterhalt, *der;* **gain** *or* **earn a ~ from sth.** sich *(Dat.)* seinen Lebensunterhalt mit etw. verdienen

liveliness ['laɪvlɪnɪs] *n., no pl.* Lebhaftigkeit, *die*

lively ['laɪvlɪ] *adj.* **a)** lebhaft; lebendig ‹*Gegenwart*›; rege ‹*Handel*›; **things start to get ~ at 9 a.m.** um 9 Uhr wird es lebhaft; **look ~** *(coll.)* sich ranhalten *(ugs.)*; **b)** *(vivid)* lebendig, anschaulich ‹*Bericht, Schilderung*›; **c)** *(joc.: exciting, dangerous, difficult)* **things were getting ~:** die Sache wurde gefährlich

liven *see* **~ up 1**

liven up [laɪvn 'ʌp] **1.** *v. t.* Leben bringen in (+ *Akk.*). **2.** *v. i.* ‹*Person:*› aufleben; **things will ~ when ...:** es wird Leben in die Bude kommen *(ugs.)*, wenn ...

liver ['lɪvə(r)] *n. (Anat., Gastr.)* Leber, *die*

liveried ['lɪvərɪd] *adj.* livriert

liverish ['lɪvərɪʃ] *adj.* **a)** *(unwell)* unwohl; **b)** *(grumpy)* mürrisch

liver: ~ salts *n. pl. (Brit.)* ≈ Magenmittel, *das;* **~ sausage** *n.* Leberwurst, *die*

livery ['lɪvərɪ] *n.* Livree, *die*

live [laɪv]: **~stock** *n. pl.* Vieh, *das;* **~ 'wire** *n. (Electr.)* stromführender Draht; *(fig.)* Energiebündel, *das (ugs.)*

livid ['lɪvɪd] *adj.* **a)** *(bluish)* bleigrau; **b)** *(Brit. coll.: furious)* fuchtig *(ugs.)*

living ['lɪvɪŋ] **1.** *n.* **a)** Leben, *das;* **b)** *(livelihood)* Lebensunterhalt, *der;* **make a ~:** seinen Lebensunterhalt verdienen; **earn one's [own] ~:** sich *(Dat.)* seinen Lebensunterhalt [selbst] verdienen; **make one's ~ out of farming** von der Landwirtschaft leben; **make a good ~:** viel verdienen; **it's a ~** *(joc.)* man kann davon leben; **c)** *(way of life)* Lebensstil, *der;* **good ~:** üppiges Leben; *(pious)* guter Lebenswandel; **high ~:** hoher Lebensstandard; **d)** *constr. as pl.* **the ~:** die Lebenden; **be still/back in the land of the ~:** noch/wieder unter den Lebenden weilen. **2.** *adj.* lebend; **~ things** Lebewesen; **within ~ memory** seit Menschengedenken

'living-room *n.* Wohnzimmer, *das*

lizard ['lɪzəd] *n.* Eidechse, *die*

llama ['lɑ:mə] *n.* Lama, *das*

lo [ləʊ] *int.* **~ and behold** *(joc.)* sieh[e] da

load [ləʊd] **1.** *n.* **a)** *(burden, weight)* Last, *die;* *(amount carried)* Ladung, *die;* **a ~ of hay** eine Ladung Heu; **barrow-~ of apples** Karre voll Äpfel; **a ~ of [old] rubbish** *or* **tripe** *(fig. coll.)* ein einziger Mist *(ugs.)*; **talk a ~ of rubbish** eine Menge Blödsinn reden *(ugs.)*; **what a ~ of rubbish!** was für ein Quatsch *(ugs.) od. (ugs. abwertend)* Schmarren!; **get a ~ of this!** *(sl.) (listen)* hör einmal gut *od.* genau zu! *(ugs.)*; *(look)* guck mal genau hin! *(ugs.)*; **b)** *(weight)* Last, *die;* *(Electr.)* Belastung, *die;* **c)** *(fig.)* Last, *die;* Bürde, *die (geh.)*; **take a ~ off sb.'s mind** jmdm. eine Last von der Seele nehmen; **that's a ~ off my mind**

damit fällt mir ein Stein vom Herzen; **d)** *usu. in pl. (coll.: plenty)* **~s of** jede Menge *od.* massenhaft *(ugs.)* ‹*Nahrungsmittel usw.*›. **2.** *v. t.* **a)** *(put ~ on)* beladen; **~ sb. with work** *(fig.)* jmdm. Arbeit auftragen *od. (ugs. abwertend)* aufhalsen; **b)** *(put as ~)* laden; **c) the dice were ~ed against him** *(fig.)* er hatte schlechte Karten; **d)** *(charge)* laden ‹*Gewehr*›; **~ a camera** einen Film [in einen Fotoapparat] einlegen; **e)** *(insert)* einlegen ‹*Film, Tonband usw.*› **(into** in + *Akk.*); **f)** *(strain)* schwer belasten; **a table ~ed with food** ein mit Speisen beladener Tisch. **3.** *v. i.* laden **(with** *Akk.*)

~ 'up *v. i.* laden **(with** *Akk.*)

loaded ['ləʊdɪd] *adj.* **a ~ question** eine suggestive Frage; **be ~** *(sl.: rich)* [schwer] Kohle haben *(salopp)*

'loading bay *n.* Ladeplatz, *der*

¹loaf [ləʊf] *n., pl.* **loaves** [ləʊvz] **a)** Brot, *das;* [Brot]laib, *der;* **a ~ of bread** ein Laib Brot; **a brown/white ~:** ein dunkles Brot/Weißbrot; **half a ~ is better than no bread** *or* **none** *(prov.)* wenig ist besser als gar nichts; **b)** *(sl.: head)* **use one's ~:** seinen Grips anstrengen *(ugs.)*

²loaf *v. i.* **~ round town/the house** in der Stadt/zu Hause herumlungern *(ugs.)*

loam [ləʊm] *n. (soil)* Lehmboden, *der*

loan [ləʊn] **1.** *n.* **a)** *(thing lent)* Leihgabe, *die;* **b)** *(lending)* **let sb. have/give sb. the ~ of sth.** jmdm. etw. leihen; **be [out] on ~** ‹*Buch, Schallplatte:*› ausgeliehen sein; **have sth. on ~ [from sb.]** etw. [von jmdm.] geliehen haben; **c)** *(money lent)* Darlehen, *das;* Kredit, *der; (public ~)* Anleihe, *die.* **2.** *v. t.* **~ sth. to sb.** jmdm. etw. leihen; etw. an jmdn. verleihen

'loan shark *n. (coll.)* Kredithai, *der (ugs. abwertend)*

loath [ləʊθ] *pred. adj.* **be ~ to do sth.** etw. ungern tun

loathe [ləʊð] *v. t.* verabscheuen; nicht ausstehen können; **he ~s eggs** er mag Eier überhaupt nicht

loathing ['ləʊðɪŋ] *n.* Abscheu, *der* **(of, for** vor + *Dat.*); **have a ~ of sth.** Abscheu vor etw. *(Dat.)* haben; etw. verabscheuen

loathsome ['ləʊðsəm] *adj.* abscheulich; widerlich; verhaßt ‹*Tätigkeit, Pflicht*›

loaves *pl. of* **¹loaf**

lob [lɒb] *v. t.,* **-bb-** in hohem Bogen werfen; *(Tennis)* lobben

lobby ['lɒbɪ] **1.** *n.* **a)** *(pressure group)* Lobby, *die;* Interessenvertretung, *die;* **b)** *(of hotel)* Eingangshalle, *die; (of theatre)* Foyer, *das.* **2.** *v. t.* zu beeinflussen suchen ‹*Abgeordnete*›. **3.** *v. i.* seinen Einfluß geltend machen; **~ for/against sth.** sich für etw. einsetzen/gegen etw. wenden

lobe [ləʊb] *n. (ear~)* Ohrläppchen, *das*

lobster ['lɒbstə(r)] *n.* Hummer, *der*

'lobster-pot *n.* Hummerkorb, *der*

local ['ləʊkl] **1.** *adj.* **a)** lokal *(bes. Zeitungsw.);* Lokal‹*teil, -nachrichten, -sender*›; Kommunal‹*politiker, -wahl, -abgaben*›; *(of this area)* hiesig; *(of that area)* dortig; ortsansässig ‹*Firma, Familie*›; ‹*Wein, Produkt, Spezialität*› [aus] der Gegend; **she's a ~ girl** sie ist von hier/dort; **~ resident** Anwohner, *der*/Anwohnerin, *die;* **~ bus** *(serving immediate area)* Nahverkehrsbus, *der;* **b)** *(Med.)* lokal ‹*Schmerzen, Entzündung*›; örtlich ‹*Betäubung*›. **2.** *n.* **a)** *(inhabitant)* Einheimische, *der/die;* **b)** *(Brit. coll.: pub)* [Stamm]kneipe, *die*

local: ~ anaes'thetic *n.* Lokalanästhetikum, *das (Med.);* **under a ~ anaesthetic** unter örtlicher Betäubung *od. (Med.)* Lokalanästhesie; **~ au'thority** *n. (Brit.)* Kommunalverwaltung, *die;* **~ call** *n. (Teleph.)* Ortsgespräch, *das;* Nahbereichsgespräch, *das (fachspr.);* **~ 'government** *n.* Kommunalverwaltung, *die;* **~ government elections/officials** Kommunalwahlen/-beamte

localise *see* **localize**

locality [lə'kælɪtɪ] *n.* Ort, *der;* Gegend, *die*

localize ['ləʊkəlaɪz] *v. t. (restrict)* eingrenzen **(to** auf + *Akk.*); lokalisieren *(bes. Politik, Med.)*

locally ['ləʊkəlɪ] *adv.* im/am Ort; in der Gegend

locate [lə'keɪt] *v. t.* **a)** *(position)* plazieren; **be ~d** liegen; gelegen sein; **b)** *(determine position of)* ausfindig machen; lokalisieren *(fachspr.);* orten *(Flugw., Seew.)*

location [lə'keɪʃn] *n.* **a)** *(position)* Lage, *die; (place)* Ort, *der; (of ship, aircraft, police car)* Position, *die; (of person, building, etc.)* Standort, *der;* **b)** *(positioning)* Positionierung, *die;* **c)** *(determination of position of)* Lokalisierung, *die;* **d)** *(Cinemat.)* Drehort, *der;* **be on ~:** bei Außenaufnahmen sein

loch [lɒx, lɒk] *n. (Scot.)* See, *der; (in Scotland)* Loch, *der; (arm of sea)* Meeresarm, *der; (in Scotland)* Loch, *der*

¹lock [lɒk] *n. (ringlet)* Locke, *die*

²lock 1. *n.* **a)** *(of door etc.)* Schloß, *das;* **under ~ and key** unter [strengem] Verschluß; **b)** *(on canal etc.)* Schleuse, *die;* **c)** *(on wheel)* Sperrvorrichtung, *die;* Sperre, *die;* **d)** *(Wrestling)* Fesselgriff, *der;* Klammergriff, *der;* **e) ~, stock, and barrel** *(fig.)* mit allem Drum und Dran *(ugs.);* **f)** *(Motor Veh.)* Lenkeinschlag, *der.* **2.** *v. t.* **a)** *(fasten)* zuschließen; abschließen; **~** *or* **shut the stable-door after the horse has bolted** *(fig.)* den Brunnen erst zudecken, wenn das Kind hineingefallen ist; **b)** *(shut)* **~ sb./sth. in sth.** jmdn./etw. in etw. *(Akk.)* [ein]schließen; **~ sb./sth. out of sth.** jmdn./etw. aus etw. aussperren; **c)** *in p.p. (joined)* **the wrestlers were ~ed in combat** die Ringer hielten sich im Fesselgriff. **3.** *v. i.* ‹*Tür, Kasten usw.:*› sich ab-/zuschließen lassen

~ a'way *v. t.* einschließen; wegschließen; einsperren ‹*Person, Tier*›

~ 'in *v. t.* einschließen; *(deliberately)* einsperren ‹*Person, Tier*›

~ 'out *v. t.* **a)** aussperren; **~ oneself out** sich aussperren; **b)** *(Industry)* aussperren ‹*Arbeiter*›

~ 'up 1. *v. i.* abschließen. **2.** *v. t.* **a)** abschließen ‹*Haus, Tür*›; **b)** *(imprison)* einsperren

locker ['lɒkə(r)] *n.* Schließfach, *das*

locket ['lɒkɪt] *n.* Medaillon, *das*

lock: ~-gate *n.* Schleusentor, *das;* **~jaw** *n. (Med.)* Kieferklemme, *die; (disease)* Wundstarrkrampf, *der;* **~-out** *n.* Aussperrung, *die;* **~smith** *n.* Schlosser, *der;* **~-up** *attrib. adj. (Brit.)* **~-up shop/garage** *Laden in einem Gebäude, in dem der Inhaber nicht wohnt/nicht unmittelbar bei der Wohnung gelegene Garage*

locomotive ['ləʊkəməʊtɪv, ləʊkə'məʊtɪv] *n.* Lokomotive, *die*

locust ['ləʊkəst] *n.* [Wander]heuschrecke, *die*

lodge [lɒdʒ] **1.** *n.* **a)** *(cottage)* Pförtner-/Gärtnerhaus, *das; (Sport)* [Jagd-/Ski]hütte, *die;* **b)** *(porter's room)* [Pförtner]loge, *die;* **c)** *(of Freemasons)* Loge, *die.* **2.** *v. t.* **a)** *(deposit formally)* einlegen ‹*Beschwerde, Protest, Berufung usw.*›; *(bring forward)* erheben ‹*Einspruch, Protest*›; einreichen ‹*Klage*›; **b)** *(house)* unterbringen; *(receive as guest)* beherbergen; bei sich unterbringen; **c)** *(leave)* ~ **sth. with sb./in a bank** *etc.* etw. bei jmdm./in einer Bank *usw.* hinterlegen *od.* deponieren; **d)** *(put, fix)* stecken; [hinein]stoßen ‹*Schwert, Messer usw.*›; **be ~d in sth.** in etw. *(Dat.)* stecken; **become ~d in sth.** ‹*Kugel, Messer:*› steckenbleiben in etw. *(Dat.)*. **3.** *v. i.* **a)** *(be paying guest)* [zur Miete] wohnen; **b)** *(enter and remain)* steckenbleiben (**in** in + *Dat.*)

lodger ['lɒdʒə(r)] *n.* Untermieter, *der*/Untermieterin, *die*

lodging ['lɒdʒɪŋ] *n.* **a)** *usu. in pl. (rented room)* [möbliertes] Zimmer; **b)** *(accommodation)* Unterkunft, *die;* **board** *or* **food and ~:** Unterkunft und Verpflegung

'**lodging-house** *n.* Pension, *die*

loft [lɒft] *n.* **a)** *(attic)* [Dach]boden, *der;* **b)** *(over stable)* Heuboden, *der*

lofty ['lɒftɪ] *adj.* **a)** *(exalted, grandiose)* hoch; hehr *(geh.);* hochfliegend ‹*Ideen*›; hochgesteckt ‹*Ziele*›; **b)** *(high)* hoch [aufragend]; hoch ‹*Flug, Raum*›; **c)** *(haughty)* hochmütig; überheblich

[1]**log** [lɒg] **1.** *n.* **a)** *(rough piece of timber)* [geschlagener] Baumstamm; *(part of tree-trunk)* Klotz, *der; (as firewood)* [Holz]scheit, *das;* **be as easy as falling off a ~:** kinderleicht sein; **sleep like a ~:** schlafen wie ein Klotz; **b)** ~[**-book**] Tagebuch, *das; (Naut.)* Logbuch, *das; (Aeronaut.)* Bordbuch, *das.* **2.** *v. t.*, **-gg-** *(record)* Buch führen über *(Akk.); (Naut.)* ins Logbuch eintragen

[2]**log, logarithm** ['lɒgərɪðm] *n.* *(Math.)* Logarithmus, *der*

log: ~-book *n.* **a)** *(Brit.: of car)* Zulassung, *die;* **b)** *see* [1]**log 1 b**; ~ '**cabin** *n.* Blockhütte, *die;* ~ '**fire** *n.* Holzfeuer, *das*

loggerheads ['lɒgəhedz] *n. pl.* **be at ~ with sb.** mit jmdm. im Clinch liegen

logging ['lɒgɪŋ] *n., no pl., no indef. art.* Holzeinschlag, *der (Forstw.)*

logic ['lɒdʒɪk] *n.* Logik, *die*

logical ['lɒdʒɪkl] *adj.* **a)** logisch; **she has a ~ mind** sie denkt logisch; **b)** *(clear-thinking)* logisch denkend; klar denkend

logically ['lɒdʒɪkəlɪ] *adv.* logisch

logistic [lə'dʒɪstɪk] *adj.* logistisch

'**log jam** *n. Stau von treibendem Holz/Flößholz;* **the talks failed to move** *or* **break the ~** *(fig.)* die Gespräche haben keinen Durchbruch gebracht

logo ['lɒgəʊ, 'ləʊgəʊ] *n., pl.* **~s** Signet, *das;* Logo, *das*

loin [lɔɪn] *n.* **a)** *in pl. (Anat.)* Lende, *die;* **b)** *(meat)* Lende, *die*

'**loincloth** *n.* Lendenschurz, *der*

loiter ['lɔɪtə(r)] *v. i.* trödeln; bummeln; *(linger suspiciously)* herumlungern; **~ with intent** sich mit gesetzwidriger Absicht herumtreiben

loll [lɒl] *v. i.* **a)** *(lounge)* sich lümmeln *(ugs. abwertend);* **b)** *(droop)* ‹*Zunge:*› heraushängen; ‹*Kopf:*› hängen

lollipop ['lɒlɪpɒp] *n.* Lutscher, *der*

lollipop: ~ man/woman *ns.* *(Brit. coll.) Mann/Frau in der Funktion eines Schülerlotsen*

lolly ['lɒlɪ] *n.* **a)** *(Brit. coll.: lollipop)* Lutscher, *der;* **ice[d] ~:** Eis am Stiel; **b)** *no pl., no indef. art. (sl.: money)* Kohle, *die (salopp)*

London ['lʌndən] **1.** *pr. n.* London *(das).* **2.** *attrib. adj.* Londoner

Londoner ['lʌndənə(r)] *pr. n.* Londoner, *der*/Londonerin, *die*

lone [ləʊn] *attrib. adj. (poet./rhet.: solitary)* einsam

loneliness ['ləʊnlɪnɪs] *n., no pl.* Einsamkeit, *die*

lonely ['ləʊnlɪ] *adj.* einsam

loner ['ləʊnə(r)] *n.* Einzelgänger, *der*/-gängerin, *die*

lonesome ['ləʊnsəm] *adj.* einsam

[1]**long** [lɒŋ] **1.** *adj.*, **~er** ['lɒŋgə(r)], **~est** ['lɒŋgɪst] **a)** lang; weit ‹*Reise, Weg*›; **be ~ in the tooth** nicht mehr der/die Jüngste sein; **take a ~ view of sth.** etw. auf lange *od.* weite Sicht sehen; **two inches/weeks ~:** zwei Zoll/Wochen lang; **b)** *(elongated)* länglich; schmal; **pull** *or* **make a ~ face** *(fig.)* ein langes Gesicht ziehen *od.* machen *(ugs.);* **c)** *(of extended duration)* lang; **~ service** *(esp. Mil.)* langjähriger Dienst; **in the '~ run** auf die Dauer; auf lange Sicht; **in the '~ term** auf lange Sicht; langfristig; **for a '~ time** lange; *(still continuing)* seit langem; **what a ~ time you've been away!** du warst aber lange [Zeit] fort!; **~ time no see!** *(coll.)* lange nicht gesehen! *(ugs.);* **d)** *(tediously lengthy)* lang[atmig]; weitschweifig; **e)** *(lasting)* lang; langjährig ‹*Gewohnheit, Freundschaft*›; **f)** klein, gering ‹*Chance*›; **g)** *(seemingly more than stated)* lang ‹*Minute, Tag, Jahre usw.*›; **h)** lang ‹*Gedächtnis*›; **have a ~ memory for sth.** etw. nicht so schnell vergessen; **i)** *(consisting of many items)* lang ‹*Liste usw.*›; hoch ‹*Zahl*›; **j)** *(Cards)* **~ suit** lange Farbe. **2.** *n.* **a)** *(long interval)* **take ~:** lange dauern; **for ~:** lange; *(since ~ ago)* seit langem; **before ~:** bald; **it is ~ since ...:** es ist lange her, daß ...; **b) the ~ and the short of it is ...:** der langen Rede kurzer Sinn ist ... **3.** *adv.*, **~er**, **~est a)** lang[e]; **as** *or* **so ~ as** solange; **you should have finished ~ before now** du hättest schon längst *od.* viel früher fertig sein sollen; **I knew her ~ before I met you** ich kenne sie schon viel länger als dich; **not ~ before that** kurz davor *od.* zuvor; **not ~ before I ...:** kurz bevor ich ...; **~ since** [schon] seit langem; **all day/night/summer ~:** den ganzen Tag/die ganze Nacht/den ganzen Sommer [über *od.* lang]; **I shan't be ~:** ich bin gleich fertig; *(departing)* bis gleich!; **don't be ~!** beeil dich!; **sb. is ~ [in** *or* **about doing sth.]** jmd. braucht lange *od.* viel Zeit[, um etw. zu tun]; **much ~er** viel länger; **not wait any/much ~er** nicht mehr länger/viel länger warten; **no ~er** nicht mehr; nicht länger ‹*warten usw.*›; **b) as** *or* **so ~ as** *(provided that)* solange; wenn

[2]**long** *v. i.* **~ for sb./sth.** sich nach jmdm./etw. sehnen; **~ for sb. to do sth.** sich *(Dat.)* [sehr] wünschen, daß jmd. etw. tut; **~ to do sth.** sich danach sehnen, etw. zu tun

long. *abbr.* **longitude** Lg.

long: ~-ago 1. *n.* längst vergangene Zeit[en]; **2.** *adj.* längst vergangen; **~-distance 1.** ['---] *adj.* Fern‹*gespräch, -verkehr usw.*›; Langstrecken‹*lauf, -läufer, -flug usw.*›; **~-distance coach** Reise- *od.* Überlandbus, *der;* **~-distance lorry-driver** Fern[last]fahrer, *der;* **2.** [-'--] *adv.* **phone ~-distance** ein Ferngespräch führen; **~ division** *see* **division f; ~-drawn[-out]** *adj.* langgezogen ‹*Schrei, Ton*›; langatmig ‹*Erklärung, Diskussion*›; **~ drink** *n.* Longdrink, *der*

longevity [lɒn'dʒevɪtɪ] *n., no pl.* Langlebigkeit, *die*

long: ~-haired *adj.* langhaarig; Langhaar‹*dackel, -katze*›; **~hand** *n.* Langschrift, *die*

longing ['lɒŋɪŋ] **1.** *n.* Verlangen, *das;* Sehnsucht, *die; (craving)* Gelüst, *das (geh.).* **2.** *adj.* sehnsüchtig

longingly ['lɒŋɪŋlɪ] *adv.* voll Sehnsucht; sehnsüchtig

longitude ['lɒŋgɪtju:d] *n. (Geog.)* [geographische] Länge; *(of a* *place)* Länge, *die;* ~ **40° E** 40° östlicher Länge

long: ~ **jump** *n. (Brit. Sport)* Weitsprung, *der;* **~-legged** *adj.* langbeinig; **~-lived** ['lɒŋlɪvd] *adj. (durable)* andauernd; *(having ~ life)* langlebig; **be ~-lived** sehr alt werden; **~-playing 'record** *n.* Langspielplatte, *die;* **~-range** *adj.* **a)** Langstrecken⟨*flugzeug, -rakete usw.*⟩; ⟨*Geschütz*⟩ mit großer Reichweite; **b)** *(relating to the future)* langfristig; **~-running** *adj.* anhaltend; Langzeit⟨*versuch*⟩; wochen-/monate-/jahrelang ⟨*Debatte, Streit usw.*⟩; lange laufend ⟨*Theaterstück*⟩; ~ **shot** *n.* **a)** *(wild guess)* reine Spekulation; **b) not by a ~ shot** bei weitem nicht; ~ **'sight** *n.* Weitsichtigkeit, *die;* **have ~ sight** weitsichtig sein; **~-sighted** [lɒŋ'saɪtɪd] *adj.* weitsichtig; *(fig.)* weitblickend; vorausschauend; **~-sleeved** ['lɒŋsli:vd] *adj.* langärmelig; **~-standing** *attrib. adj.* seit langem bestehend; langjährig ⟨*Freundschaft usw.*⟩; alt ⟨*Schulden, Rechnung, Streit*⟩; **~-suffering** *adj.* schwer geprüft; *(meek)* geduldig; **~-term** *adj.* langfristig; **~-time** *adj.* seit langem bestehend; alt ⟨*Zwist, Freund*⟩; ~ **va'cation** *n. (Brit.)* Sommer[semester]ferien *Pl.;* ~ **wave** *n. (Radio)* Langwelle, *die;* **~-wave** *adj. (Radio)* Langwellen-; **~ways** *adv.* der Länge nach; längs; **~-winded** [lɒŋ'wɪndɪd] *adj.* langatmig; weitschweifig

loo [lu:] *n. (Brit. coll.)* Klo, *das (ugs. fam.);* **go to/be on the ~:** aufs Klo gehen/auf dem Klo sein

look [lʊk] **1.** *v. i.* **a)** sehen; gucken *(ugs.);* schauen *(bes. südd., sonst geh.);* ~ **before you leap** *(prov.)* erst wägen, dann wagen *(Spr.);* ~ **the other way** *(fig.)* die Augen verschließen; **not know which way to ~:** nicht wissen, wohin man sehen soll; **b)** *(search)* nachsehen; **c)** *(face)* zugewandt sein **(to[wards]** *Dat.*); **the room ~s on to the road/into the garden** das Zimmer liegt zur Straße/zum Garten hin *od.* geht zur Straße/zum Garten; **d)** *(appear)* aussehen; ~ **as if** [so] aussehen, als ob; ~ **well/ill** gut *od.* gesund/schlecht *od.* krank aussehen; ~ **like** aussehen wie; **e)** *(seem to be)* **she ~s her age** man sieht ihr ihr Alter an; **you ~ yourself again** es scheint dir wieder gut zu gehen; **f)** ~ **[here]!** *(demanding attention)* hören Sie/hör zu!; *(protesting)* passen Sie/paß ja *od.* bloß auf!; ~ **sharp [about sth.]** *(hurry up)* sich [mit etw.] beeilen. **2.** *v. t. (ascertain by sight)* nachsehen; *in exclamation of surprise etc.* sich *(Dat.)* ansehen; ~ **what you've done!** sieh [dir mal an], was du getan *od.* angerichtet hast!; ~ **who's here!** sieh mal, wer da *od.* gekommen ist! *see also* **dagger. 3.** *n.* **a)** Blick, *der;* **get a good ~ at sb.** jmdn. gut *od.* genau sehen [können]; **have** *or* **take a ~ at sb./sth.** sich *(Dat.)* jmdn./etw. ansehen; einen Blick auf jmdn./etw. werfen; **have a ~ at a town** sich *(Dat.)* eine Stadt ansehen; **let sb. have a ~ at sth.** jmdn. etw. sehen lassen; **b)** *in sing. or pl. (person's appearance)* Aussehen, *das; (facial expression)* [Gesichts]ausdruck, *der;* **from** *or* **by the ~[s] of sb.** von jmds. Aussehen zu schließen; **good ~s** gutes Aussehen; **have good ~s** gut aussehen; **c)** *(thing's appearance)* Aussehen, *das; (Fashion)* Look, *der;* **have a neglected ~:** verwahrlost aussehen; **by the ~[s] of it** *or* **things** [so] wie es aussieht; **the house is empty, by the ~ of it** das Haus steht allem Anschein nach leer; **I don't like the ~ of this** das gefällt mir gar nicht

~ **a'bout 1.** *v. t.* ~ **about one** sich umsehen *od.* umschauen. **2.** *v. i.* sich umsehen

~ **'after** *v. t.* **a)** *(follow with one's eyes)* nachsehen (+ *Dat.*); **b)** *(attend to)* sich kümmern um; **c)** *(care for)* sorgen für; ~ **after oneself** allein zurechtkommen; für sich selbst sorgen; ~ **'after yourself!** paß auf dich auf!

~ **a'head** *v. i.* **a)** nach vorne sehen; **b)** *(fig.: plan for future)* an die Zukunft denken; vorausschauen

~ **a'round** *see* ~ **about**

~ **at** *v. t.* **a)** *(regard)* ansehen; ~ **at one's watch** auf seine Uhr sehen; **don't ~ at me like that!** sieh mich nicht so an!; **be good/not much to ~ at** nach etwas/nach nichts *od.* nicht nach viel aussehen *(ugs.);* **b)** *(examine)* sich *(Dat.)* ansehen; **c)** *(consider)* betrachten; in Betracht ziehen ⟨*Angebot*⟩

~ **a'way** *v. i.* weggucken *(ugs.);* wegsehen

~ **'back** *v. i.* **a)** sich umsehen; *(fig.: hesitate)* zurückschauen; **he's never ~ed back since then** seitdem läuft bei ihm alles bestens; **b)** *(cast one's mind back)* ~ **back [up]on** *or* **to sth.** an etw. *(Akk.)* zurückdenken

~ **'down [up]on** *v. t.* **a)** herunter-/hinuntersehen, *(ugs.)* runtergucken auf (+ *Akk.*); **b)** *(fig.: despise)* herabsehen auf (+ *Akk.*)

~ **for** *v. t.* **a)** *(expect)* erwarten; **b)** *(seek)* suchen nach; auf der Suche sein nach ⟨*neuen Ideen*⟩; ~ **for trouble** Streit suchen; *(unintentionally)* sich *(Dat.)* Ärger einhandeln

~ **'forward to** *v. t.* sich freuen auf (+ *Akk.*); ~ **forward to doing sth.** sich darauf freuen, etw. zu tun

~ **'in** *v. i.* hin-/hereinsehen; *(visit)* vorbeikommen (**on** bei)

~ **into** *v. t.* **a)** sehen in (+ *Akk.*); **b)** *(fig.: investigate)* [eingehend] untersuchen; prüfen ⟨*Beschwerde*⟩

~ **on 1.** [-'-] *v. i.* zusehen; zugucken *(ugs.).* **2.** ['--] *v. t.* ~ **on sb. as a hero** *etc.* jmdn. als Held[en] *usw.* betrachten; ~ **on sb. with distrust/suspicion** jmdn. mit Mißtrauen/Argwohn betrachten

~ **'out 1.** *v. i.* **a)** hinaus-/heraussehen (**of** aus); rausgucken *(ugs.);* **b)** *(take care)* aufpassen; **c)** *(have view)* ~ **out on sth.** ⟨*Zimmer, Wohnung usw.:*⟩ zu etw. gehen *(ugs.),* zu etw. hin liegen. **2.** *v. t. (Brit.)* [her]aussuchen

~ **'out for** *v. t. (be prepared for)* aufpassen *od.* achten auf (+ *Akk.*); sich in acht nehmen vor (+ *Dat.*) ⟨*gefährliche Person, Sturm*⟩; *(keep watching for)* Ausschau halten nach

~ **'out of** *v. t.* sehen *od. (ugs.)* gucken aus

~ **'over** *v. t.* **a)** sehen über (+ *Akk.*) ⟨*Mauer usw.*⟩; überblicken ⟨*Tal usw.*⟩; **b)** *(survey)* inspizieren, sich *(Dat.)* ansehen ⟨*Haus, Anwesen*⟩; **c)** *(scrutinize)* mustern ⟨*Person*⟩; durchsehen ⟨*Text*⟩

~ **'round** *v. i.* sich umsehen; sich umgucken *(ugs.)*

~ **through** *v. t.* **a)** ~ **through sth.** durch etw. [hindurch] sehen; **b)** *(inspect)* durchsehen ⟨*Papiere*⟩; prüfen ⟨*Antrag, Vorschlag, Aussage*⟩; **c)** *(glance through)* sich *(Dat.)* ansehen ⟨*Buch, Notizen*⟩; **d)** ~ **straight 'through sb.** *(fig.)* durch jmdn. hindurchsehen

~ **to** *v. t.* **a)** *(rely on, count upon)* ~ **to sb./sth. for sth.** etw. von jmdm./etw. erwarten; ~ **to sb./sth. to do sth.** von jmdm./etw. erwarten, daß er/es etw. tut; **b)** *(be careful about)* sorgen für; *(keep watch upon)* aufpassen auf (+ *Akk.*)

~ **'up** 1. *v. i.* **a)** aufblicken; **b)** *(improve)* besser werden; ‹*Aktien, Chancen:*› steigen; **things are ~ing up** es geht bergauf; **business is ~ing up again** das Geschäft läuft wieder besser. **2.** *v. t.* **a)** *(search for)* nachschlagen ‹*Wort*›; heraussuchen ‹*Telefonnummer, Zugverbindung usw.*›; **b)** *(coll.: visit)* ~ **sb. up** bei jmdm. reingucken *(ugs.)*; **c)** ~ **sb. up and down** jmdn. von Kopf bis Fuß mustern

~ **upon** *see* ~ **on 2**

~ **'up to** *v. t.* ~ **up to sb.** *(lit. or fig.)* zu jmdm. aufschauen *od.* aufsehen

'look-alike *n.* Doppelgänger, *der*/-gängerin, *die*

looker-'on [lʊkə(r)'ɒn] *n.* Zuschauer, *der*/Zuschauerin, *die*

'look-in *n. (opportunity)* Chance, *die;* **we didn't get a ~:** wir hatten überhaupt keine Chance

'looking-glass *n.* Spiegel, *der*

'look-out *n., pl.* **~s a)** *(keeping watch) (Naut.)* Ausschauhalten, *das; (guard)* Wache, *die;* **keep a ~** *or* **be on the ~ [for sth./sb.]** *(wanted)* [nach etw./jmdm.] Ausschau halten; *(not wanted)* [auf etw./jmdn.] aufpassen; **b)** *(observation post)* Ausguck, *der;* Beobachtungsstand, *der;* **c)** *(person)* Wache, *die; (Mil.)* Wach[t]posten, *der;* Beobachtungsposten, *der;* **d)** *(esp. Brit. fig.: prospect)* Aussichten; **that's a bad ~:** das sind schlechte Aussichten; **it's a poor/bleak** *etc.* ~ **for sb./sth.** es sieht schlecht/düster *usw.* aus für jmdn./etw.; **e)** *(concern)* **that's his [own] ~:** das ist [allein] sein Problem *od.* seine Sache

¹loom [lu:m] *n. (Weaving)* Webstuhl, *der*

²loom *v. i.* sich [bedrohlich] abzeichnen; ~ **large** [bedrohlich] auftauchen; *(fig.)* eine große Rolle spielen

~ **'up** *v. i.* ~ **up [in front of sb.]** [unmittelbar] [vor jmdm.] auftauchen

loony ['lu:nɪ] *(sl.)* **1.** *n.* Verrückte, *der/die (ugs.)*. **2.** *adj.* verrückt *(ugs.)*; irr

loop [lu:p] **1.** *n.* **a)** Schleife, *die;* **b)** *(cord)* Schlaufe, *die;* **c)** *(contraceptive coil)* Spirale, *die.* **2.** *v. t.* **a)** *(form into a ~)* zu einer Schlaufe/Öse formen; **b)** *(enclose)* umschlingen; **c)** *(fasten)* ~ **up/together** *etc.* mit einer Schlaufe hoch-/zusammenbinden *usw.;* **d)** *(Aeronaut.)* ~ **the ~:** einen Looping fliegen; loopen *(fachspr.)*

'loophole *n. (fig.)* Lücke, *die;* ~ **in the law** Gesetzeslücke, *die;* Lücke im Gesetz

loose [lu:s] **1.** *adj.* **a)** *(unrestrained)* freilaufend ‹*Tier*›; *(escaped)* ausgebrochen; **set** *or* **turn ~:** freilassen; **b)** *(not firm)* locker ‹*Zahn, Schraube, Mutter, Knopf, Messerklinge*›; **come/get/work ~:** sich lockern; *see also* **screw 1 a; c)** *(not fixed)* lose; **d)** *(not bound together)* lose; offen ‹*Haar*›; **e)** *(slack)* locker; schlaff ‹*Haut, Gewebe usw.*›; beweglich ‹*Glieder*›; **f)** *(hanging free)* lose; **be at a ~ end** *or (Amer.)* **at ~ ends** *(fig.)* beschäftigungslos sein; *(not knowing what to do with oneself)* nichts zu tun haben; nichts anzufangen wissen; **g)** *(inexact)* ungenau; schief ‹*Vergleich*›; frei ‹*Stil*›; unsauber ‹*Denken*›; **h)** *(morally lax)* liederlich ‹*Leben[swandel], Person*›; locker ‹*Moral, Lebenswandel*›; **a ~ woman** ein leichtes Mädchen. **2.** *v. t.* **a)** loslassen ‹*Hund usw.*›; **b)** *(untie)* lösen; aufmachen *(ugs.)*; **c)** ~ **[off]** abschießen ‹*Pfeil*›; abfeuern ‹*Feuerwaffe, Salve*›; abgeben ‹*Schuß, Salve*›; **d)** *(relax)* lockern; ~ **[one's] hold** loslassen

loose: ~ **'change** *see* **change 1 d;** ~ **'cover** *n. (Brit.)* Überzug, *der;* Schoner, *der;* **~-fitting** *adj.* bequem geschnitten; **~-knit** *adj.* lose zusammenhängend ‹*Organisation, Gemeinschaft usw.*›; **~-leaf** *attrib. adj.* Loseblatt-; **~-leaf file** Ringbuch, *das;* **~-limbed** ['lu:slɪmd] *adj.* gelenkig; geschmeidig; *(gawky)* schlaksig

loosely ['lu:slɪ] *adv.* **a)** *(not tightly)* locker; lose; **b)** *(not strictly)* locker ‹*gruppieren*›; lose ‹*zusammenhängen*›; frei ‹*übersetzen*›; ~ **speaking** grob gesagt

loosen ['lu:sn] **1.** *v. t.* **a)** *(make less tight etc.)* lockern; **b)** *(fig.: relax)* lockern ‹*Bestimmungen, Reglement usw.*›; ~ **sb.'s tongue** *(fig.)* jmds. Zunge lösen. **2.** *v. i. (become looser)* sich lockern

~ **up 1.** ['---] *v. t.* lockern ‹*Glieder, Muskeln*›. **2.** [--'-] *v. i.* sich auflockern; *(relax)* auftauen

loot [lu:t] **1.** *v. t.* **a)** *(plunder)* plündern; **b)** *(carry off)* rauben. **2.** *n.* **a)** [Kriegs]beute, *die;* **b)** *(sl.: money)* Zaster, *der (salopp);* Knete, *die (salopp)*

looter ['lu:tə(r)] *n.* Plünderer, *der*

lop [lɒp] *v. t.,* ~ **sth. [off** *or* **away]** etw. abhauen *od.* abhacken

lope [ləʊp] *v. i.* ‹*Hase, Kaninchen:*› springen; ‹*Wolf, Fuchs:*› laufen; ‹*Person:*› beschwingten Schrittes gehen

lopsided [lɒp'saɪdɪd] *adj.* schief; *(fig.)* einseitig

loquacious [lə'kweɪʃəs] *adj.* redselig; schwatzhaft *(abwertend)*

lord [lɔ:d] **1.** *n.* **a)** *(master)* Herr, *der;* ~ **and master** *(joc.)* Herr und Gebieter *od.* Meister *(scherzh.)*; **b) L~** *(Relig.)* Herr, *der;* **L~ God [Almighty]** unser Herr[, der allmächtige Gott]; **the L~ [God]** [Gott] der Herr; **the L~'s Prayer** das Vaterunser; **L~ only knows** *(coll.)* weiß der Himmel *(ugs.)*; **c)** *(Brit.: nobleman, or as title)* Lord, *der;* **the House of L~s** *(Brit.)* das Oberhaus; *see also* **drunk 1; d) My L~** *(Brit.) form of address (to earl, viscount)* Graf; *(to baron)* Baron; *(to bishop)* Exzellenz; *(to judge)* [mlʌd] Herr Richter. **2.** *int. (coll.)* Gott!; **oh/good L~!** du lieber Himmel *od.* Gott!; großer Gott! **3.** *v. t.* ~ **it over sb.** bei jmdm. den großen Herrn/die große Dame spielen

lordship ['lɔ:dʃɪp] *n. (title, estate)* Lordschaft, *die;* **his/your ~/their/your ~s** seine/Eure Lordschaft/ihre/Eure Lordschaften

lore [lɔ:(r)] *n.* Wissen, *das;* Kunde, *die; (body of traditions)* Überlieferung, *die; (of a people, an area)* Folklore, *die*

Lorraine [lɒ'reɪn] *pr. n.* Lothringen *(das)*

lorry ['lɒrɪ] *n. (Brit.)* Lastwagen, *der;* Lkw, *der;* Laster, *der (ugs.)*; **it fell off the back of a ~** *(joc.)* das ist mir/ihm *usw.* zugelaufen *(ugs. scherzh.)*

'lorry-driver *n. (Brit.)* Lastwagenfahrer, *der;* Lkw-Fahrer, *der*

lose [lu:z] **1.** *v. t.,* **lost** [lɒst] **a)** verlieren; kommen um, verlieren ‹*Leben, Habe*›; **sb. has nothing to ~ [by doing sth.]** es kann jmdm. nicht schaden[, wenn er etw. tut]; ~ **one's way** sich verlaufen/verfahren; **b)** *(fail to maintain)* verlieren; *(become slow by)* ‹*Uhr:*› nachgehen ‹*zwei Minuten täglich usw.*›; **c)** *(waste)* vertun ‹*Zeit*›; *(miss)* versäumen, verpassen ‹*Zeitpunkt, Gelegenheit, Ereignis*›; **d)** *(fail to obtain)* nicht bekommen ‹*Preis, Vertrag usw.*›; *(fail to hear)* nicht mitbekommen ‹*Teil einer Rede usw.*›; *(fail to catch)* verpassen, versäumen ‹*Zug, Bus*›; **the motion was lost** der Antrag kam nicht durch *od.* scheiterte; **e)** *(be defeated in)* verlieren ‹*Kampf, Spiel, Wette, Prozeß usw.*›; **f)** *(cause loss of)* ~ **sb. sth.** jmdn. um etw. bringen; **you['ve] lost me** *(fig.)* ich komme nicht mehr mit; **g)** *(get rid of)* abschütteln ⟨*Verfolger*⟩; loswerden ⟨*Erkältung*⟩; ~ **weight** abnehmen. *See also* **lost. 2.** *v. i.*, **lost a)** *suffer loss)* einen Verlust erleiden; *(in business)* Verlust machen **(on** bei); *(in match, contest)* verlieren; ~ **in freshness** an Frische verlieren; **you can't** ~ *(coll.)* du kannst nur profitieren *od.* gewinnen; **b)** *(become slow)* ⟨*Uhr:*⟩ nachgehen

~ **'out** *v. i.* verdrängt werden **(to** von)

loser ['lu:zə(r)] *n.* Verlierer, *der*/Verliererin, *die; (failure)* Versager, *der*/Versagerin, *die*

loss [lɒs] *n.* **a)** *(process)* Verlust, *der* **(of** *Gen.*); **b)** *in sing. or pl. (what is lost)* Verlust, *der;* **sell at a** ~: mit Verlust verkaufen; *see also* **cut 1 j; c)** *(state)* Verlust, *der;* **be no** ~ **to sb.** für jmdn. kein Verlust sein; **d) be at a** ~: nicht [mehr] weiterwissen; **be at a** ~ **what to do** nicht wissen, was zu tun ist; **be at a** ~ **for words/an answer** um Worte/eine Antwort verlegen sein

lost [lɒst] *adj.* **a)** verloren; ausgestorben ⟨*Kunst[fertigkeit]*⟩; **get** ~ ⟨*Person:*⟩ sich verlaufen *od.* verirren/verfahren; **get** ~! *(sl.)* verdufte! *(salopp);* **I'm** ~ *(fig.)* ich verstehe gar nichts mehr; **feel** ~ **without sb./sth.** *(fig.)* sich *(Dat.)* ohne jmdn./etw. hilflos vorkommen; *see also* **property a; b)** *(wasted)* vertan ⟨*Zeit, Gelegenheit*⟩; verschwendet ⟨*Zeit, Mühe*⟩; verpaßt, versäumt ⟨*Gelegenheit*⟩; **c)** *(not won)* verloren; aussichtslos ⟨*Sache*⟩; *see also* **all 2 d; cause 1 c; d)** ~ **in admiration** überwältigt; **be** ~ **[up]on sb.** *(unrecognized by)* bei jmdm. keine Anerkennung finden; von jmdm. nicht gewürdigt werden; **sarcasm was** ~ **on him** mit Sarkasmus konnte er nichts anfangen

lot [lɒt] *n.* **a)** *(method of choosing)* Los, *das;* **by** ~: durch das Los; **b)** *(destiny)* Los, *das;* **fall to the** ~ **of sb.** jmdm. bestimmt sein; **c)** *(item to be auctioned)* Posten, *der;* **d)** *(set of persons)* Haufen, *der;* **the** ~: [sie] alle; **'our/'your/'their** ~ *(coll.)* wir/ihr/die; **e)** *(set of things)* Menge, *die;* **divide sth. into five** ~**s** etw. in fünf Stapel/Haufen *usw.* teilen; **that's the** ~ *(coll.)* das ist alles; das wär's *(ugs.);* **f)** *(coll.: large number or quantity)* ~**s** *or* **a** ~ **of money** *etc.* viel *od.* eine Menge Geld *usw.;* ~**s of books/coins** eine Menge Bücher/Münzen; **he has a** ~ **to learn** er muß noch viel lernen; **have** ~**s to do** viel zu tun haben; **we have** ~**s of time** wir haben viel *od. (ugs.)* massenweise Zeit; ~**s** *or* **a** ~ **better** viel besser; **like sth. a** ~: etw. sehr mögen; **g)** *(for choosing)* Los, *das;* **draw/cast/throw** ~**s [for sth.]** das Los [über etw. *(Akk.)*] entscheiden lassen; [um etw.] losen; **cast/throw in one's** ~ **with sb.** sich mit jmdm. zusammentun; **draw** ~**s to determine sth.** etw. durch das Los entscheiden; **h)** *(plot of land)* Gelände, *das;* Platz, *der; (measured piece of land)* Parzelle, *die*

lotion ['ləʊʃn] *n.* Lotion, *die*

lottery ['lɒtərɪ] *n.* Lotterie, *die; (fig.)* Glücksspiel, *das*

loud [laʊd] **1.** *adj.* **a)** laut; schreiend ⟨*Reklame*⟩; lautstark ⟨*Protest, Kritik*⟩; **b)** *(flashy, conspicuous)* aufdringlich; grell, schreiend ⟨*Farbe*⟩. **2.** *adv.* laut; **laugh out** ~: laut auflachen; **say sth. out** ~: etw. aussprechen; *(fig.)* etw. laut verkünden

loud hailer ['laʊd heɪlə(r)] *n.* Megaphon, *das;* Flüstertüte, *die (ugs. scherzh.)*

loudly ['laʊdlɪ] *adv.* **a)** laut; **b)** *(flashily)* aufdringlich

loud: ~**-mouth** *n.* Großmaul, *das;* ~**-mouthed** ['laʊdmaʊðd] *adj.* großmäulig *(ugs. abwertend)*

loudness ['laʊdnɪs] *n., no pl.* **a)** Lautstärke, *die;* **b)** *(flashiness)* Aufdringlichkeit, *die*

loud'speaker n. Lautsprecher, *der*

lounge [laʊndʒ] **1.** *v. i.* ~ **[about** *or* **around]** [faul] herumliegen/-sitzen/-stehen; [faul] herumhängen *(ugs.); (in chair etc.)* sich lümmeln *(ugs.).* **2.** *n.* **a)** *(public room)* Lounge, *die; (in hotel)* Lounge, *die;* [Hotel]halle, *die; (at station)* Wartesaal, *der; (in theatre)* Foyer, *das; (at airport)* Lounge, *die;* Wartehalle, *die;* **b)** *(sitting-room)* Wohnzimmer, *das;* **c)** *(Brit.: bar)* ~ **[bar]** *see* **saloon bar**

lounger ['laʊndʒə(r)] *n.* **a)** Nichtstuer, *der;* **b)** *(sun-bed)* Liege, *die*

lour ['laʊə(r)] *v. i.* mißmutig [drein]blicken; ein finsteres Gesicht machen; *(fig.)* ⟨*Wolken, Gewitter:*⟩ sich [bedrohlich] zusammenziehen; ⟨*Himmel:*⟩ sich [bedrohlich] verfinstern

louse [laʊs] *n.* **a)** *pl.* **lice** [laɪs] Laus, *die;* **b)** *pl.* ~**s** *(sl.: person)* Ratte, *die (derb)*

lousy ['laʊzɪ] *adj.* **a)** *(infested)* verlaust; **b)** *(sl.) (disgusting)* ekelhaft; widerlich; *(very poor)* lausig *(ugs.);* mies *(ugs.);* **feel** ~: sich mies *(ugs.) od.* miserabel fühlen

lout [laʊt] *n.* Rüpel, *der;* Flegel, *der; (bumpkin)* Tolpatsch, *der (ugs.);* Tölpel, *der*

louver, louvre ['lu:və(r)] *n.* ~ **window** Jalousiefenster, *das*

lovable ['lʌvəbl] *adj.* liebenswert

love [lʌv] **1.** *n.* **a)** *(affection, sexual* ~*)* Liebe, *die* **(for** zu); **in** ~ **[with]** verliebt [in (+ *Akk.*)]; **fall in** ~ **[with]** sich verlieben [in (+ *Akk.*)]; **make** ~ **to sb.** *(have sex)* mit jmdm. schlafen; jmdn. lieben; **for** ~: aus Liebe; *(free)* unentgeltlich; umsonst; *(for pleasure)* nur zum Vergnügen *od.* Spaß; **not for** ~ **or money** um nichts in der Welt; **[Happy Christmas,]** ~ **from Beth** *(in letter)* [fröhliche Weihnachten und] herzliche Grüße von Beth; **send one's** ~ **to sb.** jmdn. grüßen lassen; **Peter sends [you] his** ~: Peter läßt [dich] grüßen; **there is no** ~ **lost between them** sie sind sich *(Dat.)* nicht grün *(ugs.);* **b)** *(devotion)* Liebe, *die* **(of, for, to[wards]** zu); ~ **of life/eating/learning** Freude am Leben/Essen/Lernen; **for the** ~ **of God** um Gottes willen; **c)** *(sweetheart)* Geliebte, *der/die;* Liebste, *der/die (veralt.);* **[my]** ~ *(coll.: form of address)* [mein] Liebling *od.* Schatz; *(to sb. less close)* mein Lieber/meine Liebe; **d)** *(Tennis)* **fifteen/thirty** ~: fünfzehn/dreißig null. **2.** *v. t.* **a)** lieben; **our/their** ~**d ones** unsere/ihre Lieben; **b)** *(like)* **I'd** ~ **a cigarette** ich hätte sehr gerne eine Zigarette; ~ **to do** *or* **doing sth.** etw. [leidenschaftlich] gern tun. **3.** *v. i.* lieben

love: ~ **affair** *n.* [Liebes]verhältnis, *das;* Liebschaft, *die;* ~**-'hate** *adj.* von Haßliebe geprägt; ~**-letter** *n.* Liebesbrief, *der;* ~**-life** *n.* Liebesleben, *das*

lovely ['lʌvlɪ] *adj.* **a)** [wunder]schön; herrlich ⟨*Tag, Essen*⟩; **b)** *(lovable)* liebenswert; **c)** *(coll.: delightful)* toll *(ugs.);* wunderbar; ~ **and warm/cool** *etc. (coll.)* schön warm/kühl *usw.*

'love-making *n. (sexual intercourse)* körperliche Liebe

lover ['lʌvə(r)] *n.* **a)** Liebhaber, *der;* Geliebte, *der; (woman)* Geliebte, *die;* **be** ~**s** ein Liebespaar sein; **b)** *(person devoted to sth.)* Liebhaber, *der*/Liebhaberin, *die;* Freund, *der*/Freundin, *die;* ~ **of the arts** Kunstliebhaber, *der*/-liebhaberin, *die;* Kunstfreund, *der*/-freundin, *die;* **dog-**~ Hundefreund, *der*/-freundin, *die*

love: ~**sick** *adj.* an Liebeskummer leidend; liebeskrank *(geh.);*

~-song *n.* Liebeslied, *das;* **~-story** *n.* Liebesgeschichte, *die*

loving *adj.* **a)** *(affectionate)* liebend; **b)** *(expressing love)* liebevoll

lovingly ['lʌvɪŋlɪ] *adv.* liebevoll; *(painstakingly)* mit viel Liebe

[1]**low** [ləʊ] **1.** *adj.* **a)** *(not reaching far up)* niedrig; niedrig, flach ‹*Absätze, Stirn*›; flach ‹*Relief*›; **b)** *(below normal level)* niedrig; tief ‹*Flug*›; flach ‹*Welle*›; tief ausgeschnitten ‹*Kleid*›; tief ‹*Ausschnitt*›; **c)** *(not elevated)* tiefliegend ‹*Wiese, Grund, Land*›; tiefhängend ‹*Wolke*›; tiefstehend ‹*Gestirne*›; tief ‹*Verbeugung*›; **d)** *(inferior)* niedrig; gering ‹*Intelligenz, Bildung*›; gewöhnlich ‹*Geschmack*›; **e)** *(not fair)* gemein; **f)** *(Cards)* niedrig; **g)** *(small in degree)* niedrig; gering ‹*Sichtweite, Wert*›; **have a ~ opinion of sb./sth.** von jmdm./etw. keine hohe Meinung haben; **h)** *(in pitch)* tief ‹*Ton, Stimme, Lage, Klang*›; *(in loudness)* leise ‹*Ton, Stimme*›; **i)** *(nearly gone)* fast verbraucht *od.* aufgebraucht; **run ~:** allmählich ausgehen *od.* zu Ende gehen. *See also* [2]**lower 1. 2.** *adv.* **a)** *(in or to a ~ position)* tief; niedrig, tief ‹*hängen*›; *see also* **high 2 a; b)** *(to a ~ level)* **prices have gone too ~:** die Preise sind zu weit gefallen; **c)** *(not loudly)* leise; **d) lay sb. ~** *(prostrate)* jmdn. niederstrecken *(geh.)*; **lie ~:** am Boden liegen; *(hide)* untertauchen. *See also* [2]**lower 2. 3.** *n.* **a)** *(Meteorol.)* Tief, *das;* **b)** Tiefststand, *der; see also* **all-time**

[2]**low** *v. i.* ‹*Kuh:*› muhen

low: ~brow *(coll.) adj.* schlicht ‹*Person*›; [geistig] anspruchslos ‹*Buch, Programm*›; **Low Countries** *pr. n. pl. (Hist.)* Niederlande *Pl.*; **~-cut** *adj.* [tief] ausgeschnitten ‹*Kleid*›; **~-down 1.** *adj. (coll.: mean)* mies *(ugs.)*; **2.** *n. (coll.)* **give [sb.]/get the ~-down on sb./sth.** [jmdm.] sagen/rauskriegen, was es mit jmdm./etw. [wirklich] auf sich hat

[1]**lower** ['ləʊə(r)] *v. t.* **a)** *(let down)* herab-/hinablassen; einholen ‹*Flagge, Segel*›; **~ oneself into** hinuntersteigen in (+ *Akk.*) ‹*Kanalschacht, Keller*›; **~ oneself into a chair** sich in einen Sessel sinken lassen; **b)** *(reduce in height)* senken ‹*Blick*›; niederschlagen ‹*Augen*›; absenken ‹*Zimmerdecke*›; auslassen ‹*Saum*›; **c)** *(lessen)* senken ‹*Preis, Miete, Zins usw.*›; **d)** *(degrade)* herabsetzen; **~ oneself to do sth.** sich so weit erniedrigen, etw. zu tun; **e)** *(weaken)* schwächen; dämpfen ‹*Licht, Stimme, Lärm*›; **~ one's voice** leiser sprechen; die Stimme senken *(geh.)*

[2]**lower 1.** *compar. adj.* **a)** unter... ‹*Nil, Themse usw., Atmosphäre*›; Unter‹*jura, -devon usw., -arm, -lippe usw.*›; Nieder‹*rhein, -kalifornien*›; **b)** *(in rank)* unter...; **~ mammals/plants** niedere Säugetiere/Pflanzen; **the ~ orders/classes** die Unterschichten/die unteren Klassen. **2.** *compar. adv.* tiefer ‹*sinken, hängen usw.*›

lower: ~ case 1. *n.* Kleinbuchstaben *Pl.*; **2.** *adj.* klein ‹*Buchstabe*›; **~ 'deck** *n. (of ship)* Unterdeck, *das; (of bus)* unteres Deck; **L~ 'Saxony** *pr. n.* Niedersachsen *(das)*

low: ~-fat *adj.* fettarm; **~-flying** *adj.* tief fliegend; **~-flying aircraft** Tiefflieger, *der;* **~-grade** *adj.* minderwertig; **~-key** *adj.* zurückhaltend; unaufdringlich ‹*Beleuchtung, Unterhaltung*›; **~land** ['ləʊlənd] **1.** *n.* Tiefland, *das;* **2.** *adj.* tiefländisch; Tiefland‹*rasse, -farm*›

lowly ['ləʊlɪ] *adj.* **a)** *(modest)* bescheiden; **b)** *(not highly evolved)* nieder...

low: ~-lying *adj.* tiefliegend; **~ point** *n.* Tiefpunkt, *der;* **~-powered** ['ləʊpaʊəd] *adj.* schwach ‹*Motor, Glühbirne*›; **~ pressure** *n. (Meteorol.)* Tiefdruck, *der;* **an area of ~ pressure** ein Tiefdruckgebiet; **~ season** *n.* Nebensaison, *die;* **~-'spirited** *adj.* niedergeschlagen; **~ 'tide** *see* **tide 1 a; ~ 'water** *n.* Niedrigwasser, *das;* **~-'water mark** *n.* Niedrigwassermarke, *die*

loyal ['lɔɪəl] *adj. (to person)* treu; *(to government etc.)* treu [ergeben]; loyal

loyalty ['lɔɪəltɪ] *n.* Treue, *die;* Loyalität, *die*

lozenge ['lɒzɪndʒ] *n.* **a)** *(tablet)* Pastille, *die;* **b)** *(diamond shape)* Raute, *die;* Rhombus, *der*

LP *abbr.* **long-playing record** LP, *die*

'L-plate *n. (Brit.)* 'L'-Schild, *das;* ≈ „Fahrschule"-Schild, *das*

LSD *abbr.* **lysergic acid diethylamide** LSD, *das*

Ltd. *abbr.* **Limited** GmbH; **... Company ~:** ...gesellschaft mbH

lubricant ['lu:brɪkənt] *n.* Schmiermittel, *das*

lubricate ['lu:brɪkeɪt] *v. t.* schmieren; einfetten ‹*Haut*›

lubrication [lu:brɪ'keɪʃn] *n.* Schmierung, *die; attrib.* Schmier‹*system, -vorrichtung*›

Lucerne [lu:'sɜ:n] *pr. n.* Luzern *(das)*; **Lake ~:** der Vierwaldstätter See

lucid ['lu:sɪd] *adj.* klar; [leicht] verständlich; einleuchtend ‹*Argumentation*›; **~ interval** *(period of sanity)* lichter Augenblick

luck [lʌk] *n.* **a)** *(good or ill fortune)* Schicksal, *das;* **as ~ would have it** wie das Schicksal es wollte; **good ~:** Glück, *das;* **bad ~:** Pech, *das;* **better ~ next time** mehr Glück beim nächsten Mal; **good ~ [to you]!** viel Glück!; alles Gute!; **good ~ to him, I say** ich wünsche ihm viel Glück; *(iron.)* na, dann viel Glück!; **just my ~:** typisch für mich; **b)** *(good fortune)* Glück, *das;* **with [any] ~:** mit ein bißchen *od.* etwas Glück; **do sth. for ~:** etw. tun, damit es einem Glück bringen soll; **be in/out of ~:** Glück/kein Glück haben; **no such ~:** schön wär's

luckily ['lʌkɪlɪ] *adv.* glücklicherweise; **~ for her** zu ihrem Glück

lucky ['lʌkɪ] *adj.* **a)** *(favoured by chance)* glücklich; **be ~ [in love/at games]** Glück [in der Liebe/im Spiel] haben; **be ~ enough to be rescued** das [große] Glück haben, gerettet zu werden; **Could you lend me £100? – 'You'll be ~!** Könntest du mir 100 Pfund leihen? – So siehst du aus!; **b)** *(favouring sb. by chance)* glücklich ‹*Umstand, Zufall, Zusammentreffen usw.*›; *see also* **escape 1 a; c)** *(bringing good luck)* Glücks‹*zahl, -tag usw.*›; **~ charm** Glücksbringer, *der;* **be born under a ~ star** ein Glückskind sein; **you can thank your ~ stars** du kannst von Glück sagen

lucrative ['lu:krətɪv] *adj.* einträglich; lukrativ

ludicrous ['lu:dɪkrəs] *adj.* lächerlich ‹*Anblick, Lohn, Argument, Vorschlag, Idee*›; lachhaft ‹*Angebot, Ausrede*›; **a ~ speed** *(low)* eine lächerliche Geschwindigkeit; *(high)* eine haarsträubende Geschwindigkeit

ludo ['lu:dəʊ] *n., no pl., no art.* Mensch-ärgere-dich-nicht[-Spiel], *das*

lug [lʌg] *v. t.,* **-gg-: a)** *(drag)* schleppen; **b)** *(force)* **~ sb. along** jmdn. mit herumschleppen *(ugs.)*

luggage ['lʌgɪdʒ] *n.* Gepäck, *das*

luggage: ~-locker *n.* [Gepäck]schließfach, *das;* **~-rack** *n.* Gepäckablage, *die;* **~ trolley** *n.* Kofferkuli, *der*

lugubrious [lu:'gu:brɪəs, lʊ'gu:-

brɪəs] *adj.* *(mournful)* kummervoll; traurig; *(dismal)* düster

lukewarm ['lu:kwɔ:m, lu:k'wɔ:m] *adj.* **a)** lauwarm; **b)** *(fig.)* lau[warm]; halbherzig

lull [lʌl] **1.** *v. t.* **a)** *(soothe)* lullen; **b)** *(fig.)* einlullen; ~ **sb. into a false sense of security** jmdn. in einer trügerischen Sicherheit wiegen. **2.** *n.* Pause, *die;* **the ~ before the storm** *(fig.)* die Ruhe vor dem Sturm

lullaby ['lʌləbaɪ] *n.* Schlaflied, *das;* Wiegenlied, *das*

lumbago [lʌm'beɪgəʊ] *n.* *(Med.)* Hexenschuß, *der;* Lumbago, *die* *(fachspr.)*

[1]**lumber** ['lʌmbə(r)] *v. i.* ‹*Person:*› schwerfällig gehen; ‹*Fahrzeug:*› rumpeln

[2]**lumber 1.** *n.* **a)** *(furniture)* Gerümpel, *das;* **b)** *(useless material)* Kram, *der* *(ugs. abwertend);* Krempel, *der (ugs. abwertend);* **c)** *(Amer.: timber)* [Bau]holz, *das.* **2.** *v. t.* *(fill up, encumber)* vollstopfen *(ugs.);* überladen ‹*Stil, Buch*›; ~ **sb. with sth./sb.** jmdm. etw./jmdn. aufhalsen *(ugs.);* **get ~ed with sth./sb.** etw./jmdn. aufgehalst kriegen *(ugs.)*

lumber: ~-jack *n.* *(Amer.)* Holzfäller, *der;* **~-room** *n.* Abstellkammer, *die;* Rumpelkammer, *die (ugs.)*

luminosity [lu:mɪ'nɒsɪtɪ] *n.* *(also Astron.)* Helligkeit, *die*

luminous ['lu:mɪnəs] *adj.* **a)** *(bright)* hell ‹*Feuer, Licht usw.*›; [hell] leuchtend; Leucht‹*anzeige, -zeiger usw.*›; ~ **paint** Leuchtfarbe, *die;* **b)** *(of light)* Leucht‹*kraft, -stärke usw.*›

[1]**lump** [lʌmp] **1.** *n.* **a)** *(shapeless mass)* Klumpen, *der;* *(of sugar, butter, etc.);* Stück, *das;* *(of wood)* Klotz, *der;* *(of dough)* Kloß, *der;* *(of bread)* Brocken, *der;* **have/get a ~ in one's throat** *(fig.)* einen Kloß im Hals haben *(ugs.);* **b)** *(swelling)* Beule, *die;* *(caused by cancer)* Knoten, *der;* **c)** *(thickset person)* Klotz, *der (ugs.);* **d) get payment in a ~:** die gesamte Summe auf einmal erhalten. **2.** *v. t.* *(mass together)* zusammentun; ~ **sth. with sth.** etw. und etw. zusammentun; ~ **sb./sth. with the rest** jmdn./etw. mit dem Rest in einen Topf werfen *(ugs.)*

~ to'gether *v. t.* zusammenfassen

[2]**lump** *v. t.* *(coll.)* sich abfinden mit; **if you don't like it you can ~ it** du mußt dich wohl oder übel damit abfinden

lump: ~ 'sugar *n.* Würfelzucker, *der;* **~ 'sum** *n.* *(covering several items)* Pauschalsumme, *die;* *(paid at once)* einmalige Pauschale

lumpy ['lʌmpɪ] *adj.* klumpig ‹*Brei, Lehm*›; ‹*Kissen, Matratze*› mit klumpiger Füllung

lunacy ['lu:nəsɪ] *n.* **a)** *(insanity)* Wahnsinn, *der;* **b)** *(mad folly)* Wahnsinn, *der (ugs.);* Irrsinn, *der*

lunar ['lu:nə(r)] *adj.* Mond-; lunar *(fachspr.)*

lunar e'clipse *n.* *(Astron.)* Mondfinsternis, *die*

lunatic ['lu:nətɪk] **1.** *adj.* **a)** *(mad)* wahnsinnig; irre *(veralt.);* *see also* **fringe 1 c; b)** *(foolish)* wahnwitzig; Wahnsinns- *(ugs.);* idiotisch *(ugs. abwertend).* **2.** *n.* Wahnsinnige, *der/die;* Irre, *der/die*

'lunatic asylum *n.* *(Hist.)* Irrenanstalt, *die (veralt., ugs. abwertend)*

lunch [lʌntʃ] **1.** *n.* Mittagessen, *das;* **have** *or* **eat [one's] ~:** zu Mittag essen. **2.** *v. i.* zu Mittag essen

'lunch-break *see* **lunch-hour**

luncheon ['lʌntʃn] *n.* *(formal)* Mittagessen, *das*

luncheon: ~ meat *n.* Frühstücksfleisch, *das;* **~ voucher** *n.* *(Brit.)* Essenmarke, *die*

lunch: ~-hour *n.* Mittagspause, *die;* **~-time** *n.* Mittagszeit, *die;* **at ~-time** mittags

lung [lʌŋ] *n.* Lunge, *die;* *(right or left)* Lungenflügel, *der;* **~s** *pl.* Lunge, *die;* **have good/weak ~s** eine gute *od.* kräftige/schwache Lunge haben

'lung cancer *n.* *(Med.)* Lungenkrebs, *der*

lunge [lʌndʒ] **1.** *n.* **a)** *(Sport)* Ausfall, *der;* **b)** *(sudden forward movement)* Sprung nach vorn. **2.** *v. i.* **a)** *(Sport)* einen Ausfall machen **(at gegen); b) ~ at sb. with a knife** jmdn. mit einem Messer angreifen

lupin ['lu:pɪn] *n.* [Edel]lupine, *die*

[1]**lurch** [lɜ:tʃ] *n.* **leave sb. in the ~:** jmdn. im Stich lassen; jmdn. hängenlassen *(ugs.)*

[2]**lurch 1.** *n.* Rucken, *das;* *(of ship)* Schlingern, *das.* **2.** *v. i.* rucken; ‹*Betrunkener:*› torkeln; ‹*Schiff:*› schlingern

lure [ljʊə(r), lʊə(r)] **1.** *v. t.* locken; **~ away from/out of/into sth.** von etw. fortlocken/aus etw. [heraus]locken/in etw. *(Akk.)* [hinein]locken. **2.** *n.* *(Hunting)* Lockvogel, *der;* *(fig.)* Lockmittel, *das*

lurid ['ljʊərɪd, 'lʊərɪd] *adj.* **a)** *(ghastly)* gespenstisch; *(highly coloured)* grell ‹*Licht, Schein, Himmel*›; **b)** *(fig.)* *(horrifying)* gräßlich; schaurig; *(sensational)* reißerisch *(abwertend)*

lurk [lɜ:k] *v. i.* **a)** lauern; ‹*Raubtier:*› auf Lauer liegen; **b)** *(fig.)* ~ **in sb.'s** *or* **at the back of sb.'s mind** ‹*Zweifel, Verdacht, Furcht:*› an jmdm. nagen

luscious ['lʌʃəs] *adj.* **a)** *(sweet in taste or smell)* köstlich [süß]; saftig [süß] ‹*Obst*›; **b)** üppig ‹*Figur, Kurven*›; knackig *(ugs.)* ‹*Mädchen*›

lush [lʌʃ] *adj.* saftig ‹*Wiese*›; grün ‹*Tal*›; üppig ‹*Vegetation*›

lust [lʌst] **1.** *n.* **a)** *(sexual drive)* Sinnenlust, *die;* sinnliche Begierde; **b)** *(passionate desire)* Gier, *die* **(for** nach). **2.** *v. i.* **~ after** [lustvoll] begehren *(geh.);* **he ~s after ...:** es gelüstet ihn nach ... *(geh.)*

lustful ['lʌstfl] *adj.* lüstern *(geh.)*

lustily ['lʌstɪlɪ] *adv.* kräftig; aus voller Kehle ‹*rufen, singen*›

lustre ['lʌstə(r)] *n.* *(Brit.)* **a)** Schimmer, *der;* [schimmernder] Glanz; **b)** *(fig.: splendour)* Glanz, *der;* **add ~ to sth.** einer Sache *(Dat.)* Glanz verleihen

lusty ['lʌstɪ] *adj.* **a)** *(healthy)* gesund; *(strong, powerful)* kräftig; **b)** *(vigorous)* herzhaft ‹*Applaus, Tritt*›; tüchtig, zupackend ‹*Arbeiter*›

lute [lu:t, lju:t] *n.* *(Mus.)* Laute, *die*

Lutheran ['lu:θərən, 'lju:θərən] **1.** *adj.* lutherisch. **2.** *n.* Lutheraner, *der*/Lutheranerin, *die*

Luxembourg, Luxemburg ['lʌksəmbɜ:g] *pr. n.* Luxemburg *(das)*

luxuriant [lʌg'zjʊərɪənt, lʌk'sjʊərɪənt] *adj.* üppig ‹*Vegetation, Farbenpracht, Blattwerk*›; voll ‹*Haar*›

luxuriate [lʌg'zjʊərɪeɪt, lʌk'sjʊərɪeɪt] *v. i.* ~ **in** sich aalen in (+ *Dat.*) ‹*Sonne, Bett usw.*›

luxurious [lʌg'zjʊərɪəs, lʌk'sjʊərɪəs] *adj.* luxuriös

luxury ['lʌkʃərɪ] **1.** *n.* **a)** Luxus, *der;* **live** *or* **lead a life of ~:** ein Leben im Luxus führen; *see also* [1]**lap; b)** *(article)* Luxusgegenstand, *der;* **luxuries** Luxus, *der;* **c)** *(sth. inessential)* Luxus, *der.* **2.** *attrib. adj.* Luxus-

LW *abbr.* *(Radio)* **long wave** LW

lying ['laɪɪŋ] **1.** *adj.* **a)** *(given to falsehood)* verlogen; ~ **scoundrel** Lügenbold, *der;* **b)** *(false, untrue)* lügnerisch; lügenhaft; erlogen ‹*Geschichte*›. **2.** *n.* Lügen, *das;* **that would be ~:** das wäre gelogen. *See also* [1]**lie 2**

lymph [lɪmf] *n.* Lymphe, *die* *(fachspr.);* Gewebsflüssigkeit, *die*

lynch [lɪntʃ] *v. t.* lynchen

'lynch law *n.* Lynchjustiz, *die*

lynx [lɪŋks] *n.* *(Zool.)* Luchs, *der*

lyre ['laɪə(r)] *n.* *(Mus.)* Lyra, *die;* Leier, *die*

lyric ['lɪrɪk] **1.** *adj.* lyrisch; ~ **poet** Lyriker, *der*/Lyrikerin, *die;* ~ **poetry** Lyrik, *die.* **2.** *n.* **a)** *(poem)* lyrisches Gedicht; **b)** *in pl.* *(of song)* Text, *der*

lyrical ['lɪrɪkl] *adj.* **a)** lyrisch; **b)** *(coll.: enthusiastic)* gefühlvoll; **become** *or* **wax** ~ **about sth.** über etw. *(Akk.)* ins Schwärmen geraten

lyricism ['lɪrɪsɪzm] *n.* Lyrismus, *der*

M

M, m [em] *n., pl.* **Ms** *or* **M's** M, m, *das*

m. *abbr.* **a) male** männl.; **b) masculine** m.; **c) married** verh.; **d) metre[s]** m; **e) milli-** m; **f) million[s]** Mill.; **g) minute[s]** Min.

MA *abbr.* **Master of Arts** M.A.; *see also* **B.Sc.**

ma [mɑ:] *n.* *(coll.)* Mama, *die;* Mutti, *die (fam.)*

mac *see* **mack**

macabre [mə'kɑ:br] *adj.* makaber

macaroni [mækə'rəʊnɪ] *n.* Makkaroni *Pl.*

macaroni 'cheese *n.* *(Brit.)* Käsemakkaroni *Pl.*

macaroon [mækə'ru:n] *n.* Makrone, *die*

¹mace [meɪs] *n.* **a)** *(Hist.: weapon)* Keule, *die;* **b)** *(staff of office)* Amtsstab, *der*

²mace *n.* *(Bot., Cookery)* Mazis, *der;* Muskatblüte, *die*

machete [mə'tʃetɪ, mə'tʃeɪtɪ] *n.* Machete, *die;* Buschmesser, *das*

machiavellian [mækɪə'velɪən] *adj.* machiavellistisch

machination [mækɪ'neɪʃn, mæʃɪ'neɪʃn] *n.* Machenschaft, *die*

machine [mə'ʃi:n] **1.** *n.* **a)** Maschine, *die;* **b)** *(bicycle)* [Fahr]rad, *das;* *(motor cycle)* Maschine, *die (ugs.);* **c)** *(computer)* Computer, *der;* **d)** *(fig.: person)* Roboter, *der;* Maschine, *die;* **e)** *(system of organization)* Apparat, *der.* **2.** *v. t.* *(make with ~)* maschinell herstellen; *(operate on with ~)* maschinell bearbeiten ‹*Werkstück*›; *(sew)* mit *od.* auf der Maschine nähen

machine: ~ **code** *n.* *(Computing)* Maschinensprache, *die;* **~-gun** *n.* Maschinengewehr, *das;* **~-made** *adj.* maschinell hergestellt; **~-minder** [mə'ʃi:nmaɪndə(r)] *n.* Maschinenwärter, *der;* **~-pistol** *n.* Maschinenpistole, *die;* **~-readable** *adj.* *(Computing)* maschinenlesbar

machinery [mə'ʃi:nərɪ] *n.* **a)** *(machines)* Maschinen *Pl.;* **b)** *(mechanism)* Mechanismus, *der;* **c)** *(organized system)* Maschinerie, *die*

machinist [mə'ʃi:nɪst] *n.* *(who makes machinery)* Maschinenbauer, *der;* *(who controls machinery)* Maschinist, *der*/Maschinistin, *die;* **[sewing-]~:** [Maschinen]näherin, *die*/-näher, *der*

machismo [mə'tʃɪzməʊ, mə'kɪzməʊ] *n., no pl.* Machismo, *der;* Männlichkeitswahn, *der*

macho ['mætʃəʊ] **1.** *n., pl.* **~s** Macho, *der.* **2.** *adj.* Macho-; **he is really ~:** er ist wirklich ein Macho

mack [mæk] *n.* *(Brit. coll.)* Regenmantel, *der*

mackerel ['mækərl] *n., pl. same or* **~s** *(Zool.)* Makrele, *die*

mackintosh ['mækɪntɒʃ] *n.* Regenmantel, *der*

macro- ['mækrəʊ] *in comb.* makro-/Makro-

macroscopic [mækrəʊ'skɒpɪk] *adj.* makroskopisch

mad [mæd] *adj.* **a)** *(insane)* geisteskrank; irr ‹*Blick, Ausdruck*›; **you must be ~!** du bist wohl verrückt! *(ugs.);* **b)** *(frenzied)* wahnsinnig; verrückt *(ugs.);* **it's one ~ rush** *(coll.)* es ist eine einzige Hetze; **drive sb. ~:** jmdn. um den Verstand bringen *od.* *(ugs.)* verrückt machen; **c)** *(foolish)* verrückt *(ugs.);* **that was a ~ thing to do** das war eine Dummheit *od.* *(ugs.)* verrückt; **d)** *(very enthusiastic)* **be/go ~ about** *or* **on sb./sth.** auf jmdn./etw. wild sein/werden *(ugs.);* **be ~ keen on sth.** *(sl.)* auf etw. *(Akk.)* ganz scharf *od.* wild sein *(ugs.);* **e)** *(coll.: annoyed)* ~ **[with** *or* **at sb.]** sauer [auf jmdn.] *(ugs.);* **f)** *(with rabies)* toll[wütig]; **[run** *etc.***] like ~:** wie wild *od.* wie ein Wilder/eine Wilde *(ugs.)* [laufen *usw.*]

Madagascan [mædə'gæskən] **1.** *adj.* madagassisch. **2.** *n.* Madagasse, *der*/Madagassin, *die*

Madagascar [mædə'gæskə(r)] *pr. n.* Madagaskar *(das)*

madam ['mædəm] *n.* **a)** *(formal address)* gnädige Frau; **M~ Chairman** Frau Vorsitzende; **Dear M~** *(in letter)* Sehr verehrte gnädige Frau; **b)** *(euphem.: woman brothel-keeper)* Bordellwirtin, *die;* Puffmutter, *die (salopp);* **c)** *(derog.: conceited, pert young woman)* Kratzbürste, *die (ugs. scherzh.)*

'madcap 1. *adj.* unbesonnen. **2.** *n.* Heißsporn, *der*

madden ['mædn] *v. t.* *(irritate)* [ver]ärgern

maddening ['mædənɪŋ] *adj.* **a)** *(irritating, tending to infuriate)* [äußerst] ärgerlich; **b)** *(tending to craze)* unerträglich

made *see* **make 1, 2**

Madeira [mə'dɪərə] **1.** *n.* Madeira[wein], *der.* **2.** *pr. n.* Madeira *(das)*

made-to-'measure *attrib. adj.* Maß-; **a ~ suit** ein Maßanzug *od.* maßgeschneiderter Anzug; *see also* **measure 1a**

'made-up *attrib. adj.* erfunden ‹*Geschichte*›

'madhouse *n.* Irrenanstalt, *die;* Irrenhaus, *das;* *(fig.)* Tollhaus, *das*

madly ['mædlɪ] *adv.* **a)** wie ein Verrückter/eine Verrückte *(ugs.);* **b)** *(coll.: passionately, extremely)* wahnsinnig *(ugs.)*

madman ['mædmən] *n., pl.* **madmen** ['mædmən] Wahnsinnige, *der;* Irre, *der*

madness ['mædnɪs] *n., no pl.* Wahnsinn, *der*

madonna [mə'dɒnə] *n.* *(Art, Relig.)* Madonna, *die*

madrigal ['mædrɪgl] *n.* *(Lit., Mus.)* Madrigal, *das*

'madwoman *n.* Wahnsinnige, *die;* Irre, *die*

maelstrom ['meɪlstrəm] *n.* *(lit. or fig.)* Ma[h]lstrom, *der;* Strudel, *der;* Sog, *der*

Mafia ['mæfɪə] *n.* Mafia, *die*

magazine [mægə'zi:n] *n.* **a)** *(periodical)* Zeitschrift, *die;* *(news ~, fashion ~, etc.)* Magazin, *das;* **b)** *(Mil.: store)* *(for arms)* Waffenkammer, *die;* *(for ammunition)* Munitionsdepot, *das;* *(for explosives)* Sprengstofflager, *das;* **c)** *(Arms, Photog.)* Magazin, *das*

magenta [mə'dʒentə] *n.* *(colour)* Magenta, *das*

maggot ['mægət] *n.* Made, *die*

magic ['mædʒɪk] **1.** *n.* **a)** *(witchcraft, lit. or fig.)* Magie, *die;* **do ~:** zaubern; **as if by ~:** wie durch Zauberei; **black ~:** Schwarze

Magie; **work like ~:** wie ein Wunder wirken; **b)** *(fig.: charm, enchantment)* Zauber, *der.* **2.** *adj.* **a)** *(of ~)* magisch ‹*Eigenschaft, Kraft*›; *(resembling ~)* zauberhaft; *(used in ~)* Zauber‹*spruch, -trank, -wort, -bann*›; **b)** *(fig.: producing surprising results)* wunderbar

magical ['mædʒɪkl] *adj.* *(of magic)* magisch; *(resembling magic)* zauberhaft; **the effect was ~:** das wirkte [wahre] Wunder

magic 'carpet *n.* fliegender Teppich

magician [mə'dʒɪʃn] *n.* *(lit. or fig.)* Magier, *der*/Magierin, *die;* *(conjurer)* Zauberer, *der*/Zauberin, *die;* **I'm not a ~:** ich kann doch nicht zaubern *(ugs.)*

magic 'wand *n.* Zauberstab, *der*

magistrate ['mædʒɪstreɪt] *n.* Friedensrichter, *der*/Friedensrichterin, *die;* **~s' court** ≈ Schiedsgericht, *das*

magnanimous [mæg'nænɪməs] *adj.*, **magnanimously** [mæg'nænɪməslɪ] *adv.* großmütig (**towards** gegen)

magnate ['mægneɪt] *n.* Magnat, *der*/Magnatin, *die*

magnesium [mæg'ni:zɪəm] *n.* *(Chem.)* Magnesium, *das*

magnet ['mægnɪt] *n.* *(lit. or fig.)* Magnet, *der*

magnetic [mæg'netɪk] *adj.* *(lit. or fig.)* magnetisch; *(fig.: very attractive)* sehr anziehend, unwiderstehlich ‹*Person*›

magnetic: ~ 'disc *see* **disc c;** **~ 'field** *n.* *(Phys.)* Magnetfeld, *das;* **~ 'pole** *n.* *(Phys.)* Magnetpol, *der;* *(Geog.)* magnetischer Pol; **~ 'tape** *n.* Magnetband, *das*

magnetise *see* **magnetize**

magnetism ['mægnɪtɪzm] *n.* **a)** *(force, lit. or fig.)* Magnetismus, *der;* **b)** *(fig.: personal charm and attraction)* Attraktivität, *die;* Anziehungskraft, *die*

magnetize ['mægnɪtaɪz] *v.t.* magnetisieren; *(fig.)* in seinen Bann schlagen

magnification [mægnɪfɪ'keɪʃn] *n.* Vergrößerung, *die*

magnificence [mæg'nɪfɪsəns] *n.*, *no pl.* *(splendour)* Prunk, *der;* Pracht, *die;* *(grandeur)* Stattlichkeit, *die;* Großartigkeit, *die;* *(beauty)* Herrlichkeit, *die;* *(lavish display)* Pracht, *die;* Üppigkeit, *die*

magnificent [mæg'nɪfɪsənt] *adj.* **a)** *(stately, sumptuously constructed or adorned)* prächtig; prachtvoll; *(sumptuous)* prunkvoll; grandios, großartig ‹*Pracht, Herrlichkeit, Anblick*›; *(beautiful)* herrlich ‹*Garten, Umgebung, Kleidung, Vorhang, Kunstwerk, Wetter, Gestalt*›; *(lavish)* üppig ‹*Freigebigkeit, Mahl*›; **b)** *(coll.: fine, excellent)* fabelhaft *(ugs.)*

magnify ['mægnɪfaɪ] *v.t.* **a)** vergrößern; **b)** *(exaggerate)* aufbauschen; übertrieben darstellen ‹*Gefahren*›

'magnifying glass *n.* Lupe, *die;* Vergrößerungsglas, *das*

magnitude ['mægnɪtju:d] *n.* **a)** *(largeness, vastness)* Ausmaß, *das;* *(of explosion, earthquake)* Stärke, *die;* **b)** *(size)* Größe, *die;* **order of ~:** Größenordnung, *die;* **problems of this ~:** Probleme dieser Größenordnung; **c)** *(importance)* Wichtigkeit, *die;* **d)** *(Astron.)* Helligkeit, *die*

magpie ['mægpaɪ] *n.* *(Ornith.)* Elster, *die*

mahogany [mə'hɒgənɪ] *n.* **a)** *(wood)* Mahagoni[holz], *das;* *attrib.* Mahagoni-; **b)** *(tree)* Mahagonibaum, *der;* **c)** *(colour)* Mahagonibraun, *das*

maid [meɪd] *n.* **a)** *(servant)* Dienstmädchen, *das;* Dienstmagd, *die (veralt.);* **b)** *(young unmarried woman, virgin)* Jungfrau, *die;* **c)** *(arch./poet./rhet.: young woman, girl)* Maid, *die (dichter. veralt.). See also* **old maid**

maiden ['meɪdn] **1.** *n.* Jungfrau, *die.* **2.** *adj.* *(first)* **~ voyage/speech** Jungfernfahrt/-rede, *die*

maiden: ~head *n.* **a)** *(virginity)* Jungfräulichkeit, *die;* **b)** *(Anat.)* Jungfernhäutchen, *das;* **~ name** *n.* Mädchenname, *der*

maid: ~ of 'honour *n.*, *pl.* **~s of honour a)** *(attendant of queen or princess)* Hof- *od.* Ehrendame, *die;* **b)** *(Amer.: chief bridesmaid)* Brautjungfer, *die;* **~servant** *n.* *(arch.)* Hausangestellte, *die;* Hausmädchen, *das*

mail [meɪl] **1.** *n.* **a)** *see* **²post 1; b)** *(vehicle carrying ~)* Postbeförderungsmittel, *das;* *(train)* Postzug, *der.* **2.** *v.t. see* **²post 2 a**

mail: ~-bag *n.* *(postman's bag)* Zustelltasche, *die;* *(sack for transporting ~)* Postsack, *der;* **~box** *n.* *(Amer.)* Briefkasten, *der;* *(slot)* Briefschlitz, *der*

mailing list ['meɪlɪŋ lɪst] *n.* Adressenliste, *die*

mail: ~man *n.* *(Amer.)* Briefträger, *der;* Postbote, *der (ugs.);* **~ order** *n.* postalische Bestellung; Mail-order, *die (Werbespr., Kaufmannsspr.);* **by ~ order** durch Bestellung *od.* Mail-order; **~-order catalogue** *n.* Versandhauskatalog, *der;* **~ shot** *n.* Versand von Werbeschriften; **~ train** *n.* Postzug, *der;* **~ van** *n.* *(Railw.)* Post- *od.* Paketwagen, *der*

maim [meɪm] *v.t.* *(mutilate)* verstümmeln; *(cripple)* zum Krüppel machen

main [meɪn] **1.** *n.* **a)** *(channel, pipe)* Hauptleitung, *die;* **~s [system]** öffentliches Versorgungsnetz; *(of electricity)* Stromnetz, *das;* **turn the gas/water off at the ~[s]** den Haupthahn [für das Gas/Wasser] abstellen; **turn the electricity off at the ~s** [den Strom] am Hauptschalter abschalten; **b)** **in the ~:** im allgemeinen; im großen und ganzen. **2.** *attrib. adj.* Haupt-; **the ~ doubt/principle** der entscheidende Zweifel/oberste Grundsatz; **the ~ thing is that ...:** die Hauptsache *od.* das Wichtigste ist, daß ...

main: ~ beam *n.* *(Motor Veh.)* **on ~ beam** aufgeblendet; **~ 'clause** *n.* *(Ling.)* Hauptsatz, *der;* **~land** ['meɪnlənd] *n.* Festland, *das;* **~ 'line** *n.* *(Railw.)* Hauptstrecke, *die;* *attrib.* **~-line station/train** Fernbahnhof/-zug, *der*

mainly ['meɪnlɪ] *adv.* hauptsächlich; in erster Linie; *(for the most part)* vorwiegend

main: ~ 'road Hauptstraße, *die;* **~spring** *n.* Hauptfeder, *die;* *(of clock, watch, etc.; also fig.)* Triebfeder, *die;* **~stay** *n.* *(Naut.)* Großstag, *das;* *(fig.)* [wichtigste] Stütze; **~stream** *n.* *(principal current)* Hauptstrom, *der;* *(fig.)* Hauptrichtung, *die;* **be in the ~stream** der Hauptrichtung angehören; **~ street** [*Brit.* -'-, *Amer.* '--] Hauptstraße, *die*

maintain [meɪn'teɪn] *v.t.* **a)** *(keep up)* aufrechterhalten; bewahren ‹*Anschein, Haltung*›; unterhalten ‹*Beziehungen, Briefwechsel*›; [beibe]halten ‹*Preise, Geschwindigkeit*›; wahren ‹*Rechte, Ruf*›; **b)** *(provide for)* **~ sb.** für jmds. Unterhalt aufkommen; **c)** *(preserve)* instand halten; warten ‹*Maschine, Gerät*›; unterhalten ‹*Straße*›; **d)** *(give aid to)* unterstützen ‹*Partei, Wohlfahrtsorganisation, Sache*›; **e)** *(assert as true)* vertreten ‹*Meinung, Lehre*›; beteuern ‹*Unschuld*›; **~ that ...:** behaupten, daß ...

maintenance ['meɪntənəns] *n.* **a)** *see* **maintain a:** Aufrechterhaltung, *die;* Bewahrung, *die;* Unterhaltung, *die;* [Beibe]halten, *das;* Wahrung, *die;* **b)** *(furnishing with means of subsistence)* Unterhaltung, *die;* **c)** *(Law: money paid to support sb.)* Unterhalt, *der;* **d)** *(preservation)* Instandhaltung, *die; (of machinery)* Wartung, *die*

maintenance: ~-free *adj.* wartungsfrei; ~ **manual** *n.* Wartungsbuch, *das*

main 'verb *n.* Hauptverb, *das*

maison[n]ette [meɪzə'net] *n.* [zweistöckige] Wohnung; Maison[n]ette, *die*

maize [meɪz] *n.* Mais, *der*

majestic [mə'dʒestɪk] *adj.* majestätisch; erhaben ‹*Erscheinung, Schönheit*›; gemessen ‹*Auftreten, Schritt*›; getragen ‹*Musik*›; *(stately)* stattlich; *(possessing grandeur)* grandios

majesty ['mædʒɪstɪ] *n.* Majestät, *die;* **Your/His/Her M~:** Eure/Seine/Ihre Majestät

major ['meɪdʒə(r)] **1.** *adj.* **a)** *attrib. (greater)* größer...; ~ **part** Großteil, *der;* **b)** *attrib. (important)* bedeutend...; *(serious)* schwer ‹*Unfall, Krankheit, Unglück, Unruhen*›; größer... ‹*Krieg, Angriff, Durchbruch*›; schwer, größer... ‹*Operation*›; **of ~ interest/importance** von größerem Interesse/von größerer Bedeutung; ~ **road** *(important)* Hauptverkehrsstraße, *die; (having priority)* Vorfahrtsstraße, *die;* **c)** *(Mus.)* Dur-; ~ **key/scale/chord** Durtonart, *die* / Durtonleiter, *die* / Durakkord, *der;* **C ~:** C-Dur; **in a ~ key** in Dur. **2.** *n.* **a)** *(Mil.)* Major, *der;* **b)** *(Amer. Univ.)* Hauptfach, *das.* **3.** *v. i. (Amer. Univ.)* ~ **in sth.** etwas als Hauptfach haben

majority [mə'dʒɒrɪtɪ] *n.* **a)** *(greater number or part)* Mehrheit, *die;* **the ~ of people think ...:** die meisten Menschen denken ...; **be in the ~:** in der Mehr- *od.* Überzahl sein; überwiegen; **b)** *(in vote)* [Stimmen]mehrheit, *die;* Majorität, *die*

majority: ~ 'rule *n.* Mehrheitsregierung, *die;* ~ **'verdict** *n.* Mehrheitsentscheid, *der*

make [meɪk] **1.** *v. t.,* **made** [meɪd] **a)** *(construct)* machen, anfertigen (**of** aus); bauen ‹*Damm, Straße, Flugzeug, Geige*›; anlegen ‹*See, Teich, Weg usw.*›; zimmern ‹*Tisch, Regal*›; basteln ‹*Spielzeug, Vogelhäuschen, Dekoration usw.*›; nähen ‹*Kleider*›; durchbrechen ‹*Türöffnung*›; *(manufacture)* herstellen; *(create)* [er]schaffen ‹*Welt*›; *(prepare)* zubereiten ‹*Mahlzeit*›; machen ‹*Frühstück, Grog*›; machen, kochen ‹*Kaffee, Tee, Marmelade*›; backen ‹*Brot, Kuchen*›; *(compose, write)* schreiben, verfassen ‹*Buch, Gedicht, Lied, Bericht*›; machen ‹*Eintrag, Zeichen, Kopie, Zusammenfassung, Testament*›; anfertigen ‹*Entwurf*›; aufsetzen ‹*Bewerbung, Schreiben, Urkunde*›; ~ **a film** einen Film drehen; ~ **a dress out of the material, ~ the material into a dress** aus dem Stoff ein Kleid machen; **a table made of wood/of the finest wood** ein Holztisch/ein Tisch aus feinstem Holz; **made in Germany** in Deutschland hergestellt; **show what one is made of** zeigen, was in einem steckt *(ugs.);* **be [simply] 'made of money** *(coll.)* im Geld [nur so] schwimmen *(ugs.);* **be 'made for sth./sb.** *(fig.: ideally suited)* wie geschaffen für etw./jmdn. sein; ~ **a bed** *(for sleeping)* ein Bett bauen *(ugs.);* ~ **the bed** *(arrange after sleeping)* das Bett machen; **have it made** *(sl.)* ausgesorgt haben *(ugs.);* **b)** *(combine into)* sich verbinden zu; bilden; **blue and yellow ~ green** aus Blau und Gelb wird Grün; **c)** *(cause to exist)* machen ‹*Ärger, Schwierigkeiten, Lärm, Aufhebens*›; ~ **enemies** sich *(Dat.)* Feinde machen *od.* schaffen; ~ **time for doing** *or* **to do sth.** sich *(Dat.)* die Zeit dazu nehmen, etw. zu tun; **d)** *(result in, amount to)* machen ‹*Unterschied, Summe*›; ergeben ‹*Resultat*›; **two and two ~ four** zwei und zwei ist *od.* macht *od.* sind vier; **they ~ a handsome pair** sie geben ein hübsches Paar ab; **qualities that ~ a man** Eigenschaften, die einen Mann ausmachen; **e)** *(establish, enact)* bilden ‹*Gegensatz*›; treffen ‹*Unterscheidung, Übereinkommen*›; ziehen ‹*Vergleich, Parallele*›; erlassen ‹*Gesetz, Haftbefehl*›; aufstellen ‹*Regeln, Behauptung*›; stellen ‹*Forderung*›; geben ‹*Bericht*›; schließen ‹*Vertrag*›; vornehmen ‹*Zahlung*›; machen ‹*Geschäft, Vorschlag, Geständnis*›; erheben ‹*Anschuldigung, Protest, Beschwerde*›; **f)** *(cause to be or become)* ~ **angry/happy/known** *etc.* wütend/glücklich/bekannt *usw.* machen; ~ **sb. captain** jmdn. zum Kapitän machen; ~ **a star of sb.** aus jmdm. einen Star machen; ~ **a friend of sb.** sich mit jmdm. anfreunden; ~ **oneself heard/respected** sich *(Dat.)* Gehör/Respekt verschaffen; ~ **oneself understood** sich verständlich machen; **shall we ~ it Tuesday then?** sagen wir also Dienstag?; **that ~s it one pound exactly** das macht genau ein Pfund; ~ **it a shorter journey by doing sth.** die Reise abkürzen, indem man etw. tut; **g)** ~ **sb. do sth.** *(cause)* jmdn. dazu bringen, etw. zu tun; *(compel)* jmdn. zwingen, etw. zu tun; ~ **sb. repeat the sentence** jmdn. den Satz wiederholen lassen; **be made to do sth.** etw. tun müssen; *(be compelled)* gezwungen werden, etw. zu tun; ~ **oneself do sth.** sich überwinden, etw. zu tun; **what ~s you think that?** wie kommst du darauf?; **h)** *(form, be counted as)* **this ~s the tenth time you've failed** das ist nun [schon] das zehnte Mal, daß du versagt hast; **will you ~ one of the party?** wirst du dabei *od. (ugs.)* mit von der Partie sein?; **i)** *(serve for)* abgeben; **this story ~s good reading** diese Geschichte ist guter Lesestoff; **j)** *(become by development or training)* **the site would ~ a good playground** der Platz würde einen guten Spielplatz abgeben; **he will ~ a good officer** aus ihm wird noch ein guter Offizier; **k)** *(gain, acquire, procure)* machen ‹*Vermögen, Profit, Verlust*›; machen *(ugs.)* ‹*Geld*›; verdienen ‹*Lebensunterhalt*›; sich *(Dat.)* erwerben ‹*Ruf*›; *(obtain as result)* kommen zu *od.* auf, herausbekommen ‹*Ergebnis, Endsumme*›; **how much did you ~?** wieviel hast du verdient?; **that ~s one pound exactly** das macht genau ein Pfund; **l)** machen ‹*Geste, Bewegung, Verbeugung*›; machen ‹*Reise, Besuch, Ausnahme, Fehler, Angebot, Entdeckung, Witz, Bemerkung*›; begehen ‹*Irrtum*›; vornehmen ‹*Änderung, Stornierung*›; vorbringen ‹*Beschwerde*›; tätigen, machen ‹*Einkäufe*›; geben ‹*Versprechen, Kommentar*›; halten ‹*Rede*›; ziehen ‹*Vergleich*›; durchführen, machen ‹*Experiment, Analyse, Inspektion*›; *(wage)* führen ‹*Krieg*›; *(accomplish)* schaffen ‹*Strecke pro Zeiteinheit*›; **m)** ~ **much of sth.** etw. betonen; ~ **little of sth.** *(play sth. down)* etw. herunterspielen; **they could ~ little of his letter** *(understand)* sie konnten mit seinem Brief nicht viel anfangen; **I don't know what to ~ of him/it** ich werde aus ihm/daraus nicht schlau *od.* klug; **what do you ~ of him?** was hältst du von ihm?; wie schätzt du ihn ein?; **n)** *(arrive at)* erreichen ‹*Bestimmungsort*›; *(coll.: catch)* [noch] kriegen *(ugs.)* ‹*Zug usw.*›; ~ **it** *(succeed in arriving)* es schaffen; **o) sth. ~s or**

breaks *or* **mars sb.** etw. entscheidet über jmds. Glück oder Verderben *(Akk.)*; ~ **sb.'s day** jmdm. einen glücklichen Tag bescheren; **p)** *(consider to be)* **What do you ~ the time? – I ~ it five past eight** Wie spät hast du es *od.* ist es bei dir? – Auf meiner Uhr ist es fünf nach acht; **q)** ~ **'do** vorliebnehmen; ~ **'do with/without sth.** mit/ohne etw. auskommen. **2.** *v.i.*, **made a)** *(proceed)* ~ **toward sth./sb.** auf etw./jmdn. zusteuern; **b)** *(act as if with intention)* ~ **to do sth.** Anstalten machen, etw. zu tun; ~ **as if** *or* **as though to do sth.** so tun, als wolle man etw. tun. **3.** *n.* **a)** *(kind of structure)* Ausführung, *die*; *(of clothes)* Machart, *die*; **b)** *(type of manufacture)* Fabrikat, *das*; *(brand)* Marke, *die*; ~ **of car** Automarke, *die*; **c) on the ~** *(sl.: intent on gain)* hinter dem Geld her *(abwertend)*

~ **for** *v.t.* **a)** *(move towards)* zusteuern auf (+ *Akk.*); zuhalten auf (+ *Akk.*); *(rush towards)* losgehen auf (+ *Akk.*); zustürzen auf (+ *Akk.*); ~ **for home** heimwärts steuern; **b)** *(be conducive to)* führen zu, herbeiführen ‹*gute Beziehungen, Erfolg, Zuversicht*›

~ **'off** *v.i.* sich davonmachen

~ **'off with** *v.t.* ~ **off with sb./sth.** sich mit jmdm./etw. [auf und] davonmachen

~ **'out 1.** *v.t.* **a)** *(write)* ausstellen ‹*Scheck, Dokument, Rechnung*›; aufstellen ‹*Liste*›; **b)** *(claim, assert)* behaupten; **you ~ me out to be a liar** du stellst mich als Lügner hin; **how do you ~ that out?** wie kommst du darauf?; *see also* **[1]case d; c)** *(understand)* verstehen; **d)** *(manage to see or hear)* ausmachen; *(manage to read)* entziffern; **e)** *(pretend)* vorgeben. **2.** *v.i.* *(coll.: ~ progress)* zurechtkommen **(at** bei)

~ **'over** *v.t.* *(transfer)* übereignen, überschreiben ‹*Geld, Geschäft, Eigentum*› **(to** *Dat.*)

~ **'up 1.** *v.t.* **a)** *(replace)* ausgleichen ‹*Fehlmenge, Verluste*›; ~ **up lost ground/time** Boden gut- *od.* wettmachen *(ugs.)*/den Zeitverlust aufholen; **b)** *(complete)* komplett machen; **c)** *(prepare, arrange)* zubereiten ‹*Arznei usw.*›; *(process material)* verarbeiten **(into** zu); **d)** *(apply cosmetics to)* schminken; ~ **up one's face/eyes** sich schminken/sich *(Dat.)* die Augen schminken; **e)** *(assemble, compile)* zusammenstellen; aufstellen ‹*Liste usw.*›; bilden ‹*ein Ganzes*›; **f)** *(invent)* erfinden; sich *(Dat.)* ausdenken; **g)** *(reconcile)* beilegen ‹*Streit, Meinungsverschiedenheit*›; **h)** *(form, constitute)* bilden; **be made up of ...:** bestehen aus ... **2.** *v.i.* **a)** *(apply cosmetics etc.)* sich schminken; **b)** *(be reconciled)* sich wieder vertragen

~ **'up for** *v.t.* **a)** *(outweigh, compensate)* wettmachen; **b)** *(~ amends for)* wiedergutmachen; **c)** ~ **up for lost time** Versäumtes nachholen *od.* *(ugs.)* wettmachen

~ **'up to** *v.t.* **a)** *(raise to, increase to)* bringen auf (+ *Akk.*); **b)** *(coll.: act flirtatiously towards)* sich heranmachen an (+ *Akk.*) *(ugs.)*; **c)** *(coll.: give compensation to)* ~ **it/this up to sb.** jmdn. dafür entschädigen

'make-believe 1. *n.* **it's only ~:** das ist bloß Phantasie. **2.** *adj.* nicht echt; **a ~ world/story** eine Scheinwelt/Phantasiegeschichte

maker ['meɪkə(r)] *n.* **a)** *(manufacturer)* Hersteller, *der*; **b) M~** *(God)* Schöpfer, *der*

make: ~shift *adj.* behelfsmäßig; **a ~shift shelter/bridge** eine Behelfsunterkunft/-brücke; **~-up** *n.* **a)** *(Cosmetics)* Make-up, *das*; *(Theatre)* Maske, *die*; **put on one's ~-up** Make-up auflegen; sich schminken; *(Theatre)* Maske machen; **b)** *(composition)* Zusammensetzung, *die*; **c)** *(character, temperament)* Veranlagung, *die*; **physical ~-up** Konstitution, *die*

making ['meɪkɪŋ] *n.* **a)** *(production)* Herstellung, *die*; **in the ~:** im Entstehen; im Werden; **be the ~ of victory/sb.'s career/sb.'s future** zum Sieg/zu jmds. Karriere führen/jmds. Zukunft sichern; **b)** *in pl.* *(qualities)* Anlagen; Voraussetzungen; **have all the ~s of sth.** alle Voraussetzungen für etw. haben; **have the ~s of a leader** über Führerqualitäten verfügen; das Zeug zum Führer haben *(ugs.)*

maladjusted [mælə'dʒʌstɪd] *adj.* *(Psych., Sociol.)* **[psychologically/socially]** ~: verhaltensgestört

maladroit [mælə'drɔɪt, 'mælədrɔɪt] *adj.* ungeschickt; taktlos ‹*Bemerkung*›

malady ['mælədɪ] *n.* Leiden, *das*; *(fig.: of society, epoch)* Übel, *das*

malaise [mə'leɪz] *n.* Unwohlsein, *das*; *(feeling of uneasiness)* Unbehagen, *das*

malaria [mə'leərɪə] *n.* Malaria, *die*

Malay [mə'leɪ] **1.** *adj.* malaiisch; **sb. is ~:** jmd. ist Malaie/Malaiin. **2.** *n.* **a)** *(person)* Malaie, *der*/Malaiin, *die*; **b)** *(language)* Malaiisch, *das*; *see also* **English 2a**

Malaya [mə'leɪə] *pr. n.* Malaya *(das)*

Malayan [mə'leɪən] *see* **Malay 1, 2a**

Malaysia [mə'leɪzɪə] *pr. n.* Malaysia *(das)*

Malaysian [mə'leɪzɪən] **1.** *adj.* malaysisch. **2.** *n.* Malaysier, *der*/Malaysierin, *die*

male [meɪl] **1.** *adj.* männlich; Männer‹*stimme, -chor, -verein*›; ~ **child / dog / cat / doctor / nurse** Junge / Rüde / Kater / Arzt / Krankenpfleger, *der*. **2.** *n.* *(person)* Mann, *der*; *(foetus, child)* Junge, *der*; *(animal)* Männchen, *das*

malevolent [mə'levələnt] *adj.* böse ‹*Macht, Tat*›; übelwollend ‹*Gott*›; boshaft, hämisch ‹*Gelächter*›; böswillig ‹*Lüge*›; boshaft ‹*Person*›

malformed [mæl'fɔ:md] *adj.* *(Med.)* mißgebildet

malfunction [mæl'fʌŋkʃn] **1.** *n.* Störung, *die*; *(Med.)* Dysfunktion, *die* *(fachspr.)*; Funktionsstörung, *die*. **2.** *v.i.* ‹*Mechanismus, System, Gerät:*› nicht richtig funktionieren; ‹*Prozeß, Vorgang:*› nicht richtig ablaufen

malice ['mælɪs] *n.* Bosheit, *die*; Böswilligkeit, *die*; **bear ~ to** *or* **towards** *or* **against sb.** jmdm. übelwollen

malicious [mə'lɪʃəs] *adj.* **a)** böse ‹*Klatsch, Tat, Person, Wort*›; böswillig ‹*Gerücht, Lüge, Verleumdung*›; boshaft ‹*Person*›; hämisch ‹*Vergnügen, Freude*›; **b)** *(Law)* böswillig ‹*Sachbeschädigung, Verleumdung*›

malign [mə'laɪn] **1.** *v.t.* *(slander)* verleumden; *(speak ill of)* schlechtmachen; ~ **sb.'s character** jmdm. Übles nachsagen. **2.** *adj.* **a)** *(injurious)* böse ‹*Macht, Geist*›; schlecht, unheilvoll ‹*Eigenschaft, Einfluß*›; **b)** *(malevolent)* böse ‹*Absicht*›; niederträchtig ‹*Motiv*›

malignant [mə'lɪgnənt] *adj.* **a)** *(Med.)* maligne *(fachspr.)*, bösartig ‹*Krankheit, Geschwür*›; ~ **cancer** Karzinom, *das* *(fachspr.)*; Krebs, *der*; **b)** *(harmful)* böse ‹*Macht*›; ungünstig ‹*Einfluß*›; **c)** *(feeling or showing ill will)* böse ‹*Geist, Zunge, Klatsch*›

malinger [mə'lɪŋgə(r)] *v.i.* simulieren

mall [mæl, mɔ:l] *n.* *(esp. Amer.: shopping centre)* Einkaufszentrum, *das*

malleable ['mælɪəbl] *adj.* formbar ‹*Material, Person*›

mallet ['mælɪt] *n.* **a)** *(hammer)* Holzhammer, *der;* Schlegel, *der; (of stonemason)* Klöpfel, *der; (of carpenter)* Klopfholz, *das;* **b)** *(Croquet)* Hammer, *der; (Polo)* Schläger, *der*

malnutrition [mælnju:'trɪʃn] *n.* Unterernährung, *die*

malpractice [mæl'præktɪs] *n. (wrongdoing)* Übeltat, *die (geh.)*

malt [mɔ:lt, mɒlt] **1.** *n.* Malz, *das.* **2.** *v. t.* mälzen ‹*Gerste*›

Malta ['mɔ:ltə, 'mɒltə] *pr. n.* Malta *(das)*

Maltese [mɔ:l'ti:z, mɒl'ti:z] **1.** *adj.* maltesisch; **sb. is ~:** jmd. ist Malteser/Malteserin. **2.** *n., pl. same* **a)** *(person)* Malteser, *der*/Malteserin, *die;* **b)** *(language)* Maltesisch *(das)*

maltreat [mæl'tri:t] *v. t.* mißhandeln

maltreatment [mæl'tri:tmənt] *n.* Mißhandlung, *die*

malt 'whisky *n.* Malzwhisky, *der*

mamma ['mæmə] *n. (coll./child lang.)* Mama, *die (fam.);* Mami, *die (fam.)*

mammal ['mæml̩] *n. (Zool.)* Säugetier, *das;* Säuger, *der*

mammoth ['mæməθ] **1.** *n. (Palaeont.)* Mammut, *das.* **2.** *adj.* Mammut-; gigantisch ‹*Vorhaben*›

man [mæn] **1.** *n., pl.* **men** [men] **a)** *no art., no pl. (human being, person)* Mensch, *der; (the human race)* der Mensch; **~ is a political animal** der Mensch ist ein politisches Wesen; **what can a ~ do?** was kann man tun?; **every ~ for himself** rette sich, wer kann; **any ~ who ...:** wer ...; jeder, der ...; **no ~:** niemand; **[all] to a ~:** allesamt; **the ~ in** *or (Amer.)* **on the street** der Mann auf der Straße; **the rights of ~:** die Menschenrechte; **b)** *(adult male, individual male)* Mann, *der;* **every ~, woman, and child** ausnahmslos jeder *od.* alle; **the [very] ~ for sth.** der richtige Mann *od.* der Richtige für etw.; **make a ~ out of sb.** *(fig.)* einen Mann aus jmdm. machen; **a ~ of property/great strength** ein vermögender/sehr kräftiger Mann; **men's clothing/outfitter** Herrenkleidung, *die*/Herrenausstatter, *der;* **be ~ enough to ...:** Manns genug sein, um zu ...; **sth. sorts out** *or* **separates the men from the boys** *(coll.)* an etw. *(Dat.)* zeigt sich, wer ein ganzer Kerl ist und wer nicht; **be one's own ~:** seine eigenen Vorstellungen haben; **men's toilet** Herrentoilette, *die;* **'Men'** „Herren"; **my [good] ~:** mein Guter; **c)** *(husband)* Mann, *der;* **be ~ and wife** verheiratet sein; **d)** *(Chess)* Figur, *die; (Draughts)* Stein, *der;* **e)** *(sl.: as int. of surprise or impatience, as mode of address)* Mensch! *(salopp);* **f)** *(type of ~)* Mann, *der;* Typ, *der;* **a ~ of the people/world/of action** ein Mann des Volkes/von Welt/der Tat; **g)** *(~servant)* Diener, *der.* **2.** *v. t.,* **-nn-** bemannen ‹*Schiff, Spill*›; besetzen ‹*Büro, Stelle usw.*›; bedienen ‹*Telefon, Geschütz*›; ‹*Soldaten:*› Stellung beziehen in (+ *Dat.*) ‹*Festung*›; mit Personal besetzen ‹*Fabrik*›

manacle ['mænəkl] **1.** *n., usu. in pl.* [Hand]fessel, *die;* Kette, *die.* **2.** *v. t.* Handfesseln anlegen (+ *Dat.*)

manage ['mænɪdʒ] **1.** *v. t.* **a)** *(handle, wield)* handhaben ‹*Werkzeug, Segel, Boot*›; bedienen ‹*Schaltbrett*›; **b)** *(conduct, organize)* durchführen ‹*Operation, Unternehmen*›; erledigen ‹*Angelegenheit*›; verwalten ‹*Geld, Grundstück*›; leiten ‹*Geschäft, Büro*›; führen ‹*Haushalt*›; **c)** *(Sport etc.: be manager of)* managen, betreuen ‹*Team, Mannschaft*›; **d)** *(cope with)* schaffen; **I couldn't ~ another apple** *(coll.)* noch einen Apfel schaffe ich nicht; **we can ~ another person in the car** einer hat noch Platz im Wagen; **e)** *(succeed in achieving)* zustandebringen ‹*Lächeln*›; **f)** *(contrive)* **~ to do sth.** *(also iron.)* es fertigbringen, etw. zu tun; **he ~d to do it** es gelang ihm, es zu tun; **I'll ~ it somehow** ich werde es schon irgendwie hinkriegen *(ugs.).* **2.** *v. i.* zurechtkommen; **~ without sth.** ohne etw. auskommen; **~ on** zurecht- *od.* auskommen mit ‹*Geld, Einkommen*›; **I can ~:** es geht; **can you ~?** geht's?; geht es?

manageable ['mænɪdʒəbl] *adj.* leicht frisierbar ‹*Haar*›; fügsam ‹*Person, Tier*›; überschaubar ‹*Größe, Menge*›; lenkbar ‹*Firma*›

management ['mænɪdʒmənt] *n.* **a)** Durchführung, *die; (of a business)* Leitung, *die;* Management, *das; (of money)* Verwaltung, *die;* **b)** *(managers)* Leitung, *die;* Management, *das; (of theatre etc.)* Direktion, *die;* **the ~:** die Geschäftsleitung

manager ['mænɪdʒə(r)] *n. (of branch of shop or bank)* Filialleiter, *der*/-leiterin, *die; (of football team)* [Chef]trainer, *der*/-trainerin, *die; (of tennis-player, boxer, pop group)* Manager, *der*/Managerin, *die; (of restaurant, shop, hotel)* Geschäftsführer, *der*/-führerin, *die; (of estate, grounds)* Verwalter, *der*/Verwalterin, *die; (of department)* Leiter, *der*/Leiterin, *die; (of theatre)* Direktor, *der*/Direktorin, *die*

manageress ['mænɪdʒəres, mænɪdʒə'res] *n. (of restaurant, shop, hotel)* Geschäftsführerin, *die; see also* **manager**

managerial [mænə'dʒɪərɪəl] *adj.* führend, leitend ‹*Stellung*›; geschäftlich ‹*Aspekt, Seite*›

managing ['mænɪdʒɪŋ] *attrib. adj.* geschäftsführend; leitend; **~ director** Geschäftsführer, *der*

¹mandarin ['mændərɪn] *n.* **~ [orange]** Mandarine, *die*

²mandarin *n.* **a) M~** *(language)* Hochchinesisch, *das;* **b)** *(party leader)* Parteiboß, *der (ugs.);* [Partei]bonze, *der (abwertend);* **c)** *(bureaucrat)* Bürokrat, *der*/Bürokratin, *die (abwertend);* Apparatschik, *der (abwertend)*

mandarine ['mændəri:n] *see* **¹mandarin**

mandate ['mændeɪt] *n. (also Polit.)* Mandat, *das*

mandatory ['mændətərɪ] *adj.* obligatorisch; **be ~:** Pflicht *od.* obligatorisch sein

mandolin, mandoline [mændə'lɪn] *n. (Mus.)* Mandoline, *die*

mane [meɪn] *n. (lit. or fig.)* Mähne, *die*

'man-eating *adj.* menschenfressend ‹*Löwe, Tiger*›; **a ~ shark** ein Menschenhai

maneuver *(Amer.) see* **manœuvre**

manful ['mænfl] *adj.* mannhaft

manfully ['mænfəlɪ] *adv.* mannhaft; wie ein Mann

manganese ['mæŋgəni:z, mæŋgə'ni:z] *n. (Chem.)* Mangan, *das*

manger ['meɪndʒə(r)] *n.* Futtertrog, *der; (Bibl.)* Krippe, *die; see also* **dog 1 a**

¹mangle ['mæŋgl] **1.** *n.* Mangel, *die.* **2.** *v. t.* mangeln ‹*Wäsche*›

²mangle *v. t.* verstümmeln, [übel] zurichten ‹*Person*›; demolieren ‹*Sache*›; verstümmeln, entstellen ‹*Zitat, Musikstück*›

mango ['mæŋgəʊ] *n., pl.* **~es** *or* **~s a)** *(tree)* Mangobaum, *der;* **b)** *(fruit)* Mango[frucht], *die*

mangrove ['mæŋgrəʊv] *n. (Bot.)* Mangrovebaum, *der*

mangy ['meɪndʒɪ] *adj.* **a)** *(Vet. Med.)* räudig; **b)** *(shabby)* schäbig ‹*Teppich, Decke, Stuhl*›

man: ~handle *v. t.* **a)** *(move by human effort)* von Hand bewegen ‹*Gegenstand*›; **b)** *(handle roughly)* grob behandeln ‹*Per-*

son〉; **~hole** *n.* Mannloch, *das;* *(in tank)* Einstiegsluke, *die;* *(to cables under pavement)* Kabelschacht, *der*

manhood ['mænhʊd] *n., no pl.* Mannesalter, *das;* *(courage)* Männlichkeit, *die*

man: ~-hour *n.* Arbeitsstunde, *die;* **~-hunt** *n.* Menschenjagd, *die;* *(for criminal)* Verbrecherjagd, *die*

mania ['meɪnɪə] *n.* **a)** *(madness)* Wahnsinn, *der;* **b)** *(enthusiasm)* Manie, *die;* **~ for detective novels** Leidenschaft für Krimis

maniac ['meɪnɪæk] **1.** *adj.* wahnsinnig; krankhaft, *(geh.)* manisch 〈*Phantasie, Verlangen*〉. **2.** *n.* **a)** *(Psych.)* Besessene, *der/die;* *(madman/-woman)* Wahnsinnige, *der/die;* **b)** *(person with passion for sth.)* Fanatiker, *der*/Fanatikerin, *die*

manicure ['mænɪkjʊə(r)] **1.** *n.* Maniküre, *die;* **give sb. a ~:** jmdn. maniküren. **2.** *v. t.* maniküren

manifest ['mænɪfest] **1.** *adj.* offenkundig; offenbar 〈*Mißverständnis*〉; sichtbar 〈*Erfolg, Fortschritt*〉; sichtlich 〈*Freude*〉. **2.** *v. t.* **a)** *(show, display)* zeigen, bekunden *(geh.)* 〈*Interesse, Mißfallen, Begeisterung, Zuneigung*〉; **b)** *(reveal)* offenbaren *(meist geh.);* **~ itself** 〈*Geist:*〉 erscheinen; 〈*Natur:*〉 sich offenbaren; 〈*Krankheit:*〉 manifest werden

manifestation [mænɪfe'steɪʃn] *n.* *(of ill-will, favour, disapproval)* Ausdruck, *der;* Bezeugung, *die;* *(appearance)* Erscheinung, *die; in pl.* Erscheinungsformen; *(visible expression, sign)* [An]zeichen, *das* **(of** von)

manifestly ['mænɪfestlɪ] *adv.* offenkundig; **it is ~ unjust that ...:** es ist ganz offensichtlich ungerecht, daß ...

manifesto [mænɪ'festəʊ] *n., pl.* **~s** Manifest, *das*

manifold ['mænɪfəʊld] **1.** *adj.* *(literary)* mannigfaltig *(geh.);* vielfältig. **2.** *(Mech. Engin.)* Verteilerrohr, *das;* **[inlet] ~:** [Ansaug]krümmer, *der;* **[exhaust] ~:** [Auspuff]krümmer, *der*

manipulate [mə'nɪpjʊleɪt] *v. t.* **a)** *(also Med.)* manipulieren; **~ sb. into doing sth.** jmdn. dahin gehend manipulieren, daß er etw. tut; **b)** *(handle)* handhaben

manipulation [mənɪpjʊ'leɪʃn] *n.* **a)** *(also Med.)* Manipulation, *die;* **b)** *(handling)* Handhabung, *die*

manipulative [mə'nɪpjʊlətɪv] *adj.* manipulativ

mankind [mæn'kaɪnd] *n.* Menschheit, *die*

manly ['mænlɪ] *adj.* männlich; *(brave)* mannhaft *(geh.)*

'man-made *adj.* künstlich 〈*See, Blumen, Schlucht*〉; vom Menschen geschaffen 〈*Gesetze*〉; *(synthetic)* Kunst〈*faser, -stoff*〉

manned [mænd] *adj.* bemannt 〈*Raumschiff usw.*〉

mannequin ['mænɪkɪn] *n.* *(person)* Mannequin, *das*

manner ['mænə(r)] *n.* **a)** *(way, fashion)* Art, *die;* Weise, *die;* *(more emphatic)* Art und Weise, *die;* **in this ~:** auf diese Art und Weise; **he acted in such a ~ as to offend her** er benahm sich so, daß sie beleidigt war; **in a ~ of speaking** mehr oder weniger; **b)** *no pl.* *(bearing)* Art, *die;* *(towards others)* Auftreten, *das;* **c)** *in pl.* *(social behaviour)* Manieren *Pl.;* Benehmen, *das;* **teach sb. some ~s** jmdm. Manieren beibringen; **that's good ~s** das gehört sich so; **that's bad ~s** das gehört sich nicht; das macht man nicht; **d)** *(artistic style)* Stil, *der;* **e)** *(type)* **all ~ of things** alles mögliche; *see also* **means c**

mannered ['mænəd] *adj.* **a)** *(showing mannerism)* manieriert; **b)** *in comb.* **... -~:** mit ... Manieren *nachgestellt;* **be well-~/bad-~:** gute/schlechte Manieren haben

mannerism ['mænərɪzm] *n.* **a)** *(addiction to a manner)* Manieriertheit, *die;* **b)** *(trick of style)* Manierismus, *der;* **c)** *(in behaviour)* Eigenart, *die*

manœuvre [mə'nu:və(r)] *(Brit.)* **1.** *n.* **a)** *(Mil., Navy)* Manöver, *das;* **be/go on ~s** im Manöver sein/ins Manöver ziehen *od.* rücken; **b)** *(deceptive movement, scheme; also of vehicle, aircraft)* Manöver, *das;* **room for ~** *(fig.)* Spielraum, *der.* **2.** *v. t.* **a)** *(Mil., Navy)* führen; dirigieren; **b)** *(bring by ~s)* manövrieren; bugsieren *(ugs.)* 〈*Sperriges*〉; **~ sb./oneself/sth. into a good position** *(fig.)* jmdn./sich/etw. in eine gute Position manövrieren; **c)** *(manipulate)* beeinflussen; **~ sb. into doing sth.** jmdn. dazu bringen, etw. zu tun. **3.** *v. i.* **a)** *(Mil., Navy)* [ein] Manöver durchführen; **b)** *(move, scheme)* manövrieren; **room to ~:** Platz zum Manövrieren; *(fig.)* Spielraum, *der*

manor ['mænə(r)] *n.* **a)** *(land)* [Land]gut, *das;* **lord/lady of the ~:** Gutsherr, *der*/Gutsherrin, *die;* **b)** *(house)* Herrenhaus, das

'manor-house *see* **manor b**

'manpower *n.* **a)** *(available power)* Arbeitspotential, *das;* *(workers)* Arbeitskräfte *Pl.;* **b)** *(Mil.)* Stärke, *die*

mansion ['mænʃn] *n.* Villa, *die;* *(of lord)* Herrenhaus, *das*

man: ~-size, ~-sized *adj.* *(suitable for a man)* 〈*Mahlzeit, Steak*〉 für einen [ganzen] Mann; *(large)* groß; **~slaughter** *n.* *(Law)* Totschlag, *der*

mantel ['mæntl]: **~piece** *n.* **a)** *(above fireplace)* Kaminsims, *der od. das;* **b)** *(around fireplace)* Kamineinfassung, *die;* **~shelf** *see* **~piece a**

mantle ['mæntl] *n.* *(cloak)* Umhang, *der;* *(fig.)* Mantel, *der;* **~ of snow** Schneedecke, *die*

'man-to-man *adj.* von Mann zu Mann *nachgestellt*

manual ['mænjʊəl] **1.** *adj.* **a)** manuell; **~ worker/labourer** Handarbeiter/Schwerarbeiter, *der;* **b)** *(not automatic)* handbetrieben; 〈*Bedienung, Kontrolle, Schaltung*〉 von Hand. **2.** *n.* *(handbook)* Handbuch, *das*

manually ['mænjʊəlɪ] *adv.* manuell; von Hand; mit der Hand

manufacture [mænjʊ'fæktʃə(r)] **1.** *n.* Herstellung, *die;* **articles of foreign/British ~:** ausländische/britische Erzeugnisse. **2.** *v. t.* *(Commerc.)* herstellen; **~d goods** Fertigprodukte

manufacturer [mænjʊ'fæktʃərə(r)] *n.* Hersteller, *der;* **'~'s recommended [retail] price'** „unverbindliche Preisempfehlung"

manure [mə'njʊə(r)] **1.** *n.* *(dung)* Dung, *der;* *(fertilizer)* Dünger, *der.* **2.** *v. t.* düngen

manuscript ['mænjʊskrɪpt] *n.* **a)** Handschrift, *die;* **b)** *(not yet printed)* Manuskript, *das*

many ['menɪ] **1.** *adj.* **a)** viele; *pred.* zahlreich; **how ~ people/books?** wie viele *od.* wieviel Leute/Bücher?; **there were as ~ as 50 of them** es waren mindestens *od.* bestimmt 50; **three accidents in as ~ days** drei Unfälle in ebenso vielen *od.* ebensoviel Tagen; **there were too ~ of them** es waren zu viele *od.* zuviel; **one is too ~/there is one too ~:** einer/eine/eins ist zuviel; **he's had one too ~** *(is drunk)* er hat einen *od.* ein Glas zuviel getrunken; **b) ~ a man** so mancher; manch einer. **2.** *n.* viele [Leute]; **~ of us** viele von uns; **a good/great ~ [of them/of the books]** eine Menge/eine ganze Reihe [von ihnen/der Bücher]

Maoist ['maʊɪst] *n.* Maoist, *der;* *attrib.* maoistisch

Maori ['maʊrɪ] **1.** *n.* **a)** *(person)* Maori, *der;* **b)** *(language)* Maori, *das.* **2.** *adj.* maorisch

map [mæp] **1.** *n.* **a)** [Land]karte, *die; (street plan)* Stadtplan, *der;* **b)** *(fig. coll.)* **off the ~:** abgelegen; **wipe off the ~:** ausradieren; **[put sth./sb.] on the ~:** [etw./jmdn.] populär [machen]. **2.** *v. t.*, **-pp-** *(make ~ of)* kartographieren; *(make survey of)* vermessen

~ 'out *v. t.* im einzelnen festlegen

maple ['meɪpl] *n.* Ahorn, *der*

'maple-leaf *n.* Ahornblatt, *das*

'map-maker *n.* Kartograph, *der*/Kartographin, *die*

mar [mɑ:(r)] *v. t.*, **-rr-** verderben; entstellen ‹*Aussehen*›; stören ‹*Veranstaltung*›

Mar. *abbr.* **March** Mrz.

marathon ['mærəθən] *n.* **a)** *(race)* Marathon[lauf], *der; attrib.* Marathon‹*läufer*›; **b)** *(fig.)* Marathon, *das; attrib.* Marathon‹*rede, -spiel, -sitzung*›

marauder [mə'rɔ:də(r)] *n.* Plünderer, *der;* Marodeur, *der (Soldatenspr.); (animal)* Räuber, *der*

marble ['mɑ:bl] *n.* **a)** *(stone)* Marmor, *der (auch fig.); attrib.* Marmor-; aus Marmor *nachgestellt;* **b)** *(toy)* Murmel, *die;* **[game of] ~s** Murmelspiel, *das;* **play ~s** murmeln; [mit] Murmeln spielen; **c)** *in pl.* **not have all** *or* **have lost one's ~s** *(sl.)* nicht alle Tassen im Schrank haben *(ugs.)*

March [mɑ:tʃ] *n.* März, *der; see also* **August; hare 1**

march 1. *n.* **a)** *(Mil., Mus.; hike)* Marsch, *der; (gait)* Marschschritt, *der;* **on the ~:** auf dem Marsch; **~ past** Vorbeimarsch, *der;* Defilee, *das;* **b)** *(in protest)* **[protest] ~:** Protestmarsch, *der;* **c)** *(progress of time, events, etc.)* Gang, *der.* **2.** *v. i. (also Mil.)* marschieren; *(fig.)* fortschreiten; **forward/quick ~!** vorwärts/im Eilschritt marsch!; **~ing orders** Marschbefehl, *der;* **give sb. his/her ~ing orders** *(fig. coll.)* jmdm. den Laufpaß geben *(ugs.)*

~ 'off 1. *v. i.* losmarschieren. **2.** *v. t.* ‹*Polizei usw.*:› abführen

marcher ['mɑ:tʃə(r)] *n.* **[protest] ~:** Demonstrant, *der*/Demonstrantin, *die*

marchioness [mɑ:ʃə'nes] *n.* Marquise, *die*

mare [meə(r)] *n.* Stute, *die; see also* **shank**

margarine [mɑ:dʒə'ri:n, mɑ:gə'ri:n], *(coll.)* **marge** [mɑ:dʒ] *ns.* Margarine, *die*

margin ['mɑ:dʒɪn] *n.* **a)** *(of page)* Rand, *der;* **notes [written] in the ~:** Randbemerkungen; **b)** *(extra amount)* Spielraum, *der;* **profit ~:** Gewinnspanne, *die;* **win by a narrow/wide ~:** knapp/mit großem Vorsprung gewinnen; **~ of error** Spielraum für mögliche Fehler; **c)** *(edge)* Rand, *der;* Saum, *der (geh.)*

marginal ['mɑ:dʒɪnl] *adj.* **a)** *(barely adequate, slight)* geringfügig; unwesentlich; **b)** *(close to limit)* marginal; *(barely profitable)* kaum rentabel; **c)** knapp ‹*Wahlergebnis*›; **~ seat/constituency** *(Brit. Polit.)* wackeliger *(ugs.) od.* nur mit knapper Mehrheit gehaltener Parlamentssitz/Wahlkreis; **d)** *(of or at the edge)* Rand‹*gebiet, -bereich usw.*›

marigold ['mærɪgəʊld] *n.* Studentenblume, *die*

marijuana (marihuana) [mærɪ'hwɑ:nə] *n.* Marihuana, *das; attrib.* Marihuana‹*zigarette*›

marina [mə'ri:nə] *n.* Marina, *die;* Jachthafen, *der*

marinade [mærɪ'neɪd] **1.** *n.* Marinade, *die.* **2.** *v. t.* marinieren

marine [mə'ri:n] **1.** *adj.* **a)** *(of the sea)* Meeres-; **b)** *(of shipping)* See‹*versicherung, -recht usw.*›; **c)** *(for use at sea)* Schiffs‹*ausrüstung, -chronometer, -kessel, -turbine usw.*›. **2.** *n. (person)* Marineinfanterist, *der;* **the M~s** die Marineinfanterie; die Marinetruppen; **tell that/it to the [horse] ~s** *(coll.)* das kannst du deiner Großmutter erzählen *(ugs.)*

mariner ['mærɪnə(r)] *n.* Seemann, *der*

marionette [mærɪə'net] *n.* Marionette, *die*

marital ['mærɪtl] *adj.* ehelich ‹*Rechte, Pflichten, Harmonie*›; Ehe‹*beratung, -glück, -krach, -krise, -probleme*›; **~ status** Familienstand, *der*

maritime ['mærɪtaɪm] *adj.* **a)** *(found near the sea)* Küsten‹*bewohner, -gebiet, -stadt, -provinz*›; **b)** *(connected with the sea)* See‹*recht, -versicherung, -volk, -wesen*›

marjoram ['mɑ:dʒərəm] *n. (Bot., Cookery)* Majoran, *der*

¹mark [mɑ:k] **1.** *n.* **a)** *(trace)* Spur, *die; (of finger, foot also)* Abdruck, *der; (stain etc.)* Fleck, *der; (scratch)* Kratzer, *der;* **dirty ~:** Schmutzfleck, *der;* **leave one's/its ~ on sth.** *(fig.)* einer Sache *(Dat.)* seinen Stempel aufdrücken; **make one's/its ~** *(fig.)* sich *(Dat.)* einen Namen machen; **b)** *(affixed sign, indication, symbol)* Zeichen, *das; (in trade names)* Typ, *der (Technik);* **distinguishing ~:** Kennzeichen, *das;* **M~ 2 version/model** Version/Modell 2; **have all the ~s of sth.** alle Anzeichen von etw. haben; **be a ~ of good taste/breeding** ein Zeichen guten Geschmacks/guter Erziehung sein; **sth. is the ~ of a good writer** an etw. *(Dat.)* erkennt man einen guten Schriftsteller; **c)** *(Sch.: grade)* Zensur, *die;* Note, *die; (Sch., Sport: unit of numerical award)* Punkt, *der;* **get good/bad/35 ~s in** *or* **for a subject** gute/schlechte Noten *od.* Zensuren/35 Punkte in einem Fach bekommen; **d)** *(line etc. to indicate position)* Markierung, *die;* **e)** *(level)* Marke, *die;* **reach the 15% ~:** die 15%-Marke erreichen; **around the 300 ~:** ungefähr 300; **f)** *(Sport: starting position)* Startlinie, *die;* **on your ~s! [get set! go!]** auf die Plätze! [Fertig! Los!]; **be quick/slow off the ~:** einen guten/schlechten Start haben; *(fig.)* fix *(ugs.)*/langsam sein; **g)** *(target, desired object)* Ziel, *das;* **hit the ~** *(fig.)* ins Schwarze treffen; **be wide of the ~** *(lit. or fig.)* danebentreffen; **be close to the ~** *(fig.)* der Sache nahekommen. **2.** *v. t.* **a)** *(stain, dirty)* Flecke[n] machen auf (+ *Dat.*); schmutzig machen; *(scratch)* zerkratzen; **b)** *(put distinguishing ~ on, signal)* kennzeichnen, markieren **(with** mit); **the bottle was ~ed 'poison'** die Flasche trug die Aufschrift „Gift"; **~ an item with its price** eine Ware auszeichnen *od.* mit einem Preisschild versehen; **ceremonies to ~ the tenth anniversary** Feierlichkeiten aus Anlaß des 10. Jahrestages; **c)** *(Sch.) (correct)* korrigieren; *(grade)* benoten; zensieren; **~ an answer wrong** eine Antwort als falsch bewerten; **d) ~ time** *(Mil.; also fig.)* auf der Stelle treten; **e)** *(characterize)* kennzeichnen; charakterisieren; **f)** *(heed)* hören auf (+ *Akk.*) ‹*Person, Wort*›; **[you] ~ my words** höre auf mich; eins kann ich dir sagen; *(as a warning)* laß dir das gesagt sein; **g)** *(Brit. Sport: keep close to)* markieren *(fachspr.),* decken ‹*Gegenspieler*›

~ 'down *v. t.* **a)** *(choose as victim, lit. or fig.)* [sich *(Dat.)*] auswählen; ausersehen *(geh.);* **b)** [im Preis] herabsetzen ‹*Ware*›; herabsetzen ‹*Preis*›

~ 'off *v. t.* abgrenzen **(from** von, gegen)

~ 'out *v. t.* **a)** *(trace out boundaries of)* markieren ‹*Spielfeld*›; **b)** *(des-*

tine) vorsehen; ‹*Schicksal:*› bestimmen, ausersehen

~ **'up** *v. t.* [im Preis] heraufsetzen ‹*Ware*›; heraufsetzen ‹*Preis*›

²mark *n. (monetary unit)* Mark, *die*

marked [mɑ:kt] *adj.* **a)** *(noticeable)* deutlich ‹*Gegensatz, Unterschied, [Ver]besserung, Veränderung*›; ausgeprägt ‹*Akzent, Merkmal, Neigung*›; **b) be a ~ man** auf der schwarzen Liste stehen *(ugs.)*

markedly ['mɑ:kɪdlɪ] *adv.* eindeutig; deutlich

marker ['mɑ:kə(r)] *n.* Markierung, *die*

'marker pen *n.* Markierstift, *der*

market ['mɑ:kɪt] **1.** *n.* **a)** Markt, *der; attrib.* Markt‹*händler, -stand*›; **at the ~:** auf dem Markt; **go to ~:** auf den Markt gehen; **b)** *(demand)* Markt, *der; (area of demand)* Absatzmarkt, *der; (persons)* Abnehmer *Pl.*; **c)** *(conditions for buying and selling, trade)* Markt, *der;* **be in the ~ for sth.** an etw. *(Dat.)* interessiert sein; **come on to the ~** ‹*neue Produkte:*› auf den Markt kommen; **put on the ~:** zum Verkauf anbieten ‹*Haus*›. **2.** *v. t.* vermarkten

market: ~ 'forces *n. pl.* Kräfte des freien Marktes; **~ 'garden** *n. (Brit.)* Gartenbaubetrieb, *der;* **~ gardener** *n. (Brit.)* Gemüseanbauer, *der/*-anbauerin, *die*

marketing ['mɑ:kɪtɪŋ] *n. (Econ.)* Marketing, *das; attrib.* Marketing-

market: ~-place *n.* Marktplatz, *der; (fig.)* Markt, *der;* **~ 'research** *n.* Marktforschung, *die*

marking ['mɑ:kɪŋ] *n.* **a)** *(identification symbol)* Markierung, *die;* Kennzeichen, *das;* **b)** *(on animal)* Zeichnung, *die;* **c)** *(Sch.) (correcting)* Korrektur, *die; (grading)* Benotung, *die;* Zensieren, *das*

'marking-ink *n.* Wäschetinte, *die*

marksman ['mɑ:ksmən] *n., pl.* **marksmen** ['mɑ:ksmən] Scharfschütze, *der*

marksmanship ['mɑ:ksmənʃɪp] *n., no pl.* Treffsicherheit, *die*

'mark-up *n. (Commerc.)* **a)** *(price increase)* Preiserhöhung, *die;* **b)** *(amount added)* Handelsspanne, *die (Kaufmannsspr.)*

marmalade ['mɑ:məleɪd] *n.* **[orange] ~:** Orangenmarmelade, *die;* **tangerine/lime ~:** Mandarinen-/Limonenmarmelade, *die*

¹maroon [mə'ru:n] **1.** *adj.* kastanienbraun. **2.** *n.* Kastanienbraun, *das*

²maroon *v. t.* **a)** *(Naut.: put ashore)* aussetzen; **b)** ‹*Flut, Hochwasser:*› von der Außenwelt abschneiden

marque [mɑ:k] *n.* Marke, *die; (of cars also)* Fabrikat, *das*

marquee [mɑ:'ki:] *n.* großes Zelt; *(for public entertainment)* Festzelt, *das*

marquess, marquis ['mɑ:kwɪs] *n.* Marquis, *der*

marriage ['mærɪdʒ] *n.* **a)** Ehe, *die* **(to** mit); **proposal** *or* **offer of ~:** Heiratsantrag, *der;* **related by ~:** verschwägert; **uncle/cousin by ~:** angeheirateter Onkel/Cousin; **b)** *(wedding)* Hochzeit, *die; (act of marrying)* Heirat, *die; (ceremony)* Trauung, *die;* **~ ceremony** Trauzeremonie, *die;* Eheschließung, *die*

marriage: ~ bureau *n.* Eheanbahnungs- *od.* Ehevermittlungsinstitut, *das;* **~ certificate** *n.* Trauschein, *der;* **~ 'guidance** *n.* Eheberatung, *die;* **~ vows** *n. pl.* Ehegelöbnis, *das (geh.)*

married ['mærɪd] **1.** *adj.* **a)** verheiratet; **~ couple** Ehepaar, *das;* **b)** *(marital)* ehelich ‹*Leben, Liebe*›; Ehe‹*leben, -name, -stand*›; **~ quarters** Verheiratetenquartiere. **2.** *n.* **young/newly ~s** Jungverheiratete

marrow ['mærəʊ] *n.* **a) [vegetable] ~:** Speisekürbis, *der;* **b)** *(Anat.)* [Knochen]mark, *das;* **to the ~** *(fig.)* durch und durch; **be chilled to the ~** *(fig.)* völlig durchgefroren sein

marry ['mærɪ] **1.** *v. t.* **a)** *(take in marriage)* heiraten; **b)** *(join in marriage)* trauen; **they were** *or* **got married last summer** sie haben letzten Sommer geheiratet; **c)** *(give in marriage)* verheiraten ‹*Kind*› **(to** mit); **d)** *(fig.: unite intimately)* verquicken; eng miteinander verbinden. **2.** *v. i.* heiraten; **~ into a [rich] family** in eine [reiche] Familie einheiraten

~ **'off** *v. t.* verheiraten ‹*Tochter*› **(to** mit)

Mars [mɑ:z] *pr. n.* **a)** *(Astron.)* Mars, *der;* **b)** *(Roman Mythol.)* Mars *(der)*

marsh [mɑ:ʃ] *n.* Sumpf, *der*

marshal ['mɑ:ʃl] **1.** *n.* **a)** *(officer of state)* [Hof]marschall, *der;* **b)** *(officer in army)* Marschall, *der;* **c)** *(Sport)* Ordner, *der.* **2.** *v. t., (Brit.)* **-ll-** *(arrange in order)* aufstellen ‹*Truppen*›; sich *(Dat.)* zurechtlegen ‹*Argumente*›; ordnen ‹*Fakten*›

'marshalling yard ['mɑ:ʃəlɪŋ jɑ:d] *n. (Railw.)* Rangierbahnhof, *der*

marsh: ~land ['mɑ:ʃlənd] *n.* Sumpfland, *das;* **~ mallow** ['mɑ:ʃ mæləʊ] *n.* **a)** *(Bot.)* Eibisch, *der;* **b)** *(confection)* Marshmallow, *das;* süßer Speck; **~'mallow** *n. (sweet)* ≈ Mohrenkopf, *der*

marshy ['mɑ:ʃɪ] *adj.* sumpfig; Sumpf‹*boden, -gebiet, -land*›

marsupial [mɑ:'sju:pɪəl, mɑ:'su:pɪəl] *(Zool.)* **1.** *adj.* Beutel‹*tier, -frosch, -mulle*›. **2.** *n.* Beuteltier, *das*

martial ['mɑ:ʃl] *adj.* kriegerisch; *see also* **court martial**

martial: ~ 'arts *n. pl. (Sport)* Kampfsportarten; **~ 'law** *n.* Kriegsrecht, *das*

martin ['mɑ:tɪn] *n. (Ornith.)* **[house-]~:** Mehlschwalbe, *die*

martyr ['mɑ:tə(r)] **1.** *n. (Relig.; also fig.)* Märtyrer, *der/*Märtyrerin, *die;* **be a ~ to rheumatism** entsetzlich unter Rheumatismus leiden; **make a ~ of oneself** den Märtyrer/die Märtyrerin spielen. **2.** *v. t.* **a)** den Märtyrertod sterben lassen; **be ~ed** den Märtyrertod sterben; **b)** *(fig.: torment)* martern *(geh.)*

martyrdom ['mɑ:tədəm] *n.* Martyrium, *das*

marvel ['mɑ:vl] **1.** *n.* Wunder, *das;* **work ~s** Wunder wirken; **be a ~ of patience** eine sagenhafte Geduld haben *(ugs.)*. **2.** *v. i., (Brit.)* **-ll-** *(literary)* **~ at sth.** über etw. *(Akk.)* staunen

marvellous ['mɑ:vələs] *adj.*, **marvellously** ['mɑ:vələslɪ] *adv.* wunderbar

marvelous, marvelously *(Amer.) see* **marvell-**

Marxism ['mɑ:ksɪzm] *n.* Marxismus, *der*

Marxist ['mɑ:ksɪst] **1.** *n.* Marxist, *der/*Marxistin, *die.* **2.** *adj.* marxistisch

marzipan ['mɑ:zɪpæn] *n.* Marzipan, *das*

mascara [mæ'skɑ:rə] *n.* Mascara, *das*

mascot ['mæskɒt] *n.* Maskottchen, *das*

masculine ['mæskjʊlɪn] *adj.* **a)** *(of men)* männlich; **b)** *(manly, manlike)* maskulin; **c)** *(Ling.)* männlich; maskulin *(fachspr.)*

masculinity [mæskjʊ'lɪnɪtɪ] *n., no pl.* Männlichkeit, *die*

mash [mæʃ] **1.** *n.* **a)** Brei, *der;* **b)** *(Brit. coll.: ~ed potatoes)* Kartoffelbrei, *der.* **2.** *v. t.* zerdrücken; zerquetschen; **~ed potatoes** Kartoffelbrei, *der*

mask [mɑ:sk] **1.** *n. (also fig., Photog.)* Maske, *die; (worn by surgeon)* Gesichtsmaske, *die;*

Mundschutz, *der.* **2.** *v. t.* **a)** *(cover with ~)* maskieren; **b)** *(fig.: disguise, conceal)* maskieren; ‹*Wolken, Bäume:*› verdecken; überdecken ‹*Geschmack*›

masochism ['mæsəkɪzm] *n.* Masochismus, *der*

masochist ['mæsəkɪst] *n.* Masochist, *der*/Masochistin, *die*

masochistic [mæsə'kɪstɪk] *adj.* masochistisch

mason ['meɪsn] *n.* **a)** *(builder)* Baumeister, *der;* Steinmetz, *der;* **b)** **M~** *(Freemason)* [Frei]maurer, *der*

masonry ['meɪsnrɪ] *n.* **a)** *(stonework)* Mauerwerk, *das;* **b)** **M~** *(of Freemasons)* [Frei]maurertum, *das*

masquerade [mæskə'reɪd, mɑ:skə'reɪd] **1.** *n. (lit. or fig.)* Maskerade, *die.* **2.** *v. i.* **~ as sb./sth.** sich als jmd./etw. ausgeben

[1]**mass** [mæs] *n. (Eccl.)* Messe, *die;* **say/hear ~:** die Messe lesen/hören; **go to** *or* **attend ~:** zur Messe gehen

[2]**mass 1.** *n.* **a)** *(solid body of matter)* Brocken, *der; (of dough, rubber)* Klumpen, *der;* **b)** *(dense aggregation of objects)* Masse, *die;* **a tangled ~ of threads** ein wirres Knäuel von Fäden; **c)** *(large number or amount of)* **a ~ of ...:** eine Unmenge von ...; **~es of ...:** massenhaft ... *(ugs.);* eine Masse ... *(ugs.);* **d)** *(unbroken expanse)* **a ~ of red** ein Meer von Rot; **be a ~ of bruises/mistakes** *(coll.)* voll blauer Flecken sein/von Fehlern nur so wimmeln; **e)** *(main portion)* Masse, *die;* **the ~es** die breite Masse; die Massen; **f)** *(Phys.)* Masse, *die;* **g)** *attrib. (for many people)* Massen-. **2.** *v. t.* **a)** anhäufen; **b)** *(Mil.)* massieren, zusammenziehen ‹*Truppen*›. **3.** *v. i.* sich ansammeln; ‹*Truppen:*› sich massieren, sich zusammenziehen; ‹*Wolken:*› sich zusammenziehen

massacre ['mæsəkə(r)] **1.** *n.* **a)** *(slaughter)* Massaker, *das;* **b)** *(coll.: defeat)* völlige Zerstörung. **2.** *v. t.* massakrieren

massage ['mæsɑ:ʒ] **1.** *n.* Massage, *die;* **~ parlour** *(often euphem.)* Massagesalon, *der.* **2.** *v. t.* massieren

mass communi'cations *n. pl.* Massenkommunikation, *die*

masseur [mæ'sɜ:(r)] *n.* Masseur, *der*

masseuse [mæ'sɜ:z] *n.* Masseurin, *die;* Masseuse, *die (oft verhüll.)*

massive ['mæsɪv] *adj. (lit. or fig.)* massiv; wuchtig ‹*Statur, Stirn*›; gewaltig ‹*Ausmaße, Aufgabe*›; enorm ‹*Schulden, Vermögen*›

mass: ~ 'media *n. pl.* Massenmedien *Pl.;* **~ 'meeting** *n.* Massenversammlung, *die; (Pol.)* Massenkundgebung, *die; (Industry)* Belegschaftsversammlung, *die;* **~-pro'duced** *adj.* serienmäßig produziert *od.* hergestellt; Massen‹*artikel*›; **~ pro'duction** *n.* Massenproduktion, *die*

mast [mɑ:st] *n. (for sail, flag, aerial, etc.)* Mast, *der;* **work** *or* **serve** *or* **sail before the ~:** als Matrose dienen

master ['mɑ:stə(r)] **1.** *n.* **a)** Herr, *der;* **be ~ of the situation/[the] ~ of one's fate** Herr der Lage/seines Schicksals sein; **be one's own ~:** sein eigener Herr sein; **b)** *(of animal, slave)* Halter, *der; (of dog)* Herrchen, *das; (of ship)* Kapitän, *der;* **be ~ in one's own house** Herr im eigenen Hause sein; **c)** *(Sch.: teacher)* Lehrer, *der;* **'French ~:** Französischlehrer, *der;* **d)** *(original of document, film, etc.)* Original, *das;* **e)** *(expert, great artist)* Meister, *der* (**at** in + *Dat.*); **be a ~ of sth.** etw. meisterhaft beherrschen; **f)** *(skilled workman)* **~ craftsman/carpenter** Handwerks-/Tischlermeister, *der;* **g)** *(Univ.)* Magister, *der;* **~ of Arts/Science** Magister Artium/rerum naturalium. **2.** *adj.* Haupt‹*strategie, -liste*›; **~ bedroom** großes Schlafzimmer; **~ tape/copy** Originalband, *das*/Original, *das;* **~ plan** Gesamtplan, *der.* **3.** *v. t.* **a)** *(learn)* erlernen; **have ~ed a language/subject** eine Sprache/ein Fach beherrschen; **b)** *(overcome)* meistern ‹*Probleme usw.*›; besiegen ‹*Feind*›; zügeln ‹*Emotionen, Gefühle*›

masterful ['mɑ:stəfl] *adj.* **a)** *(imperious)* herrisch ‹*Haltung, Ton, Person*›; **b)** *(masterly)* meisterhaft ‹*Beherrschung, Fähigkeit*›

'master-key *n.* General- *od.* Hauptschlüssel, *der*

masterly ['mɑ:stəlɪ] *adj.* meisterhaft

master: ~mind 1. *n.* führender Kopf; **2.** *v. t.* **~mind the plot/conspiracy** *etc.* der Kopf des Komplotts/der Verschwörung *usw.* sein; **~piece** *n. (work of art)* Meisterwerk, *das; (production showing masterly skill)* Meisterstück, *das;* **~ switch** *n.* Hauptschalter, *der*

mastery ['mɑ:stərɪ] *n.* **a)** *(skill)* Meisterschaft, *die;* **b)** *(knowledge)* Beherrschung, *die* (**of** *Gen.*); **c)** *(upper hand)* Oberhand, *die;* **d)** *(control)* Herrschaft, *die* (**of** über + *Akk.*)

masticate ['mæstɪkeɪt] *v. t.* zerkauen

mastiff ['mæstɪf] *n. (Zool.)* Mastiff, *der*

masturbate ['mæstəbeɪt] *v. i. & t.* masturbieren

masturbation [mæstə'beɪʃn] *n.* Masturbation, *die*

mat [mæt] *n.* **a)** *(on floor, Sport)* Matte, *die;* **pull the ~ from under sb.'s feet** *(fig.)* jmdm. den Boden unter den Füßen wegziehen; **b)** *(to protect table etc.)* Untersetzer, *der; (as decorative support)* Deckchen, *das*

[1]**match** [mætʃ] **1.** *n.* **a)** *(equal)* Ebenbürtige, *der/die;* **be no ~ for sb.** sich mit jmdm. nicht messen können; **she is more than a ~ for him** sie ist ihm mehr als gewachsen; **find** *or* **meet one's ~** *(be defeated)* seinen Meister finden; **b)** *(sb./sth. similar or appropriate)* **be a [good** *etc.*] **~ for sth.** [gut *usw.*] zu etw. passen; **c)** *(Sport)* Spiel, *das; (Football, Tennis, etc. also)* Match, *das; (Boxing)* Kampf, *der; (Athletics)* Wettkampf, *der;* **d)** *(marriage)* Heirat, *die;* **make a good ~:** eine gute Partie machen. **2.** *v. t.* **a)** *(equal)* **~ sb. at chess/in originality** es mit jmdm. im Schach/an Originalität *(Dat.)* aufnehmen [können]; **b)** *(pit)* **~ sb. with** *or* **against sb.** jmdn. jmdm. gegenüberstellen; **be ~ed against sb.** gegen jmdn. antreten; **c) be well ~ed** ‹*Mann u. Frau:*› gut zusammenpassen; ‹*Spieler, Mannschaften:*› sich *(Dat.)* ebenbürtig sein; **d)** *(harmonize with)* passen zu; **~ each other exactly** genau zueinander passen. **3.** *v. i.* *(correspond)* zusammenpassen; **with a scarf** *etc.* **to ~:** mit [dazu] passendem Schal *usw.*

~ 'up 1. *v. i.* **a)** *(correspond)* zusammenpassen; **b)** *(be equal)* **~ up to sth.** einer Sache *(Dat.)* entsprechen. **2.** *v. t.* aufeinander abstimmen ‹*Farben usw.*›; passend zusammenfügen ‹*Teile, Hälften*›

[2]**match** *n. (for lighting)* Streichholz, *das;* Zündholz, *das (südd., österr.)*

match: ~box *n.* Streichholzschachtel, *die;* **~maker** *n.* Ehestifter, *der*/Ehestifterin, *die;* **~ point** *n. (Tennis etc.)* Matchball, *der;* **~stick** *n.* Streichholz, *das;* Zündholz, *das (südd., österr.);* **~wood** *n.* **make ~wood of sth., smash sth. to ~wood** Kleinholz aus etw. machen

[1]**mate** [meɪt] **1.** *n.* **a)** Kumpel, *der*

(ugs.); (friend also) Kamerad, *der/*Kameradin, *die;* **b)** *(Naut.: officer on merchant ship)* ≈ Kapitänleutnant, *der;* **chief** *or* **first/second** ~: Erster/Zweiter Offizier; **c)** *(workman's assistant)* Gehilfe, *der;* **d)** *(Zool.) (male)* Männchen, *das; (female)* Weibchen, *das.* **2.** *v. i. (for breeding)* sich paaren. **3.** *v. t.* paaren ‹*Tiere*›; ~ **a mare and** *or* **with a stallion** eine Stute von einem Hengst decken lassen

²mate *(Chess) see* **checkmate**

material [mə'tɪərɪəl] **1.** *adj.* **a)** *(physical)* materiell; **b)** *(not spiritual)* materiell *(oft abwertend)* ‹*Person, Einstellung*›; **c)** *(relevant, important)* wesentlich. **2.** *n.* **a)** *(matter from which thing is made)* Material, *das;* **b)** *in sing. or pl. (elements)* Material, *das; (for novel, sermon also)* Stoff, *der;* **c)** *(cloth)* Stoff, *der;* **d)** *in pl.* **building/writing** ~**s** Bau-/Schreibmaterial, *das*

materialise *see* **materialize**

materialism [mə'tɪərɪəlɪzm] *n., no pl.* Materialismus, *der*

materialistic [mətɪərɪə'lɪstɪk] *adj.* materialistisch

materialize [mə'tɪərɪəlaɪz] *v. i.* **a)** ‹*Hoffnung:*› sich erfüllen; ‹*Plan, Idee:*› sich verwirklichen; ‹*Treffen, Versammlung:*› zustande kommen; **b)** *(come into view, appear)* [plötzlich] auftauchen

maternal [mə'tɜ:nl] *adj.* **a)** *(motherly)* mütterlich ‹*Liebe, Sorge, Typ*›; Mutter‹*instinkt*›; **b)** *(related)* ‹*Großeltern, Onkel, Tante*› mütterlicherseits

maternity [mə'tɜ:nɪtɪ] *n. (motherhood)* Mutterschaft, *die*

maternity: ~ **dress** *n.* Umstandskleid, *das;* ~ **home,** ~ **hospital** *ns.* Entbindungsheim, *das;* ~ **leave** *n.* Mutterschaftsurlaub, *der*

matey ['meɪtɪ] *(Brit. coll.) adj.,* **matier** ['meɪtɪə(r)], **matiest** ['meɪtɪɪst] kameradschaftlich ‹*Typ, Atmosphäre*›

math [mæθ] *(Amer. coll.) see* **maths**

mathematical [mæθɪ'mætɪkl] *adj.* mathematisch

mathematician [mæθɪmə'tɪʃn] *n.* Mathematiker, *der/*Mathematikerin, *die*

mathematics [mæθɪ'mætɪks] *n., no pl.* Mathematik, *die;* **pure/applied** ~: reine/angewandte Mathematik

maths [mæθs] *n. (Brit. coll.)* Mathe, *die (Schülerspr.)*

matinée *(Amer.:* **matinee**) ['mætɪneɪ] *n.* Matinee, *die;* Frühvorstellung, *die; (in the afternoon)* Nachmittagsvorstellung, *die*

matrices *pl. of* **matrix**

matriculate [mə'trɪkjʊleɪt] *(Univ.)* **1.** *v. t.* immatrikulieren (**in** an + *Dat.*). **2.** *v. i.* sich immatrikulieren

matrimonial [mætrɪ'məʊnɪəl] *adj.* Ehe-

matrimony ['mætrɪmənɪ] *n.* **a)** *(rite of marriage)* Eheschließung, *die;* **b)** *(married state)* Ehestand, *der;* **enter into [holy]** ~: in den [heiligen] Stand der Ehe treten *(geh.)*

matrix ['meɪtrɪks, 'mætrɪks] *n., pl.* **matrices** ['meɪtrɪsi:z, 'mætrɪsi:z] *or* ~**es** *(Math., Geol.)* Matrix, *die*

matron ['meɪtrən] *n. (in school)* ≈ Hausmutter, *die; (in hospital)* Oberin, *die;* Oberschwester, *die*

matt [mæt] *adj.* matt

matted ['mætɪd] *adj.* verfilzt

matter ['mætə(r)] **1.** *n.* **a)** *(affair)* Angelegenheit, *die;* ~**s** die Dinge; **money** ~**s** Geldangelegenheiten *od.* -fragen; **raise an important** ~: einen wichtigen Punkt ansprechen; **that's another** *or* **a different** ~ **altogether** *or* **quite another** ~: das ist etwas ganz anderes; **and to make** ~**s worse ...**: und was die Sache noch schlimmer macht/machte, ...; **b)** *(cause, occasion)* **a/no** ~ **for** *or* **of ...**: ein/kein Grund *od.* Anlaß zu ...; **it's a** ~ **of complete indifference to me** es ist mir völlig gleichgültig; **c)** *(topic)* Thema, *das;* Gegenstand, *der;* ~ **on the agenda** Punkt der Tagesordnung; **d) a** ~ **of ...** *(something that amounts to)* eine Frage (+ *Gen.*) ...; eine Sache von ...; **it's a** ~ **of taste/habit** das ist Geschmack- / Gewohnheitssache; **[only] a** ~ **of time** [nur noch] eine Frage der Zeit; **it's just a** ~ **of working harder** man muß sich ganz einfach [bei der Arbeit] mehr anstrengen; **in a** ~ **of minutes** in wenigen Minuten; **Do you know him? – Yes, as a** ~ **of fact, I do** Kennst du ihn? – Ja, ich kenne ihn tatsächlich; **e) what's the** ~**?** was ist [los]?; **is something the** ~**?** stimmt irgend etwas nicht?; ist [irgend]was *(ugs.)*?; **f) for that** ~: eigentlich; **g) no** ~**!** [das] macht nichts!; **no** ~ **how/who/what/why** *etc.* ganz gleich *od.* egal *(ugs.),* wie/wer/was/warum *usw.;* **h)** *(material, as opposed to mind, spirit, etc.)* Materie, *die;* **[in]organic/solid/vegetable** ~: [an]organische/feste/pflanzliche Stoffe. **2.** *v. i.* etwas ausmachen; **what does it** ~**?** was macht das schon?; was macht's? *(ugs.);* **what** ~**s is that ...**: worum es geht, ist ...; **not** ~ **a damn** vollkommen egal sein; **[it] doesn't** ~: [das] macht nichts *(ugs.);* **it doesn't** ~ **how/when** *etc.* es ist einerlei, wie/wann *usw.;* **does it** ~ **to you if ...?** macht es dir etwas aus, wenn ...?; **the things which** ~ **in life** [das,] worauf es im Leben ankommt

'**matter-of-fact** *adj.* sachlich; nüchtern

matting ['mætɪŋ] *n.* **coconut/straw/reed** ~: Kokos-/Stroh-/Schilfmatten

mattress ['mætrɪs] *n.* Matratze, *die*

mature [mə'tjʊə(r)] **1.** *adj.,* ~**r** [mə'tjʊərə(r)], ~**st** [mə'tjʊərɪst] reif; ausgereift ‹*Plan, Methode, Stil, Käse*›; durchgegoren ‹*Wein*›; ausgewachsen ‹*Pflanze, Tier*›; vollentwickelt ‹*Zellen*›; ~ **student** Spätstudierende, *der/die.* **2.** *v. t.* reifen lassen ‹*Frucht, Wein, Käse*›. **3.** *v. i.* **a)** ‹*Frucht, Wein, Käse usw.:*› reifen; **b)** ‹*Person:*› reifen, reifer werden

maturity [mə'tjʊərɪtɪ] *n.* Reife, *die;* **reach** ~**, come to** ~ ‹*Person:*› erwachsen werden; ‹*Tier:*› ausgewachsen sein

maudlin ['mɔ:dlɪn] *adj.* gefühlsselig

maul [mɔ:l] *v. t.* **a)** ‹*Tiger, Löwe, Bär usw.:*› Pranken-/Tatzenhiebe versetzen (+ *Dat.*); *(fig.)* malträtieren; verreißen ‹*Theaterstück, Buch*›; ‹*Boxer:*› losgehen auf (+ *Akk.*) ‹*Gegner*›; **b)** *(fondle roughly)* betatschen *(ugs.)*

mausoleum [mɔ:sə'li:əm] *n.* Mausoleum, *das*

mauve [məʊv] *adj.* mauve

maverick ['mævərɪk] *n.* Einzelgänger, *der/*Einzelgängerin, *die*

mawkish ['mɔ:kɪʃ] *adj.* rührselig

max. *abbr.* **maximum** *(adj.)* max., *(n.)* Max.

maxim ['mæksɪm] *n.* Maxime, *die*

maximize (maximise) ['mæksɪmaɪz] *v. t.* maximieren

maximum ['mæksɪməm] **1.** *n., pl.* **maxima** ['mæksɪmə] Maximum, *das.* **2.** *adj.* maximal; Maximal-; ~ **security prison** Hochsicherheitsgefängnis, *das;* ~ **temperatures today around 20°** Höchsttemperaturen am Tage um 20°

May [meɪ] *n.* Mai, *der; see also* **August**

may *v. aux., only in pres.* **may,** *neg. (coll.)* **mayn't** [meɪnt], *past* **might** [maɪt], *neg. (coll.)* **mightn't** ['maɪtnt] **a)** *expr. possibility* können; **it ~ be true** das kann stimmen; **they ~ be related** es kann sein, daß sie verwandt sind; **he ~ have missed his train** vielleicht hat er seinen Zug verpaßt; **it ~** *or* **might rain** es könnte regnen; **they might decide to stay** womöglich beschließen sie zu bleiben; **he might have been right** vielleicht hat er [ja] recht gehabt; **it's not so bad as it might have been** es hätte schlimmer kommen können; **that ~ well be** das ist durchaus möglich; **you ~ well say so** das kann man wohl sagen; **we ~** *or* **might as well go** wir könnten eigentlich ebensogut [auch] gehen; *(we are not achieving anything here)* dann können wir ja gehen; **be that as it ~**: wie dem auch sei; **b)** *expr. permission* dürfen; **you ~ go now** du kannst *od.* darfst jetzt gehen; **if I ~ say so ...**: wenn ich das sagen darf, ...; **~** *or* **might I be permitted to ...?** *(formal)* gestatten Sie, daß ...?; **~** *or* **might I ask** *(iron.)* ..., wenn ich [mal] fragen darf?; **c)** *expr. wish* mögen; **~ the best man win!** auf daß der Beste gewinnt!; **d)** *expr. request* **you might at least try [it]** du könntest es wenigstens versuchen; **e)** *used concessively* **he ~ be slow but he's accurate** mag *od.* kann sein, daß er langsam ist, aber dafür ist er auch genau; **f)** *in clauses* **so that I ~/might do sth.** damit ich etw. tun kann; **I hope he ~ succeed** ich hoffe, es gelingt ihm; **come what ~, whatever ~ happen** geschehe was will; was auch geschieht

maybe ['meɪbi:, 'meɪbɪ] *adv.* vielleicht

'May Day *n.* der Erste Mai

'Mayday *n. (distress signal)* Mayday

mayhem ['meɪhem] *n.* Chaos, *das;* **cause** *or* **create ~**: ein Chaos verursachen *od.* hervorrufen

mayn't [meɪnt] *(coll.)* = **may not;** *see* **may**

mayonnaise [meɪə'neɪz] *n.* Mayonnaise, *die*

mayor [meə(r)] *n.* Bürgermeister, *der;* **Lord M~** *(Brit.)* Lord-Mayor, *der;* ≈ Oberbürgermeister, *der*

mayoress ['meərɪs] *n. (woman mayor)* Bürgermeisterin, *die; (mayor's wife)* [Ehe]frau des Bürgermeisters

'maypole *n.* Maibaum, *der*

maze [meɪz] *n. (lit. or fig.)* Labyrinth, *das*

MC *abbr.* **a) Master of Ceremonies; b)** *(Brit.)* **Military Cross** *militärisches Verdienstkreuz*

McCoy [mə'kɔɪ] *n.* **the real ~** *(coll.)* der/die/das Echte; *(not a fake or replica)* das Original

MD *abbr.* **Doctor of Medicine** Dr. med.; *see also* **B.Sc.**

me [mɪ, *stressed* mi:] *pron.* mich; *as indirect object* mir; **bigger than/as big as me** größer als/so groß wie ich; **silly ~**: ich Dussel! *(salopp);* **who, me?** wer, ich?; **not me** ich/mich/mir nicht; **it's me** ich bin's

mead [mi:d] *n. (drink)* Met, *der*

meadow ['medəʊ] *n.* Wiese, *die;* **in the ~**: auf der Wiese

meagre (*Amer.:* **meager**) ['mi:gə(r)] *adj.* spärlich; dürftig *(auch fig.);* **a ~ attendance** eine geringe Teilnehmerzahl

[1]**meal** [mi:l] *n.* Mahlzeit, *die;* **stay for a ~**: zum Essen bleiben; **go out for a ~**: essen gehen; **make a ~ of sth.** *(fig.)* eine große Sache aus etw. machen

[2]**meal** *n. (ground grain)* Schrot[mehl], *das*

meal: ~-ticket *n.* Essenmarke, *die; (fig. coll.)* melkende Kuh *(ugs.);* **~time** *n.* Essenszeit, *die;* **at ~times** während des Essens; bei Tisch

mealy-mouthed ['mi:lɪmaʊðd] *adj. (derog.)* unaufrichtig

[1]**mean** [mi:n] *n.* Mittelweg, *der;* Mitte, *die;* **a happy ~**: der goldene Mittelweg

[2]**mean** *adj.* **a)** *(niggardly)* schäbig *(abwertend);* **b)** *(ignoble)* schäbig *(abwertend),* gemein ‹*Person, Verhalten, Gesinnung*›; **c)** *(shabby)* schäbig *(abwertend)* ‹*Haus, Wohngegend*›; armselig ‹*Verhältnisse*›; **be no ~ athlete/feat** kein schlechter Sportler/keine schlechte Leistung sein

[3]**mean** *v. t.,* **~t** [ment] **a)** *(have as one's purpose)* beabsichtigen; **~ well by** *or* **to** *or* **towards sb.** es gut mit jmdm. meinen; **I ~t him no harm** ich wollte ihm nichts Böses; **what do you ~ by [saying] that?** was willst du damit sagen?; **I ~t it** *or* **it was ~t as a joke** das sollte ein Scherz sein; **~ to do sth.** etw. tun wollen; **I ~ to be obeyed** ich verlange, daß man mir gehorcht; **I ~t to write, but forgot** ich hatte [fest] vor zu schreiben, aber habe es [dann] vergessen; **do you ~ to say that ...?** willst du damit sagen, daß ...?; **b)** *(design, destine)* **these plates are ~t to be used** diese Teller sind zum Gebrauch bestimmt *od.* sind da, um benutzt zu werden; **I ~t it to be a surprise for him** es sollte eine Überraschung für ihn sein; **they are ~t for each other** sie sind füreinander bestimmt; **I ~t you to read the letter** ich wollte, daß du den Brief liest; **be ~t to do sth.** etw. tun sollen; **c)** *(intend to convey, refer to)* meinen; **if you know** *or* **see what I ~**: du verstehst, was ich meine?; **I really ~ it, I ~ what I say** ich meine das ernst; es ist mir Ernst damit; **d)** *(signify, entail, matter)* bedeuten; **the name ~s/the instructions ~ nothing to me** der Name sagt mir nichts/ich kann mit der Anleitung nichts anfangen

meander [mɪ'ændə(r)] *v. i.* **a)** ‹*Fluß:*› sich schlängeln *od.* winden; **b)** ‹*Person:*› schlendern

meaning ['mi:nɪŋ] *n.* Bedeutung, *die; (of text etc., life)* Sinn, *der;* **this sentence has no ~**: dieser Satz ergibt keinen Sinn; **if you get my ~**: du verstehst, was ich meine?; **what's the ~ of this?** was hat [denn] das zu bedeuten?

meaningful ['mi:nɪŋfl] *adj.* sinntragend ‹*Wort, Einheit*›; *(fig.)* bedeutungsvoll ‹*Blick, Ergebnis, Folgerung*›; sinnvoll ‹*Leben, Aufgabe, Arbeit, Gespräch*›

meanness ['mi:nnɪs] *n., no pl.* **a)** *(stinginess)* Schäbigkeit, *die (abwertend);* **b)** *(baseness)* Schäbigkeit, *die (abwertend);* Gemeinheit, *die;* **c)** *(shabbiness) see* [2]**mean c**: Schäbigkeit, *die;* Armseligkeit, *die*

means [mi:nz] *n. pl.* **a)** *usu. constr. as sing. (way, method)* Möglichkeit, *die;* [Art und] Weise; **by this ~**: hierdurch; auf diese Weise; **a ~ to an end** ein Mittel zum Zweck; **we have no ~ of doing this** wir haben keine Möglichkeit, dies zu tun; **~ of transport** Transportmittel, *das;* **b)** *(resources)* Mittel *Pl.;* **live within/beyond one's ~**: seinen Verhältnissen entsprechend/über seine Verhältnisse leben; **c) Will you help me? – By all ~**: Hilfst du mir? Selbstverständlich!; **by no [manner of] ~**: ganz und gar nicht; keineswegs; **by ~ of** durch; mit [Hilfe von]

'means test *n.* Überprüfung der Bedürftigkeit

meant *see* [3]**mean**

mean: ~time 1. *n.* **in the ~time** in der Zwischenzeit; inzwischen; **2.** *adv.* inzwischen; **~while** *adv.* inzwischen

measles ['mi:zlz] *n., constr. as pl. or sing. (Med.)* Masern *Pl.; see also* **German measles**

measly ['mi:zlɪ] *adj. (coll. derog.)* pop[e]lig *(ugs. abwertend);* **a ~**

little portion eine mickrige Portion *(ugs. abwertend)*

measure ['meʒə(r)] **1.** *n.* **a)** Maß, *das;* **weights and ~s** Maße und Gewichte; **for good ~:** sicherheitshalber; *(as an extra)* zusätzlich; **give short/full ~** *(in public house)* zu wenig/vorschriftsmäßig ausschenken; **made to ~** *pred. (Brit., lit. or fig.)* maßgeschneidert; **b)** *(degree)* Menge, *die;* **in some ~:** in gewisser Hinsicht; **a ~ of freedom/responsibility** ein gewisses Maß an Freiheit/Verantwortung *(Dat.);* **c)** *(instrument or utensil for measuring)* Maß, *das; (for quantity also)* Meßglas, *das;* Meßbecher, *der; (for size also)* Meßstab, *der; (fig.)* Maßstab, *der;* **it gave us some ~ of the problems** das gab uns eine Vorstellung von den Problemen; **beyond [all] ~:** grenzenlos; über die *od.* alle Maßen *adv.;* **d)** *(Mus.: time)* Takt, *der;* **e)** *(step, law)* Maßnahme, *die; (Law: bill)* Gesetzesvorlage, *die;* **take ~s to stop/ensure sth.** Maßnahmen ergreifen *od.* treffen, um etw. zu unterbinden/sicherzustellen. **2.** *v.t.* **a)** messen ‹*Größe, Menge usw.*›; ausmessen ‹*Raum*›; **~ sb. for a suit** [bei] jmdm. Maß *od.* die Maße für einen Anzug nehmen; **b)** *(fig.: estimate)* abschätzen; **c)** *(mark off)* **~ sth. [off]** etw. abmessen. **3.** *v.i.* **a)** *(have a given size)* messen; **b)** *(take measurement[s])* Maß nehmen

~ 'out *v.t.* abmessen

~ 'up to *v.t.* entsprechen (+ *Dat.*) ‹*Maßstäben, Erwartungen*›; gewachsen sein (+ *Dat.*) ‹*Anforderungen*›

measured ['meʒəd] *adj.* rhythmisch, gleichmäßig ‹*Geräusch, Bewegung*›; gemessen ‹*Schritt, Worte, Ausdrucksweise*›

measurement ['meʒəmənt] *n.* **a)** *(act, result)* Messung, *die;* **b)** *in pl. (dimensions)* Maße *Pl.*

measuring ['meʒərɪŋ]: **~-jug** *n.* Meßbecher, *der;* **~-tape** *n.* Bandmaß, *das*

meat [mi:t] *n.* **a)** Fleisch, *das;* **b)** *(arch.: food)* **one man's ~ is another man's poison** *(prov.)* was dem einen sin Uhl, ist dem andern sin Nachtigall *(Spr.);* **c)** *(fig.)* Substanz, *die*

meat: ~ball *n.* Fleischkloß, *der;* Fleischklößchen, *das;* **~ loaf** *n.* Hackbraten, *der;* **~ 'pie** *n.* Fleischpastete, *die;* **~ safe** *n. (Brit.)* Fliegenschrank, *der*

meaty ['mi:tɪ] *adj.* fleischig; ‹*Gulasch usw.*› mit reichlich Fleisch; **have a ~ taste** nach Fleisch schmecken

mechanic [mɪ'kænɪk] *n.* Mechaniker, *der*

mechanical [mɪ'kænɪkl] *adj. (lit. or fig.)* mechanisch; **produced by ~ means** maschinell produziert

mechanical 'pencil *n. (Amer.)* Drehbleistift, *der*

mechanics [mɪ'kænɪks] *n., no pl.* **a)** Mechanik, *die;* **b)** *constr. as pl. (means of construction or operation)* Mechanismus, *der; (of writing, painting, etc.)* Technik, *die*

mechanise *see* **mechanize**

mechanism ['mekənɪzm] *n.* Mechanismus, *der*

mechanize ['mekənaɪz] *v.t.* **a)** mechanisieren; **b)** *(Mil.)* motorisieren

medal ['medl] *n.* Medaille, *die; (decoration)* Orden, *der*

medalist *(Amer.) see* **medallist**

medallion [mɪ'dæljən] *n. (large medal)* [große] Medaille

medallist ['medəlɪst] *n.* Medaillengewinner, *der*/-gewinnerin, *die (Sport)*

meddle ['medl] *v.i.* **~ with sth.** sich *(Dat.)* an etw. *(Dat.)* zu schaffen machen; **~ in sth.** sich in etw. *(Akk.)* einmischen

media ['mi:dɪə] *see* **mass media; medium 1**

mediaeval *see* **medieval**

mediate ['mi:dɪeɪt] **1.** *v.i.* vermitteln. **2.** *v.t.* **a)** *(settle)* vermitteln in (+ *Dat.*); **b)** *(bring about)* vermitteln

mediator ['mi:dɪeɪtə(r)] *n.* Vermittler, *der*/Vermittlerin, *die*

medical ['medɪkl] **1.** *adj.* medizinisch; ärztlich ‹*Behandlung*›; **~ ward** ≈ medizinische *od.* innere Abteilung. **2.** *n. (coll.) see* **medical examination**

medical: ~ certificate *n.* Attest, *das;* **~ exami'nation** *n.* ärztliche Untersuchung; **~ prac'titioner** *n.* praktischer Arzt/praktische Ärztin; Arzt/Ärztin für Allgemeinmedizin; **~ school** *n.* medizinische Hochschule; *(faculty)* medizinische Fakultät; **~ student** *n.* Medizinstudent, *der*/-studentin, *die*

medicated ['medɪkeɪtɪd] *adj.* **~ shampoo/soap** medizinisches Haarwaschmittel / medizinische Seife

medication [medɪ'keɪʃn] *n.* **a)** *(treatment)* Behandlung, *die;* Medikation, *die (Med.);* **b)** *(medicine)* Medikament, *das*

medicinal [mɪ'dɪsɪnl] *adj.* medizinisch; Arznei‹*mittel, -kohle*›; **~ qualities** Heilkräfte

medicine ['medsən, 'medɪsɪn] *n.* **a)** *no pl., no art. (science)* Medizin, *die;* **b)** *(preparation)* Medikament, *das;* Medizin, *die (veralt.);* **give sb. a dose** *or* **a taste of his/her own ~** *(fig.)* es jmdm. mit gleicher Münze heimzahlen

medicine: ~ chest *n.* Medikamentenschränkchen, *das; (in home)* Hausapotheke, *die;* **~-man** *n.* Medizinmann, *der*

medieval [medɪ'i:vl] *adj. (lit. or fig.)* mittelalterlich; **the ~ period** das Mittelalter

mediocre [mi:dɪ'əʊkə(r)] *adj.* mittelmäßig

mediocrity [mi:dɪ'ɒkrɪtɪ] *n., no pl.* Mittelmäßigkeit, *die*

meditate ['medɪteɪt] **1.** *v.t. (consider)* denken an (+ *Akk.*); erwägen; *(design)* planen. **2.** *v.i.* nachdenken, *(esp. Relig.)* meditieren ([up]on über + *Akk.*)

meditation [medɪ'teɪʃn] *n.* **a)** *(act of meditating)* Nachdenken, *das;* **b)** *(Relig.)* Meditation, *die*

Mediterranean [medɪtə'reɪnɪən] **1.** *pr. n.* **the ~:** das Mittelmeer. **2.** *adj.* mediterran *(Geogr.);* südländisch; **~ coast/countries** Mittelmeerküste, *die*/Mittelmeerländer

medium ['mi:dɪəm] **1.** *n., pl.* **media** ['mi:dɪə] *or* **~s a)** *(substance)* Medium, *das; (fig.: environment)* Umgebung, *die;* **b)** *(intermediate agency)* Mittel, *das;* **by** *or* **through the ~ of** durch; **c)** *pl.* **~s** *(Spiritualism)* Medium, *das;* **d)** *(means of communication or artistic expression)* Medium, *das;* **e)** *in pl.* **media** *(means of mass communication)* Medien *Pl.;* **f)** *(middle degree)* Mittelweg, *der; see also* **happy a. 2.** *adj.* mittler...; medium *nur präd.,* halb durchgebraten ‹*Steak*›

medium: ~size[d] *adj.* mittelgroß; **~ term** *see* **term 1 d; ~ wave** *n. (Radio)* Mittelwelle, *die*

medley ['medlɪ] *n. (forming a whole)* buntes Gemisch; *(collection of items)* Sammelsurium, *das (abwertend); (of colours)* Kunterbunt, *das*

meek [mi:k] *adj.* **a)** *(humble)* sanftmütig; **b)** *(tamely submissive)* zu nachgiebig

meet [mi:t] **1.** *v.t.,* **met** [met] **a)** *(come face to face with or into the company of)* treffen; **I have to ~ my boss at 11 a.m.** ich habe um 11 Uhr einen Termin beim Chef; **arrange to ~ sb.** sich mit jmdm. verabreden; **b)** *(go to place of arrival of)* treffen; *(collect)* abholen; **I'll ~ your train** ich hole dich vom Zug ab; **~ sb. half-way** *(fig.)*

jmdm. [auf halbem Wege] entgegenkommen; **c)** *(make the acquaintance of)* kennenlernen; **I'd like you to ~ my wife** ich möchte Sie gern meiner Frau vorstellen *od.* mit meiner Frau bekannt machen; **pleased to ~ you** [sehr] angenehm; sehr erfreut; **d)** *(reach point of contact with)* treffen auf (+ *Akk.*); **~ the eye/sb.'s eye[s]** sich den/jmds. Blicken darbieten; **~ the ear/sb.'s ears** das/jmds. Ohr treffen; **there's more to it than ~s the eye** da ist *od.* steckt mehr dahinter, als man zuerst denkt; **e)** *(experience)* stoßen auf (+ *Akk.*) ‹*Widerstand, Problem*›; ernten ‹*Gelächter, Drohungen*›; **~ [one's] death** *or* **one's end/disaster/one's fate** den Tod finden *(geh.)*/von einer Katastrophe/seinem Schicksal ereilt werden *(geh.)*; **f)** *(satisfy)* entsprechen (+ *Dat.*) ‹*Forderung, Wunsch*›; einhalten ‹*Termin, Zeitplan*›; **g)** *(pay)* decken ‹*Kosten, Auslagen*›; bezahlen ‹*Rechnung*›. **2.** *v.i.*, **met a)** *(come face to face) (by chance)* sich *(Dat.)* begegnen; *(by arrangement)* sich treffen; **we've met before** wir kennen uns bereits; **b)** *(assemble)* ‹*Komitee, Ausschuß usw.:*› tagen; **~ together** sich versammeln; **c)** *(come together)* ‹*Bahnlinien, Straßen usw.:*› aufeinandertreffen; ‹*Flüsse*› zusammenfließen

~ **'up** *v.i.* sich treffen; **~ up with sb.** *(coll.)* jmdn. treffen

~ **with** *v.t.* **a)** *(encounter)* begegnen (+ *Dat.*); **b)** *(experience)* haben ‹*Erfolg, Unfall*›; finden ‹*Zustimmung, Verständnis, Tod*›; stoßen auf (+ *Akk.*) ‹*Widerstand*›

meeting ['mi:tɪŋ] *n.* **a)** Begegnung, *die (auch fig.)*; *(by arrangement)* Treffen, *das*; **~ of minds** Verständigung, *die*; Annäherung der Standpunkte; **b)** *(assembly)* Versammlung, *die*; *(of committee, Cabinet, council, etc.)* Sitzung, *die*; *(social gathering)* Treffen, *das*; **c)** *(Sport)* Treffen, *das*; *(Racing)* Rennen, *das*

meeting: ~-place *n.* Treffpunkt, *der*; **~-point** *n.* *(of lines, roads)* Schnittpunkt, *der*; *(of rivers)* Zusammenfluß, *der*

mega- ['megə] *pref.* mega-/Mega-

megalomania [megələ'meɪnɪə] *n.* Größenwahn, *der*; Megalomanie, *die (Psych.)*

'megaphone *n.* Megaphon, *das*

'megastar *n. (coll.)* Megastar, *der (ugs.)*

melancholy ['melənkəlɪ] **1.** *n.* Melancholie, *die*; *(pensive sadness)* Schwermut, *die*. **2.** *adj.* **a)** *(gloomy, expressing sadness)* melancholisch; schwermütig; **b)** *(saddening)* deprimierend

mêlée (*Amer.*: **melee**) ['meleɪ] *n.* Handgemenge, *das*

mellow ['meləʊ] **1.** *adj.* **a)** *(softened by age or experience)* abgeklärt; **b)** *(ripe, well-matured)* reif; ausgereift ‹*Wein*›; **c)** *(genial)* freundlich; **d)** *(full and soft)* weich ‹*Stimme, Ton, Licht, Farben*›. **2.** *v.t.* reifer machen ‹*Person*›; [aus]reifen lassen ‹*Wein*›. **3.** *v.i.* ‹*Person, Obst, Wein:*› reifen; ‹*Licht, Farbe:*› weicher werden

melodious [mɪ'ləʊdɪəs] *adj.*, **melodiously** [mɪ'ləʊdɪəslɪ] *adv.* melodisch

melodrama ['melədrɑ:mə] *n. (lit. or fig.)* Melodrama, *das*

melodramatic [melədrə'mætɪk] *adj. (lit. or fig.)* melodramatisch

melody ['melədɪ] *n.* **a)** *(pleasing sound)* Gesang, *der*; **b)** *(tune)* Melodie, *die*

melon ['melən] *n.* Melone, *die*

melt [melt] **1.** *v.i.* **a)** schmelzen; *(dissolve)* sich auflösen; **~ in one's** *or* **the mouth** *(coll.)* auf der Zunge zergehen; *see also* **butter 1**; **b)** *(fig.: be softened)* dahinschmelzen *(geh.)* (**at** bei); sich erweichen lassen (**at** durch). **2.** *v.t.* **a)** schmelzen ‹*Schnee, Eis, Metall*›; *(Cookery)* zerlassen ‹*Butter*›; **b)** *(fig.: make tender)* erweichen ‹*Person, Herz*›

~ **a'way** *v.i.* ‹*Schnee, Eis:*› [weg]schmelzen; *(fig.: dwindle away)* ‹*Nebel, Dunst, Menschenmenge:*› sich auflösen; ‹*Verdacht, Mehrheit, Furcht:*› dahinschwinden *(geh.)*

~ **'down 1.** *v.i.* schmelzen. **2.** *v.t.* einschmelzen ‹*Metall, Glas*›

'melt-down *n.* Schmelzen, *das*

melting ['meltɪŋ]: **~-point** *n.* Schmelzpunkt, *der*; **~-pot** *n.* *(fig.)* Schmelztiegel, *der*; **be in the ~-pot** in rascher Veränderung begriffen sein

member ['membə(r)] *n.* **a)** Mitglied, *das*; *attrib.* Mitglieds‹*staat, -land*›; **be a ~ of the club** Mitglied des Vereins sein; **~ of the expedition** Expeditionsteilnehmer, *der*/-teilnehmerin, *die*; **~ of a/the family** Familienangehörige, *der/die*; **b) M~ [of Parliament)** *(Brit. Polit.)* Abgeordnete [des Unterhauses], *der/die*; **M~ of Congress** *(Amer. Polit.)* Kongreßabgeordnete, *der/die*

membership ['membəʃɪp] *n.* **a)** Mitgliedschaft, *die* (**of** in + *Dat.*); *attrib.* Mitglieds‹*karte, -ausweis, -beitrag*›; Mitglieder‹*liste, -verzeichnis*›; **b)** *(number of members)* Mitgliederzahl, *die*; **c)** *(body of members)* Mitglieder *Pl.*

membrane ['membreɪn] *n. (Biol.)* Membran, *die*

memento [mɪ'mentəʊ] *n., pl.* ~es *or* ~s Andenken, *das* (**of** an + *Akk.*)

memo ['meməʊ] *n., pl.* ~s *(coll.) see* **memorandum a, b**

memoirs ['memwɑ:z] *n. pl.* Memoiren *Pl.*

memorable ['memərəbl] *adj.* denkwürdig ‹*Ereignis, Gelegenheit, Tag*›; unvergeßlich ‹*Film, Aufführung*›

memorandum [memə'rændəm] *n., pl.* **memoranda** [memə'rændə] *or* ~**s a)** *(note)* Notiz, *die*; **b)** *(letter)* Mitteilung, *die*; **c)** *(Diplom.)* Memorandum, *das*

memorial [mɪ'mɔ:rɪəl] **1.** *adj.* Gedenk‹*stein, -gottesdienst, -ausstellung*›. **2.** *n.* Denkmal, *das* (**to** für)

memorize (memorise) ['meməraɪz] *v.t.* sich *(Dat.)* merken *od.* einprägen; *(learn by heart)* auswendig lernen

memory ['memərɪ] *n.* **a)** Gedächtnis, *das*; **have a good/poor ~ for faces** ein gutes/schlechtes Personengedächtnis haben; **b)** *(recollection, person or thing remembered, act of remembering)* Erinnerung, *die* (**of** an + *Akk.*); **have a vague ~ of sth.** sich nur ungenau an etw. *(Akk.)* erinnern; **it slipped** *or* **escaped my ~:** es ist mir entfallen; **from ~:** aus dem Gedächtnis *od.* Kopf; **in ~ of** zur Erinnerung an (+ *Akk.*); *attrib.* **a trip down ~ lane** eine Reise in die Vergangenheit; **c)** *(Computing)* Speicher, *der*

men *pl. of* **man**

menace ['menɪs] **1.** *v.t.* bedrohen ‹*Person*›. **2.** *n.* Plage, *die*

menacing ['menɪsɪŋ] *adj.* drohend

mend [mend] **1.** *v.t.* **a)** *(repair)* reparieren; ausbessern, flicken ‹*Kleidung, Fischernetz*›; kleben, kitten ‹*Glas, Porzellan, Sprung*›; beheben ‹*Schaden*›; beseitigen ‹*Riß*›; **b)** *(improve)* **~ one's ways** sich bessern; **~ matters** die Sache bereinigen. **2.** *v.i.* ‹*Knochen, Bein, Finger usw.:*› heilen; **has his leg ~ed yet?** ist sein Bein schon verheilt? **3.** *n.* *(in glass, china, etc.)* Kleb[e]stelle, *die*; *(in cloth)* ausgebesserte Stelle; *(repair)* Ausbesserung, *die*; **be on the ~** ‹*Person:*› auf dem Wege der Besserung sein

mendacious [men'deɪʃəs] *adj.* unwahr ‹*Bericht, Behauptung, Darstellung*›; verlogen *(abwertend)* ‹*Person, Rede, Buch*›
'**menfolk** *n. pl.* Männer
menial ['mi:nɪəl] *adj.* niedrig; untergeordnet ‹*Aufgabe*›
meningitis [menɪn'dʒaɪtɪs] *n.* *(Med.)* Meningitis, *die (fachspr.)*; Hirnhautentzündung, *die*
menopause ['menəpɔ:z] *n.* Wechseljahre *Pl.*; Klimakterium, *das (fachspr.)*
menstrual ['menstrʊəl] *adj.* *(Physiol.)* menstrual *(fachspr.)*; Menstruations-
menstruate ['menstrʊeɪt] *v.i.* *(Physiol.)* menstruieren
menstruation [menstrʊ'eɪʃn] *n.* *(Physiol.)* Menstruation, *die*
menswear ['menzweə(r)] *n., no pl.* Herrenbekleidung, *die; attrib.* Herrenbekleidungs-
mental ['mentl] *adj.* **a)** geistig; seelisch ‹*Belastung, Labilität*›; Geistes‹*zustand, -störung, -verfassung*›; ~ **process** Denkprozeß, -vorgang, *der;* **make a ~ note of sth.** sich *(Dat.)* etw. merken; **b)** *(Brit. coll.: mad)* verrückt *(salopp)*; bekloppt *(salopp)*
mental: ~ **age** *n.* geistiger Entwicklungsstand; Intelligenzalter, *das (Psych.)*; ~ **a'rithmetic** *n.* Kopfrechnen, *das;* ~ **asylum** *see* ~ **hospital;** ~ **'block** *see* **block 1 k;** ~ **'health** *n.* seelische Gesundheit; ~ **home** *n.* Nervenklinik, *die;* ~ **hospital** *n.* psychiatrische Klinik; Nervenklinik, *die;* ~ **'illness** *n.* Geisteskrankheit, *die*
mentality [men'tælɪtɪ] *n.* Mentalität, *die*
mentally ['mentəlɪ] *adv.* **a)** geistig; geistes‹*gestört, -krank*›; ~ **deficient** *or* **defective** schwachsinnig; **b)** *(inwardly)* innerlich; im Geiste; im Kopf ‹*rechnen*›
mental: ~ **patient** *n.* Geisteskranke, *der/die;* ~ **reser'vation** *n.* geheimer Vorbehalt
menthol ['menθɒl] *n.* Menthol, *das*
mention ['menʃn] **1.** *n.* Erwähnung, *die;* **there is a brief/no ~ of sth.** etw. wird kurz/nicht erwähnt; **get a ~:** erwähnt werden; **make [no] ~ of sth.** etw. [nicht] erwähnen. **2.** *v.t.* erwähnen (**to** gegenüber); ~ **as the reason for sth.** als Grund für etw. nennen; **not to ~ ...:** ganz zu schweigen von ...; **not to ~ the fact that ...:** ganz abgesehen davon, daß ...; **Thank you very much. – Don't ~ it** Vielen Dank. – Keine Ursache
mentor ['mentɔ:(r)] *n.* Mentor, *der*/Mentorin, *die*
menu ['menju:] *n.* **a)** [Speise]karte, *die;* **b)** *(fig.: diet)* Nahrung, *die;* **c)** *(Computing, Telev.)* Menü, *das*
mercenary ['mɜ:sɪnərɪ] **1.** *adj.* **a)** gewinnsüchtig; **b)** *(hired)* Söldner-. **2.** *n.* Söldner, *der*
merchandise ['mɜ:tʃəndaɪz] *n., no pl., no indef. art.* [Handels]ware, *die*
merchant ['mɜ:tʃənt] *n.* Kaufmann, *der;* **corn-/timber-~:** Getreide-/Holzhändler, *der*/-händlerin, *die*
merchant: ~ **'bank** *n.* Handelsbank, *die;* ~ **'banker** *n.* Bankier, *der (bei einer Handelsbank)*; ~ **ma'rine** *(Amer.)*, ~ **'navy** *(Brit.)* *ns.* Handelsmarine, *die;* ~ **'seaman** *n.* Matrose bei der Handelsmarine; ~ **ship** *n.* Handelsschiff, *das*
merciful ['mɜ:sɪfl] *adj.* gnädig
mercifully ['mɜ:sɪfəlɪ] *adv.* gnädig; *as sentence-modifier (fortunately)* glücklicherweise
merciless ['mɜ:sɪlɪs] *adj.*, **mercilessly** ['mɜ:sɪlɪslɪ] *adv.* gnadenlos; unbarmherzig
mercury ['mɜ:kjʊrɪ] **1.** *n.* Quecksilber, *das.* **2.** *pr. n.* **M~ a)** *(Astron.)* Merkur, *der;* **b)** *(Roman Mythol.)* Merkur *(der)*
mercy ['mɜ:sɪ] **1.** *n.* **a)** *no pl., no indef. art.* Erbarmen, *das* (**on** mit); **show sb. [no] ~:** mit jmdm. [kein] Erbarmen haben; **God's great ~:** Gottes große Barmherzigkeit; **be at the ~ of sb./sth.** jmdm./einer Sache [auf Gedeih und Verderb] ausgeliefert sein; **b)** *(instance)* glückliche Fügung; **we must be thankful** *or* **grateful for small mercies** *(coll.)* man darf [ja] nicht zuviel verlangen. **2.** *attrib. adj.* Hilfs-, Rettungs‹*einsatz, -flug*›; ~ **killings** Fälle aktiver Sterbehilfe
mere [mɪə(r)] *adj.* bloß; **he is a ~ child** er ist nur ein Kind; ~ **courage is not enough** Mut allein genügt nicht; **the ~st hint/trace of sth.** die kleinste Andeutung/Spur von etw.
merely ['mɪəlɪ] *adv.* bloß; lediglich; **not ~ ...:** nicht bloß ...
merge [mɜ:dʒ] **1.** *v.t.* **a)** zusammenschließen ‹*Firmen, Unternehmen*› (**into** zu); zusammenlegen ‹*Anteile, Abteilungen*›; **b)** *(blend gradually)* verschmelzen (**with** mit). **2.** *v.i.* **a)** ‹*Firma, Unternehmen:*› sich zusammenschließen, fusionieren (**with** mit); ‹*Abteilung:*› zusammengelegt werden (**with** mit); **b)** *(blend gradually)* ‹*Straße:*› zusammenlaufen (**with** mit); ~ **into sth.** ‹*Farbe usw.:*› in etw. *(Akk.)* übergehen
merger ['mɜ:dʒə(r)] *n.* *(of departments, parties)* Zusammenschluß, *der;* Vereinigung, *die;* *(of companies)* Fusion, *die*
meridian [mə'rɪdɪən] *n.* *(Astron., Geog.)* Meridian, *der*
meringue [mə'ræŋ] *n.* Meringe, *die;* Baiser, *das*
merit ['merɪt] **1.** *n.* **a)** *no pl.* *(worth)* Verdienst, *das;* **there is no ~ in doing that** es ist nicht [sehr] sinnvoll, das zu tun; **b)** *(good feature)* Vorzug, *der;* **on his/its ~s** nach seinen Vorzügen; **c)** *in pl.* *(rights and wrongs)* Für und Wider, *das.* **2.** *v.t.* verdienen
meritocracy [merɪ'tɒkrəsɪ] *n.* Meritokratie, *die*
mermaid ['mɜ:meɪd] *n.* Nixe, *die*
merrily ['merɪlɪ] *adv.* munter
merriment ['merɪmənt] *n., no pl.* Fröhlichkeit, *die;* **fall into fits of helpless ~:** sich vor Lachen nicht mehr halten können
merry ['merɪ] *adj.* **a)** fröhlich; **a ~ time was had by all** alle haben sich prächtig amüsiert; **the more the merrier** je mehr, desto besser; ~ **'Christmas!** frohe *od.* fröhliche Weihnachten!; **b)** *(coll.: tipsy)* beschwipst *(ugs.)*
merry: ~**-go-round** *n.* Karussell, *das;* ~**-making** *n., no pl., no indef. art.* Feiern, *das;* **the sound of ~-making** fröhlicher Festlärm
mesh [meʃ] **1.** *n.* **a)** Masche, *die;* **b)** *(netting; also fig.: network)* Geflecht, *das;* **wire ~ [fence]** Maschendraht[zaun], *der;* **c)** *in pl.* *(fig.: snare)* Maschen. **2.** *v.i.* **a)** *(Mech. Engin.)* ‹*Zahnräder:*› ineinandergreifen; ~ **with** eingreifen in *(Akk.)*; **b)** *(fig.: be harmonious)* harmonisieren (**with** mit)
mesmerize **(mesmerise)** ['mezməraɪz] *v.t.* faszinieren; erstarren lassen ‹*Tier*›
mess [mes] *n.* **a)** *(dirty/untidy state)* **[be] a ~** *or* **in a ~:** schmutzig/unaufgeräumt [sein]; **[be] a complete** *or* **in an awful ~:** in einem fürchterlichen Zustand [sein]; **what a ~!** was für ein Dreck *(ugs.)* / Durcheinander!; **look a ~:** schlimm aussehen; **your hair is a ~:** dein Haar ist ganz durcheinander; **don't make too much ~:** mach nicht zuviel Schmutz/Durcheinander; **leave a lot of ~ behind one** *(dirt)* viel Schmutz hinterlassen; *(untidiness)* eine große Unordnung hin-

terlassen; **make a ~ with sth.** mit etw. Schmutz machen; **b)** *(excreta)* **dog's/cat's ~:** Hunde-/Katzenkot, *der;* **make a ~ on the carpet** auf den Teppich machen *(ugs.);* **c)** *(bad state)* **be [in] a ~:** sich in einem schlimmen Zustand befinden; ‹*Person:*› schlimm dran sein; **get into a ~:** in Schwierigkeiten geraten; **make a ~ of** verpfuschen *(ugs.)* ‹*Arbeit, Leben, Bericht, Vertrag*›; durcheinanderbringen ‹*Pläne*›; **d)** *(eating-place)* Kantine, *die; (for officers)* Kasino, *das; (on ship)* Messe, *die;* **officers' ~:** Offizierskasino, *das*/Offiziersmesse, *die*

~ a'bout, ~ a'round 1. *v.i.* **a)** *(potter)* herumwerken; *(fool about)* herumalbern; **~ about with cars** an Autos herumbasteln *(ugs.);* **b)** *(interfere)* **~ about** *or* **around with** sich einmischen in (+ *Akk.*) ‹*Angelegenheit*›; herumspielen an (+ *Dat.*) ‹*Mechanismus, Stromkabel usw.*›. **2.** *v.t.* **~ sb. about** *or* **around** mit jmdm. nach Belieben umspringen *(abwertend)*

~ 'up *v.t.* **a)** *(make dirty)* schmutzig machen; *(make untidy)* in Unordnung bringen; **b)** *(bungle)* verpfuschen; **c)** *(interfere with)* durcheinanderbringen ‹*Plan*›

message ['mesɪdʒ] *n.* Mitteilung, *die;* Nachricht, *die;* **send/take/leave a ~:** eine Nachricht übermitteln/entgegennehmen/hinterlassen; **give sb. a ~:** jmdm. etwas ausrichten; **can I take a ~?** kann *od.* soll ich etwas ausrichten?; **get the ~** *(fig. coll.)* verstehen; es schnallen *(salopp)*

messenger ['mesɪndʒə(r)] *n.* Bote, *der*/Botin, *die*

'messenger-boy *n.* Botenjunge, *der*

Messiah [mɪ'saɪə] *n. (lit. or fig.)* Messias, *der*

Messrs ['mesəz] *n. pl.* **a)** *(in name of firm)* ≈ Fa.; **b)** *pl. of* **Mr;** *(in list of names)* **~ A, B, and C** die Herren A, B und C

'mess-up *n.* Durcheinander, *das*

messy ['mesɪ] *adj.* **a)** *(dirty)* schmutzig; *(untidy)* unordentlich; **be a ~ eater** sich beim Essen bekleckern; **b)** *(awkward)* vertrackt *(ugs.)*

met *see* **meet**

metabolism [mɪ'tæbəlɪzm] *n. (Physiol.)* Metabolismus, *der (fachspr.);* Stoffwechsel, *der*

metal ['metl] **1.** *n.* Metall, *das.* **2.** *adj.* Metall-; **be ~:** aus Metall sein. **3.** *v.t.*, *(Brit.)* **-ll-** *(Brit.: surface)* schottern ‹*Straße*›

'metal-detector *n.* Metallsuchgerät, *das*

metallic [mɪ'tælɪk] *adj.* metallisch; Metall‹*salz, -oxid*›; **have a ~ taste** nach Metall schmecken

metallurgist [mɪ'tælədʒɪst, 'metəlɜ:dʒɪst] *n.* Metallurg, *der*/Metallurgin, *die*

metallurgy [mɪ'tælədʒɪ, 'metəlɜ:dʒɪ] *n., no pl.* Metallurgie, *die*

metal: ~ polish *n.* Metallputzmittel, *das;* **~work** *n., no pl.* **a)** *(activity)* Metallbearbeitung, *die;* **b)** *(products)* Metallarbeiten; **a piece of ~work** eine Metallarbeit; **~worker** *n.* Metallarbeiter, *der*/-arbeiterin, *die*

metamorphic [metə'mɔ:fɪk] *adj. (Geol.)* metamorph ‹*Gestein*›

metamorphose [metə'mɔ:fəʊz] *v.i.* sich verwandeln (**into** in + *Akk.*)

metamorphosis [metə'mɔ:fəsɪs, metəmɔ:'fəʊsɪs] *n., pl.* **metamorphoses** [metə'mɔ:fəsi:z, metəmɔ:'fəʊsi:z] Metamorphose, *die* (**into** in + *Akk.*)

metaphor ['metəfə(r)] *n.* **a)** *no pl., no art. (stylistic device)* **[the use of] ~:** der Gebrauch von Metaphern; **b)** *(instance)* Metapher, *die;* **mixed ~:** Bildbruch, *der*

metaphorical [metə'fɒrɪkl] *adj.*, **metaphorically** [metə'fɒrɪkəlɪ] *adv.* metaphorisch

metaphysical [metə'fɪzɪkl] *adj. (Philos.)* metaphysisch

metaphysics [metə'fɪzɪks] *n., no pl. (Philos.)* Metaphysik, *die*

mete [mi:t] *v.t. (literary)* **~ out** zuteil werden lassen *(geh.)* ‹*Belohnung*› (**to** *Dat.*); auferlegen ‹*Strafe*› (**to** *Dat.*)

meteor ['mi:tɪə(r)] *n. (Astron.)* Meteor, *der*

meteoric [mi:tɪ'ɒrɪk] *adj.* **a)** *(Astron.)* Meteor‹*schweif, -tätigkeit*›; meteorisch; **b)** *(fig.)* kometenhaft

meteorite ['mi:tɪəraɪt] *n. (Astron.)* Meteorit, *der*

meteorological [mi:tɪərə'lɒdʒɪkl] *adj.* meteorologisch ‹*Instrument*›; Wetter‹*ballon, -bericht*›; **M~ Office** *(Brit.)* Meteorologisches Amt; Wetteramt, *das*

meteorologist [mi:tɪə'rɒlədʒɪst] *n.* Meteorologe, *der*/Meteorologin, *die*

meteorology [mi:tɪə'rɒlədʒɪ] *n., no pl.* Meteorologie, *die*

¹meter ['mi:tə(r)] **1.** *n.* **a)** Zähler, *der; (taking coins)* Münzzähler, *der;* **b)** *(parking-~)* Parkuhr, *die.* **2.** *v.t.* [mit einem Zähler] messen ‹*[Wasser-, Gas-, Strom]verbrauch*›

²meter *(Amer.) see* **¹,²metre**

'meter maid *n. (coll.)* Politesse, *die*

methane ['mi:θeɪn, 'meθeɪn] *n. (Chem.)* Methan, *das*

method ['meθəd] *n.* **a)** *(procedure)* Methode, *die; (process)* Verfahren, *das;* **b)** *no pl., no art. (arrangement of ideas, orderliness)* System, *das;* Systematik, *die;* **there's ~ in his madness** *(fig. joc.)* der Wahnsinn hat Methode

methodical [mɪ'θɒdɪkl] *adj.* methodisch; systematisch; **be ~:** methodisch *od.* systematisch vorgehen

methodically [mɪ'θɒdɪkəlɪ] *adv.* mit Methode; systematisch

Methodist ['meθədɪst] *n. (Relig.)* Methodist, *der*/Methodistin, *die; attrib.* Methodisten‹*kapelle, -gottesdienst, -pfarrer*›

methodology [meθə'dɒlədʒɪ] *n.* **a)** *no pl., no art. (science of method)* Methodik, *die;* Methodologie, *die;* **b)** *(methods used)* Methodik, *die*

meths [meθs] *n., no pl., no indef. art. (Brit. coll.)* [Brenn]spiritus, *der*

methylated spirit[s] [meθɪleɪtɪd 'spɪrɪt(s)] *n. [pl.]* Brennspiritus, *der;* vergällter *od.* denaturierter Alkohol *(fachspr.)*

meticulous [mɪ'tɪkjʊləs] *adj. (scrupulous)* sorgfältig; *(overscrupulous)* übergenau; **be ~ about sth.** es peinlich genau mit etw. nehmen

meticulously [mɪ'tɪkjʊləslɪ] *adv. (scrupulously)* sorgfältig; *(overscrupulously)* übergenau; **~ clean** peinlich sauber

¹metre ['mi:tə] *n. (Brit.: poetic rhythm, metrical group)* Metrum, *das*

²metre *n. (Brit.: unit)* Meter, *der od. das;* **sell cloth by the ~:** Stoff meterweise verkaufen

metric ['metrɪk] *adj.* metrisch; **~ system** metrisches System; **go ~** *(coll.)* das metrische System einführen

metrical ['metrɪkl] *adj.* metrisch

metrication [metrɪ'keɪʃn] *n.* Umstellung auf das metrische System

metronome ['metrənəʊm] *n. (Mus.)* Metronom, *das*

metropolis [mɪ'trɒpəlɪs] *n. (capital)* Hauptstadt, *die; (chief city)* Metropole, *die*

metropolitan [metrə'pɒlɪtən] *adj.* **~ New York/Tokyo** der Großraum New York/Tokio; **~ London** Großlondon *(das);* **the M~ Police** die Londoner Polizei;

~ **borough/district** *(Brit. Admin.)* Gemeinde/Bezirk im Großraum einer Großstadt

mettle ['metl] *n.* **a)** *(quality of temperament)* Wesensart, *die;* **show one's ~:** zeigen, aus welchem Holz man [geschnitzt] ist; **b)** *(spirit)* Mut, *der;* **a man of ~:** ein mutiger Mann; **be on one's ~:** zeigen müssen, was man kann

mew [mju:] **1.** *v.i.* ‹*Katze:*› miauen; ‹*Möwe:*› kreischen. **2.** *n.* *(of cat)* Miauen, *das;* *(of seagull)* Kreischen, *das*

Mexican ['meksıkən] **1.** *adj.* mexikanisch; **sb. is ~:** jmd. ist Mexikaner/Mexikanerin. **2.** *n.* Mexikaner, *der*/Mexikanerin, *die*

Mexico ['meksıkəʊ] *pr. n.* Mexiko *(das);* ~ **City** Mexiko [City] *(das)*

mg. *abbr.* **milligram[s]** mg

mi [mi:] *see* ²**me**

miaow [mı'aʊ] **1.** *v.i.* miauen. **2.** *n.* Miauen, *das*

mica ['maıkə] *n.* *(Min.)* Glimmer, *der*

mice *pl. of* **mouse**

mickey ['mıkı] *n.* *(Brit. sl.)* **take the ~ [out of sb./sth.]** jmdn./etw. durch den Kakao ziehen *(ugs.)*

micro ['maıkrəʊ] *n., pl.* **~s** *see* **microcomputer**

micro- ['maıkrəʊ] *in comb.* mikro-/Mikro-

microbe ['maıkrəʊb] *n.* *(Biol.)* Mikrobe, *die*

micro: ~bi'ology *n.* Mikrobiologie, *die;* **~chip** *n.* *(Electronics)* [Mikro]chip, *der;* **~computer** *n.* Mikrocomputer, *der;* **~dot** *n.* *(Information Sci.)* Mikrat, *das;* **~fiche** *n., pl. same or* **~fiches** Mikrofiche, *das od. der;* **~film 1.** *n.* Mikrofilm, *der;* **2.** *v.t.* auf Mikrofilm *(Akk.)* aufnehmen; **~light ['aircraft]** *n.* *(Aeronaut.)* Ultraleichtflugzeug, *das*

micrometer [maı'krɒmıtə(r)] *n.* *(Mech. Engin.)* [Fein]meßschraube, *die*

micro-'organism *n.* Mikroorganismus, *der;* Kleinstlebewesen, *das*

microphone ['maıkrəfəʊn] *n.* Mikrophon, *das*

micro'processor *n.* *(Computing)* Mikroprozessor, *der*

microscope ['maıkrəskəʊp] *n.* Mikroskop, *das*

microscopic [maıkrə'skɒpık] *adj.* **a)** mikroskopisch; **b)** *(fig.: very small)* winzig

'microwave *n.* Mikrowelle, *die;* ~ **[oven]** Mikrowellenherd, *der*

mid- [mıd] *in comb.* **in ~-air** in der Luft; **~-air collision** Zusammenstoß in der Luft; **in ~-flight/-sentence** mitten im Flug/Satz; **[in] ~-afternoon** [mitten] am Nachmittag; **~-term elections** *(Amer.)* Kongreß- und Kommunalwahlen in der Mitte der Amtszeit des Präsidenten; **~-July** Mitte Juli; **the ~-60s** die Mitte der sechziger Jahre; **a man in his ~-fifties** ein Mittfünfziger; **be in one's ~-thirties** Mitte Dreißig sein

midday ['mıddeı, mıd'deı] *n.* **a)** *(noon)* zwölf Uhr; **round about ~:** um die Mittagszeit; **b)** *(middle of day)* Mittag, *der;* *attrib.* Mittags-

middle ['mıdl] **1.** *attrib. adj.* mittler...; **the ~ one** der/die/das mittlere; ~ **point** Mittelpunkt, *der.* **2.** *n.* **a)** Mitte, *die;* *(central part)* Mittelteil, *der;* **in the ~ of the room/the table** in der Mitte des Zimmers/des Tisches; *(emphatic)* mitten im Zimmer/auf dem Tisch; **right in the ~ of Manchester** genau im Zentrum von Manchester; **in the ~ of the forest** mitten im Wald; **fold sth. down the ~:** etw. in der Mitte falten; **in the ~ of the day** mittags; **in the ~ of the morning/afternoon** mitten am Vor-/Nachmittag; **in the ~ of the night/week** mitten in der Nacht/Woche; **be in the ~ of doing sth.** *(fig.)* gerade mitten dabei sein, etw. zu tun; **b)** *(waist)* Taille, *die*

middle: ~ 'age *n.* mittleres [Lebens]alter; **~-aged** ['mıdleıdʒd] *adj.* mittleren Alters *nachgestellt;* **M~ 'Ages** *n. pl.* **the M~ Ages** das Mittelalter; **~-brow** *adj.* *(coll.)* für den [geistigen] Normalverbraucher *nachgestellt (ugs.);* **~ 'class** *n.* Mittelstand, *der;* **~-class** *adj.* bürgerlich ‹*Vorort, Einstellung, Moral, Werte*›; **~-class people** Mittelständler; **~ 'course** *n.* Mittelweg, *der;* **M~ 'East** *pr. n.* **the M~ East** der Nahe [und Mittlere] Osten; **M~ 'Eastern** *adj.* nahöstlich; des Nahen Ostens *nachgestellt;* ‹*Person*› aus dem Nahen Osten; **~ finger** *n.* Mittelfinger, *der;* **~man** *n.* *(Commerc.)* Zwischenhändler, *der*/-händlerin, *die;* *(fig.)* Vermittler, *der*/Vermittlerin, *die;* **~ name** *n.* zweiter Vorname; **~-of-the-'road** *adj.* gemäßigt; moderat; **~-of-the-road politician/politics** Politiker/Politik der Mitte; **~ school** *n.* *(Brit.)* *Schule für 9- bis 13jährige;* **~-size[d]** *adj.* mittelgroß; **~ 'way** *see* **~ course; ~weight** *n.* *(Boxing etc.)* Mittelgewicht, *das;* *(person also)* Mittelgewichtler, *der;* **M~ 'West** *pr. n.* *(Amer.)* **the M~ West** der Mittlere Westen

middling ['mıdlıŋ] **1.** *adj.* **a)** *(second-rate)* mittelmäßig; **b)** *(moderately good)* **[fair to] ~:** ganz ordentlich *(ugs.);* [ganz] passabel; **c)** *(coll.: in fairly good health)* mittelprächtig *(ugs. scherzh.).* **2.** *adv.* recht; *(only moderately)* ganz

'midfield *n.* *(Footb.)* Mittelfeld, *das;* **play in ~:** im Mittelfeld spielen; *attrib.* ~ **player** Mittelfeldspieler, *der*

midget ['mıdʒıt] **1.** *n.* **a)** *(person)* Liliputaner, *der*/Liliputanerin, *die;* Zwerg, *der*/Zwergin, *die;* **b)** *(thing)* Zwerg, *der (fig.);* *(animal)* Zwergform, *die.* **2.** *adj.* winzig; Mini‹*flugzeug, -U-Boot*›

Midland ['mıdlənd] **1.** *n.* **the ~s** *(Brit.)* Mittelengland. **2.** *adj.* **~[s]** *(Brit.)* in den Midlands *nachgestellt;* ‹*Dialekt*› der Midlands

'midnight *n.* Mitternacht, *die;* *attrib.* Mitternachts‹*stunde, -messe, -zug*›; mitternächtlich ‹*Festgelage, Feiern*›

midnight 'sun *n.* Mitternachtssonne, *die*

'midpoint *n.* Mitte, *die*

midriff ['mıdrıf] *n.* **the bulge below his ~:** die Wölbung seiner Taillengegend; **with bare ~:** nabelfrei

midst [mıdst] *n.* **in the ~ of sth.** mitten in einer Sache; **be in the ~ of doing sth.** gerade mitten dabei sein, etw. zu tun; **in our/their ~:** in unserer/ihrer Mitte

midsummer ['mıdsʌmə(r), mıd'sʌmə(r)] *n.* die [Zeit der] Sommersonnenwende; **[on] M~['s] Day** [am] Johannistag

midway ['mıdweı, mıd'weı] *adv.* auf halbem Weg[e] ‹*sich treffen, sich befinden*›; ~ **through sth.** *(fig.)* mitten in etw. *(Dat.)*

midwife ['mıdwaıf] *n., pl.* **midwives** ['mıdwaıvz] Hebamme, *die*

mid'winter *n.* die [Zeit der] Wintersonnenwende; der Mittwinter

miff [mıf] *v.t.* *(coll.)* verärgern; **be ~ed** beleidigt *od.* *(ugs.)* eingeschnappt sein

¹**might** *see* **may**

²**might** [maıt] *n.* **a)** *(force)* Gewalt, *die;* *(inner strength)* Macht, *die;* **with all one's ~:** mit aller Kraft; **b)** *(power)* Macht, *die;* ~ **is right** Macht geht vor Recht

might-have-been ['maıtəvbi:n] *n. jemand, der es zu etwas hätte bringen können;* **he is a ~:** er hat seine Chancen verpaßt

mightily ['maıtılı] *adv.* *(coll.: very)* überaus; **be ~ amused** sich köstlich amüsieren

mightn't ['maıtnt] *(coll.)* = **might not;** *see* **may**

mighty ['maɪtɪ] **1.** *adj.* **a)** *(powerful)* mächtig; gewaltig ‹*Krieger, Anstrengung*›; **b)** *(massive)* gewaltig; **c)** *(coll.: great)* riesig. *See also* **high 1e. 2.** *adv. (coll.)* verdammt *(ugs.)*
migraine ['mi:greɪn, 'maɪgreɪn] *n. (Med.)* Migräne, *die*
migrant ['maɪgrənt] **1.** *adj.* **a)** ~ **tribe** Nomadenstamm, *der;* ~ **worker** Wanderarbeiter, *der*/-arbeiterin, *die; (in EC)* Gastarbeiter, *der*/-arbeiterin, *die;* **b)** ~ **bird/fish** Zugvogel, *der*/Wanderfisch, *der.* **2.** *n.* **a)** Auswanderer, *der*/Auswanderin, *die;* **b)** *(bird)* Zugvogel, *der; (fish)* Wanderfisch, *der*
migrate [maɪ'greɪt] *v.i.* **a)** *(from rural area to town)* abwandern; *(to another country)* auswandern; *(to another place of work)* überwechseln; **b)** ‹*Vogel:*› fortziehen; ‹*Fisch:*› wandern; ~ **to the south/sea** nach Süden ziehen/zum Meer wandern
migration [maɪ'greɪʃn] *n.* **a)** *see* **migrate a:** Abwandern, *das;* Auswandern, *das;* Überwechseln, *das;* **b)** *(of birds)* Fortziehen, *das; (of fish)* Wandern, *das; (instance) (of birds)* Zug, *der; (of fish)* Wanderung, *die*
migratory ['maɪgrətərɪ, maɪ'greɪtərɪ] *adj.* **a)** ~ **tribe** Nomadenstamm, *der;* **b)** ~ **bird/fish** Zugvogel, *der*/Wanderfisch, *der*
mike [maɪk] *n. (coll.)* Mikro, *das*
Milan [mɪ'læn] *pr. n.* Mailand *(das)*
mild [maɪld] **1.** *adj.* **a)** sanft ‹*Person*›; mild ‹*Urteil, Bestrafung, Kritik*›; leicht ‹*Erkrankung, Gefühlsregung*›; gemäßigt ‹*Ausdrucksweise, Sprache*›; leicht ‹*Aufregung*› mild ‹*Wetter, Winter*›; mild, leicht ‹*Arzneimittel, Stimulans*›; **b)** *(not strong in taste)* mild. **2.** *n. schwach gehopfte englische Biersorte*
mildew ['mɪldju:] *n. (on paper, cloth, wood)* Schimmel, *der; (on plant)* Mehltau, *der*
mildly ['maɪldlɪ] *adv.* **a)** *(gently)* mild[e]; **b)** *(slightly)* ein bißchen *od.* wenig ‹*enttäuscht, bestürzt, ermutigend, begeistert*›; **c)** **to put it ~:** gelinde gesagt
mile [maɪl] *n.* **a)** Meile, *die;* ~ **after** *or* **upon** ~ *or* **~s and ~s of sand/beaches** meilenweit Sand/Strände; **~s per hour** Meilen pro Stunde; **not a million ~s from** *(joc.)* nicht allzu weit von; **b)** *(fig. coll.: great amount)* **win/miss by a ~:** haushoch gewinnen/meilenweit verfehlen; **~s better/too big** tausendmal besser/viel zu groß; **be ~s ahead of sb.** jmdm. weit voraus sein; **sb. is ~s away** *(in thought)* jmd. ist mit seinen Gedanken ganz woanders; **c)** *(race)* Meilenlauf, *der*
mileage ['maɪlɪdʒ] *n.* **a)** *(number of miles)* [Anzahl der] Meilen; **a low ~** *(on milometer)* ein niedriger Meilenstand; **b)** *(number of miles per gallon)* [Benzin]verbrauch, *der;* **what ~ do you get with your car?** wieviel verbraucht dein Auto?; **c)** *(fig.: benefit)* Nutzen, *der;* **there is no ~ in the idea** dieser Vorschlag rentiert sich nicht
'**milestone** *n. (lit. or fig.)* Meilenstein, *der*
milieu [mɪ'ljɜ:, 'mi:ljɜ:] *n., pl.* **~x** [mɪ'ljɜ:z, 'mi:ljɜ:z] *or* **~s** Milieu, *das*
militancy ['mɪlɪtənsɪ] *n., no pl.* Kampfbereitschaft, *die;* Militanz, *die*
militant ['mɪlɪtənt] **1.** *adj.* **a)** *(aggressively active)* kämpferisch; militant; **b)** *(engaged in warfare)* kriegführend. **2.** *n.* Militante, *der/die*
militarise *see* **militarize**
militarism ['mɪlɪtərɪzm] *n.* Militarismus, *der*
militarize ['mɪlɪtəraɪz] *v.t.* militarisieren
military ['mɪlɪtərɪ] **1.** *adj.* militärisch; Militär‹*regierung, -akademie, -uniform, -parade*›; ~ **service** Militärdienst, *der;* Wehrdienst, *der.* **2.** *n., constr. as sing. or pl.* **the ~:** das Militär
militate ['mɪlɪteɪt] *v.i.* ~ **against/in favour of sth.** [deutlich] gegen/für etw. sprechen; *(have effect)* sich zuungunsten/zugunsten einer Sache *(Gen.)* auswirken
militia [mɪ'lɪʃə] *n.* Miliz, *die*
milk [mɪlk] **1.** *n.* Milch, *die;* **it's no use crying over spilt ~** *(prov.)* [was] passiert ist[, ist] passiert. **2.** *v.t.* *(draw ~ from)* melken; *(fig.: get money out of)* melken *(salopp)*
milk: ~ **bar** *n.* Milchbar, *die;* **~-bottle** *n.* Milchflasche, *die;* ~ '**chocolate** *n.* Milchschokolade, *die;* **~-churn** *n.* Milchkanne, *die;* **~-float** *n. (Brit.)* Milchwagen, *der;* **~-jug** *n.* Milchkrug, *der; (with tea, coffee, etc.)* Milchkännchen, *das;* **~maid** *n.* Melkerin, *die;* **~man** ['mɪlkmən] *n., pl.* **~men** ['mɪlkmən] Milchmann, *der;* **~-powder** *n.* Milchpulver, *das;* ~ **run** *n. (fig.)* [übliche] Tour; ~ **shake** *n.* Milk-Shake, *der;* Milchshake, *der;* **~-tooth** *n.* Milchzahn, *der*
milky ['mɪlkɪ] *adj.* milchig; ~ **coffee** Milchkaffee, *der*
Milky 'Way *n.* Milchstraße, *die*
mill [mɪl] **1.** *n.* **a)** Mühle, *die;* **b)** *(factory)* Fabrik, *die; (machine)* Maschine, *die;* ~ **town** ≈ Textilstadt, *die.* **2.** *v.t.* **a)** mahlen ‹*Getreide*›; **b)** fräsen ‹*Metallgegenstand*›; rändeln ‹*Münze*›
~ **a'bout** *(Brit.),* ~ **a'round** *v.i.* durcheinanderlaufen; **a mass of people ~ing about** *or* **around in the square** eine Menschenmenge, die sich hin und her über den Platz schiebt/schob
millennium [mɪ'lenɪəm] *n., pl.* **~s** *or* **millennia** [mɪ'lenɪə] **a)** Millennium, *das;* **b)** *(Relig.)* Tausendjähriges Reich; Millennium, *das (fachspr.)*
millepede ['mɪlɪpi:d] *n. (Zool.)* Tausendfüß[l]er, *der*
miller ['mɪlə(r)] *n.* Müller, *der*
millet ['mɪlɪt] *n. (Bot.)* Hirse, *die*
milli- ['mɪlɪ] *pref.* milli-/Milli-
'**milligram** *n.* Milligramm, *das*
'**millilitre** *(Brit.; Amer.:* **milliliter**) *n.* Milliliter, *der od. das*
'**millimetre** *(Brit.; Amer.:* **millimeter**) *n.* Millimeter, *der*
milliner ['mɪlɪnə(r)] *n.* Putzmacher, *der*/-macherin, *die;* Modist, *der*/Modistin, *die*
millinery ['mɪlɪnərɪ] *n., no pl.* **a)** *(articles)* Hüte; **b)** *(business)* Hutmacherei, *die*
million ['mɪljən] **1.** *adj.* **a) a** *or* **one ~:** eine Million; **two/several ~:** zwei/mehrere Millionen; **a** *or* **one ~ and one** eine Million ein; **half a ~** eine halbe Million; **b) a ~ [and one]** *(fig.: innumerable)* tausend; **never in a ~ years** nie im Leben *(ugs.).* **2.** *n.* **a)** Million, *die;* **a** *or* **one/two ~:** eine Million/zwei Millionen; **in** *or* **by ~s** millionenweise; **a ~ and one** *etc.* eine Million einer/eine/eins; **the starving ~s** die Millionen [von] Hungerleidenden; **b)** *(indefinite amount)* **there were ~s of people** eine Unmenge Leute waren da; **he is a man/she is one in a ~:** so jemanden wie ihn/sie findet man nicht noch einmal
millionaire [mɪljə'neə(r)] *n. (lit. or fig.)* Millionär, *der*/Millionärin, *die*
millionth ['mɪljənθ] **1.** *adj.* millionst...; **a ~ part** ein Millionstel. **2.** *n. (fraction)* Millionstel, *das*
millipede *see* **millepede**
mill: **~-owner** *n.* Textilfabrikant, *der;* **~-pond** *n.* **the sea was like a ~-pond** die See war ruhig wie ein Teich; **~-race** *n.* Mühlbach, *der;* **~stone** *n.* Mühlstein,

der; **be a ~stone round sb.'s neck** *(fig.)* jmdm. ein Klotz am Bein sein *(ugs.);* **~-wheel** *n.* Mühlrad, *das*

milometer [maɪ'lɒmɪtə(r)] *n.* Meilenzähler, *der*

mime [maɪm] **1.** *n.* **a)** *(performance)* Pantomime, *die;* **b)** *no pl., no art. (art)* Pantomimik, *die;* Pantomime, *die (ugs.).* **2.** *v.i.* pantomimisch agieren. **3.** *v.t.* pantomimisch darstellen

mimic ['mɪmɪk] **1.** *n.* Imitator, *der.* **2.** *v.t.,* **-ck-: a)** nachahmen; imitieren; *(ridicule by imitating)* parodieren; **b)** *(resemble closely)* aussehen wie

mimicry ['mɪmɪkrɪ] *n., no pl.* Nachahmen, *das*

Min. *abbr.* **Minister/Ministry** Min.

min. *abbr.* **a) minute[s]** Min; **b) minimum** *(adj.)* mind., *(n.)* Min.

minaret ['mɪnəret] *n.* Minarett, *das*

mince [mɪns] **1.** *n.* Hackfleisch, *das;* Gehackte, *das.* **2.** *v.t.* **~ beef** Rindfleisch durch den [Fleisch]wolf drehen; **~d meat** Hackfleisch, *das;* **not ~ matters** die Dinge beim Namen nennen; **not ~ one's words** kein Blatt vor den Mund nehmen. **3.** *v.i.* trippeln

mince: ~meat *n.* **a)** Hackfleisch, *das;* Gehackte, *das;* **make ~meat of sb.** *(fig.)* Hackfleisch aus jmdm. machen *(ugs.);* **b)** *(sweet) süße Pastetenfüllung aus Obst, Rosinen, Gewürzen, Nierenfett usw.;* **~ 'pie** *n. mit süßem „mincemeat" gefüllte Pastete*

mincer ['mɪnsə(r)] *n.* Fleischwolf, *der*

mind [maɪnd] **1.** *n.* **a)** *(remembrance)* **bear** *or* **keep sth. in ~:** an etw. *(Akk.)* denken; etw. nicht vergessen; **have [it] in ~ to do sth.** vorhaben, etw. zu tun; **bring sth. to ~:** etw. in Erinnerung rufen; **sth. comes into sb.'s ~:** jmdm. fällt etw. ein; **it went out of my ~:** ich habe es vergessen; es ist mir entfallen; **put sth./sb. out of one's ~:** etw./jmdn. aus seinem Gedächtnis streichen; **b)** *(opinion)* **give sb. a piece of one's ~:** jmdm. gründlich die Meinung sagen; **in** *or* **to my ~:** meiner Meinung *od.* Ansicht nach; **be of one** *or* **of the same ~, be in one ~:** einer Meinung sein; **be in two ~s about sth.** [sich *(Dat.)*] unschlüssig über etw. *(Akk.)* sein; **change one's ~:** seine Meinung ändern; **have a ~ of one's own** seinen eigenen Kopf haben; **I have a good ~/half a ~ to do that** ich hätte große Lust/nicht übel Lust, das zu tun; **make up one's ~, make one's ~ up** sich entscheiden; **make up one's ~ to do sth.** sich entschließen, etw. zu tun; **read sb.'s ~:** jmds. Gedanken lesen; **c)** *(direction of thoughts)* **his ~ is on other things** er ist mit den Gedanken woanders; **give** *or* **put** *or* **turn one's ~ to** sich konzentrieren auf (+ *Akk.*) ‹*Arbeit, Aufgabe, Angelegenheit*›; **I have had sb./sth. on my ~:** jmd./etw. hat mich beschäftigt; *(worried)* ich habe mir Sorgen wegen jmdm./etw. gemacht; **she has a lot of things on her ~:** sie hat viele Sorgen; **sth. preys** *or* **weighs on sb.'s ~:** etw. macht jmdm. zu schaffen; **take sb.'s ~ off sth.** jmdn. von etw. ablenken; **keep one's ~ on sth.** sich auf etw. *(Akk.)* konzentrieren; **close one's ~ to sth.** sich einer Sache *(Dat.)* verschließen *(geh.);* **d)** *(way of thinking and feeling)* Denkweise, *die;* **frame of ~:** [seelische] Verfassung; **state of ~:** [Geistes]zustand, *der;* **be in a frame of ~ to do sth.** in der Verfassung sein, etw. zu tun; **have a logical ~:** logisch denken; **e)** *(seat of consciousness, thought, volition)* Geist, *der;* **it's all in the ~:** es ist alles nur Einstellung; **in one's ~:** im stillen; **in my ~'s eye** vor meinem geistigen Auge; im Geiste; **nothing could be further from my ~ than ...:** nichts läge mir ferner, als ...; **f)** *(intellectual powers)* Verstand, *der;* Intellekt, *der;* **have a very good ~:** einen klaren *od.* scharfen Verstand haben; **g)** *(normal mental faculties)* Verstand, *der;* **lose** *or* **go out of one's ~:** den Verstand verlieren; **be out of one's ~:** den Verstand verloren haben; **in one's right ~:** bei klarem Verstand. **2.** *v.t.* **a)** *(heed)* **don't ~ what he says** gib nichts auf sein Gerede; **~ what I say** glaub mir; **let's do it, and never ~ the expense** machen wir es doch, egal, was es kostet; **b)** *(concern oneself about)* **he ~s a lot what people think of him** es ist für ihn sehr wichtig, was die Leute von ihm denken; **I can't afford a bicycle, never ~ a car** ich kann mir kein Fahrrad leisten, geschweige denn ein Auto; **never ~ him/that** *(don't be anxious)* er/das kann dir doch egal sein *(ugs.);* **never ~ how/where ...:** es tut nichts zur Sache, wie/wo ...; **don't ~ me** nimm keine Rücksicht auf mich; *(don't let my presence disturb you)* laß dich [durch mich] nicht stören; *(iron.)* nimm bloß keine Rücksicht auf mich; **~ the doors!** Vorsicht an den Türen!; **~ one's P's and Q's** sich anständig benehmen; **c)** *usu. neg. or interrog. (object to)* **did he ~ being woken up?** hat es ihm was ausgemacht, aufgeweckt zu werden?; **would you ~ opening the door?** würdest du bitte die Tür öffnen?; **do you ~ my smoking?** stört es Sie *od.* haben Sie etwas dagegen, wenn ich rauche?; **I wouldn't ~ a walk** ich hätte nichts gegen einen Spaziergang; **d)** *(remember and take care)* **~ you don't leave anything behind** denk daran, nichts liegenlassen!; **~ how you go!** paß auf! sei vorsichtig!; *(as general farewell)* mach's gut! *(ugs.);* **~ you get this work done** sieh zu, daß du mit dieser Arbeit fertig wirst!; **e)** *(have charge of)* aufpassen auf (+ *Akk.*); **~ the shop** *or (Amer.)* **the store** *(fig.)* sich um den Laden kümmern *(ugs.).* **3.** *v.i.* **a) ~!** Vorsicht!; Achtung!; **b)** *usu. in imper. (take note)* **follow the signposts, ~, or ...:** denk daran und halte dich an die Wegweiser, sonst...; **I didn't know that, ~, or ...:** das habe ich allerdings nicht gewußt, sonst ...; **c)** *(care, object)* **do you '~?** *(may I?)* hätten Sie etwas dagegen?; *(please do not)* ich muß doch sehr bitten; **he doesn't ~ about your using the car** er hat nichts dagegen, wenn Sie den Wagen benutzen; **if you don't ~:** wenn es dir recht ist; **d)** *(give heed)* **never [you] ~** *(it's not important)* macht nichts; ist nicht schlimm; *(it's none of your business)* sei nicht so neugierig; **never ~: I can do it** schon gut – das kann ich machen; **never ~ about that now!** laß das jetzt mal [sein/liegen]!; **never ~ about him – what happened to her?** er interessiert mich nicht – was ist ihr passiert?

~ 'out *v.i.* aufpassen (**for** auf + *Akk.*); **~ out!** Vorsicht!

mind-boggling ['maɪndbɒglɪŋ] *adj. (coll.)* wahnsinnig *(ugs.)*

minded ['maɪndɪd] *adj.* **a)** *(disposed)* **be ~ to do sth.** bereit *od. (geh.)* geneigt sein, etw. zu tun; **b) mechanically ~:** technisch veranlagt; **he is not in the least politically ~:** er ist vollkommen unpolitisch

mindful ['maɪndfl] *adj.* **be ~ of sth.** *(take into account)* etw. bedenken *od.* berücksichtigen; *(give attention to)* an etw. *(Akk.)* denken

mindless ['maɪndlɪs] *adj.* geistlos, *(ugs.)* hirnlos ‹*Mensch*›; sinnlos ‹*Handlung, Gewalt*›

mind: ~-reader *see* **thought-reader; ~-set** *n.* Denkart, *die*

[1]**mine** [maɪn] **1.** *n.* **a)** *(for coal)* Bergwerk, *das; (for metal, diamonds, etc.)* Bergwerk, *das;* Mine, *die;* **go** *or* **work down the ~:** unter Tage arbeiten; **b)** *(fig.: abundant source)* unerschöpfliche Quelle; **he is a ~ of useful facts/of information** von ihm kann man eine Menge Nützliches/eine Menge erfahren; **c)** *(explosive device)* Mine, *die.* **2.** *v.t.* **a)** schürfen ‹*Gold*›; abbauen, fördern ‹*Erz, Kohle, Schiefer*›; **~ an area for ore** *etc.* in einem Gebiet Erz *usw.* abbauen *od.* fördern; **b)** *(Mil.: lay ~s in)* verminen. **3.** *v.i.* Bergbau betreiben; **~ for** *see* **2 a**

[2]**mine** *poss. pron.* **a)** *pred.* meiner/meine/mein[e]s; der/die/das meinige *(geh.);* **you do your best and I'll do ~:** du tust dein Bestes und ich auch; **those big feet of ~:** meine großen Quanten *(ugs.); see also* **hers; b)** *attrib. (arch./poet.)* mein

'**minefield** *n. (lit. or fig.)* Minenfeld, *das*

miner ['maɪnə(r)] *n.* Bergmann, *der;* Kumpel, *der (Bergmannsspr.)*

mineral ['mɪnərl] **1.** *adj.* mineralisch; Mineral‹*salz, -quelle*›. **2.** *n.* **a)** Mineral, *das;* **a country rich in ~s** ein an Bodenschätzen reiches Land; **b)** *esp. in pl. (Brit.: soft drink)* Erfrischungsgetränk, *das*

mineralogy [mɪnə'rælədʒɪ] *n.* Mineralogie, *die*

mineral: ~ oil *n.* Mineralöl, *das;* **~ water** *n.* Mineralwasser, *das*

minestrone [mɪnɪ'strəʊnɪ] *n.* *(Gastr.)* Minestrone, *die*

mine: ~sweeper *n.* Minensuchboot, *das;* **~worker** *n.* Bergmann, *der;* Kumpel, *der (Bergmannsspr.)*

mingle ['mɪŋgl] **1.** *v.t.* [ver]mischen. **2.** *v.i.* sich [ver]mischen (**with** mit); **~ with** *or* **among the crowds** sich unters Volk mischen

mingy ['mɪndʒɪ] *adj. (Brit. coll.)* mick[e]rig *(ugs.)* ‹*Gegenstand*›; knick[e]rig *(ugs.)* ‹*Person*›; lumpig *(ugs.)* ‹*Betrag*›

mini ['mɪnɪ] *n. (coll.)* **a)** *(car)* **M~, (P)** Mini, *der;* **b)** *(skirt)* Mini, *der (ugs.)*

mini- *in comb.* Mini-; Klein‹*bus, -wagen, -taxi*›

miniature ['mɪnɪtʃə(r)] **1.** *n.* **a)** *(picture)* Miniatur, *die;* **b)** *(small version)* Miniaturausgabe, *die;* **in ~:** im Kleinformat. **2.** *adj.* **a)** *(small-scale)* Miniatur-; **b)** *(smaller than normal)* Mini- *(ugs.);* Kleinst-; **~ poodle** Zwergpudel, *der;* **~ golf** Minigolf, *das;* **~ camera** Kleinstbildkamera, *die;* **~ railway** Miniaturbahn, *die*

mini: ~bus *n.* Kleinbus, *der;* **~cab** *n.* Kleintaxi, *das;* Minicar, *das;* **~computer** *n.* Minicomputer, *der*

minim ['mɪnɪm] *n. (Brit. Mus.)* halbe Note

minimal ['mɪnɪml] *adj.*, **minimally** ['mɪnɪməlɪ] *adv.* minimal

minimize (minimise) ['mɪnɪmaɪz] *v.t.* **a)** auf ein Mindestmaß reduzieren; **b)** *(understate)* bagatellisieren; verharmlosen ‹*Gefahr*›

minimum ['mɪnɪməm] **1.** *n., pl.* **minima** ['mɪnɪmə] Minimum, *das* (**of** an + *Dat.*); **keep sth. to a ~:** etw. so gering wie möglich halten; **a ~ of £5** mindestens 5 Pfund; **at the ~:** mindestens. **2.** *attrib. adj.* Mindest-; **~ temperatures tonight around 5°** nächtliche Tiefsttemperaturen um 5°; **~ wage** Mindestlohn, *der*

mining ['maɪnɪŋ] *n.* Bergbau, *der; attrib.* Bergbau-; **~ area** Bergbaugebiet, *das;* Revier, *das*

mining: ~ engineer *n.* Berg[bau]ingenieur, *der;* **~ industry** *n.* Montanindustrie, *die;* Bergbau, *der;* **~ town** *n.* Bergbaustadt, *die;* **~ village** *n.* Bergbaudorf, *das*

'**mini-roundabout** *n. (Brit.) sehr kleiner, oft nur aufs Pflaster aufgezeichneter Kreisverkehr*

'**miniskirt** *n.* Minirock, *der*

minister ['mɪnɪstə(r)] **1.** *n.* **a)** *(Polit.)* Minister, *der*/Ministerin, *die;* **M~ of State** *(Brit.)* ≈ Staatssekretär, *der*/-sekretärin, *die;* **b)** *(Eccl.)* **~ [of religion]** Geistliche, *der/die;* Pfarrer, *der*/Pfarrerin, *die.* **2.** *v.i.* **~ to sb.'s wants/needs** jmds. Wünsche/Bedürfnisse befriedigen

ministerial [mɪnɪ'stɪərɪəl] *adj.* *(Polit.)* Minister-; ministeriell

ministry ['mɪnɪstrɪ] *n.* **a)** *(Government department or building)* Ministerium, *das;* **~ official** Ministerialbeamte, *der*/-beamtin, *die;* **b)** *(profession of clergyman)* geistliches Amt; **go into** *or* **enter the ~:** Geistlicher werden

mink [mɪŋk] *n.* Nerz, *der; attrib.* **~ coat** Nerzmantel, *der*

minnow ['mɪnəʊ] *n. (Zool.)* Elritze, *die*

minor ['maɪnə(r)] **1.** *adj.* **a)** *(lesser)* kleiner...; **b)** *(unimportant)* weniger bedeutend; geringer ‹*Bedeutung*›; leicht ‹*Operation, Verletzung, Anfall*›; Neben‹*figur, -rolle*›; **~ matter** Nebensächlichkeit, *die;* **~ road** kleine Straße; **c)** *(Mus.)* Moll-; **~ key/chord** Molltonart, *die*/Mollakkord, *der;* **A ~:** a-Moll; **in a ~ key** in Moll. **2.** *n.* **a)** *(person)* Minderjährige, *der/die;* **be a ~:** minderjährig sein; **b)** *(Amer. Univ.)* Nebenfach, *das.* **3.** *v.i. (Amer.)* **~ in sth.** etw. als Nebenfach haben

minority [maɪ'nɒrɪtɪ, mɪ'nɒrɪtɪ] *n.* **a)** Minderheit, *die;* Minorität, *die;* **in the ~:** in der Minderheit; **b)** *attrib.* Minderheits‹*regierung, -bericht*›; **~ group** Minderheit, *die;* Minorität, *die;* **~ rights** Minderheitenrechte

minster ['mɪnstə(r)] *n.* Münster, *das*

minstrel ['mɪnstrl] *n.* Spielmann, *der;* fahrender Sänger

[1]**mint** [mɪnt] **1.** *n.* **a)** *(place)* Münzanstalt, *die;* Münze, *die;* **b)** *(sum of money)* **a ~ [of money]** eine schöne Stange Geld *(ugs.).* **2.** *adj.* funkelnagelneu *(ugs.);* vorzüglich ‹*Münze*› *(fachspr.);* **in ~ condition** ‹*Auto, Bild usw.*› in tadellosem Zustand. **3.** *v.t. (lit. or fig.)* prägen

[2]**mint** *n.* **a)** *(plant)* Minze, *die;* **b)** *(peppermint)* Pfefferminz, *das; attrib.* Pfefferminz-

mint 'sauce *n.* Minzsoße, *die*

minuet [mɪnjʊ'et] *n. (Mus.)* Menuett, *das*

minus ['maɪnəs] **1.** *prep.* **a)** *(with the subtraction of)* minus; weniger; *(without)* ohne; abzüglich (+ *Gen.*); **b)** *(below zero)* minus; **~ 20 degrees** 20 Grad Kälte *od.* minus 20 Grad; **c)** *(coll.: lacking)* ohne. **2.** *adj. (Math.)* negativ ‹*Wert, Menge, Größe*›; Minus‹*zeichen, -betrag*›. **3.** *n. (symbol)* Minus[zeichen], *das*

minuscule ['mɪnəskju:l] *adj.* winzig

[1]**minute** ['mɪnɪt] **1.** *n.* **a)** Minute, *die; (moment)* Moment, *der;* Augenblick, *der;* **I expect him any ~ [now]** ich erwarte ihn jeden Augenblick; **for a ~:** eine Minute/einen Moment [lang]; **in a ~** *(very soon)* gleich; **come back this ~!** komm sofort *od.* auf der Stelle zurück!; **at that very ~:** genau in diesem Augenblick; **to the ~:** auf die Minute; **up to the ~:** hochaktuell; **the ~ [that] I left** in dem Augenblick, als ich wegging; **just a ~!, wait a ~!** *(coll.)* einen Augenblick!; *(objecting)* Augenblick mal! *(ugs.);* **be five ~s' walk [away]**

fünf Minuten zu Fuß entfernt sein; **b)** *(of angle)* Minute, *die;* **c)** *in pl. (brief summary)* Protokoll, *das;* **keep** *or* **take the ~s** das Protokoll führen; **d)** *(official memorandum)* Memorandum, *das.* **2.** *v. t.* protokollieren ‹*Vernehmung, Aussage*›; zu Protokoll nehmen ‹*Bemerkung*›

²minute [maɪ'nju:t] *adj.*, **~r** [maɪ'nju:tə(r)], **~st** [maɪ'nju:tɪst] **a)** *(tiny)* winzig; **b)** *(precise)* minuziös; exakt; **with ~ care** mit peinlicher Sorgfalt

minute-hand ['mɪnɪthænd] *n.* Minutenzeiger, *der*

minutely [maɪ'nju:tlɪ] *adv.* genauestens; sorgfältigst

minx [mɪŋks] *n.* kleines Biest *(ugs.)*

miracle ['mɪrəkl] *n.* Wunder, *das;* **perform** *or* **work ~s** Wunder tun *od.* vollbringen; ‹*Mittel, Behandlung usw.:*› Wunder wirken; **economic ~:** Wirtschaftswunder, *das;* **be a ~ of ingenuity** ein Wunder an Genialität sein

miraculous [mɪ'rækjʊləs] *adj.* **a)** wunderbar; *(having ~ power)* wunderkräftig; **b)** *(surprising)* erstaunlich; unglaublich

miraculously [mɪ'rækjʊləslɪ] *adv.* **a)** auf wunderbare *od. (geh.)* wundersame Weise; **~, he escaped injury** wie durch ein Wunder blieb er unverletzt; **b)** *(surprisingly)* erstaunlicherweise

mirage ['mɪrɑ:ʒ] *n.* Fata Morgana, *die;* Luftspiegelung, *die; (fig.)* Illusion, *die*

mire ['maɪə(r)] *n.* Morast, *der*

mirror ['mɪrə(r)] **1.** *n. (lit. or fig.)* Spiegel, *der.* **2.** *v. t. (lit. or fig.)* [wider]spiegeln

mirror 'image *n.* Spiegelbild, *das*

mirth [mɜ:θ] *n.* Fröhlichkeit, *die; (laughter)* Heiterkeit, *die*

misadventure [mɪsəd'ventʃə(r)] *n.* **a)** Mißgeschick, *das;* **b)** *(Law)* **death by ~:** Tod durch Unfall

misanthropist [mɪ'zænθrəpɪst] *n.* Misanthrop, *der (geh.);* Menschenfeind, *der*

misapprehension [mɪsæprɪ'henʃn] *n.* Mißverständnis, *das;* **be under a ~:** einem Irrtum unterliegen; **have a lot of ~s about sth.** völlig falsche Vorstellungen von etw. haben

misappropriate [mɪsə'prəʊprɪeɪt] *v. t.* unterschlagen, *(Rechtsspr.)* veruntreuen ‹*Geld usw.*›

misbehave [mɪsbɪ'heɪv] *v. i. & refl.* sich schlecht benehmen

misbehaviour (*Amer.:* **misbehavior**) [mɪsbɪ'heɪvɪə(r)] *n.* schlechtes Benehmen

miscalculate [mɪs'kælkjʊleɪt] *v. t.* falsch berechnen; *(misjudge)* falsch einschätzen; **~ the distance** sich bei der Entfernung verkalkulieren

miscalculation [mɪskælkjʊ'leɪʃn] *n. (arithmetical error)* Rechenfehler, *der; (misjudgement)* Fehleinschätzung, *die*

miscarriage [mɪs'kærɪdʒ] *n.* **a)** *(Med.)* Fehlgeburt, *die;* **b)** *(Law)* **~ of justice** Justizirrtum, *der*

miscarry [mɪs'kærɪ] *v. i.* **a)** *(Med.)* eine Fehlgeburt haben; **b)** ‹*Plan, Vorhaben usw.:*› fehlschlagen

miscellaneous [mɪsə'leɪnɪəs] *adj.* **a)** *(mixed)* [kunter]bunt ‹*[Menschen]menge, Sammlung*›; **b)** *with pl. n. (of various kinds)* verschieden; verschiedenerlei

miscellany [mɪ'selənɪ] *n. (mixture)* [bunte] Sammlung; [buntes] Gemisch

mischance [mɪs'tʃɑ:ns] *n. (piece of bad luck)* unglücklicher Zufall; **by a** *or* **some ~:** durch einen unglücklichen Zufall

mischief ['mɪstʃɪf] *n.* **a)** Unsinn, *der;* Unfug, *der; (pranks)* [dumme] Streiche *Pl.; (playful malice)* Schalk, *der;* **be** *or* **get up to [some] ~:** etwas anstellen; **keep out of ~:** keinen Unfug machen; **b)** *(harm)* Schaden, *der;* **do sb./oneself a ~** *(coll.)* jmdm./sich etwas antun; **c)** *(person)* Schlawiner, *der (ugs.)*

'mischief-maker *n.* Böswillige, *der/die*

mischievous ['mɪstʃɪvəs] *adj.* **a)** spitzbübisch, schelmisch ‹*Blick, Gesichtsausdruck, Lächeln*›; **~ trick** Schabernack, *der;* **b)** *(malicious)* boshaft ‹*Person*›; böse ‹*Absicht*›; **c)** *(harmful)* bösartig ‹*Gerücht*›; böse ‹*Zeitungsartikel*›

misconceive [mɪskən'si:v] *v. t.* **be ~d** ‹*Projekt, Vorschlag, Aktion:*› schlecht konzipiert sein

misconception [mɪskən'sepʃn] *n.* falsche Vorstellung **(about** von); **be [labouring] under a ~ about sth.** sich *(Dat.)* eine falsche Vorstellung von etw. machen

misconduct [mɪs'kɒndʌkt] *n., no pl. (improper conduct)* unkorrektes Verhalten; *(Sport)* unsportliches *od.* unfaires Verhalten

misconstrue [mɪskən'stru:] *v. t.* mißverstehen; **~ sb.'s meaning** jmdn. mißverstehen

miscount [mɪs'kaʊnt] **1.** *v. i.* sich verzählen; *(when counting votes)* falsch [aus]zählen. **2.** *v. t.* falsch zählen

misdeed [mɪs'di:d] *n.* **a)** *(evil deed)* Missetat, *die (geh. veralt.);* **b)** *(crime)* Verbrechen, *das*

misdemeanour (*Amer.:* **misdemeanor**) [mɪsdɪ'mi:nə(r)] *n.* **a)** Missetat, *die (veralt., scherzh.);* **b)** *(Law)* Vergehen, *das*

misdirect [mɪsdɪ'rekt, mɪsdaɪ'rekt] *v. t.* falsch einsetzen ‹*Energien*›; in die falsche Richtung schicken ‹*nach dem Weg Fragenden*›

miser ['maɪzə(r)] *n.* Geizhals, *der;* Geizkragen, *der (ugs.)*

miserable ['mɪzərəbl] *adj.* **a)** *(unhappy)* unglücklich; erbärmlich, elend ‹*Leben[sbedingungen]*›; **feel ~:** sich elend fühlen; **b)** *(causing wretchedness)* trostlos; trist ‹*Wetter, Urlaub*›; **c)** *(contemptible, mean)* armselig; **a ~ five pounds** klägliche *od. (ugs.)* miese fünf Pfund

miserably ['mɪzərəblɪ] *adv.* **a)** unglücklich; elend, jämmerlich ‹*leben, zugrunde gehen*›; **~ poor** bettelarm; **b)** *(meanly)* miserabel, *(ugs.)* mies ‹*bezahlt*›; **c)** kläglich, jämmerlich ‹*versagen*›; völlig, total ‹*verpfuscht, unzureichend*›

miserly ['maɪzəlɪ] *adj.* geizig; armselig ‹*Portion, Essen*›

misery ['mɪzərɪ] *n.* **a)** Elend, *das;* **make sb.'s life a ~:** jmdm. das Leben zur Qual machen; **put an animal out of its ~:** ein Tier von seinen Qualen erlösen; **put sb. out of his ~** *(fig.)* jmdn. nicht länger auf die Folter spannen; **miseries** Elend, *das;* Nöte; **b)** *(coll.: discontented person)* **~[-guts]** Miesepeter, *der (ugs. abwertend)*

misfire [mɪs'faɪə(r)] *v. i.* **a)** ‹*Motor:*› eine Fehlzündung/Fehlzündungen haben; ‹*Kanone, Gewehr:*› versagen, nicht losgehen; **b)** ‹*Plan, Versuch:*› fehlschlagen; ‹*Streich, Witz:*› danebengehen

misfit ['mɪsfɪt] *n. (person)* Außenseiter, *der*/Außenseiterin, *die*

misfortune [mɪs'fɔ:tʃən, mɪs'fɔ:tʃu:n] *n.* **a)** *no pl., no art. (bad luck)* Mißgeschick, *das;* **suffer ~:** [viel] Unglück haben; **b)** *(stroke of fate)* Schicksalsschlag, *der; (unlucky incident)* Mißgeschick, *das;* **it was his ~** *or* **he had the ~ to ...:** er hatte das Pech, zu ...

misgiving [mɪs'gɪvɪŋ] *n.* Bedenken *Pl.;* Zweifel, *der;* **have some ~s about sth.** wegen einer Sache Bedenken haben

misguided [mɪs'gaɪdɪd] *adj.* töricht ‹*Person*›; unangebracht ‹*Eifer*›; unsinnig ‹*Bemühung*›

mishandle [mɪs'hændl] *v. t.* **a)** *(deal with incorrectly)* falsch behandeln ‹*Angelegenheit*›; schlecht verwalten ‹*Finanzen*›; **b)** *(handle roughly)* mißhandeln

mishap ['mɪshæp] *n.* Mißgeschick, *das;* **sb. suffers** *or* **meets with a ~:** jmdm. passiert ein Mißgeschick; **without further ~:** ohne weitere Zwischenfälle

mishear [mɪs'hɪə(r)] **1.** *v. i.,* **misheard** [mɪs'hɜ:d] sich verhören. **2.** *v. t.,* **misheard** falsch verstehen

mishit [mɪs'hɪt] *v. t.,* **-tt-, mishit** verschlagen ‹*Ball*›

mishmash ['mɪʃmæʃ] *n.* Mischmasch, *der (ugs.)* (**of** aus)

misinform [mɪsɪn'fɔ:m] *v. t.* falsch informieren

misinterpret [mɪsɪn'tɜ:prɪt] *v. t.* **a)** *(interpret wrongly)* fehlinterpretieren, falsch auslegen ‹*Text, Inschrift, Buch*›; **b)** *(make wrong inference from)* falsch deuten; mißdeuten

misinterpretation [mɪsɪntɜ:prɪ'teɪʃn] *n.* Fehlinterpretation, *die;* **be open to ~:** leicht falsch ausgelegt werden können

misjudge [mɪs'dʒʌdʒ] *v. t.* falsch einschätzen; falsch beurteilen ‹*Person*›; **~ the height/distance** sich in der Höhe/Entfernung verschätzen

misjudgement, misjudgment [mɪs'dʒʌdʒmənt] *n.* Fehleinschätzung, *die; (of person)* falsche Beurteilung; *(of distance, length, etc.)* falsche Einschätzung

mislay [mɪs'leɪ] *v. t.,* **mislaid** [mɪs'leɪd] verlegen

mislead [mɪs'li:d] *v. t.,* **misled** [mɪs'led] irreführen; täuschen; **~ sb. about sth.** jmdm. ein falsches Bild von etw. vermitteln

misleading [mɪs'li:dɪŋ] *adj.* irreführend

mismanage [mɪs'mænɪdʒ] *v. t.* herunterwirtschaften ‹*Firma*›; schlecht führen ‹*Haushalt*›; schlecht handhaben *od.* abwickeln ‹*Angelegenheit, Projekt*›

mismanagement [mɪs'mænɪdʒmənt] *n.* Mißwirtschaft, *die; (of finances)* schlechte Verwaltung; *(of matters or affairs)* schlechte Handhabung *od.* Abwicklung

misnomer [mɪs'nəʊmə(r)] *n.* unzutreffende Bezeichnung

misogynist [mɪ'sɒdʒɪnɪst] *n.* Frauenhasser, *der*

misogyny [mɪ'sɒdʒɪnɪ] *n.* Frauenhaß, *der;* Misogynie, *die (geh.)*

misplace [mɪs'pleɪs] *v. t.* an die falsche Stelle stellen/legen/setzen *usw.;* **~ one's affection/confidence** seine Zuneigung/sein Vertrauen dem Falschen/der Falschen schenken; **be ~d** *(inappropriate)* unangebracht *od.* fehl am Platz sein

misprint ['mɪsprɪnt] *n.* Druckfehler, *der*

mispronounce [mɪsprə'naʊns] *v. t.* falsch aussprechen

misquote [mɪs'kwəʊt] *v. t.* falsch zitieren

misread [mɪs'ri:d] *v. t.,* **misread** [mɪs'red] *(read wrongly)* falsch *od.* nicht richtig lesen ‹*Text, Wort, Schrift*›; *(interpret wrongly)* mißdeuten ‹*Text, Absichten*›

misrepresent [mɪsreprɪ'zent] *v. t.* falsch darstellen; verdrehen ‹*Tatsachen*›

Miss [mɪs] *n.* **a)** *(title of unmarried woman)* **~ Brown** Frau Brown; Fräulein Brown *(veralt.); (girl)* Fräulein Brown; **b)** *(title of beauty queen)* **~ France** Miß Frankreich; **c)** *(as form of address to teacher etc.)* Frau Schmidt *usw.*

miss 1. *n.* **a)** *(failure to hit or attain)* Fehlschlag, *der; (shot)* Fehlschuß, *der; (throw)* Fehlwurf, *der;* **be a ~:** danebengehen *(ugs.);* **a ~ is as good as a mile** *(prov.)* fast getroffen ist <u>au</u>ch daneben; **b) give sb./sth. a ~:** sich *(Dat.)* jmdn./etw. schenken. **2.** *v. t.* **a)** *(fail to hit, lit. or fig.)* verfehlen; **~ed!** nicht getroffen!; **the car just ~ed the tree** das Auto wäre um ein Haar gegen den Baum geprallt; **b)** *(fail to get)* nicht bekommen; *(fail to find or meet)* verpassen; **~ the goal** am Tor vorbeischießen; **c)** *(let slip)* verpassen; versäumen; **~ an opportunity** sich *(Dat.)* eine Gelegenheit entgehen lassen; **it is too good to ~** *or* **is not to be ~ed** das darf man sich *(Dat.)* [einfach] nicht entgehen lassen; **d)** *(fail to catch)* versäumen, verpassen ‹*Bus, Zug, Flugzeug*›; **~ the boat** *or* **bus** *(fig.)* den Anschluß verpassen *(fig.);* **e)** *(fail to take part in)* versäumen; **~ school** in der Schule fehlen; **f)** *(fail to see)* übersehen; *(fail to hear or understand)* nicht mitbekommen; **you can't ~ it** es ist nicht zu übersehen; **he doesn't ~ much** ihm entgeht so schnell nichts; **g)** *(feel the absence of)* vermissen; **she ~es him** er fehlt ihr; **h)** *(fail to keep or perform)* versäumen ‹*Verabredung, Vorstellung*›. **3.** *v. i.* **a)** *(not hit sth.)* nicht treffen; *(not catch sth.)* danebengreifen; **b)** ‹*Ball, Schuß usw.:*› danebengehen; **c)** ‹*Motor:*› aussetzen

~ 'out 1. *v. t.* weglassen. **2.** *v. i.* **~ out on sth.** *(coll.)* sich *(Dat.)* etw. entgehen lassen

misshapen [mɪs'ʃeɪpn] *adj.* mißgebildet; mißgestaltet

missile ['mɪsaɪl] *n.* [Wurf]geschoß, *das; (self-propelled)* Missile, *das;* Flugkörper, *der*

missile: ~ base *n.* [Raketen]abschußbasis, *die;* **~ launcher** *n.* [Raketen]abschußrampe, *die;* **~ site** *see* **~ base**

missing ['mɪsɪŋ] *adj.* vermißt; fehlend ‹*Seite, Kapitel, Teil, Hinweis, Indiz*›; **be ~** ‹*Kapitel, Wort, Seite:*› fehlen; ‹*Brille, Bleistift usw.:*› verschwunden sein; ‹*Person:*› vermißt werden; *(not be present)* nicht dasein; fehlen; **the jacket has two buttons ~:** an der Jacke fehlen zwei Knöpfe; **I am ~ £10** mir fehlen 10 Pfund; **~ person** Vermißte, *der/die;* **~ link** *(Biol.)* Missing link, *das*

mission ['mɪʃn] *n.* **a)** *(task)* Mission, *die;* Auftrag, *der;* **b)** *(journey)* Mission, *die;* **go/come on a ~ to do sth.** mit dem Auftrag reisen/kommen, etw. zu tun; **c)** *(planned operation)* Einsatz, *der;* **d)** *(vocation)* Mission, *die;* **~ in life** Lebensaufgabe, *die;* **e)** *(missionary post)* Mission[sstation], *die*

missionary ['mɪʃənərɪ] **1.** *adj.* missionarisch; Missions‹*station, -arbeit, -schrift*›. **2.** *n.* Missionar, *der*/Missionarin, *die*

missis ['mɪsɪz, 'mɪsɪs] *n. (sl./joc.: wife)* **the** *or* **my/your ~:** die *od.* meine/deine Alte *(salopp)*

misspell [mɪs'spel] *v. t., forms as* [1]**spell** falsch schreiben

misspend [mɪs'spend] *v. t., forms as* **spend** verschwenden; vergeuden

misstatement [mɪs'steɪtmənt] *n.* falsche Darstellung

mist [mɪst] *n.* **a)** *(fog)* Nebel, *der; (haze)* Dunst, *der; (on windscreen etc.)* Beschlag, *der;* **b) in the ~s of time** *or* **antiquity** *(fig.)* im Dunkel *od. (geh.)* Nebel der Vergangenheit; **c)** *(of spray, vapour, etc.)* Wolke, *die*

~ 'over *v. i.* [sich] beschlagen; **his eyes ~ed over** Tränen verschleierten seinen Blick

~ 'up *v. i.* [sich] beschlagen

mistakable [mɪ'steɪkəbl] *adj.* verwechselbar (**for** mit)

mistake [mɪ'steɪk] **1.** *n.* Fehler, *der; (misunderstanding)* Mißverständnis, *das;* **make a ~:** einen Fehler machen; *(in thinking)* sich irren; **there's some ~!** da liegt ein Irrtum *od.* Fehler vor!; **the ~ is mine** der Fehler liegt bei mir; **it is a ~ to assume that ...:** es ist ein Irrtum anzunehmen, daß ...; **by ~:** versehentlich; aus Versehen; **make no ~ about it, ...:** täusch dich nicht, ... **2.** *v. t., forms as* **take**

1 falsch verstehen; mißverstehen; ~ **sth./sb. as meaning that ...**: etw./jmdn. [fälschlicherweise] so verstehen, daß ...; ~ **x for y** x mit y verwechseln; **there is no mistaking him** man kann ihn gar nicht verwechseln; ~ **sb.'s identity** jmdn. [mit jmd. anderem] verwechseln

mistaken [mɪ'steɪkn] *adj.* **be** ~: sich täuschen; ~ **kindness/zeal** unangebrachte Freundlichkeit/unangebrachter Eifer; **or** *or* **unless I'm very much** ~: wenn mich nicht alles täuscht; **a case of** ~ **identity** eine Verwechslung

mistakenly [mɪ'steɪknlɪ] *adv.* irrtümlicherweise

mister ['mɪstə(r)] *n.* *(sl./joc.)* **hey,** ~: he, Meister *od.* Chef *(ugs.)*

mistime [mɪs'taɪm] *v.t.* einen ungünstigen Zeitpunkt wählen für; schlecht timen *(bes. Sport)*

mistletoe ['mɪsltəʊ] *n.* Mistel, *die; (sprig)* Mistelzweig, *der*

mistook *see* **mistake 2**

mistranslate [mɪstræns'leɪt] *v.t.* falsch übersetzen

mistranslation [mɪstræns'leɪʃn] *n.* falsche Übersetzung; *(error)* Übersetzungsfehler, *der*

mistreat [mɪs'tri:t] *v.t.* schlecht behandeln; *(violently)* mißhandeln

mistreatment [mɪs'tri:tmənt] *n.* schlechte Behandlung; *(violent)* Mißhandlung, *die*

mistress ['mɪstrɪs] *n.* **a)** *(of a household)* Hausherrin, *die;* **b)** *(person in control, employer)* Herrin, *die;* **she is her own** ~: sie ist ihr eigener Herr; **the dog's** ~: das Frauchen [des Hundes]; **c)** *(Brit. Sch.: teacher)* Lehrerin, *die;* '**French** ~: Französischlehrerin, *die;* **d)** *(lover)* Geliebte, *die*

mistrust [mɪs'trʌst] **1.** *v.t.* mißtrauen (+ *Dat.*). **2.** *n., no pl.* Mißtrauen, *das* (**of** gegenüber + *Dat.*)

mistrustful [mɪs'trʌstfl] *adj.* mißtrauisch; **be** ~ **of sb./sth.** jmdm./einer Sache gegenüber mißtrauisch sein

misty ['mɪstɪ] *adj.* neb[e]lig, dunstig ‹*Tag, Morgen*›; in Nebel *od.* Dunst gehüllt ‹*Berg, Hügel*›

misunderstand [mɪsʌndə'stænd] *v.t., forms as* **understand** mißverstehen; falsch verstehen; **don't** ~ **me** versteh mich nicht falsch

misunderstanding [mɪsʌndə'stændɪŋ] *n.* Mißverständnis, *das;* **there has been a** ~: da liegt ein Mißverständnis vor

misunderstood [mɪsʌndə'stʊd] *adj.* unverstanden; verkannt ‹*Künstler, Genie*›; **be** ~: kein Verständnis finden

misuse 1. [mɪs'ju:z] *v.t.* mißbrauchen; zweckentfremden ‹*Werkzeug, Gelder*›; nichts Rechtes machen aus ‹*Gelegenheit, Talent*›. **2.** [mɪs'ju:s] *n.* Mißbrauch, *der; (of funds)* Zweckentfremdung, *die*

mite [maɪt] *n.* **a)** *(Zool.)* Milbe, *die;* **b) the widow's** ~: das Scherflein der armen Witwe; **c)** *(small child)* Würmchen, *das (fam.);* **poor little** ~: armes Kleines; **d) a** ~ **too strong/outspoken** *(coll.: somewhat)* ein bißchen *od.* etwas zu stark/geradeheraus

miter *(Amer.) see* **mitre**

mitigate ['mɪtɪgeɪt] *v.t.* **a)** *(alleviate)* lindern; **b)** *(make less severe)* mildern; **mitigating circumstances** mildernde Umstände

mitigation [mɪtɪ'geɪʃn] *n. see* **mitigate**: Linderung, *die;* Milderung, *die*

mitre ['maɪtə(r)] *n.* *(Brit.)* **a)** *(Eccl.)* Mitra, *die;* **b)** *(joint)* Gehrung, *die (bes. Technik)*

mitten ['mɪtn] *n.* Fausthandschuh, *der;* Fäustling, *der; (not covering fingers)* fingerloser Handschuh

mix [mɪks] **1.** *v.t.* **a)** *(combine)* [ver]mischen; vermengen; verrühren ‹*Zutaten*›; ~ **one's drinks** alles durcheinander trinken; **b)** *(prepare by ~ing)* mischen, mixen ‹*Cocktail*›; anrühren, ansetzen ‹*Lösung, Teig*›; zubereiten ‹*Medikament*›; **c)** ~ **it [with sb.]** *(coll.)* sich [mit jmdm.] prügeln. **2.** *v.i.* **a)** *(become ~ed)* sich vermischen; **b)** *(be sociable)* Umgang mit anderen [Menschen] haben; **c)** *(be compatible)* zusammenpassen; ‹*Ideen:*› sich verbinden lassen. **3.** *n.* **a)** *(coll.: mixture)* Mischung, *die* (**of** aus); **b)** *(proportion)* [Mischungs]verhältnis, *das;* **c)** *(ready ingredients)* [gebrauchsfertige] Mischung; **[cake-]**~: Backmischung, *die*

~ **'in** *v.t.* einrühren

~ **'up** *v.t.* **a)** vermischen; verrühren ‹*Zutaten*›; **b)** *(make a muddle of)* durcheinanderbringen; *(confuse one with another)* verwechseln; **c)** *in pass. (involve)* **be/get ~ed up in sth.** in etw. *(Akk.)* verwickelt sein/werden

mixed [mɪkst] *adj.* **a)** *(diverse)* unterschiedlich ‹*Reaktionen, Kritiken*›; ~ **feelings** gemischte Gefühle; **b)** gemischt ‹*Gesellschaft*›; **a** ~ **bunch** ein bunt gemischter Haufen; **c)** *(for both sexes)* gemischt

mixed: ~ **'bag** *n.* bunte Mischung; ~ **'blessing** *n.* **be a** ~ **blessing** nicht nur Vorteile haben; **children are a** ~ **blessing** Kinder sind kein reiner Segen; ~ **'company** *n.* **in** ~ **company** in Gesellschaft von Damen [und Kindern]; ~ **'grill** *n.* Mixed grill, *der (Gastr.);* gemischte Grillplatte; ~ **'metaphor** *see* **metaphor b;** ~ **'up** *adj.* *(fig. coll.)* verwirrt, konfus ‹*Person*›; **be/feel very** ~ **up** völlig durcheinander sein; ~ **up kids** Jugendliche ohne [jeden] inneren Halt

mixer ['mɪksə(r)] *n.* *(for food)* Mixer, *der; (for concrete)* Mischmaschine, *die*

mixture ['mɪkstʃə(r)] *n.* **a)** *(mixing, being mixed)* Mischen, *das; (result)* Mischung, *die* (**of** aus); **the** ~ **as before** *(fig.)* die altbekannte Mischung; **b)** *(Motor Veh.)* Gemisch, *das*

'mix-up *n.* Durcheinander, *das; (misunderstanding)* Mißverständnis, *das*

ml. *abbr.* **millilitre[s]** ml

mm. *abbr.* **millimetre[s]** mm

mnemonic [nɪ'mɒnɪk] *n.* Gedächtnishilfe, *die;* Eselsbrücke, *die (ugs.)*

moan [məʊn] **1.** *n.* **a)** Stöhnen, *das;* **b) have a** ~ *(complain at length)* jammern; *(have a grievance)* eine Beschwerde haben. **2.** *v.i.* **a)** stöhnen (**with** vor + *Dat.*); **b)** *(complain)* jammern (**about** über + *Akk.*); ~ **at sb.** jmdm. etwas vorjammern. **3.** *v.t.* stöhnen

moat [məʊt] *n.* [Wasser]graben, *der;* **[castle]** ~: Burggraben, *der*

mob [mɒb] **1.** *n.* **a)** *(rabble)* Mob, *der (abwertend);* Pöbel, *der (abwertend);* **b)** *(sl.: associated group)* ~ **[of criminals]** Bande, *die (abwertend);* **Peter and his** ~: Peter und seine ganze Blase *(salopp);* ~ **law/rule** Gesetz/Herrschaft der Straße. **2.** *v.t.,* **-bb-: a)** *(crowd round)* belagern *(ugs.)* ‹*Schauspieler, Star*›; stürmen ‹*Kino*›; **b)** *(attack)* herfallen über (+ *Akk.*)

mobile ['məʊbaɪl] *adj.* **a)** *(able to move easily)* beweglich; *(on wheels)* fahrbar; **b)** *(accommodated in vehicle)* mobil; fahrbar; ~ **library** Fahrbücherei, *die;* ~ **canteen** Kantine auf Rädern; **c)** *(in social status)* mobil; **be upwardly** ~: sozial aufsteigen

mobile: ~ **'home** *n.* transportable Wohneinheit; *(caravan)* Wohnwagen, *der;* ~ **'phone** *n.* Mobiltelefon, *das*

mobilisation, mobilise *see* **mobiliz-**

mobility [mə'bɪlɪtɪ] *n.* **a)** *(ability to move) (of person)* Beweglichkeit, *die; (on wheels)* Fahrbarkeit, *die;* **b)** *(in social status)* Mobilität, *die*

mobilization [məʊbɪlaɪ'zeɪʃn] *n.* **a)** Mobilisierung, *die;* **b)** *(Mil.)* Mobilmachung, *die*

mobilize ['məʊbɪlaɪz] *v.t.* **a)** mobilisieren; **b)** *(Mil.)* mobil machen; *abs.* **make preparations to ~:** die Mobilmachung vorbereiten

moccasin ['mɒkəsɪn] *n.* Mokassin, *der*

mock [mɒk] **1.** *v.t.* **a)** *(subject to ridicule)* sich lustig machen über (+ *Akk.*); verspotten; **he was ~ed** man machte sich über ihn lustig; **b)** *(ridicule by imitation)* **~ sb./sth.** jmdn./etw. nachmachen[, um sich über ihn/darüber lustig zu machen]. **2.** *v.i.* **~ at sb./sth.** sich über jmdn./etw. mokieren *od.* lustig machen. **3.** *attrib. adj.* gespielt ‹*Feierlichkeit, Bescheidenheit, Ernst*›; Schein‹*kampf, -angriff*›; **~ Tudor style** Pseudotudorstil, *der;* **~ examination** simulierte Prüfung; **~-turtle soup** Mockturtlesuppe, *die (Kochk.)*

mockery ['mɒkərɪ] *n.* **a) be a ~ of justice/the truth** der Gerechtigkeit/Wahrheit *(Dat.)* hohnsprechen *(geh.);* **make a ~ of sth.** etw. zur Farce machen; **b)** *no pl., no indef. art. (derision)* Spott, *der*

mocking ['mɒkɪŋ] **1.** *adj.* spöttisch. **2.** *n.* Spott, *der*

'mocking-bird *n.* Spottdrossel, *die*

'mock-up *n.* Modell [in Originalgröße]; *(of book etc.)* Layout, *das*

mod cons [mɒd 'kɒnz] *n. pl. (Brit. coll.)* [moderner] Komfort; **have all ~:** mit allem Komfort *od. (ugs.)* allen Schikanen ausgestattet sein

mode [məʊd] *n.* **a)** *(way in which thing is done)* Art [und Weise], *die; (method of procedure)* Methode, *die; (Computing)* Betriebsart, *die;* **~ of transport** Transportmittel, *das;* **b)** *(fashion)* Mode, *die* (**for** *Gen.*)

model ['mɒdl] **1.** *n.* **a)** Modell, *das;* **b)** *(perfect example)* Muster, *das* (**of** an + *Dat.*); *(to be imitated)* Vorbild, *das;* **be a ~ of industry** ein Muster an Fleiß *(Dat.)* sein; **on the ~ of sth.** nach dem Vorbild einer Sache *(Gen.);* **c)** *(Art)* Modell, *das; (Fashion)* Model, *das;* Mannequin, *das; (male)* Dressman, *der;* **photographer's ~:** Fotomodell, *das.* **2.** *adj.* **a)** *(exemplary)* vorbildlich; Muster- *(oft iron.);* **b)** *(miniature)* Modell‹*stadt, -eisenbahn, -flugzeug*›. **3.** *v.t., (Brit.)* **-ll-: a)** modellieren; formen; **~ sth. in clay** etw. in Ton modellieren; **~ sth. after** *or* **[up]on sth.** etw. einer Sache *(Dat.)* nachbilden; **b)** *(Fashion)* vorführen ‹*Kleid, Entwurf usw.*›. **4.** *v.i., (Brit.)* **-ll-: a)** *(Fashion)* als Mannequin *od.* Model arbeiten; ‹*Mann:*› als Dressman arbeiten; *(Photog.)* als [Foto]modell arbeiten; *(Art)* Modell stehen/sitzen; **b) ~ in clay** *etc.* in Ton *usw.* modellieren

modelling (*Amer.:* **modeling**) ['mɒdəlɪŋ] *n.* **a)** *no art.* **do ~** *(Fashion)* als Mannequin *od.* Model arbeiten; ‹*Mann:*› als Dressman arbeiten; *(Photog., Art)* als Modell arbeiten; **b)** *no indef. art. (sculpturing)* Modellieren, *das;* **~ clay** Modellierton, *der*

modem ['məʊdem] *n.* Modem, *der*

moderate 1. ['mɒdərət] *adj.* **a)** gemäßigt ‹*Partei, Ansichten*›; mäßig, maßvoll ‹*Person, bes. Trinker, Esser; Forderungen*›; mäßig ‹*Begeisterung, Interesse*›; **b)** *(fairly large or good)* mittler... ‹*Größe, Menge, Wert*›; **[only] ~:** mäßig ‹*Qualität, Ernte*›; **c)** *(reasonable)* angemessen, vernünftig ‹*Preis, Summe*›; **d)** mäßig ‹*Wind*›. **2.** ['mɒdərət] *n.* Gemäßigte, *der/die.* **3.** ['mɒdəreɪt] *v.t.* mäßigen; zügeln ‹*Begeisterung*›; mildern ‹*negativen Effekt*›; **~ one's demands** seine Forderungen einschränken. **4.** ['mɒdəreit] *v.i.* nachlassen

moderately ['mɒdərətlɪ] *adv.* einigermaßen; mäßig ‹*begeistert, groß, begabt*›; **be only ~ enthusiastic about sth.** sich nicht allzu sehr *od.* übermäßig für etw. begeistern

moderation [mɒdə'reɪʃn] *n.* **a)** *(moderating)* Mäßigung, *die;* **b)** *no pl. (moderateness)* Mäßigkeit, *die;* **in ~:** mit *od.* in Maßen

modern ['mɒdn] *adj.* modern; heutig ‹*Zeit[alter], Welt, Person*›; **~ jazz** Modern Jazz, *der;* **in ~ times** in der heutigen Zeit; **~ history** neuere Geschichte; **~ languages** neuere Sprachen; *(subject of study)* Neuphilologie, *die*

modernise *see* **modernize**

modernism ['mɒdənɪzm] *n.* Modernismus, *der*

modernist ['mɒdənɪst] *n.* Modernist, *der*/Modernistin, *die*

modernity [mə'dɜ:nɪtɪ] *n.* Modernität, *die*

modernize ['mɒdənaɪz] *v.t.* modernisieren

modest ['mɒdɪst] *adj.* bescheiden; vorsichtig ‹*Schätzung*›; einfach, unauffällig ‹*Haus, Kleidung*›; **have a ~ lifestyle** bescheiden *od.* einfach leben

modestly ['mɒdɪstlɪ] *adv.* bescheiden; dezent, unauffällig ‹*sich kleiden*›

modesty ['mɒdɪstɪ] *n., no pl.* Bescheidenheit, *die;* **in all ~:** bei aller Bescheidenheit

modification [mɒdɪfɪ'keɪʃn] *n.* [Ab]änderung, *die;* Modifizierung, *die*

modifier ['mɒdɪfaɪə(r)] *n. (esp. Ling., Biol.)* Modifikator, *der*

modify ['mɒdɪfaɪ] *v.t.* **a)** *(make changes in)* [ab-, ver]ändern; modifizieren; **b)** *(tone down)* mäßigen; **~ one's position** in seiner Haltung gemäßigter werden

modular ['mɒdjʊlə(r)] *adj.* aus Elementen [zusammengesetzt]; *(in construction)* aus Baueinheiten *od.* -elementen [zusammengesetzt]; **~ construction/design** Konstruktion/Entwurf nach dem Baukastensystem; **~ unit** [Bau-, Konstruktions]element, *das*

module ['mɒdju:l] *n.* **a)** Bauelement, *das; (Electronics)* Modul, *das;* **b)** *(Educ.)* Unterrichtseinheit, *die;* **c)** *(Astronaut.)* **command ~:** Kommandoeinheit, *die*

mohair ['məʊheə(r)] *n.* Mohair, *der*

moist [mɔɪst] *adj.* feucht (**with** von)

moisten ['mɔɪsn] *v.t.* anfeuchten; feucht machen; **~ one's lips** sich *(Dat.)* die Lippen [mit der Zunge] befeuchten

moisture ['mɔɪstʃə(r)] *n.* Feuchtigkeit, *die;* **film of ~:** Feuchtigkeitsfilm, *der*

moisturizer (moisturiser) ['mɔɪstjəraɪzə(r)], **moisturizing cream** ['mɔɪstʃəraɪzɪŋ kri:m] *ns.* Feuchtigkeitscreme, *die*

molar ['məʊlə(r)] **1.** *n.* Backenzahn, *der;* Molar[zahn], *der (Anat.).* **2.** *adj.* **~ tooth** *see* **1**

molasses [mə'læsɪz] *n.* Melasse, *die*

mold *(Amer.) see* **1, 2, 3mould**

molder, molding, moldy *(Amer.) see* **mould-**

1mole [məʊl] *n. (on skin)* Leberfleck, *der; (prominent)* Muttermal, *das*

2mole *n.* **a)** *(animal)* Maulwurf, *der;* **b)** *(coll.: spy)* Maulwurf, *der (ugs.)*

molecular [mə'lekjʊlə(r)] *adj. (Phys., Chem.)* molekular

molecule ['mɒlɪkju:l, 'məʊlɪkju:l] *n. (Phys., Chem.)* Molekül, *das* '**molehill** *n.* Maulwurfshügel, *der;* **make a mountain out of a ~** *(fig.)* aus einer Mücke einen Elefanten machen *(ugs.)*

molest [mə'lest] *v.t.* belästigen; *(sexually)* [unsittlich] belästigen

moll [mɒl] *n. (coll.)* Gangsterbraut, *die*

mollify ['mɒlɪfaɪ] *v.t.* besänftigen; beschwichtigen

mollusc, *(Amer.)* **mollusk** ['mɒləsk] *n. (Zool.)* Molluske, *die (fachspr.);* Weichtier, *das*

mollycoddle ['mɒlɪkɒdl] *v.t.* [ver]hätscheln *(oft abwertend)*

molt *(Amer.) see* **moult**

molten ['məʊltn] *adj.* geschmolzen; flüssig ‹*Lava*›

mom [mɒm] *(Amer. coll.) see* **²mum**

moment ['məʊmənt] *n.* **a)** Moment, *der;* Augenblick, *der;* **barely a ~ had elapsed ...:** es war kaum eine Minute vergangen ...; **for a ~ or two** einen kurzen Augenblick; **for a few ~s** ein paar Augenblicke; **at any ~,** *(coll.)* **any ~:** jeden Augenblick *od.* Moment; **this is the ~!** dies ist der geeignete Augenblick!; **at the precise ~ she came in ...:** genau in dem Augenblick, als sie hereintrat, ...; **the ~ I get home** gleich *od.* sofort, wenn ich nach Hause komme; **one** *or* **just a** *or* **wait a ~!** einen Moment *od.* Augenblick!; **in a ~** *(instantly)* im Nu *(ugs.); (very soon)* sofort; gleich; **for a ~:** einen Moment [lang]; **not for a ~:** keinen Moment [lang]; **the ~ of truth** die Stunde der Wahrheit; **at the ~:** im Augenblick; momentan; **for the ~:** im *od.* für den Augenblick; vorläufig; **I shan't be a ~** *(I'll be back very soon)* ich bin sofort zurück; *(I have very nearly finished)* ich bin sofort soweit; **have you got a ~?** hast du mal einen Augenblick Zeit?; **be the man of the ~:** der Mann des Tages sein; **come here this ~!** komm sofort *od.* auf der Stelle her!; **b)** *(formal: importance)* **of ~:** von Bedeutung; **of little/no ~:** von geringer/ohne Bedeutung; **c)** *(Phys.)* Moment, *das*

momentarily ['məʊməntərɪlɪ] *adv.* **a)** *(for a moment)* einen Augenblick lang; **b)** *(Amer.) (at any moment)* jeden Augenblick *od.* Moment; *(in a few minutes)* in wenigen Minuten

momentary ['məʊməntərɪ] *adj.* kurz; **a ~ forgetfulness** ein Augenblick geistiger Abwesenheit

momentous [mə'mentəs] *adj.* *(important)* bedeutsam; *(of consequence)* folgenschwer; von großer Tragweite *nachgestellt*

momentum [mə'mentəm] *n., pl.* **momenta** [mə'mentə] **a)** *(impetus)* Schwung, *der;* **gather** *or* **gain ~:** schneller werden; *(fig.)* in Schwung kommen; **b)** *(Mech.)* Impuls, *der*

mommy ['mɒmɪ] *(Amer. coll.) see* **²mummy**

Mon. *abbr.* **Monday** Mo.

monarch ['mɒnək] *n.* Monarch, *der*/Monarchin, *die*

monarchist ['mɒnəkɪst] *n.* Monarchist, *der*/Monarchistin, *die*

monarchy ['mɒnəkɪ] *n.* Monarchie, *die*

monastery ['mɒnəstrɪ] *n.* [Mönchs]kloster, *das*

monastic [mə'næstɪk] *adj.* mönchisch; *(of monasteries)* klösterlich; Kloster‹*gebäude*›

Monday ['mʌndeɪ, 'mʌndɪ] **1.** *n.* Montag, *der.* **2.** *adv. (coll.)* **she comes ~s** sie kommt montags. *See also* **Friday**

monetarism ['mʌnɪtərɪzm] *n.* *(Econ.)* Monetarismus, *der*

monetarist ['mʌnɪtərɪst] **1.** *n.* *(Econ.)* Monetarist, *der*/Monetaristin, *die.* **2.** *adj.* monetaristisch

monetary ['mʌnɪtərɪ] *adj.* **a)** *(of the currency in use)* monetär; Währungs‹*politik, -system*›; **b)** *(of money)* finanziell

money ['mʌnɪ] *n.* **a)** *no pl.* Geld, *das;* **your ~ or your life!** Geld oder Leben!; **be in the ~** *(coll.)* im Geld schwimmen *(ugs.);* **there is ~ in sth.** mit etw. kann man [viel] Geld verdienen; **~ for jam** *or* **old rope** *(Brit. fig. coll.)* leicht *od.* schnell verdientes Geld; **make ~** ‹*Person:*› [viel] Geld verdienen, *(ugs.)* [das große] Geld machen; **earn good ~:** gut verdienen; **~ talks** das Geld macht's *(ugs.);* **~ makes the world go round** Geld regiert die Welt; **put ~ into sth.** Geld in etw. *(Akk.)* investieren *od. (ugs.)* hineinstecken; **have ~ to burn** *(fig. coll.)* Geld wie Heu haben *(ugs.);* **[not] be made of ~** *(fig. coll.)* [k]ein Goldesel *od.* Krösus sein; **this would only be to throw good ~ after bad** das wäre nur rausgeschmissenes *od.* rausgeworfenes Geld *(ugs.);* **for 'my ~:** wenn man mich fragt; **the best that ~ can buy** das Beste, was es für Geld gibt; **~ can't buy happiness!** Geld allein macht nicht glücklich!; *see also* **big 1 a; love 1 a; make 1 a; run 1 a; b)** *pl.* **~s** *or* **monies** ['mʌnɪz] *(sum of ~)* Geld, *das;* [Geld]betrag, *der*

money: ~-back *attrib. adj.* **~-back guarantee** Geld-zurück-Garantie, *die;* **~-bag** *n.* Geldsack, *der;* **~-box** *n.* Sparbüchse, *die;* **~-changer** ['mʌnɪtʃeɪndʒə(r)] *n.* Geldwechsler, *der;* **~-grubber** ['mʌnɪgrʌbə(r)] *n.* Raffzahn, *der (ugs.);* **~-grubbing 1.** *adj.* geldgierig *(abwertend);* **2.** *n.* Geldgier, *die (abwertend);* **~-lender** *n.* Geldverleiher, *der;* **~-maker** *n.* **be a ~-maker** ‹*Projekt, Produkt, Film:*› Geld bringen; **~-making** *adj.* gewinnbringend, einträglich ‹*Geschäft, Beschäftigung*›; **~-market** *n.* Geldmarkt, *der;* **~ order** *n.* Zahlungsanweisung, *die; (issued by Post Office)* Postanweisung, *die;* **~-spinner** *n. (Brit.)* Verkaufsschlager, *der;* **~'s-worth** *n.* **get** *or* **have one's ~'s-worth** etwas für sein Geld bekommen

Mongol ['mɒŋgl] **1.** *n.* **a)** Mongole, *der*/Mongolin, *die;* **b) m~** *(Med.)* Mongoloide, *der/die;* **she is a m~:** sie ist mongoloid. **2.** *adj.* **a)** mongolisch; **b) m~** *(Med.)* mongoloid

Mongolia [mɒŋ'gəʊlɪə] *pr. n.* Mongolei, *die*

Mongolian [mɒŋ'gəʊlɪən] **1.** *adj.* mongolisch. **2.** *n. (person)* Mongole, *der*/Mongolin, *die*

mongoose ['mɒŋgu:s] *n. (Zool.)* Indischer Mungo

mongrel ['mʌŋgrəl, 'mɒŋgrəl] *n.* **~ [dog]** Promenadenmischung, *die (scherzh., auch abwertend)*

monitor ['mɒnɪtə(r)] **1.** *n.* **a)** *(Sch.)* Aufsichtsschüler, *der*/-schülerin, *die;* **pencil/milk ~:** Bleistift-/Milchwart, *der;* **b)** *(listener)* Mithörer, *der*/Mithörerin, *die;* **c)** *(Mech. Engin., Phys., Med., Telev.)* Monitor, *der.* **2.** *v.t.* kontrollieren ‹*Strahlungsintensität*›; beobachten ‹*Wetter, Flugzeug, Bewegung*›; abhören ‹*Sendung, Telefongespräch*›

monk [mʌŋk] *n.* Mönch, *der;* **order of ~s** Mönchsorden, *der*

monkey ['mʌŋkɪ] *n.* **a)** Affe, *der;* **make a ~ of sb.** *(sl.)* jmdn. zum Gespött machen; **b)** *(mischievous child)* Schlingel, *der (scherzh.)*

monkey: ~ business *n. (coll.)* *(mischief)* Schabernack, *der; (unlawful or unfair activities)* krumme Touren *Pl. (ugs.);* **~-nut** *n.* *(Bot.)* Erdnuß, *die;* **~-wrench** *n.* Rollgabelschlüssel, *der (fachspr.);* Universalschraubenschlüssel, *der*

mono ['mɒnəʊ] *adj.* Mono‹*platte[nspieler], -wiedergabe*›

monochrome ['mɒnəkrəʊm] **1.** *n.* **in ~:** monochrom *(fachspr.);*

einfarbig; Schwarzweiß- *(Ferns.)*. **2.** *adj.* monochrom *(fachspr.)*; einfarbig; Schwarzweiß- *(Ferns.)*

monocle ['mɒnəkl] *n.* Monokel, *das;* Einglas, *das (veralt.)*

monogram ['mɒnəgræm] *n.* Monogramm, *das*

monolingual [mɒnə'lɪŋgwəl] *adj.* einsprachig

monolith ['mɒnəlɪθ] *n. (lit. or fig.)* Monolith, *der*

monolithic [mɒnə'lɪθɪk] *adj. (lit. or fig.)* monolithisch

monologue (*Amer.*: **monolog**) ['mɒnəlɒg] *n. (lit. or fig.)* Monolog, *der*

monoplane ['mɒnəpleɪn] *n. (Aeronaut.)* Eindecker, *der*

monopolize (**monopolise**) [mə'nɒpəlaɪz] *v.t. (Econ.)* monopolisieren; *(fig.)* mit Beschlag belegen; ~ **the conversation** den/die anderen nicht zu Wort kommen lassen

monopoly [mə'nɒpəlɪ] *n.* **a)** *(Econ.)* Monopol, *das* (**of** auf + *Dat.*); *(exclusive possession)* alleiniger Besitz; **you can't have a ~ of the car** du kannst das Auto nicht ständig mit Beschlag belegen; **b)** *(thing monopolized)* Monopol, *das*

monorail ['mɒnəreɪl] *n.* **a)** *(single rail)* Einschienengleis, *das;* **b)** *(vehicle)* Einschienenbahn, *die*

monosyllabic [mɒnəsɪ'læbɪk] *adj.* einsilbig ‹*Antwort, Person*›

monosyllable ['mɒnəsɪləbl] *n.* **a)** einsilbiges Wort; **b)** *(Ling.)* Einsilber, *der*

monotone ['mɒnətəʊn] *n.* gleichbleibender Ton

monotonous [mə'nɒtənəs] *adj.* eintönig; monoton

monotonously [mə'nɒtənəslɪ] *adv.* eintönig

monotony [mə'nɒtənɪ] *n.* Eintönigkeit, *die;* Monotonie, *die*

monsoon [mɒn'su:n] *n. (Geog.)* **a)** *(wind)* **summer** *or* **wet/dry ~:** Sommer-/Wintermonsun, *der;* **b)** *(season)* Regenzeit, *die*

monster ['mɒnstə(r)] **1.** *n.* **a)** Ungeheuer, *das;* Monster, *das; (huge thing)* Ungetüm, *das;* Monstrum, *das;* **what a ~!** *(in surprise or admiration)* das ist ja ungeheuer!; **b)** *(inhuman person)* Unmensch, *der; (iron.: naughty child)* Monster, *das (scherzh.)*. **2.** *attrib. adj.* riesig

monstrosity [mɒn'strɒsɪtɪ] *n.* **a)** *(outrageous thing)* Ungeheuerlichkeit, *die;* **b)** *(hideous building etc.)* Ungetüm, *das;* **c)** *(creature)* Ungeheuer, *das;* Monster, *das*

monstrous ['mɒnstrəs] *adj.* **a)** *(huge)* monströs *(geh.)*; riesig ‹*Lkw, Kuchen, Buch*›; unnatürlich groß ‹*Gemüse, Person, Baum, Pflanze*›; **b)** *(outrageous)* ungeheuerlich *(abwertend)* ‹*Vorschlag, Vorstellung, Einstellung, Entscheidung*›; **c)** *(atrocious)* scheußlich; monströs

montage [mɒn'tɑ:ʒ] *n. (Photog., Art, Radio, Film)* Montage, *die*

month [mʌnθ] *n.* Monat, *der;* **last day of the ~:** Monatsletzte, *der;* **the ~ of January** der [Monat] Januar; **for a ~/several ~s** einen Monat [lang]/mehrere Monate [lang] *od.* monatelang; **for ~s [on end]** monatelang; **every six ~s** alle sechs Monate; halbjährlich; **once every** *or* **a ~:** einmal monatlich *od.* im Monat; **in a ~['s time]** in einem Monat; **take a ~'s holiday** [sich *(Dat.)*] einen Monat Urlaub nehmen; **£10 a** *or* **per ~:** zehn Pfund im Monat; **a six-~[s]-old baby** ein sechs Monate altes *od.* sechsmonatiges Baby

monthly ['mʌnθlɪ] **1.** *adj.* monatlich; Monats‹*umsatz, -einkommen, -gehalt*›; einmonatig ‹*Abstand*›; Monats‹*zyklus, -karte*›; **three-~:** dreimonatlich; vierteljährlich; **three-~ season ticket** Dreimonats- *od.* Vierteljahreskarte, *die*. **2.** *adv.* [ein]monatlich; einmal im Monat. **3.** *n. (publication)* Monatsschrift, *die*

monument ['mɒnjʊmənt] *n.* **a)** Denkmal, *das;* **b)** *(on grave)* Grabmal, *das (geh.)*

monumental [mɒnjʊ'mentl] *adj.* **a)** *(massive)* gewaltig ‹*Skulptur*›; monumental ‹*Plastik, Gemälde, Gebäude*›; **b)** *(extremely great)* kolossal *(ugs.)*

monumentally [mɒnjʊ'mentəlɪ] *adv.* enorm ‹*stur, schlau, kreativ*›; **~ boring/stupid** sterbenslangweilig/strohdumm

moo [mu:] **1.** *n.* Muhen, *das*. **2.** *v.i.* muhen

mooch [mu:tʃ] *v.i. (sl.)* **~ about** *or* **around/along** herumschleichen *(ugs.)*/zockeln *(ugs.)*

mood [mu:d] *n.* **a)** *(state of mind)* Stimmung, *die;* **there was a [general] ~ of optimism** es herrschte allgemeiner Optimismus; **be in a good/bad ~:** [bei] guter/schlechter Laune sein; **be in a cheerful ~:** froh gelaunt sein; **be in a serious/pensive ~:** ernst/nachdenklich gestimmt sein; **be in no ~ for joking** nicht zum Scherzen aufgelegt sein; **I'm not in the ~:** ich hab' keine Lust dazu; **b)** *(fit of melancholy or bad temper)* Verstimmung, *die;* schlechte Laune; **have one's ~s** [seine] Launen haben

moody ['mu:dɪ] *adj.* **a)** *(sullen)* mißmutig; verdrossen; **b)** *(subject to moods)* launenhaft

moon [mu:n] **1.** *n.* Mond, *der;* **the ~ is full/waning/waxing** es ist Vollmond/abnehmender/zunehmender Mond; **be over the ~** *(fig. coll.)* im siebten Himmel sein *(ugs.)*; **promise sb. the ~** *(fig.)* jmdm. das Blaue vom Himmel versprechen *(ugs.)*. **2.** *v.i. (coll.)* **~ about [the house]** trübselig [im Haus] herumschleichen *(ugs.)*

moon: ~beam *n.* Mondstrahl, *der;* **~beams** Mondschein, *der;* **~light 1.** *n.* Mondlicht, *das;* Mondschein, *der;* **2.** *v.i. (coll.)* nebenberuflich abends arbeiten; **~lit** *adj.* mondbeschienen *(geh.)*

[1]**moor** [mʊə(r), mɔ:(r)] *n. (Geog.)* [Hoch]moor, *das*

[2]**moor 1.** *v.t.* festmachen; vertäuen. **2.** *v.i.* festmachen

'**moorhen** *n. (Ornith.)* [Grünfüßiges] Teichhuhn

mooring ['mʊərɪŋ, 'mɔ:rɪŋ] *n.* **a)** *usu. in pl. (means of attachment)* Vertäuung, *die;* **b)** *usu. in pl. (place)* **~[s]** Anlegestelle, *die;* **c)** *(action of making fast)* Vertäuung, *die*

moorland ['mʊələnd, 'mɔ:lənd] *n. (Geog.)* Moorland, *das*

moose [mu:s] *n., pl. same (Zool.)* Amerikanischer Elch

moot [mu:t] *adj.* umstritten; offen ‹*Frage*›; strittig ‹*Punkt*›

mop [mɒp] **1.** *n.* **a)** Mop, *der; (for washing up)* ≈ Spülbürste, *die;* **b)** **~ [of hair]** Wuschelkopf, *der*. **2.** *v.t.*, **-pp-** moppen ‹*Fußboden*›; *(wipe)* abwischen ‹*Träne, Schweiß, Stirn*›

~ 'up *v.t. (wipe up)* aufwischen ‹*Flüssigkeit*›

mope [məʊp] *v.i.* Trübsal blasen *(ugs.)*; **~ about** *or* **around** trübselig herumschleichen *(ugs.)*

moped ['məʊped] *n.* Moped, *das*

moral ['mɒrl] **1.** *adj.* **a)** moralisch; sittlich ‹*Wert*›; Moral‹*begriff, -prinzip, -vorstellung*›; Moral‹*philosoph[ie], -psychologie*›; moralisch, sittlich ‹*Verpflichtung, Pflicht*›; **be under a ~ obligation** eine moralische *od.* sittliche Pflicht haben; **b)** *(virtuous)* moralisch, sittlich ‹*Leben, Person*›. **2.** *n.* **a)** Moral, *die;* **draw the ~ from sth.** die Lehre aus etw. ziehen; **b)** *in pl. (habits)* Moral, *die*

morale [mə'rɑ:l] *n.* Moral, *die;* **low/high ~:** schlechte/gute Moral

moralise *see* **moralize**

morality [mə'rælɪtɪ] *n.* **a)** *(con-*

duct) Moral, *die;* Sittlichkeit, *die;* Moralität, *die (geh.);* **b)** *(particular system)* Ethik, *die;* **c)** *(conformity to moral principles)* Sittlichkeit, *die;* Moralität, *die (geh.)*
moralize ['mɒrəlaɪz] *v.i.* moralisieren *(geh.);* moralische Betrachtungen anstellen (**[up]on** über + *Akk.*); **do stop moralizing!** hör auf mit deinen Moralpredigten!
morally ['mɒrəlɪ] *adv.* moralisch; *(virtuously)* moralisch einwandfrei
morass [mə'ræs] *n.* Morast, *der (auch fig.);* **a ~ of paperwork** ein Wust von Papierkram *(ugs.)*
moratorium [mɒrə'tɔ:rɪəm] *n., pl.* **~s** *or* **moratoria** [mɒrə'tɔ:rɪə] **a)** [vorläufiger] Stopp (**on** für); **b)** *(authorized delay)* Moratorium, *das*
morbid ['mɔ:bɪd] *adj.* **a)** krankhaft; makaber, *(geh.)* morbid ‹*Freude, Faszination, Phantasie, Neigung*›; **b)** *(coll.: melancholy)* trübselig; **c)** *(Med.)* krankhaft ‹*Zustand, Veränderung*›
more [mɔ:(r)] **1.** *adj.* **a)** *(additional)* mehr; **would you like any** *or* **some/a few ~?** *(apples, books, etc.)* möchten Sie noch welche/ein paar?; **would you like any** *or* **some ~ apples?** möchten Sie noch Äpfel?; **would you like any** *or* **some/a little ~?** *(tea, paper, etc.)* möchten Sie noch etwas/ein wenig?; **would you like any** *or* **some ~ tea/paper?** möchten Sie noch Tee/Papier?; **I haven't any ~ [apples/tea]** ich habe keine [Äpfel]/keinen [Tee] mehr; **~ and ~:** immer mehr; **many ~ things** noch viel mehr [Dinge]; **some ~ things** noch einige Dinge; **b)** *(greater in degree)* größer; **~'s the pity** *(coll.)* leider!; **the ~ fool 'you** du bist vielleicht ein Dummkopf. **2.** *n., no pl., no indef. art.* **a)** *(greater amount or number or thing)* mehr; **~ and ~:** mehr und mehr; immer mehr; **six or ~:** mindestens sechs; **the ~ the merrier** *see* **merry a; b)** *(additional number or amount or thing)* mehr; **what is ~ ...:** außerdem ...; **and ~:** mindestens *vorangestellt;* oder mehr; **there's no need to do/say [any] ~:** da braucht nichts weiter getan/gesagt zu werden; **c) ~ than** *(coll.: exceedingly)* über‹*satt, -glücklich, -froh*›; hoch‹*erfreut, -willkommen*›. **3.** *adv.* **a)** mehr ‹*mögen, interessieren, gefallen, sich wünschen*›; *forming compar.* **a ~ interesting book** ein interessanteres Buch; **this book is ~ interesting** dieses Buch ist interessanter; **~ often** häufiger; **~ than anything [else]** vor allem; **b)** *(nearer, rather)* eher; **~ ... than ...:** eher ... als ...; **~ dead than alive** mehr tot als lebendig; **c)** *(again)* wieder; **never ~:** nie wieder *od.* mehr; **not any ~:** nicht mehr; **once ~:** noch einmal; **d) ~ and ~ ...:** mehr und mehr *od.* immer mehr ...; *with adj. or adv.* immer ... (+ *Komp.*); **become ~ and ~ absurd** immer absurder werden; **e) ~ or less** *(fairly)* mehr oder weniger; *(approximately)* annähernd; **f) ~ so** noch mehr; **the ~ so because ...:** um so mehr, als *od.* weil ...
moreish ['mɔ:rɪʃ] *adj. (coll.)* lecker
moreover [mɔ:'rəʊvə(r)] *adv.* und außerdem; zudem *(geh.)*
morgue [mɔ:g] *see* **mortuary**
moribund ['mɒrɪbʌnd] *adj. (fig.)* dem Untergang geweiht
Mormon ['mɔ:mən] *n.* Mormone, *der*/Mormonin, *die*
morning ['mɔ:nɪŋ] *n.* Morgen, *der; (as opposed to afternoon)* Vormittag, *der; attrib.* morgendlich; Morgen‹*kaffee, -spaziergang, -zeitung usw.*›; **this ~:** heute morgen *od.* früh; **tomorrow ~:** morgen früh; **during the ~:** am Morgen/Vormittag; **[early] in the ~:** am [frühen] Morgen; *(regularly)* [früh] morgens; **at one** *etc.* **in the ~ = at one a.m.** *etc. see* **a.m.; on Wednesday ~s/~:** Mittwoch morgens/[am] Mittwoch morgen *od.* früh; **one ~:** eines Morgens; **~ came** es wurde Morgen; **~s, of a ~:** morgens; **in the ~** *(coll.: next ~)* morgen früh
morning: ~-'after pill *n.* Pille [für den Morgen] danach; **~ coat** *n.* Cut[away], *der;* **~ 'service** *n. (Eccl.)* Morgenandacht, *die; (RC Ch.)* Frühmesse, *die;* **~ sickness** *n.* morgendliche Übelkeit; **~ 'star** *n.* Morgenstern, *der*
Morocco [mə'rɒkəʊ] *pr. n.* Marokko *(das)*
moron ['mɔ:rɒn] *n. (coll.)* Trottel, *der (ugs. abwertend);* Schwachkopf, *der (ugs.)*
morose [mə'rəʊs] *adj.* verdrießlich
morphine ['mɔ:fi:n] *n.* Morphin, *das (fachspr.);* Morphium, *das*
morris ['mɒrɪs]: **~ dancer** *n.* Moriskentänzer, *der;* **~ dancing** *n.* Moriskentanzen, *das*
Morse [mɔ:s] *n.* Morseschrift, *die;* Morsezeichen *Pl.*
Morse 'code *n.* Morseschrift, *die;* Morsealphabet, *das*
morsel ['mɔ:sl] *n. (of food)* Bissen, *der;* Happen, *der*
mortal ['mɔ:tl] **1.** *adj.* sterblich; *(fatal, fought to the death, intense)* tödlich (**to** für); **~ combat** ein Kampf auf Leben und Tod; **~ sin** Todsünde, *die;* **~ enemy** Todfeind, *der.* **2.** *n.* Sterbliche, *der/die*
mortality [mɔ:'tælɪtɪ] *n.* **a)** Sterblichkeit, *die;* **b)** *(number of deaths)* Sterblichkeit, *die;* Todesfälle *Pl.;* **c) ~ [rate]** Sterblichkeitsrate, *die;* Sterbeziffer, *die*
mortally ['mɔ:təlɪ] *adv.* **~ wounded** tödlich verletzt; **~ offended** zutiefst *od.* tödlich beleidigt
mortar ['mɔ:tə(r)] *n.* **a)** *(substance)* Mörtel, *der;* **b)** *(vessel)* Mörser, *der;* **c)** *(cannon)* Minenwerfer, *der;* Mörser, *der*
'mortar-board *n. (Univ.) bei bestimmten Anlässen zum Talar getragene viereckige Kopfbedeckung der Studenten und Lehrer an britischen und amerikanischen Universitäten;* ≈ Barett, *das*
mortgage ['mɔ:gɪdʒ] **1.** *n.* Hypothek, *die; attrib.* Hypotheken-; **~ repayment** Hypothekenzahlung, *die.* **2.** *v.t.* mit einer Hypothek belasten
mortice *see* **mortise**
mortician [mɔ:'tɪʃn] *n. (Amer.)* Leichenbestatter, *der*/-bestatterin, *die*
mortification [mɔ:tɪfɪ'keɪʃn] *n. (humiliation)* Beschämung, *die*
mortify ['mɔ:tɪfaɪ] *v.t.* beschämen; **he felt mortified** er empfand es als beschämend
mortise ['mɔ:tɪs] *n.* **a)** *(Woodw.)* Zapfenloch, *das;* **~ and tenon [joint]** Zapfenverbindung, *die;* Verzapfung, *die;* **b)** *attrib.* **~ lock** Steckschloß, *das*
mortuary ['mɔ:tjʊərɪ] *n.* Leichenschauhaus, *das*
mosaic [məʊ'zeɪɪk] *n. (lit. or fig.)* Mosaik, *das; attrib.* Mosaik-
Moscow ['mɒskəʊ] **1.** *pr. n.* Moskau *(das).* **2.** *attrib. adj.* Moskauer
Moslem ['mɒzləm] *see* **Muslim**
mosque [mɒsk] *n.* Moschee, *die*
mosquito [mɒs'ki:təʊ] *n., pl.* **~es** Stechmücke, *die; (in tropics)* Moskito, *der;* **~ bite** Mücken-/Moskitostich, *der*
mos'quito-net *n.* Moskitonetz, *das*
moss [mɒs] *n.* Moos, *das*
mossy ['mɒsɪ] *adj.* moosig; bemoost; moosbewachsen
most [məʊst] **1.** *adj. (in greatest number, the majority of)* die mei-

sten; *(in greatest amount)* meist...; größt... ‹*Fähigkeit, Macht, Bedarf, Geduld, Lärm*›; **make the ~ mistakes/noise** die meisten Fehler/den meisten *od.* größten Lärm machen; **~ people** die meisten Leute; **for the ~ part** größtenteils; zum größten Teil. **2.** *n.* **a)** *(greatest amount)* das meiste; **offer [the] ~ for it** das meiste *od.* am meisten dafür bieten; **pay the ~:** am meisten bezahlen; **b)** *(the greater part)* **~ of the girls** die meisten Mädchen; **~ of his friends** die meisten seiner Freunde; **~ of the poem** der größte Teil des Gedichts; **~ of the time** die meiste Zeit; *(on ~ occasions)* meistens; **~ of what he said** das meiste von dem, was er sagte; **c) make the ~ of sth., get the ~ out of sth.** etw. voll ausnützen; *(represent at its best)* das Beste aus etw. machen; **d) at [the] ~:** höchstens; **at the very ~:** allerhöchstens. **3.** *adv.* **a)** *(more than anything else)* am meisten ‹*mögen, interessieren, gefallen, sich wünschen, verlangt*›; **~ of all** am allermeisten; **b)** *forming superl.* **the ~ interesting book** das interessanteste Buch; **this book is the ~ interesting** dieses Buch ist das interessanteste; **~ often** am häufigsten; **c)** *(exceedingly)* überaus; äußerst; **~ certainly** ohne jeden Zweifel

mostly ['məʊstlı] *adv.* *(most of the time)* meistens; *(mainly)* größtenteils; hauptsächlich

MOT *see* **MOT test**

motel [məʊ'tel] *n.* Motel, *das*

motet [məʊ'tet] *n.* *(Mus.)* Motette, *die*

moth [mɒθ] *n.* Nachtfalter, *der;* *(in clothes)* Motte, *die*

moth: ~ball 1. *n.* Mottenkugel, *die;* **in ~balls** *(fig.: stored)* eingemottet ‹*Kleider, Schiff, Waffen*›; beiseite geschoben ‹*Plan, Projekt*›; **2.** *v. t.* einmotten ‹*Kleider, alte Sachen, Vorschlag, militärisches Gerät*›; beiseite schieben ‹*Plan, Projekt*›; **~-eaten** *adj.* von Motten zerfressen; *(fig.)* verstaubt; altmodisch ‹*Person*›

mother ['mʌðə(r)] **1.** *n.* Mutter, *die;* **she is a** *or* **the ~ of six [children]** sie ist Mutter von sechs Kindern; **like ~ used to make** ‹*Essen*› wie bei Muttern *(ugs.);* **M~ Superior** Äbtissin, *die;* **necessity is the ~ of invention** *(prov.)* Not macht erfinderisch *(Spr.).* **2.** *v. t.* *(over-protect)* bemuttern

'mother country *n.* Mutterland, *das*

motherhood ['mʌðəhʊd] *n.*, *no pl.* Mutterschaft, *die*

Mothering Sunday ['mʌðərıŋ sʌndı] *(Brit. Eccl.)* *see* **Mother's Day**

mother: ~-in-law *n.*, *pl.* **~s-in-law** Schwiegermutter, *die;* **~land** *n.* Vaterland, *das*

motherly ['mʌðəlı] *adj.* mütterlich; **~ love** Mutterliebe, *die*

mother: ~-of-'pearl *n.* Perlmutt, *das;* **M~'s Day** *n.* Muttertag, *der;* **~ 'tongue** *n.* Muttersprache, *die*

motif [məʊ'ti:f] *n.* Motiv, *das*

motion ['məʊʃn] **1.** *n.* **a)** *(movement)* Bewegung, *die;* Gang, *der;* **be in ~:** in Bewegung sein; sich bewegen; ‹*Maschine:*› laufen; ‹*Fahrzeug:*› fahren; **set** *or* **put sth. in ~** *(lit. or fig.)* etw. in Bewegung *od.* Gang setzen; **b)** *(gesture)* Bewegung, *die;* Wink, *der;* **c)** *(formal proposal; also Law)* Antrag, *der;* **put forward** *or* **propose a ~:** einen Antrag stellen; **d)** *(of bowels)* Stuhlgang, *der;* **have** *or* **make a ~:** Stuhlgang haben; **e)** *in sing. or pl.* *(faeces)* Stuhl, *der;* **f) go through the ~s of doing sth.** *(coll.)* *(simulate)* so tun, als ob man etw. täte; *(do superficially)* etw. pro forma tun; **go through the ~s** *(coll.)* *(simulate)* nur so tun; *(do superficially)* es nur pro forma tun. **2.** *v. t.* **~ sb. to do sth.** jmdm. bedeuten *(geh.)* *od.* winken, etw. zu tun. **3.** *v. i.* winken; **~ to sb. to come in** jmdn. hereinwinken; **~ to sb. to do sth.** jmdm. bedeuten *(geh.)*, etw. zu tun

motionless ['məʊʃnlıs] *adj.* reg[ungs]los; bewegungslos

'motion picture *n.* *(esp. Amer.)* Film, *der;* *attrib.* Film-

motivate ['məʊtıveıt] *v. t.* motivieren

motivation [məʊtı'veıʃn] *n.* **a)** *(process)* Motivierung, *die;* **b)** *(incentive)* Motivation, *die* **(for** zu); **c)** *(condition)* Motiviertheit, *die;* Motivation, *die*

motive ['məʊtıv] **1.** *n.* Motiv, *das;* Beweggrund, *der;* **the ~ for the crime** das Tatmotiv; **do sth. from ~s of kindness** etw. aus Freundlichkeit tun. **2.** *adj.* *(moving to action)* treibend ‹*Geist, Kraft*›; *(productive of motion)* Antriebs-

motley ['mɒtlı] *adj.* **a)** [bunt]gescheckt; *(multicoloured)* [kunter]bunt; **b)** *(varied)* buntgemischt; bunt ‹*Auswahl*›

motor ['məʊtə(r)] **1.** *n.* **a)** Motor, *der;* **b)** *(Brit.: ~ car)* Auto, *das.* **2.** *adj.* **a)** *(driven by engine or ~)* Motor‹*schlitten, -mäher, -jacht usw.*›; **b)** *(of ~ vehicles)* Kraftfahrzeug‹*ersatzteile, -mechaniker, -verkehr*›. **3.** *v. i.* *(Brit.)* [mit dem Auto] fahren

motor: ~ bike *(coll.)* *see* **~ cycle; ~ boat** *n.* Motorboot, *das;* **~ car** *n.* *(Brit.)* Kraftfahrzeug, *das;* Automobil, *das (geh.);* **~ caravan** *n.* *(Brit.)* Caravan, *der;* Omnibus, *der;* **~ cycle** *n.* Motorrad, *das;* Kraftrad, *das (Amtsspr.);* *attrib.* **~-cycle combination** *(Brit.)* Motorrad mit Beiwagen; **~-cyclist** *n.* Motorradfahrer, *der*/-fahrerin, *die;* **~ home** *n.* Reisemobil, *das;* **~ industry** *n.* Kraftfahrzeugindustrie, *die*

motoring ['məʊtərıŋ] *n.* *(Brit.)* Autofahren, *das;* **school of ~:** Fahrschule, *die;* *attrib.* **~ offence** Verstoß gegen die [Straßen]verkehrsordnung; **~ organisation** Automobilklub, *der*

motorise *see* **motorize**

motorist ['məʊtərıst] *n.* Autofahrer, *der*/-fahrerin, *die*

motorize ['məʊtəraız] *v. t.* motorisieren

motor: ~-racing *n.* Autorennsport, *der;* **~ scooter** *see* **scooter b; ~ show** *n.* Auto[mobil]ausstellung, *die;* **~ trade** *n.* Kraftfahrzeughandel, *der;* **~ vehicle** *n.* Kraftfahrzeug, *das;* **~way** *n.* *(Brit.)* Autobahn, *die*

MO'T test *n.* *(Brit.)* ≈ TÜV, *der*

mottled ['mɒtld] *adj.* gesprenkelt

motto ['mɒtəʊ] *n.*, *pl.* **~es a)** Motto, *das;* **my ~ is 'live and let live'** meine Devise ist „leben und leben lassen"; **b)** *(in cracker)* Spruch, *der*

¹mould [məʊld] *n.* **a)** *(earth)* Erde, *die;* **b)** *(upper soil)* [Mutter]boden, *der*

²mould 1. *n.* **a)** *(hollow)* Form, *die;* *(Metallurgy)* Kokille, *die;* *(Plastics)* Preßform, *die;* **b)** *(Cookery)* [Kuchen-/Back-/Pudding]form, *die.* **2.** *v. t.* formen **(out of, from** aus)

³mould *n.* *(Bot.)* Schimmel, *der*

moulder ['məʊldə(r)] *v. i.* **~ [away]** *(lit. or fig.)* [ver]modern

moulding ['məʊldıŋ] *n.* **a)** *(process of forming, lit. or fig.)* Formen, *das;* **b)** *(object)* Formteil, *das* **(of, in** aus); Formling, *der (fachspr.);* *(Archit.)* Zierleiste, *die;* **c)** *(wooden strip)* Leiste, *die*

mouldy ['məʊldı] *adj.* *(overgrown with mould)* schimmlig; **a ~ smell** ein Modergeruch; **go ~:** verschimmeln; schimmlig werden

moult [məʊlt] **1.** *v. t.* **a)** *(Ornith.)* verlieren ‹*Federn, Gefieder*›; **b)** *(Zool.)* verlieren ‹*Haar*›; abstreifen ‹*Haut*›; abwerfen ‹*Horn, Geweih*›. **2.** *v.i.* ‹*Vogel:*› sich mausern; ‹*Hund, Katze:*› sich haaren

mound [maʊnd] *n.* **a)** *(of earth)* Hügel, *der; (of stones)* Steinhaufen, *der;* **burial ~:** Grabhügel, *der;* **b)** *(hillock)* Anhöhe, *die;* **c)** *(heap)* Haufen, *der*

mount [maʊnt] **1.** *n.* **a)** *(mountain)* **M~ Vesuvius/Everest** der Vesuv/der Mount Everest; **b)** *(animal)* Reittier, *das; (horse)* Pferd, *das;* **c)** *(of picture, photograph)* Passepartout, *das;* **d)** *(for gem)* Fassung, *die;* **e)** *(Philat.)* [Klebe]falz, *der.* **2.** *v.t.* **a)** *(ascend)* hinaufsteigen ‹*Treppe, Leiter, Stufe*›; steigen auf (+ *Akk.*) ‹*Plattform, Kanzel*›; **b)** *(get on)* steigen auf *(Akk.)* ‹*Reittier, Fahrzeug*›; *abs.* aufsitzen; **~ the pavement** auf den Bürgersteig fahren; **c)** *(place on support)* montieren (**on** auf + *Akk.*); **d)** *(prepare)* aufstellen ‹*Maschine, Apparat*›; präparieren ‹*Exemplar*›; in ein Album einkleben ‹*Briefmarke*›; aufziehen ‹*Bild usw.*›; einfassen ‹*Edelstein usw.*›; **e)** inszenieren ‹*Stück, Show, Oper*›; organisieren ‹*Festspiele, Ausstellung*›; **f)** *(carry out)* durchführen ‹*Angriff, Operation usw.*›. **3.** *v.i.* **~ [up]** *(increase)* steigen (**to** auf + *Akk.*); **it all ~s up** es summiert sich

mountain ['maʊntɪn] *n.* **a)** *(lit. or fig.)* Berg, *der;* **in the ~s** im Gebirge; **butter/grain** *etc.* **~** *(fig.)* Butter-/Getreideberg *usw., der;* **move ~s** *(fig.)* Berge versetzen; *see also* **molehill; b)** *attrib.* Gebirgs-; **~ bike** Mountainbike, *das;* Geländefahrrad, *das*

mountaineer [maʊntɪ'nɪə(r)] *n.* Bergsteiger, *der* / Bergsteigerin, *die*

mountaineering [maʊntɪ'nɪərɪŋ] *n.* Bergsteigen, *das; attrib.* **~ expedition** Bergpartie, *die*

mountainous ['maʊntɪnəs] *adj.* **a)** gebirgig; **b)** *(huge)* riesig ‹*Gegenstand, Welle*›

mountain: ~ 'range *n.* Gebirgszug, *der;* **~ 'road** *n.* Gebirgsstraße, *die;* **~side** *n.* [Berg][ab]hang, *der;* **~ top** *n.* Berggipfel, *der*

mounted ['maʊntɪd] *adj.* *(on animal)* beritten

Mountie ['maʊntɪ] *n.* *(coll.)* Mountie, *der; berittener kanadischer Polizist*

mounting ['maʊntɪŋ] *n.* **a)** *(of performance)* Inszenierung, *die;* **b)** *(support) (Art: of drawing)* Passepartout, *das; (of engine, axle, etc.)* Aufhängung, *die*

mourn [mɔ:n] **1.** *v.i.* trauern; **~ for** *or* **over** trauern um ‹*Toten*›; nachtrauern (+ *Dat.*) ‹*Jugend, Augenlicht, Haustier*›; betrauern ‹*Verlust, Mißgeschick*›. **2.** *v.t.* betrauern; nachtrauern ‹*etw. Verlorenem*›

mourner ['mɔ:nə(r)] *n.* Trauernde, *der/die*

mournful ['mɔ:nfl] *adj.* klagend ‹*Stimme, Ton, Schrei, Geheul*›; trauervoll *(geh.)* ‹*Person*›

mourning ['mɔ:nɪŋ] *n.* **a)** *(clothes)* Trauer[kleidung], *die;* **be [dressed] in** *or* **wear/put on** *or* **go into ~:** Trauer tragen/anlegen; **b)** *(sorrowing, lamentation)* Trauer, *die*

mouse [maʊs] **1.** *n., pl.* **mice** [maɪs] **a)** Maus, *die;* **as quiet as a ~:** ganz leise; mucksmäuschenstill *(fam.)* ‹*dasitzen*›; **b)** *(fig.: timid person)* Angsthase, *der (ugs.);* **c)** *(Computing)* Maus, *die.* **2.** *v.i.* mausen

mouse: ~-hole *n.* Mauseloch, *das;* **~ trap** *n.* Mausefalle, *die*

mousse [mu:s] *n.* Mousse, *die*

moustache [mə'stɑ:ʃ] *n.* Schnurrbart, *der*

mousy ['maʊsɪ] *adj.* mattbraun ‹*Haar*›

mouth 1. [maʊθ] *n., pl.* **~s** [maʊðz] **a)** *(of person)* Mund, *der; (of animal)* Maul, *das;* **with one's ~ open** mit offenem Mund; **keep one's ~ shut** *(fig. sl.)* die *od.* seine Klappe halten *(salopp);* **put one's money where one's ~ is** *(fig. sl.)* seinen Worten Taten folgen lassen; **with one's ~ full** mit vollem Mund; **out of the ~s of babes [and sucklings]!** *(fig.)* Kindermund tut Wahrheit kund *(Spr.);* **have got many ~s to feed** viele hungrige Mäuler zu stopfen haben *(ugs.);* **take the words out of sb.'s ~:** jmdm. das Wort aus dem Mund *od.* von der Zunge nehmen; **b)** *(fig.) (entrance to harbour)* [Hafen]einfahrt, *die; (of valley, gorge, burrow, tunnel, cave)* Eingang, *der; (of bottle, cannon)* Mündung, *die;* **c)** *(of river)* Mündung, *die.* **2.** [maʊð] *v.t.* mit Lippenbewegungen sagen. **3.** [maʊð] *v.i.* lautlos die Lippen bewegen

mouthful ['maʊθfʊl] *n.* **a)** *(bite)* Mundvoll, *der; (of solid food)* Bissen, *der; (of drink)* Schluck, *der;* **b)** *(sth. difficult to say)* Zungenbrecher, *der (ugs.)*

mouth: ~-organ *n.* Mundharmonika, *die;* **~piece** *n.* **a)** *(Mus., Med.)* Mundstück, *das; (of telephone)* Sprechmuschel, *die;* **b)** *(speaker for others)* Sprachrohr, *das;* **~-to-~ resusci'tation** *n.* Wiederbelebung durch Mund-zu-Mund-Beatmung; **~wash** *n.* Mundwasser, *das;* **~-watering** *adj.* lecker

movable ['mu:vəbl] *adj.* beweglich

move [mu:v] **1.** *n.* **a)** *(change of residence)* Umzug, *der; (change of job)* Wechsel, *der;* **b)** *(action taken)* Schritt, *der; (Footb. etc.)* Spielzug, *der;* **c)** *(turn in game)* Zug, *der; (fig.)* [Schach]zug, *der;* **make a ~:** ziehen; **it's your ~:** du bist am Zug; **d) be on the ~** *(moving about)* ‹*Person:*› unterwegs sein; **e) make a ~** *(initiate action)* etwas tun *od.* unternehmen; *(coll.: leave, depart)* losziehen *(ugs.);* **make the first ~:** den Anfang machen; **make no ~:** sich nicht rühren; **make no ~ to help sb.** keine Anstalten machen, jmdm. zu helfen; **f) get a ~ on** *(coll.)* einen Zahn zulegen *(ugs.);* **get a ~ on!** *(coll.)* [mach] Tempo! *(ugs.).* **2.** *v.t.* **a)** *(change position of)* bewegen; wegräumen ‹*Hindernis, Schutt*›; *(transport)* befördern; **~ the chair over here** rück den Stuhl hier herüber!; **~ sth. to a new position** etw. an einen neuen Platz bringen; **~ house** umziehen; **~ the luggage into the building** das Gepäck ins Gebäude hineinbringen; **not ~ a muscle** sich nicht rühren; **please ~ your head [to one side]** bitte tun Sie Ihren Kopf zur Seite; **~ it!** *(coll.),* **~ yourself!** *(coll.)* Beeilung! *(ugs.);* **~ sb. to another department/job** jmdn. in eine andere Abteilung/Position versetzen; **~ police/troops into an area** Polizeikräfte/Truppen in ein Gebiet schicken; **b)** *(in game)* ziehen; **c)** *(affect)* bewegen; berühren; **~ sb. to laughter/anger** jmdn. zum Lachen bringen/jmds. Ärger erregen; **~ sb. to tears** jmdn. zu Tränen rühren; **~ sb. to pity** jmds. Mitleid erregen; **be ~d to pity** vor Mitleid gerührt sein; **be ~d by sth.** über etw. *(Akk.)* gerührt sein; **d)** *(prompt)* **~ sb. to do sth.** jmdn. dazu bewegen, etw. zu tun; **sb. is not to be ~d** jmd. läßt sich nicht erschüttern; **e)** *(propose)* beantragen ‹*Beendigung, Danksagung*›; stellen ‹*Antrag*›; **f)** *(Commerc.: sell)* absetzen. **3.** *v.i.* **a)** *(go from place to place)* sich bewegen; *(by car, bus, train)* fahren; *(on foot)* gehen; *(coll.: start, leave)* gehen; ‹*Wolken:*› ziehen (**across** über + *Akk.*); **~ with the times** *(fig.)* mit der Zeit gehen; **get moving!** beeil dich!; **start to ~** ‹*Fahrzeug:*› sich in Bewegung setzen; **nobody ~d** niemand rührte sich von der Stelle; **he has ~d to another department** er ist jetzt in einer anderen Abteilung; **Don't ~. I'll be back soon** Bleib hier *od.* Geh nicht weg. Ich bin gleich zurück; **b)** *(in games)* ziehen; **c)** *(fig.: initiate action)* handeln; aktiv werden; **~ quickly to do sth.** schnell handeln und etw. tun; **d)** *(in certain circles, part of society, part of town)* verkehren; **e)** *(change residence or accommodation)* umziehen (**to** nach); *(into flat etc.)* einziehen (**into** in + *Akk.*); *(out of town)* wegziehen (**out of** aus); *(out of flat etc.)* ausziehen (**out of** aus); **I want to ~ to London** ich will nach London ziehen; **f)** *(change posture or state)* sich bewegen; *(in order to make oneself comfortable etc.)* eine andere Haltung einnehmen; **don't ~ or I'll shoot** keine Bewegung, oder ich schieße; **g)** *(make progress)* vorankommen; **get things moving** vorankommen; **things are moving now** jetzt geht es voran; **~ towards** näherkommen (+ *Dat.*) ‹*Einigung, Höhepunkt, Kompromiß*›; **h)** *(Commerc.: be sold)* ‹*Waren:*› Absatz finden, sich absetzen lassen; **i)** *(coll.: go fast)* **that car can really ~:** der Wagen ist enorm schnell *(ugs.)*

~ a'bout 1. *v. i.* zugange sein; *(travel)* unterwegs sein. **2.** *v. t.* herumräumen ‹*Möbel, Bücher*›

~ a'long 1. *v. i.* **a)** gehen/fahren; **b)** *(make room)* Platz machen; **~ along, please!** gehen/fahren Sie bitte weiter! **2.** *v. t.* zum Weitergehen/-fahren auffordern

~ 'in 1. *v. i.* **a)** einziehen; *(to start work)* ‹*Bauarbeiter:*› kommen; **b)** *(come closer)* ‹*Truppen, Polizeikräfte:*› anrücken; **~ in on** ‹*Truppen, Polizeikräfte:*› vorrücken gegen. **2.** *v. t.* einrücken lassen ‹*Truppen, Polizeikräfte*›; hineinbringen ‹*Gepäck, Ausrüstung*›

~ 'off *v. i.* sich in Bewegung setzen

~ 'on 1. *v. i.* weitergehen/-fahren; **~ on to another question** *(fig.)* zu einer anderen Frage übergehen. **2.** *v. t.* zum Weitergehen/-fahren auffordern

~ 'out *v. i.* ausziehen (**of** aus)

~ 'over *v. i.* rücken

~ 'up *v. i.* **a)** *(in queue, hierarchy)* aufrücken; ‹*Fahrzeug:*› vorfahren; **b)** *see* **~ over**

movement ['mu:vmənt] *n.* **a)** Bewegung, *die; (of people: towards city, country, etc.)* [Ab]wanderung, *die; (trend, tendency)* Tendenz, *die;* **a ~ of the head/arm/leg** eine Kopf-/Arm-/Beinbewegung; **without ~:** bewegungslos; **b)** *in pl.* Aktivitäten *Pl.;* **keep track of sb.'s ~s** jmdn. überwachen; **c)** *(Mus.)* Satz, *der;* **d)** *(concerted action for purpose)* Bewegung, *die;* **e)** *in sing. or pl. (Mech. esp. in clock, watch)* Räderwerk, *das;* **f)** *(in price)* Preisbewegung, *die*

movie ['mu:vɪ] *n. (Amer. coll.)* Film, *der; attrib.* Film-; **the ~s** *(art form, cinema industry)* der Film

movie-goer ['mu:vɪgəʊə(r)] *n. (Amer. coll.)* Kinogänger, *der/* -gängerin, *die*

moving ['mu:vɪŋ] *adj.* **a)** beweglich; **from a ~ car** ‹*fallen, werfen, schießen*› aus einem fahrenden Auto; **b)** *(affecting)* ergreifend; bewegend

moving 'staircase *see* **escalator**

mow [məʊ] *v. t., p.p.* **mown** [məʊn] *or* **mowed** [məʊd] mähen

~ 'down *v. t.* niedermähen ‹*Soldaten*›; überfahren ‹*Fußgänger*›

mower ['məʊə(r)] *n. (for lawn)* Rasenmäher, *der*

Mozambique [məʊzəm'bi:k] *pr. n.* Mosambik *(das)*

MP *abbr.* **Member of Parliament**

m.p.g. [empi:'dʒi:] *abbr. (Motor Veh.)* **miles per gallon; do/get 34 ~** *(Brit.)* 8,31 auf 100 km [ver]brauchen

m.p.h. [empi:'eɪtʃ] *abbr.* **miles per hour; we are driving at/doing 30 ~:** wir fahren 50 [km/h]

Mr ['mɪstə(r)] *n., pl.* **Messrs** ['mesəz] *(title)* Herr; *(third person also)* Hr.; *(in an address)* Herrn; **Messrs** Hrn.; *(firm)* Fa.

Mrs ['mɪsɪz] *n., pl. same* Frau; *(third person also)* Fr.

MS *abbr.* **a) manuscript** Ms.; **b)** *(Med.)* **multiple sclerosis** MS

Ms [mɪz] *n., no pl.* Frau

M.Sc. [emes'si:] *abbr.* **Master of Science;** *see also* **B.Sc.**

Mt. *abbr.* **Mount; ~ Etna/Everest** der Ätna/der Mount Everest

much [mʌtʃ] **1.** *adj.,* **more** [mɔ:(r)], **most** [məʊst] **a)** viel; groß ‹*Erleichterung, Sorge, Dankbarkeit*›; **with ~ love** voller Liebe; **he never eats ~ breakfast/lunch** er ißt nicht viel zum Frühstück/zu Mittag; **too ~:** zuviel *indekl.;* **b) be a bit ~** *(coll.)* ein bißchen zuviel sein; *(fig.)* ein bißchen zu weit gehen. **2.** *n.; see also* **more 2; most 2;** vieles; **we don't see ~ of her any more** wir sehen sie kaum noch; **that doesn't come** *or* **amount to ~:** es kommt nicht viel dabei heraus; **he/this beer isn't up to ~** *(coll.)* mit ihm/diesem Bier ist nicht viel los *(ugs.);* **spend ~ of the day/week doing sth.** den Großteil des Tages/der Woche damit verbringen, etw. zu tun; **they have done ~ to improve the situation** sie haben viel für die Verbesserung der Situation getan; **not be ~ of a cinema-goer** *etc. (coll.)* kein großer Kinogänger *usw.* sein *(ugs.);* **it isn't ~ of a bicycle** es ist kein besonders tolles Fahrrad *(ugs.);* **not be ~ to look at** nicht sehr ansehnlich sein; **it's as ~ as she can do to get up the stairs** sie kommt gerade noch die Treppe hinauf; **I expected/thought as ~:** das habe ich erwartet/mir gedacht; **you are as ~ to blame as he is** du bist ebensosehr schuld wie er; **without so ~ as saying goodbye** ohne auch nur auf Wiedersehen zu sagen. **3.** *adv.,* **more, most a)** *modifying comparatives* viel ‹*besser*›; **~ more lively/happy/attractive** viel lebhafter/glücklicher/ attraktiver; **b)** *modifying superlatives* mit Abstand ‹*der/die/das beste, schlechteste, klügste usw.*›; **c)** *modifying passive participles and predicative adjectives* sehr; **he is ~ improved** *(in health)* es geht ihm viel besser; **d)** *modifying verbs (greatly)* sehr ‹*lieben, mögen, genießen*›; *(often)* oft ‹*sehen, treffen, besuchen*›; *(frequently)* viel; **I don't ~ like him** *or* **like him ~:** ich mag ihn nicht besonders; **not go ~ on sb./sth.** *(coll.)* nicht viel von jmdm./etw. halten; **it doesn't matter ~:** es ist nicht so wichtig; **I would ~ prefer to stay at home** ich würde viel lieber zu Hause bleiben; **~ to my surprise/annoyance, I found that ...:** zu meiner großen Überraschung/Verärgerung stellte ich fest, daß ...; **e)** *(approximately)* fast; **[pretty** *or* **very] ~ the same** fast [genau] der-/die-/dasselbe; **f) ~ as** *or* **though** *(although)* sosehr ... auch; **~ as he disliked the idea** sosehr ihm die Idee auch mißfiel; **~ as I should like to go** so gern ich auch gehen würde

muck [mʌk] *n.* **a)** *(farmyard manure)* Mist, *der;* **b)** *(coll.: anything disgusting)* Dreck, *der (ugs.); (liquid)* Brühe, *die (ugs. abwertend);* **covered in ~:** verdreckt *(ugs.)*

~ a'bout, ~ a'round *(Brit. sl.)* **1.** *v. i.* **a)** herumalbern *(ugs.);* **b)** *(tinker)* herumfummeln (**with** an + *Dat.*). **2.** *v. t.* **~ sb. about** *or* **around** jmdn. verarschen *(derb)*

~ **'in** *v.i. (coll.)* mit zugreifen *od.* mit anpacken (**with** bei)

~ **'up** *v.t.* **a)** *(Brit. coll.: bungle)* vermurksen, verbocken *(ugs.)*; ~ **it up** Mist bauen *(ugs.)*; **b)** *(make dirty)* vollschmieren *(ugs.)*; dreckig machen *(ugs.)*; einsauen *(derb)*; **c)** *(coll.: spoil)* vermasseln *(salopp)*

muckraking ['mʌkreɪkɪŋ] *n.* Skandalhascherei, *die (abwertend)*

mucky ['mʌkɪ] *adj.* dreckig *(ugs.)*

mucus ['mju:kəs] *n. (Med., Bot., Zool.)* Schleim, *der*

mud [mʌd] *n.* **a)** Schlamm, *der;* **be as clear as** ~ *(joc. iron.)* absolut unklar sein; **b)** *(fig.)* **be dragged through the** ~: in den Schmutz gezogen werden; **his name is** ~ *(coll.)* er ist unten durch *(ugs.)* (**with** bei); **sling** *or* **throw** ~ **at sb.** *(fig.)* jmdn. mit Dreck *(ugs.) od.* Schmutz bewerfen

'**mud-bath** *n. (Med.; also fig.)* Schlammbad, *das*

muddle ['mʌdl] **1.** *n.* Durcheinander, *das;* **the room is in a hopeless** ~: in dem Zimmer herrscht ein heilloses Durcheinander; **get sth. in a** ~: etw. in Unordnung bringen; etw. durcheinanderbringen; **get in[to] a** ~: durcheinanderkommen *(ugs.)*. **2.** *v.t.* ~ **[up]** durcheinanderbringen; ~ **up** *(mix up)* verwechseln (**with** mit); **be** ~**d up** *(out of order)* durcheinandergeraten sein

~ **a'long,** ~ **'on** *v.i.* vor sich *(Akk.)* hin wursteln *(ugs.)*

~ **'through** *v.i.* sich durchwursteln *(ugs.)*

muddled ['mʌdld] *adj.* benebelt ‹*Person*›; konfus ‹*Verhalten, Denken*›; verworren ‹*Situation, Information, Ideen*›

muddle-'headed *adj.* wirr

muddy ['mʌdɪ] **1.** *adj.* **a)** schlammig; **get** *or* **become** ~: verschlammen; **b)** *(turbid, dull)* trübe ‹*Flüssigkeit, Farbe*›. **2.** *v.t.* schmutzig machen; *(make turbid)* trüben ‹*Flüssigkeit*›

mud: ~**-flap** *n. (Motor Veh.)* Schmutzfänger, *der;* ~**-flat[s]** *n. [pl.] (Geog.)* Watt, *das;* ~**guard** *n.* Schutzblech, *das; (of car)* Kotflügel, *der;* ~**pack** *n.* Schlammpackung [für das Gesicht]; ~ **'pie** *n.* Kuchen *(aus Sand usw.)*

muesli ['mju:zlɪ] *n.* Müsli, *das*

[1]**muff** [mʌf] *n.* Muff, *der*

[2]**muff** *v.t.* **a)** *(bungle)* verderben; verpatzen *(ugs.)*; verhauen *(ugs.)* ‹*Examen*›; **b)** *(Theatre)* verpatzen *(ugs.)*; ~ **a line** einen Patzer machen *(ugs.)*

muffin ['mʌfɪn] *n.* Muffin, *das*

muffle ['mʌfl] *v.t.* **a)** *(envelop)* ~ **[up]** einhüllen; einmumme[l]n *(ugs.)*; **b)** dämpfen ‹*Geräusch*›; [zur Schalldämpfung] umwickeln ‹*Ruder, Trommel, Glocke*›

muffler ['mʌflə(r)] *n.* **a)** *(wrap, scarf)* Schal, *der;* **b)** *(Amer. Motor Veh.)* Schalldämpfer, *der*

[1]**mug** [mʌg] **1.** *n.* **a)** *(vessel, contents)* Becher, *der (meist mit Henkel); (for beer etc.)* Krug, *der;* **a** ~ **of milk** ein Becher Milch; **b)** *(sl.: face, mouth)* Visage, *die (salopp)*; Fresse, *die (derb)*; **c)** *(Brit. sl.: simpleton)* Schwachkopf, *der (ugs.)*; **d)** *(Brit. sl.: gullible person)* Trottel, *der (ugs. abwertend)*; Doofi, *der (ugs.)*; **that's a** '~**'s game** das ist doch Schwachsinn *(ugs.)*. **2.** *v.t.*, **-gg-** *(rob)* überfallen und berauben

[2]**mug** *(Brit. sl.: study) v.t.*, **-gg-**: ~ **up** büffeln *(ugs.)*

mugful ['mʌgfʊl] *(contents) see* [1]**mug 1 a**

mugger ['mʌgə(r)] *n.* Straßenräuber, *der*/Straßenräuberin, *die*

mugging ['mʌgɪŋ] *n.* Straßenraub, *der* (**of** an + *Dat.*)

muggins ['mʌgɪnz] *n., pl.* ~**es** *or same (coll.)* **a)** *(simpleton)* Dummkopf, *der (ugs.)*; Esel, *der (ugs.)*; **b)** *(myself, stupidly)* ich Dummkopf *(ugs.)*

muggy ['mʌgɪ] *adj.* schwül; drückend ‹*Klima, Tag, Luft*›

mulberry ['mʌlbərɪ] *n.* **a)** *(fruit)* Maulbeere, *die;* **b)** *(tree)* Maulbeerbaum, *der*

mulch [mʌltʃ] *(Agric., Hort.)* **1.** *n.* Mulch, *der.* **2.** *v.t.* mulchen

mule [mju:l] *n.* Maultier, *das; see also* **obstinate; stubborn a**

[1]**mull** [mʌl] *v.t.* ~ **over** nachdenken über (+ *Akk.*); *(in conversation)* diskutieren

[2]**mull** *v.t.* ~**ed wine** Glühwein, *der*

mullah ['mʊlə] *n. (Islam)* Mullah, *der*

multi- ['mʌltɪ] *in comb. (several)* mehr-/Mehr-; *(many)* viel-/Viel-; multi-/Multi-, poly-/Poly- *(bes. mit Fremdwörtern)*

'**multicoloured** (*Brit.; Amer.:* **multicolored**) *adj. (with several colours)* mehrfarbig, *(with many colours)* vielfarbig ‹*Gegenstand, Tier, Pflanze*›; bunt ‹*Stoff, Kleid*›

multifarious [mʌltɪ'feərɪəs] *adj.* **a)** *(having great variety)* vielgestaltig; **b)** *(many and various)* mannigfach; vielfältig

multi'lateral *adj.* mehrseitig; *(Polit.)* multilateral

multimillio'naire *n.* Multimillionär, *der*/-millionärin, *die*

multi'national *adj.* **1.** multinational. **2.** *n.* multinationaler Konzern, *der;* Multi, *der (ugs.)*

multiple ['mʌltɪpl] **1.** *adj.* **a)** *(manifold)* mehrfach; ~ **pile-up** Massenkarambolage, *die;* **b)** *(many and various)* vielerlei; vielfältig. *See also* **sclerosis. 2.** *n. (Math.)* Vielfache, *das*

multiple: ~**-'choice** *adj.* Multiple-choice-‹*Verfahren, Test, Frage*›; ~ **'store** *n. (Brit. Commerc.)* Kettenladen, *der*

multiplication [mʌltɪplɪ'keɪʃn] *n. (increase)* Vervielfachung, *die; (Math.)* Multiplikation, *die (fachspr.)*; Malnehmen, *das; attrib.* ~ **sign** Malzeichen, *das;* ~ **table** Multiplikationstabelle, *die*

multiplicity [mʌltɪ'plɪsɪtɪ] *n.* Vielfalt, *die* (**of, in** an, von + *Dat. Pl.*); *(great number)* Vielzahl, *die* (**of** von, an + *Dat.*)

multiply ['mʌltɪplaɪ] **1.** *v.t.* **a)** *(Math., also abs.)* multiplizieren *(fachspr.)*, malnehmen (**by** mit); **b)** *(increase)* vervielfachen. **2.** *v.i. (Biol.)* sich vermehren; sich fortpflanzen

'**multi-purpose** *adj.* Mehrzweck-

multi'racial *adj.* mehrrassig; gemischtrassig

'**multi-storey** *adj.* mehrstöckig; mehrgeschossig; ~ **car park/block of flats** Parkhaus/Wohnhochhaus, *das*

multitude ['mʌltɪtju:d] *n. (crowd)* Menge, *die; (great number)* Vielzahl, *die*

[1]**mum** [mʌm] *(coll.)* **1.** *int.* ~**'s the word** nicht weitersagen! **2. keep** ~: den Mund halten *(ugs.)*

[2]**mum** *n. (Brit. coll.: mother)* Mama, *die (fam.)*

mumble ['mʌmbl] **1.** *v.i.* nuscheln *(ugs.)*. **2.** *v.t.* nuscheln *(ugs.)*

mumbo-jumbo [mʌmbəʊ'dʒʌmbəʊ] *n., pl.* ~**s a)** *(meaningless ritual)* Brimborium, *das (ugs.)*; Theater, *das (ugs. abwertend)*; **b)** *(gibberish)* Kauderwelsch, *das*

[1]**mummy** ['mʌmɪ] *n.* Mumie, *die*

[2]**mummy** *n. (Brit. coll.: mother)* Mutti, *die (fam.)*; Mami, *die (fam.)*; Mama, *die (fam.)*

mumps [mʌmps] *n. sing. (Med.)* Mumps, *der*

munch [mʌntʃ] **1.** *v.t.* ~ **one's food** mampfen *(salopp)*; schmatzend kauen. **2.** *v.i.* mampfen *(salopp)*

mundane [mʌn'deɪn] *adj.* **a)** *(dull)* banal; **b)** *(worldly)* weltlich

municipal [mju:'nɪsɪpl] *adj.* gemeindlich; kommunal; Kommunal⟨*politik, -verwaltung*⟩; Gemeinde⟨*rat, -verwaltung*⟩

municipality [mju:nɪsɪ'pælɪtɪ] *n.* Gemeinde, *die*

munition [mju:'nɪʃn] *n., usu. in pl.* Kriegsmaterial, *das;* **~[s] factory** Rüstungsbetrieb, *der*

mural ['mjʊərl] *n.* Wandbild, *das; (on ceiling)* Deckengemälde, *das*

murder ['mɜ:də(r)] **1.** *n.* **a)** *(Law)* Mord, *der* (**of** an + *Dat.*); **~ investigation** Ermittlungen *Pl.* in dem/einem Mordfall; **~ hunt** Fahndung nach dem/einem Mörder; **b)** *(fig.)* **the exam/journey was ~**: die Prüfung/Reise war der glatte *od.* reine Mord *(ugs.).* **2.** *v. t.* **a)** *(kill unlawfully)* ermorden; **~ sb. with a gun/knife** jmdn. erschießen/erstechen; **b)** *(kill inhumanly)* umbringen; **c)** *(coll.: spoil)* verhunzen *(ugs.);* **d)** *(coll.: defeat)* fertigmachen *(ugs.)*

murderer ['mɜ:dərə(r)] *n.* Mörder, *der*/Mörderin, *die*

murderous ['mɜ:dərəs] *adj.* tödlich; Mord⟨*absicht, -drohung*⟩; mörderisch *(ugs.)* ⟨*Fahrweise, Kampf, Bedingung*⟩

murk [mɜ:k] *n.* Dunkelheit, *die;* Nebelnacht, *die (geh.)*

murky ['mɜ:kɪ] *adj.* **a)** *(dark)* düster; trüb ⟨*Tag, Wetter*⟩; **b)** *(dirty)* schmutzig-trüb ⟨*Wasser*⟩; **c)** *(fig.: obscure)* dunkel; unergründlich ⟨*Geheimnis, Tiefen*⟩

murmur ['mɜ:mə(r)] **1.** *n.* **a)** *(subdued sound)* Rauschen, *das; (of brook also)* Murmeln, *das (dichter.);* **b)** *(expression of discontent)* Murren, *das;* **~ of disagreement/impatience** ablehnendes/ungeduldiges Murren; **without a ~**: ohne Murren; **c)** *(soft speech)* Murmeln, *das;* **~ of approval/delight** beifälliges/freudiges Murmeln; **a ~ of voices** ein Gemurmel. **2.** *v. t.* murmeln. **3.** *v. i.* ⟨*Person:*⟩ murmeln; *(complain)* murren (**against, at** über + *Akk.*)

muscle ['mʌsl] **1.** *n.* **a)** Muskel, *der;* **not move a ~** *(fig.)* sich nicht rühren; **b)** *(tissue)* Muskeln *Pl.;* **c)** *(muscular power)* [Muskel-, Körper]kraft, *die;* Muskeln *Pl.; (fig.: force, power, influence)* Stärke, *die.* **2.** *v. i.* **~ 'in** *(coll.)* sich hineindrängen (**on** in + *Akk.*)

'muscleman *n.* Muskelmann, *der (ugs.)*

muscular ['mʌskjʊlə(r)] *adj.* **a)** *(Med.)* Muskel-; muskulär *(fachspr.);* **b)** *(sinewy)* muskulös

muscular dystrophy [mʌskjʊlə 'dɪstrəfɪ] *n. (Med.)* Muskeldystrophie, *die*

muse [mju:z] *(literary) v. i.* grübeln; [nach]sinnen *(geh.),* sinnieren (**on, about, over** über + *Akk.*)

museum [mju:'zi:əm] *n.* Museum, *das;* **~ of art** Kunstmuseum, *das*

mu'seum piece *n.* **a)** Museumsstück, *das;* **b)** *(joc. derog.)* Museumsstück, *das (ugs. iron.)*

mush [mʌʃ] *n.* **a)** *(soft pulp)* Mus, *das;* Brei, *der;* **b)** *(coll.: weak sentimentality)* Schmalz, *der (ugs.)*

mushroom ['mʌʃrʊm, 'mʌʃru:m] **1.** *n.* Pilz, *der; (edible)* [Speise]pilz, *der; (cultivated, esp. Agaricus campestris)* Champignon, *der.* **2.** *v. i. (spring up)* wie Pilze aus dem Boden schießen; **demand ~ed overnight** die Nachfrage schoß über Nacht in die Höhe

'mushroom cloud *n.* Rauchpilz, *der; (after nuclear explosion)* Atompilz, *der*

mushy ['mʌʃɪ] *adj.* **a)** *(soft)* breiig; **b)** *(coll.: feebly sentimental)* schmalzig *(abwertend)*

music ['mju:zɪk] *n.* **a)** Musik, *die;* **make ~**: Musik machen; musizieren; **piece of ~**: Musikstück, *das;* Musik, *die;* **set** *or* **put sth. to ~**: etw. vertonen *od.* in Musik setzen; **be ~ to sb.'s ears** *(fig. coll.)* Musik in jmds. Ohren sein *(ugs.); see also* **face 2 c**; **b)** *(score)* Noten *Pl.; (as merchandise also)* Musikalien *Pl.;* **sheet of ~**: Notenblatt, *das*

musical ['mju:zɪkl] **1.** *adj.* musikalisch; Musik⟨*instrument, -notation, -verein, -verständnis, -abend*⟩; Musik⟨*film, -theater*⟩. **2.** *n. (Mus., Theatre)* Musical, *das*

musical: ~ box *n. (Brit.)* Spieldose, *die;* **~ 'chairs** *n. sing.* Reise nach Jerusalem

musically ['mju:zɪkəlɪ] *adv.* musikalisch; *(melodiously)* melodisch; melodiös; **~ gifted** musikalisch [begabt]; musikbegabt

music: ~ centre *n.* Kompaktanlage, *die;* **~-hall** *n. (Brit.)* Varieté, *das; attrib.* Varieté-

musician [mju:'zɪʃn] *n.* Musiker, *der*/Musikerin, *die*

music: ~-lesson *n.* Musikstunde, *die;* **~-stand** *n.* Notenständer, *der;* **~-stool** *n.* Klavierhocker, *der;* Klavierschemel, *der;* **~-teacher** *n.* Musiklehrer, *der*/-lehrerin, *die*

musk [mʌsk] *n.* **a)** *(substance)* Moschus, *der;* **b)** *(odour)* Moschusgeruch, *der*

musket ['mʌskɪt] *n. (Hist.)* Muskete, *die*

musketeer [mʌskɪ'tɪə(r)] *n. (Hist.)* Musketier, *der*

'musk-rat *n.* **a)** *(Zool.)* Bisamratte, *die;* **b)** *(fur)* Bisam, *der*

musky ['mʌskɪ] *adj.* moschusartig ⟨*Duft, Geruch, Geschmack*⟩; Moschus⟨*duft, -parfüm*⟩

Muslim ['mʊslɪm, 'mʌzlɪm] **1.** *adj.* muslimisch *(bes. fachspr.);* moslemisch. **2.** *n.* Muslim, *der*/Muslime, *die (bes. fachspr.);* Moslem, *der*/Moslime, *die*

muslin ['mʌzlɪn] *n.* Musselin, *der*

musquash ['mʌskwɒʃ] *see* **musk-rat**

muss [mʌs] *v. t. (Amer. coll.)* verstrubbeln *(ugs.)* ⟨*Haar, Frisur*⟩

~ 'up *v. t.* durcheinanderbringen; verstrubbeln *(ugs.)* ⟨*Haar*⟩; zerknittern ⟨*Kleidung*⟩

mussel ['mʌsl] *n.* Muschel, *die*

must [məst, *stressed* mʌst] **1.** *v. aux., only in pres. and past* **must,** *neg. (coll.)* **mustn't** ['mʌsnt] **a)** *(have to)* müssen; *with negative* dürfen; **you ~ not/never do that** das darfst du nicht/nie tun; **you ~ remember ...**: du darfst nicht vergessen, ...; du mußt daran denken, ...; **you ~ listen to me!** hör mir zu!; **you ~n't do that again!** tu das [ja] nie wieder!; **I ~ get back to the office** ich muß wieder ins Büro; **~ I?** muß das sein?; **I ~ have a new dress** ich brauche ein neues Kleid; **if you '~ know** wenn du es unbedingt wissen willst; **b)** *(ought to)* müssen; *with negative* dürfen; **you ~ think about it** du solltest [unbedingt] darüber nachdenken; **I ~ not sit here drinking coffee** ich sollte *od.* dürfte eigentlich nicht hier sitzen und Kaffee trinken; **c)** *(be certain to)* müssen; **you ~ be tired** du mußt müde sein; du bist bestimmt müde; **you ~ be crazy** du bist wohl wahnsinnig!; **it ~ be about 3 o'clock** es wird wohl *od.* dürfte *od.* müßte etwa 3 Uhr sein; **it ~ have stopped raining by now** es dürfte *od.* müßte inzwischen aufgehört haben zu regnen; **there ~ have been forty of them** *(forty)* es müssen vierzig gewesen sein; *(probably about forty)* es dürften etwa vierzig gewesen sein; **d)** *expr. indignation or annoyance* **he ~ come just when ...**: er muß/mußte natürlich *od.* ausgerechnet kommen, wenn/als ... **2.** *n. (coll.)* Muß, *das;* **be a ~ for sb./sth.** ein Muß für jmdn./unerläßlich für etw. sein

mustache *see* **moustache**

mustang ['mʌstæŋ] *n.* Mustang, *der*

mustard ['mʌstəd] *n.* **a)** Senf, *der;* **~ and cress** *(Brit.)* Senfkeimlinge und Kresse; **b)** *(colour)* Senffarbe, *die; attrib.* senffarben

'**mustard gas** *n. (Chem., Mil.)* Senfgas, *das*

muster ['mʌstə(r)] **1.** *n. (Mil.)* Appell, *der;* **pass ~** *(fig.)* akzeptabel sein. **2.** *v. t.* **a)** *(summon)* versammeln; *(Mil., Naut.)* [zum Appell] antreten lassen; **b)** *(collect)* zusammenbringen; zusammenziehen ‹*Streitkräfte, Truppen*›; *(raise)* aufstellen ‹*Armee*›; ausheben ‹*Truppen*›; **c)** *(fig.: summon up)* zusammennehmen ‹*Kraft, Mut, Verstand*›; aufbringen ‹*Unterstützung*›. **3.** *v. i.* sich [ver]sammeln; ‹*Truppen:*› aufmarschieren; *(for parade)* antreten

~ 'up *v. t.* aufbringen ‹*Unterstützung, Mut, Verständnis*›; **~ up all one's courage** seinen ganzen Mut zusammennehmen

mustiness ['mʌstɪnɪs] *n., no pl. (of smell, taste)* Muffigkeit, *die*

mustn't ['mʌsnt] *(coll.)* = **must not;** *see* **must 1**

musty ['mʌstɪ] *adj.* **a)** *(smelling or tasting stale)* muffig; **b)** *(mouldy)* stockig; **c)** *(fig.)* verstaubt

mutant ['mju:tənt] *(Biol.)* **1.** *adj.* mutiert ‹*Gen, Zelle, Stamm*›. **2.** *n.* Mutante, *die*

mutate [mju:'teɪt] *(Biol.)* **1.** *v. t.* zur Mutation anregen; **be ~d** mutieren. **2.** *v. i.* mutieren

mutation [mju:'teɪʃn] *n. (Biol.)* Mutation, *die*

mute [mju:t] **1.** *adj. (dumb, silent; also Ling.)* stumm; **be ~ with rage/amazement** vor Zorn/Staunen kein Wort hervorbringen. **2.** *n.* **a)** *(dumb person)* Stumme, *der/die;* **b)** *(Mus.)* Dämpfer, *der*

muted ['mju:tɪd] *adj.* gedämpft; verhalten ‹*Kritik, Begeisterung*›

mutilate ['mju:tɪleɪt] *v. t. (lit. or fig.)* verstümmeln

mutilation [mju:tɪ'leɪʃn] *n. (lit. or fig.)* Verstümmelung, *die*

mutineer [mju:tɪ'nɪə(r)] *n.* Meuterer, *der*

mutinous ['mju:tɪnəs] *adj.* rebellisch ‹*Geist, Person*›; meuternd ‹*Mannschaft eines Schiffs, Truppen*›

mutiny ['mju:tɪnɪ] **1.** *n.* Meuterei, *die.* **2.** *v. i.* meutern

mutter ['mʌtə(r)] **1.** *v. i.* **a)** murmeln; brummeln; **b)** *(grumble)* murren **(at, about** über + *Akk.*). **2.** *v. t.* murmeln. **3.** *n.* Gemurmel, *das*

muttering ['mʌtərɪŋ] *n.* **a)** *no pl. (low speech)* Gemurmel, *das;* **b)** **~[s]** *(complaints)* Gemurre, *das*

mutton ['mʌtn] *n.* Hammelfleisch, *das;* Hammel, *der;* **~ dressed [up] as lamb** *(coll. derog.)* eine Alte, die auf jugendlich macht *(ugs.)*

mutual ['mju:tjʊəl] *adj.* **a)** gegenseitig; beiderseitig ‹*Einvernehmen, Vorteil, Bemühung*›; wechselseitig ‹*Abhängigkeit*›; **I can't bear you! – The feeling's ~:** Ich kann dich nicht riechen! – Das beruht auf Gegenseitigkeit; **to our ~ satisfaction/benefit** zu unser beider Zufriedenheit/Nutzen; **b)** *(coll.: shared)* gemeinsam ‹*Interesse, Freund, Abneigung usw.*›

mutually ['mju:tjʊəlɪ] *adv.* **a)** gegenseitig; **be ~ exclusive** einander *(geh.) od.* sich [gegenseitig] ausschließen; **~ beneficial** für beide Seiten vorteilhaft; **b)** *(coll.: in common)* gemeinsam

muzzle ['mʌzl] **1.** *n.* **a)** *(of dog)* Schnauze, *die; (of horse, cattle)* Maul, *das;* **b)** *(of gun)* Mündung, *die;* **c)** *(put over animal's mouth)* Maulkorb, *der.* **2.** *v. t.* **a)** einen Maulkorb umbinden (+ *Dat.*) ‹*Hund*›; **b)** *(fig.)* mundtot machen, einen Maulkorb anlegen *(ugs.)* (+ *Dat.*) ‹*Presse, Kritiker*›; unterdrücken ‹*Protest*›

muzzy ['mʌzɪ] *adj.* **a)** *(mentally hazy, blurred)* verschwommen; **b)** *(from intoxication)* benebelt **(with** von)

MW *abbr. (Radio)* **medium wave** MW

my [maɪ] *poss. pron. attrib.* mein; **my[, my]!, [my] oh my!** [ach du] meine Güte! *(ugs.); see also* **²her**

myopia [maɪ'əʊpɪə] *n.* Kurzsichtigkeit, *die (auch fig.);* Myopie, *die (fachspr.)*

myopic [maɪ'ɒpɪk] *adj. (lit. or fig.)* kurzsichtig

myriad ['mɪrɪəd] *(literary)* **1.** *adj.* unzählig; Myriaden von *(geh.)* ‹*Insekten, Sternen*›. **2.** *n.* Myriade, *die (geh.)*

myrrh [mɜ:(r)] *n.* Myrrhe, *die*

myself [maɪ'self] *pron.* **a)** *emphat.* selbst; **I thought so ~:** das habe ich auch gedacht; **[even] though/if I say it ~:** wenn ich es auch selbst sage; **I am quite ~ again** mir geht es wieder gut; **b)** *refl.* mich/mir; **I washed ~:** ich wusch mich; **I'm going to get ~ a car** ich werde mir ein Auto zulegen. *See also* **herself**

mysterious [mɪ'stɪərɪəs] *adj.* **a)** mysteriös; rätselhaft; geheimnisvoll ‹*Fremder, Orient*›; **b)** *(secretive)* geheimnisvoll; **be very ~ about sth.** ein großes Geheimnis aus etw. machen

mysteriously [mɪ'stɪərɪəslɪ] *adv.* auf mysteriöse *od.* rätselhafte Weise; geheimnisvoll ‹*lächeln usw.*›

mystery ['mɪstərɪ] *n.* **a)** Rätsel, *das;* **it's a ~ to me why ...:** es ist mir ein Rätsel, warum ...; **b)** *(secrecy)* Geheimnis, *das;* **shrouded in ~:** geheimnisumwittert *od.* -umwoben *(geh.);* **there's no ~ about it** das ist überhaupt kein Geheimnis

mystery: ~ tour, ~ trip *ns.* Fahrt ins Blaue *(ugs.)*

mystic ['mɪstɪk] **1.** *adj.* mystisch. **2.** *n.* Mystiker, *der/*Mystikerin, *die*

mystical ['mɪstɪkl] *adj.* mystisch

mysticism ['mɪstɪsɪzm] *n.* Mystik, *die;* Mystizismus, *der (geh.)*

mystify ['mɪstɪfaɪ] *v. t.* verwirren; **this mystifies me** das ist mir ein Rätsel *od.* rätselhaft

mystique [mɪ'sti:k] *n.* geheimnisvoller Nimbus

myth [mɪθ] *n.* Mythos, *der; (rumour)* Gerücht, *das*

mythical ['mɪθɪkl] *adj.* **a)** *(based on myth)* mythisch; **~ creatures** Sagengestalten; **b)** *(invented)* fiktiv

mythological [mɪθə'lɒdʒɪkl] *adj.* mythologisch

mythology [mɪ'θɒlədʒɪ] *n.* Mythologie, *die*

myxomatosis [mɪksəmə'təʊsɪs] *n. (Vet. Med.)* Myxomatose, *die*

N

N, n [en] *n., pl.* **Ns** *or* **N's** N, n, *das;* **for the nth** [enθ] **time** *(coll.)* zum x-ten Mal *(ugs.)*

N. *abbr.* **a) north** N; **b) northern** n.

n. *abbr.* **note** Anm.

NAAFI ['næfɪ] *abbr. (Brit.)* **Navy, Army and Air Force Institutes** *Kaufhaus für Angehörige der britischen Truppen*

nab [næb] *v. t.,* **-bb-** *(sl.)* **a)** *(arrest)* schnappen *(ugs.);* **b)** *(seize)* sich *(Dat.)* schnappen; **c)** *(steal)* klauen *(salopp);* krallen *(salopp)*

nadir ['neɪdɪə(r)] *n.* Tief[st]punkt, *der*

¹nag [næg] **1.** *v. i.,* **-gg-** nörgeln

(abwertend); ~ **at sb.** an jmdm. herumnörgeln; ~ **at sb. to do sth.** jmdm. zusetzen *(ugs.),* daß er etw. tut. **2.** *v. t.,* **-gg-** *(scold)* herumnörgeln an (+ *Dat.*) *(abwertend);* ~ **sb. about sth./to do sth.** jmdm. wegen etw. zusetzen *(ugs.)*/jmdm. zusetzen *(ugs.),* daß er etw. tut

²nag *n. (coll.: horse)* Gaul, *der*

nagging ['nægɪŋ] **1.** *adj.* **a)** *(annoying)* nörglerisch *(abwertend);* **b)** *(persistent)* quälend ‹*Angst, Sorge, Zweifel*›; bohrend ‹*Schmerz*›; **a** ~ **conscience** [quälende] Gewissensbisse *Pl.* **2.** *n.* Genörgel, *das (abwertend)*

nail [neɪl] **1.** *n.* **a)** *(on finger, toe)* Nagel, *der;* **cut one's ~s** sich *(Dat.)* die Nägel schneiden; **bite one's ~s** an den Nägeln kauen; **b)** *(metal spike)* Nagel, *der;* **be hard as ~s** *(fig.)* steinhart sein; *(fit)* topfit sein; *(unfeeling)* knallhart sein *(ugs.);* **hit the [right]** ~ **on the head** *(fig.)* den Nagel auf den Kopf treffen *(ugs.);* **be a** ~ **in sb.'s/sth.'s coffin** *(fig.)* ein Nagel zu jmds. Sarg/ein Sargnagel für etw. sein *(ugs.);* **on the** ~ *(fig. coll.)* pünktlich ‹*bezahlen, sein Geld kriegen*›. **2.** *v. t.* **a)** nageln (**to** an + *Akk.*); ~ **two planks together** zwei Bretter zusammennageln; **be ~ed to the spot/ground** *(fig.)* wie angenagelt sein *(ugs.);* **b)** *(fig.: secure, catch, engage)* an Land ziehen *(ugs.)* ‹*Vertrag, Auftrag*›; **c)** *(fig.: expose)* anprangern. *See also* **colour 1 g**

~ **'down** *v. t.* festnageln; zunageln ‹*Kiste*›

~ **'up** *v. t.* **a)** *(close)* vernageln; **b)** *(affix with ~)* annageln (**against** an + *Akk.*)

nail: **~-biting** ['neɪlbaɪtɪŋ] *adj. (fig.)* bang ‹*Minuten, Schweigen, Sorge*›; angstvoll ‹*Spannung*›; spannungsgeladen ‹*Spiel, Film*›; **~-brush** *n.* Nagelbürste, *die;* **~-clippers** *n. pl.* **[pair of]** **~-clippers** Nagelknipser, *der;* **~-file** *n.* Nagelfeile, *die;* ~ **polish** *n.* Nagellack, *der;* **~-polish remover** Nagellackentferner, *der;* **~scissors** *n. pl.* **[pair of]** **~-scissors** Nagelschere, *die;* ~ **varnish** *(Brit.) see* ~ **polish**

naïve, naive [nɑ:'i:v, naɪ'i:v] *adj.*, **naïvely, naively** [nɑ:'i:vlɪ, naɪ'i:vlɪ] *adv.* naiv

naïvety, naivety [nɑ:'i:vtɪ, naɪ'i:vtɪ] *n.* Naivität, *die*

naked ['neɪkɪd] *adj.* **a)** nackt; **strip sb.** ~: jmdn. nackt ausziehen; **b)** nackt ‹*Glühbirne*›; offen ‹*Licht, Flamme*›; **c)** *(defenceless)* wehrlos; **d)** *(plain)* nackt ‹*Tatsache, Wahrheit, Aggression, Gier, Ehrgeiz*›; **visible to** *or* **with the** ~ **eye** mit bloßem Auge zu erkennen

name [neɪm] **1.** *n.* **a)** Name, *der;* **what's your ~/the ~ of this place?** wie heißt du/dieser Ort?; **my ~ is Jack** ich heiße Jack; mein Name ist Jack; **no one of** *or* **by that** ~: niemand mit diesem Namen *od. (geh.)* dieses Namens; **last** ~: Zuname, *der,* Nachname, *der;* **the** ~ **of Edwards** der Name Edwards; **by** ~: namentlich ‹*erwähnen, aufrufen usw.*›; **know sb. by ~/by ~ only** jmdn. mit Namen/nur dem Namen nach kennen; **that's the ~ of the game** *(coll.)* darum geht es; **put one's/sb.'s** ~ **down for sth.** sich/jmdn. für etw. vormerken lassen; **put one's/sb.'s** ~ **down on the waiting-list** sich auf die Warteliste setzen lassen/jmdn. auf die Warteliste setzen; **without a penny to his** ~: ohne einen Pfennig in der Tasche; **in** ~ **[only]** [nur] auf dem Papier; **in all but** ~: im Grunde genommen; **b)** *(reputation)* Ruf, *der;* **make a** ~ **for oneself** sich *(Dat.)* einen Namen machen; **make one's/sb.'s** ~: berühmt werden/jmdn. berühmt machen; **clear one's/sb.'s** ~: seine/jmds. Unschuld beweisen; **c)** **call sb. ~s** *(abuse)* jmdn. beschimpfen; **d)** *(famous person)* Name, *der;* **many great** *or* **big ~s** viele namhafte Persönlichkeiten; viele Größen; **be a big** ~: einen großen Namen haben; **e)** *attrib.* ~ **brand** Markenartikel, *der.* **2.** *v. t.* **a)** *(give ~ to)* einen Namen geben (+ *Dat.*); ~ **sb. John** jmdn. John nennen; ~ **a ship 'Mary'** ein Schiff [auf den Namen] „Mary" taufen; ~ **sb./sth. after** *or (Amer.)* **for sb.** jmdn./etw. nach jmdm. benennen; **be ~d John** John heißen; **a man ~d Smith** ein Mann namens *od.* mit Namen Smith; **b)** *(call by right ~)* benennen; ~ **the capital of Zambia** nenne die Hauptstadt von Sambia; **c)** *(nominate)* ernennen; ~ **sb. [as] sth.** jmdn. zu etw. ernennen; ~ **one's successor** seinen Nachfolger bestimmen; **he was ~d as the winner** ihm wurde der Sieg zuerkannt; **d)** *(mention)* nennen; *(specify)* benennen; ~ **names** Namen nennen; ~ **the day** *(choose wedding-day)* den Tag der Hochzeit festsetzen; **to** ~ **but a few** um nur einige zu nennen; **we were given champagne, oysters, you** ~ **it** wir kriegten Champagner, Austern, und, und, und

name: **~-calling** *n.* Beschimpfungen *Pl.;* **~-dropping** *n.* Name-dropping, *das; Nennung bedeutender Namen, um Eindruck zu machen*

nameless ['neɪmlɪs] *adj.* **a)** *(having no name, anonymous)* namenlos; **a person who shall remain** ~: eine Person, die ungenannt bleiben soll; **b)** *(abominable)* unaussprechlich; unsäglich *(geh.);* **c)** *(inexpressible)* unbeschreiblich

namely ['neɪmlɪ] *adv.* nämlich

name: **~-plate** *n.* Namensschild, *das;* **~sake** *n.* Namensvetter, *der*/-schwester, *die*

nancy ['nænsɪ] *n. (sl.)* ~ **[boy]** Tunte, *die (salopp)*

nanny ['nænɪ] *n.* **a)** *(Brit.: nursemaid)* Kindermädchen, *das;* **b)** *(coll.: granny)* Großmama, *die (fam.);* **c)** *see* **nanny-goat**

'nanny-goat *n.* Ziege, *die;* Geiß, *die (südd., österr., schweiz., westmd.)*

nap [næp] **1.** *n.* Schläfchen, *das (ugs.);* Nickerchen, *das (fam.);* **take** *or* **have a** ~: ein Schläfchen *od.* Nickerchen machen *od.* halten. **2.** *v. i.,* **-pp-** dösen *(ugs.);* **catch sb. ~ping** *(fig.)* jmdn. überrumpeln

napalm ['neɪpɑ:m] *n.* Napalm, *das*

nape [neɪp] *n.* ~ **[of the neck]** Nacken, *der;* Genick, *das*

napkin ['næpkɪn] *n.* **a)** Serviette, *die;* **b)** *(waiter's)* Serviertuch, *das*

'napkin-ring *n.* Serviettenring, *der*

Naples ['neɪplz] *pr. n.* Neapel *(das)*

nappy ['næpɪ] *n. (Brit.)* Windel, *die*

narcissus [nɑ:'sɪsəs] *n., pl.* **narcissi** [nɑ:'sɪsaɪ] *or* **~es** *(Bot.)* Narzisse, *die*

narcotic [nɑ:'kɒtɪk] **1.** *n.* **a)** *(drug)* Rauschgift, *das;* Betäubungsmittel, *das (Rechtsw.);* **b)** *(active ingredient)* Betäubungsmittel, *das;* Narkotikum, *das (Med.).* **2.** *adj.* narkotisch; ~ **drug** Rauschgift, *das;* Betäubungsmittel, *das (Rechtsw.)*

nark [nɑ:k] *(sl.)* **1.** *n. (Brit.: informer)* Spitzel, *der (abwertend).* **2.** *v. t. (annoy)* stinken (+ *Dat.*) *(salopp);* **be ~ed [about sb./at** *or* **about sth.]** [auf jmdn./über etw. *(Akk.)*] sauer sein *(ugs.)*

narrate [nə'reɪt] *v. t.* erzählen; schildern ‹*Ereignisse*›; kommentieren ‹*Film*›

narration [nə'reɪʃn] *n.* Erzählen, *das;* Erzählung, *die; (of events)* Schilderung, *die;* Schildern, *das*

narrative ['nærətɪv] **1.** *n.* **a)** *(tale, story)* Geschichte, *die;* Erzählung, *die;* **b)** *no pl.* **be written in ~:** in der Erzählform geschrieben sein. **2.** *adj.* erzählend; Erzähl‹*kunst, -technik*›

narrator [nə'reɪtə(r)] *n.* Erzähler, *der*/Erzählerin, *die; (of film)* Kommentator, *der*/Kommentatorin, *die*

narrow ['nærəʊ] **1.** *adj.* **a)** schmal; schmal geschnitten ‹*Rock, Hose, Ärmel usw.*›; eng ‹*Tal, Gasse*›; **b)** *(limited)* eng; begrenzt, schmal ‹*Auswahl*›; **c)** *(with little margin)* knapp ‹*Sieg, Führung, Mehrheit*›; **have a ~ escape** mit knapper Not entkommen (**from** *Dat.*); **d)** *(not tolerant)* spießig *(abwertend);* engstirnig *(abwertend);* **e)** *(restricted)* eng ‹*Grenzen, Toleranzen*›; klein, begrenzt ‹*Freundeskreis*›. **2.** *v. i.* sich verschmälern; ‹*Augen, Tal:*› sich verengen; *(fig.)* [zusammen]schrumpfen; **the road ~s to one lane** die Straße wird einspurig; **'road ~s'** „Fahrbahnverengung". **3.** *v. t.* verschmälern; *(fig.)* einengen; enger fassen ‹*Definition*›; **~ one's eyes** die Augen zusammenkneifen

~ 'down 1. *v. t.* einengen, beschränken (**to** auf + *Akk.*). **2.** *v. i.* sich reduzieren (**to** auf + *Akk.*); **the choice ~s down to two possibilities** es bleiben zwei Möglichkeiten [übrig]

narrow: ~ boat *n. (Brit.) besonders schmales Binnenschiff;* **~-gauge** *adj.* schmalspurig; Schmalspur-

narrowly ['nærəʊlɪ] *adv.* **a)** *(with little width)* schmal; **b)** *(only just)* knapp; mit knapper Not ‹*entkommen*›; **he ~ escaped being run over by a car** er wäre um ein Haar *(ugs.)* überfahren worden

narrow-'minded *adj.* engstirnig *(abwertend)*

NASA ['næsə] *abbr. (Amer.)* **National Aeronautics and Space Administration** NASA, *die*

nasal ['neɪzl] *adj.* **a)** *(Anat.)* Nasen-; **b)** näselnd; **speak in a ~ voice** näseln

nastily ['nɑ:stɪlɪ] *adv.* **a)** *(disagreeably, unpleasantly)* scheußlich; **b)** *(ill-naturedly)* gemein; **behave ~:** häßlich sein; **c)** *(disgustingly)* eklig; widerlich

nasturtium [nə'stɜ:ʃəm] *n.* Kapuzinerkresse, *die*

nasty ['nɑ:stɪ] *adj.* **a)** *(disagreeable, unpleasant)* scheußlich ‹*Geruch, Geschmack, Arznei, Essen, Wetter*›; gemein ‹*Trick, Verhalten, Äußerung, Person*›; häßlich ‹*Angewohnheit*›; **that was a ~ thing to say/do** das war gemein *od.* eine Gemeinheit; **a ~ bit** *or* **piece of work** *(coll.) (man)* ein fieser Kerl *(ugs. abwertend); (woman)* ein fieses Weibsstück *(ugs. abwertend);* **b)** *(ill-natured)* böse; **be ~ to sb.** häßlich zu jmdm. sein; **c)** *(serious)* übel; böse ‹*Verletzung, Husten usw.*›; schlimm ‹*Krankheit, Husten, Verletzung*›; **she had a ~ fall** sie ist übel *od.* böse gefallen; **d)** *(disgusting)* eklig; widerlich

nation ['neɪʃn] *n.* Nation, *die; (people)* Volk, *das;* **throughout the ~:** im ganzen Land

national ['næʃənl] **1.** *adj.* national; National‹*flagge, -denkmal, -held, -theater, -tanz, -gericht, -charakter*›; Landes‹*durchschnitt, -sprache*›; Staats‹*sicherheit, -symbol*›; überregional ‹*Rundfunkstation, Zeitung*›; landesweit ‹*Streik*›. **2.** *n.* **a)** *(citizen)* Staatsbürger, *der*/-bürgerin, *die;* **foreign ~:** Ausländer, *der*/Ausländerin, *die;* **b)** *usu. in pl. (newspaper)* überregionale Zeitung

national: ~ 'anthem *n.* Nationalhymne, *die;* **~ call** *n. (Brit. Teleph.)* Inlandsgespräch, *das;* **~ 'costume** *n.* Nationaltracht, *die;* Landestracht, *die;* **N~ 'Debt** *see* **debt; ~ 'dress** *see* **~ costume; ~ 'grid** *n. (Brit. Electr.)* nationales Verbundnetz; **N~ 'Health [Service]** *n. (Brit.) staatlicher Gesundheitsdienst; attrib.* **N~ Health doctor/patient/spectacles** ≈ Kassenarzt, *der*/-patient, *der*/-brille, *die;* **~ 'holiday** *n.* Nationalfeiertag, *der; (statutory holiday)* gesetzlicher Feiertag; **N~ In'surance** *n. (Brit.)* Sozialversicherung, *die*

nationalisation, nationalise *see* **nationaliz-**

nationalism ['næʃənəlɪzm] *n.* Nationalismus, *der; (patriotism)* nationale Gesinnung

nationalist ['næʃənəlɪst] **1.** *n.* Nationalist, *der*/Nationalistin, *die.* **2.** *adj.* nationalistisch

nationalistic [næʃənə'lɪstɪk] *adj. (patriotic)* nationalistisch

nationality [næʃə'nælɪtɪ] *n.* **a)** Staatsangehörigkeit, *die;* Nationalität, *die (geh.);* **be of** *or* **have British ~:** britischer Nationalität sein *(geh.);* die britische Staatsangehörigkeit haben; **what's his ~?** welche Staatsangehörigkeit hat er?; welcher Nationalität ist er? *(geh.);* **b)** *(ethnic group)* Nationalität, *die;* Volksgruppe, *die*

nationalization [næʃənəlaɪ'zeɪʃn] *n.* Verstaatlichung, *die*

nationalize ['næʃənəlaɪz] *v. t.* verstaatlichen

nationally ['næʃənəlɪ] *adv.* als Nation; *(throughout the nation)* landesweit

national: ~ 'park *n.* Nationalpark, *der;* **~ 'service** *n. (Brit.)* Wehrdienst, *der;* **do ~ service** seinen Wehrdienst ableisten; **N~ 'Socialist** *n.* Nationalsozialist, *der*/-sozialistin, *die; attrib.* nationalsozialistisch; **N~ 'Trust** *n. (Brit.) nationale Einrichtung für Naturschutz und Denkmalpflege*

nation-wide 1. ['---] *adj.* landesweit. **2.** [--'-] *adv.* landesweit; im ganzen Land

native ['neɪtɪv] **1.** *n.* **a)** *(of specified place)* **a ~ of Britain** ein gebürtiger Brite/eine gebürtige Britin; **b)** *(indigenous person)* Eingeborene, *der/die;* **c)** *(local inhabitant)* Einheimische, *der/die;* **the ~s** die Einheimischen; **d)** *(Zool., Bot.)* **be a ~ of a place** in einem Ort beheimatet sein. **2.** *adj.* **a)** *(indigenous)* eingeboren; *(local)* einheimisch ‹*Pflanze, Tier*›; **be a ~ American** gebürtiger Amerikaner/gebürtige Amerikanerin sein; **~ inhabitant** Eingeborene/Einheimische, *der/die;* **b)** *(of one's birth)* Geburts-, Heimat‹*land, -stadt*›; Mutter‹*sprache, -sprachler*›; **he's not a ~ speaker of English** Englisch ist nicht seine Muttersprache; **c)** *(innate)* angeboren ‹*Qualitäten, Schläue*›; **d)** *(of the ~s)* Eingeborenen-; **go ~:** die Lebensweise der Eingeborenen annehmen

nativity [nə'tɪvɪtɪ] *n.* **a) the N~ [of Christ]** die Geburt Christi; **b)** *(festival)* **the N~ of Christ** das Fest der Geburt Christi; **c)** *(picture)* Geburt Christi

na'tivity play *n.* Krippenspiel, *das*

NATO, Nato ['neɪtəʊ] *abbr.* **North Atlantic Treaty Organization** NATO, *die*

natter ['nætə(r)] *(Brit. coll.)* **1.** *v. i.* quatschen *(ugs.);* quasseln *(ugs.).* **2.** *n.* **have a ~:** quatschen *(ugs.)*

natty ['nætɪ] *adj. (coll.)* schick, *(ugs.)* flott ‹*Kleidung[sstück]*›; **be a ~ dresser** immer schick *od.* flott angezogen sein

natural ['nætʃrəl] **1.** *adj.* **a)** natürlich; Natur‹*zustand, -begabung, -talent, -seide, -schwamm, -faser, -erscheinung*›; **the ~ world** die Natur[welt]; **be a ~ blonde** naturblondes Haar haben; **it is ~ for dogs to fight** es ist natürlich, daß

Hunde kämpfen; **die of** *or* **from ~ causes** eines natürlichen Todes sterben; **have a ~ tendency to ...:** naturgemäß dazu neigen, ... zu ...; **b)** *(unaffected)* natürlich ‹*Art, Lächeln, Stil*›; **c)** leiblich ‹*Eltern, Kind usw.*›; natürlich *(Rechtsspr. veralt.)* ‹*Kind*›. **2.** *n. (person)* Naturtalent, *das;* **she's a ~ for the part** die Rolle ist ihr auf den Leib geschrieben

natural: ~ 'childbirth *n.* natürliche Geburt; **~ 'death** *n.* natürlicher Tod; **~ 'gas** *see* **gas 1a; ~ 'history** *n.* **a)** *(study)* Naturkunde, *die; attrib.* Naturkunde-; naturkundlich ‹*Museum*›; **b)** *(facts)* Naturgeschichte, *die*

naturalisation, naturalise *see* **naturaliz-**

naturalist ['nætʃrəlɪst] *n.* Naturforscher, *der*/-forscherin, *die*

naturalization [nætʃrəlaɪ'zeɪʃn] *n. (admitting as citizen)* Einbürgerung, *die;* Naturalisierung, *die*

naturalize ['nætʃrəlaɪz] **1.** *v. t.* **a)** *(admit as citizen)* einbürgern; naturalisieren; **b)** naturalisieren, einbürgern ‹*Tiere, Pflanzen*›. **2.** *v. i.* eingebürgert werden

naturally ['nætʃrəlɪ] *adv.* **a)** *(by nature)* von Natur aus ‹*musikalisch, blaß, fleißig usw.*›; *(in a true-to-life way)* naturgetreu; *(with ease)* natürlich; *(in a natural manner)* auf natürliche Weise; **it comes ~ to her** es fällt ihr leicht; **lead ~ to sth.** naturgemäß zu etw. führen; **b)** *(of course)* natürlich

naturalness ['nætʃrəlnɪs] *n.* Natürlichkeit, *die*

natural: ~ re'sources *n. pl.* natürliche Ressourcen; Naturschätze *Pl.;* **~ 'science** *n.* **~ science, the ~ sciences** die Naturwissenschaften; **~ se'lection** *n. (Biol.)* natürliche Auslese

nature ['neɪtʃə(r)] *n.* **a)** Natur, *die;* **back to ~:** zurück zur Natur; **paint from ~:** nach der Natur malen; **b)** *(essential qualities)* Beschaffenheit, *die;* **in the ~ of things** naturgemäß; **c)** *(kind, sort)* Art, *die;* **things of this ~:** derartiges; Dinge dieser Art; **it's in the ~ of a command** es hat Befehlscharakter; **d)** *(character)* [Wesens]art, *die;* Wesen, *das;* **have a happy ~:** eine Frohnatur sein; **be of** *or* **have a placid ~:** eine ruhige Art haben; **have a jealous ~:** eifersüchtig sein; **it is not in her ~ to lie** es ist nicht ihre Art zu lügen; **human ~:** menschliche Natur; **it's only human ~ to ...:** es ist nur menschlich, ... zu ...

nature: ~ conservation *n.* Naturschutz, *der;* **~ cure** *n.* Naturheilverfahren, *das;* **~ lover** *n.* Naturfreund, *der*/-freundin, *die;* **~ reserve** *n.* Naturschutzgebiet, *das;* **~ study** *n.* Naturkunde, *die;* **~ trail** *n.* Naturlehrpfad, *der*

naturism ['neɪtʃərɪzm] *n. (nudism)* Naturismus, *der;* Freikörperkultur, *die*

naturist ['neɪtʃərɪst] *n. (nudist)* Naturist, *der*/Naturistin, *die;* FKK-Anhänger, *der*/FKK-Anhängerin, *die*

naught [nɔ:t] *n. (arch./dial.)* **bring to ~:** zunichte machen; **come to ~:** zunichte werden

naughtiness ['nɔ:tɪnɪs] *n.* Ungezogenheit, *die;* Unartigkeit, *die*

naughty ['nɔ:tɪ] *adj.* **a)** *(disobedient)* unartig; ungezogen; **you ~ boy/dog** du böser Junge/Hund; **b)** *(indecent)* unanständig

nausea ['nɔ:zɪə, 'nɔ:sɪə] *n.* **a)** Übelkeit, *die;* **b)** *(fig.: disgust)* Ekel, *der,* Abscheu, *der* **(with, at** vor + *Dat.*)

nauseate ['nɔ:zɪeɪt, 'nɔ:sɪeɪt] *v. t.* **a) ~ sb.** in jmdm. Übelkeit erregen; **the smell ~d him** bei dem Geruch wurde ihm übel; **b)** *(fig.: disgust)* anekeln; anwidern

nauseating ['nɔ:zɪeɪtɪŋ, 'nɔ:sɪeɪtɪŋ] *adj.* **a)** Übelkeit verursachend *od.* erregend; **b)** *(fig.: disgusting)* widerlich; ekelerregend ‹*Anblick, Geruch*›; ekelhaft ‹*Person*›

nauseous ['nɔ:zɪəs, 'nɔ:sɪəs] *adj.* **a) sb. is** *or* **feels ~:** jmdm. ist übel; **b)** *(fig.: disgusting)* widerlich

nautical ['nɔ:tɪkl] *adj.* nautisch; seemännisch ‹*Ausdruck, Können*›; **~ map** Seekarte, *die*

naval ['neɪvl] *adj.* Marine-; Flotten‹*parade, -abkommen*›; See‹*schlacht, -macht, -streitkräfte*›; ‹*Überlegenheit*› zur See; **~ ship** Kriegsschiff, *das*

naval: ~ base *n.* Flottenstützpunkt, *der;* **~ officer** *n.* Marineoffizier, *der*

nave [neɪv] *n. (Archit.)* [Mittel-, Haupt]schiff, *das*

navel ['neɪvl] *n.* Nabel, *der*

navigable ['nævɪgəbl] *adj. (suitable for ships)* schiffbar

navigate ['nævɪgeɪt] **1.** *v. t.* **a)** *(sail on)* befahren ‹*Kanal, Fluß, Gewässer*›; **b)** navigieren ‹*Schiff, Flugzeug*›. **2.** *v. i.* **a)** *(in ship, aircraft)* navigieren; **b)** *(assist driver)* den Lotsen spielen *(ugs.);* franzen *(Rallyesport);* **you drive, I'll ~:** du fährst, und ich dirigiere *od.* lotse dich

navigation [nævɪ'geɪʃn] *n.* Navigation, *die; (sailing on river etc.)* Befahren, *das; (assisting driver)* Dirigieren, *das;* Lotsen, *das;* Franzen, *das (Rallyesport)*

navi'gation lights *n. pl. (Naut.)* Lichter; *(Aeronaut.)* Kennlichter

navigator ['nævɪgeɪtə(r)] *n.* Navigator, *der*/Navigatorin, *die;* **his co-driver was acting as ~:** sein Beifahrer dirigierte *od.* lotste ihn

navvy ['nævɪ] *n. (Brit.: labourer)* Bau-/Straßenarbeiter, *der*

navy ['neɪvɪ] *n.* **a)** [Kriegs]marine, *die;* **b)** *see* **navy blue**

navy: ~ 'blue *n.* Marineblau, *das;* **~-blue** *adj.* marineblau

Nazi ['nɑ:tsɪ] **1.** *n.* **a)** Nationalsozialist, *der*/-sozialistin, *die;* Nazi, *der;* **b)** *(fig. derog.)* Faschist, *der*/Faschistin, *die;* Nazi, *der.* **2.** *adj.* **a)** nazistisch; Nazi-; **b)** *(fig. derog.)* faschistisch; Nazi-

NB *abbr.* **nota bene** NB

NCO *abbr.* **non-commissioned officer** Uffz.

NE ['nɔ:θi:st] *abbr.* **north-east** NO

near [nɪə(r)] **1.** *adv.* **a)** *(at a short distance)* nah[e]; **stand/live [quite] ~:** [ganz] in der Nähe stehen/wohnen; **come** *or* **draw ~/~er** ‹*Tag, Zeitpunkt:*› nahen/näherrücken; **get ~er together** näher zusammenrücken; **~ at hand** in Reichweite *(Dat.);* ‹*Ort*› ganz in der Nähe; **be ~ at hand** ‹*Ereignis:*› nahe bevorstehen; **so ~ and yet so far** so nah und doch so fern; **b)** *(closely)* **~ to = 2a, b, c; we were ~ to being drowned** wir wären fast *od.* beinah[e] ertrunken. **2.** *prep.* **a)** *(in space) (position)* nahe an/bei (+ *Dat.*); *(motion)* nahe an (+ *Akk.*); *(fig.)* nahe *(geh.) nachgestellt* (+ *Dat.*); in der Nähe (+ *Gen.*); **go ~ the water's edge** nahe ans Ufer gehen; **keep ~ me** halte dich *od.* bleib in meiner Nähe; **~ where ...:** in der Nähe *od.* unweit der Stelle *(Gen.),* wo ...; **move it ~er her** rücke es näher zu ihr; **don't stand so ~ the fire** geh nicht so nahe *od.* dicht an das Feuer; **when we got ~er Oxford** als wir in die Nähe von Oxford kamen; **wait till we're ~er home** warte, bis wir nicht mehr so weit von zu Hause weg sind; **don't come ~ me** komm mir nicht zu nahe; **it's ~ here** es ist hier in der Nähe; **the man ~/~est you** der Mann, der bei dir/der dir am nächsten steht; **b)** *(in quality)* **nobody comes anywhere ~ him at swimming** im Schwimmen kommt bei weitem keiner an ihn heran; **we're no ~er solving the problem** wir sind der Lösung des Problems nicht nähergekommen; **c)**

(in time) **ask me again ~er the time** frag mich, wenn der Zeitpunkt etwas näher gerückt ist, noch einmal; **it's drawing ~ Christmas** es geht auf Weihnachten zu; **come back ~er 8 o'clock** komm kurz vor 8 Uhr noch einmal zurück; **~ the end/the beginning of sth.** gegen Ende/zu Anfang einer Sache *(Gen.)*; **d)** *in comb.* Beinahe‹*unfall, -zusammenstoß, -katastrophe*›; **~-hysterical** fast hysterisch; **be in a state of ~-collapse** kurz vor dem Zusammenbruch stehen; **a ~-miracle** fast *od.* beinahe ein Wunder. **3.** *adj.* **a)** *(in space or time)* nahe; **in the ~ future** in nächster Zukunft; **the chair is ~er** der Stuhl steht näher; **our ~est neighbours** unsere nächsten Nachbarn; **b)** *(closely related)* nahe ‹*Verwandte*›; eng ‹*Freund*›; **~ and dear** lieb und teuer; **c)** *(in nature)* fast richtig ‹*Vermutung*›; groß ‹*Ähnlichkeit*›; **£30 or ~/~est offer** 30 Pfund oder nächstbestes Angebot; **this is the ~est equivalent** dies entspricht dem am ehesten; **that's the ~est you'll get to an answer** eine weitergehende Antwort wirst du nicht bekommen; **~ escape** Entkommen mit knapper Not; **round it up to the ~est penny** runde es auf den nächsthöheren Pfennigbetrag; **be a ~ miss** ‹*Schuß, Wurf:*› knapp danebengehen; **that was a ~ miss** *(escape)* das war aber knapp!; **d) the ~ side** *(Brit.) (travelling on the left/right)* die linke/rechte Seite; **e)** *(direct)* **4 miles by the ~est road** 4 Meilen auf dem kürzesten Wege. **4.** *v. t.* sich nähern (+ *Dat.*); **the building is ~ing completion** das Gebäude steht kurz vor seiner Vollendung

'nearby *adj.* nahe gelegen

Near 'East *see* **Middle East**

nearly ['nɪəlɪ] *adv.* fast; **it ~ fell over** es wäre fast umgefallen; **be ~ in tears** den Tränen nahe sein; **it is ~ six o'clock** es ist kurz vor sechs Uhr; **are you ~ ready?** bist du bald fertig?; **not ~**: nicht annähernd; bei weitem nicht

nearness ['nɪənɪs] *n., no pl. (proximity)* Nähe, *die*

near: ~-sighted *adj. (Amer.)* kurzsichtig; **~ 'thing** *n.* **that was a ~ thing!** das war knapp!

neat [ni:t] *adj.* **a)** *(tidy, clean)* sauber, ordentlich ‹*Handschrift, Arbeit*›; gepflegt ‹*Haar, Person*›; **b)** *(undiluted)* pur ‹*Getränk*›; **she drinks vodka ~**: sie trinkt Wodka pur; **c)** *(smart)* gepflegt ‹*Erscheinung, Kleidung*›; elegant, schick ‹*Anzug, Auto*›; **d)** *(deft)* geschickt; raffiniert ‹*Trick, Plan, Lösung, Gerät*›; **make a ~ job of sth./repairing sth.** etw. sehr geschickt machen/reparieren

neatly ['ni:tlɪ] *adv.* **a)** *(tidily)* ordentlich; [fein] säuberlich; **b)** *(smartly)* gepflegt; **~ groomed** äußerst gepflegt; **c)** *(deftly)* geschickt; auf raffinierte [Art und] Weise; **d)** *(briefly, clearly)* prägnant; **a ~ turned phrase** eine prägnante Formulierung

neatness ['ni:tnɪs] *n., no pl. see* **neat a, c, d**: Sauberkeit, *die;* Ordentlichkeit, *die;* Gepflegtheit, *die;* Eleganz, *die;* Geschicktheit, *die;* Raffiniertheit, *die*

nebula ['nebjʊlə] *n., pl.* **~e** ['nebjʊli:] *or* **~s** *(Astron.)* Nebel, *der*

nebulous ['nebjʊləs] *adj. (hazy)* nebelhaft, *(geh.)* nebulös ‹*Vorstellung, Werte*›; unbestimmt, vage ‹*Angst, Hoffnung*›

necessarily [nesɪ'serɪlɪ] *adv.* notwendigerweise; zwangsläufig; **it is not ~ true** es muß nicht [unbedingt] stimmen; **Do we have to do it? – Not ~**: Müssen wir es tun? – Nicht unbedingt

necessary ['nesɪsərɪ] **1.** *adj.* **a)** *(indispensable)* nötig; notwendig; unbedingt ‹*Erfordernis*›; **patience is ~ for a teacher** ein Lehrer muß Geduld haben; **it is not ~ for you to go** es ist nicht nötig *od.* notwendig, daß du gehst; **it may be ~ for him to leave** vielleicht muß er gehen; **do no more than is ~**: nur das Nötigste tun; **do everything ~** *(that must be done)* das Nötige *od.* Notwendige tun; **b)** *(inevitable)* zwangsläufig ‹*Ergebnis, Folge*›; zwingend ‹*Schluß*›; **c) a ~ evil** ein notwendiges Übel. **2.** *n.* **the necessaries of life** das Lebensnotwendige; **will you do the ~?** kümmerst du dich drum?

necessitate [nɪ'sesɪteɪt] *v. t.* erforderlich machen

necessity [nɪ'sesɪtɪ] *n.* **a)** *(power of circumstances)* Not, *die;* äußerer Zwang; **do sth. out of** *or* **from ~**: etw. notgedrungen tun; **make a virtue of ~**: aus der Not eine Tugend machen; **of ~**: notwendigerweise; **b)** *(necessary thing)* Notwendigkeit, *die;* **the necessities of life** das Lebensnotwendige; **c)** *(indispensability, imperative need)* Notwendigkeit, *die;* **there is no ~ for rudeness** es besteht keine Notwendigkeit, unhöflich zu sein; **in case of ~**: nötigenfalls; **d)** *(want)* Not, *die;* Bedürftigkeit, *die;* **be/live in ~**: Not leiden

neck [nek] **1.** *n.* **a)** Hals, *der;* **be breathing down sb.'s ~** *(fig.) (be close behind sb.)* jmdm. im Nacken sitzen *(ugs.)*; *(watch sb. closely)* jmdm. ständig auf die Finger sehen; **get it in the ~** *(coll.)* eins auf den Deckel kriegen *(ugs.)*; **give sb./be a pain in the ~** *(coll.)* jmdm. auf die Nerven *od.* den Wecker gehen *(ugs.)*; **break one's ~** *(fig. coll.)* sich *(Dat.)* den Hals brechen; **risk one's ~**: Kopf und Kragen riskieren; **save one's ~**: seinen Kopf retten; **be up to one's ~ in work** *(coll.)* bis über den Hals in Arbeit stecken *(ugs.)*; **be [in it] up to one's ~** *(coll.)* bis über den Hals drinstecken *(ugs.)*; **~ and ~**: Kopf an Kopf; **b)** *(length)* Halslänge, *die;* *(fig.)* Nasenlänge, *die;* **c)** *(cut of meat)* Hals, *der;* **d)** *(of garment)* Kragen, *der;* **that dress has a high ~**: das Kleid ist hochgeschlossen; **e)** *(narrow part)* Hals, *der.* **2.** *v. i. (sl.)* knutschen *(ugs.)*

necklace ['neklɪs] *n.* [Hals]kette, *die;* *(with jewels)* Kollier, *das*

neck: ~line *n.* [Hals]ausschnitt, *der;* **~tie** *n.* Krawatte, *die;* Binder, *der*

nectar ['nektə(r)] *n. (Bot.)* Nektar, *der;* *(delicious drink)* Göttertrank, *der (scherzh.)*

nectarine ['nektərɪn, 'nektəri:n] *n.* Nektarine, *die*

née (*Amer.:* **nee**) [neɪ] *adj.* geborene

need [ni:d] **1.** *n.* **a)** *no pl.* Notwendigkeit, *die* (**for, of** *Gen.*); *(demand)* Bedarf, *der* (**for, of** an + *Dat.*); **as the ~ arises** nach Bedarf; **if ~ arise/be** nötigenfalls; falls nötig; **there's no ~ for that** *(as answer)* [das ist] nicht nötig; **there's no ~ to do sth.** es ist nicht nötig *od.* notwendig, etw. zu tun; **there is no ~ to worry/get angry** es besteht kein Grund zur Sorge/sich zu ärgern; **be in ~ of sth.** etw. brauchen *od.* nötig haben; **there is no ~ for such behaviour** solch ein Verhalten ist unnötig; **there's no ~ for you to apologize** du brauchst dich nicht zu entschuldigen; **feel the ~ to do sth.** sich gezwungen *od.* genötigt sehen, etw. zu tun; **feel the ~ to confide in sb.** das Bedürfnis haben, sich jmdm. anzuvertrauen; **be badly in ~ of sth.** etw. dringend nötig haben; **be in ~ of repair** reparaturbedürftig sein; **have ~ of sb./sth.** jmdn./etw. brauchen *od.* nötig haben; **b)** *no pl. (emergency)* Not, *die;* **in case of ~**: im Notfall; **in times of ~**: in Notzeiten; **those in**

~: die Notleidenden *od.* Bedürftigen; *see also* **friend a; c)** *(thing)* Bedürfnis, *das.* **2.** *v. t.* **a)** *(require)* brauchen; **sth. that urgently ~s doing** etw., was dringend gemacht werden muß; **much ~ed** dringend notwendig; **that's all I ~ed!** *(iron.)* auch das noch!; das hat mir gerade noch gefehlt!; **it ~s a coat of paint** es muß gestrichen werden; **~ correction** berichtigt werden müssen; **b)** *expr. necessity* müssen; **it needs/doesn't need to be done** es muß getan werden/es braucht nicht getan zu werden; **you don't need to do that** das brauchst du nicht zu tun; **I don't ~ to be reminded** du brauchst/ihr braucht mich nicht daran zu erinnern; **it ~ed doing** es mußte getan werden; **he ~s cheering up** er muß [ein bißchen] aufgeheitert werden; **you shouldn't ~ to be told** das solltest *od.* müßtest du eigentlich wissen; **she ~s everything [to be] explained to her** man muß ihr alles erklären; **you ~ only ask** du brauchst nur zu fragen; **don't be away longer than you ~ [be]** bleib nicht länger als nötig weg; **c)** *pres.* **he ~,** *neg.* **~ not** *or (coll.)* **~n't** ['ni:dnt] *expr. desirability* müssen; *with neg.* brauchen zu; **~ I say more?** muß ich noch mehr sagen?; **I ~ hardly** *or* **hardly ~ say that ...:** ich brauche wohl kaum zu sagen, daß ...; **he ~n't be told** *(let's keep it secret)* das braucht er nicht zu wissen; **we ~n't** *or* **~ not have done it, if ...:** wir hätten es nicht zu tun brauchen, wenn ...; **that ~ not be the case** das muß nicht so sein *od.* der Fall sein

needle ['ni:dl] **1.** *n.* Nadel, *die;* **it is like looking for a ~ in a haystack** es ist, als wollte man eine Stecknadel in einem Heuhaufen finden; *see also* **pin 1 a. 2.** *v. t. (coll.)* ärgern; nerven *(ugs.);* **what's needling him?** was fuchst ihn [denn so]? *(ugs.)*

needless ['ni:dlɪs] *adj.* unnötig; *(senseless)* sinnlos; **~ to say** *or* **add, he didn't do it** überflüssig zu sagen, daß er es nicht getan hat

needlessly ['ni:dlɪslɪ] *adv.* unnötigerweise; *(senselessly)* sinnlos

'needlework *n.* Handarbeit, *die;* **do ~:** handarbeiten

needn't ['ni:dnt] *(coll.)* = **need not;** *see* **need 2 c**

needy ['ni:dɪ] *adj.* notleidend; bedürftig; **the ~:** die Notleidenden *od.* Bedürftigen

ne'er-do-well ['neədʊwel] *n.* Tunichtgut, *der*

negation [nɪ'geɪʃn] *n. (Ling.)* Negation, *die (fachspr.);* Verneinung, *die*

negative ['negətɪv] **1.** *adj.* **a)** *(also Math.)* negativ; **b)** *(Ling.)* verneint; Negations⟨*partikel*⟩; **c)** *(Electr.)* **~ pole/terminal** Minuspol, *der;* **d)** *(Photog.)* negativ; Negativ-. **2.** *n.* **a)** *(Photog.)* Negativ, *das;* **b)** *(~ statement)* negative Aussage; *(answer)* Nein, *das;* **be in the ~** ⟨*Antwort:*⟩ negativ *od.* „nein" sein

negatively ['negətɪvlɪ] *adv. (also Electr.)* negativ

neglect [nɪ'glekt] **1.** *v. t.* vernachlässigen; versäumen ⟨*Gelegenheit*⟩; unerledigt lassen, liegenlassen ⟨*Korrespondenz, Arbeit*⟩; **she ~ed to write** sie hat es versäumt zu schreiben; **not ~ doing** *or* **to do sth.** es nicht versäumen, etw. zu tun. **2.** *n.* **a)** Vernachlässigung, *die;* **be in a state of ~** ⟨*Gebäude:*⟩ verwahrlost sein; **suffer from ~:** vernachlässigt werden; **~ of duty** Pflichtvergessenheit, *die;* **b)** *(negligence)* Nachlässigkeit, *die;* Fahrlässigkeit, *die*

neglectful [nɪ'glektfl] *adj. (careless)* gleichgültig (**of** gegenüber); **be ~ of** sich nicht kümmern um

négligé, negligee ['neglɪʒeɪ] *n.* Negligé, *das*

negligence ['neglɪdʒəns] *n., no pl. (carelessness)* Nachlässigkeit, *die; (Law, Insurance, etc.)* Fahrlässigkeit, *die*

negligent ['neglɪdʒənt] *adj.* nachlässig; **be ~ about sth.** sich um etw. nicht kümmern; **be ~ of one's duties/sb.** seine Pflichten/jmdn. vernachlässigen

negligible ['neglɪdʒɪbl] *adj.* unerheblich

negotiable [nɪ'gəʊʃəbl] *adj.* **a)** *(open to discussion)* verhandlungsfähig ⟨*Forderung, Bedingungen*⟩; **b)** *(that can be got past)* zu bewältigend *nicht präd.;* zu bewältigen *nicht attr.;* passierbar ⟨*Straße, Fluß*⟩

negotiate [nɪ'gəʊʃɪeɪt] **1.** *v. i.* verhandeln (**for, on, about** über + *Akk.*); **the negotiating table** der Verhandlungstisch. **2.** *v. t.* **a)** *(arrange)* aushandeln; **b)** *(get past)* bewältigen; überwinden ⟨*Hindernis*⟩; passieren ⟨*Straße, Fluß*⟩; nehmen ⟨*Kurve*⟩; **c)** *(Commerc.) (convert into cash)* einlösen ⟨*Scheck*⟩; *(transfer)* übertragen ⟨*Wechsel, Papiere usw.*⟩

negotiation [nɪgəʊʃɪ'eɪʃn, nɪgəʊsɪ'eɪʃn] *n.* **a)** *(discussion)* Verhandlung, *die* (**for, about** über + *Akk.*); **by ~:** durch Verhandeln *od.* Verhandlungen; **enter into ~:** in Verhandlungen *(Akk.)* eintreten; **be a matter of ~:** Verhandlungssache sein; **b)** *in pl. (talks)* Verhandlungen *Pl.*

negotiator [nɪ'gəʊʃɪeɪtə(r)] *n.* Unterhändler, *der*/-händlerin, *die*

Negress ['ni:grɪs] *n.* Negerin, *die*

Negro ['ni:grəʊ] **1.** *n., pl.* **~es** Neger, *der.* **2.** *adj.* Neger-; **~ woman** Negerin, *die*

neigh [neɪ] **1.** *v. i.* wiehern. **2.** *n.* Wiehern, *das*

neighbor *etc. (Amer.) see* **neighbour** *etc.*

neighbour ['neɪbə(r)] *n.* Nachbar, *der*/Nachbarin, *die; (at table)* [Tisch]nachbar, *der*/[Tisch]nachbarin, *die; (thing)* der/die/das daneben; *(building/country)* Nachbargebäude/-land, *das;* **we're next-door ~s** wir wohnen Tür an Tür; **my next-door ~s** meine Nachbarn von nebenan

neighbourhood ['neɪbəhʊd] *n. (district)* Gegend, *die; (neighbours)* Nachbarschaft, *die; attrib.* an der *od.* um die Ecke *nachgestellt (ugs.);* **your friendly ~ bobby** *etc. (coll. joc.)* der freundliche Polizist von nebenan; **the children from the ~:** die Kinder aus der Nachbarschaft; **it was [somewhere] in the ~ of £100** es waren [so] um [die] 100 Pfund

neighbouring ['neɪbərɪŋ] *adj.* benachbart; Nachbar-; angrenzend ⟨*Felder*⟩

neighbourly ['neɪbəlɪ] *adj.* [gut]nachbarlich; *(friendly)* freundlich

neither ['naɪðə(r), ni:ðə(r)] **1.** *adj.* keiner/keine/keins der beiden; **in ~ case** in keinem Falle. **2.** *pron.* keiner/keine/keins von *od.* der beiden; **~ of them** keiner von *od.* der beiden; **Which will you have? – N~:** Welches nehmen Sie? – Keins [von beiden]. **3.** *adv. (also not)* auch nicht; **I'm not going – N~ am I** *or (sl.)* **Me ~:** Ich gehe nicht – Ich auch nicht; **if you don't go, ~ shall I** wenn du nicht gehst, gehe ich auch nicht. **4.** *conj. (not either, not on the one hand)* weder; **~ ... nor** weder ... noch; **he ~ knows nor cares** weder weiß er es, noch will er es wissen; **he ~ ate, drank, nor smoked** er aß weder, noch trank, noch rauchte er

nelly ['nelɪ] *n.* **not on your ~** *(Brit. sl.)* nie im Leben *(ugs.)*

neo- ['ni:əʊ] *in comb.* neo-/Neo-

neo'classical *adj.* klassizistisch

neolithic [ni:ə'lɪθɪk] *adj.*

(Archaeol.) neolithisch *(fachspr.)*; jungsteinzeitlich

neologism [nɪˈɒlədʒɪzm] *n.* Neubildung, *die;* Neologismus, *der (Sprachw.)*

neon [ˈniːɒn] *n. (Chem.)* Neon, *das*

neon: ~ ˈ**light** *n.* Neonlicht, *das; (fitting)* Neonlampe, *die;* ~ ˈ**sign** *n.* Neonreklame, *die*

nephew [ˈnevjuː, ˈnefjuː] *n.* Neffe, *der*

nepotism [ˈnepətɪzm] *n.* Vetternwirtschaft, *die (abwertend)*

Neptune [ˈneptjuːn] *pr. n.* **a)** *(Astron.)* Neptun, *der;* **b)** *(Roman Mythol.)* Neptun *(der)*

nerve [nɜːv] **1.** *n.* **a)** Nerv, *der;* **b)** *in pl. (fig., of mental state)* **be suffering from ~s** nervös sein; **get on sb.'s ~s** jmdm. auf die Nerven gehen *od.* fallen *(ugs.)*; **~s of steel** Nerven wie Drahtseile *(ugs.)*; **c)** *(coolness, boldness)* Kaltblütigkeit, *die;* Mut, *der;* **not have the ~ for sth.** für *od.* zu etw. nicht die Nerven haben; **lose one's ~:** die Nerven verlieren; **d)** *(coll.: audacity)* **what [a] ~!** [so eine] Frechheit!; **have the ~ to do sth.** den Nerv haben, etw. zu tun *(ugs.)*; **he's got a ~:** der hat Nerven *(ugs.)*. **2.** *v. t. (give strength or courage to)* ermutigen; **~ oneself** seinen ganzen Mut zusammennehmen

nerve: ~-centre *n. (fig.)* Schaltzentrale, *die;* **~ gas** *n.* Nervengas, *das;* **~-racking** [ˈnɜːvrækɪŋ] *adj.* nervenaufreibend

nervous [ˈnɜːvəs] *adj.* **a)** *(Anat., Med.)* Nerven-; **~ breakdown** Nervenzusammenbruch, *der;* **b)** *(having delicate nerves)* nervös; **be a ~ wreck** mit den Nerven völlig am Ende sein; **c)** *(Brit.: timid)* **be ~ of** *or* **about** Angst haben vor (+ *Dat.*); **be a ~ person** ängstlich sein

nervously [ˈnɜːvəslɪ] *adv.* nervös

nervy [ˈnɜːvɪ] *adj.* **a)** *(jerky, nervous)* nervös; unruhig; **b)** *(Amer. coll.: impudent)* unverschämt

nest [nest] **1.** *n.* **a)** *(of bird, animal, insect)* Nest, *das;* **b)** *(fig.: retreat, shelter)* Nest, *das (fig.)*; Zufluchtsort, *der;* **leave the ~:** flügge werden; **c) ~ of tables** Satz Tische. **2.** *v. i.* nisten. **3.** *v. t.* **a)** *(place as in ~)* einbetten; **b)** *(pack one inside the other)* ineinandersetzen ⟨*Töpfe usw.*⟩

ˈ**nest-egg** *n. (fig.)* Notgroschen, *der*

nestle [ˈnesl] *v. i.* **a)** *(settle oneself)* sich kuscheln; **b)** *(press oneself affectionately)* sich schmiegen **(to, up against** an + *Akk.*); **c)** *(lie half hidden)* eingebettet sein

nestling [ˈnestlɪŋ] *n.* Nestling, *der*

[1]**net** [net] **1.** *n. (lit. or fig.)* Netz, *das.* **2.** *v. t.,* **-tt-** [mit einem Netz] fangen ⟨*Tier*⟩; einfangen ⟨*Person*⟩

[2]**net 1.** *adj.* **a)** *(free from deduction)* netto; Netto⟨*einkommen, -[verkaufs]preis usw.*⟩; **b)** *(not subject to discount)* **~ price** gebundener Preis; **c)** *(excluding weight of container etc.)* netto; **~ weight** Nettogewicht, *das;* **d)** *(effective, ultimate)* End⟨*ergebnis, -effekt*⟩. **2.** *v. t.,* **-tt-** *(gain)* netto einnehmen; *(yield)* netto einbringen

net: ~ball *n.* Korbball, *der;* **~ ˈcurtain** *n.* Store [aus Gittertüll]; Tüllgardine, *die*

Netherlands [ˈneðələndz] *pr. n. sing. or pl.* Niederlande *Pl.*

net ˈprofit *n.* Reingewinn, *der*

nett *see* [2]**net**

netting [ˈnetɪŋ] *n. ([piece of] net)* Netz, *das;* **wire ~:** Drahtgeflecht, *das;* Maschendraht, *der*

nettle [ˈnetl] **1.** *n.* Nessel, *die; see also* **grasp 2 b. 2.** *v. t.* reizen; aufbringen

ˈ**network** *n.* **a)** *(of intersecting lines, electrical conductors)* Netzwerk, *das;* **b)** *(of railways etc., persons, operations)* Netz, *das;* **c)** *(of broadcasting stations)* [Sender]netz, *das; (company)* Sender, *der*

neuralgia [njʊəˈrældʒə] *n. (Med.)* Neuralgie, *die (fachspr.)*; Nervenschmerz, *der*

neurosis [njʊəˈrəʊsɪs] *n., pl.* **neuroses** [njʊəˈrəʊsiːz] Neurose, *die*

neurotic [njʊəˈrɒtɪk] **1.** *adj.* **a)** *(suffering from neurosis)* nervenkrank; **b)** *(of neurosis)* neurotisch; **c)** *(coll.: unduly anxious)* neurotisch; **don't get ~ about it** laß es nicht zu einer Neurose werden. **2.** *n.* Neurotiker, *der/* Neurotikerin, *die*

neuter [ˈnjuːtə(r)] **1.** *adj. (Ling.)* sächlich; neutral *(fachspr.)*. **2.** *v. t.* kastrieren

neutral [ˈnjuːtrl] **1.** *adj.* neutral. **2.** *n.* **a)** Neutrale, *der/die;* **b)** *(~ gear)* Leerlauf, *der*

neutralise *see* **neutralize**

neutrality [njuːˈtrælɪtɪ] *n.* Neutralität, *die*

neutralize [ˈnjuːtrəlaɪz] *v. t.* **a)** *(Chem.)* neutralisieren; **b)** *(counteract)* neutralisieren; entkräften ⟨*Argument*⟩

neutron [ˈnjuːtrɒn] *n. (Phys.)* Neutron, *das*

neutron: ~ bomb *n.* Neutronenbombe, *die;* **~ star** *n.* Neutronenstern, *der*

never [ˈnevə(r)] *adv.* **a)** nie; **the rain seemed as if it would ~ stop** der Regen schien gar nicht mehr aufhören zu wollen; **he ~ so much as apologized** er hat sich nicht einmal entschuldigt; **~, ~:** nie, nie; niemals; **he was ~ one to do sth.** es war nicht seine Art, etw. zu tun; **~ a** *(not one)* kein einziger/ keine einzige/kein einziges; **~-satisfied** unersättlich; **~-ending** endlos; **~-failing** unfehlbar; unerschöpflich ⟨*Quelle*⟩; **b)** *(coll.) expr. surprise* **you ~ believed that, did you?** du hast das doch wohl nicht geglaubt?; **well, I ~ [did]!** [na *od.* nein *od.* also] so was!; **He ate the whole turkey. – N~!** Er hat den ganzen Truthahn aufgegessen. – Nein!

never: ~-ˈ~ *n. (Brit. coll.)* Abzahlungskauf, *der;* **on the ~-~ [system]** auf Stottern *(ugs.)*; auf Raten; **~theˈless** *adv.* trotzdem; nichtsdestoweniger

new [njuː] **1.** *adj.* neu; frisch ⟨*Brot, Gemüse*⟩; neu ⟨*Kartoffeln*⟩; neu, jung ⟨*Wein*⟩; **as good as ~:** so gut wie neu; ˈ**~ boy/girl** *(lit. or fig.)* Neuling, *der;* **that's a ~ one on me** *(coll.)* das ist mir neu; *(of joke etc.)* den habe ich noch nicht gehört; **visit ~ places** unbekannte Orte besuchen; **the ~ rich** die Neureichen *(abwertend)*; **the ~ woman** *(modern)* die moderne Frau; die Frau von heute; **be like a ~ man/woman** wie neugeboren sein; **as ~:** neuwertig. **2.** *adv. (recently)* vor kurzem; frisch ⟨*gebacken, gewaschen, geschnitten*⟩; gerade erst ⟨*erblüht*⟩

new: ~-born *adj.* neugeboren; **~comer** *n.* Neuankömmling, *der; (one having no experience also)* Neuling, *der* **(to** in + *Dat.*); **~fangled** [njuːˈfæŋgld] *adj. (derog.)* neumodisch *(abwertend)*; **~-found** *adj.* neu; **~-laid** *adj.* frisch [gelegt]

newly [ˈnjuːlɪ] *adv.* neu; **~ married** seit kurzem verheiratet

ˈ**newly-wed** *n.* Jungverheiratete, *der/die*

new ˈmoon *n.* Neumond, *der*

news [njuːz] *n., no pl.* **a)** *(new information)* Nachricht, *die;* **be in the** *or* **make ~:** Schlagzeilen machen; **that's ~ to me** *(coll.)* das ist mir neu; **what's the latest ~?** was gibt es Neues?; **have you heard the ~?** hast du schon gehört?; weißt du schon das Neueste? *(ugs.)*; **have you had any ~ of him?** hast du etwas von ihm gehört?; hast du Nachricht von ihm?; **bad/good ~:** schlechte/gute

Nachrichten; **b)** *(Radio, Telev.)* Nachrichten *Pl.;* **the 10 o'clock ~:** die 10-Uhr-Nachrichten

news: ~agent *n.* Zeitungshändler, *der*/-händlerin, *die;* **~ bulletin** *n.* Nachrichten *Pl.;* **~flash** *n.* Kurzmeldung, *die;* **~caster** *n.* Nachrichtensprecher, *der*/-sprecherin, *die;* **~ 'headline** *n.* Schlagzeile, *die;* **~letter** *n.* Rundschreiben, *das;* **~paper** ['nju:speɪpə(r)] *n.* **a)** Zeitung, *die; attrib.* **~paper boy/girl** Zeitungsausträger, *der*/-austrägerin, *die;* **b)** *(material)* Zeitungspapier, *das;* **~paperman** *n.* Zeitungsmann, *der (ugs.);* Journalist, *der;* **~print** *n.* Zeitungspapier, *das;* **~reader** *n.* Nachrichtensprecher, *der*/-sprecherin, *die;* **~reel** *n.* Wochenschau, *die;* **~ room** *n.* Nachrichtenredaktion, *die;* **~-sheet** *n.* Informationsblatt, *das;* **~-stand** *n.* Zeitungskiosk, *der;* Zeitungsstand, *der;* **~ summary** *n.* Kurznachrichten *Pl.;* **~worthy** *adj.* [für die Medien] interessant; berichtenswert ‹*Ereignis*›

newsy ['nju:zɪ] *adj. (coll.)* voller Neuigkeiten *nachgestellt*

newt [nju:t] *n.* [Wasser]molch, *der*

New: ~ 'Testament *see* **testament a; new 'world** *see* **world a; new 'year** *n.* Neujahr, *das;* **over the new year** über Neujahr; **a Happy ~ Year** ein glückliches *od.* gutes neues Jahr; **bring in the ~ Year** Silvester feiern; **~ 'Year's** *(Amer.),* **~ Year's 'Day** *ns.* Neujahrstag, *der;* **~ Year's 'Eve** *n.* Silvester, *der od. das;* Neujahrsabend, *der;* **~ Zealand** [nju: 'zi:lənd] **1.** *pr. n.* Neuseeland *(das);* **2.** *attrib. adj.* neuseeländisch; **~ Zealander** [nju: 'zi:ləndə(r)] *n.* Neuseeländer, *der*/Neuseeländerin, *die*

next [nekst] **1.** *adj.* **a)** *(nearest)* nächst...; **the seat ~ to me** der Platz neben mir; **the ~ room** das Nebenzimmer; **the ~ but one** der/die/das übernächste; **~ to** *(fig.: almost)* fast; nahezu; **b)** *(in order)* nächst...; **within the ~ few days** in den nächsten Tagen; **~ month** nächsten Monat; **during the ~ year** während der nächsten zwölf Monate; **we'll come ~ May** wir kommen im Mai nächsten Jahres; **the ~ largest/larger** der/die/das nächstkleinere/nächstgrößere; **[the] ~ time** das nächste Mal; **the ~ best** der/die/das nächstbeste; **am I ~?** komme ich jetzt dran? **2.** *adv. (in the ~ place)* als nächstes; *(on the ~ occasion)* das nächste Mal; **whose name comes ~?** wessen Name kommt als nächstes *od.* nächster?; **it is my turn ~:** ich komme als nächster dran; **sit/stand ~ to sb.** neben jmdm. sitzen/stehen; **place sth. ~ to sb./sth.** etw. neben jmdn./etw. stellen; **come ~ to last** *(in race)* zweitletzter/zweitletzte werden; **come ~ to bottom** *(in exam)* der/die Zweitschlechteste sein. **3.** *n.* **a) from one day to the ~:** von einem Tag zum andern; **the week after ~:** [die] übernächste Woche; **b)** *(person)* **~ of kin** nächster/nächste Angehörige; **~ please!** der nächste, bitte!

'next-door *adj.* gleich nebenan *nachgestellt*

NHS *abbr. (Brit.)* **National Health Service**

NI *abbr. (Brit.)* **National Insurance**

nib [nɪb] *n.* Feder, *die*

nibble ['nɪbl] **1.** *v.t.* knabbern; **~ off** abknabbern. **2.** *v.i.* knabbern **(at, on** an + *Dat.*)

Nicaragua [nɪkə'rægjʊə] *pr. n.* Nicaragua *(das)*

Nicaraguan [nɪkə'rægjʊən] **1.** *adj.* nicaraguanisch; **sb. is ~:** jmd. ist Nicaraguaner/Nicaraguanerin. **2.** *n.* Nicaraguaner, *der*/Nicaraguanerin, *die*

nice [naɪs] *adj.* **a)** *(pleasing)* nett; angenehm ‹*Stimme*›; schön ‹*Wetter*›; *(iron.: disgraceful, difficult)* schön; sauber *(iron.);* **she has a ~ smile** sie lächelt so nett; **you're a ~ one, I must say** *(iron.)* du bist mir vielleicht einer; **be in a ~ mess** *(iron.)* in einem schönen Schlamassel sitzen *(ugs.);* **~ work** saubere *od.* gute Arbeit; **~ to meet you** freut mich, Sie kennenzulernen; **~ [and] warm/fast/high** schön warm/schnell/hoch; **a ~ long holiday** schöne lange Ferien; **~-looking** hübsch; gut aussehend, hübsch ‹*Person*›; **b)** *(subtle)* fein ‹*Bedeutungsunterschied*›

nicely ['naɪslɪ] *adv. (coll.)* **a)** *(well)* nett; gut ‹*arbeiten, sich benehmen, plaziert sein*›; **b)** *(all right)* gut; **he's got a new job and is doing very ~:** er hat eine neue Arbeit und kommt prima *(ugs.)* damit zurecht; **that will do ~:** das reicht völlig

nicety ['naɪsɪtɪ] *n.* **a)** *no pl. (punctiliousness)* [peinliche] Genauigkeit; **b)** *no pl. (precision, accuracy)* Feinheit, *die;* Genauigkeit, *die;* **to a ~:** perfekt ‹*arrangieren*›; sehr genau ‹*schätzen*›; **c)** *in pl. (minute distinctions)* Feinheiten

niche [nɪtʃ, ni:ʃ] *n.* **a)** *(in wall)* Nische, *die;* **b)** *(fig.: suitable place)* Platz, *der*

nick 1. *n.* **a)** *(notch)* Kerbe, *die;* **b)** *(sl.: prison)* Kittchen, *das (ugs.);* Knast, *der (salopp);* **c)** *(Brit. sl.: police station)* Wache, *die;* Revier, *das;* **d) in good/poor ~** *(coll.)* gut/nicht gut in Schuß *(ugs.);* **e) in the ~ of time** gerade noch rechtzeitig. **2.** *v.t.* **a)** *(make ~ in)* einkerben ‹*Holz*›; **~ one's chin** sich am Kinn schneiden; **b)** *(Brit. sl.) (catch)* schnappen *(ugs.); (arrest)* einlochen *(salopp);* **c)** *(Brit. sl.: steal)* klauen *(salopp);* mitgehen lassen *(ugs.)*

nickel ['nɪkl] *n.* **a)** *(metal)* Nickel, *das;* **b)** *(US coin)* Fünfcentstück, *das*

nickname ['nɪkneɪm] **1.** *n.* Spitzname, *der; (affectionate)* Koseform, *die.* **2.** *v.t.* **~ sb. ...:** jmdm. den Spitznamen ... geben; jmdn. ... taufen

nicotine ['nɪkəti:n] *n.* Nikotin, *das*

niece [ni:s] *n.* Nichte, *die*

nifty ['nɪftɪ] *adj. (sl.)* **a)** klasse *(ugs.);* flott ‹*Kleidung*›; **b)** *(clever)* geschickt; clever ‹*Plan, Idee*›

Nigeria [naɪ'dʒɪərɪə] *pr. n.* Nigeria *(das)*

Nigerian [naɪ'dʒɪərɪən] **1.** *adj.* nigerianisch; **sb. is ~:** jmd. ist Nigerianer/Nigerianerin. **2.** *n.* Nigerianer, *der*/Nigerianerin, *die*

niggardly ['nɪgədlɪ] *adj.* **a)** *(miserly)* knaus[e]rig *(ugs.);* **b)** *(given in small amounts)* armselig, kümmerlich ‹*Portion*›

nigger ['nɪgə(r)] *n. (derog.)* Nigger, *der (abwertend)*

niggle ['nɪgl] **1.** *v.i. (find fault pettily)* [herum]nörgeln *(ugs. abwertend)* **(at** an + *Dat.*). **2.** *v.t.* herumnörgeln an (+ *Dat.*)

niggling ['nɪglɪŋ] *adj.* **a)** *(petty)* belanglos; **b)** *(trivial)* nichtssagend; oberflächlich ‹*Kritik*›; krittelig ‹*Rezension, Rezensent*›

nigh [naɪ] *adv. (arch./literary/dial.)* nahe; **come** *or* **draw ~:** näherkommen; ‹*Tag, Zeitpunkt:*› nahen; **it's ~ on impossible** es ist nahezu unmöglich

night [naɪt] *n.* **a)** Nacht, *die; (evening)* Abend, *der;* **the following ~:** die Nacht/der Abend darauf; **the previous ~:** die vorausgegangene Nacht/der vorausgegangene Abend; **one ~ he came** eines Nachts/Abends kam er; **two ~s ago** vorgestern nacht/abend; **the other ~:** neulich abends/nachts; **far into the ~:** bis spät *od.* tief in die Nacht; **on Sunday ~:** Sonntag nacht/[am] Sonntag abend; **on**

Sunday ~s Sonntag abends; **[on] the ~ after/before** die Nacht danach/davor; **for the ~:** über Nacht; **late at ~:** spätabends; **take all ~** *(fig.)* den ganzen Abend brauchen; **at ~** *(in the evening, at ~fall)* abends; *(during the ~)* nachts; bei Nacht; **make a ~ of it** die Nacht durchfeiern; durchmachen *(ugs.)*; **~ and day** Tag und Nacht; **as ~ follows day** so sicher wie das Amen in der Kirche; **a ~ off** eine Nacht/ein Abend frei; **have a ~ out** *(festive evening)* [abends] ausgehen; **spend the ~ with sb.** bei jmdm. übernachten; *(implying sexual intimacy)* die Nacht mit jmdm. verbringen; **stay the ~** *or* **over ~:** über Nacht bleiben; **b)** *(darkness, lit. or fig.)* Nacht, *die;* **black as ~:** schwarz wie die Nacht; **c)** *(~fall)* Einbruch der Dunkelheit; **d)** *(~'s sleep)* **have a good/bad ~:** gut/schlecht schlafen; **have a sleepless ~:** eine schlaflose Nacht haben; **e)** *(evening of performance etc.)* Abend, *der;* **opening ~:** Premiere, *die;* **f)** *attrib.* Nacht-/Abend-

night: ~bird *n. (person)* Nachteule, *die (ugs. scherzh.)*; **~cap** *n.* **a)** Nachtmütze, *die; (woman's)* Nachthaube, *die;* **b)** *(drink)* Schlaftrunk, *der;* **~clothes** *n. pl.* Nachtwäsche, *die;* **~club** *n.* Nachtklub, *der;* Nachtlokal, *das;* **~dress** *n.* Nachthemd, *das;* **~fall** *n., no art.* Einbruch der Dunkelheit; **at/after ~fall** bei/nach Einbruch der Dunkelheit; **~gown** *n.* Nachthemd, *das*

nightie ['naɪtɪ] *n. (coll.)* Nachthemd, *das*

nightingale ['naɪtɪŋgeɪl] *n.* Nachtigall, *die*

night: ~-life *n.* Nachtleben, *das;* **~-light** *n.* Nachtlicht, *das;* **~-long 1.** *adj.* sich über die ganze Nacht hinziehend; **2.** *adv.* die ganze Nacht [lang *od.* über]

nightly ['naɪtlɪ] **1.** *adj.* nächtlich/abendlich; *(every night/evening)* allnächtlich/allabendlich. **2.** *adv. (every night)* jede Nacht; *(every evening)* jeden Abend; **twice ~** *(Theatre etc.)* zweimal pro Abend

night: ~mare *n. (lit. or fig.)* Alptraum, *der;* **~-owl** *n. (coll.: person)* Nachteule, *die (ugs. scherzh.)*; Nachtschwärmer, *der (scherzh.)*; **~ safe** *n.* Nachttresor, *der;* **~ school** *n.* Abendschule, *die;* **~ shift** *n.* Nachtschicht, *die;* **~-time** *n., no indef. art.* Nacht, *die;* **in the** *or* **at ~-time** nachts; **wait until ~-time** warten, bis es Nacht *od.* dunkel wird; **~-'watchman** *n.* Nachtwächter, *der*

nihilism ['naɪɪlɪzm, 'nɪhɪlɪzm] *n.* Nihilismus, *der*

nil [nɪl] *n.* **a)** nichts; **his chances were ~:** seine Chancen waren gleich Null; **b)** *(Sport)* null; **win one ~** *or* **by one goal to ~:** eins zu null gewinnen

Nile [naɪl] *pr. n.* Nil, *der*

nimble ['nɪmbl] *adj.* **a)** *(quick in movement)* flink; behende; beweglich ⟨*Geist*⟩; **b)** *(dextrous)* geschickt

nimbly ['nɪmblɪ] *adv.* flink ⟨*arbeiten, sich bewegen*⟩

nincompoop ['nɪŋkəmpu:p] *n.* Trottel, *der (ugs. abwertend)*

nine [naɪn] **1.** *adj.* neun; **~ times out of ten** *(fig.: nearly always)* in den weitaus meisten Fällen; **a ~ days' wonder** nur eine Eintagsfliege *(ugs.)*; *see also* **eight 1. 2.** *n.* Neun, *die;* **work from ~ to five** die übliche Arbeitszeit [von 9 bis 17 Uhr] haben; **dressed [up] to the ~s** sehr festlich gekleidet; **~-~-~, 999** *(Brit.: emergency number)* ≈ eins, eins, null; *see also* **eight 2 a, c, d**

nineteen [naɪn'ti:n] **1.** *adj.* neunzehn; *see also* **eight 1. 2.** *n.* **a)** Neunzehn, *die; see also* **eight 2 a; eighteen 2; b) talk ~ to the dozen** *(Brit.)* wie ein Wasserfall reden *(ugs.)*

nineteenth [naɪn'ti:nθ] **1.** *adj.* neunzehnt...; *see also* **eighth 1. 2.** *n. (fraction)* Neunzehntel, *das; see also* **eighth 2**

ninetieth ['naɪntɪɪθ] **1.** *adj.* neunzigst...; *see also* **eighth 1. 2.** *n. (fraction)* Neunzigstel, *das; see also* **eighth 2**

ninety ['naɪntɪ] **1.** *adj.* neunzig; *see also* **eight 1. 2.** *n.* Neunzig, *die; see also* **eight 2 a; eighty 2**

ninety: ~-'first *etc. adj.* einundneunzigst... *usw.; see also* **eighth 1; ~-'one** *etc.* **1.** *adj.* einundneunzig *usw.*; **~-nine times out of a hundred** *(fig.: nearly always)* so gut wie immer; *see also* **eight 1; 2.** *n.* Einundneunzig *usw., die; see also* **eight 2 a**

ninth [naɪnθ] **1.** *adj.* neunt...; *see also* **eighth 1. 2.** *n. (in sequence)* neunte, *der/die/das; (in rank)* Neunte, *der/die/das; (fraction)* Neuntel, *das; see also* **eighth 2**

¹nip [nɪp] **1.** *v. t.,* **-pp-: a)** zwicken; **~ sb.'s toe/sb. on the leg** jmdn. *od.* jmdm. in den Zeh/jmdn. am Bein zwicken; **b) ~ off** abzwicken; *(with scissors)* abknipsen. *See also* **bud 1. 2.** *v. i.,* **-pp-** *(Brit. sl.: step etc. quickly)* **~ in** hinein-/hereinflitzen *(ugs.)*; **~ out** hinaus-/herausflitzen *(ugs.)*. **3.** *n.* **a)** *(pinch, squeeze)* Kniff, *der; (bite)* Biß, *der;* **b)** *(coldness of air)* Kälte, *die;* **there's a ~ in the air** es ist frisch

²nip *n. (of spirits etc.)* Schlückchen, *das*

nipple ['nɪpl] *n.* **a)** *(on breast)* Brustwarze, *die;* **b)** *(of feeding-bottle)* Sauger, *der*

nippy ['nɪpɪ] *adj. (coll.)* **a)** *(nimble)* flink; spritzig ⟨*Auto*⟩; **b)** *(cold)* frisch; kühl

nit [nɪt] *n.* **a)** *(egg)* Nisse, *die;* **b)** *(sl.: stupid person)* Dussel, *der (ugs.)*; Blödmann, *der (salopp)*

nit: ~-pick *v. i.* kritteln *(abwertend)*; **~-picking** *adj. (coll.)* kleinlich *(abwertend)*

nitric acid ['naɪtrɪk æsɪd] *n. (Chem.)* Salpetersäure, *die*

nitrogen ['naɪtrədʒən] *n.* Stickstoff, *der*

nitroglycerine [naɪtrəʊ'glɪsəri:n] *n.* Nitroglyzerin, *das*

nitty-gritty [nɪtɪ'grɪtɪ] *n. (sl.)* **the ~ [of the matter]** der Kern [der Sache]; **get down to the ~:** zur Sache kommen

nitwit ['nɪtwɪt] *n. (coll.)* Trottel, *der (ugs.)*

no [nəʊ] **1.** *adj.* **a)** *(not any)* kein; **b)** *(not a)* kein; *(quite other than)* alles andere als; **she is no beauty** sie ist keine Schönheit *od.* nicht gerade eine Schönheit; **you are no friend** du bist kein [wahrer] Freund; **c)** *(hardly any)* **it's no distance from our house to the shopping centre** von unserem Haus ist es nicht weit bis zum Einkaufszentrum. **2.** *adv.* **a)** *(by no amount)* nicht; **no less [than]** nicht weniger [als]; **it is no different from before** es hat sich nichts geändert; **no more wine?** keinen Wein mehr?; **no more war!** nie wieder Krieg!; **b)** *(equivalent to negative sentence)* nein; **say/answer 'no'** nein sagen/mit Nein antworten; **I won't take 'no' for an answer** ein Nein lasse ich nicht gelten. **3.** *n., pl.* **noes** [nəʊz] Nein, *das; (vote)* Neinstimme, *die*

No. *abbr.* **number** Nr.

Noah's ark [nəʊəz 'ɑ:k] *n. (Bibl.)* die Arche Noah

nobble ['nɒbl] *v. t. (Brit. sl.)* **a)** *(durch Spritzen o. ä.)* langsam machen ⟨*Rennpferd*⟩; **b)** *(durch Bestechung o. ä.)* auf seine Seite ziehen ⟨*Person*⟩

Nobel prize [nəʊbel 'praɪz] *n.* Nobelpreis, *der*

nobility [nə'bɪlɪtɪ] *n.* **a)** *no pl.*

(character) hohe Gesinnung; Adel, *der;* **b)** *(class)* Adel, *der;* **many of the ~:** viele Adlige

noble ['nəʊbl] **1.** *adj.* **a)** *(by rank, title, or birth)* ad[e]lig; **be of ~ birth** von adliger *od.* edler Geburt sein *(geh.);* adlig sein; **b)** *(of lofty character)* edel ⟨*Gedanken, Gefühle*⟩; **~ ideals** hohe Ideale; **c)** *(showing greatness of character)* edel; hochherzig *(geh.).* **2.** *n.* Adlige, *der/die*

noble: ~man ['nəʊblmən] *n., pl.* **~men** ['nəʊblmən] Adlige, *der;* **~woman** *n.* Adlige, *die*

nobly ['nəʊblɪ] *adv.* **a)** *(with noble spirit)* edel[gesinnt]; **b)** *(generously)* edelmütig *(geh.)*

nobody ['nəʊbədɪ] *n. & pron.* niemand; keiner; *(person of no importance)* Niemand, *der*

no-'claim[s] bonus *n. (Insurance)* Schadenfreiheitsrabatt, *der*

nocturnal [nɒk'tɜ:nl] *adj.* nächtlich; nachtaktiv ⟨*Tier*⟩; **~ animal/bird** Nachttier, *das*/-vogel, *der*

nocturne ['nɒktɜ:n] *n. (Mus.)* Nocturne, *das od. die*

nod [nɒd] **1.** *v. i.,* **-dd-: a)** *(as signal)* nicken; **~ to sb.** jmdm. zunicken; **b)** *(in drowsiness)* **she sat ~ding by the fire** sie war neben dem Kamin eingenickt *(ugs.);* **her head started to ~:** sie begann einzunicken *(ugs.).* **2.** *v. t.,* **-dd-: a)** *(incline)* **~ one's head [in greeting]** [zum Gruß] mit dem Kopf nicken; **b)** *(signify by ~)* **~ approval** *or* **agreement** zustimmend nicken. **3.** *n.* [Kopf]nicken, *das*

~ 'off *v. i.* einnicken *(ugs.)*

node [nəʊd] *n. (Bot., Astron.)* Knoten, *der*

no-'go *adj.* Sperr⟨*gebiet, -zone*⟩

'no-good *adj. (coll.)* nichtsnutzig *(abwertend)*

noise [nɔɪz] *n.* **a)** *(loud outcry)* Lärm, *der;* Krach, *der;* **don't make so much ~/such a loud ~:** sei nicht so laut/mach nicht solchen Lärm *od.* Krach; **make a ~ about sth.** *(fig.: complain)* wegen etw. Krach machen *od.* schlagen *(ugs.);* **b)** *(any sound)* Geräusch, *das; (loud, harsh, unwanted)* Lärm, *der;* **c)** *(Communications)* Geräusch, *das; (hissing)* Rauschen, *das;* **d) make ~s about doing sth.** davon reden, etw. tun zu wollen

noiseless ['nɔɪzlɪs] *adj.,* **noiselessly** ['nɔɪzlɪslɪ] *adv.* **a)** *(silent[ly])* lautlos; **b)** *(making no avoidable noise)* geräuschlos

noisily ['nɔɪzɪlɪ] *adv.* laut; lärmend ⟨*spielen*⟩; geräuschvoll ⟨*stolpern, schlürfen*⟩

noisy ['nɔɪzɪ] *adj.* laut; lärmend, laut ⟨*Menschenmasse, Kinder*⟩; lautstark ⟨*Diskussion, Begrüßung*⟩; geräuschvoll ⟨*Aufbruch, Ankunft*⟩

nomad ['nəʊmæd] *n.* Nomade, *der;* **be a ~** *(fig.)* ein Nomadendasein führen

nomadic [nəʊ'mædɪk] *adj.* nomadisch; **~ tribe** Nomadenstamm, *der*

'no man's land *n.* Niemandsland, *das*

nom de plume [nɒm də 'plu:m] *n., pl.* **noms de plume** [nɒm də 'plu:m] Pseudonym, *das*

nomenclature [nə'menklətʃə(r)] *n.* Nomenklatur, *die*

nominal ['nɒmɪnl] *adj.* **a)** *(in name only)* nominell; **b)** *(virtually nothing)* äußerst gering; äußerst niedrig ⟨*Preis, Miete*⟩

nominally ['nɒmɪnəlɪ] *adv.* namentlich

nominate ['nɒmɪneɪt] *v. t.* **a)** *(propose for election)* nominieren; **b)** *(appoint to office)* ernennen

nomination [nɒmɪ'neɪʃn] *n.* **a)** *(appointment to office)* Ernennung, *die;* **b)** *(proposal for election)* Nominierung, *die*

nominative ['nɒmɪnətɪv] *(Ling.)* **1.** *adj.* Nominativ-; nominativisch; **~ case** Nominativ, *der.* **2.** *n.* Nominativ, *der*

nominee [nɒmɪ'ni:] *n. (candidate)* Kandidat, *der*/Kandidatin, *die*

non- [nɒn] *pref.* nicht-

non-ag'gression *n.* Gewaltverzicht, *der;* **~ pact** *or* **treaty** Nichtangriffspakt, *der*

non-alco'holic *adj.* alkoholfrei

non-a'ligned *adj.* blockfrei

nonchalant ['nɒnʃələnt] *adj.* nonchalant *(geh.);* unbekümmert

non-'combatant **1.** *n.* Nichtkämpfende, *der/die.* **2.** *adj.* nicht am Kampf beteiligt

non-commissioned 'officer *n.* Unteroffizier, *der*

non-committal [nɒnkə'mɪtl] *adj.* unverbindlich

noncon'formist *n.* Nonkonformist, *der*/Nonkonformistin, *die*

non-co-oper'ation *n.* Verweigerung der Kooperation

non-denominational [nɒndɪnɒmɪ'neɪʃənl] *adj.* konfessionslos

nondescript ['nɒndɪskrɪpt] *adj.* unscheinbar; undefinierbar ⟨*Farbe*⟩

none [nʌn] **1.** *pron.* kein...; **~ of them** keiner/keine/keines von ihnen; **~ of this money is mine** von diesem Geld gehört mir nichts; **~ other than ...:** kein anderer/keine andere als ... **2.** *adv.* keineswegs; **I'm ~ the wiser now** jetzt bin ich um nichts klüger; **~ the less** nichtsdestoweniger

non'entity *n.* Nichts, *das*

'non-event *n.* Reinfall, *der (ugs.);* Enttäuschung, *die*

non-existence [nɒnɪg'zɪstəns] *n., no pl.* Nichtvorhandensein, *das*

non-existent [nɒnɪg'zɪstənt] *adj.* nicht vorhanden

non-'fiction *n.* **~ [literature]** Sachliteratur, *die*

non-inter'ference, non-inter'vention *ns., no pl.* Nichteinmischung, *die*

non-'iron *adj.* bügelfrei

non-'member *n.* Nichtmitglied, *das*

non-'nuclear *adj.* Nichtnuklear-; **~ weapons** konventionelle Waffen

no-'nonsense *adj.* nüchtern

non-'party *adj.* **a)** *(not attached to a party)* parteilos; **b)** *(not related to a party)* überparteilich

nonplus [nɒn'plʌs] *v. t.,* **-ss-** verblüffen

non-'profit[-making] *adj.* nicht auf Gewinn ausgerichtet

non-'resident **1.** *adj.* *(residing elsewhere)* nicht im Haus wohnend; *(outside a country)* nicht ansässig. **2.** *n.* nicht im Haus Wohnende, *der/die; (outside a country)* Nichtansässige, *der/die;* **the bar is open to ~s** die Bar ist auch für Gäste geöffnet, die nicht im Hotel wohnen

non-re'turnable *adj.* Einweg⟨*behälter, -flasche, -[ver]packung*⟩; nicht rückzahlbar ⟨*Anzahlung*⟩

nonsense ['nɒnsəns] **1.** *n.* Unsinn, *der;* **piece of ~:** Firlefanz, *der (ugs. abwertend);* **talk ~:** Unsinn reden; **it's all a lot of ~:** das ist alles Unsinn; **make [a] ~ of sth.** etw. zur Farce machen; **make a ~ of a theory** eine Theorie in sich zusammenfallen lassen; **what's all this ~ about ...?** was soll das [dumme] Gerede über (+ *Akk.*) ...?; **stand no ~:** keinen Unfug dulden; **come along now, and no ~:** kommt jetzt, und mach keinen Unsinn. **2.** *int.* Unsinn!

nonsensical [nɒn'sensɪkl] *adj.* unsinnig

non-'slip *adj.* rutschfest

non-'smoker *n.* **a)** *(person)* Nichtraucher, *der*/Nichtraucherin, *die;* **b)** *(train compartment)* Nichtraucherabteil, *das*

non-'starter *n.* **a)** *(Sport)* Nichtstartende, *der/die;* **b)** *(fig. coll.)* Reinfall, *der (ugs.); (person)* Blindgänger, *der (fig. salopp)*

non-'stick *adj.* **~ frying-pan** *etc.*

Bratpfanne *usw.* mit Antihaftbeschichtung

non-stop 1. ['--] *adj.* durchgehend ⟨*Zug, Busverbindung*⟩; Nonstop⟨*flug, -revue*⟩. 2. [-'-] *adv.* ohne Unterbrechung ⟨*tanzen, reden, reisen, senden*⟩; nonstop, im Nonstop ⟨*fliegen, fahren*⟩

non-'violence *n., no pl.* Gewaltlosigkeit, *die*

non-'violent *adj.* gewaltlos

non-'white 1. *adj.* farbig. 2. *n.* Farbige, *der/die*

noodle ['nu:dl] *n., usu. pl. (pasta)* Nudel, *die*

nook [nʊk] *n.* Winkel, *der;* Ecke, *die;* **in every ~ and cranny** in allen Ecken und Winkeln

noon [nu:n] *n.* Mittag, *der;* zwölf Uhr [mittags]; **at/before ~:** um/vor zwölf [Uhr mittags]

'no one *pron.* **a) ~ of them** keiner/keine/keines von ihnen; **b)** *see* **nobody**

noose [nu:s] *n.* Schlinge, *die;* **put one's head in a ~** *(fig.)* den Kopf in die Schlinge stecken

nor [nə(r), *stressed* nɔ:(r)] *conj.* noch; **neither ... ~ ..., not ... ~ ...:** weder ... noch ...

norm [nɔ:m] *n.* Norm, *die*

normal ['nɔ:ml] 1. *adj.* normal. 2. *n.* **a)** *(~ value)* Normalwert, *der;* **b)** *(usual state)* normaler Stand; **everything is back to** *or* **has returned to ~:** es hat sich wieder alles normalisiert; **his temperature is above ~:** er hat erhöhte Temperatur

normalise *see* **normalize**

normality [nɔ:'mælɪtɪ] *n., no pl.* Normalität, *die*

normalize ['nɔ:məlaɪz] 1. *v.t.* normalisieren. 2. *v.i.* sich normalisieren

normally ['nɔ:məlɪ] *adv.* **a)** *(in normal way)* normal; **b)** *(ordinarily)* normalerweise

north [nɔ:θ] 1. *n.* **a)** *(direction)* Norden, *der;* **the ~:** Nord *(Met., Seew.);* **in/to[wards]/from the ~:** im/nach/von Norden; **to the ~ of** nördlich von; nördlich (+ *Gen.*); **magnetic ~:** magnetischer Nordpol; **b)** *usu.* N~ *(part lying to the ~)* Norden, *der;* **from the N~:** aus dem Norden. 2. *adj.* nördlich; Nord⟨*wind, -fenster, -küste, -grenze, -tor*⟩. 3. *adv.* nordwärts; nach Norden; **~ of** nördlich von; nördlich (+ *Gen.*)

north: N~ 'Africa *pr. n.* Nordafrika *(das);* **N~ A'merica** *pr. n.* Nordamerika *(das);* **N~ A'merican** 1. *adj.* nordamerikanisch; 2. *n.* Nordamerikaner, *der*/-amerikanerin, *die;* **~bound** *adj.* ⟨*Zug, Verkehr usw.*⟩ in Richtung Norden; **~-'east** 1. *n.* Nordosten, *der;* 2. *adj.* nordöstlich; Nordost⟨*wind, -fenster, -küste*⟩; 3. *adv.* nordostwärts; nach Nordosten; **~-'eastern** *adj.* nordöstlich

northerly ['nɔ:ðəlɪ] *adj.* **a)** *(in position or direction)* nördlich; **in a ~ direction** nach Norden; **b)** *(from the north)* ⟨*Wind*⟩ aus nördlichen Richtungen

northern ['nɔ:ðən] *adj.* nördlich; Nord⟨*grenze, -hälfte, -seite*⟩

northerner ['nɔ:ðənə(r)] *n. (male)* Nordengländer/-deutsche *usw., der; (female)* Nordengländerin/-deutsche *usw., die*

Northern: ~ 'Europe *pr. n.* Nordeuropa *(das);* **~ 'Ireland** *pr. n.* Nordirland *(das);* **n~ 'lights** *n. pl.* Nordlicht, *das*

northernmost ['nɔ:ðənməʊst] *adj.* nördlichst...

North: ~ 'German 1. *adj.* norddeutsch; 2. *n.* Norddeutsche, *der/die;* **~ 'Germany** *pr. n.* Norddeutschland *(das);* **~ Ko'rea** *pr. n.* Nordkorea *(das);* **~ of 'England** *pr. n.* Nordengland *(das); attrib.* nordenglisch; **~ 'Pole** *pr. n.* Nordpol, *der;* **~ 'Sea** *pr. n.* Nordsee, *die; attrib.* **~ Sea gas/oil** Nordseegas/-öl, *das*

northward ['nɔ:θwəd] 1. *adj.* nach Norden gerichtet; *(situated towards the north)* nördlich; **in a ~ direction** nach Norden; [in] Richtung Norden. 2. *adv.* nordwärts; **they are ~ bound** sie fahren nach *od.* [in] Richtung Norden

northwards ['nɔ:θwədz] *adv.* nordwärts

north: ~-'west 1. *n.* Nordwesten, *der;* 2. *adj.* nordwestlich; Nordwest⟨*wind, -fenster, -küste*⟩; 3. *adv.* nordwestwärts; nach Nordwesten; **~-'western** *adj.* nordwestlich

Norway ['nɔ:weɪ] *pr. n.* Norwegen *(das)*

Norwegian [nɔ:'wi:dʒn] 1. *adj.* norwegisch; **sb. is ~:** jmd. ist Norweger/Norwegerin. 2. *n.* **a)** *(person)* Norweger, *der*/Norwegerin, *die;* **b)** *(language)* Norwegisch, *das; see also* **English 2a**

Nos. *abbr.* **numbers** Nrn.

nose [nəʊz] 1. *n.* **a)** Nase, *die;* **[win] by a ~:** mit einer Nasenlänge [gewinnen]; **follow one's ~** *(fig.) (be guided by instinct)* seinem Instinkt folgen; *(go forward)* der Nase nachgehen; **get up sb.'s ~** *(sl.: annoy sb.)* jmdm. auf den Wecker gehen *(salopp);* **hold one's ~:** sich *(Dat.)* die Nase zuhalten; **pay through the ~:** tief in die Tasche greifen müssen *(ugs.);* **poke** *or* **thrust** *etc.* **one's ~ into sth.** *(fig.)* seine Nase in etw. *(Akk.)* stecken *(fig. ugs.);* **put sb.'s ~ out of joint** *(fig. coll.)* jmdn. vor den Kopf stoßen *(ugs.);* **rub sb.'s ~ in it** *(fig.)* es jmdm. ständig unter die Nase reiben *(ugs.);* **speak through one's ~:** näseln; durch die Nase sprechen; **turn up one's ~ at sth.** *(fig. coll.)* die Nase über etw. *(Akk.)* rümpfen; **under sb.'s ~** *(fig. coll.)* vor jmds. Augen *(Dat.);* **b)** *(of ship, aircraft)* Nase, *die.* 2. *v.t.* **a)** *(detect, smell out)* **~ [out]** aufspüren; **b) ~ one's way** sich *(Dat.)* vorsichtig seinen Weg bahnen. 3. *v.i. (move)* sich vorsichtig bewegen

~ about, ~ around *v.i. (coll.)* herumschnüffeln *(ugs.)*

~ out *v.t.* aufspüren

nose: ~bag *n.* Futterbeutel, *der;* **~bleed** *n.* Nasenbluten, *das;* **~dive** 1. *n.* **a)** Sturzflug, *der;* **b)** *(fig.)* Einbruch, *der;* **take a ~dive** einen Einbruch erleben; 2. *v.i.* im Sturzflug hinuntergehen

nosey *see* **nosy**

nosh [nɒʃ] *n. (esp. Brit. sl.) (snack)* Imbiß, *der; (food)* Futter, *das (salopp)*

'nosh-up *n. (Brit. sl.)* Essen, *das; (good meal)* Festessen, *das*

nostalgia [nɒ'stældʒə] *n.* Nostalgie, *die;* **~ for sth.** Sehnsucht nach etw.

nostalgic [nɒ'stældʒɪk] *adj.* nostalgisch

nostril ['nɒstrɪl] *n.* Nasenloch, *das; (of horse)* Nüster, *die*

nosy ['nəʊzɪ] *adj. (sl. derog.)* neugierig

Nosy Parker [nəʊzɪ'pɑ:kə(r)] *n.* Schnüffler, *der*/Schnüfflerin, *die (ugs. abwertend)*

not [nɒt] *adv.* **a)** nicht; **he is ~ a doctor** er ist kein Arzt; **isn't she pretty?** ist sie nicht hübsch?; **b)** *in ellipt. phrs.* nicht; **I hope ~:** hoffentlich nicht; **~ at all** überhaupt nicht; *(in polite reply to thanks)* keine Ursache; gern geschehen; **~ that [I know of]** nicht, daß [ich wüßte]; **c)** *in emphat. phrs.* **~ ... but ...:** nicht ..., sondern ...; **~ a moment** nicht ein *od.* kein einziger Augenblick; **~ a thing** gar nichts; **~ a few/everybody** nicht wenige/jeder; **~ once or** *or* **nor twice, but ...:** nicht nur ein- oder zweimal, sondern ...

notable ['nəʊtəbl] *adj.* bemerkenswert; bedeutend, angesehen ⟨*Person*⟩; **be ~ for sth.** für etw. bekannt sein

notably ['nəʊtəblɪ] *adv.* besonders

notary ['nəʊtərɪ] *n.* ~ ['**public**] Notar, *der*/Notarin, *die*

notation [nəʊ'teɪʃn] *n. (Math., Mus., Chem.)* Notation, *die (fachspr.);* Notierung, *die*

notch [nɒtʃ] **1.** *n.* Kerbe, *die; (in damaged blade)* Scharte, *die; (in belt)* Loch, *das.* **2.** *v.t.* kerben

~ '**up** *v.t.* erreichen; aufstellen ‹*Rekord*›; erringen ‹*Sieg*›

note [nəʊt] **1.** *n.* **a)** *(Mus.) (sign)* Note, *die; (key of piano)* Taste, *die; (single sound)* Ton, *der;* **strike the right** ~ ‹*Sprecher, Redner, Brief:*› den richtigen Ton treffen; **hit the wrong** ~: einen falschen Ton anschlagen; **b)** *(tone of expression)* [Unter]ton, *der;* ~ **of caution/anger** warnender/ärgerlicher [Unter]ton; **on a** ~ **of optimism, on an optimistic** ~: in optimistischem Ton; **his voice had a peevish** ~: seine Stimme klang gereizt; **a festive** ~, **a** ~ **of festivity** eine festliche Note; **c)** *(jotting)* Notiz, *die;* **take** *or* **make** ~**s** sich *(Dat.)* Notizen machen; **take** *or* **make a** ~ **of sth.** sich *(Dat.)* etw. notieren; **speak without** ~**s** frei sprechen; **d)** *(annotation, foot~)* Anmerkung, *die;* **e)** *(short letter)* [kurzer] Brief; **f)** *no pl., no art. (importance)* Bedeutung, *die;* **person/sth. of** ~: bedeutende Persönlichkeit/etw. Bedeutendes; **nothing of** ~: nichts von Bedeutung; **be of** ~: bedeutend sein; **g)** *no pl., no art. (attention)* Beachtung, *die;* **worthy of** ~: beachtenswert; **take** ~ **of sth.** *(heed)* einer Sache *(Dat.)* Beachtung schenken; *(notice)* etw. zur Kenntnis nehmen. **2.** *v.t.* **a)** *(pay attention to)* beachten; **b)** *(notice)* bemerken; **c)** *(set down)* ~ [**down**] [sich *(Dat.)*] notieren

note: ~**book** *n.* Notizbuch, *das; (for lecture* ~*s)* Kollegheft, *das;* ~**case** *n.* Brieftasche, *die*

noted ['nəʊtɪd] *adj.* bekannt, berühmt (**for** für, wegen)

note: ~**pad** *n.* Notizblock, *der;* ~**paper** *n.* Briefpapier, *das;* ~**worthy** *adj.* bemerkenswert

nothing ['nʌθɪŋ] **1.** *n.* **a)** nichts; ~ **interesting** nichts Interessantes; ~ **much** nichts Besonderes; ~ **more than** nur; ~ **more,** ~ **less** nicht mehr, nicht weniger; **I should like** ~ **more than sth./to do sth.** ich würde etw. nur zu gern haben/tun; **next to** ~: so gut wie nichts; **it's** ~ **less than suicidal to do this** es ist reiner *od.* glatter Selbstmord, dies zu tun; ~ **else than,** ~ [**else**] **but** nur; **there was** ~ [**else**] **for it but to do sth.** es blieb nichts anderes übrig, als etw. zu tun; **he is** ~ **if not active** wenn er eins ist, dann [ist er] aktiv; **there is** ~ **in it** *(in race etc.)* es ist noch nichts entschieden; *(it is untrue)* es ist nichts daran wahr; **there is** ~ '**to it** es ist kinderleicht *(fam.);* ~ **ventured** ~ **gained** *(prov.)* wer nicht wagt, der nicht gewinnt *(Spr.);* **£300 is** ~ **to him** 300 Pfund sind ein Klacks für ihn *(ugs.);* **have** [**got**] *or* **be** ~ **to do with sb./sth.** *(not concern)* nichts zu tun haben mit jmdm./etw.; **have** ~ **to do with sb./sth.** *(avoid)* jmdm./einer Sache aus dem Weg gehen; [**not**] **for** ~: [nicht] umsonst; **count** *or* **go for** ~ *(be unappreciated)* ‹*Person:*› nicht zählen; *(be profitless)* ‹*Arbeit, Bemühung:*› umsonst *od.* vergebens sein; **have** [**got**] ~ **on sb./sth.** *(be inferior to)* nicht mit jmdm./etw. zu vergleichen sein; **have** [**got**] ~ **on sb.** *(know* ~ *bad about)* nichts gegen jmdn. in der Hand haben; **have** ~ '**on** *(be naked)* nichts anhaben; *(have no engagements)* nichts vorhaben; **make** ~ **of sth.** *(make light of)* keine große Sache aus etw. machen; *(not understand)* mit etw. nichts anfangen [können]; **it means** ~ **to me** *(is not understood)* ich werde nicht klug daraus; *(is not loved)* es bedeutet mir nichts; **to say** ~ **of** ganz zu schweigen von; **b)** *(zero)* **multiply by** ~: mit null multiplizieren; **c)** *(trifling event)* Nichtigkeit, *die; (trifling person)* Nichts, *das;* Niemand, *der;* **soft** *or* **sweet** ~**s** Zärtlichkeiten *Pl.* **2.** *adv.* keineswegs; ~ **near so bad as ...**: nicht annähernd so schlecht wie ...

notice ['nəʊtɪs] **1.** *n.* **a)** Anschlag, *der;* Aushang, *der; (in newspaper)* Anzeige, *die;* **no-smoking** ~: Rauchverbotsschild, *das;* **b)** *(warning)* **give** [**sb.**] [**three days'**] ~ **of one's arrival** [jmdm.] seine Ankunft [drei Tage vorher] mitteilen; **have** [**no**] ~ [**of sth.**] [von etw.] [keine] Kenntnis haben; **at short/a moment's/ten minutes'** ~: kurzfristig/von einem Augenblick zum andern/innerhalb von zehn Minuten; **c)** *(formal notification)* Ankündigung, *die;* **until further** ~: bis auf weiteres; ~ **is given of sth.** etw. wird angekündigt; **d)** *(ending an agreement)* Kündigung, *die;* **give sb. a month's** ~: jmdm. mit einer Frist von einem Monat kündigen; **hand in one's** ~, **give** ~ *(Brit.),* **give one's** ~ *(Amer.)* kündigen; **e)** *(attention)* Beachtung, *die;* **bring sb./sth. to sb.'s** ~: jmdn. auf jmdn./etw. aufmerksam machen; **it has come to my** ~ **that ...**: ich habe bemerkt *od.* mir ist aufgefallen, daß ...; **take no** ~ **of sb./sth.** *(not observe)* jmdn./etw. nicht bemerken; *(disregard)* keine Notiz von jmdm./etw. nehmen; **take no** ~: sich nicht darum kümmern; **take** ~ **of** wahrnehmen; hören auf ‹*Rat*›; zur Kenntnis nehmen ‹*Leistung*›; **f)** *(review)* Besprechung, *die;* Rezension, *die.* **2.** *v.t.* **a)** *(perceive, take notice of)* bemerken; *abs.* **I pretended not to** ~: ich tat so, als ob ich es nicht bemerkte; **b)** *(remark upon)* erwähnen

noticeable ['nəʊtɪsəbl] *adj. (perceptible)* wahrnehmbar ‹*Fleck, Schaden, Geruch*›; merklich ‹*Verbesserung*›; spürbar ‹*Mangel*›

noticeably ['nəʊtɪsəblɪ] *adv.* sichtlich ‹*größer, kleiner*›; merklich ‹*verändern*›; spürbar ‹*kälter*›

'**notice-board** *n. (Brit.)* Anschlagtafel, *die*

notification [nəʊtɪfɪ'keɪʃn] *n.* Mitteilung, *die* (**of sb.** an jmdn.; **of sth.** über etw. *[Akk.]*)

notify ['nəʊtɪfaɪ] *v.t.* **a)** *(make known)* ankündigen; **b)** *(inform)* benachrichtigen (**of** über + *Akk.*)

notion ['nəʊʃn] *n.* **a)** Vorstellung, *die;* **not have the faintest/least** ~ **of how/what** *etc.* nicht die blasseste/geringste Ahnung haben, wie/was *usw.;* **he has no** ~ **of time** er hat kein Verhältnis zur Zeit; **b)** *(knack, inkling)* **have no** ~ **of sth.** keine Ahnung von etw. haben

notoriety [nəʊtə'raɪətɪ] *n., no pl.* traurige Berühmtheit

notorious [nə'tɔːrɪəs] *adj.* bekannt; *(infamous)* berüchtigt; notorisch ‹*Lügner*›; niederträchtig ‹*List*›; **be** *or* **have become** ~ **for sth.** wegen *od.* für etw. bekannt/berüchtigt sein

notoriously [nə'tɔːrɪəslɪ] *adv.* notorisch

notwithstanding [nɒtwɪθ'stændɪŋ, nɒtwɪð'stændɪŋ] **1.** *prep.* ungeachtet. **2.** *adv.* dennoch; dessenungeachtet. **3.** *conj.* ~ **that ...**: ungeachtet dessen, daß ...

nougat ['nuːgɑː] *n.* Nougat, *das od. der*

nought [nɔːt] *n.* Null, *die;* ~**s and crosses** *(Brit.) Spiel, bei dem innerhalb eines Feldes von Kästchen Dreierreihen von Kreisen bzw. Kreuzen zu erzielen sind*

noun [naʊn] *n. (Ling.)* Substantiv,

das; Hauptwort, *das;* Nomen, *das (fachspr.)*
nourish ['nʌrɪʃ] *v. t.* ernähren (**on** mit); *(fig.)* nähren *(geh.)*
nourishing ['nʌrɪʃɪŋ] *adj.* nahrhaft
nourishment ['nʌrɪʃmənt] *n. (food)* Nahrung, *die*
nouveau riche [nu:vəʊ 'ri:ʃ] **1.** *n., pl.* **nouveaux riches** [nu:vəʊ 'ri:ʃ] Neureiche, *der/die.* **2.** *adj.* neureich
Nov. *abbr.* **November** Nov.
novel ['nɒvl] **1.** *n.* Roman, *der.* **2.** *adj.* neuartig
novelist ['nɒvəlɪst] *n.* Romanautor, *der/*-autorin, *die*
novella [nə'velə] *n.* Novelle, *die*
novelty ['nɒvltɪ] *n.* **a) be a/no ~:** etwas/nichts Neues sein; **b)** *(newness)* Neuheit, *die;* Neuartigkeit, *die;* **c)** *(gadget)* Überraschung, *die*
November [nə'vembə(r)] *n.* November, *der; see also* **August**
novice ['nɒvɪs] *n.* **a)** *(Relig.)* Novize, *der/*Novizin, *die;* **b)** *(beginner)* Anfänger, *der/*Anfängerin, *die*
now [naʊ] **1.** *adv.* **a)** jetzt; *(nowadays)* heutzutage; *(immediately)* [jetzt] sofort; *(this time)* jetzt [schon wieder]; **just ~** *(very recently)* gerade eben; *(at this particular time)* gerade jetzt; **[every] ~ and then** *or* **again** hin und wieder; **[it's] ~ or never!** jetzt oder nie!; **b)** *(not referring to time)* **well ~:** also; **~, ~:** na, na; **~, what happened is this ...:** also, passiert ist folgendes: ...; **~ then** na *(ugs.);* **quickly ~!** nun aber schnell. **2.** *conj.* **~ [that] ...:** jetzt, wo *od.* da ... **3.** *n.* **~ is the time to do sth.** es ist jetzt an der Zeit, etw. zu tun; **before ~:** früher; **up to** *or* **until ~:** bis jetzt; **never before ~:** noch nie; **by ~:** inzwischen; **a week from ~:** [heute] in einer Woche; **between ~ and Friday** bis Freitag; **from ~ on** von jetzt an; **as of ~:** jetzt; **that's all for ~:** das ist im Augenblick alles; **bye** *etc.* **for ~!** *(coll.)* bis bald!
nowadays ['naʊədeɪz] *adv.* heutzutage
nowhere ['nəʊweə(r)] **1.** *adv.* **a)** *(in no place)* nirgends; nirgendwo; **b)** *(to no place)* nirgendwohin; **c) ~ near** *(not even nearly)* nicht annähernd. **2.** *pron.* **come from ~:** wie aus dem Nichts auftauchen; **get ~** *(make no progress)* nicht vorankommen; *(have no success)* nichts erreichen; **get sb. ~:** [jmdm.] nichts nützen
noxious ['nɒkʃəs] *adj.* giftig
nozzle ['nɒzl] *n.* Düse, *die*
nth [enθ] *see* **N, n**
nuance ['nju:ɑ̃s] *n.* Nuance, *die*
nubile ['nju:baɪl] *adj. (sexy)* sexy *(ugs.);* anziehend
nuclear ['nju:klɪə(r)] *adj.* **a)** Kern-; **b)** *(using ~ energy or weapons)* Atom-; Kern⟨*explosion, -technik*⟩; atomar ⟨*Antrieb, Gefechtskopf, Bedrohung, Gegenschlag, Wettrüsten*⟩; nuklear ⟨*Abschreckungspotential, Sprengkörper, Streitkräfte*⟩; atomgetrieben ⟨*Unterseeboot, Schiff*⟩
nuclear: ~ de'terrent *n.* atomare *od.* nukleare Abschreckung; **~ dis'armament** *n.* atomare *od.* nukleare Abrüstung; **~ 'energy** *n., no pl.* Atom- *od.* Kernenergie, *die;* **~ 'family** *n. (Sociol.)* Kernfamilie, *die;* **~ 'fission** *n.* Kernspaltung, *die;* **~-free** *adj.* atomwaffenfrei ⟨*Zone*⟩; **~ 'physics** *n.* Kernphysik, *die;* **~ 'power** *n.* **a)** Atom- *od.* Kernkraft, *die;* **b)** *(country)* Atom- *od.* Nuklearmacht, *die;* **~ 'power station** *n.* Atom- *od.* Kernkraftwerk, *das;* **~ 'warfare** *n., no pl.* Atomkrieg, *der;* **~ 'waste** *n.* Atommüll, *der*
nuclei *pl. of* **nucleus**
nucleus ['nju:klɪəs] *n., pl.* **nuclei** ['nju:klɪaɪ] Kern, *der*
nude [nju:d] **1.** *adj.* nackt; **~ figure** Akt, *der.* **2.** *n.* **a)** *(Art: figure)* Akt, *der;* **b) in the ~:** nackt
nudge [nʌdʒ] **1.** *v. t. (push gently)* anstoßen. **2.** *n.* Stoß, *der;* Puff, *der;* **give sb. a ~:** jmdn. anstoßen
nudist ['nju:dɪst] *n.* Nudist, *der/*Nudistin, *die;* FKK-Anhänger, *der/*-Anhängerin, *die*
nudity ['nju:dɪtɪ] *n.* Nacktheit, *die*
nugget ['nʌgɪt] *n. (Mining)* Klumpen, *der; (of gold)* Goldklumpen, *der;* Nugget, *das*
nuisance ['nju:səns] *n.* Ärgernis, *das;* Plage, *die;* **what a ~!** so etwas Dummes!; **make a ~ of oneself** lästig werden
null [nʌl] *adj. (Law)* **declare sth. ~ [and void]** etw. für null und nichtig erklären
nullify ['nʌlɪfaɪ] *v. t.* für null und nichtig *od.* rechtsungültig erklären ⟨*Vertrag, Testament*⟩
numb [nʌm] **1.** *adj. (without sensation)* gefühllos, taub (**with** vor + *Dat.*); *(fig.: without emotion)* benommen. **2.** *v. t.* ⟨*Kälte, Schock:*⟩ gefühllos machen; ⟨*Narkosemittel:*⟩ betäuben
number ['nʌmbə(r)] **1.** *n.* **a)** *(in series)* Nummer, *die;* **~ 3 West Street** West Street [Nr.] 3; **the ~ of sb.'s car** jmds. Autonummer; **you've got the wrong ~** *(Teleph.)* Sie sind falsch verbunden; **dial a wrong ~:** sich verwählen *(ugs.);* **~ one** *(oneself)* man selbst; *attrib.* Nummer eins *nachgestellt;* Spitzen⟨*position, -platz*⟩; **take care of** *or* **look after ~ one** an sich *(Akk.)* selbst denken; **N~ Ten [Downing Street]** *(Brit.) Amtssitz des britischen Premierministers/der britischen Premierministerin;* **sb.'s ~ is up** *(coll.)* jmds. Stunde hat geschlagen; **b)** *(esp. Math.: numeral)* Zahl, *die;* **c)** *(sum, total, quantity)* [An]zahl, *die;* **a ~ of people/things** einige Leute/Dinge; **a ~ of times/on a ~ of occasions** mehrfach *od.* -mals; **a small ~:** eine geringe [An]zahl; **large ~s** eine große [An]zahl; **in [large** *or* **great] ~s** in großer Zahl; **in a small ~ of cases** in einigen wenigen Fällen; **any ~:** beliebig viele; **on any ~ of occasions** oft[mals]; **in ~[s]** zahlenmäßig ⟨*überlegen sein, überwiegen*⟩; **d)** *(person, song, turn, edition)* Nummer, *die;* **e)** *(coll.: outfit)* Kluft, *die;* **f)** *(company)* **he was [one] of our ~:** er war einer von uns. **2.** *v. t.* **a)** *(assign ~ to)* beziffern; numerieren; **b)** *(amount to, comprise)* zählen; **the nominations ~ed ten in all** es wurden insgesamt zehn Kandidaten nominiert; **c)** *(include, regard as)* zählen, rechnen (**among, with** zu); **d) be ~ed** *(be limited)* begrenzt sein; **sb.'s days** *or* **years are ~ed** jmds. Tage sind gezählt
numberless ['nʌmbəlɪs] *adj.* unzählig; zahllos
'number-plate *n.* Nummernschild, *das*
numbness ['nʌmnɪs] *n., no pl. (caused by cold)* Gefühllosigkeit, *die;* Taubheit, *die; (caused by anaesthetic, sleeping-pill)* Betäubung, *die; (fig.: stupor)* Benommenheit, *die*
numeracy ['nju:mərəsɪ] *n.* rechnerische Fähigkeiten
numeral ['nju:mərl] *n.* Ziffer, *die; (word)* Zahlwort, *das*
numerate ['nju:mərət] *adj.* rechenkundig; **be ~:** rechnen können
numerator ['nju:məreɪtə(r)] *n. (Math.)* Zähler, *der*
numerical [nju:'merɪkl] *adj.* Zahlen⟨*wert, -folge*⟩; numerisch ⟨*Reihenfolge, Stärke*⟩; zahlenmäßig ⟨*Überlegenheit*⟩
numerous ['nju:mərəs] *adj.* zahlreich
nun [nʌn] *n.* Nonne, *die*
nunnery ['nʌnərɪ] *n.* [Nonnen]kloster, *das*
nurse [nɜ:s] **1.** *n.* Krankenschwe-

ster, *die;* **[male]** ~: Krankenpfleger, *der.* **2.** *v.t.* **a)** *(act as ~ to)* pflegen ‹*Kranke*›; ~ **sb. back to health** jmdn. gesundpflegen; **b)** *(suckle)* die Brust geben (+ *Dat.*), stillen ‹*Säugling*›; **c)** *(cradle)* vorsichtig halten; wiegen ‹*Baby*›; **d)** *(treat carefully)* ~ **gently/carefully** behutsam *od.* schonend umgehen mit. **3.** *v.i.* **a)** *(act as wet-~)* stillen; **b)** *(be a sick-~)* Krankenschwester/-pfleger sein

'nursemaid *n. (lit. or fig.)* Kindermädchen, *das*

nursery ['nɜ:sərɪ] *n.* **a)** *(room for children)* Kinderzimmer, *das;* **b)** *(crèche)* Kindertagesstätte, *die;* **c)** *see* **nursery school; d)** *(Agric.) (for plants)* Gärtnerei, *die; (for trees)* Baumschule, *die*

nursery: ~ **rhyme** *n.* Kinderreim, *der;* ~ **school** *n.* Kindergarten, *der;* **~-school teacher** *n. (female)* Kindergärtnerin, *die;* Erzieherin, *die; (male)* Erzieher, *der*

nursing ['nɜ:sɪŋ] *n., no pl., no art. (profession)* Krankenpflege, *die; attrib.* Pflege‹*personal, -beruf*›

nursing: ~ **home** *n. (Brit.) (for the aged, infirm)* Pflegeheim, *das; (for convalescents)* Genesungsheim, *das; (maternity hospital)* Entbindungsheim, *das;* ~ **'mother** *n.* stillende Mutter

nurture ['nɜ:tʃə(r)] *v.t.* **a)** *(rear)* aufziehen; **b)** *(fig.)* nähren *(geh.)*

nut [nʌt] *n.* **a)** Nuß, *die;* **be a hard** *or* **tough** ~ **[to crack]** ‹*Problem usw.*:› eine harte Nuß sein *(ugs.);* **b)** *(Mech. Engin.)* [Schrauben]mutter, *die;* **~s and bolts** *(fig.)* praktische Grundlagen; **c)** *(sl.: head)* Kürbis, *der (salopp);* **d)** *(crazy person)* Verrückte, *der/die (ugs.)*

nut: **~-case** *n. (sl.)* Verrückte, *der/die (ugs.);* **~crackers** *n. pl.* Nußknacker, *der*

nutmeg ['nʌtmeg] *n.* Muskatnuß, *die;* Muskat, *der*

nutrient ['nju:trɪənt] **1.** *adj.* **a)** *(serving as nourishment)* nahrhaft; **b)** *(providing nourishment)* Ernährungs-; Nähr‹*salze, -lösung*›. **2.** *n.* Nährstoff, *der*

nutrition [nju:'trɪʃn] *n. (nourishment, diet)* Ernährung, *die*

nutritious [nju:'trɪʃəs] *adj.* nahrhaft

nuts [nʌts] *pred. adj. (sl.)* verrückt *(ugs.)* **(about, on** nach)

'nutshell *n.* **a)** Nußschale, *die;* **b)** *(fig.)* **in a** ~: kurz gesagt

nutter ['nʌtə(r)] *n. (sl.)* Verrückte, *der/die (ugs.)*

nutty ['nʌtɪ] *adj.* **a)** *(in taste)* nussig; **b)** *(sl.: crazy)* verrückt *(ugs.)*

nuzzle ['nʌzl] *v.i. (nestle)* sich kuscheln **(up to, at, against** an + *Akk.*)

NW *abbr.* ['nɔ:θwest] **north-west** NW

nylon ['naɪlɒn] *n.* **a)** *no pl. (Textiles)* Nylon, *das; attrib.* Nylon-; **b)** *in pl. (stockings)* Nylonstrümpfe; Nylons *(ugs.)*

nymph [nɪmf] *n.* Nymphe, *die*

nymphomaniac [nɪmfə'meɪnɪæk] *n.* Nymphomanin, *die*

NZ *abbr.* **New Zealand**

O

O, o [əʊ] *n., pl.* **Os** *or* **O's a)** *(letter)* O, o, *das;* **b)** *(zero)* Null, *die*

oaf [əʊf] *n., pl.* **~s a)** *(stupid person)* Dummkopf, *der (ugs.);* **b)** *(awkward lout)* Stoffel, *der (ugs.)*

oak [əʊk] *n.* Eiche, *die; attrib.* Eichen‹*wald, -möbel, -kiste, -blatt*›

'oak-tree *n.* Eiche, *die*

OAP *abbr. (Brit.)* **old-age pensioner** Rentner, *der*/Rentnerin, *die;* ~ **club** Seniorenklub, *der*

oar [ɔ:(r)] *n.* Ruder, *das;* Riemen, *der (Sport, Seemannsspr.);* **put one's** ~ **in** *(fig. coll.)* seinen Senf dazugeben

oarsman ['ɔ:zmən] *n., pl.* **oarsmen** ['ɔ:zmən] Ruderer, *der*

oasis [əʊ'eɪsɪs] *n., pl.* **oases** [əʊ'eɪsi:z] *(lit. or fig.)* Oase, *die*

oast-house ['əʊsthaʊs] *n. (Agric., Brewing)* Hopfendarre, *die*

oat [əʊt] *n.* **~s** Hafer, *der;* **rolled ~s** Haferflocken *Pl.;* **sow one's wild ~s** *(fig.)* sich *(Dat.)* die Hörner abstoßen *(ugs.)*

oath [əʊθ] *n., pl.* **~s** [əʊðz] **a)** Eid, *der;* Schwur, *der;* **take** *or* **swear an** ~ **[on sth.] that ...:** einen Eid [auf etw. *(Akk.)*] schwören, daß ...; **b)** *(Law)* **swear** *or* **take the** ~: vereidigt werden; **on** *or* **under** ~: unter Eid; **put sb. on** *or* **under** ~: jmdn. vereidigen *od.* unter Eid nehmen; **c)** *(expletive)* Fluch, *der*

'oatmeal *n.* Hafermehl, *das*

obdurate [ɒb'djʊərət] *adj. (hardened)* unerbittlich ‹*Brutalität*›; verstockt ‹*Herz, Sünder*›; *(stubborn)* verstockt; hartnäckig ‹*Weigerung, Ablehnung*›

obedience [ə'bi:dɪəns] *n.* Gehorsam, *der;* **show** ~: gehorsam sein

obedient [ə'bi:dɪənt] *adj.* gehorsam; *(submissive)* fügsam; **be** ~ **to sb./sth.** jmdm./einer Sache gehorchen

obelisk ['ɒbəlɪsk] *n.* Obelisk, *der*

obese [əʊ'bi:s] *adj.* fett *(abwertend);* fettleibig *(bes. Med.)*

obesity [əʊ'bi:sɪtɪ] *n., no pl.* Fettheit, *die (abwertend);* Fettleibigkeit, *die (bes. Med.)*

obey [əʊ'beɪ] **1.** *v.t.* gehorchen (+ *Dat.*); ‹*Kind, Hund:*› folgen (+ *Dat.*), gehorchen (+ *Dat.*); sich halten an (+ *Akk.*) ‹*Vorschrift, Regel*›; befolgen ‹*Befehl*›. **2.** *v.i.* gehorchen

obituary [ə'bɪtjʊərɪ] **1.** *n.* Nachruf, *der* **(to, of** auf + *Akk.*); *(notice of death)* Todesanzeige, *die.* **2.** *adj.* ~ **notice/memoir** Todesanzeige, *die*/Nachruf, *der;* **the** ~ **page/column** die Todesanzeigen

object 1. ['ɒbdʒɪkt] *n.* **a)** *(thing)* Gegenstand, *der; (Philos.)* Objekt, *das;* **b)** *(purpose)* Ziel, *das;* **with this** ~ **in mind** *or* **view** mit diesem Ziel [vor Augen]; **with the** ~ **of doing sth.** in der Absicht, etw. zu tun; **c)** *(obstacle)* **money/time** *etc.* **is no** ~: Geld/Zeit *usw.* spielt keine Rolle; **d)** *(Ling.)* Objekt, *das.* **2.** [əb'dʒekt] *v.i.* **a)** *(state objection)* Einwände/einen Einwand erheben **(to** gegen); *(protest)* protestieren **(to** gegen); **b)** *(have objection or dislike)* etwas dagegen haben; ~ **to sb./sth.** etwas gegen jmdn./etw. haben; **if you don't** ~: wenn Sie nichts dagegen haben; ~ **to sb.'s doing sth.** etw. dagegen haben, daß jmd. etw. tut; **I strongly** ~ **to this tone** ich verbitte mir diesen Ton. **3.** *v.t.* [əb'dʒekt] einwenden

objection [əb'dʒekʃn] *n.* **a)** Einwand, *der;* Einspruch, *der (Amtsspr., Rechtsw.);* **raise** *or* **make an** ~ **[to sth.]** einen Einwand *od. (Rechtsw.)* Einspruch [gegen etw.] erheben; **make no** ~ **to sth.** nichts gegen etw. einzuwenden haben; **b)** *(feeling of opposition or dislike)* Abneigung, *die;* **have an/no** ~ **to sb./sth.** etw./nichts gegen jmdn./etw. haben; **have an/no** ~: etwas/nichts dagegen haben

objectionable [əb'dʒekʃənəbl] *adj.* unangenehm ‹*Anblick, Geruch*›; anstößig ‹*Wort, Benehmen*›; unausstehlich ‹*Kind*›

objective [əb'dʒektɪv] **1.** *adj. (unbiased)* objektiv. **2.** *n. (goal)* Ziel, *das*

objectively [əb'dʒektɪvlɪ] *adv.* objektiv

objectivity [ɒbdʒɪk'tɪvɪtɪ] *n., no pl.* Objektivität, *die*

'**object-lesson** *n.* Musterbeispiel, *das* (**in, on** für)

objector [əb'dʒektə(r)] *n.* Gegner, *der*/Gegnerin, *die* (**to** *Gen.*)

obligation [ɒblɪ'geɪʃn] *n.* Verpflichtung, *die; (constraint)* Zwang, *der;* **be under** *or* **have an/no ~ to do sth.** verpflichtet/nicht verpflichtet sein, etw. zu tun; **there's no ~ to buy** es besteht kein Kaufzwang

obligatory [ə'blɪgətərɪ] *adj.* obligatorisch; **make sth. ~ for sb.** etw. für jmdn. vorschreiben; **it has become ~ to do sth.** es ist zur Pflicht geworden, etw. zu tun

oblige [ə'blaɪdʒ] **1.** *v. t.* **a)** *(be binding on)* **~ sb. to do sth.** jmdm. vorschreiben, etw. zu tun; **one is ~d by law to do sth.** etw. ist gesetzlich vorgeschrieben; **b)** *(constrain, compel)* zwingen; nötigen; **you are not ~d to answer these questions** Sie sind nicht verpflichtet, diese Fragen zu beantworten; **feel ~d to do sth.** sich verpflichtet fühlen, etw. zu tun; **c)** *(be kind to)* **~ sb. by doing sth.** jmdm. den Gefallen tun und etw. tun; **~ sb. with sth.** *(help out)* jmdm. mit etw. aushelfen; **could you ~ me with a lift?** könnten Sie mich freundlicherweise mitnehmen?; **d) ~d** *(bound by gratitude)* **be much/greatly ~d to sb. [for sth.]** jmdm. [für etw.] sehr verbunden sein; **much ~d** besten Dank! **2.** *v. i.* **be always ready to ~:** immer sehr gefällig sein; **anything to ~** *(as answer)* stets zu Diensten

obliging [ə'blaɪdʒɪŋ] *adj.* entgegenkommend

oblique [ə'bli:k] **1.** *adj.* **a)** *(slanting)* schief ⟨*Gerade, Winkel*⟩; **b)** *(fig.: indirect)* indirekt ⟨*Bemerkung, Hinweis, Frage*⟩. **2.** *n.* Schrägstrich, *der*

obliquely [ə'bli:klɪ] *adv.* **a)** *(in a slanting direction)* schräg; **b)** *(fig.: indirectly)* indirekt ⟨*sich beziehen, antworten*⟩

obliterate [ə'blɪtəreɪt] *v. t.* **a)** auslöschen; **b)** *(fig.)* verschleiern ⟨*Wahrheit*⟩; auslöschen ⟨*Erinnerung*⟩; zerstreuen ⟨*Bedenken*⟩

oblivion [ə'blɪvɪən] *n., no pl. (being forgotten)* Vergessenheit, *die;* **sink** *or* **fall into ~:** in Vergessenheit geraten

oblivious [ə'blɪvɪəs] *adj.* **be ~ to** *or* **of sth.** *(be unconscious of)* sich *(Dat.)* einer Sache *(Gen.)* nicht bewußt sein; *(not notice)* etw. nicht bemerken *od.* wahrnehmen

oblong ['ɒblɒŋ] **1.** *adj.* rechteckig. **2.** *n.* Rechteck, *das*

obnoxious [əb'nɒkʃəs] *adj.* widerlich *(abwertend)*

oboe ['əʊbəʊ] *n. (Mus.)* Oboe, *die*

obscene [əb'si:n] *adj.* obszön; *(coll.: offensive)* widerlich *(abwertend);* unanständig ⟨*Profit*⟩

obscenity [əb'senɪtɪ] *n.* Obszönität, *die*

obscure [əb'skjʊə(r)] **1.** *adj.* **a)** *(unexplained)* dunkel; **for some ~ reason** aus irgendeinem verborgenen Grund; **b)** *(hard to understand)* schwer verständlich ⟨*Argument, Dichtung, Autor, Stil*⟩; unklar ⟨*Hinweis, Textstelle*⟩; **c)** *(unknown)* unbekannt ⟨*Herkunft, Schriftsteller*⟩. **2.** *v. t.* **a)** *(make indistinct)* verdunkeln; *(block)* versperren ⟨*Aussicht*⟩; *(conceal)* ⟨*Nebel:*⟩ verhüllen; **b)** *(fig.)* unverständlich machen

obscurity [əb'skjʊərɪtɪ] *n.* **a)** *no pl. (being unknown or inconspicuous)* Unbekanntheit, *die;* **sink into ~:** in Vergessenheit geraten; **in ~:** unbeachtet, unauffällig ⟨*leben*⟩; **b)** *(unintelligibleness, unintelligible thing)* Unverständlichkeit, *die;* **c)** *no pl. (darkness)* Dunkelheit, *die*

obsequious [əb'si:kwɪəs] *adj.* unterwürfig *(abwertend)*

observance [əb'zɜ:vəns] *n.* **a)** *no pl. (observing)* Beachtung, *die;* **b)** *(Relig.)* Regel, *die*

observant [əb'zɜ:vənt] *adj.* aufmerksam; **how very ~ of you!** sehr scharf beobachtet!

observation [ɒbzə'veɪʃn] *n.* **a)** *no pl.* Beobachtung, *die;* **powers of ~:** Beobachtungsgabe, *die;* **be [kept] under ~:** beobachtet werden; *(by police, detectives)* observiert *od.* überwacht werden; **b)** *(remark)* Bemerkung, *die* (**on** über + *Akk.*); **make an ~ on sth.** sich zu etw. äußern

observatory [əb'zɜ:vətərɪ] *n.* Observatorium, *das; (Astron. also)* Sternwarte, *die*

observe [əb'zɜ:v] *v. t.* **a)** *(watch)* beobachten; ⟨*Polizei, Detektiv:*⟩ observieren, überwachen; *abs.* aufpassen; *(perceive)* bemerken; **b)** *(abide by, keep)* beachten; einlegen ⟨*Schweigeminute*⟩; halten ⟨*Gelübde*⟩; feiern ⟨*Weihnachten, Jahrestag usw.*⟩; **c)** *(say)* bemerken

observer [əb'zɜ:və(r)] *n.* Beobachter, *der*/Beobachterin, *die*

obsess [əb'ses] *v. t.* **be/become ~ed with** *or* **by sb./sth.** von jmdm./etw. besessen sein/werden

obsession [əb'seʃn] *n.* **a)** *(persistent idea)* Zwangsvorstellung, *die;* **be/become an ~ with sb.** für jmdn. zur Sucht geworden sein/werden; **have an ~ with sb.** von jmdm. besessen sein; **b)** *no pl. (Psych.: condition)* Obsession, *die (fachspr.);* Besessenheit, *die*

obsessive [əb'sesɪv] *adj.* zwanghaft; obsessiv *(Psych.);* **be ~ about sth.** von etw. besessen sein; **be an ~ eater** unter Eßzwang leiden

obsolescence [ɒbsə'lesəns] *n., no pl.* Veralten, *das;* **built-in** *or* **planned ~:** geplanter Verschleiß

obsolescent [ɒbsə'lesənt] *adj.* veraltend

obsolete ['ɒbsəli:t] *adj.* veraltet; **become/have become ~:** veralten/veraltet sein

obstacle ['ɒbstəkl] *n.* Hindernis, *das* (**to** für); **put ~s in sb.'s path** *(fig.)* jmdm. Hindernisse *od.* Steine in den Weg legen

'**obstacle-race** *n.* Hindernisrennen, *das*

obstetrics [ɒb'stetrɪks] *n., no pl. (Med.)* Obstetrik, *die (fachspr.);* Geburtshilfe, *die*

obstinacy ['ɒbstɪnəsɪ] *n., no pl. see* **obstinate:** Starrsinn, *der;* Hartnäckigkeit, *die*

obstinate ['ɒbstɪnət] *adj.* starrsinnig; *(adhering to particular course of action)* hartnäckig; **be as ~ as a mule** ein sturer Bock sein *(ugs. abwertend)*

obstruct [əb'strʌkt] *v. t.* **a)** *(block)* versperren; blockieren; *(Med.)* verstopfen; behindern ⟨*Verkehr*⟩; **~ sb.'s view** jmdm. die Sicht versperren; **b)** *(fig.: impede; also Sport)* behindern

obstruction [əb'strʌkʃn] *n.* **a)** *no pl. (blocking)* Blockierung, *die; (Med.)* Verstopfung, *die; (of progress; also Sport)* Behinderung, *die;* **b)** *(obstacle)* Hindernis, *das*

obstructive [əb'strʌktɪv] *adj.* hinderlich; obstruktiv ⟨*Politik, Taktik*⟩; **be ~** ⟨*Person:*⟩ sich querlegen *(ugs.)*

obtain [əb'teɪn] *v. t.* bekommen ⟨*Ware, Information, Hilfe*⟩; erreichen, erzielen ⟨*Resultat, Wirkung*⟩; erwerben, erlangen ⟨*akademischen Grad*⟩

obtainable [əb'teɪnəbl] *adj.* erhältlich

obtrusive [əb'tru:sɪv] *adj.* aufdringlich; *(conspicuous)* auffällig

obtuse [əb'tju:s] *adj.* **a)** *(Geom.)*

stumpf ‹*Winkel*›; **b)** *(stupid)* einfältig; **he's being deliberately ~:** er stellt sich dumm

obvious ['ɒbvɪəs] *adj.* offenkundig; *(easily seen)* augenfällig; sichtlich ‹*Empfindung, innerer Zustand*›; plump ‹*Trick, Mittel*›; **she was the ~ choice** es lag nahe, daß die Wahl auf sie fiel; **the answer is ~:** die Antwort liegt auf der Hand; **the ~ thing to do is ...:** das Naheliegende ist ...; **with the ~ exception of ...:** natürlich mit Ausnahme von ...; **be ~ [to sb.] that ...:** [jmdm.] klar sein, daß ...; **that's stating the ~:** das ist nichts Neues

obviously ['ɒbvɪəslɪ] *adv.* offenkundig; sichtlich ‹*enttäuschen, überraschen usw.*›; **~, we can't expect any help** es ist klar, daß wir keine Hilfe erwarten können

occasion [ə'keɪʒn] **1.** *n.* **a)** *(opportunity)* Gelegenheit, *die;* **rise to the ~:** sich der Situation gewachsen zeigen; **b)** *(reason)* Grund, *der* **(for** zu); *(cause)* Anlaß, *der;* **should the ~ arise** falls sich die Gelegenheit ergibt; **be [an] ~ for celebration** ein Grund zum Feiern sein; **have ~ to do sth.** [eine] Gelegenheit haben, etw. zu tun; **c)** *(point in time)* Gelegenheit, *die;* **on several ~s** bei mehreren Gelegenheiten; **on that ~:** bei der Gelegenheit; damals; **on ~[s]** gelegentlich; **d)** *(special occurrence)* Anlaß, *der;* **it was quite an ~:** es war ein Ereignis; **on the ~ of** anläßlich (+ *Gen.*). **2.** *v.t.* verursachen; erregen, Anlaß geben zu ‹*Besorgnis*›

occasional [ə'keɪʒənl] *adj. (happening irregularly)* gelegentlich; vereinzelt ‹*Regenschauer*›; **take an** *or* **the ~ break** gelegentlich eine Pause machen

occasionally [ə'keɪʒənəlɪ] *adv.* gelegentlich; **[only] very ~:** gelegentlich einmal

oc'casional table *n.* Beistelltisch, *der*

occult [ɒ'kʌlt, 'ɒkʌlt] *adj. (mystical)* okkult ‹*Kunst, Wissenschaft*›; **the ~:** das Okkulte

occupant ['ɒkjʊpənt] *n.* Bewohner, *der*/Bewohnerin, *die; (of post)* Inhaber, *der*/Inhaberin, *die; (of car, bus, etc.)* Insasse, *der*/Insassin, *die*

occupation [ɒkjʊ'peɪʃn] *n.* **a)** *(of property) (tenure)* Besitz, *der; (occupancy)* Bewohnung, *die;* **b)** *(Mil.)* Okkupation, *die;* Besetzung, *die; (period)* Besatzungszeit, *die;* **c)** *(activity)* Beschäftigung, *die; (pastime)* Zeitvertreib, *der;* **d)** *(profession)* Beruf, *der;* **his ~ is civil engineering** er ist Bauingenieur [von Beruf]; **what's her ~?** was ist sie von Beruf?

occupational [ɒkjʊ'peɪʃənl] *adj.* Berufs‹*beratung, -risiko*›; betrieblich ‹*Altersversorgung*›

occupational 'therapy *n.* Beschäftigungstherapie, *die*

occupier ['ɒkjʊpaɪə(r)] *n. (Brit.)* Besitzer, *der*/Besitzerin, *die; (tenant)* Bewohner, *der*/Bewohnerin, *die*

occupy ['ɒkjʊpaɪ] *v.t.* **a)** *(Mil.; Polit. as demonstration)* besetzen; **b)** *(reside in, be a tenant of)* bewohnen; **c)** *(take up, fill)* einnehmen; besetzen ‹*Sitzplatz, Tisch*›; belegen ‹*Zimmer*›; in Anspruch nehmen ‹*Zeit, Aufmerksamkeit*›; **how did you ~ your time?** wie hast du die Zeit verbracht?; **d)** *(hold)* innehaben ‹*Stellung, Amt*›; **e)** *(busy, employ)* beschäftigen; **~ oneself [with doing sth.]** sich [mit etw.] beschäftigen; **keep sb.['s mind] occupied** jmdn. [geistig] beschäftigen

occur [ə'kɜː(r)] *v.i.*, **-rr-: a)** *(be met with)* vorkommen; ‹*Gelegenheit, Schwierigkeit, Problem:*› sich ergeben; **b)** *(happen)* ‹*Veränderung:*› eintreten; ‹*Unfall, Vorfall, Zwischenfall:*› sich ereignen; **this must not ~ again** das darf nicht wieder vorkommen; **c) ~ to sb.** *(be thought of)* jmdm. einfallen; ‹*Idee:*› jmdm. kommen; **it never ~red to me** auf den Gedanken bin ich nie gekommen

occurrence [ə'kʌrəns] *n.* **a)** *(incident)* Ereignis, *das;* **b)** *(occurring)* Vorkommen, *das;* **be of frequent ~:** häufig vorkommen

ocean ['əʊʃn] *n.* Ozean, *der;* Meer, *das*

'ocean-going *adj.* Übersee-

oceanic [əʊʃɪ'ænɪk, əʊsɪ'ænɪk] *adj.* ozeanisch; Meeres‹*tier, -klima*›; See‹*vogel, -klima*›

oceanography [əʊʃə'nɒgrəfɪ] *n.* Ozeanographie, *die;* Meereskunde, *die*

ochre (*Amer.:* **ocher**) ['əʊkə(r)] *n.* Ocker, *der od. das*

o'clock [ə'klɒk] *adv.* **it is two/six ~:** es ist zwei/sechs Uhr; **at two/six ~:** um zwei/sechs Uhr; **six ~** *attrib.* Sechs-Uhr-‹*Zug, Maschine, Nachrichten*›

Oct. *abbr.* **October** Okt.

octagon ['ɒktəgən] *n. (Geom.)* Achteck, *das;* Oktogon, *das (fachspr.)*

octane ['ɒkteɪn] *n.* Oktan, *das*

octave ['ɒktɪv] *n. (Mus.)* Oktave, *die*

October [ɒk'təʊbə(r)] *n.* Oktober, *der; see also* **August**

octopus ['ɒktəpəs] *n.* Krake, *der*

odd [ɒd] *adj.* **a)** *(extraordinary)* merkwürdig; *(strange, eccentric)* seltsam; **b)** *(surplus, spare)* übrig ‹*Stück*›; überzählig ‹*Spieler*›; restlich, übrig ‹*Silbergeld*›; **c)** *(additional)* **1,000 and ~ pounds** etwas über 1 000 Pfund; **d)** *(occasional, random)* gelegentlich; **~ job/~-job man** Gelegenheitsarbeit, *die*/-arbeiter, *der;* **e)** *(one of pair or group)* einzeln; **~ socks/gloves** *etc.* nicht zusammengehörende Socken/Handschuhe *usw.;* **be the ~ man out** *(extra person)* überzählig sein; *(thing)* nicht dazu passen; **f)** *(uneven)* ungerade ‹*Zahl, Seite, Hausnummer*›; **g)** *(plus something)* **she must be forty ~:** sie muß etwas über vierzig sein; **twelve pounds ~:** etwas mehr als zwölf Pfund

oddity ['ɒdɪtɪ] *n.* **a)** *(strangeness, peculiar trait)* Eigentümlichkeit, *die;* **b)** *(person)* Sonderling, *der;* **c)** *(object, event)* Kuriosität, *die*

oddly ['ɒdlɪ] *adv.* seltsam; merkwürdig; **~ enough** seltsamer- *od.* merkwürdigerweise

oddment ['ɒdmənt] *n.* **a)** *(left over)* [Über]rest, *der; (in sales)* Reststück, *das;* **b)** *in pl. (odds and ends)* Kleinigkeiten

oddness ['ɒdnɪs] *n., no pl.* Merkwürdigkeit, *die; (strangeness)* Seltsamkeit, *die*

'odd-numbered *adj.* ungerade

odds [ɒdz] *n. pl.* **a)** *(Betting)* Odds *pl.;* **the ~ were on Black Bess** Black Bess hatte die besten Chancen; **lay** *or* **give/take ~ of six to one in favour of/against sb./a horse** eine 6 : 1-Wette auf/gegen jmdn./ein Pferd anbieten/annehmen; **pay over the ~ for sth.** einen überhöhten Preis für etw. bezahlen; **b)** *(chances for or against)* Möglichkeit, *die; (chance for)* Aussicht, *die;* Chance, *die;* **[the] ~ are that she did it** wahrscheinlich hat sie es getan; **the ~ are against/in favour of sb./sth.** jmds. Aussichten *od.* Chancen/die Aussichten *od.* Chancen für etw. sind gering/gut; **struggle against impossible ~:** völlig chancenlos kämpfen; **c)** *(balance of advantage)* **against [all] the ~:** allen Widrigkeiten zum Trotz; **d)** *(difference)* Unterschied, *der;* **it makes no/little ~ [whether ...]** es ist völlig/ziemlich gleichgültig[, ob ...]; **what's the ~?** *(coll.)* was macht das schon?; **e)** *(variance)* **be at ~ with sb. over sth.** mit jmdm. in etw. *(Dat.)* uneinig sein; **f)** ~ **and ends** Kleinigkeiten; *(of food)* Reste

'odds-on *adj.* gut ‹*Chance, Aussicht*›; hoch, klar ‹*Favorit*›; **be ~ [favourite] to win/for sth.** klarer *od.* hoher Favorit/Favorit für etw. sein

ode [əʊd] *n.* Ode, *die* (**to** an + *Akk.*)

odious ['əʊdɪəs] *adj.* widerwärtig

odor *etc. (Amer.) see* **odour** *etc.*

odour ['əʊdə(r)] *n.* **a)** *(smell)* Geruch, *der; (fragrance)* Duft, *der;* **b)** *(fig.)* Note, *die;* **be in good/bad ~ with sb.** bei jmdm. in gutem/schlechtem Geruch stehen

odourless ['əʊdəlɪs] *adj.* geruchlos

of [əv, *stressed* ɒv] *prep.* **a)** *indicating belonging, connection, possession* **articles of clothing** Kleidungsstücke; **the brother of her father** der Bruder ihres Vaters; **a friend of mine/the vicar's** ein Freund von mir/des Pfarrers; **that dog of yours** Ihr Hund da; **it's no business of theirs** es geht sie nichts an; **where's that pencil of mine?** wo ist mein Bleistift?; **b)** *indicating starting-point* von; **within a mile of the centre** nicht weiter als eine Meile vom Zentrum entfernt; **c)** *indicating origin, cause* **it was clever of you to do that** es war klug von dir, das zu tun; **the approval of sb.** jmds. Zustimmung; **the works of Shakespeare** Shakespeares Werke; **d)** *indicating material* aus; **be made of ...:** aus ... [hergestellt] sein; **e)** *indicating closer definition, identity, or contents* **a pound of apples** ein Pfund Äpfel; **a glass of wine** ein Glas Wein; **a painting of the queen** ein Gemälde der Königin; **the city of Chicago** die Stadt Chicago; **increase of 10%** Zuwachs/Erhöhung von zehn Prozent; **battle of Hastings** Schlacht von *od.* bei Hastings; **your letter of 2 January** Ihr Brief vom 2. Januar; **be of value/interest to** von Nutzen/von Interesse *od.* interessant sein für; **the whole of ...:** der/die/das ganze ...; **f)** *indicating concern, reference* **do not speak of such things** sprich nicht von solchen Dingen; **inform sb. of sth.** jmdn. über etw. *(Akk.)* informieren; **well, what of it?** *(asked as reply)* na und?; **g)** *indicating objective relation* **his love of his father** seine Liebe zu seinem Vater; **h)** *indicating description, quality, condition* **a frown of disapproval** ein mißbilligendes Stirnrunzeln; **work of authority** maßgebendes Werk; **a boy of 14 years** ein vierzehnjähriger Junge; **i)** *indicating classification, selection* von; **the five of us** wir fünf; **the five of us went there** wir sind zu fünft hingegangen; **he of all men** *(most unsuitably)* ausgerechnet er; *(especially)* gerade er; **here of all places** ausgerechnet hier; **of an evening** *(coll.)* abends

off [ɒf] **1.** *adv.* **a)** *(away, at or to a distance)* **be a few miles ~:** wenige Meilen entfernt sein; **the lake is not far ~:** der See ist nicht weit [weg *od.* entfernt]; **Christmas is not far ~:** es ist nicht mehr lang bis Weihnachten; **some way ~:** in einiger Entfernung; **where are you ~ to?** wohin gehst du?; **I must be ~:** ich muß fort *od.* weg *od.* los; **I'm ~ now** ich gehe jetzt; **~ we go!** *(we are starting)* los *od.* ab geht's!; *(let us start)* gehen/fahren wir!; **get the lid ~:** den Deckel abbekommen; **b)** *(not in good condition)* mitgenommen; **the meat** *etc.* **is ~:** das Fleisch *usw.* ist schlecht [geworden]; **be a bit ~** *(Brit. fig.)* ein starkes Stück sein *(ugs.)*; **c) be ~** *(switched or turned ~)* ‹*Wasser, Gas, Strom:*› abgestellt sein; **the light/radio** *etc.* **is ~:** das Licht/Radio *usw.* ist aus; **put the light ~:** das Licht ausmachen; **is the gas tap ~?** ist der Gashahn zu?; **d) be ~** *(cancelled)* abgesagt sein; ‹*Verlobung:*› [auf]gelöst sein; **is Sunday's picnic ~?** fällt das Picknick am Sonntag aus?; **~ and on** immer mal wieder *(ugs.)*; **e)** *(not at work)* frei; **on my day ~:** an meinem freien Tag; **take/get/have a week** *etc.* **~:** eine Woche *usw.* Urlaub nehmen/bekommen/haben; **be ~ sick** wegen Krankheit fehlen; **f)** *(no longer available)* **soup** *etc.* **is ~:** es gibt keine Suppe *usw.* mehr; **g)** *(situated as regards money etc.)* **he is badly** *etc.* **~:** er ist schlecht *usw.* gestellt; **we'd be better ~ without him** ohne ihn wären wir besser dran; **there are many people worse ~ than you** vielen geht es schlechter als dir; **how are you ~ for food?** wieviel Eßbares hast du noch?; **be badly ~ for sth.** mit etw. knapp sein. **2.** *prep.* **a)** *(from)* von; **cut a couple of slices ~ the loaf** einige Scheiben Brot abschneiden; **b) be ~ school/work** in der Schule/am Arbeitsplatz fehlen; **c)** *(diverging from)* **get ~ the subject** [vom Thema] abschweifen; **be ~ the point** nicht zur Sache gehören; **d)** *(designed not to cover)* **~-the-shoulder** schulterfrei ‹*Kleid*›; **e)** *(having lost interest in)* **be ~ sth.** etw. leid sein *od.* haben *(ugs.)*; **be ~ one's food** keinen Appetit haben; **f)** *(leading from, not far from)* **just ~ the square** ganz in der Nähe des Platzes; **a street ~ the main road** eine Straße, die von der Hauptstraße abgeht; **g)** *(to seaward of)* vor (+ *Dat.*). **3.** *adj.* **the ~ side** *(Brit.) (when travelling on the left/right)* die rechte/linke Seite

offal ['ɒfl] *n., no pl.* Innereien *Pl.*

off: ~-beat *adj.* **a)** *(Mus.)* Off-Beat-; **b)** *(fig.: eccentric)* unkonventionell ‹*Person, Lebensweise*›; außergewöhnlich ‹*Vorlesung, Kursus*›; **~-'centre 1.** *adj.* nicht zentriert; **2.** *adv.* nicht [genau] in der Mitte; **~ chance** *see* **chance 1c**; **~ 'colour** *adj.* unwohl; **be** *or* **feel ~ colour** sich unwohl *od.* schlecht fühlen; **~-day** *n.* schlechter Tag; **~-duty** *attrib. adj.* Freizeit-; dienstfrei ‹*Zeit*›; ‹*Polizist usw.,*› der dienstfrei hat

offence [ə'fens] *n. (Brit.)* **a)** *(hurting of sb.'s feelings)* Kränkung, *die;* **I meant no ~:** ich wollte Sie/ihn *usw.* nicht kränken; **give ~:** Mißfallen erregen; **take ~:** beleidigt *od.* verärgert sein; **no ~** *(coll.)* nichts für ungut; **b)** *(transgression)* Verstoß, *der; (crime)* Delikt, *das;* Straftat, *die;* **criminal/petty ~:** strafbare Handlung/geringfügiges Vergehen

offend [ə'fend] **1.** *v.i.* verstoßen (**against** gegen). **2.** *v.t.* **~ sb.** bei jmdm. Anstoß erregen; *(hurt feelings of)* jmdn. kränken; **~ the eye** das Auge beleidigen

offender [ə'fendə(r)] *n. (against law)* Straffällige, *der/die;* Täter, *der*/Täterin, *die; (against rule)* Zuwiderhandelnde, *der/die*

offense *(Amer.) see* **offence**

offensive [ə'fensɪv] **1.** *adj.* **a)** *(aggressive)* offensiv; Angriffs- ‹*waffe, -krieg*›; **b)** *(giving offence, insulting)* ungehörig; *(indecent)* anstößig; **~ language** Beschimpfungen; **c)** *(repulsive)* widerlich; **be ~ to sb.** jmdm. zuwider sein; auf jmdn. abstoßend wirken. **2.** *n. (attack; also Sport)* Offensive, *die;* Angriff, *der;* **take the** *or* **go on the ~:** in die *od.* zur Offensive übergehen; **be on the ~:** aggressiv sein

offer ['ɒfə(r)] **1.** *v.t.* anbieten; vorbringen ‹*Entschuldigung*›; bieten ‹*Chance*›; aussprechen ‹*Beileid*›; sagen ‹*Meinung*›; unterbreiten, machen ‹*Vorschläge*›;

have something to ~: etwas zu bieten haben; **the job ~s good prospects** der Arbeitsplatz hat Zukunft; **~ resistance** Widerstand leisten; **~ to do sth.** anbieten, etw. zu tun; **~ to help** seine Hilfe anbieten. **2.** *n.* **a)** Angebot, *das;* **[have/be] on ~**: im Angebot [haben/sein]; **b)** *(marriage proposal)* Antrag, *der*

offering ['ɒfərɪŋ] *n.* Angebot, *das; (to a deity)* Opfer, *das*

offertory ['ɒfətərɪ] *n. (Eccl.)* Kollekte, *die*

off'hand 1. *adv.* **a)** *(without preparation)* auf Anhieb, aus der Hand *(ugs.)* ‹*sagen, wissen*›; **b)** *(casually)* leichthin. **2.** *adj. (casual)* beiläufig; **be ~ with sb.** zu jmdm. kurz angebunden sein

office ['ɒfɪs] *n.* **a)** Büro, *das;* **b)** *(branch of organization)* Zweigstelle, *die;* **c)** *(position with duties)* Amt, *das;* **be in/out of ~**: im/nicht mehr im Amt sein; ‹*Partei:*› an der/nicht mehr an der Regierung sein; **hold ~**: amtieren; **d)** *(government department)* **Home O~** *(Brit.)* ≈ Innenministerium, *das;* **e)** *(Eccl.: service)* Gottesdienst, *der;* **f)** *(kindness)* **[good] ~s** Hilfe, *die;* Unterstützung, *die*

office: ~-block *n.* Bürogebäude, *das;* **~ hours** *n. pl.* Dienststunden *Pl.;* **after ~ hours** nach Dienstschluß; **~ job** *n.* Bürotätigkeit, *die*

officer ['ɒfɪsə(r)] *n.* **a)** *(Army etc.)* Offizier, *der;* **b)** *(official)* Beamte, *der*/Beamtin, *die; (of club etc.)* Funktionär, *der*/Funktionärin, *die;* **c)** *(constable)* Polizeibeamte, *der/-*beamtin, *die;* **yes, ~**: jawohl, Herr Wachtmeister/Frau Wachtmeisterin

'office-worker *n.* Büroangestellte, *der/die*

official [ə'fɪʃl] **1.** *adj.* **a)** Amts‹*pflicht, -robe, -person*›; **b)** *(derived from authority, formal)* offiziell; amtlich ‹*Verlautbarung*›; regulär ‹*Streik*›; **is it ~ yet?** *(coll.)* ist das schon amtlich? **2.** *n.* Beamte, *der*/Beamtin, *die; (party, union, or sports ~)* Funktionär, *der*/Funktionärin, *die*

officialdom [ə'fɪʃldəm] *n., no pl., no art.* Beamtentum, *das*

officialese [əfɪʃə'li:z] *n., no pl. (derog.)* Behördensprache, *die*

officially [ə'fɪʃəlɪ] *adv.* offiziell

officiate [ə'fɪʃɪeɪt] *v.i.* **~ as ...**: fungieren als ...; **~ at the service** den Gottesdienst abhalten; **~ at a wedding** eine Trauung vornehmen

officious [ə'fɪʃəs] *adj.* übereifrig

offing ['ɒfɪŋ] *n.* **be in the ~** *(fig.)* bevorstehen

off: ~-'key 1. *adj.* verstimmt; **2.** *adv.* falsch ‹*singen, spielen*›; **~-licence** *n. (Brit.: premises)* ≈ Wein- und Spirituosenladen, *der;* **~-load** *v.t.* abladen; **~-load sth. on to sb.** *(fig.: get rid of)* etw. bei jmdm. loswerden; **~-peak** *attrib. adj.* **during ~-peak hours** außerhalb der Spitzenlastzeiten; **~-peak electricity** Nachtstrom, *der;* **~-putting** ['ɒfpʊtɪŋ] *adj. (Brit. coll.)* abstoßend ‹*Gesicht, Äußeres, Weg*›; abschreckend ‹*Umfang*›; **~set 1.** ['--] *n.* **~set [process]** *(Printing)* Offsetdruck, *der;* **2.** ['--, -'-] *v.t., forms as* **set 1**: ausgleichen; **~shoot** *n.* **a)** *(of plant)* Sproß, *der;* **b)** *(fig.: descendant)* Sproß, *der (geh.);* **~shore** *adj.* **a)** *(situated at sea)* küstennah; **b)** ablandig ‹*Wind*› *(Seemannsspr.);* **~'side** *adj. (Sport)* Abseits-; **be ~side** abseits *od.* im Abseits sein; **~spring** *n., pl. same (human)* Nachkommenschaft, *die; (of animal)* Junge; **~-'stage** *adv.* in den Kulissen; **go ~-stage** abgehen; **~-the-peg** *attrib. adj.* Konfektions-; von der Stange *nachgestellt;* **~ 'white** *adj.* gebrochen weiß; *(yellowish)* vergilbt

often ['ɒfn, 'ɒftn] *adv.* oft; **more ~**: häufiger; **more ~ than not** meistens; **every so ~**: gelegentlich; **once too ~**: einmal zuviel

ogle ['əʊgl] *v.i.* gaffen *(ugs. abwertend);* **~ at sb.** jmdn. angaffen *(ugs. abwertend)*

ogre ['əʊgə(r)] *n.* Oger, *der;* [menschenfressender] Riese

oh [əʊ] *int.* oh; **'oh no [you don't]!** auf keinen Fall!; **oh 'no!** o nein!; oje!; **oh 'well** na ja *(ugs.);* tja *(ugs.);* **'oh yes** oh ja; **oh 'yes?** ach ja?; **oh, 'him/'that!** *(coll.)* ach, der/das!

ohm [əʊm] *n. (Electr.)* Ohm, *das*

OHMS *abbr.* **on Her/His Majesty's Service**

oil [ɔɪl] **1.** *n.* **a)** Öl, *das;* **strike ~** *(lit.)* auf Öl stoßen; *(fig.)* das große Los ziehen; **b)** *in pl. (paints)* Ölfarben. **2.** *v.t.* ölen

oil: ~can *n.* Ölkanne, *die;* **~-colour** *n., usu. in pl.* Ölfarbe, *die;* **~ drum** *n.* Ölfaß, *das;* **~field** *n.* Ölfeld, *das;* **~-fired** ['ɔɪlfaɪəd] *adj.* ölgefeuert; ölbetrieben ‹*Zentralheizung*›; **~-lamp** *n.* Öllampe, *die;* **~-painting** *n.* Ölgemälde, *das;* **~ rig** *see* [1]**rig 1 b**; **~skin** *n.* **a)** *(material)* Öltuch, *das;* **b)** *(garment)* **put on ~skins/an ~skin** Ölzeug anziehen; **~-slick** Ölteppich, *der;* **~-tanker** *n.* Öltanker, *der;* **~ well** *n.* Ölquelle, *die*

oily ['ɔɪlɪ] *adj.* **a)** ölig ‹*Oberfläche, Hände, Lappen, Geschmack*›; Öl‹*lache, -fleck*›; ölverschmiert ‹*Gesicht, Hände*›; *(containing oil)* viel Öl enthaltend ‹*Soße*›; fettig ‹*Haut, Haar*›; **b)** *(fig.)* schmierig *(abwertend)* ‹*Kerl, Art*›; ölig ‹*Lächeln, Stimme*›

ointment ['ɔɪntmənt] *n.* Salbe, *die; see also* [1]**fly**

OK [əʊ'keɪ] *(coll.)* **1.** *adj.* in Ordnung; okay *(ugs.);* **[it's] OK by me** mir ist es recht. **2.** *adv.* gut; **be doing OK** seine Sache gut machen. **3.** *int.* okay *(ugs.);* **OK?** [ist das] klar?; okay? **4.** *n.* Zustimmung, *die;* Okay, *das (ugs.).* **5.** *v.t. (approve)* zustimmen (+ *Dat.*); **be OK'd by sb.** von jmdm. das Okay bekommen *(ugs.)*

okay [əʊ'keɪ] *see* **OK**

old [əʊld] **1.** *adj.* **a)** alt; **he is ~ enough to know better** aus diesem Alter ist er heraus; **he/she is ~ enough to be your father/mother** er/sie könnte dein Vater/deine Mutter sein; **be/seem ~ before one's time** frühzeitig gealtert sein/gealtert wirken; **be [more than] 30 years ~**: [über] 30 Jahre alt sein; **at ten years ~**: im Alter von 10 Jahren; mit 10 Jahren; **be an ~ hand** ein alter Hase sein *(ugs.);* **in the ~ days** früher; **be still working for the same ~ firm** noch immer in derselben Firma arbeiten; **b)** *in playful or friendly mention* alt *(ugs.);* **you lucky ~ so-and-so!** du bist vielleicht ein alter Glückspilz!; **I saw ~ George today** ich habe heute unsern Freund George getroffen; **good/dear ~ Harry** *(coll.)* der gute alte Harry; **have a fine ~ time** *(sl.)* sich köstlich amüsieren; **poor ~ Jim/my poor ~ arm** armer Jim/mein armer Arm *(ugs.);* **any ~ thing** *(sl.)* irgendwas *(ugs.);* **any ~ how** *(sl.)* irgendwie. **2.** *n.* **a) the ~** *constr. as pl. (~ people)* alte Menschen; **b) the knights of ~**: die Ritter früherer Zeiten

old: ~ 'age *n., no pl.* [fortgeschrittenes] Alter; **in ~ age** im [fortgeschrittenen] Alter; **~-age** *attrib. adj.* Alters‹*rente, -ruhegeld, -versicherung*›; **~-age pensioner** Rentner, *der*/Rentnerin, *die;* **~ boy** *n. (Sch.)* ehemaliger Schüler; Ehemalige, *der*

olden ['əʊldn] *adj. (literary)* **in [the] ~ days** *or* **times** in alten Zeiten

old: ~-es'tablished *adj.* alt ⟨*Tradition, Brauch*⟩; alteingesessen ⟨*Firma, Geschäft, Familie*⟩; **~-fashioned** [əʊld'fæʃnd] *adj.* altmodisch; **~ girl** *n. (Sch.)* ehemalige Schülerin; Ehemalige, *die;* **~ 'hat** *see* **hat b**

oldish ['əʊldɪʃ] *adj.* älter

old: ~ 'maid *n.* **a)** *(elderly spinster)* alte Jungfer *(abwertend);* **b)** *(fig.: fussy, prim person)* altjüngferliche Person; **~ 'man** *n.* **a)** *(sl.: superior)* **the ~ man** der Alte *(ugs.);* **b)** *(coll.: father, husband)* **the/one's ~ man** der Alte/sein Alter *(ugs.);* **~ 'master** *n. (Art)* alter Meister; **~ 'people's home** *n.* Altenheim, *das;* Altersheim, *das;* **~ 'soldier** *n.* alt[gedient]er Soldat; *(fig.)* alter Hase *(ugs.);* **Old 'Testament** *see* **testament a; ~-'timer** *n. (person with long experience)* alter Hase *(ugs.);* Oldtimer, *der (scherzh.);* **~ 'wives' tale** *n.* Ammenmärchen, *das;* Altweibermärchen, *das;* **~ 'woman** *n.* **a)** *(fig.: fussy or timid person)* altes Weib *(abwertend);* **b)** *(coll.: mother, wife)* **the/one's ~ woman** die/seine Alte *(ugs.);* **~-world** *adj.* altertümlich; altväterisch ⟨*Höflichkeit, Benehmen*⟩

O level ['əʊ levl] *n. (Brit. Sch. Hist.) Abschluß der Mittelstufe (auch in der Erwachsenenbildung als Qualifikation)*

olive ['ɒlɪv] **1.** *n.* **a)** *(tree)* Ölbaum, *der;* Olivenbaum, *der;* **b)** *(fruit)* Olive, *die.* **2.** *adj.* olivgrün

olive: ~-branch *n. (fig.)* Friedensangebot, *das;* **offer the ~-branch** ein Versöhnungs- *od.* Friedensangebot machen; **~-green** *adj.* olivgrün; **~ 'oil** *n.* Olivenöl, *das*

Olympic [ə'lɪmpɪk] *adj.* olympisch; **~ Games** Olympische Spiele; **~ champion** Olympiasieger, *der*/-siegerin, *die*

Olympics [ə'lɪmpɪks] *n. pl.* Olympiade, *die;* **Winter ~:** Winterolympiade, *die*

ombudsman ['ɒmbʊdzmən] *n., pl.* **ombudsmen** ['ɒmbʊdzmən] Ombudsmann, *der*

omelette (omelet) ['ɒmlɪt] *n. (Gastr.)* Omelett, *das*

omen ['əʊmən] *n.* Omen, *das*

ominous ['ɒmɪnəs] *adj. (of evil omen)* ominös; *(worrying)* beunruhigend

ominously ['ɒmɪnəslɪ] *adv.* bedrohlich; beunruhigend ⟨*still*⟩

omission [ə'mɪʃn] *n.* **a)** Auslassung, *die;* **b)** *(failure to act)* Unterlassung, *die*

omit [ə'mɪt] *v. t.,* **-tt-: a)** *(leave out)* weglassen; **b)** *(not perform)* versäumen; **~ to do sth.** es versäumen, etw. zu tun

omnibus ['ɒmnɪbəs] *n.* **a)** *(arch.) see* **bus 1; b)** *(book)* Sammelband, *der*

omnipotent [ɒm'nɪpətənt] *adj.* allmächtig

omniscient [ɒm'nɪsɪənt, ɒm'nɪʃɪənt] *adj.* allwissend

on [ɒn] **1.** *prep.* **a)** *(position)* auf (+ *Dat.*); *(direction)* auf (+ *Akk.*); *(attached to)* an (+ *Dat./Akk.*); **put sth. on the table** etw. auf den Tisch legen *od.* stellen; **be on the table** auf dem Tisch sein; **write sth. on the wall** etw. an die Wand schreiben; **be hanging on the wall** an der Wand hängen; **have sth. on one** etw. bei sich *(Dat.)* haben; **on the bus/train** im Bus/Zug; **be on the board/committee** im Vorstand/Ausschuß sein; **on Oxford 56767** unter der Nummer Oxford 567 67; **b)** *(with basis, motive, etc. of)* **on the evidence** auf Grund des Beweismaterials; **on the assumption/hypothesis that ...:** angenommen, ...; **c)** *in expressions of time* an ⟨*einem Abend, Tag usw.*⟩; **on Sundays** sonntags; **it's just on nine** es ist gerade 9; **on [his] arrival** bei seiner Ankunft; **on entering the room ...:** beim Betreten des Zimmers ...; **on time** *or* **schedule** pünktlich; **d)** *expr. state etc.* **be on heroin** heroinabhängig sein; **the drinks are on me** *(coll.)* die Getränke gehen auf mich; **e)** *(concerning, about)* über (+ *Akk.*). **2.** *adv.* **a) have a hat on** einen Hut aufhaben; **your hat is on crooked** dein Hut sitzt schief; **the potatoes are on** die Kartoffeln sind aufgesetzt; **b)** *(in some direction)* **face on** mit dem Gesicht voran; **on and on** immer weiter; **c)** *(switched or turned on)* **the light/radio** *etc.* **is on** das Licht/Radio *usw.* ist an; **put the light on** das Licht anmachen; **is there a gas tap on?** ist ein Gashahn aufgedreht?; **d)** *(arranged)* **is Sunday's picnic on?** findet das Picknick am Sonntag statt?; **I have nothing of importance on** ich habe nichts Wichtiges vor; **e)** *(being performed)* **what's on at the cinema?** was gibt es *od.* läuft im Kino?; **his play is currently on in London** sein Stück wird zur Zeit in London aufgeführt *od.* gespielt; **f)** *(on duty)* **come/be on** seinen Dienst antreten/Dienst haben; **g) sth. is on** *(feasible)*/**not on** etw. ist möglich/ausgeschlossen; **you're on!** *(coll.: I agree)* abgemacht!; *(making bet)* die Wette gilt!; **be on about sb./sth.** *(coll.)* [dauernd] über jmdn./etw. sprechen; **what is he on about?** was will er [sagen]?; **be on at/keep on and on at sb.** *(coll.)* jmdm. in den Ohren/dauernd in den Ohren liegen *(ugs.);* **on to, onto** auf (+ *Akk.*); **be on to sth.** *(have discovered sth.)* etw. ausfindig gemacht haben. *See also* **right 4d**

once [wʌns] **1.** *adv.* **a)** einmal; **~ a week/month** einmal die Woche/im Monat; **~ or twice** ein paarmal; einigemal; **~ again** *or* **more** noch einmal; **~ [and] for all** ein für allemal; **[every] ~ in a while** von Zeit zu Zeit; **~ an X always an X** X bleibt X; *see also* **for 1n; b)** *(multiplied by one)* ein mal; **c)** *(even for one or the first time)* je[mals]; **never/not ~:** nicht ein einziges Mal; **d)** *(formerly)* früher einmal; **~ upon a time there lived a king** es war einmal ein König; **e) at ~** *(immediately)* sofort; sogleich; *(at the same time)* gleichzeitig; **all at ~** *(all together)* alle auf einmal; *(without warning)* mit einem Mal. **2.** *conj.* sobald; **~ past the fence we are safe** wenn wir [nur] den Zaun hinter uns bringen, sind wir in Sicherheit. **3.** *n.* **[just** *or* **only] this ~:** [nur] dieses eine Mal

'once-over *n.* **give sb./sth. a/the ~:** jmdn./etw. kurz in Augenschein nehmen

'oncoming *adj.* entgegenkommend ⟨*Fahrzeug, Verkehr*⟩

one [wʌn] **1.** *adj.* **a)** *attrib.* ein; **~ thing I must say** ein[e]s muß ich sagen; **~ or two** *(fig.: a few)* ein paar; **~ more ...:** noch ein ...; **~ more time** noch einmal; **it's ~ [o'clock]** es ist eins *od.* ein Uhr; *see also* **eight 1; half 1a, 3b; quarter 1a; b)** *attrib. (single, only)* einzig; **the ~ thing** das einzige; **any ~:** irgendein; **in any ~ day/year** an einem Tag/in einem Jahr; **at any ~ time** zur gleichen Zeit; *(always)* zu jeder Zeit; **no ~:** kein; **not ~ [little] bit** überhaupt nicht; **c)** *(identical, same)* ein; **~ and the same person/thing** ein und dieselbe Person/Sache; **at ~ and the same time** gleichzeitig; *see also* **all 2a; d)** *pred. (united, unified)* **we are ~:** wir sind uns einig; **be ~ as a family/nation** eine einige Familie/Nation sein; *see also* **with a; e)** *attrib. (a particular but undefined)* **at ~ time** einmal; einst *(geh.);* **~ morning/night** eines Morgens/Nachts; ~ **day** *(on day specified)* einmal; *(at unspecified future date)* eines Tages; ~ **day soon** bald einmal; ~ **day next week** irgendwann nächste Woche; ~ **Sunday** an einem Sonntag; **f)** *attrib. contrasted with 'other'/'another'* ein; **for ~ thing** zum einen; **neither ~ thing nor the other** weder das eine noch das andere; *see also* **hand 1 n; g) in ~** *(coll.: at first attempt)* auf Anhieb; **got it in ~!** *(coll.)* [du hast es] erraten! **2.** *n.* **a)** eins; **b)** *(number, symbol)* Eins, *die; see also* **eight 2 a; c)** *(unit)* **in ~s** einzeln; **two for the price of ~:** zwei zum Preis von einem. **3.** *pron.* **a) ~ of ...:** ein... (+ *Gen.*); **~ of them/us** *etc.* einer von ihnen/uns *usw.*; **any ~ of them** jeder/jede/jedes von ihnen; **every ~ of them** jeder/jede/jedes [einzelne] von ihnen; **not ~ of them** keiner/keine/keines von ihnen; **b)** *replacing n. implied or mentioned* ein...; **big ~s and little ~s** große und kleine; **the jacket is an old ~:** die Jacke ist [schon] alt; **the older/younger ~:** der/die/das ältere/jüngere; **this is the ~ I like** den/die/das mag ich; **my husband is the tall ~ over there** mein Mann ist der große da; **you are** *or* **were the ~ who insisted on going to Scotland** du warst der-/diejenige, der/die unbedingt nach Schottland wollte; **this ~:** dieser/diese/dieses [da]; **that ~:** der/die/das [da]; **these ~s** *or* **those ~s?** *(coll.)* die [da] oder die [da]?; **these/those blue** *etc.* **~s** diese/die blauen *usw.*; **which ~?** welcher/welche/welches?; **which ~s?** welche?; **not ~:** keiner/keine/keines; *emphatic* nicht einer/eine/eines; **all but ~:** alle außer einem/einer/einem; **the last house but ~:** das vorletzte Haus; **I for ~:** ich für mein[en] Teil; **~ by ~, ~ after another** *or* **the other** einzeln; **love ~ another** sich *od.* *(geh.)* einander lieben; **be kind to ~ another** nett zueinander sein; **c)** *(contrasted with 'other'/'another')* **[the] ~ ... the other** der/die/das eine ... der/die/das andere; **d)** *(person or creature of specified kind)* **the little ~:** der/die/das Kleine; **our dear** *or* **loved ~s** unsere Lieben; **young ~** *(youngster)* Kind, *das; (young animal)* Junge, *das;* **e) [not] ~ who does** *or* **to do** *or* **for doing sth.** [nicht] der Typ, der etw. tut; **f)** *(representing people in general; also coll.: I, we)* man; *as indirect object* einem; *as direct object* einen; **~'s** sein; **wash ~'s hands** sich *(Dat.)* die Hände waschen; **g)** *(coll.: drink)* **I'll have just a little ~:** ich trinke nur einen Kleinen *(ugs.)*; **have ~ on me** ich geb dir einen aus; **h)** *(coll.: blow)* **give sb. ~ on the head/nose** jmdm. eins über den Kopf/auf die Nase geben *(ugs.)*

one: ~-armed *adj.* einarmig; **~-eyed** ['wʌnaɪd] *adj.* einäugig; **~-handed** [wʌnhændɪd] **1.** ['---] *adj.* einhändig; **2.** [-'--] *adv.* mit einer Hand; **~-legged** *adj.* einbeinig; **~-man** *attrib. adj.* Einmann⟨*boot, -betrieb usw.*⟩; **~-man band** Einmannkapelle, *die; (fig.: firm etc.)* Einmannbetrieb, *der;* **~-off** *(Brit.)* **1.** *n. (article)* Einzelstück, *das;* Einzelexemplar, *das; (operation)* einmalige Sache; **2.** *adj.* einmalig ⟨*Zahlung, Angebot, Produktion, Verkauf*⟩; Einzel⟨*stück, -modell, -anfertigung*⟩; **~-parent family** *n.* Einelternfamilie, *die;* **~-piece** *adj.* einteilig

onerous ['ɒnərəs, 'əʊnərəs] *adj.* schwer

one: ~'self *pron.* **a)** *emphat.* selbst; **as old/rich as ~self** so alt/reich wie man selbst; **be ~self** man selbst sein; **b)** *refl.* sich; *see also* **herself; ~-sided** ['wʌnsaɪdɪd] *adj.* einseitig; **~-time** *adj.* **a)** *(former)* ehemalig; **b)** *(used once only)* einmalig; **~-to-'~** *adj.* **~-to-~ relation/correspondence** hundertprozentige Parallelität; **~-to-~ teaching** Einzelunterricht, *der;* **~-track** *adj.* eingleisig; **have a ~-track mind** *(lack flexibility)* eingleisig denken; *(be obsessed by one subject)* [immer] nur eins im Kopf haben; **~-'up** *pred. adj. (coll.)* **be ~-up [on** *or* **over sb.]** *(fig.)* [jmdm.] um eine Nasenlänge voraus sein; **~-way** *adj.* **a)** in einer Richtung *nachgestellt;* Einbahn⟨*straße, -verkehr*⟩; **b)** *(single)* einfach ⟨*Fahrpreis, Fahrkarte, Flug usw.*⟩

'ongoing *adj.* aktuell ⟨*Problem, Aktivitäten, Debatte*⟩; laufend ⟨*Forschung, Projekt*⟩; andauernd ⟨*Situation*⟩

onion ['ʌnjən] *n.* Zwiebel, *die*

'onlooker *n.* Zuschauer, *der*/Zuschauerin, *die*

only ['əʊnlɪ] **1.** *attrib. adj.* **a)** einzig...; **the ~ person** der/die einzige; **my ~ regret is that ...:** ich bedaure nur, daß ...; **an ~ child** ein Einzelkind; **the ~ one/ones** der/die/das einzige/die einzigen; **the ~ thing** das einzige; **b)** *(best by far)* **the ~:** der/die/das einzig wahre; **he/she is the ~ one for me** es gibt nur ihn/sie für mich. **2.** *adv.* **a)** nur; **we had been waiting ~ 5 minutes when ...:** wir hatten erst 5 Minuten gewartet, als ...; **it's ~/~ just 6 o'clock** es ist erst 6 Uhr/gerade erst 6 Uhr vorbei; **I ~ wish I had known** wenn ich es doch nur gewußt hätte; **you ~ have** *or* **you have ~ to ask** *etc.* du brauchst nur zu fragen *usw.*; **you ~ live once** man lebt nur einmal; **~ if** nur [dann]..., wenn; **he ~ just managed it/made it** er hat es gerade so/gerade noch geschafft; **not ~ ... but also** nicht nur ... sondern auch; *see also* **if 1 d; b)** *(no longer ago than)* erst; **~ the other day/week** erst neulich *od.* kürzlich; **~ just** gerade erst; **c)** *(with no better result than)* **~ to find/discover that ...:** nur, um zu entdecken, daß ...; **d) ~ too** [sogar] ausgesprochen ⟨*froh, begierig, bereitwillig*⟩; *in context of undesirable circumstances* viel zu; **be ~ too aware of sth.** sich *(Dat.)* einer Sache *(Gen.)* voll bewußt sein; **~ too well** nur zu gut ⟨*wissen, kennen, sich erinnern*⟩; gerne ⟨*mögen*⟩. **3.** *conj.* **a)** *(but then)* nur; **b)** *(were it not for the fact that)* **~ [that] I am/he is** *etc.* **...:** ich bin/er ist *usw.* nur ...

'on-off *adj.* **~ switch** Ein-aus-Schalter

onomatopoeia [ɒnəmætə'pi:ə] *n. (Ling.)* Onomatopöie, *die*

'onset *n. (of storm)* Einsetzen, *das; (of winter)* Einbruch, *der; (of disease)* Ausbruch, *der*

'onshore *adj.* auflandig *(Seemannsspr.)* ⟨*Wind*⟩

onslaught ['ɒnslɔ:t] *n.* [heftige] Attacke *(fig.)*

onto *see* **on 2 g**

onus ['əʊnəs] *n.* Last, *die;* **the ~ is on him to do it** es ist seine Sache, es zu tun

onward[s] ['ɒnwəd(z)] *adv.* **a)** *(in space)* vorwärts; **from X ~:** von X an; **b)** *(in time)* **from that day ~:** von diesem Tag an

onyx ['ɒnɪks] *n. (Min.)* Onyx, *der*

ooze [u:z] **1.** *v. i.* **a)** *(percolate, exude)* sickern **(from** aus); *(more thickly)* quellen **(from** aus); **b)** *(become moistened)* triefen **(with** von, vor + *Dat.*). **2.** *v. t.* **a) ~ [out]** triefen von *od.* vor (+ *Dat.*); **b)** *(fig.: radiate)* ausstrahlen ⟨*Charme, Optimismus*⟩; ausströmen ⟨*Sarkasmus*⟩. **3.** *n. (mud)* Schlick, *der*

opal ['əʊpl] *n. (Min.)* Opal, *der*

opalescent [əʊpə'lesənt] *adj.* schillernd; opalisierend

opaque [əʊ'peɪk] *adj.* **a)** *(not transmitting light)* lichtundurchlässig; opak *(fachspr.)*; **b)** *(obscure)* dunkel; unverständlich

OPEC ['əʊpek] *abbr.* **Organization of Petroleum Exporting Countries** OPEC, *die*

open ['əʊpn] **1.** *adj.* **a)** offen; **with the window ~:** bei geöffnetem Fenster; **be [wide/half] ~:** [weit/halb] offenstehen; **hold the door ~ [for sb.]** [jmdm.] die Tür aufhalten; **push/pull/kick the door ~:** die Tür aufstoßen/aufziehen/eintreten; **force sth. ~:** etw. mit Gewalt öffnen; **with one's mouth ~:** mit offenem Mund; **have one's eyes ~:** die Augen geöffnet haben; **[not] be able to keep one's eyes ~:** [nicht mehr] die Augen offenhalten können; *see also* **eye 1a; b)** *(unconfined)* offen ‹*Gelände, Feuer*›; **on the ~ road** auf freier Strecke; **in the ~ air** im Freien; **c)** *(ready for business or use)* **be ~** ‹*Laden, Museum, Bank usw.:*› geöffnet sein; **'~'/'~ on Sundays'** „geöffnet"/„Sonntags geöffnet"; **d)** *(accessible)* offen; öffentlich ‹*Treffen, Rennen*›; *(available)* frei ‹*Stelle*›; freibleibend ‹*Angebot*›; **lay ~:** offenlegen ‹*Plan*›; **be ~ to the public** für die Öffentlichkeit zugänglich sein; **the offer remains ~ until the end of the month** das Angebot bleibt bestehen *od.* gilt noch bis Ende des Monats; **keep a position ~ for sb.** jmdm. eine Stelle freihalten; **e) be ~ to** *(exposed to)* ausgesetzt sein (+ *Dat.*) ‹*Wind, Sturm*›; *(receptive to)* offen sein für ‹*Ratschlag, andere Meinung, Vorschlag*›; **I hope to sell it for £1,000, but I am ~ to offers** ich möchte es für 1 000 Pfund verkaufen, aber ich lasse mit mir handeln; **lay oneself [wide] ~ to criticism** *etc.* sich der Kritik *usw.* aussetzen; **be ~ to question/doubt/argument** fraglich/zweifelhaft/umstritten sein; **f)** *(undecided)* offen; **have an ~ mind about** *or* **on sth.** einer Sache gegenüber aufgeschlossen sein; **with an ~ mind** aufgeschlossen; **leave sth. ~:** etw. offenlassen; **g)** *(undisguised, manifest)* unverhohlen ‹*Bewunderung, Haß*›; offen ‹*Verachtung, Empörung, Widerstand*›; offensichtlich ‹*Spaltung, Zwiespalt*›; **~ war/warfare** offener Krieg/Kampf; **h)** *(frank)* offen ‹*Wesen, Streit, Abstimmung, Gesicht*›; *(not secret)* öffentlich ‹*Wahl*›; **be ~ [about sth./with sb.]** [in bezug auf etw. *(Akk.)*/gegenüber jmdm.] offen sein; **i)** *(expanded, unfolded)* offen, geöffnet ‹*Pore, Regenschirm*›; aufgeblüht ‹*Blume, Knospe*›; aufgeschlagen ‹*Zeitung, Landkarte, Stadtplan*›; **sb./sth. is an ~ book [to sb.]** *(fig.)* jmd./etw. ist ein aufgeschlagenes *od.* offenes Buch [für jmdn.]. **2.** *n.* **in the ~** *(outdoors)* unter freiem Himmel; **[out] in the ~** *(fig.)* [öffentlich] bekannt; **come [out] into the ~** *(fig.) (become obvious)* herauskommen *(ugs.)*; *(speak out)* offen sprechen; **bring sth. [out] into the ~** *(fig.)* etw. an die Öffentlichkeit bringen. **3.** *v.t.* **a)** öffnen; aufmachen *(ugs.)*; **~ sth. with a key** etw. aufschließen; **b)** *(allow access to)* **~ sth. [to sb./sth.]** etw. öffnen [für jmdn./etw.]; *(fig.)* [jmdm./einer Sache] etw. öffnen; **~ sth. to the public** etw. der Öffentlichkeit *(Dat.)* zugänglich machen; **c)** *(establish)* eröffnen ‹*Konferenz, Kampagne, Diskussion, Laden*›; beginnen ‹*Verhandlungen, Krieg, Spiel*›; *(declare ~)* eröffnen ‹*Gebäude usw.*›; **~ an account** ein Konto eröffnen; **~ fire [on sb./sth.]** das Feuer [auf jmdn./etw.] eröffnen; **d)** *(unfold, spread out)* aufschlagen ‹*Zeitung, Landkarte, Stadtplan, Buch*›; aufspannen, öffnen ‹*Schirm*›; öffnen ‹*Fallschirm, Poren*›; **~ one's arms [wide]** die *od.* seine Arme [weit] ausbreiten; **e)** *(reveal, expose)* **sth. ~s new horizons/a new world to sb.** *(fig.)* etw. eröffnet jmdm. neue Horizonte/eine neue Welt; **f)** *(make more receptive)* **~ one's heart** *or* **mind to sb./sth.** sich jmdm./einer Sache öffnen; **~ sb.'s mind to sth.** jmdm. etw. nahebringen. **4.** *v.i.* **a)** sich öffnen; aufgehen; ‹*Spalt, Kluft:*› sich auftun; **'Doors ~ at 7 p.m.'** „Einlaß ab 19 Uhr"; **~ inwards/outwards** nach innen/außen aufgehen; **the door would not ~:** die Tür ging nicht auf *od.* ließ sich nicht öffnen; **his eyes ~ed wide** er riß die Augen weit auf; **~ into/on to sth.** zu etw. führen; **the kitchen ~s into the living-room** die Küche hat eine Tür zum Wohnzimmer; **b)** *(become ~ to customers)* öffnen; aufmachen *(ugs.)*; *(start trading etc.)* eröffnet werden; **the shop does not ~ on Sundays** der Laden ist sonntags geschlossen; **c)** *(make a start)* beginnen; ‹*Ausstellung:*› eröffnet werden

~ 'out 1. *v.t. (unfold)* auseinanderfalten. **2.** *v.i.* **a)** ‹*Knospe:*› sich öffnen; **b)** *(widen, expand)* **~ out into sth.** sich zu etw. erweitern

~ 'up 1. *v.t.* **a)** aufmachen *(ugs.)*; öffnen; aufschlagen ‹*Buch*›; **b)** *(form or make by cutting etc.)* machen ‹*Loch, Riß*›; **c)** *(establish, make more accessible)* eröffnen ‹*Laden, Filiale*›; erschließen ‹*neue Märkte usw.*›; **~ up a new world to sb.** jmdm. eine neue Welt erschließen. **2.** *v.i.* **a)** ‹*Blüte, Knospe:*› sich öffnen; **b)** *(be established)* ‹*Filiale:*› eröffnet werden; ‹*Firma:*› sich niederlassen; **c)** *(appear, be revealed)* entstehen; ‹*Aussichten, Möglichkeiten:*› sich eröffnen; **~ up before sb.** ‹*Blick, Aussicht:*› sich jmdm. bieten; ‹*neue Welt:*› sich vor jmdm. auftun; **d)** *(talk freely)* gesprächig werden; **~ up to sb.** sich jmdm. anvertrauen

open: ~-air *attrib. adj.* Openair‹*Konzert*›; Freiluft‹*restaurant, -aktivitäten*›; Freilicht‹*kino, -aufführung*›; ‹*Ausstellung, Markt, Versammlung*› im Freien *od.* unter freiem Himmel; **~-air [swimming-]pool** Freibad, *das*; **~-and-'shut case** *n. (coll.)* klarer Fall; **~ day** *n.* Tag der offenen Tür; **~-ended** [əʊpən'endɪd] *adj.* [am Ende] offen; *(fig.)* unbefristet ‹*Aufenthalt, Vertrag*›; Open-end-‹*Diskussion, Debatte*›; offen ‹*Frage*›

opener ['əʊpnə(r)] *n.* Öffner, *der*

open: ~-handed [əʊpn'hændɪd] *adj.* freigebig; **~-hearted** [əʊpn-hɑːtɪd] *adj.* aufrichtig ‹*Person, Mitgefühl*›; herzlich ‹*Empfang*›

opening ['əʊpnɪŋ] **1.** *n.* **a)** Öffnen, *das*; *(becoming open)* Sichöffnen, *das*; *(of crack, gap, etc.)* Entstehen, *das*; *(of exhibition, new centre)* Eröffnen, *das*; **hours** *or* **times of ~:** Öffnungszeiten; **b)** *(establishment, inauguration, ceremony)* Eröffnung, die; **~ of Parliament** Parlamentseröffnung, *die*; **c)** *(initial part)* Anfang, *der*; **d)** *(gap, aperture)* Öffnung, *die*; **e)** *(opportunity)* Möglichkeit, *die*; *(vacancy)* freie *od.* offene Stelle; **give sb. an ~ into sth.** ‹*Person:*› jmdm. den Einstieg in etw. *(Akk.)* ermöglichen; ‹*Job:*› für jmdn. ein Einstieg in etw. *(Akk.)* sein. **2.** *adj.* einleitend; **the ~ lines** *(of play, poem, etc.)* die ersten Zeilen; **~ night** *(Theatre)* Premiere, *die*; **~ move** *(Chess)* Eröffnung, *die*

opening: ~ ceremony *n.* feierliche Eröffnung (**for** *Gen.*); **~ hours** *n. pl.* Öffnungszeiten *Pl.*; **~ time** *n.* **a)** Öffnungszeit, *die*; **b) ~ times** *see* **~ hours**

openly ['əʊpnlɪ] *adv.* **a)** *(publicly)* in der Öffentlichkeit; öffentlich ‹*zugeben, verurteilen, abstreiten*›; **quite ~:** in aller Öffentlichkeit; **b)** *(frankly)* offen

open: ~ 'market *n.* offener *od.* freier Markt; **~-'minded** *adj.* aufgeschlossen (**about** für); **~-mouthed** [əʊpn'maʊðd] *adj.* mit offenem Mund; **~-necked** ['əʊpnnekt] *adj.* ‹*Hemd, Bluse*› mit offenem Kragen; ausgeschnitten ‹*Kleid, Pullover*›

openness ['əʊpnnɪs] *n., no pl.* **a)** *(receptiveness)* Empfänglichkeit, *die;* **b)** *(frankness)* Offenheit, *die*

open: ~-plan *adj.* offen angelegt ‹*Haus*›; **~-plan office** Großraumbüro, *das;* **~ 'prison** *n.* offene Anstalt; **~ 'sandwich** *n.* belegtes Brot; **~ season** *n. (Brit.)* Jagdzeit, *die; (for fish)* Fangzeit, *die;* **it is [the] ~ season for** *or* **on sth.** *(fig.)* etw. ist an der Tagesordnung; **~-top** *attrib. adj.* offen; oben offen ‹*Bus*›

[1]**opera** ['ɒpərə] *n.* **a)** Oper, *die;* **b)** *no pl.* **[the] ~:** die Oper

[2]**opera** *pl. of* **opus**

opera: ~-glasses *n. pl.* Opernglas, *das;* **~-house** *n.* Opernhaus, *das;* **~-singer** *n.* Opernsänger, *der*/-sängerin, *die*

operate ['ɒpəreɪt] **1.** *v. i.* **a)** *(be in action)* in Betrieb sein; ‹*Bus, Zug usw.:*› verkehren; *(have an effect)* sich auswirken; **b)** *(function)* arbeiten; **the torch ~s on batteries** die Taschenlampe arbeitet mit Batterien; **c)** *(perform operation)* operieren; arbeiten; **~ [on sb.]** *(Med.)* [jmdn.] operieren; **d)** *(exercise influence)* **~ [up]on sb./sth.** auf jmdn./etw. einwirken; **e)** *(follow course of conduct)* agieren; **f)** *(produce effect)* wirken; **g)** *(Mil.)* operieren. **2.** *v. t.* bedienen ‹*Maschine*›; fahren ‹*Auto*›; betreiben ‹*Unternehmen*›; unterhalten ‹*Werk, Post, Busverbindung*›

'operating-theatre *n. (Brit. Med.)* Operationssaal, *der*

operation [ɒpə'reɪʃn] *n.* **a)** *(causing to work) (of machine)* Bedienung, *die; (of factory, mine, etc.)* Betrieb, *der; (of bus service etc.)* Unterhaltung, *die;* **b)** *(way sth. works)* Arbeitsweise, *die;* **c)** *(being operative)* **come into ~** ‹*Gesetz, Gebühr usw.:*› in Kraft treten; **be in ~** ‹*Maschine, Gerät usw.:*› in Betrieb sein; ‹*Service:*› zur Verfügung stehen; ‹*Gesetz:*› in Kraft sein; **be out of ~** ‹*Maschine, Gerät usw.:*› außer Betrieb sein; **d)** *(performance)* Tätigkeit, *die;* **repeat the ~:** das Ganze [noch einmal] wiederholen; **e)** *(Med.)* Operation, *die;* **have an ~ [on one's foot]** [am Fuß] operiert werden; **f)** *(Mil.)* Einsatz, *der*

operational [ɒpə'reɪʃənl] *adj.* **a)** Einsatz‹*flugzeug, -breite*›; *(Mil.)* Einsatz-; **b)** *(esp. Mil.: ready to function)* einsatzbereit

operative ['ɒpərətɪv] **1.** *adj.* **a)** *(in operation)* **the law became ~:** das Gesetz trat in Kraft; **the scheme is fully ~:** das Programm läuft; **b)** *(most relevant)* **the ~ word is 'quietly'** die Betonung liegt auf „leise". **2.** *n.* [Fach]arbeiter, *der*/-arbeiterin, *die*

operator ['ɒpəreɪtə(r)] *n.* **a)** *(worker)* [Maschinen]bediener, *der*/-bedienerin, *die; (of crane, excavator, etc.)* Führer, *der;* **b)** *(Teleph.) (at exchange)* Vermittlung, *die; (at switchboard)* Telefonist, *der*/Telefonistin, *die;* **c)** *(coll.: shrewd person)* Schlitzohr, *das (ugs.)*

ophthalmic [ɒf'θælmɪk] *adj.* Augen-

ophthalmic op'tician *n. (Brit.)* Augenoptiker, *der*/-optikerin, *die*

opinion [ə'pɪnjən] *n.* **a)** *(belief, judgement)* Meinung, *die* (**on** über + *Akk.*, zu); Ansicht, *die* (**on** von, zu, über + *Akk.*); **his ~s on the matter/on religion** seine Meinung dazu/seine Einstellung zur Religion; **in my ~:** meiner Meinung nach; **be a matter of ~:** Ansichtssache sein; **b)** *no pl., no art. (beliefs etc. of group)* Meinung, *die* (**on** über + *Akk.*); **public ~:** die öffentliche Meinung; **c)** *(estimate)* **have a high/low ~ of sb.** eine hohe/schlechte Meinung von jmdm. haben; **d)** *(formal statement of expert)* Gutachten, *das;* **a second ~:** die Meinung eines zweiten Sachverständigen

opinionated [ə'pɪnjəneɪtɪd] *adj.* rechthaberisch

o'pinion poll *n.* Meinungsumfrage, *die*

opium ['əʊpɪəm] *n.* Opium, *das*

opponent [ə'pəʊnənt] *n.* Gegner, *der*/Gegnerin, *die*

opportune ['ɒpətju:n] *adj.* **a)** *(favourable)* günstig; **b)** *(well-timed)* zur rechten Zeit *nachgestellt*

opportunism [ɒpə'tju:nɪzm] *n., no pl.* Opportunismus, *der*

opportunist [ɒpə'tju:nɪst] *n.* Opportunist, *der*/Opportunistin, *die*

opportunity [ɒpə'tju:nɪtɪ] *n.* Gelegenheit, *die;* **have plenty of/little ~ for doing** *or* **to do sth.** reichlich/wenig Gelegenheit haben, etw. zu tun

oppose [ə'pəʊz] *v. t.* **a)** *(set oneself against)* sich wenden gegen; **b)** *(place as obstacle)* entgegenstellen (**to** *Dat.*); **c)** *(set as contrast)* gegenüberstellen (**to, against** *Dat.*)

opposed [ə'pəʊzd] *adj.* **a)** *(contrary)* gegensätzlich; entgegengesetzt; **as ~ to** im Gegensatz zu; **b)** *(hostile)* **be ~ to sth.** gegen etw. sein

opposite ['ɒpəzɪt] **1.** *adj.* **a)** *(on other or farther side)* gegenüberliegend ‹*Straßenseite, Ufer*›; entgegengesetzt ‹*Ende*›; **b)** *(contrary)* entgegengesetzt ‹*Weg, Richtung*›; **c)** *(very different in character)* entgegengesetzt, gegensätzlich ‹*Beschreibungen, Aussagen*›; **d) the ~ sex** das andere Geschlecht. **2.** *n.* Gegenteil, *das* (**of** von); **be ~s** einen Gegensatz bilden. **3.** *adv.* gegenüber; **sit ~:** auf der gegenüberliegenden Seite sitzen. **4.** *prep.* gegenüber

opposite 'number *n. (fig.)* Pendant, *das*

opposition [ɒpə'zɪʃn] *n.* **a)** *no pl. (antagonism)* Opposition, *die; (resistance)* Widerstand, *der* (**to** gegen); **in ~ to** entgegen; **b)** *(Brit. Polit.)* **the O~:** die Opposition; **[be] in ~:** in der Opposition [sein]; **c)** *(body of opponents)* Gegner *Pl.;* **d)** *(contrast, antithesis)* Gegensatz, *der* (**to** zu)

oppress [ə'pres] *v. t.* **a)** *(govern cruelly)* unterdrücken; **b)** *(fig.: weigh down)* ‹*Gefühl:*› bedrücken; ‹*Hitze:*› schwer zu schaffen machen (+ *Dat.*)

oppression [ə'preʃn] *n.* Unterdrückung, *die*

oppressive [ə'presɪv] *adj.* **a)** *(tyrannical)* repressiv; **b)** *(fig.: hard to endure)* bedrückend ‹*Ängste, Atmosphäre*›; **c)** *(fig.: hot and close)* drückend ‹*Wetter, Klima, Tag*›; **d)** *(fig.: burdensome)* drückend ‹*Steuer*›; repressiv ‹*Gesetz, Beschränkung*›

opt [ɒpt] *v. i.* sich entscheiden (**for** für); **~ to do sth.** sich dafür entscheiden, etw. zu tun; **~ out** *(not join in)* nicht mitmachen; *(cease taking part)* nicht länger mitmachen; **~ out of** nicht/nicht länger mitmachen bei

optic ['ɒptɪk] *adj. (Anat.)* Seh-

optical ['ɒptɪkl] *adj.* optisch; **~ 'fibre** Lichtleitfaser, *die*

optician [ɒp'tɪʃn] *n.* **a)** *(maker or seller of spectacles etc.)* Optiker, *der*/Optikerin, *die;* **b)** *see* **ophthalmic optician**

optics ['ɒptɪks] *n., no pl.* Optik, *die*

optima *pl. of* **optimum**
optimal ['ɒptɪml] *see* **optimum 2**
optimise *see* **optimize**
optimism ['ɒptɪmɪzm] *n., no pl.* Optimismus, *der*
optimist ['ɒptɪmɪst] *n.* Optimist, *der*/Optimistin, *die*
optimistic [ɒptɪ'mɪstɪk] *adj.* optimistisch
optimize ['ɒptimaɪz] *v. t.* optimieren
optimum ['ɒptɪməm] **1.** *n., pl.* **optima** ['ɒptɪmə] Optimum, *das.* **2.** *adj.* optimal
option ['ɒpʃn] *n.* Wahl, *die; (thing that may be chosen)* Wahlmöglichkeit, *die; (Brit. Univ., Sch.)* Wahlfach, *das;* **I have no ~ but to do sth.** mir bleibt nichts [anderes] übrig, als etw. zu tun; **keep** *or* **leave one's ~s open** sich *(Dat.)* alle Möglichkeiten offenhalten
optional ['ɒpʃənl] *adj.* nicht zwingend; **~ subject** Wahlfach, *das*
opulence ['ɒpjʊləns] *n., no pl.* Wohlstand, *der*
opulent ['ɒpjulənt] *adj.* wohlhabend ‹*Person, Aussehen*›; feudal ‹*Auto, Haus, Hotel usw.*›
opus ['əʊpəs, 'ɒpəs] *n., pl.* **opera** ['ɒpərə] *(Mus.)* Opus, *das*
or [ə(r), *stressed* ɔ:(r)] *conj.* **a)** oder; **he cannot read or write** er kann weder lesen noch schreiben; **without food or water** ohne Essen und Wasser; **[either] ... or [else] ...:** entweder ... oder [aber] ...; **b)** *introducing synonym* oder [auch]; *introducing explanation* das heißt; **or rather** beziehungsweise; **c)** *indicating uncertainty* oder; **15 or 20 minutes** 15 bis 20 Minuten; **in a day or two** in ein, zwei Tagen; **he must be ill or something** vielleicht ist er krank oder so *(ugs.);* **d)** *expr. significant afterthought* oder; **he was obviously lying – or was he?** er hat ganz offensichtlich gelogen – oder [doch nicht]?
oracle ['ɒrəkl] *n.* Orakel, *das*
oral ['ɔ:rl] **1.** *adj.* **a)** *(spoken)* mündlich ‹*Prüfung, Vereinbarung*›; mündlich überliefert ‹*Tradition*›; **b)** *(Anat.)* Mund‹*höhle, -schleimhaut*›. **2.** *n. (coll.: examination)* **the ~[s]** das Mündliche
orally ['ɔ:rəlɪ] *adv.* **a)** *(in speech)* mündlich; **b)** *(by mouth)* oral; **take ~:** einnehmen
orange ['ɒrɪndʒ] **1.** *n.* **a)** Orange, *die;* Apfelsine, *die;* **b)** *(colour)* **~[-colour]** Orange, *das.* **2.** *adj.* orange[farben]; Orangen‹*geschmack*›
orange: ~-juice *n.* Orangensaft, *der;* **~-peel** *n.* Orangenschale, *die;* **~ 'squash** *see* **squash 3a**
orang-utan [ɔ:ræŋʊ'tæn] *n. (Zool.)* Orang-Utan, *der*
oration [ə'reɪʃn] *n.* Rede, *die*
orator ['ɒrətə(r)] *n.* Redner, *der*/Rednerin, *die*
oratorio [ɒrə'tɔ:rɪəʊ] *n., pl.* **~s** *(Mus.)* Oratorium, *das*
oratory ['ɒrətərɪ] *n., no pl.* **a)** *(art)* Redekunst, *die;* **b)** *(rhetorical language)* Rhetorik, *die*
orbit ['ɔ:bɪt] **1.** *n.* **a)** *(Astron.)* [Umlauf]bahn, *die;* **b)** *(Astronaut.)* Umlaufbahn, *die;* Orbit, *der;* **put/send into ~:** in die Umlaufbahn bringen/schießen; **c)** *(fig.)* Sphäre, *die.* **2.** *v. i.* kreisen. **3.** *v. t.* umkreisen
orchard ['ɔ:tʃəd] *n.* Obstgarten, *der; (commercial)* Obstplantage, *die;* **cherry ~:** Kirschgarten, *der*
orchestra ['ɔ:kɪstrə] *n. (Mus.)* Orchester, *das*
orchestral [ɔ:'kestrl] *adj.* Orchester-
orchestrate ['ɔ:kɪstreɪt] *v. t. (Mus.; also fig.)* orchestrieren
orchestration [ɔ:kɪ'streɪʃn] *n. (Mus.)* Orchesterbearbeitung, *die*
orchid ['ɔ:kɪd] *n.* Orchidee, *die*
ordain [ɔ:'deɪn] *v. t.* **a)** *(Eccl.)* ordinieren; **b)** *(destine)* bestimmen
ordeal [ɔ:'di:l] *n.* Qual, *die* **(by** durch)
order ['ɔ:də(r)] **1.** *n.* **a)** *(sequence)* Reihenfolge, *die;* **word ~:** Wortstellung, *die;* **in ~ of importance/size/age** nach Wichtigkeit/Größe/Alter; **put sth. in ~:** etw. [in der richtigen Reihenfolge] ordnen; **keep sth. in ~:** etw. in der richtigen Reihenfolge halten; **answer the questions in ~:** die Fragen der Reihe nach beantworten; **out of ~:** nicht in der richtigen Reihenfolge; **b)** *(normal state)* Ordnung, *die;* **put** *or* **set sth./one's affairs in ~:** Ordnung in etw. bringen/seine Angelegenheiten ordnen; **be/not be in ~:** in Ordnung/nicht in Ordnung sein *(ugs.);* **be out of/in ~** *(not in/in working condition)* nicht funktionieren/funktionieren; **'out of ~'** „außer Betrieb"; **in good/bad ~:** in gutem/schlechtem Zustand; **in working ~:** betriebsfähig; **c)** *in sing. and pl. (command)* Anweisung, *die;* Anordnung, *die; (Mil.)* Befehl, *der; (Law)* Beschluß, *der;* Verfügung, *die;* **my ~s are to ..., I have ~s to ...:** ich habe Anweisung zu ...; **while following ~s** bei Befolgung der Anweisung; **act on ~s** auf Befehl handeln; **~s are ~s** Befehl ist Befehl; **court ~:** Gerichtsbeschluß, *der;* **by ~ of** auf Anordnung (+ *Gen.*); **d) in ~ to do sth.** um etw. zu tun; **in ~ that sb. should do sth.** damit jmd. etw. tut; **e)** *(Commerc.)* Auftrag, *der* **(for** über + *Akk.*); Bestellung, *die* **(for** *Gen.*); Order, *die (Kaufmannsspr.); (to waiter, ~ed goods)* Bestellung, *die;* **place an ~ [with sb.]** [jmdm.] einen Auftrag erteilen; **have sth. on ~:** etw. bestellt haben; **made to ~:** nach Maß angefertigt, maßgeschneidert ‹*Kleidung*›; **f)** *(law-abiding state)* **keep ~:** Ordnung [be]wahren; *see also* **law b; g)** *(Eccl.)* Orden, *der;* **holy ~s** heilige Weihen; **h) O~! O~!** zur Ordnung!; Ruhe bitte!; **call sb./the meeting to ~:** jmdn./die Versammlung zur Ordnung rufen; **point of ~:** Verfahrensfrage, *die;* **be in ~:** zulässig sein; *(fig.)* ‹*Forderung:*› berechtigt sein; ‹*Drink, Erklärung:*› angebracht sein; **it is in ~ for him to do that** *(fig.)* es ist in Ordnung, wenn er das tut *(ugs.);* **be out of ~** *(unacceptable)* gegen die Geschäftsordnung verstoßen; ‹*Verhalten, Handlung:*› unzulässig sein; **i)** *(kind, degree)* Klasse, *die;* Art, *die;* **j)** *(Finance)* Order, *die;* **[banker's] ~:** [Bank]anweisung, *die;* **'pay to the ~ of ...'** „zahlbar an ..." (+ *Akk.*); **k) ~ [of magnitude]** Größenordnung, *die;* **of** *or* **in the ~ of ...:** in der Größenordnung von ...; **a scoundrel of the first ~** *(fig. coll.)* ein Schurke ersten Ranges. **2.** *v. t.* **a)** *(command)* befehlen; anordnen; ‹*Richter:*› verfügen; verordnen ‹*Arznei, Ruhe usw.*›; **~ sb. to do sth.** jmdn. anweisen/*(Milit.)* jmdm. befehlen, etw. zu tun; **~ sth. [to be] done** anordnen, daß etw. getan wird; **~ sb. out of the house** jmdn. aus dem Haus weisen; **b)** *(direct the supply of)* bestellen **(from** bei); ordern ‹*Kaufmannsspr.*›; **~ in advance** vorbestellen; **c)** *(arrange)* ordnen
~ a'bout, ~ a'round *v. t. (coll.)* herumkommandieren
~ 'off *v. t. (Sport)* **~ sb. off [the field]** jmdn. vom Platz stellen
'order-form *n.* Bestellformular, *das;* Bestellschein, *der*
orderly ['ɔ:dəlɪ] **1.** *adj.* friedlich ‹*Demonstration usw.*›; diszipliniert ‹*Menge*›; *(methodical)* methodisch; ordentlich ‹*Person*›; *(tidy)* ordentlich. **2.** *n. (Mil.)* [Offiziers]bursche, *der*
ordinal ['ɔ:dɪnl] *(Math.)* **1.** *adj.* **~ number** *see* **2. 2.** *n.* Ordnungs-, Ordinalzahl, *die*

ordinance ['ɔ:dɪnəns] *n.* **a)** *(order, decree)* Verordnung, *die;* **b)** *(enactment by local authority)* Verfügung, *die;* Bestimmung, *die*

ordinarily ['ɔ:dɪnərɪlɪ] *adv.* normalerweise; gewöhnlich

ordinary ['ɔ:dɪnərɪ] *adj.* *(regular, normal)* normal ‹*Gebrauch*›; üblich ‹*Verfahren*›; *(not exceptional)* gewöhnlich; *(average)* durchschnittlich; **very ~** *(derog.)* ziemlich mittelmäßig; **~ tap-water** normales *od.* gewöhnliches Leitungswasser; **out of the ~:** außergewöhnlich; ungewöhnlich; **something/nothing out of the ~:** etwas/nichts Außergewöhnliches

ordinary level *see* **O level**

ordination [ɔ:dɪ'neɪʃn] *n.* *(Eccl.)* Ordination, *die;* Ordinierung, *die*

ordnance survey map [ɔ:dnəns 'sɜ:veɪ mæp] *n.* *(Brit.)* amtliche topographische Karte

ore [ɔ:(r)] *n.* Erz, *das*

oregano [ɒrɪ'gɑ:nəʊ] *n., no pl.* *(Cookery)* Oregano, *der*

organ ['ɔ:gən] *n.* **a)** *(Mus.)* Orgel, *die;* **b)** *(Biol.)* Organ, *das;* **c)** *(medium of communication)* Sprachrohr, *das;* *(of political party etc.)* Organ, *das*

'**organ-grinder** *n.* Drehorgelspieler, *der*/-spielerin, *die*

organic [ɔ:'gænɪk] *adj.* **a)** *(also Chem. Physiol.)* organisch; **b)** *(without chemicals)* biologisch, biodynamisch ‹*Nahrungsmittel*›; biologisch-dynamisch ‹*Ackerbau usw.*›

organisation, organise *etc. see* **organiz-**

organism ['ɔ:gənɪzm] *n.* *(Biol.)* Organismus, *der*

organist ['ɔ:gənɪst] *n.* Organist, *der*/Organistin, *die*

organization [ɔ:gənaɪ'zeɪʃn] *n.* **a)** *(organizing)* Organisation, *die;* *(of material)* Ordnung, *die;* *(of library)* Anordnung, *die;* **~ of time/work** Zeit-/Arbeitseinteilung, *die;* **b)** *(organized body, system)* Organisation, *die*

organize ['ɔ:gənaɪz] *v.t.* **a)** *(give orderly structure to)* ordnen; planen ‹*Leben*›; einteilen ‹*Arbeit, Zeit*›; *(frame, establish)* organisieren ‹*Verein, Partei, Firma, Institution*›; **I must get ~d** *(get ready)* ich muß fertig werden; **~ sb.** jmdn. an die Hand nehmen *(fig.);* **b)** *(arrange)* organisieren; **can you ~ the catering?** kümmerst du dich um die Verpflegung?

organized ['ɔ:gənaɪzd] *adj.* organisiert; geregelt ‹*Leben*›

organizer ['ɔ:gənaɪzə(r)] *n.* Organisator, *der*/Organisatorin, *die;* *(of event, festival)* Veranstalter, *der*/Veranstalterin, *die*

orgasm ['ɔ:gæzəm] *n.* Orgasmus, *der;* Höhepunkt, *der (auch fig.)*

orgy ['ɔ:dʒɪ] *n.* Orgie, *die;* **drunken ~:** Orgie unter Alkoholeinfluß

orient 1. ['ɔ:rɪənt] *n.* **the O~:** der Orient. **2.** ['ɔ:rɪent, 'ɒrɪent] *v.t.* *(fig.)* einweisen (**in** in + *Akk.*); ausrichten, abstellen (**towards** auf + *Akk.*) ‹*Programm*›; **~ oneself** sich orientieren *od.* zurechtfinden; **-~ed** -orientiert; **money-~ed** materiell orientiert

oriental [ɔ:rɪ'entl, ɒrɪ'entl] **1.** *adj.* orientalisch. **2.** *n.* Asiat, *der*/Asiatin, *die*

orientate ['ɒrɪənteɪt, 'ɔ:rɪənteɪt] *see* **orient 2**

orientation [ɒrɪən'teɪʃn, ɔ:rɪən'teɪʃn] *n.* Orientierung, *die*

orienteering [ɔ:rɪən'tɪərɪŋ, ɒrɪən'tɪərɪŋ] *n.* *(Brit.)* Orientierungsrennen, *das*

orifice ['ɒrɪfɪs] *n.* Öffnung, *die*

origin ['ɒrɪdʒɪn] *n.* *(derivation)* Abstammung, *die;* Herkunft, *die;* *(beginnings)* Anfänge *Pl.;* *(of world etc.)* Entstehung, *die;* *(source)* Ursprung, *der;* *(of belief, rumour)* Quelle, *die;* **be of humble ~, have humble ~s** bescheidener Herkunft sein; **country of ~:** Herkunftsland, *das;* **have its ~ in sth.** seinen Ursprung in etw. *(Dat.)* haben

original [ə'rɪdʒɪnl] **1.** *adj.* **a)** *(first, earliest)* ursprünglich; **the ~ inhabitants** die Ureinwohner; **b)** *(primary)* original; Original-; Ur- ‹*text, -fassung*›; eigenständig ‹*Forschung*›; *(inventive)* originell; *(creative)* schöpferisch; **an ~ painting** ein Original. **2.** *n.* Original, *das*

originality [ərɪdʒɪ'nælɪtɪ] *n.* Originalität, *die*

originally [ə'rɪdʒɪnəlɪ] *adv.* **a)** ursprünglich; **be ~ from ...:** [ursprünglich] aus ... stammen; **b)** *(in an original way)* originell ‹*schreiben usw.*›; **think ~:** originelle Gedanken haben

originate [ə'rɪdʒɪneɪt] **1.** *v.i.* **~ from** entstehen aus; **~ in** seinen Ursprung haben in (+ *Dat.*). **2.** *v.t.* schaffen; hervorbringen; *(discover)* erfinden

origination [ərɪdʒɪ'neɪʃn] *n.* Entstehung, *die*

originator [ə'rɪdʒɪneɪtə(r)] *n.* Urheber, *der*/Urheberin, *die;* *(inventor)* Erfinder, *der*/Erfinderin, *die*

ornament 1. ['ɔ:nəmənt] *n.* Schmuck-, Ziergegenstand, *der.* **2.** ['ɔ:nəment] *v.t.* verzieren

ornamental [ɔ:nə'mentl] *adj.* dekorativ; ornamental *(bes. Kunst);* Zier‹*pflanze, -naht usw.*›; **purely ~:** nur zum Schmuck *od.* zur Zierde; rein dekorativ

ornamentation [ɔ:nəmen'teɪʃn] *n., no pl.* **a)** *(ornamenting)* Ausschmückung, *die;* **b)** *(embellishment[s])* Verzierung, *die*

ornate [ɔ:'neɪt] *adj.* **a)** *(elaborately adorned)* reich verziert; prunkvoll ‹*Dekoration*›; **b)** *(style)* blumig *(abwertend);* reich ausgeschmückt ‹*Prosa*›

ornithology [ɔ:nɪ'θɒlədʒɪ] *n.* Ornithologie, *die;* Vogelkunde, *die*

orphan ['ɔ:fn] **1.** *n.* Waise, *die;* Waisenkind, *das.* **2.** *v.t.* **be ~ed** [zur] Waise werden

orphanage ['ɔ:fənɪdʒ] *n.* Waisenhaus, *das*

orthodox ['ɔ:θədɒks] *adj.* orthodox; *(conservative)* konventionell

orthodoxy ['ɔ:θədɒksɪ] *n.* Orthodoxie, *die*

orthopaedic, *(Amer.)* **orthopedic** [ɔ:θə'pi:dɪk] *adj.* orthopädisch

oscillate ['ɒsɪleɪt] *v.i.* **a)** *(swing like a pendulum)* schwingen; oszillieren *(fachspr.);* **b)** *(fig.)* schwanken

oscillation [ɒsɪ'leɪʃn] *n.* **a)** *(action)* *see* **oscillate:** Schwingen, *das;* Oszillieren, *das;* Schwanken, *das;* **b)** *(single ~)* Schwingung, *die*

osmosis [ɒz'məʊsɪs] *n., pl.* **osmoses** [ɒz'məʊsi:z] Osmose, *die*

ossify ['ɒsɪfaɪ] *v.i.* ossifizieren *(fachspr.);* verknöchern *(auch fig.)*

ostensible [ɒ'stensɪbl] *adj.* vorgeschoben; Schein-

ostensibly [ɒ'stensɪblɪ] *adv.* vorgeblich

ostentation [ɒsten'teɪʃn] *n.* Ostentation, *die (geh.);* Prahlerei, *die (abwertend);* *(showiness)* Prunk, *der*

ostentatious [ɒsten'teɪʃəs] *adj.* prunkhaft ‹*Kleidung, Schmuck*›; prahlerisch ‹*Art*›; auffällig großzügig ‹*Spende*›; **be ~ about sth.** mit etw. prunken *od. (ugs.)* protzen

osteopath ['ɒstɪəpæθ] *n.* *(Med.)* Spezialist für Knochenleiden

ostracise *see* **ostracize**

ostracism ['ɒstrəsɪzm] *n.* Ächtung, *die*

ostracize ['ɒstrəsaɪz] *v.t.* ächten

ostrich ['ɒstrɪtʃ] *n.* Strauß, *der*

other ['ʌðə(r)] **1.** *adj.* **a)** *(not the same)* ander...; **the ~ two/three** *etc.* *(the remaining)* die beiden/drei *usw.* anderen; **the ~ way**

round *or* **about** gerade umgekehrt; ~ **people's property** fremdes Eigentum; **the** ~ **one** der/die/das andere; **there is no** ~ **way** es geht nicht anders; **I know of no** ~ **way of doing it** ich weiß nicht, wie ich es sonst machen soll; **some** ~ **time** ein andermal; **b)** *(further)* **two** ~ **people/questions** noch zwei [andere *od.* weitere] Leute/Fragen; **one** ~ **thing** noch eins; **have you any** ~ **news/questions?** hast du noch weitere *od.* sonst noch Neuigkeiten/Fragen?; **c)** ~ **than** *(different from)* anders als; *(except)* außer; **any person** ~ **than yourself** jeder außer dir; **d) some writer/charity or** ~: irgendein Schriftsteller / Wohltätigkeitsverein; **some time/way or** ~: irgendwann/-wie; **something/somehow/somewhere/somebody or** ~: irgend etwas/-wie/-wo/-wer. **2.** *n.* anderer/andere/anderes; **there are six** ~**s** es sind noch sechs andere da; **tell one from the** ~: sie auseinanderhalten; **one or** ~ **of you/them** irgendwer *od.* -einer/-eine von euch/ihnen; **any** ~: irgendein anderer/-eine andere/-ein anderes; *see also* **each 2 b. 3.** *adv.* anders; ~ **than that, no real news** abgesehen davon, keine echten Neuigkeiten

otherwise ['ʌðəwaɪz] **1.** *adv.* **a)** *(in a different way)* anders; **think** ~: anders darüber denken; anderer Meinung sein; **be** ~ **engaged** anderweitige Verpflichtungen haben; **except where** ~ **stated** sofern nicht anders angegeben; **b)** *(or else)* sonst; anderenfalls; **c)** *(in other respects)* ansonsten *(ugs.);* im übrigen. **2.** *pred. adj.* anders

otter ['ɒtə(r)] *n.* [Fisch]otter, *der*

ouch [aʊtʃ] *int.* autsch

ought [ɔ:t] *v. aux. only in pres. and past* **ought,** *neg. (coll.)* **oughtn't** ['ɔ:tnt] **a) I** ~ **to do/have done it** *expr. moral duty* ich müßte es tun/hätte es tun müssen; *expr. desirability* ich sollte es tun/hätte es tun sollen; **you** ~ **to see that film** diesen Film solltest du sehen; **she** ~ **to have been a teacher** sie hätte Lehrerin werden sollen; ~ **not** *or* ~**n't you to have left by now?** müßtest du nicht schon weg sein?; **one** ~ **not to do it** man sollte es nicht tun; **he** ~ **to be hanged/in hospital** er gehört an den Galgen/ins Krankenhaus; **b)** *expr. probability* **that** ~ **to be enough** das dürfte reichen; **he** ~ **to win** er müßte [eigentlich] gewinnen; **he** ~ **to have reached Paris by now** er müßte *od.* dürfte inzwischen in Paris [angekommen] sein

oughtn't ['ɔ:tnt] *(coll.)* = **ought not**

ounce [aʊns] *n. (measure)* Unze, *die; (fig.)* **not an** ~ **of common sense** kein Fünkchen Verstand; **there is not an** ~ **of truth in it** daran ist kein Körnchen Wahrheit

our ['aʊə(r)] *poss. pron. attrib.* unser; **we have done** ~ **share** wir haben unseren Teil getan; ~ **Joe** *etc. (coll.)* unser *od. (ugs.)* uns Joe *usw.; see also* [2]**her**

ours ['aʊəz] *poss. pron. pred.* unserer/unsere/unseres; **that car is** ~: das ist unser Wagen; *see also* **hers**

ourselves [aʊə'selvz] *pron.* **a)** *emphat.* selbst; **b)** *refl.* uns. *See also* **herself**

oust [aʊst] *v. t.* **a)** *(expel, force out)* ~ **sb. from his job** jmdn. von seinem Arbeitsplatz vertreiben; ~ **the president/government from power** den Präsidenten/die Regierung entmachten; **b)** *(force out and take place of)* verdrängen; ablösen ‹*Regierung*›

out [aʊt] **1.** *adv.* **a)** *(away from place)* ~ **here/there** hier/da draußen; **'Out'** „Ausfahrt"/„Ausgang" *od.* „Aus"; **be** ~ **in the garden** draußen im Garten sein; **what's it like** ~? wie ist es draußen?; **go** ~ **shopping** *etc.* einkaufen *usw.* gehen; **be** ~ *(not at home, not in one's office, etc.)* nicht dasein; **go** ~ **in the evenings** abends aus- *od.* weggehen; **she was/stayed** ~ **all night** sie war/blieb eine/die ganze Nacht weg; **have a day** ~ **in London/at the beach** einen Tag in London/am Strand verbringen; **would you come** ~ **with me?** würdest du mit mir ausgehen?; **ten miles** ~ **from the harbour** 10 Meilen vom Hafen entfernt; **be** ~ **at sea** auf See sein; **the journey** ~: die Hinfahrt; **he is** ~ **in Africa** er ist in Afrika; **how long have you been living** ~ **here in Australia?** wie lange lebst du schon hier in Australien?; **b) be** ~ *(asleep)* weg sein *(ugs.); (drunk)* hinübersein *(ugs.); (unconscious)* bewußtlos sein; *(Boxing)* aus sein; **c)** *(no longer burning)* aus[gegangen]; **d)** *(in error)* **be 3%** ~ **in one's calculations** sich um 3% verrechnet haben; **you're a long way** ~: du hast dich gewaltig geirrt; **this is £5** ~: das stimmt um 5 Pfund nicht; **e)** *(not in fashion)* passé *(ugs.);* out *(ugs.);* **f)** *(so as to be seen or heard)* heraus; raus *(ugs.);* **there is a warrant** ~ **for his arrest** es liegt ein Haftbefehl gegen ihn vor; **say it** ~ **loud** es laut sagen; ~ **with it!** heraus *od. (ugs.)* raus damit *od.* mit der Sprache!; **their secret is** ~: ihr Geheimnis ist bekannt geworden; **[the] truth will** ~: die Wahrheit wird herauskommen; **the moon is** ~: der Mond ist zu sehen; **is the evening paper** ~ **yet?** ist die Abendausgabe schon erschienen?; **the roses are just** ~: die Rosen fangen gerade an zu blühen; **that is the best car** ~: das ist das beste Auto auf dem Markt; **g) be** ~ **for sth./to do sth.** auf etw. *(Akk.)* aussein/darauf aussein, etw. zu tun; **be** ~ **for all one can get** alles haben wollen, was man bekommen kann; **be** ~ **for trouble** Streit suchen; **they're just** ~ **to make money** sie sind nur aufs Geld aus; **h)** *(to or at an end)* **he had it finished before the day/month was** ~: er war noch am selben Tag/vor Ende des Monats damit fertig; **please hear me** ~: laß mich bitte ausreden; **Eggs? I'm afraid we're** ~: Eier? Die sind leider ausgegangen *od. (ugs.)* alle; **i) an** ~ **and** ~ **scoundrel** ein Schurke durch und durch; **an** ~ **and** ~ **disgrace** eine ungeheure Schande. *See also* **out of. 2.** *n. (way of escape)* Ausweg, *der (fig.); (excuse)* Alibi, *das*

out: ~**back** *n. (esp. Austral.)* Hinterland, *das;* ~'**bid** *v. t.,* ~**bid** überbieten; ~**board** *adj. (Naut.)* ~**board motor** Außenbordmotor, *der;* ~**break** *n.* Ausbruch, *der;* **at the** ~**break of war** bei Kriegsausbruch *od.* Ausbruch des Krieges; **an** ~**break of flu/smallpox** eine Grippe-/Pockenepidemie; ~**building** *n.* Nebengebäude, *das;* ~**burst** *n.* Ausbruch, *der;* **an** ~**burst of anger/temper** ein Zornesausbruch *(geh.) od.* Wutanfall; ~**cast** *n.* Ausgestoßene, *der/die;* **a social** ~**cast** ein Geächteter/eine Geächtete; ein Outcast *(Soziol.);* ~'**class** *v. t.* überlegen sein (+ *Dat.*); ~**come** *n.* Ergebnis, *das;* Resultat, *das;* **what was the** ~**come of your meeting?** was ist bei eurer Versammlung herausgekommen?; ~**cry** *n., no pl. (clamour)* [Aufschrei der] Empörung; [Sturm der] Entrüstung; **a public/general** ~**cry about/against sth.** allgemeine Empörung *od.* Entrüstung über etw. *(Akk.);* ~'**dated** *adj.* veraltet; überholt; antiquiert *(abwertend)* ‹*Ausdrucksweise*›; altmodisch ‹*Vorstellung, Kleidung*›; ~'**distance** *v. t.* [weit] hinter sich

(Dat.) lassen; überflügeln; ~'**do** *v. t.*, ~**doing** [aʊt'du:ɪŋ], ~**did** [aʊt'dɪd], ~**done** [aʊt'dʌn] übertreffen, überbieten (**in** an + *Dat.*); **not to be ~done [by sb.]** um nicht zurückzustehen [hinter jmdm.]; ~**door** *adj.* ~**door shoes/things** Straßenschuhe/-kleidung, *die;* **be an ~door type** gern und oft im Freien sein; **lead an ~door life** viel im Freien sein; ~**door games/pursuits** Spiele/Beschäftigungen im Freien; ~**door swimming-pool** Freibad, *das;* ~'**doors 1.** *adv.* draußen; **go ~doors** nach draußen gehen; **2.** *n.* **the [great] ~doors** die freie Natur

outer ['aʊtə(r)] *adj.* äußer...; Außen⟨*fläche, -seite, -wand, -tür, -hafen*⟩; ~ **garments** Oberbekleidung, *die*

outer 'space *n.* Weltraum, *der;* All, *das*

out: ~**fit** *n.* **a)** *(person's clothes)* Kleider *Pl.; (for fancy-dress party)* Kostüm, *das;* **b)** *(complete equipment)* Ausrüstung, *die;* Ausstattung, *die;* **c)** *(coll.: group of persons)* Haufen, *der (ugs.); (Mil.)* Haufen, *der (Soldatenspr.);* Trupp, *der;* **d)** *(coll.: organization)* Laden, *der (ugs.);* ~**fitter** *n.* Ausrüster, *der/*Ausrüsterin, *die;* **camping/sports ~fitter** Camping-/Sportgeschäft, *das;* ~**going 1.** *adj.* **a)** *(retiring from office)* [aus dem Amt] scheidend ⟨*Regierung, Präsident, Ausschuß*⟩; **b)** *(friendly)* kontaktfreudig ⟨*Person*⟩; **you should be more ~going** du solltest mehr aus dir herausgehen; **c)** *(going ~)* abgehend ⟨*Zug, Schiff*⟩; ausziehend ⟨*Mieter*⟩; ~**going flights will be delayed** bei den Abflügen wird es zu Verzögerungen kommen; **2.** *n. in pl. (expenditure)* Ausgaben *Pl.;* ~'**grow** *v. t., forms as* **grow: a)** *(leave behind)* entwachsen (+ *Dat.*); ablegen ⟨*Interesse, Schüchternheit, Vorliebe*⟩; überwinden ⟨*Ansicht, Schüchternheit*⟩; **we've ~grown all that** das alles haben wir hinter uns; **b)** *(become taller than)* größer werden als; über den Kopf wachsen (+ *Dat.*) ⟨*älterem Bruder usw.*⟩; *(grow too big for)* herauswachsen aus ⟨*Kleidung*⟩; ~**house** *n.* *(building)* Nebengebäude, *das*

outing ['aʊtɪŋ] *n.* Ausflug, *der;* **school/day's ~:** Schul-/Tagesausflug, *der;* **firm's/works ~:** Betriebsausflug, *der;* **go on an ~:** einen Ausflug machen

out: ~**landish** [aʊt'lændɪʃ] *adj.* **a)** *(looking or sounding foreign)* fremdländisch; **b)** *(bizarre)* ausgefallen; seltsam, sonderbar ⟨*Benehmen*⟩; verschroben ⟨*Ansichten*⟩; ~'**last** *v. t.* überdauern; überleben ⟨*Person, Jahrhundert*⟩; ~**law 1.** *n.* Bandit, *der/*Banditin, *die;* **2.** *v. t. (make illegal)* verbieten ⟨*Zeitung, Handlung*⟩; ~**lay** *n.* **an ~lay** Ausgaben *Pl.* (**on** für); **initial ~lay** Anschaffungskosten *Pl.;* ~**let** ['aʊtlet, 'aʊtlɪt] **a)** Ablauf, -fluß, *der;* Auslauf, -laß, *der;* **b)** *(fig.: vent)* Ventil, *das;* **c)** *(Commerc.) (market)* Absatzmarkt, *der; (shop)* Verkaufsstelle, *die;* ~**line 1.** *n.* **a)** *in sing. or pl. (line[s])* Umriß, *der;* Kontur, *die;* Silhouette, *die;* **b)** *(short account)* Grundriß, *der;* Grundzüge *Pl.; (of topic)* Übersicht, die (**of** über + *Akk.*); *(rough draft)* Entwurf, *der* (**of, for** *Gen. od.* zu); **2.** *v. t.* **a)** *(draw ~line of)* ~**line sth.** die Umrisse *od.* Konturen einer Sache zeichnen; **b) the mountain was ~lined against the sky** die Umrisse *od.* Konturen des Berges zeichneten sich gegen den Himmel ab; **c)** *(describe in general terms)* skizzieren, umreißen ⟨*Programm, Plan, Projekt*⟩; ~**live** [aʊt'lɪv] *v. t.* überleben; **it's ~lived its usefulness** es ist unbrauchbar geworden; ~**look** *n.* **a)** *(prospect)* Aussicht, *die* (**over** über + *Akk.*, **on to** auf + *Akk.*); *(fig., Meteorol.)* Aussichten *Pl.;* **b)** *(mental attitude)* Haltung, *die* (**on** gegenüber); ~**look on life** Lebensauffassung, *die;* ~**lying** *adj.* abgelegen, entlegen ⟨*Gegend, Vorort, Dorf*⟩; ~**ma'nœuvre** *v. t.* überlisten ⟨*Truppen*⟩; ausstechen, ausmanövrieren ⟨*Rivalen*⟩; ~**moded** [aʊt'məʊdɪd] *adj.* **a)** *(no longer in fashion)* altmodisch; **b)** *(obsolete)* veraltet; antiquiert *(abwertend)* ⟨*Ausdrucksweise*⟩; ~'**number** *v. t.* zahlenmäßig überlegen sein (+ *Dat.*); **they were ~numbered five to one** die anderen waren fünfmal so viele wie sie

'out of *prep.* **a)** *(from within)* aus; **go ~ the door** zur Tür hinausgehen; **fall ~ sth.** aus etw. [heraus]fallen; **b)** *(not within)* **be ~ the country** im Ausland sein; **be ~ town/the room** nicht in der Stadt/im Zimmer sein; **feel ~ it** *or* **things** sich ausgeschlossen *od.* nicht dazu gehörig fühlen; **I'm glad to be ~ it** ich bin froh, daß ich die Sache hinter mir habe; **be ~ the tournament** aus dem Turnier ausgeschieden sein; **c)** *(from among)* **one ~ every three smokers** jeder dritte Raucher; **58 ~ every 100** 58 von hundert; **pick one ~ the pile** einen/eine/eins aus dem Stapel herausgreifen; **eighth ~ ten** als Achter von zehn Teilnehmern *usw.;* **d)** *(beyond range of)* außer ⟨*Reich-/Hörweite, Sicht, Kontrolle*⟩; **e)** *(from)* aus; **get money ~ sb.** Geld aus jmdm. herausholen; **do well ~ sb./sth.** von jmdm./etw. profitieren; **f)** *(owing to)* aus ⟨*Mitleid, Furcht, Neugier usw.*⟩; **g)** *(without)* **be ~ luck** kein Glück haben; ~ **money** ohne Geld; ~ **work** ohne Arbeit; arbeitslos; **we're ~ tea** der Tee ist uns ausgegangen; **h)** *(by use of)* aus; **make a profit ~ sth.** mit etw. ein Geschäft machen; **made ~ silver** aus Silber; **i)** *(away from)* von ... entfernt; **ten miles ~ London** 10 Meilen außerhalb von London; **j)** *(beyond) see* **depth c; ordinary b**

out: ~**-of-date** *attrib. adj. (old, not relevant)* veraltet; *(old-fashioned)* altmodisch; unmodern; antiquiert *(abwertend)* ⟨*Ausdrucksweise*⟩; *(expired)* ungültig, verfallen ⟨*Karte*⟩; ~**-of-the-way** *attrib. adj. (remote)* abgelegen; entlegen; *(unusual)* ausgefallen; ~**-of-work** *attrib. adj.* arbeitslos; ~**-patient** *n.* ambulanter Patient/ambulante Patientin; ~**-patients[' department]** Poliklinik, *die;* **be an ~-patient** ambulant behandelt werden; ~'**play** *v. t. (Sport)* besser spielen als; ~**post** *n.* Außenposten, *der; (of civilization etc.; also Mil.)* Vorposten, *der;* ~**put 1.** *n.* **a)** *(amount)* Output, *der (fachspr.);* Produktion, *die; (of liquid, electricity, etc.)* Leistung, *die;* **b)** *(Computing)* Ausgabe, *die;* Output, *der (fachspr.);* **c)** *(Electr.) (energy)* [Ausgangs]leistung, *die;* Output, *der (fachspr.); (signal)* Ausgangssignal, *das;* **2.** *v. t.*, **-tt-**, ~**put** *or* ~**putted** ['aʊtpʊtɪd] *(Computing)* ausgeben ⟨*Information*⟩

outrage 1. ['aʊtreɪdʒ] *n. (deed of violence, violation of rights)* Verbrechen, *das; (during war)* Greueltat, *die; (against good taste or decency)* grober *od.* krasser Verstoß; *(upon dignity)* krasse *od.* grobe Verletzung (**upon** *Gen.*); **be an ~ against good taste/decency** den guten Geschmack/Anstand in grober *od.* krasser Weise verletzen; **an ~ against humanity** ein Verbrechen gegen die Menschheit. **2.** ['aʊtreɪdʒ, aʊt'reɪdʒ] *v. t.* **a)** empören; **be ~d at** *or* **by sth.** über etw. *(Akk.)* empört sein; **b)** *(in-*

fringe) in grober *od.* krasser Weise verstoßen gegen ⟨*Anstand, Moral*⟩

outrageous [aʊt'reɪdʒəs] *adj.* **a)** *(immoderate)* unverschämt *(ugs.)* ⟨*Forderung*⟩; unverschämt hoch ⟨*Preis, Summe*⟩; grell, schreiend ⟨*Farbe*⟩; zu auffällig ⟨*Kleidung*⟩; maßlos ⟨*Übertreibung*⟩; **it's ~!** das ist unverschämt *od.* eine Unverschämtheit!; **b)** *(grossly cruel, offensive)* ungeheuer ⟨*Grausamkeit*⟩; unverschämt ⟨*Lüge, Benehmen, Unterstellung*⟩; wüst ⟨*Schmähung*⟩; ungeheuerlich ⟨*Anklage*⟩; unerhört ⟨*Frechheit, Unhöflichkeit, Skandal*⟩; unflätig ⟨*Sprache*⟩

outrageously [aʊt'reɪdʒəslɪ] *adv.* zu auffällig, aufdringlich ⟨*sich kleiden, schminken*⟩; maßlos ⟨*übertreiben*⟩; unverschämt, schamlos ⟨*lügen, sich benehmen*⟩; fürchterlich ⟨*fluchen*⟩

out: ~rider *n. (motor-cyclist)* **[motor-cycle] ~rider** Kradbegleiter, *der*/-begleiterin, *die;* **~right 1.** [-'-] *adv.* **a)** *(altogether, entirely)* ganz, komplett ⟨*kaufen, verkaufen*⟩; *(instantaneously, on the spot)* auf der Stelle; **pay for/purchase/buy sth. ~right** sofort den ganzen Preis für etw. bezahlen; **b)** *(openly)* geradeheraus *(ugs.)*, freiheraus, rundheraus ⟨*erzählen, sagen, lachen*⟩; **2.** ['--] *adj.* ausgemacht ⟨*Unsinn, Schlechtigkeit, Unehrlichkeit*⟩; rein, pur *(ugs.)* ⟨*Arroganz, Unverschämtheit, Irrtum, Egoismus, Unsinn*⟩; glatt *(ugs.)*⟨*Ablehnung, Absage, Lüge*⟩; klar ⟨*Sieg, Niederlage, Sieger*⟩; **~'run** *v. t., forms as* **run 3: a)** *(run faster than)* schneller laufen *od.* sein als; **b)** *(escape)* entkommen (+ *Dat.*); **~'sell** *v. t., forms as* **sell 1: a)** *(be sold in greater quantities than)* sich besser verkaufen als; **b)** *(sell more than)* mehr verkaufen als; **~set** *n.* Anfang, *der;* Beginn, *der;* **at the ~set** zu Beginn *od.* Anfang; am Anfang; **from the ~set** von Anfang an; **~'shine** *v. t.*, **~shone** [aʊt'ʃɒn] *(fig.)* in den Schatten stellen

outside 1. [-'-, '--] *n.* **a)** Außenseite, *die;* **on the ~:** außen; **to/from the ~:** nach/von außen; **overtake sb. on the ~:** jmdn. außen überholen; **b)** *(external appearance)* Äußere, *das;* äußere Erscheinung; **c) at the [very] ~** *(coll.)* äußerstenfalls; höchstens. **2.** ['--] *adj.* **a)** *(of, on, nearer the ~)* äußer...; Außen⟨*wand, -mauer, -antenne, -toilette, -ansicht*⟩; **~ lane** Überholspur, *die;* **b)** *(remote)* **have only an ~ chance** nur eine sehr geringe Chance haben; **c)** fremd ⟨*Hilfe*⟩; äußer... ⟨*Einfluß*⟩; Freizeit⟨*aktivitäten, -interessen*⟩; **d)** *(greatest possible)* maximal, höchst ⟨*Schätzung*⟩; **at an ~ estimate** maximal *od.* höchstens *od.* im Höchstfall. **3.** [-'-] *adv.* **a)** *(on the ~)* draußen; *(to the ~)* nach draußen; **the world ~:** die Außenwelt; **b) ~ of** *see* **4. 4.** [-'-] *prep.* **a)** *(on outer side of)* außerhalb (+ *Gen.*); **b)** *(beyond)* außerhalb (+ *Gen.*) ⟨*Reichweite, Festival, Familie*⟩; **it's ~ the terms of the agreement** es gehört nicht zu den Bedingungen der Abmachung; **c)** *(to the ~ of)* aus ... hinaus; **go ~ the house** nach draußen gehen

outside 'broadcast *n. (Brit.)* Außenübertragung, *die*

outsider [aʊt'saɪdə(r)] *n. (Sport; also fig.)* Außenseiter, *der*

out: ~size *adj.* überdimensional; **~size clothes** Kleidung in Übergröße; **~skirts** *n. pl.* Stadtrand, *der;* **the ~skirts of the city** die Außenbezirke der Stadt; **~'smart** *v. t. (coll.)* reinlegen *(ugs.)*; **~'spoken** *adj.* freimütig ⟨*Person, Kritik, Bemerkung, Kommentar*⟩; **~spread** *adj.* ['--, *pred.* -'-] ausgebreitet; **~-'standing** *adj.* **a)** *(conspicuous)* hervorstechend ⟨*Merkmal*⟩; **b)** *(exceptional)* hervorragend ⟨*Leistung, Redner, Künstler, Dienst*⟩; überragend ⟨*Bedeutung*⟩; außergewöhnlich ⟨*Person, Mut, Fähigkeit, Geschick*⟩; **not be ~standing** nicht überragend sein; **of ~standing ability/skill** außergewöhnlich fähig/geschickt; **c)** *(not yet settled)* ausstehend ⟨*Schuld, Verbindlichkeit, Geldsumme*⟩; offen, unbezahlt ⟨*Rechnung*⟩; unerledigt ⟨*Arbeit*⟩; ungelöst ⟨*Problem*⟩; **there's £5 still ~standing** es stehen noch 5 Pfund aus; **~standingly** [aʊt'stændɪŋlɪ] *adv.* außergewöhnlich ⟨*intelligent, gut, begabt*⟩; **not ~standingly** nicht besonders; **be ~standingly good at tennis** hervorragend Tennis spielen; **~station** *n.* Außenposten, *der;* **~'stay** *v. t. (stay beyond)* überziehen ⟨*Urlaub*⟩; **~stretched** *adj.* ausgestreckt; *(spread out)* ausgebreitet; **~'strip** *v. t.* **a)** *(pass in running)* überholen; **b)** *(in competition)* überflügeln; übersteigen ⟨*Einsicht, Ressourcen, Ersparnisse*⟩; **~-tray** *n.* Ablage für Ausgänge; **~'vote** *v. t.* überstimmen

outward ['aʊtwəd] **1.** *adj.* **a)** *(external, apparent)* [rein] äußerlich; äußere ⟨*Erscheinung, Bedingung*⟩; **with an ~ show of confidence** mit einem Anstrich von Selbstsicherheit; **b)** *(going out)* Hin⟨*reise, -fracht*⟩; **~ flow of money/traffic** Kapitalabfluß, *der*/abfließender Verkehr. **2.** *adv.* nach außen ⟨*aufgehen, richten*⟩

outwardly ['aʊtwədlɪ] *adv.* nach außen hin ⟨*Gefühle zeigen*⟩; öffentlich ⟨*Loyalität erklären*⟩

outwards *see* **outward 2**

out: ~'weigh *v. t.* schwerer wiegen als; überwiegen ⟨*Nachteile*⟩; **~'wit** *v. t.*, **-tt-** überlisten

oval ['əʊvl] **1.** *adj.* oval. **2.** *n.* Oval, *das*

ovary ['əʊvərɪ] *n.* Ovarium, *das;* Eierstock, *der; (Bot.)* Ovarium, *das;* Fruchtknoten, *der*

ovation [əʊ'veɪʃn] *n.* Ovation, *die;* begeisterter Beifall; **get an ~ for sth.** Ovationen *od.* begeisterten Beifall für etw. bekommen; **a standing ~:** stehende Ovationen

oven ['ʌvn] *n.* [Back]ofen, *der;* **put sth. in the ~ for 40 minutes** etw. 40 Minuten backen; **it's like an ~ in here** hier ist es warm wie in einem Backofen

oven: ~-glove *n.* Topfhandschuh, *der;* **~-proof** *adj.* feuerfest; **~-ready** *adj.* backfertig ⟨*Pommes frites, Pastete*⟩; bratfertig ⟨*Geflügel*⟩; **~ware** *n., no pl.* feuerfestes Geschirr

over ['əʊvə(r)] **1.** *adv.* **a)** *(outward and downward)* hinüber; **b)** *(so as to cover surface)* **draw/board/cover ~:** zuziehen/-nageln/-decken; **c)** *(with motion above sth.)* **climb/look/jump ~:** hinüber- *od.* *(ugs.)* rüberklettern/-sehen/-springen; **d)** *(so as to reverse position etc.)* herum; **switch ~:** umschalten ⟨*Programm, Sender*⟩; **it rolled ~ and ~:** es rollte und rollte; **e)** *(across a space)* hinüber; *(towards speaker)* herüber; **he swam ~ to us/the other side** er schwamm zu uns herüber/hinüber zur anderen Seite; **fly ~:** vorüberfliegen; **~ here/there** *(direction)* hier herüber/dort hinüber; *(location)* hier/dort; **they are ~ [here] for the day** sie sind einen Tag hier; **ask sb. ~ [for dinner]** jmdn. [zum Essen] einladen; **f)** *(Radio)* **[come in, please,] ~:** übernehmen Sie bitte; **~ and out** Ende; **g)** *(in excess etc.)* **children of 12 and ~:** Kinder im Alter von zwölf Jahren und darüber; **be [left] ~:** übrig[geblieben] sein; **have ~:** übrig haben ⟨*Geld*⟩;

9 into 28 goes 3 and 1 ~: 28 geteilt durch neun ist gleich 3, Rest 1; **it's a bit** ~ *(in weight)* es ist ein bißchen mehr; **h)** *(from beginning to end)* von Anfang bis Ende; **say sth. twice** ~: etw. wiederholen *od.* zweimal sagen; ~ **and** ~ **[again]** immer wieder; **several times** ~: mehrmals; **i)** *(at an end)* vorbei; vorüber; **be** ~: vorbei sein; ‹*Aufführung*:› zu Ende sein; **get sth.** ~ **with** etw. hinter sich *(Akk.)* bringen; **be** ~ **and done with** erledigt sein; **j) all** ~ *(completely finished)* aus [und vorbei]; *(in or on one's whole body etc.)* überall; *(in characteristic attitude)* typisch; **I ache all** ~: mir tut alles weh; **be shaking all** ~: am ganzen Körper zittern; **embroidered all** ~ **with flowers** ganz mit Blumen bestickt; **that is him/sth. all** ~: das ist typisch für ihn/etw.; **k)** *(overleaf)* umseitig; **see** ~: siehe Rückseite. **2.** *prep.* **a)** *(above) (indicating position)* über (+ *Dat.*); *(indicating motion)* über (+ *Akk.*); **b)** *(on) (indicating position)* über (+ *Dat.*); *(indicating motion)* über (+ *Akk.*); **hit sb.** ~ **the head** jmdm. auf den Kopf schlagen; **carry a coat** ~ **one's arm** einen Mantel über dem Arm tragen; **c)** *(in or across every part of)* [überall] in (+ *Dat.*); *(to and fro upon)* über (+ *Akk.*); *(all through)* durch; **all** ~ *(in or on all parts of)* überall in (+ *Dat.*); **travel all** ~ **the country** das ganze Land bereisen; **all** ~ **Spain** überall in Spanien; **she spilt wine all** ~ **her skirt** sie hat sich *(Dat.)* Wein über den ganzen Rock geschüttet; **all** ~ **the world** in der ganzen Welt; **d)** *(round about) (indicating position)* über (+ *Dat.*); *(indicating motion)* über (+ *Akk.*); **a sense of gloom hung** ~ **him** ihn umgab eine gedrückte Stimmung; **e)** *(on account of)* wegen; **laugh** ~ **sth.** über etw. *(Akk.)* lachen; **f)** *(engaged with)* bei; **take trouble** ~ **sth.** sich *(Dat.)* mit etw. Mühe geben; **be a long time** ~ **sth.** lange für etw. brauchen; ~ **work/dinner/a cup of tea** bei der Arbeit/beim Essen/bei einer Tasse Tee; **g)** *(superior to, in charge of)* über (+ *Akk.*); **have command/authority** ~ **sb.** Befehlsgewalt über jmdn./Weisungsbefugnis gegenüber jmdm. haben; **be** ~ **sb.** *(in rank)* über jmdm. stehen; **h)** *(beyond, more than)* über (+ *Akk.*); **an increase** ~ **last year's total** eine Zunahme gegenüber der letztjährigen Gesamtmenge; **it's been** ~ **a month since ...**: es ist über einen Monat her, daß ...; ~ **and above** zusätzlich zu; **i)** *(in comparison with)* **a decrease** ~ **last year** eine Abnahme gegenüber dem letzten Jahr; **j)** *(out and down from etc.)* über (+ *Akk.*); **look** ~ **a wall** über eine Mauer sehen; **the window looks** ~ **the street** das Fenster geht zur Straße hinaus; **fall** ~ **a cliff** von einem Felsen stürzen; **k)** *(across)* über (+ *Akk.*); **the pub** ~ **the road** die Wirtschaft auf der anderen Straßenseite *od.* gegenüber; **climb** ~ **the wall** über die Mauer steigen *od.* klettern; **be** ~ **the worst** das Schlimmste hinter sich *(Dat.) od.* überstanden haben; **be** ~ **an illness** eine Krankheit überstanden haben; **l)** *(throughout, during)* über (+ *Akk.*); **stay** ~ **Christmas/the weekend/Wednesday** über Weihnachten/das Wochenende/bis Donnerstag bleiben; ~ **the summer** den Sommer über; ~ **the past years** in den letzten Jahren

over: ~-a'**bundant** *adj.* überreichlich; ~'**act 1.** *v. t.* übertrieben spielen ‹*Rolle, Theaterstück*›; chargieren ‹*Nebenrolle*›; **2.** *v. i.* übertreiben; ~**all 1.** *n.* **a)** *(Brit.: garment)* Arbeitsmantel, *der;* Arbeitskittel, *der;* **b)** *in pl.* **[pair of]** ~**alls** Overall, *der; (with a bib and strap top)* Latzhose, *die;* **2.** *adj.* **a)** *(from end to end; total)* Gesamt‹*breite, -einsparung, -klassement, -abmessung*›; **have an** ~**all majority** die absolute Mehrheit haben; **b)** *(general)* allgemein ‹*Verbesserung, Wirkung*›; **3.** ['---, --'-] *adv. (taken as a whole)* im großen und ganzen; ~-'**anxious** *adj.* übermäßig besorgt; **be** ~**-anxious to do sth.** etw. unbedingt tun wollen; ~'**awe** *v. t.* Ehrfurcht einflößen (+ *Dat.*); ‹*Waffe, Anzahl*:› einschüchtern; ~'**balance 1.** *v. i.* ‹*Person*:› das Gleichgewicht verlieren, aus dem Gleichgewicht kommen; **2.** *v. t.* aus dem Gleichgewicht bringen ‹*Person*›; ~'**bearing** *adj.* herrisch; ~'**bid** *v. t.*, ~**bid** überbieten ‹*Händler, Gegner, Gebot*›; ~**board** *adv.* über Bord; **fall** ~**board** über Bord gehen; **go** ~**board** *(fig. coll.)* ausflippen *(ugs.)* **(about** wegen); ~'**book** *v. t.* überbuchen; ~'**burden** *v. t.* *(fig.)* überlasten ‹*System, Person*› **(by** mit); ~-'**careful** *adj.* übervorsichtig; ~**cast** *adj.* trübe ‹*Wetter, Himmel, Tag*›; bewölkt ‹*Himmel, Nacht*›; bedeckt, bezogen ‹*Himmel*›; ~-'**cautious** *adj.* übervorsichtig; ~-'**charge** *v. t.* **a)** *(charge beyond reasonable price)* zuviel abnehmen *od.* abverlangen (+ *Dat.*); **b)** *(charge beyond right price)* zuviel berechnen (+ *Dat.*); ~**coat** *n.* Mantel, *der;* ~'**come 1.** *v. t., forms as* **come: a)** *(prevail* ~*)* überwinden; bezwingen ‹*Feind*›; ablegen ‹*Angewohnheit*›; widerstehen (+ *Dat.*) ‹*Versuchung*›; ‹*Schlaf*:› überkommen, übermannen; ‹*Dämpfe*:› betäuben; **b)** *in p.p. (exhausted, affected)* **he was** ~**come by grief/with emotion** Kummer/Rührung übermannte *od.* überwältigte ihn; **she was** ~**come by fear/shyness** Angst/Schüchternheit überkam *od.* überwältigte sie; **they were** ~**come with remorse** Reue befiel sie; **2.** *v. i., forms as* **come** siegen; ~-'**confidence** *n.* übersteigertes Selbstvertrauen; ~-'**confident** *adj.* übertrieben zuversichtlich; ~-'**cooked** *adj.* verkocht; ~-'**critical** *adj.* zu kritisch; **be** ~**-critical of sth.** etw. zu sehr kritisieren; ~'**crowded** *adj.* überfüllt ‹*Zug, Bus, Raum*›; übervölkert ‹*Stadt*›; ~'**crowding** *n. (of train, bus, room)* Überfüllung, *die; (of city)* Übervölkerung, *die;* ~'**do** *v. t.*, ~'**doing** [əʊvə'du:ɪŋ], ~'**did** [əʊvə'dɪd], ~'**done** [əʊvə'dʌn] **a)** *(carry to excess)* übertreiben; **b)** ~**do it** *or* **things** *(work too hard)* sich übernehmen; *(exaggerate)* es übertreiben; ~'**done** *adj.* **a)** *(exaggerated)* übertrieben; **b)** *(cooked too much)* verkocht; verbraten ‹*Fleisch*›; ~**dose 1.** ['---] *n.* Überdosis, *die;* **2.** [--'-] *v. t.* eine Überdosis geben (+ *Dat.*); **3.** ['---] *v. i.* eine Überdosis nehmen; ~**draft** *n.* Kontoüberziehung, *die;* **have an** ~**draft of £50 at the bank** sein Konto um 50 Pfund überzogen haben; **get/pay off an** ~**draft** einen Überziehungskredit erhalten/abbezahlen; ~'**draw** *v. t., forms as* **draw 1** *(Banking)* überziehen ‹*Konto*›; ~'**drawn** *adj.* überzogen ‹*Konto*›; **I am** ~**drawn [at the bank]** mein Konto ist überzogen; ~'**dress** *v. i.* sich zu fein anziehen; ~'**dressed** *adj.* zu fein angezogen; ~**drive** *n. (Motor Veh.)* Overdrive, *der;* Schongang, *der;* ~'**due** *adj.* überfällig; **the train is 15 minutes** ~**due** der Zug hat schon 15 Minuten Verspätung; ~-'**eager** *adj.* übereifrig; **be** ~**-eager to do sth.** sich übereifrig bemühen, etw. zu tun; ~'**eat** *v. i., forms as* **eat** zuviel essen; ~**eating** übermäßiges

Essen; ~-'**emphasize** *v. t.* überbetonen; ~**estimate 1.** [əʊvər'estɪmeɪt] *v. t.* überschätzen; **2.** [əʊvər'estɪmət] *n.* zu hohe Schätzung; ~-**ex'cite** *v. t.* zu sehr aufregen ⟨*Patient*⟩; **become ~-excited** ganz aufgeregt werden; ~**ex'ert** *v. refl.* sich überanstrengen; ~-**ex'pose** *v. t.* *(Photog.)* überbelichten; ~'**feed** *v. t., forms as* **feed 1** überfüttern ⟨*Tier, (fam.) Kind*⟩; ~'**fill** *v. t.* zu voll machen; ~**flow 1.** [--'-] *v. t.* **a)** *(flow ~)* laufen über (+ *Akk.*) ⟨*Rand*⟩; **b)** *(flow ~ brim of)* überlaufen aus ⟨*Tank*⟩; **a river ~flowing its banks** ein Fluß, der über die Ufer tritt; **c)** *(extend beyond limits of)* ⟨*Menge, Personen:*⟩ nicht genug Platz finden in (+ *Dat.*); **d)** *(flood)* überschwemmen ⟨*Feld*⟩; **2.** [--'-] *v. i.* **a)** *(flow ~ edge or limit)* überlaufen; **be filled/full to ~flowing** ⟨*Raum:*⟩ überfüllt sein; ⟨*Flüssigkeitsbehälter:*⟩ zum Überlaufen voll sein; **b)** *(fig.)* ⟨*Herz, Person:*⟩ überfließen *(geh.)*, überströmen (**with** vor + *Dat.*); **3.** ['---] *n.* **a)** *(what flows ~, lit. or fig.)* **the ~flow** was übergelaufen ist; **~flow of population** Bevölkerungsüberschuß, *der;* **b)** *(outlet)* **~flow [pipe]** Überlauf, *der;* ~ '**fly** *v. t., forms as* [2]**fly 2: a)** *(fly ~)* überfliegen; **b)** *(fly beyond)* hinausschießen über (+ *Akk.*) ⟨*Landebahn*⟩; ~'**full** *adj.* zu voll; übervoll; ~-'**generous** *adj.* zu *od.* übertrieben großzügig ⟨*Person*⟩; reichlich groß ⟨*Portion*⟩; ~**grown** *adj.* **a)** überwachsen, überwuchert ⟨*Beet*⟩ (**with** von); **b) he acts like an ~grown schoolboy** er führt sich auf wie ein großes Kind; ~**hang 1.** [--'-] *v. t.*, ~**hung** [əʊvə'hʌŋ] ⟨*Felsen, Stockwerk:*⟩ hinausragen über (+ *Akk.*); **2.** ['---] *n.* Überhang, *der;* ~'**hanging** *adj.* überhängend; ~-'**hasty** *adj.* vorschnell, übereilt ⟨*Urteil, Entschluß, Antwort*⟩; ~**haul 1.** [--'-] *v. t.* **a)** überholen ⟨*Auto, Schiff, Maschine, Motor*⟩; überprüfen ⟨*System*⟩; **b)** *(~take)* überholen ⟨*Fahrzeug, Person*⟩; **2.** ['---] *n.* Überholung, *die;* **need an ~haul** ⟨*Maschine:*⟩ überholt werden müssen; ⟨*System:*⟩ überarbeitet werden müssen; ~**head 1.** [--'-] *adv.* **high ~head** hoch oben; **hear a sound ~head** ein Geräusch über sich *(Dat.)* hören; **2.** ['---] *adj.* **a)** **~head wires** Hochleitung, *die;* **~head cable** Luftkabel, *das;* **~head projector** Overheadprojektor, *der;* **b)** **~head expenses/charges/costs** *(Commerc.)* Gemeinkosten *Pl.;* **3.** ['---] *n.* **~heads,** *(Amer.)* **~head** *(Commerc.)* Gemeinkosten *Pl.;* ~'**hear** *v. t., forms as* **hear 1** *(accidentally)* zufällig [mit]hören, mitbekommen ⟨*Unterhaltung, Bemerkung*⟩; *(intentionally)* belauschen ⟨*Gespräch, Personen*⟩; ~'**heat** *v. t.* überhitzen ⟨*Motor, Metall usw.*⟩; ~-**in'dulge 1.** *v. t.* zu sehr frönen *(geh.)* (+ *Dat.*) ⟨*Appetit*⟩; **~-indulge oneself** sich allzusehr gehen lassen; **2.** *v. i.* es übertreiben; ~-**in'dulgence** *n.* übermäßiger Genuß (**in** von); **~-indulgence in drink/drugs** übermäßiges Trinken/übermäßiger Drogengenuß

overjoyed [əʊvə'dʒɔɪd] *adj.* überglücklich (**at** über + *Akk.*)

over: ~**kill** *n.* *(Mil.)* Overkill, *das od. der;* **be ~kill** *(fig.)* zuviel des Guten sein; ~**land 1.** [--'-] *adv.* auf dem Landweg; **2.** ['---] *adj.* **by the ~land route** auf dem Landweg; **~land transport/journey** Beförderung/Reise auf dem Landweg; ~**lap 1.** [--'-] *v. t.* überlappen ⟨*Fläche, Dachziegel*⟩; sich überschneiden mit ⟨*Aufgabe, Datum*⟩; **2.** [--'-] *v. i.* ⟨*Flächen, Dachziegel:*⟩ sich überlappen; ⟨*Aufgaben, Daten:*⟩ sich überschneiden; ⟨*Bretter:*⟩ teilweise übereinanderliegen; **3.** ['---] *n.* Überlappung, *die;* *(of dates or tasks; between subjects, periods, etc.)* Überschneidung, *die;* ~**lay 1.** [--'-] *v. t., forms as* [2]**lay 1** *(cover)* bedecken; **2.** ['---] *n.* **a)** *(cover)* Überzug, *der;* **b)** *(transparent sheet)* Auflegefolie, *die;* ~'**leaf** *adv.* auf der Rückseite; **see diagram ~leaf** siehe das umseitige Diagramm; ~**load 1.** [--'-] *v. t.* überladen *(auch fig.)*, überlasten ⟨*Stromkreis, Lautsprecher usw.*⟩; überbelasten ⟨*Maschine, Motor, Mechanismus usw.*⟩; **2.** ['---] *n.* *(Electr.)* Überlastung, *die;* ~'**look** *v. t.* **a)** *(have view of)* ⟨*Hotel, Zimmer, Haus:*⟩ Aussicht haben *od.* bieten auf (+ *Akk.*); **house ~looking the lake** Haus mit Blick auf den See; **b)** *(not see, ignore)* übersehen; *(allow to go unpunished)* hinwegsehen über (+ *Akk.*) ⟨*Vergehen, Beleidigung*⟩; ~'**man** *v. t.* überbesetzen; ~'**manning** *n.* [personelle] Überbesetzung; ~-'**modest** *adj.* zu bescheiden; ~-'**much 1.** *adj.* allzu viel; **2.** *adv.* allzusehr

overnight 1. [--'-] *adv.* *(also fig.: suddenly)* über Nacht; **stay ~ in a hotel** in einem Hotel übernachten. **2.** ['---] *adj.* **a)** ~ **train/bus** Nachtzug, *der*/Nachtbus, *der;* ~ **stay** Übernachtung, *die;* **b)** *(fig.: sudden)* **be an ~ success** über Nacht Erfolg haben

overnight: ~ **bag** *n.* [kleine] Reisetasche; ~ **case** *n.* Handköfferchen, *das*

over: ~**pass** *see* **flyover;** ~'**pay** *v. t., forms as* **pay 2** überbezahlen; ~'**payment** *n.* Überbezahlung, *die;* ~-'**populated** *adj.* überbevölkert; ~'**power** *v. t.* überwältigen; ~**powering** [əʊvə'paʊərɪŋ] *adj.* überwältigend; durchdringend ⟨*Geruch*⟩; **the heat was ~powering** die Hitze war unerträglich; ~-**priced** [əʊvə'praɪst] *adj.* zu teuer; ~**pro'tective** *adj.* überfürsorglich (**towards** gegenüber); ~-'**qualified** *adj.* überqualifiziert; ~'**rate** *v. t.* überschätzen; **be ~rated** überschätzt werden; ⟨*Buch, Film:*⟩ überbewertet werden; ~'**reach** *v. refl.* sich übernehmen; ~-**re'act** *v. i.* unangemessen heftig reagieren (**to** auf + *Akk.*); ~-**re'action** *n.* Überreaktion, *die* (**to** auf + *Akk.*); ~**ride 1.** [--'-] *v. t., forms as* **ride 3** sich hinwegsetzen über (+ *Akk.*); **2.** ['---] *n.* **[manual] ~ride** Automatikabschaltung, *die;* ~'**riding** *adj.* vorrangig; ~**ripe** *adj.* überreif; ~'**rule** *v. t.* aufheben ⟨*Entscheidung*⟩; zurückweisen ⟨*Einwand, Appell, Forderung, Argument*⟩; **~rule sb.** jmds. Vorschlag ablehnen; ~'**run** *v. t., forms as* **run 3: a) be ~run with** überlaufen sein von ⟨*Touristen*⟩; überwuchert sein von ⟨*Unkraut*⟩; **b)** *(Mil.)* einfallen in (+ *Akk.*) ⟨*Land*⟩; überrennen ⟨*Stellungen*⟩; **c)** *(exceed)* **~run its allotted time** ⟨*Programm, Treffen, Diskussion:*⟩ länger als vorgesehen dauern; **~run [one's time]** ⟨*Dozent, Redner:*⟩ überziehen; ~**seas 1.** [--'-] *adv.* in Übersee ⟨*leben, sein, sich niederlassen*⟩; nach Übersee ⟨*gehen*⟩; **2.** ['---] *adj.* **a)** *(across the sea)* Übersee⟨*postgebühren, -handel, -telefonat*⟩; **b)** *(foreign)* Auslands⟨*hilfe, -zulage, -ausgabe, -nachrichten*⟩; ausländisch ⟨*Student*⟩; **~seas visitors** Besucher aus dem Ausland; ~'**see** *v. t., forms as* **see 1** überwachen; *(manage)* leiten ⟨*Abteilung*⟩; ~-'**sensitive** *adj.* überempfindlich; ~'**shadow** *v. t.* *(lit. or fig.)* überschatten; ~'**shoot** *v. t., forms as* **shoot 2** vorbeifahren an (+ *Akk.*) ⟨*Abzweigung*⟩; **~shoot the mark** *(fig.)* über das Ziel hinausschießen; **~shoot [the runway]** ⟨*Pilot, Flugzeug:*⟩ zu weit kommen;

~sight *n.* Versehen, *das;* **by** *or* **through an ~sight** versehentlich; aus Versehen; **~-'simplify** *v. t.* zu stark vereinfachen; **~'sleep** *v. i., forms as* **sleep 2** verschlafen; **~'spend** *v. i., forms as* **spend** zuviel [Geld] ausgeben; **~spill** *n.* Bevölkerungsüberschuß, *der; attrib.* Satelliten‹*stadt, -siedlung*›; **~'staff** *v. t.* überbesetzen; **~'state** *v. t.* übertrieben darstellen; **~'stay** *v. t.* überziehen ‹*Urlaub*›; **~'step** *v. t.* überschreiten; **~step the mark** *(fig.)* zu weit gehen; **~'stretch** *v. t.* überdehnen; *(fig.)* überfordern

overt ['əʊvət, əʊ'vɜ:t] *adj.* unverhohlen

over: ~'take *v. t.* **a)** *also abs. (esp. Brit.: pass)* überholen; **'no ~taking'** *(Brit.)* „Überholen verboten"; **b)** *(catch up)* einholen; **c)** *(fig.)* **be ~taken by events** ‹*Plan:*› von den Ereignissen überholt werden; **~'tax** *v. t.* **a)** *(demand too much tax from)* überbesteuern; **b)** *(strain)* überstrapazieren, überfordern ‹*Verstand, Geduld*›; **~tax one's strength** sich übernehmen; **~throw 1.** [--'-] *v. t., forms as* **throw 1** stürzen ‹*Regierung, Regime usw.*›; *(defeat)* schlagen, besiegen ‹*Feind*›; **2.** ['---] *n. (removal from power)* Sturz, *der;* **~time 1.** *n.* Überstunden; **work ten hours'/put in a lot of ~time** zehn/eine Menge Überstunden machen; **be on ~time** Überstunden machen; **2.** *adv.* **work ~time** Überstunden machen; **~'tire** *v. t.* übermüden; **~tone** *n. (fig.: implication)* Unterton, *der*

overture ['əʊvətjʊə(r)] *n.* **a)** *(Mus.)* Ouvertüre, *die;* **b)** *(formal proposal or offer)* Angebot, *das*

over: ~'turn 1. *v. t.* **a)** *(upset)* umstoßen; **b)** *(~throw)* umstürzen ‹*bestehende Ordnung, Vorstellung, Prinzip*›; stürzen ‹*Regierung*›; **2.** *v. i.* ‹*Auto, Boot, Kutsche:*› umkippen; ‹*Boot:*› kentern; **~-use** [əʊvə'ju:z] *v. t.* zu oft verwenden; **~'value** *v. t.* überbewerten

overweening [əʊvə'wi:nɪŋ] *adj.* maßlos ‹*Ehrgeiz, Gier, Stolz*›

'overweight *adj.* **a)** übergewichtig ‹*Person*›; **be [12 pounds] ~:** [12 Pfund] Übergewicht haben; **b) be ~** ‹*Gegenstand:*› zu schwer sein

overwhelm [əʊvə'welm] *v. t. (lit. or fig.)* überwältigen; **be ~ed with work** die Arbeit kaum bewältigen können

overwhelming [əʊvə'welmɪŋ] *adj.* überwältigend; unbändig ‹*Wut, Kraft, Verlangen, Zorn*›; unermeßlich ‹*Leid, Kummer*›; **against ~ odds** entgegen aller Wahrscheinlichkeit

over: ~'work 1. *v. t.* **a)** *(cause to work too hard)* mit Arbeit überlasten; **b)** *(fig.)* überstrapazieren ‹*Metapher, Wort usw.*›; **2.** *v. i.* sich überarbeiten; **3.** *n.* [Arbeits]überlastung, *die;* **~'wrought** *adj.* überreizt; **~-'zealous** *adj.* übereifrig

owe [əʊ] *v. t.,* **owing** ['əʊɪŋ] **a)** schulden; **~ sb. sth.** jmdm. etw. schulden; **~ it to sb. to do sth.** es jmdm. schuldig sein, etw. zu tun; **I ~ you an explanation** ich bin dir eine Erklärung schuldig; **you ~ it to yourself to take a break** du mußt dir einfach eine Pause gönnen; **can I ~ you the rest?** kann ich dir den Rest schuldig bleiben?; **~ [sb.] for sth.** [jmdm.] etw. bezahlen müssen; **I [still] ~ you for the ticket** du kriegst von mir noch das Geld für die Karte *(ugs.);* **b)** *(feel gratitude for, be indebted for)* verdanken; **~ sth. to sb.** jmdm. etw. verdanken

owing ['əʊɪŋ] *pred. adj.* ausstehend; **be ~:** ausstehen

'owing to *prep.* wegen; **~ unfortunate circumstances** auf Grund unglücklicher Umstände

owl [aʊl] *n.* Eule, *die*

own [əʊn] **1.** *adj.* eigen; **with one's ~ eyes** mit eigenen Augen; **speak from one's ~ experience** aus eigener Erfahrung sprechen; **this is all my ~ work** das habe ich alles selbst gemacht; **do one's ~ cooking/housework** selbst kochen/die Hausarbeit selbst machen; **make one's ~ clothes** seine Kleidung selbst schneidern; **a house/ideas** *etc.* **of one's ~:** ein eigenes Haus/eigene Ideen *usw.;* **have nothing of one's ~:** kein persönliches Eigentum haben; **have enough problems of one's ~:** selbst genug Probleme haben; **for reasons of his ~ ...:** aus nur ihm selbst bekannten Gründen ...; **that's where he/it comes into his/its ~** *(fig.)* da kommt er/es voll zur Geltung; **on one's/its ~** *(alone)* allein; **he's in a class of his ~** *(fig.)* er ist eine Klasse für sich; *see also* **get back 2 c; hold 1 j; man 1 b. 2.** *v. t.* besitzen; **be ~ed by sb.** jmdm. gehören; **be privately ~ed** sich in Privatbesitz befinden; **they behaved as if they ~ed the place** sie benahmen sich, als ob der Laden ihnen gehörte *(ugs.)*

~ 'up *v. i. (coll.)* ‹*Schuldiger, Täter:*› gestehen; **~ up to sth.** etw. [ein]gestehen *od.* zugeben; **~ up to having done sth.** [ein]gestehen *od.* zugeben, daß man etw. getan hat

owned [əʊnd] *adj.* **publicly ~:** gemeinde-/staatseigen; **privately ~:** in Privatbesitz *nachgestellt*

owner ['əʊnə(r)] *n.* Besitzer, *der*/Besitzerin, *die;* Eigentümer, *der*/Eigentümerin, *die; (of car also)* Halter, *der*/Halterin, *die; (of shop, hotel, firm, etc.)* Inhaber, *der*/Inhaberin, *die;* **at ~'s risk** auf eigene Gefahr

ownership ['əʊnəʃɪp] *n., no pl.* Besitz, *der;* **be under new ~** ‹*Firma, Laden, Restaurant:*› einen neuen Inhaber/eine neue Inhaberin haben

own 'goal *n. (lit. or fig.)* Eigentor, *das*

ox [ɒks] *n., pl.* **oxen** ['ɒksn] Ochse, *der*

oxidation [ɒksɪ'deɪʃn] *n. (Chem.)* Oxydation, *die*

oxide ['ɒksaɪd] *n. (Chem.)* Oxyd, *das*

oxidize (oxidise) ['ɒksɪdaɪz] *v. t. & i. (Chem.)* oxydieren

oxy-acetylene [ɒksɪə'setɪli:n] *adj.* **~ welding** Autogenschweißen, *das;* **~ torch** Schweißbrenner, *der*

oxygen ['ɒksɪdʒən] *n. (Chem.)* Sauerstoff, *der*

oyster ['ɔɪstə(r)] *n.* Auster, *die;* **the world's his ~** *(fig.)* ihm liegt die Welt zu Füßen

oz. *abbr.* **ounce[s]**

ozone ['əʊzəʊn] *n.* Ozon, *das*

ozone: ~-friendly *adj.* ozonsicher; *(not using CFCs)* FCKW-frei; **~ layer** *n.* Ozonschicht, *die;* **the hole in the ~ layer** das Ozonloch *(ugs.)*

P

P, p [pi:] *n., pl.* **Ps** *or* **P's** P, p, *das; see also* **mind 2 b**

p. *abbr.* **a) page** S.; **b)** [pi:] *(Brit.)* **penny/pence** p; **c)** *(Mus.)* **piano** p

PA *abbr.* **a) personal assistant** pers. Ass.; **b) public address: PA [system]** LS-Anlage, *die*

p.a. *abbr.* **per annum** p.a.

pace [peɪs] **1.** *n.* **a)** *(step, distance)* Schritt, *der;* **b)** *(speed)* Tempo, *das;* **slacken/quicken one's ~** *(walking)* seinen Schritt verlangsamen/beschleunigen; **at a steady/good ~:** in gleichmäßigem/zügigem Tempo; **set the ~:** das Tempo angeben *od.* bestimmen; **keep ~ [with sb./sth.]** [mit jmdm./etw.] Schritt halten; **stay** *or* **stand the ~, stay** *or* **keep with the ~** *(Sport)* das Tempo durchhalten; **c) put sb./a horse through his/its ~s** *(fig.)* jmdn./ein Pferd zeigen lassen, was er/es kann; **show one's ~s** zeigen, was man kann. **2.** *v.i.* schreiten *(geh.);* [gemessenen Schrittes] gehen; **~ up and down [the platform/room]** [auf dem Bahnsteig/im Zimmer] auf und ab gehen *od.* marschieren. **3.** *v.t.* **a)** auf- und abgehen in (+ *Dat.*); **b)** *(set the ~ for)* Schrittmacher sein für

'pace-maker *n.* **a)** *(Sport)* Schrittmacher, *der*/-macherin, *die;* **b)** *(Med.)* [Herz]schrittmacher, *der*

pacific [pə'sɪfɪk] **1.** *adj.* *(Geog.)* Pazifik‹*küste, -insel*›; **P~ Ocean** Pazifischer *od.* Stiller Ozean. **2.** *n.* **the P~:** der Pazifik

pacifism ['pæsɪfɪzm] *n., no pl., no art.* Pazifismus, *der*

pacifist ['pæsɪfɪst] *n.* Pazifist, *der*/Pazifistin, *die*

pacify ['pæsɪfaɪ] *v.t.* besänftigen; beruhigen ‹*weinendes Kind*›

pack [pæk] **1.** *n.* **a)** *(bundle)* Bündel, *das; (Mil.)* Tornister, *der; (rucksack)* Rucksack, *der;* **b)** *(derog.) (people)* Bande, *die;* **a ~ of lies/nonsense** ein Sack voll Lügen/eine Menge Unsinn; **what a ~ of lies!** alles erlogen!; **c)** *(Brit.)* **~ [of cards]** [Karten]spiel, *das;* **d)** *(wolves, wild dogs)* Rudel, *das; (hounds)* Meute, *die;* **e)** *(Cub Scouts, Brownies)* Gruppe, *die;* **f)** *(packet, set)* Schachtel, *die;* Packung, *die;* **~ of ten** Zehnerpackung, *die;* Zehnerpack, *der.* **2.** *v.t.* **a)** *(put into container)* einpacken; **~ sth. into sth.** etw. in etw. *(Akk.)* packen; **b)** *(fill)* packen; **~ one's bags** seine Koffer packen; **c)** *(cram)* vollstopfen *(ugs.);* füllen ‹*Raum, Stadion usw.*›; **d)** *(wrap)* verpacken (**in** in + *Dat. od. Akk.*); **~ed in** verpackt in (+ *Dat.*); **e)** *(sl.)* tragen, dabeihaben ‹*Waffe*›; **f) ~ [quite] a punch** *(sl.)* ganz schön zuschlagen können *(ugs.).* **3.** *v.i.* packen; **send sb. ~ing** *(fig.)* jmdn. rausschmeißen *(ugs.)*

~ a'way *v.t.* wegpacken

~ 'in *v.t. (coll.: give up)* aufstecken *(ugs.);* aufhören mit ‹*Arbeit, Spiel*›; **~ it in!** hör [doch] auf damit!

~ into *v.t.* sich drängen in (+ *Akk.*); **we all ~ed into the car** wir quetschten uns alle in das Auto *(ugs.)*

~ 'off *v.t.* *(send away)* fortschicken

~ 'up 1. *v.t.* **a)** *(package)* zusammenpacken ‹*Sachen, Werkzeug*›; packen ‹*Paket*›; **b)** *(coll.: stop)* aufhören *od. (ugs.)* Schluß machen mit; **~ up work** Feierabend machen. **2.** *v.i. (coll.)* **a)** *(give up)* aufhören; Schluß machen *(ugs.);* **b)** *(break down)* den Geist aufgeben *(ugs.)*

package ['pækɪdʒ] **1.** *n.* **a)** *(bundle; fig. coll.: transaction)* Paket, *das;* **b)** *(container)* Verpackung, *die.* **2.** *v.t. (lit. or fig.)* verpacken

package: ~ deal *n.* Paket, *das;* **~ holiday, ~ tour** *ns.* Pauschalreise, *die*

packaging ['pækɪdʒɪŋ] *n.* Verpackung, *die*

packed [pækt] *adj.* **a)** gepackt ‹*Kiste, Koffer*›; **~ meal/lunch** Eßpaket, *das*/Lunchpaket, *das;* **b)** *(crowded)* [über]voll ‹*Theater, Kino, Halle*›; **~ out** *(coll.)* gerammelt voll *(ugs.)*

packet ['pækɪt] *n.* **a)** *(package)* Päckchen, *das; (box)* Schachtel, *die;* **a ~ of cigarettes** eine Schachtel/ein Päckchen Zigaretten; **b)** *(coll.: large sum of money)* Haufen Geld *(ugs.)*

packing ['pækɪŋ] *n.* **a)** *(packaging) (material)* Verpackungsmaterial, *das; (action)* Verpacken, *das;* **including postage and ~:** einschließlich Porto und Verpackung; **b) do one's ~:** packen

'packing-case *n.* [Pack]kiste, *die*

pact [pækt] *n.* Pakt, *der;* **make a ~ with sb.** einen Pakt mit jmdm. schließen

[1]**pad** [pæd] **1.** *n.* **a)** *(cushioning material)* Polster, *das; (to protect wound)* Kompresse, *die; (Sport) (on leg)* Beinschützer, *der; (on knee)* Knieschützer, *der;* **b)** *(block of paper)* Block, *der;* **a ~ of notepaper, a [writing-]~:** ein Schreibblock; **c)** *(launching surface)* Abschußrampe, *die;* **d)** *(coll.: house, flat)* Bude, *die (ugs.).* **2.** *v.t.,* **-dd-: a)** polstern ‹*Jacke, Schulter, Stuhl*›; **b)** *(fig.: lengthen unnecessarily)* auswalzen *(ugs.)* ‹*Brief, Aufsatz usw.*› (**with** durch)

~ 'out *see* [1]**pad 2**

[2]**pad** *v.t. & i.,* **-dd-** *(walk) (in socks, slippers, etc.)* tappen; *(along path etc.)* trotten

padded ['pædɪd] *adj.* gepolstert; **~ 'cell** Gummizelle, *die;* **~ envelope** wattierter Umschlag

padding ['pædɪŋ] *n.* **a)** Polsterung, *die;* **b)** *(fig.: superfluous matter)* Füllsel, *das*

[1]**paddle** ['pædl] **1.** *n.* *(oar)* [Stech]paddel, *das.* **2.** *v.t. & i. (in canoe)* paddeln

[2]**paddle 1.** *v.i.* *(with feet)* planschen. **2.** *n.* **have a/go for a ~:** ein bißchen planschen/planschen gehen

paddle: ~-boat, ~-steamer *ns.* [Schaufel]raddampfer, *der;* **~-wheel** *n.* Schaufelrad, *das*

'paddling-pool *n.* Planschbecken, *das*

paddock ['pædək] *n.* **a)** Koppel, *die;* **b)** *(Horse-racing)* Sattelplatz, *der*

paddy ['pædɪ], **'paddy-field** *ns.* Reisfeld, *das*

'padlock 1. *n.* Vorhängeschloß, *das.* **2.** *v.t.* [mit einem Vorhängeschloß] verschließen

paediatric [pi:dɪ'ætrɪk] *adj.* *(Med.)* pädiatrisch; Kinder‹*schwester, -station*›

pagan ['peɪgən] **1.** *n.* Heide, *der*/Heidin, *die.* **2.** *adj.* heidnisch

[1]**page** [peɪdʒ] **1.** *n.* **~[-boy]** Page, *der.* **2.** *v.t. & i.* **~ [for] sb.** *(over loudspeaker)* jmdn. ausrufen; *(by pager)* jmdn. anpiepen *(ugs.)*

[2]**page** *n.* Seite, *die; (leaf, sheet of paper)* Blatt, *das;* **front/sports/fashion ~:** erste Seite/Sport-/Modeseite, *die;* **turn to the next ~:** umblättern

pageant ['pædʒənt] *n.* **a)** *(spectacle)* Schauspiel, *das;* **b)** *(play)* **historical ~:** Historienspiel, *das*

pageantry ['pædʒəntrɪ] *n.* Prachtentfaltung, *die;* Prunk, *der*

'page-boy *n.* **a)** *see* [1]**page 1; b)** *(hair-style)* Pagenkopf, der

pager ['peɪdʒə(r)] *n.* Piepser, *der (ugs.)*

pagoda [pə'gəʊdə] *n.* Pagode, *die*

paid [peɪd] **1.** *see* **pay 2, 3. 2.** *adj.* bezahlt ‹*Urlaub, Arbeit*›; **put ~ to** *(Brit. fig. coll.)* zunichte machen ‹*Hoffnung, Plan, Aussichten*›; kurzen Prozeß machen mit *(ugs.)* ‹*Person*›

'paid-up *adj.* bezahlt; **[fully] ~ member** Mitglied, das alle Beträge bezahlt hat; *(fig.)* überzeugtes Mitglied

pail [peɪl] *n.* Eimer, *der*

pain [peɪn] **1.** *n.* **a)** *no indef. art.* *(suffering)* Schmerzen; *(mental*

~*)* Qualen; **feel [some] ~, be in ~:** Schmerzen haben; **cause sb. ~** *(lit. or fig.)* jmdm. wehtun; **b)** *(instance of suffering)* Schmerz, *der;* **I have a ~ in my shoulder/knee/stomach** meine Schulter/mein Knie/Magen tut weh; **be a ~ in the neck** *see* **neck 1a; c)** *(coll.: nuisance)* Plage, *die; (sb./sth. getting on one's nerves)* Nervensäge, *die (ugs.);* **d)** *in pl. (trouble taken)* Mühe, *die;* Anstrengung, *die;* **take ~s** sich *(Dat.)* Mühe geben (**over** mit, bei); **be at ~s to do sth.** sich sehr bemühen *od.* sich *(Dat.)* große Mühe geben, etw. zu tun; **he got nothing for all his ~s** seine ganze Mühe war umsonst; **e)** *(Law)* **on** *or* **under ~ of death** bei Todesstrafe. **2.** *v. t.* schmerzen

pained [peɪnd] *adj.* gequält

painful ['peɪnfl] *adj.* **a)** *(causing pain)* schmerzhaft ‹*Krankheit, Operation, Wunde*›; **be/become ~** ‹*Körperteil:*› weh tun *od.* schmerzen; **suffer from a ~ shoulder** Schmerzen in der Schulter haben; **b)** *(distressing)* schmerzlich ‹*Gedanke, Erinnerung*›; traurig ‹*Pflicht*›; **it was ~ to watch him** es tat weh, ihm zuzusehen

painfully ['peɪnfəlɪ] *adv.* **a)** *(with great pain)* unter großen Schmerzen; **b)** *(fig.) (excessively)* über die Maßen *(geh.); (laboriously)* quälend ‹*langsam*›; ~ **obvious** nur zu offensichtlich

'**pain-killer** *n.* schmerzstillendes Mittel; Schmerzmittel, *das (ugs.)*

painless ['peɪnlɪs] *adj.* schmerzlos; *(fig.: not causing problems)* unproblematisch

painstaking ['peɪnzteɪkɪŋ] *adj.* gewissenhaft; **it is ~ work** es ist eine mühsame Arbeit; **with ~ care** mit äußerster Sorgfalt

paint [peɪnt] **1.** *n.* **a)** Farbe, *die; (on car)* Lack, *der;* **b)** *(joc.: cosmetic)* Schminke, *die.* **2.** *v. t.* **a)** *(cover, colour)* [an]streichen; **~ the town red** *(fig. sl.)* auf die Pauke hauen *(ugs.);* **b)** *(make picture of, make by ~ing)* malen; **the picture was ~ed by R.** das Bild ist von R.; **c)** *(adorn with ~ing)* bemalen ‹*Wand, Vase, Decke*›; **d) ~ a glowing/gloomy picture of sth.** *(fig.)* etw. in leuchtenden/düsteren Farben malen *od.* schildern; **e)** *(apply cosmetic to)* schminken ‹*Augen, Gesicht, Lippen*›; lackieren ‹*Nägel*›

paint: ~box *n.* Malkasten, *der;* **~brush** *n.* Pinsel, *der*

painter ['peɪntə(r)] *n.* **a)** *(artist)* Maler, *der*/Malerin, *die;* **b)** [**house-**]**~:** Maler, *der*/Malerin, *die;* Anstreicher, *der*/Anstreicherin, *die*

painting ['peɪntɪŋ] *n.* **a)** *no pl., no indef. art. (art)* Malerei, *die;* **b)** *(picture)* Gemälde, *das;* Bild, *das*

pair [peə(r)] **1.** *n.* **a)** *(set of two)* Paar, *das;* **a ~ of gloves/socks/shoes** *etc.* ein Paar Handschuhe/Socken/Schuhe *usw.;* **a** *or* **one ~ of hands/eyes** zwei Hände/Augen; **in ~s** paarweise; **the ~ of them** die beiden; **b)** *(single article)* **a ~ of pyjamas/scissors** *etc.* ein Schlafanzug/eine Schere *usw.;* **a ~ of trousers/jeans** eine Hose/Jeans; ein Paar Hosen/Jeans; **c)** *(married couple)* [Ehe]paar, *das; (mated animals)* Paar, *das;* Pärchen, *das;* **d)** *(Cards)* Pärchen, *das.* **2.** *v. t.* paaren; [paarweise] zusammenstellen

~ **'off 1.** *v. t.* zu Paaren *od.* paarweise zusammenstellen; **she was ~ed off with Alan** sie bekam Alan als Partner. **2.** *v. i.* Zweiergruppen bilden

pajamas [pə'dʒɑ:məz] *(Amer.) see* **pyjamas**

Pakistan [pɑ:kɪ'stɑ:n] *pr. n.* Pakistan *(das)*

Pakistani [pɑ:kɪ'stɑ:nɪ] **1.** *adj.* pakistanisch; **sb. is ~:** jmd. ist Pakistani. **2.** *n.* Pakistani, *der/die;* Pakistaner, *der*/Pakistanerin, *die*

pal [pæl] *(coll.) n.* Kumpel, *der (ugs.)*

palace ['pælɪs] *n.* Palast, *der*

palatable ['pælətəbl] *adj.* **a)** genießbar; trinkbar ‹*Wein*›; *(pleasant)* wohlschmeckend ‹*Speise*›; **b)** *(fig.)* annehmbar, akzeptabel ‹*Gesetz, Erhöhung, Aufführung*›

palate ['pælət] *n.* *(Anat.)* Gaumen, *der*

palatial [pə'leɪʃl] *adj.* palastartig

palaver [pə'lɑ:və(r)] *n.* *(coll.: fuss)* Umstand, *der;* Theater, *das (ugs.)*

[1]**pale** [peɪl] *n.* **be beyond the ~** ‹*Verhalten, Benehmen:*› unmöglich sein; **regard sb. as beyond the ~:** jmdn. indiskutabel finden

[2]**pale 1.** *adj.* **a)** blaß, *(esp. in illness)* fahl, *(nearly white)* bleich ‹*Gesichtsfarbe, Haut, Gesicht, Aussehen*›; **go ~:** blaß/bleich werden; **his face was ~:** er war blaß/bleich; **b)** *(light in colour)* von blasser Farbe *nachgestellt;* blaß ‹*Farbe*›; **a ~ blue/red dress** ein blaßblaues/-rotes Kleid. **2.** *v. i.* bleich/blaß werden (**at** bei); **~ into insignificance** völlig bedeutungslos werden

paleness ['peɪlnɪs] *n., no pl. (of person)* Blässe, *die*

Palestine ['pælɪstaɪn] *pr. n.* Palästina *(das)*

Palestinian [pælɪ'stɪnɪən] **1.** *adj.* palästinensisch; **sb. is ~:** jmd. ist Palästinenser/Palästinenserin. **2.** *n.* Palästinenser, *der*/Palästinenserin, *die*

palette ['pælɪt] *n.* Palette, *die*

palisade [pælɪ'seɪd] *n.* Palisade, *die;* Palisadenzaun, *der*

[1]**pall** [pɔ:l] *n.* **a)** *(over coffin)* Sargtuch, *das;* **b)** *(fig.)* Schleier, *der*

[2]**pall** *v. i.* ~ [**on sb.**] [jmdm.] langweilig werden

'**pall-bearer** *n.* Sargträger, *der*/-trägerin, *die*

pallet *n.* *(platform)* Palette, *die*

palliative ['pælɪətɪv] *n.* *(Med.)* Palliativ[um], *das (fachspr.);* Linderungsmittel, *das*

pallid ['pælɪd] *adj.* **a)** *see* [2]**pale 1a; b)** matt, blaß ‹*Farbe*›

pallor ['pælə(r)] *n.* Blässe, *die;* Fahlheit, *die*

[1]**palm** [pɑ:m] *n.* *(tree)* Palme, *die;* ~ [**branch**] *(also Eccl.)* Palmzweig, *der*

[2]**palm** *n.* Handteller, *der;* Handfläche, *die;* **have sth. in the ~ of one's hand** *(fig.)* etw. in der Hand haben

~ **'off** *v. t.* **~ sth. off on sb., ~ sb. off with sth.** jmdm. etw. andrehen *(ugs.)*

palmist ['pɑ:mɪst] *n.* Handleser, *der*/-leserin, *die*

palmistry ['pɑ:mɪstrɪ] *n., no pl.* Handlesekunst, *die*

palm: P~ 'Sunday *n.* *(Eccl.)* Palmsonntag, *der;* **~-tree** *n.* Palme, *die*

palpable ['pælpəbl] *adj.* offenkundig ‹*Lüge, Unwissenheit, Absurdität*›

palpitate ['pælpɪteɪt] *v. i.* *(pulsate)* ‹*Herz:*› palpitieren *(fachspr.),* pochen, hämmern; *(tremble)* zittern (**with** vor + *Dat.*)

palpitations [pælpɪ'teɪʃnz] *n. pl.* *(Med.)* Palpitation, *die (fachspr.);* Herzklopfen, *das*

paltry ['pɔ:ltrɪ, 'pɒltrɪ] *adj.* schäbig; armselig ‹*Auswahl*›; *(trivial)* belanglos

pampas ['pæmpəs] *n. pl.* *(Geog.)* Pampas *Pl.*

'**pampas-grass** *n.* Pampasgras, *das*

pamper ['pæmpə(r)] *v. t.* verhätscheln; ~ **oneself** sich verwöhnen

pamphlet ['pæmflɪt] *n.* *(leaflet)* Prospekt, *der; (esp. Polit.)* Flugblatt, *das; (booklet)* Broschüre, *die*

[1]**pan** [pæn] *n.* **a)** [Koch]topf, *der; (for frying)* Pfanne, *die;* **pots and ~s** Kochtöpfe; **b)** *(of scales)*

Schale, *die;* **c)** *(Brit.: of WC)* **[lavatory] ~:** Toilettenschüssel, *die*
~ 'out *v.i. (progress)* sich entwickeln
²pan *(Cinemat., Telev.)* **1. -nn-** *v.t.* schwenken. **2. -nn-** *v.i.* schwenken **(to** auf + *Akk.*); **~ning shot** Schwenk, *der*
pan- [pæn] *in comb.* pan-, Pan-
panacea [pænə'si:ə] *n.* Allheilmittel, *das*
panache [pə'næʃ] *n.* Schwung, *der;* Elan, *der*
Panama [pænə'mɑ:] **1.** *pr. n.* Panama *(das).* **2.** *n.* **p~ [hat]** Panamahut, *der*
Panama Ca'nal *pr. n.* Panamakanal, *der*
Panamanian [pænə'meɪnɪən] **1.** *adj.* panamaisch. **2.** *n.* Panamaer, *der*/Panamaerin, *die*
pancake ['pænkeɪk] *n.* Pfannkuchen, *der;* **P~ Day** *n. (Brit.)* Fastnachtsdienstag, *der*
pancreas ['pæŋkrɪəs] *n. (Anat.)* Bauchspeicheldrüse, *die;* Pankreas, *das (fachspr.)*
panda ['pændə] *n. (Zool.)* Panda, *der*
pandemonium [pændɪ'məʊnɪəm] *n.* Chaos, *das; (uproar)* Tumult, *der*
pander ['pændə(r)] *v.i.* **~ to** allzu sehr entgegenkommen (+ *Dat.*) ‹*Person, Geschmack, Instinkt*›
p. & p. *abbr. (Brit.)* **postage and packing** Porto und Verpackung
pane [peɪn] *n.* Scheibe, *die;* **window-~/~ of glass** Fenster-/Glasscheibe, *die*
panel ['pænl] *n.* **a)** *(of door, wall, etc.)* Paneel, *das;* **b)** *(esp. Telev., Radio, etc.) (quiz team)* Rateteam, *das; (in public discussion)* Podium, *das;* **c)** *(advisory body)* Gremium, *das;* Kommission, *die;* **~ of experts** Expertengremium, *das;* **d)** *(Dressmaking)* Einsatz, *der*
'panel game *n.* Ratespiel, *das*
paneling, panelist *(Amer.) see* **panell-**
panelling ['pænəlɪŋ] *n.* Täfelung, *die*
panellist ['pænəlɪst] *n. (Telev., Radio) (on quiz programme)* Mitglied des Rateteams; *(on discussion panel)* Diskussionsteilnehmer, *der*/-teilnehmerin, *die*
pang [pæŋ] *n.* **a)** *(of pain)* Stich, *der;* **b) feel ~s of conscience/guilt** Gewissensbisse haben; **feel ~s of remorse** bittere Reue empfinden
panic ['pænɪk] **1.** *n.* Panik, *die;* **be in a [state of] ~:** von Panik erfaßt sein; **hit the ~ button** *(fig. coll.)* Alarm schlagen; *(~)* durchdrehen *(ugs.).* **2.** *v.i.,* **-ck-** in Panik *(Akk.)* geraten; **don't ~!** nur keine Panik! **3.** *v.t.,* **-ck-** in Panik versetzen; **~ sb. into doing sth.** jmdn. so in Panik versetzen, daß er etw. tut
panicky ['pænɪkɪ] *adj.* von Panik bestimmt ‹*Verhalten, Handeln, Rede*›; **be ~:** in Panik sein
panic: ~ stations *n. pl. (fig. coll.)* **be at ~ stations** am Rotieren sein *(ugs.)* **(about** wegen); **~-stricken, ~-struck** *adjs.* von Panik erfaßt *od.* ergriffen
panorama [pænə'rɑ:mə] *n.* Panorama, *das; (fig.: survey)* Überblick, *der* **(of** über + *Akk.*)
panoramic [pænə'ræmɪk] *adj.* Panorama-
pansy ['pænzɪ] *n.* **a)** *(Bot.)* Stiefmütterchen, *das;* **b)** *(coll.: effeminate man)* Tunte, *die (ugs.)*
pant [pænt] *v.i.* keuchen; ‹*Hund:*› hecheln
~ for *v.t.* ringen nach ‹*Luft, Atem*›; schnappen nach ‹*Luft*›
pantechnicon [pæn'teknɪkən] *n.* **~ [van]** *(Brit.)* Möbelwagen, *der*
panther ['pænθə(r)] *n. (Zool.)* **a)** Panther, *der;* **b)** *(Amer.: puma)* Puma, *der;* Berglöwe, *der*
panties ['pæntɪz] *n. pl. (coll.)* **[pair of] ~:** Schlüpfer, *der*
pantomime ['pæntəmaɪm] *n.* **a)** *(Brit.) Märchenspiel im Varietéstil, das um Weihnachten aufgeführt wird;* **b)** *(gestures)* Pantomime, *die*
pantry ['pæntrɪ] *n.* Speisekammer, *die*
pants [pænts] *n. pl.* **a)** *(esp. Amer. coll.: trousers)* **[pair of] ~:** Hose, *die;* **catch sb. with his ~ down** *(fig. sl.)* jmdn. unvorbereitet treffen; **b)** *(Brit. coll.: underpants)* Unterhose, *die*
papacy ['peɪpəsɪ] *n.* **a)** *no pl. (office)* Papat, *der;* **b)** *(tenure)* Amtszeit als Papst; **c)** *no pl. (papal system)* Papsttum, *das*
papal ['peɪpl] *adj.* päpstlich
paper ['peɪpə(r)] **1.** *n.* **a)** *(material)* Papier, *das;* **put sth. down on ~:** etw. schriftlich festhalten *od.* niederlegen; **it looks all right on ~** *(in theory)* auf dem Papier sieht es ganz gut aus; **put pen to ~:** zur Feder greifen; **the treaty** *etc.* **isn't worth the ~ it's written on** *(coll.)* der Vertrag *usw.* ist nicht das Papier wert, auf dem er geschrieben steht; **b)** *in pl. (documents)* Dokumente; Unterlagen *Pl.; (to prove identity etc.)* Papiere *Pl.;* **c)** *(in examination) (Univ.)* Klausur, *die; (Sch.)* Arbeit, *die;* **d)** *(newspaper)* Zeitung, *die;* **daily/weekly ~:** Tages-/Wochenzeitung, *die;* **e)** *(wallpaper)* Tapete, *die;* **f)** *(wrapper)* Stück Papier; **don't scatter the ~s all over the floor** wirf das Papier nicht überall auf den Boden; **g)** *(learned article)* Referat, *das; (shorter)* Paper, *das.* **2.** *adj.* **a)** *(made of ~)* aus Papier *nachgestellt;* Papier‹*mütze, -taschentuch*›; **b)** *(theoretical)* nominell ‹*zahlenmäßige Stärke, Profit*›. **3.** *v.t.* tapezieren
~ 'over *v.t.* [mit Tapete] überkleben; **~ over the cracks** *(fig.: cover up mistakes/differences)* die Fehler/Differenzen übertünchen
paper: ~back *n.* Paperback, *das; (pocket-size)* Taschenbuch, *das;* **~ 'bag** *n.* Papiertüte, *die;* **~-boy** *n.* Zeitungsjunge, *der;* **~-chase** *n.* Schnitzeljagd, *die;* **~-clip** *n.* Büroklammer, *die; (larger)* Aktenklammer, *die;* **~ 'cup** *n.* Pappbecher, *der;* **~-girl** Zeitungsausträgerin, *die;* **~ 'handkerchief** *n.* Papiertaschentuch, *das;* **~-knife** *n.* Brieföffner, *der;* **~-mill** *n.* Papierfabrik *od.* -mühle, *die;* **~ money** *n.* Papiergeld, *das;* **~ 'napkin** *n.* Papierserviette, *die;* **~ 'plate** *n.* Pappteller, *der;* **~-round** *n.* Zeitungenaustragen, *das;* **have/do a ~-round** Zeitungen austragen; **~ servi'ette** *see* **~ napkin; ~-shop** *n.* Zeitungsgeschäft, *das;* **~-thin** *adj. (lit. or fig.)* hauchdünn; **~ 'towel** *n.* Papierhandtuch, *das;* **~weight** *n.* Briefbeschwerer, *der;* **~work** *n.* Schreibarbeit, *die*
papier mâché [pæpjeɪ 'mæʃeɪ] *n.* Papiermaché, *das;* Pappmaché, *das*
papist ['peɪpɪst] *n. (Relig. derog.)* Papist, *der*/Papistin, *die*
paprika ['pæprɪkə, pə'pri:kə] *n.* **a)** *see* **pepper 1 b; b)** *(Cookery: condiment)* Paprika, *der*
par [pɑ:(r)] *n.* **a)** *(average)* **above/below ~:** über/unter dem Durchschnitt; **feel rather below ~, not feel up to ~** *(fig.)* nicht ganz auf dem Posten *od.* Damm sein *(ugs.);* **b)** *(equality)* **be on a ~:** vergleichbar sein; **be on a ~ with sb./sth.** jmdm./einer Sache gleichkommen; **c)** *(Golf)* Par, *das;* **that's about ~ for the course** *(fig. coll.)* das ist so das Übliche
parable ['pærəbl] *n.* Gleichnis, *das;* Parabel, *die (bes. Literaturw.)*
parabola [pə'ræbələ] *n. (Geom.)* Parabel, *die*
parachute ['pærəʃu:t] **1.** *n.* **a)** Fallschirm, *der;* **b)** *(to brake air-*

craft etc.) Bremsfallschirm, *der.* **2.** *v. t.* [mit dem Fallschirm] absetzen ⟨*Person*⟩ (**into** über + *Dat.*); mit dem Fallschirm abwerfen ⟨*Vorräte*⟩. **3.** *v. i.* [mit dem Fallschirm] abspringen (**into** über + *Dat.*)

parachutist ['pærəʃu:tɪst] *n.* **[sports]** ~: Fallschirmspringer, *der*/-springerin, *die*

parade [pə'reɪd] **1.** *n.* **a)** *(display)* Zurschaustellung, *die;* **make a ~ of** zur Schau stellen ⟨*Tugend, Eigenschaft*⟩; **b)** *(Mil.: muster)* Appell, *der;* **on ~:** beim Appell; **c)** *(procession)* Umzug, *der; (of troops)* Parade, *die;* **d)** *(succession)* Reihe, *die.* **2.** *v. t.* **a)** *(display)* zur Schau stellen; vorzeigen ⟨*Person*⟩ (**before** bei); **b)** *(march through)* **~ the streets** durch die Straßen marschieren. **3.** *v. i.* paradieren; ⟨*Demonstranten:*⟩ marschieren

pa'rade-ground *n.* Exerzierplatz, *der*

paradigm ['pærədaɪm] *n. (esp. Ling.)* Paradigma, *das*

paradise ['pærədaɪs] *n.* Paradies, *das*

paradox ['pærədɒks] *n.* Paradox[on], *das*

paradoxical [pærə'dɒksɪkl] *adj.* paradox

paraffin ['pærəfɪn] *n.* **a)** *(Chem.)* Paraffin, *das;* **b)** *(Brit.: fuel)* Petroleum, *das*

paraffin: ~ stove *n.* Petroleumkocher, *der; (for heating)* Petroleumofen, *der;* **~ wax** *n.* Paraffin[wachs], *das*

paragon ['pærəgən] Muster, *das* (**of an** + *Dat.*); **~ of virtue** Tugendheld, *der*

paragraph ['pærəgrɑ:f] *n.* **a)** *(section of text)* Absatz, *der;* **b)** *(subsection of law etc.)* Paragraph, *der*

parakeet ['pærəki:t] *n. (Ornith.)* Sittich, *der*

parallel ['pærəlel] **1.** *adj.* **a)** parallel; **the railway ran ~ to the river** die Bahnlinie verlief parallel zum Fluß; **~ bars** *(Gymnastics)* Barren, *der;* **b)** *(fig.: similar)* vergleichbar; **be ~:** sich *(Dat.)* [genau] entsprechen. **2.** *n.* **a)** Parallele, *die;* **this has no ~:** dazu gibt es keine Parallele; **there is a ~ between x and y** es gibt eine Parallelität zwischen x und y; **b)** *(Electr.)* **in ~:** parallel; **c)** *(Geog.)* **~ [of latitude]** Breitenkreis, *der;* **the 42nd ~:** der 42. Breitengrad. **3.** *v. t.* gleichkommen (+ *Dat.*)

parallelogram [pærə'leləgræm] *n. (Geom.)* Parallelogramm, *das*

paralyse ['pærəlaɪz] *v. t.* lähmen; **he is ~d in both legs** seine beiden Beine sind gelähmt; *(fig.)* lahmlegen ⟨*Verkehr, Industrie*⟩; **be ~d with fright** vor Schreck wie gelähmt sein

paralysis [pə'rælɪsɪs] *n., pl.* **paralyses** [pə'rælɪsi:z] Lähmung, *die; (fig., of industry, traffic)* Lahmlegung, *die*

paralytic [pærə'lɪtɪk] *n.* Gelähmte, *der/die*

paralyze *(Amer.) see* **paralyse**

parameter [pə'ræmɪtə(r)] **a)** *(defining feature)* Faktor, *der;* **b)** *(Math.)* Parameter, *der*

paramilitary [pærə'mɪlɪtərɪ] *adj.* paramilitärisch

paramount ['pærəmaʊnt] *adj.* höchst... ⟨*Macht, Autorität, Wichtigkeit*⟩; Haupt⟨*gesichtspunkt, -überlegung*⟩

paranoia [pærə'nɔɪə] *n.* **a)** *(disorder)* Paranoia, *die (Med.);* **b)** *(tendency)* **[feeling of] ~:** krankhaftes Mißtrauen

paranormal [pærə'nɔ:ml] *adj.* paranormal; übersinnlich

parapet ['pærəpɪt, 'pærəpet] *n. (low wall or barrier)* Brüstung, *die*

paraphernalia [pærəfə'neɪlɪə] *n. sing.* **a)** *(belongings)* Utensilien *Pl.;* **b)** *(of justice, power)* Instrumentarium, *das (geh.);* Apparat, *der;* **the whole ~** *(coll.)* alles, was so dazugehört *(ugs.)*

paraphrase ['pærəfreɪz] **1.** *n.* Umschreibung, *die.* **2.** *v. t.* umschreiben

paraplegia [pærə'pli:dʒɪə] *n. (Med.)* Paraplegie, *die (fachspr.);* ≈ Querschnittslähmung, *die*

paraplegic [pærə'pli:dʒɪk] *(Med.)* **1.** *adj.* doppelseitig gelähmt; paraplegisch *(fachspr.).* **2.** *n.* doppelseitig Gelähmter/Gelähmte; Paraplegiker, *der*/Paraplegikerin, *die (fachspr.)*

parasite ['pærəsaɪt] *n. (Biol.: also fig. derog.)* Schmarotzer, *der;* Parasit, *der*

parasitic [pærə'sɪtɪk] *adj.* **a)** *(Biol.)* parasitisch; parasitär ⟨*Pilz*⟩; **be ~ on** schmarotzen an (+ *Dat.*); **b)** *(fig.)* schmarotzerisch

parasol ['pærəsɒl] *n.* Sonnenschirm, *der*

paratrooper ['pærətru:pə(r)] *n. (Mil.)* Fallschirmjäger, *der*

paratroops ['pærətru:ps] *n. pl. (Mil.)* Fallschirmjäger *Pl.*

parboil ['pɑ:bɔɪl] *v. t.* ankochen

parcel ['pɑ:sl] *n.* **a)** *(package)* Paket, *das;* **send/receive sth. by ~ post** etw. mit der Paketpost schicken/bekommen; **b) a ~ of land** ein Stück Land

~ 'out *v. t.* aufteilen ⟨*Land*⟩

~ 'up *v. t.* einwickeln

'parcel bomb *n.* Paketbombe, *die*

parched [pɑ:tʃt] *adj.* ausgedörrt ⟨*Kehle, Land, Boden*⟩; trocken ⟨*Lippen*⟩

parchment ['pɑ:tʃmənt] *n.* Pergament, *das*

pardon ['pɑ:dn] **1.** *n.* **a)** *(forgiveness)* Vergebung, *die (geh.);* Verzeihung, *die;* **b) beg sb.'s ~:** jmdn. um Entschuldigung *od. (geh.)* Verzeihung bitten; **I beg your ~:** entschuldigen *od.* verzeihen Sie bitte; *(please repeat)* wie bitte? *(auch iron.);* **beg ~** *(coll.)* Entschuldigung; Verzeihung; **~?** *(coll.)* bitte?; **~!** *(coll.)* Entschuldigung!; **c)** *(Law)* **[free] ~:** Begnadigung, *die.* **2.** *v. t.* **a)** *(forgive)* **~ sb. [for] sth.** jmdm. etw. verzeihen; **b)** *(excuse)* entschuldigen; **~ my saying so, but ...:** entschuldigen Sie bitte, daß ich es so ausdrücke, aber...; **~ 'me!** Entschuldigung!; **~ 'me?** *(Amer.)* wie bitte?; **c)** *(Law)* begnadigen

pare [peə(r)] *v. t.* **a)** *(trim)* schneiden ⟨*Finger-, Zehennägel*⟩; **b)** *(peel)* schälen ⟨*Apfel, Kartoffel*⟩

~ 'down *v. t.* reduzieren ⟨*Kosten etc.*⟩

parent ['peərənt] *n.* Elternteil, *der;* **~s** Eltern *Pl.; attrib.* Stamm⟨*firma, -organisation*⟩

parentage ['peərəntɪdʒ] *n. (lit. or fig.)* Herkunft, *die*

parental [pə'rentl] *adj.* elterlich ⟨*Gewalt*⟩; Eltern⟨*pflicht, -haus*⟩

parenthesis [pə'renθɪsɪs] *n., pl.* **parentheses** [pə'renθɪsi:z] **a)** *(bracket)* runde Klammer; Parenthese, *die (fachspr.);* **b)** *(word, clause, sentence)* Parenthese, *die (geh.);* Einschub, *der*

parenthetic [pærən'θetɪk], **parenthetical** [pærən'θetɪkl] *adj.* eingeschoben; parenthetisch *(fachspr.)*

parenthood ['peərənthʊd] *n., no pl.* Elternschaft, *die*

Paris ['pærɪs] *pr. n.* Paris *(das)*

parish ['pærɪʃ] *n.* Gemeinde, *die*

parish: ~ 'church *n.* Pfarrkirche, *die;* **~ 'council** *n. (Brit.)* Gemeinderat, *der*

parishioner [pə'rɪʃənə(r)] *n.* Gemeinde[mit]glied, *das*

parish: ~ 'priest *n.* Gemeindepfarrer, *der;* **~ 'register** *n.* Kirchenbuch, *das*

Parisian [pə'rɪzɪən] **1.** *n.* Pariser, *der*/Pariserin, *die.* **2.** *adj.* Pariser

parity ['pærɪtɪ] *n.* **a)** *(equality)* Parität, *die (geh.);* Gleichheit, *die;* **b)** *(Commerc.)* Parität, *die;* **the ~**

of sterling against the dollar die Pfund-Dollar-Parität

park [pɑ:k] **1.** *n.* **a)** Park, *der; (land kept in natural state)* Natur[schutz]park, *der;* **b)** *(sports ground)* Sportplatz, *der; (stadium)* Stadion, *das; (Baseball, Footb.)* Spielfeld, *das;* **c) amusement ~:** Vergnügungspark, *der.* **2.** *v. i.* parken; **find somewhere to ~:** einen Parkplatz finden. **3.** *v. t.* **a)** *(place, leave)* abstellen ‹*Fahrzeug*›; parken ‹*Kfz*›; **b)** *(coll.: leave, put)* deponieren ‹*scherzh.*›; **~ oneself [down]** *(sl.)* sich [hin]pflanzen *(ugs.)*

parka ['pɑ:kə] *n.* Parka, *der*

park-and-'ride *n.* Park-and-ride-System, *das; (place)* Park-and-ride-Platz, *der*

parking ['pɑ:kɪŋ] *n., no pl., no indef. art.* Parken, *das;* **'No ~'** „Parken verboten"

parking: ~ attendant *n.* Parkplatzwächter, *der*/-wächterin, *die;* **~-bay** *n.* Stellplatz, *der;* **~-light** *n.* Parklicht, *das;* Parkleuchte, *die;* **~-lot** *n. (Amer.)* Parkplatz, *der;* **~-meter** *n.* Parkuhr, *die;* **~-space** *n.* **a)** *no pl.* Parkraum, *der;* **b)** *(single space)* Platz zum Parken; Parkplatz, *der;* **~-ticket** *n.* Strafzettel [für falsches Parken]

park: ~keeper *n.* Parkwärter, *der*/-wärterin, *die;* **~land** *n.* Parklandschaft, *die*

parlance ['pɑ:ləns] *n.* **in common/legal/modern ~:** im allgemeinen/juristischen/modernen Sprachgebrauch

parliament ['pɑ:ləmənt] *n.* Parlament, *das;* **[Houses of] P~** *(Brit.)* Parlament, *das*

parliamentary [pɑ:lə'mentərɪ] *adj.* parlamentarisch; Parlaments‹*geschäfte, -reform*›

parlour *(Brit.; Amer.:* **parlor)** ['pɑ:lə(r)] *n. (dated: sitting-room)* Wohnzimmer, *das;* gute Stube *(veralt.)*

Parmesan ['pɑ:mɪzæn, pɑ:mɪ'zæn] *adj., n.* **~ [cheese]** Parmesan[käse], *der*

parochial [pə'rəʊkɪəl] *adj.* **a)** *(narrow)* krähwinklig *(abwertend);* eng ‹*Horizont*›; **be ~ in one's outlook** einen engen Horizont haben; **b)** *(Eccl.)* Gemeinde-

parody ['pærədɪ] **1.** *n.* **a)** *(humorous imitation)* Parodie, *die* **(of** auf + *Akk.*); **b)** *(feeble imitation)* Abklatsch, *der (abwertend); (of justice)* Verhöhnung, *die.* **2.** *v. t.* parodieren

parole [pə'rəʊl] **1.** *n. (conditional release)* bedingter Straferlaß *(Rechtsw.);* **he was released** *or* **let out on ~/he is on ~:** er wurde auf Bewährung entlassen. **2.** *v. t. (Law)* **~ sb.** jmdm. seine Strafe bedingt erlassen

paroxysm ['pærəksɪzm] *n.* Krampf, *der; (fit, convulsion)* Anfall, *der* **(of** von); **~ of rage/laughter** Wut-/Lachanfall, *der*

parquet ['pɑ:kɪ, 'pɑ:keɪ] *n.* **~ [flooring]** Parkett, *das;* **~ floor** Parkettfußboden, *der*

parrot ['pærət] **1.** *n.* Papagei, *der.* **2.** *v. t.* nachplappern *(abwertend);* **~ sb.** jmdm. alles nachplappern

'parrot-fashion *adv.* papageienhaft, wie ein Papagei ‹*wiederholen*›

parry ['pærɪ] *v. t. (Boxing)* abwehren ‹*Faustschlag*›; *(Fencing; also fig.)* parieren ‹*Fechthieb, Frage*›

parsimonious [pɑ:sɪ'məʊnɪəs] *adj.* sparsam; *(niggardly)* geizig

parsley ['pɑ:slɪ] *n., no pl., no indef. art.* Petersilie, *die*

parsnip ['pɑ:snɪp] *n.* Gemeiner Pastinak, *der;* Pastinake, *die*

parson ['pɑ:sn] *n. (vicar, rector)* Pfarrer, *der; (coll.: any clergyman)* Geistliche, *der*

parsonage ['pɑ:sənɪdʒ] *n.* Pfarrhaus, *das*

part [pɑ:t] **1.** *n.* **a)** Teil, *der;* **four-~:** vierteilig ‹*Serie*›; **the hottest ~ of the day** die heißesten Stunden des Tages; **accept ~ of the blame** die Schuld teilweise mit übernehmen; **for the most ~:** größtenteils; zum größten Teil; **in ~:** teilweise; **in large ~:** groß[en]teils; **in ~s** zum Teil; **~ and parcel** wesentlicher Bestandteil; **the funny ~ of it was that he ...:** das Komische daran war, daß er ...; **it's [all] ~ of the fun/job** *etc.* das gehört [mit] dazu; **be** *or* **form ~ of sth.** zu etw. gehören; **b)** *(of machine or other apparatus)* [Einzel]teil, *das;* **c)** *(share)* Anteil, *der;* **I want no ~ in this** ich möchte damit nichts zu tun haben; **d)** *(duty)* Aufgabe, *die;* **do one's ~:** seinen Teil *od.* das Seine tun; **e)** *(Theatre: character, words)* Rolle, *die;* **dress the ~** *(fig.)* die angemessene Kleidung tragen; **play a [great/considerable] ~** *(contribute)* eine [wichtige] Rolle spielen; **f)** *(Mus.)* Part, *der;* Partie, *die;* Stimme, *die;* **g)** *usu. in pl. (region)* Gegend, *die; (of continent, world)* Teil, *der;* **I am a stranger in these ~s** ich kenne mich hier nicht aus; **h)** *(side)* Partei, *die;* **take sb.'s ~:** jmds. *od.* für jmdn. Partei ergreifen; **for my ~:** für mein[en] Teil; **on my/your** *etc.* **~:** meiner-/deinerseits *usw.;* **i)** *pl. (abilities)* **a man of [many] ~s** ein [vielseitig] begabter *od.* befähigter Mann; **j)** *(Ling.)* **~ of speech** Wortart *od.* -klasse, *die;* **k) take [no] ~ [in sth.]** sich [an etw. *(Dat.)*] [nicht] beteiligen; **l) take sth. in good ~:** etw. nicht übelnehmen. **2.** *adv.* teils. **3.** *v. t.* **a)** *(divide into ~s)* teilen; scheiteln ‹*Haar*›; **b)** *(separate)* trennen. **4.** *v. i.* ‹*Menge:*› eine Gasse bilden; ‹*Wolken:*› sich teilen; ‹*Vorhang:*› sich öffnen; ‹*Seil, Tau, Kette:*› reißen; ‹*Lippen:*› sich öffnen; ‹*Wege, Personen:*› sich trennen; **~ from sb./sth.** sich von jmdm./etw. trennen; **~ with** sich trennen von ‹*Besitz, Geld*›

partake [pɑ:'teɪk] *v. i., forms as* **take 2** *(formal)* **~ of** *(eat)* zu sich nehmen ‹*Kost, Mahlzeit*›

partaken *see* **partake**

part-ex'change *n.* **accept sth. in ~ for sth.** etw. für etw. in Zahlung nehmen; **sell sth. in ~:** etw. in Zahlung geben

partial ['pɑ:ʃl] *adj.* **a)** *(biased, unfair)* voreingenommen; parteiisch ‹*Urteil*›; **b) be/not be ~ to sb./sth.** *(like/dislike)* eine Schwäche/keine besondere Vorliebe für jmdn./etw. haben; **c)** *(incomplete)* partiell ‹*Lähmung, Sonnen-, Mondfinsternis*›; teilweise ‹*Verlust, Mißerfolg*›

partiality [pɑ:ʃɪ'ælɪtɪ] *n.* **a)** *(fondness)* Vorliebe, *die;* **b)** *(bias)* Voreingenommenheit, *die*

partially ['pɑ:ʃəlɪ] *adv.* zum Teil; teilweise

participant [pɑ:'tɪsɪpənt] *n.* Beteiligte, *der/die* **(in** an + *Dat.*); *(in arranged event)* Teilnehmer, *der*/Teilnehmerin, *die* **(in** an + *Dat.*)

participate [pɑ:'tɪsɪpeɪt] *v. i. (be actively involved)* sich beteiligen **(in** an + *Dat.*); *(in arranged event)* teilnehmen **(in** an + *Dat.*)

participation [pɑ:tɪsɪ'peɪʃn] *n.* Beteiligung, *die* **(in** an + *Dat.*); *(in arranged event)* Teilnahme, *die* **(in** bei, an + *Dat.*)

participle ['pɑ:tɪsɪpl] *n. (Ling.)* Partizip, *das;* **present/past ~:** Partizip Präsens/Perfekt

particle ['pɑ:tɪkl] *n.* **a)** *(tiny portion; also Phys.)* Teilchen, *das; (of sand)* Körnchen, *das;* **b)** *(fig.) (of sense, truth)* Fünkchen, *das;* **c)** *(Ling.)* Partikel, *die*

particoloured *(Brit.; Amer.:* **particolored)** [pɑ:tɪ'kʌləd] *adj.* bunt

particular [pə'tɪkjʊlə(r)] **1.** *adj.* **a)** *(special)* besonder...; **which ~**

place do you have in mind? an welchen Ort denkst du speziell?; **here in ~:** besonders hier; **nothing/anything [in] ~:** nichts/irgend etwas Besonderes; **in his ~ case** in seinem [besonderen] Fall; **b)** *(fussy, fastidious)* genau; eigen *(landsch.)*; **I am not ~:** es ist mir gleich; **be ~ about sth.** es mit etw. genau nehmen. **2.** *n.* **a)** *in pl. (details)* Einzelheiten; Details; *(of person)* Personalien *Pl.*; *(of incident)* nähere Umstände; **b)** *(detail)* Einzelheit, *die;* Detail, *das*

particularly [pə'tıkjʊləlı] *adv.* **a)** *(especially)* besonders; **b)** *(specifically)* speziell; insbesondere

parting ['pɑ:tıŋ] **1.** *n.* **a)** *(leave-taking)* **[final] ~:** Trennung, *die;* Abschied, *der;* **b)** *(Brit.: in hair)* Scheitel, *der;* **c) ~ of the ways** *(fig.: critical point)* Scheideweg, *der;* **we came to a ~ of the ways** *(fig.)* unsere Wege trennten sich. **2.** *attrib. adj.* Abschieds-; **~ shot** Schlußbemerkung, *die*

partisan ['pɑ:tızæn] **1.** *n. (Mil.)* Partisan, *der*/Partisanin, *die.* **2.** *adj.* **a)** *(often derog.: biased)* voreingenommen, parteiisch ‹*Ansatz, Urteil, Versuch*›; Partei‹*politik, -geist*›; **b)** *(Mil.)* Partisanen-

partition [pɑ:'tıʃn] **1.** *n.* **a)** *(Polit.)* Teilung, *die;* **b)** *(room-divider)* Trennwand, *die;* **c)** *(section of hall or library)* Abteilung, *die;* Bereich, *der.* **2.** *v.t.* **a)** *(divide)* aufteilen ‹*Land, Zimmer*›; **b)** *(Polit.)* teilen ‹*Land*›

~ 'off *v.t.* abteilen ‹*Teil, Raum*›

partly ['pɑ:tlı] *adv.* zum Teil; teilweise

partner ['pɑ:tnə(r)] **1.** *n.* Partner, *der*/Partnerin, *die;* **~ in crime** Komplize, *der*/Komplizin, *die (abwertend);* **be a ~ in a firm** Teilhaber/-haberin einer Firma sein. **2.** *v.t.* **a)** *(make a ~)* **~ sb. with sb.** jmdn. mit jmdm. zusammenbringen; **b)** *(be ~ of)* **~ sb.** jmds. Partner/Partnerin sein; **~ sb. at tennis/in the dance** mit jmdm. Tennis spielen/tanzen

partnership ['pɑ:tnəʃıp] *n.* **a)** *(association)* Partnerschaft, *die;* **b)** *(Commerc.)* **business ~:** [Personen]gesellschaft, *die;* **go** *or* **enter into ~ with sb.** mit jmdm. eine [Personen]gesellschaft gründen

partook *see* **partake**

part 'payment *n.* **a)** *see* **part-exchange; b)** *(sum)* Anzahlung, *die*

partridge ['pɑ:trıdʒ] *n., pl. same or* **~s** Rebhuhn, *das*

part: ~-time 1. ['--] *adj.* Teilzeit‹*arbeit, -arbeiter*›; **he is only ~-time** er ist nur eine Teilzeitkraft; **2.** [-'-] *adv.* stundenweise, halbtags ‹*arbeiten, studieren*›; **work ~-time** als Teilzeitkraft beschäftigt sein; **~-way** *adv.* **we were ~-way through the tunnel** wir hatten ein Stück des Tunnels hinter uns; **go ~-way towards meeting sb.'s demands** jmds. Forderungen *(Dat.)* teilweise *od.* halbwegs entsprechen; **~-way through her speech** mitten in ihrer Rede

party ['pɑ:tı] *n.* **a)** *(group united in a cause etc.; Polit., Law)* Partei, *die; attrib.* Partei‹*apparat, -versammlung, -mitglied, -politik, -politiker usw.*›; **opposing ~:** Gegenpartei, *die;* **b)** *(group)* Gruppe, *die;* **a ~ of tourists** eine Touristengruppe; **c)** *(social gathering)* Party, *die;* Fete, *die (ugs.); (more formal)* Gesellschaft, *die;* **office ~:** Betriebsfest, *das;* **throw a ~** *(coll.)* eine Party schmeißen *(ugs.);* **d)** *(participant)* Beteiligte, *der/die;* **be [a] ~ in** *or* **to sth.** sich an etw. *(Dat.)* beteiligen; **parties to an agreement/a dispute** Parteien bei einem Abkommen/streitende Parteien; *see also* **third party**

party: ~ line *n.* **a)** ['---] *(Teleph.)* Gemeinschafts-, Sammelanschluß, *der;* **b)** [--'-] *(Polit.)* Parteilinie, *die;* **~ piece** *n.* **this song was my ~ piece** dieses Lied mußte ich auf jeder Gesellschaft zum besten geben; **~ 'politics** *n.* Parteipolitik, *die;* **~-wall** *n.* Mauer zum Nachbargrundstück/-gebäude

pass [pɑ:s] **1.** *n.* **a)** *(passing of an examination)* bestandene Prüfung; **get a ~ in maths** die Mathematikprüfung bestehen; **'~'** *(mark or grade)* Ausreichend, *das;* **b)** *(written permission)* Ausweis, *der; (for going into or out of a place also)* Passierschein, *der; (Mil.: for leave)* Urlaubsschein, *der; (for free transportation)* Freifahrschein, *der; (for free admission)* Freikarte, *die;* **c)** *(critical position)* Notlage, *die;* **things have come to a pretty ~ [when ...]** es muß schon weit gekommen sein[, wenn ...]; **d)** *(Football)* Paß, *der (fachspr.);* Ballabgabe, *die; (Fencing)* Ausfall, *der;* **make a ~ to a player** [den Ball] zu einem Spieler passen *(fachspr.) od.* abgeben; **e) make a ~ at sb.** *(fig. coll.: amorously)* jmdn. anmachen *(ugs.);* **f)** *(in mountains)* Paß, *der.* **2.** *v.i.* **a)** *(move onward)* ‹*Prozession:*› ziehen; ‹*Wasser:*› fließen; ‹*Gas:*› strömen; *(fig.)* ‹*Redner:*› übergehen **(to** zu); **~ further along** *or* **down the bus, please!** bitte weiter durchgehen!; **b)** *(go)* passieren; ‹*Zug, Reisender:*› fahren durch ‹*Land*›; **~ over** *(in plane)* überfliegen ‹*Ort*›; **let sb. ~:** jmdn. durchlassen *od.* passieren lassen; **c)** *(be transported, lit. or fig.)* kommen; **~ into history/oblivion** in die Geschichte eingehen/in Vergessenheit geraten; **the title/property ~es to sb.** der Titel/Besitz geht auf jmdn. über; **d)** *(change)* wechseln; **~ from one state to another** von einem Zustand in einen anderen übergehen; **e)** *(go by)* ‹*Fußgänger:*› vorbeigehen; ‹*Fahrer, Fahrzeug:*› vorbeifahren; ‹*Prozession:*› vorbeiziehen; ‹*Zeit, Sekunde:*› vergehen; *(by chance)* ‹*Person, Fahrzeug:*› vorbeikommen; **let sb./a car ~:** jmdn./ein Auto vorbeilassen *(ugs.);* **f)** *(be accepted as adequate)* durchgehen; hingehen; **let it/the matter ~:** es/die Sache durch- *od.* hingehen lassen; **g)** *(come to an end)* vorbeigehen; ‹*Fieber:*› zurückgehen; ‹*Ärger, Zorn, Sturm:*› sich legen; ‹*Gewitter, Unwetter:*› vorüberziehen; **h)** *(happen)* passieren; *(between persons)* vorfallen; **i)** *(be accepted)* durchgehen **(as** als, **for** für); **j)** *(satisfy examiner)* bestehen; **k)** *(Cards)* passen; **[I] ~!** [ich] passe! **3.** *v.t.* **a)** *(move past)* ‹*Fußgänger:*› vorbeigehen an (+ *Dat.*); ‹*Fahrer, Fahrzeug:*› vorbeifahren an (+ *Dat.*); ‹*Prozession:*› vorbeiziehen an (+ *Dat.*); **b)** *(overtake)* vorbeifahren an (+ *Dat.*) ‹*Fahrzeug, Person*›; **c)** *(cross)* überschreiten ‹*Schwelle, feindliche Linien, Grenze, Marke*›; **d)** *(reach standard in)* bestehen ‹*Prüfung*›; **e)** *(approve)* verabschieden ‹*Gesetzentwurf*›; annehmen ‹*Vorschlag*›; ‹*Zensor:*› freigeben ‹*Film, Buch, Theaterstück*›; bestehen lassen ‹*Prüfungskandidaten*›; **f)** *(be too great for)* überschreiten, übersteigen ‹*Auffassungsgabe, Verständnis*›; **g)** *(move)* bringen; **~ a thread through the eye of a needle** einen Faden durch ein Nadelöhr ziehen *od.* führen; **h)** *(Footb. etc.)* abgeben **(to** an + *Akk.*); **i)** *(spend)* verbringen ‹*Leben, Zeit, Tag*›; **j)** *(hand)* **~ sb. sth.** jmdm. etw. reichen *od.* geben; **would you ~ the salt, please?** gibst *od.* reichst du mir bitte das Salz?; **k)** *(utter)* fällen, verkünden ‹*Urteil*›; machen ‹*Bemer-*

kung⟩; **l)** *(discharge)* lassen ⟨*Wasser*⟩
~ **a'way 1.** *v. i. (euphem.: die)* die Augen schließen *od.* zumachen *(verhüll.).* **2.** *v. t.* verbringen ⟨*Zeit[raum], Abend*⟩
~ **by 1.** ['--] *v. t.* **a)** *(go past)* ⟨*Fußgänger:*⟩ vorbeigehen an (+ *Dat.*); ⟨*Fahrer, Fahrzeug:*⟩ vorbeifahren an (+ *Dat.*); ⟨*Prozession:*⟩ vorbeiziehen an (+ *Dat.*); **b)** *(omit, disregard)* übergehen. **2.** [-'-] *v. i.* ⟨*Fußgänger:*⟩ vorbeigehen; ⟨*Fahrer, Fahrzeug:*⟩ vorbeifahren; ⟨*Prozession:*⟩ vorbeiziehen
~ **'down** *see* **hand down a**
~ **for** *v. t.* durchgehen für
~ **'off 1.** *v. t. (represent falsely)* ausgeben (**as, for** als); als echt ausgeben ⟨*Fälschung*⟩. **2.** *v. i.* **a)** *(disappear gradually)* ⟨*Schock, Schmerz, Hochstimmung:*⟩ abklingen; **b)** *(take place)* verlaufen
~ **'on 1.** *v. i.* **a)** *(proceed)* fortfahren; ~ **on to sth.** zu etw. übergehen; **b)** *(euphem.: die)* die Augen schließen *(verhüll.).* **2.** *v. t.* weitergeben (**to** an + *Akk.*)
~ **'out** *v. i.* **a)** *(faint)* umkippen *(ugs.)*; **b)** *(complete military training)* seine militärische Ausbildung abschließen
~ **'over** *v. t.* übergehen
~ **'through** *v. i.* durchreisen; **be just ~ing through** nur auf der Durchreise sein
~ **'up** *v. t.* sich *(Dat.)* entgehen lassen ⟨*Gelegenheit*⟩; ablehnen ⟨*Angebot, Einladung*⟩

passable ['pɑːsəbl] *adj.* **a)** *(acceptable)* passabel; **b)** passierbar, befahrbar ⟨*Straße*⟩

passage ['pæsɪdʒ] *n.* **a)** *(going by, through, etc.) (of river)* Überquerung, *die; (of time)* [Ab-, Ver]lauf, *der; (of seasons)* Wechsel, *der;* **b)** *(transition)* Übergang, *der;* **c)** *(voyage)* Überfahrt, *die;* **d)** Gang, *der; (corridor)* Korridor, *der; (between houses)* Durchgang, *der; (in shopping precinct)* Passage, *die;* **e)** *no art., no pl. (liberty or right to pass through)* Durchreise, *die;* **f)** *(right to travel)* Passage, *die;* **work one's ~:** seine Überfahrt abarbeiten; **g)** *(part of book etc.)* Passage, *die;* **h)** *(Mus.)* Passage, *die;* Stelle, *die;* **i)** *(of a bill into law)* parlamentarische Behandlung; *(final)* Annahme, *die;* Verabschiedung, *die;* **j)** *(Anat.)* **urinary ~:** Harntrakt, *der;* **air ~s** Luft- *od.* Atemwege

'passage-way *n.* Gang, *der; (between houses)* Durchgang, *der*

pass: ~book *n.* Sparbuch, *das;* ~ **degree** *n. (Brit. Univ.)* **get a ~ degree** ein Examen ohne Prädikat bestehen

passenger ['pæsɪndʒə(r)] *n.* **a)** *(on ship)* Passagier, *der; (on plane)* Passagier, *der;* Fluggast, *der; (on train)* Reisende, *der/die; (on bus, in taxi)* Fahrgast, *der; (in car, on motor cycle)* Mitfahrer, *der*/Mitfahrerin, *die; (in front seat of car)* Beifahrer, *der*/Beifahrerin, *die;* **b)** *(coll.: ineffective member) Mensch, der von den anderen mit durchgeschleppt wird (ugs.)*

passenger: ~ aircraft *n.* Passagierflugzeug, *das;* ~ **list** *n.* Passagierliste, *die;* ~ **lounge** *n.* Warteraum, *der;* ~ **seat** *n.* Beifahrersitz, *der;* ~ **train** *n.* Zug im Personenverkehr

passer-by [pɑːsə'baɪ] *n.* Passant, *der*/Passantin, *die*

passing ['pɑːsɪŋ] **1.** *n. (of time, years)* Lauf, *der; (of winter)* Vorübergehen, *das; (of old year)* Ausklang, *der; (death)* Ende, *das;* **in ~:** beiläufig ⟨*bemerken usw.*⟩; flüchtig ⟨*begrüßen*⟩. **2.** *adj.* **a)** *(going past)* vorbeifahrend ⟨*Zug, Auto*⟩; vorbeikommend ⟨*Person*⟩; vorbeiziehend ⟨*Schatten*⟩; **b)** *(fleeting)* flüchtig ⟨*Blick*⟩; vorübergehend ⟨*Mode, Laune, Interesse*⟩; **c)** *(superficial)* flüchtig ⟨*Bekanntschaft*⟩; schnell vorübergehend ⟨*Empfindung*⟩

passion ['pæʃn] *n.* **a)** Leidenschaft, *die;* **he has a ~ for steam engines** Dampfloks sind seine Leidenschaft; er hat eine Passion für Dampfloks; **b) P~** *(Relig., Mus.)* Passion, *die*

passionate ['pæʃənət] *adj.* leidenschaftlich; **have a ~ belief in sth.** mit unbeirrbarem Eifer von etw. überzeugt sein

passion: ~-flower *n. (Bot.)* Passionsblume, *die;* **~-fruit** *n.* Passionsfrucht, *die*

passive ['pæsɪv] **1.** *adj.* **a)** passiv; widerspruchslos ⟨*Hinnahme, Annahme*⟩; ~ **smoking** passives Rauchen; **b)** *(Ling.)* Passiv-; passivisch. **2.** *n. (Ling.)* Passiv, *das*

passiveness ['pæsɪvnɪs], **passivity** [pæ'sɪvɪtɪ] *ns., no pl.* Passivität, *die*

pass: ~-mark *n.* Mindestpunktzahl, *die;* **P~over** *n.* Passah, *das*

'passport *n.* **a)** [Reise]paß, *der; attrib.* Paß-; **b)** *(fig.)* Schlüssel, *der* (**to** zu)

'password *n.* **a)** Parole, *die;* Losung, *die;* **b)** *(Computing)* Paßwort, *das*

past [pɑːst] **1.** *adj.* **a)** *pred. (over)* vorbei; vorüber; **b)** *attrib. (previous)* früher; vergangen; früher, ehemalig ⟨*Präsident, Vorsitzende usw.*⟩; **c)** *(just gone by)* letzt...; vergangen; **in the ~ few days** während der letzten Tage; **the ~ hour** die letzte *od.* vergangene Stunde; **d)** *(Ling.)* ~ **tense** Vergangenheit, *die; see also* **participle. 2.** *n.* **a)** Vergangenheit, *die; (that which happened in the ~)* Vergangene, *das;* Gewesene, *das;* **in the ~:** früher; in der Vergangenheit ⟨*leben*⟩; **be a thing of the ~:** der Vergangenheit *(Dat.)* angehören; **b)** *(Ling.)* Vergangenheit, *die.* **3.** *prep.* **a)** *(beyond in time)* nach; *(beyond in place)* hinter (+ *Dat.*); **half ~ three** halb vier; **five [minutes] ~ two** fünf [Minuten] nach zwei; **it's ~ midnight** es ist schon nach Mitternacht *od.* Mitternacht vorbei; **he is ~ sixty** er ist über sechzig; **walk ~ sb./sth.** an jmdm./etw. vorüber- *od.* vorbeigehen; **b)** *(not capable of)* **he is ~ help/caring** ihm ist nicht mehr zu helfen/es kümmert ihn nicht mehr; **be/be getting ~ it** *(coll.)* [ein bißchen] zu alt sein/allmählich zu alt werden; **I wouldn't put it ~ her to do that** ich würde es ihr schon zutrauen, daß sie das tut. **4.** *adv.* vorbei; vorüber; **hurry ~:** vorüber- *od.* vorbeieilen

pasta ['pæstə, 'pɑːstə] *n.* Nudeln *Pl.;* Teigwaren *Pl.*

paste [peɪst] **1.** *n.* **a)** Brei, *der;* **mix into a smooth/thick ~:** zu einem lockeren/dicken Brei anrühren; zu einem glatten/festen Teig anrühren ⟨*Backmischung*⟩; **b)** *(glue)* Kleister, *der;* **c)** *(of meat, fish, etc.)* Paste, *die;* **d)** *no pl., no indef. art. (imitation gems)* Straß, *der;* Similisteine *Pl.* **2.** *v. t. (fasten with glue)* kleben; ~ **sth. down/into sth.** etw. ankleben/in etw. *(Akk.)* einkleben

'pasteboard *n.* Pappe, *die;* Karton, *der*

pastel ['pæstl] **1.** *n.* **a)** *(crayon)* Pastellstift, *der;* Pastellkreide, *die;* **b)** *(drawing)* Pastellzeichnung, *die.* **2.** *adj.* pastellen; pastellfarben; Pastell⟨*farben, -töne, -zeichnung, -bild*⟩

pasteurize (pasteurise) ['pæstʃəraɪz, 'pɑːstʃəraɪz] *v. t.* pasteurisieren

pastille ['pæstɪl] *n.* Pastille, *die*

pastime ['pɑːstaɪm] *n.* Zeitvertreib, *der; (person's specific ~)* Hobby, *das;* **national ~:** Nationalsport, *der (auch iron.)*

past 'master *n. (fig.)* Meister, *der*

pastor ['pɑːstə(r)] *n.* Pfarrer, *der/*

Pfarrerin, *die;* Pastor, *der*/Pastorin, *die*

pastoral ['pɑ:stərl] *adj.* **a)** Weide-; ländlich ‹*Reiz, Idylle, Umgebung*›; **b)** *(Lit., Art, Mus.)* pastoral; **c)** *(Eccl.)* pastoral; des Pfarrers *nachgestellt;* seelsorgerisch ‹*Pflicht, Aufgabe, Leitung, Aktivitäten*›

pastry ['peɪstrɪ] *n.* **a)** *(flour-paste)* Teig, *der;* **b)** *(article of food)* Gebäckstück, *das;* **c) pastries** *collect.* [Fein]gebäck, *das*

pasture ['pɑ:stʃə(r)] *n.* **a)** *(grass)* Futter, *das;* Gras, *das;* **b)** *(land)* Weideland, *das; (piece of land)* Weide, *die;* **c)** *(fig.)* **in search of ~s new** auf der Suche nach etwas Neuem

'pasture land *n.* Weideland, *das*

[1]pasty ['pæstɪ] *n.* Pastete, *die*

[2]pasty ['peɪstɪ] *adj.* **a)** teigig; zähflüssig; **b)** *see* **pasty-faced**

pasty-faced ['peɪstɪfeɪst] *adj.* mit teigigem Gesicht *nachgestellt;* **be ~:** ein teigiges Gesicht haben

[1]pat [pæt] **1.** *n.* **a)** *(stroke, tap)* Klaps, *der;* leichter Schlag; **give sb./a dog a ~:** jmdn./einen Hund tätscheln; **give sb./a dog a ~ on the head** jmdm./einem Hund den Kopf tätscheln; **a ~ on the back** *(fig.)* eine Anerkennung; **give oneself/sb. a ~ on the back** *(fig.)* sich *(Dat.)* [selbst] auf die Schulter klopfen/jmdm. einige anerkennende Worte sagen; **b)** *(of butter)* Stückchen, *das; (of mud, clay)* Klümpchen, *das.* **2.** *v. t.*, **-tt-: a)** *(strike gently)* leicht klopfen auf (+ *Akk.*); tätscheln, *(once)* einen Klaps geben (+ *Dat.*) ‹*Person, Hund, Pferd*›; **~ sb. on the back** *(fig.)* jmdm. auf die Schulter klopfen; **b)** *(flatten)* festklopfen ‹*Sand*›; andrücken ‹*Haare*›

[2]pat 1. *adv. (ready, prepared)* **have sth. off ~:** etw. parat haben; **know sth. off ~:** etw. aus dem Effeff können *od.* beherrschen *(ugs.).* **2.** *adj. (ready)* allzu schlagfertig ‹*Antwort*›

patch [pætʃ] **1.** *n.* **a)** Stelle, *die;* **a ~ of blue sky** ein Stückchen blauer Himmel; **there were still ~es of snow** es lag vereinzelt *od.* hier und da noch Schnee; **the dog had a black ~ on its ear** der Hund hatte einen schwarzen Fleck am Ohr; **fog ~es** Nebelfelder; **in ~es** stellenweise; **go through** *or* **strike a bad/good ~** *(Brit.)* eine Pech-/Glückssträhne haben; **b)** *(on worn garment)* Flicken, *der;* **be not a ~ on sth.** *(fig. coll.)* nichts gegen etw. sein; **c)** *(on eye)* Augenklappe, *die;* **d)** *(piece of ground)* Stück Land, *das;* **potato ~:** Kartoffelacker, *der; (in garden)* Kartoffelbeet, *das;* **e)** *(area patrolled by police; also fig.)* Revier, *das.* **2.** *v. t. (apply ~ to)* flicken

~ 'up *v. t.* reparieren; zusammenflicken ‹*Segel, Buch*›; notdürftig verbinden ‹*Wunde*›; zusammenflicken *(scherzh.)* ‹*Verletzten*›; *(fig.)* beilegen ‹*Streit, Differenzen*›; kitten ‹*Ehe, Freundschaft*›

'patchwork *n.* Patchwork, *das; (fig.)* **a ~ of fields** ein bunter Teppich von Feldern

patchy ['pætʃɪ] *adj.* uneinheitlich ‹*Qualität*›; ungleichmäßig, unterschiedlich ‹*Arbeit, Aufführung, Ausstoß*›; fleckig ‹*Anstrich*›; sehr lückenhaft ‹*Wissen*›; in der Qualität unterschiedlich ‹*Film, Buch, Theaterstück*›

pâté ['pæteɪ] *n.* Pastete, *die;* **~ de foie gras** ['pæteɪ/'pɑ:teɪ də fwɑ: 'grɑ:] Gänseleberpastete, *die*

patent ['peɪtənt, 'pætənt] **1.** *adj.* **a)** patentiert; **~ medicine** Markenmedizin, *die;* **~ remedy** Spezial- *od.* Patentrezept, *das;* **b)** *(obvious)* offenkundig; offensichtlich. **2.** *n.* Patent, *das;* **~ applied for** *or* **pending** Patent angemeldet. **3.** *v. t.* patentieren lassen; **sth. has been ~ed** etw. ist patentrechtlich geschützt

patent 'leather *n.* Lackleder, *das;* **~ shoes** Lackschuhe

patently ['peɪtəntlɪ, 'pætəntlɪ] *adv.* offenkundig; offensichtlich; **~ obvious** ganz offenkundig *od.* offensichtlich

paternal [pə'tɜ:nl] *adj.* **a)** *(fatherly)* väterlich; **b)** *(related)* ‹*Großeltern, Onkel, Tante*› väterlicherseits

paternity [pə'tɜ:nɪtɪ] *n.* Vaterschaft, *die;* **~ suit** *n.* Vaterschaftsklage, *die*

path [pɑ:θ] *n., pl.* **~s** [pɑ:ðz] **a)** *(way)* Weg, *der;* Pfad, *der; (made by walking)* Trampelpfad, *der;* **keep to the ~:** auf dem Weg bleiben; **b)** *(of rocket, missile, etc.)* Bahn, *die; (of tornado)* Weg, *der;* **c)** *(fig.: course of action)* Weg, *der;* **the ~ to salvation/of virtue** der Weg des Heils/der Pfad der Tugend

pathetic [pə'θetɪk] *adj.* **a)** *(pitiful)* mitleiderregend; herzergreifend; **be a ~ sight** ein Bild des Jammers bieten; **b)** *(full of pathos)* pathetisch; **c)** *(contemptible)* armselig ‹*Entschuldigung*›; erbärmlich ‹*Darbietung, Rede, Person, Leistung*›; **you're/it's ~:** du bist ein hoffnungsloser Fall/es ist wirklich ein schwaches Bild *(ugs.)*

pathetically [pə'θetɪkəlɪ] *adv.* **a)** *(pitifully)* mitleiderregend; herzergreifend ‹*flehen*›; **b)** *(contemptibly)* erbärmlich; erschreckend ‹*wenig*›

pathological [pæθə'lɒdʒɪkl] *adj.* **a)** pathologisch; Pathologie-; **b)** *(fig.: obsessive)* krankhaft; pathologisch

pathologist [pə'θɒlədʒɪst] *n.* Pathologe, *der*/Pathologin, *die*

pathology [pə'θɒlədʒɪ] *n.* *(science)* Pathologie, *die*

pathos ['peɪθɒs] *n.* Pathos, *das*

'pathway *n.* **a)** *see* **path a; b)** *(Physiol.)* Bahn, *die;* Leitung, *die*

patience ['peɪʃəns] *n.* **a)** *no pl., no art.* Geduld, *die; (perseverance)* Ausdauer, *die;* Beharrlichkeit, *die; (forbearance)* Langmut, *die;* **with ~:** geduldig; **have endless ~:** eine Engelsgeduld haben; **lose [one's] ~ [with sth./sb.]** [mit etw./jmdm.] die Geduld verlieren; **I lost my ~:** mir riß der Geduldsfaden *(ugs.) od.* die Geduld; **b)** *(Brit. Cards)* Patience, *die*

patient ['peɪʃənt] **1.** *adj.* geduldig; *(forbearing)* langmütig; *(persevering)* beharrlich; **please be ~:** bitte hab Geduld; **remain ~:** sich in Geduld fassen. **2.** *n.* Patient, *der*/Patientin, *die*

patiently ['peɪʃəntlɪ] *adv.* geduldig; mit Geduld

patina ['pætɪnə] *n. (on bronze)* Patina, *die; (on woodwork)* Altersglanz, *der*

patio ['pætɪəʊ] *n., pl.* **~s** Veranda, *die;* Terrasse, *die*

patriarch ['peɪtrɪɑ:k] *n.* Patriarch, *der; (of tribe)* Stammesoberhaupt, *das*

patriarchal [peɪtrɪ'ɑ:kl] *adj.* patriarchalisch

patriot ['pætrɪət, 'peɪtrɪət] *n.* Patriot, *der*/Patriotin, *die*

patriotic [pætrɪ'ɒtɪk, peɪtrɪ'ɒtɪk] *adj.* patriotisch

patriotism ['pætrɪətɪzm, 'peɪtrɪətɪzm] *n.* Patriotismus, *der*

patrol [pə'trəʊl] **1.** *n.* **a)** *(of police)* Streife, *die; (of watchman)* Runde, *die; (of aircraft, ship; also Mil.)* Patrouille, *die;* **be on** *or* **keep ~** ‹*Soldat, Wächter:*› patrouillieren; **b)** *(person, group)* *(Police)* Streife, *die; (Mil.)* Patrouille, *die;* **police ~:** Polizeistreife, *die;* **c)** *(troops)* Spähtrupp, *der;* Spähpatrouille, *die.* **2.** *v. i.*, **-ll-** patrouillieren; ‹*Polizei:*› Streife laufen/fahren; ‹*Wachmann:*› seine Runde[n] machen; ‹*Flugzeug:*› Patrouille fliegen. **3.**

v. t., **-ll-** patrouillieren durch (+ *Akk.*); abpatrouillieren ⟨*Straßen, Mauer, Gegend, Lager*⟩; patrouillieren vor (+ *Dat.*) ⟨*Küste, Grenze*⟩; ⟨*Polizei:*⟩ Streife laufen/fahren in (+ *Dat.*) ⟨*Straßen, Stadtteil*⟩; ⟨*Wachmann:*⟩ seine Runde[n] machen in (+ *Dat.*)

patrol: ~ **boat** *n.* Patrouillenboot, *das;* ~ **car** *n.* Streifenwagen, *der*

patron ['peɪtrən] *n.* **a)** *(supporter)* Gönner, *der*/Gönnerin, *die; (of institution, campaign)* Schirmherr, *der*/Schirmherrin, *die;* ~ **of the arts** Kunstmäzen, *der;* **b)** *(customer) (of shop)* Kunde, *der*/Kundin, *die; (of restaurant, hotel)* Gast, *der; (of theatre, cinema)* Besucher, *der*/Besucherin, *die;* **c)** ~ **[saint]** Schutzheilige, *der/die*

patronage ['pætrənɪdʒ] *n. (support)* Gönnerschaft, *die;* Unterstützung, *die; (for campaign, institution)* Schirmherrschaft, *die*

patronise, patronising *see* **patroniz-**

patronize ['pætrənaɪz] *v. t.* **a)** *(frequent)* besuchen; **b)** *(support)* fördern; unterstützen; **c)** *(condescend to)* ~ **sb.** jmdn. von oben herab *od.* herablassend behandeln

patronizing ['pætrənaɪzɪŋ] *adj.* gönnerhaft; herablassend

patter ['pætə(r)] **1.** *n.* **a)** *(of rain)* Prasseln, *das; (of feet)* Trappeln, *das;* Getrappel, *das;* **b)** *(of salesman or comedian)* Sprüche *Pl.;* **sales** ~: Vertretersprüche *Pl.* **2.** *v. i.* **a)** ⟨*Regen, Hagel:*⟩ prasseln; ⟨*Schritte:*⟩ trappeln; **b)** *(run)* trippeln

pattern ['pætən] **1.** *n.* **a)** *(design)* Muster, *das; (on carpet, wallpaper, cloth, etc. also)* Dessin, *das;* **b)** *(form, order)* Muster, *das;* Schema, *das;* **follow a** ~: einem regelmäßigen Muster *od.* Schema folgen; **behaviour** ~: Verhaltensmuster, *das;* ~ **of thought** Denkmuster, *das;* Denkschema, *das;* ~ **of events** Ereignisfolge; **c)** *(model)* Vorlage, *die; (for sewing)* Schnittmuster, *das;* Schnitt, *der; (for knitting)* Strickanleitung, *die;* Strickmuster, *das;* **follow a** ~: nach einer Vorlage arbeiten; *(knitting)* nach einem Strickmuster stricken; **a democracy on the British** ~: eine Demokratie nach britischem Muster. **2.** *v. t. (model)* gestalten; ~ **sth. after/on sth.** etw. einer Sache *(Dat.)* nachbilden

paucity ['pɔːsɪtɪ] *n. (formal)* Mangel, *der* (**of** an + *Dat.*)

paunch [pɔːntʃ] *n.* Bauch, *der;* Wanst, *der (salopp abwertend)*

pauper ['pɔːpə(r)] *n.* Arme, *der/die*

pause [pɔːz] **1.** *n.* Pause, *die;* **without [a]** ~: ohne Pause; **an anxious** ~: ängstliches Schweigen; **give sb.** ~: jmdm. zu denken geben. **2.** *v. i.* eine Pause machen; eine Pause einlegen; ⟨*Redner:*⟩ innehalten; *(hesitate)* zögern; ~ **for reflection/thought** in Ruhe überlegen; ~ **for a rest** eine Erholungspause *od.* Ruhepause einlegen

pave [peɪv] *v. t.* pflastern; ~ **the way for** *or* **to sth.** *(fig.)* einer Sache *(Dat.)* den Weg ebnen

pavement ['peɪvmənt] *n.* **a)** *(Brit.: for pedestrians)* Bürgersteig, *der;* Gehsteig, *der;* **b)** *(Amer.: roadway)* Fahrbahn, *die*

pavilion [pə'vɪljən] *n.* **a)** Pavillon, *der;* **b)** *(Brit. Sport)* Klubhaus, *das*

paw [pɔː] **1.** *n.* **a)** Pfote, *die; (of bear, lion, tiger)* Pranke, *die;* **b)** *(coll. derog.: hand)* Pfote, *die (ugs. abwertend);* **keep your** ~**s off!** Pfoten weg! **2.** *v. t.* **a)** ⟨*Hund, Wolf:*⟩ mit der Pfote/den Pfoten berühren; ⟨*Bär, Löwe, Tiger:*⟩ mit der Pranke/den Pranken berühren; *(playfully)* tätscheln; ~ **the ground** scharren; **b)** *(coll. derog.: fondle)* befummeln *(ugs.).* **3.** *v. i.* scharren; ~ **at** mit der Pfote/den Pfoten *usw.* berühren

[1]**pawn** [pɔːn] *n.* **a)** *(Chess)* Bauer, *der;* **b)** *(fig.)* Schachfigur, *die*

[2]**pawn 1.** *n.* Pfand, *das;* **in** ~: verpfändet; versetzt. **2.** *v. t.* verpfänden; versetzen

pawn: ~**broker** *n.* Pfandleiher, *der*/-leiherin, *die;* ~**shop** *n.* Leihhaus, *das;* Pfandleihe, *die*

pay [peɪ] **1.** *n., no pl., no indef. art. (wages)* Lohn, *der; (salary)* Gehalt, *das; (of soldier)* Sold, *der;* **the** ~ **is good** die Bezahlung ist gut; **be in the** ~ **of sb./sth.** für jmdn./etw. arbeiten. **2.** *v. t.,* **paid** [peɪd] **a)** *(give money to)* bezahlen; *(fig.)* belohnen; **I paid him for the tickets** ich habe ihm das Geld für die Karten gegeben; ~ **sb. to do sth.** jmdn. dafür bezahlen, daß er etw. tut; **b)** *(hand over)* zahlen; *(~ back)* zurückzahlen; *(in instalments)* abbezahlen; ~ **the bill** die Rechnung bezahlen; ~ **sb.'s expenses** *(reimburse)* jmds. Auslagen erstatten; ~ **sb. £10** jmdm. 10 Pfund zahlen; ~ **£10 for sth.** 10 Pfund für etw. [be]zahlen; ~ **sth. into a bank account** etw. auf ein Konto ein[be]zahlen; **c)** *(yield)* einbringen, abwerfen ⟨*Dividende usw.*⟩; **this job** ~**s very little** diese Arbeit bringt sehr wenig ein; **d)** *(be profitable to)* **it would** ~ **her to do that** *(fig.)* es würde ihr nichts schaden *od.* es würde sich für sie bezahlt machen, das zu tun; **e)** ~ **the price** den Preis zahlen; **it's too high a price to** ~: das ist ein zu hoher Preis. **3.** *v. i.,* **paid a)** zahlen; ~ **for sth./sb.** etw./für jmdn. bezahlen; **sth.** ~**s for itself** etw. macht sich bezahlt; **has this been paid for?** ist das schon bezahlt?; **b)** *(yield)* sich lohnen; sich auszahlen; ⟨*Geschäft:*⟩ rentabel sein; **it** ~**s to be careful** es lohnt sich, vorsichtig zu sein; **c)** *(fig.: suffer)* büßen müssen; **if you do this you'll have to** ~ **for it later** wenn du das tust, wirst du später dafür büßen müssen

~ **'back** *v. t.* **a)** zurückzahlen; **I'll** ~ **you back later** ich gebe dir das Geld später zurück; **b)** *(fig.)* erwidern ⟨*Kompliment*⟩; sich revanchieren für ⟨*Beleidigung, Untreue*⟩; **I'll** ~ **him back** ich werde es ihm heimzahlen

~ **'in** *v. t. & i.* einzahlen

~ **'off 1.** *v. t.* auszahlen ⟨*Arbeiter*⟩; abbezahlen ⟨*Schulden*⟩; ablösen ⟨*Hypothek*⟩; befriedigen ⟨*Gläubiger*⟩; *(fig.)* abgelten ⟨*Verpflichtung*⟩. **2.** *v. i. (coll.)* sich auszahlen; sich bezahlt machen

~ **'out 1.** *v. t.* **a)** auszahlen; *(spend)* ausgeben; ~ **out large sums on sth.** hohe Beträge für etw. ausgeben; **b)** *(Naut.)* ablaufen lassen ⟨*Seil, Tau*⟩; **c)** *(coll.: punish)* ~ **sb. out for sth.** jmdm. etw. heimzahlen. **2.** *v. i.* bezahlen

~ **'up 1.** *v. t.* zurückzahlen ⟨*Schulden*⟩. **2.** *v. i.* zahlen

payable ['peɪəbl] *adj.* zahlbar; **be** ~ **to sb.** jmdm. *od.* an jmdn. zu zahlen sein; **make a cheque** ~ **to the Post Office/to sb.** einen Scheck auf die Post/auf jmds. Namen ausstellen

pay: ~**-as-you-'earn** *attrib. adj. (Brit.)* ~**-as-you-earn system/method** Quellenabzugsverfahren, *das;* ~**-as-you-earn tax system** *Steuersystem, bei dem die Lohnsteuer direkt einbehalten wird;* ~ **award** *n.* Gehaltserhöhung, *die;* ~**-bed** *n.* Privatbett, *das;* ~**-claim** *n.* Lohn-/Gehaltsforderung, *die;* ~**-day** *n.* Zahltag, *der*

PAYE *abbr. (Brit.)* **pay-as-you-earn**

payee [peɪ'iː] *n.* Zahlungsempfänger, *der*/-empfängerin, *die*

pay: ~ **envelope** *(Amer.) see* **pay-packet; ~-increase** *see* **pay-rise**

paying: ~ **'guest** *n.* zahlender Gast; **~-'in slip** *(Brit. Banking)* Einzahlungsschein, *der*

'payload *n.* Nutzlast, *die*

payment ['peɪmənt] *n.* **a)** *(act)* Zahlung, *die; (paying back)* Rückzahlung, *die; (in instalments)* Abzahlung, *die;* **in ~ [for sth.]** als Bezahlung [für etw.]; **on ~ of ...:** gegen Zahlung von ...; **b)** *(amount)* Zahlung, *die;* **make a ~:** eine Zahlung leisten; **by monthly ~s** auf Monatsraten

pay: ~ **negotiations** *n. pl.* Tarifverhandlungen; **~-off** *n. (sl.) (return)* Lohn, *der; (punishment)* Quittung, *die; (bribe)* Schmiergeld, *das (ugs. abwertend);* **~-packet** *n. (Brit.)* Lohntüte, *die;* ~ **phone** *n.* Münzfernsprecher, *der;* **~-rise** *n.* Lohn-/Gehaltserhöhung, *die;* **~-roll** *n.* Lohnliste, *die;* **have 200 workers/people on the ~-roll** 200 Arbeiter beschäftigen/Beschäftigte haben; **be on sb.'s ~-roll** für jmdn. *od.* bei jmdm. arbeiten; **~-slip** *n.* Lohnstreifen, *der*/Gehaltszettel, *der;* ~ **station** *(Amer.) see* ~ **phone;** ~ **talks** *n. pl.* Tarifverhandlungen

PC *abbr.* **a)** *(Brit.)* **police constable** Wachtm.; **b) personal computer** PC

PE *abbr.* **physical education**

pea [pi:] *n.* Erbse, *die;* **they are as like as two ~s [in a pod]** sie gleichen sich *(Dat.) od.* einander wie ein Ei dem anderen

peace [pi:s] *n.* **a)** *(freedom from war)* Frieden, *der;* **maintain/restore ~:** den Frieden bewahren/wiederherstellen; **~ talks/treaty** Friedensgespräche *Pl.*/Friedensvertrag, *der;* **make ~ [with sb.]** [mit jmdm.] Frieden schließen; **b)** *(freedom from civil disorder)* Ruhe und Ordnung; *(absence of discord)* Frieden, *der;* **in ~ [and harmony]** in [Frieden und] Eintracht; **restore ~:** Ruhe und Ordnung wiederherstellen; **bind sb. over to keep the ~:** jmdn. verwarnen, die öffentliche Ordnung zu wahren; **be at ~ [with sb./sth.]** [mit jmdm./etw.] in Frieden leben; **be at ~ with oneself** mit sich selbst im reinen sein; **make [one's] ~ [with sb.]** sich [mit jmdm.] aussöhnen; **hold one's ~:** schweigen; **c)** *(tranquillity)* Ruhe, *die;* **in ~:** in Ruhe; **leave sb. in ~:** jmdn. in Frieden *od.* in Ruhe lassen; **give sb. no ~:** jmdm. keine Ruhe lassen; **~ and quiet** Ruhe und Frieden; **d)** *(mental state)* Ruhe, *die;* **find ~:** Frieden finden; **~ of mind** Seelenfrieden, *der;* innere Ruhe; **I shall have no ~ of mind until I know it** ich werde keine ruhige Minute haben, bis ich es weiß

peaceable ['pi:səbl] *adj. (not quarrelsome)* friedfertig; friedliebend ⟨*Volk*⟩; *(calm)* friedlich

peaceably ['pi:səblɪ] *adv. (amicably)* friedlich

peaceful ['pi:sfl] *adj.* friedlich; friedfertig ⟨*Person, Volk*⟩; ruhig ⟨*Augenblick*⟩

peacefully ['pi:sfəlɪ] *adv.* friedlich; **die ~:** sanft entschlafen

peace: **~keeper** *n.* Friedenswächter, *der;* **~-keeping 1.** *adj.* ⟨*Maßnahmen, Operationen*⟩ zur Friedenssicherung; **~-keeping force** Friedenstruppe, *die;* **2.** *n.* Friedenssicherung, *die;* **~-loving** *adj.* friedliebend; **~maker** *n.* Friedensstifter, *der*/-stifterin, *die;* **~-offer** *n.* Friedensangebot, *das;* **~-offering** *n.* Friedensangebot, *das; (fig.)* Versöhnungsgeschenk, *das;* **~time** *n.* Friedenszeiten *Pl.; attrib.* Friedens⟨*produktion, -wirtschaft, -stärke*⟩

peach [pi:tʃ] *n.* **a)** Pfirsich, *der;* **b)** *(coll.)* **sb./sth. is a ~:** jmd./etw. ist spitze *od.* klasse *(ugs.);* **c)** *(colour)* Pfirsichton, *der*

'peach-tree *n.* Pfirsichbaum, *der*

'peacock *n.* Pfau, *der;* Pfauhahn, *der;* **proud/vain as a ~:** stolz/eitel wie ein Pfau

'pea-green *adj.* erbsengrün; maigrün

peak [pi:k] **1.** *n.* **a)** *(of cap)* Schirm, *der;* **b)** *(of mountain)* Gipfel, *der; (of wave)* Kamm, *der;* Krone, *die;* **c)** *(highest point)* Höhepunkt, *der;* **be at/be past its ~:** den Höhepunkt erreicht haben/den Höhepunkt überschritten haben. **2.** *attrib. adj.* Höchst-, Spitzen⟨*preise, -werte*⟩; ~ **listening/viewing period** Hauptsendezeit, *die.* **3.** *v. i.* seinen Höhepunkt erreichen

peaked [pi:kt] *adj.* ~ **cap** Schirmmütze, *die*

'peak-hour *attrib. adj.* **~-hour travel** Fahren während der Hauptverkehrszeit; **~-hour traffic** Stoßverkehr, *der*

peaky ['pi:kɪ] *adj.* kränklich; **look ~:** angeschlagen aussehen

peal [pi:l] **1.** *n.* **a)** *(ringing)* Geläut[e], *das;* Läuten, *das;* **~ of bells** Glockengeläut[e], *das;* **b)** *(set of bells)* Glockenspiel, *das;* **c) a ~ of laughter** schallendes Gelächter; **a ~ of thunder** ein Donnerschlag. **2.** *v. i.* ⟨*Glocken:*⟩ läuten

peanut ['pi:nʌt] *n.* Erdnuß, *die;* **~ butter** Erdnußbutter, *die;* **~s** *(coll.) (trivial thing)* ein Klacks *(ugs.); (money)* ein paar Kröten *(salopp);* **this is ~s compared to ...:** das ist ein Klacks gegen ...; **work for ~s** für ein Butterbrot arbeiten *(ugs.)*

pear [peə(r)] *n.* Birne, *die*

pearl [pɜ:l] *n.* Perle, *die;* **[string of] ~s** *(necklace)* Perlenkette, *die;* **~ of wisdom** *(often iron.)* Weisheit, *die*

pearl: ~ **'barley** *n.* Perlengraupen *Pl.;* **~-diver** *n.* Perlentaucher, *der*/-taucherin, *die;* **~-grey** *adj.* perlgrau; **~-oyster** *n.* Perlmuschel, *die*

pearly ['pɜ:lɪ] *adj.* perlmuttern ⟨*Glanz, Schimmer*⟩

'pear-tree *n.* Birnbaum, *der*

peasant ['pezənt] *n.* **a)** [armer] Bauer, *der;* Landarbeiter, *der;* **~ farmer** Bauer, *der;* **~ woman** Bauersfrau, *die;* **b)** *(coll. derog.) (stupid person)* Bauer, *der (ugs. abwertend); (lower-class person)* Plebejer, *der (abwertend)*

peasantry ['pezəntrɪ] *n.* Bauernschaft, *die*

pease-pudding [pi:z'pʊdɪŋ] *n.* Erbsenpudding, *der*

pea: **~-shooter** ['pi:ʃu:tə(r)] *n.* Pusterohr, *das;* ~ **'soup** *n.* Erbsensuppe, *die*

peat [pi:t] *n.* **a)** *(substance)* Torf, *der;* **b)** *(piece)* Torfstück, *das*

pebble ['pebl] *n.* Kiesel[stein], *der;* **he is/you are not the only ~ on the beach** es gibt noch andere

pebbly ['peblɪ] *adj.* steinig

peck [pek] **1.** *v. t.* **a)** hacken; picken ⟨*Körner*⟩; **the bird ~ed my finger** der Vogel pickte mir *od.* mich in den Finger; **b)** *(kiss)* flüchtig küssen. **2.** *v. i.* picken **(at** nach); **~ at one's food** in seinem Essen herumstochern. **3.** *n.* **a) the hen gave its chick a ~:** die Henne pickte *od.* hackte nach ihrem Küken; **b)** *(kiss)* flüchtiger Kuß; Küßchen, *das*

'pecking order *n.* Hackordnung, *die*

peckish ['pekɪʃ] *adj. (coll.)* hungrig; **feel/get ~:** Hunger haben/bekommen

pectin ['pektɪn] *n. (Chem.)* Pektin, *das*

peculiar [pɪ'kju:lɪə(r)] *adj.* **a)** *(strange)* seltsam; eigenartig; sonderbar; **I feel [slightly] ~:** mir ist [etwas] komisch; **b)** *(especial)* **be of ~ interest [to sb.]** [für jmdn.]

von besonderem Interesse sein; **c)** *(belonging exclusively)* eigentümlich (to *Dat.*); **this bird is ~ to South Africa** dieser Vogel kommt nur in Südafrika vor

peculiarity [pɪkju:lɪ'ærɪtɪ] *n.* **a)** *no pl., no indef. art. (unusualness)* Ausgefallenheit, *die; (of behaviour, speech)* Sonderbarkeit, *die;* **b)** *(odd trait)* Eigentümlichkeit, *die;* **c)** *(distinguishing characteristic)* [charakteristisches] Merkmal; *(special characteristic)* Besonderheit, *die*

peculiarly [pɪ'kju:lɪəlɪ] *adv.* **a)** *(strangely)* seltsam; eigenartig; sonderbar; **b)** *(especially)* besonders; **c)** *(in a way that is one's own)* **be something ~ British** etwas rein Britisches sein

pedagogic[al] [pedə'gɒgɪk(l), pedə'gɒdʒɪk(l)] *adj. (of pedagogy)* pädagogisch

pedagogy ['pedəgɒdʒɪ] *n.* Pädagogik, *die*

pedal ['pedl] **1.** *n.* Pedal, *das.* **2.** *v. i., (Brit.)* **-ll-: a)** *(work cycle ~s)* **~ [away]** in die Pedale treten; strampeln *(ugs.);* **b)** *(ride)* [mit dem Fahrrad] fahren; radeln *(ugs.);* **~ by/off** vorbeiradeln/losradeln *(ugs.)*

'pedal-bin *n.* Treteimer, *der*

pedalo ['pedələʊ] *n., pl.* **~s** Tretboot, *das*

pedant ['pedənt] *n.* Pedant, *der*/Pedantin, *die (abwertend)*

pedantic [pɪ'dæntɪk] *adj.* pedantisch *(abwertend)*

pedantry ['pedəntrɪ] *n.* Pedanterie, *die*

peddle ['pedl] *v. t.* auf der Straße verkaufen; *(from door to door)* hausieren mit; handeln mit, *(ugs.)* dealen mit ‹*Drogen, Rauschgift*›

peddler ['pedlə(r)] *see* **pedlar**

pederast ['pedəræst] *n.* Päderast, *der*

pedestal ['pedɪstl] *n.* Sockel, *der;* **put sb./sth. on a ~** *(fig.)* jmdn./etw. in den Himmel heben *(ugs.)*

pedestrian [pɪ'destrɪən] **1.** *adj. (uninspired)* trocken; langweilig. **2.** *n.* Fußgänger, *der*/-gängerin, *die*

pedestrian 'crossing *n.* Fußgängerüberweg, *der*

pedicure ['pedɪkjʊə(r)] *n.* Pediküre, *die;* **give sb. a ~:** jmdn. pediküren

pedigree ['pedɪgri:] **1.** *n.* **a)** Stammbaum, *der;* Ahnentafel, *die (geh.); (of animal)* Stammbaum, *der;* **b)** *no pl., no art. (ancient descent)* **have ~, be a man/woman of ~:** von berühmten Ahnen abstammen. **2.** *adj. (with recorded line of descent)* mit Stammbaum *nachgestellt*

pedlar ['pedlə(r)] *n.* Straßenhändler, *der*/-händlerin, *die; (from door to door)* Hausierer, *der*/Hausiererin, *die; (selling drugs)* Rauschgifthändler, *der*/-händlerin, *die;* Dealer, *der*/Dealerin, *die (ugs.)*

pee [pi:] *(coll.)* **1.** *v. i.* pinkeln *(salopp);* Pipi machen *(Kinderspr.).* **2.** *n.* **a) need/have a ~:** pinkeln müssen/pinkeln *(salopp);* **I must go for a ~:** ich muß mal eben pinkeln *(salopp);* **b)** *(urine)* Pipi, *das (Kinderspr.)*

peek [pi:k] **1.** *v. i.* gucken *(ugs.);* **no ~ing!** nicht gucken!; **~ at sb./sth.** zu jmdm./etw. hingucken. **2.** *n. (quick)* kurzer Blick; *(sly)* verstohlener Blick; **have a ~ through the keyhole** durch das Schlüsselloch gucken *(ugs.);* **take a quick ~ at sb.** kurz zu jmdm. hingucken

peel [pi:l] **1.** *v. t.* schälen; **~ the shell off an egg/the skin off a banana** ein Ei/eine Banane schälen; *see also* **eye 1 a. 2.** *v. i.* ‹*Person, Haut:*› sich schälen; ‹*Rinde, Borke:*› sich lösen; ‹*Farbe:*› abblättern. **3.** *n.* Schale, *die*

~ a'way 1. *v. t.* abschälen. **2.** *v. i.* **a)** ‹*Haut:*› sich schälen *od. (bes. nordd.)* pellen; ‹*Rinde, Borke:*› sich lösen; ‹*Farbe:*› abblättern; **b)** *(veer away)* ausscheren

~ 'back *v. t.* halb abziehen ‹*Kabelmantel, Bananenschale*›

~ 'off 1. *v. t.* abschälen; abstreifen, ausziehen ‹*Kleider*›. **2.** *v. i.* **a)** *see* **~ away 2 a; b)** *(veer away)* ausscheren; **c)** *(sl.: undress)* sich ausziehen

peeler ['pi:lə(r)] *n.* Schäler, *der;* Schälmesser, *das*

peeling ['pi:lɪŋ] *n.* Stück Schale; **~s** Schalen

¹peep [pi:p] **1.** *v. i.* ‹*Maus, Vogel:*› piep[s]en; *(squeal)* quieken. **2.** *n. (shrill sound)* Piepsen, *das; (coll.: slight utterance)* Piep[s], *der (ugs.);* **one ~ out of you and ...:** ein Pieps [von dir], und ...

²peep 1. *v. i.* **a)** *(look through narrow aperture)* gucken *(ugs.);* **b)** *(look furtively)* verstohlen gucken; **~ round** sich umgucken; **no ~ing!** nicht gucken!; **c)** *(come into view)* **~ out** [he]rausgucken; *(fig.: show itself)* zum Vorschein kommen. **2.** *n.* kurzer Blick; **take a ~ through the curtain** durch die Gardine spähen

'peep-hole *n.* Guckloch, *das*

peeping Tom [pi:pɪŋ 'tɒm] *n.* Spanner, *der (ugs.);* Voyeur, *der*

¹peer [pɪə(r)] *n.* **a)** *(Brit.: member of nobility)* **~ [of the realm]** Peer, *der;* **b)** *(equal in standing)* Gleichgestellte, *der/die;* **among her social ~s** unter ihresgleichen

²peer *v. i. (look searchingly)* forschend schauen; *(look with difficulty)* angestrengt schauen; **~ at sth./sb.** *(searchingly)* [sich *(Dat.)*] etw. genau ansehen/jmdn. forschend *od.* prüfend ansehen; *(with difficulty)* [sich *(Dat.)*] etw. angestrengt ansehen/jmdn. angestrengt ansehen; **~ into the distance** in die Ferne spähen

peerage ['pɪərɪdʒ] *n.* **a)** *no pl. (body of peers)* **the ~:** die Peers; **be raised to the ~:** in den Adelsstand erhoben werden; **b)** *(rank of peer)* Peerswürde, *die*

'peer group *n.* Peer-group, *die (Psych., Soziol.)*

peerless ['pɪəlɪs] *adj.* beispiellos

peeved [pi:vd] *adj. (coll.)* sauer *(ugs.);* **be/get ~ with sb.** auf jmdn. sauer sein/werden; **be ~ at/get ~ about sth.** über etw. *(Akk.)* sauer sein/wegen etw. sauer werden

peevish ['pi:vɪʃ] *adj. (querulous)* nörgelig *(abwertend);* quengelig *(ugs.)* ‹*Kind*›; *(showing vexation)* gereizt

peewit ['pi:wɪt] *n. (Ornith.)* Kiebitz, *der*

peg [peg] **1.** *n. (for holding together parts of framework)* Stift, *der; (for tying things to)* Pflock, *der; (for hanging things on)* Haken, *der; (clothes ~)* Wäscheklammer, *die; (for holding tent-ropes)* Hering, *der; (Mus.: for adjusting strings)* Wirbel, *der;* **off the ~** *(Brit.: ready-made)* von der Stange *(ugs.);* **take sb. down a ~ [or two]** *(fig.)* jmdm. einen Dämpfer aufsetzen *od.* geben; **a ~ to hang sth. on** *(fig.)* ein Aufhänger für etw. **2.** *v. t.,* **-gg-: a)** *(fix with ~)* mit Stiften/Pflöcken befestigen; **b)** *(Econ.: stabilize)* stabilisieren; *(support)* stützen; *(freeze)* einfrieren; **~ wages/prices/exchange rates** Löhne/Preise/Wechselkurse stabil halten

~ a'way *v. i.* schuften *(ugs.);* **[keep] ~[ging] away with sth.** nicht lockerlassen mit etw. *(ugs.)*

~ 'out 1. *v. t.* **a)** *(spread out and secure)* ausspannen ‹*Felle etc.*›; *(Brit.)* [draußen] aufhängen ‹*Wäsche*›; **b)** *(mark)* abstecken ‹*Gebiet, Fläche*›. **2.** *v. i. (sl.) (faint)* zusammenklappen *(ugs.); (die)* den Löffel abgeben *(salopp)*

pejorative [pɪ'dʒɒrətɪv] **1.** *adj.* pejorativ *(Sprachw.);* abwertend. **2.** *n.* Pejorativum, *das (Sprachw.)*

Pekingese (Pekinese) [pi:kɪ'ni:z] *n., pl. same* ~ **[dog]** Pekinese, *der*

pelican ['pelɪkən] *n. (Ornith.)* Pelikan, *der*

'**pelican crossing** *n. (Brit.)* Ampelübergang, *der*

pellet ['pelɪt] *n.* **a)** *(small ball)* Kügelchen, *das; (mass of food)* Pellet, *das (fachspr.);* **b)** *(small shot)* Schrot, *der od. das*

pell-mell [pel'mel] *adv.* **a)** *(in disorder)* durcheinander; **b)** *(headlong)* Hals über Kopf

pelmet ['pelmɪt] *n. (of wood)* Blende, *die; (of fabric)* Schabracke, *die*

[1]**pelt** [pelt] *n. (of sheep or goat)* Fell, *das; (of fur-bearing animal)* [Roh]fell, *das*

[2]**pelt 1.** *v. t. (lit. or fig.)* ~ **sb. with sth.** jmdn. mit etw. bewerfen; ~ **sb. with questions** jmdn. mit Fragen überschütten. **2.** *v. i.* **a)** ‹*Regen:*› prasseln; **it was ~ing down [with rain]** es goß wie aus Kübeln *(ugs.);* **b)** *(run fast)* rasen *(ugs.);* pesen *(ugs.).* **3.** *n.* **[at] full ~:** mit Karacho *(ugs.)*

pelvic ['pelvɪk] *adj. (Anat.)* Becken-

pelvis ['pelvɪs] *n., pl.* **pelves** ['pelvi:z] *or* **~es** *(Anat.)* Becken, *das*

[1]**pen** [pen] **1.** *n.* Pferch, *der.* **2.** *v. t.,* **-nn-: a)** *(shut up in ~)* einpferchen; **b)** ~ **sb. in a corner** jmdn. in eine Ecke drängen

~ '**in** *v. t.* **a)** einpferchen; **b)** *(fig.: restrict)* einengen

[2]**pen 1.** *n.* **a)** *(for writing)* Federhalter, *der; (fountain ~)* Füller, *der; (ball ~)* Kugelschreiber, *der;* Kuli, *der (ugs.); (felt-tip ~)* Filzstift, *der; (ball ~ or felt-tip ~)* Stift, *der;* **the ~ is mightier than the sword** *(prov.)* die Feder ist mächtiger als das Schwert; *see also* **paper 1a; b)** *(quill-feather)* Feder, *die.* **2.** *v. t.,* **-nn-** niederschreiben; ~ **a letter to/a note for sb.** jmdm. einen Brief/ein paar Worte schreiben

penal ['pi:nl] *adj.* **a)** *(of punishment)* Straf-; ~ **reform** Strafvollzugsreform, *die;* **b)** *(punishable)* strafbar ‹*Handlung, Tat*›; ~ **offence** Straftat, *die;* **c)** ~ **colony** *or* **settlement** Strafkolonie, *die*

penalize (penalise) ['pi:nəlaɪz] *v. t.* bestrafen; *(Sport)* eine Strafe verhängen gegen

penalty ['penltɪ] *n.* **a)** *(punishment)* Strafe, *die;* **the ~ for this offence is imprisonment/a fine** auf dieses Delikt steht Gefängnis/ eine Geldstrafe; **pay the ~/the ~ for sth.** *(lit. or fig.)* dafür/für etw. büßen; **on** *or* **under ~ of £200/of instant dismissal** bei einer Geldstrafe von 200 Pfund/unter Androhung *(Dat.)* der sofortigen Entlassung; **b)** *(disadvantage)* Preis, *der;* **c)** *(Sport) (Golf)* ~ **[stroke]** Strafschlag, *der; (Footb., Rugby) see* **penalty kick**

penalty: ~ **goal** *n. (Rugby)* durch einen Straftritt erzieltes Tor; ~ **kick** *n. (Footb.)* Strafstoß, *der;* Elfmeter, *der; (Rugby)* Straftritt, *der*

penance ['penəns] *n., no pl., no art.* Buße, *die;* **act of ~:** Bußübung, *die;* **undergo/do ~:** büßen/Buße tun

pence *see* **penny**

penchant ['pɑ̃ʃɑ̃] *n.* Schwäche, *die,* Vorliebe, *die* **(for** für)

pencil ['pensɪl] **1.** *n.* **a)** Bleistift, *der;* **red/coloured ~:** Rot-/Bunt- *od.* Farbstift, *der;* **write in ~:** mit Bleistift schreiben; **a ~ drawing, a drawing in ~:** eine Bleistiftzeichnung; **b)** *(cosmetic)* Stift, *der.* **2.** *v. t., (Brit.)* **-ll-: a)** *(mark)* mit Bleistift/Farbstift markieren; **b)** *(sketch)* mit Bleistift zeichnen *od.* skizzieren; **c)** *(write with ~)* mit einem Bleistift/Farbstift schreiben

~ '**in** *v. t.* **a)** *(shade with ~)* mit Bleistift [aus]schraffieren; **b)** *(note or arrange provisionally)* vorläufig notieren

pencil: ~**-case** *n.* Griffelkasten, *der; (made of a soft material)* Federmäppchen, *das;* ~**-sharpener** *n.* Bleistiftspitzer, *der*

pendant ['pendənt] *n.* Anhänger, *der*

pending ['pendɪŋ] **1.** *adj.* **a)** *(undecided)* unentschieden ‹*Angelegenheit, Sache*›; anhängig *(Rechtsspr.),* schwebend ‹*Verfahren*›; **be ~** ‹*Verfahren:*› noch anhängig sein *(Rechtsspr.),* noch schweben; ‹*Sache, Angelegenheit:*› noch unentschieden sein *od.* in der Schwebe sein; ‹*Entscheidung, Probleme:*› noch anstehen; **b)** *(about to come into existence)* bevorstehend ‹*Krieg*›; **patent ~:** Patent angemeldet. **2.** *prep. (until)* ~ **his return** bis zu seiner Rückkehr; ~ **full discussion of the matter** bis die Angelegenheit ausdiskutiert ist

'**pending-tray** *n.* Ablage für noch Unerledigtes

pendulum ['pendjʊləm] *n.* Pendel, *das*

penetrate ['penɪtreɪt] **1.** *v. t.* **a)** *(find access into)* eindringen in (+ *Akk.*); *(pass through)* durchdringen; **b)** *(permeate)* dringen in (+ *Akk.*); *(fig.)* durchdringen; ‹*Spion:*› sich einschleusen in (+ *Akk.*). **2.** *v. i.* **a)** *(make a way)* ~ **into/to sth.** in etw. *(Akk.)* eindringen/zu etw. vordringen; ~ **through sth.** durch etw. hindurch dringen; **the cold ~d through the whole house** die Kälte durchdrang das ganze Haus; **b)** *(be understood or realized)* **my hint did not ~:** mein Wink wurde nicht verstanden

penetrating ['penɪtreɪtɪŋ] *adj.* **a)** *(easily heard)* durchdringend; **b)** scharf ‹*Verstand*›; scharfsinnig ‹*Bemerkung, Kommentar, Studie*›; scharf ‹*Beobachtung*›; verstehend ‹*Blick*›

penetration [penɪ'treɪʃn] *n.* **a)** Eindringen, *das* **(of** in + *Akk.*); *(act of passing through)* Durchdringen, *das;* **b)** *no pl. (fig.: discernment)* Scharfsinn, *der;* **c)** *(act of permeating)* Durchdringen, *das; (infiltration)* Infiltration, *die;* Unterwanderung, *die;* **d)** *(seeing into sth.)* Durchdringen, *das*

'**pen-friend** *n.* Brieffreund, *der*/-freundin, *die*

penguin ['peŋgwɪn] *n.* Pinguin, *der*

penicillin [penɪ'sɪlɪn] *n. (Med.)* Penizillin, *das*

peninsula [pɪ'nɪnsjʊlə] *n.* Halbinsel, *die*

penis ['pi:nɪs] *n., pl.* **~es** *or* **penes** ['pi:ni:z] *(Anat.)* Penis, *der*

penitence ['penɪtəns] *n., no pl.* Reue, *die*

penitent ['penɪtənt] **1.** *adj.* reuevoll *(geh.);* reuig *(geh.)* ‹*Sünder*›; **be ~:** bereuen. **2.** *n.* Büßer, *der*/Büßerin, *die (Rel.)*

penitentiary [penɪ'tenʃərɪ] *n. (Amer.)* Straf[vollzugs]anstalt, *die*

pen: ~**knife** *n.* Taschenmesser, *das;* ~**-name** *n.* Schriftstellername, *der*

pennant ['penənt] *n. (Naut.: tapering flag)* Stander, *der*

penniless ['penɪlɪs] *adj.* **be ~:** keinen Pfennig Geld haben; *(fig.: be poor)* mittellos sein

penny ['penɪ] *n., pl. usu.* **pennies** ['penɪz] *(for separate coins),* **pence** [pens] *(for sum of money)* **a)** *(British coin, monetary unit)* Penny, *der;* **fifty pence** fünfzig Pence; **two/five/ten/twenty/fifty pence [piece]** Zwei-/Fünf-/Zehn-/Zwanzig-/Fünfzigpencestück, *das od.* -münze, *die; see also* **halfpenny; b) keep turning up like a bad ~** *(coll.)* immer wieder auftauchen; **the ~ has dropped** *(fig. coll.)* der

Groschen ist gefallen *(ugs.)*; **in for a ~, in for a pound** *(prov.)* wennschon, dennschon *(ugs.)*; **a pretty ~** *(coll.)* eine hübsche *od.* schöne Stange Geld *(ugs.)*; **take care of the pence** *or* **pennies, and the pounds will look after themselves** *(prov.)* spare im Kleinen, dann hast du im Großen; **a ~ for your thoughts** *(coll.)* woran denkst du [gerade]?; **sth. is two** *or* **ten a ~:** etw. gibt es wie Sand am Meer *(ugs.)*

penny: ~ 'farthing [bicycle] *n. (Brit. coll)* Hochrad, *das;* **~-pinching** ['penɪpɪnʃɪŋ] **1.** *n., no pl., no indef. art.* Pfennigfuchserei, *die (ugs.)*; **2.** *adj.* knaus[e]rig *(ugs. abwertend)*

pen: ~-pusher *n. (coll.)* Büromensch, *der; (male)* Bürohengst, *der (ugs. abwertend)*; **~-pushing** *n., no pl., no indef. art. (coll.)* Schreibkram, *der (ugs. abwertend)*

pension ['penʃn] *n.* Rente, *die; (payment to retired civil servant also)* Pension, *die;* **retire on a ~:** in *od.* auf Rente gehen *(ugs.)*; ‹*Beamter:*› in Pension gehen; **be on a ~:** eine Rente beziehen; **~ fund** Rentenfonds, *der;* **~ scheme** Rentenversicherung, *die;* **~ book** ≈ Rentenausweis, *der*

~ 'off *v.t. (discharge)* berenten *(Amtsspr.)*; auf Rente setzen *(ugs.)*; pensionieren ‹*Lehrer, Beamten*›

pensionable ['penʃənəbl] *adj.* **reach ~ age** das Rentenalter erreichen; *(as civil servant)* das Pensionsalter erreichen; **~ salary/earnings** rentenfähiges Gehalt/rentenfähiger Verdienst

pensioner ['penʃənə(r)] *n.* Rentner, *der*/Rentnerin, *die; (retired civil servant)* Pensionär, *der*/Pensionärin, *die*

pensive ['pensɪv] *adj.* **a)** *(plunged in thought)* nachdenklich; **b)** *(sorrowfully thoughtful)* schwermütig

pent [pent] *adj.* **~ in** *or* **up** eingedämmt ‹*Fluß*›; angestaut ‹*Wut, Ärger*›; *see also* **pent-up**

pentagon ['pentəgən] *n.* **a)** *(Geom.)* Fünfeck, *das;* Pentagon, *das (fachspr.)*; **b) the P~** *(Amer. Polit.)* das Pentagon

pentathlon [pen'tæθlən] *n.* *(Sport)* Fünfkampf, *der*

Pentecost ['pentɪkɒst] *n. (Relig.)* Pfingsten, *das;* Pfingstfest, *das*

pent: ~house *n.* Penthaus, *das;* Penthouse, *das;* **~-up** *attrib. adj.* angestaut ‹*Ärger, Wut*›; verhalten ‹*Freude*›; unterdrückt ‹*Sehnsucht, Gefühle*›

penultimate [pe'nʌltɪmət] *adj.* vorletzt...

penury ['penjʊərɪ] *n., no pl.* Armut, *die;* Not, *die*

peony ['pi:ənɪ] *n. (Bot.)* Pfingstrose, *die;* Päonie, *die*

people ['pi:pl] **1.** *n.* **a)** *(persons composing nation, community, etc.)* Volk, *das;* **b)** *constr. as pl. (persons forming class etc.)* Leute *Pl.;* Menschen; **city/country ~** *(inhabitants)* Stadt-/Landbewohner; *(who prefer the city/the country)* Stadt-/Landmenschen; **village ~:** Dorfbewohner; **local ~:** Einheimische; **working ~:** arbeitende Menschen; **coloured/white ~:** Farbige/Weiße; **c)** *constr. as pl. (persons not of nobility)* **the ~:** das [gemeine] Volk; **d)** *constr. as pl. (persons in general)* Menschen; Leute *Pl.; (as opposed to animals)* Menschen; **~ say he's very rich** die Leute sagen *od.* man sagt *od.* es heißt, daß er sehr reich sei; **a crowd of ~:** eine Menschenmenge; **'some ~** *(certain persons, usu. with whom the speaker disagrees)* gewisse Leute; *(you)* manche Leute; **some '~!** Leute gibt es!; **honestly, some '~!** also wirklich!; **what do you ~ think?** was denkt ihr [denn]?; **you of 'all ~ ought ...:** gerade du solltest ... **2.** *v.t.* bevölkern

peopled ['pi:pld] *adj.* bevölkert

People's Re'public *n. (Polit.)* Volksrepublik, *die;* **the ~ of China** die Volksrepublik China

pep [pep] *(coll.)* **1.** *n., no pl., no indef. art.* Schwung, *der;* Pep, *der (salopp)*. **2.** *v.t.,* **-pp-: ~ [up]** aufpeppen *(ugs.)*

pepper ['pepə(r)] **1.** *n.* **a)** Pfeffer, *der;* **b)** *(vegetable)* Paprikaschote, *die;* **red/yellow/green ~:** roter/gelber/grüner Paprika. **2.** *v.t.* **a)** *(sprinkle with ~)* pfeffern; **b)** *(pelt with missiles)* bombardieren *(ugs., auch fig.)*

pepper: ~corn *n.* Pfefferkorn, *das;* **~-mill** *n.* Pfeffermühle, *die;* **~mint** *n.* **a)** *(plant)* Pfefferminze, *die;* **b)** *(sweet)* Pfefferminz, *das;* **~-pot** *n.* Pfefferstreuer, *der*

peppery ['pepərɪ] *adj.* pfeff[e]rig; *(spicy)* scharf; *(fig.: pungent)* scharf

'pep talk *n. (coll.)* Aufmunterung, *die;* **give sb. a ~:** jmdm. ein paar aufmunternde Worte sagen

per [pə(r), *stressed* pɜ:(r)] *prep.* **a)** *(by means of)* per ‹*Post, Bahn, Schiff, Bote*›; durch ‹*Spediteur, Herrn X.*›; **b)** *(according to)* **[as] ~ sth.** wie in etw. *(Dat.)* angegeben; laut ‹*Anweisung, Preisliste*›; **c)** *(for each)* pro; **£50 ~ week** 50 Pfund in der Woche *od.* pro Woche; **fifty kilometres ~ hour** fünfzig Kilometer in der *od.* pro Stunde; **get 11 francs ~ pound** 11 Francs für ein Pfund bekommen

perambulator [pə'ræmbjʊleɪtə(r)] *(Brit. formal) see* **pram**

per annum [pər 'ænəm] *adv.* im Jahr; pro Jahr *(bes. Kaufmannsspr., ugs.)*

perceive [pə'si:v] *v.t.* **a)** *(with the mind)* spüren; bemerken; **b)** *(through the senses)* wahrnehmen; **we ~d a figure in the distance** wir erblickten in der Ferne eine Gestalt; **c)** *(regard mentally in a certain way)* wahrnehmen; **~d** vermeintlich ‹*Bedrohung, Gefahr, Wert*›

per cent (*Brit.; Amer.:* **percent**) [pə 'sent] **1.** *adv.* **ninety ~ effective** zu 90 Prozent wirksam; *see also* **hundred 1 c. 2.** *adj.* **a 5 ~ increase** ein Zuwachs von 5 Prozent; ein fünfprozentiger Zuwachs. **3.** *n.* **a)** *see* **percentage; b)** *(hundredth)* Prozent, *das*

percentage [pə'sentɪdʒ] *n.* **a)** *(rate or proportion per cent)* Prozentsatz, *der;* **a high ~ of alcohol** ein hoher Alkoholgehalt; **what ~ of 48 is 11?** wieviel Prozent von 48 sind 11?; **b)** *(proportion)* [prozentualer] Anteil

per'centage sign *n.* Prozentzeichen, *das*

perceptible [pə'septɪbl] *adj.* wahrnehmbar; **be quite ~:** ganz offensichtlich sein

perceptibly [pə'septɪblɪ] *adv.* sichtlich; sichtbar, merklich ‹*schrumpfen, welken*›

perception [pə'sepʃn] *n.* **a)** *(act)* Wahrnehmung, *die; (result)* Erkenntnis, *die;* **have keen ~s** ein stark ausgeprägtes Wahrnehmungsvermögen haben; **b)** *no pl. (faculty)* Wahrnehmungsvermögen, *das;* **c)** *(intuitive recognition)* Gespür, *das* **(of** für); *(instance)* Erfassen, *das*

perceptive [pə'septɪv] *adj.* **a)** *(discerning)* scharf ‹*Auge*›; fein ‹*Gehör, Nase, Geruchssinn*›; scharfsinnig ‹*Person*›; **b)** *(having intuitive recognition or insight)* einfühlsam ‹*Person, Zeitungsartikel, Bemerkung*›

perceptively [pə'septɪvlɪ] *adv.* **a)** *(discerningly)* mit scharfer Wahrnehmung; **b)** *(with intuitive recognition or insight)* einfühlsam

¹perch [pɜ:tʃ] *n., pl. same or* **~es** *(Zool.)* Flußbarsch, *der*

²perch 1. *n.* **a)** *(horizontal bar)* Sitzstange, *die; (for hens)* Hühnerstange, *die;* **b)** *(place to sit)* Sitzplatz, *der.* **2.** *v.i.* **a)** *(alight)* sich niederlassen; **b)** *(be supported)* sitzen. **3.** *v.t.* setzen/stellen/legen; **be ~ed** ‹*Vogel:*› sitzen; **stand ~ed on a cliff** hoch auf einer Klippe stehen; **a village ~ed on a hill** ein hoch oben auf einem Berg gelegenes Dorf

percolate ['pɜ:kəleɪt] **1.** *v.i.* **a)** *(ooze)* **~ through sth.** durch etw. [durch]sickern; **b)** *(fig.: spread gradually)* vordringen; **c)** ‹*Kaffee:*› durchlaufen. **2.** *v.t.* **a)** *(permeate)* sickern durch ‹*Gestein*›; **b)** *(fig.: penetrate)* dringen in (+ *Akk.*) ‹*Bewußtsein*›; **c)** [mit der Kaffeemaschine] machen ‹*Kaffee*›

percolator ['pɜ:kəleɪtə(r)] *n.* Kaffeemaschine, *die*

percussion [pə'kʌʃn] *n.* *(Mus.)* *(group of instruments)* Schlagzeug, *das;* **~ instrument** Schlaginstrument, *das*

peregrine ['perɪgrɪn] *n.* **~ [falcon]** *(Ornith.)* Wanderfalke, *der*

peremptory [pə'remptərɪ, 'perɪmptərɪ] *adj.* *(admitting no contradiction)* kategorisch; *(imperious)* herrisch; gebieterisch *(geh.)*

perennial [pə'renjəl] **1.** *adj.* **a)** *(lasting all year)* ganzjährig; **b)** *(lasting indefinitely)* immerwährend; ewig ‹*Jugend, Mythos, Suche*›; ungelöst ‹*Problem*›; **c)** *(Bot.)* ausdauernd. **2.** *n.* *(Bot.)* ausdauernde Pflanze

perestroika [perɪ'strɔɪkə] *n.* Perestroika, *die*

perfect 1. ['pɜ:fɪkt] *adj.* **a)** *(complete)* vollkommen; umfassend ‹*Kenntnisse, Wissen*›; **b)** *(faultless)* vollkommen; perfekt ‹*Englisch, Technik, Timing*›; tadellos ‹*Zustand*›; [absolut] gelungen ‹*Aufführung*›; lupenrein ‹*Diamant*›; *see also* [1]**practice a; c)** *(coll.: very satisfactory)* herrlich; wunderbar; **d)** *(exact)* perfekt; getreu ‹*Ebenbild, Abbild*›; *(fully what the name implies)* perfekt ‹*Gentleman, Gastgeberin*›; **e)** *(absolute)* **a ~ stranger** ein völlig Fremder; **he is a ~ stranger to me** er ist mir völlig unbekannt; **he is a ~ angel** *(coll.)***/charmer** er ist wirklich ein Engel/charmant; **I have a ~ right to stay** ich habe eindeutig *od.* durchaus das Recht zu bleiben; **f)** *(coll.: unmitigated)* absolut; **look a ~ fright/mess** wirklich zum Weglaufen/absolut verboten aussehen *(ugs.)*. **2.** [pə'fekt] *v.t.* vervollkommnen; perfektionieren

perfection [pə'fekʃn] *n., no pl.* **a)** *(making perfect)* Vervollkommnung, *die;* Perfektionierung, *die;* **b)** *(faultlessness)* Vollkommenheit, *die;* Perfektion, *die;* **to ~:** perfekt; **it/he succeeded to ~:** es war ein voller Erfolg/er war absolut erfolgreich; **c)** *(perfect person or thing)* **be ~:** perfekt sein

perfectionism [pə'fekʃənɪzm] *n., no pl.* Perfektionismus, *der*

perfectionist [pə'fekʃənɪst] *n.* Perfektionist, *der*/Perfektionistin, *die*

perfectly ['pɜ:fɪktlɪ] *adv.* **a)** *(completely)* vollkommen; völlig; **I understand that ~:** ich verstehe das vollkommen; **be ~ entitled to do sth.** durchaus berechtigt sein, etw. zu tun; **b)** *(faultlessly)* perfekt; tadellos ‹*sich verhalten*›; **c)** *(exactly)* vollkommen; exakt, genau ‹*vorhersagbar*›; **d)** *(coll.: to an unmitigated extent)* furchtbar *(ugs.)* ‹*schrecklich, schlimm, ekelhaft*›

perfidious [pə:fɪdɪəs] *adj.* perfid *(geh.)*

perforate ['pɜ:fəreɪt] *v.t.* **a)** *(make hole[s] through)* perforieren; **b)** *(make an opening into)* durchlöchern

perforation [pɜ:fə'reɪʃn] *n.* **a)** *(action of perforating)* Perforierung, *die;* **b)** *(hole)* Loch, *das;* **~s** *(line of holes esp. in paper)* Perforation, *die;* *(in sheets of stamps)* Zähnung, *die;* Perforation, *die*

perform [pə'fɔ:m] **1.** *v.t.* ausführen ‹*Befehl, Arbeit*›; erfüllen ‹*Bitte, Wunsch, Pflicht, Aufgabe*›; vollbringen ‹*[Helden]tat, Leistung*›; durchführen ‹*Operation, Experiment*›; ausfüllen ‹*Funktion*›; vollbringen ‹*Wunder*›; vorführen, zeigen ‹*Trick*›; vollziehen ‹*Trauung, Taufe, Riten*›; aufführen ‹*Theaterstück*›; vortragen, vorsingen ‹*Lied*›; vorspielen, vortragen ‹*Sonate usw.*›. **2.** *v.i.* **a)** eine Vorführung geben; *(sing)* singen; *(play)* spielen; ‹*Zauberer:*› Zaubertricks ausführen *od.* vorführen; **he ~ed very well** seine Darbietung war sehr gut; **she ~ed skilfully on the flute/piano** sie spielte mit großer Könnerschaft Flöte/Klavier; **b)** *(Theatre)* auftreten; **he ~ed very well** sein Auftritt war sehr gut; **c)** *(execute tricks)* ‹*Tier:*› Kunststücke zeigen *od.* vorführen; **train an animal to ~:** einem Tier Kunststücke beibringen; **d)** *(work, function)* ‹*Auto:*› laufen, fahren; **he ~ed all right/well [in the exam]** er machte seine Sache [in der Prüfung] ordentlich/gut

performance [pə'fɔ:məns] *n.* **a)** *(fulfilment) (of duty, task)* Erfüllung, *die;* *(of command)* Ausführung, *die;* **b)** *(carrying out)* Durchführung, *die;* **c)** *(notable feat)* Leistung, *die;* **put up a good ~:** eine gute Leistung zeigen; **d)** *(performing of play etc.)* Vorstellung, *die;* **her ~ as Desdemona** ihre Darstellung der Desdemona; **the ~ of a play/opera** die Aufführung eines Theaterstücks/einer Oper; **give a ~ of a symphony/play** eine Sinfonie/ein Stück spielen *od.* aufführen; **e)** *(achievement under test)* Leistung, *die;* **athletic ~:** die Leistung eines Sportlers; **the car has good ~:** der Wagen bringt viel Leistung; **f)** *(coll.: difficult procedure)* Theater, *das (ugs., abwertend);* Umstand, *der*

performer [pə'fɔ:mə(r)] *n.* Künstler, *der*/Künstlerin, *die*

perfume 1. ['pɜ:fju:m] *n.* **a)** *(sweet smell)* Duft, *der;* **b)** *(fluid)* Parfüm, *das.* **2.** [pə'fju:m, 'pɜ:fju:m] *v.t.* *(give sweet scent to)* mit Wohlgeruch erfüllen; *(impregnate with sweet smell)* parfümieren

perfunctory [pə'fʌŋktərɪ] *adj.* *(done for duty's sake only)* pflichtschuldig; flüchtig ‹*Erkundigung, Bemerkung*›; *(superficial)* oberflächlich ‹*Arbeit, Überprüfung*›

perhaps [pə'hæps, præps] *adv.* vielleicht; **I'll go out, ~:** ich gehe vielleicht aus; **~ so** [das] mag [ja] sein; **~ not** *(maybe this is or will not be the case)* vielleicht auch nicht; *(it might be best not to do this)* vielleicht lieber nicht

peril ['perl] *n.* Gefahr, *die;* **be in deadly ~:** in Lebensgefahr sein; **do sth. at one's ~** *(accepting risk of injury)* etw. auf eigene Gefahr tun

perilous ['perələs] *adj.* gefahrvoll; **be ~:** gefährlich sein

perilously ['perələslɪ] *adv.* gefährlich; **~ ill** todkrank

perimeter [pə'rɪmɪtə(r)] *n.* **a)** *(outer boundary)* [äußere] Begrenzung; Grenze, *die;* **at the ~ of the race-track** am Rande der Rennbahn; **b)** *(length of outline)* Umfang, *der*

period ['pɪərɪəd] **1.** *n.* **a)** *(distinct portion of history or life)* Periode, *die;* Zeit, *die;* **~s of history** geschichtliche Perioden; **at a later ~ of her life** zu einem späteren Zeitpunkt ihres Lebens; **the Classical/Romantic/Renaissance ~:** die Klassik/Romantik/Renaissance; **of the ~** *(of the time* *under discussion)* der damaligen Zeit; **b)** *(any portion of time)* Zeitraum, *der;* Zeitspanne, *die;* **over a ~ [of time]** über einen längeren Zeitraum; **within the agreed ~:** innerhalb der vereinbarten Frist; **showers and bright ~s** *(Meteorol.)* Schauer und Aufheiterungen; **c)** *(Sch.)* Stunde, *die;* **have two chemistry ~s** zwei Stunden Chemie haben; **a free ~:** eine Freistunde; **d)** *(occurrence of menstruation)* Periode, *die;* Regel[blutung], *die;* **have her/a ~:** ihre Periode *od.* Regel *od. (ugs. verhüll.)* Tage haben; **e)** *(punctuation mark)* Punkt, *der;* **f)** *(appended to statement)* **we can't pay higher wages, ~:** wir können keine höheren Löhne zahlen, da ist nichts zu machen; **g)** *(Geol.)* Periode, *die.* **2.** *adj.* zeitgenössisch ‹*Tracht, Kostüm*›; Zeit‹*roman, -stück*›; antik ‹*Möbel*›

periodic [pɪərɪ'ɒdɪk] *adj.* periodisch *od.* regelmäßig [auftretend *od.* wiederkehrend]; *(intermittent)* gelegentlich [auftretend]; vereinzelt ‹*Regenschauer*›

periodical [pɪərɪ'ɒdɪkl] **1.** *adj. see* **periodic. 2.** *n.* Zeitschrift, *die;* **weekly/monthly/quarterly ~:** Wochenzeitschrift / Monatsschrift / Vierteljahresschrift, *die*

periodically [pɪərɪ'ɒdɪkəlɪ] *adv. (at regular intervals)* regelmäßig; *(intermittently)* gelegentlich

periodic 'table *n. (Chem.)* Periodensystem, *das*

peripatetic [perɪpə'tetɪk] *adj.* **~ teacher** Lehrer, der/Lehrerin, die an mehreren Schulen unterrichtet

peripheral [pə'rɪfərl] **1.** *adj.* **a)** *(of the periphery)* ‹*Parkraum*› in Randlage; **~ road** Ringstraße, *die;* **b)** *(of minor importance)* peripher *(geh.);* marginal *(geh.);* Rand‹*problem, -erscheinung, -figur, -bemerkung*›; **c)** *(Computing)* peripher. **2.** *n. (Computing)* Peripheriegerät, *das*

periphery [pə'rɪfərɪ] *n.* **a)** *see* **circumference; b)** *(external boundary)* Begrenzung, *die; (of surface)* Außenfläche, *die;* **c)** *(outer region)* Peripherie, *die (geh.);* Rand, *der*

periscope ['perɪskəʊp] *n.* Periskop, *das*

perish ['perɪʃ] **1.** *v.i.* **a)** ‹*Person:*› umkommen; ‹*Volk, Rasse, Kultur:*› untergehen; ‹*Kraft, Energie:*› versiegen; ‹*Pflanze:*› eingehen; **~ the thought!** Gott behüte *od.* bewahre!; **b)** *(rot)* verderben; ‹*Fresken, Gemälde:*› verblassen; ‹*Gummi:*› altern. **2.** *v.t.* **a) we were ~ed [with cold]** wir waren ganz durchgefroren; **b)** *(cause to rot)* [schneller] altern lassen ‹*Gummi*›; angreifen ‹*Reifen*›

perishable ['perɪʃəbl] **1.** *adj.* [leicht] verderblich ‹*Lebensmittel, Waren*›. **2.** *n. in pl.* leicht verderbliche Güter *od.* Waren

perishing ['perɪʃɪŋ] *(coll.)* **1.** *adj.* **a)** mörderisch ‹*Wind, Kälte*›; **it's/I'm ~:** es ist bitter kalt/ich komme um vor Kälte *(ugs.);* **b)** *(Brit.: confounded)* elend; **that child is a ~ nuisance** das Kind kann einem den Nerv töten *(ugs.).* **2.** *adv.* mörderisch ‹*kalt*›

perjure ['pɜ:dʒə(r)] *v. refl. (swear to false statement)* einen Meineid leisten; *(Law: give false evidence under oath)* [unter Eid] falsch aussagen

perjury ['pɜ:dʒərɪ] *n. (swearing to false statement)* Meineid, *der; (Law: giving false evidence)* eidliche Falschaussage; **commit ~:** einen Meineid leisten/sich der eidlichen Falschaussage schuldig machen

[1]**perk** [pɜ:k] *(coll.)* **1.** *v.i.* **~ up** munter werden. **2.** *v.t.* **a) ~ up** *(restore liveliness of)* aufmuntern; **b) ~ up** *(raise briskly)* aufstellen ‹*Schwanz, Ohren*›; heben ‹*Kopf*›

[2]**perk** *n. (Brit. coll.: benefit)* [Sonder]vergünstigung, *die*

perky ['pɜ:kɪ] *adj.* **a)** *(lively)* lebhaft; munter; **b)** *(self-confident)* keck; selbstbewußt

perm [pɜ:m] **1.** *n. (permanent wave)* Dauerwelle, *die.* **2.** *v.t.* **have one's hair ~ed** sich *(Dat.)* eine Dauerwelle machen lassen; **have ~ed hair** eine Dauerwelle haben

permanence ['pɜ:mənəns] *n., no pl.* Dauerhaftigkeit, *die*

permanency ['pɜ:mənənsɪ] *n.* **a)** *no pl. see* **permanence; b)** *(condition)* Dauerzustand, *der; (job)* Dauerstellung, *die*

permanent ['pɜ:mənənt] *adj.* fest ‹*Sitz, Bestandteil, Mitglied*›; beständig, ewig ‹*Werte*›; ständig ‹*Plage, Meckern, Adresse, Kampf*›; Dauer‹*gast, -stellung, -visum*›; bleibend ‹*Folge, Schaden*›; **of ~ value** von bleibendem Wert; **sb./sth. is a ~ fixture** jmd./etw. gehört zum Inventar; **be employed on a ~ basis** fest angestellt sein

permanently ['pɜ:mənəntlɪ] *adv.* dauernd; auf Dauer ‹*verhindern, bleiben*›; fest ‹*anstellen, einstellen*›; *(repeatedly)* ständig; dauernd; **they live in France ~ now** sie leben jetzt ganz *(ugs.) od.* ständig in Frankreich; **she was ~ disabled in the accident** sie hat bei dem Unfall eine bleibende Behinderung davongetragen

permeable ['pɜ:mɪəbl] *adj.* durchlässig

permeate ['pɜ:mɪeɪt] **1.** *v.t. (get into)* dringen in (+ *Akk.*); *(pass through)* dringen durch; **be ~d with** *or* **by sth.** *(fig.)* von etw. durchdrungen sein. **2.** *v.i.* **~ through sth.** etw. durchdringen; **~ through to sb.** zu jmdm. durchdringen

permissible [pə'mɪsɪbl] *adj.* zulässig; **be ~ to** *or* **for sb.** jmdm. erlaubt sein

permission [pə'mɪʃn] *n., no indef. art.* Erlaubnis, *die; (given by official body)* Genehmigung, *die;* **ask [sb.'s] ~:** [jmdn.] um Erlaubnis bitten; **who gave you ~ to do this?** wer hat dir erlaubt, das zu tun?; **with your ~:** wenn Sie gestatten; mit Ihrer Erlaubnis; **written ~:** eine schriftliche Genehmigung

permissive [pə'mɪsɪv] *adj. (tolerant)* tolerant; großzügig; *(in relation to moral matters)* freizügig; permissiv *(geh.);* **the ~ society** die permissive Gesellschaft

permit 1. [pə'mɪt] *v.t.,* **-tt-** zulassen ‹*Berufung, Einspruch usw.*›; **~ sb. sth.** jmdm. etw. erlauben *od. (geh.)* gestatten. **2.** *v.i.,* **-tt-: a)** *(give opportunity)* es zulassen; **weather ~ting** bei entsprechendem Wetter; **b)** *(admit)* **~ of sth.** etw. erlauben *od.* gestatten; **not ~ of sth.** etw. verbieten. **3.** ['pɜ:mɪt] *n.* Genehmigung, *die; (for entering premises)* Passierschein, *der*

permutation [pɜ:mjʊ'teɪʃn] *n.* **a)** *(varying of order)* Umstellung, *die;* **b)** *(result of variation of order)* Anordnung, *die; (of series of items)* Reihenfolge, *die;* Permutation, *die (Math.)*

pernicious [pə'nɪʃəs] *adj.* verderblich; bösartig ‹*Krankheit, Person*›; schlimm, übel ‹*Angewohnheit*›

pernickety [pə'nɪkɪtɪ] *adj. (coll.)* pingelig *(ugs.)* (**about** in bezug auf + *Akk.*)

peroxide [pə'rɒksaɪd] *n.* **a)** *(Chem.)* Peroxyd, *das;* **b) [hydrogen] ~:** Wasserstoffperoxyd, *das;* **~ blonde** Wasserstoffblondine, *die*

perpendicular [pɜ:pən'dɪkjʊlə(r)] **1.** *adj.* **a)** senkrecht; lotrecht; **b)** *(very steep)* [fast] senkrecht ‹*Aufstieg, Abstieg*›; senkrecht abfallend/aufragend ‹*Kliff,*

Felswand usw.⟩; ~ **drop/slope/rock-face** Steilabfall, *der*/-hang, *der*/-wand, *die; (Geom.)* senkrecht (**to** zu); **two ~ planes/lines** zwei zueinander senkrechte Ebenen/Linien. **2.** *n.* Senkrechte, *die* (**to** zu); Lot, *das* (**to** auf + *Dat.*); **be [slightly] out of [the] ~:** [etwas] aus dem Lot sein

perpetrate ['pɜ:pɪtreɪt] *v.t.* begehen; anrichten ⟨*Schaden*⟩; verüben ⟨*Gemetzel, Greuel*⟩

perpetrator ['pɜ:pɪtreɪtə(r)] *n.* [Übel]täter, *der*/-täterin, *die;* **be the ~ of a crime/fraud/atrocity** ein Verbrechen/einen Betrug begangen haben/eine Greueltat verübt haben

perpetual [pə'petjʊəl] *adj.* **a)** *(eternal)* ewig; **b)** *(continuous)* ständig; **c)** *(coll.: repeated)* ständig; [an]dauernd; **d)** *(applicable or valid for ever)* immerwährend; ewig

perpetually [pə'petjʊəlɪ] *adv.* **a)** *(eternally)* ewig; **b)** *(continuously)* ständig; **c)** *(coll.: repeatedly)* ständig; [an]dauernd

perpetual 'motion *n., no pl., no art.* ewige Bewegung

perpetuate [pə'petjʊeɪt] *v.t.* **a)** *(preserve from oblivion)* lebendig erhalten ⟨*Andenken*⟩; unsterblich machen ⟨*Namen*⟩; aufrechterhalten ⟨*Tradition*⟩; **b)** *(make perpetual)* aufrechterhalten; erhalten ⟨*Art, Macht*⟩

perpetuity [pɜ:pɪ'tju:ɪtɪ] *n., no pl., no indef. art.* ewiger Bestand; **in** *or* **to** *or* **for ~:** für alle Ewigkeit *od.* alle Zeiten

perplex [pə'pleks] *v.t.* verwirren

perplexed [pə'plekst] *adj.* *(bewildered)* verwirrt; *(puzzled)* ratlos

perplexity [pə'pleksɪtɪ] *n. no pl.* *(bewilderment)* Verwirrung, *die; (puzzlement)* Ratlosigkeit, *die*

persecute ['pɜ:sɪkju:t] *v.t.* **a)** verfolgen; **b)** *(harass, worry)* plagen; zusetzen (+ *Dat.*)

persecution [pɜ:sɪ'kju:ʃn] *n.* **a)** Verfolgung, *die;* **b)** *(harassment)* Plagerei, *die*

perseverance [pɜ:sɪ'vɪərəns] *n.* Beharrlichkeit, *die;* Ausdauer, *die*

persevere [pɜ:sɪ'vɪə(r)] *v.i.* ausharren; **~ with** *or* **at** *or* **in sth.** bei etw. dabeibleiben; **~ in doing sth.** darauf beharren, etw. zu tun

Persia ['pɜ:ʃə] *pr. n. (Hist.)* Persien *(das)*

Persian ['pɜ:ʃn] **1.** *adj.* persisch. **2.** *n.* **a)** *(person)* Perser, *der*/Perserin, *die;* **b)** *(language)* Persisch, *das; see also* **English 2 a; c)** *see* **Persian cat**

Persian: ~ 'carpet *n.* Perser[teppich], *der;* **~ 'cat** *n.* Perserkatze, *die*

persist [pə'sɪst] *v.i.* **a)** *(continue firmly)* beharrlich sein Ziel verfolgen; nicht nachgeben; **~ in sth.** an etw. *(Dat.)* [beharrlich] festhalten; **~ in doing sth.** etw. weiterhin [beharrlich] tun; **b)** *(continue in existence)* anhalten

persistence [pə'sɪstəns] *n., no pl.* **a)** *(continuance in particular course)* Hartnäckigkeit, *die;* Beharrlichkeit, *die;* **b)** *(quality of perseverance)* Ausdauer, *die;* Zähigkeit, *die;* **c)** *(continued existence)* Fortbestehen, *das*

persistent [pə'sɪstənt] *adj.* **a)** *(continuing firmly or obstinately)* hartnäckig; **b)** *(constantly repeated)* dauernd; hartnäckig ⟨*Gerüchte*⟩; nicht nachlassend ⟨*Anstrengung, Bemühung*⟩; **~ showers** anhaltende Schauertätigkeit; **c)** *(enduring)* anhaltend

person ['pɜ:sn] *n.* **a)** Mensch, *der;* Person, *die (oft abwertend);* **a rich/sick/unemployed ~:** ein Reicher/Kranker/Arbeitsloser/eine Reiche *usw.;* **the first ~ to leave was ...:** der/die erste, der/die wegging, war ...; **what sort of ~ do you think I am?** wofür halten Sie mich eigentlich?; **in the ~ of sb.** in jmdm. *od.* jmds. Person; **in ~** *(personally)* persönlich; selbst; **b)** *(living body)* Körper, *der;* *(appearance)* [äußere] Erscheinung; Äußere, *das;* **c)** *(Ling.)* Person, *die;* **first/second/third ~:** erste/zweite/dritte Person

personable ['pɜ:sənəbl] *adj.* sympathisch

personage ['pɜ:sənɪdʒ] *n.* **a)** *(person of rank)* Persönlichkeit, *die;* **b)** *(person not known to speaker)* Person, *die*

personal ['pɜ:sənl] *adj.* persönlich; Privat⟨*angelegenheit, -leben*⟩; ⟨*Sache:*⟩ jmdm. persönlich gehören; **~ appearance** äußere Erscheinung; **~ hygiene** Körperpflege, *die;* **~ call** *(Brit. Teleph.)* Anruf mit Voranmeldung; **~ computer** Personalcomputer, *der;* **~ stereo** Walkman, *der;* **pay sb. a ~ call** jmdn. privat aufsuchen; **it's nothing ~, but ...:** nimm es bitte nicht persönlich, aber ...

personal: ~ as'sistant *n.* persönlicher Referent/persönliche Referentin; **~ column** *n.* Rubrik für private [Klein]anzeigen; **~ identifi'cation number** *n.* persönliche Identifikationsnummer

personalise *see* **personalize**

personality [pɜ:sə'nælɪtɪ] *n.* Persönlichkeit, *die;* **have a strong ~,** *(coll.)* **have lots of ~:** eine starke Persönlichkeit sein *od.* haben

personalize ['pɜ:sənəlaɪz] *v.t.* **a)** *(make personal)* persönlich gestalten; eine persönliche Note geben (+ *Dat.*); *(mark with owner's name etc.)* als persönliches Eigentum kennzeichnen; **b)** *(personify)* personifizieren

personally ['pɜ:sənəlɪ] *adv.* persönlich; **~, I see no objection** ich persönlich sehe keine Einwände

personal: ~ 'organizer *n.* Terminplaner, *der;* **~ 'property** *n.* persönliches Eigentum; **~ 'service** *n.* individueller Service; **get ~ service** individuell *od.* persönlich bedient werden

personification [pəsɒnɪfɪ'keɪʃn] *n.* Verkörperung, *die;* **be the [very] ~ of kindness** die Freundlichkeit selbst *od.* in Person sein

personify [pə'sɒnɪfaɪ] *v.t.* verkörpern; **be kindness personified, ~ kindness** die Freundlichkeit in Person sein

personnel [pɜ:sə'nel] *n.* **a)** *constr. as sing. or pl.* Belegschaft, *die; (of shop, restaurant, etc.)* Personal, *das;* **military ~:** Militärangehörige; *attrib.* Personal-; **~ manager** Personalchef, *der*/-chefin, *die;* **~ officer** Personalsachbearbeiter, *der*/-sachbearbeiterin, *die;* **b)** *no pl., no art. (department of firm)* Personalabteilung, *die*

person-to-'person *adj. (Amer. Teleph.)* **~ call** Anruf mit Voranmeldung

perspective [pə'spektɪv] *n.* **a)** Perspektive, *die; (fig.)* Blickwinkel, *der;* **throw sth. into ~:** etw. ins rechte Licht rücken; **put a different ~ on events** ein neues Licht auf die Ereignisse werfen; **b)** *(view)* Aussicht, *die; (fig.: mental view)* Ausblick, *der*

Perspex, (P) ['pɜ:speks] *n.* Plexiglas Ⓦᶻ, *das*

perspicacious [pɜ:spɪ'keɪʃəs] *adj.* scharfsinnig

perspiration [pɜ:spɪ'reɪʃn] *n.* **a)** Schweiß, *der;* **b)** *(action of perspiring)* Schwitzen, *das*

perspire [pə'spaɪə(r)] *v.i.* schwitzen; transpirieren *(geh.)*

persuadable [pə'sweɪdəbl] *adj.* leicht zu überreden; **be easily ~:** sich leicht überreden lassen

persuade [pə'sweɪd] *v.t.* **a)** *(cause to have belief)* überzeugen (**of** von); **~ oneself of sth.** sich *(Dat.)* etw. einreden; **~ oneself [that] ...:** sich *(Dat.)* einreden, daß ...; **b)**

(induce) überreden; ~ **sb. into/out of doing sth.** jmdn. [dazu] überreden, etw. zu tun/nicht zu tun

persuasion [pə'sweɪʒn] *n.* **a)** *(action of persuading)* Überzeugung, die; *(persuasiveness)* Überzeugungskraft, *die;* **it didn't take much ~:** es brauchte nicht viel Überredungskunst; **he didn't need much ~ [to have another drink]** man brauchte ihn nicht lange dazu überreden[, noch etwas zu trinken]; **b)** *(belief)* Überzeugung, *die;* **c)** *(religious belief)* Glaubensrichtung, *die; (sect)* Glaubensgemeinschaft, *die*

persuasive [pə'sweɪsɪv] *adj.*, **persuasively** [pə'sweɪsɪvlɪ] *adv.* überzeugend

persuasiveness [pə'sweɪsɪvnɪs] *n., no pl.* Überzeugungskraft, *die*

pert [pɜ:t] *adj.* **a)** *(saucy, impudent)* unverschämt; frech; **b)** *(neat)* keck ⟨*Hut, Anzug usw.*⟩; hübsch ⟨*Körper, Nase, Hinterteil*⟩

pertain [pə'teɪn] *v. i.* **a)** *(belong as part)* ~ **to** [dazu]gehören zu; **b)** *(be relevant)* ⟨*Kriterien usw.:*⟩ gelten; ~ **to** von Bedeutung sein für; **c)** *(have reference)* ~ **to sth.** etw. betreffen; mit etw. zu tun haben

pertinence ['pɜ:tɪnəns] *n., no pl.* Relevanz, *die*

pertinent ['pɜ:tɪnənt] *adj.* relevant (**to** für)

perturb [pə'tɜ:b] *v. t.* beunruhigen

Peru [pə'ru:] *pr. n.* Peru *(das)*

perusal [pə'ru:zl] *n.* Lektüre, *die; (of documents)* sorgfältiges Studium; *(fig.: action of examining) (of documents)* sorgfältige Durchsicht; **give sth. a careful ~:** etw. genau durchlesen *od.* studieren

peruse [pə'ru:z] *v. t.* genau durchlesen; *(fig.: examine)* untersuchen

Peruvian [pə'ru:vɪən] **1.** *adj.* peruanisch. **2.** *n.* Peruaner, *der/*Peruanerin, *die*

pervade [pə'veɪd] *v. t.* **a)** *(spread throughout)* durchdringen; **be ~d with** *or* **by** durchdrungen sein von; **b)** *(be rife among)* ⟨*Ansicht:*⟩ weit verbreitet sein in (+ *Dat.*)

pervasive [pə'veɪsɪv] *adj. (pervading)* durchdringend ⟨*Geruch, Feuchtigkeit, Kälte*⟩; weit verbreitet ⟨*Ansicht*⟩; sich ausbreitend ⟨*Gefühl*⟩; *(able to pervade)* alles durchdringend

perverse [pə'vɜ:s] *adj.* **a)** *(persistent in error)* uneinsichtig, verstockt ⟨*Person*⟩; borniert ⟨*Person, Argument*⟩; **b)** *(unreasonable)* verrückt

perversely [pə'vɜ:slɪ] *adv.* uneinsichtig; verstockt

perversion [pə'vɜ:ʃn] *n.* **a)** *(turning aside from proper use)* Mißbrauch, *der; (misconstruction)* Pervertierung, *die; (of words, statement)* Verdrehung, *die; (leading astray)* Verführung, *die;* **b)** *(perverted form of sth.)* Pervertierung, *die;* **c)** *(sexual)* Perversion, *die*

perversity [pə'vɜ:sɪtɪ] *n. (persistence in error)* Uneinsichtigkeit, *die;* Verstocktheit, *die*

pervert 1. [pə'vɜ:t] *v. t.* **a)** *(turn aside from proper use or nature)* pervertieren *(geh.);* beugen ⟨*Recht*⟩; untergraben ⟨*Staatsform, Demokratie*⟩; ~ **[the course of] justice** die Justiz behindern; **b)** *(misconstrue)* verfälschen; **c)** *(lead astray)* verderben. **2.** ['pɜ:vɜ:t] *n.* Perverse, *der/die*

perverted [pə'vɜ:tɪd] *adj.* **a)** *(turned aside from proper use)* pervertiert *(geh.);* **b)** *(misconstrued)* verdreht; **c)** *(led astray)* schlecht; verdorben; **d)** *(sexually)* pervers

pesky ['peskɪ] *adj. (Amer. coll.)* verdammt *(ugs.)*

pessimism ['pesɪmɪzm] *n., no pl.* Pessimismus, *der*

pessimist ['pesɪmɪst] *n.* Pessimist, *der/*Pessimistin, *die*

pessimistic [pesɪ'mɪstɪk] *adj.*, **pessimistically** [pesɪ'mɪstɪkəlɪ] *adv.* pessimistisch

pest [pest] *n. (troublesome thing)* Ärgernis, *das;* Plage, *die; (troublesome person)* Nervensäge, *die (ugs.); (destructive animal)* Schädling, *der;* **~s** *(insects)* Schädlinge; Ungeziefer, *das*

pester ['pestə(r)] *v. t.* belästigen; nerven *(ugs.);* ~ **sb. for sth.** jmdm. wegen etw. in den Ohren liegen; ~ **sb. to do sth.** jmdm. in den Ohren liegen, etw. zu tun; ~ **sb. for money** jmdn. [um Geld] anbetteln

pesticide ['pestɪsaɪd] *n.* Pestizid, *das*

pestilential [pestɪ'lenʃl] *adj.* **a)** *(fig. coll.: troublesome)* unausstehlich; **b)** *(pernicious)* verderblich

pestle ['pesl] *n.* Stößel, *der;* Pistill, *das (fachspr.)*

pet [pet] **1.** *n.* **a)** *(tame animal)* Haustier, *das;* **b)** *(darling, favourite)* Liebling, *der; (sweet person; also as term of endearment)* Schatz, *der;* **teacher's ~** *(derog.)* Liebling des Lehrers/der Lehrerin. **2.** *adj.* **a)** *(kept as ~)* zahm; **b)** *(of or for ~ animals)* Haustier-; **c)** *(favourite)* Lieblings-; **sth./sb. is sb.'s ~ aversion** *or* **hate** jmd. kann etw./jmdn. auf den Tod nicht ausstehen *(ugs.);* **d)** *(expressing fondness)* Kose⟨*form, -name*⟩. **3.** *v. t.*, **-tt-: a)** *(treat as favourite)* bevorzugen; verwöhnen; *(indulge)* verhätscheln; **b)** *(fondle)* streicheln; liebkosen. **4.** *v. i.*, **-tt-** knutschen *(ugs.);* zärtlich sein *(verhüll.)*

petal ['petl] *n.* Blütenblatt, *das*

petard [pɪ'tɑ:d] *n. (Hist.)* Petarde, *die; see also* **hoist 3**

peter ['pi:tə(r)] *v. i.* ~ **out** [allmählich] zu Ende gehen; ⟨*Wasserlauf:*⟩ versickern; ⟨*Weg:*⟩ sich verlieren; ⟨*Briefwechsel:*⟩ versanden; ⟨*Angriff:*⟩ sich totlaufen

'pet food *n.* Tierfutter, *das*

petit bourgeois [pətɪ 'bʊəʒwɑ:] *n., pl.* **petits bourgois** [pətɪ 'bʊəʒwɑ:] *(usu. derog.)* Kleinbürger, *der; attrib.* Kleinbürger-; kleinbürgerlich

petite [pə'ti:t] *adj. fem.* zierlich

petition [pə'tɪʃn] **1.** *n.* **a)** *(formal written supplication)* Petition, *die;* Eingabe, *die;* **get together** *or* **up a ~ for/against sth.** Unterschriften für/gegen etw. sammeln; **b)** *(Law)* [förmlicher] Antrag; *(for divorce)* Klage, *die.* **2.** *v. t.* eine Eingabe richten an (+ *Akk.*); ~ **sb. for sth.** jmdn. um etw. ersuchen. **3.** *v. i.* ~ **for** ersuchen um *(geh.); (present ~ for)* eine Unterschriftenliste einreichen für; ~ **for divorce** die Scheidung einreichen

petrel ['petrl] *n. (Ornith.)* Sturmvogel, *der*

petrify ['petrɪfaɪ] **1.** *v. t.* **a)** *(change into stone)* versteinern lassen; **become petrified** versteinern; **b)** *(fig.)* erstarren lassen; **be petrified with fear/shock** starr vor Angst/Schrecken sein; **be petrified by sb./sth.** vor jmdm./etw. erstarren. **2.** *v. i. (turn to stone)* versteinern; *(fig.)* erstarren

petrochemical [petrəʊ'kemɪkl] *n.* Petrochemikalie, *die*

petrol ['petrl] *n. (Brit.)* Benzin, *das;* **fill up with ~:** tanken

petrol: ~ bomb *n. (Brit.)* Benzinbombe, *die;* **~-can** *n. (Brit.)* Benzinkanister, *der;* **~-cap** *n. (Brit.)* Tankverschluß, *der*

petroleum [pɪ'trəʊlɪəm] *n.* Erdöl, *das*

petroleum 'jelly *n.* Vaseline, *die*

petrol: ~-gauge *n. (Brit.)* Benzinuhr, *die;* **~-pump** *n. (Brit.)* **a)** *(in ~-station)* Zapfsäule, *die;* **b)** *(in car, aircraft, etc.)* Benzin- *od.* Kraftstoffpumpe, *die;*

~**-station** *n. (Brit.)* Tankstelle, *die;* ~**-tank** *n. (Brit.) (in car, aircraft, etc.)* Benzintank, *der;* ~**-tanker** *n. (Brit.)* Benzintankwagen, *der*

'**pet shop** *n.* Tierhandlung, *die*

petticoat ['petɪkəʊt] *n.* Unterrock, *der*

petty ['petɪ] *adj.* **a)** *(trivial)* belanglos ‹*Detail, Sorgen*›; kleinlich ‹*Einwand, Vorschrift*›; **b)** *(minor)* Klein‹*staat, -unternehmer, -landwirt*›; klein ‹*Geschäftsmann*›; Duodez‹*fürst, -fürstentum, -staat*›; **c)** *(small-minded)* kleinlich; kleinkariert

petty: ~ '**cash** *n.* kleine Kasse; Portokasse, *die;* ~ '**officer** *n. (Navy)* ≈ [Ober]maat, *der*

petulant ['petjʊlənt] *adj.* bockig

petunia [pɪ'tju:nɪə] *n. (Bot.)* Petunie, *die*

pew [pju:] *n.* **a)** *(Eccl.)* Kirchenbank, *die;* **b)** *(coll.: seat)* [Sitz]platz, *der;* **have** *or* **take a** ~: sich platzen *(ugs. scherzh.)*

pewter ['pju:tə(r)] *n., no pl., no indef. art. (substance, vessels)* Pewter, *der;* [Hart]zinn, *das*

PG *abbr. (Brit. Cinemat.)* **Parental Guidance** ≈ bedingt jugendfrei

phallic ['fælɪk] *adj.* phallisch; ~ **symbol** Phallussymbol, *das*

phantom ['fæntəm] **1.** *n.* Phantom, *das.* **2.** *adj.* Phantom-

Pharaoh ['feərəʊ] *n.* Pharao, *der*

pharmaceutical [fɑ:mə'sju:tɪkl] **1.** *adj.* pharmazeutisch; Pharma‹*industrie, -konzern, -hersteller*›; ~ **chemist** Arzneimittelchemiker, *der*/-chemikerin, *die.* **2.** *n. in pl.* Pharmaka

pharmacist ['fɑ:məsɪst] *n.* Apotheker, *der*/Apothekerin, *die; (in research)* Pharmazeut, *der*/Pharmazeutin, *die*

pharmacology [fɑ:mə'kɒlədʒɪ] *n.* Pharmakologie, *die*

pharmacy ['fɑ:məsɪ] *n.* **a)** *no pl., no art. (preparation of drugs)* Pharmazie, *die;* **b)** *(dispensary)* Apotheke, *die*

phase [feɪz] **1.** *n.* Phase, *die; (of project, construction, history also)* Abschnitt, *der; (of illness, development also)* Stadium, *das;* **it's only** *or* **just a** ~ **[he's/she's going through]** das gibt sich [mit der Zeit] wieder *(ugs.).* **2.** *v. t.* stufenweise durchführen

~ '**in** *v. t.* stufenweise einführen

~ '**out** *v. t.* **a)** *(eliminate gradually)* nach und nach auflösen ‹*Abteilung*›; allmählich abschaffen ‹*Verfahrensweise, Methode*›; **b)** *(discontinue production of)* [langsam] auslaufen lassen

Ph.D. [pi:eɪtʃ'di:] *abbr.* **Doctor of Philosophy** Dr. phil.

pheasant ['fezənt] *n.* Fasan, *der*

phenomenal [fɪ'nɒmɪnl] *adj. (remarkable)* phänomenal; sagenhaft *(ugs.);* unwahrscheinlich *(ugs.)* ‹*Spektakel, Radau*›

phenomenon [fɪ'nɒmɪnən] *n., pl.* **phenomena** [fɪ'nɒmɪnə] Phänomen, *das*

phew [fju:] *int.* puh

phial ['faɪəl] *n.* [Medizin]fläschchen, *das;* Phiole, *die*

philanderer [fɪ'lændərə(r)] *n.* Schürzenjäger, *der (spött.)*

philanthropic [fɪlən'θrɒpɪk] *adj.* philanthropisch *(geh.);* menschenfreundlich; Wohltätigkeits‹*organisation, -verein usw.*›

philanthropist [fɪ'lænθrəpɪst] *n.* Philanthrop, *der*/Philanthropin, *die (geh.);* Menschenfreund, *der*/Menschenfreundin, *die*

philanthropy [fɪ'lænθrəpɪ] *n.* Philanthropie, *die (geh.)*

philately [fɪ'lætəlɪ] *n.* Philatelie, *die;* Briefmarkenkunde, *die*

philharmonic [fɪlhɑ:'mɒnɪk, fɪlɑ:'mɒnɪk] **1.** *adj.* philharmonisch. **2.** *n.* Philharmonie, *die*

Philippines ['fɪlɪpi:nz] *pr. n. pl.* Philippinen *Pl.*

philistine ['fɪlɪstaɪn] *n. (uncultured person)* [Kultur]banause, *der*/-banausin, *die*

philology [fɪ'lɒlədʒɪ] *n.* [historische] Sprachwissenschaft

philosopher [fɪ'lɒsəfə(r)] *n.* Philosoph, *der*/Philosophin, *die*

philosophic [fɪlə'sɒfɪk], **philosophical** [fɪlə'sɒfɪkl] *adj.* **a)** philosophisch; **b)** *(resigned, calm)* abgeklärt; gelassen

philosophize (philosophise) [fɪ'lɒsəfaɪz] *v. i.* philosophieren **(about, on** über + *Akk.*)

philosophy [fɪ'lɒsəfɪ] *n.* Philosophie, *die*

phlegm [flem] *n., no pl., no indef. art.* **a)** *(Physiol.)* Schleim, *der;* Mucus, *der (Med.);* **b)** *(coolness)* stoische Ruhe; Gleichmut, *der;* **c)** *(stolidness)* Phlegma, *das*

phlegmatic [fleg'mætɪk] *adj.* **a)** *(cool)* gleichmütig; **b)** *(stolid)* phlegmatisch

phobia ['fəʊbɪə] *n.* Phobie, *die (Psychol.);* [krankhafte] Angst

phoenix ['fi:nɪks] *n. (Mythol.)* Phönix, *der*

phone [fəʊn] *(coll.)* **1.** *n.* Telefon, *das;* **pick up/put down the** ~: [den Hörer] abnehmen/auflegen; **by** ~: telefonisch; **speak to sb. by** ~ *or* **on the** ~: mit jmdm. telefonieren. **2.** *v. i.* anrufen; **can we** ~ **from here?** können wir von hier aus telefonieren? **3.** *v. t.* anrufen; ~ **the office/home** im Büro/zu Hause anrufen

~ '**back** *v. t. & i. (make a return* ~ *call [to])* zurückrufen; *(make a further* ~ *call [to])* wieder *od.* nochmals anrufen

~ '**in 1.** *v. i.* anrufen. **2.** *v. t.* telefonisch mitteilen *od.* durchgeben

~ '**up** *v. t. & i.* anrufen

phone: ~ **book** *n.* Telefonbuch, *das;* ~ **box** *n.* Telefonzelle, *die;* ~ **call** *n.* Anruf, *der; see also* **telephone call;** ~ **card** *n.* Telefonkarte, *die;* ~**-in** *n.* ~**-in [programme]** *(Radio)* Hörersendung, *die; (Telev.)* Sendung mit Zuschaueranrufen; ~ **number** *n.* Telefonnummer, *die;* ~**-tapping** *n.* Anzapfen von Telefonleitungen

phonetic [fə'netɪk] *adj.* phonetisch

phonetics [fə'netɪks] *n.* **a)** *no pl.* Phonetik, *die;* **b)** *no pl. (phonetic script)* phonetische Umschrift; **c)** *constr. as pl.* phonetische Angaben

phoney ['fəʊnɪ] *(coll.)* **1.** *adj.*, **phonier** ['fəʊnɪə(r)], **phoniest** ['fəʊnɪɪst] **a)** *(sham)* falsch; gefälscht ‹*Brief, Dokument*›; **b)** *(fictitious)* falsch ‹*Name*›; erfunden ‹*Geschichte*›; **c)** *(fraudulent)* Schein‹*firma, -geschäft, -krieg*›; falsch, scheinbar ‹*Doktor, Diplomat, Geschäftsmann*›. **2.** *n.* **a)** *(person)* Blender, *der*/Blenderin, *die;* **this doctor is just a** ~: dieser Arzt ist ein Scharlatan; **b)** *(sham)* Fälschung, *die*

phonograph ['fəʊnəgrɑ:f] *(Amer.) see* **gramophone**

phony *see* **phoney**

phosphate ['fɒsfeɪt] *n. (Chem.)* Phosphat, *das*

phosphorescence [fɒsfə'resəns] *n.* Phosphoreszenz, *die*

phosphorescent [fɒsfə'resənt] *adj.* phosphoreszierend

phosphorus ['fɒsfərəs] *n. (Chem.)* Phosphor, *der*

photo ['fəʊtəʊ] *n., pl.* ~**s** Foto, *das; see also* **photograph 1**

photo: ~ **album** *n.* Fotoalbum, *das;* ~**copier** *n.* Fotokopiergerät, *das;* ~**copy 1.** *n.* Fotokopie, *die;* **2.** *v. t.* fotokopieren; ~**fit** *n.* Phantombild, *das*

photogenic [fəʊtə'dʒenɪk, fəʊtə'dʒi:nɪk] *adj.* fotogen

photograph ['fəʊtəgrɑ:f] **1.** *n.* Fotografie, *die;* Foto, *das;* **take a** ~ **[of sb./sth.]** [jmdn./etw.] fotografieren; ein Foto [von jmdm./etw.] machen. **2.** *v. t. & i.* fotografieren '**photograph album** *n.* Fotoalbum, *das*

photographer [fə'tɒgrəfə(r)] *n.* Fotograf, *der*/Fotografin, *die*

photographic [fəʊtə'græfɪk] *adj.* fotografisch; Foto⟨*ausrüstung, -club, -ausstellung*⟩

photography [fə'tɒgrəfɪ] *n., no pl., no indef. art.* Fotografie, *die*

photo: ~'sensitive *adj.* lichtempfindlich; **~'synthesis** *n.* *(Bot.)* Photosynthese, *die*

phrase [freɪz] **1.** *n.* **a)** *(Ling.) (idiomatic expression)* idiomatische Wendung; [Rede]wendung, *die;* **set ~:** feste [Rede]wendung; **noun/verb ~:** Nominal-/Verbalphrase, *die;* **b)** *(brief expression)* kurze Formel; *see also* **turn 1 j. 2.** *v. t.* **a)** *(express in words)* formulieren; **b)** *(Mus.)* phrasieren

'**phrase-book** *n.* Sprachführer, *der*

phraseology [freɪzɪ'ɒlədʒɪ] *n.* Ausdrucksweise, *die; (technical terms)* Terminologie, *die*

physical ['fɪzɪkl] **1.** *adj.* **a)** *(material)* physisch ⟨*Gewalt*⟩; stofflich, dinglich ⟨*Welt, Universum*⟩; **b)** *(of physics)* physikalisch; **it's a ~ impossibility** *(fig.)* es ist absolut unmöglich; **c)** *(bodily)* körperlich; physisch; **you need to take more ~ exercise** du brauchst mehr Bewegung; **d)** *(carnal, sensual)* körperlich ⟨*Liebe*⟩; sinnlich ⟨*Person, Ausstrahlung*⟩. **2.** *n.* ärztliche [Vorsorge]untersuchung; *(for joining the army)* Musterung, *die*

physical: ~ edu'cation *n.* Sport, *der;* Leibesübungen *Pl.* *(Amtsspr.);* **~ 'jerks** *n. pl.* *(coll.)* Gymnastikübungen

physically ['fɪzɪkəlɪ] *adv.* **a)** *(in accordance with physical laws)* physikalisch; **~ impossible** *(fig.)* absolut unmöglich; **b)** *(relating to the body)* körperlich; physisch; **be ~ sick** einen physischen Ekel empfinden; **~ disabled** körperbehindert

physician [fɪ'zɪʃn] *n.* Arzt, *der*/Ärztin, *die*

physicist ['fɪzɪsɪst] *n.* Physiker, *der*/Physikerin, *die*

physics ['fɪzɪks] *n., no pl.* Physik, *die*

physiological [fɪzɪə'lɒdʒɪkl] *adj.* physiologisch

physiology [fɪzɪ'ɒlədʒɪ] *n.* Physiologie, *die*

physiotherapist [fɪzɪəʊ'θerəpɪst] *n.* Physiotherapeut, *der*/-therapeutin, *die*

physiotherapy [fɪzɪəʊ'θerəpɪ] *n.* Physiotherapie, *die*

physique [fɪ'zi:k] *n.* Körperbau, *der*

pianist ['pi:ənɪst] *n.* Pianist, *der*/Pianistin, *die*

piano [pɪ'ænəʊ] *n., pl.* **~s** *(Mus.)* *(upright)* Klavier, *das; (grand)* Flügel, *der; attrib.* Klavier-; **play the ~:** Klavier spielen

piano-ac'cordion *n.* Akkordeon, *das*

piano [pɪ'ænəʊ]: **~ music** *n.* Klaviermusik, *die; (score)* Klaviernoten *Pl.;* **~-player** *n.* Klavierspieler, *der*/-spielerin, *die;* **~-stool** *n.* Klavierschemel, *der;* **~-tuner** *n.* Klavierstimmer, *der*/-stimmerin, *die*

piccolo ['pɪkələʊ] *n., pl.* **~s** *(Mus.)* Pikkoloflöte, *die;* Pikkolo, *das*

[1]**pick** [pɪk] *n.* **a)** *(for breaking up hard ground, rocks, etc.)* Spitzhacke, *die; (for breaking up ice)* [Eis]pickel, *der;* **b)** *see* **toothpick; c)** *(Mus.)* Plektrum, *das*

[2]**pick 1.** *n.* **a)** *(choice)* Wahl, *die;* **take your ~:** du hast die Wahl; **she had the ~ of several jobs** sie konnte zwischen mehreren Jobs [aus]wählen; **have [the] first ~ of sth.** als erster aus etw. auswählen dürfen; **b)** *(best part)* Elite, *die;* **the ~ of the fruit** *etc.* die besten Früchte *usw.* **2.** *v. t.* **a)** pflücken ⟨*Blumen*⟩; [ab]ernten, [ab]pflücken ⟨*Äpfel, Trauben usw.*⟩; **b)** *(select)* auswählen; aufstellen ⟨*Mannschaft*⟩; **~ the** *or* **a winner/the winning horse** auf den Sieger/das richtige *od.* siegreiche Pferd setzen; **~ one's way** sich *(Dat.)* vorsichtig [s]einen Weg suchen; **~ and choose** sich *(Dat.)* aussuchen; **~ one's time [for sth.]** den Zeitpunkt [für etw.] festlegen; **c)** *(clear of flesh)* **~ the bones [clean]** ⟨*Hund:*⟩ die Knochen [sauber] abnagen; **d) ~ sb.'s brains [about sth.]** jmdn. [über etw. *(Akk.)*] ausfragen *od.* *(ugs.)* ausquetschen; **e) ~ one's nose/teeth** in der Nase bohren/in den Zähnen [herum]stochern; **f) ~ sb.'s pocket** jmdn. bestehlen; **he had his pocket ~ed** er wurde von einem Taschendieb bestohlen; **g) ~ a lock** ein Schloß knacken *(salopp);* **h) ~ to pieces** *(fig.: criticize)* kein gutes Haar lassen an (+ *Dat.*) *(ugs.).* **3.** *v. i.* **~ and choose [too much]** [zu] wählerisch sein

~ at *v. t.* **a)** herumstochern in (+ *Dat.*) ⟨*Essen*⟩; **b)** herumspielen an (+ *Dat.*) ⟨*Pickel*⟩

~ off *v. t.* **a)** ['--] abzupfen, ablesen ⟨*Haare, Fusseln*⟩; **b)** [-'-] *(shoot one by one)* [einzeln] abschießen

~ on *v. t.* *(victimize)* es abgesehen haben auf (+ *Akk.*); **why ~ on me every time?** warum immer gerade *od.* ausgerechnet ich?; **~ on someone your own size!** leg dich doch wenigstens mit einem Gleichstarken an! *(ugs.)*

~ 'out *v. t.* **a)** *(choose)* auswählen; *(for oneself)* sich *(Dat.)* aussuchen ⟨*Kleid, Blume*⟩; **b)** *(distinguish)* ausmachen, entdecken ⟨*Detail, jmds. Gesicht in der Menge*⟩; **~ out sth. from sth.** etw. von etw. unterscheiden

~ up 1. ['--] *v. t.* **a)** *(take up)* [in die Hand] nehmen ⟨*Brief, Buch usw.*⟩; hochnehmen ⟨*Baby*⟩; [wieder] aufnehmen ⟨*Handarbeit*⟩; aufnehmen ⟨*Masche*⟩; auffinden ⟨*Fehler*⟩; *(after dropping)* aufheben; **~ sth. up from the table** etw. vom Tisch nehmen; **~ a child up in one's arms** ein Kind auf den Arm nehmen; **~ up the telephone** den [Telefon]hörer abnehmen; **~ up the pieces** *(lit. or fig.)* die Scherben aufsammeln; **b)** *(collect)* mitnehmen; *(by arrangement)* abholen **(at, from** von); *(obtain)* holen; **~ up sth. on the way home** etw. auf dem Nachhauseweg abholen; **c)** *(become infected by)* sich *(Dat.)* einfangen *od.* holen *(ugs.)* ⟨*Virus, Grippe*⟩; **d)** *(take on board)* ⟨*Bus, Autofahrer:*⟩ mitnehmen; **~ sb. up at** *or* **from the station** jmdn. vom Bahnhof abholen; **e)** *(rescue from the sea)* [aus Seenot] bergen; **f)** *(coll.: earn)* einstreichen *(ugs.);* **g)** *(coll.: make acquaintance of)* aufreißen *(ugs.);* **h)** *(find and arrest)* festnehmen; **i)** *(receive)* empfangen ⟨*Signal, Funkspruch usw.*⟩; **j)** *(obtain casually)* sich *(Dat.)* aneignen; bekommen ⟨*Sache*⟩; **~ up languages easily** mühelos Sprachen lernen; **k)** *(obtain)* auftreiben *(ugs.);* **l)** *(resume)* wieder aufnehmen ⟨*Erzählung, Gespräch*⟩; **m)** *(regain)* wiederfinden ⟨*Spur, Fährte*⟩; wieder aufnehmen ⟨*Witterung*⟩; **n)** *(pay)* **~ up the bill** *etc.* **for sth.** die Kosten *od.* die Rechnung *usw.* für etw. übernehmen. **2.** [-'-] *v. i.* ⟨*Gesundheitszustand, Befinden, Stimmung, Laune, Wetter:*⟩ sich bessern; ⟨*Person:*⟩ sich erholen; ⟨*Markt, Geschäft:*⟩ sich erholen *od.* beleben; ⟨*Gewinne:*⟩ steigen, zunehmen. **3.** *v. refl.* **~ oneself up** wieder aufstehen; *(fig.)* sich aufrappeln *(ugs.)*

'**pickaxe** *(Amer.:* '**pickax**) *see* [1]**pick a**

picket ['pɪkɪt] **1.** *n.* **a)** *(Industry)*

Streikposten, *der;* **mount a ~ [at** *or* **on a gate]** [an einem Tor] Streikposten aufstellen; **b)** *(pointed stake)* Pfahl, *der.* **2.** *v. t.* Streikposten aufstellen vor (+ *Dat.*) ‹*Fabrik, Büro usw.*›. **3.** *v. i.* Streikposten stehen

'picket line *n.* Streikpostenkette, *die*

pickings ['pɪkɪŋz] *n. pl. (gleanings)* Reste *Pl.; (things stolen)* [Aus]beute, *die; (yield)* Ausbeute, *die;* **it's easy ~:** das ist ein einträgliches Geschäft

pickle ['pɪkl] **1.** *n.* **a)** *(brine)* Salzlake, *die; (vinegar solution)* Marinade, *die;* **b)** *usu. in pl. (food)* [Mixed] Pickles *Pl.;* **c)** *(coll.: predicament)* **be in a ~:** in der Klemme sitzen *(ugs.);* **get into a ~:** in die Klemme geraten *(ugs.).* **2.** *v. t.* [in Essig *od.* sauer] einlegen ‹*Gurken, Zwiebeln, Eier*›; marinieren ‹*Hering*›

pick: ~-me-up *n.* Stärkungsmittel, *das;* **the holiday was a real ~-me-up** der Urlaub hat mir richtig gut getan; **~pocket** *n.* Taschendieb, *der*/-diebin, *die;* **~-up** *n.* **a)** *(truck)* **~-up [truck/van]** Kleinlastwagen, *der;* **b)** *(of record-player, guitar)* Tonabnehmer, *der;* **c)** *(coll.: person)* Zufallsbekanntschaft, *die*

picnic ['pɪknɪk] **1.** *n.* **a)** Picknick, *das;* **go for** *or* **on a ~:** ein Picknick machen; picknicken gehen; **have a ~:** ein Picknick machen; picknicken; **b)** *(coll.: easy task)* Kinderspiel, *das;* **be no ~:** kein Zuckerlecken *od.* Honig[sch]lecken sein. **2.** *v. i.,* **-ck-** picknicken; Picknick machen

picnic: ~ 'lunch *n.* **a)** Picknick, *das (als Mittagessen);* **b)** *(packed up)* Lunchpaket, *das;* **~ site** *n.* Picknickplatz, *der*

pictorial [pɪk'tɔ:rɪəl] *adj.* illustriert ‹*Bericht, Zeitschrift, Wochenmagazin*›; bildlich ‹*Darstellung*›

picture ['pɪktʃə(r)] **1.** *n.* **a)** Bild, *das;* **b)** *(portrait)* Porträt, *das; (photograph)* Porträtfoto, *das;* **have one's ~ painted** sich malen *od.* portraitieren lassen; **c)** *(mental image)* Vorstellung, *die;* Bild, *das;* **get a ~ of sth.** sich *(Dat.)* von etw. ein Bild machen; **give a ~ of sth.** von etw. einen Eindruck vermitteln; **present a sorry ~** *(fig.)* ein trauriges *od.* jämmerliches Bild abgeben; **look the [very] ~ of health/misery/innocence** wie das blühende Leben aussehen/ein Bild des Jammers sein/wie die Unschuld in Person aussehen; **get the ~** *(coll.)* verstehen[, worum es geht]; **I'm beginning to get the ~:** langsam *od.* allmählich verstehe *od. (ugs.)* kapiere ich; **[do you] get the ~?** verstehst du?; **put sb. in the ~:** jmdn. ins Bild setzen; **be in the ~** *(be aware)* im Bilde sein; **keep sb. in the ~:** jmdn. auf dem laufenden halten; **come** *or* **enter into the ~:** [dabei] eine Rolle spielen; **d)** *(film)* Film, *der;* **e)** *in pl. (Brit.: cinema)* Kino, *das;* **go to the ~s** ins Kino gehen; **what's on at the ~s?** was gibt's *od.* läuft im Kino?; **f)** *(delightful object)* **be a ~:** wunderschön *od. (ugs.)* ein Gedicht sein; **her face was a ~:** ihr Gesicht sprach Bände; **she looked a ~:** sie sah bildschön aus. **2.** *v. t.* **a)** *(represent)* abbilden; **b)** *(imagine)* **~ [to oneself]** sich *(Dat.)* vorstellen

picture: ~-book 1. *n.* Bilderbuch, *das;* **2.** *adj.* Bilderbuch-; **~-frame** *n.* Bilderrahmen, *der;* **~-gallery** *n.* Gemäldegalerie, *die;* **~-hook** *n.* Bilderhaken, *der;* **~ 'postcard** *n.* Ansichtskarte, *die;* **~-rail** *n.* Bilderleiste, *die*

picturesque [pɪktʃə'resk] *adj.* malerisch; pittoresk *(geh.); (vivid)* anschaulich, bildhaft ‹*Beschreibung, Erzählung*›

piddle ['pɪdl] *(coll.)* **1.** *v. i.* Pipi machen *(Kinderspr.);* pinkeln *(ugs.).* **2.** *n.* **a) have a/do one's ~:** Pipi machen *(Kinderspr.);* pinkeln *(ugs.);* **b)** *(urine)* Pipi, *das (Kinderspr.)*

pidgin ['pɪdʒɪn] *n.* Pidgin, *das*

pidgin 'English *n.* Pidgin-Englisch, *das*

pie [paɪ] *n. (of meat, fish, etc.)* Pastete, *die; (of fruit etc.)* ≈ Obstkuchen, *der;* **as sweet/nice** *etc.* **as ~** *(coll.)* superfreundlich *(ugs.);* **as easy as ~** *(coll.)* kinderleicht *(ugs.);* **have a finger in every ~** *(coll.)* überall die Finger drin haben *(ugs.);* **that's all just ~ in the sky** *(coll.)* das sind alles nur Luftschlösser

piece [pi:s] **1.** *n.* **a)** Stück, *das; (of broken glass or pottery)* Scherbe, *die; (of jigsaw puzzle, crashed aircraft, etc.)* Teil, *der; (Amer.: distance)* [kleines] Stück; **a ~ of meat** ein Stück Fleisch; **[all] in one ~:** unbeschädigt; *(fig.)* heil; wohlbehalten; **in ~s** *(broken)* kaputt *(ugs.);* zerbrochen; *(taken apart)* [in Einzelteile] zerlegt; **break into ~s, fall to ~s** zerbrechen; kaputtgehen *(ugs.);* **go [all] to ~s** *(fig.)* [völlig] die Fassung verlieren; **[all] of a ~:** aus einem Guß; **say one's ~** *(fig.)* sagen, was man zu sagen hat; **b)** *(part of set)* **~ of furniture/clothing/luggage** Möbel- / Kleidungs- / Gepäckstück, *das;* **a three-/four-~ suite** eine drei-/vierteilige Sitzgarnitur; **c)** *(enclosed area)* **a ~ of land/property** ein Stück Land/Grundstück; **d)** *(example)* **~ of luck** Glücksfall, *der;* **a fine ~ of pottery** eine sehr schöne Töpferarbeit; **fine ~ of work** hervorragende Arbeit; **he's an unpleasant ~ of work** *(fig.)* er ist ein unangenehmer Vertreter *(ugs.);* **e)** *(item)* **~ of news/gossip/information** Nachricht, *die*/Klatsch, *der*/Information, *die;* **f)** *(Chess)* Figur, *die; (Draughts, Backgammon, etc.)* Stein, *der;* **g)** *(coin)* **gold ~:** Goldstück, *das;* **a 10p ~:** ein 10-Pence-Stück; eine 10-Pence-Münze; **h)** *(article in newspaper, magazine, etc.)* Beitrag, *der;* **i)** *(literary or musical composition)* Stück, *das;* **~ of music** Musikstück, *das.* **2.** *v. t.* **~ together** *(lit. or fig.)* zusammenfügen **(from** aus)

'piecemeal *adv., adj.* stückweise

piece: ~-rate *n.* Akkordsatz, *der;* **~-work** *n., no pl.* Akkordarbeit, *die*

'pie-dish *n.* Pastetenform, *die*

pier [pɪə(r)] *n.* Pier, *der od. (Seemannsspr.) die*

pierce [pɪəs] *v. t.* **a)** *(prick)* durchbohren, durchstechen ‹*Hülle, Verkleidung, Ohrläppchen*›; *(penetrate)* sich bohren in, [ein]dringen in (+ *Akk.*) ‹*Körper, Fleisch, Herz*›; **have one's ears ~d** sich *(Dat.)* Löcher in die Ohrläppchen machen *od.* stechen lassen; **b)** *(fig.)* **the cold ~d him to the bone** die Kälte drang ihm bis ins Mark; **a scream ~d the night/silence** ein Schrei gellte durch die Nacht/zerriß die Stille

piercing ['pɪəsɪŋ] *adj.* durchdringend ‹*Stimme, Schrei, Blick*›; schneidend ‹*Sarkasmus, Kälte*›

piety ['paɪətɪ] *n., no pl.* Frömmigkeit, *die*

piffling ['pɪflɪŋ] *adj. (coll.)* lächerlich

pig [pɪg] *n.* **a)** Schwein, *das;* **~s might fly** *(iron.)* da müßte schon ein Wunder geschehen; **buy a ~ in a poke** *(fig.)* die Katze im Sack kaufen; **b)** *(coll.) (greedy person)* Vielfraß, *der (ugs.); (dirty person)* Ferkel, *das (ugs.); (unpleasant person)* Schwein, *das (derb);* **make a ~ of oneself** *(overeat)* sich *(Dat.)* den Bauch *od.* Wanst vollschlagen *(salopp)*

[1]**pigeon** ['pɪdʒɪn] *n.* Taube, *die*
[2]**pigeon** *n. (coll.: business)* **be sb.'s ~:** jmdn. angehen; **that's not my ~:** das ist nicht mein Bier *(ugs.)*
pigeon: ~-fancier *n.* Taubenfreund, *der*/-freundin, *die;* **~-hole 1.** *n.* [Ablage]fach, *das; (for letters)* Postfach, *das;* **put people in ~-holes** *(fig.)* Menschen in Schubladen einordnen; **2.** *v. t.* **a)** *(deposit)* [in die Fächer] sortieren; **b)** *(categorize)* einordnen; **~-toed 1.** *adj.* **be ~-toed** mit einwärts gerichteten Füßen gehen; **2.** *adv.* mit einwärts gerichteten Füßen
piggy ['pɪgɪ]: **~back** *n.* **give sb. a ~back** jmdn. huckepack nehmen *od.* tragen; **~ bank** *n.* Sparschwein[chen], *das*
pig: ~'headed *adj.* dickschädelig *(ugs.);* stur; **~-iron** *n.* Roheisen, *das*
piglet ['pɪglɪt] *n.* Ferkel, *das*
pigment ['pɪgmənt] **1.** *n.* Pigment, *das.* **2.** *v. t.* pigmentieren
pigmentation [pɪgmən'teɪʃn] *n.* Pigmentierung, *die*
pigmy *see* **pygmy**
pig: ~'s 'ear *n. (Brit. coll.)* **make a ~'s ear of sth.** etw. verpfuschen *od. (ugs.)* vermurksen; **~skin** *n. (leather)* Schweinsleder, *das;* **~sty** *n. (lit. or fig.)* Schweinestall, *der;* **~swill** *n.* Schweinefutter, *das; (fig. coll.: food)* Schweinefraß, *der (derb);* **~tail** *n. (plaited)* Zopf, *der;* **~tails** *(worn loose, at either side of head)* Rattenschwänzchen *Pl. (ugs.)*
[1]**pike** [paɪk] *n., pl. same (Zool.)* Hecht, *der*
[2]**pike** *n. (Arms Hist.)* Pike, *die;* Spieß, *der*
'**pikestaff** *n.* **plain as a ~:** sonnenklar *(ugs.)*
pilchard ['pɪltʃəd] *n.* Sardine, *die*
[1]**pile** [paɪl] **1.** *n.* **a)** *(heap) (of dishes, plates)* Stapel, *der; (of paper, books, letters)* Stoß, *der; (of clothes)* Haufen, *der;* **b)** *(coll.: large quantity)* Masse, *die (ugs.);* Haufen, *der (ugs.);* **a ~ of troubles/letters** eine *od. (ugs.)* jede Menge Sorgen/Briefe; **c)** *(coll.: fortune)* **make a** *or* **one's ~:** ein Vermögen machen. **2.** *v. t.* **a)** *(load)* [voll] beladen; **b)** *(heap up)* aufstapeln ⟨*Holz, Steine*⟩; aufhäufen ⟨*Abfall, Schnee*⟩; **c) ~ furniture into a van** *etc.* Möbel in einen Lieferwagen *usw.* laden
~ 'in *v. i. (get in) (seen from outside)* hineindrängen; *(seen from inside)* hereindrängen; **~ in!** [kommt] nur *od.* immer herein!; quetscht euch rein! *(ugs.)*
~ into *v. t.* drängen in (+ *Akk.*) ⟨*Stadion, Halle*⟩; drängen auf (+ *Akk.*) ⟨*Platz, Wiese*⟩; sich zwängen in (+ *Akk.*) ⟨*Auto, Zimmer, Zugabteil, Telefonzelle*⟩
~ on to *v. t.* **a) ~ logs on to the fire** Holzscheite auf das Feuer legen; **he ~d food on to my plate** er häufte mir Essen auf den Teller; **~ work on to sb.** *(fig.)* jmdm. Arbeit aufbürden; **b)** *(enter)* drängen in (+ *Akk.*) ⟨*Bus usw.*⟩
~ 'out *v. i.* nach draußen strömen *od.* drängen
~ 'up 1. *v. i.* **a)** *(accumulate)* ⟨*Waren, Post, Aufträge, Arbeit, Schnee:*⟩ sich auftürmen; ⟨*Verkehr:*⟩ sich stauen; ⟨*Schulden:*⟩ sich vermehren; ⟨*Verdacht, Eindruck, Beweise:*⟩ sich verdichten; **b)** *(crash)* aufeinander auffahren. **2.** *v. t.* aufstapeln ⟨*Steine, Bücher usw.*⟩; auftürmen ⟨*Haar, Frisur*⟩; aufhäufen ⟨*Abfall, Schnee*⟩; *(fig.)* zusammentragen ⟨*Beweise usw.*⟩; **~ up debts** sich immer mehr verschulden
[2]**pile** *n. (soft surface)* Flor, *der*
[3]**pile** *n. (stake)* Pfahl, *der*
'**pile-driver** *n.* [Pfahl]ramme, *die*
piles [paɪlz] *n. pl. (Med.)* Hämorrhoiden *Pl.*
'**pile-up** *n.* Massenkarambolage, *die*
pilfer ['pɪlfə(r)] *v. t.* stehlen; klauen *(ugs.)*
pilgrim ['pɪlgrɪm] *n.* Pilger, *der*/Pilgerin, *die;* Wallfahrer, *der*/Wallfahrerin, *die*
pilgrimage ['pɪlgrɪmɪdʒ] *n.* Pilgerfahrt, *die;* Wallfahrt, *die*
pill [pɪl] *n.* **a)** Tablette, *die;* Pille, *die (ugs.);* **be on ~s** Tabletten einnehmen müssen; **b)** *(coll.: contraceptive)* **the ~** *or* **P~:** die Pille *(ugs.);* **be on the ~:** die Pille nehmen *(ugs.);* **c)** *(fig.: unpleasant thing)* **swallow the ~:** die [bittere] Pille schlucken *(ugs.);* **sweeten the ~:** die bittere Pille versüßen *(ugs.);* **be a bitter ~ [to swallow]** eine bittere Pille *od.* bitter sein
pillage ['pɪlɪdʒ] **1.** *n.* Plünderung, *die.* **2.** *v. t.* [aus]plündern
pillar ['pɪlə(r)] *n.* **a)** *(vertical support)* Säule, *die;* **from ~ to post** *(fig.)* hin und her; **b)** *(fig.: supporter)* Stütze, *die*
'**pillar-box** *n. (Brit.)* Briefkasten, *der; attrib.* **~ red** knallrot *(ugs.)*
pillion ['pɪljən] *n.* Soziussitz, *der;* Beifahrersitz, *der;* **ride ~:** als Beifahrer/Beifahrerin *od.* auf dem Soziussitz mitfahren
pillory ['pɪlərɪ] **1.** *v. t. (lit. or fig.)* an den Pranger stellen. **2.** *n. (Hist.)* Pranger, *der*
pillow ['pɪləʊ] *n.* [Kopf]kissen, *das*
pillow: ~case, ~ slip *ns.* [Kopf]kissenbezug, *der*
pilot ['paɪlət] **1.** *n.* **a)** *(Aeronaut.)* Pilot, *der*/Pilotin, *die;* **b)** *(Naut.; also fig.: guide)* Lotse, *der.* **2.** *adj.* Pilot⟨*programm, -studie, -projekt usw.*⟩. **3.** *v. t.* **a)** *(Aeronaut.)* fliegen; **b)** *(Naut.; also fig.: guide)* lotsen
pilot: ~ boat *n.* Lotsenboot, *das;* **~-light** *n.* **a)** *(gas-burner)* Zündflamme, *die;* **b)** *(electric light)* Kontrollampe, *die*
pimento [pɪ'mentəʊ] *n., pl.* **~s** *(berry)* Piment, *der od. das*
pimp [pɪmp] *n.* Zuhälter, *der*
pimple ['pɪmpl] *n.* Pickel, *der;* Pustel, *die;* **he/his face had come out in ~s** er hat Pickel/Pickel im Gesicht bekommen
pimply ['pɪmplɪ] *adj.* pick[e]lig
PIN [pɪn] *abbr.* **~ [number]** *see* **personal identification number**
pin 1. *n.* **a)** Stecknadel, *die;* **you could have heard a ~ drop** man hätte eine Stecknadel fallen hören können; **as clean as a new ~:** blitzblank *(ugs.);* **~s and needles** *(fig.)* Kribbeln, *das;* **I had ~s and needles in my legs** *(fig.)* meine Beine kribbelten; **b)** *(peg)* Stift, *der;* **c)** *(Electr.)* Kontaktstift, *der;* **a two-/three-~ plug** ein zwei-/dreipoliger Stecker; **d) for two ~s I'd resign** es fehlt nicht mehr viel, dann kündige ich. **2.** *v. t.,* **-nn-: a)** nageln ⟨*Knochen, Bein, Hüfte*⟩; **~ a badge to one's lapel** sich *(Dat.)* ein Abzeichen ans Revers heften *od.* stecken; **~ a notice on the board** einen Zettel ans Schwarze Brett hängen *od. (ugs.)* pinnen; **~ together** mit einer Stecknadel zusammenhalten; *(Dressm.)* zusammenstecken; **b)** *(fig.)* **~ one's ears back** die Ohren spitzen *(ugs.);* **~ one's hopes on sb./sth.** seine [ganze] Hoffnung auf jmdn./etw. setzen; **~ the blame for sth. on sb.** jmdm. die Schuld an etw. *(Dat.)* zuschieben; **c)** *(seize and hold fast)* **~ sb. against the wall** jmdn. an die Wand drängen; **~ sb. to the ground** jmdn. auf den Boden drücken
~ 'down *v. t.* **a)** *(fig.: bind)* festlegen, festnageln (**to** *or* **on** auf + *Akk.*); **he's a difficult man to ~ down** man kann ihn nur schwer dazu bringen, sich [auf etwas] festzulegen; **b)** *(trap)* festhalten; **~ sb. down [to the ground]** jmdn. auf den Boden drücken; **c)** *(define exactly)* **~ sth. down in words** etw. in Worte fassen; **I**

can't quite ~ it down ich kann es nicht richtig ausmachen
~ '**up** *v.t.* aufhängen ‹*Bild, Foto*›; anschlagen ‹*Bekanntmachung, Hinweis, Liste*›; aufstecken, hochstecken ‹*Haar, Frisur*›; heften ‹*Saum, Naht*›
pinafore ['pɪnəfɔ:(r)] *n.* Schürze, *die (mit Oberteil)*
'**pin-ball** *n.* Flippern, *das*
pincers ['pɪnsəz] *n. pl.* **a) [pair of] ~:** Beiß- *od.* Kneifzange, *die;* **b)** *(of crab etc.)* Schere, *die*
pinch [pɪntʃ] **1.** *n.* **a)** *(squeezing)* Kniff, *der;* **give sb. a ~:** jmdn. kneifen; **give sb. a ~ on the arm/cheek** *etc.* jmdn. *od.* jmdm. in den Arm/die Backe *usw.* kneifen; **b)** *(fig.)* **feel the ~:** knapp bei Kasse sein *(ugs.)*; **the firm is feeling the ~:** der Firma geht es finanziell nicht gut; **at a ~:** zur Not; **if it comes to the ~:** wenn es zum Äußersten kommt; **c)** *(small amount)* Prise, *die.* **2.** *v.t.* **a)** *(grip tightly)* kneifen; **~ sb.'s cheek/bottom** jmdn. in die Wange/den Hintern *(ugs.)* kneifen; **I had to ~ myself** ich mußte mich erst mal in den Arm kneifen *(ugs.)*; **b)** *(coll.: steal)* klauen *(salopp)*; **c)** *(coll.: arrest)* sich *(Dat.)* schnappen *(ugs.)*; **get ~ed** geschnappt werden *(ugs.).* **3.** *v.i.* **a)** ‹*Schuh:*› drücken; **b)** *(be niggardly)* knausern *(ugs.)* **(on** mit)
'**pincushion** *n.* Nadelkissen, *das*
[1]**pine** [paɪn] *n.* **a)** *(tree)* Kiefer, *die;* **b)** *(wood)* Kiefernholz, *das*
[2]**pine** *v.i.* **a)** *(languish)* sich [vor Kummer] verzehren *(geh.)* **(over, about** wegen); **b)** *(long eagerly)* **~ for sb./sth.** sich nach jmdm./etw. sehnen *od. (geh.)* verzehren
~ **a'way** *v.i.* dahinkümmern
pineapple ['paɪnæpl] *n.* Ananas, *die*
pine: ~-cone *n.* Kiefernzapfen, *der;* **~-needle** *n.* Kiefernnadel, *die;* **~-tree** *n.* Kiefer, *die*
ping-pong (*Amer.:* **Ping-Pong, P**) ['pɪŋpɒŋ] *n.* Tischtennis, *das*
ping-pong: ~ ball *n.* Tischtennisball, *der;* Pingpongball, *der (ugs. veralt.)*; **~ table** *n.* Tischtennisplatte, *die*
pinion ['pɪnjən] *n. (cog-wheel)* Ritzel, *das (Technik)*; kleines Zahnrad
pink [pɪŋk] **1.** *n.* **a)** Pink, *das;* Rosa, *das;* **b) in the ~ of condition** in hervorragendem Zustand; **be in the ~** *(sl.)* kerngesund sein; **c)** *(Bot.)* [Garten]nelke, *die.* **2.** *adj.* pinkfarben, rosa ‹*Kleid, Wand*›; rosig, rosarot ‹*Himmel, Gesicht, Haut, Wangen*›
pinkie ['pɪŋkɪ] *n. (Amer., Scot.)* kleiner Finger
pinking ['pɪŋkɪŋ]: **~ scissors, ~ shears** *ns. pl.* **[pair of] ~ scissors** *or* **shears** Zackenschere, *die*
'**pin-money** *n. (for private expenditure)* Taschengeld, *das; (coll.: small sum)* Taschen- *od.* Trinkgeld, *das (ugs.)*
pinnacle ['pɪnəkl] *n.* **a)** *(Archit.)* Fiale, *die;* **b)** *(natural peak)* Gipfel, *der;* **c)** *(fig.: climax)* Höhepunkt, *der;* Gipfel, *der*
pinny ['pɪnɪ] *n. (child lang./coll.)* Schürze, *die*
pin: ~-point *v.t. (locate, define)* genau bestimmen; *(determine)* genau festlegen; **~prick** *n.* Nadelstich, *der; (fig.)* [harmlose] Stichelei; **~-stripe** *n.* Nadelstreifen, *der; (suit)* Nadelstreifenanzug, *der; attrib.* Nadelstreifen‹*anzug, -kostüm*›
pint [paɪnt] *n.* **a)** *(one-eighth of a gallon)* Pint, *das;* ≈ halber Liter; **b)** *(Brit.: quantity of liquid)* Pint, *das;* **a ~ of milk/beer** ≈ ein halber Liter Milch/Bier; **have a ~:** ein Bier trinken
'**pin-up** *(coll.) n.* **a)** *(picture) (of beautiful girl)* Pin-up[-Foto], *das; (esp. of sports, film or pop star)* Starfoto, *das;* **b)** *(beautiful girl in photograph)* Pin-up-Girl, *das*
pioneer [paɪə'nɪə(r)] **1.** *n.* Pionier, *der; (fig. also)* Wegbereiter, *der*/Wegbereiterin, *die.* **2.** *v.t.* Pionierarbeit leisten für ‹*Entwicklung, Technologie, Nutzung*›
pious ['paɪəs] *adj.* **a)** *(devout)* fromm; **b)** *(hypocritically virtuous)* heuchlerisch; scheinheilig
[1]**pip** [pɪp] *n. (seed)* Kern, *der*
[2]**pip** *n.* **a)** *(on cards, dominoes, etc.)* Auge, *das;* Punkt, *der;* **b)** *(Brit. Mil.)* Stern, *der;* **c)** *(on radar screen)* Echosignal, *das*
[3]**pip** *n. (Brit.: sound)* [kurzer] Piepston; *(time signal also)* Zeitzeichen, *das;* **when the ~s go** *(during telephone call)* wenn die Piepstöne anzeigen, daß eine neue Münze eingeworfen werden muß
[4]**pip** *n. (coll.)* **give sb. the ~:** jmdm. auf den Wecker gehen *(ugs.)*
[5]**pip** *v.t.,* **-pp-** *(Brit.) (defeat)* besiegen; schlagen; **~ sb. at the post** *(coll.)* jmdn. im Ziel abfangen; *(fig.)* jmdn. im letzten Moment ausbooten *(ugs.)*
pipe [paɪp] **1.** *n.* **a)** *(tube)* Rohr, *das;* **b)** *(Mus.)* Pfeife, *die; (flute)* Flöte, *die; (in organ)* [Orgel]pfeife, *die;* **c)** *in pl. (bagpipes)* Dudelsack, *der;* **d) [tobacco-]~:** [Tabaks]pfeife, *die;* **put that in your ~ and smoke it** schreib dir das hinter die Ohren *(ugs.).* **2.** *v.t.* **a)** *(convey by ~)* [durch ein Rohr/durch Rohre] leiten; **be ~d** ‹*Öl, Wasser:*› [durch eine Rohrleitung] fließen; ‹*Gas:*› [durch eine Rohrleitung] strömen; **b) ~d music** Hintergrundmusik, *die;* **c)** *(utter shrilly)* ‹*Vogel:*› piepsen, pfeifen; ‹*Kind:*› piepsen; **d)** *(Cookery)* spritzen. **3.** *v.i.* **a)** *(whistle)* pfeifen; **b)** ‹*Stimme:*› hell klingen, schrillen; ‹*Person:*› piepsen, mit heller *od.* schriller Stimme sprechen; ‹*Vogel:*› pfeifen, piepsen
~ '**down** *v.i. (coll.: be less noisy)* ruhig sein
~ '**up** *v.i. (begin to speak)* sich vernehmen lassen
pipe: ~-cleaner *n.* Pfeifenreiniger, *der;* **~-dream** *n.* Wunschtraum, *der;* **~line** *n.* Pipeline, *die;* **in the ~line** *(fig.)* in Vorbereitung
piper ['paɪpə(r)] *n.* **a)** Pfeifer, *der*/Pfeiferin, *die;* **he who pays the ~ calls the tune** *(prov.)* wes Brot ich ess', des Lied ich sing' *(Spr.)*; **b)** *(bagpiper)* Dudelsackspieler, *der*/-spielerin, *die*
pipette [pɪ'pet] *n. (Chem.)* Pipette, *die*
piping ['paɪpɪŋ] *n.* **a)** *(system of pipes)* Rohrleitungssystem, *das;* **b)** *(quantity of pipes)* Rohrmaterial, *das;* **c)** *(Sewing)* Paspel, *die;* **d)** *(Cookery)* Spritzgußverzierung, *die*
piping 'hot *adj.* kochendheiß
piquancy ['pi:kənsɪ] *n.* **a)** *(sharpness)* Würze, *die;* **b)** *(fig.)* Pikanterie, *die (geh.)*
piquant ['pi:kənt, 'pi:kɑ:nt] *adj. (lit. or fig.)* pikant
pique [pi:k] **1.** *v.t.* **a)** *(irritate)* verärgern; **be ~d at sb./sth.** über jmdn./etw. verärgert sein; **b)** *(wound the pride of)* kränken; **be ~d at sth.** wegen etw. gekränkt sein. **2.** *n.* **in a [fit of] ~:** verstimmt; eingeschnappt *(ugs.)*
piracy ['paɪrəsɪ] *n.* Seeräuberei, *die;* Piraterie, *die; (fig.)* Piraterie, *die*
piranha [pɪ'rɑ:nə, pɪ'rɑ:njə] *n. (Zool.)* Piranha, *der*
pirate ['paɪrət] **1.** *n.* **a)** Pirat, *der;* Seeräuber, *der; (fig.)* Schwindler, *der;* **b)** *(Radio)* [Rundfunk]pirat, *der; attrib.* **~ radio station** Piratensender, *der.* **2.** *v.t.* ausplündern ‹*Schiff*›; rauben ‹*Waren usw.*›; *(fig.)* illegal nachdrucken ‹*Buch*›; illegal pressen ‹*Schallplatte*›; illegal vervielfältigen ‹*Videoband*›; **~d edition** Raubdruck, *der*

pirouette [pɪrʊ'et] **1.** *n.* Pirouette, *die.* **2.** *v.i.* pirouettieren

Pisces ['paɪsi:z] *n., pl. same (Astrol., Astron.)* Fische *Pl.*

piss [pɪs] *(coarse)* **1.** *n.* **a)** *(urine)* Pisse, *die (derb);* **b) have a/go for a ~:** pissen/pissen gehen *(derb).* **2.** *v.i.* pissen *(derb)*

~ **'off** *(Brit. sl.)* **1.** *v.i.* sich verpissen *(salopp).* **2.** *v.t.* ankotzen *(derb)*

pissed [pɪst] *adj.* *(sl.)* voll *(salopp);* besoffen *(derb)*

pissed 'off *adj.* *(sl.)* stocksauer *(salopp)* **(with** auf + *Akk.*); **get ~ [with sb./sth.]** langsam die Schnauze voll haben [von jmdm./etw.] *(salopp)*

pistachio [pɪ'stɑ:ʃɪəʊ] *n., pl.* **~s** Pistazie, *die*

pistil ['pɪstɪl] *n.* *(Bot.)* Stempel, *der*

pistol ['pɪstl] *n.* Pistole, *die;* **hold a ~ to sb.'s head** jmdm. die Pistole an die Schläfe *od.* den Kopf setzen; *(fig.)* jmdm. die Pistole auf die Brust setzen

'pistol-shot *n.* Pistolenschuß, *der*

piston ['pɪstn] *n.* Kolben, *der*

piston: ~ **engine** *n.* Kolbenmotor, *der;* **~-rod** *n.* Kolbenstange, *die;* Pleuelstange, *die*

pit [pɪt] **1.** *n.* **a)** *(hole, mine)* Grube, *die; (natural)* Vertiefung, *die; (as trap)* Fallgrube, *die;* **[work] down the ~:** unter Tage [arbeiten] *(Bergmannsspr.);* **b) ~ of the stomach** Magengrube, *die;* **c)** *(Brit. Theatre) (for audience)* Parkett, *das;* **d)** *(Motor-racing)* Box, *die.* **2.** *v.t.,* **-tt-: a)** *(set to fight)* kämpfen lassen; **b)** *(fig.: match)* **~ sth. against sth.** etw. gegen etw. einsetzen; **~ one's wits/skill** *etc.* **against sth.** seinen Verstand/sein Können *usw.* an etw. *(Dat.)* messen; **c) be ~ted** *(have ~s)* voller Vertiefungen sein

[1]**pitch** [pɪtʃ] **1.** *n.* **a)** *(Brit.: usual place)* [Stand]platz, *der; (stand)* Stand, *der; (Sport: playing-area)* Feld, *das;* Platz, *der;* **b)** *(Mus.)* Tonhöhe, *die; (of voice)* Stimmlage, *die; (of instrument)* Tonlage, *die;* **c)** *(slope)* Neigung, *die;* **d)** *(fig.: degree, intensity)* **reach such a ~ that ...:** sich so zuspitzen, daß... **2.** *v.t.* **a)** *(erect)* aufschlagen ⟨*Zelt*⟩; **~ camp** ein/das Lager aufschlagen; **b)** *(throw)* werfen; **the horse ~ed its rider over its head** das Pferd warf den Reiter vornüber; **~ sb. out of sth.** jmdn. aus etw. hinauswerfen; **c)** *(Mus.)* anstimmen ⟨*Melodie*⟩; stimmen ⟨*Instrument*⟩; **d)** *(fig.)* **~ a programme at a particular level** ein Programm auf ein bestimmtes Niveau abstimmen; **our expectations were ~ed too high** unsere Erwartungen waren zu hoch gesteckt; **e) ~ed battle** offene [Feld]schlacht. **3.** *v.i.* *(fall)* [kopfüber] stürzen; ⟨*Schiff, Fahrzeug, Flugzeug:*⟩ mit einem Ruck nach vorn kippen; *(repeatedly)* ⟨*Schiff:*⟩ stampfen; **~ forward** vornüberstürzen

~ **'in** *v.i.* *(coll.)* loslegen *(ugs.); (begin)* sich daranmachen *(ugs.);* **~ in [and** *or* **to help]** zupacken *(ugs.)* [und helfen]; mit anpacken

~ **into** *v.t.* *(coll.)* herfallen über (+ *Akk.*); sich hermachen über (+ *Akk.*) *(ugs.)* ⟨*Essen*⟩

[2]**pitch** *n.* *(substance)* Pech, *das;* **as black as ~:** pechschwarz

pitch: **~-'black** *adj.* pechschwarz; stockdunkel *(ugs.);* **~-'dark** *adj.* stockdunkel *(ugs.);* pechfinster; **~-'darkness** *n.* tiefste Finsternis

pitched 'roof *n.* schräges Dach

[1]**pitcher** ['pɪtʃə(r)] *n.* [Henkel]krug, *der*

[2]**pitcher** *n.* *(Baseball)* Werfer, *der;* Pitcher, *der*

'pitchfork 1. *n.* *(for hay)* Heugabel, *die; (for manure)* Mistgabel, *die.* **2.** *v.t.* gabeln; **~ sb. into sth.** *(fig.)* jmdn. in etw. ⟨*Akk.*⟩ katapultieren

piteous ['pɪtɪəs] *adj.* erbärmlich; *(causing pity)* mitleiderregend; kläglich ⟨*Schrei*⟩

'pitfall *n.* Fallstrick, *der; (risk)* Gefahr, *die*

pith [pɪθ] *n.* **a)** *(in plant)* Mark, *das; (of orange etc.)* weiße Haut; **b)** *(fig.: essential part)* Kern, *der*

'pit-head *n.* ≈ Zechengelände, *das*

pithy ['pɪθɪ] *adj.* **a)** markhaltig; reich an Mark *nicht attr.;* ⟨*Orange usw.*⟩ mit dicker weißer Haut; **b)** *(fig.: full of meaning)* prägnant

pitiable ['pɪtɪəbl] *see* **pitiful**

pitiful ['pɪtɪfl] *adj.* **a)** mitleiderregend; **b)** *(contemptible)* jämmerlich *(abwertend)*

pitifully ['pɪtɪfəlɪ] *adv.* erbärmlich; jämmerlich

pitiless ['pɪtɪlɪs] *adj.,* **pitilessly** ['pɪtɪlɪslɪ] *adv.* unbarmherzig *(auch fig.);* erbarmungslos

pittance ['pɪtəns] *n.* Hungerlohn, *der (abwertend); (small allowance)* [magere] Beihilfe

pity ['pɪtɪ] **1.** *n.* **a)** *(sorrow)* Mitleid, *das;* Mitgefühl, *das;* **feel ~ for sb.** Mitgefühl für jmdn. *od.* mit jmdm. empfinden; **have/take ~ on sb.** Erbarmen mit jmdm. haben; **for ~'s sake!** um Gottes *od.* Himmels willen!; **b)** *(cause for regret)* **[what a] ~!** [wie] schade!; **it's a ~ about sb./sth.** es ist ein Jammer mit jmdm./etw. *(ugs.);* **the ~ of it is [that] ...:** das Traurige daran ist, daß ...; **more's the ~** *(coll.)* leider! **2.** *v.t.* bedauern; bemitleiden; **I ~ you** *(also contemptuously)* du tust mir leid

pitying ['pɪtɪɪŋ] *adj.,* **pityingly** ['pɪtɪɪŋlɪ] *adv.* mitleidig

pivot ['pɪvət] **1.** *n.* **a)** [Dreh]zapfen, *der;* **b)** *(fig.)* [Dreh- und] Angelpunkt, *der; (crucial point)* springender Punkt. **2.** *v.i.* sich drehen; **~ on sth.** *(fig.)* von etw. abhängen

pivotal ['pɪvətl] *adj.* *(fig.: crucial)* zentral; **~ figure** Schlüsselfigur, *die*

pixie ['pɪksɪ] *n.* Kobold, *der*

pizza ['pi:tsə] *n.* Pizza, *die*

pl. *abbr.* **plural** Pl.

placard ['plækɑ:d] *n.* Plakat, *das*

placate [plə'keɪt] *v.t.* beschwichtigen, besänftigen ⟨*Person*⟩

place [pleɪs] **1.** *n.* **a)** Ort, *der; (spot)* Stelle, *die;* Platz, *der;* **I left it in a safe ~:** ich habe es an einem sicheren Ort gelassen; **it was still in the same ~:** es war noch an derselben Stelle *od.* am selben Platz; **a ~ in the queue** ein Platz in der Schlange; **all over the ~:** überall; *(coll.: in a mess)* ganz durcheinander *(ugs.);* **from ~ to ~:** von Ort zu Ort; **in ~s** hier und da; *(in parts)* stellenweise; **find a ~ in sth.** *(be included)* in etw. *(Akk.)* eingehen; *see also* **take 1d; b)** *(fig.: rank, position)* Stellung, *die;* **put sb. in his ~:** jmdn. in seine Schranken weisen; **know one's ~:** wissen, was sich für einen gehört; **it's not my ~ to do that** es kommt mir nicht zu, das zu tun; **c)** *(building or area for specific purpose)* **a [good] ~ to park/to stop** ein [guter] Platz zum Parken/eine [gute] Stelle zum Halten; **do you know a good/cheap ~ to eat?** weißt du, wo man gut/billig essen kann?; **~ of residence** Wohnort, *der;* **~ of work** Arbeitsplatz, *der;* Arbeitsstätte, *die;* **~ of worship** Andachtsort, *der;* **d)** *(country, town)* Ort, *der;* **Paris/Italy is a great ~:** Paris ist eine tolle Stadt/Italien ist ein tolles Land *(ugs.);* **~ of birth** Geburtsort, *der;* **'go ~s** *(coll.)* herumkommen *(ugs.); (fig.)* es [im Leben] zu was bringen *(ugs.);* **e)** *(coll.: premises)* Bude, *die (ugs.); (hotel, restaurant, etc.)* Laden, *der (ugs.);* **she is at his/John's ~:** sie ist bei ihm/John; **[shall we go to] your ~ or**

mine? [gehen wir] zu dir oder zu mir?; **f)** *(seat etc.)* [Sitz]platz, *der;* **change ~s [with sb.]** [mit jmdm.] die Plätze tauschen; *(fig.)* [mit jmdm.] tauschen; **lay a/another ~**: ein/noch ein Gedeck auflegen; **is this anyone's ~?** ist dieser Platz noch frei?; **g)** *(in book etc.)* Stelle, *die;* **lose one's ~**: die Seite verschlagen *od.* verblättern; *(on page)* nicht mehr wissen, an welcher Stelle man ist; **h)** *(step, stage)* **in the first ~**: zuerst; **why didn't you say so in the first ~?** warum hast du das nicht gleich gesagt?; **in the first/second/third** *etc.* **~**: erstens/zweitens/drittens *usw.;* **i)** *(proper ~)* Platz, *der;* **everything fell into ~** *(fig.)* alles wurde klar; **a woman's ~ is in the home** eine Frau gehört ins Haus; **the clamp is properly in ~**: die Klammer sitzt richtig; **into ~**: fest‹*nageln, -schrauben, -kleben*›; **out of ~**: nicht am richtigen Platz; *(several things)* in Unordnung; *(fig.)* fehl am Platz; **take the ~ of sb.** jmds. Platz einnehmen; **j)** *(position in competition)* Platz, *der;* **take first/second** *etc.* **~**: den ersten/zweiten *usw.* Platz belegen; **k)** *(job, position, etc.)* Stelle, *die; (as pupil; in team, crew)* Platz, *der;* **l)** *(personal situation)* **what would you do in my ~?** was würden Sie an meiner Stelle tun?; **put yourself in my ~**: versetzen Sie sich in meine Lage. **2.** *v. t.* **a)** *(put) (vertically)* stellen; *(horizontally)* legen; **~ in position** richtig hinstellen/hinlegen; **~ an announcement/advertisement in a paper** eine Anzeige/ein Inserat in eine Zeitung setzen; **~ a bet** auf ein Pferd wetten; **b)** *(fig.)* **~ one's trust in sb./sth.** sein Vertrauen auf *od.* in jmdn./etw. setzen; **he ~s happiness above all other things** Glück steht für ihn an erster Stelle; **c)** *in p.p. (situated)* gelegen; **a badly ~d window** ein Fenster an einer ungünstigen Stelle; **we are well ~d for buses/shops** *etc.* wir haben es nicht weit zur Bushaltestelle/zum Einkaufen *usw.;* **how are you ~d for time/money?** *(coll.)* wie steht's mit deiner Zeit/deinem Geld?; **d)** *(find situation or home for)* unterbringen (**with** bei); **~ sb. under sb.'s care** jmdn. in jmds. Obhut geben; **e)** *(class, identify)* einordnen; einstufen; **I've seen him before but I can't ~ him** ich habe ihn schon einmal gesehen, aber ich weiß nicht, wo ich ihn unterbringen soll; **be ~d second in the race** im Rennen den zweiten Platz belegen

place: ~ card *n.* Tischkarte, *die;* **~-mat** *n.* Set, *der od. das;* **~-name** *n.* Ortsname, *der*

placenta [plə'sentə] *n., pl.* **~e** [plə'senti:] *or* **~s** *(Anat., Zool.)* Plazenta, *die (fachspr.);* Mutterkuchen, *der*

placid ['plæsɪd] *adj.* ruhig, gelassen ‹*Person*›; ruhig ‹*Wasser, Wesensart*›; *(peaceable)* friedlich, friedfertig ‹*Person*›

plagiarise *see* **plagiarize**

plagiarism ['pleɪdʒərɪzm] *n.* Plagiat, *das*

plagiarize ['pleɪdʒəraɪz] *v. t.* plagiieren

plague [pleɪg] **1.** *n.* **a)** *(esp. Hist.: epidemic)* Seuche, *die;* **the ~** *(bubonic)* die Pest; **avoid/hate sb./sth. like the ~**: jmdn./etw. wie die Pest meiden/hassen; **b)** *(infestation)* **~ of rats** Rattenplage, *die.* **2.** *v. t.* **a)** *(afflict)* plagen; quälen; **~d with** *or* **by sth.** von etw. geplagt; **b)** *(bother)* **~ sb. [with sth.]** jmdm. [mit etw.] auf die Nerven gehen *(ugs.)*

plaice [pleɪs] *n., pl. same (Zool.)* Scholle, *die*

plaid [plæd] **1.** *n.* Plaid, *das od. der.* **2.** *adj.* [bunt]kariert

plain [pleɪn] **1.** *adj.* **a)** *(clear)* klar; *(obvious)* offensichtlich; **make sth. ~ [to sb.]** [jmdm.] etw. klarmachen; **make it ~ that ...**: klarstellen, daß ...; **the reason is ~ [to see]** der Grund liegt auf der Hand; *see also* **English 2a; pikestaff; b)** *(frank, straightforward)* ehrlich; offen; schlicht ‹*Wahrheit*›; **be ~ with sb.** mit jmdm. *od.* jmdm. gegenüber offen sein; **be [all] ~ sailing** *(fig.)* [ganz] einfach sein; **c)** *(unsophisticated)* einfach; schlicht ‹*Kleidung, Frisur*›; klar ‹*Wasser*›; einfach, bescheiden ‹*Lebensstil*›; *(not lined)* unliniert ‹*Papier*›; *(not patterned)* ‹*Stoff*› ohne Muster; **d)** *(unattractive)* wenig attraktiv ‹*Mädchen*›; **e)** *(sheer)* rein; **that's ~ bad manners** das ist einfach schlechtes Benehmen. **2.** *adv.* **a)** *(clearly)* deutlich; **b)** *(simply)* einfach. **3.** *n.* **a)** Ebene, *die;* **b)** *(Knitting)* rechte Masche

plain: ~ 'chocolate *n.* halbbittere Schokolade; **~ 'clothes** *n. pl.* **in ~ clothes** in Zivil

plainly ['pleɪnlɪ] *adv.* **a)** *(clearly)* deutlich; verständlich ‹*erklären*›; **b)** *(obviously)* offensichtlich; *(undoubtedly)* eindeutig; **c)** *(frankly)* offen; **d)** *(simply, unpretentiously)* einfach; schlicht

plainness ['pleɪnnɪs] *n., no pl.* **a)** *(clearness)* Klarheit, *die;* **b)** *(frankness)* Offenheit, *die;* **c)** *(simplicity)* Schlichtheit, *die;* **d)** *(ugliness)* Unattraktivität, *die;* Unansehnlichkeit, *die*

plain-'spoken *adj.* freimütig

plaintiff ['pleɪntɪf] *n. (Law)* Kläger, *der*/Klägerin, *die*

plaintive ['pleɪntɪv] *adj.* klagend; traurig, leidend ‹*Blick*›

plait [plæt] **1.** *n. (of hair)* Zopf, *der; (of straw, ribbon, etc.)* geflochtenes Band. **2.** *v. t.* flechten

plan [plæn] **1.** *n.* Plan, *der; (for story etc.)* Konzept, *das;* Entwurf, *der; (intention)* Absicht, *die;* **~ of action** Aktionsprogramm, *das;* **have great ~s for sb.** große Pläne mit jmdm. haben; **what are your ~s for tomorrow?** was hast du morgen vor?; **[go] according to ~**: nach Plan [gehen]; planmäßig [verlaufen *od.* laufen]. **2.** *v. t.,* **-nn-** planen; *(design)* entwerfen ‹*Gebäude, Maschine*›; **~ to do sth.** planen *od.* vorhaben, etw. zu tun. **3.** *v. i.,* **-nn-** planen; **~ for sth.** Pläne für etw. machen; **~ on doing sth.** *(coll.)* vorhaben, etw. zu tun

[1]plane [pleɪn] *n.* **~[-tree]** Platane, *die*

[2]plane **1.** *n. (tool)* Hobel, *der.* **2.** *v. t.* hobeln

[3]plane *n.* **a)** *(Geom.)* Ebene, *die; (flat surface)* Fläche, *die;* **b)** *(fig.)* Ebene, *die; (moral, intellectual)* Niveau, *das;* **c)** *(aircraft)* Flugzeug, *das;* Maschine, die *(ugs.)*

planet ['plænɪt] *n.* Planet, *der*

planetarium [plænɪ'teərɪəm] *n., pl.* **~s** *or* **planetaria** [plænɪ'teərɪə] Planetarium, *das*

plank [plæŋk] *n. (piece of timber)* Brett, *das; (thicker)* Bohle, *die; (on ship)* Planke, *die;* **be as thick as two [short] ~s** *(coll.)* dumm wie Bohnenstroh sein *(ugs.)*

plankton ['plæŋktn] *n. (Biol.)* Plankton, *das*

planner ['plænə(r)] *n.* Planer, *der*/Planerin, *die*

planning ['plænɪŋ] *n.* Planen, *das;* Planung, *die;* **at the ~ stage** im Planungsstadium; **~ permission** Baugenehmigung, *die*

plant [plɑ:nt] **1.** *n.* **a)** *(Bot.)* Pflanze, *die;* **b)** *(machinery) no indef. art.* Maschinen; *(single complex)* Anlage, *die;* **c)** *(factory)* Fabrik, *die;* Werk, *das;* **d)** *(sl.: undercover agent)* Spitzel, *der;* **e)** *(sl.: thing concealed)* Untergeschobene, *das.* **2.** *v. t.* **a)** pflanzen; aussäen ‹*Samen*›; anlegen ‹*Garten usw.*›; anpflanzen ‹*Beet*›; bepflanzen

⟨*Land*⟩; ~ **a field with barley** auf einem Feld Gerste anpflanzen; **b)** *(fix)* setzen; ~ **oneself** sich hinstellen *od. (ugs.)* aufpflanzen; **c)** *(in mind)* ~ **an idea** *etc.* **in sb.'s mind/in sb.** jmdm. eine Idee *usw.* einimpfen *(ugs.) od. (geh.)* einpflanzen; **d)** *(sl.: conceal)* anbringen ⟨*Wanze*⟩; legen ⟨*Bombe*⟩; ~ **sth. on sb.** jmdm. etw. unterschieben; **e)** *(station as spy etc.)* einschmuggeln

~ '**out** *v.t.* auspflanzen ⟨*Setzlinge*⟩

plantation [plæn'teɪʃn, plɑːn'teɪʃn] *n.* **a)** *(estate)* Pflanzung, *die;* Plantage, *die;* **b)** *(group of plants)* Anpflanzung, *die*

planter ['plɑːntə(r)] *n.* **a)** Pflanzer, *der*/Pflanzerin, *die;* **b)** *(container)* Pflanzgefäß, *das*

plaque [plɑːk, plæk] *n.* **a)** *(ornamental tablet)* [Schmuck]platte, *die; (commemorating sb.)* [Gedenk]tafel, *die;* Plakette, *die (Kunstwiss.);* **b)** *(Dent.)* Plaque, *die (fachspr.);* [weißer] Zahnbelag

plasma ['plæzmə] *n.* Plasma, *das*

plaster ['plɑːstə(r)] **1.** *n.* **a)** *(for walls etc.)* [Ver]putz, *der;* **b)** ~ **[of Paris]** Gips, *der;* **have one's leg in** ~: ein Gipsbein *od.* sein Bein in Gips haben; **put sb.'s leg in** ~: jmds. Bein in Gips legen; **c)** *see* **sticking-plaster. 2.** *v.t.* **a)** verputzen ⟨*Wand*⟩; vergipsen, zugipsen ⟨*Loch, Riß*⟩; **b)** *(daub)* ~ **sth. on sth.** etw. dick auf etw. *(Akk.)* auftragen; ~**ed with mud** mit Schlamm bedeckt; **c)** *(stick on)* kleistern *(ugs.)* ⟨*Plakate, Briefmarken*⟩ (**on** auf + *Akk.*)

plaster: ~**board** *n.* Gipsplatte, *die;* ~ **cast** *n.* **a)** *(model in plaster)* Gipsabguß *od.* -abdruck, *der;* **b)** *(Med.)* Gipsverband, *der*

plastered ['plɑːstəd] *pred. adj. (sl.: drunk)* voll *(salopp);* **get** ~: sich vollaufen lassen *(salopp)*

plasterer ['plɑːstərə(r)] *n.* Gipser, *der*

plastic ['plæstɪk] **1.** *n.* Plastik, *das;* Kunststoff, *der; in pl., attrib.* Plastik-; Kunststoff-. **2.** *adj.* **a)** *(made of* ~*)* aus Plastik *od.* Kunststoff *nachgestellt;* ~ **bag** Plastiktüte, *die;* **b)** *(produced by moulding)* plastisch; **c)** *(malleable, lit. or fig.)* formbar; bildbar; **d) the** ~ **arts** die Plastik; *(including painting etc.)* die bildende Kunst; ~ **surgeon** Facharzt für plastische Chirurgie; ~ **surgery** plastische Chirurgie

Plasticine, (P) ['plæstɪsiːn] *n.* Plastilin, *das*

plate [pleɪt] **1.** *n.* **a)** *(for food)* Teller, *der; (large* ~ *for serving food)* Platte, *die;* **a** ~ **of soup/sandwiches** ein Teller Suppe/belegte Brote *od.* mit belegten Broten; **have sth. handed to one on a** ~ *(fig. coll.)* etw. auf silbernem Tablett serviert bekommen *(fig.);* **have a lot on one's** ~ *(fig. coll.)* viel am Hals *od.* um die Ohren haben *(ugs.);* **b)** *(metal* ~ *with name etc.)* Schild, *das;* **[number-]**~: Nummernschild, *das;* **c)** *no pl., no indef. art. (Brit.: tableware)* [Tafel]silber, *das;* **d)** *(for engraving, printing)* Platte, *die; (impression)* Stich, *der; (illustration)* [Bild]tafel, *die;* **e)** *(Dent.)* Gaumenplatte, *die; (coll.: denture)* [Zahn]prothese, *die;* Gebiß, *das.* **2.** *v.t.* **a)** *(coat)* plattieren; ~ **sth. [with gold/silver/chromium]** etw. vergolden / versilbern / verchromen; **b)** panzern ⟨*Schiff*⟩

plateau ['plætəʊ] *n., pl.* ~**x** ['plætəʊz] *or* ~**s** Hochebene, *die;* Plateau, *das*

plate: ~ '**glass** *n.* Flachglas, *das;* ~**-rack** *n. (Brit.)* Abtropfständer, *der;* Geschirrablage, *die*

platform ['plætfɔːm] *n.* **a)** *(Brit. Railw.)* Bahnsteig, *der;* **the train leaves from/will arrive at** ~ **4** der Zug fährt von Gleis 4 ab/in Gleis 4 ein; **b)** *(stage)* Podium, *das;* **c)** *(Polit.)* Wahlplattform, *die*

'**platform ticket** *n. (Brit.)* Bahnsteigkarte, *die*

plating ['pleɪtɪŋ] *n. (process)* Plattierung, *die; (coat)* Plattierung, *die;* Auflage, *die*

platinum ['plætɪnəm] *n.* Platin, *das*

platinum '**blonde 1.** *n.* Platinblonde, *die.* **2.** *adj.* platinblond

platitude ['plætɪtjuːd] *n.* **a)** *(trite remark)* Platitüde, *die (geh.);* Gemeinplatz, *der;* **b)** *no pl. (triteness)* Banalität, *die*

platonic [plə'tɒnɪk] *adj.* platonisch ⟨*Liebe, Freundschaft*⟩

platoon [plə'tuːn] *n. (Mil.)* Zug, *der*

plausible ['plɔːzɪbl] *adj.* plausibel; einleuchtend; glaubwürdig ⟨*Person*⟩

play [pleɪ] **1.** *n.* **a)** *(Theatre)* [Theater]stück, *das;* **put on a** ~: ein Stück aufführen; **b)** *(recreation)* Spielen, *das;* Spiel, *das;* **at** ~: beim Spielen; **say/do sth. in** ~: etw. aus *od.* im *od.* zum Spaß sagen/tun; ~ **[up]on words** Wortspiel, *das;* **c)** *(Sport)* Spiel, *das; (Amer.: manœuvre)* Spielzug, *der;* **a good piece of** ~: ein guter Spielzug; **be in/out of** ~ ⟨*Ball:*⟩ im Spiel/aus [dem Spiel] sein; **make a** ~ **for sb./sth.** *(fig. sl.)* hinter jmdm./etw. her sein *(ugs.);* es auf jmdn./etw. abgesehen haben; **d) come into** ~, **be brought** *or* **called into** ~: ins Spiel kommen; **make [great]** ~ **with sth.** viel Wesen um etw. machen; **e)** *(freedom of movement)* Spiel, *das (Technik); (fig.)* Spielraum, *der;* **give full** ~ **to one's emotions/imagination** *etc. (fig.)* seinen Gefühlen/seiner Phantasie *usw.* freien Lauf lassen; **f)** *(rapid movement)* **the** ~ **of light on water** das Spiel des Lichts auf Wasser. **2.** *v.i.* **a)** spielen; ~ **for money** um Geld spielen; **have no one to** ~ **with** niemanden zum Spielen haben; ~ **[up]on words** Wortspiele/ein Wortspiel machen; **not have much time to** ~ **with** *(coll.)* zeitlich nicht viel Spielraum haben; ~ **into sb.'s hands** *(fig.)* jmdm. in die Hand *od.* Hände arbeiten; ~ **safe** sichergehen; auf Nummer Sicher gehen *(ugs.);* ~ **for time** Zeit gewinnen wollen; **b)** *(Mus.)* spielen (**on** auf + *Dat.*). **3.** *v.t.* **a)** *(Mus.: perform on)* spielen; ~ **the violin** *etc.* Geige *usw.* spielen; ~ **sth. on the piano** *etc.* etw. auf dem Klavier *usw.* spielen; ~ **sth. by ear** etw. nach dem Gehör spielen; ~ **it by ear** *(fig.)* es dem Augenblick/der Situation überlassen; **b)** spielen ⟨*Grammophon, Tonbandgerät*⟩; abspielen ⟨*Schallplatte, Tonband*⟩; spielen lassen ⟨*Radio*⟩; **c)** *(Theatre; also fig.)* spielen; ~ **a town** in einer Stadt spielen; ~ **the fool/innocent** den Clown/Unschuldigen spielen; **d)** *(execute, practise)* ~ **a trick/joke on sb.** jmdn. hereinlegen *(ugs.)*/jmdm. einen Streich spielen; **e)** *(Sport, Cards)* spielen ⟨*Fußball, Karten, Schach usw.*⟩; spielen *od.* antreten gegen ⟨*Mannschaft, Gegner*⟩; ~ **a match** einen Wettkampf bestreiten; *(in team games)* ein Spiel machen; **he** ~**ed me at chess/squash** er war im Schach/Squash mein Gegner; ~ **it safe** auf Nummer Sicher gehen; **f)** *(Sport)* ausführen ⟨*Schlag*⟩; *(Cricket etc.)* schlagen ⟨*Ball*⟩; **g)** *(Cards)* spielen; ~ **one's cards right** *(fig.)* es richtig anfassen *(fig.);* **h)** *(coll.: gamble on)* ~ **the market** spekulieren (**in** mit *od. Wirtsch.* in + *Dat.*)

~ **a'bout** *see* ~ **around**

~ **a'long** *v.i.* mitspielen

~ **a'round** *v.i. (coll.)* spielen; ~ **around with sb./sb.'s affections/sth.** mit jmdm./jmds. Zuneigung spielen/mit etw. herumspielen *(ugs.)*

~ **at** *v.t.* spielen; **what do you think you're ~ing at?** was soll denn das?
~ **'back** *v.t.* abspielen ‹*Tonband, Aufnahme*›
~ **'down** *v.t.* herunterspielen
~ **'off 1.** *v.i.* zum Entscheidungsspiel antreten. **2.** *v.t.* ausspielen; **~ one person/firm** *etc.* **off against another** eine Person/Firma *usw.* gegen eine andere ausspielen
~ **on** *see* ~ **upon**
~ **'up 1.** *v.i.* *(coll.)* ‹*Kinder:*› nichts als Ärger machen; ‹*Auto:*› verrückt spielen; ‹*Rücken, Bein usw.:*› Schwierigkeiten machen. **2.** *v.t.* *(coll.: annoy, torment)* ärgern; ‹*Krankheit:*› zu schaffen machen (+ *Dat.*)
~ **upon** *v.t.* sich *(Dat.)* zunutze machen ‹*Gefühle, Ängste usw.*›; ~ **upon sb.'s sympathies** auf jmds. Mitgefühl *(Akk.)* spekulieren *(ugs.)*

play: ~-acting *n.* *(fig.)* Theater, *das (ugs.);* **~-back** *n.* Wiedergabe, *die;* **~boy** *n.* Playboy, *der*

played 'out *adj.* verbraucht; erschöpft ‹*Person, Tier*›; **this idea is ~:** diese Idee hat sich überlebt

player ['pleɪə(r)] *n.* **a)** Spieler, *der*/Spielerin, *die;* **b)** *(Mus.)* Musiker, *der*/Musikerin, *die;* **c)** *(actor)* Schauspieler, *der*/Schauspielerin, *die*

playful ['pleɪfl] *adj.* **a)** *(fond of playing)* spielerisch; *(frolicsome)* verspielt; **b)** *(teasing)* neckisch; *(joking)* scherzhaft

play: ~goer *n.* Theaterbesucher, *der*/-besucherin, *die;* **~ground** *n.* Spielplatz, *der; (Sch.)* Schulhof, *der;* ~ **group** *n.* Spielgruppe, *die;* **~house** *n.* **a)** *(theatre)* Schauspielhaus, *das;* **b)** *(toy house)* Spielhaus, *das*

playing: ~-card *n.* Spielkarte, *die;* **~-field** *n.* Sportplatz, *der*

play: ~mate *n.* Spielkamerad, *der*/Spielkameradin, *die;* **~-off** *n.* Entscheidungsspiel, *das;* **~-pen** *n.* Laufgitter, *das;* Laufstall, *der;* ~ **school** *n.* Kindergarten, *der;* **~thing** *n.* *(lit. or fig.)* Spielzeug, *das;* **~time** *n.* Zeit zum Spielen; **~wright** ['pleɪraɪt] *n.* Dramatiker, *der*/Dramatikerin, *die;* Stückeschreiber, *der*/-schreiberin, *die*

PLC, plc *abbr.* *(Brit.)* **public limited company** ≈ GmbH

plea [pli:] *n.* **a)** *(appeal, entreaty)* Appell, *der* (**for** zu); **make a ~ for sth.** zu etw. aufrufen; **b)** *(Law)* Verteidigungsrede, *die*

plead [pli:d] **1.** *v.i.*, **~ed** *or (esp. Amer., Scot., dial.)* **pled** [pled] **a)** *(make appeal)* inständig bitten (**for** um); *(imploringly)* flehen (**for** um); ~ **with sb. for sth./to do sth.** jmdn. inständig um etw. bitten/jmdn. inständig [darum] bitten, etw. zu tun; *(imploringly)* jmdn. um etw. anflehen/jmdn. anflehen, etw. zu tun; **b)** *(Law: put forward plea; also fig.)* plädieren; **c)** *(Law)* **how do you ~?** bekennen Sie sich schuldig? **2.** *v.t.*, **~ed** *or (esp. Amer., Scot., dial.)* **pled a)** *(beg)* inständig bitten; *(imploringly)* flehen; **b)** *(Law: offer in mitigation)* sich berufen auf (+ *Akk.*); geltend machen; *(as excuse)* sich entschuldigen mit; ~ **guilty/not guilty** *(lit. or fig.)* sich schuldig/nicht schuldig bekennen; **c)** *(present in court)* ~ **sb.'s case** *or* ~ **the case for sb.** jmds. Sache vor Gericht vertreten

pleading ['pli:dɪŋ] *adj.* flehend

pleasant ['plezənt] *adj.*, **~er** ['plezəntə(r)], **~est** ['plezəntɪst] angenehm; schön ‹*Tag, Zeit*›; nett ‹*Gesicht, Lächeln*›

pleasantry ['plezəntrɪ] *n.* *(agreeable remark)* Nettigkeit, *die;* *(humorous remark)* Scherz, *der*

please [pli:z] **1.** *v.t.* **a)** *(give pleasure to)* gefallen (+ *Dat.*); Freude machen (+ *Dat.*); **there's no pleasing her** man kann ihr nichts *od.* es ihr nicht recht machen; **she's easy to ~** *or* **easily ~d/hard to ~:** sie ist leicht/nicht leicht zufriedenzustellen; ~ **oneself** tun, was man will; ~ **yourself** ganz wie du willst; **b)** *([may it] be the will of)* gefallen; ~ **God** das gebe Gott; so Gott will. **2.** *v.i.* **a)** *(think fit)* **they come and go as they ~:** sie kommen und gehen, wie es ihnen gefällt; **do as one ~s** tun, was man will; **b)** *(give pleasure)* gefallen; **anxious** *or* **eager to ~:** bemüht, gefällig zu sein; **c) if you ~:** bitte schön; *(iron.: believe it or not)* stell dir vor. **3.** *int.* bitte; **may I have the bill, ~?** kann ich bitte zahlen?; ~ **do!** aber bitte *od.* gern!; ~ **don't** bitte nicht

pleased [pli:zd] *adj.* *(satisfied)* zufrieden (**by** mit); *(glad, happy)* erfreut (**by** über + *Akk.*); **be ~ at** *or* **about sth.** sich über etw. *(Akk.)* freuen; **be ~ with sth./sb.** mit etw./jmdm. zufrieden sein; **be ~ to do sth.** sich freuen, etw. zu tun; *see also* ¹**meet 1 c**

pleasing ['pli:zɪŋ] *adj.* gefällig; ansprechend; nett ‹*Ausblick*›

pleasurable ['pleʒərəbl] *adj.*, **pleasurably** ['pleʒərəblɪ] *adv.* angenehm

pleasure ['pleʒə(r)] *n.* **a)** *(feeling of joy)* Freude, *die;* *(sensuous enjoyment)* Vergnügen, *das;* **sth. gives sb. ~:** etw. macht jmdm. Freude; **get a lot of ~ from** *or* **out of sb./sth.** viel Freude *od.* Spaß an jmdm./etw. haben; **b)** *(gratification)* **have the ~ of doing sth.** das Vergnügen haben, etw. zu tun; **may I have the ~ [of the next dance]?** darf ich [Sie um den nächsten Tanz] bitten?; **take [a] ~ in** Vergnügen finden *od.* Spaß haben an (+ *Dat.*); **it's a ~:** gern geschehen; es war mir ein Vergnügen; **it gives me great ~ to inform you that ...** *(formal)* ich freue mich, Ihnen mitteilen zu können, daß ...; **with ~:** mit Vergnügen; gern[e]

pleasure: ~-boat *n.* Vergnügungsboot, *das;* ~ **cruise** *n.* Vergnügungsfahrt, *die*

pleat [pli:t] **1.** *n.* Falte, *die.* **2.** *v.t.* in Falten legen; fälteln

pleated ['pli:tɪd] *adj.* gefältelt; Falten‹*rock*›

plectrum ['plektrəm] *n.*, *pl.* **plectra** ['plektrə] *or* **~s** *(Mus.)* Plektrum, *das*

pled *see* **plead**

pledge [pledʒ] **1.** *n.* **a)** *(promise, vow)* Versprechen, *das;* Gelöbnis, *das (geh.);* **take** *or* **sign the ~** *(coll.)* sich zur Abstinenz verpflichten; **b)** *(as security)* Pfand, *das;* Sicherheit, *die.* **2.** *v.t.* **a)** *(promise solemnly)* versprechen; geloben ‹*Treue*›; **b)** *(bind by promise)* verpflichten; **c)** *(deposit, pawn)* verpfänden (**to** *Dat.*)

plenary ['pli:nərɪ] *adj.* Plenar‹*sitzung*›; Voll‹*versammlung*›

plenipotentiary [plenɪpə'tenʃərɪ] **1.** *adj.* *(invested with full power)* [general]bevollmächtigt ‹*Gesandte*›. **2.** *n.* [General]bevollmächtigte, *der*

plentiful ['plentɪfl] *adj.* reichlich; häufig ‹*Element, Rohstoff*›; **be ~** *or* **in ~ supply** reichlich vorhanden sein

plenty ['plentɪ] **1.** *n.*, *no pl.* ~ **of** viel; eine Menge; *(coll.: enough)* genug; **have you all got ~ of meat?** habt ihr alle reichlich Fleisch?; **we gave him ~ of warning** wir haben ihn früh genug gewarnt. **2.** *adj.* *(coll.)* reichlich vorhanden. **3.** *adv.* *(coll.)* **it's ~ large enough** es ist groß genug; **there's ~ more where this/those** *etc.* **came from** es ist noch genug da *(ugs.)*

pleurisy ['plʊərɪsɪ] *n.* *(Med.)* Pleuritis, *die (fachspr.);* Brustfellentzündung, *die*

pliable ['plaɪəbl] *adj.* biegsam;

geschmeidig ‹*Ton, Leder*›; *(fig.)* nachgiebig ‹*Charakter*›

pliers ['plaɪəz] *n. pl.* **[pair of] ~:** Zange, *die*

plight [plaɪt] *n.* Notlage, *die;* **hopeless/miserable ~:** trostloser/jämmerlicher Zustand

plimsoll ['plɪmsl] *n. (Brit.)* Turnschuh, *der*

plinth [plɪnθ] *n. (for vase, statue, etc.; of wall)* Sockel, *der*

plod [plɒd] *v.i.*, **-dd-** trotten; **~ along** dahintrotten; **~ [on] through the snow** [weiter] durch den Schnee stapfen

~ a'way *v.i. (fig.)* sich abmühen

~ 'on *v.i. (fig.)* sich weiterkämpfen

plodder ['plɒdə(r)] *n.* **he is a ~:** er arbeitet schwerfällig; *(Sch.)* er ist ein bißchen langsam

[1]**plonk** [plɒŋk] *v.t. (coll.)* **~ sth. [down]** etw. hinknallen *(ugs.)*

[2]**plonk** *n. (sl.: wine)* [billiger] Wein

plop [plɒp] **1.** *v.i.*, **-pp-** plumpsen *(ugs.)*; ‹*Regen:*› klatschen, platschen. **2.** *v.t.*, **-pp-** plumpsen lassen *(ugs.)*. **3.** *n.* Plumpsen, *das;* **with a ~:** mit einem Plumps. **4.** *adv.* plumps

plot [plɒt] **1.** *n.* **a)** *(conspiracy)* Komplott, *das;* Verschwörung, *die;* **b)** *(of play, film, novel)* Handlung, *die;* **c)** *(of ground)* Stück Land; **vegetable ~:** Gemüsebeet, *das;* **building ~:** Baugrundstück, *das.* **2.** *v.t.*, **-tt-: a)** *(plan secretly)* [heimlich] planen; **~ treason** auf Verrat sinnen *(geh.)*; **b)** *(make plan or map of)* kartieren, kartographieren ‹*Gebiet usw.*›; *(make by ~ting)* zeichnen ‹*Karte, Plan*›; **c)** *(mark on map, diagram)* **~ [down]** eintragen; einzeichnen. **3.** *v.i.*, **-tt-: ~ against sb.** sich gegen jmdn. verschwören

plotter ['plɒtə(r)] *n.* Verschwörer, *der*/Verschwörerin, *die*

plough [plaʊ] **1.** *n. (Agric.)* Pflug, *der;* **the P~** *(Astron.)* der Große Wagen *od.* Bär. **2.** *v.t.* **a)** pflügen; **~ furrows** Furchen ziehen *od.* pflügen; **b)** *(fig.)* ‹*Schiff:*› [durch]pflügen ‹*Wasserfläche*›

~ 'back *v.t.* **a)** unterpflügen; **b)** *(Finance)* reinvestieren; **~ profits** *etc.* **back into the business** *etc.* Gewinne *usw.* wieder in die Firma *usw.* stecken

~ through *v.t. (advance laboriously in)* sich kämpfen durch; *(move violently through)* rasen durch

~ 'up *v.t.* auspflügen ‹*Kartoffeln, Rüben usw.*›; zerpflügen ‹*Boden*›

ploughman ['plaʊmən] *n., pl.* **ploughmen** ['plaʊmən] Pflüger, *der;* **~'s [lunch]** *(Brit.)* *Imbiß aus Käse, Brot und Mixed Pickles*

plow *(Amer./arch.) see* **plough**

ploy [plɔɪ] *n.* Trick, *der*

pluck [plʌk] **1.** *v.t.* **a)** *(pull off, pick)* pflücken ‹*Blumen, Obst*›; **~ [out]** auszupfen ‹*Federn, Haare*›; **b)** *(pull at, twitch)* zupfen an (+ *Dat.*); zupfen ‹*Saite, Gitarre*›; **c)** *(strip of feathers)* rupfen. **2.** *v.i.* **~ at sth.** an etw. *(Dat.)* zupfen. **3.** *n.* Mut, *der;* Schneid, *der (ugs.)*

~ 'up *v.t.* **~ up [one's] courage** all seinen Mut zusammennehmen

pluckily ['plʌkɪlɪ] *adv.*, **plucky** ['plʌkɪ] *adj.* tapfer

plug [plʌg] **1.** *n.* **a)** *(filling hole)* Pfropfen, *der; (in cask)* Spund, *der;* Zapfen, *der; (stopper for basin, vessel, etc.)* Stöpsel, *der;* **b)** *(Electr.)* Stecker, *der;* **c)** *(coll.: piece of good publicity)* **give sth. a ~:** Werbung für etw. machen. **2.** *v.t.*, **-gg-: a) ~ [up]** zustopfen, verstopfen ‹*Loch usw.*›; **b)** *(coll.: advertise)* Schleichwerbung machen für; *(by presenting sth. repeatedly)* pushen *(ugs.)*

~ a'way *v.i. (coll.)* vor sich hin schuften *(ugs.)*; **~ away at sth.** sich mit etw. abschuften *(ugs.)*

~ 'in *v.t.* anschließen; **is it ~ged in?** ist der Stecker in der Steckdose *od. (ugs.)* drin?

'plug-hole *n.* Abfluß, *der*

plum [plʌm] *n.* **a)** *(fruit)* Pflaume, *die;* **b)** *(fig.)* Leckerbissen, *der; attrib.* **a ~ job/position** ein Traumjob *(ugs.)*

plumage ['plu:mɪdʒ] *n.* Gefieder, *das*

[1]**plumb** [plʌm] **1.** *v.t. (sound, measure)* [aus]loten; **~ the depths of loneliness/sorrow** die tiefsten Tiefen der Einsamkeit/Trauer erleben. **2.** *adv.* **a)** *(vertically)* senkrecht; lotrecht; **b)** *(fig.: exactly)* genau. **3.** *n.* Lot, *das;* **out of ~:** außer Lot

[2]**plumb** *v.t.* **~ in** *(connect)* fest anschließen

plumber ['plʌmə(r)] *n.* Klempner, *der;* Installateur, *der*

plumbing ['plʌmɪŋ] *n.* **a)** *(plumber's work)* Klempnerarbeiten *Pl.*; Installationsarbeiten *Pl.*; **b)** *(water-pipes)* Wasserrohre; Wasserleitungen

'plumb-line *n.* Lot, *das*

plume [plu:m] *n.* **a)** *(feather)* Feder, *die; (ornamental bunch)* Federbusch, *der;* **b) ~ of smoke/steam** Rauchwolke *od.* -fahne, *die*/Dampfwolke, *die*

plummet ['plʌmɪt] *v.i.* stürzen

plump [plʌmp] *adj.* mollig; rundlich; stämmig ‹*Arme, Beine*›; fleischig ‹*Brathuhn usw.*›; **~ cheeks** Pausbacken *Pl. (fam.)*

~ for *v.t.* **a)** *(Brit.: vote for)* stimmen für; **b)** *(choose)* sich entscheiden für

~ 'up *v.t.* aufschütteln ‹*Kissen*›; *(fatten up)* mästen

plunder ['plʌndə(r)] **1.** *v.t.* [aus]plündern ‹*Gebäude, Gebiet*›; ausplündern ‹*Person*›; rauben ‹*Sache*›. **2.** *n.* **a)** *(action)* Plünderung, *die; (spoil, booty)* Beute, *die;* **b)** *(sl.: profit)* Profit, *der*

plunge [plʌndʒ] **1.** *v.t.* **a)** *(thrust violently)* stecken; *(into liquid)* tauchen; **b)** *(fig.)* **~d in thought** in Gedanken versunken; **be ~d into darkness** in Dunkelheit getaucht sein *(geh.)*. **2.** *v.i.* **a) ~ into sth.** *(lit. or fig.)* in etw. *(Akk.)* stürzen; **~ in** sich hineinstürzen; **b)** *(descend suddenly)* ‹*Straße usw.:*› steil abfallen; **plunging neckline** tiefer Ausschnitt. **3.** *n.* Sprung, *der;* **take the ~** *(fig. coll.)* den Sprung wagen

plunger ['plʌndʒə(r)] *n.* **a)** *(part of mechanism)* [Tauch]kolben, *der;* Plunger[kolben], *der;* **b)** *(rubber suction cup)* Stampfer, *der*

pluperfect [plu:'pɜ:fɪkt] *(Ling.)* **1.** *n.* Plusquamperfekt, *das.* **2.** *adj.* **~ tense** Plusquamperfekt, *das*

plural ['plʊərl] *(Ling.)* **1.** *adj.* pluralisch; Plural-; **~ noun** Substantiv im Plural; **third person ~:** dritte Person Plural. **2.** *n.* Mehrzahl, *die;* Plural, *der*

plurality [plʊə'rælɪtɪ] *n.* **a)** *(being plural)* Pluralität, *die;* **b)** *(large number)* Vielzahl, *die*

plus [plʌs] **1.** *prep.* **a)** *(with the addition of)* plus (+ *Dat.*); *(and also)* und [zusätzlich]; **b)** *(above zero)* plus; **~ ten degrees** plus zehn Grad; zehn Grad plus. **2.** *adj.* **a)** *(additional, extra)* zusätzlich; **b)** *(at least)* **fifteen** *etc.* **~:** über fünfzehn *usw.*; **c)** *(Math.: positive)* positiv ‹*Wert, Menge, Größe*›. **3.** *n.* **a)** *(symbol)* Plus[zeichen], *das;* **b)** *(additional quantity)* Plus, *das;* **c)** *(advantage)* Pluspunkt, *der.* **4.** *conj. (sl.)* und außerdem

plush [plʌʃ] **1.** *n.* Plüsch, *der.* **2.** *adj.* Plüsch-; plüschen; *(coll.: luxurious)* feudal *(ugs.)*

Pluto ['plu:təʊ] *n.* **a)** *(Astron.)* Pluto, *der;* **b)** *(Roman Mythol.)* Pluto, *(der)*

plutonium [plu:'təʊnɪəm] *n. (Chem.)* Plutonium, *das*

[1]**ply** [plaɪ] **1.** *v.t.* **a)** *(use, wield)* gebrauchen; führen; **b)** *(work at)*

nachgehen (+ *Dat.*) ‹*Handwerk, Arbeit*›; **c)** *(supply)* ~ **sb. with sth.** jmdn. mit etw. versorgen; **d)** *(assail)* überhäufen; **e)** *(sail over)* befahren. **2.** *v. i.* **a)** *(go to and fro)* ~ **between** zwischen ‹*Orten*› [hin- und her]pendeln; *(operate on regular services)* zwischen ‹*Orten*› verkehren; **b)** *(attend regularly for custom)* seine Dienste anbieten; ~ **for customers/hire** auf Kundschaft warten

²**ply** *n.* **a)** *(of yarn, wool, etc.)* [Einzel]faden, *der; (of rope, cord, etc.)* Strang, *der; (of plywood, cloth, etc.)* Lage, *die;* Schicht, *die; see also* **three-ply**; **two-ply**; **b)** *see* **plywood**

'**plywood** *n.* Sperrholz, *das*

PM *abbr.* **Prime Minister**

p.m. [pi:'em] *adv.* nachmittags; **one** ~: ein Uhr mittags; **six/eleven** ~: sechs/elf Uhr abends

pneumatic [nju:'mætɪk] *adj.* pneumatisch; mit Druckluft betrieben *od.* arbeitend ‹*Maschine*›

pneumatic 'drill *n.* Preßluftbohrer, *der*

pneumonia [nju:'məʊnɪə] *n.* Lungenentzündung, *die;* Pneumonie, *die (Med.)*

PO *abbr.* **a) postal order** PA; **b) Post Office** PA

¹**poach** ['pəʊtʃ] **1.** *v. t.* **a)** *(catch illegally)* wildern; illegal fangen ‹*Fische*›; **b)** *(obtain unfairly)* stehlen, *(ugs.)* klauen ‹*Idee*›. **2.** *v. i.* **a)** *(catch animals illegally)* wildern; **b)** *(encroach)* ~ **[on sb.'s territory]** jmdm. ins Handwerk pfuschen

²**poach** *v. t. (Cookery)* pochieren ‹*Ei*›; dünsten, pochieren ‹*Fisch, Fleisch, Gemüse*›; ~**ed eggs** pochierte *od.* verlorene Eier

poacher ['pəʊtʃə(r)] *n.* Wilderer, *der;* Wilddieb, *der*

pocket ['pɒkɪt] **1.** *n.* **a)** Tasche, *die; (in suitcase etc.)* Seitentasche, *die; (in handbag)* [Seiten]fach, *das; (Billiards etc.)* Loch, *das;* Tasche, *die;* **be in sb.'s** ~ *(fig.)* von jmdm. abhängig sein; **b)** *(fig.: financial resources)* **it is beyond my** ~: es übersteigt meine finanziellen Möglichkeiten; **put one's hand in one's** ~: in die Tasche greifen *(ugs.)*; **be in** ~: Geld verdient haben; **be out of** ~ *(have lost money)* draufgelegt haben *(ugs.)*; zugesetzt haben; **c)** *(Mil.)* ~ **of resistance** Widerstandsnest, *das.* **2.** *adj.* Taschen‹*rechner, -uhr, -ausgabe*›. **3.** *v. t.* **a)** *(put in one's* ~*)* einstecken; **b)** *(steal)* in die eigene Tasche stecken *(ugs.)*

pocket: ~**-book** *n.* **a)** *(wallet)* Brieftasche, *die;* **b)** *(notebook)* Notizbuch, *das;* **c)** *(Amer.: paperback)* Taschenbuch, *das;* **d)** *(Amer.: handbag)* Handtasche, *die;* ~ '**handkerchief** *n.* Taschentuch, *das;* ~**-knife** *n.* Taschenmesser, *das;* ~**-money** *n.* Taschengeld, *das;* ~**-size[d]** *adj.* **a)** im Taschenformat *nachgestellt;* **b)** *(fig.: small scale)* im [Westen]taschenformat *nachgestellt (ugs. scherzh.)*

pock: ~**-mark** *n.* **a)** *(Med.)* Pockennarbe, *die;* **b)** Delle, *die; (from bullet)* Einschuß, *der;* ~**-marked** *adj.* **a)** pockennarbig ‹*Gesicht, Haut*›; **b) a wall** ~**-marked with bullets** eine mit Einschüssen übersäte Wand

pod [pɒd] *n.* **a)** *(seed-case)* Hülse, *die; (of pea)* Schote, *die;* **b)** *(in aircraft etc.) (for engine)* Gondel, *die; (for fuel)* Außentank, *der*

podgy ['pɒdʒɪ] *adj.* dicklich; pummelig *(ugs.)*, rundlich *(fam.)*, mollig ‹*Frau*›; pausbäckig, *(fam.)* rundlich ‹*Gesicht*›

podium ['pəʊdɪəm] *n., pl.* **podia** ['pəʊdɪə] *or* **-s** Podium, *das*

poem ['pəʊɪm] *n.* Gedicht, *das*

poet ['pəʊɪt] *n.* Dichter, *der;* Poet, *der (geh.)*

poetess ['pəʊɪtes] *n.* Dichterin, *die;* Poetin, *die (geh.)*

poetic [pəʊ'etɪk] *adj.*, **poetically** [pəʊ'etɪkəlɪ] *adv.* dichterisch; poetisch *(geh.)*; *see also* **justice a**; **licence 1 d**

poetry ['pəʊɪtrɪ] *n.* [Vers]dichtung, *die;* Lyrik, *die;* ~ **reading** ≈ Dichterlesung, *die*

po-faced ['pəʊfeɪst] *adj.* mit unbewegter Miene *nachgestellt*

poignancy ['pɔɪnjənsɪ] *n., no pl.* [schmerzliche] Intensität; *(of words, wit, etc.)* Schärfe, *die*

poignant ['pɔɪnjənt] *adj.* tief ‹*Bedauern, Trauer*›; überwältigend ‹*Schönheit*›; *(causing sympathy)* ergreifend, herzzerreißend ‹*Anblick, Geschichte*›

point [pɔɪnt] **1.** *n.* **a)** *(tiny mark, dot)* Punkt, *der;* **nought** ~ **two** Null Komma zwei; **b)** *(sharp end of tool, weapon, pencil, etc.)* Spitze, *die;* **come to a [sharp]** ~: spitz zulaufen; **at gun-~/knife-~**: mit vorgehaltener [Schuß]waffe/vorgehaltenem Messer; **not to put too fine a** ~ **on it** *(fig.)* um nichts zu beschönigen; **c)** *(single item)* Punkt, *der;* **agree on a** ~: in einem Punkt *od.* einer Frage übereinstimmen; **be a** ~ **of honour with sb.** für jmdn. [eine] Ehrensache sein; **d)** *(unit of scoring)* Punkt, *der;* **score ~s off sb.** *(fig.)* jmdn. an die Wand spielen; **e)** *(stage, degree)* **things have reached a** ~ **where** *or* **come to such a** ~ **that ...**: die Sache ist dahin *od.* so weit gediehen, daß ...; *(negatively)* es ist so weit gekommen, daß ...; **up to a** ~: bis zu einem gewissen Grad; **beyond a certain** ~: über einen bestimmten Punkt hinaus; **she was abrupt to the** ~ **of rudeness** sie war in einer Weise barsch, die schon an Unverschämtheit grenzte; **f)** *(moment)* Zeitpunkt, *der;* **be at/on the** ~ **of sth.** kurz vor etw. *(Dat.)* sein; einer Sache *(Dat.)* nahe sein; **be on the** ~ **of doing sth.** im Begriff sein, etw. zu tun; etw. gerade tun wollen; **g)** *(distinctive trait)* Seite, *die;* **best/strong** ~: starke Seite; Stärke, *die;* **getting up early has its ~s** frühes Aufstehen hat auch seine Vorzüge; **the** ~ *(essential thing)* das Entscheidende; **h)** *(thing to be discussed)* **that is just the** ~ *or* **the whole** ~: das ist genau der springende Punkt; **come to** *or* **get to the** ~: zur Sache *od.* zum Thema kommen; **keep** *or* **stick to the** ~: beim Thema bleiben; **be beside the** ~: unerheblich sein; keine Rolle spielen; ~ **taken** habe verstanden; **carry** *or* **make one's** ~: sich durchsetzen; **a case in** ~: ein typisches Beispiel; **make a** ~ **of doing sth.** [großen] Wert darauf legen, etw. zu tun; **make** *or* **prove a** ~: etw. beweisen; **to the** ~: sachbezogen; **more to the** ~: wichtiger; **you have a** ~ **there** da hast du recht; da ist [et]was dran *(ugs.)*; **i)** *(tip)* Spitze, *die; (Boxing)* Kinnspitze, *die;* Kinn, *das; (Ballet)* Spitze, *die;* **j)** *(of story, joke, remark)* Pointe, *die; (pungency, effect) (of literary work)* Eindringlichkeit, *die; (of remark)* Durchschlagskraft, *die;* **k)** *(purpose, value)* Zweck, *der;* Sinn, *der;* **there's no** ~ **in protesting** es hat keinen Sinn *od.* Zweck zu protestieren; **l)** *(precise place, spot)* Punkt, *der;* Stelle, *die; (Geom.)* Punkt, *der;* ~ **of contact** Berührungspunkt, *der;* ~ **of no return** Punkt, an dem es kein Zurück mehr gibt; ~ **of view** *(fig.)* Standpunkt, *der;* **m)** *(Brit.)* **[power** *or* **electric]** ~: Steckdose, *die;* **n)** *usu in pl. (Brit. Railw.)* Weiche, *die;* **o)** *usu. in pl. (Motor Veh.: contact device)* Kontakt, *der;* **p)** *(unit in competition, rationing, stocks, shares, etc.)* Punkt, *der;* **prices/the cost of living went up three ~s** die Preise/Lebenshaltungskosten sind um drei [Prozent]punkte gestiegen; **q)** *(on compass)* Strich, *der.* **2.** *v. i.* **a)** zeigen, weisen, ‹*Person auch:*› deuten (**to, at** auf + *Akk.*); **she ~ed through the window** sie zeigte aus dem Fenster; **the compass needle ~ed to the north** die Kompaßnadel zeigte *od.* wies nach Norden; **b) ~ towards** *or* **to** *(fig.)* [hin]deuten *od.* hinweisen auf (+ *Akk.*). **3.** *v. t.* **a)** *(direct)* richten ‹*Waffe, Kamera*› (**at** auf + *Akk.*); **~ one's finger at sth./sb.** mit dem Finger auf etw./jmdn. deuten *od.* zeigen *od.* weisen; **b)** *(Building)* aus-, verfugen ‹*Mauer, Steine*›

~ 'out *v. t.* hinweisen auf (+ *Akk.*); **~ sth./sb. out to sb.** jmdn. auf etw./jmdn. hinweisen *od.* aufmerksam machen; **he ~ed out the house** er zeigte das Haus; **he ~ed out my mistake** er zeigte meinen Fehler auf

point-'blank 1. *adj. (direct, lit. or fig.)* direkt; glatt ‹*Weigerung*›; **~ shot** Schuß aus kürzester Entfernung; **~ range** kürzeste Entfernung. **2.** *adv.* **a)** *(at very close range)* aus kürzester Entfernung ‹*schießen*›; **b)** *(in direct line)* direkt; *(fig.: directly)* rundheraus, *(ugs.)* geradeheraus ‹*fragen, sagen*›

pointed ['pɔɪntɪd] *adj.* **a)** spitz; **~ arch** Spitzbogen, *der;* **b)** *(fig.) (sharply expressed)* unmißverständlich; deutlich; *(ostentatious)* demonstrativ

pointer ['pɔɪntə(r)] *n.* **a)** *(indicator)* Zeiger, *der; (rod)* Zeigestock, *der;* **b) ~ [dog]** Pointer, *der;* englischer Vorstehhund

pointless ['pɔɪntlɪs] *adj. (without purpose, useless)* sinnlos; *(without force, meaningless)* belanglos ‹*Bemerkung, Geschichte*›

poise [pɔɪz] **1.** *n. (composure)* Haltung, *die; (self-confidence)* Selbstsicherheit, *die;* Selbstvertrauen, *das.* **2.** *v. t.* **a)** *in p.p.* **sit ~d on the edge of one's chair** auf der Stuhlkante balancieren; **be ~d for action** einsatzbereit sein; *see also* **poised; b)** *(balance)* balancieren

poised [pɔɪzd] *adj.* selbstsicher; *see also* **poise 2 a**

poison ['pɔɪzn] **1.** *n. (lit. or fig.)* Gift, *das;* **hate sb./sth. like ~:** jmdn./etw. wie die Pest hassen. **2.** *v. t.* **a)** vergiften; *(contaminate)* verseuchen ‹*Boden, Luft, Wasser*›; verpesten *(abwertend)* ‹*Luft*›; **b)** *(fig.)* vergiften ‹*Gedanken, Seele*›; zerstören, ruinieren ‹*Ehe, Leben*›; vergällen ‹*Freude*›; **~ sb.'s mind** jmdn. verderben *od. (geh.)* korrumpieren

poison 'gas *n.* Giftgas, *das*

poisoning ['pɔɪzənɪŋ] *n.* Vergiftung, *die; (contamination)* Verseuchung, *die*

poisonous ['pɔɪzənəs] *adj.* **a)** giftig; tödlich ‹*Dosis*›; **b)** *(fig.)* verderblich ‹*Lehre, Wirkung*›; giftig ‹*Blick, Zunge*›

poke 1. *v. t.* **a) ~ sth. [with sth.]** [mit etw.] gegen etw. stoßen; **~ sth. into sth.** etw. in etw. *(Akk.)* stoßen; **~ the fire** das Feuer schüren; **he accidentally ~d me in the eye** er stieß mir versehentlich ins Auge; **b)** *(thrust forward)* stecken ‹*Kopf*›; **~ one's head round the corner/door** um die Ecke gucken *(ugs.)*/den Kopf in die Türöffnung stecken; **c)** *(pierce)* bohren. **2.** *v. i.* **a)** *(in pond, at food, among rubbish)* [herum]stochern (**at, in, among** in + *Dat.*); **~ at sth. with a stick** *etc.* mit einem Stock *usw.* nach etw. stoßen; **b)** *(thrust itself)* sich schieben; **his elbows were poking through the sleeves** seine Ärmel hatten Löcher, aus denen die Ellbogen hervorguckten; **c)** *(pry)* schnüffeln *(ugs. abwertend).* **3.** *n. (thrust)* Stoß, *der;* **give sb. a ~ [in the ribs]** jmdm. einen [Rippen]stoß versetzen *od.* geben; **give the fire a ~:** das Feuer [an]schüren

~ a'bout, ~ a'round *v. i.* herumschnüffeln *(ugs. abwertend)*

¹poker ['pəʊkə(r)] *n.* Schürstange, *die;* Schüreisen, *das*

²poker *n. (Cards)* Poker, *das od. der*

'poker-faced *adj.* mit unbewegter Miene *nachgestellt*

poky ['pəʊkɪ] *adj.* winzig; **it's so ~ in here** es ist so eng hier drinnen

Poland ['pəʊlənd] *pr. n.* Polen *(das)*

polar ['pəʊlə(r)] *adj.* **a)** *(of pole)* polar ‹*Kaltluft, Gewässer*›; Polar‹*eis, -gebiet, -fuchs*›; **b)** *(Magn.)* polar; **c)** *(directly opposite)* [diametral] entgegengesetzt

polar 'bear *n.* Eisbär, *der*

polarisation, polarise *see* **polariz-**

polarity [pə'lærɪtɪ] *n.* **a)** *(Magn.)* Polung, *die;* Polarität, *die;* **b)** *(fig.: contrary qualities)* Gegensatz, *der*

polarization [pəʊləraɪ'zeɪʃn] *n. (Phys.)* Polarisation, *die; (fig.)* Polarisierung, *die*

polarize ['pəʊləraɪz] **1.** *v. t.* spalten; polarisieren *(geh.).* **2.** *v. i.* sich [auf]spalten

Polaroid, (P) ['pəʊlərɔɪd] *n.* **~ [camera]** Polaroidkamera, *die* Ⓦⓩ

Pole [pəʊl] *n.* Pole, *der*/Polin, *die*

¹pole *n.* **a)** *(support)* Stange, *die;* **drive sb. up the ~** *(Brit. sl.)* jmdn. zum Wahnsinn treiben *(ugs.);* **b)** *(for propelling boat)* Stake, *die (nordd.)*

²pole *n. (Astron., Geog., Magn., Electr., fig.)* Pol, *der;* **they are ~s apart** *(coll.)* zwischen ihnen liegen Welten

'polecat *n. (Zool.)* **a)** *(Brit.)* Iltis, *der;* **b)** *(Amer.) see* **skunk**

polemic [pə'lemɪk] **1.** *adj.* polemisch. **2.** *n. (discussion)* Polemik, *die; (written also)* Streitschrift, *die*

pole: ~-star *n. (Astron.)* Polarstern, *der;* **~-vault** *n.* Stabhochsprung, *der;* **~-vaulter** *n.* Stabhochspringer, *der*/-springerin, *die;* **~-vaulting** *n.* Stabhochsprung, *der;* Stabhochspringen, *das*

police [pə'li:s] **1.** *n. pl.* **a)** Polizei, *die; attrib.* Polizei‹*wagen, -hund, -schutz, -eskorte, -staat*›; **be in the ~:** bei der Polizei sein; **b)** *(members)* Polizisten *Pl.;* Polizeibeamte *Pl.; attrib.* Polizei-; **the ~ are on his trail** die Polizei ist ihm auf der Spur; **help the ~ with their enquiries** von der Polizei vernommen werden. **2.** *v. t.* [polizeilich] überwachen ‹*Gebiet, Verkehr, Fußballspiel*›; kontrollieren ‹*Gebiet, Grenze, Gewässer*›; Polizeibeamte einsetzen in (+ *Dat.*) ‹*Gebiet, Stadt usw.*›

police: ~ constable *n.* Polizist, *der*/Polizistin, *die; (rank)* Polizeihauptwachtmeister, *der;* **~ force** *n.* Polizeitruppe, *die;* **the ~ force** die Polizei; **~man** [pə'li:smən] *n., pl.* **~men** [pə'li:smən] Polizist, *der;* Polizeibeamte, *der;* **~-officer** *n.* Polizeibeamte, *der*/-beamtin, *die;* **~ station** *n.* Polizeiwache, *die;* Polizeirevier, *das;* **~woman** *n.* Polizistin, *die;* Polizeibeamtin, *die*

¹policy ['pɒlɪsɪ] *n. (method)* Handlungsweise, *die;* Vorgehensweise, *die; (overall plan)* Politik, *die;* **~ on immigration** Einwanderungspolitik, *die;* **it's bad ~ to ...:** es ist unvernünftig, ... zu ...

²policy *n. (Insurance)* Police, *die;* Versicherungsschein, *der*

polio ['pəʊlɪəʊ] *n., no pl., no art.* Polio, *die;* [spinale] Kinderlähmung

Polish ['pəʊlɪʃ] **1.** *adj.* polnisch; **sb. is ~:** jmd. ist Pole/Polin. **2.** *n.* Polnisch, *das; see also* **English 2 a**

polish ['pɒlɪʃ] **1.** *v. t.* **a)** *(make smooth)* polieren; bohnern ‹*Fußboden*›; putzen ‹*Schuhe*›; **b)** *(fig.)* ausfeilen ‹*Text, Theorie, Technik,*

Stil›; polieren ‹*Text*›; ~**ed** geschliffen ‹*Stil, Manieren*›; ausgefeilt ‹*Technik, Taktik, Plan, Satz*›. **2.** *n.* **a)** *(smoothness)* Glanz, *der;* **a table with a high ~:** ein auf Hochglanz polierter Tisch; **b)** *(substance)* Poliermittel, *das;* Politur, *die;* **c)** *(fig.)* Geschliffenheit, *die;* Schliff, *der;* **d)** *(action)* **give sth. a ~:** etw. polieren; **give the shoes a ~:** die Schuhe putzen

~ 'off *v.t. (coll.)* **a)** *(consume)* verdrücken *(ugs.);* wegputzen *(ugs.)* ‹*Essen*›; aussüffeln *(ugs.)* ‹*Getränk*›; **b)** *(complete quickly)* durchziehen *(ugs.)*

~ 'up *v.t.* **a)** *(make shiny)* polieren; **b)** *(improve)* ausfeilen ‹*Stil, Technik*›; aufpolieren ‹*[Sprach]kenntnisse*›

polite [pə'laɪt] *adj.*, **~r** [pə'laɪtə(r)], **~st** [pə'laɪtɪst] höflich; **be ~ about her dress** mach ihr ein paar Komplimente zu ihrem Kleid

politeness [pə'laɪtnɪs] *n., no pl.* Höflichkeit, *die*

politic ['pɒlɪtɪk] *adj.* klug ‹*Person, Handlung*›; opportun *(geh.)* ‹*Handlung*›; **it's not ~ to do sth.** es ist unklug *od.* nicht ratsam, etw. zu tun

political [pə'lɪtɪkl] *adj.* politisch

political: ~ a'sylum *see* **asylum; ~ 'prisoner** *n.* politischer Gefangener/politische Gefangene

politician [pɒlɪ'tɪʃn] *n.* Politiker, *der*/Politikerin, *die*

politics ['pɒlɪtɪks] *n., no pl.* **a)** *no art. (political administration)* Politik, *die; (Univ.: subject)* Politik[wissenschaft], *die;* Politologie, *die;* **b)** *no art., constr. as sing. or pl. (political affairs)* Politik, *die;* **interested/involved in ~:** politisch interessiert/engagiert; **enter ~:** in die Politik gehen; **c)** *as pl. (political principles)* Politik, *die; (of individual)* politische Einstellung; **what are his ~?** wo steht er politisch?; **practical ~:** Realpolitik, *die*

polka ['pɒlkə, 'pəʊlkə] *n.* Polka, *die*

'polka dot *n.* [großer] Tupfen

poll [pəʊl] **1.** *n.* **a)** *(voting)* Abstimmung, *die; (to elect sb.)* Wahl, *die; (result of vote)* Abstimmungsergebnis, *das*/Wahlergebnis, *das; (number of votes)* Wahlbeteiligung, *die;* **take a ~:** abstimmen lassen; eine Abstimmung durchführen; **at the ~[s]** bei den Wahlen; **go to the ~:** seine Stimme abgeben; zur Wahl gehen; **a heavy/light *or* low ~:** eine starke/geringe *od.* niedrige Wahlbeteiligung; **b)** *(survey of opinion)* Umfrage, *die.* **2.** *v.t.* **a)** *(take vote[s] of)* abstimmen/wählen lassen; **b)** *(take opinion of)* befragen; *(take survey of)* [demoskopisch] erforschen; **c)** *(obtain in ~)* erhalten ‹*Stimmen*›

pollen ['pɒlən] *n. (Bot.)* Pollen, *der;* Blütenstaub, *der*

'pollen count *n.* Pollenmenge, *die*

pollinate ['pɒlɪneɪt] *v.t. (Bot.)* bestäuben

pollination [pɒlɪ'neɪʃn] *n. (Bot.)* Bestäubung, *die*

polling ['pəʊlɪŋ]: **~-booth** *n.* Wahlkabine, *die;* **~-station** *n. (Brit.)* Wahllokal, *das*

'poll-tax *n.* Kopfsteuer, *die*

pollster ['pɒlstə(r)] *n.* Meinungsforscher, *der*/-forscherin, *die*

pollutant [pə'lu:tənt] *n.* [Umwelt]schadstoff, *der*

pollute [pə'lu:t] *v.t.* **a)** *(contaminate)* verschmutzen, verunreinigen ‹*Luft, Boden, Wasser*›; verpesten *(abwertend)* ‹*Luft*›; **b)** *(make foul)* verseuchen

pollution [pə'lu:ʃn] *n.* **a)** *(contamination)* [Umwelt]verschmutzung, *die;* **water ~:** Gewässerverschmutzung, *die;* **b)** *(polluting substance[s])* Verunreinigungen; Schadstoffe

polo ['pəʊləʊ] *n., no pl.* Polo, *das*

'polo-neck *n.* Rollkragen, *der;* **~[ed]** *attrib.* Rollkragen-

poltergeist ['pɒltəgaɪst] *n.* Klopfgeist, *der;* Poltergeist, *der*

poly ['pɒlɪ] *n., pl.* **~s** *(coll.)* Polytechnikum, *das;* ≈ TH, *die*

polyester [pɒlɪ'estə(r)] *n.* Polyester, *der*

polygamy [pə'lɪgəmɪ] *n.* Polygamie, *die (geh., fachspr.);* Mehrehe, *die;* Vielehe, *die*

polyglot ['pɒlɪglɒt] **1.** *adj.* **a)** polyglott *(geh., fachspr.);* mehrsprachig; **b)** *(speaking several languages)* polyglott *(geh.).* **2.** *n.* Polyglotte, *der/die (geh.)*

polygon ['pɒlɪgən] *n. (Geom.)* Vieleck, *das;* Polygon, *das (fachspr.)*

polymer ['pɒlɪmə(r)] *n. (Chem.)* Polymer[e], *das*

polyp ['pɒlɪp] *n. (Zool., Med.)* Polyp, *der*

polystyrene [pɒlɪ'staɪri:n] *n.* Polystyrol, *das;* **~ foam** Styropor Ⓦⓩ, *das*

polytechnic [pɒlɪ'teknɪk] *n. (Brit.)* ≈ technische Hochschule *od.* Universität

polythene ['pɒlɪθi:n] *n.* Polyäthylen, *das;* Polyethylen, *das (fachspr.); (coll.: plastic)* Plastik, *das;* **~ bag/sheet** Plastikbeutel, *der*/-folie, *die*

polyunsaturated [pɒlɪʌn'sætʃəreɪtɪd] *adj.* mehrfach ungesättigt

pomegranate ['pɒmɪgrænɪt] *n.* Granatapfel, *der*

pommel ['pʌml, 'pɒml] *n.* **a)** *(on sword)* [Schwert]knauf, *der;* **b)** *(on saddle)* Sattelknopf, *der*

'pommel-horse *n.* Seitpferd, *das*

pommy (pommie) ['pɒmɪ] *n. (Austral. and NZ sl. derog.)* Brite, *der*/Britin, *die*

pomp [pɒmp] *n.* Pomp, *der (abwertend);* Prunk, *der;* **~ and circumstance** festliches Gepränge *(geh.)*

pom-pom ['pɒmpɒm], **pompon** ['pɒmpɒn] *n. (tuft)* Pompon, *der;* **~ hat** Pudelmütze, *die*

pompous ['pɒmpəs] *adj. (self-important)* großspurig; aufgeblasen; geschwollen *(abwertend);* gespreizt *(abwertend)* ‹*Sprache*›

ponce [pɒns] *(Brit. sl.) n.* **a)** *(pimp)* Zuhälter, *der;* **b)** *(derog.: homosexual)* Schwule, *der (ugs.);* Homo, *der (ugs.)*

~ a'bout, ~ a'round *v.i. (derog.)* herumtänzeln *(ugs.)*

pond [pɒnd] *n.* Teich, *der*

ponder ['pɒndə(r)] **1.** *v.t.* nachdenken über (+ *Akk.*) ‹*Frage, Problem, Ereignis*›; bedenken ‹*Folgen*›; abwägen ‹*Vorteile, Worte*›. **2.** *v.i.* nachdenken **(over, on** über + *Akk.*)

ponderous ['pɒndərəs] *adj.* **a)** *(heavy)* schwer; **b)** *(unwieldy, laborious)* schwerfällig; umständlich ‹*Ausdrucksweise*›

pong [pɒŋ] *(Brit. coll.)* **1.** *n.* Mief, *der (ugs. abwertend).* **2.** *v.i.* miefen *(ugs. abwertend)*

pontiff ['pɒntɪf] *n.* Papst, *der*

pontificate [pɒn'tɪfɪkeɪt] *v.i.* dozieren

¹pontoon [pɒn'tu:n] *n.* **a)** *(boat)* Ponton, *der;* Prahm, *der;* **b)** *(support)* Ponton, *der*

²pontoon *n. (Brit. Cards)* Siebzehnundvier, *das*

pontoon 'bridge *n.* Pontonbrücke, *die*

pony ['pəʊnɪ] *n.* Pony, *das; see also* **shank a**

pony: ~-tail *n.* Pferdeschwanz, *der;* **~-trekking** ['pəʊnɪtrekɪŋ] *n. (Brit.)* Ponyreiten, *das*

poodle ['pu:dl] *n.* Pudel, *der;* **be sb.'s ~** *(fig.)* immer nach jmds. Pfeife tanzen

pooh [pu:] *int.* **a)** *expr. disgust* bah; bäh; pfui [Teufel]; **b)** *expr. disdain* pah

pooh-'pooh *v.t.* [als läppisch] abtun

¹pool [pu:l] *n.* **a)** *(permanent)*

Tümpel, *der;* **b)** *(temporary)* Pfütze, *die;* Lache, *die;* ~ **of blood** Blutlache, *die;* **c)** *(swimming-~)* Schwimmbecken, *das; (public swimming-~)* Schwimmbad, *das; (in house or garden)* [Swimming]pool, *der*

²pool 1. *n.* **a)** *(Gambling)* [gemeinsame Spiel]kasse; **the ~s** *(Brit.)* das Toto; **do the ~s** Toto spielen; **win the ~s** im Toto gewinnen; **b)** *(common supply)* Fonds, *der;* Topf, *der;* **a [great] ~ of experience** ein [großer] Fundus von *od.* an Erfahrung; **c)** *(game)* Pool[billard], *das.* **2.** *v. t.* zusammenlegen ⟨*Geld, Ersparnisse, Mittel, Besitz*⟩; bündeln ⟨*Anstrengungen*⟩; **they ~ed their experience** sie nutzten ihre Erfahrung gemeinsam

poor [pʊə(r)] **1.** *adj.* **a)** arm; **b)** *(inadequate)* schlecht; schwach ⟨*Rede, Spiel, Leistung, Gesundheit*⟩; dürftig ⟨*Essen, Kleidung, Unterkunft, Entschuldigung*⟩; **of ~ quality** minderer Qualität; **be ~ at maths** *etc.* schlecht *od.* schwach in Mathematik *usw.* sein; **sb. is ~ at games** Ballspiele liegen jmdm. nicht; **c)** *(paltry)* schwach ⟨*Trost*⟩; schlecht ⟨*Aussichten, Situation*⟩; *(disgusting)* mies *(ugs. abwertend);* **that's pretty ~!** das ist reichlich dürftig *od. (ugs.)* ganz schön schwach; **d)** *(unfortunate)* arm *(auch iron.);* **~ you!** du Armer/Arme!; du Ärmster/Ärmste!; **e)** *(infertile)* karg, schlecht ⟨*Boden, Land*⟩; **f)** *(spiritless, pathetic)* arm ⟨*Teufel, Dummkopf*⟩; armselig, *(abwertend)* elend ⟨*Kreatur, Stümper*⟩; **g)** *(deficient)* arm (**in** an + *Dat.*); **~ in content/ideas/vitamins** inhalts- / ideen- / vitaminarm; **h) take a ~ view of** nicht [sehr] viel halten von; für gering halten ⟨*Aussichten, Chancen*⟩. **2.** *n. pl.* **the ~:** die Armen

poorly ['pʊəlɪ] **1.** *adv.* **a)** *(scantily)* schlecht; unzureichend; **b)** *(badly)* schlecht; unbeholfen ⟨*schreiben, sprechen*⟩; **he did ~ in his exams** er war in seinen Prüfungen schlecht; **c)** *(meanly)* schlecht ⟨*leben*⟩. **2.** *pred. adj.* schlecht ⟨*aussehen, sich fühlen*⟩; **he has been ~ lately** ihm geht es in letzter Zeit schlecht

'poor: ~ man's *attrib. adj. (coll.)* des kleinen Mannes *nachgestellt;* **~ re'lation** *n.* arme Verwandte, *der/die; (fig.)* Stiefkind, *das;* **be the ~ relation** *(fig.)* im Vergleich zu etw. schlecht abschneiden

¹pop [pɒp] **1.** *v. i.,* **-pp-: a)** *(make sound)* ⟨*Korken:*⟩ knallen; ⟨*Schote, Samenkapsel:*⟩ aufplatzen, aufspringen; *(fig.)* **his eyes ~ped with amazement** er guckte wie ein Auto *(ugs.);* **b)** *(coll.: move, go quickly)* **let's ~ round to Fred's** komm, wir gehen mal eben *od.* schnell bei Fred vorbei *(ugs.);* **~ down to London** mal eben *od.* schnell nach London fahren; **you must ~ round and see us** du mußt mal vorbeikommen und uns besuchen *od.* mußt mal bei uns reingucken *(ugs.).* **2.** *v. t.,* **-pp-: a)** *(coll.: put)* **~ the meat in the fridge** das Fleisch in den Kühlschrank tun; **~ a peanut into one's mouth** [sich *(Dat.)*] eine Erdnuß in den Mund stecken; **~ one's head in at the door** den Kopf zur Tür reinstecken; **b)** *(cause to burst)* enthülsen ⟨*Erbsen, Bohnen*⟩; platzen *od. (ugs.)* knallen lassen ⟨*Luftballon*⟩; zerknallen ⟨*Papiertüte*⟩; **c) ~ the question [to sb.]** *(coll.)* jmdm. einen [Heirats]antrag machen. **3.** *n.* **a)** *(sound)* Knall, *der;* Knallen, *das;* **b)** *(coll.: drink)* Sprudel, *der; (flavoured)* Brause, *die (ugs.).* **4.** *adv.* **go ~:** knallen; peng machen *(ugs.)*

~ 'off *v. i.* **a)** *(coll.: die)* abnibbeln *(ugs., bes. nordd.);* den Löffel weglegen *od.* abgeben *(salopp);* **b)** *(move or go away)* ⟨*Person:*⟩ verschwinden, *(ugs.)* abdampfen

~ 'out *v. i.* hervorschießen aus; **~ out for a newspaper/to the shops** schnell *od.* eben mal eine Zeitung holen gehen/einkaufen gehen *(ugs.);* **he's just ~ped out for a moment** er ist nur mal kurz weggegangen *(ugs.)*

~ 'out of *v. t.* hervorschieben; **~ one's head out of the window** den Kopf zum Fenster herausstrecken; **sb.'s eyes nearly** *or* **almost ~ out of his head** *(coll.) (with surprise)* jmdm. fallen fast die Augen aus dem Kopf; *(with excitement)* jmd. *(bes. Kind)* macht große Augen

~ 'up *v. i.* **a)** *(fig.: appear)* auftauchen; **b)** *(rise up)* sich aufstellen; *(spring up)* hochspringen

²pop *(coll.)* **1.** *n. (popular music)* Popmusik, *die;* Pop, *der.* **2.** *adj.* Pop⟨*star, -musik usw.*⟩

³pop *n. (Amer. coll.: father)* Pa[pa], *der (fam.)*

pop: ~ art *n., no pl., no indef. art.* Pop-art, *die; attrib.* Pop-art-; **~ concert** *n.* Popkonzert, *das;* **~corn** *n.* Popcorn, *das*

pope [pəʊp] *n.* Papst, *der/*Päpstin, *die*

pop: ~-eyed *adj. (coll.)* großäugig; **they were ~-eyed with amazement** sie staunten Bauklötze *(salopp);* **~ festival** *n.* Popfestival, *das;* **~ group** *n.* Popgruppe, *die;* **~gun** *n.* Spielzeuggewehr, *das/* Spielzeugpistole, *die*

poplar ['pɒplə(r)] *n.* Pappel, *die*

poplin ['pɒplɪn] *n. (Textiles)* Popelin, *der;* Popeline, *der od. die*

'pop music *n.* Popmusik, *die*

popper ['pɒpə(r)] *n. (Brit. coll.)* Druckknopf, *der*

poppy ['pɒpɪ] *n.* **a)** *(Bot.)* Mohn, *der;* **opium ~:** Schlafmohn, *der;* **b)** *(Brit.: emblem)* [künstliche] Mohnblume *(als Zeichen des Gedenkens am 'Poppy Day')*

'Poppy Day *(Brit.) see* **Remembrance Day**

pop: ~ singer *n.* Popsänger, *der/*-sängerin, *die;* Schlagersänger, *der/*-sängerin, *die;* **~ song** *n.* Popsong, *der;* Schlager, *der;* **~ star** *n.* Popstar, *der;* Schlagerstar, *der*

popular ['pɒpjʊlə(r)] *adj.* **a)** *(well liked)* beliebt; populär ⟨*Entscheidung, Maßnahme*⟩; **he was a very ~ choice** mit ihm hatte man sich für einen sehr beliebten *od.* populären Mann entschieden; **be ~ with sb.** bei jmdm. beliebt sein; **b)** *(suited to the public)* volkstümlich; populär *(geh.);* **~ newspaper** Massenblatt, *das;* **c)** *(prevalent)* landläufig; allgemein ⟨*Unzufriedenheit*⟩; **d)** *(of the people)* Volks-; verbreitet ⟨*Aberglaube, Irrtum, Meinung*⟩; allgemein ⟨*Wahl, Zustimmung, Unterstützung*⟩; **by ~ request** auf allgemeinen Wunsch

popularise *see* **popularize**

popularity [pɒpjʊ'lærɪtɪ] *n., no pl.* Popularität, *die;* Beliebtheit, *die; (of decision, measure)* Popularität, *die*

popularize ['pɒpjʊləraɪz] *v. t.* **a)** *(make popular)* populär machen; **b)** *(make known)* bekannt machen; **c)** *(make understandable)* breiteren Kreisen zugänglich machen

popularly ['pɒpjʊləlɪ] *adv.* **a)** *(generally)* allgemein; landläufig; **it is ~ believed that ...:** es ist ein im Volk verbreiteter Glaube, daß ...; **b)** *(for the people)* volkstümlich; [all]gemeinverständlich

popular 'music *n.* Unterhaltungsmusik, *die*

populate ['pɒpjʊleɪt] *v. t.* bevölkern ⟨*Land, Gebiet*⟩; bewohnen ⟨*Insel, Gebiet*⟩; **heavily** *or* **densely/sparsely ~d** dicht/dünn besiedelt ⟨*Land, Gebiet usw.*⟩; dicht/dünn bevölkert ⟨*Stadt*⟩

population [pɒpjʊ'leɪʃn] *n.* Bevölkerung, *die;* **Britain has a ~ of 56 million** Großbritannien hat 56 Millionen Einwohner
popu'lation explosion *n.* Bevölkerungsexplosion, *die*
populous ['pɒpjʊləs] *adj.* dicht bevölkert
'**pop-up** *adj.* Stehauf‹*buch, -illustration*›; ~ **toaster** Toaster mit Auswerfmechanismus
porcelain ['pɔ:slɪn] *n.* Porzellan, *das; attrib.* Porzellan-
porch [pɔ:tʃ] *n.* Vordach, *das; (with side walls)* Vorbau, *der; (enclosed)* Windfang, *der; (of church etc.)* Vorhalle, *die*
porcupine ['pɔ:kjʊpaɪn] *n. (Zool.)* Stachelschwein, *das*
[1]**pore** [pɔ:(r)] *n.* Pore, *die*
[2]**pore** *v. i.* ~ **over sth.** etw. [genau] studieren; *(think deeply)* ~ **over** *or* **on sth.** über etw. *(Akk.)* [gründlich] nachdenken
pork [pɔ:k] *n.* Schweinefleisch, *das; attrib.* Schweine-; Schweins- *(bes. südd.)*
pork: ~ '**chop** *n.* Schweinekotelett, *das;* ~ '**pie** *n.* Schweinepastete, *die;* ~ '**sausage** *n.* Schweinswürstchen, *das*
porn [pɔ:n] *n., no pl. (coll.)* Pornographie, *die;* Pornos *(ugs.)*
pornographic [pɔ:nə'græfɪk] *adj.* pornographisch; Porno- *(ugs.)*
pornography [pɔ:'nɒgrəfɪ] *n.* Pornographie, *die*
porous ['pɔ:rəs] *adj.* porös ‹*Fels, Gestein, Stoff*›
porpoise ['pɔ:pəs] *n. (Zool.)* Schweinswal, *der*
porridge ['pɒrɪdʒ] *n., no pl. (food)* Porridge, *der;* [Hafer]brei, *der*
[1]**port** [pɔ:t] **1.** *n.* **a)** *(harbour)* Hafen, *der;* **come** *or* **put into ~:** [in den Hafen] einlaufen; **leave ~:** [aus dem Hafen] auslaufen; **reach ~:** den Hafen erreichen; **any ~ in a storm** *(fig. coll.)* ≈ in der Not frißt der Teufel Fliegen *(ugs.);* ~ **of call** Anlaufhafen, *der; (fig.)* Ziel, *das;* **b)** *(town)* Hafenstadt, *die;* Hafen, *der;* **c)** *(Naut., Aeronaut.: left side)* Backbord, *das;* **land to ~!** Land an Backbord! **2.** *adj. (Naut., Aeronaut.: left)* Backbord-; backbordseitig; **on the ~ bow/quarter** Backbord voraus/Backbord achteraus
[2]**port** *n. (wine)* Portwein, *der;* Port, *der (ugs.)*
portable ['pɔ:təbl] **1.** *adj.* tragbar. **2.** *n. (television)* Portable, *der; (radio)* Portable, *der;* Koffergerät, *das; (typewriter)* Portable, *die;* Koffermaschine, *die*
[1]**porter** ['pɔ:tə(r)] *n. (Brit.: doorman)* Pförtner, *der; (of hotel etc.)* Portier, *der*
[2]**porter** *n.* **a)** *(luggage-handler)* [Gepäck]träger, *der/*-trägerin, *die; (in hotel)* Hausdiener, *der;* **b)** *(Amer., Ir./Hist.: beer)* Porter, *der od. das*
portfolio [pɔ:t'fəʊlɪəʊ] *n., pl.* ~**s a)** *(list)* Portefeuille, *das;* **b)** *(Polit.)* Geschäftsbereich, *der;* Portefeuille, *das (geh.);* **c)** *(case, contents)* Mappe, *die*
porthole ['pɔ:thəʊl] *n. (Naut.)* Seitenfenster, *das; (round)* Bullauge, *das*
portion ['pɔ:ʃn] *n.* **a)** *(part)* Teil, *der; (of ticket)* Abschnitt, *der; (of inheritance)* Anteil, *der;* **b)** *(amount of food)* Portion, *die*
~ '**out** *v. t.* aufteilen (**among, between** unter + *Akk.*)
portly ['pɔ:tlɪ] *adj.* beleibt; korpulent
portmanteau [pɔ:t'mæntəʊ] *n., pl.* ~**s** *or* ~**x** [pɔ:t'mæntəʊz] Reisekoffer, *der*
portrait ['pɔ:trɪt] *n.* Porträt, *das;* Bildnis, *das (geh.); attrib.* Porträt-; **have one's ~ painted** sich porträtieren lassen
portray [pɔ:'treɪ] *v. t.* **a)** *(describe)* darstellen; schildern; **b)** *(make likeness of)* porträtieren ‹*Person*›; ‹*Schauspieler:*› darstellen ‹*Rolle, Person*›
Portugal ['pɔ:tjʊgl] *pr. n.* Portugal *(das)*
Portuguese [pɔ:tjʊ'gi:z] **1.** *adj.* portugiesisch; **sb. is ~:** jmd. ist Portugiese/Portugiesin. **2.** *n., pl. same* **a)** *(person)* Portugiese, *der/*Portugiesin, *die;* **b)** *(language)* Portugiesisch, *das; see also* **English 2 a**
pose [pəʊz] **1.** *v. t.* **a)** *(be cause of)* aufwerfen ‹*Frage, Problem*›; darstellen ‹*Bedrohung, Problem*›; bedeuten ‹*Bedrohung*›; mit sich bringen ‹*Schwierigkeiten*›; **b)** *(propound)* vorbringen; aufstellen ‹*Theorie*›; **c)** *(place)* Aufstellung nehmen lassen; posieren lassen ‹*Modell*›. **2.** *v. i.* **a)** *(assume attitude)* posieren; *(fig.)* sich geziert benehmen *(abwertend);* **b)** ~ **as** sich geben als; **he likes to ~ as an expert** er spielt gern den Experten. **3.** *n.* Haltung, *die;* Pose, *die; (fig.)* Pose, *die;* Gehabe, *das (abwertend);* **strike a ~:** eine Pose einnehmen
poser ['pəʊzə(r)] *n. (question)* knifflige Frage; *(problem)* schwieriges Problem; **that's a real ~:** das ist eine harte Nuß *(ugs.)*
posh [pɒʃ] **1.** *adj. (coll.)* vornehm; nobel *(spött.);* stinkvornehm *(salopp);* **the ~ people** die Schickeria *(ugs.).* **2.** *adv.* **talk ~:** hochgestochen reden/mit vornehmem Akzent sprechen
position [pə'zɪʃn] **1.** *n.* **a)** *(place occupied)* Platz, *der; (of player in team; of plane, ship, etc.)* Position, *die; (of hands of clock, words, stars)* Stellung, *die; (of building)* Lage, *die; (of river)* [Ver]lauf, *der;* **find one's ~ on a map** seinen Standort auf einer Karte finden; **take [up] one's ~:** seinen Platz einnehmen; **after the second lap he was in fourth ~:** nach der zweiten Runde lag er an vierter Stelle; **he finished in second ~:** er belegte den zweiten Platz; **b)** *(proper place)* **be in/out of ~:** an seinem Platz/nicht an seinem Platz sein; **c)** *(Mil.)* Stellung, *die;* **d)** *(fig.: mental attitude)* Standpunkt, *der;* Haltung, *die;* **take up a ~ on sth.** einen Standpunkt *od.* eine Haltung zu etw. einnehmen; **e)** *(fig.: situation)* **be in a good ~ [financially]** [finanziell] gut gestellt sein *od.* dastehen; **be in a ~ of strength** eine starke Position haben; **put yourself in my ~!** versetz dich [einmal] in meine Lage!; **be in a ~ to do sth.** in der Lage sein, etw. zu tun; **f)** *(rank)* Stellung, *die;* Position, *die;* **g)** *(employment)* [Arbeits]stelle, *die;* Stellung, *die;* **the ~ of assistant manager** die Stelle *od.* Position des stellvertretenden Geschäftsführers; ~ **of trust** Vertrauensstellung, *die;* Vertrauensposten, *der;* **h)** *(posture)* Haltung, *die; (during sexual intercourse)* Stellung, *die;* Position, *die;* **in a reclining ~:** zurückgelehnt. **2.** *v. t.* **a)** plazieren; aufstellen, postieren ‹*Polizisten, Wachen*›; ~ **oneself near the exit** sich in die Nähe des Ausgangs stellen/setzen; ‹*Wache, Posten usw.:*› sich in der Nähe des Ausgangs aufstellen; **b)** *(Mil.: station)* stationieren
positive ['pɒzɪtɪv] *adj.* **a)** *(definite)* eindeutig; entschieden ‹*Weigerung*›; positiv ‹*Recht*›; **in a ~ tone of voice** in bestimmtem *od.* entschiedenem Ton; **b)** *(convinced)* sicher; **Are you sure? – P~!** Bist du sicher? – Absolut [sicher]!; **I'm ~ of it** ich bin [mir] [dessen] ganz sicher; **c)** *(affirmative)* positiv; **d)** *(optimistic)* positiv; **e)** *(showing presence of sth.)* positiv ‹*Ergebnis, Befund, Test*›; **f)** *(constructive)* konstruktiv ‹*Kritik, Vorschlag*›; positiv ‹*Philo-*

sophie, Erfahrung, Denken›; **g)** *(Math.)* positiv; **h)** *(Electr.)* positiv ‹*Elektrode, Platte, Ladung, Ion*›; **i)** *as intensifier (coll.)* echt; **it would be a ~ miracle** es wäre ein echtes Wunder *od. (ugs.)* echt ein Wunder

positively ['pɒzɪtɪvlɪ] *adv.* **a)** *(constructively)* konstruktiv ‹*kritisieren*›; positiv ‹*denken*›; **b)** *(Electr.)* positiv; **c)** *(definitely)* eindeutig, entschieden ‹*sich weigern*›; **d)** *as intensifier (coll.)* echt *(ugs.)*; **it's ~ marvellous that ...:** es ist echt spitze, daß ...

posse ['pɒsɪ] *n.* **a)** *(Amer.: force with legal authority)* [Polizei]trupp, *der;* [Polizei]aufgebot, *das;* **b)** *(crowd)* Schar, *die*

possess [pə'zes] *v. t.* **a)** *(own)* besitzen; **b)** *(have as faculty or quality)* haben; **c)** *(dominate)* ‹*Furcht usw.:*› ergreifen, Besitz nehmen von ‹*Person*›; **what ~ed you/him?** *(coll.)* was ist in dich/ihn gefahren?

possessed [pə'zest] *adj. (dominated)* besessen; **like one ~:** wie ein Besessener/eine Besessene

possession [pə'zeʃn] *n.* **a)** *(thing possessed)* Besitz, *der;* **some of my ~s** einige meiner Sachen; **b)** *in pl. (property)* Besitz, *der; (territory)* Besitzungen; **c)** *(controlling)* **take ~ of** *(Mil.)* einnehmen ‹*Festung, Stadt usw.*›; besetzen ‹*Gebiet*›; **d)** *(possessing)* Besitz, *der;* **be in ~ of sth.** im Besitz einer Sache *(Gen.)* sein; **come into** *or* **get ~ of sth.** in den Besitz einer Sache *(Gen.)* gelangen; **in full ~ of one's senses** im Vollbesitz seiner geistigen Kräfte; **be in full ~ of the facts** voll im Bilde sein; **have sth. in one's ~:** im Besitz einer Sache *(Gen.)* sein; **take ~ of** in Besitz nehmen; beziehen ‹*Haus, Wohnung*›; **in ~** *(Sport)* im Ballbesitz

possessive [pə'zesɪv] *adj.* **a)** *(jealously retaining possession)* besitzergreifend; **be ~ about sth.** etw. eifersüchtig hüten; **be ~ about** *or* **towards sb.** an jmdn. Besitzansprüche stellen; **b)** *(Ling.)* possessiv; **~ adjective** Possessivadjektiv, *das;* **~ pronoun** Possessivpronomen, *das*

pos'sessive case *n.* Possessiv[us], *der*

possessor [pə'zesə(r)] *n.* Besitzer, *der*/Besitzerin, *die*

possibility [pɒsɪ'bɪlɪtɪ] *n.* **a)** Möglichkeit, *die;* **there's no ~ of his coming/agreeing** es ist ausgeschlossen, daß er kommt/zustimmt; **there's not much ~ of success** die Erfolgschancen sind nicht groß; **it's a distinct ~ that ...:** es ist gut möglich, daß ...; **b)** *in pl. (potential)* Möglichkeiten *Pl.;* **the house/subject has possibilities** aus dem Haus/Thema läßt sich etwas machen

possible ['pɒsɪbl] **1.** *adj.* **a)** möglich; **if ~:** wenn *od.* falls möglich; wenn es geht; **as ... as ~:** so ... wie möglich; möglichst ...; **all the assistance ~:** alle denkbare Unterstützung; **they made it ~ for me to be here** sie haben es mir ermöglicht, hier zu sein; **would it be ~ for me to ...?** könnte ich vielleicht ...?; **for ~ emergencies** für eventuelle Notfälle; **I'll do everything ~ to help you** ich werde mein möglichstes tun, um dir zu helfen; **we will help as far as ~:** wir werden helfen, soweit wir können; **b)** *(likely)* [durchaus *od.* gut] möglich; **c)** *(acceptable)* möglich; **there's no ~ excuse for it** dafür gibt es keine Entschuldigung; **the only ~ man for the position** der einzige Mann, der für die Stellung in Frage kommt. **2.** *n.* Anwärter, *der*/Anwärterin, *die;* Kandidat, *der*/Kandidatin, *die*

possibly ['pɒsɪblɪ] *adv.* **a)** *(by possible means)* **I cannot ~ commit myself** ich kann mich unmöglich festlegen; **how can I ~?** wie könnte ich?; **they did all they ~ could** sie haben alles Menschenmögliche getan; **as often as I ~ can** sooft ich irgend kann; **b)** *(perhaps)* möglicherweise; vielleicht; **Do you think ...? – P~:** Glaubst du ...? – Möglich[erweise] *od.* Vielleicht

[1]**post** [pəʊst] **1.** *n.* **a)** *(as support)* Pfosten, *der;* **b)** *(stake)* Pfahl, *der;* **deaf as a ~** *(coll.)* stocktaub *(ugs.)*; *see also* **pillar a; c)** *(Racing) (starting/finishing ~)* Start-/Zielpfosten, *der;* **be left at the ~:** [hoffnungslos] abgehängt werden *(ugs.)*; weit zurückbleiben; **the 'first past the ~' system** das Mehrheitswahlsystem; *see also* [5]**pip; d)** *(Sport: of goal)* Pfosten, *der.* **2.** *v. t.* **a)** *(stick up)* anschlagen, ankleben ‹*Plakat, Aufruf, Notiz, Zettel*›; **b)** *(make known)* [öffentlich] anschlagen *od.* bekanntgeben; **~ [as] missing** als vermißt melden

~ 'up *v. t.* anschlagen; ankleben; **~ up a notice** einen Anschlag machen

[2]**post 1.** *n.* **a)** *(Brit.: one dispatch of letters)* Postausgang, *der;* **by return of ~:** postwendend; **b)** *(Brit.: one collection of letters)* [Briefkasten]leerung, *die;* **c)** *(Brit.: one delivery of letters)* Post[zustellung], *die;* **in the ~:** bei der Post *(see also* **d)**; **the ~ has come** die Post ist da *od.* ist schon gekommen; **sort the ~:** die Posteingänge sortieren; **is there any ~ for me?** habe ich Post?; **d)** *no pl., no indef. art. (Brit.: official conveying)* Post, *die;* **by ~:** mit der Post; per Post; **in the ~:** in der Post *(see also* **c)**; **e)** *(~ office)* Post, *die; (~-box)* Briefkasten, *der;* **take sth. to the ~:** etw. zur Post bringen/*(to ~-box)* etw. einwerfen *od.* in den Briefkasten werfen. **2.** *v. t.* **a)** *(dispatch)* abschicken; **~ sb. sth.** jmdm. etw. schicken; **b)** *(fig. coll.)* **keep sb. ~ed [about** *or* **on sth.]** jmdn. [über etw. *(Akk.)*] auf dem laufenden halten

[3]**post 1.** *n.* **a)** *(job)* Stelle, *die;* Posten, *der;* **a teaching ~:** eine Stelle als Lehrer *od.* Lehrerstelle; **a diplomatic ~:** ein diplomatischer Posten; **b)** *(Mil.: place of duty)* Posten, *der; (fig.)* Platz, *der;* Posten, *der;* **take up one's ~** *(fig.)* seinen Platz einnehmen; **last/first ~** *(Brit. Mil.)* letzter/erster Zapfenstreich. **2.** *v. t.* **a)** *(place)* postieren; aufstellen; **b)** *(appoint)* einsetzen; **be ~ed to an embassy** an eine Botschaft versetzt werden

post- [pəʊst] *pref.* nach-/Nach-; post-/Post- *(mit Fremdwörtern)*

postage ['pəʊstɪdʒ] *n.* Porto, *das*

'postage stamp *n.* Briefmarke, *die*

postal ['pəʊstl] *adj.* Post-; postalisch ‹*Aufgabe, Einrichtung*›; *(by post)* per Post *nachgestellt*

postal: ~ district *n.* Zustellbezirk, *der;* **~ order** *n.* ≈ Postanweisung, *die;* **~ vote** *n.* Briefwahl, *die*

post: ~-bag *(Brit.) see* **mail-bag; ~-box** *n. (Brit.)* Briefkasten, *der;* **~card** *n.* Postkarte, *die;* **~ code** *n. (Brit.)* Postleitzahl, *die*

post-'date *v. t.* **a)** *(give later date to)* vordatieren; **b)** *(belong to later date than)* späteren *od.* jüngeren Datums sein als (+ *Nom.*)

poster ['pəʊstə(r)] *n.* **a)** *(placard)* Plakat, *das; (notice)* Anschlag, *der;* **b)** *(printed picture)* Plakat, *das;* Poster, *das*

posterior [pɒ'stɪərɪə(r)] *n. (joc.)* Hinterteil, *das (ugs.)*

posterity [pɒ'sterɪtɪ] *n., no pl., no art.* die Nachwelt; **go down to ~ [as sth.]** [als etw.] in die Geschichte eingehen

post-'free *(Brit.) adj., adv.* portofrei

post-'graduate 1. *adj.* Graduierten-; ⟨*College, Studiengang*⟩ für Graduierte; ~ **student** Graduierte, *der/die.* **2.** *n.* Graduierte, *der/die*

post-'haste *adv.* schnellstens

posthumous ['pɒstjʊməs] *adj.* **a)** nachgelassen, *(geh.)* postum ⟨*Buch usw.*⟩; **b)** *(occurring after death)* nachträglich; post[h]um *(geh.)*; ~ **fame** Nachruhm, *der*

post: ~man ['pəʊstmən], *pl.* **~men** ['pəʊstmən] *n.* Briefträger, *der;* Postbote, *der (ugs.)*; **~mark 1.** *n.* Poststempel, *der;* **2.** *v. t.* abstempeln; **the letter was ~marked 'Brighton'** der Brief war in Brighton abgestempelt; **~master** *n.* Postamtvorsteher, *der;* Postmeister, *der (veralt.)*

post-mortem [pəʊst'mɔ:təm] **1.** *adv.* nach dem Tode; post mortem *(fachspr.)*. **2.** *adj.* nach dem Tode eintretend; ~ **examination** Leichenschau, *die; (with dissection)* Obduktion, *die.* **3.** *n.* **a)** *(examination)* Obduktion, *die;* **b)** *(fig.)* nachträgliche Bewertung

post-natal [pəʊst'neɪtl] *adj.* nach der Geburt *nachgestellt;* postnatal *(fachspr.)*

post: ~ office *n.* **a)** *(organization)* **the P~ Office** die Post; **b)** *(place)* Postamt, *das;* Post, *die;* **~-office box** *n.* Postfach, *das;* **~-paid 1.** ['--] *adj.* frankiert; freigemacht; **2.** [-'-] *adv.* portofrei

postpone [pəʊst'pəʊn, pə'spəʊn] *v. t.* verschieben; *(for an indefinite period)* aufschieben; ~ **sth. until next week** etw. auf nächste Woche verschieben

postponement [pəʊst'pəʊnmənt, pə'spəʊnmənt] *n.* Verschiebung, *die; (for an indefinite period)* Aufschub, *der*

postpositive [pəʊst'pɒzɪtɪv] *adj. (Ling.)* nachgestellt; postpositiv *(fachspr.)*

postscript ['pəʊstskrɪpt, 'pəʊskrɪpt] *n.* Nachschrift, *die;* Postskript, *das; (fig.)* Nachtrag, *der*

postulate 1. ['pɒstjʊleɪt] *v. t. (claim as true, existent, necessary)* postulieren; ausgehen von; *(depend on)* voraussetzen; *(put forward)* aufstellen ⟨*Theorie*⟩. **2.** ['pɒstjʊlət] *n. (fundamental condition)* Postulat, *das (geh.)*; *(prerequisite)* Voraussetzung, *die;* Postulat, *das (geh.)*

posture ['pɒstʃə(r)] **1.** *n. (relative position)* [Körper]haltung, *die; (fig.: mental, political, military)* Haltung, *die.* **2.** *v. i.* posieren; *(strike a pose)* sich in Positur werfen *(ugs., leicht spött.)*

'post-war *adj.* Nachkriegs-; der Nachkriegszeit *nachgestellt*

posy ['pəʊzɪ] *n.* Sträußchen, *das*

¹pot [pɒt] **1.** *n.* **a)** *(cooking-vessel)* [Koch]topf, *der;* **go to ~** *(coll.)* den Bach runtergehen *(ugs.)*; **b)** *(container, contents)* Topf, *der; (tea~, coffee-~)* Kanne, *die;* **a ~ of tea** eine Kanne Tee; *(in café etc.)* ein Kännchen Tee; **c)** *(sl.: prize)* Preis, *der;* **d)** *(coll.: large sum)* **a ~ of/~s of** massenweise; jede Menge. **2.** *v. t.*, **-tt-: a)** *(put in container[s])* in einen Topf/in Töpfe füllen; **b)** *(put in plant-~)* ~ **[up]** eintopfen; ~ **out** austopfen; **c)** *(kill)* abschießen; abknallen *(ugs. abwertend)*; **d)** *(Brit. Billiards, Snooker)* einlochen

²pot *n. (sl.: marijuana)* Pot, *das (Jargon)*

potash ['pɒtæʃ] *n.* Kaliumkarbonat, *das;* Pottasche, *die (veralt.)*

potassium [pə'tæsɪəm] *n. (Chem.)* Kalium, *das*

potato [pə'teɪtəʊ] *n., pl.* **~es** Kartoffel, *die;* **a hot ~** *(fig. coll.)* ein heißes Eisen *(ugs.)*

'pot-belly *n.* Schmerbauch, *der (ugs.)*; Wampe, *die (ugs. abwertend)*; *(from malnutrition)* Blähbauch, *der*

potency ['pəʊtənsɪ] *n.* **a)** *(of drug)* Wirksamkeit, *die; (of alcoholic drink)* Stärke, *die; (Mil.)* Schlagkraft, *die; (of reason, argument)* Gewichtigkeit, *die; (influence)* Einfluß, *der;* Potenz, *die (geh.)*; **b)** *(of male)* **[sexual] ~:** [sexuelle] Potenz

potent ['pəʊtənt] *adj.* **a)** [hoch]wirksam ⟨*Droge*⟩; stark ⟨*Schnaps usw.*⟩; schlagkräftig ⟨*Waffe*⟩; gewichtig, schwerwiegend ⟨*Grund, Argument*⟩; wichtig, entscheidend ⟨*Faktor*⟩; stark ⟨*Motiv*⟩; *(influential)* einflußreich; potent *(geh.)*; **b)** *(sexually)* potent ⟨*Mann*⟩

potential [pə'tenʃl] **1.** *adj.* potentiell *(geh.)*; möglich. **2.** *n.* Potential, *das (geh.)*; Möglichkeiten; ~ **for growth/development** Wachstums- / Entwicklungspotential, *das;* **realize/reach one's ~:** seine Möglichkeiten ausschöpfen

pot: ~-hole 1. *n.* **a)** *(in road)* Schlagloch, *das;* **b)** *(deep cave)* [tiefe] Höhle; **2.** *v. i.* Höhlen erkunden; **~-holer** ['pɒthəʊlə(r)] *n.* [Hobby]höhlenforscher, *der/*-forscherin, *die;* **~-holing** ['pɒthəʊlɪŋ] *n.* Erkundung von Höhlen

potion ['pəʊʃn] *n.* Trank, *der*

pot 'luck *n.* **take ~:** sich überraschen lassen

pot-pourri [pəʊpʊə'ri:, pəʊ'pʊərɪ] *n.* Duftmischung, *die*

pot: ~-roast *n.* Schmorbraten, *der;* **~-shot** *n.* **a)** *(random shot)*; **take a ~-shot [at sb./sth.]** aufs Geratewohl [auf jmdn./etw.] schießen; **b)** *(fig.: critical remark)* Attacke, *die*

potted ['pɒtɪd] *adj.* **a)** *(preserved)* eingemacht; ~ **meat/fish** Fleisch-/Fischkonserven; **b)** *(planted)* Topf-; **c)** *(abridged)* kurzgefaßt

¹potter ['pɒtə(r)] *n.* Töpfer, *der*/Töpferin, *die*

²potter *v. i.* [he]rumwerkeln *(ugs.)*; ~ **round the shops** durch die Geschäfte bummeln

~ **a'bout,** ~ **a'round** *v. i.* herumwerkeln *(ugs.)* **(with** an + *Dat.*)

potter's 'wheel *n.* Töpferscheibe, *die*

pottery ['pɒtərɪ] *n.* **a)** *no pl., no indef. art. (vessels)* Töpferware, *die;* Keramik, *die; attrib.* Ton-; Keramik-; **b)** *(workshop, craft)* Töpferei, *die*

¹potty ['pɒtɪ] *adj. (Brit. sl.: crazy)* verrückt *(ugs.)* **(about, on** nach)

²potty *n. (Brit. coll.)* Töpfchen, *das*

pouch [paʊtʃ] *n.* **a)** *(small bag)* Tasche, *die;* Täschchen, *das;* **b)** *(under eye)* [Tränen]sack, *der;* **c)** *(ammunition-bag)* [Patronen]tasche, *die;* **d)** *(Zool.) (of marsupial)* Beutel, *der*

pouffe [pu:f] *n. (cushion)* Sitzpolster, *das;* Puff, *der*

poultry ['pəʊltrɪ] *n.* **a)** *constr. as pl. (birds)* Geflügel, *das;* **b)** *no pl., no indef. art. (as food)* Geflügel, *das*

pounce [paʊns] **1.** *v. i.* **a)** sich auf sein Opfer stürzen; ⟨*Raubvogel:*⟩ herabstoßen auf (+ *Akk.*); **b)** *(fig.)* ~ **[up]on/at** sich stürzen auf (+ *Akk.*). **2.** *n.* Sprung, *der;* Satz, *der*

¹pound [paʊnd] *n.* **a)** *(unit of weight)* [britisches] Pfund *(453,6 Gramm)*; **two ~s of apples** 2 Pfund Äpfel; **by the ~:** pfundweise; **b)** *(unit of currency)* Pfund, *das;* **five-~ note** Fünfpfundnote, *die;* Fünfpfundschein, *der*

²pound *n. (enclosure)* Pferch, *der; (for stray dogs)* Zwinger [für eingefangene Hunde]; *(for cars)* Abstellplatz [für polizeilich abgeschleppte Fahrzeuge]

³pound 1. *v. t.* **a)** *(crush)* zerstoßen; **b)** *(thump)* einschlagen auf (+ *Akk.*) ⟨*Person*⟩; klopfen ⟨*Fleisch*⟩; ⟨*Sturm:*⟩ heimsuchen ⟨*Gebiet, Insel*⟩; ⟨*Wellen:*⟩ klatschen auf (+ *Akk.*) ⟨*Strand,*

Ufer›, gegen *od.* an (+ *Akk.*) ‹*Felsen, Schiff*›; ‹*Geschütz:*› unter Beschuß *(Akk.)* nehmen ‹*Ziel*›; ‹*Bombenflugzeug:*› bombardieren ‹*Ziel*›; **c)** *(knock)* ~ **to pieces** ‹*Wellen:*› zertrümmern, zerschmettern ‹*Schiff*›; ‹*Geschütz, Bomben:*› in Trümmer legen ‹*Stadt, Mauern*›; **d)** *(compress)* ~ **[down]** feststampfen ‹*Erde, Boden*›. **2.** *v.i.* **a)** *(make one's way heavily)* stampfen; **b)** *(beat rapidly)* ‹*Herz:*› heftig schlagen *od.* klopfen *od.* *(geh.)* pochen

pounding ['paʊndɪŋ] *n.* **a)** *(striking) (of hammer etc.)* Schlagen, *das;* Klopfen, *das; (of artillery)* [schwerer] Beschuß; *(of waves)* Klatschen, *das;* **b)** *(of hooves, footsteps)* Stampfen, *das;* **c)** *(beating) (of heart)* Klopfen, *das;* Pochen, *das (geh.); (of music, drums)* Dröhnen, *das*

pour [pɔ:(r)] **1.** *v.t.* gießen, schütten ‹*Flüssigkeit*›; schütten ‹*Sand, Kies, Getreide usw.*›; *(into drinking-vessel)* einschenken; eingießen; *(fig.)* pumpen ‹*Geld, Geschosse*›; ~ **scorn** *or* **ridicule on sb./sth.** jmdn. mit Spott übergießen *od.* überschütten/über etw. *(Akk.)* spotten. **2.** *v.i.* **a)** *(flow)* strömen; ‹*Rauch:*› hervorquellen **(from** aus); **sweat was ~ing off the runners** den Läufern lief der Schweiß in Strömen herunter; ~ **[with rain]** in Strömen regnen; [in Strömen] gießen *(ugs.)*; **it never rains but it ~s** *(fig.)* da kommt aber auch alles zusammen; **b)** *(fig.)* strömen; ~ **in** herein-/hineinströmen; ~ **out** heraus-/hinausströmen; **letters/protests ~ed in** eine Flut von Briefen/Protesten brach herein

~ **'down** *v.i.* **it's ~ing down** es gießt [in Strömen] *(ugs.)*

~ **'forth 1.** *v.t.* von sich geben; ausschütten ‹*Kummer*›; erzählen ‹*Geschichte*›. **2.** *v.i.* ‹*Gesang, Musik usw.:*› ertönen, erklingen; ‹*Menge, Personen:*› herausströmen

~ **'off** *v.t.* abgießen

~ **'out 1.** *v.t.* eingießen, einschenken ‹*Getränk*›; ~ **out one's woes** *or* **troubles/heart to sb.** jmdm. seinen Kummer/sein Herz ausschütten. **2.** *v.i. see* ~ **2 b**

pouring ['pɔ:rɪŋ] *adj.* **a)** strömend ‹*Regen*›; **a** ~ **wet day** ein völlig verregneter Tag; **b)** *(for dispensing)* Gieß-

pout [paʊt] **1.** *v.i.* **~ing lips** Schmollippen *Pl.* **2.** *v.t.* aufwerfen, schürzen ‹*Lippen*›. **3.** *n.* Schmollmund, *der*

poverty ['pɒvətɪ] *n.* **a)** Armut, *die;* **be reduced to** ~: verarmt sein; **b)** *(fig.: deficiency)* Armut, *die* **(in** an + *Dat.*); ~ **of ideas** Ideenarmut, *die;* ~ **of imagination/intellect** Phantasielosigkeit, *die*

poverty: ~ **line** *n.* Armutsgrenze, *die;* **~-stricken** *adj.* notleidend; verarmt; *(fig.)* armselig; kümmerlich

POW *abbr.* **prisoner of war**

powder ['paʊdə(r)] **1.** *n.* **a)** Pulver, *das;* **b)** *(cosmetic)* Puder, *der.* **2.** *v.t.* **a)** pudern; **I'll just go and ~ my nose** *(euphem.)* ich muß [nur] mal verschwinden *(ugs. verhüll.)*; **b)** *(reduce to ~)* pulverisieren; zu Pulver verarbeiten ‹*Milch, Eier*›; **~ed milk** Milchpulver, *das;* Trockenmilch, *die*

'powder compact *see* ²**compact a**

powdering ['paʊdərɪŋ] *n.* [Ein]pudern, *das;* **a** ~ **of snow** eine dünne Schicht Schnee

powder: **~-keg** *n. (lit. or fig.)* Pulverfaß, *das;* **~-puff** *n.* Puderquaste, *die;* **~-room** *n.* [Damen]toilette, *die*

powdery ['paʊdərɪ] *adj.* **a)** *(like powder)* pulv[e]rig; *(in powder form)* pulverförmig; *(finer)* pud[e]rig/puderförmig; **b)** *(crumbly)* bröckelig; bröselig

power [paʊə(r)] **1.** *n.* **a)** *(ability)* Kraft, *die;* **do all in one's ~ to help sb.** alles in seiner Macht *od.* seinen Kräften Stehende tun, um jmdm. zu helfen; **b)** *(faculty)* Fähigkeit, *die;* Vermögen, *das (geh.)*; *(talent)* Begabung, *die;* Talent, *das;* **psychic ~s** übersinnliche Kräfte; **~s of persuasion** Überredungskünste; **c)** *(vigour, intensity) (of sun's rays)* Kraft, *die; (of sermon, performance)* Eindringlichkeit, *die; (solidity, physical strength)* Kraft, *die; (of a blow)* Wucht, *die;* **d)** *(authority)* Macht, *die,* Herrschaft, *die* **(over** über + *Akk.*); **she was in his ~:** sie war in seiner Gewalt; **e)** *(personal ascendancy)* **[exercise/get] ~:** Einfluß [ausüben/gewinnen] **(over** auf + *Akk.*); **f)** *(political or social ascendancy)* Macht, *die;* **hold ~:** an der Macht sein; **come into ~:** an die Macht kommen; **balance of ~:** Kräftegleichgewicht, *das;* **hold the balance of ~:** das Zünglein an der Waage sein; **g)** *(authorization)* Vollmacht, *die;* **h)** *(influential person)* Autorität, *die; (influential thing)* Machtfaktor, *der;* **be the ~ behind the throne** *(Polit.)* die graue Eminenz sein; **the ~s that be** die maßgeblichen Stellen; die da oben *(ugs.)*; **i)** *(State)* Macht, *die;* **j)** *(coll.: large amount)* Menge, *die (ugs.)*; **do sb. a ~ of good** jmdm. außerordentlich guttun; **k)** *(Math.)* Potenz, *die;* **3 to the ~ of 4** 3 hoch 4; **l)** *(mechanical, electrical)* Kraft, *die; (electric current)* Strom, *der; (of loudspeaker, engine, etc.)* Leistung, *die;* **m)** *(deity)* Macht, *die;* **the ~s of darkness** die Mächte der Finsternis. **2.** *v.t.* ‹*Treibstoff, Dampf, Strom, Gas:*› antreiben; ‹*Batterie:*› mit Energie versehen *od.* versorgen

power: **~-boat** *n.* Motorboot, *das;* ~ **cut** *n.* Stromsperre, *die;* Stromabschaltung, *die; (failure)* Stromausfall, *der;* ~ **drill** *n.* elektrische Bohrmaschine; ~ **failure** *n.* Stromausfall, *der*

powerful ['paʊəfl] *adj.* **a)** *(strong)* stark; kräftig ‹*Tritt, Schlag, Tier, Geruch, Körperbau*›; heftig ‹*Gefühl, Empfindung*›; hell, strahlend ‹*Licht*›; scharf ‹*Verstand, Geist*›; überzeugend ‹*Redner, Schauspieler*›; eindringlich ‹*Buch, Rede*›; beeindruckend ‹*Film, Darstellung*›; **b)** *(influential)*; mächtig ‹*Clique, Person, Herrscher*›; wesentlich ‹*Faktor*›

'power-house *n.* **a)** *see* **power station; b)** *(fig.)* treibende Kraft

powerless ['paʊəlɪs] *adj.* machtlos; **be ~ to do sth.** nicht die Macht haben, etw. zu tun

power: ~ **point** *n. (Brit.)* Steckdose, *die;* ~ **saw** *n.* Motorsäge, *die;* ~ **station** *n.* Kraftwerk, *das;* Elektrizitätswerk, *das*

pox [pɒks] *n.* **a)** *(disease with pocks)* Pocken *Pl.;* Blattern *Pl. (veralt.)*; **b)** *(coll.: syphilis)* Syphilis, *die;* Syph, *die od. der (salopp)*

p.p. [pi:'pi:] *abbr.* **by proxy** pp[a].

pp. *abbr.* **pages**

PR *abbr.* **a) proportional representation; b) public relations** PR; Public Relations; **PR man** Werbefachmann, *der;* PR-Mann, *der*

practicable ['præktɪkəbl] *adj.* *(feasible)* durchführbar ‹*Projekt, Idee, Plan*›; praktikabel ‹*Lösung, Plan*›

practical ['præktɪkl] *adj.* **a)** praktisch; **b)** *(inclined to action)* praktisch veranlagt ‹*Person*›; **have a ~ approach/mind** praktisch an die Dinge herangehen; **c)** *(virtual)* tatsächlich ‹*Freiheit, Organisator*›; **d)** *(feasible)* möglich ‹*Alternative*›; praktikabel ‹*Alternative, Möglichkeit*›

practicality [præktɪ'kælɪtɪ] *n.* **a)** *no pl. (of plan)* Durchführbarkeit, *die; (of person)* praktische Veranlagung; **b)** *in pl. (practical details)* **the practicalities of the situation are that ...:** die Situation sieht praktisch so aus, daß ...

practical 'joke *n.* Streich, *der;* **play ~s on sb.** jmdm. Streiche spielen

practical 'joker *n.* Witzbold, *der*

practically ['præktɪkəlɪ] *adv.* **a)** *(almost)* praktisch *(ugs.);* so gut wie; beinahe; **b)** *(in a practical manner)* praktisch

[1]**practice** ['præktɪs] *n.* **a)** *(repeated exercise)* Praxis, *die;* Übung, *die;* **put in** *or* **do some/a lot of ~:** üben/viel üben; **~ makes perfect** *(prov.)* Übung macht den Meister; **be out of ~, not be in ~:** außer Übung sein; **b)** *(spell)* Übungen *Pl.;* **piano ~:** Klavierüben, *das;* **c)** *(work or business of doctor, lawyer, etc.)* Praxis, *die; see also* **general practice; d)** *(habitual action)* übliche Praxis; Gewohnheit, *die;* **~ shows that ...:** die Erfahrung zeigt *od.* lehrt, daß ...; **good ~** *(sound procedure)* gutes Vorgehen; **e)** *(action)* Praxis, *die;* **in ~:** in der Praxis; in Wirklichkeit; **put sth. into ~:** etw. in die Praxis umsetzen; **f)** *(custom)* Gewohnheit, *die;* **regular ~:** Brauch, *der*

[2]**practice, practiced, practicing** *(Amer.) see* **practis-**

practise ['præktɪs] **1.** *v.t.* **a)** *(apply)* anwenden; praktizieren; **b)** *(be engaged in)* ausüben ‹*Beruf, Tätigkeit, Religion*›; **~ medicine** [als Arzt] praktizieren; **c)** *(exercise oneself in)* trainieren in (+ *Dat.*) ‹*Sportart*›; **~ the piano/flute** Klavier/Flöte üben. **2.** *v.i.* üben

practised ['præktɪst] *adj.* geübt ‹*Person, Auge, Blick*›; erfahren, versiert, routiniert ‹*Person*›; **with [a] ~ eye** mit geübtem Blick

practising ['præktɪsɪŋ] *adj.* praktizierend ‹*Arzt, Katholik, Anglikaner usw.*›; **~ homosexual** aktiv Homosexueller

practitioner [præk'tɪʃənə(r)] *n.* Fachmann, *der;* Praktiker, *der*/Praktikerin, *die;* **~ of the law, legal ~:** Anwalt, *der*/Anwältin, *die; see also* **general practitioner; medical practitioner**

pragmatic [præg'mætɪk] *adj.* pragmatisch

Prague [prɑ:g] *pr. n.* Prag *(das)*

prairie ['preərɪ] *n.* Grasland, *das;* Grassteppe, *die; (in North America)* Prärie, *die*

praise [preɪz] **1.** *v.t.* **a)** *(commend)* loben; *(more strongly)* rühmen; **~ sb. for doing sth.** jmdn. dafür loben, daß er etw. tut/getan hat; **b)** *(glorify)* preisen *(geh.), (dichter.)* lobpreisen ‹*Gott*›. **2.** *n.* **a)** *(approval)* Lob, *das;* **win high ~:** großes *od.* hohes Lob erhalten *od.* ernten; **sing one's own/sb.'s ~s** ein Loblied auf sich/jmdn. singen; **b)** *(worship)* Lobpreisung, *die (dichter.)*

praiseworthy ['preɪzwɜ:ðɪ] *adj.* lobenswert; löblich *(oft iron.)*

pram [præm] *n. (Brit.)* Kinderwagen, *der; (for dolls)* Puppenwagen, *der*

prance [prɑ:ns] *v.i.* **a)** ‹*Pferd:*› tänzeln; **b)** *(fig.)* stolzieren; ‹*Tänzer:*› tänzeln; **~ about** *or* **around** ‹*Kind, Tänzer:*› herumhüpfen

prank [præŋk] *n.* Streich, *der;* Schabernack, *der;* **play a ~ on sb.** jmdm. einen Streich *od.* Schabernack spielen

prattle ['prætl] **1.** *v.i.* ‹*Kleinkind:*› plappern *(ugs.);* schwafeln *(ugs. abwertend).* **2.** *n.* Geplapper, *das (ugs. abwertend);* Geschwafel, *das (ugs. abwertend)*

prawn [prɔ:n] *n.* Garnele, *die*

pray [preɪ] **1.** *v.i.* beten **(for** um); **let us ~:** lasset uns beten; **~ to God for help** Gott um Hilfe anflehen. **2.** *v.t.* **a)** *(beseech)* anflehen, flehen zu ‹*Gott, Heiligen, Jungfrau Maria*› **(for** um); **b)** *(~ to)* beten zu; **c)** *(ellipt.: I ask)* bitte

prayer [preə(r)] *n.* **a)** Gebet, *das;* **offer ~s for** beten für; **say one's ~s** beten; **b)** *no pl., no art. (praying)* Beten, *das;* **c)** *(entreaty)* inständige *od.* eindringliche Bitte

'prayerbook *n.* Gebetbuch, *das*

preach [pri:tʃ] **1.** *v.i.* predigen (**to** zu, vor + *Dat.;* **on** über + *Akk.*); *(fig.)* eine Predigt halten *(ugs.).* **2.** *v.t.* **a)** halten ‹*Predigt, Ansprache*›; predigen ‹*Evangelium, Botschaft*›; verkündigen ‹*Glauben, Lehre*›; **b)** *(advocate)* predigen *(ugs.)*

preacher ['pri:tʃə(r)] *n.* Prediger, *der*/Predigerin, *die*

preamble [pri:'æmbl] *n.* **a)** Vorbemerkung, *die;* Einleitung, *die; (to a book)* Geleitwort, *das;* **b)** *(Law)* Präambel, *die*

pre-arrange [pri:ə'reɪndʒ] *v.t.* vorher absprechen; vorher ausmachen *od.* verabreden ‹*Treffpunkt, Zeichen*›

precarious [prɪ'keərɪəs] *adj.* **a)** *(uncertain)* labil, prekär ‹*Gleichgewicht, Situation*›; gefährdet ‹*Friede*›; **make a ~ living** eine unsichere Existenz haben; **b)** *(insecure)* gefährlich ‹*Weg, Pfad*›; riskant, gefährlich ‹*Balanceakt, Leben*›; instabil *(geh.)* ‹*Bauwerk*›; unsicher ‹*Koalition*›

precaution [prɪ'kɔ:ʃn] *n.* Vorsichts-, Schutzmaßnahme, *die;* **take ~s against sth.** Vorsichts- *od.* Schutzmaßnahmen gegen etw. treffen; **do sth. as a ~:** vorsichts- *od.* sicherheitshalber etw. tun

precautionary [prɪ'kɔ:ʃənərɪ] *adj.* vorsorglich; vorbeugend; prophylaktisch *(geh., Med.);* **as a ~ measure** vorsichts- *od.* sicherheitshalber

precede [prɪ'si:d] *v.t.* **a)** *(in rank)* rangieren vor (+ *Dat.*); *(in importance)* wichtiger sein als; Vorrang haben vor (+ *Dat.*); **b)** *(in order or time)* vorangehen (+ *Dat.*); *(in vehicle)* voranfahren (+ *Dat.*); *(in time also)* vorausgehen (+ *Dat.*); **c)** *(preface, introduce)* **~ X with Y** X *(Dat.)* Y vorausschicken *od.* voranstellen

precedence ['presɪdəns] *n., no pl.* Priorität, *die (geh.)* **(over** vor + *Dat.*, gegenüber); Vorrang, *der* **(over** vor + *Dat.*)

precedent ['presɪdənt] *n.* Präzedenzfall, *der;* **set** *or* **create a ~:** ein Präzedenzfall schaffen

precept ['pri:sept] *n.* Grundsatz, *der;* Prinzip, *das*

precinct ['pri:sɪŋkt] *n.* **a)** *(traffic-free area)* **[pedestrian] ~:** Fußgängerzone, *die;* **[shopping] ~:** für den Verkehr weitgehend gesperrtes Einkaufsviertel; **b)** *(enclosed area)* Bereich, *der;* Bezirk, *der*

precious ['preʃəs] **1.** *adj.* **a)** *(costly)* wertvoll, kostbar ‹*Schmuckstück*›; Edel‹*metall, -stein*›; **b)** *(highly valued)* wertvoll, kostbar ‹*Zeit, Eigenschaft*›; **c)** *(beloved)* teuer *(geh.);* lieb; **d)** *(affected)* affektiert; **e)** *(coll.: considerable)* beträchtlich; erheblich. **2.** *adv. (coll.)* herzlich ‹*wenig, wenige*›

precipice ['presɪpɪs] *n.* Abgrund, *der*

precipitate 1. [prɪ'sɪpɪtət] *adj.* eilig ‹*Flucht*›; hastig ‹*Abreise*›; übereilt, überstürzt ‹*Tat, Entschluß, Maßnahme*›; groß, fliegend ‹*Eile*›. **2.** [prɪ'sɪpɪteɪt] *v.t.* **a)** *(throw down)* hinunterschleudern; **b)** *(hasten)* beschleunigen; *(trigger)* auslösen

precipitation [prɪsɪpɪ'teɪʃn] *n. (Meteorol.)* Niederschlag, *der*

precipitous [prɪ'sɪpɪtəs] *adj.* **a)** *(very steep)* sehr steil ‹*Schlucht, Abhang, Treppe, Weg*›; schroff ‹*Abhang, Felswand*›; **~ slope/drop** Steilhang, *der*/[steiler] Absturz; **b)** *see* **precipitate 1**

précis ['preɪsi:] **1.** *n., pl. same*

[preisi:z] Inhaltsangabe, *die;* Zusammenfassung, *die.* **2.** *v.t.* zusammenfassen

precise [prɪ'saɪs] *adj.* genau; präzise; fein ‹*Instrument*›*;* groß ‹*Genauigkeit*›*;* förmlich ‹*Art*›*;* **be [more] ~:** sich präzise[r] ausdrücken; **what are your ~ intentions?** was genau hast du vor?; ..., **to be ~:** ..., um genau zu sein; ..., genauer gesagt; **at that ~ moment** genau in dem Augenblick

precisely [prɪ'saɪslɪ] *adv.* genau; präzise ‹*antworten*›*;* **speak ~:** sich präzise ausdrücken; **that is ~ what/why ...:** genau das/deswegen ...; **what ~ do you want/mean?** was willst/meinst du eigentlich genau?; **at ~ 1.30, at 1.30 ~:** Punkt 1 Uhr 30

precision [prɪ'sɪʒn] *n., no pl.* Genauigkeit, *die; attrib.* **a ~ landing** eine Präzisionslandung

precision 'instrument *n.* Präzisions[meß]gerät, *das;* Feinmeßgerät, *das*

preclude [prɪ'klu:d] *v.t.* ausschließen ‹*Zweifel*›*;* **~ sb. from a duty/taking part** jmdn. von einer Pflicht entbinden/von der Teilnahme ausschließen

precocious [prɪ'kəʊʃəs] *adj.* frühreif ‹*Kind, Jugendlicher, Genie*›*;* altklug ‹*Äußerung*›

preconceived [pri:kən'si:vd] *adj.* vorgefaßt ‹*Ansicht, Vorstellung*›

preconception [pri:kən'sepʃn] *n.* vorgefaßte Meinung (**of** über + *Akk.*)

precondition [pri:kən'dɪʃn] *n.* Vorbedingung, *die* (**of** für)

pre-cooked [pri:'kʊkt] *adj.* vorgekocht

precursor [pri:'kɜ:sə(r)] *n.* **a)** *(of revolution, movement, etc.)* Wegbereiter, *der*/-bereiterin, *die;* **b)** *(predecessor)* Vorgänger, *der*/-gängerin, *die*

pre-date [pri:'deɪt] *v.t.* **~ sth.** ‹*Ereignis:*› einer Sache *(Dat.)* vorausgehen; ‹*Sache:*› aus der Zeit vor etw. *(Dat.)* stammen

predator ['predətə(r)] *n.* Raubtier, *das; (fish)* Raubfisch, *der*

predatory ['predətərɪ] *adj.* räuberisch; **~ animal** Raubtier, *das*

predecessor ['pri:dɪsesə(r)] *n.* **a)** *(former holder of position)* Vorgänger, *der*/-gängerin, *die;* **b)** *(preceding thing)* Vorläufer, *der*

predestination [pri:destɪ'neɪʃn] *n., no pl.* Vorherbestimmung, *die*

predestine [pri:'destɪn] *v.t.* von vornherein bestimmen (**to** zu)

predetermine [pri:dɪ'tɜ:mɪn] *v.t.* im voraus *od.* von vornherein bestimmen; ‹*Gott, Schicksal:*› vorherbestimmen

predicament [prɪ'dɪkəmənt] *n.* Dilemma, *das;* Zwangslage, *die*

predicate ['predɪkət] *n. (Ling.)* Prädikat, *das*

predicative [prɪ'dɪkətɪv] *adj.* *(Ling.)* prädikativ

predict [prɪ'dɪkt] *v.t.* voraus-, vorhersagen; prophezeien; voraus-, vorhersehen ‹*Folgen*›

predictable [prɪ'dɪktəbl] *adj.* voraus-, vorhersagbar; voraus-, vorhersehbar ‹*Folgen, Reaktion, Ereignis*›*;* berechenbar ‹*Person*›

predictably [prɪ'dɪktəblɪ] *adv.* wie voraus- *od.* vorherzusehen war

prediction [prɪ'dɪkʃn] *n.* Voraus-, Vorhersage, *die*

predilection [pri:dɪ'lekʃn] *n.* Vorliebe, *die* (**for** für)

predispose [pri:dɪ'spəʊz] *v.t.* **be ~d to do sth.** *(be willing to do sth.)* geneigt sein, etw. zu tun; *(tend to do sth.)* dazu neigen, etw. zu tun; **~ sb. in favour of sb./sth.** jmdn. für jmdn./etw. einnehmen

predisposition [pri:dɪspə'zɪʃn] *n.* Neigung, *die* (**to** zu)

predominance [prɪ'dɒmɪnəns] *n.* **a)** *(control)* Vorherrschaft, *die* (**over** über + *Akk.*); **b)** *(majority)* Überzahl, *die* (**of** von)

predominant [prɪ'dɒmɪnənt] *adj.* **a)** *(having more power)* dominierend; **b)** *(prevailing)* vorherrschend

predominantly [prɪ'dɒmɪnəntlɪ] *adv.* überwiegend

predominate [prɪ'dɒmɪneɪt] *v.i.* *(be more powerful)* dominierend sein; *(be more important)* vorherrschen; überwiegen; *(be more numerous)* in der Überzahl sein

pre-eminence [pri:'emɪnəns] *n., no pl.* Vorrangstellung, *die;* **her ~ in this field** ihre herausragende Stellung auf diesem Gebiet

pre-eminent [pri:'emɪnənt] *adj.* herausragend; **be ~:** eine herausragende Stellung einnehmen

pre-eminently [pri:'emɪnəntlɪ] *adv.* herausragend; *(mainly)* vor allem; in erster Linie

pre-empt [pri:'empt] *v.t.* *(forestall)* zuvorkommen (+ *Dat.*)

pre-emptive [pri:'emptɪv] *adj.* *(Mil.)* Präventiv‹*krieg, -maßnahme, -schlag*›

preen [pri:n] **1.** *v.t.* ‹*Vogel:*› putzen ‹*Federn, Gefieder*›. **2.** *v. refl.* ‹*Vogel:*› sich putzen; ‹*Person:*› sich herausputzen

prefab ['pri:fæb] *n. (coll.) (house)* Fertighaus, *das; (building)* Fertigbau, *der*

prefabricate [pri:'fæbrɪkeɪt] *v.t.* vorfertigen

prefabricated *adj.* [pri:'fæbrɪkeɪtɪd] vorgefertigt; **~ house/building** Fertighaus, *das*/Fertigbau, *der*

preface ['prefəs] **1.** *n.* Vorwort, *das* (**to** *Gen.*). **2.** *v.t. (introduce)* einleiten

prefect ['pri:fekt] *n. (Sch.) die Aufsicht führender älterer Schüler/führende ältere Schülerin*

prefer [prɪ'fɜ:(r)] *v.t.,* **-rr-: a)** *(like better)* vorziehen; **~ to do sth.** etw. lieber tun; es vorziehen, etw. zu tun; **~ sth. to sth.** etw. einer Sache *(Dat.)* vorziehen; **I ~ skiing to skating** ich fahre lieber Ski als Schlittschuh; **I ~ not to talk about it** darüber möchte ich lieber nicht sprechen; **I should ~ to wait** ich würde lieber warten; **he ~s blondes** er bevorzugt Blondinen; **I ~ water to wine** ich trinke lieber Wasser als Wein; **b)** *(submit)* erheben ‹*Anklage, Anschuldigungen*› (**against** gegen, **for** wegen)

preferable ['prefərəbl] *adj.* vorzuziehen *präd.;* vorzuziehend *attr.;* besser (**to** als); **the cold was ~ to the smoke** die Kälte war noch erträglicher als der Rauch

preferably ['prefərəblɪ] *adv.* am besten; *(as best liked)* am liebsten; **a piano, ~ not too expensive** ein möglichst nicht zu teures Klavier; **Wine or beer? – Wine, ~!** Wein oder Bier? – Lieber Wein!

preference ['prefərəns] *n.* **a)** *(greater liking)* Vorliebe, *die;* **for ~** *see* **preferably; have a ~ for sth. [over sth.]** etw. [einer Sache *(Dat.)*] vorziehen; **do sth. in ~ to sth. else** etw. lieber als etw. anderes tun; **b)** *(thing preferred)* **what are your ~s?** was wäre dir am liebsten?; **I have no ~:** mir ist alles gleich recht; **c)** *(favouring of one person or country)* Präferenzbehandlung, *die;* **give [one's] ~ to sb.** jmdn. bevorzugen; **give sb. ~ over others** jmdm. anderen gegenüber Vergünstigungen einräumen; **d)** *attrib. (Brit. Finance)* Vorzugs-, Prioritäts‹*obligation, -aktie*›

preferential [prefə'renʃl] *adj.* bevorzugt ‹*Behandlung*›*;* bevorrechtigt ‹*Ansprache, Stellung*›*;* **give sb. ~ treatment** jmdn. bevorzugt behandeln

prefix *(Ling.)* **1.** ['pri:fɪks, prɪ:'fɪks] *v.t.* als Präfix setzen (**to** vor + *Akk.*). **2.** ['pri:fɪks] *n.* Präfix, *das;* Vorsilbe, *die*

pregnancy ['pregnənsɪ] *n. (of*

woman) Schwangerschaft, *die; (of animal)* Trächtigkeit, *die*

'**pregnancy test** *n.* Schwangerschaftstest, *der*

pregnant ['pregnənt] *adj.* **a)** schwanger ⟨*Frau*⟩*;* trächtig ⟨*Tier*⟩*;* **she is ~ with her second child** sie erwartet ihr zweites Kind; **be six months ~:** im siebten Monat schwanger sein; **b)** *(fig.: momentous)* bedeutungsschwer *(geh.)*; **~ with meaning** bedeutungsschwanger

pre-heat [pri:'hi:t] *v.t.* vorheizen ⟨*Backofen*⟩*;* vorher erwärmen ⟨*Gas, Werkzeug*⟩

prehensile [prɪ'hensaɪl] *adj. (Zool.)* Greif⟨*-fuß, -schwanz*⟩

prehistoric [pri:hɪ'stɒrɪk] *adj.* **a)** vorgeschichtlich; prähistorisch; **b)** *(coll.) (ancient)* uralt *(ugs.); (out of date)* vorsintflutlich *(ugs.)*

prehistory [pri:'hɪstərɪ] *n.* Vorgeschichte, *die*

prejudge [pri:'dʒʌdʒ] *v.t.* **a)** *(form premature opinion about)* vorschnell *od.* voreilig urteilen über (+ *Akk.*); **b)** *(judge before trial)* im voraus beurteilen, vorverurteilen ⟨*Person*⟩*;* im voraus entscheiden ⟨*Fall*⟩

prejudice ['predʒʊdɪs] **1.** *n.* **a)** *(bias)* Vorurteil, *das;* **colour ~:** Vorurteil aufgrund der Hautfarbe; **overcome ~:** Vorurteile ablegen; **b)** *(injury)* Schaden, *der;* Nachteil, *der;* **without ~** *(Law)* unbeschadet aller Rechte. **2.** *v.t.* **a)** *(bias)* beeinflussen; **~ sb. in sb.'s favour/against sb.** jmdn. für/gegen jmdn. einnehmen; **b)** *(injure)* beeinträchtigen

prejudiced ['predʒʊdɪst] *adj.* voreingenommen (**about** gegenüber, **against** gegen); **be racially ~:** Rassenvorurteile haben

prejudicial [predʒʊ'dɪʃl] *adj.* abträglich *(geh.)* (**to** *Dat.*); nachteilig (**to** für); **be ~ to** beeinträchtigen ⟨*Anspruch, Chance, Recht*⟩*;* schaden (+ *Dat.*) ⟨*Interesse*⟩

prelate ['prelət] *n.* Prälat, *der*

prelim ['pri:lɪm] *n. (coll.: exam)* Vorprüfung, *die*

preliminary [prɪ'lɪmɪnərɪ] **1.** *adj.* Vor-; vorbereitend ⟨*Forschung, Maßnahme*⟩*;* einleitend ⟨*Kapitel, Vertragsbestimmungen*⟩*;* **~ inquiry/search** erste Nachforschung/Suche. **2.** *n., usu. in pl.* **preliminaries** Präliminarien *Pl.; (Sports)* Ausscheidungskämpfe; **as a ~ to sth.** *(as a preparation)* als Vorbereitung auf etw. *(Akk.);* **just a ~:** nur ein Vorspiel (**to** zu); **without any further preliminaries** ohne [weitere] Umschweife

prelude ['prelju:d] *n.* **a)** *(introduction)* Anfang, *der* (**to** *Gen.*); Auftakt, *der* (**to** zu); **b)** *(of play)* Vorspiel, *das* (**to** zu); **c)** *(Mus.)* Präludium, *das;* Vorspiel, *das*

pre-marital [pri:'mærɪtl] *adj.* vorehelich; **~ sex** Geschlechtsverkehr vor der Ehe

premature ['premətjʊə(r)] *adj.* **a)** *(hasty)* voreilig, übereilt ⟨*Entscheidung, Handeln*⟩*;* **b)** *(early)* früh-, vorzeitig ⟨*Altern, Ankunft, Haarausfall*⟩*;* verfrüht ⟨*Bericht, Eile, Furcht*⟩*;* **~ baby** Frühgeburt, *die;* **the baby was five weeks ~:** das Baby wurde fünf Wochen zu früh geboren

prematurely ['premətjʊəlɪ] *adv. (early)* vorzeitig; zu früh ⟨*geboren werden*⟩*; (hastily)* voreilig, übereilt ⟨*entscheiden, handeln*⟩

premeditated [pri:'medɪteɪtɪd] *adj.* vorsätzlich

premier ['premɪə(r)] *n.* Premier[minister], *der*/Premierministerin, *die*

première ['premjeə(r)] *n. (of production)* Premiere, *die;* Erstaufführung, *die; (of work)* Uraufführung, *die*

premise ['premɪs] *n.* **a)** *in pl. (building)* Gebäude, *das; (buildings and land of factory or school)* Gelände, *das; (rooms)* Räumlichkeiten *Pl.;* **on the ~s** hier/dort; *(of public house, restaurant, etc.)* im Lokal; **b)** *see* **premiss**

premiss ['premɪs] *n. (Logic)* Prämisse, *die*

premium ['pri:mɪəm] *n.* **a)** *(Insurance)* Prämie, *die;* **b)** *(reward)* Preis, *der;* Prämie, *die;* **put a ~ on sth.** *(attach special value to)* etw. [hoch ein]schätzen; **c)** *(St. Exch.)* Agio, *das;* Aufgeld, *das;* **be at a ~:** über pari stehen; *(fig.: be highly valued)* sehr gefragt sein

'**Premium Bond** *n. (Brit.)* Prämienanleihe, *die;* Losanleihe, *die*

premonition [premə'nɪʃn] *n.* **a)** *(forewarning)* Vorwarnung, *die;* **b)** *(presentiment)* Vorahnung, *die;* **feel/have a ~ of sth.** eine Vorahnung von etw. haben

preoccupation [prɪɒkjʊ'peɪʃn] *n.* Sorge, *die* (**with** um); **first** *or* **greatest** *or* **main ~:** Hauptanliegen, *das;* Hauptsorge, *die*

preoccupied [prɪ'ɒkjʊpaɪd] *adj. (lost in thought)* gedankenverloren; *(concerned)* besorgt (**with** um); *(absorbed)* beschäftigt (**with** mit)

preoccupy [prɪ'ɒkjʊpaɪ] *v.t.* beschäftigen

prep [prep] *n. (Brit. Sch. coll.)* [Haus-, Schul]aufgaben *Pl.*

pre-packaged [pri:'pækɪdʒd], **pre-packed** [pri:'pækt] *adjs.* abgepackt

prepaid *see* **prepay**

preparation [prepə'reɪʃn] *n.* **a)** Vorbereitung, *die;* **be in ~** ⟨*Publikation:*⟩ in Vorbereitung sein; **in ~ for the new baby/term** als Vorbereitung auf das neue Baby/Semester; **b)** *in pl. (things done to get ready)* Vorbereitungen *Pl.* (**for** für); **~s for war/the wedding** Kriegs-/Hochzeitsvorbereitungen; **make ~s for sth.** Vorbereitungen für etw. treffen; **c)** *(Chem., Med., Pharm.)* Präparat, *das*

preparatory [prɪ'pærətərɪ] **1.** *adj.* vorbereitend ⟨*Schritt, Maßnahme*⟩*;* Vor⟨*ermittlung, -untersuchung*⟩*;* **~ work** Vorarbeiten *Pl.* **2.** *adv.* **~ to sth.** vor etw. *(Dat.);* **~ to doing sth.** bevor man etw. tut

pre'paratory school *n.* **a)** *(Brit. Sch.) für die Aufnahme an einer Public School vorbereitende Privatschule;* **b)** *(Amer. Univ.) meist private, für die Aufnahme an einem College vorbereitende Schule*

prepare [prɪ'peə(r)] **1.** *v.t.* **a)** *(make ready)* vorbereiten; entwerfen, ausarbeiten ⟨*Plan, Rede*⟩*;* herrichten *(ugs.),* fertigmachen ⟨*Gästezimmer*⟩*; (make mentally ready, equip with necessary knowledge)* vorbereiten ⟨*Person*⟩ (**for** auf + *Akk.*); **~ oneself for a shock/the worst** sich auf einen Schock/das Schlimmste gefaßt machen; **be ~d for anything** auf alles gefaßt sein; **be ~d to do sth.** *(be willing)* bereit sein, etw. zu tun; **b)** *(make)* herstellen ⟨*Chemikalie, Metall usw.*⟩*;* zubereiten ⟨*Essen*⟩. **2.** *v.i.* sich vorbereiten (**for** auf + *Akk.*); **~ for battle/war** ⟨*Land:*⟩ zum Kampf/Krieg rüsten; **~ to do sth.** sich bereit machen etw. zu tun

prepay [pri:'peɪ] *v.t.,* **prepaid** [pri:'peɪd] im voraus [be]zahlen; *(pay postage of)* frankieren, freimachen ⟨*Brief, Paket usw.*⟩*;* **prepaid envelope** frankierter Umschlag; Freiumschlag, *der*

preponderance [prɪ'pɒndərəns] *n.* Überlegenheit, *die* (**over** über + *Akk.,* gegenüber); Übergewicht, *das*

preposition [prepə'zɪʃn] *n. (Ling.)* Präposition, *die;* Verhältniswort, *das*

prepositional [prepə'zɪʃənl] *adj. (Ling.)* präpositional; Präpositional⟨*attribut, -fall, -objekt*⟩

prepossessing [pri:pə'zesɪŋ]

adj. einnehmend, anziehend ‹*Äußeres, Erscheinung, Person, Lächeln usw.*›
preposterous [prɪ'pɒstərəs] *adj.* absurd; grotesk ‹*Äußeres, Kleidung*›
'**prep school** *(coll.) see* **preparatory school**
prerequisite [pri:'rekwɪzɪt] **1.** *n.* [Grund]voraussetzung, *die.* **2.** *adj.* unbedingt erforderlich
prerogative [prɪ'rɒgətɪv] *n.* Privileg, *das;* Vorrecht, *das*
Presbyterian [prezbɪ'tɪərɪən] **1.** *adj.* presbyterianisch. **2.** *n.* Presbyterianer, *der*/Presbyterianerin, *die*
pre-school ['pri:sku:l] *adj.* Vorschul-; ~ **years** Vorschulalter, *das*
prescribe [prɪ'skraɪb] *v. t.* **a)** *(impose)* vorschreiben; **b)** *(Med.; also fig.)* verschreiben; verordnen
prescription [prɪ'skrɪpʃn] *n.* **a)** *(prescribing)* Anordnung, *die;* Vorschreiben, *das;* **b)** *(Med.)* Rezept, *das; (medicine)* [verordnete *od.* verschriebene] Medizin; Verordnung, *die (fachspr.);* **be available only on** ~: nur auf Rezept zu bekommen sein
pre'scription charge *n.* Rezeptgebühr, *die*
prescriptive [prɪ'skrɪptɪv] *adj.* *(Ling.)* präskriptiv
presence ['prezəns] *n.* **a)** *(being present) (of person)* Gegenwart, *die;* Anwesenheit, *die; (of things)* Vorhandensein, *das;* **in the ~ of his friends** in Gegenwart *od.* Anwesenheit seiner Freunde; **make one's ~ felt** sich bemerkbar machen; **b)** *(bearing)* Auftreten, *das;* **c)** *(being represented)* Präsenz, *die;* **police ~**: Polizeipräsenz, *die;* **d)** ~ **of mind** Geistesgegenwart, *die*
[1]**present** ['prezənt] **1.** *adj.* **a)** anwesend, *(geh.)* zugegen (**at** bei); **be ~ in the air/water/in large amounts** in der Luft/im Wasser/in großen Mengen vorhanden sein; **all ~ and correct** *(joc.)* alle sind da; **all those ~**: alle Anwesenden; ~ **company excepted** Anwesende ausgenommen; **b)** *(being dealt with)* betreffend; **in the ~ case** im vorliegenden Fall; **c)** *(existing now)* gegenwärtig; jetzig, derzeitig ‹*Bischof, Chef usw.*›; **d)** *(Ling.)* ~ **tense** Präsens, *das;* Gegenwart, *die; see also* **participle. 2.** *n.* **a) the ~**: die Gegenwart; **up to the ~**: bis jetzt; bisher; **at ~**: zur Zeit; **for the ~**: vorläufig; **b)** *(Ling.)* Präsens, *das;* Gegenwart, *die*
[2]**present 1.** ['prezənt] *n. (gift)* Geschenk, *das;* **parting ~**: Abschiedsgeschenk, *das;* **make a ~ of sth. to sb., make sb. a ~ of sth.** jmdm. etw. zum Geschenk machen; *see also* **give 1b. 2.** [prɪ'zent] *v. t.* **a)** schenken; überreichen ‹*Preis, Medaille*›; ~ **sth. to sb.** *or* **sb. with sth.** jmdm. etw. schenken *od.* zum Geschenk machen; ~ **sb. with difficulties/a problem** jmdn. vor Schwierigkeiten/ein Problem stellen; **he was ~ed with an opportunity that ...**: ihm bot sich eine Gelegenheit, die ...; **b)** *(deliver)* überreichen ‹*Gesuch*› (**to** bei); vorlegen ‹*Scheck, Bericht, Rechnung*› (**to** *Dat.*); ~ **one's case** seinen Fall darlegen; **c)** *(exhibit)* zeigen; bereiten ‹*Schwierigkeit*›; aufweisen ‹*Aspekt*›; **d)** *(introduce)* vorstellen (**to** *Dat.*); **e)** *(to the public)* geben, aufführen ‹*Theaterstück*›; zeigen ‹*Film*›; moderieren ‹*Sendung*›; bringen ‹*Fernsehserie, Schauspieler in einer Rolle*›; vorstellen ‹*Produkt usw.*›; vorlegen ‹*Abhandlung*›; **f)** ~ **arms!** *(Mil.)* präsentiert das Gewehr! **3.** *v. refl.* ‹*Problem:*› auftreten; ‹*Möglichkeit:*› sich ergeben; ~ **oneself for interview/an examination** zu einem Gespräch/einer Prüfung erscheinen
presentable [prɪ'zentəbl] *adj.* ansehnlich; **the flat is not very ~ at the moment** die Wohnung ist im Augenblick nicht besonders präsentabel; **make oneself/sth. ~**: sich/etw. zurechtmachen; **I'm not ~**: ich kann mich nicht so zeigen
presentation [prezən'teɪʃn] *n.* **a)** *(giving)* Schenkung, *die; (of prize, medal, gift)* Überreichung, *die;* **b)** *(ceremony)* Verleihung, *die;* ~ **of the awards/medals** Preis-/Ordensverleihung, *die;* **c)** *(delivering) (of petition)* Überreichung, *die; (of cheque, report, account)* Vorlage, *die; (of case, position, thesis)* Darlegung, *die;* **on ~ of** gegen Vorlage (+ *Gen.*); **d)** *(exhibition)* Darstellung, *die;* **e)** *(Theatre, Radio, Telev.)* Darbietung, *die; (Theatre also)* Inszenierung, *die; (Radio, Telev. also)* Moderation, *die;* **f)** *(introduction)* Vorstellung, *die*
present-'day *adj.* heutig; zeitgemäß ‹*Einstellungen, Ansichten*›
presenter [prɪ'zentə(r)] *n. (Radio, Telev.)* Moderator, *der*/Moderatorin, *die*
presentiment [prɪ'zentɪmənt] *n.* Vorahnung, *die;* **have a ~ that ...**: vorausahnen, daß ...
presently ['prezəntlɪ] *adv.* **a)** *(soon)* bald; **b)** *(Amer., Scot.: now)* zur Zeit; derzeit
preservation [prezə'veɪʃn] *n., no pl.* **a)** *(action)* Erhaltung, *die; (of leather, wood, etc.)* Konservierung, *die;* **b)** *(state)* Erhaltungszustand, *der*
preservative [prɪ'zɜ:vətɪv] *n.* Konservierungsmittel, *das*
preserve [prɪ'zɜ:v] **1.** *n.* **a)** *in sing. or pl. (fruit)* Eingemachte, *das;* **strawberry/quince ~s** eingemachte Erdbeeren/Quitten; **b)** *(jam)* Konfitüre, *die;* **c)** *(fig.: special sphere)* Domäne, *die (geh.);* **d) wildlife/game ~**: Tierschutzgebiet, *das*/Wildpark *der.* **2.** *v. t.* **a)** *(keep safe)* schützen (**from** vor + *Dat.*); ~ **sth. from destruction** etw. vor der Zerstörung bewahren; **b)** *(maintain)* aufrechterhalten ‹*Disziplin*›; bewahren ‹*Sehfähigkeit, Brauch, Würde*›; behalten ‹*Stellung*›; wahren ‹*Anschein, Reputation*›; ~ **the peace** den Frieden bewahren *od.* erhalten; **c)** *(retain)* speichern ‹*Hitze*›; bewahren ‹*Haltung, Distanz, Humor*›; **d)** *(prepare, keep from decay)* konservieren; *(bottle)* einmachen ‹*Obst, Gemüse*›; **e)** *(keep alive)* erhalten; *(fig.)* bewahren ‹*Erinnerung, Andenken*›; **Heaven ~ us!** [Gott] bewahre!; **f)** *(care for and protect)* hegen ‹*Tierart, Wald*›; unter Schutz stellen ‹*Gewässer, Gebiet*›
pre-set [pri:'set] *v. t., forms as* **set 1** vorher einstellen
pre-shrunk [pri:'ʃrʌŋk] *adj. (Textiles)* vorgeschrumpft, vorgewaschen ‹*Jeans*›
preside [prɪ'zaɪd] *v. i.* **a)** *(at meeting etc.)* den Vorsitz haben (**at** bei); präsidieren, vorsitzen (**over** *Dat.*); **b)** *(exercise control)* ~ **over** leiten ‹*Abteilung, Organisation, Programm*›
presidency ['prezɪdənsɪ] *n.* **a)** Präsidentschaft, *die;* **b)** *(of society, legislative body)* Vorsitz, *der;* **c)** *(Univ., esp. Amer.)* Präsidentschaft, *die;* Rektorat, *das*
president ['prezɪdənt] *n.* **a)** Präsident, *der*/Präsidentin, *die;* **b)** *(of society, council, legislative body)* Vorsitzende, *der/die;* **c)** *(Univ., esp. Amer.)* Präsident, *der*/Präsidentin, *die;* Rektor, *der*/Rektorin, *die*
presidential [prezɪ'denʃl] *adj.* Präsidenten-; ~ **campaign** Präsidentschaftswahlkampf, *der*
[1]**press** [pres] **1.** *n.* **a)** *(newspapers etc.)* Presse, *die; attrib.* Presse-; der Presse *nachgestellt;* **get/have**

a good/bad ~ *(fig.)* eine gute/schlechte Presse bekommen/haben; **b)** *see* **printing-press; c)** *(printing-house)* Druckerei, *die;* **at** *or* **in [the] ~:** im Druck; **send to [the] ~:** in Druck geben; **go to [the] ~:** in Druck gehen: **d)** *(publishing firm)* Verlag, *der;* **e)** *(for flattening, compressing, etc.)* Presse, *die; (for sports racket)* Spanner, *der;* **f)** *(crowd)* Menge, *die;* **g)** *(~ing)* Druck, *der;* **give sth. a ~:** etw. drücken. **2.** *v. t.* **a)** drücken; pressen; drücken auf (+ *Akk.*) ‹*Klingel, Knopf*›; treten auf (+ *Akk.*) ‹*Gas-, Brems-, Kupplungspedal usw.*›; **~ the trigger** abdrücken; **b)** *(urge)* drängen ‹*Person*›; *(force)* aufdrängen (**[up]on** *Dat.*); *(insist on)* nachdrücklich vorbringen ‹*Forderung, Argument, Vorschlag*›; **~ sb. for an answer** jmdn. zu einer Antwort drängen; **he did not ~ the point** er ließ die Sache auf sich beruhen; **c)** *(compress)* pressen; auspressen ‹*Orangen, Saft*›; keltern ‹*Trauben, Äpfel*›; **d)** *(iron)* bügeln; **e) be ~ed for space/time/money** *(have barely enough)* zuwenig Platz/Zeit/Geld haben. **3.** *v. i.* **a)** *(exert pressure)* drücken; **the child ~ed against the railings** das Kind drückte sich gegen das Geländer; **b)** *(be urgent)* drängen; **time/sth. ~es** die Zeit drängt/etw. eilt *od.* ist dringend; **c)** *(make demand)* **~ for sth.** auf etw. *(Akk.)* drängen

~ a'head, ~ 'on *v. i. (continue activity)* [zügig] weitermachen; *(continue travelling)* [zügig] weitergehen/-fahren; **~ on with one's work** sich mit der Arbeit ranhalten *(ugs.)*

~ 'out *v. t.* auspressen; *(out of cardboard)* herausdrücken

²press *v. t.* **~ into service/use** in Dienst nehmen; einsetzen

press: ~-button *see* **push-button; ~ conference** *n.* Pressekonferenz, *die;* **~ coverage** *n.* Berichterstattung in der Presse; **~-gallery** *n.* Pressetribüne, *die;* **~-gang 1.** *n. (Hist.)* Preßgang, *der (veralt.);* **2.** *v. t. (Hist.)* pressen; zwangsrekrutieren

pressing ['presɪŋ] *adj.* **a)** *(urgent)* dringend; **b)** *(persistent)* dringlich; nachdrücklich

press: ~ release *n.* Presseinformation, *die;* **~ report** *n.* Pressebericht, *der;* **~-stud** *n. (Brit.)* Druckknopf, *der;* **~-up** *n.* Liegestütz, *der*

pressure ['preʃə(r)] **1.** *n.* **a)** *(exertion of force, amount)* Druck, *der;* **apply firm ~ to the joint** die Verbindung fest zusammendrücken; **atmospheric ~:** Luftdruck, *der;* **b)** *(oppression)* Last, *die;* Belastung, *die;* **mental ~:** psychische Belastung; **c)** *(trouble)* Druck, *der;* **~s at [one's] work** berufliche Belastungen; **d)** *(urgency)* Druck, *der; (of affairs)* Dringlichkeit, *die;* **e)** *(constraint)* Druck, *der;* Zwang, *der;* **put ~ on sb.** jmdn. unter Druck setzen; **be under a lot of ~ to do sth.** stark unter Druck gesetzt werden, etw. zu tun. *See also* **high pressure; low pressure. 2.** *v. t.* **~ sb. into doing sth.** jmdn. [dazu] drängen, etw. zu tun

pressure: ~-cooker *n.* Schnellkochtopf, *der;* **~ gauge** *n. (Motor Veh.)* Druckluftmesser, *der;* **~ group** *n.* Pressure-group, *die*

pressurize (pressurise) ['preʃəraɪz] *v. t.* **a)** *see* **pressure 2; b)** *(maintain normal pressure in)* druckfest machen, auf Normaldruck halten ‹*Flugzeugkabine*›; **~d cabin** Druckkabine, *die*

prestige [pre'sti:ʒ] **1.** *n.* Prestige, *das;* Renommee, *das.* **2.** *adj.* renommiert; Nobel‹*hotel, -gegend*›; **~ value** Prestigewert, *der*

prestigious [pre'stɪdʒəs] *adj.* angesehen

presto ['prestəʊ] *see* **hey**

presumably [prɪ'zju:məblɪ] *adv.* vermutlich; **~ he knows what he is doing** er wird schon wissen, was er tut

presume [prɪ'zju:m] **1.** *v. t.* **a)** *(venture)* **~ to do sth.** sich *(Dat.)* anmaßen, etw. zu tun; *(take the liberty)* sich *(Dat.)* erlauben, etw. zu tun; **b)** *(suppose)* annehmen; **be ~d innocent** als unschuldig gelten *od.* angesehen werden; **missing ~d dead** vermißt, wahrscheinlich *od.* mutmaßlich tot. **2.** *v. i.* sich *(Dat.)* anmaßen; **~ [up]on sth.** etw. ausnützen

presumption [prɪ'zʌmpʃn] *n.* **a)** *(arrogance)* Anmaßung, *die;* Vermessenheit, *die;* **have the ~ to do sth.** die Vermessenheit besitzen, etw. zu tun; sich *(Dat.)* anmaßen, etw. zu tun; **b)** *(assumption)* Annahme, *die;* Vermutung, *die*

presumptuous [prɪ'zʌmptjʊəs] *adj.* anmaßend; überheblich; *(impertinent)* aufdringlich

presumptuously [prɪ'zʌmptjʊəslɪ] *adv.* überheblich; *(impertinently)* aufdringlich

presuppose [pri:sə'pəʊz] *v. t. (assume, imply)* voraussetzen

pretence [prɪ'tens] *n. (Brit.)* **a)** *(pretext)* Vorwand, *der;* **under [the] ~ of helping** unter dem Vorwand zu helfen; *see also* **false pretences; b)** *no art. (make-believe, insincere behaviour)* Verstellung, *die;* **c)** *(piece of insincere behaviour)* **it is all** *or* **just a ~:** das ist alles nicht echt; **d)** *(affectation)* Affektiertheit, *die (abwertend); Unnatürlichkeit, die;* **e)** *(claim)* Anspruch, *der;* **make the/no ~ of** *or* **to sth.** Anspruch/keinen Anspruch auf etw. *(Akk.)* erheben

pretend [prɪ'tend] **1.** *v. t.* **a)** vorgeben; **she ~ed to be asleep** sie tat, als ob sie schlief[e]; **b)** *(imagine in play)* **~ to be sth.** so tun, als ob man etw. sei; **c)** *(profess falsely)* vortäuschen; *(say falsely)* vorgeben, fälschlich beteuern (**to** gegenüber); **d)** *(claim)* **not ~ to do sth.** nicht behaupten wollen, etw. zu tun. **2.** *v. i.* sich verstellen; **she's only ~ing** sie tut nur so

pretense *(Amer.) see* **pretence**

pretension [prɪ'tenʃn] *n.* **a)** *(claim)* Anspruch, *der;* **have/make ~s to great wisdom** vorgeben *od.* den Anspruch erheben, sehr klug zu sein; **b)** *(justifiable claim)* Anspruch, *der* (**to** auf + *Akk.*); **people with ~s to taste** Menschen, die Geschmack für sich in Anspruch nehmen können; **c)** *(pretentiousness)* Überheblichkeit, *die;* Anmaßung, *die; (of things: ostentation)* Protzigkeit, *die*

pretentious [prɪ'tenʃəs] *adj.* **a)** hochgestochen; wichtigtuerisch ‹*Person*›; **b)** *(ostentatious)* protzig; großspurig ‹*Person, Verhalten, Art*›

pretext ['pri:tekst] *n.* Vorwand, *der;* Ausrede, *die;* **[up]on** *or* **under the ~ of doing sth.** unter dem Vorwand *od.* mit der Entschuldigung, etw. tun zu wollen; **on the slightest ~:** mit *od.* unter dem fadenscheinigsten Vorwand

prettily ['prɪtɪlɪ] *adv.* hübsch; sehr schön ‹*singen, tanzen*›

pretty ['prɪtɪ] **1.** *adj.* **a)** *(attractive)* hübsch; nett ‹*Art*›; niedlich ‹*Geschichte, Liedchen*›; **she's not just a ~ face!** sie ist nicht nur hübsch[, sie kann auch was]!; **as ~ as a picture** bildhübsch; **not a ~ sight** *(iron.)* kein schöner Anblick; **b)** *(iron.)* hübsch, schön *(ugs. iron.).* **2.** *adv.* ziemlich; **I am ~ well** es geht mir ganz gut; **we have ~ nearly finished** wir sind so gut wie fertig; **be ~ well over/exhausted** so gut wie vorbei/erschöpft sein; **~ much the same** ziemlich unverändert; **be sitting ~** *(coll.)* sein

Schäfchen im trockenen haben *(ugs.)*
pretzel ['pretsl] *n.* Brezel, *die*
prevail [prɪ'veɪl] *v. i.* **a)** *(gain mastery)* siegen, die Oberhand gewinnen (**against, over** über + *Akk.*); ~ **[up]on sb. to do sth.** jmdn. dazu bewegen, etw. zu tun; **b)** *(predominate)* ‹*Zustand, Bedingung:*› vorherrschen; **c)** *(be current)* herrschen
prevailing [prɪ'veɪlɪŋ] *adj.* **a)** *(common)* [vor]herrschend; aktuell ‹*Mode*›; **b)** *(most frequent)* **the ~ wind is from the West** der Wind kommt vorwiegend von Westen
prevalence ['prevələns] *n., no pl.* Vorherrschen, *das; (of crime, corruption, etc.)* Überhandnehmen, *das; (of disease, malnutrition, etc.)* weite Verbreitung
prevalent ['prevələnt] *adj.* **a)** *(existing)* herrschend; weit verbreitet ‹*Krankheit*›; aktuell ‹*Trend*›; **b)** *(predominant)* vorherrschend; **be/become ~:** vorherrschen/sich durchsetzen
prevaricate [prɪ'værɪkeɪt] *v. i.* Ausflüchte machen (**over** wegen)
prevarication [prɪværɪ'keɪʃn] *n.* *(prevaricating)* Ausflüchte *Pl.*
prevent [prɪ'vent] *v. t.* *(hinder)* verhindern; verhüten; *(forestall)* vorbeugen; verhüten; ~ **sb. from doing sth.,** ~ **sb.'s doing sth.,** *(coll.)* ~ **sb. doing sth.** jmdn. daran hindern *od.* davon abhalten, etw. zu tun; **there is nothing to ~ me** nichts hindert mich daran; ~ **sb. from coming** jmdn. am Kommen hindern; **catch sb.'s arm to ~ him [from] falling** jmdn. am Arm fassen, damit er nicht fällt
prevention [prɪ'venʃn] *n.* Verhinderung, *die;* Verhütung, *die; (forestalling)* Vorbeugung, *die;* Verhütung, *die;* ~ **is better than cure** *(prov.)* Vorbeugen ist besser als Heilen *(Spr.)*
preventive [prɪ'ventɪv] *adj.* vorbeugend; präventiv *(geh.);* Präventiv‹*maßnahme, -krieg*›
preview ['pri:vju:] **1.** *n.* *(of film, play)* Voraufführung, *die; (of exhibition)* Vernissage, *die (geh.).* **2.** *v. t.* eine Vorschau sehen von ‹*Film*›
previous ['pri:vɪəs] **1.** *adj.* **a)** *(coming before)* früher ‹*Anstellung, Gelegenheit*›; ‹*Tag, Morgen, Abend, Nacht*› vorher; vorig ‹*Besitzer, Wohnsitz*›; **the ~ page** die Seite davor; **b)** *(prior)* ~ **to** vor (+ *Dat.*). **2.** *adv.* ~ **to** vor (+ *Dat.*); ~ **to being a nurse, she was ...:** bevor sie Krankenschwester wurde, war sie ...
previously ['pri:vɪəslɪ] *adv.* vorher; **two years ~:** zwei Jahre zuvor
pre-war ['pri:wɔ:(r)] *adj.* Vorkriegs-; **these houses are all ~:** diese Häuser stammen alle aus der Zeit vor dem Krieg
prey [preɪ] **1.** *n., pl. same* **a)** *(animal[s])* Beute, *die;* Beutetier, *das;* **beast/bird of ~:** Raubtier, *das*/-vogel, *der;* **b)** *(victim)* Beute, *die (geh.);* Opfer, *das.* **2.** *v. i.* ~ **[up]on** ‹*Raubtier, Raubvogel:*› schlagen; *(plunder)* ausplündern ‹*Person*›; *(exploit)* ausnutzen; ~ **[up]on sb.'s mind** jmdm. keine Ruhe lassen; ‹*Kummer, Angst:*› an jmdm. nagen
price [praɪs] **1.** *n.* **a)** *(money etc.)* Preis, *der;* **the ~ of wheat/a pint** der Weizenpreis/der Preis für ein Bier; **what is the ~ of this?** was kostet das?; **at a ~ of** zum Preis von; **sth. goes up/down in ~:** der Preis von etw. steigt/fällt; etw. steigt/fällt im Preis; **at a ~:** zum entsprechenden Preis; **b)** *(betting odds)* Eventualquote, *die;* **c)** *(value)* **be beyond ~:** [mit Geld] nicht zu bezahlen sein; **d)** *(fig.)* Preis, *der;* **he succeeded, but at a great ~:** er hatte Erfolg, mußte aber einen hohen Preis dafür bezahlen; **at/not at any ~:** um jeden/keinen Preis; **at the ~ of ruining his marriage** auf Kosten seiner Ehe; **what ~ ...?** *(Brit. sl.) (what is the chance of ...)* wie wär's mit ...?; *(... has failed)* wie steht's jetzt mit ...? *See also* **pay 2e. 2.** *v. t.* *(fix ~ of)* kalkulieren ‹*Ware*›; *(label with ~)* auszeichnen
'**price-cut** *n.* Preissenkung, *die*
priceless ['praɪslɪs] *adj.* **a)** *(invaluable)* unbezahlbar; unschätzbar ‹*Gut*›; **b)** *(coll.: amusing)* köstlich
price: ~-list *n.* Preisliste, *die;* **~-range** *n.* Preisspanne, *die;* **~-rise** *n.* Preisanstieg, *der* (**on** bei); **~-tag** *n.* Preisschild, *das*
pricey ['praɪsɪ] *adj.,* **pricier** ['praɪsɪə(r)], **priciest** ['praɪsɪɪst] *(Brit. coll.)* teuer
prick [prɪk] **1.** *v. t.* stechen; stechen in ‹*Ballon*›; aufstechen ‹*Blase*›; **he ~ed his finger with the needle** er stach sich *(Dat.)* mit der Nadel in den Finger. **2.** *v. i.* stechen. **3.** *n.* **a)** *(pain)* **[little] ~:** [leichter] Stich; **b)** *(coarse: penis)* Schwanz, *der (derb)*
~ **up 1.** *v. t.* aufrichten ‹*Ohren*›; ~ **up one's/its ears** *(listen)* die Ohren spitzen. **2.** *v. i.* ‹*Ohren:*› sich aufrichten
prickle ['prɪkl] **1.** *n.* **a)** *(thorn)* Dorn, *der;* **b)** *(Zool., Bot.)* Stachel, *der.* **2.** *v. i.* kratzen
prickly ['prɪklɪ] *adj.* **a)** *(with prickles) see* **prickle 1:** dornig; stachelig; **be ~** ‹*Pflanze:*› Dornen/Stacheln haben; **b)** *(fig.)* empfindlich
pricy *see* **pricey**
pride [praɪd] **1.** *n.* **a)** Stolz, *der; (arrogance)* Hochmut, *der;* **take** *or* **have ~ of place** die Spitzenstellung einnehmen; *(in collection etc.)* das Glanzstück sein; **take [a] ~ in sb./sth.** auf jmdn./etw. stolz sein; **b)** *(object, best one)* Stolz, *der;* **sb.'s ~ and joy** jmds. ganzer Stolz; **c)** *(of lions)* Rudel, *das.* **2.** *v. refl.* ~ **oneself [up]on sth.** *(congratulate oneself)* auf etw. *(Akk.)* stolz sein
priest [pri:st] *n.* Priester, *der; see also* **high priest**
priestess ['pri:stɪs] *n.* Priesterin, *die*
priesthood ['pri:sthʊd] *n.* *(office)* geistliches Amt; *(order of priests; priests)* Geistlichkeit, *die;* **go into the ~:** Priester werden
priestly ['pri:stlɪ] *adj.* priesterlich; Priester‹*kaste, -rolle*›
prig [prɪg] *n.* Tugendbold, *der (ugs., iron.)*
priggish ['prɪgɪʃ] *adj.* übertrieben tugendhaft
prim [prɪm] *adj.* **a)** spröde, steif ‹*Person*›; ~ **and proper** etepetete *(ugs.);* **b)** *(prudish)* zimperlich; prüde
prima facie [praɪmə 'feɪʃi:] **1.** *adv.* auf den ersten Blick. **2.** *adj.* glaubhaft klingend; ~ **evidence** *(Law)* Anscheinsbeweis, *der*
primarily ['praɪmərɪlɪ] *adv.* in erster Linie
primary ['praɪmərɪ] **1.** *adj.* **a)** *(first)* primär *(geh.);* grundlegend; ~ **source** Primärquelle, *die (geh.);* **b)** *(chief)* Haupt‹*rolle, -sorge, -ziel, -zweck*›; **of ~ importance** von höchster Bedeutung. **2.** *n.* *(Amer.: election)* Vorwahl, *die*
primary: ~ colour *see* **colour 1a;** ~ **edu'cation** *n.* Grundschulerziehung, *die;* ~ **e'lection** *n.* *(Amer.)* Vorwahl, *die;* ~ **school** *n.* Grundschule, *die; attrib.* **~-school teacher** Grundschullehrer, *der*/-lehrerin, *die*
primate ['praɪmeɪt] *n.* **a)** *(Eccl.)* Primas, *der;* **b)** *(Zool.)* Primat, *der*
¹**prime** [praɪm] **1.** *n.* **a)** Höhepunkt, *der;* Krönung, *die;* **in the ~ of life/youth** in der Blüte seiner/ihrer Jahre/der Jugend *(geh.);* **be in/past one's ~:** in den

besten Jahren sein/die besten Jahre überschritten haben; **b)** *(Math.)* Primzahl, *die.* **2.** *adj.* **a)** *(chief)* Haupt-; hauptsächlich; ~ **motive** Hauptmotiv, *das;* **be of ~ importance** von höchster Wichtigkeit sein; **b)** *(excellent)* erstklassig; vortrefflich ‹*Beispiel*›; ‹*Fleisch*› erster Güteklasse; **in ~ condition** ‹*Sportler, Tier*› in bester Verfassung; voll ausgereift ‹*Obst*›

²prime *v.t.* **a)** *(equip)* vorbereiten; ~ **sb. with information/advice** jmdn. instruieren/jmdm. Ratschläge erteilen; **well ~d** gut vorbereitet; **b)** grundieren ‹*Wand, Decke*›; **c)** füllen ‹*Pumpe*›; **d)** schärfen ‹*Sprengkörper*›

prime: ~ **'minister** *n.* Premierminister, *der*/-ministerin, *die;* ~ **'number** *n. (Math.)* Primzahl, *die*

¹primer ['praɪmə(r)] *n. (book)* Fibel, *die*

²primer *n.* **a)** *(explosive)* Zündvorrichtung, *die;* **b)** *(paint etc.)* Grundierlack, *der*

primeval [praɪ'mi:vl] *adj.* urzeitlich; Ur‹*zeiten, -wälder*›

primitive ['prɪmɪtɪv] *adj.* primitiv; *(prehistoric)* urzeitlich ‹*Mensch*›; frühzeitlich ‹*Ackerbau, Technik*›

primitively ['prɪmɪtɪvlɪ] *adv.* primitiv

primrose ['prɪmrəʊz] *n.* **a)** *(Bot.)* gelbe Schlüsselblume; **b)** *(colour)* schlüsselblumengelb

primula ['prɪmjʊlə] *n. (Bot.)* Primel, *die*

Primus, (P) ['praɪməs] *n.* ~ **[stove]** Primuskocher, *der*

prince [prɪns] *n.* **a)** *(member of royal family)* Prinz, *der;* **b)** *(rhet.: sovereign ruler)* Fürst, *der;* Monarch, *der*

Prince: ~ **'Charming** *n. (fig.)* Märchenprinz, *der;* **p~ 'consort** *n.* Prinzgemahl, *der*

princely ['prɪnslɪ] *adj. (lit. or fig.)* fürstlich; ~ **houses** Fürstenhäuser

Prince 'Regent *n.* Prinzregent, *der*

princess ['prɪnses, prɪn'ses] *n.* **a)** Prinzessin, *die;* **b)** *(wife of prince)* Fürstin, *die*

princess 'royal *n. [Titel für] älteste Tochter eines Monarchen*

principal ['prɪnsɪpl] **1.** *adj.* **a)** Haupt-; *(most important)* wichtigst...; bedeutendst...; **the ~ cause of lung cancer** die häufigste Ursache für Lungenkrebs; **b)** *(Mus.)* ~ **horn/bassoon** *etc.* erstes Horn/Fagott *usw.* **2.** *n.* **a)** *(head of school or college)* Rektor, *der*/Rektorin, *die;* **b)** *(Finance) (invested)* Kapitalbetrag, *der; (lent)* Kreditsumme, *die*

principality [prɪnsɪ'pælɪtɪ] *n.* Fürstentum, *das;* **the P~** *(Brit.)* Wales

principally ['prɪnsɪpəlɪ] *adv.* in erster Linie

principle ['prɪnsɪpl] *n.* **a)** Prinzip, *das;* **on the ~ that ...:** nach dem Grundsatz, daß ...; **be based on the ~ that ...:** auf dem Grundsatz basieren, daß ...; **basic ~:** Grundprinzip, *das;* **go back to first ~s** zu den Grundlagen zurückgehen; **in ~:** im Prinzip; **it's the ~ [of the thing]** es geht [dabei] ums Prinzip; **a man of high ~** *or* **strong ~s** ein Mann von *od.* mit hohen Prinzipien; **a matter of ~:** eine Prinzipfrage; **do sth. on ~** *or* **as a matter of ~:** etw. prinzipiell *od.* aus Prinzip tun; **b)** *(Phys.)* Lehrsatz, *der*

print [prɪnt] **1.** *n.* **a)** *(impression)* Abdruck, *der; (finger~)* Fingerabdruck, *der;* **b)** *(~ed lettering)* Gedruckte, *das; (type-face)* Druck, *der;* **clear/large ~:** deutlicher/großer Druck; **editions in large ~:** Großdruckbücher; *see also* **small print; c)** *(published or ~ed state)* **be in/out of ~** ‹*Buch:*› erhältlich/vergriffen sein; **d)** *(~ed picture or design)* Druck, *der;* **e)** *(Photog.)* Abzug, *der; (Cinemat.)* Kopie, *die;* **f)** *(Textiles) (cloth with design)* bedruckter Stoff. **2.** *v.t.* **a)** drucken ‹*Buch, Zeitschrift, Geldschein usw.*›; **b)** *(write)* in Druckschrift schreiben; **c)** *(cause to be published)* veröffentlichen ‹*Artikel, Roman, Ansichten usw.*›; **d)** *(Photog.)* abziehen; *(Cinemat.)* kopieren; **e)** *(Textiles)* bedrucken ‹*Stoff*›

~ **'out** *v.t. (Computing)* ausdrucken

printable ['prɪntəbl] *adj.* druckbar; **what he replied is not ~** *(fig.)* was er geantwortet hat, zu wiederholen, verbietet sich

printed ['prɪntɪd] *adj.* **a)** *(Printing)* gedruckt; ~ **characters** *or* **letters** Druckbuchstaben; **on the ~ page** gedruckt; **b)** *(written like print)* in Druckschrift; **c)** *(published)* veröffentlicht ‹*Artikel, Roman, Ansichten usw.*›; **d)** *(Textiles)* bedruckt ‹*Stoff*›

printed: ~ **'circuit** *n. (Electronics)* gedruckte Schaltung; ~ **matter** *n., no pl., no indef. art.* Gedruckte, *das*

printer ['prɪntə(r)] *n.* **a)** *(Printing) (worker)* Drucker, *der*/Druckerin, *die; firm of printers* Druckerei, *die;* **send sth. off to the ~'s** etw. in die Druckerei schicken; **b)** *(Computing)* Drucker, *der*

printer: ~**'s 'error** *n.* Druckfehler, *der;* ~**'s 'ink** *n.* Druckfarbe, *die*

printing ['prɪntɪŋ] *n.* **a)** Drucken, *das;* **[the]** ~ **[trade]** das Druckgewerbe; **b)** *(writing like print)* Druckschrift, *die;* **c)** *(edition)* Auflage, *die*

printing: ~ **error** *n.* Druckfehler, *der;* ~**-ink** *n.* Druckfarbe, *die;* ~**-press** *n.* Druckerpresse, *die*

'print-out *n. (Computing)* Ausdruck, *der*

prior ['praɪə(r)] **1.** *adj.* vorherig ‹*Warnung, Zustimmung, Vereinbarung usw.*›; früher ‹*Verabredung, Ehe*›; Vor‹*geschichte, -kenntnis*›; **have a** *or* **the ~ claim to sth.** ältere Rechte an etw. *(Dat.) od.* auf etw. *(Akk.)* haben. **2.** *adv.* ~ **to** vor (+ *Dat.*); ~ **to doing sth.** bevor man etw. tut/tat; ~ **to that** vorher. **3.** *n. (Eccl.)* Prior, *der*

priority [praɪ'ɒrɪtɪ] *n.* **a)** *(precedence)* Vorrang, *der; attrib.* vorrangig; **have** *or* **take ~:** Vorrang haben (**over** vor + *Dat.*); **have ~** *(on road)* Vorfahrt haben; **give ~ to sb./sth.** jmdm./einer Sache den Vorrang geben; **give top ~ to sth.** einer Sache *(Dat.)* höchste Priorität einräumen; **be listed in order of ~:** der Vorrangigkeit nach aufgeführt sein; **b)** *(matter)* vordringliche Angelegenheit; **our first ~ is to ...:** zuallererst müssen wir ...; **be high/low on the list of priorities** oben/unten auf der Prioritätenliste stehen; **get one's priorities right/wrong** seine Prioritäten richtig/falsch setzen

priory ['praɪərɪ] *n. (Eccl.)* Priorat, *das*

prise *see* **²prize**

prism ['prɪzm] *n.* Prisma, *das*

prison ['prɪzn] *n.* **a)** *(lit. or fig.)* Gefängnis, *das; attrib.* Gefängnis-; **b)** *no pl., no art. (custody)* Haft, *die;* **in ~:** im Gefängnis; **go to ~:** ins Gefängnis gehen; **send sb. to ~:** jmdn. ins Gefängnis schicken; **escape from ~:** aus dem Gefängnis ausbrechen; **let sb. out of ~:** jmdn. aus der Haft entlassen

'prison camp *n.* Gefangenenlager, *das*

prisoner ['prɪznə(r)] *n. (lit. or fig.)* Gefangene, *der/die; (accused person)* Angeklagte, *der/die;* **take/hold** *or* **keep sb. ~:** jmdn. gefangennehmen/-halten

prisoner of 'war *n.* Kriegsgefangene, *der/die;* **prisoner-of-war camp** [Kriegs]gefangenenlager, *das*

prison: ~ 'guard *n.* Gefängniswärter, *der/*-wärterin, *die;* **~ 'visitor** *n.* ≈ Gefangenenfürsorger, *der/*-fürsorgerin, *die*

pristine ['prɪsti:n, 'prɪstaɪn] *adj.* unberührt; ursprünglich ‹*Glanz, Weiße, Schönheit*›; **in ~ condition** in tadellosem Zustand

privacy ['prɪvəsɪ, 'praɪvəsɪ] *n.* **a)** *(seclusion)* Zurückgezogenheit, *die;* **in the ~ of one's [own] home** in den eigenen vier Wänden *(ugs.);* **invasion of ~/sb.'s ~:** Eindringen in die/jmds. Privatsphäre; **allow sb. no ~:** jmdm. kein Privatleben erlauben; **b)** *(confidentiality)* **in the strictest ~:** unter strengster Geheimhaltung

private ['praɪvət] **1.** *adj.* **a)** *(outside State system)* privat; Privat‹*unterricht, -schule, -industrie, -klinik, -patient, -station usw.*›; **a doctor working in ~ medicine** ein Arzt, der Privatpatienten hat; **have a ~ education** auf eine Privatschule gehen; **b)** *(belonging to individual, not public, not business)* persönlich ‹*Dinge*›; nichtöffentlich ‹*Versammlung, Sitzung*›; privat ‹*Telefongespräch, Schriftverkehr*›; Privat‹*eigentum, -wagen, -flugzeug, -strand, -parkplatz, -leben, -konto*›; '~' *(on door)* „Privat"; *(in public building)* „kein Zutritt"; *(on ~ land)* „Betreten verboten"; **for [one's own] ~ use** für den persönlichen Gebrauch; **c)** *(personal, affecting individual)* persönlich ‹*Meinung, Interesse, Überzeugung, Rache*›; privat ‹*Vereinbarung, Zweck*›; **d)** *(not for public disclosure)* geheim ‹*Verhandlung, Geschäft, Tränen*›; still ‹*Gebet, Nachdenken, Grübeln*›; persönlich ‹*Gründe*›; *(confidential)* vertraulich; **have a ~ word with sb.** jmdn. unter vier Augen sprechen; **e)** *(secluded)* still ‹*Ort*›; *(undisturbed)* ungestört; **f)** *(not in public office)* **~ citizen** *or* **individual** Privatperson, *die.* **2.** *n.* **a)** *(Brit. Mil.)* einfacher Soldat; **b)** **in ~:** privat; in kleinem Kreis ‹*feiern*›; *(confidentially)* ganz im Vertrauen; **speak to sb. in ~:** jmdn. unter vier Augen sprechen; **c)** *in pl. (coll.: genitals)* Geschlechtsteile *Pl.*

private: ~ de'tective *n.* [Privat]detektiv, *der/*-detektivin, *die;* **~ 'enterprise** *n. (Commerc.)* das freie *od.* private Unternehmertum; **~ 'eye** *(coll.) see* **~ detective**

privately ['praɪvətlɪ] *adv.* privat ‹*erziehen, zugeben*›; vertraulich ‹*jmdn. sprechen*›; insgeheim ‹*denken, glauben*›; **study ~:** private Studien betreiben; **~ owned** in Privatbesitz

private: ~ 'parts *n. pl.* Geschlechtsteile *Pl.;* **~ 'practice** *n.* Privatpraxis, *die;* **he is in ~ practice** er hat eine Privatpraxis; **~ sector** *n.* **the ~ sector [of industry]** die Privatwirtschaft

privation [praɪ'veɪʃn] *n. (lack of comforts)* Not, *die;* **suffer many ~s** viele Entbehrungen erleiden

privatisation, privatise *see* **privatiz-**

privatization [praɪvətaɪ'zeɪʃn] *n. (Econ.)* Privatisierung, *die*

privatize ['praɪvətaɪz] *v. t. (Econ.)* privatisieren

privet ['prɪvɪt] *n. (Bot.)* Liguster, *der*

privilege ['prɪvɪlɪdʒ] *n.* **a)** *(right, immunity)* Privileg, *das; collect.* Privilegien *Pl.;* **b)** *(special benefit)* Sonderrecht, *das; (honour)* Ehre, *die;* **it was a ~ to listen to him** es war ein besonderes Vergnügen, ihm zuzuhören

privileged ['prɪvɪlɪdʒd] *adj.* privilegiert; **the ~ few** die kleine Gruppe von Privilegierten; **sb. is ~ to do sth.** jmd. hat die Ehre, etw. zu tun; **be in a ~ position** eine bevorzugte Position innehaben

privy ['prɪvɪ] *adj.* **be ~ to sth.** in etw. *(Akk.)* eingeweiht sein

Privy: ~ 'Council *n. (Brit.)* Geheimer [Staats]rat; **p~ 'counsellor (p~ 'councillor)** *n. (Brit.)* Geheimer Rat

[1]**prize** [praɪz] **1.** *n.* **a)** *(reward, money)* Preis, *der;* **win** *or* **take first ~:** den ersten Preis gewinnen; **b)** *(in lottery)* Gewinn, *der;* **win sth. as a ~:** etw. gewinnen; **c)** *(fig.: something worth striving for)* Lohn, *der;* **glittering ~s** verlockender Lohn. **2.** *v. t. (value)* **~ sth. [highly]** etw. hoch schätzen; **sb.'s most ~d possessions** jmds. wertvollster Besitz. **3.** *attrib. adj.* **a)** *(~-winning)* preisgekrönt; **b)** *(awarded as ~)* **~ medal/trophy** Siegesmedaille, *die/*Siegestrophäe, *die;* **c)** *(iron.)* **~ idiot** Vollidiot, *der/*-idiotin, *die (ugs.);* **~ example** Musterbeispiel, *das (iron.)*

[2]**prize** *v. t. (force)* **~ [open]** aufstemmen; **~ the lid off a crate** eine Kiste aufstemmen; **~ information/a secret out of sb.** Informationen/ein Geheimnis aus jmdm. herauspressen

prize: ~-fight *n. (Boxing)* Preisboxkampf, *der;* **~-fighter** *n. (Boxing)* Preisboxer, *der;* **~-giving** *n. (Sch.)* Preisverleihung, *die;* **~-money** *n.* Geldpreis, *der; (Sport)* Preisgeld, *das;* **~-winner** *n.* Preisträger, *der/*-trägerin, *die; (in lottery)* Gewinner, *der/*Gewinnerin, *die;* **~-winning** *adj.* preisgekrönt; *(in lottery)* Gewinner-

[1]**pro** [prəʊ] **1.** *n. in pl.* **the ~s and cons** das Pro und Kontra. **2.** *adv.* **~ and con** pro und kontra. **3.** *prep.* für

[2]**pro 1.** *n. (Sport & Theatre coll.)* Profi, *der.* **2.** *adj.* Profi-

[3]**pro-** *pref.* pro-; **~-Communist** prokommunistisch

probability [prɒbə'bɪlɪtɪ] *n.* **a)** *(likelihood; also Math.)* Wahrscheinlichkeit, *die;* **in all ~:** aller Wahrscheinlichkeit nach; **there is little/a strong ~ that ...:** die Wahrscheinlichkeit, daß ..., ist gering/groß; **b)** *(likely event)* **the ~ is that ...:** es ist zu erwarten, daß ...; **war is becoming a ~:** der Ausbruch eines Krieges wird immer wahrscheinlicher

probable ['prɒbəbl] *adj.* wahrscheinlich; **highly ~:** höchstwahrscheinlich; **another wet summer looks ~:** es sieht ganz nach einem weiteren verregneten Sommer aus

probably ['prɒbəblɪ] *adv.* wahrscheinlich

probate ['prəʊbeɪt] *n. (Law)* gerichtliche Testamentsbestätigung

probation [prə'beɪʃn] *n.* **a)** Probezeit, *die;* **be on ~:** Probezeit haben; **b)** *(Law)* Bewährung, *die;* **be on ~:** auf Bewährung sein

probationary [prə'beɪʃənərɪ] *adj.* Probe-; **~ period** Probezeit, *die*

pro'bation officer *n.* Bewährungshelfer, *der/*-helferin, *die*

probe [prəʊb] **1.** *n.* **a)** *(investigation)* Untersuchung, *die* **(into** *Gen.*); **b)** *(Med., Electronics, Astron.)* Sonde, *die.* **2.** *v. t.* **a)** *(investigate)* erforschen; untersuchen; **b)** *(reach deeply into)* gründlich erforschen ‹*Kontinent, Weltall*›. **3.** *v. i.* **a)** *(make investigation)* forschen; **~ into a matter** einer Angelegenheit *(Dat.)* auf den Grund gehen; **b)** *(reach deeply)* vordringen **(into** in + *Akk.*)

probing ['prəʊbɪŋ] *adj. (penetrating)* gründlich; durchdringend ‹*Blick*›; **~ question** Testfrage, *die*

probity ['prəʊbɪtɪ] *n., no pl.* Rechtschaffenheit, *die*

problem ['prɒbləm] *n.* **a)** *(difficult matter)* Problem, *das; attrib.* Problem⟨*gebiet, -fall, -kind, -familie*⟩; **I find it a ~ to start** *or* **have a ~ [in] starting the car** ich habe Probleme, das Auto anzulassen; **[I see] no ~** *(coll.)* kein Problem; **what's the ~?** *(coll.)* wo fehlt's denn?; **the ~ about** *or* **with sb./sth.** das Problem mit jmdm./bei etw.; **the Northern Ireland ~:** die Nordirlandfrage; **he has a drink ~:** er hat ein Alkoholproblem; **that presents a ~:** das ist ein Problem; **b)** *(puzzle)* Rätsel, *das*

problematic [prɒblə'mætɪk], **problematical** [prɒblə'mætɪkl] *adj.* problematisch; *(doubtful)* fragwürdig

procedure [prə'si:djə(r)] *n.* **a)** *(particular course of action)* Verfahren, das; Prozedur, *die (meist abwertend);* **b)** *(way of doing sth.)* Verfahrensweise, *die;* **what is the normal ~?** wie wird das normalerweise gehandhabt?

proceed [prə'si:d] *v.i. (formal)* **a)** *(go) (on foot)* gehen; *(as or by vehicle)* fahren; *(on horseback)* reiten; *(after interruption)* weitergehen/-fahren/-reiten; **~ to business** sich geschäftlichen Dingen zuwenden; **~ to the next item on the agenda** zum nächsten Punkt der Tagesordnung übergehen; **b)** *(begin and carry on)* beginnen; *(after interruption)* fortfahren; **~ to talk/eat** *etc. (begin and carry on)* beginnen, zu sprechen/essen *usw.; (after interruption)* weitersprechen/-essen *usw.;* **~ in** *or* **with sth.** *(begin)* [mit] etw. beginnen; *(continue)* etw. fortsetzen; **c)** *(adopt course)* vorgehen; **~ discreetly with sth.** etw. diskret behandeln; **d)** *(be carried on)* ⟨*Rennen:*⟩ verlaufen; *(be under way)* ⟨*Verfahren:*⟩ laufen; *(be continued after interruption)* fortgesetzt werden; **e)** *(originate)* **~ from** *(issue from)* kommen von; *(be caused by)* herrühren von

~ against *v.t. (Law)* gerichtlich vorgehen gegen

proceeding [prə'si:dɪŋ] *n.* **a)** *(action)* Vorgehensweise, *die;* **b)** *in pl. (events)* Vorgänge; **I'll go along to watch the ~s** ich geh mal gucken, was da läuft; **c)** *in pl. (Law)* Verfahren, *das;* **legal ~s** Gerichtsverfahren, *das;* **start/take [legal] ~s** gerichtlich vorgehen (**against** gegen); **d)** *in pl. (report)* Tätigkeitsbericht, *der; (of single meeting)* Protokoll, *das*

proceeds ['prəʊsi:dz] *n. pl.* Erlös, *der* (**from** aus)

¹process ['prəʊses] **1.** *n.* **a)** *(of time or history)* Lauf, *der;* **he learnt a lot in the ~:** er lernte eine Menge dabei; **be in the ~ of doing sth.** gerade etw. tun; **be in ~:** in Gang sein; **b)** *(proceeding)* Vorgang, *der;* Prozedur, *die;* **the democratic ~:** das demokratische Verfahren; **c)** *(method)* Verfahren, *das; see also* **elimination a; d)** *(natural operation)* Prozeß, *der;* Vorgang, *der;* **~ of evolution** Evolutionsprozeß, *der.* **2.** *v.t.* verarbeiten ⟨*Rohstoff, Signal, Daten*⟩; bearbeiten ⟨*Antrag, Akte, Darlehen*⟩; *(for conservation)* behandeln ⟨*Leder, Lebensmittel*⟩; *(Photog.)* entwickeln ⟨*Film*⟩

²process [prə'ses] *v.i.* marschieren

'process cheese *(Amer.),* **'processed cheese** *ns.* Schmelzkäse, *der*

processer *see* **processor**

procession [prə'seʃn] *n.* **a)** Zug, *der; (religious)* Prozession, *die; (festive)* Umzug, *der;* **go/march/move** *etc.* **in ~:** ziehen; **funeral ~:** Trauerzug, *der;* **b)** *(fig.: series)* Reihe, *die*

processor ['prəʊsesə(r)] *n. (machine)* Prozessor, *der*

proclaim [prə'kleɪm] *v.t.* **a)** erklären ⟨*Absicht*⟩; geltend machen ⟨*Recht, Anspruch*⟩; *(declare officially)* verkünden ⟨*Amnestie*⟩; ausrufen ⟨*Republik*⟩; **~ sb./oneself King** jmdn./sich zum König ausrufen; **~ a country [to be] a republic** in einem Land die Republik ausrufen; **b)** *(reveal)* verraten; **~ sb./sth. [to be] sth.** verraten, daß jmd./etw. etw. ist

proclamation [prɒklə'meɪʃn] *n.* **a)** *(act of proclaiming)* Verkündung, *die;* Proklamation, *die (geh.); (of sovereign)* Ausrufung, *die;* **b)** *(notice)* Bekanntmachung, *die; (edict, decree)* Erlaß, *der*

proclivity [prə'klɪvɪtɪ] *n.* Neigung, *die;* **have a ~/proclivities for sth.** einen Hang zu etw. haben

procrastinate [prə'kræstɪneɪt] *v.i.* zaudern *(geh.);* **~ in doing sth.** es hinauszögern, etw. zu tun

procrastination [prəkræstɪ'neɪʃn] *n.* Saumseligkeit, *die (geh.)*

procure [prə'kjʊə(r)] **1.** *v.t.* **a)** *(obtain)* beschaffen; **~ for sb./oneself** jmdm./sich verschaffen ⟨*Arbeit, Unterkunft, Respekt, Reichtum*⟩; jmdm./sich beschaffen ⟨*Arbeit, Ware*⟩; **b)** *(bring about)* herbeiführen ⟨*Ergebnis, Wechsel, Frieden*⟩; bewirken ⟨*Freilassung*⟩; **c)** *(for sex)* beschaffen. **2.** *v.i.* Kuppelei betreiben; **procuring** Kuppelei, *die*

procurement [prə'kjʊəmənt] *n. see* **procure 1:** Beschaffung, *die;* Herbeiführung, *die;* Bewirkung, *die*

prod 1. *v.t.,* **-dd-: a)** *(poke)* stupsen *(ugs.);* stoßen mit ⟨*Stock, Finger usw.*⟩; **he ~ded the map with his finger** er stieß mit dem Finger auf die Karte; **~ sb. gently** jmdn. anstupsen *od.* leicht anstoßen; **b)** *(fig.: rouse)* antreiben; nachhelfen (+ *Dat.*) ⟨*Gedächtnis*⟩; **~ sb. into doing sth.** jmdn. drängen, etw. zu tun. **2.** *v.i.,* **-dd-** stochern. **3.** *n.* Stupser, *der;* **a ~ in the/my** *etc.* **ribs** ein Rippenstoß; **give sb. a ~:** jmdm. einen Stupser geben; *(fig.)* jmdn. auf Touren bringen

~ at *v.t.* anstupsen

prodigal ['prɒdɪgl] *adj.* verschwenderisch

prodigal 'son *n. (Bibl.; also fig. iron.)* verlorener Sohn

prodigious [prə'dɪdʒəs] *adj.* ungeheuer; unglaublich ⟨*Lügner, Dummkopf*⟩; wunderbar ⟨*Ereignis, Taten*⟩; außerordentlich ⟨*Begabung, Können*⟩; gewaltig ⟨*Fortschritt, Kraft, Energie*⟩

prodigy ['prɒdɪdʒɪ] *n.* **a)** *(gifted person)* [außergewöhnliches] Talent; **musical ~:** musikalisches Wunderkind; *see also* **child prodigy; b)** *(marvel)* Wunder, *das*

produce 1. ['prɒdju:s] *n.* Produkte *Pl.;* Erzeugnisse *Pl.;* **'~ of Spain'** „spanisches Erzeugnis". **2.** [prə'dju:s] *v.t.* **a)** *(bring forward)* erbringen ⟨*Beweis*⟩; vorlegen ⟨*Beweismaterial*⟩; beibringen ⟨*Zeugen*⟩; geben ⟨*Erklärung*⟩; vorzeigen ⟨*Paß, Fahrkarte, Papiere*⟩; herausholen ⟨*Brieftasche, Portemonnaie, Pistole*⟩; **~ sth. from one's pocket** etw. aus der Tasche ziehen; **he ~d a few coins from his pocket** er holte einige Münzen aus seiner Tasche; **she ~d a gun from her pocket** sie zog einen Revolver aus ihrer Tasche; **b)** produzieren ⟨*Show, Film*⟩; inszenieren ⟨*Theaterstück, Hörspiel, Fernsehspiel*⟩; herausgeben ⟨*Schallplatte, Buch*⟩; **well-~d** gut gemacht ⟨*Film, Theaterstück, Programm*⟩; **c)** *(manufacture)* herstellen; zubereiten ⟨*Mahlzeit*⟩; *(in nature; Agric.)* produzieren; **d)** *(create)* schreiben ⟨*Roman, Gedichte, Artikel, Aufsatz, Symphonie*⟩; schaffen ⟨*Gemälde, Skulptur, Meisterwerk*⟩; aufstellen ⟨*Theorie*⟩; **e)** *(cause)* hervorrufen; bewirken ⟨*Änderung*⟩; **f)** *(bring into being)* erzeugen; führen zu ⟨*Situation, Lage, Zustän-*

de⟩; **g)** *(yield)* erzeugen ⟨*Ware, Produkt*⟩; geben ⟨*Milch*⟩; tragen ⟨*Wolle*⟩; legen ⟨*Eier*⟩; liefern ⟨*Ernte*⟩; fördern ⟨*Metall, Kohle*⟩; abwerfen ⟨*Ertrag, Gewinn*⟩; hervorbringen ⟨*Dichter, Denker, Künstler*⟩; führen zu ⟨*Resultat*⟩; **h)** *(bear)* gebären; ⟨*Säugetier:*⟩ werfen; ⟨*Vogel, Reptil:*⟩ legen ⟨*Eier*⟩; ⟨*Fisch, Insekt:*⟩ ablegen ⟨*Eier*⟩; ⟨*Baum, Blume:*⟩ tragen ⟨*Früchte, Blüten*⟩; entwickeln ⟨*Triebe*⟩; bilden ⟨*Keime*⟩

producer [prəˈdju:sə(r)] *n.* **a)** *(Cinemat., Theatre, Radio, Telev., Econ.)* Produzent, *der*/Produzentin, *die;* **b)** *(Brit. Theatre/Radio/ Telev.)* Regisseur, *der*/Regisseurin, *die*

product [ˈprɒdʌkt] *n.* **a)** *(thing produced)* Produkt, *das; (of industrial process)* Erzeugnis, *das; (of art or intellect)* Werk, *das;* **carbon dioxide is a ~ of respiration** Kohlendioxyd entsteht bei der Atmung; **b)** *(result)* Folge, *die;* **c)** *(Math.)* Produkt, *das* (**of** aus). *See also* [1]**gross 1 d**

production [prəˈdʌkʃn] *n.* **a)** *(bringing forward) (of evidence)* Erbringung, *die; (in physical form)* Vorlage, *die; (of witness)* Beibringung, *die; (of passport etc.)* Vorzeigen, *das;* **on ~ of your passport** gegen Vorlage Ihres Passes; **b)** *(public presentation) (Cinemat.)* Produktion, *die; (Theatre)* Inszenierung, *die; (of record, book)* Herausgabe, *die;* **c)** *(action of making)* Produktion, *die; (manufacturing)* Herstellung, *die; (thing produced)* Produkt, *das;* **be in/go into ~:** in Produktion sein/gehen; **be** *or* **have gone out of ~:** nicht mehr hergestellt werden; *see also* **mass production; d)** *(thing created)* Werk, *das; (Brit. Theatre: show produced)* Inszenierung, *die;* **e)** *(causing)* Hervorrufen, *das;* **f)** *(bringing into being)* Hervorbringung, *die;* **g)** *(process of yielding)* Produktion, *die; (Mining)* Förderung, *die;* **the mine has ceased ~:** das Bergwerk hat die Förderung eingestellt; **h)** *(yield)* Ertrag, *der;* **[the] annual/ total ~ from the mine** die jährliche/gesamte Förderleistung des Bergwerks

proˈduction line *n.* Fertigungsstraße, *die*

productive [prəˈdʌktɪv] *adj.* **a)** *(producing)* **be ~** ⟨*Fabrik:*⟩ produzieren; **b)** *(producing abundantly)* ertragreich ⟨*Land, Boden, Obstbaum, Mine*⟩; leistungsfähig ⟨*Betrieb, Bauernhof*⟩; produktiv ⟨*Künstler, Komponist, Schriftsteller, Geist*⟩; **c)** *(yielding favourable results)* fruchtbar ⟨*Gespräch, Verhandlungen, Forschungsarbeit*⟩

productivity [prɒdʌkˈtɪvɪtɪ] *n.* Produktivität, *die;* **~ agreement** *or* **deal** Produktivitätsvereinbarung, *die;* **~ bonus** Leistungszulage, *die*

Prof. [prɒf] *abbr.* **Professor** Prof.

profane [prəˈfeɪn] **1.** *adj.* **a)** *(irreligious)* gotteslästerlich; **b)** *(irreverent)* respektlos ⟨*Bemerkung, Person*⟩; profan ⟨*Humor, Sprache*⟩; **c)** *(secular)* weltlich; profan. **2.** *v. t.* entweihen

profanity [prəˈfænɪtɪ] *n.* **a)** *(irreligiousness, irreligious act)* Gotteslästerung, *die;* **b)** *(irreverent behaviour, act, or utterance)* Respektlosigkeit, *die*

profess [prəˈfes] *v. t.* **a)** *(declare openly)* bekunden ⟨*Vorliebe, Abneigung, Interesse*⟩; **~ to be/do sth.** erklären, etw. zu sein/tun; **b)** *(claim)* vorgeben; geltend machen ⟨*Recht, Anspruch*⟩; **~ to be/ do sth.** behaupten, etw. zu sein/ tun; **c)** *(affirm faith in)* sich bekennen zu

professed [prəˈfest] *adj.* **a)** *(self-acknowledged)* erklärt ⟨*Marxist, Bewunderer, Absicht*⟩; **be a ~ Christian** ein bekennender Christ sein; **b)** *(alleged)* angeblich

profession [prəˈfeʃn] *n.* **a)** Beruf, *der;* **what is your ~?** was sind Sie von Beruf?; **take up/go into** *or* **enter a ~:** einen Beruf ergreifen/ in einen Beruf gehen; **she is in the legal ~:** sie ist Juristin; **be a pilot by ~:** von Beruf Pilot sein; **the [learned] ~s** Theologie, Jura und Medizin; **b)** *(body of people)* Berufsstand, *der;* **c)** *(declaration)* **~ of friendship/sympathy** Freundschafts- / Sympathiebekundung, *die;* **d)** *(Relig.: affirmation of faith)* Bekenntnis, *das* (**of** zu)

professional [prəˈfeʃənl] **1.** *adj.* **a)** *(of profession)* Berufs⟨*ausbildung, -leben*⟩; beruflich ⟨*Qualifikation, Laufbahn, Tätigkeit, Stolz, Ansehen*⟩; **~ body** Berufsorganisation, *die;* **~ advice** fachmännischer Rat; **~ standards** Leistungsniveau, *das;* **b)** *(worthy of profession) (in technical expertise)* fachmännisch; *(in attitude)* professionell; *(in experience)* routiniert; **c)** *(engaged in profession)* **~ people** Angehörige hochqualifizierter Berufe; **'apartment to let to ~ woman'** „Wohnung an berufstätige Dame zu vermieten"; **the ~ class[es]** die gehobenen Berufe; **d)** *(by profession)* gelernt; *(not amateur)* Berufs⟨*musiker, -sportler, -soldat, -fotograf*⟩; Profi⟨*sportler*⟩; **e)** *(paid)* Profi⟨*sport, -boxen, -fußball, -tennis*⟩; **go** *or* **turn ~:** Profi werden; **be in the ~ army** Berufssoldat sein; **be in the ~ theatre/on the ~ stage** beruflich am Theater/als Schauspieler arbeiten. **2.** *n. (trained person, lit. or fig.)* Fachmann, *der*/ Fachfrau, *die; (non-amateur; also Sport, Theatre)* Profi, *der*

professionalism [prəˈfeʃənəlɪzm] *n., no pl.* **a)** *(of work)* fachmännische Ausführung; *(attitude)* professionelle Einstellung; **b)** *(paid participation)* Profitum, *das*

professionally [prəˈfeʃənəlɪ] *adv.* **a)** *(in professional capacity)* geschäftlich ⟨*beraten, besuchen, konsultieren*⟩; beruflich ⟨*erfolgreich*⟩; *(in manner worthy of profession)* professionell; **be ~ trained/qualified** eine Berufsausbildung/abgeschlossene Berufsausbildung haben; **b)** *(as paid work)* berufsmäßig; **she plays tennis/the piano ~:** sie ist Tennisprofi/von Beruf Pianistin; **c)** *(by professional)* fachmännisch ⟨*leiten, betreiben*⟩; von einem Fachmann/von Fachleuten ⟨*erledigen lassen*⟩

professor [prəˈfesə(r)] *n.* **a)** *(Univ.: holder of chair)* Professor, *der*/Professorin, *die* (**of** für); **b)** *(Amer.: teacher at university)* Dozent, *der*/Dozentin, *die*

proffer [ˈprɒfə(r)] *v. t. (literary)* darbieten ⟨*Hand, Geschenk*⟩; anbieten ⟨*Frieden, Hilfe, Arm, Freundschaft*⟩; aussprechen ⟨*Dank*⟩; vorbringen ⟨*Vorschlag*⟩

proficiency [prəˈfɪʃənsɪ] *n.* Können, *das;* **degree** *or* **standard of ~:** Fertigkeit, *die;*

proˈficiency test *n.* Leistungstest, *der*

proficient [prəˈfɪʃənt] *adj.* fähig; gut ⟨*Pianist, Reiter, Skiläufer usw.*⟩; geschickt ⟨*Radfahrer, Handwerker, Lügner*⟩; **be ~ at** *or* **in maths/French** viel von Mathematik verstehen/gute Französischkenntnisse haben

profile [ˈprəʊfaɪl] *n.* **a)** *(side aspect)* Profil, *das;* **in ~:** im Profil; **b)** *(representation)* Profilbild, *das; (outline)* Umriß, *der;* **c)** *(biographical sketch)* Porträt, *das* (**of, on** *Gen.*); **d)** *(fig.)* **keep** *or* **maintain a low ~:** sich zurückhalten

profit [ˈprɒfɪt] **1.** *n.* **a)** *(Commerc.)* Gewinn, *der;* Profit, *der;* **at a ~:** mit Gewinn ⟨*verkaufen*⟩; **make a ~ from** *or* **out of sth.** mit etw.

Geld verdienen; **make [a few pence] ~ on sth.** [ein paar Pfennige] an etw. *(Dat.)* verdienen; **show a ~:** einen Gewinn verzeichnen; **yield a ~:** Gewinn abwerfen; **~-and-loss account** Gewinn-und-Verlust-Rechnung, *die;* **b)** *(advantage)* Nutzen, *der;* **there is no ~ in sth.** etw. ist zwecklos. **2.** *v.t.* **~ sb.** für jmdn. von Nutzen sein; **it did not ~ them in the end** es hat ihnen letzten Endes gar nichts gebracht. **3.** *v.i.* profitieren

~ by *v.t.* profitieren von; Nutzen ziehen aus ‹*Fehler, Erfahrung*›

~ from *v.t.* profitieren von ‹*Reise, Studium, Ratschlag*›; nutzen ‹*Gelegenheit*›

profitability [prɒfɪtə'bɪlɪtɪ] *n., no pl.* Rentabilität, *die*

profitable ['prɒfɪtəbl] *adj.* **a)** *(lucrative)* rentabel; einträglich; **b)** *(beneficial)* lohnend ‹*Unternehmung, Zeitvertreib, Kauf*›; nützlich ‹*Studium, Diskussion, Verhandlung, Nachforschungen*›

profiteer [prɒfɪ'tɪə(r)] **1.** *n.* Profitmacher, *der*/-macherin, *die.* **2.** *v.i.* sich bereichern

profiteering [prɒfɪ'tɪərɪŋ] *n.* Wucher, *der*

profit: ~-making *adj.* gewinnorientiert; **~ margin** *n.* Gewinnspanne, *die;* **~-sharing** *n.* Gewinnbeteiligung, *die; attrib.* Gewinnbeteiligungs-

profligate ['prɒflɪgət] *adj.* **a)** *(extravagant)* verschwenderisch; **b)** *(dissipated)* ausschweifend ‹*Person*›

pro forma 'invoice *n.* *(Commerc.)* Pro-Forma-Rechnung, *die*

profound [prə'faʊnd] *adj.,* **~er** [prə'faʊndə(r)], **~est** [prə'faʊndɪst] **a)** *(extreme)* tief; nachhaltig ‹*Wirkung, Einfluß, Eindruck*›; tiefgreifend ‹*Wandel, Veränderung*›; lebhaft ‹*Interesse*›; tiefempfunden ‹*Beileid, Mitgefühl*›; tiefsitzend ‹*Angst, Mißtrauen*›; völlig ‹*Unwissenheit*›; hochgradig ‹*Schwerhörigkeit*›; **it is a matter of ~ indifference to me** es ist mir völlig gleichgültig; **b)** *(penetrating)* tief; profund *(geh.)* ‹*Wissen, Erkenntnis, Werk, Kenner*›; tiefgründig ‹*Untersuchung, Abhandlung, Betrachtung*›; tiefschürfend ‹*Essay, Analyse, Forscher*›; tiefsinnig ‹*Gedicht, Buch, Schriftsteller*›; scharfsinnig ‹*Denker, Forscher*›

profoundly [prə'faʊndlɪ] *adv.* zutiefst; stark ‹*beeinflußt, mitgenommen*›; hochgradig ‹*schwerhörig*›; ungemein ‹*scharfsinnig, beschlagen, feinfühlig*›; **I am ~ indifferent about it** es ist mir völlig gleichgültig

profundity [prə'fʌndɪtɪ] *n.* **a)** *no pl.* *(extremeness)* Tiefe, *die; (of joy, sorrow, concern, change)* [großes] Ausmaß; **b)** *no pl.* *(depth of intellect)* Tiefsinnigkeit, *die; (of analysis, book)* Tiefe, *die*

profuse [prə'fju:s] *adj.* *(abundant)* verschwenderisch ‹*Fülle, Üppigkeit, Vielfalt*›; groß ‹*Dankbarkeit*›; überschwenglich ‹*Entschuldigung, Lob*›; **~ bleeding** starke Blutung

profusely [prə'fju:slɪ] *adv.* *(abundantly)* massenhaft ‹*wachsen, vorkommen*›; heftig ‹*bluten, erröten, schwitzen*›; überaus ‹*dankbar*›; überschwenglich ‹*sich entschuldigen*›

profusion [prə'fju:ʒn] *n.* ungeheure *od.* überwältigende Menge; **in ~:** in Hülle und Fülle

prognosis [prɒg'nəʊsɪs] *n., pl.* **prognoses** [prɒg'nəʊsi:z] **a)** *(Med.)* Prognose, *die;* **b)** *(prediction)* Vorhersage, *die;* Prognose, *die;* **give** *or* **make a ~ of sth.** einen Ausblick auf etw. *(Akk.)* geben

program ['prəʊgræm] **1.** *n.* **a)** *(Amer.) see* **programme 1; b)** *(Computing, Electronics)* Programm, *das.* **2.** *v.t.,* **-mm-: a)** *(Amer.) see* **programme 2; b)** *(Computing, Electronics)* programmieren; **~ming language** Programmiersprache, *die*

programer *(Amer.) see* **programmer**

programme ['prəʊgræm] **1.** *n.* **a)** *([notice of] events)* Programm, *das;* **the evening's ~:** das Abendprogramm; **what is the ~ for today?** was steht heute auf dem Programm?; **my ~ for today** mein [heutiges] Tagesprogramm; **b)** *(Radio, Telev.)* *(presentation)* Sendung, *die; (Radio: service)* Sender, *der;* Programm, *das;* **c)** *(plan, instructions for machine)* Programm, *das;* **a ~ of study** ein Studienprogramm. **2.** *v.t.* **a)** *(make ~ for)* ein Programm zusammenstellen für; **b) the tumble-drier can be ~d to operate for between 10 and 60 minutes** der Trockner kann auf 10–60 Minuten Betriebszeit eingestellt werden; **c)** *see* **program 2 b**

programmer ['prəʊgræmə(r)] *n.* *(Computing, Electronics: operator)* Programmierer, *der*/Programmiererin, *die*

progress 1. ['prəʊgres] *n.* **a)** *no pl., no indef. art.* *(onward movement)* [Vorwärts]bewegung, *die;* **our ~ has been slow** wir sind nur langsam vorangekommen; **make ~:** vorankommen; **in ~:** im Gange; **b)** *no pl., no indef. art.* *(advance)* Fortschritt, *der;* **~ of science/civilization** wissenschaftlicher/kultureller Fortschritt; **make ~:** vorankommen; ‹*Student, Patient:*› Fortschritte machen; **make good ~ [towards recovery]** ‹*Patient:*› sich gut erholen; **some ~ was made** es wurden einige Fortschritte erzielt. **2.** [prə'gres] *v.i.* **a)** *(move forward)* vorankommen; **b)** *(be carried on, develop)* Fortschritte machen; **~ towards sth.** einer Sache *(Dat.)* näherkommen. **3.** ['prəʊgres] *v.t.* vorantreiben

progression [prə'greʃn] *n.* **a)** *(development)* Fortschritt, *der* **(in** bei); **b)** *(succession)* Folge, *die;* **c)** *(Math.)* Reihe, *die*

progressive [prə'gresɪv] **1.** *adj.* **a)** *(gradual)* fortschreitend ‹*Verbesserung, Verschlechterung*›; schrittweise ‹*Reform*›; allmählich ‹*Veränderung, Herannahen, Fortschreiten, Prozeß, Besserung*›; **b)** *(worsening)* schlimmer werdend; *(Med.)* progressiv; **c)** *(favouring reform; in culture)* fortschrittlich; progressiv; **d)** *(Taxation)* gestaffelt; progressiv *(fachspr.);* **~ tax** Progressivsteuer, *die.* **2.** *n.* Progressive, *der/die*

progressively [prə'gresɪvlɪ] *adv.* *(continuously)* immer ‹*weiter, schlechter*›; *(gradually)* stetig; Schritt für Schritt ‹*reformieren*›; *(successively)* [chronologisch] fortschreitend; **move ~ towards sth.** sich immer weiter auf etw. *(Akk.)* zubewegen

'progress report *n.* Tätigkeitsbericht, *der; (fig.: news)* Lagebericht, *der*

prohibit [prə'hɪbɪt] *v.t.* **a)** *(forbid)* verbieten; **~ sb.'s doing sth., ~ sb. from doing sth.** jmdm. verbieten, etw. zu tun; **b)** *(prevent)* verhindern; **~ sb.'s doing sth., ~ sb. from doing sth.** jmdn. daran hindern, etw. zu tun

prohibition [prəʊhɪ'bɪʃn, prəʊɪ'bɪʃn] *n.* **a)** *(edict)* [gesetzliches] Verbot **(against** *Gen.*); **b)** *no pl., no art.* *(Amer. Hist.)* [gesetzliches] Alkoholverbot; **P~** *(1920–33)* die Prohibition

prohibitive [prə'hɪbɪtɪv] *adj.* unerschwinglich ‹*Preis, Miete*›; untragbar ‹*Kosten*›

prohibitively [prə'hɪbɪtɪvlɪ] *adv.* unerschwinglich ‹*hoch, teuer*›

project 1. ['prɒdʒekt] *n.* **a)** *(plan)* Plan, *der;* **b)** *(enterprise)* Projekt, *das*. **2.** [prə'dʒekt] *v.t.* **a)** werfen ‹*Schatten, Schein, Licht*›; senden ‹*Strahl*›; *(Cinemat.)* projizieren; **b)** *(make known)* vermitteln; ~ **one's own personality** seine eigene Person in den Vordergrund stellen; **c)** *(plan)* planen; **d)** *(extrapolate)* übertragen (**to** auf + *Akk.*). **3.** *v.i.* [prə'dʒekt] *(jut out)* ‹*Felsen:*› vorspringen; ‹*Zähne, Brauen:*› vorstehen; ~ **over the street** ‹*Balkon:*› über die Straße ragen. **4.** *v. refl. (transport oneself)* ~ **oneself into sth.** sich in etw. *(Akk.)* [hinein]versetzen

projectile [prə'dʒektaɪl] *n.* Geschoß, *das;* Projektil, *das (Waffent.)*

projection [prə'dʒekʃn] *n.* **a)** *(protruding thing)* Vorsprung, *der;* **b)** *(making of visible image)* Projektion, *die; (of film)* Vorführung, *die;* **c)** *(thing planned)* Plan, *der;* **d)** *(extrapolation)* Übertragung, *die;* Hochrechnung, *die (Statistik); (estimate of future possibilities)* Voraussage, *die* (**of** über + *Akk.*)

projectionist [prə'dʒekʃənɪst] *n.* *(Cinemat.)* Filmvorführer, *der/*-vorführerin, *die*

pro'jection-room *n. (Cinemat.)* Vorführraum, *der*

projector [prə'dʒektə(r)] *n.* Projektor, *der*

proletarian [prəʊlɪ'teərɪən] **1.** *adj.* proletarisch. **2.** *n.* Proletarier, *der/*Proletarierin, *die*

proletariat [prəʊlɪ'teərɪət] *n.* Proletariat, *das*

proliferate [prə'lɪfəreɪt] *v.i.* **a)** *(Biol.)* sich stark vermehren; *(Med.)* proliferieren *(fachspr.);* wuchern; **b)** *(increase, lit. or fig.)* sich ausbreiten

proliferation [prəlɪfə'reɪʃn] *n.* **a)** *(Biol.)* starke Vermehrung; *(Med.)* Proliferation, *die (fachspr.);* Wucherung, *die;* **b)** *(increase, lit. or fig.)* starke Zunahme; *(of nuclear weapons)* Proliferation, *die*

prolific [prə'lɪfɪk] *adj.* **a)** *(fertile)* fruchtbar; **b)** *(productive)* produktiv

prologue *(Amer.:* **prolog)** ['prəʊlɒg] *n.* **a)** *(introduction)* Prolog, *der* (**to** zu); **b)** *(fig.)* Vorspiel, *das* (**to** zu)

prolong [prə'lɒŋ] *v.t.* verlängern; ~ **the agony** *(fig. coll.)* die Qual [unnötig] in die Länge ziehen

prolongation [prəʊlɒŋ'geɪʃn] *n.* Verlängerung, *die*

prolonged [prə'lɒŋd] *adj.* lang; lang anhaltend ‹*Beifall*›; langgezogen ‹*Schrei*›

promenade [prɒmə'nɑ:d] **1.** *n.* *(walkway)* Promenade, *die; (Brit.: at seaside)* [Strand]promenade, *die.* **2.** *v.i.* promenieren *(geh.)*

promenade: ~ **concert** *n.* Promenadenkonzert, *das;* ~ **deck** *n.* *(Naut.)* Promenadendeck, *das*

prominence ['prɒmɪnəns] *n.* **a)** *(conspicuousness)* Auffälligkeit, *die;* **b)** *(distinction)* Bekanntheit, *die;* **come into** *or* **rise to** ~: bekannt werden; **give** ~ **to sth.** etw. in den Vordergrund stellen; **c)** *(projecting part)* Vorsprung, *der*

prominent ['prɒmɪnənt] *adj.* **a)** *(conspicuous)* auffallend; **b)** *(foremost)* herausragend; **become very** ~: sehr bekannt werden; **he was** ~ **in politics** er war ein prominenter Politiker; **c)** *(projecting)* vorspringend; vorstehend ‹*Backenknochen, Brauen*›

prominently ['prɒmɪnəntlɪ] *adv.* **a)** *(conspicuously)* auffallend; **b)** *(in forefront)* in einer führenden Rolle; **he figured** ~ **in the case** er spielte in dem Fall eine wichtige Rolle

promiscuity [prɒmɪ'skju:ɪtɪ] *n., no pl. (in sexual relations)* Promiskuität, *die (geh.)*

promiscuous [prə'mɪskjʊəs] *adj.* *(in sexual relations)* promiskuitiv; **be** ~ ‹*Person:*› den [Sexual]partner/die [Sexual]partnerin häufig wechseln; **a** ~ **man** ein Mann, der häufig die Partnerin wechselt

promiscuously [prə'mɪskjʊəslɪ] *adv. (in sexual relations)* promiskuitiv

promise ['prɒmɪs] **1.** *n.* **a)** *(assurance)* Versprechen, *das;* **sb.'s** ~**s** jmds. Versprechungen; **give** *or* **make a** ~ **[to sb.]** [jmdm.] ein Versprechen geben; **I'm not making any** ~**s** ich kann nichts versprechen; **give** *or* **make a** ~ **of sth. [to sb.]** [jmdm.] etw. versprechen; **it's a** ~: ganz bestimmt; **b)** *(guarantee)* Zusicherung, *die;* **c)** *(fig.: reason for expectation)* Hoffnung, *die;* **a painter of** *or* **with** ~: ein vielversprechender Maler; ~ **of sth.** Aussicht auf etw. *(Akk.);* **show [great]** ~: zu großen Hoffnungen berechtigen. **2.** *v.t.* **a)** *(give assurance of)* versprechen; ~ **sth. to sb.,** ~ **sb. sth.** jmdm. etw. versprechen; **b)** *(fig.: give reason for expectation of)* verheißen *(geh.);* ~ **sb. sth.** jmdm. etw. in Aussicht stellen; ~ **to do/be sth.** versprechen, etw. zu tun/zu sein; **c)** ~ **oneself sth./that one will do sth.** sich *(Dat.)* etw. vornehmen/sich vornehmen, etw. zu tun. **3.** *v.i.* **a)** ~ **well** *or* **favourably [for the future]** vielversprechend [für die Zukunft] sein; **b)** *(give assurances)* Versprechungen machen; **I can't** ~: ich kann es nicht versprechen

promising ['prɒmɪsɪŋ] *adj.* vielversprechend

promontory ['prɒməntərɪ] *n.* Vorgebirge, *das*

promote [prə'məʊt] *v.t.* **a)** *(to more senior job)* befördern; **b)** *(encourage)* fördern; **c)** *(publicize)* Werbung machen für; **d)** *(Footb.)* **be** ~**d** aufsteigen

promoter [prə'məʊtə(r)] *n.* **a)** *(who organizes and finances event)* Veranstalter, *der/*Veranstalterin, *die; (of ballet tour, pop festival, boxing-match, cycle-race also)* Promoter, *der;* **b)** *(publicizer)* Promoter, *der/*Promoterin, *die*

promotion [prə'məʊʃn] *n.* **a)** *(to more senior job)* Beförderung, *die;* **win** *or* **gain** ~: befördert werden; ~ **to [the rank of] sergeant** *etc.* Beförderung zum Unteroffizier *usw.;* **b)** *(furtherance)* Förderung, *die;* **c)** *(Sport, Theatre: event)* Veranstaltung, *die;* **d)** *(publicization)* Werbung, *die; (instance)* Werbekampagne, *die;* **sales** ~: Werbung, *die;* **e)** *(Footb.)* Aufstieg, *der;* **be sure of** ~: mit Sicherheit aufsteigen

promotional [prə'məʊʃənl] *adj.* Werbe‹*kampagne, -broschüre, -strategie usw.*›

prompt [prɒmpt] **1.** *adj.* **a)** *(ready to act)* bereitwillig; **be** ~ **in doing sth.** *or* **to do sth.** etw. unverzüglich tun; **b)** *(done readily)* sofortig; **her** ~ **answer/reaction** ihre prompte Antwort/Reaktion; **take** ~ **action** sofort handeln; **make a** ~ **decision** sich sofort entschließen; **c)** *(punctual)* pünktlich. **2.** *adv.* pünktlich; **at 6 o'clock** ~: Punkt 6 Uhr. **3.** *v.t.* **a)** *(incite)* veranlassen; ~ **sb. to sth./to do sth.** jmdn. zu etw. veranlassen/dazu veranlassen, etw. zu tun; **b)** *(supply with words; also Theatre)* soufflieren (+ *Dat.*); *(supply with answers)* vorsagen (+ *Dat.*); *(give suggestion to)* weiterhelfen (+ *Dat.*); **c)** *(inspire)* hervorrufen ‹*Kritik, Eifersucht usw.*›; provozieren ‹*Antwort*›

prompter ['prɒmptə(r)] *n.* *(Theatre)* Souffleur, *der/*Souffleuse, *die*

prompting ['prɒmptɪŋ] *n.* **a) he never needs** ~: man muß ihn nicht zweimal bitten; **b)** *(Theatre)* Soufflieren, *das*

promptly ['prɒmptlɪ] *adv.* **a)**

(quickly) prompt; **he ~ went and did the opposite** *(iron.)* er hat natürlich prompt [genau] das Gegenteil getan; **b)** *(punctually)* pünktlich; **at 8 o'clock ~, ~ at 8 o'clock** Punkt 8 Uhr; pünktlich um 8 Uhr

prone [prəʊn] *adj.* **a)** *(liable)* **be ~ to** anfällig sein für ‹*Krankheiten, Depressionen*›; neigen zu ‹*Faulheit, Meditation*›; **be ~ to do sth.** dazu neigen, etw. zu tun; **b)** *(down-facing)* **assume a ~ position on the floor** sich in Bauchlage auf den Boden legen

prong [prɒŋ] *n.* *(of fork)* Zinke, *die*

-pronged [prɒŋd] *adj. in comb.* -zinkig; **three-~ attack** *(Mil.; also fig.)* Angriff von drei Seiten

pronoun ['prəʊnaʊn] *n.* *(Ling.)* *(word replacing noun)* Pronomen, *das;* Fürwort, *das;* *(pronominal adjective)* Pronominaladjektiv, *das*

pronounce [prə'naʊns] **1.** *v.t.* **a)** *(declare formally)* verkünden; **~ judgement** das Urteil verkünden; **~ judgement on sb./sth.** über jmdn./etw. das Urteil sprechen; **~ sb./sth. [to be] sth.** jmdn./etw. für etw. erklären; **~ sb. fit for work** jmdn. für arbeitsfähig erklären; **b)** *(declare as opinion)* erklären für; **he ~d himself disgusted with it** er erklärte, er sei empört darüber; **c)** *(speak)* aussprechen ‹*Wort, Buchstaben usw.*›; **the h is not ~d** das h wird nicht gesprochen. **2.** *v.i.* **~ on sth.** zu etw. Stellung nehmen; **~ for** *or* **in favour of/against sth.** sich für/gegen etw. aussprechen

pronounced [prə'naʊnst] *adj.* **a)** *(declared)* erklärt; ausgesprochen ‹*Gegner, Autorität*›; **b)** *(marked)* ausgeprägt; **walk with** *or* **have a ~ limp** stark hinken

pronouncement [prə'naʊnsmənt] *n.* Erklärung, *die;* **make a ~ [about sth.]** eine Erklärung [zu etw.] abgeben

pronto ['prɒntəʊ] *adv.* *(sl.)* dalli *(ugs.)*; **and [do it] ~!** aber [ein bißchen] dalli! *(ugs.)*

pronunciation [prənʌnsɪ'eɪʃn] *n.* Aussprache, *die;* **what is the ~ of this word?** wie wird dieses Wort ausgesprochen?

proof [pru:f] **1.** *n.* **a)** *(fact, evidence)* Beweis, *der;* **very good ~:** sehr gute Beweise; **~ positive** eindeutige Beweise; **b)** *no pl., no indef. art.* *(Law)* Beweismaterial, *das;* **c)** *no pl.* *(proving)* **in ~ of** zum Beweis (+ *Gen.*); **d)** *no pl.* *(test, trial)* Beweis, *der;* **put a theory to the ~:** eine Theorie unter Beweis stellen; **the ~ of the pudding is in the eating** *(prov.)* Probieren geht über Studieren *(Spr.)*; **e)** *no pl., no art.* *(standard of strength)* Proof *o. Art.*; **100 ~** *(Brit.)*, **128 ~** *(Amer.)* 64 Vol.-% Alkohol; **f)** *(Printing)* Abzug, *der.* **2.** *adj.* **a)** *(impervious)* **be ~ against sth.** unempfindlich gegen etw. sein; *(fig.)* gegen etw. immun sein; **b)** *in comb.* ‹*kugel-, bruch-, einbruch-, diebes-, idioten*›sicher; ‹*schall-, wasser*›dicht; **flame-~:** nicht brennbar; **c)** hochprozentig ‹*Alkohol*›; **this liqueur is 67.4°** *(Brit.)* *or* *(Amer.)* **76.8° ~:** dieser Likör hat 38,4 Vol.-% Alkohol. **3.** *v.t.* *(Printing)* *(take ~ of)* andrucken; *(~-read)* Korrektur lesen

proof: ~-read *v.t.* *(Printing)* Korrektur lesen; **~-reader** *n.* *(Printing)* Korrektor, *der*/Korrektorin, *die;* **~-reading** *n.* *(Printing)* Korrekturlesen, *das*

prop [prɒp] **1.** *n.* *(support, lit. or fig.)* Stütze, *die;* *(Mining)* Strebe, *die.* **2.** *v.t.*, **-pp-: a)** *(support)* stützen; **the ladder was ~ped against the house** die Leiter war gegen das Haus gelehnt; **b)** *(fig.)* *see* **~ up b**

~ 'up *v.t.* **a)** *(support)* stützen; **~ oneself up on one's elbows** sich auf die Ellbogen stützen; **b)** *(fig.)* aufrichten ‹*Person*›; vor dem Konkurs bewahren ‹*Firma*›; stützen ‹*Regierung, Währung*›

propaganda [prɒpə'gændə] *n., no pl., no indef. art.* Propaganda, *die*

propagate ['prɒpəgeɪt] **1.** *v.t.* **a)** *(Hort., Bacteriol.)* vermehren **(from, by** durch); *(Breeding, Zool.)* züchten; **b)** *(spread)* verbreiten; **c)** *(Phys.)* **be ~d** sich fortpflanzen. **2.** *v.i.* **a)** *(Bot., Zool., Bacteriol.)* sich vermehren; **b)** *(spread, extend, travel)* sich ausbreiten

propagation [prɒpə'geɪʃn] *n.* **a)** *(Hort., Breeding, Bacteriol.: causing to propagate)* Züchtung, *die;* **b)** *(Bot., Zool., Bacteriol.: reproduction)* Vermehrung, *die;* **c)** *(spreading)* Verbreitung, *die;* **d)** *(Phys.)* Fortpflanzung, *die*

propagator ['prɒpəgeɪtə(r)] *n.* *(Hort.: device)* [beheizbare] Saatkiste

propane ['prəʊpeɪn] *n.* *(Chem.)* Propan, *das*

propel [prə'pel] *v.t.*, **-ll-** *(lit. or fig.)* antreiben

propeller [prə'pelə(r)] *n.* Propeller, *der*

pro'peller shaft *n.* *(Motor Veh.)* Kardanwelle, *die*

propelling 'pencil *n.* *(Brit.)* Drehbleistift, *der*

propensity [prə'pensɪtɪ] *n.* Neigung, *die;* **[have] a ~ to** *or* **towards sth.** einen Hang zu etw. [haben]; **have a ~ to do sth.** *or* **for doing sth.** dazu neigen, etw. zu tun

proper ['prɒpə(r)] **1.** *adj.* **a)** *(accurate)* richtig; wahrheitsgetreu ‹*Bericht*›; zutreffend ‹*Beschreibung*›; eigentlich ‹*Wortbedeutung*›; ursprünglich ‹*Fassung*›; **in the ~ sense** im wahrsten Sinne des Wortes; **b)** *postpos.* *(strictly so called)* im engeren Sinn *nachgestellt;* **in London ~:** in London selbst; **c)** *(genuine)* echt; richtig ‹*Wirbelsturm, Schauspieler*›; **d)** *(satisfactory)* richtig; zufriedenstellend ‹*Antwort*›; hinreichend ‹*Grund*›; **e)** *(suitable)* angemessen; *(morally fitting)* gebührend; **do sth. the ~ way** etw. richtig machen; **we must do the ~ thing by him** wir müssen ihn fair behandeln; **do as you think ~:** tu, was du für richtig hältst; **f)** *(conventionally acceptable)* gehörig; **it would not be ~ for me to ...:** es gehört sich nicht, daß ich ...; **g)** *(conventional, prim)* förmlich; **h)** *attrib.* *(coll.: thorough)* richtig; **she gave him a ~ hiding** sie gab ihm eine ordentliche Tracht Prügel; **you gave me a ~ turn** du hast mir einen ganz schönen Schrecken eingejagt. **2.** *adv.* *(coll.)* **good and ~:** gehörig; nach Strich und Faden *(ugs.)*

properly ['prɒpəlɪ] *adv.* **a)** richtig; *(rightly)* zu Recht; *(with decency)* anständig; **~ speaking** genaugenommen; **I'm not ~ authorized to do it** ich bin eigentlich nicht dazu berechtigt; **b)** *(primly)* förmlich; **c)** *(coll.: thoroughly)* total *(ugs.)*

proper ~ 'name, ~ 'noun *ns.* *(Ling.)* Eigenname, *der*

property ['prɒpətɪ] *n.* **a)** *(possession[s], ownership)* Eigentum, *das;* **lost ~:** Fundsachen *Pl.*; **lost ~ [department** *or* **office]** Fundbüro, *das;* **b)** *(estate)* Besitz, *der;* Immobilie, *die* *(fachspr.)*; **~ in London is expensive** die Immobilienpreise in London sind hoch; **c)** *(attribute)* Eigenschaft, *die;* *(effect, special power)* Wirkung, *die;* **d)** *(Cinemat., Theatre)* Requisit, *das*

prophecy ['prɒfɪsɪ] *n.* **a)** *(prediction)* Vorhersage, *die;* **b)** *(prophetic utterance)* Prophezeiung, *die;* **c)** *(prophetic faculty)* **[the**

power *or* **gift of]** ~: die Gabe der Prophetie *(geh.)*

prophesy ['prɒfɪsaɪ] **1.** *v. t. (predict)* vorhersagen; *(fig.)* prophezeien ‹*Unglück*›; *(as fortune-teller)* weissagen. **2.** *v. i.* **a)** *(foretell future)* Vorhersagen machen; **b)** *(speak as prophet)* Prophezeiungen machen

prophet ['prɒfɪt] *n. (lit. or fig.)* Prophet, *der*

prophetess ['profɪtɪs] *n.* Prophetin, *die*

prophetic [prə'fetɪk] *adj.* prophetisch

propitious [prə'pɪʃəs] *adj.* **a)** *(auspicious)* verheißungsvoll; **b)** *(favouring)* günstig; ~ **for** *or* **to sth.** günstig für etw.; ~ **for** *or* **to doing sth.** dafür geeignet, etw. zu tun

proponent [prə'pəʊnənt] *n.* Befürworter, *der*/Befürworterin, *die*

proportion [prə'pɔ:ʃn] **1.** *n.* **a)** *(portion)* Teil, *der; (in recipe)* Menge, *die;* **the ~ of deaths is high** der Anteil der Todesfälle ist hoch; **what ~ of candidates pass the exam?** wie groß ist der Anteil der erfolgreichen Prüfungskandidaten?; **b)** *(ratio)* Verhältnis, *das;* **the ~ of sth. to sth.** das Verhältnis von etw. zu etw.; **the high ~ of imports to exports** der hohe Anteil der Importe im Vergleich zu den Exporten; **in ~ [to sth.]** [einer Sache *(Dat.)*] entsprechend; **c)** *(correct relation)* Proportion, *die; (fig.)* Ausgewogenheit, *die;* **sense of ~**: Sinn für Proportionen; **be in ~ [to** *or* **with sth.]** *(lit. or fig.)* im richtigen Verhältnis [zu *od.* mit etw.] stehen; **try to keep things in ~** *(fig.)* versuchen Sie, die Dinge im richtigen Licht zu sehen; **be out of ~/all** *or* **any ~ [to** *or* **with sth.]** *(lit. or fig.)* in keinem/keinerlei Verhältnis zu etw. stehen; **get things out of ~** *(fig.)* die Dinge zu wichtig nehmen; *(worry unnecessarily)* sich *(Dat.)* zu viele Sorgen machen; **d)** *in pl. (size)* Dimension, *die;* **e)** *(Math.)* Proportion, *die;* **in direct/inverse ~**: direkt/umgekehrt proportional. **2.** *v. t. (make proportionate)* proportionieren; **~ sth. to sth.** etw. einer Sache *(Dat.)* anpassen; *see also* **proportioned**

proportional [prə'pɔ:ʃənl] *adj.* **a)** *(in proportion)* entsprechend; **be ~ to sth.** einer Sache *(Dat.)* entsprechen; **b)** *(in correct relation)* ausgewogen; **be ~ to sth.** *(lit. or fig.)* einer Sache *(Dat.)* entsprechen; **c)** *(Math.)* **be directly/indirectly ~ to sth.** einer Sache *(Dat.)* direkt/umgekehrt proportional sein

proportionally [prə'pɔ:ʃənəlɪ] *adv.* **a)** *(in proportion)* [dem]entsprechend; **b)** *(in correct relation)* proportional gesehen; **correspond/not correspond ~ to sth.** im richtigen/in keinem Verhältnis zu etw. stehen

proportional represen'tation *n. (Polit.)* Verhältniswahlsystem, *das*

proportionate [prə'pɔ:ʃənət] *adj.* **a)** *(in proportion)* entsprechend; **~ to sth.** proportional zu etw.; **b)** *(in correct relation)* ausgewogen; **~ to sth.** einer Sache *(Dat.)* entsprechend

proportioned [prə'pɔ:ʃnd] *adj.* proportioniert; **well-/ill-~**: wohlproportioniert/schlecht proportioniert

proposal [prə'pəʊzl] *n.* **a)** *(thing proposed)* Vorschlag, *der; (offer)* Angebot, *das;* **make a ~ for doing sth.** *or* **to do sth.** einen Vorschlag machen, etw. zu tun; **his ~ for improving the system** sein Vorschlag zur Verbesserung des Systems; **draw up ~s/a ~**: Pläne/einen Plan aufstellen; **b) ~ [of marriage]** [Heirats]antrag, *der*

propose [prə'pəʊz] **1.** *v. t.* **a)** *(put forward for consideration)* vorschlagen; **~ sth. to sb.** jmdm. etw. vorschlagen; **~ marriage [to sb.]** [jmdm.] einen Heiratsantrag machen; **b)** *(nominate)* **~ sb. as/for sth.** jmdn. als/für etw. vorschlagen; **c)** *(intend)* **~ doing** *or* **to do sth.** beabsichtigen, etw. zu tun; **d)** *(set up as aim)* planen. *See also* **toast 1 b. 2.** *v. i. (offer marriage)* **~ [to sb.]** jmdm. einen Heiratsantrag machen

proposition [prɒpə'zɪʃn] **1.** *n.* **a)** *(proposal)* Vorschlag, *der;* **make** *or* **put a ~ to sb.** jmdm. einen Vorschlag machen; **b)** *(statement)* Aussage, *die;* **c)** *(sl.: undertaking, problem)* Sache, *die (ugs.);* **paying ~**: lohnendes Geschäft; **d)** *(Logic)* Satz, *der;* Proposition, *die (fachspr.).* **2.** *v. t. (coll.)* jmdn. anmachen *(ugs.)*

propound [prə'paʊnd] *v. t.* darlegen; **~ a question** eine Frage aufwerfen

proprietary [prə'praɪətərɪ] *adj.* **a)** Eigentums‹*rechte-, -ansprüche usw.*›; **b)** *(patented)* Marken-; **~ brand** *or* **make of washing-powder** Markenwaschmittel, *das*

proprietary: ~ 'medicine *n.* Markenmedikament, *das;* **~ 'name, ~ 'term** *ns. (Commerc.)* Markenname, *der*

proprietor [prə'praɪətə(r)] *n.* Inhaber, *der*/Inhaberin, *die; (of newspaper)* Besitzer, *der*/Besitzerin, *die*

propriety [prə'praɪətɪ] *n.* **a)** *no pl. (decency)* Anstand, *der;* **with ~**: anständig; **breach of ~**: Verstoß gegen die guten Sitten; **b)** *no pl. (accuracy)* Richtigkeit, *die;* **with perfect ~**: völlig zu Recht

propulsion [prə'pʌlʃn] *n.* Antrieb, *der; (driving force, lit. or fig.)* Antriebskraft, *die*

prosaic [prə'zeɪɪk, prəʊ'zeɪɪk] *adj.* prosaisch *(geh.);* nüchtern

proscribe [prə'skraɪb] *v. t.* **a)** *(exile)* verbannen; *(fig.)* ächten; **b)** *(prohibit)* verbieten

prose [prəʊz] *n.* **a)** Prosa, *die; attrib.* Prosa‹*werk, -stil*›; **b)** *(Sch., Univ.)* **~ [translation]** Übersetzung in die Fremdsprache

prosecute ['prɒsɪkju:t] **1.** *v. t.* **a)** *(Law)* strafrechtlich verfolgen; **~ sb. for sth./doing sth.** jmdn. wegen etw. strafrechtlich verfolgen/jmdn. strafrechtlich verfolgen, weil er etw. tut/getan hat; **b)** *(pursue)* verfolgen; **c)** *(carry on)* ausüben. **2.** *v. i.* Anzeige erstatten

prosecution [prɒsɪ'kju:ʃn] *n.* **a)** *(Law) (bringing to trial)* [strafrechtliche] Verfolgung; *(court procedure)* Anklage, *die;* **start a ~ against sb.** Anklage gegen jmdn. erheben; **b)** *(Law: prosecuting party)* Anklage[vertretung], *die;* **the [case for the] ~**: die Anklage; **witness for the ~, ~ witness** Zeuge/Zeugin der Anklage; **~ lawyer** Staatsanwalt, *der*/-anwältin, *die;* **c)** *(pursuing)* Verfolgung, *die;* **d)** *(carrying on)* Ausübung, *die*

prosecutor ['prɒsɪkju:tə(r)] *n. (Law)* Ankläger, *der*/Anklägerin, *die;* **public ~** ≈ Generalstaatsanwalt, *der*/-anwältin, *die*

prosody ['prɒsədɪ] *n.* Verslehre, *die*

prospect 1. ['prɒspekt] *n.* **a)** *(extensive view)* Aussicht, *die* (**of** auf + *Akk.*); *(spectacle)* Anblick, *der;* **b)** *(expectation)* Erwartung, *die* (**of** hinsichtlich); **[at the] ~ of sth./doing sth.** *(mental picture, likelihood)* [bei der] Aussicht auf etw.*(Akk.)*/[darauf], etw. zu tun; **have the ~ of sth., have sth. in ~**: etw. in Aussicht haben; **c)** *in pl. (hope of success)* Zukunftsaussichten; **a man with [good] ~s** ein Mann mit Zukunft; **a job with no ~s** eine Stelle ohne Zukunft; **sb.'s ~s of sth./doing sth.** jmds. Chancen auf etw. *(Akk.)*/darauf, etw. zu tun; **the ~s for sb./sth.** die Aussichten für jmdn./etw.; **d)**

(possible customer) [möglicher] Kunde/[mögliche] Kundin; **be a good ~ for a race/the job** bei einem Rennen gute Chancen haben/ein aussichtsreicher Kandidat für den Job sein. **2.** [prə'spekt] *v.i. (explore for mineral)* prospektieren *(Bergw.)*; nach Bodenschätzen suchen; *(fig.)* Ausschau halten (**for** nach); **~ for gold** nach Gold suchen

prospective [prə'spektɪv] *adj. (expected)* voraussichtlich; zukünftig ‹*Erbe, Braut*›; potentiell ‹*Käufer, Kandidat*›

prospector [prə'spektə(r)] *n.* Prospektor, *der (Bergw.)*; *(for gold)* Goldsucher, *der*

prospectus [prə'spektəs] *n.* **a)** *(of enterprise)* Prospekt, *der (Wirtsch.)*; **b)** *(of book)* Prospekt, *der*; **c)** *(Brit. Univ.)* Studienführer, *der*

prosper ['prɒspə(r)] *v.i.* gedeihen; ‹*Geschäft*:› florieren; ‹*Kunst usw.*:› eine Blütezeit erleben; ‹*Berufstätiger*:› Erfolg haben

prosperity [prɒ'sperɪtɪ] *n., no pl.* Wohlstand, *der*

prosperous ['prɒspərəs] *adj. (flourishing)* wohlhabend; gutgehend, florierend ‹*Unternehmen*›; *(blessed with good fortune)* erfolgreich; **~ years/time** Jahre/Zeit des Wohlstands

prostate ['prɒsteɪt] *n.* **~ [gland]** *(Anat., Zool.)* Prostata, *die*; Vorsteherdrüse, *die*

prostitute ['prɒstɪtju:t] **1.** *n.* **a)** *(woman)* Prostituierte, *die*; **b)** *(man)* Strichjunge, *der (salopp)*. **2.** *v.t.* zur Prostitution anbieten; *(fig.)* prostituieren ‹*Talent, Integrität*›; **~ oneself** *(lit. or fig.)* sich prostituieren

prostitution [prɒstɪ'tju:ʃn] *n. (lit. or fig.)* Prostitution, *die*

prostrate 1. ['prɒstreɪt] *adj.* **a)** [auf dem Bauch] ausgestreckt; **b)** *(exhausted)* erschöpft; **be ~ with fever** vom Fieber geschwächt sein. **2.** [prɒ'streɪt, prə'streɪt] *v.t.* **a)** *(lay flat)* zu Boden werfen ‹*Person*›; **b)** *(overcome emotionally)* übermannen; **c)** *(exhaust)* erschöpfen; **be ~d by exhaustion** vor Erschöpfung ganz kraftlos sein. **3.** *v. refl. (throw oneself down)* **~ oneself [at sth./before sb.]** sich [vor etw./jmdm.] niederwerfen; **~ oneself at sb.'s feet** sich jmdm. zu Füßen werfen; **~ oneself [before sb.]** *(humble oneself)* sich [vor jmdm.] demütigen

protagonist [prəʊ'tægənɪst] *n.* **a)** *(advocate)* Vorkämpfer, *der*/Vorkämpferin, *die*; **b)** *(Lit./Theatre: chief character)* Protagonist, *der*/Protagonistin, *die*; *(fig.)* Hauptakteur, *der*/-akteurin, *die*

protect [prə'tekt] *v.t.* **a)** *(defend)* schützen (**from** vor + *Dat.*, **against** gegen); **~ed by law** gesetzlich geschützt; **~ sb. against** *or* **from himself/herself** jmdn. vor sich *(Dat.)* selbst schützen; **~ one's/sb.'s interests** seine/jmds. Interessen wahren; **b)** *(preserve)* unter [Natur]schutz stellen ‹*Pflanze, Tier, Gebiet*›; **~ed plants/animals** geschützte Pflanzen/Tiere; **c)** *(give legal immunity to)* schützen; **the law ~s foreign diplomats** ausländische Diplomaten genießen den Schutz der Immunität; **d)** *(Econ.)* durch Protektionismus schützen

protected 'species *n.* geschützte Art

protection [prə'tekʃn] *n.* **a)** Schutz, *der* (**from** vor + *Dat.*, **against** gegen); **under the ~ of sb./sth.** unter jmds. Schutz/dem Schutz einer Sache *(Gen.)*; **[under] police ~:** [unter] Polizeischutz; **b)** *(immunity from molestation)* Schutz, *der*; *(money paid)* Schutzgeld, *das*; **c)** *(of wildlife etc.)* Schutz, *der*; **d)** *(legal immunity)* Immunität, *die*; **e)** *(Econ.)* Schutz, *der*; *(system)* Protektionismus, *der*

protection: ~ money *n.* Schutzgeld, *das*; **~ racket** *n.* Erpresserorganisation, *die*; **run a ~ racket** die Erpressung von Schutzgeldern organisieren

protective [prə'tektɪv] *adj. (protecting)* schützend; Schutz‹*hülle, -anstrich, -vorrichtung, -maske*›; **be ~ towards sb.** fürsorglich gegenüber jmdm. sein; **~ instinct** Beschützerinstinkt, *der*; **~ clothing** Schutzkleidung, *die*

protective: ~ ar'rest, ~ 'custody *ns.* Schutzgewahrsam, *der (Amtsspr.)*; Schutzhaft, *die*

protector [prə'tektə(r)] *n.* **a)** *(person)* Beschützer, *der*/Beschützerin, *die*; **b)** *(thing)* Schutz, *der*; *in comb.* -schutz, *der*

protégé ['prɒteʒeɪ] *n.* Protegé, *der (geh.)*; Schützling, *der*

protégée ['prɒteʒeɪ] *n.* Schützling, *der*

protein ['prəʊti:n] *n. (Chem.)* Protein, *das (fachspr.)*; Eiweiß, *das*; **a high-~ diet** eine eiweißreiche Kost

protest 1. ['prəʊtest] *n.* **a)** *(remonstrance)* Beschwerde, *die*; *(Sport)* Protest, *der*; **make** *or* **lodge a ~ [against sb./sth.]** eine Beschwerde [gegen jmdn./etw.] einreichen; **b)** *(show of unwillingness, gesture of disapproval)* **~[s]** Protest, *der*; **under ~:** unter Protest; **in ~ [against sth.]** aus Protest [gegen etw.]; **c)** *no pl., no art. (dissent)* Protest, *der*; **the right of ~:** das Recht zu protestieren. **2.** [prə'test] *v.t.* **a)** *(affirm)* beteuern; **b)** *(Amer.: object to)* protestieren gegen. **3.** [prə'test] *v.i.* protestieren; *(make written or formal ~)* Protest einlegen (**to** bei); **~ about sb./sth.** gegen jmdn./etw. protestieren; **~ against being/doing sth.** dagegen protestieren, daß man etw. ist/tut

Protestant ['prɒtɪstənt] *(Relig.)* **1.** *n.* Protestant, *der*/Protestantin, *die*; Evangelische, *der/die*. **2.** *adj.* protestantisch; evangelisch

Protestantism ['prɒtɪstəntɪzm] *n., no pl., no art. (Relig.)* Protestantismus, *der*

protestation [prɒtɪ'steɪʃn] *n.* **a)** *(affirmation)* Beteuerung, *die*; **b)** *(protest)* Protest, *der*

protester [prə'testə(r)] *n. (dissenter)* Protestierende, *der/die*; *(at demonstration)* Demonstrant, *der*/Demonstrantin, *die*

protest ['prəʊtest]: **~ march** *n.* Protestmarsch, *der*; **~ marcher** *see* **marcher**; **~ song** *n.* Protestsong, *der*; **~ vote** *n.* Proteststimme, *die*

protocol ['prəʊtəkɒl] *n.* Protokoll, *das*

proton ['prəʊtɒn] *n. (Phys.)* Proton, *das*

prototype ['prəʊtətaɪp] *n.* Prototyp, *der*; **a ~ aeroplane/machine** der Prototyp eines Flugzeugs/einer Maschine

protract [prə'trækt] *v.t.* verlängern; **~ed** länger ‹*Diskussion, Krankheit, Besuch*›

protractor [prə'træktə(r)] *n. (Geom.)* Winkelmesser, *der*

protrude [prə'tru:d] **1.** *v.i.* herausragen (**from** aus); ‹*Zähne*:› vorstehen; **~ above/beneath/from behind sth.** etw. überragen/unter/hinter etw. *(Dat.)* hervorragen; **~ beyond sth.** über etw. *(Akk.)* hinausragen. **2.** *v.t.* ausstrecken ‹*Fühler*›; vorstülpen ‹*Lippen*›

protrusion [prə'tru:ʒn] *n. (projecting thing)* Vorsprung, *der*

protuberance [prə'tju:bərəns] *n. (thing)* Auswuchs, *der*

protuberant [prə'tju:bərənt] *adj.* vorstehend; hervortretend ‹*Augen*›

proud [praʊd] **1.** *adj.* **a)** stolz; **it made me [feel] really ~:** es erfüllte mich mit Stolz; **~ to do sth.** *or* **to be doing sth.** stolz darauf, etw. zu tun; ~ **of sb./sth./doing sth.** stolz auf jmdn./etw./darauf, etw. zu tun; **he is far too ~ of himself/his house** er bildet sich *(Dat.)* zu viel ein/zu viel auf sein Haus ein; **b)** *(arrogant)* hochmütig; stolz ‹*Tier*›; **I'm not too ~ to scrub floors** ich bin mir nicht zu gut zum Fußbodenschrubben. **2.** *adv. (Brit. coll.)* **do sb.** ~ *(treat generously)* jmdn. verwöhnen; *(honour greatly)* jmdm. eine Ehrung bereiten; **do oneself ~:** sich *(Dat.)* etwas Gutes tun

proudly ['praʊdlɪ] *adv.* **a)** stolz; **b)** *(arrogantly)* hochmütig

prove [pru:v] **1.** *v.t., p.p.* **~d** *or (esp. Amer., Scot., literary)* **~n** ['pru:vn] beweisen; nachweisen ‹*Identität*›; ~ **one's ability** sein Können unter Beweis stellen; **his guilt/innocence was ~d, he was ~d [to be] guilty/innocent** er wurde überführt/seine Unschuld wurde bewiesen; ~ **sb. right/wrong** ‹*Ereignis:*› jmdm. recht/unrecht geben; **be ~d wrong** *or* **to be false** ‹*Theorie, System:*› widerlegt werden; ~ **sth. to be true** beweisen, daß etw. wahr ist; ~ **one's/sb.'s case** *or* **point** beweisen, daß man recht hat/jmdm. recht geben; **it was ~d that ...:** es stellte sich heraus *od.* erwies *od.* zeigte sich, daß ... **2.** *v. refl.* ~ **oneself** sich bewähren; ~ **oneself intelligent/a good player** sich als intelligent/ als [ein] guter Spieler erweisen. **3.** *v.i. (be found to be)* sich erweisen als; ~ **[to be] unnecessary/interesting/a failure** sich als unnötig/interessant/[ein] Fehlschlag erweisen

Provence [prɒ'vɑ̃s] *pr. n.* die Provence

proverb ['prɒvɜ:b] *n.* Sprichwort, *das;* **be a ~** *(fig.)* ‹*Eigenschaft:*› sprichwörtlich sein

proverbial [prə'vɜ:bɪəl] *adj.*, **proverbially** [prə'vɜ:bɪəlɪ] *adv.* sprichwörtlich

provide [prə'vaɪd] *v.t.* **a)** *(supply)* besorgen; sorgen für; liefern ‹*Beweis*›; bereitstellen ‹*Dienst, Geld*›; **instructions are ~d with every machine** mit jeder Maschine wird eine Anleitung mitgeliefert; ~ **homes/materials/a car for sb.** jmdm. Unterkünfte/Materialien/ein Auto [zur Verfügung] stellen; ~ **sb. with money** jmdn. unterhalten; *(for journey etc.)* jmdm. Geld zur Verfügung stellen; **be [well] ~d with sth.** mit etw. [wohl]versorgt *od.* [wohl]versehen sein; ~ **oneself with sth.** sich *(Dat.)* etw. besorgen; **b)** *(stipulate)* ‹*Vertrag, Gesetz:*› vorsehen; **c) providing that** *see* **provided**

~ **for** *v.t.* **a)** *(make provision for)* vorsorgen für; Vorsorge treffen für; ‹*Plan, Gesetz:*› vorsehen ‹*Maßnahmen, Steuern*›; ‹*Schätzung:*› berücksichtigen ‹*Inflation*›; **b)** *(maintain)* sorgen für, versorgen ‹*Familie, Kind*›

provided [prə'vaɪdɪd] *conj.* ~ **[that] ...:** vorausgesetzt, [daß] ...

providence ['prɒvɪdəns] *n.* **a) [divine] ~:** die [göttliche] Vorsehung; **b) P~** *(God)* der Himmel

providential [prɒvɪ'denʃl] *adj. (opportune)* **it was ~ that ...:** es war ein Glück, daß ...

provider [prə'vaɪdə(r)] *n. (breadwinner)* Ernährer, *der*/Ernährerin, *die;* Versorger, *der*/Versorgerin, *die*

province ['prɒvɪns] *n.* **a)** *(administrative area)* Provinz, *die;* **b) the ~s** *(regions outside capital)* die Provinz *(oft abwertend);* **c)** *(sphere of action)* [Arbeits-, Tätigkeits-, Wirkungs]bereich, *der;* [Arbeits-, Tätigkeits]gebiet, *das; (area of responsibility)* Zuständigkeitsbereich, *der;* **that is not my ~:** da kenne ich mich nicht aus; *(not my responsibility)* dafür bin ich nicht zuständig

provincial [prə'vɪnʃl] **1.** *adj.* Provinz-; *(of the provinces)* Provinz-; *(typical of the provinces)* provinziell. **2.** *n.* Provinzler, *der*/Provinzlerin, *die (abwertend)*

provision [prə'vɪʒn] *n.* **a)** *(providing)* Bereitstellung, *die;* **as a** *or* **by way of ~ against ...:** zum Schutz gegen ...; ~ **of medical care** medizinische Versorgung; **make ~ for** vorsorgen *od.* Vorsorge treffen für ‹*Notfall*›; berücksichtigen ‹*Inflation*›; **make ~ for sb. in one's will** jmdn. in seinem Testament bedenken; **make ~ against sth.** Vorkehrungen zum Schutz gegen etw. treffen; **b)** *(amount available)* Vorrat, *der;* **c)** *in pl. (food)* Lebensmittel; *(for expedition also)* Proviant, *der;* **stock up with ~s** Lebensmittelvorräte anlegen; **d)** *(legal statement)* Verordnung, *die; (clause)* Bestimmung, *die*

provisional [prə'vɪʒənl] **1.** *adj.* vorläufig; provisorisch; ~ **arrangement** Provisorium, *das.* **2.** *n. in pl.* **the P~s** die provisorische IRA

provisional: P~ IR'A *n.* provisorische IRA; ~ **licence** *n.* vorläufige Fahrerlaubnis

provisionally [prə'vɪʒənəlɪ] *adv.* vorläufig; provisorisch

proviso [prə'vaɪzəʊ] *n., pl.* **~s** Vorbehalt, *der*

provocation [prɒvə'keɪʃn] *n.* Provokation, *die;* Herausforderung, *die;* **be under severe ~:** stark provoziert werden; **he loses his temper at** *or* **on the slightest** *or* **smallest ~:** er verliert die Beherrschung beim geringsten Anlaß

provocative [prə'vɒkətɪv] *adj.* provozierend; herausfordernd; *(sexually)* aufreizend; **his actions were felt to be ~:** seine Aktionen wurden als Provokation empfunden

provoke [prə'vəʊk] *v.t.* **a)** *(annoy, incite)* provozieren ‹*Person*›; reizen ‹*Person, Tier*›; *(sexually)* aufreizen; **be easily ~d** leicht reizbar sein; sich leicht provozieren lassen; ~ **sb. to anger/fury** jmdn. in Wut *(Akk.)*/zur Raserei bringen; ~ **sb. into doing sth.** jmdn. so sehr provozieren *od.* reizen, daß er etw. tut; **he was finally ~d into taking action** er ließ sich schließlich dazu hinreißen *od.* provozieren, etwas zu unternehmen; **b)** *(give rise to)* hervorrufen; erregen ‹*Ärger, Neugier, Zorn*›; auslösen ‹*Kontroverse, Krise*›; herausfordern ‹*Widerstand*›; verursachen ‹*Zwischenfall*›; Anlaß geben zu ‹*Klagen, Kritik*›

provoking [prə'vəʊkɪŋ] *adj.* provozierend; herausfordernd; **his behaviour/refusal was [very] ~:** sein Benehmen/seine Weigerung war eine [große] Provokation

prow [praʊ] *n. (Naut.)* Bug, *der*

prowess ['praʊɪs] *n.* **a)** *(valour)* Tapferkeit, *die;* **b)** *(skill)* Fähigkeiten; Können, *das;* ~ **at sports** [große] Sportlichkeit; **sexual ~:** sexuelle Leistungsfähigkeit

prowl [praʊl] **1.** *v.i.* streifen; ~ **about/around sth.** etw. durchstreifen. **2.** *v.t.* durchstreifen. **3.** *n.* Streifzug, *der;* **be on the ~:** auf einem Streifzug sein

prowler ['praʊlə(r)] *n.* **the police have warned of ~s in the area** die Polizei warnt vor verdächtigen Personen, die in der Gegend herumstreifen

proximity [prɒk'sɪmɪtɪ] *n., no pl.* Nähe, *die* **(to** zu)

proxy ['prɒksɪ] *n.* **a)** *(agency, document)* Vollmacht, *die;* Bevollmächtigung, *die;* **by ~:** durch einen Bevollmächtigten/eine Bevollmächtigte; *see also* **stand 1 g; b)** *(person)* Bevollmächtigte, *der/die; (vote)* durch einen Bevollmächtigten/eine Bevollmächtigte abgegebene Stimme; **make sb. one's ~:** jmdn. bevollmächtigen

prude [pru:d] *n.* prüder Mensch
prudence ['pru:dəns] *n., no pl.* Besonnenheit, *die;* Überlegtheit, *die;* **act with ~:** besonnen *od.* überlegt handeln
prudent ['pru:dənt] *adj.* **a)** *(careful)* besonnen ‹*Person*›; besonnen, überlegt ‹*Verhalten*›; **b)** *(circumspect)* vorsichtig; **think it more ~ to do sth.** es für klüger halten, etw. zu tun
prudish ['pru:dɪʃ] *adj.* prüde
[1]**prune** [pru:n] *n.* **a)** *(fruit)* **[dried] ~:** Back- *od.* Dörrpflaume, *die;* **b)** *(coll.: simpleton)* Trottel, *der (ugs. abwertend)*
[2]**prune** *v.t.* **a)** *(trim)* [be]schneiden; **~ back** zurückschneiden; **b)** *(lop off)* **~ [away/off]** ab- *od.* wegschneiden; **~ [out]** herausschneiden; **c)** *(fig.: reduce)* reduzieren; **~ back** Abstriche machen an (+ *Dat.*) ‹*Projekt*›
pruning-shears ['pru:nɪŋʃɪəz] *n. pl.* Gartenschere, *die;* Rosenschere, *die*
pry [praɪ] *v.i.* neugierig sein
~ a'bout *v.i.* herumschnüffeln *(ugs. abwertend) od.* -spionieren
~ into *v.t.* seine Nase stecken in (+ *Akk.*) *(ugs.)*‹*Angelegenheit*›
prying ['praɪɪŋ] *adj.* neugierig
PS *abbr.* **postscript** PS
psalm [sɑ:m] *n. (Eccl.)* Psalm, *der*
pseud [sju:d] *(coll.)* **1.** *adj.* **a)** *(pretentious)* pseudointellektuell; **b)** *see* **pseudo 1 a. 2.** *n. see* **pseudo 2**
pseudo ['sju:dəʊ] **1.** *adj.* **a)** *(sham, spurious)* unecht; **b)** *(insincere)* verlogen. **2.** *n., pl.* **~s a)** *(pretentious person)* Möchtegern, *der (ugs. spött.);* **b)** *(insincere person)* Heuchler, *der*/Heuchlerin, *die*
pseudo- *in comb.* pseudo-/Pseudo- *(fachspr., geh.)*
pseudonym ['sju:dənɪm] *n.* Pseudonym, *das*
psst, pst [pst] *int.* st
psyche ['saɪkɪ] *n.* Psyche, *die*
psychiatric [saɪkɪ'ætrɪk] *adj.* psychiatrisch
psychiatrist [saɪ'kaɪətrɪst] *n.* Psychiater, *der*/Psychiaterin, *die*
psychiatry [saɪ'kaɪətrɪ] *n.* Psychiatrie, *die*
psychic ['saɪkɪk] *adj. (having occult powers)* **be ~:** übernatürliche Fähigkeiten haben; **you must be ~** *(fig.)* du kannst wohl Gedanken lesen
psycho ['saɪkəʊ] *(coll.)* **1.** *adj.* verrückt *(ugs.).* **2.** *n., pl.* **~s** Verrückte, *der/die (ugs.)*
psycho'analyse *v.t.* psychoanalysieren *(fachspr.);* psychoanalytisch behandeln
psychoa'nalysis *n.* Psychoanalyse, *die*
psycho'analyst *n.* Psychoanalytiker, *der*/-analytikerin, *die*
psychological [saɪkə'lɒdʒɪkl] *adj.* **a)** *(of the mind)* psychisch ‹*Problem*›; psychologisch ‹*Wirkung, Druck*›; **b)** *(of psychology)* psychologisch
psychological 'warfare *n.* psychologische Kriegsführung
psychologist [saɪ'kɒlədʒɪst] *n. (also fig.)* Psychologe, *der*/Psychologin, *die*
psychology [saɪ'kɒlədʒɪ] *n.* **a)** Psychologie, *die;* **b)** *(coll.: characteristics)* Psychologie, *die (ugs.)*
psychopath ['saɪkəpæθ] *n.* Psychopath, *der*/Psychopathie, *die*
psychosis [saɪ'kəʊsɪs] *n., pl.* **psychoses** [saɪ'kəʊsi:z] Psychose, *die*
psychosomatic [saɪkəʊsə'mætɪk] *adj. (Med.)* psychosomatisch
psycho'therapy *n., no pl. (Med.)* Psychotherapie, *die*
PTO *abbr.* **please turn over** b.w.
pub [pʌb] *n. (Brit. coll.)* Kneipe, *die (ugs.); (esp. in British Isles)* Pub, *das; attrib.* Kneipen-
'pub-crawl *n. (Brit. coll.)* Zechtour, *die;* Bierreise, *die (ugs. scherzh.);* **go on a ~:** eine Zechtour machen
puberty ['pju:bətɪ] *n., no pl., no art.* Pubertät, *die;* **at ~:** in *od.* während der Pubertät
pubic ['pju:bɪk] *adj. (Anat.)* Scham-
public ['pʌblɪk] **1.** *adj.* öffentlich; **~ assembly** Volksversammlung, *die;* **a ~ danger/service** eine Gefahr für die/ein Dienst an der Allgemeinheit; **be a matter of ~ knowledge** allgemein bekannt sein; **in the ~ eye** im Blickpunkt der Öffentlichkeit; **make sth. ~:** etw. publik *(geh.) od.* bekannt machen. **2.** *n., no pl.; constr. as sing. or pl.* **a)** *(the people)* Öffentlichkeit, *die;* Allgemeinheit, *die;* **the general ~:** die Allgemeinheit; die breite Öffentlichkeit; **member of the ~:** Bürger, *der*/Bürgerin, *die;* **be open to the ~:** für den Publikumsverkehr geöffnet sein; **b)** *(section of community)* Publikum, *das; (author's readers also)* Leserschaft, *die;* **the reading ~:** das Lesepublikum; **c) in ~** *(publicly)* öffentlich; *(openly)* offen; **behave oneself in ~:** sich in der Öffentlichkeit benehmen
public-ad'dress system *n.* Lautsprecheranlage, *die*
publican ['pʌblɪkən] *n. (Brit.)* [Gast]wirt, *der*/-wirtin, *die*
publication [pʌblɪ'keɪʃn] *n. (issuing of book etc.; book etc. issued)* Veröffentlichung, *die;* Publikation, *die;* **the magazine is a weekly ~:** die Zeitschrift erscheint wöchentlich
public: ~ 'bar *n. (Brit.)* ≈ Ausschank, *der;* **~ 'company** *n. (Brit. Econ.)* Aktiengesellschaft, *die;* **~ 'figure** *n.* Persönlichkeit des öffentlichen Lebens; **~ 'footpath** *n.* öffentlicher Fußweg; **~ 'health** *n., no pl., no art.* [öffentliches] Gesundheitswesen; **~ 'holiday** *n.* gesetzlicher Feiertag; **~ 'house** *n. (Brit.)* Gastwirtschaft, *die;* Gaststätte, *die;* **~ 'interest** *n.* Interesse der Allgemeinheit
publicise *see* **publicize**
publicity [pʌb'lɪsɪtɪ] *n., no pl., no indef. art.* **a)** Publicity, *die; (advertising)* Werbung, *die;* **~ campaign** Werbekampagne, *die;* **~ material** Werbematerial, *das;* **b)** *(attention)* Publicity, *die;* Publizität, *die (geh.);* **attract ~** ‹*Vorfall:*› Aufsehen erregen
pub'licity agent *n.* Publicitymanager, *der*/-managerin, *die*
publicize ['pʌblɪsaɪz] *v.t.* publik machen ‹*Ungerechtigkeit*›; werben für, Reklame machen für ‹*Produkt, Veranstaltung*›; **well-~d** ausreichend publik gemacht
public: ~ 'library *n.* öffentliche Bücherei; **~ limited 'company** *n. (Brit.)* ≈ Aktiengesellschaft, *die*
publicly ['pʌblɪklɪ] *adv.* **a)** *(in public)* öffentlich; **b)** *(by the public)* mit öffentlichen Geldern ‹*finanzieren, subventionieren*›; **~ owned** staatseigen; staatlich
public: ~ 'nuisance *n. (Law)* Störung der öffentlichen [Sicherheit und] Ordnung; **~ o'pinion** *see* **opinion b;** **~ 'ownership** *n., no pl.* Staatseigentum, *das* (**of** an + *Dat.*); Gemeineigentum, *das* (**of** an + *Dat.*); **be taken into ~ ownership** verstaatlicht werden; **~ property** *n.* Staatsbesitz, *der;* **sth. is ~ property** *(fig.)* etw. ist allgemein bekannt; **~ 'prosecutor** *n. (Law)* Staatsanwalt, *der*/-anwältin, *die;* **~ 'purse** *see* **purse 1;** **~ re'lations** *n. pl., constr. as sing. or pl.* Public Relations *Pl.;* Öffentlichkeitsarbeit, *die; attrib.* Public-Relations-‹*Abteilung, Berater*›; **~ relations officer** Öffentlichkeitsreferent, *der*/-referentin, *die;* **~ school** *n.* **a)** *(Brit.)* Privatschule, *die; attrib.* Privatschul-; **b)** *(Scot., Amer.: school run by ~ authorities)* staatliche *od.* öffentliche Schule; **~-'spirited** *adj.*

von Gemeinsinn zeugend ⟨*Verhalten*⟩; **be a ~-spirited person** Gemeinsinn haben; **~ 'transport** *n.* öffentlicher Personenverkehr; **travel by ~ transport** mit öffentlichen Verkehrsmitteln fahren; **~ u'tility** *n.* öffentlicher Versorgungsbetrieb

publish ['pʌblɪʃ] *v. t.* **a)** ⟨*Verleger, Verlag:*⟩ verlegen ⟨*Buch, Zeitschrift, Musik usw.*⟩; ⟨*Autor:*⟩ publizieren, veröffentlichen ⟨*Text*⟩; **the book has been ~ed by a British company** das Buch ist in *od.* bei einem britischen Verlag erschienen; **b)** *(announce publicly)* verkünden; *(read out)* verlesen ⟨*Aufgebot*⟩; **c)** *(make generally known)* publik machen ⟨*Ergebnisse, Einzelheiten*⟩

publisher ['pʌblɪʃə(r)] *n.* Verleger, *der*/Verlegerin, *die;* **~[s]** *(company)* Verlag, *der;* **~s of children's books** Kinderbuchverlag, *der;* **music/scientific/magazine ~s** Musikverlag, *der*/wissenschaftlicher Verlag/Zeitschriftenverlag, *der*

publishing ['pʌblɪʃɪŋ] *n., no pl., no art.* Verlagswesen, *das; attrib.* Verlags-; **~ firm/company** Verlag, *der*

'publishing-house *n.* Verlag, *der*

puce [pju:s] **1.** *n.* Flohbraun, *das.* **2.** *adj.* flohbraun; **go ~ in the face** puterrot werden

puck [pʌk] *n. (Ice Hockey)* Puck, *der*

pucker ['pʌkə(r)] **1.** *v. t.* **~ [up]** runzeln ⟨*Brauen, Stirn*⟩; krausen, krausziehen ⟨*Stirn*⟩; kräuseln ⟨*Lippen*⟩; *(sewing)* kräuseln ⟨*Stoff*⟩. **2.** *v. i.* **~ [up]** ⟨*Gesicht:*⟩ sich in Falten legen; ⟨*Stoff:*⟩ sich kräuseln

pud [pʊd] *(coll.) see* **pudding**

pudding ['pʊdɪŋ] *n.* **a)** Pudding, *der;* **b)** *(dessert)* süße Nachspeise

pudding: ~-basin, ~-bowl *ns.* Puddingform, *die*

puddle ['pʌdl] *n.* Pfütze, *die*

puerile ['pjʊəraɪl] *adj.* kindisch *(abwertend);* infantil *(abwertend)*

Puerto Rican [pwɜ:təʊ 'ri:kən] **1.** *adj.* puertoricanisch. **2.** *n.* Puertoricaner, *der*/Puertoricanerin, *die*

Puerto Rico [pwɜ:təʊ: 'ri:kəʊ] *pr. n.* Puerto Rico *(das)*

puff [pʌf] **1.** *n.* **a)** Stoß, *der;* **~ of breath/wind** Atem-/Windstoß, *der;* **b)** *(sound of escaping vapour)* Zischen, *das;* **c)** *(quantity)* **~ of smoke** Rauchstoß, *der;* **~ of steam** Dampfwolke, *die;* **d)** *(pastry)* Blätterteigteilchen, *das;* **e) sb. runs out of ~** *(lit. or fig. coll.)* jmdm. geht die Puste aus *(ugs.).* **2.** *v. i.* **a)** ⟨*Blasebalg:*⟩ blasen; **~ [and blow]** pusten *(ugs.) od.* schnaufen [und keuchen]; **b)** *(~ cigarette smoke etc.)* paffen *(ugs.)* **(at** an + *Dat.*); **c)** *(move with ~ing)* ⟨*Person:*⟩ keuchen; ⟨*Zug, Lokomotive, Dampfer:*⟩ schnaufend fahren. **3.** *v. t.* **a)** *(blow)* pusten *(ugs.),* blasen ⟨*Rauch*⟩; stäuben ⟨*Puder*⟩; **b)** *(smoke in ~s)* paffen *(ugs.);* **c)** *(put out of breath) see* **~ out 1 b; d)** *(utter pantingly)* keuchen

~ 'out *v. t.* **a)** *(inflate)* ⟨*Wind:*⟩ blähen, bauschen ⟨*Segel*⟩; **b)** *(put out of breath)* außer Puste *(salopp) od.* Atem bringen ⟨*Person*⟩; **be ~ed out** außer Puste *(salopp) od.* Atem sein

~ 'up *v. t.* **a)** *(inflate)* aufblasen; aufpusten *(ugs.);* **b) be ~ed up** *(proud)* aufgeblasen sein

puffin ['pʌfɪn] *n. (Ornith.)* Papageientaucher, *der*

puff: ~ 'pastry *n. (Cookery)* Blätterteig, *der;* **~ 'sleeve** *n.* Puffärmel, *der*

puffy ['pʌfɪ] *adj.* verschwollen

pug [pʌg] *n.* **~[-dog]** Mops, *der*

pugnacious [pʌg'neɪʃəs] *adj. (literary)* kampflustig

'pug-nosed *adj.* stumpfnasig

puke [pju:k] *(coarse)* **1.** *v. i.* kotzen *(salopp).* **2.** *v. t.* **~ up** auskotzen *(salopp);* ausspucken *(ugs.).* **3.** *n.* Kotze, *die (salopp)*

pull [pʊl] **1.** *v. t.* **a)** *(draw, tug)* ziehen an (+ *Dat.*); ziehen ⟨*Hebel*⟩; **~ aside** beiseite ziehen; **~ sb.'s** *or* **sb. by the hair/ears/sleeve** jmdn. an den Haaren/Ohren/am Ärmel ziehen; **~ shut** zuziehen ⟨*Tür*⟩; **~ sth. over one's ears/head** sich *(Dat.)* etw. über die Ohren/den Kopf ziehen; **~ the other one** *or* **leg[, it's got bells on]** *(fig. coll.)* das kannst du einem anderen erzählen; **~ to pieces** in Stücke reißen; *(fig.: criticize severely)* zerpflücken ⟨*Argument, Artikel*⟩; **b)** *(extract)* [her]ausziehen; [heraus]ziehen ⟨*Zahn*⟩; zapfen ⟨*Bier*⟩; **c)** *(coll.: accomplish)* bringen *(ugs.);* **~ a stunt** *or* **trick** etwas Wahnsinniges tun; **d) ~ a knife/gun on sb.** ein Messer/eine Pistole ziehen und jmdn. damit bedrohen; **e) not ~ one's punches** *(fig.)* nicht zimperlich sein. **2.** *v. i.* **a)** ziehen; **'P~'** „Ziehen"; **b) ~ [to the left/right]** ⟨*Auto, Boot:*⟩ [nach links/rechts] ziehen; **c)** *(pluck)* **~ at** ziehen an (+ *Dat.*); **~ at sb.'s sleeve** jmdn. am Ärmel ziehen. **3.** *n.* **a)** Zug, *der;* Ziehen, *das; (of conflicting emotions)* Widerstreit, *der;* **give a ~ at sth.** an etw. *(Dat.)* ziehen; **b)** *no pl. (influence)* Einfluß, *der* **(with** auf + *Akk.,* bei)

~ a'head *v. i.* in Führung gehen; **~ ahead of** sich setzen vor (+ *Akk.*)

~ a'part *v. t.* **a)** *(take to pieces)* auseinandernehmen; zerlegen; **b)** *(fig.: criticize severely)* zerpflücken ⟨*Interpretation, Argumentation usw.*⟩; verreißen ⟨*Buch, [literarisches] Werk*⟩

~ a'way 1. *v. t.* wegziehen. **2.** *v. i.* anfahren; *(with effort)* anziehen

~ 'back 1. *v. i.* **a)** *(retreat)* zurücktreten; ⟨*Truppen:*⟩ sich zurückziehen; **b)** *(Sport)* [wieder]aufholen **(to** bis auf + *Akk.*). **2.** *v. t.* **a)** zurückziehen; **b)** *(Sport)* aufholen

~ 'down *v. t.* **a)** herunterziehen; **b)** *(demolish)* abreißen; **c)** *(make less)* drücken ⟨*Preis*⟩; *(weaken)* mitnehmen ⟨*Person*⟩

~ 'in 1. *v. t.* **a)** hereinziehen; zurückziehen ⟨*Beine*⟩; **b)** *(attract)* anziehen; **c)** *(coll.: detain in custody)* einkassieren *(salopp):* kassieren *(ugs.).* **2.** *v. i.* **a)** ⟨*Zug:*⟩ einfahren; **b)** *(move to side of road)* an die Seite fahren; *(stop)* anhalten; **~ in to the side of the road** an den Straßenrand fahren

~ into *v. t.* **a)** ⟨*Zug:*⟩ einfahren in (+ *Akk.*); **b)** *(move off road into)* fahren in (+ *Akk.*)

~ 'off *v. t.* **a)** *(remove)* abziehen; *(violently)* abreißen; ausziehen ⟨*Kleidungsstück, Handschuhe*⟩; **b)** *(accomplish)* an Land ziehen *(ugs.)* ⟨*Geschäft, Knüller*⟩

~ 'on *v. t.* [sich *(Dat.)*] an- *od.* überziehen; *(in a hurry)* sich werfen in (+ *Akk.*)

~ 'out 1. *v. t.* **a)** *(extract)* herausziehen; [heraus]ziehen ⟨*Zahn*⟩; **b)** *(take out of pocket etc.)* aus der Tasche ziehen; herausziehen ⟨*Messer, Pistole*⟩; [heraus]ziehen, *(scherzh.)* zücken ⟨*Brieftasche*⟩; **c)** *(withdraw)* abziehen ⟨*Truppen*⟩; herausnehmen ⟨*Spieler, Mannschaft*⟩. **2.** *v. i.* **a)** *(depart)* ⟨*Zug:*⟩ abfahren; **~ out of the station** aus dem Bahnhof ausfahren; **b)** *(away from roadside)* ausscheren; **c)** *(withdraw)* ⟨*Truppen:*⟩ abziehen **(of** aus); *(from deal, project, competition, etc.)* aussteigen *(ugs.)* **(of** aus)

~ 'over *see* **~ in 2 b**

~ 'through *v. i.* ⟨*Patient:*⟩ durchkommen

~ to'gether 1. *v. i. (fig.)* an einem *od.* am selben Strang ziehen. **2.** *v. refl.* sich zusammennehmen

~ 'up 1. *v. t.* **a)** hochziehen; **b)** ~

up a chair einen Stuhl heranziehen; **c)** [he]rausziehen ‹*Unkraut, Pflanze usw.*›; *(violently)* [he]rausreißen; **d)** *(stop)* anhalten, zum Stehen bringen ‹*Auto*›; **e)** *(reprimand)* zurechtweisen; rügen. **2.** *v. i.* **a)** *(stop)* anhalten; **b)** *(improve)* sich verbessern. **3.** *v. refl.* sich hocharbeiten

pulley ['pʊlɪ] *n.* Rolle, *die;* **set of ~s** *(tackle)* Flaschenzug, *der*

Pullman ['pʊlmən] *n.* **~ [car** *or* **coach]** Pullman[wagen], *der*

'**pull-out** *n.* **a)** *(folding portion of book etc.)* ausfaltbarer Teil; *(detachable section)* heraustrennbarer Teil; **b)** *(withdrawal)* Abzug, *der*

pullover ['pʊləʊvə(r)] *n.* Pullover, *der;* Pulli, *der (ugs.)*

pulp [pʌlp] **1.** *n.* **a)** *(of fruit)* Fruchtfleisch, *das;* **b)** *(soft mass)* Brei, *der;* **beat sb. to a ~:** jmdn. zu Brei schlagen *(salopp).* **2.** *v. t.* zerdrücken, zerstampfen ‹*Rübe*›; einstampfen ‹*Druckerzeugnis*›

pulpit ['pʊlpɪt] *n. (Eccl.)* Kanzel, *die*

pulsate [pʌl'seɪt, 'pʌlseɪt] *v. i.* **a)** *(beat, throb)* pulsieren; ‹*Herz:*› schlagen; *(fig. literary)* pulsieren; **b)** *(fig.: vibrate)* schwingen

¹pulse [pʌls] **1.** *n.* **a)** *(lit. or fig.)* Puls, *der; (single beat)* Pulsschlag, *der;* **have/keep one's finger on the ~ of sth.** die Hand am Puls einer Sache *(Gen.)* haben/auf dem laufenden über etw. *(Akk.)* bleiben; **b)** *(rhythmical recurrence)* Rhythmus, *der;* **c)** *(Electronics)* Impuls, *der.* **2.** *v. i. see* **pulsate**

²pulse *n. (variety of edible seed)* Hülsenfrucht, *die*

pulverize (pulverise) ['pʌlvəraɪz] *v. t.* **a)** *(to powder or dust)* pulverisieren; **b)** *(fig.: crush)* abservieren *(Sport)* ‹*Gegner*›; **I'll ~ you!** ich schlag' dich zu Brei! *(derb)*

puma ['pju:mə] *n. (Zool.)* Puma, *der*

pumice ['pʌmɪs] *n. (Min.)* **~[-stone]** Bimsstein, *der*

pummel ['pʌml] *v. t., (Brit.)* **-ll-** einschlagen auf (+ *Akk.*)

pump [pʌmp] **1.** *n. (machine; also fig.)* Pumpe, *die.* **2.** *v. i.* pumpen. **3.** *v. t.* **a)** pumpen; **~ bullets into sth.** Kugeln in etw. *(Akk.)* jagen *(ugs.);* **b) ~ sth. dry** etw. leerpumpen; **~ sb. for information** Auskünfte aus jmdm. herausholen; **c) ~ up** *(inflate)* aufpumpen ‹*Reifen, Fahrrad*›

pumpkin ['pʌmpkɪn] *n. (Bot.)* Kürbis, *der; attrib.* Kürbis-

pun [pʌn] *n.* Wortspiel, *das*

Punch [pʌntʃ] *n.* Punch, *der;* Hanswurst, *der;* **~ and Judy show** Kasperletheater, *das;* **be as pleased as ~:** sich freuen wie ein Schneekönig *(ugs.)*

¹punch 1. *v. t.* **a)** *(strike with fist)* boxen; **b)** *(pierce, open up)* lochen; **~ a hole** ein Loch stanzen; **~ a hole/holes in sth.** etw. lochen. **2.** *n.* **a)** *(blow)* Faustschlag, *der;* **b)** *(coll.: vigour)* Pep, *der (ugs.);* **c)** *(device for making holes) (in leather, tickets)* Lochzange, *die; (in paper)* Locher, *der. See also* **pack 2f; pull 1e**

²punch *n. (drink)* Punsch, *der*

punch: ~-ball *n. (Brit.) (ball)* Punchingball, *der; (bag)* Sandsack, *der;* **~-bowl** *n.* Bowlengefäß, *das;* Bowle, *die;* **~ card** *n. (Computing)* Lochkarte, *die;* **~-drunk** *adj. (fig.)* benommen

punched [pʌntʃt]: **~ card, ~tape** *see* **punch card, ~tape**

'**punching bag** *(Amer.) see* **punch-ball**

punch: ~ line *n.* Pointe, *die;* **~ tape** *n. (Computing)* Lochstreifen, *der;* **~-up** *n. (Brit. coll.) (fistfight, brawl)* Prügelei, *die*

punctilious [pʌŋk'tɪlɪəs] *adj.* [peinlich] korrekt; peinlich ‹*Genauigkeit*›

punctual ['pʌŋktjʊəl] *adj.* pünktlich

punctuality [pʌŋktjʊ'ælɪtɪ] *n., no pl.* Pünktlichkeit, *die*

punctuate ['pʌŋktjʊeɪt] *v. t.* interpunktieren *(fachspr.);* mit Satzzeichen versehen; *(fig.: interrupt)* unterbrechen (**with** durch)

punctuation [pʌŋktjʊ'eɪʃn] *n., no pl.* Interpunktion, *die (fachspr.);* Zeichensetzung, *die*

punctu'ation mark *n.* Satzzeichen, *das*

puncture ['pʌŋktʃə(r)] **1.** *n.* **a)** *(flat tyre)* Reifenpanne, *die;* Platte, *der (ugs.);* **b)** *(hole)* Loch, *das; (in skin)* Einstich, *der.* **2.** *v. t.* durchstechen; *(fig.)* verletzen ‹*Würde*›; **be ~d** ‹*Reifen:*› ein Loch haben, platt sein; ‹*Haut:*› einen Einstich aufweisen. **3.** *v. i.* ‹*Reifen:*› ein Loch bekommen, platt werden

pundit ['pʌndɪt] *n.* Experte, *der*/Expertin, *die*

pungent ['pʌndʒənt] *adj.* **a)** beißend, ätzend ‹*Rauch, Dämpfe*›; scharf ‹*Soße, Gewürz usw.*›; stechend riechend ‹*Gas*›; **b)** *(fig.: biting)* beißend; ätzend

punish ['pʌnɪʃ] *v. t.* **a)** bestrafen ‹*Person, Tat*›; strafen *(geh.)* ‹*Person*›; **b)** *(Boxing coll.)* schwer zusetzen (+ *Dat.*); **c)** *(coll.: tax)* auf eine harte Probe stellen; **d)** *(coll.: put under stress)* strapazieren ‹*Nerven, Bauwerk*›

punishable ['pʌnɪʃəbl] *adj.* strafbar; **it is a ~ offence to ...:** es ist strafbar, ... zu ...; **be ~ by sth.** mit etw. bestraft werden

punishing ['pʌnɪʃɪŋ] *adj.* **a)** *(Boxing coll.)* mörderisch ‹*Haken*›; **b)** *(Sport coll.)* tödlich *(Sportjargon)* ‹*Schuß, Schlag, Volley*›; **c)** *(coll.: taxing)* mörderisch *(ugs.)* ‹*Rennen, Zeitplan, Kurs*›

punishment ['pʌnɪʃmənt] *n.* **a)** *no pl. (punishing)* Bestrafung, *die;* **b)** *(penalty)* Strafe, *die;* **c)** *(coll.: rough treatment)* **take a lot of ~:** ganz schön getriezt *od.* gezwiebelt werden *(ugs.). See also* **take 1w**

punitive ['pju:nɪtɪv] *adj.* **a)** *(penal)* Straf-; **b)** *(severe)* [allzu] rigoros ‹*finanzielle Maßnahmen, Besteuerung*›; unzumutbar ‹*Steuersatz*›

punk [pʌŋk] *n.* **a)** *(Amer. sl.: worthless person)* Dreckskerl, der *(salopp);* **b)** *(Amer. coll.: young ruffian)* Rabauke, *der (ugs.);* **c)** *(admirer of ~ rock)* Punk, *der; (performer of ~ rock)* Punk[rock]er, *der*/-[rock]erin, *die;* **d)** *(music) see* **~ rock**

punk 'rock *n.* Punkrock, *der*

punnet ['pʌnɪt] *n. (Brit.)* Körbchen, *das*

punt [pʌnt] **1.** *n.* Stechkahn, *der.* **2.** *v. t.* **a)** *(propel)* staken ‹*Boot*›; *(convey)* in einem Stechkahn fahren ‹*Person*›. **3.** *v. i.* staken

punter ['pʌntə(r)] *n. (coll.)* **a)** *(gambler)* Zocker, *der*/Zockerin, *die (salopp);* **b)** *(client of prostitute)* Freier, *der (verhüll.);* **c) the ~s** *(customers)* die Leutchen *(ugs.)*

puny ['pju:nɪ] *adj.* **a)** *(undersized)* zu klein ‹*Baby, Junge*›; **b)** *(feeble)* gering ‹*Kraft*›; schwach ‹*Waffe, Person*›; **c)** *(petty)* belanglos, unerheblich ‹*Leistung, Einwand*›

pup [pʌp] *n.* **a)** *(young dog or wolf)* Welpe, *der;* **b)** *(young animal)* Junge, *das*

pupa ['pju:pə] *n., pl.* **~e** ['pju:pi:] *(Zool.)* Puppe, *die*

pupate [pju:'peɪt] *v. i. (Zool.)* sich verpuppen

pupil ['pju:pɪl] *n.* **a)** *(schoolchild, disciple)* Schüler, *der*/Schülerin, *die;* **b)** *(Anat.)* Pupille, die

puppet ['pʌpɪt] *n.* Puppe, *die; (marionette; also fig.)* Marionette, *die; attrib.* Marionetten‹*regime, -regierung*›

'**puppet-show** *n.* Puppenspiel,

das; (with marionettes) Marionettenspiel, *das*

puppy ['pʌpɪ] *n.* Hundejunge, *das;* Welpe, *der*

'**puppy fat** *n., no pl. (Brit.)* Babyspeck, *der*

purchase ['pɜ:tʃəs] **1.** *n.* **a)** *(buying)* Kauf, *der;* **make several ~s/a ~:** verschiedenes/etwas kaufen; **b)** *(thing bought)* Kauf, *der;* **c)** *no pl. (hold)* Halt, *der; (leverage)* Hebelwirkung, *die;* Hebelkraft, *die;* **get a ~:** guten *od.* festen Halt finden. **2.** *v. t.* **a)** kaufen; erwerben *(geh.);* **purchasing power** Kaufkraft, *die;* **b)** *(acquire)* erkaufen

'**purchase price** *n.* Kaufpreis, *der*

purchaser ['pɜ:tʃəsə(r)] *n.* Käufer, *der*/Käuferin, *die*

pure [pjʊə(r)] *adj. (lit. or fig.)* rein; **it is madness ~ and simple** es ist schlicht *od.* ganz einfach Wahnsinn

pure: ~-blooded *adj.* reinblütig; **~-bred** *adj.* reinrassig

purée ['pjʊəreɪ] **1.** *n.* Püree, *das;* **tomato ~:** Tomatenmark, *das.* **2.** *v. t.* pürieren

purely ['pjʊəlɪ] *adv.* **a)** *(solely)* rein; **b)** *(merely)* lediglich

purgative ['pɜ:gətɪv] *n. (medicine)* [starkes] Abführmittel

purgatory ['pɜ:gətərɪ] *n. (Relig.)* Fegefeuer, *das;* **it was ~** *(fig.)* es war eine Strafe *od.* die Hölle

purge [pɜ:dʒ] **1.** *v. t.* **a)** *(cleanse)* reinigen (**of** von); **b)** *(remove)* entfernen; **~ away** *or* **out** beseitigen; **c)** *(rid)* säubern ‹*Partei*› (**of** von); *(remove)* entfernen ‹*Person*›; **d)** *(Med.)* abführen lassen ‹*Patienten*›. **2.** *n. (clearance)* Säuberung[saktion], *die; (Polit.)* Säuberung, *die*

purification [pjʊərɪfɪ'keɪʃn] *n.* **a)** Reinigung, *die;* **b)** *(spiritual cleansing)* Läuterung, *die*

purifier ['pjʊərɪfaɪə(r)] *n. (machine)* Reinigungsapparat, *der;* Reinigungsanlage, *die*

purify ['pjʊərɪfaɪ] *v. t.* **a)** *(make pure or clear)* reinigen; **b)** *(spiritually)* reinigen; läutern

purist ['pjʊərɪst] *n.* Purist, *der*/Puristin, *die*

puritan, *(Hist.)* **Puritan** ['pjʊərɪtn] **1.** *n.* Puritaner, *der*/Puritanerin, *die.* **2.** *adj.* puritanisch

puritanical [pjʊərɪ'tænɪkl] *adj.* puritanisch

purity ['pjʊərɪtɪ] *n., no pl.* **a)** Reinheit, *die;* **b)** *(chastity)* Keuschheit, *die*

purl [pɜ:l] **1.** *n.* linke Masche. **2.** *v. t.* **~ three [stitches]** drei linke Maschen stricken; *see also* **knit 1 b**

purple ['pɜ:pl] **1.** *adj.* lila; violett; *(fig.)* überfrachtet, überladen ‹*Prosa*›; **his face went ~ with rage** vor Zorn bekam er ein hochrotes Gesicht. **2.** *n.* Lila, *das;* Violett, *das*

purport 1. [pə'pɔ:t] *v. t.* **~ to do sth.** *(profess)* [von sich] behaupten, etw. zu tun; *(be intended to seem)* den Anschein erwecken sollen, etw. zu tun; **a letter ~ing to be written by the president** ein angeblich vom Präsidenten geschriebener Brief. **2.** ['pɜ:pɔ:t] *n.* Inhalt, *der*

purpose ['pɜ:pəs] *n.* **a)** *(object)* Zweck, *der; (intention)* Absicht, *die;* **what is the ~ of doing that?** was hat es für einen Zweck, das zu tun?; **you must have had some ~ in mind** du mußt irgend etwas damit bezweckt haben; **answer** *or* **suit sb.'s ~:** jmds. Zwecken dienen *od.* entsprechen; **for a ~:** zu einem bestimmten Zweck; **for the ~ of discussing sth.** um etw. zu besprechen; **on ~:** mit Absicht; absichtlich; **for ~s of** zum Zwecke (+ *Gen.*); **b)** *(effect)* **to no ~:** ohne Erfolg; **to some/good ~:** mit einigem/gutem Erfolg; **c)** *(determination)* Entschlossenheit, *die;* **have a ~ in life** in seinem Leben einen Sinn sehen; **d)** *(intention to act)* Absicht, *die*

'**purpose-built** *adj.* [eigens] zu diesem Zweck errichtet ‹*Gebäude*›; [eigens] zu diesem Zweck hergestellt, speziell angefertigt ‹*Gerät, Bauteil*›

purposeful ['pɜ:pəsfl] *adj.* **a)** zielstrebig; *(with specific aim)* entschlossen; **b)** *(with intention)* absichtsvoll

purposely ['pɜ:pəslɪ] *adv.* absichtlich; mit Absicht

purr [pɜ:(r)] **1.** *v. i.* schnurren; *(fig.: be in satisfied mood)* strahlen. **2.** *v. t.* durch Schnurren zum Ausdruck bringen; *(fig.)* säuseln. **3.** *n.* Schnurren, *das*

purse [pɜ:s] **1.** *n. (lit. or fig.)* Portemonnaie, *das;* Geldbeutel, *der (bes. südd.);* **the public ~:** die Staatskasse. **2.** *v. t.* kräuseln, schürzen ‹*Lippen*›

purser ['pɜ:sə(r)] *n.* Zahlmeister, *der*/-meisterin, *die*

'**purse-strings** *n. pl.* Schnüre *od.* Bänder [zum Verschließen des Geldbeutels]; **hold the ~** *(fig.)* über das Geld bestimmen

pursue [pə'sju:] *v. t.* **a)** *(literary: chase, lit. or fig.)* verfolgen; **b)** *(seek after)* streben nach; suchen nach; verfolgen ‹*Ziel*›; **c)** *(look into)* nachgehen (+ *Dat.*); **d)** *(engage in)* betreiben; **e)** *(carry out)* durchführen ‹*Plan*›

pursuer [pə'sju:ə(r)] *n.* Verfolger, *der*/Verfolgerin, *die*

pursuit [pə'sju:t, pə'su:t] *n.* **a)** *(pursuing) (of person, animal, aim)* Verfolgung, *die; (of knowledge, truth, etc.)* Streben, *das* (**of** nach); *(of pleasure)* Jagd, *die* (**of** nach); **in ~ of** auf der Jagd nach ‹*Wild, Dieb usw.*›; in Ausführung (+ *Gen.*) ‹*Beschäftigung, Tätigkeit, Hobby*›; **with the police in [full] ~:** mit der Polizei [dicht] auf den Fersen; **in hot ~:** dicht auf den Fersen *(ugs.);* **b)** *(pastime)* Beschäftigung, *die;* Betätigung, *die*

purveyor [pə'veɪə(r)] *n.* Lieferant, *der*/Lieferantin, *die*

pus [pʌs] *n., no indef. art. (Med.)* Eiter, *der*

push [pʊʃ] **1.** *v. t.* **a)** schieben; *(make fall)* stoßen; schubsen *(ugs.);* **don't ~ me like that!** schieb *od.* drängel [doch] nicht so!; **~ a car** *(to start the engine)* ein Auto anschieben; **~ the door to/open** die Tür zu-/aufstoßen; **she ~ed the door instead of pulling** sie drückte gegen die Tür, statt zu ziehen; **the policemen ~ed the crowd back** die Polizisten drängten die Menge zurück; **~ sth. up the hill** etw. den Berg hinaufschieben; **~ one's way through/into/on to** *etc.* **sth.** sich *(Dat.)* einen Weg durch/in/auf *usw.* etw. *(Akk.)* bahnen; **b)** *(fig.: impel)* drängen; **~ sb. into doing sth.** jmdn. dahin bringen, daß er etw. tut; **c)** *(tax)* **~ sb. [hard]** jmdn. [stark] fordern; **~ sb. too hard/too far** jmdn. überfordern; **he ~es himself very hard** er verlangt sich *(Dat.)* sehr viel ab; **be ~ed for sth.** *(coll.: find it difficult to provide sth.)* mit etw. knapp sein; **be ~ed for money** *or* **cash** knapp bei Kasse sein *(ugs.);* **be ~ed to do sth.** *(coll.)* Mühe haben, etw. zu tun; **~ one's luck** *(coll.)* übermütig werden; **d)** *(press for sale of)* die Werbetrommel rühren für; pushen *(Werbejargon);* **e)** *(sell illegally, esp. drugs)* dealen; pushen *(Drogenjargon);* **f)** *(advance)* **~ sth. a step/stage further** etw. einen Schritt vorantreiben; **not ~ the point** die Sache auf sich beruhen lassen; **~ sth. too far** mit etw. zu weit gehen; **~ things to extremes** die Dinge *od.* es zum Äußersten *od.* auf die Spitze treiben; **g)** *(coll.)* **be ~ing sixty** *etc.*

auf die Sechzig *usw.* zugehen. **2.** *v.i.* **a)** schieben; *(in queue)* drängeln; *(at door)* drücken; **'P~'** *(on door etc.)* „Drücken"; ~ **and shove** schubsen und drängeln; ~ **at sth.** gegen etw. drücken; **b)** *(make demands)* ~ **for sth.** etw. fordern; **c)** *(make one's way)* **he ~ed between us** er drängte sich zwischen uns; ~ **through the crowd** sich durch die Menge drängeln; ~ **past** *or* **by sb.** sich an jmdm. vorbeidrängeln *od.* -drücken; **d)** *(assert oneself for one's advancement)* sich in den Vordergrund spielen. **3.** *n.* **a)** Stoß, *der;* Schubs, *der (ugs.);* **give sth. a ~:** etw. schieben *od.* stoßen; **give sb. a ~:** jmdm. einen Schubs geben *(ugs.);* jmdm. einen Stoß versetzen; **My car won't start; can you give me a ~?** Mein Auto springt nicht an. Kannst du mich anschieben?; **b)** *(effort)* Anstrengungen *Pl.; (Mil.: attack)* Vorstoß, *der;* Offensive, *die;* **c)** *(determination)* Tatkraft, *die;* Initiative, *die;* **d)** *(crisis)* **when it comes/came to the ~,** *(Amer. coll.)* **when ~ comes/came to shove** wenn es ernst wird/als es ernst wurde; **at a ~:** wenn es sein muß; **e)** *(Brit. sl.: dismissal)* **get the ~:** rausfliegen *(ugs.);* **give sb. the ~:** jmdn. rausschmeißen *(ugs.)*

~ **a'bout** *v.t.* herumschieben; *(bully)* herumkommandieren

~ **a'head** *v.i.* ‹*Armee:*› [weiter] vorstoßen; *(with plans etc.)* weitermachen; ~ **ahead with sth.** etw. vorantreiben

~ **a'round** *see* ~ **about**

~ **a'side** *v.t. (lit. or fig.)* beiseite schieben

~ **a'way** *v.t.* wegschieben

~ **'forward 1.** *v.i. see* ~ **ahead. 2.** *v.t.* vorschieben; *(Mil.)* vorstoßen; ~ **oneself forward** sich in den Vordergrund schieben

~ **in** [-'-] **1.** *v.t.* eindrücken; *(make fall into the water)* hineinstoßen. **2.** ['--] *v.i.* sich hineindrängen

~ **'off 1.** *v.i.* **a)** *(Boating)* abstoßen; **b)** *(sl.: leave)* abhauen *(salopp);* abschieben *(salopp).* **2.** *v.t.* **a)** abdrücken ‹*Deckel, Verschluß usw.*›; **b)** *(Boating)* abstoßen

~ **'on 1.** *v.i. see* ~ **ahead. 2.** *v.t.* draufdrücken ‹*Deckel, Verschluß usw.*›

~ **'out** *v.t.* hinausschieben

~ **'out of** *v.t. (force to leave)* hinausdrängen aus

~ **'over** *v.t. (make fall)* umstoßen

~ **'through** *v.t. (fig.)* durchpeitschen *(ugs.)* ‹*Gesetzesvorlage*›; durchdrücken *(ugs.)*‹*Vorschlag*›

~ **'up** *v.t.* hochschieben; *(fig.)* hochtreiben

push: ~-bike *n. (Brit. coll.)* Fahrrad, *das;* **~-button 1.** *adj.* Drucktasten‹*telefon, -radio*›; **2.** *n.* [Druck]knopf, *der;* Drucktaste, *die;* **~-chair** *n. (Brit.)* Sportwagen, *der*

pusher ['pʊʃə(r)] *n.* **a)** *(seller of drugs)* Dealer, *der (Drogenjargon);* Pusher, *der (Drogenjargon);* **b)** *(pushy person)* Streber, *der*/Streberin, *die (abwertend)*

'push-over *n. (coll.)* Kinderspiel, *das;* **he'll be a ~ for her** sie steckt ihn [glatt] in die Tasche *(ugs.)*

pushy ['pʊʃɪ] *adj. (coll.)* [übermäßig] ehrgeizig ‹*Person*›

puss [pʊs] *n. (coll.)* Mieze, *die (fam.)*

pussy ['pʊsɪ] *n. (child lang.: cat)* Miezekatze, *die (fam.);* Muschi, *die (Kinderspr.)*

pussy: ~-cat *see* **pussy; ~foot** *v.i.* [herum]schleichen; *(act cautiously)* überängstlich sein; **~ willow** *n.* Salweide, *die*

put [pʊt] **1.** *v.t.,* **-tt-, put a)** *(place)* tun; *(vertically)* stellen; *(horizontally)* legen; *(through or into narrow opening)* stecken; ~ **plates on the table** Teller auf den Tisch stellen; ~ **clean sheets on the bed** das Bett frisch beziehen; **don't ~ your elbows on the table** laß deine Ellbogen vom Tisch; **I ~ my hand on his shoulder** ich legte meine Hand auf seine Schulter; ~ **a stamp on the letter** eine Briefmarke auf den Brief kleben; ~ **salt on one's food** Salz auf sein Essen tun *od.* streuen; ~ **some more coal on the fire** Kohle nachlegen; ~ **the letter in an envelope/the letter-box** den Brief in einen Umschlag/in den Briefkasten stecken; ~ **sth. in one's pocket** etw. in die Tasche stecken; ~ **one's hands in one's pockets** die Hände in die Taschen stecken; ~ **sugar in one's tea** sich *(Dat.)* Zucker in den Tee tun; ~ **petrol in the tank** Benzin in den Tank tun *od.* füllen; ~ **the car in[to] the garage** das Auto in die Garage stellen; ~ **the cork in the bottle** die Flasche mit dem Korken verschließen; ~ **the plug in the socket** den Stecker in die Steckdose stecken; ~ **the ball into the net/over the bar** den Ball ins Netz befördern *od.* setzen/über die Latte befördern; ~ **one's arm round sb.'s waist** den Arm um jmds. Taille legen; ~ **a bandage round one's wrist** sich *(Dat.)* einen Verband ums Handgelenk legen; ~ **one's hands over one's eyes** sich *(Dat.)* die Hände auf die Augen legen; ~ **one's finger to one's lips** den *od.* seinen Finger auf die Lippen legen; ~ **the boxes one on top of the other** die Kisten übereinanderstellen; ~ **the jacket on its hanger** die Jacke auf den Bügel tun *od.* hängen; **where shall I ~ it?** wohin soll ich es tun *(ugs.)*/stellen/legen *usw.*?; wo soll ich es hintun *(ugs.)*/-stellen/-legen *usw.*?; ~ **sb. into a taxi** jmdn. in ein Taxi setzen; **we ~ our guest in Peter's room** wir haben unseren Gast in Peters Zimmer *(Dat.)* untergebracht; ~ **the baby in the pram** das Baby in den Kinderwagen legen *od. (ugs.)* stecken; **not know where to ~ oneself** *(fig.)* sehr verlegen sein/werden; ~ **it there!** *(coll.)* laß mich deine Hand schütteln!; **b)** *(cause to enter)* stoßen; ~ **a satellite into orbit** einen Satelliten in eine Umlaufbahn bringen; **c)** *(bring into specified state)* setzen; ~ **through Parliament** im Parlament durchbringen ‹*Gesetzentwurf usw.*›; **be ~ in a difficult** *etc.* **position** in eine schwierige *usw.* Lage geraten; **be ~ into power** an die Macht kommen; ~ **sb. on the committee** jmdn. in den Ausschuß schicken; ~ **sth. above** *or* **before sth.** *(fig.)* einer Sache *(Dat.)* den Vorrang vor etw. *(Dat.)* geben; **be ~ out of order** kaputtgehen *(ugs.);* ~ **sb. on to sth.** *(fig.)* jmdn. auf etw. *(Akk.)* hinweisen *od.* aufmerksam machen; ~ **sb. on to a job** *(assign)* jmdm. eine Arbeit zuweisen; **d)** *(impose)* ~ **a limit/an interpretation on sth.** etw. begrenzen *od.* beschränken/interpretieren; **e)** *(submit)* unterbreiten **(to** *Dat.***)** ‹*Vorschlag, Plan usw.*›; ~ **the situation to sb.** jmdm. die Situation darstellen; ~ **sth. to the vote** über etw. *(Akk.)* abstimmen lassen; **f)** *(cause to go or do)* ~ **sb. to work** jmdn. arbeiten lassen; **be ~ out of the game by an injury** wegen einer Verletzung nicht mehr spielen können; ~ **sb. on antibiotics** jmdn. auf Antibiotika setzen; ~ **sb. on the stage** jmdn. zur Bühne schicken; **g)** *(express)* ausdrücken; **let's ~ it like this: ...:** sagen wir so: ...; **that's one way of ~ting it** *(also iron.)* so kann man es [natürlich] auch ausdrücken; **h)** *(render)* ~ **sth. into English** etw. ins Englische übertragen *od.* übersetzen; ~ **sth. into words** etw. in Worte fassen; **i)** *(write)* schreiben; ~ **one's name on the list** seinen Namen auf die Liste setzen; ~ **a tick in the box** ein Häkchen in das Kästchen machen; ~ **one's signature to sth.** seine Unterschrift unter etw. *(Akk.)* setzen; ~ **sth. on the bill** etw. auf die Rechnung setzen; ~ **sth. on the list** *(fig.)* sich *(Dat.)* etw. [fest] vornehmen; etw. vormerken; **j)** *(imagine)* ~ **oneself in sb.'s place** *or* **situation** sich in jmds. Lage versetzen; **k)** *(invest)* ~ **money** *etc.* **into sth.** Geld *usw.* in etw. *(Akk.)* stecken; ~ **work/time/effort into sth.** Arbeit/Zeit/Energie in etw. *(Akk.)* stecken; **l)** *(stake)* setzen (**on** auf + *Akk.*); ~ **money on a horse/on sth. happening** auf ein Pferd setzen/darauf wetten, daß etw. passiert; **m)** *(estimate)* ~ **sb./sth. at** jmdn./etw. schätzen auf (+ *Akk.*); **n)** *(subject)* ~ **sb. to** jmdm. ‹*Unkosten, Mühe, Umstände*› verursachen *od.* machen; **o)** *(Athletics: throw)* stoßen ‹*Kugel*›; ~ **the shot** kugelstoßen. **2.** *v.i.*, **-tt-, put** *(Naut.)* ~ **[out] to sea** in See stechen; ~ **into port** [in den Hafen] einlaufen

~ **a'bout 1.** *v.t. (circulate)* verbreiten; in Umlauf bringen; **it was ~ about that ...:** man munkelte *(ugs.) od.* es hieß, daß ... **2.** *v.i. (Naut.)* den Kurs ändern

~ **a'cross** *v.t.* **a)** *(communicate)* vermitteln (**to** *Dat.*); **b)** *(make acceptable)* ankommen mit; *(make effective)* durchsetzen; ~ **sth. across to sb.** mit etw. bei jmdm. ankommen/etw. bei jmdm. durchsetzen

~ **a'side** *v.t.* **a)** *(disregard)* absehen von; **~ting aside the fact that ...:** wenn man von der Tatsache *od.* davon absieht, daß ...; **b)** *(save)* beiseite legen

~ **a'way** *v.t.* **a)** wegräumen; reinstellen ‹*Auto*›; *(in file)* abheften; **b)** *(save)* beiseite legen; **c)** *(coll.) (eat)* verdrücken *(ugs.)*; *(drink)* runterkippen *(ugs.)*; **d)** *(coll.: confine)* einsperren *(ugs.)*

~ **'back** *v.t.* **a)** ~ **the book back** das Buch zurücktun; ~ **the book back on the shelf** das Buch wieder ins Regal stellen; **b)** ~ **the clock back [one hour]** die Uhr [eine Stunde] zurückstellen; *see also* **clock 1 a**; **c)** *(delay)* zurückwerfen; **d)** *(postpone)* verschieben

~ **'by** *v.t.* beiseite legen; **I've got a few hundred pounds ~ by** ich habe ein paar hundert Pfund auf der hohen Kante *(ugs.)*

~ **'down 1.** *v.t.* **a)** *(set down) (vertically)* hinstellen; *(horizontally)* hinlegen; auflegen ‹*Hörer*›; ~ **sth. down on sth.** etw. auf etw. *(Akk.)* stellen/legen; ~ **down a deposit** eine Anzahlung machen; **b)** *(suppress)* niederwerfen, -schlagen ‹*Revolte, Rebellion, Aufruhr*›; **c)** *(humiliate)* herabsetzen; *(snub)* eine Abfuhr erteilen (+ *Dat.*); **d)** *(kill painlessly)* töten; **e)** *(write)* notieren; aufschreiben; ~ **sth. down in writing** etw. schriftlich niederlegen; ~ **sb.'s name down on a list** jmdn. *od.* jmds. Namen auf eine Liste setzen; ~ **sb. down for** für jmdn. reservieren ‹*Lose*›; jmdn. notieren für ‹*Dienst, Arbeit*›; jmdn. anmelden bei ‹*Schule, Verein usw.*›; **f)** *(fig.: classify)* ~ **sb./sth. down as ...:** jmdn./etw. halten für *od.* einschätzen als ...; **g)** *(attribute)* ~ **sth. down to sth.** etw. auf etw. *(Akk.)* zurückführen; **h)** *(cease to read)* weglegen, aus der Hand legen ‹*Buch*›. *See also* [3]**down 1 f. 2.** *v.i. (Aeronaut.)* niedergehen

~ **'forward** *v.t.* **a)** *(propose)* aufwarten mit; **several theories have been ~ forward to account for this** darüber gibt es verschiedene Theorien; **b)** *(nominate)* vorschlagen; **c)** ~ **the clock forward [one hour]** die Uhr [eine Stunde] vorstellen

~ **'in 1.** *v.t.* **a)** *(install)* einbauen; **b)** *(elect)* an die Regierung *od.* Macht bringen; **c)** *(enter)* melden ‹*Person*›; **d)** *(submit)* stellen ‹*Forderung, Antrag*›; einreichen ‹*Bewerbung, Antrag*›; ~ **in a claim for damages** eine Schadensersatzforderung stellen; ~ **in a plea of not guilty** sich nicht schuldig bekennen; **e)** *(devote)* aufwenden ‹*Mühe, Kraft*›; *(perform)* einlegen ‹*Sonderschicht, Überstunden*›; *(coll.: spend)* einschieben ‹*eine Stunde usw.*›; **f)** *(interpose)* einwerfen ‹*Bemerkung*›. **2.** *v.i.* ~ **in for** sich bewerben um ‹*Stellung, Posten, Vorsitz*›; beantragen ‹*Urlaub, Versetzung*›

~ **'off** *v.t.* **a)** *(postpone)* verschieben (**until** auf + *Akk.*); *(postpone engagement with)* vertrösten (**until** auf + *Akk.*); **can't you ~ her off?** kannst du ihr nicht [erst einmal] absagen?; **b)** *(switch off)* ausmachen; **c)** *(repel)* abstoßen; **don't be ~ off by his rudeness** laß dich von seiner Grobheit nicht abschrecken; ~ **sb. off sth.** jmdm. etw. verleiden; **d)** *(distract)* stören; **e)** *(fob off)* abspeisen; **f)** *(dissuade)* ~ **sb. off doing sth.** jmdn. davon abbringen, etw. zu tun

~ **'on** *v.t.* **a)** anziehen ‹*Kleidung, Hose usw.*›; aufsetzen ‹*Hut, Brille*›; draufsetzen, *(ugs.)* draufmachen ‹*Deckel, Verschluß usw.*›; *(fig.)* aufsetzen ‹*Miene, Lächeln, Gesicht*›; ~ **it on** *(coll.)* [nur] Schau machen *(ugs.)*; **his modesty is all ~ on** seine Bescheidenheit ist nur gespielt *od. (ugs.)* ist reine Schau; **b)** *(switch or turn on)* anmachen ‹*Radio, Motor, Heizung, Licht usw.*›; *(cause to heat up)* aufsetzen ‹*Wasser, Essen, Kessel, Topf*›; *(fig.: apply)* ausüben ‹*Druck*›; **c)** *(gain)* ~ **on weight/two pounds** zunehmen/zwei Pfund zunehmen; **d)** *(add)* ~ **on speed** beschleunigen; ~ **8p on [to] the price** den Preis um 8 Pence erhöhen; **e)** *(stage)* spielen ‹*Stück*›; zeigen ‹*Show, Film*›; veranstalten ‹*Ausstellung*›; *see also* **act 1 e**; **f)** *(arrange)* einsetzen ‹*Sonderzug, -bus*›; **g)** *see* ~ **forward c**; **h)** *(coll.: tease)* veräppeln *(ugs.)*

~ **'out** *v.t.* **a)** rausbringen; ~ **one's hand out** die Hand ausstrecken; *see also* **tongue a**; **b)** *(extinguish)* ausmachen ‹*Licht, Lampe*›; löschen ‹*Feuer, Brand*›; **c)** *(issue)* [he]rausgeben ‹*Buch, Zeitschrift, Broschüre, Anweisung, Erlaß*›; abgeben ‹*Stellungnahme, Erklärung*›; *(broadcast)* senden; bringen; **d)** *(annoy)* verärgern; **be ~ out** verärgert *od.* entrüstet sein; **e)** *(inconvenience)* in Verlegenheit bringen; ~ **oneself out to do sth.** die Mühe auf sich *(Akk.)* nehmen, etw. zu tun; **f)** *(make inaccurate)* verfälschen ‹*Ergebnis, Berechnung*›; **g)** *(dislocate)* verrenken; ausrenken ‹*Schulter*›

~ **'over** *see* ~ **across**

~ **'through** *v.t.* **a)** *(carry out)* durchführen ‹*Plan, Programm, Kampagne, Sanierung*›; durchbringen ‹*Gesetz, Vorschlag*›; *(complete)* zum Abschluß bringen, abschließen ‹*Geschäft usw.*›; **b)** *(Teleph.)* verbinden (**to** mit); durchstellen ‹*Gespräch*› (**to** zu). *See also* ~ **1 c**

~ **to'gether** *v.t.* zusammensetzen ‹*Bauteile, Scherben, Steine, Einzelteile, Maschine usw.*›; ordnen ‹*Gedanken*›; erstellen, ausarbeiten ‹*Begründung, Argumentation*›

~ **'up 1.** *v.t.* **a)** heben ‹*Hand*›; *(erect)* errichten ‹*Gebäude, Denkmal, Gerüst, Zaun usw.*›; bauen ‹*Haus*›; aufstellen ‹*Denkmal, Gerüst, Leinwand, Zelt*›; aufbauen ‹*Zelt, Verteidigungsanlagen*›; anbringen ‹*Schild, Notiz usw.*› (**on** an + *Dat.*); *(fig.)* aufbauen ‹*Fassade*›; abziehen ‹*Schau*›; **b)** *(display)* anschlagen; aushängen; **c)** *(offer as defence)* hochnehmen

‹*Fäuste*›; leisten ‹*Widerstand, Gegenwehr*›; ~ **up a struggle** sich wehren *od.* zur Wehr setzen; **d)** *(present for consideration)* einreichen ‹*Petition, Gesuch, Vorschlag*›; *(nominate)* aufstellen; ~ **sb. up for election** jmdn. als Kandidaten aufstellen; **e)** *(incite)* ~ **sb. up to sth.** jmdn. zu etw. anstiften; **f)** *(accommodate)* unterbringen; **g)** *(increase)* [he]raufsetzen, anheben ‹*Preis, Miete, Steuer, Zins*›; **h)** ~ **sth. up for sale** etw. zum Verkauf anbieten. **2.** *v.i.* **a)** *(be candidate)* kandidieren; sich aufstellen lassen; **b)** *(lodge)* übernachten; sich einquartieren

~ **upon** *v.t.* ausnutzen

~ **'up with** *v.t.* sich *(Dat.)* gefallen *od.* bieten lassen ‹*Beleidigung, Benehmen, Unhöflichkeit*›; sich abfinden mit ‹*Lärm, Elend, Ärger, Bedingungen*›; sich abgeben mit ‹*Person*›

'put-down *n.* Herabsetzung, *die; (snub)* Abfuhr, *die*

putrefaction [pju:trɪ'fækʃn] *n., no pl., no indef. art.* Zersetzung, *die*

putrefy ['pju:trɪfaɪ] *v.i.* sich zersetzen

putrid ['pju:trɪd] *adj.* **a)** *(rotten)* faul; **become** ~: sich zersetzen; **b)** *(of putrefaction)* faulig; ~ **smell** Fäulnisgeruch, *der*

putt [pʌt] *(Golf)* **1.** *v.i. & t.* putten. **2.** *n.* Putt, *der*

putter ['pʌtə(r)] *(Golf)* Putter, *der*

putting-green ['pʌtɪŋgri:n] *n.* **a)** *(area of grass)* Grün, *das;* **b)** *(miniature golf-course) kleiner Golfplatz nur zum Putten*

putty ['pʌtɪ] **1.** *n.* Kitt, *der.* **2.** *v.t. (fix with ~)* einkitten ‹*Fensterscheibe*›; *(fill with ~)* auskitten ‹*Risse*›

'put-up *adj.* **a** ~ **thing/job** eine abgekartete Sache/ein abgekartetes Spiel *(ugs.)*

puzzle ['pʌzl] **1.** *n.* **a)** *(problem)* Rätsel, *das; (toy)* Geduldsspiel, *das;* **b)** *(enigma)* Rätsel, *das;* **be a** ~ **to sb.** jmdm. ein Rätsel sein; **be a** ~: rätselhaft sein. **2.** *v.t.* rätselhaft *od.* ein Rätsel sein (+ *Dat.*). **3.** *v.i.* ~ **over** *or* **about sth.** sich *(Dat.)* über etw. *(Akk.)* den Kopf zerbrechen

~ **'out** *v.t.* herausfinden; ~ **out an answer to a question** eine Antwort auf eine Frage finden

puzzled ['pʌzld] *adj.* ratlos

puzzlement ['pʌzlmənt] *n., no pl.* Verwirrung, *die*

puzzling ['pʌzlɪŋ] *adj.* rätselhaft

PVC *abbr.* **polyvinyl chloride** PVC, *das*

pygmy ['pɪgmɪ] *n.* **a)** Pygmäe, *der;* **b)** *(dwarf; also fig.)* Zwerg, *der*/Zwergin, *die*

pyjamas [pɪ'dʒɑ:məz] *n. pl.* **[pair of]** ~: Schlafanzug, *der;* Pyjama, *der*

pylon ['paɪlən] *n.* Mast, *der*

pyramid ['pɪrəmɪd] *n.* Pyramide, *die*

pyre ['paɪə(r)] *n.* Scheiterhaufen, *der*

Pyrenees [pɪrə'ni:z] *pr. n. pl.* **the** ~: die Pyrenäen

Pyrex, (P) ['paɪreks] *n.* ≈ Jenaer Glas, *das* ⓦ; *attrib.* ~ **dish** feuerfeste Glasschüssel

python ['paɪθən] *n.* Python[schlange], *die*

Q

Q, q [kju:] *n., pl.* **Qs** *or* **Q's** Q, q, *das; see also* **mind 2 b**

qr. *abbr.* **quarter[s]** qr.

¹quack [kwæk] **1.** *v.i.* ‹*Ente:*› quaken. **2.** *n.* Quaken, *das*

²quack *(derog.)* **1.** *n.* Quacksalber, *der (abwertend).* **2.** *attrib. adj.* **a)** ~ **doctor** Quacksalber, *der;* **b)** Quacksalber‹*kur, -tropfen, -pillen*› *(abwertend)*

quad [kwɒd] *n. (coll.)* **a)** *(quadrangle)* Innenhof, *der;* **b)** *(quadruplet)* Vierling, *der*

quadrangle ['kwɒdræŋgl] *n. (enclosed court)* [viereckiger] Innenhof; *(with buildings)* Block, *der;* Karree, *das*

quadraphonic [kwɒdrə'fɒnɪk] *adj.* quadrophon; Quadro‹*anlage, -sound usw.*›

quadratic [kwə'drætɪk] *adj. (Math.)* quadratisch

quadrilateral [kwɒdrɪ'lætərl] *n. (Geom.)* Viereck, *das*

quadruped ['kwɒdrʊped] *n.* Vierfüßler, *der*

quadruple ['kwɒdrʊpl] **1.** *adj.* **a)** vierfach; **b)** *(four times)* viermal. **2.** *v.t.* vervierfachen ‹*Einkommen, Produktion, Profit*›. **3.** *v.i.* sich vervierfachen

quadruplet ['kwɒdrʊplɪt, kwɒ'dru:plɪt] *n.* Vierling, *der*

quagmire ['kwægmaɪə(r), kwɒgmaɪə(r)] *n.* Sumpf, *der;* Morast, *der; (fig.: complex or difficult situation)* Sumpf, *der*

¹quail [kweɪl] *n., pl. same or* ~**s** *(Ornith.)* Wachtel, *die*

²quail *v.i.* ‹*Person:*› [ver]zagen, den Mut sinken lassen; ~ **at the prospect of sth.** bei der Aussicht auf etw. *(Akk.)* verzagen

quaint [kweɪnt] *adj.* drollig; putzig *(ugs.)* ‹*Häuschen, Einrichtung*›; malerisch, pittoresk ‹*Ort*›; *(odd, strange)* kurios, seltsam ‹*Bräuche, Anblick, Begebenheit*›

quake [kweɪk] **1.** *n. (coll.)* [Erd]beben, *das.* **2.** *v.i.* beben; ‹*Sumpfboden:*› schwingen; ~ **with fear/fright** vor Angst/Schreck zittern *od.* beben

Quaker ['kweɪkə(r)] *n.* Quäker, *der*/Quäkerin, *die*

qualification [kwɒlɪfɪ'keɪʃn] *n.* **a)** *(ability)* Qualifikation, *die; (condition to be fulfilled)* Voraussetzung, *die;* **secretarial** ~**s** Ausbildung als Sekretärin; **b)** *(limitation)* Vorbehalt, *der;* **without** ~: vorbehaltlos; ohne Vorbehalt

qualified ['kwɒlɪfaɪd] *adj.* **a)** qualifiziert; *(by training)* ausgebildet; **be** ~ **for a job/to vote** die Qualifikation für eine Stelle besitzen/wahlberechtigt sein; **you are better** ~ **to judge that** du kannst das besser beurteilen; **b)** *(restricted)* nicht uneingeschränkt; **a** ~ **success** kein voller Erfolg; ~ **approval/reply** Zustimmung/Antwort unter Vorbehalt; ~ **acceptance** bedingte Annahme

qualifier ['kwɒlɪfaɪə(r)] *n.* **a)** *(restriction)* Einschränkung, *die* **(of,** *on Gen.***)**; **b)** *(person)* **be among the** ~**s** zu denen gehören, die sich qualifiziert haben; **c)** *(Sport: match)* Qualifikationsspiel, *das*

qualify ['kwɒlɪfaɪ] **1.** *v.t.* **a)** *(make competent, make officially entitled)* berechtigen **(for** zu); **b)** *(modify)* einschränken; modifizieren ‹*Meinung, Feststellung*›. **2.** *v.i.* **a)** ~ **in law/medicine** seinen [Studien]abschluß in Jura/Medizin machen; ~ **as a doctor/lawyer** sein Examen als Arzt/Anwalt machen; **b)** *(fulfil a condition)* in Frage kommen **(for** für); ~ **for admission to a university/club** die Aufnahmebedingungen einer Universität/eines Vereins erfüllen; ~ **for membership** die Bedingungen für die Mitgliedschaft erfüllen; **c)** *(Sport)* sich qualifizieren

qualifying ['kwɒlɪfaɪɪŋ] *adj.* **a)** ~ **statement** einschränkende Aussage; **b)** *(Sport)* ~ **match** Qualifikationsspiel, *das;* ~ **round/heat** Ausscheidungs- *od.* Qualifikationsrunde, *die;* **c)** ~ **examination** Zulassungsprüfung, *die*

qualitative ['kwɒlɪtətɪv] *adj.* qualitativ

quality ['kwɒlɪtɪ] **1.** *n.* **a)** Qualität, *die;* **of good/poor** *etc.* ~: von guter/schlechter *usw.* Qualität; **of the best** ~: bester Qualität; **b)** *(characteristic)* Eigenschaft, *die;* **possess the qualities of a ruler/leader** eine Führernatur sein; **c)** *(of sound, voice)* Klang, *der.* **2.** *adj.* **a)** *(excellent)* Qualitäts-; **b)** *(maintaining ~)* Qualitäts‹*prüfung, -kontrolle*›; *(denoting ~)* Güte‹*grad, -klasse, -zeichen*›

qualm [kwɑ:m, kwɔ:m] *n.* **a)** *(sudden misgiving)* ungutes Gefühl; **b)** *(scruple)* Bedenken, *das (meist Pl.)* **(over, about** gegen); **he had no ~s about borrowing money** er hatte keine Bedenken, sich *(Dat.)* Geld zu leihen

quandary ['kwɒndərɪ] *n.* Dilemma, *das;* **this demand put him in a ~**: diese Forderung brachte ihn in eine verzwickte Lage; **he was in a ~ about what to do next** er wußte nicht, was er als nächstes tun sollte

quantify ['kwɒntɪfaɪ] *v.t.* quantifizieren

quantitative ['kwɒntɪtətɪv] *adj.* quantitativ

quantity ['kwɒntɪtɪ] *n.* **a)** Quantität, *die;* **b)** *(amount, sum)* Menge, *die;* **c)** *(large amount)* [Un]menge, *die;* **d)** *(Math.)* Größe, *die;* **an unknown ~** *(fig.)* eine unbekannte Größe

'**quantity surveyor** *n.* Baukostenkalkulator, *der*/-kalkulatorin, *die*

quantum ['kwɒntəm] *n., pl.* **quanta** ['kwɒntə] *(Phys.)* Quant, *das*

quantum: ~ **jump,** ~ **leap** *ns.* *(Phys.: also fig.)* Quantumsprung, *der;* ~ **theory** *n.* *(Phys.)* Quantentheorie, *die*

quarantine ['kwɒrənti:n] **1.** *n.* Quarantäne, *die;* **be in ~**: unter Quarantäne stehen. **2.** *v.t.* unter Quarantäne stellen

quarrel ['kwɒrl] **1.** *n.* **a)** Streit, *der;* **have a ~ with sb. [about/over sth.]** sich mit jmdm. [über etw. *(Akk.)* *od.* wegen etw./um etw.] streiten; **let's not have a ~ about it** wir wollen uns nicht darüber streiten; **pick a ~ [with sb. over sth.]** [mit jmdm. wegen etw.] Streit anfangen; **b)** *(cause of complaint)* Einwand, *der* **(with** gegen); **I have no ~ with you** ich habe nichts gegen dich. **2.** *v.i., (Brit.)* **-ll-: a)** [sich] streiten **(over** um; **about** über + *Akk.,* wegen); ~ **with each other** [sich] [miteinander] streiten; *(fall out, dispute)* sich [zer]streiten **(over** um; **about** über + *Akk.,* wegen); **b)** *(find fault)* etwas auszusetzen haben **(with** an + *Dat.*); **I really can't ~ with that** daran habe ich wirklich nichts auszusetzen

quarrelsome ['kwɒrlsəm] *adj.* streitsüchtig

[1]**quarry** ['kwɒrɪ] **1.** *n.* Steinbruch, *der;* **marble ~**: Marmorbruch, *der.* **2.** *v.t.* brechen

[2]**quarry** *n.* *(prey)* Beute, *die;* *(fig.)* Opfer, *das*

quart [kwɔ:t] *n.* Quart, *das*

quarter ['kwɔ:tə(r)] **1.** *n.* **a)** Viertel, *das;* **a** *or* **one ~ of** ein Viertel (+ *Gen.*); **divide/cut sth. into ~s** etw. in vier Teile teilen/schneiden; etw. vierteln; **six and a ~**: sechseinviertel; **an hour and a ~**: eineinviertel Stunden; **a ~ [of a pound] of cheese** ein Viertel[pfund] Käse; **a ~ of a mile/an hour** eine Viertelmeile/-stunde; **b)** *(of year)* Quartal, *das;* Vierteljahr, *das;* **c)** *(point of time)* **[a] ~ to/past six** Viertel vor/nach sechs; drei Viertel sechs/Viertel sieben *(landsch.)*; **there are buses at ~ to and ~ past [the hour]** es fahren Busse um Viertel vor und Viertel nach jeder vollen Stunde; **d)** *(direction)* Richtung, *die;* **blow from all ~s** ‹*Wind:*› aus allen Richtungen wehen; **e)** *(source of supply or help)* Seite, *die;* **f)** *(area of town)* [Stadt]viertel, *das;* Quartier, *das;* **in some ~s** *(fig.)* in gewissen Kreisen; **g)** *in pl. (lodgings)* Quartier, *das (bes. Milit.)*; Unterkunft, *die;* **h)** *(Brit.: measure) (of volume)* Quarter, *der; (of weight)* ≈ Viertelzentner, *der;* **i)** *(Amer.) (school term)* Vierteljahr, *das;* *(university term)* halbes Semester; **j)** *(Astron.)* Viertel, *das;* **k)** *(mercy)* **give no ~ to sb.** jmdm. keinen Pardon *(veralt.)* gewähren *od.* geben; **l)** *(Amer.: amount, coin)* Vierteldollar, *der;* 25-Cent-Stück, *das.* **2.** *v.t.* **a)** *(divide)* vierteln; durch vier teilen ‹*Zahl, Summe*›; **b)** *(lodge)* einquartieren ‹*Soldaten*›

quarter: ~**-deck** *n.* *(Naut.)* Quarterdeck, *das;* ~**-'final** *n.* Viertelfinale, *das*

quarterly ['kwɔ:təlɪ] **1.** *adj.* vierteljährlich. **2.** *n.* Vierteljahr[e]sschrift, *die.* **3.** *adv.* vierteljährlich; alle Vierteljahre

quarter: ~**master** *n.* **a)** *(Naut.)* Quartermeister, *der;* **b)** *(Mil.)* Quartiermeister, *der (veralt.)*; ~**-note** *(Amer. Mus.) see* **crotchet**

quartet, quartette [kwɔ:'tet] *n.* *(also Mus.)* Quartett, *das*

quarto ['kwɔ:təʊ] *n., pl.* ~**s a)** *(book)* Quartband, *der;* **b)** *(size)* Quart[format], *das;* ~ **paper** Papier im Quartformat

quartz [kwɔ:ts] *n.* Quarz, *der;* ~ **clock/watch** Quarzuhr, *die*

quasar ['kweɪsɑ:(r), 'kweɪzɑ:(r)] *n.* *(Astron.)* Quasar, *der*

quash [kwɒʃ] *v.t.* **a)** *(annul, make void)* aufheben ‹*Urteil, Entscheidung*›; zurückweisen ‹*Einspruch, Klage*›; **b)** *(suppress, crush)* unterdrücken ‹*Opposition*›; niederschlagen ‹*Aufstand, Generalstreik*›

quasi- ['kweɪzaɪ, 'kwɑ:zɪ] *pref.* **a)** *(not real, seeming)* Schein-; **b)** *(half-)* Quasi-; quasi

quaver ['kweɪvə(r)] **1.** *n.* **a)** *(Brit. Mus.)* Achtelnote, *die;* **b)** *(in speech)* Zittern, *das;* Beben, *das* *(geh.)*. **2.** *v.i.* *(vibrate, tremble)* zittern

quay [ki:], '**quayside** *ns.* Kai, *der;* Kaje, *die (nordd.)*

queasy ['kwi:zɪ] *adj.* unwohl; *(uneasy)* mulmig *(ugs.)*; **a ~ feeling** ein Gefühl der Übelkeit

queen [kwi:n] *n.* **a)** *(also bee, wasp, ant)* Königin, *die;* **b)** *(Chess, Cards)* Dame, *die*

queen: ~ '**bee** *n.* Bienenkönigin, *die;* ~ '**mother** *n.* Königinmutter, *die*

queer ['kwɪə(r)] **1.** *adj.* **a)** *(strange)* sonderbar; seltsam; *(eccentric)* komisch; verschroben; **a ~ feeling** ein komisches Gefühl; **b)** *(shady, suspect)* merkwürdig; seltsam; **c)** *(out of sorts, faint)* unwohl; **I feel ~**: mir ist komisch *od.* *(ugs.)* flau; **d)** *(sl. derog.: homosexual)* schwul *(ugs.)*. **2.** *n.* *(sl. derog.: homosexual)* Schwule, *der* *(ugs.)*. **3.** *v.t.* *(sl.: spoil)* vermasseln *(salopp)*; ~ **the pitch for sb.,** ~ **sb.'s pitch** jmdm. einen Strich durch die Rechnung machen

quell [kwel] *v.t.* *(literary)* niederschlagen ‹*Aufstand, Rebellion*›; zügeln ‹*Leidenschaft, Furcht*›; überwinden ‹*Ängste, Befürchtungen*›

quench [kwentʃ] *v.t.* **a)** *(extinguish)* löschen; *(fig.)* auslöschen *(geh.)*; **b)** *(satisfy)* ~ **one's thirst** seinen Durst löschen *od.* stillen

querulous ['kwerʊləs] *adj.* gereizt; *(by nature)* reizbar

query ['kwɪərɪ] **1.** *n.* **a)** *(question)* Frage, *die;* **put/raise a ~**: eine

Frage stellen/aufwerfen; **b)** *(question mark)* Fragezeichen, *das.* **2.** *v. t.* **a)** *(call in question)* in Frage stellen ‹*Anweisung, Glaubwürdigkeit, Ergebnis usw.*›; beanstanden ‹*Rechnung, Kontoauszug*›; **b)** *(ask, inquire)* ~ **whether/if ...:** fragen, ob ...

quest [kwest] *n.* Suche, *die* (**for** nach); *(for happiness, riches, etc.)* Streben, *das* (**for** nach); **in ~ of sth.** auf der Suche nach etw.

question ['kwestʃn] **1.** *n.* **a)** Frage, *die;* **ask sb. a ~:** jmdm. eine Frage stellen; **put a ~ to sb.** an jmdn. eine Frage richten; **don't ask so many ~s!** frag nicht soviel!; **ask ~s** Fragen stellen; **and no ~s asked** ohne daß groß gefragt wird/worden ist *(ugs.);* **b)** *(doubt, objection)* Zweifel, *der* (**about** an + *Dat.*); **there is no ~ about sth.** es besteht kein Zweifel an etw. *(Dat.);* **there is no ~ [but] that ...:** es besteht kein Zweifel, daß ...; **accept/follow sth. without ~:** etwas kritiklos akzeptieren/befolgen; **not be in ~:** außer [allem] Zweifel stehen; **beyond all** *or* **without ~:** zweifellos; ohne Frage *od.* Zweifel; **c)** *(problem, concern, subject)* Frage, *die;* **sth./it is only a ~ of time** etw./es ist [nur] eine Frage der Zeit; **it is [only] a ~ of doing sth.** es geht [nur] darum, etw. zu tun; **there is no ~ of his doing that** es kann keine Rede davon sein, daß er das tut; **the ~ of sth. arises** es erhebt sich die Frage von etw.; **the person/thing in ~:** die fragliche *od.* betreffende Person/Sache; **sth./it is out of the ~:** etw./es ist ausgeschlossen; etw./es kommt nicht in Frage *(ugs.);* **the ~ is whether ...:** es geht darum, ob ...; **that is not the ~:** darum geht es nicht; **put the ~:** zur Abstimmung aufrufen (**to** *Akk.*). **2.** *v. t.* **a)** befragen; ‹*Polizei, Gericht usw.:*› vernehmen; **b)** *(throw doubt upon, raise objections to)* bezweifeln; **her goodwill cannot be ~ed** an ihrem guten Willen kann nicht gezweifelt werden

questionable ['kwestʃənəbl] *adj.* fragwürdig

questioning ['kwestʃənɪŋ] **1.** *adj.* fragend. **2.** *n.* Fragen, *das; (at examination)* Befragung, *die; (by police etc.)* Vernehmung, *die*

question: ~ mark *n. (lit. or fig.)* Fragezeichen, *das;* **~-master** *n.* Quizmaster, *der*

questionnaire [kwestʃə'neə(r)] *n.* Fragebogen, *der*

queue [kju:] **1.** *n.* Schlange, *die;* **a ~ of people/cars** eine Menschen-/Autoschlange; **stand** *or* **wait in a ~:** Schlange stehen; anstehen; **join the ~:** sich anstellen. **2.** *v. i.* **~ [up]** Schlange stehen; anstehen; *(join ~)* sich anstellen; **~ for a bus** an der Bushaltestelle Schlange stehen

quibble ['kwɪbl] **1.** *n.* **a)** *(argument)* spitzfindiges Argument; **b)** *(petty objection)* Spitzfindigkeit, *die.* **2.** *v. i.* streiten; **~ over** *or* **about sth.** über etw. *(Akk.)* streiten

quiche [ki:ʃ] *n.* Quiche, *die*

quick [kwɪk] **1.** *adj.* **a)** schnell; kurz ‹*Rede, Zusammenfassung, Pause*›; flüchtig ‹*Kuß, Blick usw.*›; **it's ~er by train** mit dem Zug geht es schneller; **'that was/'you were ~!** das ging aber schnell!; **could I have a ~ word with you?** kann ich Sie kurz einmal sprechen?; **be ~!** mach schnell! *(ugs.);* beeil[e] dich!; **b)** *(prompt to act or understand)* schnell ‹*Person*›; wach ‹*Verstand*›; aufgeweckt ‹*Kind*›; **he is very ~:** er ist sehr schnell von Begriff *(ugs.);* **be ~ to do sth.** etw. schnell tun; **be ~ to take offence** schnell *od.* leicht beleidigt sein; **she is ~ to criticize** mit Kritik ist sie schnell bei der Hand; **[have] a ~ temper** ein aufbrausendes Wesen [haben]. **2.** *adv.* schnell; **~!** [mach] schnell! **3.** *n.* empfindliches Fleisch; **bite one's nails to the ~:** die Nägel bis zum Fleisch abkauen; **be cut to the ~** *(fig.)* tief getroffen *od.* verletzt sein

quicken ['kwɪkn] **1.** *v. t.* *(make quicker)* beschleunigen. **2.** *v. i.* *(become quicker)* sich beschleunigen; schneller werden

quickie ['kwɪkɪ] *n. (coll.) (drink)* Schluck auf die Schnelle *(ugs.); (sexual intercourse)* eine Nummer auf die Schnelle *(salopp)*

'quicklime *n.* ungelöschter Kalk

quickly ['kwɪklɪ] *adv.* schnell

quickness ['kwɪknɪs] *n., no pl.* **a)** *(speed)* Schnelligkeit, *die;* **b)** *(acuteness of perception)* Schärfe, *die;* **~ of the mind** schnelle Auffassungsgabe

quick: ~sand *n.* Treibsand, *der;* **~silver** *n.* Quecksilber, *das;* **~ step** *n. (Dancing)* Quickstep, *der;* **~-tempered** *adj.* hitzig; **be ~-tempered** leicht aufbrausen; **~-witted** *adj.* geistesgegenwärtig; schlagfertig ‹*Antwort*›

quid [kwɪd] *n. (Brit. sl.) pl. same (one pound)* Pfund, *das;* **fifty ~:** fünfzig Kugeln *(salopp)*

quiet ['kwaɪət] **1.** *adj.*, **~er** ['kwaɪətə(r)], **~est** ['kwaɪətɪst] **a)** *(silent)* still; *(not loud)* leise ‹*Schritte, Musik, Stimme, Motor, Fahrzeug*›; **be ~!** *(coll.)* sei still *od.* ruhig!; **~!** Ruhe!; **keep ~:** still sein; **keep sth. ~, keep ~ about sth.** *(fig.)* etw. geheimhalten; **b)** *(peaceful, not busy)* ruhig; **c)** *(gentle)* sanft; *(peaceful)* ruhig ‹*Kind, Person*›; **d)** *(not overt, disguised)* versteckt; heimlich ‹*Groll*›; **have a ~ word with sb.** mit jmdm. unter vier Augen reden; **on the ~:** still und heimlich; **e)** *(not formal)* zwanglos; klein ‹*Feier*›; **f)** *(not showy)* dezent ‹*Farben, Muster*›; schlicht ‹*Eleganz, Stil*›. **2.** *n.* Ruhe, *die; (silence, stillness)* Stille, *die.* **3.** *v. t. see* **quieten**

quieten ['kwaɪətn] *(Brit.) v. t.* **a)** beruhigen; zur Ruhe bringen ‹*Kind, Schulklasse*›; **b)** zerstreuen ‹*Bedenken, Angst, Verdacht*›

~ 'down 1. *v. t. see* **~ a. 2.** *v. i.* sich beruhigen

quietly ['kwaɪətlɪ] *adv.* **a)** *(silently)* still; *(not loudly)* leise; **b)** *(peacefully, tranquilly)* ruhig; **be ~ drinking one's tea** in [aller] Ruhe seinen Tee trinken; **c)** *(gently)* sanft; **be ~ spoken** eine ruhige Art zu sprechen haben; **d)** *(not overtly)* insgeheim; **they settled the affair ~:** sie haben die Angelegenheit unter sich *(Dat.)* ausgemacht; **e)** *(not formally)* zwanglos; **get married ~:** im kleinen Rahmen heiraten; **f)** *(not showily)* dezent; schlicht

quietness ['kwaɪətnɪs] *n., no pl.* **a)** *(absence of noise)* Stille, *die; (of reply)* Ruhe, *die; (of car, engine)* Geräuscharmut, *die; (of footsteps)* Geräusch-, Lautlosigkeit, *die;* **b)** *(peacefulness)* Ruhe, *die*

quill [kwɪl] *n.* **a)** *see* **quill-feather; b)** *see* **quill-pen; c)** *(stem of feather)* [Feder]kiel, *der;* **d)** *(of porcupine)* Stachel, *der*

quill: ~-feather *n.* Kielfeder, *die;* **~-pen** *n.* [Feder]kiel, *der*

quilt [kwɪlt] **1.** *n.* Schlafdecke, *die;* **continental ~:** Steppdecke, *die.* **2.** *v. t.* **a)** *(cover with padded material)* wattieren; **b)** *(join like ~)* steppen

quin [kwɪn] *n. (coll.)* Fünfling, *der*

quince [kwɪns] *n.* **a)** *(fruit)* Quitte, *die;* **b)** *(tree)* Quittenbaum, *der*

quinine ['kwɪni:n, kwɪ'ni:n] *n.* Chinin, *das*

quintessence [kwɪn'tesəns] *n.* *(most perfect form)* Quintessenz, *die; (embodiment)* Inbegriff, *der*

quintet, quintette [kwɪn'tet] *n.* *(also Mus.)* Quintett, *das*

quintuplet ['kwɪntjʊplɪt, kwɪn'tju:plɪt] *n.* Fünfling, *der*
quip [kwɪp] **1.** *n.* Witzelei, *die.* **2.** *v. i.*, **-pp-** witzeln (**at** über + *Akk.*)
quirk [kwɜ:k] *n.* Marotte, *die;* **[by a] ~ of nature/fate** [durch eine] Laune der Natur/des Schicksals
quirky ['kwɜ:kɪ] *adj.* schrullig *(ugs.)*
quit [kwɪt] **1.** *pred. adj.* **be ~ of sb./sth.** jmds./einer Sache ledig sein *(geh.).* **2.** *v. t.*, **-tt-**, *(Amer.)* **quit a)** *(give up)* aufgeben; *(cease, stop)* aufhören mit; **~ doing sth.** aufhören, etw. zu tun; **b)** *(depart from)* verlassen; *(leave occupied premises)* ausziehen aus; *abs.* ausziehen; **they were given** *or* **had notice to ~ [the flat** *etc.*] ihnen wurde [die Wohnung *usw.*] gekündigt; **c)** *also abs. (from job)* kündigen
quite [kwaɪt] *adv.* **a)** *(entirely)* ganz; völlig; vollkommen; gänzlich ‹*unnötig*›; fest ‹*entschlossen*›; **not ~** *(almost)* nicht ganz; *(noticeably not)* nicht gerade; **I'm sorry – That's ~ all right** Entschuldigung – Schon gut *od.* in Ordnung; **not ~ five o'clock** noch nicht ganz 5 Uhr; **I don't need any help; I'm ~ all right, thank you** danke, es geht schon, ich komme allein zurecht; **I ~ agree/understand** ganz meine Meinung/ich verstehe schon; **~ [so]!** [ja,] genau *od.* richtig!; **that is ~ a different matter** das ist etwas ganz anderes; **~ another story/case** eine ganz andere Geschichte/ein ganz anderer Fall; **b)** *(somewhat, to some extent)* ziemlich; recht; ganz ‹*gern*›; **it was ~ an effort** es war ziemlich *od.* recht anstrengend; **that is ~ a shock/surprise** das ist ein ziemlicher Schock/eine ziemliche Überraschung; **I'd ~ like to talk to him** ich würde ganz gern mit ihm sprechen; **~ a few** ziemlich viele
quits [kwɪts] *pred. adj.* **be ~ [with sb.]** [mit jmdm.] quitt sein *(ugs.);* **call it ~** ‹*Einzelperson:*› zustimmen *od.* Ruhe geben; ‹*mehrere Personen:*› sich vertragen; **let's call it ~!** wollen wir die Sache auf sich beruhen lassen!; *(nothing owed)* sagen wir, wir sind quitt; *see also* **double 3 c**
¹quiver ['kwɪvə(r)] **1.** *v. i.* zittern (**with** vor + *Dat.*); ‹*Stimme, Lippen:*› beben *(geh.);* ‹*Lid:*› zucken. **2.** *n.* Zittern, *das; (of lips, voice also)* Beben, *das (geh.); (of eyelid)* Zucken, *das*
²quiver *n. (for arrows)* Köcher, *der*
quiz [kwɪz] **1.** *n., pl.* **~zes a)** *(Radio, Telev., etc.)* Quiz, *das;* **b)** *(questionnaire, test)* Prüfung, *die; (for pupils)* Aufgabe, *die.* **2.** *v. t.*, **-zz-** ausfragen (**about sth.** nach etw., **about sb.** über jmdn.); ‹*Polizei:*› verhören, vernehmen ‹*Verdächtige*›
quiz: ~-master *n. (Radio, Telev.)* Quizmaster, *der;* Spielleiter, *der;* **~ programme, ~ show** *ns. (Radio, Telev.)* Quizsendung, *die*
quizzical ['kwɪzɪkl] *adj.* fragend ‹*Blick, Miene*›; *(mocking)* spöttisch ‹*Lächeln*›
quoit [kɔɪt] *n. (Games)* [Gummi]ring, *der*
quoits [kɔɪts] *n., no pl. (Games)* Ringtennis, *das*
quorate ['kwɔ:rət] *adj.* beschlußfähig
quorum ['kwɔ:rəm] *n.* Quorum, *das*
quota ['kwəʊtə] *n.* **a)** *(share)* Anteil, *der;* **b)** *(quantity of goods to be produced)* Produktionsmindestquote, *die; (of work)* [Arbeits]pensum, *das;* **c)** *(maximum number)* Höchstquote, *die; (of immigrants/students permitted)* maximale Einwanderungs-/Zulassungsquote
quotation [kwəʊ'teɪʃn] *n.* **a)** Zitieren, *das; (passage)* Zitat, *das;* **b)** *(amount stated as current price)* [Börsen]kurs, *der;* [Börsen-, Kurs]notierung, *die;* **c)** *(estimate)* Kosten[vor]anschlag, *der*
quo'tation-marks *n. pl.* Anführungszeichen *Pl.*
quote [kwəʊt] **1.** *v. t.* **a)** *also abs.* zitieren (**from** aus); zitieren aus ‹*Buch, Text, Übersetzung*›; *(appeal to)* sich berufen auf (+ *Akk.*) ‹*Person, Buch, Text, Quelle*›; *(mention)* anführen ‹*Vorkommnis, Beispiel*›; **he is ~d as saying that ...:** er soll gesagt haben, daß ...; **..., and I ~, ...:** ... ich zitiere, ...; **b)** *(state price of)* angeben, nennen ‹*Preis*›; **~ sb. a price** jmdm. einen Preis nennen; **c)** *(St. Exch.)* notieren ‹*Aktie*›; **d)** *(enclose in quotation-marks)* in Anführungszeichen *(Akk.)* setzen; **..., ~, ...:** ..., Zitat, ... **2.** *n. (coll.)* **a)** *(passage)* Zitat, *das;* **b)** *(commercial quotation)* Kosten[vor]anschlag, *der;* **c)** *usu. in pl. (quotation-mark)* Anführungszeichen, *das;* Gänsefüßchen, *das (ugs.)*
quotient ['kwəʊʃnt] *n. (Math.)* Quotient, *der*

R

R, r [ɑ:(r)] *n., pl.* **Rs** *or* **R's** R, r, *das;* **the three Rs** Lesen, Schreiben und Rechnen
R. *abbr.* **a) River** Fl.; **R. Thames** die Themse; **b) Regina/Rex** Königin, *die*/König, *der*
r. *abbr.* **right** re.
rabbi ['ræbaɪ] *n.* Rabbi[ner], *der; (as title)* Rabbi, *der*
rabbit ['ræbɪt] *n.* Kaninchen, *das*
rabbit: ~-burrow, ~-hole *ns.* Kaninchenbau, *der;* **~-hutch** *n. (lit. or fig. joc.)* Kaninchenstall, *der;* **~-warren** *n.* Kaninchengehege, *das; (fig.)* Labyrinth, *das*
rabble ['ræbl] *n.* Mob, *der (abwertend);* Pöbel, *der (abwertend)*
rabid ['ræbɪd] *adj.* **a)** *([Vet.] Med.)* tollwütig ‹*Tier, Person*›; **b)** *(furious, violent)* wild ‹*Haß, Wut*›; *(extreme)* fanatisch
rabies ['reɪbi:z] *n. ([Vet.] Med.)* Tollwut, *die*
RAC *abbr. (Brit.)* **Royal Automobile Club** Königlicher Britischer Automobilklub
raccoon *see* **racoon**
¹race [reɪs] **1.** *n.* **a)** Rennen, *das;* **have a ~ [with** *or* **against sb.]** mit jmdm. um die Wette laufen/schwimmen *usw.;* **100 metres ~:** 100-m-Rennen/-Schwimmen, *das;* **b)** *in pl. (series) (for horses)* Pferderennen, *das; (for dogs)* Hunderennen, *das;* **c)** *(fig.)* **a ~ against time** ein Wettlauf mit der Zeit. **2.** *v. i.* **a)** *(in swimming, running, sailing, etc.)* um die Wette schwimmen/laufen/segeln *usw.* (**with, against** mit); **~ against time** *(fig.)* gegen die Uhr *od.* Zeit arbeiten; **b)** *(go at full or excessive speed)* ‹*Motor:*› durchdrehen; ‹*Puls:*› jagen, rasen; **c)** *(rush)* sich sehr beeilen; hetzen; *(on foot also)* rennen; jagen; **~ after sb.** jmdm. hinterherhetzen; **~ to finish sth.** sich beeilen, um etw. fertigzukriegen *(ugs.);* **~ ahead with sth.** *(hurry)* etw. im Eiltempo vorantreiben *(ugs.); (make rapid pro-*

gress) bei etw. mit Riesenschritten vorankommen *(ugs.)*. **3.** *v. t.* *(in swimming, riding, walking, running, etc.)* um die Wette schwimmen/reiten/gehen/laufen *usw.* mit; **I'll ~ you** ich mache mit dir einen Wettlauf

²race *n. (Anthrop., Biol.)* Rasse, *die;* **the human ~:** die Menschheit

race: ~course *n.* Rennbahn, *die;* **~ hatred** *n.* Rassenhaß, *der;* **~horse** *n.* Rennpferd, *das;* **~ relations** *n. pl.* Beziehung zwischen den Rassen; **~-track** *n.* Rennbahn, *die*

racial ['reɪʃl] *adj.* Rassen⟨*diskriminierung, -konflikt, -gleichheit, -vorurteil*⟩; rassisch ⟨*Gruppe, Minderheit*⟩; **~ harmony** Eintracht unter den Rassen

racialism ['reɪʃəlɪzm] *n., no pl.* Rassismus, *der*

racialist ['reɪʃəlɪst] **1.** *n.* Rassist, *der*/Rassistin, *die.* **2.** *adj.* rassistisch

racing ['reɪsɪŋ] *n., no pl., no indef. art.* **a)** *(profession, sport)* Rennsport, *der; (with horses)* Pferdesport, *der;* **b)** *(races)* Rennen *Pl.*

racing: ~-car *n.* Rennwagen, *der;* **~ driver** *n.* Rennfahrer, *der*/-fahrerin, *die*

racism ['reɪsɪzm] *n.* Rassismus, *der*

racist ['reɪsɪst] **1.** *n.* Rassist, *der*/Rassistin, *die.* **2.** *adj.* rassistisch

rack [ræk] **1.** *n.* **a)** *(for luggage in bus, train, etc.)* Ablage, *die; (for pipes, hats, toast, plates)* Ständer, *der; (on bicycle, motor cycle)* Gepäckträger, *der; (on car)* Dachgepäckträger, *der;* **b)** *(instrument of torture)* Folter[bank], *die;* **be on the ~** *(lit. or fig.)* Folterqualen leiden. **2.** *v. t.* **a)** *(lit. or fig.: torture)* quälen; plagen; **be ~ed by** *or* **with pain** *etc.* von Schmerzen *usw.* gequält und geplagt werden; **b) ~ one's brain[s]** *(fig.)* sich *(Dat.)* den Kopf zerbrechen *(ugs.)* (**for** über + *Akk.*)

¹racket ['rækɪt] *n. (Sport)* Schläger, *der; (Tennis also)* Racket, *das*

²racket *n.* **a)** *(disturbance, uproar)* Lärm, *der;* Krach, *der;* **make a ~:** Krach *od.* Lärm machen; **b)** *(dishonest scheme)* Schwindelgeschäft, *das (ugs.)*

racketeer [rækɪ'tɪə(r)] *n.* Ganove, *der; (profiteer)* Wucherer, *der*

racking ['rækɪŋ] *attrib. adj.* quälend

raconteur [rækɒn'tɜ:(r)] *n.* Geschichten-, Anekdotenerzähler, *der*/-erzählerin, *die*

racoon [rə'ku:n] *n. (Zool.)* Waschbär, *der*

racquet *see* **¹racket**

racy ['reɪsɪ] *adj.* flott *(ugs.)*, schwungvoll ⟨*Erzählweise, Stil, Sprache*⟩; schwungvoll ⟨*Rede*⟩; saftig *(ugs.)* ⟨*Humor*⟩

radar ['reɪdɑ:(r)] *n.* Radar, *das od. der*

radar: ~ operator *n.* Radartechniker, *der*/-technikerin, *die;* **~ screen** *n.* Radarschirm, *der;* **~ trap** *n.* Radarfalle, *die (ugs.)*

radial ['reɪdɪəl] **1.** *adj.* **a)** *(arranged like rays)* strahlenförmig angeordnet; strahlenförmig ⟨*Muster*⟩; **b) ~ wheel** Radialrad, *das.* **2.** *n.* Radial-, Gürtelreifen, *der*

'radial[-ply] tyre *n.* Radial-, Gürtelreifen, *der*

radiance ['reɪdɪəns], **radiancy** ['reɪdɪənsɪ] *n.* Leuchten, *das; (of sun, stars, lamp; also fig.)* Strahlen, *das*

radiant ['reɪdɪənt] *adj.* **a)** strahlend, leuchtend ⟨*Himmelskörper, Dämmerung*⟩; leuchtend ⟨*Lichtstrahl*⟩; **b)** *(fig.)* strahlend; fröhlich ⟨*Stimmung*⟩; **be ~** ⟨*Person, Augen*:⟩ strahlen (**with** vor + *Dat.*)

radiate ['reɪdɪeɪt] **1.** *v. i.* **a)** ⟨*Sonne, Sterne*:⟩ scheinen, strahlen; ⟨*Hitze, Wärme*:⟩ ausstrahlen; ⟨*Schein, Radiowellen*:⟩ ausgesendet werden, ausgehen (**from** von); **b)** *(from central point)* strahlenförmig ausgehen (**from** von). **2.** *v. t.* **a)** verbreiten, ausstrahlen ⟨*Licht, Wärme, Klang*⟩; aussenden ⟨*Strahlen, Wellen*⟩; **b)** ausstrahlen ⟨*Glück, Liebe, Gesundheit, Fröhlichkeit*⟩

radiation [reɪdɪ'eɪʃn] *n.* **a)** *(emission of energy)* Emission, *die; (of signals)* Ausstrahlung, *die;* **b)** *(energy transmitted)* Strahlung, *die;* **contaminated by ~:** strahlenverseucht; **c)** *attrib.* Strahlen⟨*therapie, -krankheit, -dosis usw.*⟩; Strahlungs⟨*intensität, -meßgerät, -niveau usw.*⟩

radiator ['reɪdɪeɪtə(r)] *n.* **a)** *(for heating a room)* [Rippen]heizkörper, *der;* Radiator, *der; (portable)* Heizgerät, *das;* **b)** *(for cooling engine)* Kühler, *der*

radical ['rædɪkl] **1.** *adj.* **a)** *(thorough, drastic; also Polit.)* radikal; drastisch, radikal ⟨*Maßnahme*⟩; umwälzend ⟨*Auswirkungen*⟩; durchgreifend ⟨*Umstrukturierung, Veränderung usw.*⟩; **a ~ cure** eine Radikalkur; **b)** *(progressive, unorthodox)* radikal; revolutionär ⟨*Stil, Design, Sprachgebrauch*⟩; **c)** *(inherent, fundamental)* grundlegend ⟨*Fehler, Unterschied*⟩. **2.** *n. (Polit.)* Radikale, *der/die*

radically ['rædɪkəlɪ] *adv.* **a)** *(thoroughly, drastically; Polit.)* radikal; **b)** *(originally, basically)* prinzipiell; **c)** *(inherently, fundamentally)* von Grund auf

radio ['reɪdɪəʊ] **1.** *n., pl.* **~s a)** *no pl., no indef. art.* Funk, *der; (for private communication)* Sprechfunk, *der;* **over the/by ~:** über/per Funk; **b)** *no pl., no indef. art. (Broadcasting)* Rundfunk, *der;* Hörfunk, *der;* **listen to the ~:** Radio hören; **on the ~:** im Radio *od.* Rundfunk; **c)** *(apparatus)* Radio, *das.* **2.** *attrib. adj.* Rundfunk-; Radio⟨*welle, -teleskop*⟩; Funk⟨*mast, -turm, -frequenz, -taxi, -telefon*⟩; **~ drama** *or* **play** Hörspiel, *das.* **3.** *v. t.* funken ⟨*Meldung, Nachricht*⟩. **4.** *v. i.* funken; eine Funkmeldung übermitteln

radio: ~'active *adj.* radioaktiv; **~ac'tivity** *n.* Radioaktivität, *die;* **~ beacon** *n.* Funkfeuer, *das;* **~-carbon dating** *n.* Radiokarbondatierung, *die;* **~-controlled** *adj.* funkgesteuert; ferngesteuert

radiography [reɪdɪ'ɒgrəfɪ] *n.* Radiographie, *die;* Röntgenographie, *die*

radiology [reɪdɪ'ɒlədʒɪ] *n., no pl.* Radiologie, *die;* Röntgenologie, *die*

radish ['rædɪʃ] *n.* Rettich, *der; (small, red)* Radieschen, *das*

radium ['reɪdɪəm] *n. (Chem.)* Radium, *das*

radius ['reɪdɪəs] *n., pl.* **radii** ['reɪdɪaɪ] *or* **~es** *(Math.)* Radius, *der; (fig.)* Umkreis, *der;* **within a ~ of 20 miles** im Umkreis von 20 Meilen

radon ['reɪdɒn] *n. (Chem.)* Radon, *das*

RAF [ɑ:reɪ'ef, *(coll.)* ræf] *abbr.* **Royal Air Force**

raffia ['ræfɪə] *n.* Raphia-, Raffiabast, *der*

raffle ['ræfl] **1.** *n.* Tombola, *die;* **~ ticket** Los, *das.* **2.** *v. t.* **~ [off]** verlosen

raft [rɑ:ft] *n.* Floß, *das*

rafter ['rɑ:ftə(r)] *n. (Building)* Sparren, *der*

¹rag [ræg] *n.* **a)** [Stoff]fetzen, *der;* [Stoff]lappen, *der;* **[all] in ~s** [ganz] zerrissen; **sb. loses his ~** *(sl.)* jmdm. reißt die Geduld; **b)** *in pl. (old and torn clothes)* Lumpen *Pl.;* **[dressed] in ~s [and tatters]** abgerissen; **go from ~s to riches** vom armen Schlucker zum

Millionär/zur Millionärin werden; **c)** *(derog.: newspaper)* Käseblatt, *das (salopp abwertend)*

²rag 1. *v. t.*, **-gg-** *(tease, play jokes on)* aufziehen; necken. **2.** *n.* **a)** *(Brit. Univ.) spaßige studentische [Wohltätigkeits]veranstaltung;* **b)** *(prank)* Ulk, *der;* Streich, *der*

ragamuffin ['rægəmʌfɪn] *n.* [zerlumptes] Gassenkind

rag: ~-and-'bone man *n. (Brit.)* Lumpensammler, *der;* **~-bag** *n. (fig.: collection)* Sammelsurium, *das (abwertend);* **~ doll** *n.* Stoffpuppe, *die*

rage [reɪdʒ] **1.** *n.* **a)** *(violent anger)* Wut, *die; (fit of anger)* Wutausbruch, *der;* **be in/fly into a ~:** in Wut *od. (ugs.)* Rage sein/geraten; **in a fit of ~:** in einem Anfall von Wut; **b)** *(vehement desire or passion)* Besessenheit, *die;* **sth. is [all] the ~:** etw. ist [ganz] groß in Mode. **2.** *v. i.* **a)** *(rave)* toben; **~ at** *or* **against sth./sb.** gegen etw./jmdn. wüten *od. (ugs.)* wettern; **b)** *(be violent, operate unchecked)* toben; ‹*Krankheit:*› wüten

ragged ['rægɪd] *adj.* **a)** zerrissen; kaputt *(ugs.);* ausgefranst ‹*Saum, Manschetten*›; **b)** *(rough, shaggy)* zottig ‹*Bart*›; **c)** *(jagged)* zerklüftet ‹*Felsen, Küste, Klippe*›; *(in tattered clothes)* abgerissen; zerlumpt

'rag trade *n. (coll.)* Modebranche, *die (ugs.)*

raid [reɪd] **1.** *n.* **a)** Einfall, *der;* Überfall, *der; (Mil.)* Überraschungsangriff *der;* **b)** *(by police)* Razzia, *die* **(on** in + *Dat.*). **2.** *v. t.* ‹*Polizei:*› eine Razzia machen auf (+ *Akk.*); ‹*Bande/Räuber/Soldaten:*› überfallen ‹*Bank/Viehherde/Land*›; ‹*Trupp, Kommando:*› stürmen ‹*feindliche Stellung*›; **~ the larder** *(joc.)* die Speisekammer plündern *(scherzh.)*

raider ['reɪdə(r)] *n. (on bank, farm)* Räuber, *der*/Räuberin, *die; (looter)* Plünderer, *der*/Plünderin, *die; (burglar)* Einbrecher, *der*/Einbrecherin, *die*

¹rail [reɪl] *n.* **a)** ‹*Kleider-, Gardinen*›stange, *die; (as part of fence) (wooden)* Latte, *die; (metal)* Stange, *die; (on ship)* Reling, *die; (as protection against contact)* Barriere, *die;* **b)** *(Railw.: of track)* Schiene, *die;* **go off the ~s** *(lit.)* entgleisen; *(fig.: depart from what is accepted)* auf die schiefe Bahn geraten; **c)** *(~way)* [Eisen]bahn, *die; attrib.* Bahn-; **by ~:** mit der Bahn; mit dem Zug

²rail *v. i.* **~ at/against sb./sth.** auf/über jmdn./etw. schimpfen

railing ['reɪlɪŋ] *n. (round garden, park)* Zaun, *der; (on sides of staircase)* Geländer, *das*

rail: ~road 1. *n. (Amer.) see* **railway; 2.** *v. t. (send or push through in haste)* **~road sb. into doing sth.** jmdn. dazu antreiben, etw. zu tun; **~road a bill through parliament** einen Gesetzentwurf im Parlament durchpeitschen *(ugs.);* **~ strike** *n.* Eisenbahnerstreik, *der*

railway ['reɪlweɪ] *n.* **a)** *(track)* Bahnlinie, *die;* Bahnstrecke, *die;* **b)** *(system)* [Eisen]bahn, *die;* **work on the ~:** bei der Bahn arbeiten

railway: ~ carriage *n.* Eisenbahnwagen, *der;* **~ crossing** *n.* Bahnübergang, *der;* **~ engine** *n.* Lokomotive, *die;* **~ line** *n.* [Eisen]bahnlinie, *die;* [Eisen]bahnstrecke, *die;* **~man** ['reɪlweɪmən] *n., pl.* **~men** ['reɪlweɪmən] Eisenbahner, *der;* **~-station** *n.* Bahnhof, *der; (smaller)* [Eisen]bahnstation, *die;* **~ worker** *n.* Bahnarbeiter, *der*

rain [reɪn] **1.** *n.* **a)** Regen, *der;* **it looks like ~:** es sieht nach Regen aus; **come ~ or shine** *(fig.)* unter allen Umständen; **b)** *(fig.: of arrows, blows, etc.)* Hagel, *der;* **c)** *in pl. (falls of ~)* **the ~s** die Regenzeit. **2.** *v. i. impers.* **it is ~ing** es regnet; **it is starting to ~:** es fängt an zu regnen. **3.** *v. t.* prasseln *od.* hageln lassen ‹*Schläge, Hiebe*›

~ 'down *v. i.* ‹*Schläge, Steine, Flüche usw.:*› niederprasseln; ‹*Schüsse, Kugeln usw.:*› niederhageln

~ 'off, *(Amer.)* **~ 'out** *v. t.* **be ~ed off** *or* **out** *(be terminated)* wegen Regen abgebrochen werden; *(be cancelled)* wegen Regen ausfallen

rainbow ['reɪnbəʊ] **1.** *n.* Regenbogen, *der;* **all the colours of the ~:** alle Regenbogenfarben. **2.** *adj.* Regenbogen‹*farben, -streifen*›; regenbogenfarbig, -farben ‹*Kleid, Blumen*›

rain: ~-check *n. (Amer. fig.)* **take a ~-check on sth.** auf etw. *(Akk.)* später wieder zurückkommen; **~coat** *n.* Regenmantel, *der;* **~drop** *n.* Regentropfen, *der;* **~fall** *n. (shower)* [Regen]schauer, *der; (quantity)* Niederschlag, *der;* **~ forest** *n.* Regenwald, *der;* **~proof 1.** *adj.* regendicht; wasserdicht; **2.** *v. t.* appretieren; **~-water** *n.* Regenwasser, *das*

rainy ['reɪnɪ] *adj.* regnerisch ‹*Tag, Wetter*›; regenreich ‹*Klima, Gebiet, Sommer, Winter*›; **~ season** Regenzeit, *die;* **keep sth. for a ~ day** *(fig.)* sich *(Dat.)* etw. für schlechte Zeiten aufheben

raise [reɪz] *v. t.* **a)** *(lift up)* heben; erhöhen ‹*Pulsfrequenz, Temperatur, Miete, Gehalt, Kosten*›; hochziehen ‹*Rolladen, Fahne, Schultern*›; aufziehen ‹*Vorhang*›; hochheben ‹*Koffer, Arm, Hand*›; **~ one's eyes to heaven** die Augen zum Himmel erheben *(geh.);* **~ one's glass to sb.** das Glas auf jmdn. erheben; **~ one's voice** die Stimme heben; **they ~d their voices** *(in anger)* sie *od.* ihre Stimmen wurden lauter; **war ~d its [ugly] head** der Krieg erhob sein [häßliches] Haupt; **b)** *(set upright, cause to stand up)* aufrichten; erheben ‹*Banner*›; aufstellen ‹*Fahnenstange, Zaun, Gerüst*›; **be ~d from the dead** von den Toten [auf]erweckt werden; **~ sb.'s spirits** jmds. Stimmung heben; **c)** *(build up, construct)* errichten ‹*Gebäude, Statue*›; erheben ‹*Forderungen, Einwände*›; entstehen lassen ‹*Vorurteile*›; *(introduce)* aufwerfen ‹*Frage*›; zur Sprache bringen, anschneiden ‹*Thema, Problem*›; *(utter)* erschallen lassen ‹*Ruf, Schrei*›; **d)** *(grow, breed, rear)* anbauen ‹*Gemüse, Getreide*›; aufziehen ‹*Vieh, [Haus]tiere*›; großziehen ‹*Familie, Kinder*›; **e)** *(bring together, procure)* aufbringen ‹*Geld, Betrag, Summe*›; aufstellen ‹*Armee, Flotte, Truppen*›; aufnehmen ‹*Hypothek, Kredit*›; **f)** *(end, cause to end)* aufheben, beenden ‹*Belagerung, Blockade*›; *(remove)* aufheben ‹*Embargo, Verbot*›; **g) ~ [merry] hell** *(coll.)* Krach schlagen *(ugs.)* **(over** wegen); **h)** *(Math.)* **~ to the fourth power** in die 4. Potenz erheben

raisin ['reɪzn] *n.* Rosine, *die*

¹rake [reɪk] **1.** *n. (Hort.)* Rechen, *der (bes. südd. u. md.);* Harke, *die (bes. nordd.).* **2.** *v. t.* **a)** harken ‹*Laub, Erde, Fußboden, Kies, Oberfläche*›; **b) ~ the fire** die Asche entfernen; **c)** *(with eyes/shots)* bestreichen

~ 'in *v. t. (coll.)* scheffeln *(ugs.)* ‹*Geld*›

~ over *v. t.* **a)** harken; **b)** *(fig.)* wieder ausgraben

~ 'up *v. t.* **a)** zusammenharken; **b)** *(fig.)* wieder ausgraben

²rake *n. (person)* Lebemann, *der*

'rake-off *n. (coll.)* [Gewinn]anteil, *der*

rakish ['reɪkɪʃ] *adj. (jaunty)* flott; keß; **wear one's hat at a ~ angle** seinen Hut frech *od.* keck aufgesetzt haben

rally ['rælɪ] **1.** *v.i.* **a)** *(come together)* sich versammeln; ~ **to the support of** *or* **the defence of,** ~ **behind** *or* **to sb.** *(fig.)* sich hinter jmdn. stellen; **b)** *(regain health)* sich wieder [ein wenig] erholen; **c)** *(reassemble)* sich [wieder] sammeln; **d)** *(increase in value after fall)* ‹*Aktie, Kurs:*› wieder anziehen, sich wieder erholen. **2.** *v.t.* **a)** *(reassemble)* wieder zusammenrufen; **b)** *(bring together)* einigen ‹*Partei, Kräfte*›; sammeln ‹*Anhänger*›; **c)** *(rouse)* aufmuntern; *(revive)* ~ **one's strength** seine [ganze] Kraft zusammennehmen. **3.** *n.* **a)** *(mass meeting)* Versammlung, *die;* **peace** ~: Friedenskundgebung, *die;* **b)** *(competition)* **[motor]** ~: Rallye, *die;* **c)** *(Tennis)* Ballwechsel, *der*

RAM [ræm] *abbr. (Computing)* **random access memory** RAM

ram [ræm] **1.** *n.* **a)** *(Zool.)* Schafbock, *der;* Widder, *der;* **b)** *see* **battering ram; c)** *(hydraulic lifting-machine)* hydraulischer Widder. **2.** *v.t.*, **-mm-: a)** *(force)* stopfen; ~ **a post into the ground** einen Pfosten in die Erde rammen; ~ **in** in etw. *(Akk.)* rammen; ~ **sth. home to sb.** jmdm. etw. deutlich vor Augen führen; **b)** *(collide with)* rammen ‹*Fahrzeug, Pfosten*›; **c)** ~ **[down]** *(beat down)* feststampfen ‹*Erde, Ton, Kies*›

ramble ['ræmbl] **1.** *n.* **[nature]** ~: Wanderung, *die.* **2.** *v.i.* **a)** *(walk)* umherstreifen **(through, in** in + *Dat.*); **b)** *(in talk)* zusammenhangloses Zeug reden *(abwertend);* **keep rambling on about sth.** sich endlos über etw. *(Akk.)* auslassen

rambler ['ræmblə(r)] *n.* **a)** Wanderer, *der*/Wanderin, *die;* **b)** *(Bot.)* Kletterrose, *die*

rambling ['ræmblɪŋ] **1.** *n.* Wandern, *das.* **2.** *adj.* **a)** *(irregularly arranged)* verschachtelt; verwinkelt ‹*Straßen*›; **b)** *(incoherent)* unzusammenhängend ‹*Erklärung, Brief*›; zerstreut ‹*Professor*›; **c)** ~ **rose** Kletterrose, *die*

ramification [ræmɪfɪ'keɪʃn] *n.* Auswirkungen

rammer ['ræmə(r)] *n.* Stampfer, *der*

ramp [ræmp] *n.* **a)** *(slope)* Rampe, *die;* **'beware** *or* **caution,** ~**!'** „Vorsicht, unebene Fahrbahn!"; **b)** *(Aeronaut.)* Gangway, *die*

rampage 1. ['ræmpeɪdʒ, ræm'peɪdʒ] *n.* Randale, *die (ugs.);* **be/go on the** ~ *(coll.)* ‹*Rowdies:*› randalieren; ‹*verärgerte Person:*› toben. **2.** [ræm'peɪdʒ] *v.i.* ‹*Rowdies:*› randalieren; ~ **about** ‹*verärgerte Person:*› toben

rampant ['ræmpənt] *adj.* zügellos ‹*Gewalt, Rassismus*›; steil ansteigend ‹*Inflation*›; üppig ‹*Wachstum*›

rampart ['ræmpɑ:t] *n.* **a)** *(walk)* Wehrgang, *der;* **b)** *(protective barrier)* Wall, *der*

'ramrod *n.* Ladestock, *der;* **as straight** *or* **stiff as a** ~ *(fig. coll.)* so steif, als ob man einen Besenstiel verschluckt hätte; stocksteif

'ramshackle *adj.* klapprig ‹*Auto*›; verkommen ‹*Gebäude*›

ran *see* **run 2, 3**

ranch [rɑ:ntʃ] *n.* Ranch, *die;* **[mink/poultry]** ~: [Nerz-/Geflügel]farm, *die*

rancher ['rɑ:ntʃə(r)] *n.* Rancher, *der*/Rancherin, *die*

rancid ['rænsɪd] *adj.* ranzig

rancour *(Brit.; Amer.:* **rancor)** ['ræŋkə(r)] *n.* [tiefe] Verbitterung

random ['rændəm] **1.** *n.* **at** ~: wahllos; willkürlich; *(aimlessly)* ziellos; **choose at** ~: aufs Geratewohl wählen. **2.** *adj.* **a)** *(unsystematic)* willkürlich ‹*Auswahl*›; **make a** ~ **guess** raten aufs Geratewohl; **b)** *(Statistics)* Zufalls-

randy ['rændɪ] *adj.* geil; scharf *(ugs.);* **feel** ~: geil sein

rang *see* **²ring 2, 3**

range [reɪndʒ] **1.** *n.* **a)** *(row)* ~ **of mountains** Bergkette, *die;* **b)** *(of subjects, interests, topics)* Palette, *die; (of musical instrument)* Tonumfang, *der; (of knowledge, voice)* Umfang, *der; (of income, department, possibility)* Bereich, *der;* **sth. is out of** *or* **beyond sb's** ~ *(lit. or fig.)* etw. ist außerhalb jmds. Reichweite; **c)** *(of telescope, missile, aircraft, etc.)* Reichweite, *die; (distance between gun and target)* Schußweite, *die;* **flying** ~: Flugbereich, *der;* **at a** ~ **of 200 metres** auf eine Entfernung von 200 Metern; **up to a** ~ **of 5 miles** bis zu einem Umkreis von 5 Meilen; **shoot at close** *or* **short/long** ~: aus kurzer/großer Entfernung schießen; **experience sth. at close** ~: etw. in unmittelbarer Nähe erleben; **d)** *(series, selection)* Kollektion, *die;* **e) [shooting]** ~: Schießstand, *der; (at funfair)* Schießbude, *die;* **f)** *(testing-site)* Versuchsgelände, *das;* **g)** *(grazing-ground)* Weide[fläche], *die.* **2.** *v.i.* **a)** *(vary within limits)* ‹*Preise, Temperaturen:*› schwanken, sich bewegen **(from ... to** zwischen [+ *Dat.*]... und); **they** ~ **in age from 3 to 12** sie sind zwischen 3 und 12 Jahre alt; **b)** *(extend)* ‹*Klippen, Gipfel, Häuser:*› sich hinziehen; **c)** *(roam)* umherziehen **(around, about** in + *Dat.*); *(fig.)* ‹*Gedanken:*› umherschweifen; **the discussion** ~**d over ...:** die Diskussion erstreckte sich auf (+ *Akk.*) ... **3.** *v.t. (arrange)* aufreihen ‹*Bücher, Tische*›; ~ **oneself against sb./sth.** *(fig.)* sich gegen jmdn./etw. zusammenschließen

'range-finder *n.* Entfernungsmesser, *der*

ranger ['reɪndʒə(r)] *n.* **a)** *(keeper)* Aufseher, *der*/Aufseherin, *die; (of forest)* Förster, *der*/Försterin, *die;* **b)** *(Amer.: law officer)* Ranger, *der; Angehöriger der berittenen Polizeitruppe*

¹rank [ræŋk] **1.** *n.* **a)** *(position in hierarchy)* Rang, *der; (Mil. also)* Dienstgrad, *der;* **be above/below sb. in** ~: einen höheren/niedrigeren Rang/Dienstgrad haben als jmd.; **b)** *(social position)* [soziale] Stellung; **people of all** ~**s** Menschen aus allen [Gesellschafts]-schichten; **c)** *(row)* Reihe, *die;* **d)** *(Brit.: taxi-stand)* [Taxen]stand, *der;* **e)** *(line of soldiers)* Reihe, *die;* **the** ~**s** *(enlisted men)* die Mannschaften und Unteroffiziere; **the** ~ **and file** die Mannschaften und Unteroffiziere; *(fig.)* die breite Masse; **close [our/their]** ~**s** die Reihen schließen; *(fig.)* sich zusammenschließen; **rise from the** ~**s** sich [aus dem Mannschaftsstand] zum Offizier hochdienen; *(fig.)* sich hocharbeiten. **2.** *v.t. (classify)* ~ **among** *or* **with** zählen *od.* rechnen zu; ~ **sth. highly** etw. hoch einstufen. **3.** *v.i.* ~ **among** *or* **with** gehören *od.* zählen zu; ~ **above/next to sb.** rangmäßig über/direkt unter jmdm. stehen

²rank *adj.* **a)** *(complete)* blank ‹*Unsinn, Frechheit*›; kraß ‹*Außenseiter, Illoyalität*›; **b)** *(stinking)* stinkend; **c)** *(rampant)* ~ **weeds** [wild] wucherndes Unkraut

rankings ['ræŋkɪŋz] *n. pl. (Sport)* Rangliste, *die*

rankle ['ræŋkl] *v.i.* **sth.** ~**s [with sb.]** etw. wurmt jmdn. *(ugs.)*

ransack ['rænsæk] *v.t.* **a)** *(search)* durchsuchen **(for** nach); **b)** *(pillage)* plündern

ransom ['rænsəm] **1.** *n.* ~ **[money]** Lösegeld, *das;* **hold to** ~: als Geisel festhalten; *(fig.)* erpressen, unter Druck *(Akk.)* setzen ‹*Regierung*›. **2.** *v.t.* **a)** *(obtain release of)* Lösegeld bezahlen für; auslösen; **b)** *(hold to* ~*)* als Geisel festhalten

ˈransom note *n.* Erpresserbrief, *der*

rant [rænt] *v. i.* ~ **[and rave]** wettern *(ugs.)* **(about** über + *Akk.*); ~ **at** anschnauzen *(ugs.)*

rap [ræp] **1.** *n.* **a)** *(sharp knock)* [energisches] Klopfen; **there was a** ~ **on** *or* **at the door** es klopfte [laut]; **give sb. a** ~ **on** *or* **over the knuckles** jmdm. auf die Finger schlagen; *(fig.)* jmdm. auf die Finger klopfen; **b)** *(sl.: blame)* **take the** ~ **[for sth.]** [für etw.] den Kopf hinhalten *(ugs.)*. **2.** *v. t.*, **-pp-** *(strike smartly)* klopfen; ~ **sb. on the knuckles** jmdm. auf die Finger klopfen. **3.** *v. i.*, **-pp-** klopfen **(on** an + *Akk.*); ~ **on the table** auf den Tisch klopfen

~ **ˈout** *v. t.* ausstoßen ‹*Befehl, Fluch*›; ~ **out a message** melden

rapacious [rəˈpeɪʃəs] *adj.* *(greedy)* habgierig

¹rape [reɪp] **1.** *n.* Vergewaltigung, *die (auch fig.);* Notzucht, *die (Rechtsspr.)*. **2.** *v. t.* vergewaltigen; notzüchtigen *(Rechtsspr.)*

²rape *n. (Bot., Agric.)* Raps, *der*

rapid [ˈræpɪd] **1.** *adj.* schnell ‹*Bewegung, Wachstum, Puls*›; rasch ‹*Folge, Bewegung, Fortschritt, Ausbreitung, Änderung*›; rapide ‹*Niedergang*›; steil ‹*Abstieg*›; reißend ‹*Gewässer, Strömung*›; stark ‹*Gefälle, Strömung*›; **there has been a** ~ **decline** es ging rapide abwärts. **2.** *n. in pl.* Stromschnellen

rapidity [rəˈpɪdɪtɪ] *n., no pl.* Schnelligkeit, *die*

rapier [ˈreɪpɪə(r)] *n. (Fencing)* Rapier, *das*

rapist [ˈreɪpɪst] *n.* Vergewaltiger, *der*

rapport [ræˈpɔː(r)] *n.* [harmonisches] Verhältnis; **have a great** ~ **with sb.** ein ausgezeichnetes Verhältnis zu jmdm. haben; **establish a** ~ **with sb.** eine Beziehung zu jmdm. aufbauen

rapt [ræpt] *adj.* gespannt ‹*Aufmerksamkeit, Miene*›; **in** ~ **contemplation** in Betrachtungen versunken

rapture [ˈræptʃə(r)] *n.* **a)** *(ecstatic delight)* **[state of]** ~: Verzückung, *die;* **b)** *in pl.* **be in** ~**s** entzückt sein **(over, about** über + *Akk.*); **go into** ~**s** [überschwenglich] schwärmen **(over, about** von)

rapturous [ˈræptʃərəs] *adj.* begeistert ‹*Applaus, Menge, Willkommen*›; verzückt ‹*Miene*›

¹rare [reə(r)] *adj.* **a)** *(uncommon)* selten; **it's** ~ **for him to do that** es kommt selten vor, daß er das tut; **b)** *(thin)* dünn ‹*Luft, Atmosphäre*›

²rare *adj. (Cookery)* englisch gebraten; nur schwach gebraten

rarebit [ˈreəbɪt] *see* **Welsh rarebit**

rarefied [ˈreərɪfaɪd] *adj.* dünn ‹*Luft*›; *(fig.)* exklusiv

rarely [ˈreəlɪ] *adv.* selten

raring [ˈreərɪŋ] *adj. (coll.)* **be** ~ **to go** kaum abwarten können, bis es losgeht

rarity [ˈreərɪtɪ] *n.* Seltenheit, *die;* Rarität, *die;* **be an object of great** ~: eine große Seltenheit sein

rascal [ˈrɑːskl] *n.* **a)** *(dishonest person)* Schuft, *der;* **b)** *(joc.: mischievous person)* Schlingel, *der (scherzh.);* Spitzbube, *der (scherzh.)*

¹rash [ræʃ] *n. (Med.)* [Haut]ausschlag, *der;* **develop a** *or* **break out** *or* **come out in a** ~: einen Ausschlag bekommen

²rash *adj.* voreilig ‹*Urteil, Entscheidung, Entschluß*›; überstürzt ‹*Versprechungen, Handlung, Erklärung*›; ungestüm ‹*Person*›

rasher [ˈræʃə(r)] *n.* ~ **[of bacon]** Speckscheibe, *die*

rasp [rɑːsp] **1.** *n.* **a)** *(tool)* Raspel, *die;* **b)** *(sound) (of metal on wood)* schneidendes Geräusch; *(of breathing)* Rasseln, *das.* **2.** *v. t.* **a)** *(scrape with* ~*)* raspeln ‹*Blech, Kante*›; **b)** *(say gratingly)* schnarren

raspberry [ˈrɑːzbərɪ] *n.* **a)** Himbeere, *die; attrib.* Himbeer‹*marmelade, -torte, -rosa, -eis*›; **b)** *(sl.: rude noise)* **blow a** ~: verächtlich prusten

rasping [ˈrɑːspɪŋ] *adj.* krächzend ‹*Husten, Stimme*›; rasselnd ‹*Geräusch*›

rat [ræt] *n.* **a)** Ratte, *die;* **brown** *or* **sewer** ~: Wanderratte, *die;* **smell a** ~ *(fig. coll.)* Lunte *od.* den Braten riechen *(ugs.);* **b)** *(coll. derog.: unpleasant person)* Ratte, *die (derb)*

ratchet [ˈrætʃɪt] *n. (Mech. Engin.) (set of teeth)* Zahnkranz, *der;* ~ **[wheel]** Klinkenrad, *das*

rate [reɪt] **1.** *n.* **a)** *(proportion)* Rate, *die;* **increase at a** ~ **of 50 a week** [um] 50 pro Woche anwachsen; ~ **of inflation/absentee** ~: Inflations-/Abwesenheitsrate, *die;* **b)** *(tariff)* Satz, *der;* **interest/taxation** ~, ~ **of interest/taxation** Zins-/Steuersatz, *der;* **c)** *(amount of money)* Gebühr, *die;* ~ **[of pay]** Lohnsatz, *der;* **letter/parcel** ~: Briefporto, *das*/Paketgebühr, *die;* **at reduced** ~: gebührenermäßigt ‹*Drucksache*›; **d)** *(speed)* Geschwindigkeit, *die;* Tempo, *das;* **at a** *or* **the** ~ **of 50 mph** mit [einer Geschwindigkeit von] 80 km/h; **at a good/fast/dangerous** ~: zügig/mit hoher Geschwindigkeit/gefährlich schnell; **e)** *(Brit.: local authority levy)* **[local** *or* **council]** ~**s** Gemeindeabgaben; **f)** *(coll.)* **at any** ~ *(at least)* zumindest; wenigstens; *(whatever happens)* auf jeden Fall; **at this** ~ **we won't get any work done** so kriegen wir gar nichts fertig *(ugs.);* **at the** ~ **you're going, ...** *(fig.)* wenn du so weitermachst, ... **2.** *v. t.* **a)** *(estimate worth of)* schätzen ‹*Vermögen*›; einschätzen ‹*Intelligenz, Leistung, Fähigkeit*›; ~ **sb./sth. highly** jmdn./etw. hoch einschätzen; **b)** *(consider)* betrachten; rechnen **(among** zu); **be** ~**d the top tennis-player in Europe** als der beste Tennisspieler Europas gelten; **c)** *(Brit.: value)* **the house is** ~**d at £100 a year** die Grundlage für die Berechnung der Gemeindeabgaben für das Haus beträgt 100 Pfund pro Jahr; **d)** *(merit)* verdienen ‹*Auszeichnung, Erwähnung*›. **3.** *v. i.* zählen **(among** zu); ~ **as** gelten als

rateable [ˈreɪtəbl] *adj. (Brit.)* ~ **value** steuerbarer Wert

ratepayer [ˈreɪtpeɪə(r)] *n. (Brit.)* Realsteuerpflichtige, *der/die;* ≈ Steuerzahler, *der*/-zahlerin, *die*

rather [ˈrɑːðə(r)] *adv.* **a)** *(by preference)* lieber; **he wanted to appear witty** ~ **than brainy** er wollte lieber geistreich als klug erscheinen; ~ **than accept bribes, he decided to resign** ehe er sich bestechen ließ, trat er lieber zurück; **b)** *(somewhat)* ziemlich ‹*gut, gelangweilt, unvorsichtig, nett, warm*›; **I** ~ **think that ...:** ich bin ziemlich sicher, daß ...; **be** ~ **better/more complicated than expected** um einiges besser/komplizierter sein als erwartet; **it is** ~ **too early** ich fürchte, es ist zu früh; **it looks** ~ **like a banana** es sieht ungefähr wie eine Banane aus; **I** ~ **like beans/him** ich esse Bohnen ganz gern/ich mag ihn recht gern; **c)** *(more truly)* vielmehr; **or** ~: beziehungsweise; [oder] genauer gesagt; **he was careless** ~ **than wicked** er war eher nachlässig als böswillig

ratify [ˈrætɪfaɪ] *v. t.* ratifizieren ‹*völkerrechtlichen Vertrag*›; bestätigen ‹*Ernennung*›; sanktionieren ‹*Vertrag, Gesetzentwurf*›

rating [ˈreɪtɪŋ] *n.* **a)** *(estimated standing)* Einschätzung, *die;* **b)** *(Radio, Telev.)* **[popularity]** ~: Einschaltquote, *die;* **be high/low in the** ~**s** eine hohe/niedrige Einschaltquote haben; **c)** *(Navy:*

rank) Dienstgrad, *der;* **d)** *(Brit. Navy: sailor)* **[naval]** ~: Mannschaftsdienstgrad, *der*

ratio ['reɪʃɪəʊ] *n., pl.* **~s** Verhältnis, *das;* **in a** *or* **the ~ of 1 to 5** im Verhältnis 1:5; **in direct ~ to** *or* **with** im gleichen Verhältnis wie; **the teacher-student ~:** das Verhältnis von Lehrern zu Schülern; **what is the ~ of men to women?** wie hoch ist der Männeranteil im Vergleich zu dem der Frauen?

ration ['ræʃn] **1.** *n.* **a)** *(daily food allowance)* [Tages]ration, *die;* **put sb. on short ~s** jmdn. auf halbe Ration setzen *(ugs.);* **b)** *(fixed allowance of food etc. for civilians)* **~[s]** Ration, *die* **(of** an + *Dat.*); **petrol/meat ~:** Benzin-/Fleischration, *die.* **2.** *v.t.* rationieren ‹*Benzin, Autos*›; Rationen zuteilen (+ *Dat.*) ‹*Person*›; **be ~ed to one glass of spirits per day** nur ein Glas Alkohol pro Tag trinken dürfen

~ 'out *v.t.* zuteilen **(to** *Dat.*); in Rationen austeilen **(to** an + *Akk.*)

rational ['ræʃənl] *adj.* *(having reason)* rational, vernunftbegabt ‹*Wesen*›; *(sensible)* vernünftig ‹*Person, Art, Politik usw.*›

rationale [ræʃə'nɑ:l] *n.* **a)** *(statement of reasons)* rationale Erklärung **(of** für); **b)** *(fundamental reason)* logische Grundlage

rationalisation, rationalise *see* **rationaliz-**

rationalization [ræʃənəlaɪ'zeɪʃn] *n.* *(Econ., Psych.)* Rationalisierung, *die*

rationalize ['ræʃənəlaɪz] **1.** *v.t.* *(Econ., Psych.)* rationalisieren. **2.** *v.i.* Scheinbegründungen finden

ration: ~ book *n.* Bezugsscheinheft, *das;* **~ card, ~ coupon** *ns.* Bezugsschein, *der*

rationing ['ræʃənɪŋ] *n.* Rationierung, *die*

rat: ~ poison *n.* Rattengift, *das;* **~ race** *n.* erbarmungsloser Konkurrenzkampf

rattle ['rætl] **1.** *v.i.* **a)** *(clatter)* ‹*Fenster, Maschinenteil, Schlüssel:*› klappern; ‹*Hagel:*› prasseln; ‹*Flaschen:*› klirren; ‹*Kette:*› rasseln; ‹*Münzen:*› klingen; **~ at the door** an der Tür rütteln; **b)** *(move)* ‹*Zug, Bus:*› rattern; ‹*Kutsche:*› rumpeln. **2.** *v.t.* **a)** *(make ~)* klappern mit ‹*Würfel, Geschirr, Dose, Münzen, Schlüsselbund*›; klirren lassen ‹*Fenster[scheiben]*›; rasseln mit ‹*Kette*›; **b)** *(sl.: disconcert)* **~ sb., get sb. ~d** jmdn. durcheinanderbringen; **don't get ~d!** reg dich nicht auf! **3.** *n.* **a)** *(of baby; Mus.)* Rassel, *die; (of sports fan)* Ratsche, *die;* **b)** *(sound)* Klappern, *das; (of hail)* Prasseln, *das; (of drums)* Schnarren, *das; (of machine-gun)* Rattern, *das; (of chains)* Rasseln, *das*

~ 'off *v.t.* *(coll.)* herunterrasseln *(ugs.)*

~ 'on *v.i.* *(coll.)* plappern *(ugs.)*

'rattlesnake *n.* Klapperschlange, *die*

ratty ['rætɪ] *adj.* *(sl.: irritable)* gereizt

raucous ['rɔ:kəs] *adj.* rauh ‹*Stimme, Lachen*›

raunchy ['rɔ:ntʃɪ] *adj.* *(suggestive)* scharf *(salopp)*

ravage ['rævɪdʒ] **1.** *v.t.* heimsuchen ‹*Gebiet, Stadt*›; so gut wie vernichten ‹*Ernte*›; schwer zeichnen ‹*Gesichtszüge*›. **2.** *n.* *in pl.* verheerende Wirkung; **the ~s of time/war** die Zeichen der Zeit/die Wunden des Krieges

rave [reɪv] **1.** *v.i.* **a)** *(talk wildly)* irrereden; **~ at** [wüst] beschimpfen; **b)** *(speak with admiration)* schwärmen **(about, over** von). **2.** *adj.* *(coll.)* [hellauf] begeistert ‹*Kritik*›

raven ['reɪvn] *n.* Rabe, *der;* Kolkrabe, *der (Zool.)*

ravenous ['rævənəs] *adj.* ausgehungert; **I'm ~:** ich habe einen Bärenhunger *(ugs.)*

ravine [rə'vi:n] *n.* Schlucht, *die; (made by river also)* Klamm, *die*

raving ['reɪvɪŋ] **1.** *n.* *in pl.* irres Gerede. **2.** *adj.* **a)** *(talking madly)* irreredend ‹*Wahnsinniger, Idiot*›; **b)** *(outstanding)* phantastisch *(ugs.)* ‹*Erfolg*›. **3.** *adv.* **be ~ mad** *(stupid)* völlig verrückt sein *(ugs.)*

ravishing ['rævɪʃɪŋ] *adj.* bildschön ‹*Anblick, Person*›; hinreißend ‹*Schönheit*›

raw [rɔ:] **1.** *adj.* **a)** *(uncooked)* roh; **b)** *(inexperienced)* unerfahren; blutig ‹*Anfänger*›; *see also* **recruit 1 a, c; c)** *(stripped of skin)* blutig ‹*Fleisch*›; offen ‹*Wunde*›; *(sore)* wund ‹*Füße*›; **d)** *(chilly)* naßkalt; **e)** *(untreated)* Roh‹*haut, -holz, -seide, -zucker, -erz, -leder*›; *(undiluted)* rein ‹*Alkohol*›; **f)** *(fig.: unpolished)* grob; **g)** *(Statistics)* unaufbereitet. **2.** *n.* **nature in the ~:** unverfälschte Natur; **touch sb. on the ~** *(Brit. coll.)* jmdn. an [s]einer verwundbaren Stelle treffen

raw ma'terial *n.* Rohstoff, *der*

¹ray [reɪ] *n.* **a)** *(lit. or fig.)* Strahl, *der;* **~ of sunshine/light** Sonnen-/Lichtstrahl, *der;* **~ of hope** Hoffnungsstrahl, *der;* **b)** *in pl.* *(radiation)* Strahlen; Strahlung, *die*

²ray *n.* *(fish)* Rochen, *der*

rayon ['reɪɒn] *n.* *(Textiles)* Reyon, *das od. der; attrib.* Reyon‹*kleid, -hemd*›

raze [reɪz] *v.t.* **~ to the ground** dem Erdboden gleichmachen

razor ['reɪzə(r)] *n.* Rasiermesser, *das;* **[electric] ~:** [elektrischer] Rasierapparat; [Elektro- *od.* Trocken]rasierer, *der (ugs.)*

razor: ~-blade *n.* Rasierklinge, *die;* **~-edge** *n.* Rasierschneide, *die;* **be** *or* **stand on a ~-edge** *or* **~'s edge** *(fig. coll.)* sich auf einer Gratwanderung befinden; **~-sharp** *adj.* sehr scharf ‹*Messer*›; *(fig.)* messerscharf ‹*Verstand, Intellekt*›; scharfsinnig ‹*Person*›

RC *abbr.* **Roman Catholic** r.-k.; röm.-kath.

Rd. *abbr.* **road** Str.

RE *abbr.* *(Brit.)* **Religious Education** Religionslehre, *die*

re [ri:] *prep.* *(coll.)* über (+ *Akk.*)

're [ə(r)] *(coll.)* = **are;** *see* **be**

reach [ri:tʃ] **1.** *v.t.* **a)** *(arrive at)* erreichen; ankommen *od.* eintreffen in (+ *Dat.*) ‹*Stadt, Land*›; erzielen ‹*Übereinstimmung, Übereinkunft*›; kommen zu ‹*Entscheidung, Entschluß; Ausgang, Eingang*›; **be easily ~ed** leicht erreichbar *od.* zu erreichen sein **(by** mit); **not a sound ~ed our ears** kein Laut drang an unsere Ohren; **have you ~ed page 45 yet?** bist du schon auf Seite 45 [angelangt]?; **you can ~ her at this number/by radio** du kannst sie unter dieser Nummer/über Funk erreichen; **b)** *(extend to)* ‹*Straße:*› führen bis zu; ‹*Leiter, Haar:*› reichen bis zu; **c)** *(pass)* **~ me that book** reich mir das Buch herüber. **2.** *v.i.* **a)** *(stretch out hand)* **~ for sth.** nach etw. greifen; **how high can you ~?** wie hoch kannst du langen?; **b)** *(be long/tall enough)* **sth. will/won't ~:** etw. ist/ist nicht lang genug; **he can't ~ up to the top shelf** er kann das oberste Regal nicht [mit der Hand] erreichen; **will it ~ as far as ...?** wird es bis zu ... reichen? **can you ~?** kannst *od.* kommst du dran? *(ugs.);* **c)** *(go as far as)* ‹*Wasser, Gebäude, Besitz:*› reichen **([up] to** bis [hinauf] zu). **3.** *n.* **a)** *(extent of ~ing)* Reichweite, *die;* **be within easy ~ [of a place]** [von einem Ort aus] leicht erreichbar sein; **be above sb.'s ~:** zu hoch für jmdn. sein; **keep sth. out of ~ of sb.** etw. unerreichbar für jmdn. aufbewahren; **keep sth. within easy ~:** etw. in greifbarer Nähe aufbewahren; **be within/beyond the ~**

of sb. in/außer jmds. Reichweite sein; *(fig.)* für jmdn. im/nicht im Bereich des Möglichen liegen; *(financially)* für jmdn. erschwinglich/unerschwinglich sein; **b)** *(expanse)* Abschnitt, *der*

~ **'down 1.** *v. i.* den Arm nach unten ausstrecken; ~ **down to sth.** *(be long enough)* bis zu etw. [hinunter]reichen. **2.** *v. t.* hinunterholen; *(to receiving speaker)* herunterreichen

~ **'out 1.** *v. t.* *(stretch out)* ausstrecken ‹*Fuß, Bein, Hand, Arm*› (**for** nach). **2.** *v. i.* die Hand ausstrecken (**for** nach); ~ **out for,** ~ **out to grasp** ‹*Person, Hand:*› greifen nach

reachable ['ri:tʃəbl] *adj.* erreichbar

react [rɪ'ækt] *v. i.* **a)** *(respond)* reagieren (**to** auf + *Akk.*); **b)** *(act in opposition)* sich widersetzen (**against** *Dat.*); **c)** *(Chem., Phys.)* reagieren

reaction [rɪ'ækʃn] *n.* Reaktion, *die* (**to** auf + *Akk.*); ~ **against sth.** Widerstand gegen etw.; **action and** ~: Wirkung und Gegenwirkung; **what was his** ~**?** wie hat er reagiert?; **there was a favourable** ~ **to the proposal** der Vorschlag ist positiv aufgenommen worden

reactionary [rɪ'ækʃənərɪ] *(Polit.)* **1.** *adj.* reaktionär. **2.** *n.* Reaktionär, *der*/Reaktionärin, *die*

reactor [rɪ'æktə(r)] *n.* **[nuclear]** ~: Kernreaktor, *der*

read [ri:d] **1.** *v. t.*, **read** [red] **a)** lesen; ~ **sb. sth.,** ~ **sth. to sb.** jmdm. etwas vorlesen; *see also* **take 1 u; b)** *(show a reading of)* anzeigen; **c)** *(interpret)* deuten; ~ **sb.'s hand** jmdm. aus der Hand lesen; ~ **sb.'s mind** *or* **thoughts** jmds. Gedanken lesen; ~ **sth. into sth.** etw. in etw. *(Akk.)* hineinlesen; **d)** *(Brit. Univ.: study)* studieren. **2.** *v. i.*, **read** [red] **a)** lesen; ~ **to sb.** jmdm. vorlesen; **b)** *(convey meaning)* lauten; **the contract** ~**s as follows** der Vertrag hat folgenden Wortlaut; **c)** *(affect reader)* sich lesen. **3.** *n.* **a) have a quiet** ~: in Ruhe lesen; **b)** *(Brit. coll.: reading matter)* **be a good** ~: sich gut lesen. **4.** [red] *adj.* **widely** *or* **deeply** ~: sehr belesen ‹*Person*›; **the most widely** ~ **book/author** das meistgelesene Buch/der meistgelesene Autor

~ **'back** *v. t.* wiederholen; noch einmal vorlesen

~ **'off** *v. t.* durchlesen; *(from meter, board)* ablesen ‹*Zahl, Stand*›

~ **'out** *v. t.* laut vorlesen

~ **'over,** ~ **'through** *v. t.* durchlesen

~ **'up** *v. t.* sich informieren (**on** über + *Akk.*)

readable ['ri:dəbl] *adj.* **a)** *(pleasant to read)* lesenswert; **b)** *(legible)* leserlich

readdress [ri:ə'dres] *v. t.* umadressieren

reader ['ri:də(r)] *n.* **a)** Leser, *der*/Leserin, *die;* **be a slow/good/great** ~ **[of sth.]** [etw.] langsam/gut/gern lesen; **b)** *(who reads aloud)* Vorlesende, *der/die;* **c)** *(textbook)* Lehrbuch, *das; (to learn to read, containing original texts)* Lesebuch, *das;* **d)** *(Brit. Univ.)* ≈ Assistenzprofessor, *der*/-professorin, *die* (**in** für)

readership ['ri:dəʃɪp] *n.* *(number or type of readers)* Leserschaft, *die;* Leserkreis, *der;* **what is the** ~ **of the paper?** wie groß ist die Leserschaft der Zeitung?

readily ['redɪlɪ] *adv.* **a)** *(willingly)* bereitwillig; **b)** *(without difficulty)* ohne weiteres

readiness ['redɪnɪs] *n., no pl.* Bereitschaft, *die;* ~ **to learn** Lernbereitschaft, *die;* **have/be in** ~ **[for sth.]** [für etw.] bereithalten/bereit sein

reading ['ri:dɪŋ] *n.* **a)** Lesen, *das;* **b)** *(matter to be read)* Lektüre, *die;* **make interesting/be good/dull** ~: interessant/gut/langweilig zu lesen sein; **c)** *(figure shown)* Anzeige, *die;* **d)** *(recital)* Lesung, *die* (**from** aus); **e)** *(interpretation)* [Aus]deutung, *die;* **f)** *(Parl.)* **[first/second/third]** ~: [erste/zweite/dritte] Lesung

reading: ~-glasses *n. pl.* Lesebrille, *die;* ~ **knowledge** *n.* **have a** ~ **knowledge of a language** Texte in einer Sprache lesen können; **~-lamp, ~-light** *ns.* Leselampe, *die;* **~-list** *n.* Literaturliste, *die;* ~ **matter** *n., no pl., no indef. art.* Lesestoff, *der;* Lektüre, *die;* **~-room** *n.* Lesesaal, *der*

readjust [ri:ə'dʒʌst] **1.** *v. t.* neu einstellen. **2.** *v. i.* ~ **to** sich wieder gewöhnen an (+ *Akk.*) ‹*Leben*›

ready ['redɪ] **1.** *adj.* **a)** *(prepared)* fertig; **be** ~ **to do sth.** bereit sein, etw. zu tun; **I'm not** ~ **to go to the cinema yet** ich kann jetzt noch nicht ins Kino gehen; **the troops are** ~ **to march/for battle** die Truppen sind marsch-/gefechtsbereit; **be** ~ **for work/school** zur Arbeit/für die Schule bereit sein; *(about to leave)* für die Arbeit/Schule fertig sein; **be** ~ **to leave** aufbruchsbereit sein; **be** ~ **for sb.** bereit sein, sich jmdm. zu stellen; **be** ~ **for anything** auf alles vorbereitet sein; **get** ~ **to go** sich zum Aufbruch bereit machen; ~**, set** *or* **steady, go!** Achtung, fertig, los!; **b)** *(willing)* bereit; **c)** *(prompt)* schnell; **have** ~**, be** ~ **with** parat haben, nicht verlegen sein um ‹*Antwort, Ausrede, Vorschlag*›; **d)** *(likely)* im Begriff; **be** ~ **to cry** den Tränen nahe sein; **e)** *(within reach)* griffbereit ‹*Fahrkarte, Taschenlampe, Waffe*›; **have your tickets** ~**!** halten Sie Ihre Fahrkarten bitte bereit! **2.** *adv.* fertig; ~ **cooked** vorgekocht. **3.** *n.* **at the** ~: schußbereit, im Anschlag ‹*Schußwaffe*›

ready: ~ **'cash** *see* ~ **money; ~-'made** *adj.* **a)** Konfektions‹*anzug, -kleidung*›; **~-made curtains** Fertiggardinen; **b)** *(fig.)* vorgefertigt; ~ **'money** *n.* **a)** *(cash)* Bargeld, *das;* **b)** *(immediate payment)* **for** ~ **money** gegen bar; ~ **reckoner** [redɪ 'rekənə(r)] *n.* Berechnungstabelle, *die; (for conversion)* Umrechnungstabelle, *die;* **~-to-eat** *adj.* Fertig‹*mahlzeit, -dessert*›; **~-to-'wear** *adj.* Konfektions‹*anzug, -kleidung*›

reaffirm [ri:ə'fɜ:m] *v. t.* [erneut] bekräftigen

real [rɪəl] *adj.* **a)** *(actually existing)* real ‹*Gestalt, Ereignis, Lebewesen*›; wirklich ‹*Macht*›; **b)** *(genuine)* echt ‹*Interesse, Gold, Seide*›; **c)** *(true)* wahr ‹*Grund, Freund, Name, Glück*›; echt ‹*Mitleid, Vergnügen, Sieg*›; **the** ~ **thing** *(genuine article)* der/die/das Echte; **be [not] the** ~ **thing** [un]echt sein; **d)** *(Econ.)* real; Real-; **in** ~ **terms** real ‹*sinken, steigen*›; **e) be for** ~ *(sl.)* echt sein; ‹*Angebot, Drohung:*› ernst gemeint sein

real: ~ **'ale** *n. (Brit.)* echtes Ale; ~ **'coffee** *n.* Bohnenkaffee, *der;* ~ **e'state** *n. (Law)* Immobilien *Pl.*

realisation, realise *see* **realiz-**

realism ['rɪəlɪzm] *n.* Realismus, *der*

realist ['rɪəlɪst] *n.* Realist, *der*/Realistin, *die*

realistic [rɪə'lɪstɪk] *adj.* realistisch; **be** ~ **about sth.** etw. realistisch sehen

reality [rɪ'ælɪtɪ] *n.* **a)** *no pl.* Realität, *die;* **bring sb. back to** ~: jmdn. in die Realität zurückholen; **in** ~: in Wirklichkeit; **b)** *no pl.* *(resemblance to original)* Naturtreue, *die*

realization [rɪəlaɪ'zeɪʃn] *n.* **a)** *(understanding)* Erkenntnis, *die;* **b)** *(becoming real)* Verwirkli-

chung, *die;* **c)** *(Finance: act of selling)* Realisierung, *die*
realize ['rɪəlaɪz] *v.t.* **a)** *(be aware of)* bemerken; realisieren; erkennen ‹*Fehler*›; ~ **[that] ...:** merken, daß ...; **I didn't** ~ *(abs.)* ich habe es nicht gewußt/*(had not noticed)* bemerkt; **b)** *(make happen)* verwirklichen; **c)** *(Finance: sell for cash)* realisieren *(fachspr.)*; in Geld *(Akk.)* umsetzen; **d)** *(fetch as price or profit)* erbringen ‹*Summe, Gewinn, Preis*›
real: ~ **'life** *n* das wirkliche Leben; die Realität; ~**-life** *attrib. adj.* real
really ['rɪəlɪ] *adv.* wirklich; **I don't** ~/~ **don't know what to do now** ich weiß eigentlich/wirklich nicht, was ich jetzt tun soll; **not** ~**:** eigentlich nicht; **that's not** ~ **a problem** das ist eigentlich kein Problem; **[well,]** ~**!** [also] so was!; ~**?** wirklich?; tatsächlich?
realm [relm] *n.* [König]reich, *das;* **be within/beyond the** ~**s of possibility** im/nicht im Bereich des Möglichen liegen
reap [ri:p] *v.t.* **a)** *(cut)* schneiden ‹*Getreide*›; **b)** *(gather in)* einfahren ‹*Getreide, Ernte*›; **c)** *(fig.)* ernten ‹*Ruhm, Lob*›; erhalten ‹*Belohnung*›; erzielen ‹*Gewinn*›
reappear [ri:ə'pɪə(r)] *v.i.* wieder auftauchen; *(come back)* [wieder] zurückkommen
reappearance [ri:ə'pɪərəns] *n.* Wiederauftauchen, *das*
reappraisal [ri:ə'preɪzl] *n.* Neubewertung, *die*
reappraise [ri:ə'preɪz] *v.t.* neu bewerten
[1]**rear** [rɪə(r)] **1.** *n.* **a)** *(back part)* hinterer Teil; **at** *or (Amer.)* **in the** ~ **of** im hinteren Teil (+ *Gen.*); **b)** *(back)* Rückseite, *die;* **be in** *or* **bring up the** ~**:** den Schluß bilden; **to the** ~ **of the house there is ...:** hinter dem Haus ist ...; **go round to the** ~ **of the house** hinter das Haus gehen; **c)** *(Mil.)* rückwärtiger Teil; **d)** *(coll.: buttocks)* Hintern, *der (ugs.)*. **2.** *adj.* hinter ... ‹*Eingang, Tür, Blinklicht*›; Hinter‹*achse, -rad*›
[2]**rear 1.** *v.t.* **a)** großziehen ‹*Kind, Familie*›; halten ‹*Vieh*›; hegen ‹*Wild*›; **b)** *(lift up)* heben ‹*Kopf*›; ~ **its ugly head** *(fig.)* seine häßliche Fratze zeigen. **2.** *v.i.* ‹*Pferd:*› sich aufbäumen
rear: ~ **'door** *n. (Motor Veh.)* Fondtür, *die;* Hintertür, *die;* ~**guard** *n. (Mil.)* Nachhut, *die;* ~**-light** *n.* Rücklicht, *das*
rearm [ri'ɑ:m] **1.** *v.i.* wiederaufrüsten. **2.** *v.t.* wiederaufrüsten ‹*Land*›; wiederbewaffnen/*(give more modern arms to)* neu bewaffnen *od.* ausrüsten ‹*Truppen*›
rearmament [ri:'ɑ:məmənt] *n.* Wiederbewaffnung, *die; (of country also)* Wiederaufrüstung, *die*
rearrange [ri:ə'reɪndʒ] *v.t.* umräumen ‹*Möbel, Zimmer*›; verlegen ‹*Treffen, Spiel*› **(for** auf + *Akk.*); ändern ‹*Anordnung, Programm*›
rearrangement [ri:ə'reɪndʒmənt] *n. see* **rearrange:** Umräumen, *das;* Verlegung, *die;* Änderung, *die*
rear-view 'mirror *n.* Rückspiegel, *der*
reason ['ri:zn] **1.** *n.* **a)** *(cause)* Grund, *der;* **there is [no/every]** ~ **to assume** *or* **believe that ...:** es besteht [kein/ein guter] Grund zu der Annahme, daß ...; **have no** ~ **to complain** *or* **for complaint** sich nicht beklagen können; **for that [very]** ~**:** aus [eben] diesem Grund; **no particular** ~ *(as answer)* einfach so; **all the more** ~ **for doing sth.** ein Grund mehr, etw. zu tun; **for no obvious** ~**:** aus keinem ersichtlichen Grund; **for the [simple]** ~ **that ...:** [einfach,] weil ...; **by** ~ **of** wegen; aufgrund; **with** ~**:** aus gutem Grund; **b)** *no pl., no art. (power to understand; sense; Philos.)* Vernunft, *die; (sanity)* gesunder Verstand; **lose one's** ~**:** den Verstand verlieren; **you can have anything within** ~**:** du kannst alles haben, solange es im Rahmen bleibt; **stand to** ~**:** unzweifelhaft sein; **not listen to** ~**:** sich *(Dat.)* nichts sagen lassen; **see** ~**:** zur Einsicht kommen. **2.** *v.i.* **a)** schlußfolgern **(from** aus); **b)** ~ **with** diskutieren mit **(about, on** über + *Akk.*); **you can't** ~ **with her** mit ihr kann man nicht vernünftig reden. **3.** *v.t.* schlußfolgern; **ours not to** ~ **why** es ist nicht unsere Sache, nach dem Warum zu fragen
~ **'out** *v.t.* sich *(Dat.)* überlegen
reasonable ['ri:zənəbl] *adj.* **a)** vernünftig; angemessen, vernünftig ‹*Forderung*›; **b)** *(inexpensive)* günstig; **it's a** ~ **price** das ist ein vernünftiger Preis; **c)** *(fair)* passabel ‹*Leistung, Wein*›; **d)** *(within limits)* realistisch ‹*Chancen, Angebot*›
reasonably ['ri:zənəblɪ] *adv.* **a)** *(within reason)* vernünftig; **b)** *(moderately)* ~ **priced** preisgünstig; **c)** *(rather)* ganz ‹*gut*›; ziemlich ‹*gesund*›
reasoned ['ri:znd] *adj.* durchdacht
reasoning ['ri:zənɪŋ] *n.* logisches Denken; *(argumentation)* Argumentation, *die*
reassurance [ri:ə'ʃʊərəns] *n.* **a)** *(calming)* **give sb.** ~**:** jmdn. beruhigen; **b)** *(confirmation in opinion)* Bestätigung, *die; in pl.* [wiederholte] Versicherungen
reassure [ri:ə'ʃʊə(r)] *v.t.* **a)** *(calm fears of)* beruhigen; **b)** *(confirm in opinion)* bestätigen
reassuring [ri:ə'ʃʊərɪŋ] *adj.* beruhigend
[1]**rebate** ['ri:beɪt] *n.* **a)** *(refund)* Rückzahlung, *die;* **b)** *(discount)* Rabatt, *der* **(on** auf + *Akk.*)
[2]**rebate** *n. (groove)* Falz, *der; (to receive edge of door or window)* Anschlag, *der*
rebel 1. ['rebl] *n.* Rebell, *der*/Rebellin, *die.* **2.** *attrib. adj.* **a)** *(of rebels)* Rebellen-; **b)** *(refusing obedience to ruler)* rebellisch; aufständisch. **3.** [rɪ'bel] *v.i.,* **-ll-** rebellieren
rebellion [rɪ'beljən] *n.* Rebellion, *die;* **rise [up] in** ~**:** rebellieren
rebellious [rɪ'beljəs] *adj. (defiant)* rebellisch; aufsässig
rebirth [ri:'bɜ:θ] *n.* **a)** Wiedergeburt, *die;* **b)** *(revival)* Wiederaufleben, *das*
reborn [ri:'bɔ:n] *adj.* wiedergeboren; **be** ~**:** wiedergeboren werden
rebound 1. [rɪ'baʊnd] *v.i.* **a)** *(spring back)* abprallen **(from** von); **b)** *(have adverse effect)* zurückfallen **(upon** auf + *Akk.*). **2.** ['ri:baʊnd] *n.* **a)** *(recoil)* Abprall, *der;* **b)** *(fig.: emotional reaction)* **marry sb. on the** ~**:** in seiner Enttäuschung jmdn. heiraten
rebuff [rɪ'bʌf] **1.** *n.* [schroffe] Abweisung; **be met with a** ~**:** auf Ablehnung stoßen. **2.** *v.t.* [schroff] zurückweisen
rebuild [ri:'bɪld] *v.t.,* **rebuilt** [ri:'bɪlt] *(lit. or fig.)* wieder aufbauen; *(make extensive changes to)* umbauen
rebuke [rɪ'bju:k] **1.** *v.t.* tadeln, rügen **(for** wegen); ~ **sb. for doing sth.** jmdn. zurechtweisen, weil er etwas tut/getan hat. **2.** *n.* Rüge, *die;* Zurechtweisung, *die*
rebut [rɪ'bʌt] *v.t.,* **-tt-** *(formal)* widerlegen
rebuttal [rɪ'bʌtl] *n. (Law)* Widerlegung, *die*
recalcitrant [rɪ'kælsɪtrənt] *adj.* aufsässig ‹*Person*›
recall 1. [rɪ'kɔ:l] *v.t.* **a)** *(remember)* sich erinnern an (+ *Akk.*); **b)** *(serve as reminder of)* erinnern an (+ *Akk.*); ~ **sth. to sb.** jmdn. an etw. *(Akk.)* erinnern; **c)** *(summon back)* zurückrufen ‹*Soldat,*

fehlerhaftes Produkt⟩; zurückfordern ⟨*Buch*⟩; **d)** abberufen ⟨*Botschafter, Delegation*⟩ (**from** aus). **2.** [rɪ'kɔ:l, 'ri:kɔ:l] *n.* **a)** *(ability to remember)* **[powers of]** ~: Erinnerungsvermögen, *das;* Gedächtnis, *das;* **b)** *(possibility of annulling)* **beyond** *or* **past** ~: unwiderruflich; **c)** *(summons back)* Rückruf, *der; (to active duty)* Wiedereinberufung, *die*

recant [ri:'kænt] **1.** *v.i.* [öffentlich] widerrufen. **2.** *v.t.* widerrufen

recap ['ri:kæp] *(coll.)* **1.** *v.t. & i.*, **-pp-** rekapitulieren; kurz zusammenfassen. **2.** *n.* Zusammenfassung, *die*

recapitulate [ri:kə'pɪtjʊleɪt] *v.t. & i.* rekapitulieren; kurz zusammenfassen

recapitulation [ri:kəpɪtjʊ'leɪʃn] *n.* Zusammenfassung, *die*

recapture [ri:'kæptʃə(r)] *v.t.* **a)** *(capture again)* wieder ergreifen ⟨*Gefangenen*⟩; wieder einfangen ⟨*Tier*⟩; zurückerobern ⟨*Stadt*⟩; **b)** *(re-create)* wieder lebendig werden lassen ⟨*Atmosphäre*⟩

recede [ri'si:d] *v.i.* **a)** ⟨*Hochwasser, Flut:*⟩ zurückgehen; ⟨*Küste:*⟩ zurückweichen; **his hair is beginning to** ~: er bekommt eine Stirnglatze; **b)** *(be left at increasing distance)* ~ **[into the distance]** in der Ferne verschwinden

receding [rɪ'si:dɪŋ] *adj.* fliehend ⟨*Kinn, Stirn*⟩; zurückgehend ⟨*Flut, Hochwasser*⟩

receipt [rɪ'si:t] *n.* **a)** Empfang, *der;* **please acknowledge** ~ **of this letter/order** bestätigen Sie bitte den Empfang dieses Briefes/dieser Bestellung; **be in** ~ **of** *(formal)* erhalten haben ⟨*Brief*⟩; **b)** *(written acknowledgement)* Empfangsbestätigung, *die;* Quittung, *die;* **c)** *in pl. (amount received)* Einnahmen (**from** aus)

receive [rɪ'si:v] *v.t.* **a)** *(get)* erhalten; beziehen ⟨*Gehalt, Rente*⟩; verliehen bekommen ⟨*akademischer Grad*⟩; **'payment ~d with thanks'** „Betrag dankend erhalten"; **she ~d a lot of attention/sympathy [from him]** es wurde ihr [von ihm] viel Aufmerksamkeit/Verständnis entgegengebracht; ~ **[fatal] injuries** [tödlich] verletzt werden; ~ **30 days [imprisonment]** 30 Tage Gefängnis bekommen; ~ **the sacraments/holy communion** *(Relig.)* das Abendmahl/die heilige Kommunion empfangen; **b)** *(accept)* entgegennehmen ⟨*Bukett, Lieferung*⟩; *(submit to)* über sich *(Akk.)* ergehen lassen; **be convicted for receiving [stolen goods]** *(Law)* der Hehlerei überführt werden; **c)** *(serve as receptacle for)* aufnehmen; **d)** *(greet)* reagieren auf *(Akk.)*, aufnehmen ⟨*Angebot, Nachricht, Theaterstück, Roman*⟩; empfangen ⟨*Person*⟩; **e)** *(entertain)* empfangen ⟨*Botschafter, Delegation, Nachbarn, Gast*⟩; **f)** *(Radio, Telev.)* empfangen ⟨*Sender, Signal*⟩; **are you receiving me?** können Sie mich hören?

receiver [rɪ'si:və(r)] *n.* **a)** Empfänger, *der*/Empfängerin, *die;* **b)** *(Teleph.)* [Telefon]hörer, *der;* **c)** *(Radio, Telev.)* Empfänger, *der;* Receiver, *der (Technik);* **d)** **[official]** ~ *(Law: for property of bankrupt)* [gerichtlich bestellter/bestellte] Konkursverwalter/-verwalterin; **e)** *(of stolen goods)* Hehler, *der*/Hehlerin, *die*

recent ['ri:sənt] *adj.* **a)** *(not long past)* jüngst ⟨*Ereignisse, Wahlen, Vergangenheit usw.*⟩; **the** ~ **closure of the factory** die kürzlich erfolgte Schließung der Fabrik; **at our** ~ **meeting** als wir uns kürzlich *od.* vor kurzem trafen; **a ~/more ~ survey** eine neuere Untersuchung; **at our most** ~ **meeting** bei unserer letzten Begegnung; **b)** *(not long established)* Neu⟨*auflage, -anschaffung, -erscheinung*⟩

recently ['ri:səntlɪ] *adv. (a short time ago)* neulich; kürzlich; vor kurzem; *(in the recent past)* in der letzten Zeit; **until ~/until quite** ~: bis vor kurzem/bis vor ganz kurzer Zeit; ~ **we've been following a different policy** seit kurzem verfolgen wir eine andere Politik; **as** ~ **as last year** *(last year still)* noch letztes Jahr; **as** ~ **as this morning** *(not until this morning)* [gerade] erst heute morgen

receptacle [rɪ:septəkl] *n.* Behälter, *der;* Gefäß, *das*

reception [rɪ'sepʃn] *n.* **a)** *(welcome) (of person)* Empfang, *der;* Aufnahme, *die; (of play, speech)* Aufnahme, *die;* **meet with a cool** ~: kühl aufgenommen werden; **give sb. a warm** ~: jmdn. herzlich empfangen; **b)** *(party)* Empfang, *der;* **hold** *or* **give a** ~: einen Empfang geben; **c)** *no art. (Brit.: foyer)* die Rezeption; **d)** *no art. (Radio, Telev.)* der Empfang

re'ception desk *n.* Rezeption, *die*

receptionist [rɪ'sepʃənɪst] *n.* *(in hotel)* Empfangschef, *der*/-dame, *die; (at doctor's, dentist's)* Sprechstundenhilfe, *die; (with firm)* Empfangssekretärin, *die*

receptive [rɪ'septɪv] *adj.* aufgeschlossen, empfänglich (**to** für); **have a** ~ **mind** aufgeschlossen sein

recess [rɪ'ses, 'ri:ses] *n.* **a)** *(alcove)* Nische, *die;* **b)** *(Brit. Parl.; Amer.: short vacation)* Ferien *Pl.; (Amer. Sch.: between classes)* Pause, *die*

recession [rɪ'seʃn] *n.* **a)** *(Econ.: decline)* Rezession, *die;* **b)** *(receding)* Zurückgehen, *das*

recharge [ri:'tʃɑ:dʒ] *v.t.* aufladen ⟨*Batterie*⟩; nachladen ⟨*Waffe*⟩

rechargeable [ri:'tʃɑ:dʒəbl] *adj.* wiederaufladbar

recipe ['resɪpɪ] *n. (lit. or fig.)* Rezept, *das;* ~ **for success** Erfolgsrezept, *das;* **it's a** ~ **for disaster** damit ist die Katastrophe vorprogrammiert

recipient [rɪ'sɪpɪənt] *n.* Empfänger, *der*/Empfängerin, *die*

reciprocal [rɪ'sɪprəkl] *adj.* gegenseitig ⟨*Abkommen, Zuneigung, Hilfe*⟩

reciprocate [rɪ'sɪprəkeɪt] **1.** *v.t.* austauschen ⟨*Versprechen*⟩; erwidern ⟨*Gruß, Lächeln, Abneigung, Annäherungsversuch*⟩; sich revanchieren für ⟨*Hilfe*⟩. **2.** *v.i. (respond)* sich revanchieren

recital [rɪ'saɪtl] *n.* **a)** *(performance)* [Solisten]konzert, *das; (of literature also)* Rezitation, *die;* **b)** *(detailed account)* Schilderung, *die*

recitation [resɪ'teɪʃn] *n.* Rezitation, *die*

recite [rɪ'saɪt] **1.** *v.t.* **a)** *(speak from memory)* rezitieren ⟨*Passage, Gedicht*⟩; **b)** *(give list of)* aufzählen. **2.** *v.i.* rezitieren

reckless ['reklɪs] *adj.* unbesonnen; rücksichtslos ⟨*Fahrweise*⟩; tollkühn ⟨*Fluchtversuch*⟩

reckon ['rekn] **1.** *v.t.* **a)** *(work out)* ausrechnen ⟨*Kosten, Lohn, Ausgaben*⟩; bestimmen ⟨*Position*⟩; **b)** *(conclude)* schätzen; **I** ~ **you're lucky to be alive** ich glaube, du kannst von Glück sagen, daß du noch lebst!; **I** ~ **to arrive** *or* **I shall arrive there by 8.30** ich nehme an, daß ich [spätestens] halb neun dort bin; **I usually** ~ **to arrive there by 8.30** in der Regel bin ich [spätestens] halb neun dort; **c)** *(consider)* halten (**as** für); **be ~ed as** *or* **to be sth.** als etw. gelten; **d)** *(arrive at as total)* kommen auf (+ *Akk.*). **2.** *v.i.* rechnen

~ **'in** *v.t.* [mit] einrechnen

~ **on** *see* ~ **upon**

~ **'up 1.** *v.t.* zusammenzählen. **2.** *v.i.* ~ **up with sb.** mit jmdm. abrechnen

~ **upon** *v. t.* **a)** *(rely on)* zählen auf (+ *Akk.*); **b)** *(expect)* rechnen mit

~ **with** *v. i.* **a)** *(take into account)* rechnen mit ‹*Hindernis, Möglichkeit*›; **he is a man to be ~ed with** er ist ein Mann, den man nicht unterschätzen sollte; **b)** *(deal with)* abrechnen mit

~ **without** *v. i.* nicht rechnen mit

reckoning ['rekniŋ] *n.* **a)** *(calculation)* Berechnung, *die;* **by my ~:** nach meiner Rechnung; **day of ~** *(fig.)* Tag der Abrechnung; *(moment of truth)* Stunde der Wahrheit; **be [wildly] out in one's ~:** sich [gehörig] verrechnet haben; **b)** *(bill)* Rechnung, *die*

reclaim [rɪ'kleɪm] **1.** *v. t.* **a)** urbar machen ‹*Land, Wüste*›; ~ **land from the sea** dem Meer Land abgewinnen; **b)** *(for reuse)* zur Wiederverwertung sammeln; wiederverwenden ‹*Rohstoff*›. **2.** *n.* **be past** *or* **beyond ~:** unwiederbringlich verloren sein

reclamation [reklə'meɪʃn] *n.* Urbarmachung, *die;* **land ~:** Landgewinnung, *die*

recline [rɪ'klaɪn] **1.** *v. i.* **a)** *(lean back)* sich zurücklehnen; **the chair ~s** die Rückenlehne des Sessels läßt sich [nach hinten] verstellen; **reclining seat** *(in car)* Liegesitz, *der;* **b)** *(be lying down)* liegen. **2.** *v. t.* [nach hinten] lehnen

recluse [rɪ'klu:s] *n.* Einsiedler, *der*/Einsiedlerin, *die*

recognisable, recognise *see* **recogniz-**

recognition [rekəg'nɪʃn] *n.* **a)** *no pl., no art.* Wiedererkennen, *das;* **he's changed beyond all ~:** er ist nicht mehr wiederzuerkennen; **b)** *(acceptance, acknowledgement)* Anerkennung, *die;* **achieve/receive ~:** Anerkennung finden; **in ~ of** als Anerkennung für

recognizable ['rekəgnaɪzəbl] *adj.* erkennbar; deutlich ‹*Unterschied*›; **be ~:** wiederzuerkennen sein

recognize ['rekəgnaɪz] *v. t.* **a)** *(know again)* wiedererkennen (**by** an + *Dat.*, **from** durch); **b)** *(acknowledge)* erkennen; anerkennen ‹*Gültigkeit, Land, Methode, Leistung, Bedeutung, Dienst*›; **be ~d as** angesehen werden *od.* gelten als; **c)** *(admit)* zugeben; **d)** *(identify nature of)* erkennen; ~ **sb. to be a fraud** erkennen, daß jmd. ein Betrüger ist

recoil 1. [rɪ'kɔɪl] *v. i.* **a)** *(shrink back)* zurückfahren; ~ **from an idea** vor einem Gedanken zurückschrecken; **b)** ‹*Waffe:*› einen Rückstoß haben. **2.** ['ri:kɔɪl, rɪ'kɔɪl] *n.* Rückstoß, *der*

recollect [rekə'lekt] *v. t.* sich erinnern an (+ *Akk.*); ~ **meeting sb.** sich daran erinnern, jmdn. getroffen zu haben

recollection [rekə'lekʃn] *n.* Erinnerung, *die;* **have a/no ~ of sth.** sich an etw. *(Akk.)* erinnern/nicht erinnern können

recommend [rekə'mend] *v. t.* **a)** empfehlen; ~ **sb. to do sth.** jmdm. empfehlen, etw. zu tun; **b)** *(make acceptable)* sprechen für; **the plan has little/nothing to ~ it** es spricht wenig/nichts für den Plan

recommendation [rekəmen'deɪʃn] *n.* Empfehlung, *die;* **on sb.'s ~:** auf jmds. Empfehlung *(Akk.)*

recompense ['rekəmpens] *(formal)* **1.** *v. t.* **a)** *(reward)* belohnen; **b)** *(make amends to)* entschädigen. **2.** *n., no art., no pl.* **a)** *(reward)* Anerkennung, *die;* **b)** *(compensation)* Entschädigung, *die*

reconcile ['rekənsaɪl] *v. t.* **a)** *(restore to friendship)* versöhnen; **become ~d** sich versöhnen; **b)** *(resign oneself)* ~ **oneself** *or* **become/be ~d to sth.** sich mit etw. versöhnen; **c)** *(make compatible)* in Einklang bringen ‹*Vorstellungen, Überzeugungen*›; *(show to be compatible)* miteinander vereinen; **d)** *(settle)* beilegen ‹*Meinungsverschiedenheit*›

reconciliation [rekənsɪlɪ'eɪʃn] *n.* **a)** *(restoring to friendship)* Versöhnung, *die;* **b)** *(making compatible)* Harmonisierung, *die*

recondition [ri:kən'dɪʃn] *v. t.* [generalüberholen; **~ed engine** Austauschmotor, *der*

reconnaissance [rɪ'kɒnɪsəns] *n., no pl., no def. art. (Mil.)* Aufklärung, *die; (of area)* Erkundung, *die;* **the plane was on ~:** das Flugzeug war auf einem Aufklärungsflug; *attrib.* ~ **aircraft** Aufklärungsflugzeug, *das*

reconnoitre (*Brit.; Amer.:* **reconnoiter**) [rekə'nɔɪtə(r)] **1.** *v. t. (esp. Mil.)* auskundschaften; erkunden ‹*Gelände*›; *(fig.)* erkunden. **2.** *v. i. (esp. Mil.)* auf Erkundung [aus]gehen; *(fig.)* sich umsehen

reconsider [ri:kən'sɪdə(r)] *v. t.* [noch einmal] überdenken; ~ **a case** einen Fall von neuem aufrollen; *abs.* **there is still time to ~:** du kannst es dir/wir können es uns *usw.* immer noch überlegen

reconstruct [ri:kən'strʌkt] *v. t.* *(build again)* wieder aufbauen ‹*Stadt, Gebäude*›; neu errichten ‹*Gerüst*›; rekonstruieren ‹*Anlage*›; *(fig.)* rekonstruieren

reconstruction [ri:kən'strʌkʃn] *n.* **a)** *(process)* Wiederaufbau, *der;* **b)** *(thing reconstructed)* Rekonstruktion, *die*

record 1. [rɪ'kɔ:d] *v. t.* **a)** aufzeichnen; ~ **a new LP** eine neue LP aufnehmen; ~ **sth. in a book/painting** etw. in einem Buch/auf einem Gemälde festhalten; **b)** *(register officially)* dokumentieren; protokollieren ‹*Verhandlung*›. **2.** *v. i.* aufzeichnen; *(on tape)* Tonbandaufnahmen/eine Tonbandaufnahme machen. **3.** ['rekɔ:d] *n.* **a)** **be on ~** ‹*Prozeß, Verhandlung, Besprechung:*› protokolliert sein; **there is no such case on ~:** ein solcher Fall ist nicht dokumentiert; **it is on ~ that ...:** es ist dokumentiert, daß ...; **have sth. on ~:** etw. dokumentiert haben; **put sth. on ~:** etw. schriftlich festhalten; **b)** *(report)* Protokoll, *das; (Law: official report)* [Gerichts]akte, *die;* **c)** *(document)* Dokument, *das; (piece of evidence)* Zeugnis, *das;* Beleg, *der;* **medical ~s** medizinische Unterlagen; **for the ~:** für das Protokoll; **just for the ~:** der Vollständigkeit halber; *(iron.)* nur der Ordnung halber; **[strictly] off the ~:** [ganz] inoffiziell; **get** *or* **keep** *or* **put** *or* **set the ~ straight** keine Mißverständnisse aufkommen lassen; **d)** *(disc for gramophone)* [Schall]platte, *die;* **e)** *(facts of sb.'s/sth.'s past)* Ruf, *der;* **have a good ~ [of achievements]** gute Leistungen vorweisen können; **have a [criminal/police] ~:** vorbestraft sein; **f)** *(best performance)* Rekord, *der;* **set a ~:** einen Rekord aufstellen; **break** *or* **beat the ~:** den Rekord brechen. **4.** *attrib. adj.* Rekord-

'record-breaking *adj.* Rekord-

recorded [rɪ'kɔ:dɪd] *adj.* aufgezeichnet ‹*Film, Konzert, Rede*›; überliefert ‹*Ereignis, Geschichte*›; bespielt ‹*Band*›; ~ **music** Musikaufnahmen

recorded de'livery *n. (Brit. Post)* eingeschriebene Sendung *(ohne Versicherung)*

recorder [rɪ'kɔ:də(r)] *n.* **a)** *(instrument/apparatus)* Aufzeichnungsgerät, *das;* **b)** *see* **tape recorder; c)** *(Mus.)* Blockflöte, *die*

'record-holder *n. (Sport)* Rekordhalter, *der*/-halterin, *die*

recording [rɪ'kɔ:dɪŋ] *n. (what is recorded)* Aufnahme, *die; (to be*

heard or seen later) Aufzeichnung, *die*
recording: ~ **head** *n.* Aufnahmekopf, *der;* ~ **session** *n.* Aufnahme, *die;* ~ **studio** *n.* Tonstudio, *das*
record ['rekɔ:d]: ~ **library** *n.* Phonothek, *die;* ~**-player** *n.* Plattenspieler, *der;* ~ **sleeve** *n.* Plattenhülle, *die;* Plattencover, *das;* ~ **token** *n.* [Schall]plattengutschein, *der*
recount [rɪ'kaʊnt] *v. t. (tell)* erzählen
re-count 1. [ri:'kaʊnt] *v. t. (count again)* [noch einmal] nachzählen. **2.** ['ri:kaʊnt] *n.* Nachzählung, *die;* **have a** ~: nachzählen
recoup [rɪ'ku:p] *v. t. (regain)* ausgleichen ‹*Verlust*›; [wieder] hereinbekommen ‹*[Geld]einsatz*›
recourse [rɪ'kɔ:s] *n.* **a)** *(resort)* Zufluchtnahme, *die;* **have** ~ **to sb./sth.** bei jmdm./zu etw. Zuflucht nehmen; **b)** *(person or thing resorted to)* Zuflucht, *die*
recover [rɪ'kʌvə(r)] **1.** *v. t.* **a)** *(regain)* zurückerobern; **b)** *(find again)* wiederfinden ‹*Verlorenes, Fährte, Spur*›; **c)** *(retrieve)* zurückbekommen; bergen ‹*Wrack*›; **d)** *(make up for)* aufholen ‹*verlorene Zeit*›; **e)** *(acquire again)* wiedergewinnen ‹*Vertrauen*›; wiederfinden ‹*Gleichgewicht, innere Ruhe usw.*›; ~ **consciousness** das Bewußtsein wiedererlangen; ~ **one's senses** *(lit. or fig.)* wieder zur Besinnung kommen; ~ **one's sight** sein Sehvermögen wiedergewinnen; ~ **one's breath** wieder zu Atem kommen; **f)** *(reclaim)* ~ **land from the sea** dem Meer Land abgewinnen; ~ **metal from scrap** Metall aus Schrott gewinnen; **g)** *(Law)* erheben ‹*Steuer, Abgabe*›; erhalten ‹*Schadenersatz, Schmerzensgeld*›. **2.** *v. i.* ~ **from sth.** sich von etw. [wieder] erholen; **how long will it take him to** ~**?** wann wird er wieder gesund sein?; **be [completely** *or* **fully** ~**ed** [völlig] wiederhergestellt sein
re-cover [ri:'kʌvə(r)] *v. t.* neu beziehen ‹*Sessel, Schirm usw.*›
recovery [rɪ'kʌvərɪ] *n.* **a)** *(after illness)* Erholung, *die;* **make a quick/good** ~: sich schnell/gut erholen; **he is past** ~: für ihn gibt es keine Hoffnung mehr; **b)** *(of sth. lost)* Wiederfinden, *das;* **c)** *(of raw materials)* Rückgewinnung, *die*
recreation [rekrɪ'eɪʃn] *n. (means of entertainment)* Freizeitbeschäftigung, *die;* Hobby, *das;* **for** *or* **as a** ~: zur Freizeitbeschäftigung *od.* Entspannung
recreational [rekrɪ'eɪʃənl] *adj.* Freizeit‹*wert, -möglichkeiten, -gelände*›; Erholungs‹*gebiet*›
recre'ation ground *n.* Freizeitgelände, *das*
recrimination [rɪkrɪmɪ'neɪʃn] *n.* Gegenbeschuldigung, *die; (counter-accusation)* **[mutual]** ~**s** [gegenseitige] Beschuldigungen
recruit [rɪ'kru:t] **1.** *n.* **a)** *(Mil.)* Rekrut, *der;* **a raw** ~: ein frisch Eingezogener; **b)** *(new member)* neues Mitglied; **c) [raw]** ~ *(fig.: novice)* blutiger Anfänger. **2.** *v. t.* **a)** *(Mil.: enlist)* anwerben; *(into society, party, etc.)* werben ‹*Mitglied*›; **b)** *(select for appointment)* neu einstellen
recruitment [ri'kru:tmənt] *n.* **a)** *(Mil.)* Anwerbung, *die; (for membership)* ~ **of members** Mitgliederwerbung, *die;* **b)** *(process of selecting for appointment)* Neueinstellung, *die*
recta *pl. of* **rectum**
rectangle ['rektæŋgl] *n.* Rechteck, *das*
rectangular [rek'tæŋgjʊlə(r)] *adj.* rechteckig
rectify ['rektɪfaɪ] *v. t.* korrigieren ‹*Fehler, Berechnung, Kurs*›; richtigstellen ‹*Bemerkung, Sachverhalt*›; Abhilfe schaffen (+ *Dat.*) ‹*Mangel, Mißstand*›
rector ['rektə(r)] *n.* **a)** Pfarrer, *der;* **b)** *(Univ.)* Rektor, *der*/Rektorin, *die*
rectory ['rektərɪ] *n.* Pfarrhaus, *das*
rectum ['rektəm] *n., pl.* ~**s** *or* **recta** ['rektə] *(Anat.)* Mastdarm, *der;* Rektum, *das (fachspr.)*
recuperate [rɪ'kju:pəreɪt] **1.** *v. i.* sich erholen. **2.** *v. t.* wiederherstellen ‹*Gesundheit*›
recuperation [rɪkju:pə'reɪʃn] *n.* Erholung, *die*
recur [rɪ'kɜ:(r)] *v. i.*, **-rr-: a)** sich wiederholen; ‹*Krankheit, Beschwerden usw.:*› wiederkehren; ‹*Problem, Symptom:*› wieder auftreten; **b)** *(return to one's mind)* ‹*Gedanke, Furcht, Gefühl:*› wiederkehren; **c)** *(Math.)* **2.3** ~**ring** 2 Komma 3 Periode
recurrence [rɪ'kʌrəns] *n.* Wiederholung, *die; (of illness, complaint)* Wiederkehr, *die; (of problem, symptom)* Wiederauftreten, *das*
recurrent [rɪ'kʌrənt] *adj.* immer wiederkehrend; wiederholt ‹*Hinweis, Bezugnahme*›
recycle [ri:'saɪkl] *v. t. (reuse)* wiederverwerten ‹*Papier, Glas, Abfall*›; *(convert)* wiederaufbereiten ‹*Metall, Brauchwasser, Abfall*›
recycling [ri:'saɪklɪŋ] *n.* Recycling, *das;* Wiederaufbereitung, *die*
red [red] **1.** *adj.* **a)** rot; Rot‹*wild, -buche*›; rotglühend ‹*Feuer, Lava usw.*›; **go** ~ **with shame** rot vor Scham werden; **go** ~ **in the face** rot werden; **as** ~ **as a beetroot** puterrot; rot wie eine Tomate *(ugs. scherzh.)*; **her eyes were** ~ **with crying** sie hatte rotgeweinte Augen; *see also* **paint 2 a; see 2 a; b) Red** *(Soviet Russian)* rot, kommunistisch ‹*Soldat, Propaganda*›; **the Red Army** die Rote Armee. **2.** *n.* **a)** *(colour, traffic light)* Rot, *das;* **underline sth. in** ~: etw. rot unterstreichen; **b)** *(debt)* **[be] in the** ~: in den roten Zahlen [sein]; **c) Red** *(communist)* Rote, *der/die*
red: ~**-blooded** ['redblʌdɪd] *adj.* heißblütig; ~**brick** *adj. (Brit.)* weniger traditionsreich ‹*Universität*›; ~ **'carpet** *n. (lit. or fig.)* roter Teppich; **Red 'Cross** *n.* Rotes Kreuz; ~ **'currant** *n.* [rote] Johannisbeere
redden ['redn] **1.** *v. i.* ‹*Gesicht, Himmel:*› sich röten; ‹*Person:*› rot werden, erröten; ‹*Blätter, Wasser:*› sich rot färben. **2.** *v. t.* rot färben; röten *(geh.)*
reddish ['redɪʃ] *adj.* rötlich
redecorate [ri:'dekəreɪt] *v. t.* renovieren; *(with wallpaper)* neu tapezieren; *(with paint)* neu streichen
redeem [rɪ'di:m] *v. t.* **a)** *(regain)* wiederherstellen ‹*Ehre, Gesundheit*›; wiedergewinnen ‹*Position*›; **b)** *(buy back)* tilgen ‹*Hypothek*›; [wieder] einlösen ‹*Pfand*›; abzahlen ‹*Grundstück*›; **c)** *(convert)* einlösen ‹*Gutschein, Coupon*›; **d)** *(make amends for)* ausgleichen, wettmachen ‹*Fehler, Schuld usw.*›; **he has one** ~**ing feature** man muß ihm eins zugute halten; **e)** *(repay)* abzahlen ‹*Schuld, Kredit*›; **f)** *(save)* retten; **g)** ~ **oneself** sich freikaufen
Redeemer [rɪ'di:mə(r)] *n. (Relig.)* Erlöser, *der;* Heiland, *der*
redemption [rɪ'dempʃn] *n.* **a)** *(of pawned goods)* Einlösen, *das;* Rückkauf, *der;* **b)** *(of tokens, trading stamps, stocks, etc.)* Einlösen, *das;* **c)** *(of mortgage, debt)* Tilgung, *die; (of land)* Abzahlung, *die;* **d)** *(of person, country)* Befreiung, *die;* **he's past** *or* **beyond** ~: für ihn gibt es keine Rettung mehr; **e)** *(deliverance from sin)* Erlösung, *die*
redeploy [ri:dɪ'plɔɪ] *v. t.* umsta-

tionieren ‹*Truppen, Raketen*›; woanders einsetzen ‹*Arbeitskräfte*›

red: ~**-eyed** *adj.* **be ~-eyed** ['redaɪd] rote Augen haben; ~**-faced** ['redfeɪst] *adj.* rotgesichtig; **be ~-faced** *(with rage/embarrassment)* ein [hoch]rotes Gesicht haben/vor Verlegenheit rot werden; ~**-haired** *adj.* rothaarig; ~**-'handed** *adj.* **catch sb. ~-handed** jmdn. auf frischer Tat ertappen; ~**head** *n.* Rotschopf, *der (ugs.);* Rothaarige, *der/die;* ~**-headed** *adj.* rothaarig; ~ **'herring** *n. (fig.)* Ablenkungsmanöver, *das; (in thriller, historical research)* falsche Fährte; ~**-hot** *adj.* **a)** [rot]glühend; **b)** *(fig.)* glühend ‹*Anhänger, Gläubiger*›; heiß ‹*Blondine, Thema, Musik*›; brandaktuell ‹*Nachricht*›

redid *see* **redo**

Red 'Indian *(Brit.)* **1.** *n.* Indianer, *der*/Indianerin, *die.* **2.** *adj.* Indianer-

redirect [ri:daɪ'rekt, ri:dɪ'rekt] *v.t.* nachsenden ‹*Post, Brief usw.*›; umleiten ‹*Verkehr*›; weiterleiten (**to** an + *Akk.*) ‹*Anfrage*›

rediscover [ri:dɪ'skʌvə(r)] *v.t.* wiederentdecken

redistribute [ri:dɪ'strɪbju:t] *v.t.* umverteilen

red: ~**-'letter day** *n. (memorable day)* im Kalender rot anzustreichender Tag; großer Tag; ~ **'light** *n.* **a)** [rotes] Warnlicht; *(of traffic-lights)* rote [Verkehrs]ampel; **b)** *(fig.)* Warnzeichen, *das;* ~**-'light district** *n.* Amüsierviertel, *das;* Strich, *der (salopp)*

redness ['rednɪs] *n., no pl. (of face, skin, eyes, sky)* Röte, *die; (of blood, fire, rose, dress, light)* rote Farbe

redo [ri:'du:] *v.t.*, **redoes** [ri:'dʌz], **redoing** [ri:'du:ɪŋ], **redid** [ri:'dɪd], **redone** [ri:'dʌn] *(do again)* wiederholen ‹*Prüfung, Spiel, Test*›; neu frisieren ‹*Haare*›; erneuern ‹*Make-up, Lidschatten*›; noch einmal machen ‹*Bett, Hausaufgabe*›; überarbeiten ‹*Aufsatz, Übersetzung, Komposition*›

redone *see* **redo**

redouble [ri:'dʌbl] **1.** *v.t.* verdoppeln. **2.** *v.i.* sich verdoppeln

redoubtable [rɪ'daʊtəbl] *adj.* ehrfurchtgebietend ‹*Person*›; gewaltig ‹*Aufgabe, Pflicht usw.*›; gefürchtet ‹*Gegner, Krieger*›

red: ~ **'pepper** *n.* **a)** *see* **cayenne; b)** *(vegetable) see* **pepper 1 b;** ~ **'rag** *n. (fig.)* rotes Tuch (**to** für); **be like a ~ rag to a bull [to sb.]** wie ein rotes Tuch [auf jmdn.] wirken

redress [rɪ'dres] **1.** *n. (reparation, correction)* Entschädigung, *die;* **seek ~ for sth.** eine Entschädigung für etw. verlangen; **seek [legal] ~:** auf Schadenersatz klagen; **have no ~:** keine Entschädigung erhalten; *(Law)* keinen Rechtsanspruch auf Entschädigung haben. **2.** *v.t.* **a)** *(adjust again)* ins Gleichgewicht bringen; ~ **the balance** das Gleichgewicht wiederherstellen; **b)** *(set right, rectify)* wiedergutmachen ‹*Unrecht*›; ausgleichen ‹*Ungerechtigkeiten*›; abhelfen (+ *Dat.*) ‹*Beschwerden, Mißbrauch*›

Red 'Sea *pr. n.* Rote Meer, *das*

red: ~**skin** *see* **Red Indian 1;** ~ **'squirrel** *n.* Eichhörnchen, *das;* ~ **'tape** *n. (fig.)* [unnötige] Bürokratie

reduce [rɪ'dju:s] *v.t.* **a)** *(diminish)* senken ‹*Preis, Gebühr, Fieber, Aufwendungen, Blutdruck usw.*›; verbilligen ‹*Ware*›; reduzieren ‹*Geschwindigkeit, Gewicht, Anzahl, Menge, Preis*›; **at ~d prices** zu herabgesetzten Preisen; **b)** ~ **to despair/silence/tears** in Verzweiflung stürzen/verstummen lassen/zum Weinen bringen; ~ **sb. to begging** jmdn. an den Bettelstab bringen; **be ~d to starvation** hungern müssen

reduction [rɪ'dʌkʃn] *n.* **a)** *(amount, process) (in price, costs, wages, rates, speed, etc.)* Senkung, *die* (**in** *Gen.*); *(in numbers, output, etc.)* Verringerung, *die* (**in** *Gen.*); ~ **in prices/wages** Preis-/Lohnsenkung, *die;* **there is a ~ on all furniture** alle Möbel sind im Preis heruntergesetzt; **a ~ of £10** ein Preisnachlaß von 10 Pfund; **b)** *(smaller copy)* Verkleinerung, *die*

redundancy [rɪ'dʌndənsɪ] *n.* **a)** *(Brit.)* Arbeitslosigkeit, *die;* **redundancies** Entlassungen; **b)** *(being more than needed)* Überfluß, *der*

re'dundancy payment *n.* Abfindung, *die*

redundant [rɪ'dʌndənt] *adj.* **a)** *(Brit.: now unemployed)* arbeitslos; **be made** *or* **become ~:** den Arbeitsplatz verlieren; **make ~:** entlassen; **b)** *(more than needed)* überflüssig

red 'wine *n.* Rotwein, *der*

reed [ri:d] *n.* **a)** *(Bot.)* Schilf[rohr], *das;* Ried, *das;* **b)** *(Mus.: part of instrument)* Rohrblatt, *das*

'reed instrument *n. (Mus.)* Rohrblattinstrument, *das*

reef [ri:f] *n. (ridge)* Riff, *das*

'reef-knot *n.* Kreuzknoten, *der*

reek [ri:k] **1.** *n.* Geruch, *der;* Gestank, *der (abwertend).* **2.** *v.i.* riechen, *(abwertend)* stinken (**of** nach)

reel [ri:l] **1.** *n.* **a)** *(roller, cylinder)* ‹*Papier-, Schlauch-, Garn-, Angel*›rolle, *die;* ‹*Film-, Tonband-, Garn*›spule, *die;* **b)** *(quantity)* Rolle, *die;* **c)** *(dance, music)* Reel, *der.* **2.** *v.t.* ~ **[up]** *(wind on)* aufspulen. **3.** *v.i.* **a)** *(be in a whirl)* sich drehen; **his head was ~ing** in seinem Kopf drehte sich alles; **b)** *(sway)* torkeln; *(fig.: be shaken)* taumeln

~ **'in** *v.t.* an Land ziehen ‹*Fisch*›

~ **'off** *v.t. (say rapidly)* herunterleiern *(ugs. abwertend),* hersagen ‹*Geschichte*›; *(without apparent effort)* abspulen *(ugs.)* ‹*Gedicht, Namen, Einzelheiten*›

re-elect [ri:ɪ'lekt] *v.t.* wiederwählen

re-election [ri:ɪ'lekʃn] *n.* Wiederwahl, *die*

re-enact [ri:ɪ'nækt] *v.t.* nachspielen ‹*Szene, Schlacht*›; ~ **a crime** den Hergang eines Verbrechens nachspielen

re-enter [ri:'entə(r)] **1.** *v.i.* **a)** wieder eintreten; **b)** *(for race, exam, etc.)* wieder antreten. **2.** *v.t.* wieder betreten ‹*Raum, Gebäude*›; wieder eintreffen in (+ *Dat.*) ‹*Ortschaft*›; wieder einreisen in (+ *Akk.*) ‹*Land*›; wiedereintreten in (+ *Akk.*) ‹*Erdatmosphäre*›

re-entry [ri:'entrɪ] *n.* Wiedereintreten, *das; (into country)* Wiedereinreise, *die; (of spacecraft)* Wiedereintritt, *der*

ref [ref] *n. (Sport coll.)* Schiri, *der (Sportjargon); (Boxing)* Ringrichter, *der*

ref. *abbr.* **our/your ~:** unser/Ihr Zeichen

refashion [ri:'fæʃn] *v.t.* umgestalten

refectory [rɪ'fektərɪ] *n.* Mensa, *die*

refer [rɪ'fɜ:(r)] **1.** *v.i.*, **-rr-: a)** ~ **to** *(allude to)* sich beziehen auf (+ *Akk.*) ‹*Buch, Person usw.*›; *(speak of)* sprechen von ‹*Person, Problem, Ereignis usw.*›; **b)** ~ **to** *(apply to, relate to)* betreffen; ‹*Beschreibung:*› sich beziehen auf (+ *Akk.*); **does that remark ~ to me?** gilt diese Bemerkung mir?; **c)** ~ **to** *(consult, cite as proof)* konsultieren *(geh.);* nachsehen in (+ *Dat.*). **2.** *v.t.*, **-rr-** *(send on to)* ~ **sb./sth. to sb./sth.** jmdn./etw. an jmdn./auf etw. *(Akk.)* verweisen; ~ **a patient to a specialist** einen Patienten an einen Facharzt überweisen; ~ **sb. to a paragraph/**

an article jmdn. auf einen Absatz/Artikel aufmerksam machen

referee [refə'ri:] **1.** *n.* **a)** *(Sport: umpire)* Schiedsrichter, *der*/-richterin, *die; (Boxing)* Ringrichter, *der; (Wrestling)* Kampfrichter, *der;* **b)** *(Brit.) see* **reference e; c)** *(person who assesses)* Gutachter, *der*/Gutachterin, *die.* **2.** *v.t. (Sport: umpire)* als Schiedsrichter/-richterin leiten; ~ **a football game** ein Fußballspiel pfeifen *od.* leiten. **3.** *v.i. (Sport: umpire)* Schiedsrichter/-richterin sein

reference ['refrəns] *n.* **a)** *(allusion)* Hinweis, *der* (**to** auf + *Akk.*); **make [several] ~[s] to sth.** sich [mehrfach] auf etw. *(Akk.)* beziehen; **make no ~ to sth.** etw. nicht ansprechen; **b)** *(note directing reader)* Verweis, *der* (**to** auf + *Akk.*); **c)** *(cited book, passage)* Quellenangabe, *die;* **d)** *(testimonial)* Zeugnis, *das;* Referenz, *die;* **character ~:** persönliche Referenzen; **give sb. a good ~:** jmdm. ein gutes Zeugnis ausstellen; **e)** *(person willing to testify)* Referenz, *die;* **quote sb. as one's ~:** jmdn. als Referenz angeben; **f)** *(act of referring)* Konsultation, *die* (**to** *Gen.*) *(geh.);* ~ **to a dictionary/map** Nachschlagen in einem Wörterbuch/Nachsehen auf einer Karte; **work of ~:** Nachschlagewerk, *das*

reference: ~ book *n.* Nachschlagewerk, *das;* ~ **mark** *n.* Verweiszeichen, *das*

referendum [refə'rendəm] *n., pl.* **~s** *or* **referenda** [refə'rendə] Volksentscheid, *der;* Referendum, *das*

refill 1. [ri:'fıl] *v.t.* nachfüllen ‹*Glas, Feuerzeug*›; neu füllen ‹*Kissen*›; mit einer neuen Füllung versehen ‹*Zahn*›; ~ **the glasses** nachschenken. **2.** ['ri:fıl] *n.* **a)** *(cartridge)* [Nachfüll]patrone, *die; (for ball-pen)* Ersatzmine, *die;* **b) can I have a ~?** *(coll.)* gießt du mir noch einmal nach?

refine [rı'faın] **1.** *v.t.* **a)** *(purify)* raffinieren; **b)** *(make cultured)* kultivieren; **c)** *(improve)* verbessern; verfeinern ‹*Stil, Technik*›. **2.** *v.i.* **a)** *(become pure)* rein werden; **b)** *(become more cultured)* sich verfeinern

refined [rı'faınd] *adj.* **a)** *(purified)* raffiniert; Fein‹*kupfer, -silber usw.*›; ~ **sugar** [Zucker]raffinade, *die;* **b)** *(cultured)* kultiviert

refinement [rı'faınmənt] *n.* **a)** *(purifying)* Raffination, *die;* **b)** *(fineness of feeling, elegance)* Kultiviertheit, *die;* **c)** *(improvement)* Verbesserung, *die;* Weiterentwicklung, *die* (**[up]on** *Gen.*)

refinery [rı'faınərı] *n.* Raffinerie, *die*

refit 1. [ri:'fıt] *v.t.,* **-tt-** überholen; reparieren; *(equip with new things)* neu ausstatten. **2.** *(renew supplies or equipment)* sich neu ausrüsten. **3.** ['ri:fıt] *n.* Überholung, *die; (with supplies or equipment)* Neuausstattung, *die*

reflate [ri:'fleıt] *v.t. (Econ.)* ankurbeln ‹*Wirtschaft, Konjunktur*›

reflation [ri:'fleıʃn] *n. (Econ.)* Reflation, *die*

reflect [rı'flekt] **1.** *v.t.* **a)** *(throw back)* reflektieren; **b)** *(reproduce)* spiegeln; *(fig.)* widerspiegeln ‹*Ansichten, Gefühle, Werte*›; **be ~ed** sich spiegeln; **c)** *(contemplate)* nachdenken über (+ *Akk.*). **2.** *v.i. (meditate)* nachdenken

~ **[up]on** *v.t. (consider, contemplate)* nachdenken über (+ *Akk.*); abwägen ‹*Konsequenzen*›

reflection [rı'flekʃn] *n.* **a)** *(of light etc.)* Reflexion, *die; (by surface of water etc.)* Spiegelung, *die;* **b)** *(reflected light, heat, or colour)* Reflexion, *die; (image, lit. or fig.)* Spiegelbild, *das;* **c)** *(meditation, consideration)* Nachdenken, *das* (**upon** über + *Akk.*); **be lost in ~:** in Gedanken versunken sein; **on ~:** bei weiterem Nachdenken; **on ~, I think ...:** wenn ich mir das recht überlege, [so] glaube ich ...; **d)** *(remark)* Reflexion, die *(geh.),* Betrachtung, *die* (**on** über + *Akk.*)

reflective [rı'flektıv] *adj.* **a)** reflektierend; **be ~:** reflektieren; **b)** *(thoughtful)* nachdenklich

reflector [rı'flektə(r)] *n.* **a)** Rückstrahler, *der;* **b)** *(telescope)* Reflektor, *der*

reflex ['ri:fleks] **1.** *n. (Physiol.)* Reflex, *der.* **2.** *adj. (by reflection)* Reflex-

reflex: ~ action *n. (Physiol.)* Reflexhandlung, *die;* ~ **camera** *n. (Photog.)* Spiegelreflexkamera, *die*

reflexive [rı'fleksıv] *adj. (Ling.)* reflexiv

reflex re'action *n. (Physiol.; also fig.)* Reflexreaktion, *die*

refloat [ri:'fləʊt] *v.t.* [wieder] flottmachen ‹*Schiff*›; *(fig.)* wieder flüssig machen *(ugs.)*

reform [rı'fɔ:m] **1.** *v.t.* **a)** *(make better)* bessern ‹*Person*›; reformieren ‹*Institution*›; **b)** *(abolish)* ~ **sth.** mit etw. aufräumen. **2.** *v.i.* sich bessern. **3.** *n. (of person)* Besserung, *die; (in a system)* Reform, *die* (**in** *Gen.*)

re-form [ri:'fɔ:m] **1.** *v.t.* neu gründen ‹*Gesellschaft usw.*›. **2.** *v.i.* sich neu bilden; ‹*Band, Gesellschaft:*› neu gegründet werden

reformation [refə'meıʃn] *n. (of person, character)* Wandlung, *die* (**in** + *Gen.*); **the R~** *(Hist.)* die Reformation

reformed [rı'fɔ:md] *adj.* gewandelt; **he's a ~ character** er hat sich positiv verändert

reformer [rı'fɔ:mə(r)] *n.* **[political] ~:** Reformpolitiker, *der*/-politikerin, *die*

refraction [rı'frækʃn] *n. (Phys.)* Brechung, *die*

refractor [rı'fræktə(r)] *n. (telescope)* Refraktor, *der*

refractory [rı'fræktərı] *adj.* **a)** *(stubborn)* störrisch; widerspenstig; **b)** *(heat-resistant)* hitzebeständig

[1]**refrain** [rı'freın] *n.* Refrain, *der*

[2]**refrain** *v.i.* ~ **from doing sth.** es unterlassen, etw. zu tun; **'please ~ from smoking'** „bitte nicht rauchen"; **he ~ed from comment** er enthielt sich jeden Kommentars *(geh.)*

refresh [rı'freʃ] *v.t.* **a)** erfrischen; *(with food and/or drink)* stärken; ~ **oneself** *(with rest)* sich ausruhen; *(with food and/or drink)* sich stärken; **b)** auffrischen ‹*Wissen*›; **let me ~ your memory** lassen Sie mich Ihrem Gedächtnis nachhelfen

refreshing [rı'freʃıŋ] *adj.* **a)** wohltuend ‹*Abwechslung, Ruhe*›; erfrischend ‹*Brise, Schlaf, Getränk*›; **b)** *(interesting)* erfrischend

refreshment [rı'freʃmənt] *n.* Erfrischung, *die*

refreshment: ~ room *n.* Imbißstube, *die;* ~ **stall** *n.* Erfrischungsstand, *der*

refrigerate [rı'frıdʒəreıt] *v.t.* **a)** kühl lagern ‹*Lebensmittel*›; **b)** *(chill)* kühlen; *(freeze)* einfrieren; **c)** *(make cool)* abkühlen ‹*Luft*›

refrigeration [rıfrıdʒə'reıʃn] *n.* kühle Lagerung; *(chilling)* Kühlung, *die; (freezing)* Einfrieren, *das*

refrigerator [rı'frıdʒəreıtə(r)] *n.* Kühlschrank, *der*

refuel [ri:'fju:əl], *(Brit.)* **-ll-: 1.** *v.t.* auftanken. **2.** *v.i.* [auf]tanken

refuge ['refju:dʒ] *n.* Zuflucht, *die;* **take ~ in** Schutz *od.* Zuflucht suchen in (+ *Dat.*) (**from** vor + *Dat.*); **women's ~:** Frauenhaus, *das*

refugee [refjʊ'dʒi:] *n.* Flüchtling, *der;* **economic ~:** Wirtschaftsflüchtling, *der*

refu'gee camp *n.* Flüchtlingslager, *das*
refund 1. [ri:'fʌnd] *v. t. (pay back)* zurückzahlen ⟨*Geld, Schulden*⟩; erstatten ⟨*Kosten*⟩. **2.** ['ri:fʌnd] *n.* Rückzahlung, *die; (of expenses)* [Rück]erstattung, *die;* **obtain a ~ of sth.** etw. zurückbekommen
refurbish [ri:'fɜ:bɪʃ] *v. t.* renovieren ⟨*Haus*⟩; aufarbeiten ⟨*Kleidung*⟩; aufpolieren ⟨*Möbel*⟩
refusal [rɪ'fju:zl] *n.* Ablehnung, *die; (after a period of time)* Absage, *die; (of admittance, entry, permission)* Verweigerung, *die;* **~ to do sth.** Weigerung, etw. zu tun; **have/get [the] first ~ on sth.** das Vorkaufsrecht für etw. haben/eingeräumt bekommen
[1]**refuse** [rɪ'fju:z] **1.** *v. t.* **a)** ablehnen; abweisen ⟨*Heiratsantrag*⟩; verweigern ⟨*Nahrung, Befehl, Bewilligung, Zutritt, Einreise, Erlaubnis*⟩; **~ sb. admittance/entry/permission** jmdm. den Zutritt/die Einreise/die Erlaubnis verweigern; **~ to do sth.** sich weigern, etw. zu tun; **b)** *(not oblige)* abweisen ⟨*Person*⟩; **c)** ⟨*Pferd:*⟩ verweigern ⟨*Hindernis*⟩. **2.** *v. i.* **a)** ablehnen; *(after request)* sich weigern; **b)** ⟨*Pferd:*⟩ verweigern
[2]**refuse** ['refju:s] *n.* Müll, *der;* Abfall, *der*
refuse ['refju:s]: **~ collection** *n.* Müllabfuhr, *die;* **~ collector** *n.* Müllwerker, *der;* **~ disposal** *n.* Abfallbeseitigung, *die;* **~ heap** *n.* Müllhaufen, *der*
refute [rɪ'fju:t] *v. t.* widerlegen
regain [rɪ'geɪn] *v. t.* zurückgewinnen ⟨*Zuversicht, Vertrauen, Augenlicht*⟩; zurückerobern ⟨*Gebiet*⟩; **~ control of sth.** etw. wieder unter Kontrolle bringen; *see also* **consciousness a**
regal ['ri:gl] *adj.* **a)** *(magnificent, stately)* majestätisch ⟨*Person, Baum, Art, Tier, Würde*⟩; groß ⟨*Luxus*⟩; **b)** *(royal)* königlich
regale [rɪ'geɪl] *v. t. (entertain)* verwöhnen (**with, on** mit); **~ sb. with stories** jmdn. mit Geschichten unterhalten
regalia [rɪ'geɪlɪə] *n. pl.* **a)** *(of royalty)* Krönungsinsignien; **b)** *(of order)* Ordensinsignien
regard [rɪ'gɑ:d] **1.** *v. t.* **a)** *(gaze upon)* betrachten; **b)** *(give heed to)* beachten ⟨*jmds. Worte, Rat*⟩; Rücksicht nehmen auf (+ *Akk.*) ⟨*Wunsch, Gesundheit, jmds. Recht*⟩; **c)** *(fig.: look upon)* betrachten; **~ sb. kindly/warmly** jmdm. freundlich gesinnt/herzlich zugetan sein; **~ sb. with envy/scorn** neidisch auf jmdn. sein/jmdn. verachten; **~ sb. as a friend/fool** jmdn. als Freund betrachten/für einen Dummkopf halten; **be ~ed as** gelten als; **~ sth. as wrong** etw. für falsch halten; **d)** *(concern, have relation to)* betreffen; berücksichtigen ⟨*Tatsachen*⟩; **as ~s sb./sth., ~ing sb./sth.** was jmdn./etw. angeht *od.* betrifft. **2.** *n.* **a)** *(attention)* Beachtung, *die;* **pay ~ to/have ~ to** *or* **for sb./sth.** jmdm./etw. Beachtung schenken; **without ~ to** ohne Rücksicht auf (+ *Akk.*); **b)** *(esteem, kindly feeling)* Achtung, *die;* **hold sb./sth. in high/low ~, have** *or* **show a high/low ~ for sb./sth.** jmdn./etw. sehr schätzen/geringschätzen; **c)** *in pl.* Grüße; **send one's ~s** grüßen lassen; **give her my ~s** grüße sie von mir; **with kind[est] ~s** mit herzlich[st]en Grüßen; **d)** *(relation, respect)* Beziehung, *die;* **in this ~:** in dieser Beziehung *od.* Hinsicht; **in** *or* **with ~ to sb./sth.** in bezug auf jmdn./etw.
regarding [rɪ'gɑ:dɪŋ] *see* **regard 1 d**
regardless [rɪ'gɑ:dlɪs] **1.** *adj.* **~ of sth.** ungeachtet *od.* trotz einer Sache *(Gen.);* **~ of the cost** ohne Rücksicht auf die Kosten. **2.** *adv.* trotzdem; **carry on ~:** trotzdem weitermachen
regatta [rɪ'gætə] *n.* Regatta, *die*
regenerate [rɪ'dʒenəreɪt] *v. t.* **a)** *(generate again, re-create)* regenerieren *(bes. Chemie, Biol.);* **b)** *(improve, reform)* erneuern ⟨*Kirche, Gesellschaft*⟩; **feel ~d** sich wie neugeboren fühlen
regeneration [rɪdʒenə'reɪʃn] *n.* **a)** *(re-creation)* Neuentstehung, *die; (fig.: revival, renaissance)* Wiederbelebung, *die; (of church, society)* Erneuerung, *die;* **b)** *(Biol.: regrowth)* Regeneration, *die (fachspr.);* Neubildung, *die*
regent ['ri:dʒənt] *n.* Regent, *der*/Regentin, *die*
reggae ['regeɪ] *n. (Mus.)* Reggae, *der*
regime, régime [reɪ'ʒi:m] *n. (system)* [Regierungs]system, *das; (derog.)* Regime, *das; (fig.)* bestehende Ordnung
regiment 1. ['redʒɪmənt, 'redʒmənt] *n.* **a)** *(Mil.: organizational unit)* Regiment, *das;* **parachute ~:** Luftlanderegiment, *das;* **b)** *(Mil.: operational unit)* Abteilung, *die;* **tank ~:** Panzerabteilung, *die.* **2.** ['redʒɪmənt, 'redʒɪment] *v. t. (organize)* reglementieren
regimental [redʒɪ'mentl] *(Mil.)* *adj.* Regiments⟨*kleidung, -vorräte*⟩
regimentation [redʒɪmən'teɪʃn, redʒɪmen'teɪʃn] *n.* Reglementierung, *die*
region ['ri:dʒn] *n.* **a)** *(area)* Gebiet, *das;* **b)** *(administrative division)* Bezirk, *der;* **administrative ~:** Verwaltungsbezirk, *der;* **Strathclyde R~:** Bezirk Strathclyde; **c)** *(fig.: sphere)* Bereich, *der;* Gebiet, *das;* **in the ~ of two tons** ungefähr zwei Tonnen
regional ['ri:dʒənl] *adj.* regional ⟨*System, Akzent, Förderung*⟩; Regional⟨*planung, -fernsehen, -programm, -ausschuß*⟩
register ['redʒɪstə(r)] **1.** *n. (book, list)* Register, *das; (at school)* Klassenbuch, *das;* **parish/hotel/marriage ~:** Kirchen-/Fremden-/Heiratsbuch, *das;* **~ of births, deaths and marriages** Personenstandsbuch, *das;* **medical ~:** Ärzteregister, *das;* **electoral ~:** Wählerverzeichnis, *das.* **2.** *v. t.* **a)** *(set down)* schriftlich festhalten ⟨*Name, Zahl, Detail*⟩; **b)** *(enter)* registrieren ⟨*Geburt, Heirat, Todesfall, Patent*⟩; *(cause to be entered)* registrieren lassen; eintragen ⟨*Warenzeichen, Firma, Verein*⟩; anmelden ⟨*Auto, Patent*⟩; *abs. (at hotel)* sich ins Fremdenbuch eintragen; **~ [oneself] with the police** sich polizeilich anmelden; **c)** *(enrol)* anmelden; *(Univ.)* einschreiben; immatrikulieren; *(as voter)* eintragen (**on** in + *Akk.*) ⟨*Person*⟩; *abs. (as student)* sich einschreiben *od.* immatrikulieren; *(in list of voters)* sich ins Wählerverzeichnis eintragen lassen; **d)** *(Post)* eingeschrieben versenden; **have sth. ~ed** etw. einschreiben lassen; **e)** zum Ausdruck bringen ⟨*Entsetzen, Überraschung*⟩; **~ a protest** Protest anmelden. **3.** *v. i. (make impression)* einen Eindruck machen (**on, with** auf + *Akk.*); **it didn't ~ with him** er hat das nicht registriert
registered ['redʒɪstəd] *adj.* [ins Standesregister] eingetragen ⟨*Taufe, Heirat*⟩; [ins Handelsregister] eingetragen ⟨*Firma*⟩; eingeschrieben, immatrikuliert ⟨*Student*⟩; eingeschrieben ⟨*Brief, Post, Päckchen*⟩; **~ disabled** ≈ Behinderter/Behinderte mit Schwerbehindertenausweis; **~ trade mark** Warenzeichen; **by ~ post** per Einschreiben
registrar ['redʒɪstrɑ:(r), redʒɪ'strɑ:(r)] *n.* **a)** *(Univ.)* ≈ Kanzler, *der*/Kanzlerin, *die; (public official)*Standesbeamte, *der*/-beam-

tin, *die;* **b)** *(Med.) Arzt/Ärztin in der klinischen Fachausbildung*
registration [redʒɪ'streɪʃn] *n. (act of registering)* Registrierung, *die; (enrolment)* Anmeldung, *die; (of students)* Einschreibung, *die;* Immatrikulation, *die; (of voters)* Eintragung ins Wählerverzeichnis
registration: ~ document *n. (Brit.)* Kraftfahrzeugbrief, *der;* **~ number** *n. (Motor Veh.)* amtliches *od.* polizeiliches Kennzeichen
registry ['redʒɪstrɪ] *n.* **~ [office]** Standesamt, *das; attrib.* standesamtlich ‹*Trauung*›; **be married in a ~ [office]** sich standesamtlich trauen lassen
regret [rɪ'gret] **1.** *v. t.,* **-tt-: a)** *(feel sorrow for loss of)* nachtrauern (+ *Dat.*); **b)** *(be sorry for)* bedauern; **~ having done sth.** es bedauern, daß man etw. getan hat; **it is to be ~ted that ...:** es ist bedauerlich, daß ...; **I ~ to say that ...:** ich muß leider sagen, daß ... **2.** *n.* Bedauern, *das;* **much to my ~:** zu meinem großen Bedauern; **have no ~s** nichts bereuen; **send one's ~s** *(polite refusal)* sich entschuldigen lassen
regretfully [rɪ'gretfəlɪ] *adv.* mit Bedauern
regrettable [rɪ'gretəbl] adj. bedauerlich
regrettably [rɪ'gretəblɪ] *adv.* bedauerlicherweise; bedauerlich ‹*teuer*›
regroup [ri:'gru:p] **1.** *v. t.* **a)** umgruppieren; **b)** *(Mil.: reorganize)* neu formieren ‹*Truppen*›. **2.** *v. i.* **a)** *(form a new group)* sich neu gruppieren; **b)** *(Mil.)* sich neu formieren
regular ['regjʊlə(r)] **1.** *adj.* **a)** *(recurring uniformly, habitual)* regelmäßig; geregelt ‹*Arbeit*›; fest ‹*Anstellung, Reihenfolge*›; **~ customer** Stammkunde, *der*/-kundin, *die;* **our ~ postman** unser [gewohnter] Briefträger; **get ~ work** ‹*Freiberufler:*› regelmäßig Aufträge bekommen; **have** *or* **lead a ~ life** ein geregeltes Leben führen; **b)** *(evenly arranged, symmetrical)* regelmäßig; **c)** *(properly qualified)* ausgebildet; **~ soldiers** Berufssoldaten; **d)** *(Ling.)* regelmäßig; **e)** *(coll.: thorough)* richtig *(ugs.).* **2.** *n.* **a)** *(coll.: ~ customer, visitor, etc.)* Stammkunde, *der*/-kundin, *die; (in pub)* Stammgast, *der;* **b)** *(soldier)* Berufssoldat, *der*
regularise *see* **regularize**
regularity [regjʊ'lærɪtɪ] *n.* Regelmäßigkeit, *die*
regularize ['regjʊləraɪz] *v. t.* **a)** *(make regular)* regeln; *(by law)* gesetzlich regeln *od.* festlegen; **b)** *(make steady)* stabilisieren ‹*Atmung, Puls, Spannung*›
regularly ['regjʊləlɪ] *adv.* **a)** *(at fixed times)* regelmäßig; *(constantly)* ständig; **b)** *(steadily)* gleichmäßig; **c)** *(symmetrically)* regelmäßig ‹*bauen, anlegen*›
regulate ['regjʊleɪt] *v. t.* **a)** *(control)* regeln; *(subject to restriction)* begrenzen; **b)** *(adjust)* regulieren; einstellen ‹*Apparat, Maschine*›; [richtig ein]stellen ‹*Uhr*›
regulation [regʊ'leɪʃn] *n.* **a)** *(regulating)* Regelung, *die; (of quantity, speed)* Regulierung, *die; (of machine)* Einstellen, *das;* **b)** *(rule)* Vorschrift, *die;* **be against ~s** vorschriftswidrig sein; **c)** *attrib.* vorschriftsmäßig ‹*Kleidung*›
regulator ['regjʊleɪtə(r)] *n. (device)* Regler, *der; (of clock, watch)* Gangregler, *der*
rehabilitate [ri:hə'bɪlɪteɪt] *v. t.* rehabilitieren; **~ [back into society]** wieder [in die Gesellschaft] eingliedern
rehabilitation [ri:həbɪlɪ'teɪʃn] *n.* Rehabilitation, *die;* **~ [in society]** Wiedereingliederung [in die Gesellschaft]
rehash 1. [ri:'hæʃ] *v. t.* aufwärmen. **2.** ['ri:hæʃ] *n. (restatement)* Aufguß, *der (abwertend)*
rehearsal [rɪ'hɜ:sl] *n. (Theatre, Mus., etc.)* Probe, *die; see also* **dress rehearsal**
rehearse [rɪ'hɜ:s] *v. t. (Theatre, Mus., etc.)* proben
re-heat [ri:'hi:t] *v. t.* wieder erwärmen; aufwärmen ‹*Essen*›
rehouse [ri:'haʊz] *v. t.* umquartieren
reign [reɪn] **1.** *n.* Herrschaft, *die; (of monarch also)* Regentschaft, *die;* **in the ~ of King Charles** während der Regentschaft König Karls. **2.** *v. i.* **a)** *(hold office)* herrschen (**over** über + *Akk.*); **~ing champion** amtierender Meister/amtierende Meisterin; **b)** *(prevail)* herrschen
reimburse [ri:ɪm'bɜ:s] *v. t.* [zurück]erstatten ‹*[Un]kosten, Spesen*›; entschädigen ‹*Person*›; **~ sb. for** jmdm. [zurück]erstatten ‹*[Un]kosten, Spesen*›; jmdm. ersetzen ‹*Verlust*›
reimbursement [ri:ɪm'bɜ:smənt] *n.* Rückzahlung, *die; (of expenses)* Erstattung, *die*
rein [reɪn] *n.* **a)** Zügel, *der;* **keep a child on ~s** ein Kind am Laufgurt führen; **b)** *(fig.)* Zügel, *der;* **hold the ~s** die Zügel in der Hand haben; **keep a tight ~ on** an der Kandare halten ‹*Person*›; im Zaum halten ‹*Gefühle*›; *see also* **free 1 a**
~ 'in *v. t. (check, lit. or fig.)* zügeln
reincarnation [ri:ɪnkɑ:'neɪʃn] *n. (Relig.)* Reinkarnation, *die;* Wiedergeburt, *die*
reindeer ['reɪndɪə(r)] *n., pl. same* Ren[tier], *das*
reinforce [ri:ɪn'fɔ:s] *v. t.* verstärken ‹*Truppen, Festung, Stoff*›; erhöhen ‹*Anzahl*›; untermauern ‹*Argument*›; bestätigen ‹*Behauptung*›; **~ sb.'s opinion** jmdn. in seiner Meinung bestärken; **~d concrete** Stahlbeton, *der*
reinforcement [ri:ɪn'fɔ:smənt] *n.* **a)** Verstärkung, *die; (of numbers)* Zunahme, *die; (of argument)* Untermauerung, *die;* **b) ~[s]** *(additional men etc.)* Verstärkung, *die*
reinstate [ri:ɪn'steɪt] *v. t. (in job)* wieder einstellen
reinterpret [ri:ɪn'tɜ:prɪt] *v. t. (interpret afresh)* noch einmal interpretieren; *(give different interpretation)* neu interpretieren
reinvest [ri:ɪn'vest] *v. t.* reinvestieren *(fachspr.);* wieder anlegen ‹*Kapital*›
reissue [ri:'ɪʃu:, ri:'ɪsju:] **1.** *v. t.* neu herausbringen. **2.** *n.* Neuauflage, *die*
reiterate [ri:'ɪtəreɪt] *v. t.* wiederholen
reject 1. [rɪ'dʒekt] *v. t.* **a)** ablehnen; abweisen ‹*Freier*›; zurückweisen ‹*Bitte, Annäherungsversuch*›; **b)** *(Med.)* nicht vertragen ‹*Nahrung, Medizin*›; abstoßen ‹*Transplantat*›. **2.** ['ri:dʒekt] *n. (thing)* Ausschuß, *der*
rejection [rɪ'dʒekʃn] *n.* **a)** *see* **reject 1 a:** Ablehnung, *die;* Abweisung, *die;* Zurückweisung, *die;* **parental ~:** Ablehnung durch die Eltern; **b)** *(Med.)* Abstoßung, *die*
rejoice [rɪ'dʒɔɪs] *v. i.* **a)** *(feel great joy)* sich freuen (**over, at** über + *Akk.*); **b)** *(make merry)* feiern
rejoicing [rɪ'dʒɔɪsɪŋ] *n.* **a) [sounds of] ~:** Jubel, *der;* **b)** *in pl. (celebrations)* Feier, *die*
[1]**rejoin** [rɪ'dʒɔɪn] *v. t. (reply)* erwidern (**to** auf + *Akk.*)
[2]**rejoin** [ri:'dʒɔɪn] *v. t.* **a)** *(join again)* wieder stoßen zu ‹*Regiment*›; wieder eintreten in (+ *Akk.*) ‹*Partei, Verein*›; **~ one's ship** wieder an Bord gehen; **b)** ‹*Verkehrsteilnehmer:*› wieder kommen auf (+ *Akk.*) ‹*Straße, Autobahn*›; ‹*Straße:*› wieder [ein]münden in (+ *Akk.*) ‹*Straße, Autobahn*›

rejoinder [rɪ'dʒɔɪndə(r)] *n.* Erwiderung, *die* (**to** auf + *Akk.*)
rejuvenate [rɪ'dʒu:vəneɪt] *v. t.* verjüngen ‹*Person, Haut*›
rekindle [ri:'kɪndl] *v. t.* **a)** *(relight)* wieder anfachen; **b)** *(fig.: reawaken)* wieder entfachen ‹*Liebe, Leidenschaft*›; wieder aufleben lassen ‹*Sehnsucht, Verlangen, Hoffnung*›
relapse [rɪ'læps] **1.** *v. i.* ‹*Kranker:*› einen Rückfall bekommen; **~ into** zurückfallen in (+ *Akk.*) ‹*Götzendienst, Barbarei*›; **~ into silence/lethargy** wieder in Schweigen/Lethargie verfallen. **2.** *n.* Rückfall, *der* (**into** in + *Akk.*)
relate [rɪ'leɪt] **1.** *v. t.* **a)** *(tell)* erzählen ‹*Geschichte*›; erzählen von ‹*Abenteuer*›; **b)** *(bring into relation)* in Zusammenhang bringen (**to, with** mit); **c)** *(establish relation or connection between)* einen Zusammenhang herstellen zwischen. **2.** *v. i.* **a) ~ to** *(have reference)* ‹*Behauptung, Frage, Angelegenheit:*› in Zusammenhang stehen mit; betreffen ‹*Person*›; **b) ~ to** *(feel involved or connected with)* eine Beziehung haben zu
related [rɪ'leɪtɪd] *adj.* **a)** *(by kinship or marriage)* verwandt (**to** mit); **~ by marriage** verschwägert; **b)** *(connected)* miteinander in Zusammenhang stehend; verwandt ‹*Sprache, Begriff, Spezies, Fach*›
relation [rɪ'leɪʃn] *n.* **a)** *(connection)* Beziehung, *die;* Zusammenhang, *der* (**of ... and** zwischen ... und); **be out of all ~ to** in keinem Verhältnis stehen zu ‹*Kosten, geleisteter Arbeit*›; **in** *or* **with ~ to** in bezug auf (+ *Akk.*); *see also* [2]**bear 1 c**; **b)** *in pl. (dealings) (with parents, police)* Verhältnis, *das* (**with** zu); *(with country)* Beziehungen (**with** zu, mit); *(sexual intercourse)* intime Beziehungen (**with** zu); **c)** *(kin, relative)* Verwandte, *der/die;* **what ~ is he to you?** wie ist er mit dir verwandt?; **is she any ~ [to you]?** ist sie mit dir verwandt?
relationship [rɪ'leɪʃnʃɪp] *n.* **a)** *(mutual tie)* Beziehung, *die* (**with** zu); **have a good/bad ~ with sb.** zu jmdm. ein gutes/schlechtes Verhältnis haben; **doctor-patient ~:** Verhältnis zwischen Arzt und Patient; **b)** *(kinship)* Verwandtschaftsverhältnis, *das;* **c)** *(connection)* Beziehung, *die;* **d)** *(sexual)* Verhältnis, *das*
relative ['relətɪv] **1.** *n.* Verwandte, *der/die.* **2.** *adj.* relativ; *(comparative)* jeweilig; **the ~ costs of a and b** die Kostenrelation zwischen a und b; **with ~ calmness** relativ gelassen; **be ~ to sth.** sich nach etw. richten; **a large population ~ to the town's size** eine im Verhältnis zur Größe der Stadt beachtliche Einwohnerzahl
relative 'clause *n. (Ling.)* Relativsatz, *der*
relatively ['relətɪvlɪ] *adv.* relativ; verhältnismäßig
relative 'pronoun *n. (Ling.)* Relativpronomen, *das*
relativity [relə'tɪvɪtɪ] *n. (Phys.)* Relativität, *die;* **~ theory, the theory of ~:** die Relativitätstheorie
relax [rɪ'læks] **1.** *v. t.* **a)** *(make less tense)* entspannen ‹*Muskel, Körper[teil]*›; lockern ‹*Muskel, Feder, Griff*›; *(fig.)* lockern; **b)** *(make less strict)* lockern ‹*Gesetz, Disziplin, Sitten*›; **c)** *(slacken)* nachlassen in (+ *Dat.*) ‹*Bemühungen, Aufmerksamkeit*›; verlangsamen ‹*Tempo*›. **2.** *v. i.* **a)** *(become less tense)* sich entspannen; **b)** *(slacken)* nachlassen (**in** in + *Dat.*); **c)** *(become less stern)* sich mäßigen (**in** in + *Dat.*); **d)** *(cease effort)* sich entspannen; ausspannen; *(stop worrying, calm down)* sich beruhigen
relaxation [ri:læk'seɪʃn] *n.* **a)** *(recreation)* Freizeitbeschäftigung, *die;* **play tennis as a ~:** zur Entspannung Tennis spielen; **b)** *(cessation of effort)* Erholung, *die* (**from** von); **c)** *(reduction of tension; lit. or fig.)* Lockerung, *die*
relaxed [rɪ'lækst] *adj.* **a)** *(informal, not anxious)* entspannt, gelöst ‹*Atmosphäre, Lächeln, Gefühl, Person*›; **b)** *(not strict or exact)* gelockert ‹*Regel, Beschränkung*›
relaxing [rɪ'læksɪŋ] *adj.* entspannend; erholsam; **have a ~ bath** zur Entspannung ein Bad nehmen
relay 1. ['ri:leɪ] *n.* **a)** *(gang)* Schicht, *die;* **work in ~s** schichtweise arbeiten; **b)** *(race)* Staffel, *die;* **c)** *(Electr.)* Relais, *das;* **d)** *(Radio, Telev.)* **radio ~:** Richtfunkverbindung, *die;* **~ station** Relaisstation, *die;* **e)** *(transmission)* Übertragung, *die.* **2.** [ri:'leɪ, 'ri:leɪ] *v. t.* **a)** *(pass on)* weiterleiten; **~ a message to sb. that ...:** jmdm. ausrichten *od.* mitteilen, daß ...; **b)** *(Radio, Telev., Teleph.)* übertragen
'relay race *n. (Running)* Staffellauf, *der;* *(Swimming)* Staffelschwimmen, *das*
release [rɪ'li:s] **1.** *v. t.* **a)** *(free)* freilassen ‹*Tier, Häftling, Sklaven*›; *(from jail)* entlassen (**from** aus); *(from bondage, trap)* befreien (**from** aus); *(from pain)* erlösen (**from** von); *(from promise, obligation, vow)* entbinden (**from** von); **b)** *(let go, let fall)* loslassen; lösen ‹*Handbremse*›; ausklinken ‹*Bombe*›; **~ one's hold** *or* **grip on sth.** etw. loslassen; **c)** *(make known)* veröffentlichen ‹*Erklärung, Nachricht*›; *(issue)* herausbringen ‹*Film, Schallplatte, Produkt*›. **2.** *n.* **a)** *(act of freeing) see* **1 a:** Freilassung, *die;* Entlassung, *die;* Befreiung, *die;* Erlösung, *die;* Entbindung, *die;* **b)** *(of published item)* Veröffentlichung, *die;* **when does the film go out on general ~?** wann kommt der Film in die Kinos?; **a new ~ by Bob Dylan** eine neue Platte *od.* eine Neuveröffentlichung von Bob Dylan; **c)** *(handle, lever, button)* Auslöser, *der*
relegate ['relɪgeɪt] *v. t.* **a) ~ sb. to the position** *or* **status of ...:** jmdn. zu ... degradieren; **b)** *(Sport)* absteigen lassen; **be ~d** absteigen (**to** in + *Akk.*)
relegation [relɪ'geɪʃn] *n.* **a)** Degradierung, *die;* **her ~ to the position of ...:** ihre Degradierung zu ...; **b)** *(Sport)* Abstieg, *der*
relent [rɪ'lent] *v. i.* sich erweichen lassen; *(yield to compassion)* Mitleid zeigen; ‹*Wetter:*› besser werden
relentless [rɪ'lentlɪs] *adj.* unerbittlich; schonungslos ‹*Kritik, Heftigkeit*›
relevance ['relɪvəns] *n.* Relevanz, die (**to** für)
relevant ['relɪvənt] *adj.* relevant (**to** für); wichtig ‹*Information, Dokument*›; entsprechend ‹*Formular*›
reliability [rɪlaɪə'bɪlɪtɪ] *n., no pl.* Zuverlässigkeit, *die*
reliable [rɪ'laɪəbl] *adj.* zuverlässig
reliably [rɪ'laɪəblɪ] *adv.* zuverlässig; **I am ~ informed that ...:** ich habe aus zuverlässiger Quelle erfahren, daß ...
reliance [rɪ'laɪəns] *n. (trust, confidence)* Vertrauen, *das* (**in** zu, **on** auf + *Akk.*)
reliant [rɪ'laɪənt] *adj.* **be ~ on sb./sth.** auf jmdn./etw. angewiesen sein
relic ['relɪk] *n.* **a)** *(Relig.)* Reliquie, *die;* **b)** *(surviving trace)* Überbleibsel, *das (ugs.);* Relikt, *das*
[1]**relief** [rɪ'li:f] *n.* **a)** *(alleviation, deliverance)* Erleichterung, *die;* **give**

or **bring [sb.] ~ [from pain]** [jmdm.] [Schmerz]linderung verschaffen; **breathe** *or* **heave a sigh of ~:** erleichtert aufatmen; **what a ~!, that's a ~!** da bin ich aber erleichtert!; **b)** *(assistance)* Hilfe, *die; (financial state assistance)* Sozialhilfe, *die; attrib.* Hilfs⟨*fond, -organisation, -komitee*⟩; **c)** *(replacement of person)* Ablösung, *die; attrib.* **~ driver** ablösender Fahrer

²**relief** *n.* **a)** *(Art)* **works in ~:** Reliefarbeiten; **high/low ~:** Hoch-/Flachrelief, *das;* **b)** *(a sculpture)* Relief, *das;* **c) stand out in strong ~ against sth.** sich scharf gegen etw. abheben; *(fig.)* in krassem Gegensatz zu etw. stehen

relief: ~ bus *n.* Entlastungsbus, *der; (as replacement)* Ersatzbus, *der;* **~ map** *n.* Reliefkarte, *die;* **~ road** *n.* Entlastungsstraße, *die*

relieve [rɪ'li:v] *v.t.* **a)** *(lessen, mitigate)* lindern; verringern ⟨*Dampfdruck, Anspannung*⟩; unterbrechen ⟨*Eintönigkeit*⟩; erleichtern ⟨*Gewissen*⟩; *(remove)* abbauen ⟨*Anspannung*⟩; stillen ⟨*Schmerzen*⟩; *(remove or lessen monotony of)* auflockern; **I am ~d to hear that ...:** es erleichtert mich zu hören, daß ...; **b)** *(release from duty)* ablösen ⟨*Wache, Truppen*⟩; **c) ~ sb.** *(of task, duty)* jmdn. entbinden **(of** von); *(of responsibility, load)* jmdm. abnehmen **(of** *Akk.*); *(from debt)* jmdm. erlassen **(from** *Akk.*); *(of burden, duty; from sorrow, worry)* jmdn. befreien **(of, from** von); **d) ~ oneself** *(empty the bladder or bowels)* sich erleichtern *(verhüll.);* **e)** *(release from a post)* entbinden **(of, from** von); *(dismiss)* entheben *(geh.)* **(of, from** *Gen.*)

religion [rɪ'lɪdʒn] *n.* Religion, *die;* **freedom of ~:** Glaubensfreiheit, *die;* **what is your ~?** welcher Religion gehörst du an?

religious [rɪ'lɪdʒəs] *adj.* **a)** *(pious)* religiös; fromm; **b)** *(concerned with religion)* Glaubens⟨*freiheit, -eifer*⟩; Religions⟨*freiheit, -unterricht, -kenntnisse*⟩; religiös ⟨*Überzeugung, Zentrum*⟩; **c)** *(of monastic order)* religiös ⟨*Orden*⟩; **~ community** Ordensgemeinschaft, *die;* **d)** *(scrupulous)* peinlich ⟨*Sorgfalt, Genauigkeit*⟩

religiously [rɪ'lɪdʒəslɪ] *adv.* **a)** *(piously, reverently)* inbrünstig ⟨*beten*⟩; ehrfürchtig ⟨*verehren, niederknien*⟩; **b)** *(conscientiously)* gewissenhaft ⟨*durchsehen, verbessern*⟩; peinlich genau ⟨*saubermachen, verbessern*⟩

relinquish [rɪ'lɪŋkwɪʃ] *v.t.* **a)** *(give up, abandon)* aufgeben; ablassen von ⟨*Glaube*⟩; verzichten auf (+ *Akk.*) ⟨*Recht, Anspruch, Macht*⟩; aufgeben ⟨*Anspruch, Stelle, Arbeit, Besitz*⟩; **~ the right/one's claim to sth.** auf sein Recht/seinen Anspruch auf etw. *(Akk.)* verzichten; **b) ~ one's hold** *or* **grip on sb./sth.** jmdn./etw. loslassen

relish ['relɪʃ] **1.** *n.* **a)** *(liking)* Vorliebe, *die;* **do sth. with [great] ~:** etw. mit [großem] Genuß tun; **he takes [great] ~ in doing sth.** es bereitet ihm [große] Freude, etw. zu tun; **b)** *(condiment)* Relish, *das (Kochk.).* **2.** *v.t.* genießen; reizvoll finden ⟨*Gedanke, Vorstellung*⟩

reload [ri:'ləʊd] *v.t.* nachladen ⟨*Schußwaffe*⟩; **~ the camera** einen neuen Film einlegen

relocate [ri:lə'keɪt] *v.t.* verlegen ⟨*Fabrik, Büro*⟩

reluctance [rɪ'lʌktəns] *n., no pl.* Widerwille, *der;* Abneigung, *die;* **have a [great] ~ to do sth.** etw. nur mit Widerwillen tun

reluctant [rɪ'lʌktənt] *adj.* unwillig; **be ~ to do sth.** etw. nur ungern *od.* widerstrebend tun

rely [rɪ'laɪ] *v.i.* **a)** *(have trust)* sich verlassen **([up]on** auf + *Akk.*); **b)** *(be dependent)* angewiesen sein **([up]on** auf + *Akk.*); **[have to] ~ on sb. to help** darauf angewiesen sein, daß jmd. hilft

remain [rɪ'meɪn] *v.i.* **a)** *(be left over)* übrigbleiben; **all that ~ed for me to do was to ...:** ich mußte *od.* brauchte nur noch ...; **nothing ~s but to thank you all** es bleibt mir nur, Ihnen allen zu danken; **b)** *(stay)* bleiben; **~ behind** noch dableiben; **~ in sb.'s memory** jmdm. im Gedächtnis bleiben; **c)** *(continue to be)* bleiben; **it ~s to be seen** das bleibt abzuwarten *od.* wird sich zeigen; **the fact ~s that ...:** das ändert nichts an der Tatsache *od.* daran, daß ...

remainder [rɪ'meɪndə(r)] *n.* **a)** *(sb. or sth. left over; also Math.)* Rest, *der;* **b)** *(remaining stock)* Restposten, *der*

remaining [rɪ'meɪnɪŋ] *adj.* restlich; übrig; **spend one's ~ years ...:** seinen Lebensabend ... verbringen

remains [rɪ'meɪnz] *n. pl.* **a)** *(leftover part)* Reste; **b)** *(corpse)* sterbliche [Über]reste *(verhüll.);* **c)** *(relics)* Relikte; Reste; **Roman ~:** Relikte aus der Römerzeit

remake ['ri:meɪk] *n. (Cinemat.)* Remake, *das (fachspr.);* Neuverfilmung, *die*

remand [rɪ'mɑ:nd] **1.** *v.t.* **~ sb. [in custody]** jmdn. in Untersuchungshaft behalten; **be ~ed in custody/on bail** in Untersuchungshaft bleiben müssen/gegen Kaution aus der Untersuchungshaft entlassen werden. **2.** *n.* **[period of] ~:** Untersuchungshaft, *die;* **place** *or* **put sb. on ~:** jmdn. in Untersuchungshaft nehmen; **be on ~:** in Untersuchungshaft sein; **be held on ~:** in Untersuchungshaft bleiben müssen

re'mand centre *n. (Brit.) Untersuchungsgefängnis für jugendliche Straftäter zwischen 14 und 21 Jahren*

remark [rɪ'mɑ:k] **1.** *v.t.* bemerken **(to** gegenüber). **2.** *v.i.* eine Bemerkung machen **([up]on** zu, über + *Akk.*). **3.** *n. (comment)* Bemerkung, *die* **(on** über + *Akk.*); **make a ~:** eine Bemerkung machen **(about, at** über + *Akk.*)

remarkable [rɪ'mɑ:kəbl] *adj.* **a)** *(notable)* bemerkenswert; **b)** *(extraordinary)* außergewöhnlich

remarkably [rɪ'mɑ:kəblɪ] *adv.* **a)** *(notably)* bemerkenswert; **b)** *(exceptionally)* außergewöhnlich

remarry [ri:'mærɪ] *v.i. & t.* wieder heiraten

remedial [rɪ'mi:dɪəl] *adj.* **a)** *(affording a remedy)* Heil⟨*behandlung, -wirkung*⟩; *(intended to remedy deficiency etc.)* rehabilitierend ⟨*Maßnahme*⟩; **take ~ action** Hilfsmaßnahmen ergreifen; **b)** *(Educ.)* Förder-; **classes in ~ reading** Förderunterricht im Lesen

remedy ['remɪdɪ] **1.** *n.* **a)** *(cure)* [Heil]mittel, *das* **(for** gegen); **cough/herbal ~:** Husten-/Kräutermittel, *das;* **cold/flu ~:** Mittel gegen Erkältung/Grippe; **b)** *(means of counteracting)* [Gegen]mittel, *das* **(for** gegen). **2.** *v.t.* beheben ⟨*Sprachfehler, Problem*⟩; ausgleichen ⟨*Kurzsichtigkeit*⟩; retten ⟨*Situation*⟩; **the situation cannot be remedied** die Situation ist nicht zu retten

remember [rɪ'membə(r)] *v.t.* **a)** *(keep in memory)* denken an (+ *Akk.*); *(bring to mind)* sich erinnern an (+ *Akk.*); **don't you ~ me?** erinnern Sie sich nicht an mich?; **~ who/where you are!** vergiß nicht, wer/wo du bist; **I can't ~ the word I want** das Wort, das ich brauche, fällt mir gerade nicht ein; **I ~ed to bring the book** ich habe daran gedacht, das Buch mitzubringen; **I can never ~ her name** ich kann mir ihren Namen einfach nicht merken; **if I ~**

correctly *(abs.)* wenn ich mich recht erinnere; **an evening to ~:** ein unvergeßlicher Abend; **b)** *(convey greetings from)* grüßen; **~ me to them** grüße sie von mir; **she asked to be ~ed to you** sie läßt dich grüßen

remembrance [rɪ'membrəns] *n.* Gedenken, *das;* **in ~ of sb.** zu jmds. Gedächtnis; zum Gedenken an jmdn.

Remembrance: ~ Day, ~ Sunday *ns. (Brit.)* ≈ Volkstrauertag, *der*

remind [rɪ'maɪnd] *v.t.* erinnern (**of** an + *Akk.*); **~ sb. to do sth.** jmdn. daran erinnern, etw. zu tun; **that ~s me, ...:** dabei fällt mir ein, ...; **you are ~ed that ...:** beachten Sie bitte, daß ...

reminder [rɪ'maɪndə(r)] *n.* Erinnerung, *die* (**of** an + *Akk.*); *(mnemonic)* Gedächtnishilfe *od.* -stütze, *die;* **give sb. a ~ that ...:** jmdn. daran erinnern, daß ...; **serve as/be a ~ of sth.** an etw. *(Akk.)* erinnern

reminisce [remɪ'nɪs] *v.i.* sich in Erinnerungen *(Dat.)* ergehen (**about** an + *Akk.*)

reminiscence [remɪ'nɪsəns] *n.* Erinnerung, *die* (**of** an + *Akk.*)

reminiscent [remɪ'nɪsənt] *adj.* **~ of sth.** an etw. *(Akk.)* erinnernd; **be ~ of sth.** an etw. *(Akk.)* erinnern

remiss [rɪ'mɪs] *adj.* nachlässig (**of** von)

remission [rɪ'mɪʃn] *n.* **a)** *(of sins)* Vergebung, *die;* **b)** *(of debt, punishment)* Erlaß, *der;* **c)** *(prison sentence)* Straferlaß, *der;* **he gained one year's ~:** ihm ist ein Jahr erlassen worden; **d)** *(Med.)* Remission, *die;* **go into ~:** remittieren

remit [rɪ'mɪt] *v.t.,* **-tt-: a)** *(pardon)* vergeben ‹*Sünde, Beleidigung usw.*›; **b)** *(cancel)* erlassen ‹*Steuer, Gebühr usw.*›; **c)** *(send)* überweisen ‹*Geld*›

remittance [rɪ'mɪtəns] *n.* Überweisung, *die*

remnant ['remnənt] *n.* Rest, *der;* *(trace)* Überrest, *der*

remold *(Amer.) see* **remould**

remonstrance [rɪ'mɒnstrəns] *n.* Protest, *der* (**with, against** gegen)

remonstrate ['remənstreɪt] *v.i.* protestieren (**against** gegen); **~ with sb.** jmdm. Vorhaltungen machen (**on** wegen)

remorse [rɪ'mɔ:s] *n.* Reue, *die* (**for, about** über + *Akk.*); **without ~** *(merciless)* erbarmungslos

remorseful [rɪ'mɔ:sfl] *adj.* reuig; reuevoll *(geh.)*

remorseless [rɪ'mɔ:slɪs] *adj.* **a)** *(merciless)* erbarmungslos ‹*Grausamkeit, Barbarei*›; **b)** *(relentless)* unerbittlich ‹*Schicksal, Logik*›

remote [rɪ'məʊt] *adj.,* **~r** [rɪ'məʊtə(r)], **~st** [rɪ'məʊtɪst] **a)** *(far apart)* entfernt; **b)** *(far off)* fern ‹*Vergangenheit, Zukunft, Zeit*›; früh ‹*Altertum*›; abgelegen, *(geh.)* entlegen ‹*Ort, Gebiet*›; **~ from** *(lit. or fig.)* weit entfernt von; **c)** *(not closely related)* entfernt, weitläufig ‹*Vorfahr, Nachkomme, Verwandte*›; **d)** *(slight)* gering ‹*Chance, Möglichkeit*›

remote: ~ con'trol *n. (of vehicle)* Fernlenkung, *die;* Fernsteuerung, *die; (of apparatus)* Fernbedienung, *die;* **~-control[led]** *adj.* ferngesteuert; ferngelenkt; fernbedient ‹*Anlage*›

remotely [rɪ'məʊtlɪ] *adv.* **a)** *(distantly)* entfernt, weitläufig ‹*verwandt*›; **b)** *(slightly)* **they are not [even] ~ alike** sie haben [aber auch] nicht die entfernteste Ähnlichkeit [miteinander]; **it is ~ conceivable that ...:** es ist nicht völlig auszuschließen, daß ...

remould 1. [ri:'məʊld] *v.t.* *(refashion)* ummodeln, umgestalten (**into** zu); *(Motor Veh.)* runderneuern ‹*Reifen*›. **2.** ['ri:məʊld] *n.* *(Motor Veh.)* runderneuerter Reifen

remount [ri:'maʊnt] *v.i.* *(on horse)* wieder aufsitzen; *(on bicycle)* wieder aufs Fahrrad steigen

removable [rɪ'mu:vəbl] *adj.* abnehmbar; entfernbar ‹*Fleck, Trennwand*›; herausnehmbar ‹*Futter*›

removal [rɪ'mu:vl] *n.* **a)** *(taking away)* Entfernung, *die; (of traces)* Beseitigung, *die;* **b)** *(dismissal)* Entlassung, *die;* **the minister's ~ from office** die Entfernung des Ministers aus dem Amt; **c)** *see* **remove 1 c:** Beseitigung, *die;* Vertreibung, *die;* Zerstreuung, *die;* **d)** *(transfer of furniture)* Umzug, *der;* **'Smith & Co., R~s'** „Smith & Co., Spedition"

removal: ~ firm *n.* Spedition, *die;* **~ man** *n.* Möbelpacker, *der;* **~ van** *n.* Möbelwagen, *der*

remove [rɪ'mu:v] **1.** *v.t.* **a)** *(take away)* entfernen; streichen ‹*Buchpassage*›; wegnehmen, wegräumen ‹*Papiere, Ordner usw.*›; abräumen ‹*Geschirr*›; beseitigen ‹*Spur*›; *(take off)* abnehmen; ausziehen ‹*Kleidungsstück*›; **she ~d her/the child's coat** sie legte ihren Mantel ab/sie zog dem Kind den Mantel aus; **~ one's make-up** sich abschminken; **the parents ~d the child from the school** die Eltern nahmen das Kind von der Schule; **b)** *(dismiss)* entlassen; **~ sb. from office/his post** jmdn. aus dem Amt/von seinem Posten entfernen; **c)** *(eradicate)* beseitigen ‹*Gefahr, Hindernis, Problem, Zweifel*›; vertreiben ‹*Angst*›; zerstreuen ‹*Verdacht, Befürchtungen*›; **d)** *in p.p.* *(remote)* **be entirely ~d from politics/everyday life** gar nichts mit Politik zu tun haben/völlig lebensfremd sein. **2.** *v.i. (formal)* [um]ziehen. **3.** *n.* **be but one ~ from** nur noch einen Schritt entfernt sein von; **at one ~:** auf Distanz (**from** gegenüber)

remover [rɪ'mu:və(r)] *n.* **a)** *(of paint/varnish/hair/rust)* Farb-/Lack-/Haar-/Rostentferner, *der;* **b)** *(removal man)* Möbelpacker, *der*

remuneration [rɪmju:nə'reɪʃn] *n.* Bezahlung, *die;* Entlohnung, *die;* *(reward)* Belohnung, *die*

Renaissance [rə'neɪsəns, rɪ'neɪsəns] *n., no pl. (Hist.)* Renaissance, *die*

rename [ri:'neɪm] *v.t.* umbenennen; umtaufen ‹*Schiff*›

render ['rendə(r)] *v.t.* **a)** *(show, give)* leisten ‹*Gehorsam, Hilfe*›; erweisen ‹*Ehre, Achtung, Respekt, Dienst*›; bieten, gewähren ‹*Schutz*›; **~ a service to sb., ~ sb. a service** jmdm. einen Dienst erweisen; **b)** *(pay)* entrichten ‹*Tribut, Steuern, Abgaben*›; **c)** *(represent, reproduce)* wiedergeben, spielen ‹*Musik, Szene*›; *(translate)* übersetzen (**by** mit); **~ a text into another language** einen Text in eine andere Sprache übertragen

rendering ['rendərɪŋ] *n. see* **render c:** Wiedergabe, *die;* Spielen, *das;* Übersetzung, *die;* Übertragung, *die*

rendezvous ['rɒndɪvu:, 'rɒndeɪvu:] *n., pl. same* ['rɒndɪvu:z, 'rɒndeɪvu:z] **a)** *(meeting-place)* Treffpunkt, *der;* **b)** *(meeting)* Rendezvous, *das (veralt.);* Verabredung, *die;* **c)** *(Astronaut.)* Rendezvous, *das*

rendition [ren'dɪʃn] *see* **rendering**

renegade ['renɪgeɪd] **1.** *n.* Abtrünnige, *der/die.* **2.** *adj.* abtrünnig

renew [rɪ'nju:] *v.t.* **a)** *(restore, regenerate, recover)* erneuern; wieder wecken *od.* wachrufen ‹*Gefühle*›; wiederherstellen ‹*Kraft*›; **b)** *(replace)* erneuern; auffüllen ‹*Vorrat*›; ausbessern ‹*Kleidungs-*

stück⟩; **c)** *(begin again)* erneuern ⟨*Bekanntschaft*⟩; fortsetzen ⟨*Angriff, Bemühungen*⟩; **d)** *(repeat)* wiederholen ⟨*Aussage, Beschuldigung*⟩; **e)** *(extend)* erneuern, verlängern ⟨*Vertrag, Genehmigung, Ausweis etc.*⟩; ~ **a library book** ⟨*Bibliothekar/Benutzer:*⟩ ein Buch [aus der Bücherei] verlängern/verlängern lassen

renewable [rɪ'nju:əbl] *adj.* regenerationsfähig ⟨*Energiequelle*⟩; verlängerbar ⟨*Vertrag, Genehmigung, Ausweis*⟩

renewal [rɪ'nju:əl] *n.* Erneuerung, *die; (of contract, passport etc. also)* Verlängerung, *die; (of attack)* Wiederaufnahme, *die; (of library book)* Verlängerung der Leihfrist

renounce [rɪ'naʊns] *v. t.* **a)** *(abandon)* verzichten auf (+ *Akk.*); **b)** *(refuse to recognize)* aufkündigen ⟨*Vertrag, Freundschaft*⟩; aufgeben ⟨*Grundsatz, Plan, Versuch*⟩; verstoßen ⟨*Person*⟩; ~ **the devil/one's faith** dem Teufel/seinem Glauben abschwören

renovate ['renəveɪt] *v. t.* renovieren ⟨*Gebäude*⟩; restaurieren ⟨*Möbel, Gemälde*⟩

renovation [renə'veɪʃn] *n.* Renovierung, *die; (of furniture etc.)* Restaurierung, *die*

renown [rɪ'naʊn] Renommee, *das;* Ansehen, *das;* **of [great]** ~: von hohem Ansehen

renowned [rɪ'naʊnd] *adj.* berühmt (**for** wegen, für)

rent [rent] **1.** *n. (for house, flat, etc.)* Miete, *die; (for land)* Pacht, *die.* **2.** *v. t.* **a)** *(use)* mieten ⟨*Haus, Wohnung usw.*⟩; pachten ⟨*Land*⟩; mieten ⟨*Auto, Gerät*⟩; **b)** *(let)* vermieten ⟨*Haus, Wohnung, Auto etc.*⟩ (**to** *Dat.*, an + *Akk.*); verpachten ⟨*Land*⟩ (**to** *Dat.*, an + *Akk.*)

~ **'out** *see* **rent 2 b**

rental ['rentl] *n.* **a)** *(from houses etc.)* Miete, *die; (from land)* Pacht, *die;* **b)** *see* **rent 2:** Mietung, *die;* Pachtung, *die; (letting)* Vermietung, *die;* Verpachtung, *die;* **car** ~: Autoverleih, *der*

rent: ~**-controlled** *adj.* mietpreisgebunden; ~**-free** *adj.* mietfrei; ~ **rebate** *n.* Mietermäßigung, *die*

renunciation [rɪnʌnsɪ'eɪʃn] *n.* **a)** *see* **renounce 1 a, b:** Verzicht, *der;* Aufkündigung, *die;* Aufgabe, *die;* Verstoßung, *die;* **b)** *(self-denial)* Selbstverleugnung, *die*

reopen [rɪ'əʊpn] **1.** *v. t.* **a)** *(open again)* wieder öffnen; wieder aufmachen; wiedereröffnen ⟨*Geschäft, Lokal usw.*⟩; **b)** *(return to)* wiederaufnehmen ⟨*Diskussion, Verhandlung, Feindseligkeiten*⟩; wiederaufnehmen, wieder aufrollen ⟨*Fall*⟩; zurückkommen auf (+ *Akk.*) ⟨*Angelegenheit*⟩. **2.** *v. i.* ⟨*Geschäft, Lokal usw.:*⟩ wieder öffnen; wiedereröffnet werden; ⟨*Verhandlungen, Unterricht:*⟩ wieder beginnen

reorder [ri:'ɔ:də(r)] *v. t.* **a)** *(Commerc.: order again)* nachbestellen ⟨*Ware*⟩; *(after theft, loss)* neu bestellen; **b)** *(rearrange)* umordnen

reorganisation, reorganise *see* **reorganiz-**

reorganization [ri:ɔ:gənaɪ'zeɪʃn] *n.* Umorganisation, *die; (of time, work)* Neueinteilung, *die; (of text)* Neugliederung, *die*

reorganize [ri:'ɔ:gənaɪz] *v. t.* umorganisieren; neu einteilen ⟨*Zeit, Arbeit*⟩; neu gliedern ⟨*Aufsatz, Referat*⟩

reorient [ri:'ɔ:rɪent, ri:'ɒrɪent], **reorientate** [ri:'ɔ:rɪənteɪt] *v. t.* neu ausrichten; ~ **sb.** jmdm. eine neue Orientierung geben

reorientation [ri:ɔ:rɪən'teɪʃn] *n.* Neuorientierung, *die*

¹rep [rep] *n. (coll.: representative)* Vertreter, *der*/Vertreterin, *die*

²rep *n. (Theatre coll.)* Repertoiretheater, *das;* **be in** ~: an einem Repertoiretheater spielen

repaid *see* **repay**

repaint [ri:'peɪnt] *v. t.* neu streichen ⟨*Gebäude, Wand, Tür usw.*⟩; neu lackieren ⟨*Auto*⟩

repair [rɪ'peə(r)] **1.** *v. t.* **a)** *(restore, mend)* reparieren; ausbessern ⟨*Kleidung, Straße*⟩; **b)** *(remedy)* wiedergutmachen ⟨*Schaden, Fehler*⟩; beheben ⟨*Schaden, Mangel*⟩. **2.** *n.* **a)** *(restoring, renovation)* Reparatur, *die;* **be beyond** ~: sich nicht mehr reparieren lassen; **be in need of** ~: reparaturbedürftig sein; **b)** *no pl., no art. (condition)* **be in good/bad** ~: in gutem/schlechtem Zustand sein

repairable [rɪ'peərəbl] *adj.* reparabel

reparation [repə'reɪʃn] *n.* **a)** *(making amends)* Wiedergutmachung, *die;* **b)** *(compensation)* Entschädigung, *die;* ~**s** *(for war damage)* Reparationen; **make** ~ **[for sth.]** [für etw.] Ersatz leisten

repartee [repɑ:'ti:] *n.* **a)** *(skill in making retorts)* Schlagfertigkeit, *die;* **be good at** ~: schlagfertig sein; **b)** *(conversation)* von [Geist und] Schlagfertigkeit sprühende Unterhaltung

repatriate [ri:'pætrɪeɪt] *v. t.* repatriieren

repatriation [ri:pætrɪ'eɪʃn] *n.* Repatriierung, *die*

repay [ri:'peɪ] **1.** *v. t.*, **repaid** [ri:'peɪd] **a)** *(pay back)* zurückzahlen ⟨*Schulden usw.*⟩; erstatten ⟨*Spesen*⟩; **b)** *(return)* erwidern ⟨*Besuch, Freundlichkeit*⟩; **c)** *(give in recompense)* ~ **sb. for sth.** jmdm. etw. vergelten. **2.** *v. i.*, **repaid** Rückzahlungen leisten

repayable [ri:'peɪəbl] *adj.* rückzahlbar

repayment [ri:'peɪmənt] *n.* **a)** *(paying back)* Rückzahlung, *die;* **b)** *(reward)* Lohn, *der* (**for** für)

repeal [rɪ'pi:l] **1.** *v. t.* aufheben ⟨*Gesetz, Erlaß usw.*⟩. **2.** *n.* Aufhebung, *die*

repeat [rɪ'pi:t] **1.** *n.* **a)** Wiederholung, *die; (Radio, TV also)* Wiederholungssendung, *die;* **do a** ~ **of sth.** etw. wiederholen; **b)** *(Commerc.)* Nachbestellung, *die.* **2.** *v. t.* **a)** *(say, do, broadcast again)* wiederholen; **please** ~ **after me: ...:** sprich/sprecht/sprechen Sie mir bitte nach: ...; **b)** *(recite)* aufsagen ⟨*Gedicht, Strophe, Text*⟩; **c)** *(report)* weitererzählen (**to** *Dat.*). **3.** *v. i. (Math.: recur)* ⟨*Zahl:*⟩ periodisch sein

repeat: ~ **'order** *n. (Commerc.)* Nachbestellung, *die;* ~ **per'formance** *n.* Wiederholungsvorstellung, *die (Theater)*

repel [rɪ'pel] *v. t.*, **-ll-: a)** *(drive back)* abwehren ⟨*Feind, Annäherungsversuch usw.*⟩; abstoßen ⟨*Feuchtigkeit, elektrische Ladung, Magnetpol*⟩; **b)** *(be repulsive to)* abstoßen

repellent [rɪ'pelənt] *adj.* **a)** *(repugnant)* abstoßend; **b)** *(repelling)* **water-**~: wasserabstoßend

repent [rɪ'pent] *v. i.* bereuen (**of** *Akk.*)

repentance [rɪ'pentəns] *n.* Reue, *die*

repentant [rɪ'pentənt] *adj.* reuig; reuevoll *(geh.);* reumütig *(öfter scherzh.)*

repercussion [ri:pə'kʌʃn] *n. usu. in pl.* Auswirkung, *die* (**[up]on** auf + *Akk.*)

repertoire ['repətwɑ:(r)] *n. (Mus., Theatre)* Repertoire, *das* (**of** an + *Dat.*, von)

repertory ['repətərɪ] *n.* **a)** *see* **repertoire; b)** *(Theatre)* Repertoiretheater, *das*

'repertory company *n.* Repertoiretheater, *das*

repetition [repɪ'tɪʃn] *n.* Wiederholung, *die*

repetitious [repɪ'tɪʃəs] *adj.* sich immer wiederholend *attr.*

repetitive [rɪ'petɪtɪv] *adj.* eintönig; **sth. is ~**: etw. bietet keine Abwechslung

rephrase [ri:'freɪz] *v. t.* umformulieren; **I'll ~ that** ich will es anders ausdrücken

replace [rɪ'pleɪs] *v. t.* **a)** *(vertically)* zurückstellen; *(horizontally)* zurücklegen; wieder einordnen ‹*Karteikarte*›; [wieder] auflegen ‹*Telefonhörer*›; **b)** *(take place of, provide substitute for)* ersetzen; **~ A with** *or* **by B** A durch B ersetzen; **c)** *(renew)* ersetzen ‹*Gestohlenes usw.*›; austauschen, auswechseln ‹*Maschinen[teile] usw.*›

replacement [rɪ'pleɪsmənt] *n.* **a)** *see* **replace a**: Zurückstellen, *das;* Zurücklegen, *das;* Wiedereinordnen, *das;* Auflegen, *das;* **b)** *(provision of substitute for)* Ersatz, *der;* Ersetzen, *das; attrib.* Ersatz-; **c)** *(substitute)* Ersatz, *der;* **~ [part]** Ersatzteil, *das*

replay 1. [ri:'pleɪ] *v. t.* wiederholen ‹*Spiel*›; nochmals abspielen ‹*Tonband usw.*›. **2.** ['ri:pleɪ] *n.* Wiederholung, *die; (match)* Wiederholungsspiel, *das*

replenish [rɪ'plenɪʃ] *v. t.* [wieder] auffüllen

replete [rɪ'pli:t] *adj. (filled)* reich (**with** an + *Dat.*)

replica ['replɪkə] *n.* Nachbildung, *die; (of work of art)* Kopie, *die*

reply [rɪ'plaɪ] **1.** *v. i.* **~ [to sb./sth.]** [jmdm./auf etw. *(Akk.)*] antworten. **2.** *v. t.* **~ that ...**: antworten, daß ... **3.** *n.* Antwort, *die* (**to** auf + *Akk.*); **in/by way of ~**: als Antwort; **in ~ to your letter** in Beantwortung Ihres Schreibens *(Amtsspr.)*

report [rɪ'pɔ:t] **1.** *v. t.* **a)** *(relate)* berichten/*(in writing)* einen Bericht schreiben über (+ *Akk.*) ‹*Ereignis usw.*›; *(state formally also)* melden; **sb. is/was ~ed to be ...**: jmd. soll ... sein/gewesen sein; **~ sb. missing** jmdn. als vermißt melden; **b)** *(repeat)* übermitteln (**to** *Dat.*) ‹*Botschaft*›; wiedergeben (**to** *Dat.*) ‹*Worte, Sinn*›; **he is ~ed as having said that ...**: er soll gesagt haben, daß ...; **c)** *(name or notify to authorities)* melden (**to** *Dat.*); *(for prosecution)* anzeigen (**to** bei). **2.** *v. i.* **a)** Bericht erstatten (**on** über + *Akk.*); berichten (**on** über + *Akk.*); *(Radio, Telev.)* **[this is] John Tally ~ing [from Delhi]** John Tally berichtet [aus Delhi]; **b)** *(present oneself)* sich melden (**to** bei); **~ for duty** sich zum Dienst melden; **~ sick** sich krank melden; **c)** *(be responsible)* **~ to sb.** jmdm. unterstehen. **3.** *n.* **a)** *(account)* Bericht, *der* (**on, about** über + *Akk.*); *(in newspaper etc. also)* Reportage, *die* (**on** über + *Akk.*); **make a ~**: einen Bericht abfassen; **b)** *(Sch.)* Zeugnis, *das;* **c)** *(sound)* Knall, *der;* **d)** *(rumour)* Gerücht, *das*

~ 'back *v. i.* **a)** *(present oneself again)* sich zurückmelden (**for** zu); **b)** *(give a ~)* Bericht erstatten (**to** *Dat.*)

reported 'speech *n. (Ling.)* indirekte Rede

reporter [rɪ'pɔ:tə(r)] *n. (Radio, Telev., Journ.)* Reporter, *der*/Reporterin, *die;* Berichterstatter, *der*/-erstatterin, *die*

repose [rɪ'pəʊz] *(literary)* **1.** *n. (rest, respite)* Ruhe, *die.* **2.** *v. i. (lie)* ruhen

reprehensible [reprɪ'hensɪbl] *adj.* tadelnswert; sträflich

represent [reprɪ'zent] *v. t.* **a)** *(symbolize)* verkörpern; **b)** *(denote, depict, present)* darstellen (**as** als); *(Theatre also)* spielen; **c)** *(correspond to)* entsprechen (+ *Dat.*); **d)** *(be specimen of, act for)* vertreten

representation [reprɪzen'teɪʃn] *n.* **a)** *(depicting, image)* Darstellung, *die;* **b)** *(acting for sb.)* Vertretung, *die;* **c)** *(protest)* Protest, *der;* **make ~s to sb.** bei jmdm. Protest einlegen

representative [reprɪ'zentətɪv] **1.** *n.* **a)** *(member, agent, deputy)* Vertreter, *der*/Vertreterin, *die; (firm's agent, deputy also)* Repräsentant, *der*/Repräsentantin, *die;* **b) R~** *(Amer. Polit.)* Abgeordneter/Abgeordnete im Repräsentantenhaus; **House of R~s** Repräsentantenhaus, *das.* **2.** *adj.* **a)** *(typical)* repräsentativ (**of** für); **b)** *(consisting of deputies)* Abgeordneten‹*versammlung, -kammer usw.*›; **c)** *(Polit.: based on representation)* repräsentativ; Repräsentativ‹*system, -verfassung*›; **d)** **be ~ of** *(portray)* darstellen; *(symbolize)* symbolisieren; ‹*Person:*› verkörpern

repress [rɪ'pres] *v. t.* **a)** unterdrücken ‹*Aufruhr, Gefühle, Lachen usw.*›; **b)** *(Psych.)* verdrängen ‹*Gefühle*› (**from** aus)

repressed [rɪ'prest] *adj.* unterdrückt; *(Psych.)* verdrängt

repression [rɪ'preʃn] *n.* Unterdrückung, *die; (Psych.)* Verdrängung, *die*

repressive [rɪ'presɪv] *adj.* repressiv

reprieve [rɪ'pri:v] **1.** *v. t.* **~ sb.** *(postpone execution)* jmdm. Strafaufschub gewähren; *(remit execution)* jmdn. begnadigen; *(fig.)* verschonen. **2.** *n.* Strafaufschub, *der* (**of** für); Begnadigung, *die; (fig.)* Gnadenfrist, *die*

reprimand ['reprɪmɑ:nd] **1.** *n.* Tadel, *der;* Verweis, *der.* **2.** *v. t.* tadeln; einen Verweis erteilen (+ *Dat.*)

reprint 1. [ri:'prɪnt] *v. t.* **a)** *(print again)* wieder abdrucken; **b)** *(make ~ of)* nachdrucken. **2.** ['ri:prɪnt] *n. (book ~ed)* Nachdruck, *der*

reprisal [rɪ'praɪzl] *n.* Vergeltungsakt, *der* (**for** gegen)

reproach [rɪ'prəʊtʃ] **1.** *v. t.* **~ sb.** jmdm. Vorwürfe machen; **~ sb. with** *or* **for sth.** jmdm. etw. vorwerfen *od.* zum Vorwurf machen; **have nothing to ~ oneself for** *or* **with** sich *(Dat.)* nichts vorzuwerfen haben. **2.** *n.* **a)** *(rebuke)* Vorwurf, *der;* **be above** *or* **beyond ~**: über jeden Vorwurf erhaben sein; **look of ~**: vorwurfsvoller Blick; **b)** *(disgrace)* Schande, *die* (**to** für)

reproachful [rɪ'prəʊtʃfl] *adj.* vorwurfsvoll

reprobate ['reprəbeɪt] *n.* Halunke, *der*

reprocess [ri:'prəʊses] *v. t.* wiederaufbereiten

reproduce [ri:prə'dju:s] **1.** *v. t.* wiedergeben; reproduzieren *(Druckw.)* ‹*Bilder usw.*›. **2.** *v. i.* *(multiply)* sich fortpflanzen; sich vermehren

reproduction [ri:prə'dʌkʃn] *n.* **a)** Wiedergabe, *die;* Reproduktion, *die (Druckw.);* **~ of sound** Tonwiedergabe, *die;* **b)** *(producing offspring)* Fortpflanzung, *die;* **c)** *(copy)* Reproduktion, *die; attrib.* **~ furniture** Stilmöbel *Pl.*

reproof [rɪ'pru:f] *n.* Tadel, *der*

reprove [rɪ'pru:v] *v. t.* tadeln ‹*Verhalten usw.*›; tadeln, zurechtweisen ‹*Person*›

reptile ['reptaɪl] *n.* Reptil, *das;* Kriechtier, *das*

reptilian [rep'tɪljən] *adj.* reptilartig; *(of reptile)* Reptilien‹*knochen, -schädel*›

republic [rɪ'pʌblɪk] *n.* Republik, *die*

republican [rɪ'pʌblɪkən] **1.** *adj.* **a)** republikanisch; **b)** *(Amer. Polit.)* **R~ Party** Republikanische Partei. **2.** *n.* **R~** *(Amer. Polit.)* Republikaner, *der*/Republikanerin, *die*

repudiate [rɪ'pju:dɪeɪt] *v. t.* **a)** *(deny)* zurückweisen ‹*Anschuldigung usw.*›; *(reject)* nicht anerkennen ‹*Autorität, Vertrag usw.*›; **b)** *(disown)* verstoßen ‹*Person*›

repugnance [rɪ'pʌgnəns] *n.* Abscheu, *der* (**to[wards]** vor + *Dat.*)
repugnant [rɪ'pʌgnənt] *adj.* widerlich; abstoßend; **be ~ to sb.** jmdm. widerlich sein
repulse [rɪ'pʌls] *v.t. (lit. or fig.)* abwehren
repulsion [rɪ'pʌlʃn] *n.* **a)** *(disgust)* Widerwille, *der* (**towards** gegen); **b)** *(Phys.)* Repulsion, *die*
repulsive [rɪ'pʌlsɪv] *adj. (disgusting)* abstoßend; widerwärtig
reputable ['repjʊtəbl] *adj.* angesehen ‹*Person, Familie, Beruf, Zeitung usw.*›; anständig ‹*Verhalten*›; seriös ‹*Firma*›
reputation [repjʊ'teɪʃn] *n.* **a)** Ruf, *der;* **have a ~ for doing/being sth.** in dem Ruf stehen, etw. zu tun/sein; **what sort of ~ do they have?** wie ist ihr Ruf?; **b)** *(good name)* Name, *der;* **c)** *(bad name)* schlechter Ruf
repute [rɪ'pju:t] **1.** *v.t. in pass.* **be ~d [to be] sth.** als etw. gelten; **she is ~d to have/make ...:** man sagt, daß sie ... hat/macht. **2.** *n.* Ruf, *der;* Ansehen, *das;* **hold sb./sth. in high ~:** von jmdm./etw. eine hohe Meinung haben; jmdn./etw. hochschätzen *(geh.)*
reputedly [rɪ'pju:tɪdlɪ] *adv.* angeblich; vermeintlich
request [rɪ'kwest] **1.** *v.t.* bitten; **~ sth. of** *or* **from sb.** jmdn. um etw. bitten; **~ a record** einen Plattenwunsch äußern; **~ sb. to do sth.** jmdn. [darum] bitten, etw. zu tun; **'You are ~ed not to smoke'** „Bitte nicht rauchen". **2.** *n.* Bitte, *die* (**for** um); **at sb.'s ~:** auf jmds. Bitte *od.* Wunsch *(Akk.)* [hin]; **I have one ~ to make of you** ich habe eine Bitte an Sie; **by** *or* **on ~:** auf Wunsch; **record ~s** *(Radio)* Plattenwünsche *Pl.*
re'quest stop *n. (Brit.)* Bedarfshaltestelle, *die*
requiem ['rekwɪem] *n.* Requiem, *das*
require [rɪ'kwaɪə(r)] *v.t.* **a)** *(need, wish to have)* brauchen; benötigen; erfordern ‹*Tun, Verhalten*›; **a catalogue/guide is available if ~d** bei Bedarf ist ein Katalog erhältlich/auf Wunsch steht ein Führer zur Verfügung; **is there anything else you ~?** brauchen/ *(want)* wünschen Sie außerdem noch etwas?; **b)** *(order, demand)* verlangen (**of** von); **~ sb. to do sth., ~ of sb. that he does sth.** von jmdm. verlangen, daß er etw. tut; **be ~d to do sth.** etw. tun müssen *od.* sollen
requirement [rɪ'kwaɪəmənt] *n.* **a)** *(need)* Bedarf, *der;* **meet the ~s** den Bedarf decken; **meet sb.'s ~s** jmds. Wünschen entsprechen; **b)** *(condition)* Erfordernis, *das; (for a job)* Voraussetzung, *die;* **fulfil sb.'s ~s** jmds. Anforderungen *(Dat.)* genügen
requisite ['rekwɪzɪt] **1.** *adj.* notwendig (**to, for** für). **2.** *n.* **toilet/travel ~s** Toiletten-/Reiseartikel *Pl.*
requisition [rekwɪ'zɪʃn] **1.** *n.* **a)** *(esp. Law: demand)* Aufforderung, *die;* **b)** *(order for sth.)* Anforderung, *die* (**for** *Gen.*); *(by force if necessary)* Beschlagnahmung, *die* (**for** *Gen.*). **2.** *v.t.* anfordern; *(by force if necessary)* beschlagnahmen
reran *see* **rerun 1**
reread [ri:'ri:d] *v.t.*, **reread** [ri:'red] wieder *od.* nochmals lesen
re-route [ri:'ru:t] *v.t.*, **~ing** umleiten
rerun 1. [ri:'rʌn] *v.t., forms as* **run 3** wiederholen ‹*Rennen*›; wieder auf- *od.* vorführen ‹*Film*›; wieder abspielen ‹*Tonband*›. **2.** ['ri:rʌn] *n. see* **1:** Wiederholung, *die;* Wiederaufführung, *die*
resat *see* **resit 1**
reschedule [ri:'ʃedju:l, ri:'skedju:l] *v.t.* zeitlich neu festlegen ‹*Veranstaltung, Flug, Programm usw.*›; **the flight will be ~d for 5 o'clock** der Flug wird auf 5 Uhr verlegt
rescind [rɪ'sɪnd] *v.t.* für ungültig erklären
rescue ['reskju:] **1.** *v.t.* retten (**from** aus); *(set free)* befreien (**from** aus); **~ sb. from drowning** jmdn. vorm Ertrinken retten. **2.** *n. see* **1:** Rettung, *die;* Befreiung, *die; attrib.* Rettungs‹*dienst, -versuch, -mannschaft, -aktion*›; **go/come to the/sb.'s ~:** jmdm. zu Hilfe kommen
rescuer ['reskju:ə(r)] *n.* Retter, *der*/Retterin, *die*
research [rɪ'sɜ:tʃ, 'ri:sɜ:tʃ] **1.** *n.* **a)** *(scientific study)* Forschung, *die* (**into, on** über + *Akk.*); **do ~ in biochemistry** auf dem Gebiet der Biochemie forschen; **piece of ~:** Forschungsarbeit, *die; (investigation)* Untersuchung, *die;* **b)** *(inquiry)* Nachforschung, *die* (**into** über + *Akk.*). **2.** *v.i.* forschen; **~ into sth.** etw. erforschen *od.* untersuchen; *(esp. Univ.)* über etw. *(Akk.)* forschen. **3.** *v.t.* erforschen; untersuchen; recherchieren ‹*Buch usw.*›
research assistant [-'- ---, '-- ---] *n.* wissenschaftlicher Assistent/ wissenschaftliche Assistentin
researcher [rɪ'sɜ:tʃə(r), 'ri:sɜ:tʃə(r)] *n.* Forscher, *der*/Forscherin, *die*
research: ~ fellowship *n.* Forschungsstipendium, *das;* **~ student** *n.* ≈ Doktorand, *der*/Doktorandin, *die;* **~ worker** *n. mit Nachforschungen beauftragte Person;* ≈ Rechercheur, *der*/Rechercheurin, *die*
reselect [ri:sɪ'lekt] *v.t. (Parl.)* wieder aufstellen ‹*Abgeordneten*›
resemblance [rɪ'zembləns] *n.* Ähnlichkeit, *die* (**to** mit, **between** zwischen + *Dat.*); **bear a faint/strong/no ~ to ...:** eine geringe/ starke/keine Ähnlichkeit mit ... haben
resemble [rɪ'zembl] *v.t.* ähneln, gleichen (+ *Dat.*)
resent [rɪ'zent] *v.t.* übelnehmen; **she ~ed his success** sie mißgönnte ihm seinen Erfolg; **she ~ed his having won** sie ärgerte sich darüber, daß er gewonnen hatte
resentful [rɪ'zentfl] *adj.* übelnehmerisch, nachtragend ‹*Person, Art, Verhalten*›; grollend *(geh.)* ‹*Blick*›; **feel ~ about sth.** etw. übelnehmen; **be ~ of sb.'s success** jmdm. seinen Erfolg mißgönnen
resentment [rɪ'zentmənt] *n., no pl.* Groll, *der (geh.);* **feel ~ towards** *or* **against sb.** einen Groll auf jmdn. haben
reservation [rezə'veɪʃn] *n.* **a)** Reservierung, *die;* **[seat] ~:** [Platz]-reservierung, *die;* **b)** *(doubt, objection)* Vorbehalt, *der* (**about** gegen); Bedenken (**about** bezüglich + *Gen.*); **without ~:** ohne Vorbehalt; vorbehaltlos; **with ~s** mit [gewissen] Vorbehalten; **c)** *see* **central reservation**
reserve [rɪ'zɜ:v] **1.** *v.t.* **a)** *(secure)* reservieren lassen ‹*Tisch, Platz, Zimmer*›; *(set aside)* reservieren; **~ the right to do sth.** sich *(Dat.)* [das Recht] vorbehalten, etw. zu tun; **all rights ~d** alle Rechte vorbehalten; **b)** *in pass. (be kept)* **be ~d for sb.** ‹*Funktion, Tätigkeit:*› jmdm. vorbehalten sein; **c)** *(postpone)* **~ judgement** sein Urteil aufschieben. **2.** *n.* **a)** *(extra amount)* Reserve, *die* (**of an** + *Dat.*); *(Banking also)* Rücklage, *die;* **~s of energy/strength** Energie-/Kraftreserven; **keep sth. in ~:** etw. in Reserve halten; **b)** *in sing. or pl. (Mil.) (troops)* Reserve, *die;* **c)** *(Sport)* Reservespieler, *der*/-spielerin, *die;* **the R~s** die Reserve; **d)** *(restriction)* Vorbehalt, *der;* **without ~:** ohne Vorbehalt; vorbehaltlos; **e)** *(reticence)* Reserve, *die;* Zurückhaltung, *die*

reserved [rɪ'zɜ:vd] *adj.* **a)** *(reticent)* reserviert; zurückhaltend; **b)** *(booked)* reserviert

reservist [rɪ'zɜ:vɪst] *n. (Mil.)* Reservist, *der*

reservoir ['rezəvwɑ:(r)] *n.* **a)** *([artificial] lake)* Reservoir, *das;* **b)** *(container)* Behälter, *der;* Speicher, *der*

resettle [ri:'setl] *v.t.* umsiedeln ‹*Flüchtlinge usw.*› (**in** in + *Akk.*)

reshape [ri:'ʃeɪp] *v.t.* **a)** *(give new form to)* umgestalten; umstellen ‹*Politik*›; **b)** *(remould)* umformen

reshuffle [ri:'ʃʌfl] **1.** *v.t.* **a)** *(reorganize)* umbilden ‹*Kabinett usw.*›; **b)** *(Cards)* neu mischen. **2.** *n.* Umbildung, *die;* **Cabinet ~:** Kabinettsumbildung, *die*

reside [rɪ'zaɪd] *v.i. (formal)* **a)** *(dwell)* wohnen; wohnhaft sein *(Amtsspr.);* ‹*Monarch, Präsident usw.:*› residieren; **b)** *(be vested, present)* liegen (**in** bei)

residence ['rezɪdəns] *n.* **a)** *(abode)* Wohnsitz, *der; (house)* Wohnhaus, *das; (mansion)* Villa, *die; (of a head of state or church, an ambassador)* Residenz, *die;* **the President's official ~:** der offizielle Wohnsitz des Präsidenten; **b)** *(residing)* Aufenthalt, *der;* **take up ~ in Rome** seinen Wohnsitz in Rom nehmen; **be in ~** ‹*König, Präsident usw.:*› [an seinem offiziellen Wohnsitz] anwesend sein

'residence permit *n.* Aufenthaltsgenehmigung, *die*

resident ['rezɪdənt] **1.** *adj.* **a)** *(residing)* wohnhaft; **he is ~ in England** er hat seinen Wohnsitz in England; **b)** *(living in)* im Haus wohnend ‹*Haushälterin*›; Anstalts‹*arzt, -geistlicher*›. **2.** *n.* *(inhabitant)* Bewohner, *der*/Bewohnerin, *die; (in a town etc. also)* Einwohner, *der*/Einwohnerin, *die; (at hotel)* Hotelgast, *der;* **'access/parking for ~s only'** „Anlieger frei"/„Parken nur für Anlieger"

residential [rezɪ'denʃl] *adj.* **a)** Wohn‹*gebiet, -siedlung, -straße*›; **b) ~ course** *Kurs, dessen Teilnehmer am Ort wohnen*

residual [rɪ'zɪdjʊəl] *adj.* zurückgeblieben; noch vorhanden

residue ['rezɪdju:] *n.* **a)** *(remainder)* Rest, *der;* **b)** *(Chem.)* Rückstand, *der*

resign [rɪ'zaɪn] **1.** *v.t. (hand over)* zurücktreten von ‹*Amt*›; verzichten auf (+ *Akk.*) ‹*Recht, Anspruch*›; **~ one's job/post** seine Stelle/Stellung kündigen. **2.** *v. refl.* **~ oneself to sth.** sich mit etw. abfinden. **3.** *v.i.* ‹*Arbeitnehmer:*› kündigen; ‹*Regierungsbeamter:*› zurücktreten (**from** von); ‹*Vorsitzender:*› zurücktreten, sein Amt niederlegen

resignation [rezɪg'neɪʃn] *n.* **a)** *see* **resign 3:** Kündigung, *die;* Rücktritt, *der;* **give in** *or* **tender one's ~:** seine Kündigung/seinen Rücktritt einreichen; **b)** *(being resigned)* Ergebenheit, *die* (**to** in + *Akk.*)

resigned [rɪ'zaɪnd] *adj.* resigniert; **be ~ to sth.** sich mit etw. abgefunden haben

resilience [rɪ'zɪlɪəns] *n., no pl.* **a)** *(elasticity)* Elastizität, *die;* **b)** *(fig.)* Unverwüstlichkeit, *die*

resilient [rɪ'zɪlɪənt] *adj.* **a)** *(elastic)* elastisch; **b)** *(fig.)* unverwüstlich; **be ~:** sich nicht [so leicht] unterkriegen lassen

resin ['rezɪn] *n. (Bot.)* Harz, *das*

resist [rɪ'zɪst] **1.** *v.t.* **a)** *(withstand action of)* standhalten (+ *Dat.*) ‹*Frost, Hitze, Feuchtigkeit usw.*›; **b)** *(oppose, repel)* sich widersetzen (+ *Dat.*) ‹*Maßnahme, Festnahme, Plan usw.*›; widerstehen (+ *Dat.*) ‹*Versuchung, jmds. Charme*›; Widerstand leisten gegen ‹*Angriff, Feind*›; sich wehren gegen ‹*Veränderung, Einfluß*›. **2.** *v.i. see* **1b:** sich widersetzen; widerstehen; Widerstand leisten; sich wehren

resistance [rɪ'zɪstəns] *n.* **a)** *(resisting, opposing force)* Widerstand, *der* (**to** gegen); **make** *or* **offer no ~ [to sb./sth.]** [jmdm./einer Sache] keinen Widerstand leisten; **b)** *(Biol., Med.)* Widerstandskraft, *die* (**to** gegen); **c)** *(against occupation)* Widerstand, *der*

resistant [rɪ'zɪstənt] *adj.* **a)** *(having power to resist)* widerstandsfähig (**to** gegen); **heat-/water-/rust-~:** hitze-/wasser-/rostbeständig; **b)** *(Med., Biol.)* resistent (**to** gegen)

resit 1. [ri:'sɪt] *v.t.,* **-tt-, resat** [ri:'sæt] wiederholen ‹*Prüfung*›. **2.** ['ri:sɪt] *n.* Wiederholungsprüfung, *die*

resolute ['rezəlu:t] *adj.* resolut, energisch ‹*Person*›; entschlossen ‹*Tat*›; entschieden ‹*Antwort, Weigerung*›

resolution [rezə'lu:ʃn] *n.* **a)** *(decision)* Entschließung, *die; (Polit. also)* Resolution, *die;* **b)** *(resolve)* Vorsatz, *der;* **make a ~:** einen Vorsatz fassen; **make a ~ to do sth.** den Vorsatz fassen, etw. zu tun; **New Year['s] ~s** gute Vorsätze fürs neue Jahr; **c)** *no pl. (firmness)* Entschlossenheit, *die;* **d)** *no pl. (solving) see* **resolve 1 a, b:** Beseitigung, *die;* Ausräumung, *die;* Lösung, *die*

resolve [rɪ'zɒlv] **1.** *v.t.* **a)** *(dispel)* beseitigen, ausräumen ‹*Schwierigkeit, Zweifel, Unklarheit*›; **b)** *(explain)* lösen ‹*Problem, Rätsel*›; **c)** *(decide)* beschließen; **d)** *(settle)* beilegen ‹*Streit*›; klären ‹*Streitpunkt*›; regeln ‹*Angelegenheit*›. **2.** *v.i. (decide)* **~ [up]on sth./doing sth.** sich zu etw. entschließen/sich [dazu] entschließen, etw. zu tun. **3.** *n.* Vorsatz, *der;* **make a ~ to do sth.** den Vorsatz fassen, etw. zu tun

resolved [rɪ'zɒlvd] *pred. adj.* **~ [to do sth.]** entschlossen[, etw. zu tun]

resonance ['rezənəns] *n.* Resonanz, *die; (of voice)* voller Klang; *(fig.)* Widerhall, *der*

resonant ['rezənənt] *adj.* **a)** hallend ‹*Echo, Ton, Klang*›; volltönend ‹*Stimme*›; **b)** ‹*Raum, Körper:*› mit viel Resonanz

resort [rɪ'zɔ:t] **1.** *n.* **a)** *(resource, recourse)* Ausweg, *der;* **you were my last ~:** du warst meine letzte Rettung *(ugs.);* **as a last ~:** als letzter Ausweg; **b)** *(place frequented)* Aufenthalt[sort], *der;* **[holiday] ~:** Urlaubsort, *der;* Ferienort, *der;* **ski/health ~:** Skiurlaubs-/Kurort, *der;* **seaside ~:** Seebad, *das.* **2.** *v.i.* **~ to sth./sb.** zu etw. greifen/sich an jmdn. wenden (**for** um); **~ to violence** Gewalt anwenden; **~ to stealing/shouting** *etc.* sich aufs Stehlen/Schreien *usw.* verlegen

resound [rɪ'zaʊnd] *v.i.* **a)** *(ring)* widerhallen (**with** von); **b)** *(produce echo)* hallen

resounding [rɪ'zaʊndɪŋ] *adj.* hallend ‹*Lärm, Schreie*›; schallend ‹*Gelächter, Stimme*›; überwältigend ‹*Sieg, Erfolg*›; gewaltig ‹*Niederlage, Mißerfolg*›

resource [rɪ'sɔ:s, rɪ'zɔ:s] *n.* **a)** *usu. in pl. (stock)* Mittel *Pl.;* Ressource, *die;* **financial/mineral ~s** Geldmittel *Pl.*/Bodenschätze *Pl.;* **b)** *usu. pl. (Amer.: asset)* Aktivposten, *der;* **c)** *(expedient)* Ausweg, *der;* **be left to one's own ~s** sich *(Dat.)* selbst überlassen sein; *see also* **throw 1 b**

resourceful [rɪ'sɔ:sfl, rɪ'zɔ:sfl] *adj.* findig ‹*Person*›; einfallsreich ‹*Plan*›

respect [rɪ'spekt] **1.** *n.* **a)** *(esteem)* Respekt, *der* (**for** vor + *Dat.*); Achtung, *die* (**for** vor + *Dat.*); **show ~ for sb./sth.** Respekt vor jmdm./etw. zeigen; **hold sb. in [high** *or* **great] ~:** jmdn. [sehr]

achten; **treat sb./sth. with ~:** jmdm./etw. mit Respekt *od.* Achtung begegnen/etw. mit Vorsicht behandeln; **with [all due] ~, ...:** bei allem Respekt, ...; **b)** *(consideration)* Rücksicht, *die* **(for** auf + *Akk.*); **c)** *(aspect)* Beziehung, *die;* Hinsicht, *die;* **in all/many/some ~s** in jeder/vieler/mancher Beziehung *od.* Hinsicht; **d)** *(reference)* Bezug, *der;* **with ~ to ...:** in bezug auf ... *(Akk.);* was ... [an]betrifft; **e)** *in pl.* **pay one's ~s to sb.** *(formal)* jmdm. seine Aufwartung machen *(veralt.).* **2.** *v. t.* respektieren; achten; **~ sb.'s feelings** auf jmds. Gefühle Rücksicht nehmen

respectability [rɪspektəˈbɪlɪtɪ] *n., no pl. see* **respectable a:** Ansehen, *das;* Ehrbarkeit, *die (geh.)*

respectable [rɪˈspektəbl] *adj.* **a)** *(of good character)* angesehen ⟨*Bürger usw.*⟩; ehrenwert ⟨*Motive*⟩; *(decent)* ehrbar *(geh.)* ⟨*Leute, Kaufmann, Hausfrau*⟩; **b)** *(presentable)* anständig, respektabel ⟨*Beschäftigung usw.*⟩; vornehm, gut ⟨*Adresse*⟩; ordentlich, *(that one can be seen in)* vorzeigbar *(ugs.)* ⟨*Kleidung*⟩; **c)** *(considerable)* beachtlich ⟨*Summe*⟩

respectably [rɪˈspektəblɪ] *adv.* anständig ⟨*sich benehmen*⟩; ordentlich ⟨*gekleidet*⟩

respectful [rɪˈspektfl] *adj.* respektvoll **(to[wards]** gegenüber)

respecting [rɪˈspektɪŋ] *prep.* bezüglich; hinsichtlich

respective [rɪˈspektɪv] *adj.* jeweilig

respectively [rɪˈspektɪvlɪ] *adv.* beziehungsweise; **he and I contributed £10 and £1 ~:** er und ich steuerten 10 bzw. 1 Pfund bei

respiration [respɪˈreɪʃn] *n. (one breath)* Atemzug, *der; (breathing)* Atmung, *die*

respirator [ˈrespɪreɪtə(r)] *n.* **a)** *(protecting device)* Atemschutzgerät, *das;* **b)** *(Med.)* Respirator, *der*

respiratory [ˈrespərətərɪ, rɪˈspɪrətərɪ] *adj.* Atem⟨*geräusch, -wege*⟩; Atmungs⟨*system, -organ*⟩

respite [ˈrespaɪt] *n.* **a)** *(delay)* Aufschub, *der;* **b)** *(interval of relief)* Ruhepause, *die;* **without ~:** ohne Pause *od.* Unterbrechung

resplendent [rɪˈsplendənt] *adj.* prächtig

respond [rɪˈspɒnd] **1.** *v. i.* **a)** *(answer)* antworten **(to** auf + *Akk.*); **~ to sb.'s greeting** jmds. Gruß erwidern; **b)** *(react)* reagieren **(to** auf + *Akk.*); ⟨*Patient, Bremsen, Lenkung usw.:*⟩ ansprechen **(to** auf + *Akk.*); **they ~ed very generously to this appeal** der Aufruf fand bei ihnen ein großes Echo. **2.** *v. t.* antworten; erwidern

response [rɪˈspɒns] *n.* **a)** *(answer)* Antwort, *die* **(to** auf + *Akk.*); **in ~ [to]** als Antwort [auf (+ *Akk.*)]; **in ~ to your letter** in Beantwortung Ihres Schreibens *(Papierdt.);* **make no ~:** nicht antworten; **b)** *(reaction)* Reaktion, *die;* **make no ~ to sth.** auf etw. *(Akk.)* nicht reagieren

responsibility [rɪspɒnsɪˈbɪlɪtɪ] *n.* **a)** *no pl., no indef. art. (being responsible)* Verantwortung, *die;* **take** *or* **accept/claim [full] ~ [for sth.]** die [volle] Verantwortung [für etw.] übernehmen; **do sth. on one's own ~:** etw. in eigener Verantwortung tun; *(at one's own risk)* etw. auf eigene Verantwortung tun; **b)** *(duty)* Verpflichtung, *die;* **that's ˈyour ~:** dafür bist du verantwortlich

responsible [rɪˈspɒnsɪbl] *adj.* **a)** verantwortlich **(for** für); **hold sb. ~ for sth.** jmdn. für etw. verantwortlich machen; **be ~ to sb. [for sth.]** jmdm. gegenüber [für etw.] verantwortlich sein; **be ~ for sth.** ⟨*Person:*⟩ für etw. verantwortlich sein; ⟨*Sache:*⟩ die Ursache für etw. sein; **b)** verantwortlich, verantwortungsvoll ⟨*Stellung, Tätigkeit, Aufgabe*⟩; **c)** *(trustworthy)* verantwortungsvoll, verantwortungsbewußt ⟨*Person*⟩

responsive [rɪˈspɒnsɪv] *adj.* aufgeschlossen ⟨*Person*⟩; **be ~ to sth.** auf etw. *(Akk.)* reagieren

[1]rest [rest] **1.** *v. i.* **a)** *(lie, lit. or fig.)* ruhen; **~ on** ruhen auf (+ *Dat.*); *(fig.)* ⟨*Argumentation:*⟩ sich stützen auf (+ *Akk.*); ⟨*Ruf:*⟩ beruhen auf (+ *Dat.*); **~ against sth.** an etw. *(Dat.)* lehnen; **b)** *(take repose)* ruhen; sich ausruhen **(from** von); *(pause)* eine Pause machen *od.* einlegen; **I won't ~ until ...:** ich werde nicht ruhen noch rasten, bis ...; **tell sb. to ~** ⟨*Arzt:*⟩ jmdm. Ruhe verordnen; **c)** *(be left)* **let the matter ~:** die Sache ruhenlassen; **~ assured that ...:** seien Sie versichert, daß ...; **d) ~ with sb.** ⟨*Verantwortung, Entscheidung, Schuld:*⟩ bei jmdm. liegen. **2.** *v. t.* **a)** *(place for support)* **~ sth. against sth.** etw. an etw. *(Akk.)* lehnen; **~ sth. on sth.** *(lit. or fig.)* etw. auf etw. *(Akk.)* stützen; **b)** *(give relief to)* ausruhen lassen ⟨*Pferd, Person*⟩; ausruhen ⟨*Augen*⟩; schonen ⟨*Stimme, Körperteil*⟩. **3.** *n.* **a)** *(repose)* Ruhe, *die;* **get a good night's ~:** sich ordentlich ausschlafen; **be at ~** *(euphem.: be dead)* ruhen *(geh.);* **lay to ~** *(euphem.: bury)* zur letzten Ruhe betten *(geh. verhüll.);* **b)** *(freedom from exertion)* Ruhe[pause], *die;* Erholung, *die* **(from** von); **take a ~:** sich ausruhen **(from** von); **tell sb. to take a ~** ⟨*Arzt:*⟩ jmdm. Ruhe verordnen; **set sb.'s mind at ~:** jmdn. beruhigen **(about** hinsichtlich); **c)** *(pause)* **have** *or* **take a ~:** [eine] Pause machen; **give sb./sth. a ~:** ausruhen lassen ⟨*Person, Nutztier*⟩; *(fig.)* ruhenlassen ⟨*Thema, Angelegenheit*⟩; **d)** *(stationary position)* **at ~:** in Ruhe; **come to ~:** zum Stehen kommen; *(have final position)* landen; **e)** *(Mus.)* Pause, *die*

[2]rest *n. (remainder)* **the ~:** der Rest; **we'll do the ~:** alles Übrige erledigen wir; **the ~ of her clothes** ihre übrigen Kleider; **she's no different from the ~:** sie ist nicht besser als die anderen; **and [all] the ~ of it** und so weiter; **for the ~:** im übrigen; sonst

restart [riːˈstɑːt] *v. t.* **a)** *(start again)* wieder anstellen ⟨*Maschine*⟩; wieder anlassen ⟨*Auto, Motor*⟩; **b)** *(resume)* wiederaufnehmen ⟨*Verhandlungen, Berufstätigkeit*⟩; fortsetzen ⟨*Spiel*⟩; neu starten ⟨*Rennen*⟩

restate [riːˈsteɪt] *v. t.* *(express again)* noch einmal darlegen; *(express differently)* anders darlegen

restaurant [ˈrestərɔ̃, ˈrestərɒnt] *n.* Restaurant, *das*

ˈrestaurant car *n. (Brit. Railw.)* Speisewagen, *der*

rest: ~-cure *n. (Med.)* Erholungskur, *die;* **~-day** *n.* Ruhetag, *der*

rested [ˈrestɪd] *adj.* ausgeruht

restful [ˈrestfl] *adj.* **a)** *(free from disturbance)* ruhig ⟨*Tag, Woche, Ort*⟩; **b)** *(conducive to rest)* beruhigend

ˈrest-home *n.* Pflegeheim, *das*

restive [ˈrestɪv] *adj.* **a)** *(restless)* unruhig; **b)** *(unmanageable)* aufsässig ⟨*Einwohner, Bevölkerung*⟩

restless [ˈrestlɪs] *adj.* unruhig ⟨*Nacht, Schlaf, Bewegung*⟩; ruhelos ⟨*Person, Sehnsucht*⟩

restock [riːˈstɒk] *v. t.* **a) ~ a shop** das Lager eines Geschäfts wieder auffüllen; **b)** wieder besetzen ⟨*Fluß, Teich*⟩

restoration [restəˈreɪʃn] *n.* **a)** *(restoring) (of peace, health)* Wiederherstellung, *die; (of a work of art, building, etc.)* Restaurierung, *die;* Restauration, *die (fachspr.);* **b)** *(giving back)* Rückgabe, *die;* **c)**

(re-establishment) Wiedereinführung, *die;* **the R~** *(Brit. Hist.)* die Restauration

restorative [rɪ'stɒrətɪv, rɪ'stɔ:rətɪv] *adj.* stärkend; aufbauend

restore [rɪ'stɔ:(r)] *v. t.* **a)** *(bring to original state)* restaurieren ‹*Bauwerk, Kunstwerk usw.*›; konjizieren ‹*Text, Satz*› *(Literaturw.);* ~ **sb. to health** jmds. Gesundheit wiederherstellen; **his strength was ~d** er kam wieder zu Kräften; **b)** *(give back)* zurückgeben; **c)** *(reinstate)* wiedereinsetzen (**to** in + *Akk.*); ~ **sb. to power** jmdn. wieder an die Macht bringen; **d)** *(re-establish)* wiederherstellen ‹*Ordnung, Ruhe, Vertrauen*›

restorer [rɪ'stɔ:rə(r)] *n.* *(Art, Archit.: person)* Restaurator, *der*/ Restauratorin, *die*

restrain [rɪ'streɪn] *v. t.* zurückhalten ‹*Gefühl, Lachen, Drang, Person*›; bändigen ‹*unartiges Kind, Tier*›; ~ **sb./oneself from doing sth.** jmdn. davon abhalten/sich zurückhalten, etw. zu tun; ~ **yourself!** beherrsch dich!

restrained [rɪ'streɪnd] *adj.* zurückhaltend ‹*Wesen, Kritik*›; verhalten ‹*Blick, Geste, Gefühl*›; beherrscht ‹*Reaktion, Worte*›

restraint [rɪ'streɪnt] *n.* **a)** *(restriction)* Einschränkung, *die;* **without ~:** ungehindert; **b)** *(reserve)* Zurückhaltung, *die;* **c)** *(moderation)* Unaufdringlichkeit, *die; (self-control)* Selbstbeherrschung, *die;* **without ~:** ungehemmt

restrict [rɪ'strɪkt] *v. t.* beschränken (**to** auf + *Akk.*); ‹*Kleidung:*› be-, einengen

restricted [rɪ'strɪktɪd] *adj.* **a)** *(limited)* beschränkt; begrenzt; ~ **diet** Diät, *die;* **b)** *(subject to restriction)* Sperr‹*gebiet*›; begrenzt ‹*Zulassung, Aufnahme, Anwendbarkeit*›; **be ~ to doing sth.** sich darauf beschränken müssen, etw. zu tun

restricted 'area *n.* **a)** Sperrgebiet, *das;* **b)** *(Brit.: with speed limit) Gebiet mit Geschwindigkeitsbeschränkung*

restriction [rɪ'strɪkʃn] *n.* Beschränkung, *die;* Einschränkung, *die* (**on** *Gen.*); **without ~:** ohne Einschränkung; **put** *or* **place** *or* **impose ~s on sth.** etw. einschränken; **speed/weight ~:** Geschwindigkeits- / Gewichtsbeschränkung, *die*

restrictive [rɪ'strɪktɪv] *adj.* restriktiv; einschränkend *nicht präd.*

'rest-room *n. (esp. Amer.)* Toilette, *die*

restructure [ri:'strʌktʃə(r)] *v. t.* umstrukturieren

restyle [ri:'staɪl] *v. t.* neu stylen

result [rɪ'zʌlt] **1.** *v. i.* **a)** *(follow)* ~ **from sth.** die Folge einer Sache *(Gen.)* sein; von etw. herrühren; *(future)* aus etw. resultieren; **b)** *(end)* ~ **in sth.** in etw. *(Dat.)* resultieren; zu etw. führen; **the game ~ed in a draw** das Spiel endete mit einem Unentschieden; ~ **in sb.'s doing sth.** zur Folge haben, daß jmd. etw. tut. **2.** *n.* Ergebnis, *das;* Resultat, *das;* **be the ~ of sth.** die Folge einer Sache *(Gen.)* sein; **as a ~ [of this]** infolgedessen; **without ~:** ergebnislos

resultant [rɪ'zʌltənt] *attrib. adj.* daraus resultierend

resume [rɪ'zju:m] **1.** *v. t.* **a)** *(begin again)* wiederaufnehmen; fortsetzen ‹*Reise*›; **b)** *(get back)* wieder-, zurückgewinnen; wieder übernehmen ‹*Kommando*›. **2.** *v. i.* weitermachen; ‹*Parlament:*› die Sitzung fortsetzen; ‹*Unterricht:*› wieder beginnen

résumé ['rezʊmeɪ] *n.* *(summary)* Zusammenfassung, *die*

resumption [rɪ'zʌmpʃn] *n.* **a)** *see* **resume 1 a:** Wiederaufnahme, *die;* Fortsetzung, *die;* **b)** *see* **resume 1 b:** Wieder-, Zurückgewinnung, *die;* Wiederübernahme, *die*

resurface [ri:'sɜ:fɪs] **1.** *v. t.* ~ **a road** den Belag einer Straße erneuern. **2.** *v. i. (lit. or fig.)* wieder auftauchen

resurrection [rezə'rekʃn] *n.* *(Relig.)* Auferstehung, *die;* **the R~:** die Auferstehung Christi

resuscitate [rɪ'sʌsɪteɪt] *v. t. (lit. or fig.)* wiederbeleben

retail 1. ['ri:teɪl] *n.* Einzelhandel, *der.* **2.** *adj.* Einzel‹*handel*›; Einzelhandels‹*geschäft, -preis*›; [End]verkaufs‹*preis*›. **3.** *adv.* **buy/sell ~:** en détail kaufen/verkaufen *(Kaufmannsspr.).* **4.** *v. t.* ['ri:teɪl, rɪ'teɪl] *(sell)* [im Einzelhandel] verkaufen. **5.** ['ri:teɪl, rɪ'teɪl] *v. i.* im Einzelhandel verkauft werden (**at, for** für)

retailer ['ri:teɪlə(r)] *n.* Einzelhändler, *der/*-händlerin, *die*

retail 'price index *n. (Brit.)* Preisindex des Einzelhandels

retain [rɪ'teɪn] *v. t.* **a)** *(keep)* behalten; sich *(Dat.)* bewahren ‹*Witz, Fähigkeit*›; ein-, zurückbehalten ‹*Gelder*›; gespeichert lassen ‹*Information*›; ~ **power** ‹*Partei:*› an der Macht bleiben; ~ **control [of sth.]** die Kontrolle [über etw. *(Akk.)*] behalten; **b)** *(keep in place)* ‹*Damm:*› stauen/‹*Deich:*› zurückhalten/‹*Gefäß:*› halten ‹*Wasser*›; ~ **sth. in position** etw. in der richtigen Position halten; **c)** *(secure services of)* beauftragen ‹*Anwalt*›; **d)** *(not forget)* behalten, sich *(Dat.)* merken ‹*Gedanke, Tatsache*›

retainer [rɪ'teɪnə(r)] *n.* *(fee)* Honorarvorschuß, *der*

retake [ri:'teɪk] *v. t., forms as* **take 1, 2: a)** *(recapture)* wieder einnehmen ‹*Stadt, Festung*›; **b)** *(take again)* wiederholen ‹*Prüfung, Strafstoß*›

retaliate [rɪ'tælɪeɪt] *v. i.* Vergeltung üben (**against** an + *Dat.*); ‹*Truppen:*› zurückschlagen; kontern (**against** *Akk.*) ‹*Maßnahme, Kritik*›

retaliation [rɪtælɪ'eɪʃn] *n.* *(in war, fight)* Vergeltung, *die;* Gegenschlag, *der; (in argument etc.)* Konter, *der (ugs.);* Konterschlag, *der;* **in ~ for** als Vergeltung für

retard [rɪ'tɑ:d] *v. t.* verzögern; retardieren *(bes. Physiol., Psych.)*

retarded [rɪ'tɑ:dɪd] *adj. (Psychol.)* **[mentally]** ~**:** [geistig] zurückgeblieben

retch [retʃ, ri:tʃ] **1.** *v. i.* würgen. **2.** *n.* Würgen, *das*

retell [ri:'tel] *v. t.,* **retold** [ri:'təʊld] nacherzählen; *(tell again)* noch einmal erzählen

retentive [rɪ'tentɪv] *adj.* gut ‹*Gedächtnis*›

reticence ['retɪsəns] *n., no pl.* Zurückhaltung, *die*

reticent ['retɪsənt] *adj.* zurückhaltend (**on, about** in bezug auf + *Akk.*)

retina ['retɪnə] *n., pl.* **~s** *or* **~e** ['retɪni:] *(Anat.)* Retina, *die (fachspr.);* Netzhaut, *die*

retinue ['retɪnju:] *n.* Gefolge, *das*

retire [rɪ'taɪə(r)] **1.** *v. i.* **a)** *(give up work or position)* ausscheiden (**from** aus); ‹*Angestellter, Arbeiter:*› in Rente gehen; ‹*Beamter, Militär:*› in Pension *od.* den Ruhestand gehen; ‹*Selbständiger:*› sich zur Ruhe setzen; **b)** *(withdraw)* sich zurückziehen (**to** in + *Akk.*); *(Sport)* aufgeben; ~ **[to bed]** sich [zum Schlafen] zurückziehen. **2.** *v. t.* aus Altersgründen entlassen; pensionieren, in den Ruhestand versetzen ‹*Beamten, Militär*›

retired [rɪ'taɪəd] *adj.* aus dem Berufsleben ausgeschieden ‹*Angestellter, Arbeiter, Selbständiger*›; ‹*Beamter, Soldat:*› im Ruhestand, pensioniert; **be ~:** nicht mehr arbeiten; ‹*Angestellter, Arbeiter:*› Rentner/Rentnerin *od.* in Rente sein; ‹*Beamter, Soldat:*› im Ruhestand *od.* pensioniert sein

retirement [rɪ'taɪəmənt] *n.* **a)** *(leaving work)* Ausscheiden aus dem Arbeitsleben; **b)** *no art. (period)* Ruhestand, *der;* **take early** ~ ‹*Selbständiger:*› sich vorzeitig zur Ruhe setzen; ‹*Angestellter, Arbeiter:*› vorzeitig in Rente gehen; ‹*Beamter, Militär:*› sich vorzeitig pensionieren lassen; **c)** *(withdrawing)* Rückzug, *der* (**to, into** in + *Akk.*)

retirement: ~ **age** *n.* Altersgrenze, *die; (of employees also)* Rentenalter, *das;* ~ **pension** *n. (for employees)* [Alters]rente, *die; (for civil servants, servicemen)* Pension, *die*

retiring [rɪ'taɪərɪŋ] *adj. (shy)* zurückhaltend

retold *see* **retell**

retook *see* **retake**

[1]**retort** [rɪ'tɔ:t] **1.** *n.* Entgegnung, *die,* Erwiderung, *die* (**to** auf + *Akk.*). **2.** *v.t.* entgegnen. **3.** *v.i.* scharf antworten

[2]**retort** *n. (Chem., Industry)* Retorte, *die*

retrace [rɪ'treɪs] *v.t.* **a)** *(trace back)* zurückverfolgen; **b)** *(trace again)* nachvollziehen ‹*Entwicklung*›; **c)** *(go back over)* zurückgehen; ~ **one's steps** denselben Weg noch einmal zurückgehen

retract [rɪ'trækt] **1.** *v.t.* **a)** *(withdraw)* zurücknehmen; **b)** *(Aeronaut.)* einziehen, einfahren ‹*Fahrgestell*›; **c)** *(draw back)* zurückziehen; einziehen ‹*Fühler, Krallen*›. **2.** *v.i.* **a)** *(Aeronaut.)* ‹*Fahrgestell:*› einziehbar *od.* einfahrbar sein; **b)** *(be drawn back)* ‹*Fühler, Krallen:*› eingezogen werden

retraction [rɪ'trækʃn] *n. (withdrawing)* Zurücknahme, *die*

retread ['ri:tred] *n. (Motor Veh.)* runderneuerter Reifen

retreat [rɪ'tri:t] **1.** *n.* **a)** *(withdrawal; also Mil. or fig.)* Rückzug, *der;* **beat a** ~: den Rückzug antreten; *(fig.)* das Feld räumen; **b)** *(place of seclusion)* Zuflucht, *die;* Zufluchtsort, *der; (hiding-place also)* Unterschlupf, *der.* **2.** *v.i. (withdraw; also Mil. or fig.)* sich zurückziehen; *(in fear)* zurückweichen; ~ **within oneself** sich in sich *(Akk.)* selbst zurückziehen

retrench [rɪ'trentʃ] **1.** *v.t.* senken ‹*Ausgaben, Lohn*›. **2.** *v.i.* sich einschränken

retrial [ri:'traɪəl] *n. (Law)* Wiederaufnahmeverfahren, *das*

retribution [retrɪ'bju:ʃn] *n.* Vergeltung, *die;* **in** ~ **for** zur Vergeltung für

retrieval [rɪ'tri:vl] *n.* **a)** *(setting right) (of situation)* Rettung, *die; (of mistake)* Wiedergutmachung, *die;* **beyond** *or* **past** ~: hoffnungslos; **b)** *(rescue)* Rettung, *die; (from wreckage)* Bergung, *die* (**from** aus); **c)** *(recovery) (of letter)* Zurückholen, *das; (of ball)* Wiederholen, *das;* **d)** *(Computing)* Wiederauffinden, *das*

retrieve [rɪ'tri:v] *v.t.* **a)** *(set right)* wiedergutmachen ‹*Fehler*›; retten ‹*Situation*›; **b)** *(rescue)* retten (**from** aus); *(from wreckage)* bergen (**from** aus); **c)** *(recover)* zurückholen ‹*Brief*›; wiederholen ‹*Ball*›; wiederbekommen ‹*Geld*›; **d)** *(Computing)* wiederauffinden ‹*Information*›; **e)** *(fetch)* ‹*Hund:*› apportieren

retriever [rɪ'tri:və(r)] *n.* Apportierhund, *der; (breed)* Retriever, *der*

retrograde ['retrəgreɪd] *adj.* rückschrittlich ‹*Idee, Politik, Maßnahme*›; ~ **step** *(fig.)* Rückschritt, *der*

retro-rocket ['retrəʊrɒkɪt] *n. (Astronaut.)* Bremsrakete, *die*

retrospect ['retrəspekt] *n.* **in** ~: im nachhinein

retrospective [retrə'spektɪv] *adj.* **a)** retrospektiv *(geh.);* **take a** ~ **look at sth.** Rückschau auf etw. *(Akk.)* halten *(geh.);* **b)** *(applying to the past)* rückwirkend ‹*Lohnerhöhung, Gesetz, Vertragsänderung*›

retrospectively [retrə'spektɪvlɪ] *adv. (so as to apply to the past)* rückwirkend

return [rɪ'tɜ:n] **1.** *v.i.* **a)** *(come back)* zurückkommen; zurückkehren *(geh.); (go back)* zurückgehen; zurückkehren *(geh.); (go back by vehicle)* zurückfahren; zurückkehren *(geh.);* ~ **home** wieder nach Hause kommen/gehen/fahren/zurückkehren; ~ **to work** *(after holiday or strike)* die Arbeit wieder aufnehmen; **b)** *(revert)* ~ **to a subject** auf ein Thema zurückkommen. **2.** *v.t.* **a)** *(bring back)* zurückbringen; zurückgeben ‹*geliehenen/gestohlenen Gegenstand, gekaufte Ware*›; [wieder] zurückschicken ‹*unzustellbaren Brief*›; *(hand back, refuse)* zurückweisen ‹*Scheck*›; ~**ed with thanks** mit Dank zurück; '~ **to sender'** *(on letter)* „zurück an Absender"; **b)** *(restore)* ~ **sth. to its original state** *or* **condition** etw. wieder in seinen ursprünglichen Zustand versetzen; **c)** *(yield)* abwerfen ‹*Gewinn*›; **d)** *(give back sth. similar)* erwidern ‹*Besuch, Gruß, Liebe, Gewehrfeuer*›; sich revanchieren für *(ugs.)* ‹*Freundlichkeit, Gefallen*›; zurückgeben ‹*Schlag*›; **e)** *(elect)* wählen ‹*Kandidaten*›; ~ **sb. to Parliament** jmdn. ins Parlament wählen; **f)** *(Sport)* zurückschlagen ‹*Ball*›; *(throw back)* zurückwerfen; **g)** *(answer)* erwidern; entgegnen; **h)** *(declare)* ~ **a verdict of guilty/not guilty** ‹*Geschworene:*› auf „schuldig"/„nicht schuldig" erkennen. **3.** *n.* **a)** *(coming back)* Rückkehr, *die; (to home)* Heimkehr, *die;* ~ **to health** Genesung, *die (geh.);* **many happy** ~**s [of the day]!** herzlichen Glückwunsch [zum Geburtstag]!; **b) by** ~ **[of post]** postwendend; **c)** *(ticket)* Rückfahrkarte, *die;* **single or** ~**?** einfach oder hin und zurück?; **d)** *(proceeds)* ~**[s]** Ertrag, Gewinn, *der* (**on, from** aus); ~ **on capital** Kapitalgewinn, *der;* **e)** *(bringing back)* Zurückbringen, *das; (of property, goods, book)* Rückgabe, *die* (**to** an + *Akk.*); **f)** *(giving back of sth. similar)* Erwiderung, *die;* **receive/get sth. in** ~ **[for sth.]** etw. [für etw.] bekommen

returnable [rɪ'tɜ:nəbl] *adj.* Mehrweg‹*behälter, -flasche usw.*›; rückzahlbar ‹*Gebühr, Kaution*›

return: ~ '**fare** *n.* Preis für eine Rückfahrkarte/*(for flight)* einen Rückflugschein; **what is the** ~ **fare?** wieviel kostet eine Rückfahrkarte/ein Rückflugschein?; ~ '**flight** *n.* Rückflug, *der; (both ways)* Hin- und Rückflug, *der*

re'turning officer *n. (Brit. Parl.)* Wahlleiter, *der*/-leiterin, *die*

return: ~ '**journey** *n.* Rückreise, *die;* Rückfahrt, *die; (both ways)* Hin- und Rückfahrt, *die;* ~ '**match** *n.* Rückspiel, *das;* ~ '**ticket** *n. (Brit.)* Rückfahrkarte, *die; (for flight)* Rückflugschein, *der*

retype [ri:'taɪp] *v.t.* neu tippen

reunification [ri:ju:nɪfɪ'keɪʃn] *n.* Wiedervereinigung, *die*

reunion [ri:'ju:njən] *n.* **a)** *(gathering)* Treffen, *das;* **b)** *(reuniting)* Wiedersehen, *das;* **c)** *(reunited state)* Wiedervereinigung, *die*

reunite [ri:jʊ'naɪt] **1.** *v.t.* wieder zusammenführen; **a** ~**d Germany** ein wiedervereinigtes Deutschland. **2.** *v.i.* sich wieder zusammenschließen

reuse [ri:'ju:z] *v.t.* wiederverwenden

Rev. ['revərənd, *(coll.)* rev] *abbr.* **Reverend** Rev.

rev [rev] *(coll.)* **1.** *n., usu. in pl.* Umdrehung, *die;* Tour, *die (Tech-*

nikjargon). **2.** *v.i.*, **-vv-** mit hoher Drehzahl *od.* hochtourig laufen. **3.** *v.t.*, **-vv-** hochdrehen *(Technikjargon); (noisily)* aufheulen lassen ‹*Motor*›
~ 'up 1. *v.i.* ‹*Motor:*› hochgejagt werden *(Technikjargon).* **2.** *v.t.* hochjagen *(Technikjargon);* aufheulen lassen ‹*Motor[rad]*›
revaluation [ri:væljʊ'eɪʃn] *n. (Econ.)* Aufwertung, *die*
revalue [ri:'vælju:] *v.t. (Econ.)* aufwerten ‹*Währung*›
revamp [ri:'væmp] *(coll.) v.t.* renovieren ‹*Zimmer, Gebäude*›; [wieder] aufmöbeln *od.* aufpolieren ‹*Schrank, Auto usw.*›; neu bearbeiten ‹*Stück, Musical usw.*›
reveal [rɪ'vi:l] *v.t.* enthüllen *(geh.);* verraten; offenbaren *(geh., Theol.),* [offen] zeigen ‹*Gefühle*›; **be ~ed** ‹*Wahrheit:*› ans Licht kommen; **~ one's identity** seine Identität preisgeben *(geh.);* **~ sb. to be sth.** jmdn. als etw. enthüllen *(geh.)*
revealing [rɪ'vi:lɪŋ] *adj.* aufschlußreich ‹*Darstellung, Dokument*›; verräterisch ‹*Bemerkung, Versprecher*›; offenherzig *(scherzh.)* ‹*Kleid, Bluse usw.*›
reveille [rɪ'vælɪ] *n. (Mil.)* Wecksignal, *das*
revel ['revl] **1.** *v.i.*, *(Brit.)* **-ll-: a)** *(take delight)* genießen (**in** *Akk.*); **~ in doing sth.** es [richtig] genießen, etw. zu tun; **b)** *(carouse)* feiern. **2.** *n. usu pl.* Feiern, *das;* Feierei, *die (ugs.)*
revelation [revə'leɪʃn] *n.* **a)** Enthüllung, *die;* **be a ~:** einem die Augen öffnen; **be a ~ to sb.** jmdm. die Augen öffnen; **b)** *(Relig.)* Offenbarung, *die*
reveller ['revələ(r)] *n.* Feiernde, *der/die*
revelry ['revəlrɪ] *n.* Feiern, *das;* Feierei, *die (ugs.)*
revenge [rɪ'vendʒ] **1.** *v.t.* rächen ‹*Person, Tat*›; sich rächen für ‹*Tat*›; **~ oneself** *or* **be ~d [on sb.] [for sth.]** sich [für etw.] [an jmdm.] rächen. **2.** *n.* Rache, *die;* **[desire for] ~:** Rachsucht, *die (geh.);* **take ~** *or* **have one's ~ [on sb.] [for sth.]** Rache [an jmdm.] [für etw.] nehmen *od. (geh.)* üben; **in ~ for sth.** als Rache für etw.
revenue ['revənju:] *n.* **a)** *(State's income)* **[national/state] ~:** Staatseinnahmen; öffentliche Einnahmen; **b) ~[s]** *(income)* Einnahmen; Einkünfte *Pl.*
reverberate [rɪ'vɜ:bəreɪt] *v.i.* ‹*Geräusch, Musik:*› widerhallen
reverberation [rɪvɜ:bə'reɪʃn] *n.* **~[s]** Widerhall, *der*
revere [rɪ'vɪə(r)] *v.t.* verehren
reverence ['revərəns] *n. (revering)* Verehrung, *die;* Ehrfurcht, *die*
reverend ['revərənd] **1.** *adj.* ehrwürdig; **the R~ John Wilson** Hochwürden John Wilson. **2.** *n. (coll.)* Pfarrer, *der*
reverent ['revərənt] *adj.* ehrfürchtig
reverie ['revərɪ] *n.* Träumerei, *die;* **fall into a ~:** in Träumereien *(Akk.)* versinken
reversal [rɪ'vɜ:sl] *n.* Umkehrung, *die*
reverse [rɪ'vɜ:s] **1.** *adj.* entgegengesetzt ‹*Richtung*›; Rück‹*seite*›; umgekehrt ‹*Reihenfolge*›. **2.** *n.* **a)** *(contrary)* Gegenteil, *das;* **quite the ~!** ganz im Gegenteil!; **b)** *(Motor Veh.)* Rückwärtsgang, *der;* **in ~:** im Rückwärtsgang; **put the car into ~, go into ~:** den Rückwärtsgang einlegen; **c)** *(defeat)* Rückschlag, *der.* **3.** *v.t.* **a)** *(turn around)* umkehren ‹*Reihenfolge, Wortstellung, Bewegung, Richtung*›; grundlegend revidieren ‹*Politik*›; **~ the charge[s]** *(Brit.)* ein R-Gespräch anmelden; **b)** *(cause to move backwards)* zurücksetzen; **c)** *(revoke)* aufheben ‹*Urteil, Anordnung*›; rückgängig machen ‹*Maßnahme*›. **4.** *v.i.* zurücksetzen; rückwärts fahren
reverse 'gear *n. (Motor Veh.)* Rückwärtsgang, *der; see also* **gear 1 a**
reversible [rɪ'vɜ:sɪbl] *adj.* **a)** umkehrbar, *(fachspr.)* reversibel ‹*Vorgang*›; *(capable of being revoked)* aufhebbar ‹*Entscheidung, Anordnung*›; **b)** *(having two usable sides)* beidseitig verwendbar ‹*Stoff*›; beidseitig tragbar ‹*Kleidungsstück*›
re'versing light *n.* Rückfahrscheinwerfer, *der*
revert [rɪ'vɜ:t] *v.i.* **a)** *(recur, return)* zurückkommen (**to** auf + *Akk.*), wieder aufgreifen (**to** *Akk.*) ‹*Thema, Angelegenheit, Frage*›; ‹*Gedanken:*› zurückkehren *(geh.)* (**to** zu); **to ~ to ...:** um wieder auf ... *(Akk.)* zurückzukommen; **b)** *(Law)* ‹*Eigentum:*› zurückfallen, *(Rechtsspr.)* heimfallen (**to** an + *Akk.*)
review [rɪ'vju:] **1.** *n.* **a)** *(survey)* Übersicht, *die* (**of** über + *Akk.*); *(of past events)* Rückschau, *die* (**of** auf + *Akk.*); **be a ~ of sth.** einen Überblick *od.* eine Übersicht über etw. *(Akk.)* geben; **b)** *(re-examination)* [nochmalige] Überprüfung; *(of salary)* Revision, *die;* **be under ~** ‹*Vereinbarung, Lage:*› nochmals geprüft werden; **c)** *(of book, play, etc.)* Besprechung, *die;* Kritik, *die;* Rezension, *die;* **d)** *(periodical)* Zeitschrift, *die;* **e)** *(Mil.)* Inspektion, *die.* **2.** *v.t.* **a)** *(survey)* untersuchen; prüfen; **b)** *(re-examine)* überprüfen; **c)** *(Mil.)* inspizieren; mustern; **d)** *(write a criticism of)* besprechen; rezensieren; **e)** *(Law)* überprüfen
revile [rɪ'vaɪl] *v.t.* schmähen *(geh.)*
revise [rɪ'vaɪz] *v.t.* **a)** *(amend)* revidieren ‹*Urteil, Gesetz, Vorschlag*›; **b)** *(check over)* durchsehen ‹*Manuskript, Text, Notizen*›; **c)** *(reread)* noch einmal durchlesen ‹*Notizen*›; *abs.* lernen; **~ one's maths** Mathe *(ugs.)* wiederholen
revision [rɪ'vɪʒn] *n.* **a)** *(amending)* Revision, *die;* **in need of ~:** revisionsbedürftig; **b)** *(checking over)* Durchsicht, *die;* **c)** *(amended version)* [Neu]bearbeitung, *die;* überarbeitete *od.* revidierte Fassung; **d)** *(rereading)* Wiederholung, *die*
revisit [ri:'vɪzɪt] *v.t.* wieder besuchen
revitalize (revitalise) [ri:'vaɪtəlaɪz] *v.t.* neu beleben
revival [rɪ'vaɪvl] *n.* **a)** *(making active again)* Wieder- *od.* Neubelebung, *die;* **b)** *(Theatre)* Wiederaufführung, *die;* Revival, *das;* **c)** *(Relig.: awakening)* Erweckung, *die;* **d)** *(restoration)* Wiederherstellung, *die;* Regenerierung, *die (geh.); (to consciousness or life; also fig.)* Wiederbelebung, *die*
revive [rɪ'vaɪv] **1.** *v.i.* **a)** *(come back to consciousness)* wieder zu sich kommen; **b)** *(be revitalized)* wieder aufleben; zu neuem Leben erwachen; ‹*Geschäft:*› sich wieder beleben. **2.** *v.t.* **a)** *(restore to consciousness)* wiederbeleben; **b)** *(restore to healthy state)* wieder auf die Beine bringen ‹*Person*›; *(strengthen, reawaken)* wieder wecken ‹*Wunsch, Interesse, Ehrgeiz*›; **~ sb.'s hopes** jmdn. neue Hoffnung schöpfen lassen; **c)** *(make active again)* wieder aufleben lassen; **d)** *(Theatre)* wieder auf die Bühne bringen
revoke [rɪ'vəʊk] *v.t.* aufheben ‹*Erlaß, Privileg, Entscheidung*›; zurückziehen ‹*Auftrag*›; widerrufen ‹*Befehl, Erlaubnis, Genehmigung*›; zurücknehmen ‹*Versprechen*›
revolt [rɪ'vəʊlt] **1.** *v.i.* **a)** *(rebel)* revoltieren, aufbegehren *(geh.)*

(**against** gegen); **b)** *(feel revulsion)* sich sträuben (**at, against, from** gegen). **2.** *v. t.* mit Abscheu erfüllen. **3.** *n. (rebelling)* Aufruhr, *der;* Rebellion, *die; (rising)* Revolte, *die (auch fig.);* Aufstand, *der;* **be** *or* **rise in** ~: revoltieren; aufbegehren *(geh.)*

revolting [rɪ'vəʊltɪŋ] *adj. (repulsive)* abscheulich; scheußlich ‹*Gedanke, Wetter*›; widerlich ‹*Person*›

revolution [revə'lu:ʃn] *n.* **a)** *(lit. or fig.)* Revolution, *die;* **b)** *(single turn)* Umdrehung, *die;* **number of** ~**s** Drehzahl, *die*

revolutionary [revə'lu:ʃənərɪ] **1.** *adj.* **a)** *(Polit.)* revolutionär; **b)** *(involving great changes)* revolutionär; umwälzend; *(pioneering)* bahnbrechend. **2.** *n.* Revolutionär, *der*/Revolutionärin, *die*

revolutionize (revolutionise) [revə'lu:ʃənaɪz] *v. t.* grundlegend verändern; revolutionieren ‹*Gesellschaft, Technik*›

revolve [rɪ'vɒlv] **1.** *v. t.* drehen. **2.** *v. i.* sich drehen (**round, about, on** um); **everything** ~**s around her** sie ist der Mittelpunkt[, um den sich alles dreht]

revolver [rɪ'vɒlvə(r)] *n.* [Trommel]revolver, *der*

revolving [rɪ'vɒlvɪŋ] *attrib. adj.* drehbar; Dreh‹*stuhl, -bühne, -tür*›

revue [rɪ'vju:] *n.* Kabarett, *das; (musical show)* Revue, *die*

revulsion [rɪ'vʌlʃn] *n. (feeling)* Abscheu, *der* (**at** vor + *Dat.*, gegen)

reward [rɪ'wɔ:d] **1.** *n.* Belohnung, *die; (for kindness)* Dank, *der;* Lohn, *der; (recognition of merit etc.)* Auszeichnung, *die;* **offer a** ~ **of £100** 100 Pfund Belohnung aussetzen. **2.** *v. t.* belohnen

rewarding [rɪ'wɔ:dɪŋ] *adj.* lohnend ‹*Zeitvertreib, Beschäftigung*›; **be** ~**/financially** ~: sich lohnen/einträglich sein

rewind [ri:'waɪnd] *v. t.*, **rewound** [ri:'waʊnd] **a)** *(wind again)* wieder aufziehen ‹*Uhr*›; **b)** *(wind back)* zurückspulen ‹*Film, Band*›

rewire [ri:'waɪə(r)] *v. t.* mit neuen Leitungen versehen

reword [ri:'wɜ:d] *v. t.* umformulieren; neu formulieren

rewrite [ri:'raɪt] *v. t.*, **rewrote** [ri:'rəʊt], **rewritten** [ri:'rɪtn] *(write again)* noch einmal [neu] schreiben; *(write differently)* umschreiben

rhapsody ['ræpsədɪ] *n.* **a)** *(Mus.)* Rhapsodie, *die;* **b)** *(ecstatic utterance)* Schwärmerei, *die*

rhesus ['ri:səs]: ~ **factor** *n. (Med.)* Rhesusfaktor, *der;* ~ **monkey** *n.* Rhesusaffe, *der*

rhetoric ['retərɪk] *n.* **a)** [**art of**] ~: Redekunst, *die;* Rhetorik, *die;* **b)** *(derog.)* Phrasen *(abwertend)*

rhetorical [rɪ'tɒrɪkl] *adj.* **a)** rhetorisch ‹*Frage, Diskurs*›; **b)** *(derog.)* phrasenhaft *(abwertend)*

rheumatic [ru:'mætɪk] **1.** *adj.* rheumatisch. **2.** *n.* **a)** *in pl. (coll.)* Rheuma, *das (ugs.);* **b)** *(person)* Rheumatiker, *der*/Rheumatikerin, *die;* Rheumakranke, *der/die*

rheumatism ['ru:mətɪzm] *n. (Med.)* Rheumatismus, *der;* Rheuma, *das (ugs.)*

rheumatoid arthritis [ru:mətɔɪd ɑ:'θraɪtɪs] *n. (Med.)* chronischer Gelenkrheumatismus

Rhine [raɪn] *pr. n.* Rhein, *der*

rhino ['raɪnəʊ] *n., pl. same or* ~**s** *(coll.),* **rhinoceros** [raɪ'nɒsərəs] *n., pl. same or* ~**es** Nashorn, *das;* Rhinozeros, *das*

rhododendron [rəʊdə'dendrən] *n. (Bot.)* Rhododendron, *der;* Alpenrose, *die*

rhubarb ['ru:bɑ:b] *n.* Rhabarber, *der*

rhyme [raɪm] **1.** *n.* **a)** Reim, *der;* **without** ~ **or reason** ohne Sinn und Verstand; **b)** *(short poem)* Reim, *der; (rhyming verse)* gereimte Verse; **c)** *(rhyming word)* Reimwort, *das.* **2.** *v. i.* sich reimen (**with** auf + *Akk.*). **3.** *v. t.* reimen

rhythm [rɪðm] *n.* Rhythmus, *der*

rhythmic ['rɪðmɪk], **rhythmical** ['rɪðmɪkl] *adj.* rhythmisch; gleichmäßig

rib [rɪb] **1.** *n.* **a)** *(Anat.)* Rippe, *die;* **b)** ~[**s**] *(joint of meat)* Rippenstück, *das;* **c)** *(supporting piece) (of insect's wing)* Ader, *die; (of feather)* Kiel, *der;* Schaft, *der; (of leaf, in knitting)* Rippe, *die.* **2.** *v. t.*, **-bb-** *(coll.)* aufziehen *(ugs.)*

ribald ['rɪbəld] *adj.* zotig; schmutzig ‹*Lachen*›; unanständig ‹*Ausdrücke*›; *(irreverent)* anzüglich

ribbon [rɪbn] *n.* **a)** *(band for hair, dress, etc.)* Band, *das; (on typewriter)* [Farb]band, *das; (on medal)* [Ordens]band, *das;* **b)** *(fig.: strip)* Streifen, *der*

rice [raɪs] *n.* Reis, *der*

rice: ~**-paper** *n.* Reispapier, *das;* ~ '**pudding** *n.* Reispudding, *der*

rich [rɪtʃ] **1.** *adj.* **a)** *(wealthy)* reich; **b)** *(having great resources)* reich (**in** an + *Dat.*); *(fertile)* fruchtbar ‹*Land, Boden*›; **oil-**~: ölreich; ~ **in vitamins/lime** vitamin-/kalkreich; **c)** *(splendid)* prachtvoll; prächtig; reich ‹*Ausstattung*›; **d)** *(containing much fat, oil, eggs, etc.)* gehaltvoll; *(indigestible)* schwer ‹*Essen*›; **e)** *(deep, full)* voll[tönend] ‹*Stimme*›; voll ‹*Ton*›; satt ‹*Farbe, Farbton*›; voll ‹*Geschmack*›; **f)** *(valuable)* reich *(geh.)* ‹*Geschenke, Opfergaben*›; **g)** *(amusing)* köstlich; **that's** ~! köstlich!; *(iron.)* das ist stark! *(ugs.).* **2.** *n. pl.* **the** ~: die Reichen

riches ['rɪtʃɪz] *n. pl.* Reichtum, *der*

richly ['rɪtʃlɪ] *adv.* **a)** *(splendidly)* reich; üppig ‹*ausgestattet*›; prächtig ‹*gekleidet*›; ~ **ornamented** reichverziert; **b)** *(fully)* voll und ganz; ~ **deserved** wohlverdient

richness ['rɪtʃnɪs] *n., no pl.* **a)** *(elaborateness)* Pracht, *die;* Prächtigkeit, *die;* **b)** *(of food)* Reichhaltigkeit, *die;* **c)** *(fullness) (of voice)* voller Klang; *(of colour)* Sattheit, *die;* **d)** *(great resources)* Reichtum, *der* (**in** an + *Dat.*); *(of soil)* Fruchtbarkeit, *die*

rickets ['rɪkɪts] *n., constr. as sing. or pl. (Med.)* Rachitis, *die*

rickety ['rɪkɪtɪ] *adj.* wack[e]lig ‹*Tisch, Stuhl usw.*›; klapp[e]rig ‹*Auto*›

rickshaw ['rɪkʃɔ:] *n.* Rikscha, *die*

ricochet ['rɪkəʃeɪ] **1.** *n.* **a)** Abprallen, *das;* **b)** *(hit)* Abpraller, *der.* **2.** *v. i.*, ~**ed** ['rɪkəʃeɪd] *or* ~**ted** ['rɪkeʃetɪd] abprallen (**off** von)

rid [rɪd] *v. t.*, **-dd-, rid:** ~ **sth. of sth.** etw. von etw. befreien; ~ **oneself of sb./sth.** sich von jmdm./etw. befreien; sich jmds./einer Sache entledigen *(geh.);* **be** ~ **of sb./sth.** jmdn./etw. los sein *(ugs.);* **get** ~ **of sb./sth.** jmdn./etw. loswerden

riddance ['rɪdəns] *n.* **good** ~ [**to bad rubbish**]! zum Glück *od.* Gott sei Dank ist er/es *usw.* weg!

ridden *see* **ride 2, 3**

[1]**riddle** ['rɪdl] *n.* Rätsel, *das;* **tell sb. a** ~: jmdm. ein Rätsel aufgeben

[2]**riddle** *v. t. (fill with holes)* durchlöchern; ~**d with bullets** von Kugeln durchsiebt; ~**d with corruption** *(fig.)* von Korruption durchsetzt

ride [raɪd] **1.** *n.* **a)** *(journey) (on horseback)* [Aus]ritt, *der; (in vehicle, at fair)* Fahrt, die; ~ **in a train/coach** Zug-/Busfahrt, *die;* **go for a** ~: ausreiten; **go for a** [**bi**]**cycle** ~: radfahren; *(longer distance)* eine Radtour machen; **go for a** ~ [**in the car**] [mit dem Auto] wegfahren; **have a** ~ **in a train/taxi/on the merry-go-round** mit dem Zug/Taxi/Karussell fahren; **give sb. a ~:** jmdn. mitnehmen; **take sb. for a ~:** jmdn. spazierenfahren; *(fig. sl.: deceive)* jmdn. reinlegen *(ugs.)*; **b)** *(quality of ~)* Fahrkomfort, *der.* **2.** *v.i.*, **rode** [rəʊd], **ridden** ['rɪdn] **a)** *(travel) (on horse)* reiten; *(on bicycle, in vehicle; Amer.: in elevator)* fahren; **~ to town on one's bike/in one's car/on the train** mit dem Rad/Auto/Zug in die Stadt fahren; **b)** *(float)* **~ at anchor** vor Anker liegen *od. (Seemannsspr.)* reiten; **c)** *(be carried)* reiten; rittlings sitzen; **'X ~s again'** *(fig.)* „X ist wieder da"; **be riding high** *(fig.)* Oberwasser haben *(ugs.)*; **let sth. ~** *(fig.)* etw. auf sich beruhen lassen. **3.** *v.t.*, **rode, ridden a)** *(~ on)* reiten ‹*Pferd usw.*›; fahren mit ‹*Fahrrad*›; **learn to ~ a bicycle** radfahren lernen; **b)** *(traverse) (on horseback)* reiten; *(on cycle)* fahren

~ a'way, ~ 'off *v.i.* wegreiten/wegfahren

~ 'out *v.t.* abreiten *(Seemannsspr.)* ‹*Sturm*›; *(fig.)* überstehen

~ 'up *v.i.* **a) ~ up [to sth.]** ‹*Reiter:*› an etw. *(Akk.)* heranreiten; ‹*Fahrer:*› an etw. *(Akk.)* heranfahren; **b) the skirt rode up over her knees** *(fig.)* der Rock rutschte über ihr Knie

rider ['raɪdə(r)] *n.* **a)** Reiter, *der*/Reiterin, *die; (of cycle, motorcycle)* Fahrer, *der*/Fahrerin, *die;* **b)** *(addition)* Zusatz, *der;* **add a ~:** einen Zusatz machen

ridge [rɪdʒ] *n.* **a)** *(of roof)* First, *der; (of nose)* Rücken, *der;* **b)** *(long hilltop)* Grat, *der;* Kamm, *der;* **~ of mountains** Gebirgskamm, *der;* **c)** *(Meteorol.)* **~ [of high pressure]** langgestrecktes Hoch

ridicule ['rɪdɪkju:l] **1.** *n.* Spott, *der;* **hold sb./sth. up to ~:** jmdn./etw. der Lächerlichkeit preisgeben. **2.** *v.t.* verspotten; spotten über (+ *Akk.*)

ridiculous [rɪ'dɪkjʊləs] *adj.* lächerlich; **don't be ~!** sei nicht albern!; **make oneself [look] ~:** sich lächerlich machen

riding ['raɪdɪŋ] *n.* Reiten, *das*

riding: ~-breeches *n. pl.* Reithose, *die;* **~ lesson** *n.* Reitstunde, *die;* **~-school** *n.* Reitschule, *die*

rife [raɪf] *pred. adj. (widespread)* weit verbreitet; **rumours were ~:** es gingen Gerüchte um

riff-raff ['rɪfræf] *n.* Gesindel, *das*

rifle ['raɪfl] **1.** *n.* Gewehr, *das; (hunting-~)* Büchse, *die.* **2.** *v.t. (ransack)* durchwühlen; *(pillage)* plündern. **3.** *v.i.* **~ through sth.** etw. durchwühlen

rifle: ~-range *n.* Schießstand, *der;* Schießplatz, *der;* **~-shot** *n.* Gewehrschuß, *der*

rift [rɪft] *n.* **a)** *(dispute)* Unstimmigkeit, *die;* **b)** *(cleft)* Spalte, *die*

¹rig [rɪg] **1.** *n.* **a)** *(Naut.)* Takelung, *die;* **b)** *(for oil well)* [Öl]förderturm, *der; (off shore)* Förderinsel, *die;* **drilling ~:** Bohrturm, *der; (off shore)* Bohrinsel, *die.* **2.** *v.t.*, **-gg-** *(Naut.)* auftakeln

~ 'out *v.t.* ausstaffieren

~ 'up *v.t.* aufbauen

²rig *v.t.*, **-gg-** *(falsify)* fälschen ‹*Wahl*›; verfälschen, *(geh.)* manipulieren ‹*[Wahl]ergebnis*›

rigging ['rɪgɪŋ] *n. (Naut.)* Takelung, *die*

right [raɪt] **1.** *adj.* **a)** *(just, morally good)* richtig; **it is only ~ [and proper] to do sth./that sb. should do sth.** es ist nur recht und billig, etw. zu tun/daß jmd. etw. tut; **b)** *(correct, true)* richtig; **~ enough** völlig richtig; **you're [quite] ~:** du hast [völlig] recht; **too ~!** *(coll.)* allerdings!; **how ~ you are!** wie recht du hast!; **be ~ in sth.** recht mit etw. haben; **let's get it ~ this time!** machen wir es diesmal besser!; **is that clock ~?** geht die Uhr da richtig?; **have you got the ~ fare?** haben Sie das Fahrgeld passend?; **put** *or* **set ~:** richtigstellen ‹*Irrtum*›; wiedergutmachen ‹*Unrecht*›; berichtigen ‹*Fehler*›; bereinigen ‹*Mißverständnis*›; wieder in Ordnung bringen ‹*Situation, Angelegenheit, Gerät*›; **put** *or* **set sb. ~:** jmdn. berichtigen *od.* korrigieren; **~ [you are]!**, *(Brit.)* **~ oh!** *(coll.)* okay! *(ugs.)*; alles klar! *(ugs.)*; **that's ~:** ja[wohl]; so ist es; **is that ~?** stimmt das?; *(indeed?)* aha!; **[am I] ~?** nicht [wahr]?; oder [nicht]? *(ugs.)*; *see also* **all 3**; **c)** *(preferable, most suitable)* richtig; recht; **do sth. the ~ way** etw. richtig machen; **say/do the ~ thing** das Richtige sagen/tun; **d)** *(sound, sane)* richtig; **not be quite ~ in the head** nicht ganz richtig [im Kopf] sein; **as ~ as rain** *(coll.) (in health)* gesund wie ein Fisch im Wasser; *(satisfactory)* in bester Ordnung; **put sb. ~** *(restore to health)* jmdn. [wieder] auf die Beine bringen; *see also* **mind 1 g**; **e) you're a ~ one!** *(coll.)* du bist mir der/die Richtige!; **f)** *(opposite of left)* recht...; **on the ~ side** auf der rechten Seite; rechts; *see also* **turn 1 c**; **be sb.'s ~ arm** *(fig.)* jmds. rechte Hand sein; **g) R~** *(Polit.)* recht... *See also* **right side.** **2.** *v.t.* **a)** *(correct)* berichtigen; richtigstellen; **b)** *(restore to upright position)* [wieder] aufrichten; ‹*Boot usw.:*› **~ itself** sich [von selbst] [wieder] aufrichten; *(fig.: come to proper state)* ‹*Mangel:*› sich [von selbst] geben. **3.** *n.* **a)** *(fair claim, authority)* Recht, *das;* Anrecht, *das;* **have a/no ~ to sth.** ein/kein Anrecht *od.* Recht auf etw. *(Akk.)* haben; **have a** *or* **the/no ~ to do sth.** das/kein Recht haben, etw. zu tun; **by ~ of** auf Grund (+ *Gen.*); **belong to sb. as of** *or* **by ~:** jmds. rechtmäßiges Eigentum sein; **what ~ has he [got] to do that?** mit welchem Recht tut er das?; **in one's own ~:** aus eigenem Recht; **the ~ to work/life** das Recht auf Arbeit/Leben; **~ of way** *(~ to pass across)* Wegerecht, *das; (path)* öffentlicher Weg; *(precedence)* Vorfahrtsrecht, *das;* **who has the ~ of way?** wer hat Vorfahrt?; **be within one's ~s to do sth.** etw. mit [Fug und] Recht tun können; **b)** *(what is just)* Recht, *das;* **~ is on our side** das Recht ist auf unserer Seite; **by ~[s]** von Rechts wegen; **do ~:** sich richtig verhalten; richtig handeln; **do ~ to do sth.** recht daran tun, etw. zu tun; **in the ~:** im Recht; **c)** *(~-hand side)* rechte Seite; **move to the ~:** nach rechts rücken; **on** *or* **to the ~ [of sb./sth.]** rechts [von jmdm./etw.]; **on** *or* **to my ~, to the ~ of me** rechts von mir; zu meiner Rechten; **drive on the ~:** rechts fahren; **d)** *(Polit.)* **the R~:** die Rechte; **be on the R~ of the party** dem rechten Flügel der Partei angehören; **e)** *in pl. (proper state)* **set** *or* **put sth. to ~s** etw. in Ordnung bringen; **f)** *(in marching) see* **²left 3 d**; **g)** *(Boxing)* Rechte, *die.* **4.** *adv.* **a)** *(properly, correctly, justly)* richtig ‹*machen, raten, halten*›; **go ~** *(succeed)* klappen *(ugs.)*; **nothing is going ~ for me today** bei mir klappt heute nichts *(ugs.)*; **b)** *(to the side opposite left)* nach rechts; **~ of the road** rechts von der Straße; **c)** *(all the way)* bis ganz; *(completely)* ganz; völlig; **~ through the summer** den ganzen Sommer hindurch; **~ round the house** ums ganze Haus [herum]; **rotten ~ through** durch und durch verfault; **d)** *(exactly)* genau; **~ in the middle of sth.** mitten in etw. *(Dat./Akk.)*; **~ now** im Moment; jetzt sofort, gleich ‹*handeln*›; **~ at the beginning** gleich am Anfang;

~ **on!** *(coll.) (approving)* recht so!; so ist's recht!; *(agreeing)* genau!; ganz recht!; **e)** *(straight)* direkt; genau; **go ~ on [the way one is going]** [weiter] geradeaus gehen *od.* fahren; **f)** *(coll.: immediately)* ~ **[away/off]** sofort; gleich; **g)** *(arch./dial.: very)* sehr

right: ~ angle *n.* rechter Winkel; **at ~ angles to ...:** rechtwinklig zu ...; im rechten Winkel zu ...; **~-angled** *adj.* rechtwinklig

righteous ['raɪtʃəs] *adj.* **a)** *(upright)* rechtschaffen, *(bibl.)* gerecht ‹*Person*›; gerecht ‹*Gott*›; **b)** *(morally justifiable)* gerecht ‹*Sache, Zorn*›

rightful ['raɪtfl] *adj.* **a)** *(fair)* gerecht ‹*Sache, Strafe*›; berechtigt ‹*Forderung, Anspruch*›; **b)** *(entitled)* rechtmäßig ‹*Besitzer, Herrscher, Erbe, Anteil*›

right: ~ 'hand *n.* **a)** rechte Hand; Rechte, *die;* **b)** *(~ side)* **on** *or* **at sb.'s ~ hand** zu jmds. Rechten; rechts von jmdm.; **~-hand** *adj.* recht...; **~-hand bend** Rechtskurve, *die; see also* **drive 1g; ~-handed** [raɪt'hændɪd] **1.** *adj.* **a)** rechtshändig; ‹*Werkzeug*› für Rechtshänder; **be ~-handed** ‹*Person:*› Rechtshänder/Rechtshänderin sein; **b)** *(turning to ~)* rechtsgängig, rechtsdrehend ‹*Schraube, Gewinde*›; **2.** *adv.* rechtshändig; mit der rechten Hand; **~-hand 'man** *n. (chief assistant)* rechte Hand

rightly ['raɪtlɪ] *adv.* **a)** *(fairly, correctly)* richtig; **do ~:** richtig handeln; **..., and ~ so ...,** und zwar zu Recht; **~ or wrongly, ...:** ob es nun richtig ist/war oder nicht, ...; **b)** *(fitly)* zu Recht

right-'minded *adj.* gerecht denkend

righto ['raɪtəʊ, raɪ'təʊ] *int. (Brit.)* okay *(ugs.);* alles klar *(ugs.)*

right: ~ side *n.* **a)** *(of fabric)* Oberseite, *die;* **b) be on the ~ side of fifty** noch keine fünfzig sein; **[the] ~ side out/up** richtig herum; **~ 'wing** *n.* rechter Flügel; **~-wing** *adj.* **a)** *(Sport)* Rechtsaußen‹*spieler, -position*›; **b)** *(Polit.)* recht...; rechtsgerichtet; Rechts‹*intellektueller, -extremist, -radikalismus*›; **~-winger** *n.* **a)** *(Sport)* Rechtsaußen, *der;* **b)** *(Polit.)* Angehöriger/Angehörige des rechten Flügels

rigid ['rɪdʒɪd] *adj.* **a)** starr; *(stiff)* steif; *(hard)* hart; *(firm)* fest; **b)** *(fig.: harsh, inflexible)* streng ‹*Person*›; unbeugsam ‹*Haltung, System*›

rigidity [rɪ'dʒɪdɪtɪ] *n., no pl. see* **rigid: a)** Starrheit, *die;* Steifheit, *die;* Härte, *die;* Festigkeit, *die;* **b)** Strenge, *die*

rigidly ['rɪdʒɪdlɪ] *adv.* **a)** starr; **b)** *(harshly, inflexibly)* [allzu] streng; peinlich ‹*korrekt*›; rigoros ‹*beschränken*›

rigmarole ['rɪgmərəʊl] *n. (derog.)* **a)** *(long story)* langatmiges Geschwafel *(ugs. abwertend);* **b)** *(complex procedure)* Zirkus, *der (ugs. abwertend)*

rigor *(Amer.) see* **rigour**

rigor mortis [rɪgə 'mɔːtɪs] *n. (Med.)* Totenstarre, *die;* Rigor mortis, *der (fachspr.)*

rigorous ['rɪgərəs] *adj.* **a)** *(strict)* streng; rigoros ‹*Methode, Maßnahme, Beschränkung, Strenge*›; **b)** *(marked by extremes)* hart ‹*Leben, Bedingungen*›; **c)** *(precise)* peinlich ‹*Genauigkeit, Beachtung*›; exakt ‹*Analyse*›; streng ‹*Beurteilung, Maßstab*›; schlüssig ‹*Argumentation*›

rigour ['rɪgə(r)] *n. (Brit.)* **a)** *(strictness)* Strenge, *die;* **b)** *(of life, conditions, etc.)* Härte, *die;* Strenge, *die;* **the ~s of sth.** die Unbilden *(geh.)* einer Sache *(Gen.);* **c)** *(precision)* Stringenz, *die (geh.); (of argument)* Schlüssigkeit, *die*

rile [raɪl] *v.t. (coll.)* ärgern; **get/feel ~d** sich ärgern

rim [rɪm] *n.* Rand, *der; (of wheel)* Felge, *die*

rimless ['rɪmlɪs] *adj.* randlos

rind [raɪnd] *n. (of fruit)* Schale, *die; (of cheese)* Rinde, *die; (of bacon)* Schwarte, *die*

¹ring [rɪŋ] **1.** *n.* **a)** Ring, *der;* **b)** *(Horse-racing, Boxing)* Ring, *der; (in circus)* Manege, *die;* **c)** *(group)* Ring, *der; (gang)* Bande, *die; (controlling prices)* Kartell, *das;* **d)** *(circle)* Kreis, *der;* **make** *or* **run ~s [a]round sb.** *(fig.)* jmdn. in die Tasche stecken *(ugs.).* **2.** *v.t.* **a)** *(surround)* umringen; einkreisen ‹*Wort, Buchstaben usw.*›; **b)** *(Brit.: put ~ on leg of)* beringen ‹*Vogel*›

²ring 1. *n.* **a)** *(act of sounding bell)* Läuten, *das;* Klingeln, *das;* **there's a ~ at the door** es hat geklingelt; **give two ~s** zweimal läuten *od.* klingeln; **b)** *(Brit. coll.: telephone call)* Anruf, *der;* **give sb. a ~:** jmdn. anrufen; **c)** *(resonance; fig.: impression)* Klang, *der; (fig.)* **have the ~ of truth** glaubhaft klingen. **2.** *v.i.,* **rang** [ræŋ], **rung** [rʌŋ] **a)** *(sound clearly)* [er]schallen; ‹*Hammer:*› [er]dröhnen; **b)** *(be sounded)* ‹*Glocke, Klingel, Telefon:*› läuten; ‹*Kasse, Telefon, Wecker:*› klingeln; **the doorbell rang** die Türklingel ging; es klingelte; **c)** *(~ bell)* läuten (**for** nach); **please ~ for attention** bitte läuten; **d)** *(Brit.: make telephone call)* anrufen; **e)** *(resound)* **~ in sb.'s ears** jmdm. in den Ohren klingen; **~ true/false** *(fig.)* glaubhaft/unglaubhaft klingen; **f)** *(hum)* summen; *(loudly)* dröhnen; **my ears are ~ing** mir dröhnen die Ohren. **3.** *v.t.,* **rang, rung a)** läuten ‹*Glocke*›; **~ the [door]bell** läuten; klingeln; **it ~s a bell** *(fig. coll.)* es kommt mir [irgendwie] bekannt vor; **b)** *(Brit.: telephone)* anrufen

~ 'back *v.t. & i. (Brit.)* **a)** *(again)* wieder anrufen; **b)** *(in return)* zurückrufen

~ 'in *v.i. (Brit.)* anrufen

~ 'off *v.i. (Brit.)* auflegen; abhängen

~ 'out 1. *v.i.* ertönen. **2.** *v.t.* ausläuten

~ round *(Brit.)* **1.** [-'-] *v.i.* herumtelefonieren. **2.** ['--] *v.t.* herumtelefonieren bei

~ 'up *v.t.* **a)** *(Brit.: telephone)* anrufen; **b)** *(record on cash register)* [ein]tippen; bongen *(ugs.)*

ring: ~-a-~-o'-'roses *n.* Ringelreihen, *der;* **~ binder** *n.* Ringbuch, *das*

ringed [rɪŋd] *adj.* beringt

'ring-finger *n.* Ringfinger, *der*

ringing ['rɪŋɪŋ] **1.** *adj. (clear and full)* schallend ‹*Stimme, Gelächter*›; *(sonorous)* klangvoll, volltönend ‹*Stimme, Lachen, Lied*›; *(resounding)* dröhnend ‹*Schlag*›. **2.** *n.* **a)** *(sounding, sound)* Läuten, *das;* **b)** *(Brit. Teleph.)* **~ tone** Freiton, *der*

'ringleader *n.* Anführer, *der*/Anführerin, *die*

ringlet ['rɪŋlɪt] *n.* [Ringel]löckchen, *das*

ring: ~master *n.* Dresseur, *der;* **~-pull** *adj.* **~-pull can** Aufreißdose, *die;* Ring-Pull-Dose, *die;* **~ road** *n.* Ringstraße, *die*

rink [rɪŋk] *n. (for ice-skating)* Eisbahn, *die; (for roller-skating)* Rollschuhbahn, *die*

rinse [rɪns] **1.** *v.t.* **a)** *(wash out)* ausspülen ‹*Mund, Gefäß usw.*›; **b)** *(put through water)* [aus]spülen ‹*Wäsche usw.*›; abspülen ‹*Hände, Geschirr*›. **2.** *n. (rinsing)* Spülen, *das;* Spülung, *die*

~ a'way *v.t.* wegspülen

~ 'out *v.t.* **a)** *(wash with clean water)* ausspülen ‹*Wäsche, Mund, Behälter*›; **b)** *(remove by washing)* [her]ausspülen

riot ['raɪət] **1.** *n.* **a)** *(violent disturbance)* Aufruhr, *der;* **~s** Unruhen

Pl.; Aufstand, *der;* **b)** *(noisy or uncontrolled behaviour)* Krawall, *der;* Tumult, *der;* **run ~:** randalieren; **let one's imagination run ~:** seiner Phantasie freien Lauf lassen; **c)** *(coll.: amusing thing or person)* **be a ~:** zum Piepen sein *(ugs.).* **2.** *v.i.* einen Aufstand machen

'riot act *n.* **read sb. the ~** *(fig. coll.)* jmdm. die Leviten lesen

rioter ['raɪətə(r)] *n.* Aufrührer, *der*

riotous ['raɪətəs] *adj.* **a)** *(turbulent)* aufrührerisch ‹*Menge*›; tumultartig ‹*Vorgang*›; **b)** *(dissolute)* ausschweifend; **c)** *(unrestrained)* wild

riotously ['raɪətəslɪ] *adv.* **~ funny** *(coll.)* urkomisch; zum Schreien *präd. (ugs.)*

riot: ~ police *n.* Bereitschaftspolizei, *die;* **~ shield** *n.* Schutzschild, *der*

RIP *abbr.* **rest in peace** R.I.P.

rip [rɪp] **1.** *n. (tear)* Riß, *der.* **2.** *v.t.,* **-pp-: a)** *(make tear in)* zerreißen; **~ open** aufreißen; *(with knife)* aufschlitzen; **~ one's skirt on sth.** sich *(Dat.)* an etw. *(Dat.)* das Kleid einreißen; **b)** *(make by tearing)* reißen ‹*Loch*›. **3.** *v.i.,* **-pp-: a)** *(split)* [ein]reißen; **b) let ~** *(coll.)* loslegen *(ugs.)*

~ a'part *v.t. (tear apart)* auseinanderreißen; zerreißen; *(destroy)* demolieren

~ into *v.t. (fig.: attack verbally)* jmdm. ins Gesicht springen *(ugs.)*

~ 'off *v.t.* **a)** *(remove from)* reißen von; *(remove)* abreißen; herunterreißen ‹*Maske, Kleidungsstück*›; **b)** *(sl.: defraud)* übers Ohr hauen *(ugs.);* bescheißen *(derb)*

~ 'out *v.t.* herausreißen **(of** aus)

~ 'up *v.t.* zerreißen; kaputtreißen *(ugs.);* **~ up an agreement** *(fig.)* aus einer Vereinbarung einfach wieder aussteigen *(ugs.)*

'rip-cord *n.* Reißleine, *die*

ripe [raɪp] *adj.* reif **(for** zu); ausgereift ‹*Käse, Wein, Plan*›; **the time is ~ for doing sth.** es ist an der Zeit, etw. zu tun; **~ old age** hohes Alter

ripen ['raɪpn] **1.** *v.t.* zur Reife bringen; *(fig.)* reifen lassen *(geh.).* **2.** *v.i. (lit. or fig.)* reifen; **~ into sth.** *(fig.)* zu etw. reifen *(geh.)*

'rip-off *n. (sl.)* Nepp, *der (ugs. abwertend)*

riposte [rɪ'pɒst] **1.** *n. (retort)* [rasche] Entgegnung *od. (geh.)* Replik. **2.** *v.i. (retort)* [rasch] antworten

ripple ['rɪpl] **1.** *n.* kleine Welle; **a ~ of applause** kurzer Beifall. **2.** *v.i.* **a)** ‹*See:*› sich kräuseln; ‹*Welle:*› plätschern; **b)** *(sound)* erklingen. **3.** *v.t.* kräuseln

'rip-roaring *adj.* wahnsinnig *(ugs.);* Wahnsinns- *(ugs.)*

rise [raɪz] **1.** *n.* **a)** *(going up) (of sun etc.)* Aufgang, *der; (Theatre: of curtain)* Aufgehen, *das; (advancement)* Aufstieg, *der;* **b)** *(emergence)* Aufkommen, *das;* **c)** *(increase) (in value, price, cost)* Steigerung, *die; (St. Exch.: in shares)* Hausse, *die; (in population, temperature)* Zunahme, *die;* **d)** *(Brit.)* **[pay] ~** *(in wages)* Lohnerhöhung, *die; (in salary)* Gehaltserhöhung, *die;* **e)** *(hill)* Anhöhe, *die;* Erhebung, *die;* **f)** *(origin)* Ursprung, *der;* **give ~ to** führen zu; ‹*Ereignis:*› Anlaß geben zu ‹*Spekulation*›; **g) get** *or* **take a ~ out of sb.** *(fig.: make fun of)* sich über jmdn. lustig machen. **2.** *v.i.,* **rose** [rəʊz], **risen** ['rɪzn] **a)** *(go up)* aufsteigen; **~ [up] into the air** ‹*Rauch:*› aufsteigen, in die Höhe steigen; ‹*Ballon, Vogel, Flugzeug:*› sich in die Luft erheben; **b)** *(come up)* ‹*Sonne, Mond:*› aufgehen; ‹*Blase:*› aufsteigen; **c)** *(reach higher level)* steigen; ‹*Stimme:*› höher werden; **d)** *(extend upward)* aufragen; sich erheben; ‹*Weg, Straße:*› ansteigen; **~ to 2,000 metres** ‹*Berg:*› 2000 m hoch aufragen; **e)** *(advance)* ‹*Person:*› aufsteigen, aufrücken; **~ to be the director** zum Direktor aufsteigen; **~ in the world** voran- *od.* weiterkommen; **f)** *(increase)* steigen; ‹*Stimme:*› lauter werden; ‹*Wind, Sturm:*› auffrischen, stärker werden; **g)** *(Cookery)* ‹*Teig, Kuchen:*› aufgehen; **h)** ‹*Stimmung, Moral:*› steigen; **i)** *(come to surface)* ‹*Fisch:*› steigen; **~ to the bait** *(fig.)* sich ködern lassen *(ugs.);* **j)** *(Theatre)* ‹*Vorhang:*› aufgehen, sich heben; **k)** *(rebel, cease to be quiet)* ‹*Person:*› aufbegehren *(geh.),* sich erheben; **l)** *(get up)* **~ [to one's feet]** aufstehen; **~ on its hind legs** ‹*Pferd:*› steigen; **m)** *(adjourn)* ‹*Parlament:*› in die Ferien gehen, die Sitzungsperiode beenden; *(end a session)* die Sitzung beenden; **n)** *(come to life again)* auferstehen; **o)** *(have origin)* ‹*Fluß:*› entspringen

~ to *see* **occasion 1a**

~ 'up *v.i.* **a)** *(get up)* aufstehen; sich erheben; **b)** *(advance)* aufsteigen; *(in level)* ansteigen; **c)** *(rebel)* **~ up [in revolt]** aufbegehren *(geh.);* sich erheben; **d)** ‹*Berg:*› aufragen; **~ up to 2,000 metres** 2000 m hoch aufragen

riser ['raɪzə(r)] *n.* **early ~:** Frühaufsteher, *der*/Frühaufsteherin, *die;* **late ~:** Spätaufsteher, *der*/Spätaufsteherin, *die*

rising ['raɪzɪŋ] **1.** *n.* **a)** *(of sun, moon, star)* Aufgang, *der;* **b)** *(getting up)* Aufstehen, *das;* **c)** *(revolt)* Aufstand, *der.* **2.** *adj.* **a)** aufgehend ‹*Sonne, Mond, Stern*›; **b)** *(increasing)* steigend ‹*Kosten, Temperatur*›; *(fig.)* wachsend ‹*Entrüstung, Wut, Ärger, Bedeutung*›; **c)** steigend ‹*Wasser, Flut*›; **d) the ~ generation** die heranwachsende Generation; **e)** *(advancing in standing)* aufstrebend; **f)** *(sloping upwards)* ansteigend

rising 'damp *n.* aufsteigende Feuchtigkeit

risk [rɪsk] **1.** *n.* **a)** *(hazard)* Gefahr, *die; (chance taken)* Risiko, *das;* **there is a/no ~ of sb.'s doing sth.** *or* **that sb. will do sth.** es besteht die/keine Gefahr, daß jmd. etw. tut; **at one's own ~:** auf eigene Gefahr *od.* eigenes Risiko; **put at ~:** gefährden; in Gefahr bringen; **run the ~ of doing sth.** Gefahr laufen, etw. zu tun; *(knowingly)* es riskieren, etw. zu tun; **take the ~ of doing sth.** es riskieren, etw. zu tun; das Risiko eingehen, etw. zu tun; **b)** *(Insurance)* **he is a poor/good ~:** bei ihm ist das Risiko groß/gering. **2.** *v.t.* riskieren; wagen ‹*Sprung, Kampf*›; **you'll ~ losing your job** du riskierst es, deinen Job zu verlieren; **I'll ~ it!** ich lasse es drauf ankommen; ich riskiere es; **~ one's life** sein Leben riskieren; *(thoughtlessly)* sein Leben aufs Spiel setzen

risky ['rɪskɪ] *adj.* gefährlich; riskant, gewagt ‹*Experiment, Unternehmen, Projekt*›

risqué ['rɪskeɪ] *adj.* gewagt; nicht ganz salonfähig

rissole ['rɪsəʊl] *n.* Rissole, *die*

rite [raɪt] *n.* Ritus, *der*

ritual ['rɪtʃʊəl] **1.** *adj.* rituell; Ritual‹*mord, -tötung*›. **2.** *n. (act)* Ritual, *das*

rival ['raɪvl] **1.** *n.* **a)** *(competitor)* Rivale, *der*/Rivalin, *die;* **~s in love** Nebenbuhler; **business ~s** Konkurrenten; **b)** *(equal)* **have no ~/~s** seines-/ihresgleichen suchen; **without ~[s]** konkurrenzlos. **2.** *v.t., (Brit.)* **-ll-** gleichkommen (+ *Dat.*); nicht nachstehen (+ *Dat.*). **3.** *adj.* rivalisierend ‹*Gruppen*›; konkurrierend ‹*Forderungen*›; Konkurrenz‹*unternehmen usw.*›

rivalry ['raɪvlrɪ] *n.* Rivalität, *die (geh.);* **business ~:** Wettbewerb, *der*

river ['rɪvə(r)] *n.* **a)** Fluß, *der; (large)* Strom, *der;* **the ~ Thames** *(Brit.),* **the Thames ~** *(Amer.)* die Themse; **sell sb. down the ~** *(fig. coll.)* jmdn. verschaukeln *(ugs.);* **b)** *(fig.)* Strom, *der*

river: ~-bed *n.* Flußbett, *das;* **~side 1.** *n.* Flußufer, *das;* **on** *or* **by the ~side** am Fluß; **2.** *attrib. adj.* am Fluß gelegen; am Fluß *nachgestellt*

rivet ['rɪvɪt] **1.** *n.* Niete, *die;* Niet, *der od. das (Technik).* **2.** *v.t.* **a)** [ver]nieten; **~ sth. together** etw. zusammennieten; **b)** *(fig.: hold firmly)* fesseln ‹*Person, Aufmerksamkeit, Blick*›; **be ~ed to the spot** wie angenagelt [da]stehen *(ugs.)*

riveting ['rɪvɪtɪŋ] *adj.* fesselnd

RN *abbr. (Brit.)* **Royal Navy** Königl. Mar.

RNLI *abbr. (Brit.)* **Royal National Lifeboat Institution** *Königliches Institut für Rettungsboote*

road [rəʊd] *n.* **a)** Straße, *die;* **the Birmingham/London ~:** die Straße nach Birmingham/London; *(name of ~/street)* **London/Shelley R~:** Londoner Straße/Shelleystraße; **'~ up'** „Straßenarbeiten"; **across** *or* **over the ~ [from us]** [bei uns *od. (geh.)* uns *(Dat.)*] gegenüber; **by ~** *(by car/bus)* per Auto/Bus; *(by lorry/truck)* per Lkw; **off the ~** *(being repaired)* in der Werkstatt; in Reparatur; **one for the ~** *(coll.)* ein Glas zum Abschied; **be on the ~:** auf Reisen *od.* unterwegs sein; ‹*Theaterensemble usw.:*› auf Tournee *od. (ugs.)* Tour *(Dat.)* sein; **put a vehicle on the ~:** ein Fahrzeug in Betrieb nehmen; **b)** *(means of access)* Weg, *der;* **set sb. on the ~ to ruin** jmdn. ins Verderben führen; **be on the right ~:** auf dem richtigen Weg sein; **end of the ~** *(destination)* Ziel, *das; (limit)* Ende, *das;* **c)** *(one's way)* Weg, *der;* **get in sb.'s ~** *(coll.)* jmdm. in die Quere kommen *(ugs.);* **get out of my ~!** *(coll.)* geh mir aus dem Weg!; **d)** *(Amer.) see* **railway; e)** *(Mining)* Strecke, *die;* **f)** *usu. in pl. (Naut.)* Reede, *die*

road: ~ accident *n.* Verkehrsunfall, *der;* **~ atlas** *n.* Autoatlas, *der;* **~-block** *n.* Straßensperre, *die;* **~ haulage** *n.* Gütertransport auf der Straße; **~-hog** *n.* Verkehrsrowdy, *der (abwertend);* **~-holding** *n. (Brit. Motor Veh.)* Straßenlage, *die;* **~-manager** *n.* Roadmanager, *der;* **~-map** *n.* Straßenkarte, *die;* **~ safety** *n.* Verkehrssicherheit, *die;* **~ sense** *n.* Gespür für Verkehrssituationen; **~side 1.** *n.* Straßenrand, *der;* **at** *or* **by/along the ~side** am Straßenrand; an/entlang der Straße; **2.** *adj.* ‹*Gasthaus usw.*› am Straßenrand, an der Straße; **~ sign** *n.* Straßenschild, *das (ugs.);* Verkehrszeichen, *das;* **~-sweeper** *n.* Straßenkehrer, *der/*-kehrerin, *die (bes. südd.);* Straßenfeger, *der/*-fegerin, *die (bes. nordd.);* **~ test** *n.* Fahrtest, *der;* **~-test** *v.t.* einem Fahrtest unterziehen; **~ transport** *n.* Personen- und Güterbeförderung auf der Straße; **~-user** *n.* Verkehrsteilnehmer, *der/*-teilnehmerin, *die;* **~way** *n.* Fahrbahn, *die;* **~-works** *n. pl.* Straßenbauarbeiten *Pl.;* **'~-works'** „Baustelle"; **~worthy** *adj.* fahrtüchtig ‹*Fahrzeug*›

roam [rəʊm] **1.** *v.i.* umherstreifen; herumstreifen *(ugs.);* ‹*Nomade:*› wandern; *(stray)* ‹*Tier:*› streunen. **2.** *v.t.* streifen durch; durchstreifen *(geh.)*

~ a'bout, ~ a'round 1. *v.i.* herumstreifen *(ugs.);* umherstreifen. **2.** *v.t.* herumstreifen in (+ *Dat.*) *(ugs.);* durchstreifen *(geh.)*

roar [rɔ:(r)] **1.** *n. (of wild beast)* Brüllen, *das;* Gebrüll, *das; (of water, applause)* Tosen, *das;* Getose, *das; (of avalanche, guns)* Donner, *der; (of machine, traffic)* Dröhnen, *das;* Getöse, *das;* **~s/a ~ [of laughter]** dröhnendes *od.* brüllendes Gelächter. **2.** *v.i.* **a)** *(cry loudly)* brüllen **(with** vor + *Dat.*); **~ [with laughter]** [vor Lachen] brüllen; **b)** ‹*Motor:*› dröhnen; ‹*Artillerie:*› donnern; *(blaze up)* ‹*Feuer:*› bullern *(ugs.).* **3.** *v.t.* brüllen

roaring ['rɔ:rɪŋ] **1.** *adj.* **a)** dröhnend ‹*Motor, Donner*›; tosend ‹*Meer*›; brüllend ‹*Löwe*›; **b)** *(blazing loudly)* bullernd *(ugs.)* ‹*Feuer*›; **c)** *(riotous)* **a ~ success** ein Bombenerfolg *(ugs.);* **the ~ twenties** die wilden zwanziger Jahre; die Roaring Twenties; **d)** *(brisk)* **do a ~ trade** ein Bombengeschäft machen. **2.** *adv.* **~ drunk** sternhagelvoll *(salopp)*

roast [rəʊst] **1.** *v.t.* braten; rösten ‹*Kaffeebohnen, Erdnüsse, Mandeln, Kastanien*›. **2.** *attrib. adj.* gebraten ‹*Fleisch, Ente usw.*›; Brat‹*hähnchen, -kartoffeln*›; Röst‹*kastanien*›; **eat ~ duck/pork/beef** Enten-/Schweine-/Rinderbraten essen; **~ beef** *(sirloin)* Roastbeef, *das.* **3.** *n.* Braten, *der*

rob [rɒb] *v.t.,* **-bb-** ausrauben ‹*Bank, Safe, Kasse*›; berauben ‹*Person*›; *abs.* rauben; **~ sb. of sth.** jmdm. etw. rauben *od.* stehlen; *(deprive of what is due)* jmdn. um etw. bringen *od.* betrügen; *(withhold sth. from)* jmdm. etw. vorenthalten; **be ~bed** bestohlen werden; *(by force)* beraubt werden

robber ['rɒbə(r)] *n.* Räuber, *der/* Räuberin, *die*

robbery ['rɒbərɪ] *n.* Raub, *der;* **robberies** Raubüberfälle

robe [rəʊb] *n.* **a)** *(ceremonial garment)* Gewand, *das (geh.); (of judge, vicar)* Talar, *der;* **~ of office** Amtstracht, *die;* **b)** *(long garment)* [langes Über]gewand; **c)** *(dressing-gown)* Morgenrock, *der;* **beach ~:** Bademantel, *der*

robin ['rɒbɪn] *n. (Ornith.)* **~ [redbreast]** Rotkehlchen, *das*

robot ['rəʊbɒt] *n.* Roboter, *der*

robust [rəʊ'bʌst] *adj.* robust ‹*Person, Gesundheit*›; kräftig ‹*Person, Gestalt, Körperbau*›; widerstandsfähig ‹*Pflanze*›; robust ‹*Fahrzeug, Maschine, Möbel*›; stabil ‹*Haus*›

[1]**rock** [rɒk] *n.* **a)** *(piece of ~)* Fels, *der;* **be as solid as a ~** *(fig.)* absolut zuverlässig sein; **b)** *(large ~, hill)* Felsen, *der;* Fels, *der (geh.);* **c)** *(substance)* Fels, *der; (esp. Geol.)* Gestein, *das;* **d)** *(boulder)* Felsbrocken, *der; (Amer.: stone)* Stein, *der;* Steinbrocken, *der;* **'danger, falling ~s'** „Achtung *od.* Vorsicht, Steinschlag!"; „Steinschlaggefahr!"; **e)** *no pl., no indef. art. (hard sweet)* **stick of ~:** Zuckerstange, *die;* **f)** *(fig.: support)* Stütze, *die;* Rückhalt, *der; (of society)* Fundament, *das;* **g)** **be on the ~s** *(fig. coll.: have failed)* ‹*Ehe, Firma:*› kaputt sein *(ugs.);* **h)** **on the ~s** *(with ice cubes)* mit Eis *od.* on the rocks

[2]**rock 1.** *v.t.* **a)** *(move to and fro)* wiegen; *(in cradle)* schaukeln; wiegen; **b)** *(shake)* erschüttern; *(fig.)* erschüttern ‹*Person*›; **~ the boat** *(fig. coll.)* Trouble machen *(ugs.).* **2.** *v.i.* **a)** *(move to and fro)* sich wiegen; schaukeln; **b)** *(sway)* schwanken; wanken; **c)** *(dance)* **~ and roll** Rock and Roll tanzen. **3.** *n. (music)* Rock, *der; attrib.* Rock-; **~ and** *or* **'n' roll [music]** Rock and Roll, *der;* Rock 'n' Roll, *der*

rock: ~-'bottom *(coll.)* **1.** *adj.* **~-bottom prices** Schleuderpreise *(ugs.);* **2.** *n.* **reach** *or* **touch ~-bottom** ‹*Handel, Währung, Preis usw.:*› in den Keller fallen *od.* sinken *(ugs.);* **her spirits reached ~-bottom** ihre Stimmung war auf

dem Tiefpunkt [angelangt]; ~-**climbing** *n.* [Fels]klettern, *das*
rocker ['rɒkə(r)] *n.* **a)** *(Brit.: gang member)* Rocker, *der;* **b) be off one's ~** *(fig. sl.)* übergeschnappt *od.* durchgedreht sein *(ugs.)*
rockery ['rɒkərɪ] *n.* Steingarten, *der*
rocket ['rɒkɪt] **1.** *n.* **a)** Rakete, *die;* **b)** *(Brit. sl.: reprimand)* **give sb. a ~:** jmdm. eine Zigarre verpassen *(ugs.).* **2.** *v.i.* ⟨*Preise:*⟩ in die Höhe schnellen
rocket: ~ engine *n.* Raketentriebwerk, *das;* **~ flight** *n.* Raketenflug, *der;* **~-launcher** *n.* Raketenwerfer, *der;* **~-powered, ~-propelled** *adjs.* raketengetrieben
rock: ~ face *n.* Felswand, *die;* **~fall** *n.* Steinschlag, *der;* **~-garden** *n.* Steingarten, *der;* **~-hard** *adj.* steinhart
Rockies ['rɒkɪz] *pr. n. pl.* **the ~:** die Rocky Mountains
rocking: ~-chair *n.* Schaukelstuhl, *der;* **~-horse** *n.* Schaukelpferd, *das*
rock: ~like *adj.* felsartig; felsenfest ⟨*Glaube usw.*⟩; **~-plant** *n.* Felsenpflanze, *die; (Hort.)* Steingartengewächs, *das*
rocky ['rɒkɪ] *adj.* **a)** *(coll.: unsteady)* wackelig *(ugs.);* **b)** *(full or consisting of rocks)* felsig; **c) the R~ Mountains** *see* **Rockies**
rococo [rə'kəʊkəʊ] *adj.* Rokoko-
rod [rɒd] *n.* **a)** Stange, *die;* **b)** *(shorter)* Stab, *der;* **c)** *(for punishing)* Stock, *der;* Rute, *die;* **rule with a ~ of iron** *(fig.)* mit eiserner Faust *od.* Rute regieren; **spare the ~ and spoil the child** wer die Rute schont, verdirbt das Kind
rode *see* **ride 2, 3**
rodent ['rəʊdənt] *n.* Nagetier, *das*
rodeo ['rəʊdɪəʊ, rə'deɪəʊ] *n., pl.* **~s** Rodeo, *der od. das*
[1]**roe** [rəʊ] *n.* *(of fish)* **[hard] ~:** Rogen, *der;* **[soft] ~:** Milch, *die*
[2]**roe** *n.* *(deer)* Reh, *das*
roe: ~buck *n.* Rehbock, *der;* **~-deer** *n.* Reh, *das*
roger ['rɒdʒə(r)] *int.* *(message received)* verstanden
rogue [rəʊg] *n.* **a)** Gauner, *der (abwertend);* **~s' gallery** *(Police)* Verbrecheralbum, *das;* **b)** *(joc.: mischievous child)* Spitzbube, *der (scherzh.);* **c)** *(dangerous animal)* **~ [buffalo/elephant** *etc.*] bösartiger Einzelgänger
roguish ['rəʊgɪʃ] *adj.* **a)** gaunerhaft; **b)** *(mischievous)* spitzbübisch
role, rôle [rəʊl] *n.* Rolle, *die*
role: ~-playing *n.* Rollenspiel, *das;* Rollenverhalten, *das;* **~ reversal** *n.* Rollentausch, *der*
[1]**roll** [rəʊl] *n.* **a)** Rolle, *die; (of cloth, tobacco, etc.)* Ballen, *der; (of fat on body)* Wulst, *der;* **~ of film** Rolle Film; **b)** *(of bread etc.)* **[bread] ~:** Brötchen, *das;* **egg/ham ~:** Eier-/Schinkenbrötchen, *das;* **c)** *(document)* [Schrift]rolle, *die;* **d)** *(register, catalogue)* Liste, *die;* Verzeichnis, *das;* **~ of honour** Gedenktafel [für die Gefallenen]; **e)** *(Mil., Sch.: list of names)* Liste, *die;* **schools with falling ~s** Schulen mit sinkenden Schülerzahlen; **call the ~:** die Anwesenheit feststellen
[2]**roll 1.** *n.* **a)** *(of drum)* Wirbel, *der; (of thunder)* Rollen, *das;* **b)** *(motion)* Rollen, *das;* **c)** *(single movement)* Rolle, *die; (of dice)* Wurf, *der.* **2.** *v.t.* **a)** *(move, send)* rollen; *(between surfaces)* drehen; **b)** *(shape by ~ing)* rollen; **~ a cigarette** eine Zigarette rollen *od.* drehen; **~ one's own** [selbst] drehen; **~ snow/wool into a ball** einen Schneeball formen/Wolle zu einem Knäuel aufwickeln; **[all] ~ed into one** *(fig.)* in einem; **~ oneself/itself into a ball** sich zusammenrollen; **c)** *(flatten)* walzen ⟨*Rasen, Metall usw.*⟩; ausrollen ⟨*Teig*⟩; **d) ~ one's eyes** die Augen rollen; **e) ~ one's r's** das r rollen. **3.** *v.i.* **a)** *(move by turning over)* rollen; **heads will ~** *(fig.)* es werden Köpfe rollen; **b)** *(operate)* ⟨*Maschine:*⟩ laufen; ⟨*Presse:*⟩ sich drehen; *(on wheels)* rollen; **c)** *(wallow, sway, walk)* sich wälzen; **d)** *(Naut.)* ⟨*Schiff:*⟩ rollen, schlingern; **e)** *(revolve)* ⟨*Augen:*⟩ sich [ver]drehen; **f)** *(flow, go forward)* sich wälzen *(fig.);* ⟨*Wolken:*⟩ ziehen; ⟨*Tränen:*⟩ rollen; **g)** ⟨*Donner:*⟩ rollen; ⟨*Trommel:*⟩ dröhnen
~ a'bout *v.i.* herumrollen; ⟨*Schiff:*⟩ schlingern, rollen; ⟨*Kind, Hund usw.:*⟩ sich wälzen; **be ~ing about with laughter** sich vor Lachen wälzen
~ a'way 1. *v.i.* ⟨*Ball:*⟩ wegrollen; ⟨*Nebel, Wolken:*⟩ sich verziehen. **2.** *v.t.* wegrollen
~ 'back *v.t.* **a)** zurückrollen; **b)** *(cause to retreat)* zurückschlagen ⟨*Feinde, Truppen*⟩
~ 'by *v.i.* vorbeirollen; ⟨*Zeit:*⟩ vergehen; **the years ~ed by** die Jahre zogen ins Land
~ 'in *v.i. (coll.)* ⟨*Briefe, Geschenke, Geldbeträge:*⟩ eingehen; **~ in an hour late** mit einer Stunde Verspätung aufkreuzen *(salopp)*
~ 'on 1. *v.t.* mit einer Rolle auftragen ⟨*Farbe*⟩. **2.** *v.i.* **a)** *(pass by)* ⟨*Jahre:*⟩ vergehen; **b)** *(Brit. coll.)* **~ on Saturday!** wenn doch schon Samstag wäre!
~ 'out 1. *v.t.* **a)** *(make flat and smooth)* auswalzen ⟨*Metall*⟩; ausrollen ⟨*Teig, Teppich*⟩; **b)** *(bring out)* herausbringen. **2.** *v.i.* heraus-/hinausrollen
~ 'over 1. *v.i.* ⟨*Person:*⟩ sich umdrehen, *(to make room)* sich zur Seite rollen; **~ over [and over]** ⟨*Auto:*⟩ sich [immer wieder] überschlagen; **the dog ~ed over on to its back** der Hund rollte sich auf den Rücken. **2.** *v.t.* herumdrehen; *(with effort)* herumwälzen
~ 'up 1. *v.t.* aufrollen ⟨*Teppich, Maßband*⟩; zusammenrollen ⟨*Regenschirm, Landkarte, Dokument usw.*⟩; hochkrempeln ⟨*Hose*⟩; *see also* **sleeve a. 2.** *v.i.* **a)** *(curl up)* sich zusammenrollen; **b)** *(arrive)* aufkreuzen *(salopp);* **~ up! ~ up!** hereinspaziert!
roll: ~-bar *n. (Motor Veh.)* Überrollbügel, *der;* **~-call** *n.* Aufrufen aller Namen; *(Mil.)* Zählappell, *der*
roller ['rəʊlə(r)] *n.* **a)** *(heavy, for pressing, smoothing road, lawn, etc.)* Walze, *die; (smaller, for towel, painting, pastry)* Rolle, *die;* **b)** *(for hair)* Lockenwickler, *der;* **put one's hair in ~s** sich *(Dat.)* die Haare aufdrehen; **c)** *(wave)* Roller, *der (Meeresk.)*
roller: ~ blind *n.* Rouleau, *das;* Rollo, *das;* **~-coaster** *n.* Achterbahn, *die;* **~-skate 1.** *n.* Rollschuh, *der;* **2.** *v.i.* Rollschuh laufen; **~-skating** *n.* Rollschuhlaufen, *das; attrib.* **~-skating rink** Rollschuhbahn, *die;* **~ towel** *n.* *auf einer Rolle hängendes endloses Handtuch*
'roll film *n.* Rollfilm, *der*
rolling ['rəʊlɪŋ] *adj.* **a)** *(moving from side to side)* rollend ⟨*Augen*⟩; schwankend ⟨*Gang*⟩; schlingernd ⟨*Schiff*⟩; **b)** *(undulating)* wogend ⟨*See*⟩; wellig ⟨*Gelände*⟩; **~ hills** sanfte Hügel
rolling: ~-mill *n.* Walzwerk, *das;* **~-pin** *n. (Cookery)* Teigrolle, *die;* Nudelholz, *das;* **~-stock** *n.* *(Brit. Railw.)* Fahrzeugbestand, *der;* rollendes Material *(fachspr.);* **~ 'stone** *n. (fig.)* unsteter Mensch; **a ~ stone gathers no moss** *(prov.)* wer ein unstetes Leben führt, bringt es zu nichts
ROM [rɒm] *abbr. (Computing)* **read-only memory** ROM
Roman ['rəʊmən] **1.** *n.* Römer, *der*/Römerin, *die.* **2.** *adj.* römisch; **~ road** Römerstraße, *die*
Roman: ~ 'alphabet *n.* lateini-

sches Alphabet; ~ **'Catholic 1.** *adj.* römisch-katholisch; **2.** *n.* Katholik, *der*/Katholikin, *die;* **sb. is a ~ Catholic** jmd. ist römisch-katholisch

romance [rəʊ'mæns] **1.** *n.* **a)** *(love affair)* Romanze, *die;* **b)** *(love-story)* [romantische] Liebesgeschichte; **c)** *(romantic quality)* Romantik, *die;* **d)** *(Lit.) (medieval tale)* Romanze, *die; (improbable tale)* phantastische Geschichte; **e)** *(make-believe)* Phantasterei, *die;* **f) R~** *(Ling.)* Romanisch, *das.* **2.** *adj.* **R~** *(Ling.)* romanisch; **R~ languages and literature** *(subject)* Romanistik, *die*

Romanesque [rəʊmə'nesk] *n.* *(Art, Archit.)* Romanik, *die*

Romania [rəʊ'meɪnɪə] *pr. n.* Rumänien *(das)*

Romanian [rəʊ'meɪnɪən] **1.** *adj.* rumänisch; **sb. is ~:** jmd. ist Rumäne/Rumänin. **2.** *n.* **a)** *(person)* Rumäne, *der*/Rumänin, *die;* **b)** *(language)* Rumänisch, *das; see also* **English 2 a**

Roman 'numeral *n.* römische Ziffer

romantic [rəʊ'mæntɪk] **1.** *adj.* **a)** *(emotional)* romantisch; ~ **fiction** *(love-stories)* Liebesromane; **b)** **R~** *(Lit., Art)* romantisch; der Romantik *nachgestellt.* **2.** *n.* **R~** *(Lit., Art, Mus.)* Romantiker, *der*/Romantikerin, *die*

romanticise *see* **romanticize**

Romanticism [rəʊ'mæntɪsɪzm] *n.* *(Lit., Art, Mus.)* Romantik, *die*

romanticize [rəʊ'mæntɪsaɪz] *v. t.* romantisieren

Romany ['rəʊmənɪ] **1.** *n.* **a)** *(gypsy)* Rom, *der;* **the Romanies** die Roma; **b)** *(language)* Romani, *das.* **2.** *adj.* **a)** Roma-; **b)** *(Ling.)* Romani-

Rome [rəʊm] *pr. n.* Rom *(das);* ~ **was not built in a day** *(prov.)* Rom ist nicht an einem Tag erbaut worden *(Spr.)*

romp [rɒmp] **1.** *v. i.* **a)** [herum]tollen; **b)** *(coll.: win, succeed, etc. easily)* ~ **home** *or* **in** spielend gewinnen; ~ **through sth.** etw. spielend schaffen. **2.** *n.* Tollerei, *die;* **have a ~:** [herum]tollen

roof [ru:f] **1.** *n.* **a)** Dach, *das;* **under one ~:** unter einem Dach; **have a ~ over one's head** ein Dach über dem Kopf haben; **go through the ~** ‹*Preise:*› kraß in die Höhe steigen; **sb. goes through** *or* **hits the ~** *(fig. coll.)* jmd. geht an die Decke *(ugs.);* **b)** *(Anat.)* ~ **of the mouth** Gaumen, *der.* **2.** *v. t.* bedachen; ~ **in** *or* **over** überdachen

roofless ['ru:flɪs] *adj.* dachlos

roof: ~-rack *n.* Dachgepäckträger, *der;* **~-top** *n.* Dach, *das;* **shout sth. from the ~-tops** *(fig.)* etw. in die Welt hinausrufen

¹rook [rʊk] **1.** *n.* *(Ornith.)* Saatkrähe, *die.* **2.** *v. t.* *(charge extortionately)* neppen *(ugs. abwertend)*

²rook *n.* *(Chess)* Turm, *der*

rookery ['rʊkərɪ] *n.* Saatkrähenkolonie, *die*

room [ru:m, rʊm] *n.* **a)** *(in building)* Zimmer, *das; (esp. without furniture)* Raum, *der; (large ~, for function)* Saal, *der;* **leave the ~** *(coll.: go to lavatory)* austreten *(ugs.);* **b)** *no pl., no indef. art.* *(space)* Platz, *der;* **give sb. ~:** jmdm. Platz machen; **give sb. ~ to do sth.** *(fig.)* jmdm. die Freiheit lassen, etw. zu tun; **make ~ [for sb./sth.]** [jmdm./einer Sache] Platz machen; **c)** *(scope)* **there is still ~ for improvement in his work** seine Arbeit ist noch verbesserungsfähig; *see also* **manœuvre 1 b, 3 b; d)** *in pl.* *(apartments, lodgings)* Wohnung, *die;* **'~s to let'** „Zimmer zu vermieten". *See also* **cat a**

-roomed [ru:md, rʊmd] *adj.* *in comb.* **a three-~ flat** eine Dreizimmerwohnung; **a one-~/four-~ building** ein Haus mit einem Zimmer/vier Zimmern

room: ~-mate *n.* Zimmergenosse, *der*/-genossin, *die;* Stubenkamerad, *der (Milit.);* ~ **service** *n.* Zimmerservice, *der;* ~ **temperature** *n.* Zimmertemperatur, *die*

roomy ['ru:mɪ] *adj.* geräumig

roost [ru:st] **1.** *n.* Schlafplatz, *der;* *(perch)* [Sitz]stange, *die;* **come home to ~** *(fig.)* jmdm. heimgezahlt werden; *see also* **rule 2 b. 2.** *v. i.* ‹*Vogel:*› sich [zum Schlafen] niederlassen

rooster ['ru:stə(r)] *n.* *(Amer.)* Hahn, *der*

¹root [ru:t] **1.** *n.* **a)** Wurzel, *die;* **pull sth. up by the ~s** etw. mit den Wurzeln ausreißen; *(fig.)* etw. mit der Wurzel ausrotten; **put down ~s/strike** *or* **take ~** *(lit. or fig.)* Wurzeln schlagen; **have ~s** verwurzelt sein; **b)** *(source)* Wurzel, *die; (basis)* Grundlage, *die;* **have its ~s in sth.** einer Sache *(Dat.)* entspringen; **get at** *or* **to the ~[s] of things** den Dingen auf den Grund kommen; **be at the ~ of the matter** der Kern der Sache sein. **2.** *v. t.* ~ **a plant firmly** eine Pflanze fest einpflanzen; **have ~ed itself in sth.** *(fig.)* in etw. *(Dat.)* verwurzelt sein; **stand ~ed to the spot** wie angewurzelt dastehen. **3.** *v. i.* ‹*Pflanze:*› wurzeln, anwachsen

~ **'out** *v. t.* ausrotten; ausmerzen

²root *v. i.* **a)** *(turn up ground)* wühlen **(for** nach); **b)** *(coll.)* ~ **for** *(cheer)* anfeuern; *(wish for success of)* Stimmung machen für

'root crop[s] *n. [pl.]* Hackfrüchte *Pl.*

rooted ['ru:tɪd] *adj.* eingewurzelt

rootless ['ru:tlɪs] *adj.* wurzellos

rope [rəʊp] *n.* **a)** *(cord)* Seil, *das;* Tau, *das;* **b)** *(Amer.: lasso)* Lasso, *das;* **c)** *(for hanging sb.)* **the ~:** der Strang; *(fig.: death penalty)* die Todesstrafe; **d)** *in pl.* *(Boxing)* **the ~s** die Seile; **be on the ~s** *(lit. or fig.)* in den Seilen hängen; **e)** *in pl.* **learn the ~s** lernen, sich zurechtzufinden; *(at work)* sich einarbeiten; **know the ~s** sich auskennen; **show sb. the ~s** jmdn. mit allem vertraut machen

~ **'in** *v. t.* **a)** mit einem Seil/mit Seilen absperren ‹*Gebiet*›; **b)** *(fig.)* einspannen *(ugs.);* **how did you get ~d in to that?** warum hast du dich dazu breitschlagen lassen? *(ugs.)*

~ **'off** *v. t.* [mit einem Seil/mit Seilen] absperren

~ **to'gether** *v. t.* *(Mount.)* aneinanderseilen

rope: ~-'ladder *n.* Strickleiter, *die;* **~way** *n.* Seilbahn, *die*

ropy ['rəʊpɪ] *adj.* *(coll.)* *(poor)* schäbig; *(in a bad state)* mitgenommen; **you look a bit ~:** du siehst ziemlich kaputt aus

rosary ['rəʊzərɪ] *n.* *(Relig.)* Rosenkranz, *der*

¹rose [rəʊz] **1.** *n.* **a)** *(plant, flower)* Rose, *die;* **no bed of ~s** *(fig.)* kein Honigschlecken; **it's not all ~s** es ist nicht alles [so] rosig; **everything's [coming up] ~s** alles ist bestens; **b)** *(colour)* Rosa, *das.* **2.** *adj.* rosa[farben]

²rose *see* **rise 2**

rose: ~-bed *n.* Rosenbeet, *das;* **~-bud** *n.* Rosenknospe, *die;* **~-bush** *n.* Rosenstrauch, *der;* **~-coloured** *adj.* *(lit. or fig.)* rosarot; **see things through ~-coloured spectacles** die Dinge durch eine rosarote Brille sehen; **~-hip** *n.* *(Bot.)* Hagebutte, *die;* **~-hip tea** Hagebuttentee, *der*

rosemary ['rəʊzmərɪ] *n.* *(Bot.)* Rosmarin, *der*

rose: ~-petal *n.* Rosen[blüten]blatt, *das;* **~-tinted** *see* **rose-coloured**

rosette [rəʊ'zet] *n.* Rosette, *die*

roster ['rɒstə(r)] *n.* Dienstplan, *der*

rostrum ['rɒstrəm] *n., pl.* **rostra** ['rɒstrə] *or* ~**s** *(platform)* Podium, *das; (desk)* Rednerpult, *das*

rosy ['rəʊzɪ] *adj.* **a)** rosig; **b)** *(fig.)* rosig ‹*Zukunft, Aussichten*›; **paint a ~ picture of sth.** etw. in den rosigsten Farben schildern

rot [rɒt] **1.** *n.* **a)** *see* **2a:** Verrottung, *die;* Fäulnis, *die;* Verwesung, *die; (fig.: deterioration)* Verfall, *der;* **stop the ~** *(fig.)* dem Verfall Einhalt gebieten; **the ~ has set in** *(fig.)* der Verfall hat eingesetzt; *see also* **dry rot; b)** *(sl.: nonsense)* Quark, *der (salopp);* ~! Blödsinn! *(ugs.).* **2.** *v.i.,* **-tt-: a)** *(decay)* verrotten; ‹*Fleisch, Gemüse, Obst:*› verfaulen; ‹*Leiche:*› verwesen; ‹*Holz:*› faulen; ‹*Zähne:*› schlecht werden; **b)** *(fig.: go to ruin)* verrotten. **3.** *v.t.,* **-tt-** verrotten lassen; verfaulen lassen ‹*Fleisch, Gemüse, Obst*›; faulen lassen ‹*Holz*›; verwesen lassen ‹*Leiche*›; zerstören ‹*Zähne*›

~ **a'way** *v.i.* verfaulen; ‹*Leiche:*› verwesen; ‹*Holz:*› faulen

rota ['rəʊtə] *n. (Brit.)* **a)** *(order of rotation)* Turnus, *der;* **b)** *(list of persons)* [Arbeits]plan, *der*

rotary ['rəʊtərɪ] *adj.* **a)** *(acting by rotation)* rotierend; Rotations-; ~ **engine** Drehkolbenmotor, *der;* **b) R~:** Rotarier-; **R~ Club** Rotary-Club, *der*

rotate [rəʊ'teɪt] **1.** *v.i. (revolve)* rotieren; sich drehen; ~ **on an axis** sich um eine Achse drehen. **2.** *v.t.* **a)** *(cause to revolve)* in Rotation versetzen; **b)** *(alternate)* abwechselnd erledigen ‹*Aufgaben*›; abwechselnd erfüllen ‹*Pflichten*›; ~ **[the] crops** Fruchtwechselwirtschaft betreiben

rotation [rəʊ'teɪʃn] *n.* **a)** Rotation, *die,* Drehung, *die* (**about** um); **b)** *(succession)* turnusmäßiger Wechsel; *(in political office)* Rotation, *die;* ~ **of crops** Fruchtfolge, *die;* **by** ~**:** im Turnus

rote [rəʊt] *n.* **by** ~**:** auswendig ‹*lernen, aufsagen*›

rotisserie [rəʊ'tɪsərɪ] *n.* **a)** *(restaurant)* Rotisserie, *die;* **b)** *(appliance)* Grill, *der*

rotor ['rəʊtə(r)] *n.* Rotor, *der (Technik)*

rotten ['rɒtn] **1.** *adj.,* ~**er** ['rɒtənə(r)], ~**est** ['rɒtənɪst] **a)** *(decayed)* verrottet; verwest ‹*Leiche*›; verrottet ‹*Holz*›; verfault ‹*Obst, Gemüse, Fleisch*›; faul ‹*Ei, Zähne*›; *(rusted)* verrostet; ~ **to the core** *(fig.)* verdorben bis ins Mark; völlig verrottet ‹*System, Gesellschaft*›; **b)** *(corrupt)* verdorben; verkommen; **c)** *(sl.: bad)* mies *(ugs.);* **feel ~** *(ill)* sich mies fühlen *(ugs.); (have a bad conscience)* ein schlechtes Gewissen haben; ~ **luck** saumäßiges Pech *(salopp).* **2.** *adv. (sl.)* saumäßig *(salopp);* **hurt/stink something ~:** saumäßig weh tun/stinken *(salopp);* **spoilt ~:** ganz schön verwöhnt *(ugs.)*

rotund [rəʊ'tʌnd] *adj.* **a)** *(round)* rund; **b)** *(plump)* rundlich

rouble ['ru:bl] *n.* Rubel, *der*

rouge [ru:ʒ] *n. (cosmetic powder)* Rouge, *das*

rough [rʌf] **1.** *adj.* **a)** *(coarse, uneven)* rauh; holp[e]rig ‹*Straße usw.*›; uneben ‹*Gelände*›; aufgewühlt ‹*Wasser*›; **b)** *(violent)* rauh, roh ‹*Person, Worte, Behandlung, Benehmen*›; rauh ‹*Gegend*›; **c)** *(harsh to the senses)* rauh; kratzig ‹*Geschmack, Getränk*›; sauer ‹*Apfelwein*›; **d)** *(trying)* hart; **this is ~ on him** das ist hart für ihn; **have a ~ time** es schwer haben; **give sb. a ~ time** es jmdm. schwer machen; **e)** *(fig.: lacking finish, polish)* derb; rauh ‹*Empfang*›; unbeholfen ‹*Stil*›; ungeschliffen ‹*Benehmen, Sprache*›; **he has a few ~ edges** *(fig.)* er ist ein wenig ungeschliffen; **f)** *(rudimentary)* primitiv ‹*Unterkunft, Leben*›; *(approximate)* grob ‹*Skizze, Schätzung, Einteilung, Übersetzung*›; vag ‹*Vorstellung*›; ~ **notes** stichwortartige Notizen; ~ **draft** Rohentwurf, *der;* ~ **paper/notebook** Konzeptpapier, *das*/Kladde, *die;* **g)** *(coll.: ill)* angeschlagen *(ugs.).* **2.** *n.* **a)** *(Golf)* Rough, *das;* **b) take the ~ with the smooth** die Dinge nehmen, wie sie kommen; **c)** *(unfinished state)* **[be] in ~:** [sich] im Rohzustand [befinden]. **3.** *adv.* rauh ‹*spielen*›; scharf ‹*reiten*›; **sleep ~:** im Freien schlafen. **4.** *v.t.* ~ **it** primitiv leben

~ **'out** *v.t.* [grob] entwerfen

~ **'up** *v.t. (sl.: deal roughly with)* anrempeln *(ugs.)*

roughage ['rʌfɪdʒ] *n.* Ballaststoffe *Pl. (Med.)*

rough: ~**-and-ready** *adj.* **a)** *(not elaborate)* provisorisch; skizzenhaft ‹*Beschreibung*›; behelfsmäßig ‹*Hütte, Methode*›; **b)** *(not refined)* rauhbeinig *(ugs.)* ‹*Person*›; ~**-and-'tumble** *n.* [wildes] Handgemenge; [wilde] Rauferei; ~ **'copy** *n.* **a)** *(original draft)* [erster] Entwurf; Konzept, *das;* **b)** *(simplified copy)* grobe Skizze; ~ **'diamond** *n. (fig.)* ungehobelter, aber guter Mensch; **he's a [bit of a] ~ diamond** er ist rauh, aber herzlich

roughen ['rʌfn] **1.** *v.t.* aufrauhen ‹*Oberfläche*›; rauh machen ‹*Hände*›. **2.** *v.i.* rauh werden

rough: ~ **house** *n. (sl.)* Keilerei, *die (ugs.);* ~ **'justice** *n.* ziemlich willkürliche Urteile; ~ **'luck** *n.* Pech, *das*

roughly ['rʌflɪ] *adv.* **a)** *(violently)* roh; grob; **b)** *(crudely)* leidlich; grob ‹*skizzieren, bearbeiten, bauen*›; **c)** *(approximately)* ungefähr; grob ‹*geschätzt*›

roughness ['rʌfnɪs] *n.* **a)** *no pl.* Rauheit, *die; (unevenness)* Unebenheit, *die;* **b)** *no pl. (sharpness) (of wine, fruit juice)* Säure, *die; (of voice)* Rauheit, *die;* **c)** *no pl. (violence)* Roheit, *die;* **the ~ of the area** die Häufigkeit von Gewalttaten in der Gegend; **d)** *(rough place or part)* unausgefeilte Stelle

rough: ~ **shod** *adj.* **ride ~shod over sb./sth.** jmdn./etw. mit Füßen treten; ~ **stuff** *n. (sl.)* Zoff, *der (salopp)*

roulette [ru:'let] *n.* Roulette, *das*

round [raʊnd] **1.** *adj.* rund; rundlich ‹*Arme*›; ~ **cheeks** Pausbacken *Pl. (fam.);* **in ~ figures, it will cost £1,000** rund gerechnet wird es 1 000 Pfund kosten. **2.** *n.* **a)** *(recurring series)* Serie, *die;* ~ **of talks/negotiations** Gesprächs-/Verhandlungsrunde, *die;* **the daily ~:** der Alltag; **b)** *(charge of ammunition)* Ladung, *die;* **50 ~s [of ammunition]** 50 Schuß Munition; **fire five ~s** fünf Schüsse abfeuern; **c)** *(division of game or contest)* Runde, *die;* **d)** *(burst)* ~ **of applause** Beifallssturm, *der;* ~**s of cheers** Hochrufe; **e)** ~ **[of drinks]** Runde, *die;* **f)** *(regular calls)* Runde, *die;* Tour, *die;* **the doctor is on her ~ at present** Frau Doktor macht gerade Hausbesuche; **go [on]** *or* **make one's ~s** ‹*Posten, Wächter usw.:*› seine Runde machen *od.* gehen; ‹*Krankenhausarzt:*› Visite machen; **do** *or* **go the ~s** ‹*Person, Gerücht usw.:*› die Runde machen *(ugs.);* **g)** *(Golf)* Runde, *die;* **a ~ of golf** eine Runde Golf; **h)** *(slice)* **a ~ of bread/toast** eine Scheibe Brot/Toast. **3.** *adv.* **a) all the year ~:** das ganze Jahr hindurch; **the third time ~:** beim dritten Mal; **have a wall all ~:** von einer Mauer eingeschlossen sein; **have a look ~:** sich umsehen; **b)** *(in girth)* **be [all of] ten feet ~:** einen Umfang von [mindestens] zehn Fuß haben; **c)** *(from one point, place, person, etc. to another)* **he asked ~ among his friends** er fragte seine Freunde; **d)** *(by indirect* *way)* herum; **walk ~**: außen herum gehen; **go a/the long way ~**: einen weiten Umweg machen; **e)** *(here)* hier; *(there)* dort; **I'll go ~ tomorrow** ich gehe morgen hin; **call ~ any time!** kommen Sie doch jederzeit vorbei!; **ask sb. ~ [for a drink]** jmdn. [zu einem Gläschen zu sich] einladen; *see also* **clock 1 a. 4.** *prep.* **a)** um [... herum]; **a tour ~ the world** eine Weltreise; **travel ~ England** durch England reisen; **she had a blanket ~ her** sie hatte eine Decke um sich geschlungen; **right ~ the lake** um den ganzen See herum; **be ~ the back of the house** hinter dem Haus sein; **run ~ the streets** durch die Straßen rennen; **walk** *etc.* **~ and ~ sth.** immer wieder um etw. herumgehen *usw.*; **we looked ~ the shops** wir sahen uns in den Geschäften um; **b)** *(in various directions from)* um [... herum]; rund um ‹*einen Ort*›; **look ~ one** um sich schauen; **do you live ~ here?** wohnst du [hier] in der Nähe?; **if you're ever ~ this way** wenn du hier in der Nähe bist. **5.** *v. t.* **a)** *(give ~ shape to)* rund machen; runden ‹*Lippen, Rücken*›; **b)** *(state as ~ number)* runden **(to** auf + *Akk.*); **c)** *(go ~)* umfahren/umgehen *usw.*; **~ a bend** um eine Kurve fahren/gehen/kommen *usw.*

~ 'down *v. t.* abrunden ‹*Zahl*› **(to** auf + *Akk.*)

~ 'off *v. t. (also fig.: complete)* abrunden

~ on *v. t.* anfahren

~ 'up *v. t.* **a)** *(gather, collect together)* verhaften ‹*Verdächtige*›; zusammentreiben ‹*Vieh*›; beschaffen, *(ugs.)* auftreiben ‹*Geld*›; **b)** *(to ~ figure)* aufrunden **(to** auf + *Akk.*)

round: ~ a'bout 1. *adv.* **a)** *(on all sides)* ringsum; **the villages ~ about** die umliegenden Dörfer; **b)** *(indirectly)* auf Umwegen; **c)** *(approximately)* rund; **~ about 2,500 people** um die *od.* rund 2 500 Leute; **2.** *prep.* rund um; **~about 1.** *n.* **a)** *(Brit.: road junction)* Verkehrskreisel, *der;* **b)** *(Brit.: merry-go-round)* Karussell, *das;* **it's swings and ~abouts** es gleicht sich aus; **2.** *adj.* **a)** *(meandering)* **a [very] ~about way** ein [sehr] umständlicher Weg; **the taxi went a ~about way** das Taxi machte einen Umweg; **b)** *(fig.: indirect)* umständlich

rounders ['raʊndəz] *n. sing. (Brit.)* Rounders, *das;* Rundball, *das (dem Baseball ähnliches Spiel)*

round: ~-eyed ['raʊndaɪd] *adj.* mit großen Augen *nachgestellt;* **be ~-eyed with amazement** große Augen machen; **~-faced** ['raʊndfeɪst] *adj.* pausbäckig *(fam.);* **R~head** *n. (Brit. Hist.)* Rundkopf, *der*

roundly ['raʊndlɪ] *adv.* entschieden

roundness ['raʊndnɪs] *n., no pl.* Rundheit, *die; (of figure)* Rundlichkeit, *die*

round: ~ 'number *n.* runde Zahl; **~-shouldered** [raʊnd-'ʃəʊldəd] *adj.* ‹*Person*› mit einem Rundrücken; **be ~-shouldered** einen Rundrücken haben; **~-the-'clock** *adj.* rund um die Uhr *nachgestellt;* **~ 'trip** *n.* **a)** Rundreise, *die;* **b)** *(Amer.: return trip)* Hin- und Rückfahrt, *die;* **~-up** *n.* **a)** *(gathering-in) (of persons)* Einfangen, *das; (arrest)* Verhaftung, *die; (of animals)* Zusammentreiben, *das;* **b)** *(summary)* Zusammenfassung, *die*

rouse [raʊz] *v. t.* **a)** *(awaken, lit. or fig.)* wecken **(from** aus); **~ oneself** aufwachen; *(overcome indolence)* sich aufraffen; **~ sb./oneself to action** jmdn. zur Tat anstacheln/sich zur Tat aufraffen; **b)** *(provoke)* reizen; **he is terrible when ~d** er ist furchtbar, wenn man ihn reizt; **~ sb. to anger** jmdn. in Wut bringen; **c)** *(cause)* wecken; hervorrufen, auslösen ‹*Empörung, Beschuldigungen*›

rousing ['raʊzɪŋ] *adj.* mitreißend ‹*Lied*›; leidenschaftlich ‹*Rede*›; stürmisch ‹*Beifall*›

¹rout [raʊt] **1.** *n. (disorderly retreat)* [wilde] Flucht; *(disastrous defeat)* verheerende Niederlage; **put to ~**: in die Flucht schlagen. **2.** *v. t.* aufreiben ‹*Feind, Truppen*›; vernichtend schlagen ‹*Gegner*›

²rout *v. i. (root)* wühlen

~ 'out *v. t.* herausjagen; **~ sb. out of sth.** jmdn. aus etw. jagen

route [ru:t, *Mil. also:* raʊt] **1.** *n. (course)* Route, *die;* Weg, *der;* **shipping ~**: Schiffahrtsstraße, *die.* **2.** *v. t.*, **~ing** fahren lassen ‹*Fahrzeug*›; führen ‹*Linie*›; **the train is ~d through** *or* **via Crewe** der Zug fährt über Crewe

'route march *n. (Mil.)* Übungsmarsch, *der*

routine [ru:'ti:n] **1.** *n.* **a)** *(regular procedure; Computing)* Routine, *die;* **b)** *(coll.) (set speech)* Platte, *die (ugs.); (formula)* Spruch, *der;* **c)** *(Theatre)* Nummer, *die; (Dancing, Skating)* Figur, *die; (Gymnastics)* Übung, *die.* **2.** *adj.* routinemäßig; Routine‹*arbeit, -untersuchung usw.*›

rove [rəʊv] **1.** *v. i.* ziehen; ‹*Blick:*› schweifen *(geh.);* **~ [about]** herumziehen; **have a roving eye** den Frauen/Männern schöne Augen machen. **2.** *v. t.* streifen durch; durchstreifen *(geh.);* ‹*Blick:*› durchschweifen ‹*Raum*›

¹row [raʊ] *(coll.)* **1.** *n.* **a)** *(noise)* Krach, *der;* **make a ~**: Krach machen; *(protest)* Rabatz machen *(ugs.);* **b)** *(quarrel)* Krach, *der (ugs.);* **have/start a ~**: Krach haben/anfangen *(ugs.).* **2.** *v. i.* sich streiten

²row [rəʊ] *n.* **a)** Reihe, *die;* **in a ~**: in einer Reihe; *(coll.: in succession)* nacheinander; **b)** *(line of numbers etc.)* Zeile, *die;* **c)** *(terrace)* **~ [of houses]** [Häuser]zeile, *die;* [Häuser]reihe, *die*

³row [rəʊ] **1.** *v. i.* rudern. **2.** *v. t.* rudern; **~ sb. across** jmdn. hinüberrudern. **3.** *n.* **go for a ~**: rudern gehen

rowan ['rəʊən, 'raʊən] *n.* **~[-tree]** Eberesche, *die*

row-boat ['rəʊbəʊt] *n. (Amer.)* Ruderboot, *das*

rowdy ['raʊdɪ] **1.** *adj.* rowdyhaft *(abwertend);* **~ adolescents** jugendliche Rowdys *(abwertend);* **~ scenes** tumultartige Szenen; **the party was ~**: auf der Party ging es laut zu. **2.** *n.* Krawallmacher, *der;* Rabauke, *der*

rowing ['rəʊɪŋ] *n., no pl.* Rudern, *das*

rowing: ~-boat *n. (Brit.)* Ruderboot, *das;* **~-club** *n.* Ruderklub, *der*

royal ['rɔɪəl] **1.** *adj.* königlich. **2.** *n. (coll.)* Mitglied der Königsfamilie; **the ~s** die Königsfamilie

royal: R~ 'Air Force *n. (Brit.)* Königliche Luftwaffe; **~ 'blue** *n. (Brit.)* Königsblau, *das;* **~ 'family** *n.* königliche Familie

royalist ['rɔɪəlɪst] *n.* Royalist, *der*/Royalistin, *die; attrib.* Royalisten-; royalistisch

Royal 'Navy *n. (Brit.)* Königliche Kriegsmarine

royalty ['rɔɪəltɪ] *n.* **a)** *(payment)* Tantieme, *die* **(on** für); **b)** *collect. (royal persons)* Mitglieder des Königshauses; **c)** *no pl., no art. (member of royal family)* ein Mitglied der königlichen Familie

r.p.m. [ɑ:pi:'em] *abbr.* **revolutions per minute** U.p.M.

RSPCA *abbr. (Brit.)* **Royal Society for the Prevention of Cruelty to Animals** *britischer Tierschutzverein*

rub [rʌb] **1.** *v. t.*, **-bb-:** reiben **(on,**

against an + *Dat.*); *(with ointment etc.)* einreiben; *(to remove dirt etc.)* abreiben; *(to dry)* trockenreiben; *(with sandpaper)* [ab]schmirgeln; ~ **sth. off sth.** etw. von etw. reiben; ~ **one's hands** sich *(Dat.)* die Hände reiben; ~ **shoulders** *or* **elbows with sb.** *(fig.)* Tuchfühlung mit jmdm. haben; ~ **two things together** zwei Dinge aneinanderreiben; ~ **sth. through a sieve** etw. durch ein Sieb streichen. **2.** *v. i.*, **-bb-: a)** *(exercise friction)* reiben (**[up]on, against** an + *Dat.*); **b)** *(get frayed)* sich abreiben. **3.** *n.* Reiben, *das;* **give it a quick** ~**:** reib es kurz ab; **there's the** ~ *(fig.)* da liegt der Haken [dabei] *(ugs.)*

~ **'down** *v. t.* **a)** *(prepare)* abschmirgeln; **b)** *(dry)* abreiben

~ **'in** *v. t.* einreiben; **there's no need to** *or* **don't** ~ **it in** *(fig.)* reib es mir nicht [dauernd] unter die Nase *(ugs.)*

~ **'off 1.** *v. t.* wegreiben; wegwischen. **2.** *v. i. (lit. or fig.)* abfärben (**on** auf + *Akk.*)

~ **'out 1.** *v. t.* ausreiben; *(from paper)* ausradieren. **2.** *v. i.* sich ausreiben/*(from paper)* sich ausradieren lassen

~ **'up** *v. t.* **a)** *(polish)* blank reiben; wienern *(ugs.)*; **b)** ~ **sb. up the wrong way** *(fig.)* jmdm. auf den Schlips treten *(ugs.)*

[1]**rubber** ['rʌbə(r)] *n.* **a)** Gummi, *das od. der; attrib.* Gummi-; **b)** *(eraser)* Radiergummi, *der*

[2]**rubber** *n. (Cards)* Robber, *der*

rubber: ~ **'band** *n.* Gummiband, *das;* ~ **plant** *n. (Bot.)* Gummibaum, *der;* ~ **'stamp** *n.* Gummistempel, *der;* ~**-'stamp** *v. t. (fig.: approve)* absegnen *(ugs. scherzh.)*

rubbery ['rʌbərı] *adj.* gummiartig; *(tough)* zäh; **be tough and** ~**:** zäh wie Gummi sein

rubbish ['rʌbıʃ] **1.** *n., no pl., no indef. art.* **a)** *(refuse)* Abfall, *der;* Abfälle; *(to be collected and dumped)* Müll, *der;* **b)** *(worthless material)* Plunder, *der (ugs. abwertend);* **be** ~**:** nichts taugen; **c)** *(nonsense)* Quatsch, *der (ugs.);* Blödsinn, *der (ugs.);* **what** ~**!** was für ein Quatsch *od.* Schmarren! **2.** *int.* Quatsch *(ugs. abwertend)*

rubbish: ~**-bin** *n.* Abfall-/Mülleimer, *der; (in factory)* Abfall-/Mülltonne, *die;* ~**-chute** *n.* Müllschlucker, *der;* ~**-dump** *n.* Müllkippe, *die;* ~**-heap** *n.* Müllhaufen, *der; (in garden)* Abfallhaufen, *der;* ~**-tip** *n.* Müllabladeplatz, *der*

rubbishy ['rʌbıʃı] *adj.* mies *(ugs.);* ~ **newspaper** Käseblatt, *das (salopp abwertend)*

rubble ['rʌbl] *n. (from damaged building)* Trümmer *Pl.; (Geol. also)* Schutt, *der;* **reduce sth. to** ~**:** etw. in Schutt und Asche legen

ruby ['ru:bı] **1.** *n.* **a)** *(precious stone)* Rubin, *der;* **b)** *(colour)* Rubinrot, *das.* **2.** *adj.* **a)** *(red)* rubinfarben; rubinrot; **b)** Rubin*(ring, -brosche usw.)*

ruby: ~**-red** *adj.* rubinrot; ~ **'wedding** *n.* Rubinhochzeit, *die*

RUC *abbr.* **Royal Ulster Constabulary** *nordirische Polizei*

ruck [rʌk], *(Brit.)* **ruckle** ['rʌkl] **1.** *n. (crease)* Falte, *die.* **2.** *v. i.* ~ **up** hochrutschen

rucksack ['rʌksæk, 'rʊksæk] *n.* Rucksack, *der*

ruckus ['rʌkəs] *n.*, **ructions** ['rʌkʃnz] *n. pl. (coll.)* Rabatz, *der (ugs.)*

rudder ['rʌdə(r)] *n.* Ruder, *das*

ruddy ['rʌdı] *adj.* **a)** *(reddish)* rötlich; **b)** *(Brit. sl. euphem.: bloody)* verdammt *(salopp)*

rude [ru:d] *adj.* **a)** *(impolite)* unhöflich; *(stronger)* rüde; **say** ~ **things** *or* **be** ~ **about sb.** in ungehöriger Weise von jmdm. sprechen; **be** ~ **to sb.** zu jmdm. grob unhöflich sein/jmdn. rüde behandeln; **b)** *(abrupt)* unsanft; ~ **awakening** böses *od. (geh.)* jähes Erwachen; **c)** *(obscene)* unanständig

rudeness ['ru:dnıs] *n., no pl. (bad manners)* ungehöriges *od.* rüdes Benehmen

rudimentary [ru:dı'mentərı] *adj. (elementary)* elementar; primitiv *(Gebäude);* ~ **knowledge** Grundkenntnisse *Pl.*

rudiments ['ru:dımənts] *n. pl.* **a)** *(first principles)* Grundzüge *Pl.;* Grundlagen *Pl.;* **b)** *(imperfect beginning)* [erster] Ansatz

rueful ['ru:fl] *adj.*, **ruefully** ['ru:fəlı] *adv.* reumütig; reuig

ruff [rʌf] *n.* Halskrause, *die*

ruffian ['rʌfıən] *n.* Rohling, *der (abwertend);* **gang of** ~**s** Schlägerbande, *die*

ruffle ['rʌfl] **1.** *v. t.* **a)** *(disturb smoothness of)* kräuseln; ~ **sb.'s hair** jmdm. durch die Haare fahren; **b)** *(upset)* aus der Fassung bringen; **be easily** ~**d** leicht aus der Fassung geraten. **2.** *n. (frill)* Rüsche, *die*

~ **'up** *v. t.* sträuben *(Gefieder)*

rug [rʌg] *n.* **a)** *(for floor)* [kleiner, dicker] Teppich; **Persian** ~**:** Perserbrücke, *die;* **pull the** ~ **[out] from under sb.** *(fig.)* jmdm. den Boden unter den Füßen wegziehen; **b)** *(wrap, blanket)* [dicke] Wolldecke

Rugby ['rʌgbı] *n.* Rugby, *das*

Rugby: ~ **ball** *n.* Rugbyball, *der;* ~ **tackle** *n.* tiefes Fassen *(Rugby);* **the policeman brought him down with a** ~ **tackle** der Polizist warf sich auf ihn und riß ihn zu Boden

rugged ['rʌgıd] *adj.* **a)** *(sturdy)* robust; **b)** *(involving hardship)* hart *(Test);* **c)** *(unpolished)* rauh; **with** ~ **good looks** gutaussehend mit markanten Gesichtszügen; **d)** *(uneven)* zerklüftet; unwegsam *(Land, Anstieg);* zerfurcht *(Gesicht)*

rugger ['rʌgə(r)] *n. (Brit. coll.)* Rugby, *das*

ruin ['ru:ın] **1.** *n.* **a)** *no pl., no indef. art. (decay)* Verfall, *der;* **go to** *or* **fall into rack and** ~ *(Gebäude:)* völlig verfallen; *(Garten:)* völlig verwahrlosen; **b)** *no pl., no indef. art. (downfall)* Ruin, *der;* ~ **stared her in the face** sie stand vor dem Ruin; **c)** *in sing. or pl. (remains)* Ruine, *die;* **in** ~**s** in Trümmern; **d)** *(cause of* ~*)* Ruin, *der;* Untergang, *der;* **you'll be the** ~ **of me** du ruinierst mich [noch]. **2.** *v. t.* ruinieren; verderben *(Urlaub, Abend);* zunichte machen *(Aussichten, Möglichkeiten usw.)*

ruined ['ru:ınd] *adj.* **a)** *(reduced to ruins)* verfallen; ~ **town** Ruinenstadt, *die;* **a** ~ **castle/palace/church** eine Burg-/Palast-/Kirchenruine; **b)** *(brought to ruin)* ruiniert; **c)** *(spoilt)* verdorben

ruinous ['ru:ınəs] *adj.* ruinös; katastrophal *(Wirkung);* **be** ~ **to sb./sth.** jmdn./etw. ruinieren

rule [ru:l] **1.** *n.* **a)** *(principle)* Regel, *die;* **the** ~**s of the game** *(lit. or fig.)* die Spielregeln; **stick to** *or* **play by the** ~**s** *(lit. or fig.)* sich an die Spielregeln halten; ~**s and regulations** Regeln und Vorschriften; **be against the** ~**s** regelwidrig sein; *(fig.)* gegen die Spielregeln verstoßen; **as a** ~**:** in der Regel; ~ **of thumb** Faustregel, *die;* **b)** *(custom)* Regel, *die;* **the** ~ **of the house is that ...:** in diesem Haus ist es üblich, daß ...; **c)** *no pl. (government)* Herrschaft, *die* (**over** über + *Akk.*); **the** ~ **of law** die Autorität des Gesetzes; **d)** *(graduated measure)* Maß, *das; (tape)* Bandmaß, *das; (folding)* Zollstock, *der.* **2.** *v. t.* **a)** *(control)* beherrschen; **b)** *(be the ruler of)* regieren; *(Monarch, Diktator usw.:)* herrschen über (+ *Akk.*); ~ **the roost [in the house]** Herr im Hause sein; **c)** *(give as decision)* entscheiden; ~ **a motion out of order** einen Antrag nicht zulassen; **d)** *(draw)* ziehen ‹*Linie*›; *(draw lines on)* linieren ‹*Papier*›. **3.** *v.i.* **a)** *(govern)* herrschen; **b)** *(decide, declare formally)* entscheiden (**against** gegen; **in favour of** für); ~ **on a matter** in einer Sache entscheiden

~ **'off 1.** *v.t.* mit einem Strich abtrennen. **2.** *v.i.* eine Schlußstrich ziehen

~ **'out** *v.t.* **a)** *(exclude, eliminate)* ausschließen; **b)** *(prevent)* unmöglich machen

ruled [ru:ld] *adj.* liniert ‹*Papier*›

ruler ['ru:lə(r)] *n.* **a)** *(person)* Herrscher, *der*/Herrscherin, *die;* **b)** *(for drawing or measuring)* Lineal, *das*

ruling ['ru:lɪŋ] **1.** *n. (decision)* Entscheidung, *die.* **2.** *adj.* **a)** *(predominating)* herrschend ‹*Meinung*›; vorherrschend ‹*Charakterzug*›; **b)** *(governing, reigning)* herrschend ‹*Klasse*›; regierend ‹*Partei*›; amtierend ‹*Regierung*›

rum [rʌm] *n.* Rum, *der*

¹rumble ['rʌmbl] **1.** *n.* Grollen, *das; (of heavy vehicle)* Rumpeln, *das (ugs.)* **2.** *v.i.* **a)** grollen; ‹*Magen:*› knurren; **b)** *(go with rumbling noise)* rumpeln *(ugs.)*

²rumble *v.t. (coll.: understand)* spitzkriegen *(ugs.)* ‹*Sache*›; auf die Schliche kommen (+ *Dat.*) ‹*Person*›

ruminate ['ru:mɪneɪt] *v.i.* **a)** ~ **over** *or* **about** *or* **on sth.** über etw. *(Akk.)* nachsinnen *(geh.) od.* grübeln; **b)** *(Zool.)* wiederkäuen

rummage ['rʌmɪdʒ] **1.** *v.i.* wühlen *(ugs.);* kramen *(ugs.);* ~ **about** *or* **around** herumkramen *(ugs.).* **2.** *n.* **have a** ~ **through sth.** etw. durchwühlen *od.* durchstöbern

'rummage sale *(esp. Amer.) see* **jumble sale**

rummy ['rʌmɪ] *n. (Cards)* Rommé, *das*

rumour *(Brit.; Amer.:* **rumor***)* ['ru:mə(r)] **1.** *n. (unverified story)* Gerücht, *das;* **there is a** ~ **that** *or* ~ **has it that ...:** es geht das Gerücht, daß ... **2.** *v.t.* **sb. is ~ed to have done sth., it is ~ed that sb. has done sth.** man munkelt *(ugs.) od.* es geht das Gerücht, daß jmd. etw. getan hat

rump [rʌmp] *n.* **a)** *(buttocks)* Hinterteil, *das (ugs.);* **b)** *(remnant)* Rest, *der*

'rump steak *n.* Rumpsteak, *das*

rumpus ['rʌmpəs] *n., no pl. (coll.)* Krach, *der (ugs.);* Spektakel, *der (ugs.);* **kick up** *or* **make a** ~**:** einen Spektakel veranstalten *(ugs.)*

run [rʌn] **1.** *n.* **a)** Lauf, *der;* **go for a** ~ **before breakfast** vor dem Frühstück einen Lauf machen; **make a late** ~ *(Sport or fig.)* zum Endspurt ansetzen; **come towards sb./start off at a** ~**:** jmdm. entgegenlaufen/losrennen; **I've had a good** ~ **for my money** ich bin auf meine Kosten gekommen; **on the** ~**:** auf der Flucht; **b)** *(trip in vehicle)* Fahrt, *die; (for pleasure)* Ausflug, *der;* **on the** ~ **down to Cornwall** auf der Fahrt nach Cornwall; **go for a** ~ **[in the car]** einen [Auto]ausflug machen; **c) she has had a long** ~ **of success** sie war lange [Zeit] erfolgreich; **have a long** ~ ‹*Stück, Show:*› viele Aufführungen erleben; **d)** *(succession)* Serie, *die; (Cards)* Sequenz, *die;* **a** ~ **of victories** eine Siegesserie; **e)** *(tendency)* Ablauf, *der;* **the general** ~ **of things/events** der Lauf der Dinge/der Gang der Ereignisse; **f)** *(regular route)* Strecke, *die;* **g)** *(Cricket, Baseball)* Lauf, *der;* Run, *der;* **h)** *(quantity produced) (of book)* Auflage, *die;* **production** ~**:** Ausstoß, *der (Wirtsch.);* **i)** *(demand)* Run, *der* (**on** auf + *Akk.*); **j) the ~s** *(sl.: diarrhoea)* Durchmarsch, *der (salopp);* **k)** *(unrestricted use)* **give sb. the** ~ **of sth.** jmdm. etw. zu seiner freien Verfügung überlassen; **have the** ~ **of sth.** etw. zu seiner freien Verfügung haben; **l)** *(animal enclosure)* Auslauf, *der.* **2.** *v.i.*, **-nn-, ran** [ræn], **run a)** laufen; *(fast also)* rennen; ~ **for the bus** laufen *od.* rennen, um den Bus zu kriegen *(ugs.);* ~ **to help sb.** jmdm. zu Hilfe eilen; **b)** *(compete)* laufen; **c)** *(hurry)* laufen; **don't** ~ **to me when things go wrong** komm mir nicht angelaufen, wenn etwas schiefgeht *(ugs.);* ~ **to meet sb.** jmdm. entgegenlaufen; **d)** *(roll)* laufen; ‹*Ball, Kugel:*› rollen, laufen; **e)** *(slide)* laufen; ‹*Schlitten, [Schiebe]tür:*› gleiten; **f)** *(revolve)* ‹*Rad, Maschine:*› laufen; **g)** *(flee)* davonlaufen; **h)** *(operate on a schedule)* fahren; ~ **between two places** ‹*Zug, Bus:*› zwischen zwei Orten verkehren; **the train is ~ning late** der Zug hat Verspätung; **the train doesn't** ~ **on Sundays** der Zug verkehrt nicht an Sonntagen; **i)** *(pass cursorily)* ~ **through** überfliegen ‹*Text*›; ~ **through one's head** *or* **mind** ‹*Gedanken, Ideen:*› einem durch den Kopf gehen; ~ **through the various possibilities** die verschiedenen Möglichkeiten durchspielen; **j)** *(flow)* laufen; ‹*Fluß:*› fließen; ~ **dry** ‹*Fluß:*› austrocknen; ‹*Quelle:*› versiegen; ~ **low** *or* **short** knapp werden; ausgehen; **k)** *(be current)* ‹*Vertrag, Theaterstück:*› laufen; **l)** *(be present)* ~ **through sth.** sich durch etw. ziehen; ~ **in the family** ‹*Eigenschaft, Begabung:*› in der Familie liegen; **m)** *(function)* laufen; **keep/leave the engine ~ning** den Motor laufen lassen/nicht abstellen; **the machine ~s on batteries/oil** *etc.* die Maschine läuft mit Batterien/Öl *usw.;* **n)** *(have a course)* ‹*Straße, Bahnlinie:*› verlaufen; **o)** *(have wording)* lauten; ‹*Geschichte:*› gehen *(fig.);* **p)** *(have certain level)* **inflation is ~ning at 15%** die Inflationsrate beläuft sich auf *od.* beträgt 15%; **q)** *(seek election)* kandidieren; ~ **for mayor** für das Amt des Bürgermeisters kandidieren; **r)** *(spread quickly)* **a shiver ran down my spine** ein Schau[d]er *(geh.)* lief mir den Rücken hinunter; **s)** *(spread undesirably)* ‹*Butter, Eis:*› zerlaufen; *(in washing)* ‹*Farben:*› auslaufen; **t)** *(ladder)* ‹*Strumpf:*› Laufmaschen bekommen. **3.** *v.t.*, **-nn-, ran, run a)** *(cause to move)* laufen lassen; *(drive)* fahren; ~ **one's hand/fingers through/along** *or* **over sth.** mit der Hand/den Fingern durch etw. fahren/über etw. *(Akk.)* streichen; ~ **an** *or* **one's eye along** *or* **down** *or* **over sth.** *(fig.)* etw. überfliegen; **b)** *(cause to flow)* [ein]laufen lassen; ~ **a bath** ein Bad einlaufen lassen; **c)** *(organize, manage)* führen, leiten ‹*Geschäft usw.*›; durchführen ‹*Experiment*›; veranstalten ‹*Wettbewerb*›; führen ‹*Leben*›; **d)** *(operate)* bedienen ‹*Maschine*›; verkehren lassen ‹*Verkehrsmittel*›; einsetzen ‹*Sonderbus, -zug*›; laufen lassen ‹*Motor*›; abspielen ‹*Tonband*›; ~ **forward/back** vorwärts-/zurückspulen ‹*Film, Tonband*›; **e)** *(own and use)* sich *(Dat.)* halten ‹*Auto*›; **this car is expensive to** ~**:** dieses Auto ist im Unterhalt sehr teuer; ~ **a car with defective brakes** ein Auto mit defekten Bremsen fahren; **f)** *(take for journey)* fahren; **I'll** ~ **you into town** ich fahre *od.* bringe dich in die Stadt; **g)** *(pursue)* jagen; ~ **sb. hard** *or* **close** jmdm. auf den Fersen sein *od.* sitzen *(ugs.);* **be** ~ **off one's feet** alle Hände voll zu tun haben *(ugs.); (in business)* Hochbetrieb haben *(ugs.); see also* **earth 1d; h)** *(complete)* laufen ‹*Rennen, Marathon, Strecke*›; ~

messages/errands Botengänge machen; **i)** ~ **a fever/a temperature** Fieber/erhöhte Temperatur haben; **j)** *(publish)* bringen *(ugs.)* ‹*Bericht, Artikel usw.*›

~ **a'bout** *v.i.* **a)** *(bustle)* hin- und herlaufen; **b)** *(play without restraint)* herumtollen; herumspringen *(ugs.)*

~ **a'cross** *v.t.* ~ **across sb.** jmdn. treffen; jmdm. über den Weg laufen; ~ **across sth.** auf etw. *(Akk.)* stoßen

~ **after** *v.t.* hinterherlaufen (+ *Dat.*)

~ **a'long** *v.i.* *(coll.: depart)* sich trollen *(ugs.)*

~ **a'round** **1.** *v.i.* **a)** ~ **around with sb.** sich mit jmdm. herumtreiben; **b)** *see* **run about a**; **c)** *see* **run about b.** **2.** *v.t.* herumfahren

~ **a'way** *v.i.* **a)** *(flee)* weglaufen; fortlaufen; **b)** *(abscond)* ~ **away [from home/from the children's home]** [von zu Hause/aus dem Kinderheim] weglaufen; **c)** *(elope)* ~ **away with sb./together** mit jmdm./zusammen durchbrennen *(ugs.)*; **d)** *(bolt)* ‹*Pferd:*› durchgehen; **e)** ‹*Wasser:*› ablaufen

~ **a'way with** *v.t.* **a)** *(coll.: steal)* abhauen mit *(salopp)*; **b)** *(fig.: win)* ~ **away with the top prize/all the trophies** den 1. Preis/alle Trophäen erringen; **c)** *(fig.: be misled by)* **don't** ~ **away with the idea that ...:** glaub bloß nicht, daß ...; **he let his imagination/enthusiasm** ~ **away with him** seine Phantasie/Begeisterung ist mit ihm durchgegangen

~ **'down** **1.** *v.t.* **a)** *(collide with)* überfahren; **b)** *(criticize)* heruntermachen *(ugs.)*; herabsetzen; **c)** *(cause to diminish)* abbauen; verringern ‹*Produktion*›; **d)** *(cause to lose power)* leer machen ‹*Batterie*›. **2.** *v.i.* **a)** hin-/herunterlaufen/-rennen/-fahren; **b)** *(decline)* sich verringern; **c)** *(lose power)* ausgehen; ‹*Batterie:*› leer werden; ‹*Uhr, Spielzeug:*› ablaufen

~ **'in** *v.t.* **a)** *(prepare for use)* einfahren ‹*Auto*›; sich einlaufen lassen ‹*Maschine*›; **b)** *(coll.: arrest)* hoppnehmen *(salopp)*

~ **into** *v.t.* **a)** ~ **into a telegraph pole/tree** gegen einen Telegrafenmast/Baum fahren; **b)** *(cause to collide with)* ~ **one's car into a tree** seinen Wagen gegen einen Baum fahren; **c)** *(fig.: meet)* ~ **into sb.** jmdm. in die Arme laufen *(ugs.)*; **d)** *(be faced with)* stoßen auf (+ *Akk.*) ‹*Schwierigkeiten, Widerstand, Probleme usw.*›; **e)** *(enter)* geraten in (+ *Akk.*) ‹*Sturm, schlechtes Wetter, Schulden*›; **his debts** ~ **into thousands** seine Schulden gehen in die Tausende

~ **'off** **1.** *v.i. see* ~ **away a, c.** **2.** *v.t.* **a)** *(compose rapidly)* hinwerfen ‹*ein paar Zeilen, Verse, Notizen*›; zu Papier bringen ‹*Brief*›; **b)** *(produce on machine)* abziehen ‹*Kopien, Handzettel usw.*›; **c)** *(cause to drain away)* ablaufen lassen

~ **'off with** *v.t.* **a)** *(coll.: steal)* abhauen mit *(salopp)*; **b)** = ~ **away with** *see* ~ **away c**; **c)** *see* ~ **away with b**

~ **'on** *v.i.* weitergehen; ‹*Krankheit:*› fortschreiten

~ **'out** *v.i.* **a)** hin-/herauslaufen/-rennen; **b)** *(become exhausted)* ‹*Vorräte, Bestände:*› zu Ende gehen; ‹*Geduld:*› sich erschöpfen; **we have** ~ **out** wir haben keinen/keine/keines mehr; *(sold everything)* wir sind ausverkauft; **c)** *(expire)* ‹*Vertrag:*› ablaufen

~ **'out of** *v.t.* **sb.** ~**s out of sth.** jmdm. geht etw. aus; **I'm** ~**ning out of patience** meine Geduld geht zu Ende; **we're** ~**ning out of time** uns wird die Zeit allmählich knapp

~ **over** **1.** ['---] *v.t.* *(knock down)* überfahren. **2.** [-'--] *v.i.* überlaufen

~ **through** *v.t.* **a)** ['--] abspielen ‹*Tonband, Film*›; **b)** ['--] *(rehearse)* durchspielen ‹*Theaterstück*›; **c)** [-'-] *(pierce right through)* ~ **sb. through with sth.** jmdn. mit etw. durchbohren. *See also* ~ **2 i, l**

~ **to** *v.t.* **a)** *(amount to)* umfassen; ‹*Geldsumme, Kosten:*› sich belaufen auf (+ *Akk.*); **b)** *(be sufficient for)* **sth. will** ~ **to sth.** etw. reicht für etw.; **c)** *(afford)* **sb. can** ~ **to sth.** jmd. kann sich *(Dat.)* etw. leisten

~ **'up** **1.** *v.i.* hinlaufen; **come** ~**ning up** herangelaufen kommen. **2.** *v.t.* **a)** *(hoist)* hissen ‹*Fahne*›; **b)** *(make quickly)* rasch nähen ‹*Kleidungsstück*›; **c)** *(allow to accumulate)* ~ **up debts** Schulden zusammenkommen lassen

~ **'up against** *v.t.* stoßen auf (+ *Akk.*) ‹*Probleme, Widerstand usw.*›

run: ~**-around** *n.* *(coll.)* **give sb. the** ~**-around** jmdn. an der Nase herumführen *(ugs.)*; ~**away** **1.** *n.* Ausreißer, *der*/Ausreißerin, *die* *(ugs.)*; **2.** *attrib. adj.* **a)** *(out of control)* durchgegangen ‹*Pferd*›; außer Kontrolle geraten ‹*Fahrzeug, Preise*›; *(fig.)* galoppierend ‹*Inflation*›; **b)** *(outstanding)* überwältigend ‹*Erfolg*›; triumphal ‹*Sieg*›; ~**-down** **1.** ['--] *n.* *(coll.: briefing)* Übersicht, *die* **(on** über + *Akk.*); **2.** [-'-] *adj.* *(tired)* mitgenommen

[1]**rung** [rʌŋ] *n.* *(of ladder)* Sprosse, *die*

[2]**rung** *see* [2]**ring 2, 3**

runner ['rʌnə(r)] *n.* **a)** Läufer, *der*/Läuferin, *die*; **b)** *(horse in race)* **eight** ~**s were in the race** acht Pferde liefen beim Rennen; **c)** *(messenger)* Bote, *der*; **d)** **curtain** ~**:** Gardinenröllchen, *das*; **e)** *(part on which sth. slides)* Kufe, *die*; *(groove)* Laufschiene, *die*; **f)** *(carpet)* Läufer, *der*

runner: ~ **bean** *n.* *(Brit.)* Stangenbohne, *die*; ~**-'up** *n.* Zweite, *der/die*; **the** ~**s-up** die Plazierten

running ['rʌnɪŋ] **1.** *n.* **a)** *(management)* Leitung, *die*; **b)** *(action)* Laufen, *das*; *(jogging)* Jogging, *das*; **make the** ~ *(in competition)* an der Spitze liegen; *(fig.: have the initiative)* den Ton angeben; **in/out of the** ~**:** im/aus dem Rennen; **c)** *(of engine, machine)* Laufen, *das*. **2.** *adj.* **a)** *(continuous)* ständig; fortlaufend ‹*Erklärungen*›; **have** *or* **fight a** ~ **battle** *(fig.)* ständig im Streit liegen; **b)** *(in succession)* hintereinander; **win for the third year** ~**:** schon drei Jahre hintereinander gewinnen

running: ~ **'commentary** *n.* *(Broadcasting; also fig.)* Live-Kommentar, *der*; ~ **costs** *n. pl.* Betriebskosten *Pl.*; ~ **'jump** *n.* **you can [go and] take a** ~ **jump** *(fig. sl.)* du kannst mir den Buckel herunterrutschen *(ugs.)*; ~ **re'pairs** *n. pl.* laufende Reparaturen; ~**-shoe** *n.* Rennschuh, *der*; ~**-shorts** *n. pl.* Sporthose, *die*; ~ **'sore** *n.* nässende Wunde; *(fig.)* schwärende Wunde; ~ **'water** *n.* **a)** *(in stream)* fließendes Gewässer; **b)** *(available through pipe)* fließendes Wasser

runny ['rʌnɪ] *adj.* **a)** *(secreting mucus)* laufend ‹*Nase*›; **b)** *(excessively liquid)* zerlaufend; zu dünn ‹*Farbe, Marmelade*›

run: ~**-of-the-'mill** *adj.* ganz gewöhnlich; ~**-through** *n.* **a)** *(cursory reading)* **give a text a [quick]** ~**-through** einen Text [kurz] überfliegen; **b)** *(rapid summary)* Überblick, *der* **(of** über + *Akk.*); **c)** *(rehearsal)* Durchlaufprobe, *die*; ~**-up** *n.* **a)** *(approach to an event)* **during** *or* **in the** ~**-up to an event** im Vorfeld *(fig.)* eines Ereignisses; **b)** *(Sport)* Anlauf, *der*; **take a** ~**-up** Anlauf nehmen

runway ['rʌnweɪ] *n.* *(for take-off)* Startbahn, *die;* *(for landing)* Landebahn, *die*

rupture ['rʌptʃə(r)] **1.** *n.* **a)** *(lit. or fig.)* Bruch, *der;* **b)** *(Med.)* Ruptur, *die.* **2.** *v. t.* **a)** *(burst)* aufreißen; **a ~d appendix/spleen** ein geplatzter Blinddarm/eine gerissene Milz; **b) ~ oneself** sich *(Dat.)* einen Bruch zuziehen *od.* heben

rural ['rʊərl] *adj.* ländlich; **~ life** Landleben, *das*

ruse [ru:z] *n.* List, *die*

[1]**rush** [rʌʃ] *n.* *(Bot.)* Binse, *die*

[2]**rush 1.** *n.* **a)** *(rapid moving forward)* **make a ~ for sth.** sich auf etw. *(Akk.)* stürzen; **the holiday ~:** der [hektische] Urlaubsverkehr; **b)** *(hurry)* Eile, *die;* **what's all the ~?** wozu diese Hast?; **be in a [great] ~:** in [großer] Eile sein; es [sehr] eilig haben; **c)** *(surging)* Anwandlung, *die* **(of** von); **a ~ of blood [to the head]** *(fig. coll.)* eine [plötzliche] Anwandlung; **d)** *(period of great activity)* Hochbetrieb, *der;* **there is a ~ on** es herrscht Hochbetrieb *(ugs.);* **a ~ of new orders** eine Flut von neuen Aufträgen; **e)** *(heavy demand)* Ansturm, *der* **(for, on** auf + *Akk.*). **2.** *v. t.* **a)** *(convey rapidly)* **~ sb./sth. somewhere** jmdn./etw. auf schnellstem Wege irgendwohin bringen; **~ through Parliament** im Parlament durchpeitschen *(ugs. abwertend)* ‹*Gesetz*›; **be ~ed** *(have to hurry)* in Eile sein; **b)** *(cause to act hastily)* **~ sb. into doing sth.** jmdn. dazu drängen, etw. zu tun; **she hates to be ~ed** sie kann es nicht ausstehen, wenn sie sich [ab]hetzen muß; **c)** *(perform quickly)* auf die Schnelle erledigen; *(perform too quickly)* **~ it** zu schnell machen; **d)** *(Mil. or fig.: charge)* stürmen; überrumpeln ‹*feindliche Gruppe*›. **3.** *v. i.* **a)** *(move quickly)* eilen; ‹*Hund, Pferd:*› laufen; **she ~ed into the room** sie stürzte ins Zimmer; **~ through Customs/the exit** durch den Zoll/Ausgang stürmen; **b)** *(hurry unduly)* sich zu sehr beeilen; **don't ~!** nur keine Eile!; **c)** *(flow rapidly)* stürzen; **~ past** vorbeistürzen; **d) the blood ~ed to his face** das Blut schoß ihm ins Gesicht

~ a'bout, ~ a'round *v. i.* herumhetzen

~ into *v. t.* **~ into sth.** in etw. *(Akk.)* hin-/hereinstürzen; *(fig.)* sich in etw. *(Akk.)* stürzen/etw. überstürzt tun

~ 'up *v. i.* angestürzt kommen

rush: ~-hour *n.* Stoßzeit, *die; attrib.* **~-hour traffic** Berufsverkehr, *der;* **~ job** *n.* eilige Arbeit; **~ mat** *n.* Binsenmatte, *die;* **~ order** *n.* Eilauftrag, *der;* dringende Bestellung

rusk [rʌsk] *n.* Zwieback, *der*

russet ['rʌsɪt] **1.** *n.* *(reddish-brown)* Rotbraun, *das.* **2.** *adj.* rotbraun

Russia ['rʌʃə] *pr. n.* Rußland *(das)*

Russian ['rʌʃn] **1.** *adj.* russisch; **sb. is ~:** jmd. ist Russe/Russin. **2.** *n.* **a)** *(person)* Russe, *der*/Russin, *die;* **b)** *(language)* Russisch, *das; see also* **English 2 a**

rust [rʌst] **1.** *n., no pl., no indef. art.* Rost, *der.* **2.** *v. i.* rosten. **3.** *v. t.* [ver]rosten lassen

~ 'through *v. i.* durchrosten

rustic ['rʌstɪk] *adj.* **a)** *(of the country)* ländlich; **b)** *(unrefined)* bäurisch *(abwertend);* **c)** *(roughly built)* rustikal ‹*Mobiliar*›

rustle ['rʌsl] **1.** *n.* Rascheln, *das.* **2.** *v. i.* rascheln. **3.** *v. t.* **a)** rascheln lassen; rascheln mit ‹*Papieren*›; **b)** *(Amer.: steal)* stehlen

~ 'up *v. t.* *(coll.: produce)* auftreiben *(ugs.);* zusammenzaubern *(fig.)* ‹*Mahlzeit*›

rustler ['rʌslə(r)] *n.* *(Amer.)* Viehdieb, *der*

'rust-proof 1. *adj.* rostfrei. **2.** *v. t.* rostfrei *od.* rostbeständig machen

rusty ['rʌstɪ] *adj.* **a)** *(rusted)* rostig; **b)** *(fig.: impaired by neglect)* eingerostet; **I am a bit ~:** ich bin ein bißchen aus der Übung

[1]**rut** [rʌt] *n.* **a)** *(track)* Spurrille, *die;* **b)** *(fig.: established procedure)* **get into a ~:** in einen gewissen Trott verfallen; **be in a ~:** aus dem [Alltags]trott nicht mehr herauskommen

[2]**rut** *n.* *(sexual excitement)* Brunst, *die;* *(of roe-deer, stag, etc.)* Brunft, *die (Jägersprache)*

ruthless ['ru:θlɪs] *adj.,* **ruthlessly** ['ru:θlɪslɪ] *adv.* rücksichtslos

rutted ['rʌtɪd] *adj.* zerfurcht

rye [raɪ] *n.* **a)** *(cereal)* Roggen, *der;* **b) ~ [whisky]** Roggenwhisky, *der;* Rye, *der*

'rye bread *n.* Roggenbrot, *das*

S

S, s [es] *n., pl.* Ss *or* S's ['esɪs] S, s, *das*

S. *abbr.* **a) south** S; **b) southern** s.; **c) Saint** St.

s. *abbr.* **second[s]** Sek.

sabbath ['sæbəθ] *n.* **a)** *(Jewish)* Sabbat, *der;* **b)** *(Christian)* Sonntag, *der*

sabbatical [sə'bætɪkl] **1.** *adj.* **~ term/year** Forschungssemester/-jahr, *das.* **2.** *n.* Forschungsurlaub, *der*

saber *(Amer.) see* **sabre**

sable ['seɪbl] *n.* *(Zool., also fur)* Zobel, *der*

sabotage ['sæbətɑ:ʒ] **1.** *n.* *(lit. or fig.)* Sabotage, *die;* **act of ~:** Sabotageakt, *der.* **2.** *v. t.* einen Sabotageakt verüben auf (+ *Akk.*); *(fig.)* sabotieren ‹*Pläne usw.*›

saboteur [sæbə'tɜ:(r)] *n.* Saboteur, *der*

sabre ['seɪbə(r)] *n.* *(Brit.)* Säbel, *der*

saccharin ['sækərɪn] *n.* Saccharin, *das*

sachet ['sæʃeɪ] *n.* **a)** *(small packet)* *(for shampoo etc.)* Beutel, *der;* *(cushion-shaped)* Kissen, *das;* **b)** *(bag for scenting clothes)* Duftkissen, *das*

[1]**sack** [sæk] **1.** *n.* **a)** Sack, *der;* **a ~ of potatoes** ein Sack Kartoffeln; **b)** *(coll.: dismissal)* Rausschmiß, *der (ugs.);* **get the ~:** rausgeschmissen werden *(ugs.);* **give sb. the ~:** jmdn. rausschmeißen *(ugs.);* **c) hit the ~** *(sl.)* sich in die Falle hauen *(salopp).* **2.** *v. t.* *(coll.: dismiss)* rausschmeißen *(ugs.)* **(for** wegen)

[2]**sack 1.** *v. t.* *(loot)* plündern. **2.** *n.* Plünderung, *die*

sacking ['sækɪŋ] *n.* *(coll.: dismissal)* Rausschmiß, *der (ugs.)*

'sack race *n.* Sackhüpfen, *das*

sacrament ['sækrəmənt] *n.* Sakrament, *das;* **the Holy S~** *(in the Eucharist)* das Allerheiligste

sacred ['seɪkrɪd] *adj.* heilig; geheiligt ‹*Tradition*›; geistlich

⟨*Musik, Dichtung*⟩; **is nothing ~?** *(iron.)* scheut man denn vor nichts mehr zurück?
sacred 'cow *n. (lit. or fig.)* heilige Kuh
sacrifice ['sækrɪfaɪs] **1.** *n.* **a)** *(giving up valued thing)* Opferung, *die; (of principles)* Preisgabe, *die; (of pride, possessions)* Aufgabe, *die;* **make ~s** Opfer bringen; **b)** *(offering to deity)* Opfer, *das;* **c)** *(Games: deliberate incurring of loss)* Opfern, *das.* **2.** *v. t. (give up, offer as ~)* opfern; **~ oneself/sth. to sth.** sich/etw. einer Sache *(Dat.)* opfern
sacrificial [sækrɪ'fɪʃl] *adj.* Opfer-
sacrilege ['sækrɪlɪdʒ] *n., no pl.* **[act of] ~:** Sakrileg, *das*
sacrilegious [sækrɪ'lɪdʒəs] *adj.* sakrilegisch; *(fig.)* frevelhaft
sacrosanct ['sækrəsæŋkt] *adj. (lit. or fig.)* sakrosankt
sad [sæd] *adj.* **a)** *(sorrowful)* traurig **(at, about** über + *Akk.*); **feel ~, be in a ~ mood** traurig sein; **b)** *(causing grief)* traurig; schmerzlich ⟨*Tod, Verlust*⟩; **~ to say, ...:** bedauerlicherweise ...; leider ...; **c)** *(derog./joc.: deplorably bad)* traurig
sadden ['sædn] *v. t.* traurig stimmen; **be deeply ~ed** tieftraurig sein; **I was ~ed to see that ...:** es betrübte mich, zu sehen, daß ...
saddle ['sædl] **1.** *n.* **a)** *(seat for rider)* Sattel, *der;* **be in the ~** *(fig.)* das Heft in der Hand haben *(geh.);* **b)** *(ridge between summits)* [Berg]sattel, *der.* **2.** *v. t.* **a)** satteln ⟨*Pferd usw.*⟩; **b)** *(fig.)* **~ sb. with sth.** jmdm. etw. aufbürden *(geh.)*
saddle: ~-bag *n.* Satteltasche, *die;* **~ sore** *n.* Sattelwunde, *die;* **~-sore** *adj.* **be ~-sore** wund vom Reiten/Radfahren sein
sadism ['seɪdɪzm] *n.* Sadismus, *der*
sadist ['seɪdɪst] *n.* Sadist, *der*/Sadistin, *die*
sadistic [sə'dɪstɪk] *adj.*, **sadistically** [sə'dɪstɪkəlɪ] *adv.* sadistisch
sadly ['sædlɪ] *adv.* **a)** *(with sorrow)* traurig; **b)** *(unfortunately)* leider; **c)** *(deplorably)* erbärmlich *(abwertend)*
sadness ['sædnɪs] *n., no pl.* Traurigkeit, *die* **(at, about** über + *Akk.*)
s.a.e. [eseɪ'i:] *abbr.* **stamped addressed envelope** adressierter Freiumschlag
safari [sə'fɑ:rɪ] *n.* Safari, *die;* **be/go on ~:** auf Safari sein/gehen
sa'fari park *n.* Safaripark, *der*
safe [seɪf] **1.** *n.* Safe, *der;* Geldschrank, *der.* **2.** *adj.* **a)** *(out of danger)* sicher **(from** vor + *Dat.*); **he's ~:** er ist in Sicherheit; **make sth. ~ from sth.** etw. gegen etw. sichern; **~ and sound** sicher und wohlbehalten; **b)** *(free from danger)* ungefährlich; sicher ⟨*Ort, Hafen*⟩; **better ~ than sorry** Vorsicht ist besser als Nachsicht *(ugs.);* **wish sb. a ~ journey** jmdm. eine gute Reise wünschen; **is the car ~ to drive?** ist der Wagen verkehrssicher?; **to be on the ~ side** zur Sicherheit; **c)** *(unlikely to produce controversy)* sicher; bewährt *(iron.)* ⟨*Klischee*⟩; **it is ~ to say [that ...]** man kann mit einiger Sicherheit sagen[, daß ...]; **d)** *(reliable)* sicher ⟨*Methode, Investition, Stelle*⟩; naheliegend ⟨*Vermutung*⟩; **e)** *(secure)* **your secrets will be ~ with me** deine Geheimnisse sind bei mir gut aufgehoben. *See also* **play 2a, 3e**
safe: ~ 'bet *n.* **it is a ~ bet he will be there** man kann darauf wetten, daß er dort ist; **~-breaker** *n.* Geldschrankknacker, *der (ugs.);* **~ 'conduct** *n.* freies *od.* sicheres Geleit; **~ de'posit** *n.* Tresor, *der; attrib.* **~-deposit box** *(at the bank)* Banksafe, *der;* **~guard 1.** *n.* Schutz, *der;* **2.** *v. t.* schützen; **~guard sb.'s future/interests** jmds. Zukunft sichern/Interessen wahren; **~ 'keeping** *n.* sichere Obhut *(geh.); (of thing)* [sichere] Aufbewahrung
safely ['seɪflɪ] *adv.* **a)** *(without harm)* sicher; **did the parcel arrive ~?** ist das Paket heil angekommen?; **b)** *(securely)* sicher; **be ~ behind bars** [in sicherem Gewahrsam] hinter Schloß und Riegel sein; **c)** *(with certainty)* **one can ~ say [that] she will come** man kann mit ziemlicher Sicherheit sagen, daß sie kommt
safety ['seɪftɪ] *n.* **a)** *(being out of danger)* Sicherheit, *die;* **b)** *(lack of danger)* Ungefährlichkeit, *die; (of a machine)* Betriebssicherheit, *die;* **there is ~ in numbers** zu mehreren ist man sicherer; **a ~ first policy** eine Politik der Vorsicht; **c)** *attrib.* Sicherheits⟨*netz, -kette, -faktor, -maßnahmen, -vorrichtungen, -lampe*⟩
safety: ~-belt *n.* Sicherheitsgurt, *der;* **~-catch** *n. (of door)* Sicherheitsverriegelung, *die; (of gun)* Sicherungshebel, *der;* **~ helmet** *n.* Schutzhelm, *der;* **~ margin** *n.* Spielraum, *der;* **~ match** *n.* Sicherheitszündholz, *das;* **~-pin** *n.* Sicherheitsnadel, *die;* **~ razor** *n.* Rasierapparat, *der;* **~-valve** *n.* Sicherheitsventil, *das (Technik); (fig.)* Ventil, *das (fig.)*
saffron ['sæfrən] **1.** *n.* Safran, *der.* **2.** *adj.* safrangelb
sag [sæg] **1.** *v. i.*, **-gg-: a)** *(have downward bulge)* durchhängen; **b)** *(sink)* sich senken; absacken *(ugs.);* ⟨*Gebäude:*⟩ [in sich *(Akk.)*] zusammensacken *(ugs.);* ⟨*Schultern:*⟩ herabhängen; ⟨*Brüste:*⟩ hängen; *(fig.: decline)* ⟨*Mut, Stimmung:*⟩ sinken. **2.** *n.* **a)** *(amount that rope etc. ~s)* Durchhang, *der;* **b)** *(sinking)* **there was a ~ in the seat** der Sitz war durchgesessen
saga ['sɑ:gə] *n.* **a)** *(story of adventure)* Heldenepos, *das (fig.); (medieval narrative)* Saga, *die (Literaturw.);* **b)** *(coll.: long involved story)* [ganzer] Roman *(fig.)*
sagacious [sə'geɪʃəs] *adj.* klug
¹sage [seɪdʒ] *n. (Bot.)* Salbei, *der od. die*
²sage 1. *n.* Weise, *der.* **2.** *adj.* weise
Sagittarius [sædʒɪ'teərɪəs] *n. (Astrol., Astron.)* der Schütze
sago ['seɪgəʊ] *n., pl.* **~s** Sago, *der*
Sahara [sə'hɑ:rə] *pr. n.* **the ~ [Desert]** die [Wüste] Sahara
said *see* **say 1**
sail [seɪl] **1.** *n.* **a)** *(voyage in ~ing vessel)* Segelfahrt, *die;* **go for a ~:** eine Segelfahrt machen; **set ~** *(begin voyage)* losfahren **(for** nach); **b)** *(piece of canvas)* Segel, *das.* **2.** *v. i.* **a)** *(travel on water)* fahren; *(in ~ing boat)* segeln; **b)** *(start voyage)* auslaufen **(for** nach); in See stechen; **c)** *(glide in air)* segeln; **d)** *(fig.: be thrown)* segeln *(ugs.);* **e)** *(move smoothly)* gleiten; **f)** *(fig. coll.: pass easily)* **~ through an examination** eine Prüfung spielend schaffen. **3.** *v. t.* **a)** steuern ⟨*Boot, Schiff*⟩; segeln mit ⟨*Segeljacht, -schiff*⟩; **b)** *(travel across)* durchfahren, befahren ⟨*Meer*⟩
sail: ~board *n.* Surfbrett, *das (zum Windsurfen);* **~-boarding** *see* **windsurfing; ~boat** *n. (Amer.)* Segelboot, *das*
sailing ['seɪlɪŋ] *n.* **a)** *(handling a boat)* Segeln, *das;* **b)** *(departure from a port)* Abfahrt, *die;* **there are regular ~s from here across to the island** von hier fahren regelmäßig Schiffe hinüber zur Insel
sailing: ~ boat *n.* Segelboot, *das;* **~ ship, ~ vessel** *ns.* Segelschiff, *das*
sailor ['seɪlə(r)] *n.* Seemann, *der; (in navy)* Matrose, *der;* **be a good/**

bad ~ *(not get seasick/get seasick)* seefest/nicht seefest sein
saint 1. [sənt] *adj.* **S~ Michael/ Helena** der heilige Michael/die heilige Helena; Sankt Michael/ Helena; ~ **Michael's [Church]** die Michaelskirche. **2.** [seɪnt] *n.* Heilige, *der/die;* **make** *or* **declare sb. a** ~ *(RC Ch.)* jmdn. heiligsprechen; **be as patient as a** ~: eine Engelsgeduld haben
saintly ['seɪntlɪ] *adj.* heilig
¹sake [seɪk] *n.* **for the** ~ **of** um ... *(Gen.)* willen; **for my** *etc.* ~: um meinetwillen *usw.;* mir *usw.* zuliebe; **for your/its own** ~: um deiner/seiner selbst willen; **for the** ~ **of a few pounds** wegen ein paar Pfund; **for Christ's** *or* **God's** *or* **goodness'** *or* **Heaven's** *or (coll.)* **Pete's** ~: um Gottes *od.* Himmels willen; **for old times'** ~: um der schönen Erinnerung willen
²sake ['sɑːkɪ] *n. (drink)* Sake, *der*
salacious [sə'leɪʃəs] *adj.* **a)** *(lustful)* lüstern; **b)** *(inciting sexual desire)* pornographisch
salad ['sæləd] *n.* Salat, *der*
salad: ~ **cream** *n.* ≈ Mayonnaise, *die;* ~**-dressing** *n.* Dressing, *das;* Salatsoße, *die;* ~**-servers** ['sælədsɜːvəz] *n. pl.* Salatbesteck, *das*
salami [sə'lɑːmɪ] *n.* Salami, *die*
salaried ['sælərɪd] *adj.* **a)** *(receiving salary)* Gehalt beziehend; ~ **employee** Angestellte, *der/die;* **b)** ~ **post** Stelle mit festem Gehalt
salary ['sælərɪ] *n.* Gehalt, *das; attrib.* ~ **increase** Gehaltserhöhung, *die*
sale [seɪl] *n.* **a)** *(selling)* Verkauf, *der;* **[up] for** ~: zu verkaufen; **put up** *or* **offer for** ~: zum Verkauf anbieten; **on** ~ **at your chemist's** in Ihrer Apotheke erhältlich; **offer** *etc.* **sth. on a** ~ **or return basis** etw. auf Kommissionsbasis anbieten *usw.;* **b)** *(instance of selling)* Verkauf, *der;* **c)** *in pl., no art. (amount sold)* Verkaufszahlen *Pl.* **(of** für); Absatz, *der;* **d)** *(disposal at reduced prices)* Ausverkauf, *der;* **clearance/end-of-season** ~: Räumungs-/Schlußverkauf, *der*
saleable ['seɪləbl] *adj.* verkäuflich; **be [highly]** ~: sich [gut] verkaufen lassen
'sale-room *n. (Brit.)* Auktionsraum, *der*
sales: ~ **assistant** *(Brit.),* ~ **clerk** *(Amer.) ns.* Verkäufer, *der*/Verkäuferin, *die;* ~ **department** *n.* Verkaufsabteilung, *die;* ~**girl,** ~**lady** *ns.* Verkäuferin, *die;* ~**man** ['seɪlzmən] *n., pl.* ~**men** ['seɪlzmən] Verkäufer, *der;* ~ **manager** *n.* Verkaufsleiter, *der/*-leiterin, *die;* Sales-manager, *der;* ~ **patter,** ~ **pitch** *ns.* Verkaufsargumentation, *die;* ~ **rep** *(coll.),* ~ **representative** *ns.* [Handels]vertreter, *der/*-vertreterin, *die;* ~ **talk** *see* ~ **patter;** ~**woman** *n.* Verkäuferin, *die*
salient ['seɪlɪənt] *adj.* *(striking)* auffallend; ins Auge springend; hervorstechend ⟨*Charakterzug*⟩; **the** ~ **points of a speech** die herausragenden Punkte einer Rede
saline ['seɪlaɪn] *adj.* salzig
saliva [sə'laɪvə] *n.* Speichel, *der*
salivate ['sælɪveɪt] *v. i.* speicheln
sallow ['sæləʊ] *adj.* blaßgelb
sally *n.* **a)** *(Mil.: sortie)* Ausfall, *der;* **b)** *(excursion)* Ausflug, *der*
salmon ['sæmən] **1.** *n., pl. same* Lachs, *der.* **2.** *adj. (colour)* lachsfarben; lachsrosa ⟨*Farbton*⟩
'salmon-pink 1. *n.* lachsrosa Farbton. **2.** *adj.* lachsfarben
salon ['sælɔ̃] *n.* Salon, *der*
saloon [sə'luːn] *n.* **a)** *(public room in ship, hotel, etc.)* Salon, *der;* **dining** ~: Speisesaal, *der;* **b)** *(Brit.: motor car)* Limousine, *die;* **c)** *(Amer.: bar)* Saloon, *der*
saloon: ~ **bar** *n. (Brit.) separater Teil eines Pubs mit mehr Komfort;* ~ **'car** *see* **saloon b**
SALT [sɔːlt, sɒlt] *abbr.* **Strategic Arms Limitation Talks/Treaty** SALT
salt 1. *n.* **a)** *(for food etc.; also Chem.)* **[common]** ~: [Koch]salz, *das;* **rub** ~ **in[to] the wound** *(fig.)* Salz in die Wunde streuen; **take sth. with a grain** *or* **pinch of** ~ *(fig.)* etw. cum grano salis *(geh.) od.* nicht ganz wörtlich nehmen; **be the** ~ **of the earth** *(fig.)* anständig und rechtschaffen sein; **b)** *in pl. (medicine)* Salz, *das;* **like a dose of** ~**s** *(sl.)* in Null Komma nichts *(ugs.).* **2.** *adj.* **a)** *(containing or tasting of* ~*)* salzig; *(preserved with* ~*)* gepökelt ⟨*Fleisch*⟩; gesalzen ⟨*Butter*⟩; **b)** *(bitter)* salzig ⟨*Tränen*⟩. **3.** *v. t. (add* ~ *to)* salzen; *(fig.)* würzen
~ **a'way** *v. t. (coll.)* auf die hohe Kante legen *(ugs.)*
salt: ~**-cellar** *n. (open)* Salzfaß, *das; (sprinkler)* Salzstreuer, *der;* ~**-spoon** *n.* Salzlöffelchen, *das;* ~ **'water** *n.* Salzwasser, *das;* ~**-water** *adj.* Salzwasser-
salty ['sɔːltɪ, 'sɒltɪ] *adj.* salzig
salubrious [sə'luːbrɪəs] *adj.* gesund; **not a very** ~ **area** *(fig.)* ein etwas zweifelhaftes Viertel
salutary ['sæljʊtərɪ] *adj.* heilsam ⟨*Wirkung, Einfluß, Schock*⟩
salute [sə'luːt] **1.** *v. t.* **a)** *(Mil., Navy)* ~ **sb.** jmdn. [militärisch] grüßen; *(fig.: pay tribute to)* sich vor jmdm. verneigen; **b)** *(greet)* grüßen. **2.** *v. i. (Mil., Navy)* [militärisch] grüßen. **3.** *n. (Mil., Navy)* Salut, *der;* militärischer Gruß; **fire a seven-gun** ~: sieben Schuß Salut abfeuern
salvage ['sælvɪdʒ] **1.** *n.* **a)** *(rescue of property)* Bergung, *die; attrib.* Bergungs⟨*arbeiten, -aktion*⟩; **b)** *(rescued property)* Bergegut, *das; (for recycling)* Sammelgut, *das.* **2.** *v. t.* **a)** *(rescue)* bergen; retten *(auch fig.)* **(from** von); **b)** *(save for recycling)* für die Wiederverwendung sammeln
salvation [sæl'veɪʃn] *n.* **a)** *no art. (Relig.)* Erlösung, *die;* **b)** *(means of preservation)* Rettung, *die*
Salvation 'Army *n.* Heilsarmee, *die*
salvo ['sælvəʊ] *n., pl.* ~**es** *or* ~**s** *(of guns)* Salve, *die*
Samaritan [sə'mærɪtən] *n.* **good** ~: [barmherziger] Samariter; **the** ~**s** *(organization)* ≈ die Telefonseelsorge
same [seɪm] **1.** *adj.* **the** ~: der/die/das gleiche; **the** ~ **[thing]** *(identical)* der-/die-/dasselbe; **the** ~ **afternoon/evening** *(of* ~ *day)* schon am Nachmittag/Abend; **she seemed just the** ~ **[as ever] to me** sie schien mir unverändert *od.* immer noch die alte; **one and the** ~ **person/man** ein und dieselbe Person/ein und derselbe Mann; **the very** ~: genau der/die/das; ebenderselbe/-dieselbe/-dasselbe; **much the** ~ **as** fast genauso wie. **2.** *pron.* **the** ~, *(coll.)* ~ *(the* ~ *thing)* der-/die-/dasselbe; **they look [exactly] the** ~: sie sehen gleich aus; **more of the** ~: noch mehr davon; **and the** ~ **to you!** *(also iron.)* danke gleichfalls; **[the]** ~ **again** das gleiche noch einmal; **I feel bored – S~ here** *(coll.)* Ich langweile mich – Dito. **3.** *adv.* **[the]** ~ **as you do** genau wie du; **the** ~ **as before** genau wie vorher; **all** *or* **just the** ~: trotzdem; nichtsdestotrotz *(ugs., oft scherzh.);* **think the** ~ **of/feel the** ~ **towards** dasselbe halten von/empfinden für
sameness ['seɪmnɪs] *n., no pl.* Gleichheit, *die*
Samoa [sə'məʊə] *pr. n.* Samoa *(das)*
sample ['sɑːmpl] **1.** *n.* **a)** *(representative portion)* Auswahl, *die; (in opinion research, statistics)* Querschnitt, *der;* Sample, *das;* **b)** *(example)* [Muster]beispiel, *das; (specimen)* Probe, *die;* **[commer-**

cial] ~: Muster, *das; attrib.* Probe⟨*exemplar, -seite*⟩. **2.** *v.t.* probieren; ~ **the pleasures of country life** die Freuden des Landlebens kosten *(geh.)*

sanatorium [sænəˈtɔːrɪəm] *n., pl.* ~**s** *or* **sanatoria** [sænəˈtɔːrɪə] *(clinic)* Sanatorium, *das*

sanctify [ˈsæŋktɪfaɪ] *v.t.* **a)** heiligen; **b)** *(consecrate)* weihen; heiligen *(bes. bibl.)*

sanctimonious [sæŋktɪˈməʊnɪəs] *adj.*, **sanctimoniously** [sæŋktɪˈməʊnɪəslɪ] *adv.* scheinheilig

sanction [ˈsæŋkʃn] **1.** *n.* **a)** *(official approval)* Sanktion, *die;* **give one's ~ to sth.** seine Erlaubnis für etw. geben; **b)** *(Polit.: penalty; Law: punishment)* Sanktion, *die.* **2.** *v.t.* sanktionieren

sanctity [ˈsæŋktɪtɪ] *n., no pl.* Heiligkeit, *die*

sanctuary [ˈsæŋktʃʊərɪ] *n.* **a)** *(holy place)* Heiligtum, *das;* **b)** *(part of church)* Altarraum, *der;* **c)** *(place of refuge)* Zufluchtsort, *der;* **d)** *(for animals or plants)* Naturschutzgebiet, *das;* **e) take ~:** Zuflucht suchen

sand [sænd] **1.** *n.* **a)** Sand, *der;* **have** *or* **keep** *or* **bury one's head in the ~** *(fig.)* den Kopf in den Sand stecken; **b)** *in pl. (expanse)* Sandbank, *die; (beach)* Sandstrand, *der.* **2.** *v.t.* **a)** *(sprinkle)* ~ **the road** die Straße mit Sand streuen; **b)** *(polish)* ~ **sth. down** etw. [ab]schmirgeln

sandal [ˈsændl] *n.* Sandale, *die*

sand: ~bag *n.* Sandsack, *der;* **~bank** *n.* Sandbank, *die;* **~-blast** *v.t.* sandstrahlen *(Technik);* **~-box** *n. (Amer.)* Sandkasten, *der;* **~boy** *n.* **be happy as a ~boy** glücklich und zufrieden sein; **~-castle** *n.* Sandburg, *die;* **~-dune** *n.* Düne, *die*

sander [ˈsændə(r)] *n.* Sandpapierschleifmaschine, *die*

sand: ~paper 1. *n.* Sandpapier, *das;* **2.** *v.t.* [mit Sandpapier] [ab]schmirgeln; **~-pit** *n.* Sandkasten, *der;* **~stone** *n.* Sandstein, *der;* **~storm** *n.* Sandsturm, *der;* ~ **trap** *n. (Amer. Golf)* Bunker, *der*

sandwich [ˈsændwɪdʒ, ˈsændwɪtʃ] **1.** *n.* Sandwich, *der od. das;* ≈ [zusammengeklapptes] belegtes Brot; **cheese ~:** Käsebrot, *das.* **2.** *v.t.* einschieben (**between** zwischen + *Akk.;* **into** in + *Akk.*); **be ~ed between other people/cars** zwischen andere Personen gequetscht werden/Autos eingeklemmt sein

ˈsandwich course *n. Ausbildung mit abwechselnd theoretischem und praktischem Unterricht*

sandy [ˈsændɪ] *adj.* **a)** sandig; Sand⟨*boden, -strand*⟩; **b)** *(yellowish-red)* rotblond ⟨*Haar*⟩

sane [seɪn] *adj.* **a)** geistig gesund; **b)** *(sensible)* vernünftig

sang *see* **sing**

sanguine [ˈsæŋgwɪn] *adj. (confident)* zuversichtlich (**about** was ... betrifft); heiter ⟨*Temperament*⟩

sanitary [ˈsænɪtərɪ] *adj.* sanitär ⟨*Verhältnisse, Anlagen*⟩; gesundheitlich ⟨*Gesichtspunkt, Problem*⟩; Gesundheits⟨*behörde*⟩; hygienisch ⟨*Küche, Krankenhaus, Gewohnheit*⟩

sanitary: ~ **napkin** *(Amer.),* ~ **towel** *(Brit.) ns.* Damenbinde, *die*

sanitation [sænɪˈteɪʃn] *n., no pl.* **a)** *(drainage, refuse disposal)* Kanalisation und Abfallbeseitigung; **b)** *(hygiene)* Hygiene, *die*

sanity [ˈsænɪtɪ] *n.* **a)** *(mental health)* geistige Gesundheit; **lose one's ~:** den Verstand verlieren; **fear for/doubt sb.'s ~:** um jmds. Zurechnungsfähigkeit fürchten/an jmds. Verstand *(Dat.)* zweifeln; **b)** *(good sense)* Vernünftigkeit, *die;* **restore ~ to the proceedings** die Veranstaltung wieder in vernünftige Bahnen lenken

sank *see* **sink 2, 3**

Santa [ˈsæntə] *(coll.),* **Santa Claus** [ˈsæntə klɔːz] *n.* Weihnachtsmann, *der*

sap [sæp] **1.** *n.* Saft, *der; (fig.: vital spirit)* belebende Kraft. **2.** *v.t.,* **-pp-** *(fig.: exhaust vigour of)* zehren an (+ *Dat.*)

sapling [ˈsæplɪŋ] *n.* junger Baum

sapper [ˈsæpə(r)] *n. (Brit. Mil.)* Pionier, *der*

sapphire [ˈsæfaɪə(r)] *n.* Saphir, *der; attrib.* ~ **blue** saphirblau; ~ **ring** Saphirring, *der*

sarcasm [ˈsɑːkæzm] *n.* Sarkasmus, *der*

sarcastic [sɑːˈkæstɪk] *adj.*, **sarcastically** [sɑːˈkæstɪkəlɪ] *adv.* sarkastisch

sardine [sɑːˈdiːn] *n. (Zool.)* Sardine, *die; (Gastr.)* [Öl]sardine, *die;* **like ~s** *(fig.)* wie die Ölsardinen

Sardinia [sɑːˈdɪnɪə] *pr. n.* Sardinien *(das)*

Sardinian [sɑːˈdɪnɪən] **1.** *n.* **a)** *(person)* Sarde, *der*/Sardin, *die;* Sardinier, *der*/Sardinierin, *die;* **b)** *(language)* Sardisch, *das.* **2.** *adj.* sardisch

sardonic [sɑːˈdɒnɪk] *adj.* höhnisch ⟨*Bemerkung*⟩; sardonisch ⟨*Lachen, Lächeln*⟩

sari [ˈsɑːrɪ] *n.* Sari, *der*

sarong [səˈrɒŋ] *n.* Sarong, *der*

¹sash [sæʃ] *n.* Schärpe, *die*

²sash *n.* **a)** *(of window)* Fensterrahmen, *der;* **b)** *(window)* Schiebefenster, *das*

sash-ˈwindow *n.* Schiebefenster, *das*

sat *see* **sit**

Sat. *abbr.* **Saturday** Sa.

Satan [ˈseɪtən] *pr. n.* Satan, *der*

satanic [səˈtænɪk] *adj.* satanisch; teuflisch

satchel [ˈsætʃl] *n.* [Schul]ranzen, *der*

sate [seɪt] *v.t. (literary)* **a)** *(gratify)* stillen ⟨*Hunger, Verlangen*⟩; zufriedenstellen ⟨*Person*⟩; **b)** *(cloy)* übersättigen ⟨*Lust, Verlangen*⟩; **become ~d with/be ~d by sth.** einer Sache *(Gen.)* überdrüssig werden/sein

satellite [ˈsætəlaɪt] *n. (Astronaut., Astron.; also country)* Satellit, *der;* **by ~:** über Satellit

satellite: ~ **ˈbroadcasting** *n., no pl., no art.* Satellitenfunk, *der;* ~ **dish** *n.* Parabolantenne, *die;* ~ **town** *n.* Satelliten- *od.* Trabantenstadt, *die*

satiate [ˈseɪʃɪeɪt] *see* **sate**

satin [ˈsætɪn] **1.** *n.* Satin, *der.* **2.** *attrib. adj.* **a)** *(made of ~)* Satin-; **b)** *(like ~)* seidig

satire [ˈsætaɪə(r)] *n.* Satire, *die* (**on** auf + *Akk.*)

satirical [səˈtɪrɪkl] *adj.*, **satirically** [səˈtɪrɪkəlɪ] *adv.* satirisch

satirise *see* **satirize**

satirist [ˈsætɪrɪst] *n.* Satiriker, *der*/Satirikerin, *die*

satirize [ˈsætɪraɪz] *v.t.* **a)** *(write satire on)* satirisch darstellen; **b)** *(describe satirically)* ⟨*Buch, Film usw.:*⟩ eine Satire sein auf (+ *Akk.*)

satisfaction [sætɪsˈfækʃn] *n.* **a)** *no pl. (act)* Befriedigung, *die;* **b)** *no pl. (feeling of gratification)* Befriedigung, *die* (**at, with** über + *Akk.*); Genugtuung, *die* (**at, with** über + *Akk.*); **job ~:** Befriedigung in der Arbeit; **what ~ can it give you?** was befriedigt dich daran?; **c)** *no pl. (gratified state)* **meet with sb.'s** *or* **give sb. [complete] ~:** jmdn. [in jeder Weise] zufriedenstellen; **to sb.'s ~, to the ~ of sb.** zu jmds. Zufriedenheit; **d)** *(instance of gratification)* Befriedigung, *die;* **it is a great ~ to me that ...:** es erfüllt mich mit großer Befriedigung, daß ...; **have the ~ of doing sth.** das Vergnügen haben, etw. zu tun

satisfactory [sætɪsˈfæktərɪ] *adj.* zufriedenstellend; angemessen

⟨*Bezahlung*⟩; '~' *(as school mark)* „ausreichend"
satisfied ['sætɪsfaɪd] *adj.* **a)** *(contented)* zufrieden; **be ~ with doing sth.** sich damit begnügen, etw. zu tun; **b)** *(convinced)* überzeugt (**of** von); **be ~ that ...:** [davon] überzeugt sein, daß ...
satisfy ['sætɪsfaɪ] *v. t.* **a)** *(content)* befriedigen; zufriedenstellen ⟨*Kunden, Publikum*⟩; entsprechen (+ *Dat.*) ⟨*Vorliebe, Empfinden, Meinung, Zeitgeist*⟩; erfüllen ⟨*Hoffnung, Erwartung*⟩; **b)** *(rid of want)* befriedigen; *(put an end to)* stillen ⟨*Hunger, Durst*⟩; *(make replete)* sättigen; **c)** *(convince)* ~ **sb.** [**of sth.**] jmdn. [von etw.] überzeugen; ~ **oneself of** *or* **as to** sich überzeugen von ⟨*Wahrheit, Ehrlichkeit*⟩; sich *(Dat.)* Gewißheit verschaffen über (+ *Akk.*) ⟨*Motiv*⟩; **d)** *(adequately deal with)* ausräumen ⟨*Einwand, Zweifel*⟩; erfüllen ⟨*Bitte, Forderung, Bedingung*⟩; **e)** *(fulfil)* erfüllen ⟨*Vertrag, Verpflichtung, Forderung*⟩
satisfying ['sætɪsfaɪɪŋ] *adj.* befriedigend; zufriedenstellend ⟨*Antwort, Lösung, Leistung*⟩
satsuma [sæt'su:mə] *n.* Satsuma, *die*
saturate ['sætʃəreɪt, 'sætjʊreɪt] *v. t.* **a)** *(soak)* durchnässen; [mit Feuchtigkeit durch]tränken ⟨*Boden, Erde*⟩; **b)** *(fill to capacity)* auslasten; sättigen ⟨*Markt*⟩; **c)** *(Phys., Chem.)* sättigen
saturated ['sætʃəreɪtɪd, 'sætjʊreɪtɪd] *adj.* **a)** *(soaked)* durchnäßt; völlig naß ⟨*Boden*⟩; **b)** *(imbued)* durchdrungen (**with, in** von); **c)** *(filled to capacity)* ausgelastet; gesättigt ⟨*Markt*⟩; **d)** *(Phys., Chem.)* gesättigt ⟨*Lösung, Verbindung, Fett*⟩
saturation point [sætʃə'reɪʃn pɔɪnt, sætjʊ'reɪʃn pɔɪnt] *n.* *(limit of capacity)* [Ober]grenze, *die;* *(of market; Phys.)* Sättigungspunkt, *der*
Saturday ['sætədeɪ, 'sætədɪ] **1.** *n.* Sonnabend, *der;* Samstag, *der.* **2.** *adv.* *(coll.)* **he comes ~s** er kommt sonnabends *od.* samstags. *See also* **Friday**
Saturn ['sætən] *pr. n.* **a)** *(Astron.)* Saturn, *der;* **b)** *(Roman Mythol.)* Saturn *(der)*
sauce [sɔ:s] **1.** *n.* **a)** Soße, *die;* **b)** *(impudence)* Frechheit, *die* **2.** *v. t.* *(coll.)* frech sein zu
sauce: ~-boat *n.* Sauciere, *die;* **~pan** ['sɔ:spən] *n.* Kochtopf, *der;* *(with straight handle)* [Stiel]kasserolle, *die*
saucer ['sɔ:sə(r)] *n.* Untertasse, *die;* **their eyes were like ~s** *(fig.)* sie machten große Augen *(ugs.)*
saucy ['sɔ:sɪ] *adj.* **a)** *(rude)* frech; **b)** *(pert, jaunty)* keck
Saudi Arabia [saʊdɪ ə'reɪbɪə] *pr. n.* Saudi-Arabien *(das)*
Saudi-Arabian [saʊdɪə'reɪbɪən] **1.** *adj.* saudiarabisch. **2.** *n.* Saudi[araber], *der*/-araberin, *die*
sauna ['sɔ:nə] *n.* Sauna, *die;* **have** *or* **take a ~:** saunieren; ein Saunabad nehmen
saunter ['sɔ:ntə(r)] **1.** *v. i.* schlendern. **2.** *n.* *(stroll)* Bummel, *der* *(ugs.);* *(leisurely pace)* Schlenderschritt, *der*
sausage ['sɒsɪdʒ] *n.* Wurst, *die;* *(smaller)* Würstchen, *das;* **not a ~** *(fig. sl.)* gar nix *(ugs.)*
sausage: ~-dog *n.* *(Brit. coll.)* Dackel, *der;* **~-meat** *n.* Wurstmasse, *die;* ~ **'roll** *n.* *Blätterteig mit Wurstfüllung*
sauté ['səʊteɪ] *(Cookery)* **1.** *adj.* sautiert *(fachspr.);* kurz [an]gebraten; **~ potatoes** ≈ Bratkartoffeln. **2.** *n.* Sauté, *das.* **3.** *v. t.*, **~d** *or* **~ed** ['səʊteɪd] sautieren *(fachspr.);* kurz [an]braten
savage ['sævɪdʒ] **1.** *adj.* **a)** *(uncivilized)* primitiv; wild ⟨*Volksstamm*⟩; unzivilisiert ⟨*Land*⟩; **b)** *(fierce)* brutal; wild ⟨*Tier*⟩; scharf ⟨*Hund*⟩; jähzornig ⟨*Temperament*⟩; **make a ~ attack on sb.** brutal über jmdn. herfallen; *(fig.)* jmdn. schonungslos angreifen. **2.** *n.* **a)** *(uncivilized person)* Wilde, *der/die (veralt.);* **b)** *(barbarous or uncultivated person)* Barbar, *der*/Barbarin, *die (abwertend).* **3.** *v. t.* ⟨*Hund:*⟩ anfallen ⟨*Kind usw.*⟩
savagery ['sævɪdʒrɪ] *n., no pl.* *(ferocity)* Brutalität, *die*
savannah (savanna) [sə'vænə] *n.* *(Geog.)* Savanne, *die*
save [seɪv] **1.** *v. t.* **a)** *(rescue)* retten (**from** vor + *Dat.*); **please, ~ me!** bitte helfen Sie mir!; **~ sb. from the clutches of the enemy/from making a mistake** jmdn. aus den Klauen des Feindes retten/davor bewahren, daß er einen Fehler macht; **~ oneself from falling** sich [beim Hinfallen] fangen; **~ the day** die Situation retten; **b)** *(keep undamaged)* schonen ⟨*Kleidung, Möbelstück*⟩; **c) God ~ the King/Queen** *etc.* Gott behüte *od.* beschütze den König/die Königin *usw.;* **d)** *(Theol.)* retten ⟨*Sünder, Seele, Menschen*⟩; **be past saving** nicht mehr zu retten sein; **e)** *(put aside)* aufheben; sparen ⟨*Geld*⟩; sammeln ⟨*Rabattmarken, Briefmarken*⟩; *(conserve)* sparsam umgehen mit ⟨*Geldmitteln, Kräften, Wasser*⟩; **~ money for a rainy day** *(fig.)* einen Notgroschen zurücklegen; **~ oneself** sich schonen; seine Kräfte sparen; **~ one's breath** sich *(Dat.)* seine Worte sparen; **~ a seat for sb.** jmdm. einen Platz freihalten; **f)** *(make unnecessary)* sparen ⟨*Geld, Zeit, Energie*⟩; **~ sb./oneself sth.** jmdm./sich etw. ersparen; **~ sb./oneself doing sth.** *or* **having to do sth.** es jmdm./sich ersparen, etw. tun zu müssen; **g)** *(avoid losing)* nicht verlieren ⟨*Satz, Karte, Stich*⟩; *(Sport)* abwehren ⟨*Schuß, Ball*⟩; verhindern ⟨*Tor*⟩. **2.** *v. i.* **a)** *(put money by)* sparen; **~ with a building society** bei einer Bausparkasse sparen; **b)** *(avoid waste)* sparen (**on** *Akk.*); **~ on food** am Essen sparen; **c)** *(Sport)* ⟨*Torwart:*⟩ halten. **3.** *n.* *(Sport)* Abwehr, *die;* Parade, *die (fachspr.);* **make a ~** ⟨*Torwart:*⟩ halten. **4.** *prep.* *(arch./poet./rhet.)* mit Ausnahme (+ *Gen.*); **~ for sth.** von etw. abgesehen
~ 'up 1. *v. t.* sparen; sammeln, sparen ⟨*Marken, Gutscheine usw.*⟩. **2.** *v. i.* sparen (**for** für, auf + *Akk.*)
save-as-you-'earn *n.* *(Brit.)* *Sparen durch regelmäßige Abbuchung eines bestimmten Betrages vom Lohn-/Gehaltskonto*
saveloy ['sævəlɔɪ] Zervelatwurst, *die*
saver ['seɪvə(r)] *n.* **a)** *(of money)* Sparer, *der*/Sparerin, *die;* **b)** *in comb.* *(device)* **sth. is a time-~/labour-~/money-~:** etw. spart Zeit/Arbeit/Geld
saving ['seɪvɪŋ] **1.** *n.* **a)** *in pl.* *(money saved)* Ersparnisse *Pl.;* **have money put by in ~s** Geld zurückgelegt haben; **b)** *(rescue; also Theol.)* Rettung, *die;* **c)** *(instance of economy)* Ersparnis, *die.* **2.** *adj.* *in comb.* ⟨*kosten-, benzin*⟩sparend. **3.** *prep.* bis auf (+ *Akk.*)
saving 'grace *n.* versöhnender Zug
'savings bank *n.* Sparkasse, *die*
saviour (*Amer.:* **savior**) ['seɪvjə(r)] *n.* **a)** Retter, *der*/Retterin, *die;* **b)** *(Relig.)* **our/the S~:** unser/der Heiland
savor, savory *(Amer.)* *see* **savour, savoury**
savour ['seɪvə(r)] *(Brit.)* **1.** *n.* **a)** *(flavour)* Geschmack, *der;* *(fig.)* Charakter, *der;* **b)** *(trace)* **a ~ of sth.** ein Hauch *od.* Anflug von etw.; **c)** *(enjoyable quality)* Reiz,

der. **2.** *v. t. (lit. or fig., literary)* genießen

savoury ['seɪvərɪ] *(Brit.)* **1.** *adj.* **a)** *(not sweet)* pikant; *(having salt flavour)* salzig; **b)** *(appetizing)* appetitanregend. **2.** *n.* [pikantes] Häppchen

¹saw [sɔ:] **1.** *n.* Säge, *die.* **2.** *v. t., p.p.* **~n** [sɔ:n] *or* **~ed** [zer]sägen; *(make with ~)* sägen; **~ in half** in der Mitte durchsägen. **3.** *v. i., p.p.* **~n** *or* **~ed** sägen; **~ through sth.** etw. durchsägen

~ 'down *v. t.* umsägen ‹*Baum*›

~ 'off *v. t.* absägen

~ 'up *v. t.* zersägen (**into** in + *Akk.*)

²saw *see* **see**

saw: ~dust *n.* Sägemehl, *das;* **~mill** *n.* Sägemühle, *die*

'sawn-off *adj. (Brit.)* abgesägt; ‹*Gewehr*› mit abgesägtem Lauf

Saxon ['sæksn] **1.** *n.* **a)** Sachse, *der*/Sächsin, *die;* **b)** *(Ling.)* Sächsisch, *das.* **2.** *adj.* **a)** sächsisch; **b)** *(Ling.)* sächsisch

Saxony ['sæksənɪ] *pr. n.* Sachsen *(das)*

saxophone ['sæksəfəʊn] *n.* *(Mus.)* Saxophon, *das*

saxophonist [sæk'sɒfənɪst, 'sæksəfəʊnɪst] *n.* Saxophonist, *der*/Saxophonistin, *die*

say [seɪ] **1.** *v. t., pres. t.* **he ~s** [sez], *p.t. & p.p.* **said** [sed] **a)** sagen; **~ sth. out loud** etw. aussprechen *od.* laut sagen; **he said something about going out** er hat etwas von Ausgehen gesagt; **what more can I ~?** was soll ich da noch [groß] sagen?; **it ~s a lot** *or* **much** *or* **something for sb./sth. that ...:** es spricht sehr für jmdn./etw., daß ...; **have a lot/not much to ~ for oneself** viel reden/nicht viel von sich geben; **to ~ nothing of** *(quite apart from)* ganz zu schweigen von; mal ganz abgesehen von; **that is to ~:** das heißt; **having said that, that said** *(nevertheless)* abgesehen davon; **when all is said and done** letzten Endes; **you can ~ 'that again, you 'said it** *(coll.)* das kannst du laut sagen *(ugs.);* **you don't '~ [so]** *(coll.)* was du nicht sagst *(ugs.);* **~s 'you** *(sl.)* wer's glaubt, wird selig *(ugs. scherzh.);* **I'll ~ [it is]!** *(coll.: it certainly is)* und wie!; **don't let** *or* **never let it be said [that] ...:** niemand soll sagen können, [daß] ...; **I can't ~ [that] I like cats/the idea** ich kann nicht gerade sagen *od.* behaupten, daß ich Katzen mag/die Idee gut finde; **[well,] I 'must ~:** also, ich muß schon sagen; **I should ~ so/not** ich glaube schon/nicht; *(emphatic)* bestimmt/bestimmt nicht; **what have you got to ~ for yourself?** was haben Sie zu Ihren Gunsten zu sagen?; **there's something to be said on both sides/either side** man kann für beide Seiten/jede Seite Argumente anführen; **I can't ~ fairer than that** ein besseres Angebot kann ich nicht machen; **and so ~ all of us** der Meinung sind wir auch; **what do** *or* **would you ~ to sb./sth.?** *(think about)* was hältst du von jmdm./etw.?; was würdest du zu jmdm./etw. sagen?; **what I'm trying to ~ is this** was ich sagen will, ist folgendes; **~ nothing to sb.** *(fig.)* ‹*Musik, Kunst:*› jmdm. nichts bedeuten; **which/that is not ~ing much** *or* **a lot** was nicht viel heißen will/das will nicht viel heißen; **b)** *(recite, repeat, speak words of)* sprechen ‹*Gebet, Text*›; aufsagen ‹*Einmaleins, Gedicht*›; **c)** *(have specified wording or reading)* sagen; ‹*Zeitung:*› schreiben; ‹*Uhr:*› zeigen ‹*Uhrzeit*›; **the Bible ~s** *or* **it ~s in the Bible [that] ...:** in der Bibel heißt es, daß ...; **a sign ~ing ...:** ein Schild mit der Aufschrift ...; **what does it ~ here?** was steht hier?; **d)** *in pass.* **she is said to be clever/have done it** man sagt, sie sei klug/habe es getan. **2.** *v. i., forms as* **1: a)** *(speak)* sagen; **I ~!** *(Brit.)* *(seeking attention)* Entschuldigung!; *(admiring)* Donnerwetter!; **b)** *in imper. (Amer.)* Mensch! **3.** *n.* **a)** *(share in decision)* **have a** *or* **some ~:** ein Mitspracherecht haben (**in** bei); **have no ~:** nichts zu sagen haben; **b)** *(power of decision)* **the [final] ~:** das letzte Wort (**in** bei); **c)** *(what one has to ~)* **have one's ~:** seine Meinung sagen; *(chance to speak)* **get one's** *or* **have a ~:** zu Wort kommen

SAYE *abbr. (Brit.)* **save-as-you-earn**

saying ['seɪɪŋ] *n.* **a)** *(maxim)* Redensart, *die;* **there is a ~ that ...:** wie es [im Sprichwort/in der Maxime] heißt, ...; **as the ~ goes** wie es so schön heißt; **b)** *(remark)* Ausspruch, *der;* **c) there is no ~ what/why ...:** man kann nicht sagen, was/warum ...; **go without ~:** sich von selbst verstehen

'say-so *n.* **a)** *(power of decision)* **on/without sb.'s ~:** auf/ohne jmds. Anweisung *(Akk.);* **b)** *(assertion)* **I won't believe it just on your ~:** das glaube ich dir nicht einfach so

scab [skæb] *n.* **a)** *(over wound or sore)* [Wund]schorf, *der;* **b)** *no pl. (skin-disease)* Räude, *die;* **c)** *(derog.: strike-breaker)* Streikbrecher, *der*/-brecherin, *die*

scabbard ['skæbəd] *n.* Scheide, *die*

scaffold ['skæfəld] *n. (for execution)* Schafott, *das;* **go to the ~:** aufs Schafott kommen

scaffolding ['skæfəldɪŋ] *n., no pl.* Gerüst, *das; (materials)* Gerüstmaterial, *das;* **be surrounded by ~:** eingerüstet sein *(Bauw.)*

scald [skɔ:ld, skɒld] **1.** *n.* Verbrühung, *die.* **2.** *v. t.* **a)** verbrühen; **~ oneself** *or* **one's skin** sich verbrühen; **~ing hot** brühheiß; **b)** *(Cookery)* erhitzen ‹*Milch*›; **c)** *(clean with boiling water)* auskochen

¹scale [skeɪl] *n.* **a)** *(of fish, reptile)* Schuppe, *die;* **b)** *no pl. (deposit) (in kettles, boilers, etc.)* Kesselstein, *der; (on teeth)* Zahnstein, *der*

²scale *n.* **a)** *in sing. or pl. (weighing instrument)* **~[s]** Waage, *die;* **a pair** *or* **set of ~s** eine Waage; **bathroom/kitchen ~[s]** Personen-/Küchenwaage, *die;* **the ~s are evenly balanced** *(fig.)* die Chancen sind ausgewogen; **b)** *(dish of balance)* Waagschale, *die;* **tip** *or* **turn the ~[s]** *(fig.)* den Ausschlag geben

³scale 1. *n.* **a)** *(series of degrees)* Skala, *die;* **the social ~:** die gesellschaftliche Stufenleiter; **b)** *(Mus.)* Tonleiter, *die;* **c)** *(dimensions)* Ausmaß, *das;* **on a grand ~:** im großen Stil; **on a commercial ~:** gewerbsmäßig; **plan on a large ~:** in großem Rahmen planen; **on an international ~:** auf internationaler Ebene; ‹*Katastrophe*› von internationalem Außmaß; **d)** *(ratio of reduction)* Maßstab, *der; attrib.* maßstab[s]gerecht ‹*Modell, Zeichnung*›; **a map with a ~ of 1:250,000** eine Karte im Maßstab 1:250 000; **to ~:** maßstab[s]gerecht; **be out of ~:** im Maßstab nicht passen (**with** zu); **e)** *(indication) (on map, plan)* Maßstab, *der; (on thermometer, ruler, exposure meter)* [Anzeige]skala, *die; (instrument)* Meßstab, *der.* **2.** *v. t.* **a)** *(climb, clamber up)* ersteigen ‹*Festung, Mauer, Leiter, Gipfel*›; erklettern ‹*Felswand, Leiter, Gipfel*›; **b)** [ab]stufen, staffeln ‹*Fahrpreise*›; maßstab[s]gerecht anfertigen ‹*Zeichnung*›; **~ production to demand** die Produktion an die Nachfrage anpassen

~ '**down** *v.t.* [entsprechend] drosseln ⟨*Produktion*⟩; [entsprechende] Abstriche machen an (+ *Dat.*) ⟨*Ideen*⟩; **we ~d down our plans** wir haben bei unseren Planungen Abstriche gemacht

~ '**up** *v.t.* [entsprechend] vergrößern ⟨*Umfang, Ausmaß*⟩; hochfahren ⟨*Produktion, Plan*⟩; **we ~d up our plans** wir haben im größeren Maßstab neu geplant

scallop ['skæləp, 'skɒləp] *n.* **a)** *in pl. (ornamental edging)* Feston, *das;* Bogenkante, *die;* **b)** *(Zool.)* Kammuschel, *die; (Gastr.)* Jakobsmuschel, *die*

scalp [skælp] **1.** *n.* **a)** Kopfhaut, *die;* **b)** *(war-trophy)* Skalp, *der; (fig.)* Trophäe, *die;* **be after sb.'s ~** *(fig.)* jmdm. an den Kragen wollen. **2.** *v.t.* skalpieren

scalpel ['skælpl] *n. (Med.)* Skalpell, *das*

scaly ['skeɪlɪ] *adj.* schuppig

scam [skæm] *n. (Amer. sl.)* Masche, *die (ugs.)*

scamp [skæmp] *n. (derog./joc.)* Spitzbube, *der (abwertend/fam.)*

scamper ['skæmpə(r)] *v.i.* ⟨*Person:*⟩ flitzen; ⟨*Tier:*⟩ huschen; *(hop)* hoppeln; **~ down the stairs** die Treppe hinunterflitzen

scan [skæn] **1.** *v.t.*, **-nn-: a)** *(examine intensely)* [genau] studieren; *(search thoroughly, lit. or fig.)* absuchen (**for** nach); **b)** *(look over cursorily)* flüchtig ansehen; überfliegen ⟨*Zeitung, Liste usw.*⟩ (**for** auf der Suche nach); **c)** ⟨*Radar:*⟩ [mittels Strahlen] abtasten ⟨*Luftraum*⟩; ⟨*Flugsicherung:*⟩ [mittels Radar] überwachen ⟨*Luftraum*⟩; **d)** *(Med.)* szintigraphisch untersuchen ⟨*Körper, Organ*⟩. **2.** *v.i.*, **-nn-** ⟨*Vers[zeile]:*⟩ das richtige Versmaß haben. **3.** *n.* **a)** *(thorough search)* Absuchen, *das;* **b)** *(quick look)* **[cursory] ~:** flüchtiger Blick; **c)** *(examination by beam)* Durchleuchtung, *die;* **d)** *(Med.)* szintigraphische Untersuchung; **body-/brain-~:** Ganzkörper-/Gehirnscan, *der*

scandal ['skændl] *n.* **a)** Skandal, *der* (**about/of** um); *(story)* Skandalgeschichte, *die;* **cause *or* create a ~:** einen Skandal verursachen; **b)** *(outrage)* Empörung, *die;* **c)** *no art. (damage to reputation)* Schande, *die;* **be untouched by ~:** einen makellosen Ruf haben; **d)** *(malicious gossip)* Klatsch, *der (ugs.); (in newspapers etc.)* Skandalgeschichten

scandalize (scandalise) ['skændəlaɪz] *v.t.* schockieren

scandalous ['skændələs] *adj.* skandalös; schockierend ⟨*Bemerkung*⟩; Skandal⟨*blatt, -geschichte, -bericht*⟩; **how ~!** unerhört!

Scandinavia [skændɪ'neɪvɪə] *pr. n.* Skandinavien *(das)*

Scandinavian [skændɪ'neɪvɪən] **1.** *adj.* skandinavisch; **sb. is ~:** jmd. ist Skandinavier/Skandinavierin. **2.** *n.* **a)** *(person)* Skandinavier, *der*/Skandinavierin, *die;* **b)** *(Ling.)* skandinavische Sprachen

scanner ['skænə(r)] *n.* **a)** *(to detect radioactivity)* Geigerzähler, *der;* **b)** *(radar aerial)* Radarantenne, *die;* **c)** *(Med.)* [Szinti]scanner, *der*

scant [skænt] *adj. (arch./literary)* karg *(geh.)* ⟨*Lob, Lohn*⟩; wenig ⟨*Rücksicht*⟩; **pay sb./sth. ~ attention** jmdn./etw. kaum beachten

scantily ['skæntɪlɪ] *adv.* kärglich; spärlich ⟨*bekleidet*⟩

scanty ['skæntɪ] *adj.* spärlich; knapp ⟨*Bikini*⟩; nur wenig ⟨*Vergnügen, Spaß*⟩

scapegoat ['skeɪpgəʊt] *n.* Sündenbock, *der;* **make sb. a ~:** jmdn. zum Sündenbock machen

scar [skɑː(r)] **1.** *n. (lit. or fig.)* Narbe, *die;* **bear the ~s of sth.** *(fig.)* von etw. gezeichnet sein. **2.** *v.t.*, **-rr-: ~ sb./sb.'s face** bei jmdm./in jmds. Gesicht *(Dat.)* Narben hinterlassen; **~ sb. for life** *(fig.)* jmdn. für sein ganzes Leben zeichnen

scarce [skeəs] *adj.* **a)** *(insufficient)* knapp; **b)** *(rare)* selten; **make oneself ~** *(coll.)* sich aus dem Staub machen *(ugs.)*

scarcely ['skeəslɪ] *adv.* kaum; **there was ~ a drop of wine left** es war fast kein Tropfen Wein mehr da; **~ [ever]** kaum [jemals]; **it is ~ likely** es ist wenig wahrscheinlich

scarcity ['skeəsɪtɪ] *n.* **a)** *(short supply)* Knappheit, *die* (**of** an + *Dat.*); **~ of teachers** Lehrermangel, *der;* **food ~:** Lebensmittelknappheit, *die;* **b)** *no pl. (rareness)* Seltenheit, *die*

scare [skeə(r)] **1.** *n.* **a)** *(sensation of fear)* Schreck[en], *der;* **give sb. a ~:** jmdm. einen Schreck[en] einjagen; **b)** *(general alarm; panic)* [allgemeine] Hysterie; **bomb ~:** Bombendrohung, *die; attrib.* **~ story** Schauergeschichte, *die.* **2.** *v.t.* **a)** *(frighten)* Angst machen (+ *Dat.*); *(startle)* erschrecken; **~ sb. into doing sth.** jmdn. dazu bringen, etw. [aus Angst] zu tun; **horror films ~ the pants off me** *(coll.)* bei Horrorfilmen habe ich immer eine wahnsinnige Angst *(ugs.);* **b)** *(drive away)* verscheuchen ⟨*Vögel*⟩. **3.** *v.i.* erschrecken (**at** bei); **~ easily** sich leicht erschrecken lassen

~ **a'way,** ~ **'off** *v.t.* verscheuchen

'scarecrow *n. (lit. or fig.)* Vogelscheuche, *die*

scared [skeəd] *adj.* verängstigt ⟨*Gesicht, Stimme*⟩; **be/feel [very] ~:** [große] Angst haben; **be ~ of sb./sth.** vor jmdm./etw. Angst haben; **be ~ of doing/to do sth.** sich nicht [ge]trauen, etw. zu tun

scaremonger ['skeəmʌŋgə(r)] *n.* Panikmacher, *der*/-macherin, *die (abwertend)*

scarf [skɑːf] *n., pl.* **~s *or* scarves** [skɑːvz] Schal, *der; (triangular/square piece of fine material)* Halstuch, *das; (worn over hair)* Kopftuch, *das; (worn over shoulders)* Schultertuch, *das*

scarlet ['skɑːlɪt] **1.** *n.* Scharlach, *der;* Scharlachrot, *das.* **2.** *adj.* scharlachrot; **I turned ~:** ich wurde puterrot

scarlet 'fever *n. (Med.)* Scharlach, *der*

scarper ['skɑːpə(r)] *v.i. (Brit. sl.)* abhauen *(salopp);* sich aus dem Staub machen *(ugs.)*

scarves *see* **scarf**

scary ['skeərɪ] *adj. (coll.: frightening)* furchterregend ⟨*Anblick*⟩; schaurig ⟨*Film, Geschichte*⟩; angsteinflößend ⟨*Person, Gesicht*⟩; **it was ~ to listen to** beim Zuhören konnte man richtig Angst kriegen *(ugs.)*

scathing ['skeɪðɪŋ] *adj.* beißend ⟨*Spott, Kritik*⟩; scharf ⟨*Angriff*⟩; bissig ⟨*Person, Humor, Bemerkung*⟩; **be ~ about sth.** etw. heruntermachen

scatter ['skætə(r)] **1.** *v.t.* **a)** vertreiben; zerstreuen, auseinandertreiben ⟨*Menge*⟩; **b)** *(distribute irregularly)* verstreuen; ausstreuen ⟨*Samen*⟩. **2.** *v.i.* sich auflösen; ⟨*Menge:*⟩ sich zerstreuen; *(in fear)* auseinanderstieben

scatter: ~-brain *n.* zerstreuter Mensch; Schussel, *der (ugs.);* **~-brained** ['skætəbreɪnd] *adj.* zerstreut; schusselig *(ugs.)*

scattered ['skætəd] *adj.* verstreut; vereinzelt ⟨*Fälle, Anzeichen, Regenschauer*⟩; **thinly ~ population** verstreut lebende Bevölkerung

scatty ['skætɪ] *adj. (Brit. sl.)* dußlig *(salopp);* **drive sb. ~:** jmdn. verrückt machen *(ugs.)*

scavenge ['skævɪndʒ] **1.** *v.t.* **a)** sich *(Dat.)* holen; **b)** *(search)* durchstöbern (**for** nach); absuchen ⟨*Strand*⟩; fleddern ⟨*Leiche*⟩. **2.** *v.i.* **~ for sth.** nach etw. suchen

scavenger ['skævɪndʒə(r)] *n.*

(animal) Aasfresser, *der; (fig. derog.: person)* Aasgeier, *der (ugs. abwertend)*

scenario [sɪ'nɑːrɪəʊ, sɪ'neərɪəʊ] *n., pl.* **~s** *(also fig.)* Szenario, *das*

scene [siːn] *n.* **a)** *(place of event)* Schauplatz, *der; (in novel, play, etc.)* Ort der Handlung; **~ of the crime** Ort des Verbrechens; Tatort, *der;* **b)** *(portion of play, film, or book)* Szene, *die; (division of act)* Auftritt, *der;* **love/trial ~:** Liebes-/Gerichtsszene, *die;* **c)** *(display of passion, anger, jealousy)* Szene, *die;* **create** *or* **make a ~:** eine Szene machen; **d)** *(view)* Anblick, *der; (as depicted)* Aussicht, *die;* **change of ~:** Tapetenwechsel, *der (ugs.);* **e)** *(place of action)* Ort des Geschehens; **arrive** *or* **come on the ~:** auftauchen; **f)** *(field of action)* **the political/drug/artistic ~:** die politische/Drogen-/Kunstszene; **the social ~:** das gesellschaftliche Leben; **g)** *(sl.: area of interest)* **what's your ~?** worauf stehst du? *(ugs.);* **that's not my ~:** das ist nicht mein Fall *(ugs.);* **h)** *(Theatre: set)* Bühnenbild, *das;* **behind the ~s** *(lit. or fig.)* hinter den Kulissen; **set the ~ [for sb.]** *(fig.)* [jmdm.] die Ausgangssituation darlegen

scenery ['siːnərɪ] *n., no pl.* **a)** *(Theatre)* Bühnenbild, *das;* **b)** *(landscape)* Landschaft, *die; (picturesque)* [malerische] Landschaft; **change of ~:** Tapetenwechsel, *der (ugs.)*

scenic ['siːnɪk] *adj. (with fine natural scenery)* landschaftlich schön; **a ~ drive** eine Fahrt durch schöne Landschaft; **~ railway** Berg-und-Tal-Bahn, *die*

scent [sent] **1.** *n.* **a)** *(smell)* Duft, *der; (fig.)* [Vor]ahnung, *die;* **catch the ~ of sth.** den Duft von etw. in die Nase bekommen; **b)** *(Hunting; also fig.: trail)* Fährte, *die;* **be on the ~ of sb./sth.** *(fig.)* jmdm./einer Sache auf der Spur sein; **put** *or* **throw sb. off the ~** *(fig.)* jmdn. auf eine falsche Fährte bringen; **c)** *(Brit.: perfume)* Parfüm, *das;* **d)** *(sense of smell)* Geruchssinn, *der; (fig.: power to detect)* Spürsinn, *der.* **2.** *v. t.* **a)** *(lit. or fig.)* wittern; **b)** *(apply perfume to)* parfümieren

scented ['sentɪd] *adj.* **a)** *(having smell)* duftend; **be ~** ⟨*Blume:*⟩ duften; **b)** *(perfumed)* parfümiert

scepter *(Amer.) see* **sceptre**

sceptic ['skeptɪk] *n.* Skeptiker, *der*/Skeptikerin, *die; (with religious doubts)* [Glaubens]zweifler, *der*/-zweiflerin, *die*

sceptical ['skeptɪkl] *adj.* skeptisch; **be ~ about** *or* **of sb./sth.** jmdm./einer Sache skeptisch gegenüberstehen

scepticism ['skeptɪsɪzm] *n.* Skepsis, *die; (Philos.)* Skeptizismus, *der; (religious doubt)* Glaubenszweifel *Pl.*

sceptre ['septə(r)] *n. (Brit.; lit. or fig.)* Zepter, *das*

schedule ['ʃedjuːl, 'skedjuːl] **1.** *n.* **a)** *(list)* Tabelle, *die; (for event, festival)* Programm, *das;* **b)** *(plan of procedure)* Zeitplan, *der;* **we are working to a tight ~:** unsere Termine sind sehr eng; **c)** *(set of tasks)* Terminplan, *der;* Programm, *das;* **work/study ~:** Arbeits-/Studienplan, *der;* **d)** *(tabulated statement)* Aufstellung, *die;* **e)** *(time stated in plan)* **on ~:** programmgemäß; **arrive on ~:** pünktlich ankommen. **2.** *v. t.* **a)** *(make plan of)* zeitlich planen; **be ~d for Thursday** für Donnerstag geplant sein; **b)** *(make timetable of)* einen Fahrplan aufstellen für; *(include in timetable)* in den Fahrplan aufnehmen

scheduled ['ʃedjuːld, 'skedjuːld] *adj. (according to timetable)* [fahr]planmäßig ⟨*Zug, Halt*⟩; flugplanmäßig ⟨*Zwischenlandung*⟩; **~ flight** Linienflug, *der*

schematic [skɪ'mætɪk, skiː'mætɪk] *adj.* schematisch

scheme [skiːm] **1.** *n.* **a)** *(arrangement)* Anordnung, *die;* **b)** *(table of classification, outline)* Schema, *das;* **c)** *(plan)* Programm, *das; (project)* Projekt, *das;* **pension ~:** Altersversorgung, *die;* **d)** *(dishonest plan)* Intrige, *die.* **2.** *v. i.* Pläne schmieden

scheming ['skiːmɪŋ] **1.** *n., no pl., no indef. art.* Winkelzüge *Pl.;* Machenschaften *Pl.* **2.** *adj.* intrigant

schizophrenia [skɪtsə'friːnɪə] *n. (Psych.)* Schizophrenie, *die*

schizophrenic [skɪtsə'frenɪk, skɪtsə'friːnɪk] *(Psych.; also fig. coll.)* **1.** *adj.* schizophren. **2.** *n.* Schizophrene, *der/die*

scholar ['skɒlə(r)] *n.* **a)** *(learned person)* Gelehrte, *der/die;* **literary ~:** Literaturwissenschaftler *der*/-wissenschaftlerin, *die;* **Shakespeare[an] ~:** Shakespeare-Forscher, *der*/-Forscherin, *die;* **b)** *(one who learns)* Schüler, *der*/Schülerin, *die;* **c)** *(holder of scholarship)* Stipendiat, *der*/Stipendiatin, *die*

scholarly ['skɒləlɪ] *adj.* wissenschaftlich; *(having much learning)* gelehrt

scholarship ['skɒləʃɪp] *n.* **a)** *(payment for education)* Stipendium, *das;* **b)** *no pl. (scholarly work)* Gelehrsamkeit, *die (geh.); (methods)* Wissenschaftlichkeit, *die;* **c)** *no pl. (body of learning)* **literary/linguistic/historical ~:** Literatur- / Sprach- / Geschichtswissenschaft, *die*

[1]school [skuːl] **1.** *n.* **a)** Schule, *die; (Amer.: university, college)* Hochschule, *die; attrib.* Schul-; **be at** *or* **in ~:** in der Schule sein; *(attend ~)* zur Schule gehen; **to/from ~:** zur/von *od.* aus der Schule; **go to ~:** zur Schule gehen; **have time off ~:** schulfrei haben; **there will be no ~ today** heute ist keine Schule; **b)** *attrib.* Schul⟨*aufsatz, -bus, -jahr, -system*⟩; **~ holidays** Schulferien *Pl.;* **~ exchange** Schüleraustausch, *der;* **the ~ term** die Schulzeit; **c)** *(disciples)* Schule, *die;* **~ of thought** Lehrmeinung, *die;* **d)** *(Brit.: group of gamblers)* Runde, *die.* **2.** *v. t. (train)* erziehen; dressieren ⟨*Pferd*⟩; **~ sb. in sth.** jmdn. in etw. *(Akk.)* unterweisen *(geh.)*

[2]school *n. (of fish)* Schwarm, *der;* Schule, *die (Zool.)*

school: ~ age *n.* Schulalter, *das;* **children of ~ age** Kinder im schulpflichtigen Alter; **~boy** *n.* Schüler, *der; (with reference to behaviour)* Schuljunge, *der;* **~child** *n.* Schulkind, *das;* **~-days** *n. pl.* Schulzeit, *die;* **~-friend** *n.* Schulfreund, *der*/-freundin, *die;* **~girl** *n.* Schülerin, *die; (with reference to behaviour)* Schulmädchen, *das*

schooling ['skuːlɪŋ] *n.* Schulbildung, *die;* **he has had little ~:** er hat keine richtige Schulbildung gehabt

school: ~-leaver ['skuːlliːvə(r)] *n. (Brit.)* Schulabgänger, *der*/-abgängerin, *die;* **~-'leaving age** *n. (Brit.)* Schulabgangsalter, *das;* **~master** *n.* Lehrer, *der;* **~mistress** *n.* Lehrerin, *die;* **~room** *n.* Schulzimmer, *das;* **~teacher** *n.* Lehrer, *der*/Lehrerin, *die;* **~-work** *n.* Schularbeiten *Pl.*

schooner ['skuːnə(r)] *n.* **a)** *(Naut.)* Schoner, *der;* **b)** *(Brit.: sherry glass)* [hohes] Sherryglas

sciatica [saɪ'ætɪkə] *n. (Med.)* Ischias, *die (fachspr. der od. das)*

science ['saɪəns] *n.* **a)** *no pl., no art.* Wissenschaft, *die;* **applied/pure ~:** angewandte/reine Wissenschaft; **b)** *(branch of knowledge)* Wissenschaft, *die;* **c) [natural] ~:** Naturwissenschaften; *attrib.* naturwissenschaftlich

⟨*Buch, Labor*⟩; **d)** *(technique, expert's skill)* Kunst, *die*
science: ~ **'fiction** *n.* Science-fiction, *die;* ~ **park** *n.* Technologiepark, *der*
scientific [saɪən'tɪfɪk] *adj.* **a)** wissenschaftlich; *(of natural science)* naturwissenschaftlich; **b)** *(using technical skill)* technisch gut ⟨*Boxer, Schauspieler, Tennis*⟩
scientist ['saɪəntɪst] *n.* Wissenschaftler, *der*/Wissenschaftlerin, *die; (in physical or natural science)* Naturwissenschaftler, *der*/-wissenschaftlerin, *die;* **biological/social/computer** ~**s** Biologen/Soziologen/Informatiker
Scillies ['sɪlɪz], **Scilly Isles** ['sɪlɪ aɪlz] *pr. n. pl.* Scilly-Inseln *Pl.*
scimitar ['sɪmɪtə(r)] *n.* Krummsäbel, *der*
scintillating ['sɪntɪleɪtɪŋ] *adj. (fig.)* geistsprühend
scissors ['sɪzəz] *n. pl.* **[pair of]** ~: Schere, *die;* **be a** ~**-and-paste job** ⟨*Buch, Werk:*⟩ [aus anderen Werken] zusammengeschrieben sein
'scissors kick *n. (Swimming)* Scherenschlag, *der*
sclerosis [sklɪə'rəʊsɪs] *n., pl.* **scleroses** [sklɪə'rəʊsi:z] *(Med.)* Sklerose, *die;* **disseminated** *or* **multiple** ~: multiple Sklerose
¹scoff [skɒf] *v. i. (mock)* spotten; ~**ing remarks** spöttische Bemerkungen; ~ **at sb./sth.** sich über jmdn./etw. lustig machen
²scoff *(sl.)* **1.** *v. t. (eat greedily)* verschlingen. **2.** *v. i.* sich [*(Dat.)* den Bauch] vollschlagen *(salopp)*
scold [skəʊld] **1.** *v. t.* schelten *(geh.);* ausschimpfen **(for** wegen); **she** ~**ed him for coming late** sie schimpfte ihn aus *od.* schalt ihn, weil er zu spät kam. **2.** *v. i.* schimpfen; ~**ing wife** zänkische Ehefrau
scolding ['skəʊldɪŋ] *n.* Schimpfen, *das; (instance)* Schelte, *die (geh.);* Schimpfe, *die (ugs.);* **get a** ~: ausgeschimpft werden
scone [skɒn, skəʊn] *n. weicher, oft zum Tee gegessener kleiner Kuchen*
scoop [sku:p] **1.** *n.* **a)** *(shovel)* Schaufel, *die;* **b)** *(ladle, ladleful)* Schöpflöffel, *der;* [Schöpf]kelle, *die;* **c)** *(for ice-cream, mashed potatoes)* Portionierer, *der; (of ice-cream)* Kugel, *die;* **d)** *(Journ.)* Knüller, *der (ugs.);* Scoop, *der (fachspr.).* **2.** *v. t.* **a)** *(lift)* schaufeln ⟨*Kohlen, Zucker*⟩; *(with ladle)* schöpfen ⟨*Flüssigkeit, Schaum*⟩; **b)** *(secure)* erzielen ⟨*Gewinn*⟩; hereinholen *(ugs.)* ⟨*Auftrag*⟩; *(in a lottery, bet)* gewinnen ⟨*Vermögen*⟩; **c)** *(Journ.)* ausstechen
~ **'out** *v. t.* **a)** *(hollow out)* aushöhlen; schaufeln ⟨*Loch, Graben*⟩; **b)** *(remove)* [her]ausschöpfen ⟨*Flüssigkeit*⟩; auslöffeln ⟨*Fruchtfleisch*⟩; *(with a knife)* herausschneiden ⟨*Fruchtfleisch, Gehäuse*⟩; *(excavate)* ausbaggern ⟨*Erde*⟩
~ **'up** *v. t.* schöpfen ⟨*Wasser, Suppe*⟩; schaufeln ⟨*Erde*⟩; aufschaufeln ⟨*Kohlen, Kies*⟩
scoot [sku:t] *v. i. (coll.)* rasen; *(to escape)* die Kurve kratzen *(ugs.)*
scooter ['sku:tə(r)] *n.* **a)** *(toy)* Roller, *der;* **b) [motor]** ~: [Motor]roller, *der*
scope [skəʊp] *n., no indef. art.* **a)** Bereich, *der; (of person's activities)* Betätigungsfeld, *das; (of person's job)* Aufgabenbereich, *der; (of department etc.)* Zuständigkeitsbereich, *der;* Zuständigkeit, *die; (of discussion, meeting, negotiations, investigations, etc.)* Rahmen, *der;* **that is a subject beyond my** ~: das fällt nicht in meine Sparte; *(beyond my grasp)* das ist mir zu hoch; **b)** *(opportunity)* Entfaltungsmöglichkeiten *Pl.;* **give ample** ~ **for new ideas** weiten Raum für neue Ideen bieten
scorch [skɔ:tʃ] **1.** *v. t.* verbrennen; versengen. **2.** *v. i.* versengt werden; verbrennen. **3.** *n.* versengte Stelle; Brandfleck, *der*
scorcher ['skɔ:tʃə(r)] *n. (Brit. coll.: hot day)* **what a** ~**!** ist das eine Affenhitze heute!
scorching ['skɔ:tʃɪŋ] *adj.* glühend heiß; sengend, glühend ⟨*Hitze*⟩
score [skɔ:(r)] **1.** *n.* **a)** *(points)* [Spiel]stand, *der; (made by one player)* Punktzahl, *die;* **What's the** ~**? – The** ~ **was 4–1 at half-time** Wie steht es? – Der Halbzeitstand war 4 : 1; **final** ~: Endstand, *der;* **keep [the]** ~: zählen; **know the** ~ *(fig. coll.)* wissen, was Sache ist *od.* was läuft *(salopp);* **b)** *(Mus.)* Partitur, *die; (Film)* [Film]musik, *die;* **c)** *pl.* ~ *or* ~**s** *(group of 20)* zwanzig; **d)** *in pl. (great numbers)* ~**s [and** ~**s] of** zig *(ugs.);* Dutzende [von]; ~**s of times** zigmal *(ugs.);* **e)** *(notch)* Kerbe, *die; (weal)* Striemen, *der;* **f) pay off** *or* **settle an old** ~ *(fig.)* eine alte Rechnung begleichen; **g)** *(reason)* Grund, *der;* **on that** ~: was das betrifft *od.* angeht; diesbezüglich. **2.** *v. t.* **a)** *(win)* erzielen ⟨*Erfolg, Punkt, Treffer usw.*⟩; ~ **a direct hit on sth.** ⟨*Person:*⟩ einen Volltreffer landen; ⟨*Bombe:*⟩ etw. voll treffen; **they** ~**d a success** sie konnten einen Erfolg [für sich] verbuchen; ~ **a goal** ein Tor schießen/werfen; **b)** *(make notch/notches in)* einkerben; **c)** *(be worth)* zählen; **d)** *(Mus.)* setzen; *(orchestrate)* orchestrieren ⟨*Musikstück*⟩. **3.** *v. i.* **a)** *(make* ~*)* Punkte/einen Punkt erzielen *od. (ugs.)* machen; punkten *(bes. Boxen); (~ goal/goals)* ein Tor/Tore schießen/werfen; ~ **high** *or* **well** *(in test etc.)* eine hohe Punktzahl erreichen *od.* erzielen; **b)** *(keep* ~*)* aufschreiben; anschreiben; **c)** *(secure advantage)* die besseren Karten haben **(over** gegenüber, im Vergleich zu)
~ **'out,** ~ **'through** *v. t.* durchstreichen; ausstreichen
score: ~**-board** *n.* Anzeigetafel, *die;* ~**-card** *n. (Sport)* Anschreibekarte, *die; (Golf)* Scorekarte, *die*
scorer ['skɔ:rə(r)] *n.* **a)** *(recorder of score)* Anschreiber, *der*/Anschreiberin, *die;* **b)** *(Footb.)* Torschütze, *der*/-schützin, *die*
scorn [skɔ:n] **1.** *n., no pl., no indef. art.* Verachtung, *die;* **with** ~: mit *od.* voll[er] Verachtung; verachtungsvoll. **2.** *v. t.* **a)** *(hold in contempt)* verachten; **b)** *(refuse)* in den Wind schlagen ⟨*Rat*⟩; ausschlagen ⟨*Angebot*⟩; ~ **doing** *or* **to do sth.** es für unter seiner Würde halten, etw. zu tun
scornful ['skɔ:nfl] *adj.* verächtlich ⟨*Lächeln, Blick*⟩; **be** ~ **of sth.** für etw. nur Verachtung haben
Scorpio ['skɔ:pɪəʊ] *n. (Astrol., Astron.)* der Skorpion
scorpion ['skɔ:pɪən] *n. (Zool.)* Skorpion, *der*
Scot [skɒt] *n.* Schotte, *der*/Schottin, *die*
Scotch [skɒtʃ] **1.** *adj.* **a)** *(of Scotland) see* **Scottish; b)** *(Ling.) see* **Scots 1 b. 2.** *n.* **a)** *(whisky)* Scotch, *der;* schottischer Whisky; **b)** *(Ling.) see* **Scots 2; c)** *constr. as pl.* **the** ~: die Schotten
scotch *v. t.* **a)** *(frustrate)* zunichte machen ⟨*Plan*⟩; **b)** *(put an end to)* den Boden entziehen (+ *Dat.*) ⟨*Gerücht, Darstellung*⟩
Scotch: ~ **'egg** *n. (Gastr.) hartgekochtes Ei in Wurstbrät;* ~ **'mist** *n.* dichter Nieselregen; ~ **tape, (P)** *n. (Amer.)* ≈ Tesafilm, *der* Ⓦᶻ; ~ **'terrier** *n.* Scotch[terrier], *der;* ~ **'whisky** *n.* schottischer Whisky
scot-'free *pred. adj.* ungeschoren; **get off/go/escape** ~: ungeschoren davonkommen

Scotland ['skɒtlənd] *pr. n.* Schottland *(das)*

Scots [skɒts] **1.** *adj.* **a)** *(esp. Scot.) see* **Scottish**; **b)** *(Ling.)* schottisch. **2.** *n. (dialect)* Schottisch, *das*

Scots: ~**man** ['skɒtsmən] *n., pl.* ~**men** ['skɒtsmən] Schotte, *der;* ~**woman** *n.* Schottin, *die*

Scottish ['skɒtɪʃ] *adj.* schottisch

scoundrel ['skaʊndrl] *n.* Schuft, *der (abwertend); (villain)* Schurke, *der (abwertend)*

[1]**scour** [skaʊə(r)] *v.t.* **a)** scheuern ‹*Topf, Metall*›; ~ **out** ausscheuern ‹*Topf*›; **b)** *(clear out)* ~ **[out]** durchspülen ‹*Rohr*›; **c)** *(remove by rubbing)* [ab]scheuern

[2]**scour** *v.t. (search)* durchkämmen **(for** nach)

scourer ['skaʊərə(r)] *n.* Topfreiniger, *der;* Topfkratzer, *der*

scourge [skɜːdʒ] **1.** *n. (lit. or fig.)* Geißel, *die.* **2.** *v.t.* **a)** *(whip)* geißeln; **b)** *(afflict)* heimsuchen

scout [skaʊt] **1.** *n.* **a) [Boy] S~:** Pfadfinder, *der;* **b)** *(Mil. etc.: sent to get information)* Späher, *der/* Späherin, *die;* Kundschafter, *der/* Kundschafterin, *die.* **2.** *v.i.* auf Erkundung gehen; ~ **for sb./sth.** nach jmdm./etw. Ausschau halten: **be ~ing for talent** auf Talentsuche sein

~ **a'bout,** ~ **a'round** *v.i.* sich umsehen **(for** nach)

'scout leader *n.* Pfadfinderführer, *der*

scowl [skaʊl] **1.** *v.i.* ein mürrisches Gesicht machen; ~ **at sb.** jmdn. mürrisch ansehen. **2.** *n.* mürrischer [Gesichts]ausdruck

scrabble [skræbl] **1.** *v.i.* ‹*Maus, Hund:*› scharren, kratzen. **2.** *n.* **S~, (P)** Scrabble, *das*

scram [skræm] *v.i.,* **-mm-** *(sl.)* abhauen *(salopp);* verschwinden *(ugs.)*

scramble ['skræmbl] **1.** *v.i.* **a)** *(clamber)* klettern; kraxeln *(ugs.);* **b)** *(move hastily)* hasten *(geh.);* rennen *(ugs.);* ~ **for sth.** um etw. rangeln; ‹*Kinder:*› sich um etw. balgen; **c)** *(Air Force)* [im Alarmfalle] aufsteigen. **2.** *v.t.* **a)** *(Cookery)* ~ **some eggs** Rührei[er] machen; *see also* **scrambled egg; b)** *(Teleph., Radio)* verschlüsseln; **c)** *(mix together)* [ver]mischen; **d)** ~ **the ball away** *(Footb.)* den Ball [irgendwie] wegschlagen. **3.** *n.* **a)** *(struggle)* Gerangel, *das* **(for** um); **b)** *(climb)* Kletterpartie, *die (ugs.)*

scrambled egg [skræmbld 'eg] *n. (Gastr.)* Rührei, *das*

[1]**scrap** [skræp] **1.** *n.* **a)** *(fragment) (of paper, conversation)* Fetzen, *der; (of food)* Bissen, *der;* ~ **of paper** Stück Papier; *(small, torn)* Papierfetzen, *der;* **b)** *in pl. (odds and ends) (of food)* Reste *Pl.; (of language)* Brocken *Pl.;* **a few ~s of information/news** ein paar bruchstückhafte Informationen/Nachrichten; **c)** *(smallest amount)* **not a ~ of** kein bißchen; *(of sympathy, truth also)* nicht ein Fünkchen; **not a ~ of evidence** nicht die Spur eines Beweises; **d)** *no pl., no indef. art. (waste metal)* Schrott, *der; attrib.* ~ **metal** Schrott, *der;* Altmetall, *das;* **e)** *no pl., no indef. art. (rubbish)* Abfall, *der;* **they are ~:** das ist Abfall *od.* sind Abfälle. **2.** *v.t.,* **-pp-** wegwerfen; wegschmeißen *(ugs.); (send for ~)* verschrotten; *(fig.)* aufgeben ‹*Plan, Projekt usw.*›; **you can ~ that idea right away** die Idee kannst du gleich vergessen *(ugs.)*

[2]**scrap** *(coll.)* **1.** *n. (fight)* Rauferei, *die;* Klopperei, *die (ugs.).* **2.** *v.i.,* **-pp-** sich raufen **(with** mit)

'scrap-book *n.* [Sammel]album, *das*

scrape [skreɪp] **1.** *v.t.* **a)** *(make smooth)* schaben ‹*Häute, Möhren, Kartoffeln usw.*›; abziehen ‹*Holz*›; *(damage)* verkratzen, verschrammen ‹*Fußboden, Auto*›; schürfen ‹*Körperteil*›; **b)** *(remove)* [ab]schaben, [ab]kratzen ‹*Farbe, Schmutz, Rost*› **(off, from** von); **c)** *(draw along)* schleifen; **d)** *(remove dirt from)* abstreifen ‹*Schuhe, Stiefel*›; **e)** *(draw back)* straff kämmen ‹*Haar*›; **f)** *(excavate)* scharren ‹*Loch*›; **g)** *(accumulate by care with money)* ~ **together/up** *(raise)* zusammenkratzen *(ugs.); (save up)* zusammensparen; **h)** ~ **together/up** *(amass by scraping)* zusammenscharren ‹*Sand, Kies*›; **i)** *(leave no food on or in)* abkratzen ‹*Teller*›; auskratzen ‹*Schüssel*›. *See also* **barrel a. 2.** *v.i.* **a)** *(make with scraping sound)* schleifen; **b)** *(rub)* streifen **(against, over** *Akk.*); **c)** ~ **past each other** ‹*Autos:*› haarscharf aneinander vorbeifahren; **d) bow and ~:** katzbuckeln *(abwertend). See also* **scrimp. 3.** *n.* **a)** *(act, sound)* Kratzen, *das* **(against** an + *Dat.*); Schaben, das **(against** an + *Dat.*); **b)** *(predicament)* Schwulitäten *Pl. (ugs.);* **be in a/get into a ~:** in Schwulitäten sein/kommen; **get sb. out of a ~:** jmdm. aus der Bredouille *od.* Patsche helfen *(ugs.);* **c)** *(~d place)* Kratzer, *der (ugs.);* Schramme, *die*

~ **a'long** *v.i. (fig.)* sich über Wasser halten **(on** mit)

~ **a'way** *v.t.* abkratzen; abschaben

~ **'by** *see* ~ **along**

~ **'out** *v.t.* **a)** *(excavate)* buddeln *(ugs.);* scharren; **b)** *(clean)* auskratzen; ausschaben

~ **through 1.** ['--] *v.t.* **a)** sich zwängen durch; **b)** *(fig.: just succeed in passing)* mit Hängen und Würgen kommen durch ‹*Prüfung*›. **2.** [-'-] *v.i.* **a)** sich durchzwängen; **b)** *(fig.: just succeed in passing examination)* mit Hängen und Würgen durchkommen

scraper ['skreɪpə(r)] *n.* **a)** *(for shoes)* Kratzeisen, *das; (grid)* Abtreter, *der;* Abstreifer, *der;* **b)** *(hand tool, kitchen utensil)* Schaber, *der; (for clearing snow)* Schneescharre, *die; (decorator's)* Spachtel, *der; (for removing ice from car windows)* [Eis]kratzer, *der*

scrap: ~**-heap** *n.* Schutthaufen, *der;* Müllhaufen, *der;* ~ **merchant** *n.* Schrotthändler, *der/* -händlerin, *die;* ~ **'paper** *n.* Schmierpapier, *das*

scrappy ['skræpɪ] *adj.* **a)** *(not complete)* lückenhaft ‹*Bericht, Bildung usw.*›; **b)** *(made up of bits or scraps)* zusammengestoppelt *(abwertend)*

'scrap-yard *n.* Schrottplatz, *der;* **be sent to the ~:** verschrottet werden

scratch [skrætʃ] **1.** *v.t.* **a)** *(score surface of)* zerkratzen; verkratzen; *(score skin of)* kratzen; ~ **the surface [of sth.]** ‹*Geschoß usw.:*› [etw.] streifen; **he has only ~ed the surface [of the problem]** er hat das Problem nur oberflächlich gestreift; **b)** *(get ~ on)* ~ **oneself/one's hands** *etc.* sich schrammen/sich *(Dat.)* die Hände *usw.* zerkratzen *od.* [zer]schrammen *od.* ritzen; **c)** *(scrape without marking)* kratzen; kratzen an (+ *Dat.*) ‹*Insektenstich usw.*›; ~ **oneself/one's arm** *etc.* sich kratzen/sich *(Dat.)* den Arm *usw. od.* am Arm *usw.* kratzen; *abs.* ‹*Person:*› sich kratzen; ~ **one's head** sich am Kopf kratzen; ~ **one's head [over sth.]** *(fig.)* sich *(Dat.)* den Kopf über etw. *(Akk.)* zerbrechen; **you ~ my back and I'll ~ yours** *(fig. coll.)* eine Hand wäscht die andere *(Spr.);* **d)** *(form)* kratzen, ritzen ‹*Buchstaben etc.*›; kratzen, scharren ‹*Loch*› **(in** in + *Akk.*); ~ **a living** sich schlecht und recht ernähren **(from** von); **e)** *(erase from list)* streichen **(from** aus); *(withdraw* *from competition)* von der Starter- *od.* Teilnehmerliste streichen ‹*Rennpferd, Athleten*›. **2.** *v. i.* **a)** kratzen; **b)** *(scrape)* ‹*Huhn:*› kratzen, scharren. **3.** *n.* **a)** *(mark, wound; coll.: trifling wound)* Kratzer, *der (ugs.)*; Schramme, *die;* **b)** *(sound)* Kratzen, *das* (**at** an + *Dat.*); Kratzgeräusch, *das;* **c) have a [good]** ~: sich [ordentlich] kratzen; **d) start from** ~ *(fig.)* bei Null anfangen *(ugs.)*; **be up to** ~ ‹*Arbeit, Leistung:*› nichts zu wünschen übriglassen; ‹*Person:*› in Form *od. (ugs.)* auf Zack sein; **bring sth. up to** ~: etw. auf Vordermann *(scherzh.)* bringen. **4.** *adj. (collected haphazardly)* bunt zusammengewürfelt

~ **a'bout,** ~ **a'round** *v. i.* scharren; *(fig.: search)* suchen (**for** nach)

~ **'off** *v. t.* abkratzen

~ **'out** *v. t.* **a)** *(score out)* aus-, durchstreichen ‹*Name, Wort*›; **b)** *(gouge out)* auskratzen ‹*Auge*›

scratchy ['skrætʃɪ] *adj.* **a)** kratzig [klingend] ‹*Schallplatte*›; **b)** kratzig ‹*Wolle, Kleidungsstück*›

scrawl [skrɔ:l] **1.** *v. t.* hinkritzeln; ~ **sth. on sth.** etw. auf etw. *(Akk.)* kritzeln. **2.** *v. i.* kritzeln. **3.** *n. (piece of writing)* Gekritzel, *das; (handwriting)* Klaue, *die (salopp abwertend)*

~ **'out** *v. t.* wegstreichen ‹*Wort*›

scrawny ['skrɔ:nɪ] *adj. (derog.)* hager, dürr ‹*Hals, Person*›; mager ‹*Vieh*›

scream [skri:m] **1.** *v. i.* **a)** schreien (**with** vor + *Dat.*); ~ **at sb.** jmdn. anschreien; **b)** ‹*Vogel, Affe:*› schreien; ‹*Sirene, Triebwerk:*› heulen; ‹*Reifen:*› quietschen; ‹*Säge:*› kreischen. **2.** *v. t.* schreien. **3.** *n.* **a)** Schrei, *der; (of siren or jet engine)* Heulen, *das;* ~**s of laughter/pain** gellendes Gelächter/Schmerzensschreie; **b)** *(sl.: comical person or thing)* **be a** ~: zum Schreien sein *(ugs.)*

scree [skri:] *n.* ~**[s]** *(stones)* Schutt, *der;* Geröll, *das;* Schotter, *der*

screech [skri:tʃ] **1.** *v. i.* ‹*Kind, Eule:*› kreischen, schreien; ‹*Bremsen:*› quietschen, kreischen; ~ **to a halt, come to a** ~**ing halt** ‹*Auto:*› quietschend *od.* kreischend zum Stehen kommen. **2.** *v. t.* kreischen. **3.** *n. (cry)* Schrei, *der;* Kreischen, *das;* **give a** ~ **of laughter** gellend auflachen

screen [skri:n] **1.** *n.* **a)** *(partition)* Trennwand, *die; (piece of furniture)* Wandschirm, *der;* **b)** *(sth. that conceals from view)* Sichtschutz, *der; (of trees, persons, fog)* Wand, *die; (of persons)* Mauer, *die; (of secrecy)* Wand, *die;* Mauer, *die;* **c)** *(on which pictures are projected)* Leinwand, *die; (of computer; radar* ~*)* **[TV]** ~: [Fernseh]schirm, *der;* Bildschirm, *der;* **the** ~ *(Cinemat.)* die Leinwand; **d)** *(Phys.)* [Schutz]schirm, *der; (Electr.)* Abschirmung, *die;* **e)** *(Motor Veh.) see* **windscreen; f)** *(Amer.: netting to exclude insects)* Fliegendraht, *der;* Fliegengitter, *das.* **2.** *v. t.* **a)** *(shelter)* schützen (**from** vor + *Dat.*); *(conceal)* verdecken; ~ **one's eyes from the sun** seine Augen vor der Sonne schützen *od. (geh.)* gegen die Sonne beschirmen; ~ **sth. from sb.** etw. jmds. Blicken entziehen; **b)** *(show)* vorführen, zeigen ‹*Dias, Film*›; **c)** *(check) (for disease)* untersuchen (**for** auf + *Akk.*); *(for loyalty etc.)* unter die Lupe nehmen

~ **'off** *v. t.* abteilen ‹*Teil eines Raums*›; [mit einem Wandschirm] abtrennen ‹*Bett*›

screen: ~**-play** *n. (Cinemat.)* Drehbuch, *das;* ~**-printing** *n. (Textiles)* Gewebefilmdruck, *der*

screw [skru:] **1.** *n.* **a)** Schraube, *die;* **he has a** ~ **loose** *(coll. joc.)* bei ihm ist eine Schraube locker *od.* lose *(salopp)*; **put the** ~**[s] on sb.** *(fig. coll.)* jmdm. [die] Daumenschrauben anlegen *(ugs.)*; **b)** *(Naut., Aeronaut.)* Schraube, *die;* **c)** *(sl.: prison warder)* Wachtel, *die (salopp)*; **d)** *(coarse) (copulation)* Fick, *der (vulg.)*; Nummer, *die (derb)*; *(partner in copulation)* Ficker, *der*/Fickerin, *die (vulg.)*; **have a** ~: ficken *(vulg.)*; vögeln *(vulg.)*. **2.** *v. t.* **a)** *(fasten)* schrauben (**to** an + *Akk.*); ~ **down** festschrauben; **have one's head** ~**ed on [straight** *or* **the right way** *or* **properly]** *(coll.)* ein vernünftiger Mensch sein; **b)** *(turn)* schrauben ‹*Schraubverschluß usw.*›; ~ **a piece of paper into a ball** ein Stück Papier zu einer Kugel zusammendrehen; **c)** *(sl.: extort)* [raus]quetschen *(salopp)* ‹*Geld, Geständnis*› (**out of** aus); **d)** *(coarse: copulate with)* ‹*Mann:*› ficken *(vulg.)*, vögeln *(vulg.)*; ‹*Frau:*› ficken mit *(vulg.)*, vögeln mit *(vulg.)*. **3.** *v. i.* **a)** *(revolve)* sich schrauben lassen; sich drehen lassen; **b)** *(coarse: copulate)* ficken *(vulg.)*; vögeln *(vulg.)*

~ **'up** *v. t.* **a)** *(crumple up)* zusammenknüllen ‹*Blatt Papier*›; **b)** verziehen ‹*Gesicht*›; zusammenkneifen ‹*Augen, Mund*›; **c)** *(sl.: bungle)* vermurksen *(ugs.)*; vermasseln *(salopp)*; ~ **it/things up** Mist bauen *(salopp)*; **d)** ~ **up one's courage** sich *(Dat.)* ein Herz fassen

screw: ~**ball** *(Amer. sl.)* **1.** *n.* Spinner, der/Spinnerin, *die (ugs. abwertend)*; **2.** *adj.* spleenig; ~**-cap** *n.* Schraubdeckel, *der;* Schraubverschluß, *der;* ~**driver** *n.* Schraubenzieher, *der*

'screwed-up *adj. (fig. coll.)* neurotisch

screw-top *see* **screw-cap**

screwy ['skru:ɪ] *adj. (sl.: eccentric)* spinnig *(ugs. abwertend)*

scribble ['skrɪbl] **1.** *v. t.* **a)** *(write hastily)* hinkritzeln ‹*Zeilen, Nachricht*›; **b)** *(draw carelessly or meaninglessly)* kritzeln ‹*Skizze, Muster*›. **2.** *v. i.* kritzeln. **3.** *n.* Gekritzel, *das (abwertend)*

scribbler ['skrɪblə(r)] *n. (joc. derog.)* Schreiberling, *der (abwertend); (of poems also)* Dichterling, *der (abwertend)*

scribe [skraɪb] *n.* **a)** *(producer of manuscripts)* Schreiber, *der;* Skriptor, *der; (copyist)* Abschreiber, *der;* Kopist, *der;* **b)** *(Bibl.: theologian)* Schriftgelehrte, *der*

scrimmage ['skrɪmɪdʒ] *n.* Gerangel, *das*

scrimp [skrɪmp] *v. i.* knausern *(ugs.)*; knapsen *(ugs.)*; ~ **and save** *or* **scrape** knapsen und knausern *(ugs.)*

script [skrɪpt] *n.* **a)** *(handwriting)* Handschrift, *die;* **b)** *(of play)* Regiebuch, *das; (of film)*; [Dreh]buch, *das;* Skript, *das (fachspr.)*; **c)** *(for broadcaster)* Skript, *das;* Manuskript, *das;* **d)** *(system of writing)* Schrift, *die*

scripture ['skrɪptʃə(r)] *n.* **a)** *(Relig.: sacred book)* heilige Schrift; **[Holy] S**~**, the [Holy] S**~**s** *(Christian Relig.)* die [Heilige] Schrift; *attrib.* Bibel‹*text, -stunde*›; **b)** *no pl., no art. (Sch.)* Religion, *die*

'script-writer *n. (of film)* Drehbuchautor, *der*/-autorin, *die*

scroll [skrəʊl] **1.** *n.* **a)** *(roll)* Rolle, *die;* **b)** *(Archit.)* Volute, *die;* Schnecke, *die.* **2.** *v. t. (Computing)* verschieben; scrollen *(fachspr.)*

scrounge [skraʊndʒ] *(coll.)* **1.** *v. t.* schnorren *(ugs.)* (**off, from** von); ~ **things** schnorren. **2.** *v. i.* schnorren *(ugs.)* (**from** bei)

scrounger ['skraʊndʒə(r)] *n. (coll.)* Schnorrer, *der*/Schnorrerin, *die (ugs. abwertend)*

¹scrub [skrʌb] **1.** *v. t.,* **-bb-: a)** *(rub)* schrubben *(ugs.)*; scheuern;

b) *(coll.: cancel, scrap)* zurücknehmen ‹*Befehl*›; sausenlassen, schießenlassen *(salopp)* ‹*Plan, Projekt*›; wegschmeißen *(ugs.)* ‹*Brief*›. **2.** *v.i.*, **-bb-** schrubben *(ugs.)*; scheuern. **3.** *n.* **give sth. a ~:** etw. schrubben *(ugs.) od.* scheuern

[2]**scrub** *n.* *(brushwood)* Buschwerk, *das;* Strauchwerk, *das;* Gesträuch, *das;* *(area of brushwood)* Buschland, *das*

scrubber ['skrʌbə(r)] *n.* *(sl.: immoral woman)* Flittchen, *das (ugs. abwertend);* Nutte, *die (abwertend)*

'**scrubbing-brush,** *(Amer.)* '**scrub-brush** *ns.* Scheuerbürste, *die*

scrubby ['skrʌbɪ] *adj.* **a)** *(bristly)* stoppelig ‹*Kinn*›; stachelig, borstig ‹*Bart*›; **b)** *(stunted)* krüppelhaft ‹*Büsche, Sträucher*›

[1]**scruff** [skrʌf] *n.* **by the ~ of the neck** beim *od.* am Genick

[2]**scruff** *n.* *(Brit. coll.)* *(man)* vergammelter Typ *(ugs.);* *(woman, girl)* Schlampe, *die (abwertend)*

scruffy ['skrʌfɪ] *adj.* vergammelt *(ugs.);* heruntergekommen ‹*Gegend, Haus*›

scrum [skrʌm] *n.* *(Rugby)* Gedränge, *das*

scrumptious ['skrʌmʃəs] *adj.* *(coll.)* lecker ‹*Essen*›

scruple ['skru:pl] *n., usu. pl.* Skrupel, *der;* Bedenken, *das;* **a person with no ~s** ein gewissen- *od.* skrupelloser Mensch; **have no ~s about doing sth.** keine Bedenken *od.* Skrupel haben, etw. zu tun

scrupulous ['skru:pjʊləs] *adj.* gewissenhaft ‹*Person*›; unbedingt ‹*Ehrlichkeit*›; peinlich ‹*Sorgfalt*›; **pay ~ attention to sth.** peinlich auf etw. *(Akk.)* achten

scrupulously ['skru:pjʊləslɪ] *adv.* peinlich ‹*sauber, genau*›; **~ honest** auf unbedingte Ehrlichkeit bedacht

scrutinize (scrutinise) ['skru:tɪnaɪz] *v.t.* [genau] untersuchen ‹*Gegenstand, Forschungsgegenstand*›; [über]prüfen ‹*Rechnung, Paß, Fahrkarte*›; mustern ‹*Miene, Person*›

scrutiny ['skru:tɪnɪ] *n.* *(critical gaze)* prüfender Blick; *(close examination)* *(of recruit)* Musterung, *die;* *(of bill, passport)* [Über]prüfung, *die;* **bear ~:** einer [genauen] Prüfung standhalten

scuba ['sku:bə, 'skju:bə] *n.* *(Sport)* Regenerationstauchgerät, *das;* *attrib.* Geräte‹*tauchen*›; [Geräte]tauch‹*ausrüstung*›

scuff [skʌf] **1.** *v.t.* **a)** *(graze)* streifen; **~ one's shoe against sth.** etw. mit dem Schuh streifen; **b)** *(mark by grazing)* verkratzen, verschrammen ‹*Schuhe, Fußboden*›. **2.** *n.* Kratzer, *der;* Kratzspur, *die;* Schramme, *die*

scuffle ['skʌfl] **1.** *n.* Handgreiflichkeiten *Pl.;* Tätlichkeiten *Pl.;* **a ~ broke out** es kam zu Handgreiflichkeiten *od.* Tätlichkeiten. **2.** *v.i.* handgreiflich *od.* tätlich werden (**with** gegen)

scull [skʌl] **1.** *n.* *(oar)* Skull, *das.* **2.** *v.t.* skullen; rudern. **3.** *v.i.* skullen

scullery ['skʌlərɪ] *n.* Spülküche, *die*

sculpt [skʌlpt] *v.t. & i.* *(coll.)* bildhauern *(ugs.)*

sculptor ['skʌlptə(r)] *n.* Bildhauer, *der*/-hauerin, *die*

sculptress ['skʌlptrɪs] *n.* Bildhauerin, *die*

sculpture ['skʌlptʃə(r)] **1.** *n.* **a)** *(art)* Bildhauerei, *die;* **b)** *(piece of work)* Skulptur, *die;* Plastik, *die;* *(pieces collectively)* Skulpturen; Plastiken. **2.** *v.t.* **a)** *(represent)* skulpt[ur]ieren *(geh.);* bildhauerisch darstellen; **b)** *(shape)* formen (**into** zu)

scum [skʌm] *n.* **a)** Schmutzschicht, *die;* *(film)* Schmutzfilm, *der;* **a ring of ~ around the bath** ein Schmutzrand in der Badewanne; **b)** *no pl., no indef. art.* *(fig. derog.)* Abschaum, *der (abwertend);* Auswurf, *der (abwertend);* **the ~ of the earth/of humanity** der Abschaum der Menschheit

scupper ['skʌpə(r)] *v.t.* **a)** *(Brit. coll.)* über den Haufen werfen *(ugs.)* ‹*Plan*›; **we're ~ed if the police arrive** wenn die Polizei kommt, sind wir erledigt; **b)** *(sink)* versenken ‹*Schiff*›

scurrilous ['skʌrɪləs] *adj.* niederträchtig

scurry ['skʌrɪ] *v.i.* huschen; flitzen *(ugs.)*

scurvy ['skɜ:vɪ] *n.* *(Med.)* Skorbut, *der*

[1]**scuttle** ['skʌtl] *n.* *(coal-box)* Kohlenfüller, *der*

[2]**scuttle** *(Naut.)* *v.t.* versenken

[3]**scuttle** *v.i.* *(scurry)* rennen; flitzen *(ugs.);* ‹*Maus, Krabbe:*› huschen; **she ~d off** sie huschte davon

scythe [saɪð] **1.** *n.* Sense, *die.* **2.** *v.t.* [mit der Sense] mähen

SDI *abbr.* **strategic defence initiative** SDI

SDP *abbr.* *(Brit. Polit.)* **Social Democratic Party**

SE [saʊθ'i:st] *abbr.* **south-east** SO

sea [si:] **a)** Meer, *das;* **the ~:** das Meer; die See; **by ~:** mit dem Schiff; **by the ~:** am Meer; an der See; **at ~:** auf See *(Dat.);* **be all at ~** *(fig.)* nicht mehr weiter wissen; **go to ~:** in See stechen; *(become sailor)* zur See gehen *(ugs.);* **put [out] to ~:** in See *(Akk.)* gehen *od.* stechen; auslaufen; **b)** *(specific tract of water)* Meer, *das;* **the seven ~s** *(literary/poet.)* die sieben [Welt]meere; **c)** *(freshwater lake)* See, *der;* **d)** *in sing. or pl.* *(state of ~)* See, *die;* *(wave)* Welle, *die;* Woge, *die (geh.);* See, *die (Seemannsspr.);* **e)** *(fig.: vast quantity)* Meer, *das;* *(of drink)* Strom, *der;* **f)** *attrib.* *(of or on the ~)* See‹*wasser, -schlacht, -karte, -wind*›; Meer‹*ungeheuer, -wasser, -salz usw.*›; Meeres‹*grund, -küste usw.*›; *(in names of marine fauna or flora)* See‹*maus, -anemone, -löwe usw.*›; Meer‹*brasse, -neunauge usw.*›

sea: ~ '**air** *n.* Seeluft, *die;* ~ **a'nemone** *n.* *(Zool.)* Seeanemone, *die;* Seerose, *die;* **~-'bed** *n.* Meeresboden *der;* Meeresgrund, *der;* **~-bird** *n.* Seevogel, *der;* ~ **breeze** *n.* *(Meteorol.)* Seewind, *der;* Seebrise, *die;* **~faring** ['si:feərɪŋ] *adj.* **~faring man** Seemann, *der;* **~faring nation** Seefahrernation, *die;* seefahrende Nation; **~food** *n.* Meeresfrüchte *Pl.;* *attrib.* Fisch‹*restaurant*›; ~ **front** *n.* *unmittelbar am Meer gelegene Straße[n] einer Seestadt;* **a walk along the ~ front** ein Spaziergang am Wasser *od.* auf der Uferpromenade; **~going** *adj.* *(for crossing ~)* seegehend; **~-going yacht** Hochseejacht, *die;* **~-green 1.** [-'-] *n.* Seegrün, *das;* Meergrün, *das;* **2.** ['--] *adj.* seegrün; meergrün; **~-gull** *n.* [See]möwe, *die;* **~-horse** *n.* *(Zool.)* Seepferdchen, *das*

[1]**seal** [si:l] *n.* *(Zool.)* Robbe, *die;* **[common] ~:** [Gemeiner] Seehund

[2]**seal 1.** *n.* **a)** *(piece of wax, lead, etc., stamp, impression)* Siegel, *das;* *(lead ~ also)* Plombe, *die;* *(stamp also)* Siegelstempel, *der;* Petschaft, *das;* *(impression also)* Siegelabdruck, *der;* **b) set the ~ on** *(fig.)* zementieren (+ *Akk.*); **gain the ~ of respectability** sich *(Dat.)* großes Ansehen erwerben; **have the ~ of official approval** offiziell gebilligt werden; **c)** *(to close aperture)* Abdichtung, *die.* **2.** *v.t.* **a)** *(stamp with ~, affix ~ to)* siegeln ‹*Dokument*›; *(fasten with ~)* verplomben, plombieren ‹*Tür,*

Stromzähler⟩; **b)** *(close securely)* abdichten ⟨*Behälter, Rohr usw.*⟩; zukleben ⟨*Umschlag, Paket*⟩; [zum Verschließen der Poren] kurz anbraten ⟨*Fleisch*⟩; **my lips are ~ed** *(fig.)* meine Lippen sind versiegelt; **c)** *(stop up)* verschließen; abdichten ⟨*Leck*⟩; verschmieren ⟨*Riß*⟩; **d)** *(decide)* besiegeln ⟨*Geschäft, Abmachung, jmds. Schicksal*⟩

~ 'in *v.t.* bewahren ⟨*Geschmack*⟩; am Austreten hindern ⟨*Fleischsaft*⟩

~ 'off *v.t.* abriegeln

~ 'up *see* **~ 2b, c**

sealant ['si:lənt] *n.* Dichtungsmaterial, *das*

sea: ~-legs *n.pl.* Seebeine *Pl. (Seemannsspr.)*; **get** *or* **find one's ~-legs** sich *(Dat.)* Seebeine wachsen lassen *(Seemannsspr.)*; **~-level** *n.* Meeresspiegel, *der (fachspr.)*; **200 feet above/below ~-level** 200 Fuß über/unter dem Meeresspiegel *od.* über/unter Meereshöhe *od. (fachspr.)* Normalnull; **at ~-level** auf Meereshöhe *(Dat.)*

'sealing-wax *n.* Siegellack, *der;* Siegelwachs, *das*

'sea-lion *n. (Zool.)* Seelöwe, *der*

seam [si:m] *n.* **a)** *(line of joining)* Naht, *die;* **come apart at the ~s** aus den Nähten gehen; *(fig. coll.: fail)* zusammenbrechen; **burst at the ~s** *(fig.)* aus den *od.* allen Nähten platzen *(ugs.)*; **b)** *(fissure)* Spalt, *der;* Spalte, *die;* **c)** *(Mining)* Flöz, *das; (Geol.: stratum)* Schicht, *die*

seaman ['si:mən] *n., pl.* **seamen** ['si:mən] **a)** *(sailor)* Matrose, *der;* **b)** *(expert in navigation etc.)* Seemann, *der*

seamanship ['si:mənʃɪp] *n., no pl.* seemännisches Geschick; Seemannschaft, *die (fachspr.)*

'sea-mark *n. (Naut.)* Seezeichen, *das*

seamed [si:md] *adj.* **~ stockings** Strümpfe mit Naht

'sea mist *n.* Küstennebel, *der*

seamless ['si:mlɪs] *adj.* nahtlos

seamstress ['semstrɪs] *n.* Näherin, *die*

seamy ['si:mɪ] *adj.* **the ~ side [of life** *etc.*] *(fig.)* die Schattenseite[n] [des Lebens *usw.*]

seance ['seɪɑ̃ns], **séance** ['seɪɑ̃s] *n.* Séance, *die (fachspr.)*; spiritistische Sitzung

sea: ~plane *n.* Wasserflugzeug, *das;* **~port** *n.* Seehafen, *der;* **~ power** *n.* Seemacht, *die*

sear [sɪə(r)] *v.t.* verbrennen; versengen

search [sɜ:tʃ] **1.** *v.t.* durchsuchen (**for** nach); absuchen ⟨*Gebiet, Fläche*⟩ (**for** nach); prüfen *od.* musternd blicken in (+ *Akk.*) ⟨*Gesicht*⟩; *(fig.: probe)* erforschen ⟨*Herz, Gewissen*⟩; suchen in (+ *Dat.*), durchstöbern *(ugs.)* ⟨*Gedächtnis*⟩ (**for** nach); **~ me!** *(coll.)* keine Ahnung! **2.** *v.i.* suchen. **3.** *n.* Suche, *die* (**for** nach); *(of building, room, etc.)* Durchsuchung, *die;* **make a ~ for** suchen nach ⟨*Waffen, Drogen, Diebesgut*⟩; **in ~ of sb./sth.** auf der Suche nach jmdm./etw.

~ for *v.t.* suchen [nach]

~ 'out *v.t.* heraussuchen; aufspüren ⟨*Person mit unbekanntem Aufenthalt*⟩

~ through *v.t.* durchsuchen; durchsehen ⟨*Buch*⟩

searching ['sɜ:tʃɪŋ] *adj.* prüfend, forschend ⟨*Blick*⟩; bohrend ⟨*Frage*⟩; *(thorough)* eingehend ⟨*Untersuchung*⟩

search: ~light *n.* **a)** *(lamp)* Suchscheinwerfer, *der;* **b)** *(beam of light)* Scheinwerferlicht, *das (auch fig.)*; *(fig.)* Rampenlicht, *das;* **~-party** *n.* Suchtrupp, *der;* Suchmannschaft, *die;* **~-warrant** *n. (Law)* Durchsuchungsbefehl, *der*

searing ['sɪərɪŋ] *adj.* sengend ⟨*Hitze*⟩; brennend ⟨*Schmerz*⟩

sea: ~-salt *n.* Meersalz, *das;* Seesalz, *das;* **~scape** ['si:skeɪp] *n. (Art: picture)* Seestück, *das;* Marine, *die;* **S~ Scout** *n. (Brit.)* Seepfadfinder, *der*/-pfadfinderin, *die;* **~ shell** *n.* Muschel[schale], *die;* **~-shore** *n.* *(land near ~)* [Meeres]küste, *die; (beach)* Strand, *der;* **walk along the ~-shore** am Meer/Strand entlanggehen; **~sick** *adj.* seekrank; **~sickness** *n., no pl.* Seekrankheit, *die;* **~side** *n., no pl.* [Meeres]küste, *die;* **by/to/at the ~side** am/ans/am Meer; an der/an die/an der See; *attrib.* **~side town** Seestadt, *die*

season ['si:zn] **1.** *n.* **a)** *(time of the year)* Jahreszeit, *die;* **dry/rainy ~:** Trocken-/Regenzeit, *die;* **b)** *(time of breeding) (for mammals)* Tragezeit, *die; (for birds)* Brutzeit, *die; (time of flourishing)* Blüte[zeit], *die; (time when animal is hunted)* Jagdzeit, *die;* **nesting ~:** Nistzeit, *die;* Brut[zeit], *die; see also* **close season; open season; c)** *(time devoted to specified, social activity)* Saison, *die;* **harvest/opera ~:** Erntezeit, die/Opernsaison, *die;* **football ~:** Fußballsaison, *die;* **holiday** *or (Amer.)* **vacation ~:** Urlaubszeit, *die;* Ferienzeit, *die;* **tourist ~:** Touristensaison, *die;* Reisezeit, *die;* **'the ~'s greetings'** „ein frohes Weihnachtsfest und ein glückliches neues Jahr"; **d) raspberries are in/out of** *or* **not in ~:** jetzt ist die/nicht die Saison *od.* Zeit für Himbeeren; **e)** *(ticket) see* **season-ticket; f)** *(Theatre, Cinemat.)* Spielzeit, *die. See also* **high season; low season; silly 1a. 2.** *v.t.* **a)** *(lit. or fig.)* würzen ⟨*Fleisch, Rede*⟩; **b)** *(mature)* ablagern lassen ⟨*Holz*⟩; **~ed** erfahren ⟨*Wahlkämpfer, Soldat, Reisender*⟩

seasonable ['si:zənəbl] *adj.* der Jahreszeit gemäß

seasonal ['si:zənl] *adj.* Saison⟨*arbeit, -geschäft*⟩; saisonabhängig ⟨*Preise*⟩

seasoning ['si:zənɪŋ] *n.* **a)** *(Cookery)* Gewürze *Pl.*; Würze, *die;* **b)** *(fig.)* Würze, *die*

'season-ticket *n.* Dauerkarte, *die; (for one year/month)* Jahres-/Monatskarte, *die*

seat [si:t] **1. a)** *(thing for sitting on)* Sitzgelegenheit, *die; (in vehicle, cinema, etc.)* Sitz, *der; (of toilet)* [Klosett]brille, *die (ugs.)*; **b)** *(place)* Platz, *der; (in vehicle)* [Sitz]platz, *der;* **have** *or* **take a ~:** sich [hin]setzen; Platz nehmen *(geh.)*; **take one's ~ at table** sich zu Tisch setzen; **c)** *(part of chair)* Sitzfläche, *die;* **d)** *(buttocks)* Gesäß, *das; (of trousers)* Sitz, *der;* Hosenboden, *der;* **by the ~ of one's pants** *(coll. fig.)* nach Gefühl; **e)** *(site)* Sitz, *der; (of disease also)* Herd, *der (Med.)*; *(of learning)* Stätte, *die (geh.)*; *(of trouble)* Quelle, *die;* **~ of the fire** Brandherd, *der;* **f)** *(right to sit in Parliament etc.)* Sitz, *der;* Mandat, *das;* **be elected to a ~ in Parliament** ins Parlament gewählt werden. **2.** *v.t.* **a)** *(cause to sit)* setzen; *(accommodate at table etc.)* unterbringen; *(ask to sit)* ⟨*Platzanweiser:*⟩ einen Platz anweisen (+ *Dat.*); **~ oneself** sich setzen; **b)** *(have ~s for)* Sitzplätze bieten (+ *Dat.*); **~ 500 people** 500 Sitzplätze haben; **the car ~s five comfortably** in dem Auto haben fünf Personen bequem Platz

'seat-belt *n. (Motor Veh., Aeronaut.)* Sicherheitsgurt, *der;* **fasten one's ~:** sich anschnallen; den Gurt anlegen; **wear a ~:** angeschnallt sein; *(during journey)* angeschnallt fahren

seated ['si:tɪd] *adj.* sitzend; **remain ~:** sitzen bleiben; **be ~** *(formal)* Platz nehmen *(geh.)*

-seater ['sɪ:tə(r)] *adj. in comb.* -sitzig; **two-~ [car]** Zweisitzer, *der*

seating ['si:tɪŋ] *n., no pl., no indef. art.* **a)** *(seats)* Sitzplätze; Sitzgelegenheiten; **b)** *attrib.* Sitz⟨*ordnung, -plan*⟩; **the ~ arrangements** die Sitzordnung

sea: ~-urchin *n. (Zool.)* Seeigel, *der;* **~-wall** *n.* Strandmauer, *die; (dike)* Deich, *der*

seaward ['si:wəd] **1.** *adj.* seewärtig ⟨*Kurs, Wind*⟩. **2.** *adv.* seewärts; nach See zu

sea: ~-water *n.* Meerwasser, *das;* Seewasser, *das;* **~weed** *n.* [See]tang, *der;* **~worthy** *adj.* seetüchtig

sec [sek] *(coll.) see* **¹second 2 b**

Sec. *abbr.* **Secretary** Sekr.

sec. *abbr.* **second[s]** Sek.

secateurs [sekə'tɜ:z, 'sekətɜ:z] *n. pl. (Brit.)* Gartenschere, *die;* Rosenschere, *die*

secede [sɪ'si:d] *v.i. (Polit./Eccl./formal)* sich abspalten (**from** von)

secession [sɪ'seʃn] *n. (Polit./Eccl./formal)* Abspaltung, *die*

secluded [sɪ'klu:dɪd] *adj.* **a)** *(hidden from view)* versteckt; *(somewhat isolated)* abgelegen; **b)** *(solitary)* zurückgezogen ⟨*Leben*⟩

seclusion [sɪ'klu:ʒn] *n., no pl.* **a)** *(keeping from company)* Absonderung, *die; (being kept from company)* Abgesondertheit, *die;* **b)** *(privacy of life)* Zurückgezogenheit, *die;* **c)** *(remoteness)* Abgelegenheit, *die*

¹second ['sekənd] **1.** *adj.* zweit...; zweitwichtigst... ⟨*Stadt, Hafen usw.*⟩; **~ largest/highest** *etc.* zweitgrößt.../-höchst... *usw.*; **every ~ week** jede zweite Woche; **~ to none** unübertroffen. **2.** *n.* **a)** *(unit of time or angle)* Sekunde, *die;* **b)** *(coll.: moment)* Sekunde, *die (ugs.);* **wait a few ~s** einen Moment warten; **in a ~** *(immediately)* sofort *(ugs.); (very quickly)* im Nu *(ugs.);* **just a ~!** *(coll.)* einen Moment!; **c)** *(additional person or thing)* **a ~:** noch einer/eine/eins; **d) the ~** *(in sequence)* der/die/das zweite; *(in rank)* der/die/das Zweite; **be the ~ to arrive** als zweiter/zweite ankommen; **e)** *(in duel, boxing)* Sekundant, *der*/Sekundantin, *die;* **f)** *in pl. (helping of food)* zweite Portion; *(~ course)* zweiter Gang; **g)** *(day)* **the ~ of May** der zweite Mai; **the ~ [of the month]** der Zweite [des Monats]; **h)** *in pl. (goods of ~ quality)* Waren zweiter Wahl; **i)** *(Brit. Univ.)* ≈ Gut, *das;* ≈ Zwei, *die.* **3.** *v.t. (support)* unterstützen ⟨*Antrag, Nominierung*⟩; **I'll ~ that!** *(coll.)* dem schließe ich mich an!

²second [sɪ'kɒnd] *v.t. (transfer)* vorübergehend versetzen

secondary ['sekəndərɪ] *adj.* **a)** *(of less importance)* zweitrangig; sekundär *(geh.);* Neben⟨*akzent, -sache*⟩; *(derived from sth. primary)* weiterverarbeitend ⟨*Industrie*⟩; **~ literature** Sekundärliteratur, *die;* **be ~ to sth.** einer Sache *(Dat.)* untergeordnet sein; **b)** *(indirectly caused)* sekundär *(geh., Med., Biol.)*

secondary: ~ education *n.* höhere Schule; *(result)* höhere Schulbildung; **~ 'modern [school]** *n. (Brit. Hist.)* ≈ Mittelschule, *die (veralt.);* Realschule, *die;* **~ school** *n.* höhere *od.* weiterführende Schule

second: ~-best 1. ['---] *adj.* zweitbest...; **2.** [--'-] *adv.* **come off ~-best** den kürzeren ziehen *(ugs.);* **3.** [--'-] *n., no pl.* Zweitbeste, *der/die/das;* **~ 'childhood** *see* **childhood; ~ 'class** *n.* **a)** zweite Kategorie; **b)** *(Transport, Post)* zweite Klasse; **c)** *(Brit. Univ.)* ≈ Gut, *das;* Zwei, *die;* **~-class 1.** ['---] *adj.* **a)** *(of lower class)* zweiter Klasse *nachgestellt;* Zweiter-Klasse-⟨*Fahrkarte, Abteil, Passagier, Post, Brief usw.*⟩; **~-class stamp** *Briefmarke für einen Zweiter-Klasse-Brief;* **b)** *(of inferior class)* zweitklassig *(abwertend);* **~-class citizen** Bürger zweiter Klasse; **2.** [--'-] *adv.* zweiter Klasse ⟨*fahren*⟩; **send a letter ~-class** einen Brief mit Zweiter-Klasse-Post schicken; **~ 'cousin** *see* **cousin**

seconder ['sekəndə(r)] *n.* Befürworter, *der*/-worterin, *die*

second: ~ gear *n., no pl. (Motor Veh.)* zweiter Gang; *see also* **gear 1 a; ~ hand** *n. (Horol.)* Sekundenzeiger, *der;* **~-hand 1.** ['---] *adj.* **a)** gebraucht ⟨*Kleidung, Auto usw.*⟩; antiquarisch ⟨*Buch*⟩; **~-hand car** Gebrauchtwagen, *der;* **b)** *(selling used goods)* Gebrauchtwaren-; Secondhand⟨*laden*⟩; **c)** *(obtained from sb. else)* ⟨*Nachrichten, Bericht*⟩ aus zweiter Hand; **2.** [--'-] *adv.* aus zweiter Hand *(auch fig.);* **~ 'home** *n.* Zweitwohnung, *die; (holiday house)* Ferienhaus, *das*

secondly ['sekəndlɪ] *adv.* zweitens

secondment [sɪ'kɒndmənt] *n. (Brit.)* **a)** *(of official)* vorübergehende Versetzung; **b)** *(Mil.)* Abstellung, *die*

second: ~ name *n.* Nachname, *der;* Zuname, *der;* **~ 'nature** *n., no pl., no art. (coll.)* zweite Natur; **become/be ~ nature to sb.** jmdm. zur zweiten Natur werden/geworden sein; jmdm. in Fleisch und Blut *(Akk.)* übergehen/übergegangen sein; **~-'rate** *adj.* zweitklassig; **~s hand** *see* **~ hand; ~ sight** *see* **sight 1 a; ~ 'thoughts** *n. pl.* **have ~ thoughts** es sich *(Dat.)* anders überlegen (**about** mit); **we've had ~ thoughts about buying the house** wir wollen das Haus nun doch nicht kaufen; **but on ~ thoughts I think I will** wenn ich mir's [noch mal] überlege, werde ich es doch tun

secrecy ['si:krɪsɪ] *n.* **a)** *(keeping of secret)* Geheimhaltung, *die;* **with great ~:** in aller Heimlichkeit *od.* ganz im geheimen; **b)** *(secretiveness)* Heimlichtuerei, *die (abwertend);* **c)** *(unrevealed state)* Heimlichkeit, *die;* **in ~:** im geheimen

secret ['si:krɪt] **1.** *adj.* **a)** *(kept private)* geheim; Geheim⟨*fach, -tür, -abkommen, -kode*⟩; **keep sth. ~:** etw. geheimhalten (**from** vor + *Dat.*); **b)** *(acting in ~)* heimlich ⟨*Trinker, Liebhaber, Bewunderer*⟩. **2.** *n.* **a)** Geheimnis, *das;* **make no ~ of sth.** kein Geheimnis aus etw. machen; *(not conceal feelings, opinion)* kein[en] Hehl aus etw. machen; **keep ~s/a ~:** schweigen *(fig.);* den Mund halten *(ugs.);* **can you keep a ~?** kannst du schweigen?; **keep ~s from sb.** Geheimnisse vor jmdm. haben; **be in the ~:** eingeweiht sein; **open ~:** offenes Geheimnis; **b) in ~:** im geheimen; heimlich

secret 'agent *n.* Geheimagent, *der*/-agentin, *die*

secretarial [sekrə'teərɪəl] *adj.* Sekretariats⟨*personal*⟩; Sekretärinnen⟨*kursus, -tätigkeit*⟩; ⟨*Arbeit*⟩ als Sekretärin

secretariat [sekrə'teərɪət] *n.* Sekretariat, *das*

secretary ['sekrətərɪ] *n.* Sekretär, *der*/Sekretärin, *die; (of company)* Schriftführer, *der*/-führerin, *die*

Secretary: ~-'General *n., pl.* **Secretaries-General** Generalsekretär, *der*/-sekretärin, *die;* **~ of 'State** *n.* **a)** *(Brit. Polit.)* Minister, *der*/Ministerin, *die;* **b)** *(Amer. Polit.)* Außenminister, *der*/-ministerin, *die*

secret 'ballot *n.* geheime Abstimmung

secrete [sɪ'kri:t] *v.t.* **a)** *(Physiol.)* absondern; **b)** *(formal/literary: hide)* verbergen

secretion [sɪ'kri:ʃn] *n.* **a)** *(Phys-*

iol.) Absonderung, *die; (substance also)* Sekret, *das (fachspr.);* **b)** *(formal/literary: concealing)* Verbergen, *das*

secretive ['si:krɪtɪv] *adj.* verschlossen ‹*Person*›; geheimnisvoll ‹*Lächeln*›; **be ~:** heimlich tun *(abwertend),* geheimnisvoll tun **(about** mit); **she was being very ~ about something** sie versuchte, irgend etwas zu verheimlichen

secretly ['si:krɪtlɪ] *adv.* heimlich; insgeheim ‹*etw. glauben*›

Secret: ~ Po'lice *n.* Geheimpolizei, *die;* **~ 'Service** *n.* Geheimdienst, *der;* **s~ so'ciety** *n.* Geheimbund, *der*

sect [sekt] *n.* Sekte, *die*

sectarian [sek'teərɪən] *adj.* konfessionell; konfessionell motiviert ‹*Handlungen*›; konfessionell ausgerichtet ‹*Erziehung*›; Konfessions‹*krieg, -streit*›

section ['sekʃn] *n.* **a)** *(part cut off)* Abschnitt, *der;* Stück, *das; (part of divided whole)* Teil, *der; (of railway track)* Teilstück, *das;* [Strecken]abschnitt, *der;* **b)** *(of firm)* Abteilung, *die; (of organization etc.)* Sektion, *die; (of orchestra or band)* Gruppe, *die;* **c)** *(component part)* [Einzel]teil, *das;* [Bau]element, *das;* **d)** *(of chapter, book)* Abschnitt, *der; (of statute, act)* Paragraph, *der;* **e)** *(part of community)* Gruppe, *die*

sectional ['sekʃənl] *adj.* Gruppen‹*interessen*›; partikular ‹*Interessen*›; ‹*Auseinandersetzung*› zwischen den Bevölkerungsgruppen

sector ['sektə(r)] *n.* **a)** Sektor, *der;* **the leisure/industrial ~:** der Freizeitsektor/der Bereich der Industrie; **b)** *(Mil.: area)* Kampfabschnitt, *der;* Gefechtsabschnitt, *der*

secular ['sekjʊlə(r)] *adj.* säkular *(geh.);* weltlich ‹*Angelegenheit, Schule, Musik, Gericht*›

secure [sɪ'kjʊə(r)] **1.** *adj.* **a)** *(safe)* sicher; **~ against burglars/fire** gegen Einbruch/Feuer geschützt; einbruch-/feuersicher; **make sth. ~ from attack/enemies** etw. gegen Angriffe/Feinde sichern; **b)** *(firmly fastened)* fest; **be ~** ‹*Ladung:*› gesichert sein; ‹*Riegel, Tür:*› fest zu sein; ‹*Tür:*› ver- *od.* zugeriegelt sein; ‹*Schraube:*› fest sein *od.* sitzen; **make sth. ~:** etw. sichern; **c)** *(untroubled)* sicher, gesichert ‹*Existenz*›; **feel ~:** sich sicher *od.* geborgen fühlen; **~ in the knowledge that ...:** in dem sicheren Bewußtsein, daß ... **2.** *v. t.* **a)** *(obtain)* sichern **(for** *Dat.*); beschaffen ‹*Auftrag*› **(for** *Dat.*); *(for oneself)* sich *(Dat.)* sichern; **b)** *(confine)* fesseln ‹*Gefangenen*›; *(in container)* einschließen ‹*Wertsachen*›; *(fasten firmly)* sichern, fest zumachen ‹*Fenster, Tür*›; festmachen ‹*Boot*› **(to** an + *Dat.*); **c)** *(guarantee)* absichern ‹*Darlehen*›

securely [sɪ'kjʊəlɪ] *adv.* **a)** *(firmly)* fest ‹*verriegeln, zumachen*›; sicher ‹*befestigen*›; **b)** *(safely)* sicher ‹*untergebracht sein*›

security [sɪ'kjʊərɪtɪ] *n.* **a)** *(safety)* Sicherheit, *die;* **~ [measures]** Sicherheitsmaßnahmen; **national ~:** nationale Sicherheit; **b)** *(thing that guarantees)* Sicherheit, *die;* Gewähr, *die;* **c)** *usu. in pl. (Finance)* Wertpapier, *das;* **securities** Wertpapiere; **d) emotional ~:** emotionale Sicherheit; **he needs the ~ of a good home** er braucht die Geborgenheit eines guten Zuhauses

security: ~ check *n.* Sicherheitskontrolle, *die;* **S~ Council** *n. (Polit.)* Sicherheitsrat, *der;* **~ forces** *n. pl.* Sicherheitskräfte *Pl.;* **~ guard** *n.* Wächter, *der*/Wächterin, *die;* **~ officer** *n.* Sicherheitsbeauftragte, *der/die*

sedan [sɪ'dæn] *n.* **a)** *(Hist.: chair)* Sänfte, *die;* **b)** *(Amer. Motor Veh.)* Limousine, *die*

se'dan-chair *see* **sedan a**

sedate [sɪ'deɪt] **1.** *adj.* bedächtig; gesetzt ‹*alte Dame*›; ruhig ‹*Kind*›; gemächlich ‹*Tempo, Leben, Auto*›. **2.** *v. t. (Med.)* sedieren *(fachspr.);* ruhigstellen

sedation [sɪ'deɪʃn] *n. (Med.)* Sedation, *die (fachspr.);* Ruhigstellung, *die*

sedative ['sedətɪv] **1.** *n. (Med.)* Sedativum, *das (fachspr.);* Beruhigungsmittel, *das.* **2.** *adj.* **a)** *(Med.)* sedativ *(fachspr.);* **b)** *(fig.: calming)* beruhigend ‹*Wirkung*›

sedentary ['sedəntərɪ] *adj.* sitzend ‹*Haltung, Lebensweise, Tätigkeit*›; **lead a ~ life** eine sitzende Lebensweise haben

sediment ['sedɪmənt] *n.* **a)** *(matter)* Ablagerung, *die;* Ablagerungen *Pl.;* **b)** *(lees)* Bodensatz, *der; (of wine also)* Depot, *das (fachspr.)*

sedition [sɪ'dɪʃn] *n.* Aufruhr, *der*

seditious [sɪ'dɪʃəs] *adj.* aufrührerisch; staatsgefährdend ‹*Delikt*›

seduce [sɪ'dju:s] *v. t.* **a)** *(sexually)* verführen; **b)** *(lead astray)* verführen; *(distract)* ablenken **(away from** von); **~ sb. into doing sth.** jmdn. dazu verführen *od.* verleiten, etw. zu tun

seducer [sɪ'dju:sə(r)] *n.* Verführer, *der*

seduction [sɪ'dʌkʃn] *n.* **a)** *(sexual)* Verführung, *die;* **b)** *(leading astray)* Verführung, *die* **(into** zu)

seductive [sɪ'dʌktɪv] *adj.* verführerisch; verlockend ‹*Angebot*›

see [si:] **1.** *v. t.,* **saw** [sɔ:], **seen** [si:n] **a)** sehen; **let sb. ~ sth.** *(show)* jmdm. etw. zeigen; **let me ~:** laß mich mal sehen; **I saw her fall** *or* **falling** ich habe sie fallen sehen; **he was ~n to leave** *or* **~n leaving the building** er ist beim Verlassen des Gebäudes gesehen worden; **I'll believe it when I ~ it** das will ich erst mal sehen; **they saw it happen** sie haben gesehen, wie es passiert ist; **can you ~ that house over there?** siehst du das Haus da drüben?; **be worth ~ing** sehenswert sein; sich lohnen *(ugs.);* **~ the light** *(fig.: undergo conversion)* das Licht schauen *(geh.);* **I saw the light** *(I realized my error etc.)* mir ging ein Licht auf *(ugs.);* **'~ things** Halluzinationen haben; **I must be ~ing things** *(joc.)* ich glaub', ich seh' nicht richtig; **~ the sights/town** sich *(Dat.)* die Sehenswürdigkeiten/Stadt ansehen; **~ one's way [clear] to do** *or* **to doing sth.** es einrichten, etw. zu tun; **b)** *(watch)* sehen; **let's ~ a film** sehen wir uns *(Dat.)* einen Film an!; **c)** *(meet [with])* sehen; treffen; *(meet socially)* zusammenkommen mit; sich treffen mit; **I'll ~ you there/at 5** wir sehen uns dort/um 5; **~ you!** *(coll.),* **[I'll] be ~ing you!** *(coll.)* bis bald! *(ugs.);* **~ you on Saturday/soon** bis Samstag/bald; *see also* **[1]long 1 c; d)** *(speak to)* sprechen ‹*Person*› **(about** wegen); *(pay visit to)* gehen zu, *(geh.)* aufsuchen ‹*Arzt, Anwalt usw.*›; *(receive)* empfangen; **the doctor will ~ you now** Herr/Frau Doktor läßt bitten; **whom would you like to ~?** wen möchten Sie sprechen?; zu wem möchten Sie?; **e)** *(discern mentally)* sehen; **I ~ it all!** jetzt ist mir alles klar; **I can ~ it's difficult for you** ich verstehe, daß es nicht leicht für dich ist; **I ~ what you mean** ich verstehe [was du meinst]; **~ what I mean?** siehst du?; **I saw that it was a mistake** mir war klar, daß es ein Fehler war; **I don't ~ the point of it** ich sehe keinen Sinn darin; **he didn't ~ the joke** er fand es [gar] nicht lustig; *(did not understand)* er hat den Witz nicht verstanden; **I can't think what she ~s in him** ich weiß nicht, was sie an ihm findet; **f)** *(consider)* sehen; **let me ~ what I can do** [ich will] mal sehen, was ich tun kann; **g)** *(foresee)* sehen; **I can ~ I'm going to be busy** ich sehe [es] schon [kommen], daß ich beschäftigt sein werde; **I can ~ it won't be easy** ich sehe schon, daß es nicht einfach sein wird; **h)** *(find out)* feststellen; *(by looking)* nachsehen; **that remains to be ~n** das wird man sehen; **~ if you can read this** guck mal, ob du das hier lesen kannst *(ugs.)*; **i)** *(take view of)* sehen; betrachten; **~ things as sb. does** jmds. Ansichten teilen; **try to ~ it my way** versuche es doch mal aus meiner Sicht zu sehen; **as I ~ it** meines Erachtens; **j)** *(learn)* sehen; **I ~ from your letter that ...:** ich entnehme Ihrem Brief, daß ...; **k)** *(make sure)* **~ [that] ...:** zusehen *od.* darauf achten, daß ...; **l)** *usu. in imper. (look at)* einsehen ‹*Buch*›; **~ below/p. 15** siehe unten/S. 15; **m)** *(experience, be witness of)* erleben; **live to ~ sth.** etw. miterleben; **now I've ~n everything!** *(iron.)* hat man so etwas schon erlebt *od.* gesehen!; **we shall ~:** wir werden [ja/schon] sehen; **he will not** *or* **never ~ 50 again** er ist [bestimmt] über 50; **n)** *(imagine)* sich *(Dat.)* vorstellen; **~ sb./oneself doing sth.** sich vorstellen, daß jmd./man etw. tut; **I can ~ it now – ...:** ich sehe es schon bildhaft vor mir – ...; **o)** *(contemplate)* mit ansehen; zusehen bei; **[stand by and] ~ sb. doing sth.** [tatenlos] zusehen *od.* es [tatenlos] mit ansehen, wie jmd. etw. tut; **p)** *(escort)* begleiten, bringen **(to** [bis] zu); **q)** *(consent willingly to)* einsehen; **not ~ oneself doing sth.** es nicht einsehen, daß man etw. tut. **2.** *v. i.*, **saw, seen a)** *(discern objects)* sehen; **~ for yourself!** sieh doch selbst!; **as far as the eye can ~:** soweit das Auge reicht; **~ red** rotsehen *(ugs.)*; **b)** *(make sure)* nachsehen; **c)** *(reflect)* überlegen; **let me ~:** laß mich überlegen; warte mal ['n Moment] *(ugs.)*; **d) I ~:** ich verstehe; aha *(ugs.)*; ach so *(ugs.)*; **you ~:** weißt du/wißt ihr/wissen Sie; **there you are, you ~!** Siehst du? Ich hab's doch gesagt!; **as far as I can ~:** soweit ich das *od.* es beurteilen kann

~ about *v. t.* sich kümmern um; **I've come to ~ about the room/cooker** ich komme wegen des Zimmers/des Herdes

~ into *v. t.* **a)** *(gain view into)* [hinein]sehen in (+ *Akk.*); [rein]gucken *(ugs.)* in (+ *Akk.*); **b)** *(fig.: investigate)* nachgehen, auf den Grund gehen (+ *Dat.*) ‹*Angelegenheit, Klage*›

~ 'off *v. t.* **a)** *(say farewell to)* verabschieden; **b)** *(chase away)* vertreiben

~ 'out 1. *v. i.* hinaussehen; rausgucken *(ugs.)*. **2.** *v. t.* **a)** *(remain till end of)* ‹*Zuschauer:*› sich *(Dat.)* zu Ende ansehen ‹*Spiel*›; ableisten ‹*Amtsperiode*›; ‹*Patient:*› überleben ‹*Zeitraum*›; **b)** *(escort from premises)* hinausbegleiten **(of** aus); **~ oneself out** allein hinausfinden

~ over, ~ round *v. t.* besichtigen

~ through *v. t.* **a)** ['--] hindurchsehen durch; durchgucken *(ugs.)* durch; *(fig.)* durchschauen; **b)** [-'-] *(not abandon)* zu Ende *od.* zum Abschluß bringen; **c)** [-'-] *(be sufficient for)* **~ sb. through** jmdm. reichen

~ to *v. t.* sich kümmern um; **I'll ~ to that** dafür werde ich sorgen; **~ to it that ...:** dafür sorgen, daß ...

seed [si:d] **1.** *n.* **a)** *(grain)* Samen, *der;* Samenkorn, *das; (of grape etc.)* Kern, *der; (for birds)* Korn, *das;* **b)** *no pl., no indef. art. (~s collectively)* Samen[körner] *Pl.*; *(for sowing)* Saatgut, *das;* Saat, *die;* **grass-~:** Grassamen *Pl.*; **go** *or* **run to ~:** Samen bilden; ‹*Salat:*› [in Samen] schießen; *(fig.)* herunterkommen *(ugs.)*; **c)** *(fig.: beginning)* Saat, *die;* **sow [the] ~s of doubt/discord** Zweifel aufkommen lassen/Zwietracht säen; **d)** *(Sport coll.)* gesetzter Spieler/gesetzte Spielerin. **2.** *v. t.* **a)** *(place ~s in)* besäen; **b)** *(Sport)* setzen ‹*Spieler*›; **be ~ed number one** als Nummer eins gesetzt werden/sein; **c)** *(lit. or fig.: sprinkle [as] with ~)* besäen. **3.** *v. i. (produce ~s)* Samen bilden

seed: ~-bed *n.* **a)** *(Hort.)* [Saat]beet, *das;* **b)** *(fig.: place of development)* Grundlage, *die;* **~-cake** *n.* Kümmelkuchen, *der;* **~-corn** *n.* Saatgetreide, *das;* Saatkorn, *das*

seedless ['si:dlɪs] *adj.* kernlos

seedling ['si:dlɪŋ] *n.* Sämling, *der*

seedy ['si:dɪ] *adj.* **a)** *(coll.: unwell)* **feel ~:** sich [leicht] angeschlagen fühlen; **b)** *(shabby)* schäbig, *(ugs. abwertend)* vergammelt ‹*Aussehen, Kleidung*›; heruntergekommen ‹*Stadtteil*›; **c)** *(disreputable)* zweifelhaft ‹*Person*›

seeing ['si:ɪŋ] **1.** *conj.* **~ [that] ...:** da ...; wo ... *(ugs.)*. **2.** *n.* **~ is believing** so was glaubt man erst, wenn man es gesehen hat

seek [si:k] *v. t.*, **sought** [sɔ:t] **a)** suchen; anstreben ‹*Posten, Amt*›; sich bemühen um ‹*Anerkennung, Freundschaft, Interview, Einstellung*›; *(try to reach)* aufsuchen; **b)** *(literary/formal: attempt)* suchen *(geh.)*; versuchen; **~ to do sth.** suchen, etw. zu tun *(geh.)*

~ after *v. t.* suchen nach; **be much sought after** sehr gesucht sein

~ for *v. t.* suchen nach

~ 'out *v. t.* ausfindig machen ‹*Sache, Ort*›; aufsuchen, kommen zu ‹*Personen*›

seem [si:m] *v. i.* **a)** *(appear [to be])* scheinen; **you ~ tired** du wirkst müde; **she ~s nice** sie scheint nett zu sein; **it's not quite what it ~s** es ist nicht ganz das, was es [zunächst] zu sein scheint; **it ~s like only yesterday** es ist, als wäre es erst gestern gewesen; **he ~s certain to win** es sieht ganz so aus, als würde er gewinnen; **what ~s to be the trouble?** wo fehlt's denn? **I ~ to recall having seen him before** ich glaube mich zu erinnern, ihn schon einmal gesehen zu haben; **it ~s [that] ...:** anscheinend ...; **it would ~ to be ...:** es scheint ja wohl ... zu sein; **so it ~s** *or* **would ~:** so will es scheinen; **b) sb. can't ~ to do sth.** *(coll.)* jmd. scheint etw. nicht tun zu können; **she doesn't ~ to notice such things** *(coll.)* so was merkt sie irgendwie nicht *(ugs.)*

seeming ['si:mɪŋ] *adj.* scheinbar

seemingly ['si:mɪŋlɪ] *adv.* **a)** *(evidently)* offensichtlich; **b)** *(to outward appearance)* scheinbar

seemly ['si:mlɪ] *adj.* anständig; **it isn't ~ to praise oneself** es gehört sich nicht, sich selbst zu loben

seen *see* **see**

seep [si:p] *v. i.* **~ [away]** [ab]sickern; **~ in through sth.** durch etw. hineinsickern; **~ out of sth.** aus etw. heraussickern

seer [sɪə(r)] *n.* *(prophet)* Seher, *der*/Seherin, *die*

'see-saw 1. *n.* **a)** *(plank)* Wippe, *die;* **b)** *no art. (game)* Wippen, *das;* **c)** *(fig.: contest)* Auf und Ab, *das.* **2.** *v. i.* ‹*Weg, Straße:*› auf und ab führen; ‹*Deck:*› [auf und ab] schaukeln

seethe [si:ð] *v. i.* **a)** *(surge)* ‹*Wellen, Meer:*› branden; ‹*Straßen usw.:*› wimmeln **(with** von); *(bubble or foam as if boiling)* schäumen; **b)** *(fig.: be agitated)* **~ [with anger/inwardly]** vor Wut/innerlich schäumen

'see-through *adj.* durchsichtig

segment ['segmənt] *n. (of orange, pineapple)* Scheibe, *die; (of cake, pear)* Stück, *das; (of worm, skull, limb)* Segment, *das; (of economy, market)* Bereich, *der*

segregate ['segrɪgeɪt] *v. t.* **a)** trennen; isolieren ‹*Kranke*›; aussondern ‹*Forschungsgebiet*›; **b)** *(racially)* segregieren *(geh.)*; absondern

segregation [segrɪ'geɪʃn] *n., no pl.* **a)** Trennung, *die;* **b)** **[racial] ~:** Rassentrennung, *die*

seismic ['saɪzmɪk] *adj.* seismisch

seize [si:z] *v. t.* **a)** ergreifen; **~ power** die Macht ergreifen; **~ sb. by the arm/collar/shoulder** jmdn. am Arm/Kragen/an der Schulter packen; **~ the opportunity** *or* **occasion/moment [to do sth.]** die Gelegenheit ergreifen/den günstigen Augenblick nutzen [und etw. tun]; **~ any/a** *or* **the chance [to do sth.]** jede/die Gelegenheit nutzen[, um etw. zu tun]; **be ~d with remorse/panic** von Gewissensbissen geplagt/von Panik ergriffen werden; **b)** *(capture)* gefangennehmen ‹*Person*›; kapern ‹*Schiff*›; mit Gewalt übernehmen ‹*Flugzeug, Gebäude*›; einnehmen ‹*Festung, Brücke*›; **c)** *(confiscate)* beschlagnahmen

~ on *v. t.* sich *(Dat.)* vornehmen ‹*Einzelheit, Aspekt, Schwachpunkt*›; aufgreifen ‹*Idee, Vorschlag*›; ergreifen ‹*Chance*›

~ 'up *v. i.* sich festfressen; ‹*Verkehr:*› zusammenbrechen, zum Erliegen kommen

~ upon *see* **~ on**

seizure ['si:ʒə(r)] *n.* **a)** *(capturing)* Gefangennahme, *die; (of ship)* Kapern, *das; (of aircraft, building)* Übernahme, *die; (of fortress, bridge)* Einnahme, *die;* **~ of power** Machtergreifung, *die;* **b)** *(confiscation)* Beschlagnahme, *die;* **c)** *(Med.: attack)* Anfall, *der*

seldom ['seldəm] *adv.* selten; **~ or never** so gut wie nie; **~, if ever** fast nie; äußerst selten

select [sɪ'lekt] **1.** *adj.* **a)** *(carefully chosen)* ausgewählt; **b)** *(exclusive)* exklusiv. **2.** *v. t.* auswählen; **~ one's own apples** sich *(Dat.)* die Äpfel selbst aussuchen

select com'mittee *n.* Sonderkommission, *die*

selection [sɪ'lekʃn] *n.* **a)** *(what is selected [from])* Auswahl, *die* (**of** an + *Dat.*, **from** aus); *(person)* Wahl, *die;* **a ~ from ...** *(Mus.)* eine Auswahl aus ...; **make a ~** *(one)* eine Wahl treffen; *(several)* eine Auswahl treffen; **~s from the best writers** ausgewählte Werke der besten Schriftsteller; **b)** *(act of choosing)* [Aus]wahl, *die;* **~ committee** Auswahlkomitee, *das;* **c)** *(being chosen)* Wahl, *die;* **his ~ as president** seine Wahl zum Präsidenten

selective [sɪ'lektɪv] *adj. (using selection)* selektiv; *(careful in one's choice)* wählerisch

selectively [sɪ'lektɪvlɪ] *adv.* selektiv; **shop ~:** gezielt einkaufen

selector [sɪ'lektə(r)] *n.* **a)** *(person who selects team)* Mannschaftsaufsteller, *der*/-aufstellerin, *die;* **b)** *(knob)* Schaltknopf, *der; (switch)* Wahlschalter, *der*

self [self] *n., pl.* **selves** [selvz] *(person's essence)* Selbst, *das (geh.);* Ich, *das;* **be one's usual ~:** man selbst sein; **be back to one's former** *or* **old ~ [again]** wieder der/die alte sein; **one's better ~:** sein besseres Ich

self- *in comb.* selbst-/Selbst-

self: ~-ad'dressed *adj.* **a ~-addressed envelope** ein adressierter Rückumschlag; **~-ad'hesive** *adj.* selbstklebend; **~-ap'pointed** *adj.* selbsternannt; **~-as'surance** *n., no pl.* Selbstbewußtsein, *das;* Selbstsicherheit, *die;* **~-as'sured** *adj.* selbstsicher; selbstbewußt; **~-'catering 1.** *adj.* mit Selbstversorgung *nachgestellt;* **2.** *n.* Selbstversorgung, *die;* **~-con'fessed** *adj.* erklärt; **~-'confidence** *n., no pl.* Selbstvertrauen, *das;* **~-'confident** *adj.*, **~-'confidently** *adv.* selbstsicher; **~-'conscious** *adj.* **a)** *(ill at ease)* unsicher; **b)** *(deliberate)* reflektiert ‹*Prosa, Stil*›; **~-'consciousness** *n.* **a)** Unsicherheit, *die;* **b)** *(deliberateness)* Reflektiertheit, *die;* **~-con'tained** *adj.* **a)** *(not dependent)* selbstgenügsam; **b)** *(Brit.: complete in itself)* abgeschlossen ‹*Wohnung*›; **~-contra'dictory** *adj.* mit sich selbst in Widerspruch; **~-con'trol** *n., no pl.* Selbstbeherrschung, *die;* **~-con'trolled** *adj.* voller Selbstbeherrschung *nachgestellt;* **~-'critical** *adj.* selbstkritisch; **~-de'ception** *n.* Selbsttäuschung, *die;* **~-de'feating** *adj.* unsinnig; zwecklos; **~-de'fence** *n., no pl., no indef. art.* Notwehr, *die;* Selbstverteidigung, *die;* **in ~-defence** aus Notwehr; **~-de'structive** *adj.* selbstzerstörerisch; **~-'discipline** *n., no pl.* Selbstdisziplin, *die;* **~-drive** *adj.* **~-drive hire [company]** Autovermietung, *die;* **~-drive vehicle** Mietwagen, *der;* **~-ef'facing** *adj.* zurückhaltend; **~-em'ployed** *adj.* selbständig; **~-employed man/woman** Selbständige, *der/die;* **~-e'steem** *n.* Selbstachtung, *die;* **~-'evident** *adj.*, **~-'evidently** *adv.* offenkundig; **~-ex'planatory** *adj.* ohne weiteres verständlich; **be ~-explanatory** für sich selbst sprechen; **~-ex'pression** *n., no pl., no indef. art.* Selbstdarstellung, *die;* **~-'governing** *adj.* selbstverwaltet; **~-'help** *n., no pl.* Selbsthilfe, *die;* **~-im'portance** *n., no pl.* Selbstgefälligkeit, *die; (arrogant and pompous bearing)* Selbstherrlichkeit, *die;* **~-im'portant** *adj.* selbstgefällig; *(arrogant and pompous)* selbstherrlich; **~-im'posed** *adj.* selbstauferlegt; **~-in'dulgence** *n.* Maßlosigkeit, *die;* **~-in'dulgent** *adj.* maßlos; **~-in'flicted** *adj.* selbst beigebracht ‹*Wunde*›; selbst auferlegt ‹*Strafe*›; **~-'interest** *n.* Eigeninteresse, *das*

selfish ['selfɪʃ] *adj.* egoistisch; selbstsüchtig

selfishness ['selfɪʃnɪs] *n., no pl.* Egoismus, *der;* Selbstsucht, *die*

selfless ['selflɪs] *adj.* selbstlos

self: ~-'loading *adj.* mit Selbstladevorrichtung *nachgestellt;* **~-'locking** *adj.* selbstschließend; **~-made** *adj.* selbstgemacht; **a ~-made man** ein Selfmademan; **she is a ~-made woman** sie hat sich aus eigener Kraft hochgearbeitet; **~-o'pinionated** *adj.* **a)** *(conceited)* eingebildet; von sich eingenommen; **b)** *(obstinate)* starrköpfig; rechthaberisch; **~-'pity** *n., no pl.* Selbstmitleid, *das;* **~-'portrait** *n.* Selbstporträt, *das;* **~-pos'sessed** *adj.* selbstbeherrscht; **~-preser'vation** *n., no pl., no indef. art.* Selbsterhaltung, *die;* **~-'raising flour** *n. (Brit.)* mit Backpulver versetztes Mehl; **~-re'liant** *adj.* selbstbewußt; selbstsicher; **~-re'spect** *n., no pl.* Selbstachtung, *die;* **~-re'specting** *adj.* mit Selbstachtung *nachgestellt;* **no ~-respecting person ...:** niemand, der etwas auf sich hält, ...; **~-re'straint** *n., no pl.* Selbstbeherrschung, *die;* **~-'righteous** *adj.* selbstgerecht; **~-'righteousness** *n., no pl.* Selbstgerechtigkeit, *die;* **~-'sacrifice** *n.* Selbstaufopferung, *die;* **~-'sacrificing** *adj.* [sich] aufopfernd ‹*Mutter, Vater*›; aufopfernd ‹*Liebe*›; **~same** *adj.* **the ~same** der-/die-/dasselbe;

~-'**satisfied** *adj.* selbstzufrieden; *(smug)* selbstgefällig; ~-**seeking** *adj.* selbstsüchtig; ~-'**service** *n.* Selbstbedienung, *die; attrib.* Selbstbedienungs-; ~-**styled** *adj.* selbsternannt; ~-**suf'ficiency** *n.* Unabhängigkeit, *die; (of country)* Autarkie, *die;* ~-**suf'ficient** *adj. (independent)* unabhängig; autark ‹*Land*›; selbständig ‹*Person*›; ~-**sup'porting** *adj.* sich selbst tragend ‹*Unternehmen, Verein*›; finanziell unabhängig ‹*Person*›; **the club is ~-supporting** der Verein trägt sich selbst; ~-'**tapping** *adj.* selbstschneidend ‹*Schraube*›; ~-'**taught** *adj.* autodidaktisch; selbsterlernt ‹*Fertigkeiten*›; **~-taught person** Autodidakt, *der*/Autodidaktin, *die;* **she is ~-taught** sie ist Autodidaktin; ~-'**will** *n., no pl.* Eigensinn, *der;* ~-**willed** [self'wɪld] *adj.* eigensinnig

sell [sel] **1.** *v. t.,* **sold** [səʊld] **a)** verkaufen; **the shop ~s groceries** in dem Laden gibt es Lebensmittel [zu kaufen]; ~ **sth. to sb.,** ~ **sb. sth.** jmdm. etw. verkaufen; ~ **by ...** *(on package)* ≈ mindestens haltbar bis ...; **b)** *(betray)* verraten; **c)** *(offer dishonourably)* verkaufen; verhökern *(ugs. abwertend);* ~ **oneself/one's soul** sich/seine Seele verkaufen (**to** *Dat.*); **d)** *(sl.: cheat, disappoint)* verraten; anschmieren *(salopp);* **I've been sold!, sold again!** ich bin [wieder] der/die Dumme! *(ugs.);* **e)** *(gain acceptance for)* ~ **sb. as ...:** jmdn. als ... verkaufen *(ugs.);* ~ **sth. to sb.** jmdn. für etw. gewinnen; ~ **sb. the idea of doing sth.** jmdn. für den Gedanken gewinnen, etw. zu tun; **f)** ~ **sb. on sth.** *(coll.: make enthusiastic)* jmdn. für etw. begeistern *od.* erwärmen; **be sold on sth.** *(coll.)* von etw. begeistert sein. **2.** *v. i.,* **sold a)** sich verkaufen [lassen]; ‹*Person:*› verkaufen; **the book sold 5,000 copies in a week** in einer Woche wurden 5 000 Exemplare des Buches verkauft; **b)** ~ **at** *or* **for** kosten. *See also* **river 1 a**

~ '**off** *v. t.* verkaufen; abstoßen ‹*Anteile, Aktien*›

~ '**out 1.** *v. t.* **a)** ausverkaufen; **the play/performance was sold out** das Stück/die Aufführung war ausverkauft; **b)** *(coll.: betray)* verpfeifen *(ugs.).* **2.** *v. i.* **a) we have** *or* **are sold out** wir sind ausverkauft; **b)** *(coll.: betray one's cause)* ~ **out to sb./sth.** zu jmdm./etw. überlaufen

~ '**out of** *v. t.* **we have** *or* **are sold out of sth.** etw. ist ausverkauft

~ '**up** *v. t. (Brit.)* verkaufen; *abs.* sein Hab und Gut verkaufen

'**sell-by date** *n.* ≈ Mindesthaltbarkeitsdatum, *das*

seller ['selə(r)] *n.* **a)** Verkäufer, *der*/Verkäuferin, *die;* **a ~'s** *or* **~s' market** ein Verkäufermarkt; **b)** *(product)* **be a slow/bad ~:** sich nur langsam/schlecht verkaufen; **be a good ~:** sich gut verkaufen

selling ['selɪŋ] *n.* **a)** *(act, occupation)* Verkaufen, *das;* **b)** *(salesmanship)* Verkauf, *der*

selling: ~-**point** *n.* **a [good] ~-point** ein Verkaufsargument; *(fig.)* ein Pluspunkt; ~ **price** *n.* Verkaufspreis, *der*

Sellotape, (P) ['seləʊteɪp] *n., no pl., no indef. art.* ≈ Tesafilm, *der* Ⓦz

sellotape *v. t.* mit Klebeband kleben

'**sell-out** *n.* **a)** *(event)* **be a ~:** ausverkauft sein; **b)** *(coll.: betrayal)* Verrat, *der*

selves *pl. of* **self**

semantic [sɪ'mæntɪk] *adj.* semantisch

semantics [sɪ'mæntɪks] *n., no pl.* Semantik, *die*

semaphore ['seməfɔ:(r)] **1.** *n. (system)* Winken, *das.* **2.** *v. i.* ~ **to sb.** jmdm. ein Winksignal übermitteln

semblance ['sembləns] *n.* Anschein, *der;* **without a ~ of regret/a smile** ohne das geringste Zeichen von Bedauern/den Anflug eines Lächelns; **bring some ~ of order to sth.** wenigstens den Anschein von Ordnung in etw. *(Akk.)* bringen

semen ['si:men] *n. (Physiol.)* Samen, *der;* Sperma, *das*

semester [sɪ'mestə(r)] *n. (Univ.)* Semester, *das*

semi ['semɪ] *n. (Brit. coll.: house)* Doppelhaushälfte, *die*

semi- *in comb.* halb-/Halb-

semi: ~-**bold** *adj. (Printing)* halbfett; ~**breve** *n. (Brit. Mus.)* ganze Note; ~**circle** *n.* Halbkreis, *der;* ~'**circular** *adj.* halbkreisförmig; ~'**colon** *n.* Semikolon, *das;* ~**conductor** *n. (Phys.)* Halbleiter, *der;* ~-'**conscious** *adj.* halb bewußtlos; ~-**de'tached 1.** *adj.* **the house is ~-detached** es ist eine Doppelhaushälfte; **a ~-detached house** eine Doppelhaushälfte; **2.** *n. (Brit.: house)* Doppelhaushälfte, *die;* ~'**final** *n.* Halbfinale, *das;* **in the ~finals** im Halbfinale; ~'**finalist** *n.* Halbfinalteilnehmer, *der*/-teilnehmerin, *die;* Halbfinalist, *der*/-finalistin, *die;* ~-**literate** *adj.* **be ~-literate** kaum lesen und schreiben können

seminal ['semɪnl] *adj. (strongly influencing later developments)* schöpferisch

seminar ['semɪnɑ:(r)] *n.* Seminar, *das*

semi: ~-**precious** *adj.* ~-**precious stone** Halbedelstein, *der;* ~**quaver** *n. (Brit. Mus.)* Sechzehntelnote, *die;* ~-**skilled** *adj.* angelernt; ~**tone** *n. (Mus.)* Halbton, *der*

semolina [semə'li:nə] *n.* Grieß, *der*

Sen. *abbr.* **a) Senator** Sen.; **b) Senior** sen.

senate ['senət] *n.* Senat, *der*

senator ['senətə(r)] *n.* Senator, *der*/Senatorin, *die*

send [send] *v. t.,* **sent** [sent] **a)** *(cause to go)* schicken; senden *(geh.);* ~ **sb. to boarding-school/university** jmdn. ins Internat/auf die Universität schicken; ~ **sb. on a course/tour** jmdn. in einen Kurs/auf eine Tour schicken; **she ~s her best wishes/love** sie läßt grüßen/herzlich grüßen; ~ **[sb.] apologies/congratulations** sich [bei jmdm.] entschuldigen lassen/[jmdm.] seine Glückwünsche übermitteln; ~ **sb. home/to bed** jmdn. nach Hause/ins Bett schicken; *see also* **word 1 f; b)** *(propel)* ~ **a rocket into space** eine Rakete in den Weltraum schießen; ~ **up clouds of dust** Staubwolken aufwirbeln; ~ **sb. sprawling/reeling** jmdn. zu Boden strecken/ins Wanken bringen; *see also* **²fly 1 c; c)** ~ **sb. mad** *or* **crazy** jmdn. verrückt machen *(ugs.);* ~ **sb. to sleep** jmdn. zum Einschlafen bringen

~ **a'head** *v. t.* vorausschicken

~ **a'way 1.** *v. t.* wegschicken. **2.** *v. i.* ~ **away [to sb.] for sth.** etw. [bei jmdm.] anfordern

~ '**back** *v. t.* **a)** *(return)* zurückschicken; **b)** *(because of dissatisfaction)* zurückgehen lassen ‹*Speise, Getränk*›; *(by post)* zurückschicken ‹*Ware*›

~ '**down** *v. t.* [hinunter]schicken

~ **for** *v. t.* **a)** *(tell to come)* holen lassen; rufen ‹*Polizei, Arzt, Krankenwagen*›; **b)** *(order from elsewhere)* anfordern

~ '**in** *v. t.* einschicken

~ '**off 1.** *v. t.* **a)** *(dispatch)* abschicken ‹*Sache*›; losschicken *(ugs.)*‹*Person*›; **b)** *(bid farewell to)* verabschieden; **c)** *(Sport)* vom Platz stellen (**for** wegen). **2.** *v. i.*

~ **off for sth. [to sb.]** etw. [von jmdm.] anfordern
~ **'on** *v.t.* **a)** *(forward)* nachsenden ‹*Post*›; **b)** *(cause to go ahead)* ~ **on [ahead]** vorausschicken; **c)** *(cause to participate)* ~ **a player on** einen Spieler einsetzen
~ **'out 1.** *v.t.* **a)** *(issue)* verschicken; **b)** *(emit)* aussenden ‹*Hilferuf, Nachricht*›; abgeben ‹*Hitze*›; senden ‹*Lichtstrahlen*›; ausstoßen ‹*Rauch*›; **c)** *(dispatch)* schicken; ~ **sb. out for sth.** jmdn. schicken, um etw. zu besorgen; **d)** *(order to leave)* hinausschicken. **2.** *v.i.* ~ **out for sth.** etw. besorgen *od.* holen lassen
~ **'up** *v.t.* **a)** *(Brit. coll.: parody)* parodieren; **b)** *(cause to rise)* steigen lassen ‹*Ballon*›; hochtreiben ‹*Preis, Kosten, Temperatur*›; ~ **sb.'s temperature up** *(fig. joc.)* jmdn. zum Kochen bringen *(ugs.)*

sender ['sendə(r)] *n.* *(of goods)* Lieferant, *der*/Lieferantin, *die*; *(of letter)* Absender, *der*/Absenderin, *die*

send: ~-off *n.* Verabschiedung, *die*; **~-up** *n.* *(Brit. coll.: parody)* Parodie, *die*; **do a ~-up of sb./sth.** jmdn./etw. parodieren

senile ['si:naɪl] *adj.* senil; *(physically)* altersschwach

senility [sɪ'nɪlɪtɪ] *n.*, *no pl.* Senilität, *die*; *(physical infirmity)* Altersschwäche, *die*

senior ['si:nɪə(r)] **1.** *adj.* **a)** *(older)* älter; **be ~ to sb.** älter als jmd. sein; **b)** *(of higher rank)* höher ‹*Rang, Beamter, Stellung*›; leitend ‹*Angestellter, Stellung*›; *(longest-serving)* ältest ...; **someone ~:** jemand in höherer Stellung; ~ **management** Geschäftsleitung, *die*; **be ~ to sb.** eine höhere Stellung als jmd. haben; **c)** *appended to name (the elder)* **Mr Smith S~:** Mr. Smith senior; **d)** *(Amer. Sch., Univ.)* ~ **class** Abschlußklasse, *die*; ~ **year** letztes Jahr vor der Abschlußprüfung. **2.** *n.* *(older person)* Ältere, *der/die*; *(person of higher rank)* Vorgesetzte, *der/die*; **be sb.'s ~ [by six years]** *or* **[six years] sb.'s ~:** [sechs Jahre] älter als jmd. sein

senior 'citizen *n.* Senior, *der*/Seniorin, *die*

seniority [si:nɪ'ɒrɪtɪ] *n.* **a)** *(superior age)* Alter, *das*; **b)** *(priority in length of service)* höheres Dienstalter; **c)** *(superior rank)* höherer Rang

senior: ~ 'officer *n.* höherer Beamter/höhere Beamtin; *(Mil.)* ranghöchster Offizier; ~ **'partner** *n.* Seniorpartner, *der*/-partnerin, *die*; ~ **school** *(Brit.)* *see* **secondary school**

sensation [sen'seɪʃn] *n.* **a)** *(feeling)* Gefühl, *das*; ~ **of giddiness** Schwindelgefühl, *das*; **have a ~ of falling** das Gefühl haben zu fallen; **b)** *(person, event)* Sensation, *die*; **a great ~:** ein großes Ereignis; **c)** *(excitement)* Aufsehen, *das*

sensational [sen'seɪʃənl] *adj.* **a)** *(spectacular)* aufsehenerregend; sensationell; **b)** *(arousing intense response)* reißerisch *(abwertend)*; Sensations‹*blatt, -presse*›; **c)** *(phenomenal)* phänomenal

sense [sens] **1.** *n.* **a)** *(faculty of perception)* Sinn, *der*; ~ **of smell/touch/taste** Geruchs-/Tast-/Geschmackssinn, *der*; **come to one's ~s** das Bewußtsein wiedererlangen; **b)** *in pl.* *(normal state of mind)* Verstand, *der*; **have taken leave of one's ~s** den Verstand verloren haben; **come to one's ~s** zur Vernunft kommen; **bring sb. to his ~s** jmdn. zur Vernunft *od.* Besinnung bringen; **c)** *(consciousness)* Gefühl, *das*; ~ **of responsibility/guilt** Verantwortungs-/Schuldgefühl, *das*; **out of a ~ of duty** aus Pflichtgefühl; **d)** *(practical wisdom)* Verstand, *der*; **there's a lot of ~ in what he's saying** was er sagt, klingt sehr vernünftig; **have the ~ to do sth.** so vernünftig sein, etw. zu tun; **what is the ~ of** *or* **in doing that?** was hat man davon *od.* wozu soll es gut sein, das zu tun?; **talk ~:** vernünftig reden; **now you are talking ~:** jetzt wirst du vernünftig; **see ~:** zur Vernunft kommen; **make sb. see ~:** jmdn. zur Vernunft bringen; *see also* **common sense; good 11; e)** *(meaning)* Sinn, *der*; *(of word)* Bedeutung, *die*; **in the strict** *or* **literal ~:** im strengen *od.* wörtlichen Sinn; **in every ~ [of the word]** in jeder Hinsicht; **in some ~:** irgendwie; **in a** *or* **one ~:** in gewisser Hinsicht *od.* Weise; **make ~:** einen Sinn ergeben; **her arguments do not make ~ to me** ihre Argumente leuchten mir nicht ein; **it does not make ~ to do that** es ist Unsinn *od.* unvernünftig, das zu tun; **it makes [a lot of] ~** *(is [very] reasonable)* es ist [sehr] sinnvoll; **it all makes ~ to me now** jetzt verstehe ich alles; **it just doesn't make ~:** es ergibt einfach keinen Sinn; **make ~ of sth.** etw. verstehen. **2.** *v.t.* spüren; ‹*Tier:*› wittern

senseless ['senslɪs] *adj.* **a)** *(unconscious)* bewußtlos; **b)** *(foolish)* unvernünftig; dumm; **c)** *(purposeless)* unsinnig ‹*Argument*›; sinnlos ‹*Diskussion, Vergeudung*›

sensibilities [sensɪ'bɪlɪtɪz] *n. pl.* *(susceptibility)* Empfindlichkeit, *die*; **her ~ are easily wounded** sie ist sehr schnell gekränkt

sensible ['sensɪbl] *adj.* **a)** *(reasonable)* vernünftig; **b)** *(practical)* praktisch; zweckmäßig; fest ‹*Schuhe*›

sensibly ['sensɪblɪ] *adv.* **a)** *(reasonably)* vernünftig; besonnen; ~ **enough, he refused** er war so vernünftig abzulehnen; **b)** *(practically)* zweckmäßig

sensitise *see* **sensitize**

sensitive ['sensɪtɪv] *adj.* **a)** *(responsive)* empfindlich **(to** gegen**)**; ~ **to light** lichtempfindlich; **b)** *(easily upset)* empfindlich; sensibel; **be ~ to sth.** empfindlich auf etw. *(Akk.)* reagieren; **sb. is ~ about sth.** etw. ist bei jmdm. ein wunder Punkt; **c)** heikel ‹*Thema, Diskussion*›; **d)** *(perceptive)* einfühlsam

sensitivity [sensɪ'tɪvɪtɪ] *n.* *see* **sensitive:** Empfindlichkeit, *die*; Sensibilität, *die*; Heikelkeit, *die*; Einfühlsamkeit, *die*; ~ **to light** Lichtempfindlichkeit, *die*

sensitize ['sensɪtaɪz] *v.t.* sensibilisieren **(to** für**)**

sensor ['sensə(r)] *n.* Sensor, *der*

sensory ['sensərɪ] *adj.* sensorisch; Sinnes‹*wahrnehmung, -organ*›

sensual ['sensjʊəl, 'senʃʊəl] *adj.* sinnlich; lustvoll ‹*Leben*›; Sinnen‹*freude, -genuß*›

sensuous ['sensjʊəs] *adj.* sinnlich

sent *see* **send**

sentence ['sentəns] **1.** *n.* **a)** *(decision of lawcourt)* [Straf]urteil, *das*; *(fig.)* Strafe, *die*; **pass ~ [on sb.]** [jmdm.] das Urteil verkünden; **be under ~ of death** zum Tode verurteilt sein; **b)** *(Ling.)* Satz, *der*. **2.** *v.t.* *(lit. or fig.)* verurteilen **(to** zu**)**

'sentence-modifier *n.* *(Ling.)* Satzpartikel, *die*

sentiment ['sentɪmənt] *n.* **a)** *(mental feeling)* Gefühl, *das*; **b)** *(emotion conveyed in art)* Empfindung, *die*; **c)** *no pl.* *(emotional weakness)* Sentimentalität, *die*

sentimental [sentɪ'mentl] *adj.* sentimental; **sth. has ~ value [for sb.]** jmd. hängt an etw. *(Dat.)*; **for ~ reasons** aus Sentimentalität

sentinel ['sentɪnl] *n.* *(lit. or fig.)* Wache, *die*

sentry ['sentrɪ] *n.* *(lit. or fig.)* Wache, *die*

'sentry-box *n.* Wachhäuschen, *das*

separable ['sepərəbl] *adj.* **a)** trennbar; zerlegbar ‹*Werkzeug, Gerät*›; **b)** *(Ling.)* trennbar

separate 1. ['sepərət] *adj.* verschieden ‹*Fragen, Probleme*›; getrennt ‹*Konten, Betten*›; gesondert ‹*Teil*›; separat ‹*Eingang, Toilette, Blatt Papier, Abteil*›; Sonder‹*vereinbarung*›; *(one's own, individual)* eigen ‹*Zimmer, Identität, Organisation*›; **lead ~ lives** getrennt leben; **go ~ ways** getrennte Wege gehen; **keep two things ~:** zwei Dinge auseinanderhalten; **keep issue A ~ from issue B** Frage A und Frage B getrennt behandeln. **2.** ['sepəreɪt] *v.t.* trennen; **they are ~d** *(no longer live together)* sie leben getrennt. **3.** ['sepəreɪt] *v.i.* **a)** *(disperse)* sich trennen; **b)** ‹*Ehepaar:*› sich trennen

separately ['sepərətlɪ] *adv.* getrennt

separates ['sepərəts] *n. pl. (Fashion)* Separates; einzelne Kleidungsstücke [die man kombinieren kann]

separation [sepə'reɪʃn] *n.* Trennung, *die*

separatist ['sepərətɪst] *n.* Separatist, *der*/Separatistin, *die; attrib.* **~ movement** Separatistenbewegung, *die*

sepia ['si:pɪə] *n.* **a)** *(pigment)* Sepia, *die;* **b)** *(colour)* Sepiabraun, *das*

Sept. *abbr.* **September** Sept.

September [sep'tembə(r)] *n.* September, *der; see also* **August**

septic ['septɪk] *adj.* septisch

septic 'tank *n.* Faulraum, *der*

sepulchre (*Amer.:* **sepulcher**) ['seplkə(r)] *n.* Grab, *das*

sequel ['si:kwl] *n.* **a)** *(consequence, result)* Folge, *die* **(to** von); **b)** *(continuation)* Fortsetzung, *die;* **there was a tragic ~:** es gab ein tragisches Nachspiel

sequence ['si:kwəns] *n.* **a)** *(succession)* Reihenfolge, *die;* **rapid/logical ~:** rasche/logische Abfolge; **b)** *(part of film)* Sequenz, *die*

sequestered [sɪ'kwestəd] *adj.* abgelegen; ‹*Leben*› in Abgeschiedenheit

sequin ['si:kwɪn] *n.* Paillette, *die*

serenade [serə'neɪd] **1.** *n. (Mus.)* Ständchen, *das.* **2.** *v.t. (Mus.)* **~ sb.** jmdm. ein Ständchen bringen

serene [sɪ'ri:n] *adj.,* **~r** [sɪ'ri:nə(r)], **~st** [sɪ'ri:nɪst] **a)** *(placid)* ruhig; gelassen; **b)** *(calm)* klar ‹*Wetter, Himmel*›; *(unruffled)* unbewegt ‹*See, Wasser usw.*›

serenity [sɪ'renɪtɪ, sə'renɪtɪ] *n., no pl.* **a)** *(placidity)* Gelassenheit, *die;* **b)** *(of clear weather)* Klarheit, *die*

serge [sɜ:dʒ] *n. (Textiles)* Serge, *die*

sergeant ['sɑ:dʒənt] *n.* **a)** *(Mil.)* Unteroffizier, *der;* **b)** *(police officer)* ≈ Polizeimeister, *der*

sergeant-'major ≈ [Ober]stabsfeldwebel, *der*

serial ['sɪərɪəl] *n.* Fortsetzungsgeschichte, *die; (on radio, television)* Serie, *die*

serialize ['sɪərɪəlaɪz] *v.t.* in Fortsetzungen veröffentlichen; *(on radio, television)* in Fortsetzungen *od.* als Serie senden

'serial number *n.* Seriennummer, *die*

series ['sɪəri:z, 'sɪərɪz] *n., pl. same* **a)** *(sequence)* Reihe, *die;* **a ~ of events/misfortunes** eine Folge von Ereignissen/Mißgeschicken; **b)** *(set of successive issues)* Serie, *die;* **radio/TV ~:** Hörfunkreihe/Fernsehserie, *die;* **~ of programmes** Sendereihe, *die;* **c)** *(set of books)* Reihe, *die;* **d)** *(group of stamps, games, etc.)* Serie, *die;* **e)** *(Electr.)* **in ~:** in Reihe

serious ['sɪərɪəs] *adj.* **a)** *(earnest)* ernst; **~ music** ernste Musik; **a ~ play** ein ernstes Stück; **b)** *(important, grave)* ernst ‹*Angelegenheit, Lage, Problem, Zustand*›; ernsthaft ‹*Frage, Einwand, Kandidat*›; gravierend ‹*Änderung*›; schwer ‹*Krankheit, Unfall, Fehler, Verstoß, Niederlage*›; ernstzunehmend ‹*Rivale*›; ernstlich ‹*Gefahr, Bedrohung*›; bedenklich ‹*Verschlechterung, Mangel*›; schwerwiegend ‹*Vorwurf*›; **things are/sth. is getting ~:** die Lage spitzt sich zu/etw. nimmt ernste Ausmaße an; **there is a ~ danger that ...:** es besteht ernste Gefahr, daß ...; **c)** *(in earnest)* **are you ~?** ist das dein Ernst?; **be ~ about sth./doing sth.** etw. ernst nehmen/ernsthaft tun wollen; **is he ~ about her?** meint er es ernst mit ihr?

seriously ['sɪərɪəslɪ] *adv.* **a)** *(earnestly)* ernst[haft]; **quite ~, ...:** ganz im Ernst, ...; **take sth./sb. ~:** etw./jmdn. ernst nehmen; **b)** *(severely)* ernstlich; schwer ‹*verletzt*›

seriousness ['sɪərɪəsnɪs] *n., no pl.* **a)** *(earnestness)* Ernst, *der;* Ernsthaftigkeit, *die;* **in all ~:** ganz im Ernst; **b)** *(gravity)* Schwere, *die; (of situation)* Ernst, *der*

sermon ['sɜ:mən] *n. (Relig.)* Predigt, *die;* **give a ~:** eine Predigt halten

serpent ['sɜ:pənt] *n.* **a)** *(snake)* Schlange, *die;* **b)** *(fig.: treacherous person)* falsche Schlange

SERPS [sɜ:ps] *abbr. (Brit.)* **State earnings-related pension scheme** staatliche einkommensbezogene Rentenversicherung

serrated [se'reɪtɪd] *adj.* gezackt; **~ knife** Sägemesser, *das*

serum ['sɪərəm] *n., pl.* **sera** ['sɪərə] *or* **~s** *(Physiol.)* Serum, *das*

servant ['sɜ:vənt] *n.* Diener, *der*/Dienerin, *die; (female also)* Dienstmädchen, *das;* **keep** *or* **have ~s** Bedienstete haben

'servant-girl *n.* Dienstmädchen, *das*

serve [sɜ:v] **1.** *v.t.* **a)** *(work for)* dienen (+ *Dat.*); **b)** *(be useful to)* dienlich sein (+ *Dat.*); **this car ~d us well** dieses Auto hat uns gute Dienste getan; **if my memory ~s me right** wenn mich mein Gedächtnis nicht täuscht; **c)** *(meet needs of)* nutzen (+ *Dat.*); **~ a/no purpose** einen Zweck erfüllen/keinen Zweck haben; **~ its purpose** *or* **turn** seinen Zweck erfüllen; **d)** *(go through period of)* durchlaufen ‹*Lehre*›; absitzen, verbüßen ‹*Haftstrafe*›; **~ [one's] time** *(undergo apprenticeship)* seine Lehrzeit durchmachen; *(undergo imprisonment)* seine Zeit absitzen; **e)** *(dish up)* servieren; *(pour out)* einschenken **(to** *Dat.*); **dinner is ~d** das Essen ist aufgetragen; **f)** *(render obedience to)* dienen (+ *Dat.*) ‹*Gott, König, Land*›; **g)** *(attend)* bedienen; **are you being ~d?** werden Sie schon bedient?; **h)** *(supply)* versorgen; **~s three** *(in recipe)* für drei Personen *od.* Portionen; **i)** *(provide with food)* bedienen; **j)** *(make legal delivery of)* zustellen; **~ a summons on sb.** jmdn. vorladen; **he has been ~d notice to quit** ihm ist gekündigt worden; **k)** *(Tennis etc.)* aufschlagen; **~ an ace** ein As schlagen; **l) ~[s]** *or* **it ~s him right!** *(coll.)* [das] geschieht ihm recht! **2.** *v.i.* **a)** *(do service)* dienen; **~ as chairman** das Amt des Vorsitzenden innehaben; **~ on a jury** Geschworener/Geschworene sein; **b)** *(be employed; be soldier etc.)* dienen; **c)** *(be of use)* **~ to do sth.** dazu dienen, etw. zu tun; **~ to show sth.** etw. zeigen; **~ for** *or* **as** dienen als; **d)** *(~ food)* **shall I ~?** soll ich auftragen?; **e)** *(attend in shop etc.)* bedienen; **f)** *(Eccl.)* ministrieren; **g)** *(Tennis etc.)* aufschlagen; **it's your turn to ~:** du hast Aufschlag. **3.** *n. see* **service 1 h**

~ '**up** *v. t.* **a)** *(put before eaters)* servieren; **b)** *(offer for consideration)* auftischen *(ugs.)*

service ['sɜ:vɪs] **1.** *n.* **a)** *(doing of work for employer etc.)* Dienst, *der;* **give good ~:** gute Dienste leisten; **do ~ as sth.** als etw. dienen; **he died in the ~ of his country** er starb in Pflichterfüllung für sein Vaterland; **b)** *(sth. done to help others)* **do sb. a ~:** jmdm. einen guten Dienst erweisen; ~s Dienste; *(Econ.)* Dienstleistungen; **[in recognition of her] ~s to the hospital/state** [in Anerkennung ihrer] Verdienste um das Krankenhaus/den Staat; **c)** *(Eccl.)* Gottesdienst, *der;* **d)** *(act of attending to customer)* Service, *der; (in shop, garage, etc.)* Bedienung, *die;* **e)** *(system of transport)* Verbindung, *die;* **there is no bus ~ on Sundays** sonntags verkehren keine Busse; **the number 325 bus ~:** die Buslinie Nr. 325; **f)** *(provision of maintenance)* **[after-sale** *or* **follow-up] ~:** Kundendienst, *der;* **take one's car in for a ~:** sein Auto zur Inspektion bringen; **g)** *no pl., no art. (operation)* Betrieb, *der;* **bring into ~:** in Betrieb nehmen; **out of ~:** außer Betrieb; **take out of ~:** außer Betrieb setzen; **go** *or* **come into ~:** in Betrieb genommen werden; **h)** *(Tennis etc.)* Aufschlag, *der;* **whose ~ is it?** wer hat Aufschlag?; **i)** *(crockery set)* Service, *das;* **dessert/tea ~:** Dessert-/Tee-Service, *das;* **j)** *(assistance)* **can I be of ~ [to you]?** kann ich Ihnen behilflich sein?; **k)** *(person's behalf)* **in his ~:** in seinem Auftrag; **I'm at your ~:** ich stehe zu Ihren Diensten; **l) the consular ~:** der Konsulatsdienst; **BBC World S~:** BBC Weltsender; **public ~:** öffentlicher Dienst; **m)** *in pl. (Brit.: public supply)* Versorgungseinrichtungen; **n)** *(Mil.)* **the [armed** *or* **fighting] ~s** die Streitkräfte; **in the ~s** beim Militär; **o)** *(being servant)* **be in/go into ~:** in Stellung sein/gehen *(veralt.)* (**with** bei). **2.** *v. t.* **a)** *(provide maintenance for)* warten ‹*Wagen, Waschmaschine, Heizung*›; **take one's car to be ~d** sein Auto zur Inspektion bringen; **b)** *(pay interest on)* Zinsen zahlen für ‹*Schulden*›

serviceable ['sɜ:vɪsəbl] *adj.* **a)** *(useful)* nützlich; **b)** *(durable)* haltbar

service: ~ **area** *n. (for motorists' needs)* Raststätte, *die;* ~ **charge** *n. (in restaurant)* Bedienungsgeld, *das; (of bank)* Bearbeitungsgebühr, *die;* ~ **industry** *n.* Dienstleistungsbetrieb, *der;* ~**man** *n. (in armed services)* Militärangehörige, *der;* ~ **station** *n.* Tankstelle, *die*

serviette [sɜ:vɪ'et] *n. (Brit.)* Serviette, *die*

servile ['sɜ:vaɪl] *adj.* unterwürfig; erbärmlich ‹*Furcht, Unterwürfigkeit*›

serving ['sɜ:vɪŋ] *n.* Portion, *die*

serving: ~ **dish** *n.* Servierschüssel, *die;* ~ **hatch** *n.* Durchreiche, *die;* ~**-spoon** *n.* Vorlegelöffel, *der*

servitude ['sɜ:vɪtju:d] *n., no pl. (lit. or fig.)* Knechtschaft, *die*

session ['seʃn] *n.* **a)** *(meeting)* Sitzung, *die;* **discussion ~:** Diskussionsrunde, *die;* **be in ~:** tagen; **b)** *(period spent)* Sitzung, *die; (by several people)* Treffen, *das;* **recording ~:** Aufnahme, *die*

set [set] **1.** *v. t.,* **-tt-, set a)** *(put) (horizontally)* legen; *(vertically)* stellen; ~ **sb. ashore** jmdn. an Land setzen; ~ **the proposals before the board** *(fig.)* dem Vorstand die Vorschläge unterbreiten *od.* vorlegen; ~ **sth. against sth.** *(balance)* etw. einer Sache *(Dat.)* gegenüberstellen; **b)** *(apply)* setzen; ~ **pen to paper** etwas zu Papier bringen; ~ **a match to sth.** ein Streichholz an etw. *(Akk.)* halten; **c)** *(adjust)* einstellen (**at** auf + *Akk.*); aufstellen ‹*Falle*›; stellen ‹*Uhr*›; ~ **the alarm for 5.30 a.m.** den Wecker auf 5.30 Uhr stellen; **d) be ~** *(have location of action)* ‹*Buch, Film:*› spielen; ~ **a book/film in Australia** ein Buch/einen Film in Australien spielen lassen; **e)** *(specify)* festlegen ‹*Bedingungen*›; festsetzen ‹*Termin, Ort usw.*› (**for** auf + *Akk.*); ~ **the interest rate at 10%** die Zinsen auf 10% festsetzen; ~ **limits** Grenzen setzen; **f)** *(bring into specified state)* ~ **sth./things right** *or* **in order** etw./die Dinge in Ordnung bringen; ~ **sb. laughing** jmdn. zum Lachen bringen; ~ **a dog barking** einen Hund anschlagen lassen; ~ **sb. thinking that ...:** jmdn. auf den Gedanken bringen, daß ...; **the news ~ me thinking** die Nachricht machte mich nachdenklich; **g)** *(put forward)* stellen ‹*Frage, Aufgabe*›; aufgeben ‹*Hausaufgabe*›; vorschreiben ‹*Textbuch, Lektüre*›; *(compose)* zusammenstellen ‹*Rätsel, Fragen*›; ~ **sb. a task/problem** jmdm. eine Aufgabe stellen/jmdn. vor ein Problem stellen; ~ **[sb./oneself] a target** [jmdm./sich] ein Ziel setzen; **h)** *(turn to solid)* fest werden lassen; **is the jelly ~ yet?** ist das Gelee schon fest?; **i)** *(lay for meal)* decken ‹*Tisch*›; auflegen ‹*Gedeck*›; **j)** *(establish)* aufstellen ‹*Rekord, Richtlinien*›; **k)** *(Med.: put into place)* [ein]richten; einrenken ‹*verrenktes Gelenk*›; **l)** *(fix)* legen ‹*Haare*›; ~ **eyes on sb./sth.** jmdn./etw. sehen; **m)** *(Printing)* setzen; **n)** ~ **sb. in charge of sth.** jmdn. mit etw. betrauen; ~ **a dog on sb.** einen Hund auf jmdn. hetzen; ~ **the police after sb.** die Polizei auf jmdn. hetzen; ~ **sb. against sb.** jmdn. gegen jmdn. aufbringen; **o) be ~ on a hill** ‹*Haus:*› auf einem Hügel stehen. **2.** *v. i.,* **-tt-, set a)** *(solidify)* fest werden; **has the jelly ~ yet?** ist das Gelee schon fest?; **b)** *(go down)* ‹*Sonne, Mond:*› untergehen. **3.** *n.* **a)** *(group)* Satz, *der;* ~ **[of two]** Paar, *das;* **a ~ of chairs** eine Sitzgruppe; **a complete ~ of Dickens' novels** eine Gesamtausgabe der Romane von Dickens; **chess ~:** Schachspiel, *das;* **b)** *see* **service 1 i; c)** *(section of society)* Kreis, *der;* **racing ~:** Rennsportfreunde *od.* -fans; **the fast ~:** die Lebewelt; **d)** *(Math.)* Menge, *die;* **e)** ~ **[of teeth]** Gebiß, *das;* **f)** *(radio or TV receiver)* Gerät, *das;* Apparat, *der;* **g)** *(Tennis)* Satz, *der;* **h)** *(of hair)* Frisieren, *das;* Einlegen, *das;* **i)** *(Theatre: built-up scenery)* Szenenaufbau, *der;* **j)** *(area of performance) (of film)* Drehort, *der; (of play)* Bühne, *die;* **on the ~** *(for film)* bei den Dreharbeiten; *(for play)* bei den Proben. **4.** *adj.* **a)** *(fixed)* starr ‹*Linie, Gewohnheit, Blick, Lächeln*›; fest ‹*Absichten, Zielvorstellungen, Zeitpunkt*›; **be ~ in one's ways** *or* **habits** in seinen Gewohnheiten festgefahren sein; **deep-~ eyes** tiefliegende Augen; **b)** *(assigned for study)* vorgeschrieben ‹*Buch, Text*›; **be a ~ book** Pflichtlektüre sein; **c)** *(according to fixed menu)* ~ **meal** *or* **menu** Menü, *das;* **d)** *(ready)* **sth. is ~ to increase** etw. wird bald steigen; **be/get ~ for sth.** zu etw. bereit sein/sich zu etw. fertigmachen; **be/get ~ to leave** bereit sein/sich fertigmachen zum Aufbruch; **all ~?** *(coll.)* alles klar *od.* fertig?; **be all ~ to do sth.** bereit sein, etw. zu tun; **e)** *(determined)* **be ~ on sth./doing sth.** zu etw. entschlossen sein/entschlossen sein, etw. zu tun; **be [dead] ~ against sth.** [absolut] gegen etw. sein

~ **about** *v.t.* **a)** *(begin purposefully)* ~ **about sth.** sich an etw. *(Akk.)* machen; etw. in Angriff nehmen; ~ **about doing sth.** sich daranmachen, etw. zu tun; **b)** *(coll.: attack)* herfallen über (+ *Akk.*)

~ **a'part** *v.t.* **a)** *(reserve)* reservieren; einplanen ⟨*Zeit*⟩; **b)** *(make different)* abheben **(from** von)

~ **a'side** *v.t.* **a)** *(put to one side)* beiseite legen ⟨*Buch, Zeitung, Strickzeug*⟩; beiseite stellen ⟨*Stuhl, Glas usw.*⟩; unterbrechen ⟨*Arbeit, Tätigkeit*⟩; außer acht lassen ⟨*Frage*⟩; *(postpone)* aufschieben ⟨*Arbeit*⟩; **b)** *(cancel)* aufheben ⟨*Urteil, Entscheidung*⟩; **c)** *(pay no attention to)* außer acht lassen ⟨*Unterschiede, Formalitäten*⟩; **d)** *(reserve)* aufheben ⟨*Essen, Zutaten*⟩; einplanen ⟨*Minute, Zeit*⟩; beiseite legen ⟨*Geld*⟩; *(save for customer)* zurücklegen ⟨*Ware*⟩

~ **'back** *v.t.* **a)** *(hinder progress of)* behindern ⟨*Fortschritt*⟩; aufhalten ⟨*Entwicklung*⟩; zurückwerfen ⟨*Projekt, Programm*⟩; **b)** *(coll.: be an expense to)* ~ **sb. back a fair amount** jmdn. eine hübsche Summe kosten; **c)** *(place at a distance)* zurücksetzen; **the house is ~ back some distance from the road** das Haus steht in einiger Entfernung von der Straße; **d)** *(postpone)* verschieben ⟨*Termin*⟩ **(to** auf + *Akk.*)

~ **'by** *see* ~ **aside a, d**

~ **'down** *v.t.* **a)** absetzen ⟨*Fahrgast, Ladung*⟩; **the bus will ~ you down there** du kannst dort aus dem Bus aussteigen; **b)** *(record on paper)* niederschreiben; **c)** *(place on surface)* absetzen; abstellen

~ **'forth 1.** *v.i. (begin journey)* aufbrechen; ~ **forth on a journey** eine Reise antreten. **2.** *v.t. (present)* darstellen ⟨*Zahlen, Kosten*⟩; darlegen ⟨*Programm, Ziel, Politik*⟩

~ **'in** *v.i.* ⟨*Dunkelheit, Regen, Reaktion, Verfall:*⟩ einsetzen

~ **'off 1.** *v.i.* *(begin journey)* aufbrechen; *(start to move)* loslaufen; ⟨*Zug:*⟩ losfahren; ~ **off for work** sich auf den Weg zur Arbeit machen. **2.** *v.t.* **a)** *(show to advantage)* hervorheben; **b)** *(start)* führen zu; auslösen ⟨*Reaktion, Alarmanlage*⟩; **c)** *(cause to explode)* explodieren lassen; abbrennen ⟨*Feuerwerk*⟩; **d)** *(counterbalance)* ausgleichen; ~ **sth. off against sth.** etw. einer Sache *(Dat.)* gegenüberstellen

~ **on** *v.t. (attack)* überfallen

~ **'out 1.** *v.i.* **a)** *(begin journey)* aufbrechen; **b)** ~ **out to do sth.** sich *(Dat.)* vornehmen, etw. zu tun. **2.** *v.t.* **a)** *(present)* darlegen ⟨*Gedanke, Argument*⟩; auslegen ⟨*Waren*⟩; ausbreiten ⟨*Geschenke*⟩; aufstellen ⟨*Schachfiguren*⟩; **b)** *(state, specify)* darlegen ⟨*Bedingungen, Einwände, Vorschriften*⟩

~ **'to** *v.i.* **a)** *(begin vigorously)* sich daranmachen; **b)** *(begin to fight)* loslegen *(ugs.)*

~ **'up 1.** *v.t.* **a)** *(erect)* errichten ⟨*Straßensperre, Denkmal*⟩; aufstellen ⟨*Kamera*⟩; aufbauen ⟨*Zelt, Spieltisch*⟩; **b)** *(establish)* bilden ⟨*Regierung usw.*⟩; gründen ⟨*Gesellschaft, Organisation, Orden*⟩; aufbauen ⟨*Kontrollsystem, Verteidigung*⟩; einleiten ⟨*Untersuchung*⟩; einrichten ⟨*Büro*⟩; ~ **oneself up in business** ein Geschäft aufmachen; ~ **sb. up in business** jmdm. die Gründung eines eigenen Geschäfts ermöglichen; **c)** *(begin to utter)* anstimmen; **the class ~ up such a din** die Klasse veranstaltete einen solchen Lärm; **d)** *(coll.: make stronger)* stärken; **a good breakfast should ~ you up for the day** ein gutes Frühstück gibt dir Kraft für den ganzen Tag; **e)** *(achieve)* aufstellen ⟨*Rekord, Zeit*⟩; **f)** *(provide adequately)* ~ **sb. up with sth.** jmdn. mit etw. versorgen; **g)** *(place in view)* anbringen ⟨*Schild, Warnung*⟩; **h)** *(prepare)* vorbereiten ⟨*Experiment*⟩; betriebsbereit machen ⟨*Maschine*⟩. **2.** *v.i.* ~ **up in business** ein Geschäft aufmachen; ~ **up as a dentist** sich als Zahnarzt niederlassen

set: ~**-back** *n.* Rückschlag, *der;* Niederlage, *die;* ~ **phrase** *n.* feste Wendung; Phrase, *die;* ~ **'piece** *n. (Footb.)* Standardsituation, *die;* ~ **point** *n. (Tennis etc.)* Satzball, *der*

settee [se'ti:] *n.* Sofa, *das*

setting ['setɪŋ] *n.* **a)** *(Mus.)* Vertonung, *die;* **b)** *(frame for jewel)* Fassung, *die;* **c)** *(surroundings)* Rahmen, *der;* *(of novel etc.)* Schauplatz, *der;* **d)** *(plates and cutlery)* Gedeck, *das*

settle ['setl] **1.** *v.t.* **a)** *(place)* *(horizontally)* [sorgfältig] legen; *(vertically)* [sorgfältig] stellen; *(at an angle)* [sorgfältig] lehnen; **he ~d himself comfortably on the couch** er machte es sich *(Dat.)* auf der Couch bequem; **b)** *(establish)* *(in house or business)* unterbringen; *(in country or colony)* ansiedeln ⟨*Volk*⟩; **c)** *(determine, resolve)* aushandeln, sich einigen auf ⟨*Preis*⟩; beilegen ⟨*Streit, Konflikt, Meinungsverschiedenheit*⟩; beseitigen, ausräumen ⟨*Zweifel, Bedenken*⟩; entscheiden ⟨*Frage, Spiel*⟩; regeln, in Ordnung bringen ⟨*Angelegenheit*⟩; **nothing has been ~d as yet** es ist noch nichts entschieden; **that ~s it** dann ist ja alles klar *(ugs.)*; *expr. exasperation* jetzt reicht's! *(ugs.)*; ~ **a case out of court** sich außergerichtlich vergleichen; ~ **one's affairs** seine Angelegenheiten in Ordnung bringen; seinen Nachlaß regeln; **d)** *(deal with, dispose of)* fertig werden mit; **e)** bezahlen, *(geh.)* begleichen ⟨*Rechnung, Betrag*⟩; erfüllen ⟨*Forderung, Anspruch*⟩; ausgleichen ⟨*Konto*⟩; **f)** *(cause to sink)* sich absetzen lassen ⟨*Bodensatz, Sand, Sediment*⟩; **a shower will ~ the dust** ein Schauer wird den Staub binden; **g)** *(calm)* beruhigen ⟨*Nerven, Magen*⟩; **h)** *(colonize)* besiedeln; **i)** *(bestow)* ~ **money/property on sb.** jmdm. Geld/Besitz übereignen. **2.** *v.i.* **a)** *(become established)* sich niederlassen; *(as colonist)* sich ansiedeln; **b)** *(end dispute)* sich einigen; **c)** *(pay what is owed)* abrechnen; **d)** *(in chair etc.)* sich niederlassen; *(to work etc.)* sich konzentrieren **(to** auf + *Akk.*); *(into way of life etc.)* sich gewöhnen **(into** an + *Akk.*); **the snow/dust ~d on the ground** der Schnee blieb liegen/der Staub setzte sich [am Boden] ab; **darkness/silence/fog ~d over the village** Dunkelheit/Stille/Nebel legte *od.* senkte sich über das Dorf; **e)** *(subside)* ⟨*Haus, Fundament, Boden:*⟩ sich senken; ⟨*Sediment:*⟩ sich ablagern; **f)** *(be digested)* ⟨*Essen:*⟩ sich setzen; *(become calm)* ⟨*Magen:*⟩ sich beruhigen; **g)** *(become clear)* ⟨*Wein, Bier:*⟩ sich klären

~ **'back** *v.i.* **a)** *(relax)* sich zurücklehnen **(in** in + *Dat.*); **b)** ~ **back into one's routine** sich wieder in die Alltagsroutine hineinfinden

~ **'down 1.** *v.i.* **a)** *(make oneself comfortable)* sich niederlassen **(in** in + *Dat.*); ~ **down for the night** sich schlafen *od.* zur Ruhe legen; **b)** *(become established in a town or house)* seßhaft *od.* heimisch werden; ~ **down in a job** *(find permanent work)* eine feste Anstellung finden; *(get used to a job)* sich einarbeiten; **c)** *(marry)* **it's about time he ~d down** er sollte allmählich häuslich werden [und heiraten]; **d)** *(calm down)* ⟨*Person:*⟩ sich beruhigen; ⟨*Lärm,*

Aufregung:⟩ sich legen; ~ **down to work** richtig mit der Arbeit anfangen. **2.** *v.t.* **a)** *(make comfortable)* ~ **oneself down** sich [gemütlich] hinsetzen; ~ **the baby down for the night/to sleep** das Baby schlafen legen; **b)** *(calm down)* beruhigen

~ **for** *v.t.* **a)** *(agree to)* sich zufriedengeben mit; **b)** *(decide on)* sich entscheiden für

~ **'in** *v.i. (in new home)* sich einleben; *(in new job or school)* sich eingewöhnen

~ **on** *v.t.* **a)** *(decide on)* sich entscheiden für; **b)** *(agree on)* sich einigen auf (+ *Akk.*)

~ **'up** *v.i.* abrechnen; ~ **up with the waiter** beim Kellner bezahlen

~ **with** *v.t.* ~ **with sb.** *(pay agreed amount to sb.)* jmdm. eine Abfindung zahlen; *(pay all the money owed to sb.)* bei jmdm. seine Rechnung begleichen

settled ['setld] *adj.* vorausbestimmt ⟨*Zukunft*⟩; beständig ⟨*Wetter*⟩; geregelt ⟨*Lebensweise*⟩; **I don't feel ~ in this house/job** ich kann mich in diesem Haus nicht heimisch fühlen/in diese Arbeit nicht hineinfinden

settlement ['setlmənt] *n.* **a)** Entscheidung, *die; (of price)* Einigung, *die; (of argument, conflict, etc.)* Beilegung, *die; (of problem)* Lösung, *die; (of question)* Klärung, *die; (of affairs)* Regelung, *die; (of court case)* Vergleich, *der;* **reach a ~:** zu einer Einigung kommen; **reach a ~ out of court** sich außergerichtlich vergleichen; **b)** *(of bill, account, etc.)* Bezahlung, *die;* Begleichung, *die;* **c)** *(Law: bestowal)* Zuwendung, *die; (in will)* Legat, *das (fachspr.);* Vermächtnis, *das;* **d)** *(colony)* Siedlung, *die; (colonization)* Besiedlung, *die*

settler ['setlə(r)] *n.* Siedler, *der/* Siedlerin, *die*

set: ~-to ['settu:] *n., pl.* **~-tos** Streit, *der;* **~-tos** Streitereien; *(with fists)* Prügeleien; **have a ~-to** Streit haben; *(with fists)* sich prügeln; **~-up** *n. (coll.)* **a)** *(organization)* System, *das; (structure)* Aufbau, *der;* **b)** *(situation)* Zustand, *der;* **what's the ~-up here?** wie läuft das hier? *(ugs.)*

seven ['sevn] **1.** *adj.* sieben; *see also* **eight 1. 2.** *n.* Sieben, *die; see also* **eight 2 a, c, d**

seventeen [sevn'ti:n] **1.** *adj.* siebzehn; *see also* **eight 1. 2.** *n.* Siebzehn, *die; see also* **eight 2 a; eighteen 2**

seventeenth [sevn'ti:nθ] **1.** *adj.* siebzehnt...; *see also* **eighth 1. 2.** *n. (fraction)* Siebzehntel, *das; see also* **eighth 2**

seventh ['sevnθ] **1.** *adj.* sieb[en]t...; *see also* **eighth 1. 2.** *n. (in sequence)* sieb[en]te, *der/die/das; (in rank)* Sieb[en]te, *der/die/das; (fraction)* Sieb[en]tel, *das; see also* **eighth 2**

seventieth ['sevntɪɪθ] **1.** *adj.* siebzigst...; *see also* **eighth 1. 2.** *n. (fraction)* Siebzigstel, *das; see also* **eighth 2**

seventy ['sevntɪ] **1.** *adj.* siebzig; *see also* **eight 1. 2.** *n.* Siebzig, *die; see also* **eight 2 a; eighty 2**

seventy: ~-eight *n. (record)* Achtundsiebziger[platte], *die;* **~-first** *etc. adj.* einundsiebzigst... *usw.; see also* **eighth 1; ~-one** *etc.* **1.** *adj.* einundsiebzig *usw.; see also* **eight 1; 2.** *n.* Einundsiebzig *usw., die; see also* **eight 2 a**

sever ['sevə(r)] *v.t.* **a)** *(cut)* durchtrennen; *(fig.: break off)* abbrechen ⟨*Beziehungen, Verbindung*⟩; **b)** *(separate with force)* abtrennen; *(with axe etc.)* abhacken

several ['sevrl] **1.** *adv.* **a)** *(a few)* mehrere; einige; ~ **times** mehrmals; mehrere *od.* einige Male; ~ **more copies** noch einige Exemplare mehr; **b)** *(separate, diverse)* verschieden. **2.** *pron.* einige; ~ **of us** einige von uns; ~ **of the buildings** einige *od.* mehrere [der] Gebäude

severance ['sevərəns] *n. (of diplomatic relations)* Abbruch, *der; (of communications)* Unterbrechung, *die; attrib.* ~ **pay** Abfindung, *die*

severe [sɪ'vɪə(r)] *adj.*, **~r** [sɪ'vɪərə(r)], **~st** [sɪ'vɪərɪst] **a)** *(strict)* streng; hart ⟨*Urteil, Strafe, Kritik*⟩; **be ~ on** *or* **with sb.** streng mit jmdm. sein *od.* umgehen; **b)** *(violent, extreme)* streng ⟨*Frost, Winter*⟩; schwer ⟨*Sturm, Dürre, Verlust, Behinderung, Verletzung*⟩; rauh ⟨*Wetter*⟩; heftig ⟨*Anfall, Schmerz*⟩; **c)** *(making great demands)* hart ⟨*Test, Prüfung, Konkurrenz*⟩; **d)** *(serious, not slight)* bedrohlich ⟨*Mangel, Knappheit*⟩; heftig, stark ⟨*Blutung*⟩; schwer ⟨*Krankheit*⟩; **e)** *(unadorned)* streng ⟨*Stil, Schönheit, Dekor*⟩

severely [sɪ'vɪəlɪ] *adv.* hart; hart, streng ⟨*bestrafen*⟩; schwer ⟨*verletzt, behindert*⟩; **be ~ critical of sth.** etw. scharf kritisieren

severity [sɪ'verɪtɪ] *n.* Strenge, *die; (of drought, shortage)* großes Ausmaß; *(of criticism)* Schärfe, *die*

sew [səʊ] **1.** *v.t., p.p.* **~n** [səʊn] *or* **~ed** [səʊd] nähen; ~ **together** zusammennähen ⟨*Stoff, Leder usw.*⟩. **2.** *v.i., p.p.* **~n** *or* **~ed** nähen

~ **'on** *v.t.* annähen ⟨*Knopf*⟩; aufnähen ⟨*Abzeichen, Band*⟩

~ **'up** *v.t.* **a)** nähen ⟨*Saum, Naht, Wunde*⟩; **they ~ed me up after the operation** *(coll.)* nach der Operation haben sie mich wieder zugenäht; **b)** *(Brit. fig. coll.: settle, arrange)* **be ~n up** unter Dach und Fach sein; *(completely organized)* durchorganisiert sein

sewage ['sju:ɪdʒ, 'su:ɪdʒ] *n.* Abwasser, *das*

sewage: ~ disposal *n.* Abwasserbeseitigung, *die;* **~ farm** *n.*, **~ works** *n. sing., pl. same* Kläranlage, *die*

sewer ['sju:ə(r), 'su:ə(r)] *n. (tunnel)* Abwasserkanal, *der; (pipe)* Abwasserleitung, *die*

sewerage ['sju:ərɪdʒ, 'su:ərɪdʒ] *n.* **a)** *(system of sewers)* Kanalisation, *die;* **b)** *no pl. (removal of sewage)* Abwasserbeseitigung, *die;* **c)** *(sewage)* Abwasser, *das*

sewing ['səʊɪŋ] *n.* Näharbeit, *die*

sewing: ~-basket *n.* Nähkorb, *der;* **~-machine** *n.* Nähmaschine, *die*

sewn *see* **sew**

sex [seks] **1.** *n.* **a)** Geschlecht, *das;* **what ~ is the baby/puppy?** welches Geschlecht hat das Baby/ der Welpe?; **b)** *(sexuality; coll.: intercourse)* Sex, *der (ugs.);* **have ~ with sb.** *(coll.)* mit jmdm. schlafen *(verhüll.);* Sex mit jmdm. haben *(salopp).* **2.** *attrib. adj.* Geschlechts⟨*organ, -trieb*⟩; Sexual⟨*verbrechen, -trieb, -instinkt*⟩

sex: ~ act *n.* Geschlechtsakt, *der;* **~ appeal** *n.* Sex Appeal, *der;* **~ discrimination** *n.* sexuelle Diskriminierung; **~ education** *n.* Sexualerziehung, *die*

sexily ['seksɪlɪ] *adv.* aufreizend, *(ugs.)* sexy ⟨*sprechen, lächeln*⟩

sexism ['seksɪzm] *n., no pl.* Sexismus, *der*

sexist ['seksɪst] **1.** *n.* Sexist, *der/* Sexistin, *die.* **2.** *adj.* sexistisch

sex: ~ life *n.* Geschlechtsleben, *das;* **~ maniac** *n.* Triebverbrecher, *der;* **~ symbol** *n.* Sexidol, *das*

sexual ['seksjʊəl, 'sekʃʊəl] *adj.* sexuell; geschlechtlich, sexuell ⟨*Anziehung, Erregung, Verlangen, Diskriminierung*⟩; ~ **maturity/ behaviour** Geschlechtsreife, *die/* Sexualverhalten, *das*

sexual 'intercourse *n., no pl., no indef. art.* Geschlechtsverkehr, *der*

sexuality [seksjʊ'ælɪtɪ, sekʃʊ'ælɪtɪ] *n., no pl.* Sexualität, *die*
sexy ['seksɪ] *adj.* sexy *(ugs.)*; erotisch ‹*Film, Buch, Gemälde*›
Seychelles [seɪ'ʃelz] *pr. n.* Seychellen *Pl.*
shabbily ['ʃæbɪlɪ] *adv.*, **shabby** ['ʃæbɪ] *adj.* schäbig
shack [ʃæk] **1.** *n.* Hütte, *die.* **2.** *v. i. (coll.)* ~ **up with sb.** mit jmdm. zusammenziehen
shackle ['ʃækl] **1.** *n., usu. in pl. (lit. or fig.)* Fessel, *die.* **2.** *v. t. (lit. or fig.)* anketten (**to** an + *Akk.*)
shade [ʃeɪd] **1.** *n.* **a)** Schatten, *der;* **put sb./sth. in[to] the ~** *(fig.)* jmdn./etw. in den Schatten stellen; **38[°C] in the ~:** 38° im Schatten; **b)** *(colour)* Ton, *der; (fig.)* Schattierung, *die;* **various ~s of purple** verschiedene Violettöne; **~s of meaning** Bedeutungsnuancen *od.* -schattierungen; **c)** *(eye-shield)* [Augen]schirm, *der; (lamp~)* [Lampen]schirm, *der.* **2.** *v. t.* **a)** *(screen)* beschatten *(geh.)*; Schatten geben (+ *Dat.*); **be ~d from the sun** vor Sonneneinstrahlung geschützt sein; ~ **one's eyes with one's hand** die Hand schützend über die Augen halten; **b)** abdunkeln ‹*Fenster, Lampe, Licht*›; **c)** *(just defeat)* knapp überbieten. **3.** *v. i. (lit. or fig.)* übergehen (**into** in + *Akk.*)
~ **'in** *v. t.* [ab]schattieren
shading [ʃeɪdɪŋ] *n.* Schattierung, *die; (protection from light)* Lichtschutz, *der*
shadow ['ʃædəʊ] **1.** *n.* **a)** Schatten, *der;* **cast a ~ over** *(lit. or fig.)* einen Schatten werfen auf (+ *Akk.*); **be in sb.'s ~** *(fig.)* in jmds. Schatten stehen; **be afraid of one's own ~** *(fig.)* sich vor seinem eigenen Schatten fürchten; **b)** *(slightest trace)* **without a ~ of doubt** ohne den Schatten eines Zweifels; **catch at** *or* **chase after ~s** einem Phantom *od. (geh.)* Schatten nachjagen; **c)** *(ghost, lit. or fig.)* Schatten, *der;* **d) S~** *attrib. (Brit. Polit.)* ‹*Minister, Kanzler*› im Schattenkabinett; **S~ Cabinet** Schattenkabinett, *das.* **2.** *v. t.* **a)** *(darken)* überschatten; **b)** *(follow secretly)* beschatten
shadowy ['ʃædəʊɪ] *adj.* **a)** *(not distinct)* schattenhaft; schemenhaft *(geh.)*; **b)** *(full of shade)* schattig
shady ['ʃeɪdɪ] *adj.* **a)** *(giving shade)* schattenspendend *(geh.)*; *(situated in shade)* schattig; **b)** *(disreputable)* zwielichtig
shaft [ʃɑ:ft] *n.* **a)** *(of tool, golf club, spear)* Schaft, *der;* **b)** *(Mech. Engin.)* Welle, *die;* **c)** *(of cart or carriage)* Deichsel, *die:* **d)** *(of mine, lift, etc.)* Schacht, *der*
shaggy ['ʃægɪ] *adj.* **a)** *(hairy)* zottelig; **b)** *(unkempt)* struppig
shaggy-'dog story *n. endlos langer Witz ohne richtige Pointe*
Shah [ʃɑ:] *n.* Schah, *der*
shake [ʃeɪk] **1.** *n.* Schütteln, *das;* **give sb./sth. a ~:** jmdn./etw. schütteln; **with a ~ of the head** mit einem Kopfschütteln; **be no great ~s** *(coll.)* nicht gerade umwerfend sein *(ugs.).* **2.** *v. t.*, **shook** [ʃʊk], **shaken** ['ʃeɪkn] *or (arch./coll.)* **shook a)** *(move violently)* schütteln; **the dog shook itself** der Hund schüttelte sich; **be ~n to pieces** völlig durchgeschüttelt werden; ~ **one's fist/a stick at sb.** jmdm. mit der Faust/einem Stock drohen; **'~ [well] before using'** „vor Gebrauch [gut] schütteln!"; ~ **hands** sich *(Dat.) od.* einander die Hand geben *od.* schütteln; ~ **sb. by the hand** jmdm. die Hand schütteln *od.* drücken; **b)** *(cause to tremble)* erschüttern ‹*Gebäude usw.*›; ~ **one's head [over sth.]** [über etw. *(Akk.)*] den Kopf schütteln; **c)** *(weaken)* erschüttern; ~ **sb.'s faith in sth./sb.** jmds. Glauben an etw./jmdn. erschüttern; **d)** *(agitate)* erschüttern; **she was badly ~n by the news of his death** die Nachricht von seinem Tod erschütterte sie sehr; **he failed his exam – that shook him!** er hat die Prüfung nicht bestanden – das war ein Schock für ihn!; ~ **sb.'s composure** jmdn. aus dem Gleichgewicht bringen. **3.** *v. i.*, **shook, shaken** *or (arch./coll.)* **shook a)** *(tremble)* wackeln; ‹*Boden, Stimme:*› beben; ‹*Hand:*› zittern; ~ **[all over] with cold/fear** [am ganzen Leib] vor Kälte/Angst schlottern; ~ **like a leaf** wie Espenlaub zittern; ~ **with emotion** vor Erregung beben; ~ **in one's shoes** *(coll.)* vor Angst schlottern; **b)** *(coll.: ~ hands)* sich *(Dat.)* die Hand geben; **let's ~ on it!** schlag ein!; Hand drauf!
~ **'off** *v. t. (lit. or fig.)* abschütteln
~ **'out** *v. t.* ausschütteln; *(spread out)* ausbreiten
~ **'up** *v. t.* **a)** *(mix)* schütteln; **b)** aufschütteln ‹*Kissen*›; **c)** *(discompose)* einen Schrecken einjagen (+ *Dat.*); **she felt pretty ~n up** sie hatte einen ziemlichen Schrecken bekommen; **d)** *(rouse to activity)* aufrütteln; **e)** *(coll.: reorganize)* umkrempeln *(ugs.)*
shaken *see* **shake 2, 3**
'shake-up *n. (coll.: reorganization)* **give sth. a [good] ~:** etw. [total] umkrempeln *(ugs.)*; **sth. needs a ~:** etw. muß [mal] umgekrempelt werden *(ugs.)*
shaky ['ʃeɪkɪ] *adj.* **a)** *(unsteady)* wack[e]lig ‹*Möbelstück, Leiter*›; zittrig ‹*Hand, Stimme, Bewegung, Greis*›; **feel ~:** sich zittrig fühlen; **be ~ on one's legs** wacklig auf den Beinen sein *(ugs.)*; **b)** *(unreliable)* **his German is rather ~:** sein Deutsch steht auf wackligen Füßen *(ugs.)*
shall [ʃl, *stressed* ʃæl] *v.aux. only in pres.* **shall,** *neg. (coll.)* **shan't** [ʃɑ:nt], *past* **should** [ʃəd, *stressed* ʃʊd], *neg. (coll.)* **shouldn't** ['ʃʊdnt] **a)** *expr. simple future* werden; **b) should** *expr. conditional* würde/würdest/würden/würdet; **I should have been killed if I had let go** ich wäre getötet worden, wenn ich losgelassen hätte; **c)** *expr. command* **the committee ~ not be disturbed** der Ausschuß darf nicht gestört werden; **d)** *expr. will or intention* **what ~ we do?** was sollen wir tun?; **let's go in, ~ we?** gehen wir doch hinein, oder?; **I'll buy six, ~ I?** ich kaufe 6 [Stück], ja?; **you ~ pay for this!** das sollst du mir büßen!; **we should be safe now** jetzt dürften wir in Sicherheit sein; **he shouldn't do things like that!** er sollte so etwas nicht tun!; **oh, you shouldn't have!** *expr. gratitude* das wäre doch nicht nötig gewesen!; **you should be more careful** du solltest vorsichtiger *od.* sorgfältiger sein; **e)** *in conditional clause* **if we should be defeated** falls wir unterliegen [sollten]; **I should hope so** ich hoffe es; *(indignant)* das möchte ich hoffen!; **f)** *in tentative assertion* **I should like to disagree with you on that point** in dem Punkt *od.* da möchte ich dir widersprechen; **I should say it is time we went home** ich würde sagen *od.* ich glaube, es ist Zeit, daß wir nach Hause gehen
shallot [ʃə'lɒt] *n.* Schalotte, *die*
shallow ['ʃæləʊ] *adj.* seicht ‹*Wasser, Fluß*›; flach ‹*Schüssel, Teller, Wasser*›; *(fig.)* seicht *(abwertend)* ‹*Unterhaltung, Gerede*›; flach *(abwertend)* ‹*Person, Denker, Geist*›; platt *(abwertend)* ‹*Argument, Verallgemeinerung*›
sham [ʃæm] **1.** *adj.* unecht; imitiert ‹*Leder, Holz, Pelz, Stein*›. **2.** *n. (pretence)* Heuchelei, *die; (person)* Heuchler, *der*/Heuchlerin, *die;* **their marriage is only a ~:** ihre Ehe besteht nur auf dem Papier. **3.** *v. t.*, **-mm-** vortäuschen;

simulieren; ~ **dead/ill** sich tot/ krank stellen. **4.** *v.i.*, **-mm-** simulieren; sich verstellen

shamble ['ʃæmbl] **1.** *v.i.* schlurfen. **2.** *n.* Schlurfen, *das*

shambles ['ʃæmblz] *n. sing. (coll.: mess)* Chaos, *das;* **the house/room was a ~:** das Haus/ Zimmer glich einem Schlachtfeld; **the economy is in a ~:** in der Wirtschaft herrschen chaotische Zustände

shame [ʃeɪm] **1.** *n.* **a)** Scham, *die;* **feel ~/no ~ for what one did** sich schämen/sich nicht schämen für das, was man getan hat; **hang one's head in** *or* **for ~:** beschämt den Kopf senken; **blush with ~:** vor Scham erröten; **have no [sense of] ~:** kein[erlei] Schamgefühl besitzen; **have you no ~?** schämst du dich nicht?; **to my ~ I must confess ...:** ich muß zu meiner Schande gestehen ...; **b)** *(state of disgrace)* Schande, *die;* **~ on you!** du solltest dich schämen; **put sb./ sth. to ~:** jmdn. beschämen/etw. in den Schatten stellen; **c) what a ~!** *(bad luck)* so ein Pech!; *(pity)* wie schade!; **it is a crying** *or* **terrible** *or* **great ~:** es ist eine wahre Schande. **2.** *v.t.* beschämen; **~ sb. into doing/out of doing sth.** jmdn. dazu bringen, daß er sich schämt und etw. tut/nicht tut

'shamefaced *adj.* betreten; **have a ~ look, look ~:** betreten dreinblicken

shameful ['ʃeɪmfl] *adj.* beschämend

shameless ['ʃeɪmlɪs] *adj.* schamlos

shampoo [ʃæm'pu:] **1.** *v.t.* schamponieren ‹*Haar, Teppich, Polster*›. **2.** *n.* Shampoo[n], *das;* **carpet ~:** Teppichschaum, *der;* **have a ~ and set** sich *(Dat.)* die Haare waschen und [ein]legen lassen; **give one's hair a ~:** sich *(Dat.)* die Haare waschen

shamrock ['ʃæmrɒk] *n.* Klee, *der; (emblem of Ireland)* Shamrock, *der*

shandy ['ʃændɪ] *n.* Bier mit Limonade; Radler, *der (bes. südd.)*

shank [ʃæŋk] *n.* **a)** *(of person)* Unterschenkel, *der;* **[go] on S~s's mare** *or* **pony** auf Schusters Rappen [reisen] *(scherzh.);* **b)** *(of horse)* Vordermittelfuß, *der*

shan't [ʃɑ:nt] *(coll.)* = **shall not**

[1]**shanty** ['ʃæntɪ] *n. (hut)* [armselige] Hütte

[2]**shanty** *n. (song)* Shanty, *das;* Seemannslied, *das*

'shanty town *n.* Elendsviertel, *das*

shape [ʃeɪp] **1.** *v.t.* **a)** *(create, form)* formen; bearbeiten ‹*Holz, Stein*› **(into** zu); **b)** *(adapt, direct)* prägen, formen ‹*Charakter, Person*›; [entscheidend] beeinflussen ‹*Gang der Geschichte, Leben, Zukunft, Gesellschaft*›. **2.** *v.i.* sich entwickeln. **3.** *n.* **a)** *(external form, outline)* Form, *die;* **spherical/rectangular in ~:** kugelförmig/rechteckig; **take ~** ‹*Konstruktion, Skulptur:*› Gestalt annehmen *(see also* **c**); **b)** *(appearance)* Gestalt, *die;* **in the ~ of a woman** in Gestalt einer Frau; **c)** *(specific form)* Form, *die;* Gestalt, *die;* **take ~** ‹*Plan, Vorhaben:*› Gestalt *od.* feste Formen annehmen *(see also* **a**); **get one's ideas into ~:** seine Gedanken sammeln; **knock sth. into ~:** etw. wieder in Form bringen; **in all ~s and sizes, in every ~ and size** in allen Formen und Größen; **the ~ of things to come** die Dinge, die da kommen sollen/sollten; **d)** *(condition)* Form, *die (bes. Sport);* **do yoga to keep in ~:** Yoga machen, um in Form zu bleiben; **be in good/bad ~:** gut/schlecht in Form sein; **e)** *(person seen, ghost)* Gestalt, *die*

~ **'up** *v.i.* sich entwickeln; **how's the new editor shaping up?** wie macht sich der neue Redakteur?

shaped [ʃeɪpt] *adj.* geformt; **be ~ like a pear** die Form einer Birne haben

shapeless ['ʃeɪplɪs] *adj.* formlos; unförmig ‹*Kleid, Person*›

shapely ['ʃeɪplɪ] *adj.* wohlgeformt ‹*Beine, Busen*›; gut ‹*Figur*›

share [ʃeə(r)] **1.** *n.* **a)** *(portion)* Teil, *der od. das; (part one is entitled to)* **[fair] ~:** Anteil, *der;* **he had a large ~ in bringing it about** er hatte großen Anteil daran, daß es zustande kam; **pay one's ~ of the bill** seinen Teil der Rechnung bezahlen; **have a ~ in the profits** am Gewinn beteiligt sein; **do more than one's [fair] ~ of the work** mehr als seinen Teil zur Arbeit beitragen; **each had his ~ of the cake** jeder bekam seinen Teil vom Kuchen ab; **have more than one's [fair] ~ of the blame/attention** mehr Schuld zugewiesen bekommen/mehr Beachtung finden, als man verdient; **she had her ~ of luck/bad luck** sie hat aber auch Glück/Pech gehabt; **take one's ~ of the responsibility** seinen Teil Verantwortung tragen; **take one's ~ of the blame** seinen Teil Schuld auf sich *(Akk.)* nehmen; **go ~s** teilen; **b)** *(part-ownership of property)* [Geschäfts]anteil, *der; (part of company's capital)* Aktie, *die;* **hold ~s in a company** *(Brit.)* Anteile *od.* Aktien einer Gesellschaft besitzen. **2.** *v.t.* teilen; gemeinsam tragen ‹*Verantwortung*›; **~ the same birthday/surname** am gleichen Tag Geburtstag/den gleichen Nachnamen haben. **3.** *v.i.* **~ in** teilnehmen an (+ *Dat.*); beteiligt sein an (+ *Dat.*) ‹*Gewinn, Planung*›; teilen ‹*Freude, Erfahrung*›

~ **'out** *v.t.* aufteilen **(among** unter + *Akk.*)

share: ~ certificate *n.* Aktienurkunde, *die;* **~holder** *n.* Aktionär, *der*/Aktionärin, *die;* **~-out** *n.* Aufteilung, *die*

shark [ʃɑ:k] *n.* **a)** Hai[fisch], *der;* **b)** *(fig.: swindler)* gerissener Geschäftemacher; **property ~:** Grundstückshai, *der (ugs. abwertend); see also* **loan shark**

sharp [ʃɑ:p] **1.** *adj.* **a)** *(with fine edge)* scharf; *(with fine point)* spitz ‹*Nadel, Bleistift, Giebel, Gipfel*›; **b)** *(clear-cut)* scharf ‹*Umriß, Kontrast, Bild, Gesichtszüge, Linie*›; deutlich ‹*Unterscheidung*›; präzise ‹*Eindruck*›; **c)** *(abrupt, angular)* scharf ‹*Kurve, Winkel*›; steil, schroff ‹*Abhang*›; stark ‹*Gefälle*›; **a ~ rise/fall in prices** ein jäher Preisanstieg/ Preissturz; **d)** *(intense)* groß ‹*Appetit, Hunger[gefühl]*›; *(acid, pungent)* scharf ‹*Würze, Geschmack, Sauce, Käse*›; sauer ‹*Apfel*›; herb ‹*Wein*›; *(shrill, piercing)* schrill ‹*Schrei, Pfiff*›; *(biting)* scharf ‹*Wind, Frost, Luft*›; *(sudden, severe)* heftig ‹*Schmerz, Anfall, Krampf, Kampf*›; *(harsh, acrimonious)* scharf ‹*Protest, Tadel, Ton, Stimme, Zunge, Worte*›; **a short ~ struggle** ein kurzer, heftiger Kampf; **e)** *(acute, quick)* scharf ‹*Augen, Verstand, Gehör, Ohr, Beobachtungsgabe, Intelligenz, Geruchssinn*›; aufgeweckt ‹*Kind*›; scharfsinnig ‹*Bemerkung*›; begabt ‹*Schüler, Student*›; **that was pretty ~!** das war ganz schön clever!; **keep a ~ look-out for the police!** halt die Augen offen, falls die Polizei kommt!; **f)** *(derog.: artful, dishonest, quick to take advantage)* gerissen; **g)** *(Mus.)* [um einen Halbton] erhöht ‹*Note*›; **F/G/C** *etc.* **~:** fis, Fis/gis, Gis/cis, Cis *usw.*, *das.* **2.** *adv.* **a)** *(punctually)* **at six o'clock ~:** Punkt sechs Uhr; **b)** *(suddenly)* scharf ‹*bremsen*›; plötzlich ‹*anhalten*›; **turn ~ right** scharf nach rechts abbiegen; **c) look ~!** halt dich ran! *(ugs.);* **d)**

(Mus.) zu hoch ⟨*singen, spielen*⟩. **3.** *n. (Mus.)* erhöhter Ton; *(symbol)* Kreuz, *das;* Erhöhungszeichen, *das*

sharpen [ˈʃɑːpn] *v.t.* schärfen *(auch fig.);* [an]spitzen ⟨*Bleistift*⟩; *(fig.)* anregen ⟨*Appetit*⟩

sharpener [ˈʃɑːpnə(r)] *n. (for pencil)* Bleistiftspitzer, *der;* Spitzer, *der (ugs.); (for tools)* Abziehstein, *der;* Schleifstein, *der*

sharp-eyed [ˈʃɑːpaɪd] *adj.* scharfäugig; **be ~:** scharfe Augen haben

sharpish [ˈʃɑːpɪʃ] *adv. (coll.) (quickly)* rasch; *(promptly)* unverzüglich; sofort

sharply [ˈʃɑːplɪ] *adv.* **a)** *(acutely)* spitz; **~ angled** spitzwinklig; **b)** *(clearly)* scharf ⟨*voneinander unterschieden, kontrastierend, umrissen*⟩; **c)** *(abruptly)* scharf ⟨*bremsen, abbiegen*⟩; steil, schroff ⟨*abfallen*⟩; **d)** *(acidly)* scharf ⟨*gewürzt*⟩; *(harshly)* in scharfem Ton ⟨*antworten*⟩; **~ worded letter** Brief in scharfem Ton; **e)** *(quickly)* schnell, rasch ⟨*denken, handeln*⟩

sharp: ~shooter *n.* Scharfschütze, *der;* **~-witted** [ˈʃɑːpwɪtɪd] *adj.* scharfsinnig

shatter [ˈʃætə(r)] **1.** *v.t.* **a)** *(smash)* zertrümmern; **b)** *(destroy)* zerschlagen ⟨*Hoffnungen*⟩; **c)** *(coll.: greatly upset)* schwer mitnehmen. **2.** *v.i.* zerbrechen; zerspringen

shattered [ˈʃætəd] *adj.* **a)** zerbrochen, zersprungen ⟨*Scheibe, Glas, Fenster*⟩; *(fig.)* zerstört ⟨*Hoffnungen*⟩; zerrüttet ⟨*Nerven*⟩; **b)** *(coll.: greatly upset)* **she was ~ by the news** die Nachricht hat sie schwer mitgenommen; **I'm ~!** ich bin ganz erschüttert!; **c)** *(Brit. coll.: exhausted)* **I'm ~:** ich bin [völlig] kaputt *(ugs.)*

shattering [ˈʃætərɪŋ] *adj.* **a)** *(ruinously destructive)* verheerend ⟨*Wirkung, Explosion*⟩; vernichtend ⟨*Schlag, Niederlage*⟩; **b)** *(coll.: very upsetting)* erschütternd

shave [ʃeɪv] **1.** *v.t.* **a)** rasieren; abrasieren ⟨*Haare*⟩; **he ~d his beard** er hat sich *(Dat.)* den Bart abrasiert; **b)** *(graze)* ⟨*Auto:*⟩ streifen. **2.** *v.i.* **a)** sich rasieren; **b)** *(scrape)* **~ past sth.** etw. [leicht] streifen. **3.** *n.* **a)** Rasur, *die;* **have** *or* **get a ~:** sich rasieren; **a clean** *or* **close ~:** eine Glattrasur; **b)** **close ~** *(fig.) see* **close 1f**

~ ˈoff *v.t.* abrasieren ⟨*Bart, Haare*⟩

shaven [ˈʃeɪvn] *adj.* rasiert; [kahl]geschoren ⟨*Kopf*⟩

shaver [ˈʃeɪvə(r)] *n.* Rasierapparat, *der;* Rasierer, *der (ugs.)*

ˈshaver point *n.* Anschluß *od.* Steckdose für den Rasierapparat

shaving [ˈʃeɪvɪŋ] *n.* **a)** *(action)* Rasieren, *das;* **b)** *in pl. (of wood, metal, etc.)* Späne

shaving: ~-brush *n.* Rasierpinsel, *der;* **~-cream** *n.* Rasiercreme, *die;* **~-foam** *n.* Rasierschaum, *der*

shawl [ʃɔːl] *n.* Schultertuch, *das*

she [ʃɪ, *stressed* ʃiː] *pron.* sie; *referring to personified things or animals which correspond to German masculines/neuters* er/es; **it was ~** *(formal)* sie war es; *see also* **[1,2]her; hers; herself**

she- [ʃiː] *in comb.* weiblich; **~-ass/-bear** Eselin, *die*/Bärin, *die*

sheaf [ʃiːf] *n., pl.* **sheaves** [ʃiːvz] *(of corn etc.)* Garbe, *die; (of paper, arrows, etc.)* Bündel, *das*

shear [ʃɪə(r)] *v.t., p.p.* **shorn** [ʃɔːn] *or* **~ed** *(clip)* scheren

~ ˈoff 1. *v.t.* abtrennen. **2.** *v.i.* abscheren *(Technik)*

shears [ʃɪəz] *n. pl.* **[pair of] ~** *(große)* Schere, *die;* **garden ~:** Heckenschere, *die*

sheath [ʃiːθ] *n., pl.* **~s** [ʃiːðz, ʃiːθs] **a)** *(for sword etc.)* Scheide, *die;* **b)** *(condom)* Gummischutz, *der;* **c)** *(Electr.)* Mantel, *der*

ˈsheath-knife *n.* Fahrtenmesser, *das*

sheaves *pl. of* **sheaf**

[1]shed [ʃed] *v.t.,* **-dd-, shed a)** *(part with)* verlieren; abwerfen, verlieren ⟨*Laub, Geweih*⟩; abstreifen ⟨*Haut, Hülle, Badehose*⟩; ausziehen ⟨*Kleidung*⟩; **the snake is ~ding its skin** die Schlange häutet sich; **you should ~ a few pounds** du solltest ein paar Pfund abspecken *(salopp);* **b)** vergießen ⟨*Blut, Tränen*⟩; **don't ~ any tears over him** seinetwegen solltest du keine Tränen vergießen; **c)** *(dispense)* verbreiten ⟨*Wärme, Licht*⟩; *see also* **[1]light 1f; d)** *(fig.: cast off)* abschütteln ⟨*Sorgen, Bürde*⟩

[2]shed *n.* Schuppen, *der*

she'd [ʃɪd, *stressed* ʃiːd] **a)** = **she had; b)** = **she would**

sheen [ʃiːn] *n.* Glanz, *der*

sheep [ʃiːp] *n., pl. same* Schaf, *das;* **separate the ~ from the goats** *(fig.)* die Böcke von den Schafen trennen; **count ~** *(fig.)* Schäfchen zählen *(fam.);* **follow sb. like ~:** jmdm. wie eine Schafherde folgen

ˈsheep-dog *n.* Hütehund, *der;* Schäferhund, *der;* **Old English S~:** Bobtail, *der*

sheepish [ˈʃiːpɪʃ] *adj. (awkwardly self-conscious)* verlegen; *(embarrassed)* kleinlaut; **he felt a bit ~** *(foolish)* es war ihm ein bißchen peinlich

ˈsheepskin *n.* Schaffell, *das; attrib.* **~ [jacket]** Schaffelljacke, *die*

sheer [ʃɪə(r)] *adj.* **a)** *attrib. (mere, absolute)* rein; blank ⟨*Unsinn, Gewalt*⟩; **by ~ chance** rein zufällig; **the ~ insolence of it!** so eine Frechheit!; **only by ~ hard work** nur durch harte Arbeit; **b)** *(perpendicular)* schroff ⟨*Felsen, Abfall*⟩; steil ⟨*Felsen, Abfall, Aufstieg*⟩; **c)** *(finely woven)* hauchfein

sheet [ʃiːt] *n.* **a)** Laken, *das; (for covering mattress)* Bettuch, *das;* Laken, *das;* **put clean ~s on the bed** das Bett frisch beziehen; **between the ~s** *(in bed)* im Bett; **b)** *(of thin metal, plastic)* Folie, *die; (of iron, tin)* Blech, *das; (of glass, of thicker metal, plastic)* Platte, *die; (of stamps)* Bogen, *der; (of paper)* Bogen, *der;* Blatt, *das;* **a ~ of paper** ein Papierbogen; ein Bogen *od.* Blatt Papier; **start with/have a clean ~** *(fig.)* ganz neu beginnen/eine reine Weste haben *(ugs.); attrib.* **~ glass/metal/iron** Flachglas, *das*/Blech, *das*/Eisenblech, *das;* **c)** *(wide expanse)* ⟨*Eis-, Lava-, Nebel*⟩decke, *die;* **a huge ~ of flame** ein Flammenmeer; **the rain was coming down in ~s** es regnete in Strömen

sheet: ~ lightning *n. (Meteorol.)* Flächenblitz, *der;* **~ music** *n.* Notenblätter

sheik, sheikh [ʃeɪk, ʃiːk] *ns.* Scheich, *der*

shelf [ʃelf] *n., pl.* **shelves** [ʃelvz] *(flat board)* Brett, *das;* Bord, *das; (compartment)* Fach, *das; (set of shelves)* Regal, *das;* **~ of books** Bücherbrett, *das;* **be left on the ~** *(fig.)* sitzengeblieben sein *(ugs.);* **be put on the ~** *(fig.)* aufs Abstellgleis geschoben werden *(ugs.)*

shelf: ~-life *n.* Lagerfähigkeit, *die;* **~-room, ~-space** *ns.* Stellfläche [im Regal]; **give sth. ~-room** *or* **~-space** sich *(Dat.)* etw. ins Regal stellen

shell [ʃel] **1.** *n.* **a)** *(casing)* Schale, *die; (of turtle, tortoise)* Panzer, *der; (of snail)* Haus, *das; (of pea)* Schote, *die;* Hülse, *die;* **collect ~s on the beach** am Strand Muscheln sammeln; **come out of one's ~** *(fig.)* aus sich herausgehen; **retire** *or* **go into one's ~** *(fig.)* sich in sein Schneckenhaus zurückziehen *(ugs.);* **b)** *(pastry case)* Teighülle, *die;* **c)** *(Mil.: bomb)* Granate, *die;* **d)** *(of unfinished building)* Rohbau, *der; (of ruined building)*

Ruine, *die;* **e)** *(Motor Veh.)* Aufbau, *der;* Karosserie, *die; (after fire, at breaker's, etc.)* [Karosserie]gerippe, *das.* **2.** *v.t.* **a)** *(take out of ~)* schälen; knacken, schälen ‹*Nuß*›; enthülsen, *(nordd.)* palen ‹*Erbsen*›; **b)** *(Mil.)* [mit Artillerie] beschießen

~ **'out** *v.t. & i. (sl.)* blechen *(ugs.)* **(on** für)

she'll [ʃɪl, *stressed* ʃi:l] = **she will**

shell: ~fish *n., pl. same* **a)** Schal[en]tier, *das; (oyster, clam)* Muschel, *die; (crustacean)* Krebstier, *das;* **b)** *in pl. (Gastr.)* Meeresfrüchte *Pl.;* **~-shock** *n. (Psych.)* Kriegsneurose, *die;* **~-shocked** *adj.* **be ~-shocked** eine Kriegsneurose haben; *(fig.)* niedergeschmettert sein

shelter ['ʃeltə(r)] **1.** *n.* **a)** *(shield)* Schutz, *der* **(against** vor + *Dat.,* gegen); **bomb** *or* **air-raid ~:** Luftschutzraum, *der;* **get under ~:** sich unterstellen; **b)** *no pl. (place of safety)* Zuflucht, *die;* **we needed food and ~:** wir brauchten etwas zu essen und eine Unterkunft; **look for ~ for the night** eine Unterkunft für die Nacht suchen; **offer** *or* **give sb. ~, provide ~ for sb.** jmdm. Zuflucht gewähren *od.* bieten; **take ~ [from a storm]** [vor einem Sturm] Schutz suchen; **seek/reach ~:** Schutz *od.* Zuflucht suchen/finden. **2.** *v.t.* schützen **(from** vor + *Dat.*); Unterschlupf gewähren (+ *Dat.*) ‹*Flüchtling*›; ~ **sb. from blame/harm** jmdn. decken/gegen alle Gefahren schützen. **3.** *v.i.* Schutz *od.* Zuflucht suchen **(from** vor + *Dat.*); **this is a good place to ~:** hier ist man gut geschützt

sheltered ['ʃeltəd] *adj.* geschützt ‹*Platz, Tal*›; behütet ‹*Leben*›

shelve [ʃelv] **1.** *v.t. (put on ~s)* ins Regal stellen; *(fig.) (abandon)* ad acta *od. (ugs.)* zu den Akten legen; *(defer)* auf Eis legen *(ugs.).* **2.** *v.i.* ~ **away/off/out into** ‹*Berg, Boden, Ebene:*› abfallen nach

shelves *pl. of* **shelf**

shepherd ['ʃepəd] **1.** *n.* Schäfer, *der;* Schafhirt, *der.* **2.** *v.t.* hüten; *(fig.)* führen

shepherdess ['ʃepədɪs] *n.* Schäferin, *die;* Schafhirtin, *die*

shepherd's 'pie *n. (Gastr.) Auflauf aus Hackfleisch mit einer Schicht Kartoffelbrei darüber*

sherbet ['ʃɜ:bət] *n.* **a)** *(fruit juice; also Amer.: water-ice)* Sorbet[t], *der od. das;* **b)** *(effervescent drink)* Brauselimonade, *die; (powder)* Brausepulver, *das*

sheriff ['ʃerɪf] *n.* Sheriff, *der*

sherry ['ʃerɪ] *n.* Sherry, *der*

she's [ʃɪz, *stressed* ʃi:z] **a)** = **she is; b) she has**

Shetland Islands ['ʃetlənd aɪləndz] *pr. n. pl.* Shetlandinseln *Pl.;* Shetlands *Pl.*

Shetland 'pony *n.* Shetlandpony, *das*

Shetlands ['ʃetləndz] *pr. n. pl.* Shetlands *Pl.*

shield [ʃi:ld] **1.** *n.* **a)** *(piece of armour)* Schild, *der;* **b)** *(in machinery etc.)* Schutz, *der;* **radiation ~:** Strahlenschutz, *der;* **c)** *(fig.: person or thing that protects)* Schild, *der (geh.);* **d)** *(Sport: trophy)* Trophäe, *die (in Form eines Schildes).* **2.** *v.t.* **a)** *(protect)* schützen **(from** vor + *Dat.*); **b)** *(conceal)* decken ‹*Schuldigen*›; ~ **sb. from the truth** die Wahrheit von jmdm. fernhalten

shier, shiest *see* [1]**shy 1**

shift [ʃɪft] **1.** *v.t.* **a)** *(move)* verrücken, umstellen ‹*Möbel*›; wegnehmen ‹*Arm, Hand, Fuß*›; wegräumen ‹*Schutt*›; entfernen ‹*Schmutz, Fleck*›; *(to another floor, room, or place)* verlegen ‹*Büro, Patienten, Schauplatz*›; ~ **one's weight to the other foot** sein Gewicht auf den anderen Fuß verlagern; ~ **the responsibility/blame on to sb.** *(fig.)* die Verantwortung/Schuld auf jmdn. schieben; **b)** *(Amer. Motor Veh.)* ~ **gears** schalten. **2.** *v.i.* **a)** ‹*Wind:*› drehen **(to** nach); ‹*Ladung:*› verrutschen; ~ **uneasily in one's chair** unruhig auf dem Stuhl hin und her rutschen; **b)** *(manage)* ~ **for oneself** selbst für sich sorgen; **c)** *(sl.: move quickly)* rasen; **this new Porsche really ~s** der neue Porsche geht ab wie eine Rakete *(ugs.);* **d)** *(Amer. Motor Veh.: change gear)* schalten; ~ **down into second gear** in den zweiten Gang runterschalten *(ugs.).* **3.** *n.* **a) a ~ in emphasis** eine Verlagerung des Akzents; **a ~ in values/public opinion** ein Wandel der Wertvorstellungen/ein Umschwung der öffentlichen Meinung; **a ~ towards/away from liberalism** eine Hinwendung zum/Abwendung vom Liberalismus; **b)** *(for work)* Schicht, *die;* **eight-hour/late ~:** Achtstunden-/Spätschicht, *die;* **do** *or* **work the late ~:** Spätschicht haben; **work in ~s** Schichtarbeit machen; **c) make ~ with/without sth.** sich *(Dat.)* mit/ohne etw. behelfen; **d)** *(of typewriter)* Umschaltung, *die;* **e)** *(Amer. Motor Veh.: gear-change)* Schaltung, *die*

shifty ['ʃɪftɪ] *adj.* verschlagen *(abwertend)*

shilling ['ʃɪlɪŋ] *n. (Hist.)* Shilling, *der*

shilly-shally ['ʃɪlɪʃælɪ] *v.i.* zaudern; **stop ~ing!** entschließ dich endlich!

shimmer ['ʃɪmə(r)] **1.** *v.i.* schimmern. **2.** *n.* Schimmer, *der*

shin [ʃɪn] **1.** *n.* Schienbein, *das.* **2.** *v.i.,* **-nn-:** ~ **up/down a tree** *etc.* einen Baum *usw.* hinauf-/hinunterklettern

'shin-bone *n.* Schienbein, *das*

shindig ['ʃɪndɪg] *n. (coll.)* **a)** *see* **shindy; b)** *(party)* Fete, *die (ugs.)*

shindy ['ʃɪndɪ] *n. (brawl)* Rauferei, *die; (row)* Streit, *der; (noise)* Krach, *der*

shine [ʃaɪn] **1.** *v.i.,* **shone** [ʃɒn] **a)** ‹*Lampe, Licht, Stern:*› leuchten; ‹*Sonne:*› scheinen; *(reflect light)* glänzen; ‹*Mond:*› scheinen; **his face shone with happiness/excitement** *(fig.)* er strahlte vor Glück/sein Gesicht glühte vor Aufregung; **b)** *(fig.: be brilliant)* glänzen; **a shining example/light** ein leuchtendes Beispiel/eine Leuchte; ~ **at sport** im Sport glänzen. **2.** *v.t.* **a)** *p.t. & p.p.* **shone** leuchten lassen; ~ **a light on sth./in sb.'s eyes** etw. anleuchten/jmdm. in die Augen leuchten; **b)** *p.t. & p.p.* **~d** *(clean and polish)* putzen; *(make shiny)* polieren. **3.** *n., no pl.* **a)** *(brightness)* Schein, *der;* Licht, *das;* **b)** *(polish)* Glanz, *der;* **have a ~** ‹*Oberfläche:*› glänzen; **take the ~ off sth.** *(fig.: spoil sth.)* einen Schatten auf etw. *(Akk.)* werfen; **c) take a ~ to sb./sth.** *(coll.)* Gefallen an jmdm./etw. finden

shingle ['ʃɪŋgl] *n., no pl., no indef. art. (pebbles)* Kies, *der*

shingles ['ʃɪŋglz] *n. sing. (Med.)* Gürtelrose, *die*

shin: ~-guard, ~-pad *ns.* Schienbeinschutz, *der*

shiny ['ʃaɪnɪ] *adj.* glänzend

ship [ʃɪp] **1.** *n.* Schiff, *das.* **2.** *v.t.,* **-pp-** *(take on board)* einschiffen, an Bord bringen ‹*Vorräte, Ladung, Passagiere*›; *(transport by sea)* verschiffen ‹*Auto, Truppen*›; *(send by train, road, or air)* verschicken, versenden ‹*Waren*›

~ **'out** *v.t.* verschiffen ‹*Ladung, Güter*›

ship: ~builder *n.* Schiff[s]bauer, *der;* **~building** *n., no pl., no indef. art.* Schiffbau, *der*

shipment ['ʃɪpmənt] *n.* **a)** Versand, *der; (by sea)* Verschiffung, *die;* **b)** *(amount)* Sendung, *die*

'shipowner *n.* Schiffseigentümer, *der/*-eigentümerin, *die; (of* *several ships)* Reeder, *der*/Reederin, *die*

shipper ['ʃɪpə(r)] *n.* *(merchant)* Spediteur, *der*/Spediteurin, *die; (company)* Spedition, *die*

shipping ['ʃɪpɪŋ] *n.* **a)** *no pl., no indef. art. (ships)* Schiffe *Pl.; (traffic)* Schiffahrt, *die;* Schiffsverkehr, *der;* **all ~:** alle Schiffe/der ganze Schiffsverkehr; **closed to ~:** für Schiffe/für die Schiffahrt gesperrt; **b)** *(transporting)* Versand, *der*

'**shipping forecast** *n.* Seewetterbericht, *der*

ship: ~shape *pred. adj.* in bester Ordnung; **get sth. ~shape** etw. in Ordnung bringen; **~wreck 1.** *n. (lit. or fig.)* Schiffbruch, *der;* **2.** *v.t.* **be ~wrecked** Schiffbruch erleiden; *(fig.: be ruined)* ‹*Hoffnung:*› sich zerschlagen haben; ‹*Karriere:*› gescheitert sein; **~yard** *n.* [Schiffs]werft, *die*

shire ['ʃaɪə(r)] *n.* *(county)* Grafschaft, *die*

'**shire-horse** *n. bes. in Mittelengland gezüchtetes schweres Zugpferd*

shirk [ʃɜ:k] *v.t.* sich entziehen (+ *Dat.*) ‹*Pflicht, Verantwortung*›; **~ one's job/doing sth.** sich vor der Arbeit drücken/sich davor drücken *(ugs.),* etw. zu tun

shirker ['ʃɜ:kə(r)] *n.* Drückeberger, *der (ugs. abwertend)*

shirt [ʃɜ:t] *n.* **[man's] ~:** [Herren- *od.* Ober]hemd, *das;* **[woman's or lady's] ~:** Hemdbluse, *die;* **sports/rugby/football ~:** Trikot/Rugby-/Fußballtrikot, *das;* **keep your ~ on!** *(fig. sl.)* [nur] ruhig Blut! *(ugs.)*

'**shirt-sleeve** *n.* Hemdsärmel, *der;* **work in one's ~s** in Hemdsärmeln arbeiten

shit [ʃɪt] *(coarse)* **1.** *v.i.,* **-tt-, shitted** *or* **shit** *or* **shat** [ʃæt] scheißen *(derb);* **~ in one's pants** sich *(Dat.)* in die Hose[n] scheißen. **2.** *v. refl.,* **-tt-, shitted** *or* **shit** *or* **shat** sich *(Dat.)* in die Hose[n] scheißen *(derb).* **3.** *int.* Scheiße *(derb).* **4.** *n.* **a)** *(excrement)* Scheiße, *die (derb);* **have** *(Brit.) or (Amer.)* **take a ~:** scheißen *(derb);* **b)** *(person)* Scheißkerl, *der; (nonsense)* Scheiß, *der (salopp abwertend);* **don't give me that ~:** erzähl mir nicht so einen Scheiß! *(salopp);* **I don't give a ~ [about it]** das ist mir scheißegal *(salopp);* **be up ~ creek [without a paddle]** *(fig.)* bis zum Hals in der Scheiße stecken *(derb)*

shiver ['ʃɪvə(r)] **1.** *v.i.* *(tremble)* zittern (**with** vor + *Dat.*). **2.** *n.* *(trembling, lit. or fig.)* Schau[d]er, *der (geh.);* **~ of cold/fear** Kälte-/Angstschauer, *der;* **send ~s** *or* **a ~ up** *or* **[up and] down sb.'s back** *or* **spine** jmdm. [einen] Schauder über den Rücken jagen; **give sb. the ~s** *(fig.)* jmdn. schaudern lassen

shivery ['ʃɪvərɪ] *adj.* verfroren ‹*Person*›

shoal [ʃəʊl] *n.* *(of fish)* Schwarm, *der*

shock [ʃɒk] **1.** *n.* **a)** Schock, *der;* **I got the ~ of my life** ich erschrak zu Tode; **come as a ~ to sb.** ein Schock für jmdn. sein; **give sb. a ~:** jmdm. einen Schock versetzen; **he's in for a [nasty] ~!** er wird eine böse Überraschung erleben!; **b)** *(violent impact)* Erschütterung, *die* (**of** durch); **c)** *(Electr.)* Schlag, *der;* **d)** *(Med.)* Schock, *der;* **be in [a state of] ~:** unter Schock[wirkung] stehen; **[electric] ~:** Elektroschock, *der.* **2.** *v.t.* **a) ~ sb. [deeply]** ein [schwerer] Schock für jmdn. sein; **b)** *(scandalize)* schockieren; **I'm not easily ~ed** mich schockiert so leicht nichts; **be ~ed by sth.** über etw. *(Akk.)* schockiert sein

shock absorber ['ʃɒk æbsɔ:bə(r), 'ʃɒk æbzɔ:bə(r)] *n.* Stoßdämpfer, *der*

shocking ['ʃɒkɪŋ] *adj.* **a)** schockierend; **b)** *(coll.: very bad)* fürchterlich *(ugs.);* **what a ~ thing to say!** wie kann man nur so etwas sagen! *(ugs.)*

shock: ~proof *adj.* stoßfest ‹*Uhr, Kiste*›; erschütterungsfest ‹*Gebäude*›; **~ therapy, ~ treatment** *ns. (Med.)* Schocktherapie, *die;* Schockbehandlung, *die*

shod *see* **shoe 2**

shoddy ['ʃɒdɪ] *adj.* schäbig *(abwertend); (poorly done, poor in quality)* minderwertig ‹*Arbeit, Stoff, Artikel*›

shoe [ʃu:] **1.** *n.* **a)** Schuh, *der;* **I shouldn't like to be in his ~s** *(fig.)* ich möchte nicht in seiner Haut stecken *(ugs.);* **put oneself into sb.'s ~s** *(fig.)* sich in jmds. Lage *(Akk.)* versetzen; **sb. shakes in his ~s** jmdm. schlottern die Knie; **b)** *(of horse)* [Huf]eisen, *das.* **2.** *v.t.,* **~ing** ['ʃu:ɪŋ], **shod** [ʃɒd] beschlagen ‹*Pferd*›

shoe: ~ bar *n.* Schnellschusterei, *die;* **~horn** *n.* Schuhlöffel, *der;* **~-lace** *n.* Schnürsenkel, *der;* Schuhband, *das;* **~maker** *n.* Schuhmacher, *der;* Schuster, *der;* **~making** *n., no pl.* Schuhmacherei, *die;* **~-polish** *n.* Schuhcreme, *die;* **~-shop** *n.* Schuhgeschäft, *das;* **~-string** *n.* **a)** *see* **~-lace; b)** *(coll.: small amount)* **on a ~-string** mit ganz wenig Geld; *attrib.* **a ~-string budget** ein minimaler Etat

shone *see* **shine 1, 2**

shoo [ʃu:] **1.** *int.* sch. **2.** *v.t.* scheuchen; **~ away** fort- *od.* wegscheuchen

shook *see* **shake 2, 3**

shoot [ʃu:t] **1.** *v.i.,* **shot** [ʃɒt] **a)** schießen (**at** auf + *Akk.*); **~ to kill** ‹*Polizei:*› scharf schießen; **b)** *(move rapidly)* schießen *(ugs.);* **~ past sb./down the stairs** an jmdm. vorbeischießen/die Treppe hinunterschießen *(ugs.);* **pain shot through/up his arm** ein Schmerz schoß durch seinen Arm/seinen Arm hinauf; **c)** *(Bot.)* austreiben; **d)** *(Sport)* schießen. **2.** *v.t.,* **shot a)** *(wound)* anschießen; *(kill)* erschießen; *(hunt)* schießen; **~ sb. dead** jmdn. erschießen *od. (ugs.)* totschießen; **you'll get shot for this** *(fig.)* du kannst dein Testament machen *(ugs.);* **he ought to be shot** *(fig.)* der gehört aufgehängt *(ugs.);* **b)** schießen mit ‹*Bogen, Munition, Pistole*›; abschießen ‹*Pfeil, Kugel*› (**at** auf + *Akk.*); **c)** *(sl.: inject)* schießen *(Drogenjargon)* ‹*Heroin, Kokain*›; **d)** *(send out)* zuwerfen ‹*Lächeln, Blick*› (**at** *Dat.*); [aus]treiben ‹*Knospen, Schößlinge*›; **e)** *(Sport)* schießen ‹*Tor, Ball, Puck*›; *(Basketball)* werfen ‹*Korb*›; **f)** *(push, slide)* vorschieben ‹*Riegel*›; **g)** *(Cinemat.)* drehen ‹*Film, Szene*›; **h)** *(pass swiftly over, under, etc.)* durchfahren ‹*Stromschnelle*›; unterfahren ‹*Brücke*›; **~ the lights** *(sl.)* eine rote Ampel überfahren. **3.** *n.* **a)** *(Bot.)* Trieb, *der;* **b)** *(~ing-party, -expedition, -practice, -land)* Jagd, *die;* **the whole [bang] ~** *(sl.)* der ganze Kram *od.* Krempel *(ugs. abwertend)*

~ a'head *v.i.* vorpreschen; **~ ahead of sb.** jmdn. blitzschnell hinter sich *(Dat.)* lassen

~ 'down *v.t.* niederschießen ‹*Person*›; abschießen ‹*Flugzeug*›; *(fig.)* entkräften ‹*Argument*›

~ 'off *v.i.* losschießen *(ugs.)*

~ 'out 1. *v.i.* hervorschießen; **the dog shot out of the gate** der Hund schoß aus dem Tor heraus *(ugs.).* **2.** *v.t.* herausschleudern; **~ it out** *(sl.)* sich schießen

~ 'up *v.i.* in die Höhe schießen; ‹*Preise, Temperatur, Kosten, Pulsfrequenz:*› in die Höhe schnellen

shooting ['ʃu:tɪŋ] *n.* **a)** Schießerei, *die;* **two more ~s were reported** Meldungen zufolge wurden erneut zwei Menschen von Schüssen getroffen; **b)** *(Sport)* Schießen, *das;* **rifle ~:** Gewehrschießen, *das;* **c)** *(Hunting)* **go ~:** auf die Jagd gehen; **d)** *(Cinemat.)* Dreharbeiten *Pl.*

shooting: ~-gallery *n.* Schießstand, *der; (at fun-fair)* Schießbude, *die;* **~-match** *n.* **a)** Wettschießen, *das;* **b) the whole ~-match** *(sl.)* der ganze Kram *od.* Krempel *(ugs. abwertend);* **~-range** *n.* Schießstand, *der;* **~ 'star** *n.* Sternschnuppe, *die*

'shoot-out *n.* Schießerei, *die*

shop [ʃɒp] **1.** *n.* **a)** *(premises)* Laden, *der;* Geschäft, *das;* **go to the ~s** einkaufen gehen; **keep a ~:** einen Laden *od.* ein Geschäft haben; **keep [the] ~ for sb.** jmdn. im Laden *od.* Geschäft vertreten; **all over the ~** *(fig. sl.)* überall; **b)** *(business)* **set up ~:** ein Geschäft eröffnen; *(as a lawyer, dentist, etc.)* eine Praxis aufmachen; **shut up ~:** das Geschäft schließen; **talk ~:** fachsimpeln *(ugs.);* **c)** *(workshop)* Werkstatt, *die.* **2.** *v. i.,* **-pp-** einkaufen; **go ~ping** einkaufen gehen; **~** *or* **go ~ping for shoes** Schuhe kaufen gehen. **3.** *v. t.,* **-pp-** *(Brit. sl.)* verpfeifen

~ a'round *v. i.* sich umsehen **(for** nach)

shop: ~ assistant *n. (Brit.)* Verkäufer, *der*/Verkäuferin, *die;* **~-'floor** *n.* **a)** *(place)* Produktion, *die (ugs.);* **b)** *(workers)* **the ~-floor** die Arbeiter; *attrib.* Arbeiter-; **~keeper** *n.* Ladenbesitzer, *der*/-besitzerin, *die;* **~-lifter** ['ʃɒplɪftə(r)] *n.* Ladendieb, *der*/-diebin, *die;* **~-lifting** *n., no pl., no indef. art.* Ladendiebstahl, *der;* **~owner** *see* **~keeper**

shopper ['ʃɒpə(r)] *n.* **a)** *(person)* Käufer, *der*/Käuferin, *die;* **b)** *(wheeled bag)* Einkaufsroller, *der*

shopping ['ʃɒpɪŋ] *n., no pl., no indef. art.* **a)** *(buying goods)* Einkaufen, *das;* **do the/one's ~:** einkaufen/[seine] Einkäufe machen; **b)** *(items bought)* Einkäufe *Pl.*

shopping: ~-bag *n.* Einkaufstasche, *die;* **~-basket** *n.* Einkaufskorb, *der;* **~ centre** *n.* Einkaufszentrum, *das;* **~-list** *n.* Einkaufszettel, *der; (fig.)* Wunschliste, *die;* **~ mall** *n.* Einkaufszentrum, *das;* **~ street** *n.* Geschäftsstraße, *die;* **~ trolley** *n.* Einkaufswagen, *der*

shop: ~-soiled *adj. (Brit.)* *(slightly damaged)* leicht beschädigt; *(slightly dirty)* angeschmutzt; **~-steward** *n.* [gewerkschaftlicher] Vertrauensmann/[gewerkschaftliche] Vertrauensfrau; **~-'window** *n.* Schaufenster, *das*

¹shore [ʃɔː(r)] *n.* Ufer, *das; (coast)* Küste, *die; (beach)* Strand, *der;* **on the ~:** am Ufer/an der Küste/am Strand; **on the ~[s] of Lake Garda** am Ufer des Gardasees; **off ~:** vor der Küste; **be on ~** ‹*Seemann:*› an Land sein

²shore *v. t. (support)* abstützen ‹*Tunnel*›

~ 'up *v. t. (support)* abstützen ‹*Mauer, Haus*›; *(fig.)* stützen ‹*Preis, Währung, Wirtschaft*›

shorn *see* **shear**

short [ʃɔːt] **1.** *adj.* **a)** kurz; **a ~ time** *or* **while ago/later** vor kurzem/kurze Zeit später; **for a ~ time** *or* **while** eine kleine Weile; ein [kleines] Weilchen; **a ~ time before he left** kurz bevor er ging; **a ~ time** *or* **while before/after sth.** kurz vor/nach etw. *(Dat.);* **in a ~ time** *or* **while** *(soon)* bald; in Kürze; **within a ~ [space of] time** innerhalb kurzer Zeit; **in the ~ run** *or* **term** kurzfristig; kurzzeitig; **wear one's hair/skirts ~:** seine Haare kurz tragen/kurze Röcke tragen; **b)** *(not tall)* klein ‹*Person, Wuchs*›; niedrig ‹*Gebäude, Baum, Schornstein*›; **c)** *(not far-reaching)* kurz ‹*Wurf, Schuß, Gedächtnis*›; **d)** *(deficient, scanty)* knapp; **be in ~ supply** knapp sein; **good doctors are in ~ supply** gute Ärzte sind rar *od. (ugs.)* sind Mangelware; **be [far/not far] ~ of a record** einen Rekord [bei weitem] nicht erreichen/[knapp] verfehlen; **sb./sth. is so much/so many ~:** jmdm./einer Sache fehlt soundsoviel/fehlen soundsoviele; **sb. is ~ of sth.** jmdm. fehlt es an etw. *(Dat.);* **time is getting/is ~:** die Zeit wird/ist knapp; **keep sb. ~ [of sth.]** jmdn. [mit etw.] kurzhalten; **[have to] go ~ [of sth.]** [an etw. *(Dat.)*] Mangel leiden [müssen]; **she is ~ of milk today** sie hat heute nicht genug Milch; **the firm is ~ of staff** die Firma hat zu wenig Arbeitskräfte; **be ~ [of cash]** knapp [bei Kasse] sein *(ugs.);* **he is just ~ of six feet/not far ~ of 60** er ist knapp sechs Fuß [groß]/sechzig [Jahre alt]; **it is nothing ~ of miraculous** es ist ein ausgesprochenes Wunder; **e)** *(brief, concise)* kurz; **a ~ history of Wales** eine kurzgefaßte Geschichte von Wales; **the ~ answer is ...:** um es kurz zu machen: die Antwort ist ...; **~ and sweet** *(iron.)* kurz und schmerzlos *(ugs.);* **in ~, ...:** kurz, ... ; **f)** *(curt, uncivil)* kurz angebunden; barsch; **g)** *(Cookery)* mürbe ‹*Teig*›; **h) sell oneself ~** *(fig.)* sein Licht unter den Scheffel stellen; **sell sb./sth. ~** *(fig.)* jmdn./etw. unterschätzen. **2.** *adv.* **a)** *(abruptly)* plötzlich; **stop ~:** plötzlich abbrechen; ‹*Musik, Gespräch:*› jäh *(geh.)* abbrechen; **stop ~ at sth.** über etw. *(Akk.)* nicht hinausgehen; **stop sb. ~:** jmdm. ins Wort fallen; **pull up ~:** plötzlich anhalten; **bring** *or* **pull sb. up ~:** jmdn. stutzen lassen; **b)** *(curtly)* kurz angebunden; barsch; **c)** *(before the expected place or time)* **jump/land ~:** zu kurz springen/zu früh landen *(ugs.);* **~ of sth.** vor etw. *(Dat.);* **stop ~ of the line** vor der Linie stehen-/liegenbleiben; **the bomb dropped ~ [of its target]** die Bombe fiel vor das Ziel; **fall** *or* **come [far/considerably] ~ of sth.** etw. [bei weitem] nicht erreichen; **stop ~ of sth.** *(fig.)* vor etw. zurückschrecken; **stop ~ of doing sth.** davor zurückschrecken, etw. zu tun; **d) nothing ~ of a catastrophe/miracle can ...:** nur eine Katastrophe/ein Wunder kann ...; **~ of locking him in, how can I keep him from going out?** wie kann ich ihn daran hindern auszugehen – es sei denn ich schlösse ihn ein? **3.** *n.* **a)** *(Electr. coll.)* Kurze, *der (ugs.);* **b)** *(coll.: drink)* Schnaps, *der (ugs.).* **4.** *v. t.* *(Electr. coll.)* kurzschließen. **5.** *v. i. (Electr. coll.)* einen Kurzschluß kriegen *(ugs.)*

shortage ['ʃɔːtɪdʒ] *n.* Mangel, *der* **(of** an + *Dat.*); **~ of fruit/teachers** Obstknappheit, *die*/Lehrermangel, *der*

short: ~bread, ~cake *ns.* Shortbread, *das; Keks aus Butterteig;* **~-'change** *v. t.* zu wenig [Wechselgeld] herausgeben (+ *Dat.*); *(fig.)* übers Ohr hauen *(ugs.);* **~ 'circuit** *n. (Electr.)* Kurzschluß, *der;* **~-'circuit** *(Electr.)* **1.** *v. t.* kurzschließen; *(fig.)* umgehen; **2.** *v. i.* einen Kurzschluß bekommen; **~coming** *n., usu. in pl.* Unzulänglichkeit, *die;* **~ 'cut** *n.* Abkürzung, *die;* **take a ~ cut** *(lit. or fig.)* eine Abkürzung machen; **be a ~ cut to sth.** *(fig.)* den Weg zu etw. abkürzen; **~ division** *see* **division f; ~ 'drink** *n.* hochprozentiges Getränk

shorten ['ʃɔːtn] **1.** *v. i. (become shorter)* kürzer werden. **2.** *v. t.* *(make shorter)* kürzen; *(curtail)* verkürzen ‹*Besuch, Wartezeit, Inkubationszeit*›

short: ~fall *n.* Fehlmenge, *die; (in budget, financial resources)* Defizit, *das;* **~-haired** *adj.* kurzhaarig; **~hand** *n.* Kurzschrift, *die;* Stenographie, *die;* **write ~hand** stenographieren; **that's ~hand for ...** *(fig.)* das ist eine Kurzformel für ...; *see also* **typist**

shortish ['ʃɔ:tɪʃ] *adj.* ziemlich kurz; ziemlich klein ‹*Person*›

short: ~-legged *adj.* kurzbeinig; **~ list** *n. (Brit.)* engere Auswahl; **be on/put sb. on the ~ list** in der engeren Auswahl sein/jmdn. in die engere Auswahl nehmen; **~-list** *v. t.* in die engere Auswahl nehmen; **~-lived** ['ʃɔ:tlɪvd] *adj.* kurzlebig

shortly ['ʃɔ:tlɪ] *adv.* **a)** *(soon)* in Kürze; gleich *(ugs.);* **~ before/after sth.** kurz vor/nach etw.; **b)** *(briefly)* kurz; **c)** *(curtly)* kurz angebunden; in barschem Ton

shortness ['ʃɔ:tnɪs] *n., no pl.* **a)** Kürze, *die;* **b)** *(of person)* Kleinheit, *die;* geringe Körpergröße; **c)** *(scarcity, lack)* Knappheit, *die* **(of** an + *Dat.*); **d)** *(curtness)* Barschheit, *die*

'**short: ~ 'pastry** *n. (Cookery)* Mürbeteig, *der;* **~-range** *adj.* **a)** Kurzstrecken‹*flugzeug, -rakete usw.*›; **b)** *(relating to ~ future period)* kurzfristig

shorts [ʃɔ:ts] *n. pl.* **a)** *(trousers)* kurze Hose[n]; Shorts *Pl.; (in sports)* Sporthose, *die;* **b)** *(Amer.: underpants)* Unterhose, *die*

short: ~ 'sight *n., no pl., no art.* Kurzsichtigkeit, *die;* **have ~ sight** kurzsichtig sein; **~-sighted** [ʃɔ:t'saɪtɪd] *adj. (lit. or fig.)* kurzsichtig; **~sleeved** ['ʃɔ:tsli:vd] *adj.* kurzärm[e]lig; **~-staffed** [ʃɔ:t'stɑ:ft] *adj.* **be [very] ~-staffed** [viel] zu wenig Personal haben; **~ 'story** *n. (Lit.)* Short story, *die;* Kurzgeschichte, *die;* **~ 'temper** *n.* **have a ~ temper** aufbrausend *od.* cholerisch sein; **~-'tempered** *adj.* aufbrausend; cholerisch; **~-term** *adj.* kurzfristig; *(provisional)* vorläufig ‹*Lösung, Antwort*›; **~ 'wave** *n. (Radio)* Kurzwelle, *die;* **~-wave** *adj. (Radio)* Kurzwellen-

shot [ʃɒt] **1.** *n.* **a)** *(discharge of gun)* Schuß, *der; (firing of rocket)* Abschuß, *der;* Start, *der;* **fire a ~ [at sb./sth.]** einen Schuß [auf jmdn./etw.] abgeben; **like a ~** *(fig.)* wie der Blitz *(ugs.);* **I'd do it like a ~:** ich würde es auf der Stelle tun; **have a ~ at sth./at doing sth.** *(fig.)* etw. versuchen/versuchen, etw. zu tun; *see also* **dark 2 b; long shot; b)** *(Athletics)* Kugel, *die;* **put the ~:** die Kugel stoßen; kugelstoßen; **[putting] the ~:** Kugelstoßen, *das;* **c)** *(Sport: stroke, kick, throw, Archery, Shooting)* Schuß, *der;* **d)** *(Photog.)* Aufnahme, *die; (Cinemat.)* Einstellung, *die;* **do** *or* **film interior/location ~s** *(Cinemat.)* Innenaufnahmen machen/am Originalschauplatz drehen; **e)** *(injection)* Spritze, *die; (of drug)* Schuß, *der (Jargon);* **be a ~ in the arm for sb./sth.** *(fig.)* jmdm./einer Sache Aufschwung geben. **2.** *see* **shoot 1, 2. 3.** *adj.* **be/get ~ of sb./sth.** *(sl.)* jmdn./etw. los sein/loswerden

shot: ~gun *n.* [Schrot]flinte, *die;* **~gun wedding/marriage** *(fig. coll.)* Mußheirat/Mußehe, *die (ugs.);* **~-put** *n., no pl., no indef. art. (Athletics)* Kugelstoßen, *das;* **~-putter** ['ʃɒtpʊtə(r)] *n. (Athletics)* Kugelstoßer, *der*/-stoßerin, *die*

should *see* **shall**

shoulder ['ʃəʊldə(r)] **1.** *n.* **a)** Schulter, *die;* **~ to ~** *(lit. or fig.)* Schulter an Schulter; **straight from the ~** *(fig.)* unverblümt; **cry on sb.'s ~** *(fig.)* sich bei jmdm. ausweinen; **give sb. the cold '~:** jmdn. schneiden; **b)** *in pl. (upper part of back)* Schultern *Pl.; (of garment)* Schulterpartie, *die;* **lie** *or* **rest/fall on sb.'s ~s** *(fig.)* auf jmds. Schultern *(Dat.)* lasten/jmdm. aufgebürdet werden; **he has broad ~s** *(fig.: is able to take responsibility)* er hat einen breiten Rücken; **c)** *(Anat.: ~-joint)* Schultergelenk, *das;* **d)** *(Gastr.)* Bug, *der;* Schulter, *die;* **~ of lamb** Lammschulter, *die;* **e)** *(Road Constr.)* Randstreifen, *der;* Seitenstreifen, *der; see also* **hard shoulder. 2.** *v. t.* **a)** *(push with ~)* rempeln; **~ one's way through the crowd** sich rempelnd einen Weg durch die Menge bahnen; **b)** *(take on one's ~s)* schultern; *(fig.)* übernehmen ‹*Verantwortung, Aufgabe*›

~ a'side *v. t.* beiseite rempeln; *(fig.)* beiseite schieben

shoulder: ~-bag *n.* Umhängetasche, *die;* **~-blade** *n.* Schulterblatt, *das;* **~-strap** *n.* **a)** *(on ~ of garment)* Schulterklappe, *die;* **b)** *(on bag)* Tragriemen, *der; (suspending a garment)* Träger, *der*

shouldn't ['ʃʊdnt] *(coll.)* = **should not;** *see* **shall**

shout [ʃaʊt] **1.** *n.* Ruf, *der; (inarticulate)* Schrei, *der;* **warning ~, ~ of alarm** Warnruf, *der*/-schrei, *der;* **~ of joy/rage** Freuden-/Wutschrei, *der;* **~ of encouragement/approval** Anfeuerungs-/Beifallsruf, *der.* **2.** *v. i.* schreien; **~ with laughter/pain** vor Lachen/Schmerzen schreien; **~ with** *or* **for joy** vor Freude schreien; **~ at sb.** *(abusively)* jmdn. anschreien; **~ for sb./sth.** nach jmdm./etw. schreien; **~ for help** um Hilfe schreien *od.* rufen. **3.** *v. t.* schreien; **~ abuse** pöbeln; **~ abuse at sb.** jmdn. anpöbeln

~ 'down *v. t.* **~ sb. down** *(prevent from being heard)* jmdn. niederschreien

~ 'out 1. *v. i.* aufschreien **2.** *v. t.* [laut] rufen

shouting ['ʃaʊtɪŋ] *n. (act)* Schreien, *das; (shouts)* Geschrei, *das;* **it's all over but** *or* **bar the ~** *(fig.)* das Rennen ist im Grunde schon gelaufen *(ugs.)*

shove [ʃʌv] **1.** *n.* Stoß, *der.* **2.** *v. t.* **a)** stoßen; schubsen *(ugs.);* **b)** *(use force to propel)* schieben; **c)** *(coll.: put)* tun. **3.** *v. i.* drängen; drängeln *(ugs.);* **~ through the crowd** *(coll.)* sich durch die Menge drängeln *(ugs.). See also* **push 2 a, 3 d**

~ a'way *v. t. (coll.)* wegschubsen *(ugs.)*

shovel ['ʃʌvl] **1.** *n.* Schaufel, *die; (machine)* Bagger, *der.* **2.** *v. t., (Brit.)* **-ll-: a)** schaufeln; **b)** *(fig.)* **~ food into one's mouth** Essen in sich *(Akk.)* reinschaufeln *od.* -stopfen *(ugs.)*

shovelful ['ʃʌvlfʊl] *n.* **a ~ of earth** *etc.* eine Schaufel Erde *usw.*

show [ʃəʊ] **1.** *n.* **a)** *(display)* Pracht, *die;* **a ~ of flowers/colour** eine Blumen-/Farbenpracht; **~ of force/strength** *etc.* Demonstration der Macht/Stärke *usw.;* **be on ~:** ausgestellt sein; **put sth. on ~:** etw. ausstellen; **b)** *(exhibition)* Ausstellung, *die;* Schau, *die;* **dog ~:** Hundeschau, *die;* **c)** *(entertainment, performance)* Show, *die; (Theatre)* Vorstellung, *die; (Radio, Telev.)* [Unterhaltungs]sendung, *die;* **d)** *(coll.: effort)* **it's a poor ~:** das ist ein schwaches Bild; **put up a good/poor ~:** eine gute/schlechte Figur machen; **good ~!** gut [gemacht]!; **e)** *(sl.: undertaking, business)* **it's his ~:** er ist der Boß *(ugs.);* **run the ~:** der Boß sein *(ugs.);* **give the [whole] ~ away** alles ausquatschen *(salopp);* **f)** *(outward appearance)* Anschein, *der;* **make a great ~ of friendliness** ungeheuer freundlich tun; **make** *or* **put on a [great] ~ of doing sth.** sich *(Dat.)* [angestrengt] den Anschein geben, etw. zu tun; **be for ~:** reine Angeberei sein *(ugs.)*; **do sth. just for ~:** etw. nur aus Prestigegründen tun. **2.** *v. t., p.p.* **~n** [ʃəʊn] *or* **~ed a)** *(allow or cause to be seen)* zeigen; vorzeigen ⟨*Paß, Fahrschein usw.*⟩; **~ sb. sth., ~ sth. to sb.** jmdm. etw. zeigen; **have nothing/something to ~ for it** [dabei] nichts/etwas zum Vorzeigen haben; **that dress ~s your petticoat** bei diesem Kleid sieht man deinen Unterrock; **this material does not ~ the dirt** auf diesem Material sieht man den Schmutz nicht; *see also* **colour 1 e; sign 1 e; b)** *(manifest, give evidence of)* zeigen; beweisen ⟨*Mut, Entschlossenheit, Urteilsvermögen usw.*⟩; **~ hesitation** zaudern; **he is ~ing his age** man sieht ihm sein Alter an; **c) ~ [sb.] kindness/mercy** freundlich [zu jmdm.] sein/Erbarmen [mit jmdm.] haben; **~ mercy on** *or* **to sb.** Erbarmen mit jmdm. haben; **d)** *(indicate)* zeigen ⟨*Gefühl, Freude usw.*⟩; ⟨*Thermometer, Uhr usw.:*⟩ anzeigen; **as ~n in the illustration** wie die Abbildung zeigt; **frontiers are ~n by blue lines and towns are ~n in red** Grenzen sind durch blaue Linien und Städte sind rot gekennzeichnet; **the accounts ~ a profit** die Bücher weisen einen Gewinn aus; **the firm ~s a profit/loss** die Firma macht Gewinn/Verlust; **e)** *(demonstrate, prove)* zeigen; **~ sb. that ...:** jmdm. beweisen, daß ...; **it all/just goes to ~ that ...:** das beweist nur, daß ...; **it all goes to ~, doesn't it?** das beweist es doch, oder?; **I'll ~ you/him** *etc.*! ich werd's dir/ihm *usw.* schon zeigen!; **~ sb. who's boss** jmdm. zeigen, wer das Sagen hat; **f)** *(conduct)* führen; **~ sb. over** *or* **round the house/to his place** jmdn. durchs Haus/an seinen Platz führen. **3.** *v. i., p.p.* **~n** *or* **~ed a)** *(be visible)* sichtbar *od.* zu sehen sein; **he was angry/bored, and it ~ed** er war wütend/langweilte sich, und man sah es [ihm an]; **his age is beginning to ~:** man sieht ihm sein Alter allmählich an; **b)** *(be ~n)* ⟨*Film:*⟩ laufen; ⟨*Künstler:*⟩ ausstellen; **'Gandhi' – now ~ing in the West End** „Gandhi" – Jetzt im West End; **c)** *(make sth. known)* **time will ~:** man wird es [ja] sehen

~ 'in *v. t.* hineinführen/hereinführen

~ 'off 1. *v. t.* **a)** *(display)* **~ sth./sb. off** etw./jmdn. vorführen *od.* vorzeigen; *(in order to impress)* mit etw./jmdm. prahlen *od. (ugs.)* angeben; **b)** *(display to advantage)* zur Geltung bringen. **2.** *v. i.* angeben *(ugs.)*; prahlen

~ 'out *v. t.* hinausführen

~ 'round *v. t.* herumführen

~ 'through *v. i.* durchscheinen

~ 'up 1. *v. t.* **a)** *(make visible)* deutlich sichtbar machen; aufdecken ⟨*Betrug*⟩; **b)** *(coll.: embarrass)* blamieren. **2.** *v. i.* **a)** *(be easily visible)* [deutlich] zu sehen *od.* erkennen sein; *(fig.)* sich zeigen; **b)** *(coll.: arrive)* sich blicken lassen *(ugs.)*; auftauchen

show: ~ biz ['ʃəʊbɪz] *(sl.)*, **~ business** *ns., no pl., no art.* Schaugeschäft, *das;* Showbusineß, *das;* **~-case** *n.* Vitrine, *die;* Schaukasten, *der; (fig.)* Schaufenster, *das;* **~down** *n. (fig.)* Kraftprobe, *die;* **have a ~down [with sb.]** sich [mit jmdm.] auseinandersetzen

shower ['ʃaʊə(r)] **1.** *n.* **a)** Schauer, *der;* **~ of rain/sleet/hail** Regen-/Schneeregen-/Hagelschauer, *der;* **a ~ of confetti/sparks/stones** ein Konfettiregen/Funkenregen/Steinhagel; **b)** *(for washing)* Dusche, *die; attrib.* Dusch-; **have** *or* **take a [cold/quick] ~:** [kalt/schnell] duschen; **be under the ~:** unter der Dusche stehen; **c)** *(Amer.: party)* **~ [party]** Geschenkparty, *die (für eine Braut, bei der sie Aussteuergegenstände geschenkt bekommt).* **2.** *v. t.* **a) ~ sth. over** *or* **on sb., ~ sb. with sth.** jmdn. mit etw. überschütten; **b)** *(fig.: lavish)* **~ sth. [up]on sb., ~ sb. with sth.** jmdn. mit etw. überhäufen. **3.** *v. i.* **a)** *(fall in ~s)* **~ down [up]on sb.** ⟨*Wasser, Konfetti:*⟩ auf jmdn. herabregnen; ⟨*Steine, Verwünschungen:*⟩ auf jmdn. niederhageln; **b)** *(have a ~)* duschen

'shower: ~-curtain *n.* Duschvorhang, *der;* **~ gel** *n.* Duschgel, *das;* **~-proof** *adj.* [bedingt] regendicht

showery ['ʃaʊərɪ] *adj.* **it is ~:** es gibt immer wieder kurze Schauer; **a cold and ~ day** ein kalter Tag mit häufigen Schauern

show: ~girl *n.* Showgirl, *das;* **~ground** *n.* Ausstellungsgelände, *das;* **~ house** *n.* Musterhaus, *das*

showing ['ʃəʊɪŋ] *n.* **a)** *(of film)* Vorführung, *die; (of television programme)* Sendung, *die;* **b)** *(evidence)* **on this ~:** demnach; **on any ~:** wie man es auch [dreht und] wendet; **c)** *(quality of performance)* Leistung, *die;* **make a good/poor** *etc.* **~:** eine gute/schwache *usw.* Leistung zeigen; **on this ~:** bei dieser Leistung

show: ~-jumper *n. (Sport)* **a)** *(person)* Springreiter, *der*/-reiterin, *die;* **b)** *(horse)* Springpferd, *das;* **~-jumping** *n. (Sport)* Springreiten, *das;* **~man** ['ʃəʊmən] *n., pl.* **~men** ['ʃəʊmən] **a)** *(proprietor of fairground booth etc.)* Schausteller, *der;* **b)** *(effective presenter)* Showman, *der*

shown *see* **show 2, 3**

show: ~-off *n. (coll.)* Angeber, *der*/Angeberin, *die (ugs.)*; **don't be such a ~-off** gib nicht so an!; **~-piece** *n. (of exhibition, collection)* Schaustück, *das; (highlight)* Paradestück, *das;* **~-place** *n.* Attraktion, *die;* **~room** *n.* Ausstellungsraum, *der;* **~ trial** *n.* Schauprozeß, *der*

showy ['ʃəʊɪ] *adj.* **a)** *(gaudy, ostentatious)* protzig *(ugs.)*; **b)** *(striking)* prächtig ⟨*Farben*⟩; [farben]prächtig ⟨*Blumen*⟩

shrank *see* **shrink**

shrapnel *n.* ['ʃræpnl] *n. (Mil.: fragments)* Bomben-/Granatsplitter

shred [ʃred] *n.* Fetzen, *der;* **not a ~ of evidence** keine Spur eines Beweises; **tear** *etc.* **sth. to ~s** etw. in Fetzen reißen *usw.*; **tear a theory/an argument to ~s** eine Theorie/eine Argumentation zerpflücken; **our clothes were in ~s** unsere Kleidung war zerfetzt

shrew [ʃru:] *n.* **a)** *(Zool.)* Spitzmaus, *die;* **b)** *(woman)* Beißzange, *die (salopp)*

shrewd [ʃru:d] *adj.* scharfsinnig ⟨*Person*⟩; klug ⟨*Entscheidung, Investition, Schritt, Geschäftsmann*⟩; genau ⟨*Schätzung, Einschätzung*⟩; treffsicher ⟨*Urteilsvermögen*⟩

shriek [ʃri:k] **1.** *n.* [Auf]schrei, *der;* **give a ~:** [auf]schreien; **give a ~ of horror/fear** *etc.* einen Schrei des Entsetzens/der Angst *usw.* ausstoßen. **2.** *v. i.* [auf]schreien; **~ with horror/fear** *etc.* vor Entsetzen/Angst *usw.* [auf]schreien

shrift [ʃrɪft] *n.* **give sb. short ~:** jmdn. kurz abfertigen *(ugs.)*; **get short ~ [from sb.]** [von jmdm.] kurz abgefertigt werden *(ugs.)*

shrill [ʃrɪl] *adj.*, **shrilly** ['ʃrɪlɪ] *adv.* schrill

shrimp [ʃrɪmp] *n., pl.* **~s** *or* **~** *(Zool.)* Garnele, *die;* Krabbe, *die (ugs.)*; *(Gastr.)* Krabbe, *die; attrib.* Garnelen-/Krabben-

shrine [ʃraɪn] *n.* Heiligtum, *das; (tomb)* Grab, *das; (casket)* Schrein, *der (veralt.)*; *(casket* *holding sacred relics)* Reliquienschrein, *der*

shrink [ʃrɪŋk] **1.** *v.i.*, **shrank** [ʃræŋk], **shrunk** [ʃrʌŋk] **a)** *(grow smaller)* schrumpfen; ‹*Person:*› kleiner werden; ‹*Kleidung, Stoff:*› einlaufen; ‹*Holz:*› sich zusammenziehen; ‹*Handel, Einkünfte:*› zurückgehen; **b)** *(recoil)* sich zusammenkauern; ~ **from sb./sth.** vor jmdm. zurückweichen/vor etw. *(Dat.)* zurückschrecken; ~ **from doing sth.** sich scheuen, etw. zu tun. **2.** *v.t.*, **shrank, shrunk** sich zusammenziehen lassen ‹*Holz*›; einlaufen lassen ‹*Textilien*›

~ **a'way** *v.i.* **a)** *(recoil)* zurückweichen (**from** vor + *Dat.*); **b)** *(grow smaller)* zusammenschrumpfen

~ **'back** *v.i.* zurückweichen (**from** vor + *Dat.*); ~ **back from sth./doing sth.** *(fig.)* vor etw. *(Dat.)* zurückschrecken/sich scheuen, etw. zu tun

shrinkage ['ʃrɪŋkɪdʒ] *n.* **a)** *(act) (of clothing, material)* Einlaufen, *das; (of income, trade, etc.)* Rückgang, *der;* **b)** *(degree)* Schrumpfung, *die*

shrink: ~-proof, ~-resistant *adjs.* schrumpffrei; **~-wrap** *v.t.* in einer Schrumpffolie verpacken

shrivel ['ʃrɪvl] **1.** *v.t., (Brit.)* **-ll-:** ~ **[up]** schrump[e]lig machen; runzlig machen ‹*Haut, Gesicht*›; welk werden lassen ‹*Pflanze, Blume*›. **2.** *v.i., (Brit.)* **-ll-:** ~ **[up]** verschrumpeln; ‹*Haut, Gesicht:*› runzlig werden; ‹*Pflanze, Blume:*› welk werden; ‹*Ballon:*› zusammenschrumpfen

shroud [ʃraʊd] **1.** *n.* **a)** Leichentuch, *das (veralt.);* **b)** *(fig.: of fog, mystery, etc.)* Schleier, *der.* **2.** *v.t.* einhüllen; ~ **sth. in sth.** etw. in etw. *(Akk.)* hüllen

Shrove Tuesday [ʃrəʊv 'tju:zdeɪ, ʃrəʊv 'tju:zdɪ] *n.* Fastnachtsdienstag, *der*

shrub [ʃrʌb] *n.* Strauch, *der*

shrubbery ['ʃrʌbərɪ] *n.* **a)** Gesträuch, *das;* **b)** *(shrubs collectively)* Sträucher

shrug [ʃrʌg] **1.** *n.* ~ **[of one's** *or* **the shoulders]** Achselzucken, *das;* **give a ~ [of one's** *or* **the shoulders]** die *od.* mit den Achseln zucken. **2.** *v.t. & i.*, **-gg-:** ~ **[one's shoulders]** die *od.* mit den Achseln zucken

~ **'off** *v.t.* in den Wind schlagen; ~ **sth. off as unimportant** etw. als unwichtig abtun

shrunk *see* **shrink**

shrunken ['ʃrʌŋkn] *adj.* verhutzelt *(ugs.)* ‹*Person*›; schrumpelig, verschrumpelt ‹*Äpfel*›; ~ **head** Schrumpfkopf, *der*

shudder ['ʃʌdə(r)] **1.** *v.i.* **a)** *(shiver)* zittern (**with** vor + *Dat.*); **sb. ~s to think of sth.** jmdn. schaudert bei dem Gedanken an etw. *(Akk.);* **b)** *(vibrate)* zittern; ~ **to a halt** zitternd zum Stehen kommen. **2.** *n.* **a)** *(shivering)* Zittern, *das;* Schauder, *der;* **sb. has/gets the ~s** *(coll.)* jmdn. schaudert; **it gives me the ~s to think of it** *(coll.)* mich schaudert, wenn ich daran denke; **b)** *(vibration)* Zittern, *das*

shuffle ['ʃʌfl] **1.** *n.* **a)** Schlurfen, *das;* **walk with a ~:** schlurfend gehen; schlurfen; **b)** *(Cards)* Mischen, *das;* **give the cards a [good] ~:** die Karten [gut] mischen; **c)** *(fig.: change)* Umbildung, *die;* **cabinet ~:** Kabinettsumbildung, *die.* **2.** *v.t.* **a)** *(rearrange)* umbilden ‹*Kabinett*›; neu verteilen ‹*Aufgaben*›; sortieren ‹*Schriftstücke usw.*›; *(mix up)* durcheinanderbringen; **b)** *(Cards)* mischen; **c) he ~s his feet when he walks** er schlurft beim Gehen. **3.** *v.i.* **a)** *(Cards)* mischen; **b)** *(move, walk)* schlurfen; **c)** *(shift one's position)* herumrutschen

shun [ʃʌn] *v.t.*, **-nn-** meiden

shunt [ʃʌnt] *v.t. (Railw.)* rangieren; ~ **[off]** *(fig.)* abschieben

shush [ʃʊʃ] **1.** *int. see* **hush 3.** **2.** *v.t.* zum Schweigen bringen

shut [ʃʌt] **1.** *v.t.*, **-tt-, shut a)** zumachen; schließen; ~ **sth. to sb./sth.** etw. für jmdn./etw. schließen; ~ **a road to traffic** eine Straße für den Verkehr sperren; ~ **the door on sb.** jmdm. die Tür vor der Nase zuschlagen *(ugs.);* ~ **the door on sth.** *(fig.)* die Möglichkeit einer Sache *(Gen.)* verbauen; ~ **one's eyes to sth.** *(fig.)* seine Augen vor etw. *(Dat.)* verschließen; *(choose to ignore sth.)* über etw. *(Akk.)* hinwegsehen; ~ **one's ears to sth.** *(fig.)* die Ohren vor etw. *(Dat.)* verschließen; ~ *or* **lock the stable door after the horse has bolted** *see* [2]**lock 2 a; b)** *(confine)* ~ **sb./an animal in[to] sth.** jmdn./ein Tier in etw. *(Akk.)* sperren; ~ **oneself in[to] a room** sich in einem Zimmer einschließen; **c)** *(exclude)* ~ **sb./an animal out of sth.** jmdn./ein Tier aus etw. aussperren; **d)** *(trap)* ~ **one's finger/coat in a door** sich *(Dat.)* den Finger/Mantel in einer Tür einklemmen; **e)** *(fold up)* schließen, zumachen ‹*Buch, Hand*›; zusammenklappen ‹*Klappmesser, Fächer*›. **2.** *v.i.*, **-tt-, shut** schließen; ‹*Laden:*› schließen, zumachen; ‹*Blüte:*› sich schließen; **the door/case won't ~:** die Tür/der Koffer geht nicht zu *od.* schließt nicht; **the door ~ on/after him** die Tür schloß sich vor/hinter ihm. **3.** *adj.* zu; geschlossen; **we are ~ on Saturdays/for lunch** wir haben samstags/über Mittag geschlossen *od.* zu; **keep sth. ~:** etw. geschlossen halten *od.* zu lassen

~ **a'way** *v.t.* wegschließen; **keep sth. ~ away safely** etw. unter sicherem Verschluß halten

~ **'down 1.** *v.t.* **a)** schließen, zumachen ‹*Deckel, Fenster*›; **b)** *(end operation of)* stillegen; abschalten ‹*Kernreaktor*›; einstellen ‹*Aktivitäten*›; *(Radio, Telev.)* einstellen ‹*Sendebetrieb*›. **2.** *v.i. (cease working)* ‹*Laden, Fabrik:*› geschlossen werden; ‹*Zeitung, Sendebetrieb:*› eingestellt werden

~ **'in** *v.t.* **a)** *(keep in)* einschließen; **b)** *(encircle)* umschließen; **feel ~ in** sich eingeschlossen fühlen

~ **'off** *v.t.* **a)** *(stop)* unterbrechen ‹*Strom, Fluß*›; abstellen ‹*Motor, Maschine, Gerät*›; **b)** *(isolate)* absperren; ~ **sb. off from sb./sth.** jmdn. von jmdm./etw. abschneiden; ~ **oneself off from sb./sth.** sich gegen jmdn./etw. abkapseln

~ **'out** *v.t.* aussperren; versperren ‹*Aussicht*›; *(exclude from view)* verdecken; *(prevent)* ausschließen ‹*Gefahr, Möglichkeit*›; **the tree ~s out the light** der Baum nimmt das Licht weg

~ **'up 1.** *v.t.* **a)** *(close)* abschließen; zuschließen; ~ **up [the/one's] house** das/sein Haus [sicher] abschließen; *see also* **shop 1 b; b)** *(put away)* einschließen ‹*Dokumente, Wertsachen usw.*›; einsperren ‹*Tier, Person*›; ~ **sth. up in sth.** etw. in etw. *(Akk.)* schließen; ~ **sb. up in an asylum** jmdn. in eine Anstalt sperren; **c)** *(reduce to silence)* zum Schweigen bringen. **2.** *v.i. (coll.: be quiet)* den Mund halten *(ugs.);* ~ **up!** halt den Mund! *(ugs.)*

'shut-eye *n. (coll.)* **get** *or* **have some** *or* **a bit of ~:** ein Nickerchen halten *(fam.)*

shutter ['ʃʌtə(r)] *n.* **a)** Laden, *der; (of window)* Fensterladen, *der;* **put up the ~s** *(fig.: cease business)* zumachen; schließen; **b)** *(Photog.)* Verschluß, *der;* ~ **release** Auslöser, *der;* ~ **speed** Verschlußzeit, *die*

shuttle ['ʃʌtl] **1.** *n.* **a)** *(in loom, sewing-machine)* Schiffchen, *das;* **b)** *(Transport) (service)* Pendelverkehr, *der; (bus)* Pendelbus, *der; (aircraft)* Pendelmaschine, *die; (train)* Pendelzug, *der; see also* **space shuttle. 2.** *v.t.* ~ **sth. backwards and forwards** etw. hin und her schicken; ~ **passengers about** Passagiere hin und her fahren. **3.** *v.i.* pendeln

shuttle: ~cock *n.* Federball, *der;* ~ **service** *n.* Pendelverkehr, *der*

[1]**shy** [ʃaɪ] **1.** *adj.*, **~er** *or* **shier** ['ʃaɪə(r)], **~est** *or* **shiest** ['ʃaɪɪst] scheu; *(diffident)* schüchtern; **don't be** ~: sei nicht [so] schüchtern!; **feel** ~ **about doing sth.** sich genieren, etw. zu tun; **be** ~ **of doing sth.** Hemmungen haben, etw. zu tun. **2.** *v.i.* scheuen (**at** vor + *Dat.*)

~ **a'way** *v.i.* ~ **away from sth.** ⟨*Pferd:*⟩ vor etw. *(Dat.)* scheuen; ~ **away from sth./doing sth.** *(fig.)* etw. scheuen/sich scheuen, etw. zu tun

[2]**shy 1.** *v.t. (throw)* ~ **sth. at sth./sb.** etw. auf etw./jmdn. schmeißen *(ugs.).* **2.** *v.i.* schmeißen *(ugs.)* (**at** nach)

shyly ['ʃaɪlɪ] *adv.* scheu; *(diffidently)* schüchtern

shyness ['ʃaɪnɪs] *n., no pl.* Scheuheit, *die; (diffidence)* Schüchternheit, *die*

Siamese [saɪə'mi:z] **1.** *adj.* siamesisch. **2.** *n., pl. same (Zool.)* Siamese, *der*

Siamese: ~ **'cat** *n.* Siamkatze, *die;* ~ **'twins** *n. pl.* siamesische Zwillinge

Siberia [saɪ'bɪərɪə] *pr. n.* Sibirien *(das)*

sibling ['sɪblɪŋ] *n. (male)* Bruder, *der; (female)* Schwester, *die; in pl.* Geschwister *Pl.*

sic [sɪk] *adv.* sic

Sicily ['sɪsɪlɪ] *pr. n.* Sizilien *(das)*

sick [sɪk] **1.** *adj.* **a)** *(ill)* krank; **be** ~ **with sth.** etw. haben; **go** ~, **fall** *or (coll.)* **take** ~: krank werden; **be off** ~: krank [gemeldet] sein; **sb. is** ~ **at** *or* **to his/her stomach** *(Amer.)* jmdm. ist [es] schlecht *od.* übel; **b)** *(Brit.: vomiting or about to vomit)* **be** ~: sich erbrechen; **be** ~ **over sb./sth.** sich über jmdn./etw. erbrechen; **I think I'm going to be** ~: ich glaube, ich muß [mich er]brechen; **a** ~ **feeling** ein Übelkeitsgefühl; **sb. gets/feels** ~: jmdm. wird/ist [es] übel *od.* schlecht; **he felt** ~ **with fear** ihm war vor Angst [ganz] übel; **sth. makes sb.** ~: von etw. wird [es] jmdm. schlecht *od.* übel *(see also* **d)**; **c)** *(sickly)* elend ⟨*Aussehen*⟩; leidend ⟨*Blick*⟩; **d)** *(fig.)* **worried** ~: krank vor Sorgen; **be/get** ~ **of sb./sth.** jmdn./etw. satt haben/allmählich satt haben; **be** ~ **and tired** *or* ~ **to death of sb./sth.** *(coll.)* von jmdm./etw. die Nase [gestrichen] voll haben *(ugs.);* **be** ~ **of the sight/sound of sb./sth.** *(coll.)* jmdn./etw. nicht mehr sehen/hören können; **be** ~ **of doing sth.** es satt haben, etw. zu tun; **make sb.** ~ *(disgust)* jmdn. anekeln; *(coll.: make envious)* jmdn. ganz neidisch machen *(see also* **b)**; **e)** *(deranged)* pervers; *(morally corrupt)* krank ⟨*Gesellschaft*⟩; *(morbid)* makaber ⟨*Witz, Humor, Phantasie*⟩. **2.** *n. pl.* **the** ~: die Kranken

sick: ~-bay *see* [2]**bay b; ~-bed** *n.* Krankenbett, *das*

sicken ['sɪkn] **1.** *v.i.* **a)** *(become ill)* krank werden; **be ~ing for something/the measles** *(Brit.)* krank werden *od. (ugs.)* etwas ausbrüten/[die] Masern bekommen; **b)** *(feel nausea or disgust)* ~ **at sth.** sich vor etw. *(Dat.)* ekeln; ~ **of sth./of doing sth.** einer Sache *(Gen.)* überdrüssig sein/es überdrüssig sein, etw. zu tun. **2.** *v.t.* **a)** **sth. ~s sb.** bei etw. wird jmdm. übel; **b)** *(disgust)* anwidern

sickening ['sɪknɪŋ] *adj.* **a)** ekelerregend; widerlich ⟨*Anblick, Geruch*⟩; **b)** *(coll.: infuriating)* unerträglich; **it's really** ~: es kann einen krank machen

sickle ['sɪkl] *n.* Sichel, *die*

sick: ~-leave *n.* Urlaub wegen Krankheit, *der;* **be on ~-leave** ≈ krank geschrieben sein; **~-list** *n.* Liste der Kranken, *die;* **on the ~-list** krank [gemeldet/geschrieben]

sickly ['sɪklɪ] *adj.* **a)** *(ailing)* kränklich; **b)** *(weak, faint)* schwach; matt ⟨*Lächeln*⟩; kraftlos ⟨*Sonne*⟩; fahl ⟨*Licht*⟩; blaß ⟨*Hautfarbe, Gesicht*⟩; **c)** *(nauseating)* ekelhaft; widerlich; *(mawkish)* süßlich

sickness ['sɪknɪs] *n.* **a)** *no art. (being ill)* Krankheit, *die;* **b)** *(disease; also fig.)* Krankheit, *die;* **c)** *(nausea)* Übelkeit, *die; (vomiting)* Erbrechen, *das*

sick: ~-pay *n.* Entgeltfortzahlung im Krankheitsfalle; *(paid by insurance)* Krankengeld, *das;* **~-room** *n.* Krankenzimmer, *das*

side [saɪd] **1.** *n.* **a)** *(also Geom.)* Seite, *die;* **this** ~ **up** oben; **lie on its** ~: auf der Seite liegen; **on both ~s** auf beiden Seiten; **b)** *(of animal or person)* Seite, *die;* **sleep on one's right/left** ~: auf der rechten/linken Seite schlafen; ~ **of mutton/beef/pork** Hammel-/Rinder-/ Schweinehälfte, *die;* ~ **of bacon** Speckseite, *die;* **split one's ~s [laughing]** *(fig.)* vor Lachen platzen; **walk/stand** ~ **by** ~: nebeneinander gehen/stehen; **work/fight** *etc.* ~ **by** ~ **[with sb.]** Seite an Seite [mit jmdm.] arbeiten/kämpfen *usw.;* **c)** *(part away from the centre)* Seite, *die;* **the eastern** ~ **of the town** der Ostteil der Stadt; **the ~s of sb.'s mouth** jmds. Mundwinkel; **right[-hand]/left[-hand]** ~: rechte/linke Seite; **on the right[-hand]/left[-hand]** ~ **of the road** auf der rechten/linken Straßenseite; **from** ~ **to** ~ *(right across)* quer hinüber; *(alternately each way)* von einer Seite auf die andere *od.* zur anderen; **to one** ~: zur Seite; **on one** ~: an der Seite; **stand on** *or* **to one** ~: an *od.* auf der Seite stehen; **on the** ~ *(fig.: in addition to regular work or income)* nebenbei; nebenher; **d)** *(space beside person or thing)* Seite, *die;* **at** *or* **by sb.'s** ~: an jmds. Seite *(Dat.);* neben jmdm.; **at** *or* **by the** ~ **of the car** beim *od.* am Auto; **at** *or* **by the** ~ **of the road/lake/grave** an der Straße/am See/am Grab; **on all ~s** *or* **every** ~: von allen Seiten ⟨*umzingelt, kritisiert*⟩; **e)** *(in relation to dividing line)* Seite, *die;* **[on] either** ~ **of** beiderseits, auf beiden Seiten (+ *Gen.*); **[to** *or* **on] one** ~ **of** neben (+ *Dat.*); **this/the other** ~ **of** *(with regard to space)* diesseits/jenseits (+ *Gen.*); *(with regard to time)* vor/nach (+ *Dat.*); **he is this** ~ **of fifty** er ist unter fünfzig; *see also* **right side; wrong side; f)** *(aspect)* Seite, *die;* **there are two ~s to every question** alles hat seine zwei Seiten; **look on the bright/gloomy** ~ **[of things]** die Dinge von der angenehmen/düsteren Seite sehen; **be on the high/expensive** *etc.* ~: [etwas] hoch/teuer *usw.* sein; **g)** *(opposing group or position)* Seite, *die;* Partei, *die; (Sport: team)* Mannschaft, *die;* **put sb.'s** ~: jmds. Seite vertreten; **be on the winning** ~ *(fig.)* auf der Seite der Gewinner stehen; **let the** ~ **down** *(fig.)* versagen; **change ~s** zur anderen Seite überwechseln; **time is on sb.'s** ~: die Zeit arbeitet für jmdn.; **take sb.'s** ~: sich auf jmds. Seite stellen; **take ~s [with/against sb.]** [für/gegen jmdn.] Partei ergreifen; **h)** *(of family)* Seite, *die;* **on one's/sb.'s father's/mother's** ~: väterlicher-/mütterlicherseits. **2.** *v.i.* ~ **with sb.** sich auf jmds. Seite *(Akk.)* stellen; ~ **against sb.** sich gegen jmdn. stellen. **3.** *adj.* seitlich; Seiten-

side: ~board *n.* Anrichte, *die;* **~boards** *(coll.),* **~burns** *ns. pl.* **a)** *(hair on cheeks)* Backenbart, *der;* **b)** *(hair in front of the ears)* Koteletten *Pl.;* **~-car** *n.* Beiwagen, *der;* **~-dish** *n.* Beilage, *die;* **~-door** *n.* Seitentür, *die;* **~-effect** *n.* Nebenwirkung, *die;* **~-entrance** *n.* Seiteneingang, *der;* **~-exit** *n.* Seitenausgang, *der;* **~-glance** *n. (lit. or fig.)* Seitenblick, *der* (**at** auf + *Akk.*); **~-kick** *n. (coll.)* Kumpan, *der;* **~light** *n. (Motor Veh.)* Begrenzungsleuchte, *die;* **drive on ~lights** mit Standlicht fahren; **~line** *n.* **a)** *(goods)* Nebensortiment, *das;* **b)** *(occupation)* Nebenbeschäftigung, *die;* **c)** *in pl. (Sport)* Begrenzungslinien; **on the ~lines** *(outside play area/track etc.)* am Spielfeldrand/am Rande der Bahn *usw.;* **remain on the ~lines** *(fig.)* sich [aus allem] heraushalten; **~-road** *n.* Seitenstraße, *die;* **~-saddle 1.** *n.* Damensattel, *der;* **2.** *adv.* **ride ~-saddle** im Damensattel reiten; **~-salad** *n.* Salat [als Beilage]; **~-show** *n.* Nebenattraktion, *die;* **~-splitting** *adj.* zwerchfellerschütternd; **~-step 1.** *n.* Schritt zur Seite, *der;* **2.** *v.t. (lit. or fig.)* ausweichen (+ *Dat.*); **~-street** *n.* Seitenstraße, *die;* **~-table** *n.* Beistelltisch, *der;* **~track** *v.t.* **get ~tracked** abgelenkt werden; **~walk** *(Amer.) see* **pavement a;** **~ways** ['saɪdweɪz] **1.** *adv.* seitwärts; **look at sb./sth. ~ways** jmdn./etw. von der Seite ansehen; **2.** *adj.* seitlich; **~-whiskers** *n. pl.* Backenbart, *der;* **~ wind** *n.* Seitenwind, *der*

siding ['saɪdɪŋ] *n. (Railw.)* Abstellgleis, *das;* Rangiergleis, *das*

sidle ['saɪdl] *v.i.* schleichen; **~ up to sb.** [sich] zu jmdm. schleichen

siege [si:dʒ] *n. (Mil.)* Belagerung, *die; (by police)* Umstellung, *die;* **be under ~** *(lit. or fig.)* belagert sein; *(by police)* umstellt sein; **lay ~ to sth.** *(lit. or fig.)* etw. belagern

siesta [sɪ'estə] *n.* Siesta, *die;* **have** *or* **take a ~:** [eine] Siesta halten *od.* machen

sieve [sɪv] **1.** *n.* Sieb, *das;* **have a head** *or* **memory like a ~** *(coll.)* ein Gedächtnis wie ein Sieb haben *(ugs.).* **2.** *v.t.* sieben

sift [sɪft] **1.** *v.t.* sieben; *(fig.: examine closely)* unter die Lupe nehmen; **~ sth. from sth.** etw. von etw. trennen. **2.** *v.i.* **~ through** durchsehen ‹*Briefe, Dokumente usw.*›; durchsuchen ‹*Trümmer, Asche, Habseligkeiten usw.*›

~ 'out *v.t. (lit. or fig.)* aussieben; **~ out sth. from sth.** etw. aus etw. heraussieben; *(fig.)* etw. von etw. trennen

sigh [saɪ] **1.** *n.* Seufzer, *der;* **give** *or* **breathe** *or* **utter** *or* **heave a ~:** einen Seufzer ausstoßen *od.* tun; **~ of relief/contentment** Seufzer der Erleichterung/Zufriedenheit. **2.** *v.i.* seufzen; **~ with relief/contentment** *etc.* vor Erleichterung/Zufriedenheit *usw. od.* erleichtert/zufrieden *usw.* seufzen; **~ for sth./sb.** *(fig.)* sich nach etw./jmdm. sehnen. **3.** *v.t.* seufzen

sight [saɪt] **1.** *n.* **a)** *(faculty)* Sehvermögen, *das;* **loss of ~:** Verlust des Sehvermögens; **second ~:** das Zweite Gesicht; **near ~** *see* **short sight;** **by ~:** mit dem Gesichtssinn *od.* den Augen; **know sb. by ~:** jmdn. vom Sehen kennen; *see also* **long sight; short sight; b)** *(act of seeing)* Anblick, *der;* **at [the] ~ of sb./blood** bei jmds. Anblick/beim Anblick von Blut; **catch ~ of sb./sth.** *(lit. or fig.)* jmdn./etw. erblicken; **lose ~ of sb./sth.** *(lit. or fig.)* jmdn./etw. aus dem Auge *od.* den Augen verlieren; **play sth. at ~:** etw. vom Blatt spielen; **shoot sb. at** *or* **on ~:** jmdn. gleich [bei seinem Erscheinen] erschießen; **at first ~:** auf den ersten Blick; **love at first ~:** Liebe auf den ersten Blick; **c)** *(spectacle)* Anblick, *der;* **be a sorry ~:** einen traurigen Anblick *od.* ein trauriges Bild bieten; **it is a ~ to see** *or* **to behold** *or* **worth seeing** das muß man gesehen haben; **a ~ for sore eyes** eine Augenweide; **be/look a [real] ~** *(coll.) (amusing)* [vollkommen] unmöglich aussehen *(ugs.); (horrible)* böse *od.* schlimm aussehen; **d)** *in pl. (noteworthy features)* Sehenswürdigkeiten *Pl.;* **see the ~s** sich *(Dat.)* die Sehenswürdigkeiten ansehen; **e)** *(range)* Sichtweite, *die;* **in ~** *(lit. or fig.)* in Sicht; **come into ~:** in Sicht kommen; **keep sb./sth. in ~** *(lit. or fig.)* jmdn./etw. im Auge behalten; **within** *or* **in ~ of sb./sth.** *(able to see)* in jmds. Sichtweite *(Dat.)*/in Sichtweite einer Sache; **out of sb.'s ~:** außerhalb jmds. Sichtweite; **be out of ~:** außer Sicht sein; *(sl.: be excellent)* wahnsinnig sein *(ugs.);* **keep** *or* **stay out of [sb.'s] ~:** sich [von jmdm.] nicht sehen lassen; **keep sb./sth. out of ~:** jmdn./etw. niemanden sehen lassen; **keep sth./sb. out of sb.'s ~:** jmdn. etw./jmdn. nicht sehen lassen; **not let sb./sth. out of one's ~:** jmdn./etw. nicht aus den Augen lassen; **out of ~, out of mind** *(prov.)* aus den Augen, aus dem Sinn; **f)** *(device for aiming)* Visier, *das;* **~s** Visiervorrichtung, *die;* **set/have [set] one's ~s on sth.** *(fig.)* etw. anpeilen; **set one's ~s [too] high** *(fig.)* seine Ziele [zu] hoch stecken; **lower/raise one's ~s** *(fig.)* zurückstecken/sich *(Dat.)* ein höheres Ziel setzen. **2.** *v.t.* sichten ‹*Land, Schiff, Flugzeug, Wrack*›; sehen ‹*Entflohenen, Vermißten*›; antreffen ‹*seltenes Tier, seltene Pflanze*›

sighted ['saɪtɪd] *adj.* sehend; **partially ~:** [hochgradig] sehbehindert

sight: ~-read *v.t. & i. (Mus.)* ‹*Pianist usw.:*› vom Blatt spielen; ‹*Sänger:*› vom Blatt singen; **~seeing** *n.* Sightseeing, *das (Touristikjargon);* **go ~seeing** Besichtigungen machen; **~-seer** ['saɪtsi:ə(r)] *n.* Tourist *(der die Sehenswürdigkeiten besichtigt)*

sign [saɪn] **1.** *n.* **a)** *(symbol, gesture, signal, mark)* Zeichen, *das;* **b)** *(Astrol.)* **~ [of the zodiac]** [Tierkreis]zeichen, *das;* Sternzeichen, *das;* **what ~ are you?** welches Tierkreiszeichen *od.* Sternzeichen bist du?; **sb.'s birth ~:** jmds. Tierkreiszeichen; **c)** *(notice)* Schild, *das;* **[direction] ~:** Wegweiser, *der;* **[advertising] ~:** Reklameschild, *das;* Reklame, *die; (illuminated, flashing)* Leuchtreklame, *die;* **danger ~** *(lit. or fig.)* Gefahrenzeichen, *das;* **d)** *(outside shop etc.) see* **signboard; e)** *(indication)* Zeichen, *das; (of future event)* Anzeichen, *das;* **there is little/no/every ~ of sth./that ...:** wenig/nichts/alles deutet auf etw. *(Akk.)* hin *od.* deutet darauf hin, daß ...; **show [no] ~s of fatigue/strain/improvement** *etc.* [keine] Anzeichen der Müdigkeit / Anstrengung / Besserung *usw.* zeigen *od.* erkennen lassen; **the carpet showed little/some ~[s] of wear** der Teppich wirkte kaum/etwas abgenutzt; **as a ~ of** als Zeichen (+ *Gen.*); **do sth. as a ~ of sth.** etw. zum Zeichen einer Sache *(Gen.)* tun; **at the first** *or* **slightest ~ of sth.** schon beim geringsten Anzeichen von etw.; **there was no ~ of him/the car anywhere** er/der Wagen war nirgends zu sehen; **there was no ~ of life** keine Menschenseele war zu sehen; **~ of the times** Zeichen der Zeit. **2.** *v.t.* **a)** *(write one's name etc. on)* unterschreiben; ⟨*Künstler, Autor:*⟩ signieren ⟨*Werk*⟩; **b)** ~ **one's name** [mit seinem Namen] unterschreiben; ~ **oneself R.A. Smith** mit R.A. Smith unterschreiben. **3.** *v.i.* *(write one's name)* unterschreiben; ~ **for sth.** *(acknowledge receipt of sth.)* den Empfang einer Sache *(Gen.)* bestätigen

~ **a'way** *v.t.* abtreten ⟨*Eigentum*⟩; verzichten auf ⟨*Recht, Freiheit usw.*⟩

~ **'off 1.** *v.i.* **a)** *(cease employment)* kündigen; **b)** *(at end of shift etc.)* sich [zum Feierabend *usw.*] abmelden; **c)** *(Radio)* sich verabschieden. **2.** *v.t.* kündigen

~ **'on 1.** *v.t.* einstellen ⟨*Arbeiter*⟩; verpflichten ⟨*Fußballspieler*⟩; anwerben ⟨*Soldaten*⟩; anheuern, anmustern ⟨*Seeleute*⟩. **2.** *v.i.* **a)** sich verpflichten (**with** bei); **b)** ~ **on [for the dole]** sich arbeitslos melden; stempeln gehen *(ugs. veralt.)*

~ **'out 1.** *v.t.* ~ **books out from the library** Bücher als [aus der Bibliothek] entliehen eintragen. **2.** *v.i.* sich [schriftlich] abmelden; ⟨*Hotelgast:*⟩ abreisen

~ **'over** *v.t.* überschreiben ⟨*Immobilien*⟩; übertragen ⟨*Rechte*⟩

~ **'up 1.** *v.t.* *(engage)* [vertraglich] verpflichten; einstellen ⟨*Arbeiter*⟩; aufnehmen ⟨*Mitglied*⟩; einschreiben ⟨*Kursteilnehmer*⟩. **2.** *v.i.* sich [vertraglich] verpflichten (**with** bei); *(join a course etc.)* sich einschreiben

signal ['sɪgnl] **1.** *n.* Signal, *das;* **a ~ for sth./to sb.** ein Zeichen zu etw./für jmdn.; **at a ~ from the headmaster** auf ein Zeichen des Direktors; **the ~ was against us/at red** *(Railw.)* das Signal zeigte „halt"/stand auf Rot; **hand ~s** *(Motor Veh.)* Handzeichen; **radio ~:** Funkspruch, *der.* **2.** *v.i.*, *(Brit.)* **-ll-** signalisieren; Signale geben; ⟨*Kraftfahrer:*⟩ blinken; *(using hand etc. ~s)* anzeigen; ~ **for assistance** ein Hilfesignal geben; ~ **to sb. [to do sth.]** jmdm. ein Zeichen geben[, etw. zu tun]. **3.** *v.t.*, *(Brit.)* **-ll-: a)** *(lit. or fig.)* signalisieren; ~ **sb. [to do sth.]** jmdm. ein Zeichen geben[, etw. zu tun]; **the driver ~led that he was turning right** der Fahrer zeigte an, daß er [nach] rechts abbiegen wollte; **b)** *(Radio etc.)* funken; [über Funk] durchgeben. **4.** *adj.* außergewöhnlich

signal: ~-box *n.* *(Railw.)* Stellwerk, *das;* **~man** ['sɪgnlmən] *n.*, *pl.* **~men** ['sɪgnlmən] *(Brit. Railw.)* Stellwerkswärter, *der*

signatory ['sɪgnətərɪ] *n.* *(person)* Unterzeichner, *der;* *(party)* vertragschließende Partei; *(state)* Signatarstaat, *der*

signature ['sɪgnətʃə(r)] *n.* **a)** Unterschrift, *die;* *(on painting)* Signatur, *die;* **put one's ~ to sth.** seine Unterschrift unter etw. *(Akk.)* setzen; **b)** *(Mus.)* *see* **key signature; time signature**

'signature tune *n.* *(Radio, Telev.)* Erkennungsmelodie, *die*

'signboard *n.* Schild, *das;* *(advertising)* Reklameschild, *das*

signet-ring ['sɪgnɪt rɪŋ] *n.* Siegelring, *der*

significance [sɪg'nɪfɪkəns] *n.* *(meaning, importance)* Bedeutung, *die;* **be of [no] ~:** [nicht] von Bedeutung sein; **a matter of great/little/no ~:** eine [sehr] wichtige/ziemlich unwichtige/völlig unwichtige Angelegenheit

significant [sɪg'nɪfɪkənt] *adj.* **a)** *(noteworthy, important)* bedeutend; **b)** *(full of meaning)* bedeutsam

significantly [sɪg'nɪfɪkəntlɪ] *adv.* **a)** *(meaningfully)* bedeutungsvoll; *as sentence-modifier* ~ **[enough]** bedeutsamerweise; **b)** *(notably)* bedeutend; signifikant *(geh., fachspr.)*

signify ['sɪgnɪfaɪ] *v.t.* **a)** *(indicate, mean)* bedeuten; **b)** *(communicate, make known)* kundtun *(geh.)*; zum Ausdruck bringen

sign: ~ language *n.* Zeichensprache, *die;* **~post 1.** *n.* *(lit. or fig.)* Wegweiser, *der;* **2.** *v.t.* ausschildern ⟨*Route, Umleitungsstrecke usw.*⟩; mit Wegweisern versehen ⟨*Straße*⟩; **~-writer** *n.* Schildermaler, *der*

silage ['saɪlɪdʒ] *n.* *(Agric.)* Silage, *die;* Gärfutter, *das*

silence ['saɪləns] **1.** *n.* Schweigen, *das;* *(keeping a secret)* Verschwiegenheit, *die;* *(taciturnity)* Schweigsamkeit, *die;* *(stillness)* Stille, *die;* **there was ~:** es herrschte Schweigen/Stille; **~!** Ruhe!; **in ~:** schweigend; **call for ~:** um Ruhe bitten; **keep ~** *(lit. or fig.)* schweigen; **break the ~:** die Stille unterbrechen; *(be the first to speak)* das Schweigen brechen; **break one's ~** *(lit. or fig.)* sein Schweigen brechen; **a minute's ~:** eine Schweigeminute. **2.** *v.t.* zum Schweigen bringen; *(fig.)* ersticken ⟨*Zweifel, Ängste, Proteste*⟩; mundtot machen ⟨*Gegner, Zeugen*⟩

silencer ['saɪlənsə(r)] *n.* *(Brit. Motor Veh., Arms)* Schalldämpfer, *der*

silent ['saɪlənt] *adj.* **a)** stumm; *(noiseless)* unhörbar; *(still)* still; **be ~** *(say nothing)* schweigen; *(be still)* still sein; *(not be working)* ⟨*Maschine:*⟩ stillstehen; ⟨*Waffen:*⟩ schweigen; **fall ~:** verstummen; **keep** *or* **remain ~** *(lit. or fig.)* schweigen; ⟨*jmd., der verhört wird:*⟩ beharrlich schweigen; **b)** *(taciturn)* schweigsam; **c)** *(Ling.)* stumm; **d)** *(Cinemat.)* ~ **film** Stummfilm, *der*

silently ['saɪləntlɪ] *adv.* schweigend; stumm ⟨*weinen, beten*⟩; *(noiselessly)* lautlos

silent ma'jority *n.* schweigende Mehrheit

Silesia [saɪ'li:ʃə] *pr. n.* Schlesien *(das)*

Silesian [saɪ'li:ʃn] **1.** *adj.* schlesisch. **2.** *n.* **a)** *(person)* Schlesier, *der*/Schlesierin, *die;* **b)** *(dialect)* Schlesisch, *das*

silhouette [sɪlʊ'et] **1.** *n.* **a)** *(picture)* Schattenriß, *der;* **b)** *(appearance against the light)* Silhouette, *die.* **2.** *v.t.* **be ~d against sth.** sich als Silhouette gegen etw. abheben

silicon ['sɪlɪkən] *n.* *(Chem.)* Silicium, *das;* Silizium, *das;* ~ **chip** Siliciumchip, *der;* Siliziumchip, *der*

silicone ['sɪlɪkəʊn] *n.* *(Chem.)* Silikon, *das*

silk [sɪlk] **1.** *n.* **a)** Seide, *die;* **take ~** *(Brit. Law)* Kronanwalt werden; **b)** *in pl.* *(garments)* seidene Kleider *od.* Kleidungsstücke; **c)** *(of spider etc.)* [Spinnen]faden, *der;* **d)** *(Brit. Law coll.)* Kronanwalt, *der.* **2.** *attrib. adj.* seiden; Seiden-

silken ['sɪlkn] *adj.* seiden; Seiden-

silk: ~-screen printing *see* **screen-printing; ~worm** *n.* *(Zool.)* Seidenraupe, *die*

silky ['sɪlkɪ] *adj.* seidig

sill [sɪl] *n.* *(of door)* [Tür]schwelle, *die;* *(of window)* Fensterbank, *die*

silliness ['sɪlɪnɪs] *n.*, *no pl.* Dummheit, *die;* Blödheit, *die* *(ugs.)*

silly ['sɪlɪ] **1.** *adj.* dumm; blöd[e] *(ugs.)*; *(imprudent, unwise)* töricht; unklug; *(childish)* albern; **the ~ season** *(Journ.)* die Sauregurkenzeit; **a ~ thing** *(a foolish action)* etwas Dummes *od.* *(ugs.)* Blödes; *(a trivial matter)* eine blödsinnige Kleinigkeit *(ugs.)*; **it/that was a ~ thing to do** es/das war dumm *od.* *(ugs.)* blöd; **I was scared ~:** mir rutschte das Herz in die Hose *(ugs.)*. **2.** *n.* *(coll.)* Dummchen, *das;* Dummerchen, *das* *(fam.)*

silly-billy ['sɪlɪbɪlɪ] *n. (coll.)* Kindskopf, *der*

silo ['saɪləʊ] *n., pl.* **~s a)** *(Agric.)* Silo, *der;* **b)** *(Mil.)* **[missile] ~:** [Raketen]silo, *der*

silt [sɪlt] **1.** *n.* Schlamm, *der;* Schlick, *der.* **2.** *v.t.* **~ up** verschlämmen. **3.** *v.i.* **~ up** verschlammen

silver ['sɪlvə(r)] **1.** *n.* **a)** *no pl., no indef. art.* Silber, *das;* **b)** *(colour, medal, vessels, cutlery)* Silber, *das; (cutlery of other material)* Besteck, *das;* **c)** *no pl., no indef. art. (coins)* Silbermünzen *Pl.;* Silber, *das (ugs.).* **2.** *attrib. adj.* silbern; Silber‹*pokal, -münze*›; *see also* **spoon. 3.** *v.t. (coat with ~)* versilbern; *(coat with amalgam)* verspiegeln ‹*Glas*›

silver: ~ 'birch *n. (Bot.)* Weißbirke, *die;* **~-coloured** *adj.* silberfarben; silberfarbig; **~-haired** *adj.* silberhaarig *(geh.);* **~ 'medal** *n.* Silbermedaille, *die;* **~ 'medallist** *n.* Silbermedaillengewinner, *der*/-gewinnerin, *die;* **~ 'paper** *n.* Silberpapier, *das;* **~ 'plate** *n., no pl., no indef. art.* versilberte Ware; *(coating)* Silberauflage, *die;* **~smith** *n.* Silberschmied, *der*/-schmiedin, *die;* **~ware** *n., no pl.* Silber, *das;* **~ 'wedding** *n.* Silberhochzeit, *die;* silberne Hochzeit

silvery ['sɪlvərɪ] *adj. (silver-coloured)* silbrig; *(clear-sounding)* silbern *(dichter.);* silbrig *(geh.)*

similar ['sɪmɪlə(r)] *adj.* ähnlich (**to** *Dat.*); **some flour and a ~ amount of sugar** etwas Mehl und ungefähr die gleiche Menge Zucker; **of ~ size/colour** *etc.* von ähnlicher Größe/Farbe *usw.;* **be ~ in size/appearance** *etc.* **[to sb./sth.]** eine ähnliche Größe/ein ähnliches Aussehen haben [wie jmd./etw.]; **look/taste/smell** *etc.* **~ [to sth.]** ähnlich aussehen/schmecken/riechen *usw.* [wie etw.]; **the two brothers look very ~:** die beiden Brüder sehen sich *(Dat.)* sehr ähnlich

similarity [sɪmɪ'lærɪtɪ] *n.* Ähnlichkeit, *die* (**to** mit)

similarly ['sɪmɪləlɪ] *adv.* ähnlich; *(to the same degree)* ebenso; *as sentence-modifier* ebensogut

simile ['sɪmɪlɪ] *n. (Lit.)* Vergleich, *der*

simmer ['sɪmə(r)] **1.** *v.i.* **a)** *(Cookery)* ‹*Flüssigkeit:*› sieden; **allow the fish to ~ for ten minutes** den Fisch zehn Minuten ziehen lassen; **b)** *(fig.)* gären; **let things ~:** die Dinge sich entwickeln lassen; **~ with rage/excitement** eine Wut haben/innerlich ganz aufgeregt sein. **2.** *v.t. (Cookery)* köcheln lassen ‹*Suppe, Soße usw.*›; ziehen lassen ‹*Fisch, Klöße usw.*›

~ 'down *v.i.* sich abregen *(ugs.)*

simper ['sɪmpə(r)] *v.i.* affektiert *od.* gekünstelt lächeln

simple ['sɪmpl] *adj.* **a)** *(not compound, not complicated)* einfach; *(not elaborate)* schlicht ‹*Mobiliar, Schönheit, Kunstwerk, Kleidung*›; **the ~ life** das einfache Leben; **b)** *(unqualified, absolute)* einfach; simpel; **it was a ~ misunderstanding** es war [ganz] einfach ein Mißverständnis; **it is a ~ fact that ...:** es ist [ganz] einfach eine Tatsache *od.* eine simple Tatsache, daß ...; **c)** *(easy)* einfach; **it's [not] as ~ as that** so einfach ist das [nicht]; **d)** *(unsophisticated)* schlicht; *(foolish)* dumm; einfältig

'simple-minded *adj.* **a)** *(unsophisticated)* schlicht; **b)** *(feeble-minded)* debil

simpleton ['sɪmpltən] *n.* Einfaltspinsel, *der (ugs.)*

simplicity [sɪm'plɪsɪtɪ] *n., no pl.* Einfachheit, *die; (unpretentiousness, lack of sophistication)* Schlichtheit, *die*

simplification [sɪmplɪfɪ'keɪʃn] *n.* Vereinfachung, *die*

simplify ['sɪmplɪfaɪ] *v.t.* vereinfachen; **~ matters** die Sache vereinfachen

simplistic [sɪm'plɪstɪk] *adj.* [all]zu simpel

simply ['sɪmplɪ] *adv.* **a)** *(in an uncomplicated manner)* einfach; *(in an unsophisticated manner)* schlicht; **live/eat ~:** einfach leben/essen; **b)** *(absolutely)* einfach; **c)** *(categorically, without good reason, without asking)* einfach; *(merely)* nur; **it ~ isn't true** es ist einfach nicht wahr; **you ~ must see that film** du mußt den Film einfach sehen; **I was ~ trying to help** ich wollte nur helfen; **quite ~:** ganz einfach; **~ because ...:** einfach weil ...

simulate ['sɪmjʊleɪt] *v.t.* **a)** *(feign)* vortäuschen; heucheln ‹*Reue, Entrüstung, Begeisterung*›; simulieren, vortäuschen ‹*Krankheit*›; **b)** *(mimic)* nachahmen; **c)** simulieren ‹*Bedingungen, Wetter, Umwelt usw.*›

simulated ['sɪmjʊleɪtɪd] *adj.* **a)** *(feigned)* vorgetäuscht; geheuchelt; **b)** *(artificial)* imitiert ‹*Leder, Pelz usw.*›; **c)** simuliert ‹*Bedingungen, Wetter, Umwelt usw.*›

simulator ['sɪmjʊleɪtə(r)] *n.* Simulator, *der*

simultaneous [sɪml'teɪnɪəs] *adj.* gleichzeitig (**with** mit); simultan *(fachspr., geh.);* **be ~:** gleichzeitig/simultan erfolgen

simultaneous interpre'tation *n.* Simultandolmetschen, *das*

simultaneously [sɪml'teɪnɪəslɪ] *adv.* gleichzeitig

sin [sɪn] **1.** *n.* Sünde, *die;* **live in ~** *(coll.)* in Sünde leben *(veralt., scherzh.);* **[as] miserable as ~:** todunglücklich; **for my ~s** *(joc.)* um meiner Missetaten willen *(scherzh.).* **2.** *v.i.,* **-nn-** sündigen; **~ against sb./God** an jmdm./Gott *od.* gegen jmdn./Gott sündigen

since [sɪns] **1.** *adv.* seitdem; **he has ~ remarried, he has remarried ~:** er hat danach wieder geheiratet; **long ~:** vor langer Zeit; **not long ~:** vor nicht allzulanger Zeit; **he is long ~ dead** er ist seit langem tot. **2.** *prep.* seit; **~ seeing you ...:** seit ich dich gesehen habe; **~ then/that time** inzwischen; **~ when?** seit wann? **3.** *conj.* **a)** seit; **it is a long time/so long/not so long ~ ...:** es ist lange/so lange/gar nicht lange her, daß ...; **how long is it ~ he left you?** wie lange ist es her, daß er dich verlassen hat?; **b)** *(seeing that, as)* da

sincere [sɪn'sɪə(r)] *adj.,* **~r** [sɪn'sɪərə(r)], **~st** [sɪn'sɪərɪst] aufrichtig; herzlich ‹*Grüße, Glückwünsche usw.*›; wahr ‹*Freund*›

sincerely [sɪn'sɪəlɪ] *adv.* aufrichtig; **yours ~** *(in letter)* mit freundlichen Grüßen

sincerity [sɪn'serɪtɪ] *n., no pl.* Aufrichtigkeit, *die*

sine [saɪn] *n. (Math.)* Sinus, *der*

sinecure ['sɪnɪkjʊə(r), 'saɪnɪkjʊə(r)] *n.* Pfründe, *die*

sinew ['sɪnju:] *n. (Anat.)* Sehne, *die*

sinewy ['sɪnju:ɪ] *adj.* sehnig; *(fig.: vigorous)* kraftvoll

sinful ['sɪnfl] *adj.* sündig; *(reprehensible)* sündhaft; **it is ~ to ...:** es ist eine Sünde, ... zu ...

sing [sɪŋ] **1.** *v.i.,* **sang** [sæŋ], **sung** [sʌŋ] singen; *(fig.)* ‹*Kessel, Wind:*› singen. **2.** *v.t.,* **sang, sung** singen; **~ sb. a song** *or* **a song for sb.** jmdm. ein Lied vorsingen; **~ sb. to sleep** jmdn. in den Schlaf singen

~ a'long *v.i.* mitsingen

~ 'out 1. *v.i.* **a)** *(~ loudly)* [laut *od.* aus voller Kehle] singen; **b)** *(call out)* [laut] rufen; **~ out for sb./sth.** nach jmdm./etw. rufen. **2.** *v.t.* rufen; schreien

~ 'up *v.i.* lauter singen

Singapore [sɪŋgə'pɔ:(r)] *pr. n.* Singapur *(das)*

singe [sɪndʒ] 1. *v.t.*, ~ing ansengen; versengen. 2. *v.i.*, ~ing [ver]sengen

singer ['sɪŋə(r)] *n.* Sänger, *der*/ Sängerin, *die*

singing ['sɪŋɪŋ] *n.*, *no pl.* *(lit. or fig.: of kettle, wind)* Singen, *das;* **the ~ of the birds** der Gesang der Vögel

single ['sɪŋgl] 1. *adj.* **a)** einfach; einzig ‹*Ziel, Hoffnung*›; *(for one person)* Einzel‹*bett, -zimmer*›; einfach ‹*Größe*›; *(without the other one of a pair)* einzeln; **speak with a ~ voice** *(fig.)* mit einer Stimme sprechen; **~ sheet** Betttuch für ein Einzelbett; **~ ticket** *(Brit.)* einfache Fahrkarte; **~ fare** *(Brit.)* Preis für [die] einfache Fahrt; **b)** *(one by itself)* einzig; *(isolated)* einzeln; **one ~ ...:** ein einziger/eine einzige/ein einziges ...; **at a** *or* **one ~ blow** *or* **stroke** mit einem Schlag; **c)** *(unmarried)* ledig; **a ~ man/woman/~ people** ein Lediger/eine Ledige/Ledige; **~ parent** alleinerziehender Elternteil; **d)** *(separate, individual)* einzeln; **every ~ one** jeder/jede/ jedes einzelne; **every ~ time/day** aber auch jedesmal/jeden Tag; **not a ~ one** kein einziger/keine einzige/kein einziges; **not a ~ word** kein einziges Wort; **not/ never for a ~ minute** *or* **moment** keinen [einzigen] Augenblick [lang]. 2. *n.* **a)** *(Brit.: ticket)* einfache Fahrkarte; **[a] ~/two ~s to Manchester, please** einmal/zweimal einfach nach Manchester, bitte; **b)** *(record)* Single, *die;* **c)** *in pl.* *(Golf)* Single, *das;* *(Tennis)* Einzel, *das;* **men's/women's** *or* **ladies' ~s** Herren-/Dameneinzel, *das*

~ **out** *v.t.* aussondern; *(be distinctive quality of)* auszeichnen (**from** vor + *Dat.*); **~ sb./sth. out as/for sth.** jmdn./etw. als etw./für etw. auswählen; **~ sb. out for promotion/special attention** jmdn. für eine Beförderung vorsehen/ sich mit jmdm. besonders befassen

single: ~ **cream** *n.* [einfache] Sahne; **~-decker** 1. *n.* **be a ~-decker** ‹*Bus, Straßenbahn:*› nur ein Deck haben; 2. *adj.* **~-decker bus/tram** Bus/Straßenbahn mit [nur] einem Deck; **~ [European] market** *n.* [europäischer] Binnenmarkt; **~-handed** 1. ['----] *adj.* Einhand‹*segeln, -segler*›; **his ~-handed efforts to get a new hospital** seine einsamen Bemühungen um ein neues Krankenhaus; 2. [--'--] *adv.* allein; **sail round the world ~-handed** als Einhandsegler um die Welt fahren; **~-lens 'reflex [camera]** *n.* *(Photog.)* einäugige Spiegelreflexkamera; **~-line** *adj.* einspurig; **~-minded** *adj.* zielstrebig

singleness ['sɪŋglnɪs] *n.*, *no pl.* **~ of purpose** Zielstrebigkeit, *die*

'singles bar *n.* Singlekneipe, *die*

singlet ['sɪŋglɪt] *n.* *(Brit.)* *(vest)* Unterhemd, *das;* *(Sport)* Trikot, *das*

singly ['sɪŋglɪ] *adv.* **a)** einzeln; **b)** *(by oneself)* allein

'singsong *n.* *(Brit.)* **have a ~:** gemeinsam singen

singular ['sɪŋgjʊlə(r)] 1. *adj.* **a)** *(Ling.)* singularisch; Singular-; **~ noun** Substantiv im Singular; **first person ~:** erste Person Singular; **b)** *(individual)* einzeln; *(unique)* einmalig; einzigartig; **c)** *(extraordinary)* einmalig; einzigartig. 2. *n.* *(Ling.)* Einzahl, *die;* Singular, *der*

singularity [sɪŋgjʊ'lærɪtɪ] *n.*, *no pl.* Eigenartigkeit, *die;* Sonderbarkeit, *die*

singularly ['sɪŋgjʊləlɪ] *adv.* *(extraordinarily)* außerordentlich; einmalig ‹*schön*›; *(strangely)* seltsam

sinister ['sɪnɪstə(r)] *adj.* **a)** *(of evil omen)* unheilverkündend; **b)** *(suggestive of malice)* finster; *(wicked)* übel

sink [sɪŋk] 1. *n.* Spülbecken, *das;* Spüle, *die;* **pour sth. down the ~:** etw. in den Ausguß schütten. 2. *v.i.*, **sank** [sæŋk] *or* **sunk** [sʌŋk], **sunk a)** sinken; **leave sb. to ~ or swim** *(fig.)* jmdn. seinem Schicksal überlassen; **b) ~ into** *(become immersed in)* sinken in (+ *Akk.*); versinken in (+ *Dat.*); *(penetrate)* eindringen in (+ *Akk.*); *(fig.: be absorbed into)* dringen in (+ *Akk.*) ‹*Bewußtsein*›; **~ into an armchair/the cushions** in einen Sessel/die Kissen sinken; **~ into a deep sleep/a coma** in einen tiefen Schlaf/in ein Koma sinken *(geh.)*; **be sunk in thought/despair** in Gedanken/in Verzweiflung *(Akk.)* versunken sein; **c)** *(come to lower level or pitch)* sinken; *(fig.: fail)* ‹*Moral, Hoffnung:*› sinken; **sb.'s heart ~s/spirits ~:** jmds. Stimmung sinkt; **~ to one's knees** auf die *od.* seine Knie sinken; **d)** *(fall)* ‹*Preis, Temperatur, Währung, Produktion usw.:*› sinken; **~ in value** im Wert sinken. 3. *v.t.*, **sank** *or* **sunk, sunk a)** versenken; *(cause failure of)* zunichte machen; **be sunk** *(fig. coll.: have failed)* aufgeschmissen sein *(ugs.)*; **~ one's differences** seine Streitigkeiten begraben; **b)** *(lower)* senken; *(Golf)* ins Loch schlagen ‹*Ball*›; **c)** *(dig)* niederbringen; *(recess)* versenken; *(embed)* stoßen ‹*Schwert, Messer*›; graben *(geh.)* ‹*Zähne, Klauen*›

~ **'in** *v.i.* **a)** *(become immersed)* einsinken; *(penetrate)* eindringen; **b)** *(fig.: be absorbed into the mind)* jmdm. ins Bewußtsein dringen; ‹*Warnung, Lektion:*› verstanden werden

sinking ['sɪŋkɪŋ] 1. *adj.* **a)** sinkend; **b)** *(declining)* untergehend ‹*Sonne*›; **c)** *(falling in value)* sinkend; **d) with a ~ heart** *(fig.)* beklommen; resigniert. 2. *n.* **a)** *(of ship)* *(deliberate)* Versenkung, *die;* *(accidental)* Sinken, *das;* Untergang, *der;* *(of well)* Niederbringung, *die;* **b)** *attrib.* **a ~ feeling** *(fig.)* ein flaues Gefühl [im Magen]

'sink unit *n.* Spüle, *die*

sinner ['sɪnə(r)] *n.* Sünder, *der*/ Sünderin, *die*

sinuous ['sɪnjʊəs] *adj.* gewunden; sich schlängelnd ‹*Schlange*›; *(lithe)* geschmeidig ‹*Körper, Bewegungen*›

sinus ['saɪnəs] *n.* *(Anat.)* Sinus, *der (fachspr.)*

sip [sɪp] 1. *v.t.*, **-pp-:** **~ [up]** schlürfen. 2. *v.i.*, **-pp-:** **~ at/from sth.** an etw. *(Dat.)* nippen. 3. *n.* Schlückchen, *das*

siphon ['saɪfn] 1. *n.* Siphon, *der.* 2. *v.t.* [durch einen Saugheber] laufen lassen; **~ sth. from a tank** etw. [mit einem Saugheber] aus einem Tank ablassen

~ **'off** *v.t.* [mit einem Saugheber] ablassen; *(fig.: transfer)* abzweigen

sir [sɜ:(r)] *n.* **a)** *(formal address)* der Herr; *(to teacher)* Herr Meier/Schmidt *usw.*; **no '~!** keinesfalls!; von wegen! *(ugs.)*; **yes '~!** allerdings; **Sir!** *(Mil.)* Herr Oberst/Leutnant *usw.*!; *(yes)* jawohl, Herr Oberst/Leutnant *usw.*!; **b)** *(in letter)* **Dear Sir** Sehr geehrter Herr; **Dear Sirs** Sehr geehrte [Damen und] Herren; **Dear Sir or Madam** Sehr geehrte Dame/Sehr geehrter Herr; **c) Sir** [sə(r)] *(title of knight etc.)* Sir

sire ['saɪə(r)] 1. *n.* Vatertier, *das.* 2. *v.t.* zeugen

siren ['saɪrən] *n.* **a)** Sirene, *die;* **b)** *(temptress)* Sirene, *die (geh.)*

sirloin ['sɜ:lɔɪn] *n.* **a)** *(Brit.: upper part of loin of beef)* Roastbeef, *das;* **a ~ of beef** ein Stück Roastbeef; **~ steak** Rumpsteak, *das;* **b)** *(Amer.)* Rumpsteak, *das*

sisal ['saɪsl] *n.* *(fibre)* Sisal, *der*
sissy ['sɪsɪ] **1.** *n.* Waschlappen, *der.* **2.** *adj.* feige
sister ['sɪstə(r)] *n.* **a)** Schwester, *die;* **b)** *(fellow member of trade union)* Kollegin, *die;* **c)** *(Brit.: senior nurse)* Oberschwester, *die*
'**sister-in-law** *n., pl.* **sisters-in-law** Schwägerin, *die*
sisterly ['sɪstəlɪ] *adj.* schwesterlich
sit [sɪt] **1.** *v.i.,* **-tt-, sat** [sæt] **a)** *(become seated)* sich setzen; ~ **on** *or* **in a chair/in an armchair** sich auf einen Stuhl/in einen Sessel setzen; ~ **by** *or* **with sb.** sich zu jmdm. setzen; ~ **over there!** setz dich dort drüben hin!; **b)** *(be seated)* sitzen; **don't just ~ there!** sitz nicht einfach rum! *(ugs.);* ~ **in judgement on** *or* **over sb./sth.** über jmdn./etw. zu Gericht sitzen; ~ **still!** sitz ruhig *od.* still!; ~ **tight** *(coll.)* ruhig sitzen bleiben; *(fig.: stay in hiding)* sich nicht fortrühren; **c)** ~ **for one's portrait** Porträt sitzen; **d)** *(take a test)* ~ **for sth.** die Prüfung für etw. machen; **e)** *(be in session)* tagen; **f)** *(be on perch or nest)* sitzen. **2.** *v.t.,* **-tt-, sat a)** *(cause to be seated, place)* setzen; **b)** *(Brit.)* ~ **an examination** eine Prüfung machen
~ **a'bout,** ~ **a'round** *v.i.* herumsitzen *(ugs.)*
~ '**back** *v.i.* **a)** sich zurücklehnen; **b)** *(fig.: do nothing)* sich im Sessel zurücklehnen *(fig.)*
~ '**down 1.** *v.i.* **a)** *(become seated)* sich setzen **(on/in** auf/in + *Akk.*); **b)** *(be seated)* sitzen; **take sth. ~ting down** *(fig.)* etw. auf sich *(Dat.)* sitzen lassen. **2.** *v.t.* ~ **sb. down** *(invite to ~)* jmdn. Platz nehmen lassen; *(help to ~)* jmdm. helfen, sich zu setzen
~ '**in** *v.i.* **a)** *(occupy place as protest)* ein Sit-in veranstalten; **b)** ~ **in on** *(be present at)* teilnehmen an (+ *Dat.*); dabeisein bei
~ **on** *v.t.* **a)** *(serve as member of)* sitzen in (+ *Dat.*) ‹*Ausschuß usw.*›; **b)** *(coll.: delay)* in der Schublade liegen lassen *(fig. ugs.);* auf die lange Bank schieben *(ugs.)* ‹*Entscheidung*›; **c)** *(coll.: repress)* unterdrücken; **d)** *(fig.: hold on to)* festhalten
~ '**up 1.** *v.i.* **a)** *(rise)* sich aufsetzen; **b)** *(be sitting erect)* [aufrecht] sitzen; **c)** *(not slouch)* gerade sitzen; ~ **up straight!** sitz gerade!; ~ **up [and take notice]** *(fig. coll.)* aufhorchen; **d)** *(delay going to bed)* aufbleiben; ~ **up [waiting] for sb.** aufbleiben und auf jmdn. warten; ~ **up with sb.** bei jmdm. Nachtwache halten. **2.** *v.t.* aufsetzen
~ **upon** *see* ~ **on**
sitcom ['sɪtkɒm] *(coll.) see* **situation comedy**
'**sit-down 1.** *n.* **have a ~:** sich setzen. **2.** *adj.* ~ **demonstration** Sitzblockade, *die;* ~ **strike** Sitzstreik, *der*
site [saɪt] **1.** *n.* **a)** *(land)* Grundstück, *das;* **b)** *(location)* Sitz, *der; (of new factory etc.)* Standort, *der.* **2.** *v.t.* *(locate)* stationieren ‹*Raketen*›; ~ **a factory in London** London als Standort einer Fabrik wählen; **be ~d** gelegen sein
'**sit-in** *n.* Sit-in, *das*
siting ['saɪtɪŋ] *n.* Standortwahl, *die* **(of** für); *(position)* Lage, *die; (of missiles)* Stationierung, *die*
sitter ['sɪtə(r)] *n.* *(artist's model)* Modell, *das*
sitting ['sɪtɪŋ] *n.* Sitzung, *die;* **in one** *or* **at a** ~ *(fig.)* in einem Zug[e]
sitting: ~ '**duck** *n.* *(fig.)* leichtes Ziel; **~-room** *n.* Wohnzimmer, *das;* ~ '**target** *see* ~ **duck;** ~ '**tenant** *n.* **he is/was the ~ tenant** er ist/war der jetzige/damalige Mieter; **there is a ~ tenant** es ist ein Mieter vorhanden
situate ['sɪtjʊeɪt] *v.t.* legen
situated ['sɪtjʊeɪtɪd] *adj.* **a)** gelegen; **be ~:** liegen; **a badly ~ house** ein Haus in schlechter *od.* ungünstiger Lage; **b) be well/badly ~ financially** finanziell gut/schlecht gestellt sein
situation [sɪtjʊ'eɪʃn] *n.* **a)** *(location)* Lage, *die;* **b)** *(circumstances)* Situation, *die;* **be in the happy ~ of being able to do sth.** in der glücklichen Lage sein, etw. tun zu können; **what's the ~?** wie steht's?; **c)** *(job)* Stelle, *die*
situation '**comedy** *n.* Situationskomödie, *die (Serie von Radio- oder Fernsehkomödien mit unverbundenen Episoden bei gleichbleibenden Rollen)*
six [sɪks] **1.** *adj.* sechs; **be ~ feet** *or* **foot under** *(coll.)* unter der Erde liegen; **it is ~ of one and half-a-dozen of the other** *(coll.)* das ist Jacke wie Hose *(ugs.); see also* **eight 1. 2.** *n.* Sechs, *die;* **be at ~es and sevens** sich in einem heillosen Durcheinander befinden; *(on an issue or matter)* heillos zerstritten sein **(on** über + *Akk.*); *see also* **eight 2 a, c, d; hit 1 i**
six: ~-footer [sɪks'fʊtə(r)] *n.* *(person)* Zwei-Meter-Mann, *der/* -Frau, *die;* **~-pack** *n.* Sechserpack, *der;* **~pence** ['sɪkspəns] *n.* *(Brit. Hist.: coin)* Sixpence, *der;* **~-shooter** *n.* sechsschüssiger Revolver
sixteen [sɪks'ti:n, 'sɪksti:n] **1.** *adj.* sechzehn; *see also* **eight 1. 2.** *n.* Sechzehn, *die; see also* **eight 2 a, d; eighteen 2**
sixteenth [sɪks'ti:nθ] **1.** *adj.* sechzehnt...; *see also* **eighth 1. 2.** *n.* *(fraction)* Sechzehntel, *das; see also* **eighth 2**
six'teenth-note *n.* *(Amer. Mus.)* Sechzehntelnote, *die*
sixth [sɪksθ] **1.** *adj.* sechst...; *see also* **eighth 1. 2.** *n.* *(in sequence)* sechste, *der/die/das; (in rank)* Sechste, *der/die/das; (fraction)* Sechstel, *das; see also* **eighth 2**
'**sixth form** *n.* *(Brit. Sch.)* ≈ zwölfte/dreizehnte Klasse
sixtieth ['sɪkstɪɪθ] **1.** *adj.* sechzigst...; *see also* **eighth 1. 2.** *n.* *(fraction)* Sechzigstel, *das; see also* **eighth 2**
sixty ['sɪkstɪ] **1.** *adj.* sechzig; *see also* **eight 1. 2.** *n.* Sechzig, *die; see also* **eight 2 a; eighty 2**
sixty: ~-'first *etc. adj.* einundsechzigst... *usw.; see also* **eighth 1; ~-'one** *etc.* **1.** *adj.* einundsechzig *usw.; see also* **eight 1; 2.** *n.* Einundsechzig *usw., die; see also* **eight 2 a**
[1]**size** [saɪz] *n.* **a)** Größe, *die; (fig. of problem, project)* Umfang, *der;* Ausmaß, *das;* **reach full ~:** auswachsen; **be quite a ~:** ziemlich groß sein; **be twice the ~ of sth.** zweimal so groß wie etw. sein; **who can afford a car that ~?** wer kann sich *(Dat.)* einen so großen Wagen leisten?; **what ~ [of] box do you want?** welche Größe soll die [gewünschte] Schachtel haben?; **be small in ~:** klein sein; **be the ~ of sth.** so groß wie etw. sein; **that's [about] the ~ of it** *(fig. coll.)* so sieht die Sache aus *(ugs.);* **try sth. for ~:** etw. [wegen der Größe] anprobieren; *(fig.)* es einmal mit etw. versuchen; **what ~?** wie groß?; **b)** *(graded class)* Größe, *die; (of paper)* Format, *das;* **collar/waist ~:** Kragen-/Taillenweite, *die;* **take a ~ 7 shoe, take ~ 7 in shoes** Schuhgröße 7 haben
~ '**up** *v.t.* taxieren ‹*Lage*›
[2]**size** *n.* Leim, *der; (for textiles)* Schlichte, *die*
sizeable ['saɪzəbl] *adj.* ziemlich groß; beträchtlich ‹*Summe, Wissen, Einfluß, Unterschied*›; ansehnlich ‹*Betrag*›
sizzle ['sɪzl] **1.** *v.i.* zischen. **2.** *n.* Zischen, *das*
[1]**skate** [skeɪt] *n.* *(Zool.)* Rochen, *der*

[2]**skate** 1. *n.* *(ice-~)* Schlittschuh, *der;* *(roller-~)* Rollschuh, *der;* **get one's ~s on** *(Brit. fig. coll.)* sich beeilen. 2. *v. i.* *(ice-~)* Schlittschuh laufen; *(roller-~)* Rollschuh laufen; **~ on thin ice** *(fig.)* sich auf dünnem Eis bewegen; *(put oneself in danger)* sich auf dünnes Eis begeben

~ over, ~ round *v. t.* *(fig.)* *(avoid)* hinweggehen über (+ *Akk.*) ‹*Frage, Problem*›; *(touch lightly on)* [nur] streifen

skate: ~board 1. *n.* Skateboard, *das;* Rollerbrett, *das;* 2. *v. i.* Skateboard fahren; **~boarding** *n., no pl.* Skateboardfahren, *das*

skater ['skeɪtə(r)] *n.* *(ice-~)* Eisläufer, *der*/-läuferin, *die;* *(roller-~)* Rollschuhläufer, *der*/-läuferin, *die*

skating ['skeɪtɪŋ] *n., no pl.* *(ice-~)* Schlittschuhlaufen, *das;* *(roller-~)* Rollschuhlaufen, *das*

'**skating-rink** *n.* **a)** *(ice)* Eisbahn, *die;* Eisfläche, *die;* **b)** *(for roller-skating)* Rollschuhbahn, *die*

skeleton ['skelɪtn] *n.* Skelett, *das;* Gerippe, *das;* **have a ~ in the cupboard** *(Brit.)* *or* *(Amer.)* **closet** *(fig.)* eine Leiche im Keller haben *(ugs.)*

skeleton: ~ crew *n.* Stammbesatzung, *die;* **~ key** *n.* Dietrich, *der;* **~ service** *n.* **provide a ~ service** den Betrieb notdürftig aufrechterhalten; **there were buses running, but it was only a ~ service** es fuhren zwar Busse, aber nur einige wenige; **~ staff** *n.* Minimalbesetzung, *die*

skeptic *etc.* *(Amer.)* *see* **sceptic** *etc.*

sketch [sketʃ] 1. *n.* **a)** *(drawing)* Skizze, *die;* **do** *or* **make a ~:** eine Skizze anfertigen; **b)** *(play)* Sketch, *der;* **c)** *(Lit.)* Skizze, *die.* 2. *v. t.* *(lit. or fig.)* skizzieren

~ 'in *v. t.* **a)** *(draw)* einzeichnen; **b)** *(fig.: outline)* skizzieren

~ 'out *v. t.* *(lit. or fig.)* [in groben Umrissen] skizzieren

sketch: ~-book *n.* Skizzenbuch, *das;* **~ map** *n.* Faustskizze, *die;* **~-pad** *n.* Skizzenblock, *der*

sketchy ['sketʃɪ] *adj.* **a)** skizzenhaft; **b)** *(incomplete)* lückenhaft ‹*Information, Bericht*›; **c)** *(inadequate)* unzureichend

skew [skju:] 1. *adj.* schräg; schief ‹*Gesicht*›. 2. *n.* **on the ~:** schräg ‹*überqueren*›; schief ‹*tragen, aufsetzen*›; **the picture is [hanging] on the ~:** das Bild hängt schief. 3. *v. i.* **~ round** sich drehen

skewer ['skju:ə(r)] 1. *n.* [Brat]spieß, *der.* 2. *v. t.* aufspießen

skew-'whiff *(Brit. coll.)* *see* **askew**

ski [ski:] 1. *n.* **a)** Ski, *der;* **b)** *(on vehicle)* Kufe, *die.* 2. *v. i.* Ski laufen *od.* fahren

'**ski boot** *n.* Skistiefel, *der*

skid [skɪd] 1. *v. i.,* **-dd-: a)** schlittern; *(from one side to the other; spinning round)* schleudern; **b)** *(on foot)* rutschen. 2. *n.* **a)** Schlittern/Schleudern, *das;* **go into a ~:** ins Schlittern/Schleudern geraten; **b)** *(Aeronaut.)* Gleitkufe, *die*

skid: ~-marks *n. pl.* Schleuderspur, *die;* **~ row** [skɪd 'rəʊ] *n.* *(Amer.)* Pennerviertel, *das* *(salopp abwertend);* **end up on ~ row** *(coll.)* als Penner enden *(salopp abwertend)*

skier ['ski:ə(r)] *n.* Skiläufer, *der*/-läuferin, *die;* Skifahrer, *der*/-fahrerin, *die*

skiing ['ski:ɪŋ] *n., no pl.* Skilaufen, *das;* Skifahren, *das;* *(Sport)* Skisport, *der*

ski: ~-jump *n.* **a)** *(slope)* Sprungschanze, *die;* **b)** *(leap)* Skisprung, *der;* **~-jumper** *n.* Skispringer, *der*/-springerin, *die;* **~-jumping** *n., no pl.* Skispringen, *das*

skilful ['skɪlfl] *adj.* **a)** *(having skill)* geschickt; gewandt ‹*Redner*›; gut ‹*Beobachter, Lehrer*›; **b)** *(well executed)* geschickt; kunstvoll ‹*Gemälde, Plastik, Roman, Komposition*›; *(expert)* fachgerecht ‹*Beurteilung*›; kunstgerecht ausgeführt ‹*Operation*›

skilfully ['skɪlfəlɪ] *adv.* *see* **skilful b:** geschickt; kunstvoll; fachgerecht; kunstgerecht

'**ski-lift** *n.* Skilift, *der*

skill [skɪl] *n.* **a)** *(expertness)* Geschick, *das,* Fertigkeit, *die* (**at, in** in + *Dat.*); *(of artist)* Können, *das;* **b)** *(technique)* Fertigkeit, *die;* *(of weaving, bricklaying)* Technik, *die;* Kunst, *die;* **c)** *in pl.* *(abilities)* Fähigkeiten; **office ~s** Büroerfahrung, *die;* **language ~s** Sprachkenntnisse, *die*

skilled [skɪld] *adj.* **a)** *see* **skilful a;** **b)** *(requiring skill)* qualifiziert ‹*Arbeit, Tätigkeit*›; **~ trade** Ausbildungsberuf, *der;* **c)** *(trained)* ausgebildet; *(experienced)* erfahren; **be ~ in diplomacy/sewing** ein guter Diplomat sein/gut nähen können

skillful, skillfully *(Amer.)* *see* **skilful, skilfully**

skim [skɪm] 1. *v. t.,* **-mm-: a)** *(remove)* abschöpfen; abrahmen ‹*Milch*›; **b)** *(touch in passing)* streifen; **c)** *(pass closely over)* **~ sth.** dicht über etw. *(Akk.)* fliegen; **d)** *(scan briefly)* *see* **~ through.** 2. *v. i.,* **-mm-** segeln

~ 'off *v. t.* **a)** abschöpfen; **b)** *(fig.)* *see* **cream off**

~ through *v. t.* überfliegen ‹*Buch, Zeitung*›

skimmed 'milk, skim 'milk *n.* entrahmte Milch

skimp [skɪmp] 1. *v. t.* sparen an (+ *Dat.*); **he did the work badly, ~ing it** er schluderte bei seiner Arbeit *(ugs.).* 2. *v. i.* sparen (**with, on** an + *Dat.*); **he had to ~ on food/clothes** er mußte am Essen/an der Kleidung sparen

skimpy ['skɪmpɪ] *adj.* sparsam; karg ‹*Mahl*›; kärglich ‹*Leben*›; winzig ‹*Badeanzug*›; [zu] knapp ‹*Anzug*›; spärlich ‹*Wissen*›

skin [skɪn] 1. *n.* **a)** Haut, *die;* **be all** *or* **just ~ and bone** *(fig.)* nur Haut und Knochen sein *(ugs.);* **be soaked** *or* **wet to the ~:** bis auf die Haut durchnäßt sein; **by** *or* **with the ~ of one's teeth** mit knapper Not; **get under sb.'s ~** *(fig. coll.)* *(irritate sb.)* jmdm. auf die Nerven gehen *od.* fallen *(ugs.);* *(fascinate or enchant sb.)* jmdm. unter die Haut gehen *(ugs.);* **have a thick/thin ~** *(fig.)* ein dickes Fell haben *(ugs.)*/dünnhäutig sein; **jump out of one's ~** *(fig.)* aus dem Häuschen geraten *(ugs.);* **save one's ~** *(fig.)* seine Haut retten *(ugs.);* **it's no ~ off my/his** *etc.* **nose** *(coll.)* das braucht mich/ihn *usw.* nicht zu jucken *(ugs.);* **b)** *(hide)* Haut, *die;* **c)** *(fur)* Fell, *das;* **d)** *(peel)* Schale, *die;* *(of onion, peach also)* Haut, *die;* **e)** *(sausage-casing)* Haut, *die;* **f)** *(on milk)* Haut, *die.* 2. *v. t.,* **-nn-** häuten; schälen ‹*Frucht*›; **~ sb. alive** *(fig. coll.)* Hackfleisch aus jmdm. machen *(ugs.);* *see also* **eye 1 a**

skin: ~-cream *n.* Hautcreme, *die;* **~-'deep** *adj.* *(fig.)* oberflächlich; *see also* **beauty a;** **~-diver** *n.* Taucher, *der*/Taucherin, *die;* **~-diving** *n., no pl.* Tauchen, *das;* **~flint** *n.* Geizhals, *der* *(abwertend)*

skinful ['skɪnfʊl] *n.* *(coll.)* **have had a ~:** voll sein *(salopp)*

skin: ~-graft *n.* Hauttransplantation, *die;* **~head** *n.* *(Brit.)* Skinhead, *der*

skinny ['skɪnɪ] *adj.* mager

skint [skɪnt] *adj.* *(Brit. coll.)* bankrott; **be ~:** blank *od.* pleite sein *(ugs.)*

'**skin-tight** *adj.* hauteng

[1]**skip** [skɪp] 1. *v. i.,* **-pp-: a)** hüpfen; **b)** *(use skipping-rope)* seilspringen; **c)** *(change quickly)* springen *(fig.);* **d)** *(make omis-*

sions) überspringen. **2.** *v.t.*, **-pp-: a)** *(omit)* überspringen; *(in mentioning names)* übergehen; **my heart ~ped a beat** *(fig.)* mir stockte das Herz; **b)** *(coll.: miss)* schwänzen *(ugs.)* ‹*Schule usw.*›; liegenlassen ‹*Hausarbeit*›; **~ breakfast/lunch** *etc.* das Frühstück/Mittagessen *usw.* auslassen. **3.** *n.* Hüpfer, *der;* Hopser, *der (ugs.)*

~ a'bout, ~ a'round *v.i.* **a)** herumhüpfen; **b) he did not stay with his subject but ~ped about** er hielt sich nicht an sein Thema, sondern sprang von einem Gegenstand zum anderen *od.* nächsten

~ over *see* **~ 2a**

~ through *v.t.* **a)** *see* **skim through; b)** *(make short work of)* [rasch] durchziehen *(ugs.);* herunterschnurren *(ugs.)* ‹*Vorlesung*›

²skip *n.* *(Building)* Container, *der*

ski: ~ pass *n.* Skipaß, *der;* **~ pole** *n.* Skistock, *der*

skipper ['skɪpə(r)] *n.* **a)** *(Naut.)* Kapitän, *der; (of yacht)* Skipper, *der (Seglerjargon);* **b)** *(Aeronaut.)* [Flug]kapitän, *der;* **c)** *(Sport)* [Mannschafts]kapitän, *der*

'skipping-rope *(Brit.),* **'skip-rope** *(Amer.) ns.* Sprungseil, *das;* Springseil, *das*

'ski resort *n.* Skiurlaubsort, *der*

skirmish ['skɜːmɪʃ] *n.* **a)** *(fight)* Rangelei, *die (ugs.); (of troops, armies)* Gefecht, *das (Milit.);* **b)** *(fig.: argument)* Auseinandersetzung, *die*

skirt [skɜːt] **1.** *n.* Rock, *der.* **2.** *v.t.* herumgehen um. **3.** *v.i.* **~ along sth.** an etw. *(Dat.)* entlanggehen/-fahren/-reiten *usw.*

~ round *v.t.* herumgehen um; *(fig.)* umgehen; ausweichen (+ *Dat.*)

skirting ['skɜːtɪŋ] *n.* **~[-board]** *(Brit.)* Fußleiste, *die*

ski: ~-run *n.* Skihang, *der; (prepared)* [Ski]piste, *die;* **~ stick** *n.* Skistock, *der*

skit [skɪt] *n.* parodistischer Sketch (**on** über + *Akk.*)

skittish ['skɪtɪʃ] *adj.* **a)** *(nervous)* nervös ‹*Pferd*›; *(inclined to shy)* schreckhaft ‹*Pferd*›; **b)** *(lively)* ausgelassen; aufgekratzt *(ugs.)*

skittle ['skɪtl] *n.* **a)** Kegel, *der;* **b)** *in pl., constr. as sing. (game)* Kegeln, *das*

skive [skaɪv] *v.i.* *(Brit. sl.)* sich drücken *(ugs.)*

~ 'off *(Brit. sl.)* **1.** *v.i.* sich verdrücken *(ugs.).* **2.** *v.t.* schwänzen *(ugs.)*

skulk [skʌlk] *v.i.* **a)** *(lurk)* lauern; **b)** *(move stealthily)* schleichen

~ 'off *v.i.* sich fortschleichen

skull [skʌl] *n.* **a)** *(Anat.)* Schädel, *der;* **b)** *(as object)* Totenschädel, *der; (representation)* Totenkopf, *der*

skull and cross-bones [skʌl ən 'krɒsbəʊnz] *n.* Totenkopf, *der (mit gekreuzten Knochen); (flag)* Totenkopfflagge, *die*

skunk [skʌŋk] *n.* *(Zool.)* Stinktier, *das*

sky [skaɪ] *n.* Himmel, *der;* **in the ~:** am Himmel; **praise sb./sth. to the skies** jmdn./etw. in den Himmel heben *(ugs.);* **the ~'s the limit** *(fig.)* da gibt es [praktisch] keine Grenze

sky: ~-blue 1. *adj.* himmelblau; **2.** *n.* Himmelblau, *das;* **~-diver** *n.* Fallschirmspringer, *der*/-springerin, *die;* **~-diving** *n.* Fallschirmspringen, *das (als Sport);* Fallschirmsport, *der;* **~-high 1.** *adj.* himmelhoch; astronomisch *(ugs.)* ‹*Preise usw.*›; **2.** *adv.* hoch in die Luft ‹*werfen, steigen usw.*›; **go ~-high** ‹*Preise usw.:*› in astronomische Höhen klettern *(ugs.);* **~lark 1.** *n.* *(Ornith.)* [Feld]lerche, *die;* **2.** *v.i.* **~lark [about** *or* **around]** herumalbern *(ugs.);* **~light** *n.* Dachfenster, *das;* **~line** *n.* Silhouette, *die; (characteristic of certain town)* Skyline, *die;* **~scraper** *n.* Wolkenkratzer, *der*

slab [slæb] *n.* **a)** *(flat stone etc.)* Platte, *die;* **b)** *(thick slice)* [dicke] Scheibe; *(of cake)* [dickes] Stück; *(of chocolate, toffee)* Tafel, *die*

slack [slæk] **1.** *adj.* **a)** *(lax)* nachlässig; schlampig *(ugs. abwertend);* **be ~ about** *or* **in** *or* **with sth.** in bezug auf etw. *(Akk.)* nachlässig sein; **b)** *(loose)* schlaff; locker ‹*Verband, Strumpfband*›; **c)** *(sluggish)* schlaff; schwach ‹*Wind, Flut*›; **d)** *(Commerc.: not busy)* flau. **2.** *n.* **there's too much ~ in the rope** das Seil ist zu locker *od.* nicht straff genug; **take in** *or* **up the ~:** das Seil/die Schnur *usw.* straffen. **3.** *v.i.* *(coll.)* bummeln *(ugs.)*

slacken ['slækn] **1.** *v.i.* **a)** *(loosen)* sich lockern; ‹*Seil:*› schlaff werden; **b)** *(diminish)* nachlassen; ‹*Geschwindigkeit:*› sich verringern; ‹*Schritt:*› sich verlangsamen. **2.** *v.t.* **a)** *(loosen)* lockern; **b)** *(diminish)* verringern; verlangsamen ‹*Schritt*›

~ 'off 1. *v.i.* **a)** *(loosen) see* **~ 1a; b)** *(diminish) see* **~ 1b; c)** *(relax)* es etwas langsamer angehen lassen *(ugs.).* **2.** *v.t.* **a)** *(loosen) see* **~ 2a; b)** *(diminish) see* **~ 2b**

slacker ['slækə(r)] *n.* *(derog.)* Faulenzer, *der*

slackness ['slæknɪs] *n., no pl.* **a)** *(negligence)* Nachlässigkeit, *die;* **b)** *(idleness)* Bummelei, *die (ugs.);* **c)** *(looseness)* Schlaffheit, *die;* **d)** *(of market, trade)* Flaute, *die*

slacks [slæks] *n. pl.* **[pair of] ~:** lange Hose; Slacks *Pl. (Mode)*

slag [slæg] *n.* *(Metallurgy)* Schlacke, *die*

'slag-heap *n.* *(Mining)* Schlackenhalde, *die*

slain *see* **slay a**

slake [sleɪk] *v.t.* stillen

slalom ['slɑːləm] *n.* *(Skiing)* Slalom, *der*

¹slam [slæm] **1.** *v.t.*, **-mm-: a)** *(shut)* zuschlagen; zuknallen *(ugs.);* **~ the door in sb.'s face** jmdm. die Tür vor der Nase zuschlagen; **b)** *(put violently)* knallen *(ugs.);* **~ one's foot on the brake** *(coll.)* auf die Bremse steigen *(ugs.).* **2.** *v.i.*, **-mm-: a)** zuschlagen; zuknallen *(ugs.);* **b)** *(move violently)* stürmen; **the car ~med against** *or* **into the wall** das Auto knallte *(ugs.)* gegen die Mauer. **3.** *n.* Knall, *der*

~ 'on *v.t.* *(coll.)* **~ on the brakes** auf die Bremse latschen *(salopp)*

²slam *n.* **a)** *(Cards)* Schlemm, *der;* **grand ~:** großer Schlemm; **b)** *(Sport)* **achieve the grand ~:** alle [wichtigen] Meistertitel gewinnen; *(Tennis)* den Grand Slam gewinnen

slander ['slɑːndə(r)] **1.** *n.* Verleumdung, *die* (**on** *Gen.*). **2.** *v.t.* verleumden; schädigen ‹*Ruf*›

slanderous ['slɑːndərəs] *adj.* verleumderisch

slang [slæŋ] *n.* Slang, *der;* ‹*Theater-, Soldaten-, Juristen*›jargon, *der; attrib.* Slang‹*wort, -ausdruck*›

'slanging-match *n.* gegenseitige [lautstarke] Beschimpfung

slangy ['slæŋɪ] *adj.* Slang‹*ausdruck, -wort*›; salopp ‹*Wortwahl, Redeweise*›

slant [slɑːnt] **1.** *v.i.* ‹*Fläche:*› sich neigen; ‹*Linie:*› schräg verlaufen. **2.** *v.t.* **a)** abschrägen; schräg zeichnen ‹*Linie*›; **b)** *(fig.: bias)* [so] hinbiegen *(ugs.)* ‹*Meldung, Bemerkung*›. **3.** *n.* **a)** Schräge, *die;* **be on a** *or* **the ~:** schräg sein; **b)** *(fig.: bias)* Tendenz, *die;* Färbung, *die;* **have a left-wing ~** ‹*Bericht:*› links gefärbt sein

slanted ['slɑːntɪd] *adj.* *(fig.)* gefärbt; **a ~ question** eine Suggestivfrage

slanting ['slɑːntɪŋ] *adj.* schräg

slap [slæp] **1.** *v.t.*, **-pp-: a)** schlagen; ~ **sb. on the face/arm/hand** jmdm. ins Gesicht/auf den Arm/ auf die Hand schlagen; ~ **sb.'s face** *or* **sb. in** *or* **on the face** jmdn. ohrfeigen; ~ **sb. on the back** jmdm. auf die Schulter klopfen; **she deserves to be ~ped on the back** *(fig.)* sie verdient Beifall; Hut ab vor ihr! *(ugs.)*; **b)** *(put forcefully)* knallen *(ugs.)*; ~ **a fine on sb.** *(fig.)* jmdm. eine Geldstrafe aufbrummen *(ugs.)*. **2.** *v.i.*, **-pp-** schlagen; klatschen. **3.** *n.* Schlag, *der;* **give sb. a ~:** jmdn. [mit der flachen Hand] schlagen; **a ~ in the face** *(lit. or fig.)* ein Schlag ins Gesicht; **give sb. a ~ on the back** *(lit. or fig.)* jmdm. auf die Schulter klopfen; **a ~ on the back for sb./sth.** *(fig.)* eine Anerkennung für jmdn./etw. **4.** *adv.* voll; **run ~ into sb.** *(lit. or fig.)* mit jmdm. zusammenprallen; **hit sb. ~ in the eye/face** *etc.* jmdn. mit voller Wucht ins Auge/Gesicht *usw.*treffen; ~ **in the middle** genau in der Mitte

~ **'down** *v.t.* **a)** *(lay forcefully)* hinknallen *(ugs.)*; ~ **sth. down on sth.** etw. auf etw. *(Akk.)* knallen *(ugs.)*; **b)** *(coll.: reprimand)* ~ **sb. down** jmdm. eins auf den Deckel geben *(ugs.)*; **be ~ ped down** eins auf den Deckel kriegen *(ugs.)*

~ **'on** *v.t.* **a)** *(coll.: apply hastily)* draufklatschen *(ugs.)* ‹*Farbe, Make-up*›; zuschnappen lassen ‹*Handschellen*›; **b)** *(coll.: impose)* draufschlagen *(ugs.)*. *See also* ~ **1a**

slap: ~-bang *adv.* **the table was ~-bang in the middle of the room** der Tisch stand einfach mitten im Zimmer; **~dash 1.** *adv.* ruck, zuck *(ugs.)*; **2.** *adj.* schludrig *(ugs. abwertend)*; **in a ~dash fashion** schludrig; **her essay is ~dash** ihr Aufsatz ist hingeschludert *(ugs. abwertend)*; **be ~dash in one's work** bei der Arbeit schludern *(ugs. abwertend)*; **~-happy** *adj. (coll.: cheerfully casual)* unbekümmert; **~-up** *attrib. adj. (sl.)* ‹*Essen, Diner*› mit allen Schikanen *(ugs.)*

slash [slæʃ] **1.** *v.i.* ~ **at sb./sth. with a knife** auf jmdn./etw. mit einem Messer losgehen. **2.** *v.t.* **a)** *(make gashes in)* aufschlitzen; **b)** *(fig.: reduce sharply)* [drastisch] reduzieren; [drastisch] kürzen ‹*Etat, Gehalt, Umfang*›; ~ **costs by one million** die Kosten um eine Million reduzieren. **3.** *n.* **a)** *(~ing stroke)* Hieb, *der;* **b)** *(wound)* Schnitt, *der*

slat [slæt] *n.* Leiste, *die; (of wood in bedstead, fence)* Latte, *die; (in Venetian blind)* Lamelle, *die*

slate [sleɪt] **1.** *n.* **a)** *(Geol.)* Schiefer, *der;* **b)** *(Building)* Schieferplatte, *die;* **c)** *(writing-surface)* Schiefertafel, *die;* **put sth. on the ~** *(Brit. coll.)* etw. anschreiben *(ugs.)*; **wipe the ~ clean** *(fig.)* einen Schlußstrich ziehen. **2.** *attrib. adj.* Schiefer-. **3.** *v.t. (Brit. coll.: criticize)* in der Luft zerreißen *(ugs.)* **(for** wegen)

slate: ~-coloured *adj.* schieferfarben; **~-grey 1.** *n.* Schiefergrau, *das;* **2.** *adj.* schiefergrau

slaughter ['slɔ:tə(r)] **1.** *n.* **a)** *(killing for food)* Schlachten, *das;* Schlachtung, *die; see also* **lamb 1a; b)** *(massacre)* Abschlachten, *das; (in battle)* Gemetzel, *das.* **2.** *v.t.* **a)** *(kill for food)* schlachten; **b)** *(massacre)* abschlachten; niedermetzeln *(abwertend)*; **c)** *(coll.: defeat)* fertigmachen *(salopp)*

'slaughterhouse *see* **abattoir**

Slav [slɑ:v] *n.* Slawe, *der*/Slawin, *die*

slave [sleɪv] **1.** *n.* **a)** Sklave, *der*/ Sklavin, *die;* **b)** *(fig.)* **be a ~ of** *or* **to sth.** Sklave von etw. sein; **be a ~ to sb.** jmdm. verfallen sein. **2.** *v.i.* ~ **[away]** schuften *(ugs.)*; sich abplagen *od. (salopp)* abrackern **(at** mit); ~ **over a hot stove all day** den ganzen Tag am Herd stehen *(ugs.)*

slave: ~-driver *n.* **a)** Sklavenaufseher, *der;* **b)** *(fig.)* Sklaventreiber, *der*/-treiberin, *die (abwertend)*; ~ **'labour** *n.* Sklavenarbeit, *die; (fig.)* Ausbeutung, *die*

slavery ['sleɪvərɪ] *n., no pl.* **a)** Sklaverei, *die;* **b)** *(drudgery)* Sklavenarbeit, *die;* Sklaverei, *die*

slavish ['sleɪvɪʃ] *adj.* sklavisch

Slavonic [slə'vɒnɪk] **1.** *adj.* slawisch. **2.** *n.* Slawisch, *das*

slay [sleɪ] *v.t.* **a) slew** [slu:], **slain** [sleɪn] *(literary)* ermorden; *(with sword, club also)* erschlagen; **b)** **~ed, ~ed** *(coll.: amuse greatly)* **he/his jokes ~ed me** über ihn/seine Witze hätte ich mich totlachen können *(ugs.)*

sleazy ['sli:zɪ] *adj.* schäbig *(abwertend)*; heruntergekommen *(ugs.)* ‹*Person*›; *(disreputable)* anrüchig

sled [sled] *(Amer.)*, **sledge** [sledʒ] *ns.* Schlitten, *der*

'sledge-hammer *n.* Vorschlagshammer, *der*

sleek [sli:k] *adj.* **a)** *(glossy)* seidig ‹*Fell, Haar, Pelz*›;‹*Tier*› mit seidigem Fell; **b) the ~ lines of the car** die schnittige Form des Wagens

sleep [sli:p] **1.** *n.* Schlaf, *der;* **get some ~:** schlafen; **get/go to ~:** einschlafen; **go to ~!** schlaf jetzt!; **not lose [any] ~ over sth.** *(fig.)* wegen etw. keine schlaflose Nacht haben; **put an animal to ~** *(euphem.)* ein Tier einschläfern; **talk in one's ~:** im Schlaf sprechen; **walk in one's ~:** schlafwandeln; **I can/could do it in my ~** *(fig.)* ich kann/könnte es im Schlaf; **get** *or* **have a good night's ~:** [sich] gründlich ausschlafen; **have a ~:** schlafen. **2.** *v.i.*, **slept** [slept] schlafen; ~ **late** lange schlafen; ausschlafen; ~ **like a log** *or* **top** wie ein Stein schlafen *(ugs.)*; ~ **tight!** *(coll.)* schlaf gut! **3.** *v.t.*, **slept** schlafen lassen; **the hotel ~s 80** das Hotel hat 80 Betten

~ **a'round** *v.i. (coll.)* herumschlafen *(ugs.)*

~ **'in** *v.i.* im Bett bleiben

~ **'off** *v.t.* ausschlafen; ~ **it off** seinen Rausch ausschlafen

~ **on 1.** *v.i.* [-'-] weiterschlafen. **2.** *v.t.* ['--] überschlafen

~ **through** *v.t.* ~ **through the noise/alarm** trotz des Lärms/ Weckerklingelns [weiter]schlafen

~ **together** *v.i. (coll. euphem.)* miteinander schlafen

~ **with** *v.t.* ~ **with sb.** *(coll. euphem.)* mit jmdm. schlafen

sleeper ['sli:pə(r)] *n.* **a)** Schläfer, *der;* **be a heavy/light ~:** einen tiefen/leichten Schlaf haben; **b)** *(Brit. Railw.: support)* Schwelle, *die;* **c)** *(Railw.) (coach)* Schlafwagen, *der; (berth)* Schlafwagenplatz, *der; (overnight train)* **[night] ~:** Nachtzug mit Schlafwagen

sleeping ['sli:pɪŋ] *adj. (lit. or fig.)* schlafend; **let ~ dogs lie** *(fig.)* keine schlafenden Hunde wecken

sleeping: ~-bag *n.* Schlafsack, *der;* **~-car** *n. (Railw.)* Schlafwagen, *der;* ~ **'partner** *n. (Commerc.)* stiller Teilhaber; **~-pill, ~-tablet** *ns.* Schlaftablette, *die*

sleepless ['sli:plɪs] *adj.* schlaflos

sleep: ~-walk *v.i.* schlafwandeln; **~-walker** *n.* Schlafwandler, *der*/-wandlerin, *die*

sleepy ['sli:pɪ] *adj.* **a)** *(drowsy)* schläfrig; **b)** *(sluggish)* schwerfällig; *(unobservant)* schlafmützig *(ugs. abwertend)*; **c)** *(peaceful)* verschlafen ‹*Dorf, Stadt usw.*›

sleet [sli:t] **1.** *n., no indef. art.* Schneeregen, *der.* **2.** *v.i. impers.* **it was ~ing** es gab Schneeregen

sleeve [sli:v] *n.* **a)** Ärmel, *der;* **have sth. up one's ~** *(fig.)* etw. in petto haben *(ugs.)*; **roll up one's ~s** *(lit. or fig.)* die Ärmel hochkrempeln *(ugs.)*; **b)** *(record-cover)* Hülle, *die*

sleeveless ['sli:vlɪs] *adj.* ärmellos

sleigh [sleɪ] *n.* Schlitten, *der*

'**sleigh-ride** *n.* Schlittenfahrt, *die*

sleight of hand [slaɪt əv 'hænd] *n.* Fingerfertigkeit, *die*

slender ['slendə(r)] *adj.* **a)** *(slim)* schlank; schmal ‹*Buch, Band*›; **b)** *(meagre)* mager ‹*Einkommen, Kost*›; gering ‹*Chance, Mittel, Vorräte, Hoffnung, Kenntnis*›; schwach ‹*Entschuldigung, Argument, Grund*›

slept *see* **sleep 2**

sleuth [slu:θ] *n.* Detektiv, *der*

[1]**slew** [slu:] **1.** *v.i.* ~ **to the side/left** sich [schnell] seitwärts/nach links drehen; ‹*Kran:*› seitwärts/nach links schwenken. **2.** *v.t.* schwenken ‹*Kran*›

[2]**slew** *see* **slay a**

slice [slaɪs] **1.** *n.* **a)** *(cut portion)* Scheibe, *die; (of apple, melon, peach, apricot, cake, pie)* Stück, *das;* **a ~ of life** ein Ausschnitt aus dem Leben; *see also* **cake 1 a; b)** *(share)* Teil, *der; (allotted part of profits, money)* Anteil, *der;* **c)** *(utensil)* [Braten]wender, *der.* **2.** *v.t.* **a)** in Scheiben schneiden; in Stücke schneiden ‹*Bohnen, Apfel, Pfirsich, Kuchen usw.*›; **b)** *(Golf)* slicen; *(Tennis)* unterschneiden; slicen. **3.** *v.i.* schneiden; ~ **through** durchschn**e**iden; durchpfl**ü**gen ‹*Wellen, Meer*›

~ '**off** *v.t.* abschneiden

~ '**up** *v.t.* aufschneiden; *(fig.: divide)* aufteilen

sliced [slaɪst] *adj. (cut into slices)* aufgeschnitten; kleingeschnitten ‹*Gemüse*›; ~ **bread** Schnittbrot, *das;* **the greatest thing since ~ bread** *(coll. joc.)* der/die/das Größte seit der Erfindung der Bratkartoffel *(ugs. scherzh.)*

slick [slɪk] **1.** *adj. (coll.)* **a)** *(dextrous)* professionell; **b)** *(pretentiously dextrous)* clever *(ugs.)*. **2.** *n.* **[oil-]~**: Ölteppich, *der*

slid *see* **slide 1, 2**

slide [slaɪd] **1.** *v.i.*, **slid** [slɪd] **a)** rutschen; ‹*Kolben, Schublade, Feder:*› gleiten; ~ **down sth.** etw. hinunterrutschen; **b)** *(glide over ice)* schlittern; **c)** *(move smoothly)* gleiten; **d)** *(fig.: take its own course)* **let sth./things ~**: etw./die Dinge schleifen lassen *(fig.)*. **2.** *v.t.*, **slid a)** schieben; **b)** *(place unobtrusively)* gleiten lassen. **3.** *n.* **a)** *(Photog.)* Dia[positiv], *das;* **b)** *(chute) (in children's playground)* Rutschbahn, *die; (for goods etc.)* Rutsche, *die;* **c)** *see* **hair-slide; d)** *(fig.: decline)* **the ~ in the value of the pound** das Abgleiten des Pfundes; **e)** *(for microscope)* Objektträger, *der*

slide: ~ **film** *n. (Photog.)* Diafilm, *der;* ~ **projector** *n.* Diaprojektor, *der;* ~**-rule** *n. (Math.)* Rechenschieber, *der*

sliding ['slaɪdɪŋ]: ~ '**door** *n.* Schiebetür, *die;* ~ '**roof** *n.* Schiebedach, *das;* ~ '**scale** *n.* ~ **scale [of fees]** gleitende [Gebühren]skala; ~ **seat** *n. (Rowing)* Rollsitz, *der*

slight [slaɪt] **1.** *adj.* **a)** leicht; schwach ‹*Hoffnung, Aussichten, Wirkung*›; gedämpft ‹*Optimismus*›; gering ‹*Bedeutung*›; **the ~est thing makes her nervous** die kleinste Kleinigkeit macht sie nervös; **b)** *(scanty)* oberflächlich; **c)** *(slender)* zierlich; *(weedy)* schmächtig; *(flimsy)* zerbrechlich; **d) not in the ~est** nicht im geringsten; **not the ~est ...**: nicht der/die/das geringste ...; **I haven't the ~est idea** ich habe nicht die leiseste Ahnung. **2.** *v.t.* *(disparage)* herabsetzen; *(be discourteous or disrespectful to)* brüskieren; *(ignore)* ignorieren. **3.** *n. (on sb.'s character, reputation, good name)* Verunglimpfung, *die* (**on** *Gen.*); *(on sb.'s abilities)* Herabsetzung, *die* (**on** *Gen.*); *(lack of courtesy)* Affront, *der* (**on** gegen)

slightly ['slaɪtlɪ] *adv.* **a)** ein bißchen; leicht ‹*verletzen, riechen nach, ansteigen*›; flüchtig ‹*jmdn. kennen*›; oberflächlich ‹*etw. kennen*›; **b)** ~ **built** *(slender)* zierlich; *(weedy)* schmächtig

slim [slɪm] **1.** *adj.* **a)** schlank; schmal ‹*Band, Buch*›; **b)** *(meagre)* mager; schwach ‹*Entschuldigung, Aussicht, Hoffnung*›; gering ‹*Gewinn, Chancen*›. **2.** *v.i.*, **-mm-** abnehmen. **3.** *v.t.*, **-mm-** schlanker machen; *(fig.: decrease)* kürzen ‹*Budget*›; verschlanken *(Jargon)* ‹*Produktion*›

~ '**down 1.** *v.i.* abnehmen; schlanker werden. **2.** *v.t. see* ~ **3**

slime [slaɪm] *n.* Schlick, *der; (mucus, viscous matter)* Schleim, *der*

'**slimline** *adj.* schlank; schlank geschnitten ‹*Kleid*›; kalorienarm ‹*Lebensmittel*›

slimmer ['slɪmə(r)] *n. (Brit.) jmd., der etwas für die schlanke Linie tut*

slimming ['slɪmɪŋ] **1.** *n.* **a)** Abnehmen, *das; attrib.* Schlankheits-; **b)** *(fig.: reduction of budget)* Kürzung, *die.* **2.** *adj.* schlank machend ‹*Lebensmittel*›; **be ~**: schlank machen

slimy ['slaɪmɪ] *adj.* schleimig; schlickig ‹*Schlamm*›

sling [slɪŋ] **1.** *n.* **a)** *(weapon)* Schleuder, *die;* **b)** *(Med.)* Schlinge, *die;* **c)** *(carrying-belt)* Tragriemen, *der; (for carrying babies)* Tragehöschen, *das.* **2.** *v.t.*, **slung** [slʌŋ] **a)** *(hurl from ~)* schleudern; **b)** *(coll.: throw)* schmeißen *(ugs.)*; **she slung him his coat** sie schmiß ihm seinen Mantel zu *(ugs.)*

~ **a'way** *v.t. (coll.)* wegschmeißen *(ugs.)*

~ '**out** *v.t. (coll.)* **a)** *(throw out)* ~ **sb. out** jmdn. rausschmeißen *(ugs.)*; **b)** *see* ~ **away**

'**slingshot** *(Amer.) see* **catapult 1**

slink [slɪŋk] *v.i.*, **slunk** [slʌŋk] schleichen

~ **a'way,** ~ **off** *v.i.* davonschleichen; sich fortstehlen

slinky ['slɪŋkɪ] *adj.* aufreizend; hauteng ‹*Kleidung*›

slip [slɪp] **1.** *v.i.*, **-pp-: a)** *(slide)* rutschen; ‹*Messer:*› abrutschen; *(and fall)* ausrutschen; **b)** *(escape)* schlüpfen; **let a chance/opportunity ~**: sich *(Dat.)* eine Chance/Gelegenheit entgehen lassen; **let [it] ~ that ...**: verraten, daß ...; **c)** *(go)* ~ **to the butcher's** [rasch] zum Fleischer rüberspringen *(ugs.)*; ~ **from the room** aus dem Zimmer schlüpfen; **d)** *(move smoothly)* gleiten; **everything ~ped into place** *(fig.)* alles fügte sich zusammen; **e)** *(make mistake)* einen [Flüchtigkeits]fehler machen; **f)** *(deteriorate)* nachlassen; ‹*Moral, Niveau, Ansehen:*› sinken. **2.** *v.t.*, **-pp-: a)** stecken; ~ **the dress over one's head** das Kleid über den Kopf streifen; ~ **sb. sth.** jmdm. etw. zustecken; **b)** *(escape from)* entwischen (+ *Dat.*); **the dog ~ped its collar** der Hund streifte sein Halsband ab; **the boat ~ped its mooring** das Boot löste sich aus seiner Verankerung; ~ **sb.'s attention** jmds. Aufmerksamkeit *(Dat.)* entgehen; ~ **sb.'s memory** *or* **mind** jmdm. entfallen; **c)** *(release)* loslassen. **3.** *n.* **a)** *(fall)* **after his ~**: nachdem er ausgerutscht [und gestürzt] war; **a ~ on these steps could be nasty** auf diesen Stufen auszurutschen könnte schlimme Folgen haben; **b)** *(mistake)* Versehen, *das;* Ausrutscher, *der (ugs.)*; **a ~ of the tongue/pen** ein Versprecher/Schreibfehler; **make a ~**: einen Fehler machen; **c)** *(underwear)* Unterrock, *der;* **d)** *(pillowcase)* [Kopf]kissenbezug, *der;* **e)** *(strip)* ~ **of metal/plastic**

Metall-/Plastikstreifen, *der;* **f)** *(piece of paper)* ‹*Einzahlungs-, Wett*›schein, *der;* ~ **[of paper]** Zettel, *der;* **g) give sb. the ~** *(escape)* jmdm. entwischen *(ugs.); (avoid)* jmdm. ausweichen; **h) a ~ of a girl** ein zierliches Mädchen

~ **a'cross** *v.i.* rüberspringen *(ugs.)*

~ **a'way** *v.i.* **a)** ‹*Person:*› sich fortschleichen; **b)** *(pass quickly)* ‹*Zeit, Tage, Wochen usw.:*› verfliegen

~ **'back** *v.i.* zurückschleichen; *(very quickly)* zurücksausen; ~ **back into unconsciousness** wieder das Bewußtsein verlieren

~ **be'hind** *v.i.* zurückfallen; *(with one's work)* in Rückstand geraten

~ **'by** *v.i.* **a)** *(pass unnoticed)* vorbeischleichen; ‹*Fehler:*› durchrutschen *(ugs.);* **b)** *see* ~ **away b**

~ **'down** *v.i.* runterrutschen *(ugs.);* ‹*Getränk:*› die Kehle runterlaufen *(ugs.)*

~ **'in 1.** *v.i.* sich hineinschleichen; *(enter briefly)* [kurz] reinkommen *(ugs.); (enter unnoticed)* ‹*Fehler:*› sich einschleichen. **2.** *v.t.* einfließen lassen ‹*Bemerkung*›

~ **into** *v.t.* **a)** *(put on)* schlüpfen in (+ *Akk.*) ‹*Kleidungsstück*›; **b)** *(lapse into)* verfallen in (+ *Akk.*)

~ **'off 1.** *v.i.* **a)** *(slide down)* runterrutschen *(ugs.);* **b)** *see* ~ **away a. 2.** *v.t.* abstreifen ‹*Schmuck, Bezug, Handschuh*›; schlüpfen aus ‹*Kleid, Hose, Schuh*›; ausziehen ‹*Strumpf, Handschuh*›

~ **'on** *v.t.* überstreifen ‹*Bezug, Handschuh, Ring*›; schlüpfen in (+ *Akk.*) ‹*Kleid, Hose, Schuh*›; anziehen ‹*Strumpf, Handschuh*›; anlegen ‹*Schmuck*›

~ **'out** *v.i.* **a)** *(leave)* sich hinausschleichen; ~ **out to the butcher's** zum Fleischer rüberspringen *(ugs.);* **b)** *(be revealed)* **it ~ped out** es ist mir/dir/ihm *usw.* herausgerutscht

~ **'over** *v.i.* **a)** *(fall)* ausrutschen; **b)** *see* ~ **across**

~ **'past** *see* ~ **by**

~ **'through** *v.i.* durchschlüpfen; ‹*Fehler:*› durchrutschen *(ugs.)*

~ **'up** *v.i. (coll.)* einen Schnitzer machen *(ugs.)* **(on, over** bei)

slip: ~-case *n.* Schuber, *der;* **~-cover** *n. (for unused furniture)* Schutzüberzug, *der;* **~-knot** *n. (easily undone knot)* Slipstek, *der;* **~-on 1.** *adj.* **~-on shoes** Slipper; **2.** *n. (shoe)* Slipper, *der*

slipper ['slɪpə(r)] *n.* Hausschuh, *der*

slippery ['slɪpərɪ] *adj.* **a)** schlüpfrig; glitschig; **be on a ~ path** *or* **slope** *(fig.)* auf einem verhängnisvollen Weg sein; **b)** *(elusive)* schlüpfrig; glitschig; wendig ‹*Spieler*›; *(shifty)* aalglatt *(abwertend);* **he is a ~ customer** er ist aalglatt *(abwertend);* **c)** *(fig.: delicate)* heikel ‹*Thema, Fall*›

slippy ['slɪpɪ] *(coll.) see* **slippery a**

slip: ~-road *n. (Brit.) (for approach)* Zufahrtsstraße, *die; (to motorway)* Auffahrt, *die; (for leaving)* Ausfahrt, *die;* **~shod** *adj.* schlampig *(ugs. abwertend); (fig.: careless, unsystematic)* schludrig *(ugs. abwertend);* **~stream** *n.* **a)** *(of car, motor cycle)* Fahrtwind, *der; (Racing)* Windschatten, *der;* **b)** *(of propeller)* Propellerwind, *der; (of ship; also Brit. fig.)* Kielwasser, *das;* **~way** *(Shipb.)* Helling, *die od. der*

slit [slɪt] **1.** *n.* Schlitz, *der.* **2.** *v.t.,* **-tt-, slit** aufschlitzen; ~ **sb.'s throat** jmdm. die Kehle durchschneiden

slither ['slɪðə(r)] *v.i.* rutschen; *(on ice, polished floor also)* schlittern

sliver ['slɪvə(r)] *n.* **a)** *(thin slice of food)* dünne Scheibe; **b)** *(splinter)* Splitter, *der;* ~ **of wood/glass/bone** Holz-/Glas-/Knochensplitter, *der*

slob [slɒb] *n. (coll.)* Schwein, *das (derb);* **lazy ~:** fauler Sack *(salopp abwertend);* **fat ~:** Fettsack, *der (salopp abwertend)*

slobber ['slɒbə(r)] *v.i.* sabbern *(ugs.);* ~ **over sb./sth.** jmdn./etw. besabbern; *(fig.)* von jmdm./etw. schwärmen

sloe [sləʊ] *n. (Bot.)* Schlehe, *die*

slog [slɒg] **1.** *v.t.,* **-gg-** dreschen *(ugs.)* ‹*Ball*›; *(in boxing, fight)* voll treffen. **2.** *v.i.,* **-gg-: a)** *(hit)* draufschlagen *(ugs.);* **b)** *(fig.: work doggedly)* sich abplagen; schuften *(ugs.); (for school, exams)* büffeln *(ugs.);* **c)** *(walk doggedly)* sich schleppen. **3.** *n.* **a)** *(hit)* [wuchtiger] Schlag; **give sb./sth. a ~:** jmdm./einer Sache einen wuchtigen Schlag versetzen; **b)** *(hard work)* Plackerei, *die (ugs.)*

~ **at** *v.t.* **a)** *(hit)* ~ **at sb./sth.** auf jmdn./etw. eindreschen *(ugs.);* **b)** *(work hard at)* sich abplagen mit

~ **a'way** *v.i.* sich abplagen **(at** mit)

~ **'out** *v.t. (coll.)* ~ **it out** es [bis zum Ende] durchstehen; ~ **one's guts out** sich kaputtarbeiten *(ugs.)*

slogan ['sləʊgən] *n.* **a)** *(striking phrase)* Slogan, *der; (advertising ~)* Werbeslogan, *der;* Werbespruch, *der;* **b)** *(motto)* Wahlspruch, *der; (in political campaign)* [Wahl]slogan, *der*

slogger ['slɒgə(r)] *n.* **a)** *(hitter)* **be a [real] ~:** immer nur draufschlagen *(ugs.);* **b)** *(hard worker)* Arbeitstier, *das (fig.)*

slop [slɒp] **1.** *v.i.,* **-pp-** schwappen **(out of, from** aus). **2.** *v.t.,* **-pp-** schwappen; *(intentionally)* kippen; klatschen *(ugs.)* ‹*Farbe an die Wand*›

~ **a'bout,** ~ **a'round** *v.i.* herumschwappen *(ugs.)*

~ **'over** *v.i. (splash over)* überschwappen

slope [sləʊp] **1.** *n.* **a)** *(slant)* Neigung, *die; (of river)* Gefälle, *das;* **the roof was at a ~ of 45°** das Dach hatte eine Neigung von 45°; **be on a** *or* **the ~:** geneigt sein; **b)** *(slanting ground)* Hang, *der;* **c)** *(Skiing)* Piste, *die.* **2.** *v.i. (slant)* sich neigen; ‹*Wand, Mauer:*› schief sein; ‹*Boden, Garten:*› abschüssig sein; ~ **upwards/downwards** ‹*Straße:*› ansteigen/abfallen

~ **a'way** *v.i.* abfallen

~ **down** *v.i.* sich hinabneigen

~ **'off** *v.i. (sl.)* sich verdrücken *(ugs.)*

sloppy ['slɒpɪ] *adj.* **a)** *(careless)* schlud[e]rig *(ugs. abwertend);* **b)** *(untidy)* unordentlich; schlampig *(ugs. abwertend)*

slosh [slɒʃ] **1.** *v.i.* platschen *(ugs.);* ‹*Flüssigkeit:*› schwappen. **2.** *v.t. (coll.: pour clumsily)* schwappen

slot [slɒt] **1.** *n.* **a)** *(hole)* Schlitz, *der;* **b)** *(groove)* Nut, *die;* **c)** *(coll.: position)* Platz, *der;* **d)** *(coll.: in schedule)* Termin, *der; (Radio, Telev.)* Sendezeit, *die.* **2.** *v.t.,* **-tt-:** ~ **sth. into place/sth.** etw. einfügen/in etw. *(Akk.)* einfügen. **3.** *v.i.,* **-tt-:** ~ **into place/sth.** *(lit. or fig.)* sich einfügen/in etw. *(Akk.)* einfügen; **everything ~ted into place** *(fig.)* alles fügte sich zusammen

~ **'in 1.** *v.t.* einfügen. **2.** *v.i. (lit. or fig.)* sich einfügen

~ **to'gether 1.** *v.t.* zusammenfügen. **2.** *v.i. (lit. or fig.)* sich zusammenfügen

sloth [sləʊθ] *n.* **a)** *no pl. (lethargy)* Trägheit, *die;* **b)** *(Zool.)* Faultier, *das*

slothful ['sləʊθfl] *adj.* träge; schwerfällig ‹*Anstrengungen, Versuche*›

'slot-machine *n.* **a)** *(vending-machine)* Automat, *der;* **b)** *(Amer.) see* **fruit machine**

slouch [slaʊtʃ] **1.** *n.* **a)** *(posture)* schlaffe Haltung; **b)** *(sl.: lazy person)* Faulpelz, *der;* **be no ~ at sth.** etwas loshaben in etw. *(Dat.).* **2.**

v. i. **a)** sich schlecht halten; **don't ~!** halte dich gerade!; **b)** *(be ungainly)* sich herumflegeln *(ugs. abwertend)*
slovenly ['slʌvnlɪ] *adj.* schlampig *(ugs.)*; schlud[e]rig *(ugs.)*
slow [sləʊ] **1.** *adj.* **a)** langsam; **~ but sure** langsam, aber zuverlässig; **b)** *(gradual)* langsam; langwierig ‹*Suche, Arbeit*›; **get off to a ~ start** beim Start langsam wegkommen; ‹*Aufruf, Produkt:*› zunächst nur wenig Anklang finden; **make ~ progress [in** *or* **at** *or* **with sth.]** nur langsam [mit etw.] vorankommen; **c) be ~ [by ten minutes], be [ten minutes] ~** ‹*Uhr:*› [zehn Minuten] nachgehen; **d)** *(preventing quick motion)* nur langsam befahrbar ‹*Strecke, Straße, Belag*›; **e)** *(tardy)* **[not] be ~ to do sth.** [nicht] zögern, etw. zu tun; **f)** *(not easily roused)* **be ~ to anger/to take offence** sich nicht leicht ärgern/beleidigen lassen; **g)** *(dull-witted)* schwerfällig; langsam; *see also* **uptake; h)** *(burning feebly)* schwach; **in a ~ oven** bei schwacher Hitze [im Backofen]; **i)** *(uninteresting)* langweilig; **j)** *(Commerc.)* flau ‹*Geschäft*›. **2.** *adv.* langsam; **'~'** „langsam fahren!"; **go ~:** langsam fahren; *(Brit. Industry)* langsam arbeiten. **3.** *v. i.* langsamer werden; **~ to a halt** anhalten; ‹*Zug:*› zum Stehen kommen. **4.** *v. t.* **~ a train/car** die Geschwindigkeit eines Zuges/Wagens verringern
~ 'down 1. *v. i.* **a)** langsamer werden; seine Geschwindigkeit verringern; *(in working/speaking)* langsamer arbeiten/sprechen; ‹*Produktion, Geburten-/Sterbeziffer, Inflation[srate]:*› sinken; **b)** *(reduce pace of living)* langsamer machen *(ugs.)*. **2.** *v. t.* verlangsamen; **the driver ~ed the car/train down** der Autofahrer/Lokomotivführer fuhr langsamer
~ 'up *see* **~ down**
slow: ~coach *n.* Trödler, *der*/Trödlerin, *die (ugs. abwertend)*; **~-down** *n.* Verlangsamung, *die* **(in** *Gen.***)**; *(in birth, death, inflation rate, output, production, number)* Sinken, *das* **(in** *Gen.***)**
slowly ['sləʊlɪ] *adv.* langsam; **~ but surely** langsam, aber sicher
slow: ~ 'motion *n. (Cinemat.)* Zeitlupe, *die;* **in ~ motion** in Zeitlupe; *attrib.* **~ motion replay** Zeitlupenwiederholung, *die;* **~-moving** *adj.* sich langsam fortbewegend
slowness ['sləʊnɪs] *n., no pl.* **a)** Langsamkeit, *die;* **b)** *(gradualness)* Langsamkeit, *die; (of search, work)* Langwierigkeit, *die;* **c)** *(slackness)* Zögern, *das;* **his ~ to react** *or* **in reacting** sein zögerndes Reagieren; **d)** *(stupidity)* Schwerfälligkeit, *die;* **~ [of comprehension/mind/wit]** Begriffsstutzigkeit, *die (abwertend)*; **e)** *(dullness)* Langweiligkeit, *die*
slow: ~-witted [sləʊ'wɪtɪd] *adj.* [geistig] schwerfällig; **~-worm** *n. (Zool.)* Blindschleiche, *die*
SLR *abbr. (Photog.)* **single-lens reflex**
sludge [slʌdʒ] *n.* **a)** *(mud)* Matsch, *der (ugs.)*; Schlamm, *der;* **b)** *(sediment)* [schlammiger] Bodensatz
¹slug [slʌg] *n.* **a)** *(Zool.)* Nacktschnecke, *die;* **b)** *(bullet)* [rohe] Gewehrkugel; **c)** *(Amer.: tot of liquor)* **a ~ of whisky/rum** *etc.* ein Schluck Whisky/Rum *usw.*
²slug *(Amer.: hit)* **1.** *v. t.*, **-gg-** niederschlagen. **2.** *n.* [harter] Schlag
sluggish ['slʌgɪʃ] *adj.* träge; schleppend ‹*Gang, Schritt*›; schwerfällig ‹*Reaktion, Vorstellungskraft*›; *(Commerc.)* flau; schleppend ‹*Nachfrage, Geschäftsgang*›
sluice [slu:s] **1.** *n. (Hydraulic Engin.)* Schütz, *das.* **2.** *v. t.* **~ [down]** *(with hose)* abspritzen; *(with bucket)* übergießen
slum [slʌm] **1.** *n.* Slum, *der; (single house or apartment)* Elendsquartier, *das.* **2.** *v. t.*, **-mm-: ~ it** wie arme Leute leben; *(fig.)* sich unters [gemeine] Volk mischen
slumber ['slʌmbə(r)] *(poet./rhet.)* **1.** *n. (lit. or fig.)* **~[s]** Schlummer, *der (geh.)*; **fall into a light/long ~:** in leichten/tiefen Schlummer sinken. **2.** *v. i. (lit. or fig.)* schlummern *(geh.)*
slump [slʌmp] **1.** *n.* Sturz, *der (fig.)*; *(in demand, investment, sales, production)* starker Rückgang **(in** *Gen.***)**; *(economic depression)* Depression, *die (Wirtsch.)*; *(in morale, support, popularity)* Nachlassen, *das* **(in** *Gen.***)**. **2.** *v. i.* **a)** *(Commerc.)* stark zurückgehen; ‹*Preise, Kurse:*› stürzen *(fig.)*; **b)** *(be diminished)* ‹*Popularität, Moral, Unterstützung usw.:*› nachlassen; **c)** *(collapse)* fallen; **they found him ~ed over the table** sie fanden ihn über dem Tisch zusammengesunken
slung *see* **sling 2**
slunk *see* **slink**
slur [slɜ:(r)] **1.** *v. t.*, **-rr-: ~ one's words/speech** undeutlich sprechen; **~red speech** undeutliche Aussprache. **2.** *n. (insult)* Beleidigung, *die* **(on** für); **cast a ~ on sb./sth.** jmdn./etw. verunglimpfen *(geh.)*; **it's a ~ on his reputation** es schmälert seinen Ruf
slurp [slɜ:p] *(coll.)* **1.** *v. t.* **~ [up]** schlürfen. **2.** *n.* Schlürfen, *das*
slush [slʌʃ] *n.* **a)** *(thawing snow)* Schneematsch, *der;* **b)** *(fig. derog.: sentiment)* sentimentaler Kitsch
'slush fund *n.* Fonds für Bestechungsgelder
slushy ['slʌʃɪ] *adj.* **a)** *(wet)* matschig; **b)** *(derog.: sloppy)* sentimental
slut [slʌt] *n.* Schlampe, *die (ugs. abwertend)*
sly [slaɪ] **1.** *adj.* **a)** *(crafty)* schlau; gerissen *(ugs.)* ‹*Geschäftsmann, Schachzug, Trick*›; verschlagen *(abwertend)* ‹*Blick*›; **he is a ~ one** *or* **customer** das ist ein ganz Gerissener *od.* Schlauer *(ugs.)*; **b)** *(secretive)* heimlichtuerisch; verschlagen *(abwertend)* ‹*Rivale*›; **c)** *(knowing)* vielsagend ‹*Blick, Lächeln*›. **2.** *n.* **on the ~:** heimlich
¹smack [smæk] **1.** *n.* **a)** *(sound)* Klatsch, *der;* **b)** *(blow)* Schlag, *der; (on child's bottom)* Klaps, *der (ugs.)*. **2.** *v. t.* **a)** *(slap)* [mit der flachen Hand] schlagen; *(lightly)* einen Klaps *(ugs.)* geben (+ *Dat.*); **~ sb.'s face/bottom/hand** jmdn. ohrfeigen/jmdm. eins hintendrauf geben *(ugs.)*/jmdm. eins auf die Hand geben *(ugs.)*; **b) ~ one's lips** [mit den Lippen] schmatzen. **3.** *v. i.* **~ into the net/wall** ins Netz/gegen die Mauer knallen *(ugs.)*. **4.** *adv.* **a)** *(coll.: with a ~)* **go ~ into a lamp-post** gegen einen Laternenpfahl knallen *(ugs.)*; **b)** *(exactly)* direkt
²smack *v. i.* **~ of** schmecken nach; *(fig.)* riechen nach *(ugs.)*
small [smɔ:l] **1.** *adj.* **a)** *(in size)* klein; gering ‹*Wirkung, Appetit, Fähigkeit*›; schmal ‹*Taille, Handgelenk*›; dünn ‹*Stimme*›; **it's a ~ world** die Welt ist klein; **b)** *attrib. (~-scale)* klein; Klein‹*aktionär, -sparer, -händler, -betrieb, -bauer*›; **c)** *(young, not fully grown)* klein; **d)** *(of the ~er kind)* klein; **~ letter** Kleinbuchstabe, *der;* **feel ~** *(fig.)* sich *(Dat.)* ganz klein vorkommen; **make sb. feel/look ~** *(fig.)* jmdn. beschämen/ein schlechtes Licht auf jmdn. werfen; **e)** *(not much)* wenig; **demand for/interest in the product was ~:** die Nachfrage nach/das Interesse an dem Produkt war gering; **[it's] ~ wonder** [es ist] kein

Wunder; **f)** *(trifling)* klein; **we have a few ~ matters/points/problems to clear up before ...:** es sind noch ein paar Kleinigkeiten zu klären, bevor ...; **g)** *(minor)* unbedeutend; **great and ~:** hoch und niedrig; **h)** *(petty)* kleinlich *(abwertend);* **have a ~ mind** ein Kleinkrämer sein *(abwertend).* **2.** *n. (Anat.)* **~ of the back** Kreuz, *das.* **3.** *adv.* klein

small: ~ 'ad *n. (coll.)* Kleinanzeige, *die;* **~ 'change** *n., no pl., no indef. art.* Kleingeld, *das;* **~holder** *n. (Brit. Agric.)* Kleinbauer, *der*/-bäuerin, *die;* **~holding** *n. (Brit. Agric.)* landwirtschaftlicher Kleinbetrieb

smallish ['smɔ:lɪʃ] *adj.* ziemlich klein/gering; ziemlich schmal ‹*Taille*›

small: ~-'minded *adj.* kleinlich; engstirnig, kleingeistig ‹*Einstellung*›; **~pox** *n. (Med.)* Pocken *Pl.;* **~ 'print** *n. (lit. or fig.)* Kleingedruckte, *das;* **~-scale** *attrib. adj.* in kleinem Maßstab *nachgestellt;* klein ‹*Konflikt, Unternehmer*›; Klein‹*betrieb, -bauer, -gärtner*›; **~ 'screen** *n. (Telev.)* Bildschirm, *der;* **~-size[d]** *adj.* klein; **~ talk** *n.* leichte Unterhaltung; *(at parties)* Smalltalk, *der;* **make ~ talk [with sb.]** [mit jmdm.] Konversation machen; **~-time** *attrib. adj. (coll.)* Schmalspur- *(ugs. abwertend);* **~-time crook** kleiner Ganove *(ugs. abwertend)*

smarmy ['smɑ:mɪ] *adj.* kriecherisch, schmeichlerisch ‹*Stimme*›

smart [smɑ:t] **1.** *adj.* **a)** *(clever)* clever; *(ingenious)* raffiniert; *(accomplished)* hervorragend; **b)** *(neat)* schick; schön ‹*Haus, Garten, Auto*›; **c)** *attrib. (fashionable)* elegant; smart; **the ~ set** die elegante Welt; die Schickeria; **d)** *(vigorous)* hart ‹*Schlag, Gefecht*›; scharf ‹*Zurechtweisung, Schritt*›; **e)** *(prompt)* flink; **look ~:** sich beeilen. **2.** *v.i.* schmerzen; **~ under sth.** *(fig.)* unter etw. *(Dat.)* leiden

smart alec[k], smart alick [smɑ:t 'ælɪk] *(coll. derog.)* **1.** *n.* Besserwisser, *der (abwertend).* **2.** *attrib. adj.* neunmalklug; besserwisserisch *(abwertend)*

'smart-arse, *(Amer.)* **'smart-ass** *ns. (sl.)* Klugscheißer, *der (salopp abwertend)*

smarten ['smɑ:tn] **1.** *v.t.* **a)** *(make spruce)* herrichten; **he ~ed his hair/clothes** er brachte sein Haar/seine Kleidung in Ordnung *(ugs.);* **~ oneself** *(tidy up)* sich zurechtmachen; *(dress up)* sich herrichten; *(improve appearance in general)* auf sein Äußeres achten; **b)** *(accelerate)* **~ one's pace** seinen Schritt/seine Schritte beschleunigen. **2.** *v.i.* **the pace ~ed** das Tempo beschleunigte sich

~ 'up 1. *v.t.* **a)** *see* **~ 1 a; b)** *(fig.)* **~ up one's ideas** sich am Riemen reißen *(ugs.).* **2.** *v.i. (tidy up)* sich zurechtmachen; *(improve appearance in general)* auf sein Äußeres achten

smartly ['smɑ:tlɪ] *adv.* **a)** *(cleverly)* clever; **b)** *(neatly)* schmuck ‹*[an]gestrichen*›; smart, flott ‹*gekleidet, geschnitten*›; **c)** *(fashionably)* vornehm; **d)** *(vigorously)* hart; *(sharply)* scharf ‹*zurechtweisen*›; hart ‹*anpacken*›; **e)** *(promptly)* sofort; auf der Stelle

smartness ['smɑ:tnɪs] *n., no pl.* **a)** *(cleverness)* Cleverneß, *die;* **b)** *(neatness)* Gepflegtheit, *die;* **~ [of appearance]** ansprechendes Äußeres

smash [smæʃ] **1.** *v.t.* **a)** *(break)* zerschlagen; **~ sth. to pieces** etw. zerschmettern; **b)** *(defeat)* zerschlagen ‹*Rebellion, Revolution, Opposition*›; zerschmettern ‹*Feind*›; *(in games)* vernichtend schlagen; klar verbessern ‹*Rekord*›; **c)** *(hit hard)* **~ sb. in the face/mouth** jmdm. [hart] ins Gesicht/auf den Mund schlagen; **d)** *(Tennis)* schmettern; **e)** *(propel forcefully)* schmettern; **he ~ed the car into a wall** er knallte *(ugs.)* mit dem Wagen gegen eine Mauer. **2.** *v.i.* **a)** *(shatter)* zerbrechen; **b)** *(crash)* krachen; **the cars ~ed into each other** die Wagen krachten zusammen *(salopp).* **3.** *n.* **a)** *(sound)* Krachen, *das; (of glass)* Klirren, *das;* **b)** *(collision) see* **smash-up**

~ 'down *v.t.* einschlagen ‹*Tür*›

~ 'in *v.t.* zerschmettern; eindrücken ‹*Rippen, Motorhaube, Kotflügel*›; einschlagen ‹*Fenster, Tür, Schädel*›; **~ sb.'s face in** *(coll.)* jmdm. die Fresse polieren *(derb)*

~ 'up 1. *v.t.* zertrümmern. **2.** *v.i.* zerschellen; ‹*Auto:*› zertrümmert werden

smash-and-'grab [raid] *n. (coll.)* Schaufenstereinbruch, *der*

smasher ['smæʃə(r)] *n. (coll.)* **be a ~:** [ganz] große Klasse sein *(ugs.)*

smash 'hit *n. (coll.) (film, play)* Kassenschlager, *der (ugs.); (song, record)* Riesenhit, *der (ugs.)*

smashing ['smæʃɪŋ] *adj. (coll.)* toll *(ugs.);* klasse *(ugs.)*

'smash-up *n.* schwerer Zusammenstoß

smatter ['smætə(r)], **smattering** ['smætərɪŋ] *ns.* oberflächliche Kenntnisse; *(feeble)* Halbwissen, *das (abwertend);* **have a ~ of German** ein paar Brocken Deutsch können

smear [smɪə(r)] **1.** *v.t.* **a)** *(daub)* beschmieren; *(put on or over)* schmieren; **~ cream/ointment over one's body/face** sich *(Dat.)* den Körper/das Gesicht mit Creme/Salbe einreiben; **~ed with blood** blutbeschmiert *od.* -verschmiert; **b)** *(smudge)* verwischen; verschmieren; **c)** *(fig.: defame)* in den Schmutz ziehen. **2.** *n.* **a)** *(blotch)* [Schmutz]fleck, *der;* **b)** *(fig.: defamation)* **a ~ on him/his [good] name** eine Beschmutzung seiner Person/seines [guten] Namens

smear: ~ tactics *n. pl.* schmutzige Mittel; **~ test** *n. (Med.)* Abstrich, *der*

smell [smel] **1.** *n.* **a)** *no pl., no art.* **have a good/bad sense of ~:** einen guten/schlechten Geruchssinn haben; **b)** *(odour)* Geruch, *der* **(of** nach); *(pleasant also)* Duft, *der* **(of** nach); **a ~ of burning/gas** ein Brand-/ Gasgeruch; **there was a ~ of coffee** es duftete nach Kaffee; **sth. has a nice/strong** *etc.* **~ [to it]** etw. riecht angenehm/stark *usw.;* **c)** *(stink)* Gestank, *der.* **2.** *v.t.*, **smelt** [smelt] *or* **~ed a)** *(perceive)* riechen; *(fig.)* wittern; **I can ~ burning/gas** es riecht brandig/ nach Gas; **I could ~ trouble** *(fig.)* es roch nach Ärger; **b)** *(inhale ~ of)* riechen an (+ *Dat.*). **3.** *v.i.*, **smelt** *or* **~ed a)** *(emit ~)* riechen; *(pleasantly also)* duften; **b)** *(recall ~; fig.: suggest)* **~ of sth.** nach etw. riechen; **c)** *(stink)* riechen; **his breath ~s** er riecht aus dem Mund

~ 'out *v.t. (lit. or fig.)* aufspüren

smelling salts ['smelɪŋ sɔ:lts, 'smelɪŋ sɒlts] *n. pl.* Riechsalz, *das*

smelly ['smelɪ] *adj.* stinkend *(abwertend);* **be ~:** stinken *(abwertend)*

¹smelt [smelt] *v.t. (Metallurgy)* **a)** *(melt)* verhütten ‹*Erz*›; **b)** *(refine)* erschmelzen ‹*Metall*›

²smelt *see* **smell 2, 3**

smile [smaɪl] **1.** *n.* Lächeln, *das;* **a ~ of joy/satisfaction** ein freudiges/befriedigtes Lächeln; **be all ~s** über das ganze Gesicht strahlen; **break into a ~:** [plötzlich] zu lächeln beginnen; **give sb. a ~:** jmdn. anlächeln; **raise a few ~s** zum Lächeln anregen; **take that ~ off your face!** hör auf zu grinsen!; **with a ~:** mit einem Lächeln

[auf den Lippen]; lächelnd. **2.** *v. i.* lächeln; **make sb. ~:** jmdn. zum Lachen bringen; **keep smiling** *(fig.: not despair)* das Lachen nicht verlernen *(fig.)*; **keep smiling!** Kopf hoch!; ~ **at sth.** *(lit. or fig.)* über etw. *(Akk.)* lächeln; ~ **with delight/pleasure** vor Freude strahlen; **Fortune ~d on us** das Glück lachte uns *(veralt.)*
smirk [smɜ:k] **1.** *v. i.* grinsen. **2.** *n.* Grinsen, *das*
smite [smaɪt] *v. t.*, **smote** [sməʊt], **smitten** ['smɪtn] *(arch./literary)* **a)** *(strike)* schlagen (**on** auf, an + *Akk.*); **b)** *(afflict)* **be smitten by** *or* **with desire/terror/the plague** von Verlangen/Schrecken ergriffen/ mit der Pest geschlagen sein *(geh.)*; **be smitten by** *or* **with sb./ sb.'s charms** jmdm./jmds. Zauber erlegen sein
smith [smɪθ] *n.* Schmied, *der*
smithy ['smɪðɪ] *n.* Schmiede, *die*
smitten *see* **smite**
smock [smɒk] *n.* [Arbeits]kittel, *der*
smog [smɒg] *n.* Smog, *der*
smoke [sməʊk] **1.** *n.* **a)** Rauch, *der;* **go up in ~:** in Rauch [und Flammen] aufgehen; *(fig.)* in Rauch aufgehen; **[there is] no ~ without fire** *(prov.)* kein Rauch ohne Flamme *(Spr.)*; **b)** *(act of smoking tobacco)* **a ~ would be nice just now** jetzt würde ich gern eine rauchen. **2.** *v. i.* **a)** *(~ tobacco)* rauchen; ~ **like a chimney** rauchen wie ein Schlot *(ugs.)*; **b)** *(emit ~)* rauchen; *(burn imperfectly)* qualmen; *(emit vapour)* dampfen. **3.** *v. t.* **a)** rauchen; **b)** *(darken)* schwärzen ‹*Glas*›; ‹*Petroleumlampe:*› verräuchern ‹*Wand, Decke*›; **c)** räuchern ‹*Fleisch, Fisch*›
~ **'out** *v. t.* ausräuchern; *(fig.: discover)* aufspüren ‹*Verbrecher*›
'smoke-bomb *n.* Rauchbombe, *die*
smoked [sməʊkt] *adj. (Cookery)* geräuchert
'smoke-detector *n.* Rauchmelder, *der*
smokeless ['sməʊklɪs] *adj.* rauchlos; rauchfrei ‹*Zone*›
smoker ['sməʊkə(r)] *n.* **a)** Raucher, *der*/Raucherin, *die;* **be a heavy ~:** ein starker Raucher/eine starke Raucherin sein; **b)** *see* **smoking-compartment**
smoke: ~-screen *n.* [künstliche] Nebelwand; *(fig.)* Vernebelung, *die* (**for** *Gen.*); **~-signal** *n.* Rauchzeichen, *das;* Rauchsignal, *das*
smoking ['sməʊkɪŋ] *n.* **a)** *(act)* Rauchen, *das;* **'no ~'** „Rauchen verboten"; **b)** *no art. (seating area)* **[do you want to sit in] ~ or non-~?** [möchten Sie für] Raucher oder Nichtraucher?
'smoking-compartment *n. (Railw.)* Raucherabteil, *das*
smoky ['sməʊkɪ] *adj. (emitting smoke)* rauchend; qualmend; *(smoke-filled, smoke-stained)* verräuchert; *(coloured or tasting like smoke)* rauchig
smolder *(Amer.) see* **smoulder**
smooth [smu:ð] **1.** *adj.* **a)** *(even)* glatt; eben ‹*Straße, Weg*›; **as ~ as glass/silk** spiegelglatt/glatt wie Seide; **be worn ~** ‹*Treppenstufe:*› abgetreten sein; ‹*Reifen:*› abgefahren sein; ‹*Fels, Stein:*› glattgeschliffen sein; **this razor gives a ~ shave** dieser Rasierapparat rasiert glatt; **b)** *(mild)* weich; **as ~ as velvet** *(fig.)* samtweich; **c)** *(fluent)* flüssig; geschliffen ‹*Stil, Diktion*›; **d)** *(not jerky)* geschmeidig ‹*Bewegung*›; ruhig ‹*Fahrt, Flug, Lauf einer Maschine, Bewegung, Atmung*›; weich ‹*Start, Landung, Autofahren, Schalten*›; **e)** *(without problems)* reibungslos; **the change-over was fairly ~:** der Wechsel ging ziemlich reibungslos vonstatten; **f)** *(derog.: suave)* glatt; *(~-tongued)* glattzüngig *(geh. abwertend)*; **he is a ~ operator** er ist gewieft; **g)** *(coll.: elegant)* schick; **h)** *(skilful)* geschickt; souverän. **2.** *v. t.* glätten; glattstreichen, glätten ‹*Stoff, Tuch, Papier*›; glattstreichen ‹*Haar*›; *(with sandpaper)* glattschleifen, glätten ‹*Holz*›; *(fig.: soothe)* besänftigen
~ **'down** *v. t.* glattstreichen ‹*Haar*›; *(fig.)* schlichten ‹*Streit*›
smoothly ['smu:ðlɪ] *adv.* **a)** *(evenly)* glatt; **b)** *(not jerkily)* geschmeidig ‹*sich bewegen*›; reibungslos ‹*funktionieren*›; ruhig ‹*atmen, fließen, fahren*›; weich ‹*starten, landen, schalten*›; **a ~ running engine** *(Motor Veh.)* ein rund laufender Motor; **c)** *(without problems)* reibungslos; glatt; **d)** *(derog.: suavely)* aalglatt *(abwertend)*; glattzüngig *(geh. abwertend)* ‹*sprechen*›; **e)** *(coll.: elegantly)* schick; **f)** *(skilfully)* geschickt; souverän
smoothness ['smu:ðnɪs] *n., no pl.* **a)** *(evenness)* Glätte, *die;* **b)** *(mildness)* Weichheit, *die;* **c)** *(of movement)* Geschmeidigkeit, *die; (of machine operation, breathing)* Gleichmäßigkeit, *die;* **d)** *(lack of problems)* Reibungslosigkeit, *die;* **e)** *(derog.: suavity)* Glätte, *die* *(abwertend)*; **f)** *(coll.: elegance)* Schick, *der;* **g)** *(skill)* Geschicklichkeit, *die;* Souveränität, *die*
smote *see* **smite**
smother ['smʌðə(r)] *v. t.* **a)** *(stifle, extinguish)* ersticken; **b)** *(overwhelm)* überschütten (**with,** in mit); ~ **sb. with kisses** jmdn. mit seinen Küssen [fast] ersticken; **c)** *(fig.: suppress)* unterdrücken ‹*Kichern, Gähnen, Wahrheit*›; ersticken ‹*Kritik, Gerücht, Schluchzen, Gelächter, Schreie*›; dämpfen ‹*Lärm*›
smoulder ['sməʊldə(r)] *v. i.* **a)** schwelen; **b)** *(fig.)* ‹*Haß, Rebellion:*› schwelen; ‹*Liebe:*› glimmen *(geh.)*; **she was ~ing with rage** Zorn schwelte in ihr
smudge [smʌdʒ] **1.** *v. t.* **a)** *(blur)* verwischen; **b)** *(smear)* schmieren; **c)** *(make smear on)* verschmieren. **2.** *v. i.* ‹*Füller, Tinte, Farbe:*› schmieren. **3.** *n.* **a)** *(smear)* Fleck, *der; (fig.)* Schandfleck, *der;* **b)** *(blur)* Schmierage, *die (ugs.)*
smug [smʌg] *adj.* selbstgefällig *(abwertend)*
smuggle ['smʌgl] *v. t.* schmuggeln
~ **'in** *v. t.* einschmuggeln; hinein-/ hereinschmuggeln ‹*Person*›
~ **'out** *v. t.* hinaus-/herausschmuggeln
smuggler ['smʌglə(r)] *n.* Schmuggler, *der*/Schmugglerin, *die*
smuggling ['smʌglɪŋ] *n.* Schmuggel, *der;* Schmuggeln, *das*
smut [smʌt] *n.* Rußflocke, *die; (smudge)* Rußfleck, *der*
smutty ['smʌtɪ] *adj.* **a)** *(dirty)* verschmutzt; **b)** *(lewd)* schmutzig *(abwertend)*
snack [snæk] *n.* Imbiß, *der;* Snack, *der;* **have a [quick] ~:** [rasch] eine Kleinigkeit essen *(ugs.)*
'snack-bar *n.* Schnellimbiß, *der;* Snackbar, *die*
snag [snæg] **1.** *n.* **a)** *(jagged point)* Zacke, *die;* **b)** *(problem)* Haken, *der;* **what's the ~?** wo klemmt es [denn]? *(ugs.)*; **hit a ~, run up against a ~:** auf ein Problem *od.* eine Schwierigkeit stoßen; **there's a ~ in it** die Sache hat einen Haken. **2.** *v. t.*, **-gg-: I've ~ged my coat** mein Mantel hat sich verfangen
snail [sneɪl] *n.* Schnecke, *die;* **at [a] ~'s pace** im Schneckentempo *(ugs.)*
snake [sneɪk] *n.* **a)** Schlange, *die;* **b)** *(derog.)* ~ **[in the grass]** *(woman)* [falsche] Schlange;

(man) falscher Kerl *od.* *(ugs.)* Hund

snake: **~-bite** *n.* Schlangenbiß, *der;* **~-charmer** ['sneɪktʃɑːmə(r)] *n.* Schlangenbeschwörer, *der;* **~-skin** *n.* Schlangenleder, *das*

snap [snæp] **1.** *v. t.*, **-pp-:** **a)** *(break)* zerbrechen; **~ sth. in two** *or* **in half** etw. in zwei Stücke brechen; **b)** **~ one's fingers** mit den Fingern schnalzen; **c)** *(move with snapping sound)* **~ sth. home** *or* **into place** etw. einrasten *od.* einschnappen lassen; **~ shut** zuschnappen lassen ‹*Portemonnaie, Schloß*›; zuklappen ‹*Buch, Zigarettendose, Etui*›; **d)** *(take photograph of)* knipsen; **e)** *(say in sharp manner)* fauchen; *(speak crisply or curtly)* bellen. **2.** *v. i.*, **-pp-:** **a)** *(break)* brechen; **b)** *(fig.: give way under strain)* ausrasten *(ugs.)*; **my patience has finally ~ped** nun ist mir der Geduldsfaden aber gerissen; **c)** *(make as if to bite)* [zu]schnappen; **d)** *(move smartly)* **~ into action** loslegen *(ugs.)*; **~ to attention** strammstehen; **e)** **~ shut** zuschnappen; ‹*Kiefer:*› zusammenklappen; ‹*Mund:*› zuklappen; **f)** *(speak sharply)* fauchen. **3.** *n.* **a)** *(sound)* Knacken, *das;* **b)** *(Photog.)* Schnappschuß, *der;* **c)** *(Brit. Cards)* Schnippschnapp[schnurr], *das.* **4.** *attrib. adj.* *(spontaneous)* spontan. **5.** *int.* *(Brit. Cards)* schnapp

~ at *v. t.* **a)** *(bite)* **~ at sb./sth.** nach jmdm./etw. schnappen; **~ at sb.'s heels** jmdm. auf den Fersen sein; **b)** *(speak sharply to)* anfauchen *(ugs.)*

~ 'off **1.** *v. i.* abbrechen; abknicken ‹*Zweig, Antenne*›. **2.** *v. t.* **a)** *(break)* abbrechen; **b)** *(bite)* abbeißen; **~ sb.'s head off** *(fig.)* jmdm. den Kopf abreißen *(fig.)*

~ 'out *v. t.* bellen ‹*Befehl, Anweisung*›

~ 'out of *v. t.* abwerfen; sich befreien von ‹*Gefühl, Stimmung, Komplex*›; **~ out of it!** *(coll.)* hör auf damit!; *(wake up)* wach auf!

~ 'up *v. t.* **a)** *(pick up)* [sich *(Dat.)*] schnappen; **b)** *(fig. coll.: seize)* [sich *(Dat.)*] schnappen *(ugs.)*; **~ up a bargain/an offer** bei einem Angebot [sofort] zugreifen *od.* *(salopp)* zuschlagen; **the tickets were ~ped up immediately** die Karten waren sofort weg

'snapdragon *n.* *(Bot.)* Löwenmäulchen, *das*

snappy ['snæpɪ] *adj.* **a)** *(lively)* lebhaft; **b)** *(smart)* schick; **be a ~ dresser** sich flott *od.* schick kleiden; **c)** *(coll.)* **look ~!, make it ~!** ein bißchen dalli! *(ugs.)*

'snapshot *n.* *(Photog.)* Schnappschuß, *der*

snare [sneə(r)] **1.** *n.* **a)** *(trap)* Schlinge, *die;* Falle, *die (auch fig.)*; **b)** *(temptation)* Fallstrick, *der.* **2.** *v. t.* [mit einer Schlinge] fangen ‹*Tier*›

[1]snarl [snɑːl] **1.** *v. i.* **a)** *(growl)* ‹*Hund:*› knurren; **b)** *(speak)* knurren. **2.** *n.* Knurren, *das*

[2]snarl **1.** *n.* *(tangle)* Knoten, *der.* **2.** *v. t.* verheddern *(ugs.)*. **3.** *v. i.* sich verheddern *(ugs.)*

~ 'up **1.** *v. t.* *(confuse)* durcheinanderbringen; *(bring to a halt)* zum Erliegen bringen; **get ~ed up in the traffic** im Verkehr steckenbleiben. **2.** *v. i.* ‹*Verkehr:*› stocken; ‹*Wolle:*› sich verheddern *(ugs.)*

'snarl-up *n.* Stau, *der;* Stockung, *die*

snatch [snætʃ] **1.** *v. t.* **a)** *(grab)* schnappen; **~ a bite to eat** [schnell] einen Bissen zu sich nehmen; **~ a rest** sich *(Dat.)* eine Ruhepause verschaffen; **~ some sleep** ein bißchen schlafen; **~ sth. from sth.** etw. schnell von etw. nehmen; *(very abruptly)* etw. von etw. reißen; **~ sth. from sb.** jmdm. etw. wegreißen; **b)** *(steal)* klauen *(ugs.)*; *(kidnap)* kidnappen. **2.** *v. i.* einfach zugreifen. **3.** *n.* **a)** **make a ~ at sb./sth.** nach jmdm./etw. greifen; **b)** *(Brit. sl.: robbery)* Raub, *der;* **c)** *(sl.: kidnap)* Kidnapping, *das;* **d)** *(fragment)* **~es of talk/conversation** Gesprächsfetzen *od.* -brocken *Pl.*

~ a'way *v. t.* [schnell] wegziehen (**from** *Dat.*); **~ sth. away from sb.** jmdm. etw. wegreißen

~ 'up *v. t.* [sich *(Dat.)*] schnappen

sneak [sniːk] **1.** *v. t.* **a)** *(take)* stibitzen *(fam.)*; **b)** *(fig.)* **~ a look at sb./sth.** nach jmdm./etw. schielen; **c)** *(bring)* **~ sth./sb. into a place** etw./jmdn. in einen Ort schmuggeln. **2.** *v. i.* **a)** *(Brit. Sch. sl.: tell tales)* petzen *(Schülerspr.)*; **b)** *(move furtively)* schleichen. **3.** *attrib. adj.* **a)** *(without warning)* **~ attack/raid** Überraschungsangriff, *der;* **b)** **a ~ preview of the film** eine inoffizielle Vorpremiere des Films. **4.** *n.* *(Brit. Sch. sl.)* Petze, *die (Schülerspr.)*

~ a'way *v. i.* [sich] fortschleichen; sich davonmachen

~ 'in **1.** *v. i.* **a)** *(enter stealthily)* sich hineinschleichen; *(fig.)* sich einschleichen; **b)** *(win narrowly)* knapp siegen. **2.** *v. t.* *(bring in)* einschmuggeln *(ugs.)*

~ 'out *v. i.* [sich] hinausschleichen

~ 'out of *(Amer.: avoid)* **~ out of sth./doing sth.** sich vor etw. *(Dat.)* drücken *(ugs.)*/sich davor drücken *(ugs.)*, etw. zu tun

sneaking ['sniːkɪŋ] *attrib. adj.* heimlich; leise ‹*Verdacht*›

'sneak-thief *n.* Einschleichdieb, *der*

sneaky ['sniːkɪ] **a)** *(underhand)* hinterhältig; **b)** **have a ~ feeling that ...:** so ein leises Gefühl haben, daß ...

sneer [snɪə(r)] *v. i.* **a)** *(smile scornfully)* spöttisch *od.* höhnisch lächeln/grinsen; hohnlächeln; **b)** *(speak scornfully)* höhnen *(geh.)*; spotten

~ at *v. t.* **a)** *(smile scornfully at)* höhnisch anlächeln/angrinsen; **b)** *(express scorn for)* verhöhnen *(geh.)*; spotten über (+ *Akk.*)

sneeze [sniːz] **1.** *v. i.* niesen; **not to be ~d at** *(fig. coll.)* nicht zu verachten *(ugs.)*. **2.** *n.* Niesen, *das*

snide [snaɪd] *adj.* abfällig

sniff [snɪf] **1.** *n.* Schnüffeln, *das;* Schnuppern, *das;* *(with running nose, while crying)* Schniefen, *das;* *(contemptuous)* Naserümpfen, *das;* **have a ~ at sth.** an etw. *(Dat.)* riechen *od.* schnuppern. **2.** *v. i.* schniefen; die Nase hochziehen; *(to detect a smell)* schnuppern; *(to express contempt)* die Nase rümpfen. **3.** *v. t.* *(smell)* riechen *od.* schnuppern an (+ *Dat.*) ‹*Essen, Getränk, Blume, Parfüm, Wein*›

~ at *v. t.* **a)** schnuppern *od.* riechen an (+ *Dat.*) ‹*Blume, Essen*›; **b)** *(show contempt for)* die Nase rümpfen über (+ *Akk.*); **not to be ~ed at** *(fig. coll.)* nicht zu verachten *(ugs.)*

snigger ['snɪgə(r)] **1.** *v. i.* [boshaft] kichern. **2.** *n.* [boshaftes] Kichern

snip [snɪp] **1.** *v. t.*, **-pp-** schnippeln *(ugs.)*, schneiden ‹*Loch*›; schnippeln *(ugs.)* *od.* schneiden an (+ *Dat.*) ‹*Tuch, Haaren, Hecke*›; *(cut off)* abschnippeln *(ugs.)*; abschneiden. **2.** *v. i.*, **-pp-** schnippeln *(ugs.)*; schneiden. **3.** *n.* **a)** *(Brit. sl.: good bargain)* Schnäppchen, *das (ugs.)*; **b)** *(cut)* Schnitt, *der;* Schnipser, *der (ugs.)*

snipe [snaɪp] *v. i.* *(Mil.)* aus dem Hinterhalt schießen

~ at *v. t.* **a)** *(Mil.)* aus dem Hinterhalt beschießen; **b)** *(fig.: make snide comments about)* anschießen *(ugs.)*

sniper ['snaɪpə(r)] *n.* Heckenschütze, *der*

snippet ['snɪpɪt] *n.* **a)** *(piece)* Schnipsel, *der od. das;* **b)** *(of in-*

formation in newspaper) Notiz, *die; (of knowledge)* Bruchstück, *das; (from a book)* Passage, *die; (of conversation)* Gesprächsfetzen, *der;* **useful ~s of information** nützliche Hinweise

snivel ['snɪvl] *v.i., (Brit.)* **-ll-** *(sniff, sob)* schniefen; schnüffeln *(ugs.)*

snob [snɒb] *n.* Snob, *der; attrib.* **~ appeal** *or* **value** Snob-Appeal, *der*

snobbery ['snɒbərɪ] *n.* Snobismus, *der*

snobbish ['snɒbɪʃ] *adj.* snobistisch

snooker ['snu:kə(r)] *n., no pl., no indef. art.* Snooker [Pool], *das;* Taschenbillard, *das*

snoop [snu:p] *(coll.) v.i.* schnüffeln *(ugs.);* **~ about** *or* **around** herumschnüffeln *(ugs.)*

snooper ['snu:pə(r)] *n. (coll.)* Schnüffler, *der*/Schnüfflerin, *die (ugs.)*

snootily ['snu:tɪlɪ] *adv.,* **snooty** ['snu:tɪ] *adj. (coll.)* hochnäsig *(ugs.)*

snooze [snu:z] *(coll.)* **1.** *v.i.* dösen *(ugs.).* **2.** *n.* Nickerchen, *das (fam.);* **have a ~:** ein Nickerchen machen

snore [snɔ:(r)] **1.** *v.i.* schnarchen. **2.** *n.* Schnarcher, *der (ugs.);* **~s** Schnarchen, *das*

snorkel ['snɔ:kl] **1.** *n.* Schnorchel, *der.* **2.** *v.i., (Brit.)* **-ll-** schnorcheln

snort [snɔ:t] **1.** *v.i.* schnauben **(with, in** vor + *Dat.*); **~ with laughter** vor Lachen prusten. **2.** *v.t.* schnauben. **3.** *n.* Schnauben, *das;* **with a ~ of rage** wutschnaubend

snot [snɒt] *n. (sl.)* Rotz, *der (derb)*

snotty ['snɒtɪ] *adj. (sl.)* **a)** *see* **snooty; b)** *(running with nasal mucus)* rotznäsig *(salopp);* **~ child/nose** Rotznase, *die;* **~ handkerchief** Rotzfahne, *die (salopp)*

snout [snaʊt] *n. (nose)* Schnauze, *die; (of pig, ant-eater)* Rüssel, *der*

snow [snəʊ] **1.** *n.* **a)** *no indef. art.* Schnee, *der;* **b)** *in pl. (falls)* Schneefälle *Pl.;* **c)** *(on TV screen etc.)* Schnee, *der.* **2.** *v.i. impers.* **it ~s** *or* **is ~ing** es schneit

~ **'in** *v.t.* **they are ~ed in** sie sind eingeschneit

~ **'under** *v.t.* **be ~ed under** *(with work)* erdrückt werden; *(with presents, letters)* überschüttet werden

snow: ~ball 1. *n.* Schneeball, *der; attrib.* **have a ~ball effect** eine Kettenreaktion auslösen; **2.** *v.i.* **a)** Schneebälle werfen; **b)** *(fig.: increase greatly)* lawinenartig zunehmen; **~bound** *adj.* eingeschneit; **~-covered** *adj.* schneebedeckt; **~-drift** *n.* Schneeverwehung, *die;* Schneewehe, *die;* **~drop** *n.* Schneeglöckchen, *das;* **~fall** *n.* Schneefall, *der;* **~flake** *n.* Schneeflocke, *die;* **~man** *n.* Schneemann, *der;* **~-plough** *n.* Schneepflug, *der;* **~storm** *n.* Schneesturm, *der;* **~-white** *adj.* schneeweiß

snowy ['snəʊɪ] *adj.* **a)** schneereich ‹*Gegend, Monat*›; schneebedeckt ‹*Berge*›; **b)** *(white)* schneeweiß

snub [snʌb] **1.** *v.t.,* **-bb-: a)** *(rebuff)* brüskieren; vor den Kopf stoßen; **b)** *(reprove)* zurechtweisen; *(insult)* beleidigen; **c)** *(reject)* ablehnen. **2.** *n.* Abfuhr, *die*

snub: ~ 'nose *n.* Stupsnase, *die;* **~-nosed** ['snʌbnəʊzd] *adj.* stupsnasig

[1]snuff [snʌf] *n.* Schnupftabak, *der;* **take a pinch of ~:** eine Prise schnupfen

[2]snuff *v.t.* beschneiden, putzen ‹*Kerze*›; **~ it** *(sl.: die)* ins Gras beißen *(salopp)*

~ **'out** *v.t.* **a)** *(extinguish)* löschen ‹*Kerze*›; **b)** *(fig.: put an end to)* zerstören; zunichte machen ‹*Hoffnung*›

snuffle ['snʌfl] *v.i. (sniff)* schnüffeln **(at** an + *Dat.*); *(with cold, after crying)* schniefen

snug [snʌg] *adj.* **a)** *(cosy)* gemütlich, behaglich ‹*Haus, Zimmer, Bett*›; *(warm)* mollig warm ‹*Zimmer, Mantel, Bett*›; **b)** *(sheltered)* geschützt; **c)** *(close-fitting)* **be a ~ fit** genau passen; ‹*Kleidung:*› wie angegossen passen

so [səʊ] **1.** *adv.* **a)** *(by that amount)* so; **as winter draws near, so it gets darker** je näher der Winter rückt, desto dunkler wird es; **as fast as the water poured in, so we bailed it out** in dem Maße, wie das Wasser eindrang, schöpften wir es heraus; **so ... as** so ... wie; **there is nothing so fine as ...:** es gibt nichts Schöneres als ...; **not so [very] difficult/easy** *etc.* nicht so schwer/leicht *usw.;* **so beautiful a present** so ein schönes Geschenk; ein so schönes Geschenk; **so far** bis hierher; *(until now)* bisher; bis jetzt; *(to such a distance)* so weit; **and so on [and so forth]** und so weiter und so fort; **so long!** bis dann *od.* nachher! *(ugs.);* **so many** so viele; *(unspecified number)* soundso viele; **so much** so viel; *(unspecified amount)* soundsoviel; **the villages are all so much alike** die Dörfer gleichen sich alle so sehr; **so much for him/his plans** *(that is all)* das wär's, was ihn/seine Pläne angeht; **so much for my hopes** und ich habe mir solche Hoffnungen gemacht; **so much the better** um so besser; **not so much ... as** weniger ... als [eher]; **not so much as** *(not even)* [noch] nicht einmal; **b)** *(in that manner)* so; **so be it** einverstanden; **this being so** da dem so ist *(geh.);* **it so happened that he was not there** er war [zufällig] gerade nicht da; **c)** *(to such a degree)* so; **this answer so provoked him that ...:** diese Antwort provozierte ihn so *od.* derart, daß ...; **so much so that ...:** so sehr, daß ...; das geht/ging so weit, daß ...; **d)** *(with the intent)* **so as to** um ... zu; **~ [that]** damit; **e)** *(emphatically)* so; **I'm so glad/tired!** ich bin ja so froh/müde!; **so kind of you!** wirklich nett von Ihnen!; **so sorry!** *(coll.)* Entschuldigung!; Verzeihung!; **f)** *(indeed)* **It's a rainbow! – So it is!** Es ist ein Regenbogen! – Ja, wirklich!; **you said it was good, and so it was** du sagtest, es sei gut, und so war es auch; **is that so?** so? *(ugs.);* wirklich?; **and so he did** und das machte/tat er [dann] auch; **it may be so, possibly so** [das ist] möglich; **g)** *(likewise)* **so am/have/would/could/will/do I** ich auch; **h)** *(thus)* so; **and so it was that ...:** und so geschah es, daß ...; **not so!** nein, nein! **2.** *pron.* **he suggested that I should take the train, and if I had done so, ...:** er riet mir, den Zug zu nehmen, und wenn ich es getan hätte, ...; **I'm afraid so** leider ja; ich fürchte schon; **the teacher said so** der Lehrer hat es gesagt; **I suppose so** ich nehme an *(ugs.); expr. reluctant agreement* wenn es sein muß; *granting grudging permission* von mir aus; **I told you so** ich habe es dir [doch] gesagt; **he is a man of the world, so to say** *or* **speak** es menschelt sozusagen bei ihm; **it will take a week or so** es wird so *(ugs.) od.* etwa eine Woche dauern; **there were twenty or so people** es waren so *(ugs.)* um die zwanzig Leute da; **very much so** in der Tat; allerdings. **3.** *conj. (therefore)* daher; **so 'that's what he meant** das hat er also gemeint; **so what is the answer?** wie lautet also die Antwort?; **so 'there you are!** da bist du also!; **so there you 'are!** ich habe also recht!; **so that's 'that** *(coll.) (it's done)* [al]so, das war's *(ugs.); (it's over)* das war's also *(ugs.); (everything has been taken care of)* das wär's dann *(ugs.)*; **so 'there!** [und] fertig!; [und damit] basta! *(ugs.)*; **so you see ...:** du siehst also ...; **so?** na und?

soak [səʊk] **1.** *v. t.* **a)** *(steep)* einweichen ‹*Wäsche in Lauge*›; eintauchen ‹*Brot in Milch*›; ~ **oneself in the sun** sich in der Sonne aalen *(ugs.)*; **b)** *(wet)* naß machen; durchnässen; durchtränken ‹*Erde*›; ~ **sb. from head to foot** jmdn. von Kopf bis Fuß durchnässen; **a rag ~ed in petrol** ein mit Benzin getränkter Lappen; **~ed in sweat** schweißgebadet. **2.** *v. i.* **a)** *(steep)* **put sth. in sth. to ~:** etw. in etw. *(Dat.)* einweichen; **lie ~ing in the bath** ‹*Person:*› sich im Bad durchweichen lassen; **b)** *(drain)* ‹*Feuchtigkeit, Nässe:*› sickern; ~ **away** wegsickern. **3.** *n.* **give sth. a [good] ~:** etw. [gründlich] einweichen

~ **'in** *v. i.* **a)** *(seep in)* einsickern; eindringen; **b)** *(fig.)* **let the atmosphere ~ in** die Atmosphäre auf sich *(Akk.)* einwirken lassen

~ **into** *v. t.* sickern in (+ *Akk.*); ‹*Tinte usw.:*› einziehen in (+ *Akk.*)

~ **through 1.** *v. t.* **a)** ['--] *(penetrate)* ‹*Flüssigkeit, Strahlen:*› dringen durch; ‹*Regenwasser, Blut:*› sickern durch; **b)** [-'-] *(drench)* durchnässen. **2.** *v. i.* [-'-] durchdringen

~ **'up** *v. t.* **a)** *(absorb)* aufsaugen; ~ **up the sunshine** in der Sonne baden; **b)** *(fig.)* aufnehmen ‹*Atmosphäre*›; aufnehmen, in sich *(Akk.)* aufsaugen ‹*Wissen usw.*›

soaking ['səʊkɪŋ] **1.** *n.* *(drenching)* **need a [good] ~** ‹*Garten:*› [gut] gewässert werden müssen; **get a ~:** eine Dusche abbekommen; **give sb./sth. a ~:** jmdn./etw. naß machen. **2.** *adv.* ~ **wet** völlig durchnäßt; klatsch- *od.* patschnaß *(ugs.)*. **3.** *adj.* **a)** *(drenched)* naß [bis auf die Haut]; patschnaß *(ugs.)*; **be ~** ‹*Kleidung:*› völlig durchnäßt sein; **b)** *(saturating)* alles durchnässend ‹*Strom, Regen*›

'so-and-so *n., pl.* **~'s a)** *(person not named)* [Herr/Frau] Soundso; *(thing not named)* Dings, *das*; **b)** *(coll.: contemptible person)* Biest, *das (ugs.)*

soap [səʊp] **1.** *n., no indef. art.* **a)** Seife, *die*; **a bar** *or* **tablet of ~:** ein Stück Seife; **b)** *(coll.) see* **soap opera. 2.** *v. t.* ~ **[down]** einseifen

soap: ~-box *n.* **a)** *(packing-box)* Seifenschachtel, *die*; **b)** *(stand)* ≈ Apfelsinenkiste, *die*; **get on one's ~-box** *(fig.)* laut seine Meinung äußern; Volksreden halten; **c)** *(cart)* Seifenkiste, *die*; ~ **bubble** *n.* Seifenblase, *die*; **~-dish** *n.* Seifenschale, *die*; ~ **flakes** *n. pl.* Seifenflocken *Pl.*; ~ **opera** *n.* *(Telev., Radio)* Seifenoper, *die (ugs.)*; ~ **powder** *n.* Seifenpulver, *das*; **~suds** *n. pl.* Seifenschaum, *der*

soapy ['səʊpɪ] *adj.* seifig; ~ **water** Seifenlauge, *die*

soar [sɔ:(r)] *v. i.* **a)** *(fly up)* aufsteigen; *(hover in the air)* segeln; **b)** *(extend)* ~ **into the sky** in den Himmel ragen; **c)** *(fig.: rise rapidly)* steil ansteigen; ‹*Preise, Kosten usw.:*› in die Höhe schießen *(ugs.)*; **my hopes have ~ed again** ich schöpfe wieder große Hoffnung

soaring ['sɔ:rɪŋ] *attrib. adj.* **a)** *(flying)* segelnd; [hoch am Himmel] schwebend; **b)** *(fig.: rising rapidly)* sprunghaft ansteigend; galoppierend ‹*Preise, Inflation, Kosten*›

sob [sɒb] **1.** *v. i.*, **-bb-** schluchzen (**with** vor + *Dat.*). **2.** *v. t.*, **-bb-** schluchzen. **3.** *n.* Schluchzer, *der*

~ **'out** *v. t.* schluchzen; ~ **one's heart out** bitterlich weinen

sober ['səʊbə(r)] *adj.* **a)** *(not drunk)* nüchtern; **as ~ as a judge** stocknüchtern; **b)** *(moderate)* solide; **c)** *(solemn)* ernst

~ **'down** *v. i.* ruhig werden; **he has ~ed down a lot** er ist wesentlich vernünftiger geworden

~ **'up 1.** *v. i.* nüchtern werden; ausnüchtern. **2.** *v. t.* ausnüchtern

sobering ['səʊbərɪŋ] *adj.* ernüchternd

sobriety [sə'braɪətɪ] *n., no pl., no indef. art.* **a)** *(not being drunk)* Nüchternheit, *die*; **b)** *(moderation)* Bescheidenheit, *die*

so-called ['səʊkɔ:ld] *adj.* sogenannt; *(alleged)* angeblich

soccer ['sɒkə(r)] *n.* *(coll.)* Fußball, *der*

sociable ['səʊʃəbl] *adj.* gesellig; **he did it just to be ~:** er hat es nur getan, um nicht ungesellig zu sein

social ['səʊʃl] *adj.* **a)** sozial; gesellschaftlich; ~ **welfare** Fürsorge, *die*; **b)** *(of ~ life)* gesellschaftlich; gesellig ‹*Abend, Beisammensein*›; ~ **behaviour** Benehmen in Gesellschaft

social: ~ 'class *n.* Gesellschaftsschicht, *die*; [Gesellschafts]klasse, *die*; ~ **'climber** *n.* Emporkömmling, *der (abwertend)*; [sozialer] Aufsteiger *(ugs.)*; ~ **club** *n.* Klub für geselliges Beisammensein; **S~ 'Democrat** *n.* *(Polit.)* Sozialdemokrat, *der*/-demokratin, *die*; **S~ Demo'cratic Party** *n.* *(Brit. Polit.)* Sozialdemokratische Partei; ~ **'history** *n.* Sozialgeschichte, *die*

socialise *see* **socialize**

socialism ['səʊʃəlɪzm] *n.* Sozialismus, *der*

socialist ['səʊʃəlɪst] **1.** *n.* Sozialist, *der*/Sozialistin, *die*. **2.** *adj.* sozialistisch

socialize ['səʊʃəlaɪz] *v. i.* geselligen Umgang pflegen; ~ **with sb.** *(chat)* sich mit jmdm. unterhalten

'social life *n.* gesellschaftliches Leben; **a place with plenty of ~:** ein Ort, wo etwas los ist *(ugs.)*; **not have much ~** ‹*Person:*› nicht viel ausgehen

socially ['səʊʃəlɪ] *adv.* **meet ~:** sich privat treffen; ~ **deprived** sozial benachteiligt

social: ~ 'science *n.* Sozialwissenschaften *Pl.*; Gesellschaftswissenschaften *Pl.*; ~ **se'curity** *n.* soziale Sicherheit; ~ **'service** *n.* *(service provided by the government)* staatliche Sozialleistung; ~ **studies** *n.* *(Educ.)* Gemeinschaftskunde, *die*; ~ **system** *n.* Gesellschaftssystem, *das*; ~ **work** *n.* Sozialarbeit, *die*; ~ **worker** *n.* Sozialarbeiter, *der*/-arbeiterin, *die*

society [sə'saɪətɪ] **1.** *n.* **a)** Gesellschaft, *die*; **high ~:** High-Society, *die*; **b)** *(club, association)* Verein, *der*; *(Commerc.)* Gesellschaft, *die*; *(group of persons with common beliefs, aims, interests, etc.)* Gemeinschaft, *die*. **2.** *attrib. adj.* **a)** *(of high ~)* Gesellschafts-; [High-]Society-; **she is a ~ hostess** sie gibt Feste für die [gehobene] Gesellschaft; **b)** *(of club or association)* Vereins-, Klub‹*vorsitzender, -treffen, -ausflug usw.*›

sociologist [səʊsɪ'ɒlədʒɪst] *n.* Soziologe, *der*/Soziologin, *die*

sociology [səʊsɪ'ɒlədʒɪ] *n.* Soziologie, *die*

[1]**sock** [sɒk] *n., pl.* **~s** *or* *(Commerc./coll.)* **sox** [sɒks] Socke, *die*; Socken, *der (südd., österr., schweiz.)*; *(ankle ~, esp. for children also)* Söckchen, *das*; **pull one's ~s up** *(Brit. fig. coll.)* sich am Riemen reißen *(ugs.)*; **put a ~ in it!** *(Brit. sl.)* halt die Klappe! *(salopp)*

[2]**sock** *v. t.* *(coll.: hit)* schlagen; hauen *(ugs.)*

socket ['sɒkɪt] *n.* **a)** *(Anat.)* *(of eye)* Höhle, *die*; *(of joint)* Pfanne, *die*; **b)** *(Electr.)* Steckdose, *die*; *(receiving a bulb)* Fassung, *die*; **c)** *(for attachment)* Fassung, *die*

[1]sod [sɒd] *n. (turf)* Sode, *die*

[2]sod *(sl.)* **1.** *n. (bastard, swine)* Sau, *die (derb);* **the poor old ~:** das arme Schwein *(salopp).* **2.** *v. t.,* **-dd-: ~ that/you!** verdammter Mist/scher dich zum Teufel! *(ugs.)*

~ **'off** *v. i. imper. (sl.)* verpiß dich *(derb)*

soda ['səʊdə] *n.* **a)** *(sodium compound)* Soda, *die od. das;* **b)** *(drink)* Soda[wasser], *das;* **whisky and ~:** Whisky mit Soda

'soda-water *n.* Soda[wasser], *das*

sodden ['sɒdn] *adj.* durchnäßt **(with** von)

sodium ['səʊdɪəm] *n. (Chem.)* Natrium, *das*

sodium 'chloride *n. (Chem.)* Natriumchlorid, *das*

sodomy ['sɒdəmɪ] *n.* Analverkehr, *der*

sofa ['səʊfə] *n.* Sofa, *das; attrib.* ~ **bed** Bettcouch, *die*

soft [sɒft] *adj.* **a)** weich; zart, weich ⟨*Haut*⟩; **the ground is ~:** der Boden ist aufgeweicht; *(Sport)* der Boden ist schwer; **as ~ as butter** weich wie Butter; butterweich; ~ **ice-cream** Soft-Eis, *das;* ~ **toys** Stofftiere; **b)** *(mild)* sanft; mild ⟨*Klima*⟩; zart ⟨*Duft*⟩; **c)** *(compassionate)* **have a ~ spot for sb./sth.** eine Vorliebe *od.* Schwäche für jmdn./etw. haben; **d)** *(delicate)* sanft ⟨*Augen*⟩; weich ⟨*Farbe, Licht*⟩; **e)** *(quiet)* leise; sanft; **f)** *(gentle)* sanft; **be ~ on** *or* **with sb.** *(coll.: be unusually lenient with)* mit jmdm. sanft umgehen; **g)** *(sl.: easy)* bequem, *(ugs.)* locker ⟨*Job, Leben*⟩; **have a ~ job** eine ruhige Kugel schieben *(ugs.);* **h)** *(compliant)* nachgiebig; **i)** *(too indulgent)* zu nachsichtig; zu lasch *(ugs.)*

soft: ~-boiled *adj.* weichgekocht ⟨*Ei*⟩; **~-centred** *adj.* ⟨*Praline usw.*⟩ mit weicher Füllung; ~ **currency** *n. (Econ.)* weiche Währung; ~ **drink** *n.* alkoholfreies Getränk; ~ **drug** *n.* weiche Droge

soften ['sɒfn] **1.** *v. i.* weicher werden. **2.** *v. t.* weich klopfen ⟨*Fleisch*⟩; aufweichen ⟨*Boden*⟩; dämpfen ⟨*Beleuchtung*⟩; mildern ⟨*Farbe, Farbton*⟩; enthärten ⟨*Wasser*⟩; ~ **the blow** *(fig.)* den Schock mildern

~ **'up** *v. t.* weichklopfen *(ugs.)* ⟨*Boxgegner*⟩; aufweichen ⟨*Verteidigungsanlagen*⟩; *(verbally)* milder stimmen

softener ['sɒfənə(r)] *n.* **a)** *(for water)* [Wasser]enthärter, *der;* **b)** *(for fabrics)* Weichspülmittel, *das;* Weichspüler, *der*

soft: ~ fruit *n.* Beerenobst, *das;* **~-hearted** [sɒft'hɑːtɪd] *adj.* weichherzig

softie *see* **softy**

softly ['sɒftlɪ] *adv.* **a)** *(quietly)* leise ⟨*sprechen, singen, gehen*⟩; **b)** *(gently)* sanft; **speak ~:** mit sanfter Stimme sprechen

softness ['sɒftnɪs] *n., no pl. see* **soft 1: a)** Weichheit, *die;* Zartheit, *die;* **b)** Sanftheit, *die;* Milde, *die;* Zartheit, *die;* **c)** *(delicacy)* Sanftheit, *die;* Weichheit, *die;* **d)** *(of voice, music, etc.)* Gedämpftheit, *die;* **e)** *(gentleness)* Sanftheit, *die;* **f)** *(leniency)* Nachsichtigkeit, *die;* Laschheit, *die (ugs.)*

soft: ~ option *n.* Weg des geringsten Widerstandes; **~-'pedal** *v. i. (tone down)* herunterspielen; ~ **'porn** *(coll.),* ~ **por'nography** *ns.* Softpornographie, *die;* ~ **'sell** *n.* **give sb. the ~ sell** jmdn. auf die sanfte Tour *(ugs.)* zum Kauf zu bewegen versuchen; ~ **'soap** *n.* **a)** *(cleanser)* Schmierseife, *die;* **b)** *(fig.: flattery)* Schmeichelei, *die;* **use ~ soap** schmeicheln; schöntun *(ugs.);* **~-'soap** *v. t.* **~-soap sb.** jmdm. Honig um den Bart schmieren *(ugs.);* **~-spoken** *adj.* leise sprechend ⟨*Person*⟩; **be ~-spoken** leise sprechen; **~ware** *n., no pl., no indef. art. (Computing)* Software, *die;* **~wood** *n.* Weichholz, *das; attrib.* Weichholz-

softy ['sɒftɪ] *n.* **a)** *(coll.: weakling)* Weichling, *der;* Waschlappen, *der (ugs.);* **b)** *(sentimental person)* **be a ~:** sentimental sein

soggy ['sɒgɪ] *adj.* aufgeweicht ⟨*Boden*⟩; durchnäßt ⟨*Kleider*⟩; matschig ⟨*Salat*⟩; nicht durchgebacken, *(landsch.)* glitschig ⟨*Brot, Kuchen*⟩

[1]soil [sɔɪl] *n.* **a)** *(earth)* Erde, *die;* Boden, *der;* **b)** *(ground)* Boden, *der;* **on British/foreign ~:** auf britischem Boden/im Ausland *od.* *(geh.)* in der Fremde

[2]soil *v. t. (lit. or fig.)* beschmutzen

soiled [sɔɪld] *adj.* schmutzig ⟨*Wäsche, Windel*⟩; gebraucht ⟨*Damenbinde*⟩

sojourn ['sɒdʒɜːn, 'sɒdʒən] *(literary)* **1.** *v. i.* verweilen *(geh.);* weilen *(geh.)* **(at** in + *Dat.*). **2.** *n.* Aufenthalt, *der*

solace ['sɒləs] *n.* Trost, *der;* **take** *or* **find ~ in sth.** Trost in etw. *(Dat.)* finden; sich mit etw. trösten; **turn to sb./sth. for ~:** bei jmdm./etw. Trost suchen

solar ['səʊlə(r)] *adj.* Sonnen-

solar: ~ cell *n.* Sonnenzelle, *die;* Solarzelle, *die;* ~ **e'clipse** *n.* *(Astron.)* Sonnenfinsternis, *die;* ~ **energy** *n.* Solarenergie, *die;* Sonnenenergie, *die*

solarium [sə'leərɪəm] *n., pl.* **solaria** [sə'leərɪə] Solarium, *das*

solar: ~ plexus [səʊlə 'pleksəs] *n. (Anat.)* Solarplexus, *der;* Sonnengeflecht, *das;* ~ **'power** *n.* Sonnenenergie, *die;* **~-powered** *adj.* mit Sonnenenergie betrieben; ~ **system** *n. (Astron.)* Sonnensystem, *das*

sold *see* **sell**

solder ['səʊldə(r), 'sɒldə(r)] **1.** *n.* Lot, *das (Technik).* **2.** *v. t.* löten

'soldering-iron *n.* Lötkolben, *der*

soldier ['səʊldʒə(r)] *n.* Soldat, *der;* ~ **of fortune** Glücksritter, *der (abwertend); (mercenary)* Söldner, *der*

~ **'on** *v. i. (coll.)* weitermachen

[1]sole [səʊl] **1.** *n. (Anat.; of shoe)* Sohle, *die.* **2.** *v. t.* [be]sohlen

[2]sole *n. (fish)* Seezunge, *die*

[3]sole *adj.* einzig; alleinig ⟨*Verantwortung, Erbe, Recht*⟩; Allein⟨*erbe, -eigentümer*⟩; **he is the ~ judge of whether ...:** er allein urteilt darüber, ob .../entscheidet, ob ...

solely ['səʊllɪ] *adv.* einzig und allein; ausschließlich; ~ **because ...:** nur [deswegen], weil ...; einzig und allein, weil ...

solemn ['sɒləm] *adj.* feierlich; ernst ⟨*Anlaß, Gespräch*⟩

solemnity [sə'lemnɪtɪ] *n.* **a)** *no pl.* Feierlichkeit, *die;* **b)** *(rite)* Feierlichkeit, *die*

solenoid ['səʊlənɔɪd] *n.* Zylinderspule, *die; (converting energy)* Magnetspule, *die*

solicit [sə'lɪsɪt] **1.** *v. t.* **a)** *(appeal for)* werben um ⟨*Wählerstimmen, Unterstützung*⟩; **b)** *(appeal to)* ~ **sb. for sth.** bei jmdm. um etw. werben; **c)** *(make sexual offer to)* ~ **sb.** sich jmdm. anbieten. **2.** *v. i.* **a)** *(make request)* ~ **for sth.** um etw. bitten *od. (geh.)* ersuchen; *(in a petition)* etw. [mit einer Eingabe] fordern; **b)** *(Commerc.)* ~ **for sth.** um etw. werben; **c)** *(offer illicit sex)* ~ **[for custom]** sich anbieten

solicitor [sə'lɪsɪtə(r)] *n. (Brit.: lawyer)* Rechtsanwalt, *der/*-anwältin, *die (der/die nicht vor höheren Gerichten auftritt)*

solicitous [sə'lɪsɪtəs] *adj.* **a)** *(eager)* **be ~ of sth.** um etw. bemüht sein; **be ~ to do sth.** [darum] bemüht sein, etw. zu tun; **b)** *(anxious)* besorgt; ~ **about sb./sth.** um jmdn./etw. besorgt

solid ['sɒlɪd] **1.** *adj.* **a)** *(rigid)* fest; **freeze/be frozen ~:** [fest] gefrieren/gefroren sein; **set ~:** fest werden; **b)** *(of the same substance all through)* massiv; **~ silver** massives Silber; **~ gold** reines Gold; **~ tyre** Vollgummireifen, *der;* **be packed ~** *(coll.)* gerammelt voll sein *(ugs.);* **c)** *(well-built)* stabil; solide gebaut ‹*Haus, Mauer usw.*›; **have a ~ majority** *(Polit.)* eine solide Mehrheit haben; **d)** *(reliable)* verläßlich, zuverlässig ‹*Freund, Helfer, Verbündeter*›; fest ‹*Stütze*›; **e)** *(complete)* ganz; **a good ~ meal** eine kräftige Mahlzeit; **f)** *(sound)* stichhaltig ‹*Argument, Grund*›; solide ‹*Arbeiter, Finanzlage, Firma*›; solide, gediegen ‹*Komfort, Grundlage*›; **g)** *(Geom.: having three dimensions)* dreidimensional; räumlich. **2.** *n.* **a)** *(substance)* fester Körper; **b)** *in pl. (food)* feste Nahrung

solidarity [sɒlɪ'dærɪtɪ] *n., no pl.* Solidarität, *die*

solidify [sə'lɪdɪfaɪ] **1.** *v. t.* verfestigen. **2.** *v. i. (become solid)* hart *od.* fest werden; erstarren; ‹*Flüssigkeit, Lava:*› erstarren

solidity [sə'lɪdɪtɪ] *n., no pl. see* **solid 1: a)** Festigkeit, *die;* **b)** Massivität, *die;* **c)** Stabilität, *die;* **d)** *(of reasons, argument)* Stichhaltigkeit, *die*

solidly ['sɒlɪdlɪ] *adv.* **a)** *(firmly)* stabil; **b)** *(compactly)* **a ~ built person** ein kräftig gebauter Mensch; **c)** *(ceaselessly)* pausenlos; **d)** *(whole-heartedly)* **be ~ behind sb./sth.** uneingeschränkt hinter jmdm./einer Sache stehen

'solid-state *adj. (Phys.)* Festkörper‹*physik, -geräte, -schaltung*›

soliloquy [sə'lɪləkwɪ] *n.* Monolog, *der; (talking to oneself)* Selbstgespräch, *das*

solitaire [sɒlɪ'teə(r)] *n.* **a)** *(gem)* Solitär, *der;* **b)** *(ring)* Solitärring, *der;* **c)** *(game)* Solitär, *das*

solitary ['sɒlɪtərɪ] *adj.* einsam; **a ~ existence/life** ein Einsiedlerdasein/-leben

solitude ['sɒlɪtju:d] *n.* Einsamkeit, *die*

solo ['səʊləʊ] **1.** *n., pl.* **~s a)** *(Mus.)* Solo, *das;* **b)** *(Cards)* **~ [whist]** Solo[-whist], *das.* **2.** *adj.* **a)** *(Mus.)* Solo‹*spiel, -part, -tanz, -instrument*›; **b)** *(unaccompanied)* **~ flight** Alleinflug, *der.* **3.** *adv. (unaccompanied)* solo ‹*singen, spielen, tanzen usw.*›; **go/fly ~** *(Aeronaut.)* einen Alleinflug machen

soloist ['səʊləʊɪst] *n. (Mus.)* Solist, *der*/Solistin, *die*

solstice ['sɒlstɪs] *n.* Sonnenwende, *die;* **summer/winter ~:** Sommer-/Wintersonnenwende, *die*

soluble ['sɒljʊbl] *adj.* **a)** *(esp. Chem.)* löslich; solubel *(fachspr.);* **~ in water, water-~:** wasserlöslich; **b)** *(solvable)* lösbar

solution [sə'lu:ʃn, sə'lju:ʃn] *n.* **a)** *(esp. Chem.)* Lösung, *die;* **b)** *([result of] solving)* Lösung, *die* **(to** *Gen.***)**; **find a ~ to sth.** eine Lösung für etw. finden; etw. lösen

solve [sɒlv] *v. t.* lösen

solvent ['sɒlvənt] **1.** *adj.* **a)** *(Chem.: dissolving)* lösend: **b)** *(Finance)* solvent. **2.** *n. (Chem.)* Lösungsmittel, *das* **(of, for** für); **~ abuse** Mißbrauch von Lösungsmitteln als Rauschmittel

sombre (*Amer.:* **somber**) ['sɒmbə(r)] *adj.* dunkel; düster ‹*Atmosphäre, Stimmung*›

some [səm, *stressed* sʌm] **1.** *adj.* **a)** *(one or other)* [irgend]ein; **~ fool** irgendein Dummkopf *(ugs.);* **~ day** eines Tages; **~ shop/book or other** irgendein Laden/Buch; **~ person or other** irgend jemand; irgendwer; **b)** *(a considerable quantity of)* einig...; etlich... *(ugs. verstärkend);* **speak at ~ length/wait for ~ time** ziemlich lang[e] sprechen/warten; **~ time/weeks/days/years ago** vor einiger Zeit/vor einigen Wochen/Tagen/Jahren; **~ time soon** bald [einmal]; **c)** *(a small quantity of)* ein bißchen; **would you like ~ wine?** möchten Sie [etwas] Wein?; **do ~ shopping/reading** einkaufen/lesen; **d)** *(to a certain extent)* **~ guide** eine gewisse Orientierungshilfe; **that is ~ proof** das ist [doch] gewissermaßen ein Beweis; **e) this is ~ war/poem/car!** *(sl.)* das ist vielleicht ein Krieg/Gedicht/Wagen! *(ugs.);* **f)** *(approximately)* etwa; ungefähr. **2.** *pron.* einig...; **she only ate ~ of it** sie hat es nur teilweise aufgegessen; **~ of her ideas are good** sie hat einige gute Ideen; **~ of the greatest music** einige der größten Werke der Musik; **~ say ...:** manche sagen ...; **~ ..., others ...:** manche ..., andere ...; die einen ..., andere ...; **... and then ~:** und noch einige/einiges mehr. **3.** *adv. (coll.: in ~ degree)* ein bißchen; etwas; **~ more** noch ein bißchen

somebody ['sʌmbədɪ] *n. & pron.* jemand; **~ or other** irgend jemand; *(important person)* **be ~:** jemand *od.* etwas sein

'somehow *adv.* **~ [or other]** irgendwie

someone ['sʌmwən, *stressed* 'sʌmwʌn] *pron. see* **somebody**

'someplace *(Amer. coll.) see* **somewhere**

somersault ['sʌməsɔ:lt, 'sʌməsɒlt] **1.** *n.* Purzelbaum, *der (ugs.);* Salto, *der (Sport);* **turn a ~:** einen Purzelbaum schlagen *(ugs.);* einen Salto springen *(Sport);* **the car ~ed [into a ditch]** das Auto überschlug sich [und landete in einem Graben]. **2.** *v. i.* einen Purzelbaum schlagen *(ugs.);* einen Salto springen *(Sport)*

'something *n. & pron.* **a)** *(some thing)* etwas; **~ new/old/good/bad** etwas Neues/Altes/Gutes/Schlechtes; **b)** *(some unspecified thing)* [irgend] etwas; **~ or other** irgend etwas; **c)** *(some quantity of a thing)* etwas; **there is ~ in what you say** was du sagst, hat etwas für sich; an dem, was du sagst, ist etwas dran *(ugs.);* **he has ~ about him** er hat etwas Besonderes an sich *(Dat.);* **d)** *(impressive or important thing, person, etc.)* **the party was quite ~:** die Party war spitze *(ugs.);* **e) or ~** *see* [1]**or c; f) ~ like** etwa wie; **that's ~ like it** das ist schon besser; **g) ~ of an expert/a specialist** so etwas wie ein Fachmann/Spezialist; **see ~ of sb.** jmdn. sehen

'sometime 1. *adj.* ehemalig. **2.** *adv.* irgendwann

'sometimes *adv.* manchmal; **~ ..., at other times ...:** manchmal ..., manchmal ...

'somewhat *adv. (rather)* irgendwie; ziemlich

'somewhere 1. *adv.* **a)** *(in a place)* irgendwo; **~ about** *or* **around thirty [years old]** [so *(ugs.)*] um die dreißig [Jahre alt]; **~ between five and ten** [so *(ugs.)*] zwischen fünf und zehn; **b)** *(to a place)* irgendwohin; **get ~** *(coll.) (in life)* es zu etwas bringen; *(in a task)* weiterkommen. **2.** *n.* **look for ~ to stay** sich nach einer Unterkunft umsehen; **she prefers ~ hot for her holidays** in den Ferien fährt sie am liebsten irgendwohin, wo es heiß ist

son [sʌn] *n.* Sohn, *der; (as address)* **[my] ~:** mein Sohn; **~ and heir** Sohn und Erbe; **the Son [of God]** *(Relig.)* der Sohn [Gottes]

sonar ['səʊnɑ:(r)] *n.* Sonar, *der*

sonata [sə'nɑ:tə] *n. (Mus.)* Sonate, *die*

song [sɒŋ] *n.* **a)** Lied, *das; (esp. political ballad, pop ~)* Song, *der;* **b)** *no pl. (singing)* Gesang, *der;* **on ~** *(fig. coll.)* in Spitzenform; **break** *or* **burst into ~:** ein Lied anstimmen; **for a ~:** für einen Apfel und ein Ei *(ugs.)*; **a ~ and dance** *(Brit. coll.: fuss; Amer. coll.: rigmarole)* viel *od.* großes Trara *(ugs.)*; **c)** *(bird cry)* Gesang, *der; (of cuckoo)* Ruf, *der*

song: ~bird *n.* Singvogel, *der;* **~-book** *n.* Liederbuch, *das;* **~-writer** *n.* Songschreiber, *der/* -schreiberin, *die*

sonic ['sɒnɪk] *attrib. adj.* Schall-

sonic 'boom *n.* Überschallknall, *der*

'son-in-law *n., pl.* **sons-in-law** Schwiegersohn, *der*

sonnet ['sɒnɪt] *n.* Sonett, *das*

sonny ['sʌnɪ] *n. (coll.)* Kleiner *(der);* kleiner Mann *(ugs.)*

sonorous ['sɒnərəs] *adj.* volltönend; sonor ‹*Stimme*›; klangvoll ‹*Instrument, Sprache*›

soon [su:n] **a)** bald; *(quickly)* schnell; **b)** *(early)* früh; **how ~ will it be ready?** wann ist es denn fertig?; **none too ~:** keinen Augenblick zu früh; **no ~er said than done** gesagt, getan; **~er said than done** leichter gesagt als getan; **no ~er had I arrived than ...:** kaum war ich angekommen, da ...; **~er or later** früher oder später; **the ~er [...] the better** *(coll.)* je früher *od.* eher [...], desto besser; **c) we'll set off just as ~ as he arrives** sobald er ankommt, machen wir uns auf den Weg; **as ~ as possible** so bald wie möglich; **d)** *(willingly)* **just as ~ [as ...]** genauso gern [wie ...]; **she would ~er die than ...:** sie würde lieber sterben, als ...; **they would kill you as ~ as look at you** *(coll.)* sie würden dich auf der Stelle umbringen; **~er you than me** lieber du als ich

soot [sʊt] *n.* Ruß, *der*

soothe [su:ð] *v. t.* **a)** *(calm)* beruhigen; beschwichtigen ‹*Gefühle*›; **b)** *(make less severe)* mildern; lindern ‹*Schmerz*›

soothing ['su:ðɪŋ] *adj.* beruhigend; wohltuend ‹*Bad, Creme, Massage*›

sooty ['sʊtɪ] *adj.* verrußt; rußig

sop [sɒp] **1.** *n.* **a)** *(piece of bread)* Stück eingeweichtes Brot; **b)** *(fig.)* Beschwichtigungsmittel, *das.* **2.** *v. i.*, **-pp-: be ~ping [wet]** völlig durchnäßt sein

sophisticated [sə'fɪstɪkeɪtɪd] *adj.* **a)** *(cultured)* kultiviert; gepflegt ‹*Restaurant, Küche*›; anspruchsvoll ‹*Roman, Autor, Unterhaltung, Stil*›; **b)** *(elaborate)* ausgeklügelt ‹*Autozubehör*›; differenziert, subtil ‹*Argument, System, Ansatz*›; hochentwickelt ‹*Technik, Elektronik, Software, Geräte*›

sophistication [səfɪstɪ'keɪʃn] *n.* **a)** *(refinement)* Kultiviertheit, *die; (of argument)* Differenziertheit, *die; (of style, manner)* Subtilität, *die;* **b)** *(advanced methods, state)* hoher Entwicklungsstand [der Technik]; **era of technical ~:** Zeitalter hochentwickelter Technik

sophomore ['sɒfəmɔ:(r)] *n. (Amer. Sch./Univ.) Student/Studentin einer High-School bzw. Universität im zweiten Studienjahr*

soporific [sɒpə'rɪfɪk] *adj.* schläfrig ‹*Person*›; einschläfernd ‹*Wirkung, Rede*›

soppy ['sɒpɪ] *adj. (Brit. coll.: sentimental)* rührselig; sentimental ‹*Person*›

soprano [sə'prɑ:nəʊ] *n., pl.* **~s** *or* **soprani** [sə'prɑ:ni:] *(Mus.) (voice, singer, part)* Sopran, *der; (female singer also)* Sopranistin, *die*

sorcerer ['sɔ:sərə(r)] *n.* Zauberer, *der*

sorcery ['sɔ:sərɪ] *n.* Zauberei, *die*

sordid ['sɔ:dɪd] *adj.* **a)** *(base)* dreckig *(abwertend);* unehrenhaft, unlauter ‹*Motiv*›; unerfreulich ‹*Detail, Geschichte*›; *(greedy)* schmutzig ‹*Geschäft*›; **b)** *(squalid)* schmutzig; schäbig ‹*Wohnung, Verhältnisse*›; heruntergekommen ‹*Stadtviertel*›

sore [sɔ:(r)] **1.** *adj.* **a)** weh; *(inflamed or injured)* wund; **sb. has a ~ back/foot/arm** *etc.* jmdm. tut der Rücken/Fuß/Arm *usw.* weh; **~ point** *or* **spot** *(fig.)* wunder Punkt; **b)** *(irritated)* verärgert; sauer *(ugs.).* **2.** *n.* wunde Stelle

sorely ['sɔ:lɪ] *adv.* sehr; dringend ‹*nötig, benötigt*›; **be ~ in need of sth.** etw. dringend brauchen; **~ tempted** stark versucht

soreness ['sɔ:nɪs] *n.* Schmerz, *der*

sorrow ['sɒrəʊ] *n.* **a)** Kummer, *der;* Leid, *das;* **feel [great] ~ that ...:** es [sehr] bedauern, daß ...; **cause sb. [great] ~:** jmdm. [großen] Kummer bereiten; **b)** *(misfortune)* Sorge, *die;* **he has had many ~s** er hat viel [Schweres] durchgemacht; *see also* **drown 2 b**

sorrowful ['sɒrəʊfl, 'sɒrəfl] *adj.* betrübt ‹*Person*›; traurig ‹*Anlaß, Lächeln, Herz*›

sorry ['sɒrɪ] *adj.* **sb. is ~ to do sth.** jmdm. tut es leid, etw. tun zu müssen; **I am ~ to disappoint you** ich muß dich leider enttäuschen; **sb. is ~ that ...:** es tut jmdm. leid, daß ...; **sb. is ~ about sth.** jmdm. tut etw. leid; **~, but ...** *(coll.)* tut mir leid, aber ...; **I'm ~** *(won't change my mind)* tut mir leid; **~ I'm late** *(coll.)* Entschuldigung, daß ich zu spät komme; **I'm ~ to say** leider; **I can't say [that] I'm ~!** ich bin nicht gerade traurig darüber; **sb. is** *or* **feels ~ for sb./sth.** jmd. tut jmdm. leid/jmd. bedauert etw.; **you'll be ~:** das wird dir noch leid tun; **feel ~ for oneself** *(coll.)* sich selbst bemitleiden; sich *(Dat.)* leid tun; **~!** Entschuldigung!; **~?** wie bitte?; **~ about that!** *(coll.)* tut mir leid!

sort [sɔ:t] **1.** *n.* **a)** Art, *die; (type)* Sorte, *die;* **cakes of several ~s** verschiedene Kuchensorten; **a new ~ of bicycle** ein neuartiges Fahrrad; **people of every/that ~:** Menschen jeden/diesen Schlages; **it takes all ~s [to make a world]** *(coll.)* es gibt so'ne und solche *(ugs.);* **all ~s of ...:** alle möglichen ...; **support sb. in all ~s of ways** jmdn. auf vielerlei Art und Weise unterstützen; **she is just/not my ~:** sie ist genau/nicht mein Typ *(ugs.);* **what ~ of [a] person do you think I am?** für wen hältst du mich?; **you'll do nothing of the ~:** das kommt gar nicht in Frage; **~ of** *(coll.)* irgendwie; *(more or less)* mehr oder weniger; *(to some extent)* ziemlich *(ugs.);* **nothing of the ~:** nichts dergleichen; **or something of the ~:** oder so [etwas ähnliches] *(ugs.);* **he is a doctor/footballer of a ~** *or* **of ~s** *(derog.)* er nennt sich Arzt/Fußballspieler; **we don't mix with people of that ~:** mit solchen Leuten wollen wir nichts zu tun haben; **he/she is a good ~** *(coll.)* er/sie ist schon in Ordnung *(ugs.);* **b) be out of ~s** nicht in Form sein; *(be irritable)* schlecht gelaunt sein. **2.** *v. t.* sortieren

~ 'out *v. t.* **a)** *(arrange)* sortieren; **b)** *(settle)* klären; schlichten ‹*Streit*›; beenden ‹*Verwirrung*›; **it will ~ itself out** es wird schon in Ordnung kommen; **c)** *(organize)* durchorganisieren; auf Vordermann bringen *(ugs.);* **things have ~ed themselves out** die Dinge haben sich eingerenkt; **d)** *(sl.: punish)* **~ sb. out** jmdm. zeigen, wo's langgeht *(ugs.);* **e)** *(select)* aussuchen; wählen

'sort code *n. (Banking)* Bankleitzahl, *die*

sortie ['sɔ:ti:, 'sɔ:tɪ] *n. (Mil.; also fig.)* **a)** Ausfall, *der;* **b)** *(flight)* Einsatz, *der*

SOS [esəʊ'es] *n.* SOS, *das*

'so so, 'so-so *adj., adv.* so lala *(ugs.)*

soufflé ['su:fleɪ] *n. (Gastr.)* Soufflé, *das;* **~ dish** Souffléform, *die*

sought *see* **seek**

soul [səʊl] *n.* **a)** Seele, *die;* **sell one's ~ for sth.** *(fig.)* seine Seele für etw. verkaufen; **bare one's ~ to sb.** jmdm. sein Herz ausschütten; **b)** *(person)* Seele, *die;* **not a ~:** keine Menschenseele; **the poor little ~:** das arme kleine Ding

'**soul-destroying** *adj.* **a)** *(boring)* nervtötend; geisttötend; **b)** *(depressing)* deprimierend

soulful ['səʊlfl] *adj.* gefühlvoll; *(sad)* schwermütig

soul: ~ mate *n.* Seelenverwandte, *der/die;* **~-searching** *n.* Gewissenskampf, *der*

[1]**sound** [saʊnd] **1.** *adj.* **a)** *(healthy)* gesund; intakt ‹*Gebäude, Mauerwerk*›; gut ‹*Frucht, Obst, Holz, Boden*›; **of ~ mind** im Vollbesitz der geistigen Kräfte; **the building was structurally ~:** das Gebäude hatte eine gesunde Bausubstanz; **b)** *(well-founded)* vernünftig ‹*Argument, Rat*›; klug ‹*Wahl*›; **it makes ~ sense** es ist sehr vernünftig; **c)** *(Finance: secure)* gesund, solide ‹*Basis*›; klug ‹*Investition*›; **d)** *(competent, reliable)* solide ‹*Spieler*›; **have a ~ character** charakterfest sein; **e)** *(undisturbed)* tief, gesund ‹*Schlaf*›; **f)** *(thorough)* gehörig *(ugs.)* ‹*Niederlage, Tracht Prügel*›; gekonnt ‹*Leistung*›. **2.** *adv.* fest, tief ‹*schlafen*›

[2]**sound 1.** *n.* **a)** *(Phys.)* Schall, *der;* **b)** *(noise)* Laut, *der; (of wind, sea, car, footsteps, breaking glass or twigs)* Geräusch, *das; (of voices, laughter, bell)* Klang, der; **do sth. without a ~:** etw. lautlos tun; **c)** *(Radio, Telev., Cinemat.)* Ton, *der;* **loss of ~:** Tonausfall, *der;* **d)** *(music)* Klang, *der;* **e)** *(Phonet.: articulation)* Laut, *der;* **f)** *(fig.: impression)* **I like the ~ of your plan** ich finde, Ihr Plan hört sich gut an; **I don't like the ~ of this** das hört sich nicht gut an. **2.** *v. i.* **a)** *(seem)* klingen; **it ~s as if .../like ...:** es klingt, als .../wie ...; **it ~s to me from what you have said that ...:** was du gesagt hast, klingt für mich so, als ob ...; **that ~s [like] a good idea to me** ich finde, die Idee hört sich gut an; **~s good to me!** klingt gut! *(ugs.);* gute Idee! *(ugs.);* **b)** *(emit ~)* [er]tönen. **3.** *v. t.* **a)** *(cause to emit ~)* ertönen lassen; **~ the trumpet** trompeten; in die Trompete blasen; **b)** *(utter)* **~ a note of caution** zur Vorsicht mahnen; **c)** *(pronounce)* aussprechen

~ 'off *v. i.* tönen *(ugs.),* schwadronieren **(on, about** von)

[3]**sound** *n.* *(strait)* Sund, *der;* Meerenge, *die*

[4]**sound** *v. t.* **a)** *(Naut.: fathom)* ausloten; sondieren; **b)** *(fig.: test) see* **~ out**

~ 'out *v. t.* ausfragen ‹*Person*›; sondieren *(geh.),* herausbekommen ‹*Sache*›; **~ sb. out on sth.** bei jmdm. wegen etw. vorfühlen

sound: ~ barrier *n.* Schallmauer, *die;* **~ broadcasting** *n.* Hörfunk, *der;* **~ effect** *n.* Geräuscheffekt, *der;* **~ engineer** *n.* *(Radio, Telev., Cinemat.)* Toningenieur, *der/*-ingenieurin, *die*

sounding ['saʊndɪŋ] *n.* **a)** *(Naut.: measurement)* Lotung, *die;* **take ~s** Lotungen vornehmen; loten; **b)** *(fig.)* Sondierung, *die (geh.);* **carry out ~s of public opinion/of interested parties** die öffentliche Meinung sondieren/mit den Beteiligten Sondierungsgespräche führen

soundless ['saʊndlɪs] *adj.* lautlos; stumm, tonlos ‹*Sprache, Gebet*›

soundly ['saʊndlɪ] *adv.* **a)** *(solidly)* stabil, solide ‹*bauen*›; **b)** *(well)* vernünftig ‹*argumentieren, urteilen, investieren*›; **c)** *(deeply)* tief, fest ‹*schlafen*›; **d)** *(thoroughly)* ordentlich *(ugs.)* ‹*verhauen*›; vernichtend ‹*schlagen, besiegen*›

soundness ['saʊndnɪs] *n., no pl.* **a)** *(of mind, body)* Gesundheit, *die; (of construction, structure)* Solidität, *die;* **b)** *(of argument)* Stichhaltigkeit, *die; (of policy, views)* Vernünftigkeit, *die;* **c)** *(of sleep)* Tiefe, *die;* **d)** *(competence, reliability)* Solidität, *die;* **e)** *(solvency)* wirtschaftliche Gesundheit; Solvenz, *die*

sound: ~-proof 1. *adj.* schalldicht; **2.** *v. t.* schalldicht machen; **~ recorder** *n.* Tonaufnahmegerät, *das;* **~-track** *n.* *(Cinemat.)* Soundtrack, *der;* **~-wave** *n.* *(Phys.)* Schallwelle, *die*

soup [su:p] *n.* Suppe, *die;* **be/land in the ~** *(fig. sl.)* in der Patsche sitzen/landen *(ugs.)*

souped-up ['su:ptʌp] *attrib. adj.* *(Motor Veh. coll.)* frisiert *(ugs.)*

soup: ~-plate *n.* Suppenteller, *der;* **~-spoon** *n.* Suppenlöffel, *der*

sour [saʊə(r)] **1.** *adj.* **a)** *(having acid taste)* sauer; **b)** *(morose)* griesgrämig *(abwertend);* säuerlich ‹*Blick*›; **c)** *(unpleasant)* bitter; **when things go ~:** wenn man *od.* einem alles leid wird. **2.** *v. t.* **a)** versauern lassen; sauer machen; **b)** *(fig.: spoil)* verbauen ‹*Karriere*›; trüben ‹*Beziehung*›; **c)** *(fig.: make gloomy)* verbittern. **3.** *v. i.* ‹*Beziehungen:*› sich trüben

source [sɔ:s] *n.* Quelle, *die;* **~ of income/infection** Einkommensquelle, *die/*Infektionsherd, *der;* **locate the ~ of a leak** *(lit. or fig.)* feststellen, wo eine undichte Stelle ist; **the whole thing is a ~ of some embarrassment to us** das Ganze ist für uns ziemlich unangenehm; **at ~:** an der Quelle

'**sourpuss** *n. (sl.) (male)* Miesepeter, der *(ugs.); (female)* miesepetrige Ziege *(ugs.)*

souse [saʊs] *v. t.* eintauchen

south [saʊθ] **1.** *n.* **a)** *(direction)* Süden, *der;* **the ~:** Süd *(Met., Seew.);* **in/to[wards]/from the ~:** im/nach/von Süden; **to the ~ of** südlich von; südlich (+ *Gen.*); **b)** *usu.* **S~** *(part lying to the ~)* Süden, *der;* **from the S~:** aus dem Süden. **2.** *adj.* südlich; Süd‹*küste, -wind, -grenze, -tor*›. **3.** *adv.* südwärts; nach Süden; **~ of** südlich von; südlich (+ *Gen.*)

South: ~ 'Africa *pr. n.* Südafrika *(das);* **~ 'African 1.** *adj.* südafrikanisch; **2.** *n.* Südafrikaner, *der/*-afrikanerin, *die;* **~ A'merica** *pr. n.* Südamerika *(das);* **~ A'merican 1.** *adj.* südamerikanisch; **2.** *n.* Südamerikaner, *der/*-amerikanerin, *die*

south: ~bound *adj.* ‹*Zug, Verkehr usw.*› in Richtung Süden; **~-'east 1.** *n.* Südosten, *der;* **2.** *adj.* südöstlich; Südost‹*wind, -küste*›; **3.** *adv.* südostwärts; nach Südosten; **~-'eastern** *adj.* südöstlich

southerly ['sʌðəlɪ] *adj.* **a)** *(in position or direction)* südlich; **in a ~ direction** nach Süden; **b)** *(from the south)* ‹*Wind*› aus südlichen Richtungen

southern ['sʌðən] *adj.* südlich; Süd‹*grenze, -hälfte, -seite*›; südländisch ‹*Temperament*›; **~ Spain** Südspanien; das südliche Spanien; **~ Africa** das südliche Afrika

southerner [sʌðənə(r)] *n. (male)* Südengländer / -franzose / -italiener *usw., der; (female)* Südengländerin / -französin / -italienerin *usw., die*

Southern 'Europe pr. n. Südeuropa, *(das)*

southernmost [sʌðənməʊst] *adj.* südlichst...

South: ~ 'German 1. *adj.* süddeutsch; **2.** *n.* Süddeutsche, *der/die;* **~ 'Germany** *pr. n.* Süddeutschland *(das);* **~ Ko'rea** *pr. n.* Südkorea *(das);* **~ of 'Eng-**

land *pr. n.* Südengland *(das); attrib.* südenglisch; ~ **'Pole** *pr. n.* Südpol, *der;* ~ **'Seas** *pr. n. pl.* Südsee, *die*

southward ['saʊθwəd] **1.** *adj.* nach Süden gerichtet; *(situated towards the south)* südlich; **in a ~ direction** nach Süden; [in] Richtung Süden. **2.** *adv.* südwärts; **they are ~ bound** sie fahren nach *od.* [in] Richtung Süden

southwards ['saʊθwədz] *adv.* südwärts

south: ~-'west 1. *n.* Südwesten, *der;* **2.** *adj.* südwestlich; Südwest⟨*wind, -küste*⟩; **3.** *adv.* südwestwärts; nach Südwesten; **S~-West 'Africa** *pr. n.* Südwestafrika, *(das);* **~-'western** *adj.* südwestlich

souvenir [su:və'nɪə(r)] *n. (of holiday)* Andenken, *das;* Souvenir, *das* (**of** aus); *(of wedding-day, one's youth, etc.)* Andenken, *das* (**of** an + *Akk.*)

sou'wester [saʊ'westə(r)] *n.* **a)** *(hat)* Südwester, *der;* **b)** *(coat)* Ölhaut, *die*

sovereign ['sɒvrɪn] *n.* **a)** *(ruler)* Souverän, *der;* **b)** *(Brit. Hist.: coin)* Sovereign, *der;* 20-Shilling-Münze, *die*

sovereignty ['sɒvrɪntɪ] *n.* Souveränität, *die;* Oberhoheit, *die*

Soviet ['səʊvɪət, 'sɒvɪət] **1.** *adj.* sowjetisch; Sowjet⟨*bürger, -literatur, -kultur, -ideologie*⟩. **2.** *n.* Sowjet, *der*

Soviet 'Union *pr. n.* Sowjetunion, *die*

¹sow [səʊ] *v. t., p.p.* **sown** [səʊn] *or* **~ed** [səʊd] **a)** *(plant)* [aus]säen; **b)** *(plant with seed)* einsäen, besäen ⟨*Feld, Boden*⟩; **c)** *(cover thickly)* spicken *(ugs.)*

²sow [saʊ] *n. (female pig)* Sau, *die*

sown *see* **¹sow**

soya [bean] ['sɔɪə (bi:n)], **soy bean** ['sɔɪ bi:n] *ns.* **a)** *(plant)* Soja[bohne], *die;* **b)** *(seed)* Sojabohne, *die*

soy sauce ['sɔɪ sɔ:s] *n.* Sojasoße, *die*

sozzled ['sɒzld] *adj. (sl.)* voll *(ugs.)*

spa [spɑ:] *n.* **a)** *(place)* Bad, *das;* Badeort, *der;* **b)** *(spring)* Mineralquelle, *die*

space [speɪs] **1.** *n.* **a)** Raum, *der;* **stare into ~:** in die Luft *od.* ins Leere starren; **b)** *(interval between points)* Platz, *der;* **clear a ~:** Platz schaffen; **c)** *(area on page)* Platz, *der;* **d) the wide open ~s** das weite, flache Land; **e)** *(Astron.)* Weltraum, *der; see also* **outer space; f)** *(blank between words)* Zwischenraum, *der;* **g)** *(interval of time)* Zeitraum, *der;* **in the ~ of a minute/an hour** *etc.* innerhalb einer Minute/Stunde *usw.* **2.** *v. t.* **the posts are ~d at intervals of one metre** die Pfosten sind im Abstand von einem Meter aufgestellt

~ **'out** *v. t.* verteilen

space: ~ age *n.* [Welt]raumzeitalter, *das;* Zeitalter der Raumfahrt, *das;* **~-bar** *n.* Leertaste, *die;* **~craft** *n.* Raumfahrzeug, *das; (unmanned)* Raumsonde, *die;* ~ **flight** *n.* **a)** *(a journey through ~)* [Welt]raumflug, *der;* **b)** *see* ~ **travel; ~-heater** *n.* Heizgerät, *das;* **~man** *n.* Raumfahrer, *der*/-fahrerin, *die;* **~-saving** *adj.* platzsparend; **~ship** *n.* Raumschiff, *das;* ~ **shuttle** *n.* Raumfähre, *die;* Raumtransporter, *der;* ~ **station** *n.* [Welt]raumstation, *die;* **~suit** *n.* Raumanzug, *der;* ~ **travel** *n.* Raumfahrt, *die;* ~ **walk** *n.* Spaziergang im All

spacing ['speɪsɪŋ] *n.* Zwischenraum, *der; (Printing)* Sperrungen; Spationierung, *die (Druckw.);* **single/double ~** *(on typewriter)* einfacher/doppelter Zeilenabstand

spacious ['speɪʃəs] *adj.* **a)** *(vast in area)* weitläufig ⟨*Garten, Park, Ländereien*⟩; **b)** *(roomy)* geräumig ⟨*Raum*⟩; breit ⟨*Straße*⟩

spade [speɪd] *n.* **a)** Spaten, *der;* **call a ~ a ~:** das Kind beim [rechten] Namen nennen *(ugs.);* **b)** *(Cards)* Pik, *das; see also* **club 1 d**

'spadework *n. (preliminary work)* Vorarbeit, *die*

spaghetti [spə'getɪ] *n.* Spaghetti *Pl.*

Spain [speɪn] *pr. n.* Spanien *(das)*

Spam, (P) [spæm] *n.* Frühstücksfleisch, *das*

¹span [spæn] **1.** *n.* **a)** *(full extent)* Spanne, *die;* ~ **of life/time** Lebens-/Zeitspanne, *die;* **b)** *(of bridge)* Spannweite, *die.* **2.** *v. t.,* **-nn-** überspannen ⟨*Fluß*⟩; umfassen ⟨*Zeitraum*⟩

²span *see* **spick**

spangle ['spæŋgl] **1.** *n. see* **sequin. 2.** *v. t.* **~d with stars/buttercups** von glitzernden Sternen/mit leuchtenden Butterblumen übersät

Spaniard ['spænjəd] *n.* Spanier, *der*/Spanierin, *die*

spaniel ['spænjəl] *n.* Spaniel, *der*

Spanish ['spænɪʃ] **1.** *adj.* spanisch; **sb. is ~:** jmd. ist Spanier/Spanierin. **2.** *n.* **a)** *(language)* Spanisch, *das; see also* **English 2 a; b)** *constr. as pl. (people)* Spanier

spank [spæŋk] **1.** *n.* ≈ Klaps, *der (ugs.).* **2.** *v. t.* ~ **sb.** jmdm. den Hintern versohlen *(ugs.);* **get ~ed** den Hintern voll kriegen *(ugs.)*

spanking ['spæŋkɪŋ] *n.* Tracht Prügel, *die (ugs.)*

spanner ['spænə(r)] *n. (Brit.)* Schraubenschlüssel, *der;* **put** *or* **throw a ~ in the works** *(fig.)* Sand ins Getriebe streuen

¹spar [spɑ:(r)] *v. i.,* **-rr-** *(Boxing)* sparren

²spar *n. (pole)* Rundholz, *das;* Spiere, *die (Seemannsspr.)*

spare [speə(r)] **1.** *adj.* **a)** *(not in use)* übrig; ~ **time/moment** Freizeit, *die*/freier Augenblick; **there is one ~ seat** ein Platz ist noch frei; **are there any ~ tickets for Friday?** gibt es noch Karten für Freitag?; **b)** *(for use when needed)* zusätzlich, Extra⟨*bett, -tasse*⟩; ~ **room** Gästezimmer, *das;* **go ~** *(Brit. sl.: be very angry)* durchdrehen *(salopp).* **2.** *n.* Ersatzteil, *das*/-reifen, *der usw.* **3.** *v. t.* **a)** *(do without)* entbehren; **can you ~ me a moment?** hast du einen Augenblick Zeit für mich?; **we arrived with ten minutes to ~:** wir kamen zehn Minuten früher an; **b)** *(not inflict on)* ~ **sb. sth.** jmdm. etw. ersparen; **c)** *(not hurt)* [ver]schonen; **d)** *(fail to use)* **not ~ any expense/pains** *or* **efforts** keine Kosten/Mühe scheuen; **no expense ~d** an nichts gespart. *See also* **rod c**

spare: ~ 'part *n.* Ersatzteil, *das;* ~ **'tyre** *n.* **a)** Reserve-, Ersatzreifen, *der;* **b)** *(Brit. fig. coll.)* Rettungsring, *der (ugs.);* ~ **'wheel** *n.* Ersatzrad, *das*

sparing ['speərɪŋ] *adj.* sparsam; **be ~ of sth./in the use of sth.** mit etw. sparsam umgehen

spark [spɑ:k] **1.** *n.* **a)** Funke, *der;* **the ~s [begin to] fly** *(fig.)* es funkt *(ugs.);* **a ~ of generosity/decency** *(fig.)* ein Funke[n] Großzügigkeit/Anstand; **b)** *(in sparking-plug)* Zündfunke[n], *der (Kfz-W.);* **c) a bright ~** *(clever person; also iron.)* ein schlauer Kopf. **2.** *v. t. see* **~ off**

~ **'off** *v. t.* **a)** *(cause to explode)* zünden; **b)** *(fig.: start)* auslösen

'sparking-plug *n. (Brit. Motor Veh.)* Zündkerze, *die*

sparkle ['spɑ:kl] **1.** *v. i.* **a)** *(flash)* ⟨*Tautropfen:*⟩ glitzern; ⟨*Augen:*⟩ funkeln, sprühen; **b)** *(perform brilliantly)* glänzen; **c)** *(be lively)* sprühen (**with** vor + *Dat.*). **2.** *n.* Glitzern, *das;* Funkeln, *das*

sparkler ['spɑ:klə(r)] *n.* Wunderkerze, *die*
sparkling ['spɑ:klɪŋ] *adj.* **a)** *(flashing)* glitzernd ‹*Stein, Diamant*›; **b)** *(bright)* funkelnd ‹*Augen*›; **c)** *(brilliant)* glänzend ‹*Schauspiel, Aufführung, Rede*›
sparkling 'wine *n.* Schaumwein, *der*
'spark-plug *see* **sparking-plug**
'sparring partner *n. (Boxing)* Sparringspartner, *der*
sparrow ['spærəʊ] *n.* Sperling, *der;* Spatz, *der*
sparse [spɑ:s] *adj.* spärlich; dünn ‹*Besiedlung*›
Spartan ['spɑ:tn] **1.** *adj.* spartanisch. **2.** *n.* Spartaner, *der*/Spartanerin, *die*
spasm ['spæzm] *n.* **a)** Krampf, *der;* Spasmus, *der (Med.);* **b)** *(convulsive movement)* Anfall, *der;* **c)** *(coll.)* **a ~ of activity** plötzliche fieberhafte Aktivität
spasmodic [spæz'mɒdɪk] *adj.* **a)** *(marked by spasms)* krampfartig; **b)** *(intermittent)* sporadisch ‹*Anwachsen, Bemühungen*›
spastic ['spæstɪk] *(Med.)* **1.** *n.* Spastiker, *der*/Spastikerin, *die.* **2.** *adj.* spastisch [gelähmt]
spat *see* **[1]spit 1, 2**
spate [speɪt] *n.* **a)** *(flood)* **the river/waterfall is in [full] ~:** der Fluß/Wasserfall führt Hochwasser; **b)** *(fig.: large amount)* **a ~ of sth.** eine Flut von etw.; **a ~ of burglaries** eine Einbruchsserie
spatial ['speɪʃl] *adj.* räumlich
spatter ['spætə(r)] *v.t.* spritzen; **~ sb./sth. with sth.** jmdn./etw. mit etw. bespritzen
spatula ['spætjʊlə] *n.* Spachtel, *die*
spawn [spɔ:n] *(Zool.)* **1.** *v.t.* ablegen ‹*Eier*›; *(fig.)* hervorbringen. **2.** *v.i.* laichen. **3.** *n., constr. as sing. or pl.* Laich, *der*
spay [speɪ] *v.t.* sterilisieren ‹*Katze, Hündin*›
speak [spi:k] **1.** *v.i.,* **spoke** [spəʊk], **spoken** ['spəʊkn] **a)** sprechen; **~ [with sb.] on** *or* **about sth.** [mit jmdm.] über etwas *(Akk.)* sprechen; **~ for/against sth.** sich für/gegen etw. aussprechen; **sth. ~s well for sb.** etw. spricht für jmdn.; **b)** *(on telephone)* **Is Mr Grant there? – S~ing!** Ist Mister Grant da? – Am Apparat!; **Mr Grant ~ing** *(when connected to caller)* Grant hier; hier ist Grant; **who is ~ing, please?** wer ist am Apparat, bitte?; mit wem spreche ich, bitte? **2.** *v.t.,* **spoke, spoken a)** *(utter)* sprechen ‹*Satz, Wort, Sprache*›; **b)** *(make known)* sagen ‹*Wahrheit*›; **~ one's opinion/mind** seine Meinung sagen/sagen, was man denkt; **c)** *(convey without words)* **sth. ~s volumes** etw. spricht Bände
~ for *v.t.* sprechen für; **~ for yourself!** das ist [nur] deine Meinung!; **~ for itself/themselves** für sich selbst sprechen; **sth. is spoken for** *(reserved)* etw. ist schon vergeben
~ of *v.t.* sprechen von; **~ing of Mary** da wir gerade von Mary sprechen; apropos Mary; **nothing to ~ of** nichts Besonderes *od.* Nennenswertes
~ 'out *v.i.* seine Meinung sagen; **~ out against sth.** sich gegen etw. aussprechen
~ to *v.t.* **a)** *(address)* sprechen mit; reden mit; **I know him to ~ to** ich kenne ihn [nur] flüchtig; **b)** *(request action from)* **~ to sb. about sth.** mit jmdm. wegen einer Sache *od.* über etw. *(Akk.)* sprechen; **c)** *(coll.: reprove)* **~ to sb.** sich mit jmdm. unterhalten *(verhüllend)*
~ 'up *v.i.* **a)** *(~ more loudly)* lauter sprechen; **b)** *see* **~ out**
speaker ['spi:kə(r)] *n.* **a)** *(in public)* Redner, *der*/Rednerin, *die;* **b)** *(of a language)* Sprecher, *der*/Sprecherin, *die;* **be a French ~, be a ~ of French** Französisch sprechen; **c)** S~ *(Polit.)* Sprecher, *der;* ≈ Parlamentspräsident, *der;* **d)** *see* **loudspeaker**
speaking ['spi:kɪŋ] **1.** *n. (talking)* Sprechen, *das;* **a good ~ voice** eine gute Sprechstimme; **not be on ~ terms with sb.** nicht [mehr] mit jmdm. reden; **~ clock** *(Brit.)* telefonische Zeitansage. **2.** *adv.* **strictly/roughly/generally/legally ~:** genaugenommen/grob gesagt/im allgemeinen/aus juristischer Sicht
spear [spɪə(r)] **1.** *n.* Speer, *der.* **2.** *v.t.* aufspießen
spear: ~head 1. *n. (fig.)* Speerspitze, *die; (Mil.)* Angriffsspitze, *die;* **2.** *v.t. (fig.)* **~head sth.** etw. anführen; **~mint** *n.* Grüne Minze
special ['speʃl] *adj.* speziell; besonder ...; Sonder‹*korrespondent, -zug, -mission usw.*›; **nobody ~:** niemand Besonderer; **a ~ occasion** ein besonderer Anlaß
special: S~ Branch *n. (Brit. Police) Abteilung der britischen Polizei, deren Aufgabe die Wahrung der inneren Sicherheit ist;* ≈ Sicherheitsdienst, *der;* **~ 'case** *n.* Sonderfall, *der;* **~ correspondent** *n.* Sonderkorrespondent, *der*/-korrespondentin, *die;* Sonderberichterstatter, *der*/-berichterstatterin, *die;* **~ de'livery** *n. (Post)* Eilzustellung, *die;* **~ e'dition** *n.* Sonderausgabe, *die;* **~ effects** *n. pl. (Cinemat.)* Special effects *Pl. (fachspr.);* Spezialeffekte
specialise *see* **specialize**
specialist ['speʃəlɪst] *n.* **a)** Spezialist, *der*/Spezialistin, *die* (**in** für); Fachmann, *der*/Fachfrau, *die* (**in** für); **~ knowledge** Fachwissen, *das;* **b)** *(Med.)* Facharzt, *der*/-ärztin, *die*
speciality [speʃɪ'ælɪtɪ] *n.* Spezialität, *die*
specialize ['speʃəlaɪz] *v.i.* sich spezialisieren (**in** auf + *Akk.*)
specially ['speʃəlɪ] *adv.* **a)** speziell; **make sth. ~:** etw. speziell *od.* extra anfertigen; **~ made/chosen for me** eigens für mich gemacht/ausgewählt; **a ~ made wheelchair/lift** ein spezieller Rollstuhl/Lift; **b)** *(especially)* besonders
special 'offer *n.* Sonderangebot, *das;* **on ~:** im Sonderangebot
specialty ['speʃltɪ] *(esp. Amer.) see* **speciality**
species ['spi:ʃi:z, 'spi:si:z] *n., pl. same* **a)** *(Biol.)* Spezies, *die (fachspr.);* Art, *die;* **b)** *(sort)* Art, *die*
specific [spɪ'sɪfɪk] *adj.* deutlich, klar ‹*Aussage*›; bestimmt ‹*Ziel, Grund*›; **make a ~ request** einen bestimmten Wunsch äußern; **could you be more ~?** kannst du dich genauer ausdrücken?
specifically [spɪ'sɪfɪkəlɪ] *adv.* ausdrücklich; eigens; extra *(ugs.)*
specification [spesɪfɪ'keɪʃn] *n.* **a)** *often pl. (details)* technische Daten; *(instructions)* Konstruktionsplan, *der; (for building)* Baubeschreibung, *die;* **b)** *(specifying)* Spezifizierung, *die (geh.);* **c) [patent] ~:** Patentschrift, *die*
specify ['spesɪfaɪ] *v.t.* ausdrücklich sagen; ausdrücklich nennen ‹*Namen*›; **as specified above** wie oben aufgeführt; **unless otherwise specified** wenn nicht anders angegeben
specimen ['spesɪmən] *n.* **a)** *(example)* Exemplar, *das;* **a ~ of his handwriting** eine Schriftprobe von ihm; **~ signature** Unterschriftsprobe, *die;* **b)** *(sample)* Probe, *die;* **a ~ of his urine was required** es wurde eine Urinprobe von ihm benötigt; **c)** *(coll./derog.: type)* Marke, *die (salopp)*
specious ['spi:ʃəs] *adj.* **a ~ argument** ein nur scheinbar treffendes

Argument; **a ~ pretence/appearance of honesty** ein Anschein von Ehrlichkeit

speck [spek] *n.* **a)** *(spot)* Fleck, *der; (of paint also)* Spritzer, *der;* **b)** *(particle)* Teilchen, *das;* **~ of soot/dust** Rußflocke, *die*/Staubkörnchen, *das*

specs [speks] *n. pl. (coll.: spectacles)* Brille, *die*

spectacle ['spektəkl] *n.* **a)** *in pl.* **[pair of] ~s** Brille, *die;* **b)** *(public show)* Spektakel, *das;* **c)** *(object of attention)* Anblick, *der;* Schauspiel, *das;* **make a ~ of oneself** sich unmöglich aufführen

'**spectacle case** *n.* Brillenetui, *das*

spectacular [spek'tækjʊlə(r)] **1.** *adj.* spektakulär. **2.** *n.* Spektakel, *das*

spectator [spek'teɪtə(r)] *n.* Zuschauer, *der*/Zuschauerin, *die*

spec'tator sport *n.* Publikumssport, *der*

specter *(Amer.) see* **spectre**

spectra *pl. of* **spectrum**

spectral ['spektrl] *adj. (Phys.)* spektral

spectre ['spektə(r)] *n. (Brit.)* **a)** *(apparition)* Gespenst, *das;* **b)** *(disturbing image)* Schreckgespenst, *das*

spectrum ['spektrəm] *n., pl.* **spectra** ['spektrə] Spektrum, *das;* **~ of opinion** Meinungsspektrum, *das*

speculate ['spekjʊleɪt] *v. i.* spekulieren **(about, on** über + *Akk.*); Vermutungen *od.* Spekulationen anstellen **(about, on** über + *Akk.*); **~ on the Stock Exchange/in rubber** an der Börse/mit *od. (Wirtsch. Jargon)* in Gummi spekulieren

speculation [spekjʊ'leɪʃn] *n.* Spekulation, *die* **(over** über + *Akk.*)

speculative ['spekjʊlətɪv] *adj.* spekulativ

speculator ['spekjʊleɪtə(r)] *n.* Spekulant, *der*/Spekulantin, *die*

sped *see* **speed 2, 3**

speech [spi:tʃ] *n.* **a)** *(public address)* Rede, *die;* **make** *or* **deliver** *or* **give a ~:** eine Rede halten; **b)** *(faculty of speaking)* Sprache, *die;* **c)** *(act of speaking)* Sprechen, *das;* Sprache, *die;* **d)** *(manner of speaking)* Sprache, *die;* Sprechweise, *die;* **his ~ was slurred** er sprach undeutlich

speech: ~-day *n. (Brit. Sch.)* jährliches Schulfest; **~ defect** *n.* Sprachfehler, *der*

speechless ['spi:tʃlɪs] *adj.* sprachlos **(with** vor + *Dat.*)

speed [spi:d] **1.** *n.* **a)** Geschwindigkeit, *die;* Schnelligkeit, *die;* **at full** *or* **top ~:** mit Höchstgeschwindigkeit; mit Vollgas *(ugs.);* **pick up ~:** schneller werden; **at a ~ of eighty miles an hour** mit einer Geschwindigkeit von achtzig Meilen in der Stunde; **at ~:** mit hoher Geschwindigkeit; **b)** *(gear)* Gang, *der;* **a five-~ gearbox** eine 5-Gang-Schaltung; **c)** *(Photog.) (of film etc.)* Lichtempfindlichkeit, *die; (of lens)* **[shutter] ~:** Belichtungszeit, *die.* **2.** *v. i.,* **sped** [sped] *or* **~ed a)** *(go fast)* schnell fahren; rasen *(ugs.);* **b)** *p. t. & p.p.* **~ed** *(go too fast)* zu schnell fahren; rasen *(ugs.).* **3.** *v. t.,* **sped** *or* **~ed: ~ sb. on his/her way** jmdn. verabschieden

~ 'off *v. i.* davonbrausen

~ 'up 1. *v. t.,* **~ed up** beschleunigen; **~ up the work** die Arbeit vorantreiben; *(one's own work)* sich mit der Arbeit beeilen. **2.** *v. i.,* **~ed up** sich beeilen

speed: ~boat *n.* Rennboot, *das;* **~ bump, ~ hump** *ns.* Bodenschwelle, *die*

speeding ['spi:dɪŋ] *n. (going too fast)* zu schnelles Fahren; Rasen, *das (ugs. abwertend);* Geschwindigkeitsüberschreitung, *die (Verkehrsw.)*

'**speed limit** *n.* Tempolimit, *das;* Geschwindigkeitsbeschränkung, *die (Verkehrsw.)*

speedometer [spi:'dɒmɪtə(r)] *n.* Tachometer, *der od. das*

speed: ~ trap *n.* Geschwindigkeitskontrolle, *die; (with radar)* Radarfalle, *die (ugs.);* **~way** *n.* **a)** *(motor-cycle racing)* Speedwayrennen, *das;* **b)** *(race-track)* Speedwaybahn, *die*

speedy ['spi:dɪ] *adj.* schnell; umgehend, prompt ‹*Antwort*›

[1]**spell** [spel] **1.** *v. t.,* **~ed** *or (Brit.)* **spelt** [spelt] **a)** schreiben; *(aloud)* buchstabieren; **b)** *(form)* **what do these letters/what does b-a-t ~?** welches Wort ergeben diese Buchstaben/die Buchstaben b-a-t?; **c)** *(fig.: have as result)* bedeuten; **that ~s trouble** das bedeutet nichts Gutes. **2.** *v. i.,* **~ed** *or (Brit.)* **spelt** *(say)* buchstabieren; *(write)* richtig schreiben; **he can't ~:** er kann keine Rechtschreibung *(ugs.)*

~ 'out, ~ 'over *v. t.* **a)** *(read letter by letter)* [langsam] buchstabieren; **b)** *(fig.: explain precisely)* genau erklären; genau darlegen

[2]**spell** *n.* Weile, *die;* **a ~ of overseas service** eine Zeitlang Dienst in Übersee; **on Sunday it will be cloudy with some sunny ~s** am Sonntag wolkig mit sonnigen Abschnitten; **a cold ~:** eine Kälteperiode; **a long ~ when ...:** eine lange Zeit, während der ...

[3]**spell** *n.* **a)** *(words used as a charm)* Zauberspruch, *der;* **cast a ~ over** *or* **on sb./sth., put a ~ on sb./sth.** jmdn./etw. verzaubern; **b)** *(fascination)* Zauber, *der;* **break the ~:** den Bann brechen; **be under a ~:** unter einem Bann stehen

'**spellbound** *adj.* verzaubert; **he can hold his readers ~:** er kann seine Leser in seinem Bann halten

spelling ['spelɪŋ] *n.* **a)** Rechtschreibung, *die;* **b)** *(sequence of letters)* Schreibweise, *die*

spelling: ~-bee *n.* Rechtschreib[e]wettbewerb, *der;* **~ checker** *n.* Rechtschreibprogramm, *das;* **~ mistake** *n.* Rechtschreibfehler, *der*

spelt *see* [1]**spell**

spend [spend] *v. t.,* **spent** [spent] **a)** *(pay out)* ausgeben; **~ money like water** *or (coll.)* **as if it's going out of fashion** sein *od.* das Geld mit beiden Händen ausgeben *od.* hinauswerfen *(ugs.);* **it was money well spent** es hat sich ausgezahlt; **~ a penny** *(fig. coll.)* mal verschwinden [müssen] *(ugs.);* **b)** *(use)* aufwenden **(on** für)

'**spending money** *n.* **a)** *(Amer.) see* **pocket-money; b)** *(Brit.: sum intended for spending)* verfügbares Geld

'**spendthrift** *n.* Verschwender, *der*/Verschwenderin, *die*

spent 1. *see* **spend. 2.** *adj.* **a)** *(used up)* verbraucht; **~ cartridge** leere Geschoßhülse; **b)** *(drained of energy)* erschöpft; ausgelaugt; **a ~ force** *(fig.)* eine Kraft, die sich erschöpft hat

sperm ['spɜ:m] *n., pl.* **~s** *or same (Biol.) (semen)* Sperma, *der*

'**sperm whale** *n.* Pottwal, *der*

spew [spju:] **1.** *v. t.* spucken. **2.** *v. i.* sich ergießen

~ 'out 1. *v. t.* erbrechen, [aus]spucken ‹*Gegessenes*›; ‹*Vulkan:*› spucken, speien ‹*Lava*›. **2.** *v. i.* sich ergießen **(of, from** aus)

sphere [sfɪə(r)] *n.* **a)** *(field of action)* Bereich, *der;* Sphäre, *die (geh.);* **be distinguished in many ~s** sich auf vielen Gebieten ausgezeichnet haben; **that's outside my ~:** das gehört nicht zu meinem Tätigkeitsbereich; **~ of influence** Einflußbereich, *der;* **b)** *(Geom.)* Kugel, *die*

spherical ['sferɪkl] *adj.* kugelförmig

spice [spaɪs] **1.** *n.* **a)** Gewürz, *das; (collectively)* Gewürze *Pl.; attrib.* Gewürz-; **b)** *(fig.: excitement)* Würze, *die;* **the ~ of life** die Würze des Lebens. **2.** *v. t.* würzen

spick [spɪk] *adj.* **~ and span** blitzblank *od.* -sauber *(ugs.)*

spicy ['spaɪsɪ] *adj.* pikant; würzig

spider ['spaɪdə(r)] *n.* Spinne, *die*

'spider's web (*Amer.*: **'spider web**) Spinnennetz, *das; (fig.)* Netz, *das*

spidery ['spaɪdərɪ] *adj.* spinnenförmig; krakelig *(ugs.)*‹*Schrift*›

spike [spaɪk] **1.** *n.* **a)** Stachel, *der; (of running-shoe)* Spike, *der;* **b)** *in pl. (shoes)* Spikes *Pl.* **2.** *v. t.* **a) ~ sb.'s guns** *(fig.)* jmdm. einen Strich durch die Rechnung machen *(ugs.);* **b)** *(coll.: add spirits or drugs to)* **sb. ~d his drink** jmd. hat ihm etwas in seinen Drink getan

spiky ['spaɪkɪ] *adj.* **a)** *(like a spike)* spitz [zulaufend]; stachelig ‹*Haare*›; **b)** *(having spikes)* stach[e]lig

spill [spɪl] **1.** *v. t.*, **spilt** [spɪlt] *or* **~ed a)** verschütten ‹*Flüssigkeit*›; **~ sth. on sth.** etw. auf etw. *(Akk.)* schütten; **b)** *(sl.: divulge)* ausquatschen *(salopp);* **~ the beans [to sb.]** [jmdm. gegenüber] aus der Schule plaudern; **not ~ the beans [to sb.]** [jmdm. gegenüber] dichthalten *(ugs.). See also* **milk 1. 2.** *v. i.*, **spilt** *or* **~ed** überlaufen. **3.** *n. (fall)* Sturz, *der*

~ 'over *v. i.* überlaufen; *(fig.)* überquellen; ‹*Unruhen:*› sich ausbreiten

spillage ['spɪlɪdʒ] *n.* **a)** *(act)* Verschütten, *das;* **~ of oil** *(from tanker)* das Auslaufen von Öl; **b)** *(quantity)* Verschüttete, *das*

spilt *see* **spill 1, 2**

spin [spɪn] **1.** *v. t.*, **-nn-, spun** [spʌn] **a)** spinnen; **~ a yarn** *(fig.)* ein Garn spinnen *(bes. Seemannsspr.);* fabulieren; **b)** *(in washing-machine etc.)* schleudern; **c)** *(cause to whirl round)* [schnell] drehen; wirbeln [lassen]; **~ a coin** eine Münze kreiseln lassen; *(toss)* eine Münze werfen. **2.** *v. i.*, **-nn-, spun** sich drehen; **my head is ~ning** *(fig.) (from noise)* mir brummt der Schädel *(ugs.); (from many impressions)* mir schwirrt der Kopf. **3.** *n.* **a)** *(whirl)* **give sth. a ~:** etw. in Drehung versetzen; **b)** *(Aeronaut.)* Trudeln, *das;* **c)** *(Sport: revolving motion)* Effet, *der;* Spin, *der;* **d)** *(outing)* **go for a ~:** einen Ausflug machen; **a ~ in the car** eine Spritztour mit dem Auto

~ 'out *v. t.* **a)** *(prolong)* in die Länge ziehen; **b)** *(use sparingly)* **~ one's money out until pay-day** sein Geld bis zum Zahltag strecken

~ 'round 1. *v. i.* sich drehen; ‹*Person:*› sich [schnell] umdrehen. **2.** *v. t.* [schnell] drehen

spinach ['spɪnɪdʒ] *n.* Spinat, *der*

spinal ['spaɪnl] *adj. (Anat.)* Wirbelsäulen-; Rückgrat[s]-

spinal: ~ 'column *n.* Wirbelsäule, *die;* **~ 'cord** *n.* Rückenmark, *das*

spindle ['spɪndl] *n.* Spindel, *die*

spindly ['spɪndlɪ] *adj.* spindeldürr ‹*Person, Beine, Arme*›

spin-'drier *n.* Wäscheschleuder, *die*

spin-'dry *v. t.* schleudern

spine [spaɪn] *n.* **a)** *(backbone)* Wirbelsäule, *die;* **b)** *(Bot., Zool.)* Stachel, *der;* **c)** *(of book)* Buchrücken, *der*

spine: ~-chiller *n.* Schocker, *der (ugs.);* **~-chilling** *adj.* gruselig

spineless ['spaɪnlɪs] *adj. (fig.)* rückgratlos

spinney ['spɪnɪ] *n. (Brit.)* Gehölz, *das*

spinning: ~-top *n.* Kreisel, *der;* **~-wheel** *n.* Spinnrad, *das*

'spin-off *n.* Nebenprodukt, *das*

spinster ['spɪnstə(r)] *n.* **a)** ledige Frau; Junggesellin, *die;* **b)** *(derog.: old maid)* alte Jungfer *(abwertend)*

spiny ['spaɪnɪ] *adj.* dornig; stachelig

spiral ['spaɪrl] **1.** *adj.* spiralförmig; spiralig; **~ spring** Spiralfeder, *die.* **2.** *n.* Spirale, *die;* **the ~ of rising prices and wages** die Lohn-Preis-Spirale. **3.** *v. i. (Brit.)* **-ll-** ‹*Weg:*› sich hochwinden; ‹*Kosten, Profite:*› in die Höhe klettern; ‹*Rauch:*› in einer Spirale aufsteigen

spiral 'staircase *n.* Wendeltreppe, *die*

spire [spaɪə(r)] *n.* Turmspitze, *die*

spirit ['spɪrɪt] **1.** *n.* **a)** *in pl. (distilled liquor)* Spirituosen *Pl.;* **b)** *(mental attitude)* Geisteshaltung, *die;* **in the right/wrong ~:** mit der richtigen/falschen Einstellung; **take sth. in the wrong ~:** etw. falsch auffassen; **take sth. in the ~ in which it is meant** etw. so auffassen, wie es gemeint ist; **enter into the ~ of sth.** innerlich bei einer Sache [beteiligt] sein *od.* dabeisein; **c)** *(courage)* Mut, *der;* **d)** *(vital principle, soul, inner qualities)* Geist, *der;* **in ~:** innerlich; im Geiste; **be with sb. in ~:** in Gedanken od. im Geist[e] bei jmdm. sein; **e)** *(real meaning)* Geist, *der;* Sinn, *der;* **f)** *(mental tendency)* Geist, *der; (mood)* Stimmung, *die;* **the ~ of the age** *or* **times** der Zeitgeist; **g) high ~s** gehobene Stimmung; gute Laune; **in poor** *or* **low ~s** niedergedrückt; **h)** *(liquid obtained by distillation)* Spiritus, *der; (purified alcohol)* reiner Alkohol. **2.** *v. t.* **~ away, ~ off** verschwinden lassen

spirited ['spɪrɪtɪd] *adj.* **a)** beherzt ‹*Angriff, Versuch, Antwort, Verteidigung*›; lebhaft ‹*Antwort*›; **b) low-/proud-~:** niedergedrückt/stolz; **high-~:** ausgelassen; temperamentvoll ‹*Pferd*›; **mean-~:** gemein

'spirit-level *n.* Wasserwaage, *die*

spiritual ['spɪrɪtʃʊəl] **1.** *adj.* spirituell *(geh.);* **his ~ home** seine geistige Heimat. **2.** *n.* **[Negro] ~:** [Negro] Spiritual, *das*

spiritualism ['spɪrɪtʃʊəlɪzm] *n.* Spiritismus, *der*

spiritualist ['spɪrɪtʃʊəlɪst] *n.* Spiritist, *der*/Spiritistin, *die*

[1]spit [spɪt] **1.** *v. i.*, **-tt-, spat** [spæt] *or* **spit a)** spucken; **he spat in his enemy's face** er spuckte seinem Feind ins Gesicht; **b)** *(make angry noise)* fauchen; **~ at sb.** jmdn. anfauchen; **c)** *(rain lightly)* **~ [down]** tröpfeln *(ugs.).* **2.** *v. t.*, **-tt-, spat** *or* **spit a)** spucken; **b)** *(fig.: utter angrily)* **~ defiance at sb.** jmdn. trotzig anfauchen. **3.** *n.* Spucke, *die;* **~ and polish** *(cleaning work)* Putzen und Reinigen; Wienern, *das*

~ 'out *v. t.* ausspucken; **she spat out the words** sie spuckte die Worte nur so aus; **~ it out!** *(fig. coll.)* spuck es aus! *(ugs.)*

[2]spit *n.* **a)** *(point of land)* Halbinsel, *die;* **b)** *(reef)* Riff, *das; (shoal)* Untiefe, *die; (sandbank)* Sandbank, *die;* **c)** *(for roasting meat)* Spieß, *der*

spite [spaɪt] **1.** *n.* **a)** *(malice)* Boshaftigkeit, *die;* **b) in ~ of** trotz; **in ~ of oneself** obwohl man es eigentlich nicht will. **2.** *v. t.* ärgern; **cut off one's nose to ~ one's face** sich *(Dat. od. Akk.)* ins eigene Fleisch schneiden

spiteful ['spaɪtfl] *adj.*, **spitefully** ['spaɪtfəlɪ] *adv.* boshaft; gehässig *(abwertend)*

spitting 'image *n.* **be the ~ of sb.** jmdm. wie aus dem Gesicht geschnitten sein

spittle ['spɪtl] *n.* Spucke, *die;* Speichel, *der*

spiv [spɪv] *n. (Brit. sl.)* **a)** *(person living by his wits)* smarter kleiner Geschäftemacher; **b)** *(black-market dealer)* Schwarzhändler, *der;* Schieber, *der (ugs.)*

splash [splæʃ] **1.** *v.t.* **a)** spritzen; ~ **sb./sth. with sth.** jmdn./etw. mit etw. bespritzen; **b)** *(Journ.)* als Aufmacher bringen ‹*Story usw.*›. **2.** *v.i.* **a)** *(fly about in drops)* spritzen; **b)** *(cause liquid to fly about)* [umher]spritzen; **c)** *(move with ~ing)* platschen *(ugs.)*. **3.** *n.* **a)** Spritzen, *das;* **hit the water with a ~:** ins Wasser platschen *(ugs.)*; **make a [big] ~** *(fig.)* Furore machen; **b)** *(liquid)* Spritzer, *der;* **c)** *(noise)* Plätschern, *das*

~ **a'bout** *v.i.* herumspritzen *(ugs.)*; [herum]planschen

~ **'out** *v.i. (coll.)* ~ **out on sth.** für etw. umbekümmert Geld ausgeben

splay [spleɪ] **1.** *v.t.* **a)** *(spread)* ~ **[out]** spreizen; **b)** *(construct with divergent sides)* ausschrägen. **2.** *v.i.* ‹*Linien:*› [schräg] auseinanderlaufen; ‹*Tischbeine, Stuhlbeine:*› schräg nach außen gehen; ‹*Finger, Zehen:*› gespreizt sein

spleen [spli:n] *n.* Milz, *die*

splendid ['splendɪd] *adj. (excellent, outstanding)* großartig; *(beautiful)* herrlich; *(sumptuous, magnificent)* prächtig

splendour (*Brit.; Amer.:* **splendor**) ['splendə(r)] *n.* **a)** *(magnificence)* Pracht, *die;* **b)** *(brightness)* Glanz, *der*

splice [splaɪs] *v.t.* **a)** *(join ends of by interweaving)* verspleißen *(Seemannsspr.)*; **b)** *(join in overlapping position)* [an den Enden überlappend] zusammenfügen; zusammenkleben ‹*Filmstreifen usw.*›

splint [splɪnt] *n.* Schiene, *die;* **put sb.'s arm in a ~:** jmds. Arm schienen

splinter ['splɪntə(r)] *n.* Splitter, *der*

'splinter group *n.* Splittergruppe, *die*

split [splɪt] **1.** *n.* **a)** *(tear)* Riß, *der;* **b)** *(division into parts)* [Auf]teilung, *die;* **c)** *(fig.: rift)* Spaltung, *die;* **a ~ between Moscow and her allies** ein Bruch zwischen Moskau und seinen Verbündeten; **d)** *(Gymnastics, Skating)* **the ~s** *or (Amer.)* **~:** Spagat, *der od. das;* **do the ~s** Spagat machen. **2.** *adj.* gespalten; **be ~ on a question** [sich *(Dat.)*] in einer Frage uneins sein. **3.** *v.t.*, **-tt-, split a)** *(tear)* zerreißen; **b)** *(divide)* teilen; spalten ‹*Holz*›; ~ **persons/things into groups** Personen/Dinge in Gruppen *(Akk.)* aufteilen *od.* einteilen; ~ **the difference** sich in der Mitte treffen; ~ **hairs** *(fig.)* Haare spalten; **c)** *(divide into disagreeing parties)* spalten; **d)** *(remove by breaking)* ~ **[off** *or* **away]** abbrechen. **4.** *v.i.*, **-tt-, split a)** *(break into parts)* ‹*Holz:*› splittern; ‹*Stoff, Seil:*› reißen; **b)** *(divide into parts)* sich teilen; ‹*Gruppe:*› sich spalten; ‹*zwei Personen:*› sich trennen; **c)** *(be removed by breaking)* ~ **from** absplittern von; ~ **apart** zersplittern; **d)** *(sl.: depart)* abhauen *(ugs.)*

~ **a'way** *v.i.* absplittern; ~ **away from** absplittern von; ‹*Parteiflügel, Gruppierung:*› sich abspalten von

~ **'off 1.** *v.t.* abspalten. **2.** *v.i. see* ~ **away**

~ **on** *v.t. (sl.)* ~ **on sb. [to sb.]** jmdn. [bei jmdm.] verpfeifen *(ugs.)*

~ **'open 1.** *v.i.* aufbrechen. **2.** *v.t.* öffnen ‹*Nuß, Schote*›; **he ~ his head open** er hat sich *(Dat.)* den Kopf aufgeschlagen

~ **'up 1.** *v.t.* aufteilen. **2.** *v.i. (coll.)* sich trennen; ~ **up with sb.** sich von jmdm. trennen; mit jmdm. Schluß machen *(ugs.)*

split: ~ **in'finitive** *n. (Ling.) Konstruktion im Englischen, bei der zwischen Infinitivkonjunktion und Infinitiv ein Adverb eingeschoben wird;* **~-level** *adj.* mit Zwischengeschoß; auf zwei Ebenen; ~ **perso'nality** *n.* gespaltene Persönlichkeit *(Psych.)*; ~ **'second** *n.* **in a ~ second** im Bruchteil einer Sekunde; **~-second timing** [zeitliche] Abstimmung auf die Sekunde genau

splitting ['splɪtɪŋ] *adj.* **a ~ headache** rasende Kopfschmerzen

splutter ['splʌtə(r)] **1.** *v.i.* ‹*Feuer, Gaslampe:*› flackern; ‹*Fett:*› spritzen; ‹*Person:*› prusten; ‹*Motor:*› stottern; ~ **with rage/indignation** vor Wut/Entrüstung schnauben. **2.** *v.t.* stottern ‹*Worte*›

spoil [spɔɪl] **1.** *v.t.*, **~t** [spɔɪlt] *or* **~ed a)** *(impair)* verderben; ruinieren ‹*Leben*›; **the news ~t his dinner/evening** die Nachricht verdarb ihm das Essen/den Abend; **~t ballot papers** ungültige Stimmzettel; **b)** *(injure character of)* verderben *(geh.)*; verziehen ‹*Kind*›; ~ **sb. for sth.** jmdn. für etw. zu anspruchsvoll machen; **c)** *(pamper)* verwöhnen; **be ~t for choice** die Qual der Wahl haben. **2.** *v.i.*, **~t** *or* **~ed a)** *(go bad)* verderben; **b) be ~ing for a fight/for trouble** Streit/Ärger suchen. **3.** *n. (plunder)* **~[s]** Beute, *die*

'spoil-sport *n.* Spielverderber, *der*/-verderberin, *die*

spoilt 1. *see* **spoil 1, 2. 2.** *adj.* verzogen ‹*Kind*›

[1]**spoke** [spəʊk] *n.* Speiche, *die;* **put a ~ in sb.'s wheel** *(fig.)* jmdm. einen Knüppel zwischen die Beine werfen

[2]**spoke, spoken** *see* **speak**

spokesman ['spəʊksmən] *n., pl.* **spokesmen** ['spəʊksmən] Sprecher, *der*

spokesperson ['spəʊkspɜ:sn] *n.* Sprecher, *der*/Sprecherin, *die*

spokeswoman ['spəʊkswʊmən] *n.* Sprecherin, *die*

sponge [spʌndʒ] **1.** *n.* **a)** Schwamm, *der;* **b)** *see* **sponge-cake; sponge pudding. 2.** *v.t.* **a)** *see* **cadge 1; b)** *(wipe)* mit einem Schwamm waschen

~ **off** *v.t.* **a)** [-'-] *(wipe off)* mit einem Schwamm abwischen; *(wash off)* mit einem Schwamm abwaschen; **b)** ['--] *see* ~ **on**

~ **on** *v.t.* ~ **on sb.** bei *od.* von jmdm. schnorren

sponge: **~-bag** *n. (Brit.)* Kulturbeutel, *der;* **~-cake** *n.* Biskuitkuchen, *der;* ~ **'pudding** *n.* Schwammpudding, *der (Kochk.)*

sponger ['spʌndʒə(r)] *n.* Schmarotzer, *der*/Schmarotzerin, *die;* Schnorrer, *der*/Schnorrerin, *die*

spongy ['spʌndʒɪ] *adj.* schwammig

sponsor ['spɒnsə(r)] **1.** *n. (firm paying for event, one donating to charitable event)* Sponsor, *der.* **2.** *v.t.* **a)** sponsern ‹*Wohlfahrtsverband, Teilnehmer, Programm, Veranstaltung*›; **b)** *(support in election)* unterstützen ‹*Kandidaten*›; ~ **sb.** jmds. Kandidatur unterstützen

sponsored ['spɒnsəd] *adj.* gesponsert; finanziell gefördert; ~ **run** *als Wohltätigkeitsveranstaltung durchgeführter Dauerlauf mit gesponserten Teilnehmern*

sponsorship ['spɒnsəʃɪp] *n.* **a)** *(financial support)* Sponsorschaft, *die;* **b)** *(support of candidate)* Unterstützung, *die;* **the party's ~ of sb.** die Unterstützung von jmds. Kandidatur durch die Partei

spontaneous [spɒn'teɪnɪəs] *adj.* spontan; **make a ~ offer of sth.** spontan etw. anbieten

spontaneous com'bustion *n.* Selbstentzündung, *die*

spoof [spu:f] *(coll.) n.* Veralberung, *die* (**of, on** von); Parodie, *die* (**of, on** auf + *Akk.*)

spook [spu:k] *n. (joc.)* Geist, *der;* Gespenst, *das*

spooky ['spu:kɪ] *adj.* gespenstisch

spool [spu:l] *n.* Spule, *die*

spoon [spu:n] *n.* Löffel, *der;* **be born with a silver ~ in one's mouth** mit einem goldenen *od.* silbernen Löffel im Mund geboren werden

spoonerism ['spu:nərɪzm] *n. witziges Vertauschen der Anfangsbuchstaben o.ä. von zwei oder mehr Wörtern (wie bei „Leichenzehrer" für „Zeichenlehrer")*

'**spoon-feed** *v.t.* mit dem Löffel füttern; **~ sb.** *(fig.)* jmdm. alles vorkauen *(ugs.)*

spoonful ['spu:nfʊl] *n.* **a ~ of sugar** ein Löffel [voll] Zucker

sporadic [spə'rædɪk] *adj.* sporadisch; vereinzelt ‹*Schauer, Schüsse*›

spore [spɔ:(r)] *n.* Spore, *die*

sport [spɔ:t] **1.** *n.* **a)** *(pastime)* Sport, *der;* **~s** Sportarten; **team/winter/water/indoor ~:** Mannschafts- / Winter- / Wasser- / Hallensport, *der;* **b)** *no pl., no art. (collectively)* Sport, *der;* **go in for ~, do ~:** Sport treiben; **c)** *in pl. (Brit.)* **[athletic] ~s** Athletik, *die;* **S~s Day** *(Sch.)* Sportfest, *das;* **d)** *no pl., no art. (fun)* Spaß, *der;* **do/say sth. in ~:** etw. im *od.* zum Scherz tun/sagen; **e)** *(coll.: easy-going person)* **be a [real] ~:** ein prima Kerl sein *(ugs.);* [schwer] in Ordnung sein *(ugs.);* **be a ~!** sei kein Spielverderber!; **Aunt Joan is a real ~:** Tante Joan ist echt *(ugs.)* in Ordnung; **be a good/bad ~** *(in games)* ein guter/schlechter Verlierer sein. **2.** *v.t.* stolz tragen ‹*Kleidungsstück*›

sporting ['spɔ:tɪŋ] *adj.* **a)** *(interested in sport)* sportlich; **b)** *(generous)* großzügig; *(fair)* fair; anständig; **give sb. a ~ chance** jmdm. eine [faire] Chance geben; **c)** *(relating to sport)* Sport-

sports: ~ car *n.* Sportwagen, *der;* **~ field** *n.* Sportplatz, *der;* **~ jacket** *n.* sportlicher Sakko; **~man** ['spɔ:tsmən] *n., pl.* **~men** ['spɔ:tsmən] Sportler, *der*

sportsmanship ['spɔ:tsmənʃɪp] *n., no pl.* **a)** *(fairness)* [sportliche] Fairneß; **b)** *(skill)* sportliche Leistung

sports: ~ programme *n. (Radio, Telev.)* Sportsendung, *die;* **~wear** *n., no pl.* Sport[be]kleidung, *die;* **~woman** *n.* Sportlerin, *die*

sporty ['spɔ:tɪ] *adj.* **a)** *(coll.: sport-loving)* sportlich; **the whole family is ~:** die ganze Familie ist sportbegeistert; **b)** *(jaunty)* sportlich ‹*Aussehen*›; **wear one's hat at a ~ angle** seinen Hut flott aufgesetzt haben

spot [spɒt] **1.** *n.* **a)** *(precise place)* Stelle, *die;* **on this ~:** an dieser Stelle; **on the ~** *(fig.) (instantly)* auf der Stelle; **be on the ~** *(be present)* zur Stelle sein; **be in/get into/get out of a [tight] ~** *(fig. coll.)* in der Klemme sitzen/in die Klemme geraten/sich aus einer brenzligen Lage befreien *(ugs.);* **put sb. on the ~** *(fig. coll.: cause difficulties for sb.)* jmdn. in Verlegenheit bringen; **b)** *(inhabited place)* Ort, *der;* **a nice ~ on the Moselle** ein hübscher Flecken an der Mosel; **c)** *(suitable area)* Platz, *der;* **holiday/sun ~:** Ferienort, *der*/Ferienort [mit Sonnengarantie]; **picnic ~:** Picknickplatz, *der;* **d)** *(dot)* Tupfen, *der;* Tupfer, *der; (larger)* Flecken, *der;* **knock ~s off sb.** *(fig. coll.)* jmdn. in die Pfanne hauen *(ugs.);* **see ~s before one's eyes** Sterne sehen *(ugs.);* **e)** *(stain)* **~ [of blood/grease/ink]** [Blut-/Fett-/Tinten]fleck, *der;* **f)** *(Brit. coll.: small amount)* **do a ~ of work/sewing** ein bißchen arbeiten/nähen; **how about a ~ of lunch?** wie wär's mit einem Bissen zu Mittag?; **have** *or* **be in a ~ of bother** *or* **trouble** etwas Ärger haben; **g)** *(drop)* **a ~** *or* **a few ~s of rain** ein paar Regentropfen; **h)** *(area on body)* [Körper]stelle, *die;* **have a weak ~** *(fig.)* eine Schwachstelle haben; *see also* **sore 1a; i)** *(Telev. coll.: position in programme)* Sendezeit, *die;* **the 7 o'clock ~:** das Siebenuhrprogramm; **j)** *(Med.)* Pickel, *der;* **heat ~:** Hitzebläschen, *das;* **break out in ~s** Ausschlag bekommen; **k)** *(on dice, dominoes)* Punkt, *der;* **l)** *(spotlight)* Spot, *der.* **2.** *v.t.,* **-tt-: a)** *(detect)* entdecken; identifizieren ‹*Verbrecher*›; erkennen ‹*Gefahr*›; **b)** *(take note of)* erkennen ‹*Flugzeugtyp, Vogel, Talent*›; **go train-/plane-~ting** Zug-/Flugzeugtypen bestimmen; **c)** *(coll.: pick out)* tippen auf (+ *Akk.*) *(ugs.)* ‹*Sieger, Gewinner usw.*›; **d)** *(stain)* beflecken; *(with ink or paint)* beklecksen; *(with mud)* beschmutzen

spot: ~ '**check** *n. (test made immediately)* sofortige Überprüfung (**on** *Gen.*); *(test made on randomly selected subject)* Stichprobe, *die;* **~-check** *v.t.* stichprobenweise überprüfen; **~ lamp** *n.* Spotlight, *das*

spotless ['spɒtlɪs] *adj.* **a)** *(unstained)* fleckenlos; **her house is absolutely ~** *(fig.)* ihr Haus ist makellos sauber; **b)** *(fig.: blameless)* mustergültig; untadelig ‹*Charakter*›

spot: ~light 1. *n.* **a)** *(Theatre)* [Bühnen]scheinwerfer, *der;* **b)** *(Motor Veh.)* Scheinwerfer, *der;* **c)** *(fig.: attention)* **be in the ~light** im Rampenlicht [der Öffentlichkeit] stehen; **keep out of the ~light** sich von der Öffentlichkeit fernhalten; **2.** *v.t.,* **~lighted** *or* **~lit a)** *(Theatre)* [mit dem Scheinwerfer] anstrahlen; **b)** *(fig.: highlight)* in den Blickpunkt der Öffentlichkeit bringen; **~-'on** *(coll.)* **1.** *adj.* goldrichtig *(ugs.);* **I was ~-on** ich lag genau richtig *(ugs.);* **your estimate was ~-on** mit deiner Schätzung hast du ins Schwarze getroffen; **2.** *adv.* haargenau *(ugs.)*

spotted ['spɒtɪd] *adj.* **a)** gepunktet; **b)** *(Zool.)* **~ woodpecker/hyena** Buntspecht, *der*/Tüpfelhyäne, *die*

spotty ['spɒtɪ] *adj.* **a)** *(spotted)* gefleckt; *(stained)* fleckig; **b)** *(pimply)* picklig; **be ~:** viele Pickel haben

spouse [spaʊz, spaʊs] *n.* [Ehe]gatte, *der*/-gattin, *die; (joc.)* Angetraute, *der/die;* Gemahl, *der*/Gemahlin, *die*

spout [spaʊt] **1.** *n.* Schnabel, *der; (of water-pump)* [Auslauf]rohr, *das; (of tap)* Ausflußrohr, *das;* **be up the ~** *(sl.: ruined)* im Eimer sein *(ugs.).* **2.** *v.t.* **a)** *(discharge)* ausstoßen ‹*Wasser, Lava, Öl*›; **b)** *(declaim)* deklamieren ‹*Verse*›; *(rattle off)* herunterrasseln *(ugs.)* ‹*Zahlen, Fakten usw.*›; **~ nonsense** Unsinn verzapfen *(ugs.).* **3.** *v.i.* **a)** *(gush)* schießen (**from** aus); **b)** *(declaim)* schwadronieren *(abwertend);* schwafeln *(ugs. abwertend)*

~ '**out** *v.i.* herausströmen; **~ out of sth.** aus etw. strömen

sprain [spreɪn] **1.** *v.t.* verstauchen; **~ one's ankle/wrist** sich *(Dat.)* den Knöchel/das Handgelenk verstauchen. **2.** *n.* Verstauchung, *die*

sprang *see* **spring 2, 3**

sprawl [sprɔ:l] **1.** *n.* **a)** *(slump)* **lie in a ~:** ausgestreckt [da]liegen; **b)** *(straggle)* verstreute Ansammlung. **2.** *v.i.* **a)** *(spread oneself)* sich ausstrecken; **b)** *(fall)* der Länge nach hinfallen; **c)** *(straggle)* sich ausbreiten

sprawling ['sprɔ:lɪŋ] *attrib. adj.* **a)** *(extended)* ausgestreckt [liegend]; **b)** *(falling)* der Länge nach hinfallend; **c)** *(straggling)* verstreut liegend ‹*Gebäude*›; wuchernd ‹*Großstadt*›

[1]**spray** [spreɪ] *n.* **a)** *(bouquet)* Strauß, *der;* **b)** *(branch)* Zweig, *der; (of palm or fern)* Wedel, *der*

[2]**spray 1.** *v. t.* **a)** *(in a stream)* spritzen; *(in a mist)* sprühen ‹*Parfum, Farbe, Spray*›; **they ~ed the general's car with bullets** sie durchsiebten den Wagen des Generals mit Kugeln; **b)** *(treat)* besprühen ‹*Haar, Haut, Pflanze*›; spritzen ‹*Nutzpflanzen*›. **2.** *v. i.* spritzen. **3.** *n.* **a)** *(drops)* Sprühnebel, *der;* **b)** *(liquid)* Spray, *der od. das;* **c)** *(container)* Spraydose, *die; (in gardening)* Spritze, *die;* **hair/throat ~:** Haar-/Rachenspray, *der od. das*

~ on [to] *v. t.* **~ sth. on [to] sth.** etw. mit etw. besprühen

'**spray can** *n.* Spraydose, *die*

spread [spred] **1.** *v. t.*, **spread a)** ausbreiten ‹*Tuch, Landkarte*› (**on** auf + *Dat.*); streichen ‹*Butter, Farbe, Marmelade*›; **b)** *(cover)* **~ a roll with marmalade/butter** ein Brötchen mit Marmelade/Butter bestreichen; **the sofa was ~ with a blanket** auf dem Sofa lag eine Decke [ausgebreitet]; **c)** *(fig.: display)* **a magnificent view was ~ before us** uns *(Dat.)* bot sich eine herrliche Aussicht; **d)** *(extend range of)* verbreiten; **e)** *(distribute)* verteilen; *(untidily)* verstreuen; streuen ‹*Dünger*›; verbreiten ‹*Zerstörung, Angst, Niedergeschlagenheit*›; **f)** *(make known)* verbreiten; **~ the word** *(tell news)* es weitersagen; **g)** *(separate)* ausbreiten ‹*Arme*›. **2.** *v. i.*, **spread a)** sich ausbreiten; **a smile ~ across** *or* **over his face** ein Lächeln breitete sich *(geh.)* über sein Gesicht; **margarine ~s easily** Margarine läßt sich leicht streichen; **~ like wildfire** sich in *od.* mit Windeseile verbreiten; **b)** *(scatter, disperse)* sich verteilen; **the odour ~s through the room** der Geruch breitet sich im ganzen Zimmer aus; **c)** *(circulate)* ‹*Neuigkeiten, Gerücht, Kenntnis usw.*:› sich verbreiten. **3.** *n.* **a)** *(expanse)* Fläche, *die;* **b)** *(span) (of tree)* Kronendurchmesser, *der; (of wings)* Spann[weite], *die;* **c)** *(breadth)* **have a wide ~** ‹*Interessen, Ansichten*:› breit gefächert sein; **d)** *(extension)* Verbreitung, *die; (of city, urbanization, poverty)* Ausbreitung, *die;* **e)** *(diffusion)* Ausbreitung, *die; (of learning, knowledge)* Verbreitung, *die;* Vermittlung, *die;* **f)** *(distribution)* Verteilung, *die;* **g)** *(coll.: meal)* Festessen, *das;* **h)** *(paste)* Brotaufstrich, *der;* ‹*Rindfleisch-, Lachs*›paste, *die;* ‹*Käse-, Erdnuß-, Schokoladen*›krem, *die*

~ a'bout, ~ a'round *v. t.* **a)** verbreiten ‹*Neuigkeiten, Gerücht usw.*›; **b)** *(strew)* verstreuen

~ 'out 1. *v. t.* **a)** *(extend)* ausbreiten ‹*Arme*›; **b)** *(space out)* verteilen ‹*Soldaten, Tänzer, Pfosten*›; legen ‹*Karten*›; ausbreiten ‹*Papiere*›. **2.** *v. i.* sich verteilen; ‹*Soldaten*:› ausschwärmen

'**spreadsheet** *n. (Computing)* Arbeitsblatt, *das*

spree [spri:] *n.* **a)** *(spending ~)* Einkaufsorgie, *die (ugs.);* **go on a shopping ~:** ganz groß einkaufen gehen; **b) be/go out on the ~** *(coll.)* einen draufmachen *(ugs.)*

sprig [sprɪg] *n.* Zweig, *der*

sprightly ['spraɪtlɪ] *adj.* munter

spring [sprɪŋ] **1.** *n.* **a)** *(season)* Frühling, *der;* **in ~ 1969, in the ~ of 1969** im Frühjahr 1969; **in early/late ~:** zu Anfang/Ende des Frühjahrs; **last/next ~:** letzten/nächsten Frühling; **in [the] ~:** im Frühling *od.* Frühjahr; **b)** *(source, lit. or fig.)* Quelle, *die;* **c)** *(Mech.)* Feder, *die;* **~s** *(vehicle suspension)* Federung, *die;* **d)** *(jump)* Sprung, *der;* **make a ~ at sb./at an animal** sich auf jmdn./ein Tier stürzen; **e)** *(elasticity)* Elastizität, *die;* **walk with a ~ in one's step** mit beschwingten Schritten gehen. **2.** *v. i.*, **sprang** [spræŋ] *or (Amer.)* **sprung** [sprʌŋ], **sprung a)** *(jump)* springen; **~ [up] from sth.** von etw. aufspringen; **~ to one's feet** aufspringen; **~ to sb.'s assistance/defence** jmdm. beispringen; **~ to life** *(fig.)* [plötzlich] zum Leben erwachen; **b)** *(arise)* entspringen (**from** *Dat.*); ‹*Saat, Hoffnung*:› keimen; **~ to fame** über Nacht bekannt werden; **~ to mind** jmdm. einfallen; **c)** *(recoil)* **~ back into position** zurückschnellen; **~ to** *or* **shut** ‹*Tür, Falle, Deckel*:› zuschnappen. **3.** *v. t.*, **sprang** *or (Amer.)* **sprung, sprung a)** *(make known suddenly)* **~ a new idea on sb.** jmdn. mit einer neuen Idee überfallen; **~ a surprise on sb.** jmdn. überraschen; **b)** aufspringen lassen ‹*Schloß*›; zuschnappen lassen ‹*Falle*›; **c)** *(sl.: set free)* herausholen (**from** aus)

~ 'back *v. i.* zurückschnellen

~ from *v. t.* **a)** *(appear from)* [plötzlich] herkommen; **b)** *(originate from)* herrühren von; ‹*Person*:› abstammen von

~ 'up *v. i.* ‹*Wind, Zweifel*:› aufkommen; ‹*Gebäude*:› aus dem Boden wachsen; ‹*Pflanze*:› aus dem Boden schießen; ‹*Organisation, Freundschaft*:› entstehen

spring: ~ 'binder *n.* Klemmappe, *die;* **~board** *n. (Sport; also fig.)* Sprungbrett, *das; (in circus)* Schleuderbrett, *das;* **~ 'chicken** *n.* **a)** *(fowl)* junges Hähnchen; **b)** *(fig.: person)* **be no ~ chicken** nicht mehr der/die Jüngste sein *(ugs.);* **~-'clean 1.** *n.* [großer] Hausputz; *(in spring)* Frühjahrsputz, *der;* **2.** *v. t.* **~-clean [the whole house]** [großen] Hausputz/Frühjahrsputz machen; **~-loaded** *adj.* mit Sprungfeder *nachgestellt;* **~ 'onion** *n.* Frühlingszwiebel, *die;* **~ 'tide** *n.* Springflut, *die;* **~time** *n.* Frühling, *der*

springy ['sprɪŋɪ] *adj.* elastisch; federnd ‹*Schritt, Brett, Boden*›

sprinkle ['sprɪŋkl] *v. t.* streuen; sprengen ‹*Flüssigkeit*›; **~ sth. over/on sth.** etw. über/auf etw. *(Akk.)* streuen/sprengen; **~ sth. with sth.** etw. mit etw. bestreuen/besprengen

sprinkler ['sprɪŋklə(r)] *n.* **a)** *(Hort.: for watering)* Sprinkler, *der; (Agric.)* Regner, *der;* **b)** *(fire extinguisher)* **~s, ~ system** Sprinkleranlage, *die*

sprinkling ['sprɪŋklɪŋ] *n.* **a ~ of snow/sugar/dust** eine dünne Schneedecke / Zucker- / Staubschicht; **there was only a ~ of holidaymakers on the beach** nur ein paar vereinzelte Urlauber waren am Strand

sprint [sprɪnt] **1.** *v. t. & i.* rennen; sprinten *(bes. Sport);* spurten *(bes. Sport).* **2.** *n.* Sprint, *der (bes. Sport);* **the hundred-metres ~:** der Hundertmeterlauf

sprinter ['sprɪntə(r)] *n.* Sprinter, *der*/Sprinterin, *die*

spritzer ['sprɪtsə(r)] *n. (Amer.)* Schorle, *die*

sprout [spraʊt] **1.** *n.* **a)** *in pl. (coll.) see* **Brussels sprouts; b)** *(Bot.) see* **shoot 3a. 2.** *v. i.* **a)** *(lit. or fig.)* sprießen *(geh.);* **b)** *(grow)* emporschießen; **c)** *(fig.)* ‹*Gebäude*:› wie Pilze aus dem Boden schießen. **3.** *v. t.* [aus]treiben ‹*Blüten, Knospen*›; sich *(Dat.)* wachsen lassen ‹*Bart*›

spruce [spru:s] **1.** *adj.* gepflegt; **look ~:** adrett aussehen. **2.** *n. (Bot.)* Fichte, *die.* **3.** *v. t.* **~ up** verschönern; **~ the house up** das [ganze] Haus aufräumen und putzen; **get ~d up** sich feinmachen *(ugs.)*

sprung [sprʌŋ] **1.** *see* **spring 2, 3. 2.** *attrib. adj.* gefedert

spry [spraɪ] *adj.* rege

spud [spʌd] *n. (Brit. sl.: potato)* Kartoffel, *die*

spun *see* **spin 1, 2**

spur [spɜ:(r)] **1.** *n.* **a)** Sporn, *der;* **b)** *(fig.: stimulus)* Ansporn, *der* (**to** für); **on the ~ of the moment** ganz spontan. **2.** *v. t.*, **-rr-: a)** die Sporen geben (+ *Dat.*) ‹*Pferd*›; **b)** *(fig.: incite)* anspornen; **~ sb. [on] to sth./to do sth.** jmdn. zu etw. anspornen/anspornen, etw. zu tun; **c)** *(fig.: stimulate)* hervorrufen; in Gang setzen ‹*Aktivität*›; erregen ‹*Interesse*›

spurious ['spjʊərɪəs] *adj.* unaufrichtig ‹*Charakter, Verhalten*›; gespielt ‹*Gefühl, Interesse*›; zweifelhaft ‹*Anspruch, Vergnügen*›; falsch ‹*Name, Münze*›

spurn [spɜ:n] *v. t.* zurückweisen; abweisen; ausschlagen ‹*Angebot, Gelegenheit*›

[1]**spurt** [spɜ:t] **1.** *n.* Spurt, *der (bes. Sport);* **final ~:** Endspurt, *der;* **there was a ~ of activity** es brach kurzzeitig lebhafte Aktivität aus; **in a sudden ~ of energy** in einem plötzlichen Anfall von Energie; **put on a ~:** einen Spurt einlegen. **2.** *v. i.* spurten *(bes. Sport)*

[2]**spurt 1.** *v. i.* **~ out [from** *or* **of]** herausspritzen [aus]; **~ from** spritzen aus. **2.** *v. t.* **the wound ~ed blood** aus der Wunde spritzte Blut. **3.** *n.* Strahl, *der*

spy [spaɪ] **1.** *n.* **a)** *(secret agent)* Spion, *der*/Spionin, *die;* **b)** *(watcher)* Spion, *der*/Spionin, *die;* Schnüffler, *der*/Schnüfflerin, *die (abwertend);* **~ in the sky/cab** *(coll.)* Spionagesatellit, *der*/Fahrt[en]schreiber, *der.* **2.** *v. t. (literary)* ausmachen. **3.** *v. i. (watch closely)* [herum]spionieren; *(practise espionage)* Spionage treiben; **~ on sb./a country** jmdm. nachspionieren/gegen ein Land spionieren

~ 'out *v. t.* aufspüren; ausspionieren ‹*Feind, feindliche Stellung*›; **~ out the land** *(lit. or fig.)* die Lage erkunden

spy: ~-ring *n.* Spionagering, *der;* **~ satellite** *n.* Spionagesatellit, *der*

sq., Sq. *abbr.* **square, Square**

squabble ['skwɒbl] **1.** *n.* Streit, *der;* **petty ~s** kleine Streitereien; **have a ~ [with sb. about sth.]** sich [mit jmdm. wegen einer Sache] streiten. **2.** *v. i.* sich zanken (**over, about** wegen)

squad [skwɒd] *n.* **a)** *(Mil.)* Gruppe, *die;* Trupp, *der;* **b)** *(group)* Mannschaft, *die;* **c)** *(Police)* **Drug/Fraud S~:** Rauschgift-/Betrugsdezernat, *das*

squadron ['skwɒdrən] *n.* **a)** *(Mil.) (of tanks)* Bataillon, *das; (of cavalry)* Schwadron, *die;* **b)** *(Navy)* Geschwader, *das;* **c)** *(Air Force)* Staffel, *die*

squalid ['skwɒlɪd] *adj.* **a)** *(dirty)* [abstoßend] schmutzig; **b)** *(poor)* schäbig; armselig; **c)** *(fig.: sordid)* abstoßend

squall [skwɔ:l] *n. (gust)* Bö, *die*

squalor ['skwɒlə(r)] *n., no pl.* Schmutz, *der;* **live in ~:** in Schmutz und Elend leben

squander ['skwɒndə(r)] *v. t.* vergeuden ‹*Talent, Zeit, Geld*›; verschleudern ‹*Ersparnisse, Vermögen*›; nicht nutzen ‹*Chance, Gelegenheit*›

square [skweə(r)] **1.** *n.* **a)** *(Geom.)* Quadrat, *das;* **b)** *(object, arrangement)* Quadrat, *das;* **carpet ~:** Teppichfliese, *die;* **c)** *(on board in game)* Feld, *das;* **be** *or* **go back to ~ one** *(fig. coll.)* wieder von vorn anfangen müssen; **d)** *(open area)* Platz, *der;* **e)** *(scarf)* [quadratisches] Tuch; **silk ~:** Seidentuch, *das;* **f)** *(Mil.: drill area)* Kasernenhof, *der;* **g)** *(Math.: product)* Quadrat, *das;* **h)** *(sl.: old-fashioned person)* Spießer, *der*/Spießerin, *die (abwertend).* **2.** *adj.* **a)** quadratisch; **b) a ~ foot/mile/metre** *etc.* ein Quadratfuß/eine Quadratmeile/ein Quadratmeter *usw.;* **a foot ~:** ein Fuß im Quadrat; **c)** *(right-angled)* rechtwink[e]lig; **~ with** *or* **to** im rechten Winkel zu; **d)** *(stocky)* gedrungen ‹*Statur, Gestalt*›; **e)** *(in outline)* rechteckig; eckig ‹*Schultern, Kinn*›; **f)** *(quits)* quitt *(ugs.);* **be [all] ~:** [völlig] quitt sein *(ugs.);* ‹*Spieler:*› gleich stehen; ‹*Spiel:*› unentschieden stehen. **3.** *adv.* breit ‹*sitzen*›; **put sth. ~ in the middle of sth.** etw. mitten auf etw. *(Akk.)* stellen; **the ball hit him ~ on the head** der Ball traf ihn genau am Kopf. **4.** *v. t.* **a)** *(make right-angled)* rechtwinklig machen; vierkantig zuschneiden ‹*Holz*›; **b)** *(place ~ly)* **~ one's shoulders** seine Schultern straffen; **c)** *(divide into ~s)* in Karos einteilen; **~d paper** kariertes Papier; **d)** *(Math.: multiply)* quadrieren; **3 ~d is 9** 3 [im] Quadrat ist 9; 3 hoch 2 ist 9; **e)** *(reconcile)* **~ sth. with sth.** etw. mit etw. in Einklang bringen; **f) ~ it with sb.** *(coll.: get sb.'s approval)* es mit jmdm. klären. **5.** *v. i. (be consistent)* übereinstimmen; **sth. does not ~ with sth.** etw. steht nicht im Einklang mit etw.; **it just does not ~:** hier stimmt doch etwas nicht

~ 'up *v. i. (settle up)* abrechnen

square: ~ 'brackets *n. pl.* eckige Klammern; **~ 'deal** *n.* faires Geschäft; **get a ~ deal** kein schlechtes Geschäft machen

squarely ['skweəlɪ] *adv.* fest ‹*ansehen*›; genau ‹*treffen*›; aufrecht ‹*sitzen*›

square: ~ 'meal *n.* anständige Mahlzeit *(ugs.);* **~ 'root** *n. (Math.)* Quadratwurzel, *die;* **~-root sign** [Quadrat]wurzelzeichen, *das*

[1]**squash** [skwɒʃ] **1.** *v. t.* **a)** *(crush)* zerquetschen; **~ sth. flat** etw. platt drücken; **b)** *(compress)* pressen; **~ in/up** eindrücken/zusammendrücken ‹*Gegenstand*›; **~ sb./sth. into sth.** jmdn./etw. in etw. *(Akk.)* [hinein]zwängen; **c)** *(put down)* niederschlagen ‹*Aufstand*›; zunichte machen ‹*Hoffnung, Traum*›; **d)** *(coll.: dismiss)* ablehnen ‹*Vorschlag, Plan*›; **e)** *(coll.: silence)* zum Schweigen bringen. **2.** *v. i.* sich quetschen; **~ in** sich hineinquetschen; **we ~ed up** wir drängten uns zusammen. **3.** *n.* **a)** *(Brit.: drink)* Fruchtsaftgetränk, *das;* **orange/lemon ~:** Orangen- / Limonensaftgetränk, *das;* **b)** *(Sport)* **~ [rackets]** Squash, *das*

[2]**squash** *n. (gourd)* [Speise]kürbis, *der*

squat [skwɒt] **1.** *v. i.*, **-tt-: a)** *(crouch)* hocken; *(crouch down)* sich hocken; **b)** *(coll.: sit)* sitzen; *(sit down)* sich setzen; **c)** *(coll.: occupy property) (house)* eine Hausbesetzung machen; *(land)* eine Landbesetzung machen; **~ in a house/on land** ein Haus besetzen/Land besetzen. **2.** *adj.* rundlich; untersetzt

~ 'down *v. i.* sich [nieder]hocken; *(on seat)* sich hinsetzen

squatter ['skwɒtə(r)] *n. (illegal occupier)* Besetzer, *der*/Besetzerin, *die; (of house also)* Hausbesetzer, *der*/-besetzerin, *die*

squaw [skwɔ:] *n.* Squaw, *die*

squawk [skwɔ:k] **1.** *v. i.* ‹*Hahn, Krähe, Rabe:*› krähen; ‹*Huhn:*› kreischen; *(complain)* ‹*Person:*› keifen *(abwertend).* **2.** *n.* **~[s]** *(of crow, cockerel, raven)* Krähen, *das; (of hen)* Kreischen, *das*

squeak [skwi:k] **1.** *n.* **a)** *(of animal)* Quieken, *das;* **b)** *(of hinge, door, brake, shoe, etc.)* Quietschen, *das;* **c)** *(coll.: escape)* **have a narrow ~:** gerade noch [mit dem Leben] davonkommen. **2.** *v. i.* **a)** ‹*Tier:*› quieken; **b)** ‹*Scharnier, Tür, Bremse, Schuh usw.:*› quietschen

squeaky ['skwi:kɪ] *adj.* quietschend; schrill ‹*Stimme*›; **be ~ clean** blitzsauber sein *(ugs.)*; *(fig.)* eine blütenweiße Weste haben *(fig. ugs.)*

squeal [skwi:l] **1.** *v. i.* **a)** ~ **with pain/in fear** ‹*Person:*› vor Schmerz/Angst aufschreien; ‹*Tier:*› vor Schmerz/Angst laut quieken; ~ **with laughter/in excitement** vor Lachen/Aufregung kreischen; **b)** ‹*Bremsen, Räder:*› kreischen; ‹*Reifen:*› quietschen; **c)** *(sl.: protest)* ~ **[in protest]** lauthals protestieren. **2.** *v. t.* kreischen. **3.** *n.* Kreischen, *das; (of tyres)* Quietschen, *das; (of animal)* Quieken, *das*

squeamish ['skwi:mɪʃ] *adj.* **be ~:** zartbesaitet sein; **this film is not for the ~:** dieser Film ist nichts für zarte Gemüter

squeegee [skwi:'dʒi:] *n. (for floor)* [Boden]wischer, *der; (for window)* [Fenster]wischer, *der*

squeeze [skwi:z] **1.** *n.* **a)** *(pressing)* Druck, *der;* **it only takes a gentle ~:** man braucht nur leicht zu drücken; **give sth. a [small] ~:** etw. [leicht] drücken; **b)** *(small quantity)* **a ~ of juice/washing-up liquid** ein Spritzer Saft/Spülmittel; **c)** *(crush)* Gedränge, *das.* **2.** *v. t.* **a)** *(press)* drücken; drücken auf (+ *Akk.*) ‹*Tube, Plastikflasche*›; kneten ‹*Ton, Knetmasse*›; ausdrücken ‹*Schwamm, Wäsche, Pickel*›; *(to get juice)* auspressen ‹*Früchte, Obst*›; ~ **sb.'s hand** jmdm. die Hand drücken; ~ **the trigger** auf den Abzug drücken; **b)** *(extract)* drücken **(out of** aus); ~ **out sth.** etw. herausdrücken; **c)** *(force)* zwängen; ~ **one's way past/into/out of sth.** sich an etw. *(Dat.)* vorbei-/in etw. *(Akk.)* hinein-/aus etw. herauszwängen; **d)** *(fig. coll.)* ~ **sth. from sb.** etw. aus jmdm. herauspressen. **3.** *v. i.* ~ **past sb./sth.** sich an jmdm./etw. vorbeidrängen; ~ **between two persons** sich zwischen zwei Personen *(Dat.)* durchdrängen; ~ **together** sich zusammendrängen

~ **'in 1.** *v. t.* **a)** reinquetschen; **b)** *(fig.: fit in)* einschieben. **2.** *v. i.* sich hineinzwängen

squelch [skweltʃ] *v. i.* **a)** *(make sucking sound)* quatschen *(ugs.)*; **b)** *(go over wet ground)* patschen

squib [skwɪb] *n. (firework)* Knallfrosch, *der;* **damp ~** *(fig.)* Reinfall, *der*

squid [skwɪd] *n. (Zool., Gastr.)* Kalmar, *der*

squidgy ['skwɪdʒɪ] *adj. (Brit. coll.)* durchweicht; matschig *(ugs.)*

squiggle ['skwɪgl] *n.* Schnörkel, *der*

squint [skwɪnt] **1.** *n.* **a)** *(Med.)* Schielen, *das;* **have a ~:** schielen; **b)** *(stealthy look)* Schielen, *das (ugs.)*; **c)** *(coll.: glance)* kurzer Blick; **have** *or* **take a ~ at** einen Blick werfen auf (+ *Akk.*); überfliegen ‹*Text, Zeitung*›. **2.** *v. i.* **a)** *(Med.)* schielen; **b)** *(with half-closed eyes)* blinzeln; die Augen zusammenkneifen; **c)** *(obliquely)* ~ **through a gap** durch eine Lücke lugen; **d)** *(coll.: glance)* ~ **at** einen [kurzen] Blick werfen auf (+ *Akk.*); überfliegen ‹*Zeitung, Text*›

squire ['skwaɪə(r)] *(country gentleman)* Squire, *der;* ≈ Gutsherr, *der*

squirm [skwɜ:m] *v. i.* **a)** *see* **wriggle 1; b)** *(fig.: show unease)* sich winden **(with** vor)

squirrel ['skwɪrl] *n. (Zool.)* Eichhörnchen, *das*

squirt [skwɜ:t] **1.** *v. t.* spritzen; sprühen ‹*Spray, Puder*›; ~ **sth. at sb.** jmdn. mit etw. bespritzen/besprühen; ~ **sb. in the eye/face [with sth.]** jmdm. [etw.] ins Auge/Gesicht spritzen/sprühen; ~ **oneself with water/deodorant** sich mit Wasser bespritzen/mit Deodorant besprühen. **2.** *v. i.* spritzen. **3.** *n.* Spritzer, *der*

Sr. *abbr.* **Senior** sen.

Sri Lanka [sri: 'læŋkə] *pr. n.* Sri Lanka *(das)*

Sri Lankan [sri: 'læŋkən] **1.** *adj.* srilankisch. **2.** *n.* Srilanker, *der*/Srilankerin, *die*

St *abbr.* **Saint** St.

St. *abbr.* **Street** Str.

st. *abbr. (Brit.: unit of weight)* **stone**

stab [stæb] **1.** *v. t.*, **-bb-** stechen; ~ **sb. in the chest** jmdm. in die Brust stechen. **2.** *v. i.*, **-bb-: a)** *(pierce)* stechen; **b)** *(thrust)* zustechen; ~ **at sb.** nach jmdm. stechen. **3.** *n.* **a)** *(act)* Stich, *der;* **b)** *(coll.: attempt)* **make** *or* **have a ~ [at it]** [es] probieren

stabbing ['stæbɪŋ] **1.** *n.* Messerstecherei, *die.* **2.** *attrib. adj.* stechend ‹*Schmerz*›

stability [stə'bɪlɪtɪ] *n., no pl.* Stabilität, *die*

stabilize ['steɪbɪlaɪz] **1.** *v. t.* stabilisieren. **2.** *v. i.* sich stabilisieren

stable ['steɪbl] **1.** *adj.* **a)** *(steady)* stabil; **a ~ family background** geordnete Familienverhältnisse; **b)** *(resolute)* gefestigt ‹*Person, Charakter*›. **2.** *n.* Stall, *der.* **3.** *v. t. (put in ~)* in den Stall bringen; *(keep in ~)* **the pony was ~d at a nearby farm** das Pony war im Stall eines nahegelegenen Bauernhofes untergebracht

staccato [stə'kɑ:təʊ] *(Mus.)* **1.** *adj.* staccato gesetzt; *(fig.)* abgehackt ‹*Sprache*›. **2.** *adv.* staccato

stack [stæk] **1.** *n.* **a)** *(of hay etc.)* Schober, *der (südd., österr.)*; Feim, *der (nordd., md.)*; **b)** *(pile)* Stoß, *der;* Stapel, *der;* **place sth. in ~s** etw. [auf]stapeln; **c)** *(coll.: large amount)* Haufen, *der (ugs.)*; **a ~ of work/money** ein Haufen Arbeit/Geld; **have a ~ of things to do** einen Haufen zu tun haben *(ugs.)*; **d) [chimney-]~:** Schornstein, *der.* **2.** *v. t.* **a)** *(pile)* ~ **[up]** [auf]stapeln; ~ **logs in a pile** Holz zu einem Stoß aufschichten; **b)** *(arrange fraudulently)* ~ **the cards** beim Mischen betrügen; **the odds** *or* **cards** *or* **chips are ~ed against sb.** *(fig.)* jmd. hat schlechte Karten *(fig. ugs.)*

~ **'up** *see* ~ **2 a**

stadium ['steɪdɪəm] *n.* Stadion, *das*

staff [stɑ:f] **1.** *n.* **a)** *(stick)* Stock, *der;* **b)** *constr. as pl. (personnel)* Personal, *das;* **editorial ~:** Redaktion, *die;* **the ~ of the firm** die Betriebsangehörigen; die Belegschaft [der Firma]; **c)** *constr. as pl. (of school)* Lehrerkollegium, *das;* Lehrkörper, *der (Amtsspr.)*; *(of university or college)* Dozentenschaft, *die;* **d)** *pl.* **staves** [steɪvz] *(Mus.)* Liniensystem, *das.* **2.** *v. t.* mit Personal ausstatten

staff: ~ **meeting** *n.* [Lehrer]konferenz, *die;* ~ **nurse** *n. (Brit.)* Zweitschwester, *die/Krankenpfleger in der Stellung einer Zweitschwester;* **~-room** *n.* Lehrerzimmer, *das*

stag [stæg] *n.* Hirsch, *der*

stage [steɪdʒ] **1.** *n.* **a)** *(Theatre)* Bühne, *die;* **down/up ~** *(position)* vorne/hinten auf der Bühne; *(direction)* nach vorn/nach hinten; **b)** *(fig.)* **the ~:** das Theater; **go on the ~:** zur Bühne *od.* zum Theater gehen; **c)** *(part of process)* Stadium, *das;* Phase, *die;* **be at a late/critical ~:** sich in einer späten/kritischen Phase befinden; **at this ~:** in diesem Stadium; **do sth. in** *or* **by ~s** etw. abschnittsweise *od.* nach und nach tun; **in the final ~s** in der Schlußphase; **d)** *(raised platform)* Gerüst, *das;* **e)** *(of microscope)* Mikroskoptisch, *der;* **f)** *(fig.: scene)* Bühne, *die;* **set the ~ for sb./sth.** jmdm. den Weg ebnen/etw. in die Wege leiten; **g)** *(distance)* Etappe, *die.* **2.** *v. t.* **a)** *(present)* inszenieren; **b)** *(ar-*

range) veranstalten ⟨*Wettkampf, Ausstellung*⟩; ausrichten ⟨*Veranstaltung*⟩; organisieren ⟨*Streik*⟩; bewerkstelligen ⟨*Rückzug*⟩

stage: ~-coach *n.* Postkutsche, *die;* ~ **'door** *n. (Theatre)* Bühneneingang, *der;* ~ **fright** *n. (Theatre)* Lampenfieber, *das;* **~-manage** *v. t.* **a)** *(Theatre)* als Inspizient/Inspizientin mitwirken bei ⟨*Inszenierung*⟩; **b)** *(fig.)* veranstalten; inszenieren ⟨*Revolte usw.*⟩; **~-manager** *n. (Theatre)* Inspizient, der/Inspizientin, *die;* **~-struck** *adj.* theaterbesessen; ~ **whisper** *n.* Beiseitesprechen, *das;* **in a ~ whisper** beiseite

stagger ['stægə(r)] **1.** *v. i.* schwanken; torkeln *(ugs.).* **2.** *v. t.* **a)** *(astonish)* die Sprache verschlagen (+ *Dat.*); **b)** versetzt anordnen; **~ed junction** versetzt angelegte Kreuzung

staggering ['stægərıŋ] *adj.* erschütternd ⟨*Schlag, Schock, Verlust*⟩; schwindelerregend ⟨*Höhe, Menge*⟩; folgenschwer ⟨*Auswirkung, Bedeutung*⟩; zutiefst beunruhigend ⟨*Nachricht*⟩

stagnant ['stægnənt] *adj.* **a)** *(motionless)* stehend ⟨*Gewässer*⟩; **the water is ~:** das Wasser steht; **b)** *(fig.: lifeless)* abgestumpft ⟨*Geist, Seele*⟩; stagnierend ⟨*Wirtschaft*⟩; dumpf ⟨*Leben*⟩; **the economy is ~:** die Wirtschaft stagniert

stagnate [stæg'neıt] *v. i.* **a)** ⟨*Wasser usw.:*⟩ abstehen; **b)** *(fig.)* ⟨*Wirtschaft, Geschäft:*⟩ stagnieren; ⟨*Geist:*⟩ in Lethargie verfallen; ⟨*Person:*⟩ abstumpfen

stagnation [stæg'neıʃn] *n., no pl.* **a)** *(of water etc.)* Stehen, *das;* **b)** *(fig.)* Stagnation, *die*

'stag night *n. Zechabend des Bräutigams mit seinen Freunden kurz vor seiner Hochzeit*

staid [steıd] *adj.* **a)** *(steady, sedate)* gesetzt; **b)** *(serious)* bieder

stain [steın] **1.** *v. t.* **a)** *(discolour)* verfärben; *(make ~s on)* Flecken hinterlassen auf (+ *Dat.*); **b)** *(fig.: damage)* beflecken; besudeln *(geh. abwertend);* **c)** *(colour)* färben; beizen ⟨*Holz*⟩. **2.** *n.* **a)** *(discoloration)* Fleck, *der;* **b)** *(fig.: blemish)* Schandfleck, *der*

stained 'glass *n.* farbiges Glas; Farbglas, *das;* ~ **'window** Fenster mit Glasmalerei

stainless ['steınlıs] *adj.* **a)** *(spotless)* fleckenlos; **b)** *(non-rusting)* rostfrei

stainless 'steel *n.* Edelstahl, *der*

stair [steə(r)] *n.* ~s *or (arch./Scot.) same* Treppe, *die*

stair: ~-carpet *see* **carpet 1 a;** **~case** *n.* Treppenhaus, *das; (one flight)* Treppe, *die;* **on the ~case** auf der Treppe; **~way** *n.* **a)** *(access via ~s)* Treppenaufgang, *der;* **b)** *(~-case)* Treppe, *die;* **~-well** *n.* Treppenhaus, *das*

stake [steık] **1.** *n.* **a)** *(pointed stick)* Pfahl, *der;* **b)** *(wager)* Einsatz, *der;* **be at ~:** auf dem Spiel stehen. **2.** *v. t.* **a)** *(secure)* [an einem Pfahl/an Pfählen] anbinden; **b)** *(wager)* setzen (**on** auf + *Akk.*); **c)** *(risk)* aufs Spiel setzen (**on** für)

~ **'out** *v. t.* **a)** *(mark out)* mit Pfählen begrenzen; **b)** *(fig.: claim)* beanspruchen; **c)** *(Amer. coll.: observe)* überwachen

stalactite ['stæləktaıt] *n. (Geol.)* Stalaktit, *der*

stalagmite ['stæləgmaıt] *n. (Geol.)* Stalagmit, *der*

stale [steıl] *adj.* alt; muffig; abgestanden ⟨*Luft*⟩; alt[backen] ⟨*Brot*⟩; schal ⟨*Bier, Wein usw.*⟩; *(fig.)* abgedroschen ⟨*Witz, Trick*⟩; überholt ⟨*Nachricht*⟩

'stalemate *n. (Chess; also fig.)* Patt, *das*

[1]**stalk** [stɔ:k] *v. t.* sich heranpirschen an (+ *Akk.*)

[2]**stalk** *n. (Bot.) (main stem)* Stengel, *der; (of leaf, flower, fruit)* Stiel, *der*

[1]**stall** [stɔ:l] **1.** *n.* **a)** Stand, *der;* **b)** *(for horse)* Box, *die; (for cow)* Stand, *der;* **c)** *(Eccl.: seat)* Stuhl, *der;* **the choir ~s** das Chorgestühl; **d)** *in pl. (Brit. Theatre: seats)* **[front] ~s** Parkett, *das.* **2.** *v. t.* abwürgen *(ugs.)* ⟨*Motor*⟩. **3.** *v. i.* ⟨*Motor:*⟩ stehenbleiben

[2]**stall 1.** *v. i.* ausweichen. **2.** *v. t.* blockieren ⟨*Gesetz, Fortschritt*⟩; aufhalten ⟨*Feind, Fortschritt*⟩

stallion ['stæljən] *n.* Hengst, *der*

stalwart ['stɔ:lwət] *adj.* **a)** *(sturdy)* stämmig; **b)** *attrib. (fig.: determined)* entschieden; entschlossen ⟨*Kämpfer*⟩; *(loyal)* treu; getreu *(geh.)*

stamen ['steımen, 'steımən] *n. (Bot.)* Staubblatt, *das*

stamina ['stæmınə] *n.* **a)** *(physical staying-power)* Ausdauer, *die;* **b)** *(endurance)* Durchhaltevermögen, *das*

stammer ['stæmə(r)] **1.** *v. i.* stottern. **2.** *v. t.* stammeln. **3.** *n.* Stottern, *das*

stamp [stæmp] **1.** *v. t.* **a)** *(impress, imprint sth. on)* [ab]stempeln; ~ **sth. on sth.** etw. auf etw. *(Akk.)* [auf]stempeln; **b)** ~ **one's foot/feet** mit dem Fuß/den Füßen stampfen; ~ **the floor** *or* **ground [in anger/with rage]** [ärgerlich/wütend] auf den Boden stampfen; **c)** *(put postage ~ on)* frankieren; freimachen *(Postw.);* **~ed addressed envelope** frankierter Rückumschlag; **d)** *(mentally)* **become** *or* **be ~ed on sb.['s memory** *or* **mind]** sich jmdm. fest einprägen. **2.** *v. i.* aufstampfen. **3.** *n.* **a)** Marke, *die; (postage ~)* Briefmarke, *die;* **b)** *(instrument for ~ing, mark)* Stempel, *der;* **c)** *(fig.: characteristic)* **bear the ~ of genius/greatness** Genialität/Größe erkennen lassen

~ **on** *v. t.* **a)** zertreten ⟨*Insekt, Dose*⟩; zertrampeln ⟨*Blumen*⟩; ~ **on sb.'s foot** jmdm. auf den Fuß treten; **b)** *(suppress)* durchgreifen gegen. *See also* ~ **1 a, d**

~ **'out** *v. t.* **a)** *(eliminate)* ausmerzen; ersticken ⟨*Aufstand, Feuer*⟩; niederwalzen ⟨*Opposition, Widerstand*⟩; **b)** *(cut out)* [aus]stanzen

stamp: ~-album *n.* Briefmarkenalbum, *das;* **~-collecting** *n.* Briefmarkensammeln, *das;* **~-collection** *n.* Briefmarkensammlung, *die;* **~-collector** *n.* Briefmarkensammler, *der*/-sammlerin, *die*

stampede [stæm'pi:d] *n.* Stampede, *die*

'stamp-machine *n.* Briefmarkenautomat, *der*

stanch [stɑ:ntʃ, stɔ:ntʃ] *v. t.* **a)** *(stop flow of)* stillen ⟨*Blut*⟩; **b)** *(stop flow from)* abbinden ⟨*Wunde*⟩

stand [stænd] **1.** *v. i.*, **stood** [stʊd] **a)** stehen; ~ **in a line** *or* **row** sich in einer Reihe aufstellen; *(be ~ing)* in einer Reihe stehen; **we stood talking** wir standen da und unterhielten uns; **b)** *(have height)* **he ~s six feet tall/the tree ~s 30 feet high** er ist sechs Fuß groß/der Baum ist 30 Fuß hoch; **c)** *(be at level)* ⟨*Aktien, Währung, Thermometer:*⟩ stehen (**at** auf + *Dat.*); ⟨*Fonds:*⟩ sich belaufen (**at** auf + *Akk.*); ⟨*Absatz, Export usw.:*⟩ liegen (**at** bei); **d)** *(hold good)* bestehenbleiben; **my offer/promise still ~s** mein Angebot/Versprechen gilt nach wie vor; **e)** *(find oneself, be)* ~ **convicted of treachery** wegen Verrats verurteilt sein; **as it ~s, as things ~:** wie die Dinge [jetzt] liegen; **the law as it ~s** das bestehende *od.* gültige Recht; **I'd like to know where I ~** *(fig.)* ich möchte wissen, wo ich dran bin; ~ **in need of sth.** einer Sache *(Gen.)* dringend bedürfen; **f)** *(be candidate)* kandidieren (**for** für); ~ **in an election** bei einer Wahl kandidieren; ~ **as a Liberal/Conservative** für die Liberalen/Konservativen kandidieren; ~ **for Parliament** *(Brit.)* für einen Parlamentssitz kandidieren; **g)** ~ **proxy for sb.** jmdn. vertreten; **h)** *(place oneself)* sich stellen; ~ **in the way of sth.** *(fig.)* einer Sache *(Dat.)* im Weg stehen; **[not]** ~ **in sb.'s way** *(fig.)* jmdm. [keine] Steine in den Weg legen; **i)** *(be likely)* ~ **to win** *or* **gain/lose sth.** etw. gewinnen/verlieren können. **2.** *v. t.*, **stood a)** *(set in position)* stellen; ~ **sth. on end/upside down** etw. hochkant/auf den Kopf stellen; **b)** *(endure)* ertragen; vertragen ‹*Klima*›; **I can't** ~ **the heat/noise** ich halte die Hitze/den Lärm nicht aus; **I cannot** ~ **[the sight of] him/her** ich kann ihn/sie nicht ausstehen; **he can't** ~ **the pressure/strain/stress** er ist dem Druck/den Strapazen/dem Streß nicht gewachsen; **I can't** ~ **it any longer!** ich halte es nicht mehr aus!; *see also* **time 1a**; **c)** *(undergo)* ausgesetzt sein (+ *Dat.*); ~ **trial [for sth.]** [wegen etw.] vor Gericht stehen; **d)** *(buy)* ~ **sb. sth.** jmdm. etw. ausgeben *od.* spendieren *(ugs.)*. **3.** *n.* **a)** *(support)* Ständer, *der;* **b)** *(stall; at exhibition)* Stand, *der;* **c)** *(raised structure, grand~)* Tribüne, *die;* **d)** *(resistance)* Widerstand, *der;* **take** *or* **make a** ~ *(fig.)* klar Stellung beziehen **(for/against/on** für/gegen/zu); **e)** *(~ing-place for taxi, bus, etc.)* Stand, *der*

~ **a'bout,** ~ **a'round** *v. i.* herumstehen

~ **a'side** *v. i.* zur Seite treten; Platz machen

~ **'back** *v. i.* **a)** ~ **[well] back [from sth.]** [ein gutes Stück] [von etw.] entfernt stehen; **b)** *see* ~ **aside**; **c)** *(fig.: distance oneself)* zurücktreten; **d)** *(fig.: withdraw)* ~ **back from sth.** sich aus einer Sache heraushalten

~ **between** *v. t.* **sth.** ~**s between sb. and sth.** *(fig.)* etw. steht jmdm. bei etw. im Wege

~ **by 1.** [-'-] *v. i.* **a)** *(remain apart)* abseits stehen; **b)** *(be near)* daneben stehen; **c)** *(be ready)* sich zur Verfügung halten. **2.** ['--] *v. t.* **a)** *(support)* ~ **by sb./one another** jmdm./sich [gegenseitig] *od.* *(geh.)* einander beistehen; **b)** *(adhere to)* ~ **by sth.** zu etw. stehen; ~ **by the terms of a contract** einen Vertrag einhalten; ~ **by a promise** ein Versprechen halten

~ **'down** *v. i.* *(withdraw, retire)* verzichten; ~ **down in favour of a person** zugunsten einer Person *(Gen.)* zurücktreten

~ **for** *v. t.* **a)** *(signify)* bedeuten; **b)** *(coll.: tolerate)* sich bieten lassen

~ **'in** *v. i.* *(deputize)* aushelfen; ~ **in for sb.** für jmdn. einspringen

~ **'out** *v. i.* **a)** *(be prominent)* herausragen; ~ **out against** *or* **in contrast to sth.** sich gegen etw. abheben; ~ **out a mile** nicht zu übersehen sein; ‹*Grund, Antwort:*› [klar] auf der Hand liegen; **b)** *(be outstanding)* herausragen **(from** aus)

~ **over** *v. t.* beaufsichtigen

~ **to'gether** *v. i.* zusammenstehen; *(for a photograph)* sich [gemeinsam] aufstellen; *(fig.)* zusammenhalten

~ **'up 1.** *v. i.* **a)** *(rise)* aufstehen; **b)** *(be upright)* stehen; ~ **up straight** sich aufrecht hinstellen; **c)** *(be valid)* gelten; Gültigkeit haben; **d)** ~ **up well [in comparison with sb./sth.]** [im Vergleich zu jmdm./etw.] gut abschneiden. **2.** *v. t.* **a)** *(put upright)* aufstellen; [wieder] hinstellen ‹*Fahrrad, Stuhl usw.*›; **b)** *(coll.: fail to keep date with)* ~ **sb. up** jmdn. versetzen *(ugs.)*

~ **'up for** *v. t.* ~ **up for sb./sth.** für jmdn./etw. Partei ergreifen; sich für jmdn./etw. stark machen

~ **'up to** *v. t.* **a)** *(face steadfastly)* ~ **up to sb.** sich jmdm. entgegenstellen; jmdm. die Stirn bieten; ~ **up to sth.** sich einer Sache *(Dat.)* stellen; **b)** *(survive intact under)* ~ **up to sth.** einer Sache *(Dat.)* standhalten; ~ **up to wear and tear** eine starke Beanspruchung aushalten

stand-a'lone *adj.* *(Computing)* selbständig

standard ['stændəd] **1.** *n.* **a)** *(norm)* Maßstab, *der;* **safety** ~**s** Sicherheitsnormen; **above/below/up to** ~**:** überdurchschnittlich [gut]/unter dem Durchschnitt/der Norm entsprechend; **b)** *(degree)* Niveau, *das;* **set a high/low** ~ **in** *or* **of sth.** hohe/niedrige Ansprüche an etw. *(Akk.)* stellen; ~ **of living** Lebensstandard, *der;* **c)** *in pl.* *(moral principles)* Prinzipien; **d)** *(flag)* Standarte, *die.* **2.** *adj.* **a)** *(conforming to ~)* Standard-; *(used as reference)* Normal-; **b)** *(widely used)* normal; **be** ~ **procedure** Vorschrift sein; **be fitted with sth. as** ~**:** serienmäßig mit etw. ausgerüstet sein; **a** ~ **letter** ein Schemabrief *(Bürow.)*; **be** ~ **practice** allgemein üblich sein

standard: ~**-bearer** *n.* **a)** *(Mil.: flag-bearer)* Standartenträger, *der;* **b)** *(fig.: leader)* Vorkämpfer, *der*/-kämpferin, *die;* **S**~ **'English** *n.* Standardenglisch, *das*

standardize (standardise) ['stændədaɪz] *v. t.* standardisieren

'standard lamp *n.* *(Brit.)* Stehlampe, *die*

'stand-by 1. *n., pl.* ~**s a)** *(reserve)* **[act] as a** ~**:** als Ersatz [bereitstehen]; **be on** ~ ‹*Polizei, Feuerwehr, Truppen:*› einsatzbereit sein; **b)** *(resource)* Rückhalt, *der;* **sth. is a good** ~**:** auf etw. *(Akk.)* kann man jederzeit zurückgreifen. **2.** *attrib. adj.* Ersatz-; ~ **ticket/passenger** Stand-by-Ticket, *das*/-Passagier, *der*

'stand-in 1. *n.* Ersatz, *der;* *(in theatre, film)* Ersatzdarsteller, *der*/-darstellerin, *die.* **2.** *attrib. adj.* Ersatz-

standing ['stændɪŋ] **1.** *n.* **a)** *(repute)* Ansehen, *das;* **be of** *or* **have [a] high** ~**:** ein hohes Ansehen genießen; **what is his** ~**?** welchen Rang bekleidet er?; **b)** *(service)* **be an MP of twenty years'** ~**:** seit zwanzig Jahren [ununterbrochen] dem Parlament angehören; **c)** *(duration)* **of long/short** ~**:** von langer/kurzer Dauer. **2.** *adj.* **a)** *(erect)* stehend; **after the storm there was scarcely a tree still** ~**:** nach dem Sturm stand kaum mehr ein Baum; **leave sb.** ~ *(lit. or fig.: progress much faster)* jmdn. weit hinter sich *(Dat.)* lassen; **b)** *attrib.* *(established)* fest ‹*Regel, Brauch*›; **he has a** ~ **excuse** er bringt immer die gleiche Entschuldigung; **c)** *attrib.* *(permanent)* stehend ‹*Heer*›

standing: ~ **com'mittee** *n.* ständiger Ausschuß; ~ **'order** *n.* Dauerauftrag, *der;* *(for regular supply)* Abonnement, *das;* ~ **o'vation** *n.* stürmischer Beifall; stehende Ovation *(geh.)*; ~**-room** *n., no pl., no indef. art.* Stehplätze

stand-offish [stænd'ɒfɪʃ] *adj.* reserviert

stand: ~**-pipe** *n.* *(for water-supply)* Standrohr, *das;* ~**point** *n.* **a)** *(observation point)* Standort, *der;* **b)** *(fig.: viewpoint)* Standpunkt, *der;* ~**still** *n.* Stillstand, *der;* **be at a** ~**still** stillstehen; **come to a** ~**still** zum Stehen kommen; ‹*Verhandlungen:*› zum Stillstand kommen; **the traffic/production came to a** ~**still** der Verkehr/die Produktion kam zum Erliegen; **bring to a** ~**still** zum Stehen bringen; zum Erliegen bringen ‹*Produktion*›; ~**-up** *adj.* ~**-up fight** Schlägerei, *die*

stank *see* **stink 1**

stanza ['stænzə] *n.* *(Pros.)* Strophe, *die*

¹staple ['steɪpl] **1.** *n. (for fastening paper)* [Heft]klammer, *die.* **2.** *v. t.* heften (**on to** an + *Akk.*)

²staple *attrib. adj.* **a)** *(principal)* Grund-; **b)** *(Commerc.: important)* grundlegend

stapler ['steɪplə(r)] *n.* [Draht]hefter, *der*

star [stɑː(r)] **1.** *n.* **a)** Stern, *der;* **three/four ~ hotel** Drei-/Vier-Sterne-Hotel, *das;* **two/four ~ [petrol]** Normal-/Super[benzin], *das;* **the S~s and Stripes** *(Amer.)* das Sternenbanner; **b)** *(prominent person)* Star, *der;* **c)** *(asterisk)* Stern, *der;* Sternchen, *das;* **d)** *(Astrol.)* Stern, *der;* **read one's/the ~s** sein/das Horoskop lesen. **2.** *attrib. adj.* Star-; **~ pupil** bester Schüler/beste Schülerin; **~ turn** *or* **attraction** Hauptattraktion, *die.* **3.** *v. t.*, **-rr-** *(feature as ~)* **~ring Humphrey Bogart and Lauren Bacall** mit Humphrey Bogart und Lauren Bacall in den Hauptrollen. **4.** *v. i.*, **-rr-: ~ in a film/play/TV series** in einem Film/einem Stück/einer Fernsehserie die Hauptrolle spielen

starboard ['stɑːbəd] *(Naut., Aeronaut.)* **1.** *n.* Steuerbord, *das;* **land to ~!** Land an Steuerbord! **2.** *adj.* steuerbord-; steuerbordseitig; **on the ~ bow/quarter** Steuerbord voraus/achteraus

starch [stɑːtʃ] **1.** *n.* Stärke, *die.* **2.** *v. t.* stärken

starchy ['stɑːtʃɪ] *adj.* **a)** stärkehaltig ‹*Nahrungsmittel*›; **b)** *(fig.: prim)* steif

stardom ['stɑːdəm] *n.* Starruhm, *der*

stare [steə(r)] **1.** *v. i.* **a)** *(gaze)* starren; **~ in surprise/amazement** überrascht/erstaunt starren; **~ at sb./sth.** jmdn./etw. anstarren; **b)** *(have fixed gaze)* starr blicken. **2.** *v. t.* **~ sb. in the face** jmdn. [feindselig] fixieren; *(fig.)* jmdm. ins Auge springen; **ruin was staring him in the face** ihm drohte der Ruin. **3.** *n.* Starren, *das;* **fix sb. with a [curious/malevolent] ~:** jmdn. [neugierig/böse] anstarren

~ 'down, ~ 'out *v. t.* **~ sb. down** *or* **out** jmdn. so lange anstarren, bis er/sie die Augen abwendet

'starfish *n.* Seestern, *der*

staring ['steərɪŋ] *attrib. adj.* starrend ‹*Augen*›; **with ~ eyes** mit starrem Blick; **be stark ~ mad** *(fig. coll.)* völlig verrückt sein *(ugs.)*.

stark [stɑːk] **1.** *adj.* **a)** *(bleak)* öde; spröde ‹*Schönheit, Dichtung*›; **b)** *(obvious)* scharf umrissen; nackt ‹*Wahrheit*›; scharf ‹*Kontrast, Umriß*›; kraß ‹*Unterschied, Gegensatz, Realismus*›; **c)** *(extreme)* schier ‹*Entsetzen, Dummheit*›; nackt ‹*Armut, Angst*›. **2.** *adv.* völlig; **~ naked** splitternackt *(ugs.); see also* **staring**

'starlight *n., no pl.* Sternenlicht, *das*

starling ['stɑːlɪŋ] *n.* *(Ornith.)* [Gemeiner] Star

'starlit *adj.* sternhell

starry ['stɑːrɪ] *adj.* sternklar ‹*Himmel, Nacht*›; sternenübersät ‹*Himmel*›

'starry-eyed *adj.* blauäugig *(fig.)*

'star-studded *adj.* ‹*Show, Film, Besetzung*› mit großem Staraufgebot

start [stɑːt] **1.** *v. i.* **a)** *(begin)* anfangen; beginnen *(oft geh.)*; **~ on sth.** etw. beginnen; **~ with sth./sb.** bei *od.* mit etw./jmdm. anfangen; **prices ~ at ten dollars** die Preise beginnen bei zehn Dollar; **~ at the beginning** am Anfang beginnen; **to ~ with** zuerst *od.* zunächst einmal; **~ing from next month** ab nächsten Monat; **b)** *(set out)* aufbrechen; **c)** *(make sudden movement)* aufschrecken; **~ with pain/surprise** vor Schmerz/Überraschung auffahren; **~ from one's chair** von seinem Stuhl hochfahren; **d)** *(begin to function)* anlaufen; ‹*Auto, Motor usw.*:› anspringen. **2.** *v. t.* **a)** *(begin)* beginnen [mit]; **~ school** in die Schule kommen; **~ work** mit der Arbeit beginnen (**on** an + *Dat.*); *(after leaving school)* zu arbeiten anfangen; **~ doing** *or* **to do sth.** [damit] anfangen, etw. zu tun; **b)** *(cause)* auslösen; anfangen ‹*Streit, Schlägerei*›; legen ‹*Brand*›; *(accidentally)* verursachen ‹*Brand*›; **c)** *(set up)* ins Leben rufen ‹*Organisation, Projekt*›; aufmachen ‹*Laden, Geschäft*›; gründen ‹*Verein, Firma, Zeitung*›; **d)** *(switch on)* einschalten; starten, anlassen ‹*Motor, Auto*›; **e) ~ sb. doing sth.** jmdn. anfangen lassen, etw. zu tun; **~ sb. drinking/coughing/laughing** jmdn. zum Trinken/Husten/Lachen bringen; **~ sb. on a diet** jmdn. auf Diät *(Akk.)* setzen; **~ sb. in business/a trade** jmdm. die Gründung eines Geschäfts ermöglichen/jmdn. in ein Handwerk einführen; **f)** *(Sport)* **~ a race** ein Rennen starten; **~ a football match** ein Fußballspiel anpfeifen. **3.** *n.* **a)** Anfang, *der;* Beginn, *der;* *(of race)* Start, *der;* **from the ~:** von Anfang an; **from ~ to finish** von Anfang bis Ende; **at the ~:** am Anfang; **at the ~ of the war/day** bei Kriegsbeginn/zum Tagesanfang; **make a ~:** anfangen (**on, with** mit); *(on journey)* aufbrechen; **get off to** *or* **make a good/slow/poor ~:** einen guten/langsamen/schlechten Start haben; **for a ~** *(coll.)* zunächst einmal; **b)** *(Sport: ~ing-place)* Start, *der;* **c)** *(Sport: advantage)* Vorsprung, *der;* **give sb. [a] 60 metres ~:** jmdm. eine Vorgabe von 60 Metern geben; **have a ~ over** *or* **on sb./sth.** *(fig.)* einen Vorsprung vor jmdm./etw. haben; **d)** *(jump)* **she remembered** *or* **realized with a ~ that ...:** sie schreckte zusammen, als ihr einfiel, daß ...; **give sb. [a] ~:** jmdm. einen Schreck einjagen

~ 'off 1. *v. i.* **a)** *see* **set off 1; b)** *(coll.: begin action)* **~ off by showing sth.** zu Beginn etw. zeigen; **c) ~ off with** *or* **on sth.** *(begin on)* mit etw. beginnen. **2.** *v. t.* **a) ~ sb. off on a task/job** jmdn. in eine Aufgabe/Arbeit einweisen; **b)** *see* **set off 2 b**

~ 'out *v. i.* **a)** *see* **set out 1; b)** *see* **set off 1**

~ 'up 1. *v. i.* **a)** *see* **jump up; b)** *(be set going)* starten; ‹*Motor*:› anspringen; **c)** *(begin to work)* **~ up in engineering/insurance** als Ingenieur/in der Versicherungsbranche anfangen. **2.** *v. t.* **a)** beginnen ‹*Gespräch*›; gründen ‹*Geschäft, Firma*›; schließen ‹*Freundschaft*›; **b)** starten ‹*Fahrzeug, Motor*›

starter ['stɑːtə(r)] *n.* **a)** *(Sport: signaller)* Starter, *der;* **b)** *(Sport: entrant)* Starter, *der*/Starterin, *die;* *(horse)* startendes Pferd; **c)** *(Motor Veh.)* **~ [motor]** Anlasser, *der;* **d)** *(initial action)* Anfang, *der;* **as a ~:** zuerst; **e)** *(hors d'œuvre etc.)* Vorspeise, *die*

starting: ~-block *n.* *(Athletics)* Startblock, *der;* **~-line** *n.* Startlinie, *die;* **~-point** *n.* *(lit. or fig.)* Ausgangspunkt, *der;* *(for solving a problem)* Ansatzpunkt, *der*

startle ['stɑːtl] *v. t.* erschrecken; **be ~d by sth.** über etw. *(Akk.)* erschrecken

startling ['stɑːtlɪŋ] *adj.* erstaunlich; überraschend ‹*Nachricht*›

starvation [stɑː'veɪʃn] *n.* Verhungern, *das;* **die of** *or* **from/suffer from ~:** verhungern/hungern *od.* Hunger leiden; **be** *or* **live on a ~ diet** fast am Verhungern sein; **~ wages** Hungerlohn, *der*

starve [stɑːv] **1.** *v. i.* **a)** *(die of hunger)* **~ [to death]** verhungern; **b)** *(suffer hunger)* hungern; **c) be**

starving *(coll.: feel hungry)* am Verhungern sein *(ugs.).* **2.** *v. t.* **a)** *(kill by starving)* ~ **sb. [to death]** jmdn. verhungern lassen; **b)** *(deprive of food)* hungern lassen; **c)** *(deprive)* **we were ~d of knowledge** uns *(Dat.)* wurde [viel] Wissen vorenthalten; **feel ~d of affection** unter einem Mangel an Zuneigung leiden

~ **'out** *v. t.* aushungern

'star wars *n. pl.* Krieg der Sterne

state [steɪt] **1.** *n.* **a)** *(condition)* Zustand, *der;* ~ **of the economy** Wirtschaftslage, *die;* **the ~ of play** *(Sport)* der Spielstand; **the ~ of play in the negotiations/debate** *(fig.)* der [gegenwärtige] Stand der Verhandlungen/Debatte; **the ~ of things in general** die allgemeine Lage; **the ~ of the nation** die Lage der Nation; **a ~ of war exists** es herrscht Kriegszustand; **be in a ~ of excitement/sadness/anxiety** aufgeregt/traurig/ängstlich sein; **b)** *(mess)* **what a ~ you're in!** wie siehst du denn aus!; **c)** *(anxiety)* **be in a ~** *(be in a panic)* aufgeregt sein; *(be excited)* ganz aus dem Häuschen sein *(ugs.);* **get into a ~** *(coll.)* Zustände kriegen *(ugs.);* **don't get into a ~!** reg dich nicht auf! *(ugs.);* **d)** *(nation)* Staat, *der;* **[affairs] of S~:** Staats[geschäfte]; **e)** *(federal ~) (of Germany, Austria)* Land, *das; (of America)* Staat, *der;* **the [United] S~s** *sing.* die [Vereinigten] Staaten; **f)** S~ *(civil government)* Staat, *der;* **g)** *(pomp)* Prunk, *der;* **in ~:** in vollem Staat; **lie in ~:** aufgebahrt sein. **2.** *attrib. adj.* **a)** *(of nation or federal ~)* staatlich; Staats‹*bank, -sicherheit, -geheimnis, -dienst*›; ~ **education** staatliches Erziehungswesen; **b)** *(ceremonial)* Staats-. **3.** *v. t.* **a)** *(express)* erklären; *(fully or clearly)* darlegen; äußern ‹*Meinung*›; angeben ‹*Alter usw.*›; **'please ~ full particulars'** „bitte genaue Angaben machen"; **b)** *(specify)* festlegen; **at ~d intervals** in genau festgelegten Abständen

'State Department *n. (Amer. Polit.)* Außenministerium, *das*

stateless ['steɪtlɪs] *adj.* staatenlos; ~ **person** Staatenlose, *der/die*

stately ['steɪtlɪ] *adj.* majestätisch; stattlich ‹*Körperbau, Erscheinung, Gebäude*›; hochtrabend ‹*Stil*›; feierlich ‹*Prozession*›; **at a ~ pace** gemessenen Schrittes

stately 'home *n. (Brit.)* Herrensitz, *der; (grander)* Schloß, *das*

statement ['steɪtmənt] *n.* **a)** *(stating, account, thing stated)* Aussage, *die; (declaration)* Erklärung, *die; (allegation)* Behauptung, *die;* **make a ~** ‹*Zeuge:*› eine Aussage machen; ‹*Politiker:*› eine Erklärung abgeben **(on** zu); **b)** *(Finance: report)* **[bank] ~:** Kontoauszug, *der*

state: ~-of-the-'art *adj.* auf dem neuesten Stand der Technik *nachgestellt;* **~-owned** *adj.* staatlich; in Staatsbesitz *nachgestellt;* **S~ school** *n. (Brit.)* staatliche Schule; **S~side** *adv. (Amer. coll.)* **be/work/travel S~side** in den Staaten *(ugs.)* sein/arbeiten/in die Staaten *(ugs.)* reisen

statesman ['steɪtsmən] *n., pl.* **statesmen** ['steɪtsmən] Staatsmann, *der*

static ['stætɪk] **1.** *adj.* **a)** *(Phys.)* statisch; **b)** *(not moving)* statisch; *(not changing)* konstant ‹*Umweltbedingungen*›. **2.** *n. (atmospherics)* atmosphärische Störungen

station ['steɪʃn] **1.** *n.* **a)** *(position)* Position, *die;* **b)** *(establishment)* Station, *die;* **c)** *see* **railway-station; d)** *(status)* Rang, *der.* **2.** *v. t.* **a)** *(assign position to)* stationieren; abstellen ‹*Auto*›; aufstellen ‹*Wache*›; **b)** *(place)* stellen; ~ **oneself** sich aufstellen

stationary ['steɪʃənərɪ] *adj.* **a)** *(not moving)* stehend; **be ~:** stehen; **the traffic was ~:** der Verkehr war zum Erliegen gekommen; **b)** *(fixed)* stationär

stationer ['steɪʃənə(r)] *n.* Schreibwarenhändler, *der/*-händlerin, *die;* **~'s [shop]** Schreibwarengeschäft, *das*

stationery ['steɪʃənərɪ] *n.* **a)** *(writing-materials)* Schreibwaren *Pl.;* **b)** *(writing-paper)* Briefpapier, *das*

station: ~-master *n. (Railw.)* Stationsvorsteher, *der/*-vorsteherin, *die;* **~-wagon** *n. (Amer.)* Kombi[wagen], *der*

statistical [stə'tɪstɪkl] *attrib. adj.,* **statistically** [stə'tɪstɪkəlɪ] *adv.* statistisch

statistician [stætɪ'stɪʃn] *n.* Statistiker, *der/*Statistikerin, *die*

statistics [stə'tɪstɪks] *n.* **a)** *as pl. (facts)* Statistik, *die;* **b)** *no pl. (science)* Statistik, *die*

statue ['stætʃu:, 'stætju:] *n.* Statue, *die*

statuette [stætʃʊ'et, stætjʊ'et] *n.* Statuette, *die*

stature ['stætʃə(r)] *n.* **a)** *(body height)* Statur, *die;* **b)** *(fig.: standing)* Format, *das;* **a person of [some] ~:** eine [recht] bedeutende Persönlichkeit

status ['steɪtəs] *n.* **a)** *(position)* Rang, *der;* **rise in ~:** an Ansehen gewinnen; **social ~:** [gesellschaftlicher] Status; **equality of ~ [with sb.]** Gleichstellung [mit jmdm.]; **financial ~:** finanzielle *od.* wirtschaftliche Lage; **b)** *(superior position)* Status, *der*

status: ~ quo [steɪtəs 'kwəʊ] *n.* Status quo, *der;* **~ symbol** *n.* Statussymbol, *das*

statute ['stætju:t] *n.* **a)** *(Law)* Gesetz, *das;* **by ~:** per Gesetz; **b)** *in pl. (rules)* Statut, *das;* Satzung, *die*

'statute-book *n. (Law)* Gesetzbuch, *das*

statutory ['stætjʊtərɪ] *adj.* **a)** *(Law)* gesetzlich ‹*Feiertag, Bestimmung, Erfordernis, Erbe*›; gesetzlich vorgeschrieben ‹*Strafe*›; gesetzlich festgeschrieben ‹*Löhne, Zinssatz*›; gesetzlich festgelegt ‹*Voraussetzung, Sätze, Zeit*›; ~ **law** kodifiziertes Recht; ~ **rights** [gesetzliche] Rechte; **b)** *(relating to the statutes of an institution)* Satzungs‹*bestimmungen*›; von der Satzung vorgesehen ‹*Geldbuße usw.*›

[1]staunch [stɔ:ntʃ, stɑ:ntʃ] *adj.* treu ‹*Freund, Anhänger*›; streitbar ‹*Kämpfer, Anhänger*›; überzeugt ‹*Katholik, Demokrat usw.*›; unerschütterlich ‹*Mut, Hingabe, Glaube*›; standhaft ‹*Herz*›

[2]staunch *see* **stanch**

stave [steɪv] **1.** *n. (Mus.) see* **staff 1 d. 2.** *v. t.,* **~d** *or* **stove** [stəʊv] ein Loch schlagen in (+ *Akk.*)

~ **'in** *v. t. (crush)* eindrücken ‹*Karosserie, Tür, Rippen*›; einschlagen ‹*Kopf, Kiste*›; *(break hole in)* ein Loch schlagen in (+ *Akk.*)

~ **'off** *v. t.,* **~d off** abwenden; abwehren ‹*Angriff*›; verhindern ‹*Krankheit*›; stillen ‹*Hunger, Durst*›

stay [steɪ] **1.** *n.* **a)** Aufenthalt, *der; (visit)* Besuch, *der;* **during her ~ with us** während sie bei uns zu Besuch war; **come/go for a short ~ with sb.** jmdn. kurz besuchen; **have a week's ~ in London** eine Woche in London verbringen; **b)** *(Law)* ~ **[of execution]** Aussetzung [der Vollstreckung]; *(fig.)* Galgenfrist, *die.* **2.** *v. i.* **a)** *(remain)* bleiben; **be here to ~, have come to ~:** sich fest eingebürgert haben; ‹*Arbeitslosigkeit, Inflation:*› zum Dauerzustand geworden sein; ‹*Modeartikel:*› in Mode bleiben; ~ **for** *or* **to dinner/for the party** zum Essen/zur Party bleiben; ~ **put** *(coll.)*‹*Ball, Haar:*› liegen bleiben; ‹*Hut:*› fest sitzen;

⟨*Bild:*⟩ hängen bleiben; ⟨*Person:*⟩ bleiben[, wo man ist]; **b)** *(dwell temporarily)* wohnen; ~ **abroad** im Ausland leben; ~ **the night in a hotel** die Nacht in einem Hotel verbringen; ~ **at sb.'s** *or* **with sb. for the weekend** das Wochenende bei jmdm. verbringen; **c)** *(Sport)* durchhalten. **3.** *v. t.* **a)** *(arch./literary: stop)* aufhalten; ~ **sb.'s hand** *(fig.)* jmdn. zurückhalten; **b)** *(endure)* ~ **the course** *or* **distance** die [ganze] Strecke durchhalten; *(fig.)* durchhalten

~ **a'way** *v. i.* **a)** *(not attend)* ~ **away [from sth.]** [von etw.] wegbleiben; [einer Sache *(Dat.)*] fernbleiben; ~ **away from school/a meeting** nicht zur Schule/zu einem Treffen gehen/kommen; **b)** *(~ distant)* **he ~ed well away from the wall** er hielt sich ein gutes Stück von der Wand entfernt

~ **'back** *v. i.* **a)** *(not approach)* zurückbleiben; **b)** *see* ~ **behind**

~ **be'hind** *v. i.* zurückbleiben; **have to ~ behind [after school]** nachsitzen müssen

~ **'down** *v. i.* **a)** *(remain lowered)* unten bleiben; **b)** *(not increase)* stabil bleiben; **c)** *(Educ.: not go to higher form)* sitzenbleiben *(ugs.)*

~ **'in** *v. i.* **a)** *(remain in position)* halten; **will these creases ~ in?** bleiben diese Falten [drin *(ugs.)*]?; **this passage [of the book] should ~ in** diese Passage sollte nicht gestrichen werden; **b)** *(remain at home)* zu Hause bleiben

~ **off** *v. t.* ~ **off the bottle/off drugs** die Finger vom Alkohol/von Drogen lassen *(ugs.)*

~ **'on** *v. i.* **a)** *(remain in place)* ⟨*Hut, Perücke, Kopftuch:*⟩ sitzen bleiben; ⟨*falsche Wimpern, Aufkleber:*⟩ haften; ⟨*Deckel, Rad:*⟩ halten; **b)** *(remain in operation)* angeschaltet bleiben; anbleiben *(ugs.)*; **c)** *(remain present)* noch [da]bleiben; ~ **on at school** auf der Schule bleiben; ~ **on as chairman** Vorsitzender bleiben

~ **'out** *v. i.* **a)** *(not go home)* wegbleiben *(ugs.)*; nicht nach Hause kommen/gehen; **don't ~ out late!** komm nicht zu spät nach Hause!; **b)** *(remain outside)* draußen bleiben; **c)** *(fig.)* ~ **out of sb.'s way** jmdm. aus dem Wege gehen; **d)** *(remain on strike)* ~ **out [on strike]** im Ausstand bleiben

~ **'up** *v. i.* **a)** *(not go to bed)* aufbleiben; **b)** *(remain in position)* ⟨*Pfosten, Gebäude:*⟩ stehenbleiben; ⟨*Plakat:*⟩ hängen bleiben; ⟨*Flugzeug, Haare:*⟩ oben bleiben

'stay-at-home 1. *n.* häuslicher Mensch. **2.** *attrib. adj.* häuslich

'staying-power *n.* Durchhaltevermögen, *das*

STD *abbr. (Brit. Teleph.)* **subscriber trunk dialling** Selbstwählfernverkehr, *der;* ~ **code** Vorwahl[nummer], *die*

stead [sted] *n., no pl., no art.* **a)** **in sb.'s ~:** an jmds. Stelle; **b)** **stand sb. in good ~:** jmdm. zustatten kommen; **that car has stood her in good ~:** dieser Wagen hat ihr gute Dienste geleistet

steadfast ['stedfəst, 'stedfɑ:st] *adj.* standhaft; zuverlässig ⟨*Freund*⟩; fest ⟨*Entschluß*⟩; unverwandt ⟨*Blick*⟩; unerschütterlich ⟨*Glaube*⟩; unverbrüchlich *(geh.)* ⟨*Freundschaft, Treue*⟩

steadily ['stedɪlɪ] *adv.* **a)** *(stably)* fest; festen Schrittes ⟨*gehen*⟩; sicher ⟨*radfahren*⟩; **b)** *(without faltering)* fest ⟨*[an]blicken*⟩; **c)** *(continuously)* stetig; ohne Unterbrechung ⟨*arbeiten, marschieren*⟩; **it was raining ~:** es hat ununterbrochen geregnet; **progress ~:** stetige Fortschritte machen; **d)** *(firmly)* standhaft ⟨*sich weigern*⟩; fest ⟨*glauben*⟩; **e)** *(reliably)* zuverlässig

steady ['stedɪ] **1.** *adj.* **a)** *(stable)* stabil; *(not wobbling)* standfest; **as ~ as a rock** völlig standfest ⟨*Leiter, Tisch*⟩; völlig stabil ⟨*Boot*⟩; ganz ruhig ⟨*Hand*⟩; **be ~ on one's feet** *or* **legs/bicycle** sicher auf den Beinen sein/sicher auf seinem Fahrrad fahren; **hold** *or* **keep the ladder ~:** die Leiter festhalten; ~ **as she goes!** *(coll.)* immer so weiter!; **b)** *(still)* ruhig; **turn a ~ gaze** *or* **look on sb.** jmdn. fest ansehen; **c)** *(regular, constant)* stetig; gleichmäßig ⟨*Tempo*⟩, stabil ⟨*Preis, Lohn*⟩; gleichbleibend ⟨*Temperatur*⟩; beständig ⟨*Klima, Summen, Lärm*⟩; **we had ~ rain/drizzle** wir hatten Dauerregen/es nieselte [bei uns] ständig; **~!** Vorsicht!; *(to dog, horse)* ruhig!; ~ **on!** langsam! *(ugs.)*; **d)** *(invariable)* unerschütterlich; beständig ⟨*Wesensart*⟩; standhaft ⟨*Weigerung*⟩; fest ⟨*Charakter, Glaube*⟩; **e)** *(enduring)* **a ~ job** eine feste Stelle; **a ~ boy-friend/girl-friend** ein fester Freund/eine feste Freundin *(ugs.)*. **2.** *v. t.* festhalten ⟨*Leiter*⟩; beruhigen ⟨*Pferd, Nerven*⟩; ruhig halten ⟨*Boot, Flugzeug*⟩; **she steadied herself against the table/with a stick** sie hielt sich am Tisch fest/stützte sich mit einem Stock. **3.** *v. i.* ⟨*Preise:*⟩ sich stabilisieren; ⟨*Geschwindigkeit:*⟩ sich mäßigen. **4.** *adv.* **go ~ with sth.** mit etw. vorsichtig sein; **go ~ with sb.** *(coll.)* mit jmdm. gehen *(ugs.)*

steak [steɪk] *n.* Steak, *das; (of ham, bacon, gammon, salmon, etc.)* Scheibe, *die;* **a chicken/turkey/veal ~:** ein Hähnchen-/Puten-/Kalbsschnitzel; ~ **and kidney pie/pudding** Rindfleisch-Nieren-Pastete, *die*

steal [sti:l] **1.** *v. t.,* **stole** [stəʊl], **stolen** ['stəʊln] **a)** stehlen (**from** *Dat.*); ~ **sb.'s boy-friend/girl-friend** jmdm. den Freund/die Freundin ausspannen *(ugs.)*; **she was the star of the play, but the little dog stole the show** *(fig.)* sie war der Star des Stückes, aber der kleine Hund stahl ihr die Schau; **b)** *(get slyly)* rauben *(geh. scherzh.)* ⟨*Kuß, Umarmung*⟩; entlocken ⟨*Worte, Interview*⟩; sich *(Dat.)* genehmigen *(ugs. scherzh.)* ⟨*Nickerchen*⟩; ~ **a glance [at sb./sth.]** jmdm. einen verstohlenen Blick zuwerfen/einen verstohlenen Blick auf etw. *(Akk.)* werfen; **c)** *(fig.: win)* **she stole my heart** sie eroberte mein Herz. **2.** *v. i.,* **stole, stolen a)** stehlen; ~ **from sb.** jmdn. bestehlen; ~ **from the till/supermarket** aus der Kasse/im Supermarkt stehlen; **b)** *(move furtively)* sich stehlen; ~ **in/out/up** sich hinein-/hinaus-/hinaufstehlen; ~ **up [on sb./sth.]** sich [an jmdn./etw.] heranschleichen

~ **a'way** *v. i.* sich fortstehlen

stealth [stelθ] *n.* Heimlichkeit, *die;* **by ~:** heimlich

stealthy ['stelθɪ] *adj.* heimlich; verstohlen ⟨*Blick, Bewegung, Tun*⟩

steam [sti:m] **1.** *n., no pl., no indef. art.* Dampf, *der;* **the window was covered with ~:** das Fenster war beschlagen; **get up ~:** Dampf aufmachen; **let off ~** *(fig.)* Dampf ablassen *(ugs.)*; **run out of ~:** keinen Dampf mehr haben; *(fig.)* den Schwung verlieren; **under one's own ~** *(fig.)* aus eigener Kraft. **2.** *v. t.* **a)** *(Cookery)* dämpfen; dünsten; **~ed pudding** gedämpfter Pudding; **b)** ~ **open an envelope** einen Umschlag mit [heißem] Wasserdampf öffnen. **3.** *v. i.* dampfen; **~ing hot** dampfend heiß

~ **'up 1.** *v. t.* **a)** beschlagen lassen; **be ~ed up** beschlagen sein; **b)** *(fig. coll.)* **be/get [all] ~ed up** [total] ausrasten *(ugs.)* **2.** *v. i.* beschlagen

steam: ~boat *n.* Dampfschiff, *das;* ~ **engine** *n.* **a)** *(Railw.)*

Dampflok[omotive], *die;* **b)** *(stationary engine)* Dampf[kraft]maschine, *die*

steamer ['sti:mə(r)] *n.* **a)** *(Naut.)* Dampfer, *der;* **b)** *(Cookery)* Dämpfer, *der*

steam: ~ **iron** *n.* Dampfbügeleisen, *das;* ~**roller 1.** *n.* Dampfwalze, *die;* **2.** *v.t.* [mit der Dampfwalze] walzen; ~**ship** *n.* Dampfschiff, *das;* ~ **train** *n.* Dampfzug, *der*

steamy ['sti:mɪ] *adj.* **a)** dunstig; feucht ‹*Hitze*›; beschlagen ‹*Glas*›; **b)** *(coll.: erotic)* heiß

steel [sti:l] **1.** *n.* Stahl, *der;* **as hard as** ~: stahlhart. **2.** *attrib. adj.* stählern; Stahl‹*helm, -block, -platte*›. **3.** *v.t.* ~ **oneself for/against sth.** sich für/gegen etw. wappnen *(geh.);* ~ **oneself to do sth.** allen Mut zusammennehmen, um etw. zu tun

steel: ~ '**band** *n.* *(Mus.)* Steelband, *die;* ~ **gui'tar** *n.* *(Mus.)* Hawaiigitarre, *die;* ~**worker** *n.* Stahlarbeiter, *der*/-arbeiterin, *die;* ~**works** *n. sing., pl. same* Stahlwerk, *das*

steely ['sti:lɪ] *adj.* **a)** *(strong)* stählern; **b)** *(resolute)* eisern; **c)** *(severe)* steinern

[1]**steep** [sti:p] *adj.* **a)** steil; **b)** *(rapid)* stark ‹*Preissenkung*›; steil ‹*Preisanstieg*›; **c)** *(coll.: excessive)* happig *(ugs.);* **the bill is [a bit]** ~: die Rechnung ist [ziemlich] gesalzen *(ugs.);* **be a bit** ~: ein bißchen zu weit gehen

[2]**steep** *v.t.* **a)** *(soak)* einweichen; **b)** *(bathe)* baden

steeped [sti:pt] *adj.* durchdrungen (**in** von); **a place** ~ **in history/tradition** ein geschichtsträchtiger/von der Tradition durchdrungener Ort

steepen ['sti:pn] *v.i.* steil[er] werden

steeple ['sti:pl] *n.* Kirchturm, *der*

steeple: ~**chase** *n.* *(Sport)* **a)** *(horse-race)* Steeplechase, *die;* Hindernisrennen, *das;* **b)** *(Athletics)* Hindernislauf, *der;* ~**jack** *n.* *Arbeiter, der/Arbeiterin, die Reparaturarbeiten an Kaminen, Kirchtürmen usw. ausführt*

steeply ['sti:plɪ] *adv.* steil ‹*ansteigen, abfallen*›

[1]**steer** [stɪə(r)] **1.** *v.t.* **a)** steuern; lenken; **this car is easy to** ~: dieser Wagen ist leicht lenkbar; **b)** *(direct)* ~ **a** *or* **one's way through ...:** steuern durch ...; ~ **a** *or* **one's course for a place** auf einen Ort zusteuern; *(in ship, plane, etc.)* Kurs auf einen Ort nehmen; **c)** *(guide movement of)* führen, lotsen ‹*Person*›; ~ **sb./the conversation towards/away from a subject** jmdn./das Gespräch auf ein Thema lenken/von einem Thema ablenken. **2.** *v.i.* steuern; ~ **clear of sb./sth.** *(fig. coll.)* jmdm./einer Sache aus dem Weg[e] gehen; ~ **for sth.** etw. ansteuern

[2]**steer** *n.* *(Zool.)* junger Ochse

steering ['stɪərɪŋ] *n.* **a)** *(Motor Veh.)* Lenkung, *die;* **b)** *(Naut.)* Ruder, *das;* Steuerung, *die*

steering: ~**-column** *n.* *(Motor Veh.)* Lenksäule, *die;* ~ **committee** *n.* Lenkungsausschuß, *der;* ~**-lock** *n.* *(Motor Veh.)* Lenkradschloß, *das;* ~**-wheel** *n.* **a)** *(Motor Veh.)* Lenkrad, *das;* **b)** *(Naut.)* Steuerrad, *das*

[1]**stem** [stem] **1.** *n.* **a)** *(Bot.) (of tree, shrub)* Stamm, *der;* *(of flower, leaf, fruit)* Stiel, *der;* **b)** *(of glass)* Stiel, *der;* **c)** *(of tobacco-pipe)* Pfeifenrohr, *das;* **d)** *(Ling.)* Stamm, *der.* **2.** *v.i.,* **-mm-:** ~ **from sth.** auf etw. *(Akk.)* zurückzuführen sein

[2]**stem** *v.t.,* **-mm-** *(check, dam up)* aufhalten; eindämmen ‹*Flut*›; stillen ‹*Blutung, Wunde*›; *(fig.)* Einhalt gebieten (+ *Dat.*) *(geh.);* stoppen ‹*Redefluß*›

stench [stentʃ] *n.* Gestank, *der* *(abwertend)*

stencil ['stensl] **1.** *n.* **a)** ~ **[-plate]** Schablone, *die;* **b)** *(for duplicating)* Matrize, *die;* **c)** *(~led pattern/lettering)* schabloniertes Muster/schablonierte Schrift. **2.** *v.t., (Brit.)* **-ll-** mit einer Schablone zeichnen; schablonieren

step [step] **1.** *n.* **a)** *(movement, distance)* Schritt, *der;* **at every** ~: mit jedem Schritt; **watch sb.'s every** ~ *(fig.)* jmdn. auf Schritt und Tritt überwachen; **take a** ~ **towards/away from sb.** einen Schritt auf jmdn. zugehen/von jmdm. wegtreten; **take a** ~ **back/sideways/forward** einen Schritt zurücktreten/zur Seite treten/nach vorn treten; **a** ~ **forward/back** *(fig.)* ein Schritt nach vorn/zurück; **a** ~ **in the right/wrong direction** *(fig.)* ein Schritt in die richtige/falsche Richtung; **mind** *or* **watch your** ~! *(lit. or fig.)* paß auf!; **I can't walk another** ~: ich kann keinen Schritt mehr gehen; **b)** *(stair)* Stufe, *die;* *(on vehicle)* Tritt, *der;* **a flight of** ~**s** eine Treppe; **[pair of]** ~**s** *(ladder)* Stehleiter, *die;* *(small)* Trittleiter, *die;* **c) follow** *or* **walk in sb.'s** ~**s** *(fig.)* in jmds. Fußstapfen treten; **d)** *(short distance)* **it's only a** ~ **to my house** es sind nur ein paar Schritte bis zu mir; **e) be in** ~: im Schritt sein; *(with music, in dancing)* im Takt sein; **be in/out of** ~ **with sth.** *(fig.)* mit etw. Schritt/nicht Schritt halten; **be out of** ~: aus dem Schritt geraten sein; *(with music, in dancing)* nicht im Takt sein; **f)** *(action)* Schritt, *der;* **take** ~**s to do sth.** Schritte unternehmen, um etw. zu tun; **g)** ~ **by** ~: Schritt für Schritt; **what is the next** ~? wie geht es weiter?; **h)** *(grade)* Stufe, *die.* **2.** *v.i.,* **-pp-** treten; ~ **lightly** *or* **softly** leise auftreten; ~ **inside** eintreten; **please** ~ **inside for a moment** kommen Sie bitte auf einen Augenblick herein; ~ **into sb.'s shoes** *(fig.)* an jmds. Stelle treten; ~ **on sth.** *(on the ground)* auf etw. *(Akk.)* treten; ~ **on [to]** steigen auf (+ *Akk.*); steigen in (+ *Akk.*) ‹*Fahrzeug, Flugzeug*›; ~ **on it** *(coll.)* auf die Tube drücken *(ugs.);* ~ **on sb.'s toes** *(lit. or fig.)* jmdm. auf die Füße treten; ~ **out of one's dress/trousers** aus seinem Kleid/seiner Hose steigen *(ugs.);* ~ **over sb./sth.** über jmdn./etw. steigen

~ **a'side** *v.i.* **a)** zur Seite treten; **b)** *(fig.: resign)* seine Stellung räumen

~ '**back** *v.i.* zurücktreten; ~ **back in fright/surprise** vor Schreck/Überraschung [einen Schritt] zurückweichen

~ '**down** *v.i.* **a)** ~ **down from the train/into the boat** aus dem Zug/in das Boot steigen; **b)** *(fig.)* see **stand down**

~ '**forward** *v.i.* **a)** [einen Schritt] vortreten; **b)** *(fig.: present oneself)* sich melden; **would somebody like to** ~ **forward and help with the trick?** würde jemand gern nach vorn kommen und bei dem Trick assistieren?

~ '**in** *v.i.* **a)** eintreten; *(into vehicle)* einsteigen; *(into pool)* hineinsteigen; **b)** *(fig.) (take sb.'s place)* einspringen; *(intervene)* eingreifen

~ '**off 1.** *v.i.* *(from vehicle)* aussteigen; *(from a height)* hinabspringen. **2.** *v.t.* *(get off)* steigen aus ‹*Fahrzeug*›; treten von ‹*Bürgersteig*›

~ '**out** *v.i.* hinausgehen; **the car/boat stopped and she** ~**ped out** der Wagen/das Boot hielt an und sie stieg aus

~ '**up 1.** *v.i.* **a)** *(ascend)* hinaufsteigen; ~ **up into** [ein]steigen in (+ *Akk.*) ‹*Fahrzeug*›; ~ **up on to** steigen auf (+ *Akk.*) ‹*Podest, Tisch*›; **b)** *(approach)* ~ **right up!** treten Sie näher!; ~ **up to sb.** zu jmdm. treten; **c)** *(increase)* zuneh-

men. **2.** *v.t.* erhöhen; intensivieren ⟨*Wahlkampf*⟩; verstärken ⟨*Anstrengungen*⟩; verschärfen ⟨*Sicherheitsmaßnahmen, Streik*⟩
step: ~brother *n.* Stiefbruder, *der;* **~child** *n.* Stiefkind, *das;* **~daughter** *n.* Stieftochter, *die;* **~father** *n.* Stiefvater, *der;* **~-ladder** *n.* Stehleiter, *die;* **~mother** *n.* Stiefmutter, *die*
steppe [step] *n.* *(Geog.)* Steppe, *die*
'stepping-stone *n.* Trittstein, *der;* *(fig.)* Sprungbrett, *das* (to für, in)
step: ~sister *n.* Stiefschwester, *die;* **~son** *n.* Stiefsohn, *der*
stereo ['sterɪəʊ] **1.** *n., pl.* **~s a)** *(equipment)* Stereoanlage, *die;* **b)** *no pl. (sound reproduction)* Stereo, *das.* **2.** *adj.* stereo; Stereo⟨*effekt, -aufnahme, -platte*⟩
stereophonic [sterɪə'fɒnɪk] *adj.* stereophon
stereoscopic [sterɪə'skɒpɪk] *adj.* stereoskopisch
stereotype ['sterɪətaɪp] **1.** *n.* Stereotyp, *das (Psych.);* Klischee, *das.* **2.** *v.t.* in ein Klischee zwängen; **~d** stereotyp ⟨*Redensart, Frage, Vorstellung*⟩; klischeehaft ⟨*Sprache, Denkweise*⟩
sterile ['steraɪl] *adj.* **a)** *(germ-free)* steril; **b)** *(barren, lit. or fig.)* steril; *(fig.)* nutzlos ⟨*Tätigkeit*⟩; fruchtlos ⟨*Diskussion, Gespräch*⟩
sterilize (sterilise) ['sterɪlaɪz] *v.t.* sterilisieren
sterling ['stɜ:lɪŋ] **1.** *n., no pl., no indef. art.* Sterling, *der;* **five pounds ~:** fünf Pfund Sterling; **in ~:** in Pfund [Sterling]. **2.** *attrib. adj.* **a) ~ silver** Sterlingsilber, *das;* **b)** *(fig.)* gediegen; **do ~ work** erstklassige Arbeit leisten
¹stern [stɜ:n] *adj.* streng; hart ⟨*Strafe*⟩; ernst ⟨*Warnung*⟩
²stern *n.* *(Naut.)* Heck, *das*
sternly ['stɜ:nlɪ] *adv.* streng; ernsthaft ⟨*warnen*⟩; in strengem Ton ⟨*sprechen*⟩
steroid ['stɪərɔɪd, 'sterɔɪd] *n.* *(Chem.)* Steroid, *das*
stet [stet] *(Printing) v.i. imper.* bleibt
stethoscope ['steθəskəʊp] *n.* *(Med.)* Stethoskop, *das*
stetson ['stetsn] *n.* Stetson[hut], *der*
stevedore ['sti:vədɔ:(r)] *n.* *(Naut.)* Schauermann, *der*
stew [stju:] **1.** *n.* *(Gastr.)* Eintopf, *der;* **Irish ~:** Irish-Stew, *das.* **2.** *v.t.* *(Cookery)* schmoren [lassen]; **~ apples/plums** Apfel-/Pflaumenkompott kochen. **3.** *v.i.* *(Cookery)* schmoren; ⟨*Obst:*⟩ gedünstet werden; **~ [in one's own juice]** *(fig.)* [im eigenen Saft] schmoren *(ugs.)*
steward ['stju:əd] *n.* **a)** *(on ship, plane)* Steward, *der;* **b)** *(at public meeting, ball, etc.)* Ordner, *der*/Ordnerin, *die;* **~s** *(of race)* Rennleitung, *die;* **c)** *(estate manager)* Verwalter, *der*/Verwalterin, *die*
stewardess ['stju:ədɪs] *n.* Stewardeß, *die*
'stewing steak *n., no pl., no indef. art.* [Rinder]schmorfleisch, *das*
stick [stɪk] **1.** *v.t.,* **stuck** [stʌk] **a)** *(thrust point of)* stecken; **~ sth. in[to] sth.** mit etw. in etw. *(Akk.)* stechen; **get stuck into sb./sth./a meal** *(sl.: begin action)* jmdm. eine Abreibung verpassen/sich in etw. *(Akk.)* reinknien/tüchtig reinhauen *(salopp);* **b)** *(impale)* spießen; **~ sth. [up]on sth.** etw. auf etw. *(Akk.)* [auf]spießen; **c)** *(coll.: put)* stecken; **he stuck a feather in his hat** er steckte sich *(Dat.)* eine Feder an den Hut; **~ a picture on the wall/a vase on the shelf** ein Bild an die Wand hängen/eine Vase aufs Regal stellen; **~ sth. in the kitchen** etw. in die Küche tun *(ugs.);* **~ one on sb.** *(sl.: hit)* jmdm. eine langen *(ugs.);* **you know where you can ~ that!, [you can] ~ it!** *(sl.)* das kannst du dir sonstwohin stecken!; **d)** *(with glue etc.)* kleben; **e)** *(make immobile)* **the car is stuck in the mud** das Auto ist im Schlamm steckengeblieben; **the door is stuck** die Tür klemmt [fest]; **f)** *(puzzle)* **be stuck for an answer/for ideas** um eine Antwort/um Ideen verlegen sein; **Can you help me with this problem? I'm stuck** Kannst du mir bei diesem Problem helfen? Ich komme nicht weiter; **g)** *(cover)* **~ sth. with pins/needles** Stecknadeln/Nadeln in etw. *(Akk.)* stecken; **h)** *(Brit. coll.: tolerate)* **~ it** durchhalten; **she can't ~ him** sie kann ihn nicht riechen *(salopp);* **i)** *(coll.)* **be stuck with sth.** *(have to accept)* sich mit etw. herumschlagen müssen *(ugs.);* **be stuck with sb.** jmdn. am *od.* auf dem Hals haben *(ugs.).* **2.** *v.i.,* **stuck a)** *(be fixed by point)* stecken; **b)** *(adhere)* kleben; **~ to sth.** an etw. *(Dat.)* kleben; **~ in the/sb.'s mind** *(fig.)* im/jmdm. im Gedächtnis haftenbleiben; **c)** *(become immobile)* ⟨*Auto, Räder:*⟩ steckenbleiben; ⟨*Schublade, Tür, Griff, Bremse:*⟩ klemmen; ⟨*Schlüssel:*⟩ feststecken; **~ fast** ⟨*Auto, Rad:*⟩ feststecken; ⟨*Reißverschluß, Tür, Schublade:*⟩ festklemmen; **the record is stuck** die Platte ist hängengeblieben; **d)** *(protrude)* **a letter stuck from his pocket** ein Brief schaute ihm aus der Tasche. **3.** *n.* **a)** *([cut] shoot of tree, piece of wood; also for punishment)* Stock, *der;* *(staff)* [Holz]stab, *der;* *(walking-~)* Spazierstock, *der;* *(for handicapped person)* Krückstock, *der;* **b)** *(Hockey etc.)* Schläger, *der;* **c)** *(long piece)* **a ~ of chalk/shaving-soap** ein Stück Kreide/Rasierseife; **a ~ of rock/celery/rhubarb** eine Zuckerstange/eine Stange Sellerie/Rhabarber; **d)** *no pl., no art. (coll.: criticism)* **get** *or* **take [some] ~:** viel einstecken müssen; **give sb. [some] ~:** jmdn. zusammenstauchen *(ugs.)*
~ a'bout, ~ a'round *v.i. (coll.)* dableiben; *(wait)* warten
~ at *v.t.* **~ at one's books/studying** fleißig Bücher wälzen/studieren; **~ 'at it** *(coll.)* dranbleiben *(ugs.)*
~ by *v.t. (fig.)* stehen zu
~ 'down *v.t.* festkleben; zukleben ⟨*Umschlag*⟩
~ 'in *v.t.* **a)** *(jab in)* hineinstechen ⟨*Spritze, Nadel*⟩; anstecken ⟨*Hutnadel*⟩; **b)** *(glue in)* einkleben; **c)** *(coll.: put in)* hineinstecken
~ 'on 1. *v.t.* **a)** *(glue on)* aufkleben ⟨*Briefmarke, Etikett*⟩; ankleben ⟨*Tapete*⟩; **b)** *(attach by pin etc.)* anstecken. **2.** *v.i.* kleben[bleiben]
~ 'out 1. *v.t.* **a)** herausstrecken ⟨*Brust, Zunge*⟩; ausstrecken ⟨*Arm, Bein*⟩; **b) ~ it out** *(coll.)* durchhalten; ausharren. **2.** *v.i.* **a)** *(project)* ⟨*Brust, Bauch:*⟩ vorstehen; ⟨*steifes Kleid:*⟩ abstehen; ⟨*Nagel, Ast:*⟩ herausstehen; **his ears ~ out** er hat abstehende Ohren; **b)** *(fig.: be obvious)* sich abheben; **~ out a mile** *(sl.)* [klar] auf der Hand liegen; **~ out like a sore thumb** *(coll.)* ins Auge springen
~ to *v.t.* **a)** *(be faithful to)* halten zu ⟨*Person*⟩; halten ⟨*Versprechen*⟩; bleiben bei ⟨*Entscheidung, Meinung*⟩; treu bleiben (+ *Dat.*) ⟨*Idealen, Grundsätzen*⟩; **b)** *(not deviate from)* sich halten an (+ *Akk.*) ⟨*Plan, Text, Original*⟩; bleiben an (+ *Dat.*) ⟨*Arbeit*⟩; bleiben bei ⟨*Wahrheit, Thema*⟩; **~ to business** bei der Sache bleiben; **~ to the point** beim Thema bleiben
~ to'gether 1. *v.t.* zusammenkleben. **2.** *v.i.* **a)** *(adhere together)* zusammenkleben; **b)** *(fig.: remain united)* zusammenhalten
~ 'up 1. *v.t.* **a)** *(seal)* zukleben; **b)**

(coll.: put up, raise) strecken, recken ‹*Kopf, Hals*›; anschlagen ‹*Bekanntmachung, Poster*›; aufschlagen ‹*Zelt*›; hinbauen, -setzen ‹*Häuser*›; raufsetzen *(ugs.)* ‹*Preise*›; ~ **up one's hand** die Hand heben; ~ **'em up!** *(sl.)* Pfoten hoch! *(salopp)*; **c)** *(sl.: rob)* ausrauben; **d) stuck up** *(conceited)* eingebildet. **2.** *v.i.* **a)** ‹*Haar, Kragen:*› hochstehen; ‹*Nagel, Pflasterstein:*› hervorstehen; **b)** ~ **up for sb./sth.** für jmdn./etw. eintreten; ~ **up for yourself!** setz dich zur Wehr!

~ **with** *v.t. (coll.)* **a)** *(keep contact with)* ~ **with the leaders** sich an der Spitze halten *(bes. Sport)*; ~ **'with it!** bleib dran! *(ugs.)*; **b)** *(remain faithful to)* bleiben bei ‹*Gruppe, Partei*›; halten zu ‹*Freund*›

sticker ['stɪkə(r)] *n.* Aufkleber, *der*

'sticking-plaster *n. (Med.)* Heftpflaster, *das*

'stick-in-the-mud 1. *n. (person lacking initiative)* Trantüte, *die (ugs. abwertend)*; *(unprogressive person)* Spießer, *der (abwertend)*. **2.** *adj. (lacking in initiative)* schlafmützig *(ugs. abwertend)*; *(unprogressive)* spießig *(abwertend)*

stickleback ['stɪklbæk] *n. (Zool.)* Stichling, *der*

stickler ['stɪklə(r)] *n.* **be a ~ for tidiness/authority** es mit der Sauberkeit sehr genau nehmen/in puncto Autorität keinen Spaß verstehen

stick: ~-on *adj.* selbstklebend: **~-up** *n. (sl.)* bewaffneter Raubüberfall

sticky ['stɪkɪ] *adj.* **a)** klebrig; feucht ‹*Farbe, gestrichener/gewaschener Gegenstand*›; zäh ‹*Teig, Brei, Mischung*›; ~ **label** Aufkleber, *der;* ~ **tape** Klebestreifen, *der;* **b)** *(humid)* schwül ‹*Klima, Luft*›; feucht ‹*Haut*›; **c)** *(sl.: unpleasant)* vertrackt *(ugs.)*; heikel; **a ~ situation** eine brenzlige Lage

stiff [stɪf] *adj.* **a)** *(rigid)* steif; hart ‹*Bürste, Stock*›; **be frozen ~:** steif vor Kälte sein; ‹*Wäsche, Körper[teile]:*› steif gefroren sein; **b)** *(intense, severe)* hartnäckig; schroff ‹*Absage*›; kräftig ‹*Standpauke*›; ~ **competition** scharfe Konkurrenz; **c)** *(formal)* steif; förmlich ‹*Brief, Stil*›; **d)** *(difficult)* hart ‹*Test*›; schwer ‹*Frage, Prüfung*›; steil ‹*Abstieg, Anstieg*›; **be ~ going** *(fig. coll.)* harte Arbeit sein; **e)** stark, *(Seemannsspr.)* steif ‹*Wind, Brise*›; **f)** *(not bending, not working freely, aching)* steif ‹*Gelenk, Gliedmaßen, Nacken, Person*›; schwergängig ‹*Angel, Kolben, Gelenk*›; **g)** *(coll.: excessive)* saftig *(ugs.)* ‹*Preis, Strafe*›; **h)** *(strong)* steif *(ugs.)* ‹*Drink*›; stark ‹*Dosis, Medizin*›; **i)** *(thick)* zäh[flüssig]; **j)** *(coll.)* **be bored/scared/worried ~:** sich zu Tode langweilen/eine wahnsinnige Angst haben *(ugs.)*/sich *(Dat.)* furchtbare *(ugs.)* Sorgen machen

stiffen ['stɪfn] **1.** *v.t.* **a)** steif machen; stärken ‹*Kragen*›; versteifen ‹*Material*›; zäh[flüssig]er machen ‹*Paste, Teig*›; **b)** *(fig.: bolster)* verstärken ‹*Widerstand*›; stärken ‹*Moral, Entschlossenheit*›. **2.** *v.i.* **a)** ‹*Person:*› erstarren; **b)** ‹*Wind, Brise:*› steifer werden *(Seemannsspr.)*, auffrischen; **c)** *(become thicker)* ‹*Teig:*› steifer werden; ‹*Mischung:*› zäher werden; **d)** *(fig.: become more resolute)* sich verstärken

stiffness ['stɪfnɪs] *n., no pl.* **a)** *(rigidity, formality)* Steifheit, *die; (of letter, language)* Förmlichkeit, *die;* **b)** *(intensity)* Härte, *die;* **c)** *(difficulty)* Schwierigkeit, *die;* **d)** *(of wind)* Stärke, *die;* Steifheit, *die (Seemannsspr.)*; **e)** *(lack of suppleness)* Steifheit, *die; (of hinge, piston)* geringe Beweglichkeit; **f)** *(coll.: excessiveness) (of punishment)* Strenge, *die; (of demand, price)* Überzogenheit, *die;* **g)** *(thick consistency)* Zähheit, *die*

stifle ['staɪfl] **1.** *v.t.* ersticken; *(fig.: suppress)* unterdrücken; ersticken ‹*Widerstand, Aufstand, Schrei*›; **we were ~d by the heat** wir erstickten fast vor Hitze. **2.** *v.i.* ersticken

stifling ['staɪflɪŋ] *adj.* stickig; drückend ‹*Hitze*›; *(fig.)* einengend ‹*Atmosphäre*›; erdrückend ‹*Einfluß, Herrschaft*›

stigma ['stɪgmə] *n., pl.* **~s** *or* **~ta** ['stɪgmətə, stɪg'mɑːtə] Stigma, *das (geh.)*; Makel, *der (geh.)*

stile [staɪl] *n.* Zauntritt, *der;* Trittleiter, *die*

stiletto [stɪ'letəʊ] *n., pl.* **~s** *or* **~es** **a)** *(dagger)* Stilett, *das;* **b)** ~ **[heel]** Stöckelabsatz, *der*

¹still [stɪl] **1.** *adj.* **a)** *pred.* still; **be ~:** [still] stehen; ‹*Fahne:*› sich nicht bewegen; ‹*Hand:*› ruhig sein; **hold** *or* **keep sth. ~:** etw. ruhig halten; **hold** *or* **keep a ladder/horse ~:** eine Leiter/ein Pferd festhalten; **hold ~!** halt still!; **keep** *or* **stay ~:** stillhalten; *(not change posture)* ruhig bleiben; ‹*Pferd:*› stillstehen; ‹*Gegenstand:*› liegenbleiben; **sit ~:** stillsitzen; **stand ~:** stillstehen; ‹*Uhr:*› stehen; ‹*Arbeit:*› ruhen; *(stop)* stehenbleiben; **b)** *(calm)* ruhig; **c)** *(without sound)* still; ruhig; **d)** *(not sparkling)* nicht moussierend ‹*Wein*›; still ‹*Mineralwasser*›; **e)** *(hushed)* leise. **2.** *adv.* **a)** *(without change)* noch; *expr. surprise or annoyance* immer noch; **drink your tea while it is ~ hot** trink deinen Tee, solange er [noch] heiß ist; **b)** *(nevertheless)* trotzdem; **~, what can you do about it?** aber was kann man dagegen tun?; **c)** *with comparative (even)* noch; **become fatter ~** *or* **~ fatter** noch *od.* immer dicker werden; **better/worse ~** *as sentence-modifier* besser/schlimmer noch. **3.** *n. (Photog.)* Fotografie, *die*

²still *n.* Destillierapparat, *der*

still: ~born *adj.* totgeboren; **the child was ~born** das Kind war eine Totgeburt *od.* kam tot zur Welt; ~ **life** *n., pl.* ~ **lifes** *or* **lives** *(Art)* Stilleben, *das*

stillness ['stɪlnɪs] *n., no pl.* **a)** *(motionlessness)* Bewegungslosigkeit, *die;* **b)** *(quietness)* Stille, *die*

stilt [stɪlt] *n.* Stelze, *die*

stilted ['stɪltɪd] *adj.* gestelzt; gespreizt

stimulant ['stɪmjʊlənt] **1.** *attrib. adj. (Med.)* stimulierend. **2.** *n. (lit. or fig.)* Stimulans, *das;* Anregungsmittel, *das*

stimulate ['stɪmjʊleɪt] *v.t.* **a)** anregen; stimulieren *(geh.)*; beleben ‹*Körper*›; *(sexually)* erregen; **b)** *(fig.)* anregen ‹*Geist, Diskussion, Appetit*›; hervorrufen ‹*Reaktion*›; wecken ‹*Interesse, Neugier*›; beleben ‹*Wirtschaft, Wachstum, Markt, Absatz*›

stimulation [stɪmjʊ'leɪʃn] *n.* **a)** Anregung, *die;* Stimulierung, *die (geh.)*; *(sexual)* Erregung, *die;* **b)** *(fig.)* Anregung, *die; (of reaction)* Hervorrufen, *das; (of interest, curiosity)* Wecken, *das; (of economy, market, growth, sales)* Belebung, *die*

stimulus ['stɪmjʊləs] *n., pl.* **stimuli** ['stɪmjʊlaɪ] **a)** *(spur)* Ansporn, *der* **(to** zu); **b)** *(rousing effect)* Anregung, *die*

sting [stɪŋ] **1.** *n.* **a)** *(wounding)* Stich, *der; (by jellyfish, nettles)* Verbrennung, *die;* **b)** *(pain)* Stechen, *das;* stechender Schmerz; *(from ointment, cane, whip, wind, rash)* Brennen, *das;* **a ~ in the tail** *(fig.)* ein Pferdefuß; **take the ~ out of sth.** *(fig.)* einer Sache

(Dat.) den Stachel nehmen *(geh.)*; **c)** *(Zool.)* [Gift]stachel, *der;* **d)** *(fraud)* Ding, *das (ugs.); (police operation)* Operation, *die.* **2.** *v. t.*, **stung** [stʌŋ] **a)** *(wound)* stechen; **a bee stung [him on] his arm** eine Biene stach ihm in den Arm; **a jellyfish stung me/my leg** ich habe mich/mein Bein an einer Qualle verbrannt; **b)** *(cause pain to)* **the smoke/the wind stung my eyes** der Rauch/der Wind brannte mir in den Augen; **c)** *(hurt mentally)* tief treffen; [zutiefst] verletzen; **~ing** scharf ‹*Vorwürfe, Anklagen, Kritik*›; **d)** *(incite)* **~ sb. into sth./doing sth.** jmdn. zu etw. anstacheln/dazu anstacheln, etw. zu tun; **e)** *(sl.: swindle)* übers Ohr hauen *(ugs.)*. **3.** *v. i.*, **stung a)** *(feel pain)* brennen; **b)** *(have ~)* stechen

'stinging-nettle *n. (Bot.)* Brennessel, *die*

stingy ['stɪndʒɪ] *adj.* geizig; knaus[e]rig *(ugs.)*; kümmerlich ‹*Spende, Portion, Mahlzeit*›

stink [stɪŋk] **1.** *v. i.*, **stank** [stæŋk] *or* **stunk** [stʌŋk], **stunk a)** stinken (**of** nach); *(fig.)* ‹*Angelegenheit, Korruption:*› zum Himmel stinken; **b)** *(fig.: be repulsive)* **sth. ~s** an etw. (+ *Dat.*) stinkt etwas *(ugs.)*. **2.** *n.* **a)** *(bad smell)* Gestank, *der;* **b)** *(coll.: fuss)* Stunk, *der (ugs.);* **kick up** *or* **raise a ~ about sth.** wegen etw. Stunk machen *(ugs.)*

'stink-bomb *n.* Stinkbombe, *die*

stint [stɪnt] **1.** *v. t.* kurzhalten; **~ oneself [of sth.]** sich [mit etw.] einschränken. **2.** *v. i.* **~ on sth.** an etw. *(Dat.)* sparen. **3.** *n.* **a)** *(allotted amount)* [Arbeits]pensum, *das;* **each of us did a ~ at the wheel** jeder von uns saß eine Zeitlang am Steuer; **b)** *(limitation)* **without ~:** uneingeschränkt

stipulate ['stɪpjʊleɪt] *(demand)* fordern; verlangen; *(lay down)* festlegen; *(insist on)* sich *(Dat.)* ausbedingen

stipulation [stɪpjʊ'leɪʃn] *n.* **a)** *(condition)* Bedingung, *die;* **on** *or* **with the ~ that ...:** unter der Bedingung, daß ...; **b)** *(act) see* **stipulate:** Forderung, *die;* Festlegung, *die;* Ausbedingung, *die*

stir [stɜ:(r)] **1.** *v. t.*, **-rr-: a)** *(mix)* rühren; umrühren ‹*Tee, Kaffee*›; **~ sth. into sth.** etw. in etw. *(Akk.)* [ein]rühren; **b)** *(move)* bewegen; **c)** *(fig.: arouse)* bewegen; wecken ‹*Neugier, Interesse, Gefühle, Phantasie*›. **2.** *v. i.*, **-rr-** *(move)* sich rühren; *(in sleep, breeze)* sich bewegen; **without ~ring** regungslos. **3.** *n., no pl.* Aufregung, *die; (bustle, activity)* Betriebsamkeit, *die;* **cause** *or* **create a [big** *or* **great] ~:** [großes] Aufsehen erregen

~ 'in *v. t.* einrühren

~ 'up *v. t.* **a)** *(disturb)* aufrühren; **b)** *(fig.: arouse, provoke)* wecken ‹*Neugier, Interesse, Leidenschaft*›; aufrütteln ‹*Anhänger, Gefolgsleute*›; entfachen ‹*Liebe, Haß, Streit, Zorn, Revolution*›; schüren ‹*Haß, Feindseligkeit*›

'stir-fry *v. t. (Cookery)* unter Rühren schnell braten

stirring ['stɜ:rɪŋ] *adj.* bewegend ‹*Musik, Theaterstück, Poesie*›; spannend ‹*Roman, Geschichte*›; mitreißend ‹*Auftritt, Rede, Marsch*›; bewegt ‹*Zeiten*›

stirrup ['stɪrəp] *n. (Riding)* Steigbügel, *der*

stitch [stɪtʃ] **1.** *n.* **a)** *(Sewing: pass of needle)* Stich, *der;* **b)** *(result of needle movement) (Knitting, Crocheting)* Masche, *die; (Sewing, Embroidery)* Stich, *der;* **drop a ~** *(Knitting)* eine Masche fallenlassen; **c)** *(coll.: piece of clothing)* **not have a ~ on** splitter[faser]nackt *(ugs.)* sein; **d)** *(pain)* **[have] a ~ [in the side]** Seitenstechen [haben]; **e)** *(coll.)* **be in ~es** sich kugeln vor Lachen *(ugs.)*; **f)** *(Med.: to sew up wound)* Stich, *der;* **~es** Naht, *die;* **he had his ~es taken out** ihm wurden die Fäden gezogen. **2.** *v. t.* nähen; *(Embroidery)* sticken. **3.** *v. i.* nähen; *(Embroidery)* sticken

~ 'on *v. t.* annähen ‹*Knopf*›; aufnähen ‹*Flicken, Borte*›

~ 'up *v. t.* nähen; zusammennähen ‹*Stoffteile*›; vernähen ‹*Loch, Riß, Wunde*›

stoat [stəʊt] *n.* Hermelin, *das*

stock [stɒk] **1.** *n.* **a)** *(origin, family, breed)* Abstammung, *die;* **be** *or* **come of farming/French ~:** bäuerlicher/französischer Herkunft sein; **b)** *(supply, store)* Vorrat, *der; (in shop etc.)* Warenbestand, *der;* **our ~s of food/sherry** unsere Lebensmittelvorräte/unser Vorrat an Sherry *(Dat.)*; **be in ~/out of ~:** vorrätig/nicht vorrätig sein; **have sth. in ~:** etw. auf *od. (Kaufmannsspr.)* am Lager haben; **keep sth. in ~** *(have available as a general policy)* etw. führen; **take ~:** Inventur machen; *(fig.)* Bilanz ziehen; **take ~ of sth.** *(fig.)* über etw. *(Akk.)* Bilanz ziehen; **take ~ of one's situation/prospects** seine Situation/seine Zukunftsaussichten bestimmen; **c)** *(Cookery)* Brühe, *die;* **d)** *(Finance)* Wertpapiere; *(shares)* Aktien; **sb.'s ~ is high/low** *(fig.)* jmds. Aktien stehen gut/schlecht *(fig.)*; **e)** *(Hort.)* Stamm, *der; (for grafting)* Unterlage, *die;* **f)** *(handle)* Griff, *der; (of gun)* Schaft, *der;* **g)** *(Agric.)* Vieh, *das;* **h)** *(raw material)* [Roh]material, *das;* **[film] ~:** Filmmaterial, *das.* **2.** *v. t.* **a)** *(supply with ~)* beliefern; **~ a pond/river/lake with fish** einen Teich/Fluß/See mit Fischen besetzen; **b)** *(Commerc.: keep in ~)* auf *od. (fachspr.)* am Lager haben; führen. **3.** *attrib. adj.* **a)** *(Commerc.)* vorrätig; **a ~ size/model** eine Standardgröße/ein Standardmodell; **b)** *(fig.: trite, unoriginal)* abgedroschen *(ugs.)*; **~ character** Standardrolle, *die*

~ 'up 1. *v. i.* **~ up [with sth.]** sich *(Dat.)* einen Vorrat an etw. *(Dat.)* anlegen; **~ up on sth.** seine Vorräte an etw. *(Dat.)* auffüllen. **2.** *v. t.* auffüllen; mit Fischen besetzen ‹*Teich, Fluß, See*›

stockade [stɒ'keɪd] *n.* Palisade, *die*

stock: ~broker *n. (Finance)* Effektenmakler, *der*/-maklerin, *die;* **~broking** ['stɒkbrəʊkɪŋ] *n., no pl. (Finance)* Effektenhandel, *der;* **~ cube** *n. (Cookery)* Brühwürfel, *der;* **~ exchange** *n. (Finance)* Börse, *die;* **the S~ Exchange** *(Brit.)* die [Londoner] Börse

stocking ['stɒkɪŋ] *n.* Strumpf, *der;* **in one's ~[ed] feet** in Strümpfen; **hang up one's ~:** den Strumpf für den Weihnachtsmann aufhängen

stocking: ~-filler *(Brit.)*, **~-stuffer** *(Amer.) ns.* **a)** *kleines Geschenk, das in den für den Weihnachtsmann aufgehängten Strumpf gesteckt wird;* **b)** zusätzliche Kleinigkeit *(als Weihnachtsgeschenk)*

stock-in-'trade *n.* Inventar, *das; (workman's tools)* Handwerkszeug, *das; (fig.: resource)* [festes] Repertoire; **be the ~ of sb.** zu jmds. festem Repertoire gehören

stockist ['stɒkɪst] *n. (Brit. Commerc.)* Fachhändler/-händlerin [mit größerem Warenlager]

stock: ~-market *n. (Finance)* **a)** *see* **stock exchange; b)** *(trading)* Börsengeschäft, *das;* **~pile 1.** *n.* Vorrat, *der; (of weapons)* Arsenal, *das;* **2.** *v. t.* horten; anhäufen ‹*Waffen*›; **~-pot** *n. (Cookery)* Suppentopf, *der;* **~-room** *n.* Lager, *das;* **~-'still** *pred. adj.* bewegungslos; **stand ~-still** regungslos [da]stehen; **~-taking** *n. (Commerc.)* Inventur, *die;* **closed**

for ~-**taking** wegen Inventur geschlossen

stocky ['stɒkɪ] *adj.* stämmig

stockyard [stɒkjɑ:d] *n.* Viehhof, *der*

stodgy ['stɒdʒɪ] *adj.* pappig [und schwerverdaulich] ‹*Essen*›

stoic ['stəʊɪk] **1.** *n.* **a)** S~ *(Philos.)* Stoiker, *der;* **b)** *(impassive person)* Stoiker, *der*/Stoikerin, *die.* **2.** *adj.* **a)** S~ *(Philos.)* stoisch; **b)** stoisch ‹*Person, Ablehnung, Antwort usw.*›

stoical ['stəʊɪkl] *adj.* stoisch

stoke [stəʊk] *v.t.* heizen ‹*Ofen, Kessel*›; unterhalten ‹*Feuer*›

~ **'up 1.** *v.t.* aufheizen ‹*Kessel, Ofen, Dampfmaschine*›. **2.** *v.i.* *(coll.: feed oneself)* sich vollstopfen *(ugs.)*

stoker ['stəʊkə(r)] *n.* Heizer, *der*/Heizerin, *die*

stole *see* **steal**

stolen ['stəʊln] **1.** *see* **steal. 2.** *attrib. adj.* heimlich ‹*Vergnügen, Kuß*›; verstohlen ‹*Blick*›; ~ **goods** Diebesgut, *das;* **receiver of ~ goods** Hehler, *der*/Hehlerin, *die*

stolid ['stɒlɪd] *adj.* stur *(ugs.);* unbeirrbar ‹*Entschlossenheit*›; hartnäckig ‹*Schweigen, Weigerung, Gleichgültigkeit*›

stomach ['stʌmək] **1.** *n.* **a)** *(Anat., Zool.)* Magen, *der;* **on an empty ~:** mit leerem Magen ‹*arbeiten, fahren, weggehen*›; auf nüchternen Magen ‹*Alkohol trinken, Medizin einnehmen*›; **on a full ~:** mit vollem Magen; **turn sb.'s ~:** jmdm. den Magen umdrehen *(ugs.);* **b)** *(abdomen, paunch)* Bauch, *der;* **have a pain in one's ~:** Bauchschmerzen haben; **c)** **have the/no ~ [for sth.]** *(wish/not wish to eat)* Appetit/keinen Appetit [auf etw. *(Akk.)*] haben; *(fig.: courage)* Mut/keinen Mut [zu etw.] haben. **2.** *v.t.* **a)** *(eat, drink)* herunterbekommen *(ugs.); (keep down)* bei sich behalten; **b)** *(fig.: tolerate)* ausstehen; akzeptieren ‹*Vorstellung, Vorgehen, Rat*›

stomach: ~-ache *n.* Magenschmerzen *Pl.;* **have a ~-ache** Magenschmerzen haben; ~ **upset** *n.* Magenverstimmung, *die*

stone [stəʊn] **1.** *n.* **a)** *(also Med., Bot.)* Stein, *der;* **[as] hard as [a] ~:** steinhart; **throw ~s/a ~ at sb.** jmdn. mit Steinen bewerfen/einen Stein auf jmdn. werfen; **only a ~'s throw [away]** *(fig.)* nur einen Steinwurf weit entfernt; **leave no ~ unturned** *(fig.)* Himmel und Hölle in Bewegung setzen; **b)** *(gem)* [Edel]stein, *der;* **c)** *pl. same (Brit.: weight unit) Gewicht von 6,35 kg.* **2.** *adj.* steinern; Stein‹*hütte, -kreuz, -mauer, -brücke*›. **3.** *v.t.* **a)** mit Steinen bewerfen; ~ **me!,** ~ **the crows!** *(sl.)* mich laust der Affe! *(ugs.);* **b)** entsteinen ‹*Obst*›

stone: S~ Age *n. (Archaeol.)* Steinzeit, *die; attrib.* Steinzeit-; **~-cold 1.** *adj.* eiskalt; **2.** *adv.* **~-cold sober** stocknüchtern

stoned [stəʊnd] *adj. (sl.)* stoned *(Drogenjargon); (drunk)* voll zu *(salopp)*

stone: ~-'dead *pred. adj.* mausetot *(fam.);* **kill sth. ~-dead** *(fig.)* etw. völlig zunichte machen; **~-'deaf** *adj.* stocktaub *(ugs.);* **~'wall** *(Brit.)* **1.** *v.i.* mauern *(fig.);* **2.** *v.t.* **~wall sth.** bei etw. mauern; **~walling** ['stəʊnwɔ:lɪŋ] *n. (Brit.)* **~walling [tactics]** Hinhaltetaktik, *die;* **~ware** *n., no pl.* Steingut, *das; attrib.* ‹*Krug, Vase*› aus Steingut

stony ['stəʊnɪ] *adj.* **a)** *(full of stones)* steinig; **b)** *(like stone)* steinartig; **c)** *(hostile)* steinern *(geh.)* ‹*Blick, Miene*›; frostig ‹*Person, Empfang, Schweigen*›

stood *see* **stand 1, 2**

stool [stu:l] *n.* Hocker, *der;* **fall between two ~s** *(fig.)* sich zwischen zwei Stühle setzen

stoop [stu:p] **1.** *v.i.* **a)** ~ **[down]** sich bücken; ~ **over sth.** sich über etw. *(Akk.)* beugen; **he'd ~ to anything to get his way** *(fig.)* ihm ist jedes Mittel recht[, um sein Ziel zu erreichen]; ~ **to do sth.** *(fig.)* sich dazu erniedrigen, etw. zu tun; **b)** *(have ~)* gebeugt gehen. **2.** *v.t.* beugen; **~ed with old age** vom Alter gebeugt. **3.** *n.* gebeugte Haltung; **have a/walk with a ~:** einen krummen Rücken haben/gebeugt gehen

stop [stɒp] **1.** *v.t.*, **-pp-: a)** *(not let move further)* anhalten ‹*Person, Fahrzeug*›; aufhalten ‹*Fortschritt, Verkehr, Feind*›; verstummen lassen *(geh.)* ‹*Gerücht, Geschichte, Lüge*›; ‹*Tormann:*› halten ‹*Ball*›; **she ~ped her car** sie hielt an; ~ **thief!** haltet den Dieb!; **there's no ~ping sb.** jmd. läßt sich nicht aufhalten; **b)** *(not let continue)* unterbrechen ‹*Redner, Spiel, Gespräch, Vorstellung*›; beenden ‹*Krieg, Gespräch, Treffen, Spiel, Versuch, Arbeit*›; stillen ‹*Blutung*›; stoppen ‹*Produktion, Uhr, Streik, Inflation*›; einstellen ‹*Handel, Zahlung, Lieferung, Besuche, Subskriptionen, Bemühungen*›; abstellen ‹*Strom, Gas, Wasser, Mißstände*›; beseitigen ‹*Schmerz*›; ~ **that/that nonsense/that noise!** hör damit/mit diesem Unsinn/diesem Lärm auf!; **bad light ~ped play** *(Sport)* das Spiel wurde wegen schlechter Lichtverhältnisse abgebrochen; ~ **the show** *(fig.)* Furore machen; **just you try and ~ me!** versuch doch, mich daran zu hindern!; ~ **smoking/crying** aufhören zu rauchen/weinen; **never ~ doing sth.** etw. unaufhörlich tun; ~ **it!** hör auf [damit]!; *(in more peremptory tone)* Schluß damit!; ~ **oneself** sich zurückhalten; **I couldn't ~ myself** ich konnte nicht anders; **c)** *(not let happen)* verhindern ‹*Verbrechen, Unfall*›; **he tried to ~ us parking** er versuchte uns am Parken zu hindern; **he phoned his mother to ~ her [from] worrying** er rief seine Mutter an, damit sie sich keine Sorgen machte; ~ **sth. [from] happening** verhindern, daß etw. geschieht; **d)** *(cause to cease working)* abstellen ‹*Maschine usw.*›; ‹*Streikende:*› stillegen ‹*Betrieb*›; **e)** *(block up)* zustopfen ‹*Loch, Öffnung, Riß, Ohren*›; verschließen ‹*Wasserhahn, Rohr, Schlauch, Flasche*›; **f)** *(withhold)* streichen; ~ **[payment of] a cheque** einen Scheck sperren lassen. **2.** *v.i.*, **-pp-: a)** *(not extend further)* aufhören; ‹*Straße, Treppe:*› enden; ‹*Ton:*› verstummen; ‹*Ärger:*› verfliegen; ‹*Schmerz:*› abklingen; ‹*Zahlungen, Lieferungen:*› eingestellt werden; **b)** *(not move or operate further)* ‹*Fahrzeug, Fahrer:*› halten; ‹*Maschine, Motor:*› stillstehen; ‹*Uhr, Fußgänger, Herz:*› stehenbleiben; **he ~ped in the middle of the sentence** er unterbrach sich mitten im Satz; **he never ~s to think [before he acts]** er denkt nie nach [bevor er handelt]; **~!** halt!; ~ **at nothing** vor nichts zurückschrecken; ~ **dead** plötzlich stehenbleiben; ‹*Redner:*› abbrechen. **c)** *(coll.: stay)* bleiben; ~ **at a hotel/at a friend's house/with sb.** in einem Hotel/im Hause eines Freundes/bei jmdm. wohnen. **3.** *n.* **a)** *(halt)* Halt, *der;* **there will be two ~s for coffee on the way** es wird unterwegs zweimal zum Kaffeetrinken angehalten; **this train goes to London with only two ~s** dieser Zug fährt mit nur zwei Zwischenhalten nach London; **bring to a ~:** zum Stehen bringen ‹*Fahrzeug*›; zum Erliegen bringen ‹*Verkehr*›; unterbrechen ‹*Arbeit, Diskussion, Treffen*›; **come to**

a ~: stehenbleiben; ‹*Fahrzeug:*› zum Stehen kommen; ‹*Gespräch:*› abbrechen; ‹*Arbeit, Verkehr:*› zum Erliegen kommen; ‹*Vorlesung:*› abgebrochen werden; **make a ~ at** *or* **in a place** in einem Ort haltmachen; **put a ~ to** abstellen ‹*Mißstände, Unsinn*›; unterbinden ‹*Versuche*›; aus der Welt schaffen ‹*Gerücht*›; **put a ~ on a cheque** einen Scheck sperren lassen; **without a ~:** ohne Halt ‹*fahren, fliegen*›; ohne anzuhalten ‹*gehen, laufen*›; ununterbrochen ‹*arbeiten, reden*›; **b)** *(place)* Haltestelle, *die;* **the ship's first ~ is Cairo** der erste Hafen, den das Schiff anläuft, ist Kairo; **the plane's first ~ is Frankfurt** die erste Zwischenlandung des Flugzeuges ist in Frankfurt; **c)** *(Brit.: punctuation-mark)* Satzzeichen, *das; see also* **full stop a; d)** *(in telegram)* stop

~ **be'hind** *(coll.) see* **stay behind**

~ **'by** *(Amer.)* **1.** *v.i.* vorbeischauen *(ugs.).* **2.** *v.t.* ~ **by sb.'s house** *or* **place [and have a drink]** bei jmdm. [auf einen Drink] vorbeischauen *(ugs.)*

~ **'off** *v.i.* einen Zwischenaufenthalt einlegen

~ **'out** *v.i. (coll.)* **a)** draußen bleiben; **b)** *(remain on strike)* ‹*Arbeiter:*› weiterstreiken *(ugs.)*

~ **'over** *v.i.* einen Zwischenaufenthalt machen; *(remain for the night)* übernachten **(at** bei)

~ **'up 1.** *v.t. see* ~ **1e. 2.** *v.i. (coll.) see* **stay up a**

stop: **~cock** *n.* Abstellhahn, *der;* Absperrhahn, *der (Technik);* **~gap** *n.* Notbehelf, *der; (scheme, measure, plan, person)* Notlösung, *die; attrib.* behelfsmäßig; **a ~gap measure** eine Behelfsmaßnahme; **~-'go** *n. (Brit.)* Hin und Her, *das; (boom and recession)* Auf und Ab, *das;* **~-light** *n.* **a)** *(red traffic-light)* rotes Licht; **b)** *(Motor Veh.)* Bremslicht, *das;* **~over** *n.* Stopover, *der;* Zwischenaufenthalt, *der; (of aircraft)* Zwischenlandung, *die*

stoppage ['stɒpɪdʒ] *n.* **a)** *(halt)* Stillstand, *der; (strike)* Streik, *der;* **b)** *(cancellation)* Sperrung, *die; (of delivery)* Einstellung, *die;* **c)** *(deduction)* Abzug, *der*

stopper ['stɒpə(r)] **1.** *n.* Stöpsel, *der;* Pfropfen, *der.* **2.** *v.t.* zustöpseln

stop: **~-press** *n. (Brit. Journ.)* letzte Meldung/Meldungen; **~ sign** *n.* Stoppschild, *das;* **~-signal** *n.* Haltesignal, *das;* **~-watch** *n.* Stoppuhr, *die*

storage ['stɔ:rɪdʒ] *n., no pl., no indef. art. (storing)* Lagerung, *die; (of furniture)* Einlagerung, *die; (of films, books, documents)* Aufbewahrung, *die; (of data, water, electricity)* Speicherung, *die*

storage: ~ **heater** *n.* [Nacht]speicherofen, *der;* ~ **space** *n.* Lagerraum, *der; (in house)* Platz [zum Aufbewahren]; ~ **tank** *n.* Sammelbehälter, *der*

store [stɔ:(r)] **1.** *n.* **a)** *(Amer.: shop)* Laden, *der;* **b)** *in sing. or pl. (Brit.: large general shop)* Kaufhaus, *das;* **c)** *(warehouse)* Lager, *das; (for valuables)* Depot, *das; (for books, films, documents)* Magazin, *das;* **put sth. in ~:** etw. [bei einer Spedition] einlagern; **d)** *(stock)* Vorrat, *der* **(of** an + *Dat.*); **get in** *or* **lay in a ~ of sth.** einen Vorrat an etw. *(Dat.)* anlegen; **be** *or* **lie in ~ for sb.** jmdn. erwarten; **have a surprise in ~ for sb.** eine Überraschung für jmdn. [auf Lager] haben; **who knows what the future has in ~?** wer weiß, was die Zukunft mit sich bringt?; **e)** *in pl. (supplies)* Vorräte; **the ~s** *(place)* das [Vorrats]lager; **f) lay** *or* **put** *or* **set [great] ~ by** *or* **on sth.** [großen] Wert auf etw. *(Akk.)* legen. **2.** *v.t.* **a)** *(put in ~)* einlagern; speichern ‹*Getreide, Energie, Wissen*›; einspeichern ‹*Daten*›; ablegen ‹*Papiere, Dokumente*›; **b)** *(leave for storage)* unterbringen; **c)** *(hold)* aufnehmen; speichern ‹*Energie, Daten*›

~ **a'way** *v.t.* lagern; ablegen ‹*Akten*›; ~ **things away in a trunk/at a friend's house** Sachen in einer Truhe verstauen/bei einem Freund aufbewahren

~ **'up** *v.t.* speichern; ~ **up provisions/food/nuts** sich *(Dat.)* Vorräte / Lebensmittelvorräte / einen Vorrat an Nüssen anlegen; **you're only storing up trouble for yourself** du handelst dir nur immer mehr Schwierigkeiten ein

store: ~ **detective** *n.* Kaufhausdetektiv, *der;* **~house** *n.* Lager[haus], *das;* **sb. is a ~house of knowledge/information [about angling]** jmd. ist ein wandelndes Lexikon[, was das Angeln betrifft]; **the book is a real ~house of facts [about Germany]** das Buch ist eine wahre Fundgrube [für jeden, der sich über Deutschland orientieren will]; **~keeper** *n.* **a)** *(in charge of ~s)* Lagerist, *der*/Lageristin, *die; (Mil.)* Verwalter der Materialausgabe; **b)** *(Amer.: shopkeeper)* Besitzer eines Einzelhandelsgeschäftes; **~-room** *n.* Lagerraum, *der*

storey ['stɔ:rɪ] *n.* Stockwerk, *das;* Geschoß, *das;* **a five-~ house** ein fünfgeschossiges Haus; **third-~ window** Fenster im zweiten Stock[werk]

stork [stɔ:k] *n.* Storch, *der*

storm [stɔ:m] **1.** *n.* **a)** Unwetter, *das; (thunder~)* Gewitter, *das;* **the night of the ~:** die Sturmnacht; **a ~ in a teacup** *(fig.)* ein Sturm im Wasserglas; **b)** *(fig.: dispute)* Sturm der Entrüstung; **c)** *(fig.: outburst) (of applause, protest, indignation, criticism)* Sturm, *der; (of abuse)* Flut, *die;* **d)** *(Mil.: attack)* Sturm, *der;* **take sb./sth. by ~:** jmdn. überrumpeln/etw. im Sturm nehmen. **2.** *v.i.* **a)** stürmen; **he ~ed in** er kam hereingestürmt; **b)** *(talk violently)* toben; ~ **at sb.** jmdn. andonnern *(ugs.).* **3.** *v.t. (Mil.)* stürmen

'storm-cloud *n. (Meteorol.)* Gewitterwolke, *die*

stormy ['stɔ:mɪ] *adj.* **a)** stürmisch; hitzig ‹*Auseinandersetzung*›; **b)** *(indicating storms)* auf Sturm hindeutend; **be** *or* **look ~:** nach Sturm aussehen

[1]**story** ['stɔ:rɪ] *n.* **a)** *(account of events)* Geschichte, *die;* **give the ~ of sth.** etw. schildern *od.* darstellen; **it is quite another ~ now** *(fig.)* jetzt sieht alles ganz anders aus; **the [old,] old ~, the same old ~** *(fig.)* das alte Lied *(ugs.);* **tall ~:** unglaubliche Geschichte; **that's [a bit of] a tall ~!** das ist ein bißchen dick aufgetragen! *(ugs.);* **that's a different ~** *(fig.)* das ist etwas ganz anderes; **that's his ~ [and he's sticking to it]** er bleibt bei dem, was er gesagt hat; **that's only 'half the ~:** das ist noch nicht alles; **the ~ goes that ...:** man erzählt sich, daß ...; **that's not the whole ~:** das ist noch nicht alles; **to cut** *or* **make a long ~ short, ...:** kurz [gesagt], ...; **b)** *(narrative)* Geschichte, *die;* **that's the ~ of my life!** *(fig.)* das ist mein ewiges Problem!; **c)** *(news item)* Bericht, *der;* Story, *die (ugs.);* **d)** *(plot)* Story, *die;* **e)** *(set of [interesting] facts)* **the objects in the room have a ~:** die Gegenstände in dem Zimmer haben ihre eigene Geschichte; **f)** *(coll./child lang.: lie)* Märchen, *das;* **tell stories** Märchen erzählen

[2]**story** *(Amer.) see* **storey**

story: **~-book 1.** *n.* Geschichtenbuch, *das; (with fairy-tales)* Märchenbuch, *das;* **2.** *attrib. adj.* Bilderbuch-; **~-book world** Märchenwelt, *die;* **~-teller** *n.* **a)** *(narrator)* [Geschichten]erzähler, *der*/-erzählerin, *die;* **b)** *(writer)* Erzähler, *der*/Erzählerin, *die;* **c)** *(raconteur)* Anekdotenerzähler, *der*/-erzählerin, *die;* **she's a wonderful ~-teller** sie kann wundervoll erzählen

stout [staʊt] **1.** *adj.* **a)** *(strong)* fest; stabil ‹*Boot, Werkzeug, Messer, Zaun*›; dick ‹*Tür, Mauer, Damm, Stock, Papier*›; robust ‹*Material, Kleidung*›; stark ‹*Seil, Abwehr*›; kräftig ‹*Pflanze, Pferd, Pfeiler*›; **b)** *(fat)* beleibt; **c)** *(brave, staunch)* unverzagt; heftig ‹*Widerstand, Opposition*›; entschieden ‹*Ablehnung*›; stark ‹*Gegner*›; fest ‹*Glaube*›; **a ~ heart** ein festes Herz. **2.** *n. (drink)* Stout, *der*

stout-hearted ['staʊthɑ:tɪd] *adj.* beherzt; unerschrocken

stoutly ['staʊtlɪ] *adv.* **a)** *(strongly)* stabil ‹*gebaut, gezimmert*›; **~ made** solide, robust ‹*Schuhwerk*›; stark ‹*Seil*›; **~ built** stämmig; kräftig ‹*Tier*›; stabil ‹*Haus, Zaun, Tor*›; dick ‹*Tür, Mauer*›; **b)** *(staunchly)* beherzt; hartnäckig ‹*behaupten, ablehnen, widerstehen*›; fest ‹*glauben*›

[1]**stove** [stəʊv] *n.* Ofen, *der; (for cooking)* Herd, *der;* **electric ~:** Elektroherd, *der*

[2]**stove** *see* **stave 2**

stow [stəʊ] *v. t.* **a)** *(put into place)* packen (**into** in + *Akk.*); verstauen (**into** in + *Dat.*); *(Naut.)* stauen; **b)** *(fill)* vollpacken; vollstopfen *(ugs.); (Naut.)* befrachten

~ a'way 1. *v. t.* verwahren. **2.** *v. i.* als blinder Passagier reisen

'stowaway *n.* blinder Passagier

straddle ['strædl] *v. t.* **~** *or* **sit straddling a fence/chair** rittlings auf einem Zaun/Stuhl sitzen; **~** *or* **stand straddling a ditch** mit gespreizten Beinen über einem Graben stehen; **his legs ~d the chair/brook** er saß rittlings auf dem Stuhl/stand mit gespreizten Beinen über dem Bach; **their farm ~s the border** ihre Farm liegt beiderseits der Grenze; **the bridge ~s the river/road** die Brücke überspannt den Fluß/die Straße

straggle ['strægl] *v. i.* **a)** *(trail)* **~ [along] behind the others** den anderen hinterherzockeln *(ugs.);* **b)** *(spread in irregular way)* ‹*Dorf, Stadt:*› sich ausbreiten; ‹*Häuser, Bäume:*› verstreut stehen; **c)** *(grow untidily)* ‹*Pflanze:*› wuchern; ‹*Haar, Bart:*› zottig wachsen

straggler ['stræglə(r)] *n.* Nachzügler, *der*

straggling ['stræglɪŋ] *adj.* **a)** *(trailing)* nachzockelnd *(ugs.);* **b)** *(irregular)* verstreut ‹*Häuser*›; ungeordnet ‹*Reihe*›; unregelmäßig ‹*Baumreihe, Schrift*›; weiträumig angelegt ‹*Stadt, Gebäude*›; **c)** *(long and untidy)* wuchernd; zottig ‹*Haar, Bart*›

straggly ['stræglɪ] *see* **straggling c**

straight [streɪt] **1.** *adj.* **a)** gerade; aufrecht ‹*Haltung*›; glatt ‹*Haar*›; **in a ~ line** in gerader Linie; **b)** *(not having been bent)* ausgestreckt ‹*Arm, Bein*›; durchgedrückt ‹*Knie*›; **c)** *(not misshapen)* gerade ‹*Bein*›; **d)** *(Fashion)* gerade geschnitten; **e)** *(undiluted, unmodified)* unvermischt; **have** *or* **drink whisky/gin ~:** Whisky/Gin pur trinken; **a ~ choice** eine klare Wahl; **f)** *(successive)* fortlaufend; **win in ~ sets** *(Tennis)* ohne Satzverlust gewinnen; **the team had ten ~ wins** die Mannschaft hat zehn Spiele hintereinander gewonnen; **~ As** *(Amer.)* lauter Einsen; **g)** *(undeviating)* direkt ‹*Blick, Schlag, Schuß, Paß, Ball, Weg*›; **h)** *(candid)* geradlinig ‹*Person*›; ehrlich ‹*Antwort*›; klar ‹*Abfuhr, Weigerung, Verurteilung*›; unmißverständlich ‹*Rat*›; **~ dealings/speaking** direkte Verhandlungen/unverblümte Sprache; **a ~ answer to a ~ question** eine klare Antwort auf eine klare Frage; **he did some ~ talking with her** er sprach sich mit ihr offen aus; **be ~ with sb.** zu jmdm. offen sein; **i)** *(Theatre)* ernst; *(not avant-garde)* konventionell; **j)** *(in good order, not askew)* **the accounts are ~:** die Bücher sind in Ordnung; **the picture is ~:** das Bild hängt gerade; **is my hair/tie ~?** sitzt meine Frisur/Krawatte [richtig]?; **is my hat [on] ~?** sitzt mein Hut [richtig]?; **put ~:** geradeziehen ‹*Krawatte*›; gerade aufsetzen ‹*Hut*›; gerade hängen ‹*Bild*›; aufräumen ‹*Zimmer, Sachen*›; richtigstellen ‹*Fehler, Mißverständnis*›; **put things ~:** alles in Ordnung bringen; **put things ~ with sb.** mit jmdm. alles klären; **get sth. ~** *(fig.)* etw. genau *od.* richtig verstehen; **let's get it** *or* **things** *or* **the facts ~:** wir sollten alles genau klären; **get this ~!** merk dir das [ein für allemal]!; **put sb. ~:** jmdn. aufklären; **put** *or* **set the record ~:** die Sache *od.* das richtigstellen. **2.** *adv.* **a)** *(in a ~ line)* gerade; **she came ~ at me** sie kam geradewegs auf mich zu; **~ opposite** genau gegenüber; **head ~ for the wall** genau auf die Mauer zusteuern; **go ~** *(fig.: give up crime)* ein bürgerliches Leben führen; **b)** *(directly)* geradewegs; **~ after** sofort nach; **come ~ to the point** direkt *od.* gleich zur Sache kommen; **look sb. ~ in the eye** jmdm. direkt in die Augen blicken; **~ ahead** *or* **on** immer geradeaus; **they went ~ ahead and did it** sie taten es sofort; **c)** *(honestly, frankly)* aufrichtig; **give it to me ~:** sei ganz offen zu mir!; **he came ~ out with it** er sagte es ohne Umschweife; **I told him ~ [out] that ...:** ich sagte [es] ihm ins Gesicht, daß ...; **play ~ with sb.** mit jmdm. ein ehrliches Spiel spielen; **d)** *(upright)* gerade ‹*sitzen, stehen, wachsen*›; **e)** *(accurately)* zielsicher; **he can't shoot [very] ~:** er ist nicht [sehr] zielsicher; **f)** *(clearly)* klar ‹*sehen, denken*›. **3.** *n.* (**~** *stretch)* gerade Strecke; *(Sport)* Gerade, *die;* **final** *or* **home** *or* **finishing ~** *(Sport; also fig.)* Zielgerade, *die*

straight a'way *adv. (coll.)* sofort; gleich

straighten ['streɪtn] **1.** *v. t.* **a)** geradeziehen ‹*Kabel, Teppich, Seil*›; geradebiegen ‹*Draht*›; glätten ‹*Falte, Kleidung, Haare*›; geradehalten ‹*Rücken*›; strecken ‹*Beine, Arme*›; gerade hängen ‹*Bild*›; **b)** *(put in order)* aufräumen; einrichten ‹*neue Wohnung*›; in Ordnung bringen ‹*Geschäftsbücher, Finanzen*›. **2.** *v. i.* gerade werden

~ 'out 1. *v. t.* **a)** geradebiegen ‹*Draht*›; geradeziehen ‹*Seil, Kabel*›; glätten ‹*Decke, Teppich*›; begradigen ‹*Fluß, Straße*›; **b)** *(put in order, clear up)* klären; aus der Welt schaffen ‹*Mißverständnis, Meinungsverschiedenheit*›; in Ordnung bringen ‹*Angelegenheit*›; berichtigen ‹*Fehler*›. **2.** *v. i.* gerade werden

~ 'up 1. *v. t. see* **tidy up 2. 2.** *v. i.* sich aufrichten

straight: ~ 'face *n.* unbewegtes Gesicht; **with a ~ face** ohne eine Miene zu verziehen; **keep a ~ face** keine Miene verziehen; **~-faced** ['streɪtfeɪst] *adj.* mit unbewegter Miene *nachgestellt;* **be ~-faced** keine Miene verziehen; **~'forward** *adj.* **a)** *(frank)* freimütig; geradlinig ‹*Politik*›; schlicht ‹*Stil, Sprache, Erzählung, Bericht*›; klar ‹*Anweisung, Vorstellungen*›; **have a ~forward approach to a problem** ein Problem direkt angehen; **b)** *(simple)* einfach; eindeutig ‹*Lage*›; ~ 'off *adv. (coll.)* schlankweg *(ugs.)*

[1]**strain** [streɪn] **1.** *n.* **a)** *(pull)* Belastung, *die; (on rope)* Spannung, *die;* **put a ~ on sb./sth.** jmdn./etw. belasten; **b)** *(extreme physical or mental tension)* Streß, *der;* **feel the ~:** die Anstrengung spüren; **stand** *or* **take the ~:** die Belastung *od.* den Streß aushalten; **place sb. under [a] great ~:** jmdn. einer starken Belastung aussetzen; **be under [a great deal of] ~:** unter großem Streß stehen; **c)** *(person, thing)* **be a ~ on sb./sth.** jmdn./etw. belasten; eine Belastung für jmdn./etw. sein; **find sth. a ~:** etw. als Belastung empfinden; **d)** *(injury) (muscular)* Zerrung, *die; (over~ on heart, back, etc.)* Überanstrengung, *die;* **e)** *in sing. or pl. (burst of music)* Klänge; *(burst of poetry)* Vers, *der;* Zeile, *die.* **2.** *v. t.* **a)** *(over-exert)* überanstrengen; zerren ‹*Muskel*›; überbeanspruchen ‹*Geduld, Loyalität usw.*›; **b)** *(stretch tightly)* [fest] spannen; **c)** *(exert to maximum)* ~ **oneself/sb./sth.** das Letzte aus sich/jmdm./etw. herausholen; ~ **one's ears/eyes/voice** seine Ohren/Augen/Stimme anstrengen; ~ **oneself to do sth.** sich nach Kräften bemühen, etw. zu tun; **d)** *(use beyond proper limits)* verzerren ‹*Wahrheit, Lehre, Tatsachen*›; überbeanspruchen ‹*Geduld, Wohlwollen*›; **e)** *(filter)* durchseihen; seihen (**through** durch); ~ **[the water from] the vegetables** das Gemüse abgießen. **3.** *v. i. (strive intensely)* sich anstrengen; ~ **at sth.** an etw. *(Dat.)* zerren; ~ **at the leash** an der Leine zerren; *(fig.)* es kaum erwarten können; ~ **after sth.** sich mit aller Gewalt um etw. bemühen

~ **a'way,** ~ **'off** *v. t.* abseihen; abgießen ‹*Wasser*›

~ **'out** *v. t.* [her]ausfiltern

[2]**strain** *n.* **a)** *(breed)* Rasse, *die; (of plants)* Sorte, *die; (of virus)* Art, *die;* **b)** *no pl. (tendency)* Neigung, *die* (**of** zu); Hang, *der* (**of** zu); **a cruel ~:** ein grausamer Zug

strained [streɪnd] *adj.* gezwungen ‹*Lächeln*›; künstlich ‹*Humor, Witz*›; gewagt ‹*Interpretation*›; ~ **relations** gespannte Beziehungen

strainer ['streɪnə(r)] *n.* Sieb, *das*

strait [streɪt] *n.* **a)** *in sing. or pl. (Geog.)* [Wasser]straße, *die;* Meerenge, *die;* **b)** *in pl. (bad situation)* Schwierigkeiten

strait: ~**-jacket** *n. (lit. or fig.)* Zwangsjacke, *die;* ~**-laced** [streɪt'leɪst] *adj. (fig.)* puritanisch

[1]**strand** [strænd] *n. (thread)* Faden, *der; (of wire)* Litze, *die (Elektrot.); (of rope)* Strang, *der; (of beads, pearls, flowers, etc.)* Kette, *die; (of hair)* Strähne, *die*

[2]**strand** *v. t.* **a)** *(leave behind)* trocken setzen; **be [left] ~ed** *(fig.)* seinem Schicksal überlassen sein; *(be stuck)* festsitzen; **the strike left them ~ed in England** wegen des Streiks saßen sie in England fest; **b)** *(wash ashore)* an Land spülen ‹*Leiche, Wrackteile*›; *(run aground)* auf Grund setzen ‹*Schiff*›

strange [streɪndʒ] *adj.* **a)** *(peculiar)* seltsam; sonderbar; merkwürdig; **feel [very] ~:** sich [ganz] komisch fühlen; **it feels ~ to do sth.** es ist ein merkwürdiges *od.* komisches Gefühl, wenn man etw. tut; ~ **to say** seltsamerweise; **b)** *(alien, unfamiliar)* fremd; ~ **to sb.** jmdm. fremd; **c)** *(unaccustomed)* ~ **to sth.** nicht vertraut mit etw.; **feel ~:** sich nicht zu Hause fühlen

strangely ['streɪndʒlɪ] *adv.* seltsam; merkwürdig; ~ **enough, ...:** seltsamerweise ...

stranger ['streɪndʒə(r)] *n.* **a)** *(foreigner, unknown person)* Fremde, *der/die;* **he is a/no ~ to me** er ist mir nicht bekannt/ist mir bekannt; **hello, ~:** hallo, lange nicht gesehen; **b)** *(one lacking certain experience)* **be a/no ~ to sth.** etw. nicht gewöhnt/etw. gewöhnt sein; **he is no ~ to this sort of work** diese Arbeit ist ihm nicht fremd; **he is a ~ here/to the town** er ist hier/in der Stadt fremd; **be a/no ~ to Oxford** Oxford gar nicht/[recht gut] kennen

strangle ['stræŋgl] *v. t.* erdrosseln; erwürgen

'**stranglehold** *n. (lit. or fig.)* Würgegriff, *der;* **have a ~ on sb./sth.** jmdn./etw. im Würgegriff haben

strangulation [stræŋgjʊ'leɪʃn] *n.* Erdrosseln, *das;* Erwürgen, *das*

strap [stræp] **1.** *n.* **a)** *(leather strip)* Riemen, *der; (textile strip)* Band, *das; (shoulder-~)* Träger, *der; (for watch)* Armband, *das;* **b)** *(to grasp in vehicle)* Halteriemen, *der.* **2.** *v. t.,* **-pp-:** ~ **[into position/down]** festschnallen; ~ **oneself in** sich anschnallen

~ **'up** *v. t.* zuschnallen

'**straphanger** *n.* stehender Fahrgast

strapless ['stræplɪs] *adj.* trägerlos

strapping ['stræpɪŋ] *adj.* stramm

Strasburg ['stræzbɜ:g] *pr. n.* Straßburg *(das)*

strata *pl. of* **stratum**

stratagem ['strætədʒəm] *n.* *(trick)* [Kriegs]list, *die*

strategic [strə'ti:dʒɪk] *adj.* **a)** strategisch; **b)** *(of great military importance)* strategisch wichtig; *(necessary to plan)* bedeutsam ‹*Element, Faktor*›

strategist ['strætɪdʒɪst] *n.* Stratege, *der*/Strategin, *die*

strategy ['strætɪdʒɪ] *n.* Strategie, *die; (fig. also)* Taktik, *die;* **it was bad ~** *(fig.)* es war taktisch *od.* strategisch unklug

stratosphere ['strætəsfɪə(r)] *n.* Stratosphäre, *die*

stratum ['strɑ:təm, 'streɪtəm] *n., pl.* **strata** ['strɑ:tə, 'streɪtə] Schicht, *die*

straw [strɔ:] *n.* **a)** *no pl. (stalks of grain)* Stroh, *das;* **b)** *(single stalk)* Strohhalm, *der;* **clutch** *or* **grasp at ~s** *(fig. coll.)* sich an einen Strohhalm klammern; **be the last ~, be the ~ that broke the camel's back** *(coll.)* das Faß zum Überlaufen bringen; **that's the last** *or* **final ~:** jetzt reicht's aber; **draw ~s [for sth.]** Hölzchen [um etw.] ziehen; **pick the short ~** *(fig.)* das schlechtere Los ziehen; **c) [drinking-]~:** Trinkhalm, *der;* Strohhalm, *der*

strawberry ['strɔ:bərɪ] *n.* Erdbeere, *die*

straw: ~ **boss** *n. (Amer.)* Vorarbeiter, *der;* ~**-coloured** *adj.* strohgelb; ~ **'hat** *n.* Strohhut, *der*

stray [streɪ] **1.** *v. i.* **a)** *(wander)* streunen; *(fig.: in thought etc.)* abschweifen (**into** in + *Akk.*); ~ **[away] from** sich absondern von; **the child had ~ed from his parents** das Kind war seinen Eltern weggelaufen; ~ **into enemy territory** sich auf feindliches Gebiet verirren; **b)** *(deviate)* abweichen (**from** von); **have ~ed** sich verirrt haben; ~ **from the point/from** *or* **off the road** vom Thema/von der Straße abkommen. **2.** *n. (animal)* streunendes Tier; *(without owner)* herrenloses Tier. **3.** *adj.* **a)** streunend; *(without owner)* herrenlos; *(out of proper place)* verirrt; **b)** *(occasional, isolated)* vereinzelt

streak [stri:k] **1.** *n.* **a)** *(narrow line)* Streifen, *der; (in hair)* Strähne, *die;* ~ **of lightning** Blitzstrahl, *der;* **like a ~ [of lightning]** [schnell] wie der Blitz *(ugs.);* wie ein geölter Blitz *(ugs.);* **b)** *(fig.: element)* **have a jealous/cruel ~:** zur Eifersucht/Grausamkeit neigen; **have a ~ of meanness/jealousy** eine geizige/eifersüchtige Ader haben; **c)** *(fig.: spell)* ~ **of good/bad luck, lucky/unlucky ~:**

Glücks-/Pechsträhne, *die;* **be on a** *or* **have a winning/losing ~:** eine Glücks-/Pechsträhne haben. **2.** *v. t.* streifen; **~ sth. with green** etw. mit grünen Streifen versehen; **hair ~ed with grey** Haar mit grauen Strähnen; **~ed with paint/tears/mud** farb-/tränenverschmiert/dreckbeschmiert. **3.** *v. i.* **a)** *(move rapidly)* flitzen *(ugs.);* **b)** *(coll.: run naked)* blitzen *(ugs.);* flitzen *(ugs.)*

streaker ['stri:kə(r)] *n. (coll.)* Blitzer, *der*/Blitzerin, *die (ugs.);* Flitzer, *der*/Flitzerin, *die (ugs.)*

streaky ['stri:kɪ] *adj.* streifig; gestreift ‹*Muster, Fell*›

streaky 'bacon *n.* durchwachsener Speck

stream [stri:m] **1.** *n.* **a)** *(of flowing water)* Wasserlauf, *der; (brook)* Bach, *der;* **b)** *(flow, large quantity)* Strom, *der; (of abuse, excuses, words)* Schwall, *der;* **~s** *or* **a ~ of applications** eine Flut von Bewerbungen; **in ~s** in Strömen; **the children rushed in ~s/in a ~ through the school gates** die Kinder strömten durch die Schultore; **c)** *(current)* Strömung, *die; (fig.)* Trend, *der;* **against/with the ~ of sth.** *(fig.)* gegen den/mit dem Strom einer Sache; **go against/ with the ~** ‹*Person:*› gegen den/ mit dem Strom schwimmen; **d)** *(Brit. Educ.)* Parallelzug, *der;* **e) be/go on ~** *(Industry)* in Betrieb sein/den Betrieb aufnehmen. **2.** *v. i.* **a)** *(flow)* strömen; ‹*Sonnenlicht:*› fluten; **tears ~ed down her face** Tränen strömten ihr über das Gesicht; **b) my eyes ~ed** mir tränten die Augen. **3.** *v. t.* **his nose was ~ing blood** Blut floß ihm aus der Nase

~ 'in *v. i.* hereinströmen/hineinströmen

~ 'out *v. i.* herausströmen/hinausströmen

~ 'past *v. i.* vorbeiströmen

~ 'through *v. i.* hindurchströmen

streamer ['stri:mə(r)] *n.* Band, *das; (of paper)* Luftschlange, *die*

'streamline *v. t.* **a)** [eine] Stromlinienform geben (+ *Dat.*); **be ~d** eine Stromlinienform haben; **b)** *(simplify)* rationalisieren; *(reduce)* einschränken

street [stri:t] *n.* Straße, *die;* **in** *(Brit.) or* **on ... Street** in der ...straße; **in the ~:** auf der Straße; **be on the ~[s]** *(be published)* ‹*Zeitung:*› draußen sein; *(have no place to live)* auf der Straße liegen *(ugs.);* **keep the youngsters off the ~s** dafür sorgen, daß sich die Jugendlichen nicht auf der Straße herumtreiben; **~s ahead [of sb./sth.]** *(coll.)* um Längen besser [als jmd./etw.] *(ugs.);* **be [right] up sb.'s ~** *(coll.)* jmds. Fall sein *(ugs.)*

street: ~car *n. (Amer.)* Straßenbahn, *die;* Tram, *die (südd., österr., schweiz.);* **~ credibility** *n.* [glaubwürdiges] Image; **~ crime** *n., no pl., no indef. art.* Straßenkriminalität, *die;* **~ door** *n.* [vordere] Haustür; **~ furniture** *n.: Gegenstände wie Straßenlaternen, Abfallkörbe, Telefonzellen, Verkehrszeichen usw.;* **~-lamp, ~-light** *ns.* Straßenlaterne, *die;* **~-lighting** *n.* Straßenbeleuchtung, *die;* **~-map** *n.* Stadtplan, *der;* **~-market** *n.* Markt, *der;* **~-plan** *see* **~-map; ~-sweeper** *n.* **a)** *(person)* Straßenfeger, *der*/-fegerin, *die (bes. nordd.);* Straßenkehrer, *der*/ -kehrerin, *die (bes. südd.);* **b)** *(machine)* Kehrmaschine, *die; (vehicle)* Straßenkehrmaschine, *die;* **~ value** *n.* Straßenverkaufswert, *der;* **~ vendor** *n.* Straßenhändler, *der*/-händlerin, *die;* **~-wise** *adj. (coll.)* **be ~-wise** wissen, wo es langgeht

strength [streŋθ] *n.* **a)** Stärke, *die; (power)* Kraft, *die; (of argument)* [Überzeugungs]kraft, *die; (of poison, medicine)* Wirksamkeit, *die; (of legal evidence)* [Beweis]kraft, *die; (resistance of material, building, etc.)* Stabilität, *die;* **not know one's own ~:** nicht wissen, wie stark man ist; **give sb. ~:** jmdn. stärken; jmdm. Kraft geben; **go from ~ to ~:** immer erfolgreicher werden; **on the ~ of sth./that** auf Grund einer Sache *(Gen.)*/dessen; **b)** *(proportion present)* Stärke, *die; (full complement)* **be below ~/up to ~:** weniger als/etwa die volle Stärke haben; **in [full] ~:** in voller Stärke; **the police were there in ~:** ein starkes Polizeiaufgebot war da

strengthen ['streŋθən, 'streŋkθən] **1.** *v. t. (give power to)* stärken; *(reinforce, intensify, increase in number)* verstärken; erhöhen ‹*Anteil*›; *(make more effective)* unterstützen; **~ sb.'s resolve** jmdn. in seinem Entschluß bestärken; **~ sb.'s hand** *(fig.)* jmds. Position stärken. **2.** *v. i.* stärker werden

strenuous ['strenjʊəs] *adj.* **a)** *(energetic)* energisch; gewaltig ‹*Anstrengung*›; **b)** *(requiring exertion)* anstrengend

stress [stres] **1.** *n.* **a)** *(strain)* Streß, *der;* **be under ~:** unter Streß *(Dat.)* stehen; **b)** *(emphasis)* Betonung, *die;* Nachdruck, *der;* **lay** *or* **place** *or* **put [a] ~ on sth.** auf etw. *(Akk.)* Wert *od.* Gewicht legen; **c)** *(accentuation)* Betonung, *die;* **put the/a ~ on sth.** etw. betonen. **2.** *v. t.* **a)** *(emphasize)* betonen; Wert legen auf (+ *Akk.*) ‹*richtige Ernährung, gutes Benehmen, Sport usw.*›; **~ [the point] that ...:** darauf hinweisen, daß ...; **b)** *(Ling.)* betonen ‹*Silbe, Vokal usw.*›

'stress-mark *n.* Betonungszeichen, *das*

stretch [stretʃ] **1.** *v. t.* **a)** *(lengthen, extend)* strecken ‹*Arm, Hand*›; recken ‹*Hals*›; dehnen ‹*Gummiband*›; *(spread)* ausbreiten ‹*Decke*›; *(tighten)* spannen; **he lay ~ed out on the ground** er lag ausgestreckt auf dem Boden; **~ one's legs** *(by walking)* sich *(Dat.)* die Beine vertreten; **b)** *(widen)* dehnen; **~ [out of shape]** ausweiten ‹*Schuhe, Jacke*›; **c)** *(fig.: make the most of)* ausschöpfen ‹*Reserve*›; fordern ‹*Person, Begabung*›; **d)** *(fig.: extend beyond proper limit)* überschreiten ‹*Befugnis, Grenzen des Anstands*›; strapazieren *(ugs.)* ‹*Geduld*›; es nicht so genau nehmen mit ‹*Gesetz, Bestimmung, Begriff, Grundsätzen*›; **~ a point** großzügig sein; **~ the truth** ‹*Aussage:*› nicht ganz der Wahrheit entsprechen; **he's certainly ~ing the truth there** er nimmt es hier mit der Wahrheit nicht so genau; **we're a bit ~ed at the moment** wir sind zur Zeit ziemlich überlastet; **~ it/things** den Bogen überspannen. **2.** *v. i.* **a)** *(extend in length)* sich dehnen; ‹*Person, Tier:*› sich strecken; **b)** *(have specified length)* sich ausdehnen; **~ from A to B** sich von A bis B erstrecken; **c) ~ to sth.** *(be sufficient for)* für etw. reichen; **could you ~ to £10?** hast du vielleicht sogar 10 Pfund? **3.** *v. refl.* sich strecken. **4.** *n.* **a)** *(lengthening, drawing out)* **have a ~:** sich strecken; **give sth. a ~:** etw. dehnen; **b)** *(exertion)* **by no ~ of the imagination** auch mit viel Phantasie nicht; **at a ~** *(fig.)* wenn es sein muß (*see also* **d**); **at full ~:** auf Hochtouren; **c)** *(expanse, length)* Abschnitt, *der;* **a ~ of road/open country** ein Stück Straße/freies Gelände; **d)** *(period)* **for a ~:** eine Zeitlang; **a four-hour ~:** eine [Zeit]spanne von vier Stunden; **at a ~:** ohne Unterbrechung (*see also* **b**). **5.** *adj.* dehnbar; Stretch‹*hose, -gewebe*›

~ 'out 1. *v.t.* **a)** [aus]strecken ‹*Arm, Bein*›; ausbreiten ‹*Decke*›; auseinanderziehen ‹*Seil*›; ~ **oneself out** sich [lang] ausstrecken; **b)** *(eke out)* ~ **sth. out** mit etw. reichen. **2.** *v.i.* **a)** *(~ one's hands out, lit. or fig.)* die Hände ausstrecken **(to** nach); **b)** *(extend)* sich ausdehnen

stretcher ['stretʃə(r)] *n. (for carrying a person)* [Trag]bahre, *die*

'**stretcher-bearer** *n.* [Kranken]träger, *der*

strew [stru:] *v.t., p.p.* **~ed** [stru:d] *or* **~n** [stru:n] **a)** *(scatter)* streuen ‹*Blumen, Sand usw.*›; **clothes were ~n about the room** Kleider lagen im ganzen Zimmer verstreut herum; **b)** *(cover, lit. or fig.)* bestreuen; **the grass was ~n with litter** [überall] auf dem Gras war Abfall verstreut

stricken ['strɪkn] *adj. (afflicted)* heimgesucht; havariert ‹*Schiff, Flugzeug*›; *(showing affliction)* schmerzerfüllt; **be ~ with fever** von Fieber geschüttelt werden; ~ **with fear/grief** angsterfüllt/gramgebeugt

strict [strɪkt] *adj.* **a)** *(firm)* streng; strenggläubig ‹*Katholik, Moslem usw.*›; **in ~ confidence** streng vertraulich; **b)** *(precise)* streng; genau ‹*Übersetzung*›; **in the ~ sense [of the word]** im strengen Sinn[e] [des Wortes]

strictly ['strɪktlɪ] *adv.* streng; ~ **no smoking** Rauchen streng[stens] verboten; ~ **[speaking]** strenggenommen

strictness ['strɪktnɪs] *n., no pl.* **a)** *(firmness)* Strenge, *die;* **b)** *(precision)* Genauigkeit, *die*

stricture ['strɪktʃə(r)] *n. usu. in pl. (critical remark)* **~[s]** [scharfe *od.* heftige] Kritik

stride [straɪd] **1.** *n.* Schritt, *der;* **make ~s [towards sth.]** *(fig.)* [in Richtung auf etw. *(Akk.)*] Fortschritte machen; **get into one's ~:** seinen Rhythmus finden; *(fig.)* in Fahrt *od.* Schwung kommen; **put sb. off his ~** *(fig.)* jmdn. aus dem Konzept bringen; **take sth. in one's ~** *(fig.)* mit etw. gut fertig werden. **2.** *v.i.*, **strode** [strəʊd], **stridden** ['strɪdn] [mit großen Schritten] gehen; *(solemnly)* schreiten *(geh.)*

~ '**out** *v.i.* ausschreiten *(geh.)*

strident ['straɪdənt] *adj.* schrill ‹*Stimme, Blech[bläser]*›; *(fig.)* grell ‹*Farbe*›; schrill ‹*Protest, Ton*›

strife [straɪf] *n., no pl., no indef. art.* Streit, *der*

strike [straɪk] **1.** *n.* **a)** *(Industry)* Streik, *der;* Ausstand, *der;* **be on/go [out]** *or* **come out on ~:** in den Streik getreten sein/in den Streik treten; **b)** *(Finance, Mining, Oil Industry)* Treffer, *der (fig. ugs.);* **make a ~:** sein Glück machen; *(Mining)* fündig werden; **c)** *(sudden success)* **[lucky] ~:** Glückstreffer, *der;* **d)** *(act of hitting)* Schlag, *der;* **e)** *(Mil.)* Angriff, *der* **(at** auf + *Akk.*). **2.** *v.t.*, **struck** [strʌk], **struck** *or (arch.)* **stricken** ['strɪkn] **a)** *(hit)* schlagen; ‹*Schlag, Geschoß:*› treffen ‹*Ziel*›; ‹*Blitz:*› [ein]schlagen in (+ *Akk.*), treffen; *(afflict)* treffen; ‹*Epidemie, Seuche, Katastrophe usw.:*› heimsuchen; ~ **one's head on** *or* **against the wall** mit dem Kopf gegen die Wand schlagen; **the car struck a pedestrian** das Auto erfaßte einen Fußgänger; **the ship struck the rocks** das Schiff lief auf die Felsen; **b)** *(delete)* streichen **(from, off** aus); **c)** *(deliver)* ~ **two punches** zweimal zuschlagen; ~ **sb. a blow** jmdm. einen Schlag versetzen; **who struck [the] first blow?** wer hat zuerst geschlagen?; ~ **a blow against sb./against** *or* **to sth.** *(fig.)* jmdm./einer Sache einen Schlag versetzen; ~ **a blow for sth.** *(fig.)* eine Lanze für etw. brechen; **d)** *(produce by hitting flint)* schlagen ‹*Funken*›; *(ignite)* anzünden ‹*Streichholz*›; **e)** *(chime)* schlagen; **f)** *(Mus.)* anschlagen ‹*Töne auf dem Klavier*›; anzupfen, anreißen ‹*Töne auf der Gitarre*›; *(fig.)* anschlagen ‹*Ton*›; **g)** *(impress)* beeindrucken; ~ **sb. as [being] silly** jmdm. dumm zu sein scheinen *od.* dumm erscheinen; **it ~s sb. that ...:** es scheint jmdm., daß...; **how does it ~ you?** was hältst du davon?; **h)** *(occur to)* einfallen (+ *Dat.*); **i)** *(cause to become)* **a heart attack struck him dead** er erlag einem Herzanfall; **be struck blind/dumb** erblinden/verstummen; **j)** *(attack)* überfallen; *(Mil.)* angreifen; **k)** *(encounter)* begegnen (+ *Dat.*); **l)** *(Mining)* stoßen auf (+ *Akk.*); ~ **gold** auf Gold stoßen; *(fig.)* einen Glückstreffer landen *(ugs.)* **(in** mit); **m)** *(reach)* stoßen auf (+ *Akk.*) ‹*Hauptstraße, Weg, Fluß*›; **n)** *(adopt)* einnehmen ‹*[Geistes]haltung*›; **o)** *(take down)* einholen ‹*Segel, Flagge*›; abbrechen ‹*Zelt, Lager*›. **3.** *v.i.*, **struck, struck** *or (arch.)* **stricken a)** *(deliver a blow)* zuschlagen; ‹*Pfeil:*› treffen; ‹*Blitz:*› einschlagen; ‹*Unheil, Katastrophe, Krise, Leid:*› hereinbrechen *(geh.)*; *(collide)* zusammenstoßen; *(hit)* schlagen **(against** gegen, **[up]on** auf + *Akk.*); **b)** *(ignite)* zünden; **c)** *(chime)* schlagen; **d)** *(Industry)* streiken; **e)** *(attack; also Mil.)* zuschlagen *(fig.)*; **f)** *(make a find) (Mining)* fündig werden; ~ **lucky** Glück haben; **g)** *(direct course)* ~ **south** *etc.* sich nach Süden *usw.* wenden

~ **at** *v.t.* schlagen nach; *(fig.)* einen Schlag versetzen (+ *Dat.*); rütteln an (+ *Dat.*) ‹*Grundfesten*›

~ '**back** *v.i. (lit. or fig.)* zurückschlagen; ~ **back at sb./sth.** sich gegen jmdn./etw. zur Wehr setzen

~ '**down** *v.t.* niederschlagen; *(fig.)* niederwerfen *(geh.)*

~ '**off** *v.t.* **a)** *(remove)* abschlagen; **b)** *(remove from membership)* streichen ‹*Namen*›; *(from professional body)* die Zulassung/Approbation entziehen (+ *Dat.*)

~ '**out 1.** *v.t. (delete)* streichen. **2.** *v.i.* **a)** *(hit out)* zuschlagen; ~ **out at sb./sth.** nach jmdm./etw. schlagen; *(fig.)* jmdn./etw. scharf angreifen; **b)** *(set out, lit. or fig.)* aufbrechen; ~ **out in a new direction** *(fig.)* etwas Neues anfangen

~ **through** *v.t.* durchstreichen

~ **up** *v.t.* **a)** *(start)* beginnen ‹*Unterhaltung*›; anknüpfen ‹*Bekanntschaft*›; schließen ‹*Freundschaft*›; **b)** *(begin to play)* anstimmen

strike: ~ **action** *n.* Streikaktionen; ~ **ballot** *n.* Urabstimmung, *die;* ~ **benefit** *see* ~ **pay;** **~-breaker** *n.* Streikbrecher, *der/*-brecherin, *die;* **~-force** *see* **striking-force;** ~ **pay** *n.* Streikgeld, *das*

striker ['straɪkə(r)] *n.* **a)** *(worker on strike)* Streikende, *der/die;* **b)** *(Footb.)* Stürmer, *der/*Stürmerin, *die*

striking ['straɪkɪŋ] *adj.* auffallend; erstaunlich ‹*Ähnlichkeit, Unterschied*›; bemerkenswert ‹*Idee*›; schlagend ‹*Beispiel*›

striking: ~-distance *n.* Reichweite, *die;* **within easy ~-distance of a town** *(fig.)* in unmittelbarer Nähe einer Stadt; **~-force** *n. (Mil., Police)* Einsatzkommando, *das*

string [strɪŋ] **1.** *n.* **a)** *(thin cord)* Schnur, *die; (to tie up parcels etc. also)* Bindfaden, *der; (ribbon)* Band, *das;* **how long is a piece of ~?** *(fig.)* wie weit ist der Himmel?; **[have/keep sb.] on a ~:** [jmdn.] an der Leine *(ugs.) od.* am Gängelband [haben/halten]; **pull [a few** *or* **some] ~s** *(fig.)* seine Beziehungen spielen lassen; **there are ~s attached** *(fig.)* es sind Bedingungen/es ist eine Bedingung damit verknüpft; **without ~s, with no ~s attached** ohne Bedingung[en]; **b)** *(of bow)* Sehne, *die; (of racket, musical instrument)* Saite, *die;* **have another ~ to one's bow** *(fig.)* noch ein Eisen im Feuer haben *(ugs.);* **c)** *in pl. (Mus.) (instruments)* Streichinstrumente; *(players)* Streicher; **~ quartet/orchestra** Streichquartett/-orchester, *das;* **d)** *(series, sequence)* Kette, *die; (procession)* Zug, *der.* **2.** *v. t.,* **strung** [strʌŋ] **a)** bespannen ‹*Tennisschläger, Bogen, Gitarre usw.*›; **b)** *(thread)* auffädeln; aufziehen

~ a'long *(coll.)* **1.** *v. i.* sich anschließen; **~ along with sb.** mit jmdm. mitgehen; *(have relationship)* mit jmdm. gehen *(ugs.).* **2.** *v. t. (deceive)* an der Nase herumführen *(ugs.)*

~ 'out 1. *v. t.* verstreuen. **2.** *v. i. (in space)* sich verteilen

~ to'gether *v. t. (on a thread)* auffädeln; aufziehen; *(by tying)* zusammenbinden; miteinander verknüpfen ‹*Worte*›

~ 'up *v. t.* **a)** *(hang up)* aufhängen ‹*Lampions, Papiergirlanden*›; **b)** *(coll.: kill by hanging)* aufhängen *(ugs.);* **c)** *(make tense)* **strung up** angespannt

string: ~ 'bag *n.* [Einkaufs]netz, *das;* **~ band** *n. (Mus.)* Streichorchester, *das*

stringed [strɪŋd] *attrib. adj. (Mus.)* Saiten-

stringent ['strɪndʒənt] *adj.* **a)** *(strict)* streng ‹*Bestimmung, Gesetz, Maßnahme, Test*›; **b)** *(tight)* angespannt ‹*Finanzlage*›

string 'vest *n.* Netzhemd, *das*

stringy ['strɪŋɪ] *adj.* **a)** *(fibrous)* faserig; **b)** *(resembling string)* dünn ‹*Haar*›; faserig ‹*Gewebe*›

¹strip [strɪp] **1.** *v. t.,* **-pp-: a)** ausziehen ‹*Person*›; leerräumen, ausräumen ‹*Haus, Schrank, Regal*›; abziehen ‹*Bett*›; entrinden ‹*Baum*›; abbeizen ‹*Möbel, Türen*›; ausschlachten, *(dismantle)* auseinandernehmen ‹*Maschine, Auto*›; **~ped to the waist** mit nacktem Oberkörper; **~ sb. of sth.** jmdn. einer Sache *(Gen.)* berauben *(geh.);* **~ sb. of his rank/title/medals/decorations/office** jmdm. seinen Rang/Titel/seine Medaillen/Auszeichnungen aberkennen/jmdn. seines Amtes entkleiden *(geh.);* **~ the walls** die Tapeten entfernen; **b)** *(remove)* entfernen **(from, off** von); abziehen ‹*Laken*›; abstreifen ‹*Hülle*›. **2.** *v. i.,* **-pp-** sich ausziehen; **~ to the waist/[down] to one's underwear** den Oberkörper freimachen/sich bis auf die Unterwäsche ausziehen

~ 'down 1. *v. t.* **a)** *(dismantle)* auseinandernehmen; **b)** *(undress)* ausziehen. **2.** *v. i.* sich ausziehen

~ 'off 1. *v. t.* **a)** abreißen; abschälen ‹*Rinde*›; abziehen ‹*Tapete*›; **~ sth. off sth.** etw. von etw. abreißen/abschälen/abziehen; **b)** ausziehen ‹*Kleidung*›. **2.** *v. i.* sich ausziehen

²strip *n.* **a)** *(narrow piece)* Streifen, *der;* **a ~ of land** ein schmales Stück *od.* Streifen Land; **tear sb. off a ~, tear a ~ off sb.** *(Brit. sl.)* jmdm. den Marsch blasen *(ugs.);* **b)** *see* **strip cartoon**

'strip cartoon *n.* Comic[strip], *der*

stripe [straɪp] *n.* **a)** Streifen, *der;* **b)** *(Mil.)* [Ärmel]streifen, *der. See also* **star 1 a**

striped [straɪpt] *adj.* gestreift; Streifen‹*muster*›

strip: ~ light *n.* Neonröhre, *die;* **~ lighting** *n.* Neonbeleuchtung, *die;* Neonlicht, *das*

stripling ['strɪplɪŋ] *n.* Jüngelchen, *das*

'stripped pine *n.* abgebeizte Kiefer

stripper ['strɪpə(r)] *n.* **a)** *(solvent)* Farbentferner, *der; (for wallpaper)* Tapetenlöser, *der; (tool)* Kratzer, *der;* **b)** *(strip-tease performer)* Stripper, *der*/Stripperin, *die (ugs.)*

strip: ~-show *n.* Strip-Show, *die;* **~-'tease** *n.* Striptease, *der*

stripy ['straɪpɪ] *adj.* gestreift ‹*Fell, Blazer*›; Streifen‹*muster, -stoff*›

strive [straɪv] *v. i.,* **strove** [strəʊv], **striven** ['strɪvn] **a)** *(endeavour)* sich bemühen; **~ to do sth.** bestrebt sein *(geh.) od.* sich bemühen, etw. zu tun; **~ after** *or* **for sth.** nach etw. streben; **b)** *(contend)* kämpfen **(for** um)

strobe [strəʊb] *n. (coll.)* Stroboskoplicht, *das*

strode *see* **stride 2**

¹stroke [strəʊk] *n.* **a)** *(act of striking)* Hieb, *der;* Schlag, *der; (of sword, axe)* Hieb, *der;* **finishing ~** *(lit. or fig.)* Todesstoß, *der;* **b)** *(Med.)* Schlaganfall, *der;* **paralytic/apoplectic ~:** paralytischer/apoplektischer Anfall; **c)** *(sudden impact)* **~ of lightning** Blitzschlag, *der;* **by a ~ of fate/fortune** durch eine Fügung des Schicksals/einen [glücklichen] Zufall; **~ of [good] luck** Glücksfall, *der;* **have a ~ of bad/[good] luck** Pech/Glück haben; **d)** *(single effort)* Streich, *der; (skilful effort)* Schachzug, *der;* **at a** *or* **one ~:** auf einen Schlag *od.* Streich; **not do a ~ [of work]** keinen [Hand]schlag tun; **~ of genius** genialer Einfall; **e)** *(of pendulum, heart, wings, oar)* Schlag, *der; (in swimming)* Zug, *der;* **f)** *(Billiards etc.)* Stoß, *der; (Tennis, Cricket, Golf, Rowing)* Schlag, *der;* **g)** *(mark, line)* Strich, *der; (of handwriting; also fig.: detail)* Zug, *der;* (*symbol* /) Schrägstrich, *der;* **h)** *(sound of clock)* Schlag, *der;* **on the ~ of nine** Punkt neun [Uhr]

²stroke 1. *v. t.* streicheln; **~ sth. over/across sth.** mit etw. über etw. *(Akk.)* streichen; **~ sth. back** etw. zurückstreichen. **2.** *n.* **give sb./sth. a ~:** jmdn./etw. streicheln

stroll [strəʊl] **1.** *v. i. (saunter)* spazierengehen; **~ into sth.** in etw. *(Akk.)* schlendern. **2.** *n.* Spaziergang, *der;* **go for a ~:** einen Spaziergang machen

~ a'long *v. i.* daherspazieren *od.* -schlendern

~'on *v. i.* weiterschlendern

stroller ['strəʊlə(r)] *n. (pushchair)* Sportwagen, *der*

strong [strɒŋ] **1.** *adj.,* **~er** ['strɒŋgə(r)], **~est** ['strɒŋgɪst] **a)** *(resistant)* stark; gefestigt ‹*Ehe*›; stabil ‹*Möbel*›; solide, fest ‹*Fundament, Schuhe*›; streng ‹*Vorschriften, Vorkehrungen*›; robust ‹*Konstitution, Magen, Stoff, Porzellan*›; **you have to have a ~ stomach** *(fig.)* man muß einiges vertragen können; **b)** *(powerful)* stark, kräftig ‹*Person, Tier*›; kräftig ‹*Arme, Beine, Muskeln, Tritt, Schlag, Zähne*›; stark ‹*Linse, Brille, Strom, Magnet*›; gut ‹*Augen*›; **as ~ as a horse** *or* **an ox** *(fig.)* bärenstark *(ugs.);* **a man of ~ character** ein charakterstarker Mann; **c)** *(effective)* stark ‹*Regierung, Herrscher, Wille*›; streng ‹*Disziplin, Lehrer*›; gut ‹*Gedächtnis, Schüler*›; fähig ‹*Redner, Mathematiker*›; *(formidable)* stark ‹*Gegner, Kombination*›; aussichtsreich ‹*Kandidat*›; *(powerful in resources)* reich ‹*Nation, Land*›; leistungsfähig ‹*Wirtschaft*›; stark ‹*Besetzung, Delegation, Truppe, Kontingent usw.*›; **sb.'s ~ point** jmds. Stärke; **d)** *(convincing)* gut, handfest ‹*Grund, Beispiel, Argument*›; **there is a ~ possibility that ...:** es ist sehr wahrscheinlich, daß ...; **e)** *(vigorous, moving forcefully)* stark; voll ‹*Unterstützung*›; fest

⟨*Überzeugung*⟩; kraftvoll ⟨*Stil*⟩; *(fervent)* glühend ⟨*Anhänger, Verfechter einer Sache*⟩; **take ~ measures/action** energisch vorgehen; **f)** *(affecting the senses)* stark; kräftig, stark ⟨*Geruch, Geschmack, Stimme*⟩; markant ⟨*Gesichtszüge*⟩; *(pungent)* streng ⟨*Geruch, Geschmack*⟩; kräftig ⟨*Käse*⟩; **g)** *(concentrated)* stark; kräftig ⟨*Farbe*⟩; **I need a ~ drink** ich muß mir erst mal einen genehmigen *(ugs.)*; **h)** *(emphatic)* stark ⟨*Ausdruck, Protest*⟩; heftig ⟨*Worte, Wortwechsel*⟩. **2.** *adv.* stark; **sb. is going ~:** es geht jmdm. gut; **they are still going ~** *(after years of marriage)* mit ihnen geht es noch immer gut; *(after hours of work)* sie sind noch immer eifrig dabei

strong: ~ 'arm *n., no pl.* Muskelkraft, *die; attrib.* **~-arm methods** brutale Methoden; **~-box** *n.* Kassette, *die;* **~hold** *n.* Festung, *die; (fig.)* Hochburg, *die;* **~ 'language** *n., no pl., no indef. art.* derbe Ausdrucksweise; **use ~ language** sich derb ausdrücken

strongly ['strɒŋlɪ] *adv.* **a)** stark; fest ⟨*etabliert*⟩; solide ⟨*gearbeitet*⟩; **~ built** solide gebaut; *(in body)* kräftig gebaut; **b)** *(powerfully)* stark; **c)** *(convincingly)* überzeugend ⟨*darlegen*⟩; **d)** *(vigorously)* energisch ⟨*protestieren, bestreiten*⟩; nachdrücklich ⟨*unterstützen*⟩; dringend ⟨*raten*⟩; fest ⟨*glauben*⟩; **I feel ~ about it** es ist mir sehr ernst damit; es liegt mir sehr am Herzen; **I ~ suspect that ...:** ich habe den starken Verdacht, daß ...

strong: ~ man *n.* Muskelmann, *der (ugs.);* **~-'minded** *adj.* [seelisch] robust; *(determined)* willensstark; **~ point** *n. (fortified position)* Stützpunkt, *der; see also* **~ 1c;** **~-room** *n.* Tresorraum, *der;* Stahlkammer, *die*

strontium ['strɒntɪəm] *n. (Chem.)* Strontium, *das*

stroppy ['strɒpɪ] *adj. (Brit. sl.)* pampig *(salopp)*

strove *see* **strive**

struck *see* **strike 2, 3**

structural ['strʌktʃərl] *adj.* baulich; Bau⟨*material*⟩; tragend ⟨*Wand, Säule, Balken*⟩; Konstruktions⟨*fehler*⟩

structure ['strʌktʃə(r)] **1.** *n.* **a)** Struktur, *die;* Aufbau, *der; (Mus.)* Kompositionsweise, *die; (manner of construction)* Bauweise, *die;* Struktur, *die;* **b)** *(something constructed)* Konstruktion, *die; (building)* Bauwerk, *das; (complex whole; also Biol.)* Struktur, *die.* **2.** *v.t.* strukturieren; regeln ⟨*Leben*⟩; aufbauen ⟨*literarisches Werk*⟩; *(construct)* konstruieren; bauen

struggle ['strʌgl] **1.** *v.i.* **a)** *(try with difficulty)* kämpfen; **~ to do sth.** sich abmühen, etw. zu tun; **~ for a place/a better world** um einen Platz/für eine bessere Welt kämpfen; **~ against** *or* **with sb./sth.** mit jmdm./etw. *od.* gegen jmdn./etw. kämpfen; **~ with sth.** *(try to cope)* sich mit etw. quälen; mit etw. kämpfen; **b)** *(proceed with difficulty)* sich quälen; *(into tight dress, through narrow opening)* sich zwängen; **I ~d past** ich kämpfte mich vorbei; **c)** *(physically)* kämpfen; *(resist)* sich wehren; **~ free** freikommen; sich befreien; **d)** *(be in difficulties)* kämpfen *(fig.);* **after three laps I was struggling** nach drei Runden hatte ich zu kämpfen. **2.** *n.* **a)** *(exertion)* **with a ~:** mit Mühe; **it was a long ~:** es kostete viel Mühe; **have a [hard] ~ to do sth.** [große] Mühe haben, etw. zu tun; **the ~ for freedom** der Kampf für die Freiheit; **b)** *(physical fight)* Kampf, *der;* **the ~ against** *or* **with sb./sth.** der Kampf gegen *od.* mit jmdm./etw.; **the ~ for influence/power** der Kampf um Einfluß/die Macht; **surrender without a ~:** kampflos aufgeben

strum [strʌm] **1.** *v.i.,* **-mm-** klimpern *(ugs.)* **(on** auf + *Dat.*). **2.** *v.t.,* **-mm-** klimpern *(ugs.)* auf (+ *Dat.*)

strung *see* **string 2**

¹strut [strʌt] **1.** *v.i.,* **-tt-** *(walk)* stolzieren. **2.** *n.* stolzierender Gang

²strut *n. (support)* Strebe, *die*

stub [stʌb] **1.** *n.* **a)** *(short remaining portion)* Stummel, *der; (of cigarette)* Kippe, *die;* **~ of pencil** Bleistiftstummel, *der;* **b)** *(counterfoil)* Abschnitt, *der; (of ticket)* Abriß, *der.* **2.** *v.t.,* **-bb-: a) ~ one's toe [against** *or* **on sth.]** sich *(Dat.)* den Zeh [an etw. *(Dat.)*] stoßen; **b)** ausdrücken ⟨*Zigarette usw.*⟩; *(with one's foot)* austreten ⟨*Zigarette usw.*⟩

~ 'out *v.t.* ausdrücken

stubble ['stʌbl] *n., no pl.* Stoppeln *Pl.*

stubborn ['stʌbən] *adj.* **a)** *(obstinate)* starrköpfig *(abwertend);* dickköpfig *(ugs.);* störrisch ⟨*Tier, Gesicht, Haltung*⟩; hartnäckig ⟨*Vorurteil*⟩; **be ~ in insisting on sth.** stur *(ugs. abwertend)* auf etw. *(Dat.)* beharren; **[as] ~ as a mule** störrisch wie ein Maulesel *(ugs.);* **b)** *(resolute)* hartnäckig; fest ⟨*Mut, Entschlossenheit, Treue*⟩; **c)** *(intractable)* störrisch *(fig.);* vertrackt *(ugs.)* ⟨*Problem*⟩

stubbornness ['stʌbənnɪs] *n., no pl.* **a)** *(obstinacy)* Starrköpfigkeit, *die;* **b)** *(resolution, intractability)* Hartnäckigkeit, *die*

stucco ['stʌkəʊ] *n., pl.* **~es** *(fine plaster)* Stuck, *der; (coarse plaster)* Putz, *der*

stuck *see* **stick 1, 2**

'stuck up *see* **stick up 1d**

¹stud [stʌd] **1.** *n.* **a)** *(nail)* Beschlagnagel, *der; (on clothes)* Niete, *die; (on boot)* Stollen, *der; (marker in road)* Nagel, *der (Verkehrsw.);* **b)** *(for ear)* Ohrstecker, *der.* **2.** *v.t.,* **-dd-** *(set with ~s)* beschlagen; *(be scattered over)* verstreut sein über (+ *Akk.*); **~ded with flowers/stars** *etc.* mit Blumen/Sternen *usw.* übersät

²stud *n.* **a)** *(Breeding)* Gestüt, *das;* **b)** *(stallion)* Zuchthengst, *der*

student ['stju:dənt] *n.* Student, *der*/Studentin, *die; (in school or training establishment)* Schüler, *der*/Schülerin, *die;* **be a ~ of sth.** etw. studieren; **~ of medicine** Student/Studentin der Medizin; Medizinstudent, *der*/-studentin, *die; attrib.* **~ days** Studenten-/Schulzeit, *die;* **~ driver** *(Amer.)* Fahrschüler, *der*/-schülerin, *die;* **~ nurse** Lernschwester, *die*/Pflegeschüler, *der;* **be a ~ doctor/teacher** ein medizinisches Praktikum/Schulpraktikum machen

'stud-farm *n.* Gestüt, *das*

studied ['stʌdɪd] *adj.* **a)** *(thoughtful)* [wohl]überlegt; **b)** *(intentional)* gewollt; gesucht ⟨*Stil, Ausdrucksweise*⟩

studio ['stju:dɪəʊ] *n., pl.* **~s a)** *(photographer's or painter's workroom)* Atelier, *das; (workshop for the performing arts)* Studio, *das;* **b)** *(Cinemat., Radio, Telev.)* Studio, *das*

studio: ~ apartment *(Amer.),* **~ flat** *(Brit.) ns.* Einzimmerwohnung, *die*

studious ['stju:dɪəs] *adj.* *(assiduous in study)* lerneifrig; gelehrt ⟨*Beschäftigung, Buch, Aussehen, Atmosphäre*⟩

study ['stʌdɪ] **1.** *n.* **a)** Studium, *das;* Lernen, *das;* **the ~ of mathematics/law** das Studium der Mathematik/der Rechtswissenschaft; **[books on] African/Social Studies** *(Educ./Univ.)* [Bücher zur] Afrikanistik/Sozialwissenschaft; **graduate studies** *(Educ./Univ.)* Graduiertenstudium, *das;*

b) *(piece of work)* **a ~ of** *or* **on sth.** eine Studie über etw. *(Akk.)*; **studies are being carried out** zur Zeit werden Untersuchungen durchgeführt; **c) a ~ in sth.** ein Musterbeispiel *(fig.)* für etw.; **his face was a ~!** sein Gesicht war sehenswert!; **d)** *(Art)* Studie, *die; (Mus.)* Etüde, *die;* Übung, *die; (Lit., Theatre)* Studie, *die* **(in, of** über + *Akk.*); **e)** *(room)* Arbeitszimmer, *das.* **2.** *v. t.* **a)** studieren; *(at school)* lernen; **b)** *(scrutinize)* studieren; **c)** *(read attentively)* studieren ‹*Fahrplan*›; sich *(Dat.)* [sorgfältig] durchlesen ‹*Prüfungsfragen, Bericht*›. **3.** *v. i.* lernen; *(at university)* studieren; **~ to be a doctor/teach French** Medizin studieren/Französisch für das Lehramt studieren

stuff [stʌf] **1.** *n.* **a)** *no pl., no indef. art. (material[s])* Zeug, *das (ugs.)*; **be made of sterner ~:** aus härterem Stoff gemacht sein *(fig.)*; **the ~ that dreams/heroes are made of** der Stoff, aus dem die Träume sind/Helden gemacht sind *(fig.)*; **plastic is useful ~:** Plastik ist eine nützliche Sache; **b)** *no pl., no indef. art. (activity, knowledge)* **do painting or drawing, ~ like that** malen oder zeichnen oder so was *(ugs.)*; **do one's ~** *(coll.)* seine Sache machen; **know one's ~** *(coll.: be knowledgeable)* sich auskennen; *(know one's job)* seine Sache verstehen; **that's the ~!** *(coll.)* so ist's richtig! **2.** *v. t.* **a)** stopfen; zustopfen ‹*Loch, Ohren*›; *(in taxidermy)* ausstopfen; *(Cookery)* füllen; **~ sth. with** *or* **full of sth.** etw. mit etw. vollstopfen *(ugs.)*; **[go and] get ~ed!** *(sl.)* hau ab! *(ugs.)*; **~ oneself** *(sl.)* sich vollstopfen *(ugs.)*; **~ one's face** *(sl.)* sich *(Dat.)* den Bauch vollstopfen *(ugs.)*; **~ ballot boxes** *(Amer.: insert bogus votes)* Stimmen fälschen; **b)** *(sl.)* **~ him!** zum Teufel mit ihm!; **~ it!** Scheiß drauf! *(derb)*; **he can ~ it!** er kann mich mal! *(derb)*. **3.** *v. i.* sich vollstopfen *(ugs.)*

stuffing ['stʌfɪŋ] *n.* **a)** *(material)* Füllmaterial, *das;* **a ~ of horsehair** eine Füllung aus Roßhaar; **knock the ~ out of sb.** *(fig. coll.)* jmdn. umhauen *(ugs.)*; **b)** *(Cookery)* Füllung, *die*

stuffy ['stʌfɪ] *adj.* **a)** *(stifling)* stickig ‹*Zimmer, Atmosphäre*›; **b)** *(congested)* verstopft; **c)** *(coll.: prim)* spießig **(about** gegenüber)

stultify ['stʌltɪfaɪ] *v. t.* lähmen; **have a ~ing effect on sth.** sich lähmend auf etw. *(Akk.)* auswirken; **~ing boredom/monotony** lähmende Langeweile/Monotonie

stumble ['stʌmbl] *v. i.* **a)** stolpern **(over** über + *Akk.*); **b)** *(falter)* stocken; **~ over sth./through life** über etw. *(Akk.)*/durchs Leben stolpern; **c) ~ across** *or* **[up]on sb./sth.** *(find by chance)* über jmdn. stolpern *(fig. ugs.)*/auf etw. *(Akk.)* stoßen

stumbling-block ['stʌmblɪŋblɒk] *n.* Stolperstein, *der*

stump [stʌmp] **1.** *n.* **a)** *(of tree, branch, tooth)* Stumpf, *der; (of cigar, pencil, limb, tail, etc.)* Stummel, *der;* **b)** *(Cricket)* Stab, *der.* **2.** *v. t. (confound)* verwirren; durcheinanderbringen; **be ~ed** ratlos sein; **be ~ed for an answer** um eine Antwort verlegen sein. **3.** *v. i. (walk stiffly)* stapfen; *(walk noisily)* trampeln

stumpy ['stʌmpɪ] *adj.* gedrungen; **~ tail** Stummelschwanz, *der*

stun [stʌn] *v. t.*, **-nn-: a)** *(knock senseless)* betäuben; **be ~ned** *(unconscious)* bewußtlos sein; *(dazed)* benommen sein; **b)** *(fig.)* **be ~ned at** *or* **by sth.** von etw. wie betäubt sein

stung *see* **sting 2, 3**

stunk *see* **stink 1**

stunning ['stʌnɪŋ] *adj. (coll.)* **a)** *(splendid)* hinreißend; umwerfend *(ugs.)*; **b)** *(causing insensibility)* wuchtig ‹*Schlag*›; **c)** *(shocking)* bestürzend ‹*Nachricht*›; *(amazing)* sensationell

[1]**stunt** [stʌnt] *v. t.* hemmen, beeinträchtigen ‹*Wachstum, Entwicklung*›; **~ed trees** verkümmerte Bäume

[2]**stunt** *n.* halsbrecherisches Kunststück; *(Cinemat.)* Stunt, *der; (Advertising)* [Werbe]gag, *der*

stupendous [stju:'pendəs] *adj.* gewaltig; außergewöhnlich ‹*Schönheit, Intelligenz, Talent*›; großartig ‹*Urlaub, Schauspieler*›

stupid ['stju:pɪd] *adj.*, **~er** ['stju:pɪdə(r)], **~est** ['stju:pɪdɪst] *(slow-witted, unintelligent)* dumm; einfältig ‹*Person, Aussehen*›; *(ridiculous)* lächerlich; *(pointless)* dumm *(ugs.)* ‹*Witz, Geschichte, Gedanke*›; *expr. rejection or irritation* blöd *(ugs.)*; **it would be ~ to do sth.** es wäre töricht, etw. zu tun

stupidity [stju:'pɪdɪtɪ] *n.* Dummheit, *die; (of action also)* Torheit, *die*

stupidly ['stju:pɪdlɪ] *adv.* dumm; **~ [enough], I have ...:** dummerweise habe ich ...

stupor ['stju:pə(r)] *n.* Benommenheit, *die;* **in a drunken ~:** sinnlos betrunken

sturdy ['stɜ:dɪ] *adj. (robust)* stabil ‹*Haus, Stuhl, Schiff*›; kräftig ‹*Rasse, Pflanze, Pferd, Kind*›; kräftig [gebaut] ‹*Person*›; *(thickset)* stämmig ‹*Person*›; *(strong)* stämmig ‹*Beine, Arme*›; *(sound)* solide; *(resolute)* stark ‹*Gegner, Verfechter, Widerstand*›

sturgeon ['stɜ:dʒən] *n.* Stör, *der*

stutter ['stʌtə(r)] **1.** *v. i.* stottern; ‹*Gewehr:*› tacken. **2.** *n.* Stottern, *das;* **have a bad ~:** stark stottern

[1]**sty** [staɪ] *see* **pigsty**

[2]**sty, stye** [staɪ] *n. (Med.)* Gerstenkorn, *das*

style [staɪl] **1.** *n.* **a)** Stil, *der; (in conversation)* Ton, *der; (in performance)* Art, *die; (of habitual behaviour)* Art, *die;* **that's the ~!** so ist es richtig!; **be bad** *or* **not good ~:** schlechter *od.* kein guter Stil sein; **it's not my ~ [to do that]** das ist nicht mein Stil; **dress in the latest/modern ~:** sich nach der neuesten/neuen Mode kleiden; **cook in the French ~:** französisch kochen; **b)** *(superior way of living, behaving, etc.)* Stil, *der;* **in ~:** stilvoll; *(on a grand scale)* im großen Stil; **in the grand ~:** im großen Stil; **c)** *(sort)* Art, *die;* **~ of music** Musikrichtung, *die;* **d)** *(pattern)* Art, *die; (of clothes)* Machart, *die; (hair-~)* Frisur, *die.* **2.** *v. t. (design)* entwerfen; **elegantly ~d clothes** elegant geschnittene Kleidung

styli *pl. of* **stylus**

stylish ['staɪlɪʃ] *adj.* stilvoll; elegant ‹*Kleidung, Auto, Hotel, Person*›

stylist ['staɪlɪst] *n.* Designer, *der*/Designerin, *die; (hair-~)* Haarstilist, *der*/-stilistin, *die*

stylistic [staɪ'lɪstɪk] *adj.* stilistisch; Stil‹*mittel, -merkmale*›

stylus ['staɪləs] *n., pl.* **styli** ['staɪlaɪ] *or* **~es** *(of record-player)* [Abtast]nadel, *die*

styptic ['stɪptɪk] *adj.* blutstillend

suave [swɑ:v] *adj.* gewandt

sub [sʌb] *n. (coll.)* **a)** *(subscription)* Abo, *das (ugs.)*; **b)** *(esp. Sport: substitute)* Ersatz, *der;* **c)** *see* **sub-editor**

sub- *pref.* unter-; *(mit Fremdwörtern meist)* sub-

'**subcommittee** *n.* Unterausschuß, *der*

sub'conscious *(Psych.)* **1.** *adj.* unterbewußt; **~ mind** Unterbewußtsein, *das.* **2.** *n.* Unterbewußtsein, *das*

sub'consciously *adv. (Psych.)* unterbewußt

'**subcontinent** *n. (Geog.)* Subkontinent, *der*

subcon'tract 1. *v. t. (accept under secondary contract)* als Subunternehmer übernehmen; *(offer under secondary contract)* an Subunternehmer/an einen Subunternehmer vergeben; ~ **a job to sb.** eine Arbeit an jmdn. [in einem Untervertrag] vergeben. **2.** *v. i. (accept secondary contract)* als Subunternehmer arbeiten; *(offer secondary contract)* Aufträge an Subunternehmer/an einen Subunternehmer vergeben

'subcontractor *n.* Subunternehmer, *der*/-unternehmerin, *die*

subdivide ['sʌbdɪvaɪd, sʌbdɪ'vaɪd] **1.** *v. t. (further divide)* erneut teilen; *(divide into parts)* unterteilen. **2.** *v. i.* ~ **into sth.** sich in etw. *(Akk.)* teilen

subdivision ['sʌbdɪvɪʒn, sʌbdɪ'vɪʒn] *n. (subdividing)* erneute Teilung; *(subordinate division)* Unterabteilung, *die;* ~ **[of sth.] into sth.** Unterteilung [einer Sache *(Gen.)*] in etw. *(Akk.)*

subdue [səb'dju:] *v. t. (conquer)* besiegen; unterwerfen; *(bring under control)* bändigen ‹*Kind, Tier*›; ruhigstellen ‹*Patienten*›; unter Kontrolle bringen ‹*Demonstranten usw.*›; bezähmen ‹*Gefühle, zornige Person*›; *(reduce in intensity)* dämpfen ‹*Zorn, Heftigkeit, gute Laune, Lärm, Licht*›; abkühlen *(fig.)* ‹*Leidenschaft*›; verblassen lassen ‹*Farben*›

subdued [səb'dju:d] *adj.* gedämpft; **he seemed rather** ~: er schien ziemlich gedämpfter Stimmung zu sein

sub-'editor *n. (Journ., Publishing)* **a)** *(assistant editor)* Mitherausgeber, *der*/Mitherausgeberin, *die;* **b)** *(Brit.: one who prepares material)* Redaktionsassistent, *der*/-assistentin, *die*

'subgroup *n.* Untergruppe, *die*

'subheading *n.* **a)** *(subordinate division)* Unterabschnitt, *der;* **b)** *(subordinate title)* Untertitel, *der*

sub'human *adj.* unmenschlich

subject 1. ['sʌbdʒɪkt] *n.* **a)** *(citizen)* Staatsbürger, *der*/-bürgerin, *die; (in relation to monarch)* Untertan, *der*/Untertanin, *die;* **b)** *(topic)* Thema, *das; (department of study)* Fach, *das; (area of knowledge)* Fach[gebiet], *das; (Art)* Motiv, *das; (Mus.)* Thema, *das;* **be the ~ of an investigation** Gegenstand einer Untersuchung sein; **on the ~ of money** über das Thema Geld ‹*reden usw.*›; beim Thema Geld ‹*sein, bleiben*›; **change the ~:** das Thema wechseln; **c) be a ~ for sth.** *(cause sth.)* zu etw. Anlaß geben; **d)** *(Ling., Logic, Philos.)* Subjekt, *das.* **2.** ['sʌbdʒɪkt] *adj.* **a)** *(conditional)* **be ~ to sth.** von etw. abhängig sein *od.* abhängen; **sth. is ~ to alteration** etw. kann geändert werden; **b)** *(prone)* **be ~ to** anfällig sein für ‹*Krankheit*›; neigen zu ‹*Melancholie*›; **c)** *(dependent)* abhängig; ~ **to** *(dependent on)* untertan (+ *Dat.*) ‹*König usw.*›; unterworfen (+ *Dat.*) ‹*Verfassung, Gesetz, Krone*›; untergeben (+ *Dat.*) ‹*Dienstherrn*›. **3.** ['sʌbdʒɪkt] *adv.* ~ **to sth.** vorbehaltlich einer Sache *(Gen.)*. **4.** [səb'dʒekt] *v. t.* **a)** *(subjugate, make submissive)* unterwerfen (**to** *Dat.*); **b)** *(expose)* ~ **sb./sth. to sth.** jmdn./etw. einer Sache *(Dat.)* aussetzen; ~ **sb. to torture** jmdn. der Folter unterwerfen; ~ **sth. to chemical analysis** etw. einer chemischen Analyse unterziehen

subjective [səb'dʒektɪv] *adj.* **a)** subjektiv; **b)** *(Ling.)* Subjekt-

'subject-matter *n., no pl., no indef. art.* Gegenstand, *der*

subjugate ['sʌbdʒʊgeɪt] *v. t.* unterjochen (**to** unter + *Akk.*)

subjunctive [səb'dʒʌŋktɪv] *(Ling.)* **1.** *adj.* konjunktivisch; Konjunktiv-; ~ **mood** Konjunktiv, *der.* **2.** *n.* Konjunktiv, *der;* **past/present ~:** Konjunktiv II *od.* Präteritum/Konjunktiv I *od.* Präsens

sub'let *v. t.,* **-tt-, sublet** untervermieten

sublime [sə'blaɪm] *adj.,* **~r** [sə'blaɪmə(r)], **~st** [sə'blaɪmɪst] *(exalted)* erhaben; *(iron.)* vollendet *(fig. iron.)* ‹*Chaos*›

subliminal [sʌb'lɪmɪnl] *adj.* ~ **advertising** unterschwellige Werbung

sub-ma'chine-gun *n.* Maschinenpistole, *die*

submarine [sʌbmə'ri:n, 'sʌbməri:n] **1.** *n.* Unterseeboot, *das;* U-Boot, *das.* **2.** *adj.* Unterwasser-; unterseeisch *(Geol.)*

submerge [səb'mɜ:dʒ] **1.** *v. t.* **a)** *(place under water)* ~ **sth. [in the water]** etw. eintauchen *od.* ins Wasser tauchen; **b)** *(inundate)* ‹*Wasser:*› überschwemmen; **be ~d in water** unter Wasser stehen. **2.** *v. i.* abtauchen *(Seemannsspr.)*

submission [səb'mɪʃn] *n.* **a)** *(surrender)* Unterwerfung, *die* (**to** unter + *Akk.*); **force/frighten sb. into ~:** jmdn. zwingen, sich zu unterwerfen/jmdm. durch Einschüchterung seinen Willen aufzwingen; **b)** *no pl., no art. (meekness)* Unterwerfung, *die;* **c)** *(presentation)* Einreichung, *die* (**to** bei); *(thing put forward)* Einsendung, *die; (by witness)* Aussage, *die*

submissive [səb'mɪsɪv] *adj.* gehorsam; unterwürfig *(abwertend);* **be ~ to sb./sth.** sich jmdm./einer Sache unterwerfen

submit [səb'mɪt] **1.** *v. t.,* **-tt-: a)** *(present)* einreichen; vorbringen ‹*Vorschlag*›; abgeben ‹*[Doktor]arbeit usw.*›; ~ **sth. to sb.** jmdm. etw. vorlegen; ~ **sth. to scrutiny/investigation** etw. einer Prüfung/Untersuchung unterziehen; ~ **one's entry to a competition** seine Teilnehmerkarte *usw.* für ein Preisausschreiben einsenden; ~ **that ...** *(suggest, argue)* behaupten, daß ...; **b)** *(surrender)* ~ **oneself to sb./sth.** sich jmdm./einer Sache unterwerfen; **c)** *(subject)* ~ **sth. to heat** etw. der Hitze *(Dat.)* aussetzen; ~ **oneself to sth.** sich einer Sache *(Dat.)* unterziehen. **2.** *v. i.,* **-tt-: a)** *(surrender)* aufgeben; sich unterwerfen (**to** *Dat.*); **b)** *(defer)* ~ **to sb./sth.** sich jmdm./einer Sache beugen; **c)** *(agree to undergo)* ~ **to sth.** sich einer Sache *(Dat.)* aussetzen

sub'normal *adj.* unterdurchschnittlich; subnormal *(Med.)*

subordinate 1. [sə'bɔ:dɪnət] *adj.* *(inferior)* untergeordnet; *(lower-ranking)* rangniedriger; *(secondary)* zweitrangig; **be ~ to sb./sth.** jmdm./einer Sache untergeordnet sein. **2.** [sə'bɔ:dɪnət] *n.* Untergebene, *der/die.* **3.** [sə'bɔ:dɪneɪt] *v. t. (render subject)* unterordnen (**to** *Dat.*)

'subroutine *n. (Computing)* Unterprogramm, *das;* Subroutine, *die*

subscribe [səb'skraɪb] **1.** *v. t. ([promise to] contribute)* ~ **sth.** zusichern, etw. zu spenden; **be ~d** als Spende zugesichert worden sein. **2.** *v. i.* **a)** *(express support)* ~ **to sth.** sich einer Sache *(Dat.)* anschließen; **b)** *([promise to] make contribution)* ~ **to** *or* **for sth.** eine Spende für etw. zusichern

subscriber [səb'skraɪbə(r)] *n.* **a)** *(one who assents)* Befürworter, *der*/Befürworterin, *die* (**to** *Gen.*); **b)** *(contributor)* Spender, *der*/Spenderin, *die* (**of, to** für); *(to a newspaper etc.)* Abonnent, *der*/Abonnentin, *die* (**to** *Gen.*); **c)** *(Teleph.)* Fernsprechkunde, *der*/-kundin, *die*

subscription [səb'skrɪpʃn] *n.* **a)** *(thing subscribed)* Spendenbeitrag, *der* (**to** für); *(membership fee)* Mitgliedsbeitrag, *der* (**to**

für); *(prepayment for newspaper etc.)* Abonnement, *das* (**to** *Gen.*); **[buy] by ~:** im Abonnement [beziehen]; **b)** *(act of subscribing money)* Spende, *die;* **[be built] by ~:** mit Spenden [gebaut werden]

'subsection *n.* Unterabschnitt, *der*

subsequent ['sʌbsɪkwənt] *adj.* folgend; nachfolgend ‹*Kind*›; später ‹*Gelegenheit*›; **~ events** spätere *od.* die folgenden Ereignisse

subservient [səb'sɜ:vɪənt] *adj.* **a)** *(merely instrumental)* dienend; **be ~ to sb./sth.** jmdm./einer Sache dienen; **b)** *(subordinate)* untergeordnet (**to** *Dat.*); **c)** *(obsequious)* unterwürfig; servil *(abwertend)*

subside [səb'saɪd] *v.i.* **a)** *(sink to lower level)* ‹*Wasser, Flut, Fluß:*› sinken; ‹*Boden, Haus:*› sich senken; ‹*Schwellung:*› zurückgehen; **b)** *(abate)* nachlassen; **~ into** verfallen in (+ *Akk.*) ‹*Untätigkeit, Schweigen usw.*›

subsidence [səb'saɪdəns, 'sʌbsɪdəns] *n.* **a)** *(sinking) (of ground, structure)* Senkung, *die; (of liquid)* Sinken, *das;* **b)** *(abatement)* Nachlassen, *das*

subsidiary [səb'sɪdɪərɪ] **1.** *adj.* **a)** *(auxiliary)* unterstützend; subsidiär *(fachspr.);* untergeordnet ‹*Funktion, Stellung*›; Neben‹*fach, -aspekt*›; **~ to sth.** einer Sache *(Dat.)* untergeordnet; *(secondary)* gegenüber einer Sache zweitrangig; **b)** *(Commerc.)* **~ company** *see* **2. 2.** *n. (Commerc.)* Tochtergesellschaft, *die*

subsidize (subsidise) ['sʌbsɪdaɪz] *v.t.* subventionieren; finanziell unterstützen ‹*Person*›

subsidy ['sʌbsɪdɪ] *n.* Subvention, *die;* **receive a ~:** subventioniert werden

subsist [səb'sɪst] *v.i. (keep oneself alive)* existieren; **~ on sth.** von etw. leben

subsistence [səb'sɪstəns] *n.* **a)** *(subsisting)* [Über]leben, *das;* **be enough for a bare ~:** gerade genug zum [Über]leben sein; ‹*Einkommen:*› das Existenzminimum sein; **b) [means of] ~:** Lebensgrundlage, *die*

subsistence: ~ allowance *n.* Außendienstzulage, *die;* **~ level** *n.* Existenzminimum, *das;* **live at ~ level** gerade genug zum Leben haben

sub'sonic *adj.* Unterschall-

substance ['sʌbstəns] *n.* **a)** Stoff, *der;* Substanz, *die;* **b)** *no pl. (solidity)* Substanz, *die;* **c)** *no pl. (content) (of book etc.)* Inhalt, *der;* **there is no ~ in his claim/the rumour** seine Behauptung/das Gerücht entbehrt jeder Grundlage; **d)** *no pl. (essence)* Kern, *der;* **in ~:** im wesentlichen

sub'standard *adj.* **a)** unzulänglich; **the printing/recording was ~:** der Druck/die Aufnahme war nicht zufriedenstellend; **b)** *(Ling.)* nicht standardsprachlich

substantial [səb'stænʃl] *adj.* **a)** *(considerable)* beträchtlich; erheblich ‹*Zugeständnis, Verbesserung*›; größer... ‹*Darlehen*›; **b)** gehaltvoll ‹*Essen, Nahrung*›; **c)** *(solid in structure)* solide, stabil ‹*Möbel*›; solide ‹*Haus*›; kräftig ‹*Körperbau*›; wesentlich ‹*Unterschied, Argument*›

substantially [səb'stænʃəlɪ] *adv.* **a)** *(considerably)* wesentlich; **b)** *(solidly)* **~ built** solide gebaut ‹*Haus usw.*›; kräftig gebaut ‹*Person*›; **c)** *(essentially)* im wesentlichen; **~ free from sth.** weitgehend frei von etw.

substitute ['sʌbstɪtju:t] **1.** *n.* **a)** **~[s]** Ersatz, *der;* **~s for rubber** Ersatzstoffe für Gummi; **coffee ~:** Kaffee-Ersatz, *der;* **there is no ~ for real ale/hard work** es geht nichts über das echte englische Bier/über harte Arbeit; **b)** *(Sport)* Ersatzspieler, *der*/-spielerin, *die.* **2.** *adj.* Ersatz-; **a ~ teacher/secretary** *etc.* eine Vertretung. **3.** *v.t.* **~ A for B** B durch A ersetzen; **~ oil for butter** statt Butter Öl nehmen; **~ a striker for a midfield player** einen Mittelfeldspieler gegen einen Stürmer auswechseln *od.* austauschen. **4.** *v.i.* **~ for sb.** jmdn. vertreten; für jmdn. einspringen; *(Sport)* für jmdn. ins Spiel kommen

substitution [sʌbstɪ'tju:ʃn] *n.* Ersetzung, *die; (Sport)* Spielerwechsel, *der;* **~ of A for B** Verwendung von A statt B; **make a ~** *(Sport)* [einen Spieler] auswechseln

'subtenant *n.* Untermieter, *der*/-mieterin, *die; (of land, farm, shop)* Unterpächter, *der*/-pächterin, *die*

subterranean [sʌbtə'reɪnɪən] *adj.* unterirdisch

'subtitle 1. *n.* Untertitel, *der.* **2.** *v.t.* untertiteln; **the book is ~d ...:** das Buch hat den Untertitel ...

subtle ['sʌtl] *adj.*, **~r** ['sʌtlə(r)], **~st** ['sʌtlɪst] **a)** zart ‹*Duft, Dunst, Parfüm*›; fein ‹*Geschmack, Aroma*›; **b)** *(elusive)* subtil *(geh.);* fein ‹*Unterschied*›; unaufdringlich ‹*Charme*›; **c)** *(refined)* fein ‹*Ironie, Humor*›; zart ‹*Hinweis*›; subtil *(geh.)* ‹*Scherz*›; **d)** *(perceptive)* feinsinnig ‹*Beobachter, Kritiker*›; fein ‹*Intellekt*›

subtlety ['sʌtltɪ] *n., no pl. see* **subtle:** Zartheit, *die;* Feinheit, *die;* Subtilität, *die (geh.)*

subtly ['sʌtlɪ] *adv.* auf subtile Weise *(geh.);* zart ‹*hinweisen auf, andeuten*›; **~ flavoured/perfumed** von feinem Geschmack *nachgestellt*/zart duftend

'subtotal *n.* Zwischensumme, *die*

subtract [səb'trækt] *v.t.* abziehen (**from** von); subtrahieren (**from** von)

subtraction [səb'trækʃn] *n.* Subtraktion, *die*

suburb ['sʌbɜ:b] *n.* Vorort, *der;* **live in the ~s** am Stadtrand leben

suburban [sə'bɜ:bən] *adj.* **a)** *(of suburbs)* Vorort-; ‹*Leben, Haus*› am Stadtrand; **~ spread** *or* **sprawl** eintönige, endlose Vororte; **b)** *(derog.: limited in outlook)* spießig *(abwertend)*

suburbia [sə'bɜ:bɪə] *n. (derog.)* die [eintönigen] Vororte

subversive [səb'vɜ:sɪv] **1.** *adj.* subversiv. **2.** *n.* Subversive, *der/die*

'subway *n.* **a)** *(passage)* Unterführung, *die;* **b)** *(Amer.: railway)* Untergrundbahn, *die;* U-Bahn, *die (ugs.)*

sub'zero *adj.* **~ temperatures/conditions** Temperaturen unter Null

succeed [sək'si:d] **1.** *v.i.* **a)** *(achieve aim)* Erfolg haben; **sb. ~s in sth.** jmdm. gelingt etw.; jmd. schafft etw.; **sb. ~s in doing sth.** es gelingt jmdm., etw. zu tun; **~ in business/college** geschäftlich/im Studium erfolgreich sein; **I ~ed in passing the test** ich habe die Prüfung mit Erfolg *od.* erfolgreich abgelegt; **the plan did not ~:** der Plan ist gescheitert; **b)** *(come next)* die Nachfolge antreten; **~ to an office/the throne** die Nachfolge in einem Amt/die Thronfolge antreten; **~ to a title/an estate** einen Titel/ein Gut erben. **2.** *v.t.* ablösen ‹*Monarchen, Beamten*›; **~ sb. [in a post]** jmds. Nachfolge [in einem Amt] antreten

success [sək'ses] *n.* Erfolg, *der;* **meet with ~:** Erfolg haben; erfolgreich sein; **make a ~ of sth.** bei etw. Erfolg haben

successful [sək'sesfl] *adj.* erfolgreich; **be ~ in sth./doing sth.** Erfolg bei etw. haben/dabei haben, etw. zu tun; **she made a ~ attempt on the record** der Rekordversuch ist ihr gelungen

successfully [sək'sesfəlɪ] *adv.* erfolgreich

succession [sək'seʃn] *n.* **a)** Folge, *die;* **four games/years** *etc.* **in ~:** vier Spiele/Jahre *usw.* hintereinander; **in close ~** *(in space)* dicht hintereinander; *(in time)* kurz hintereinander; **b)** *(series)* Serie, *die;* **a ~ of losses/visitors** eine Verlust-/Besucherserie; **c)** *(right of succeeding to the throne etc.)* Erbfolge, *die;* **he is second in ~:** er ist Zweiter in der Erbfolge; **in ~ to his uncle** als Nachfolger seines Onkels

successive [sək'sesɪv] *adj.* aufeinanderfolgend; **five ~ games/jobs** fünf Spiele/Stellungen hintereinander

successively [sək'sesɪvlɪ] *adv.* hintereinander

successor [sək'sesə(r)] *n.* Nachfolger, *der*/Nachfolgerin, *die;* **sb.'s ~, the ~ to sb.** jmds. Nachfolger; **the ~ to the throne** der Nachfolger auf dem Thron

suc'cess story *n.* Erfolgsstory, *die (ugs.)*

succinct [sək'sɪŋkt] *adj. (terse)* knapp; *(clear, to the point)* prägnant

succinctly [sək'sɪŋktlɪ] *adv. (tersely)* in knappen Worten; *(clearly)* prägnant

succinctness [sək'sɪŋktnɪs] *n., no pl. (terseness)* Knappheit, *die; (clarity)* Prägnanz, *die*

succulent ['sʌkjʊlənt] **1.** *adj.* **a)** saftig ‹*Pfirsich, Steak usw.*›; **b)** *(Bot.)* sukkulent; fleischig; **~ plants** Sukkulenten. **2.** *n. (Bot.)* Sukkulente, *die;* Fettpflanze, *die*

succumb [sə'kʌm] *v.i.* **a)** *(be forced to give way)* unterliegen; **~ to sth.** einer Sache *(Dat.)* erliegen; **~ to temptation** der Versuchung erliegen; **~ to pressure** dem Druck nachgeben; **b)** *(die)* **~ [to one's illness/wounds** *etc.***]** seiner Krankheit/seinen Verletzungen *usw.* erliegen

such [sʌtʃ] **1.** *adj., no compar. or superl.* **a)** *(of that kind)* solch ...; **~ a person** solch *od. (ugs.)* so ein Mensch; ein solcher Mensch; **~ a book** solch *od. (ugs.)* so ein Buch; ein solches Buch; **~ people** solche Leute; **~ things** so etwas; **symphonies and other ~ compositions** Sinfonien und andere Kompositionen dieser Art; **or some ~ thing** oder so etwas; oder etwas in der Art; **I said no ~ thing** ich habe nichts dergleichen gesagt; **you'll do no ~ thing** das wirst du nicht tun; **there is no ~ bird** solch einen *od.* einen solchen Vogel gibt es nicht; **experiences ~ as these** solche *od.* derartige Erfahrungen; **there is no ~ thing as a unicorn** Einhörner gibt es gar nicht; **~ writers as Eliot and Fry** Schriftsteller wie Eliot und Fry; **I will take ~ steps as I think necessary** ich werde die Schritte unternehmen, die ich für notwendig halte; **at ~ a time** zu einer solchen Zeit; **at ~ a moment as this** in einem Augenblick wie diesem; *(disapproving)* gerade jetzt; **in ~ a case** in einem solchen *od. (ugs.)* so einem Fall; **for** *or* **on ~ an occasion** zu einem solchen Anlaß; **~ a one as he/she is impossible to replace** jemand wie er/sie ist unersetzlich; **b)** *(so great)* solch ...; derartig; **I got ~ a fright that ...:** ich bekam einen derartigen *od. (ugs.)* so einen Schrecken, daß ...; **~ was the force of the explosion that ...:** die Explosion war so stark, daß ...; **to ~ an extent** dermaßen; **c)** *with adj.* so; **~ a big house** ein so großes Haus; **she has ~ lovely blue eyes** sie hat so schöne blaue Augen; **~ a long time** so lange. **2.** *pron.* **as ~:** als solcher/solche/solches; *(strictly speaking)* im Grunde genommen; an sich; **~ is life** so ist das Leben; **~ as** wie [zum Beispiel]

such-and-such ['sʌtʃənsʌtʃ] **1.** *adj.* **in ~ a place at ~ a time** an dem und dem Ort um die und die Zeit; **Mr ~:** Herr Sowieso. **2.** *pron.* der und der/die und die/das und das

suchlike ['sʌtʃlaɪk] *pron. (coll.)* derlei

suck [sʌk] **1.** *v.t.* saugen **(out of** aus); lutschen ‹*Bonbon*›; **~ one's thumb** am Daumen lutschen. **2.** *v.i.* ‹*Baby:*› saugen; **~ at sth.** an etw. *(Dat.)* saugen; **~ at a lollipop** an einem Lutscher lecken

~ **'down** *v.t.* hinunterziehen; ‹*Strudel:*› in die Tiefe ziehen

~ **'in** *v.t.* einsaugen; ‹*Strudel:*› in die Tiefe ziehen

~ **'under** *v.t.* in die Tiefe ziehen

~ **'up 1.** *v.t.* aufsaugen ‹*Staub, Feuchtigkeit*›; *(with a straw)* einsaugen. **2.** *v.i.* **~ up to sb.** *(sl.)* jmdm. in den Hintern kriechen *(salopp)*

sucker ['sʌkə(r)] *n.* **a)** *(suction pad)* Saugfuß, *der; (Zool.)* Saugnapf, *der;* **b)** *(one attracted)* **be a ~ for sb./sth.** eine Schwäche für jmdn./etw. haben; **c)** *(sl.: dupe)* Dumme, *der/die;* **poor ~:** armer Trottel

suckle ['sʌkl] **1.** *v.t.* säugen. **2.** *v.i.* [an der Brust] trinken

suction ['sʌkʃn] *n.* **a)** *(sucking)* Absaugen, *das; (force)* Saugwirkung, *die;* **b)** *(of air, currents, etc.)* Sog, *der*

Sudan [su:'dɑ:n] *pr. n.* **[the] ~:** [der] Sudan

sudden ['sʌdn] **1.** *adj.* **a)** *(unexpected)* plötzlich; **I had a ~ thought** auf einmal *od.* plötzlich fiel mir etwas ein; **b)** *(abrupt, without warning)* jäh ‹*Abgrund, Übergang, Ruck*›; **there was a ~ bend in the road** plötzlich machte die Straße eine Biegung. **2.** *n.* **all of a ~:** plötzlich

sudden 'death *attrib. adj. (Sport coll.)* **a ~ play-off** ein Stichentscheid; *(Footb.: using penalties)* ein Elfmeterschießen

suddenly ['sʌdnlɪ] *adv.* plötzlich

suddenness ['sʌdnnɪs] *n., no pl.* Plötzlichkeit, *die*

suds [sʌdz] *n. pl.* **[soap-]~:** [Seifen]lauge, *die; (froth)* Schaum, *der*

sue [su:, sju:] *(Law)* **1.** *v.t.* verklagen **(for** auf + *Akk.*). **2.** *v.i.* klagen **(for** auf + *Akk.*)

suede [sweɪd] *n.* Wildleder, *das*

suet ['su:ɪt, 'sju:ɪt] *n.* Talg, *der*

Suez ['sʊɪz, 'sju:ɪz] *pr. n.* Suez *(das);* **~ Canal** Suez-Kanal, *der*

suffer ['sʌfə(r)] **1.** *v.t.* **a)** *(undergo)* erleiden ‹*Verlust, Unrecht, Schmerz, Niederlage*›; durchmachen, erleben ‹*Schweres, Kummer*›; dulden ‹*Unverschämtheit*›; **the dollar ~ed further losses against the yen** der Dollar mußte weitere Einbußen gegenüber dem Yen hinnehmen; **b)** *(tolerate)* dulden; **not ~ fools gladly** mit dummen Leuten keine Geduld haben. **2.** *v.i.* leiden; **~ for sth.** *(for a cause)* für etw. leiden; *(to make amends)* für etw. büßen

~ **from** *v.t.* leiden unter (+ *Dat.*); leiden an (+ *Dat.*) ‹*Krankheit*›; **~ from shock** unter Schock[wirkung] stehen; **~ from faulty planning** an falscher Planung kranken

sufferance ['sʌfərəns] *n.* Duldung, *die;* **he remains here on ~ only** er ist hier bloß geduldet

sufferer ['sʌfərə(r)] *n.* Betroffene, *der/die; (from disease)* Leidende, *der/die*

suffering ['sʌfərɪŋ] *n.* Leiden, *das;* **her ~s are now at an end** sie hat jetzt ausgelitten *(geh.)*

suffice [sə'faɪs] **1.** *v.i.* genügen; **~ it to say: ...:** nur soviel sei gesagt: ... **2.** *v.t.* genügen (+ *Dat.*); reichen für

sufficiency [sə'fɪʃənsɪ] *n., no pl.* Zulänglichkeit, *die*

sufficient [sə'fɪʃənt] *adj.* genug; ~ **money/food** genug Geld/genug zu essen; **be** ~: genügen; ~ **reason** Grund genug; **have you had** ~? *(food, drink)* haben Sie schon genug?

sufficiently [sə'fɪʃəntlɪ] *adv.* genug; *(adequately)* ausreichend; ~ **large** groß genug; **a** ~ **large number** eine genügend große Zahl

suffix ['sʌfɪks] *n. (Ling.)* Suffix, *das (fachspr.);* Nachsilbe, *die*

suffocate ['sʌfəkeɪt] **1.** *v.t.* ersticken; **he was ~d by the smoke** der Rauch erstickte ihn; er erstickte an dem Rauch. **2.** *v.i.* ersticken

suffocation [sʌfə'keɪʃn] *n.* Erstickung, *die;* **a feeling of** ~: das Gefühl, zu ersticken

suffrage ['sʌfrɪdʒ] *n.* Wahlrecht, *das;* **female** ~: das Frauenwahlrecht

suffragette [sʌfrə'dʒet] *n. (Hist.)* Frauenrechtlerin, *die;* Suffragette, *die*

sugar ['ʃʊgə(r)] **1.** *n.* Zucker, *der;* **two ~s, please** *(spoonfuls)* zwei Löffel Zucker, bitte; *(lumps)* zwei Stück Zucker, bitte. **2.** *v.t.* zuckern; *(fig.)* versüßen

sugar: ~ **basin** *see* **~-bowl;** **~-beet** *n.* Zuckerrübe, *die;* **~-bowl** *n.* Zuckerschale, *die; (covered)* Zuckerdose, *die;* **~-cane** *n.* Zuckerrohr, *das;* **~-coated** *adj.* gezuckert; mit Zucker überzogen ‹*Dragee usw.*›; **~-lump** *n.* Zuckerstück, *das; (when counted)* Stück Zucker, *das*

sugary ['ʃʊgərɪ] *adj.* süß; *(fig.)* süßlich ‹*Lächeln, Stimme, Musik*›

suggest [sə'dʒest] **1.** *v.t.* **a)** *(propose)* vorschlagen; ~ **sth. to sb.** jmdm. etw. vorschlagen; **he ~ed going to the cinema** er schlug vor, ins Kino zu gehen; **b)** *(assert)* **are you trying to** ~ **that he is lying?** wollen Sie damit sagen, daß er lügt?; **he ~ed that the calculation was incorrect** er sagte, die Rechnung sei falsch; **I** ~ **that ...** *(Law)* ich unterstelle, daß ...; **c)** *(make one think of)* suggerieren; ‹*Symptome, Tatsachen:*› schließen lassen auf (+ *Akk.*). **2.** *v. refl.* ~ **itself [to sb.]** ‹*Möglichkeiten, Ausweg:*› sich [jmdm.] anbieten; ‹*Gedanke:*› sich [jmdm.] aufdrängen

suggestion [sə'dʒestʃn] *n.* **a)** Vorschlag, *der;* **at** *or* **on sb.'s** ~: auf jmds. Vorschlag *(Akk.);* **b)** *(insinuation)* Andeutungen *Pl.;* **there is no** ~ **that he co-operated with the kidnappers** niemand unterstellt, daß er mit den Entführern zusammengearbeitet hat; **what a** ~! wie kann man so etwas nur sagen!; **c)** *(fig.: trace)* Spur, *die*

suggestive [sə'dʒestɪv] *adj.* **a)** suggestiv *(geh.);* **be** ~ **of sth.** auf etw. *(Akk.)* schließen lassen; **b)** *(risqué)* anzüglich; gewagt; zweideutig ‹*Scherze, Lieder*›

suicidal [su:ɪ'saɪdl, sju:ɪ'saɪdl] *adj.* **a)** selbstmörderisch ‹*Akt, Absicht*›; suizidal *(fachspr.)* ‹*Verhalten, Patient*›; ~ **tendencies** eine Neigung zum Selbstmord; **I felt** *or* **was quite** ~: ich hätte mich am liebsten gleich umgebracht; **b)** *(dangerous)* selbstmörderisch ‹*Fahrweise, Verhalten usw.*›

suicide ['su:ɪsaɪd, 'sju:ɪsaɪd] *n.* Selbstmord, *der (auch fig.);* Suizid, *der (fachspr.)*

suicide: ~ **attempt** *n.* Selbstmordversuch, *der;* ~ **pact** *n.* Selbstmordpakt, *der*

suit [su:t, sju:t] **1.** *n.* **a)** *(for men)* Anzug, *der; (for women)* Kostüm, *das;* **a three-piece-~:** ein dreiteiliger Anzug; ein Dreiteiler; **buy [oneself] a new** ~ **of clothes** sich neu einkleiden; **b)** *(Law)* ~ **[at law]** Prozeß, *der;* [Gerichts]verfahren, *das;* **c)** *(Cards)* Farbe, *die;* **follow** ~: Farbe bedienen; *(fig.)* das Gleiche tun. **2.** *v.t.* **a)** anpassen **(to** *Dat.*); **b) be ~ed [to sth./one another]** [zu etw./zueinander] passen; **he is not at all ~ed to marriage** er eignet sich überhaupt nicht für die Ehe; **they are ill/well ~ed** sie passen schlecht/gut zueinander; **c)** *(satisfy needs of)* passen (+ *Dat.*); recht sein (+ *Dat.*); **does the climate** ~ **you/your health?** bekommt Ihnen das Klima?; **dried fruit/asparagus does not** ~ **me** ich vertrage kein Trockenobst/keinen Spargel; **d)** *(go well with)* passen zu; **does this hat** ~ **me?** steht mir dieser Hut?; **black ~s her** Schwarz steht ihr gut. **3.** *v. refl.* ~ **oneself** tun, was man will; ~ **yourself!** [ganz] wie du willst!

suitability [su:tə'bɪlɪtɪ, sju:tə'bɪlɪtɪ] *n., no pl.* Eignung, *die* **(for** für); *(of clothing, remark; for an occasion)* Angemessenheit, *die* **(for** für); **his** ~ **as a teacher** seine Eignung zum *od.* als Lehrer

suitable ['su:təbl, 'sju:təbl] *adj.* geeignet; *(for an occasion)* angemessen ‹*Kleidung*›; angebracht ‹*Bemerkung*›; *(matching, convenient)* passend; **this girl-friend is not** ~ **for him** diese Freundin paßt nicht zu ihm; **Monday is the most** ~ **day [for me]** Montag paßt [mir] am besten

suitably ['su:təblɪ, 'sju:təblɪ] *adv.* angemessen; gehörig ‹*entrüstet*›; gebührend ‹*beeindruckt*›; entsprechend ‹*gekleidet*›

'**suitcase** *n.* Koffer, *der;* **live out of a** ~: aus dem Koffer leben

suite [swi:t] *n.* **a)** *(of furniture)* Garnitur, *die;* **three-piece** ~: Polstergarnitur, *die;* **bedroom** ~: Schlafzimmereinrichtung, *die;* **b)** *(of rooms)* Suite, *die;* **c)** *(Mus.)* Suite, *die*

suitor ['su:tə(r), 'sju:tə(r)] *n.* Freier, *der*

sulfate, sulfide, sulfur, sulfuric *(Amer.) see* **sulph-**

sulk [sʌlk] **1.** *n., usu. in pl.* **have a** ~ *or* **the ~s, be in** *or* **have a fit of the ~s** eingeschnappt sein *(ugs.);* schmollen. **2.** *v.i.* schmollen

sulky ['sʌlkɪ] *adj.* schmollend; eingeschnappt *(ugs.)*

sullen ['sʌlən] *adj.* mürrisch; verdrießlich; *(fig.)* düster ‹*Himmel*›

sully ['sʌlɪ] *v.t. (formal)* besudeln *(geh.)*

sulphate ['sʌlfeɪt] *n.* Sulfat, *das*

sulphide ['sʌlfaɪd] *n.* Sulfid, *das*

sulphur ['sʌlfə(r)] *n.* Schwefel, *der*

sulphuric [sʌl'fjʊərɪk] *adj.* ~ **acid** Schwefelsäure, *die*

sultan ['sʌltən] *n.* Sultan, *der*

sultana [sʌl'tɑ:nə] *n. (raisin)* Sultanine, *die*

sultry ['sʌltrɪ] *adj.* schwül ‹*Wetter, Tag, Atmosphäre*›; *(fig.: sensual)* sinnlich; schwül ‹*Schönheit*›

sum [sʌm] **1.** *n.* **a)** *(total amount, lit. or fig.)* Summe, *die* **(of** aus); ~ **[total]** Ergebnis, *das;* **that was the** ~ **total of our achievements** *or* **of what we achieved** das war alles, was wir erreicht haben; **b)** *(amount of money)* Summe, *die;* **a cheque for this** ~: ein Scheck über diesen Betrag; **c)** *(Arithmetic)* Rechenaufgabe, *die;* **do ~s** rechnen; **she is good at ~s** sie kann gut rechnen; sie ist gut im Rechnen. **2.** *v.t.,* **-mm-** addieren

~ '**up 1.** *v.t.* **a)** zusammenfassen; **b)** *(Brit.: assess)* einschätzen; **this ~med him up perfectly** damit war er treffend charakterisiert. **2.** *v.i.* ein Fazit ziehen; ‹*Richter:*› resümieren; **in ~ming up, I should like to ...:** zusammenfassend möchte ich ...

summarily ['sʌmərɪlɪ] *adv.* **a)** *(shortly)* knapp; **b)** *(without formalities or delay)* summarisch; ~ **dismissed** fristlos entlassen; ~ **convicted** *(Law)* im summarischen Verfahren verurteilt

summarize (summarise) ['sʌməraɪz] *v.t.* zusammenfassen

summary ['sʌmərɪ] **1.** *adj.* **a)** knapp; **b)** *(without formalities or delay)* summarisch; fristlos ‹*Entlassung*›; ~ **justice** Schnelljustiz, *die.* **2.** *n.* Zusammenfassung, *die*

summer ['sʌmə(r)] **1.** *n.* Sommer, *der;* **in [the]** ~: im Sommer; **in early/late** ~: im Früh-/Spätsommer; **last/next** ~: letzten/nächsten Sommer; **a ~'s day/night** ein Sommertag/eine Sommernacht; **in the ~ of 1983, in ~ 1983** im Sommer 1983; **two ~s ago we went to France** im Sommer vor zwei Jahren waren wir in Frankreich. **2.** *attrib. adj.* Sommer-

summer: **~-house** *n.* [Garten]laube, *die;* ~ **school** *n.* Sommerkurs, *der;* ~ **term** *n.* Sommerhalbjahr, *das;* **S~ Time** *n.* *(Brit.: system)* die Sommerzeit; **~time** *n.* *(season)* Sommer, *der;* **in [the] ~time** im Sommer

summery ['sʌmərɪ] *adj.* sommerlich

summing-'up *n.* Zusammenfassung, *die*

summit ['sʌmɪt] *n.* **a)** *(peak, lit. or fig.)* Gipfel, *der;* **b)** *(discussion)* Gipfel, *der;* ~ **conference/meeting** Gipfelkonferenz, *die*/-treffen, *das*

summon ['sʌmən] *v.t.* **a)** *(call upon)* rufen **(to** zu); holen ‹*Hilfe*›; zusammenrufen ‹*Aktionäre*›; **b)** *(call by authority)* zu sich zitieren; einberufen ‹*Parlament*›; **c)** *(Law: to court)* vorladen ‹*Angeklagten, Zeugen*›

~ **'up** *v.t.* aufbringen ‹*Mut, Kräfte, Energie, Begeisterung*›

summons ['sʌmənz] *n.* *(Law)* Vorladung, *die;* **serve a ~ on sb.** jmdm. eine Vorladung zustellen

sump [sʌmp] *n.* *(Brit. Motor Veh.)* Ölwanne, *die*

sumptuous ['sʌmptjʊəs] *adj.* üppig; luxuriös ‹*Einband, Möbel, Kleidung*›

sun [sʌn] **1.** *n.* Sonne, *die;* **catch the** ~ *(be in a sunny position)* viel Sonne abbekommen; *(get ~burnt)* einen Sonnenbrand bekommen; **a touch of the** ~: ein leichter Sonnenstich; **under the** ~ *(fig.)* auf der Welt. **2.** *v. refl.*, **-nn-** sich sonnen

Sun. *abbr.* **Sunday** So.

sun: **~bathe** *v.i.* sonnenbaden; **~bather** *n.* Sonnenbadende, *der/die;* **~bathing** *n.* Sonnenbaden, *das;* **~beam** *n.* Sonnenstrahl, *der;* **~-bed** *n.* *(with UV lamp)* Sonnenbank, *die;* *(in garden etc.)* Gartenliege, *die;* **~-blind** *n.* Markise, *die;* **~burn** *n.* Sonnenbrand, *der;* **~burnt** *adj.* **a)** *(suffering from ~burn)* **be ~burnt** einen Sonnenbrand haben; **have a ~burnt back/face** einen Sonnenbrand auf dem Rücken/im Gesicht haben; **get badly ~burnt** einen schlimmen Sonnenbrand bekommen; **b)** *(tanned)* sonnenverbrannt ‹*Person, Gesicht usw.*›

sundae ['sʌndeɪ, 'sʌndɪ] *n.* **[ice-cream]** ~: Eisbecher, *der*

Sunday ['sʌndeɪ, 'sʌndɪ] **1.** *n.* Sonntag, *der.* **2.** *adv.* *(coll.)* **she comes ~s** sie kommt sonntags. *See also* **Friday**

'Sunday school *n.* Sonntagsschule, *die*

sun: **~dial** *n.* Sonnenuhr, *die;* **~down** *see* **sunset**

sundry ['sʌndrɪ] **1.** *adj.* verschieden; ~ **articles** verschiedene *od.* diverse Artikel. **2.** *n. in pl.* Verschiedenes; Diverses; *see also* **all 2 a**

'sunflower *n.* Sonnenblume, *die;* ~ **seeds** Sonnenblumenkerne

sung *see* **sing**

sun: **~-glasses** *n. pl.* Sonnenbrille, *die;* **~-hat** *n.* Sonnenhut, *der*

sunk *see* **sink 2, 3**

sunken ['sʌŋkn] *adj.* versunken ‹*Schatz*›; gesunken ‹*Schiff*›; eingefallen ‹*Augen, Wangen*›; tieferliegend ‹*Garten, Zimmer*›; in den Boden eingelassen ‹*Badewanne*›

'sun-lamp *n.* Höhensonne, *die*

sunless ['sʌnlɪs] *adj.* ‹*Ecke, Stelle, Tal*› wo die Sonne nie hinkommt; trübe ‹*Tag*›

sun: **~light** *n.* Sonnenlicht, *das;* **~lit** *adj.* sonnenbeschienen ‹*Landschaft*›; sonnig ‹*Zimmer, Garten*›; ~ **lounge** *n.* Veranda, *die*

sunny ['sʌnɪ] **a)** sonnig; ~ **intervals** Aufheiterungen; **the ~ side of the house/street** die Sonnenseite des Hauses/der Straße; ~ **side up** ‹*Spiegelei*› mit dem Gelben nach oben; **b)** *(cheery)* fröhlich ‹*Wesen, Lächeln*›

sun: **~ray** *n.* Sonnenstrahl, *der;* **~rise** *n.* Sonnenaufgang, *der;* **at ~rise** bei Sonnenaufgang; *attrib.* **~rise industry** Zukunftsindustrie, *die;* **~-roof** *n.* *(Motor Veh.)* Schiebedach, *das;* **~set** *n.* Sonnenuntergang, *der;* **at ~set** bei Sonnenuntergang; **~shade** *n.* Sonnenschirm, *der;* *(awning)* Markise, *die;* **~shine** *n.* Sonnenschein, *der;* **~shine roof** *see* **sun-roof;** **~stroke** *n.* Sonnenstich, *der;* **suffer from/get ~stroke** einen Sonnenstich haben/bekommen; **~-tan** *n.* [Sonnen]bräune, *die;* **get a ~-tan** braun werden; **~-tan lotion** *n.* Sonnencreme, *die;* **~-tanned** *adj.* braun[gebrannt]; sonnengebräunt *(geh.);* **~-tan oil** *n.* Sonnenöl, *das;* **~-trap** *n.* sonniges Plätzchen; **~-up** *(Amer.)* *see* **~-rise**

super ['su:pə(r)] *adj.* *(Brit. coll.)* super *(ugs.)*

superannuation [su:pərænjʊ'eɪʃn] *n.* **a)** ~ **[contribution/payment]** Beitrag zur Rentenversicherung; **b)** *(pension)* Rente, *die*

superb [sʊ'pɜ:b, sju:'pɜ:b] *adj.* einzigartig; erstklassig ‹*Essen, Zustand*›

supercilious [su:pə'sɪlɪəs] *adj.* hochnäsig

supercomputer ['su:pəkəmpju:tə(r)] *n.* Supercomputer, *der*

superficial [su:pə'fɪʃl] *adj.* *(also fig.)* oberflächlich; leicht ‹*Änderung, Schaden*›; äußerlich ‹*Ähnlichkeit*›

superfluous [sʊ'pɜ:flʊəs, sju:'pɜ:flʊəs] *adj.* überflüssig

superglue ['su:pəglu:] *n.* Sekundenkleber, *der*

supergrass ['su:pəgrɑ:s] *n.* Superspitzel, *der (abwertend)*

superhighway ['su:pəhaɪweɪ] *n.* *(Amer.)* Autobahn, *die*

superhuman [su:pə'hju:mən] *adj.* übermenschlich

superimpose [su:pərɪm'pəʊz] *v.t.* aufbringen ‹*Schicht usw.*›; aufkopieren ‹*Bild*›

superintend [su:pərɪn'tend] *v.t.* überwachen; beaufsichtigen

superintendent [su:pərɪn'tendənt] *n.* *(Brit. Police)* Kommissar, *der*/Kommissarin, *die;* *(Amer. Police)* [Polizei]präsident, *der*/-präsidentin, *die*

superior [su:'pɪərɪə(r), sju:'pɪərɪə(r), sʊ'pɪərɪə(r)] **1.** *adj.* **a)** *(of higher quality)* besonders gut ‹*Restaurant, Qualität, Stoff*›; überlegen ‹*handwerkliches Können, Technik, Intelligenz*›; **b)** *(having higher rank)* höher... ‹*Stellung, Rang, Gericht*›; **be ~ to sb.** einen höheren Rang als jmd. haben. **2.** *n.* **a)** *(sb. higher in rank)* Vorgesetzte, *der/die;* **b)** *(sb. better)* Überlegene, *der/die*

superiority [su:pɪərɪ'ɒrɪtɪ, sju:pɪərɪ'ɒrɪtɪ, sʊpɪərɪ'ɒrɪtɪ] *n.* Überlegenheit, *die* **(to** über + *Akk.*); *(of goods)* besondere Qualität

superlative [su:'pɜ:lətɪv, sju:'pɜ:lətɪv] **1.** *adj.* **a)** unübertrefflich; **b)** *(Ling.)* superlativisch; **a ~ adjective/adverb** ein Adjektiv/Adverb im Superlativ. **2.** *n.* *(Ling.)* Superlativ, *der*

supermarket ['su:pəma:kɪt] *n.* Supermarkt, *der*

supernatural [su:pə'nætʃərl] *adj.* übernatürlich

superpower ['su:pəpaʊə(r)] *n.* *(Polit.)* Supermacht, *die*

supersede [su:pə'si:d] *v. t.* ablösen (**by** durch); **old ~d ideas** alte, überholte Vorstellungen

supersonic [su:pə'sɒnɪk] *adj.* Überschall-; **go ~:** die Schallmauer durchbrechen

superstar ['su:pəsta:(r)] *n.* Superstar, *der*

superstition [su:pə'stɪʃn] *n.* *(lit. or fig.)* Aberglaube, *der;* **~s** abergläubische Vorstellungen

superstitious [su:pə'stɪʃəs] *adj.* abergläubisch

superstore ['su:pəstɔ:(r)] *n.* Großmarkt, *der*

superstructure ['su:pəstrʌktʃə(r)] *n.* **a)** Aufbau, *der;* **b)** *(Sociol.)* Überbau, *der*

supertanker ['su:pətæŋkə(r)] *n.* Supertanker, *der*

supervise ['su:pəvaɪz] *v. t.* beaufsichtigen

supervision [su:pə'vɪʒn] *n.* Aufsicht, *die*

supervisor ['su:pəvaɪzə(r)] *n.* Aufseher, *der*/Aufseherin, *die;* *(for Ph.D. thesis)* Doktorvater, *der;* **office ~:** Bürovorsteher, *der*/-vorsteherin, *die*

supervisory ['su:pəvaɪzərɪ] *adj.* Aufsichts-

supper ['sʌpə(r)] *n.* Abendessen, *das;* *(simpler meal)* Abendbrot, *das;* **have** *or* **eat [one's] ~:** zu Abend essen; **be at** *or* **eating** *or* **having [one's] ~:** beim Abendessen/Abendbrot sein; **The Last S~:** das [letzte] Abendmahl

'**supper-time** *n.* Abendbrotzeit, *die;* **it's ~:** es ist Zeit zum Abendessen

supplant [sə'pla:nt] *v. t.* ablösen, ersetzen (**by** durch); ausstechen ‹*Widersacher, Rivalen*›

supple ['sʌpl] *adj.* geschmeidig

supplement 1. ['sʌplɪmənt] *n.* **a)** Ergänzung, *die* (**to** + *Gen.*); *(addition)* Zusatz, *der;* **b)** *(of book)* Nachtrag, *der;* *(separate volume)* Supplement, *das;* Nachtragsband, *der;* *(of newspaper)* Beilage, *die;* **c)** *(to fare etc.)* Zuschlag, *der.* **2.** ['sʌplɪmənt, sʌplɪ'ment] *v. t.* ergänzen

supplementary [sʌplɪ'mentərɪ] *adj.* zusätzlich; Zusatz‹*rente, -frage*›; **~ fare/charge** Zuschlag, *der*

supplier [sə'plaɪə(r)] *n.* *(Commerc.)* Lieferant, *der*/Lieferantin, *die*

supply [sə'plaɪ] **1.** *v. t.* **a)** liefern ‹*Waren usw.*›; sorgen für ‹*Unterkunft*›; zur Verfügung stellen ‹*Lehrmittel, Arbeitskleidung usw.*›; beliefern ‹*Kunden, Geschäft*›; versorgen ‹*System*›; **~ sth. to sb., ~ sb. with sth.** jmdn. mit etw. versorgen/*(Commerc.)* beliefern; **b)** *(make good)* erfüllen ‹*Nachfrage, Bedarf*›; abhelfen (+ *Dat.*) ‹*Mangel*›. **2.** *n.* **a)** *(stock)* Vorräte *Pl.;* **a large ~ of food** große Lebensmittelvorräte; **military/medical supplies** militärischer/medizinischer Nachschub; **~ and demand** *(Econ.)* Angebot und Nachfrage; **b)** *(provision)* Versorgung, *die* (**of** mit); **their gas ~ was cut off** ihnen ist das Gas abgestellt worden; **the blood ~ to the brain** die Versorgung des Gehirns mit Blut; **c) ~ [teacher]** Vertretung, *die.* **3.** *attrib.* Versorgungs‹*schiff, -netz, -basis, -lager usw.*›; **~ lines** Nachschubwege

support [sə'pɔ:t] **1.** *v. t.* **a)** *(hold up)* stützen ‹*Mauer, Verletzten*›; *(bear weight of)* tragen ‹*Dach*›; **b)** *(give strength to)* stärken; **c)** unterstützen ‹*Politik, Verein*›; *(Footb.)* **~ Spurs** Spurs-Fan sein; **d)** *(give money to)* unterstützen; spenden für; **e)** *(provide for)* ernähren ‹*Familie, sich selbst*›; **f)** *(bring facts to confirm)* stützen ‹*Theorie, Anspruch, Behauptung*›; *(speak in favour of)* befürworten ‹*Streik, Maßnahme*›. **2.** *n.* **a)** Unterstützung, *die;* **give ~ to sb./sth.** jmdn./etw. unterstützen; **in ~:** zur Unterstützung; **speak in ~ of sb./sth.** jmdn. unterstützen/etw. befürworten; **b)** *(sb./sth. that ~s)* Stütze, *die;* **hold on to sb./sth. for ~:** sich an jmdm./etw. festhalten

supporter [sə'pɔ:tə(r)] *n.* Anhänger, *der*/Anhängerin, *die;* **a football ~:** ein Fußballfan; **~s of a strike** Befürworter eines Streiks

sup'porters' club *n.* *(Sport)* Fanclub, *der*

supporting [sə'pɔ:tɪŋ] *adj.* *(Cinemat., Theatre)* **~ role** Nebenrolle, *die;* **~ actor/actress** Schauspieler/-spielerin in einer Nebenrolle; **~ film** Vorfilm, *der*

supportive [sə'pɔ:tɪv] *adj.* hilfreich; **be very ~ [to sb.]** [jmdm.] eine große Hilfe *od.* Stütze sein

suppose [sə'pəʊz] *v. t.* **a)** *(assume)* annehmen; **~** *or* **supposing [that] he ...:** angenommen, [daß] er ...; **always supposing that ...:** immer vorausgesetzt, daß ...; **~ we wait until tomorrow** wir könnten eigentlich bis morgen warten; **b)** *(presume)* vermuten; **I ~d she was in Glasgow** ich vermutete sie in Glasgow; **I don't ~ you have an onion to spare?** Sie haben wohl nicht zufällig eine Zwiebel übrig?; **we're not going to manage it, are we? – I ~ not** wir werden es wohl nicht schaffen – ich glaube kaum; **I ~ so** ich nehme es an; *(doubtfully)* ja, vermutlich; *(more confidently)* ich glaube schon; **c) be ~d to do/be sth.** *(be generally believed to do/be sth.)* etw. tun/sein sollen; **cats are ~d to have nine lives** Katzen sollen angeblich neun Leben haben; **d)** *(allow)* **you are not ~d to do that** das darfst du nicht; **I'm not ~d to be here** ich dürfte eigentlich gar nicht hier sein; **e)** *(presuppose)* voraussetzen

supposedly [sə'pəʊzɪdlɪ] *adv.* angeblich

supposition [sʌpə'zɪʃn] *n.* Annahme, *die;* Vermutung, *die;* **be based on ~:** auf Annahmen *od.* Vermutungen beruhen

suppress [sə'pres] *v. t.* unterdrücken

suppression [sə'preʃn] *n.* Unterdrückung, *die*

supremacy [su:'preməsɪ, sju:'preməsɪ] *n., no pl.* **a)** *(supreme authority)* Souveränität, *die;* **b)** *(superiority)* Überlegenheit, *die*

supreme [su:'pri:m, sju:'pri:m] *adj.* höchst...

Supt. *abbr.* **Superintendent**

surcharge ['sɜ:tʃa:dʒ] *n.* Zuschlag, *der*

sure [ʃʊə(r)] **1.** *adj.* **a)** *(confident)* sicher; **be ~ of sth.** sich *(Dat.)* einer Sache *(Gen.)* sicher sein; **~ of oneself** selbstsicher; **don't be too ~:** da wäre ich mir nicht so sicher; **b)** *(safe)* sicher; **be on ~r ground** *(lit. or fig.)* sich auf festerem Boden befinden; **c)** *(certain)* sicher; **you're ~ to be welcome** Sie werden ganz sicher *od.* bestimmt willkommen sein; **it's ~ to rain** es wird bestimmt regnen; **don't worry, it's ~ to turn out well** keine Sorge, es wird schon alles gutgehen; **he is ~ to ask questions about the incident** er wird auf jeden Fall Fragen zu dem Vorfall stellen; **d)** *(undoubtedly true)* sicher; **to be ~** *expr. concession* natürlich; *expr. surprise* wirklich!; tatsächlich!; **for ~** *(coll.: without doubt)* auf jeden Fall; **e) make ~ [of sth.]** sich [einer Sache] vergewissern; *(check)* [etw.] nachprüfen; **you'd better make ~ of a seat** *or* **that you have a seat** du solltest dir einen Platz sichern; **make** *or*

be ~ you do it, be ~ to do it *(do not fail to do it)* sieh zu, daß du es tust; *(do not forget)* vergiß nicht, es zu tun; **be ~ you finish the work by tomorrow** machen Sie die Arbeit auf jeden Fall bis morgen fertig; **f)** *(reliable)* sicher ‹*Zeichen*›; zuverlässig ‹*Freund, Bote, Heilmittel*›; **a ~ winner** ein todsicherer Tip *(ugs.)*. **2.** *adv.* **a) as ~ as ~ can be** *(coll.)* so sicher wie das Amen in der Kirche; **as ~ as I'm standing here** so wahr ich hier stehe; **~ enough** tatsächlich; **b)** *(Amer. coll.: certainly)* wirklich; echt *(ugs.)*. **3.** *int.* **~!, ~ thing!** *(Amer.)* na klar! *(ugs.)*

sure: ~-fire *attrib. adj. (Amer. coll.)* todsicher; **~-footed** ['ʃʊəfʊtɪd] *adj. (lit. or fig.)* trittsicher

surely ['ʃʊəlɪ] *adv.* **1. a)** *as sentence-modifier* doch; **~ we've met before?** wir kennen uns doch, oder?; **~ you are not going out in this snowstorm?** du willst doch wohl in dem Schneesturm nicht rausgehen?; **b)** *(steadily)* sicher; **slowly but ~:** langsam, aber sicher; **c)** *(certainly)* sicherlich; **the plan will ~ fail** der Plan wird garantiert scheitern. **2.** *int. (Amer.)* natürlich; selbstverständlich

surf [sɜ:f] *n.* Brandung, *die*

surface ['sɜ:fɪs] **1.** *n.* **a)** *no pl.* Oberfläche, *die;* **outer ~:** Außenfläche, *die;* **the earth's ~:** die Erdoberfläche; **the ~ of the lake** die Seeoberfläche; **on the ~:** an der Oberfläche; *(Mining)* über Tage; **b)** *(outward appearance)* Oberfläche, *die;* **on the ~:** oberflächlich betrachtet; **come to the ~:** an die Oberfläche kommen; ‹*Taucher, Unterseeboot:*› auftauchen; *(fig.)* ans Licht kommen *(fig.)*. **2.** *v.i.* auftauchen; *(fig.)* hochkommen

surface: ~ area *n.* Oberfläche, *die;* **~ mail** *n.* gewöhnliche Post *(die auf dem Land- bzw. Seeweg befördert wird)*

'surfboard *n.* Surfbrett, *das*

surfeit ['sɜ:fɪt] *n.* Übermaß, das

surfer ['sɜ:fə(r)] *n.* Surfer, *der*/Surferin, *die*

surfing ['sɜ:fɪŋ] *n.* Surfen, *das*

surge [sɜ:dʒ] **1.** *v.i.* ‹*Wellen:*› branden; ‹*Fluten, Menschenmenge:*› sich wälzen; ‹*elektrischer Strom:*› ansteigen; **the crowd ~d forward** die Menschenmenge drängte sich nach vorn. **2.** *n.* **a)** *(of the sea)* Branden, *das;* **b)** *(of crowd)* Sichwälzen, *das; (of electric current)* Anstieg, *der*

~ 'up *v.i.* aufsteigen; ‹*Gefühl:*› aufwallen

surgeon ['sɜ:dʒən] *n.* Chirurg, *der*/Chirurgin, *die*

surgery ['sɜ:dʒərɪ] *n.* **a)** *no pl., no indef. art.* Chirurgie, *die;* **need ~:** operiert werden müssen; **undergo ~:** sich einer Operation *(Dat.)* unterziehen; **b)** *(Brit.: place)* Praxis, *die;* **doctor's/dental ~:** Arzt-/Zahnarztpraxis, *die;* **c)** *(Brit.: time; session)* Sprechstunde, *die;* **when is his ~?** wann hat er Sprechstunde?; **hold a ~** *(Brit. coll.)* ‹*Abgeordneter, Anwalt usw.:*› eine Sprechstunde abhalten

surgical ['sɜ:dʒɪkl] *adj.* chirurgisch; **~ treatment** Operation, *die*/Operationen

surly ['sɜ:lɪ] *adj.* mürrisch; verdrießlich

surmise [sə'maɪz] **1.** *n.* Vermutung, *die;* Mutmaßung, *die.* **2.** *v.t.* mutmaßen

surmount [sə'maʊnt] *v.t.* überwinden ‹*Hindernis, Schwierigkeiten*›

surname ['sɜ:neɪm] *n.* Nachname, *der;* Zuname, *der*

surpass [sə'pɑ:s] *v.t.* übertreffen (**in** an + *Dat.*); **~ oneself** sich selbst übertreffen; **sth. ~es [sb.'s] comprehension** etw. ist [jmdm.] unbegreiflich

surplice ['sɜ:plɪs] *n. (Eccl.)* Chorhemd, *das*

surplus ['sɜ:pləs] **1.** *n.* Überschuß, *der* (**of** an + *Dat.*); **army ~ store/boots** Laden für Restbestände/Schuhe aus Restbeständen der Armee. **2.** *adj.* überschüssig; **be ~ to sb.'s requirements** von jmdm. nicht benötigt werden; **~ stocks** Überschüsse *Pl.*

surprise [sə'praɪz] **1.** *n.* **a)** Überraschung, *die;* **take sb. by ~:** jmdn. überrumpeln; **give sb. a ~:** jmdn. erschrecken; **to my great ~, much to my ~:** zu meiner großen Überraschung; sehr zu meiner Überraschung; **it came as a ~ to us** es war für uns eine Überraschung; **~, ~!** *(iron.)* sieh mal einer an! *(spött.)*; **b)** *attrib.* überraschend, unerwartet ‹*Besuch*›; **a ~ attack/defeat** ein Überraschungsangriff/eine überraschende Niederlage; **it's to be a ~ party** die Party soll eine Überraschung sein. **2.** *v.t.* überraschen; überrumpeln ‹*Feind*›; **I shouldn't be ~d if ...:** es würde mich nicht wundern, wenn ...; **be ~d at sb./sth.** sich über jmdn./etw. wundern

surprising [sə'praɪzɪŋ] *adj.* überraschend; **it's hardly ~ that ...:** es ist kaum verwunderlich, daß ...

surreal [sə'ri:əl] *adj.* surrealistisch

surrealism [sə'ri:əlɪzm] *n., no pl.* Surrealismus, *der*

surrealist [sə'ri:əlɪst] **1.** *n.* Surrealist, *der*/Surrealistin, *die.* **2.** *adj.* surrealistisch

surrender [sə'rendə(r)] **1.** *n.* **a)** *(submitting to enemy)* Kapitulation, *die;* **b)** *(giving up possession)* Aufgabe, *die; (of insurance policy)* Rückkauf, *der; (of firearms)* Abgabe, *die.* **2.** *v.i.* kapitulieren; **~ to despair** sich der Verzweiflung überlassen. **3.** *v.t. (give up possession of)* aufgeben; preisgeben ‹*Freiheit, Privileg*›; niederlegen ‹*Amt*›; abgeben, aushändigen ‹*Wertgegenstände*›. **4.** *v. refl.* sich hingeben (**to** + *Dat.*)

surreptitious [sʌrəp'tɪʃəs] *adj.* heimlich; verstohlen ‹*Blick*›

surrogate ['sʌrəgət] *n. (substitute)* Ersatz, *der*

surrogate 'mother *n.* Leihmutter, *die*

surround [sə'raʊnd] *v.t.* **a)** *(come or be all round)* umringen; ‹*Truppen, Heer:*› umzingeln ‹*Stadt, Feind*›; **b)** *(enclose, encircle)* umgeben; **be ~ed by** *or* **with sth.** von etw. umgeben sein

surrounding [sə'raʊndɪŋ] *adj.* umliegend ‹*Dörfer*›; **~ area** Umgebung, *die;* **the ~ countryside** die [Landschaft in der] Umgebung

surroundings [sə'raʊndɪŋz] *n. pl.* Umgebung, *die*

surtax ['sɜ:tæks] *n.* Ergänzungsabgabe *od.* -steuer, *die*

surveillance [sə'veɪləns] *n.* Überwachung, *die;* **keep sb. under ~:** jmdn. überwachen; **be under ~:** überwacht werden

survey 1. [sə'veɪ] *v.t.* **a)** *(take general view of)* betrachten; *(from high point)* überblicken ‹*Landschaft, Umgebung*›; **b)** *(examine)* inspizieren ‹*Gebäude usw.*›; **c)** *(assess)* bewerten ‹*Situation, Problem usw.*›. **2.** ['sɜ:veɪ] *n.* **a)** *(general view, critical inspection)* Überblick, *der* (**of** über + *Akk.*); **b)** *(by opinion poll)* Umfrage, *die; (by research)* Untersuchung, *die;* **conduct a ~ into sth.** eine Umfrage zu etw. veranstalten/etw. untersuchen; **c)** *(Surv.)* Vermessung, *die;* **d)** *(building inspection)* Inspektion, *die*

surveying [sə'veɪɪŋ] *n.* **a)** Landvermessung, *die;* **b)** *(Constr.)* Abstecken, *das*

surveyor [sə'veɪə(r)] *n.* **a)** *(of building)* Gutachter, *der*/Gutachterin, *die;* **b)** *(of land)* Landvermesser, *der*/-vermesserin, *die*

survival [sə'vaɪvl] *n., no pl.* Überleben, *das; (of tradition)* Fortbestand, *der; (of building)* Erhaltung, *die;* **fight for ~:** Existenzkampf, *der;* **the ~ of the fittest** *(Biol.)* [das] Überleben der Stärkeren

survive [sə'vaɪv] **1.** *v. t.* überleben. **2.** *v. i.* ‹*Person:*› überleben; ‹*Schriften, Gebäude, Traditionen:*› erhalten bleiben

survivor [sə'vaɪvə(r)] *n.* Überlebende, *der/die;* **he's a ~:** er ist nicht unterzukriegen

susceptibility [səseptɪ'bɪlɪtɪ] *n. (to flattery, persuasion, etc.)* Empfänglichkeit, *die* **(to** für); *(to illness, injury, etc.)* Anfälligkeit, *die* **(to** für)

susceptible [sə'septɪbl] *adj.* **a)** *(sensitive) (to flattery, persuasion, etc.)* empfänglich **(to** für); *(to illness, injury, etc.)* anfällig **(to** für); **b)** *(easily influenced)* empfindsam; beeindruckbar

suspect 1. [sə'spekt] *v. t.* **a)** *(imagine to be likely)* vermuten; **~ the worst** das Schlimmste befürchten; **~ sb. to be sth., ~ that sb. is sth.** glauben *od.* vermuten, daß jmd. etw. ist; **b)** *(mentally accuse)* verdächtigen; **~ sb. of sth./of doing sth.** jmdn. einer Sache verdächtigen/jmdn. verdächtigen, etw. zu tun; **~ed of drug-trafficking** des Drogenhandels verdächtig; **c)** *(mistrust)* bezweifeln ‹*Echtheit*›; **~ sb.'s motives** jmds. Beweggründen mit Argwohn gegenüberstehen. **2.** ['sʌspekt] *adj.* fragwürdig; suspekt *(geh.);* verdächtig ‹*Stoff, Paket, Fahrzeug*›. **3.** ['sʌspekt] *n.* Verdächtige, *der/die;* **a murder ~:** ein Mordverdächtiger/eine Mordverdächtige

suspected [sə'spektɪd] *adj.* verdächtig; **~ smallpox cases, ~ cases of smallpox** Fälle mit Verdacht auf Pocken

suspend [sə'spend] *v. t.* **a)** *(hang up)* [auf]hängen; **be ~ed [from sth.]** [von etw.] [herab]hängen; **b)** *(stop, defer)* suspendieren ‹*Rechte*›; [vorübergehend] einstellen ‹*Zugverkehr, Kampfhandlungen*›; **~ judgement** sich des Urteils enthalten; **c)** *(remove from work etc.)* ausschließen **(from** von); sperren ‹*Sportler*›; vom Unterricht ausschließen ‹*Schüler*›; **~ sb. from duty [pending an inquiry]** jmdn. [während einer schwebenden Untersuchung] vom Dienst suspendieren

suspended 'sentence *n. (Law)* Strafe mit Bewährung; **he was given a two-year ~:** er erhielt zwei Jahre Haft auf Bewährung

suspender belt [sə'spendə belt] *n. (Brit.)* Strumpfbandgürtel, *der*

suspenders [sə'spendəz] *n. pl.* **a)** *(Brit.: for stockings)* Strumpfbänder *od.* -halter; **b)** *(Amer.: for trousers)* Hosenträger

suspense [sə'spens] *n.* Spannung, *die;* **the ~ is killing me** *(joc.)* ich bin gespannt wie ein Regenschirm *(ugs. scherzh.);* **keep sb. in ~:** jmdn. auf die Folter spannen

suspension [sə'spenʃn] *n.* **a)** *(action of debarring)* Ausschluß, *der; (from office)* Suspendierung, *die; (Sport)* Sperrung, *die;* **be under ~** ‹*Schüler:*› [zeitweilig] vom Unterricht ausgeschlossen sein; ‹*Sportler:*› [zeitweilig] gesperrt sein; **b)** *(temporary cessation)* Suspendierung, *die; (of train service, hostilities)* [vorübergehende] Einstellung; **c)** *(Motor Veh.)* Federung, die

su'spension bridge *n.* Hängebrücke, *die*

suspicion [sə'spɪʃn] *n.* **a)** *(uneasy feeling)* Mißtrauen, *das* **(of** gegenüber); *(more specific)* Verdacht, *der; (unconfirmed belief)* Ahnung, *die;* Verdacht, *der;* **have a ~ that ...:** den Verdacht haben, daß ...; **I have my ~s about him** er kommt mir verdächtig vor; **b)** *(suspecting)* Verdacht, *der* **(of** auf + *Akk.*); **on ~ of theft/murder** *etc.* wegen Verdachts auf Diebstahl/Mordverdachts *usw.;* **lay oneself open to ~:** sich verdächtig machen; **be under ~:** verdächtigt werden

suspicious [sə'spɪʃəs] *adj.* **a)** *(tending to suspect)* mißtrauisch **(of** gegenüber); **be ~ of sb./sth.** jmdm./einer Sache mißtrauen; **b)** *(arousing suspicion)* verdächtig

suspiciously [sə'spɪʃəslɪ] *adv.* **a)** *(as to arouse suspicion)* verdächtig; **look ~ like sth.** verdächtig nach etw. aussehen; **b)** *(warily)* mißtrauisch

suss out [sʌs 'aʊt] *v. t. (Brit. sl.)* checken *(ugs.);* spannen *(ugs.)*

sustain [sə'steɪn] *v. t.* **a)** *(withstand)* widerstehen (+ *Dat.*) ‹*Druck*›; standhalten (+ *Dat.*) ‹*Angriff*›; tragen ‹*Gewicht*›; **b)** *(support, uphold)* aufrechterhalten; **~ an objection** einem Einwand stattgeben; **c)** *(suffer)* erleiden ‹*Niederlage, Verlust, Verletzung*›; **~ damage** Schaden nehmen; **d)** *(maintain)* bestreiten ‹*Unterhaltung*›; bewahren ‹*Interesse*›

sustained [sə'steɪnd] *adj. (prolonged)* länger ...; anhaltend ‹*Beifall*›; ausdauernd ‹*Anstrengung*›

sustenance ['sʌstɪnəns] *n.* **a)** *(nourishment, food)* Nahrung, *die;* **b)** *(nourishing quality)* Nährwert, *der*

SW *abbr.* **a)** [saʊθ'west] **south-west** SW; **b)** *(Radio)* **short wave** KW

swab [swɒb] *n.* **a)** *(Med.: absorbent pad)* Tupfer, *der;* **b)** *(Med.: specimen)* Abstrich, *der*

Swabia ['sweɪbɪə] *pr. n.* Schwaben *(das)*

swagger ['swægə(r)] **1.** *v. i.* **a)** *(walk with a ~)* großspurig stolzieren; **b)** *(boast)* angeben *(ugs.).* **2.** *n. see* **1: a)** großspuriges Stolzieren; **b)** Angeberei, *die (ugs.)*

Swahili [swɑː'hiːlɪ, swə'hiːlɪ] **1.** *adj.* Swahili-. **2.** *n.* Swahili, *das; see also* **English 2 a**

[1]**swallow** ['swɒləʊ] **1.** *v. t.* **a)** schlucken; *(by mistake)* verschlucken ‹*Fischgräte; fig.: Wort, Silbe*›; **~ the bait** *(fig.)* den Köder schlucken *(ugs.);* **b)** *(repress)* hinunterschlucken *(ugs.)* ‹*Stolz, Ärger*›; **~ one's words** [demütig] zurücknehmen, was man gesagt hat; **c)** *(believe)* schlucken *(ugs.),* glauben ‹*Geschichte, Erklärung*›; **d)** *(put up with)* schlucken *(ugs.)* ‹*Beleidigung, Unrecht*›. **2.** *v. i.* schlucken

~ 'up *v. t.* **a)** *(make disappear)* verschlucken; schlucken ‹*kleinere Betriebe, Gebiete*›; **I wished the earth would ~ me up** ich wäre am liebsten vor Scham in den Boden versunken; **b)** *(exhaust, consume)* auffressen; verschlingen ‹*große Summen*›

[2]**swallow** *n. (Ornith.)* Schwalbe, *die*

swam *see* **swim 1, 2**

swamp [swɒmp] **1.** *n.* Sumpf, *der.* **2.** *v. t.* **a)** *(flood)* überschwemmen; **b)** *(overwhelm)* **be ~ed with letters/applications/work** mit Briefen/Bewerbungen überschwemmt werden/bis über den Hals in Arbeit stecken *(ugs.)*

swampy ['swɒmpɪ] *adj.* sumpfig

swan [swɒn] *n.* Schwan, *der*

swank [swæŋk] *v. i. (coll.)* angeben *(ugs.)* **(about** mit)

swanky ['swæŋkɪ] *adj. (coll.)* protzig *(ugs.)*

'swansong *n. (fig.)* Schwanengesang, *der*

swap [swɒp] **1.** *v. t.,* **-pp-** tauschen **(for** gegen); austauschen ‹*Erfahrungen, Erinnerungen*›; **~ places [with sb.]** [mit jmdm.] den Platz *od.* die Plätze tauschen. **2.** *v. i.,* **-pp-** tauschen. **3.** *n.* Tausch, *der;* **do a ~ [with sb.]** [mit jmdm.] tauschen

swarm [swɔ:m] **1.** *n.* **a)** Schwarm, *der;* **~ [of bees]** Bienenschwarm, *der;* **b)** *in pl. (great numbers)* **~s of tourists/children** Scharen von Touristen/Kindern. **2.** *v.i.* **a)** *(move in a ~)* schwärmen; **b)** *(teem)* wimmeln **(with** von)

swarthy ['swɔ:ðɪ] *adj.* dunkel ⟨*Gesichtsfarbe*⟩; dunkelhäutig ⟨*Person*⟩

swastika ['swɒstɪkə] *n. (of Nazis)* Hakenkreuz, *das*

swat [swɒt] *v.t.*, **-tt-** totschlagen ⟨*Fliege, Wespe*⟩

swathe [sweɪð] *v.t.* [ein]hüllen

swatter ['swɒtə(r)] *n.* Klatsche, *die*

sway [sweɪ] **1.** *v.i.* [hin und her] schwanken; *(gently)* sich wiegen. **2.** *v.t.* **a)** wiegen ⟨*Kopf, Hüften, Zweig, Wipfel*⟩; hin und her schwanken lassen ⟨*Baum, Mast, Antenne*⟩; **b)** *(have influence over)* beeinflussen; *(persuade)* überreden. **3.** *n.* Herrschaft, *die;* **have sb. under one's ~, hold ~ over sb.** über jmdn. herrschen

swear [sweə(r)] **1.** *v.t.*, **swore** [swɔ:(r)], **sworn** [swɔ:n] **a)** schwören ⟨*Eid usw.*⟩; **I could have sworn [that] it was him** ich hätte schwören können, daß er es war; **b)** *(administer oath to)* vereidigen ⟨*Zeugen*⟩; **~ sb. to secrecy** jmdn. auf Geheimhaltung einschwören. **2.** *v.i.*, **swore, sworn a)** *(use ~-words)* fluchen; **b) ~ to sth.** *(be certain of)* etw. beschwören; einen Eid auf etw. *(Akk.)* ablegen; **I wouldn't like to ~ to it** *(coll.)* ich will es nicht beschwören; **c)** *(take oath)* schwören, einen Eid ablegen **(on** auf + *Akk.*)

~ at *v.t.* beschimpfen

~ by *v.t. (coll.: have confidence in)* schwören auf (+ *Akk.*)

~ in *v.t.* vereidigen ⟨*Geschworenen, Zeugen*⟩

'swear-word *n.* Kraftausdruck, *der;* Fluch, *der;* **use ~s** fluchen

sweat [swet] **1.** *n.* **a)** Schweiß, *der;* **in** *or* **by the ~ of one's brow** im Schweiße seines Angesichtes; **I came** *or* **broke out in a ~:** mir brach der [Angst]schweiß aus; **don't get in such a ~!** reg dich nicht so auf!; **b)** *(drudgery)* Plagerei, *die;* Plackerei, *die (ugs.);* **no ~!** *(coll.)* kein Problem! *(ugs.).* **2.** *v.i.*, **~ed** *or (Amer.)* **~: a)** *(perspire)* schwitzen; **~ like a pig** *(coll.)* schwitzen wie die Sau *(salopp);* **~ with fear** vor Angst schwitzen; **b)** *(fig.: suffer)* **he made me sit outside ~ing** er ließ mich draußen sitzen und schmoren *(ugs.);* **c)** *(drudge)* sich placken *(ugs.).* **3.** *v.t.* **a) ~ blood** *(fig.)* Blut und Wasser schwitzen *(ugs.);* **b) ~ it out** *(coll.)* durchhalten; ausharren

sweated labour [swetɪd 'leɪbə(r)] *n.* unterbezahlte [Schwer]arbeit

sweater ['swetə(r)] *n.* Pullover, *der*

sweat: ~-shirt *n.* Sweatshirt, *das;* **~-shop** *n.* ausbeuterische [kleine] Klitsche *(ugs.)*

sweaty ['swetɪ] *adj. (moist with sweat)* schweißig; schweißnaß

Swede [swi:d] *n.* Schwede, *der*/Schwedin, *die*

swede *n.* Kohlrübe, *die*

Sweden ['swi:dn] *pr. n.* Schweden *(das)*

Swedish ['swi:dɪʃ] **1.** *adj.* schwedisch; **sb. is ~:** jmd. ist Schwede/Schwedin. **2.** *n.* Schwedisch, *das; see also* **English 2 a**

sweep [swi:p] **1.** *v.t.*, **swept** [swept] **a)** fegen *(bes. nordd.);* kehren *(bes. südd.);* **~ the board, ~ all before one** *(fig.: win all awards)* auf der ganzen Linie siegen; **b)** *(move with force)* fegen; **the current swept the logs along** die Strömung riß die Hölzer mit; **c)** *(traverse swiftly)* **~ the hillside/plain** ⟨*Wind:*⟩ über die Hügel/Ebene fegen; **~ the country** ⟨*Epidemie, Mode:*⟩ das Land überrollen; ⟨*Feuer:*⟩ durch das Land fegen. **2.** *v.i.*, **swept a)** *(clean)* fegen *(bes. nordd.);* kehren *(bes. südd.);* **b)** *(go fast, in stately manner)* ⟨*Vogel:*⟩ gleiten; ⟨*Person, Auto:*⟩ rauschen; ⟨*Wind usw.:*⟩ fegen; **c)** *(extend)* sich erstrecken; **the road ~s to the left** die Straße macht einen großen Bogen nach links; **his glance swept from left to right** sein Blick glitt von links nach rechts. **3.** *n.* **a)** *(cleaning)* **give sth. a ~:** etw. fegen *(bes. nordd.);* etw. kehren *(bes. südd.);* **make a clean ~** *(fig.: get rid of everything)* gründlich aufräumen; **b)** *see* **chimney-sweep; c)** *(coll.) see* **sweepstake; d)** *(motion of arm)* ausholende Bewegung; **e)** *(stretch)* **a wide/an open ~ of country** ein weiter Landstrich; **f)** *(curve of road, river)* Bogen, *der;* **the wide ~ of the bay** die geschwungene Kurve der Bucht

~ a'side *v.t.* **a)** *(dismiss)* beiseite schieben ⟨*Einwand, Zweifel*⟩; **b)** *(push aside)* wegfegen; beiseite fegen

~ a'way *v.t.* fortreißen; *(fig.)* hinwegfegen *(geh.)* ⟨*Traditionen*⟩; *(abolish)* aufräumen mit ⟨*Privilegien, Korruption*⟩

~ 'by *v.i.* vorbeirauschen

~ 'down *v.i.* **the hills ~ down to the sea** die Berge fallen in sanftem Bogen zum Meer hinab

~ 'in *v.i. (enter majestically)* einziehen

~ 'out *v.t.* ausfegen *(bes. nordd.);* auskehren *(bes. südd.)*

~ 'up 1. *v.t.* zusammenfegen *(bes. nordd.);* zusammenkehren *(bes. südd.).* **2.** *v.i.* angerauscht kommen

sweeper ['swi:pə(r)] *n.* **[road] ~** *(person)* Straßenfeger, *der (bes. nordd.);* Straßenkehrer, *der (bes. südd.); (machine)* Straßenkehrmaschine, *die*

sweeping ['swi:pɪŋ] *adj.* **a)** *(without limitations)* pauschal; **b)** *(far-reaching)* weitreichend ⟨*Einsparung*⟩; umfassend ⟨*Reform*⟩; durchschlagend ⟨*Sieg, Erfolg*⟩; umwälzend ⟨*Veränderung*⟩

'sweepstake *n.* **a)** *(race, contest)* Sweepstake[rennen], *das;* **b)** *(lottery) Pferdetoto, bei dem sich die Gewinnsumme aus den Einsätzen zusammensetzt*

sweet [swi:t] **1.** *adj.* **a)** *(to taste)* süß; **~ tea** gesüßter Tee; **have a ~ tooth** gern Süßes mögen; **b)** *(lovely)* süß; reizend ⟨*Wesen, Gesicht, Mädchen, Kleid*⟩; **~ dreams!** träum[e]/träumt süß!; **how ~ of you!** wie nett *od.* lieb von dir!; **go one's own ~ way** machen, was einem paßt; **c)** *(fragrant)* süß; frisch ⟨*Atem*⟩; **d)** *(musical)* süß *(geh.);* lieblich ⟨*Stimme, Musik, Klang*⟩. **2.** *n.* **a)** *(Brit.: piece of confectionery)* Bonbon, *das od. der; (with chocolate, fudge, etc.)* Süßigkeit, *die;* **b)** *(Brit.: dessert)* Nachtisch, *der;* Dessert, *das;* **for ~:** zum Nachtisch *od.* Dessert

sweet: ~-and-'sour *attrib. adj.* süßsauer; **~ corn** *n.* Zuckermais, *der*

sweeten ['swi:tn] *v.t.* **a)** *(add sugar etc. to)* süßen; **b)** *(add fragrance to)* süß machen; versüßen; *(remove bad smell of)* reinigen ⟨*Luft, Atem*⟩; **c)** *(make agreeable)* versüßen ⟨*Leben, Abend*⟩; milde stimmen ⟨*Person*⟩

sweetener ['swi:tnə(r)] *n.* **a)** Süßstoff, *der;* **b)** *(bribe)* kleine Aufmerksamkeit *(iron.)*

'sweetheart *n.* Schatz, *der;* Liebling, *der*

sweetie ['swi:tɪ] *(Brit. child lang.) see* **sweet 2 a**

sweetness ['swi:tnɪs] *n., no pl.* **a)**

Süße, *die;* **b)** *(fragrance)* süßer Duft; **c)** *(melodiousness)* Süße, *die (geh.)*

sweet: ~ '**pea** *n. (Bot.)* Wicke, *die;* ~-**shop** *n. (Brit.)* Süßwarengeschäft, *das;* ~-**smelling** *adj.* süß [duftend]; ~-**tempered** ['swi:ttempəd] *adj.* sanftmütig

swell [swel] **1.** *v. t.*, **~ed, swollen** ['swəʊlən] *or* **~ed a)** *(increase in size, height)* anschwellen lassen; aufquellen lassen ⟨*Holz*⟩; **b)** *(increase amount of)* anschwellen lassen; vergrößern; ~ **the ranks [of participants]** die Zahl der Teilnehmer vergrößern; **c)** blähen ⟨*Segel*⟩. **2.** *v. i.*, **~ed, swollen** *or* **~ed a)** *(expand)* ⟨*Körperteil:*⟩ anschwellen; ⟨*Segel:*⟩ sich blähen; ⟨*Material:*⟩ aufquellen; **b)** *(increase in amount)* ⟨*Anzahl:*⟩ zunehmen; **c)** *(become louder)* anschwellen (**[in]to** zu). **3.** *n. (of sea)* Dünung, *die*

swelling ['swelɪŋ] **1.** *n.* Schwellung, *die (Med.).* **2.** *adj. (growing larger, louder)* anschwellend

swelter ['sweltə(r)] *v. i.* [vor Hitze] [fast] vergehen; ~ **in the heat** in der Hitze schmoren *(ugs.);* **~ing** glühend heiß ⟨*Tag, Wetter*⟩; **~ing heat** Bruthitze, *die*

swept *see* **sweep 1, 2**

swerve [swɜ:v] **1.** *v. i. (deviate)* einen Bogen *od. (ugs.)* Schlenker machen; ~ **to the right/left** nach rechts/links [aus]schwenken. **2.** *n. (divergence from course)* Bogen, *der;* Schlenker, *der (ugs.)*

swift [swɪft] **1.** *adj.* schnell; flink, schnell ⟨*Bewegung*⟩; ~ **action** rasches Handeln; ~ **retribution** prompte Bestrafung. **2.** *n. (Ornith.)* Mauersegler, *der*

swiftly ['swɪftlɪ] *adv.* schnell; *(soon)* bald

swiftness ['swɪftnɪs] *n.* Schnelligkeit, *die;* ~ **of action** schnelles *od.* rasches Handeln

swig [swɪg] *(coll.)* **1.** *v. t.*, **-gg-** schlucken *(ugs.);* [herunter]kippen *(ugs.).* **2.** *v. i.*, **-gg-** [hastig] trinken. **3.** *n.* Schluck, *der:* **have/take a ~ [of beer** *etc.*] einen tüchtigen Schluck [Bier *usw.*] trinken/nehmen

swill [swɪl] **1.** *v. t.* **a)** *(rinse)* ~ **[out]** [aus]spülen; **b)** *(derog.: drink greedily)* hinunterspülen *(ugs.).* **2.** *n.* **give sth. a ~ [out]/down** etw. [aus]spülen/abspülen

swim [swɪm] **1.** *v. i.*, **-mm-, swam** [swæm], **swum** [swʌm] **a)** schwimmen; ~ **with/against the tide/stream** *(fig.)* mit dem/gegen den Strom schwimmen; **b)** *(fig.: be flooded, overflow)* ~ **with** *or* **in sth.** in etw. *(Dat.)* schwimmen; **the deck was ~ming with water** das Deck stand unter Wasser; **c)** *(appear to whirl)* ~ **[before sb.'s eyes]** [vor jmds. Augen] verschwimmen; **d)** *(have dizzy sensation)* **my head was ~ming** mir war schwindelig. *See also* **sink 2 a. 2.** *v. t.*, **-mm-, swam, swum** schwimmen ⟨*Strecke*⟩; durchschwimmen ⟨*Fluß, See*⟩. **3.** *n.* **a) have a/go for a ~:** schwimmen/schwimmen gehen; **b) be in the ~ [of things]** mitten im Geschehen sein

swimmer ['swɪmə(r)] *n.* Schwimmer, *der*/Schwimmerin, *die;* **be a good/poor ~:** gut/schlecht schwimmen können

swimming ['swɪmɪŋ] *n.* Schwimmen, *das*

swimming: ~-**baths** *n. pl.* Schwimmbad, *das;* ~-**costume** *n.* Badeanzug, *der;* ~-**lesson** *n.* Schwimmstunde, *die;* **~-lessons** Schwimmunterricht, *der;* ~-**pool** *n.* Schwimmbecken, *das; (in house or garden)* Swimmingpool, *der; (building)* Schwimmbad, *das;* ~-**trunks** *n. pl.* Badehose, *die*

'**swim-suit** *n.* Badeanzug, *der*

swindle ['swɪndl] **1.** *v. t.* betrügen; ~ **sb. out of sth.** jmdn. um etw. betrügen; *(take by persuasion)* jmdm. etw. abschwindeln. **2.** *n.* Schwindel, *der;* Betrug, *der*

swindler ['swɪndlə(r)] *n.* Schwindler, *der*/Schwindlerin, *die*

swine [swaɪn] *n., pl. same* **a)** *(Amer./formal/Zool.)* Schwein, *das;* **b)** *(derog.: contemptible person)* Schwein, *das (abwertend)*

swing [swɪŋ] **1.** *n.* **a)** *(apparatus)* Schaukel, *die;* **b)** *(spell of ~ing)* Schaukeln, *das;* **c)** *(Sport: strike, blow)* Schlag, *der; (Boxing)* Schwinger, *der; (Golf)* Schwung, *der;* **take a ~ at sb./sth.** zum Schlag gegen jmdn./auf etw. *(Akk.)* ausholen; **d)** *(of suspended object)* Schwingen, *das;* **in full ~** *(fig.)* in vollem Gang[e]; **e)** *(steady movement)* Rhythmus, *der;* **the party went with a ~:** auf der Party herrschte eine tolle Stimmung *(ugs.);* **get into/be in the ~ of things** *or* **it** richtig reinkommen/richtig drin sein *(ugs.);* **f)** *(Mus.)* Swing, *der;* **g)** *(shift)* Schwankung, *die; (of public opinion)* Wende, *die; (amount of change in votes)* Abwanderung, *die.* **2.** *v. i.*, **swung** [swʌŋ] **a)** *(turn on axis, sway)* schwingen; *(in wind)* schaukeln; ~ **open** ⟨*Tür:*⟩ aufgehen; **b)** *(go in sweeping curve)* schwenken; ~ **from sb.'s arm/a tree** an jmds. Arm/einem Baum schwingen *(geh.) od.* baumeln; **c)** ~ **into action** *(fig.)* loslegen *(ugs.);* **d)** *(move oneself by ~ing)* sich schwingen; **the car swung out of the drive** der Wagen schwenkte aus der Einfahrt; **e)** *(sl.: be executed by hanging)* baumeln *(salopp);* **he'll ~ for it** dafür wird er baumeln. **3.** *v. t.*, **swung a)** schwingen; *(rock)* schaukeln; ~ **sth. round and round** etw. kreisen *od.* im Kreise wirbeln lassen; **b)** *(cause to face in another direction)* schwenken; **he swung the car off the road/into the road** er schwenkte [mit dem Auto] von der Straße ab/in die Straße ein; **c)** *(have influence on)* umschlagen lassen ⟨*öffentliche Meinung*⟩; ~ **the elections** den Ausgang der Wahlen entscheiden; **what swung it for me ...:** was für mich den Ausschlag gab ...

~ '**round** *v. i.* sich schnell umdrehen (**on** nach); *(in surprise)* herumfahren

swing: ~ **bridge** *n.* Drehbrücke, *die;* ~-'**door** *n.* Pendeltür, *die*

swingeing ['swɪndʒɪŋ] *adj. (Brit.)* hart ⟨*Schlag*⟩; *(fig.)* drastisch ⟨*Kürzung, Maßnahme*⟩; scharf ⟨*Attacke*⟩

swinging ['swɪŋɪŋ] *adj.* **a)** schwingend; **b)** *(rhythmical)* [stark] rhythmisch; **c)** *(sl.: lively)* wild *(ugs.);* swingend *(ugs.)*

swipe [swaɪp] *(coll.)* **1.** *v. i.* ~ **at** eindreschen auf (+ *Akk.*) *(ugs.).* **2.** *v. t.* **a)** *(hit hard)* knallen *(ugs.);* **b)** *(sl.: steal)* klauen *(ugs.).* **3.** *n.* **take a wild ~ at sth.** wild auf etw. *(Akk.)* losschlagen

swirl [swɜ:l] **1.** *v. i.* wirbeln. **2.** *v. t.* umherwirbeln. **3.** *n. (spiralling shape)* Spirale, *die*

swish [swɪʃ] **1.** *v. t.* schlagen mit ⟨*Schwanz*⟩; sausen lassen ⟨*Stock*⟩. **2.** *v. i.* zischen. **3.** *n.* Zischen, *das.* **4.** *adj. (coll.)* schick *(ugs.)*

Swiss [swɪs] **1.** *adj.* Schweizer; schweizerisch; **sb. is ~:** jmd. ist Schweizer/Schweizerin. **2.** *n.* Schweizer, *der*/Schweizerin, *die;* **the ~** *pl.* die Schweizer

Swiss: ~ '**German** **1.** *adj.* schweizerdeutsch; **2.** *n.* Schweizerdeutsch, *das;* ~ '**roll** *n.* Biskuitrolle, *die*

switch [swɪtʃ] **1.** *n.* **a)** *(esp. Electr.)* Schalter, *der;* **b)** *(Amer. Railw.)* Weiche, *die;* **c)** *(change with another)* Wechsel, *der;* **d)** *(flexible shoot, whip)* Gerte, *die.* **2.** *v. t.* **a)** *(change)* ~ **sth. [over] to sth.**

etw. auf etw. *(Akk.)* umstellen *od.* *(Electr.)* umschalten; ~ **the conversation to another topic** das Gespräch auf ein anderes Thema lenken; **b)** *(exchange)* tauschen. **3.** *v. i.* wechseln; ~ **[over] to sth.** auf etw. *(Akk.)* umstellen *od.* *(Electr.)* umschalten

~ **a'round 1.** *v. t.* umstellen ‹*Möbel, Dienstplan*›. **2.** *v. i.* [die Stellung] wechseln

~ **'off** *v. t. & i.* ausschalten; *(also fig. coll.)* abschalten

~ **'on 1.** *v. t.* einschalten; anschalten. **2.** *v. i.* sich anschalten

~ **'over 1.** *v. t. see* ~ **2 a. 2.** *v. i. see* ~ **3**

~ **'round** *see* ~ **around**

~ **'through** *v. t.* durchstellen ‹*Telefongespräch, Anrufer*›

switch: ~back *n. (roller-coaster)* Achterbahn, *die;* **~board** *n. (Teleph.)* [Telefon]zentrale, *die;* Vermittlung, *die;* **~board operator** Telefonist, *der*/Telefonistin, *die*

Switzerland ['swɪtsələnd] *pr. n.* die Schweiz

swivel ['swɪvl] **1.** *n.* Drehgelenk, *das.* **2.** *v. i., (Brit.)* **-ll-** sich drehen. **3.** *v. t., (Brit.)* **-ll-** drehen

'swivel chair *n.* Drehstuhl, *der*

swollen ['swəʊlən] **1.** *see* **swell 1, 2. 2.** *adj.* geschwollen; angeschwollen ‹*Fluß*›; **have a ~ head** *(fig.)* sehr eingebildet *od.* von sich eingenommen sein

'swollen-headed *adj.* eingebildet

swoon [swu:n] *(literary)* **1.** *v. i.* **a)** *(faint)* ohnmächtig werden; **b)** *(go into ecstasies)* ~ **over sb./sth.** von jmdm./etw. schwärmen. **2.** *n.* Ohnmacht, *die*

swoop [swu:p] **1.** *n.* **a)** *(downward plunge)* Sturzflug, *der;* **b)** *(coll.: raid)* Razzia, *die.* **2.** *v. i. (plunge suddenly)* herabstoßen; *(pounce)* ~ **on sb.** sich auf jmdn. stürzen; *(to attack)* gegen jmdn. einen Schlag führen; **the police ~ed on several addresses** die Polizei führte in mehreren Wohnungen Razzien durch

swop *see* **swap**

sword [sɔ:d] *n.* Schwert, *das*

'swordfish *n.* Schwertfisch, *der*

swore *see* **swear**

sworn [swɔ:n] **1.** *see* **swear. 2.** *attrib. adj.* **a)** *(bound by an oath)* verschworen ‹*Freund*›; ~ **enemy** Todfeind, *der;* **b)** *(certified by oath)* beeidigt; ~ **evidence** Aussage unter Eid; ~ **affidavit/statement** eidesstattliche Versicherung/eidliche Erklärung

swot [swɒt] *(Brit. coll.)* **1.** *n.* Streber, *der*/Streberin, *die (abwertend).* **2.** *v. i.,* **-tt-** büffeln *(ugs.)*

~ **'up** *v. t.* büffeln *(ugs.)*

swum *see* **swim 1, 2**

swung *see* **swing 2, 3**

sycamore ['sɪkəmɔ:(r)] *n.* Bergahorn, *der; (Amer.: plane-tree)* Platane, *die*

sycophant ['sɪkəfænt] *n.* Kriecher, *der;* Schranze, *die*

syllable ['sɪləbl] *n. (lit. or fig.)* Silbe, *die;* **in words of one ~** *(fig.)* mit [sehr] einfachen Worten

syllabus ['sɪləbəs] *n., pl.* **~es** *or* **syllabi** ['sɪləbaɪ] Lehrplan, *der; (for exam)* Studienplan, *der*

symbiosis [sɪmbɪ'əʊsɪs] *n., pl.* **symbioses** [sɪmbɪ'əʊsi:z] *(Biol.; also fig.)* Symbiose, *die*

symbiotic [sɪmbɪ'ɒtɪk] *adj.* symbiotisch

symbol ['sɪmbl] *n.* Symbol, *das* **(of** für)

symbolic [sɪm'bɒlɪk], **symbolical** [sɪm'bɒlɪkl] *adj.* symbolisch

symbolise *see* **symbolize**

symbolism ['sɪmbəlɪzm] *n.* Symbolik, *die*

symbolize ['sɪmbəlaɪz] *v. t.* symbolisieren

symmetrical [sɪ'metrɪkl] *adj.*, **symmetrically** [sɪ'metrɪkəlɪ] *adv.* symmetrisch

symmetry ['sɪmɪtrɪ] *n.* Symmetrie, *die*

sympathetic [sɪmpə'θetɪk] *adj.* **a)** *(showing pity)* mitfühlend; *(understanding)* verständnisvoll; **b)** *(favourably inclined)* wohlgesinnt; geneigt ‹*Leser*›; **be ~ to a cause/to new ideas** einer Sache wohlwollend gegenüberstehen/für neue Ideen empfänglich *od.* zugänglich sein; **give sb. a ~ hearing** ein offenes Ohr für jmdn. haben; **he is not at all ~ to this idea** er ist von dieser Idee ganz und gar nicht angetan

sympathise, sympathiser *see* **sympathiz-**

sympathize ['sɪmpəθaɪz] *v. i.* **a)** *(feel or express sympathy)* ~ **with sb.** mit jmdm. [mit]fühlen *od.* Mitleid haben; *(by speaking)* sein Mitgefühl mit jmdm. äußern; **I do ~:** es tut mir wirklich leid; **b)** ~ **with** *(have understanding for)* Verständnis haben für ‹*jmds. Not, Denkweise usw.*›; *(Polit.: share ideas of)* sympathisieren mit ‹*Partei usw.*›

sympathizer ['sɪmpəθaɪzə(r)] *n.* Sympathisant, *der*/Sympathisantin, *die*

sympathy ['sɪmpəθɪ] *n.* **a)** *(sharing feelings of another)* Mitgefühl, *das;* **in deepest ~:** mit aufrichtigem Beileid; **b)** *(agreement in opinion or emotion)* Sympathie, *die;* **my sympathies are with Schmidt** ich bin auf Schmidts Seite; **be in/out of ~ with sth.** mit etw. sympathisieren/nicht sympathisieren; **come out** *or* **strike in ~ with sb.** mit jmdm. in einen Sympathiestreik treten

'sympathy strike *n.* Sympathiestreik, *der*

symphonic [sɪm'fɒnɪk] *adj.* sinfonisch; symphonisch

symphony ['sɪmfənɪ] *n.* Sinfonie, *die*

'symphony orchestra *n.* Sinfonieorchester, *das*

symposium [sɪm'pəʊzɪəm] *n., pl.* **symposia** [sɪm'pəʊzɪə] Symposion, *das;* Symposium, *das*

symptom ['sɪmptəm] *n. (Med.; also fig.)* Symptom, *das*

symptomatic [sɪmptə'mætɪk] *adj. (Med.; also fig.)* symptomatisch **(of** für)

synagogue (*Amer.:* **synagog**) ['sɪnəgɒg] *n.* Synagoge, *die*

sync, synch [sɪŋk] *n. (coll.)* **in/out of ~:** synchron/nicht synchron

synchromesh ['sɪŋkrəmeʃ] *n.* *(Motor Veh.)* ~ **[gearbox]** Synchrongetriebe, *das;* **there is ~ on all gears** alle Gänge sind synchronisiert

synchronize (synchronise) ['sɪŋkrənaɪz] *v. t.* **a)** synchronisieren ‹*Vorgänge, Maschinen, Bild und Ton*›; **b)** *(set to same time)* gleichstellen ‹*Uhren*›; **we'd better ~ [our] watches** wir sollten Uhrenvergleich machen

syndicate ['sɪndɪkət] *n.* **a)** *(for business, in organized crime)* Syndikat, *das;* **b)** *(in newspapers)* *Presseagentur, die Beiträge ankauft und an eine od. mehrere Zeitungen vertreibt*

syndrome ['sɪndrəʊm] *n. (Med.; also fig.)* Syndrom, *das*

synod ['sɪnəd] *n.* Synode, *die*

synonym ['sɪnənɪm] *n. (Ling.)* Synonym, *das*

synonymous [sɪ'nɒnɪməs] *adj.* **a)** *(Ling.)* synonym **(with** mit); **b)** ~ **with** *(fig.: suggestive of, linked with)* gleichbedeutend mit

synonymy [sɪ'nɒnəmɪ] *n. (Ling.)* Synonymie, *die*

synopsis [sɪ'nɒpsɪs] *n., pl.* **synopses** [sɪ'nɒpsi:z] Inhaltsangabe, *die*

syntactic [sɪn'tæktɪk] *adj. (Ling.)* syntaktisch

syntax ['sɪntæks] *n. (Ling.)* Syntax, *die*

synthesis ['sɪnθɪsɪs] *n., pl.* **syntheses** ['sɪnθɪsi:z] Synthese, *die*

synthesise, synthesiser *see* **synthesiz-**

synthesize ['sɪnθɪsaɪz] *v. t.* **a)** *(form into a whole)* zur Synthese bringen; **b)** *(Chem.)* synthetisieren; **c)** *(Electronics)* ~ **speech** Sprache elektronisch generieren

synthesizer ['sɪnθɪsaɪzə(r)] *n.* *(Mus.)* Synthesizer, *der*

synthetic [sɪn'θetɪk] **1.** *adj.* synthetisch. **2.** *n.* Kunststoff, *der;* ~**s** *(Textiles)* Synthetics

syphilis ['sɪfɪlɪs] *n.* *(Med.)* Syphilis, *die*

syphon *see* **siphon**

Syria ['sɪrɪə] *pr. n.* Syrien *(das)*

Syrian ['sɪrɪən] **1.** *adj.* syrisch; **sb. is** ~: jmd. ist Syrer/Syrerin. **2.** *n.* Syrer, *der*/Syrerin, *die*

syringe [sɪ'rɪndʒ] **1.** *n.* Spritze, *die; see also* **hypodermic 1. 2.** *v. t.* spritzen; ausspritzen ‹*Ohr*›

syrup ['sɪrəp] *n.* Sirup, *der;* **cough** ~: Hustensaft, *der*

system ['sɪstəm] *n.* **a)** *(lit. or fig.)* System, *das; (of roads, railways also)* Netz, *das;* **root** ~ *(Bot.)* Wurzelgeflecht, *das;* **b)** *(Anat., Zool.: body)* Körper, *der; (part)* **digestive/muscular/nervous** ~: Verdauungsapparat, *der*/Muskulatur, *die*/Nervensystem, *das;* **get sth. out of one's** ~ *(fig.)* etw. loswerden; *(by talking)* sich *(Dat.)* etw. von der Seele reden

systematic [sɪstə'mætɪk] *adj.*, **systematically** [sɪstə'mætɪkəlɪ] *adv.* systematisch

systematize (systematise) ['sɪstəmətaɪz] *v. t.* systematisieren **(into** zu)

systemic [sɪ'stemɪk] *adj.* *(Biol.)* systemisch

'**systems analyst** *n.* Systemanalytiker, *der*/-analytikerin, *die*

T

T, t [ti:] *n., pl.* **Ts** *or* **T's** T, t, *das;* **to a T** ganz genau; haargenau; **T-junction** Einmündung, *die (in eine Vorfahrtsstraße);* **T-bone steak** T-bone-Steak, *das;* **T-shirt** T-shirt, *das*

ta [tɑ:] *int.* *(Brit. coll.)* danke

[1]**tab** [tæb] *n.* **a)** *(projecting flap)* Zunge, *die; (label)* Schildchen, *das; (on clothing)* Etikett, *das; (with name)* Namensschild, *das; (on file [card])* Reiter, *der;* **b)** *(Amer. coll.: bill)* Rechnung, *die;* **pick up the** ~: die Zeche bezahlen; **c) keep** ~**s** *or* **a** ~ **on sb./sth.** *(watch)* jmdn./etw. [genau] beobachten

[2]**tab** *see* **tabulator**

tabby ['tæbɪ] *n.* **a)** ~ **[cat]** Tigerkatze, *die;* **b)** *(female cat)* [weibliche] Katze; Kätzin, *die*

table ['teɪbl] **1.** *n.* **a)** Tisch, *der;* **at** ~: bei Tisch; **sit down at** ~: sich zu Tisch setzen; **after two whiskies he was under the** ~ *(coll.)* nach zwei Whisky lag er unter dem Tisch *(ugs.);* **drink sb. under the** ~: jmdn. unter den Tisch trinken *(ugs.);* **get sb./get round the** ~: jmdn. an einen Tisch bringen/ sich an einen Tisch setzen; **turn the** ~**s [on sb.]** *(fig.)* [jmdm. gegenüber] den Spieß umdrehen *od.* umkehren; *see also* [2]**lay 1 e; b)** *(list)* Tabelle, *die;* ~ **of contents** Inhaltsverzeichnis, *das;* **learn one's** ~**s** das Einmaleins lernen; **say one's nine times** ~: die Neunerreihe aufsagen. **2.** *v. t.* einbringen; auf den Tisch legen *(ugs.)*

tableau ['tæbləʊ] *n., pl.* ~**x** ['tæbləʊz] *(lit. or fig.)* Tableau, *das*

table: ~**-cloth** *n.* Tischdecke, *die;* Tischtuch, *das;* ~**-knife** *n.* Messer, *das;* ~**-lamp** *n.* Tischlampe, *die;* ~ **manners** *n. pl.* Tischmanieren *Pl.;* ~**-mat** *n.* Set, *das;* ~ **salt** *n.* Tafelsalz, *das;* ~**spoon** *n.* Servierlöffel, *der;* ~**spoonful** *n.* Servierlöffel[voll], *der*

tablet ['tæblɪt] *n.* **a)** *(pill)* Tablette, *die;* **b)** *(of soap)* Stück, *das;* **c)** *(stone slab)* Tafel, *die*

table: ~ **tennis** *n.* *(Sport)* Tischtennis, *das;* ~ **tennis bat** Tischtennisschläger, *der;* ~**ware** *n., no pl.* Geschirr, Besteck und Gläser; ~ **wine** *n.* Tischwein, *der*

tabloid ['tæblɔɪd] *n.* *(kleinformatige, bebilderte)* Boulevardzeitung; **the** ~**s** *(derog.)* die Boulevardpresse

taboo, tabu [tə'bu:] **1.** *n.* Tabu, *das.* **2.** *adj.* tabuisiert; Tabu‹*wort*›; **be** ~: tabu sein

tabulate ['tæbjʊleɪt] *v. t.* tabellarisch darstellen; tabellarisieren

tabulation [tæbjʊ'leɪʃn] *n.* tabellarische Aufstellung; Tabellarisierung, *die*

tabulator ['tæbjʊleɪtə(r)] *n.* Tabulator, *der*

tachograph ['tækəgrɑ:f] *n.* *(Motor Veh.)* Fahrt[en]schreiber, *der*

tacit ['tæsɪt] *adj.*, **tacitly** ['tæsɪtlɪ] *adv.* stillschweigend

taciturn ['tæsɪtɜ:n] *adj.* schweigsam; wortkarg

tack [tæk] **1.** *n.* **a)** *(small nail)* kleiner Nagel; **b)** *(temporary stitch)* Heftstich, *der;* **c)** *(Naut.: direction of vessel; also fig.)* Kurs, *der;* **on the right/wrong** ~ *(fig.)* auf dem richtigen/falschen Weg *od.* Kurs; **change one's** ~, **try another** ~ *(fig.)* einen anderen Kurs einschlagen. **2.** *v. t.* **a)** *(stitch loosely)* heften; **b)** *(nail)* festnageln. **3.** *v. i.* *(Naut.)* kreuzen

~ **'on** *v. t.* anhängen **(to** an + *Akk.*)

tackle ['tækl] **1.** *v. t.* **a)** angehen, in Angriff nehmen ‹*Problem usw.*›; ~ **sb. about/on/over sth.** jmdn. auf etw. *(Akk.)* ansprechen; *(ask for sth.)* jmdn. um etw. angehen; **b)** *(Sport)* angreifen ‹*Spieler*›; *(Amer. Footb.; Rugby)* fassen. **2.** *n.* **a)** *(equipment)* Ausrüstung, *die;* **b)** *(Sport)* Angriff, *der; (sliding* ~*)* Tackling, *das; (Amer. Footb.; Rugby)* Fassen und Halten

tacky ['tækɪ] *adj.* *(sticky)* klebrig

tact [tækt] *n.* Takt, *der;* **he has no** ~: er hat kein Taktgefühl

tactful ['tæktfl] *adj.*, **tactfully** ['tæktfəlɪ] *adv.* taktvoll

tactic ['tæktɪk] *n.* Taktik, *die*

tactical ['tæktɪkl] *adj.* taktisch ‹*Fehler, Manöver, Rückzug*›; ~ **voting** taktische Stimmabgabe

tactics ['tæktɪks] *n. pl.* Taktik, *die*

tactless ['tæktlɪs] *adj.* taktlos

tactlessly ['tæktlɪslɪ] *adv.* taktlos; *as sentence-modifier* taktloserweise

tadpole ['tædpəʊl] *n.* Kaulquappe, *die*

[1]**tag** [tæg] **1.** *n.* **a)** *(label)* Schild, *das; (on clothes)* Etikett, *das; (on animal's ear)* Ohrmarke, *die;* **b)** *(loop)* Schlaufe, *die;* **c)** *(stock phrase)* Zitat, *das;* geflügeltes Wort. **2.** *v. t.*, **-gg-** *(attach)* anhängen **(to** an + *Akk.*); ~ **together** aneinanderhängen; zusammenheften ‹*Blätter*›. **3.** *v. i.*, **-gg-:** ~ **behind** [nach]folgen; ~ **after sb.** hinter jmdm. hertrotteln *(ugs.)*

~ **a'long** *v. i.* hinterherlaufen; **do you mind if I** ~ **along?** darf ich mich anschließen?

~ **'on** *v. t.* anhängen **(to** an + *Akk.*)

[2]**tag** *n.* *(game)* Fangen, *das*

tail [teɪl] **1.** *n.* **a)** Schwanz, *der;* **b)** *(fig.)* **have sb./sth. on one's** ~ *(coll.)* jmdn./etw. auf den Fersen haben *(ugs.)*; **be/keep on sb.'s ~** *(coll.)* jmdm. auf den Fersen sein/ bleiben *(ugs.)*; **with one's ~ between one's legs** mit eingezogenem Schwanz *(ugs.)*; **sb. has his ~ up** jmd. ist übermütig; **turn ~ [and run]** Fersengeld geben *(ugs.)*; die Flucht ergreifen; **c)** *(of comet)* Schweif, *der;* **d) [shirt-]~:** Hemdzipfel, *der (ugs.)*; **e)** *(of man's coat)* Schoß, *der;* **f)** *in pl. (man's evening dress)* Frack, *der;* **g)** *in pl. (on coin)* **~s [it is]** Zahl; *see also* **head 1 e. 2.** *v. t.* **a)** *(remove stalks of)* **top and ~ gooseberries** Stachelbeeren putzen; **b)** *(sl.: follow)* beschatten

~ 'away *see* **~ off**

~ 'back *v. i.* sich stauen

~ 'off *v. i.* **a)** *(decrease)* zurückgehen; **b)** *(fade into silence)* verstummen

tail: ~back *n. (Brit.)* Rückstau, *der;* **~board** *n.* hintere Bordwand; **~ coat** *n.* Frack, *der;* **~-end** *n. (hindmost end)* Schwanz, *der; (fig.)* Ende, *das;* **~-gate** *n. (Motor Veh.)* Heckklappe, *die;* **~-lamp** *(esp. Amer.)*, **~-light** *ns.* Rück- *od.* Schlußlicht, *das*

tailor ['teɪlə(r)] **1.** *n.* Schneider, *der*/Schneiderin, *die; see also* **baker. 2.** *v. t.* **a)** schneidern; **b)** *(fig.)* **~ed to** *or* **for sb./sth.** für jmdn./etw. maßgeschneidert; **~ed to sb.'s needs** auf jmds. Bedürfnisse zugeschnitten

'tailor-made *adj. (lit. or fig.)* maßgeschneidert

tail: ~plane *n. (Aeronaut.)* Höhenleitwerk, *das;* **~ wind** *n.* Rückenwind, *der*

taint [teɪnt] **1.** *n.* Makel, *der.* **2.** *v. t.* verderben; beflecken ⟨*Ruf*⟩; **be ~ed with sth.** mit etw. behaftet sein *(geh.)*

Taiwan [taɪ'wɑːn] *pr. n.* Taiwan *(das)*

take [teɪk] **1.** *v. t.*, **took** [tʊk], **taken** ['teɪkn] **a)** *(get hold of, grasp, seize)* nehmen; **~ sb.'s arm** jmds. Arm nehmen; **~ sb. by the hand/arm** jmdn. bei der Hand/am Arm nehmen; **b)** *(capture)* einnehmen ⟨*Stadt, Festung*⟩; machen ⟨*Gefangenen*⟩; *(chess)* schlagen; nehmen; **c)** *(gain, earn)* ⟨*Laden:*⟩ einbringen; ⟨*Film, Stück:*⟩ einspielen; *(win)* gewinnen ⟨*Satz, Spiel, Preis, Titel*⟩; erzielen ⟨*Punkte*⟩; *(Cards)* machen ⟨*Stich*⟩; **~ first/second** *etc.* **place** den ersten/zweiten *usw.* Platz belegen; *(fig.)* an erster/zweiter *usw.* Stelle kommen; **~ the biscuit** *(Brit. coll.) or (coll.)* **cake** *(fig.)* alle/alles übertreffen; **d)** *(assume possession of)* nehmen; *(~ away with one)* mitnehmen; *(steal)* mitnehmen *(verhüll.)*; *(obtain by purchase)* kaufen, *(by rent)* mieten ⟨*Auto, Wohnung, Haus*⟩; nehmen ⟨*Klavier-, Deutsch-, Fahrstunden*⟩; mitmachen ⟨*Tanzkurs*⟩; *(buy regularly)* nehmen; lesen ⟨*Zeitung, Zeitschrift*⟩; *(subscribe to)* beziehen; *(obtain)* erwerben ⟨*akademischen Grad*⟩; *(form a relationship with)* sich *(Dat.)* nehmen ⟨*Frau, Geliebten usw.*⟩; **that woman took my purse** die Frau hat mir meinen Geldbeutel gestohlen; **he took his degree at Sussex University** er hat sein Examen an der Universität von Sussex gemacht; **~ place** stattfinden; *(spontaneously)* sich ereignen; ⟨*Wandlung:*⟩ sich vollziehen; **I'll ~ this handbag/the curry, please** ich nehme diese Handtasche/das Curry; **who has ~n my pencil?** wer hat meinen Bleistift weggenommen?; **e)** *(avail oneself of, use)* nehmen; machen ⟨*Pause, Ferien, Nickerchen*⟩; nehmen ⟨*Beispiel, Zitat usw.*⟩ **(from** aus); **~ the opportunity to do/of doing sth.** die Gelegenheit dazu benutzen, etw. zu tun; **~ the car/bus into town** mit dem Auto/Bus in die Stadt fahren; **~ two eggs** *etc. (in recipe)* man nehme zwei Eier *usw.*; **~ all the time you want** nimm dir ruhig Zeit; **[let's] ~ a more recent example/my sister [for example]** nehmen wir ein Beispiel neueren Datums/einmal meine Schwester; **f)** *(carry, guide, convey)* bringen; **~ sb.'s shoes to the mender['s]/sb.'s coat to the cleaner's** jmds. Schuhe zum Schuster/ jmds. Mantel in die Reinigung bringen; **~ a message to sb.** jmdm. eine Nachricht überbringen; **~ sb. to school/hospital** jmdn. zur Schule/ins Krankenhaus bringen; **~ sb. to visit sb.** jmdn. zu Besuch bei jmdm. mitnehmen; **~ sb. to the zoo/cinema/ to dinner** mit jmdm. in den Zoo/ ins Kino/zum Abendessen gehen; **~ sb. into one's home/house** jmdn. bei sich aufnehmen; **the road ~s you/the story ~s us to London** die Straße führt nach/die Erzählung führt uns nach London; **his ability will ~ him far/to the top** mit seinen Fähigkeiten wird er es weit bringen/wird er ganz nach oben kommen; **~ sb./ sth. with one** jmdn./etw. mitnehmen; **~ home** mit nach Hause nehmen; *(earn)* nach Hause bringen ⟨*Geld*⟩; *(accompany)* nach Hause bringen *od.* begleiten; *(to meet one's parents etc.)* mit nach Hause bringen; **~ sb. through/ over sth.** *(fig.)* mit jmdm. etw. durchgehen; **~ in hand** *(begin)* in Angriff nehmen; *(assume responsibility for)* sich kümmern um; **~ sb. into partnership [with one]/into the business** jmdn. zu seinem Teilhaber machen/in sein Geschäft aufnehmen; **~ a stick** *etc.* **to sb.** den Stock *usw.* bei jmdm. gebrauchen; **~ sth. to pieces** *or* **bits** etw. auseinandernehmen; **you can/can't ~ sb. anywhere** *(fig. coll.)* man kann jmdn. überallhin/nirgendwohin mitnehmen; **you can't ~ it 'with you** *(coll.)* man kann es ja nicht mitnehmen; **g)** *(remove)* nehmen; *(deduct)* abziehen; **~ sth./sb. from sb.** jmdm. etw./jmdn. wegnehmen; **I took the parcel from her** ich nahm ihr das Paket ab; **~ all the fun/hard work out of sth.** einem alle Freude an etw. *(Dat.)* nehmen/einem die schwere Arbeit bei etw. ersparen; **h) sb. ~s courage from sth.** etw. macht jmdm. Mut; *see also* **heart a; i) be ~n ill** *or (coll.)* **sick** krank werden; **j)** *(make)* machen ⟨*Foto, Kopie*⟩; *(photograph)* aufnehmen; **k)** *(perform, execute)* aufnehmen ⟨*Brief, Diktat*⟩; machen ⟨*Prüfung, Sprung, Spaziergang, Reise, Umfrage*⟩; durchführen ⟨*Befragung, Volkszählung*⟩; ablegen ⟨*Gelübde, Eid*⟩; übernehmen ⟨*Rolle, Part*⟩; treffen ⟨*Entscheidung*⟩; **~ a fail/ tumble** stürzen/straucheln; **~ a step forward/backward** einen Schritt vor-/zurücktreten; **~ a turn for the better/worse** eine Wende zum Besseren/Schlechteren nehmen; **l)** *(negotiate)* nehmen ⟨*Zaun, Mauer, Hürde, Kurve, Hindernis*⟩; **m)** *(conduct)* halten ⟨*Gottesdienst, Andacht, Unterricht*⟩; **Ms X ~s us for maths** in Mathe haben wir Frau X; **n)** *(be taught)* **~ Latin at school** in der Schule Latein haben; **o)** *(consume)* trinken ⟨*Tee, Kaffee, Kognak usw.*⟩; nehmen ⟨*Zucker, Milch, Überdosis, Tabletten, Medizin*⟩; **~ sugar in one's tea** den Tee mit Zucker trinken; **what can I ~ for a cold?** was kann ich gegen eine Erkältung nehmen?; **to be ~n three times a day** dreimal täglich einzunehmen; **not to be ~n [internally]** nicht zur innerlichen Anwendung; **p)** *(occupy)* einnehmen ⟨*Sitz im Parlament*⟩; übernehmen, antreten ⟨*Amt*⟩; **~ sb.'s seat** sich auf jmds. Platz setzen; **is that/this seat ~n?** ist da/hier noch frei?; **q)** *(need, require)* brauchen ‹*Platz, Zeit*›; haben ‹*Kleider-, Schuhgröße usw.*›; *(Ling.)* haben ‹*Objekt, Plural-s*›; gebraucht werden mit ‹*Kasus*›; **this verb ~s 'sein'** dieses Verb wird mit „sein" konjugiert; **the wound will ~ some time to heal** es braucht einige Zeit, bis die Wunde geheilt ist; **the ticket-machine ~s 20p and 50p coins** der Fahrkartenautomat nimmt 20-Pence- und 50-Pence-Stücke; **as long as it ~s** so lange wie nötig; **sth. ~s an hour/a year/all day** etw. dauert eine Stunde/ein Jahr/einen ganzen Tag; **it ~s an hour** *etc.* **to do sth.** es dauert eine Stunde *usw.*, [um] etw. zu tun; **sb. ~s** *or* **it ~s sb. a long time/an hour** *etc.* **to do sth.** jmd. braucht lange/eine Stunde *usw.*, um etw. zu tun; **what took you so long?** was hast du denn so lange gemacht?; **~ a lot of work/effort/courage** viel Arbeit/Mühe/Mut kosten; **have [got] what it ~s** das Zeug dazu haben; **it will ~ [quite] a lot of explaining** es wird schwer zu erklären sein; **that story of his ~s some believing** die Geschichte, die er da erzählt, ist kaum zu glauben; **it ~s a thief to know a thief** nur ein Dieb kennt einen Dieb; **it ~s all sorts [to make a world]** es gibt solche und solche; **r)** *(contain, hold)* fassen; *(support)* tragen; **s)** *(ascertain and record)* notieren ‹*Namen, Adresse, Autonummer usw.*›; fühlen ‹*Puls*›; messen ‹*Temperatur, Größe usw.*›; **~ the minutes of a meeting** bei einer Sitzung [das] Protokoll führen; **t)** *(apprehend, grasp)* **~ sb.'s meaning/drift** verstehen, was jmd. meint; **~ sb.'s point** jmds. Standpunkt verstehen; **~ it [that] ...:** annehmen, [daß] ...; **can I ~ it that ...?** kann ich davon ausgehen, daß ...?; **~ sth. to mean sth.** etw. so verstehen, daß ...; **what do you ~ that to mean?** wie verstehen Sie das?; **~ sth. as settled/as a compliment/refusal** etw. als erledigt betrachten/als eine Ablehnung/ein Kompliment auffassen; **~ sb./sth. for/to be sth.** jmdn./etw. für etw. halten; **what do you ~ me for?** wofür halten Sie mich?; **u)** *(treat or react to in a specified manner)* aufnehmen; **~ sth. like a man** etw. wie ein Mann nehmen; **~ sth. well/badly/hard** etw. gut/schlecht/nur schwer verkraften; **sb. ~s sth. very badly/hard** etw. trifft jmdn. sehr; **~ sth. calmly** *or* **coolly** etw. gelassen [auf- *od.* hin]nehmen; **~ sth. as read** etw. als bekannt voraussetzen; **you can/may ~ it as read that ...:** du kannst sicher sein, daß ...; **taking it all in all, taking one thing with another** alles in allem; **v)** *(accept)* annehmen; **~ money** *etc.* **[from sb./for sth.]** Geld *usw.* [von jmdm./für etw.] [an]nehmen; **will you ~ £500 for the car?** wollen Sie den Wagen für 500 Pfund verkaufen?; **[you can] ~ it or leave it** entweder du bist damit einverstanden, oder du läßt es bleiben; **~ the hint** den Wink verstehen; **~ sb.'s word for it** sich auf jmdn. *od.* jmds. Wort[e] verlassen; **you don't have to ~ my word for it** du brauchst es mir nicht zu glauben; **~ things as they come, ~ it as it comes** es nehmen, wie es kommt; **w)** *(receive, submit to)* einstecken [müssen] ‹*Schlag, Tritt, Stoß*›; *(Boxing)* nehmen [müssen] ‹*Schlag*›; *(endure, tolerate)* aushalten; vertragen ‹*Klima, Alkohol, Kaffee, Knoblauch*›; verwinden ‹*Schock*›; *(put up with)* sich *(Dat.)* gefallen lassen [müssen] ‹*Kritik, Grobheit*›; **~ one's punishment bravely** seine Strafe tapfer ertragen; **~ no nonsense** sich *(Dat.)* nichts bieten lassen; **~ 'that!** nimm das!; **~ it** *(coll.)* es verkraften; *(referring to criticism, abuse)* damit fertigwerden; **x)** *(adopt, choose)* ergreifen ‹*Maßnahmen*›; unternehmen ‹*Schritte*›; einschlagen ‹*Weg*›; sich entschließen zu ‹*Schritt, Handlungsweise*›; **~ the wrong road** die falsche Straße nehmen; **~ a firm** *etc.* **stand [with sb./on** *or* **over sth.]** jmdm. gegenüber/hinsichtlich einer Sache nicht nachgeben; **~ the easy way out** die einfachste Lösung wählen; **y)** *(receive, accommodate)* [an]nehmen ‹*Bewerber, Schüler*›; aufnehmen ‹*Gäste*›; **the city ~s its name from its founder** die Stadt ist nach ihrem Gründer benannt; **z)** *(swindle)* **he was ~n for £500 by the con-man** *(coll.)* der Schwindler hat ihm 500 Pfund abgeknöpft *(ugs.)*; **aa) be ~n with sb./sth.** von jmdm./etw. angetan sein. **2.** *v. i.*, **took, taken a)** *(be successful, effective)* ‹*Transplantat:*› vom Körper angenommen werden; ‹*Impfung:*› anschlagen; ‹*Pfropfreis:*› anwachsen; ‹*Sämling, Pflanze:*› angehen; ‹*Feuer:*› zu brennen beginnen; ‹*Fisch:*› [an]beißen; **b)** *(detract)* **~ from sth.** etw. schmälern. **3.** *n. (Telev., Cinemat.)* Einstellung, *die;* Take, *der od. das (fachspr.)*

~ after *v. t.* **~ after sb.** *(resemble)* jmdm. ähnlich sein; *(~ as one's example)* es jmdm. gleichtun

~ a'long *v. t.* mitnehmen

~ a'part *see* **apart b**

~ a'round *v. t.* **a)** *(~ with one)* überallhin mitnehmen; **b)** *(show around)* herumführen

~ a'side *see* **aside 1**

~ a'way *v. t.* **a)** *(remove)* wegnehmen; *(to a distance)* mitnehmen; **~ sth. away from sb.** jmdm. etw. abnehmen; **~ sb.'s licence/passport away** jmdm. den Führerschein/Paß abnehmen; **to ~ away** ‹*Pizza, Snack usw.*› zum Mitnehmen; **~ away sb.'s rights/privileges/freedom** jmdm. seine Rechte/Privilegien/die Freiheit nehmen; **~ sb. away** jmdn. wegbringen; ‹*Polizei:*› jmdn. abführen; **~ him away!** schafft ihn fort!; hinweg mit ihm! *(geh.)*; **~ a child away from its parents/home/from school** ein Kind den Eltern wegnehmen/aus seiner häuslichen Umgebung herausreißen/aus der Schule nehmen; **b)** *(Math.: deduct)* abziehen

~ a'way from *v. t.* schmälern

~ 'back *v. t.* **a)** *(retract, have back)* zurücknehmen; wieder einstellen ‹*Arbeitnehmer*›; wieder [bei sich] aufnehmen ‹*Ehepartner*›; *(reclaim)* sich *(Dat.)* wiedergeben lassen; **b)** *(return)* zurückbringen; *(~ somewhere again)* wieder bringen ‹*Person*›; *(carry or convey back)* wieder mitnehmen; **that ~s me back [to my childhood]** das weckt bei mir [Kindheits]erinnerungen

~ 'down *v. t.* **a)** *(carry or lead down)* hinunterbringen; **this path ~s you down to the harbour** auf diesem Weg kommen Sie zum Hafen [hinunter]; **b)** *(lower or lift down)* abnehmen ‹*Bild, Ankündigung, Weihnachtsschmuck*›; einholen ‹*Fahne*›; herunterziehen, herunterlassen ‹*Hose*›; **~ a box down from a shelf** eine Schachtel von einem Regal herunternehmen; **c)** *(dismantle)* abreißen; abbauen ‹*Gerüst, Zelt*›; **d)** *(write down)* aufnehmen ‹*Brief, Personalien*›; aufschreiben ‹*Autonummer*›; mitschreiben ‹*Vortrag*›

~ 'in *v. t.* **a)** *(convey to a place)* hinbringen; *(conduct)* hineinführen ‹*Gast*›; *(coll.: ~ for repair or service)* wegbringen *(ugs.)* ‹*Auto, Gerät usw.*›; **~ sb. in [in the car]** jmdn. [mit dem Auto] reinfahren *(ugs.)*; **I took the car in** ich fuhr mit dem Auto rein *(ugs.)*; **b)** *(bring indoors)* hereinholen; **c)** *(receive, admit)* aufnehmen; *(for payment)* vermieten an (+ *Akk.*); [auf]nehmen ‹*[Kur]gäste*›; ~ **in lodgers** ‹*Haus-, Wohnungseigentümer:*› Zimmer vermieten; **d)** *(make narrower)* enger machen ‹*Kleidungsstück*›; **e)** *(include, comprise)* einbeziehen; **f)** *(coll.: visit)* mitnehmen *(ugs.)*; **our tour took in most of the main sights** auf unserer Rundfahrt haben wir die wichtigsten Sehenswürdigkeiten besichtigt; **g)** *(understand, grasp)* begreifen; überblicken, erfassen ‹*Lage*›; **h)** *(observe)* erfassen; *(watch, listen to)* mitbekommen; **i)** *(deceive)* einwickeln *(salopp)*; **be ~n in [by sb./sth.]** sich [von jmdm./durch etw.] einwickeln lassen *(salopp)*

~ **'off 1.** *v. t.* **a)** abnehmen ‹*Deckel, Hut, Bild, Hörer, Tischtuch, Verband*›; ausziehen ‹*Schuhe, Handschuhe*›; ablegen ‹*Hut, Mantel, Schmuck*›; ~ **the cover off a pillow/bed** ein Kissen abziehen/ein Bett abdecken; ~ **a parcel off sb.** jmdm. ein Paket abnehmen; ~ **your hands off me!** faß mich nicht an!; **b)** *(transfer from)* übernehmen ‹*Passagiere, Besatzung, Fracht*›; ~ **sb. off sth.** jmdn. von etw. holen; *(withdraw from job, assignment, etc.)* jmdm. etw. entziehen; ~ **sth. off a list/the menu** etw. von einer Liste streichen/von der Speisekarte nehmen; ~ **a train/bus off a route** einen Zug/Bus vom Fahrplan streichen; **c)** *(cut off)* abtrennen; *(with saw)* absägen; *(with knife, scissors, etc.)* abschneiden; *(amputate)* abnehmen; **d)** *(lead, conduct)* ~ **sb. off to hospital/prison** jmdn. ins Krankenhaus/Gefängnis bringen; **e)** *(deduct)* abziehen; ~ **sth. off sth.** etw. von etw. abziehen; ~ **£10 off the price** den Preis um zehn Pfund reduzieren; **f)** ~ **off weight/a few pounds** *(lose weight)* abnehmen/einige Pfund abnehmen; **g)** *(have free)* ~ **a day** *etc.* **off** sich *(Dat.)* einen Tag *usw.* frei nehmen *(ugs.)*; ~ **time off [work** *or* **from work]** sich *(Dat.)* frei nehmen; **h)** *(mimic)* nachmachen *(ugs.)*. **2.** *v. i.* **a)** *(Aeronaut.)* starten; **b)** *(Sport)* ‹*Springer, Pferd:*› abspringen

~ **'on** *v. t.* **a)** *(undertake)* übernehmen; annehmen ‹*Herausforderung, Wette usw.*›; auf sich *(Akk.)* nehmen ‹*Bürde*›; *(accept responsibility for)* sich einlassen auf (+ *Akk.*) ‹*Person*›; sich *(Dat.)* aufbürden *od.* aufladen ‹*Sache*›; **b)** *(enrol, employ)* einstellen; aufnehmen ‹*Schüler, Studenten*›; annehmen ‹*Privatschüler*›; **c)** *(acquire, assume)* annehmen ‹*Farbe, Form, Ausdruck, Ausmaße*›; erhalten ‹*Bedeutung*›; **d)** *(accept as opponent)* sich auf eine Auseinandersetzung einlassen mit; es aufnehmen mit; den Kampf aufnehmen mit ‹*Regierung, Gesetz*›; **e)** *(~ on board)* aufnehmen

~ **'out** *v. t.* **a)** *(remove)* herausnehmen; ziehen ‹*Zahn*›; ~ **sth. out of sth.**, ~ **out sth. from sth.** etw. aus etw. [heraus]nehmen; ~ **it/a lot out of sb.** *(fig.)* jmdn. mitnehmen/sehr mitnehmen; **b)** *(destroy)* zerstören; *(fig.)* *(Footb. etc.)* ausschalten; *(kill)* töten; **c)** *(withdraw)* abheben ‹*Geld*›; **d)** *(deduct)* abziehen (**of** von); **e)** *(go out with)* ~ **sb. out** mit jmdm. ausgehen; ~ **sb. out for a walk/drive** mit jmdm. einen Spaziergang/eine Spazierfahrt machen; ~ **sb. out to** *or* **for lunch/out to the cinema** jmdn. zum Mittagessen/ins Kino einladen; ~ **the dog out [for a walk]** den Hund ausführen; **f)** *(get issued)* erwerben; erhalten; abschließen ‹*Versicherung*›; ausleihen ‹*Bücher*›; aufgeben ‹*Anzeige*›; ~ **out a subscription to sth.** etw. abonnieren; **g)** ~ **it/sth. out on sb./sth.** seine Wut/etw. an jmdm./etw. auslassen

~ **'over 1.** *v. t.* **a)** *(assume control of)* übernehmen; ~ **sth. over from sb.** etw. von jmdm. übernehmen; ~ **sb./sth. over** *(fig.)* von jmdm./etw. Besitz ergreifen; **b)** *(carry or transport over)* ~ **sb./sth. over to sb./sb.'s flat/Guildford** jmdn./etw. zu jmdm./in jmds. Wohnung/nach Guildford bringen *od.* *(ugs.)* rüberbringen. **2.** *v. i.* übernehmen; ‹*Manager, Firmenleiter:*› die Geschäfte übernehmen; ‹*Regierung, Präsident:*› die Amtsgeschäfte übernehmen; ‹*Beifahrer:*› das Steuer übernehmen; **the night nurse ~s over at 10 p.m.** um zehn Uhr [abends] tritt die Nachtschwester ihren Dienst an

~ **'round** *v. t.* **a)** *(carry, deliver)* vorbeibringen; **b)** *(show around)* [herum]führen; ~ **sb. round the factory** jmdn. durch die Fabrik führen

~ **to** *v. i.* **a)** *(get into habit of)* ~ **to doing sth.** anfangen, etw. zu tun; es sich *(Dat.)* angewöhnen, etw. zu tun; ~ **to drugs/gambling/crime** zu Drogen greifen/dem Spiel/der Kriminalität verfallen; **b)** *(escape to)* sich flüchten in (+ *Akk.*); **c)** *(develop a liking for)* sich hingezogen fühlen zu ‹*Person*›; sich erwärmen für ‹*Sache*›; *(adapt oneself to)* sich gewöhnen an (+ *Akk.*)

~ **'up 1.** *v. t.* **a)** *(lift up)* hochheben; *(pick up)* aufheben; aufnehmen ‹*Staub, Partikel*›; herausnehmen ‹*Pflanzen*›; herausreißen ‹*Schienenstrang, Dielen*›; aufreißen ‹*Straße*›; **b)** *(move up)* weiter nach oben rücken; *(shorten)* kürzer machen; **c)** *(carry or lead up)* ~ **sb./sth. up** jmdn./etw. hinaufbringen (**to** zu); ~ **sth. up to sb.** jmdm. etw. hinaufbringen; **d)** *(absorb)* aufnehmen; **e)** *(wind up)* aufwickeln; **f)** *(occupy, engage)* beanspruchen; **I'm sorry to have ~n up so much of your time** es tut mir leid, Ihre Zeit so lange in Anspruch genommen zu haben; **g)** ergreifen ‹*Beruf*›; anfangen ‹*Jogging, Tennis, Schach, Gitarre*›; ~ **up a musical instrument** ein Instrument zu spielen beginnen; ~ **up German/a hobby** anfangen, Deutsch zu lernen/sich *(Dat.)* ein Hobby zulegen; **h)** *(start, adopt)* aufnehmen ‹*Arbeit, Kampf*›; antreten ‹*Stelle*›; übernehmen ‹*Pflicht, Funktion*›; einnehmen ‹*Haltung, Position*›; eintreten für, sich einsetzen für ‹*Sache*›; ~ **up a/one's position** ‹*Polizeiposten, Politiker:*› Position beziehen; **i)** *(accept)* annehmen; aufnehmen ‹*Idee, Vorschlag, Kredit, Geld*›; kaufen ‹*Aktien*›; **j)** *(raise, pursue further)* aufgreifen; ~ **sth. up with sb.** sich in einer Sache an jmdn. wenden; **k)** ~ **sb. up [on sth.]** *(accept)* jmdn. [in bezug auf etw. *(Akk.)*] beim Wort nehmen. **2.** *v. i.* **a)** *(coll.: become friendly)* ~ **up with sb.** sich mit jmdm. einlassen; **b)** ~ **up where sb./sth. has left off** da einsetzen, wo jmd./etw. aufgehört hat

~ **upon** *v. t.* ~ **upon oneself** auf sich *(Akk.)* nehmen ‹*Aufgabe, Pflicht, Verantwortung*›; ~ **it upon oneself to do sth.** es auf sich *(Akk.)* nehmen, etw. zu tun; *(in an interfering way)* sich *(Dat.)* herausnehmen *(ugs.)*, etw. zu tun

'take-away *n.* *(restaurant)* Restaurant mit Straßenverkauf; *(meal)* Essen zum Mitnehmen; *attrib.* ‹*Restaurant*› mit Straßenverkauf; ‹*Essen, Mahlzeit*› zum Mitnehmen

taken *see* **take 1, 2**

take: ~-off *n.* **a)** *(Sport)* Absprung, *der;* **b)** *(Aeronaut.)* Start, *der;* Take-off, *das (fachspr.);* **c)** *(coll.: caricature)* Parodie, *die;* **~-over** *n. (Commerc.)* Übernahme, *die;* **~-over bid** Übernahmeangebot, *das*

taker ['teɪkə(r)] *n.* **there were no ~s [for the offer]** niemand hat [das Angebot] angenommen

taking ['teɪkɪŋ] *n.* **a)** *in pl. (amount taken)* Einnahmen; **b)** *(seizure)* Einnahme, *die;* **c) they are yours/his** *etc.* **for the ~:** du kannst/er kann *usw.* sie haben; **victory was his for the ~:** sein Sieg war so gut wie sicher

talc [tælk] *n.* Talkum, *das*

talcum ['tælkəm] *n.* Talkumpuder, *der;* Talkum, *das; (as cosmetic)* **~ [powder]** Körperpuder, *der*

tale [teɪl] *n.* **a)** *(story)* Erzählung, *die;* Geschichte, *die* **(of** von, **about** über + *Akk.*); **b)** *(piece of gossip)* Geschichte, *die (ugs.)*

talent ['tælənt] *n.* **a)** *(ability)* Talent, *das;* **have [great/no** *etc.*] **~ [for sth.]** [viel/kein *usw.*] Talent [zu *od.* für etw.] haben; **have a ~ for music** musikalisches Talent haben; **have a [great] ~ for doing sth.** das Talent haben, etw. zu tun; **b)** *(people with ability)* Talente; Begabungen

talented ['tæləntɪd] *adj.* talentiert

talk [tɔ:k] **1.** *n.* **a)** *(discussion)* Gespräch, *das;* **have a ~ [with sb.] [about sth.]** [mit jmdm.] [über etw. *(Akk.)*] reden *od.* sprechen; **have a long ~ on the phone** lange miteinander telefonieren; **could I have a ~ with you?** könnte ich Sie einmal sprechen?; **have** *or* **hold ~s [with sb.]** [mit jmdm.] Gespräche führen; **b)** *(speech, lecture)* Vortrag, *der;* **give a ~/a series of ~s [on sth./sb.]** einen Vortrag/eine Vortragsreihe [über etw./jmdn.] halten; **c)** *no pl. (form of communication)* Sprache, *die;* **d)** *no pl. (talking)* Gerede, *das (abwertend);* **there's too much ~ [of ...]** es wird zuviel [von ...] geredet; **be the ~ of the town/neighbourhood** *etc.* Stadtgespräch/das Thema in der Nachbarschaft *usw.* sein. **2.** *v. i.* **a)** *(speak)* sprechen, reden **(with, to** mit); *(lecture)* sprechen; *(converse)* sich unterhalten; *(have ~s)* Gespräche führen; *(gossip)* reden; **be ~ing in German** deutsch sprechen; **love to hear oneself ~:** sich gern reden hören; **we must ~:** wir müssen miteinander reden; **~ on the phone** telefonieren; **keep sb. ~ing** jmdn. in ein [längeres] Gespräch verwickeln; **now you're ~ing!** *(coll.)* das hört sich schon besser an; **that's no way to ~/~ to your uncle** das darfst du nicht sagen/so darfst du aber nicht mit deinem Onkel reden!; **it's easy for you/him** *etc.* **to ~:** du hast/er hat *usw.* gut reden; **look who's ~ing** *(iron.)* das mußt du gerade sagen; **you can** *(iron.) or* **can't ~!** sei du nur ganz still!; **could I ~ to you for a moment?** könnte ich Sie einen Augenblick sprechen?; **get ~ing [to sb.]** [mit jmdm.] ins Gespräch kommen; **~ to oneself** Selbstgespräche führen; **~ of** *or* **about sb./sth.** über jmdn./etw. reden; **everyone's ~ing about him/his divorce** er/seine Scheidung ist in aller Munde; **~ of** *or* **about doing sth.** davon reden, etw. zu tun; **[not] know what one is ~ing about** [gar nicht] wissen, wovon man redet; **[not] know what sb. is ~ing about** [nicht] wissen, was jmd. meint *od.* wovon jmd. spricht; **~ing of holidays** *etc.* da wir [gerade] vom Urlaub *usw.* sprechen; **b)** *(have power of speech)* sprechen; **c)** *(betray secrets)* reden; **make sb. ~:** jmdn. zum Reden bringen. **3.** *v. t.* **a)** *(utter, express)* **~ [a load of] nonsense** [eine Menge] Unsinn *od.* *(ugs.)* Stuß reden; **b)** *(discuss)* **~ politics/music** *etc.* über Politik/Musik *usw.* reden; **c)** *(use)* sprechen ‹*Sprache, Dialekt usw.*›; **d) ~ oneself hoarse** sich heiser reden; **~ oneself** *or* **one's way out of trouble** sich aus Schwierigkeiten herausreden; **~ sb. into/out of sth.** jmdn. zu etw. überreden/jmdm. etw. ausreden

~ 'down 1. *v. t. (silence)* in Grund und Boden reden. **2.** *v. i.* **~ down to sb.** von oben herab *od.* herablassend mit jmdm. reden

~ 'over *v. t.* **a) ~ sth. over [with sb.]** etw. [mit jmdm.] besprechen; **b)** *(persuade)* **~ sb. over** jmdn. überreden

~ 'round *v. t.* **a)** *(persuade)* **~ sb. round** jmdn. überreden; **b)** *(skirt)* **~ round sth.** um etw. herumreden *(ugs.)*

~ 'through *v. t.* **~ sb. through sth.** etw. mit jmdm. durchgehen *od.* durchsprechen; **~ sth. through** etw. durchsprechen

talkative ['tɔ:kətɪv] *adj.* gesprächig; geschwätzig *(abwertend)*

'talked-of *attrib. adj.* **much ~:** viel diskutiert ‹*Buch, Stück, Projekt*›; **a much ~ actor/artist** ein Schauspieler/Künstler, der in aller Munde ist

talker ['tɔ:kə(r)] *n.* **a)** Redner, *der*/Rednerin, *die;* **b)** *(one who talks but does not act)* Schwätzer, *der*/Schwätzerin, *die*

talking ['tɔ:kɪŋ] **1.** *n.* Reden, *das;* **do [all] the ~:** das Gespräch dominieren; **let me do the ~:** überlaß lieber mir das Reden. **2.** *adj.* sprechend

talking: ~ point *n.* Gesprächsthema, *das;* **~-shop** *n. (derog.)* Quasselbude, *die (ugs. abwertend)*

'talk-show *n.* Talk-Show, *die*

tall [tɔ:l] *adj.* **a)** hoch; groß ‹*Person, Tier*›; **grow ~:** groß werden; wachsen; **b)** *(coll.: excessive)* **a ~ tale** eine unglaubwürdige Geschichte; **that's a ~ order** das ist ziemlich viel verlangt; *see also* [1]**story a**

'tallboy *n.* Doppelkommode, *die;* Tallboy, *der*

tallow ['tæləʊ] *n.* Talg, *der*

tally ['tælɪ] **1.** *n. (record)* **sb.'s ~ is 18 goals** jmd. kann 18 Tore für sich verbuchen; **keep a [daily] ~ of sth.** [täglich] über etw. *(Akk.)* Buch führen. **2.** *v. i.* übereinstimmen

talon ['tælən] *n.* Klaue, *die;* **~s** *(fig.: long fingernails)* Krallen *(ugs. abwertend)*

tambourine [tæmbə'ri:n] *n.* *(Mus.)* Tamburin, *das*

tame [teɪm] **1.** *adj.* **a)** zahm; *(joc.)* hauseigen ‹*Anarchist, Genie usw.*›; **grow/become ~:** zahm werden; **b)** *(spiritless)* lahm *(ugs.),* lustlos ‹*Einwilligung, Anerkennung, Kampagne, Versuch*›; zahm *(ugs.)* ‹*Besprechung, Kritik*›; **c)** *(dull)* wenig aufregend; lasch ‹*Stil*›. **2.** *v. t. (lit. or fig.)* zähmen

tamper ['tæmpə(r)] *v. i.* **~ with** sich *(Dat.)* zu schaffen machen an (+ *Dat.*)

tampon ['tæmpɒn] *n.* Tampon, *der*

tan [tæn] **1.** *v. t.,* **-nn-: a)** gerben; **b)** *(bronze)* ‹*Sonne:*› bräunen; ‹*Person:*› braun werden lassen ‹*Körperteil*›; **c)** *(sl.: beat)* das Fell gerben *(salopp)* (+ *Dat.*). **2.** *v. i.,* **-nn-** braun werden. **3.** *n.* **a)** *(colour)* Gelbbraun, *das;* **b)** *(sun-~)* Bräune, *die;* **have/get a ~:** braun sein/werden. **4.** *adj.* gelbbraun

tandem ['tændəm] *n.* Tandem, *das;* **~ bicycle** Tandem, *das;* **coupled/harnessed in ~:** hintereinandergekoppelt/-gespannt

tang [tæŋ] *n. (taste/smell)* **[sharp] ~:** scharfer Geschmack/Geruch; **[spicy/salty] ~:** würziger/salziger Geschmack/Geruch

tangent ['tændʒənt] *n.* *(Math.)*

Tangente, *die;* **go** *or* **fly off at a ~** *(fig.)* plötzlich vom Thema abschweifen

tangerine [tændʒə'ri:n] *n.* **a)** *(fruit)* ~ **[orange]** Tangerine, *die;* **b)** *(colour)* Orangerot, *das*

tangible ['tændʒɪbl] *adj.* **a)** *(perceptible by touch)* fühlbar; **b)** *(fig.: real, definite)* greifbar; spürbar, merklich ‹*Unterschied, Verbesserung*›; handfest ‹*Beweis*›

tangle ['tæŋgl] **1.** *n.* Gewirr, *das; (in hair)* Verfilzung, *die; (fig.: dispute)* Auseinandersetzung, *die;* **be in a ~:** sich verheddert haben *(ugs.);* ‹*Haar:*› sich verfilzt haben; *(fig.)* ‹*Angelegenheiten:*› in Unordnung *(Dat.)* sein; ‹*Person:*› verwirrt sein. **2.** *v.t.* verheddern *(ugs.);* verfilzen ‹*Haar*›

~ **'up** *v.t.* verheddern *(ugs.);* verfilzen ‹*Haar*›; **become** *or* **get ~d up** sich verheddern *(ugs.)*

~ **with** *v.t. (coll.)* ~ **with sb.** sich mit jmdm. anlegen

tangled ['tæŋgld] *adj.* verheddert *(ugs.);* verfilzt ‹*Haar*›; *(confused, complicated)* verworren; verwickelt ‹*Angelegenheit*›

tango ['tæŋgəʊ] *n., pl.* **~s** Tango, *der*

tangy ['tæŋɪ] *adj.* scharf; *(spicy)* würzig; *(salty)* salzig

tank [tæŋk] *n.* **a)** Tank, *der; (for fish etc.)* Aquarium, *das; (for rain-water)* Auffangbecken, *das;* **fill the ~** *(with petrol)* volltanken; **b)** *(Mil.)* Panzer, *der*

~ **'up 1.** *v.i. (get fuel)* auftanken. **2.** *v.t.* auftanken; **get ~ed up** *(sl.: drunk)* sich volltanken *(salopp)*

tankard ['tæŋkəd] *n.* Krug, *der;* **a ~ of beer** *etc.* ein Krug Bier *usw.*

tanker ['tæŋkə(r)] *n.* *(ship)* Tanker, *der;* Tankschiff, *das; (vehicle)* Tank[last]wagen, *der*

tanned [tænd] *adj.* braungebrannt

tanner ['tænə(r)] *n.* *(person)* Gerber, *der*/Gerberin, *die*

tannery ['tænərɪ] *n.* Gerberei, *die*

Tannoy, (P) ['tænɔɪ] *n.* Lautsprecher, *der;* **over** *or* **on the ~:** über Lautsprecher

tantalise, tantalising *see* **tantaliz-**

tantalize ['tæntəlaɪz] *v.t.* reizen; *(tease also)* zappeln lassen *(ugs.); (with promises)* [falsche] Hoffnungen wecken bei

tantalizing ['tæntəlaɪzɪŋ] *adj.* verlockend; **a ~ puzzle** ein Rätsel, das einen nicht losläßt

tantamount ['tæntəmaʊnt] *pred. adj.* **be ~ to sth.** gleichbedeutend mit etw. sein; einer Sache *(Dat.)* gleichkommen

tantrum ['tæntrəm] *n.* Wutanfall, *der; (of child)* Trotzanfall, *der;* **be in a ~:** einen Wutanfall/Trotzanfall haben; **get into/throw a ~:** einen Wutanfall/Trotzanfall bekommen

Tanzania [tænzə'ni:ə] *pr. n.* Tansania *(das)*

[1]**tap** [tæp] **1.** *n.* **a)** Hahn, *der; (on barrel, cask)* [Zapf]hahn, *der;* **hot/cold[-water] ~:** Warm-/Kaltwasserhahn, *der;* **on ~:** vom Faß *nachgestellt;* **be on ~** *(fig.)* zur Verfügung stehen; **have on ~** *(fig.)* zur Verfügung haben ‹*Geld, Mittel*›; an der Hand haben ‹*Experten*›; **b)** *(plug)* Zapfen, *der;* Spund, *der.* **2.** *v.t.,* **-pp-: a)** *(make use of)* erschließen ‹*Reserven, Ressourcen, Bezirk, Markt, Land, Einnahmequelle*›; **b)** *(Teleph.: intercept)* abhören; anzapfen *(ugs.)*

[2]**tap 1.** *v.t.,* **-pp-** *(strike lightly)* klopfen an (+ *Akk.*); *(on upper surface)* klopfen auf (+ *Akk.*); **~ one's fingers on the table** *(repeatedly)* mit den Fingern auf den Tisch trommeln; **~ one's foot** mit dem Fuß auf den Boden klopfen; **~ one's foot to the music** mit dem Fuß den Takt schlagen; **~ sb. on the shoulder** jmdm. auf die Schulter klopfen/*(more lightly)* tippen. **2.** *v.i.,* **-pp-: ~ at/on sth.** an etw. *(Akk.)* klopfen; *(on upper surface)* auf etw. *(Akk.)* klopfen. **3.** *n.* Klopfen, *das; (given to naughty child)* Klaps, *der (ugs.);* **there was a ~ at/on the door** es klopfte an die Tür; **I felt a ~ on my shoulder** jemand klopfte/*(more lightly)* tippte mir auf die Schulter

~ **'in** *v.t.* einklopfen ‹*Nagel usw.*›

~ **'out** *v.t.* **a)** *(knock out)* ausklopfen ‹*Pfeife*›; herausklopfen ‹*Nagel, Keil*›; **b)** klopfen ‹*Rhythmus, Takt*›; *(in Morse)* morsen ‹*Nachricht*›; *(on typewriter)* tippen *(ugs.)*

tap: ~-dance 1. *n.* Step[tanz], *der;* **2.** *v.i.* Step tanzen; steppen; **~-dancer** *n.* Steptänzer, *der*/-tänzerin, *die;* **~-dancing** *n.* Steptanz, *der;* Steppen, *das*

tape [teɪp] **1.** *n.* **a)** Band, *das;* **adhesive/***(coll.)* **sticky ~:** Klebstreifen, *der;* Klebeband, *das;* **b)** *(Sport)* Zielband, *das;* **c)** *(for recording)* [Ton]band, *das* (**of** mit); **[have sth.] on ~:** [etw.] auf Band *(Dat.)* [haben]; **put/record sth. on ~, make a ~ of sth.** etw. auf Band *(Akk.)* aufnehmen; **blank ~:** unbespieltes Band; **d) [paper] ~:** Papierstreifen, *der; (punched with holes)* Lochstreifen, *der.* **2.** *v.t.* **a)** *(record on ~)* [auf Band *(Akk.)*] aufnehmen; **b)** *(bind with ~)* [mit Klebeband *od.* Klebstreifen] zukleben ‹*Paket*›; kleben ‹*Einband, eingerissene Seite*›; **c) have got sb./sth. ~d** *(sl.)* jmdn. durchschaut haben/etw. im Griff *od.* unter Kontrolle haben

~ **to'gether** *v.t.* [mit Klebeband] zusammenkleben

~ **'up** *v.t.* [mit Klebeband] zukleben; [mit Klebeband] zusammenkleben ‹*zerrissene Seite, zerbrochene Pfeife usw.*›

tape: ~ cassette *n.* Tonbandkassette, *die;* **~ deck** *n.* Tapedeck, *das;* **~-measure** *n.* Bandmaß, *das; (for measuring garments etc.)* [Zenti]metermaß, *das;* **~-player** *n.* Tonband[wiedergabe]gerät, *das*

taper ['teɪpə(r)] **1.** *v.t.* sich verjüngen lassen; **~ [to a point]** spitz zulaufen lassen; **be ~ed** sich verjüngen; *(to a point)* spitz zulaufen. **2.** *v.i.* sich verjüngen; **~ [to a point]** spitz zulaufen. **3.** *n.* **[wax] ~:** Wachsstock, *der*

~ **away** *see* **~ off 2**

~ **'off 1.** *v.t. see* **~ 1. 2.** *v.i.* **a)** *see* **~ 2; b)** *(fig.: decrease gradually)* zurückgehen

tape: ~-record ['teɪprɪkɔ:d] *v.t.* [auf Tonband *(Akk.)*] aufnehmen *od.* aufzeichnen; **~ recorder** *n.* Tonbandgerät, *das;* **~ recording** *n.* Tonbandaufnahme, *die*

tapered ['teɪpəd], **tapering** ['teɪpərɪŋ] *adjs.* sich verjüngend; *(to a point)* spitz zulaufend

tapestry ['tæpɪstrɪ] *n.* Gobelingewebe, *das; (wall-hanging)* Bildteppich, *der;* Tapisserie, *die*

'tapeworm *n.* Bandwurm, *der*

tapioca [tæpɪ'əʊkə] *n.* Tapioka, *die*

'tap-water *n.* Leitungswasser, *das*

tar [tɑ:(r)] **1.** *n.* Teer, *der;* **high-~/low-~ cigarette** Zigarette mit hohem/niedrigem Teergehalt. **2.** *v.t.,* **-rr-** teeren; **they are ~red with the same brush** *or* **stick** *(fig.)* der eine ist nicht besser als der andere

tardy ['tɑ:dɪ] *adj.* **a)** *(slow)* [zögernd] langsam; **b)** *(late)* spät; *(too late)* zu spät

target ['tɑ:gɪt] **1.** *n.* **a)** *(lit. or fig.)* Ziel, *das;* **hit/miss the/one's/its ~:** [das Ziel] treffen/das Ziel verfehlen; **set oneself a ~** *(fig.)* sich *(Dat.)* ein Ziel setzen *od.* stecken; **set oneself a ~ of £5,000** sich *(Dat.)* 5 000 Pfund zum Ziel setzen; **set sb. a ~ of six months** jmdm. eine Frist von sechs Monaten setzen; **reach one's ~** *(fig.)* sein Ziel erreichen; **be on/off** *or* **not on ~** ‹*Geschoß, Schuß:*› treffen/danebengehen; **be on ~** *(fig.)* ‹*Sparer, Sammler:*› auf dem Wege dahin sein[, sein Ziel zu erreichen]; **be on ~ for sth.** *(lit. or fig.)* auf etw. *(Akk.)* zusteuern; **be above/below ~** *(fig.)* das Ziel über-/unterschritten haben; **b)** *(Sport)* Zielscheibe, *die.* **2.** *v. t.* **a)** *(Mil.)* angreifen; **b)** *(fig.)* zielen auf ‹*Käufergruppe*›; **be ~ed on sth.** auf etw. *(Akk.)* gerichtet sein; **be ~ed on** *or* **at sth.** *(fig.)* auf etw. *(Akk.)* abzielen

'target practice *n., no art.* Schießübungen

tariff ['tærɪf] *n.* **a)** *(tax)* Zoll, *der; (table or scale of customs duties)* Zolltarif, *der;* **[import] ~:** Einfuhr- *od.* Importzoll, *der;* **b)** *(list of charges)* Tarif, *der*

Tarmac, tarmac ['tɑ:mæk] **1.** *n.* **(P) a)** Makadam, *der (Bauw.);* **b)** *(at airport)* Rollbahn, *die.* **2.** *v. t.,* **-ck-** makadamisieren *(Bauw.)*

tarnish ['tɑ:nɪʃ] **1.** *v. t.* stumpf werden lassen ‹*Metall*›; *(fig.)* beflecken ‹*Ruf, Namen*›. **2.** *v. i.* ‹*Metall:*› stumpf werden, anlaufen. **3.** *n. (discolouring film)* Beschlag, *der;* Überzug, *der*

tarnished ['tɑ:nɪʃt] *adj.* stumpf ‹*Metall*›; *(fig.)* befleckt ‹*Ruf, Name, Image*›

tarpaulin [tɑ:'pɔ:lɪn] *n.* Persenning, *die*

tarry ['tɑ:rɪ] *adj.* teerig; teerverschmiert ‹*Hand, Kleidung*›

[1]**tart** [tɑ:t] *adj.* herb; sauer ‹*Obst usw.*›; *(fig.)* scharfzüngig

[2]**tart** *n.* **a)** *(Brit.) (filled pie)* ≈ Obstkuchen, *der; (small pastry)* Obsttörtchen, *das;* **jam ~:** Marmeladentörtchen, *das;* **b)** *(sl.: prostitute)* Nutte, *die (salopp)*

~ 'up *v. t. (Brit. coll.)* **~ oneself up, get ~ed up** sich auftakeln *(ugs.);* **~ a pub/restaurant up** *(fig.)* eine Kneipe/ein Lokal aufmotzen *(ugs.)*

tartan ['tɑ:tən] **1.** *n.* Schotten[stoff], *der; (pattern)* **the Stewart ~:** der Stewart *(Textilw.); das Schottenmuster des Stewart-Clans.* **2.** *adj.* **a)** Schotten‹*rock, -jacke*›; **~ plaid/rug** Tartan, *der;* **b) T~ track (P)** Tartanbahn, *die*

tartare ['tɑ:tɑ:(r)] *adj.* **~ sauce, sauce ~** *see* **tartar sauce**

tartar sauce [tɑ:tə'sɔ:s] *n.* *(Gastr.)* Remoulade[nsoße], *die*

task [tɑ:sk] *n.* Aufgabe, *die;* **set sb. the ~ of doing sth.** jmdm. auftragen, etw. zu tun; **set oneself the ~ of doing sth.** es sich *(Dat.)* zur Aufgabe machen, etw. zu tun; **carry out/perform a ~:** eine Aufgabe erfüllen; **take sb. to ~:** jmdm. eine Lektion erteilen

task: ~ force, ~ group *ns. (sent out)* Sonderkommando, *das; (set up)* Sonderkommission, *die;* **~master** *n.* **a hard ~master** ein strenger Vorgesetzter; *(teacher)* ein strenger Lehrmeister

Tasmania [tæz'meɪnɪə] *pr. n.* Tasmanien *(das)*

tassel ['tæsl] *n.* Quaste, *die*

taste [teɪst] **1.** *v. t.* **a)** schmecken; *(try a little)* probieren; kosten; **b)** *(recognize flavour of)* [heraus]schmecken; **c)** *(fig.: experience)* kosten *(geh.)* ‹*Macht, Freiheit, [Miß]erfolg, Glück, Niederlage*›. **2.** *v. i.* **a)** *(have sense of flavour)* schmecken; **b)** *(have certain flavour)* schmecken **(of** nach); **not ~ of anything** nach nichts schmecken. **3.** *n.* **a)** *(flavour)* Geschmack, *der;* **to ~:** nach Geschmack ‹*verdünnen*›; **this dish has no ~:** dieses Gericht schmeckt nach nichts; **there's a ~ of garlic in sth.** etw. schmeckt nach Knoblauch; **leave a nasty/bad** *etc.* **~ in the mouth** *(lit. or fig.)* einen unangenehmen/üblen *usw.* Nachgeschmack hinterlassen; **b)** *(sense)* **[sense of] ~:** Geschmack[ssinn], *der;* **c)** *(discernment)* Geschmack, *der;* **~ in art/music** Kunst-/Musikgeschmack, *der;* **have good ~ in clothes** sich geschmackvoll kleiden; **it would be bad ~ to do that** es wäre geschmacklos, das zu tun; **in good/bad ~:** geschmackvoll/geschmacklos; **d)** *(sample, lit. or fig.)* Kostprobe, *die;* **have a ~ of** probieren ‹*Speise, Getränk*›; kennenlernen ‹*Freiheit, jmds. Jähzorn, Arroganz*›; **give sb. a ~ of sth.** *(lit. or fig.)* jmdm. eine Kostprobe einer Sache *(Gen.)* geben; **e)** *(liking)* Geschmack, *der* **(in** für); **have a/no ~ for sth.** an etw. *(Dat.)* Geschmack/keinen Geschmack finden; **have expensive ~s in clothes** *etc.* eine Vorliebe für teure Kleidung *usw.* haben; **be/not be to sb.'s ~:** nach jmds./nicht nach jmds. Geschmack sein

'taste-bud *n.* Geschmacksknospe, *die*

tasteful ['teɪstfl] *adj.,* **tastefully** ['teɪstfəlɪ] *adv.* geschmackvoll

tasteless ['teɪstlɪs] *adj.* geschmacklos

tasty ['teɪstɪ] *adj.* lecker

tat *see* [2]**tit**

ta-ta [tæ'tɑ:] *int. (coll.)* tschüs *(ugs.)*

tattered ['tætəd] *adj.* zerlumpt ‹*Kleidung, Person*›; zerrissen ‹*Segel*›; zerfleddert ‹*Buch, Zeitschrift*›; *(fig.)* ramponiert *(ugs.)* ‹*Ruf*›

tatters ['tætəz] *n.pl.* Fetzen; **be in ~:** in Fetzen sein; *(fig.)* ‹*Karriere, Leben:*› ruiniert sein

tattoo [tə'tu:] **1.** *v. t.* tätowieren; **~ sth. on sb.'s arm** jmdm. etw. auf den Arm tätowieren. **2.** *n.* Tätowierung, *die*

tattooed [tə'tu:d] *adj.* tätowiert

tatty ['tætɪ] *adj. (coll.)* schäbig *(abwertend);* zerfleddert ‹*Zeitschrift, Buch*›; *(inferior)* mies *(ugs.)* ‹*Publikation, Firma*›; *(threadbare)* billig ‹*Ausrede*›

taught *see* **teach**

taunt [tɔ:nt] **1.** *v. t.* verspotten **(about** wegen); **~ sb. with being a weakling** jmdn. als Schwächling verspotten. **2.** *n.* spöttische Bemerkung

taunting ['tɔ:ntɪŋ] *n.* Spott, *der*

Taurus ['tɔ:rəs] *n. (Astrol., Astron.)* der Stier

taut [tɔ:t] *adj.* **a)** *(tight)* straff ‹*Seil, Kabel, Saite*›; gespannt ‹*Muskel*›; **b)** *(fig.: tense)* angespannt ‹*Nerven, Ausdruck*›; **c)** *(fig.: concise)* knapp ‹*Stil*›

tautology [tɔ:'tɒlədʒɪ] *n.* Tautologie, *die*

tavern ['tævən] *n. (literary)* Schenke, *die*

tawdry ['tɔ:drɪ] *adj.* billig und geschmacklos; *(fig.)* zweifelhaft

tawny ['tɔ:nɪ] *adj.* gelbbraun

tax [tæks] **1.** *n.* **a)** Steuer, *die;* **pay 20% in ~ [on sth.]** 20% Steuern [für etw.] zahlen; **before/after ~:** vor Steuern/nach Abzug der Steuern; **free of ~:** steuerfrei; *(after ~, ~ paid)* nach Abzug der Steuern; netto; **b)** *(fig.: burden)* Belastung, *die* **(on** für). **2.** *v. t.* **a)** *(impose ~ on)* besteuern; *(pay ~ on)* versteuern ‹*Einkommen*›; **b)** *(make demands on)* strapazieren ‹*Mittel, Kräfte, Geduld usw.*›; **c)** *(accuse)* beschuldigen, bezichtigen **(with** *Gen.*)

taxable ['tæksəbl] *adj.* steuerpflichtig

'tax allowance *n.* Steuerfreibetrag, *der*

taxation [tæk'seɪʃn] *n. (imposition of taxes)* Besteuerung, *die; (taxes payable)* Steuern

tax: ~ avoidance *n.* Steuerminderung, *die;* **~ bill** *n.* Steuerbescheid, *der; (amount)* Steuerschuld, *die;* **~ bracket** *n.* Stufe im Steuertarif; **~-collector** *n.* Finanzbeamte, *der*/-beamtin, *die;* **~-deductible** *adj.* steuerabzugsfähig; [steuerlich] absetzbar; ~ **demand** *n.* Steuerforderung, *die;* ~ **disc** *n.* Steuerplakette, *die;* ~ **evasion** *n.* Steuerhinterziehung, *die;* ~ **exile** *n.* **a)** *(person)* Steuerflüchtling, *der;* **b)** *(place)* Steueroase, *die (ugs.);* ~-**free 1.** *adj.* steuerfrei; *(after payment of ~)* Netto-; ~-**free allowance** Steuerfreibetrag, *der;* **2.** *adv.* steuerfrei; *(after payment of ~)* netto; ~ **haven** *n.* Steueroase, *die (ugs.)*

taxi ['tæksɪ] **1.** *n.* Taxi, *das;* **go by ~:** mit dem Taxi fahren. **2.** *v.i.*, **~ing** *or* **taxying** ['tæksɪɪŋ] *(Aeronaut.)* ‹*Flugzeug:*› rollen

'**taxi-cab** *see* **taxi 1**

taxidermist ['tæksɪdɜ:mɪst] *n.* Präparator, *der*/Präparatorin, *die*

'**taxi-driver** *n.* Taxifahrer, *der*/-fahrerin, *die*

taxing ['tæksɪŋ] *adj.* strapaziös, anstrengend ‹*Arbeit, Rolle, Reise*›; schwierig ‹*Problem*›

'**tax inspector** *n.* Steuerinspektor, *der*/-inspektorin, *die*

taxi: ~-**rank** *(Brit.), (Amer.)* ~ **stand** *ns.* Taxistand, *der*

tax: ~**man** *n. (coll.)* Finanzbeamte, *der*/-beamtin, *die;* **a letter from the ~man** ein Brief vom Finanzamt; ~ **office** *n.* Finanzamt, *das;* ~**payer** *n.* Steuerzahler, *der*/-zahlerin, *die;* ~**paying** *attrib. adj.* Steuern zahlend...; ~ **return** *n.* Steuererklärung, *die*

TB *abbr.* **tuberculosis** Tb, *die*

tea [ti:] *n.* **a)** Tee, *der;* **[not] be sb.'s cup of ~** *(fig. coll.)* [nicht] jmds. Fall sein *(ugs.);* **b)** *(meal)* **[high] ~:** Abendessen, *das;* **afternoon ~:** [Nachmittags]tee, *der*

tea: ~-**bag** *n.* Teebeutel, *der;* ~-**break** *n. (Brit.)* Teepause, *die;* ~-**caddy** *n.* Teebüchse, *die;* ~**cake** *n.* **a)** *(Brit.: sweet bread bun)* ≈ Rosinenbrötchen, *das;* **b)** *(Amer.: sweet cake)* Keks, *der;* ~**cakes** Teegebäck, *das*

teach [ti:tʃ] **1.** *v.t.*, **taught** [tɔ:t] unterrichten; *(at university)* lehren; ~ **music** *etc.* **to sb.,** ~ **sb. music** *etc.* jmdn. in Musik *usw.* unterrichten; ~ **oneself** es sich *(Dat.)* selbst beibringen; ~ **sb./oneself/an animal sth.** jmdm./sich/einem Tier etw. beibringen; ~ **sb./oneself/an animal to do sth.** jmdm./sich/einem Tier beibringen, etw. zu tun; ~ **sb. to ride/to play the piano** jmdm. das Reiten/Klavierspielen beibringen; **I'll/that'll ~ you** *etc.* **to do that!** *(coll. iron.)* ich werde/das wird dich *usw.* lehren, das zu tun! *(iron.);* **that'll ~ him/you** *etc.*! *(coll. iron.)* das hat er/hast du *usw.* nun davon! *(iron.);* ~ **sb. how/that ...:** jmdm. beibringen, wie/daß ...; ‹*Bibel, Erfahrung:*› jmdn. lehren, wie/daß ... **2.** *v.i.*, **taught** unterrichten

teacher ['ti:tʃə(r)] *n.* Lehrer, *der*/Lehrerin, *die;* **she's a university/evening-class ~:** sie lehrt an der Universität/unterrichtet an der Abendschule; **kindergarten ~:** ≈ Vorschullehrer, *der*/-lehrerin, *die;* **geography/music ~:** Geographie-/Musiklehrer, *der*/Geographie-/Musiklehrerin, *die*

'**teacher training college** *n.* ≈ pädagogische Hochschule, *die*

'**tea-chest** *n.* Teekiste, *die*

teaching ['ti:tʃɪŋ] *n.* **a)** *(act)* Unterrichten, *das* **(of** von); **the ~ of languages, language ~:** der Sprachunterricht; **b)** *no pl., no art. (profession)* Lehrberuf, *der*

teaching: ~ **aid** *n.* Lehr- *od.* Unterrichtsmittel, *das;* ~ **hospital** *n.* Ausbildungskrankenhaus, *das*

tea: ~-**cloth** *n. (for drying)* Geschirrtuch, *das;* ~-**cosy** *n.* Teewärmer, *der;* ~**cup** *n.* Teetasse, *die;* ~**cupful** *n.* Tasse, *die;* **a ~cupful of flour** eine Tasse Mehl

teak [ti:k] *n.* Teak[holz], *das; attrib.* Teak[holz]‹*öl, -furnier, -möbel*›

tea: ~-**kettle** *n.* Teekessel, *der;* ~-**lady** *n. Frau, die in einer Firma, Behörde o. ä. den Pausentee usw. zubereitet;* ~-**leaf** *n.* Teeblatt, *das*

team [ti:m] *n.* **a)** *(group)* Team, *das; (Sport also)* Mannschaft, *die;* **a football/cricket ~:** eine Fußball-/Kricketmannschaft; **a ~ of scientists** eine Gruppe *od.* ein Team von Wissenschaftlern; **make a good ~:** ein gutes Team *od.* Gespann sein; **work as a ~:** im Team zusammenarbeiten; **b)** *(draught animals)* Gespann, *das*

~ '**up 1.** *v.t.* zusammenbringen. **2.** *v.i.* sich zusammentun *(ugs.)*

'**tea-maker** *n. (device)* Teemaschine, *die*

team: ~ **effort** *n.* Team- *od.* Gemeinschaftsarbeit, *die;* **a great ~ effort** eine großartige Gemeinschaftsleistung; ~ **game** *n.* Mannschaftsspiel, *das;* ~-**leader** *n.* Gruppenleiter, *der*/-leiterin, *die;* ~-**mate** *n.* Mannschaftskamerad, *der*/-kameradin, *die;* ~-**member** *n.* Mitglied des Teams/der Mannschaft/der Gruppe; ~ '**spirit** *n.* Teamgeist, *der; (Sport also)* Mannschaftsgeist, *der*

teamster ['ti:mstə(r)] *n. (Amer.)* Lkw-Fahrer, *der*/-Fahrerin, *die*

'**team-work** *n.* Teamarbeit, *die*

tea: ~-**party** *n.* Teegesellschaft, *die;* ~**pot** *n.* Teekanne, *die*

[1]**tear** [teə(r)] **1.** *n.* Riß, *der; see also* **wear 1 a. 2.** *v.t.*, **tore** [tɔ:(r)], **torn** [tɔ:n] **a)** *(rip, lit. or fig.)* zerreißen; *(pull apart)* auseinanderreißen; *(damage)* aufreißen; ~ **open** aufreißen ‹*Brief, Schachtel, Paket*›; ~ **one's dress [on a nail]** sich *(Dat.)* das Kleid [an einem Nagel] aufreißen; ~ **a hole/gash in sth.** ein Loch/eine klaffende Wunde in etw. *(Akk.)* reißen; ~ **sth. in half** *or* **in two** etw. entzweireißen; ~ **to shreds** *or* **pieces** *(lit.)* zerfetzen; in Stücke reißen ‹*Flagge, Kleidung, Person*›; ~ **to shreds** *(fig.) (destroy)* ruinieren ‹*Ruf, Leumund*›; zerrütten ‹*Nerven*›; zunichte machen ‹*Argument, Alibi*›; auseinandernehmen *(salopp)* ‹*Mannschaft*›; *(criticize)* verreißen *(ugs.);* **a country torn by war** ein durch Krieg zerrissenes Land; **be torn between two things/people/between x and y** zwischen zwei Dingen/Personen/x und y hin- und hergerissen sein; **that's torn it** *(Brit. fig. coll.)* das hat alles vermasselt *(salopp);* **b)** *(remove with force)* reißen; ~ **sth. out of** *or* **from sb.'s hands** jmdm. etw. aus der Hand reißen; ~ **one's hair** *(fig.)* sich *(Dat.)* die Haare raufen *(ugs.).* **3.** *v.i.*, **tore, torn a)** *(rip)* [zer]reißen; **it ~s along the perforation** es läßt sich entlang der Perforation abreißen; ~ **in half** *or* **in two** entzweireißen; durchreißen; **b)** *(move hurriedly)* rasen *(ugs.);* ~ **past** vorbeirasen *(ugs.);* ~ **along the street** die Straße hinunterrasen *(ugs.);* ~ **off** losrasen *(ugs.)*

~ **apart** *v.t. (lit. or fig.)* auseinanderreißen; *(coll.: criticize)* zerreißen *(ugs.)*

~ **at** *v.t.* zerren an (+ *Dat.*)

~ **a'way** *v.t.* wegreißen; abreißen ‹*Tapete, Verpackung*›; ~ **sb./oneself away [from sb./sth.]** *(fig.)* jmdn./sich [von jmdm./etw.] loseisen *(ugs.);* ~ **oneself away [from a sight/book]** *(fig.)* sich [von einem Anblick/Buch] losreißen

~ '**down** *v.t.* herunterreißen; niederreißen ‹*Zaun, Mauer*›; abreißen ‹*Gebäude*›

~ **into** *v.t.* ‹*Geschoß:*› ein Loch reißen in (+ *Akk.*); ‹*Säge:*› sich [hinein]fressen in (+ *Akk.*); ‹*Raubtier:*› zerfleischen; *(fig.: tell off, criticize)* heftig angreifen

~ **'off** *v.t.* abreißen; *see also* ~ **3b**
~ **'out** *v.t.* herausreißen; ausreißen ‹*Baum*›; *see also* ~ **2b**
~ **'up** *v.t.* **a)** *(remove)* aufreißen ‹*Straße, Bürgersteig*›; herausreißen ‹*Zaun, Pflanze*›; ausreißen ‹*Baum*›; **b)** *(destroy)* zerreißen; *(fig.)* für null und nichtig erklären ‹*Vertrag, Abkommen*›

²tear [tɪə(r)] *n.* Träne, *die;* **there were ~s in her eyes** sie hatte *od.* ihr standen Tränen in den Augen; **with ~s in one's eyes** mit Tränen in den Augen; **burst into ~s** in Tränen ausbrechen; **move sb. to ~s** jmdn. zu Tränen rühren; **bore sb. to ~s** jmdn. zu Tode langweilen; **be in ~s** in Tränen aufgelöst sein

tearaway ['teərəweɪ] *n.* Rabauke, *der (ugs.)*

tear-drop ['tɪədrɒp] *n.* Träne, *die*

tearful ['tɪəfl] *adj.* *(crying)* weinend; tränenreich ‹*Versöhnung, Abschied, Anlaß*›; **she was looking very ~:** sie sah sehr verweint aus; *(about to cry)* sie schien den Tränen nahe

tear-gas ['tɪəgæs] *n.* Tränengas, *das*

tearing ['teərɪŋ] *adj.* **a)** reißend ‹*Geräusch*›; **b)** *(coll.: violent)* rasend; **be in a ~ hurry** schrecklich in Eile sein

tear-off ['teərɒf] *attrib. adj.* ~ **calendar** Abreißkalender, *der*

tea: **~-room** *n.* Teestube, *die;* ≈ Café, *das;* **~-rose** *n.* Teerose, *die*

tease [ti:z] **1.** *v.t.* necken; ~ **sb. [about sth.]** jmdn. [mit etw.] aufziehen *(ugs.);* jmdn. [wegen etw.] verspotten; **he's only teasing you** er macht nur Spaß *(ugs.);* **stop teasing the dog** hör auf, den Hund zu ärgern. **2.** *v.i.* seine Späße machen; **I'm only teasing** ich mache nur Spaß

teasel ['ti:zl] *n.* *(Bot.)* Karde, *die*

teaser ['ti:zə(r)] *n.* *(coll.: puzzle)* **brain-~:** Denk[sport]aufgabe, *die;* **be a [real] ~** *(fig.)* eine harte Nuß sein *(ugs.)*

tea: **~-service, ~-set** *ns.* Tee-Service, *das;* **~-shop** *(Brit.) see* **tea-room**

teasing ['ti:zɪŋ] *adj.* neckend

tea: **~spoon** *n.* Teelöffel, *der;* **~spoonful** *n.* Teelöffel, *der;* **a ~spoonful** ein Teelöffel [voll]; **~-strainer** *n.* Teesieb, *das*

teat [ti:t] *n.* **a)** *(nipple)* Zitze, *die;* **b)** *(of rubber or plastic)* Sauger, *der*

tea: **~-table** *n.* Teetisch, *der;* **~-things** *n. pl. (coll.)* Teegeschirr, *das;* **~-time** *n.* Teezeit, *die;* **~-towel** *n.* Geschirrtuch, *das;* **~-trolley,** *(Amer.)* **~-wagon** *ns.* Teewagen, *der*

teazel, teazle ['ti:zl] *see* **teasel**

technical ['teknɪkl] *adj.* **a)** technisch ‹*Problem, Detail, Daten, Fortschritt*›; *(of particular science, art, etc.)* fachlich; Fach‹*kenntnis, -sprache, -begriff, -wörterbuch*›; *(of the execution of a work of art)* technisch ‹*Fertigkeit, Schwierigkeit*›; ~ **expertise/expert** Sachkenntnis, *die*/Fachmann, *der;* ~ **college/school** Fachhochschule, *die*/Fachschule, *die;* **explain sth. without being** *or* **getting too ~:** etw. erklären, ohne sich zu fachsprachlich auszudrücken; ~ **hitch** technisches Problem; ~ **term** Fachbegriff, *der;* Fachausdruck, *der;* Fachterminus, *der;* **for ~ reasons** aus technischen Gründen; **b)** *(Law)* formaljuristisch; **c)** ~ **knock-out** *(Boxing)* technischer K.o.

technical 'drawing *n., no pl., no art. (Brit.)* technisches Zeichnen

technicality [teknɪ'kælɪtɪ] *n.* *(technical expression)* Fachausdruck, *der;* *(technical distinction)* technisches Detail; *(technical point)* technische Frage; **be acquitted on a ~** *(Law)* auf Grund eines Formfehlers freigesprochen werden

technician [tek'nɪʃn] *n.* Techniker, *der*/Technikerin, *die*

technique [tek'ni:k] *n.* Technik, *die;* *(procedure)* Methode, *die*

technological [teknə'lɒdʒɪkl] *adj. see* **technology:** technisch; technologisch

technologist [tek'nɒlədʒɪst] *n.* Technologe, *der*/Technologin, *die;* ‹*Lebensmittel-, Erdöl-*›techniker, *der*/-technikerin, *die*

technology [tek'nɒlədʒɪ] *n.* Technik, *die;* *(application of science)* Technologie, *die;* **science and ~:** Wissenschaft und Technik; **college of ~:** Fachhochschule für Technik

teddy ['tedɪ] *n.* ~ **[bear]** Teddy[bär], *der*

tedious ['ti:dɪəs] *adj.* langwierig ‹*Reise, Arbeit*›; *(uninteresting)* langweilig

tee [ti:] *n.* *(Golf)* Tee, *das*
~ **'off** *v.i.* *(Golf)* abschlagen

teem [ti:m] *v.i.* wimmeln **(with** von)

teenage ['ti:neɪdʒ], **teenaged** ['ti:neɪdʒd] *attrib. adj.* im Teenageralter *nachgestellt*

teenager ['ti:neɪdʒə(r)] *n.* Teenager, *der;* *(loosely)* Jugendliche, *der/die*

teens [ti:nz] *n. pl.* Teenagerjahre; **be out of/in one's ~:** aus den Teenagerjahren heraussein/in den Teenagerjahren sein

'tee-shirt *n.* T-shirt, *das*

teeter ['ti:tə(r)] *v.i.* wanken; ~ **on the edge** *or* **brink of sth.** schwankend am Rande einer Sache *(Gen.)* stehen; *(fig.)* am Rande einer Sache stehen

teeth *pl. of* **tooth**

teething troubles ['ti:ðɪŋ trʌblz] *n.pl.* Beschwerden während des Zahnens; **have ~** *(fig.)* ‹*Person, Vorhaben:*› Anfangsschwierigkeiten haben; ‹*Maschine usw.:*› Kinderkrankheiten haben

teetotal [ti:'təʊtl] *adj.* abstinent lebend; alkoholfrei ‹*Restaurant, Hotel, Feier*›; **sb. is ~:** jmd. ist Abstinenzler/Abstinenzlerin

teetotaller [ti:'təʊtələ(r)] *n.* Abstinenzler, *der*/Abstinenzlerin, *die*

Tel., tel. *abbr.* **telephone** Tel.

telecommunication [telɪkəmju:nɪ'keɪʃn] *n.* **a)** *(long-distance communication)* Fernmeldeverkehr, *der; attrib.* Fernmelde-; **b)** *in pl. (science)* Fernmelde- *od.* Nachrichtentechnik, *die; attrib.* Fernmelde- *od.* Nachrichten‹*techniker, -satellit*›

telegram ['telɪgræm] *n.* Telegramm, *das;* **by ~:** telegrafisch

telegraph ['telɪgrɑ:f] *n.* Telegraf, *der; attrib.* Telegrafen-; ~ **pole** Telegrafenmast, *der*

telepathic [telɪ'pæθɪk] *adj.* telepathisch; **be ~:** telepathische Fähigkeiten haben

telepathy [tɪ'lepəθɪ] *n.* Telepathie, *die*

telephone ['telɪfəʊn] **1.** *n.* Telefon, *das; attrib.* Telefon-; **[public] ~:** öffentlicher Fernsprecher *(Amtsspr.);* [öffentliches] Telefon; **answer the ~:** Anrufe entgegennehmen; *(on one occasion)* ans Telefon gehen; *(speak)* sich melden; **by ~:** telefonisch; **over** *or* **on the ~:** am Telefon; **speak** *or* **talk to sb. on the** *or* **by ~:** mit jmdm. telefonieren; **be on the ~** *(be connected to the system)* Telefon haben; *(be speaking)* telefonieren (**to** mit); **it's your sister on the ~:** deine Schwester ist am Apparat; **get on the ~ to sb.** jmdn. anrufen; **get sb. on the ~:** jmdn. telefonisch erreichen; **be wanted on the ~:** am Telefon verlangt werden; *attrib.* ~ **answering machine** Anrufbeantworter, *der.* **2.** *v.t.* anrufen; telefonisch übermitteln ‹*Nachricht, Ergebnis usw.*› (**to** *Dat.*); ~ **the office/~**

home im Büro/zu Hause anrufen. **3.** *v.i.* anrufen; ~ **for a taxi/the doctor** nach einem Taxi/dem Arzt telefonieren; **can we ~ from here?** können wir von hier aus telefonieren?

telephone: ~ **book** *n.* Telefonbuch, *das;* ~ **booth,** *(Brit.)* ~-**box** *ns.* Telefonzelle, *die;* ~ **call** *n.* Telefonanruf, *der;* Telefongespräch, *das;* **make a ~ call** ein Telefongespräch führen; **have** *or* **receive a ~ call** einen Anruf erhalten; **there was a ~ call for you** es hat jemand für Sie angerufen; **international ~ call** Auslandsgespräch, *das;* ~ **directory** *n.* Telefonverzeichnis, *das;* Telefonbuch, *das;* ~ **exchange** *n.* Fernmeldeamt, *das;* ~ **kiosk** *n.* Telefonzelle, *die;* ~ **line** *n.* Telefonleitung, *die;* ~ **message** *n.* telefonische Nachricht; ~ **number** *n.* Telefonnummer, *die;* ~ **operator** *n.* Telegrafist, *der*/Telegrafistin, *die;* ~ **receiver** *n.* Telefonhörer, *der*

telephoto [telɪ'fəʊtəʊ] *adj.* *(Photog.)* telefotografisch; ~ **lens** Teleobjektiv, *das*

teleprinter ['telɪprɪntə(r)] *n.* Fernschreiber, *der*

telescope ['telɪskəʊp] **1.** *n.* Teleskop, *das;* Fernrohr, *das.* **2.** *v.t.* zusammenschieben ⟨*Antenne, Rohr*⟩; ineinanderschieben ⟨*Abschnitte, Waggons*⟩; *(fig.)* komprimieren (**into** zu)

telescopic [telɪ'skɒpɪk] *adj.* *(collapsible)* ausziehbar; Teleskop⟨*antenne, -mast*⟩; ~ **umbrella** Taschenschirm, *der*

teletext ['telɪtekst] *n.* Teletext, *der*

televise ['telɪvaɪz] *v.t.* im Fernsehen senden *od.* übertragen

television ['telɪvɪʒn, telɪ'vɪʒn] *n.* **a)** *no pl., no art.* das Fernsehen; **colour/black and white ~:** das Farb- / Schwarzweißfernsehen; **we have ten hours of ~ a day** bei uns gibt es täglich 10 Stunden Fernsehprogramm; **live ~:** Live-Sendungen [im Fernsehen]; **on ~:** im Fernsehen; **what's on ~?** was läuft *od.* gibt's im Fernsehen?; **watch ~:** fernsehen; **b)** *(~ set)* Fernsehapparat, *der;* Fernseher, *der (ugs.);* **portable ~:** tragbares Fernsehgerät

television: ~ **advertising** *n.* Fernsehwerbung, *die;* ~ **aerial** *n.* Fernsehantenne, *die;* ~ **camera** *n.* Fernsehkamera, *die;* ~ **channel** *n.* [Fernseh]kanal, *der;* ~ **licence** *n.* *(Brit.)* Fernsehgenehmigung, *die (die jährlich gegen Zahlung der Gebühren erneuert wird); attrib.* ~ **licence fee** Fernsehgebühren *Pl.;* ~ **personality** *n.* Fernsehgröße, *die (ugs.);* ~ **programme** *n.* Fernsehsendung, *die; (sequence)* Fernsehprogramm, *das;* **my favourite ~ programme** meine Lieblingssendung im Fernsehen; ~ **screen** *n.* Bildschirm, *der;* ~ **serial** *n.* Fernsehserie, *die;* ~ **set** *n.* Fernsehgerät, *das;* ~ **studio** *n.* Fernsehstudio, *das;* ~ **viewer** *n.* Fernsehzuschauer, *der*/-zuschauerin, *die*

Telex, telex ['teleks] **1.** *n.* Telex, *das;* **by ~:** über Telex. **2.** *v.t.* ein Telex schicken (+ *Dat.*) ⟨*Person, Firma*⟩; telexen ⟨*Nachricht*⟩

tell [tel] **1.** *v.t.,* **told** [təʊld] **a)** *(make known)* sagen ⟨*Name, Adresse, Alter*⟩; *(give account of)* erzählen ⟨*Neuigkeit, Sorgen*⟩; anvertrauen ⟨*Geheimnis*⟩; ~ **sb. sth.** *or* **sth. to sb.** jmdm. etw. sagen/erzählen/anvertrauen; ~ **sb. the way to the station** jmdm. den Weg zum Bahnhof beschreiben; ~ **sb. the time** jmdm. sagen, wie spät es ist; jmdm. die Uhrzeit sagen; ~ **all** auspacken *(ugs.);* ~ **sb. [something] about sb./sth.** jmdm. [etwas] von jmdm./etw. erzählen; ~ **sb. nothing/all about what happened** jmdm. nichts davon/alles erzählen, was passiert ist; **will you ~ him [that] I will come?** sag ihm bitte, daß ich kommen werde; **they ~ me/us [that]...** *(according to them)* man sagt, daß ...; **I'll ~ you what I'll do** weißt du, was ich machen werde?; ~ **everyone/** *(coll.)* **the world [that/how** *etc.*] jedem/*(ugs.)* aller Welt erzählen[, daß/wie *usw.*]; **I cannot ~ you how ...** *(cannot express how ...)* ich kann dir gar nicht sagen, wie ...; **I couldn't ~ you** *(I don't know)* das kann ich nicht sagen; **I can ~ you,...** *(I can assure you)* ich kann dir sagen, ...; **... , I can ~ you ...** , das kann ich dir sagen; **you can't ~ me [that] ...** *(it can't be true that ...)* du kannst mir doch nicht erzählen, daß ...; **you can't ~ him anything** *(he won't accept advice)* er läßt sich *(Dat.)* ja nichts sagen; *(he is well-informed)* ihm kannst du nichts erzählen; **let me ~ you** *(let me assure you)* ... , das kann ich dir sagen; **let me ~ you that ...:** ich kann dir versichern, daß ...; **... , I ~ you** *or* **I'm ~ing you** ..., das sage ich dir; **you're ~ing 'me!** *(coll.)* wem sagst du das! *(ugs.);* **I don't need to ~ you [that] ...:** ich brauche dir wohl nicht extra zu sagen, daß ...; **be told sth. by sb.** etw. von jmdm. erfahren; **I was told that ...:** mir wurde gesagt, daß ...; **so I've been told** *(I know that)* [das] habe ich schon gehört; **... or so I've been/I'm told** ... , wie ich gehört habe/höre; **no, don't ~ me, let me guess** [nein,] sag's nicht, laß mich raten; **don't ~ me [that] ...** *(expressing incredulity, dismay, etc.)* jetzt sag bloß nicht, [daß] ...; **you aren't trying** *or* **don't mean to ~ me [that] ...?** du wirst doch nicht sagen wollen, daß ...?; **b)** *(relate, lit. or fig.)* erzählen; **sth. ~s its own story** *or* **tale** *(needs no comment)* etw. spricht für sich selbst; ~ **a different story** *or* **tale** *(reveal the truth)* eine andere Sprache sprechen *(fig.);* **live** *or* **survive to ~ the tale** überleben; ~ **tales [about sb.]** *(gossip; reveal secret)* [über jmdn.] tratschen *(ugs. abwertend);* ~ **tales [to sb.]** *(report)* andere/einen anderen [bei jmdm.] anschwärzen; [bei jmdm.] petzen *(Schülerspr. abwertend);* ~ **tales** *(lie)* Lügengeschichten erzählen; **c)** *(instruct)* sagen; ~ **sb. [not] to do sth.** jmdm. sagen, daß er etw. [nicht] tun soll; jmdm. sagen, er soll[e] etw. [nicht] tun; ~ **sb. what to do** jmdm. sagen, was er tun soll; **do as** *or* **what I ~ you** tu, was ich dir sage; **do as you are told** tu, was man dir sagt; **d)** *(determine)* feststellen; *(see, recognize)* erkennen (**by** an + *Dat.*); *(with reference to the future)* [vorher]sagen; ~ **the difference [between ...]** den Unterschied [zwischen ...] erkennen *od.* feststellen; **it's impossible/difficult to ~ [if/what** *etc.*] es ist unmöglich/schwer zu sagen[, ob/was *usw.*]; **it's easy to ~ whether ...:** es läßt sich leicht sagen, ob ...; **you never can ~ how/what** *etc.* man weiß nie, wie/was *usw.;* **e)** *(distinguish)* unterscheiden; **f)** *(utter)* sagen; **g) all told** insgesamt. **2.** *v.i.,* **told a)** *(determine)* **how can you ~?** wie kann man das feststellen *od.* wissen?; **it's difficult** *or* **hard to ~:** das ist schwer zu sagen; **how can one ~?, how can** *or* **do you ~?** woran kann man das erkennen?; **as far as one/I can ~, ...:** wie es aussieht, ...; **you never can ~:** man kann nie wissen; **who can ~?** wer kann das sagen *od.* will das wissen?; **b)** *(give information)* erzählen (**of, about** von); *(give evidence)* ~ **of sth.** von etw. Zeugnis geben *od.* ablegen; **c)** *(reveal secret)* es verraten; **time [alone] will**

~: das wird sich [erst noch] zeigen; **d)** *(produce an effect)* sich auswirken; ‹*Wort, Fausthieb, Schuß:*› sitzen; ~ **in favour of sb.** *or* **in sb.'s favour** sich zu jmds. Gunsten auswirken; ~ **against sb./sth.** sich nachteilig für jmdn./auf etw. *(Akk.)* auswirken

~ **a'part** *v. t.* auseinanderhalten

~ **'off** *v. t.* *(coll.: scold)* ~ **sb. off [for sth.]** jmdn. [für *od.* wegen etw.] ausschimpfen

~ **on** *v. t.* **a)** *(affect)* ~ **on sb./sth.** sich bei jmdm. bemerkbar machen/sich [nachteilig] auf etw. *(Akk.)* auswirken; **b)** *(coll.: inform against)* ~ **on sb.** jmdn. verpetzen *(Schülerspr. abwertend)*

teller ['telə(r)] *n.* **a)** *(in bank) see* **cashier; b)** *(counter of votes)* Stimmenzähler, *der*/-zählerin, *die*

telling ['telɪŋ] **1.** *adj.* *(effective, striking)* schlagend ‹*Argument, Antwort*›; wirkungsvoll ‹*Worte, Phrase, Stil*›; ~ **blow** *(Boxing)* Wirkungstreffer, *der; (fig.)* empfindlicher Schlag; **with ~ effect** mit durchschlagender Wirkung. **2.** *n.* Erzählen, *das;* **he did not need any ~, he needed no ~:** dazu brauchte man ihn nicht lange *od.* eigens aufzufordern; **that would be ~:** damit würde ich ein Geheimnis verraten; **there's no ~ what/how ...:** man weiß nie, was/wie ...

telling-'off *n.* *(coll.)* Standpauke, *die (ugs.);* **give sb. a ~:** jmdn. ausschimpfen (**for** wegen); **get a ~:** Schimpfe kriegen *(ugs.)*

'tell-tale *n.* Klatschmaul, *das (ugs. abwertend);* Petze, *die (Schülerspr. abwertend); attrib.* vielsagend ‹*Blick, Lächeln*›; verräterisch ‹*Röte, Fleck, Glanz, Zucken, Zeichen*›

telly ['telɪ] *n.* *(Brit. coll.)* Fernseher, *der (ugs.);* Glotze, *die (salopp);* **watch ~:** Fernsehen gucken *(ugs.);* **what's on [the] ~?** was kommt im Fernsehen?

temp [temp] *(Brit. coll.)* **1.** *n.* Zeitarbeitskraft, *die.* **2.** *v. i.* Zeitarbeit machen

temper ['tempə(r)] **1.** *n.* **a)** *(disposition)* Naturell, *das;* **be in a good/bad ~:** gute/schlechte Laune haben; gut/schlecht gelaunt sein; **be in a foul** *or* **filthy ~:** eine miese Laune haben *(ugs.);* **keep/lose one's ~:** sich beherrschen/die Beherrschung verlieren; **lose one's ~ with sb.** die Beherrschung bei jmdm. verlieren; **control one's ~:** sich beherrschen; **b)** *(anger)* **fit/outburst of ~:** Wutanfall, *der*/-ausbruch, *der;* **have a ~:** jähzornig sein; **be in/get into a ~:** wütend sein/werden (**over** wegen). **2.** *v. t.* mäßigen; mildern ‹*Trostlosigkeit, Strenge, Kritik*›

temperament ['temprəmənt] *n.* *(nature)* Veranlagung, *die;* Natur, *die; (disposition)* Temperament, *das;* **have an artistic ~:** künstlerisch veranlagt sein

temperamental [temprə'mentl] *adj.* launisch *(abwertend);* launenhaft; **be a bit ~** *(fig. coll.)* ‹*Auto, Maschine:*› seine Mucken haben *(ugs.)*

temperance ['tempərəns] *n.* **a)** *(moderation)* Mäßigung, *die; (in one's eating, drinking)* Mäßigkeit, *die;* **b)** *(total abstinence)* Abstinenz, *die*

temperate ['tempərət] *adj.* gemäßigt

temperature ['temprɪtʃə(r)] *n.* **a)** Temperatur, *die;* **what is the ~?** wieviel Grad sind es?; **the ~ is below/above ...:** die Temperatur liegt unter/über ... *(Dat.);* **at high/low ~s** bei hohen/niedrigen Temperaturen; **b)** *(Med.)* Temperatur, *die;* **have** *or* **run a ~** *(coll.)* Temperatur *od.* Fieber haben; **have a slight/high ~:** leichtes/hohes Fieber haben; **take sb.'s ~:** jmds. [Körper]temperatur messen

template ['templɪt] *n.* Schablone, *die*

[1]**temple** ['templ] *n.* Tempel, *der*

[2]**temple** *n.* *(Anat.)* Schläfe, *die*

tempo ['tempəʊ] *n., pl.* **~s** *or* **tempi** ['tempi:] **a)** *(fig.: pace)* **the ~ of life in the town** der Rhythmus der Stadt; **the campaign ~ stepped up** der Wahlkampf ging in die heiße Phase über; **b)** *(Mus.: speed)* Tempo, *das*

temporal ['tempərl] *adj.* *(of this life)* irdisch; *(secular)* weltlich

temporarily ['tempərərɪlɪ] *adv.* vorübergehend

temporary ['tempərərɪ] **1.** *adj.* vorübergehend; provisorisch ‹*Gebäude, Büro*›; ~ **worker** Aushilfe, *die.* **2.** *n.* Aushilfe, *die;* Aushilfskraft, *die*

temporize (temporise) ['tempəraɪz] *v. i.* **a)** *(adopt indecisive policy)* sich nicht festlegen; **b)** *(act so as to gain time)* sich abwartend verhalten

tempt [tempt] *v. t.* **a)** *(attract)* ~ **sb. out/into the town** jmdn. hinauslocken/in die Stadt locken; ~ **sb. to do sth.** in jmdm. den Wunsch wecken, etw. zu tun; **b)** *(cause to have strong urge)* ~ **sb. to do sth.** jmdn. geneigt machen, etw. zu tun; **be ~ed to do sth.** versucht sein, etw. zu tun; **c)** *(entice)* verführen; **be ~ed into doing sth.** sich dazu verleiten lassen, etw. zu tun; ~ **sb. away from sth.** jmdn. von etw. weglocken; **don't ~ me!** verleite mich nicht!; **d)** *(provoke)* herausfordern; ~ **fate** *or* **providence** das Schicksal herausfordern

temptation [temp'teɪʃn] *n.* **a)** *no pl.* *(attracting)* Verlockung, *die; (being attracted)* Versuchung, *die; (enticing)* Verführung, *die* (**into** zu); *(being enticed)* Versuchung, *die (geh.);* **feel a ~ to do sth.** versucht sein, etw. zu tun; **give in to [the] ~:** der Versuchung erliegen; **b)** *(thing)* Verlockung, *die* (**to** zu)

tempting ['temptɪŋ] *adj.* verlockend; verführerisch

ten [ten] **1.** *adj.* zehn; *see also* **eight 1. 2.** *n.* **a)** *(number, symbol)* Zehn, *die;* **b)** *(set of ~)* Zehnerpackung, *die;* **c) bet sb. ~ to one that ...** *(fig.)* jede Wette halten, daß ... *(ugs.). See also* **eight 2 a, c, d**

tenable ['tenəbl] *adj.* **a)** haltbar; *(fig.)* haltbar ‹*Theorie, Annahme*›; vertretbar ‹*Standpunkt*›; **b)** ~ **for five years** auf fünf Jahre befristet ‹*Arbeitsverhältnis, Stelle*›

tenacious [tɪ'neɪʃəs] *adj.* **a)** *(holding fast)* hartnäckig haftend ‹*Dornen, Samen*›; **b)** *(resolute)* hartnäckig; **be ~:** sich hartnäckig halten

tenacity [tɪ'næsɪtɪ] *n., no pl.* Hartnäckigkeit, *die*

tenancy ['tenənsɪ] *n.* **a)** *(of flat, residential building)* Mietverhältnis, *das; (of farm, shop)* Pachtverhältnis, *das;* **have ~ of a flat** eine Wohnung gemietet haben; **b)** *(period)* Mietdauer, *die*

tenant ['tenənt] *n.* **a)** *(of flat, residential building)* Mieter, *der*/Mieterin, *die; (of farm, shop)* Pächter, *der*/Pächterin, *die;* **b)** *(occupant)* Bewohner, *der*/Bewohnerin, *die*

[1]**tend** [tend] *v. i.* ~ **to do sth.** dazu neigen *od.* tendieren, etw. zu tun; ~ **to sth.** zu etw. neigen; **it ~s to get quite cold there at nights** es wird dort nachts oft sehr kalt; **he ~s to get upset if ...:** er regt sich leicht auf, wenn ...; **this ~s to suggest that ...:** dies deutet darauf hin, daß ...

[2]**tend** *v. t.* sich kümmern um; hüten ‹*Schafe*›; bedienen ‹*Maschine*›; **rice has to be ~ed carefully** Reis erfordert sorgfältige Pflege

tendency ['tendənsɪ] *n.* *(inclination)* Tendenz, *die;* **have a ~ to do sth.** dazu neigen, etw. zu tun; **there is a ~ for everyone to get**

complacent die Leute neigen dazu, selbstzufrieden zu werden
tendentious [ten'denʃəs] *adj.* tendenziös
[1]**tender** ['tendə(r)] *adj.* **a)** *(not tough)* zart; **b)** *(sensitive)* empfindlich; ~ **spot** *(fig.)* wunder Punkt; **c)** *(loving)* zärtlich; liebevoll; **d) be of ~ age** *or* **years** noch sehr jung sein; **at a ~ age** in jungen Jahren; **at the ~ age of twelve** im zarten Alter von zwölf Jahren
[2]**tender** *n. (Naut.)* Tender, *der*
[3]**tender 1.** *v.t.* **a)** *(present)* einreichen ‹*Rücktritt*›; vorbringen ‹*Entschuldigung*›; **b)** *(offer as payment)* anbieten; **please ~ exact fare** bitte den genauen Betrag bereithalten. **2.** *v.i.* ~ **for sth.** ein Angebot für etw. einreichen. **3.** *n.* Angebot, *das;* **put in a ~:** ein Angebot einreichen; **put sth. out to ~:** etw. ausschreiben
tender-hearted ['tendəhɑ:tɪd] *adj.* weichherzig
tenderize (tenderise) ['tendəraɪz] *v.t. (Cookery)* zart machen; *(by beating)* weich klopfen
tenderly ['tendəlɪ] *adv.* **a)** *(gently)* behutsam ‹*behandeln*›; **b)** *(lovingly)* zärtlich
tenderness ['tendənɪs] *n., no pl.* **a)** *(of meat etc.)* Zartheit, *die;* **b)** *(loving quality)* Zärtlichkeit, *die;* **c)** *(delicacy)* Empfindlichkeit, *die*
tendon ['tendən] *n. (Anat.)* Sehne, *die;* **Achilles ~:** Achillessehne, *die*
tendril ['tendrɪl] *n.* Ranke, *die*
tenement ['tenɪmənt] *n.* **a)** *(Scot.: house containing several dwellings)* Mietshaus, *das;* Mietskaserne, *die (abwertend);* **b)** *(Amer.: house containing several apartments)* ~**[-house]** Mietshaus, *das*
Tenerife [tenə'ri:f] *pr. n.* Teneriffa *(das)*
tenet ['tenɪt] *n.* Grundsatz, *der*
ten-gallon 'hat *n.* Cowboyhut, *der*
tenner ['tenə(r)] *n. (coll.) (Brit.)* Zehnpfundschein, *der;* Zehner, *der (ugs.); (Amer.)* Zehndollarschein, *der;* Zehner, *der (ugs.)*
tennis ['tenɪs] *n., no pl.* Tennis, *das*
tennis: ~-ball *n.* Tennisball, *der;* **~-club** *n.* Tennisverein, *der;* **~-court** *n. (for lawn ~)* Tennisplatz, *der; (for indoor ~)* Tennishalle, *die;* ~ **'elbow** *n., no pl., no art. (Med.)* Tennisell[en]bogen, *der;* **~-match** *n.* Tennismatch, *das;* Tennisspiel, *das;* **~-player** *n.* Tennisspieler, *der/* -spielerin, *die;* **~-racket** *n.* Tennisschläger, *der*
tenon ['tenən] *n. (Woodw.)* Zapfen, *der; see also* **mortise a**
tenor ['tenə(r)] *n.* **a)** *(Mus.: voice, singer, part)* Tenor, *der;* ~ **voice** Tenorstimme, *die;* **b)** *(of argument, speech)* Tenor, *der*
tenpenny ['tenpənɪ] *adj.* für zehn Pence *nachgestellt*
tenpenny 'piece *n. (Brit.)* Zehnpencemünze, *die*
tenpin bowling [tenpɪn 'bəʊlɪŋ] *n.* Bowling, *das*
[1]**tense** [tens] *n. (Ling.)* Zeit, *die;* **in the present/future** *etc.* ~: im Präsens/Futur *usw.*
[2]**tense 1.** *adj.* **a)** *(taut; showing nervous tension)* gespannt; **a ~ silence** eine [an]gespannte Stille; **b)** *(causing nervous tension)* spannungsgeladen. **2.** *v.i.* **sb. ~s** jmds. Muskeln spannen sich an. **3.** *v.t.* anspannen
~ **'up** *v.i.* ‹*Muskeln:*› sich anspannen; ‹*Person:*› sich verkrampfen
tension ['tenʃn] *n.* **a)** *(latent hostility)* Spannung, *die;* ~ **between the police and the people is on the increase** die Spannungen zwischen Polizei und Bevölkerung wachsen; **there is a lot of ~ between them** zwischen ihnen herrscht ein gespanntes Verhältnis; **racial ~:** Rassenspannungen *Pl.;* **b)** *(mental strain)* Anspannung, *die;* **c)** *no pl. (of violin string, tennis-racket)* Spannung, *die*
tent [tent] *n.* Zelt, *das*
tentacle ['tentəkl] *n. (Zool., Bot.)* Tentakel, *der od. das*
tentative ['tentətɪv] *adj.* **a)** *(not definite)* vorläufig; **make a ~ suggestion** einen Vorschlag in den Raum stellen; **say a ~ 'yes'** vorläufig „ja" sagen; **b)** *(hesitant)* zaghaft
tentatively ['tentətɪvlɪ] *adv.* **a)** *(not definitely)* vorläufig; **b)** *(hesitantly)* zaghaft
tenterhooks ['tentəhʊks] *n. pl.* **be on ~:** [wie] auf glühenden Kohlen sitzen; **keep sb. on ~:** jmdn. auf die Folter spannen
tenth [tenθ] **1.** *adj.* zehnt...; *see also* **eighth 1. 2.** *n. (in sequence)* zehnte, *der/die/das; (in rank)* Zehnte, *der/die/das; (fraction)* Zehntel, *das; see also* **eighth 2**
'tent-peg *n.* Zeltpflock, *der;* Hering, *der*
tenuous ['tenjʊəs] *adj.* dünn ‹*Faden*›; zart ‹*Spinnwebe*›; *(fig.)* dünn ‹*Atmosphäre*›; dürftig ‹*Argument*›; unbegründet ‹*Anspruch*›
tenure ['tenjə(r)] *n.* **a)** *(right, title)* Besitztitel, *der;* **b)** *(possession)* Besitz, *der;* **c)** *(period)* ~ **[of office]** Amtszeit, *die;* **d)** *(permanent appointment)* Dauerstellung, *die*
tepid ['tepɪd] *adj.* lauwarm
term [tɜ:m] **1.** *n.* **a)** *(word expressing definite concept)* [Fach]begriff, *der;* **legal/medical ~:** juristischer/medizinischer Fachausdruck; ~ **of reproach** Vorwurf, *der;* **in ~s of money/politics** unter finanziellem/politischem Aspekt; **b)** *in pl. (conditions)* Bedingungen; **he does everything on his own ~s** er tut alles, wie er es für richtig hält; **come to** *or* **make ~s [with sb.]** sich [mit jmdm.] einigen; **come to ~s [with each other]** sich einigen; **come to ~s with sth.** *(be able to accept sth.)* mit etw. zurechtkommen; *(resign oneself to sth.)* sich mit etw. abfinden; **~s of reference** *(Brit.)* Aufgabenbereich, *der;* **c)** *in pl. (charges)* Konditionen; **their ~s are ...:** sie verlangen ...; **hire-purchase on easy ~s** Ratenkauf zu günstigen Bedingungen; **d) in the short/long/medium ~:** kurz-/lang-/mittelfristig; **e)** *(Sch.)* Halbjahr, *das; (Univ.: one of two/three/four divisions per year)* Semester, *das*/Trimester, *das*/Quartal, *das;* **during ~:** während des Halbjahres/Semesters *usw.;* **out of ~:** in den Ferien; **end of ~:** Halbjahres-/Semesterende *usw.;* **f)** *(limited period)* Zeitraum, *der; (period of tenure)* ~ **[of office]** Amtszeit, *die;* **g)** *(period of imprisonment)* Haftzeit, *die;* **h)** *in pl. (mode of expression)* Worte; **praise in the highest ~s** in den höchsten Tönen loben; **i)** *in pl. (relations)* **be on good/poor/friendly ~s with sb.** mit jmdm. auf gutem/schlechtem/freundschaftlichem Fuß stehen. **2.** *v.t.* nennen
terminal ['tɜ:mɪnl] **1.** *n.* **a)** *(Electr.)* Anschluß, *der; (of battery)* Pol, *der;* **b)** *(for train or bus)* Bahnhof, *der; (for airline passengers)* Terminal, *der od. das;* **c)** *(Teleph., Computing)* Terminal, *das.* **2.** *adj.* **a)** End‹*bahnhof, -station*›; **b)** *(Med.)* unheilbar; **have a ~ illness** unheilbar krank sein; **a ~ case** ein hoffnungsloser Fall
terminate ['tɜ:mɪneɪt] **1.** *v.t.* **a)** beenden; **the contract was ~d** der Vertrag wurde gelöst; **b)** *(Med.)* unterbrechen ‹*Schwangerschaft*›. **2.** *v.i.* enden; ‹*Vertrag:*› ablaufen
termination [tɜ:mɪ'neɪʃn] *n.* **a)** *no pl. (coming to an end)* Ende, *das; (of lease)* Ablauf, *der;* **b)** *no pl. (bringing to an end)* Beendigung, *die; (of a marriage)* Auflösung,

die; **c)** *(Med.)* Schwangerschaftsabbruch, *der*
terminology [tɜ:mɪ'nɒlədʒɪ] *n.* Terminologie, *die*
terminus ['tɜ:mɪnəs] *n., pl.* ~**es** *or* **termini** ['tɜ:mɪnaɪ] *(of bus, train, etc.)* Endstation, *die*
termite ['tɜ:maɪt] *n. (Zool.)* Termite, *die*
tern [tɜ:n] *n. (Ornith.)* Seeschwalbe, *die*
terrace ['terəs, 'terɪs] *n.* **a)** *(row of houses)* Häuserreihe, *die;* **b)** *(adjacent to house; Agric.: on hillside)* Terrasse, *die;* **c)** *in pl. (Footb.)* Ränge
terraced house ['terəsthaʊs, 'terɪsthaʊs], '**terrace-house** *ns.* Reihenhaus, *das*
terracotta [terə'kɒtə] *n., no pl., no indef. art.* Terrakotta, *die*
terra firma [terə 'fɜ:mə] *n., no pl., no art.* fester Boden
terrain [te'reɪn] *n.* Gelände, *das;* Terrain, *das (bes. Milit.)*
terrapin ['terəpɪn] *n. (Zool.)* Sumpfschildkröte, *die*
terrestrial [tə'restrɪəl, tɪ'restrɪəl] *adj.* terrestrisch; Erd⟨*satellit, -bevölkerung*⟩
terrible ['terɪbl] *adj.* **a)** *(coll.: very great or bad)* schrecklich *(ugs.);* fürchterlich *(ugs.);* **I feel ~ about doing it** es tut mir schrecklich leid, es zu tun; **b)** *(coll.: incompetent)* schlecht; **be ~ at maths/tennis/carpentry** in Mathe schlecht sein/schlecht Tennis spielen/ein schlechter Tischler sein; **c)** *(causing terror)* furchtbar
terribly ['terɪblɪ] *adv.* **a)** *(coll.: very)* unheimlich *(ugs.);* furchtbar *(ugs.);* **b)** *(coll.: appallingly)* furchtbar *(ugs.);* **c)** *(coll.: incompetently)* schlecht; **d)** *(fearfully)* auf erschreckende Weise
terrier ['terɪə(r)] *n.* Terrier, *der*
terrific [tə'rɪfɪk] *adj. (coll.)* **a)** *(great, intense)* irrsinnig *(ugs.);* Wahnsinns- *(ugs.);* unwahrscheinlich *(ugs.);* **b)** *(magnificent)* sagenhaft *(ugs.);* **c)** *(highly expert)* klasse *(ugs.);* toll *(ugs.);* **be ~ at sth.** in etw. *(Dat.)* Spitze sein *(ugs.);* **a ~ singer** ein Spitzensänger/eine Spitzensängerin *(ugs.)*
terrify ['terɪfaɪ] *v.t.* **a)** *(fill with terror)* angst machen (+ *Dat.*); **terrified** verängstigt; **b)** *(coll.: make very anxious)* angst machen (+ *Dat.*); **be terrified that ...:** Angst haben, daß ...; **c)** *(scare)* Angst einjagen (+ *Dat.*)
terrifying ['terɪfaɪɪŋ] *adj.* **a)** *(causing terror)* entsetzlich ⟨*Erlebnis, Film, Buch, Theaterstück*⟩; erschreckend ⟨*Klarheit, Gedanke*⟩; furchterregend ⟨*Anblick*⟩; beängstigend ⟨*Geschwindigkeit, Neigungswinkel*⟩; **b)** *(formidable)* furchterregend; beängstigend ⟨*Gelehrsamkeit, Intensität*⟩
territorial [terɪ'tɔ:rɪəl] *adj.* territorial; Gebiets⟨*anspruch, -hoheit usw.*⟩; Hoheits⟨*gebiet*⟩
territorial: T~ 'Army *n. (Brit. Mil.)* Territorialarmee, *die;* ~ '**waters** *n. pl.* Hoheitsgewässer
territory ['terɪtərɪ] *n.* **a)** *(Polit.)* Staatsgebiet, *das;* Hoheitsgebiet, *das;* **b)** *(fig.: area of knowledge or action)* Gebiet, *das;* **c)** *(of commercial traveller etc.)* Bezirk, *der;* **d)** *(large tract of land)* Region, *die;* Gebiet, *das*
terror ['terə(r)] *n.* **a)** *(extreme fear)* [panische] Angst; **in ~:** in panischer Angst; **b)** *(person or thing causing ~)* Schrecken, *der;* **c)** **[holy] ~** *(troublesome person)* Plage, *die*
terrorise *see* **terrorize**
terrorism ['terərɪzm] *n.* Terrorismus, *der; (terrorist acts)* Terror, *der;* **acts of ~:** Terrorakte
terrorist ['terərɪst] *n.* Terrorist, *der*/Terroristin, *die; attrib.* Terror⟨*gruppe, -organisation*⟩
terrorize ['terəraɪz] *v.t.* **a)** *(frighten)* in [Angst und] Schrecken versetzen; **b)** *(coerce by terrorism)* terrorisieren; *(intimidate)* durch Terror[akte] einschüchtern
terror: ~-stricken, ~-struck *adjs.* zu Tode erschrocken
terse [tɜ:s] *adj.* **a)** *(concise)* kurz und bündig; **b)** *(curt)* knapp
tertiary ['tɜ:ʃərɪ] *adj. (of third order or rank)* tertiär
Terylene, (P) ['terɪli:n] *n.* Terylen, *das* Ⓦ
test [test] **1.** *n.* **a)** *(examination) (Sch.)* Klassenarbeit, *die; (Univ.)* Klausur, *die; (short examination)* Test, *der;* **put sb./sth. to the ~:** jmdn./etw. erproben; **b)** *(critical inspection, analysis)* Test, *der;* **c)** *(basis for evaluation)* Prüfstein, *der;* **d)** *(Cricket)* Test Match, *das.* **2.** *v.t.* **a)** *(examine, analyse)* untersuchen ⟨*Wasser, Gehör, Augen*⟩; testen ⟨*Gehör, Augen*⟩; prüfen ⟨*Schüler*⟩; überprüfen ⟨*Hypothese, Aussage, Leitungen*⟩; **~ sb. for Aids** jmdn. auf Aids untersuchen; **b)** *(try severely)* auf die Probe stellen
~ out *v.t.* ausprobieren ⟨*neue Produkte*⟩ (**on** an + *Dat.*); erproben ⟨*Theorie, Idee*⟩
testament ['testəmənt] *n.* **a) Old/New T~** *(Bibl.)* Altes/Neues Testament; **b)** *see* [2]**will 1 b**
test: ~ ban treaty *n.* [Atom]teststopp-Abkommen, *das;* **~ card** *n. (Telev.)* Testbild, *das;* ~ '**case** *n. (Law)* Musterprozeß, *der;* **~ drive** *n.* Probefahrt, *die;* **~-drive** *v.t.* probefahren
tester ['testə(r)] *n.* Prüfer, *der*/Prüferin, *die; (device)* Prüfgerät, *das*
'**test flight** *n.* Testflug, *der;* Erprobungsflug, *der*
testicle ['testɪkl] *n. (Anat., Zool.)* Testikel, *der (fachspr.);* Hoden, *der*
testify ['testɪfaɪ] **1.** *v.i.* **a)** **~ to sth.** etw. bezeugen; **~ to sb.'s high intelligence** jmdm. große Intelligenz bescheinigen; **b)** *(Law)* **~ against sb./before sth.** gegen jmdn./vor etw. *(Dat.)* aussagen. **2.** *v.t.* **a)** *(declare)* bestätigen; **b)** *(be evidence of)* beweisen
testimonial [testɪ'məʊnɪəl] *n. (certificate of character)* Zeugnis, *das; (recommendation)* Referenz, *die*
testimony ['testɪmənɪ] *n.* **a)** *(witness)* Aussage, *die;* **b)** *(Law)* [Zeugen]aussage, *die;* **c)** *no pl. (statements)* Angaben
test: ~ match *n. (Sport)* Test Match, *das;* **~ paper** *n.* **a)** *(Educ.)* Übungsarbeit, *die; (Univ.)* Übungsklausur, *die;* **b)** *(Chem.)* Indikatorpapier, *das;* **~-piece** *n.* Pflicht[übung], *die; (Mus.)* Pflichtstück, *das;* **~ pilot** *n. (Aeronaut.)* Testpilot, *der*/-pilotin, *die;* **~-tube** *n. (Chem., Biol.)* Reagenzglas, *das; attrib.* **~-tube baby** *(coll.)* Retortenbaby, *das (ugs.)*
testy ['testɪ] *adj.* leicht reizbar ⟨*Person*⟩; gereizt ⟨*Antwort*⟩
tetanus ['tetənəs] *n. (Med.)* Tetanus, *der (fachspr.);* [Wund]starrkrampf, *der*
tetchy ['tetʃɪ] *adj.* leicht reizbar; *(on single occasion)* gereizt
tête-à-tête [teɪtɑ:'teɪt] *n.* Tête-à-tête, *das (veralt.);* Gespräch unter vier Augen
tether ['teðə(r)] **1.** *n. (chain)* Kette, *die; (rope)* Strick, *der;* **be at the end of one's ~:** am Ende [seiner Kraft] sein. **2.** *v.t.* anbinden (**to** an)
Teutonic [tju:'tɒnɪk] *adj.* **a)** *(Germanic)* germanisch; **b)** *(with German characteristics)* teutonisch *(abwertend, auch scherzh.)*
text [tekst] *n.* **a)** Text, *der;* **they couldn't agree on the ~ of the agreement** sie konnten sich über den Wortlaut des Vertrages nicht einigen; **b)** *(passage of Scripture)* Bibelstelle, *die*

'textbook *n. (Educ.)* Lehrbuch, *das*
textile ['tekstaɪl] *n.* Stoff, *der;* ~**s** Textilien *Pl.*
textual ['tekstjʊəl] *adj.* textlich
texture ['tekstʃə(r)] *n.* **a)** Beschaffenheit, *die; (of fabric, material)* Struktur, *die; (of food)* Konsistenz, *die;* **have a smooth** ~: sich glatt anfühlen; **b)** *(of prose, music, etc.)* Textur, *die (geh.)*
Th. *abbr.* **Thursday** Do.
Thai [taɪ] **1.** *adj.* thailändisch. **2.** *n.* **a)** *pl.* ~**s** *or same* Thai, *der/die;* Thailänder, *der*/Thailänderin, *die;* **b)** *(language)* Thai, *das*
Thailand ['taɪlænd] *pr. n.* Thailand *(das)*
Thames [temz] *pr. n.* Themse, *die*
than [ðən, *stressed* ðæn] *conj.* **a)** *(in comparison)* als; **I know you better** ~ **[I do] him** ich kenne dich besser als ihn; **I know you better** ~ **he [does]** ich kenne dich besser als er; **you are taller** ~ **he [is]** *or (coll.)* **him** du bist größer als er; **b)** *(introducing statement of difference)* als
thank [θæŋk] *v. t.* ~ **sb. [for sth.]** jmdm. [für etw.] danken; sich bei jmdm. [für etw.] bedanken; **have sb./sth. to** ~ **for sth.** jmdm./einer Sache etw. zu verdanken haben; **have [only] oneself to** ~ **for sth.** etw. sich *(Dat.)* selbst zuzuschreiben haben; **he won't** ~ **you for that** *(iron.)* er wird dir dafür nicht gerade dankbar sein; ~ **God** *or* **goodness** *or* **heaven[s]** Gott sei Dank; **[I]** ~ **you** danke; *(slightly formal)* vielen Dank; **no,** ~ **you** nein, danke; **yes,** ~ **you** ja, bitte; danke, ja; ~ **you very much [indeed]** vielen herzlichen Dank; ~**ing 'you** *(coll.)* danke; ~ **you for nothing!** *(iron.)* danke bestens!; **I will** ~ **you to do as you are told** *(iron.)* ich wäre dir sehr verbunden, wenn du tätest, was man dir sagt
thankful ['θæŋkfl] *adj.* dankbar; **I am just** ~ **that it's all over** ich bin nur froh, daß das jetzt alles vorüber ist
thankfully ['θæŋkfəlɪ] *adv.* **a)** *(gratefully)* dankbar; **b)** *(as sentence-modifier: fortunately)* glücklicherweise
thankless ['θæŋklɪs] *adj.* undankbar ⟨*Aufgabe, Person*⟩
thanks [θæŋks] *n. pl.* **a)** *(gratitude)* Dank, *der;* **accept sth. with** ~: etw. dankend annehmen; **that's all the** ~ **one gets** das ist nun der Dank dafür!; **give** ~ **[to God]** dem Herrn danken; ~ **to** *(with the help of)* dank; *(on account of the bad influence of)* wegen; ~ **to you** dank deiner; *(reproachfully)* deinetwegen; **it is small** *or* **no** ~ **to him that we won** ihm haben wir es jedenfalls nicht zu verdanken, daß wir gewonnen haben; **b)** *(as formula expressing gratitude)* danke; **no,** ~: nein, danke; **yes,** ~: ja, bitte; ~ **awfully** *or* **a lot** *or* **very much, many** ~ *(coll.)* vielen *od.* tausend Dank
Thanksgiving ['θæŋksgɪvɪŋ] *n.* ~ **[Day]** *(Amer.)* [amerikanisches] Erntedankfest; Thanksgiving Day, *der*
'thank-you *n. (coll.)* Dankeschön, *das;* **a warm** *or* **hearty** ~: ein herzliches Dankeschön; *attrib.* ~ **letter** Dankbrief, *der;* **give sb. a** ~ **present** jmdm. zum Dank etwas schenken
that 1. [ðæt] *adj., pl.* **those** [ðəʊz] **a)** dieser/diese/dieses; **b)** *expr. strong feeling* der/die/das; **never will I forget** ~ **day** den Tag werde ich nie vergessen; **c)** *(coupled or contrasted with 'this')* der/die/das [da]. **2.** [ðæt] *pron., pl.* **those a)** der/die/das; **who is** ~ **in the garden?** wer ist das [da] im Garten?; **what bird is** ~**?** was für ein Vogel ist das?; **I know all** ~: ich weiß das alles; **and [all]** ~: und so weiter; **like** ~: *(of the kind or in the way mentioned, of* ~ *character)* so; **[just] like '**~ *(without effort, thought)* einfach so; **don't be like** ~**!** sei doch nicht so; **don't talk like** ~: hör auf, so zu reden; **he is 'like** ~: so ist er eben; ~ **is [to say]** *introducing explanation* das heißt; *introducing reservation* das heißt; genauer gesagt; **if they'd have me,** ~ **is** das heißt, wenn sie mich nehmen; **'**~**'s more like it** *(of suggestion, news)* das hört sich schon besser an; *(of action, work)* das sieht schon besser aus; ~**'s right!** *expr. approval* gut *od.* recht so; *(iron.)* nur so weiter!; *(coll.: expr. assent)* jawohl; ~**'s a good** *etc.* **boy/girl** das ist lieb [von dir, mein Junge/Mädchen]; *(with request)* sei so lieb *usw.;* ~ **will do** das reicht; **sb./sth. is not as ... as all '**~ *(coll.)* so ... ist jmd./etw. nun auch wieder nicht; **[so]** ~**'s '**~ *(it's finished)* so, das wär's; *(it's settled)* so ist es nun mal; **you are not going to the party, and** ~**'s '**~**!** du gehst nicht zu der Party, und damit Schluß!; **b)** *(Brit.: person spoken to)* **who is** ~**?** wer ist da?; *(behind wall etc.)* wer ist denn da?; *(on telephone)* wer ist am Apparat?; **who was** ~**?** wer war das? **3.** [ðət] *rel. pron., pl. same* der/die/das; **the people** ~ **you got it from** die Leute, von denen du es bekommen hast; **the box** ~ **you put the apples in** die Kiste, in die du die Äpfel getan hast; **is he the man** ~ **you saw last night?** ist das der Mann, den Sie gestern abend gesehen haben?; **everyone** ~ **I know** jeder, den ich kenne; **this is all [the money]** ~ **I have** das ist alles [Geld], was ich habe. **4.** [ðæt] *adv. (coll.)* so; **he may be daft, but he's not [all] '**~ **daft** er mag ja blöd sein, aber so blöd [ist er] auch wieder nicht. **5.** [ðət] *rel. adv.* der/die/das; **at the speed** ~ **he was going** bei der Geschwindigkeit, die er hatte; **the day** ~ **I first met her** der Tag, an dem ich sie zum ersten Mal sah. **6.** [ðət, *stressed* ðæt] *conj.* **a)** *introducing statement; expr. result, reason or cause* daß; **b)** *expr. purpose* **[in order]** ~: damit
thatch [θætʃ] **1.** *n. (of straw)* Strohdach, *das; (of reeds)* Schilf- *od.* Reetdach, *das; (of palm-leaves)* Palmblattdach, *das; (material)* Stroh, *das*/Schilf, *das*/ Palmblätter; *(roofing)* Dachbedeckung, *die.* **2.** *v. t.* mit Stroh/ Schilf/Palmblättern decken
thatched [θætʃt] *adj.* strohgedeckt/schilf- *od.* reetgedeckt; gedeckt ⟨*Dach*⟩*;* Stroh-/Schilf- *od.* Reet⟨*dach*⟩
Thatcherism ['θætʃərɪzm] *n. (Polit.)* Thatcherismus, *der*
thaw [θɔː] **1.** *n.* **a)** *(warmth)* Tauwetter, *das;* **b)** *(act of* ~*ing)* **after the** ~: nachdem es getaut hat/ hatte; **c)** *(fig.)* Tauwetter, *das;* Tauwetterperiode, *die.* **2.** *v. i.* **a)** *(melt)* auftauen; **b)** *(become warm enough to melt ice etc.)* tauen; **c)** *(fig.: become less aloof or hostile)* auftauen. **3.** *v. t.* **a)** *(cause to melt)* auftauen; **b)** *(fig.: cause to be less aloof or hostile)* auftauen; entspannen ⟨*Atmosphäre*⟩
~ **'out** *see* ~ **2, 3**
the [*before vowel* ðɪ, *before consonant* ðə, *when stressed* ðiː] **1.** *def. art.* **a)** der/die/das; **all** ~ **doors** alle Türen; **play** ~ **piano** Klavier spielen; **if you want a quick survey, this is** ~ **book** für einen raschen Überblick ist dies das richtige Buch; **it's** *or* **there's only** ~ **one** es ist nur dieser/diese/dieses eine; **he lives in** ~ **district** er wohnt in dieser Gegend; **£5** ~ **square metre/**~ **gallon/**~ **kilogram** 5 Pfund der Quadratmeter/ die Gallone/das Kilogramm; **14 miles to** ~ **gallon** 14 Meilen auf eine Gallone; ≈ 20 l auf 100 km;

a scale of one mile to ~ inch ein Maßstab von 1:63 360; **b)** *(denoting one best known)* **it is '~ restaurant in this town** das ist das Restaurant in dieser Stadt; **red is '~ colour this year** Rot ist in diesem Jahr die Farbe; **c)** *with names of diseases* **have got ~ toothache/measles** *(coll.)* Zahnschmerzen/die Masern haben; **d)** *(Brit. coll.: my, our, etc.)* mein/unser *usw.* **2.** *adv.* **~ more I practise ~ better I play** je mehr ich übe, desto *od.* um so besser spiele ich; **so much ~ worse for sb./sth.** um so schlimmer für jmdn./etw.

theatre (*Amer.*: **theater**) ['θɪətə(r)] *n.* **a)** Theater, *das;* **at the ~:** im Theater; **go to the ~:** ins Theater gehen; **b)** *(lecture ~)* Hörsaal, *der;* **c)** *(Brit. Med.)* *see* **operating-theatre; d)** *(dramatic art)* **the ~:** das Theater; **e)** *(scene of action)* Schauplatz, *der; (of war)* Kriegsschauplatz, *der*

'theatre-goer *n.* Theaterbesucher, *der*/-besucherin, *die*

theatrical [θɪ'ætrɪkl] *adj.* **a)** schauspielerisch; **a ~ company** eine Schauspiel- *od.* Theatertruppe; **b)** *(showy)* theatralisch ‹*Benehmen, Verbeugung, Person*›

thee [ði:] *pron. (arch./poet./dial.)* dich; *as indirect object* dir; *(Relig.: God)* Dich/Dir

theft [θeft] *n.* Diebstahl, *der;* **~ of cars** Autodiebstahl, *der*

their [ðeə(r)] *poss. pron. attrib.* **a)** ihr; *see also* **²her; our; b)** *(coll.: his or her)* **who has forgotten ~ ticket?** wer hat seine Karte vergessen?

theirs [ðeəz] *poss. pron. pred.* ihrer/ihre/ihres; *see also* **hers; ours**

them [ðəm, *stressed* ðem] *pron.* **a)** sie; *as indirect object* ihnen; *see also* **¹her; b)** *(coll.: him/her)* ihn/sie

theme [θi:m] *n.* **a)** *(of speaker, writer, or thinker)* Gegenstand, *der;* Thema, *das;* **b)** *(Mus.)* Thema, *das;* Leitmotiv, *das*

theme: ~ music *n.* Titelmelodie, *die;* **~ park** *n. Freizeitpark, dessen Attraktionen und Einrichtungen auf ein bestimmtes Thema bezogen sind;* **~ song** *n.* **a)** *see* **~ music; b)** *see* **signature tune; ~ tune** *see* **signature tune**

themselves [ðəm'selvz] *pron.* **a)** *emphat.* selbst; **the results ~ were ...:** die Ergebnisse an sich waren ...; **b)** *refl.* sich ‹*waschen usw.*›; sich selbst ‹*die Schuld geben, regieren*›. *See also* **herself**

then [ðen] **1.** *adv.* **a)** *(at that time)* damals; **~ and there** auf der Stelle; *see also* **now 1 a, b; b)** *(after that)* dann; **~ [again]** *(and also)* außerdem; **but ~** *(after all)* aber schließlich; **c)** *(in that case)* dann; **~ why didn't you say so?** warum hast du dann nichts gesagt?; **hurry up, ~:** dann beeil dich aber; **but ~ again** aber andererseits; **d)** *expr. grudging or impatient concession* dann eben; **well, take it, ~:** dann nimm es eben; **e)** *(accordingly)* [dann] also. **2.** *n.* **before ~:** vorher; davor; **by ~:** bis dahin; **from ~ on** von da an; **till ~:** bis dahin; **oh, we should get there long before ~:** ach, bis dahin sind wir längst dort; **since ~:** seitdem. **3.** *adj.* damalig

theodolite [θɪ'ɒdəlaɪt] *n. (Surv.)* Theodolit, *der*

theologian [θi:ə'ləʊdʒɪən] *n.* Theologe, *der*/Theologin, *die*

theological [θi:ə'lɒdʒɪkl] *adj.* theologisch; Theologie‹*student, -dozent*›

theology [θɪ'ɒlədʒɪ] *n.* **a)** *no pl., no indef. art.* Theologie, *die;* **b)** *(religious system)* Glaubenslehre, *die*

theoretical [θɪə'retɪkl] *adj.* theoretisch; **your arguments are only ~:** deine Argumentation ist reine Theorie

theoretically [θi:ə'retɪkəlɪ] *adv.* theoretisch

theorise *see* **theorize**

theorist ['θɪərɪst] *n.* Theoretiker, *der*/Theoretikerin, *die*

theorize ['θɪəraɪz] *v. i.* theoretisieren

theory ['θɪərɪ] *n. (also Math.)* Theorie, *die;* **~ of evolution/music** Evolutions-/Musiktheorie, *die;* **in ~:** theoretisch; **have a ~ that ...:** die Theorie vertreten, daß ...

therapeutic [θerə'pju:tɪk] *adj.* therapeutisch; *(curative)* therapeutisch wirksam

therapist ['θerəpɪst] *n. (Med.)* Therapeut, *der*/Therapeutin, *die*

therapy ['θerəpɪ] *n. (Med., Psych.)* Therapie, *die;* [Heil]behandlung, *die*

there [ðeə(r)] **1.** *adv.* **a)** *(in/at that place)* da; dort; *(fairly close)* da; **sb. has been ~ before** *(fig. coll.)* jmd. weiß Bescheid; **~ or ~a'bouts** so ungefähr; **be down/in/up ~:** da unten/drin/oben sein; **~ goes ...:** da geht/fährt *usw.* ...; **are you ~?** *(on telephone)* sind Sie noch da *od.* *(ugs.)* dran?; **~ and then** auf der Stelle; **b)** *(calling attention)* **hello** *or* **hi ~!** hallo!; **you ~!** Sie da!; **move along ~!** weitergehen!; **~'s a good** *etc.* **boy/girl** das ist lieb [von dir, mein Junge/Mädchen]; **c)** *(in that respect)* da; **so ~:** und damit basta *(ugs.)*; **~ you are wrong** da irrst du dich; **~, it is a loose wire** da haben wir's – ein loser Draht; **~ it is** *(nothing can be done about it)* da kann man nichts machen; **~ you are** *(giving sth.)* [da,] bitte schön (*see also* **2 b**); **d)** *(to that place)* dahin, dorthin ‹*gehen, gelangen, fahren, rücken, stellen*›; **we got ~ and back in two hours** wir brauchten für Hin- und Rückweg [nur] zwei Stunden; **down/in/up ~:** dort hinunter/hinein/hinauf; **get ~ first** jmdm./den anderen zuvorkommen; **get ~** *(fig.)* *(achieve)* es [schon] schaffen; *(understand)* es verstehen; **e)** [ðə(r), *stressed* ðeə(r)] *as introductory function-word* da; **was ~ anything in it?** war da irgend etwas drin? *(ugs.)*; **~ is enough food** es gibt genug zu essen; **~ are many kinds of ...:** es gibt viele Arten von ...; **~ were four of them** sie waren zu viert; **~ was once an old woman who ...:** es war einmal eine alte Frau, die ...; **~ was no beer left** es gab kein Bier mehr; **~ appears to be some error** da scheint ein Irrtum unterlaufen zu sein; **~'s no time for that now** dafür haben wir/habe ich jetzt keine Zeit; **... if ever ~ was one** ... wie er/sie/es im Buche steht; **what is ~ for supper?** was gibt's zum Abendessen? **2.** *int.* **a)** *(to soothe child etc.)* **~, ~:** na, na *(ugs.)*; **b)** *expr. triumph or dismay* **~ [you are]!** da, siehst du! (*see also* **1 c**); **~, you've dropped it!** da, jetzt hast du es doch fallen lassen! **3.** *n.* da, dort; **near ~:** da *od.* dort in der Nähe

there: ~abouts ['ðeərəbaʊts] *adv.* **a)** *(near that place)* da [in der Nähe]; **the locals ~abouts** die Leute, die dort wohnen; **b)** *(near that number)* **two litres or ~abouts** zwei Liter [so] ungefähr; *see also* **there 1 a; ~by** [ðeə'baɪ, 'ðeəbaɪ] *adv.* dadurch; **~fore** *adv.* deshalb; also; **~u'pon** *adv.* **a)** *(soon after that)* kurz darauf; **b)** *(in consequence of that)* daraufhin

thermal ['θɜ:ml] **1.** *adj.* thermisch ‹*Erscheinung, Anforderungen*›; Wärme‹*dämmung, -strahlung*›; **~ underwear** kälteisolierende Unterwäsche. **2.** *n.* Thermik, *die*

thermodynamics [θɜ:mədaɪ'næmɪks] *n., no pl. (Phys.)* Thermodynamik, *die*

thermometer [θə'mɒmɪtə(r)] *n.* Thermometer, *das*

Thermos, thermos, (P) ['θɜ:məs] *n.* ~ **[flask/jug/bottle]** Thermosflasche, *die* Ⓦⓩ

thermostat ['θɜ:məstæt] *n.* Thermostat, *der*

thesaurus [θɪ'sɔ:rəs] *n., pl.* **thesauri** [θɪ'sɔ:rɪ] *or* **~es** Thesaurus, *der*

these *pl. of* **this 1, 2**

thesis ['θi:sɪs] *n., pl.* **theses** ['θi:si:z] **a)** *(proposition)* These, *die;* **b)** *(dissertation)* Dissertation, *die,* Doktorarbeit, *die* (**on** über + *Akk.*)

they [ðeɪ] *pron.* **a)** sie; **b)** *(people in general)* man; **c)** *(coll.: he or she)* **everyone thinks ~ know best** jeder denkt, er weiß es am besten; **d)** *(those in authority)* sie; die *(ugs.). See also* **their; theirs; them; themselves**

they'd [ðeɪd] **a)** = **they would; b)** = **they had**

they'll [ðeɪl] = **they will**

they're [ðeə(r)] = **they are**

they've [ðeɪv] = **they have**

thick [θɪk] **1.** *adj.* **a)** dick; breit, dick ‹*Linie*›; **that's** *or* **it's a bit ~!** *(Brit. fig. coll.)* das ist ein starkes Stück! *(ugs.);* **a rope two inches ~, a two-inch ~ rope** ein zwei Zoll starkes *od.* dickes Seil; **b)** *(dense)* dicht ‹*Haar, Nebel, Wolken, Gestrüpp usw.*›; **c)** *(filled)* **~ with** voll von; **air ~ with fog and smoke** von Nebel und Rauch erfüllte Luft; **d)** steif ‹*Gallerte*›; dickflüssig ‹*Sahne*›; dick ‹*Suppe, Schlamm, Brei, Kleister*›; **e)** *(stupid)* dumm; **you're just plain ~:** du bist ganz einfach doof *(salopp);* **[as] ~ as two short planks** *(coll.)* dumm wie Bohnenstroh *(ugs.);* **f)** *(coll.: intimate)* **be very ~ with sb.** mit jmdm. dick befreundet sein *(ugs.);* **be [as] ~ as thieves** dicke Freunde sein *(ugs.).* **2.** *n., no pl., no indef. art.* **in the ~ of** mitten in (+ *Dat.*); **in the ~ of it** *or* **things** mitten drin; **stay with sb./stick together through ~ and thin** mit jmdm./zusammen durch dick und dünn gehen. **3.** *adv.* **job offers/complaints came in ~ and fast** es kam eine Flut von Stellenangeboten/Beschwerden

thick 'ear *n. (Brit. sl.)* **give sb. a ~:** jmdm. ein paar hinter die Ohren geben *(ugs.)*

thicken [θɪkn] **1.** *v.t.* dicker machen; eindicken ‹*Sauce*›. **2.** *v.i.* **a)** dicker werden; **b)** *(become dense)* ‹*Nebel:*› dichter werden; **c)** *(become blurred)* **his speech ~ed** er bekam eine schwere Zunge *(geh.);* **d)** *(become complex)* **the plot ~s!** die Sache wird kompliziert!; *(iron.)* die Sache wird langsam interessant!

thicket ['θɪkɪt] *n.* Dickicht, *das*

thick: ~head *n.* Dummkopf, *der;* **~headed** *adj.* dumm

thickly ['θɪklɪ] *adv.* **a)** *(in a thick layer)* dick; **b)** *(densely, abundantly)* dicht

thickness ['θɪknɪs] *n.* **a)** Dicke, *die;* **be two metres in ~:** zwei Meter dick sein; **b)** *no pl. (denseness)* Dichte, die; *(of hair)* Fülle, *die;* **c)** *no pl. (of jelly)* Steifheit, *die; (of cream)* Dickflüssigkeit, *die; (of soup, porridge, glue)* Dicke, *die;* **d)** *(layer)* Lage, *die*

thick: ~set *adj. (stocky)* gedrungen; **~-skinned** *adj. (fig.)* unsensibel; dickfellig *(ugs. abwertend)*

thief [θi:f] *n., pl.* **thieves** [θi:vz] Dieb, *der*/Diebin, *die*

thieve [θi:v] *v.i.* stehlen

thigh [θaɪ] *n.* **a)** *(Anat.)* Oberschenkel, *der;* **b)** *(Zool.)* Schenkel, *der*

thigh: ~-bone *n. (Anat.)* Oberschenkelknochen, *der;* **~-boot** *n.* Kanonenstiefel, *der;* Schaftstiefel, *der*

thimble ['θɪmbl] *n.* Fingerhut, *der*

thimbleful ['θɪmblfʊl] *n.* Fingerhut [voll], *der*

thin [θɪn] **1.** *adj.* **a)** *(of small thickness or diameter)* dünn; **b)** *(not fat)* dünn; **a tall, ~ man** ein großer, hagerer Mann; **as ~ as a rake** *or* **lath** spindeldürr; **c)** *(narrow)* schmal ‹*Baumreihe*›; dünn ‹*Linie*›; **d)** *(sparse)* dünn, schütter ‹*Haar*›; fein ‹*Regen, Dunst*›; spärlich ‹*Publikum, Besuch*›; gering ‹*Beteiligung*›; dünn ‹*Luft*›; **he is already ~ on top** *or* **going ~ on top** bei ihm lichtet es sich oben schon; **be ~ on the ground** *(fig.)* dünn gesät sein; **vanish** *or* **disappear into ~ air** *(fig.)* sich in Luft auflösen; **e)** *(sl.: wretched)* enttäuschend, unbefriedigend ‹*Zeit*›. *See also* **thick 2. 2.** *adv.* dünn. **3.** *v.t.,* **-nn-: a)** *(make less deep or broad)* dünner machen; **b)** *(make less dense, dilute)* verdünnen; **c)** *(reduce in number)* dezimieren. **4.** *v.i.,* **-nn-** ‹*Haar, Nebel:*› sich lichten; ‹*Menschenmenge:*› sich zerstreuen

~ 'out *v.i.* ‹*Menschenmenge:*› sich verlaufen; ‹*Verkehr:*› abnehmen; ‹*Häuser:*› spärlicher werden

thine [ðaɪn] *poss. pron. (arch./poet./dial.)* **a)** *pred.* deiner/deine/dein[e]s; der/die/das deinige *(geh.); see also* **hers; b)** *attrib.* dein

thing [θɪŋ] *n.* **a)** *(inanimate object)* Sache, *die;* Ding, *das;* **what's that ~ in your hand?** was hast du da in der Hand?; **be a rare ~:** etwas Seltenes sein; **neither one ~ nor the other** weder das eine noch das andere; **not a ~:** überhaupt *od.* gar nichts; **b)** *(action)* **that was a foolish ~ to do** das war eine große Dummheit; **it was the right ~ to do** es war das einzig Richtige; **we can't do a ~ about it** wir können nichts dagegen tun; **do ~s to sb./sth.** *(fig. coll.)* auf jmdn./etw. eine enorme Wirkung haben *(ugs.);* **c)** *(fact)* [Tat]sache, *die;* **a ~ which is well known to everybody** eine allgemein bekannte Tatsache; **it's a strange ~ that ...:** es ist seltsam, daß ...; **for one ~, you don't have enough money[, for another ~ ...]** zunächst einmal hast du nicht genügend Geld [, außerdem ...]; **the best/worst ~ about the situation/her** das Beste/Schlimmste an der Situation/an ihr; **know/learn a ~ or two about sth./sb.** sich mit etw./jmdm. auskennen/einiges über etw. *(Akk.)* lernen/über jmdn. erfahren; **the [only] ~ is that ...:** die Sache ist [nur] die, daß ...; **d)** *(idea)* **say the first ~ that comes into one's head** das sagen, was einem gerade so einfällt; **what a ~ to say!** wie kann man nur so etwas sagen!; **have a ~ about sb./sth.** *(coll.) (be obsessed about)* auf jmdn./etw. abfahren *(salopp); (be prejudiced about)* etwas gegen jmdn./etw. haben; *(be afraid of or repulsed by)* einen Horror vor jmdm./etw. haben *(ugs.);* **e)** *(task)* **she has a reputation for getting ~s done** sie ist für ihre Tatkraft bekannt; **a big ~ to undertake** ein großes Unterfangen; **f)** *(affair)* Sache, *die;* Angelegenheit, *die;* **make a mess of ~s** alles vermasseln *(salopp);* **make a [big] ~ of sth.** *(regard as essential)* auf etw. besonderen Wert legen; *(get excited about)* sich über etw. *(Akk.)* aufregen; **it's one ~ after another** es kommt eins zum anderen; **g)** *(circumstance)* **take ~s too seriously** alles zu ernst nehmen; **how are ~s?** wie geht's [dir]?; **as ~s stand [with me]** so wie die Dinge [bei mir] liegen; **it's just one of those ~s** *(coll.)* so was kommt schon mal vor *(ugs.);* **h)** *(individual, creature)* Ding, *das;* **she is in hospital, poor ~:** sie ist im Krankenhaus, das arme Ding; **you spiteful ~!** du [gemeines] Biest!; **i)** *in pl. (personal belongings, outer clothing)* Sachen; **wash up the dinner ~s** das Geschirr vom Abendessen abwaschen; **j)** *in pl.* *(matters)* **an expert/authority on ~s historical** ein Fachmann/eine Autorität in geschichtlichen Fragen; **k)** *(product of work)* Sache, *die;* **the latest ~ in hats** der letzte Schrei in der Hutmode; **l)** *(special interest)* **what's your ~?** was machst du gerne?; **do one's own ~** *(coll.)* sich selbst verwirklichen; **m)** *(coll.: sth. remarkable)* **now 'there's a ~!** das ist ja ein Ding! *(ugs.);* **n) the ~** *(what is proper or needed or important)* das Richtige; **blue jeans are the ~ among teenagers** Bluejeans sind der Hit *(ugs.)* unter den Teenagern; **but the ~ is, will she come in fact?** aber die Frage ist, wird sie auch tatsächlich kommen?

thingamy ['θɪŋəmɪ], **thingumabob** ['θɪŋəməbɒb], **thingumajig** ['θɪŋəmədʒɪg], **thingummy** ['θɪŋəmɪ], **thingy** ['θɪŋɪ] *ns. (coll.)* Dings, *der/die/das (salopp);* Dingsbums, *der/die/das (ugs.)*

think [θɪŋk] **1.** *v.t.,* **thought** [θɔ:t] **a)** *(consider)* meinen; **we ~ [that] he will come** wir denken *od.* glauben, daß er kommt; **we do not ~ it probable** wir halten es nicht für wahrscheinlich; **he is thought to be a fraud** man hält ihn für einen Betrüger; **what do you ~?** was meinst du?; **what do you ~ of** *or* **about him/it?** was hältst du von ihm/davon?; **I thought to myself ...:** ich dachte mir [im stillen]; **that's what 'they ~!** das meinen die!; **..., don't you ~?** ..., findest *od.* meinst du nicht auch?; **where do you ~ you are?** was glaubst du eigentlich, wo du bist?; **who does he/she ~ he/she is?** für wen *od.* wofür hält er/sie sich eigentlich?; **you** *or* **one** *or* **anyone would ~ that ...:** man sollte [doch] eigentlich annehmen, daß ...; **I ~ not** ich glaube nicht; **I should '~ so/~ 'not!** *(indignant)* das will ich meinen/das will ich nicht hoffen; **I thought as much** *or* **so** das habe ich mir schon gedacht; **I ~ so** ich glaube schon; **do you really ~ so?** findest du wirklich? **I wouldn't ~ so** das glaube ich kaum; **yes, I ~ so too** ja, das finde ich auch *(ugs.);* **I should ~ not!** *(no!)* auf keinen Fall; **that'll be great fun, I 'don't ~** *(coll. iron.)* das kann ja lustig werden *(ugs. iron.);* **to ~ [that] he should treat me like this!** man sollte es nicht für möglich halten, daß er mich so behandelt!; **I wouldn't have thought it possible** ich hätte das nicht für möglich gehalten; **b)** *(coll.: remember)* **~ to do sth.** daran denken, etw. zu tun; **c)** *(imagine)* sich *(Dat.)* vorstellen. **2.** *v.i.,* **thought a)** [nach]denken; **we want to make the students ~:** wir möchten die Studenten zum Denken bringen; **I need time to ~:** ich muß es mir erst überlegen; **I've been ~ing** ich habe nachgedacht; **~ in German** *etc.* deutsch *usw.* denken; **it makes you ~:** es macht *od.* stimmt einen nachdenklich; **just ~!** stell dir das mal vor!; **~ for oneself** sich *(Dat.)* seine eigene Meinung bilden; **~ [to oneself] ...:** sich *(Dat.)* im stillen denken ...; **let me ~:** laß [mich] mal nachdenken *od.* überlegen; **you'd better ~ again!** da hast du dich aber geschnitten! *(ugs.);* **~ twice** es sich *(Dat.)* zweimal überlegen; **this made her ~ twice** das gab ihr zu denken; **~ twice about doing sth.** es sich *(Dat.)* zweimal überlegen, ob man etw. tut; **b)** *(have intention)* **I ~ I'll try** ich glaube *od.* denke, ich werde es versuchen; **we ~ we'll enter for the regatta** wir haben vor, an der Regatta teilzunehmen. **3.** *n. (coll.)* **have a [good] ~:** es sich *(Dat.)* gut überlegen; **have a ~ about that!** denk mal drüber nach! *(ugs.);* **you have [got] another ~ coming!** da irrst du dich aber gewaltig!

~ about *v.t.* **a)** *(consider)* nachdenken über (+ *Akk.*); **what are you ~ing about?** woran *od.* was denkst du [gerade]?; **give sb. something to ~ about** jmdm. etwas geben, worüber er/sie nachdenken kann; *(to worry about)* jmdm. zu denken geben; **it doesn't bear ~ing about** man darf gar nicht daran denken; **b)** *(consider practicability of)* sich *(Dat.)* durch den Kopf gehen lassen; sich *(Dat.)* überlegen; **it's worth ~ing about** es ist überlegenswert

~ a'head *v.i.* vorausdenken

~ 'back to *v.t.* sich zurückerinnern an (+ *Akk.*)

~ of *v.t.* **a)** *(consider)* denken an (+ *Akk.*); **... but I can't ~ of everything at once!** ... aber ich habe schließlich auch nur einen Kopf!; **he ~s of everything** er denkt einfach an alles; **he never ~s of anyone but himself** er denkt immer nur an sich; **[just] ~** *or* **to ~ of it!** man stelle sich *(Dat.) od.* stell dir das bloß vor!; **[now I] come to ~ of it, ...:** wenn ich es mir recht überlege, ...; **b)** *(be aware of in the mind)* denken an (+ *Akk.*); **c)** *(consider the possibility of)* denken an (+ *Akk.*); **be ~ing of resigning** sich mit dem Gedanken tragen, zurückzutreten; **I don't know what she was ~ing of!** ich weiß nicht, was sie sich dabei gedacht hat!; **d)** *(choose from what one knows)* **I want you to ~ of a word beginning with B** überlege dir ein Wort, das mit B beginnt; **~ of a number, double it and ...:** denk dir eine Zahl, verdoppele sie und ...; **e)** *(have as idea)* **we'll ~ of something** wir werden uns etwas einfallen lassen; **can you ~ of anyone who ...?** fällt dir jemand ein, der ...?; **we're still trying to ~ of a suitable title for the book** wir suchen noch immer einen passenden Titel für das Buch; **what 'will they ~ of next?** was werden sie sich *(Dat.)* wohl [sonst] noch alles einfallen lassen?; **f)** *(remember)* sich erinnern an (+ *Akk.*); **I just can't ~ of her name** ich komme einfach nicht auf ihren Namen; **g) ~ little/nothing of sb./sth.** *(consider contemptible)* wenig/nichts von jmdm./etw. halten; **~ little/nothing of doing sth.** *(consider insignificant)* wenig/nichts dabei finden, etw. zu tun; **~ much** *or* **a lot** *or* **well** *or* **highly of sb./sth.** viel von jmdm./etw. halten; **not ~ much of sb./sth.** nicht viel von jmdm./etw. halten

~ 'out *v.t.* **a)** *(consider carefully)* durchdenken; **b)** *(devise)* sich *(Dat.)* ausdenken

~ 'over *v.t.* sich *(Dat.)* überlegen; überdenken; **I will ~ it over** ich lasse es mir durch den Kopf gehen

~ 'through *v.t.* [gründlich] durchdenken

~ 'up *v.t. (coll.)* sich *(Dat.)* ausdenken

thinker ['θɪŋkə(r)] *n.* Denker, *der*/Denkerin, *die*

thinking ['θɪŋkɪŋ] **1.** *n.* **in modern ~ ...:** nach heutiger Auffassung ...; **what is your ~ on this question?** wie ist deine Meinung zu dieser Frage? **2.** *attrib. adj.* [vernünftig] denkend

'thinking-cap *n.* **put on one's ~:** scharf nachdenken; seinen Geist anstrengen

'think-tank *n.* Beraterstab, *der*

thinly ['θɪnlɪ] **a)** *adv.* dünn; **b)** *(sparsely)* spärlich ⟨*bevölkert, bewaldet*⟩; dünn ⟨*besiedelt*⟩; **c)** *(inadequately)* leicht bekleidet; *(fig.)* dürftig ⟨*verschleiert, verkleidet*⟩

thinner ['θɪnə(r)] **1.** *adj., adv. com-*

par. of **thin 1, 2. 2.** *n.* **~[s]** Verdünner, *der;* Verdünnungsmittel, *das*
'thin-skinned *adj. (fig.)* empfindlich; dünnhäutig *(geh.)*
third [θɜ:d] **1.** *adj.* dritt...; **the ~ finger** der Ringfinger; **~ largest/highest** *etc.* drittgrößt.../-höchst... *usw.;* **every ~ week** jede dritte Woche; **a ~ part** *or* **share** ein Drittel. **2.** *n.* **a)** *(in sequence)* dritte, *der/die/das; (in rank)* Dritte, *der/die/das; (fraction)* Drittel, *das;* **be the ~ to arrive** als dritter/dritte ankommen; **b)** *(day)* **the ~ of May** der dritte Mai; **the ~ [of the month]** der Dritte [des Monats]
third 'gear *n., no pl. (Motor Veh.)* dritter Gang; *see also* **gear 1 a**
thirdly ['θɜ:dlɪ] *adv.* drittens
third: ~ 'party *n.* Dritte, *der/die;* dritte Person; *attrib.* **~-party insurance** Haftpflichtversicherung, *die;* **be covered by ~-party insurance** haftpflichtversichert sein; **~ 'person** *n.* **a)** *see* **~ party; b)** *see* **person c; ~-rate** *adj.* drittklassig; **T~ 'World** *n.* dritte Welt; **countries of the T~ World, T~ World countries** Länder der dritten Welt
thirst [θɜ:st] **1.** *n.* Durst, *der;* **die of ~:** verdursten; *(fig.: be very thirsty)* vor Durst sterben *(ugs.);* **~ for knowledge** Wissensdurst, *der.* **2.** *v. i.* **~ for revenge/knowledge** nach Rache/Wissen dürsten *(geh.)*
thirsty ['θɜ:stɪ] *adj.* **a)** durstig; **be ~:** Durst haben; **sb. is ~ for sth.** *(fig.)* jmd. *od.* jmdn. dürstet nach etw. *(dichter.);* **b)** *(coll.: causing thirst)* durstig machend; **this is ~ work** diese Arbeit macht durstig
thirteen [θɜ:'ti:n] **1.** *adj.* dreizehn; *see also* **eight 1. 2.** *n.* Dreizehn, *die; see also* **eight 2 a, d; eighteen 2**
thirteenth [θɜ:'ti:nθ] **1.** *adj.* dreizehnt...; *see also* **eighth 1. 2.** *n.* **a)** *(fraction)* Dreizehntel, *das;* **b) Friday the ~:** Freitag, der Dreizehnte. *See also* **eighth 2**
thirtieth ['θɜ:tɪɪθ] **1.** *adj.* dreißigst...; *see also* **eighth 1. 2.** *n.* *(fraction)* Dreißigstel, *das; see also* **eighth 2**
thirty ['θɜ:tɪ] **1.** *adj.* dreißig; *see also* **eight 1. 2.** *n.* Dreißig, *die; see also* **eight 2 a; eighty 2**
thirty: ~-'first *etc. adj.* einunddreißigst... *usw.; see also* **eighth 1; ~-'one** *etc.* **1.** *adj.* einunddreißig *usw.; see also* **eight 1; 2.** *n.* Einunddreißig *usw., die; see also* **eight 2 a**
this [ðɪs] **1.** *adj., pl.* **these** [ði:z] **a)** dieser/diese/dieses; *(with less emphasis)* der/die/das; **at ~ time** zu dieser Zeit; **before ~ time** vorher; zuvor; **these days** heut[zutag]e; **I'll say ~ much/I can tell you ~ much ...:** soviel kann ich sagen/soviel kann ich dir verraten ...; **all ~ week** die[se] ganze Woche; **by ~ time** inzwischen; mittlerweile; **~ morning/evening** *etc.* heute morgen/abend *usw.;* **these last three weeks** die letzten drei Wochen; **~ Monday** *(to come)* nächsten Montag; **b)** *(coll.: previously unspecified)* **they dug ~ great big trench** sie hoben einen riesigen Graben aus; **I was in the pub when ~ fellow came up to me** ich war in der Kneipe, als [so] einer *od.* so'n Typ auf mich zukam *(ugs.). See also* **that 1 c. 2.** *pron., pl.* **these a) what's ~?** was ist [denn] das?; **what is all ~?** was soll das alles?; **what flower is ~?** was ist das für eine Blume?; **fold it like ~!** falte es so!; **I knew all ~ before** ich wußte dies *od.* das alles schon vorher; **~ is not fair!** das ist nicht fair!; **what's all ~ about Jan and Angela separating?** stimmt das, daß Jan und Angela sich trennen wollen?; **b)** *(the present)* **before ~:** bis jetzt; **c)** *(Brit. Teleph.: person speaking)* **~ is Andy [speaking]** hier [spricht *od.* ist] Andy; **d)** *(Amer. Teleph.: person spoken to)* **who did you say ~ was?** wer ist am Apparat?; mit wem spreche ich, bitte?; **e) ~ and that** dies und das; **~, that, and the other** alles mögliche. **3.** *adv. (coll.)* so; **~ much** so viel
thistle ['θɪsl] *n.* Distel, *die*
thong [θɒŋ] *n.* [Leder]riemen, *der*
thorax ['θɔ:ræks] *n., pl.* **thoraces** ['θɔ:rəsi:z] *or* **~es** *(Anat., Zool.)* Thorax, *der*
thorn [θɔ:n] *n.* **a)** *(part of plant)* Dorn, *der;* **b)** *(plant)* Dornenstrauch, *der;* **c) a ~ in the flesh** *or* **side/in sb.'s flesh** *or* **side** ein Pfahl im Fleische/im Fleische für jmdn.
thorny ['θɔ:nɪ] *adj.* **a)** dornig; **b)** *(fig.: difficult)* heikel; dornenreich ⟨*Weg*⟩
thorough ['θʌrə] *adj.* **a)** gründlich; durchgreifend ⟨*Reform*⟩; genau ⟨*Beschreibung, Anweisung*⟩; **b)** *(downright)* ausgemacht ⟨*Halunke, Nervensäge*⟩
thorough: ~bred 1. *adj.* **a)** reinrassig ⟨*Tier*⟩; vollblütig ⟨*Pferd*⟩; **b)** *(fig.)* rassig ⟨*Sportwagen*⟩; **2.** *n.* reinrassiges Tier; *(horse)* Rassepferd, *das; (Horse-racing)* Vollblut, *das;* **~fare** *n.* Durchfahrtsstraße, *die;* **'no ~fare'** „Durchfahrt verboten"; *(on foot)* „kein Durchgang"; **~going** *adj.* **a)** *see* **thorough a; b)** *(extreme)* radikal ⟨*Konservative, Sozialist*⟩
thoroughly ['θʌrəlɪ] *adv.* gründlich ⟨*untersuchen, prüfen*⟩; gehörig ⟨*müde, erschöpft*⟩; so richtig ⟨*genießen*⟩; ausgesprochen ⟨*langweilig*⟩; zutiefst ⟨*beschämt*⟩; *(completely)* völlig ⟨*durchnäßt, verzogen*⟩; total ⟨*verdorben, verwöhnt*⟩; **be ~ fed up with sth.** *(sl.)* von etw. die Nase gestrichen voll haben *(ugs.);* **be ~ delighted with sth.** sich außerordentlich über etw. *(Akk.)* freuen
thoroughness ['θʌrənɪs] *n., no pl.* Gründlichkeit, *die*
those *see* **that 1, 2**
thou [ðaʊ] *pron. (arch./poet./dial.)* du; *(Relig.: God)* Du
though [ðəʊ] **1.** *(conj.)* **a)** *(despite the fact that)* obwohl; **late ~ it was** obwohl es so spät war; **the car, ~ powerful, is also economical** der Wagen ist zwar stark, aber [zugleich] auch wirtschaftlich; **b)** *(but nevertheless)* aber; **a slow ~ certain method** eine langsame, aber *od.* wenn auch sichere Methode; **c)** *(even if)* **[even] ~:** auch wenn; **as ~ = as if** *see* **if 1 a; d)** *(and yet)* **~ you never know** obwohl man nie weiß; **she read on, ~ not to the very end** sie las weiter, wenn auch nicht bis ganz zum Schluß. **2.** *adv. (coll.)* trotzdem; **I like him ~:** ich mag ihn aber [trotzdem]; **you don't know him, ~:** aber du kennst ihn nicht
thought [θɔ:t] **1.** *see* **think 1, 2. 2.** *n.* **a)** *no pl.* Denken, *das;* **[lost] in ~:** in Gedanken [verloren *od.* versunken]; **b)** *no pl., no art. (reflection)* Überlegung, *die;* Nachdenken, *das;* **after serious ~:** nach reiflicher Überlegung; **c)** *(consideration)* Rücksicht, *die* (**for** auf + *Akk.*); **he has no ~ for others** er nimmt keine Rücksicht auf andere; **give [plenty of] ~ to sth., give sth. [plenty of] ~:** [reiflich] über etw. *(Akk.)* nachdenken; **he never gave the matter a moment's ~:** er dachte keinen Augenblick daran; **d)** *(idea, conception)* Gedanke, *der;* **I've just had a ~!** mir ist gerade ein [guter] Gedanke gekommen; **it's the ~ that counts** der gute Wille zählt; **at the [very] ~ of sth./of doing sth./that ...:** beim [bloßen] Gedanken an etw. *(Akk.)*/daran, etw. zu tun/, daß ...; **that's** *or* **there's a ~!** das ist aber eine [gute] Idee!; **she is [constantly] in his**

~s er muß ständig an sie denken; **e)** *in pl. (opinion)* Gedanken; **I'll tell you my ~s on the matter** ich sage dir, wie ich darüber denke; **f)** *(intention)* **have no ~ of doing sth.** überhaupt nicht daran denken, etw. zu tun; **give up all ~[s] of sth./doing sth.** sich *(Dat.)* etw. aus dem Kopf schlagen/es sich *(Dat.)* aus dem Kopf schlagen, etw. zu tun; **nothing was further from my ~s** nicht im Traum hätte ich daran gedacht

thoughtful ['θɔ:tfl] *adj.* **a)** *(meditative)* nachdenklich; **b)** *(considerate)* rücksichtsvoll; *(helpful)* aufmerksam; **c)** *(showing original thought)* gedankenreich; *(well thought out)* [gut] durchdacht; wohlüberlegt ‹*Bemerkung*›

thoughtfully ['θɔ:tfəlɪ] *adv.* **a)** *(meditatively)* nachdenklich; **b)** *(considerately)* rücksichtsvollerweise; **c)** *(in a well-thought-out manner)* **a ~ written article** ein gut durchdachter Artikel

thoughtless ['θɔ:tlɪs] *adj.* **a)** gedankenlos; **~ of the danger, ...:** ohne an die Gefahr zu denken ...; **b)** *(inconsiderate)* rücksichtslos

thoughtlessly ['θɔ:tlɪslɪ] *adv.* **a)** gedankenlos; **b)** *(inconsiderately)* aus Rücksichtslosigkeit

thought: ~-provoking *adj.* **be ~-provoking** nachdenklich stimmen; **~-reader** *n.* Gedankenleser, *der*/-leserin, *die*

thousand ['θaʊznd] **1.** *adj.* **a)** tausend; **a** *or* **one ~:** eintausend; **two/several ~:** zweitausend/mehrere tausend; **one and a half ~:** [ein]tausendfünfhundert; **a** *or* **one ~ and one** [ein]tausend[und]eins; **a** *or* **one ~ and one people** [ein]tausendundeine Person; **b) a ~ [and one]** *(fig.: innumerable)* tausend *(ugs.)*; **a ~ thanks** tausend Dank. *See also* **eight 1. 2.** *n.* **a)** *(number)* tausend; **a** *or* **one/two ~:** ein-/zweitausend; **a ~ and one** [ein]tausend[und]eins; **a ~-to-one chance** eine Chance von tausend zu eins; **b)** *(symbol, written figure)* Tausend, *die; (in adding numbers by columns)* Tausender, *der (Math.); (set or group)* Tausend, *das;* **c)** *(indefinite amount)* **~s** Tausende. *See also* **eight 2 a**

thousandth ['θaʊzndθ] **1.** *adj.* tausendst...; **a ~ part** ein Tausendstel. **2.** *n. (fraction)* Tausendstel, *das; (in sequence)* tausendste, *der/die/das; (in rank)* Tausendste, *der/die/das. See also* **eighth 2**

thrash [θræʃ] *v.t.* **a)** *(beat)* [ver]prügeln; **b)** *(defeat)* vernichtend schlagen; **c)** *see* **thresh**

~ **'out** *v.t.* ausdiskutieren ‹*Problem, Frage*›; ausarbeiten ‹*Plan*›

thrashing ['θræʃɪŋ] *n.* **a)** *(beating)* Prügel *Pl.;* **b)** *(defeat)* Schlappe, *die*

thread [θred] **1.** *n.* **a)** Faden, *der;* **b)** *(fig.)* **hang by a ~** *(be in a precarious state)* an einem [dünnen *od.* seidenen] Faden hängen; *(depend on sth. still in doubt)* auf Messers Schneide stehen; **lose the ~:** den Faden verlieren; **take** *or* **pick up the ~ of the conversation** den Gesprächsfaden wieder aufnehmen; **c)** *(of screw)* Gewinde, *das.* **2.** *v.t.* **a)** *(pass ~ through)* einfädeln; auffädeln ‹*Perlen*›; **b) ~ one's way through sth.** *(lit. or fig.)* sich durch etw. schlängeln

'threadbare *adj.* abgenutzt; abgetragen ‹*Kleidung*›; *(fig.)* abgedroschen ‹*Argument*› *(ugs.)*

threat [θret] *n.* Drohung, *die;* **make a ~ against sb.** jmdm. drohen; **under ~ of** unter Androhung von; **at the slightest ~ of sth.** wenn etw. auch nur ganz entfernt droht

threaten ['θretn] *v.t.* **a)** *(use threats towards)* bedrohen; **~ sb. with prosecution/a beating** jmdm. Verfolgung/Schläge androhen; **b)** *(announce one's intention)* **~ to do sth.** damit drohen, etw. zu tun; **the fire ~ed to engulf the whole village** *(fig.)* das Feuer drohte das ganze Dorf einzuschließen; **~ to commit suicide/to resign** mit Selbstmord/dem Rücktritt drohen; **c)** drohen mit ‹*Gewalt, Repressalien, Rache usw.*›; **the sky ~s rain** am Himmel hängen drohende Regenwolken

threatening ['θretnɪŋ] *adj.* drohend; **~ letter** Drohbrief, *der*

three [θri:] **1.** *adj.* drei; *see also* **eight 1; R b. 2.** *n.* **a)** *(number, symbol)* Drei, *die;* **b)** *(set of ~ people)* Dreiergruppe, *die;* **the ~ [of them]** die Drei. *See also* **eight 2 a, c, d**

three: ~-dimensional [θri:dɪ'menʃənl, θri:daɪ'menʃənl] *adj.* dreidimensional; **~fold** *adj., adv.* dreifach; **a ~-fold increase** ein Anstieg auf das Dreifache; **~penny bit** ['θrepənɪ bɪt] *n. (Brit. Hist.)* Dreipencestück, *das;* **~-piece** *adj. see* **piece 1 b; ~-pin** *adj. see* **pin 1 c; ~-ply** *adj.* dreilagig ‹*Holz*›; dreifädig ‹*Wolle, Zwirn*›; **~-quarter** *adj.* dreiviertel; **~-quarters 1.** *n.* **a)** drei Viertel *pl.* **(of + *Gen.*)**; **~-quarters of an hour** eine Dreiviertelstunde; **b)** *attrib.* Dreiviertel‹*mehrheit usw.*›; **2.** *adv.* dreiviertel ‹*voll*›; zu drei Vierteln ‹*fertig*›

threesome ['θri:səm] *n.* Dreigespann, *das;* Trio, *das;* **go as a ~:** zu dritt gehen

three: ~-storey *adj.* dreistöckig; **~-way adaptor** *n. (Electr.)* Dreifachstecker, *der;* **~-wheeler** [θri:'wi:lə(r)] *n.* Dreirad, *das (Kfz-W.)*

thresh [θreʃ] *v.t. (Agric.)* dreschen

threshold ['θreʃəʊld] *n. (lit. or fig.)* Schwelle, *die;* **be on the ~ of sth.** *(fig.)* an der Schwelle einer Sache *(Gen.)* stehen

threw *see* **throw 1**

thrift [θrɪft] *n.* **a)** *no pl.* Sparsamkeit, *die;* **b)** *(Bot.)* Grasnelke, *die*

'thrift account *n. (Amer.)* Sparkonto, *das*

thrifty ['θrɪftɪ] *adj.* sparsam

thrill [θrɪl] **1.** *v.t. (excite)* faszinieren; *(delight)* begeistern; **be ~ed by/with sth.** von etw. fasziniert/begeistert sein. **2.** *n.* **a)** *(wave of emotion)* Erregung, *die;* **a ~ of joy/pleasure** freudige Erregung; **a ~ of excitement/anticipation** prickelnde Erregung/Vorfreude; **b)** *(exciting experience)* aufregendes Erlebnis; **sb. gets a ~ out of sth.** etw. erregt jmdn.; **cheap ~s** anspruchsloser Nervenkitzel *(ugs.)*

thriller ['θrɪlə(r)] *n.* Thriller, *der*

thrilling ['θrɪlɪŋ] *adj.* aufregend; spannend ‹*Buch, Film, Theaterstück, Geschichte*›; packend ‹*Ereignis*›; mitreißend ‹*Musik*›; prickelnd ‹*Gefühl*›

thrive [θraɪv] *v.i.*, **~d** *or* **throve** [θrəʊv], **~d** *or* **thriven** ['θrɪvn] **a)** *(grow vigorously)* wachsen und gedeihen; **b)** *(prosper)* aufblühen **(on** bei**)**; **business is thriving** das Geschäft floriert; **c)** *(grow rich)* reich werden

throat [θrəʊt] *n.* **a)** *(outside and inside of neck)* Hals, *der; (esp. inside)* Kehle, *die;* **look down sb.'s ~:** jmdm. in den Hals *od.* Rachen schauen; **a [sore] ~:** Halsschmerzen; **cut sb.'s ~:** jmdm. die Kehle durchschneiden; **cut one's own ~** *(fig.)* sich *(Dat.)* ins eigene Fleisch schneiden; **ram** *or* **thrust sth. down sb.'s ~** *(fig.)* jmdm. etw. aufzwingen; **b)** *(of bottle, vase)* Hals, *der*

throaty ['θrəʊtɪ] *adj.* **a)** *(from the throat)* kehlig; **b)** *(hoarse)* heiser

throb [θrɒb] **1.** *v.i.*, **-bb-: a)** *(palpitate, pulsate)* pochen; **b)** ‹*Motor, Artillerie:*› dröhnen. **2.** *n. see* **1:** Pochen, *das;* Dröhnen, *das*

throes [θrəʊz] *n. pl.* Qual, *die;* **be in the ~ of sth.** *(fig.)* mitten in etw. *(Dat.)* stecken *(ugs.)*

thrombosis [θrɒm'bəʊsɪs] *n., pl.* **thromboses** [θrɒm'bəʊsi:z] *(Med.)* Thrombose, *die*

throne [θrəʊn] *n.* Thron, *der;* **succeed to the ~:** die Thronfolge antreten

throng [θrɒŋ] **1.** *n.* [Menschen]menge, *die.* **2.** *v. i.* strömen **(into** in + *Akk.*); *(press)* sich drängen. **3.** *v. t.* sich drängen in (+ *Dat.*)

throttle ['θrɒtl] **1.** *n. (Mech. Engin.)* **~[-valve]** Drosselklappe, *die;* **~[-pedal]** *(Motor Veh.)* Gas[pedal], *das;* **~[-lever]** Gashebel, *der;* **at full ~** *(Motor Veh.)* mit Vollgas. **2.** *v. t.* erdrosseln; *(fig.)* ersticken

through [θru:] **1.** *prep.* **a)** durch; *(fig.)* **search/read ~ sth.** etw. durchsuchen/durchlesen; **live ~ sth.** *(survive)* etw. überleben; *(experience)* etw. erleben; **b)** *(Amer.: up to and including)* bis [einschließlich]; **c)** *(by reason of)* durch; infolge von ‹*Vernachlässigung, Einflüssen*›; **it was all ~ you that we were late** es war nur deine Schuld, daß wir zu spät gekommen sind; **it happened ~ no fault of yours** es geschah nicht durch deine Schuld. **2.** *adv.* **a) let sb. ~:** jmdn. durchlassen; **be ~ with a piece of work/with sb.** mit einer Arbeit fertig/mit jmdm. fertig *(ugs.)* sein; **b)** *(Teleph.)* **be ~:** durch sein *(ugs.);* **be ~ to sb.** mit jmdm. verbunden sein. **3.** *attrib. adj.* durchgehend ‹*Zug*›; **~ coach** *or* **carriage** Kurswagen, *der* **(for** nach); **~ traffic** Durchgangsverkehr, *der;* **'no ~ road'** „keine Durchfahrt[sstraße]"

through: ~'out 1. *prep.* **~out the war/period** den ganzen Krieg/die ganze Zeit hindurch; **spread ~out the country** sich im ganzen Land verbreiten; **2.** *adv. (entirely)* ganz; *(always)* stets; die ganze Zeit [hindurch]; **~way** *n. (Amer.: expressway)* Schnellstraße, *die*

throve *see* **thrive**

throw [θrəʊ] **1.** *v. t.,* **threw** [θru:], **thrown** [θrəʊn] **a)** werfen; **~ sth. to sb.** jmdm. etw. zuwerfen; **~ sth. at sb.** etw. nach jmdm. werfen; **~ me that towel, please** wirf mal bitte das Handtuch rüber *(ugs.);* **~ a punch/punches** zuschlagen; **~ a left/right** eine Linke/Rechte schlagen; **~ oneself on one's knees/to the floor/into a chair** sich auf die Knie/zu Boden/in einen Sessel werfen; **~ oneself at sb.** sich auf jmdn. werfen; *(fig.)* sich jmdm. an den Hals werfen *(ugs.);* **b)** *(fig.)* **~ sb. out of work/into prison** jmdn. entlassen *od. (ugs.)* hinauswerfen/ins Gefängnis werfen *(geh.);* **be ~n upon one's own resources** selbst für sich aufkommen müssen; **~ oneself into a task** sich in eine Arbeit *(Akk.)* stürzen; **~ sth. into disarray** etw. durcheinanderbringen; **c)** *(bring to the ground)* zu Boden werfen ‹*Ringer, Gegner*›; *(unseat)* abwerfen ‹*Reiter*›; **d)** *(coll.: disconcert)* ‹*Frage:*› aus der Fassung bringen; **e)** *(Pottery)* drehen; **f)** *also abs. (Games)* werfen; **~ [the/a dice]** würfeln. **2.** *n.* Wurf, *der*

~ a'bout *v. t.* herumwerfen *(ugs.);* **~ one's money about** *(fig.)* mit Geld um sich werfen; *see also* **weight 1 a**

~ a'round *v. t.* **a)** *see* **~ about; b)** *(surround with)* **~ a cordon around an area** ein Gebiet abriegeln

~ a'way *v. t.* **a)** *(get rid of, waste)* wegwerfen; *(discard)* abwerfen ‹*Spielkarte*›; **~ away money on sth.** Geld für etw. wegwerfen; **~ oneself away on sb.** sich an jmdn. wegwerfen; **b)** *(lose by neglect)* verschenken ‹*Vorteil, Vorsprung, Spiel usw.*›

~ 'back *v. t.* **a)** *(return, repulse)* zurückwerfen; **b)** zurückschlagen ‹*Bettuch, Vorhang, Teppich*›; zurückwerfen ‹*Kopf*›

~ 'down *v. t.* **~ down [on the ground]** auf den Boden werfen; **it's ~ing it down** *(coll.)* es gießt [wie aus Eimern] *(ugs.)*

~ 'in *v. t.* **a)** *(include as free extra)* [gratis] dazugeben; **with ... ~n in** mit ... als Zugabe; **b)** *(interpose)* einstreuen ‹*Bemerkung*›; **c)** *(Footb., Rugby)* einwerfen; **d) ~ one's hand in** *(fig.: withdraw)* aufgeben

~ 'off *v. t.* **a)** *(discard)* ablegen ‹*Maske, Verkleidung*›; von sich werfen ‹*Kleider*›; *(get rid of)* loswerden ‹*Erkältung, lästige Person*›; **b)** *(perform or write casually)* [mühelos] hinwerfen ‹*Rede, Gedicht usw.*›

~ 'on 1. *v. t.* sich werfen in ‹*Kleider*›. **2.** *v. refl.* **~ oneself [up]on sb.** sich auf jmdn. stürzen

~ 'out *v. t.* **a)** *(discard)* wegwerfen; **b)** *(expel)* **~ sb. out [of sth.]** jmdn. [aus etw.] hinauswerfen *(ugs.);* **~ sb. out of work** jmdn. hinauswerfen *(ugs.);* **c)** *(refuse)* verwerfen ‹*Plan usw.*›; **d)** *(put forward tentatively)* in den Raum stellen ‹*Vorschläge*›; **e) ~ out one's chest** die Brust herausdrücken; **f)** *(confuse)* durcheinanderbringen; aus dem Konzept bringen ‹*Sprecher*›

~ to'gether *v. t.* **a)** *(assemble hastily)* zusammenhauen *(ugs.);* zusammenwerfen ‹*Ideen, Zutaten*›; herzaubern ‹*Essen*›; zusammenschustern *(ugs. abwertend)* ‹*Aufsatz, Artikel*›; zusammenschreiben ‹*Buch, Artikel, Rede*›; **b)** *(bring together)* zusammenwürfeln

~ 'up 1. *v. t.* **a)** *(lift quickly)* hochwerfen ‹*Arme, Hände*›; [plötzlich] hochschieben ‹*Fenster*›; **b)** *(erect quickly)* hochziehen *(salopp)* ‹*Gebäude*›; **c)** *(give up)* hinwerfen *(ugs.)* ‹*Arbeit*›; aufgeben ‹*Versuch*›; abbrechen ‹*Laufbahn, Ausbildung*›; **d)** *(produce)* hervorbringen ‹*Führer, Ideen usw.*›; **e)** *(coll.: vomit)* ausspucken *(ugs.).* **2.** *v. i. (coll.: vomit)* brechen *(ugs.)*

throw: ~-away 1. *adj.* **a)** *(disposable)* Wegwerf-; Einweg-; **b)** *(underemphasized)* beiläufig [gesprochen] ‹*Bemerkung*›; **2.** *n.* Wegwerfartikel, *der; (bottle)* Einwegflasche, *die;* **~-back** *n.* Rückkehr, *die* **(to** zu)

thrower ['θrəʊə(r)] *n.* Werfer, *der*/Werferin, *die*

'throw-in *n. (Footb., Rugby)* Einwurf, *der*

thrown *see* **throw 1**

thru [θru:] *(Amer.) see* **through**

thrush [θrʌʃ] *n. (Ornith.)* Drossel, *die*

thrust [θrʌst] **1.** *v. t.,* **thrust a)** *(push suddenly)* stoßen; **he ~ his fist into my face** er stieß mir seine Faust ins Gesicht; *(fig.)* **~ aside** beiseite schieben; in den Wind schlagen ‹*Warnungen*›; **~ extra work [up]on sb.** jmdm. zusätzliche Arbeit aufbürden; **fame was ~ upon her** sie wurde unversehens berühmt; **b) ~ one's way through/into/out of sth.** sich durch/in/aus etw. drängen. **2.** *n.* **a)** *(sudden push)* Stoß, *der;* **b)** *(gist)* Stoßrichtung, *die;* **c)** *(Mil.: advance)* Vorstoß, *der;* **d)** *(force of jet engine)* Schub, *der*

thud [θʌd] **1.** *v. i.,* **-dd-** dumpf schlagen; **~ to the floor/ground** dumpf [auf dem Fußboden/Boden] aufschlagen. **2.** *n.* dumpfer Schlag

thug [θʌg] *n.* Schläger, *der*

thuggery ['θʌgərɪ] *n., no pl.* Schlägerunwesen, *das*

thuggish ['θʌgɪʃ] *adj.* aggressiv ‹*Verhalten, Fußballfan*›; **~ lout/youth** jugendlicher Schläger

thumb [θʌm] **1.** *n.* Daumen, *der;* **give sb. the ~s down on a**

proposal/idea jmds. Vorschlag/ Idee ablehnen; **get the ~s down** ‹*Idee*:› verworfen werden; ‹*Kandidat*:› abgelehnt werden; **get the ~s up** ‹*Person, Projekt*:› akzeptiert werden; **have ten ~s, be all ~s** zwei linke Hände haben *(ugs.)*; **have sb. under one's ~**: jmdn. unter der Fuchtel haben *(ugs.)*; **be under sb.'s ~**: unter jmds. Fuchtel stehen. **2.** *v. t.* **a)** ~ **a lift** einem Autofahrer winken, um sich mitnehmen zu lassen; *(hitch-hike)* per Anhalter fahren; **b)** *(turn over)* [mit dem Daumen] durchblättern ‹*Buch*›; [mit dem Daumen] umblättern ‹*Seiten*›; **c)** ~ **one's nose [at sb.]** [jmdm.] eine lange Nase machen

~ **through** *v. t.* [mit dem Daumen] durchblättern ‹*Buch*›

thumb: ~-index *n.* Daumenregister, *das;* **~-nail** *n.* Daumennagel, *der; attrib.* **~-nail sketch** *(Art)* Miniaturportrait, *das; (fig.: brief description)* kurze Beschreibung; **~-tack** *n. (Amer.)* Reißzwecke, *die*

thump [θʌmp] **1.** *v. t.* [mit Wucht] schlagen. **2.** *v. i.* **a)** hämmern **(at, on** gegen); ‹*Herz*:› heftig pochen; **b)** *(move noisily)* ~ **around** herumpoltern. **3.** *n.* **a)** *(blow)* Schlag, *der;* **b)** *(dull sound)* Bums, *der (ugs.);* dumpfer Schlag

thunder ['θʌndə(r)] **1.** *n.* **a)** *no pl., no indef. art.* Donner, *der;* **roll/ crash of ~**: Donnerrollen, *das*/ -schlag, *der;* **b) steal sb.'s ~** *(fig.)* jmdm. die Schau stehlen *(ugs.).* **2.** *v. i.* donnern

thunder: ~bolt *n.* Blitzschlag [mit Donner]; *(from God)* Blitzstrahl, *der (geh.);* **come as something of a ~bolt** wie ein Blitz einschlagen; **~clap** *n.* Donnerschlag, *der;* **~-cloud** *n.* Gewitterwolke, *die*

thunderous ['θʌndərəs] *adj.* donnernd

thunder: ~storm *n.* Gewitter, *das;* **~struck** *adj.* **be ~struck** wie vom Donner gerührt sein

thundery ['θʌndərɪ] *adj.* gewittrig; **it looks ~**: es sieht nach Gewitter aus

Thurs. *abbr.* **Thursday** Do.

Thursday ['θɜ:zdeɪ, 'θɜ:zdɪ] **1.** *n.* Donnerstag, *der.* **2.** *adv. (coll.)* **she comes ~s** sie kommt donnerstags. *See also* **Friday**

thus [ðʌs] *adv.* **a)** *(in the way indicated)* so; *(thereby)* dadurch; **b)** *(accordingly)* deshalb; daher; **c)** *(to this extent)* ~ **much/far** so viel/so weit

thwart [θwɔ:t] *v. t.* durchkreuzen ‹*Pläne, Absichten*›; vereiteln ‹*Versuch*›; ~ **sb.** jmdm. einen Strich durch die Rechnung machen

thy [ðaɪ] *poss. pron. attrib. (arch./ poet./dial.)* dein; *see also* **²her**

thyme [taɪm] *n. (Bot.)* Thymian, *der*

thyroid ['θaɪrɔɪd] *n.* ~ **[gland]** *(Anat., Zool.)* Schilddrüse, *die*

tiara [tɪ'ɑ:rə] *n.* Diadem, *das*

Tibet [tɪ'bet] *pr. n.* Tibet *(das)*

Tibetan [tɪ'betn] **1.** *adj.* tibetisch; **sb. is ~**: jmd. ist Tibeter/Tibeterin. **2.** *n.* **a)** *(person)* Tibeter, *der*/Tibeterin, *die;* **b)** *(language)* Tibetisch, *das*

¹tick [tɪk] **1.** *v. i.* ticken; **what makes sb. ~** *(fig.)* worauf jmd. anspricht. **2.** *v. t.* **a)** mit einem Häkchen versehen; **b)** *see* ~ **off a. 3.** *n.* **a)** *(of clock etc.)* Ticken, *das;* **b)** *(Brit. coll.: moment)* Sekunde, *die;* **half a ~!, just a ~!** Momentchen! *(ugs.);* **I'll be with you in a ~** *or* **two ~s** ich komme gleich; **c)** *(mark)* Häkchen, *das;* **put a ~ against your preference** kennzeichnen Sie das, was Sie bevorzugen, mit einem Häkchen

~ **a'way** *v. i.* [weiter] ticken; **the minutes ~ed away** die Minuten verstrichen

~ **'off** *v. t.* **a)** *(cross off)* abhaken; **b)** *(coll.: reprimand)* rüffeln *(ugs.)*

~ **'over** *v. i.* **a)** *(Motor Veh.)* im Leerlauf laufen; ~ **over noisily/ too slowly/too fast** im Leerlauf [zu] laut/zu langsam/zu schnell drehen; **b)** *(fig.)* ~ **over [nicely]** *(progress satisfactorily)* ganz gut laufen *(ugs.)*

²tick *n. (insect)* Lausfliege, *die*

³tick *n. (coll.: credit)* **buy on ~**: auf Pump kaufen *(salopp);* **can I have it on ~?** kann ich das anschreiben lassen?

ticker-tape ['tɪkəteɪp] *n. (Amer.)* [Papier]streifen, *der (aus dem Fernschreiber)*

ticket ['tɪkɪt] *n.* **a)** Karte, *die; (for concert, theatre, cinema, exhibition)* [Eintritts]karte, *die; (for public transport)* Fahrschein, *der; (of cardboard)* Fahrkarte, *die; (for aeroplane)* Flugschein, *der;* Ticket, *das; (of lottery, raffle)* Los, *das; (for library)* Ausweis, *der;* **price ~**: Preisschild, *das;* **[parking] ~** *(notification of traffic offence)* Strafmandat, *das;* Strafzettel, *der (ugs.);* **b)** *(Amer. Polit.: list of candidates)* [Wahl]liste, *die;* **run on the Democratic/Republican ~**: für die Demokraten/Republikaner kandidieren

ticket: ~-collector *n. (on train)* Schaffner, *der*/Schaffnerin, *die; (on station)* Fahrkartenkontrolleur, *der*/-kontrolleurin, *die;* **~-holder** *n. (at concert, theatre, cinema, exhibition)* Besitzer/Besitzerin einer Eintrittskarte; **~-inspector** *n.* Fahrkartenkontrolleur, *der*/-kontrolleurin, *die;* **~-office** *n.* Kartenschalter, *der; (for public transport)* Fahrkartenschalter, *der; (for advance booking)* Kartenvorverkaufsstelle, *die*

ticking-'off *n. (sl.)* Rüffel, *der (ugs.)*

tickle ['tɪkl] **1.** *v. t.* **a)** *(touch lightly)* kitzeln; **b)** *(amuse)* **be ~d by sth.** sich über etw. *(Akk.)* amüsieren; **be ~d pink about sth.** *(coll.)* sich wahnsinnig über etw. *(Akk.)* freuen *(ugs.);* ~ **sb.'s fancy** jmdn. reizen. **2.** *v. i.* kitzeln

ticklish ['tɪklɪʃ] *adj. (lit. or fig.)* kitzlig

tick: ~-over *n. (Motor Veh.)* Leerlauf, *der;* **~-tock** ['tɪktɒk] *n.* Ticktack, *das*

tidal ['taɪdl] *adj.* Gezeiten-; ~ **power-station** Gezeitenkraftwerk, *das*

'tidal wave *n.* Flutwelle, *die*

tiddler ['tɪdlə(r)] *n. (Brit. coll./ child lang.)* **a)** *(fish)* Fischchen, *das;* **b)** *(child)* Kleine, *das*

tiddly-wink ['tɪdlɪwɪŋk] *n.* **a)** *(counter)* farbiges Plättchen; **b) ~s** *sing. (game)* Flohhüpfen, *das*

tide [taɪd] **1.** *n.* **a)** *(rise or fall of sea)* Tide, *die (nordd., bes. Seemannsspr.);* **high ~**: Flut, *die;* **low ~**: Ebbe, *die;* **the ~s** die Gezeiten; **sail on the next ~**: mit der nächsten Flut auslaufen; **the ~ is in/out** es ist Flut/Ebbe; **when the ~ is in/out** bei Flut/Ebbe; *see also* **turn 1 g; b)** *(fig.: trend)* Trend, *der;* **go with/against the ~**: mit dem/gegen den Strom schwimmen; *see also* **turn 3 c. 2.** *v. t.* ~ **sb. over** jmdm. über die Runden helfen *(ugs.);* ~ **sb. over a difficult period** jmdm. über eine schwierige Zeit hinweghelfen

'tide-mark *n.* **a)** Flutmarke, *die;* **b)** *(Brit. coll.: line on body, bath, etc.)* Schmutzrand, *der*

tidily ['taɪdɪlɪ] *adv.* ordentlich; *(clearly)* übersichtlich ‹*präsentieren, gestalten*›

tidiness ['taɪdɪnɪs] *n., no pl.* Ordentlichkeit, *die*

tidings ['taɪdɪŋz] *n. pl. (literary)* Kunde, *die (geh.)*

tidy ['taɪdɪ] **1.** *adj.* **a)** *(neat)* ordentlich; aufgeräumt ‹*Zimmer, Schreibtisch*›; **make oneself/a room ~**: sich zurechtmachen/ein Zimmer aufräumen; **b)** *(coll.: considerable)* ordentlich *(ugs.);* **a**

~ **sum** *or* **penny** ein hübsches Sümmchen *(ugs.).* **2.** *v.t.* aufräumen ‹*Zimmer*›; ~ **oneself** sich zurechtmachen

~ **a'way** *v.t.* wegräumen

~ **'up 1.** *v.i.* aufräumen. **2.** *v.t.* aufräumen; in Ordnung bringen ‹*Text*›

tie [taɪ] **1.** *v.t.*, **tying** ['taɪɪŋ] **a)** binden (**to** an + *Akk.*, **into** zu); ~ **the prisoner's legs together** dem Gefangenen die Beine zusammenbinden; ~ **an apron round you[r waist]** binde dir eine Schürze um; ~ **a knot** einen Knoten machen; **b)** *(Sport: gain equal score in)* ~ **the match** unentschieden spielen; **c)** *(restrict)* binden (**to** an + *Akk.*). **2.** *v.i.*, **tying a)** *(be fastened)* **it won't** ~: es läßt sich nicht binden; **it ~s at the back** es wird hinten gebunden; **b)** *(have equal scores, votes, etc.)* ~ **for second place in the competition/election** mit gleicher Punktzahl den zweiten Platz im Wettbewerb/mit gleicher Stimmenzahl den zweiten Platz bei der Wahl erreichen; ~ **6:6** mit 6:6 ein Unentschieden erreichen. **3.** *n.* **a)** Krawatte, *die;* **b)** *(cord etc. for fastening)* Band, *das;* **c)** *(fig.) (bond)* Band, *das; (restriction)* Bindung, *die;* **d)** *(equality) (of scores)* Punktgleichheit, *die; (of votes)* Patt, *das;* Stimmengleichheit, *die;* **there was a ~ for third place** zwei Teilnehmer landeten punktgleich auf dem dritten Platz; **end in** *or* **be a** ~: unentschieden *od.* mit einem Unentschieden enden; **e)** *(Sport: match)* Begegnung, *die*

~ **'back** *v.t.* zurückbinden

~ **'down** *v.t.* **a)** *(fasten)* festbinden; **b)** *(fig.: restrict)* binden; **be ~d down by sth.** durch etw. gebunden *od.* eingeschränkt sein; ~ **sb. down to a time/a schedule** jmdn. auf eine Zeit/einen Zeitplan festlegen

~ **'in 1.** *v.i.* ~ **in with sth.** zu etw. passen. **2.** *v.t.* ~ **sth. in with sth.** etw. mit etw. abstimmen

~ **'up** *v.t.* **a)** *(bind)* festbinden; festmachen ‹*Boot*›; ~ **up a parcel with string** ein Paket verschnüren; **b)** *(complete arrangements for)* abschließen; **c)** *(make unavailable)* fest anlegen ‹*Geld*›; **d)** *see* ~ **in 2; e)** *(keep busy)* beschäftigen

tie: ~-break, ~-breaker *ns.* Tie-Break, *der od. das;* **~-clip** *n.* Krawattenhalter, *der;* **~-on** *adj.* Anhänge-; **~-on label** Anhänger, *der;* **~-pin** *n.* Krawattennadel, *die*

tier [tɪə(r)] *n.* **a)** *(row)* Rang, *der;* **b)** *(unit)* Stufe, *die*

tiger ['taɪgə(r)] *n.* **a)** *(Zool.)* Tiger, *der;* **paper** ~ *(fig.)* Papiertiger, *der;* **b)** *(fierce or energetic person)* Kämpfernatur, *die*

tight [taɪt] **1.** *adj.* **a)** *(firm)* fest; fest angezogen ‹*Schraube, Mutter*›; festsitzend ‹*Deckel, Korken*›; **be very** ~: sehr fest sitzen; **the drawer/window is** ~: die Schublade/das Fenster klemmt; **b)** *(close-fitting)* eng ‹*Kleid, Hose, Schuh usw.*›; **this shoe is rather [too]** ~ *or* **a rather ~ fit** dieser Schuh ist etwas zu eng; **c)** *(impermeable)* ~ **seal/joint** dichter Verschluß/dichte Fuge; **d)** *(taut)* straff; **a ~ feeling in one's chest** ein Gefühl der Beklemmung *od.* Enge in der Brust; **e)** *(with little space)* knapp; gedrängt ‹*Programm*›; **f)** *(difficult to negotiate)* **a ~ corner** eine enge Kurve; **be in/get oneself into a ~ corner** *or (coll.)* **spot [over sth.]** *(fig.)* [wegen etw.] in der Klemme sein/in die Klemme geraten *(ugs.);* **g)** *(strict)* streng ‹*Kontrolle, Disziplin*›; straff ‹*Organisation*›; **h)** *(coll.: stingy)* knauserig *(ugs.);* **i)** *(coll.: drunk)* voll *(salopp);* **get** ~: sich vollaufen lassen *(salopp).* **2.** *adv.* **a)** *(firmly)* fest; **hold ~!** halt dich fest!; **b)** *(so as to leave no space)* [ganz] voll. **3.** *n. in pl.* **a)** *(Brit.)* **[pair of] ~s** Strumpfhose, *die;* **b)** *(of dancer etc.)* Trikothose, *die*

tighten ['taɪtn] **1.** *v.t.* **a)** [fest] anziehen ‹*Knoten, Schraube, Mutter usw.*›; straffziehen ‹*Seil, Schnur*›; anspannen ‹*Muskeln*›; verstärken ‹*Griff*›; ~ **one's belt** *(fig.)* den Gürtel enger schnallen *(ugs.);* **b)** *(make stricter)* verschärfen ‹*Kontrolle, Gesetz, Vorschrift*›. **2.** *v.i.* sich spannen; ‹*Knoten:*› sich zusammenziehen

~ **'up 1.** *v.t.* **a)** anziehen; *(retighten)* nachziehen; **b)** *(make stricter)* verschärfen ‹*Gesetze, Bestimmungen, Kontrollen*›; ~ **up security** die Sicherheitsmaßnahmen verschärfen. **2.** *v.i.* härter durchgreifen; ~ **up on security/drunken driving** die Sicherheitsmaßnahmen verschärfen/bei Trunkenheit am Steuer schärfer durchgreifen

tight: ~-fisted [taɪt'fɪstɪd] *adj.* geizig; **~-fitting** *adj.* enganliegend ‹*Pullover, Trikot*›; **~-lipped** ['taɪtlɪpt] *adj.* **a)** *(without emotion)* mit zusammengepreßten Lippen *nachgestellt;* **b)** *(silent)* verschwiegen

tightness ['taɪtnɪs] *n., no pl.* **a)** *(firmness)* Festigkeit, *die;* **b)** *(closeness of fit)* enger Sitz; **c)** *(lack of leakage)* Dichtheit, *die;* **d)** *(tautness)* Straffheit, *die;* **e)** *(strictness of control or discipline)* Strenge, *die*

'tightrope *n.* Drahtseil, *das; attrib.* ~ **walker** Seiltänzer, *der*/-tänzerin, *die*

tigress ['taɪgrɪs] *n. (Zool.)* Tigerin, *die*

tile [taɪl] **1.** *n.* **a)** *(on roof)* Ziegel, *der; (on floor, wall)* Fliese, *die; (on stove; also esp. designer ~)* Kachel, *die;* **spend the night on the ~s** *(fig. sl.)* die ganze Nacht durchsumpfen *(salopp);* **b)** *(Games)* Spielstein, *der.* **2.** *v.t.* [mit Ziegeln] decken ‹*Dach*›; fliesen ‹*Wand, Fußboden*›; kacheln ‹*Wand*›; **~d roof** Ziegeldach, *das;* **~d floor** Fliesenboden, *der*

tiling ['taɪlɪŋ] *n., no pl., no indef. art.* **a)** *(fixing tiles) (on roof)* [Dach]decken, *das; (on floor)* Fliesen[legen], *das; (on wall)* Kacheln, *das;* Fliesen, *das;* **b)** *(set of tiles) see* **tile 1 a**: Ziegel/Kacheln/Fliesen

[1]**till** [tɪl] *v.t. (Agric.)* bestellen

[2]**till 1.** *prep.* bis; *(followed by article + noun)* bis zu; **not [...]** ~: erst; *see also* **until 1. 2.** *conj.* bis; *see also* **until 2**

[3]**till** *n.* Kasse, *die;* **at the** ~: an der Kasse; **have/put one's hand** *or* **fingers in the** ~ *(fig.)* in die Kasse greifen

tiller ['tɪlə(r)] *n. (Naut.)* Pinne, *die (Seemannsspr.)*

tilt [tɪlt] **1.** *v.i.* kippen. **2.** *v.t.* kippen; neigen ‹*Kopf*›. **3.** *n.* **a)** Schräglage, *die;* **a 45° ~**: eine Neigung *od.* ein Neigungswinkel von 45°; **b) [at] full ~**: mit voller Wucht

timber ['tɪmbə(r)] *n.* **a)** *no pl. (wood for building)* [Bau]holz, *das;* **b)** *(type of wood)* Holzart, *die;* Holz, *das;* **c)** *no pl., no indef. art. (trees)* Wald, *der;* **d)** *(beam, piece of wood)* Balken, *der; (Naut.)* Spant, *das*

timbre ['tæmbə(r), 'tæbr] *n. (Mus.)* Timbre, *das*

time [taɪm] **1.** *n.* **a)** *no pl., no art.* Zeit, *die;* **the greatest composer of all** ~: der größte Komponist aller Zeiten; **for all** ~: für immer [und ewig]; **past/present/future** ~: Vergangenheit, *die*/Gegenwart, *die*/Zukunft, *die;* **stand the test of** ~: die Zeit überdauern; sich bewähren; **in [the course of] ~, as ~ goes on/went on** mit der Zeit; im Laufe der Zeit; **as old as** ~: uralt; ~ **will tell** *or* **show** die Zukunft wird es zeigen; **at this point** *or* **moment in ~:** zum gegenwärtigen Zeitpunkt; **~ flies** die Zeit vergeht [wie] im Fluge; **work against ~:** unter Zeitdruck arbeiten; **in ~, with ~** *(sooner or later)* mit der Zeit; **b)** *(interval, available or allotted period)* Zeit, *die;* **in a week's/month's/year's ~:** in einer Woche/in einem Monat/Jahr; **there is ~ for that** dafür ist *od.* haben wir noch Zeit; **it takes me all my ~ to do it** es beansprucht meine ganze Zeit, es zu tun; **give one's ~ to sth.** einer Sache *(Dat.)* seine Zeit opfern; **waste of ~:** Zeitverschwendung, *die;* **spend [most of one's/a lot of] ~ on sth./[in] doing sth.** [die meiste/viel] Zeit mit etw. zubringen/damit verbringen, etw. zu tun; **I have been waiting for some/a long ~:** ich warte schon seit einiger Zeit/schon lange; **she will be there for [quite] some ~:** sie wird ziemlich lange dort sein; **be pressed for ~:** keine Zeit haben; *(have to finish quickly)* in Zeitnot sein; **pass the ~:** sich *(Dat.)* die Zeit vertreiben; **length of ~:** Zeit[dauer], *die;* **make ~ for sb./sth.** sich *(Dat.)* für jmdn./etw. Zeit nehmen; **a short ~ ago** vor kurzem; **that's a long ~ ago** das ist schon lange her; **in one's own ~:** in seiner Freizeit; *(whenever one wishes)* wann man will; **take one's ~ [over sth.]** sich *(Dat.)* [für etw.] Zeit lassen; *(be slow)* sich *(Dat.)* Zeit [mit etw.] lassen; **~ is money** *(prov.)* Zeit ist Geld *(Spr.);* **in [good] ~** *(not late)* rechtzeitig; **all the** *or* **this ~:** die ganze Zeit; *(without ceasing)* ständig; **in [less than** *or* **next to] 'no ~:** innerhalb kürzester Zeit; im Nu *od.* Handumdrehen; **in 'half the ~:** in der Hälfte der Zeit; **'half the ~** *(coll.: as often as not)* fast immer; **it will take [some] ~:** es wird einige Zeit dauern; **have the/no ~:** Zeit/keine Zeit haben; **have no ~ for sb./sth.** für jmdn./etw. ist einem seine Zeit zu schade; **there is no ~ to lose** *or* **be lost** es ist keine Zeit zu verlieren; **lose no ~ in doing sth.** *(not delay)* etw. unverzüglich tun; **do ~** *(sl.)* eine Strafe absitzen *(ugs.);* **in my '~** *(heyday)* zu meiner Zeit *(ugs.); (in the course of my life)* im Laufe meines Lebens; **in 'my ~** *(period at a place)* zu meiner Zeit *(ugs.);* **~ off** *or* **out** freie Zeit; **get/take ~ off** frei bekommen/sich *(Dat.)* frei nehmen *(ugs.);* **T~!** *(Boxing)* Stop!; Time!; *(Brit.: in pub)* Feierabend!; **c)** *no pl. (moment or period destined for purpose)* Zeit, *die;* **harvest/Christmas ~:** Ernte-/Weihnachtszeit, *die;* **there is a ~ and place for everything** alles zu seiner Zeit; **now is the ~ to do it** jetzt ist die richtige Zeit, es zu tun; **~ for lunch** Zeit zum Mittagessen; **it is ~ to go** es wird Zeit zu gehen; **and not before ~:** und es wurde auch Zeit; **when the ~ comes/came** wenn es so weit ist/als es so weit war; **on ~** *(punctually)* pünktlich; **ahead of ~:** zu früh ⟨*ankommen*⟩; vorzeitig ⟨*fertig werden*⟩; **all in good ~:** alles zu seiner Zeit; *see also* **be 2a; d)** *in sing. or pl. (circumstances)* Zeit, *die;* **~s are good/bad/have changed** die Zeiten sind gut/schlecht/haben sich verändert; **have a good ~:** Spaß haben *(ugs.);* sich amüsieren; **have a hard ~ [of it]** eine schwere Zeit durchmachen; **e)** *(associated with events or person[s])* Zeit, *die;* **in ~ of peace/war** in Friedens-/Kriegszeiten; **in Tudor/ancient ~s** zur Zeit der Tudors/der Antike; **in former/modern ~s** früher/heutzutage; **ahead of** *or* **before one's/its ~:** seiner Zeit voraus; **at 'one ~** *(previously)* früher; **f)** *(occasion)* Mal, *das;* **for the first ~:** zum ersten Mal; **next ~ you come** wenn du das nächste Mal kommst; **ten/a hundred/a thousand ~s** zehn- / hundert- / tausendmal; **many ~s** sehr oft; **many's the ~ [that] ..., many a ~ ...:** viele Male ...; **at all ~s** jederzeit; **at ~s** gelegentlich; **from ~ to ~:** von Zeit zu Zeit; **at other ~s** sonst; **at one ~ or another** irgendwann einmal; **at a ~ like this/that** unter diesen/solchen Umständen; **at the** *or* **that ~** *(in the past)* damals; **at one ~, at [one and] the same ~** *(simultaneously)* gleichzeitig; **at the same ~** *(nevertheless)* gleichwohl; **between ~s** zwischendurch; **~ and [~] again, ~ after ~:** immer [und immer] wieder; **pay sb. £6 a ~:** jmdm. für jedes Mal 6 Pfund zahlen; **one at a ~:** einzeln; **two at a ~:** jeweils zwei; **for hours/weeks at a ~:** stundenlang/wochenlang [ohne Unterbrechung]; **g)** *(point in day etc.)* [Uhr]zeit, *die;* **at the same ~ every morning** jeden Morgen um dieselbe Zeit; **what ~ is it?, what is the ~?** wie spät ist es?; **have you [got] the ~?** kannst du mir sagen, wie spät es ist?; **tell the ~** *(read a clock)* die Uhr lesen; **~ of day** Tageszeit, *die;* **[at this] ~ of [the] year** [um diese] Jahreszeit; **at this ~ of [the] night** zu dieser Nachtstunde; **pass the ~ of day** *(coll.)* ein paar Worte wechseln **by this/that ~:** inzwischen; **by the ~ [that] we arrived** bis wir hinkamen; **[by] this ~ tomorrow** morgen um diese Zeit; **keep good ~** ⟨*Uhr:*⟩ genau *od.* richtig gehen; **h)** *(amount)* Zeit, *die;* **make good ~** gut vorwärtskommen; **[your] ~'s up!** deine Zeit ist um *(ugs.) od* abgelaufen; **i)** *(multiplication)* mal; **three ~s four** drei mal vier **four ~s the size of/higher than sth.** viermal so groß wie/höher als etw.; **j)** *(Mus.) (duration of note)* Zeitdauer, *die; (measure)* Takt, *der;* **in three-four ~:** im Dreivierteltakt; **keep in ~ with the music** den Takt halten; **out of ~/in ~:** aus dem/im Takt; **keep ~ with sth.** bei etw. den Takt [ein]halten. **2.** *v. t.* **a)** *(do at correct ~)* zeitlich abstimmen; **be well/ill ~d** zur richtigen/falschen Zeit kommen; **b)** *(set to operate at correct ~)* justieren *(Technik);* einstellen; **c)** *(arrange ~ of arrival/departure of)* **the bus is ~d to connect with the train** der Bus hat einen direkten Anschluß an den Zug; **be ~d to take 90 minutes** fahrplanmäßig 90 Minuten dauern; **d)** *(measure ~ taken by)* stoppen; **~ an egg** auf die richtige Kochdauer für ein Ei achten

time: ~ bomb *n. (lit. or fig.)* Zeitbombe, *die;* **~-consuming** *adj* **a)** *(taking ~)* zeitaufwendig; **b)** *(wasteful of ~)* zeitraubend; **~ exposure** *n. (Photog.)* Zeitaufnahme, *die;* **~-honoured** *adj.* altehrwürdig *(geh.);* althergebracht ⟨*Brauch, Vorstellung*⟩ **~-keeping** *n. (at work)* Einhaltung der Arbeitsstunden; **~-lag** *n.* zeitliche Verzögerung; **~-limit** *n.* Frist, *die;* **put a ~-limit on sth.** eine Frist für etw. setzen **~-lock** *n.* Zeitschloß, *das*

timely ['taımlı] *adj.* rechtzeitig

'timepiece *n.* Chronometer, *das*

timer ['taımə(r)] *n. (device)* Kurzzeitmesser, *der; (with switch)* Schaltuhr, *die*

time: ~-scale *n.* Zeitskala, *die;* **~-share 1.** *attrib. adj.* **~-share apartment** *Ferienwohnung, an der man einen Besitzanteil hat, der es einem erlaubt, eine bestimmte Zeit pro Jahr in dieser Wohnung zu verbringen;* **2.** *n. see* **~-sharing b; ~-sharing** *n., no pl., no art.* **a)** *(Computing)* Time-sharing, *das;* **b)** *(joint ownership)* *Eigentum an einer Ferienwohnung o. ä., das für eine festgelegte Zeit des Jahres gilt;* Time-sharing, *das*

(Wirtsch.); ~**-signal** *n.* Zeitzeichen, *das;* ~ **signature** *n.* *(Mus.)* Taktbezeichnung, *die;* ~**-switch** *n.* Zeitschalter, *der;* ~**table** *n.* **a)** *(scheme of work)* Zeitplan, *der; (Educ.)* Stundenplan, *der;* **b)** *(Transport)* Fahrplan, *der;* ~**-travel** *n.* Reise durch die Zeit; ~**-zone** *n.* Zeitzone, *die*

timid ['tɪmɪd] *adj.* **a)** scheu ‹*Tier, Vogel*›; **b)** *(fearful)* ängstlich ‹*Person, Miene, Worte*›; **c)** *(lacking boldness)* zaghaft; *(shy)* schüchtern

timing ['taɪmɪŋ] *n., no pl.* **a) that was perfect** ~! *(as sb. arrives)* du kommst gerade im richtigen Augenblick!; **b)** *(Theatre)* Timing, *das*

timpani ['tɪmpəni:] *n. pl. (Mus.)* Kesselpauke, *die;* Timpani *(fachspr.)*

tin [tɪn] **1.** *n.* **a)** *(metal)* Zinn, *das;* ~**[-plate]** Weißblech, *das;* **b)** *(Cookery)* **cooking** ~**s** Back- und Bratformen; **c)** *(Brit.: for preserving)* [Konserven]dose, *die;* **a** ~ **of peas** eine Dose Erbsen; **d)** *(with separate or hinged lid)* Dose, *die;* **bread** ~: Brotkasten, *der.* **2.** *v. t.,* **-nn-** *(Brit.)* zu Konserven verarbeiten

tinder ['tɪndə(r)] *n.* Zunder, *der*

tin 'foil *n., no pl.* Stanniol, *das; (aluminium foil)* Alufolie, *die*

tinge [tɪndʒ] **1.** *v. t.,* ~**ing** ['tɪndʒɪŋ] tönen; **a white curtain** ~**d with pink** ein weißer, ins Zartrosa gehender Vorhang; **her black hair was** ~**d with grey** ihr schwarzes Haar war graumeliert; *(fig.)* **her admiration was** ~**d with envy** ihre Bewunderung war nicht ganz frei von Neid. **2.** *n.* [leichte] Färbung; *(fig.)* Hauch, *der;* **a** ~ **of red in the sky** eine leicht rötliche Färbung des Himmels; **white with a** ~ **of blue** weiß mit einem Stich ins Bläuliche

tingle ['tɪŋgl] **1.** *v. i.* kribbeln. **2.** *n.* Kribbeln, *das;* **feel a** ~ **of excitement** vor Aufregung ganz kribbelig sein *(ugs.)*

tinker ['tɪŋkə(r)] **1.** *n.* Kesselflicker, *der.* **2.** *v. i.* ~ **with sth.** an etw. *(Dat.)* herumbasteln *(ugs.)*/*(incompetently; also fig.)* herumpfuschen *(ugs.)*

tinkle ['tɪŋkl] **1.** *n.* Klingeln, *das; (of coins)* Klimpern, *das.* **2.** *v. i.* ‹*Glocke:*› klingeln; ‹*Münzen:*› klimpern

'tin mine *n.* Zinnbergwerk, *das*

tinned [tɪnd] *adj. (Brit.)* Dosen-; **be** ~: aus der Dose sein

tin: ~**-opener** *n. (Brit.)* Dosen-, Büchsenöffner, *der;* ~ **'plate** *n.* Weißblech, *das;* ~**pot** *attrib. adj. (derog.)* schäbig; ~**pot town** Kaff, *das (ugs.);* ~**pot dictator** Operettendiktator, *der*

tinsel ['tɪnsl] *n.* Lametta, *das*

tint [tɪnt] **1.** *n.* Farbton, *der.* **2.** *v. t.* tönen; kolorieren ‹*Zeichnung, Stich*›

tiny ['taɪnɪ] *adj.* winzig; **a** ~ **bit better** *(coll.)* ein klein wenig besser

¹tip [tɪp] **1.** *n.* *(end, point)* Spitze, *die;* **the** ~ **of his nose/finger/toe** seine Nasen-/Finger-/Zehenspitze; **on the** ~**s of one's toes** auf Zehenspitzen; **from** ~ **to toe** vom Scheitel bis zur Sohle; **it is on the** ~ **of my tongue** es liegt mir auf der Zunge. **2.** *v. t.,* **-pp-:** ~ **sth. [with stone/brass]** etw. mit einer [Stein-/Messing]spitze versehen

²tip 1. *v. i.,* **-pp-** *(lean, fall)* kippen; ~ **over** umkippen. **2.** *v. t.,* **-pp-: a)** *(make tilt)* kippen; ~ **the balance** *(fig.)* den Ausschlag geben; *see also* **²scale 1 b; b)** *(make overturn)* umkippen; *(Brit.: discharge)* kippen; **c)** *(mention as likely winner etc.)* voraussagen ‹*Sieger*›; ~ **sb. to win** auf jmds. Sieg tippen; **be** ~**ped for the Presidency/a post** als Favorit für die Präsidentschaftswahlen/einen Posten genannt werden; **d)** *(sl.: give)* geben; ~ **sb. the wink** *(fig.)* jmdm. Bescheid sagen; *(~ sb. off)* jmdm. einen Tip geben *(ugs.);* **e)** *(give money to)* ~ **sb. [20p]** jmdm. [20 Pence] Trinkgeld geben. **3.** *n.* **a)** *(money)* Trinkgeld, *das;* **as a** ~: als Trinkgeld; **b)** *(special information)* Hinweis, *der;* Tip, *der (ugs.); (advice)* Rat, *der;* **hot** ~: heißer Tip; **c)** *(Brit.: place for refuse)* Müllkippe, *die;* **d)** *(derog.: untidy place)* Schweinestall, *der*

~ **'off** *v. t.* ~ **sb. off** jmdm. einen Hinweis *od. (ugs.)* Tip geben

'tip-off *n.* Hinweis, *der*

tipple ['tɪpl] **1.** *v. i.* trinken. **2.** *n.* *(coll.: drink)* **have a** ~: einen trinken *(ugs.);* **what's your** ~**?** was trinken Sie?

tippler ['tɪplə(r)] *n.* Trinker, *der*/Trinkerin, *die*

tipsy ['tɪpsɪ] *adj. (coll.)* angeheitert; beschwipst *(ugs.)*

tip: ~**toe 1.** *v. i.* auf Zehenspitzen gehen; *(walk quietly)* sich schleichen *od.* stehlen; **2.** *adv.* auf Zehenspitzen; **3.** *n.* **on** ~**toe[s]** auf Zehenspitzen; **stand on** ~**toe** sich auf die Zehenspitzen stellen; ~**top** *adj. (coll.)* ausgezeichnet; tipptopp *(ugs.);* **be in** ~**top condition** in einem Topzustand/‹*Person:*› in Topform sein; ~**-up seat** *n.* Klappsitz, *der*

tirade [taɪ'reɪd, tɪ'reɪd] *n.* Tirade, *die (geh.)*

¹tire ['taɪə(r)] *(Amer.) see* **tyre**

²tire 1. *v. t.* ermüden. **2.** *v. i.* müde werden; ermüden; ~ **of sth./doing sth.** einer Sache *(Gen.)* überdrüssig werden/es müde werden *(geh.),* etw. zu tun

~ **'out** *v. t.* erschöpfen; ~ **oneself out doing sth.** etw. bis zur Erschöpfung tun

tired ['taɪəd] *adj.* **a)** *(weary)* müde; **b)** *(fed up)* **be** ~ **of sth./doing sth.** etw. satt haben/es satt haben *od. (geh.)* es müde sein, etw. zu tun; **get** *or* **grow** ~ **of sb./sth.** jmds./einer Sache überdrüssig werden; **c)** *(fig.: hackneyed)* abgegriffen; abgedroschen *(ugs.)*

tiredness ['taɪədnɪs] *n., no pl.* Müdigkeit, *die*

tireless ['taɪəlɪs] *adj.* unermüdlich

tiresome ['taɪəsəm] *adj.* **a)** *(wearisome)* mühsam; **b)** *(annoying)* lästig; **how** ~**!** so ein Ärger!

tiring ['taɪərɪŋ] *adj.* ermüdend; anstrengend ‹*Tag, Person*›

tissue ['tɪʃu:, 'tɪsju:] *n.* **a)** *(woven fabric; also Biol.)* Gewebe, *das;* **b)** *(absorbent paper)* **[paper]** ~: Papiertuch, *das; (handkerchief)* Papiertaschentuch, *das;* **c)** *(for wrapping)* ~ **[paper]** Seidenpapier, *das;* **d)** *(fig.: web)* Geflecht, *das;* ~ **of lies** Lügengewebe, *das*

¹tit [tɪt] *n. (Ornith.)* Meise, *die*

²tit *n.* **it's** ~ **for tat** wie du mir, so ich dir

'titbit *n.* **a)** *(food)* Häppchen, *das (ugs.);* **b)** *(piece of news)* Neuigkeit, *die*

titch [tɪtʃ] *n. (coll.)* Knirps, *der (ugs.)*

titchy ['tɪtʃɪ] *adj. (coll.)* klitzeklein *(ugs.)*

title ['taɪtl] *n.* **a)** *(of book etc.)* Titel, *der; (of article, chapter)* Überschrift, *die;* **the flyweight** ~ *(Sport)* der Titel im Fliegengewicht; **the** ~**s** *(Cinemat., Telev.)* der Vorspann; **b)** *(of person)* Titel, *der; (of nobility)* [Adels]titel, *der; (of organization)* Name, *der;* **c)** *(Law: recognized claim)* Rechtsanspruch, *der* (**to** auf + *Akk.*)

titled ['taɪtld] *adj.* adlig

title: ~**-page** *n.* Titelseite, *die;* ~**-role** *n.* Titelrolle, *die*

titter ['tɪtə(r)] **1.** *v. i.* kichern. **2.** *n.* ~**[s]** Kichern, *das*

tittle-tattle ['tɪtltætl] *n.* Klatsch, *der (ugs. abwertend)*

tizzy ['tɪzɪ] *n. (sl.)* **be in a/get into a ~:** durchdrehen *(ugs.)* **(over** wegen)

T-junction *see* T

TNT *abbr.* **trinitrotoluene** TNT, *das*

to 1. [*before vowel* tʊ, *before consonant* tə, *stressed* tu:] *prep.* **a)** *(in the direction of and reaching)* zu; *(with name of place)* nach; **go to work/to the theatre** zur Arbeit/ins Theater gehen; **to Paris/France** nach Paris/Frankreich; **go from town to town** von Stadt zu Stadt ziehen; **throw the ball to me** wirf mir den Ball zu; **b)** *(towards a condition or quality)* zu; **appoint sb. to a post** jmdn. auf einen Posten berufen; **c)** *(as far as)* bis zu; **from London to Edinburgh** von London [bis] nach Edinburgh; **increase from 10% to 20%** von 10% auf 20% steigen; **d)** *(next to, facing)* **with one's back to the wall** mit dem Rücken zur Wand; **e)** *(implying comparison, ratio, etc.)* **[compared] to** verglichen mit; im Vergleich zu; **3 is to 4 as 6 is to 8** 3 verhält sich zu 4 wie 6 zu 8; **it's ten to one he does sth.** die Chancen stehen zehn zu eins, daß er etw. tut; **f)** *introducing relationship or indirect object* **to sb./sth.** jmdm./einer Sache *(Dat.)*; **lend/explain** *etc.* **sth. to sb.** jmdm. etw. leihen/erklären *usw.*; **speak to sb.** mit jmdm. sprechen; **relate to sth.** sich auf etw. *(Akk.)* beziehen; **to me** *(in my opinion)* meiner Meinung nach; **secretary to the Minister** Sekretär des Ministers; **a room to oneself** ein eigenes Zimmer; **get four apples to the pound** vier Äpfel je Pfund bekommen; **that's all there is to it** mehr ist nicht dazu zu sagen; **what's that to you?** was geht das dich an?; **g)** *(until)* bis; **to the end** bis zum Ende; **to this day** bis heute; **five [minutes] to eight** fünf [Minuten] vor acht; **h)** *with infinitive of a verb* zu; *expressing purpose, or after* **too** um [...] zu; **want to know** wissen wollen; **do sth. to annoy sb.** etw. tun, um jmdn. zu ärgern; **too young to marry** zu jung, um zu heiraten; zu jung zum Heiraten; **too hot to drink** zu heiß zum Trinken; **to rebel is pointless** es ist sinnlos zu rebellieren; **he woke to find himself in a strange room** er erwachte und fand sich in einem fremden Zimmer wieder; **to use a technical term** um einen Fachausdruck zu gebrauchen; **i)** *as substitute for infinitive* **he would have phoned but forgot to** er hätte angerufen, aber er vergaß es; **she didn't want to go there, but she had to** sie wollte nicht hingehen, aber sie mußte. **2.** [tu:] *adv.* **a)** *(just not shut)* **be to** ‹*Tür, Fenster:*› angelehnt sein; **push a door to** eine Tür anlehnen; **b) to and fro** hin und her

toad [təʊd] *n. (Zool.; fig. derog.)* Kröte, *die*

toad: ~-in-the-hole *n. (Gastr.) Würstchen, in einen Teig eingebacken;* **~stool** *n.* Giftpilz, *der; (Bot.)* Schirmpilz, *der*

toady ['təʊdɪ] **1.** *n.* Kriecher, *der.* **2.** *v.i.* **~ [to sb.]** [vor jmdm.] kriechen *(abwertend)*

toast [təʊst] **1.** *n.* **a)** *no pl., no indef. art.* Toast, *der;* **a piece of ~:** eine Scheibe Toast; **cheese/egg on ~:** Toast mit Käse/Ei; **as warm as ~** *(fig.)* schön warm *(ugs.)*; **b)** *(call to drink)* Toast, *der;* **drink/propose a ~ to sb./sth.** auf jmdn./etw. trinken/einen Toast auf jmdn./etw. ausbringen; **be the ~ of the town** von der ganzen Stadt gefeiert werden. **2.** *v.t.* **a)** rösten; toasten ‹*Brot*›; **b)** *(drink in honour of)* trinken auf (+ *Akk.*)

toaster ['təʊstə(r)] *n.* Toaster, *der*

tobacco [tə'bækəʊ] *n., pl.* **~s** Tabak, *der*

tobacconist [tə'bækənɪst] *n.* Tabak[waren]händler, *der*/-händlerin, *die; see also* **baker**

toboggan [tə'bɒgən] **1.** *n.* Schlitten, *der;* Toboggan, *der.* **2.** *v.i.* Schlitten fahren

today [tə'deɪ] **1.** *n.* heute; **~'s newspaper** die Zeitung von heute. **2.** *adv.* heute; **a week/fortnight [from] ~:** heute in einer Woche/in vierzehn Tagen; **a year [ago] ~:** heute vor einem Jahr; **early ~:** heute früh; **later [on] ~:** später [am Tage]; **earlier ~:** heute vor wenigen Stunden

toddle ['tɒdl] *v.i.* **a)** *(with tottering steps)* mit wackligen Schritten gehen; wackeln *(ugs.)*; **b)** *(coll.: leave)* **~ [off]** sich verziehen *(ugs.)*

toddler ['tɒdlə(r)] *n.* ≈ Kleinkind, *das*

to-do [tə'du:] *n.* Getue, *das (ugs.)*

toe [təʊ] **1.** *n.* **a)** *(Anat.)* Zeh, *der;* Zehe, *die;* **be on one's ~s** *(fig.)* auf Zack sein *(ugs.)*; **keep sb. on his/her ~s** *(fig.)* jmdn. in Trab halten *(ugs.)*; **b)** *(of footwear)* Spitze, *die;* **at the ~:** an den Zehen; **c)** *(Zool.)* Zeh, *der.* **2.** *v.t.,* **~ing** *(fig.)* **~ the line** *or (Amer.)* **mark** sich einordnen; **refuse to ~ the line** aus der Reihe tanzen; **~ the party line** linientreu sein

toe-: ~-cap *n.* Vorderkappe, *die; (of boot)* Stiefelkappe, *die;* **~-hold** *n.* Tritt, *der; (fig.)* **gain a ~-hold** einen Fuß in die Tür bekommen; **~-nail** *n.* Zeh[en]nagel, *der*

toffee ['tɒfɪ] *n.* **a)** Karamel, *der;* **b)** *(Brit.: piece)* Toffee, *das;* Sahnebonbon, *das;* **c) sb. can't do sth. for ~** *(fig. sl.)* jmd. kann etw. nicht für fünf Pfennig tun *(ugs.)*

toffee: ~-apple *n. mit Karamel überzogener Apfel am Stiel;* **~-nosed** *adj. (Brit. sl.)* hochnäsig

toga ['təʊgə] *n. (Roman Ant.)* Toga, *die*

together [tə'geðə(r)] *adv.* **a)** *(in or into company)* zusammen; **sit down ~:** sich zusammensetzen; **gather ~:** sich [ver]sammeln; **taken all ~:** alle zusammengenommen; **~ with** zusammen mit; **b)** *(simultaneously)* gleichzeitig; **all ~ now!** jetzt alle zusammen *od.* im Chor!; **c)** *(one with another)* miteinander; **put them ~ to compare them** halte sie nebeneinander, um sie zu vergleichen; **d)** *(without interruption)* **for weeks/days/hours ~:** wochen-/tage-/stundenlang; **for three days ~:** drei Tage hintereinander

togetherness [tə'geðənɪs] *n., no pl.* Zusammengehörigkeit, *die*

toggle ['tɒgl] *n.* Knebelknopf, *der*

toil [tɔɪl] **1.** *v.i.* **a)** *(work laboriously)* schwer arbeiten; sich abarbeiten; **~ at/over sth.** sich mit etw. abplagen/abmühen; **~ through a book** sich mühsam durch ein Buch arbeiten; **b)** *(move laboriously)* sich schleppen. **2.** *n.* [harte] Arbeit

toilet ['tɔɪlɪt] *n.* Toilette, *die;* **down the ~:** in die Toilette; **go to the ~:** auf die Toilette gehen

toilet: ~-bag *n.* Kulturbeutel, *der;* **~-bowl** *n.* Toilettenbecken, *das;* Klosettbecken, *das;* **~-paper** *n.* Toilettenpapier, *das*

toiletries ['tɔɪlɪtrɪz] *n.pl.* Körperpflegemittel; Toilettenartikel

toilet: ~-roll *n.* Rolle Toilettenpapier; **~-roll holder** *n.* Toilettenpapierhalter, *der;* **~-seat** *n.* Klosettbrille, *die (ugs.)*; **~ tissue** *see* **toilet-paper; ~ water** *n.* Toilettenwasser, *das;* Eau de Toilette, *das*

toing and froing [tu:ɪŋ ən 'frəʊɪŋ] *n.* Hin und Her, *das*

token ['təʊkn] **1.** *n.* **a)** *(voucher)* Gutschein, *der;* **b)** *(counter, disc)* Marke, *die;* **c)** *(sign)* Zeichen, *das; (evidence)* Beweis, *der;* **as a** *or* **in ~ of sth.** als Zeichen/zum Beweis einer Sache; **d) by the same** *or* **this ~:** ebenso. **2.** *attrib.*

adj. symbolisch ⟨*Preis*⟩; nominal *(Wirtsch.)* ⟨*Lohnerhöhung, Miete*⟩; **a ~ woman on the staff** eine Alibifrau als Mitarbeiterin; **offer** *or* **put up ~ resistance** pro forma Widerstand leisten

Tokyo ['təʊkjəʊ] *pr. n.* Tokio *(das)*

told *see* **tell**

tolerable ['tɒlərəbl] *adj.* **a)** *(endurable)* erträglich **(to, for** für); **b)** *(fairly good)* leidlich; annehmbar

tolerably ['tɒlərəblɪ] *adv.* leidlich; annehmbar; einigermaßen ⟨*gut, richtig*⟩

tolerance ['tɒlərəns] *n.* Toleranz, *die* **(for, towards,** gegen[über])

tolerant ['tɒlərənt] *adj.* tolerant **(of, towards** gegen[über])

tolerate ['tɒləreɪt] *v. t.* **a)** dulden; tolerieren *(geh.)*; **b)** *(put up with)* **~ sb./sth.** sich mit jmdm./etw. abfinden; **~ one another** sich [gegenseitig] akzeptieren; **c)** *(sustain)* ertragen ⟨*Schmerzen, Hitze, Lärm*⟩

toleration [tɒlə'reɪʃn] *n.* Tolerierung, *die (geh.)*; **religious ~:** religiöse Toleranz

[1]**toll** [təʊl] *n.* **a)** *(tax, duty)* Gebühr, *die; (for road)* [Straßen]gebühr, *die;* Maut, *die (bes. österr.)*; **b)** *(damage etc. incurred)* Aufwand, *der;* **take** *or* **exact a /its ~ of sth.** einen Tribut an etw. *(Dat.)* fordern *(fig.)*

[2]**toll 1.** *v. t.* läuten; ⟨*Turmuhr:*⟩ schlagen ⟨*Stunde*⟩. **2.** *v. i.* läuten

toll: ~-bridge *n.* gebührenpflichtige Brücke; Mautbrücke, *die (bes. österr.)*; **~ call** *n. (Amer. Teleph.)* gebührenpflichtiges Gespräch; **~-road** *n.* gebührenpflichtige Straße; Mautstraße, *die (bes. österr.)*

tom [tɒm] *n.* **a) any** *or* **every Tom, Dick, and Harry** Hinz und Kunz *(ugs. abwertend)*; **b)** *(cat)* Kater, *der.* *See also* **peeping Tom**

tomahawk ['tɒməhɔ:k] *n.* Tomahawk, *der*

tomato [tə'mɑ:təʊ] *n., pl.* **~es** Tomate, *die*

tomato: ~-juice *n.* Tomatensaft, *der;* **~ 'ketchup** *n.* Tomatenketchup, *der od. das;* **~ 'purée** *n.* Tomatenmark, *das;* **~ 'sauce** *n.* **a)** Tomatensoße, *die;* **b)** *see* **~ ketchup; ~ 'soup** *n.* Tomatensuppe, *die*

tomb [tu:m] *n.* **a)** *(grave)* Grab, *das;* **b)** *(monument)* Grabmal, *das*

tombola [tɒm'bəʊlə] *n.* Tombola, *die*

'tomboy *n.* Wildfang, *der*

'tombstone *n.* Grabstein, *der;* Grabmal, *das*

'tom-cat *n.* Kater, *der*

tome [təʊm] *n.* dicker Band; Wälzer, *der (ugs.)*

tomfoolery [tɒm'fu:lərɪ] *n.* Blödsinn, *der (ugs.)*

tommy-gun ['tɒmɪgʌn] *n.* Maschinenpistole, *die*

tomorrow [tə'mɒrəʊ] **1.** *n.* **a)** morgen; **~ morning/afternoon/evening/night** morgen früh *od.* vormittag / nachmittag / abend / nacht; **~ is another day** *(prov.)* morgen ist auch [noch] ein Tag *(Spr.)*; **b)** *(the future)* Morgen, *das;* **who knows what ~ will bring?** wer weiß, was die Zukunft bringt?; **like there's no ~** *(coll.)* als ginge morgen die Welt unter. **2.** *adv.* morgen; **a week/month [from] ~:** morgen in einer Woche/in einem Monat; **a year [ago] ~:** morgen vor einem Jahr; **[I'll] see you ~!** *(coll.)* bis morgen!; **never put off till ~ what you can do today** *(prov.)* was du heute kannst besorgen, das verschiebe nicht auf morgen *(Spr.)*; **the day after ~:** übermorgen; **this time ~:** morgen um diese Zeit; **~ afternoon/morning** morgen nachmittag/früh; **~ evening** *or* **night** morgen abend

'tom-tom *n. (Mus.)* Tomtom, *das*

ton [tʌn] *n.* **a)** Tonne, *die;* **a five-~ lorry** ein Lastwagen von fünf Tonnen [Leergewicht]; ein Fünftonner *(ugs.)*; **metric ~:** metrische Tonne; **two ~[s] of coal** zwei Tonnen Kohle; **b)** *(fig. coll.: a lot)* **it weighs [half] a ~:** es ist zentnerschwer *(fig.)*; **~s [of food/people/reasons** *etc.*] haufenweise *(ugs.)* [Essen/Leute/Gründe *usw.*]

tone [təʊn] **1.** *n.* **a)** *(sound)* Klang, *der; (Teleph.)* Ton, *der;* **b)** *(style of speaking)* Ton, *der;* **don't speak to me in that ~ [of voice]** sprich mit mir nicht in diesem Ton; **in an angry** *etc.* **~, in angry** *etc.* **~s** in ärgerlichem *usw.* Ton; **in a ~ of reproach/anger** *etc.* in vorwurfsvollem/wütendem *usw.* Ton; **c)** *(tint, shade)* [Farb]ton, *der;* **~s of blue** Blautöne; blaue Töne; **grey with a blue ~:** bläulichgrau; **d)** *(style of writing)* [Grund]stimmung, *die; (of letter)* Ton, *der;* **e)** *(Mus.) (note)* Ton, *der; (quality of sound)* Klang, *der; (Brit.: interval)* Intervall, *das;* **f)** *(fig.: character)* Stimmung, *die;* **give a serious/flippant ~ to sth.** einer Sache *(Dat.)* eine ernsthafte/frivole Note verleihen; **lower/raise the ~ of sth.** das Niveau einer Sache *(Gen.)* senken/erhöhen; **set the ~:** den Ton angeben; **set the ~ of** *or* **for sth.** für etw. bestimmend sein; **g)** *(Art: general effect of colour)* Farbgebung, *die;* Kolorit, *das;* **h)** *(degree of brightness)* Schattierung, *die;* Nuancierung, *die;* **bright ~:** Helligkeit, *die;* **i)** *(Photog.)* Ton, *der.* **2.** *v. i. see* **~ in**

~ 'down *v. t.* **a)** *(Art)* [ab]dämpfen ⟨*Farbe*⟩; **~ a painting down** die Farben eines Bildes abdämpfen; **b)** *(fig.: soften)* mäßigen ⟨*Sprache*⟩; abschwächen ⟨*Verbalattacke, Forderung*⟩

~'in *v. i.* farblich harmonieren

tone: ~-arm *n.* Tonarm, *der;* **~ control** *n. (device)* Klangregler, *der;* **~-'deaf** *adj.* ohne musikalisches Gehör *(Musik)*; **be ~-deaf** kein musikalisches Gehör haben

tongs [tɒŋz] *n. pl.* **[pair of] ~:** Zange, *die*

tongue [tʌŋ] *n.* **a)** Zunge, *die;* **bite one's ~** *(lit. or fig.)* sich auf die Zunge beißen; **put out your ~, please** strecken Sie [bitte] mal Ihre Zunge heraus!; **put** *or* **stick one's ~ out [at sb.]** [jmdm.] die Zunge herausstrecken; **with one's ~ hanging out** mit [heraus]hängender Zunge; **he made the remark ~ in cheek** *(fig.)* er meinte die Bemerkung nicht ernst; **hold one's ~** *(fig.)* stillschweigen; **b)** *(meat)* Zunge, *die;* **c)** *(manner or power of speech)* **find/lose one's ~:** seine Sprache wiederfinden/die Sprache verlieren; **get one's ~ round sth.** etw. aussprechen; **have a sharp** *etc.* **~:** eine scharfe *usw.* Zunge haben; **d)** *(language)* Sprache, *die;* **e)** *(of shoe)* Zunge, *die;* **f)** *(promontory)* **~ [of land]** Landzunge, *die;* **g)** *(of buckle)* Dorn, *der*

tongue: ~-in-'cheek *adj.* nicht ernst gemeint; *(ironical)* ironisch; *see also* **tongue a; ~-tied** *adj.* schüchtern; gehemmt; **be ~-tied [with** *or* **by fear/embarrassment** *etc.*] [vor Angst/Verlegenheit *usw.*] kein Wort herausbringen; **~-twister** *n.* Zungenbrecher, *der (ugs.)*

tonic ['tɒnɪk] **1.** *n.* **a)** *(Med.)* Tonikum, *das;* **it was as good as a ~:** es hat mir/ihm *usw.* richtig gutgetan; **b)** *(fig.: invigorating influence)* Wohltat, *die (geh.)*; **c)** *(~ water)* Tonic, *das;* **gin** *etc.* **and ~:** Gin *usw.* [mit] Tonic; **d)** *(Mus.)* Tonika, *die.* **2.** *adj.* **a)** *(Med.)* kräftigend; *(fig.)* wohltuend ⟨*Wirkung*⟩; **b)** *(Mus.)* tonisch

'tonic water *n.* Tonic[wasser], *das*

tonight [tə'naɪt] **1.** *n.* **a)** *(this evening)* heute abend; **~ has been such**

fun heute abend war es so lustig; **after** ~: nach dem heutigen Abend; **I enjoyed** ~: es war ein schöner Abend; ~**'s [news]paper** die heutige Abendzeitung; ~**'s performance** die heutige [Abend]vorstellung; ~**'s the night!** heute abend ist es soweit!; ~**'s weather will be cold** heute abend wird es kalt; **b)** *(this or the coming night)* heute nacht; ~ **will be colder** heute nacht wird es kälter werden. **2.** *adv.* **a)** *(this evening)* heute abend; **b)** *(during this or the coming night)* heute nacht; **[I'll] see you** ~! bis heute abend!

tonne ['tʌn] *n.* [metrische] Tonne

tonsil ['tɒnsl] *n. (Anat.)* [Gaumen]mandel, *die;* **have one's** ~**s out** sich *(Dat.)* die Mandeln herausnehmen lassen

tonsillitis [tɒnsə'laɪtɪs] *n. (Med.)* Mandelentzündung, *die*

too [tu:] *adv.* **a)** *(excessively)* zu; **far** *or* **much** ~ **much** viel zu viel; ~ **much** zuviel; **I've had** ~ **much to eat/drink** ich habe zuviel gegessen/getrunken; **but not** ~ **much, please** aber bitte nicht allzuviel; **the problem/he was** ~ **much for her** sie war der Aufgabe/ihm nicht gewachsen; **things are getting** ~ **much for me** es wird mir allmählich zu viel; **this is** '~ **much!** *(indignantly)* jetzt reicht's!; **she's/that's just** '~ **much** *(intolerable)* sie ist/das ist zuviel! *(ugs.)*; *(sl.: wonderful)* sie ist/das ist echt spitze *(ugs.)*; ~ **difficult a task** eine zu schwierige Aufgabe; **none** ~ *or* **not any** ~ **easy** nicht allzu leicht; *(less than one had expected)* gar nicht so leicht; **he is none** ~ *or* **not any** ~ **clever/quick** *etc.* er ist nicht der Schlauste/Schnellste *usw.;* **none** ~ **soon** keinen Augenblick zu früh; *see also* **all 3; good 1b, e; many 1a; much 1a; only 2d; b)** *(also)* auch; **she can sing, and play the piano,** ~: sie kann singen und auch *od.* außerdem Klavier spielen; **c)** *(coll.: very)* besonders; **I'm not feeling** ~ **good** mir geht es nicht besonders [gut]; **I'm not** ~ **sure** ich bin mir nicht ganz sicher; **not** ~ **pleased** nicht gerade erfreut; **d)** *(moreover)* **he lost in twenty moves, and to an amateur** ~: er verlor in zwanzig Zügen, und noch dazu gegen einen Amateur; **there was frost last night, and in May/Spain** ~! es hat letzte Nacht gefroren, und das im Mai/in Spanien!

took *see* **take 1, 2**

tool [tu:l] *n.* **a)** Werkzeug, *das;* *(garden* ~*)* Gerät, *das;* **set of** ~**s** Werkzeug, *das; see also* [3]**down 4c; b)** *(machine)* Werkzeugmaschine, *die;* **c)** *(Mech. Engin.: lathe* ~*)* Meißel, *der;* **d)** *(fig.: means)* [Hilfs]mittel, *das;* **pen and paper are the writer's basic** ~**s** Feder und Papier sind das wichtigste Handwerkszeug des Schriftstellers; **the** ~**s of the trade** das Handwerkszeug; das Rüstzeug; **e)** *(fig.: person)* Werkzeug, *das*

tool: ~**-bag** *n.* Werkzeugtasche, *die;* ~**-box,** ~**-case** *ns.* Werkzeugkasten, *der;* ~**-kit** *n. (Brit.)* Werkzeugsatz, *der; (more general)* Werkzeug, *das; (for vehicle)* **is there a** ~**-kit?** gibt es Bordwerkzeug?; ~**-shed** *n.* Geräteschuppen, *der*

toot [tu:t] **1.** *v. t.* **the driver** ~**ed his horn** der Fahrer hupte. **2.** *v. i. (on car etc. horn)* hupen. **3.** *n.* Tuten, *das;* **give a** ~ **on one's/its horn** ⟨*Autofahrer/Auto:*⟩ hupen

tooth [tu:θ] *n., pl.* **teeth** [ti:θ] **a)** Zahn, *der;* **say sth. between one's teeth** etw. mit zusammengebissenen Zähnen hervorstoßen; **have a** ~ **out/filled** sich *(Dat.)* einen Zahn ziehen/füllen lassen; **armed to the teeth** bis an die Zähne bewaffnet; ~ **and nail** verbissen ⟨*kämpfen, bekämpfen*⟩; **get one's teeth into sth.** *(fig.)* etw. in Angriff nehmen; **show one's teeth** ⟨*Hund:*⟩ die Zähne fletschen; *(fig.)* die Zähne zeigen *(ugs.)*; **b)** *(of rake, fork, comb)* Zinke, *die; (of cog-wheel, saw, comb)* Zahn, *der*

tooth: ~**ache** *n.* Zahnschmerzen *Pl.;* Zahnweh, *das (ugs.)*; ~**-brush** *n.* Zahnbürste, *die*

toothed [tu:θt] *adj.* **a)** *(Mech. Engin.)* gezähnt; ~ **wheel** Zahnrad; **b)** *(Bot.)* gezähnt; **c)** *in comb. (having teeth)* **sharp-**~ ⟨*Tier*⟩ mit scharfen Zähnen

toothless ['tu:θlɪs] *adj.* zahnlos

tooth: ~**-mug** *n.* Zahnputzbecher, *der;* ~**paste** *n.* Zahnpasta, *die;* ~**pick** *n.* Zahnstocher, *der;* ~**-powder** *n.* Zahnpulver, *das*

toothy ['tu:θɪ] *adj.* **give sb. a** ~ **smile** jmdn. mit entblößten Zähnen anlächeln; **he is a bit** ~: er hat ein ziemliches Pferdegebiß *(ugs.)*

[1]**top** [tɒp] **1.** *n.* **a)** *(highest part)* Spitze, *die; (of table)* Platte, *die; (of bench seat)* Sitzfläche, *die; (*~ *floor)* oberstes Stockwerk; *(flat roof, roof garden)* Dach, *das; (rim of glass, bottle, etc.)* Rand, *der; (*~ *end)* oberes Ende; *(of tree)* Spitze, *die;* Wipfel, *der;* **a cake with a cherry on** ~: ein Kuchen mit einer Kirsche [oben]drauf; **at the** ~: oben; **at the** ~ **of the building/hill/pile/stairs** oben im Gebäude/[oben] auf dem Hügel/[oben] auf dem Stapel/oben an der Treppe; **be at/get to** *or* **reach the** ~ **[of the ladder** *or* **tree]** *(fig.)* auf der obersten Sprosse [der Leiter] stehen/die oberste Sprosse [der Leiter] erreichen *(fig.)*; **be/get on** ~ **of a situation/subject** eine Situation/eine Materie im Griff haben/in den Griff bekommen; **don't let it get on** ~ **of you** *(fig.)* laß dich davon nicht unterkriegen! *(ugs.)*; **he put it on [the]** ~ **of the pile** er legte es [oben] auf den Stapel; **on** ~ **of one another** *or* **each other** aufeinander; **on** ~ **of sth.** *(fig.: in addition)* zusätzlich zu etw.; **on** ~ **of everything else** zu alledem noch; **come/be on** ~ **of sth.** *(be additional)* zu etw. [hinzu]kommen; **on** ~ **of the world** *(fig.)* überglücklich; **be/go thin on** ~: licht auf dem Kopf sein/werden; **be on** ~: ganz oben sein/liegen; **come out on** ~ *(be successful)* Erfolg haben; *(win)* gewinnen; **get to the** ~ *(fig.)* eine Spitzenposition erringen; ganz nach oben kommen *(ugs.)*; **from** ~ **to toe** von Kopf bis Fuß; **be over the** ~: übertrieben *od.* überzogen sein; **he searched the house from** ~ **to bottom** er durchsuchte das Haus von oben bis unten; **b)** *(highest rank)* Spitze, *die;* **the man at the** ~: der [oberste] Chef *od. (ugs.)* Boß; ~ **of the table** *(Sport)* Tabellenspitze, *die;* **[at the]** ~ **of the agenda is ...:** ganz oben auf der Tagesordnung steht ...; **be [at the]** ~ **of the class** der/die Klassenbeste sein; ~ **of the bill** *(Theatre)* Zugpferd, *das;* **c)** *(of vegetable)* Kraut, *das;* **d)** *(upper surface)* Oberfläche, *die; (of cupboard, wardrobe, chest)* Oberseite, *die;* **on [the]** ~ **of sth.** [oben] auf etw. *(position: Dat./direction: Akk.)*; **cut the** ~ **off an egg** ein Ei köpfen; **they climbed to the** ~ **of the hill/slope** sie kletterten auf den Hügel/den Hang hinauf; **e)** *(folding roof)* Verdeck, *das;* **f)** *(upper deck of bus, boat)* Oberdeck, *das;* **g)** *(cap of pen)* [Verschluß]kappe, *die;* **h)** *(cream on milk)* Sahne, *die;* Rahm, *der (regional, bes. südd., österr., schweiz.)*; **i)** *(upper part of page)* oberer Teil; **at the** ~ **[of the page]** oben [auf der/die Seite]; **j)** *(upper garment)* Oberteil, *das;* **k)** *(turn-down of sock)* Umschlag, *der;* **l)** *(head end)*

Kopf, *der; (of street)* oberes Ende; **m)** *(utmost)* Gipfel, *der;* **shout/talk at the ~ of one's voice** aus vollem Halse schreien/so laut wie möglich sprechen; **n) be the ~s** *(coll.) (the best)* der/die/das Größte sein *(ugs.); (marvellous)* spitze sein *(ugs.);* **o)** *(surface)* Oberfläche, *die;* **p)** *(lid)* Deckel, *der; (of bottle, glass, jar, etc.)* Stöpsel, *der;* **q)** *(Brit. Motor Veh.)* **in ~:** im größten Gang. **2.** *adj.* oberst...; höchst... ‹*Ton, Preis*›; **~ end** oberes Ende; **the/a ~ award** die höchste/eine hohe Auszeichnung; **the/a ~ chess-player** der beste Schachspieler/einer der besten Schachspieler *od.* ein Spitzenschachspieler; **~ scientists/actors** *etc.* hochkarätige Wissenschaftler/Schauspieler *usw.*; **~ sportsman/job/politician** Spitzensportler, *der*/Spitzenposition, *die*/Spitzenpolitiker, *der;* **the ~ pupil/school/marks** der beste Schüler/die beste Schule/die besten Noten; **~ manager/management** Topmanager/-management; **a ~ speed of 100 m.p.h.** eine Spitzen- *od.* Höchstgeschwindigkeit von 100 Meilen pro Stunde; **go at ~ speed** mit Spitzen- *od.* Höchstgeschwindigkeit fahren; **be/come ~ [in a subject]** [in einem Fach] der/die Beste sein/werden; **give sth. ~ priority** einer Sache *(Dat.)* höchste Priorität einräumen; **have a record in the ~ ten** eine Platte in den Top-ten haben; **in the ~ left/right corner** in der linken/rechten oberen Ecke; **on the ~ floor** im obersten Stockwerk; **the ~ people** *(in society)* die Spitzen der Gesellschaft; *(in a particular field)* die besten Leute. **3.** *v.t.*, **-pp-: a)** *(cover)* **the hills were ~ped with** *or* **by snow** die Hügelspitzen waren schneebedeckt; **b)** *(Hort.: cut ~ off)* stutzen ‹*Pflanze*›; kappen ‹*Baum*›; **c)** *(be taller than)* überragen; **d)** *(surpass, excel)* übertreffen; **exports have ~ped [the] £40 million [mark/level]** die Exporte haben die [Grenze von] 40 Millionen Pfund überschritten; **to ~ it all** [noch] obendrein; **e)** *(head)* anführen; **~ the bill** *(Theatre)* das Zugpferd sein

~ 'off 1. *v.t. (coll.)* beschließen. **2.** *v.i. (coll.)* schließen

~ 'up *(Brit. coll.)* **1.** *v.t.* auffüllen ‹*Batterie, Tank, Flasche, Glas*›; **~ up the petrol/oil/water** Benzin/Öl/Wasser nachfüllen; **~ up sb.'s drink** jmdm. nachschenken. **2.** *v.i. (fill one's tank up)* volltanken; *(fill one's glass up)* sich nachschenken

[2]**top** *n. (toy)* Kreisel, *der*

topaz ['təʊpæz] *n. (Min.)* Topas, *der*

top: ~ 'brass *see* **brass 1g; ~ coat** *n.* **a)** *(overcoat)* Überzieher, *der;* Mantel, *der;* **b)** *(of paint)* Deckanstrich, *der;* **~ copy** *n.* Original, *das;* **~ 'dog** *n. (fig. sl.)* Boß, *der (ugs.);* **~-flight** *attrib. adj.* erstrangig; Spitzen‹*sportler, -politiker*›; **~ 'hat** *n.* Zylinder[hut], *der;* **~-heavy** *adj.* oberlastig; kopflastig ‹*Baum, Pflanze, Bürokratie*›

topic ['tɒpɪk] *n.* Thema, *das;* **~ of debate/conversation** Diskussions-/Gesprächsthema, *das*

topical ['tɒpɪkl] *adj.* aktuell

topicality [tɒpɪ'kælɪtɪ] *n., no pl.* Aktualität, *die*

topically ['tɒpɪkəlɪ] *adv.* mit aktuellem Bezug

topless ['tɒplɪs] *adj.* **a) a ~ statue/column** eine Statue/Säule mit fehlendem oberem Teil; **b) a ~ dress/swimsuit** ein busenfreies Kleid/ein Oben-ohne-Badeanzug; **c)** *(bare-breasted)* barbusig; **~ girl/waitress** Oben-ohne-Mädchen, *das*/-Bedienung, *die*

'top-level *attrib. adj.* Gipfel‹*treffen, -konferenz*›; Spitzen‹*politiker, -funktionär*›; **~ discussions** Diskussionen auf höchster Ebene

topmost ['tɒpməʊst, 'tɒpməst] *adj.* oberst... ‹*Schicht, Stufe*›; höchst... ‹*Gipfel, Beamte, Note*›

top-'notch *adj. (coll.)* phantastisch *(ugs.)*

topography [tə'pɒgrəfɪ] *n.* **a)** Topographie, *die;* **b)** *(features)* örtliche *od. (geh.)* topographische Gegebenheiten

topping ['tɒpɪŋ] *n. (Cookery)* Überzug, *der*

topple ['tɒpl] **1.** *v.i.* fallen; **the tower/pile ~d to the ground** der Turm/Stapel fiel um *od.* kippte um; **~ [from power]** *(fig.)* stürzen. **2.** *v.t.* stürzen; **~ sb./a government [from power]** ‹*Gegner:*› jmdn./eine Regierung stürzen; ‹*Skandal, Abstimmung:*› jmdn./eine Regierung zu Fall bringen

~ 'down *v.i.* hinab-/herabfallen

~ 'over *v.i.* ‹*Turm, Stapel, Baum, Auto:*› umstürzen, umfallen; ‹*Vase, Ohnmächtiger:*› umfallen

top: ~-quality *attrib. adj.* [qualitativ] hochwertig; **~-ranking** *attrib. adj.* Spitzen‹*funktionär, -beamter, -politiker, -sportler, -orchester, -delegierter*›; hochrangig ‹*Offizier*›; erstrangig ‹*Autor, Schauspieler*›; führend ‹*Wissenschaftler*›; **~ 'secret** *adj.* streng geheim; **~side** *n. (joint of beef)* Oberschale, *die;* **~soil** *n. (Agric.)* Mutterboden, *der; (of field)* [Acker]krume, *die*

topsy-turvy [tɒpsɪ'tɜ:vɪ] **1.** *adv.* verkehrtrum *(ugs.);* auf dem Kopf *(ugs.)* ‹*stehen, liegen*›; **turn sth. ~** *(lit. or fig.)* etw. auf den Kopf stellen *(ugs.).* **2.** *adj.* chaotisch; *(fig.)* **a world where things are all ~:** eine Welt, in der alles auf dem Kopf steht

'top-up *n. (coll.)* Auffüllung, *die;* **give the tank/oil a ~:** den Tank auffüllen/Öl nachfüllen; **would you like a ~?** soll ich dir noch mal nachgießen?

torch [tɔ:tʃ] *n.* **a) [electric] ~** *(Brit.)* Taschenlampe, *die;* **b)** *(blowlamp) (for welding)* Schweißbrenner, *der; (for soldering)* Lötlampe, *die; (for cutting)* Schneidbrenner, *der*

tore *see* [1]**tear 2, 3**

toreador ['tɒrɪədɔ:(r)] *n.* Toreador, *der*

torment 1. ['tɔ:ment] *n.* Qual, *die;* **be in ~:** Qualen ausstehen. **2.** [tɔ:'ment] *v.t.* **a)** quälen; peinigen; **be ~ed by** *or* **with sth.** von etw. gequält werden; **b)** *(tease, worry)* quälen

torn *see* [1]**tear 2, 3**

tornado [tɔ:'neɪdəʊ] *n., pl.* **~es** Wirbelsturm, *der; (in North America)* Tornado, *der*

torpedo [tɔ:'pi:dəʊ] **1.** *n., pl.* **~es** Torpedo, *der.* **2.** *v.t. (auch fig.)* torpedieren

torpedo: ~-boat *n.* Torpedoboot, *das;* **~-tube** *n.* Torpedorohr, *das*

torpid ['tɔ:pɪd] *adj.* träge

torpor [tɔ:pə(r)] *n.* Trägheit, *die*

torque [tɔ:k] *n. (Mech.)* Drehmoment, *das*

torrent ['tɒrənt] *n.* **a)** reißender Bach; *(stream having steep course)* Sturzbach, *der;* **mountain ~:** reißender Gebirgsbach; **a ~ of rain** ein Regenguß; **the rain came down in ~s** es regnete in Strömen; **b)** *(fig.: violent flow)* Flut, *die;* Schwall, *der*

torrential [tə'renʃl] *adj.* **a)** reißend ‹*Gebirgsbach, Fluten*›; wolkenbruchartig ‹*Regen, Schauer*›; **b)** *(fig.)* überwältigend; gewaltig

torrid ['tɒrɪd] *adj.* **a)** *(intensely hot)* glutheiß; **the ~ heat of the desert** die Gluthitze der Wüste; **b)** *(fig.: intense, ardent)* glühend *(geh.);* ‹*Liebesszene*› voller Leidenschaft

torso ['tɔ:səʊ] *n., pl.* **~s a)** *(Art)*

Torso, *der;* **b)** *(human trunk)* Rumpf, *der;* **bare ~:** nackter Oberkörper

tortoise ['tɔ:təs] *n.* Schildkröte, *die*

tortoiseshell ['tɔ:təsʃel] *n.* Schildpatt, *das; attrib.* Schildpatt-

tortoiseshell 'cat *n.* Katze mit Schildpattzeichnung

tortuous ['tɔ:tjʊəs] *adj.* **a)** *(full of twists and turns)* verschlungen ‹*Weg*›; gewunden ‹*Flußlauf*›; **b)** *(fig.: circuitous)* umständlich; verworren ‹*Argumentation, Denken, Sprache*›

torture ['tɔ:tʃə(r)] **1.** *n.* **a)** Folter, *die;* **the ~ of sb.** jmds. Folterung; **instrument of ~:** Folterwerkzeug, *das;* Folterinstrument, *das;* **b)** *(fig.: agony)* Qual, *die;* **it was ~:** es war eine Tortur. **2.** *v. t.* foltern; *(fig.)* quälen

'torture-chamber *n.* Folterkammer, *die*

torturer ['tɔ:tʃərə(r)] *n.* Folterer, *der*/Folterin, *die*

Tory ['tɔ:rɪ] *(Brit. Polit. coll.)* **1.** *n.* Tory, *der.* **2.** *adj.* Tory-

toss [tɒs] **1.** *v. t.* **a)** *(throw upwards)* hochwerfen; **~ a pancake** einen Pfannkuchen [durch Hochwerfen] wenden; **b)** *(throw casually)* werfen; schmeißen *(ugs.);* **~ it over!** *(coll.)* schmeiß es/ihn/sie rüber *(ugs.);* **~ sth. to sb.** jmdm. etw. zuwerfen; **c) ~ a coin** eine Münze werfen; **~ sb. for sth.** mit jmdm. durch Hochwerfen einer Münze um etw. losen; **d) be ~ed by a bull/horse** von einem Stier auf die Hörner genommen werden/von einem Pferd abgeworfen werden; **e)** *(move about)* hin und her werfen; **f)** *(Cookery: mix gently)* wenden; **~ a salad in oil** einen Salat mit Öl anmachen. **2.** *v. i.* **a)** *(be restless in bed)* sich hin und her werfen; **~ and turn** sich [schlaflos] im Bett wälzen; **b)** ‹*Schiff, Boot:*› hin und her geworfen werden; **c)** *(~ coin)* eine Münze werfen; **~ for sth.** mit einer Münze um etw. losen. **3.** *n.* **a)** *(of coin)* **~ of a coin** Hochwerfen einer Münze; **argue the ~** *(fig.)* die Entscheidung nicht akzeptieren wollen; **lose/win the ~:** bei der Auslosung verlieren/gewinnen; *(Footb.)* die Seitenwahl verlieren/gewinnen; **b) give a contemptuous/proud ~ of the head** den Kopf verächtlich/stolz in den Nacken werfen; **c)** *(throw)* Wurf, *der*

~ about, ~ around 1. *v. i.* **a)** *(be restless in bed)* sich [schlaflos] im Bett wälzen; **b)** *see* **~ 2 b. 2.** *v. t.* **~ sth. around** *or* **about** etw. herumwerfen; *(fig.)* etw. in die Debatte werfen

~ a'side *v. t.* **a)** *(throw to one side)* hinwerfen; **b)** *(fig.: reject, abandon)* beiseite schieben

~ a'way *v. t.* wegwerfen

~ 'back *v. t.* zurückwerfen ‹*Kopf, Haar*›; runterkippen *(ugs.)* ‹*Getränk*›

~ 'out *v. t.* **a)** *(throw out)* **~ sth. out** etw. wegwerfen *od. (ugs.)* -schmeißen; **b)** *(fig.: reject)* [kurzerhand] ablehnen

~ 'up 1. *v. i.* eine Münze werfen; **~ up for sth.** mit einer Münze um etw. losen. **2.** *v. t.* *(throw)* hochwerfen; in die Luft werfen

'toss-up *n.* **a)** *(tossing of coin)* Hochwerfen einer Münze; **b)** *(even chance)* **it is a ~ [whether ...]** es ist noch ganz ungewiß[, ob ...]

[1]**tot** [tɒt] *n. (coll.)* **a)** *(small child)* kleines Kind; Wicht, der *(fam.);* **tiny ~:** kleiner Wicht; **b)** *(dram of liquor)* Gläschen, *das*

[2]**tot** *(coll.)* **1.** *v. t.,* **-tt-:** **~ 'up** zusammenziehen *(ugs.).* **2.** *v. i.,* **-tt-:** **~ 'up** sich summieren; sich [zusammen]läppern *(ugs.);* **that ~s up to £5** das macht zusammen 5 Pfund *(ugs.)*

total ['təʊtl] **1.** *adj.* **a)** *(comprising the whole)* gesamt; Gesamt‹*gewicht, -wert, -bevölkerung usw.*›; **what are your ~ debts?** wieviel Schulden hast du insgesamt?; **a ~ increase of £100** eine Steigerung von insgesamt 100 Pfund; **b)** *(absolute)* völlig *nicht präd.;* **be in ~ ignorance of sth.** von etw. überhaupt *od.* absolut nichts wissen; **a ~ beginner** ein absoluter Anfänger; **~ nonsense** totaler Unsinn; **have a ~ lack of interest in sth.** sich für etw. absolut nicht interessieren. **2.** *n. (number)* Gesamtzahl, *die; (amount)* Gesamtbetrag, *der; (result of addition)* Summe, *die;* **a ~ of 200/£200** *etc.* insgesamt 200/200 Pfund *usw.;* **in ~:** insgesamt. **3.** *v. t., (Brit.)* **-ll-:** **a)** *(add up)* addieren, zusammenzählen ‹*Zahlen, Posten, Beträge*›; **b)** *(amount to)* [insgesamt] betragen

~ 'up 1. *v. t.* addieren; zusammenzählen. **2.** *v. i.* **~ up to sth.** sich auf etw. *(Akk.)* belaufen

total e'clipse *n. (Astron.)* totale Finsternis

totalitarian [təʊtælɪ'teərɪən] *adj. (Polit.)* totalitär

totality [tə'tælɪtɪ] *n.* Gesamtheit, *die*

totally ['təʊtəlɪ] *adv.* völlig

total: ~ re'call *n.* **have [the power of] ~ recall** ein absolutes Erinnerungsvermögen haben; **~ 'war** *n.* totaler Krieg

tote [təʊt] *v. t. (coll.)* schleppen

'tote bag *n.* ≈ Reisetasche, *die*

totem ['təʊtəm] *n.* Totem, *das (Völkerk.)*

'totem-pole *n.* Totempfahl, *der (Völkerk.)*

totter ['tɒtə(r)] *v. i.* wanken; taumeln

toucan ['tu:kən] *n. (Ornith.)* Tukan, *der*

touch [tʌtʃ] **1.** *v. t.* **a)** *(lit. or fig.)* berühren; *(inspect by ~ing)* betasten; **~ the sky** *(fig.)* an den Himmel stoßen; **~ sb. on the shoulder** jmdm. auf die Schulter tippen; **~ A to B** B mit A berühren; **~ glasses** anstoßen; **b)** *(harm, interfere with)* anrühren; **the police can't ~ you [for it]** die Polizei kann dich nicht [dafür] belangen; **c)** *(fig.: rival)* **~ sth.** an etw. *(Akk.)* heranreichen; **d)** *(affect emotionally)* rühren; **e)** *(concern oneself with)* anrühren; **f) ~ sb. for a loan/£5** *(sl.)* jmdn. anpumpen *(salopp)*/um 5 Pfund anpumpen *od.* anhauen *(salopp).* **2.** *v. i.* sich berühren; ‹*Grundstücke:*› aneinanderstoßen; **don't ~!** nicht anfassen!; **'please do not ~'** „bitte nicht berühren!" **3.** *n.* **a)** Berührung, *die;* **at a ~:** bei bloßer Berührung; **be soft/warm** *etc.* **to the ~:** sich weich/ warm *usw.* anfühlen; **b)** *no pl., no art. (faculty)* **[sense of] ~:** Tastsinn, *der;* **c)** *(small amount)* **a ~ of salt/pepper** *etc.* eine Spur Salz/Pfeffer *usw.;* **a ~ of irony/sadness** *etc.* ein Anflug von Ironie/Traurigkeit *usw.;* **have a ~ of rheumatism** ein bißchen Rheuma haben; **a ~** *(slightly)* ein [ganz] kleines bißchen; **d)** *(game of tag)* Fangen, *das;* **e)** *(Art: stroke)* Strich, *der; (fig.)* Detail, *das;* **to mention it in such a way was a clever/subtle ~:** es auf eine solche Weise zu erwähnen, war ein schlauer/raffinierter Einfall; **add** *or* **put the final ~es to sth.** einer Sache *(Dat.)* den letzten Schliff geben; **f)** *(manner, style) (on keyboard instrument, typewriter)* Anschlag, *der; (of writer, sculptor)* Stil, *der;* **a personal ~:** eine persönliche *od.* individuelle Note; **lose one's ~:** seinen Schwung verlieren; *(Sport)* seine Form verlieren; **g)** *(communication)* **be in/out of ~ [with sb.]** [mit jmdm.] Kontakt/keinen Kontakt haben; **I shall be in ~ with them** ich werde mit ihnen Kontakt aufnehmen; **be in/**

out of ~ with sth. über etw. (+ *Akk.*) auf dem laufenden/nicht auf dem laufenden sein; **get in ~ [with sb.]** mit jmdm. Kontakt/Verbindung aufnehmen; **keep in ~ [with sb.]** [mit jmdm.] in Verbindung *od.* Kontakt bleiben; **keep in ~!** laß von dir hören!; **keep in ~ with sth.** sich über etw. *(Akk.)* auf dem laufenden halten; **lose ~ with sb.** den Kontakt zu jmdm. verlieren; **we have lost ~:** wir haben keinen Kontakt mehr [zueinander]; **have lost ~ with sth.** über etw. *(Akk.)* nicht mehr auf dem laufenden sein; **put sb. in ~ with sb.** jmdn. mit jmdm. zusammenbringen; **h)** *(Footb., Rugby: part of field)* Aus, *das;* Mark, *die (Rugby);* **in ~:** im Aus; **i)** *(sl.)* **be an easy** *or* **a soft ~** *(be a person who gives money readily)* leicht rumzukriegen sein *(ugs.)*

~ 'down *v. i.* **a)** *(Rugby)* den Ball niederlegen; *(Amer. Footb.)* den Ball hinter die Grundlinie bringen; **b)** ‹*Flugzeug:*› aufsetzen; *(land)* landen

~ on *v. t.* **a)** *(treat briefly)* ansprechen; **b)** *(verge on)* grenzen an (+ *Akk.*)

~ 'up *v. t.* **a)** *(improve)* ausbessern; **b)** *(sl.: fondle)* befummeln *(ugs.)*

~ upon *see* **~ on**

touch: ~-and-'go *adj.* prekär ‹*Situation*›; **it is ~-and-go [whether ...]** es steht auf des Messers Schneide[, ob ...]; **~down** *n.* **a)** *(Amer. Footb.)* Touchdown, *der;* **b)** *(Aeronaut.)* Landung, *die*

touched [tʌtʃt] *pred. adj.* **a)** *(moved)* gerührt; **b)** *(coll.: mad)* meschugge *(salopp)*

touching ['tʌtʃɪŋ] *adj.* rührend; *(moving)* bewegend; ergreifend

touch: ~-line *n. (Footb., Rugby)* Seitenlinie, *die;* **~-paper** *n.* Zündpapier, *das; (on firework)* Papierlunte, *die;* **~stone** *n. (fig.)* Prüfstein, *der;* **~-type** *v. i.* blindschreiben; **~-typing** *n.* Blindschreiben, *das*

touchy ['tʌtʃɪ] *adj.* empfindlich ‹*Person*›; heikel ‹*Thema, Sache*›

tough [tʌf] **1.** *adj.* **a)** fest ‹*Material, Stoff, Leder, Metall, Werkstoff*›; zäh ‹*Fleisch; fachspr.: Werkstoff, Metall, Kunststoff*›; widerstandsfähig ‹*Straßenbelag, Bodenbelag, Gummi, Glas, Haut*›; strapazierfähig ‹*Kleidung, Stoff, Schuhe*›; **b)** *(hardy, unyielding)* zäh ‹*Person*›; **a ~ customer** *(coll.)* ein harter Brocken *(ugs.);* **c)** *(difficult, trying)* schwierig; vertrackt *(ugs.)* ‹*Problem*›; hart ‹*Kampf, Wettkampf*›; strapaziös ‹*Reise*›; schwer ‹*Zeit*›; **we had a ~ time** wir haben viel durchgemacht; **d)** *(severe, harsh)* hart; **get ~** *(coll.)* andere Saiten aufziehen; **e)** *(coll.: unfortunate, hard)* **~ luck** Pech, *das;* **that's ~ [luck]** so'n Pech! *(ugs.);* **be ~ on sb.** hart für jmdn. sein. **2.** *n.* Rowdy, *der (abwertend)*

toughen ['tʌfn] *v. t.* größere Festigkeit geben (+ *Dat.*); zäher machen *(fachspr.)* ‹*Werkstoff, Metall, Kunststoff*›; abhärten, *(geh.)* stählen ‹*Person, Körper*›; verschärfen ‹*Gesetz, Widerstand*›

~ 'up *v. t.* abhärten; stählen *(geh.);* verschärfen ‹*Gesetz, Verbrechensbekämpfung*›

toughness ['tʌfnɪs] *n., no pl. see* **tough 1 a:** Festigkeit, *die;* Zäheit, *die;* Zähigkeit, *die (fachspr.);* Widerstandsfähigkeit, *die;* Strapazierfähigkeit, *die*

toupee, toupet ['tu:peɪ] *n.* Toupet, *das*

tour [tʊə(r)] **1.** *n.* **a)** [Rund]reise, *die;* Tour, *die (ugs.);* **a ~ of** *or* **through Europe** eine Reise durch Europa; eine Europareise; **a world ~/round-the-world ~:** eine Weltreise/Reise um die Welt; **a walking/cycling ~:** eine Wanderung/[Fahr]radtour; **b)** *(Theatre, Sport)* Tournee, *die;* Tour, *die (Jargon);* **be/go on ~:** auf Tournee/Tour sein/gehen; **c)** *(excursion, inspection) (of museum, palace, house)* Besichtigung, *die;* **go on/make/do a ~ of** besichtigen ‹*Museum, Haus, Schloß usw.*›; **a ~ of the countryside/the city/the factory** ein Ausflug in die Umgebung/eine Besichtigungstour durch die Stadt/ein Rundgang durch die Fabrik; **d) ~ [of duty]** Dienstzeit, *die.* **2.** *v. i.* **a) ~/go ~ing in** *or* **through a country** eine Reise *od. (ugs.)* Tour durch ein Land machen; **be ~ing in a country** auf einer Reise *od. (ugs.)* Tour durch ein Land sein; **b)** *(Theatre, Sport, exhibition)* eine Tournee *od. (Jargon)* Tour machen; *(be on ~)* auf Tournee *od. (Jargon)* Tour sein; touren *(Jargon); (go on ~)* auf Tournee *od. (Jargon)* Tour gehen. **3.** *v. t.* **a)** besichtigen ‹*Stadt, Gebäude, Museum*›; **~ a country/region** eine Reise *od. (ugs.)* Tour durch ein Land/Gebiet machen; **~ an area on foot/by bicycle** eine Wanderung/Radtour durch eine Gegend machen; **b)** *(Theatre, Sport)* **~ a country/the provinces** eine Tournee *od. (Jargon)* Tour durch das Land/die Provinz machen

tourer ['tʊərə(r)] *n. (Motor Veh.)* Kabriolimousine, *die*

tourism ['tʊərɪzm] *n., no pl., no indef. art.* **a)** Tourismus, *der;* **b)** *(operation of tours)* Touristik, *die*

tourist ['tʊərɪst] **1.** *n.* Tourist, *der*/Touristin, *die.* **2.** *attrib. adj.* Touristen-; **special ~ rates** ermäßigte Preise für Touristen

tourist: ~ attraction *n.* Touristenattraktion, *die;* **~ board** *n. (Brit.)* Amt für Fremdenverkehrswesen; **~ guide** *n.* **a)** *(person)* Touristenführer, *der*/-führerin, *die;* **b)** *(book)* Reiseführer, *der* **(to, of** von); **~ hotel** *n.* Touristenhotel, *das;* **~ industry** *n.* **a)** *(business)* Tourismusindustrie, *die;* **b)** *(firms)* Touristik[branche], *die;* **~ infor'mation centre, ~ office** *ns.* Fremdenverkehrsbüro, *das;* Touristeninformation, *die (ugs.);* **~ season** *n.* Touristensaison, *die;* **~ trade** *see* **~ industry**

touristy ['tʊərɪstɪ] *adj. (derog.)* auf Tourismus getrimmt *(ugs.);* Touristen‹*stadt, -nest, -gegend*› *(ugs. abwertend)*

tournament ['tʊənəmənt] *n. (Hist.; Sport)* Turnier, *das*

tourniquet ['tʊənɪkeɪ] *n. (Med.)* Tourniquet, *das*

'tour operator *n.* Reiseveranstalter, *der*/-veranstalterin, *die*

tousle ['taʊzl] *v. t.* zerzausen

tout [taʊt] **1.** *v. i.* **~ [for business/custom/orders]** Kunden anreißen *(ugs.) od.* werben; **~ for customers/buyers** Kunden/Käufer anreißen *(ugs.) od.* werben. **2.** *n.* Anreißer, *der*/Anreißerin, *die (ugs.);* Kundenwerber, *der*/-werberin, *die;* **ticket ~:** Kartenschwarzhändler, *der*/-händlerin, *die*

tow [təʊ] **1.** *v. t.* schleppen; ziehen ‹*Anhänger, Wasserskiläufer, Handwagen*›; **he ~ed my car to get it started** er hat meinen Wagen angeschleppt. **2.** *n.* Schleppen, *das;* **My car's broken down. – Do you want a ~?** Mein Wagen ist stehengeblieben. – Soll ich Sie [ab]schleppen?; **give a boat/car a ~:** ein Boot/einen Wagen schleppen; **have sth. in** *or* **on ~:** etw. im Schlepp[tau] haben; **have sb. in ~** *(fig.)* jmdn. im Schlepptau haben *(ugs.);* **take a boat/car in ~:** ein Boot/einen Wagen in Schlepp nehmen

~ a'way *v. t.* abschleppen

toward [tə'wɔ:d], **towards** [tə'wɔ:dz] *prep.* **a)** *(in direction of)* **~ sb./sth.** auf jmdn./etw. zu; **the ship sailed ~ France/the open sea**

das Schiff fuhr in Richtung Frankreich/offenes Meer; ~ **[the] town** in Richtung [auf die] Stadt; **point ~ the north** nach Norden zeigen; **turn ~ sb.** sich zu jmdm. umdrehen; **sit/stand with one's back [turned] ~ sth.** mit dem Rücken zu etw. sitzen/stehen; **the country was drifting ~ war/economic chaos** das Land trieb dem Krieg/wirtschaftlichem Chaos zu; **b)** *(in relation to)* gegenüber; **feel sth. ~ sb.** jmdm. gegenüber etw. empfinden; **be fair/unfair** *etc.* **~ sb.** jmdm. gegenüber *od.* zu jmdm. fair/unfair *usw.* sein; **feel angry/sympathetic ~ sb.** böse auf jmdn. sein/Verständnis für jmdn. haben; **c)** *(for)* **a contribution ~ sth.** ein Beitrag zu etw.; **save up ~ a car/one's holidays** auf *od.* für einen Wagen/für seine Ferien sparen; **proposals ~ solving a problem** Vorschläge zur Lösung eines Problems; **work together ~ a solution** gemeinsam auf eine Lösung hinarbeiten; **d)** *(near)* gegen; **~ the end of May/of the year** *etc.* [gegen] Ende Mai/des Jahres

towel ['taʊəl] **1.** *n.* Handtuch, *das;* **throw in the ~** *(Boxing; also fig.)* das Handtuch werfen. **2.** *v. t., (Brit.)* **-ll-** abtrocknen; **~ oneself** sich abtrocknen

'towel-rail *n.* Handtuchhalter, *der*

tower ['taʊə(r)] **1.** *n.* **a)** Turm, *der;* **b)** *(fortress)* Festung, *die;* Wehrturm, *der;* **the T~ [of London]** der Tower [von London]; **c) be a ~ of strength [to sb.]** *(fig.)* [jmdm.] ein fester Rückhalt sein. **2.** *v. i.* in die Höhe ragen

~ above, ~ over *v. t.* **~ above** *or* **over sb./sth.** *(lit. or fig.)* jmdn./etw. überragen

'tower block *n.* Hochhaus, *das*

towering ['taʊərɪŋ] *attrib. adj.* **a)** hoch aufragend; **b)** *(fig.)* herausragend ‹*Leistung, Gestalt*›; **c)** *(fig.: violent, intense)* blind ‹*Wut*›; maßlos ‹*Ehrgeiz, Stolz*›

town [taʊn] *n.* **a)** Stadt, *die;* **the ~ of Cambridge** die Stadt Cambridge; **in [the] ~:** in der Stadt; **the ~** *(people)* die Stadt; **on the outskirts/in the centre of ~:** in den Randbezirken der Stadt/in der Stadtmitte *od.* Innenstadt; **go [up] to ~:** in die Stadt fahren; **be in/out of ~:** in der Stadt/nicht in der Stadt sein; **the best coffee/tea/cake** *etc.* **in ~:** der beste Kaffee/Tee/Kuchen *usw.* in der Stadt; **go out/have a night on the ~** *(coll.)* [in die Stadt gehen und] einen draufmachen *(ugs.);* **go to ~** *(fig. coll.)* in die vollen gehen **(on** bei) *(ugs.);* **b)** *(business or shopping centre)* Stadt, *die;* **in ~:** in der Stadt; **go into ~:** in die Stadt gehen/fahren

town: ~ 'centre *n.* Stadtmitte, *die;* Stadtzentrum, *das;* **~ 'clerk** *n.* ≈ [Ober]stadtdirektor, *der*/-direktorin, *die;* **~ 'council** *n.* *(Brit.)* Stadtrat, *der;* **~ 'councillor** *n.* *(Brit.)* Stadtrat, *der*/-rätin, *die;* **~ 'hall** *n.* Rathaus, *das;* **~ house** *n.* **a)** *(residence in ~)* Stadthaus, *das;* **b)** *(terrace-house)* Reihenhaus, *das;* **~ 'planning** *n.* Stadtplanung, *die*

tow: ~-path *n.* Leinpfad, *der;* **~-rope** *n.* Abschleppseil, *das*

toxic ['tɒksɪk] *adj.* giftig; toxisch *(fachspr.)*

toy [tɔɪ] **1.** *n. (lit. or fig.)* Spielzeug, *das;* **~s** Spielzeug, *das;* Spielwaren *Pl. (Wirtsch.).* **2.** *adj.* **a)** Spielzeug-; **b)** *(Breeding)* Zwerg-. **3.** *v. i.* **~ with the idea of doing sth.** mit dem Gedanken spielen, etw. zu tun; **~ with one's food** *(nibble at)* in seinem Essen herumstochern

toy: ~-boy *n. (coll.)* Gespiele, *der* *(scherzh.);* **~shop** *n.* Spielwarengeschäft, *das;* **~ 'soldier** *n.* Spielzeugsoldat, *der*

¹trace [treɪs] **1.** *v. t.* **a)** *(copy)* durchpausen; abpausen; **~ sth. on to sth.** etw. auf etw. *(Akk.)* pausen; **b)** *(delineate)* zeichnen ‹*Form, Linie*›; malen ‹*Buchstaben, Wort*›; *(fig.)* entwerfen; **she ~d our route on the map with her finger** sie zeichnete unsere Route mit dem Finger auf der Landkarte nach; **c)** *(follow track of)* folgen (+ *Dat.*); verfolgen; **~ a river to its source** einen Fluß [bis] zur Quelle zurückverfolgen; **the police ~d him to Spain** die Polizei spürte ihn in Spanien auf; **d)** *(observe, find)* finden; **~ a connection** einen Zusammenhang sehen. **2.** *n.* Spur, *die;* **there is no ~ of your letter in our records** in unseren Aufzeichnungen findet sich kein Hinweis auf Ihr Schreiben; **I can't find any ~ of him/it** *(cannot locate)* ich kann ihn/es nirgends finden; **lose [all] ~ of sb.** jmdn. [völlig] aus den Augen verlieren; **sink without ~:** sinken, ohne eine Spur zu hinterlassen; *(fig.)* in der Versenkung verschwinden *(ugs.);* ‹*bekannte Persönlichkeit:*› von der Bildfläche verschwinden *(ugs.)*

~ 'back *v. t.* zurückverfolgen

~ 'out *see* **~ 1 b**

²trace *n.* Strang, *der;* **kick over the ~s** *(fig.)* über die Stränge schlagen *(ugs.)*

traceable ['treɪsəbl] *adj.* **a) sth. is ~ to sth./through sth.** etw. läßt sich bis zu etw./durch etw. hindurch zurückverfolgen; **b)** *(discoverable)* auffindbar

'trace element *n. (Chem.)* Spurenelement, *das*

tracer ['treɪsə(r)] *n. (Mil.)* Leuchtspurgeschoß, *das*

trachea [trə'ki:ə] *n., pl.* **~e** [trə'ki:i:] *(Anat.)* Trachea, *die* *(fachspr.);* Luftröhre, *die*

tracing ['treɪsɪŋ] *n.* **a)** *(action)* [Durch]pausen, *das;* [Ab]pausen, *das;* **b)** *(copy)* Pause, *die*

'tracing-paper *n.* Pauspapier, *das*

track [træk] **1.** *n.* **a)** Spur, *die;* *(of wild animal)* Fährte, *die;* **~s** *(footprints)* [Fuß]spuren; *(of animal also)* Fährte, *die;* **cover one's ~s** *(fig.)* seine Spur verwischen; **be on sb.'s ~:** jmdm. auf der Spur sein; *(fig.: in possession of clue to sb.'s plans)* jmdm. auf die Schliche gekommen sein; **be on the right/wrong ~** *(fig.)* auf der richtigen/falschen Spur sein; **keep ~ of sb./sth.** jmdn./etw. im Auge behalten; **lose ~ of sb./sth.** jmdn./etw. aus den Augen verlieren; **make ~s** *(coll.) (depart)* sich auf die Socken machen *(ugs.);* **stop [dead] in one's ~s** *(coll.)* auf der Stelle stehenbleiben; **b)** *(path)* [unbefestigter] Weg; *(footpath)* Pfad, *der;* *(fig.)* Weg, *der;* **c)** *(Sport)* Bahn, *die;* **cycling/greyhound ~:** Radrennbahn, *die*/Windhundrennbahn, *die;* **circuit of the ~:** Bahnrunde, *die;* **d)** *(Railw.)* Gleis, *das;* **single/double ~:** eingleisige/zweigleisige Strecke; **e)** *(course taken)* Route, *die;* *(of rocket, satellite, comet, missile, hurricane, etc.)* Bahn, *die;* **f)** *(of tank, tractor, etc.)* Kette, *die;* **g)** *(section of record)* Stück, *das;* **h)** *see* **sound-track. 2.** *v. t.* **~ an animal** die Spur/Fährte eines Tieres verfolgen; **the police ~ed him [to Paris]** die Polizei folgte seiner Spur [bis nach Paris]; **~ a rocket/satellite** die Bahn einer Rakete/eines Satelliten verfolgen

~ 'down *v. t.* aufspüren

tracker ['trækə(r)] *n.* **a)** Fährtensucher, *der;* **b) ~[dog]** Spürhund, *der*

track: ~ events *n. pl. (Athletics)* Laufwettbewerbe; **~ record** *n.* *(fig.)* **his ~ record is good, he has a good ~ record** er hat gute Leistungen vorzuweisen; **~ shoe** *n.*

Rennschuh, *der;* ~ **suit** *n.* Trainingsanzug, *der*

[1]**tract** [trækt] *n.* **a)** *(area)* Gebiet, *das;* **b)** *(Anat.)* Trakt, *der*

[2]**tract** *n.* *(pamphlet)* [Flug]schrift, *die;* Traktat, *der (veralt.)*

traction ['trækʃn] *n., no pl., no indef. art.* **a)** *(drawing along)* Traktion, *die (fachspr.);* Ziehen, *das;* **b)** *(grip of tyre etc.)* Haftung, *die;* **c)** *(Med.)* Zug, *der;* **in** ~: im Zug- *od.* Streckverband

'**traction engine** *n.* Zugmaschine, *die*

tractor ['træktə(r)] *n.* Traktor, *der*

trade [treɪd] **1.** *n.* **a)** *(line of business)* Gewerbe, *das;* **the wool/furniture/hotel** ~: die Woll-/Möbel-/Hotelbranche; **the retail/wholesale** ~: der Einzel-/Großhandel; **he's a butcher/lawyer/baker** *etc.* **by** ~: er ist von Beruf Metzger / Rechtsanwalt / Bäcker *usw.;* **trick of the** ~: einschlägiger Trick; **b)** *no pl., no indef. art (commerce)* Handel, *der;* **be bad/good for** ~: schlecht/gut fürs Geschäft sein; **foreign** ~: Außenhandel, *der;* **c)** *no pl. (business done)* Geschäft, *das;* *(between countries)* Handel, *der;* **do a good/roaring** ~ **[in sth.]** ein gutes Geschäft/ein Riesengeschäft [mit etw.] machen; **d)** *(craft)* Handwerk, *das;* **e)** *no pl., no indef. art. (persons)* **the** ~: die Branche; **f)** *in pl. (Meteorol.)* Passat, *der.* **2.** *v.i.* **a)** *(buy and sell)* Handel treiben; ~ **as a wholesale/retail dealer** ein Großhandels- / Einzelhandelsgeschäft betreiben; ~ **in sth.** in *od.* mit etw. *(Dat.)* handeln; **b)** *(have an exchange)* tauschen; ~ **with sb. for sth.** jmdm. etw. abhandeln. **3.** *v.t.* **a)** tauschen; austauschen ‹*Waren, Grüße, Informationen, Geheimnisse*›; sich *(Dat.)* sagen ‹*Beleidigungen*›; **b)** ~ **sth. for sth.** etw. gegen etw. tauschen; ~ **an old car** *etc.* **for a new one** einen alten Wagen *usw.* für einen neuen in Zahlung geben

~ '**in** *v.t.* in Zahlung geben; einlösen ‹*Gutschein, Kupon usw.*›

~ '**off** *v.t. (coll.)* ~ **sth. off for sth.** etw. gegen etw. tauschen

~ **on** *v.t. (fig.)* ~ **on sth.** aus etw. Kapital schlagen; sich *(Dat.)* etw. zunutze machen

~ '**up** *v.i.* sich verbessern

~ **upon** *see* ~ **on**

trade: ~ **balance** *n. (Econ.)* Handelsbilanz, *die;* ~ **cycle** *n. (Brit. Econ.)* Konjunkturzyklus, *der;* ~ **deficit** *n. (Econ.)* Handelsbilanzdefizit, *das;* ~ '**discount** *n.* Branchenrabatt, *der;* ~ **fair** *n.* [Fach]messe, *die;* ~ **gap** *see* ~ **deficit;** ~ **journal** *n.* Fachzeitschrift, *die;* ~ **mark** *n.* **a)** Warenzeichen, *das;* **b)** *(fig.)* **leave one's** ~ **mark on sth.** einer Sache *(Dat.)* seinen Stempel aufdrücken; ~ **name** *n.* **a)** *(name used in the* ~*)* Fachbezeichnung, *die;* **b)** *(proprietary name)* Markenname, *der;* **c)** *(name of business)* Firmenname, *der;* ~ **price** *n.* Einkaufspreis, *der*

trader ['treɪdə(r)] *n.* **a)** Händler, *der*/Händlerin, *die;* **b)** *(Naut.)* Handelsschiff, *das*

trade: ~ **route** *n.* Handelsweg, *der;* Handelsstraße, *die;* ~ '**secret** *n.* Geschäftsgeheimnis, *das;* ~**sman** ['treɪdzmən] *n., pl.* ~**smen** ['treɪdzmən] **a)** *(shopkeeper)* [Einzel]händler, *der;* Ladeninhaber, *der;* ~**smen's entrance** Lieferanteneingang, *der;* **b)** *(craftsman)* Handwerker, *der;* ~**speople** ['treɪdzpi:pl] *n. pl.* **a)** *(shopkeepers)* [Einzel]händler; Ladeninhaber; **b)** *(craft workers)* Handwerker; ~**s'** '**union** *see* ~ **union; T~s Union** '**Congress** *pr. n. (Brit.)* Gewerkschaftsbund, *der;* ~ '**union** *n.* Gewerkschaft, *die; attrib.* Gewerkschafts-; ~-'**unionism** [treɪd'ju:nɪənɪzm, treɪd'ju:njənɪzm] *n., no pl.* Gewerkschaftswesen, *das;* ~-'**unionist** *n.* Gewerkschaft[l]er, *der*/Gewerkschaft[l]erin, *die;* ~ **wind** *n. (Meteorol.)* Passatwind, *der*

trading ['treɪdɪŋ] *n.* Handel, *der*

trading: ~ **estate** *n. (Brit.)* Gewerbegebiet, *das;* ~ **hours** *n.pl.* Geschäftszeit, *die;* ~ **stamp** *n.* Rabattmarke, *die*

tradition [trə'dɪʃn] *n.* Tradition, *die;* *(story)* [mündliche] Überlieferung; **family** ~: Familientradition, *die;* **in the best** ~**[s]** nach bester Tradition; **break with** ~: mit der Tradition brechen

traditional [trə'dɪʃənl] *adj.* traditionell; mündlich überliefert ‹*Geschichte*›; herkömmlich ‹*Erziehung, Einrichtung, Methode*›; überkommen ‹*Brauch, Sitte, Werte, Moral*›; **it is** ~ **to do sth.** es ist Tradition, etw. zu tun

traditionally [trə'dɪʃənəlɪ] *adv.* *(in a traditional manner)* traditionell; *(by tradition)* traditionell[erweise]

traffic ['træfɪk] **1.** *n., no pl.* **a)** *no indef. art.* Verkehr, *der;* ~ **is heavy/light** es herrscht starker/geringer Verkehr; **b)** *(trade)* Handel, *der;* ~ **in drugs/arms** Drogen-/Waffenhandel, *der;* **c)** *(amount of business)* Verkehr, *der.* **2.** *v.i.,* **-ck-** Geschäfte machen; ~ **in sth.** mit etw. handeln *od.* Handel treiben; *(fig.)* mit etw. schachern *(abwertend)*

traffic: ~ **circle** *n. (Amer.)* Kreisverkehr, *der;* ~ **cone** *n.* Pylon, *der;* Leitkegel, *der;* ~ **hold-up** *see* ~ **jam;** ~ **island** *n.* Verkehrsinsel, *die;* ~ **jam** *n.* [Verkehrs]stau, *der*

trafficker ['træfɪkə(r)] *n.* Händler, *der*/Händlerin, *die*

traffic: ~ **lights** *n.pl.* [Verkehrs]ampel, *die;* ~ **police** *n.* Verkehrspolizei, *die;* ~ **policeman** *n.* Verkehrspolizist, *der;* ~ **sign** *n.* Verkehrszeichen, *das;* ~ **signals** *see* ~ **lights;** ~ **warden** *n.* *(Brit.)* Hilfspolizist, *der;* *(woman)* Hilfspolizistin, *die;* Politesse, *die*

tragedy ['trædʒɪdɪ] *n.* **a)** *(sad event or fact)* Tragödie, *die;* *(sad story)* tragische Geschichte; **the** ~ **[of it] is that ...:** das Tragische [daran] ist, daß ...; **b)** *(accident)* Tragödie, *die;* **c)** *(Theatre)* Tragödie, *die;* Trauerspiel, *das*

tragic ['trædʒɪk] *adj.* **a)** tragisch; **a** ~ **waste of talent/money** eine schlimme Vergeudung von Talenten/Geldverschwendung; **b)** *attrib.* *(Theatre)* tragisch; ~ **actor/actress** Tragöde, *der*/Tragödin, *die*

tragically ['trædʒɪkəlɪ] *adv.* tragisch; **their predictions have been** ~ **fulfilled** ihre Prophezeiungen haben sich auf tragische Weise erfüllt

trail [treɪl] **1.** *n.* **a)** Spur, *die;* *(of meteor)* Schweif, *der;* **a** ~ **of blood** eine Blutspur; ~ **of smoke/dust** Rauch-/Staubfahne, *die;* **b)** *(Hunting)* Spur, *die;* Fährte, *die;* **be on the** ~ **of an animal** der Fährte eines Tieres folgen; **be/get on sb.'s** ~ *(lit. or fig.)* jmdm. auf der Spur *od.* Fährte sein/jmdm. auf die Spur *od.* Fährte kommen; **be hard** *or* **hot on the** ~ **of sb.** *(lit. or fig.)* jmdm. dicht auf den Fersen sein *(ugs.);* **c)** *(path)* Pfad, *der;* *(wagon* ~*)* Weg, *der.* **2.** *v.t.* **a)** *(pursue)* verfolgen; *(shadow)* beschatten; ~ **sb./an animal to a place** jmdm./einem Tier bis zu einem Ort folgen; **b)** *(drag)* ~ **sth. [after** *or* **behind one]** etw. hinter sich *(Dat.)* herziehen; ~ **sth. on the ground** etw. über den Boden schleifen lassen. **3.** *v.i.* **a)** *(be dragged)* schleifen; **b)** *(hang loosely)* herabhängen; **c)** *(walk wearily etc.)* trotten; *(lag)* hintertrotten; **d)** *(Sport: be losing)* zurückliegen; **e)** *(creep)* ⟨*Pflanze:*⟩ kriechen

~ **a'way** *see* ~ **off**

~ **be'hind** *v. i.* hinterhertrödeln *(ugs.)*

~ **'off** *v. i.* **his voice/shout ~ed off into a whisper/into silence** seine Stimme/sein Schreien wurde schwächer, bis er schließlich nur noch flüsterte/bis er schließlich ganz verstummte

trailer ['treɪlə(r)] *n.* **a)** *(Motor Veh.)* Anhänger, *der; (boat ~ also)* Trailer, *der; (Amer.: caravan)* Wohnanhänger, *der;* **b)** *(Cinemat., Telev.)* Trailer, *der*

train [treɪn] **1.** *v. t.* **a)** ausbilden (**in** in + *Dat.*); erziehen ⟨*Kind*⟩; abrichten ⟨*Hund*⟩; dressieren ⟨*Tier*⟩; schulen ⟨*Geist, Auge, Ohr*⟩; bilden ⟨*Charakter*⟩; ~ **sb. as a teacher/soldier/engineer** jmdn. zum Lehrer/Soldaten/Ingenieur ausbilden; **he/she has been well/badly/fully ~ed** er/sie besitzt eine gute/schlechte/umfassende Ausbildung; **b)** *(Sport)* trainieren; ~ **oneself** trainieren; **c)** *(teach and accustom)* ~ **an animal to do sth./to sth.** einem Tier beibringen, etw. zu tun/etw. beibringen; ~ **oneself to do sth.** sich dazu erziehen, etw. zu tun; ~ **a child to do sth./to sth.** ein Kind dazu erziehen, etw. zu tun/zu etw. erziehen; ~ **sb. to use a machine** jmdn. in der Bedienung einer Maschine schulen; **d)** *(Hort.)* ziehen; erziehen *(fachspr.);* **e)** *(aim)* richten (**on** auf + *Akk.*). **2.** *v. i.* **a)** eine Ausbildung machen; **he is ~ing as** *or* **to be a teacher/doctor/engineer** er macht eine Lehrer- / Arzt- / Ingenieursausbildung; **b)** *(Sport)* trainieren. **3.** *n.* **a)** *(Railw.)* Zug, *der;* **go** *or* **travel by ~:** mit dem Zug *od.* der Bahn fahren; **on the ~:** im Zug; **which is the ~ for Oxford?** welcher Zug fährt nach Oxford?; **b)** *(of skirt etc.)* Schleppe, *die;* **c)** ~ **of thought** Gedankengang, *der*

'train-driver *n.* Lokomotivführer, *der/*-führerin, *die*

trained [treɪnd] *adj.* ausgebildet ⟨*Arbeiter, Lehrer, Arzt, Stimme*⟩; abgerichtet ⟨*Hund*⟩; dressiert ⟨*Tier*⟩; geschult ⟨*Geist, Auge, Ohr*⟩

trainee [treɪ'ni:] *n.* Auszubildende, *der/die; (business management ~)* Trainee, *der/die; (in academic or technical professions)* Praktikant, *der/*Praktikantin, *die*

trainer ['treɪnə(r)] *n.* **a)** *(Sport)* [Konditions]trainer, *der/*-trainerin, *die;* **b)** *in pl.* Trainingsschuhe

'train fare *n.* Fahrpreis, *der*

training ['treɪnɪŋ] *n., no pl.* **a)** Ausbildung, *die;* **b)** *(Sport)* Training, *das;* **be in ~** *(train)* trainieren; im Training sein; *(be fit)* in [guter] Form sein; **be out of ~:** außer Form sein

training: ~ **college** *n.* berufsbildende Schule; **~-course** *n.* Lehrgang, *der;* ~ **scheme** *n.* Ausbildungsprogramm, *das;* **be on a ~ scheme** an einem Ausbildungsprogramm teilnehmen; ~ **shoes** *n. pl.* Trainingsschuhe

train: ~ **journey** *n.* Bahnfahrt, *die; (long)* Bahnreise, *die;* ~ **service** *n.* Zugverbindung, *die;* [Eisen]bahnverbindung, *die;* ~ **set** *n.* [Modell]eisenbahn, *die;* **~-spotter** *n. jmd., der als Hobby die Nummern von Lokomotiven aufschreibt;* **~-spotting** *n., no pl., no indef. art.: das Aufschreiben von Lokomotivnummern als Hobby;* ~ **station** *n. (Amer.)* Bahnhof, *der*

trait [treɪ] *n.* Eigenschaft, *die*

traitor ['treɪtə(r)] *n.* Verräter, *der/*Verräterin, *die;* **turn ~:** zum Verräter/zur Verräterin werden

trajectory [trə'dʒektərɪ] *n. (Phys.)* [Flug]bahn, *die*

tram [træm] *n. (Brit.)* Straßenbahn, *die;* **go by ~:** mit der Straßenbahn fahren; **on the ~:** in der Straßenbahn

'tramlines *n. pl. (Brit.)* **a)** Straßenbahnschienen; **b)** *(fig.: rigid principles)* starre Vorschriften; **c)** *(Tennis coll.)* Korridor, *der*

trammel ['træml] *v. t., (Brit.)* **-ll-** einengen

tramp [træmp] **1.** *n.* **a)** *(vagrant)* Landstreicher, *der/*-streicherin, *die; (in city)* Stadtstreicher, *der/*-streicherin, *die;* **b)** *(sound of steps)* Schritte; *(of horses)* Getrappel, *das; (of elephants)* Trampeln, *das;* **c)** *(walk)* [Fuß]marsch, *der.* **2.** *v. i.* **a)** *(tread heavily)* trampeln; **b)** *(walk)* marschieren. **3.** *v. t.* **a)** ~ **one's way** trotten; **b)** durchwandern; *(with no particular destination)* durchstreifen

trample ['træmpl] **1.** *v. t.* zertrampeln; ~ **sth. into the ground** etw. in den Boden treten; **he was ~d to death by elephants** er wurde von Elefanten zu Tode getrampelt. **2.** *v. i.* trampeln

~ **on** *v. t.* herumtrampeln auf (+ *Dat.*); ~ **on sb./sth./sb.'s feelings** *(fig.)* jmdn./etw./jmds. Gefühle mit Füßen treten

trampoline ['træmpəli:n] **1.** *n.* Trampolin, *das.* **2.** *v. i.* Trampolin springen

trance [trɑ:ns] *n.* Trance, *die; (half-conscious state, hypnotic state)* tranceartiger Zustand; **be** *or* **lie in a ~:** in Trance/in einem tranceartigen Zustand sein; **fall** *or* **go into a ~:** in Trance/in einen tranceartigen Zustand fallen; **put** *or* **send sb. into a ~:** jmdn. in Trance/in einen tranceartigen Zustand versetzen; **she's been walking about in a ~ all day** sie ist den ganzen Tag wie in Trance herumgelaufen

tranquil ['træŋkwɪl] *adj.* ruhig; friedlich ⟨*Stimmung, Szene*⟩

tranquillity [træŋ'kwɪlɪtɪ] *n.* Ruhe, *die*

tranquillizer ['træŋkwɪlaɪzə(r)] *n. (Med.)* Tranquilizer, *der;* Beruhigungsmittel, *das*

transact [træn'sækt] *v. t.* ~ **business [with sb.]** [mit jmdm.] Geschäfte tätigen *(Kaufmannsspr., Papierdt.)*

transaction [træn'sækʃn] *n.* **a)** *(doing of business)* **after the ~ of their business** nachdem sie das Geschäftliche erledigt hatten; **b)** *(piece of business)* Geschäft, *das; (financial)* Transaktion, *die*

transatlantic [trænsət'læntɪk] *adj.* **a)** *(Brit.: American)* transatlantisch; amerikanisch; **b)** *(Amer.: European)* transatlantisch; europäisch; **c)** *(crossing the Atlantic)* transatlantisch; **a ~ voyage** eine Reise über den Atlantik

transcend [træn'send] *v. t. (be beyond range of)* übersteigen; hinausgehen über ⟨*Grenzen*⟩; *(Philos.)* transzendieren

transcendental [trænsen'dentl] *adj. (Philos.)* transzendental

transcontinental [trænskɒntɪ'nentl] *adj.* transkontinental

transcribe [træn'skraɪb] *v. t. (copy in writing)* abschreiben; aufschreiben ⟨*mündliche Überlieferung*⟩; mitschreiben ⟨*Rede*⟩; ~ **a tape/a taped interview** von einem Tonband/von der Tonbandaufzeichnung eines Interviews eine Niederschrift anfertigen

transcript ['trænskrɪpt] *n.* Abschrift, *die; (of trial, interview, speech, conference)* Protokoll, *das; (of tape, taped material)* Niederschrift, *die*

transept ['trænsept] *n. (Eccl. Archit.)* Querschiff, *das;* **north/south ~:** nördlicher/südlicher Kreuzarm

transfer 1. [træns'fɜ:(r)] *v. t.,* **-rr-:** **a)** *(move)* verlegen (**to** nach); überweisen ⟨*Geld*⟩ (**to** auf + *Akk.*); transferieren ⟨*große Geldsumme*⟩; übertragen ⟨*Befugnis,*

Macht⟩ (to *Dat.*); ~ **a prisoner to a different gaol** einen Gefangenen in ein anderes Gefängnis verlegen *od.* überführen; ~ **one's allegiance [from sb.] to sb.** [von jmdm.] zu jmdm. überwechseln; **b)** übereignen ⟨*Gegenstand, Grundbesitz*⟩ (to *Dat.*); **c)** versetzen ⟨*Arbeiter, Angestellte, Schüler*⟩; *(Footb.)* transferieren; **d)** übertragen ⟨*Bedeutung, Sinn*⟩. **2.** [træns'fɜ:(r)] *v.i.*, **-rr-: a)** *(change to continue journey)* umsteigen; ~ **from Heathrow to Gatwick** zum Weiterflug *od.* Umsteigen von Heathrow nach Gatwick fahren; **b)** *(move to another place or group)* wechseln; ⟨*Firma:*⟩ übersiedeln. **3.** ['trænsfɜ:(r)] *n.* **a)** *(moving)* Verlegung, *die; (of powers)* Übertragung, *die* (to an + *Akk.*); *(of money)* Überweisung, *die; (of large sums)* Transfer, *der (Wirtsch.)*; **b)** *(of employee, pupil)* Versetzung, *die; (Footb.)* Transfer, *der;* **c)** *(Amer.: ticket)* Umsteigefahrkarte, *die;* **d)** *(picture)* Abziehbild, *das*

transferable [træns'fɜ:rəbl, 'trænsfərəbl] *adj.* übertragbar

transference ['trænsfərəns] *n.* Übertragung, *die*

transfigure [træns'fɪgə(r)] *v.t.* verklären

transform [træns'fɔ:m] *v.t.* verwandeln; ~ **heat into energy** Wärme in Energie umwandeln

transformation [trænsfə'meɪʃn] *n.* Verwandlung, *die; (of heat into energy)* Umwandlung, *die*

transformer [træns'fɔ:mə(r)] *n.* *(Electr.)* Transformator, *der*

transfusion [træns'fju:ʒn] *n.* *(Med.)* Transfusion, *die;* Übertragung, *die*

transient ['trænzɪənt] *adj.* kurzlebig; vergänglich

transistor [træn'sɪstə(r)] *n.* **a)** ~ **[radio]** Transistor, *der;* Transistorradio, *das;* **b)** *(Electronics)* Transistor, *der*

transit ['trænsɪt] *n.* **a) passengers in ~:** Transitreisende; Durchreisende; **be in ~:** auf der Durchreise sein; **b)** *(conveyance)* Transport, *der;* **goods in ~ from London to Hull** Waren auf dem Transport von London nach Hull

transition [træn'sɪʒn, træn'zɪʃn] *n.* Übergang, *der; (sudden change)* Wechsel, *der*

transitional [træn'sɪʒənl, træn'zɪʃənl] *adj.* Übergangs-

transitive ['trænsɪtɪv] *adj. (Ling.)* transitiv

'**transit lounge** *n.* Transithalle, *die;* Transitlounge, *die*

transitory ['trænsɪtərɪ] *adj.* vergänglich; *(fleeting)* flüchtig

transit: ~ **passenger** *n.* Transitpassagier, *der;* ~ **visa** *n.* Transitvisum, *das*

translate [træns'leɪt] **1.** *v.t.* **a)** übersetzen; ~ **a novel from English into German** einen Roman aus dem Englischen ins Deutsche übersetzen; ~ **'Abgeordneter' as 'Deputy'** „Abgeordneter" mit „Deputy" übersetzen; **b)** *(convert)* ~ **words into action[s]** Worte in die Tat/in Taten umsetzen. **2.** *v.i.* sich übersetzen lassen

translation [træns'leɪʃn] *n.* Übersetzung, *die;* **his works are available in ~:** seine Werke liegen in Übersetzung *od.* übersetzt vor; **read sth. in ~:** etw. in der Übersetzung lesen

translator [træns'leɪtə(r)] *n.* Übersetzer, *der*/Übersetzerin, *die*

translucent [træns'lu:sənt] *adj.* **a)** *(partly transparent)* durchscheinend; **b)** *(transparent)* durchsichtig

transmission [træns'mɪʃn] *n.* **a)** *see* **transmit a:** Übersendung, *die;* Übertragung, *die;* Überlieferung, *die;* [Weiter]vererbung, *die;* **b)** *(Radio, Telev.)* Ausstrahlung, *die; (via satellite also; by wire)* Übertragung, *die;* **c)** *(Motor Veh.) (drive)* Antrieb, *der; (gearbox)* Getriebe, *das;* **manual/automatic ~:** Schalt-/Automatikgetriebe, *das*

transmit [træns'mɪt] *v.t.*, **-tt-: a)** *(pass on)* übersenden ⟨*Nachricht*⟩; übertragen ⟨*Krankheit*⟩; überliefern ⟨*Wissen, Kenntnisse*⟩; *(genetically)* [weiter]vererben ⟨*Eigenschaft*⟩; **b)** durchlassen ⟨*Licht*⟩; übertragen ⟨*Druck, Schall*⟩; leiten ⟨*Wärme, Elektrizität*⟩; **c)** *(Radio, Telev.)* ausstrahlen; *(via satellite also; by wire)* übertragen

transmitter [træns'mɪtə(r)] *n.* Sender, *der*

transmute [træns'mju:t] *v.t.* umwandeln

transparency [træns'pærənsɪ] *n.* **a)** Durchsichtigkeit, *die; (fig. also)* Durchschaubarkeit, *die;* Fadenscheinigkeit, *die (abwertend);* **b)** *(Photog.)* Transparent, *das; (slide)* Dia, *das*

transparent [træns'pærənt] *adj.* durchsichtig; *(fig.) (obvious)* offenkundig; *(easily understood)* klar

transpire [træn'spaɪə(r)] *v.i.* **a)** *(coll.: happen)* passieren; **b)** *(come to be known)* sich herausstellen; **she had not, it ~d, seen the letter** sie hatte, so stellte sich heraus, den Brief nicht gesehen

transplant 1. [træns'plɑ:nt] *v.t.* **a)** *(Med.)* transplantieren *(fachspr.)*, verpflanzen ⟨*Organ, Gewebe*⟩; **b)** *(plant in another place)* umpflanzen. **2.** ['trænsplɑ:nt] *n.* *(Med.) (operation)* Transplantation, *die (fachspr.);* Verpflanzung, *die; (thing ~ed)* Transplantat, *das (fachspr.)*

transport 1. [træn'spɔ:t] *v.t.* **a)** *(convey)* transportieren; befördern; **b)** *(literary: affect with emotion)* anrühren; anwandeln *(geh.)*; **~ed with joy** von Freude überkommen. **2.** ['trænspɔ:t] *n.* **a)** *(conveyance)* Transport, *der;* Beförderung, *die; attrib.* Transport-; Beförderungs-; **b)** *(means of conveyance)* Verkehrsmittel, *das; (for persons also)* Fortbewegungsmittel, *das;* ~ **was provided** für die Beförderung wurde gesorgt; **be without ~:** kein [eigenes] Fahrzeug haben; **Ministry of T~:** Verkehrsministerium, *das;* **c)** *(vehement emotion)* Ausbruch, *der;* **be in/send sb. into ~s of joy** außer sich vor Freude sein/jmdn. in helles Entzücken versetzen

transportable [træn'spɔ:təbl] *adj.* transportabel

transportation [trænspə'teɪʃn] *n.* **a)** *(conveying)* Transport, *der;* Beförderung, *die;* ~ **by air/sea/road/rail** Luft-/See-/Straßen-/Bahntransport, *der;* **b)** *(Amer.) see* **transport 2 b**

'**transport café** *n. (Brit.)* Fernfahrerlokal, *das*

transporter [træn'spɔ:tə(r)] *n.* Transporter, *der*

transpose [træns'pəʊz] *v.t.* **a)** *(cause to change places)* vertauschen; **b)** *(change order of)* umstellen; **c)** *(Mus.)* transponieren

transverse ['trænsvɜ:s] *adj.* querliegend; Quer⟨*balken, -lage, -streifen, -verstrebung*⟩; ~ **section** Querschnitt, *der*

transvestite [træns'vestaɪt] *n.* *(Psych.)* Transvestit, *der*

trap [træp] **1.** *n.* **a)** *(lit. or fig.)* Falle, *die;* **set a ~ for an animal** eine Falle für ein Tier legen *od.* [auf]stellen; **set a ~ for sb.** *(fig.)* jmdm. eine Falle stellen; **fall into a/sb.'s ~** *(fig.)* in die/jmdm. in die Falle gehen; **b)** *(sl.: mouth)* Klappe, *die (salopp);* Fresse, *die (derb);* **shut your ~!, keep your ~ shut!** halt die Klappe *(salopp) od. (derb)* Fresse!; **c)** *(carriage) (leichter zweirädriger)* Einspänner. **2.** *v.t.*, **-pp-: a)** *(catch)* [in *od.* mit einer Falle] fangen ⟨*Tier*⟩; *(fig.)* in eine Falle locken ⟨*Person*⟩; **be ~ped** *(fig.)* in eine Falle gehen/in der Falle sitzen; **be ~ped in a cave/by the tide** in einer Höhle festsitzen/von der Flut abgeschnitten sein; **she ~ped him into contradicting himself** sie brachte ihn durch eine List dazu, sich zu widersprechen; **b)** *(confine)* einschließen; *(immobilize)* einklemmen ⟨*Person, Körperteil*⟩; **~ one's finger/foot** sich *(Dat.)* den Finger/Fuß einklemmen; **c)** *(entangle)* verstricken

'trapdoor *n.* Falltür, *die*

trapeze [trə'pi:z] *n.* Trapez, *das*

tra'peze artist *n.* Trapezkünstler, *der/*-künstlerin, *die*

trapper ['træpə(r)] *n.* Fallensteller, *der;* *(in North America)* Trapper, *der*

trappings ['træpɪŋz] *n. pl.* **a)** [äußere] Zeichen; *(of power, high office)* Insignien; **b)** *(ornamental harness)* ≈ Schabracke, *die*

trash [træʃ] *n., no pl., no indef. art.* **a)** *(rubbish)* Abfall, *der;* **b)** *(badly made thing)* Mist, *der (ugs. abwertend); (bad literature)* Schund, *der (ugs. abwertend);* **be [just] ~:** nichts taugen

'trashcan *n. (Amer.)* Mülltonne, *die*

trashy ['træʃɪ] *adj.* minderwertig; Schund⟨*literatur, -roman*⟩

trauma ['trɔ:mə] *n., pl.* **~ta** ['trɔ:mətə] *or* **~s** Trauma, *das (fachspr.); (injury also)* Verletzung, *die; (shock also)* Schock, *der*

traumatic [trɔ:'mætɪk] *adj.* **a)** *(Med.)* traumatisch; **b)** *(coll.: devastating)* furchtbar

travel ['trævl] **1.** *n.* Reisen, *das; attrib.* Reise-; **be off on one's ~s** verreist sein; **if you see him on your ~s, ...** *(joc.)* wenn er dir über den Weg läuft, ... **2.** *v. i., (Brit.)* **-ll-: a)** *(make a journey)* reisen; *(go in vehicle)* fahren; **~ a lot** viel reisen; **b)** *(coll.: withstand long journey)* **~ [well]** ⟨*Ware:*⟩ lange Transporte vertragen; **~ badly** ⟨*Ware:*⟩ lange Transporte nicht vertragen; **c)** *(work as travelling sales representative)* reisen; Vertreter/Vertreterin sein; **~ in stationery** in Schreibwaren reisen *(Kaufmannsspr.);* **d)** *(move)* sich bewegen; ⟨*Blick, Schmerz:*⟩ wandern; ⟨*Tier:*⟩ sich fortbewegen; ⟨*Licht, Schall:*⟩ sich ausbreiten; **e)** *(coll.: move briskly)* kacheln *(ugs.);* **that car can really ~:** das Auto zieht ganz schön ab *(ugs.).* **3.** *v. t., (Brit.)* **-ll-** zurücklegen ⟨*Strecke, Entfernung*⟩; bereisen ⟨*Bezirk*⟩; benutzen, passieren ⟨*Weg, Straße*⟩; **we had ~led 10 miles** wir waren 10 Meilen gefahren

~ a'bout, ~ a'round 1. *v. i.* umherreisen. **2.** *v. t.* **~ about** *or* **around the country** durchs Land reisen *od.* fahren

travel: ~ agency *n.* Reisebüro, *das;* **~ agent** *n.* Reisebürokaufmann, *der/*-kauffrau, *die*

travelator ['trævəleɪtə(r)] *n.* Fahr- *od.* Rollsteig, *der*

travel: ~ brochure *n.* Reiseprospekt, *der;* **~ bureau** *n.* Reisebüro, *das*

traveled, traveler, traveling *(Amer.) see* **travell-**

travelled ['trævld] *adj. (Brit.)* **be much ~** ⟨*Person:*⟩ weit gereist sein; **be well ~** ⟨*Weg, Straße:*⟩ viel befahren sein

traveller ['trævlə(r)] *n. (Brit.)* **a)** Reisende, *der/die;* **be a poor ~:** das Reisen nicht [gut] vertragen; **b)** *(sales representative)* Vertreter, *der/*Vertreterin, *die;* **c)** *in pl. (gypsies etc.)* fahrendes Volk

'traveller's cheque *n.* Reisescheck, *der*

travelling ['trævlɪŋ] *adj. (Brit.)* Wander⟨*zirkus, -ausstellung*⟩

travelling: ~-bag *n.* Reisetasche, *die;* **~ clock** *n.* Reisewecker, *der;* **~ expenses** *n. pl.* Reisekosten *Pl.;* **~ 'salesman** *n.* Vertreter, *der*

travelogue (*Amer.:* **travelog**) ['trævəlɒg] *n.* Reisebericht, *der*

travel: ~-sick *adj.* reisekrank; **~-sickness** *n., no pl.* Reisekrankheit, *die;* **~-sickness pill** *n.* Tablette gegen Reisekrankheit

traverse ['trævəs, trə'vɜ:s] **1.** *v. t.* **a)** überqueren ⟨*Gebirge*⟩; durchqueren ⟨*Gebiet*⟩; **b)** *(Mountaineering)* traversieren. **2.** *n. (Mountaineering)* Traversierung, *die*

travesty ['trævɪstɪ] **1.** *n.* **a)** *(parody)* Karikatur, *die;* **be a ~ [of justice]** ein Hohn [auf die Gerechtigkeit] sein; **b)** *(Lit.: burlesque)* Travestie, *die (fachspr.).* **2.** *v. t.* ins Lächerliche ziehen

trawl [trɔ:l] **1.** *v. i.* mit dem Grundnetz fischen. **2.** *n.* **~[-net]** Grund[schlepp]netz, *das*

trawler ['trɔ:lə(r)] *n. (vessel)* [Fisch]trawler, *der*

trawlerman ['trɔ:ləmən] *n., pl.* **trawlermen** ['trɔ:ləmən] ≈ Hochseefischer, *der*

tray [treɪ] *n.* **a)** Tablett, *das;* **b)** *(for correspondence)* Ablagekorb, *der*

treacherous ['tretʃərəs] *adj.* **a)** treulos ⟨*Person*⟩; heimtückisch ⟨*Intrige, Feind*⟩; **b)** *(deceptive)* tückisch; **the ice looks pretty ~:** das Eis sieht nicht sehr vertrauenerweckend aus

treachery ['tretʃərɪ] *n.* Verrat, *der;* **act of ~:** Verrat, *der*

treacle ['tri:kl] *n. (Brit.)* **a)** *(golden syrup)* Sirup, *der;* **b)** *see* **molasses**

tread [tred] **1.** *n.* **a)** *(of tyre, shoe, boot, etc.)* Lauffläche, *die;* **2 millimetres of tread on a tyre** 2 Millimeter Profil auf einem Reifen; **b)** *(manner of walking)* Gang, *der; (sound of walking)* Schritt, *der;* **walk with a springy/catlike ~:** einen federnden/katzenhaften Gang haben; **c)** *(of staircase)* [Tritt]stufe, *die.* **2.** *v. i.,* **trod** [trɒd], **trodden** ['trɒdn] *or* **trod** treten **(in/on** in/auf + *Akk.*); *(walk)* gehen; **~ carefully** *(fig.)* behutsam vorgehen; **~ on sb.'s toes** *(lit. or fig.)* jmdm. auf die Füße treten; **~ dirt into the carpet/all over the house** Schmutz in den Teppich treten/im ganzen Haus herumtreten. **3.** *v. t.,* **trod, trodden** *or* **trod a)** *(walk on)* treten auf (+ *Akk.*); stampfen ⟨*Weintrauben*⟩; *(fig.)* gehen ⟨*Weg*⟩; **be trodden underfoot** mit Füßen getreten werden; **~ water** *(Swimming)* Wasser treten; **b)** *(make by walking or ~ing)* austreten ⟨*Weg*⟩

~ 'down *v. t.* festtreten ⟨*Erde*⟩; *(destroy)* zertreten ⟨*Blume, Beet*⟩

~ 'in *v. t.* festtreten

treadle ['tredl] *n.* Tritt, *der*

'treadmill *n. (lit. or fig.)* Tretmühle, *die*

treason ['tri:zn] *n.* **[high] ~:** Hochverrat, *der*

treasure ['treʒə(r)] **1.** *n.* **a)** Schatz, *der;* Kostbarkeit, *die;* **art ~s** Kunstschätze; **b)** *no pl., no indef. art. (riches)* Schätze; **buried ~:** ein vergrabener Schatz; **c)** *(coll.: valued person)* Schatz, *der (ugs.).* **2.** *v. t.* in Ehren halten; die Erinnerung bewahren an (+ *Dat.*); **I'll always ~ this moment** ich werde diesen Augenblick niemals vergessen

treasure: ~-house *n.* Schatzkammer, *die; (fig.)* [wahre] Fundgrube; **~-hunt** *n.* Schatzsuche, *die*

treasurer ['treʒərə(r)] *n.* **a)** *(of club, society)* Kassenwart, *der/*-wartin, *die; (of club, party)* Schatzmeister, *der/*-meisterin, *die; (of company)* Leiter/Leiterin der Finanzabteilung; **b)** *(local government official)* Leiter/Leiterin der Finanzverwaltung

treasure trove ['treʒə trəʊv] *n.* Schatz, *der; (fig.: valuable source)* [wahre] Fundgrube

treasury ['treʒərɪ] *n.* **a)** *(as book-title)* Schatzkästchen, *das;* **b)** *(government department)* **the T~:** das Finanzministerium

treat [tri:t] **1.** *n.* **a)** [besonderes] Vergnügen; *(sth. to eat)* [besonderer] Leckerbissen; **what a ~ [it is] to do/not to have to do that!** welch ein Genuß *od.* eine Wohltat, das zu tun/nicht tun zu müssen!; **have a ~ in store for sb.** noch eine besondere Freude für jmdn. auf Lager haben; **there was a ~ in store for them** auf sie wartete noch eine besondere Freude; **go down a ~** *(coll.)* ‹*Essen, Getränk:*› prima schmecken *(ugs.);* **work a ~** *(coll.)*‹*Maschine:*› prima arbeiten *(ugs.);* ‹*Plan:*› prima funktionieren *(ugs.);* **b)** *(entertainment) Vergnügen, für dessen Kosten jmd. anderes aufkommt;* **lay on a special ~ for sb.** jmdm. etwas Besonderes bieten; **c)** *(act of ~ing)* Einladung, *die.* **2.** *v. t.* **a)** *(act towards)* behandeln; **~ sth. as a joke** etw. als Witz nehmen; **~ sth. with contempt** für etw. nur Verachtung haben; **b)** *(Med.)* behandeln; **~ sb. for sth.** jmdn. wegen etw. behandeln; *(before confirmation of diagnosis)* jmdn. auf etw. *(Akk.)* behandeln; **c)** *(apply process to)* behandeln ‹*Material, Stoff, Metall, Leder*›; klären ‹*Abwässer*›; **d)** *(handle in literature etc.)* behandeln; **e)** *(provide with at own expense)* einladen; **~ sb. to sth.** jmdm. etw. spendieren; **~ oneself to a holiday/a new hat** sich *(Dat.)* Urlaub gönnen/sich *(Dat.)* einen neuen Hut leisten

treatise ['tri:tɪs, 'tri:tɪz] *n.* Abhandlung, *die*

treatment ['tri:tmənt] *n.* **a)** Behandlung, *die;* **his ~ of the staff/you** die Art, wie er das Personal/dich behandelt; **give sb. the [full] ~** *(coll.) (treat cruelly/harshly)* jmdn. in die Mangel nehmen *(salopp); (entertain on a lavish scale)* jmdn. verwöhnen; **b)** *(Med.)* Behandlung, *die;* **need immediate medical ~:** sofort ärztlich behandelt werden müssen; **c)** *(processing)* Behandlung, *die; (of sewage)* Klärung, *die*

treaty ['tri:tɪ] *n.* [Staats]vertrag, *der;* **make** *or* **sign a ~:** einen Vertrag schließen

treble ['trebl] **1.** *adj.* **a)** dreifach; **~ the amount compared to ...:** dreimal so viel wie ...; **sell sth. for ~ the price** etw. dreimal so teuer verkaufen; **b)** *(Brit. Mus.)* **~ voice** Sopranstimme, *die.* **2.** *n.* **a)** *(~ quantity)* Dreifache, *das;* **b)** *(Mus.)* **he is a ~/is singing the ~:** er singt Sopran/den Sopran. **3.** *v. t.* verdreifachen; **be ~d** ‹*Wert einer Aktie usw.:*› sich verdreifachen. **4.** *v. i.* sich verdreifachen

'**treble clef** *n. (Mus.)* Violinschlüssel, *der*

tree [tri:] *n.* Baum, *der;* **not grow on ~s** *(fig.)* nicht [einfach] vom Himmel fallen

'**tree-house** *n.* Baumhaus, *das*

treeless ['tri:lɪs] *adj.* baumlos

tree: ~-lined *adj.* von Bäumen gesäumt; **~-top** *n.* [Baum]wipfel, *der;* **~-trunk** *n.* Baumstamm, *der*

trek [trek] **1.** *v.i.*, **-kk-** ziehen (**across** durch). **2.** *n.* [schwierige] Reise

trellis ['trelɪs] *n.* Gitter, *das; (for plants)* Spalier, *das*

tremble ['trembl] **1.** *v.i.* zittern (**with** vor + *Dat.*); **~ for sb./sth.** *(fig.)* um jmdn./etw. zittern; **I ~ to think what .../at the thought** *(fig.)* mir wird bange, wenn ich daran denke, was .../wenn ich daran denke. **2.** *n.* Zittern, *das;* **be all of a ~** *(coll.)* am ganzen Körper zittern; **there was a ~ in her voice** ihre Stimme zitterte

trembling ['tremblɪŋ] **1.** *adj.* zitternd. **2.** *n.* Zittern, *das*

tremendous [trɪ'mendəs] *adj.* **a)** *(immense)* gewaltig; enorm ‹*Fähigkeiten*›; **b)** *(coll.: wonderful)* großartig

tremolo ['tremələʊ] *n., pl.* **~s** *(Mus.)* Tremolo, *das*

tremor ['tremə(r)] *n.* **a)** Zittern, *das;* **feel a ~ of delight/fear** freudig erregt sein/vor Angst zittern; **b) [earth] ~** *(Geol.)* leichtes Erdbeben

tremulous ['tremjʊləs] *adj.* **a)** *(trembling)* zitternd; **be ~:** zittern; **b)** *(timid)* zaghaft ‹*Lächeln*›; ängstlich ‹*Person*›

trench [trentʃ] *n.* Graben, *der; (Geog.)* [Tiefsee]graben, *der; (Mil.)* Schützengraben, *der*

trenchant ['trentʃənt] *adj.* deutlich, energisch ‹*Kritik, Sprache*›; energisch ‹*Verteidiger, Kritiker, Politik*›; prägnant ‹*Stil*›

'**trench coat** *n. (Mil.)* Wettermantel, *der; (Fashion)* Trenchcoat, *der*

trend [trend] **1.** *n.* **a)** Trend, *der;* **population ~s** die Bevölkerungsentwicklung; **upward ~:** steigende Tendenz; **b)** *(fashion)* Mode, *die;* [Mode]trend, *der;* **set the ~:** den Trend bestimmen. **2.** *v.i.* **a)** *(take a course)* verlaufen; **b)** *(fig.: move)* sich entwickeln; **~ upward** steigen

'**trend-setter** *n.* Trendsetter, *der*

trendy ['trendɪ] *(Brit. coll.)* **1.** *adj.* modisch; Schickimicki‹*kneipe, -wohngegend*› *(ugs.);* fortschrittlich-modern ‹*Geistlicher, Lehrer*›. **2.** *n.* Schickimicki, *der (ugs.)*

trepidation [trepɪ'deɪʃn] *n.* Beklommenheit, *die;* **with some ~, not without ~:** ziemlich beklommen

trespass ['trespəs] **1.** *v. i.* **~ on** unerlaubt betreten ‹*Grundstück*›; eingreifen in (+ *Akk.*) ‹*jmds. Rechte*›; **'no ~ing'** „Betreten verboten!"; **~ on sb.'s time/privacy** *(fig.)* jmds. Zeit über Gebühr in Anspruch nehmen/jmds. Privatsphäre verletzen. **2.** *n. (Law)* Hausfriedensbruch, *der*

trespasser ['trespəsə(r)] *n.* Unbefugte, *der/die;* **'~s will be prosecuted'** „Betreten verboten, Zuwiderhandlungen werden verfolgt"

tress [tres] *n. (literary/arch.)* Haarstrang, *der; (curly)* Locke, *die*

trestle ['tresl] *n.* **a)** [Auflager]bock, *der;* **b) ~[-table]** Tapeziertisch, *der*

trial ['traɪəl] *n.* **a)** *(Law)* [Gerichts]verfahren, *das;* **be on ~ [for murder]** [wegen Mordes] angeklagt sein; **go on ~ [for one's life]** [wegen eines Verbrechens, auf das die Todesstrafe steht,] vor Gericht gestellt werden; **bring sb. to ~, put sb. on ~:** jmdm. den Prozeß machen (**for** wegen); **the case was brought to ~:** der Fall wurde vor Gericht verhandelt; **b)** *(testing)* Test, *der;* **be given ~s** getestet werden; **employ sb. on ~:** jmdn. probeweise einstellen; **give sth. a ~:** etw. ausprobieren; **[by] ~ and error** [durch] Ausprobieren; **~ of strength** Kraftprobe, *die;* **c)** *(trouble)* Prüfung, *die (geh.);* Problem, *das;* **find sth. a ~:** etw. als lästig empfinden; **be a ~ to sb.** jmdm. zu schaffen machen; **d)** *(Sport) (competition)* Prüfung, *die; (for selection)* Testspiel, *das. See also* **tribulation a**

trial: ~ pack *n.* Probepackung, *die;* **~ 'run** *n.* **a)** *(of car)* Testfahrt, *die; (of machine)* Probelauf, *der;* **b)** *(fig.)* Probelauf, *der;* **give sth. a ~ run** etw. testen

triangle ['traɪæŋgl] *n.* **a)** Dreieck, *das;* **b)** *(Mus.)* Triangel, *das od. der*

triangular [traɪ'æŋgjʊlə(r)] *adj.* dreieckig; dreiseitig ‹*Pyramide*›

tribal ['traɪbl] *adj.* Stammes-

tribalism ['traɪbəlɪzm] *n.* Tribalismus, *der (fachspr.)*

tribe [traɪb] *n.* Stamm, *der*

tribesman ['traɪbzmən] *n., pl.* **tribesmen** ['traɪbzmən] Stammesangehörige, *der*

tribulation [trɪbjʊ'leɪʃn] *n.* **a)** *(great affliction)* Kummer, *der;* **trials and ~s** Probleme und Sorgen; **b)** *(cause of trouble etc.)* **be a ~ to sb.** jmdm. zur Last fallen

tribunal [traɪ'bju:nl, trɪ'bju:nl] *n.* **a)** Schiedsgericht, *das; (court of justice)* Gericht, *das;* **b)** *(fig.)* Tribunal, *das*

tribune ['trɪbju:n] *n. (platform)* [Redner]tribüne, *die*

tributary ['trɪbjʊtərɪ] *n. (river)* Nebenfluß, *der*

tribute ['trɪbju:t] *n.* Tribut, *der* (to an + *Akk.*); **pay ~ to sb./sth.** jmdm./einer Sache den schuldigen Tribut zollen *(geh.);* **floral ~s** Blumen [als Zeichen der Anerkennung]; *(to deceased person)* Blumen und Kränze; **as a ~ to his work** zur Würdigung seiner Arbeit; **she is a ~ to her trainer** sie macht ihrem Trainer alle Ehre

trice [traɪs] *n.* **in a ~:** im Handumdrehen

trick [trɪk] **1.** *n.* **a)** Trick, *der;* **I suspect some ~:** es könnte ein Trick sein; **it was all a ~:** das war [alles] nur Bluff; **it was such a shabby ~ [to play on her]** es war [ihr gegenüber] eine derartige Gemeinheit *od.* dermaßen gemein; **b)** *(feat of skill etc.)* Kunststück, *das;* **try every ~ in the book** es mit allen Tricks probieren; **he never misses a ~** *(fig.)* ihm entgeht nichts; **that should do the ~** *(coll.)* damit dürfte es klappen *(ugs.);* **c)** *(knack)* **get** *or* **find the ~ [of doing sth.]** den Dreh finden[, wie man etw. tut]; **d) how's ~s?** *(coll.)* was macht die Kunst? *(ugs.);* **e)** *(mannerism)* Eigenart, *die;* **have a ~ of doing sth.** die Eigenart haben, etw. zu tun; **f)** *(prank)* Streich, *der;* **play a ~ on sb.** jmdm. einen Streich spielen; **be up to one's [old] ~s again** immer noch auf dieselbe Tour reisen *(ugs.);* **~ or treat** Trick-or-Treat, *das (Kinderspiel);* **g)** *(illusion)* **~ of vision/lighting/the light** Augentäuschung, *die;* **h)** *(Cards)* Stich, *der.* **2.** *v.t.* täuschen; hereinlegen; **~ sb. into doing sth.** jmdn. mit einem Trick *od.* einer List dazu bringen, etw. zu tun; **~ sb. out of/into sth.** jmdm. etw. ablisten. **3.** *adj.* **~ photograph** Trickaufnahme, *die;* **~ photography** Trickfotografie, *die;* **~ question** Fangfrage, *die*

~ 'out, ~ 'up *v.t.* schmücken

trickery ['trɪkərɪ] *n.* [Hinter]list, *die;* **piece of ~:** List, *die;* Trick, *der*

trickle ['trɪkl] **1.** *n.* Rinnsal, *das (geh.)* (of von); **in a ~:** als Rinnsal; **a ~ of rain ran down the window** Regenwasser rann am Fenster hinunter; **there was a ~ of people leaving the room** *(fig.)* einige wenige Menschen verließen nacheinander den Raum. **2.** *v.i.* rinnen; *(in drops)* tröpfeln; *(fig.)* ⟨*Ball:*⟩ langsam rollen; **~ out** ⟨*Zuschauer:*⟩ nach und nach [hinaus]gehen; **~ through** *or* **out** ⟨*Informationen:*⟩ durchsickern

trickster ['trɪkstə(r)] *n.* Schwindler, *der*/Schwindlerin, *die*

tricky ['trɪkɪ] *adj.* verzwickt *(ugs.)*

tricycle ['traɪsɪkl] *n.* Dreirad, *das*

trident ['traɪdənt] *n.* dreizackiger Fischspeer; *(held by Britannia, Neptune, etc.)* Dreizack, *der*

tried *see* **try 2, 3**

trier [traɪə(r)] *n.* **he's a real ~:** er wirft die Flinte nicht so schnell ins Korn

trifle ['traɪfl] *n.* **a)** *(Brit. Gastr.)* Trifle, *das;* **b)** *(thing of slight value)* Kleinigkeit, *die;* **the merest ~:** die geringste Kleinigkeit; **it's only a ~:** es ist nichts Besonderes; **c) a ~ tired/angry** *etc.* ein bißchen müde/böse *usw.*

~ with *v.i.* spielen mit ⟨*jmds. Gefühlen*⟩; nicht ernst genug nehmen ⟨*Person*⟩; **he is not a person you can ~ with** er läßt nicht mit sich spaßen

trifling ['traɪflɪŋ] *adj.* unbedeutend ⟨*Angelegenheit, Irrtum*⟩; lächerlich ⟨*Gedanke*⟩; gering ⟨*Gefahr, Wert*⟩; [lächerlich] gering ⟨*Summe*⟩

trigger ['trɪgə(r)] **1.** *n.* **a)** *(of gun)* Abzug, *der; (of machine)* Drücker, *der;* **pull the ~:** abdrücken; **be quick on the ~** *(fig.)* prompt reagieren; **b)** *(that sets off reaction)* Auslöser, *der.* **2.** *v.t.* **~ [off]** auslösen

'trigger-happy *adj.* schießwütig

trigonometry [trɪgə'nɒmɪtrɪ] *n.* *(Math.)* Trigonometrie, *die*

trike [traɪk] *n. (coll.)* Dreirad, *das*

trilateral [traɪ'lætərl] **1.** *adj. (having three sides)* dreiseitig; *(involving three parties also)* trilateral *(geh.).* **2.** *n.* Dreieck, *das*

trilby ['trɪlbɪ] *n. (Brit.)* **~ [hat]** Klapprandhut, *der*

trill [trɪl] *n.* **a)** Trillern, *das;* **b)** *(Mus.)* Triller, *der*

trillion ['trɪlɪən] *n.* **a)** *(million million)* Billion, *die;* **b)** *(Brit.: million million million)* Trillion, *die*

trilogy ['trɪlədʒɪ] *n.* Trilogie, *die*

trim [trɪm] **1.** *v.t.,* **-mm-: a)** schneiden ⟨*Hecke*⟩; [nach]schneiden ⟨*Haar*⟩; beschneiden *(auch fig.)* ⟨*Papier, Hecke, Docht, Budget*⟩; **~ £100 off** *or* **from a budget** ein Budget um 100 Pfund kürzen; **b)** *(ornament)* besetzen (**with** mit). **2.** *adj.* proper; gepflegt ⟨*Garten*⟩; **keep sth. ~:** etw. in Ordnung halten. **3.** *n.* **a)** *(state of adjustment)* Bereitschaft, *die;* **be in fine physical ~:** in guter körperlicher Verfassung sein; **get/be in ~** *(healthy)* sich trimmen/in Form *od.* fit sein; **b)** *(cut)* Nachschneiden, *das;* **my hair needs a ~:** ich muß mir die Haare nachschneiden lassen; **give a hedge a ~:** eine Hecke nachschneiden; **just a ~, please** *(said to hairdresser)* nur nachschneiden, bitte

~ 'down *v.t. (fig.)* verringern; **her figure needed ~ming down** sie mußte etwas für ihre Figur tun

~ 'off *v.t.* abschneiden; *(fig.)* abnehmen

trimmer ['trɪmə(r)] *n.* Schneider, *der;* **hedge-~:** Heckenschere, *die*

trimming ['trɪmɪŋ] *n.* **a)** *(decorative addition)* Verzierung, *die;* **lace ~s** Spitzenbesatz, *der;* **b)** *in pl. (coll.: accompaniments) (for main dish)* Beilagen; *(extra fittings on car)* Extras; **with all the ~s** mit allem Drum und Dran *(ugs.);* **c)** *in pl. (pieces cut off)* Abfall, *der (vom Zuschneiden); (of meat)* abgeschnittene Stücke

Trinidad ['trɪnɪdæd] *pr. n.* Trinidad *(das)*

Trinidadian [trɪnɪ'dædɪən] **1.** *adj.* trinidadisch; **sb. is ~:** jmd. ist Trinidader/Trinidaderin. **2.** *n.* Trinidader, *der*/Trinidaderin, *die*

Trinity ['trɪnɪtɪ] *n.* **a)** *(Theol.)* **the [Holy] ~:** die [Heilige] Dreifaltigkeit *od.* Dreieinigkeit *od.* Trinität; **b)** *(Eccl.)* **~ [Sunday]** Dreifaltigkeitssonntag, *der*

trinket ['trɪŋkɪt] *n.* kleines, billiges Schmuckstück; *(on bracelet)* Anhänger, *der*

trio ['tri:əʊ] *n., pl.* **~s** *(also Mus.)* Trio, *das*

trip [trɪp] **1.** *n.* **a)** *(journey)* Reise, *die;* Trip, *der (ugs.); (shorter)* Ausflug, *der;* Trip, *der (ugs.);* **two ~s were necessary to transport everything** zwei Fahrten waren nötig, um alles zu transportieren; **make a ~ to London** nach London fahren; **b)** *(coll.: drug-induced hallucinations)* Trip, *der (Jargon);* **[good/bad] ~ on LSD** [guter/schlechter] LSD-Trip. **2.** *v.i.,* **-pp-: a)** *(stumble)* stolpern (**on** über + *Akk.*); **b)** *(coll.: hallucin-*

ate while on drugs) ~ **[on LSD]** auf einem [LSD-]Trip sein; **c)** *(walk etc. with light steps)* trippeln. **3.** *v. t.*, **-pp-** *see* ~ **up 2 a**
~ **over** *v. t.* stolpern über (+ *Akk.*)
~ **'up 1.** *v. i.* **a)** *(stumble)* stolpern; **b)** *(fig.: make a mistake)* einen Fehler machen. **2.** *v. t.* **a)** *(cause to stumble)* stolpern lassen; **b)** *(cause to make a mistake)* aufs Glatteis führen *(fig.)*
tripe [traɪp] *n.* **a)** Kaldaunen; **b)** *(sl.: rubbish)* Quatsch, *der (ugs. abwertend)*
triple ['trɪpl] **1.** *adj.* **a)** dreifach; **b)** *(three times greater than)* ~ **the ...**: der/die/das dreifache ...; **at ~ the speed** mit der dreifachen Geschwindigkeit *od.* dreimal so schnell; ~ **the number of machines** dreimal so viele Maschinen. **2.** *n.* Dreifache, *das.* **3.** *v. i.* sich verdreifachen. **4.** *v. t.* verdreifachen
'triple jump *n.* Dreisprung, *der*
triplet ['trɪplɪt] *n.* Drilling, *der*
triplicate ['trɪplɪkət] **1.** *adj.* dreifach. **2.** *n.* **in** ~: in dreifacher Ausfertigung
trip 'mileage recorder *n.* *(Motor Veh.)* Tageskilometerzähler, *der*
tripod ['traɪpɒd] *n.* Dreibein, *das;* [dreibeiniges] Stativ
tripper ['trɪpə(r)] *n.* *(Brit.)* Ausflügler, *der*/Ausflüglerin, *die*
'trip-wire *n.* Stolperdraht, *der*
trite [traɪt] *adj.* banal
triumph ['traɪəmf, 'traɪʌmf] **1.** *n.* Triumph, *der* (**over** über + *Akk.*). **2.** *v. i.* triumphieren (**over** über + *Akk.*)
triumphant [traɪ'ʌmfənt] *adj.* **a)** *(victorious)* siegreich; **b)** *(exulting)* triumphierend ⟨*Blick*⟩; ~ **shouts** Triumphgeschrei, *das*
trivia ['trɪvɪə] *n. pl.* Belanglosigkeiten
trivial ['trɪvɪəl] *adj.* belanglos; trivial *(geh.)*
triviality [trɪvɪ'ælɪtɪ] *n.* Belanglosigkeit, *die;* Trivialität, *die (geh.)*
trivialize (trivialise) ['trɪvɪəlaɪz] *v. t.* trivialisieren *(geh.)*
trod, trodden *see* **tread 2, 3**
trolley ['trɒlɪ] *n.* *(Brit.)* **a)** *(on rails)* Draisine, *die;* **b)** *(for serving food)* Servierwagen, *der;* **c)** **[supermarket]** ~: Einkaufswagen, *der*
'trolley bus *n.* *(Brit.)* Oberleitungsomnibus, *der*
trombone [trɒm'bəʊn, 'trɒmbəʊn] *n.* Posaune, *die*
troop [tru:p] **1.** *n.* **a)** *in pl.* Truppen; **our best ~s** unsere besten Soldaten; **b)** *(of cavalry)* Schwadron, *die; (artillery and armour)* Batterie, *die;* **c)** *(assembled company)* Schar, *die.* **2.** *v. i.* strömen; *(in an orderly fashion)* marschieren; ~ **in/out** hinein-/hinausströmen
'troop-carrier *n.* Truppentransporter, *der*
trooper ['tru:pə(r)] *n.* **a)** *(soldier)* einfacher Soldat; **swear like a** ~ *(coll.)* wie ein Fuhrmann fluchen *(ugs.);* **b)** *(Amer.: policeman)* Polizist, *der*
'troop-ship *n.* *(Mil.)* Truppentransporter, *der*
trophy ['trəʊfɪ] *n.* Trophäe, *die*
tropic ['trɒpɪk] *n.* **the T~s** *(Geog.)* die Tropen; **the ~ of Cancer/Capricorn** *(Astron., Geog.)* der Wendekreis des Krebses/Steinbocks
tropical ['trɒpɪkl] *adj.* tropisch; Tropen⟨*krankheit, -kleidung*⟩
tropical: ~ **'medicine** *n.* Tropenmedizin, *die;* ~ **'rain forest** *n.* tropischer Regenwald
trot [trɒt] **1.** *n.* *(coll.)* **on the** ~ *(in succession)* hintereinander; **every weekend for five weeks on the** ~: an fünf Wochenenden hintereinander; **be on the** ~: auf Trab sein *(ugs.).* **2.** *v. i.*, **-tt-** traben. **3.** *v. t.*, **-tt-** traben lassen ⟨*Pferd*⟩
~ **'out** *v. t.* *(fig. coll.)* **a)** *(produce for approval)* vorführen; **b)** *(produce unthinkingly)* kommen mit *(ugs.)*
trotter ['trɒtə(r)] *n.* Fuß, *der;* **pigs' ~s** *(Cookery)* Schweinsfüße
trouble ['trʌbl] **1.** *n.* **a)** Ärger, *der;* Schwierigkeiten *Pl.;* **have ~ with sb./sth.** mit jmdm./etw. Ärger haben; **put one's ~s behind one** seine Probleme vergessen; **be out of** ~: aus den Schwierigkeiten heraussein; **keep out of** ~: nicht [wieder] in Schwierigkeiten kommen; **in** ~: in Schwierigkeiten; **be in ~ with the police** Ärger mit der Polizei haben; **be in serious** *or* **real** *or* **a lot of ~ [over sth.]** [wegen einer Sache] in ernsten *od.* großen Schwierigkeiten sein; **get sb. into** ~: jmdn. in Schwierigkeiten bringen; **get a girl into** ~ *(coll.)* einem Mädchen ein Kind machen *(ugs.);* **get into ~ [over sth.]** [wegen einer Sache] in Schwierigkeiten geraten; **get into ~ with the law** mit dem Gesetz in Konflikt geraten; **there'll be ~ [if ...]** es wird Ärger geben[, wenn ...]; **what's** *or* **what seems to be the ~?** was ist denn?; was ist los? *(ugs.); (doctor's question to patient)* wo fehlt's denn?; **you are asking for** ~ *(coll.)* du machst dir nur selber Schwierigkeiten; **that's asking for** ~ *(coll.)* das muß ja Ärger geben; **make** *or* **cause** ~ *(cause disturbance)* Ärger machen (**about** wegen); *(cause disagreement)* Zwietracht säen; **b)** *(faulty operation)* Probleme; **engine/clutch/brake** ~: Probleme mit dem Motor/der Kupplung/der Bremse; **c)** *(disease)* **suffer from heart/liver** ~: Probleme mit dem Herz/der Leber haben; **she's got some ~ with her back** ihr Rücken macht ihr zu schaffen; **d)** *(cause of vexation etc.)* Problem, *das;* **half the** ~ *(fig.)* das größte Problem; **your ~ is that ...**: dein Fehler ist, daß ...; **e)** *(inconvenience)* Mühe, *die;* **it's more ~ than it's worth** es lohnt sich nicht; **I don't want to put you to any** ~: ich möchte Ihnen keine Umstände machen; **not worth the** ~: nicht der Mühe wert; **give sb. no** ~: jmdm. keine Mühe machen; **take the ~ to do sth., go to the ~ of doing sth.** sich *(Dat.)* die Mühe machen, etw. zu tun; **go to** *or* **take a lot of/some** ~: sich *(Dat.)* sehr viel/viel Mühe geben; **please don't go to a lot of** ~: bitte machen Sie sich *(Dat.)* nicht allzuviel Umstände; **of course I'll help you – [it's] no ~ at all** natürlich helfe ich dir – das macht keine Umstände *od.* das ist nicht der Rede wert; **f)** *(source of inconvenience)* **be a ~ [to sb.]** jmdm. zur Last fallen; **he won't be any** ~: er wird [Ihnen] keine Schwierigkeiten machen; **g)** *in sing. or pl. (unrest)* Unruhen. **2.** *v. t.* **a)** *(agitate)* beunruhigen; **don't let it ~ you** mach dir deswegen keine Sorgen; **b)** *(inconvenience)* stören; **[I'm] sorry to ~ you** bitte entschuldigen Sie die Störung. **3.** *v. i.* **a)** *(be disturbed)* sich *(Dat.)* Sorgen machen (**over** um); **don't ~ about it** mach dir deswegen keine Gedanken; **b)** *(make an effort)* sich bemühen; **don't ~ to explain/to get up** du brauchst mir gar nichts zu erklären/bitte bleiben Sie sitzen
troubled ['trʌbld] *adj.* **a)** *(worried)* besorgt; **what are you so ~ about?** was macht dir denn solche Sorgen?; **b)** *(restless)* unruhig; **c)** *(agitated)* aufgewühlt; unruhig ⟨*Zeit*⟩; bewegt ⟨*Geschichte*⟩
trouble: **~-free** *adj.* problemlos; **~-maker** *n.* Unruhestifter, *der*/-stifterin, *die;* **~-shooter** *n.* *jemand, der Störungen od. Probleme findet und beseitigt; (in disputes)* Vermittler, *der*/Vermittlerin, *die*

troublesome ['trʌblsəm] *adj.* schwierig; lästig ‹*Krankheit*›
'**trouble-spot** *n.* **a)** Unruheherd, *der;* **b)** *(in machine)* Schwachstelle, *die*
trough [trɒf] *n.* **a)** Trog, *der;* **a drinking-~:** ein Wassertrog; **b)** *(between waves)* Wellental, *das;* **c)** *(Meteorol.)* Trog, *der;* **a ~ of low pressure** eine Tiefdruckrinne
troupe [tru:p] *n.* Truppe, *die*
trouser ['traʊzə]: **~-leg** *n.* Hosenbein, *das;* **~ pocket** *n.* Hosentasche, *die*
trousers ['traʊzəz] *n. pl.* **[pair of] ~:** Hose, *die;* Hosen *Pl.;* **wear the ~** *(fig.)* die Hosen anhaben *(ugs.)*
'**trouser suit** *n. (Brit.)* Hosenanzug, *der*
trousseau ['tru:səʊ] *n., pl.* **~s** *or* **~x** ['tru:səʊz] Aussteuer, *die*
trout [traʊt] *n., pl. same* Forelle, *die*
trowel ['traʊəl] *n.* **a)** Kelle, *die;* **lay it on with a ~** *(fig.)* [es] dick auftragen *(ugs. abwertend);* **b)** *(Hort.)* Pflanzkelle, *die*
truancy ['tru:ənsɪ] *n.* Schuleschwänzen, *das (ugs.);* unentschuldigtes Fernbleiben vom Unterricht
truant ['tru:ənt] *n.* [Schul]schwänzer, *der*/-schwänzerin, *die (ugs.);* **play ~:** [die Schule] schwänzen *(ugs.)*
truce [tru:s] *n.* Waffenstillstand, *der;* **call a ~:** einen Waffenstillstand schließen
truck [trʌk] *n.* **a)** *(road vehicle)* Last[kraft]wagen, *der;* Lkw, *der;* **b)** *(Brit. Railw.: wagon)* offener Güterwagen
'**truck driver** *n.* Lastwagenfahrer, *der*/-fahrerin, *die; (long-distance)* Fernfahrer, *der*/-fahrerin, *die*
trucker ['trʌkə(r)] *n.* **a)** *(Amer.: market-gardener)* Gemüsegärtner, *der*/-gärtnerin, *die;* **b)** *(esp. Amer.) see* **truck driver**
'**truck-load** *n.* Wagenladung, *die*
truculent ['trʌkjʊlənt] *adj.* aufsässig
trudge [trʌdʒ] *v.i.* trotten; *(through mud, snow, etc.)* stapfen
true [tru:] **1.** *adj.,* **~r** ['tru:ə(r)], **~st** ['tru:ɪst] **a)** *(in accordance with fact)* wahr; wahrheitsgetreu ‹*Bericht, Beschreibung*›; **is it ~ that ...?** stimmt es, daß ...?; **[only] too ~:** nur zu wahr; **that is too good to be ~:** das ist zu schön, um wahr zu sein; **[that's] ~ [enough]** [das] stimmt; **come ~** ‹*Traum, Wunsch:*› Wirklichkeit werden, wahr werden; ‹*Befürchtung, Prophezeihung:*› sich bewahrheiten; **b)** richtig ‹*Vorteil, Einschätzung*›; *(rightly so called)* eigentlich; **the frog is not a ~ reptile** der Frosch ist kein echtes Reptil; **c)** *(not sham)* wahr; echt, wahr ‹*Freund, Freundschaft, Christ*›; **that's not a ~ antique** das ist keine echte Antiquität; **d)** *(accurately conforming)* getreu ‹*Wiedergabe*›; **be ~ to sth.** einer Sache *(Dat.)* genau entsprechen; **~ to type** typisch; **~ to life** lebensecht; **e)** *(loyal)* treu; **~ to one's word** *or* **promise** getreu seinem Versprechen; **f)** *(in correct position)* gerade ‹*Pfosten*›. **2.** *n.* **out of [the] ~:** schief ‹*Mauer, Pfosten, Räder*›. **3.** *adv.* **a)** *(truthfully)* aufrichtig ‹*lieben*›; **tell me ~:** sag mir die Wahrheit; **b)** *(accurately)* gerade; genau ‹*zielen*›
true: ~-blue 1. *adj.* in der Wolle gefärbt; **2.** *n.* Hundertfünfzigprozentige, *der/die (abwertend);* **~-life** *adj.* aus dem Leben gegriffen ‹*Geschichte, Drama*›
truffle ['trʌfl] *n.* Trüffel, *die od. (ugs.) der*
truism ['tru:ɪzm] *n.* Binsenweisheit, *die*
truly ['tru:lɪ] *adv.* **a)** *(genuinely)* wirklich; **be ~ grateful** wirklich sehr *od.* aufrichtig dankbar sein; **b)** *(accurately)* zutreffend, richtig ‹*darstellen, sagen*›; *see also* **yours b**
trump [trʌmp] *(Cards)* **1.** *n.* Trumpf, *der;* **turn up ~s** *(Brit. coll.) (turn out better than expected)* doch noch ein voller Erfolg werden; *(do the right thing)* die Situation retten; **hold all the ~s** *(fig.)* alle Trümpfe in der Hand haben *od.* halten. **2.** *v.t.* übertrumpfen
~ up *v.t. (coll.)* konstruieren; **~ed up charge** falsche Beschuldigung
'**trump card** *n. (lit. or fig.)* Trumpf, *der;* **play one's ~** *(lit. or fig.)* seinen [größten *od.* stärksten] Trumpf ausspielen
trumpet ['trʌmpɪt] **1.** *n. (Mus., Bot.)* Trompete, *die.* **2.** *v.t. & i.* trompeten
trumpeter ['trʌmpɪtə(r)] *n.* Trompeter, *der*/Trompeterin, *die*
truncate [trʌŋ'keɪt] *v.t.* **a)** stutzen ‹*Baum, Spitze*›; **b)** *(fig.)* kürzen
truncheon ['trʌntʃn] *n.* Schlagstock, *der*
trundle ['trʌndl] *v.t. & i.* rollen
trunk [trʌŋk] *n.* **a)** *(of elephant etc.)* Rüssel, *der;* **b)** *(large box)* Schrankkoffer, *der;* **c)** *(of tree)* Stamm, *der;* **d)** *(of human or animal body)* Rumpf, *der;* **e)** *(Amer.: of car)* Kofferraum, *der;* **f)** *in pl. (Brit.: shorts)* Unterhose, *die;* **[swimming] ~s** Badehose, *die*
'**trunk: ~-call** *n.* Ferngespräch, *das;* **~-line** *n. (Railw.)* Hauptstrecke, *die; (Teleph.)* Fernleitung, *die;* **~-road** *n. (Brit.)* Fernstraße, *die*
truss [trʌs] **1.** *n.* **a)** *(of roof etc.)* Gebälk, *das;* **b)** *(Med.)* Bruchband, *das.* **2.** *v.t.* **~ [up]** fesseln
trust [trʌst] **1.** *n.* **a)** *(firm belief)* Vertrauen, *das;* **place** *or* **put one's ~ in sb./sth.** sein Vertrauen auf *od.* in jmdn./etw. setzen; **have [every] ~ in sb./sth.** [volles] Vertrauen zu jmdm./etw. haben; **b)** *(reliance)* **take sth. on ~:** etw. einfach glauben; **c)** *(organization managed by trustees)* Treuhandgesellschaft, *die;* **[charitable] ~:** Stiftung, *die;* **d)** *(body of trustees)* Treuhänder *Pl.; (of charitable ~)* [Stiftungs]beirat, *der;* Kuratorium, *das;* **e)** *(organized association of companies)* Trust, *der;* **f)** *(responsibility)* **position of ~:** Vertrauensstellung, *die;* **g)** *(obligation)* Verpflichtung, *die;* **h)** *(Law)* **hold in ~:** treuhänderisch verwalten. **2.** *v.t.* **a)** *(rely on)* trauen (+ *Dat.*); vertrauen (+ *Dat.*) ‹*Person*›; **not ~ sb. an inch** jmdm. nicht über den Weg trauen; **he/what he says is not to be ~ed** er ist nicht vertrauenswürdig/auf das, was er sagt, kann man sich nicht verlassen; **~ sb. with sth.** jmdm. etw. anvertrauen; **~ 'you/'him!** *etc. (coll. iron.)* typisch!; **~ 'him to get it wrong!** er muß natürlich einen Fehler machen!; **b)** *(hope)* hoffen; **I ~ he is not hurt?** er ist doch hoffentlich nicht verletzt? **3.** *v.i.* **a) ~ to** sich verlassen auf (+ *Akk.*); **b)** *(believe)* **~ in sb./sth.** auf jmdn./etw. vertrauen
trustee [trʌ'sti:] *n.* **a)** *(person holding property in trust; also fig.)* Treuhänder, *der*/Treuhänderin, *die;* **b)** *(one appointed to manage institution)* Kurator, *der*/Kuratorin, *die*
trustful ['trʌstfl] *adj.* vertrauensvoll
'**trust fund** *n.* Treuhandvermögen, *das*
trusting ['trʌstɪŋ] *adj.* vertrauensvoll
trustworthy ['trʌstwɜ:ðɪ] *adj.* vertrauenswürdig
truth [tru:θ] *n., pl.* **~s** [tru:ðz, tru:θs] **a)** *no pl.* Wahrheit, *die;* **there is some/not a word of** *or* **no ~ in that** es ist etwas Wahres/kein wahres Wort/nichts Wahres daran; **in ~** *(literary)* wahrlich *(geh.);* **b)** *(what is true)* Wahrheit, *die;*

(principle) Grundsatz, *der;* **tell the [whole] ~:** die [ganze] Wahrheit sagen; **the ~ is that I forgot** um ehrlich zu sein, ich habe es vergessen; **to tell the ~, ~ to tell** ehrlich gesagt

truthful ['tru:θfl] *adj.* ehrlich; wahrheitsgetreu ⟨*Darstellung, Schilderung*⟩; **be ~ about sth.** die Wahrheit über etw. *(Akk.)* sagen

try [traɪ] **1.** *n.* **a)** *(attempt)* Versuch, *der;* **have a ~ at sth./doing sth.** etw. versuchen/versuchen, etw. zu tun; **give sb./sth. a ~:** jmdm. eine Chance geben/etw. einmal ausprobieren; **I'll give him another ~** *(ask him again for help, a favour, etc.)* ich versuche es noch einmal bei ihm; *(give him another chance)* ich versuche es noch einmal mit ihm; *(on telephone)* ich versuche noch einmal, ihn zu erreichen; **give it a ~:** es versuchen; **b)** *(Rugby)* Versuch, *der.* **2.** *v.t.* **a)** *(attempt, make effort)* versuchen; **it's ~ing to rain** es tröpfelt ein wenig; **do ~ to be on time** bitte versuche, pünktlich zu sein; **it's no use ~ing to do sth.** es hat keinen Zweck zu versuchen, etw. zu tun; **I've given up ~ing to do sth.** ich versuche schon gar nicht mehr, etw. zu tun; **~ one's best** sein Bestes tun; **b)** *(test usefulness of)* probieren; **if the stain is difficult to remove, ~ soap and water** wenn der Fleck schwer zu entfernen ist, versuche *od.* probiere es doch mal mit Wasser und Seife; **I've tried all the bookshops for this book** ich habe in allen Buchhandlungen versucht, dieses Buch zu bekommen; **~ one's hand at sth.** sich an etw. *(Dat.)* versuchen; **~ shaking it!** probier es mal mit Schütteln!; **I'll ~ anything once** ich probiere alles einmal aus; **c)** *(test)* auf die Probe stellen ⟨*Fähigkeit, Kraft, Mut, Geduld*⟩; **~ the door/window [to see if it's locked]** versuchen, die Tür/das Fenster zu öffnen[, um zu sehen, ob sie/es verschlossen ist]; **d)** *(Law.: take to trial)* **~ a case** einen Fall verhandeln; **~ sb. [for sth.]** jmdn. [wegen einer Sache] vor Gericht stellen; jmdm. [wegen einer Sache] den Prozeß machen; **he was tried for murder** er stand wegen Mordes vor Gericht; **he was tried before a jury** er wurde vor ein Schwurgericht gestellt. **3.** *v.i.* es versuchen; **she wasn't even ~ing** sie hat sich *(Dat.)* überhaupt keine Mühe gegeben *od.* es gar nicht erst versucht; **it was not for want of ~ing** es lag nicht daran, daß er/sie *usw.* sich nicht bemüht hätte; **~ and do sth.** *(coll.)* versuchen, etw. zu tun; **~ hard/harder** sich *(Dat.)* viel/mehr Mühe geben

~ for *v.t.* *(compete for)* sich bemühen um ⟨*Arbeitsstelle, Stipendium*⟩; kämpfen um ⟨*Sieg im Sport*⟩

~ 'on *v.t.* **a)** anprobieren ⟨*Kleidungsstück*⟩; **b)** *(Brit. coll.)* **~ it on** provozieren

~ 'out *v.t.* **~ sth./sb. out** etw. ausprobieren/jmdm. eine Chance geben

trying ['traɪɪŋ] *adj.* **a)** *(testing)* schwierig; **b)** *(difficult to endure)* anstrengend

'try-out *n.* Erprobung, *die;* **give sth. a ~:** etw. ausprobieren

tsar [zɑ:(r)] *n.* *(Hist.)* Zar, *der*

tsetse [fly] ['tsetsɪ (flaɪ)] *n.* Tsetsefliege, *die*

'T-shirt *n.* T-Shirt, *das*

'T-square *n.* Kreuzwinkel, *der*

Tu. *abbr.* **Tuesday** Di.

tub [tʌb] *n.* **a)** Kübel, *der;* **b)** *(for ice-cream etc.)* Becher, *der;* **c)** *(Brit. coll.: bath)* Bad, *das*

tuba ['tju:bə] *n.* *(Mus.)* Tuba, *die*

tubby ['tʌbɪ] *adj.* rundlich; pummelig *(ugs.),* rundlich ⟨*Kind*⟩

tube [tju:b] *n.* **a)** *(for conveying liquids etc.)* Rohr, *das;* **b)** *(small cylinder)* Tube, *die;* *(for sweets, tablets)* Röhrchen, *das;* **c)** *(Anat., Zool.)* Röhre, *die;* **d)** *(cathode-ray ~)* Röhre, *die;* *(coll.: television)* **watch the ~:** vor der Röhre sitzen *(ugs.);* **e)** *(Amer.: thermionic valve)* Röhre, *die;* **f)** *(Brit. coll.: underground railway)* U-Bahn, *die*

tuber ['tju:bə(r)] *n.* *(Bot.)* Knolle, *die*

tuberculosis [tju:bɜ:kjʊ'ləʊsɪs] *n., no pl.* *(Med.)* Tuberkulose, *die*

tube: ~ station *n.* *(Brit. coll.)* U-Bahnhof, *der;* **~ train** *n.* *(Brit. coll.)* U-Bahn-Zug, *der*

tubing ['tju:bɪŋ] *n.* Rohre *Pl.*

tubular ['tju:bjʊlə(r) *adj.* *(tube-shaped)* rohrförmig

tubular 'bells *n. pl.* Glockenspiel, *das*

TUC *abbr.* *(Brit.)* **Trades Union Congress**

tuck [tʌk] **1.** *v.t.* stecken; **he ~ed his legs under him** er schlug die Beine unter. **2.** *n.* *(in fabric)* *(for decoration)* Biese, *die;* *(to shorten or tighten)* Abnäher, *der*

~ 'in 1. *v.t.* hineinstecken; **~ in the blankets** die Decken an den Seiten feststecken; **~ your shirt in!** steck dein Hemd in die Hose! **2.** *v.i.* *(coll.)* zulangen *(ugs.)*

~ into *v.i.* *(coll.: eat)* **~ into sth.** sich *(Dat.)* etw. schmecken lassen

~ 'up *v.t.* **a)** hochkrempeln ⟨*Ärmel, Hose*⟩; hochnehmen ⟨*Rock*⟩; **b)** *(cover snugly)* zudecken; **be ~ed up [in bed]** zugedeckt [im Bett] sein

'tuck-shop *n.* *(Brit. Sch.)* *Laden für Erfrischungen, Süßigkeiten usw. in einer Schule*

Tudor ['tju:də(r)] *(Brit. Hist.)* **1.** *n.* Tudor, *der/die.* **2.** *attrib. adj.* Tudor-

Tue., Tues. *abbrs.* **Tuesday** Di.

Tuesday ['tju:zdeɪ, 'tju:zdɪ] **1.** *n.* Dienstag, *der.* **2.** *adv.* *(coll.)* **she comes ~s** sie kommt dienstags. *See also* **Friday**

tuft [tʌft] *n.* Büschel, *das;* **~ of grass/hair** Gras-/Haarbüschel, *das*

tufted ['tʌftɪd] *adj.* büschelig; **~ carpet** Tuftingteppich, *der*

tug [tʌg] **1.** *n.* **a)** Ruck, *der;* **he felt a ~ on the fishing-line** er spürte, wie etwas an der Angel zog; **~ of love [battle]** *(coll.)* *Streit bei der Ehescheidung, wem das Kind zugesprochen wird;* **~ of war** *(lit. or fig.)* Tauziehen, *das;* **b) ~ [boat]** Schlepper, *der.* **2.** *v.t.,* **-gg-** ziehen; schleppen ⟨*Boot*⟩. **3.** *v.i.,* **-gg-** zerren **(at** an + *Dat.)*

tuition [tju:'ɪʃn] *n.* Unterricht, *der*

tulip ['tju:lɪp] *n.* Tulpe, *die*

tumble ['tʌmbl] **1.** *v.i.* **a)** *(fall suddenly)* stürzen; fallen; **~ off sth.** von etw. fallen; **b)** *(move in headlong fashion)* stürzen; **c)** ⟨*Preise usw.:*⟩ fallen. **2.** *v.t.* *(fling headlong)* schleudern. **3.** *n.* Sturz, *der*

~ on *v.t.* *(chance on)* stolpern über (+ *Akk.*)

~ 'over *v.i.* hinfallen; ⟨*Kartenhaus:*⟩ umfallen

~ to *v.t.* *(Brit. coll.)* durchschauen

tumble: ~down *adj.* verfallen; **~-drier** *n.* Wäschetrockner, *der;* **~-dry** *v.t.* im Automaten trocknen

tumbler ['tʌmblə(r)] *n.* *(glass)* *(short)* Whiskyglas, *das;* *(long)* Wasserglas, *das*

tummy ['tʌmɪ] *n.* *(child lang./coll.)* Bäuchlein, *das;* **I've got an upset ~:** ich habe mir den Magen verdorben

'tummy-ache *n.* *(child lang./coll.)* Bauchweh, *das*

tumour (*Brit.; Amer.:* **tumor**) ['tju:mə(r)] *n.* Tumor, *der*

tumult ['tju:mʌlt] *n.* *(commotion, uproar)* Tumult, *der;* **be in ~:** sich in Aufruhr befinden

tuna ['tju:nə] *n., pl. same or* **~s a)** *(fish)* Thunfisch, *der;* **b)** *(as food)* **~[-fish]** Thunfisch, *der*

tune [tju:n] **1.** *n.* **a)** *(melody)* Melodie, *die;* **change one's ~, sing another** *or* **a different ~** *(fig.) (behave differently)* sein Verhalten ändern; *(assume different tone)* einen anderen Ton anschlagen; **call the ~:** den Ton angeben; **b)** *(correct pitch)* **sing in/out of ~:** richtig/falsch singen; **be in/out of ~** ‹*Instrument:*› richtig gestimmt/verstimmt sein; **c)** *(fig.: agreement)* **be in/out of ~ with sth.** mit etw. in Einklang/nicht in Einklang stehen; **d)** *(amount)* **to the ~ of [£50,000]** sage und schreibe [50 000 Pfund]. **2.** *v. t.* **a)** *(Mus.: put in ~)* stimmen; **b)** *(Radio, Telev.)* einstellen (**to** auf + *Akk.*); **stay ~d!** bleiben Sie auf dieser Welle!; **c)** einstellen ‹*Motor, Vergaser*›; *(for more power)* frisieren ‹*Motor, Auto*›

~ 'in *v. i. (Radio, Telev.)* **~ in to a station** einen Sender einstellen; **~ in to** *(fig.)* sich einstellen auf (+ *Akk.*)

~ 'up 1. *v. i.* [die Instrumente] stimmen. **2.** *v. t.* einstellen

tuneful ['tju:nfl] *adj.* melodisch

tuneless ['tju:nlɪs] *adj.* unmelodisch

tuner ['tju:nə(r)] *n.* **a)** *(Mus.)* Stimmer, *der*/Stimmerin, *die;* **b)** *(knob etc.)* Einstellknopf, *der;* Tuner, *der (Technik);* **c)** *(radio)* Tuner, *der*

tungsten ['tʌŋstən] *n.* Wolfram, *das*

tunic ['tju:nɪk] *n. (of soldier, policeman)* Uniformjacke, *die; (of schoolgirl)* Kittel, *der*

tuning ['tju:nɪŋ] *n.* **a)** *(Mus.)* Stimmen, *das;* **b)** *(Radio)* Einstellen, *das;* **c)** *(Motor Veh.)* Einstellen, *das; (to increase power)* Frisieren, *das;* Tuning, *das;* **the engine needs ~:** der Motor muß eingestellt werden

'tuning-fork *n. (Mus.)* Stimmgabel, *die*

Tunisia [tju:'nɪzɪə] *pr. n.* Tunesien *(das)*

tunnel ['tʌnl] **1.** *n.* Tunnel, *der; (dug by animal)* Gang, *der.* **2.** *v. i., (Brit.)* **-ll-** einen Tunnel graben; **~ under sth.** etw. untertunneln; **~ through sth.** durch etw. *(Akk.)* einen Tunnel graben

'tunnel vision *n.* Röhrengesichtsfeld, *das (Med.); (fig.)* enges Blickfeld

turban ['tɜ:bən] *n.* Turban, *der*

turbid ['tɜ:bɪd] *adj.* *(muddy)* trüb[e]

turbine ['tɜ:baɪn] *n.* Turbine, *die*

turbo ['tɜ:bəʊ] *n. (coll.)* Turbo, *der*

turbo: ~-charged *adj.* mit Turbolader *nachgestellt;* **~-charger** *n.* Turbolader, *der*

turbot ['tɜ:bət] *n. (Zool.)* Steinbutt, *der*

turbulence ['tɜ:bjʊləns] *n., no pl.* **a)** *(agitation)* Aufgewühltheit, *die; (fig.)* Aufruhr, *der; (unruliness)* Unruhe, *die;* **b)** *(Phys.)* Turbulenz, *die*

turbulent ['tɜ:bjʊlənt] *adj.* **a)** aufgewühlt ‹*Gedanken, Leidenschaften, Wellen*›; turbulent ‹*Herrschaft, Kindheit*›; ungestüm ‹*Menge*›; aufrührerisch ‹*Stadt, Mob*›; **b)** *(Phys.)* turbulent

tureen [tjʊə'ri:n] *n.* Terrine, *die*

turf [tɜ:f] *n., pl.* **~s** *or* **turves** [tɜ:vz] **a)** *no pl. (covering of grass etc.)* Rasen, *der;* **b)** *(cut patch of grass)* [abgestochenes] Rasenstück; Sode, *die (bes. nordd.);* **lay ~:** Fertigrasen verlegen; **c) the ~** *(racecourse)* der Turf *(Pferdesport);* die Rennbahn; *(horse-racing)* der Pferderennsport

~ 'out *v. t. (sl.)* rausschmeißen *(ugs.)*

~ 'over *v. t.* mit Rasenstücken bedecken

'turf accountant *n.* Buchmacher, *der*

Turk [tɜ:k] *n.* Türke, *der*/Türkin, *die*

Turkey ['tɜ:kɪ] *pr. n.* die Türkei

turkey *n. (fowl)* Truthahn, *der*/Truthenne, *die; (esp. as food)* Puter, *der*/Pute, *die*

Turkish ['tɜ:kɪʃ] **1.** *adj.* türkisch; **sb. is ~:** jmd. ist Türke/Türkin. **2.** *n.* Türkisch, *das; see also* **English 2 a**

Turkish: ~ 'bath *n.* türkisches Bad; **~ de'light** *n. mit Puderzucker bestreutes, gelatinehaltiges Konfekt;* Lokum, *das*

turmeric ['tɜ:mərɪk] *n.* Gelbwurzel, *die; (spice)* Kurkuma, *die*

turmoil ['tɜ:mɔɪl] *n.* Aufruhr, *der;* [wildes] Durcheinander

turn [tɜ:n] **1.** *n.* **a) it is sb.'s ~ to do sth.** jmd. ist an der Reihe, etw. zu tun; **it's your ~ [next]** du bist als nächster/nächste dran *(ugs.) od.* an der Reihe; **wait one's ~:** warten, bis man an der Reihe ist; **your ~ will come** du kommst auch [noch] an die Reihe; **by ~s** abwechselnd; **he gave it to her, and she in ~ passed it on to me** er gab es ihr, und sie wiederum reichte es an mich weiter; **in one's ~:** wiederum; **out of ~** *(before or after one's ~)* außer der Reihe; *(fig.)* an der falschen Stelle ‹*lachen*›; **excuse me if I'm talking out of ~** *(fig.)* entschuldige, wenn ich etwas Unpassendes sage; **take [it in] ~s** sich abwechseln; **take ~s at doing sth., take it in ~s to do sth.** etw. abwechselnd tun; **b)** *(rotary motion)* Drehung, *die;* **give the handle a ~:** den Griff [herum]drehen; **[done] to a ~:** genau richtig [zubereitet]; **c)** *(change of direction)* Wende, *die;* **take a ~ to the right/left, do** *or* **make** *or* **take a right/left ~:** nach rechts/links abbiegen; **'no left/right ~'** „links/rechts abbiegen verboten!"; **the ~ of the year/century** die Jahres-/Jahrhundertwende; **a ~ of fortune** eine Schicksalswende; **take a favourable ~** *(fig.)* sich zum Guten wenden; **d)** *(deflection)* Biegung, *die;* **e)** *(bend)* Kurve, *die; (corner)* Ecke, *die;* **at every ~** *(fig.) (constantly)* ständig; **f)** *(short performance on stage etc.)* Nummer, *die;* **do one's ~:** auftreten; **g)** *(change of tide)* **~ of the tide** Gezeitenwechsel, *der;* **h)** *(character)* **be of a mechanical/speculative ~:** technisch begabt sein/einen Hang zum Spekulativen haben; **i)** *(literary: formation)* Rundung, *die;* **j)** *(form of expression)* **an elegant ~ of speech/phrase** eine elegante Ausdrucksweise; **k)** *(service)* **do sb. a good/bad ~:** jmdm. einen guten/schlechten Dienst erweisen; **one good ~ deserves another** *(prov.)* hilfst du mir, so helf ich dir; **l)** *(coll.: fright)* **give sb. quite a ~:** jmdm. einen gehörigen Schrecken einjagen *(ugs.).* **2.** *v. t.* **a)** *(make revolve)* drehen; **~ the tap** am Wasserhahn drehen; **~ the key in the lock** den Schlüssel im Schloß herumdrehen; **b)** *(reverse)* umdrehen; wenden ‹*Pfannkuchen, Matratze, Auto, Heu, Teppich*›; umgraben ‹*Erde*›; **~ sth. upside down** *or* **on its head** *(lit. or fig.)* etw. auf den Kopf stellen; **~ sth. back to front** die Vorderseite einer Sache nach hinten drehen; **~ sth. inside out** etw. nach außen stülpen *od.* drehen; **~ the page** umblättern; **c)** *(give new direction to)* drehen, wenden ‹*Kopf*›; **she could still ~ heads** die Leute drehten sich immer noch nach ihr um; **~ a hose/gun on sb./sth.** einen Schlauch/ein Gewehr auf jmdn./etw. richten; **~ one's attention/mind to sth.** sich/seine Gedanken einer Sache *(Dat.)* zuwenden; **~ one's thoughts to a subject** sich [in Gedanken] mit einem Thema beschäftigen; **~ a car into a road** [mit einem Auto] in eine Straße einbiegen; **~ the tide [of sth.]** [bei etw.] den Ausschlag geben; ~ **sb. from his purpose** jmdn. von seinem Vorhaben abbringen; **d)** *(send)* ~ **sb. loose on sb./sth.** jmdn. auf jmdn./etw. loslassen; ~ **sb. from one's door/off one's land** jmdn. von seiner Tür/von seinem Land verjagen; ~ **a dog on sb.** einen Hund auf jmdn. hetzen; **e)** *(cause to become)* verwandeln; **the cigarette smoke has ~ed the walls yellow** der Zigarettenrauch hat die Wände vergilben lassen; ~ **the lights low** das Licht dämpfen; ~ **a play/book into a film** ein Theaterstück/Buch verfilmen; **the thought ~ed him pale** der Gedanke ließ ihn erbleichen *(geh.)*; **f)** *(make sour)* sauer werden lassen ‹*Milch*›; **g)** ~ **sb.'s stomach** jmdm. den Magen umdrehen; **h)** ~ **sb.'s head** *(make conceited)* jmdm. zu Kopf steigen; **i)** *(shape in lathe)* drechseln ‹*Holz*›; drehen ‹*Metall*›; **j)** drehen ‹*Pirouette*›; schlagen ‹*Rad, Purzelbaum*›; **k)** *(reach the age of)* ~ **40** 40 [Jahre alt] werden; **l) it's just ~ed 12 o'clock/quarter past 4** es ist gerade 12 Uhr/viertel nach vier vorbei. **3.** *v.i.* **a)** *(revolve)* sich drehen; ‹*Wasserhahn, Schlüssel*:› sich drehen lassen; **the earth ~s on its axis** die Erde dreht sich um ihre Achse; **he couldn't get the key to ~:** er konnte den Schlüssel nicht drehen; **b)** *(reverse direction)* ‹*Person*:› sich herumdrehen; ‹*Auto*:› wenden; **the car ~ed upside down** das Auto überschlug sich; ~ **back to front** sich von hinten nach vorne drehen; **c)** *(take new direction)* sich wenden; *(~ round)* sich umdrehen; **his thoughts/attention ~ed to her** er wandte ihr seine Gedanken/Aufmerksamkeit zu; **left/right ~!** *(Mil.)* links/rechts um!; ~ **into a road/away from the river** in eine Straße einbiegen/vom Fluß abbiegen; ~ **to the left** nach links abbiegen/‹*Schiff, Flugzeug*:› abdrehen; ~ **up/down a street** in eine Straße einbiegen; ~ **towards home** den Heimweg einschlagen; **when the tide ~s** wenn die Ebbe/Flut kommt; **not know where** *or* **which way to ~** *(fig.)* keinen Ausweg [mehr] wissen; **my luck has ~ed** *(fig.)* mein Glück hat sich gewendet; **d)** *(become)* werden; ~ **traitor/statesman/Muslim** zum Verräter/zum Staatsmann/Moslem werden; ~ **[in]to sth.** zu etw. werden; *(be transformed)* sich in etw. *(Akk.)* verwandeln; **her face ~ed green** sie wurde [ganz] grün im Gesicht; **e)** *(change colour)* ‹*Laub*:› sich [ver]färben; **f)** *(become sour)* ‹*Milch*:› sauer werden; **g) my stomach ~s** mir dreht sich der Magen um *(ugs.)*

~ **a'bout 1.** *v.i.* sich umdrehen; ‹*Kompanie*:› kehrtmachen; *(fig.)* eine Kehrtwendung machen. **2.** *v.t.* wenden ‹*Auto, Boot usw.*›

~ **against** *v.t.* **a)** ~ **against sb.** sich gegen jmdn. wenden; ~ **sb. against sb.** jmdn. gegen jmdn. aufbringen; **b) they ~ed his own arguments against him** sie verwendeten seine eigenen Argumente gegen ihn

~ **a'way 1.** *v.i.* sich abwenden; ~ **away from sth.** *(fig.)* sich von etw. abwenden. **2.** *v.t.* **a)** *(avert)* abwenden; **b)** *(send away)* wegschicken

~ **'back 1.** *v.i.* **a)** *(retreat, lit. or fig.)* umkehren; kehrtmachen *(ugs.)*; **there can be no ~ing back** es gibt kein Zurück *od.* keinen Weg zurück; **b)** *(in book etc.)* zurückgehen. **2.** *v.t.* **a)** *(cause to retreat)* zurückweisen; zurückschlagen ‹*Feind*›; **b)** *(fold back)* zurückschlagen ‹*Bettdecke, Teppich*›; herunterschlagen ‹*Kragen*›

~ **'down** *v.t.* **a)** herunterschlagen ‹*Kragen, Hutkrempe*›; umknicken ‹*Buchseite*›; [nach unten] umschlagen ‹*Laken*›; **b)** *(reduce level of)* niedriger stellen ‹*Heizung, Kochplatte*›; dämpfen ‹*Licht*›; herunterdrehen ‹*Gas, Heizung*›; leiser stellen ‹*Ton, Radio, Fernseher*›; **c)** *(reject, refuse)* ablehnen; abweisen ‹*Bewerber, Kandidaten usw.*›

~ **'in 1.** *v.t.* **a)** *(fold inwards)* nach innen drehen; **b)** *(hand in)* abgeben; **c)** *(surrender)* [der Polizei] übergeben; **d)** *(register)* hinlegen *(ugs.)* ‹*Auftritt, Leistung*›; **e)** *(coll.: give up)* aufstecken *(ugs.)* ‹*Arbeit*›; hinschmeißen *(salopp)* ‹*Arbeit, Dienstabzeichen*›. **2.** *v.i.* **a)** *(incline inwards)* nach innen gebogen sein; *(narrow)* sich verjüngen; **b)** *(enter)* einbiegen; **c)** *(coll.: go to bed)* in die Falle gehen *(salopp)*

~ **'off 1.** *v.t.* **a)** abschalten; abstellen ‹*Wasser, Gas*›; zudrehen ‹*Wasserhahn*›; **b)** *(coll.: cause to lose interest)* anwidern. **2.** *v.i.* abbiegen

~ **on 1.** *v.t.* **a)** [-'-] anschalten; aufdrehen ‹*Wasserhahn, Gas*›; **b)** [-'-] *(coll.: cause to take interest)* anmachen *(ugs.)*; **c)** ['--] *(be based on)* ‹*Argument*:› beruhen auf (+ *Dat.*); ‹*Gespräch, Diskussion*› sich drehen um *(ugs.)*; **d)** ['--] *(become hostile towards)* sich wenden gegen; *(attack)* angreifen. **2.** [-'-] *v.i.* einschalten

~ **'out 1.** *v.t.* **a)** *(expel)* hinauswerfen *(ugs.)*; ~ **sb. out of a room/out into the street** jmdn. aus einem Zimmer weisen *od. (ugs.)* werfen/auf die Straße werfen *od.* setzen; **b)** *(switch off)* ausschalten; abdrehen ‹*Gas*›; **c)** *(incline outwards)* nach außen drehen ‹*Füße, Zehen*›; **d)** *(equip)* ausstaffieren; **e)** *(produce)* produzieren; hervorbringen ‹*Fachkräfte, Spezialisten*›; *(in great quantities)* ausstoßen; **f)** *(Brit.) (empty)* ausräumen; ausschütten ‹*Büchse*›; leeren ‹*Inhalt eines Koffers, einer Büchse*›; stürzen ‹*Götterspeise, Pudding*› **(on to** auf + *Akk.*); *(clean)* [gründlich] aufräumen; *(get rid of)* wegwerfen; ~ **out one's pockets** seine Taschen umdrehen. **2.** *v.i.* **a)** *(prove to be)* **sb./sth. ~s out to be sth.** jmd./etw. stellt sich als jmd./etw. heraus *od.* erweist sich als jmd./etw.; **it ~s out that ...:** es stellt sich heraus, daß ...; **as it ~ed out, as things ~ed out** wie sich [nachher] herausstellte; **b)** *(come to be eventually)* **the day ~ed out wet** der Tag wurde regnerisch; **see how things ~ out** sehen, wie sich die Dinge entwickeln; ~ **out to be sth.** sich zu etw. entwickeln; **everything ~ed out well/all right in the end** alles endete gut; **c)** *(end)* **the story ~ed out happily** die Geschichte ging gut aus; **d)** *(appear)* ‹*Menge, Fans usw.*:› erscheinen; **he ~s out every Saturday to watch his team** er kommt jeden Samstag, um seine Mannschaft zu sehen

~ **'over 1.** *v.t.* **a)** *(cause to fall over)* umwerfen; **b)** *(expose the other side of)* umdrehen; umgraben ‹*Erde*›; **c)** drehen ‹*Motor*›; **d)** ~ **sth. over [in one's mind]** sich *(Dat.)* etw. hin und her überlegen; **e)** *(hand over)* übergeben **(to** *Dat.*) ‹*Betrieb, Amt*›. **2.** *v.i.* **a)** *(tip over)* umkippen; ‹*Boot*:› kentern, umschlagen; ‹*Auto, Flugzeug*:› sich überschlagen; **b)** *(from one side to the other)* sich umdrehen; ~ **over on to one's back** sich auf den Rücken drehen; **c)** ‹*Motor*:› laufen; **d) my stomach ~ed over at the thought of it** beim Gedanken daran drehte sich mir der Magen um *(ugs.)*; **e)** *(~ a page)* weiterblättern

~ **'round 1.** *v.i.* **a)** sich umdrehen; ~ **round and go back the same way** umkehren und denselben Weg zurückgehen; **b)** *(rotate)* sich drehen; **c)** ~ **round and do sth.** *(fig.)* plötzlich etw. tun. **2.** *v.t.* **a)** *(unload and reload)* be- und entladen ‹*Frachtschiff*›; abfertigen ‹*Passagierschiff*›; **b)** *see* ~ **about 2; c)** *(reverse)* umdrehen; auf den Kopf stellen *(ugs.)* ‹*Theorie, Argument*›

~ **to** *v.t.* **a)** *(set about)* ~ **to work** an die Arbeit gehen; **b)** *(go to for help etc.)* ~ **to sb./sth.** sich an jmdn. wenden/etw. zu Hilfe nehmen; ~ **to sb. for money** jmdn. um Geld bitten; ~ **to sb. for comfort/help/advice** bei jmdm. Trost/Hilfe/Rat suchen; ~ **to drugs** zu Drogen greifen; ~ **to drink/one's work** *(seeking consolation)* sich in den Alkohol/seine Arbeit flüchten; **c)** *(go on to consider next)* ~ **to a subject/topic** sich einem Thema zuwenden; *see also* ~ **2c**

~ **'up 1.** *v.i.* **a)** *(make one's appearance)* erscheinen; aufkreuzen *(ugs.)*; **b)** *(happen)* passieren; geschehen; **c)** *(present itself)* auftauchen; ‹*Gelegenheit:*› sich bieten; **something is sure to** ~ **up** irgend etwas wird sich schon finden; **d)** *(be found)* sich finden. **2.** *v.t.* **a)** *(dig up)* freilegen; *(fig.)* ans Licht bringen; **I ~ed up a lot of interesting information** ich habe viele interessante Informationen aufgetrieben; **b)** hochschlagen ‹*Kragen, Hutkrempe*›; **her nose is ~ed up** sie hat eine Stupsnase; **c)** lauter stellen, *(ugs.)* aufdrehen ‹*Ton, Fernseher, Radio*›; aufdrehen ‹*Wasser, Heizung, Gas*›; heller machen ‹*Licht*›; **d)** ~ **it up!** *(Brit. sl.)* hör auf damit!

~ **upon** *see* ~ **on 1c, d**

turning ['tɜ:nɪŋ] *n. (off road)* Abzweigung, *die; (fig.)* Kreuzweg, *der (geh.)*; **take the second ~ to the left** die zweite Abzweigung nach links nehmen

turning: ~-circle *n. (Motor Veh.)* Wendekreis, *der;* **~-point** *n.* Wendepunkt, *der*

turnip ['tɜ:nɪp] *n.* Kohlrübe, *die;* Steckrübe, *die*

turn: ~-off *n.* **a)** *(turning)* Abzweigung, *die; (off motorway)* Ausfahrt, *die;* **b)** *(coll.: repellent person or thing)* **be a ~-off** abstoßend sein; **be a ~-off for sb.** jmdn. abstoßen; **~-on** *n. (coll.)* **be a ~-on [for sb.]** [jmdn.] anmachen *(ugs.)*; **~-out** *n.* **a)** *(~ing out for duty)* Einsatz, *der;* Ausrücken, *das;* **b)** *(number voting)* **~-out [of voters]** Wahlbeteiligung, *die;* **c)** *(number assembled)* Beteiligung, *die* (**for** an + *Dat.*); **there was a large ~-out of fans at the airport** eine große Zahl von Fans war zum Flughafen gekommen; **~over** *n.* **a)** *(tart etc.)* **apple/apricot ~over** Apfel-/Aprikosentasche, *die;* **b)** *(Commerc.) (of business, money)* Umsatz, *der; (of stock)* Umschlag, *der;* **c)** *(of staff)* Fluktuation, *die;* **~pike** *n.* **a)** *(Brit. Hist.: toll road)* gebührenpflichtige Straße; **b)** *(Amer.: expressway)* gebührenpflichtige Autobahn; **~stile** *n.* Drehkreuz, *das;* **~table** *n.* Plattenteller, *der;* **~-up** *n.* **a)** *(Brit. Fashion)* Aufschlag, *der;* **b)** *(Brit. coll.: unexpected event)* **a ~-up [for the book]** eine Riesenüberraschung *(ugs.)*

turpentine ['tɜ:pntaɪn] *n.* **a)** *(resin)* Terpentin, *das;* **b) [oil of]** ~: Terpentin, *das (ugs.)*; Terpentinöl, *das; attrib.* ~ **substitute** Terpentinersatz, *der*

turps [tɜ:ps] *n. (coll.)* Terpentin, *das (ugs.)*

turquoise ['tɜ:kwɔɪz] **1.** *n.* **a)** Türkis, *der;* **b)** *(colour)* Türkis, *das.* **2.** *adj.* türkis[farben]

turquoise: ~ **'blue** *n.* Türkisblau, *das;* ~ **'green** *n.* Türkisgrün, *das*

turret ['tʌrɪt] *n.* **a)** *(Archit.)* Türmchen, *das;* **b)** *(of tank etc.)* [Geschütz]turm, *der*

turtle [tɜ:tl] *n.* **a)** *(marine reptile)* Meeresschildkröte, *die;* **b)** *(Amer.: freshwater reptile)* Wasserschildkröte, *die;* **c) turn** ~ ‹*Schiff, Boot:*› kentern

turtle: ~-dove *n.* Turteltaube, *die;* **~-neck** *n.* Stehbundkragen, *der; attrib.* **~-neck pullover** Pullover mit Stehbund

turves *see* **turf**

tusk [tʌsk] *n. (of elephant)* Stoßzahn, *der; (of boar, walrus)* Hauer, *der*

tussle ['tʌsl] **1.** *n.* Gerangel, *das (ugs.)*. **2.** *v.i.* sich balgen; *(fig.)* sich auseinandersetzen (**about** wegen)

tussock ['tʌsək] *n.* [Gras]büschel, *das*

tut[-tut] [tʌt('tʌt)] *int.* na[, na]

tutor ['tju:tə(r)] **1.** *n.* **a) [private]** ~: [Privat]lehrer, *der*/-lehrerin, *die; (for extra help)* Nachhilfelehrer, *der*/-lehrerin, *die;* **b)** *(Brit. Univ.)* ≈ Tutor, *der.* **2.** *v.t.* ~ **sb.** *(teach privately)* jmdm. Privatstunden geben; *(give extra lessons to)* jmdm. Nachhilfestunden geben

tutorial [tju:tɔ:rɪəl] *n. (Brit. Univ.)* ≈ Kolloquium, *das*

tuxedo [tʌk'si:dəʊ] *n., pl.* **~s** *or* **~es** *(Amer.)* Smoking, *der*

TV [ti:'vi:] *n.* **a)** *(television)* Fernsehen, *das; attrib.* Fernseh-; **b)** *(television set)* Fernseher, *der (ugs.)*

twaddle ['twɒdl] *n.* Gewäsch, *das (ugs.)*

twang [twæŋ] **1.** *v.i.* ‹*Bogen:*› mit vibrierendem Ton zurückschnellen. **2.** *v.t.* zupfen ‹*Saite*›; ~ **a guitar** auf einer Gitarre [herum]klimpern *(ugs.)*. **3.** *n. (tone of voice)* **[nasal]** ~: Näseln, *das*

tweak [twi:k] **1.** *v.t.* ~ **sb. in the arm,** ~ **sb.'s arm** jmdn. in den Arm kneifen; ~ **sb.'s ear** jmdn. am Ohr ziehen. **2.** *n.* Kneifen, *das*

twee [twi:] *adj.,* **tweer** ['twi:ə(r)], **tweest** ['twi:ɪst] *(Brit. derog.)* geziert ‹*Wesen, Art, Ausdrucksweise*›; kitschig ‹*Stil, Bild*›; niedlich, putzig ‹*Kleidung, Dorf*›

tweed [twi:d] *n.* **a)** *(fabric)* Tweed, *der; attrib.* Tweed-; **b)** *in pl. (clothes)* Tweedkleidung, *die*

tweet [twi:t] **1.** *n.* Zwitschern, *das.* **2.** *v.i.* zwitschern

tweeter ['twi:tə(r)] *n.* Hochtonlautsprecher, *der*

tweezers ['twi:zəz] *n. pl.* **[pair of]** ~: Pinzette, *die*

twelfth [twelfθ] **1.** *adj.* zwölft...; *see also* **eighth 1. 2.** *n. (fraction)* Zwölftel, *das; see also* **eighth 2**

twelve [twelv] **1.** *adj.* zwölf; ~ **noon** [zwölf Uhr] Mittag; ~ **midnight** [zwölf Uhr] Mitternacht; *see also* **eight 1. 2.** *n.* Zwölf, *die; see also* **eight 2a, d**

twentieth ['twentɪɪθ] **1.** *adj.* zwanzigst...; *see also* **eighth 1. 2.** *n. (fraction)* Zwanzigstel, *das; see also* **eighth 2**

twenty ['twentɪ] **1.** *adj.* zwanzig; *see also* **eight 1. 2.** *n.* Zwanzig, *die; see also* **eight 2a; eighty 2**

twenty: ~-first *etc. adj.* einundzwanzigst... *usw.; see also* **eighth 1; ~-four-hour** *see* **hour a; ~-one** *etc.* **1.** *adj.* einundzwanzig *usw.; see also* **eight 1; 2.** *n.* Einundzwanzig *usw., die; see also* **eight 2a**

twerp [twɜ:p] *n. (sl.) (male)* Blödmann, *der (derb); (female)* blöde Kuh *(derb)*

twice [twaɪs] *adv.* **a)** *(two times)* zweimal; **she didn't have to be asked ~!** da brauchte man sie nicht zweimal zu fragen!; ~ **a year** zweimal im Jahr; **b)** *(doubly)* doppelt; ~ **as strong** *etc.* doppelt so stark *usw.*; **he's ~ her age** er ist doppelt so alt wie sie; *see also* **think 2a**

twiddle ['twɪdl] **1.** *v.t.* herumdrehen an (+ *Dat.*) *(ugs.)*; ~ **one's thumbs** *(lit. or fig.)* Däumchen drehen *(ugs.)*. **2.** *v.i.* ~ **with sth.**

mit etw. spielen; an etw. *(Dat.)* herumfummeln *(ugs.)*
[1]**twig** [twɪg] *n.* Zweig, *der*
[2]**twig** *(coll.)* **1.** *v. t.* **-gg-: a)** *(understand)* kapieren *(ugs.)*; **b)** *(notice)* mitkriegen *(ugs.)*. **2.** *v. i.* **-gg-: a)** *(understand)* es kapieren *(ugs.)*; **b)** *(notice)* es mitkriegen *(ugs.)*
twilight ['twaɪlaɪt] *n.* **a)** *(evening light)* Dämmerlicht, *das;* Zwielicht, *das;* **b)** *(period of half-light)* Dämmerung, *die*
twin [twɪn] **1.** *attrib. adj.* **a)** Zwillings-; ~ **brother/sister** Zwillingsbruder, *der*/-schwester, *die;* **b)** *(forming a pair)* Doppel-; doppelt ‹*Problem, Verantwortung*›; **c)** *(Bot.)* paarig; **d)** Doppel‹*vergaser, -propeller, -schraube usw.*›. **2.** *n.* **a)** Zwilling, *der;* **his ~:** sein Zwillingsbruder/seine Zwillingsschwester; **b)** *(exact counterpart)* Gegenstück, *das.* **3.** *v. t.*, **-nn-** eng verbinden; **Bottrop is ~ned with Blackpool** Bottrop und Blackpool sind Partnerstädte
twin: ~ **'bed** *n.* eines von zwei [gleichen] Einzelbetten; ~ **beds** zwei Einzelbetten; **~-bedded** ['twɪnbedɪd] *adj.* **a ~-bedded room** ein Zweibettzimmer
twine [twaɪn] **1.** *n.* Bindfaden, *der; (thicker)* Kordel, *die; (for nets)* Garn, *das.* **2.** *v. t.* **a)** *(form by twisting strands together)* [zusammen]drehen; **b)** *(form by interlacing)* winden *(geh.)* ‹*Kranz, Girlande*›. **3.** *v. i.* sich winden **(about, around** um)
twin-engined ['twɪnendʒɪnd] *adj.* zweimotorig
twinge [twɪndʒ] *n.* Stechen, *das;* **a ~ of toothache/pain** ein stechender Zahnschmerz/ein stechender Schmerz; **~[s] of remorse/conscience** *(fig.)* Gewissensbisse
twinkle ['twɪŋkl] **1.** *v. i.* ‹*Sterne, Augen:*› funkeln, blitzen **(with** vor + *Dat.*). **2.** *n.* **a) in a ~:** im Handumdrehen; **b)** *(sparkle of the eyes)* Funkeln, *das;* **'...', she said with a ~ in her eye** „...", sagte sie augenzwinkernd
twinkling ['twɪŋklɪŋ] *n.* **in a ~, in the ~ of an eye** im Handumdrehen
twin: ~set *n.* *(Brit.)* Twinset, *das;* ~ **town** *n.* *(Brit.)* Partnerstadt, *die;* **~-tub** *n.* halbautomatische Waschmaschine *(mit separater Schleuder)*
twirl [twɜ:l] **1.** *v. t.* **a)** *(spin)* [schnell] drehen; **b)** *(twiddle)* zwirbeln ‹*Schnurrbart*›; drehen ‹*Haar*›. **2.** *v. i.* wirbeln **(around** über + *Akk.*); **sb. ~s around** jmd. wirbelt herum. **3.** *n.* [Herum]wirbeln, *das*
twist [twɪst] **1.** *v. t.* **a)** *(distort)* verdrehen ‹*Worte, Bedeutung*›; ~ **out of shape** verbiegen; ~ **one's ankle** sich *(Dat.)* den Knöchel verrenken; **her face was ~ed with pain** ihr Gesicht war schmerzverzerrt; ~ **sb.'s arm** jmdm. den Arm umdrehen; *(fig.)* jmdm. [die] Daumenschrauben anlegen *(scherzh.)*; **b)** *(wind about one another)* flechten ‹*Blumen, Haare*› **(into** zu); **c)** *(rotate)* drehen; *(back and forth)* hin und her drehen; **d)** *(interweave)* verweben; **e)** *(give spiral form to)* drehen **(into** zu). **2.** *v. i.* **a)** sich winden; ~ **and turn** sich drehen und winden; ~ **around sth.** sich um etw. winden; **b)** *(take ~ed position)* sich winden. **3.** *n.* **a)** *(thread etc.)* Zwirn, *der;* **b)** ~ **of lemon/orange** Zitronen-/Orangenscheibe, *die;* **c)** *(~ing)* Drehung, *die;* **d)** *(unexpected occurrence)* überraschende Wendung; ~ **of fate** Laune des Schicksals; **e)** *(peculiar tendency)* **give a ~ to sth.** etw. verdrehen; **f) round the ~:** = **round the bend** *see* **bend 1 a**
~ **'off 1.** *v. t.* abdrehen. **2.** *v. i.* **the cap ~s off** der Verschluß läßt sich abdrehen
twisted ['twɪstɪd] *adj.* verbogen; *(fig.)* verdreht *(ugs. abwertend)* ‹*Geist*›; verquer ‹*Humor*›
twit [twɪt] *n.* *(Brit. sl.)* Trottel, *der (ugs.)*
twitch [twɪtʃ] **1.** *v. t.* **a)** zupfen; **b)** zucken mit ‹*Nase, Schwanz*›; wackeln mit ‹*Ohr*›. **2.** *v. i.* **a)** *(pull sharply)* zupfen **(at** an + *Dat.*); **b)** ‹*Mund, Lippen, Hand, Nase:*› zucken. **3.** *n.* Zucken, *das*
twitter ['twɪtə(r)] **1.** *n.* *(chirping)* Zwitschern, *das;* Gezwitscher, *das.* **2.** *v. i.* zwitschern; ‹*Person:*› schnattern *(ugs.)*
two [tu:] **1.** *adj.* zwei; **a box/shirt or ~:** ein, zwei Schachteln/Hemden; ein oder zwei Schachteln/Hemden; *see also* **eight 1. 2.** *n.* Zwei, *die;* **the ~:** die beiden; die zwei; **just the ~ of us** nur wir zwei *od.* beiden; **put ~ and ~ together** *(fig.)* zwei und zwei zusammenzählen; **cut/break in ~:** zweiteilen/entzweibrechen; ~ **and ~,** ~ **by** ~ *(~ at a time)* [zu] zwei und zwei; zu zweien; **that makes ~ of us** *(coll.)* mir geht's/ging's genauso *(ugs.)*; *see also* **eight 2 a, c, d;** [1]**game 1 a; penny b**
two: ~-bit *adj.* *(Amer.: of poor quality)* mies *(ugs.)*; **~-dimensional** [tu:dɪ'menʃənl, tu:daɪ'menʃənl] *adj.* zweidimensional; **~-door** *attrib. adj.* zweitürig ‹*Auto*›; **~-edged** *adj.* *(lit. or fig.)* zweischneidig; **~-faced** ['tu:feɪst] *adj.* *(fig.)* falsch *(abwertend)*
twofold ['tu:fəʊld] *adj.* **a)** zweifach; **be ~:** zweifacher Art *od.* Natur sein; **b)** *(double)* **a ~ increase** ein Anstieg auf das Doppelte
two: ~-'handed *adj.* **a)** *(having ~ hands)* zweihändig; **b)** *(requiring both hands)* beidhändig; **~-party system** *n.* Zweiparteiensystem, *das;* **~pence** ['tʌpəns] *n.* *(Brit.)* zwei Pence; **~-piece 1.** *n.* Zweiteiler, *der;* **2.** *adj.* zweiteilig; **~-pin** *see* **pin 1 c;** **~-ply** *adj.* zweilagig ‹*Holz*›; zweifädig ‹*Wolle, Zwirn*›; **~-seater 1.** ['-'--] *n.* Zweisitzer, *der;* **2.** ['---] *attrib. adj.* zweisitzig
twosome ['tu:səm] *n.* **a)** Paar, *das;* **b)** *(Golf)* Zweier, *der*
two: ~-storey *adj.* zweigeschossig; **~-stroke** *adj.* *(Mech. Engin.)* Zweitakt‹*motor, -verfahren*›; **~-time** *v. t.* *(sl.)* **~-time sb.** *(be unfaithful)* jmdm. fremdgehen; **~-tone** *adj.* **a)** *(in colour)* zweifarbig; **b)** *(in sound)* Zweiklang-; **~-up-~-down** *n.* kleines [Reihen]haus; **~-way** *adj.* **a)** *(in both directions)* zweibahnig *(Verkehrsw.)*; **'~-way traffic ahead'** „Achtung Gegenverkehr"; **b) ~-way mirror** Einwegspiegel, *der;* **~-wheeler** ['tu:wi:lə(r)] *n.* Zweirad, *das*
tycoon [taɪ'ku:n] *n.* Magnat, *der;* Tycoon, *der*
type [taɪp] **1.** *n.* **a)** Art, *die; (person)* Typ, *der;* **what ~ of car ...?** was für ein Auto ...?; **he's not the ~ to let people down** er ist nicht der Typ, der andere im Stich läßt; **he is a different ~ of person** er ist eine andere Art Mensch *od.* ein anderer Typ; **books of this ~:** derartige Bücher; **b)** *(coll.: character)* Type, *die (ugs.)*; **c)** *(Printing)* Drucktype, *die;* **be in small/italic ~:** kleingedruckt/kursiv gedruckt sein. **2.** *v. t.* [mit der Maschine] schreiben; tippen *(ugs.)*; **~d letter** maschinegeschriebener Brief. **3.** *v. i.* maschineschreiben
~ **'in** *v. t.* eintippen *(ugs.)*
~ **'out** *v. t.* [mit der Schreibmaschine] abschreiben; abtippen *(ugs.)*
-type [taɪp] *in comb.* -artig
type: ~-cast *v. t.* [auf eine bestimmte Rolle] festlegen; **~-face** *n.* Schriftbild, *das;* **~script** *n.* maschine[n]geschriebene Fas-

sung; Typoscript, *das;* **~set** *v. t. (Printing)* setzen; **~setter** *n.* [Schrift]setzer, *der*/[Schrift]setzerin, *die*

'typewriter *n.* Schreibmaschine, *die;* **~ ribbon** Farbband, *das*

'typewritten *adj.* maschine[n]geschrieben; mit der [Schreib]maschine geschrieben

typhoid ['taɪfɔɪd] *n. (Med.)* **~ [fever]** Typhus, *der*

typhoon [taɪ'fu:n] *n.* Taifun, *der*

typhus ['taifəs] *n. (Med.)* Fleckfieber, *das*

typical ['tɪpɪkl] *adj.* typisch (**of** für); **that's just ~!** [das ist mal wieder] typisch! *(ugs.)*

typically ['tɪpɪklɪ] *adv.* typischerweise; **~ she turned up late** wie üblich kam sie zu spät

typify ['tɪpɪfaɪ] *v. t.* **a)** *(represent)* [symbolhaft] darstellen; **b)** *(be an example of)* **~ sth.** als typisches Beispiel für etw. dienen

typing ['taɪpɪŋ] *n.* Maschineschreiben, *das;* **his ~ is excellent** er kann sehr gut maschineschreiben

typing: ~ error *n.* Tippfehler, *der (ugs.);* **~ pool** *n.* Schreibzentrale, *die*

typist ['taɪpɪst] *n.* Schreibkraft, *die;* **shorthand ~:** Stenotypist, *der*/-typistin, *die*

typographic [taɪpə'græfɪk], **typographical** [taɪpə'græfɪkl] *adj.* typographisch; **~ error** Setzfehler, *der*

typography [taɪ'pɒgrəfɪ] *n.* Typographie, *die*

tyrannical [tɪ'rænɪkl, taɪ'rænɪkl] *adj.* tyrannisch

tyrannize (tyrannise) ['tɪrənaɪz] *v. t.* ‹*Chef, Vater, Ehemann:*› tyrannisieren; ‹*Herrscher:*› als Tyrann herrschen über (+ *Akk.*)

tyranny ['tɪrənɪ] *n.* Tyrannei, *die*

tyrant ['taɪrənt] *n. (lit. or fig.)* Tyrann, *der*

tyre ['taɪə(r)] *n.* Reifen, *der*

tyre: ~-gauge *n.* Reifendruckprüfer, *der;* **~ pressure** *n.* Reifendruck, *der*

Tyrol [tɪ'rəʊl] *pr. n.* Tirol *(das)*

Tyrolean [tɪrə'li:ən] *adj.* Tiroler

tzar *see* **tsar**

U

U, u [ju:] *n., pl.* **Us** *or* **U's** U, u, *das*

'U-bend *n.* U-Rohr, *das;* Knie, *das (ugs.)*

ubiquitous [ju:'bɪkwɪtəs] *adj.* allgegenwärtig

'U-boat *n. (Hist.)* [deutsches] U-Boot

udder ['ʌdə(r)] *n.* Euter, *das*

UFO ['ju:fəʊ] *n., pl.* **~s** Ufo, *das*

Uganda [ju:'gændə] *pr. n.* Uganda *(das)*

Ugandan [ju:'gændən] **1.** *adj.* ugandisch; **sb. is ~:** jmd. ist Ugander/Uganderin. **2.** *n.* Ugander, *der*/Uganderin, *die*

ugh [ʌh, ʊh, ɜ:h] *int.* bah

ugliness ['ʌglɪnɪs] *n., no pl.* Häßlichkeit, *die*

ugly ['ʌglɪ] *adj.* **a)** *(in appearance, morally)* häßlich; **~ duckling** *(fig.)* häßliches Entlein *(ugs. scherzh.);* **as ~ as sin** *(coll.)* potthäßlich *(ugs.);* häßlich wie die Nacht; **b)** *(nasty)* übel ‹*Wunde, Laune, Szene usw.*›; **have an ~ temper** übellaunig sein

UHF *abbr.* **ultra-high frequency** UHF

UHT *abbr.* **ultra-high temperature** ultrahocherhitzt; **UHT milk** H-Milch, *die*

UK *abbr.* **United Kingdom**

Ukraine [ju:'kreɪn] *pr. n.* Ukraine, *die*

ulcer ['ʌlsə(r)] *n.* Geschwür, *das;* **mouth ~[s]** Aphthe, *die (Med.)*

Ulster ['ʌlstə(r)] *pr. n.* Ulster *(das)*

ulterior [ʌl'tɪərɪə(r)] *adj.* hintergründig; geheim; **~ motive/thought** Hintergedanke, *der*

ultimate ['ʌltɪmət] **1.** *attrib. adj.* **a)** *(final)* letzt...; *(eventual)* endgültig ‹*Sieg*›; letztendlich ‹*Rettung*›; größt... ‹*Opfer*›; **~ result/goal/decision** Endergebnis, *das*/Endziel, *das*/endgültige Entscheidung; **the ~ deterrent** das äußerste Abschreckungsmittel; **b)** *(fundamental)* tiefst... ‹*Grundlage, Wahrheit*›; **the ~ origin** der eigentliche Ursprung. **2.** *n.* **the ~** *(maximum)* das absolute Maximum; *(minimum)* das absolute Minimum; **the ~ in comfort/luxury/style/fashion** der Gipfel an Bequemlichkeit/Luxus/das Exzellenteste an Stil/in der Mode

ultimately ['ʌltɪmətlɪ] *adv.* **a)** *(in the end)* schließlich; **b)** *(in the last analysis)* letzten Endes; *(basically)* im Grunde [genommen]

ultimatum [ʌltɪ'meɪtəm] *n., pl.* **~s** *or* **ultimata** [ʌltɪ'meɪtə] Ultimatum, *das;* **give sb. an ~:** jmdm. ein Ultimatum stellen

ultra- ['ʌltrə] *in comb.* ultra‹*konservativ, -modern*›; hyper‹*modern, -modisch*›

ultra'sonic *adj.* Ultraschall-

'ultrasound *n., no pl.* Ultraschall, *der*

ultra'violet *adj. (Phys.)* ultraviolett ‹*Strahlen, Licht*›; UV- ‹*Lampe, Filter*›

umbilical cord [ʌm'bɪlɪkl kɔ:d] *n.* Nabelschnur, *die*

umbrage ['ʌmbrɪdʒ] *n., no pl., no indef. art.* **take ~ [at** *or* **over sth.]** [an etw. (+ *Dat.*)] Anstoß nehmen

umbrella [ʌm'brelə] *n.* **a)** [Regen]schirm, *der;* **put up an ~:** einen Schirm aufspannen; **b)** *(fig.: protection)* Schutz, *der;* **c)** *attrib.* **an ~ organization** eine Dachorganisation

umlaut ['ʊmlaʊt] *n.* **a)** *(vowel change)* Umlaut, *der;* **b)** *(mark)* Umlautzeichen, *das*

umpire ['ʌmpaɪə(r)] *n.* Schiedsrichter, *der*/-richterin, *die*

umpteen [ʌmp'ti:n] *adj. (coll.)* zig *(ugs.);* x *(ugs.)*

umpteenth [ʌmp'ti:nθ] *adj. (coll.)* zigst... *(ugs.);* **for the ~ time** zum zigsten *or* x-sten Mal *(ugs.)*

UN *abbr.* **United Nations** UN[O], *die*

unabashed [ʌnə'bæʃt] *adj.* ungeniert; *(without shame)* schamlos; *(undaunted)* unerschrocken ‹*Kämpfer*›

unabated [ʌnə'beɪtɪd] *adj.* unvermindert

unable [ʌn'eɪbl] *pred. adj.* **be ~ to do sth.** nicht in der Lage sein, etw. zu tun; etw. nicht tun können; **he wanted to attend but was ~ to** er wollte kommen, aber er war dazu nicht in der Lage

unabridged [ʌnə'brɪdʒd] *adj.* ungekürzt

unacceptable [ʌnək'septəbl] *adj.* unannehmbar; **[be] not ~:** durchaus akzeptabel [sein]

unaccompanied [ʌnə'kʌmpənɪd] *adj.* ohne Begleitung ‹*reisen, singen*›; unbegleitet ‹*Gepäck,*

Chor⟩; (on aircraft etc.) ~ **minor** alleinreisendes Kind

unaccountable [ʌnə'kaʊntəbl] *adj.* unerklärlich

unaccountably [ʌnə'kaʊntəblɪ] *adv.* unerklärlicherweise; *with adj.* unerklärlich

unaccounted [ʌnə'kaʊntɪd] *adj.* ~ **for** unauffindbar; **several passengers are still** ~ **for** einige Passagiere werden noch vermißt

unaccustomed [ʌnə'kʌstəmd] *adj.* ungewohnt; **be** ~ **to sth.** etw. *(Akk.)* nicht gewöhnt sein

unacquainted [ʌnə'kweɪntɪd] *adj.* **be [completely]** ~ **with sth.** mit etw. [überhaupt] nicht vertraut sein

unadulterated [ʌnə'dʌltəreɪtɪd] *adj.* **a)** *(pure)* unverfälscht; rein *⟨Wasser, Wein⟩;* **b)** *(utter)* völlig

unadventurous [ʌnəd'ventʃərəs] *adj.* bieder *⟨Person⟩;* ereignislos *⟨Leben⟩;* einfallslos *⟨Inszenierung, Buch, usw.⟩*

unaffected [ʌnə'fektɪd] *adj.* **a)** *(not affected)* unberührt; *(Med.)* nicht angegriffen *⟨Organ⟩;* **the area was** ~ **by the strike** die Gegend war vom Streik nicht betroffen; **b)** *(natural)* natürlich; ungekünstelt

unafraid [ʌnə'freɪd] *adj.* **be** ~ **[of sb./sth.]** keine Angst [vor jmdm./etw.] haben

unaided [ʌn'eɪdɪd] *adj.* ohne fremde Hilfe; **walk** ~: ohne Hilfe gehen

unalterable [ʌn'ɔ:ltərəbl, ʌn'ɒltərəbl] *adj.* unabänderlich *⟨Gesetz, Schicksal⟩;* unverrückbar *⟨Entschluß⟩*

unaltered [ʌn'ɔ:ltəd, ʌn'ɒltəd] *adj.* unverändert

unambiguous [ʌnæm'bɪgjʊəs] *adj.* unzweideutig

unambitious [ʌnæm'bɪʃəs] *adj.* *⟨Person⟩* ohne Ehrgeiz; **be** ~: keinen Ehrgeiz haben

un-American [ʌnə'merɪkn] *adj.* **a)** unamerikanisch; **b)** *(anti-American)* antiamerikanisch; ~ **activities** unamerikanische Umtriebe

unanimity [ju:nə'nɪmɪtɪ] *n., no pl.* Einmütigkeit, *die*

unanimous [ju:'nænɪməs] *adj.* einstimmig; **be** ~ **in doing sth.** etw. einmütig tun; **be** ~ **in rejecting** *or* **in their** *etc.* **rejection of sth.** etw. einmütig ablehnen

unanimously [ju:'nænɪməslɪ] *adv.* einstimmig

unanswerable [ʌn'ɑ:nsərəbl] *adj.* unbeantwortbar *⟨Frage⟩;* unwiderlegbar *⟨Argument⟩*

unanswered [ʌn'ɑ:nsəd] *adj.* unbeantwortet; **go** ~, **be left** ~: unbeantwortet bleiben

unappetizing [ʌn'æpɪtaɪzɪŋ] *adj.* unappetitlich

unapproachable [ʌnə'prəʊtʃəbl] *adj.* unzugänglich

unarmed [ʌn'ɑ:md] *adj.* unbewaffnet; ~ **combat** Kampf ohne Waffen

unashamed [ʌnə'ʃeɪmd] *adj.* schamlos; *(not embarrassed)* ungeniert; unverhohlen *⟨Individualist⟩*

unasked [ʌn'ɑ:skt] *adj.* **a)** *(uninvited)* ungebeten; **b)** *(not asked for)* ~ **[for]** ungefragt

unassailable [ʌnə'seɪləbl] *adj.* **a)** *(not open to assault)* uneinnehmbar; **an** ~ **lead** ein nicht aufzuholender Vorsprung; **b)** *(irrefutable)* unwiderlegbar

unassisted [ʌnə'sɪstɪd] *see* **unaided**

unassuming [ʌnə'sju:mɪŋ] *adj.* bescheiden; unprätentiös *(geh.)*

unattached [ʌnə'tætʃt] *adj.* **a)** *(not fixed)* nicht befestigt; **b)** *(without a partner)* ungebunden

unattended [ʌnə'tendɪd] *adj.* **a)** ~ **to** *(not dealt with)* unerledigt, unbearbeitet *⟨Post, Angelegenheit⟩;* nicht bedient *⟨Kunde⟩;* nicht behandelt *⟨Patient, Wunde⟩;* **leave a customer/patient** ~ **to** einen Kunden nicht bedienen/einen Patienten nicht behandeln; **b)** *(not supervised)* unbewacht *⟨Parkplatz, Gepäck⟩*

unattractive [ʌnə'træktɪv] *adj.* unattraktiv; unschön *⟨Ort, Merkmal⟩;* wenig verlockend *⟨Angebot, Vorschlag⟩*

unauthorized [ʌn'ɔ:θəraɪzd] *adj.* unbefugt; nicht autorisiert *⟨Biographie⟩;* nicht genehmigt *⟨Demonstration⟩;* **no entry for** ~ **persons** Zutritt für Unbefugte verboten

unavailable [ʌnə'veɪləbl] *adj.* nicht erhältlich *⟨Ware⟩;* **be** ~ **for comment** zu einer Stellungnahme nicht zur Verfügung stehen

unavoidable [ʌnə'vɔɪdəbl] *adj.* unvermeidlich; ~ **delays** unvermeidbare Verzögerungen

unavoidably [ʌnə'vɔɪdəblɪ] *adv.* **we were** ~ **delayed** unsere Verspätung ließ sich nicht vermeiden; **he has been** ~ **detained** er konnte nicht verhindern, daß er aufgehalten wurde

unaware [ʌnə'weə(r)] *adj.* **be** ~ **of sth.** sich *(Dat.)* einer Sache *(Gen.)* nicht bewußt sein

unawares [ʌnə'weəz] *adv.* unerwartet; **come upon sb./catch sb.** ~: jmdn. überraschen

unbalanced [ʌn'bælənst] *adj.* **a)** unausgewogen; **b)** *(mentally* ~*)* unausgeglichen

unbearable [ʌn'beərəbl] *adj.*, **unbearably** [ʌn'beərəblɪ] *adv.* unerträglich

unbeatable [ʌn'bi:təbl] *adj.* unschlagbar *(ugs.)*

unbeaten [ʌn'bi:tn] *adj.* **a)** *(not defeated)* ungeschlagen; **they lost their** ~ **record** ihre Siegesserie endete; **b)** *(not surpassed)* unerreicht; **this record is still** ~: dieser Rekord ist immer noch ungebrochen

unbecoming [ʌnbɪ'kʌmɪŋ] *adj.* *(improper)* unschicklich *(geh.)*

unbelievable [ʌnbɪ'li:vəbl] *adj.* **a)** *(hardly believable)* unglaublich; **b)** *(tremendous)* unwahrscheinlich *⟨Hunger, Durst⟩*

unbelievably [ʌnbɪ'li:vəblɪ] *adv.* *as intensifier* unglaublich

unbeliever [ʌnbɪ'li:və(r)] *n.* Ungläubige, *der/die*

unbiased, unbiassed [ʌn'baɪəst] *adj.* unvoreingenommen

unblemished [ʌn'blemɪʃt] *adj.* makellos *⟨Haut, Lack, Ruf⟩;* unbefleckt *(geh.)* *⟨Ehre⟩*

unblock [ʌn'blɒk] *v.t.* frei machen *od.* bekommen; **remain ~ed** frei bleiben

unbolt [ʌn'bəʊlt] *v.t.* aufriegeln *⟨Tür, Tor⟩*

unborn [ʌn'bɔ:n, *attrib.* 'ʌnbɔ:n] *adj.* ungeboren

unbound [ʌn'baʊnd] *adj.* **a)** *(not tied)* offen *⟨Haar⟩;* **b)** ungebunden *⟨Buch⟩*

unbounded [ʌn'baʊndɪd] *adj.* **a)** *(unchecked)* uneingeschränkt *⟨Freiheit⟩;* unkontrolliert *⟨Gefühl⟩;* **b)** *(unlimited)* grenzenlos

unbreakable [ʌn'breɪkəbl] *adj.* unzerbrechlich

un-British [ʌn'brɪtɪʃ] *adj.* unbritisch

unbroken [ʌn'brəʊkn] *adj.* **a)** *(undamaged)* heil; unbeschädigt; **b)** *(not interrupted)* ununterbrochen; ~ **sleep/peace/silence** ungestörter Schlaf/Friede/durch nichts unterbrochene Stille

unbuckle [ʌn'bʌkl] *v.t.* aufschnallen

unburden [ʌn'bɜ:dn] *v.t.* *(literary)* befreien *⟨Gewissen⟩;* ~ **oneself/one's heart [to sb.]** [jmdm.] sein Herz ausschütten; ~ **oneself of sth.** sich von etw. befreien

unbusinesslike [ʌn'bɪznɪslaɪk] *adj.* **he is** ~, **he has an** ~ **approach** er geht nicht wie ein Geschäftsmann an die Dinge heran

unbutton [ʌn'bʌtn] *v.t.* aufknöpfen

unbuttoned [ʌn'bʌtnd] *adj.* *(lit. or fig.)* aufgeknöpft; offen

uncalled-for [ʌn'kɔ:ldfɔ:(r)] *adj.* unangebracht

uncanny [ʌn'kænɪ] *adj.* **a)** *(seemingly supernatural)* unheimlich; **b)** *(mysterious)* verblüffend

uncap [ʌn'kæp] *v. t.*, **-pp-** öffnen ⟨*Flasche*⟩

uncared-for [ʌn'keədfɔ:(r)] *adj.* vernachlässigt

uncaring [ʌn'keərɪŋ] *adj.* gleichgültig

uncarpeted [ʌn'kɑ:pɪtɪd] *adj.* teppichlos

unceasing [ʌn'si:sɪŋ] *adj.* unaufhörlich

uncensored [ʌn'sensəd] *adj.* unzensiert

unceremonious [ʌnserɪ'məʊnɪəs] *adj.* **a)** *(informal)* formlos; **b)** *(abrupt)* brüsk

unceremoniously [ʌnserɪ'məʊnɪəslɪ] *adv.* ohne Umschweife

uncertain [ʌn'sɜ:tn] *adj.* **a)** *(not sure)* **be ~ [whether ...]** sich *(Dat.)* nicht sicher sein[, ob ...]; **b)** *(not clear)* ungewiß ⟨*Ergebnis, Zukunft, Schicksal*⟩; **of ~ age/origin** unbestimmten Alters/unbestimmter Herkunft; **it is still ~ whether ...:** es ist noch ungewiß, ob ...; **it is ~ who was the inventor** der Erfinder ist nicht [genau] bekannt; **c)** *(unsteady)* unsicher ⟨*Schritte*⟩; **d)** *(changeable)* unbeständig ⟨*Charakter, Wetter*⟩; **e)** *(ambiguous)* vage; **in no ~ terms** ganz eindeutig

uncertainly [ʌn'sɜ:tnlɪ] *adv.* **a)** *(without definite aim)* ziellos; **b)** *(without confidence)* unsicher

uncertainty [ʌn'sɜ:tntɪ] *n.* **a)** *no pl. (doubtfulness)* Ungewißheit, *die;* **there is some ~ about it** es ist etwas ungewiß; **b)** *(doubtful point)* Unklarheit, *die;* **c)** *no pl. (hesitation)* Unsicherheit, *die*

unchanged [ʌn'tʃeɪndʒd] *adj.* unverändert

unchanging [ʌn'tʃeɪndʒɪŋ] *adj.* unveränderlich

uncharacteristic [ʌnkærɪktə'rɪstɪk] *adj.* uncharakteristisch **(of** für); ungewohnt ⟨*Grobheit, Schärfe*⟩

uncharitable [ʌn'tʃærɪtəbl] *adj.*, **uncharitably** [ʌn'tʃærɪtəblɪ] *adv.* lieblos

unchecked [ʌn'tʃekt] *adj.* **a)** *(not examined)* ungeprüft; **b)** *(unrestrained)* ungehindert; nicht eingedämmt ⟨*Epidemie, Inflation*⟩; **sth. goes ~:** gegen etw. wird nichts getan

uncivil [ʌn'sɪvɪl, ʌn'sɪvl] *adj.* unhöflich

uncivilized [ʌn'sɪvɪlaɪzd] *adj.* unzivilisiert

unclaimed [ʌn'kleɪmd] *adj.* herrenlos; nicht abgeholt ⟨*Brief, Preis*⟩; **the money is still ~:** bis jetzt hat niemand Anspruch auf das Geld erhoben

unclassified [ʌn'klæsɪfaɪd] *adj.* nicht klassifiziert; *(not subject to security classification)* nicht geheim

uncle ['ʌŋkl] *n.* Onkel, *der*

unclean [ʌn'kli:n] *adj.* unrein

unclothed [ʌn'kləʊðd] *adj.* unbekleidet

uncluttered [ʌn'klʌtəd] *adj.* ordentlich

uncoil [ʌn'kɔɪl] **1.** *v. t.* abwickeln. **2.** *v. refl.* sich abwickeln; ⟨*Schlange:*⟩ sich strecken

uncomfortable [ʌn'kʌmfətəbl] *adj.* **a)** *(causing physical discomfort)* unbequem; **b)** *(feeling discomfort)* **be ~:** sich unbehaglich fühlen; **c)** *(uneasy, disconcerting)* unangenehm; peinlich ⟨*Stille*⟩; **his gaze made me ~:** sein Blick war mir unangenehm

uncomfortably [ʌn'kʌmfətəblɪ] *adv.* **a)** unbequem; **b)** *(uneasily)* unbehaglich; **be** *or* **feel ~ aware of sth.** sich *(Dat.)* einer Sache *(Gen.)* peinlich bewußt sein

uncommitted [ʌnkə'mɪtɪd] *adj.* unbeteiligt

uncommon [ʌn'kɒmən] *adj.* ungewöhnlich; **it is not ~ for him to be found there** es ist [ganz und gar] nicht ungewöhnlich, daß man ihn dort findet

uncommunicative [ʌnkə'mju:nɪkətɪv] *adj.* verschlossen

uncomplaining [ʌnkəm'pleɪnɪŋ] *adj.*, **uncomplainingly** [ʌnkəm'pleɪnɪŋlɪ] *adv.* klaglos

uncompleted [ʌnkəm'pli:tɪd] *adj.* unvollendet

uncomplicated [ʌn'kɒmplɪkeɪtɪd] *adj.* unkompliziert

uncomplimentary [ʌnkɒmplɪ'mentərɪ] *adj.* wenig schmeichelhaft

uncomprehending [ʌnkɒmprɪ'hendɪŋ] *adj.* verständnislos

uncompromising [ʌn'kɒmprəmaɪzɪŋ] *adj.* kompromißlos

unconcealed [ʌnkən'si:ld] *adj.* unverhohlen

unconcern [ʌnkən'sɜ:n] *n.*, *no pl.* Gleichgültigkeit, *die*

unconcerned [ʌnkən'sɜ:nd] *adj.* gleichgültig; *(free from anxiety)* unbekümmert; **sb. is ~ about sb./sth.** jmdm. ist jmd./etw. gleichgültig

unconditional [ʌnkən'dɪʃənl] *adj.* bedingungslos ⟨*Kapitulation*⟩; kategorisch ⟨*Ablehnung*⟩; ⟨*Versprechen*⟩ ohne Vorbehalte

unconfirmed [ʌnkən'fɜ:md] *adj.* unbestätigt

uncongenial [ʌnkən'dʒi:nɪəl] *adj.* unsympathisch ⟨*Person*⟩; **I find the work ~:** die Arbeit sagt mir nicht zu *od.* liegt mir nicht

unconnected [ʌnkə'nektɪd] *adj.* **a)** nicht verbunden; **~ with any party** nicht parteigebunden; **b)** *(disjointed, isolated)* zusammenhanglos

unconscious [ʌn'kɒnʃəs] **1.** *adj.* **a)** *(Med.: senseless)* bewußtlos; **b)** *(unaware)* **be ~ of sth.** sich einer Sache *(Gen.)* nicht bewußt sein; **she was ~ of the tragedy** sie wußte nichts von der Tragödie; **c)** *(not intended; Psych.)* unbewußt. **2.** *n.* Unbewußte, *das*

unconsciously [ʌn'kɒnʃəslɪ] *adv.* unbewußt

unconsciousness [ʌn'kɒnʃəsnɪs] *n.*, *no pl.* *(loss of consciousness)* Bewußtlosigkeit, *die*

uncontaminated [ʌnkən'tæmɪneɪtɪd] *adj.* unverschmutzt, nicht verseucht **(with** von)

uncontested [ʌnkən'testɪd] *adj.* unangefochten; **it was an ~ election** bei der Wahl gab es keinen Gegenkandidaten

uncontrollable [ʌnkən'trəʊləbl] *adj.* unkontrollierbar

uncontrolled [ʌnkən'trəʊld] *adj.* unkontrolliert

uncontroversial [ʌnkɒntrə'vɜ:ʃl] *adj.* nicht kontrovers; **be ~:** keinerlei Widerspruch hervorrufen

unconventional [ʌnkən'venʃənl] *adj.*, **unconventionally** [ʌnkən'venʃənəlɪ] *adv.* unkonventionell

unconvinced [ʌnkən'vɪnst] *adj.* nicht überzeugt

unconvincing [ʌnkən'vɪnsɪŋ] *adj.*, **unconvincingly** [ʌnkən'vɪnsɪŋlɪ] *adv.* nicht überzeugend

uncooked [ʌn'kʊkt] *adj.* roh

uncooperative [ʌnkəʊ'ɒpərətɪv] *adj.* unkooperativ; wenig entgegenkommend; *(unhelpful)* wenig hilfsbereit

uncoordinated [ʌnkəʊ'ɔ:dɪneɪtɪd] *adj.* unkoordiniert

uncork [ʌn'kɔ:k] *v. t.* entkorken

uncountable [ʌn'kaʊntəbl] *adj.* *(Ling.)* unzählbar

uncouth [ʌn'ku:θ] *adj.* **a)** *(lacking refinement)* ungeschliffen; ungehobelt ⟨*Person, Benehmen*⟩; grob ⟨*Bemerkung, Sprache*⟩; **b)** *(boorish)* unkultiviert; flegelhaft *(abwertend)*

uncover [ʌn'kʌvə(r)] *v. t.* **a)**

(remove cover from) aufdecken; freilegen ⟨*Wunde, Begrabenes*⟩; **b)** *(disclose)* aufdecken ⟨*Skandal, Verschwörung, Wahrheit*⟩

uncritical [ʌn'krɪtɪkl] *adj.* unkritisch

uncrossed [ʌn'krɒst] *adj.* **an ~ cheque/postal order** ein Barscheck/Postbarscheck

uncrowded [ʌn'kraʊdɪd] *adj.* nicht überlaufen

uncurl [ʌn'kɜ:l] **1.** *v.t.* auseinanderrollen. **2.** *v. refl.* sich strecken. **3.** *v.i.* sich auseinanderollen

uncut [ʌn'kʌt] *adj.* nicht geschnitten ⟨*Gras, Haare usw.*⟩; nicht gemäht ⟨*Rasen*⟩; ungeschliffen ⟨*Edelstein*⟩

undamaged [ʌn'dæmɪdʒd] *adj.* unbeschädigt

undated [ʌn'deɪtɪd] *adj.* undatiert

undaunted [ʌn'dɔ:ntɪd] *adj.* unverzagt; **~ by threats** durch Drohungen nicht eingeschüchtert

undecided [ʌndɪ'saɪdɪd] *adj.* **a)** *(not settled)* nicht entschieden; **b)** *(hesitant)* unentschlossen; **be ~ whether to do sth.** sich *(Dat.)* noch unschlüssig sein, ob man etw. tun soll

undefeated [ʌndɪ'fi:tɪd] *adj.* ungeschlagen ⟨*Mannschaft*⟩; unbesiegt ⟨*Heer*⟩

undefined [ʌndɪ'faɪnd] *adj.* nicht definiert; *(indefinite)* unbestimmt

undelivered [ʌndɪ'lɪvəd] *adj.* nicht zugestellt ⟨*Postsendung*⟩; *(on letter)* **if ~:** wenn unzustellbar

undemanding [ʌndɪ'mɑ:ndɪŋ] *adj.* anspruchslos

undemocratic [ʌndemə'krætɪk] *adj.* undemokratisch

undemonstrative [ʌndɪ'mɒnstrətɪv] *adj.* zurückhaltend

undeniable [ʌndɪ'naɪəbl] *adj.* unbestreitbar

undeniably [ʌndɪ'naɪəblɪ] *adv.* unbestreitbar

under ['ʌndə(r)] **1.** *prep.* **a)** *(underneath, below) (indicating position)* unter (+ *Dat.*); *(indicating motion)* unter (+ *Akk.*); **from ~ the table/bed** unter dem Tisch/Bett hervor; **b)** *(undergoing)* **~ treatment** in Behandlung; **~ repair** in Reparatur; **fields ~ cultivation** bebaute Felder; *see also* **discussion b; influence 1; pain 1 e; c)** *(in conditions of)* bei ⟨*Streß, hohen Temperaturen usw.*⟩; **d)** *(subject to)* unter (+ *Dat.*); **~ the doctor, ~ doctor's orders** in ärztlicher Behandlung; **e)** *(in accordance with)* **~ the terms of the contract/agreement** nach den Bestimmungen des Vertrags/Abkommens; **f)** *(with the use of)* unter (+ *Dat.*); **~ an assumed name** unter falschem Namen; **g)** *(less than)* unter (+ *Dat.*); **for ~ five pounds** für weniger als fünf Pfund; *see also* **age 1 a. 2.** *adv.* **a)** *(in or to a lower or subordinate position)* darunter; **stay ~** *(~ water)* unter Wasser bleiben; *see also* **go under; b)** *(in/into a state of unconsciousness)* **be ~/put sb. ~:** in Narkose liegen/jmdn. in Narkose versetzen

under: ~arm *adj.* **a)** *(Cricket, etc.)* ⟨*Aufschlag, Wurf:*⟩ von unten; **b)** *(in armpit)* Achsel⟨*haare, -schweiß*⟩; **~carriage** *n.* Fahrwerk, *das;* **~'charge** *v.t.* **~charge sb. [by several pounds]** jmdm. [einige Pfund] zuwenig berechnen; **~clothes** *n. pl.*, **~clothing** *see* **~wear; ~coat** *n.* **a)** *(layer of paint)* Grundierung, *die;* **b)** *(paint)* Grundierfarbe, *die;* **~'cooked** *adj.* zu kurz gekocht/gebraten; noch nicht gar; **~cover** *adj. (disguised)* getarnt; *(secret)* verdeckt; *(engaged in international spying)* geheim[dienstlich]; **~cover agent** Untergrund-/Geheimagent, *der;* **~current** *n.* Unterströmung, *die; (fig.: ~lying feeling)* Unterton, *der;* **~'cut** *v.t.*, **~cut** unterbieten; **~de'veloped** *adj.* unterentwickelt; **~de'velopment** *n.*, *no pl.* Unterentwicklung, *die;* **~dog** *n.* **a)** *(in fight, match)* Unterlegene, *der/die;* **b)** *(fig.: disadvantaged person)* Benachteiligte, *der/die;* **~'done** *adj.* halbgar; **I don't like my steak ~done** ich habe mein Steak gern gut durchgebraten; **~em'ployed** *adj.* unterbeschäftigt; **~em'ployment** *n.* Unterbeschäftigung, *die;* **~estimate 1.** [ʌndər'estɪmeɪt] *v.t.* unterschätzen; **2.** [ʌndər'estɪmət] *n.* Unterschätzung, *die;* **~ex'pose** *v.t. (Photog.)* unterbelichten; **~ex'posure** *n. (Photog.)* Unterbelichtung, *die;* **~'fed** *adj.* unterernährt; **~-'fives** *n.pl.* Kinder unter fünf Jahren; **~'floor heating** *n.* [Fuß]bodenheizung, *die;* **~'foot** *adv.* am Boden; **it's rough/muddy ~foot** der Boden ist uneben/matschig; **be trampled ~foot** mit Füßen zertrampelt werden; *(fig.)* wie der letzte Dreck behandelt werden *(salopp);* **~'go** *v.t., forms as* **go 1** durchmachen ⟨*schlimme Zeiten*⟩; ertragen ⟨*Demütigung*⟩; **~go treatment/an operation** sich einer Behandlung/Operation unterziehen; **~go a change** sich verändern; **~'graduate** *n.* **~graduate [student]** Student/Studentin vor der ersten Prüfung; **~ground 1.** [--'-] *adv.* **a)** *(beneath surface of ground)* unter der Erde; *(Mining)* unter Tage; **an explosion ~ground** eine unterirdische Explosion; **b)** *(fig.) (in hiding)* im Untergrund; *(into hiding)* in den Untergrund; **go ~ground** untertauchen; in den Untergrund gehen; **2.** ['---] *adj.* **a)** unterirdisch ⟨*Höhle, See*⟩; **~ground railway** Untergrundbahn, *die;* **~ground car-park** Tiefgarage, *die;* **b)** *(fig.: secret)* **~ground organization/movement/press** Untergrundorganisation/-bewegung/-presse; **3.** *n.* **a)** *(railway)* U-Bahn, *die;* **b)** *(clandestine movement)* Untergrund, *der;* Untergrundbewegung, *die;* **~growth** *n.* Unterholz, *das;* **~hand, ~handed** *adjs.* **a)** *(secret)* heimlich; **b)** *(crafty)* hinterhältig; **~in'sured** *adj.* unterversichert; **[1]~'lay** *see* **~lie; [2]~lay** *n.* Unterlage, *die;* **~'lie** *v.t., forms as* **[2]lie 2: a)** *(lie ~)* **~lie sth.** unter etw. *(Dat.)* liegen; **b)** *(fig.: be [at] the basis of)* **~lie sth.** einer Sache *(Dat.)* zugrundeliegen; **~lying cause of sth.** eigentliche Ursache für etw.; **~line 1.** [--'-] *v.t. (lit. or fig.)* unterstreichen; **2.** ['---] *n.* Unterstreichung, *die*

underling ['ʌndəlɪŋ] *n. (derog.)* Untergebene, *der/die*

under: ~'lying *see* **~lie; ~'manned** *adj.* [personell] unterbesetzt; **~manning** *n.* [personelle] Unterbesetzung, *die;* **~mentioned** *adj. (Brit.)* untengenannt; untenerwähnt; **~'mine** *v.t.* **a)** unterhöhlen; ⟨*Wasser:*⟩ unterspülen; **b)** *(fig.: weaken)* untergraben; erschüttern ⟨*Vertrauen*⟩; unterminieren ⟨*Autorität*⟩; schwächen ⟨*Gesundheit*⟩

underneath [ʌndə'ni:θ] **1.** *prep. (indicating position)* unter (+ *Dat.*); *(indicating motion)* unter (+ *Akk.*); **from ~ the bed** unter dem Bett hervor. **2.** *adv.* darunter. **3.** *n.* Unterseite, *die*

under: ~'nourished *adj.* unterernährt; **~'paid** *adj.* unterbezahlt; **~pants** *n.pl.* Unterhose, *die;* Unterhosen *Pl.;* **~pass** *n.* Unterführung, *die;* **~'pay** *v.t., forms as* **pay 2** unterbezahlen; **~'payment** *n.* Unterbezahlung, *die;* **~'pin** *v.t.* [ab]stützen; *(fig.)* untermauern; **~play** *v.t. (play down)* herunterspielen; **~'privileged** *adj.* unterprivilegiert; **~'rate** *v.t.* unterschätzen; **be ~rated** [allgemein] unterschätzt werden; **~'score** *see* **~line 1;**

~score *see* **~line 2; ~secretary** *n.* **a)** *(esp. Amer.: assistant to secretary)* Unterstaatssekretär, *der;* **b)** *(Brit.)* **Parliamentary U~secretary** [parlamentarischer] Staatssekretär, *der;* **~'sell** *v. t., forms as* **sell 1: a)** *(set at lower price than)* [im Preis] unterbieten; **b)** *(present inadequately)* nicht genug anpreisen; **~-side** *n.* Unterseite, *die;* **~signed** *adj. (esp. Law)* **the ~signed** der/die Unterzeichnete/*(pl.)* die Unterzeichneten *(Papierdt.);* **~sized** *adj.* unter Normalgröße *nachgestellt;* [ziemlich] klein geraten ‹*Mensch, Tier*›; **~skirt** *n.* Unterrock, *der;* **~sold** *see* **~sell; ~'staffed** *adj.* unterbesetzt; **be ~staffed** an Personalmangel leiden

understand [ʌndə'stænd] **1.** *v. t.,* **understood** [ʌndə'stʊd] **a)** verstehen; **~ sth. by sth.** etw. unter etw. *(Dat.)* verstehen; **~ mathematics** mathematisches Verständnis haben; **is that understood?** ist das klar?; **make oneself understood** sich verständlich machen; **b)** *(have heard)* gehört haben; **I ~ him to be a distant relation** ich glaube, er ist ein entfernter Verwandter; **c)** *(take as implied)* **~ sth. from sb.'s words** etw. aus jmds. Worten entnehmen; **it was understood that ...:** es wurde allgemein angenommen, daß ...; **do I ~ that ...?** gehe ich recht in der Annahme, daß ...? *See also* **give 1 e; make 1 f. 2.** *v. i.,* **understood a)** *(have understanding)* verstehen; **~ about sth.** etwas von etw. verstehen; **I quite ~:** ich verstehe schon; **b)** *(gather, hear)* **if I ~ correctly** wenn ich mich nicht irre; **he is, I ~, no longer here** er ist, wie ich höre, nicht mehr hier

understandable [ʌndə'stændəbl] *adj.* verständlich

understandably [ʌndə'stændəblɪ] *adv.* verständlicherweise

understanding [ʌndə'stændɪŋ] **1.** *adj. (able to sympathize)* verständnisvoll; **you could be a bit more ~:** du könntest etwas mehr Verständnis zeigen. **2.** *n.* **a)** *(agreement)* Verständigung, *die;* **reach an ~ with sb.** sich mit jmdm. verständigen; **the good ~ between them** das gute Einverständnis zwischen ihnen; **have a secret ~ with sb.** eine geheime Vereinbarung mit jmdm. haben; **on the ~ that ...:** unter der Voraussetzung, daß ...; **on the clear ~ that ...** *(condition)* unter der ausdrücklichen Bedingung, daß ...; **b)** *(intelligence)* Verstand, *der;* **c)** *(insight, comprehension)* Verständnis, *das* **(of, for** für); **beyond ~:** unbegreiflich; **my ~ of the matter is that she has won** so wie ich es verstehe, hat sie gewonnen

under: ~'state *v. t.* **a)** herunterspielen; **~state the case** untertreiben; **b)** *(represent inadequately)* zu gering veranschlagen; **~'statement** *n. (avoidance of emphasis)* Untertreibung, *die;* Understatement, *das;* **~'stocked** *adj.* unterversorgt; **~'stood** *see* **understand; ~study** *n.* Ersatzspieler, *der*/-spielerin, *die;* zweite Besetzung; **~'take** *v. t., forms as* **take 1: a)** *(set about)* unternehmen; **~take a task** eine Aufgabe übernehmen; **~take to do sth.** sich verpflichten, etw. zu tun; **b)** *(guarantee)* **~take sth./that ...:** sich für etw. verbürgen/sich dafür verbürgen, daß ...; **~taker** *n.* Leichenbestatter, *der*/-bestatterin, *die;* **[firm of] ~takers** Bestattungsunternehmen, *das;* **~taking** *n.* **a)** *no pl. (taking on) (of task)* Übernehmen, *das; (of journey etc.)* Unternehmen, *das;* **b)** *(task)* Aufgabe, *die;* **a dangerous ~taking** ein gefährliches Unterfangen; **c)** *(business)* Unternehmen, *das;* Betrieb, *der;* **d)** *(pledge)* Versprechen, *das;* **give an ~taking that .../to do sth.** zusichern, daß .../sich verpflichten, etw. zu tun; **~tone** *n.* **a)** *(low voice)* **in ~tones** *or* **an ~tone** in gedämpftem Ton; **b)** *(~current)* **~tone of criticism** kritischer Unterton; **c)** *(subdued colour)* Tönung, *die;* **~took** *see* **~take; ~tow** *n.* Unterströmung, *die;* **~-'used** *adj.* nicht voll genutzt; **~'value** *v. t.* unterbewerten; **~vest** *n.* Unterhemd, *das;* **~water 1.** ['----] *attrib. adj.* Unterwasser-; **2.** [--'--] *adv.* unter Wasser; **~wear** *n., no pl., no indef. art.* Unterwäsche, *die;* **~'weight** *adj.* untergewichtig; **~went** *see* **~go; ~world** *n. (lit. or fig.)* Unterwelt, *die;* **~'write** *v. t., forms as* **write** *(accept liability for)* [als Versicherer] unterzeichnen; **~write a risk** ein Risiko versichern; **~writer** *n. (of insurance policy)* Versicherer, *der; (of stock issue)* Garant, *der*/Garantin, *die;* **~written** *see* **~write**

undeserved [ʌndɪ'zɜ:vd] *adj.* unverdient

undeserving [ʌndɪ'zɜ:vɪŋ] *adj.* unwürdig **(of** *Gen.*)

undesirability [ʌndɪzaɪərə'bɪlɪtɪ] *n., no pl.* Unerwünschtheit, *die*

undesirable [ʌndɪ'zaɪərəbl] *adj.* unerwünscht; **it is ~ that ...:** es ist nicht wünschenswert, daß ...

undesirably [ʌndɪ'zaɪərəblɪ] *adv.* unerwünscht

undetectable [ʌndɪ'tektəbl] *adj.* nicht nachweisbar

undetected [ʌndɪ'tektɪd] *adj.* unentdeckt; **go** *or* **pass ~:** unentdeckt bleiben

undeterred [ʌndɪ'tɜ:d] *adj.* nicht entmutigt **(by** durch); **remain ~:** sich nicht abschrecken lassen; **continue ~:** unbeirrt weitermachen

undeveloped [ʌndɪ'veləpt] *adj.* **a)** *(immature)* nicht voll ausgebildet; **b)** *(Photog.)* nicht entwickelt; **c)** *(not built on)* nicht bebaut

undid *see* **undo**

undies ['ʌndɪz] *n. pl. (coll.)* Unterwäsche, *die*

undignified [ʌn'dɪgnɪfaɪd] *adj.* würdelos; **consider it ~ to do sth.** es für unter seiner Würde halten, etw. zu tun

undiplomatic [ʌndɪplə'mætɪk] *adj.* undiplomatisch

undipped [ʌn'dɪpt] *adj.* nicht abgeblendet ‹*Scheinwerfer*›

undisciplined [ʌn'dɪsɪplɪnd] *adj.* undiszipliniert

undisclosed [ʌndɪs'kləʊzd] *adj.* geheim; **an ~ sum** ein nicht genannter Betrag

undiscoverable [ʌndɪ'skʌvərəbl] *adj.* nicht feststellbar

undiscovered [ʌndɪ'skʌvəd] *adj.* unentdeckt

undiscriminating [ʌndɪ'skrɪmɪneɪtɪŋ] *adj.* unkritisch; *(undemanding)* anspruchslos

undisguised [ʌndɪs'gaɪzd] *adj.* unverhohlen

undismayed [ʌndɪs'meɪd] *see* **undeterred**

undisputed [ʌndɪ'spju:tɪd] *adj.* unbestritten ‹*Fertigkeit, Kompetenz*›; unangefochten ‹*Führer, Autorität*›

undistinguished [ʌndɪ'stɪŋgwɪʃt] *adj.* mittelmäßig; *(ordinary)* gewöhnlich

undisturbed [ʌndɪ'stɜ:bd] *adj.* **a)** *(untouched)* unberührt; **b)** *(not interrupted)* ungestört; **c)** *(not worried)* ungerührt

undivided [ʌndɪ'vaɪdɪd] *adj.* ungeteilt ‹*Sympathie, Aufmerksamkeit*›; uneingeschränkt ‹*Loyalität*›

undo [ʌn'du:] **1.** *v. t.,* **undoes** [ʌn'dʌz], **undoing** [ʌn'du:ɪŋ], **undid** [ʌn'dɪd], **undone** [ʌn'dʌn] **a)** *(unfasten)* aufmachen; **b)** *(cancel)* ungeschehen machen. **2.** *v. i.,*

forms as **1: ~ at the back** ‹*Kleid usw.*:› hinten aufgemacht werden
undoing [ʌn'du:ɪŋ] *n., no pl., no indef. art.* **be sb.'s ~:** jmds. Verderben sein
undone [ʌn'dʌn] *adj.* **a)** *(not accomplished)* unerledigt; **leave the work** *or* **job ~:** die Arbeit liegen lassen; **b)** *(not fastened)* offen
undoubted [ʌn'daʊtɪd] *adj.* unzweifelhaft
undoubtedly [ʌn'daʊtɪdlɪ] *adv.* zweifellos
undreamed-of [ʌn'dri:mdɒv], **undreamt-of** [ʌn'dremtɒv] *adjs.* *(unheard-of)* unerhört; *(unimaginable)* unvorstellbar; ungeahnt ‹*Reichtum*›; **such a thing was ~:** an so etwas hätte man nicht im Traum gedacht
undress [ʌn'dres] **1.** *v.t.* ausziehen; entkleiden *(geh.)*; **get ~ed** sich ausziehen; **can he ~ himself?** kann er sich selbst ausziehen? **2.** *v.i.* sich ausziehen. **3.** *n., no pl., no art.* **in a state of ~:** halbbekleidet
undressed [ʌn'drest] *adj.* **a)** *(not clothed)* unbekleidet; *(no longer clothed)* ausgezogen; *(not yet clothed)* nicht angezogen; **b)** *(unfinished)* unbearbeitet ‹*Stein, Holz*›; ungegerbt ‹*Leder, Haut*›
undrinkable [ʌn'drɪŋkəbl] *adj.* nicht trinkbar; ungenießbar
undue [ʌn'dju:] *attrib. adj.* übertrieben; übermäßig; unangemessen hoch ‹*Gewinn*›; unberechtigt ‹*Optimismus*›
undulating [ʌndjʊ'leɪtɪŋ] *adj.* Wellen‹*linie, -bewegung*›; **~ country/hills** sanfte Hügellandschaft; **~ road** auf- und abführende Straße
undulation ['ʌndjʊleɪʃn] *n.* **a)** *(wavy motion)* Wellenbewegung, *die*; **b)** *(wavy line)* Wellenlinie, *die*
unduly [ʌn'dju:lɪ] *adv.* übermäßig; übertrieben ‹*ängstlich*›; unangemessen ‹*hoch*›; **not ~ worried** nicht besonders beunruhigt
unearned ['ʌnɜ:nd] *adj.* unverdient; **~ income** Kapitalertrag, *der*
unearth [ʌn'ɜ:θ] *v.t.* **a)** *(dig up)* ausgraben; **b)** *(fig.: discover)* aufdecken; zutage fördern
unearthly [ʌn'ɜ:θlɪ] *adj.* **a)** *(mysterious)* unheimlich; **b)** *(coll.: terrible)* **~ din** Höllenlärm, *der* *(ugs.)*; **at an ~ hour** in aller Herrgottsfrühe
unease [ʌn'i:z] *see* **uneasiness**
uneasily [ʌn'i:zɪlɪ] *adv.* **a)** *(anxiously)* mit Unbehagen; **b)** *(with embarrassment)* **be ~ aware of sth.** sich *(Dat.)* einer Sache *(Gen.)* peinlich bewußt sein; **c)** *(restlessly)* unruhig ‹*schlafen, sitzen*›
uneasiness [ʌn'i:zɪnɪs] *n., no pl.* **a)** *(anxiety)* [ängstliches] Unbehagen; **b)** *(restlessness)* Unruhe, *die*
uneasy [ʌn'i:zɪ] *adj.* **a)** *(anxious)* besorgt; **be ~ about sth.** sich wegen etw. Sorgen machen; **he felt ~:** ihm war unbehaglich zumute; **b)** *(restless)* unruhig ‹*Schlaf*›; **c)** *(disturbing)* quälend ‹*Verdacht*›; **~ conscience** schlechtes Gewissen
uneatable [ʌn'i:təbl] *adj.* ungenießbar
uneaten [ʌn'i:tn] *adj.* ungegessen
uneconomic [ʌni:kə'nɒmɪk, ʌnekə'nɒmɪk] *adj.* unrentabel
uneconomical [ʌni:kə'nɒmɪkl, ʌnekə'nɒmɪkl] *adj.* verschwenderisch ‹*Person*›; **~ [to run]** unwirtschaftlich
uneducated [ʌn'edjʊkeɪtɪd] *adj.* ungebildet
unemotional [ʌnɪ'məʊʃənl] *adj.* emotionslos; nüchtern
unemployable [ʌnɪm'plɔɪəbl] *adj.* als Arbeitskraft ungeeignet
unemployed [ʌnɪm'plɔɪd] **1.** *adj.* **a)** *(out of work)* arbeitslos; **b)** *(with nothing to do)* beschäftigungslos. **2.** *n. pl.* **the ~:** die Arbeitslosen
unemployment [ʌnɪm'plɔɪmənt] *n., no pl., no indef. art.* Arbeitslosigkeit, *die*; *(number unemployed)* Arbeitslosenzahl, *die*
unemployment: ~ benefit *n.* Arbeitslosengeld, *das*; **~ figures** *n. pl.* Arbeitslosenzahl, *die*
unending [ʌn'endɪŋ] *adj.* endlos; ewig ‹*Fortschritt*›; **her ordeal seemed ~:** ihre Qualen schienen nie enden zu wollen
unenterprising [ʌn'entəpraɪzɪŋ] *adj.* wenig unternehmungslustig; **an ~ person** eine Person ohne Unternehmungsgeist
unenthusiastic [ʌnɪnθju:zɪ'æstɪk, ʌnɪnθu:zɪ'æstɪk] *adj.* wenig begeistert **(about** von); distanziert ‹*Buchkritik*›
unenviable [ʌn'envɪəbl] *adj.* wenig beneidenswert
unequal [ʌn'i:kwl] *adj.* **a)** *(not equal)* unterschiedlich; ungleich ‹*Kampf*›; **b)** *(inadequate)* **be ~** *or* **show oneself ~ to sth.** einer Sache *(Dat.)* nicht gewachsen sein; **c)** *(of varying quality)* ungleichmäßig
unequalled *(Amer.:* **unequaled**) [ʌn'i:kwld] *adj.* unerreicht; unübertroffen
unequivocal [ʌnɪ'kwɪvəkl] *adj.* eindeutig
unerring [ʌn'ɜ:rɪŋ] *adj.* untrüglich ‹*Instinkt, Geschmack*›; unbedingt ‹*Treffsicherheit*›; mathematisch ‹*Genauigkeit*›; unfehlbar ‹*Instinkt*›
UNESCO [ju:'neskəʊ] *abbr.* **United Nations Educational, Scientific and Cultural Organization** UNESCO, *die*
unethical [ʌn'eθɪkl] *adj.* unmoralisch
uneven [ʌn'i:vn] *adj.* **a)** *(not smooth)* uneben; **b)** *(not uniform)* ungleichmäßig; unregelmäßig ‹*Pulsschlag*›; unausgeglichen ‹*Temperament*›
unevenly [ʌn'i:vnlɪ] *adv.* ungleichmäßig
uneventful [ʌnɪ'ventfl] *adj.* **a)** *(quiet)* ereignislos; ruhig ‹*Leben*›; **b)** *(normal)* ‹*Fahrt, Landung*› ohne Zwischenfälle
unexceptional [ʌnɪk'sepʃənl] *adj.* alltäglich; *(average)* durchschnittlich
unexciting [ʌnɪk'saɪtɪŋ] *adj.* wenig aufregend; *(boring)* langweilig
unexpected [ʌnɪk'spektɪd] *adj.* unerwartet; **this news was entirely ~:** diese Nachricht kam völlig unerwartet
unexpectedly [ʌnɪk'spektɪdlɪ] *adv.* unerwartet
unexplained [ʌnɪk'spleɪnd] *adj.* ungeklärt; unentschuldigt ‹*Abwesenheit*›
unexploded [ʌnɪk'spləʊdɪd] *adj.* nicht explodiert *od.* detoniert
unexplored [ʌnɪk'splɔ:d] *adj.* unerforscht
unexposed [ʌnɪk'spəʊzd] *adj.* **a)** *(not brought to light)* unaufgeklärt; nicht entlarvt ‹*Verbrecher*›; **b)** *(Photog.)* unbelichtet
unexpressive [ʌnɪk'spresɪv] *adj.* ausdruckslos
unfailing [ʌn'feɪlɪŋ] *adj.* unerschöpflich
unfailingly [ʌn'feɪlɪŋlɪ] *adv.* stets
unfair [ʌn'feə(r)] *adj.* unfair; ungerecht, unfair ‹*Kritik, Urteil*›; unlauter ‹*Wettbewerb*›; ungerecht ‹*Strafe*›; **an ~ share** ein ungerechtfertigt hoher Anteil; **be ~ to sb.** jmdm. gegenüber ungerecht sein
unfairly [ʌn'feəlɪ] *adv.* **a)** *(unjustly)* ungerecht; unfair ‹*spielen*›; **b)** *(unreasonably)* zu Unrecht
unfairness [ʌn'feənɪs] *n., no pl.* Ungerechtigkeit, *die*
unfaithful [ʌn'feɪθfl] *adj.* untreu; **~ to sb./sth.** jmdm./einer Sache untreu
unfamiliar [ʌnfə'mɪljə(r)] *adj.* **a)**

(strange) unbekannt; fremd ⟨*Stadt*⟩; ungewohnt ⟨*Arbeit, Tätigkeit*⟩; **b)** *(not well acquainted)* nicht vertraut; **be ~ with sth.** sich mit etw. nicht auskennen

unfamiliarity [ʌnfəmɪlɪ'ærɪtɪ] *n., no pl.* **a)** *(strangeness)* Fremdheit, *die; (of activity)* Ungewohntheit, *die;* **b) ~ with sth.** *(poor knowledge of)* Unvertrautheit mit etw.

unfashionable [ʌn'fæʃənəbl] *adj.* unmodern ⟨*Kleidung*⟩; nicht eben schick ⟨*Wohngegend*⟩; **become ~:** aus der Mode kommen; **a view now ~:** eine jetzt überholte Ansicht

unfasten [ʌn'fɑ:sn] *v.t.* **a)** öffnen; **b)** *(detach)* lösen

unfathomable [ʌn'fæðəməbl] *adj.* **a)** *(incomprehensible)* unergründlich; **b)** *(immeasurable)* unermeßlich

unfavorable, unfavorably *(Amer.) see* **unfavourable, unfavourably**

unfavourable [ʌn'feɪvərəbl] *adj.* **a)** *(negative)* ungünstig; unfreundlich ⟨*Kommentar, Reaktion*⟩; negativ ⟨*Kritik, Antwort*⟩; **b)** *(tending to make difficult)* ungünstig **(to, for** für)

unfavourably [ʌn'feɪvərəblɪ] *adv.* ungünstig; **be ~ disposed towards sb./sth.** jmdm./etw. gegenüber ablehnend eingestellt sein

unfeeling [ʌn'fi:lɪŋ] *adj. (unsympathetic)* gefühllos

unfinished [ʌn'fɪnɪʃt] *adj.* **a)** *(not completed)* unvollendet ⟨*Gedicht, Werk*⟩; unerledigt ⟨*Arbeit, Geschäft*⟩; **b)** *(in rough state)* unbearbeitet

unfit [ʌn'fɪt] **1.** *adj.* **a)** *(unsuitable)* ungeeignet; **b)** *(not physically fit)* nicht fit; **~ for military service** [wehrdienst]untauglich. **2.** *v.t.*, **-tt-** untauglich machen; *see also* **unfitted**

unfitness [ʌn'fɪtnɪs] *n., no pl.* **a)** *(unsuitability)* fehlende Eignung; **b)** *(poor physical condition)* **[state of] ~:** schlechte körperliche Verfassung

unfitted [ʌn'fɪtɪd] *adj. (unsuited)* ungeeignet

unflagging [ʌn'flægɪŋ] *adj.* unermüdlich

unflappable [ʌn'flæpəbl] *adj. (coll.)* unerschütterlich; **an ~ person** jemand, der sich durch nichts aus der Ruhe bringen läßt

unflattering [ʌn'flætərɪŋ] *adj.* wenig schmeichelhaft

unflinching [ʌn'flɪntʃɪŋ] *adj.* unerschrocken; unbeirrbar ⟨*Entschlossenheit*⟩

unfold [ʌn'fəʊld] **1.** *v.t.* entfalten; ausbreiten ⟨*Zeitung, Landkarte*⟩; **~ one's arms** die Arme ausstrecken. **2.** *v.i.* **a)** *(open out)* ⟨*Knospe:*⟩ sich öffnen; ⟨*Flügel:*⟩ sich entfalten; **b)** *(develop)* sich entwickeln; ⟨*Geheimnis:*⟩ sich aufklären; **as the story ~ed** im weiteren Verlauf der Geschichte

unforeseeable [ʌnfɔ:'si:əbl] *adj.* unvorhersehbar; **be ~:** nicht vorauszusehen sein

unforeseen [ʌnfɔ:'si:n] *adj.* unvorhergesehen

unforgettable [ʌnfə'getəbl] *adj.* unvergeßlich

unforgivable [ʌnfə'gɪvəbl] *adj.* unverzeihlich

unforgiving [ʌnfə'gɪvɪŋ] *adj.* nachtragend

unformed [ʌn'fɔ:md] *adj.* unausgereift

unfortunate [ʌn'fɔ:tʃʊnət, ʌn'fɔ:tʃənət] *adj.* **a)** *(unlucky)* unglücklich; *(unfavourable)* ungünstig ⟨*Tag, Zeit*⟩; **the poor ~ woman** die arme bedauernswerte Frau; **be ~ [enough] to do sth.** das Pech haben, etw. zu tun; **b)** *(regrettable)* bedauerlich

unfortunately [ʌn'fɔ:tʃʊnətlɪ, ʌn'fɔ:tʃənətlɪ] *adv.* leider

unfounded [ʌn'faʊndɪd] *adj. (fig.)* unbegründet; **the rumours are totally ~:** die Gerüchte entbehren jeder Grundlage

unfreeze [ʌn'fri:z] *v.t. & i.*, **unfroze** [ʌn'frəʊz], **unfrozen** [ʌn'frəʊzn] auftauen

unfriendly [ʌn'frendlɪ] *adj.* unfreundlich; feindlich ⟨*Staat*⟩

unfulfilled [ʌnfʊl'fɪld] *adj.* **a)** unerfüllt ⟨*Person*⟩; **b)** *(not carried out)* unerledigt

unfurl [ʌn'fɜ:l] *v.t.* aufrollen; losmachen ⟨*Segel*⟩

unfurnished [ʌn'fɜ:nɪʃt] *adj.* unmöbliert

ungainly [ʌn'geɪnlɪ] *adj.* unbeholfen; ungelenk

ungentlemanly [ʌn'dʒentlmənlɪ] *adj.* unfein; *(impolite)* unhöflich; **it is ~:** es gehört sich nicht für einen Gentleman

ungodly [ʌn'gɒdlɪ] *adj.* **a)** gottlos; **b)** *(coll.: outrageous)* unchristlich *(ugs.)*

ungracious [ʌn'greɪʃəs] *adj.* unhöflich; *(tactless)* taktlos

ungrammatical [ʌngrə'mætɪkl] *adj.* ungrammatisch

ungrateful [ʌn'greɪtfl] *adj.* undankbar

ungrudging [ʌn'grʌdʒɪŋ] *adj.* bereitwillig; *(generous)* großzügig; herzlich ⟨*Gastfreundschaft*⟩; neidlos ⟨*Bewunderung*⟩

unguarded [ʌn'gɑ:dɪd] *adj.* **a)** *(not guarded)* unbewacht; **b)** *(incautious)* unvorsichtig

unhappily [ʌn'hæpɪlɪ] *adv.* **a)** *(unfortunately)* unglücklicherweise; leider; **b)** *(without happiness)* unglücklich

unhappiness [ʌn'hæpɪnɪs] *n., no pl.* Bekümmertheit, *die;* **he has been the cause of much ~ to her** er hat ihr viel Kummer gemacht

unhappy [ʌn'hæpɪ] *adj.* **a)** unglücklich; *(not content)* unzufrieden **(about** mit); **be** *or* **feel ~ about doing sth.** Bedenken haben, etw. zu tun; **b)** *(unfortunate)* unglückselig ⟨*Zeit, Zufall*⟩; unglücklich ⟨*Zusammenstellung, Wahl*⟩

unharmed [ʌn'hɑ:md] *adj.* unbeschädigt; *(uninjured)* unverletzt

unhealthy [ʌn'helθɪ] *adj.* **a)** *(not in good health, harmful to health)* ungesund; **b)** *(unwholesome)* ungesund, krankhaft ⟨*Gier*⟩; schädlich ⟨*Einfluß*⟩; schlecht ⟨*Angewohnheit*⟩

unheard [ʌn'hɜ:d] *adj.* **~-of** *(unknown)* [gänzlich] unbekannt; *(unprecedented)* beispiellos

unheeded [ʌn'hi:dɪd] *adj.* unbeachtet; **go ~:** nicht beachtet werden; ⟨*Gebet, Wunsch:*⟩ nicht erhört werden

unhelpful [ʌn'helpfl] *adj.* wenig hilfsbereit ⟨*Person*⟩; ⟨*Bemerkung, Kritik*⟩ die einem nicht weiterhilft

unholy [ʌn'həʊlɪ] *adj.* **a)** *(wicked)* unheilig ⟨*Allianz*⟩; **b)** *(coll.: dreadful)* fürchterlich ⟨*Krawall, Durcheinander*⟩

unhook [ʌn'hʊk] *v.t.* vom Haken nehmen; aufhaken ⟨*Kleid*⟩; loshaken ⟨*Tor*⟩

unhoped-for [ʌn'həʊptfɔ:(r)] *adj.* unverhofft

unhurried [ʌn'hʌrɪd] *adj.*, **unhurriedly** [ʌn'hʌrɪdlɪ] *adv.* gemächlich

unhurt [ʌn'hɜ:t] *adj.* unverletzt

unhygienic [ʌnhaɪ'dʒi:nɪk] *adj.* unhygienisch

unicorn ['ju:nɪkɔ:n] *n. (Mythol.)* Einhorn, *das*

unidentified [ʌnaɪ'dentɪfaɪd] *adj.* nicht identifiziert; **~ flying object** unbekanntes Flugobjekt

unification [ju:nɪfɪ'keɪʃn] *n.* Einigung, *die; (of system)* Vereinheitlichung, *die*

uniform ['ju:nɪfɔ:m] **1.** *adj. (the same for all)* einheitlich; *(unvarying)* gleichbleibend ⟨*Strömung, Temperatur, Qualität*⟩; gleichmäßig ⟨*Tempo*⟩; **be ~ in shape/size/appearance** die gleiche Form/Größe/das gleiche Aussehen haben. **2.** *n.* Uniform, *die;* **in/out of ~:** in/ohne Uniform; **be in/out of ~:** Uniform/keine Uniform tragen

uniformed ['ju:nɪfɔ:md] *adj.* uniformiert

uniformity [ju:nɪ'fɔ:mɪtɪ] *n.* Einheitlichkeit, *die*

uniformly ['ju:nɪfɔ:mlɪ] *adv.* **a)** *(without variation)* einheitlich; **b)** *(equally)* gleichmäßig

unify ['ju:nɪfaɪ] *v.t.* einigen ‹*Volk, Land*›; vereinheitlichen ‹*System*›

unilateral [ju:nɪ'lætərl] *adj.* einseitig

unimaginable [ʌnɪ'mædʒɪnəbl] *adj.* unvorstellbar

unimaginative [ʌnɪ'mædʒɪnətɪv] *adj.*, **unimaginatively** [ʌnɪ'mædʒɪnətɪvlɪ] *adv.* phantasielos

unimpaired [ʌnɪm'peəd] *adj.* unbeeinträchtigt

unimportance [ʌnɪm'pɔ:təns] *n., no pl.* Unwichtigkeit, *die;* Bedeutungslosigkeit, *die*

unimportant [ʌnɪm'pɔ:tənt] *adj.* unwichtig; bedeutungslos

unimpressed [ʌnɪm'prest] *adj.* nicht beeindruckt

unimpressive [ʌnɪm'presɪv] *adj.* nicht eindrucksvoll; unscheinbar ‹*Gebäude*›

uninformed [ʌnɪn'fɔ:md] *adj.* **a)** *(not informed)* uninformiert; **b)** *(based on ignorance)* auf Unkenntnis beruhend ‹*Urteil, Ansicht*›; **~ guess** reine Vermutung

uninhabitable [ʌnɪn'hæbɪtəbl] *adj.* unbewohnbar

uninhabited [ʌnɪn'hæbɪtɪd] *adj.* unbewohnt

uninhibited [ʌnɪn'hɪbɪtɪd] *adj.* ungehemmt; ohne Hemmungen *nachgestellt*

uninitiated [ʌnɪ'nɪʃɪeɪtɪd] *adj.* uneingeweiht; **~ in the mysteries** nicht in die Geheimnisse eingeweiht; **the ~:** Außenstehende

uninjured [ʌn'ɪndʒəd] *adj.* unverletzt

uninspired [ʌnɪn'spaɪəd] *adj.* einfallslos; **I am/feel ~:** mir fehlt die Inspiration

uninspiring [ʌnɪn'spaɪərɪŋ] *adj.* langweilig

unintelligent [ʌnɪn'telɪdʒənt] *adj.* nicht intelligent

unintelligible [ʌnɪn'telɪdʒɪbl] *adj.* unverständlich

unintended [ʌnɪn'tendɪd] *adj.* unbeabsichtigt

unintentional [ʌnɪn'tenʃənl] *adj.*, **unintentionally** [ʌnɪn'tenʃənəlɪ] *adv.* unabsichtlich

uninterested [ʌn'ɪntrestɪd, ʌn'ɪntrɪstɪd] *adj.* desinteressiert (**in** an + *Dat.*)

uninteresting [ʌn'ɪntrestɪŋ, ʌn'ɪntrɪstɪŋ] *adj.* uninteressant

uninterrupted [ʌnɪntə'rʌptɪd] *adj.* **a)** *(continuous)* ununterbrochen; nicht unterbrochen; **b)** *(not disturbed)* ungestört

uninvited [ʌnɪn'vaɪtɪd] *adj.* ungeladen

uninviting [ʌnɪn'vaɪtɪŋ] *adj.* wenig verlockend; wenig einladend ‹*Ort, Wetter*›

union ['ju:nɪən, 'ju:njən] *n.* **a)** *(trade ~)* Gewerkschaft, *die;* **b)** *(political unit)* Union, *die*

unionist ['ju:nɪənɪst, 'ju:njənɪst] *n.* **a)** *(member of trade union)* Gewerkschafter, *der*/Gewerkschafterin, *die;* *(advocate of trade unions)* Gewerkschaftsanhänger, *der*/-anhängerin, *die;* **b)** **U~** *(Polit.)* Unionist, *der*/Unionistin, *die*

Union 'Jack *n.* *(Brit.)* Union Jack, *der*

unique [ju:'ni:k] *adj.* *(unparalleled)* einzigartig; *(not repeated)* einmalig ‹*Gelegenheit, Angebot*›; **this problem is ~ to our society** dieses Problem gibt es nur in unserer Gesellschaft

uniquely [ju:'ni:klɪ] *adv.* **a)** *(exclusively)* einzig und allein; **b)** *(to a unique degree)* einzigartig; einmalig ‹*talentiert, begabt*›

unisex ['ju:nɪseks] *adj.* Unisex‹*mantel, -kleidung*›; **~ hairdresser** Damen-und-Herren-Frisör

unison ['ju:nɪsən] *n.* **a)** *(Mus.)* Unisono, *das;* **in ~:** unisono; einstimmig; **act in ~** *(fig.)* vereint handeln; **b)** *(concord)* Einmütigkeit, *die*

unit ['ju:nɪt] *n.* **a)** *(element, group; also Mil.)* Einheit, *die;* *(in complex mechanism)* Element, *das;* **armoured ~** *(Mil.)* Panzereinheit, *die;* **b)** *(in adding numbers by columns)* Einer, *der* *(Math.)*; **c)** *(quantity chosen as standard)* [Maß]einheit, *die;* *(of gas, electricity)* Einheit, *die;* **~ of length/monetary ~:** Längen-/Währungseinheit, *die;* **d)** *(piece of furniture)* Element, *das;* **kitchen ~:** Küchenelement, *das*

unite [ju:'naɪt] **1.** *v.t.* vereinigen; verbinden ‹*Einzelteile*›; ein[ig]en ‹*Partei, Mitglieder*›. **2.** *v.i.* *(join together)* sich vereinigen; ‹*Elemente:*› sich verbinden

united [ju:'naɪtɪd] *adj.* **a)** *(harmonious)* einig; **a ~ front** eine geschlossene Front; **b)** *(combined)* vereint *(geh.)*; gemeinsam

United: ~ 'Kingdom *pr. n.* Vereinigtes Königreich [Großbritannien und Nordirland]; **~ 'Nations** *pr. n. sing.* Vereinte Nationen *Pl.*; **~ 'States** *see* **state 1 e**; **~ States of A'merica** *n. sing.* Vereinigte Staaten von Amerika

unit: ~ 'furniture *n.* Anbaumöbel *Pl.*; **~ 'trust** *n.* *(Brit. Finance)* ≈ Investmentfonds, *der*

unity ['ju:nɪtɪ] *n.* **a)** *(state of being united)* Einheit, *die;* **their ~ of purpose** die Gemeinsamkeit ihres Wollens; **b)** *(Math.)* Einselement, *das*

universal [ju:nɪ'vɜ:sl] *adj.* **a)** *(prevailing everywhere)* allgemein; allgemeingültig ‹*Regel, Wahrheit*›; **become ~:** sich allgemein verbreiten; **b)** universal ‹*Bildung, Wissen*›; **c)** *(common to all members of a class)* universell

universally [ju:nɪ'vɜ:səlɪ] *adv.* allgemein

universe ['ju:nɪvɜ:s] *n.* Universum, *das;* *(world; fig.: mankind)* Welt, *die*

university [ju:nɪ:'vɜ:sɪtɪ] *n.* Universität, *die;* *attrib.* Universitäts-; **go to ~:** auf die *od.* zur Universität gehen; **at ~:** an der Universität

unjust [ʌn'dʒʌst] *adj.* ungerecht (**to** *Dat.* + gegenüber)

unjustifiable [ʌn'dʒʌstɪfaɪəbl] *adj.* ungerechtfertigt; **be ~:** nicht zu rechtfertigen sein

unjustifiably [ʌn'dʒʌstɪfaɪəblɪ] *adv.* ungerechtfertigterweise

unjustified [ʌn'dʒʌstɪfaɪd] *adj.* ungerechtfertigt

unjustly [ʌn'dʒʌstlɪ] ungerechterweise; zu Unrecht

unkempt [ʌn'kempt] *adj.* **a)** *(dishevelled)* ungekämmt ‹*Haare*›; **b)** *(untidy)* ungepflegt

unkind [ʌn'kaɪnd] *adj.* unfreundlich; **be ~ to sb./animals** jmdn./Tiere schlecht behandeln

unkindly [ʌn'kaɪndlɪ] *adv.* unfreundlich

unkindness [ʌn'kaɪndnɪs] *n.* Unfreundlichkeit, *die*

unknowing [ʌn'nəʊɪŋ] *see* **unwitting**

unknowingly [ʌn'nəʊɪŋlɪ] *see* **unwittingly**

unknown [ʌn'nəʊn] **1.** *adj.* unbekannt; **sb./sth. is ~ to sb.** jmd./etw. ist jmdm. nicht bekannt; **a drug ~ to us** ein uns unbekanntes Heilmittel; **it is ~/not ~ for him to do such a thing** es ist nie vorgekommen/ist schon vorgekommen, daß er so etwas getan hat; **the U~ Soldier** *or* **Warrior** der Unbekannte Soldat; **~ strengths/reserves** *(unsuspected)* ungeahnte Kräfte/Reserven. **2.** *adv.* **~ to sb.** ohne daß jmd. davon weiß/wußte. **3.** *n.* **the ~**: das Unbekannte; **journey/voyage into the ~** *(lit. or fig.)* Reise in unbekannte Regionen

unlace [ʌn'leɪs] *v. t.* aufschnüren

unladylike [ʌn'leɪdɪlaɪk] *adj.* nicht sehr damenhaft; **very ~**: gar nicht damenhaft

unlawful [ʌn'lɔ:fl] *adj.* ungesetzlich; gesetzwidrig; **~ possession of firearms/drugs** illegaler Waffen-/Drogenbesitz

unleaded [ʌn'ledɪd] *adj.* bleifrei ‹*Benzin*›

unless ['ʌn'les, ən'les] *conj.* es sei denn; wenn ... nicht; **I shall not do it ~ I am paid for it** ich werde es nur tun, wenn ich dafür bezahlt werde; **I shall expect you tomorrow ~ I hear from you/hear to the contrary** falls *od.* sofern ich nichts von dir/nichts Gegenteiliges höre, erwarte ich dich morgen; **I might go, but not ~ I'm asked to** vielleicht gehe ich, aber nur, wenn man mich darum bittet; **~ I'm [very much] mistaken** wenn ich mich nicht [sehr] irre *od.* täusche; **~ otherwise indicated** *or* **stated** wenn nicht anders angegeben

unliberated [ʌn'lɪbəreɪtɪd] *adj.* nicht emanzipiert ‹*Frau*›; unfrei ‹*Massen, Land*›

unlicensed [ʌn'laɪsənst] *adj.* ‹*Händler, Makler, Buchmacher*› ohne Konzession; nicht angemeldet ‹*Fernsehgerät, Auto*›; **~ premises** Gaststättenbetrieb ohne [Schank]konzession

unlike [ʌn'laɪk] **1.** *adj.* nicht ähnlich; unähnlich; *(unequal)* **~ signs** *(Math.)* ungleiche Vorzeichen; **~ poles** *(Phys.)* ungleiche Pole; **they are ~**: sie sind sich *(Dat.)* nicht ähnlich. **2.** *prep.* **be ~ sb./sth.** jmdm./einer Sache nicht ähnlich sein; **be not ~ sb./sth.** jmdm./etw. nicht unähnlich sein *od.* ganz ähnlich sein; **sth. is ~ sb.** *(not characteristic of)* etw. sieht jmdm. gar nicht ähnlich *(ugs.)*; etw. ist für jmdn. nicht typisch; **it is ~ him to be late** es ist sonst nicht seine Art, zu spät zu kommen; **~ her brother, she likes walking** im Gegensatz zu ihrem Bruder geht sie gern spazieren

unlikelihood [ʌn'laɪklɪhʊd] *n., no pl.* Unwahrscheinlichkeit, *die*

unlikely [ʌn'laɪklɪ] *adj.* **be ~ to do sth.** etw. wahrscheinlich nicht tun; **in the ~ event that ...**: sollte der unwahrscheinliche Fall eintreten, daß ...; **he's ~ to be chosen for the part/post** er wird die Rolle/Stelle kaum bekommen

unlimited [ʌn'lɪmɪtɪd] *adj.* unbegrenzt; grenzenlos, unendlich ‹*Himmel, Meer, Geduld*›

[1]unlined [ʌn'laɪnd] *adj. (without lining)* ungefüttert ‹*Kleidung, Briefumschlag*›

[2]unlined *adj. (without lines)* unliniert ‹*Papier*›

unlisted [ʌn'lɪstɪd] *adj.* **~ [telephone] number** Geheimnummer, *die*

unlit [ʌn'lɪt] *adj.* unbeleuchtet ‹*Straße, Korridor, Zimmer*›; nicht angezündet ‹*Lampe, Kamin, Kerze*›

unload [ʌn'ləʊd] **1.** *v. t.* **a)** entladen ‹*Lastwagen, Waggon*›; löschen ‹*Schiff, Schiffsladung*›; ausladen ‹*Gepäck*›; **b)** *(dispose of; Commerc.: sell off, dump)* abstoßen ‹*Aktien, Wertpapiere*›; **~ sb./sth. on [to] sb.** *(fig.)* jmdn./etw. bei jmdm. abladen. **2.** *v. i.* ‹*Schiff:*› gelöscht werden; ‹*Lastwagen:*› entladen werden

unloaded [ʌn'ləʊdɪd] *adj.* nicht geladen ‹*Gewehr, Pistole*›

unlock [ʌn'lɒk] *v. t.* aufschließen; lösen ‹*Rad, Taste*›; **~ed** unverschlossen ‹*Tür, Tor*›; **leave the door ~ed when you go out** schließ die Tür nicht ab, wenn du gehst; **the gate was left ~ed** das Tor war nicht abgeschlossen

unloose [ʌn'lu:s] *see* **loose 2**

unlovable [ʌn'lʌvəbl] *adj.* wenig liebenswert

unloved [ʌn'lʌvd] *adj.* ungeliebt

unluckily [ʌn'lʌkɪlɪ] *adv.* unglücklich; *as sentence-modifier* unglücklicherweise: **~ for him/her** *etc.* zu seinem/ihrem *usw.* Pech

unlucky [ʌn'lʌkɪ] *adj.* **a)** unglücklich; *(not successful)* glücklos; **be [very/really] ~**: [großes/wirkliches] Pech haben; **b)** *(bringing bad luck)* **an ~ date/number** ein Unglückstag/eine Unglückszahl; **an ~ sign/omen** ein schlechtes Zeichen/Omen; **be ~**: Unglück bringen

unmade [ʌn'meɪd] *adj.* ungemacht ‹*Bett*›; unbefestigt ‹*Straße*›

unmanageable [ʌn'mænɪdʒəbl] *adj.* **a)** *(difficult to control)* widerspenstig ‹*Kind, Pferd, Haare*›; **b)** *(unwieldy)* sperrig

unmanly [ʌn'mænlɪ] *adj.* unmännlich

unmanned [ʌn'mænd] *adj.* unbemannt ‹*Leuchtturm, Raumschiff, Bahnübergang*›; *(with nobody in attendance)* nicht besetzt ‹*Schalter, Rezeption*›; unbewacht ‹*Posten, Eingang*›

unmarked [ʌn'mɑ:kt] *adj.* **a)** *(without markings)* ‹*Schachtel, Kiste*› ohne Aufschrift; nicht gezeichnet ‹*Wäschestück*›; anonym ‹*Grab*›; **an ~ police car** ein Zivilfahrzeug der Polizei; **b)** *(not spoilt by marks)* fleckenlos ‹*Fußboden, Oberfläche*›; makellos ‹*Haut, Pfirsich, Apfel*›; **c)** *(not corrected)* unkorrigiert ‹*Klassenarbeit*›; **d)** *(not noticed)* unbemerkt; **e)** *(Sport)* ungedeckt ‹*Spieler*›

unmarried [ʌn'mærɪd] *adj.* unverheiratet; ledig; **~ mother** ledige Mutter

unmask [ʌn'mɑ:sk] *v. t.* **~ sb.** jmdm. die Maske entreißen; *(fig.)* jmdn. entlarven (**as** als)

unmatched [ʌn'mætʃt] *adj.* **be ~ [for sth.]** [in etw. *(Dat.)*] unübertroffen sein

unmentionable [ʌn'menʃənəbl] *adj.* unaussprechlich ‹*Sünde, Verbrechen*›

unmerciful [ʌn'mɜ:sɪfl] *adj.* erbarmungslos; unbarmherzig

unmerited [ʌn'merɪtɪd] *adj.* unverdient

unmetalled [ʌn'metld] *adj.* *(Brit.)* unbefestigt ‹*Straße*›

unmethodical [ʌnmɪ'θɒdɪkl] *adj.* unmethodisch

unmindful [ʌn'maɪndfl] *adj.* **be ~ of sth.** etw. nicht beachten

unmistakable [ʌnmɪ'steɪkəbl] *adj.* deutlich; klar ‹*Beweis*›; unverwechselbar ‹*Handschrift, Stimme*›

unmistakably [ʌnmɪ'steɪkəblɪ] *adv.* unverkennbar

unmitigated [ʌn'mɪtɪgeɪtɪd] *adj* vollkommen ‹*Unsinn*›; **an ~ scoundrel** ein Erzschurke; **be an ~ disaster** *(coll.)* eine einzige Katastrophe sein *(ugs.)*

unmoved [ʌn'mu:vd] *adj.* unbewegt; ungerührt; **be/remain ~ by sb.'s pleas** sich von jmds. Bitten nicht rühren lassen

unmusical [ʌn'mju:zɪkl] *adj.* unmusikalisch ‹*Person*›

unnamed [ʌn'neɪmd] *adj.* **a)** *(unidentified)* [namentlich] nicht genannt ‹*Ort, Person, Medizin*›; ungenannt ‹*Wohltäter*›; **b)** *(having no name)* namenlos ‹*Findling*›

unnatural [ʌn'nætʃrəl] *adj.* **a)** unnatürlich; *(abnormal)* nicht normal; *(perverted)* widernatürlich; **b)** *(affected)* unnatürlich; gekünstelt

unnaturally [ʌn'nætʃrəlɪ] *adv.* unnatürlich; **not ~**: natürlich

unnecessarily [ʌn'nesɪsərɪlɪ] *adv.* **a)** unnötig[erweise] ‹*sich ärgern, sich aufregen, sich sorgen*›; **spend money/time ~**: unnötig

Geld/Zeit aufwenden; **b)** *(excessively)* unnötig ⟨*streng, kompliziert*⟩

unnecessary [ʌn'nesəsərɪ] *adj.* unnötig; **it is ~ for sb. to do sth.** es ist unnötig *od.* es muß nicht sein, daß jmd. etw. tut

unnerve [ʌn'nɜ:v] *v. t.* entnerven

unnerving [ʌn'nɜ:vɪŋ] *adj.* entnervend; zermürbend ⟨*Warten*⟩; nervenaufreibend ⟨*Erlebnis*⟩

unnoticed [ʌn'nəʊtɪst] *adj.* unbemerkt; **pass** *or* **go ~:** unbemerkt bleiben

UNO ['ju:nəʊ] *abbr.* **United Nations Organization** UNO, *die*

unobjectionable [ʌnəb'dʒekʃənəbl] *adj.* gefällig; **sth./sb. is ~:** gegen etw./jmdn. gibt es nichts einzuwenden

unobservant [ʌnəb'zɜ:vənt] *adj.* unaufmerksam

unobserved [ʌnəb'zɜ:vd] *adj.* unbeobachtet

unobstructed [ʌnəb'strʌktɪd] *adj.* frei ⟨*Weg, Rohr, Ausgang*⟩; ungehindert ⟨*Vormarsch, Durchfahrt*⟩

unobtainable [ʌnəb'teɪnəbl] *adj.* nicht erhältlich; **number ~** *(Teleph.)* kein Anschluß unter dieser Nummer

unobtrusive [ʌnəb'tru:sɪv] *adj.* unaufdringlich ⟨*Geste, Bemerkung, Muster, Farbe*⟩; unauffällig ⟨*Riß, Bewegung*⟩

unoccupied [ʌn'ɒkjʊpaɪd] *adj.* **a)** *(empty)* unbesetzt; unbewohnt ⟨*Haus, Wohnung, Raum*⟩; **b)** *(not busy)* unbeschäftigt; **~ moments** freie Augenblicke

unofficial [ʌnə'fɪʃl] *adj.* inoffiziell; **an ~ strike** ein wilder Streik; **take ~ action** einen wilden Streik durchführen

unofficially [ʌnə'fɪʃəlɪ] *adv.* inoffiziell

unopened [ʌn'əʊpnd] *adj.* ungeöffnet; noch nicht aufgegangen ⟨*Knospe, Blüte*⟩

unopposed [ʌnə'pəʊzd] *adj.* unangefochten ⟨*Kandidat, Wahlsieger*⟩; ungehindert ⟨*Vormarsch*⟩

unorganized [ʌn'ɔ:gənaɪzd] *adj.* unsystematisch ⟨*Arbeitsweise*⟩; konfus ⟨*Essay, Person*⟩; ungeordnet ⟨*Struktur, Leben*⟩

unoriginal [ʌnə'rɪdʒɪnl] *adj.* unoriginell

unorthodox [ʌn'ɔ:θədɒks] *adj.* unorthodox *(geh.)*

unpack [ʌn'pæk] *v. t. & i.* auspacken

unpaid [ʌn'peɪd] *adj.* **a)** *(not yet paid)* unbezahlt; nicht bezahlt; **~ for** nicht bezahlt; **b)** *(not providing or receiving a salary)* unbezahlt ⟨*Arbeit, Stelle, Freiwilliger usw.*⟩; *(honorary)* ehrenamtlich; **~ leave** unbezahlter Urlaub

unpalatable [ʌn'pælətəbl] *adj.* ungenießbar; *(fig.)* unverdaulich ⟨*Tatsache, Wahrheit*⟩

unparalleled [ʌn'pærəleld] *adj.* beispiellos; unvergleichlich ⟨*Schönheit*⟩

unpardonable [ʌn'pɑ:dənəbl] unverzeihlich

unpatriotic [ʌnpætrɪ'ɒtɪk, ʌnpeɪtrɪ'ɒtɪk] *adj.* unpatriotisch

unpaved [ʌn'peɪvd] *adj.* ungepflastert

unperceptive [ʌnpə'septɪv] *adj.* unaufmerksam; nicht sehr tiefgründig ⟨*Bemerkung*⟩

unperturbed [ʌnpə'tɜ:bd] *adj.* **he was ~ by the prospect of ...:** die Aussicht auf ... beunruhigte ihn nicht; **remain ~:** sich nicht aus der Ruhe bringen lassen

unpick [ʌn'pɪk] *v. t.* auftrennen

unpin [ʌn'pɪn] *v. t.*, **-nn-** abnehmen ⟨*Zettel, Brosche*⟩

unplaced [ʌn'pleɪst] *adj. (Sport)* unplaziert

unplanned [ʌn'plænd] *adj.* nicht geplant; ungeplant

unpleasant [ʌn'plezənt] *adj.* unangenehm; unfreundlich ⟨*Bemerkung*⟩; böse ⟨*Lächeln*⟩; **she can be really ~:** sie kann sehr unangenehm werden; **be ~ with sb.** zu jmdm. unfreundlich sein

unpleasantly [ʌn'plezəntlɪ] *adv.* unangenehm; böse ⟨*lächeln*⟩; unfreundlich ⟨*antworten*⟩

unpleasantness [ʌn'plezəntnɪs] *n.* **a)** *no pl. (unpleasant nature)* Unerfreulichkeit, *die; (of person)* Unfreundlichkeit, *die;* **b)** *(bad feeling, quarrel)* Verstimmung, *die*

unplug [ʌn'plʌg] *v. t.*, **-gg-** *(Electr.: disconnect)* **~ a radio/a television set** den Stecker eines Radio-/Fernsehgeräts herausziehen

unpolished [ʌn'pɒlɪʃt] *adj.* unpoliert ⟨*Holz, Marmor, Schuhe, Reis*⟩; *(fig.)* ungeschliffen ⟨*Person, Manieren, Sprache*⟩

unpolluted [ʌnpə'lu:tɪd] sauber ⟨*Wasser, Fluß, Umwelt*⟩

unpopular [ʌn'pɒpjʊlə(r)] *adj.* unbeliebt ⟨*Lehrer, Regierung usw.*⟩; unpopulär ⟨*Maßnahme, Politik*⟩; **be ~ with sb.** *(not liked)* ⟨*Person:*⟩ bei jmdm. unbeliebt sein; ⟨*Maßnahme, Steuern:*⟩ bei jmdm. unpopulär sein; **I'm rather ~ with my wife at the moment** meine Frau ist auf mich zur Zeit ziemlich schlecht zu sprechen

unpopularity [ʌnpɒpjʊ'lærɪtɪ] *n., no pl. see* **unpopular:** Unbeliebtheit, *die* (**with** bei); Unpopularität, *die* (**with** bei)

unprecedented [ʌn'presɪdentɪd] *adj.* beispiellos; [noch] nie dagewesen

unpredictable [ʌnprɪ'dɪktəbl] *adj.* unberechenbar ⟨*Person, Charakter, Wetter*⟩

unprejudiced [ʌn'predʒʊdɪst] *adj.* unvoreingenommen

unpremeditated [ʌnprɪ'medɪteɪtɪd] *adj.* nicht vorsätzlich ⟨*Verbrechen*⟩; nicht geplant ⟨*Angriff, Tat*⟩

unprepared [ʌnprɪ'peəd] *adj.* **a)** *(not yet prepared)* nicht vorbereitet ⟨*Zimmer, Mahlzeit*⟩; **be [not] ~ for sth.** auf etw. *(Akk.)* [nicht] unvorbereitet sein; **b)** *(improvised)* Stegreif⟨*rede, -erklärung*⟩

unprepossessing [ʌnpri:pə'zesɪŋ] *adj.* wenig attraktiv; wenig einnehmend ⟨*Aussehen, Person*⟩

unpretentious [ʌnprɪ'tenʃəs] *adj.* unprätentiös *(geh.)*; einfach ⟨*Wein, Stil, Haus*⟩; bescheiden ⟨*Person*⟩

unprincipled [ʌn'prɪnsɪpld] *adj.* skrupellos

unprintable [ʌn'prɪntəbl] *adj. (lit. or fig.)* nicht druckreif

unproductive [ʌnprə'dʌktɪv] *adj.* unfruchtbar ⟨*Boden, Gegend*⟩; fruchtlos ⟨*Diskussion, Anstrengung, Nachforschung*⟩; unproduktiv ⟨*Zeit, Arbeit, Kapital*⟩

unprofessional [ʌnprə'feʃənl] *adj.* **a)** *(contrary to standards)* standeswidrig; **b)** *(amateurish)* unfachmännisch; stümperhaft

unprofitable [ʌn'prɒfɪtəbl] *adj.* unrentabel ⟨*Zeche, Investition, Geschäft*⟩; wenig einträglich ⟨*Arbeit*⟩; *(fig.)* fruchtlos

unpromising [ʌn'prɒmɪsɪŋ] *adj.* nicht sehr vielversprechend

unpronounceable [ʌnprə'naʊnsəbl] *adj.* unaussprechbar

unprotected [ʌnprə'tektɪd] *adj.* ungeschützt (**against** vor + *Dat.*, gegen); nicht geschützt ⟨*Art, Tier*⟩; **an ~ machine** eine Maschine ohne Schutzvorrichtung[en]; **~ sex** ungeschützter Geschlechtsverkehr

unproved [ʌn'pru:vd], **unproven** [ʌn'pru:vn] *adj.* **a)** *(not proved)* unbewiesen; **b)** *(untested)* ungeprüft

unprovoked [ʌnprə'vəʊkt] *adj.* grundlos

unpublished [ʌn'pʌblɪʃt] *adj.* unveröffentlicht

unpunished [ʌn'pʌnɪʃt] *adj.* ungesühnt ⟨*Verbrechen*⟩; unbestraft ⟨*Verbrecher*⟩; **go ~:** ohne Strafe bleiben; ⟨*Verbrecher:*⟩ straffrei ausgehen

unqualified [ʌn'kwɒlɪfaɪd] *adj.* **a)** *(lacking qualifications)* unqualifiziert; ⟨*Arzt*⟩ ohne Abschluß; **be ~ for sth.** für etw. nicht qualifiziert sein; **be ~ to do sth.** nicht dafür qualifiziert sein, etw. zu tun; **b)** *(absolute)* uneingeschränkt ⟨*Zustimmung*⟩; rein ⟨*Freude, Vergnügen*⟩; voll ⟨*Erfolg*⟩

unquestionable [ʌn'kwestʃənəbl] *adj.* unbezweifelbar ⟨*Tatsache, Beweis*⟩; unbestreitbar ⟨*Recht, Fähigkeiten, Ehrlichkeit*⟩; unanfechtbar ⟨*Autorität*⟩

unquestionably [ʌn'kwestʃənəblɪ] *adv.* zweifellos; ohne Frage

unquestioned [ʌn'kwestʃənd] *adj.* unangefochten ⟨*Fähigkeit, Macht, Autorität, Recht*⟩; unbestritten ⟨*Talent*⟩; **his ability/loyalty is ~:** seine Fähigkeit/Loyalität steht außer Frage

unquestioning [ʌn'kwestʃənɪŋ] *adj.*, **unquestioningly** [ʌn'kwestʃənɪŋlɪ] *adv.* bedingungslos; blind

unravel [ʌn'rævl] **1.** *v. t.*, *(Brit.)* **-ll-** entwirren; *(undo)* aufziehen; *(fig.)* **~ a mystery/the truth/a plot** ein Geheimnis enträtseln/die Wahrheit aufdecken/ein Komplott aufdecken. **2.** *v. i.*, *(Brit.)* **-ll-** aufgehen; sich aufziehen

unread [ʌn'red] *adj.* ungelesen

unreadable [ʌn'ri:dəbl] *adj.* **a)** *(illegible)* unleserlich; **b)** *(too difficult, boring, etc.)* unlesbar

unready [ʌn'redɪ] *adj.* nicht bereit; **the country is ~ for war** das Land ist für einen Krieg nicht gerüstet

unreal [ʌn'rɪəl] *adj.* unwirklich

unrealistic [ʌnrɪə'lɪstɪk] *adj.* unrealistisch

unreality [ʌnrɪ'ælɪtɪ] *n.*, *no pl.* Unwirklichkeit, *die*

unreasonable [ʌn'ri:zənəbl] *adj.* unvernünftig; übertrieben ⟨*Ansprüche, Forderung*⟩; übertrieben [hoch] ⟨*Preis, Kosten*⟩

unrecognizable [ʌn'rekəgnaɪzəbl] *adj.* **be [absolutely** *or* **quite] ~:** [überhaupt] nicht wiederzuerkennen sein

unrecognized [ʌn'rekəgnaɪzd] *adj.* **a)** *(not identified)* unerkannt; **b)** *(not officially recognized)* nicht anerkannt; **c)** *(not appreciated)* nicht [gebührend] gewürdigt ⟨*Talent, Genie*⟩; nicht [genügend] beachtet ⟨*Gefahr, Tatsache*⟩

unrecorded [ʌnrɪ'kɔ:dɪd] *adj.* **a)** *(not documented)* nicht [dokumentarisch] belegt; **b)** *(not recorded)* nicht aufgezeichnet; unbespielt, leer ⟨*Tonband, Kassette*⟩

unreel [ʌn'ri:l] **1.** *v. t.* abwickeln; abspulen ⟨*Film, Tonband*⟩. **2.** *v. i.* sich abwickeln; sich abspulen

unrefined [ʌnrɪ'faɪnd] *adj.* **a)** *(not refined)* nicht raffiniert; ungebleicht *(Mehl)*; **b)** *(fig.)* unkultiviert, ungeschliffen ⟨*Geschmack, Manieren, Person, Sprache*⟩

unrelated [ʌnrɪ'leɪtɪd] *adj.* unzusammenhängend; **be ~** *(not connected)* nicht miteinander zusammenhängen; *(not related by family)* nicht [miteinander] verwandt sein; **be ~ to sth.** mit etw. in keinem Zusammenhang stehen

unrelenting [ʌnrɪ'lentɪŋ] *adj.* unvermindert, nicht nachlassend ⟨*Hitze, Kälte, Regen*⟩; unerbittlich ⟨*Kampf, Opposition, Verfolgung, Haß*⟩; unnachgiebig ⟨*Entschlossenheit, Ehrgeiz*⟩

unreliable [ʌnrɪ'laɪəbl] *adj.* unzuverlässig

unremitting [ʌnrɪ'mɪtɪŋ] *adj.* nicht nachlassend; unermüdlich ⟨*Anstrengung, Versuche, Sorge*⟩; beharrlich ⟨*Kampf*⟩

unrepeatable [ʌnrɪ'pi:təbl] *adj.* **a)** *(unique)* einzigartig; einmalig ⟨*Angebot, Preis*⟩; **b)** *(not fit to be repeated)* **sth. is ~:** etw. ist nicht zitierfähig

unrepresentative [ʌnreprɪ'zentətɪv] *adj.* nicht repräsentativ (**of** für); *(Polit.)* nicht demokratisch gewählt ⟨*Regierung, Führer*⟩

unrequited [ʌnrɪ'kwaɪtɪd] *adj.* unerwidert

unreserved [ʌnrɪ'zɜ:vd] *adj.* **a)** *(not booked)* nicht reserviert; **b)** *(full, without any reservations)* uneingeschränkt ⟨*Zustimmung, Aufnahme, Entschuldigung usw.*⟩

unresponsive [ʌnrɪ'spɒnsɪv] *adj.* **be ~:** nicht reagieren (**to** auf + *Akk.*); **an ~ audience** ein teilnahmsloses Publikum

unrest [ʌn'rest] *n.* Unruhen *Pl.*

unrestrained [ʌnrɪ'streɪnd] *adj.* uneingeschränkt ⟨*Freude, Begeisterung, Wachstum, Überfluß*⟩; unbeherrscht ⟨*Gefühlsäußerung, Wut, Gewalt*⟩; unkontrolliert ⟨*Entwicklung, Wachstum*⟩; ungeniert ⟨*Sprache, Benehmen*⟩

unrestricted [ʌnrɪ'strɪktɪd] *adj.* unbeschränkt; uneingeschränkt; frei ⟨*Sicht*⟩; **have ~ use of sth.** etw. uneingeschränkt nutzen [dürfen]

unrewarded [ʌnrɪ'wɔ:dɪd] *adj.* **go ~:** keine Belohnung bekommen; ⟨*Tat, Mühe:*⟩ nicht belohnt werden

unrewarding [ʌnrɪ'wɔ:dɪŋ] *adj.* unbefriedigend; undankbar ⟨*Aufgabe*⟩

unripe [ʌn'raɪp] *adj.* unreif

unrivalled (*Amer.:* **unrivaled**) [ʌn'raɪvld] *adj.* unvergleichlich; beispiellos; unübertroffen ⟨*Ruf, Luxus, Erfahrung, Könnerschaft*⟩; **our goods are ~ in** *or* **for quality** unsere Waren sind in ihrer Qualität konkurrenzlos *od.* unerreicht

unroadworthy [ʌn'rəʊdwɜ:ðɪ] *adj.* nicht verkehrssicher ⟨*Fahrzeug*⟩

unroll [ʌn'rəʊl] **1.** *v. t.* aufrollen. **2.** *v. i.* sich aufrollen; *(fig.)* ⟨*Geschichte, Handlung:*⟩ sich entrollen

unromantic [ʌnrə'mæntɪk] *adj.* unromantisch

unruffled [ʌn'rʌfld] *adj.* ruhig; glatt ⟨*Gewässer, Haar, Feder*⟩

unruled [ʌn'ru:ld] unliniert ⟨*Papier*⟩

unruliness [ʌn'ru:lɪnɪs] *n.* Ungebärdigkeit, *die*

unruly [ʌn'ru:lɪ] *adj.* ungebärdig ⟨*Person, Benehmen*⟩; widerspenstig ⟨*Haar, Person, Benehmen*⟩

unsafe [ʌn'seɪf] *adj.* nicht sicher ⟨*Leiter, Konstruktion*⟩; baufällig ⟨*Gebäude*⟩; nicht verkehrssicher ⟨*Fahrzeug*⟩; gefährlich ⟨*Maschine, Leitungen, Spielzeug*⟩; **the food is ~ to eat** das Essen ist ungenießbar; **feel ~:** sich unsicher fühlen; **it is ~ to do that** es ist gefährlich, das zu tun

unsaid [ʌn'sed] *adj.* ungesagt; unausgesprochen; **leave sth. ~:** etw. ungesagt lassen

unsaleable [ʌn'seɪləbl] *adj.* unverkäuflich

unsatisfactory [ʌnsætɪs'fæktərɪ] *adj.* unbefriedigend; nicht befriedigend; schlecht ⟨*Service, Hotel*⟩; mangelhaft ⟨*schulische Leistung*⟩

unsatisfied [ʌn'sætɪsfaɪd] *adj.* unzufrieden; unerfüllt ⟨*Wunsch, Bedürfnis*⟩; nicht befriedigt ⟨*Wunsch, Bedürfnis, Neugier, Nachfrage*⟩; nicht gestillt ⟨*Hunger, Neugier, Appetit*⟩; **leave sb. ~:** jmdn. nicht befriedigen

unsatisfying [ʌn'sætɪsfaɪɪŋ] *adj.* unbefriedigend

unsavoury (*Amer.:* **unsavory**) [ʌn'seɪvərɪ] *adj.* unangenehm ⟨*Geruch, Geschmack, Mahlzeit*⟩; zwielichtig ⟨*Charakter, Person*⟩; zweifelhaft ⟨*Ruf, Geschäfte, Angelegenheit*⟩; unerfreulich ⟨*Einzelheiten*⟩

unscathed [ʌn'skeɪðd] *adj.* unversehrt ⟨*Person*⟩; unbeschädigt ⟨*Sache*⟩

unscented [ʌn'sentɪd] *adj.* nicht parfümiert ‹*Seife, Shampoo*›
unscheduled [ʌn'ʃedju:ld] *adj.* außerplanmäßig
unscientific [ʌnsaɪən'tɪfɪk] *adj.* unwissenschaftlich ‹*Methode, Buch, Ansatz*›
unscramble [ʌn'skræmbl] *v. t.* (*lit. or fig.*) entwirren; (*Teleph.: decode*) entschlüsseln
unscratched [ʌn'skrætʃt] *adj.* (*unhurt*) unverletzt
unscrew [ʌn'skru:] **1.** *v. t.* ab- *od.* losschrauben ‹*Regal, Deckel usw.*›; herausdrehen ‹*Schraube*›. **2.** *v. i.* ‹*Brett, Verschluß:*› sich abschrauben lassen; ‹*Schraube:*› sich lösen *od.* abschrauben lassen; **come ~ed** sich lösen
unscrupulous [ʌn'skru:pjʊləs] *adj.* skrupellos
unseal [ʌn'si:l] *v. t.* (*break seal of*) entsiegeln; (*open*) öffnen ‹*Brief, Paket, Behälter*›
unsealed [ʌn'si:ld] *adj.* offen; unverschlossen
unseasoned [ʌn'si:znd] *adj.* (*not flavoured*) ungewürzt; (*not matured*) nicht abgelagert ‹*Holz*›; unerfahren ‹*Soldat*›
unseaworthy [ʌn'si:wɜ:ðɪ] *adj.* nicht seetüchtig
unseemly [ʌn'si:mlɪ] *adj.* unschicklich; ungehörig ‹*Benehmen*›; ungebührlich ‹*Eile, Benehmen*›
unseen [ʌn'si:n] *adj.* **a)** (*not seen*) ungesehen; unbekannt ‹*Text*›; **b)** (*invisible*) unsichtbar
unselfconscious [ʌnself'kɒnʃəs] *adj.*, unbefangen
unselfish [ʌn'selfɪʃ] *adj.* selbstlos
unserviceable [ʌn'sɜ:vɪsəbl] *adj.* unbrauchbar
unsettle [ʌn'setl] *v. t.* durcheinanderbringen; verwirren ‹*menschlichen Geist*›; stören ‹*Friede*›; verstören ‹*Kind, Tier*›
unsettled [ʌn'setld] *adj.* **a)** (*changeable*) wechselhaft; (*fig.*) unstet (*geh.*), ruhelos ‹*Leben*›; unsicher ‹*Zukunft*›; **b)** (*upset*) verstimmt ‹*Magen*›; gestört ‹*Verdauung*›; unruhig ‹*Zeit, Land*›; **c)** (*open to further discussion*) ungeklärt ‹*Angelegenheit, Frage*›
unsettling [ʌn'setlɪŋ] *adj.* störend ‹*Vorfall, Einfluß*›; beunruhigend ‹*Nachricht*›; **have an ~ effect on sb.** jmdn. aus dem Gleichgewicht bringen
unshak[e]able [ʌn'ʃeɪkəbl] *adj.* unerschütterlich
unshaken [ʌn'ʃeɪkn] *adj.* **be ~:** nicht erschüttert sein
unshaven [ʌn'ʃeɪvn] *adj.* unrasiert
unsightly [ʌn'saɪtlɪ] *adj.* unschön
unsigned [ʌn'saɪnd] *adj.* nicht unterzeichnet ‹*Brief, Dokument*›; unsigniert ‹*Gemälde*›
unskilful [ʌn'skɪlfl] *adj.* ungeschickt
unskilled [ʌn'skɪld] *adj.* **a)** (*lacking skills*) ungeschickt; stümperhaft; **b)** (*without special training*) ungelernt ‹*Arbeiter*›; **c)** (*done without skill*) schlecht; stümperhaft; **d)** keine besonderen Fertigkeiten erfordernd ‹*Arbeit*›
unskillful (*Amer.*) *see* **unskilful**
unslept-in [ʌn'sleptɪn] *adj.* **the bed was ~:** in dem Bett hatte niemand geschlafen
unsociable [ʌn'səʊʃəbl] *adj.* ungesellig
unsocial [ʌn'səʊʃl] *adj.* ungesellig; **at this ~ hour** (*joc.*) zu dieser unchristlichen Tageszeit; **work ~ hours** nachts/sonn- und feiertags arbeiten
unsold [ʌn'səʊld] *adj.* unverkauft
unsolicited [ʌnsə'lɪsɪtɪd] *adj.* nicht angefordert *od.* erbeten; nicht bestellt ‹*Waren*›; unverlangt eingesandt ‹*Manuskript*›
unsolved [ʌn'sɒlvd] *adj.* ungelöst; unaufgeklärt ‹*Verbrechen*›
unsophisticated [ʌnsə'fɪstɪkeɪtɪd] *adj.* schlicht, einfach ‹*Person, Geschmack, Vergnügen, Spiel*›; unkompliziert ‹*Maschine, Küche, Methode*›; einfach ‹*Wein*›
unsound [ʌn'saʊnd] *adj.* **a)** (*diseased*) nicht gesund; krank; **b)** (*defective*) baufällig ‹*Gebäude*›; **structurally ~:** baufällig; **c)** (*ill-founded*) wenig stichhaltig; anfechtbar ‹*Gesetz*›; nicht vertretbar ‹*Ansichten, Methoden*›; **d)** (*unreliable*) unzuverlässig; **the firm is financially ~:** die Firma steht finanziell auf schwachen Füßen; **e) of ~ mind** unzurechnungsfähig
unsparing [ʌn'speərɪŋ] *adj.* **a)** (*lavish*) großzügig; **give sb. one's ~ help/support** jmdm. seine volle Hilfe/Unterstützung geben; **be ~ of** *or* **in sth.** mit etw. nicht geizen; **b)** (*merciless*) schonungslos
unspeakable [ʌn'spi:kəbl] *adj.* unbeschreiblich; (*indescribably bad*) unsäglich
unspecified [ʌn'spesɪfaɪd] *adj.* nicht näher bezeichnet; nicht genannt ‹*Anzahl, Summe*›
unspectacular [ʌnspek'tækjʊlə(r)] *adj.* wenig eindrucksvoll
unspoiled [ʌn'spɔɪld], **unspoilt** [ʌn'spɔɪlt] *adj.* unverdorben; unberührt ‹*Dorf, Landschaft*›
unspoken [ʌn'spəʊkn] *adj.* ungesagt; (*tacit*) unausgesprochen; stillschweigend ‹*Übereinkunft*›; **be left ~:** ungesagt bleiben
unstable [ʌn'steɪbl] *adj.* nicht stabil; instabil (*geh.*); labil ‹*Wirtschaft, Beziehungen, Verhältnisse*›; **[mentally/emotionally] ~:** [psychisch] labil
unsteady [ʌn'stedɪ] *adj.* unsicher; wechselhaft ‹*Entwicklung*›; ungleichmäßig ‹*Flamme, Rhythmus*›; wackelig ‹*Leiter, Stuhl, Tisch, Konstruktion*›; **be ~ on one's feet** unsicher auf den Beinen sein
unstinting [ʌn'stɪntɪŋ] *adj.* großzügig; **be ~ in sth.** mit etw. nicht geizen; **be ~ in one's efforts** keine Mühe scheuen
unstuck [ʌn'stʌk] *adj.* **come ~:** sich lösen; ‹*Briefumschlag:*› aufgehen; (*fig. coll.: come to grief, fail*) ‹*Person:*› baden gehen (*ugs.*) (**over** mit); ‹*Projekt, Plan, Theorie, Geschäft:*› in die Binsen gehen (*ugs.*)
unsubtle [ʌn'sʌtl] *adj.* plump
unsuccessful [ʌnsək'sesfl] *adj.* erfolglos; **be ~:** keinen Erfolg haben; **the operation was ~:** die Operation hatte keinen Erfolg *od.* mißlang; **be ~ in an examination** eine Prüfung nicht bestehen
unsuccessfully [ʌnsək'sesfəlɪ] *adv.* erfolglos; vergebens ‹*versuchen*›
unsuitability [ʌnsu:tə'bɪlɪtɪ, ʌnsju:tə'bɪlɪtɪ] *n., no pl.* Ungeeignetsein, *das;* (*for job*) mangelnde Eignung
unsuitable [ʌn'su:təbl, ʌn'sju:təbl] *adj.* ungeeignet; **~ clothes** (*for weather, activity*) unzweckmäßige Kleider; (*for occasion, age*) unpassende Kleider; **be ~ for sb./sth.** für jmdn./etw. ungeeignet sein
unsuitably [ʌn'su:təblɪ, ʌn'sju:təblɪ] *adv.* unpassend
unsuited [ʌn'su:tɪd, ʌn'sju:tɪd] *adj.* ungeeignet; **be ~ for** *or* **to sb./sth.** für jmdn./etw. ungeeignet sein; ‹*Verhalten, Sprache:*› für jmdn./etw. unpassend sein
unsung [ʌn'sʌŋ] *adj.* unbesungen ‹*Held, Tat*›
unsure [ʌn'ʃʊə(r)] *adj.* unsicher; **be ~ about sb./sth.** sich (*Dat.*) über jmdn./etw. nicht im klaren sein; **be ~ whether to do sth.** sich (*Dat.*) nicht sicher sein, ob man etw. tun soll; **be ~ of sb./sth.** sich (*Dat.*) jmds./einer Sache nicht sicher sein; **be ~ of a date/of one's facts** ein Datum nicht genau wissen/seine Fakten nicht genau kennen; **be ~ of oneself** unsicher sein

unsurprisingly [ʌnsə'praɪzɪŋlɪ] *adv.* wie zu erwarten war
unsuspected [ʌnsə'spektɪd] *adj.* ungeahnt ‹*Talent, Kräfte, Stärke, Tiefe, Charme*›; unvermutet ‹*Defekt, Leck, Ergebnis, Folge*›
unsuspecting [ʌnsə'spektɪŋ] *adj.*, **unsuspectingly** [ʌnsə'spektɪŋlɪ] *adv.* nichtsahnend
unsweetened [ʌn'swi:tnd] *adj.* ungesüßt
unswerving [ʌn'swɜ:vɪŋ] *adj.* unerschütterlich ‹*Glaube, Treue*›; unbeirrbar ‹*Entschlossenheit*›
unsympathetic [ʌnsɪmpə'θetɪk] *adj.* **a)** wenig mitfühlend; **be ~:** kein Mitgefühl zeigen; **be ~ to sth./not ~ to sth.** kein Verständnis/durchaus Verständnis für etw. haben; **b)** *(unlikeable)* unsympathisch
unsystematic [ʌnsɪstə'mætɪk] *adj.* unsystematisch
untamed [ʌn'teɪmd] *adj. (lit. or fig.)* ungezähmt; wild
untangle [ʌn'tæŋgl] *v.t. (lit. or fig.)* entwirren
untenable [ʌn'tenəbl] *adj.* unhaltbar
untested [ʌn'testɪd] *adj.* nicht erprobt
unthinkable [ʌn'θɪŋkəbl] *adj.* unvorstellbar
unthinking [ʌn'θɪŋkɪŋ] *adj.* gedankenlos
untidiness [ʌn'taɪdɪnɪs] *n., no pl. see* **untidy:** Ungepflegtheit, *die;* Unaufgeräumtheit, *die;*
untidy [ʌn'taɪdɪ] *adj.* ungepflegt ‹*Äußeres, Person, Garten*›; unaufgeräumt ‹*Zimmer*›
untie [ʌn'taɪ] *v.t.*, **untying** [ʌn'taɪɪŋ] aufknüpfen, aufknoten ‹*Faden, Seil, Paket*›; aufbinden ‹*Knoten, Schnürsenkel*›; losbinden ‹*Pferd, Boot, Seil vom Pfosten*›; **~ sb./sb.'s hands** jmdn./jmds. Hände von den Fesseln befreien
until [ən'tɪl] **1.** *prep.* bis; *(followed by article + noun)* bis zu; **~ [the] evening/the end** bis zum Abend/bis zum Ende; **~ his death/retirement** bis zu seinem Tod/seiner Pensionierung; **~ next week** bis nächste Woche; **~ then** *or* **that time** bis dahin *od.* dann; **not ~:** erst; **not ~ Christmas/the summer/his birthday** erst an Weihnachten/im Sommer/an seinem Geburtstag. **2.** *conj.* bis; **~ you find the key, we shall not be able to get in** solange du den Schlüssel nicht findest, kommen wir nicht hinein; **I did not know ~ you told me** ich wußte das nicht, bis du es mir gesagt hast
untimely [ʌn'taɪmlɪ] *adj.* **a)** *(inopportune)* ungelegen; *(inappropriate)* unpassend; **be ~:** ungelegen kommen/unpassend sein; **b)** *(premature)* vorzeitig; allzu früh ‹*Tod*›
untiring [ʌn'taɪərɪŋ] *adj.* unermüdlich; **be ~ in one's efforts to do sth.** sich unermüdlich bemühen, etw. zu tun
untold [ʌn'təʊld] *adj.* **a)** *(immeasurable)* unbeschreiblich; unsagbar ‹*Elend*›; unermeßlich ‹*Reichtümer, Anzahl*›; **b)** *(countless)* unzählig; **c)** *(not related)* nicht erzählt
untouchable [ʌn'tʌtʃəbl] **1.** *adj. (beyond reach)* unberührbar; **sth. is ~:** etw. kann nicht berührt werden. **2.** *n.* Unberührbare, *der/die*
untouched [ʌn'tʌtʃt] *adj.* **a)** *(not handled, untasted)* unberührt; **leave sth. ~:** etw. nicht anrühren; **b)** *(not changed)* unverändert; **c)** *(not affected)* unberührt
untoward [ʌntə'wɔ:d, ʌn'təʊəd] *adj.* **a)** *(unfavourable)* ungünstig; unglücklich ‹*Unfall*›; **b)** *(unseemly)* ungehörig
untraceable [ʌn'treɪsəbl] *adj.* unauffindbar
untrained [ʌn'treɪnd] *adj.* unausgebildet; ungelernt ‹*Arbeitskräfte*›; nicht dressiert ‹*Tier*›; **to the ~ eye/ear** dem ungeschulten Auge/Ohr
untranslatable [ʌntræns'leɪtəbl] *adj.* unübersetzbar
untried [ʌn'traɪd] *adj.* **a)** *(not tested)* unerprobt; **leave nothing ~:** nichts unversucht lassen; **b)** *(Law)* nicht vor Gericht gestellt ‹*Person*›; nicht verhandelt ‹*Fall*›
untroubled [ʌn'trʌbld] *adj.* ungestört ‹*Schlaf, Ruhe*›; sorglos ‹*Gesicht, Geist*›
untrue [ʌn'tru:] *adj.* **a)** *(false)* unwahr; **that's ~:** das ist nicht wahr; **b)** *(unfaithful)* **~ to sb./sth.** jmdm./etw. untreu
untrustworthy [ʌn'trʌstwɜ:ðɪ] *adj.* unzuverlässig
untruth [ʌn'tru:θ] *n., pl.* **~s** [ʌn'tru:ðz, ʌn'tru:θs] Unwahrheit, *die*
untruthful [ʌn'tru:θfl] *adj.* verlogen *(abwertend);* **an ~ story** eine Lügengeschichte *(abwertend)*
untuneful [ʌn'tju:nfl] *adj.* unmelodisch
unusable [ʌn'ju:zəbl] *adj.* unbrauchbar
[1]**unused** [ʌn'ju:zd] *adj.* *(new, fresh)* unbenutzt; *(not utilized)* ungenutzt; ungestempelt ‹*Briefmarke*›
[2]**unused** [ʌn'ju:st] *adj.* *(unaccustomed)* **be ~ to sth./to doing sth.** etw. *(Akk.)* nicht gewohnt sein/nicht gewohnt sein, etw. zu tun
unusual [ʌn'ju:ʒʊəl] *adj.* ungewöhnlich; *(exceptional)* außergewöhnlich; **an ~ number of ...:** eine ungewöhnlich große Zahl von ...; **it is ~ for him to do that** er tut das gewöhnlich nicht; **it is not ~ for her to do that** es ist durchaus nicht ungewöhnlich, daß sie das tut
unusually [ʌn'ju:ʒʊəlɪ] *adv.* ungewöhnlich
unutterable [ʌn'ʌtərəbl] *adj.*, **unutterably** [ʌn'ʌtərəblɪ] *adv.* unsäglich
unvarying [ʌn'veərɪɪŋ] *adj.* gleichbleibend
unveil [ʌn'veɪl] *v.t.* entschleiern ‹*Gesicht*›; enthüllen ‹*Statue, Gedenktafel*›; *(fig.)* vorstellen ‹*neues Auto, Produkt, Modell*›; veröffentlichen, *(geh.)* enthüllen ‹*Plan, Projekt*›
unversed [ʌn'vɜ:st] *adj.* nicht bewandert **(in** in + *Dat.*)
unvoiced [ʌn'vɔɪst] *adj.* **a)** unausgesprochen ‹*Ansichten, Gefühle, Zweifel*›; **b)** *(Phonet.)* stimmlos
unwaged [ʌn'weɪdʒd] *adj.* arbeitslos
unwanted [ʌn'wɒntɪd] *adj.* unerwünscht; **one's ~ clothes/books** die Kleider/Bücher, die man nicht mehr [haben] will
unwarranted [ʌn'wɒrəntɪd] *adj.* ungerechtfertigt
unwary [ʌn'weərɪ] *adj.* unvorsichtig; unüberlegt ‹*Tat, Schritt*›
unwashed [ʌn'wɒʃt] *adj.* ungewaschen ‹*Person, Kleidung*›; ungespült ‹*Geschirr*›
unwavering [ʌn'weɪvərɪŋ] *adj.* fest ‹*Blick*›; *(fig.: firm, resolute)* unerschütterlich
unwelcome [ʌn'welkəm] *adj.* unwillkommen; ungebeten ‹*Besucher*›
unwell [ʌn'wel] *adj.* unwohl; **look ~:** nicht wohl *od.* gut aussehen; **he feels ~** *(feels poorly)* er fühlt sich nicht wohl; *(feels sick)* ihm ist [es] schlecht *od.* übel
unwholesome [ʌn'həʊlsəm] *adj. (lit. or fig.)* ungesund
unwieldy [ʌn'wi:ldɪ] *adj.* unhandlich ‹*Werkzeug, Waffe*›; sperrig ‹*Karton, Form, Paket*›
unwilling [ʌn'wɪlɪŋ] *adj.* widerwillig ‹*Partner, Unterstützung, Zustimmung*›; unfreiwillig ‹*Helfer*›; **be ~ to do sth.** etw. nicht tun wollen
unwillingly [ʌn'wɪlɪŋlɪ] *adv.* widerwillig
unwillingness [ʌn'wɪlɪŋnɪs] *n., no pl.* Widerwille, *der;* **~ to help/**

listen mangelnde Bereitschaft zu helfen/zuzuhören

unwind [ʌn'waɪnd] **1.** *v.t.*, **unwound** [ʌn'waʊnd] abwickeln; abspulen ‹*Film*›. **2.** *v.i.*, **unwound a)** *(unreel)* sich abwickeln; **b)** *(fig.: unfold)* sich entwickeln; **c)** *(coll.: relax)* sich entspannen

unwise [ʌn'waɪz] *adj.* unklug

unwitting [ʌn'wɪtɪŋ] *adj.* ahnungslos ‹*Opfer*›; unwissentlich ‹*Komplize, Urheber*›; unbeabsichtigt ‹*Fehler, Handlung*›; ungewollt ‹*Beleidigung*›

unwittingly [ʌn'wɪtɪŋlɪ] *adv.* unwissentlich; unabsichtlich ‹*beleidigen*›

unwonted [ʌn'wəʊntɪd] *adj.* ungewohnt

unworkable [ʌn'wɜːkəbl] *adj.* unbrauchbar ‹*Material*›; nicht abbaubar ‹*Flöz*›; *(fig.: impracticable)* unbrauchbar ‹*System*›; undurchführbar ‹*Plan, Projekt*›

unworldly [ʌn'wɜːldlɪ] *adj.* weltabgewandt; *(naïve, not worldly-wise)* weltfremd

unworn [ʌn'wɔːn] *adj.* **a)** *(new)* ungetragen ‹*Kleidung*›; **b)** *(not damaged)* nicht abgetreten ‹*Teppich, Treppe*›; nicht abgetragen ‹*Kleidungsstück*›; nicht abgefahren ‹*Reifen*›

unworried [ʌn'wʌrɪd] *adj.* unbekümmert; **she was completely ~ by it** sie machte sich *(Dat.)* keine Sorgen darum

unworthy [ʌn'wɜːðɪ] *adj.* unwürdig; **be [not] ~ of sth.** einer Sache nicht [un]würdig sein; **be ~ of sb./sth.** ‹*Verhalten, Einstellung usw.*:› einer Person/Sache *(Gen.)* unwürdig sein

unwrap [ʌn'ræp] *v.t.*, **-pp-** auswickeln; abwickeln ‹*Bandage*›

unwritten [ʌn'rɪtn] *adj.* ungeschrieben; nicht schriftlich festgehalten ‹*Märchen, Lied, Vertrag, Verfassung*›; unbeschrieben ‹*Papier, Seite*›

unyielding [ʌn'jiːldɪŋ] *adj.* hart; *(fig.)* unnachgiebig; unerschütterlich ‹*Mut*›

unzip [ʌn'zɪp] **1.** *v.t.*, **-pp-** öffnen ‹*Reißverschluß*›; **~ a dress/bag** *etc.* den Reißverschluß eines Kleides/einer Tasche *usw.* öffnen. **2.** *v.i.*, **-pp-**: **the dress ~s at the back** das Kleid hat hinten einen Reißverschluß

up [ʌp] **1.** *adv.* **a)** *(to higher place)* nach oben; *(in lift)* aufwärts; **[right] up to sth.** *(lit. or fig.)* [ganz] bis zu etw. hinauf; **the bird flew up to the roof** der Vogel flog aufs Dach [hinauf]; **up into the air** in die Luft [hinauf] ...; **climb up on sth./climb up to the top of sth.** auf etw. *(Akk.)* [hinauf]steigen/bis zur Spitze einer Sache hinaufsteigen; **the way up [to sth.]** der Weg hinauf [zu etw.]; **on the way up** *(lit. or fig.)* auf dem Weg nach oben; **up here/there** hier herauf/dort hinauf; **high/higher up** hoch/höher hinauf; **farther up** weiter hinauf; **half-way/a long/little way up** den halben Weg/ein weites/kurzes Stück hinauf; **come on up!** komm [hier/weiter] herauf!; **up it** *etc.* **comes/goes** herauf kommt/hinauf geht es *usw.*; **up you go!** rauf mit dir! *(ugs.)*; **b)** *(to upstairs, northwards)* rauf *(bes. ugs.)*; herauf/hinauf *(bes. schriftsprachlich)*; nach oben; **come up from London to Edinburgh** von London nach Edinburgh [he]raufkommen; **c)** *(to place regarded as more important)* **go up to Leeds from the country** vom Land in die Stadt Leeds *od.* nach Leeds fahren; **d)** *(Brit.: to capital)* rein *(bes. ugs.)*; herein/hinein *(bes. schriftsprachlich)*; **go up to town** *or* **London** nach London gehen/fahren; **get up to London from Reading** von Reading nach London [he]reinfahren; **e)** *(in higher place, upstairs, in north)* oben; **up here/there** hier/da oben; **high up** hoch oben; **an order from high up** *(fig.)* ein Befehl von ganz oben *(ugs.)*; **higher up in the mountains** weiter oben in den Bergen; **the picture should be higher up** das Bild müßte höher hängen; **farther up** weiter oben; **half-way/a long/little way up** auf halbem Weg nach oben/ein gutes/kurzes Stück weiter oben; **10 metres up** 10 Meter hoch; **live four floors** *or* **storeys up** im vierten Stockwerk wohnen; **his flat is on the next floor up** seine Wohnung ist ein Stockwerk höher; **up north** oben im Norden *(ugs.)*; **f)** *(erect)* hoch; **keep your head up** halte den Kopf hoch; *see also* **chin**; **g)** *(out of bed)* **be up** aufsein; **up and about** auf den Beinen; **h)** *(in place regarded as more important; Brit.: in capital)* **up in town** *or* **London/Leeds** in London/Leeds; **i)** *(in price, value, amount)* **prices have gone/are up** die Preise sind gestiegen; **butter is up [by ...]** Butter ist [...] teurer; **j)** *(including higher limit)* **up to** bis ... hinauf; **up to midday/up to £2** bis zum Mittag/bis zu 2 Pfund; **k)** *(in position of gain)* **we're £300 up on last year** wir liegen 300 Pfund über dem letzten Jahr; **the takings were £500 up on the previous month** die Einnahmen lagen 500 Pfund über denen des Vormonats; **l)** *(ahead)* **be three points/games/goals up** *(Sport)* mit drei Punkten/Spielen/Toren vorn liegen; **be three points up on sb.** drei Punkte vor jmdm. sein *od.* liegen; **m)** *(as far as)* **up to sth.** bis zu etw.; **she is up to Chapter 3** sie ist bis zum dritten Kapitel gekommen *od.* ist beim dritten Kapitel; **up to here/there** bis hier[hin]/bis dorthin; **I've had it up to here** *(coll.)* mir steht es bis hier [hin] *(ugs.)*; **up to now/then/that time/last week** bis jetzt/damals/zu jener Zeit/zur letzten Woche; **n) up to** *(comparable with)* **be up to expectation[s]** den Erwartungen entsprechen; **his last opera is not up to his others** seine neueste Oper reicht an seine früheren nicht heran; **o) up to** *(capable of)* **[not] be/feel up to sth.** einer Sache *(Dat.)* [nicht] gewachsen sein/sich einer Sache *(Dat.)* [nicht] gewachsen fühlen; **[not] be/feel up to doing sth.** [nicht] in der Lage sein/sich nicht in der Lage fühlen, etw. zu tun; **not be up to much** nicht viel taugen; **p) up to** *(derog.: doing)* **be up to sth.** etw. anstellen *(ugs.)*; **what is he up to?** was hat er [bloß] vor?; **q) it is [not] up to sb. to do sth.** *(sb.'s duty)* es ist [nicht] jmds. Sache, etw. zu tun; **it is up to us to help them** es ist unsere Pflicht, ihnen zu helfen; **now it's up to him to do something** nun liegt es bei *od.* an ihm, etwas zu tun; **it's/that's up to you** *(is for you to decide)* es/das hängt von dir ab; *(concerns only you)* es/das ist deine Sache; **r)** *(close)* **up against sb./sth.** an jmdm./etw. ‹*lehnen*›; an jmdn./etw. ‹*stellen*›; **sit up against the wall** mit dem Rücken zur *od.* an der Wand sitzen; **s)** *(confronted by)* **be up against a problem/difficulty** *etc.* *(coll.)* vor einem Problem/einer Schwierigkeit *usw.* stehen; **be up against a tough opponent** es mit einem harten Gegner zu tun haben; **be up against it** in großen Schwierigkeiten stecken; **t) up and down** *(upwards and downwards)* hinauf und hinunter; *(to and fro)* auf und ab; **be up and down** *(coll.: variable)* Hochs und Tiefs haben; **u)** *(facing upwards)* **'this side/way up'** *(on box etc.)* „[hier] oben"; **turn sth. this/the other side/way up** diese/die andere Seite einer Sache nach oben drehen; **the right/wrong way up** richtig/verkehrt *od.* falsch herum; **v)** *(finished, at an end)* abgelaufen; **time is up** die Zeit ist abgelaufen. **2.** *prep.* **a)** *(upwards along, from bottom to top)* rauf *(bes. ugs.)*; herauf/hinauf *(bes. schriftsprachlich)*; **walk up sth.** etw. hinaufgehen; **up hill and down dale** bergauf und bergab; **b)** *(upwards through)* **force a liquid up a pipe** eine Flüssigkeit durch eine Röhre nach oben pressen; **c)** *(upwards over)* **up sth.** etw. *(Akk.)* hinauf; **ivy grew up the wall** Efeu wuchs die Mauer hinauf; **d)** *(along)* **come up the street** die Straße herauf- *od.* entlangkommen; **walk up and down the platform** auf dem Bahnsteig auf und ab gehen; **e)** *(at or in higher position in or on)* [weiter] oben; **further up the ladder/coast** weiter oben auf der Leiter/an der Küste; **f)** *(from bottom to top along)* **up the side of a house** an der Seite eines Hauses hinauf. **3.** *adj.* **a)** *(directed upwards)* aufwärts führend ‹*Rohr, Kabel*›; ‹*Rolltreppe*› nach oben; nach oben gerichtet ‹*Kolbenhub*›; **up train/line** *(Railw.)* Zug/Gleis Richtung Stadt; **b)** *(well-informed)* **be up in a subject/on the news** in einem Fach auf der Höhe [der Zeit] sein/über alle Neuigkeiten Bescheid wissen *od.* gut informiert sein; **c)** *(coll.: ready)* **tea['s]/grub['s] up!** Tee/Essen ist fertig!; **d)** *(coll.: amiss)* **what's up?** was ist los? *(ugs.)*; **something is up** irgendwas ist los *(ugs.)*. **4.** *n. in pl.* **the ups and downs** *(lit. or fig.)* das Auf und Ab; *(fig.)* die Höhen und Tiefen. **5.** *v. i.*, **-pp-** *(coll.)* **up and leave/resign** einfach abhauen *(ugs.)*/kündigen; **he ups and says ...:** da sagt er doch [ur]plötzlich ... **6.** *v. t.*, **-pp-** *(coll.) (increase)* erhöhen; *(raise up)* heben

'up-and-coming *adj. (coll.)* aufstrebend

'up-and-up *n. (coll.)* **be on the ~:** auf dem aufsteigenden Ast sein *(ugs.)*

'upbeat *n. (Mus.)* Auftakt, *der*

upbringing ['ʌpbrɪŋɪŋ] *n.* Erziehung, *die*

update 1. [ʌp'deɪt] *v. t. (bring up to date)* aktualisieren; auf den aktuellen Stand bringen; *(modernize)* modernisieren. **2.** ['ʌpdeɪt] *n.* Lagebericht, *der* (**on** zu); *(~d version)* Neuausgabe, *die*

up-'end *v. t. (lit. or fig.)* auf den Kopf stellen

up 'front *adv. (coll.)* **a)** *(at the front)* vorne; **b)** *(as down payment)* im voraus

upgrade [ʌp'greɪd] *v. t.* **a)** *(raise)* befördern ‹*Beschäftigte*›; aufwerten ‹*Stellung*›; **b)** *(improve)* verbessern

upheaval [ʌp'hi:vl] *n.* Aufruhr, *der; (commotion, disturbance)* Durcheinander, *das;* **an emotional ~:** ein Aufruhr der Gefühle

upheld *see* **uphold**

up'hill 1. *adj.* bergauf führend ‹*Weg, Pfad*›; ‹*Fahrt, Reise*› bergauf; *(fig.)* **an ~ task/struggle** eine mühselige Aufgabe/ein harter Kampf. **2.** *adv.* bergauf; **it's ~ all the way** es geht immer bergauf; *(fig.)* es ist ein mühseliges Geschäft

up'hold *v. t.*, **upheld a)** *(support)* unterstützen; hochhalten, wahren ‹*Tradition, Ehre*›; schützen ‹*Verfassung*›; **b)** *(confirm)* aufrechterhalten ‹*Forderung, Einwand*›; anerkennen ‹*Einwand, Beschwerde*›

upholster [ʌp'həʊlstə(r)] *v. t.* polstern

upholstery [ʌp'həʊlstərɪ] *n.* **a)** *(craft)* Polster[er]handwerk, *das;* **b)** *(padding)* Polsterung, *die; (cover also)* Bezug, *der; attrib.* Polster-

'upkeep *n.* Unterhalt, *der*

'up-market *adj.* exklusiv ‹*Waren, Hotel, Geschäft*›; Luxus‹*güter, -hotel, -restaurant*›; **go ~:** exklusiver [und teurer] werden

upon [ə'pɒn] *prep.* **a)** *(indicating direction)* auf (+ *Akk.*); *(indicating position)* auf (+ *Dat.*); **b)** *see* **on 1 a, b; a house ~ the river bank** ein Haus am Flußufer

upper ['ʌpə(r)] **1.** *compar. adj.* **a)** ober... ‹*Nil, Themse usw., Atmosphäre*›; Ober‹*grenze, -lippe, -arm usw., -schlesien, -österreich usw., -kreide, -devon usw.*›; *(Mus.)* hoch ‹*Tonlage, Noten*›; **~ circle** *(Theatre)* oberer Rang; **have/get/gain the ~ hand [of sb./sth.]** die Oberhand [über jmdn./etw.] haben/erhalten/gewinnen; **b)** *(in rank)* ober...; **the ~ ranks/echelons of the civil service/Army** die oberen *od.* höheren Ränge des Beamtentums/der Armee; **~ class[es]** Oberschicht, *die;* **the ~ crust** *(coll.)* die oberen Zehntausend. **2.** *n.* Oberteil, *das;* **'leather ~s'** „Obermaterial Leder"

upper: ~ case 1. *n.* Großbuchstaben; **2.** *adj.* groß ‹*Buchstabe*›; **~-class** *adj.* Oberschicht-; **~-class people/family/accent** Leute/Familie aus der Oberschicht/Akzent der Oberschicht; **~ 'deck** *n. (of ship, bus)* Oberdeck, das

uppermost ['ʌpəməʊst] **1.** *adj.* oberst...; **be ~ in sb.'s mind** jmdn. am meisten beschäftigen. **2.** *adv.* ganz oben; obenauf; **come ~** *(fig.)* an erster Stelle stehen

uppity ['ʌpɪtɪ] *adj. (coll.)* hochnäsig *(ugs.)*; **get ~:** sich aufblasen

upright ['ʌpraɪt] **1.** *adj.* **a)** aufrecht; steil ‹*Schrift*›; **a chair with an ~ back** ein Stuhl mit einem geraden Rücken[teil]; **~ piano** Klavier, *das;* **set/stand/hold sth. ~:** etw. aufrecht hinstellen/halten; **stand ~:** aufrecht stehen; **sit ~:** aufrecht sitzen; **hold oneself ~:** sich geradehalten; *see also* **bolt 4; b)** *(fig.: honourable)* aufrecht. **2. a)** *n.* seitliche Leiste; *(of ladder)* Holm, *der;* **b)** *(piano)* Klavier, *das*

'uprising *n.* Aufstand, *der*

up-river *see* **upstream**

'uproar *n.* Aufruhr, *der;* Tumult, *der;* **be in [an] ~:** in Aufruhr sein

uproarious [ʌp'rɔ:rɪəs] *adj.* zum Schreien komisch *(ugs.)* ‹*Witz, Anblick, Komödie*›; schallend ‹*Gelächter*›

up'root *v. t.* [her]ausreißen; ‹*Sturm:*› entwurzeln; **~ sb.** jmdn. aus der gewohnten Umgebung herausreißen; **people were ~ed by the war** die Menschen wurden durch den Krieg entwurzelt

upset 1. [ʌp'set] *v. t.*, **-tt-, upset a)** *(overturn)* umkippen; *(accidentally)* umstoßen ‹*Tasse, Vase, Milch usw.*›; **~ sth. over sth.** etw. über etw. *(Akk.)* kippen; **b)** *(distress)* erschüttern; mitnehmen *(ugs.)*; *(disturb the composure or temper of)* aus der Fassung bringen; *(shock, make angry, excite)* aufregen; **don't let it ~ you** nimm es nicht so schwer; **~ oneself** sich aufregen; **c)** *(make ill)* **sth. ~s sb.** etw. bekommt jmdm. nicht; **d)** *(disorganize)* stören; durcheinanderbringen ‹*Plan, Berechnung, Arrangement*›. **2.** *v. i.*, **-tt-, upset** umkippen. **3.** *adj.* **a)** *(overturned)* umgekippt; **b)** *(distressed)* bestürzt; *(agitated)* aufgeregt; *(unhappy)* unglücklich; *(put out)* aufgebracht; verärgert; *(offended)* gekränkt; **be ~ [about sth.]** *(be distressed)* [über etw. *(Akk.)*] bestürzt sein; *(be angry)* sich [über etw. *(Akk.)*] ärgern; **we were very ~ to hear of his illness** die Nachricht von seiner Krankheit ist uns sehr nahegegangen; **get ~ [about/over sth.]** sich [über etw. *(Akk.)*] aufregen; **there's no point in getting ~ about it** es hat keinen Sinn, sich darüber aufzuregen; **c)** ['--] *(disordered)* **an ~ stomach** ein verdorbener Magen. **4.** ['ʌpset] *n.* **a)** *(overturning)* Umkippen, *das;* **b)** *(agitation)* Aufregung, *die; (shock)* Schock, *der; (annoyance)* Verärgerung, *die;* **c)** *(slight quarrel)* Mißstimmung, *die;* **d)** *(slight illness)* Unpäßlichkeit, *die;* **digestive/stomach ~:** Verdauungsstörung, *die*/Magenverstimmung, *die;* **e)** *(disturbance)* Zwischenfall, *der; (confusion, upheaval)* Aufruhr, *der;* **f)** *(surprising result)* Überraschung, *die*

up'setting *adj.* erschütternd; *(sad)* traurig; bestürzend; *(annoying)* ärgerlich; **my mother found the obscene language ~:** meine Mutter fand die obszöne Sprache anstößig

'upshot *n.* Ergebnis, *das*

upside 'down 1. *adv.* verkehrt herum; **turn sth. ~** *(lit. or fig.)* etw. auf den Kopf stellen. **2.** *adj.* auf dem Kopf stehend ‹*Bild*›; **be ~:** auf dem Kopf stehen; **the acrobat hung ~:** der Akrobat hing mit dem Kopf nach unten *od.* kopfüber

up'stage *v.t.* **~ sb.** *(fig. coll.)* jmdm. die Schau stehlen *(ugs.)*

upstairs 1. [-'-] *adv.* nach oben ‹*gehen, kommen*›; oben ‹*sein, wohnen*›. **2.** ['--] *adj.* im Obergeschoß *nachgestellt.* **3.** [-'-] *n.* Obergeschoß, *das*

up'standing *adj.* **a)** *(strong and healthy)* stattlich; **b)** *(honest)* aufrichtig; aufrecht

'upstart *n.* Emporkömmling, *der*

'upstate *(Amer.)* **1.** *adj.* **~ New York** nördlicher Teil des Staates New York; **an ~ town** eine Stadt im nördlichen Teil des Staates. **2.** *adv.* **live ~:** im nördlichen Teil des Staates leben; **go/travel ~:** in den nördlichen Teil des Staates fahren/reisen

upstream 1. [-'-] *adv.* flußaufwärts. **2.** ['--] *adj.* flußaufwärts gelegen ‹*Ort*›

'upsurge *n.* Aufwallen, *das (geh);* **she felt an ~ of tenderness** sie fühlte Zärtlichkeit in sich *(Dat.)* aufwallen

'uptake *n.* **be quick/slow on** *or* **in the ~** *(coll.)* schnell begreifen/schwer von Begriff sein *(ugs.)*

uptight [ʌp'taɪt, 'ʌptaɪt] *adj. (coll.)* *(tense)* nervös **(about** wegen); *(touchy, angry)* sauer *(ugs.)* **(about** wegen)

up to 'date *pred. adj.* **be/keep [very] ~:** auf dem [aller]neusten Stand sein/bleiben; **keep/bring sth. ~:** etw. auf dem neusten Stand halten/auf den neusten Stand bringen

up-to-'date *attrib. adj. (current)* aktuell; *(modern)* modern; aktuell ‹*Mode*›

up-to-the-'minute *adj.* hochaktuell

'upturn *n.* Aufschwung, *der* **(in** + *Gen.*); **an ~ in prices** ein Anstieg der Preise

'upturned *adj.* **a)** *(upside-down)* umgedreht; **b)** *(turned upwards)* hochgeschlagen ‹*Rand, Krempe*›; nach oben gerichtet ‹*Gesicht, Auge*›; **~ nose** Stupsnase, *die*

upward ['ʌpwəd] **1.** *adj.* nach oben *nachgestellt;* nach oben gerichtet; **~ movement/trend** *(lit. or fig.)* Aufwärtsbewegung, *die*/-trend, *der;* **~ gradient** *or* **slope** Steigung, *die.* **2.** *adv.* aufwärts ‹*sich bewegen*›; nach oben ‹*sehen, gehen*›; *see also* **face up[ward]**

upwards ['ʌpwədz] *adv.* **a)** *see* **upward 2; b) ~ of** mehr als; über; **they cost £200 and ~:** sie kosten 200 Pfund und darüber

Urals ['jʊərlz] *pr. n. pl.* Ural, *der*

uranium [jʊə'reɪnɪəm] *n. (Chem.)* Uran, *das*

Uranus ['jʊərənəs, jʊə'reɪnəs] *pr. n. (Astron.)* Uranus, *der*

urban ['3:bn] *adj.* städtisch; Stadt‹*gebiet, -bevölkerung, -planung, -sanierung, -guerilla*›; **~ life** Leben in der Stadt

urbane [3:'beɪn] *adj.* weltmännisch

urchin ['3:tʃɪn] *n.* Range, *die; (boy)* Strolch, *der*

urge [3:dʒ] **1.** *v.t.* **a) ~ sb. to do sth.** jmdn. drängen, etw. zu tun; **~ sb. to sth.** jmdn. zu etw. drängen; **we ~d him to reconsider** wir rieten ihm dringend, es sich *(Dat.)* noch einmal zu überlegen; **~ sth. [on** *or* **upon sb.]** [jmdn.] zu etw. drängen; **~ caution/patience [on** *or* **upon sb.]** [jmdn.] zur Vorsicht/Geduld mahnen; **~ on** *or* **upon sb. the need for sth./for doing sth.** jmdm. die Notwendigkeit einer Sache/die Notwendigkeit, etw. zu tun, ans Herz legen; **~ that sth. [should] be done** darauf dringen, daß etw. getan wird; **b)** *(drive on)* [an]treiben; **~ forward/onward** vorwärts treiben; *(fig.)* treiben. **2.** *n.* Trieb, *der;* **have/feel an/the ~ to do sth.** den Drang verspüren, etw. zu tun; **resist the ~ to do sth.** dem [inneren] Drang widerstehen, etw. zu tun

~ 'on *v.t.* antreiben; *(hasten)* vorantreiben; *(encourage)* anfeuern; **~d on by hunger/ambition** vom Hunger/Ehrgeiz getrieben

urgency ['3:dʒənsɪ] *n., no pl.* Dringlichkeit, *die;* **there is no ~:** es eilt nicht *od.* ist nicht dringend; **be of the utmost ~:** äußerst dringend sein; **a matter of great ~:** eine sehr dringende Angelegenheit

urgent ['3:dʒənt] *adj.* dringend; *(to be dealt with immediately)* eilig; **be in ~ need of sth.** etw. dringend brauchen; **give ~ consideration to sth.** etw. vordringlich in Betracht ziehen; **on ~ business** in dringenden Geschäften; **it is ~:** es eilt

urgently ['3:dʒəntlɪ] *adv.* dringend; *(without delay)* eilig

urinal [jʊə'raɪnl, 'jʊərɪnl] *n. (fitting)* Urinal, *das;* **[public] ~:** [öffentliche] Herrentoilette; Pissoir, *das*

urinate ['jʊərɪneɪt] *v.t.* urinieren

urine ['jʊərɪn] *n.* Urin, *der;* Harn, *der*

urn [3:n] *n.* **a) tea/coffee ~:** Tee-/Kaffeemaschine, *die;* **b)** *(vessel)* Urne, *die*

Uruguay ['jʊərəgwaɪ] *pr. n.* Uruguay *(das)*

Uruguayan [jʊərə'gwaɪən] **1.** *adj.* uruguayisch; **sb is ~:** jmd. ist Uruguayer/Uruguayerin. **2.** *n.* Uruguayer, *der*/Uruguayerin, *die*

US *abbr.* **United States** USA; *attrib.* US-

us [əs, *stressed* ʌs] *pron.* **a)** uns; **it's us** wir sind's *(ugs.);* **b)** *(sl.: me)* **give us a clue/kiss!** gib mir 'nen Tip/Kuß! *(ugs.)*

USA *abbr.* **United States of America** USA; *attrib.* der USA *nachgestellt*

usable ['ju:zəbl] *adj.* brauchbar; gebräuchlich ‹*Wort*›

USAF *abbr.* **United States Air Force** Luftwaffe der Vereinigten Staaten

usage ['ju:zɪdʒ, 'ju:sɪdʒ] *n.* **a)** Brauch, *der;* Gepflogenheit, *die (geh.);* **be in common ~:** allgemein gebräuchlich sein; **b)** *(Ling.: use of language)* Sprachgebrauch, *der;* **~ [of a word]** Verwendung [eines Wortes]; **in American** *etc.* **~:** im amerikanischen *usw.* Sprachgebrauch; **c)** *(treatment)* Behandlung, *die;* **have rough ~:** schlecht behandelt werden

use 1. [ju:s] *n.* **a)** Gebrauch, *der; (of dictionary, calculator, room)* Benutzung, *die; (of word, expression; of pesticide, garlic, herb, spice)* Verwendung, *die; (of name, title)* Führung, *die; (of alcohol, drugs)* Konsum, *der;* **the ~ of brutal means/methods** die Anwendung brutaler Mittel/Methoden; **the ~ of troops/tear-**

gas/violence der Einsatz von Truppen/Tränengas/die Gewaltanwendung; **constant/rough ~:** dauernder Gebrauch/schlechte Behandlung; **[not] be in ~:** [nicht] in Gebrauch sein; **be no longer in ~:** nicht mehr verwendet werden; **be in daily** *etc.* **~:** täglich *usw.* in Gebrauch *od.* Benutzung sein; **come into ~:** in Gebrauch kommen; **go/fall out of ~:** außer Gebrauch kommen; **instructions/directions for ~:** Gebrauchsanweisung, *die;* **ready for [immediate] ~:** [sofort] gebrauchsfertig; **batteries for ~ in** *or* **with watches** Batterien [speziell] für Armbanduhren; **a course for ~ in schools** ein Kurs für die Schule *od.* zur Verwendung im Schulunterricht; **for the ~ of sb.** für jmdn.; **for personal/private ~:** für den persönlichen Gebrauch/den Privatgebrauch; **for external ~ only** nur zur äußerlichen Anwendung; **for ~ in an emergency/only in case of fire** für den Notfall/nur bei Feuer zu benutzen; **with ~:** durch den Gebrauch; **with careful** *etc.* **~:** bei sorgsamer *usw.* Behandlung; **make ~ of sb./sth.** jmdn./etw. gebrauchen/*(exploit)* ausnutzen; **make the best ~ of sth./it** das Beste aus etw./daraus machen; **make good ~ of, turn** *or* **put to good ~:** gut nutzen ‹*Zeit, Talent, Geld*›; **put sth. to ~:** etw. verwenden; **b)** *(utility, usefulness)* Nutzen, *der;* **these tools/clothes will be of ~ to sb.** dieses Werkzeug wird/diese Kleider werden für jmdn. von Nutzen sein; **is it of [any] ~?** ist das [irgendwie] zu gebrauchen *od.* von Nutzen?; **can I be of any ~ to you?** kann ich dir irgendwie helfen?; **be [of] no ~ [to sb.]** [jmdm.] nichts nützen; **he is [of] no ~ in a crisis/as a manager** er ist in einer Krise/als Manager zu nichts nütze *od. (ugs.)* nicht zu gebrauchen; **it's no ~ [doing that]** es hat keinen Zweck *od.* Sinn[, das zu tun]; **you're/that's a fat lot of ~** *(coll. iron.)* du bist ja eine schöne Hilfe/davon haben wir aber was *(ugs. iron.);* **what's the ~ of that/of doing that?** was nützt das/was nützt es, das zu tun?; **what's the ~?** was nützt es?; **oh well, what's the ~!** ach, was soll's schon! *(ugs.);* **c)** *(purpose)* Verwendung, *die;* Verwendungszweck, *der;* **have its/one's ~s** seinen Nutzen haben; **have/find a ~ for sth./sb.** für etw./jmdn. Verwendung haben/finden; **have no/not much ~ for sth./sb.** etw./jmdn. nicht/kaum brauchen; **put sth. to a good/a new ~:** etw. sinnvoll/auf neu[artig]e Weise verwenden; **d)** *(right or power of using)* **have the ~ of sth.** etw. benutzen können; **[have the] ~ of kitchen and bathroom** Küchen- und Badbenutzung [haben]; **let sb. have** *or* **give sb. the ~ of sth.** jmdn. etw. benutzen lassen. **2.** [ju:z] *v.t.* **a)** benutzen; nutzen ‹*Gelegenheit*›; anwenden ‹*Gewalt*›; einsetzen ‹*Tränengas, Wasserwerfer*›; in Anspruch nehmen ‹*Firma, Agentur, Agenten, Dienstleistung*›; nutzen ‹*Zeit, Gelegenheit, Talent, Erfahrung*›; führen ‹*Namen, Titel*›; **do you know how to ~ this tool?** kannst du mit diesem Werkzeug umgehen?; **anything you say may be ~d in evidence** was Sie sagen, kann vor Gericht verwendet werden; **~ sb.'s name [as a reference]** sich [als Empfehlung] auf jmdn. berufen; **I could ~ the money/a drink** *(coll.)* ich könnte das Geld brauchen/einen Drink vertragen *(ugs.);* **~ one's time to do sth.** seine Zeit dazu nutzen, etw. zu tun; **b)** *(consume as material)* verwenden; **~ gas/oil for heating** mit Gas/Öl heizen; **the camera ~s 35 mm film** für die Kamera braucht man einen 35-mm-Film; **'~ sparingly'** „sparsam verwenden!"; **c)** *(take habitually)* **~ drugs/heroin** *etc.* Drogen/Heroin *usw.* nehmen; **d)** *(employ in speaking or writing)* benutzen; gebrauchen; verwenden; **~ strong language** Kraftausdrücke gebrauchen; **e)** *(exercise, apply)* Gebrauch machen von ‹*Autorität, Einfluß, Können, Menschenverstand*›; **~ diplomacy/tact [in one's dealings** *etc.* **with sb.]** [bei jmdm.] diplomatisch vorgehen/[zu jmdm.] taktvoll sein; **he ~d all his strength** er wandte seine ganze Kraft auf; **~ a method/tactics** eine Methode anwenden/nach einer [bestimmten] Taktik vorgehen; **f)** *(take advantage of)* **~ sb.** jmdn. ausnutzen; **g)** *(treat)* behandeln; **~ sb./sth. well/badly** jmdn./etw. gut/schlecht behandeln; **h) ~d to** ['ju:st tə] *(formerly)* **I ~d to live in London/work in a factory** früher habe ich in London gelebt/in einer Fabrik gearbeitet; **he ~d to be very shy** er war früher sehr schüchtern; **my mother always ~d to say ...:** meine Mutter hat immer gesagt *od.* pflegte zu sagen ...; **this ~d to be my room** das war [früher] mein Zimmer; **things aren't what they ~d to be** es ist nichts mehr so wie früher; **he smokes much more than he ~d to** er raucht viel mehr als früher; **I ~d not** *or* **I did not ~** *or (coll.)* **I didn't ~** *or (coll.)* **I ~[d]n't to smoke** früher habe ich nicht geraucht

~ 'up *v.t.* aufbrauchen; verwenden ‹*[Essens]reste*›; verbrauchen, erschöpfen ‹*Kraft, Geld, Energie*›; **~ up a dozen eggs** ein Dutzend Eier verbrauchen

used 1. *adj.* **a)** [ju:zd] *(no longer new)* gebraucht; benutzt ‹*Handtuch, Teller*›; gestempelt ‹*Briefmarke*›; **~ car** Gebrauchtwagen, *der;* **~-car salesman** Gebrauchtwagenhändler, *der;* **b)** [ju:st] *(accustomed)* **~ to sth.** [an] etw. *(Akk.)* gewöhnt; etw. gewohnt; **be/get ~ to sb./sth.** [an] jmdn./etw. gewöhnt sein/sich an jmdn./etw. gewöhnen; **I'm not ~ to this kind of treatment** *or* **to being treated in this way** ich bin eine solche Behandlung nicht gewohnt; ich bin es nicht gewohnt, so behandelt zu werden; **you'll soon be ~ to it** du wirst dich bald *od.* schnell daran gewöhnen; **[not] be ~ to sb. doing sth./to having sb. do sth.** [es] [nicht] gewohnt sein, daß jmd. etw. tut; **she was ~ to getting up early** sie war daran gewöhnt, früh aufzustehen. **2.** [ju:st] *see* **use 2 h**

useful ['ju:sfl] *adj.* **a)** nützlich; praktisch ‹*Werkzeug, Gerät, Auto*›; brauchbar ‹*Rat, Idee, Wörterbuch*›; hilfreich ‹*Gespräch, Rat, Idee*›; **he is a ~ person to know** es ist nützlich, ihn zu kennen; **this is ~ to know** das ist gut zu wissen; **be ~ to sb.** jmdm. *od.* für jmdn. nützlich sein; jmdm. nützen; **sb. finds sth. ~:** etw. nützt jmdm.; **make oneself ~:** sich nützlich machen; **b)** *(coll.: worthwhile)* ordentlich *(ugs.);* ansehnlich ‹*Betrag, Stück, Arbeit*›

usefulness ['ju:sflnɪs] *n., no pl.* Nützlichkeit, *die;* Brauchbarkeit, *die*

useless ['ju:slɪs] *adj.* unbrauchbar ‹*Werkzeug, Gerät, Rat, Vorschlag, Idee, Material*›; nutzlos ‹*Wissen, Information, Fakten, Protest, Anstrengung, Kampf*›; vergeblich ‹*Anstrengung, Maßnahme, Kampf, Klage*›; zwecklos ‹*Widerstand, Protest, Argumentieren*›; **be ~ to sb.** jmdm. nichts nützen; **feel ~:** sich nutzlos fühlen; **it's ~ to do that** *or* **doing that** es hat keinen Zweck *od.* Sinn, das zu tun

uselessly ['ju:slɪslɪ] *adv.* unnütz,

sinnlos ⟨*verschwenden, aufwenden*⟩; vergeblich ⟨*kämpfen, protestieren*⟩

user ['ju:zə(r)] *n.* Benutzer, *der*/ Benutzerin, *die; (of drugs, alcohol)* Konsument, *der*/Konsumentin, *die; (of coal, electricity, gas)* Verbraucher, *der*/Verbraucherin, *die; (of telephone)* Kunde, *der*/Kundin, *die*

'**user-friendly** *adj.* benutzerfreundlich

usher ['ʌʃə(r)] **1.** *n. (in court)* Gerichtsdiener, *der; (at cinema, theatre, church)* Platzanweiser, *der*/-anweiserin, *die.* **2.** *v.t.* führen; geleiten *(geh.)*; ~ **sb. to his seat** jmdn. an seinen Platz führen

~ '**in** *v.t.* ~ **sb. in** jmdn. hineinführen *od. (geh.)* -geleiten; ~ **sth. in** *(fig.)* etw. einläuten

~ '**out** *v.t.* hinausführen *od. (geh.)* -geleiten

usherette [ʌʃə'ret] *n.* Platzanweiserin, *die*

USN *abbr.* **United States Navy** Marine der Vereinigten Staaten

USS *abbr.* **United States Ship** *Schiff aus den Vereinigten Staaten*

USSR *abbr.* **Union of Soviet Socialist Republics** UdSSR, *die; attrib.* der UdSSR *nachgestellt*

usual ['ju:ʒʊəl] *adj.* üblich; **be ~ for sb.** bei jmdm. üblich sein; **it is ~ for sb. to do sth.** es ist üblich, daß jmd. etw. tut; **[no] better/bigger/more** *etc.* **than ~:** [nicht] besser/größer/mehr *usw.* als gewöhnlich *od.* üblich; **as [is] ~:** wie üblich; **as is ~ in such cases** wie in solchen Fällen üblich

usually ['ju:ʒʊəlɪ] *adv.* gewöhnlich; normalerweise; **more than ~ tired** *etc.* noch müder *usw.* als üblich; ganz ungewöhnlich müde *usw.*

usurp [ju:'zɜ:p] *v.t.* sich *(Dat.)* widerrechtlich aneignen ⟨*Titel, Recht, Position*⟩; usurpieren *(geh.)* ⟨*Macht, Thron*⟩

usury ['ju:ʒərɪ] *n.* Wucher, *der*

utensil [ju:'tensɪl] *n.* Utensil, *das;* **writing ~s** Schreibutensilien

uterus ['ju:tərəs] *n., pl.* **uteri** ['ju:təraɪ] *(Anat.)* Gebärmutter, *die;* Uterus, *der (fachspr.)*

utilisation, utilise *see* **utiliz-**

utilitarian [ju:tɪlɪ'teərɪən] *adj.* **a)** *(functional)* funktionell; utilitär ⟨*Ziele*⟩; **b)** *(Philos.)* utilitaristisch

utility [ju:'tɪlɪtɪ] *n.* **a)** Nutzen, *der;* **b)** *see* **public utility**

u'**tility room** *n. Raum, in den [größere] Haushaltsgeräte (z.B. Waschmaschine) installiert sind*

utilization [ju:tɪlaɪ'zeɪʃn] *n.* Nutzung, *die*

utilize ['ju:tɪlaɪz] *v.t.* nutzen

utmost ['ʌtməʊst] **1.** *adj.* äußerst...; tiefst... ⟨*Verachtung*⟩; höchst... ⟨*Verehrung, Gefahr*⟩; größt... ⟨*Höflichkeit, Eleganz, Einfachheit, Geschwindigkeit*⟩; **of [the] ~ importance** von äußerster Wichtigkeit; **with the ~ caution** mit größter *od.* äußerster Vorsicht. **2.** *n.* Äußerste, *das;* **do** *or* **try one's ~ to do sth.** mit allen Mitteln versuchen, etw. zu tun

Utopia [ju:'təʊpɪə] *n. (place)* Utopia *(das)*

Utopian [ju:'təʊpɪən] *adj.* utopisch

[1]**utter** ['ʌtə(r)] *attrib. adj.* vollkommen, völlig ⟨*Chaos, Verwirrung, Fehlschlag, Einsamkeit, Unsinn*⟩; ungeheuer ⟨*Elend, Dummheit, Glück, Schönheit*⟩; größt... ⟨*Freude, Vergnügen*⟩; ~ **fool** Vollidiot, *der (ugs.)*

[2]**utter** *v.t.* **a)** von sich geben ⟨*Schrei, Seufzer, Ächzen*⟩; **b)** *(say)* sagen ⟨*Wahrheit, Wort*⟩; schwören ⟨*Eid*⟩; äußern ⟨*Drohung*⟩; zum Ausdruck bringen ⟨*Gefühle*⟩; **the last words he ~ed** die letzten Worte, die er sprach

utterance ['ʌtərəns] *n. (spoken words)* Worte *Pl.*; *(Ling.)* [sprachliche] Äußerung; *(sentence)* Satz, *der*

utterly ['ʌtəlɪ] *adv.* völlig; vollkommen; restlos ⟨*elend, deprimiert*⟩; absolut ⟨*entzückend, bezaubernd*⟩; hinreißend ⟨*schön*⟩; äußerst ⟨*dumm, lächerlich*⟩; aus tiefster Seele ⟨*verabscheuen, ablehnen, bereuen*⟩

uttermost ['ʌtəməʊst] *see* **utmost**

'**U-turn** *n.* Wende [um 180°]; **the driver/car made a ~:** der Fahrer/Wagen wendete; **make a ~ [on sth.]** *(fig.)* eine Kehrtwendung [bei etw.] vollziehen *od.* machen

UV *abbr.* **ultraviolet** UV

V

V, v [vi:] *n., pl.* **Vs** *or* **V's** V, v, *das*

v. *abbr.* **a)** ['vɜ:səs, vi:] **versus** gg.; **b) very**

vacancy ['veɪkənsɪ] *n.* **a)** *(job)* freie Stelle; **fill a ~:** eine [freie] Stelle besetzen; **have a ~:** eine freie Stelle *od.* Stelle frei haben; **'vacancies'** *(notice outside factory)* „Stellen frei"; *(in newspaper)* „Stellenangebote"; **b)** *(unoccupied room)* freies Zimmer; **have a ~:** ein Zimmer frei haben; **'vacancies'** „Zimmer frei"; **'no vacancies'** „belegt"; **c)** *no pl. (of look, mind, etc.)* Leere, *die*

vacant ['veɪkənt] *adj.* **a)** *(not occupied)* frei; **'~'** *(on door of toilet)* „frei"; **'situations ~'** „Stellenangebote"; **b)** *(mentally inactive)* leer

vacate [və'keɪt] *v.t.* räumen ⟨*Gebäude, Büro, Wohnung*⟩; aufgeben ⟨*Stelle, Amt*⟩

vacation [və'keɪʃn] **1.** *n.* **a)** *(Brit. Law, Univ.: recess)* Ferien *Pl.*; **b)** *(Amer.) see* **holiday 1 b. 2.** *v.i. (Amer.)* ~ **[at/in a place]** [an einem Ort] Urlaub machen

vaccinate ['væksɪneɪt] *v.t. (Med.)* impfen

vaccination [væksɪ'neɪʃn] *n. (Med.)* Impfung, *die; attrib.* Impf-; **have a ~:** geimpft werden

vaccine ['væksi:n, 'væksɪn] *n.* Impfstoff, *der*

vacillate ['væsɪleɪt] *v.i. (lit. or fig.)* schwanken

vacuum ['vækjʊəm] **1.** *n.* **a)** *pl.* **vacua** ['vækjʊə] *or* **~s** *(Phys.; also fig.)* Vakuum, *das;* **live in a ~** *(lit. or fig.)* im luftleeren Raum leben; **b)** *pl.* **~s** *(coll.: ~ cleaner)* Sauger, *der (ugs.)*. **2.** *v.t.& i.* [staub]saugen

vacuum: ~ cleaner *n.* Staubsauger, *der;* **~ flask** *n. (Brit.)* Thermosflasche, *die;* **~-packed** *adj.* vakuumverpackt

vagabond ['vægəbɒnd] *n.* Landstreicher, *der*/Landstreicherin, *die (oft abwertend)*; Vagabund, *der*/Vagabundin, *die (veralt.)*

vagaries ['veɪgərɪz] *n. pl. (lit. or fig.)* Launen *Pl.*

vagina [və'dʒaɪnə] *n., pl.* **~e** [və'dʒaɪni:] *or* **~s** *(Anat.)* Scheide, *die;* Vagina, *die (fachspr.)*

vagrant ['veɪgrənt] *n.* Landstreicher, *der*/Landstreicherin, *die (oft abwertend); (in cities)* Stadtstreicher, *der*/Stadtstreicherin, *die*

vague [veɪg] *adj.* vage; verschwommen, undeutlich ⟨*Form, Umriß*⟩; undefinierbar ⟨*Farbe*⟩; *(absent-minded)* geistesabwesend; *(inattentive)* unkonzentriert; **not have the ~st idea** *or* **notion** nicht die blasseste *od.* leiseste Ahnung haben; **be ~ about**

sth. etw. nur vag[e] andeuten; *(in understanding)* nur eine vage Vorstellung von etw. haben

vaguely ['veɪglɪ] *adv.* vage; entfernt ⟨*bekannt sein, erinnern an*⟩; schwach ⟨*sich erinnern*⟩; **he was ~ alarmed/disappointed** er war irgendwie beunruhigt/enttäuscht

vain [veɪn] *adj.* **a)** *(conceited)* eitel; **b)** *(useless)* leer ⟨*Drohung, Versprechen, Worte*⟩; eitel *(geh.)* ⟨*Vergnügungen*⟩; vergeblich ⟨*Hoffnung, Erwartung, Versuch*⟩; **in ~:** vergeblich; vergebens

vainly ['veɪnlɪ] *adv.* vergebens; vergeblich

vale [veɪl] *n. (arch./poet.)* Tal, *das*

valentine ['væləntaɪn] *n.* **a)** *jmd., dem man am Valentinstag einen Gruß schickt;* **b)** ~ **[card]** Grußkarte zum Valentinstag; **c) St. V~'s Day** Valentinstag, *der*

valet ['vælɪt, 'væleɪ] *n.* **a)** Kammerdiener, *der;* **b)** ~ **service** Reinigungs[- und Reparatur]service

valiant ['væljənt] *adj.* tapfer; kühn *(geh.);* tapfer ⟨*Versuch*⟩; **he made a ~ effort to disguise his disappointment** er versuchte tapfer, seine Enttäuschung zu verbergen

valiantly ['væljəntlɪ] *adv.* tapfer; kühn *(geh.)*

valid ['vælɪd] *adj.* **a)** *(legally acceptable)* gültig; berechtigt ⟨*Anspruch*⟩; *(legally ~)* rechtsgültig; *(having legal force)* rechtskräftig; bindend ⟨*Vertrag*⟩; **a ~ claim** ein Rechtsanspruch (**to** auf + *Akk.*); **b)** *(justifiable)* stichhaltig ⟨*Argument, Einwand, Theorie*⟩; triftig ⟨*Grund*⟩; zuverlässig ⟨*Methode*⟩; begründet ⟨*Entschuldigung, Einwand*⟩

validate ['vælɪdeɪt] *v.t.* rechtskräftig machen ⟨*Anspruch, Vertrag, Testament*⟩; bestätigen, beweisen ⟨*Hypothese, Theorie*⟩; für gültig erklären ⟨*Wahl*⟩

validity [və'lɪdɪtɪ] *n., no pl.* **a)** *(of ticket, document)* Gültigkeit, *die; (of claim, contract, marriage, etc.)* Rechtsgültigkeit, *die;* **b)** *(of argument, excuse, objection, theory)* Stichhaltigkeit, *die; (of reason)* Triftigkeit, *die; (of method)* Zuverlässigkeit, *die*

valley ['vælɪ] *n. (lit. or fig.)* Tal, *das*

valour *(Amer.:* **valor***)* ['vælə(r)] *n.* Tapferkeit, *die;* **fight with ~:** tapfer kämpfen

valuable ['væljʊəbl] **1.** *adj.* wertvoll; **be ~ to sb.** für jmdn. wertvoll sein. **2.** *n., in pl.* Wertgegenstände; Wertsachen

valuation [væljʊ'eɪʃn] *n.* Schätzung, *die;* **make/get a ~ of sth.** etw. schätzen/etw. schätzen lassen

value ['vælju:] **1.** *n.* **a)** Wert, *der;* **be of great/little/some/no ~ [to sb.]** [für jmdn.] von großem/geringem/einigem/keinerlei Nutzen sein; **be of [no] practical ~ to sb.** für jmdn. von [keinerlei] praktischem Nutzen sein; **set** *or* **put a high/low ~ on sth.** etw. hoch/niedrig einschätzen; **attach great ~ to sth.** einer Sache *(Dat.)* große Wichtigkeit beimessen; **what would be the ~ of it?** was ist es wohl wert?; **know the ~ of sth.** wissen, was etw. wert ist; **sth./nothing of ~:** etw./nichts Wertvolles; **an object of ~:** ein Wertgegenstand; **items of great/little/no ~:** sehr wertvolle/nicht sonderlich wertvolle/wertlose Gegenstände; **be of great/little/no** *etc.* **~:** viel/wenig/nichts *usw.* wert sein; **increase** *or* **go up in ~:** an Wert gewinnen; wertvoller werden; **decline** *or* **decrease** *or* **fall** *or* **go down in ~:** an Wert verlieren; **put a ~ on sth.** den Wert einer Sache schätzen; **sth. to the ~ of ...:** etw. im Werte von ...; **be good/poor etc. ~ [for money]** seinen Preis wert/nicht wert sein; **get [good]/poor ~ [for money]** etwas/nicht viel für sein Geld bekommen; **b)** *in pl. (principles)* Werte; Wertvorstellungen; **c)** *(Math.)* [Zahlen]wert, *der.* **2.** *v.t.* **a)** *(appreciate)* schätzen; **if you ~ your life** wenn dir dein Leben lieb ist; **b)** *(put price on)* schätzen, taxieren (**at** auf + *Akk.*)

value added 'tax *n. (Brit.)* Mehrwertsteuer, *die*

valued ['vælju:d] *adj.* geschätzt ⟨*Freund, Kollege, Kunde*⟩; wertvoll ⟨*Rat, Hilfe*⟩

'value-judgement *n.* Werturteil, *das*

valueless ['væljʊlɪs] *adj.* wertlos

valve [vælv] *n.* **a)** Ventil, *das;* **b)** *(Anat., Zool.)* Klappe, *die*

vamoose [və'mu:s] *v.i. (Amer. sl.)* verduften *(ugs.)*

vampire ['væmpaɪə(r)] *n.* Vampir, *der*

¹van [væn] *n.* **a) [delivery] ~:** Lieferwagen, *der;* **baker's/laundry ~:** Bäckerauto, *das*/Wäschereiauto, *das (ugs.);* **b)** *(Brit. Railw.)* [geschlossener] Wagen

²van *n. (foremost part)* Vorhut, *die; (fig.: leaders of movement, opinion)* Vorkämpfer *Pl.;* **be in the ~ of a movement/the attack** zu den Vorkämpfern einer Bewegung gehören/den Angriff anführen

vandal ['vændl] *n.* **a)** Rowdy, *der;* **~-proof** unzerstörbar; **b)** *(Hist.)* **V~:** Wandale, *der;* Vandale, *der*

vandalise *see* **vandalize**

vandalism ['vændəlɪzm] *n.* Wandalismus, *der;* Vandalismus, *der*

vandalize ['vændəlaɪz] *v.t. (destroy)* [mutwillig] zerstören; *(damage)* [mutwillig] beschädigen

vane [veɪn] *n.* **a)** *(weathercock) (in shape of arrow)* Wetterfahne, *die; (in shape of cock)* Wetterhahn, *der;* **b)** *(blade)* Blatt, *das*

vanguard ['vængɑ:d] *n.* **a)** *(Mil., Navy)* Vorhut, *die;* **b)** *(fig.: leaders)* Vorreiter; *(of literary, artistic, etc. movement)* Avantgarde, *die;* **be in the ~ of progress/a movement** an der Spitze des Fortschritts/einer Bewegung stehen

vanilla [və'nɪlə] **1.** *n.* Vanille, *die.* **2.** *adj.* Vanille-

vanish ['vænɪʃ] *v.i.* **a)** *(disappear; coll.: leave quickly)* verschwinden; **~ from sight** verschwinden; **~ into the distance** in der Ferne verschwinden; *see also* **thin 1d; b)** *(cease to exist)* ⟨*Gebäude:*⟩ verschwinden; ⟨*Sitte, Tradition:*⟩ untergehen; ⟨*Zweifel, Bedenken:*⟩ sich auflösen; ⟨*Hoffnung, Chancen:*⟩ schwinden

vanishing ['vænɪʃɪŋ]: **~-cream** *n.* Feuchtigkeitscreme, *die;* **~-point** *n. (Art, Math.)* Fluchtpunkt, *der; (fig.)* Nullpunkt, *der*

vanity ['vænɪtɪ] *n.* **a)** *(pride, conceit)* Eitelkeit, *die;* **b)** *(worthlessness)* Nichtigkeit, *die*

'vanity bag *n.* Kosmetiktäschchen, *das*

vanquish ['væŋkwɪʃ] *v.t. (literary)* bezwingen

vantage-point ['vɑ:ntɪdʒ pɔɪnt] *n.* Aussichtspunkt, *der; (fig.)* **his ~ as director** der Überblick, den er als Direktor hat/hatte

vapid ['væpɪd] *adj.* schal ⟨*Geschmack*⟩; geistlos ⟨*Gerede, Bemerkungen*⟩

vapor *(Amer.) see* **vapour**

vaporize (vaporise) ['veɪpəraɪz] *v.t. & i.* verdampfen

vapour ['veɪpə(r)] *n. (Brit.)* **a)** Dampf, *der; (mist)* Dunst, *der;* **~s** *(rising from the ground)* Schwaden; **b)** *(Phys.)* Dampf, *der*

'vapour trail *n. (Aeronaut.)* Kondensstreifen, *der*

variable ['veərɪəbl] *adj.* **a)** *(alterable)* veränderbar; **be ~:** verändert werden können; ⟨*Gerät:*⟩ eingestellt werden können; **b)** *(inconsistent, changeable)* unbeständig ⟨*Wetter, Wind, Strömung, Stimmung, Leistung*⟩; wechselhaft ⟨*Wetter, Launen, Schicksal, Qualität, Erfolg*⟩; schwankend ⟨*Kosten*⟩; **c)** *(Astron., Math.)* veränderlich; variabel

variance ['veərıəns] *n.* Uneinigkeit, *die;* **be at ~:** [sich *(Dat.)*] uneinig sein (**on** über + *Akk.*); ⟨*Theorien, Meinungen, Philosophien usw.:*⟩ nicht übereinstimmen; **be at ~ with sb./sth.** [sich *(Dat.)*] mit jmdm. uneinig sein/mit etw. nicht übereinstimmen

variant ['veərıənt] **1.** *attrib. adj.* verschieden. **2.** *n.* Variante, *die*

variation [veərı'eıʃn] *n.* **a)** *(varying)* Veränderung, *die; (in style, diet, routine, programme)* Abwechslung, *die; (difference)* Unterschied, *der;* **be subject to ~** ⟨*Preise:*⟩ Schwankungen unterworfen sein; ⟨*Regeln:*⟩ Änderungen unterworfen sein; **b)** *(variant)* Variante, *die* (**of, on** *Gen.*); **c)** *(Mus., Biol., Ballet, Math.)* Variation, *die*

varicose vein [værıkəʊs 'veın] *n.* *(Med.)* Krampfader, *die*

varied ['veərıd] *adj.* *(differing)* unterschiedlich; *(marked by variation)* abwechslungsreich ⟨*Land, Diät, Leben*⟩; vielseitig ⟨*Arbeit, Stil, Sammlung*⟩; vielgestaltig ⟨*Landschaft*⟩; bunt ⟨*Mischung*⟩

variegated ['veərıgeıtıd] *adj.* *(Bot.)* mehrfarbig; panaschiert ⟨*grüne Blätter*⟩

variety [və'raıətı] *n.* **a)** *(diversity)* Vielfältigkeit, *die; (in style, diet, routine, programme)* Abwechslung, *die;* **add** *or* **give ~ to sth.** etw. abwechslungsreicher gestalten; **for the sake of ~:** zur Abwechslung; **b)** *(assortment)* Auswahl, *die* (**of** an + *Dat.*, von); **for a ~ of reasons** aus verschiedenen Gründen; **a wide ~ of birds/flowers** viele verschiedene Vogelarten/Blumen; **c)** *(Theatre)* Varieté, *das;* **d)** *(form)* Art, *die; (of fruit, vegetable, cigarette)* Sorte, *die;* **e)** *(Biol.)* Unterart, *die;* Varietät, *die (fachspr.); (cultivated)* Züchtung, *die;* Rasse, *die*

variety: ~ act *n.* Varieténummer, *die;* **~ artist** *n.* *(Theatre)* Varietékünstler, *der*/-künstlerin, *die; (Telev.)* Showstar, *der;* **~ show** *n.* **a)** *(Theatre)* Varieté, *das;* **b)** *(Telev.) (varietéähnliche)* Show; **~ theatre** *n.* Varieté[theater], *das*

various ['veərıəs] *adj.* **a)** *pred. (different)* verschieden; unterschiedlich; *(manifold)* vielfältig; **b)** *attrib. (several)* verschiedene

variously ['veərıəslı] *adv.* unterschiedlich

varmint ['vɑ:mınt] *n.* *(Amer./dial.) (animal)* Biest, *das (ugs.); (person)* Halunke, *der; (child)* Racker, *der (fam.)*

varnish ['vɑ:nıʃ] **1.** *n.* **a)** Lack, *der; (transparent)* Lasur, *die;* **b)** *(Art)* Firnis, *der;* **c)** *(Ceramics)* Glasur, *die;* **d)** *(glossiness, lit. or fig.)* Glanz, *der.* **2.** *v.t.* **a)** lackieren; *(with transparent ~)* lasieren; **b)** *(Art)* firnissen; **c)** *(Ceramics)* glasieren; **d)** *(fig.: gloss over)* beschönigen; übertünchen ⟨*Fehler, Verbrechen, Laster*⟩

vary ['veərı] **1.** *v.t.* verändern; ändern ⟨*Bestimmungen, Programm, Methode, Verhalten, Stil, Route, Kurs*⟩; abwandeln ⟨*Rezept, Muster*⟩; *(add variety to)* abwechslungsreicher gestalten. **2.** *v.i.* *(become different)* sich ändern; ⟨*Preis, Nachfrage, Qualität, Temperatur:*⟩ schwanken; *(be different)* unterschiedlich sein; *(between extremes)* wechseln; *(deviate)* abweichen; **~ in weight/size/shape/colour** *etc.* im Gewicht/in der Größe/Form/Farbe variieren (**from ... to ...:** zwischen ... + *Dat.* und ... + *Dat.*); **opinions ~ on this point** die Meinungen gehen in diesem Punkt auseinander

varying ['veərııŋ] *attrib. adj.* wechselnd; wechselhaft, veränderlich ⟨*Wetter*⟩; *(different)* unterschiedlich

vase [vɑ:z] *n.* Vase, *die*

vasectomy [və'sektəmı] *n.* *(Med.)* Vasektomie, *die*

Vaseline, (P) ['væsəli:n] *n., no pl., no indef. art.* Vaseline, *die*

vassal ['væsl] *n.* *(Hist.)* Vasall, *der*/Vasallin, *die*

vast [vɑ:st] *adj.* **a)** *(huge)* riesig; weit ⟨*Fläche, Meer, Kontinent, Welt[raum]*⟩; umfangreich ⟨*Sammlung*⟩; **b)** *(coll.: great)* enorm; Riesen⟨*menge, -summe, -fehler*⟩; unermeßlich ⟨*Reichtümer*⟩; überwältigend ⟨*Mehrheit*⟩; **a ~ amount of time/money** enorm viel Zeit/viel Geld

vastly ['vɑ:stlı] *adv.* *(coll.)* enorm; weitaus ⟨*besser*⟩; weit ⟨*überlegen, unterlegen*⟩; gewaltig ⟨*sich verbessern, irren, überschätzen, unterschätzen*⟩; köstlich ⟨*sich amüsieren*⟩

vastness ['vɑ:stnıs] *n., no pl.* **a)** *(hugeness)* [immense *od.* ungeheure] Weite; *(of building, crowd, army)* [immense *od.* ungeheure] Größe; *(of collection etc.)* [riesiger] Umfang; **b)** *(greatness)* [immenses] Ausmaß

VAT [vi:eı'ti:, væt] *abbr.* **value added tax** MwSt.

vat [væt] *n.* Bottich, *der; (in paper-making)* Bütte, *die*

Vatican ['vætıkən] *pr. n.* Vatikan, *der*

¹vault [vɔ:lt, vɒlt] *n.* **a)** *(Archit.)* Gewölbe, *das;* **b)** *(cellar)* [Gewölbe]keller, *der;* **c)** *(in bank)* Tresorraum, *der;* **d)** *(tomb)* Gruft, *die*

²vault 1. *v.i.* *(leap)* sich schwingen. **2.** *v.t.* sich schwingen über (+ *Akk.*); *(Gymnastics)* springen über (+ *Akk.*). **3.** *n.* Sprung, *der*

vaulted ['vɔ:ltıd, 'vɒltıd] *adj.* *(Archit.)* gewölbt

vaunt [vɔ:nt] *(literary) v.t.* sich brüsten mit; **much ~ed** vielgepriesen *od.* -gerühmt

VCR *abbr.* **video cassette recorder**

VD [vi:'di:] *n.* Geschlechtskrankheit, *die;* **get** *or* **catch ~:** sich *(Dat.)* eine Geschlechtskrankheit zuziehen

VDU *abbr.* **visual display unit**

veal [vi:l] *n., no pl.* Kalb[fleisch], *das; attrib.* Kalbs-; **roast ~:** Kalbsbraten, *der*

vector ['vektə(r)] *n.* *(Math.)* Vektor, *der*

veer [vıə(r)] *v.i.* **a)** ⟨*Wind:*⟩ [sich] im Uhrzeigersinn drehen; ⟨*Schiff, Flugzeug:*⟩ abdrehen; ⟨*Auto:*⟩ ausscheren; **~ off course/off the road** *(unintentionally)* vom Kurs/von der Straße abkommen; *(intentionally)* vom Kurs abdrehen/von der Straße abbiegen; **~ out of control** außer Kontrolle geraten und ins Schleudern kommen; **b)** *(fig.: change)* schwanken (**from ... to ...:** zwischen ... + *Dat.* und ... + *Dat.*); **~ from one extreme to the other** ⟨*Person:*⟩ von einem Extrem ins andere fallen; **~ to the left** *(in politics)* auf Linkskurs umschwenken

~ a'way, ~ 'off *v.i.* ⟨*Schiff, Flugzeug:*⟩ abdrehen; ⟨*Auto:*⟩ ausscheren; ⟨*Fahrer, Straße:*⟩ abbiegen

~ 'round 1. *v.i.* drehen; *(through 180°)* wenden. **2.** *v.t.* wenden

veg [vedʒ] *n., pl. same (coll.)* Gemüse, *das;* **meat and two ~:** Fleisch mit Kartoffeln und Gemüse

vegan ['vi:gən] **1.** *n.* Veganer, *der*/Veganerin, *die.* **2.** *adj.* vegan

vegetable ['vedʒıtəbl] *n.* **a)** Gemüse, *das;* **spring/summer/winter ~:** Frühjahrs-/Sommer-/Wintergemüse, *das;* **fresh ~s** frisches Gemüse; **green ~s** Grüngemüse, *das;* **meat and two ~s** Fleisch mit Kartoffeln und Gemüse; *attrib.* Gemüse⟨*suppe, -extrakt*⟩; **b)** *(fig.)* **become/be a ~** *(through injury etc.)* nur noch [dahin]vegetieren

vegetable: ~ **dish** *n.* **a)** *(food)* Gemüsegericht, *das;* **b)** *(bowl)* Gemüseschüssel, *die;* ~ **dye** *n.* Pflanzenfarbe, *die;* ~ **garden** *n.* Gemüsegarten, *der;* ~ **knife** *n.* Küchenmesser, *das;* ~ **oil** *n.* *(Cookery)* Pflanzenöl, *das*
vegetarian [vedʒɪ'teərɪən] **1.** *n.* Vegetarier, *der*/Vegetarierin, *die.* **2.** *adj.* vegetarisch
vegetarianism [vedʒɪ'teərɪənɪzm] *n., no pl., no indef. art.* Vegetarismus, *der*
vegetate ['vedʒɪteɪt] *v. i.* *(as result of injury or illness)* nur noch [dahin]vegetieren
vegetation [vedʒɪ'teɪʃn] *n., no pl.* *(plants)* Vegetation, *die*
vehemence ['vi:əməns] *n., no pl.* Heftigkeit, *die;* Vehemenz, *die;* **with ~:** heftig; vehement
vehement ['vi:əmənt] *adj.* heftig; vehement; leidenschaftlich ‹*Gefühle, Rede*›; stark ‹*Wunsch, Abneigung*›; hitzig ‹*Debatte*›
vehemently ['vi:əməntlɪ] *adv.* heftig; vehement
vehicle ['vi:ɪkl] *n.* **a)** Fahrzeug, *das;* **b)** *(fig.: medium)* Vehikel, *das*
veil [veɪl] **1.** *n.* **a)** Schleier, *der;* **take the ~** *(Relig.)* den Schleier nehmen *(geh.);* **b) beyond the ~** *(fig.)* im Jenseits; **c)** *(fig.: obscuring medium)* Schleier, *der;* ~ **of mist/clouds** Dunst-/Wolkenschleier, *der;* **draw a ~ over sth.** den Mantel des Schweigens über etw. *(Akk.)* breiten. **2.** *v. t.* **a)** verschleiern; **b)** *(fig.: cover)* verhüllen; *(conceal)* verbergen ‹*Gefühle, Motive*› **(with, in** hinter + *Dat.*); verschleiern ‹*Fakten, Wahrheit, Bedeutung*›
veiled [veɪld] *adj.* **a)** verschleiert; **b)** *(fig.: covert)* versteckt ‹*Groll, Drohung*›; verhüllt ‹*Anspielung*›
vein [veɪn] *n.* **a)** Vene, *die;* *(in popular use: any blood-vessel)* Ader, *die;* **b)** *(Geol., Mining, Zool.)* Ader, *die;* **c)** *(Bot.)* Blattrippe, *die;* Ader, *die;* **d)** *(streak)* Ader, *die;* ~**s** *(in wood, marble)* Maserung, *die;* **e)** *(fig.: character, tendency)* Zug, *der;* **a ~ of melancholy/humour** ein melancholischer/humorvoller Zug; **f)** *(fig.)* *(mood)* Stimmung, *die;* *(style)* Art, *die;* **be in a happy/sad ~:** frohgelaunt/traurig gestimmt sein; **in a similar ~:** vergleichbarer Art
Velcro, (P) ['velkrəʊ] *n., no pl., no indef. art.* Klettverschluß, *der* Ⓦⓩ
vellum ['veləm] *n.* Pergament, *das*
velocity [vɪ'lɒsɪtɪ] *n.* Geschwindigkeit, *die;* ~ **of the wind, wind ~:** Windgeschwindigkeit, *die;* ~ **of light** *(Phys.)* Lichtgeschwindigkeit, *die*
velvet ['velvɪt] **1.** *n.* Samt, *der;* **[as] smooth as ~:** weich wie Samt; samtweich. **2.** *adj.* aus Samt *nachgestellt;* Samt-; *(soft as ~)* samten; samtweich
velveteen [velvi'ti:n] **1.** *n.* Baumwollsamt, *der;* Velveton, *der* *(fachspr.).* **2.** *adj.* aus Baumwollsamt *nachgestellt;* Velveton-
velvety ['velvɪtɪ] *adj.* samtig; samtweich
venal ['vi:nl] *adj.* käuflich, korrupt ‹*Person*›; korrupt ‹*Verhalten, Praktiken*›; eigennützig ‹*Interessen, Motive, Dienste*›
vendetta [ven'detə] *n.* **a)** Hetzkampagne, *die;* *(feud)* Fehde, *die;* **b)** *(killings)* Blutrache, *die*
vending-machine ['vendɪŋ məʃi:n] *n.* [Verkaufs]automat, *der*
vendor ['vendə(r), 'vendɔ:(r)] *n.* *(esp. Law)* Verkäufer, *der*/Verkäuferin, *die*
veneer [vɪ'nɪə(r)] *n.* **a)** Furnier, *das;* *(layer in plywood)* Furnierblatt, *das;* **b)** *(fig.: disguise)* Tünche, *die;* **beneath a ~ of respectability** hinter einer Fassade der Wohlanständigkeit
venerable ['venərəbl] *adj.* ehrwürdig; heilig ‹*Reliquien*›
venerate ['venəreɪt] *v. t.* verehren; hochachten; in Ehren halten ‹*jmds. Andenken, Traditionen, heilige Orte*›
veneration [venə'reɪʃn] *n.* **a)** *(reverence)* Ehrfurcht, *die* **(of, for** vor + *Dat.*); **b)** *(venerating, being venerated)* Verehrung, *die* **(of** für)
venereal disease [vɪ'nɪərɪəl dɪzi:z] *n.* *(Med.)* Geschlechtskrankheit, *die;* venerische Krankheit *(fachspr.)*
venetian blind [vɪ'ni:ʃn blaɪnd] *n.* Jalousie, *die*
Venezuela [venɪ'zweɪlə] *pr. n.* Venezuela *(das)*
Venezuelan [venɪ'zweɪlən] **1.** *adj.* venezolanisch; **sb. is ~:** jmd. ist Venezolaner/Venezolanerin. **2.** *n.* Venezolaner, *der*/Venezolanerin, *die*
vengeance ['vendʒəns] *n.* **a)** Rache, *die;* Vergeltung, *die;* **take ~ [up]on sb. [for sth.]** sich an jmdm. [für etw.] rächen; **b) with a ~** *(coll.)* gewaltig *(ugs.);* **go to work with a ~** *(coll.)* sich tüchtig ins Zeug legen *(ugs.)*
vengeful ['vendʒfl] *adj.* rachedurstig *(geh.);* rachsüchtig *(geh.)*
venial ['vi:nɪəl] *adj.* **a)** *(pardonable)* verzeihlich; entschuldbar; **b)** *(Theol.)* läßlich ‹*Sünde*›
Venice ['venɪs] *pr. n.* Venedig *(das)*
venison ['venɪsn, 'venɪzn] *n., no pl.* Hirsch[fleisch], *das;* *(of roe deer)* Reh[fleisch], *das;* **roast ~:** Hirsch-/Rehbraten, *der*
venom ['venəm] *n.* **a)** *(Zool.)* Gift, *das;* **b)** *(fig.)* Boshaftigkeit, *die;* Gehässigkeit, *die*
venomous ['venəməs] *adj.* **a)** *(Zool.)* giftig; Gift‹*schlange, -stachel*›; **b)** *(fig.)* giftig *(ugs.);* boshaft
[1]**vent** [vent] **1.** *n.* **a)** *(for gas, liquid to escape)* Öffnung, *die;* **b)** *(flue)* [Rauch]abzug, *der;* **c)** *(Geol.)* [Vulkan]schlot, *der;* **d)** *(fig.: for emotions)* Ventil, *das (fig.);* **give ~ to** Luft machen (+ *Dat.*) ‹*Ärger, Wut*›; freien Lauf lassen (+ *Dat.*) ‹*Gefühlen*›. **2.** *v. t.* *(fig.)* freien Lauf lassen (+ *Dat.*) ‹*Kummer, Schmerz*›; Luft machen (+ *Dat.*) ‹*Ärger, Wut*›; ~ **one's anger on sb.** seinen Ärger an jmdm. auslassen *od.* abreagieren
[2]**vent** *n.* *(in garment)* Schlitz, *der*
ventilate ['ventɪleɪt] *v. t.* **a)** lüften; *(by permanent installation)* belüften; **b)** *(fig.)* *(submit to public consideration)* [offen] erörtern; *(voice)* kundtun, äußern ‹*Meinung*›; vorbringen ‹*Beschwerden*›
ventilation [ventɪ'leɪʃn] *n.* **a)** *no pl.* Belüftung, *die;* **b)** *no pl.* *(installation)* Lüftung, *die*
ventilator ['ventɪleɪtə(r)] *n.* **a)** Lüftung[svorrichtung], *die;* *(fan)* Ventilator, *der;* **b)** *(Med.)* Beatmungsgerät, *das*
ventriloquism [ven'trɪləkwɪzm] *n., no pl.* Bauchreden, *das*
ventriloquist [ven'trɪləkwɪst] *n.* Bauchredner, *der*/-rednerin, *die*
venture ['ventʃə(r)] **1.** *n.* **a)** Unternehmung, *die;* **their ~ into space/the unknown** ihre Reise in den Weltraum/ins Unbekannte; **a new ~ in sth.** ein neuer Vorstoß in etw. *(Dat.);* **I can't lose much by the ~:** ich kann bei dem Versuch nicht viel verlieren; **b)** *(Commerc.)* Unternehmung, *die;* **a successful ~:** ein erfolgreiches Geschäft; **a new publishing ~:** ein neues verlegerisches Vorhaben *od.* Projekt. **2.** *v. i.* **a)** *(dare)* wagen; **if I might ~ to suggest ...:** wenn Sie [mir] gestatten, möchte ich vorschlagen ...; **may I ~ to ask ...:** darf ich mir erlauben, zu fragen ...; **b)** *(dare to go)* sich wagen; ~ **further into the cave** sich weiter *od.* tiefer in die Höhle vorwagen; ~ **out of doors** sich vor die Tür wagen. **3.** *v. t.* **a)** wagen ‹*Bitte, Bemerkung, Blick, Vermu-*

tung⟩; zu äußern wagen ⟨*Ansicht*⟩; sich *(Dat.)* erlauben ⟨*Frage, Scherz, Bemerkung*⟩; ~ **an explanation for sth.** etw. zu erklären versuchen; **if I might ~ a suggestion** wenn ich mir einen Vorschlag erlauben darf; **b)** *(risk, stake)* aufs Spiel setzen ⟨*Leben, Ruf, Vermögen, Glück*⟩; setzen ⟨*Wettsumme*⟩ (**on** auf + *Akk.*); *see also* **nothing 1 a**

~ **'forth** *(literary) see* ~ **out**

~ **on** *v. t.* sich einlassen auf (+ *Akk.*); sich wagen an (+ *Akk.*) ⟨*Aufgabe*⟩; sich wagen auf (+ *Akk.*) ⟨*Reise*⟩

~ **'out** *v. i.* sich hinauswagen

~ **upon** *see* ~ **on**

venue ['venju:] *n. (Sport)* [Austragungs]ort, *der; (Mus., Theatre)* [Veranstaltungs]ort, *der; (meeting-place)* Treffpunkt, *der*

Venus ['vi:nəs] *pr. n.* **a)** *(Astron.)* Venus, *die;* **b)** *(Roman Mythol.)* Venus *(die)*

veracity [və'ræsıtı] *n., no pl.* Wahrheitstreue, *die*

veranda[h] [və'rændə] *n.* Veranda, *die*

verb [vɜ:b] *n. (Ling.)* Verb, *das*

verbal ['vɜ:bl] *adj.* **a)** *(relating to words)* sprachlich; **his skills are ~:** seine Fähigkeiten liegen auf sprachlichem Gebiet; **b)** *(oral)* mündlich; verbal, mündlich ⟨*Bekenntnis, Anerkennung, Protest*⟩; **c)** *(Ling.)* verbal

verbally ['vɜ:bəlı] *adv.* **a)** *(regarding words)* sprachlich; mit Worten, verbal ⟨*beschreiben*⟩; **b)** *(orally)* mündlich; verbal, mündlich ⟨*protestieren*⟩

verbal 'noun *n. (Ling.)* Verbalsubstantiv, *das*

verbatim [və'beıtım] **1.** *adv.* im Wortlaut ⟨*veröffentlichen*⟩; [wort]wörtlich ⟨*sagen, abschreiben, zitieren*⟩. **2.** *adj.* wortgetreu; [wort]wörtlich

verbiage ['vɜ:bııdʒ] *n., no pl., no indef. art.* **a)** *(wordiness)* Geschwätzigkeit, *die;* **b)** *(words)* Geschwätz, *das*

verbose [və'bəʊs] *adj.* geschwätzig; weitschweifig ⟨*Roman, Vortrag, Autor*⟩; langatmig ⟨*Rede, Redner, Stil*⟩

verdant ['vɜ:dənt] *adj. (literary)* [saft]grün

verdict ['vɜ:dıkt] *n.* **a)** *(Law)* Urteil, *das;* [Urteils]spruch, *der;* **open ~:** *Feststellung eines gewaltsamen Todes ohne Nennung der Ursache (bei einer gerichtlichen Untersuchung);* ~ **of guilty/not guilty** Schuld-/Freispruch, *der;* **reach a ~:** zu einem Urteil kommen; **b)** *(judgement)* Urteil, *das* (**on** über + *Akk.*); *(decision)* Entscheidung, *die;* **the ~ of the electors** die Entscheidung der Wähler; **give** *or* **pass a/one's ~ [on sb./sth.]** ein/sein Urteil [über jmdn./etw.] abgeben

verge [vɜ:dʒ] *n.* **a)** *(grass edging)* Rasensaum, *der; (on road)* Bankette, *die;* **'keep off the ~'** „Bankette nicht befahrbar"; **b)** *(brink, border, lit. or fig.)* Rand, *der; (fig.: point at which something begins)* Schwelle, *die;* **be on the ~ of economic collapse/of war** am Rand des wirtschaftlichen Zusammenbruchs/an der Schwelle des Krieges stehen; **be on the ~ of despair/tears/a breakthrough** der Verzweiflung/den Tränen/dem Durchbruch nahe sein; **be on the ~ of doing sth.** kurz davor stehen, etw. zu tun; **bring sb./sth. to the ~ of sth.** jmdn./etw. an den Rand von etw. bringen

~ **on** *v. t.* [an]grenzen an (+ *Akk.*); **be verging on 70** an die 70 sein; **an estate verging on four acres** *(fig.)* ein Grundstück von fast vier Morgen [Größe]; **be verging on tears/madness** den Tränen/dem Wahnsinn nahe sein

verger ['vɜ:dʒə(r)] *n. (Eccl.)* Küster, *der*

verifiable ['verıfaıəbl] *adj.* nachprüfbar

verification [verıfı'keıʃn] *n.* **a)** *(check)* Überprüfung, *die;* **b)** *see* **verify b:** Bestätigung, *die;* Bekräftigung, *die;* Nachweis, *der;* **c)** *(bearing out)* Bestätigung, *die*

verify ['verıfaı] *v. t.* **a)** *(check)* überprüfen; prüfen ⟨*Bücher*⟩; **ring sb. up to ~ the news** jmdn. anrufen, um sich *(Dat.)* die [Richtigkeit der] Nachricht bestätigen zu lassen; **b)** *(confirm)* bestätigen ⟨*Vermutung, Diagnose*⟩; bekräftigen ⟨*Anspruch, Forderung*⟩; nachweisen ⟨*Identität*⟩; **c)** *(bear out)* bestätigen; beweisen ⟨*Theorie*⟩

veritable ['verıtəbl] *adj. (literary)* richtig; wahr, richtig ⟨*Engel, Genie*⟩; wahr ⟨*Wunder*⟩

vermilion [və'mıljən] **1.** *n. (colour)* Zinnoberrot. **2.** *adj.* zinnoberrot

vermin ['vɜ:mın] *n., no pl., no indef. art.* Ungeziefer, *das; (fig. derog.)* Pack, *das (abwertend);* Abschaum, *der (abwertend)*

vermouth ['vɜ:məθ, və'mu:θ] *n.* Wermut[wein], *der*

vernacular [və'nækjʊlə(r)] **1.** *adj. (native)* landessprachlich; ⟨*Predigt, Zeitung*⟩ in der Landessprache; *(not learned or technical)* volkstümlich; *(in dialect)* mundartlich. **2.** *n.* **a)** *(native language)* Landessprache, *die; (dialect)* Dialekt, *der;* **b)** *(jargon)* Sprache, *die; (of a profession or group)* Jargon, *der*

versatile ['vɜ:sətaıl] *adj.* vielseitig; *(mentally)* flexibel; *(having many uses)* vielseitig verwendbar

versatility [vɜ:sə'tılıtı] *n., no pl.* Vielseitigkeit, *die; (mental)* Flexibilität, *die; (variety of uses)* vielseitige Verwendbarkeit

verse [vɜ:s] *n.* **a)** *(line)* Vers, *der;* **b)** *(stanza)* Strophe, *die;* **of** *or* **in** *or* **with five ~s** fünfstrophig; **c)** *no pl., no indef. art. (poetry)* Lyrik, *die;* **write some ~:** einige Verse schreiben; **piece of ~:** Gedicht, *das;* **written in ~:** in Versform; **put sth. into ~:** etw. in Verse fassen; **d)** *(in Bible)* Vers, *der*

versed [vɜ:st] *adj.* **be [well] ~ in sth.** sich in etw. *(Dat.)* [gut] auskennen

version ['vɜ:ʃn] *n.* Version, *die; (in another language)* Übersetzung, *die; (in another form also)* Fassung, *die; (of vehicle, machine, tool)* Modell, *das*

versus ['vɜ:səs] *prep.* gegen

vertebra ['vɜ:tıbrə] *n., pl.* **~e** ['vɜ:tıbri:] *(Anat.)* Wirbel, *der;* **~e** *(backbone)* Wirbelsäule, *die*

vertebrate ['vɜ:tıbrət, 'vɜ:tıbreıt] *(Zool.)* **1.** *adj.* Wirbel⟨*tier*⟩; Wirbeltier⟨*skelett, -fossilien*⟩. **2.** *n.* Wirbeltier, *das*

vertex ['vɜ:teks] *n., pl.* **vertices** ['vɜ:tısi:z] *or* **~es** *(highest point)* Gipfel, *der; (of tower, turret)* Spitze, *die; (Archit.: of dome, arch)* Scheitel[punkt], *der*

vertical ['vɜ:tıkl] **1.** *adj.* senkrecht; senkrecht aufragend *od.* abfallend ⟨*Klippe*⟩; **be ~:** senkrecht stehen. **2.** *n.* senkrechte *od.* vertikale Linie; **be out of [the] ~:** nicht im *od.* außer Lot sein

vertically ['vɜ:tıkəlı] *adv.* senkrecht; vertikal

vertical: ~ **'take-off** *n. (Aeronaut.)* Senkrechtstart, *der;* ~ **'take-off aircraft** *n. (Aeronaut.)* Senkrechtstarter, *der*

vertices *pl. of* **vertex**

vertigo ['vɜ:tıgəʊ] *n., pl.* **~s** Schwindel, *der;* Vertigo, *die (Med.)*

verve [vɜ:v] *n.* Schwung, *der; (of artist, orchestra's playing, sports team's play)* Temperament, *das*

very ['verı] **1.** *attrib. adj.* **a)** *(precise, exact)* genau; **on the ~ day when ...:** genau am [selben] Tag, an dem ...; **you're the ~ person I wanted to see** genau dich wollte ich sehen; **at the ~ moment when ...:** im selben Augenblick, als ...; **in the ~ centre** genau in der Mitte; **the ~ thing** genau das Richtige; **b)** *(extreme)* **at the ~ back/front** ganz hinten/vorn; **at the ~ edge of the cliff** ganz am Rand der Klippe; **at the ~ end/beginning** ganz am Ende/Anfang; **from the ~ beginning** von Anfang an; **only a ~ little** nur ein ganz kleines bißchen; **c)** *(mere)* bloß ‹*Gedanke*›; **at the ~ thought** allein schon beim Gedanken; **the ~ fact of his presence** allein schon seine Anwesenheit; **d)** *(absolute)* absolut ‹*Minimum, Maximum*›; **do one's ~ best** *or* **utmost** sein menschenmöglichstes tun; **the ~ most I can offer is ...:** ich kann allerhöchstens ... anbieten; **it's the ~ least** das ist das Allermindeste; **£50 at the ~ most** allerhöchstens 50 Pfund; **be the ~ first to arrive** als allererster ankommen; **for the ~ last time** zum allerletzten Mal; **e)** *emphat.* **before their ~ eyes** vor ihren Augen; **be caught in the ~ act** auf frischer Tat ertappt werden; **under sb.'s ~ nose** *(fig. coll.)* direkt vor jmds. Augen *(Dat.)*. **2.** *adv.* **a)** *(extremely)* sehr; **it's ~ near** es ist ganz in der Nähe; **in the ~ near future** in allernächster Zukunft; **it's ~ possible that ...:** es ist sehr gut möglich, daß ...; **~ probably** höchstwahrscheinlich; **she's ~/so ~ thin** sie ist sehr dünn/so dünn; **how ~ rude [of him]!** das ist aber unhöflich [von ihm]!; **[yes,] ~ much [so]** [ja,] sehr; **~ much prettier/better** [sehr] viel hübscher/besser; **not ~ much** nicht sehr; **~ little** [nur] sehr wenig ‹*verstehen, essen*›; **thank you [~,] ~ much** [vielen,] vielen Dank; **not ~ big** *(not extremely big)* nicht sehr groß; *(not at all big)* nicht gerade groß; **b)** *(absolutely)* aller‹*best..., -letzt..., -leichtest...*›; **at the ~ latest** allerspätestens; **the ~ last thing I expected** das, womit ich am allerwenigsten gerechnet hatte; **c)** *(precisely)* **the ~ same one** genau der-/die-/dasselbe; **that is the ~ word he used** das ist genau das Wort, das er gebrauchte; **in his ~ next sentence/breath** schon im nächsten Satz/Atemzug; **d) ~ good** *(accepting)* sehr wohl; *(agreeing)* sehr schön; **~ well** *expr. reluctant consent* also gut; na schön; **that's all ~ well, but ...:** das ist ja alles schön und gut, aber ...

vespers ['vespəz] *n. constr. as sing. or pl. (Eccl.)* Vesper, *die*

vessel ['vesl] *n.* **a)** *(receptacle; also Anat., Bot.)* Gefäß, *das;* **[drinking-]~:** Trinkgefäß, *das; see also* **blood-vessel; b)** *(Naut.)* Schiff, *das*

vest [vest] **1.** *n. (Brit.: undergarment)* Unterhemd, *das; (woman's)* Hemd, *das.* **2.** *v.t.* **~ sb. with sth., ~ sth. in sb.** jmdm. etw. verleihen; **be ~ed in sb.** jmdm. übertragen sein; *see also* **vested**

vested ['vestɪd] *adj.* **~ interest/right** wohlerworbener Anspruch; *(established by law)* gesetzlicher Anspruch; **~ interests** *(groups of persons)* Interessengruppen; **have a ~ interest in sth.** *(fig.)* ein persönliches Interesse an etw. *(Dat.)* haben

vestibule ['vestɪbju:l] *n.* **a)** *(indoors)* [Eingangs]halle, *die;* **b)** *(external porch)* Vorhalle, *die;* **c)** *(Amer. Railw.)* Vorraum, *der*

vestige ['vestɪdʒ] *n.* Spur, *die;* **not a ~ of truth/honour** kein Fünkchen Wahrheit/Ehre

vestment ['vestmənt] *n.* [Priester]gewand, *das; (worn on special occasions)* Ornat, *das*

vestry ['vestrɪ] *n. (Eccl.)* Sakristei, *die*

vet [vet] **1.** *n.* Tierarzt, *der/*-ärztin, *die.* **2.** *v.t.*, **-tt-** überprüfen; **~ an article for errors** einen Artikel auf Fehler [hin] durchsehen

veteran ['vetərən] **1.** *n.* Veteran, *der/*Veteranin, *die.* **2.** *attrib. adj.* altgedient ‹*Offizier, Politiker, Schauspieler*›

veteran 'car *n. (Brit.)* Veteran, *der*

veterinarian [vetərɪ'neərɪən] *(Amer.) see* **veterinary surgeon**

veterinary ['vetərɪnərɪ] *attrib. adj.* tiermedizinisch; veterinär; **~ science/medicine** Veterinär- *od.* Tiermedizin, *die*

veterinary 'surgeon *n. (Brit.)* Tierarzt, *der/*-ärztin, *die*

veto ['vi:təʊ] **1.** *n., pl.* **~es a) [power** *or* **right of] ~:** Veto[recht], *das;* **b)** *(rejection, prohibition)* Veto, *das* **(on** gegen, **from** von seiten); **put a** *or* **one's ~ on sth.** sein Veto gegen etw. einlegen. **2.** *v.t.* sein Veto einlegen gegen

vex [veks] *v.t.* [ver]ärgern; *(cause to worry)* beunruhigen; *(dissatisfy, disappoint)* bekümmern; **be ~ed with sb.** sich über jmdn. ärgern

vexation [vek'seɪʃn] *n.* **a)** *(act of harassing)* Belästigung, *die;* **b)** *(state of irritation)* Verärgerung, *die* **(with, at** über + *Akk.*); *(state of worry)* Beunruhigung, *die; (dissatisfaction, disappointment)* Kummer, *der;* **c)** *(annoying thing)* Ärgernis, *das* **(to, for** für)

vexatious [vek'seɪʃəs] *adj.* ärgerlich; unausstehlich ‹*Person*›

vexed [vekst] *adj. (annoyed)* verärgert **(by** über + *Akk.*); *(distressed)* bekümmert **(by** über + *Akk.*)

vexing ['veksɪŋ] *adj.* lästig ‹*Angelegenheit, Problem, Sorgen*›; ärgerlich ‹*Zwickmühle*›

VG *abbr.* **very good**

VHF *abbr.* **Very High Frequency** UKW

via ['vaɪə] *prep.* über (+ *Akk.*) ‹*Ort, Sender, Telefon*›; auf (+ *Dat.*) ‹*Weg*›; durch ‹*Eingang, Schornstein, Person*›; per ‹*Post*›

viability [vaɪə'bɪlɪtɪ] *n., no pl.* **a)** *(of foetus, animal, plant)* Lebensfähigkeit, *die;* **b)** *(fig.) (of state, company)* Lebensfähigkeit, *die; (feasibility)* Realisierbarkeit, *die*

viable ['vaɪəbl] *adj.* **a)** *(capable of maintaining life)* lebensfähig; **b)** *(fig.)* lebensfähig ‹*Staat, Firma*›; *(feasible)* realisierbar

viaduct ['vaɪədʌkt] *n.* Viadukt, *das od. der*

vibrant ['vaɪbrənt] *adj.* pulsierend ‹*Leben*›; lebensprühend ‹*Atmosphäre*›; dynamisch ‹*Kraft*›; lebhaft ‹*Farbe, Rot*›

vibrate [vaɪ'breɪt] **1.** *v.i.* **a)** vibrieren; *(under strong impact)* beben; **b)** *(resound)* [nach]klingen; **c)** *(Phys.)* schwingen; ‹*Glocke:*› vibrieren; **d)** *(thrill)* ‹*Stimme, Körper:*› vibrieren **(with** vor + *Dat.*). **2.** *v.t.* vibrieren lassen; zum Schwingen bringen ‹*Saite*›

vibration [vaɪ'breɪʃn] *n.* **a)** *(vibrating)* Vibrationen; *(visible)* Vibrieren, *das; (under strong impact)* Beben, *das;* **send ~s** *or* **a ~ through sth.** ‹*Erdstoß:*› etw. erzittern lassen; **b)** *(Phys.)* Schwingung, *die;* **c)** *in pl. (fig.)* **get some ~s** etwas spüren; **I get good ~s from this place** dieser Ort hat eine wohltuende Ausstrahlung

vibrato [vɪ'brɑ:təʊ] *n., pl.* **~s** *(Mus.)* Vibrato, *das*

vicar ['vɪkə(r)] *n.* Pfarrer, *der*

vicarage ['vɪkərɪdʒ] *n.* Pfarrhaus, *das*

vicarious [vɪ'keərɪəs] *adj.* **a)** *(delegated)* Stellvertreter-; **b)** *(done for another)* stellvertretend; **c)** *(experienced through another)* nachempfunden ‹*Freude, Erregung usw.*›; **~ [sexual] satisfaction** Ersatzbefriedigung, *die*

¹vice [vaɪs] *n.* **a)** Laster, *das;* **a life/den of ~:** ein Lasterleben/eine Lasterhöhle; **b)** *(character or behaviour defect)* Fehler, *der*

²vice *n. (Brit.: tool)* Schraubstock, *der*

vice- *pref.* Vize-

vice: ~-'chairman *n.* stellvertretender Vorsitzender; **~-'chairmanship** *n.* Amt des/der stellvertretenden Vorsitzenden; **~-'chancellor** *n. (Univ.)* Vizekanzler, *der*/Vizekanzlerin, *die*

'vicelike *adj.* eisern ‹*Griff*›; fest ‹*Umklammerung*›

vice: ~-'presidency *n.* Amt des Vizepräsidenten/der Vizepräsidentin; **~-'president** *n.* Vizepräsident, *der*/-präsidentin, *die*; **~ squad** *n. (Police)* Sittenpolizei, *die*

vice versa [vaɪsɪ 'vɜ:sə] *adv.* umgekehrt

vicinity [vɪ'sɪnɪtɪ] *n.* **a)** *(neighbourhood)* Umgebung, *die*; **in our ~:** nicht weit von uns [entfernt]; **in the immediate ~:** ganz in der Nähe; **in the ~ [of a place]** in der Nähe [eines Ortes]; **in the ~ of 50** *(fig.)* so um die 50; **b)** *no pl. (nearness)* Nähe, *die*

vicious ['vɪʃəs] *adj.* **a)** *(malicious, spiteful)* böse; boshaft ‹*Äußerung*›; böswillig ‹*Versuch, Kritik*›; bösartig ‹*Äußerung, Tier*›; **b)** *(wicked)* skrupellos ‹*Tyrann, Verbrecher*›; schlecht ‹*Person*›; **c)** *(violent, severe)* brutal; unerträglich ‹*Wetter, Schmerz*›

vicious 'circle *n.* Teufelskreis, *der*

viciously ['vɪʃəslɪ] *adv.* **a)** *(maliciously, spitefully)* boshaft; auf gehässige Weise ‹*kritisieren*›; **b)** *(violently, severely)* brutal

vicissitude [vɪ'sɪsɪtju:d] *n.* steter Wandel; **~s** *(fickleness)* Unbeständigkeit, *die*; **the ~s of life** die Wechselfälle des Lebens

victim ['vɪktɪm] *n. (also dupe, Relig.)* Opfer, *das*; *(of sarcasm, abuse)* Zielscheibe, *die (fig.)*; **be the ~ of sb.'s anger/envy/policy** unter jmds. Zorn/Neid/Politik *(Dat.)* zu leiden haben; **fall [a] ~ to sth.** das Opfer einer Sache *(Gen.)* werden; **fall ~ to famine** der Hungersnot *(Dat.)* zum Opfer fallen

victimisation, victimise *see* **victimiz-**

victimization [vɪktɪmaɪ'zeɪʃn] *n.* Schikanierung, *die*; *(selective punishment)* gezielte Bestrafung

victimize ['vɪktɪmaɪz] *v. t.* **a)** *(make a victim)* schikanieren; **be ~d [by sb.]** unter jmdm. zu leiden haben; **b)** *(punish selectively)* gezielt bestrafen

victor ['vɪktə(r)] *n. (rhet.)* Sieger, *der*/Siegerin, *die*

Victorian [vɪk'tɔ:rɪən] **1.** *adj.* viktorianisch. **2.** *n.* Viktorianer, *der*/Viktorianerin, *die*

victorious [vɪk'tɔ:rɪəs] *adj.* siegreich; **be ~ over sb./sth.** über jmdn./etw. siegreich bleiben; **be ~ in one's struggle** aus seinem Kampf siegreich hervorgehen

victory ['vɪktərɪ] *n.* Sieg, *der* (**over** über + *Akk.*); *attrib.* Sieges-; **achieve ~:** den Sieg erringen; **be sure of ~:** der sichere Sieger sein

victuals ['vɪtlz] *n. pl. (dated)* Eßwaren *Pl.*; *(of fort, ship, for journey)* Proviant, *der*

video ['vɪdɪəʊ] **1.** *adj.* Video‹*recorder, -kassette, -kopf*›. **2.** *n., pl.* **~s a)** *(~ recorder)* Videorecorder, *der*; *(~ film, ~tape, ~ recording)* Video, *das (ugs.)*; **have sth. on ~:** etw. auf Video haben *(ugs.)*; **b)** *(visual element of TV broadcasts)* Bild, *das.* **3.** *v. t. see* **videotape 2**

video: ~ camera *n.* Videokamera, *die*; **~ cas'sette** *n.* Videokassette, *die*; **~ cas'sette recorder** *n.* Videokassettenrecorder, *der*; **~ disc** *n.* Bild- *od.* Videoplatte, *die*; **~ film** *n.* Videofilm, *der*; **~ game** *n.* Videospiel, *das*; **~ 'nasty** *n.* Horrorvideo, *das*; **~ recorder** *n.* Videorecorder, *der*; **~ recording** *n.* Videoaufnahme, *die*; **~tape 1.** *n.* Videoband, *das*; **2.** *v. t.* [auf Videoband *(Akk.)*] aufnehmen; **~ telephone** *n.* Bildtelefon, *das*; **~text** *n.* Bildschirmtext, *der*; *(teletext)* Videotext, *der*

vie [vaɪ] *v. i.*, **vying** ['vaɪɪŋ] **~ [with sb.] for sth.** [mit jmdm.] um etw. wetteifern

Vienna [vɪ'enə] **1.** *pr. n.* Wien *(das).* **2.** *attrib. adj.* Wiener

Viennese [vɪə'ni:z] **1.** *adj.* Wiener; **sb. is ~:** jmd. ist Wiener/Wienerin. **2.** *n., pl. same* Wiener, *der*/Wienerin, *die*

Vietnam [vɪet'næm] *pr. n.* **a)** Vietnam *(das)*; **b) ~ [War]** Vietnamkrieg, *der*

Vietnamese [vɪetnə'mi:z] **1.** *adj.* vietnamesisch. **2.** *n., pl. same* **a)** *(person)* Vietnamese, *der*/Vietnamesin, *die*; **b)** *(language)* Vietnamesisch, *das*

view [vju:] **1.** *n.* **a)** *(range of vision)* Sicht, *die*; **get a good ~ of sth.** etw. gut sehen können; **have a clear/distant ~ of sth.** etw. deutlich/in der Ferne sehen können; **be out of/in ~:** nicht zu sehen/zu sehen sein; **come into ~:** in Sicht kommen; **our hotel has a good ~ of the sea** von unserem Hotel aus kann man das Meer gut sehen; **b)** *(what is seen)* Aussicht, *die*; **the ~s from here** die Aussicht von hier; **a room with a ~:** ein Zimmer mit Aussicht; **c)** *(picture)* Ansicht, *die*; **photographic ~:** Foto, *das*; **d)** *(opinion)* Ansicht, *die*; **what is your ~** *or* **are your ~s on this?** was meinst du dazu?; **don't you have any ~[s] about it?** hast du keine Meinung dazu?; **the general/majority ~ is that ...:** die Allgemeinheit/Mehrheit ist der Ansicht, daß ...; **take a favourable ~ of sth.** etw. billigen; **have** *or* **hold ~s about** *or* **on sth.** eine Meinung über etw. *(Akk.)* haben; **hold** *or* **take the ~ that ...:** der Ansicht sein, daß ...; **in my ~:** meiner Ansicht nach; **in sb.'s ~:** nach jmds. Ansicht; **I take a different ~:** ich bin anderer Ansicht; **take a critical/grave/optimistic ~ of sth.** etw. kritisch/ernst/optimistisch beurteilen; **e) be on ~** ‹*Waren, Haus:*› besichtigt werden können; ‹*Bauplan:*› [zur Einsicht] ausliegen; **in ~ of sth.** *(fig.)* angesichts einer Sache; **with a ~ to** *or* **with a** *or* **the ~ of doing sth.** in der Absicht, etw. zu tun; **with a ~ to sth.** *(fig.)* mit etw. im Auge; **with this in ~:** in Anbetracht dessen; *see also* **point 11**; **f)** *(survey)* Betrachtung, *die*; *(of house, site)* Besichtigung, *die.* **2.** *v. t.* **a)** *(look at)* sich *(Dat.)* ansehen; **b)** *(consider)* betrachten; beurteilen ‹*Situation, Problem*›; **~ed in this light ...:** so gesehen ...; **I ~ the matter differently** ich sehe das anders; **c)** *(inspect)* besichtigen; **ask to ~ sth.** darum bitten, etw. besichtigen zu dürfen. **3.** *v. i. (Telev.)* fernsehen

viewdata ['vju:deɪtə] *n. (Teleph.)* Bildschirmtextsystem, *das*

viewer ['vju:ə(r)] *n.* **a)** *(Telev.)* [Fernseh]zuschauer, *der*/-zuschauerin, *die*; **b)** *(Photog.) (for cine film)* Filmbetrachter, *der*; *(for slides)* Diabetrachter, *der*

'viewfinder *n. (Photog.)* Sucher, *der*

viewing ['vju:ɪŋ] *n.* **a)** *(Telev.)* Fernsehen, *das*; **~ figures** Einschaltquoten; **at peak ~ time** zur besten Sendezeit; **b)** *(of house, at auction, etc.)* Besichtigung, *die*

viewpoint ['vju:pɔɪnt] *n.* Standpunkt, *der*; Sehweise, *die*; **from a general/the political ~ ...:** allgemein/politisch gesehen ...

vigil ['vɪdʒɪl] *n.* Wachen, *das*; **keep ~ [over sb.]** [bei jmdm.] wachen

vigilance ['vɪdʒɪləns] *n., no pl.* Wachsamkeit, *die*

vigilant ['vɪdʒɪlənt] *adj.* wachsam; **be ~ for sth.** auf etw. *(Akk.)* achten

vigilante [vɪdʒɪ'læntɪ] *n.* Mitglied einer/der Bürgerwehr; ~ **group** Bürgerwehr, *die*

vignette [vi:'njet] *n. (Lit.)* Skizze, *die*

vigor *(Amer.) see* **vigour**

vigorous ['vɪgərəs] *adj.* kraftvoll; kräftig ⟨*Person, Tier, Stoß, Pflanze, Wachstum, Trieb*⟩; robust ⟨*Gesundheit*⟩; leidenschaftlich ⟨*Verteidigung, Befürworter*⟩; heftig ⟨*Nicken, Attacke, Kritik, Protest*⟩; intensiv ⟨*Gymnastik, Denksport*⟩; energisch ⟨*Versuch, Anstrengung, Leugnen, Maßnahme*⟩; schwungvoll ⟨*Rede*⟩

vigour ['vɪgə(r)] *n. (Brit.)* **a)** *(of person, animal, sexuality)* Vitalität, *die; (of limbs, body)* Kraft, *die; (of health)* Robustheit, *die; (of argument, struggle, protest, denial, attack, criticism)* Heftigkeit, *die; (of performance, speech)* Schwung, *der; (of words, style, mind, intellect)* Lebendigkeit, *die;* **with** ~: schwungvoll ⟨*musizieren, reden, singen, schauspielern*⟩; kräftig ⟨*reiben, schrubben, drücken, ziehen*⟩; **b)** *(Bot.)* Wuchskraft, *die*

Viking ['vaɪkɪŋ] *n. (Hist.)* Wikinger, *der*/Wikingerin, *die; attrib.* Wikinger-

vile [vaɪl] *adj.* **a)** *(base)* verwerflich *(geh.)*; abscheulich ⟨*Charakter, Verbrechen*⟩; gemein ⟨*Verleumdung*⟩; vulgär ⟨*Sprache*⟩; *(repulsive)* widerwärtig; **b)** *(coll.: very unpleasant)* scheußlich *(ugs.)*

villa ['vɪlə] *n.* **a)** *(holiday house)* **[holiday]** ~: Ferienhaus, *das;* **b)** *(country house)* **[country]** ~: Landhaus, *das*

village ['vɪlɪdʒ] *n.* Dorf, *das; attrib.* Dorf⟨*leben, -kneipe usw.*⟩

village: ~ **'green** *n.* Dorfwiese, *die;* ~**'idiot** *n.* Dorftrottel, *der*

villager ['vɪlɪdʒə(r)] *n.* Dorfbewohner, *der*/-bewohnerin, *die*

villain ['vɪlən] *n.* **a)** *(scoundrel)* Verbrecher, *der;* **b)** ~ **[of the piece]** *(Theatre; also fig.)* Bösewicht, *der;* **c)** *(coll.: rascal)* [kleiner] Halunke *(scherzh.)*

villainous ['vɪlənəs] *adj.* **a)** gemein; abscheulich; **b)** *(coll.: very bad)* scheußlich *(ugs.)*

villainy ['vɪlənɪ] *n.* Gemeinheit, *die*

vindicate ['vɪndɪkeɪt] *v. t.* **a)** *(justify, establish)* verteidigen, rechtfertigen ⟨*Person, Meinung, Handeln, Verhalten, Anspruch, Politik*⟩; beweisen ⟨*Behauptung*⟩; *(confirm)* bestätigen ⟨*Recht, Meinung, Urteil, Theorie*⟩; **b)** *(exonerate)* rehabilitieren

vindication [vɪndɪ'keɪʃn] *n. see* **vindicate: a)** Verteidigung, *die;* Rechtfertigung, *die;* Beweis, *der* (**of** für); Bestätigung, *die;* **be a** ~ **of sth.** etw. rechtfertigen/verteidigen/beweisen/bestätigen; **in** ~ **of his claim/conduct** *etc.* zur Rechtfertigung seines Anspruchs/Benehmens *usw.*; **b)** Rehabilitierung, *die*

vindictive [vɪn'dɪktɪv] *adj.* nachtragend ⟨*Person*⟩; unversöhnlich ⟨*Stimmung*⟩; ~ **act/move/attack** Racheakt, *der (geh.)*

vine [vaɪn] *n.* **a)** Weinrebe, *die;* **b)** *(stem of trailer or climber)* Ranke, *die*

vinegar ['vɪnɪgə(r)] *n.* Essig, *der;* **[as] sour as** ~: sehr sauer

vineyard ['vɪnjɑ:d, 'vɪnjəd] *n.* Weinberg, *der*

vintage ['vɪntɪdʒ] **1.** *n.* **a)** *(season's wine)* Jahrgang, *der; (season's grapes)* Traubenernte, *die;* **last/this year's** ~: der letzte/dieser Jahrgang; **the 1981** ~**/a 1983** ~: der 81er/ein 83er; **b)** *(fig.: particular period)* Jahrgang, *der; (of car, machine)* Baujahr, *das;* **c)** *(grape-harvest; season)* Weinlese, *die.* **2.** *adj.* erlesen ⟨*Wein, Sekt, Whisky*⟩; herrlich ⟨*Komödie, Melodie*⟩; brillant ⟨*Leistung, Interpretation*⟩; *(old-fashioned)* alt ⟨*Modell*⟩; altmodisch ⟨*Stil*⟩

vintage 'car *n. (Brit.)* [zwischen 1917 und 1930 gebauter] Oldtimer

vinyl ['vaɪnɪl] *n.* Vinyl, *das*

viola [vɪ'əʊlə] *n. (Mus.)* Bratsche, *die*

violate ['vaɪəleɪt] *v. t.* **a)** verletzen; brechen ⟨*Vertrag, Versprechen, Gesetz*⟩; verstoßen gegen ⟨*Regel, Vorschrift, Prinzipien, Bestimmungen*⟩; verletzen ⟨*Vorschrift*⟩; stören ⟨*Ruhe, Frieden*⟩; verschandeln ⟨*Wälder, Landschaft*⟩; **b)** *(profane)* schänden; entheiligen ⟨*Sabbat*⟩; **c)** *(rape)* vergewaltigen; schänden *(veralt.)*

violation [vaɪə'leɪʃn] *n. see* **violate: a)** Verletzung, *die;* Bruch, *der;* Verstoß, *der* (**of** gegen); Störung, *die;* Verschandelung, *die;* **traffic** ~: Verkehrsdelikt, *das;* **be/act in** ~ **of** verletzen/brechen/verstoßen gegen; **b)** Schändung, *die;* Entheiligung, *die;* **c)** Vergewaltigung, *die;* Schändung, *die (veralt.)*

violence ['vaɪələns] *n., no pl.* **a)** *(intensity, force)* Heftigkeit, *die; (of blow, waterfall)* Wucht, *die; (of temper)* Ungestüm, *das; (of contrast)* Kraßheit, *die;* **b)** *(brutality)* Gewalt, *die; (at public event)* Gewalttätigkeiten; **by** *or* **with** ~: mit Gewalt; **resort to** *or* **use** ~: Gewalt anwenden; **c)** *(Law)* Gewalt, *die;* **threaten sb. with** ~: jmdm. Gewalt androhen; **act/crime of** ~: Gewalttat, *die*/Gewaltverbrechen, *das;* **robbery with** ~: [bewaffneter] Raubüberfall

violent ['vaɪələnt] *adj.* gewalttätig; heftig ⟨*Schlag, Attacke, Leidenschaft, Auseinandersetzung, Erschütterung, Reaktion, Schmerzen, Wind*⟩; wuchtig ⟨*Schlag, Stoß*⟩; schwer ⟨*Schock*⟩; krass ⟨*Gegensatz, Kontrast*⟩; grell ⟨*Farbe*⟩; knall⟨*rot, -grün usw.*⟩; Gewalt⟨*verbrecher, -tat*⟩; **don't be so** ~: sei nicht so aggressiv; **he has a** ~ **temper** er neigt zum Jähzorn; ~ **death** gewaltsamer *od.* unnatürlicher Tod

violet ['vaɪələt] **1.** *n.* **a)** Veilchen, *das;* **shrinking** ~ *(fig.)* schüchternes Pflänzchen *(ugs.)*; **b)** *(colour)* Violett, *das.* **2.** *adj.* violett

violin [vaɪə'lɪn] *n. (Mus.)* Violine, *die;* Geige, *die*

vio'lin case *n.* Geigenkasten, *der*

violinist [vaɪə'lɪnɪst], **vio'lin player** *ns. (Mus.)* Geiger, *der*/Geigerin, *die*

VIP [vi:aɪ'pi:] *n.* Prominente, *der/die;* **the** ~**s** die Prominenz

viper ['vaɪpə(r)] *n.* **a)** *(Zool.)* Viper, *die;* **b)** *(fig.)* Schlange, *die (abwertend)*

VIP: ~ **lounge** *n.* VIP-Halle, *die;* ~ **treatment** *n.* Vorzugsbehandlung, *die;* **give sb.** ~ **treatment** jmdn. mit allen Ehren behandeln

viral ['vaɪərl] *adj. (Med.)* Virus-

virgin ['vɜ:dʒɪn] **1.** *n.* **a)** Jungfrau, *die;* **she/he is still a** ~: sie ist noch Jungfrau/er ist noch unschuldig; **b) the [Blessed] V**~ **[Mary]** *(Relig.)* die [Heilige] Jungfrau [Maria]. **2.** *adj.* **a)** *(chaste)* jungfräulich; **b)** *(untouched, unspoiled)* unberührt ⟨*Land, Wälder*⟩; jungfräulich ⟨*Schnee*⟩; makellos ⟨*Weiß*⟩

virginal ['vɜ:dʒɪnl] *adj.* jungfräulich

virginity [vɜ:'dʒɪnɪtɪ] *n.* Unschuld, *die; (of girl also)* Jungfräulichkeit, *die*

Virgo ['vɜ:gəʊ] *n., pl.* ~**s** *(Astrol., Astron.)* die Jungfrau

virile ['vɪraɪl] *adj.* **a)** *(masculine)* männlich; maskulin *(geh.)*; **b)** *(sexually potent)* viril; **c)** *(fig.: forceful, vigorous)* kraftvoll

virility [vɪ'rɪlɪtɪ] *n.* **a)** Männlichkeit, *die;* **b)** *(sexual potency)* Virilität, *die;* Manneskraft, *die;* **c)** *(fig.)* kraftvoller Schwung

virtual ['vɜ:tjʊəl] *adj.* **a** ~ **...:** so

gut wie ein/eine ...; praktisch ein/eine ... *(ugs.)*; **he is the ~ head of the business** er ist quasi der Chef des Geschäfts *(ugs.)*; **the whole day was a ~ disaster** der ganze Tag war geradezu eine Katastrophe; **the traffic came to a ~ standstill** der Verkehr kam praktisch zum Stillstand *(ugs.)*

virtually ['vɜ:tjʊəlɪ] *adv.* so gut wie; praktisch *(ugs.)*

virtue ['vɜ:tju:] *n.* **a)** *(moral excellence)* Tugend, *die; (chastity)* Tugendhaftigkeit, *die;* **~ is its own reward** *(prov.)* die Tugend trägt ihren Lohn in sich selbst; **b)** *(advantage)* Vorteil, *der;* Vorzug, *der;* **what is the ~ in that?** welchen Vorteil hat das?; **there's no ~ in doing that** es bringt keinen Vorteil, das zu tun; **c) by ~ of** auf Grund, aufgrund (+ *Gen.*)

virtuoso [vɜ:tjʊ'əʊzəʊ] *n., pl.* **virtuosi** [vɜ:tjʊ'əʊzi:] *or* **~s** Virtuose, *der*/Virtuosin, *die; attrib.* virtuos ⟨*Spiel, Aufführung*⟩; **a ~ performer** ein Virtuose/eine Virtuosin

virtuous ['vɜ:tjʊəs] *adj.* rechtschaffen ⟨*Person*⟩; tugendhaft ⟨*Leben*⟩; **if you're feeling ~ you can ...** *(iron.)* wenn du etwas Gutes tun willst, kannst du ...

virulent ['vɪrʊlənt, 'vɪrjʊlənt] *adj.* **a)** *(Med.)* virulent; starkwirkend ⟨*Gift*⟩; **b)** *(fig.: malignant)* heftig; scharf ⟨*Angriff*⟩

virus ['vaɪərəs] *n.* Virus, *der, (fachspr.) das;* **a ~ infection** eine Virusinfektion

visa ['vi:zə] *n.* Visum, *das*

vis-à-vis [vi:zɑ:'vi:] *prep.* **a)** *(in relation to)* bezüglich (+ *Gen.*); **b)** *(facing)* gegenüber; **c)** *(compared with)* im Vergleich zu

viscount ['vaɪkaʊnt] *n.* Viscount, *der*

viscountess ['vaɪkaʊntɪs] *n.* Viscountess, *die*

viscous ['vɪskəs] *adj.* dickflüssig

vise *(Amer.) see* [2]**vice**

visibility [vɪzɪ'bɪlɪtɪ] *n., no pl.* **a)** *(being visible)* Sichtbarkeit, *die;* **b)** *(range of vision)* Sicht, *die; (Meteorol.)* Sichtweite, *die;* **reduce ~ to ten metres** die Sichtweite auf zehn Meter verringern

visible ['vɪzɪbl] *adj.* **a)** *(also Econ.)* sichtbar; **be ~ to the naked eye** mit bloßem Auge erkennbar sein; **~ to observers in X** für Beobachter in X zu sehen; **highly ~** *(fig.)* unübersehbar; **b)** *(apparent)* erkennbar; **with ~ impatience** mit sichtlicher Ungeduld

visibly ['vɪzɪblɪ] *adv.* sichtlich

vision ['vɪʒn] *n.* **a)** *(sight)* Sehkraft, *die;* **[range of] ~:** Sichtweite, *die;* **[field of] ~:** Sehfeld, *das;* **b)** *(dream)* Vision, *die;* Gesicht, *das (geh.); (person seen in dream)* Phantom, *das;* **c)** *usu. pl. (imaginings)* Phantasien; Phantasiebilder; **have ~s of sth.** von etw. phantasieren; *(more specific)* sich *(Dat.)* etw. ausmalen; **have ~s of having to do sth.** kommen sehen, daß man etw. tun muß; **d)** *(insight, foresight)* Weitblick, *der;* **a man/woman of ~:** ein Mann/eine Frau mit Weitblick; **e)** *(Telev.)* Bild, *das;* **in sound and ~:** in Ton und Bild

visionary ['vɪʒənərɪ] **1.** *adj.* **a)** *(imaginative)* phantasievoll; *(fanciful)* phantastisch; **b)** *(imagined)* eingebildet; **c)** *(seeing visions)* visionär. **2.** *n.* Visionär, *der*/Visionärin, *die*

visit ['vɪzɪt] **1.** *v. t.* **a)** besuchen; aufsuchen ⟨*Arzt*⟩; **b)** *(dated: afflict)* heimsuchen. **2.** *v. i.* einen Besuch/Besuche machen; **I'm only ~ing** ich bin nur zu Besuch; **be ~ing with sb.** *(Amer.)* bei jmdm. zu Besuch sein. **3.** *n.* Besuch, *der;* **pay** *or* **make a ~ to sb.** jmdm. einen Besuch abstatten; **pay a ~** *(coll.: go to the toilet)* aufs Klo gehen *(ugs.)*; **have** *or* **receive a ~ [from sb.]** [von jmdm.] besucht werden; **a ~ to a** *or* **the theatre/a museum** ein Theater-/Museumsbesuch; **a ~ to the British Museum** ein Besuch des Britischen Museums; **a ~ to Rome/the USA** ein Besuch *od.* Aufenthalt in Rom/in den USA; **a home ~ by the doctor [to sb.]** ein Hausbesuch des Arztes [bei jmdm.]

visiting: ~-card *n. (lit. or fig.)* Visitenkarte, *die;* **~ hours** *n. pl.* Besuchszeiten; **~ team** *n. (Sport)* Gastmannschaft, *die*

visitor ['vɪzɪtə(r)] *n.* Besucher, *der*/Besucherin, *die; (to hotel, beach, etc.)* Gast, *der;* **have ~s/a ~:** Besuch haben

'visitors' book *n.* Gästebuch, *das*

visor ['vaɪzə(r)] *n.* **a)** *(of helmet)* Visier, *das;* **b)** *(eye-shade, peak of cap)* Schirm, *der;* **c)** *(Motor Veh.)* **[sun] ~:** Blendschirm, *der*

vista ['vɪstə] *n.* **a)** *(view)* [Aus]blick, *der* **(of** auf + *Akk.*); *(long, narrow view)* Perspektive, *die;* **b)** *(fig.)* **open up new ~s** neue Perspektiven eröffnen

visual ['vɪzjʊəl, 'vɪʒjʊəl] *adj.* **a)** *(related to vision)* Seh⟨*nerv, -organ*⟩; **~ sense** Gesichtssinn, *der;* **b)** *(attained by sight)* visuell; optisch ⟨*Eindruck, Darstellung*⟩; bildlich ⟨*Vorstellungsvermögen*⟩; **the ~ arts** die bildenden und darstellenden Künste; **~ display** *(Computing)* Sichtanzeige, *die*

visual: ~ aids *n. pl.* Anschauungsmaterial, *das;* **~ dis'play unit** *n.* Bildschirmgerät, *das*

visualize ['vɪzjʊəlaɪz, 'vɪʒjʊəlaɪz] *v. t.* **a)** *(imagine)* sich *(Dat.)* vorstellen; **b)** *(envisage, foresee)* voraussehen

visually ['vɪzjʊəlɪ, 'vɪʒjʊəlɪ] *adv.* **a)** *(with regard to vision)* optisch; **b)** *(by visual means)* bildlich

vital ['vaɪtl] *adj.* **a)** *(essential to life)* lebenswichtig; **b)** *(essential)* unbedingt notwendig; *(crucial)* entscheidend, ausschlaggebend ⟨*Frage, Entschluß*⟩ **(to** für); **it is of ~ importance** *or* **~ that you ...:** es ist von entscheidender Bedeutung, daß Sie ...; **c)** *(full of life)* lebendig, kraftvoll ⟨*Stil*⟩; vital ⟨*Person*⟩

vitality [vaɪ'tælɪtɪ] *n., no pl.* **a)** *(ability to sustain life)* Lebenskraft, *die;* **b)** *(liveliness)* Vitalität, *die; (of style, language)* Lebendigkeit, *die; (energy)* Energie, *die;* **c)** *(fig.: of institution, organization, etc.)* Dauerhaftigkeit, *die*

vital sta'tistics *n. pl. (coll.: of woman)* Maße; **her ~ are 34-26-34** sie hat die Maße 34/26/34

vitamin ['vɪtəmɪn, 'vaɪtəmɪn] *n.* Vitamin, *das;* **~ C** Vitamin C

vitamin: ~ deficiency *n.* Vitaminmangel, *der;* **~ pill** *n.* Vitamintablette, *die*

vitiate ['vɪʃɪeɪt] *v. t.* **a)** *(impair quality of, corrupt)* beeinträchtigen; **b)** *(invalidate)* zunichte machen; hinfällig machen ⟨*Vereinbarung, Vertrag*⟩

vitreous ['vɪtrɪəs] *adj. (glasslike)* glasartig; **~ china** Halbporzellan, *das*

vitriolic [vɪtrɪ'ɒlɪk] *adj.* ätzend; giftig ⟨*Bemerkung*⟩; geharnischt ⟨*Attacke, Rede*⟩

viva ['vaɪvə] *(Brit. Univ. coll.) n.* Mündliche, *das (ugs.)*

vivacious [vɪ'veɪʃəs] *adj.* lebhaft; lebendig ⟨*Stil*⟩; munter ⟨*Lachen, Lächeln*⟩

viva voce [vaɪvə 'vəʊtsɪ, vaɪvə 'vəʊsɪ] *(Univ.)* **1.** *adv., adj.* mündlich. **2.** *n.* mündliche Prüfung; *(doctoral)* Rigorosum, *das*

vivid ['vɪvɪd] *adj.* **a)** *(bright)* strahlend ⟨*Helligkeit*⟩; hell ⟨*Blitz*⟩; lebhaft ⟨*Farbe*⟩; **b)** *(clear, lifelike)* lebendig ⟨*Schilderung*⟩; lebhaft ⟨*Phantasie, Erinnerung*⟩; **c)** *(intense)* kraftvoll ⟨*Töne*⟩

vividness ['vɪvɪdnɪs] *n., no pl.* **a)** *(brightness)* Helligkeit, *die;* **b)**

(liveliness, realism) Lebhaftigkeit, *die; (of description)* Lebendigkeit, *die*

vivisection [vɪvɪ'sekʃn] *n.* Vivisektion, *die (fachspr.)*

vixen ['vɪksn] *n.* Füchsin, *die*

viz [vɪz] *adv.* d. h.

'**V-neck** *n.* V-Ausschnitt, *der*

'**V-necked** *adj.* ‹*Pullover, Kleid*› mit V-Ausschnitt

vocabulary [və'kæbjʊlərɪ] *n.* **a)** *(list)* Vokabelverzeichnis, *das;* **learn ~:** Vokabeln lernen; **~ book** Vokabelheft, *das;* **~ test** Vokabeltest, *der;* **b)** *(language of particular field)* Vokabular, *das;* **c)** *(range of language)* Wortschatz, *der*

vocal ['vəʊkl] **1.** *adj.* **a)** *(concerned with voice)* stimmlich; **b)** *(expressing oneself freely)* gesprächig; lautstark ‹*Minderheit, Gruppe, Protest*›. **2.** *n. in sing. or pl. (Mus.)* Vokalpartie, *die;* Vocal, *das (fachspr.)*

'**vocal cords** *n. pl.* Stimmbänder

vocalist ['vəʊkəlɪst] *n.* Sänger, *der*/Sängerin, *die (bei einer Band od. Combo)*

vocal: ~ **music** *n.* Vokalmusik, *die;* ~ **score** *n. (Mus.)* Vokalpartitur, *die*

vocation [və'keɪʃn] *n.* **a)** *(call to career; also Relig.)* Berufung, *die;* **he felt no ~ for the ministry** er fühlte sich nicht zum Geistlichen berufen; **b)** *(special aptitude)* Begabung, *die* **(for** für)

vocational [və'keɪʃənl] *adj.* berufsbezogen

vocational: ~ **college** *n.* Berufsschule, *die;* ~ **guidance** *n.* Berufsberatung, *die;* ~ **training** *n.* berufliche Bildung

vociferous [və'sɪfərəs] *adj.* *(noisy)* laut; krakeelend *(ugs.)* ‹*Zwischenrufer usw.*›; *(insistent)* lautstark ‹*Forderung, Protest*›

vodka ['vɒdkə] *n.* Wodka, *der*

vogue [vəʊg] *n.* Mode, *die;* **the ~ for large hats** die Mode mit den großen Hüten; **be in/come into ~:** in Mode sein/kommen; **go out of ~:** aus der Mode kommen

voice [vɔɪs] **1.** *n.* **a)** *(lit. or fig.)* Stimme, *die;* **in a firm/loud ~:** mit fester/lauter Stimme; **lose one's ~:** die Stimme verlieren; **make one's ~ heard** sich verständlich machen; *(fig.)* sich *(Dat.)* Gehör verschaffen; **b)** *(expression)* **give ~ to sth.** einer Sache *(Dat.)* Ausdruck geben; **c) with one ~:** einstimmig; **d)** *(Mus.)* Stimme, *die;* **[singing] ~:** Singstimme, *die;* **e)** *(Ling.)* **the active/passive ~:** das Aktiv/Passiv. **2.** *v. t.* **a)** *(express)* zum Ausdruck bringen ‹*Meinung*›; **b)** *esp. in p.p. (Phonet.)* stimmhaft aussprechen; **a ~d consonant** ein stimmhafter Konsonant

voice: ~**-box** *n.* Kehlkopf, *der;* ~**-over** *n.* Begleitkommentar, *der*

void [vɔɪd] **1.** *adj.* **a)** *(empty)* leer; **b)** *(invalid)* ungültig. **2.** *n. (empty space)* Nichts, *das; (fig.)* **there was an aching ~ in her heart** sie spürte im Innern ein schmerzliches Gefühl der Leere

vol. *abbr.* **volume** Bd.

volatile ['vɒlətaɪl] *adj.* **a)** *(Chem.)* flüchtig; **b)** *(fig.) (lively)* impulsiv; *(changeable)* unbeständig ‹*Person, Laune*›; *(likely to erupt)* explosiv ‹*Temperament*›; brisant ‹*Lage*›

vol-au-vent ['vɒləʊvɑ̃] *n. (Gastr.)* Pastete, *die*

volcanic [vɒl'kænɪk] *adj.* **a)** vulkanisch; **~ eruption** Vulkanausbruch, *der;* **b)** *(fig.: violent)* leidenschaftlich

volcano [vɒl'keɪnəʊ] *n., pl.* **~es** Vulkan, *der*

vole [vəʊl] *n.* Wühlmaus, *die;* **field ~:** Feldmaus, *die*

Volga ['vɒlgə] *pr. n.* Wolga, *die*

volition [və'lɪʃn] *n.* Wille, *der;* **of one's own ~:** aus eigenem Willen; freiwillig

volley ['vɒlɪ] *n.* **a)** *(of missiles)* Salve, *die;* **a ~ of stones/arrows** ein Hagel von Steinen/Pfeilen; ein Stein-/Pfeilhagel; **b)** *(fig.)* **a ~ of oaths/curses** eine Schimpfkanonade; **direct a ~ of questions at sb.** jmdn. mit Fragen bombardieren; **c)** *(Tennis)* Volley, *der*

'**volleyball** *n.* Volleyball, *der*

volt [vəʊlt] *n. (Electr.)* Volt, *das*

voltage ['vəʊltɪdʒ] *n. (Electr.)* Spannung, *die;* **high/low ~:** Hoch-/Niederspannung, *die*

volte-face [vɒlt'fæs] *n. (fig.)* Kehrtwendung, *die*

voluble ['vɒljʊbl] *adj.* redselig *(abwertend);* wortreich ‹*Rede*›

volume ['vɒlju:m] *n.* **a)** *(book, set of periodicals)* Band, *der; (on periodical)* **V~ II no. 3** Jahrgang II, Nr. 3; *see also* **speak 2 c; b)** *(loudness)* Lautstärke, *die; (of voice)* Volumen, *das;* **turn the ~ up/down** das Radio *usw.* lauter/leiser stellen; **c)** *(amount of space)* Rauminhalt, *der;* Volumen, *das; (amount of substance)* Teil, *der;* **d)** *(amount, quantity) (of sales etc.)* Volumen, *das;* **~ of traffic/passenger travel** Verkehrs-/Passagieraufkommen, *das*

'**volume control** *n.* Lautstärkeregelung, *die; (device)* Lautstärkeregler, *der*

voluminous [və'lju:mɪnəs, və'lu:mɪnəs] *adj.* **a)** *(great in quantity)* voluminös *(geh.);* sehr umfangreich; **b)** *(bulky, loose)* weit ‹*Kleider*›; voluminös *(geh.)* ‹*Tasche usw.*›

voluntarily ['vɒləntərɪlɪ] *adv.* freiwillig

voluntary ['vɒləntərɪ] *adj.* freiwillig; **~ organizations** Freiwilligenverbände

volunteer [vɒlən'tɪə(r)] **1.** *n.* Freiwillige, *der/die;* **any ~s?** Freiwillige vor!; *attrib.* **~ army/force** Freiwilligenheer, *das*/Freiwilligenverband, *der.* **2.** *v. t. (offer)* anbieten ‹*Hilfe, Dienste*›; herausrücken mit *(ugs.)* ‹*Informationen, Neuigkeiten*›; **~ advice** unerbetene Ratschläge erteilen. **3.** *v. i.* sich [freiwillig] melden; **~ to do** *or* **for the shopping** sich zum Einkaufen bereiterklären

voluptuous [və'lʌptjʊəs] *adj.* üppig ‹*Figur, Kurven, Blondine*›; aufreizend ‹*Bewegungen*›; sinnlich ‹*Mund*›

vomit ['vɒmɪt] **1.** *v. t.* erbrechen. **2.** *v. i.* sich übergeben; [sich] erbrechen. **3.** *n.* Erbrochene, *das*

voodoo ['vu:du:] *n. (witchcraft)* Wodu, *der*

voracious [və'reɪʃəs] *adj.* **a)** *(ravenous)* gefräßig ‹*Person, Tier*›; unbändig ‹*Appetit*›; **b)** *(fig.: insatiable)* unersättlich ‹*Lust, Leser*›

vortex ['vɔ:teks] *n., pl.* **vortices** ['vɔ:tɪsi:z] *or* **~es** *(whirlpool, whirlwind)* Wirbel, *der; (eddying current; also fig.: whirl)* Strudel, *der*

vote [vəʊt] **1.** *n.* **a)** *(individual ~)* Stimme, *die;* **a majority of ~s** eine Stimmenmehrheit; **my ~ goes to X, X has my ~** *(fig. coll.)* ich stimme *od.* bin für X; **b)** *(act of voting)* Abstimmung, *die;* **take a ~ on sth.** über etw. *(Akk.)* abstimmen; **c)** *(right to ~)* **have/be given** *or* **get the ~:** das Stimmrecht haben/bekommen; **d)** *(collective)* Stimmen; *(result)* Abstimmungsergebnis, *das;* **the ~ in favour of capital punishment** die Stimmenzahl für die Todesstrafe; **e)** *(expression of opinion)* Votum, *das;* **give sb. a ~ of confidence/no confidence** jmdm. sein Vertrauen/Mißtrauen aussprechen; **~ of confidence/no confidence** Vertrauens-/Mißtrauensvotum, *das;* **propose a ~ of thanks** eine Dankadresse halten. **2.** *v. i.* abstimmen; *(in election)* wählen; **~ for/against** stimmen für/gegen; **~ for**

Smith! wählen Sie Smith!; ~ **on a motion** über einen Antrag abstimmen; ~ **to do sth.** beschließen, etw. zu tun; ~ **by ballot/[a] show of hands** mit Stimmzetteln/durch Handzeichen abstimmen; ~ **Conservative/Labour** *etc.* die Konservativen/Labour *usw.* wählen. **3.** *v.t.* **a)** *(elect)* ~ **sb. Chairman/President** *etc.* jmdn. zum Vorsitzenden/Präsidenten *usw.* wählen; *(approve)* ~ **a sum of money for sth.** einen Betrag für etw. bewilligen; **b)** *(coll.: pronounce)* bezeichnen; ~ **sth. a success/failure** etw. als Erfolg/Mißerfolg bezeichnen
~ **'down** *v.t.* niederstimmen
~ **'in** *v.t.* wählen ‹*Partei, Regierung*›
~ **'out** *v.t.* abwählen
voter ['vəʊtə(r)] *n.* Wähler, *der*/Wählerin, *die*
voting ['vəʊtɪŋ] *n.* Abstimmen, *das; (in election)* Wählen, *das;* **the ~ was 220 for, 165 against** das Ergebnis der Abstimmung war 220 [Stimmen] dafür, 165 dagegen
'voting system *n.* Wahlsystem, *das*
vouch [vaʊtʃ] **1.** *v.t.* ~ **that ...:** sich dafür verbürgen, daß ... **2.** *v.i.* ~ **for sb./sth.** sich für jmdn./etw. verbürgen
voucher ['vaʊtʃə(r)] *n.* Gutschein, *der;* Voucher, *der (Touristmus)*
vow [vaʊ] **1.** *n.* Gelöbnis, *das; (Relig.)* Gelübde, *das.* **2.** *v.t.* ~ **sth./to do sth.** etw. geloben/geloben, etw. zu tun; ~ **to take revenge on sb.** jmdm. Rache schwören
vowel ['vaʊəl] *n.* Vokal, *der;* Selbstlaut, *der;* ~ **sound** Vokallaut, *der*
voyage ['vɔɪɪdʒ] **1.** *n.* Reise, *die; (sea ~)* Seereise, *die;* **outward/homeward ~, ~ out/home** Hin-/Rückreise, *die;* **he was on a ~ of discovery** *(lit. or fig.)* er war auf einer Entdeckungsreise. **2.** *v.i.* *(literary)* reisen
vroom [vru:m, vrʊm] *int.* brumm
vs *abbr.* **versus** gg.
'V-shaped *adj.* V-förmig
'V-sign *n.* **a)** *(sign for victory)* Siegeszeichen, *das;* **b)** *(gesture of abuse, contempt) Zeichen, das „Du kannst mich mal!" signalisiert*
VSO *abbr.* **Voluntary Service Overseas**
VTO[L] ['vi:tɒl] *abbr. (Aeronaut.)* **vertical take-off [and landing]** Senkrechtstart [und -landung]
vulgar ['vʌlgə(r)] *adj.* vulgär; ordinär ‹*Person, Benehmen, Witz, Film*›; geschmacklos ‹*Kleidung*›
vulgarity [vʌl'gærɪtɪ] *n., no pl.* Vulgarität, *die; (of clothing)* Geschmacklosigkeit, *die*
vulnerability [vʌlnərə'bɪlɪtɪ] *n., no pl.* **a)** Angreifbarkeit, *die; (to criticism, temptation)* Anfälligkeit, *die* **(to** für); **b)** *(to injury)* Empfindlichkeit, *die* **(to** gegen); Schutzlosigkeit, *die; (emotional)* Verletzlichkeit, *die*
vulnerable ['vʌlnərəbl] *adj.* **a)** *(exposed to danger)* angreifbar; **a ~ spot/point** ein schwacher Punkt; **be ~ to sth.** für etw. anfällig sein; **be ~ to attack/in a ~ position** leicht angreifbar sein; **b)** *(susceptible to injury)* empfindlich **(to** gegen); *(without protection)* schutzlos; ~ **to infection** anfällig für Infektionen
vulture ['vʌltʃə(r)] *n. (lit. or fig.)* Geier, *der*
vying *see* **vie**

W, w ['dʌblju:] *n., pl.* **Ws** *or* **W's** W, w, *das*
W. *abbr.* **a) watt[s]** W; **b) west** W.; **c) western** w.
wad [wɒd] *n.* **a)** *(material)* Knäuel, *das; (smaller)* Pfropfen, *der;* **a ~ of cotton wool** ein Wattebausch; **b)** *(of papers)* Bündel, *das;* **~s of money** bündelweise Geld
wadding ['wɒdɪŋ] *n. (lining)* Futter, *das; (for packing)* Füllmaterial, *das;* Füllsel *Pl.*
waddle ['wɒdl] **1.** *v.i.* watscheln. **2.** *n.* watschelnder Gang
wade [weɪd] *v.i.* waten; *(in snow, sand)* stapfen
~ **'in** *v.i. (fig. coll.)* [gleich] losgehen; *(tackle task)* sich hineinknien *(ugs.)*
~ **into** *v.t. (fig. coll.)* losgehen auf (+ *Akk.*)
~ **through** *v.t.* **a)** waten durch; stapfen durch ‹*Schnee, Unkraut*›; **b)** *(fig. coll.)* durchackern *(ugs.)* ‹*Manuskript, Buch*›
wafer ['weɪfə(r)] *n.* **a)** Waffel, *die; (very thin)* Oblate, *die;* **b)** *(Eccl.)* Hostie, *die;* **c)** *(Electronics)* Wafer, *der*
'wafer-thin *adj.* hauchdünn
[1]waffle ['wɒfl] *n. (Gastr.)* Waffel, *die*
[2]waffle *(Brit. coll.: talk)* **1.** *v.i.* schwafeln *(ugs. abwertend);* faseln *(ugs. abwertend).* **2.** *n.* Geschwafel, *das (ugs. abwertend);* Faselei, *die (ugs. abwertend)*
waft [wɒft, wɑ:ft] **1.** *v.t.* wehen. **2.** *v.i.* ‹*Geruch, Duft:*› ziehen, *(with perceptible air-movement)* wehen
[1]wag [wæg] **1.** *v.t.,* **-gg-** ‹*Hund:*› wedeln mit ‹*Schwanz*›; ‹*Vogel:*› wippen mit ‹*Schwanz*›; ‹*Person:*› schütteln ‹*Kopf*›; ~ **one's finger at sb.** jmdm. mit dem Finger drohen. **2.** *v.i.,* **-gg-** ‹*Schwanz:*› wedeln/*(of bird)* wippen; **her tongue never stops ~ging** ihre Zunge steht niemals still. **3.** *n. (of dog's tail)* Wedeln, *das* **(of** mit); *(of bird's tail)* Wippen, *das* **(of** mit)
[2]wag *n. (facetious person)* Witzbold, *der (ugs.)*
wage [weɪdʒ] **1.** *n. in sing. or pl.* Lohn, *der.* **2.** *v.t.* führen ‹*Krieg, Feldzug*›; ~ **war on** *or* **against crime** *(fig.)* gegen das Verbrechen zu Felde ziehen
wage: ~-claim *n.* Lohnforderung, *die;* **~-earner** *n.* Lohnempfänger, *der/*-empfängerin, *die;* ~ **freeze** *n.* Lohnstopp, *der;* ~ **increase** *n.* Lohnerhöhung, *die;* ~ **packet** *n.* Lohntüte, *die*
wager ['weɪdʒə(r)] *(dated/formal)* **1.** *n.* Wette, *die;* **a ~ of £50** eine Wette um 50 Pfund. **2.** *v.t.* wetten; *(on a horse)* setzen; ~ **one's life/one's whole fortune on sth.** seinen Kopf/sein ganzes Vermögen auf etw. *(Akk.)* verwetten; **I ~ you £10 that ...:** ich wette mit dir um 10 Pfund, daß ... **3.** *v.i.* wetten; **he's there by now, I'll ~:** ich möchte wetten, daß er inzwischen da ist
waggle ['wægl] *(coll.)* **1.** *v.t.* ~ **its tail** ‹*Hund:*› mit dem Schwanz wedeln; ‹*Vogel:*› mit dem Schwanz wippen. **2.** *v.i.* hin und her schlagen
waggon *(Brit.),* **wagon** ['wægən] *n.* **a)** *(horse-drawn)* Wagen, *der;* **covered ~:** Planwagen, *der;* **b)** *(Amer.: motor vehicle)* Wagen, *der;* **c) go/be on the ~** *(go/be teetotal)* keinen Tropfen mehr/keinen Tropfen anrühren; **d)** *(Brit. Railw.)* Wagen, *der;* Waggon, *der (volkst.)*
'wagtail *n. (Ornith.)* Bachstelze, *die*

waif [weɪf] *n.* verlassenes Kind; **~s and strays** obdachlose Kinder

wail [weɪl] **1.** *v. i.* **a)** *(lament)* klagen *(geh.)* (**for** um); jammern *(ugs.)* (**for** um); ‹*Kind:*› heulen; **b)** *(fig.)* ‹*Wind, Sirene:*› heulen. **2.** *n.* **a)** *(cry)* klagender Schrei; **~s** Geheul, *das;* **b)** *(fig.: of wind etc.)* Heulen, *das;* Geheul, *das*

waist [weɪst] *n.* **a)** *(part of body or garment)* Taille, *die;* **tight round the ~:** eng in der Taille; **b)** *(Amer.)* *(blouse)* Bluse, *die;* *(bodice)* Mieder, *das*

'**waistband** *n.* Gürtelbund, *der;* *(of trousers)* [Hosen]bund, *der;* *(of skirt)* [Rock]bund, *der*

waistcoat ['weɪskəʊt, 'weɪstkəʊt] *n. (Brit.)* Weste, *die*

'**waist-deep 1.** *adj.* bis zur Taille reichend; **be ~:** einem bis zur Taille reichen. **2.** *adv.* bis zur Taille

'**waistline** *n.* Taille, *die;* **be bad for the ~:** schlecht für die schlanke Linie sein

wait [weɪt] **1.** *v. i.* **a)** warten; **~ [for] an hour** eine Stunde warten; **~ a moment** Moment mal; **keep sb. ~ing, make sb. ~:** jmdn. warten lassen; **~ to see sth. happen** darauf warten, daß etw. passiert; **'repairs [done]/keys cut while you ~'** „Reparatur-/Schlüsselschnelldienst"; **sth. is still ~ing to be done** etw. muß noch gemacht werden; **~ and see** abwarten[, was passiert]; **sth. can/can't** *or* **won't ~:** etw. kann/kann nicht warten; **I can't ~ to do sth.** *(am eager)* ich kann es kaum erwarten, etw. zu tun; **[just] you ~!** warte mal ab!; *(as threat)* warte nur!; **b) ~ at** *or* *(Amer.)* **on table** servieren; ‹*Ober:*› kellnern *(ugs.).* **2.** *v. t.* *(await)* warten auf (+ *Akk.*); **~ one's chance/opportunity** auf eine [günstige] Gelegenheit warten; **~ one's turn** warten, bis man dran ist *od.* drankommt. **3.** *n.* **a)** *(act, time)* **after a long/short ~:** nach langer/kurzer Wartezeit; **have a long/short ~ for sth.** lange/nicht lange auf etw. *(Akk.)* warten müssen; **b)** *(watching for enemy)* **lie in ~:** im Hinterhalt liegen; **lie in ~ for sb./sth.** jmdm./einer Sache auflauern

~ a'bout, ~ a'round *v. i.* herumstehen

~ be'hind *v. i.* noch hier-/dableiben; **~ behind for sb.** auf jmdn. warten

~ for *v. t.* warten auf (+ *Akk.*); **~ for the rain to stop** warten, bis der Regen aufhört; **~ for it!** warte/wartet!; *(to create suspense before saying something surprising)* warte ab!

~ 'in *v. i.* zu Hause warten (**for** auf + *Akk.*)

~ on *v. t. (serve)* bedienen

~ 'up *v. i.* aufbleiben (**for** wegen)

waiter ['weɪtə(r)] *n.* Kellner, *der;* **~!** Herr Ober!

waiting ['weɪtɪŋ] *n.* **a)** Warten, *das;* **b)** *no pl., no art. (working as waiter)* Servieren, *das;* Kellnern, *das (ugs.)*

waiting: ~ game *n.* Hinhaltetaktik, *die;* **play a ~ game** erst einmal abwarten; **~-list** *n.* Warteliste, *die;* **~-room** *n.* Wartezimmer, *das;* *(at railway or bus station)* Warteraum, *der;* *(larger)* Wartesaal, *der*

waitress ['weɪtrɪs] *n.* Serviererin, *die;* **~!** Fräulein! *(veralt.)*

waive [weɪv] *v. t.* verzichten auf (+ *Akk.*); nicht vollstrecken ‹*Strafe*›; nicht anwenden ‹*Regel*›

[1]**wake** [weɪk] **1.** *v. i.*, **woke** [wəʊk] *or (arch.)* **~d, woken** ['wəʊkn] *or (arch.)* **~d** aufwachen; *(fig.)* ‹*Natur, Gefühle:*› erwachen; **I woke to the sound of soft music** beim Aufwachen hörte ich leise Musik; **~ to sth.** *(fig.: realize)* etw. erkennen; sich *(Dat.)* einer Sache *(Gen.)* bewußt werden. **2.** *v. t.* **woke** *or (arch.)* **~d, woken** *or (arch.)* **~d** wecken; *(fig.)* erwecken *(geh.)* ‹*die Natur, Erinnerungen*›; wecken ‹*Erinnerungen*›; **be quiet, you'll ~ your baby brother** sei still, sonst wacht dein Brüderchen auf! **3.** *n. (Ir.: watch by corpse)* Totenwache, *die*

~ 'up 1. *v. i. (lit. or fig.)* aufwachen; **~ up!** wach auf!; *(fig.: pay attention)* paß besser auf!; **~ up to sth.** *(fig.: realize)* etw. erkennen. **2.** *v. t.* **a)** *(rouse from sleep)* wecken; **b)** *(fig.: enliven)* wachrütteln; Leben bringen in (+ *Akk.*) ‹*Stadt*›; **you need to ~ your ideas up a bit** du müßtest dich ein bißchen zusammenreißen

[2]**wake** *n.* **a)** *(water)* Kielwasser, *das;* **b)** *(air)* Turbulenz, *die;* **c)** *(fig.)* **in the ~ of sth./sb.** im Gefolge von etw./in jmds. Gefolge; **follow in the ~ of sb./sth.** jmdm./einer Sache folgen; **bring sth. in its ~:** etw. zur Folge haben

wakeful ['weɪkfl] *adj.* **a)** *(sleepless)* schlaflos ‹*Nacht*›; **b)** *(vigilant)* wachsam

waken ['weɪkn] **1.** *v. t.* **a)** wecken; **b)** *(fig.: arouse)* wecken ‹*Interesse, Gefühl*›; erregen ‹*Zorn*›. **2.** *v. i. see* [1]**wake 1**

waking ['weɪkɪŋ] *adj.* **in one's ~ hours** den ganzen Tag; von früh bis spät; **spend all one's ~ hours [on] doing sth.** etw. von früh bis spät tun

Wales [weɪlz] *pr. n.* Wales *(das)*

walk [wɔ:k] **1.** *v. i.* **a)** laufen; *(as opposed to running)* gehen; *(as opposed to driving)* zu Fuß gehen; **you can ~ there in five minutes** es sind nur 5 Minuten zu Fuß bis dorthin; **'~'/'don't ~'** *(Amer.: at pedestrian lights)* „gehen"/„warten"; **~ on crutches/with a stick** an Krücken/am Stock gehen; **learn to ~:** laufen lernen; **~ tall** *(fig.)* erhobenen Hauptes gehen *(fig.);* **b)** *(exercise)* gehen; marschieren *(ugs.).* **2.** *v. t.* **a)** entlanggehen; ablaufen ‹*Strecke, Weg*›; durchwandern ‹*Gebiet*›; **~ the streets** durch die Straßen gehen/*(aimlessly)* laufen; *(as prostitute)* auf den Strich gehen *(ugs.);* **b)** *(cause to ~; lead)* führen; ausführen ‹*Hund*›; **~ sb. off his/her feet** jmdn. [bis zur Erschöpfung] durch die Gegend schleifen *(ugs.);* **c)** *(accompany)* bringen; **he ~ed his girl-friend home** er brachte seine Freundin nach Hause. **3.** *n.* **a)** Spaziergang, *der;* **go [out] for** *or* **take** *or* **have a ~:** einen Spaziergang machen; **take sb./the dog for a ~:** jmdn./den Hund spazierenführen; **a ten-mile ~:** eine Wanderung von zehn Meilen; *(distance)* **ten minutes' ~ from here** zehn Minuten zu Fuß von hier; **b)** *(gait)* Gang, *der;* *(characteristic)* normale Gangart; **c)** *(Sport: race)* Wettbewerb im Gehen; **the 10,000 metres ~:** das 10 000-m-Gehen; **d)** *(path, route)* [Spazier]weg, *der;* **e) people from all ~s of life** Leute aus den verschiedensten gesellschaftlichen Gruppierungen

~ a'bout *v. i.* herumlaufen

~ a'way *v. i.* **a)** weggehen; **she was lucky to ~ away from the accident** sie hatte großes Glück, den Unfallort unverletzt verlassen zu können; **b)** *(fig.)* **he tried to ~ away from the problem** *(ignore it)* er versuchte, dem Problem aus dem Weg zu gehen; **~ away with sth.** *(coll.)* *(win easily)* etw. spielend leicht gewinnen; *(steal)* sich mit etw. davonmachen *(ugs.)*

~ 'in *v. i.* **a)** *(enter)* hereinkommen/hineingehen; reinkommen/-gehen *(ugs.);* **'please ~ in'** „[bitte] eintreten, ohne zu klopfen"; **b)** *(enter without permission)* hinein-/hereinspazieren

~ into *v. t.* **a)** *(enter)* betreten; treten in (+ *Akk.*) ‹*Pfütze*›; *(without permission)* eindringen in (+ *Akk.*) ‹*Haus*›; **b)** *(hit by accident)* laufen gegen ‹*Pfosten, Laternenpfahl*›; ~ **into sb.** mit jmdm. zusammenstoßen; ~ **into a trap** *(lit. or fig.)* in eine Falle gehen; **c)** *(coll.: come easily into)* **she ~ed into the top job** ihr ist der Topjob einfach zugefallen

~ **'off 1.** *v.i.* **a)** *(leave)* weggehen; verschwinden; **b)** ~ **off with sth.** *(coll.)* sich mit etw. davonmachen *(ugs.)*; ~ **off with all the prizes** alle Preise einheimsen *(ugs.)*. **2.** *v.t.* ~ **off a hangover** einen Spaziergang machen, um seinen Kater loszuwerden

~ **'on** *v.i.* **a)** *(go further)* weitergehen; ~ **on!** *(to horse)* hü!; **b)** *(go on stage)* auf die Bühne kommen

~ **'out** *v.i.* **a)** *(leave)* hinausgehen; rausgehen *(ugs.)*; **b)** *(Mil.: leave barracks)* ausgehen; **c)** *(leave in protest)* aus Protest den Saal verlassen; *(leave organization)* austreten; **d)** *(go on strike)* in den Streik *od.* Ausstand treten

~ **'out of** *v.t.* **a)** *(leave)* gehen aus; **b)** *(leave in protest)* verlassen ‹*Saal, Versammlung*›

~ **'out on** *v.t. (coll.)* verlassen; sitzenlassen *(ugs.)* ‹*Frau, Mann*›

~ **'over** *v.t.* ~ **[all] over sb.** jmdn. fertigmachen *(ugs.)*

~ **'up** *v.i.* **a)** *(approach)* sich nähern; ~ **up to sb.** zu jmdm. hingehen; **he ~ed up to me** er kam zu mir [heran]; **b)** *(ascend)* hochlaufen; nach oben laufen

'walkabout *n.* **a)** *(through crowds)* Bad in der Menge *(scherzh.)*; **go on a ~:** sich unters Volk mischen; **b)** *(Austral.: in bush)* Buschwanderung, *die*

walker ['wɔ:kə(r)] *n.* Spaziergänger, *der*/-gängerin, *die; (in race)* Geher, *der*/Geherin, *die; (rambler, hiker)* Wanderer, *der*/Wanderin, *die*

walkie-talkie [wɔ:kɪ'tɔ:kɪ] *n.* Walkie-talkie, *das*

walking ['wɔ:kɪŋ] **1.** *attrib. adj.* **a ~ dictionary/encyclopaedia** *(joc.)* ein wandelndes Wörterbuch/Konversationslexikon. **2.** *n., no pl., no art.* [Spazieren]gehen, *das;* Laufen, *das; attrib.* **at ~ pace** im Schrittempo; **be within ~ distance** zu Fuß zu erreichen sein

walking: ~ **holiday** *n.* Wanderurlaub, *der;* ~ **shoe** *n.* Wanderschuh, *der;* ~**-stick** *n.* Spazierstock, *der;* ~**-tour** *n.* Wanderung, *die*

Walkman, (P) ['wɔ:kmən] *n., pl.* **~s** Walkman, *der* Ⓦⓩ

walk: ~-on part *n. (Theatre)* Statistenrolle, *die;* **~-out** *n.* Arbeitsniederlegung, *die;* **~-over** *n. (fig.: easy victory)* Spaziergang, *der (ugs.);* **~way** *n.* Fußweg, *der; (over machinery etc.)* Laufsteg, *der*

wall [wɔ:l] **1.** *n.* **a)** *(of building, part of structure)* Wand, *die; (external, also free-standing)* Mauer, *die;* **town/garden ~:** Stadt-/Gartenmauer, *die;* **the south ~ of the house** die Südwand des Hauses; **a concrete ~:** eine Betonwand/-mauer; **the Great W~ of China** die Chinesische Mauer; **the Berlin W~** *(Hist.)* die [Berliner] Mauer; **b)** *(internal)* Wand, *die;* **be hanging on the ~:** an der Wand hängen; **hang a picture on the ~:** ein Bild an die Wand hängen; **drive** *or* **send sb. up the ~** *(fig. coll.)* jmdn. auf die Palme bringen *(ugs.);* **go up the ~** *(fig. coll.)* die Wände hochgehen *(ugs.);* **c)** *(Mount., Min.)* Wand, *die; (fig.)* Mauer, *die;* **a ~ of water/fire** *(fig.)* eine Wasser-/Feuerwand; **the North W~ of the Eiger** die Eigernordwand; *(fig.)* **a ~ of silence/prejudice** eine Mauer des Schweigens/von Vorurteilen. **2.** *v.t.* **[be] ~ed** von einer Mauer/Mauern umgeben [sein]; **X is a ~ed city/town** X hat eine Stadtmauer

~ **'in** *v.t.* mit einer Mauer umgeben; *(fig.)* umzingeln

~ **'off** *v.t.* abteilen

~ **'up** *v.t.* zumauern; einmauern ‹*Person*›

wallaby ['wɒləbɪ] *n. (Zool.)* Wallaby, *das*

wall: ~ **bars** *n. pl.* Sprossenwand, *die;* **~-cupboard** *n.* Hängeschrank, *der*

wallet ['wɒlɪt] *n.* Brieftasche, *die; (for cheque card etc.)* Etui, *das*

wall: ~flower *n.* **a)** *(Bot.)* Goldlack, *der;* **b)** *(coll.: person)* Mauerblümchen, *das (ugs.);* **~-hanging** *n.* Wandbehang, *der;* **~-light** *n.* Wandlampe, *die;* ~ **map** *n.* Wandkarte, *die*

wallop ['wɒləp] *(coll.)* **1.** *v.t. (hit)* schlagen. **2.** *n.* Schlag, *der;* **give sb./sth. a ~:** auf jmdn./etw. draufhauen *(ugs.)*

wallow ['wɒləʊ] *v.i.* **a)** *(roll around)* sich wälzen; ‹*Schiff:*› schlingern; *(in mud also)* sich suhlen; **b)** *(fig.: take delight)* schwelgen (**in** in + *Dat.*)

wall: ~-painting *n.* Wandgemälde, *das;* **~paper 1.** *n.* Tapete, *die;* **2.** *v.t.* tapezieren; ~ **socket** *n. (Electr.)* Wandsteckdose, *die;* **~-to-~** *adj. (covering floor)* **~-to-~ carpeting** Teppichboden, *der;* **~-unit** *n.* Hängeelement, *das*

wally ['wɒlɪ] *n. (Brit. sl.)* Blödmann, *der (salopp)*

walnut ['wɔ:lnʌt] *n.* **a)** *(nut)* Walnuß, *die;* **b)** *(tree)* [Wal]nußbaum, *der;* **c)** *(wood)* Nußbaumholz, *das*

walrus ['wɔ:lrəs, 'wɒlrəs] *n.* Walroß, *das*

waltz [wɔ:lts, wɔ:ls, wɒlts, wɒls] **1.** *n.* Walzer, *der.* **2.** *v.i.* Walzer tanzen

wan [wɒn] *adj.* fahl *(geh.);* bleich; ~ **smile** mattes Lächeln

wand [wɒnd] *n.* Stab, *der*

wander ['wɒndə(r)] **1.** *v.i.* **a)** *(go aimlessly)* umherirren; *(walk slowly)* bummeln; **she ~ed over to me** sie kam zu mir herüber; **b)** *(stray)* ‹*Katze:*› streunen; ‹*Schafe:*› sich verlaufen; **c)** *(fig.: stray from subject)* abschweifen. **2.** *v.t.* wandern durch. **3.** *n. (coll.: walk)* Spaziergang, *der;* **I'll go for** *or* **take a ~ round** *or* **through the town** ich werd' mal einen Bummel durch die Stadt machen

~ **a'bout** *v.i.* sich herumtreiben

~ **a'long** *v.i.* dahintrotten; ‹*Fahrzeug:*› dahinzockeln *(ugs.)*

~ **'in** *v.i.* hineinspazieren; *(towards speaker)* hereinspaziert kommen

~ **'off** *v.i.* **a)** *(stray)* weggehen; ‹*Kind:*› sich selbständig machen *(scherzh.);* **b)** *(coll.: go away)* sich davonmachen *(ugs.)*

wanderer ['wɒndərə(r)] *n.* Streuner, *der*/Streunerin, *die*

wane [weɪn] *v.i.* ‹*Mond:*› abnehmen; ‹*Kraft, Einfluß, Macht:*› schwinden, abnehmen; ‹*Ruf, Ruhm:*› verblassen; *see also* ²**wax**

wangle ['wæŋgl] *(coll.)* **1.** *v.t. (get by devious means)* organisieren *(ugs.)* ‹*Karte, Einladung*›; ~ **sth. out of sb.** jmdm. etw. abluchsen *(ugs.).* **2.** *n.* Kniff, *der;* **by a ~:** durch Schiebung *(ugs.)*

want [wɒnt] **1.** *v.t.* **a)** *(desire)* wollen; **I ~ my mummy** ich will zu meiner Mama; **I ~ it done by tonight** ich will, daß es bis heute abend fertig wird; **I don't ~ there to be any misunderstanding** ich will *od.* möchte nicht, daß da ein Mißverständnis aufkommt; **b)** *(require, need)* brauchen; **'W~ed – cook for small family'** „Koch/Köchin für kleine Familie gesucht"; **you're ~ed on the phone** du wirst am Telefon verlangt; **feel ~ed** das Gefühl haben, gebraucht zu werden; **the windows ~ painting** die Fenster müßten gestrichen werden; **you ~ to be [more] careful** *(ought to be)* du solltest vorsichtig[er] sein; **c) ~ed [by the police]** [polizeilich] gesucht (**for** wegen); **d)** *(lack)* **sb./sth. ~s sth.** jmdm./einer Sache fehlt es an etw. *(Dat.)*. **2.** *n.* **a)** *no pl. (lack)* Mangel, *der* (**of** an + *Dat.*); **there is no ~ of ...**: es fehlt nicht an ... *(Dat.)*; **for ~ of sth.** aus Mangel an etw. *(Dat.)*; **for ~ of a better word** in Ermangelung eines besseren Ausdrucks; **b)** *no pl. (need)* Not, *die;* **suffer ~**: Not leiden; **c)** *(desire)* Bedürfnis, *das;* **we can supply all your ~s** wir können alles liefern, was Sie brauchen; **~ ad** *(Amer.)* Kaufgesuch, *das*

~ for *v. t.* **sb. ~s for nothing** *or* **doesn't ~ for anything** jmdm. fehlt es an nichts

wanting ['wɒntɪŋ] *adj.* **be ~**: fehlen; **sb./sth. is ~ in sth.** jmdm./einer Sache fehlt es an etw. *(Dat.)*; **be found ~**: für unzureichend befunden werden

wanton ['wɒntən] *adj.* **a)** *(dated: licentious)* lüstern; wollüstig ⟨*Person, Gedanken, Benehmen*⟩; **b)** *(wilful)* mutwillig ⟨*Beschädigung, Grausamkeit, Verschwendung*⟩; leichtfertig ⟨*Vernachlässigung*⟩

war [wɔ:(r)] *n.* **a)** Krieg, *der;* **between the ~s** zwischen den Weltkriegen; **declare ~**: den Krieg erklären (**on** *Dat.*); **be at ~**: sich im Krieg befinden; **make ~**: Krieg führen (**on** gegen); **go to ~**: in den Krieg ziehen (**against** gegen); **b)** *(science)* Kriegführung, *die;* **c)** *(fig.: conflict)* Krieg, *der;* **price ~**: Preiskrieg, *der;* **~ of nerves** Nervenkrieg, *der;* **d)** *(fig.: fight, campaign)* Kampf, *der* (**on, against** gegen); **declare ~ on poverty** der Armut den Kampf ansagen

warble ['wɔ:bl] *v. t. & i.* trällern

warbler ['wɔ:blə(r)] *n. (Ornith.)* Grasmücke, *die*

war: ~ correspondent *n.* Kriegsberichterstatter, *der/*-berichterstatterin, *die;* **~ crime** *n.* Kriegsverbrechen, *das;* **~ criminal** *n.* Kriegsverbrecher, *der/*-verbrecherin, *die;* **~-cry** *n.* **a)** Kriegsruf, *der;* **b)** *(slogan)* Schlachtruf, *der*

ward [wɔ:d] *n.* **a)** *(in hospital)* Station, *die; (single room)* Krankensaal, *der;* **geriatric/maternity ~**: geriatrische Abteilung/Entbindungsstation, *die;* **she's in W~ 3** sie liegt auf Station 3; **b)** *(minor)* Mündel, *das od. die;* **c)** *(electoral division)* Wahlbezirk, *der*

~ 'off *v. t.* **a)** *(prevent)* abwehren; schützen vor (+ *Dat.*) ⟨*Erkältung, Depressionen*⟩; abwenden ⟨*Gefahr*⟩; **b)** *(keep at distance)* sich *(Dat.)* vom Leibe halten ⟨*Verehrer*⟩

warden ['wɔ:dn] *n.* **a)** *(president, governor)* Direktor, *der/*Direktorin, *die; (of college, school)* Rektor, *der/*Rektorin, *die; (of hostel, sheltered housing)* Heimleiter, *der/*-leiterin, *die; (of youth hostel)* Herbergsvater, *der/*-mutter, *die;* **b)** *(supervisor)* Aufseher, *der/*Aufseherin, *die*

warder ['wɔ:də(r)] *n. (Brit.)* Wärter, *der;* Aufseher, *der*

wardrobe ['wɔ:drəʊb] *n.* **a)** *(piece of furniture)* Kleiderschrank, *der;* **b)** *(stock of clothes)* Garderobe, *die; (in theatre)* Kostüme *Pl.*

ware [weə(r)] *n.* **a)** *(pottery)* Steinzeug, *das;* **Delft ~**: Delfter Keramik; **b)** *in pl. (goods)* Ware, *die*

warehouse 1. ['weəhaʊs] *n.* Lagerhaus, *das; (part of building)* Lager, *das; (Brit.: retail or wholesale store)* Großmarkt, *der.* **2.** ['weəhaʊs, 'weəhaʊz] *v. t.* einlagern

warfare ['wɔ:feə(r)] *n. (lit. or fig.)* Krieg, *der;* **in modern ~**: in der modernen Kriegführung; **economic ~**: Wirtschaftskrieg, *der*

war: ~-game *n.* Kriegsspiel, *das;* **~head** *n.* Sprengkopf, *der;* **~-horse** *n. (Hist., fig.)* Schlachtroß, *das;* **~like** *adj. (bellicose)* kriegerisch

warm [wɔ:m] **1.** *adj.* **a)** warm; **come inside and get ~**: komm rein und wärm dich auf; **I am very ~ from running** mir ist sehr warm vom Rennen; **it's ~ work** bei der Arbeit kommt man ins Schwitzen; **keep sb.'s food ~**: jmdm. das Essen warm halten; **keep a seat/job ~ for sb.** *(fig.)* jmdm. einen Platz/eine Stellung freihalten; **b)** *(enthusiastic)* herzlich ⟨*Grüße, Dank*⟩; eng ⟨*Freundschaft*⟩; lebhaft ⟨*Interesse*⟩; begeistert ⟨*Unterstützung, Applaus*⟩; **c)** *(cordial, sympathetic)* warm ⟨*Herz, Wesen, Gefühl*⟩; herzlich ⟨*Lächeln*⟩; echt empfunden ⟨*Hochachtung*⟩; **d)** *(passionate)* heiß ⟨*Temperament, Küsse*⟩; **e)** *(unpleasant)* ungemütlich; **he left when things began to get too ~ for him** er ging, als ihm die Sache zu ungemütlich wurde; **f)** *(recent)* heiß ⟨*Spur*⟩; **g)** *(in games: close)* **you're getting ~!** warm! **2.** *v. t.* wärmen; warm machen ⟨*Flüssigkeit*⟩; **~ one's hands** sich *(Dat.)* die Hände wärmen. **3.** *v. i.* **a) ~ to sb./sth.** *(come to like)* sich für jmdn./etw. erwärmen; **the speaker ~ed to his subject** der Redner steigerte sich in sein Thema hinein; **b)** *(get ~er)* warm werden

~ 'up 1. *v. i.* **a)** *(get ~)* warm werden; ⟨*Motor:*⟩ warmlaufen; **b)** *(prepare)* ⟨*Sportler:*⟩ sich aufwärmen; **c)** *(fig.: become animated)* warm werden; ⟨*Party:*⟩ in Schwung kommen; ⟨*Publikum:*⟩ in Stimmung kommen. **2.** *v. t.* aufwärmen ⟨*Speisen*⟩; erwärmen ⟨*Raum, Zimmer*⟩; warmlaufen lassen ⟨*Motor*⟩; *(fig.)* in Stimmung bringen ⟨*Publikum*⟩

warm-blooded ['wɔ:mblʌdɪd] *adj.* warmblütig ⟨*Tier*⟩; **~ animals** Warmblüter

'war memorial *n.* Kriegerdenkmal, *das*

warm-hearted ['wɔ:mhɑ:tɪd] *adj.* herzlich; warmherzig ⟨*Person*⟩

warmly ['wɔ:mlɪ] *adv.* **a)** *(to maintain warmth)* warm; **b)** *(enthusiastically)* herzlich ⟨*willkommen heißen, gratulieren, begrüßen, grüßen, danken*⟩; wärmstens ⟨*empfehlen*⟩; begeistert ⟨*sprechen von, applaudieren*⟩

warmonger ['wɔ:mʌŋgə(r)] *n.* Kriegshetzer, *der/*-hetzerin, *die*

warmth [wɔ:mθ] *n.* **a)** *(state of being warm; also of colour)* Wärme, *die;* **b)** *(enthusiasm, affection, cordiality)* Herzlichkeit, *die;* Wärme, *die*

warn [wɔ:n] *v. t.* **a)** *(inform, give notice)* warnen (**against, of, about** vor + *Dat.*); **~ sb. that ...**: jmdn. darauf hinweisen, daß ...; **you have been ~ed!** ich habe/wir haben dich gewarnt!; **~ sb. not to do sth.** jmdn. davor warnen, etw. zu tun; **b)** *(admonish)* ermahnen; *(officially)* abmahnen

~ 'off *v. t.* warnen; **~ sb. off doing sth.** jmdn. davor warnen, etw. zu tun

warning ['wɔ:nɪŋ] **1.** *n.* **a)** *(advance notice)* Vorwarnung, *die;* **we had no ~ of their arrival** sie kamen ohne Vorwarnung; **give sb. plenty of/a few days' ~**: jmdm. rechtzeitig/ein paar Tage vorher Bescheid sagen; **b)** *(lesson)* Lehre, *die;* **let that be a ~ to you** laß dir/laßt euch das eine Warnung sein; **c)** *(caution)* Verwarnung, *die; (less official)* Warnung, *die.* **2.** *attrib. adj.* Warn⟨*schild, -zeichen, -signal usw.*⟩; **~ light/shot** Warnleuchte, *die/*-schuß, *der;* **~ notice** Warnung, *die;* **a ~ look/gesture** ein warnender Blick/eine warnende Geste

warp [wɔ:p] **1.** *v. i. (become bent)*

sich verbiegen; ‹*Holz, Schallplatte:*› sich verziehen. **2.** *v. t.* **a)** verbiegen; **b)** *(fig.: pervert)* verformen; verbiegen; **~ed** getrübt ‹*Urteilsvermögen*›; pervertiert ‹*Denken, Gehirn*›. **3.** *n. (Weaving)* Kettfaden, *der*

war: ~-paint *n. (also fig. coll. joc.)* Kriegsbemalung, *die;* **~-path** *n.* Kriegspfad, *der;* **be on the ~-path** auf dem Kriegspfad sein; *(fig.)* in Rage sein; **~-plane** *n.* Kampfflugzeug, *das*

warrant ['wɒrənt] **1.** *n. (for sb.'s arrest)* Haftbefehl, *der;* **[search] ~:** Durchsuchungsbefehl, *der.* **2.** *v. t. (justify)* rechtfertigen

warranty ['wɒrəntɪ] *n. (Law)* Garantie, *die;* **it is still under ~:** es steht noch unter Garantie

warren ['wɒrn] *n.* **a)** *see* **rabbit-warren; b)** *(fig.: maze)* Labyrinth, *das*

warring ['wɔ:rɪŋ] *attrib. adj.* kriegführend

warrior ['wɒrɪə(r)] *n. (esp. literary)* Krieger, *der (geh.)*

Warsaw ['wɔ:sɔ:] **1.** *pr. n.* Warschau *(das).* **2.** *attrib. adj.* Warschauer; **~ Pact** *(Hist.)* Warschauer Pakt

'warship *n.* Kriegsschiff, *das*

wart [wɔ:t] *n.* Warze, *die;* **~s and all** *(fig.)* schonungslos; ungeschminkt [bis ins kleinste Detail]

'wart-hog *n.* Warzenschwein, *das*

'wartime *n.* **a)** Kriegszeit, *die;* **in** *or* **during ~:** während des Krieges; im Krieg; **b)** *attrib.* Kriegs‹*rationierung, -evakuierung usw.*›; **~ England** [das] England während des Krieges

wary ['weərɪ] *adj.* vorsichtig; *(suspicious)* mißtrauisch (**of** gegenüber); **be ~ of** *or* **about doing sth.** sich davor hüten, etw. zu tun; **be ~ of sb./sth.** sich vor jmdm./etw. in acht nehmen

was *see* **be**

wash [wɒʃ] **1.** *v. t.* **a)** waschen; **~ oneself/one's hands** *(also euphem.)***/face/hair** sich waschen/ sich *(Dat.)* die Hände *(auch verhüll.)*/das Gesicht/die Haare waschen; **~ the dishes** abwaschen; [Geschirr] spülen; **~ the floor** den Fußboden aufwischen *od.* feucht wischen; **~ one's hands of sb./sth.** mit jmdm./etw. nichts mehr zu tun haben wollen; **b)** *(remove)* waschen ‹*Fleck*› (**out of** aus); abwaschen ‹*Schmutz*› (**off** von); **c)** *(by licking)* putzen; **the cat ~ed its fur** die Katze putzte sich *(Dat.)* das Fell; **d)** *(carry along)* spülen; **be ~ed downstream** von der Strömung mitgerissen werden. **2.** *v. i.* **a)** sich waschen; **b)** *(clean clothes)* waschen; **c)** ‹*Stoff, Kleidungsstück, Handtuch:*› sich waschen lassen; **that won't ~** *(fig. coll.)* das zieht nicht *(ugs.).* **3.** *n.* **a) give sb./sth. a [good] ~:** jmdn./etw. [gründlich] waschen; **the baby/ car needs a ~** *or (coll.)* **could do with a ~:** das Kind/Auto müßte mal gewaschen werden; **b)** *(laundering)* Wäsche, *die;* **it is in the ~:** es ist in der Wäsche; **it'll all come out in the ~** *(fig. coll.)* das wird sich alles klären; **c)** *(of ship, aircraft, etc.)* Sog, *der;* **d)** *(lotion)* Waschlotion, *die;* **a ~ for disinfecting the mouth** ein desinfizierendes Mundwasser

~ a'way *v. t.* **a)** wegspülen; **b) ~ a stain/the mud away** einen Fleck/ den Schmutz auswaschen

~ 'down *v. t.* **a)** *(with a hose)* abspritzen ‹*Auto, Deck, Hof*›; *(with soap and water)* abwaschen; aufwaschen ‹*Fußboden*›; **b)** *(help to go down)* runterspülen *(ugs.)*

~ 'off 1. *v. t.* **~ sth. off** etw. abwaschen. **2.** *v. i.* abgehen; *(from fabric etc.)* herausgehen

~ 'out *v. t. (clean)* auswaschen ‹*Kleidungsstück*›; ausscheuern ‹*Topf*›; ausspülen ‹*Mund*›; **~ dirt/marks out of clothes** Schmutz/Flecken aus Kleidern [her]auswaschen

~ 'over *v. t. (fig. coll.: not affect)* **~ over sb.** ‹*Streit, Lärm, Unruhe usw.:*› jmdn. gar nicht berühren

~ 'up 1. *v. t.* **a)** *(Brit.: clean)* **~ the dishes up** das Geschirr abwaschen *od.* spülen; **b)** *(carry to shore)* anspülen ‹*Leiche, Strandgut, Wrackteile usw.*›. **2.** *v. i.* abwaschen; spülen

washable ['wɒʃəbl] *adj.* waschbar ‹*Stoff*›; abwaschbar ‹*Tapete, Farbe*›

wash: ~-and-'wear *adj.* bügelfrei; **~-basin** *n.* Waschbecken, *das;* **~-day** *n.* Waschtag, *der*

washed-'out *adj.* **a)** *attrib.* verwaschen ‹*Farbe, Kleidungsstück*›; **b)** *(fig.: exhausted)* abgespannt; mitgenommen

washer ['wɒʃə(r)] *n. (Mech. Engin.)* Unterlegscheibe, *die; (of tap)* Dichtungsring, *der;* Dichtungsscheibe, *die*

washing ['wɒʃɪŋ] *n., no pl., no indef. art.* **a)** *(clothes to be washed)* Wäsche, *die;* **b)** *(cleansing)* Waschen, *das;* **do the ~:** waschen

washing: ~-machine *n.* Waschmaschine, *die;* **~-powder** *n.* Waschpulver, *das;* **~-'up** *n. (Brit.)* Abwasch, *der;* **do the ~-up** den Abwasch machen; abwaschen; **~-'up liquid** *n.* Spülmittel, *das;* **~-'up machine** *see* **dishwasher**

wash: ~-out *n.* **a)** *(sl.: failure)* Pleite, *die (ugs.);* Reinfall, *der (ugs.);* **b)** *(sl.: useless person)* Niete, *die (salopp abwertend);* **~room** *n. (Amer.)* WC, *das*

wasn't ['wɒznt] *(coll.)* = **was not;** *see* **be**

wasp [wɒsp] *n.* Wespe, *die*

waspish ['wɒspɪʃ] *adj.* bissig

wastage ['weɪstɪdʒ] *n.* **a)** *(loss by wear etc.)* Schwund, *der;* **b) [natural] ~** *(Admin.)* ≈ natürliche Fluktuation

waste [weɪst] **1.** *n.* **a)** *(useless remains)* Abfall, *der;* **kitchen ~:** Küchenabfälle *Pl.;* **b)** *(extravagant use)* Verschwendung, *die;* Vergeudung, *die;* **it's a ~ of time/ money/energy** das ist Zeit-/ Geld-/Energieverschwendung; **it would be a ~ of effort** das wäre vergeudete Mühe; **go** *or* **run to ~:** vergeudet werden. **2.** *v. t.* **a)** *(squander)* verschwenden; vergeuden (**on** auf + *Akk.*, an + *Akk.*); **he is ~d on an audience like that** für ein solches Publikum ist er zu schade; **all his efforts were ~d** all seine Mühe war umsonst; **don't ~ my time!** stehlen Sie mir nicht die Zeit!; **~ not, want not** *(prov.)* spare in der Zeit, so hast du in der Not *(Spr.);* **b) be ~d** *(reduced)* ‹*Vorräte, Bevölkerung:*› abnehmen, schrumpfen; **c)** *(cause to shrink)* aufzehren ‹*Kräfte*›; auszehren ‹*Körper*›. **3.** *v. i.* dahinschwinden; *(gradually)* im Schwinden begriffen sein. **4.** *adj.* **a)** *(not wanted)* **~ material** Abfall, *der;* **~ food** Essensreste *Pl.;* **~ water** Abwasser, *das;* **b) lay ~ sth.** etw. verwüsten

~ a'way *v. i.* immer mehr abmagern

waste: ~-basket *see* **waste-paper basket; ~ disposal** *n.* Abfallbeseitigung, *die;* Entsorgung, *die (Amtsspr.)*

wasteful ['weɪstfl] *adj.* **a)** *(extravagant)* verschwenderisch; **too much ~ expenditure** zuviel Geldverschwendung; **b)** *(causing waste)* unwirtschaftlich; **be ~ of sth.** etw. vergeuden

wastefulness ['weɪstflnɪs] *n., no pl.* **a)** *(extravagance)* Verschwendung, *die; (character trait)* Verschwendungssucht, *die;* **b)** *(of manufacturing process)* Unwirtschaftlichkeit, *die*

waste: ~land *n. (not cultivated)*

Ödland, *das; (not built on)* unbebautes Land; *(fig.)* Einöde, *die;* ~ **'paper** *n.* Papierabfall, *der;* ~-**'paper basket** *n.* Papierkorb, *der;* ~-**pipe** *n.* Abflußrohr, *das*

watch [wɒtʃ] **1.** *n.* **a)** **[wrist-/pocket-]**~: [Armband-/Taschen]uhr, *die;* **b)** *(constant attention)* Wache, *die;* **keep** ~: Wache halten; **keep [a]** ~ **for sb./sth.** auf jmdn./etw. achten *od.* aufpassen; **keep [a]** ~ **for enemy aircraft** nach feindlichen Flugzeugen Ausschau halten; **keep a close** ~ **on the time** genau auf die Zeit achten; **they kept a** ~ **on all his activities** sie überwachten alle seine Aktivitäten; **the police were on the** ~ **for car thieves** die Polizei hielt nach Autodieben Ausschau; **c)** *(Naut.)* Wache, *die;* **the officer of the** ~: der wachhabende Offizier. **2.** *v. i.* **a)** *(wait)* ~ **for sb./sth.** auf jmdn./etw. warten; **b)** *(keep* ~*)* Wache stehen. **3.** *v. t.* **a)** *(observe)* sich *(Dat.)* ansehen ⟨*Sportveranstaltung, Fernsehsendung*⟩; ~ **[the] television** *or* **TV** fernsehen; Fernsehen gucken *(ugs.);* ~ **sth. [on television** *or* **TV]** sich *(Dat.)* etw. [im Fernsehen] ansehen; ~ **sb. do** *or* **doing sth.** zusehen, wie jmd. etw. tut; **we are being** ~**ed** wir werden beobachtet; ~ **one's weight** auf sein Gewicht achten; **b)** *(be careful of, look after)* achten auf (+ *Akk.*); ~ **your manners!** *(coll.)* benimm dich!; ~ **your language!** *(coll.)* drück dich bitte etwas gepflegter *od.* nicht so ordinär aus!; ~ **him, he's an awkward customer** *(coll.)* paß/paßt auf, er ist mit Vorsicht zu genießen *(ugs.);* ~ **how you go/drive** paß auf!/fahr vorsichtig!; ~ **it** *or* **oneself** sich vorsehen; **[just]** ~ **it [or you'll be in trouble]!** paß bloß auf[, sonst gibt's Ärger]! *(ugs.);* **c)** *(look out for)* warten auf (+ *Akk.*)

~ **'out** *v. i.* **a)** *(be careful)* sich vorsehen; aufpassen; **W**~ **out! There's a car coming!** Vorsicht! Da kommt ein Auto!; **b)** *(look out)* ~ **out for sb./sth.** auf jmdn./etw. achten; *(wait)* auf jmdn./etw. warten

~ **'over** *v. t.* sich kümmern um; in Obhut nehmen ⟨*Wertgegenstand*⟩; ⟨*Gott, Schutzengel:*⟩ wachen über (+ *Akk.*)

'watch-dog *n.* Wachhund, *der;* *(fig.)* Wächter, *der;* Aufpasser, *der (ugs.)*

watchful ['wɒtʃfl] *adj.* wachsam; **be** ~ **for sth.** vor etw. *(Dat.)* auf der Hut sein; **keep a** ~ **eye on sb./sth.** ein wachsames Auge auf jmdn./etw. haben

watch: ~**maker** *n.* Uhrmacher, *der*/Uhrmacherin, *die;* ~**man** ['wɒtʃmən] *n., pl.* ~**men** ['wɒtʃmən] Wachmann, *der;* ~-**strap** *n.* [Uhr]armband, *das;* ~-**tower** *n.* Wachturm, *der;* ~**word** *n.* Parole, *die*

water ['wɔːtə(r)] **1.** *n.* **a)** Wasser, *das;* **be under** ~ ⟨*Straße, Sportplatz usw.:*⟩ unter Wasser stehen; **the island across** *or* **over the** ~: die Insel drüben; **send/carry sth. by** ~: etw. auf dem Wasserweg versenden/befördern; **be in deep** ~ *(fig.)* in großen Schwierigkeiten sein; **get [oneself] into deep** ~ *(fig.)* sich in große Schwierigkeiten bringen; **on the** ~ *(in boat etc.)* auf dem Wasser; **pour** *or* **throw cold** ~ **on sth.** *(fig.)* einer Sache *(Dat.)* einen Dämpfer aufsetzen; ~ **under the bridge** *or* **over the dam** *(fig.)* Schnee von gestern *(fig.);* **b)** *in pl. (part of the sea etc.)* Gewässer *Pl.;* **c)** *in pl. (mineral* ~ *at spa etc.)* Heilquelle, *die;* Brunnen, *der;* **take** *or* **drink the** ~**s** eine Brunnenkur machen; **d) of the first** ~: reinsten Wassers; **a genius of the first** ~: ein Genie ersten Ranges. **2.** *v. t.* **a)** bewässern ⟨*Land*⟩; wässern ⟨*Pflanzen*⟩; ~ **the flowers** die Blumen [be]gießen; **b)** *(adulterate)* verwässern ⟨*Wein, Bier usw.*⟩; **c)** ⟨*Fluß:*⟩ bewässern ⟨*Land*⟩; **d)** *(give drink of* ~ *to)* tränken ⟨*Tier, Vieh*⟩. **3.** *v. i.* **a)** ⟨*Augen:*⟩ tränen; **b) my mouth was** ~**ing as ...**: mir lief das Wasser im Munde zusammen, als ...; **the very thought of it made my mouth** ~: allein bei dem Gedanken lief mir das Wasser im Munde zusammen

~ **'down** *v. t. (lit. or fig.)* verwässern

water: ~-**bottle** *n.* Wasserflasche, *die;* ~-**butt** *n.* Regentonne, *die;* ~-**cannon** *n.* Wasserwerfer, *der;* ~-**closet** *n.* Toilette, *die;* WC, *das;* ~-**colour** *n.* **a)** *(paint)* Wasserfarbe, *die;* **b)** *(picture)* Aquarell, *das;* ~**course** *n.* *(stream etc.)* Wasserlauf, *der;* *(bed)* Flußbett, *das;* ~**cress** *n.* Brunnenkresse, *die;* ~-**diviner** ['wɔːtədɪvaɪnə(r)] *n.* [Wünschel]rutengänger, *der*/-gängerin, *die;* ~**fall** *n.* Wasserfall, *der;* ~**front** *n.* Ufer, *das;* **down on the** ~**front** unten am Wasser; *attrib.* **a** ~**front restaurant** ein Restaurant am Wasser; ~-**heater** *n.* Heißwassergerät, *das*

watering ['wɔːtərɪŋ] *n.* Bewässerung, *die;* *(of flowers, houseplants)* Gießen, *das;* **give the plants a thorough** ~: die Pflanzen gut wässern *od.* gießen

'watering-can *n.* Gießkanne, *die*

water: ~-**level** *n.* **a)** *(in reservoir etc.)* Wasserstand, *der;* Pegelstand, *der;* **b)** *(below which ground is saturated)* Grundwasserspiegel, *der;* ~-**lily** *n.* Seerose, *die;* ~-**line** *n.* *(Naut.)* Wasserlinie, *die;* ~**logged** ['wɔːtəlɒgd] *adj.* vollgesogen ⟨*Holz*⟩; ⟨*Boot*⟩ voll Wasser; naß, feucht ⟨*Boden*⟩; aufgeweicht ⟨*Sportplatz*⟩; ~-**main** *n.* Hauptwasserleitung, *die;* **a burst** ~-**main** ein Wasserrohrbruch; ~**mark 1.** *n.* Wasserzeichen, *das;* **2.** *v. t.* mit Wasserzeichen versehen; ~-**melon** *n.* Wassermelone, *die;* ~ **meter** *n.* Wasseruhr, *die;* ~-**mill** *n.* Wassermühle, *die;* ~-**pipe** *n.* **a)** Wasserrohr, *das;* **b)** *(for smoking)* Wasserpfeife, *die;* ~-**pistol** *n.* Wasserpistole, *die;* ~ **polo** *n.* Wasserball, *der;* ~-**power** *n.* Wasserkraft, *die;* ~**proof 1.** *adj.* wasserdicht; wasserfest ⟨*Farbe*⟩; **2.** *n.* Regenhaut, *die;* *(raincoat)* Regenmantel, *der;* **3.** *v. t.* wasserdicht machen; imprägnieren ⟨*Stoff*⟩; wetterfest machen ⟨*Holzzaun, Gartenmöbel*⟩; ~-**rate** *n.* Wassergeld, *das;* **the** ~-**rates** die Wassergebühren; ~-**repellent** *adj.* wasserabstoßend; ~-**resistant** *adj.* wasserundurchlässig; wasserfest ⟨*Farbe*⟩; ~**shed** *n.* *(fig.: turning-point)* Wendepunkt, *der;* ~-**ski 1.** *n.* Wasserski, *der;* **2.** *v. i.* Wasserski laufen; ~-**skiing** *n., no pl., no art.* Wasserskilaufen, *das;* ~-**softener** *n.* Wasserenthärter, *der;* ~-**supply** *n.* **a)** *no pl., no indef. art. (providing)* Wasserversorgung, *die;* **b)** *(stored drinking* ~*)* Trinkwasser, *das;* *(amount)* [Trink]wasservorrat, *der;* ~-**table** *n.* Grundwasserspiegel, *der;* ~ **tap** *n.* Wasserhahn, *der;* ~**tight** *adj. (lit. or fig.)* wasserdicht; ~-**tower** *n.* Wasserturm, *der;* ~ **vapour** *n.* Wasserdampf, *der;* ~**way** *n.* Wasserstraße, *die;* **inland** ~**ways** Binnenwasserstraßen; ~-**wheel** *n.* Wasserrad, *das;* *(used to raise* ~*)* Schöpfrad, *das;* ~-**wings** *n. pl.* Schwimmflügel; ~**works** *n.* **a)** *sing., pl. same (system)* Wasserversorgungssystem, *das;* *(building)* Wasserwerk, *das;* **b)** *pl. (sl.: tears)* **turn on the** ~**works** losheulen *(ugs.);* **c)** *pl. (coll.: urinary system)* Blase, *die*

watery ['wɔːtərɪ] *adj.* wäßrig, wässerig ‹*Essen, Suppe*›; feucht ‹*Augen*›; dünn ‹*Getränk*›

watt [wɒt] *n. (Electr., Phys.)* Watt, *das*

wave [weɪv] **1.** *n.* **a)** *(lit. or fig.)* Welle, *die;* Woge, *die (geh.);* **his hair has a natural ~ in it** sein Haar ist von Natur aus wellig; **a ~ of enthusiasm/pain** eine Welle der Begeisterung/des Schmerzes; **~s of attackers** Angriffswellen; **b)** *(gesture)* **give sb. a ~:** jmdm. zuwinken; **with a ~ of one's hand** mit einem Winken. **2.** *v.i.* **a)** ‹*Fahne, Flagge, Wimpel:*› wehen; ‹*Baum, Gras, Korn:*› sich wiegen; ‹*Kornfeld:*› wogen; **b)** *(gesture with hand)* winken; **~ at** *or* **to sb.** jmdm. winken. **3.** *v.t.* schwenken; *(brandish)* schwingen ‹*Schwert, Säbel*›; **~ one's hand at** *or* **to sb.** jmdm. winken; **she ~d her umbrella angrily at him** sie drohte ihm wütend mit dem Regenschirm; **stop waving that rifle/those scissors around** hör auf, mit dem Gewehr/der Schere herumzufuchteln *(ugs.);* **~ sb. on/over** jmdn. weiter-/herüberwinken; **~ goodbye to sb.** jmdm. zum Abschied zuwinken

~ a'side *v.t.* **a)** *(refuse to accept)* abtun ‹*Zweifel, Einwand*›; **b)** *(signal to move aside)* **I tried to speak but she ~d me aside** ich wollte reden, aber sie winkte ab

~ a'way *v.t.* wegwinken

~ 'down *v.t.* [durch Winken] anhalten

~ 'off *v.t.* **~ sb. off** jmdm. nachwinken

wave: ~band *n.* Wellenbereich, *der;* **~length** *n. (Radio, Telev., Phys.; also fig.)* Wellenlänge, *die;* **be on sb.'s ~length** *(fig.)* die gleiche Wellenlänge wie jmd. haben; **~ power** *n.* Wellenkraft, *die*

waver ['weɪvə(r)] *v.i.* **a)** *(begin to give way)* wanken; **start** *or* **begin to ~:** ins Wanken geraten; **b)** *(be irresolute)* schwanken (**between** zwischen + *Dat.*)

wavy ['weɪvɪ] *adj.* **a)** *(undulating)* wellig; wogend ‹*Gras*›; **b)** *(forming wave-like curves)* geschlängelt; **~ line** Schlangenlinie, *die*

¹wax [wæks] **1.** *n.* **a)** Wachs, *das;* **be [like] ~ in sb.'s hands** [wie] Wachs in jmds. Händen sein; **b)** *(in ear)* Schmalz, *das.* **2.** *adj.* Wachs-. **3.** *v.t.* wachsen; wichsen ‹*Schnurrbart*›

²wax *v.i. (increase)* ‹*Mond:*› zunehmen; **~ and wane** *(fig.)* zu- und abnehmen

way [weɪ] **1.** *n.* **a)** *(road etc., lit. or fig.)* Weg, *der;* **across** *or* **over the ~:** gegenüber; **b)** *(route)* Weg, *der;* **ask the** *or* **one's ~:** nach dem Weg fragen; **ask the ~ to ...:** fragen *od.* sich erkundigen, wo es nach ... geht; **pick one's ~:** sich *(Dat.)* einen Weg suchen; **lead the ~:** vorausgehen; *(fig.: show how to do sth.)* es vormachen; **find the** *or* **one's ~ in/out** den Eingang/Ausgang finden; **find a ~ out** *(fig.)* einen Ausweg finden; **I'll take the letter to the post office – it's on my ~:** ich bringe den Brief zur Post – sie liegt auf meinem Weg; **'W~ In/Out'** „Ein-/Ausgang"; **go to Italy by ~ of Switzerland** über die Schweiz nach Italien fahren; **there's no ~ out** *(fig.)* es gibt keinen Ausweg; **the ~ back/down/up** der Weg zurück/nach unten/nach oben; **go one's own ~/their separate ~s** *(fig.)* eigene/getrennte Wege gehen; **be going sb.'s ~** *(coll.)* denselben Weg wie jmd. haben; **things are really going my ~ at the moment** *(fig.)* im Moment läuft [bei mir] alles so, wie ich es mir vorgestellt habe; **money came his ~:** er kam zu Geld; **many offers came his ~:** er kriegte viele Angebote; **go out of one's ~ to collect sth. for sb.** einen Umweg machen, um etw. für jmdn. abzuholen; **go out of one's ~ to be helpful** sich *(Dat.)* besondere Mühe geben, hilfsbereit zu sein; **out of the ~:** abgelegen; **c)** *(method)* Art und Weise, *die;* **there is a right ~ and a wrong ~ of doing it** es gibt einen richtigen und einen falschen Weg, es zu tun; **that is not the ~ to do it** so macht man das nicht; **do it this ~:** mach es so; **do it my ~:** mach es wie ich; **I don't like the ~ she smiles** mir gefällt ihr Lächeln nicht; **that's no ~ to speak to a lady** so spricht man nicht mit einer Dame; **he has a strange ~ of talking** er hat eine seltsame Sprechweise *od.* Art zu sprechen; **from** *or* **by the ~ [that] she looked at me, I knew that there was something wrong** an ihrem Blick konnte ich erkennen, daß etwas nicht stimmte; **find a** *or* **some ~ of doing sth.** einen Weg finden, etw. zu tun; **there are no two ~s about it** da gibt es gar keinen Zweifel; **Are you going to give me that money? – No ~!** *(coll.)* Gibst du mir das Geld? – Nichts da! *(ugs.);* **there was no ~ he would change his stand** er würde auf gar keinen Fall seinen Standpunkt ändern; **no ~ is he coming with us** es kommt überhaupt nicht in Frage, daß er mit uns kommt; **one ~ or another** irgendwie; **~s and means [to do sth.** *or* **of doing sth.]** Mittel und Wege, etw. zu tun; **be built** *or* **made that ~** *(fig. coll.)* so gestrickt sein *(fig. ugs.);* **be that ~** *(coll.)* so sein; **better that ~:** besser so; **either ~:** so oder so; **d)** *(desired course of action)* Wille, *der;* **get** *or* **have one's [own] ~, have it one's [own] ~:** seinen Willen kriegen; **all right, have it your own ~[, then]!** na gut *od.* schön, du sollst deinen Willen haben!; **e)** *in sing. or (Amer. coll.) pl. (distance between two points)* Stück, *das;* **a little ~:** ein kleines Stück[chen]; *(fig.)* ein klein[es] bißchen; **it's a long ~ off** *or* **a long ~ from here** es ist ein ganzes Stück von hier aus; es ist weit weg von hier; **the summer holidays are only a little ~ away** bis zu den Sommerferien ist es nicht mehr lange; **there's [still] some ~ to go yet** es ist noch ein ganzes Stück; *(fig.)* es dauert noch ein Weilchen; **I went a little/a long/some ~ to meet him** ich bin ihm ein kleines/ganzes/ziemliches Stück entgegengegangen/-gefahren *usw.,* um mich mit ihm zu treffen; *(fig.)* ich bin ihm etwas/sehr/ziemlich entgegengekommen; **have gone/come a long ~** *(fig.)* es weit gebracht haben; **go a long ~ toward sth./doing sth.** viel zu etw. beitragen/viel dazu beitragen, etw. zu tun; **a little kindness goes a long ~:** ein bißchen Freundlichkeit ist viel wert *od.* hilft viel; **all the ~:** den ganzen Weg; **go all the ~ [with sb.]** *(fig.)* [jmdm.] in jeder Hinsicht zustimmen; *(sl.: have full sexual intercourse)* es [mit jmdm.] richtig machen *(salopp);* **f)** *(room for progress)* Weg, *der;* **block the ~:** den Weg versperren; **leave the ~ open for sth.** *(fig.)* etw. möglich machen; **clear the ~ [for sth.]** *(lit. or fig.)* [einer Sache *(Dat.)*] den Weg freimachen; **be in sb.'s** *or* **the ~:** [jmdm.] im Weg sein; **get in sb.'s ~** *(lit. or fig.)* jmdm. im Wege stehen; **put difficulties/obstacles in sb.'s ~** *(fig.)* jmdm. Schwierigkeiten bereiten/Hindernisse in den Weg legen; **make ~ for sth.** für etw. Platz schaffen *od. (fig.)* machen; **make ~!** Platz da!; **[get] out of the/my ~!** [geh] aus dem Weg!; **move one's car out of the ~:** seinen Wagen aus dem Weg fahren; **get sth. out of the ~** *(settle sth.)* etw. erledigen; **g)** *(journey)* **on his**

~ to the office/London auf dem Weg ins Büro/nach London; **on the ~ out to Singapore** auf dem Hinweg/der Hinfahrt/dem Hinflug nach Singapur; **on the ~ back from Nigeria** auf dem Rückweg/der Rückfahrt/dem Rückflug von Nigeria; **she is just on the** *or* **her ~ in/out** sie kommt/geht gerade; **be on the ~ out** *(fig. coll.) (be losing popularity)* passé sein *(ugs.); (be reaching end of life)* ‹*Hund, Auto, Person*:› es nicht mehr lange machen *(ugs.);* **we stopped on the ~ to have lunch** wir hielten unterwegs zum Mittagessen an; **on her ~ home** auf dem Nachhauseweg; **they're on their ~**: sie sind unterwegs; **on the ~ there** auf dem Hinweg; **[be] on your ~!** nun geh schon!; **by the ~**: übrigens; **all this is by the ~**: das alles nur nebenbei; **h)** *(specific direction)* Richtung, *die;* **she went this/that/the other ~**: sie ist in diese/die/die andere Richtung gegangen; **look this ~, please** sieh/seht bitte hierher!; **he wouldn't look my ~**: er hat nicht zu mir herübergesehen; **I will call next time I'm [down] your ~**: wenn ich das nächste Mal in deiner Gegend bin, komme ich [bei dir] vorbei; **look the other ~** *(lit. or fig.)* weggucken; **the other ~ about** *or* **round** andersherum; **this/which ~ round** so/wie herum; **stand sth. the right/wrong ~ up** etw. richtig/falsch herum stellen; **turn sth. the right ~ round** etw. richtig herum drehen; **'this ~ up'** „hier oben"; **i)** *(advance)* Weg, *der;* **fight/push** *etc.* **one's ~ through** sich durchkämpfen/-drängen; **be under ~** ‹*Person*:› aufgebrochen sein; ‹*Fahrzeug*:› abgefahren sein; *(fig.: be in progress)* ‹*Besprechung, Verhandlung, Tagung*:› im Gange sein; **get sth. under ~** *(fig.)* etw. in Gang bringen; **get under ~**: wegkommen; **make one's ~ to Oxford/the station** nach Oxford/zum Bahnhof gehen/fahren; **Do you need a lift? – No, I'll make my own ~**: Soll ich dich mitnehmen? – Nein, ich komme alleine; **make one's [own] ~ in the world** seinen Weg gehen *(fig.);* **make** *or* **pay its ~**: ohne Verlust arbeiten; **pay one's ~**: für sich selbst aufkommen; **j)** *(respect)* Hinsicht, *die;* **in [exactly] the same ~**: [ganz] genauso; **in some ~s** in gewisser Hinsicht; **in one ~**: auf eine Art; **not in any ~**: in keiner Weise; **in every ~**: in jeder Hinsicht; **in a ~**: auf eine Art; **in more ~s than one** auf mehr als eine Art; **in no ~**: auf keinen Fall; durchaus nicht; **k)** *(state)* Verfassung, *die;* **in a bad ~**: schlecht; **they are in a very bad ~**: es geht ihnen sehr schlecht; **either ~**: so oder so; **in a small ~**: in bescheidenem Rahmen; **by ~ of** *(as a kind of)* als; *(for the purpose of)* um ... zu; **by ~ of illustration / greeting / apology / introduction** zur Illustration / Begrüßung / Entschuldigung/ Einführung; **l)** *(custom)* Art, *die;* **get into/out of the ~ of doing sth.** sich *(Dat.)* etw. an-/abgewöhnen; **he has a ~ of leaving his bills unpaid** es ist so seine Art, seine Rechnungen nicht zu bezahlen; **in its ~**: auf seine/ihre Art; **~ of life** Lebensstil, *der;* **change one's ~s** sich ändern; **~ of thinking** Denkungsart, *die;* **to my ~ of thinking** meiner Meinung nach; **m)** *(normal course of events)* **be the ~**: so *od.* üblich sein; **that is always the ~**: das ist immer so; **n)** *(ability to charm sb. or attain one's object)* **he has a ~ with him** er hat so eine Art; **she has a ~ with children/animals** sie kann mit Kindern/Tieren gut umgehen; **o)** *(specific manner)* Eigenart, *die;* **fall into bad ~s** schlechte [An]gewohnheiten annehmen; **p)** *(ordinary course)* Rahmen, *der;* **q)** *in pl. (parts)* Teile; **split sth. three ~s** etw. in drei Teile teilen. **2.** *adv.* weit; **~ off/ahead/above** weit weg von/weit voraus/weit über; **~ back** *(coll.)* vor langer Zeit; **~ back in the early fifties/before the war** vor langer Zeit, Anfang der fünfziger Jahre/vor dem Krieg; **~ up in the clouds** hoch oben in den Wolken; **he was ~ out with his guess, his guess was ~ out** er lag mit seiner Schätzung gewaltig daneben; **~ down south/in the valley** tief [unten] im Süden/Tal

way: ~'lay *v. t., forms as* [2]**lay 1: a)** *(ambush)* überfallen; **b)** *(stop for conversation)* abfangen; **~-'out** *adj. (coll.)* verrückt *(ugs.);* irre *(salopp);* **~side** *n.* Wegrand, *der;* **fall by the ~side** *(fig.)* auf der Strecke bleiben *(ugs.); attrib.* **~side flowers/inns** Blumen/Gasthöfe am Wegrand

wayward ['weɪwəd] *adj.* eigenwillig; ungezügelt ‹*Talent*›

WC *abbr.* **water-closet** WC, *das*

we [wɪ, *stressed* wi:] *pl. pron.* wir; **how are we feeling today?** *(coll.)* wie geht's uns denn heute? *(ugs.); see also* **our; ours; ourselves; us**

weak [wi:k] *adj.* **a)** *(lit. or fig.)* schwach; matt ‹*Lächeln*›; schwach ausgeprägt ‹*Kinn*›; jämmerlich ‹*Kapitulation*›; *(easily led)* labil ‹*Charakter, Person*›; **go/feel ~ at the knees** weiche Knie kriegen/haben; **the ~er sex** das schwache Geschlecht; **~ eyes** *or* **sight** schlechte Augen; **a ~ stomach** ein empfindlicher Magen; **be ~ in the head** schwachsinnig sein; **his French/maths is rather ~, he's rather ~ in French/maths** in Französisch/Mathematik ist er ziemlich schwach; **sb.'s ~ side** *or* **point** jmds. schwache Seite *od.* schwacher Punkt *od.* Schwachpunkt; **he has only a ~ case** seine Sache steht auf schwachen Füßen; **b)** *(watery)* schwach ‹*Kaffee, Tee*›; wäßrig, wässerig ‹*Suppe*›; dünn ‹*Bier, Suppe, Kaffee, Tee*›

weaken ['wi:kn] **1.** *v.t.* schwächen; beeinträchtigen ‹*Augen*›; entkräften, schwächen ‹*Argument*›; lockern ‹*Griff*›; **be ~ed by stress/too much work** durch Streß/zuviel Arbeit angegriffen werden. **2.** *v.i.* ‹*Kraft, Entschlossenheit*:› nachlassen; **the patient was visibly ~ing** der Patient wurde sichtlich schwächer

weak-kneed ['wi:kni:d] *adj.* **a)** **be ~**: weiche Knie haben **(with** vor + *Dat.*); **b)** *(fig.)* feige

weakling ['wi:klɪŋ] *n.* Schwächling, *der*

'weak-minded *adj.* **a)** *(lacking strength of purpose)* entschlußlos; unentschlossen; **b)** *(mentally deficient)* schwachsinnig

weakness ['wi:knɪs] *n.* Schwäche, *die; (in argument, defence)* schwacher Punkt

'weak-willed *adj.* willensschwach

weal [wi:l] *n. (ridge on flesh)* Striemen, *der*

wealth [welθ] *n., no pl.* **a)** *(abundance)* Fülle, *die;* **a great ~ of detail** große Detailfülle; **b)** *(riches, being rich)* Reichtum, *der*

wealthy ['welθɪ] **1.** *adj.* reich. **2.** *n. pl.* **the ~**: die Reichen

wean [wi:n] *v.t.* abstillen; entwöhnen ‹*Tier*›; **~ sb. [away] from sth.** *(fig.)* jmdm. etw. abgewöhnen

weapon ['wepən] *n. (lit. or fig.)* Waffe, *die*

weaponry ['wepənrɪ] *n.* Waffen *Pl.*

wear [weə(r)] **1.** *n., no pl., no indef. art.* **a)** *(rubbing)* **~ [and tear]** Verschleiß, *der;* Abnutzung, *die;* **show signs of ~**: Verschleiß- *od.*

Abnutzungserscheinungen aufweisen; **the worse for ~:** abgetragen ⟨*Kleider*⟩; abgelaufen ⟨*Schuhe*⟩; abgenutzt ⟨*Teppich, Sessel, Möbel*⟩; **feel the worse for ~:** sich angeschlagen fühlen *(ugs.)*; **b)** *(clothes, use of clothes)* Kleidung, *die;* **clothes for everyday ~:** Alltagskleidung, *die;* **a jacket for casual ~:** ein Freizeitsakko; **children's/ladies' ~:** Kinder-/Damen[be]kleidung, *die.* **2.** *v.t.*, **wore** [wɔ:(r)], **worn** [wɔ:n] **a)** tragen ⟨*Kleidung, Schmuck, Bart, Brille, Perücke, Abzeichen*⟩; **I haven't a thing to ~:** ich habe überhaupt nichts anzuziehen; **what size shoes do you ~?** welche Schuhgröße haben Sie?; **~ one's hair long** lange Haare tragen; **b)** abtragen ⟨*Kleidungsstück*⟩; abtreten, abnutzen ⟨*Teppich*⟩; **be worn [smooth]** ⟨*Stufen:*⟩ ausgetreten sein; ⟨*Gestein:*⟩ ausgewaschen sein; ⟨*Gesicht:*⟩ abgehärmt sein; **a [badly] worn tyre** ein [stark] abgefahrener Reifen; **c)** *(make by rubbing)* scheuern; **the water had worn a channel in the rock** das Wasser hatte sich durch den Felsen gefressen; **d)** *(exhaust)* erschöpfen; **e)** *(coll.: accept)* **I won't ~ that!** das nehme ich dir/ihm *usw.* nicht ab! *(ugs.).* **3.** *v.i.*, **wore, worn a)** ⟨*Kante, Saum, Kleider:*⟩ sich durchscheuern; ⟨*Absätze, Schuhsohlen:*⟩ sich ablaufen; ⟨*Teppich:*⟩ sich abnutzen; **~ thin** *(fig.)* ⟨*Freundschaft, Stil:*⟩ verflachen, oberflächlicher werden; ⟨*Witz, Ausrede:*⟩ schon reichlich alt sein; **b)** *(endure rubbing)* ⟨*Material, Stoff:*⟩ halten; *(fig.)* sich halten; **~ well/badly** sich gut/schlecht tragen

~ a'way 1. *v.t.* abschleifen ⟨*Kanten, Grate*⟩; **be worn away** ⟨*Stufen:*⟩ ausgetreten werden; ⟨*Inschrift:*⟩ verwittern. **2.** *v.i.* sich abnutzen; ⟨*Gestein:*⟩ verwittern; ⟨*Schuhabsätze:*⟩ sich ablaufen; *(fig.: weaken, lessen)* dahinschwinden

~ 'down 1. *v.t.* **a) be worn down** ⟨*Stufen:*⟩ ausgetreten werden; ⟨*Absätze:*⟩ sich ablaufen; ⟨*Reifen:*⟩ sich abfahren; ⟨*Berge:*⟩ abgetragen werden; **b)** *(fig.)* **~ down sb.'s resistance/defence/opposition** jmds. Widerstand/Verteidigung/Opposition zermürben; **~ sb. down** jmdn. zermürben; **worn down with hard work** abgearbeitet. **2.** *v.i.* ⟨*Absätze:*⟩ sich ablaufen; ⟨*Reifen:*⟩ sich abfahren

~ 'off *v.i.* ⟨*Auflage, Schicht:*⟩ abgehen; ⟨*Muster:*⟩ sich verlieren; *(fig.: pass away gradually)* sich legen; ⟨*Wirkung, Schmerz:*⟩ nachlassen

~ 'out 1. *v.t.* **a)** *(make useless)* aufbrauchen; ablaufen ⟨*Schuhe*⟩; auftragen ⟨*Kleidungsstück*⟩; **b)** *(fig.: exhaust)* kaputtmachen *(ugs.)*; **~ oneself out** sich kaputtmachen *(ugs.)*; **be worn out** kaputt sein *(ugs.).* **2.** *v.i.* *(become unusable)* kaputtgehen; **his patience finally wore out** seine Geduld war schließlich erschöpft

~ 'through 1. *v.i.* sich durchscheuern; **my trousers have worn through at the knee** meine Hose ist an den Knien durchgescheuert. **2.** *v.t.* durchscheuern

wearisome ['wɪərɪsəm] *adj. (lit. or fig.)* ermüdend

weary ['wɪərɪ] **1.** *adj.* **a)** *(tired)* müde; **b)** *(bored, impatient)* **be ~ of sth.** einer Sache *(Gen.)* überdrüssig sein; etw. satt haben *(ugs.)*; **c)** *(tiring)* ermüdend. **2.** *v.t.* **be wearied by sth.** durch etw. erschöpft sein. **3.** *v.i.* **~ of sth./sb.** einer Sache/jmds. überdrüssig werden

weasel ['wi:zl] *n.* Wiesel, *das*

weather ['weðə(r)] **1.** *n.* Wetter, *das;* **what's the ~ like?** wie ist das Wetter?; **the ~ has turned cooler** es ist kühler geworden; **he goes out in all ~s** er geht bei jedem Wetter hinaus; **he is feeling under the ~** *(fig.)* er ist [zur Zeit] nicht ganz auf dem Posten; **make heavy ~ of sth.** *(fig.)* sich mit etw. schwertun. **2.** *attrib. adj.* **keep a ~ eye on sth.** ein wachsames Auge auf etw. *(Akk.)* haben. **3.** *v.t.* **a)** *(expose to open air)* auswittern ⟨*Kalk, Holz*⟩; **b) be ~ed** ⟨*Gesicht:*⟩ wettergegerbt sein; **c)** *(wear away)* verwittern lassen ⟨*Gestein*⟩; **d)** *(come safely through)* abwettern ⟨*Sturm*⟩; *(fig.)* durchstehen ⟨*schwere Zeit*⟩. **4.** *v.i.* **a)** *(be discoloured)* ⟨*Holz, Farbe:*⟩ verblassen; *(wear away)* **~ [away]** ⟨*Gestein:*⟩ verwittern; **b)** *(survive exposure)* wetterfest sein

weather: ~-beaten *adj.* wettergegerbt ⟨*Gesicht, Haut*⟩; verwittert ⟨*Felsen, Gebäude*⟩; **~-chart** *n.* Wetterkarte, *die;* **~cock** *n.* Wetterhahn, *der;* **~ forecast** *n.* Wettervorhersage, *die;* **~man** *n.* Meteorologe, *der;* **~-map** *n.* Wetterkarte, *die;* **~proof 1.** *adj.* wetterfest; **2.** *v.t.* wetterfest machen; **~-report** *n.* Wetterbericht, *der;* **~ satellite** *n.* Wettersatellit, *der;* **~-vane** *n.* Wetterfahne, *die*

¹weave [wi:v] **1.** *n. (Textiles)* Bindung, *die.* **2.** *v.t.*, **wove** [wəʊv], **woven** ['wəʊvn] **a)** *(intertwine)* weben ⟨*[Baum]wolle, Garn, Fäden*⟩; **~ sth. into sth.** etw. zu etw. verweben; **~ flowers into wreaths** aus Blumen Kränze flechten; **b)** *(make by weaving)* weben ⟨*Textilien*⟩; flechten ⟨*Girlande, Korb, Kranz*⟩; **c)** *(fig.)* einflechten ⟨*Nebenhandlung, Thema usw.*⟩ **(into** in + *Akk.*); **d)** *(fig.: contrive)* ausspinnen ⟨*Geschichte*⟩. **3.** *v.i.*, **wove, woven** *(make fabric by weaving)* weben

²weave *v.i.* **a)** *(move repeatedly from side to side)* torkeln; **b)** *(take devious course)* sich schlängeln

weaver ['wi:və(r)] *n.* Weber, *der*/Weberin, *die*

web [web] *n.* **a)** Netz, *das;* **spider's ~:** Spinnennetz, *das;* **b)** *(woven fabric)* Gewebe, *das;* *(fig.)* Gespinst, *das;* **a ~ of lies/intrigue** ein Gespinst von Lügen/Intrigen

webbing ['webɪŋ] *n.* Gurtstoff, *der*

web: ~-foot *n.* Schwimmfuß, *der;* Ruderfuß, *der (Zool.)*; **~-footed** *adj.* schwimmfüßig; **~-'offset** *n. (Printing)* Rollenoffset[druck], *der*

wed [wed] *(rhet.)* **1.** *v.t.*, **-dd-** *(marry)* heiraten; ehelichen *(veralt., scherzh.)*; *(perform wedding ceremony for)* trauen ⟨*Brautpaar*⟩. **2.** *v.i.* heiraten; sich vermählen *(geh.)*

we'd [wɪd, *stressed* wi:d] **a) = we had; b) = we would**

Wed. *abbr.* **Wednesday** Mi.

wedded ['wedɪd] *adj.* **a)** *(married)* angetraut; **a ~ couple** ein getrautes Paar; **b)** *(of marriage)* **~ life** Eheleben, *das;* **~ bliss** Eheglück, *das;* **c)** *(fig.: devoted)* **be ~ to an idea/a dogma/a party** sich einer Idee/einem Dogma/einer Partei verschrieben haben; **be ~ to the view that ...:** immer noch davon überzeugt sein, daß ...; **d)** *(fig.: united)* vereint **(to** mit)

wedding ['wedɪŋ] *n.* Hochzeit, *die;* **have a registry office/a church ~:** sich standesamtlich/kirchlich trauen lassen; standesamtlich/kirchlich heiraten

wedding: ~ anniversary *n.* Hochzeitstag, *der;* **~ breakfast** *n.* Hochzeitsessen, *das;* **~-cake** *n.* Hochzeitskuchen, *der;* **~-day** *n.* Hochzeitstag, *der;* **~-dress** *n.* Brautkleid, *das;* **~-night** *n.* Hochzeitsnacht, *die;* **~ present** *n.* Hochzeitsgeschenk, *das;* **~-ring** *n.* Ehering, *der;* Trauring, *der*

wedge [wedʒ] **1.** *n.* **a)** Keil, *der;*

it's the thin end of the ~ *(fig.)* so fängt es immer an; **b) a ~ of cake** ein Stück Torte; **a ~ of cheese** eine Ecke Käse. **2.** *v. t.* **a)** *(fasten)* verkeilen; ~ **a door/window open** eine Tür/ein Fenster festklemmen, damit sie/es offen bleibt; **b)** *(pack tightly)* verkeilen; **there were five of them ~d together in the back of the car** sie saßen zu fünft eingezwängt *od.* zusammengepfercht hinten im Wagen

'wedge-shaped *adj.* keilförmig

wedlock ['wedlɒk] *n. (literary)* Ehe, *die;* Ehebund, *der (geh.);* **born in/out of ~:** ehelich/unehelich geboren

Wednesday ['wenzdeɪ, 'wenzdɪ] **1.** *n.* Mittwoch, *der.* **2.** *adv. (coll.)* **she comes ~s** sie kommt mittwochs. *See also* **Friday**

¹wee [wi:] *adj.* **a)** *(child lang./Scot.)* klein; lütt *(nordd.);* **b)** *(coll.: extremely small)* **a ~ bit** ein ganz klein bißchen *(ugs.)*

²wee *see* **wee-wee**

weed [wi:d] **1.** *n.* **a)** Unkraut, *das;* **~s** Unkräuter; Unkraut, *das;* **it's only a ~:** das ist bloß Unkraut; **b)** *(weakly person)* Kümmerling, *der.* **2.** *v. t.* jäten. **3.** *v. i.* [Unkraut] jäten

~ **'out** *v. t. (fig.)* aussieben

weeding ['wi:dɪŋ] *n., no pl., no indef. art.* [Unkraut]jäten, *das;* **do the/some ~:** Unkraut jäten

'weed-killer *n.* Unkrautvertilgungsmittel, *das*

weedy ['wi:dɪ] *adj.* **a)** von Unkraut überwachsen; **b)** *(scrawny)* spillerig *(ugs.);* schmächtig

week [wi:k] *n.* Woche, *die;* **what day of the ~ is it today?** was für ein Wochentag ist heute?; **he was away for a ~:** er war [für] eine Woche weg; **I haven't seen you for ~s** ich habe dich seit Wochen nicht gesehen; **~s ago** vor Wochen; **three times a ~:** dreimal die *od.* in der Woche; **£40 a** *or* **per ~:** 40 Pfund die *od.* in der *od.* pro Woche; **a ~'s leave/rest** eine Woche Urlaub/Pause; **for several ~s** mehrere Wochen lang; wochenlang; **once a ~, every ~:** einmal die Woche *od.* in der Woche; einmal wöchentlich; **~ in ~ out** Woche für Woche; **in a ~['s time]** in einer Woche; **in two ~s[' time]** in zwei Wochen; in vierzehn Tagen; **take a ~'s holiday** [sich *(Dat.)*] eine Woche Urlaub nehmen; **from ~ to ~, ~ by ~:** Woche für *od.* um Woche; **a three-~ period** ein Zeitraum von drei Wochen; **a two-~ visit** ein zweiwöchiger Besuch; **a six-~[s]-old baby** ein sechs Wochen altes *od.* sechswöchiges Baby; **a ~ [from] today/from** *or* **on Monday, today/Monday ~:** heute/Montag in einer Woche; **a ~ ago today/Sunday** heute/Sonntag vor einer Woche; **tomorrow ~:** morgen in einer Woche; **in** *or* **during the ~:** während der Woche; **42-hour/five-day ~:** 42-Stunden-Woche, *die*/Fünftagewoche, *die*

week: ~day *n.* Werktag, *der;* Wochentag, *der;* **on ~days** werktags; wochentags; **~end** [-'-, '--] *n.* Wochenende, *das;* **at the ~end** am Wochenende; **at** *or (Amer.)* **on ~ends** an Wochenenden; **go/be away for the ~end** übers Wochenende wegfahren/weg sein; **~-long** *adj.* einwöchig

weekly ['wi:klɪ] **1.** *adj.* wöchentlich; **~ wages** Wochenlohn, *der;* **a ~ season-ticket/magazine** eine Wochenkarte/Wochenzeitschrift; **at ~ intervals** wöchentlich; einmal pro Woche; **three-~:** dreiwöchentlich. **2.** *adv.* wöchentlich; einmal die Woche *od.* in der Woche. **3.** *n. (newspaper)* Wochenzeitung, *die; (magazine)* Wochenzeitschrift, *die*

'week-night *n.* **on a ~:** abends an einem Werktag; **on ~s** werktags abends

weep [wi:p] **1.** *v. i.,* **wept** [wept] **a)** weinen; **~ with** *or* **for joy/rage** vor Freude/Zorn weinen; **~ for sb./sth.** um jmdn./etw. weinen; **it makes you want to ~:** man könnte weinen; **b)** ‹*Wunde:*› nässen. **2.** *v. t.,* **wept a)** weinen ‹*Tränen*›; **b)** *(lament over)* beweinen

weepie ['wi:pɪ] *n. (coll.)* Schmachtfetzen, *der*

weeping 'willow *n.* Trauerweide, *die*

weevil ['wi:vɪl] *n.* Rüsselkäfer, *der*

'wee-wee *(coll.)* **1.** *n.* Pipi, *das (ugs.);* **do a ~:** Pipi machen. **2.** *v. i.* Pipi machen *(ugs.)*

weft [weft] *n.* **a)** *(set of threads)* Schuß, *der;* **b)** *(yarn)* Schußfaden, *der*

weigh [weɪ] **1.** *v. t.* **a)** *(find weight of)* wiegen; **the shop assistant was ~ing the fruit for her** die Verkäuferin wog ihr das Obst ab; **b)** *(estimate value of)* abwägen; **c)** *(consider)* abwägen; **~ in one's mind whether ...:** sich *(Dat.)* überlegen, ob ...; **~ the consequences of one's actions** sich *(Dat.)* die Folgen seines Handelns klarmachen; **d)** *(balance in one's hand)* wiegen; **e)** *(have the weight of)* wiegen; **it ~s very little** es wiegt sehr wenig; **a steak ~ing two pounds** ein zwei Pfund schweres Steak. **2.** *v. i.* **a) ~ [very] heavy/light** [sehr] viel/wenig wiegen; **b)** *(be important)* **~ with sb.** bei jmdm. Gewicht haben; **~ in sb.'s favour** für jmdn. sprechen

~ **a'gainst** *v. t. (fig.)* sprechen gegen; **~ heavily against sb.** sehr *od.* stark gegen jmdn. sprechen

~ **'down** *v. t.* **a)** *(cause to sag)* **fruit ~ed down the branches of the tree** die Äste des Baumes bogen sich unter der Last der Früchte; **be ~ed down by packages** mit Paketen schwer beladen sein; **b)** *(cause to be anxious or depressed)* niederdrücken; **~ed down with cares** bedrückt von Sorgen

~ **'in** *v. i. (Sport)* sich wiegen lassen; **~ in at 200 kg** 200 kg auf die Waage bringen

~ **on** *v. t.* lasten auf (+ *Dat.*); **~ [heavily] on sb.'s mind** jmdm. [schwer] auf der Seele liegen

~ **'out** *v. t.* abwiegen

~ **'up** *v. t.* abwägen; sich *(Dat.)* eine Meinung bilden über (+ *Akk.*) ‹*Person*›

weigh: ~bridge *n.* Brückenwaage, *die;* **~-in** *n. (Sport)* Wiegen, *das*

'weighing-machine *n.* Waage, *die*

weight [weɪt] **1.** *n.* **a)** Gewicht, *das;* **what is your ~?** wieviel wiegen Sie?; **be under/over ~:** zuwenig/zuviel wiegen; **throw one's ~ about** *or* **around** *(fig. coll.)* sich wichtig machen; **pull one's ~** *(do one's fair share)* sich voll einsetzen; **~s and measures** Maße und Gewichte; **lift ~s** Lasten heben; **b)** *(Athletics)* Kugel, *die;* **c)** *(surface density of cloth etc.)* Qualität, *die;* **d)** *(fig.: heavy burden)* Last, *die;* **it would be a ~ off my mind if ...:** mir würde ein Stein vom Herzen fallen, wenn ...; **e)** *(importance)* Gewicht, *das;* **give due ~ to sth.** einer Sache *(Dat.)* die nötige Beachtung schenken; **carry ~:** ins Gewicht fallen; **his opinion carries no ~ with me** seine Meinung ist für mich unbedeutend; **f)** *(preponderance)* Übergewicht, *das;* **the ~ of evidence is against him** praktisch alle Beweise sprechen gegen ihn; **~ of numbers** zahlenmäßiges Übergewicht. **2.** *v. t.* **a)** *(add ~ to)* beschweren; **b)** *(hold with ~)* **~ [down]** beschweren; *(fig.)* belasten

weightlessness ['weɪtlɪsnɪs] *n.* Schwerelosigkeit, *die*

weight: ~-lifter *n.* Gewichtheber, *der*/-heberin, *die;* **~-lifting**

n., no pl., no indef. art. Gewichtheben, *das;* **~-watcher** *n.* Schlankheitsbewußte, *der/die*

weighty ['weɪtɪ] *adj.* **a)** *(heavy)* schwer; **b)** *(important)* gewichtig

weir [wɪə(r)] *n.* Wehr, *das*

weird [wɪəd] *adj. (coll.: odd)* bizarr; verrückt *(ugs.)*

welcome ['welkəm] **1.** *int.* willkommen; **~ home/to England!** willkommen zu Hause/in England!; **~ aboard!** willkommen an Bord! **2.** *n.* **a)** Willkommen, *das;* **outstay** *or* **overstay one's ~:** zu lange bleiben; **give sb. a warm ~:** jmdn. herzlich willkommen heißen; **b)** *(reception)* Empfang, *der;* **give a proposal a warm ~:** einen Vorschlag zustimmend aufnehmen; **give sb. a warm ~** *(iron.)* jmdn. gebührend empfangen *(iron.);* **receive a rather cool ~:** ziemlich kühl empfangen werden. **3.** *v.t.* begrüßen; willkommen heißen *(geh.).* **4.** *adj.* **a)** willkommen; gefällig ⟨*Anblick*⟩; **make sb. [feel] ~:** jmdm. das Gefühl geben *od.* vermitteln, willkommen zu sein; **b)** *pred.* **you're ~ to take it** du kannst es gern nehmen; **you're ~** *(it was no trouble)* gern geschehen!; keine Ursache!; **if you want to stay here for the night you are more than ~:** wenn Sie die Nacht über hier bleiben möchten, sind Sie herzlich willkommen

welcoming ['welkəmɪŋ] *adj.* einladend; **a ~ cup of tea awaited us** zur Begrüßung erwartete uns eine Tasse Tee

weld [weld] *v.t.* **a)** *(unite)* verschweißen; *(repair, make, or attach by ~ing)* schweißen (**[on]to** an + *Akk.*); **~ two pipes together** zwei Rohre zusammenschweißen; **b)** *(fig.: unite closely)* zusammenschweißen (**into** zu); **~ two elements together** zwei Elemente zusammenschweißen

welder ['weldə(r)] *n.* **a)** *(person)* Schweißer, *der*/Schweißerin, *die;* **b)** *(machine)* Schweißgerät, *das*

welfare ['welfeə(r)] *n.* **a)** *(health and prosperity)* Wohl, *das;* **b)** *(social work; payments etc.)* Sozialhilfe, *die*

welfare: W~ 'State *n.* Wohlfahrtsstaat, *der;* **~ work** *n.* Sozialarbeit, *die;* **~ worker** *n.* Sozialarbeiter, *der*/-arbeiterin, *die*

[1]**well** [wel] *n.* **a)** *(water ~, mineral spring)* Brunnen, *der;* **b)** *(Archit.)* Schacht, *der; (of staircase)* Treppenloch, *das*

[2]**well 1.** *int.* **a)** *expr. astonishment* mein Gott; meine Güte; nanu; **~, ~!** sieh mal einer an!; **b)** *expr. relief* mein Gott; **c)** *expr. concession* na ja; **~ then, let's say no more about it** schon gut, reden wir nicht mehr davon; **d)** *expr. resumption* nun; **~ [then], who was it?** nun, wer war's?; **e)** *expr. qualified recognition of point* **~[, but] ...:** na ja, aber ...; ja schon, aber ...; **f)** *expr. resignation* **[oh] ~:** nun denn; **ah ~:** na ja; **g)** *expr. expectation* **~ [then]?** na? **2.** *adv.,* **better** ['betə(r)], **best** [best] **a)** *(satisfactorily)* gut; **the business is doing ~:** das Geschäft geht gut; **do ~ for oneself** Erfolg haben; **do ~ out of sth.** mit etw. ein gutes Geschäft machen; **the patient is doing ~:** dem Patienten geht es gut; **a ~ situated house** ein günstig gelegenes Haus; **you did ~ to come** gut, daß du gekommen bist; **~ done!** großartig!; **didn't he do ~!** hat er sich nicht gut geschlagen?; **you would do ~ to ...:** Sie täten gut daran, zu ...; **come off ~:** gut abschneiden; **you're ~ out of it** es ist gut, daß du damit nichts mehr zu tun hast; **b)** *(thoroughly)* gründlich ⟨*trocknen, polieren, schütteln*⟩; tüchtig ⟨*verprügeln*⟩; genau ⟨*beobachten*⟩; gewissenhaft ⟨*urteilen*⟩; **be ~ able to do sth.** durchaus *od.* sehr wohl in der Lage sein, etw. zu tun; **I'm ~ aware of what has been going on** mir ist sehr wohl klar *od.* bewußt, was sich abgespielt hat; **let** *or* **leave ~ alone** sich zufrieden geben; **be ~ pleased** sehr erfreut sein; **~ out of sight** *(very far off)* völlig außer Sichtweite (**of** *Gen.*); **we arrived ~ before the performance began** wir kamen eine ganze Zeit vor Beginn der Vorstellung; **be ~ in with sb.** bei jmdm. gut angeschrieben sein; **~ and truly** vollkommen; **I know only too ~ how/what** *etc.* **...:** ich weiß nur zu gut, wie/was *usw.* ...; **c)** *(considerably)* weit; **it was ~ on into the afternoon** es war schon spät am Nachmittag; **he is ~ past** *or* **over retiring age** er hat schon längst das Rentenalter erreicht; **he is ~ past** *or* **over forty** er ist weit über vierzig; **be ~ away** *(lit. or fig.)* einen guten Vorsprung haben; *(coll.: be drunk)* ziemlich benebelt sein *(ugs.);* **d)** *(approvingly, kindly)* gut, anständig ⟨*jmdn. behandeln*⟩; **think ~ of sb./sth.** eine gute Meinung von jmdm./etw. haben; **speak ~ of sb./sth.** sich positiv über jmdn./etw. äußern; **wish sb. ~:** jmdm. alles Gute wünschen; **e)** *(in all likelihood)* sehr wohl; **f)** *(easily)* ohne weiteres; **you cannot very ~ refuse their help** du kannst ihre Hilfe nicht ohne weiteres *od.* nicht gut ausschlagen; **g) as ~** *(in addition)* auch; ebenfalls; *(as much, not less truly)* genauso; ebenso; *(with equal reason)* genausogut; ebensogut; *(advisable)* ratsam; *(equally ~)* genauso gut; **Coming for a drink? – I might as ~:** Kommst du mit, einen trinken? – Warum nicht?; **you might as ~ go** du kannst ruhig gehen; **that is [just] as ~** *(not regrettable)* um so besser; **it was just as ~ that I had ...:** zum Glück hatte ich ...; **A as ~ as B:** B und auch [noch] A; **she can sing as ~ as dance** sie kann singen und auch tanzen; **as ~ as helping** *or (coll.)* **help me, she continued her own work** sie half mir und machte dabei noch mit ihrer eigenen Arbeit weiter. **3.** *adj.* **a)** *(in good health)* gesund; **How are you feeling now? – Quite ~, thank you** Wie fühlen Sie sich jetzt? – Ganz gut, danke; **look ~:** gut *od.* gesund aussehen; **I am perfectly ~:** ich fühle mich bestens; **get ~ soon!** gute Besserung!; **he hasn't been very ~ lately** es geht ihm in letzter Zeit nicht sehr gut; **feel ~:** sich wohl fühlen; **make sb. ~:** jmdn. gesund machen; **b)** *pred. (satisfactory)* **I am very ~ where I am** ich bin hier sehr zufrieden; **all's ~:** es ist alles in Ordnung; **all's ~ that ends ~** *(prov.)* Ende gut, alles gut; **all is not ~ with sb./sth.** mit jmdm./etw. ist etwas nicht in Ordnung; **[that's all] ~ and good** [das ist alles] gut und schön; **all being ~:** wenn alles gutgeht; **c)** *pred. (advisable)* ratsam

we'll [wɪl, *stressed* wi:l] = **we will**

well: ~-advised *see* **advised; ~-aimed** *adj.* gezielt ⟨*Schuß, Tritt, Stoß, Schlag*⟩; **~-behaved** *adj. see* **behave 1a; ~-being** *n.* Wohl, *das;* **~-bred** *adj. (having good manners)* anständig; **~-built** *adj.* ⟨*Person*⟩ mit guter Figur; **be ~-built** eine gute Figur haben; **~-chosen** *adj.* wohlgesetzt ⟨*Worte*⟩; wohlüberlegt ⟨*Bemerkungen*⟩; **~-defined** *adj.* klar definiert; **~-deserved** *adj.* wohlverdient ⟨*Lob, Ruhe*⟩; verdient ⟨*Belohnung, Prügel*⟩; **~-done** *adj. (Cookery)* durchgebraten; durch *nicht attr.;* **~-dressed** *adj.* gutgekleidet *präd. getrennt geschrieben;* **~-earned** *adj.* wohlverdient; **~-educated** *adj.* gebildet ⟨*Person, Benehmen*⟩; **~-equipped** *adj.* gut ausgestattet ‹*Büro, Studio, Krankenwagen*›; gut ausgerüstet ‹*Polizei, Armee, Expedition, Flugzeug*›; **~-established** *adj.* bewährt; **~-fed** *adj.* wohlgenährt; **~-founded** *adj.* [wohl] fundiert; **~-groomed** *adj.* gepflegt; **~-heeled** ['welhi:ld] *adj.* *(coll.)* gutbetucht *(ugs.)* *präd. getrennt geschrieben*

wellies ['welɪz] *n. pl. (Brit. coll.)* Gummistiefel

'well-informed *adj.* gutunterrichtet *präd. getrennt geschrieben*

wellington ['welɪŋtən] *n.* ~ **[boot]** Gummistiefel, *der*

well: ~-intentioned ['welɪntenʃənd] *adj.* gutgemeint *präd. getrennt geschrieben;* **~-judged** *adj.* gut gezielt; **~-kept** *adj.* gepflegt; in gutem Zustand *nachgestellt;* wohlgehütet ‹*Geheimnis*›; **~-known** *adj.* **a)** *(known to many)* bekannt; **b)** *(known thoroughly)* vertraut; **~-loved** *adj.* beliebt; **~ made** *adj.* *(skilfully manufactured)* gut [gearbeitet]; **~-mannered** *see* **mannered b; ~ marked** *adj.* gut gekennzeichnet ‹*Strecke, Wanderung*›; **~-meaning** *adj.* wohlmeinend; **be ~-meaning** es gut meinen; **~-meant** *adj.* gutgemeint; **~nigh** *adv. (rhet.)* nahezu; **~ off** *adj.* **a)** *(rich)* wohlhabend; **sb. is ~ off** jmdm. geht es [finanziell] gut; **b) be ~ off for sth.** *(provided with)* mit etw. gut versorgt sein; **c)** *(favourably situated)* **she is perfectly ~ off** es geht ihr ausgezeichnet; **~-preserved** *adj.* gut erhalten ‹*Holz, Mumie, (scherzh.) Achtzigjährige usw.*›; **~-read** ['welred] *adj.* belesen; **~-spent** *adj.* sinnvoll verbracht ‹*Zeit*›; vernünftig ausgegeben ‹*Geld*›; **~-thought-out** *adj.* gut durchdacht; **~-thumbed** *adj.* zerlesen ‹*Buch*›; **~-timed** *adj.* zeitlich gut gewählt; **~-to-do** *adj.* wohlhabend; **~-wisher** *n.* Sympathisant, *der*/Sympathisantin, *die;* **cards and gifts from ~-wishers** Kartengrüße und Geschenke; **~-worn** *adj.* abgetragen ‹*Kleidungsstück*›; abgenutzt ‹*Teppich*›; ausgetreten ‹*Pfad*›; abgedroschen ‹*Redensart, Ausdruck*›

Welsh [welʃ] **1.** *adj.* walisisch; **sb. is ~:** jmd. ist Waliser/Waliserin. **2.** *n.* **a)** *(language)* Walisisch, *das; see also* **English 2a; b)** *pl.* **the ~:** die Waliser

welsh *v. i. (leave without paying)* sich davonmachen, ohne zu bezahlen; **~ on** *v. t. (coll.)* **~ on sb./sth.** jmdn. sitzen lassen/sich um etw. herumdrücken *(ugs.)*

Welsh: ~man ['welʃmən] *n., pl.* **~men** ['welʃmən] Waliser, *der;* **~ 'rabbit, ~ 'rarebit** *ns.* Käsetoast, *der;* **~woman** *n.* Waliserin, *die*

welter ['weltə(r)] **1.** *v. i.* sich wälzen. **2.** *n.* Chaos, *das;* **a ~ of foam** eine schäumende Flut; **a ~ of emotions** ein Sturm von Gefühlen

'welterweight *n. (Boxing etc.)* Weltergewicht, *das; (person also)* Weltergewichtler, *der*

wend [wend] *v. t. (literary/arch.)* **~ one's way homewards** sich auf den Heimweg machen; **they ~ed their way back towards the village** sie machten sich auf den Weg zurück ins Dorf

Wendy house ['wendɪ haʊs] *n.* Spielhaus, *das*

went *see* **go 1, 2**

wept *see* **weep**

were *see* **be**

we're [wɪə(r)] = **we are**

weren't [wɜ:nt] *(coll.)* = **were not;** *see* **be**

werewolf ['wɪəwʊlf, 'weəwʊlf] *n., pl.* **werewolves** ['wɪəwʊlvz, 'weəwʊlvz], **werwolf** ['wɜ:wʊlf] *n., pl.* **werwolves** ['wɜ:wʊlvz] *(Mythol.)* Werwolf, *der*

west [west] **1.** *n.* **a)** *(direction)* Westen, *der;* **the ~:** West *(Met., Seew.);* **in/to[wards]/from the ~:** im/nach/von Westen; **to the ~ of** westlich von; westlich (+ *Gen.*); **b)** *usu.* **W~** *(also Polit.)* Westen, *der;* **from the W~:** aus dem Westen. **2.** *adj.* westlich; West‹*küste, -wind, -grenze, -tor*›. **3.** *adv.* westwärts; nach Westen; **~ of** westlich von; westlich (+ *Gen.*); **go ~** *(fig. sl.: be killed or wrecked or lost)* hopsgehen *(salopp)*

West: ~ 'Africa *pr. n.* Westafrika *(das);* **~ 'Bank** *pr. n.* **the ~ Bank** *(of the Jordan)* das Westjordanland; **~ Ber'lin** *pr. n. (Hist.)* West-Berlin *(das);* **w~bound** *adj.* ‹*Zug, Verkehr usw.*› in Richtung Westen; **~ Country** *n. (Brit.)* Westengland, *das;* **~ 'End** *n. (Brit.)* Westend, *das*

westerly ['westəlɪ] *adj.* **a)** *(in position or direction)* westlich; **in a ~ direction** nach Westen; **b)** *(from the west)* ‹*Wind*› aus westlichen Richtungen

western ['westən] **1.** *adj.* westlich; West‹*grenze, -hälfte, -seite*›; **~ Germany** Westdeutschland, *das.* **2.** *n.* Western, *der*

westerner ['westənə(r)] *n.* Abendländer, *der*/Abendländerin, *die*

Western 'Europe *pr. n.* Westeuropa *(das)*

westernize (westernise) ['westənaɪz] *v. t.* verwestlichen

westernmost ['westənməʊst] *adj.* westlichst...

West: ~ 'German *(Hist.)* **1.** *adj.* westdeutsch; **2.** *n.* Westdeutsche, *der/die;* **~ 'Germany** *pr. n. (Hist.)* Westdeutschland *(das);* **~ 'Indian 1.** *adj.* westindisch; **2.** *n.* Westinder, *der*/-inderin, *die;* **~ Indies** *pr. n. pl.* Westindische Inseln

Westphalia [west'feɪlɪə] *pr. n.* Westfalen *(das)*

Westphalian [west'feɪlɪən] **1.** *adj.* westfälisch. **2.** *n.* Westfale, *der*/Westfalin, *die*

westward ['westwəd] **1.** *adj.* nach Westen gerichtet; *(situated towards the west)* westlich; **in a ~ direction** nach Westen; [in] Richtung Westen. **2.** *adv.* westwärts; **they are ~ bound** sie fahren nach *od.* [in] Richtung Westen. **3.** *n.* Westen, *der*

westwards ['westwədz] *adv.* westwärts

wet [wet] **1.** *adj.* **a)** naß; **~ with tears** tränenfeucht; **~ behind the ears** *(fig.)* feucht hinter den Ohren *(ugs.);* **~ to the skin, ~ through** naß bis auf die Haut; **b)** *(rainy)* regnerisch; feucht ‹*Klima*›; **c)** *(recently applied)* frisch ‹*Farbe*›; **'~ paint'** „frisch gestrichen"; **d)** *(sl.: feeble)* schlapp *(ugs.);* schlappschwänzig *(salopp).* **2.** *v. t.,* **-tt-, wet** *or* **wetted a)** befeuchten; **b)** *(urinate on)* **~ one's bed/pants** das Bett/sich *(Dat.)* die Hosen naß machen. **3.** *n.* **a)** *(moisture)* Feuchtigkeit, *die;* **b)** *(rainy weather)* Regenwetter, *das; (rainy conditions)* Nässe, *die;* **in the ~:** im Regen; **c)** *(sl.: feeble person)* Flasche, *die (salopp abwertend);* **d)** *(Brit. Polit. coll.)* Schlappschwanz, *der (salopp abwertend)*

wet: ~back *n. (Amer. coll.)* illegaler mexikanischer Einwanderer; **~-nurse 1.** *n.* Amme, *die;* **2.** *v. t. (fig. derog.)* bemuttern; **~ suit** *n.* Tauchanzug, *der*

we've [wɪv, *stressed* wi:v] = **we have**

whack [wæk] **1.** *v. t. (coll.: strike heavily)* hauen *(ugs.).* **2.** *n.* **a)** *(coll.: heavy blow)* Schlag, *der;* **give sb. a ~ on the bottom** jmdm. eins auf den Hintern geben *(ugs.);* **b)** *(sl.: share)* Anteil, *der*

whale [weɪl] *n., pl.* **~s** *or same* **a)**

(Zool.) Wal, *der;* Walfisch, *der (volkst.);* **b)** *no pl. (coll.)* **we had a ~ of a [good] time** wir haben uns bombig *(ugs.)* amüsiert

wham [wæm] **1.** *int.* wumm. **2.** *v.t.*, **-mm-: ~ sb.** jmdm. einen Schlag versetzen

wharf [wɔ:f] *n.*, *pl.* **wharves** [wɔ:vz] *or* **~s** Kai, *der;* Kaje, *die (nordd.)*

what [wɒt] **1.** *interrog. adj.* **a)** *asking for selection* welch...; **~ book did you choose?** welches Buch hast du ausgesucht?; **b)** *asking for statement of amount* wieviel; *with pl. n.* wie viele; **~ men/money has he?** wie viele Leute/wieviel Geld hat er?; **I know ~ time it starts** ich weiß, um wieviel Uhr es anfängt; **~ more can I do/say?** was kann ich sonst noch tun/sagen?; **~ more do you want?** was willst du [noch] mehr?; **c)** *asking for statement of kind* was für; **~ kind of man is he?** was für ein Mensch ist er?; **~ good** *or* **use is it?** wozu soll das gut sein? **2.** *excl. adj.* **a)** *(how great)* was für; **~ impudence** *or* **cheek/luck!** was für eine Unverschämtheit *od.* Frechheit/was für ein Glück!; **b)** *before adj. and n. (to ~ extent)* was für. **3.** *rel. adj.* **we can dispose of ~ difficulties there are remaining** wir können die verbleibenden Schwierigkeiten ausräumen; **lend me ~ money you can** leih mir soviel Geld, wie du kannst; **I will give you ~ help I can** ich werde dir helfen, so gut ich kann. **4.** *adv.* **a)** *(to ~ extent)* **~ do I care?** was kümmert's mich?; **~ does it matter?** was macht's?; **b) ~ with ...:** wenn man an ... denkt; **~ with changing jobs and moving house I haven't had time to do any studying** da ich eine neue Stellung angetreten habe und umgezogen bin, hatte ich keine Zeit zum Lernen. **5.** *interrog. pron.* **a)** *(~ thing)* was; **~ is your name?** wie heißt du/heißen Sie?; **~ about ...?** *(is there any news of ...?, ~ will become of ...?)* was ist mit ...?; **~ about a game of chess?** wie wär's mit einer Partie Schach?; **~ to do?** was tun?; **~-d'you-[ma-] call-him/-her/-it, ~'s-his/-her/-its-name** wie heißt er/sie/es noch; **~ for?** wozu?; **and/or ~ 'have you** und/oder was sonst noch [alles]; **~ if ...?** was ist, wenn ...?; **~ is it** *etc.* **like?** wie ist es *usw.*?; **~ not** wer weiß was alles; **~ 'of it?** was ist dabei?; was soll [schon] dabeisein?; **~ do you say** *or (Amer.)* **~ say we have a rest?** was hältst du davon, wenn wir mal Pause machen?; wie wär's mit einer Pause?; **[I'll] tell you ~:** weißt du, was; paß mal auf; **[and] ~ then?** [na] und?; **or ~?** oder was?; **so ~?** na und?; **b)** *asking for confirmation* **~?** wie?; was? *(ugs.);* **you did ~?** was hast du gemacht?; **c)** *in rhet. questions equivalent to neg. statement* **~ is the use in trying/the point of going on?** wozu [groß] versuchen/weitermachen? **6.** *rel. pron. (that which)* was; **do ~ I tell you** tu, was ich dir sage; **~ little I know/remember** das bißchen, das ich weiß/an das ich mich erinnere; **this is ~ I mean: ...:** ich meine Folgendes: ...; **tell sb. ~ to do** *or* **~ he can do with sth.** *(coll. iron.)* jmdm. sagen, wo er sich *(Dat.)* etw. hinstecken kann *(salopp);* **~ is more** außerdem; **the weather being ~ it is ...:** so, wie es mit dem Wetter aussieht, ...; **for ~ it is** in seiner Art. **7.** *excl. pron.* was; **~ she must have suffered!** wie sie gelitten haben muß!

whatever [wɒt'evə(r)] **1.** *adj.* **a)** *rel. adj.* **~ measures we take** welche Maßnahmen wir auch immer ergreifen; **~ materials you will need** alle Materialien, die du vielleicht brauchst; **b)** *(notwithstanding which)* was für ... auch immer; **~ problems you encounter** auf welche Probleme Sie auch stoßen [mögen]; **c)** *(at all)* überhaupt; **I can't see anyone ~:** ich kann überhaupt niemanden sehen. **2.** *pron.* **a)** *rel. pron.* was für ... [auch immer]; **do ~ you like** mach, was du willst; **b)** *(notwithstanding anything)* was auch [immer]; **~ happens, ...:** was auch geschieht, ...; **c) or ~:** oder was auch immer; oder sonst was *(ugs.)*

'whatnot *n. (coll.: indefinite thing)* Dingsbums, *das*

whatsit ['wɒtsɪt] *n. (coll.) (thing)* Dingsbums, *das (ugs.); (peron)* Dingsda, *der (ugs.)*

wheat [wi:t] *n., no pl., no indef. art.* Weizen, *der*

wheedle ['wi:dl] *v.t.* **a)** *(coax)* **~ sb. into doing sth.** jmdm. so lange gut zureden, bis er etw. tut; **b)** *(get by cajoling)* sich *(Dat.)* verschaffen; **~ sth. out of sb.** jmdm. etw. abschwatzen *(ugs.)*

wheel [wi:l] **1.** *n.* **a)** Rad, *das;* **[potter's] ~:** Töpferscheibe, *die;* **[roulette] ~:** Roulett, *das;* **reinvent the ~** *(fig.)* sich mit Problemen aufhalten, die längst gelöst sind; **put** *or* **set the ~s in motion** *(fig.)* die Sache in Gang setzen; **the ~s of bureaucracy turn slowly** *(fig.)* die Mühlen der Bürokratie mahlen langsam; **b)** *(for steering) (Motor Veh.)* Lenkrad, *das; (Naut.)* Steuerrad, *das;* **at** *or* **behind the ~** *(of car)* am *od.* hinterm Steuer; *(of ship; also fig.)* am Ruder; **c)** *(Mil.: drill movement)* Schwenkung, *die;* **left/right ~:** Links-/Rechtsschwenkung, *die.* **2.** *v.t.* **a)** *(turn round)* wenden; **b)** *(push)* schieben; **~ oneself** *(in a ~chair)* fahren. **3.** *v.i.* **a)** kehrtmachen; **b) ~ and deal** mauscheln

~ a'bout, ~ a'round 1. *v.t.* herumdrehen; wenden ⟨*Pferd*⟩. **2.** *v.i.* kehrtmachen; *(face the other way)* sich umdrehen

~ 'in *v.t.* hinein-/hereinschieben

~ 'out *v.t.* hinaus-/herausschieben; **~ sb. out** *(fig. derog.)* jmdn. vorführen

~ 'round *see* **~ about**

wheel: ~barrow *n.* Schubkarre, *die;* Schubkarren, *der;* **~base** *n. (Motor Veh., Railw.)* Radstand, *der;* **~chair** *n.* Rollstuhl, *der;* **~-clamp** *n.* Parkkralle, *die*

-wheeled [wi:ld] *adj. in comb.* ⟨*vier-, sechs-, acht*⟩räd[e]rig

wheeler-dealer [wi:lə'di:lə(r)] *n.* Mauschler, *der*/Mauschlerin, *die (ugs.); (financial)* Geschäftemacher, *der*/-macherin, *die*

wheelie ['wi:lɪ] *n. (sl.) Fahren auf dem Hinterrad;* Wheelie, *das;* **do a ~/do ~s** auf dem Hinterrad fahren; ein Wheelie/Wheelies fahren

wheeling and 'dealing *n.* Mauschelei, *die (ugs.); (shady deals)* undurchsichtige Geschäfte

wheeze [wi:z] **1.** *v.i.* schnaufen; keuchen. **2.** *n.* Schnaufen, *das;* Keuchen, *das*

wheezy ['wi:zɪ] *adj. (coll.)* pfeifend, keuchend ⟨*Atem, Stimme*⟩

whelk [welk] *n. (Zool.)* Wellhornschnecke, *die*

when [wen] **1.** *adv.* **a)** *(at what time)* wann; **say ~** *(coll.: pouring drink)* sag halt; **that was ~ I intervened** das war der Moment, wo ich eingriff; **b)** *(at which)* **the time ~ ...:** die Zeit, zu der *od. (ugs.)* wo/*(with past tense)* als ...; **the day ~ ...:** der Tag, an dem *od. (ugs.)* wo/*(with past tense)* als ...; **do you remember [the time] ~ we ...:** erinnerst du dich daran, wie wir ... **2.** *conj.* **a)** *(at the time that)* als; *(with present or future tense)* wenn; **~ [I was] young** als ich jung war; in meiner Jugend; **~ in doubt** im Zweifelsfall; **~ cleaning the gun** beim Putzen des Gewehrs; **~ speaking French** wenn ich/sie *usw.* Französisch spreche/spricht *usw.;* **b)** *(whereas)* **why do you go abroad ~ it's cheaper here?** warum fährst du ins Ausland, wo es doch hier billiger ist?; **I received only £5 ~ I should have got £10** ich bekam nur 5 Pfund, hätte aber 10 Pfund bekommen sollen; **c)** *(considering that)* wenn; **how can I finish it ~ you won't help?** wie soll ich es fertigmachen, wenn du nicht hilfst?; **d)** *(and at that moment)* als. **3.** *pron.* **by/till ~ ...?;** bis wann ...?; **from/since ~ ...?** ab/seit wann ...?; **~ are we invited for?** für wann sind wir eingeladen?; **but that was yesterday, since ~ things have changed** aber das war gestern, und inzwischen hat sich manches geändert

whence [wens] *(arch./literary)* **1.** *adv.* woher; **the village ~ comes the famous cheese** das Dorf, aus dem der berühmte Käse kommt. **2.** *conj. (to the place from which)* dorthin, woher

whenever [wen'evə(r)] **1.** *adv.* **a)** wann immer; **or ~:** oder wann immer; **b)** *(coll.)* = **when ever** *see* **ever e. 2.** *conj.* jedesmal wenn

where [weə(r)] **1.** *adv.* **a)** *(in or at what place)* wo; **~ shall we sit?** wo wollen wir sitzen *od.* uns hinsetzen?; wohin wollen wir uns setzen?; **~ was I?** *(fig.)* wo war ich stehengeblieben?; **~ did Orwell say/write that?** wo *od.* an welcher Stelle sagt/schreibt Orwell das? **~ is the harm in it/the sense of it?** was macht das schon/welchen *od.* was für einen Sinn hat das?; **this is ~ I was born** hier bin ich geboren; **b)** *(from what place)* woher; **~ did you get that information?** wo hast du das erfahren?; **c)** *(to what place, to which)* wohin; **~ shall I put it?** wohin soll ich es legen?; wo soll ich es hinlegen?; **~ do we go from here?** *(fig.)* was tun wir jetzt *od.* als nächstes?; **d)** *(in what respect)* inwiefern; **I don't know ~ they differ/I've gone wrong** ich weiß nicht, worin sie sich unterscheiden/wo ich den Fehler gemacht habe; **that is ~ you are wrong** in diesem Punkt irrst du dich; **e)** *(in which)* wo; **in the box ~ I keep my tools** in der Kiste, worin *od.* in der ich mein Werkzeug habe; **f)** *(in what situation)* wo; **~ will/would they be if ...?** was wird/würde aus ihnen, wenn ...? **2.** *conj.* wo; **~ uncertain, leave blank** bei Unsicherheit [bitte] freilassen. **3.** *pron.* **near/not far from ~ it happened** nahe der Stelle/nicht weit von der *od.* unweit der Stelle, wo es passiert ist; **from ~ I'm standing** von meinem Standort [aus]; **they continued from ~ they left off** sie machten da weiter, wo sie aufgehört hatten; **~ [...] from?** woher [...]?; von wo [...]?; **~ [...] to?** wohin [...]?; **~ are you going to?** wohin gehst du?; wo gehst du hin?

whereabouts 1. [weərə'baʊts] *adv. (in what place)* wo; *(to what place)* wohin. **2.** [weərə'baʊts] *pron.* **~ are you from?** woher kommst du? **3.** ['weərəbaʊts] *n., constr. as sing. or pl. (of thing)* Verbleib, *der; (of person)* Aufenthalt[sort], *der*

where: ~'as *conj.* während; **he is very quiet, ~as she is an extrovert** er ist sehr ruhig, sie dagegen ist eher extravertiert; **~'by** *adv. (by which)* mit dem/der/denen; mit dessen/deren Hilfe; **~upon** [weərə'pɒn] *adv.* worauf

wherever [weər'evə(r)] **1.** *adv.* **a)** *(in whatever place)* wo immer; **sit ~ you like** setz dich, wohin du magst; **or ~:** oder wo immer; oder sonstwo *(ugs.);* **b)** *(to whatever place)* wohin immer; **I shall go ~ I like** ich gehe, wohin ich will; **or ~:** oder wohin immer; oder sonstwohin *(ugs.);* **c)** *(coll.: where ever)* **~ in the world have you been?** wo in aller Welt hast du bloß gesteckt? **2.** *conj.* **a)** *(in every place that)* überall [da], wo; **do it ~ possible** tun Sie es, wo *od.* wenn [irgend] möglich; **b)** *(to every place that)* wohin auch; **~ he went** wohin er auch ging. **3.** *pron.* wo ... auch; **~ you're going to** wo du auch hingehst; wohin du auch gehst

whet [wet] *v. t.,* **-tt-: a)** *(sharpen)* wetzen; **b)** *(fig.: stimulate)* anregen ⟨*Appetit*⟩

whether ['weðə(r)] *conj.* ob; **I don't know ~ to go [or not]** ich weiß nicht, ob ich gehen soll [oder nicht]; **the question [of] ~ to do it [or not]** die Frage, ob man es tun soll [oder nicht]; **~ you like it or not, I'm going** ob es dir paßt oder nicht, ich gehe

whey [weɪ] *n., no pl., no indef. art.* Molke, *die*

which [wɪtʃ] **1.** *adj.* **a)** *interrog.* welch...; **~ one** welcher/welche/welches; **~ ones** welche; **~ one of you did it?** wer von euch hat es getan?; **~ way** *(how)* wie; *(in ~ direction)* wohin; **b)** *rel.* welch... *(geh.);* **he usually comes at one o'clock, at ~ time I'm having lunch/by ~ time I've finished** er kommt immer um ein Uhr; dann esse ich gerade zu Mittag/bis dahin bin ich schon fertig. **2.** *pron.* **a)** *interrog.* welcher/welche/welches; **~ of you?** wer von euch?; **~ is ~?** welcher/welche/welches ist welcher/welche/welches?; **b)** *rel.* der/die/das; welcher/welche/welches *(veralt.); referring to a clause* was; **of ~:** dessen/deren; **everything ~ I predicted** alles, was ich vorausgesagt habe; **the crime of ~ you accuse him** das Verbrechen, dessen Sie ihn anklagen; **I intervened, after ~ they calmed down** ich griff ein, worauf[hin] sie sich beruhigten; **Our Father, ~ art in Heaven** *(Rel.)* Vater unser, der du bist im Himmel

whichever [wɪtʃ'evə(r)] **1.** *adj.* **a)** *(any ... that)* der *od.* derjenige, der/die *od.* diejenige, die/das *od.* dasjenige, das/die *od.* diejenigen, die; **go ~ way you want** es ist egal, welchen Weg du nimmst; **take ~ apple/apples you wish** nimm den Apfel, den du willst/die Äpfel, die du willst; **..., ~ period is the longer** ... , je nachdem, welches der längere Zeitraum ist; **b)** *(no matter which/who/whom)* welche/welcher/welches ... auch; **~ way you go** welchen Weg du auch nimmst. **2.** *pron.* **a)** *(any one[s] that)* der *od.* derjenige, der/die *od.* diejenige, die/das *od.* dasjenige, das/die *od.* diejenigen, die; **~ of you/the children wins will get a prize** wer von euch gewinnt/das Kind, das gewinnt, bekommt einen Preis; **b)** *(no matter which one[s])* welcher/welche/welches ... auch; **~ of them comes/come** wer von ihnen auch kommt; **c)** *(coll.: which ever)* **~ could it be?** welcher/welche/welches könnte das nur sein?

whiff [wɪf] *n.* **a)** *(smell)* [leichter] Geruch; *(puff, breath)* Hauch, *der;* **~s of smoke** Rauchwölkchen; **b)** *(fig.: trace)* Hauch, *der;* **the faintest ~ of sentiment** der leiseste Anflug von Sentimentalität

while [waɪl] **1.** *n.* Weile, *die;* **quite a** *or* **quite some ~, a good ~:** eine ganze Weile; ziemlich lange; **[for] a ~:** eine Weile; **where have you been all the** *or* **this ~?** wo warst du die ganze Zeit?; **a long ~:** lange; **for a little** *or* **short ~:** eine kleine Weile; **stay a little ~ [longer]** bleib noch ein Weilchen; **in a little** *or* **short ~:** gleich; **be worth [sb.'s] ~:** sich [für jmdn.] lohnen; **make sth. worth sb.'s ~:** jmdn. für etw. entsprechend belohnen; **once in a ~:** von Zeit zu Zeit [mal]; hin und wieder [mal].

2. *conj.* **a)** während; *(as long as)* solange; ~ **in London he took piano lessons** als er in London war, nahm er Klavierstunden; **b)** *(although)* obgleich; **c)** *(whereas)* während

~ **away** *v. t.* ~ **away the time** sich *(Dat.)* die Zeit vertreiben (**by, with** mit)

whilst [waɪlst] *(Brit.) see* **while 2**

whim [wɪm] *n. (mood)* Laune, *die*

whimper ['wɪmpə(r)] **1.** *n.* ~**[s]** Wimmern, *das; (of dog etc.)* Winseln, *das.* **2.** *v. i.* wimmern; ‹*Hund:*› winseln

whimsical ['wɪmzɪkl] *adj.* launenhaft; *(odd, fanciful)* spleenig

whine [waɪn] **1.** *v. i.* **a)** *(make moaning sound)* heulen; ‹*Hund:*› jaulen; ‹*Baby:*› quengeln *(ugs.);* **b)** *(complain)* jammern; **he's been whining to the boss about it** er hat dem Chef darüber etwas vorgejammert. **2.** *n.* **a)** *(sound)* Heulen, *das; (esp. of dog)* Jaulen, *das;* **b)** *(complaint)* ~**[s]** Gejammer, *das*

whinge ['wɪndʒ] *coll.* **1.** *v. i.*, ~**ing** *see* **whine 1 b.** **2.** *n. see* **whine 2 b**

whip [wɪp] **1.** *n.* **a)** Peitsche, *die;* **b)** *(Brit. Parl.: official)* Einpeitscher, *der*/Einpeitscherin, *die (Jargon);* Fraktionsgeschäftsführer, *der*/-führerin, *die (Amtsspr.);* **c)** *(Brit. Parl.: notice)* **[three-line] ~:** *[verbindliche] Aufforderung zur Teilnahme an einer Plenarsitzung [wegen einer wichtigen Abstimmung];* **resign the ~:** aus der Fraktion austreten. **2.** *v. t.*, **-pp-:** **a)** *(lash)* peitschen; **the rider ~ped his horse** der Reiter gab seinem Pferd die Peitsche; **b)** *(Cookery)* schlagen; **c)** *(move quickly)* reißen ‹*Gegenstand*›; **she ~ped it out of my hand** sie riß es mir aus der Hand; **d)** *(sl.: defeat)* auseinandernehmen *(salopp bes. Sport);* **e)** *(sl.: steal)* klauen *(ugs.).* **3.** *v. i.*, **-pp-:** **a)** *(move quickly)* flitzen *(ugs.);* ~ **through a book in no time** ein Buch in Null Komma nichts durchlesen *(ugs.);* **b)** *(lash)* peitschen

~ **a'way** *v. t.* wegreißen (**from** *Dat.*)

~ **'out** *v. t.* [blitzschnell] herausziehen

~ **'up** *v. t.* **a)** *(snatch up)* [blitz]schnell aufheben; **b)** *(Cookery)* [kräftig] schlagen; **c)** *(arouse)* aufpeitschen ‹*Wellen*›; *(fig.)* anheizen *(ugs.),* anfachen *(geh.)* ‹*Emotionen, Interesse*›; schüren ‹*Haß, Unzufriedenheit*›; **d)** *(coll.: make quickly)* schnell hinzaubern ‹*Gericht, Essen*›

whip: ~cord *n.* **a)** *(cord)* Peitschenschnur, *die;* **b)** *(fabric)* Whipcord, *der;* ~ **hand** *n.* **have** *or* **hold the ~ hand [of** *or* **over sb.]** *(fig.)* die Oberhand [über jmdn.] haben; **~lash** *n.* **a)** Peitschenriemen, *der;* **b)** *(Med.)* **~lash [injury]** Peitschenschlagverletzung, *die;* Schleudertrauma, *das*

whipped 'cream *n.* Schlagsahne, *die*

whippet ['wɪpɪt] *n.* Whippet, *der*

whipping ['wɪpɪŋ] *n. (flogging)* Schlagen [mit der Peitsche]; *(as form of punishment)* Prügelstrafe, *die; (flagellation)* Geißelung, *die;* **give sb. a ~:** jmdn. auspeitschen; *(sl.: defeat)* jmdm. eins überbraten *(salopp)*

'whip-round *n. (Brit. coll.)* Sammlung, *die;* **have** *or* **hold a ~-round [for sb./sth.]** [für jmdn./etw.] den Hut herumgehen lassen *(ugs.)*

whirl [wɜ:l] **1.** *v. t.* **a)** *(rotate)* [im Kreis] herumwirbeln; **b)** *(fling)* schleudern; *(with circling motion)* wirbeln ‹*Blätter, Schneeflocken usw.*›; **c)** *(convey rapidly)* in Windeseile fahren. **2.** *v. i.* **a)** *(rotate)* wirbeln; **b)** *(move swiftly)* sausen; *(with circling motion)* wirbeln; **c)** *(fig.: reel)* **everything/the room ~ed about me** mir drehte sich alles/das Zimmer drehte sich vor meinen Augen. **3.** *n.* **a)** Wirbeln, *das;* **she was** *or* **her thoughts were** *or* **her head was in a ~** *(fig.)* ihr schwirrte der Kopf; **b)** *(bustle)* Trubel, *der*

~ **a'bout,** ~ **a'round** *v. t. & i.* herumwirbeln

~ **a'way,** ~**'off** **1.** *v. t.* in Windeseile wegfahren. **2.** *v. i.* lossausen

~ **'round** **1.** *v. t.* [im Kreis] herumwirbeln. **2.** *v. i.* [im Kreis] herumwirbeln; ‹*Rad, Rotor, Strudel:*› wirbeln

whirl: ~pool *n.* Strudel, *der; (bathing pool)* Whirlpool, *der;* **~wind** *n.* **a)** Wirbelwind, *der; (stronger)* Wirbelsturm, *der;* **b)** *(fig.: tumult)* Wirbel, *der;* Trubel, *der; attrib.* **~wind romance** heftige Romanze

whirr [wɜ:(r)] **1.** *v. i.* surren; ‹*Heuschrecke, Grille usw.:*› zirpen; ‹*Flügel eines Vogels, Propeller:*› schwirren. **2.** *n. see* **1:** Surren, *das;* Zirpen, *das;* Schwirren, *das*

whisk [wɪsk] **1.** *n.* **a)** Wedel, *der;* **b)** *(Cookery)* Schneebesen, *der; (part of mixer)* Rührbesen, *der.* **2.** *v. t.* **a)** *(Cookery)* [mit dem Schnee-/Rührbesen] schlagen; **b)** *(convey rapidly)* in Windeseile bringen

~ **a'way** *v. t.* **a)** *(flap away)* wegscheuchen; **b)** *(remove suddenly)* ~ **sth. away [from sb.]** [jmdm.] etw. [plötzlich] wegreißen; **c)** *(convey rapidly)* in Windeseile wegbringen

~ **'off** *v. t.* **a)** *(flap off) see* ~ **away a;** **b)** *(remove suddenly)* [plötzlich] wegreißen; ~ **one's coat off** seinen Mantel von sich werfen; **c)** *see* ~ **away c**

~ **'up** *see* ~ **2 a**

whisker ['wɪskə(r)] *n.* **a)** *in pl. (hair on man's cheek)* Backenbart, *der;* **b)** *(Zool.) (of cat, mouse, rat)* Schnurrhaar, *das; (of walrus)* Bartborste, *die;* **c)** *(fig. coll.: small distance)* **be within a ~ of sth./doing sth.** kurz vor etw. *(Dat.)* stehen/kurz davorstehen, etw. zu tun; **win by a ~:** ganz knapp gewinnen

whiskey *(Amer., Ir.),* **whisky** ['wɪskɪ] *n.* Whisky, *der; (Irish or American ~)* Whiskey, *der*

whisper ['wɪspə(r)] **1.** *v. i.* **a)** flüstern; ~ **to sb.** jmdm. etwas zuflüstern; ~ **to me so that no one else will hear** flüster es mir ins Ohr, damit es niemand [anders] hört; **b)** *(speak secretly)* tuscheln; ~ **against sb.** über jmdn. tuscheln; **c)** *(rustle)* [leise] rauschen; säuseln *(geh.);* flüstern *(poet.).* **2.** *v. t.* **a)** flüstern; ~ **sth. to sb./in sb.'s ear** jmdm. etw. zuflüstern/ins Ohr flüstern; **b)** *(rumour)* [hinter vorgehaltener Hand] erzählen; **the story is being ~ed about the village that ...:** im Dorf macht die Geschichte die Runde, daß ...; **it is ~ed that ...:** man munkelt, daß ... *(ugs.).* **3.** *n.* **a)** *(~ed speech)* Flüstern, *das;* **in a ~, in ~s** im Flüsterton; **b)** *(~ed remark)* **their ~s** ihr Geflüster; **c)** *(rumour)* Gerücht, *das;* **there were ~s that ...:** es gab Gerüchte, daß ...

whist [wɪst] *n. (Cards)* Whist, *das*

whistle ['wɪsl] **1.** *v. i.* pfeifen; ~ **at a girl** hinter einem Mädchen herpfeifen; **the spectators ~d at the referee** die Zuschauer pfiffen den Schiedsrichter aus; ~ **to sb.** jmdm. pfeifen; ~ **for sth.** nach etw. pfeifen; ~ **in the dark** *(fig.)* seine Angst verdrängen; **you can ~ for it!** *(fig. coll.)* da kannst du lange warten! **2.** *v. t.* **a)** pfeifen; **b)** *(summon)* her[bei]pfeifen; **he ~d his dog and it came running** er pfiff seinem Hund, und er kam angelaufen. **3.** *n.* **a)** *(sound)* Pfiff, *der; (act of whistling)* Pfeifen, *das;* **he gave a ~ of surprise** er ließ ein überraschtes Pfeifen vernehmen; **b)** *(instrument)* Pfeife, *die;*

penny *or* **tin** ~: Blechflöte, *die;* **the referee blew his** ~: der Schiedsrichter pfiff; **[as] clean/clear as a** ~ *(fig.)* blitzsauber/absolut frei; **blow the** ~ **on sb./sth.** *(fig.)* jmdn./etw. auffliegen lassen *(ugs.)*

'**whistle-stop** *n. (Amer.)* **a)** *(Railw.) (small town)* kleines Nest *(ugs.) (an einer Bahnlinie); (station)* Bedarfshaltepunkt, *der;* **b)** *(Polit.) kurzer Auftritt eines Politikers während einer Wahlkampfreise; (rapid visit)* Stippvisite, *die; attrib.* **~-stop tour/campaign** Reise mit vielen Kurzaufenthalten/Wahlkampf[reise] mit vielen kurzen Auftritten *od.* Terminen

whistling '**kettle** *n.* Pfeifkessel, *der*

whit [wɪt] *n., no pl., no def. art. (arch./literary)* **no** ~, **not a** ~: kein bißchen

white [waɪt] **1.** *adj.* **a)** weiß; **[as]** ~ **as snow** schneeweiß; **he prefers his coffee** ~ *(Brit.)* er trinkt seinen Kaffee am liebsten mit Milch; **b)** *(pale)* weiß; *(through illness)* blaß; bleich; *(through fear or rage)* bleich; weiß; **[as]** ~ **as chalk** *or* **a sheet** kreidebleich; **c)** *(light-skinned)* weiß; ~ **people** Weiße *Pl.* **2.** *n.* **a)** *(colour)* Weiß, *das;* **b)** *(of egg)* Eiweiß, *das;* **c)** *(of eye)* Weiße, *das;* **the ~s of their eyes** das Weiße in ihren Augen; **d)** **W**~ *(person)* Weiße, *der/die;* **e)** *(~ clothes)* **dressed in** ~: weiß gekleidet; **~s** weißer Dreß

white: ~bait *n., pl. same: junger Hering/junge Sprotte o. ä.;* ~ '**bread** *n.* Weißbrot, *das;* ~ '**coffee** *n. (Brit.)* Kaffee mit Milch; ~-'**collar** *adj.* **~-collar worker** Angestellte, *der/die;* **~-collar union** Angestelltengewerkschaft, *die;* ~ '**elephant** *n. (fig.)* nutzloser Besitz; **be a** ~ **elephant** ‹*Gebäude, Einkaufszentrum usw.:*› reine Geldverschwendung sein; *attrib.* **a** ~ **elephant stall** *eine Bude, an der Sachen angeboten werden, die deren ehemalige Besitzer gern loswerden wollen;* **~-faced** ['waɪtfeɪsd] *adj.* [kreide]bleich; **W~hall** *pr. n. (Brit. Polit.: Government)* Whitehall *(das);* ~ '**hope** *n.* Hoffnungsträger, *der*/Hoffnungsträgerin, *die;* ~ '**horse** *n.* **a)** Schimmel, *der;* **b)** *in pl. (on waves)* Schaumkronen; ~-'**hot** *adj.* weißglühend; *(fig.)* glühend; **W**~ **House** *pr. n. (Amer. Polit.)* **the W~ House** das Weiße Haus; ~ **man** *n. (Anthrop.)* Weiße, *der;* **the** ~ **man** *(~ people)* der weiße Mann

whiten ['waɪtn] **1.** *v. t.* weiß machen; weißen ‹*Wand, Schuhe*›. **2.** *v. i.* **a)** *(become white)* weiß werden; **b)** *(turn pale)* [kreide]weiß werden

whiteness ['waɪtnɪs] *n., no pl.* **a)** Weiß, *das;* **b)** *(paleness)* Blässe, *die*

white: W~ '**Paper** *n. (Brit.) öffentliches Diskussionspapier über Vorhaben der Regierung;* ~ '**sauce** *n.* weiße *od.* helle Soße; ~ '**slave** *n.* weiße Sklavin; *Opfer des Mädchenhandels; attrib.* ~ **slave trade** *or* **traffic** Mädchenhandel, *der;* ~ '**spirit** *n. (Chem.)* Terpentin[öl]ersatz, *der;* ~ '**stick** *n.* Blindenstock, *der;* ~ '**sugar** *n.* weißer Zucker; **~wash 1.** *n.* **a)** [weiße] Tünche; *(fig.)* Schönfärberei, *die;* **the report is a ~wash of the Government** der Bericht versucht, die Regierung reinzuwaschen; **b)** *(defeat)* Zu-Null-Niederlage, *die;* **2.** *v. t.* **a)** [weiß] tünchen; **b)** *(defeat)* zu Null schlagen; ~ '**wedding** *n.* Hochzeit in Weiß; **have a** ~ **wedding** in Weiß heiraten; ~ '**wine** *n.* Weißwein, *der;* ~ **woman** *n. (Anthrop.)* Weiße, *die*

whither ['wɪðə(r)] *(arch./rhet.)* **1.** *adv.* wohin; ~ **democracy/Ulster?** *(fig. rhet.)* wohin *od. (geh.)* quo vadis, Demokratie/Ulster? **2.** *conj.* dorthin *od.* dahin, wohin

whitish ['waɪtɪʃ] *adj.* weißlich

Whit Monday [wɪt 'mʌndeɪ, wɪt 'mʌndɪ] *n.* Pfingstmontag, *der*

Whitsun ['wɪtsn] *n.* Pfingsten, *das od. Pl.;* **at** ~: zu *od.* an Pfingsten; **next/last** ~: nächste/letzte Pfingsten

Whit Sunday [wɪt 'sʌndeɪ, wɪt 'sʌndɪ] *n.* Pfingstsonntag, *der*

whittle ['wɪtl] **1.** *v. t.* schnitzen an (+ *Dat.*); ~ **a stick to a point** einen Stock anspitzen. **2.** *v. i.* ~ **at sth.** an etw. *(Dat.)* [herum]schnitzen

~ **a'way,** ~ '**down** *v. t. (fig.)* **a)** *(completely)* auffressen ‹*Gewinn, Geldmittel usw.*›; ~ **away sb.'s rights/power** jmdm. nach und nach alle Rechte/alle Macht nehmen; **b)** *(partly)* allmählich reduzieren ‹*Anzahl, Team, Gewinn, Verlust*›; verkürzen ‹*Liste*›

Whit [wɪt]: ~ **week** *n.* Pfingstwoche, *die;* ~ **week'end** *n.* Pfingstwochenende, *das*

whiz, whizz [wɪz] **1.** *v. i.,* **-zz-** zischen. **2.** *n.* Zischen, *das*

~ '**past** *v. i.* vorbeizischen; ‹*Vogel:*› vorbeischießen

'**whiz[z]-kid** *n. (coll.)* Senkrechtstarter, *der;* **he is a financial** ~: er macht eine steile Karriere als Finanzmann

who [hʊ, *stressed* hu:] *pron.* **a)** *interrog.* wer; *(coll.: whom)* wen; *(coll.: to whom)* wem; ~ **are you talking about?** *(coll.)* von wem *od.* über wen sprichst du?; **I don't know ~'s ~ in the firm yet** ich kenne die Leute in der Firma noch nicht richtig; ~ **am 'I to object/argue** *etc.*? wie könnte ich Einwände erheben/etwas dagegen sagen *usw.*?; ~ **would have thought it?** *(rhet.)* wer hätte das gedacht!; **b)** *rel.* der/die/das; *pl.* die; *(coll.: whom)* den/die/das; *(coll.: to whom)* dem/der/denen; **any person/he/those** ~ **...:** wer ...; **they** ~ **...:** diejenigen, die *od.* welche ...; **everybody** ~ **...:** jeder, der ...; **I/you** ~ **...:** ich, der ich/du, der du ...; **the man** ~ **I met last week/~ you were speaking to** *(coll.)* der Mann, den ich letzte Woche getroffen habe/mit dem du gesprochen hast

whoa [wəʊ] *int.* brr

who'd [hʊd, *stressed* hu:d] **a)** = **who had; b)** = **who would**

whodun[n]it [hu:'dʌnɪt] *n. (coll.)* Krimi, *der (ugs.)*

whoever [hu:'evə(r)] *pron.* **a)** wer [immer]; ~ **comes will be welcome** jeder, der kommt, ist willkommen; **b)** *(no matter who)* wer ... auch; ~ **you may be** wer Sie auch sind; **c)** *(coll.: who ever)* ~ **could it be?** wer könnte das nur sein?

whole [həʊl] **1.** *adj.* **a)** ganz; **that's the** ~ **point [of the exercise]** das ist der ganze Zweck der Übung *(ugs.);* **the** ~ **lot [of them]** [sie] alle; **a** ~ **lot of people** eine ganze Menge Leute; **b)** *(intact)* ganz; **roast sth.** ~: etw. im ganzen braten; **c)** *(undiminished)* ganz; **three** ~ **hours** drei volle Stunden. **2.** *n.* **a) the** ~: das Ganze; **the** ~ **of my money/the village/London** mein ganzes *od.* gesamtes Geld/das ganze Dorf/ganz London; **he spent the** ~ **of that year/of Easter abroad** er war jenes Jahr/zu Ostern die ganze Zeit im Ausland; **the** ~ **of Shakespeare** *or* **of Shakespeare's works** Shakespeares gesamte Werke; **b)** *(total of parts)* Ganze, *das;* **as a** ~: als Ganzes; **sell sth. as a** ~: etw. im ganzen verkaufen; **on the** ~: im großen und ganzen

whole: ~food *n.* Vollwertkost, *die;* **~-hearted** [həʊl'hɑ:tɪd] *adj.* herzlich ‹*Dank, Dankbarkeit, Glückwünsche*›; tiefempfunden ‹*Dankbarkeit, Reue*›; rückhaltlos ‹*Unterstützung, Hingabe,*

Ergebenheit⟩; **~meal** *adj.* Vollkorn-; **~ note** *n. (Amer. Mus.)* ganze Note; **~ 'number** *n. (Math.)* ganze Zahl; **~sale 1.** *adj.* **a)** *(Commerc.)* Großhandels-; **~sale dealer** *or* **merchant** Großhändler, *der*/-händlerin, *die;* **the ~sale trade** der Großhandel; **b)** *(fig.: on a large scale)* massenhaft; Massen-; **in a ~sale way** massenweise; **c)** *(fig.: indiscriminate)* pauschal; **2.** *adv.* **a)** *(Commerc.)* en gros ⟨*[ein]kaufen, verkaufen*⟩; im Großhandel ⟨*[ein]kaufen*⟩; *(at ~sale price)* zum Einkaufs- *od.* Großhandelspreis; **b)** *(fig.: on a large scale)* massenweise; **c)** *(fig.: indiscriminately)* pauschal

wholesaler ['həʊlseɪlə(r)] *n. (Commerc.)* Grossist, *der*/Grossistin, *die (fachspr.);* Großhändler, *der*/-händlerin, *die*

wholesome ['həʊlsəm] *adj.* gesund

whole 'wheat *n.* Vollweizen, *der*

who'll [hʊl, *stressed* hu:l] = **who will**

wholly ['həʊllɪ] *adv.* völlig; durch und durch ⟨*böse*⟩

whom [hu:m] *pron.* **a)** *interrog.* wen; *as indirect object* wem; **to ~/of ~ did you speak?** mit wem/von wem haben Sie gesprochen?; **b)** *rel.* den/die/das; *pl.* die; *as indirect object* dem/der/dem; *pl.* denen; **the children, the mother of ~ ...:** die Kinder, deren Mutter ...; **five children, all of ~ are coming** fünf Kinder, die alle mitkommen; **ten candidates, only the best of ~ ...:** zehn Kandidaten, von denen nur die besten ...

whoop [wu:p] **1.** *v.i.* [aufgeregt] schreien; *(with joy, excitement)* juchzen *(ugs.);* jauchzen. **2.** *v.t.* **~ it up** *(coll.)* die Sau rauslassen *(salopp); (Amer.: stir up enthusiasm)* Stimmung machen. **3.** *n.* [aufgeregter] Schrei; *(of joy, excitement)* Juchzer, *der (ugs.);* Jauchzer, *der*

whoopee 1. [wʊ'pi:] *int.* juhu. **2.** ['wʊpi:] *n.* **make ~** *(coll.)* die Sau rauslassen *(ugs.)*

whooping cough ['hu:pɪŋ kɒf] *n. (Med.)* Keuchhusten, *der*

whoosh [wʊʃ] **1.** *v.i.* brausen; ⟨*Rakete, Geschoß:*⟩ zischen. **2.** *n.* Brausen, *das; (of rocket, projectile)* Zischen, *das;* **with a [loud] ~:** [laut] brausend/zischend

whopper ['wɒpə(r)] *n. (coll.)* **a)** Riese, *der;* **a ~ of a marrow/fish** ein Riesending von einem Kürbis/Fisch *(ugs.);* **b)** *(lie)* faustdicke Lüge

whopping ['wɒpɪŋ] *(coll.)* **1.** *adj.* riesig; Riesen- *(ugs.);* faustdick ⟨*Lüge*⟩. **2.** *adv.* **~ great** *see* **1**

whore [hɔ:(r)] *(derog.)* **1.** *n.* **a)** *(prostitute)* Hure, *die;* **b)** *(loose woman)* Flittchen, *das.* **2.** *v.i.* **~ [around]** [herum]huren

whorl [wɔ:l] *n. (Bot.)* Wirtel, *der;* Quirl, *der*

who's [hu:z] **a)** = **who is; b)** = **who has**

whose [hu:z] *pron.* **a)** *interrog.* wessen; **b)** *rel.* dessen/deren/dessen; *pl.* deren; **the people ~ house this is** die Leute, denen dieses Haus gehört

whosever [hu:z'evə(r)] *pron.* wessen ... auch; **~ it is, ...:** wem er/sie/es auch gehört, ...

who've [hʊv, *stressed* hu:v] = **who have**

why [waɪ] **1.** *adv.* **a)** *(for what reason)* warum; *(for what purpose)* wozu; **~ is that?** warum das?; **and this/that is ~ I believe ...:** und darum glaube ich ...; **~ not buy it, if you like it?** kauf es dir doch, wenn es dir gefällt; **~ do we need another car?** wozu brauchen wir noch ein Auto?; **b)** *(on account of which)* **the reason ~ he did it** der Grund, aus dem *od.* warum er es tat; **I can see no reason ~ not** ich wüßte nicht, warum nicht. **2.** *int.* **~, certainly/of course!** aber sicher!; **~, if it isn't Jack!** aber das ist ja Jack!

WI *abbr.* **a) West Indies; b)** *(Brit.)* **Women's Institute**

wick [wɪk] *n.* Docht, *der;* **get on sb.'s ~** *(fig. sl.)* jmdm. auf den Keks gehen *(salopp)*

wicked ['wɪkɪd] **1.** *adj.* **a)** *(evil)* böse; schlecht ⟨*Charakter, Person, Welt*⟩; niederträchtig ⟨*Gedanken, Plan, Verhalten*⟩; schändlich ⟨*Gesetz, Buch*⟩; **the ~ villain** der Schurke; der Bösewicht *(veralt.);* **it was ~ of you to torment the poor cat** es war gemein von dir, die arme Katze zu quälen; **b)** *(vicious)* boshaft ⟨*Zunge*⟩; **c)** *(coll.: scandalous)* himmelschreiend; sündhaft *(ugs.)* ⟨*Preis*⟩; **it's ~ how he's been treated** wie man ihn behandelt hat, das schreit zum Himmel; **it's a ~ shame** es ist eine wahre Schande. **2.** *n. pl.* **the ~:** die Bösen

wickedly ['wɪkɪdlɪ] *adv.* **a)** *(evilly)* niederträchtig; *as sentence-modifier* niederträchtigerweise; **b)** *(coll.: scandalously)* himmelschreiend; sündhaft *(ugs.)* ⟨*teuer*⟩

wickedness ['wɪkɪdnɪs] *n.* **a)** *no pl. see* **wicked a:** Bosheit, *die;* Schlechtigkeit, *die;* Niederträchtigkeit, *die;* Schändlichkeit, *die;* **b)** *(evil act)* Niederträchtigkeit, *die;* **c)** *no pl. (viciousness)* Boshaftigkeit, *die;* **d)** *no pl. (coll.: scandalousness)* Schändlichkeit, *die;* **the ~ of this waste** so eine himmelschreiende Verschwendung

wicker ['wɪkə(r)] *n.* Korbgeflecht, *das; attrib.* Korb⟨*waren, -möbel, -stuhl*⟩; geflochten ⟨*Korb, Matte*⟩

'wickerwork *n.* **a)** *(material)* Korbgeflecht, *das;* **b)** *(articles)* Korbwaren

wicket ['wɪkɪt] *n. (Cricket) (stumps)* Tor, *das;* Wicket, *das; (central area of pitch)* Wurfbahn, *die;* **at the ~:** [als Schlagmann] auf dem Spielfeld; **keep ~:** als Torwächter spielen

'wicket-keeper *n. (Cricket)* Torwächter, *der*/-wächterin, *die;* Wicketkeeper, *der*

wide [waɪd] **1.** *adj.* **a)** *(broad)* breit; groß ⟨*Unterschied, Abstand, Winkel, Loch*⟩; weit ⟨*Kleidung*⟩; **allow** *or* **leave a ~ margin** *(fig.)* viel Spielraum lassen; **three feet ~:** drei Fuß breit; **b)** *(extensive)* weit; umfassend ⟨*Lektüre, Wissen, Kenntnisse*⟩; weitreichend ⟨*Einfluß*⟩; vielseitig ⟨*Interessen*⟩; groß ⟨*Vielfalt, Bekanntheit, Berühmtheit*⟩; reichhaltig ⟨*Auswahl, Sortiment*⟩; breit ⟨*Publizität*⟩; **have ~ appeal** weite Kreise ansprechen; **the ~ world** die weite Welt; **c)** *(liberal)* großzügig; **d)** *(fully open)* weit geöffnet; **e)** *(off target)* **be ~ of sth.** etw. verfehlen; **be ~ of the mark** *(fig.)* ⟨*Annahme, Bemerkung:*⟩ nicht zutreffen; **you're ~ of the mark** *(fig.)* du liegst falsch *(ugs.).* **2.** *adv.* **a)** *(fully)* weit; **~ awake** hellwach; *(fig. coll.)* gewitzt; **I'm ~ awake to your tricks** ich durchschaue deine Tricks; **b)** *(off target)* **shoot ~:** danebenschießen; **fall ~ of the target, go ~:** das Ziel verfehlen; **aim ~/~ of sth.** daneben/neben etw. *(Akk.)* zielen

-wide *in comb.* **city-/county-~:** in der ganzen Stadt/Grafschaft *nachgestellt*

wide: ~-angle 'lens *n. (Photog.)* Weitwinkelobjektiv, *das;* **~-eyed** ['waɪdaɪd] *adj. (surprised)* mit großen Augen *nachgestellt*

widely ['waɪdlɪ] *adv.* **a)** *(over a wide area)* weit ⟨*verbreitet, gestreut*⟩; locker, in großen Abständen ⟨*verteilt*⟩; **he has travelled ~ in Europe** er ist in Europa viel gereist; **a ~ read man** ein [sehr] belesener Mann; **b)** *(by many people)* weithin ⟨*bekannt, akzeptiert*⟩; **a ~**

held view eine weitverbreitete Ansicht; **it is ~ rumoured that ...:** allgemein wird gemunkelt *(ugs.),* daß...; **c)** *(in a wide sense)* im weiten Sinne ‹*gebraucht*›; weit ‹*interpretiert*›; **d)** *(greatly)* stark, erheblich ‹*sich unterscheiden*›; sehr ‹*verschieden, unterschiedlich*›

widen ['waɪdn] **1.** *v.t.* verbreitern; *(fig.)* erweitern. **2.** *v.i.* sich verbreitern; breiter werden; *(fig.)* sich erweitern; ‹*Interessen:*› vielfältiger werden

~ **'out** *v.i.* sich verbreitern; breiter werden; **~ out into sth.** sich zu etw. erweitern

wide: ~-open *attrib. adj.*, ~ **'open** *pred. adj.* weit aufstehend *od.* geöffnet ‹*Fenster, Tür*›; weit aufgerissen ‹*Mund, Augen*›; **the ~-open spaces of North America** die Weite der nordamerikanischen Landschaft; **be ~ open** ‹*Fenster, Tür:*› weit offenstehen; **be ~ open to attack/criticism** Angriffen/der Kritik ausgesetzt sein; **lay** *or* **leave oneself/sb. ~ open to sth.** sich/jmdn. einer Sache *(Dat.)* schutzlos preisgeben; **the contest is still ~ open** der Wettbewerb *od.* der Ausgang des Wettbewerbs ist noch völlig offen; **~-ranging** ['waɪdreɪndʒɪŋ] *adj.* weitgehend ‹*Maßnahme, Veränderung*›; weitreichend ‹*Auswirkungen*›; ausführlich ‹*Diskussion, Gespräch*›; **~spread** *adj.* weitverbreitet *präd. getrennt geschrieben* ‹*Art, Ansicht*›; groß ‹*Nachfrage, Beliebtheit*›; von vielen geteilt ‹*Sympathie*›; **become ~spread** sich [weit] ausbreiten; **there was a ~spread demand for reform** Reformen wurden allgemein gefordert

widow ['wɪdəʊ] **1.** *n.* Witwe, *die;* **be left/made a ~:** zur Witwe werden. **2.** *v.t.* zur Witwe machen ‹*Frau*›; zum Witwer machen ‹*Mann*›; **be ~ed** zur Witwe/zum Witwer werden (**by** durch)

widowed ['wɪdəʊd] *adj.* verwitwet

widower ['wɪdəʊə(r)] *n.* Witwer, *der*

width [wɪdθ] *n.* **a)** *(measurement)* Breite, *die; (of garment)* Weite, *die;* **what is the ~ of ...?** wie breit/weit ist ...?; **be half a metre in ~:** einen halben Meter breit/weit sein; **b)** *(large scope)* großer Umfang; *(of definition)* Weite, *die; (of interests)* Vielseitigkeit, *die;* **c)** *(piece of material)* Bahn, *die*

widthways ['wɪdθweɪz], **widthwise** ['wɪdθwaɪz] *adv.* in der Breite

wield [wi:ld] *v.t. (literary)* führen *(geh.); (fig.)* ausüben ‹*Macht, Einfluß usw.*›; **~ a stick/sword** einen Stock/ein Schwert schwingen

wiener ['vi:nə(r)] *n. (Amer.)* Würstchen, *das*

wife [waɪf] *n., pl.* **wives** [waɪvz] Frau, *die;* **make sb. one's ~:** jmdn. zur Frau nehmen; **lawful wedded ~** *(Eccl.)* rechtmäßig angetraute Frau; **old wives' tale** Ammenmärchen, *das*

wife-swapping ['waɪfswɒpɪŋ] *n. (coll.)* Partnertausch, *der*

wig [wɪg] *n.* Perücke, *die*

wiggle ['wɪgl] *(coll.)* **1.** *v.t.* hin und her bewegen. **2.** *v.i.* wackeln; *(move)* sich schlängeln; **~ into sth.** sich in etw. *(Akk.)* zwängen. **3.** *n.* Wackeln, *das*

wigwam ['wɪgwæm] *n.* Wigwam, *der*

wild [waɪld] **1.** *adj.* **a)** wildlebend ‹*Tier*›; wildwachsend ‹*Pflanze*›; **grow ~:** wild wachsen; **~ beast** wildes Tier; **b)** *(rough)* unzivilisiert; *(bleak)* wild ‹*Landschaft, Gegend*›; **c)** *(unrestrained)* wild; ungezügelt; wild, wüst ‹*Bursche, Unordnung, Durcheinander*›; **run ~** ‹*Pferd, Hund:*› frei herumlaufen; ‹*Kind:*› herumtoben; ‹*Pflanzen:*› wuchern; **let one's imagination run ~:** seiner Phantasie freien Lauf lassen; **d)** *(stormy)* stürmisch; tobend ‹*Wellen*›; **e)** rasend ‹*Wut, Zorn, Eifersucht, Beifall*›; unbändig ‹*Freude, Wut, Zorn, Schmerz*›; wild ‹*Erregung, Zorn, Geschrei*›; panisch ‹*Angst*›; irr ‹*Blick*›; **be/become ~ [with sth.]** [vor etw. *(Dat.)*] außer sich *(Dat.)* sein/außer sich *(Akk.)* geraten; **send** *or* **drive sb. ~:** jmdn. rasend vor Erregung machen; **f)** *(coll.: very keen)* **be ~ about sb./sth.** wild auf jmdn./etw. sein; **I'm not ~ about it** ich bin nicht wild darauf *(ugs.);* **g)** *(coll.: angry)* wütend; **be ~ with** *or* **at sb.** eine Wut auf jmdn. haben; **h)** *(reckless)* ungezielt ‹*Schuß, Schlag*›; unbedacht ‹*Verhalten, Versprechen, Gerede*›; aus der Luft gegriffen ‹*Anschuldigungen, Behauptungen*›; maßlos ‹*Übertreibung*›; irrwitzig ‹*Plan, Idee, Versuch, Hoffnung*›. **2.** *n.* **the ~[s]** die Wildnis; **see an animal in the ~:** ein Tier in freier Wildbahn sehen; **in the ~s** *(coll.)* in der Pampa *(ugs.);* **the call of the ~:** der Ruf der Wildnis

wild 'cat *n. (Zool.)* Wildkatze, *die;* **~cat** *attrib. adj.* **~cat strike** wilder Streik

wilderness ['wɪldənɪs] *n.* Wildnis, *die; (desert)* Wüste, *die;* **be in the ~** *(Polit.)* alle Bedeutung verloren haben

wild: ~-eyed ['waɪldaɪd] *adj.* mit irrem Blick *nachgestellt;* **~fire** *see* **spread 2a; ~fowl** *n., pl. same* Federwild, *das; (Cookery)* Wildgeflügel, *das;* **~-'goose chase** *n. (fig.: hopeless quest)* aussichtslose Suche; **send sb. on a ~-goose chase** jmdn. einem Phantom nachjagen lassen; **~ 'horse** *n.* Wildpferd, *das;* **~ horses would not drag it from me** *(fig.)* eher beiße ich mir die Zunge ab[, als daß ich es erzähle]; **~life** *n., no pl., no indef. art.* die Tier- und Pflanzenwelt; die Natur; *attrib.* **~life park/reserve/sanctuary** Naturpark, *der*/-reservat, *das*/-schutzgebiet, *das*

wildly ['waɪldlɪ] *adv.* **a)** *(unrestrainedly)* wild; **run ~ all over the house** ‹*Kinder:*› wie wild im ganzen Haus herumtoben; **b)** *(stormily)* wild; **the wind blew ~:** der Wind blies heftig; **c)** *(excitedly)* rasend ‹*eifersüchtig*›; unbändig ‹*verliebt, sich freuen, sich verlieben*›; wild ‹*schreien, applaudieren*›; erregt ‹*diskutieren*›; **I'm not ~ interested in it** *(iron.)* ich interessiere mich nicht übermäßig dafür; **be ~ excited about sth.** über etw. *(Akk.)* ganz aus dem Häuschen sein *(ugs.);* **he looked ~ about him** er blickte irr um sich; **d)** *(recklessly)* aufs Geratewohl; maßlos ‹*übertreiben*›; wirr ‹*daherreden, denken*›; **~ inaccurate** völlig ungenau

wildness ['waɪldnɪs] *n., no pl.* **a)** *(bleakness)* Wildheit, *die;* **b)** *(lack of restraint)* Wildheit, *die;* **after the ~ of his youth** nach seiner wilden *od.* stürmischen Jugend; **c)** *(storminess)* **the ~ of the weather/sea** das stürmische Wetter/die stürmische See; **d)** *(of promise, words)* Unbedachtheit, *die; (of scheme, attempt, idea, hope, quest)* Irrwitzigkeit, *die*

Wild 'West *pr. n.* Wilder Westen

wile [waɪl] *n.* List, *die;* Schlich, *der*

wilful ['wɪlfl] *adj.* **a)** *(deliberate)* vorsätzlich; bewußt ‹*Täuschung*›; **b)** *(obstinate)* starrsinnig

¹will [wɪl] **1.** *v.t., only in pres.* **will,** *neg. (coll.)* **won't** [wəʊnt], *past* **would** [wʊd], *neg. (coll.)* **wouldn't** [wʊdnt] **a)** *(consent to)* wollen; **They won't help me. W~/Would you?** Sie wollen mir nicht helfen. Bist du bereit?; **you ~ help her, won't you?** du hilfst ihr doch *od.* du wirst ihr doch helfen, nicht wahr?; **the car won't start** das Auto will nicht anspringen *od.* springt nicht an; **~/would you pass the salt, please?** gibst du bitte mal das Salz rüber?/würdest du bitte mal das Salz rübergeben?; **~/would you come in?** kommen Sie doch herein; **now just listen, ~ you!** jetzt hör/hört gefälligst zu!; **~ you be quiet!** willst du/wollt ihr wohl ruhig sein!; **b)** *(be accustomed to)* pflegen; **he ~ sit there hour after hour** er pflegt dort stundenlang zu sitzen; *(emphatic)* **children '~ make a noise** Kinder machen [eben] Lärm; **..., as young people '~:** ..., wie alle jungen Leute [es tun]; **he '~ insist on doing it** er besteht unbedingt darauf, es zu tun; **it 'would have to rain** natürlich mußte es regnen; **c)** *(wish)* wollen; **~ you have some more cake?** möchtest *od.* willst du noch etwas Kuchen?; **do as/what you ~:** mach, was du willst; **call it what [ever] you ~:** nenn es, wie du willst; **would to God that ...:** wollte Gott, daß ...; **d)** *(be able to)* **the box ~ hold 5 lb. of tea** in die Kiste gehen 5 Pfund Tee; **the theatre ~ seat 800** das Theater hat 800 Sitzplätze. **2.** *v. aux., forms as* **1: a)** *expr. simple future* werden; **this time tomorrow he ~ be in Oxford** morgen um diese Zeit ist er in Oxford; **tomorrow he ~ have been here a month** morgen ist er einen Monat hier; **one more cherry, and I ~ have eaten a pound** noch eine Kirsche und ich habe ein Pfund gegessen; **b)** *expr. intention* **I promise I won't do it again** ich verspreche, ich mach's nicht noch mal; **You won't do that, ~ you? – Oh yes, I ~!** Du machst es doch nicht, oder? – Doch[, ich mach's]!; **~ do** *(coll.)* wird gemacht; mach ich *(ugs.)*; **c)** *in conditional clause* **if he tried, he would succeed** wenn er es versuchen würde, würde er es erreichen; **he would like/would have liked to see her** er würde sie gerne sehen/er hätte sie gerne gesehen; **d)** *(request)* **~ you please tidy up** würdest du bitte aufräumen?

[2]**will 1.** *n.* **a)** *(faculty)* Wille, *der;* **freedom of the ~:** Willensfreiheit, *die;* **have a ~ of one's own** [s]einen eigenen Willen haben; **an iron ~, a ~ of iron** ein eiserner Wille; **b)** *(Law: testament)* Testament, *das;* **c)** *(desire)* **at ~:** nach Belieben; **~ to live** Lebenswille, *der;* **you must have the ~ to win** du mußt gewinnen wollen; **against one's/sb.'s ~:** gegen seinen/jmds. Willen; **of one's own [free] ~:** aus freien Stücken; **do sth. with a ~:** etw. mit großem Eifer *od.* Elan tun; **where there's a ~ there's a way** *(prov.)* wo ein Wille ist, ist auch ein Weg; **d)** *(disposition)* **with the best ~ in the world** bei allem Wohlwollen; *in neg. clause* beim besten Willen. **2.** *v. t.* durch Willenskraft erzwingen; **~ oneself to do sth.** sich zwingen, etw. zu tun; **~ sb. to win** jmds. Sieg mit aller Kraft herbeiwünschen

willies ['wɪlɪz] *n. pl. (sl.)* **sb. gets the ~:** jmdm. wird ganz anders *(ugs.)*; **it gives me the ~:** dabei wird mir ganz anders *(ugs.)*

willing ['wɪlɪŋ] **1.** *adj.* **a)** willig; **ready and ~:** bereit; **be ~ to do sth.** bereit sein, etw. zu tun; **I'm ~ to believe you're right** ich will gerne glauben, daß du recht hast; **b)** *attrib. (readily offered)* willig; **she gave ~ assistance/help** sie half bereitwillig; **lend a ~ hand** bereitwillig helfen. **2.** *n.* **show ~:** guten Willen zeigen

willingly ['wɪlɪŋlɪ] *adv.* **a)** *(with pleasure)* gern[e]; **b)** *(voluntarily)* freiwillig; **they did not come ~:** sie kamen nur widerstrebend

willingness ['wɪlɪŋnɪs] *n., no pl.* Bereitschaft, *die*

will-o'-the-wisp [wɪləðə'wɪsp] *n.* **a)** Irrlicht, *das;* **b)** *(fig.)* Schimäre, *die*

willow ['wɪləʊ] *n.* Weide, *die*

'**willow-pattern** *n.* Weidenmuster, *das*

'**will-power** *n.* Willenskraft, *die*

willy-nilly [wɪlɪ'nɪlɪ] *adv.* wohl oder übel ‹*etw. tun müssen*›

wilt [wɪlt] *v. i.* **a)** *(Bot.: wither)* welk werden; welken; **b)** *(fig.)* ‹*Person:*› schlapp werden, *(ugs.)* abschlaffen; ‹*Interesse, Begeisterung:*› abflauen; ‹*Energie, Kraft*› dahinschwinden

wily ['waɪlɪ] *adj.* listig; gewieft ‹*Person*›; raffiniert ‹*Trick, Argumentation, Plan usw.*›

wimp [wɪmp] *n. (coll. derog.)* Schlappschwanz, *der (ugs.)*

win [wɪn] **1.** *v. t.*, **-nn-, won** [wʌn] **a)** gewinnen; bekommen ‹*Stipendium, Auftrag, Vertrag, Recht*›; ernten ‹*Beifall, Dank*›; **~ the long jump** im Weitsprung gewinnen; **~ an argument/debate** aus einem Streit/einer Debatte als Sieger hervorgehen; **~ promotion** befördert werden; **~ sb. sth.** jmdm. etw. einbringen; **~ a reputation [for oneself]** sich *(Dat.)* einen Ruf erwerben *od.* einen Namen machen; **~ sth. from** *or* **off sb.** jmdm. etw. abnehmen; **you can't ~ them all** *(coll.)*, **you ~ some, you lose some** *(coll.)* man kann nicht immer Glück haben; **b)** *(sl.: steal)* organisieren *(ugs.)*; **c) ~ one's way into sb.'s heart/affections** jmds. Herz/Zuneigung gewinnen. **2.** *v. i.*, **-nn-, won** gewinnen; *(in battle)* siegen; **you ~** *(have defeated me)* du hast gewonnen *(ugs.)*; **~ or lose** wie es auch ausgeht/ausgehen würde; **you can't ~** *(coll.)* da hat man keine Chance *(ugs.)*. **3.** *n.* Sieg, *der;* **have a ~:** gewinnen

~ 'back *v. t.* zurückgewinnen

~ 'out *v. i. (coll.)* **~ out [over sb./sth.]** sich [gegen jmdn./etw.] durchsetzen

~ 'over, ~ 'round *v. t.* bekehren; *(to one's side)* auf seine Seite bringen; *(convince)* überzeugen; **~ sb. over** *or* **round to a plan/to one's point of view** jmdn. für einen Plan gewinnen/zu seiner Ansicht bekehren *od.* von seiner Ansicht überzeugen

~ 'through *v. i.* Erfolg haben; **~ through to the next round** die nächste Runde erreichen

wince [wɪns] *v. i.* zusammenzucken (**at** bei); **he ~d under the pain/the insult** der Schmerz/die Beleidigung ließ ihn zusammenzucken

winch [wɪntʃ] **1.** *n.* Winde, *die.* **2.** *v. t.* winden; mit einer Winde ziehen; **~ up** hochwinden

[1]**wind** [wɪnd] **1.** *n.* **a)** Wind, *der;* **be in the ~** *(fig.)* in der Luft liegen; **see how** *or* **which way the ~ blows** *or* **lies** *(fig.)* sehen, woher der Wind weht; **sail close to** *or* **near the ~:** hart am Wind segeln; *(fig.)* sich hart an der Grenze des Erlaubten bewegen; **take the ~ out of sb.'s sails** *(fig.)* jmdm. den Wind aus den Segeln nehmen; **the ~[s] of change** ein frischer Wind *(fig.)*; **b)** *no pl. (Mus.) (stream of air) (in organ)* Wind, *der; (in other instruments)* Luftstrom, *der; (instruments)* Bläser; **c) get ~ of sth.** *(fig.)* Wind von etw. bekommen; **d)** *no pl., no indef. art. (flatulence)* Blähungen; **break ~:** eine Blähung abgehen lassen; **get/have the ~ up** *(sl.)* Manschetten *(ugs.) od.* Schiß *(salopp)* kriegen/haben; **put the ~ up sb.** *(sl.)* jmdm. Schiß machen *(salopp)*; **e)** *(breath)* **lose/have lost one's ~:** außer Atem kommen/sein; **recover** *or* **get one's ~:** wieder zu Atem kommen; **get one's second ~** *(lit. or fig.)* sich wieder steigern. **2.** *v. t.* außer Atem brin-

gen; **the blow ~ed him** der Schlag nahm ihm den Atem; **be ~ed** außer Atem sein; **he was ~ed by the blow to his stomach** nach dem Schlag in die Magengrube schnappte er nach Luft

[2]**wind** [waɪnd] **1.** *v.i.*, **wound** [waʊnd] **a)** *(curve)* sich winden; *(move)* sich schlängeln; **b)** *(coil)* sich wickeln. **2.** *v.t.*, **wound a)** *(coil)* wickeln; *(on to reel)* spulen; **~ wool into a ball** Wolle zu einem Knäuel aufwickeln; **~ sth. off sth./on [to] sth.** etw. von etw. [ab]wickeln/auf etw. *(Akk.)* [auf]wickeln; **~ sb. round one's finger** jmdn. um den Finger wickeln *(ugs.)*; **b)** *(with key etc.)* aufziehen ‹*Uhr*›; **c) ~ one's/its way** sich winden; sich schlängeln; **d)** *(coil into ball)* zu einem Knäuel/zu Knäueln aufwickeln; **e)** *(surround)* wickeln; **he wound the injured arm in a piece of cloth** er umwickelte den verletzten Arm mit einem Tuch; **f)** *(winch)* winden; **~ sth. with a winch** etw. mit einer Winde ziehen. **3.** *n.* **a)** *(curve)* Windung, *die;* **b)** *(turn)* Umdrehung, *die;* **give sth. a ~:** etw. aufziehen

~ 'back *v.t. & i.* zurückspulen

~ 'down 1. *v.t.* **a)** *(lower)* mit einer Winde herunter-/hinunterlassen; herunterdrehen ‹*Autofenster*›; **b)** *(fig.: reduce gradually)* einschränken; drosseln ‹*Produktion*›; *(and cease)* allmählich einstellen. **2.** *v.i.* *(lose momentum)* ablaufen; *(fig.)* ‹*Produktion:*› zurückgehen; *(cease)* auslaufen

~ 'on *v.t. & i.* weiterspulen

~ 'up 1. *v.t.* **a)** *(raise)* hochwinden; *(winch up)* [mit einer Winde] hochziehen; hochdrehen ‹*Autofenster*›; **b)** *(coil)* aufwickeln; **c)** *(with key etc.)* aufziehen ‹*Uhr*›; **d)** *(make tense)* aufregen; erregen; **get wound up** sich aufregen; sich erregen; **e)** *(coll.: annoy deliberately)* auf die Palme bringen *(ugs.)*; **f)** *(conclude)* beschließen ‹*Debatte, Rede*›; **g)** *(Finance, Law)* auflösen; einstellen ‹*Aktivitäten*›; **~ up one's affairs** seine Angelegenheiten in Ordnung bringen. **2.** *v.i.* **a)** *(conclude)* schließen; **he wound up for the Government** er sprach als letzter Redner aus dem Regierungslager; **b)** *(coll.: end up)* **~ up in prison/hospital** [zum Schluß] im Gefängnis/Krankenhaus landen *(ugs.)*

wind [wɪnd]: **~bag** *n.* *(derog.)* Schwätzer, *der*/Schwätzerin, *die;* **~-blown** *adj.* vom Wind zerzaust ‹*Haar*›; **~-break** *n.* Windschutz, *der;* **~breaker** *(Amer.)*, **~-cheater** *(Brit.)* *ns.* Windjacke, *die;* **~-chill factor** *n.* Wind-chill-Index, *der (Meteor.)*

winder ['waɪndə(r)] *n.* *(of watch)* Krone, *die;* *(of clock, toy)* Aufziehschraube, *die;* *(key)* Schlüssel, *der*

wind [wɪnd]: **~fall** *n.* **a)** Stück Fallobst; *(apple)* Fallapfel, *der;* **~s** Fallobst, *das;* **b)** *(fig.)* warmer Regen *(ugs.)*; **~ farm** *n.* Windpark, *der;* Windfarm, *die*

winding ['waɪndɪŋ] **1.** *attrib. adj.* gewunden. **2.** *n.* **a)** *in pl.* *(of road, river)* Windungen; **b)** *(Electr.)* Wicklung, *die*

wind instrument ['wɪnd ɪnstrəmənt] *n.* *(Mus.)* Blasinstrument, *das*

windlass ['wɪndləs] *n.* Winde, *die*

windmill ['wɪndmɪl] *n.* **a)** Windmühle, *die;* *(to drive generator, water pump, etc.)* Windrad, *das;* **b)** *(toy)* Windrädchen, *das*

window ['wɪndəʊ] *n.* **a)** Fenster, *das;* **break a ~:** eine Fensterscheibe zerbrechen; ‹*Einbrecher:*› eine Fensterscheibe einschlagen; **go out of the ~** *(fig. coll.)* den Bach runtergehen *(ugs.)*; **b)** *(fig.: means of observation)* **a ~ on the West/world** ein Fenster zum Westen/zur Welt; **c)** *(for display of goods)* [Schau]fenster, *das;* **d)** *(for issue of tickets etc.)* Schalter, *der;* **e)** *(Computing)* Fenster, *das*

window: ~-box *n.* Blumenkasten, *der;* **~-cleaner** *n.* Fensterputzer, *der*/-putzerin, *die;* **~-cleaning** *n.* Fensterputzen, *das;* **~-frame** *n.* Fensterrahmen, *der;* **~-ledge** *n.* *(inside)* Fensterbank, *die;* *(outside)* Fenstersims, *der od. das;* **~-pane** *n.* Fensterscheibe, *die;* **~-shopper** *n.* Schaufensterbummler, *der*/ -bummlerin, *die;* **~-shopping** *n.* Schaufensterbummeln, *das;* **go ~-shopping** einen Schaufensterbummel machen; **~-sill** *see* **~-ledge**

wind [wɪnd]: **~pipe** *n.* *(Anat.)* Luftröhre, *die;* **~ power** *n.* Windkraft, *die;* **~ pump** *n.* Windpumpe, *die;* **~screen,** *(Amer.)* **~shield** *ns.* *(Motor Veh.)* Windschutzscheibe, *die;* **~screen-/~shield-wiper** Scheibenwischer, *der;* **~surfer** *n.* Windsurfer, *der;* **~surfing** *n.* *(Sport)* Windsurfen, *das;* **~-swept** *adj.* windgepeitscht; vom Wind zerzaust ‹*Person, Haare*›; **~-tunnel** *n.* *(Aeronaut.)* Windkanal, *der*

windward ['wɪndwəd] **1.** *adj.* **~ side** Windseite, *die;* **in a ~ direction** gegen den Wind. **2.** *adv.* gegen den Wind. **3.** *n.* Windseite, *die;* **sail to ~:** gegen den Wind segeln

windy ['wɪndɪ] *adj.* windig ‹*Tag, Ort, Wetter*›

wine [waɪn] **1.** *n.* Wein, *der.* **2.** *v.t.* **~ and 'dine** in großem Stil *od.* *(ugs.)* groß bewirten

wine: ~-bar *n.* Weinstube, *die;* **~bottle** *n.* Weinflasche, *die;* **~-cellar** *n.* [Wein]keller, *der;* **~glass** *n.* Weinglas, *das;* **~-list** *n.* Weinkarte, *die;* **~-taster** *n.* Weinverkoster, *der*/-verkosterin, *die;* **~-tasting** ['waɪnteɪstɪŋ] *n.* Weinprobe, *die;* **~-'vinegar** *n.* Weinessig, *der;* **~-waiter** *n.* Weinkellner, *der*

wing [wɪŋ] *n.* **a)** *(Ornith., Archit., Sport)* Flügel, *der;* **on the ~:** im Fluge; **spread** *or* **stretch one's ~s** *(fig.)* sich auf eigene Füße stellen; **take sb. under one's ~:** jmdn. unter seine Fittiche nehmen; **b)** *(Aeronaut.)* [Trag]flügel, *der;* Tragfläche, *die;* **c)** *in pl.* *(Theatre)* Kulissen; **wait in the ~s** *(fig.)* auf seine Chance warten; **d)** *(Brit. Motor Veh.)* Kotflügel, *der*

winged [wɪŋd] *adj.* geflügelt

winger ['wɪŋə(r)] *n.* *(Sport)* Außenstürmer, *der*/-stürmerin, *die;* Flügel, *der*

wing: ~-mirror *n.* *(Brit. Motor Veh.)* Außenspiegel, *der;* **~-span, ~-spread** *ns.* [Flügel]spannweite, *die;* **~-tip** *n.* Flügelspitze, *die*

wink [wɪŋk] **1.** *v.i.* **a)** *(blink)* blinzeln; *(as signal)* zwinkern; **~ at sb.** jmdm. zuzwinkern; **b)** *(twinkle, flash)* blinken. **2.** *v.t.* **~ one's eye/eyes** blinzeln; *(as signal)* zwinkern; **~ one's eye at sb.** jmdm. zuzwinkern. **3.** *n.* **a)** Blinzeln, *das;* *(signal)* Zwinkern, *das;* **give sb. a [secret/sly/knowing** *etc.***] ~:** jmdm. [heimlich/verschmitzt/wissend *usw.*] zuzwinkern; *see also* [2]**tip 2d**; **b) not get a ~ of sleep, not sleep a ~:** kein Auge zutun; *see also* **forty 1**

winker ['wɪŋkə(r)] *n.* *(Motor Veh.)* Blinker, *der*

winkle ['wɪŋkl] **1.** *n.* Strandschnecke, *die.* **2.** *v.t.* **~ out** herausholen, *(ugs.)* rauspfriemeln ‹*Gegenstand, Substanz*›; herausholen ‹*Person, Tier*›; **~ sth. out of sb.** *(fig.)* etw. aus jmdm. rauskriegen *(ugs.)*

winner ['wɪnə(r)] *n.* **a)** Sieger, *der*/Siegerin, *die;* *(of competition or prize)* Gewinner, *der*/Gewinnerin, *die; (winning shot)* Siegestreffer, *der; (winning goal)* Siegestor, *das;* **b)** *(successful thing)* Erfolg, *der; (successful play, product)* Renner, *der (ugs.);* Hit, *der (ugs.);* **you're on [to] a ~ with this idea/book** *(coll.)* diese Idee/dieses Buch wird garantiert ein Renner *od.* Hit *(ugs.)*

winning ['wɪnɪŋ] *adj.* **a)** *attrib.* siegreich; **~ team** siegreiche Mannschaft; Siegermannschaft, *die;* **the ~ captain** der Kapitän der Siegermannschaft; **b)** *attrib. (bringing victory)* den Sieg bringend; **~ number** Gewinnzahl, *die;* **c)** *(charming)* einnehmend; gewinnend ⟨*Lächeln*⟩

'**winning-post** *n. (Sport)* Zielpfosten, *der*

winnings ['wɪnɪŋz] *n. pl.* Gewinn, *der*

winter ['wɪntə(r)] **1.** *n.* Winter, *der;* **in [the] ~:** im Winter; **last/next ~:** letzten/nächsten Winter; **the ~ of 1947–8** *or* **of 1947** der Winter 1947–48 *od.* [des Jahres] 1947; **~'s day** Wintertag, *der.* **2.** *attrib. adj.* Winter-. **3.** *v.i.* den Winter verbringen; ⟨*Truppe, Tier:*⟩ überwintern

winter: ~ 'sport *n.* **a)** *usu. in pl.* Wintersport, *der;* **b)** *(particular sport)* Wintersportart, *die;* **~time** *n.* Winter[s]zeit, *die;* **in [the] ~time** im Winter

wintry ['wɪntrɪ] *adj.* **a)** winterlich; rauh ⟨*Klima*⟩; kalt ⟨*Wind*⟩; **~ shower** Schneegestöber, *das;* **cold and ~:** winterlich kalt; **b)** *(fig.)* frostig ⟨*Lächeln*⟩

wipe [waɪp] *v.t.* **a)** abwischen; [auf]wischen ⟨*Fußboden*⟩; *(dry)* abtrocknen; **~ one's mouth** sich *(Dat.)* den Mund abwischen; **~ one's brow/eyes/nose** sich *(Dat.)* die Stirn wischen/die Tränen abwischen/die Nase abwischen; **~ one's feet/shoes** [sich *(Dat.)*] die Füße/Schuhe abtreten; **~ sb./sth. clean/dry** jmdn./etw. abwischen/abtrocknen; **b)** *(get rid of)* [ab]wischen; löschen ⟨*Bandaufnahme*⟩; **~ one's/sb.'s tears/the tears from one's/sb.'s eyes** sich/jmdm. die Tränen abwischen/aus den Augen wischen

~ a'way *v.t.* wegwischen; **~ away a tear** sich *(Dat.)* eine Träne abwischen

~ 'down *v.t.* abwischen; *(dry)* abtrocknen

~ 'off *v.t.* **a)** *(remove)* wegwischen; löschen ⟨*Bandaufnahme*⟩; **b)** *(pay off)* zurückzahlen ⟨*Schulden*⟩; ablösen ⟨*Hypothek*⟩

~ 'out *v.t.* **a)** *(clean)* auswischen; **b)** *(remove)* wegwischen; *(erase)* auslöschen; **c)** *(cancel)* tilgen; zunichte machen ⟨*Vorteil, Gewinn usw.*⟩; **d)** *(destroy, abolish)* ausrotten ⟨*Rasse, Tierart, Feinde*⟩; ausmerzen ⟨*Seuche, Korruption, Terrorismus*⟩; **e)** *(coll.: murder)* aus dem Weg räumen

~ 'over *v.t.* wischen über (+ *Akk.*)

~ 'up *v.t.* **a)** aufwischen; **b)** *(dry)* abtrocknen

wiper ['waɪpə(r)] *n. (Motor Veh.)* Wischer, *der*

wire [waɪə(r)] **1.** *n.* **a)** Draht, *der;* **b)** *(barrier)* Drahtverhau, *der od. das; (fence)* Drahtzaun, *der;* **c)** *(Electr., Teleph.)* Leitung, *die;* **a piece** *or* **length of ~:** ein Stück [Leitungs]draht; **telephone/telegraph ~:** Telefon-/Telegrafenleitung, *die;* **get one's** *or* **the ~s crossed** *(fig.)* auf der Leitung stehen *(ugs.);* **d)** *(coll.: telegram)* Telegramm, *das.* **2.** *v.t.* **a)** *(fasten with ~)* mit Draht zusammenbinden; *(stiffen with ~)* mit Draht versteifen; **~ sth. together** etw. mit Draht verbinden; **b)** *(Electr.)* **~ sth. to sth.** etw. an etw. *(Akk.)* anschließen; **~ a house** *(lay wiring circuits)* in einem Haus die Stromleitungen legen; **c)** *(coll.: telegraph)* **~ sb.** jmdm. *od.* an jmdn. telegrafieren; **~ money** Geld telegrafisch überweisen

~ 'up *v.t. (Electr.)* anschließen **(to** an + *Akk.*)

wire: ~ 'brush *n.* Drahtbürste, *die;* **~-cutters** *n. pl.* Drahtschneider, *der*

wireless ['waɪəlɪs] **1.** *adj. (Brit.) see* **radio 2. 2.** *n.* **a)** *(Brit.)* Radio, *das;* **b)** *(telegraphy)* Funk, *der;* **by ~:** über Funk *(Akk.)*

wire: ~ 'netting *see* **netting; ~ 'rope** *n.* Drahtseil, *das;* **~-strippers** *n. pl.* Abisolierzange, *die;* **~-tapping** *see* **phone-tapping**

wiring ['waɪərɪŋ] *n., no pl., no indef. art. (Electr.)* [elektrische] Leitungen

wisdom ['wɪzdəm] *n., no pl.* **a)** Weisheit, *die;* **worldly ~:** Weltklugheit, *die;* **b)** *(prudence)* Klugheit, *die;* **where is the ~ of such a move/in doing that?** was für einen Sinn hat solch ein Schritt/hat es, das zu tun?; **words of ~:** weise Worte; *(advice)* weise Ratschläge

'**wisdom tooth** *n.* Weisheitszahn, *der*

wise [waɪz] *adj.* **a)** weise; vernünftig ⟨*Meinung*⟩; **be ~ after the event** so tun, als hätte man es immer schon gewußt; **b)** *(prudent)* klug ⟨*Vorgehensweise*⟩; vernünftig ⟨*Lebensweise, Praktik*⟩; **c)** *(informed)* **be none the** *or* **no/not much ~r** kein bißchen *od.* nicht/nicht viel klüger als vorher sein; **without anyone's being [any] the ~r** ohne daß es jemand merkt; **d)** *(coll.: aware)* **be ~ to sb./sth.** jmdn./etw. kennen; **get ~ to sb./sb.'s tricks** jmdm. auf die Schliche kommen; **get ~ to sth.** etw. spitzkriegen *(ugs.);* **put sb. ~:** jmdm. die Augen öffnen; **put sb. ~ to sth.** jmdn. über etw. *(Akk.)* aufklären; **put sb. ~ to sb.** jmdm., was jmdn. betrifft, die Augen öffnen

~ 'up *(Amer. sl.)* **1.** *v.t.* **~ sb. up [to sth.]** jmdn. [über etw.] aufklären. **2.** *v.i.* **~ up to sth.** sich *(Dat.)* über etw. klarwerden; **~ up to sb./sb.'s tricks** jmdm. auf die Schliche kommen

-wise *adv. in comb.* **a)** *(in the direction of)* **length~:** der Länge nach; **clock~:** im Uhrzeigersinn; **b)** *(coll.: as regards)* -mäßig; was ... betrifft; **weather~:** wettermäßig; was das Wetter betrifft; **health~:** in puncto Gesundheit; gesundheitlich

wise: ~crack *(coll.)* **1.** *n.* witzige Bemerkung; **2.** *v.i.* witzeln; **~ guy** *n. (coll.)* Klugscheißer, *der (salopp abwertend)*

wisely ['waɪzlɪ] *adv.* weise; *(prudently)* klug; *as sentence-modifier* klugerweise

wish [wɪʃ] **1.** *v.t.* **a)** *(desire, hope)* wünschen; **I ~ I was** *or* **were rich** ich wollte *od. (geh.)* wünschte, ich wäre reich; **I do ~ he would come** wenn er nur kommen würde; **I ~ you would shut up** es wäre mir lieb, wenn du den Mund hieltest; **'~ you were here'** *(on postcard)* „schade, daß du nicht hier bist"; **b)** *with inf. (want)* wünschen *(geh.);* **I ~ to go** ich möchte *od.* will gehen; **I ~ you to stay** ich möchte *od.* will, daß du bleibst; **c)** *(say that one hopes sb. will have sth.)* wünschen; **~ sb. luck/success** *etc.* jmdm. Glück/Erfolg *usw.* wünschen; **~ sb. good morning/a happy birthday** jmdm. guten Morgen sagen/zum Geburtstag gratulieren; **~ sb. ill/well** jmdm. [etwas] Schlechtes/alles Gute wünschen; **d)** *(coll.: foist)* **~ sb./sth. on sb.** jmdm. jmdn./etw. aufhalsen *(ugs.).* **2.** *v.i.* wünschen; **come on, ~!** nun, wünsch dir was!; **~ for sth.** sich *(Dat.)* etw. wünschen; **what more could one ~ for?** was will man mehr?; **they have everything they could**

possibly ~ for sie haben alles, was sie sich *(Dat.)* nur wünschen können. **3.** *n.* **a)** Wunsch, *der;* **her ~ is that ...:** es ist ihr Wunsch *od.* sie wünscht, daß ...; **I have no [great/ particular] ~ to go** ich habe keine [große/besondere] Lust zu gehen; **make a ~:** sich *(Dat.)* etwas wünschen; **with best/[all] good ~es, with every good ~:** mit den besten/allen guten Wünschen **(on, for** zu); **b)** *(thing desired)* **get** *or* **have one's ~:** seinen Wunsch erfüllt bekommen; **at last he has [got] his ~:** endlich ist sein Wunsch in Erfüllung gegangen

~ a'way *v. t.* wegwünschen

'wishbone *n. (Ornith.)* Gabelbein, *das*

wishful ['wɪʃfl] *adj.* sehnsuchtsvoll *(geh.)* ⟨*Blick, Verlangen*⟩; **~ thinking** Wunschdenken, *das*

'wishing-well *n.* Wunschbrunnen, *der*

wishy-washy ['wɪʃɪwɒʃɪ] *adj.* labberig *(ugs.); (fig.)* lasch

wisp [wɪsp] *n. (of straw)* Büschel, *das;* **~ of hair** Haarsträhne, *die;* **~ of cloud/smoke** Wolkenfetzen, *der*/Rauchfahne, *die*

wistful ['wɪstfl] *adj.* wehmütig; melancholisch ⟨*Person, Typ*⟩; traurig ⟨*Augen*⟩

¹wit [wɪt] *n.* **a)** *(humour)* Witz, *der;* **have a ready ~:** schlagfertig sein; **b)** *(intelligence)* Geist, *der;* **battle of ~s** intellektueller Schlagabtausch; **be at one's ~'s** *or* **~s' end** sich *(Dat.)* keinen Rat mehr wissen; **collect** *or* **gather one's ~s** zu sich kommen; **drive sb. out of his/ her ~s** jmdn. um den Verstand bringen; **frighten** *or* **scare sb. out of his/her ~s** jmdm. Todesangst einjagen; **be frightened** *or* **scared out of one's ~s** Todesangst haben; **have/keep one's ~s about one** auf Draht sein *(ugs.)*/nicht den Kopf verlieren; **c)** *(person)* geistreicher Mensch

²wit *v. i.* **to ~:** nämlich

witch [wɪtʃ] *n. (lit. or fig.)* Hexe, *die*

witch: ~craft *n., no pl.* Hexerei, *die;* **~-doctor** *n.* Medizinmann, *der;* **~-hunt** *n. (lit. or fig.)* Hexenjagd, *die* **(for** auf + *Akk.*)

with [wɪð] *prep.* **a)** mit; **put sth. ~ sth.** etw. zu etw. stellen/legen; **have no pen to write ~:** nichts zum Schreiben haben; **I'll be ~ you in a minute** ich komme gleich; **be ~ it** *(coll.)* up to date sein; **not be ~ sb.** *(coll.: fail to understand)* jmdm. nicht folgen können; **I'm not ~ you** *(coll.)* ich komme nicht mit; **be one ~ sb./sth.** mit jmdm./etw. eins sein; **b)** *(in the care or possession of)* bei; **I have no money ~ me** ich habe kein Geld dabei *od.* bei mir; **c)** *(owing to)* vor (+ *Dat.*); **tremble ~ fear** vor Angst zittern; **d)** *(displaying)* mit; **~ courage** mutig; **handle ~ care** vorsichtig behandeln; **e)** *(while having)* bei; **sleep ~ the window open** bei offenem Fenster schlafen; **speak ~ one's mouth full** mit vollem Mund sprechen; **f)** *(in regard to)* **be patient ~ sb.** mit jmdm. geduldig sein; **what do you want ~ me?** was wollen Sie von mir?; **how are things ~ you?** wie geht es dir?; **what can he want ~ it?** was mag er damit vorhaben?; **g)** *(at the same time as, in the same way as)* mit; **~ that** damit; **h)** *(employed by)* bei; **i)** *(despite)* trotz; *see also* **²will 1 d**

with'draw 1. *v. t., forms as* **draw 1: a)** *(pull back, retract)* zurückziehen; **b)** *(remove)* nehmen *(fig.)* **(from** aus); abziehen ⟨*Truppen*⟩ **(from** aus). **2.** *v. i., forms as* **draw 1** sich zurückziehen

withdrawal [wɪð'drɔ:əl] *n.* **a)** Zurücknahme, *die;* **b)** *(removal) (of privilege)* Entzug, *der; (of troops)* Abzug, *der; (of money)* Abhebung, *die*

with'drawal slip *n.* Auszahlungsschein, *der*

with'drawn *adj. (unsociable)* verschlossen

wither ['wɪðə(r)] **1.** *v. t.* verdorren lassen. **2.** *v. i.* [ver]welken

~ a'way *v. i. (lit. or fig.)* dahinwelken *(geh.)*

~ 'up *v. i.* [ver]welken

withered ['wɪðəd] *adj.* verwelkt ⟨*Gras, Pflanze*⟩; verkrüppelt ⟨*Gliedmaße*⟩

withering ['wɪðərɪŋ] *adj.* vernichtend ⟨*Blick, Bemerkung*⟩; sengend ⟨*Hitze*⟩

with'hold *v. t., forms as* **²hold: a)** *(refuse to grant)* verweigern; versagen *(geh.);* **b)** *(hold back)* verschweigen ⟨*Wahrheit*⟩; **~ sth. from sb.** jmdm. etw. vorenthalten

within [wɪ'ðɪn] *prep.* **a)** *(on the inside of)* innerhalb; **b)** *(not beyond)* im Rahmen (+ *Gen.*); **stay/be ~ the law** den Boden des Gesetzes nicht verlassen; **c)** *(not farther off than)* **~ eight miles of sth.** acht Meilen im Umkreis von etw.; **we were ~ eight miles of our destination when ...:** wir waren kaum noch acht Meilen von unserem Ziel entfernt, als ...; **d)** *(subject to)* innerhalb; **work ~ certain conditions** unter bestimmten Bedingungen arbeiten; **e)** *(in a time no longer than)* innerhalb; binnen; **within an/the hour** innerhalb einer Stunde

without [wɪ'ðaʊt] *prep.* ohne; **~ doing sth.** ohne etw. zu tun; **can you do it ~ his knowing?** kannst du das machen, ohne daß er davon weiß?

with'stand *v. t.,* **withstood** [wɪθ'stʊd] standhalten (+ *Dat.*); aushalten ⟨*Beanspruchung, hohe Temperaturen*⟩

witless ['wɪtlɪs] *adj.* **a)** *(foolish)* töricht; **b)** *(insane)* geistesgestört; **c)** *(dull-witted)* beschränkt

witness ['wɪtnɪs] **1.** *n.* **a)** Zeuge, *der*/Zeugin, *die* **(of, to** *Gen.*); **b)** *see* **eyewitness; c)** *no pl. (evidence)* Zeugnis, *das (geh.);* **bear ~ to** *or* **of sth.** ⟨*Person:*⟩ etw. bezeugen; *(fig.)* von etw. zeugen. **2.** *v. t.* **a)** *(see)* **~ sth.** Zeuge/Zeugin einer Sache *(Gen.)* sein; **they have ~ed many changes** sie haben viele Veränderungen erlebt; **b)** *(attest genuineness of)* bestätigen ⟨*Unterschrift, Echtheit eines Dokuments*⟩

witness: ~-box *(Brit.),* **~-stand** *(Amer.) ns.* Zeugenstand, *der*

witter ['wɪtə(r)] *v. i. (Brit. coll.)* **~ [on]** quatschen *(ugs. abwertend)*

witticism ['wɪtɪsɪzm] *n.* Witzelei, *die*

wittingly ['wɪtɪŋlɪ] *adv.* wissentlich

witty ['wɪtɪ] *adj.* **a)** witzig; **b)** *(possessing wit)* geistreich ⟨*Person*⟩

wives *pl. of* **wife**

wizard ['wɪzəd] *n.* **a)** *(sorcerer)* Zauberer, *der;* **b)** *(very skilled person)* Genie, *das* **(at** in + *Dat.*)

wizened ['wɪzənd] *adj.* runz[e]lig

wobble ['wɒbl] *v. i.* **a)** *(rock)* wackeln; ⟨*Kompaßnadel:*⟩ zittern; **b)** *(go unsteadily)* wackeln *(ugs.)*

wobbly ['wɒblɪ] *adj.* wack[e]lig; zitt[e]rig ⟨*Schrift, Hand, Stimme*⟩

woe [wəʊ] *n. (arch./literary/joc.)* **a)** *(distress)* Jammer, *der;* **a tale of ~:** eine jammervolle Geschichte; **~ betide you!** wehe dir!; **b)** *in pl. (troubles)* Jammer, *der*

woebegone ['wəʊbɪgɒn] *adj.* jammervoll

woeful ['wəʊfl] *adj.* beklagenswert

wog [wɒg] *n. (sl. derog.)* Kanake, *der (ugs. abwertend)*

woke, woken *see* **¹wake 1, 2**

wolf [wʊlf] **1.** *n., pl.* **wolves** [wʊlvz] *(Zool.)* Wolf, *der;* **keep the ~ from the door** *(fig.)* den größten Hunger stillen; **be a ~ in sheep's clothing** *(fig.)* ein Wolf im Schafspelz sein. **2.** *v.t.* ~ **[down]** verschlingen

'wolf-whistle 1. *n.* anerkennender Pfiff. **2.** *v.i.* anerkennend pfeifen

wolves *pl. of* **wolf 1**

woman ['wʊmən] *n., pl.* **women** ['wɪmɪn] **a)** Frau, *die;* **women and children first** Frauen und Kinder zuerst; **a ~'s work is never done** eine Frau hat immer etwas zu tun; **that's ~'s work** das ist Frauenarbeit; **women's page** Frauenseite, *die;* **women's [toilet]** Damen[toilette], *die;* **the other ~:** die Geliebte; **b)** *attrib. (female)* weiblich; ~ **friend** Freundin, *die;* ~ **doctor** Ärztin, *die;* **a ~ driver** eine Frau am Steuer; **c)** *no pl.* **[the]** ~ *(an average ~)* die Frau

womanhood ['wʊmənhʊd] *n., no pl.* Weiblichkeit, *die*

womanizer ['wʊmənaɪzə(r)] *n.* Schürzenjäger, *der*

womanliness ['wʊmənlɪnɪs] *n., no pl.* Fraulichkeit, *die*

womanly ['wʊmənlɪ] *adj.* fraulich; weiblich

womb [wu:m] *n. (Anat.)* Gebärmutter, *die;* **in her ~:** in ihrem Leib *(geh.)*

wombat ['wɒmbæt] *n. (Zool.)* Wombat, *der*

women *pl. of* **woman**

Women: w~folk *n. pl.* Frauen; **~'s 'Institute** *n. (Brit.) britischer Frauenverband;* **~'s 'Lib** *(coll.) see* **~'s Liberation; ~'s Libber** [wɪmɪnz 'lɪbə(r)] *n. (coll.)* Emanze, *die (ugs. abwertend);* Frauenrechtlerin, *die;* **~'s Libe'ration** *n.* die Frauenbewegung; **w~'s 'rights** *n. pl.* die Rechte der Frau

won *see* **win 1, 2**

wonder ['wʌndə(r)] **1.** *n.* **a)** *(extraordinary thing)* Wunder, *das;* **do** *or* **work ~s** Wunder tun *od.* wirken; *(fig.)* Wunder wirken; **~s will never cease** *(iron.)* Wunder über Wunder!; **small** *or* **what** *or* **[it is] no ~ [that] ...:** [es ist] kein Wunder, daß ...; **the ~ is, ...:** das Erstaunliche ist, ...; **b)** *(marvellously successful person)* Wunderkind, *das; (marvellously successful thing)* Wunderding, *das;* **boy/girl ~:** Wunderkind, *das;* **the seven ~s of the world** die Sieben Weltwunder; **c)** *no pl. (feeling)* Staunen, *das;* **be lost in ~:** in Staunen versunken sein. **2.** *adj.* Wunder-. **3.** *v.i.* sich wundern; staunen (**at** über + *Akk.*); **that's not to be ~ed at** darüber braucht man sich nicht zu wundern; **I shouldn't ~ [if ...]** *(coll.)* es würde mich nicht wundern[, wenn ...]. **4.** *v.t.* **a)** sich fragen; **I ~ what the time is** wieviel Uhr mag es wohl sein?; **I was ~ing what to do** ich habe mir überlegt, was ich tun soll; **I ~ whether I might open the window** dürfte ich vielleicht das Fenster öffnen?; **she ~ed if ...** *(enquired)* sie fragte, ob ...; **I ~ if you'd mind if ...?** würde es Ihnen etwas ausmachen, wenn ...?; **b)** *(be surprised to find)* ~ **[that] ...:** sich wundern, daß ...

wonderful ['wʌndəfl] *adj.* wunderbar; wundervoll

wonderfully ['wʌndəfəlɪ] *adv.* wunderbar

wondering ['wʌndərɪŋ] *adj.* staunend

'wonderland *n.* **a)** *(wonderful place)* Paradies, *das;* **b)** *(fairyland)* Wunderland, *das*

wonderment ['wʌndəmənt] *n., no pl.* Verwunderung, *die*

wonky ['wɒŋkɪ] *adj. (Brit. sl.)* wack[e]lig; *(crooked)* schief

wont [wəʊnt] **1.** *pred. adj. (arch./literary)* gewohnt; **as he was ~ to say** wie er zu sagen pflegte. **2.** *n. (literary)* Gepflogenheit, *die (geh.);* **as was her ~:** wie sie zu tun pflegte

won't [wəʊnt] *(coll.)* = **will not;** *see* [1]**will**

woo [wu:] *v.t.* **a)** *(literary: court)* ~ **sb.** um jmdn. werben *(geh.);* **b)** *(seek to win)* umwerben ‹*Kunden, Wähler*›; ~ **away** abwerben ‹*Arbeitskräfte*›

wood [wʊd] *n.* **a)** *in sing. or pl. (area with trees)* Wald, *der;* **sb. cannot see the ~ for the trees** *(fig.)* jmd. sieht den Wald vor [lauter] Bäumen nicht *(scherzh.);* **be out of the ~** *(Brit.) or (Amer.)* **~s** *(fig.)* über den Berg sein *(ugs.);* **b)** *(substance, material)* Holz, *das;* **touch ~** *(Brit.),* **knock [on] ~** *(Amer.)* unberufen!

wooded ['wʊdɪd] *adj.* bewaldet

wooden ['wʊdn] *adj.* **a)** hölzern ‹*Brücke, Spielzeug*›; Holz‹*haus, -brücke, -bein, -griff, -spielzeug*›; **b)** *(fig.: stiff)* hölzern

wooden 'spoon *n. (fig.)* Trostpreis, *der (für den Letzten eines Wettbewerbs, oft in ironischer Weise überreicht)*

wood: ~land ['wʊdlənd] *n.* Waldland, *das;* Wald, *der;* **~-louse** *n. (Zool.)* Kellerassel, *die;* **~pecker** *n.* Specht, *der;* **~-pigeon** *n.* Ringeltaube, *die;* **~-screw** *n.* Holzschraube, *die;* **~-shed** *n.* Holzschuppen, *der;* **~-wind** *n. (Mus.)* Holzblasinstrument, *das;* **the ~-wind [section]** die Holzbläser; **~-wind instrument** Holzblasinstrument, *das;* **~work** *n., no pl.* **a)** *(making things out of ~)* Arbeiten mit Holz; **b)** *(things made of ~)* Holzarbeit[en]; **~worm** *n., no pl., no art.* Holzwurm; **it's got ~worm** da ist der Holzwurm drin *(ugs.)*

woody ['wʊdɪ] *adj.* **a)** *(well-wooded)* waldreich; **b)** *(consisting of wood)* holzig ‹*Pflanze[nteil], Wurzel*›; Holz‹*stamm*›

woof [wʊf] *n.* [dumpfes] Bellen; **~ ~! went the dog** wau, wau! bellte der Hund

woofer ['wʊfə(r)] *n.* Baß[lautsprecher], *der*

wool [wʊl] *n.* **a)** Wolle, *die; attrib.* Woll-; **pull the ~ over sb.'s eyes** jmdm. etwas vormachen *(ugs.);* **b)** *(garments)* Wolle, *die*

woollen (*Amer.:* **woolen**) ['wʊlən] **1.** *adj.* wollen. **2.** *n. in pl. (garments)* Wollsachen *Pl.*

woolly ['wʊlɪ] **1.** *adj.* **a)** wollig; Woll‹*pullover, -mütze*›; **b)** *(confused)* verschwommen. **2.** *n. (coll.)* **a)** *(Brit.: knitted garment)* **[winter] woollies** [Winter]wollsachen *Pl. (ugs.);* **a ~:** ein Wollpullover/eine Wolljacke; **b)** *in pl. (Amer.: undergarments)* wollene Unterwäsche

word [wɜ:d] **1.** *n.* **a)** Wort, *das;* **~s cannot describe it** mit Worten läßt sich das nicht beschreiben; **in a** *or* **one ~** *(fig.)* mit einem Wort; **[not] in so many ~s** [nicht] ausdrücklich; **in other ~s** mit anderen Worten; **not a ~ of sth.** kein Wort von etw.; **bad luck/drunk is not the ~ for it** Pech/betrunken ist gar kein Ausdruck dafür *(ugs.);* **that's not the ~ I would have used** das ist gar kein Ausdruck *(ugs.);* **put sth. into ~s** etw. in Worte fassen; **~ for ~:** Wort für Wort; **without a** *or* **one/another ~:** ohne ein/ein weiteres Wort; **too funny** *etc.* **for ~s** unsagbar komisch *usw.; see also* **fail 2e; play 1b, 2a; b)** *(thing said)* Wort, *das;* **hard ~s** harte Worte; **exchange** *or* **have ~s** einen Wortwechsel haben; **a man of few ~s** ein Mann von wenig Worten; **have a ~ [with sb.] about sth.** [mit jmdm.] über etw. *(Akk.)* sprechen; **could I have a ~ [with you]?** kann ich dich mal sprechen?; **have ~s with sb.** sich mit jmdm. streiten; **say a few ~s** ein paar Worte sprechen; **suit the action to the ~:** seinen Worten Taten folgen lassen; **it's his ~ against mine** sein Wort steht gegen meins; **take sb. at his/her ~:** jmdn. beim Wort nehmen; **~ of command/advice**

Kommando, *das*/Rat, *der;* **at a ~ of command** auf Befehl; **the W~ [of God]** *(Bible)* das Wort [Gottes]; **put in a good ~ for sb. [with sb.]** [bei jmdm.] ein [gutes] Wort für jmdn. einlegen; **c)** *(promise)* Wort, *das;* **doubt sb.'s ~:** jmds. Wort in Zweifel ziehen; **give [sb.] one's ~:** jmdm. sein Wort geben; **keep/break one's ~:** sein Wort halten/brechen; **upon my ~!** *(dated)* meiner Treu! *(veralt.);* **my ~!** meine Güte!; *see also* **take 1 v; d)** *no pl. (speaking)* Wort, *das;* **by ~ of mouth** durch mündliche Mitteilung; **e)** *in pl. (text of song, spoken by actor)* Text, *der;* **f)** *no pl., no indef. art. (news)* Nachricht, *die;* **~ had just reached them** die Nachricht hatte sie gerade erreicht; **~ has it** *or* **the ~ is [that] ...:** es geht das Gerücht, daß ...; **~ went round that ...:** es ging das Gerücht, daß ...; **send/leave ~ that/of when ...:** Nachricht geben/eine Nachricht hinterlassen, daß/wenn ...; **is there any ~ from her?** hat sie schon von sich hören lassen?; **g)** *(command)* Kommando, *das;* **just say the ~:** sag nur ein Wort; **at the ~ 'run', you run!** bei dem Wort „rennen" rennst du!; **h)** *(password)* Parole, *die (Milit.);* **give the ~:** die Parole sagen. **2.** *v. t.* formulieren

'word-game *n.* Buchstabenspiel, *das*

wording ['wɜ:dɪŋ] *n.* Formulierung, *die;* Wortwahl, *die;* **the exact ~ of the contract** der genaue Wortlaut des Vertrages

word: ~ order *n. (Ling.)* Wortstellung, *die;* **~-'perfect** *adj.* **be ~-perfect** seinen Text beherrschen; **~ processing** *n.* Textverarbeitung, *die;* **~ processor** *n.* Textverarbeitungssystem, *das*

wordy ['wɜ:dɪ] *adj.* weitschweifig

wore *see* **wear 2, 3**

work [wɜ:k] **1.** *n.* **a)** *no pl., no indef. art.* Arbeit, *die;* **at ~** *(engaged in ~ing)* bei der Arbeit; *(fig.: operating)* am Werk (*see also* **e**); **be at ~ on sth.** an etw. *(Dat.)* arbeiten; *(fig.)* auf etw. *(Akk.)* wirken; **set to ~** ‹*Person:*› sich an die Arbeit machen; **set sb. to ~:** jmdn. an die Arbeit schicken; **all ~ and no play** immer nur arbeiten; **have one's ~ cut out** viel zu tun haben; sich ranhalten müssen *(ugs.);* **b)** *(thing made or achieved)* Werk, *das;* **a good day's ~:** eine gute Tagesleistung; **do a good day's ~:** ein tüchtiges Stück Arbeit hinter sich bringen; **is that all your own ~?** hast du das alles selbst gemacht?; **~ of art** Kunstwerk, *das;* **c)** *(book, piece of music)* Werk, *das;* **a ~ of reference/literature/art** ein Nachschlagewerk/literarisches Werk/Kunstwerk; **d)** *in pl. (of author or composer)* Werke; **e)** *(employment)* Arbeit, *die,* **out of ~:** arbeitslos; ohne Arbeit; **be in ~:** eine Stelle haben; **go out to ~:** arbeiten gehen; **at ~** *(place of employment)* auf der Arbeit (*see also* **a**); **from ~:** von der Arbeit; **f)** *in pl., usu. constr. as sing. (factory)* Werk, *das;* **g)** *in pl. (Mil.)* Werke; Befestigungen; **h)** *in pl. (operations of building etc.)* Arbeiten; **i)** *in pl. (machine's operative parts)* Werk, *das;* **j)** *in pl. (sl.: all that can be included)* **the [whole/full] ~s** der ganze Kram *(ugs.);* **give sb. the ~s** *(fig.) (give sb. the best possible treatment)* jmdm. richtig verwöhnen *(ugs.); (give sb. the worst possible treatment)* jmdn. fertigmachen *(salopp).* **2.** *v. i.,* **~ed** *or (arch./literary)* **wrought** [rɔ:t] **a)** arbeiten; **~ to rule** Dienst nach Vorschrift machen; **~ for a cause** *etc.* für eine Sache *usw.* arbeiten; **~ against sth.** *(impede)* einer Sache *(Dat.)* entgegenstehen; **b)** *(function effectively)* funktionieren; ‹*Charme:*› wirken (**on** auf + *Akk.*); **make the washing-machine/television ~:** die Waschmaschine/den Fernsehapparat in Ordnung bringen; **c)** ‹*Rad, Getriebe, Kette:*› laufen; **d)** *(be craftsman)* **~ in a material** mit *od.* *(fachspr.)* in einem Material arbeiten; **e)** ‹*Faktoren, Einflüsse:*› wirken (**on** auf + *Akk.*); **~ against** arbeiten gegen; *see also* **~ on; f)** *(make its/one's way)* sich schieben; **~ loose** sich lockern; **~ round to a question** *(fig.)* sich zu einer Frage vorarbeiten. **3.** *v. t.,* **~ed** *or (arch./literary)* **wrought a)** *(operate)* bedienen ‹*Maschine*›; fahren ‹*Schiff*›; betätigen ‹*Bremse*›; **~ed by electricity** elektrisch betrieben; **b)** *(get labour from)* arbeiten lassen; **c)** *(get material from)* ausbeuten ‹*Steinbruch, Grube*›; **d)** *(operate in or on)* ‹*Vertreter:*› bereisen; **e)** *(control)* steuern; **f)** *(effect)* bewirken ‹*Änderung*›; wirken ‹*Wunder*›; **~ it** *or* **things so that ...** *(sl.)* es deichseln, daß ... *(ugs.);* **g)** *(cause to go gradually)* führen; **~ one's way up/into sth.** sich hocharbeiten/in etw. *(Akk.)* hineinarbeiten; **h)** *(get gradually)* bringen; **~ oneself into a position** sich in eine Position hocharbeiten; **i)** *(knead, stir)* **~ sth. into sth.** etw. zu etw. verarbeiten; *(mix in)* etw. unter etw. *(Akk.)* rühren; **j)** *(gradually excite)* **~ oneself into a state/a rage** sich aufregen/in einen Wutanfall hineinsteigern; **k)** *(make by needle~ etc.)* arbeiten; aufsticken ‹*Muster*› (**on** auf + *Akk.*); **l)** *(purchase, obtain with labour)* abarbeiten; *(fig.)* **~ one's keep** für sein Geld etwas leisten; **she ~ed her way through college** sie hat sich *(Dat.)* ihr Studium selbst verdient; *see also* **passage f**

~ a'way *v. i.* **~ away [at sth.]** [an etw. *(Dat.)*] arbeiten

~ 'in *v. t. (include)* hineinbringen; *(mix in)* hineinrühren; *(rub in)* einreiben

~ 'off *v. t.* **a)** *(get rid of)* loswerden; abreagieren ‹*Wut*›; **~ sth. off on sb./sth.** etw. an jmdm./etw. auslassen; **~ off some excess energy** überschüssige Energie loswerden; **b)** *(pay off)* abtragen ‹*Schuld*›

~ on 1. ['--] *v. t.* **a)** *(expend effort on)* **~ on sth.** an etw. *(Dat.)* arbeiten; **b)** *(use as basis)* **~ on sth.** von etw. ausgehen; **c)** *(try to persuade)* **~ on sb.** jmdn. bearbeiten *(ugs.).* **2.** [-'-] *v. i.* weiterarbeiten

~ 'out 1. *v. t.* **a)** *(find by calculation)* ausrechnen; **b)** *(solve)* lösen ‹*Problem, Rechenaufgabe*›; **c)** *(resolve)* **~ things out with sb./for oneself** die Angelegenheit mit jmdm./sich selbst ausmachen; **d)** *(devise)* ausarbeiten ‹*Plan, Strategie*›; **e)** *(make out)* herausfinden; *(understand)* verstehen; **I can't ~ him out** ich werde aus ihm nicht klug; **f)** *(Mining: exhaust)* ausbeuten. **2.** *v. i.* **a)** *(be calculated)* **sth. ~s out at £250/[an increase of] 22%** etw. ergibt 250 Pfund/bedeutet [eine Steigerung von] 22%; **b)** *(give definite result)* ‹*Gleichung, Rechnung:*› aufgehen; **c)** *(have result)* laufen; **things ~ed out [well] in the end** es ist schließlich doch alles gutgegangen; **things didn't ~ out the way we planned** es kam ganz anders, als wir geplant hatten

~ through *v. t.* durcharbeiten

~ towards *v. t. (lit. or fig.)* hinarbeiten auf (+ *Akk.*)

~ 'up 1. *v. t.* **a)** *(develop)* verarbeiten (**into** zu); *(create)* erarbeiten; **b)** *(excite)* aufpeitschen ‹*Menge*›; **get ~ed up** sich aufregen; **~ oneself up into a rage/fury** sich in einen Wutanfall/in Raserei hineinsteigern. **2.** *v. i.* **a)** **~ up to sth.** ‹*Musik:*› sich zu etw. steigern; ‹*Geschichte, Film:*› auf etw.

(Akk.) zusteuern; **I'll have to ~ up to it** ich muß darauf hinarbeiten; **b)** ‹*Rock usw.*:› sich hochschieben

workable ['wɜ:kəbl] *adj.* **a)** *(capable of being worked)* bebaubar ‹*Land*›; abbauwürdig ‹*Mine*›; **be ~** ‹*Mörtel*:› sich verarbeiten lassen; ‹*Stahl*:› sich bearbeiten lassen; ‹*Mine*:› sich ausbeuten lassen; **b)** *(feasible)* durchführbar

workaday ['wɜ:kədeɪ] *adj.* alltäglich

workaholic [wɜ:kə'hɒlɪk] *n.* *(coll.)* arbeitswütiger Mensch; Workaholic, *der (Psych.); attrib.* arbeitswütig

work: ~-bench *n.* Werkbank, *die; (of tailor, glazier)* Arbeitstisch, *der;* **~-box** *n.* Nähkasten, *der;* **~day** *n.* Werktag, *der*

worker ['wɜ:kə(r)] *n.* **a)** Arbeiter, *der*/Arbeiterin, *die;* **b)** *(Zool.)* Arbeiterin, *die*

work: ~force *n.* Belegschaft, *die;* **~-horse** *n.* *(lit. or fig.)* Arbeitspferd, *das;* **~house** *n.* *(Brit. Hist., Amer.)* Arbeitshaus, *das*

working ['wɜ:kɪŋ] **1.** *n.* **a)** Arbeiten, *das;* **b)** *(way sth. works)* Arbeitsweise, *die;* **I cannot follow the ~s of his mind** ich kann seinen Gedankengängen nicht folgen. **2.** *attrib. adj.* **a)** handlungsfähig ‹*Mehrheit*›; ‹*Entwurf, Vereinbarung*› als Ausgangspunkt; **b)** *(in employment)* arbeitend; werktätig; **~ man** *(labourer)* Arbeiter, *der*

working: ~ 'breakfast *n.* Arbeitsfrühstück, *das;* **~ 'class** *n.* Arbeiterklasse, *die;* **~-class** *adj.* der Arbeiterklasse *nachgestellt;* **sb. is ~-class** jmd. gehört zur Arbeiterklasse; **~ clothes** *n. pl.* Arbeitskleidung, *die;* **~ 'day** *n.* **a)** *(portion of the day)* Arbeitstag, *der;* **b)** *(day when work is done) see* **workday;** **~ 'hours** *n. pl.* Arbeitszeit, *die;* **~ 'knowledge** *n.* ausreichende Kenntnisse **(of** in + *Dat.*); **sb. with a ~ knowledge of these machines** jmd., der im Umgang mit diesen Maschinen erfahren ist; **~ 'lunch** *n.* Arbeitsessen, *das;* **~ 'model** *n.* funktionsfähiges Modell; **~ 'mother** *n.* berufstätige Mutter; **~ 'order** *n.* betriebsfähiger Zustand; **be in good ~ order** betriebsbereit sein; **~ party** *n.* *(Brit.)* Arbeitsgruppe, *die;* **~ 'week** *n.* Arbeitswoche, *die;* **a 35-hour ~ week** eine 35-Stunden-Woche; **~ 'wife** *n.* berufstätige Ehefrau; **~ 'woman** *n.* berufstätige Frau

work: ~-load *n.* Arbeitslast, *die;* **~man** ['wɜ:kmən] *n., pl.* **~men** ['wɜ:kmən] Arbeiter, *der;* **council ~man** städtischer Arbeiter; **a bad ~man blames his tools** *(prov.)* ein schlechter Handwerker schimpft über sein Werkzeug

workmanlike ['wɜ:kmənlaɪk] *adj.* fachmännisch

workmanship ['wɜ:kmənʃɪp] *n., no pl.* **a)** *(person's skill)* handwerkliches Können; **b)** *(quality of execution)* Kunstfertigkeit, *die*

work: ~mate *n.* *(Brit.)* Arbeitskollege, *der*/-kollegin, *die;* **~-out** *n.* [Fitneß]training, *das;* **have a good ~-out** hart trainieren; **go for a ~-out** zum [Fitneß]training gehen; **~-people** *n. pl.* Arbeiter; **~ permit** *n.* Arbeitserlaubnis, *die;* **~-place** *n.* Arbeitsplatz, *der;* **~sheet** *n.* **a)** *(recording ~ done etc.)* Arbeitszettel, *der;* **b)** *(for student)* Formular mit Prüfungsfragen; **~shop** *n.* **a)** *(room)* Werkstatt, *die; (building)* Werk, *das;* **b)** *(meeting)* Workshop, *der;* Arbeitstreffen, *das;* **drama ~shop** Theaterworkshop, *der;* **~-shy** *adj.* arbeitsscheu; **~-station** *n.* *(Computing)* graphischer Arbeitsplatz; ≈ Terminal, *das;* **~ surface, ~top** *ns.* Arbeitsplatte, *die;* **~-to-'rule** *n.* Dienst nach Vorschrift

world [wɜ:ld] *n.* **a)** Welt, *die;* **go/sail round the ~:** eine Weltreise machen/die Welt umsegeln; **money makes the ~ go round** Geld regiert die Welt; **it's the same the ~ over** es ist doch überall das gleiche; **the eyes of the ~ are on them** die Welt blickt auf sie; **[all] the ~ over, all over the ~:** in *od.* auf der ganzen Welt; **lead the ~ [in sth.]** [in etw. *(Dat.)*] führend in der Welt sein; **the Old/New W~:** die Alte/Neue Welt; **who/what in the ~ was it?** wer/was in aller Welt war es? *(ugs.);* **how in the ~ was it that ...?** wie in aller Welt *(ugs.)* war es möglich, daß ...?; **nothing in the ~ would persuade me** um nichts in der Welt ließe ich mich überreden; **not for anything in the ~:** um nichts in der Welt; **look for all the ~ as if ...:** geradezu aussehen, als ob ...; **in a ~ of one's own** in einer anderen Welt *(fig.);* **not do sth. for the whole ~:** etw. um alles in der Welt nicht tun; **be all the ~ to sb.** jmdm. das Wichtigste/Liebste auf der Welt sein; **think the ~ of sb.** große Stücke auf jmdn. halten *(ugs.);* **all alone in the ~:** ganz allein auf der Welt; **sb. is not long for this ~:** jmds. Tage sind gezählt; **out of this ~** *(fig. coll.)* phantastisch *(ugs.);* **come into the ~:** auf die Welt kommen; **the best of all possible ~s** die beste aller Welten; **get the best of both ~s** am meisten profitieren; **the ~'s end, the end of the ~:** das Ende der Welt; **it's not the end of the ~** *(iron.)* davon geht die Welt nicht unter *(ugs.);* **know/have seen a lot of the ~:** die Welt kennen/viel von der Welt gesehen haben; **see the ~:** die Welt kennenlernen; **a man/woman of the ~:** ein Mann/eine Frau mit Welterfahrung; **go up/come down in the ~:** [gesellschaftlich] aufsteigen/absteigen; *attrib.* **~ politics** Weltpolitik, *die;* **b)** *(domain)* **the literary/sporting/animal ~:** die literarische Welt *(geh.)*/die Welt *(geh.)* des Sports/die Tierwelt; **the ~ of letters/art/sport** die Welt *(geh.)* der Literatur/Kunst/des Sports; **c)** *(vast amount)* **it will do him a** *or* **the ~ of good** es wird ihm unendlich guttun; **a ~ of difference** ein weltweiter Unterschied; **a ~ away from sth.** Welten von etw. entfernt; **they are ~s apart in their views** ihre Ansichten sind Welten voneinander entfernt

world: W~ Bank *n.* Weltbank, *die;* **~-beater** *n.* **be a ~-beater** zur Spitzenklasse gehören; **~ 'champion** *n.* Weltmeister, *der*/-meisterin, *die;* **W~ 'Cup** *n.* *(Sport)* Worldcup, *der;* **~-famous** *adj.* weltberühmt

worldly ['wɜ:ldlɪ] *adj.* weltlich; weltlich eingestellt ‹*Person*›

world: ~ 'power *n.* Weltmacht, *die;* **~ 'record** *n.* Weltrekord, *der; attrib.* **~-record holder** Weltrekordhalter, *der*/-halterin, *die;* **~-shaking** *adj.* welterschütternd; **~ 'war** *n.* Weltkrieg, *der;* **the First/Second W~ War, W~ War I/II** der erste/zweite Weltkrieg; der 1./2. Weltkrieg; **~-wide 1.** ['--] *adj.* weltweit *nicht präd.;* **2.** [-'-] *adv.* weltweit

worm [wɜ:m] **1.** *n.* **a)** Wurm, *der;* **[even] a ~ will turn** *(prov.)* auch der Wurm krümmt sich, wenn er getreten wird *(Spr.);* **b)** *in pl. (intestinal parasites)* Würmer. **2.** *v. t.* **a) ~ oneself into sb.'s favour** sich in jmds. Gunst schleichen; **b)** *(draw by crafty persistence)* **~ sth. out of sb.** etw. aus jmdm. herausbringen *(ugs.).* **3.** *v. i.* sich winden

worm: ~-eaten *adj.* wurmstichig; *(fig.)* vom Zahn der Zeit angenagt; **~'s-eye-'view** *n.* Froschperspektive, *die (auch fig.)*

worn *see* **wear 2, 3**

ˈ**worn-out** *attrib. adj.* abgetragen ⟨*Kleidungsstück*⟩; abgenutzt ⟨*Teppich*⟩; abgedroschen ⟨*Redensart, Ausdruck*⟩; erschöpft, *(ugs.)* erledigt ⟨*Person*⟩

worried [ˈwʌrɪd] *adj.* besorgt; **give sb. a ~ look** jmdn. besorgt ansehen; **you had me ~:** ich habe mir [deinetwegen] Sorgen gemacht; **don't look so ~!** schau nicht so bekümmert drein!; **~ sick** krank vor Sorge; **be very ~:** sich *(Dat.)* große Sorgen machen

worrier [ˈwʌrɪə(r)] *n.* **be too much of a ~:** sich *(Dat.)* immer [zuviel] Sorgen machen; **he's a [real] ~:** er macht sich *(Dat.)* um alles Sorgen

worry [ˈwʌrɪ] **1.** *v.t.* **a)** beunruhigen; **it worries me to death to think that ...:** ich sorge mich zu Tode, wenn ich [daran] denke, daß ...; **~ oneself [about sth.]** sich *(Dat.)* um etw. Sorgen machen; **b)** *(bother)* stören; **c) ~ a bone** ⟨*Hund usw.:*⟩ an einem Knochen [herum]nagen. **2.** *v.i.* sich *(Dat.)* Sorgen machen; sich sorgen; **~ about sth.** sich *(Dat.)* um etw. Sorgen machen; **don't ~ about it** mach dir deswegen keine Sorgen!; ˈ**I should ~** *(coll. iron.)* was kümmert mich das?; **not to ~** *(coll.)* kein Problem *(ugs.)*. **3.** *n.* Sorge, *die;* **sth. is the least of sb.'s worries** etw. ist jmds. geringste Sorge; **it must be a great ~ to you** es muß dir große Sorgen bereiten

worrying [ˈwʌrɪɪŋ] **1.** *adj. (full of worry)* sorgenvoll ⟨*Zeit, Woche usw.*⟩; **it is a ~ time for her** sie hat zur Zeit große Sorgen. **2.** *n.* **~ only makes everything worse** sich *(Dat.)* Sorgen zu machen macht alles nur noch schlimmer

worse [wɜːs] **1.** *adj. compar. of* **bad 1** schlechter; schlimmer ⟨*Schmerz, Krankheit, Benehmen*⟩; **things could not/could be ~:** es kann nicht mehr schlimmer kommen/es könnte schlimmer sein; **the food is bad, and the service ~:** das Essen ist schlecht und die Bedienung noch schlechter; **he's getting ~:** mit ihm wird es schlimmer; *(his health)* ihm geht es schlechter; **be ~ than useless** ⟨*Sache:*⟩ mehr als unbrauchbar sein; ⟨*Person:*⟩ ein hoffnungsloser Fall sein; **sb. is [none] the ~ for sth.** jmdm. geht es wegen etw. [nicht] schlechter; **~ and ~:** immer schlechter/schlimmer; **to make matters ~, ...:** zu allem Übel ...; **it could have been ~:** es hätte schlimmer sein *od.* kommen können; **~ luck!** so ein Pech!; *see also* **drink 1c; wear 1a. 2.** *adv. compar. of* **badly** schlechter; schlimmer, schlechter ⟨*sich benehmen*⟩; **~ and ~:** immer schlechter/schlimmer; *see also* **better 3a; off 1g. 3.** *n.* Schlimmeres; **she might do ~ than settle for that job** es wäre bestimmt kein Fehler, wenn sie sich für die Stelle entschiede; **go from bad to ~:** immer schlimmer werden; **or ~:** oder noch Schlimmeres; **~ still** schlimmer noch; **a change for the ~:** eine Wende zum Schlechteren; **take a turn for the ~:** sich verschlechtern; ⟨*Krankheit:*⟩ sich verschlimmern; **nobody will think any the ~ of you** niemand wird deswegen schlechter von dir denken; **there is ~ to come** es kommt noch schlimmer; *see also* **worst 3a**

worsen [ˈwɜːsn] **1.** *v.t.* verschlechtern; verschlimmern ⟨*Knappheit*⟩. **2.** *v.i.* sich verschlechtern; ⟨*Hungersnot, Sturm, Problem:*⟩ sich verschlimmern

worship [ˈwɜːʃɪp] **1.** *v.t., (Brit.)* **-pp-: a)** verehren, anbeten ⟨*Gott, Götter*⟩; **b)** *(idolize)* abgöttisch verehren. **2.** *v.i., (Brit.)* **-pp-** am Gottesdienst teilnehmen. **3.** *n.* **a)** Anbetung, *die; (service)* Gottesdienst, *der;* **b) Your/His W~:** *Anrede für Richter, Bürgermeister;* ≈ Euer/seine Ehren

worshipper (*Amer.:* **worshiper**) [ˈwɜːʃɪpə(r)] *n.* **a)** *(in church)* Gottesdienstbesucher, *der*/-besucherin, *die;* **b)** *(of deity)* Anbeter, *der*/Anbeterin, *die*

worst [wɜːst] **1.** *adj. superl. of* **bad 1** *see* **worse 1:** schlechtest.../schlimmst...; **be ~:** am schlechtesten/schlimmsten sein; **the ~ thing about it was ...:** das Schlimmste daran war ... **2.** *adv. superl. of* **badly** am schlimmsten; am schlechtesten ⟨*gekleidet*⟩. **3.** *n.* **a) [the] ~:** der/die/das Schlimmste; **prepare for the ~:** sich auf das Schlimmste gefaßt machen; **at ~, at the [very] ~:** schlimmstenfalls; im [aller]schlimmsten Fall[e]; **get** *or* **have the ~ of it** *(be defeated)* geschlagen werden; *(suffer the most)* am meisten zu leiden haben; **if the ~** *or* **it comes to the ~** *(Brit.)*, **if worse comes to ~** *(Amer.)* wenn es zum Schlimmsten kommt; **do your ~:** mach, was du willst!; **let him do his ~:** er soll machen, was er will; **b)** *(what is of poorest quality)* Schlechteste, *der/die/das*

worsted [ˈwʊstɪd] *n. (Textiles)* Kammgarn, *das*

worth [wɜːθ] **1.** *adj.* **a)** *(of value equivalent to)* wert; **it's ~/not ~ £80** es ist 80 Pfund wert/80 Pfund ist es nicht wert; **it is not ~ much** *or* **a lot [to sb.]** es ist [jmdm.] nicht viel wert; **be ~ the money** das Geld wert sein; **not ~ a penny** keinen Pfennig wert *(ugs.)*; **for what it is ~:** was immer auch davon zu halten ist; **b)** *(worthy of)* **is it ~ hearing/the effort?** ist es hörenswert/der Mühe wert?; **is it ~ doing?** lohnt es sich?; **if it's ~ doing, it's ~ doing well** wenn schon, denn schon; **it isn't ~ it** es lohnt sich nicht; **it's ~ a try** es ist einen Versuch wert; **it would be [well] ~ it** *(coll.)* es würde sich [sehr] lohnen; **be well ~ sth.** durchaus *od.* sehr wohl etw. wert sein; **c) be ~ sth.** *(possess)* etw. wert sein *(ugs.)*; **run/cycle for all one is ~** *(coll.)* rennen/fahren, was man kann. **2.** *n.* **a)** *(equivalent of money etc. in commodity)* **ten pounds' ~ of petrol** Benzin für zehn Pfund; *(more formal)* Benzin im Wert von zehn Pfund; **b)** *(value, excellence)* Wert, *der*

worthless [ˈwɜːθlɪs] *adj.* **a)** *(valueless)* wertlos; **b)** *(despicable)* nichtswürdig

ˈ**worthwhile** *attrib. adj.* lohnend; *see also* **while 1**

worthy [ˈwɜːðɪ] **1.** *adj.* **a)** *(adequate, estimable)* würdig; verdienstvoll ⟨*Tat*⟩; angemessen ⟨*Belohnung*⟩; **~ of the occasion** dem Anlaß angemessen; **b)** *(deserving)* würdig; verdienstvoll ⟨*Sache, Organisation*⟩; **be ~ of the name** den Namen verdienen; **~ of note/mention** erwähnenswert. **2.** *n.* **local worthies** *(joc.)* örtliche Honoratioren

would *see* [1]**will**

would-be [ˈwʊdbiː] *attrib. adj.* **a ~ philosopher** ein Möchtegernphilosoph; **a ~ aggressor** ein möglicher Aggressor

wouldn't [ˈwʊdnt] *(coll.)* = **would not**; *see* [1]**will**

[1]**wound** [wuːnd] **1.** *n. (lit. or fig.)* Wunde, *die;* **a war ~:** eine Kriegsverletzung. **2.** *v.t.* verwunden; *(fig.)* verletzen; **be ~ed in the thigh/arm** am Oberschenkel/Arm verwundet werden

[2]**wound** *see* [2]**wind 1, 2**

wove, woven *see* [1]**weave 2, 3**

wow [waʊ] **1.** *int.* hoi. **2.** *n. (sl.)* **be a ~:** eine Wucht sein *(salopp)*. **3.** *v.t. (sl.)* umhauen *(ugs.)*

WP *abbr.* **word processor**

wraith [reɪθ] *n.* Gespenst, *das*

wrangle [ˈræŋgl] **1.** *v.i.* [sich] streiten. **2.** *n.* Streit, *der*

wrap [ræp] **1.** *v.t.*, **-pp-: a)** einwickeln; *(fig.)* hüllen; **~ped** abgepackt ‹*Brot usw.*›; **~ sth. in paper/cotton wool** etw. in Papier/Watte [ein]wickeln; **~ sth. [a]round sth.** etw. um etw. wickeln; **b)** schlingen ‹*Schal, Handtuch usw.*› **(about, round** um). **2.** *n.* Umschlag[e]tuch, *das; under* **~s** *(fig.)* unter Verschluß; **keep sth. under ~s** etw. geheimhalten

~ 'up 1. *v.t.* **a)** *see* **wrap 1; wrapped up; b)** *(fig.: conclude)* abschließen. **2.** *v.i.* sich warm einpacken *(ugs.)*

wrapped up [ræpt 'ʌp] *adj.* **be ~ in one's work** in seine Arbeit völlig versunken sein; **a country whose prosperity is ~ in its shipping** ein Land, dessen Reichtum eng mit seiner Schiffahrt verknüpft ist

wrapper ['ræpə(r)] *n.* **a)** *(around newspaper etc.)* Streifband, *das (Postw.);* **b)** *(around sweet etc.)* **sweet-/toffee-~[s]** Bonbonpapier, *das;* **c)** *(of book) see* **jacket c**

wrapping ['ræpɪŋ] *n.* Verpackung, *die;* **~s** Verpackung, *die; (fig.)* Hülle, *die (dichter.)*

'wrapping-paper *n. (strong paper)* Packpapier, *das; (decorative paper)* Geschenkpapier, *das*

wrath [rɒθ] *n. (poet./rhet.)* Zorn, *der*

wreak [ri:k] *v.t.* **a)** *(inflict)* **~ vengeance on sb.** an jmdm. Rache nehmen; **b)** *(vent)* auslassen ‹*Wut, Ärger*› **(on** an + *Dat.*); **c)** *(cause)* anrichten ‹*Verwüstung, Unheil*›

wreath [ri:θ] *n.*, *pl.* **wreaths** [ri:ðz, ri:θs] Kranz, *der;* **a ~ of smoke** ein Ring aus Rauch

wreathe [ri:ð] *v.t.* **a)** *(encircle)* umkränzen; **her face was ~d in smiles** ein Lächeln umspielte ihre Lippen; **b)** *(make by interweaving)* flechten; winden *(geh.)*

wreck [rek] **1.** *n.* **a)** *(destruction)* Schiffbruch, *der; (fig.)* Zerstörung, *die;* **b)** *(ship)* Wrack, *das;* **c)** *(broken remains, lit. or fig.)* Wrack, *das;* **she was a physical/mental ~:** sie war körperlich/geistig ein Wrack; **I feel/you look a ~** *(coll.)* ich fühle mich kaputt *(ugs.)*/du siehst kaputt aus *(ugs.)*. **2.** *v.t.* **a)** *(destroy)* ruinieren; zu Schrott fahren ‹*Auto*›; **be ~ed** *(shipwrecked)* ‹*Schiff, Person:*› Schiffbruch erleiden; **b)** *(fig.: ruin)* zerstören; ruinieren ‹*Gesundheit, Urlaub*›

wreckage ['rekɪdʒ] *n.* Wrackteile; *(fig.)* Trümmer *Pl.*

Wren [ren] *n. (Brit.) Angehörige des weiblichen Marinedienstes;* **join the ~s** in den weiblichen Marinedienst eintreten

wren *n.* Zaunkönig, *der*

wrench [rentʃ] **1.** *n.* **a)** *(tool)* verstellbarer Schraubenschlüssel; **b)** *(Amer.) see* **spanner; c)** *(violent twist)* Verrenkung, *die;* **d)** *(fig.)* **be a great ~ [for sb.]** sehr schmerzhaft für jmdn. sein. **2.** *v.t.* **a)** *(tug violently)* reißen; **~ at sth.** an etw. *(Dat.)* reißen; **~ sth. round/off/open** etw. herum-/ab-/aufreißen; **~ sth. from sb.** jmdm. etw. entreißen; **b)** *(injure by twisting)* **~ one's ankle** *etc.* sich *(Dat.)* den Knöchel *usw.* verrenken

wrest [rest] *v.t.* **~ sth. from sb./sb.'s grasp** *(lit. or fig.)* jmdm./jmds. Griff etw. entreißen *od. (geh.)* entwinden; **~ sth. from sth.** einer Sache *(Dat.)* etw. abringen

wrestle ['resl] **1.** *n. (hard struggle)* Ringen, *das.* **2.** *v.i.* **a)** ringen; **b)** *(fig.: grapple)* sich abmühen; **~ with one's conscience** mit seinem Gewissen ringen

wrestler ['reslə(r)] *n.* Ringer, *der*/Ringerin, *die*

wrestling ['reslɪŋ] *n., no pl., no indef. art.* Ringen, *das*

wretch [retʃ] *n.* Kreatur, *die; (joc.: child)* Gör, *das*

wretched ['retʃɪd] *adj.* **a)** *(miserable)* unglücklich; **feel ~ about sb./sth.** *(be embarrassed)* über jmdn./etw. todunglücklich sein; **feel ~** *(be very unwell)* sich elend fühlen; **b)** *(coll.: damned)* elend *(abwertend);* **c)** *(very bad)* erbärmlich; miserabel ‹*Wetter*›; **d)** *(causing discomfort)* schrecklich ‹*Reise, Erfahrung, Zeit*›

wretchedness ['retʃɪdnɪs] *n., no pl.* **a)** *(misery)* Elend, *das;* **b)** *(badness)* Erbärmlichkeit, *die*

wriggle ['rɪgl] **1.** *v.i.* **a)** sich winden; ‹*Fisch:*› zappeln; **b)** *(make one's/its way by wriggling)* sich schlängeln; **~ out of a difficulty** *etc. (fig.)* sich aus einer schwierigen Situation *usw.* herauswinden. **2.** *v.t.* **~ one's way** *(lit. or fig.)* sich schlängeln

wring [rɪŋ] *v.t.*, **wrung** [rʌŋ] **a)** wringen; **~ out** auswringen; **~ the water out of the towels** das Wasser aus den Handtüchern wringen; **b)** *(squeeze forcibly)* **~ sb.'s hand** jmdm. fest die Hand drücken; *(twist forcibly)* **~ one's hands** die Hände ringen *(geh.);* **~ the neck of an animal** einem Tier den Hals umdrehen; **I could have wrung his neck** *(fig.)* ich hätte ihm den Hals umdrehen können; **c)** *(extract)* wringen; **~ sth. from** *or* **out of sb.** *(fig.)* jmdm. etw. abpressen

wringer ['rɪŋə(r)] *n.* Wringmaschine, *die*

wringing 'wet *adj.* tropfnaß

wrinkle ['rɪŋkl] *n.* Falte, *die; (in paper)* Knick, *der*

wrinkled ['rɪŋkld] *adj.* runz[e]lig

wrinkly ['rɪŋklɪ] **1.** *adj.* runz[e]lig. **2.** *n. (sl.)* Grufti, *der (ugs.)*

wrist [rɪst] *n.* Handgelenk, *das*

'wrist-watch *n.* Armbanduhr, *die*

¹writ [rɪt] *n. (Law)* Verfügung, *die*

²writ *see* **write 2a**

write [raɪt] **1.** *v.i.*, **wrote** [rəʊt], **written** ['rɪtn] schreiben; **~ to sb./a firm** jmdm./an eine Firma schreiben. **2.** *v.t.*, **wrote, written a)** schreiben; ausschreiben ‹*Scheck*›; **the written language** die Schriftsprache; **written applications** schriftliche Anträge; **the paper had been written all over** das Papier war ganz vollgeschrieben; **be written into the contract** [ausdrücklich] im Vertrag stehen; **~ sb. into/out of a serial** für jmdn. eine Rolle in einer Serie schreiben/jmdm. einen Abgang aus einer Serie verschaffen; **writ large** *(fig.)* im Großformat *(fig.);* **b)** *(Amer./Commerc./coll.: ~ letter to)* anschreiben; **c)** *in pass. (fig.: be apparent)* **sb. has sth. written in his face** jmdm. steht etw. im Gesicht geschrieben; **guilt was written all over her face** die Schuld stand ihr ins Gesicht geschrieben

~ a'way *v.i.* **~ away for sth.** etw. [schriftlich] anfordern

~ 'back *v.i.* zurückschreiben

~ 'down *v.t.* aufschreiben

~ 'in *v.i.* hinschreiben *(ugs.); (include)* hineinschreiben; **~ in for sth.** etw. [schriftlich] anfordern

~ 'off 1. *v.t.* **a)** *(compose with ease)* herunterschreiben *(ugs.);* **b)** *(cancel)* abschreiben ‹*Schulden, Verlust*›; *(fig.)* **~ sb. off [as a failure** *etc.*] jmdn. [als Versager] abschreiben *(ugs.);* **c)** *(destroy)* zu Schrott fahren. **2.** *v.i. see* **~ away**

~ 'out *v.t.* **a)** ausschreiben ‹*Scheck*›; schreiben ‹*Rezept*›; **b)** *(~ in final form)* ausarbeiten; *(~ in full)* ausschreiben

~ 'up *v.t.* **a)** *(praise)* eine gute Kritik schreiben über (+ *Akk.*); **b)** *(~ account of)* einen Bericht schreiben über (+ *Akk.*); *(~ in full)* aufarbeiten; **c)** *(bring up to date)* auf den neuesten Stand bringen

'write-off *n. (person)* Versager, *der*/Versagerin, *die; (vehicle)* Totalschaden, *der*

writer ['raɪtə(r)] *n.* **a)** *(author)* Schriftsteller, *der*/Schriftstellerin, *die; (of letter, article)* Schreiber, *der*/Schreiberin, *die;* Verfasser, *der*/Verfasserin, *die; (of lyrics, advertisements)* Texter, *der*/Texterin, *die; (of music)* Komponist, *der*/Komponistin, *die;* **be a ~:** Schriftsteller/Schriftstellerin sein; **b) be a good/bad ~** *(as to handwriting)* eine gute/schlechte Schrift haben

'write-up *n.* Bericht, *der; (by critic)* Kritik, *die;* **get a good ~:** gut besprochen werden

writhe [raɪð] *v.i. (lit. or fig.)* sich winden; **he/it makes me ~** *(with embarrassment)* er/es bringt mich in ziemliche Verlegenheit; *(with disgust)* er/es ist mir zuwider

writing ['raɪtɪŋ] *n.* **a)** Schreiben, *das;* **put sth. in ~:** etw. schriftlich machen *(ugs.);* **b)** *(handwriting)* Schrift, *die;* **c)** *(something written)* Schrift, *die;* **the ~ on the wall** *(fig.)* das Menetekel an der Wand

writing: ~-case *n.* Schreibmappe, *die;* **~-pad** *n.* Schreibblock, *der;* **~-paper** *n.* Schreibpapier, *das;* Briefpapier, *das*

written *see* **write**

wrong [rɒŋ] **1.** *adj.* **a)** *(morally bad)* unrecht *(geh.); (unfair)* ungerecht; **you were ~ to be so angry** es war nicht richtig von dir, so ärgerlich zu sein; **b)** *(mistaken)* falsch; **be ~** ‹*Person:*› sich irren; **I was ~ about you** ich habe mich in dir geirrt; **the clock is ~:** die Uhr geht falsch; **the clock is ~ by ten minutes** *(fast/slow)* die Uhr geht 10 Minuten vor/nach; **how ~ can you be** *or* **get!** wie man sich irren kann!; **c)** *(not suitable)* falsch; **give the ~ answer** eine falsche Antwort geben; **say/do the ~ thing** das Falsche sagen/tun; **be the ~ person for the job** für die Stelle ungeeignet sein; **take the ~ turning** falsch abbiegen; **get hold of the ~ end of the stick** *(fig.)* alles völlig falsch verstehen; **[the] ~ way round** verkehrt herum; **d)** *(out of order)* nicht in Ordnung; **there's something ~ here/with him** hier/mit ihm stimmt etwas nicht; **there's nothing ~:** es ist alles in Ordnung; **what's ~?** ist etwas nicht in Ordnung? *See also* **wrong side. 2.** *adv.* falsch; **get it ~:** es falsch *od.* verkehrt machen; *(misunderstand)* sich irren; **I got the answer ~ again** meine Antwort war wieder falsch; **get sb. ~:** jmdn. falsch verstehen; **go ~** *(take ~ path)* sich verlaufen; *(fig.)* ‹*Person:*› vom rechten Weg abkommen *(fig. geh.);* ‹*Maschine, Mechanismus:*› kaputtgehen *(ugs.);* ‹*Angelegenheit:*› danebengehen *(ugs.).* **3.** *n.* Unrecht, *das;* **two ~s don't make a right** das gibt nur ein Unrecht mehr; **do ~:** unrecht tun; **she can do no ~:** sie kann überhaupt nichts Unrechtes tun; **be in the ~:** im Unrecht sein. **4.** *v.t.* **~ sb.** jmdn. ungerecht behandeln

wrong: ~doer *n.* Übeltäter, *der*/-täterin, *die;* Missetäter, *der*/-täterin, *die (geh.);* **~doing** *n.* **a)** *no pl., no indef. art.* Missetaten *(geh.);* **b)** *(instance)* Missetat, *die (geh.);* **~-'foot** *v.t.* **a)** *(Sport)* **~-foot sb.** jmdn. auf dem falschen Fuß erwischen *(Sportjargon);* **b)** *(fig. coll.)* unvorbereitet treffen

wrongful ['rɒŋfl] *adj.* **a)** *(unfair)* unrecht *(geh.);* **b)** *(unlawful)* rechtswidrig

wrongfully ['rɒŋfəlɪ] *adv.* **a)** *(unfairly)* unrecht *(geh.)* ‹*handeln*›; zu Unrecht ‹*beschuldigen*›; **b)** *(unlawfully)* rechtswidrig

wrongly ['rɒŋlɪ] *adv.* **a)** *(inappropriately, incorrectly)* falsch; **b)** *(mistakenly)* zu Unrecht; **I believed, ~, that ...:** ich habe fälschlicherweise geglaubt, daß ...; **c)** *see* **wrongfully a**

'wrong side *n.* **a)** *(of fabric)* linke Seite; **[the] ~ out/up** verkehrt herum; **b) be on the ~ of thirty** die dreißig überschritten haben; **get on the ~ of sb./the law** *(fig.)* jmdn. falsch anfassen/mit dem Gesetz in Konflikt geraten; *see also* **bed 1 a**

wrote *see* **write**

wrought *see* **work 2**

wrought 'iron *n.* Schmiedeeisen, *das; attrib.* schmiedeeisern ‹*Tor, Zaun*›

wrung *see* **wring**

wry [raɪ] *adj.*, **~er** *or* **wrier** ['raɪə(r)], **~est** *or* **wriest** ['raɪɪst] ironisch ‹*Blick*›; fein ‹*Humor, Witz*›

wryly ['raɪlɪ] *adv.* ironisch ‹*blicken, sagen*›; schief ‹*lächeln*›

wt. *abbr.* **weight** Gew.

WW *abbr. (Amer.)* **World War** WK

X, x [eks] *n., pl.* **Xs** *or* **X's** ['eksɪz] **a)** *(letter)* X, x, *das;* **b)** *(Math.)* x; **c)** *(unknown person)* **Mr X** Herr X; **d) x marks the spot** die Stelle ist durch ein Kreuz markiert

xenon ['zenɒn] *n. (Chem.)* Xenon, *das*

Xerox, (P), xerox ['zɪərɒks, 'zerɒks] **1.** *n.* **a)** *(process)* Xerographie, *die (Druckw.);* **b)** *(copy)* Xerokopie, *die.* **2. xerox** *v.t.* xerokopieren

Xmas ['krɪsməs, 'eksməs] *n. (coll.)* Weihnachten, *das*

'X-ray 1. *n.* **a)** *in pl.* Röntgenstrahlen *Pl.;* **b)** *(picture)* Röntgenaufnahme, *die;* **c)** *attrib.* Röntgen-. **2.** *v.t.* röntgen; durchleuchten ‹*Gepäck*›

xylophone ['zaɪləfəʊn] *n. (Mus.)* Xylophon, *das*

Y

Y, y [waɪ] *n., pl.* **Ys** *or* **Y's a)** *(letter)* Y, y, *das;* **b)** *(Math.)* y

yacht [jɒt] **1.** *n.* **a)** *(for racing)* Segelboot, *das;* Segeljacht, *die;* **b)** *(for pleasure travel etc.)* Jacht, *die.* **2.** *v.i.* segeln

'yacht-club *n.* Jachtklub, *der*

yachting ['jɒtɪŋ] *n., no pl., no art.* Segeln, *das*

yachtsman ['jɒtsmən] *n., pl.* **yachtsmen** ['jɒtsmən] Segler, *der*

Yank [jæŋk] *n. (Brit. coll.: American)* Yankee, *der;* Ami, *der (ugs.)*

yank *(coll.)* **1.** *v.t.* reißen an

(+ *Dat.*); ~ **sth. off/out** etw. ab-/ausreißen. **2.** *n.* Reißen, *das*

Yankee ['jæŋkɪ] *see* **Yank**

yap [jæp] *v.i.*, **-pp-** kläffen

[1]**yard** [jɑ:d] *n.* Yard, *das;* **by the ~:** ≈ meterweise; *(fig.)* am laufenden Band *(ugs.);* **have a face a ~ long** ein Gesicht wie drei Tage Regenwetter machen

[2]**yard** *n.* **a)** *(attached to building)* Hof, *der;* **in the ~:** auf dem Hof; **b)** *(for manufacture)* Werkstatt, *die; (for storage)* Lager, *das; (ship~)* Werft, *die;* **builder's ~:** Bauhof, *der;* **c)** *(Amer.: garden)* Garten, *der*

'**yardstick** *n.* *(fig.: standard)* Maßstab, *der*

yarn [jɑ:n] *n.* **a)** *(thread)* Garn, *das;* **b)** *(coll.: story)* Geschichte, *die; (of sailor)* [Seemanns]garn, *das*

yawn [jɔ:n] **1.** *n.* Gähnen, *das;* **give a [long] ~:** [herzhaft] gähnen. **2.** *v.i.* **a)** gähnen; **b)** *(fig.)* ‹*Abgrund, Kluft, Spalte:*› gähnen *(geh.)*

yawning ['jɔ:nɪŋ] *adj.* gähnend *(auch fig. geh.)*

yd[s]. *abbr.* **yard[s]** Yd[s].

[1]**ye** [ji:] *pron.* *(arch./poet./dial./joc.)* Ihr *(veralt.); (as direct or indirect object)* Euch *(veralt.)*

[2]**ye** *adj.* *(pseudo-arch.)* = **the**

yea [jeɪ] *adv.* *(arch.)* ja

yeah [jeə] *adv.* *(coll.)* ja; **[oh] ~?** [ach] ja?

year [jɪə(r)] *n.* **a)** Jahr, *das;* **she gets £10,000 a ~:** sie verdient 10 000 Pfund im Jahr; **~ in ~ out** jahrein, jahraus; **~ after ~:** Jahr für *od.* um Jahr; **all [the] ~ round** das ganze Jahr hindurch; **in a ~['s time]** in einem Jahr; **once a ~, once every ~:** einmal im Jahr; **Christian** *or* **Church** *or* **ecclesiastical ~** *(Eccl.)* Kirchenjahr, *das;* liturgisches Jahr *(kath. Kirche);* **a ten-~-old** ein Zehnjähriger/eine Zehnjährige; **a ten-~[s]-old child** ein zehn Jahre altes Kind; **in her thirtieth ~:** in ihrem 30. Lebensjahr; **financial** *or* **fiscal** *or* **tax ~:** Finanz- *od.* Rechnungsjahr, *das;* **calendar** *or* **civil ~:** Kalenderjahr, *das;* **school ~:** Schuljahr, *das;* **for a ~ and a day** ein Jahr und einen Tag [lang]; **a ~ [from] today** *etc.* heute *usw.* in einem Jahr; **a ~ [ago] today** *etc.* heute *usw.* vor einem Jahr; **... of the ~** *(best)* ... des Jahres; **she looks ~s older** sie sieht um Jahre älter aus; **take ~s off sb./sb.'s life** jmdn. um Jahre jünger/älter machen; *see also* [1]**by 1 m; from b; b)** *(group of students)* Jahrgang, *der;* **first-~ student** Student/Studentin im ersten Jahr; **c)** *in pl. (age)* **be getting on/be well on in ~s** in die Jahre kommen/in vorgerücktem Alter sein *(geh.)*

year: ~-book *n.* Jahrbuch, *das;* **~-long** *adj.* *(lasting a ~)* einjährig; *(lasting the whole ~)* ganzjährig

yearly ['jɪəlɪ] **1.** *adj.* **a)** *(annual)* jährlich; **ten-~:** zehnjährig; **at twice-~ intervals** zweimal im Jahr; **b)** *(lasting a year)* Einjahres‹*vertrag, -abonnement*›. **2.** *adv.* jährlich

yearn [jɜ:n] *v.i.* **~ for** *or* **after sth./for sb./to do sth.** sich nach etw./jmdm. sehnen/sich danach sehnen, etw. zu tun

yearning ['jɜ:nɪŋ] *n.* Sehnsucht, *die*

'**year-round** *adj.* ganzjährig

yeast [ji:st] *n.* Hefe, *die*

yell [jel] **1.** *n.* gellender Schrei; **let out a ~:** einen Schrei ausstoßen. **2.** *v.t. & i.* [gellend] schreien

yellow ['jeləʊ] **1.** *adj.* **a)** gelb; flachsblond ‹*Haar*›; golden ‹*Getreide*›; vergilbt ‹*Papier*›; **b)** *(fig. coll.: cowardly)* feige. **2.** *n.* Gelb, *das.* **3.** *v.t. & i.* vergilben

yellowish ['jeləʊɪʃ] *adj.* gelblich

yellow: ~ 'line *n.* *(Brit.)* gelbe [Markierungs]linie; **I'm on double ~ lines** ich stehe im Parkverbot; **~ 'pages, (P)** *n. pl.* gelbe Seiten; Branchenverzeichnis, *das*

yelp [jelp] **1.** *v.i.* aufheulen *(ugs.);* ‹*Hund:*› jaulen. **2.** *n.* Heulen, *das; (of dog)* Jaulen, *das*

[1]**yen** [jen] *n., pl. same (Japanese currency)* Yen, *der*

[2]**yen** *n.* *(coll.: longing)* Drang, *der* **(for** nach); **sb. has a ~ to do sth.** es drängt jmdn. danach, etw. zu tun

yeoman ['jəʊmən] *n., pl.* **yeomen** ['jəʊmən] **a)** *(with small estate)* Kleinbauer, *der;* **b)** *(Hist.: freeholder)* Freisasse, *der*

yep [jep] *int.* *(Amer. coll.)* ja

yes [jes] **1.** *adv.* ja; *(in contradiction)* doch; **~, sir** jawohl!; **~?** *(indeed?)* ach ja?; *(what do you want?)* ja?; *(to customer)* ja, bitte?; **say '~'** ja sagen; **say ~ to a proposal** einem Vorschlag zustimmen; **~ and no** ja und nein. **2.** *n., pl.* **~es** Ja, *das*

'**yes-man** *n.* *(coll. derog.)* Jasager, *der (abwertend)*

yesterday ['jestədeɪ, 'jestədɪ] **1.** *n.* gestern; **the day before ~:** vorgestern; **~'s paper** die gestrige Zeitung; die Zeitung von gestern; **~ morning/afternoon/evening/night** gestern vormittag/nachmittag/abend/nacht; **a week [from] ~:** gestern in einer Woche; **~ evening's concert** das Konzert gestern abend *od.* am gestrigen Abend. **2.** *adv.* **a)** gestern; **the day before ~:** vorgestern; **~ morning/afternoon/evening/night** gestern vormittag / nachmittag / abend / nacht; **b)** *(in the recent past)* gestern; *see also* **born 1**

yet [jet] **1.** *adv.* **a)** *(still)* noch; **have ~ to reach sth.** etw. erst noch erreichen müssen; **much ~ remains to be done** noch bleibt viel zu tun; *see also* **as 5; b)** *(hitherto)* bisher; **the play is his best ~:** das Stück ist sein bisher bestes; **c)** *neg. or interrog.* **not [just] ~:** [jetzt] noch nicht; **never ~:** noch nie; **need you go just ~?** mußt du [jetzt] schon gehen?; **you haven't seen anything** *or (coll.)* **ain't seen nothing ~:** das ist noch gar nichts; **d)** *(before all is over)* doch noch; **he could win ~:** er könnte noch gewinnen; **e)** *with compar. (even)* noch; **f)** *(nevertheless)* doch; **g)** *(again)* noch; **~ again** noch einmal; **she has never voted for that party, nor ~ intends to** sie hat nie für diese Partei gestimmt, und sie hat es auch nicht vor. **2.** *conj.* doch; **a faint ~ unmistakable smell** ein schwacher, aber unverkennbarer Geruch

yew [ju:] *n.* **~ [-tree]** Eibe, *die*

YHA *abbr.* *(Brit.)* **Youth Hostels Association** Jugendherbergsverband, *der*

Yiddish ['jɪdɪʃ] **1.** *adj.* jiddisch. **2.** *n.* Jiddisch, *das; see also* **English 2 a**

yield [ji:ld] **1.** *v.t.* **a)** *(give)* bringen; hervorbringen ‹*Ernte*›; tragen ‹*Obst*›; abwerfen ‹*Gewinn*›; ergeben ‹*Resultat, Informationen*›; **b)** *(surrender)* übergeben ‹*Festung*›; lassen ‹*Vortritt*›; abtreten ‹*Besitz*› **(to** an + *Akk.*); **~ the point** [in diesem Punkt] nachgeben; **~ a point to sb.** jmdm. in einem Punkt nachgeben. **2.** *v.i.* **a)** *(surrender)* sich unterwerfen; **~ to threats/temptation** Drohungen *(Dat.)* nachgeben/der Versuchung *(Dat.)* erliegen; **~ to persuasion/sb.'s entreaties** sich überreden lassen/jmds. Bitten *(Dat.)* nachgeben; **b)** *(give right of way)* Vorfahrt gewähren. **3.** *n.* **a)** Ertrag, *der;* **b)** *(return on investment)* Zins[ertrag], *der;* **a 10% ~:** 10% Zinsen

yippee ['jɪpi:, jɪ'pi:] *int.* hurra

yob [jɒb], **yobbo** ['jɒbəʊ] *ns., pl.* **~s** *(Brit. sl.)* Rowdy, *der*

yodel ['jəʊdl] **1.** *v.i. & t., (Brit.)* **-ll-** jodeln. **2.** *n.* Jodeln, *das*

yoga ['jəʊgə] *n.* Joga, *der od. das*
yoghurt, yogurt ['jɒgət] *n.* Joghurt, *der od. das*
yoke [jəʊk] **1.** *n.* **a)** *(for animal)* Joch, *das;* **b)** *(for person)* [Trag]joch, *das;* **c)** *(of garment)* Sattel, *der (Textilw.).* **2.** *v.t.* **a)** ins Joch spannen ‹*Tier*›; ~ **an animal to sth.** ein Tier vor etw. *(Akk.)* spannen; **b)** *(fig.: couple)* verbinden
yokel ['jəʊkl] *n.* *(derog.)* [Bauern]tölpel, *der*
yolk [jəʊk] *n.* Dotter, *der;* Eigelb, *das*
yonder ['jɒndə(r)] *(literary)* **1.** *adj.* ~ **tree/peasant** jener Baum/Bauer dort *(geh.).* **2.** *adv.* dort drüben
Yorkshire pudding [jɔ:kʃɪə 'pʊdɪŋ, jɔ:kʃə 'pʊdɪŋ] *n.* *(Gastr.)* Yorkshirepudding, *der*
you [jʊ, *stressed* 'ju:] *pron.* **a)** *sing./pl.* du/ihr; *in polite address sing. or pl.* Sie; *as direct object* dich/euch/Sie; *as indirect object* dir/euch/Ihnen; *refl.* dich/dir/euch; *in polite address* sich; **it was ~:** du warst/ihr wart/Sie waren es; **~-know-what/-who** du weißt/ihr wißt/Sie wissen schon, was/wer/wen/wem; **b)** *(one)* man; **smoking is bad for you** Rauchen ist ungesund. *See also* **your; yours; yourself; yourselves**
you'd [jʊd, *stressed* ju:d] **a)** = **you had; b)** = **you would**
you'll [jʊl, *stressed* ju:l] **a)** = **you will; b)** = **you shall**
young [jʌŋ] **1.** *adj.*, **~er** ['jʌŋgə(r)], **~est** ['jʌŋgɪst] **a)** *(lit. or fig.)* jung; **a very ~ child** ein ganz kleines Kind; **the ~ boys** die [kleinen] Jungen; **~ at heart** im Herzen jung geblieben; **sb. is not getting any ~er** jmd. wird auch nicht jünger; **you're only ~ once** man ist nur einmal jung; **the night is still ~:** die Nacht ist jung; **~ Jones** der junge Jones *(ugs.);* **b)** *(characteristic of youth)* jugendlich; **~ love/fashion** junge Liebe/Mode. **2.** *n. pl.* *(of animals)* Junge; *(of humans)* Kinder; **with ~:** trächtig; **the ~** *(~ people)* die jungen Leute; **~ and old** jung und alt
'**young days** *n. pl.* Jugendjahre *Pl.;* **in my ~:** in meiner Jugend[zeit]
youngish ['jʌŋgɪʃ] *adj.* ziemlich jung
young: ~ 'lady *n.* **a)** junge Dame; **b)** *(girl-friend)* Freundin, *die;* **~ 'man** *n.* **a)** junger Mann; **b)** *(boy-friend)* Freund, *der*
youngster ['jʌŋstə(r)] *n.* **a)** *(child)* Kleine, *der/die/das;* **b)** *(young person)* Jugendliche, *der/die;* **you're just a ~ compared with me** im Vergleich zu mir bist du noch jung
young 'woman *n.* **a)** junge Frau; **b)** *(girl-friend)* Freundin, *die*
your [jə(r), *stressed* jʊə(r), jɔ:(r)] *poss. pron. attrib.* *(of you, sing./pl.)* dein/euer; *in polite address* Ihr. *See also* **²her**
you're [jə(r), *stressed* jʊə(r), jɔ:(r)] = **you are**
yours [jʊəz, jɔ:z] *poss. pron. pred.* **a)** *(to or of you, sing.)* deiner/deine/dein[e]s; *(to or of you, pl.)* eurer/eure/eures; *in polite address* Ihrer/Ihre/Ihr[e]s; **what's ~?** *(coll.)* was nimmst du/nehmen Sie?; *see also* **hers; ours; b)** *(your letter)* Ihr Brief; *(Commerc.)* Ihr Schreiben; **c)** *(ending letter)* ~ **[obediently]** Ihr [sehr ergebener *(geh.)*]; ~ **truly** in alter Verbundenheit Dein/Deine; *(in business letter)* mit freundlichen Grüßen; *(joc.: I)* meine Wenigkeit *(scherzh.); see also* **faithfully c; sincerely**
yourself [jə'self, *stressed* jʊə'self, jɔ:'self] *pron.* **a)** *emphat.* selbst; **for ~:** für dich/*in polite address* Sie selbst; **you must do sth. for ~:** du mußt selbst etw. tun; **relax and be ~:** entspann dich und gib dich ganz natürlich; **b)** *refl.* dich/dir; *in polite address* sich. *See also* **herself; myself**
yourselves [jə'selfs, *stressed* jʊə'selvz, jɔ:'selvz] *pron.* **a)** *emphat.* selbst; **for ~:** für euch/*in polite address* Sie selbst; **b)** *refl.* euch/sich. *See also* **herself**
youth [ju:θ] *n.* **a)** *no pl., no art.* Jugend, *die;* **b)** *pl.* **~s** [ju:ðz] *(young man)* Jugendliche, *der;* **c)** *constr. as pl.* *(young people)* Jugend, *die*
youth: ~ centre *n.* Jugendzentrum, *das;* **~ club** *n.* Jugendklub, *der*
youthful ['ju:θfl] *adj.* jugendlich
'**youth hostel** *n.* Jugendherberge, *die*
you've [jʊv, *stressed* ju:v] = **you have**
Yo-Yo, (P) ['jəʊjəʊ] *n., pl.* **~s** Jo-Jo, *das*
Yugoslav ['ju:gəslɑ:v] *see* **Yugoslavian**
Yugoslavia [ju:gə'slɑ:vɪə] *pr. n.* Jugoslawien *(das)*
Yugoslavian [ju:gə'slɑ:vɪən] **1.** *adj.* jugoslawisch; **sb. is ~:** jmd. ist Jugoslawe/Jugoslawin. **2.** *n.* Jugoslawe, *der*/Jugoslawin, *die*
yuk [jʌk] *int.* *(sl.)* bäh; äks
yule [ju:l], '**yule-tide** *ns.* *(arch.)* Weihnachtszeit, *die*
yummy ['jʌmɪ] *(coll.)* **1.** *adj.* lecker. **2.** *int.* *(child lang.)* lecker, lecker
yuppie ['jʌpɪ] *n.* *(coll.)* Yuppie, *der*

Z

Z, z [zed] *n., pl.* **Zs** *or* **Z's a)** *(letter)* Z, z, *das;* **b)** *(Math.)* z
Zaire [zɑ:'ɪə(r)] *pr. n.* Zaire *(das)*
Zambia ['zæmbɪə] *pr. n.* Sambia *(das)*
Zambian ['zæmbɪən] **1.** *adj.* sambisch. **2.** *n.* Sambier, *der*/Sambierin, *die*
zeal [zi:l] *n., no pl.* **a)** *(fervour)* Eifer, *der;* **b)** *(hearty endeavour)* Hingabe, *die*
zealous ['zeləs] *adj.* **a)** *(fervent)* glühend *(geh.)* ‹*Verehrer*›; begeistert ‹*Fan*›; **b)** *(eager)* eifrig
zebra ['zebrə] *n.* Zebra, *das*
zebra 'crossing *n.* *(Brit.)* Zebrastreifen, *der*
zed [zed] *(Brit.),* **zee** [zi:] *(Amer.)* *ns.* Zett, *das*
zenith ['zenɪθ] *n.* Zenit, *der*
zero ['zɪərəʊ] *n., pl.* **~s a)** *(nought)* Null, *die;* **b)** *(fig.: nil)* null; **her chances are ~:** ihre Aussichten sind gleich Null *(ugs.);* **c)** *(starting point of scale; of temperature)* Null, *die;* **in ~ gravity** im Zustand der Schwerelosigkeit; **absolute ~** *(Phys.)* absoluter Nullpunkt; **d) ~ [hour]** die Stunde X
zero-'rated *adj.* **~ goods** nicht mehrwertsteuerpflichtige Güter
zest [zest] *n.* **a)** *(lit. or fig.)* Würze, *die;* **add a ~ to the dish** das Gericht würzig machen; **add ~ and life to sth.** etw. beleben; **b)** *(gusto)* Begeisterung, *die;* **~ for living** Lebenslust, *die*
zigzag ['zɪgzæg] **1.** *adj.* zickzackförmig; Zickzack‹*muster, -anordnung*›; **~ line** Zickzacklinie, *die.* **2.** *adv.* zickzack. **3.** *n.* Zickzacklinie, *die*
zilch [zɪltʃ] *n., no pl., no art.* *(esp. Amer. sl.)* rein *od.* reineweg gar nichts *(ugs.);* **be ~:** gleich Null sein *(ugs.)*

Zimbabwe [zɪm'bɑːbwɪ] *pr. n.* Simbabwe *(das)*

Zimbabwean [zɪm'bɑːbwɪən] **1.** *adj.* simbabwisch. **2.** *n.* Simbabwer, *der*/Simbabwerin, *die*

zinc [zɪŋk] *n.* Zink, *das*

Zionism ['zaɪənɪzm] *n., no pl.* Zionismus, *der*

Zionist ['zaɪənɪst] *n.* Zionist, *der*/Zionistin, *die*

zip [zɪp] **1.** *n.* **a)** Reißverschluß, *der;* **b)** *(fig.: energy, vigour)* Schwung, *der.* **2.** *v.t.,* **-pp-: a)** *(close)* ~ **[up] sth.** den Reißverschluß an etw. *(Dat.)* zuziehen *od.* zumachen; ~ **sb. up** jmdm. den Reißverschluß zuziehen *od.* zumachen; **b)** ~ **[up]** *(enclose)* [durch Schließen des Reißverschlusses] einpacken *(ugs.).* **3.** *v.i.,* **-pp-: a)** *(fasten)* ~ **[up]** mit Reißverschluß geschlossen werden; **the dress ~s up [at the back/side]** das Kleid hat [hinten/seitlich] einen Reißverschluß; **b)** *(move fast)* sausen

zip: ~-bag *n.* Tasche mit Reißverschluß; **Zip code** *n. (Amer.)* Postleitzahl, *die;* **~-fastener** *see* ~ **1 a**

zipper ['zɪpə(r)] *see* **zip 1 a**

zit [zɪt] *n. (esp. Amer. sl.)* Pickel, *der*

zither ['zɪðə(r)] *n. (Mus.)* Zither, *die*

zodiac ['zəʊdɪæk] *n. (Astron.)* Tierkreis, *der;* **sign of the ~** *(Astrol.)* Tierkreiszeichen, *das;* Sternzeichen, *das*

zombie (*Amer.:* **zombi**) ['zɒmbɪ] *n. (lit. or fig.)* Zombie, *der*

zone [zəʊn] *n.* Zone, *die;* **[time] ~:** Zeitzone, *die*

zonked [zɒŋkt] *adj. (sl.)* **be ~** *(by drugs)* stoned sein *(Drogenjargon); (by alcohol)* zusein *(salopp); (be tired)* erschlagen sein *(ugs.)*

zoo [zuː] *n.* Zoo, *der*

'zoo-keeper *n.* Zoowärter, *der*/-wärterin, *die*

zoological [zəʊə'lɒdʒɪkl] *adj.* zoologisch

zoological 'garden[s] *n.* zoologischer Garten

zoologist [zəʊ'ɒlədʒɪst] *n.* Zoologe, *der*/Zoologin, *die*

zoology [zəʊ'ɒlədʒɪ] *n.* Zoologie, *die*

zoom [zuːm] *v.i.* rauschen; **we ~ed along on our bicycles** wir sausten auf unseren Fahrrädern daher

~ **'in** *v.i.* **a)** *(Cinemat., Telev.)* zoomen *(fachspr.);* nahe heranfahren; ~ **in on sth.** auf etw. *(Akk.)* zoomen *(fachspr.);* etw. nahe heranholen; **b)** ~ **in on sth.** *(fig.)* sich auf etw. *(Akk.)* konzentrieren

'zoom lens *n. (Photog.)* Zoomobjektiv, *das;* Gummilinse, *die (ugs.)*

zucchini [zʊ'kiːnɪ] *n., pl. same or* **~s** *(esp. Amer.)* Zucchino, *der*

Zurich ['zjʊərɪk] **1.** *pr. n.* Zürich *(das).* **2.** *attrib. adj.* **a)** *(of canton)* des Kantons Zürich *nachgestellt;* **b)** *(of city)* Züricher; Zürcher *(schweiz.)*

A

a, A [a:] **das; ~, ~ a)** *(Buchstabe)* a/A; **kleines a** small a; **großes A** capital A; **das A und O** *(fig.)* the essential thing/things (*Gen.* for); **von A bis Z** *(fig. ugs.)* from beginning to end; **wer A sagt, muß auch B sagen** *(fig.)* if one starts a thing, one must go through with it; **b)** *(Musik)* [key of] A

ä, Ä [ɛ:] **das; ~, ~:** a umlaut

a *Abk.* **Ar, Are**

à [a] *Präp. mit Nom., Akk. (Kaufmannsspr.)* **zehn Marken à 50 Pfennig** ten stamps at 50 pfennigs each; **zehn Kisten à zwölf Flaschen** ten cases of twelve bottles each

A *Abk.* **a) Autobahn; b) Ampere** A

Aa [a'|a] **das; ~** *(Kinderspr.)* poo[-poo] *(child lang.)*; **Aa machen** do poo-poo *or* big jobs *(child lang.)*; do a big job *(Amer. child lang.)*

[1]**AA** *Abk.* **Anonyme Alkoholiker** AA

[2]**AA** [a:'|a:] **das; ~** *Abk.* **Auswärtiges Amt**

Aal [a:l] **der; ~[e]s, ~e** eel; **glatt wie ein ~ sein** be as slippery as an eel; **sich [drehen und] winden** *od.* **krümmen wie ein ~:** twist and turn like an eel

aalen *refl. V. (ugs.)* **sich am Strand/in der Sonne ~:** lie stretched out on the beach/in the sun; **er ging an den Strand, um sich in der Sonne zu ~:** he went to the beach to stretch out in the sun

aal·glatt *(abwertend)* **1.** *Adj.* slippery; **~ sein** be as slippery as an eel. **2.** *adv.* smoothly

Aas [a:s] **das; ~es, ~e** *od.* **Äser** ['ɛ:zɐ] **a)** *o. Pl.* carrion *no art.*; **b)** *Pl.* **~e** [rotting] carcass; **c)** *Pl.* **Äser** *(salopp) (abwertend)* swine; *(anerkennend)* devil; **kein ~:** not one damned person

aasen *itr. V. (ugs., bes. nordd.)* **mit etw. ~:** be wasteful with sth.

Aas-: ~fliege die *(Zool.)* blowfly; **~fresser der** *(Zool.)* carrion-eater; scavenger; **~geier der** vulture; **wie [die] ~geier** *(abwertend)* like vultures

aasig *Adv. (ugs.)* **~ kalt** damned cold; **es tut ~ weh** it hurts like mad *(sl.)*

ab [ap] **1.** *Präp. mit Dat.* **a)** *(zeitlich)* from; **ab 1980** as from 1980; **Jugendliche ab 16 Jahren** young people over the age of 16; **ab [dem] 3. April** from the 3rd of April; **b)** *(bes. Kaufmannsspr.: räumlich)* ex; **ab Werk** ex works; **ab Frankfurt fliegen** fly from Frankfurt; **c)** *([Rang]folge)* from ... on[wards]; **ab 20 DM** from 20 DM [upwards]. **2.** *Adv.* **a)** *(weg)* off; away; **nicht weit ab vom Weg** not far [away] from the path; **an der Kreuzung links ab** turn off left at the junction; **b)** *(ugs.: Aufforderung)* off; away; **ab nach Hause** get off home; **ab die Post** *(fig.)* off you/we *etc.* go; **ab nach Kassel** *(fig.)* it's off and away; **c)** *(milit. Kommando)* **Gewehr ab!** order arms!; **d) ab und zu** *od. (norddt.)* **an** now and then; from time to time; *s. auch* **absein; auf 3e, f; von 1a, b**

Abakus ['a(:)bakʊs] **der; ~, ~:** abacus

ab|ändern *tr. V.* alter; change; amend ⟨*text*⟩

ab|arbeiten 1. *tr. V.* **a)** *(abgelten)* work for ⟨*meal*⟩; work off ⟨*debt, amount*⟩; **b)** *(abnutzen)* wear out [with work]; *s. auch* **abgearbeitet. 2.** *refl. V.* slave [away]; work like a slave

Ab·art die variety

ab·artig *Adj.* deviant; abnormal

Ab·artigkeit die abnormality; deviancy

ab|asten *refl. V. (ugs.)* slave [away]; **sich mit etw. ~:** heave sth. around

Abb. *Abk.* **Abbildung** Fig.

Ab·bau der a) dismantling; *(von Zelten, Lagern)* striking; **b)** *(Senkung)* reduction; **c)** *s.* **abbauen 1d:** cutback (*Gen.* in); pruning; **der ~ von Vorurteilen** the breaking down of prejudices; **d)** *(Chemie, Biol.)* breakdown; **e)** *(Bergbau) s.* **abbauen 1f:** mining; quarrying; working

ab|bauen 1. *tr. V.* **a)** dismantle; strike ⟨*tent, camp*⟩; dismantle, take down ⟨*scaffolding*⟩; **b)** *(senken)* reduce ⟨*wages*⟩; **c)** *(beseitigen)* gradually remove; break down ⟨*prejudices, inhibitions*⟩; **d)** *(verringern)* cut back ⟨*staff*⟩; prune ⟨*jobs*⟩; **e)** *(Chemie, Biol.)* break down ⟨*carbohydrates, alcohol*⟩; **f)** *(Bergbau)* mine ⟨*coal, gold*⟩; quarry ⟨*stone*⟩; work ⟨*seam*⟩. **2.** *itr. V.* fade; slow down; **körperlich ~:** decay physically

ab|beißen 1. *unr. tr. V.* bite off. **2.** *unr. itr. V.* have a bite

ab|beizen *tr. V. (Handw.)* strip ⟨*wooden object*⟩

ab|bekommen *unr. tr. V.* **a)** get; **sie hat keinen Mann ~** *(ugs.)* she didn't catch herself a husband; **b)** *(hinnehmen müssen)* **einen Schlag/ein paar Kratzer ~:** get hit/get a few scratches; **etwas ~** *(getroffen werden)* get *or* be hit; *(verletzt werden)* get *or* be hurt; **der Wagen hat nichts ~:** the car wasn't damaged; **c)** *(entfernen können)* get ⟨*paint, lid, chain*⟩ off

ab|berufen *unr. tr. V.* recall ⟨*ambassador, envoy*⟩ **(aus, von** from)

Ab·berufung die recall

ab|bestellen *tr. V.* cancel

Ab·bestellung die cancellation

ab|betteln *tr. V. (ugs.)* **jmdm. etw. ~:** beg sth. from sb.

ab|bezahlen *tr. V.* pay off

ab|biegen 1. *unr. itr. V.; mit sein* turn off; **links ~:** turn [off] left. **2.** *unr. tr. V.* **a)** bend ⟨*rod, metal sheet, etc.*⟩; **b)** *(ugs.: abwenden)* get out of *(coll.)* ⟨*obligation*⟩; head off *(coll.)* ⟨*row*⟩

Abbiege·spur die turning-off lane

Ạb·bild das *(eines Menschen)* likeness; *(eines Gegenstandes)* copy; *(fig.)* portrayal

ạb|bilden *tr. V.* copy; reproduce ⟨*object, picture*⟩; portray, depict ⟨*person*⟩; depict ⟨*landscape*⟩; *(fig.)* portray; depict

Ạb·bildung die a) *(Bild)* illustration; *(Schaubild)* diagram; **die ~ einer Frau** a/the picture of a woman; **~ 5** *(in einem Buch)* figure *or* fig. 5; **b)** *o. Pl. (das Abbilden)* reproduction; *(fig.)* portrayal

ạb|binden *unr. tr. V.* **a)** *(losbinden)* untie; undo; **eine Schnur ~:** untie a piece of string; **b)** *(abschnüren)* put a tourniquet on ⟨*artery, arm, leg, etc.*⟩; tie ⟨*umbilical cord*⟩

Ạb·bitte die *(geh.)* **jmdm. ~ leisten** *od.* **tun** ask sb.'s pardon

ạb|blasen *unr. tr. V. (ugs.: absagen)* call off ⟨*enterprise, party*⟩

ạb|blättern *itr. V.; mit sein* flake off

ạb|bleiben *unr. itr. V.; mit sein (ugs., bes. nordd.)* **wo ist er/es nur abgeblieben?** where has he/it got to *(Brit.) or (Amer.)* gone?; where can he/it be?

ạb|blenden 1. *tr. V.* dip *(Brit.)*, dim *(Amer.)* ⟨*headlights*⟩. **2.** *itr. V.* dip *(Brit.) or (Amer.)* dim one's headlights

Ạbblend·licht das; *o. Pl.* dipped *(Brit.) or (Amer.)* dimmed beam; **mit ~ fahren** drive on dipped *or* dimmed headlights

ạb|blitzen *itr. V.; mit sein (ugs.)* **sie ließ alle Verehrer ~:** she gave all her admirers the brush-off; **bei jmdm. [mit etw.] ~:** fail to get anywhere [with sth.] with sb. *(coll.)*

ạb|blocken *tr. V. (Sport, fig.)* block

ạb|brausen 1. *tr. V. s.* **abduschen. 2.** *itr. V.; mit sein (ugs.)* roar off

ạb|brechen 1. *unr. tr. V.* **a)** break off; *(durchbrechen)* break ⟨*needle, pencil*⟩; **sich** *(Dat.)* **einen Fingernagel/Zahn ~:** break a fingernail/a tooth; **b)** *(abbauen)* strike ⟨*tent, camp*⟩; **c)** *(abreißen)* demolish, pull down ⟨*building, tower*⟩; **d)** *(beenden)* break off ⟨*negotiations, [diplomatic] relations, discussion, connection, activity*⟩; *(vorzeitig, wider Erwarten)* cut short ⟨*conversation, studies, holiday, activity*⟩; **den Kampf ~** *(Boxen)* stop the fight. **2.** *unr. itr. V.* **a)** *mit sein (entzweigehen)* break [off]; **b)** *(aufhören)* break off; **c)** *mit sein (beendet werden)* **die Verbindung brach ab** the connection was cut off. **3.** *unr. refl. V.* **sich** *(Dat.)* **einen/keinen ~** *(salopp)* put/not put oneself out

ạb|bremsen 1. *tr. V.* **a)** brake; **den Wagen ~:** slow the car down; **b)** retard ⟨*motion*⟩; break ⟨*fall*⟩. **2.** *itr. V.* brake; apply the brakes

ạb|brennen 1. *unr. itr. V.; mit sein* **a)** be burned down; ⟨*farm*⟩ be burned out; **das Haus ist abgebrannt** the house has burned down; **wir sind schon zweimal abgebrannt** we've been burned out twice already; *s. auch* **Grundmauer; b)** *(sich aufbrauchen)* ⟨*fuse*⟩ burn away; ⟨*candle*⟩ burn down; **abgebrannte Streichhölzer** used *or* burnt matches. **2.** *unr. tr. V.* **a)** let off ⟨*firework*⟩; **b)** burn down ⟨*building*⟩

ạb|bringen *unr. tr. V.* **jmdn. von etw. ~:** make sb. give up sth.; **jmdn. vom Kurs ~:** make sb. change course; **jmdn. von der Fährte ~:** throw sb. off the scent; **jmdn. davon ~, etw. zu tun** stop sb. doing sth.; *(durch Worte)* dissuade sb, from doing sth.; **jmdn. vom Thema ~:** get sb. away from the subject

ạb|bröckeln *itr. V.; mit sein (auch fig.)* crumble away; ⟨*price, exchange rate*⟩ decline gradually

Ạb·bruch der a) *o. Pl. (Abriß)* demolition; pulling down; **b)** *(Beendigung)* breaking-off; *(Boxen)* stopping; **c) einer Sache** *(Dat.)* **[keinen] ~ tun** do [no] harm to sth.

ạbbruch-, Ạbbruch-: ~firma die demolition firm; **~haus das** condemned house; **~reif** *Adj.* ripe for demolition *postpos.*

ạb|brummen *tr. V. (ugs.)* do *(coll.)* ⟨*prison sentence, time in prison*⟩

ạb|buchen *tr. V.* ⟨*bank*⟩ debit (von to); ⟨*creditor*⟩ claim by direct debit (von to); **etw. ~ lassen** *(durch die Bank)* pay sth. by standing order; *(durch Gläubiger)* pay sth. by direct debit

Ạb·buchung die debiting

ạb|bürsten *tr. V.* **a)** brush off; **etw. von etw. ~:** brush sth. off sth.; **jmdm. die Haare/den Schmutz ~:** brush the hairs/the dirt off sb.; **b)** *(säubern)* brush ⟨*garment*⟩

ạb|büßen *tr. V.* serve [out] ⟨*prison sentence*⟩

Abc [a(:)be(:)'tse:] **das; ~ a)** ABC; **b)** *(fig.: Grundlagen)* ABC; fundamentals *pl.*

ạb|checken *tr. V.* check

Abc-Schütze der child just starting school

ABC-Waffen *Pl.* ABC weapons

ạb|dampfen *itr. V.; mit sein (ugs.: abfahren)* set off

ạb|danken *itr. V.* ⟨*monarch, ruler*⟩ abdicate; ⟨*government, minister*⟩ resign

Ạbdankung die; ~, ~en *(eines Herrschers)* abdication; *(eines Ministers, einer Regierung)* resignation

ạb|decken 1. *tr. V.* **a)** open up; uncover ⟨*container*⟩; ⟨*gale*⟩ take the roof/roofs off ⟨*house*⟩, take the tiles off ⟨*roof*⟩; **b)** *(herunternehmen, -reißen)* take off; remove; **c)** *(abräumen)* clear ⟨*table*⟩; clear away ⟨*dishes*⟩; **d)** *(schützen)* cover ⟨*person*⟩; *(Schach)* defend; **e)** *(Sport)* mark ⟨*player*⟩; **f)** *(bezahlen, ausgleichen)* cover; meet ⟨*need, demand*⟩

Ạbdecker der; ~s, ~ *(veralt.)* knacker *(Brit.)*

Abdeckerei die; ~, ~en *(veralt.)* knacker's yard *(Brit.)*

Ạb·deckung die covering

ạb|dichten *tr. V.* seal; seal, stop up ⟨*hole, crack, gap, etc.*⟩; plug ⟨*leak*⟩; *(gegen Zugluft)* draughtproof ⟨*door, window*⟩

ạb|drängen *tr. V.* push away; force away; drive ⟨*animal*⟩ away; **jmdn. von etw. ~:** push sb. away from sth.; **einen Spieler vom Ball ~** *(Fußball)* force a player off the ball

ạb|drehen 1. *tr. V.* **a)** *(ausschalten)* turn off; turn *or* switch off ⟨*light, lamp, electricity, fire, radio*⟩; **den Hahn ~** *(fig.)* turn off the supply; **b)** *(abtrennen)* twist off; *s. auch* **Gurgel; c)** *(abschrauben)* screw off ⟨*lid, top*⟩. **2.** *itr. V.; meist mit sein (die Richtung ändern)* turn off

ạb|driften *itr. V.; mit sein (Seew.)* be blown; make leeway *(Naut.)*

[1]**Ạb·druck der;** *Pl.* **Ạbdrücke** mark; imprint; *(Finger~)* fingermark; *(Fuß~)* footprint; footmark; *(Wachs~)* impression; *(Gips~)* cast

[2]**Ạb·druck der;** *Pl.* **~e a)** *o. Pl. (Vorgang)* printing; **b)** *(Ergebnis) (einer Graphik)* print

ạb|drucken *tr. V.* print; *(veröffentlichen)* publish

ạb|drücken 1. *itr. V.* pull the trigger; shoot; **auf jmdn./etw. ~:** shoot *or* fire at sb./sth. **2.** *tr. V.* **a)** *(abfeuern)* fire ⟨*revolver, gun*⟩; **b)** *(zudrücken)* constrict; **c) jmdm. die Luft ~:** stop sb. breathing. **3.** *refl. V.* **a) sich [in etw.** *(Dat.)***] ~:** make marks [in sth.]; ⟨*track*⟩ be imprinted [in sth.]; **b) sich [mit**

dem Fuß] ~: push oneself away with one's foot

ab|dunkeln *tr. V.* darken ⟨*room*⟩; dim ⟨*light*⟩

ab|duschen *tr. V.* **sich/jmdn. [kalt/warm]** ~: take/give sb. a [cold/hot] shower; **sich/jmdm. den Rücken** ~: shower one's/sb.'s back

ab|ebben *itr. V.; mit sein* recede; abate

abend ['a:bn̩t] *Adv.* **heute/morgen/gestern** ~: this/tomorrow/yesterday evening; tonight/tomorrow night/last night *(coll.)*; **Sonntag** ~: [on] Sunday evening *or (coll.)* night; **was gibt es heute** ~ **[zu essen]?** what's for dinner/supper?

Abend der; ~s, ~e a) evening; **es wird** ~: evening is drawing in; **guten** ~! good evening; **eines [schönen] ~s** one evening; **am [frühen/späten]** ~: [early/late] in the evening; **am** ~ **vorher** *od.* **zuvor** the evening *or (coll.)* night before; the previous evening; **bis zum [späten]** ~: until [late in the] evening; *(als Frist)* by [late] evening; **am selben/nächsten** ~: the same/following evening *or (coll.)* night; **gegen** ~: towards evening; **zu** ~ **essen** have dinner; *(allgemeiner)* have one's evening meal; *s. auch* **heilig b; Tag a; b)** *(Geselligkeit)* evening; *(Kultur~)* soirée; **ein bunter** ~: a social [evening] *or (coll.)* night

abend-, Abend-: ~**an·zug der** dinner dress; evening suit; ~**blatt das** evening [news] paper; ~**brot das** evening meal; supper; **wann gibt es ~brot?** when's supper?; ~**dämmerung die** [evening] twilight; ~**essen das** dinner; ~**füllend** *Adj.* occupying a whole evening *postpos., not pred.*; **ein ~füllendes Programm** a full evening's programme; ~**gebet das** evening prayers *pl.*; *(von Kindern)* bedtime prayers *pl.*; ~**kasse die** box-office *(open on the evening of the performance)*; ~**kleid das** evening dress *or* gown; ~**kurs[us] der** evening class *or* course; ~**land das**; *o. Pl.* West; Occident *(literary)*; ~**ländisch** [-lɛndɪʃ] *Adj.* Western; Occidental *(literary)*

abendlich *Adj.; nicht präd.* evening; ⟨*quiet, coolness*⟩ of the evening; **die ~en Straßen der Stadt** the streets of the town at evening

Abend·mahl das; *o. Pl.* **a)** *(christl. Rel.)* Communion; **das ~ nehmen** receive Communion; **b)** *(N. T.)* Last Supper

Abend-: ~**mahl·zeit die** evening meal; ~**nachrichten** *Pl.* evening news *sing.*; ~**programm das** evening programmes *pl.*; ~**rot das** red glow of the sunset sky

abends *Adv.* in the evenings; **sechs Uhr** ~: six o'clock in the evening; **Montag** *od.* **montags** ~: [on] Monday evenings; **von morgens bis** ~: from morning to night; **spät** ~: late in the evening

Abend-: ~**schule die** night school; evening classes *pl.*; ~**schüler der** student at evening classes; ~**sonne die** evening sun; ~**stern der** evening star; ~**stunde die** evening hour; **in den frühen/späten ~stunden** early/late in the evening; ~**vorstellung die** evening performance; ~**zeitung die** *s.* ~**blatt**

Abenteuer ['a:bn̩tɔyɐ] **das; ~s, ~ a)** *(auch fig.)* adventure; **b)** *(Unternehmen)* venture; **c)** *(Liebesaffäre)* affair

abenteuerlich *Adj.* **a)** *(riskant)* risky; hazardous; **b)** *(bizarr)* bizarre

Abenteuer-: ~**roman der** adventure novel; ~**spielplatz der** adventure playground

Abenteurer ['a:bn̩tɔyrɐ] **der; ~s, ~:** adventurer

Abenteurerin die; ~, ~nen adventuress

aber ['a:bɐ] **1.** *Konj.* but; **wir ~ ...:** we, however, ...; ~ **trotzdem** but in spite of that; **oder** ~: or else; ~ **warum denn?** but why?; **das stimmt ~ nicht** but that's not right. **2.** *Adv. (veralt.: wieder)*; ~ **und abermals** again and again; time and again; **tausend und ~ tausend** thousands upon thousands. **3.** *Partikel* **das ist ~ schön!** why, isn't that nice!; ~ **ja/nein!** why, yes/no! ~ **natürlich!** but *or* why of course!; **das ist ~ auch zu dumm** it's just 'too stupid *or (Amer.)* dumb; **du bist ~ groß!** aren't you tall!; ~, ~! now, now!

Aber·glaube[n] der a) superstition; **b)** *(Vorurteil)* myth

aber·gläubisch [-glɔybɪʃ] *Adj.* superstitious

Aber·hunderte *Pl. (geh.)* hundreds [upon hundreds]

ab|erkennen *unr. tr. V.* **jmdm. ein Recht** ~: revoke sb.'s right; *(Sport)* **jmdm. den Sieg/Titel** ~: disallow sb.'s victory/strip sb. of his/her title

abermalig *Adj.; nicht präd.* renewed

abermals ['a:bɐma:ls] *Adv.* once again; once more

ab|ernten *tr. V.* finish the harvesting of; finish harvesting *or* picking ⟨*fruit*⟩

Aber·tausende *Pl. (geh.)* thousands [upon thousands]

aber·witzig *Adj.* crazy

ab|essen *unr. tr. V.* **a) etw. [von etw.]** ~: eat sth. off [sth.]; **b)** *(leer essen)* clear ⟨*plate, table*⟩; **abgegessene Teller** empty plates

Abf. *Abk.* **Abfahrt** dep.

ab|fahren 1. *unr. itr. V.; mit sein* **a)** *(wegfahren)* leave; depart; **wo fährt der Zug nach Paris ab?** where does the Paris train leave from?; **b)** *(hinunterfahren)* drive down; *(Skisport)* ski *or* go down; **c)** *(salopp)* **auf jmdn./etw. [voll]** ~: be mad about sb./sth.; **d)** *(salopp)* **jmdn. ~ lassen** tell sb. where he/she can go *(sl.)*; **bei jmdm. ganz schön** ~: get absolutely nowhere with sb. *(coll.)*. **2.** *unr. tr. V.* **a)** take away; **b)** *(abnutzen)* wear out; **abgefahrene Reifen** worn tyres; **c)** *auch mit sein (entlangfahren)* drive the whole length of ⟨*street, route*⟩; drive through ⟨*district*⟩

Ab·fahrt die a) departure; **b)** *(Skisport)* descent; *(Strecke)* run; **c)** *(Autobahn~)* exit

abfahrt·bereit *s.* **abfahrbereit**

Abfahrts-: ~**lauf der** *(Skisport)* downhill [racing]; ~**läufer der** *(Skisport)* downhill racer

Abfahrt[s]·zeit die time of departure; departure time

Ab·fall der a) rubbish, *(Amer.)* garbage *or* trash *no indef. art., no pl.*; *(Fleisch~)* offal *no indef. art., no pl.*; *(Industrie~)* waste *no indef. art.*; *(auf der Straße)* litter *no indef. art., no pl.*; **b)** *o. Pl. (Rückgang)* drop (*Gen.*, **in** + *Dat.* in)

Abfall·eimer der rubbish *or* waste bin; trash *or* garbage can *(Amer.)*; *(auf der Straße)* litter bin; trash *or* litter basket *or* can *(Amer.)*

ab|fallen *unr. itr. V.; mit sein* **a)** *(ugs.)* **wieviel fällt für jeden ab?** what will each person's share be?; **für dich wird auch eine Kleinigkeit** ~: you'll get something out of it too; **b)** *(herunterfallen)* fall off; **von jmdm.** ~ *(fig.)* leave sb. **c)** *(sich lossagen)* ⟨*country*⟩ secede; **vom Glauben/von jmdm.** ~: desert the faith/sb.; **d)** *(nachlassen)* drop; **e)** *(bes. Sport: zurückfallen)* drop *or* fall back; **f)** *(sich senken)* ⟨*land, hillside, road*⟩ drop away; slope; **g)** *(im Vergleich)* **gegenüber jmdm./etw.** *od.* **gegen jmdn./etw. stark** ~: be markedly inferior to sb./sth.

ạb·fällig 1. *Adj.* disparaging; derogatory. **2.** *adv.* **sich ~ über jmdn. äußern** make disparaging *or* derogatory remarks about sb.

Ạbfall·produkt das *(auch fig.)* by-product

ạb|fangen *unr. tr. V.* **a)** intercept ⟨*agent, message, aircraft*⟩; **b)** *(auf-, anhalten)* catch; **c)** *(abwehren)* repel ⟨*charge, assault*⟩; ward off ⟨*blow, attack*⟩; *(fig.)* stop ⟨*development*⟩; cushion ⟨*impact*⟩; *(unter Kontrolle bringen)* get ⟨*vehicle, aircraft*⟩ under control

ạb|färben *itr. V.* ⟨*colour, garment, etc.*⟩ run; **auf jmdn./etw. ~** *(fig.)* rub off on sb./sth.

ạb|fassen *tr. V.* write ⟨*report, letter, etc.*⟩; draw up ⟨*will*⟩

ạb|fegen *tr. V.* **a)** brush off; **etw. von etw. ~:** brush sth. off sth.; **b)** *(säubern)* **etw. ~:** brush sth. clean

ạb|feilen *tr. V.* **a)** *(entfernen)* file off; **b)** *(verkürzen, glätten)* file down

ạb|fertigen *tr. V.* **a)** handle, dispatch ⟨*mail*⟩; deal with ⟨*applicant, application*⟩; deal with, handle ⟨*passengers*⟩; serve ⟨*customer*⟩; clear ⟨*ship*⟩ for sailing; clear ⟨*aircraft*⟩ for take-off; clear ⟨*lorry*⟩ for departure; *(kontrollieren)* clear; check; **b)** *(ugs.: unfreundlich behandeln)* **jmdn. [grob/barsch] ~:** [roughly/rudely] turn sb. away

Ạb·fertigung die *s.* **abfertigen a:** handling; dispatching; serving; clearing for sailing/take-off/departure; *(Kontrolle)* clearance; checking

ạb|feuern *tr. V.* fire; **Schüsse/eine Kanone [auf jmdn./etw.] ~:** fire shots/a cannon [at sb./sth.]

ạb|finden 1. *unr. tr. V.* **jmdn. mit etw. ~:** compensate sb. with sth.; **seine Gläubiger ~:** settle with one's creditors; **er wurde großzügig abgefunden** he received a generous settlement. **2.** *unr. refl. V.* **sich ~:** resign oneself; **sich mit etw. ~:** come to terms with sth.

Ạbfindung die; ~, ~en a) *(Summe)* settlement; **eine ~ zahlen** make a settlement; **b)** *(Vorgang) (Entschädigung)* compensation; *(von Gläubigern)* paying-off

ạb|flachen ['apflaxn̩] *tr. V.* flatten [out]

ạb|flauen *itr. V.; mit sein* die down; subside; ⟨*interest, conversation*⟩ flag; ⟨*business*⟩ become slack

ạb|fliegen 1. *unr. itr. V.; mit sein* ⟨*person*⟩ leave [by aeroplane]; ⟨*aircraft*⟩ take off; ⟨*bird*⟩ fly off *or* away; **die Maschine nach Brüssel fliegt um 13[30] Uhr ab** the plane for Brussels leaves at 13.30. **2.** *unr. tr. V.* fly over ⟨*district*⟩; fly along ⟨*road*⟩

ạb|fließen *unr. itr. V.; mit sein* **a)** flow off; *(wegfließen)* flow away; **aus etw. ~:** drain away from sth.; **von etw. ~** run off sth.; **b)** *(sich leeren)* empty

Ạb·flug der departure

Ạbflug·zeit die departure time

Ạb·fluß der a) drain; *(von Gewässern)* outlet; *(Rohr)* drain-pipe; *(für Abwasser)* waste-pipe; **b)** *o. Pl. (das Abfließen)* draining away

Ạbfluß·rohr das outlet pipe

Ạb·folge die sequence

ạb|fragen *tr. V.* **a)** test; **jmdn.** *od.* **jmdm. die Vokabeln ~:** test sb. on his/her vocabulary; **b)** *(DV)* retrieve, read out ⟨*data*⟩; interrogate ⟨*measuring-instrument, store*⟩

ạb|fressen *unr. tr. V.* **a) etw. [von etw.] ~:** eat sth. off [sth.]; **b)** *(leer fressen)* strip ⟨*tree, stem, etc.*⟩ bare

ạb|frieren 1. *unr. itr. V.; mit sein* **die Ohren froren ihm ab** he lost his ears through frostbite. **2.** *unr. refl. V.* **sich** *(Dat.)* **etw. ~:** lose sth. by frostbite; **sich** *(Dat.)* **einen ~** *(ugs.)* freeze to death *(coll.)*

ạb|frottieren *tr. V.* rub down [thoroughly]

ạb|fühlen *tr. V.* feel

Ạbfuhr die; ~, ~en a) removal; **b) jmdm. eine ~ erteilen** rebuff sb.; turn sb. down; **sich eine ~ holen** be rebuffed *or* turned down

ạb|führen 1. *tr. V.* **a)** *(nach Festnahme)* take away; **b)** *(zahlen)* pay out; pay ⟨*taxes*⟩; **c)** *auch itr. (abbringen)* take away. **2.** *itr. V. (für Stuhlgang sorgen)* be a laxative; have a laxative effect; **ein ~des Mittel** a laxative

Ạbführ·mittel das laxative; *(stärker)* purgative

ạb|füllen *tr. V.* fill ⟨*sack, bottle, barrel*⟩; **Wein in Flaschen ~:** bottle wine; **Bier in Dosen ~:** can beer

ạb|füttern *tr. V.* line ⟨*jacket, coat*⟩

Ạb·gabe die a) handing in; *(eines Briefes, Pakets, Telegramms)* delivery; *(eines Gesuchs, Antrags)* submission; **b)** *(Steuer, Gebühr)* tax; *(auf Produkte)* duty; *(Gemeinde~)* rate; *(Beitrag)* contribution; **c)** *(Ausstrahlung)* release; emission; **d)** *(Sport: Abspiel)* pass; **e)** *o. Pl. (das Abfeuern)* firing; **f)** *(von Erklärungen)* giving; *(von Urteilen, Aussagen)* making; *(Stimm~)* casting

ạbgabe·pflichtig *Adj.* ⟨*person, business, trade*⟩ liable to tax; ⟨*product*⟩ subject to duty

Ạb·gang der a) leaving; departure; *(Abfahrt)* departure; *(Theater)* exit; *(fig.)* departure; **sich einen guten ~ verschaffen** *(fig.)* make a good exit; **b)** *(jmd., der ausscheidet)* departure; *(Schule)* leaver; **c)** *(bes. Amtsspr.: Todesfall)* death; **d)** *(Turnen)* dismount; **e)** *o. Pl. (Ausscheidung)* passing; *(von Eiter, Würmern)* discharge; **f)** *(Med.: Fehlgeburt)* miscarriage; **g)** *o. Pl. (Absendung)* dispatch

Ạbgangs·zeugnis das *(Schulw.)* ≈ leaving certificate

Ạb·gas das exhaust; **~e** exhaust fumes

ạbgearbeitet *Adj.* work-worn ⟨*hands*⟩

ạb|geben 1. *unr. tr. V.* **a)** *(aushändigen)* hand over; deliver ⟨*letter, parcel, telegram*⟩; hand in, submit ⟨*application*⟩; hand in ⟨*school work*⟩; **etw. bei jmdm. ~:** deliver sth. *or* hand sth. over to sb.; **den Mantel an der Garderobe ~:** leave one's coat in the cloakroom; **b)** *auch itr. (abtreten)* **jmdm. [etwas] von etw. ~:** let sb. have some of sth.; **den Vorsitz/die Spitze ~:** give up the chair/the leadership; **einen Punkt/Satz/eine Runde ~** *(Sport)* drop a point/set/round; **c)** *(abfeuern)* fire; **d)** *(ausstrahlen)* emit ⟨*radiation*⟩; radiate ⟨*heat*⟩; give off ⟨*gas*⟩; transmit ⟨*radio message*⟩; **e)** make ⟨*judgement, statement*⟩; cast ⟨*vote*⟩; **seine Stimme für jmdn. ~:** cast one's vote in favour of sb.; vote for sb.; **f)** *(fungieren als)* make; **eine traurige Figur ~:** cut a sorry figure; **g)** *(verkaufen)* sell; *(zu niedrigem Preis)* sell off; **gebrauchte Skier billig abzugeben** second-hand skis for sale cheap; **h)** *auch itr. (Sport: abspielen)* pass. **2.** *unr. refl. V. (sich befassen)* **sich mit jmdm./etw. ~:** spend time on sb./sth.; *(geringschätzig)* waste one's time on sb./sth.

ạb·gebrannt *Adj.; nicht attr. (ugs.)* broke *(coll.)*

ạbgebrüht *Adj. (ugs.)* hardened

ạb·gedroschen *Adj. (ugs.)* hackneyed; well-worn; trite

abgefeimt ['apgəfaimt] *Adj.* infernally cunning ⟨*villain, rogue*⟩; villainous ⟨*scheme*⟩

ạb·gegriffen *Adj.* **a)** *(abgenutzt)* battered; **b)** *(fig.: abgedroschen)* hackneyed; well-worn

ạbgehackt 1. *Adj.* clipped ⟨*speech*⟩. **2.** *adv.* ⟨*speak*⟩ in short bursts

ab·gehangen *Adj.* hung; **gut ~es Fleisch** well-hung meat

abgehärmt *Adj.* careworn; haggard

abgehärtet *Adj.* *(körperlich)* tough; *(seelisch)* callous

ab|gehen *unr. itr. V.; mit sein* **a)** *(sich entfernen)* leave; go away *or* off; *(Theater)* exit; go off; **b)** *(ausscheiden)* leave; **von der Schule ~:** leave school; **c)** *(abfahren)* ‹*train, ship, bus*› leave, depart; **d)** *(abgeschickt werden)* ‹*message, letter*› be sent [off]; **e)** *(abzweigen)* branch off; *(in andere Richtung)* turn off; **f)** *(sich lösen)* come off; ‹*spot, stain*› come out; ‹*avalanche*› come down; **g)** *(Turnen)* dismount; **h)** *(abgerechnet werden)* **von etw. ~:** have to be deducted from sth.; **i)** *(fehlen)* **jmdm. geht etw. [völlig] ab** sb. is [totally] lacking in sth.; **j) ihm ging einer ab** *(derb)* he shot his load *(coarse)*

abgehetzt *Adj.* exhausted; *(außer Atem)* breathless

abgekämpft *Adj.* worn out; exhausted

abgekartet *Adj.* *(ugs.)* prearranged; **von vornherein ~:** set up in advance; **eine ~e Sache** *od.* **ein ~es Spiel sein** be rigged in advance

abgeklärt *Adj.* serene

ab·gelegen *Adj.* remote; *(einsam)* isolated; out-of-the-way ‹*district*›; *(abgeschieden)* secluded

ab|gelten *unr. tr. V.* satisfy, settle ‹*claim*›

abgemagert *Adj.* emaciated; wasted

ab·geneigt *Adj.* **einer Sache** *(Dat.)* **~ sein** be averse to sth.; **[nicht] ~ sein, etw. zu tun** [not] be averse to doing sth.

abgenutzt *Adj.* worn ‹*tyre, chair, handle*›; well-used ‹*implement*›

Abgeordnete der/die; *adj. Dekl.* member [of parliament]; *(z. B. in Frankreich)* deputy

ab·gerissen *Adj.* ragged

Ab·gesandte der/die emissary

Ab·gesang der a) *(Abschied)* **ein ~ auf etw.** *(Akk.)* a farewell to sth.; **b)** *(geh.: letztes Werk)* swansong

abgeschabt *Adj.* shabby; worn

ab·geschieden *Adj.* secluded; *(abgelegen)* isolated

ab·geschlagen *Adj.* *(Sport)* [well] beaten; **~ auf dem neunten Tabellenplatz** in lowly ninth place

ab·geschlossen *Adj.* **a)** *(abgesondert)* secluded; solitary; **b)** *(in sich geschlossen)* enclosed; self-contained ‹*flat*›

abgeschmackt ['abgəʃmakt] *Adj.* tasteless

ab·geschnitten *Adj.* isolated; **von der Außenwelt ~:** cut off from the outside world

ab·gesehen *Adv.:* **~ von jmdm./etw.** apart from sb./sth.; **~ davon, daß ...:** apart from the fact that ...

ab·gespannt *Adj.* weary; exhausted

ab·gestanden *Adj.* **a)** *(schal)* flat; **b)** *(verbraucht)* stale

ab·gestorben *Adj.* dead ‹*branch, tree*›; numb ‹*fingers, legs, etc.*›

abgestumpft *Adj.* apathetic and insensitive ‹*person*›; deadened ‹*conscience, perception*›

abgetakelt *Adj.* *(salopp)* faded

ab·getragen *Adj.* well-worn

ab·getreten *Adj.* worn down

abgewetzt *Adj.* well-worn; battered ‹*suitcase etc.*›

ab|gewinnen *unr. tr. V.* **a) jmdm. etw. ~:** get sth. out of sb.; win sth. from sb.; **einer Sache** *(Dat.)* **etw. ~:** win *or* gain sth. from sth.; **b)** *(fig.)* **ich kann ihm/dem nichts ~:** he/it does not do anything for me *(coll.)*; *s. auch* **Geschmack a**

ab·gewogen *Adj.* carefully weighed; balanced ‹*judgement*›; carefully considered ‹*account*›

ab|gewöhnen 1. *tr. V.* **jmdm. etw. ~:** make sb. give up *or* stop sth. **2.** *refl. V.* **sich** *(Dat.)* **etw. ~:** give up *or* stop sth.; **zum Abgewöhnen [sein]** *(ugs.)* [be] awful

abgezehrt *Adj.* emaciated

abgezirkelt *Adj.* measured out

ab|gießen *unr. tr. V.* pour away ‹*liquid*›; drain ‹*potatoes*›

Ab·glanz der distant echo; pale reflection

Ab·gott der idol

abgöttisch ['apgœtɪʃ] **1.** *Adj.* idolatrous. **2.** *adv.* **jmdn. ~ verehren/lieben** idolize sb.

ab|graben *unr. tr. V.* dig out; *s. auch* **Wasser a**

ab|grasen *tr. V.* **a)** graze away ‹*pasture*›; **b)** *(ugs.: absuchen)* **etw. nach etw. ~:** comb *or* scour sth. for sth.

ab|grenzen *tr. V.* **a)** bound; **etw. gegen** *od.* **von etw. ~:** separate sth. from sth.; **b)** *(unterscheiden)* differentiate; distinguish; **sich von jmdm. ~:** differentiate oneself from sb.

Abgrenzung die; ~, ~en a) boundary; **b)** *(Unterscheidung)* differentiation

Ab·grund der a) *(Schlucht)* abyss; chasm; *(Abhang)* precipice; **b)** *(fig. geh.)* dark abyss; **die Abgründe der menschlichen Seele** the depths of the human soul

abgründig ['apgrʏndɪç] *(geh.)* *Adj.* inscrutable ‹*smile*›; hidden ‹*meaning*›; dark ‹*secret*›

abgrund·tief *Adj.* out and out

ab|gucken *(ugs.)* **1.** *tr., auch itr. V.* **[bei** *od.* **von] jmdm. etw. ~:** learn sth. by watching sb. **2.** *itr., auch tr. V.* *(abschreiben)* **[bei jmdm.] ~:** copy [from sb.]; copy [sb. else's] work

Ab·guß der cast

ab|haben *(ugs.)* *unr. tr. V.* **etwas/ein Stück** *usw.* **[von etw.] ~:** have some/a piece *etc.* [of sth.]

ab|hacken *tr. V.* chop off

ab|haken *tr. V.* tick off; check off *(Amer.)*

ab|halten *unr. tr. V.* **a) jmdn./etw. [von jmdm./etw.] ~:** keep sb./sth. off [sb./sth.]; **b) jmdn. davon ~, etw. zu tun** stop sb. doing sth.; prevent sb. from doing sth.; **c)** *(durchführen)* hold ‹*elections, meeting, referendum*›

ab|handeln *tr. V.* **a) jmdm. etw. ~:** do a deal with sb. for sth.; **b)** *(darstellen)* treat; deal with

abhanden [ap'handn̩] *Adv.* **in ~ kommen** get lost; go astray; **etw. kommt jmdm. ~:** sb. loses sth.

Ab·handlung die treatise

Ab·hang der slope; incline

¹ab|hängen *unr. itr. V.* **von jmdm./etw. ~:** depend on sb./sth.; **davon hängt sehr viel für mich ab** a lot depends on it for me

²ab|hängen 1. *tr. V.* **a)** *(abnehmen)* take down; **ein Bild von der Wand ~:** take a picture [down] off the wall; **b)** *(abkuppeln)* uncouple; **c)** *(ugs.: abschütteln)* shake off *(coll.)* ‹*pursuer, competitor*›. **2.** *itr. V.* *(den Hörer auflegen)* hang up

abhängig ['aphɛŋɪç] *Adj.* **a) von jmdm./etw. ~ sein** *(bedingt)* depend on sb./sth.; *(angewiesen)* be dependent on sb./sth.; **b)** *(süchtig)* addicted **(von** to); **c)** *(Sprachw.)* indirect *or* reported ‹*speech*›; subordinate ‹*clause*›

Abhängigkeit die; ~, ~en a) dependence; **in ~ von jmdm./etw. geraten** become dependent on sb./sth.; **b)** *(Sucht)* addiction **(von** to)

ab|härten *tr. V.* harden

ab|hauen 1. *unr. tr. V.* **a)** *(abtrennen)* chop off; **b)** *Prät. nur* haute ab *(abschlagen)* knock off. **2.** *unr. itr. V.; mit sein; Prät. nur* haute ab *(salopp: verschwinden)* beat it *(sl.)*; **hau ab!** get lost! *(sl.)*

ạb|heben 1. *unr. tr., auch itr. V.* **a)** lift off ⟨*lid, cover, etc.*⟩; **[den Hörer] ~:** answer [the telephone]; **b)** *(Kartenspiel) (teilen)* cut [the pack]; *(nehmen)* draw ⟨*card*⟩; **c)** *(von einem Konto)* withdraw ⟨*money*⟩. **2.** *unr. itr. V.* ⟨*balloon*⟩ rise; ⟨*aircraft, bird*⟩ take off; ⟨*rocket*⟩ lift off. **3.** *unr. refl. V.* stand out; contrast; **sich von** *od.* **gegen etw./von jmdm. ~:** stand out against *or* contrast with sth./sb.

ạb|heften *tr. V.* file

ạb|helfen *unr. itr. V.* **einem Mißstand ~:** put an end to an abuse; **dem ist leicht abzuhelfen** that is easily remedied

ạb|hetzen 1. *tr. V.* ride ⟨*horse etc.*⟩ to exhaustion. **2.** *refl. V.* rush *or* dash [around]

Ạb·hilfe die; *o. Pl.* action to improve matters; **~ schaffen** find a remedy; put things right

ạb|hobeln *tr. V.* plane down

ạb|holen *tr. V.* collect, pick up ⟨thing⟩; pick up, fetch ⟨*person*⟩; **ich hole Sie am Bahnhof ab** I'll pick you up at the station

ạb|holzen *tr. V.* fell ⟨*trees*⟩; clear ⟨*area*⟩ [of trees]

ạb|horchen *tr. V.* sound ⟨*chest, lungs*⟩; **jmdn. ~:** sound sb.'s chest/lungs

ạb|hören *tr. V.* **a)** *(abfragen)* **jmdm.** *od.* **jmdn. Vokabeln ~:** test sb.'s vocabulary [orally]; **b)** *(heimlich anhören)* listen to; **c)** *(überwachen)* tap ⟨*telephone, telephone conversation*⟩; bug *(coll.)* ⟨*conversation, premises*⟩; **jmdn. ~:** tap sb.'s telephone; **d)** *s.* **abhorchen**

Ạbhör·gerät das listening device; bug *(coll.)*

ạbhör·sicher *Adj.* bug-proof *(coll.)*; tap-proof ⟨*telephone*⟩

ạb|hungern *tr. V.* take off, lose ⟨*weight*⟩

Abi ['abi] **das; ~s, ~s** *(Schülerspr.) s.* **Abitur**

Abitur [abi'tu:ɐ̯] **das; ~s, ~e** Abitur *(school-leaving examination at grammar school needed for entry to higher education)*; ≈ A levels *(Brit.)*; **sein** *od.* **das ~ machen** do *or* take one's Abitur

Abiturient [abitu'ri̯ɛnt] **der; ~en, ~en, Abituriẹntin die; ~, ~nen** *sb. who is taking/has passed the 'Abitur'*

Abitụr·zeugnis das Abitur certificate

ạb|jagen *tr. V.* **jmdm. etw. ~:** finally get sth. away from sb.

Abk. *Abk.* **Abkürzung** abbr.

ạb|kämmen *tr. V. (absuchen)* comb, scour (**nach** for)

ạb|kanzeln *tr. V. (ugs.)* **jmdn. ~:** give sb. a dressing down; reprimand sb.

ạb|kapseln *tr. V.* encapsulate; **sich gegen die Umwelt ~** *(fig.)* isolate oneself from one's surroundings

ạb|kaufen *tr. V.* **a) jmdm. etw. ~:** buy sth. from sb.; **b)** *(ugs.: glauben)* **das kaufe ich dir nicht ab** I'm not buying that story *(coll.)*

[1]**ạb|kehren** *tr. V.* turn away; **sich [von jmdm./etw.] ~:** turn away [from sb./sth.]; **die uns abgekehrte Seite des Mondes/des Schiffes** the far side of the moon/the ship

[2]**ạb|kehren** *tr. V. s.* **abfegen**

ạb|kippen 1. *tr., auch itr. V. (abladen)* tip out; dump ⟨*refuse*⟩. **2.** *itr. V.; mit sein (herunterfallen)* tip over

ạb|klappern *tr. V. (ugs.)* trudge round ⟨*town, district*⟩; **alle Läden nach etw. ~:** do the rounds of all the shops looking for sth.

Ạb·klatsch der *(abwertend)* pale imitation; poor copy

ạb|klemmen *tr. V.* **a)** *(zusammenpressen)* clamp; **b)** *(lösen)* disconnect

ạb|klingen *unr. itr. V.; mit sein* **a)** *(leiser werden)* grow fainter; **b)** *(nachlassen)* subside; die away

ạb|klopfen *tr. V.* **a)** knock *or* tap off; **jmdm. etw. [von der Jacke] ~:** tap sth. off sb.['s jacket]; **b)** *(säubern)* knock/tap the dirt/snow/crumbs *etc.* off; **sich** *(Dat.)* **die Hände ~:** clap one's hands together to knock the flour/powder *etc.* off; **c)** *(untersuchen)* tap

ạb|knallen *tr. V. (salopp)* shoot down; gun down

ạb|knicken 1. *tr. V.* **a)** *(abbrechen)* snap *or* break off; **b)** *(knicken)* bend. **2.** *itr. V.; mit sein* **a)** *(abbrechen)* snap; break; **b)** *(einknicken)* bend over

ạb|knöpfen *tr. V.* **a)** unbutton; **b)** *(salopp)* **jmdm. Geld ~:** get money out of sb.

ạb|knutschen *tr. V. (ugs.)* **a)** kiss and fondle; **b)** *(sexuell)* **jmdn. ~/sich mit jmdm. ~:** smooch *(coll.) or (sl.)* neck with sb.; **sich ~:** smooch *(coll.)*; neck *(sl.)*

ạb|kochen 1. *tr. V.* boil. **2.** *itr. V. (im Freien kochen)* cook in the open air

ạb|kommandieren *tr. V.* detail; *(fig.)* detail; send; **jmdn. zum Dienst/zu einer Einheit ~:** detail sb. for duty/to a unit; **jmdn. an die Front ~:** send sb. to the front

ạb|kommen *unr. itr. V.; mit sein* **a) vom Weg ~:** lose one's way; **vom Kurs ~:** go off course; **von der Fahrbahn ~:** leave the road; **b)** *(abschweifen)* digress; **vom Thema ~:** stray from the topic; digress; **c) von etw. ~** *(etw. aufgeben)* give sth. up; **von einem Plan ~:** abandon *or* give up a plan; **d)** *(aus der Mode, außer Gebrauch kommen)* ⟨*method, clothes*⟩ go out of fashion; ⟨*tradition*⟩ disappear

Ạb·kommen das; ~s, ~: agreement; **ein ~ [über etw.** *(Akk.)***] schließen** come to an agreement [on sth.]

abkömmlich ['apkœmlɪç] *Adj.* free; available

ạb|können *unr. tr. V. (nordd.: mögen)* stand; *(vertragen)* take

ạb|koppeln *tr. V.* uncouple

ạb|kratzen 1. *tr. V.* **a)** *(entfernen) (mit den Fingern)* scratch off; *(mit einem Werkzeug)* scrape off; **b)** *(säubern)* scrape [clean]. **2.** *itr. V.; mit sein (derb)* croak *(sl.)*; snuff it *(sl.)*

ạb|kriegen *tr. V. (ugs.) s.* **abbekommen**

ạb|kühlen 1. *tr. V.* cool down; **jmds. Eifer ~** *(fig.)* dampen sb.'s ardour; **jmdn. ~** *(fig.)* cool sb. off. **2.** *itr. V.; meist mit sein* cool down; get cooler; **es hat stark abgekühlt** *(Met.)* it has become a lot cooler. **3.** *refl. V.* cool down; get cooler

Ạb·kühlung die cooling

ạb|kuppeln *tr. V. s.* **abkoppeln**

ạb|kürzen *tr., itr. V.* **a)** *(räumlich)* shorten; **den Weg ~:** take a shorter route; **b)** *(zeitlich)* cut short; **c)** *(kürzer schreiben)* abbreviate (**mit** to)

Ạb·kürzung die a) *(Weg)* short cut; **b)** *(Wort)* abbreviation; **c)** *(das Abkürzen)* cutting short; **zur ~ des Verfahrens** to shorten the procedure

ạb|küssen *tr. V.* cover with kisses

ạb|laden *unr. tr., itr. V.* **a)** unload, off-load ⟨*case, sack, barrel, goods, vehicle*⟩; dump, unload ⟨*gravel, sand, rubble*⟩; **seine Sorgen bei jmdm. ~** *(fig.)* unburden oneself to sb.

Ạb·lage die a) *(Vorrichtung)* storage place; **b)** *(Bürow.)* filing

ạb|lagern 1. *tr. V.* **a)** *(absetzen)* deposit; **b)** *(deponieren)* dump. **2.** *refl. V.* be deposited

Ạb·lagerung die a) deposit; **b)** *(das Absetzen)* deposition; **c)** *(das Deponieren)* dumping

Ablaß ['aplas] **der; Ạblasses, Ạblässe** *(kath. Rel.)* indulgence

ạb|lassen 1. *unr. tr. V.* **a)** *(ablaufen lassen)* let out (**aus** of); **b)** *(ausströmen lassen)* let off

⟨*steam*⟩*;* let out ⟨*air*⟩*;* **c)** *(leeren)* empty. **2.** *unr. itr. V.* **a)** *(aufgeben)* **von etw. ~:** give sth. up; **b) von jmdm./etw. ~:** leave sb./sth. alone

Ablativ ['ablati:f] **der; ~s, ~e** *(Sprachw.)* ablative

Ab·lauf der a) *(Verlauf)* course; **der ~ der Ereignisse** the course of events; **b)** *(einer Veranstaltung)* passing *or* going off; **c)** *o. Pl. (Ende)* **nach ~ eines Jahres** after a year; **nach ~ einer Frist** at the end of a period of time

ab|laufen 1. *unr. itr. V.; mit sein* **a)** *(abfließen)* flow away; *(herausfließen)* run *or* flow out; **b)** *(herabfließen)* run down; **von/an etw.** *(Dat.)* **~:** run off sth.; **c)** *(verlaufen)* pass *or* go off; **gut abgelaufen sein** have gone *or* passed off well; **d)** ⟨*alarm clock*⟩ run down; ⟨*parking meter*⟩ expire; **e)** ⟨*period, contract, passport*⟩ expire; **f) ~ lassen** play ⟨*tape*⟩*;* run ⟨*film*⟩ through. **2.** *unr. tr. V.* **a)** *auch mit sein (entlanglaufen)* walk all along; go over ⟨*area*⟩ on foot; *(schnell)* run all along; **b)** *(abnutzen)* wear down

Ab·leben das; *o. Pl. (geh.)* decease; demise

ab|lecken *tr. V.* **a)** lick off; **b)** *(säubern)* lick clean; **sich** *(Dat.)* **die Finger ~:** lick one's fingers

ab|legen 1. *tr. V.* **a)** *(niederlegen)* lay *or* put down; lay ⟨*egg*⟩*;* **b)** *(Bürow.)* file; **c)** *(nicht mehr tragen)* stop wearing; **abgelegte Kleidung** old clothes *pl.;* cast-offs *pl.;* **d)** *(aufgeben)* give up ⟨*habit*⟩*;* lose ⟨*shyness*⟩*;* put aside ⟨*arrogance*⟩*;* **e)** *(machen, leisten)* swear ⟨*oath*⟩*;* sit ⟨*examination*⟩*;* make ⟨*confession*⟩*; s. auch* **Bekenntnis a, Rechenschaft. 2.** *tr., itr. V.* **a)** *(ausziehen)* take off; **möchten Sie ~?** would you like to take your coat off?; **b)** *(Kartenspiel) (abwerfen)* discard; *(auflegen)* put down. **3.** *itr. V. (Seemannsspr.: losfahren)* **[vom Kai] ~:** cast off

Ableger der; ~s, ~ a) *(Bot.)* layer; **b)** *(Steckling)* cutting

ab|lehnen 1. *tr. V.* **a)** *(zurückweisen)* decline; decline, turn down ⟨*money, invitation, position*⟩*;* reject ⟨*suggestion, applicant*⟩*;* **b)** *(nicht genehmigen)* turn down; reject; reject, throw out ⟨*bill*⟩*;* **c)** *(verweigern)* **es ~, etw. zu tun** refuse to do sth.; **d)** *(mißbilligen)* disapprove of; reject. **2.** *itr. V.* decline; **sie haben ohne Begründung abgelehnt** *(nicht genehmigt)* they rejected it/them without giving any reason

ablehnend 1. *Adj.* negative ⟨*reply, attitude*⟩*;* **ein ~er Bescheid** a rejection. **2.** *adv.* **einer Sache** *(Dat.)* **~ gegenüberstehen** take a negative view of sth.; **sich ~ zu etw. äußern** voice one's opposition to sth.

Ablehnung die; ~, ~en a) *(Zurückweisung)* rejection; **auf ~ stoßen** meet with opposition; **b)** *(Mißbilligung)* disapproval; **auf ~ stoßen** meet with disapproval

ab|leiten 1. *tr. V.* **a)** divert; **b)** *(herleiten; auch Sprachw., Math.)* **etw. aus/von etw. ~:** derive sth. from sth.; **c)** *(Math.: differenzieren)* differentiate ⟨*function*⟩. **2.** *refl. V. (sich herleiten)* **sich aus/ von etw. ~:** derive *or* be derived from sth.

Ab·leitung die a) *(das Ableiten; auch Math., Sprachw.)* derivation; **b)** *(Sprachw.: Wort; Math.: Ergebnis des Differenzierens)* derivative

ab|lenken 1. *tr. V.* **a)** *(weglenken)* deflect; **den Verdacht von sich ~** *(fig.)* divert suspicion from oneself; **b)** *auch itr. (abbringen)* **jmdn. von etw. ~:** distract sb. from sth.; **alles, was ablenkt** everything that is distracting; **c)** *auch itr. (zerstreuen)* divert; **das lenkt dich davon ab** that'll take your mind off it. **2.** *itr. V.* **[vom Thema] ~:** change the subject

Ab·lenkung die a) *(Richtungsänderung)* deflection; **b)** *(Störung)* distraction; **c)** *(Zerstreuung)* diversion

Ablenkungs·manöver das diversion[ary tactic]

¹ab|lesen *unr. tr. V.* **a)** pick off; **b)** *(säubern)* pick clean; groom ⟨*coat*⟩

²ab|lesen 1. *unr. tr., itr. V.* **a)** read ⟨*speech, lecture*⟩*;* **werden Sie frei sprechen oder ~?** will you be talking from notes or reading your speech?; **b)** *(feststellen, prüfen)* check ⟨*time, speed, temperature*⟩*;* **[das Gas/den Strom] ~:** read the gas/electricity meter; **die Temperatur auf dem** *od.* **am Thermometer ~:** read off the temperature on the thermometer; **das Thermometer/den Tacho ~:** read the thermometer/speedo. **2.** *unr. tr. V. (erkennen)* see; **etw. an etw.** *(Dat.)* **~:** see sth. from sth.; **jmdm. jeden Wunsch von den Augen ~:** read sb.'s every wish in his/her eyes

ab|leugnen *tr. V.* deny

ab|lichten *tr. V.* **a)** *(fotokopieren)* photocopy; **b)** *(fotografieren)* take a photograph of

ab|liefern *tr. V.* deliver ⟨*goods*⟩*;* hand in ⟨*manuscript, examination paper, weapon, etc.*⟩*; (fig. ugs.)* take/bring ⟨*person*⟩ **(in/auf +** *Dat.,* **bei** to)

ab|löschen 1. *tr. V.* **a)** *(trocknen)* blot ⟨*ink, letter, etc.*⟩*;* **b)** *(abwischen)* wipe ⟨*blackboard*⟩*;* wipe out ⟨*writing*⟩

ab|lösen 1. *tr. V.* **a)** relieve; take over from; *(ersetzen)* replace; **sich** *od.* **einander ~:** take turns; **b)** *(lösen)* **etw. [von etw.] ~:** get sth. off [sth.]; remove sth. [from sth.]; **c)** *(verhüll.: entlassen)* remove from office. **2.** *refl. V. (sich lösen)* ⟨*retina*⟩ become detached; **sich von etw. ~:** come off sth.

Ablöse·summe die *(Sport)* transfer fee

Ab·lösung die a) *(eines Postens)* changing; *(Ersetzung)* replacement; **ich schicke Ihnen jemanden zur ~:** I'll send someone to relieve you; **b)** *(Ersatz)* relief; **c)** *(der Netzhaut)* detachment; **d)** *(verhüll.: Entlassung)* removal

ab|luchsen ['aplʊksn̩] *tr. V. (salopp)* **jmdm. etw. ~:** get *or (sl.)* wangle sth. out of sb.

ab|lutschen *tr. V.* **a) etw. [von etw.] ~:** suck sth. off [sth.]; **b)** *(säubern)* suck clean; **ein abgelutschter Bonbon** a half-sucked sweet

ab|machen *tr. V.* **a)** *(ugs.)* take off; take down ⟨*sign, rope*⟩*;* **etw. von etw. ~:** take sth. off sth.; **b)** *(vereinbaren)* agree; arrange; **abgemacht, wir kommen mit!** all right, we'll come; **c)** *(klären)* sort out; **das muß er mit sich selbst ~:** that's something he'll have to sort out by himself

Abmachung die; ~, ~en agreement; arrangement; **eine ~ [mit jmdm.] treffen** come to an agreement *or* arrangement [with sb.]

ab|magern *itr. V.; mit sein* become thin; *(absichtlich)* slim; **bis auf die Knochen ~:** become a mere skeleton

Abmagerungs·kur die reducing diet; **eine ~ machen** go on a diet

ab|mähen *tr. V.* mow

ab|malen *tr. V.* paint a picture of **(aus, von** from)

Ab·marsch der departure

abmarsch·bereit *Adj.* ready to depart; *(Milit.)* ready to march

ab|marschieren *itr. V.; mit sein* depart; *(Milit.)* march off

ab|melden *tr. V.* **a) sich/jmdn. ~:** report that one/sb. is leaving; *(bei Wegzug) notify the authorities that one/sb. is moving from an address;* **sich [bei jmdm.] vom Dienst**

~: report absent from duty [to sb.]; **b) ein Auto** ~: cancel a car's registration; **c)** *(ugs.)* **[bei jmdm.] abgemeldet sein** no longer be of interest [to sb.]; **der ist jetzt bei mir abgemeldet** I want nothing more to do with him

Ab·meldung die a) *(beim Weggehen)* report that one is leaving; **b)** *(beim Wegzug) registration of a move with the authorities at one's old address;* **c) die** ~ **eines Autos** the cancellation of a car's registration

ab|messen *unr. tr. V.* measure; *(fig.)* measure; assess

Ab·messung die *meist Pl. (Dimension)* dimension; measurement

ab|mildern *tr. V.* **a)** break, cushion ‹*fall, impact*›; **b)** *(fig.: abschwächen)* tone down; take the edge off

ab|montieren *tr. V.* take off, remove ‹*part, wheel*›; dismantle ‹*machine, equipment*›

ab|mühen *refl. V.* **sich [mit jmdm./etw.]** ~: toil [for sb.'s benefit/with sth.]; **sie mühte sich mit dem schweren Koffer ab** she struggled with the heavy suitcase

ab|murksen [-mʊrksn̩] *tr. V. (salopp)* do in *(sl.)*

ab|nagen *tr. V.* **etw. [von etw.]** ~: gnaw sth. off [sth.]; **einen Knochen** ~: gnaw a bone

Abnahme ['apna:mə] **die;** ~, ~**n a)** *o. Pl. (das Entfernen)* removal; **b)** *(Verminderung)* decrease; decline; **c)** *(Kauf)* purchasing; **bei** ~ **größerer Mengen** when large quantities are purchased; **d)** *(Prüfung) (einer Strecke, eines Gebäudes)* inspection and approval; *(eines Fahrzeugs)* testing and passing; *(Freigabe)* passing

ab|nehmen 1. *unr. tr. V.* **a)** *(entfernen)* take off; remove; take down ‹*picture, curtain, lamp*›; **jmdm. das Bein** ~: take sb.'s leg off; **b)** *(übernehmen)* take; **jmdm. den Koffer** ~: take sb.'s suitcase [from him/her]; **kann/darf ich Ihnen etwas** ~**?** can/may I carry something for you?; **jmdm. seine Sorgen** ~: relieve sb. of his/her worries; **c)** *(entgegennehmen)* **jmdm. ein Versprechen/einen Eid** ~: make sb. give a promise/swear an oath; **jmdm. die Beichte** ~: hear sb.'s confession; **eine Prüfung** ~: conduct an examination; **d)** *(prüfen)* inspect and approve; test and pass ‹*vehicle*›; **e)** *(wegnehmen)* take away ‹*driving licence, passport*›; **jmdm. etw.** ~: take sth. off sb.; **f)** *(abverlangen)* **jmdm. etw.** ~: charge sb. sth.; **g)** *(abkaufen)* **jmdm. etw.** ~: buy sth. from sb.; **h)** *(ugs.: glauben)* **das nehme ich dir/ihm** *usw.* **nicht ab** I won't buy that *(coll.);* **i)** *(beim Telefon)* pick up ‹*receiver*›; answer ‹*telephone*›; **j)** *auch itr. (Handarb.)* decrease. **2.** *unr. itr. V.* **a)** *(ans Telefon gehen)* answer the telephone; **es nimmt niemand ab** there is no answer; **b)** *(Gewicht verlieren)* lose weight; **sechs Kilo** ~: lose six kilos; **c)** *(sich verringern)* decrease; drop; ‹*attention, interest*› flag; ‹*brightness*› diminish; ‹*moon*› wane; **wir haben** ~**den Mond** there is a waning moon

Abnehmer der; ~**s,** ~: buyer

Ab·neigung die dislike, aversion **(gegen** for)

abnorm [ap'nɔrm] **1.** *Adj.* abnormal. **2.** *adv.* abnormally

ab|nutzen, *(landsch.:)* **ab|nützen 1.** *tr. V.* wear out. **2.** *refl. V.* wear out; become worn; **das Material nutzt sich rasch ab** the material wears very quickly

Ab·nutzung, *(landsch.:)* **Abnützung die;** ~: wear [and tear] *no indef. art.*

Abonnement [abɔnə'mã:] **das;** ~**s,** ~**s** subscription (*Gen.* to); *(Theater)* subscription ticket

Abonnent [abɔ'nɛnt] **der;** ~**en,** ~**en** subscriber (*Gen.* to); *(Theater)* season-ticket holder

abonnieren [abɔ'ni:rən] **1.** *tr. V.* subscribe to; *(Theater)* get a season-ticket for. **2.** *itr. V. (bes. schweiz.)* **abonniert sein auf** (+ *Akk.*) have a subscription to ‹*newspaper, magazine, concerts*›; *(Theater)* have a season-ticket for; *(fig.)* get as a matter of course

ab|ordnen *tr. V.* send; **jmdn. als Delegierten** ~: delegate sb.; **jmdn. zu einer Konferenz** ~: delegate sb. to a conference

Ab·ordnung die delegation

Abort [a'bɔrt] **der;** ~**[e]s,** ~**e** *(veralt., noch fachspr.)* lavatory

[2]**Abort der;** ~**s,** ~**e** *(Med.)* **a)** *(Fehlgeburt)* miscarriage; **b)** *(Abtreibung)* abortion

ab|packen *tr. V.* pack; wrap ‹*bread*›; **abgepacktes Obst** packaged fruit

ab|passen *tr. V.* **a)** *(abwarten)* wait for; **b)** *(aufhalten)* catch

ab|pausen *tr. V.* trace

ab|perlen *itr. V.; mit sein* **von etw.** ~: roll off sth.

ab|pfeifen *(Sport)* **1.** *unr. itr. V.* blow the whistle. **2.** *unr. tr. V.* **a)** *(unterbrechen)* [blow the whistle to] stop; **b)** *(beenden)* blow the whistle for the end of ‹*match, game, half*›

Ab·pfiff der *(Sport)* final whistle; *(Halbzeit~)* half-time whistle

ab|pflücken *tr. V.* pick

ab|placken *(ugs.),* **ab|plagen** *refl. V.* slave away; flog oneself to death *(coll.);* **sich mit etw./jmdm.** ~: slave away at sth./for sb.'s benefit

ab|platzen *itr. V.; mit sein* ‹*lacquer, enamel, plaster*› flake off; ‹*button*› fly off

ab|prallen *itr. V.; mit sein* rebound; bounce off; ‹*missile*› ricochet; **an** *od.* **von etw.** *(Dat.)* ~: rebound/ricochet off sth.; **an jmdm.** ~ *(fig.)* bounce off sb.

ab|pumpen *tr. V.* pump out; extract ‹*milk*› by breast-pump

ab|putzen *tr. V. (ugs.)* **a)** wipe; **jmdm./sich das Gesicht** *usw.* ~: clean sb.'s/one's face *etc.;* **b)** *(entfernen)* **etw. von jmdm./etw.** ~: wipe sth. off sb./sth.

ab|quälen *refl. V.* **sich [mit etw.]** ~: struggle [with sth.]; **sich** *(Dat.)* **einen Brief** ~ *(ugs.)* force oneself to write a letter

ab|quetschen *tr. V.* **jmdm. einen Arm/ein Bein** ~: crush sb.'s arm/leg

ab|rackern *refl. V. (ugs.)* slave [away]; flog oneself to death *(coll.);* **sich mit etw.** ~: slave away at sth.

Abrakadabra [a:braka'da:bra] **das;** ~**s** abracadabra

ab|rasieren *tr. V.* shave off; **jmdm./sich den Bart** ~: shave off sb.'s/one's beard

ab|raten *unr. itr., tr. V.* **jmdm. von etw.** ~: advise sb. against sth.; **jmdm. [davon]** ~, **etw. zu tun** advise sb. not to do sth. *or* against doing sth.

Ab·raum der; *o. Pl. (Bergbau)* overburden

ab|räumen *tr. V.* **a)** clear away; **b)** *(leer machen)* clear ‹*table*›

ab|rauschen *itr. V.; mit sein (ugs.) (schnell)* rush off; *(auffällig)* sweep off

ab|reagieren 1. *tr. V.* work off; **seine Wut an jmdm.** ~: take one's anger out on sb. **2.** *refl. V.* work off one's feelings

ab|rechnen 1. *itr. V.* **a)** cash up; **b) mit jmdm.** ~ *(fig.)* call sb. to account. **2.** *tr. V.* **a) die Kasse** ~: reckon up the till; total the cash *or* register *(Amer.);* **seine Spesen** ~: claim one's expenses; **b)** *(abziehen)* deduct

Ab·rechnung die a) *(Schlußrechnung)* cashing up *no art.;* **die**

Kellnerin machte die ~: the waitress was cashing up; **b)** *(Aufstellung)* statement; *(Kaufmannsspr.: Bilanz)* balance; *(Dokument)* balance-sheet; **c)** *(Vergeltung)* reckoning; **d)** *(Abzug)* deduction; **nach ~ der Unkosten** after deducting expenses

Ab·rede die a) arrangement; agreement; **b) etw. in ~ stellen** deny sth.

ab|regen *refl. V. (ugs.)* calm down; **reg dich ab!** cool it! *(coll.);* calm down!

ab|reiben *unr. tr. V.* **a)** rub off; **etw. von etw. ~:** rub sth. off sth.; **b)** *(säubern)* rub; **[sich** *(Dat.)***] die Hände an der Hose ~:** rub one's hands on one's trousers; **c)** *(frottieren)* rub down

Ab·reibung die *(ugs.: Prügel)* hiding *(coll.);* licking *(Amer. coll.)*

Ab·reise die departure **(nach** for); **bei meiner ~:** when I left/leave

ab|reisen *itr. V.; mit sein* leave **(nach** for)

ab|reißen 1. *unr. tr. V.* **a)** *(entfernen)* tear off; tear down ‹*poster, notice*›; pull off ‹*button*›; break off ‹*thread*›; *s. auch* **Kopf a; b)** *(niederreißen)* demolish, pull down ‹*building*›. **2.** *unr. itr. V.; mit sein* **a)** *(sich lösen)* fly off; ‹*shoe-lace*› break off; **b)** *(aufhören)* come to an end; ‹*connection, contact*› be broken off

Abreiß·kalender der tear-off calendar

ab|richten *tr. V.* train

ab|riegeln *tr., itr. V.* **a)** *(zusperren)* **[die Tür] ~:** bolt the door; **b)** *(absperren)* seal *or* cordon off ‹*area*›

ab|ringen *unr. tr. V.* **jmdm. etw. ~:** extract sth. from sb.; **sich** *(Dat.)* **ein Lächeln ~:** force a smile

Ab·riß der a) *o. Pl.: s.* **abreißen 1 b:** demolition; pulling down; **b)** *(knappe Darstellung)* outline

ab|rollen 1. *tr. V.* unwind; **sich ~:** unwind [itself]. **2.** *itr. V.; mit sein* **a)** unwind [itself]; **b)** *(vonstatten gehen)* go off; ‹*events*› unfold

ab|rücken 1. *tr. V. (wegschieben)* move away. **2.** *itr. V.; mit sein* **a)** move away; **von jmdm./etw. ~:** move away from sb./sth.; **b)** *(Milit.)* move out; **c)** *(ugs.: sich entfernen)* clear out *(coll.)*

Ab·ruf der *in* **auf ~:** on call; **sich auf ~ bereithalten** be on call

abruf·bereit *Adj.* on call *postpos.*

ab|rufen *unr. tr. V.* **a)** summon, call ‹*person*›; **er wurde ins Jenseits/aus diesem Leben abgerufen** *(geh. verhüll.)* he was taken from us; **b)** *(DV)* retrieve; **c)** *(Kaufmannsspr.)* **etw. ~:** ask for sth. to be delivered; **d)** *(Finanzw.)* withdraw

ab|runden *tr. V.* **a)** round off; **abgerundete Ecken** rounded corners; **b)** *(auf eine runde Zahl bringen)* round up/down **(auf +** *Akk.* to); **etw. nach oben/unten ~:** round sth. up/down; **ein Betrag von abgerundet 27,50 Mark** a rounded [up/down] sum of 27.50 marks; **c)** *(vervollkommnen)* round off; complete

Ab·rundung die a) rounding off; **b)** *(von Zahlen)* rounding up/down; **c)** *(Vervollkommnung)* rounding off

ab|rupfen *tr. V.* pull off

abrupt [ap'rʊpt] **1.** *Adj.* abrupt. **2.** *adv.* abruptly

ab|rüsten *itr., tr. V.* disarm

Ab·rüstung die; ~: disarmament

ab|rutschen *itr. V.; mit sein* **a)** *(abgleiten)* slip; **von etw. ~:** slip off sth.; **sie ist mit dem Messer abgerutscht** her knife slipped; **b)** *(nach unten rutschen)* slide down; ‹*earth*› subside; ‹*snow*› give way; *(fig.)* ‹*pupil, competitor, etc.*› slip **(auf +** *Akk.* to)

Abs. *Abk.* **a) Absatz c** para; **b) Absender**

ab|säbeln *tr. V. (ugs.)* hack off

ab·sacken *itr. V.; mit sein (ugs.)* **a)** fall; ‹*ground*› subside; ‹*aircraft*› lose altitude; **b)** *(fig.)* go downhill

Ab·sage die *(auf eine Einladung)* refusal; *(auf eine Bewerbung)* rejection; **jmdm. eine ~ erteilen** reject sb.

ab|sagen 1. *tr. V.* cancel; withdraw ‹*participation, co-operation*›. **2.** *itr. V.* **jmdm. ~:** tell sb. one cannot come; put sb. off *(coll.);* **telefonisch ~:** ring to say one cannot come

ab|sägen *tr. V.* **a)** saw off; **b)** *(ugs.)* **jmdn. ~:** get rid of sb.

ab|sahnen *(ugs.)* **1.** *itr. V.* make a killing *(coll.).* **2.** *tr. V.* **100 000 Mark ~:** pocket 100,000 marks

Ab·satz der a) *(am Schuh)* heel; **b)** *(Textunterbrechung)* break; **einen ~ machen** make a break; start a new line; **c)** *(Abschnitt)* paragraph; **d)** *(Kaufmannsspr.)* sales *pl.;* **~ finden** sell; *s. auch* **reißend; e)** *(einer Innentreppe)* landing; *(zwischen Geschossen)* half-landing; **f)** *(Mauer~)* ledge

absatz-, Absatz-: ~markt der *(Kaufmannsspr.)* market; **~trick der** *(Fußball)* clever back-heel; **~weise** *Adv.* paragraph by paragraph

ab|saufen *unr. itr. V.; mit sein* **a)** *(salopp: untergehen)* go to the bottom; **b)** *(derb: ertrinken)* drown; **c)** *(ugs.)* ‹*engine, car*› flood; **d)** *(salopp: sich mit Wasser füllen)* flood; **abgesoffen sein** be under water; be flooded

ab|saugen *tr. V.* **a)** suck away; **etw. aus/von etw. ~:** suck sth. out of/off sth.; **b)** *(säubern)* hoover *(Brit. coll.);* vacuum

ab|schaben *tr. V.* **a)** scrape off; **b)** *(säubern)* scrape [clean]; *s. auch* **abgeschabt**

ab|schaffen 1. *tr. V.* **a)** *(beseitigen)* abolish ‹*regulation, institution*›; repeal ‹*law*›; put an end to ‹*injustice, abuse*›; **er möchte alle Flugzeuge ~:** he'd like to do away with aeroplanes completely; **b)** *(sich trennen von)* get rid of. **2.** *refl. V. (südd., schweiz.)* slave away; work oneself hard

Ab·schaffung die abolition; *(von Gesetzen)* repeal; *(von Unrecht, Mißstand)* ending

ab|schälen 1. *tr. V.* **a)** *(lösen)* peel off; **etw. von etw. ~:** peel sth. off sth.; **b)** *(befreien von)* bark ‹*tree*›. **2.** *refl. V. (sich lösen)* peel off; **die Haut schält sich ab** the skin is peeling

ab|schalten 1. *tr., itr. V. (ausschalten)* switch off; turn off; shut down ‹*power-station*›. **2.** *itr. V. (fig. ugs.)* switch off

ab|schätzen *tr. V.* estimate; size up ‹*person, possibilities*›

abschätzig ['ap-ʃɛtsɪç] **1.** *Adj.* derogatory; disparaging. **2.** *adv.* derogatorily; disparagingly

Ab·schaum der; *o. Pl. (abwertend)* scum; dregs *pl.*

ab|scheiden *unr. tr. V. (Chemie)* precipitate; *(Physiol.)* secrete

Ab·scheu der; ~s, *(selten:)* **die; ~:** detestation; abhorrence; **einen ~ vor jmdm./etw. haben** detest *or* abhor sb./sth.

ab|scheuern *tr. V.* **a)** scrub off; **b)** *(säubern)* scrub; **c)** *(beschädigen)* graze ‹*skin*›; wear away ‹*cloth*›

abscheulich [ap'ʃɔylɪç] **1.** *Adj.* **a)** *(widerwärtig)* disgusting, awful ‹*smell, taste*›; repulsive, awful ‹*sight*›; **b)** *(verwerflich, schändlich)* disgraceful ‹*behaviour*›; abominable ‹*crime*›. **2.** *adv.* **a)** disgracefully; abominably; **b)** *(ugs.: sehr)* **~ frieren** freeze [half] to death *(coll.);* **das tut ~ weh** it hurts like hell *(coll.);* **~ kalt/scharf** terribly cold/sharp *(coll.)*

Abscheulichkeit die; ~: *s.* **abscheulich 1:** disgustingness; awfulness; repulsiveness; disgracefulness; abominableness

ab|schicken *tr. V.* send [off], post ‹*letter, parcel*›; dispatch, send [off], post ‹*goods, money*›

ab|schieben 1. *unr. tr. V.* **a)** push *or* shove away; **das Bett von der Wand ~:** push *or* shove the bed away from the wall; **b)** *(abwälzen)* shift; **die Verantwortung/Schuld auf jmdn. ~:** shift [the] responsibility/the blame on to sb.; **c)** *(Rechtsw.: ausweisen)* deport; **jmdn. über die Grenze ~:** put sb. over the border; **d)** *(ugs.: entfernen)* get rid of; **jmdn. in ein Heim ~:** shove sb. into a home *(coll.).* **2.** *unr. itr. V.; mit sein (salopp: weggehen)* push off *(sl.);* shove off *(coll.)*

Ab·schiebung die *(Rechtsw.)* deportation

Abschied ['ap-ʃi:t] **der; ~[e]s, ~e a)** *(Trennung)* parting **(von** from); farewell **(von** to); **[von jmdm./etw.] ~ nehmen** say goodbye [to sb./sth.]; take one's leave [of sb./sth.]; **beim ~:** at parting; when saying goodbye; **sich zum ~ die Hände schütteln** shake hands on parting; **jmdm. zum ~ zuwinken** wave goodbye to sb.; **b)** *(geh.: Entlassung)* resignation; **seinen ~ nehmen** *(geh.)* resign; ‹*officer*› resign one's commission

Abschieds-: ~besuch der farewell visit; **~brief der** farewell letter; **~feier die** farewell ceremony; *(Party)* farewell *or* leaving party; **~geschenk das** farewell *or* parting gift; **~gruß der** goodbye; farewell; **~kuß der** goodbye *or* parting kiss; **~schmerz der;** *o. Pl.* sorrow at parting

ab|schießen *unr. tr. V.* **a)** loose, fire ‹*arrow*›; fire ‹*rifle, pistol, cannon, missile*›; launch ‹*spacecraft*›; **b)** *(töten)* take; **c)** *(ugs.: entfernen)* kick *or* throw ‹*person*› out; **d)** *(von sich geben)* fire off ‹*question*›; shoot ‹*glance*›; **e)** *(zerstören)* shoot down ‹*aeroplane*›; put ‹*tank*› out of action; **f)** *(wegreißen)* shoot off ‹*arm, leg, etc.*›

ab|schinden *unr. refl. V.* **sich ~:** work *or (Brit. coll.)* flog oneself to death

ab|schirmen *tr. V.* **a)** *(schützen)* shield; **jmdn./sich von der** *od.* **gegen die Umwelt ~:** screen sb./oneself off from the outside world; **b)** *(abhalten)* screen off ‹*light, radiation*›

ab|schlachten *tr. V.* slaughter

ab|schlaffen ['ap-ʃlafn̩] *itr. V.; mit sein (ugs.)* wilt; sag; **ein abgeschlaffter Typ** a lackadaisical fellow; **er saß abgeschlafft im Sessel** he sat limply in his chair; **geistig ~:** lose one's intellectual vigour

Ab·schlag der a) *(Kaufmannsspr.)* reduction; discount; **b)** *(Teilzahlung)* interim payment; *(Vorschuß)* advance; **c)** *(Fußball)* goalkeeper's kick out

ab|schlagen *unr. tr. V.* **a)** knock off; *(mit dem Beil, Schwert usw.)* chop off; **b)** *(ablehnen)* refuse; **jmdm. etw. ~:** refuse *or* deny sb. sth.; **c)** *(abwehren)* beat *or* fend off; **d)** *(zerlegen)* dismantle; strike ‹*tent*›; **e)** *auch itr. (Fußball)* **[den Ball] ~:** kick the ball out

abschlägig ['apʃlɛ:gɪç] *(Amtsspr.)* **1.** *Adj.* negative; **ein ~er Bescheid** a refusal *or* rejection. **2.** *adv.* **jmdn. ~ bescheiden** refuse sb.

Abschlag[s]·zahlung die *s.* **Abschlag b**

ab|schlecken *tr. V. (österr., südd.) s.* **ablecken**

ab|schleifen 1. *unr. tr. V.* **a)** *(entfernen) (von Holz)* sand off; *(von Metall, Glas usw.)* grind off; **b)** *(glätten)* sand down ‹*wood*›; grind down ‹*metal, glass, etc.*›; smooth down ‹*broken tooth*›. **2.** *unr. refl. V. (sich abnutzen)* wear away; **das schleift sich noch ab** *(fig.)* that will wear off in time

Abschlepp·dienst der *(Kfz-W.)* breakdown recovery service; tow[ing] service *(Amer.)*

ab|schleppen 1. *tr. V.* **a)** tow away; *(schleppen)* tow; **b)** *(salopp: mitnehmen)* **jmdn. ~:** drag sb. off. **2.** *refl. V. (ugs.: schwer tragen)* **sich mit etw. ~:** break one's back carrying sth. *(fig.)*

Abschlepp-: ~seil das towrope; *(aus Draht)* towing cable; **~stange die** tow-bar; **~wagen der** breakdown vehicle; tow truck *(Amer.)*

ab|schließen 1. *unr. tr. V.* **a)** *auch itr. (zuschließen)* lock ‹*door, gate, cupboard*›; lock [up] ‹*house, flat, room, park*›; **b)** *(verschließen)* seal; **etw. luftdicht ~:** seal sth. hermetically; **c)** *(begrenzen)* border; **d)** *(zum Abschluß bringen)* bring to an end; conclude; **sein Studium ~:** finish one's studies; **e)** *(vereinbaren)* strike ‹*bargain, deal*›; make ‹*purchase*›; enter into ‹*agreement*›; **Geschäfte ~:** conclude deals; *(im Handel)* do business; *s. auch* **Versicherung b; Wette. 2.** *unr. itr. V.* **a)** *(begrenzt sein)* be bordered **(mit** by); **b)** *(aufhören, enden)* end; **~d sagte er ...:** in conclusion he said ...; **c) mit jmdm./etw. abgeschlossen haben** have finished with sb./sth.

Ab·schluß der a) *(Verschluß)* seal; **b)** *(abschließender Teil)* edge; **c)** *(Beendigung)* conclusion; end; **vor ~ der Arbeiten** before the completion of the work; **etw. zum ~ bringen** finish sth.; bring sth. to an end *or* a conclusion; **zum ~ unseres Programms** to end our programme; **d)** *(ugs.: ~zeugnis)* **einen/keinen ~ haben** *(Hochschulw.)* ≈ have a/have no degree *or (Amer.)* diploma; *(Schulw.)* ≈ have some/no GCSE passes *(Brit.)*; *(Lehre)* have/not have finished one's apprenticeship; **e)** *(Kaufmannsspr.: Schlußrechnung)* balancing; **f)** *(Kaufmannsspr.: geschäftliche Vereinbarung)* business deal; **g)** *(eines Geschäfts, Vertrags)* conclusion

Abschluß-: ~prüfung die *(Hochschulw.)* final examination; finals *pl.; (Schulw.)* leaving *or (Amer.)* final examination; **~zeugnis das** *(Schulw.)* ≈ leaving certificate *(Brit.);* ≈ diploma *(Amer.)*

ab|schmecken *tr. V.* **a)** *(kosten)* taste; try; **b)** *(würzen)* season

ab|schmieren 1. *tr. V.* **a)** *(Technik)* grease; **b)** *(ugs. abwertend: abschreiben)* copy; ‹*child in school*› crib **(von, bei** from). **2.** *itr. V.* **a)** *(ugs. abwertend)* crib **(von, bei** from); **b)** *(Fliegerspr.)* sideslip

ab|schminken *tr. V.* **a) jmdn./sich ~:** remove sb.'s/one's make-up; **b) sich** *(Dat.)* **etw. ~** *(salopp)* get sth. out of one's head

ab|schmirgeln *tr. V.;* **a)** *(polieren)* rub down with emery; *(mit Sandpapier)* sand down; **b)** *(entfernen)* rub off with emery; *(mit Sandpapier)* sand off

ab|schnallen *tr. V.* unfasten; **[sich** *(Dat.)***] den Tornister ~:** take off one's knapsack; **sich ~:** unfasten one's seat-belt

ab|schneiden 1. *unr. tr. V.* **a)** *(abtrennen)* cut off; cut down ‹*sth. hanging*›; **etw. von etw. ~:** cut sth. off sth.; **sich** *(Dat.)* **den Finger ~:** cut one's finger off; **sich** *(Dat.)* **eine Scheibe Brot ~:** cut oneself a slice of bread; *s. auch* **Scheibe b; b)** *(kürzer schneiden)* cut; **jmdm./sich die Haare ~:** cut sb.'s/one's hair; **ein Kleid [ein Stück] ~:** cut [a piece] off a dress; **c) jmdm. den Weg ~:** take a short cut to get ahead of sb.; **d)** *(trennen, isolieren)* cut off. **2.** *unr. itr. V.* **bei etw. gut/schlecht ~:** do well/badly in sth.

Ab·schnitt der a) *(Kapitel)* section; **b)** *(Milit.: Gebiet, Gelände)* sector; **c)** *(Zeitspanne)* phase; **d)** *(eines Formulars)* [detachable] portion

ab|schnüren *tr. V.* constrict; *(als medizinische Maßnahme)* apply a tourniquet to; **jmdm. die Luft/das Blut ~:** stop sb. from breathing/restrict sb.'s circulation

ab|schöpfen *tr. V.* skim off; *(fig.)* siphon off ‹profits›; **den Rahm ~** *(fig.)* cream off the best

ab|schrauben *tr. V.* unscrew [and remove]

ab|schrecken 1. *tr. V.* **a)** *(abhalten)* deter; **b)** *(fernhalten)* scare off; **c)** *(Kochk.)* pour cold water over; put ‹*boiled eggs*› into cold water. **2.** *itr. V.* act as a deterrent

abschreckend 1. *Adj.* **a)** *(warnend)* deterrent; **ein ~es Beispiel für alle Raucher** a warning to all smokers; **b)** *(abstoßend)* repulsive. **2.** *adv.* **~ wirken** have a deterrent effect; **~ häßlich** repulsively ugly

Abschreckung die; ~: deterrence

ab|schreiben 1. *unr. tr. V.* **a)** *(kopieren)* copy out; **sich** *(Dat.)* **etw. ~:** copy sth. down; *(aus einem Buch, einer Zeitung usw.)* copy sth. out; **b) etw. von** *od.* **bei jmdm. ~** *(in der Schule)* copy sth. from *or* off sb.; *(als Plagiator)* plagiarize sth. from sb.; **c)** *(Wirtsch.)* amortize, write down **(mit** by); **d)** *(ugs.: verlorengeben)* write off; **jmdn. abgeschrieben haben** have written sb. off. **2.** *unr. itr. V.* **bei** *od.* **von jmdm. ~** *(in der Schule)* copy off sb.; *(als Plagiator)* copy from sb.

Ab·schreibung die *(Wirtsch.)* amortization

Ab·schrift die copy

ab|schrubben *tr. V.* *(ugs.)* **a)** scrub away *or* off; **sich/jmdm. den Schmutz ~:** scrub the dirt off oneself/sb.; **b)** *(säubern)* scrub; **sich/jmdm. den Rücken ~:** scrub one's/sb.'s back [down]; **sich/jmdn. ~:** scrub oneself/sb. [down]

ab|schürfen *tr. V.* **sich** *(Dat.)* **die Knie/die Ellenbogen ~:** graze one's knees/one's elbows; **sich** *(Dat.)* **die Haut ~:** chafe the skin

Ab·schuß der a) *(eines Flugzeugs)* shooting down; *(eines Panzers)* putting out of action; **b)** *(von Wild)* shooting; **c)** *(das Abfeuern) (von Geschossen, Torpedos)* firing; *(in den Weltraum)* launching

abschüssig ['ap-ʃʏsɪç] *Adj.* downward sloping ‹*land*›; **die Straße ist sehr ~:** the road goes steeply downhill

Abschuß·rampe die launch pad; launching pad

ab|schütteln *tr. V.* **a)** shake down ‹*fruit*›; **[sich** *(Dat.)***] den Staub/den Schnee [vom Mantel] ~:** shake off the dust/the snow [from one's coat]; **b)** *(fig.)* shake off ‹*pursuer, marker, yoke of tyranny etc.*›

ab|schwächen 1. *tr. V.* **a)** tone down, moderate ‹*statement, criticism*›; **b)** *(verringern)* lessen ‹*effect, impression*›; cushion ‹*blow, impact*›. **2.** *refl. V.* ‹*interest, demand*› wane; **der Preisauftrieb schwächt sich ab** price increases are slowing down

Ab·schwächung die; ~, ~en a) *(Milderung)* toning down, moderation; *(abgemilderte Form)* attenuation; **b)** *(eines Aufpralls, Stoßes usw.)* cushioning; **c)** *(das Nachlassen)* waning; *(eines Hochs, Tiefs)* weakening; *(zahlenmäßig)* drop (*Gen.* in)

ab|schwatzen *tr. V.* **jmdm. etw. ~:** talk sb. into giving one sth.

ab|schweifen *itr. V.; mit sein* digress

Abschweifung die; ~, ~en digression

ab|schwellen *unr. itr. V.; mit sein* go down

ab|schwenken *itr. V.; mit sein* turn aside; **links/rechts ~** *(abbiegen)* turn left/right; *(die Richtung allmählich ändern)* bear to the left/right

ab|schwören *unr. itr. V.* **dem Teufel/seinem Glauben ~:** renounce the Devil/one's faith; **dem Alkohol/Laster ~:** forswear *or* swear off alcohol/vice

Ab·schwung der a) *(Turnen)* dismount; **beim ~:** when dismounting; **b)** *(Wirtsch.)* downward trend

ab|segeln *itr. V.; mit sein* sail away; **von Kiel ~:** sail from Kiel

ab|segnen *tr. V.* *(ugs. scherzh.)* sanction

absehbar *Adj.* foreseeable; **in ~er Zeit** within the foreseeable future; **etw. ist noch gar nicht ~:** sth. cannot yet be predicted; **auf** *od.* **für ~e Zeit** for the foreseeable future; **nicht ~:** unforeseeable

ab|sehen 1. *unr. tr. V.* **a)** *(voraussehen)* predict; foresee ‹*event*›; **b) es auf etw.** *(Akk.)* **abgesehen haben** be after sth.; **er hat es auf sie abgesehen** he's got his eye on her; **er hat es darauf abgesehen, uns zu ärgern** he's out to annoy us; **der Chef hat es auf ihn abgesehen** the boss has got it in for him. **2.** *unr. itr. V.* **a)** *(nicht beachten)* **von etw. ~:** leave aside *or* ignore sth.; *s. auch* **abgesehen; b)** *(verzichten)* **von etw. ~:** refrain from sth.; **von einer Anzeige/Klage ~:** not report sth./not press charges

ab|seifen *tr. V.* **jmdn./sich ~:** soap sb./oneself down

ab|seilen 1. *tr. V.* lower [with a rope]. **2.** *refl. V.* *(Bergsteigen)* abseil

ab|sein *unr. itr. V.; mit sein (Zusschr. nur im Inf. u. Part.)* **[an etw.** *(Dat.)***] ~** *(ugs.: abgegangen sein)* have come off [sth.]

abseits ['apzaits] **1.** *Präp. mit Gen.* away from. **2.** *Adv.* **a)** *(entfernt)* far away; **etwas ~:** a little way away; **b)** *(Ballspiele)* **~ sein** *od.* **stehen** be offside

Abseits das; ~, ~ a) *(Ballspiele)* **das war ein klares ~:** that was clearly offside; **im ~ stehen** be offside; **der Spieler lief ins ~:** the player put himself offside; **b)** *(fig.)* **im ~ stehen** have been pushed out into the cold; **ins ~ geraten** be pushed out into the cold

ab|senden *unr. od. regelm. tr. V.* dispatch

Ab·sender der sender; *(Anschrift)* sender's address

ab|senken 1. *refl. V.* **sich [zum See/Fluß hin] ~:** slope [down to the lake/river]. **2.** *tr. V.* lower

ab|servieren 1. *itr. V.* clear away. **2.** *tr. V.* **a) ein Gedeck/den Tisch ~:** clear away a cover/clear the table; **b)** *(salopp: absetzen, kaltstellen)* throw out

absetzbar *Adj.* **a)** *(Steuerw.)* **[steuerlich] ~:** [tax-]deductible; **b)** *(verkäuflich)* saleable; **c)** *s.* **absetzen 1 d: er ist nicht ~:** he cannot be dismissed/be removed from office

ab|setzen 1. *tr. V.* **a)** *(abnehmen)* take off; **b)** *(hinstellen)* put down ‹*glass, bag, suitcase*›; **c)** *(aussteigen lassen)* **jmdn. ~** *(im öffentlichen Verkehr)* put sb. down; let sb. out *(Amer.)*; *(im privaten Verkehr)* drop sb. [off]; **d)** *(entlassen)* dismiss ‹*minister, official*›; remove ‹*chancellor, judge*› from office; depose ‹*king, emperor*›; **e)** *(ablagern)* deposit; **f)** *(absagen)* drop; call off ‹*strike, football match*›; **g)** *(nicht mehr anwenden)* discontinue ‹*treatment, therapy*›;

stop taking ⟨*medicine, drug*⟩; **h)** *(von den Lippen nehmen)* take ⟨*glass, trumpet*⟩ from one's lips; **i)** *(verkaufen)* sell; **j)** *(Steuerw.)* **etw. [von der Steuer] ~:** deduct sth. [from tax]. **2.** *refl. V.* **a)** *(sich ablagern)* be deposited; ⟨*dust*⟩ settle; ⟨*particles in suspension*⟩ settle out; **b)** *(sich distanzieren)* **sich von etw. ~:** distance oneself from sth.; **c)** *(sich unterscheiden) s.* **abheben 3; d)** *(ugs.: sich davonmachen)* get away

Absetzung die; ~, ~en *s.* **absetzen 1d:** dismissal; removal; deposition

ab|sichern 1. *tr. V.* **a)** make safe; **b)** *(fig.)* substantiate ⟨*argument, conclusions*⟩; validate ⟨*result*⟩. **2.** *refl. V.* safeguard oneself

Ab·sicht die; ~, ~en intention; **die ~ haben, etw. zu tun** plan *or* intend to do sth.; **etw. mit ~ tun** do sth. intentionally *or* deliberately; **etw. ohne** *od.* **nicht mit ~ tun** do sth. unintentionally; **in der besten ~:** with the best of intentions; **in betrügerischer ~:** with intent to deceive

ab·sichtlich 1. *Adj.* intentional; deliberate. **2.** *adv.* intentionally; deliberately

ab|singen *unr. tr. V.* **a) etw. vom Blatt ~:** sing sth. at sight; **b) unter Absingen** *(Dat.)* **der Nationalhymne** singing the national anthem

ab|sinken *unr. itr. V.; mit sein* sink; *(fig.)* decline; ⟨*temperature, blood pressure*⟩ drop

Absinth [ap'zɪnt] **der; ~[e]s, ~e** absinth[e]

ab|sitzen 1. *unr. tr. V. (hinter sich bringen)* sit through; sit out ⟨*hours of duty etc.*⟩; *(im Gefängnis)* serve; **zehn Jahre ~:** serve *or (coll.)* do ten years. **2.** *unr. itr. V. mit sein* dismount **(von** from)

absolut [apzo'lu:t] **1.** *Adj.* absolute. **2.** *adv.* absolutely

Absolution [apzolu'tsi̯o:n] **die; ~, ~en** *(kath. Rel.)* absolution; **jmdm. die ~ erteilen** give sb. absolution

Absolutismus der; ~ *(hist.)* absolutism *no art.*

absolutistisch 1. *Adj.* absolutist. **2.** *adv.* in an absolutist manner

Absolvent [apzɔl'vɛnt] **der; ~en, ~en, Absolventin die; ~, ~nen** *(einer Schule)* one who has taken the leaving *or (Amer.)* final examination; *(einer Akademie, Hochschule)* graduate

absolvieren [apzɔl'vi:rən] *tr. V.* **a)** complete; **das Gymnasium ~:** complete a grammar-school education; **b)** *(erledigen, verrichten)* put in ⟨*hours*⟩; do ⟨*performance, route, task*⟩; make ⟨*visit*⟩

Absolvierung die; ~: completion

ab·sonderlich *Adj.* strange; odd

ab|sondern 1. *tr. V.* **a)** isolate ⟨*patient*⟩; separate ⟨*prisoner*⟩; **b)** *(Physiol.)* secrete. **2.** *refl. V.* isolate oneself

Absonderung die; ~, ~en a) isolation; **b)** *(Physiol.)* secretion

absorbieren *tr. V. (fachspr., fig. geh.)* absorb

Absorption [apzɔrp'tsi̯o:n] **die; ~** *(fachspr., fig. geh.)* absorption

ab|spalten *unr. od. regelm. refl. V.* split off *or* away

ab|spannen *tr. V.* unharness ⟨*horse*⟩; unhitch ⟨*wagon*⟩; unyoke ⟨*oxen*⟩

Abspannung die; ~, ~en *(Ermüdung)* weariness; fatigue

ab|sparen *refl. V. in* **sich** *(Dat.)* **etw. vom Munde ~:** scrimp and save for sth.

ab|speisen *tr. V.* **jmdn. mit etw. ~:** fob sb. off with sth.

abspenstig ['ap-ʃpɛnstɪç] *Adj.; nicht attr.* **jmdm. etw. ~ machen** get sb. to part with sth.; **jmdm. den Freund/die Freundin ~ machen** steal sb.'s boy-/girl-friend

ab|sperren 1. *tr. V.* **a)** seal off; close off; **b) jmdm. das Gas/das Wasser/den Strom ~:** cut off sb.'s gas/water/electricity; **c)** *(österr., südd.: abschließen)* lock ⟨*door*⟩. **2.** *itr. V. (österr., südd.)* lock up

Ab·sperrung die a) sealing off; closing off; **b)** *(Sperre)* barrier

Ab·spiel das *(Ballspiele)* **a)** *(das Abspielen)* passing; **b)** *(Schuß)* pass

ab|spielen 1. *tr. V.* **a)** play ⟨*record, tape*⟩; **b) vom Blatt ~:** play ⟨*piece of music*⟩ at sight; **c)** *(Ballspiele)* pass. **2.** *refl. V.* take place. **3.** *itr. V. (Ballspiele)* pass; **an jmdn. ~:** pass [the ball] to sb.

ab|splittern *itr. V.; mit sein* ⟨*wood*⟩ splinter off; ⟨*lacquer, paint*⟩ flake off

Ab·sprache die agreement; arrangement; **eine ~ treffen** come to an agreement *or* make an arrangement; **nach ~ mit** by arrangement with

ab|sprechen 1. *unr. tr. V.* **a)** *(aberkennen)* **jmdm. etw. ~:** deprive sb. of sth.; **b)** *(ableugnen)* **jmdm. etw. ~:** deny that sb. has sth.; **jmdm. das Recht auf etw.** *(Akk.)* **~:** deny sb.'s right to sth.; **jmdm. das Recht ~, etw. zu tun** deny sb. the right to do sth.; **c)** *(vereinbaren)* arrange. **2.** *unr. refl. V.* come to *or* reach an agreement

ab|spreizen *tr. V.* stretch out ⟨*arm, leg*⟩ sideways; splay out ⟨*finger, toe*⟩

ab|sprengen *tr. V.* split off

ab|springen *unr. itr. V.; mit sein* **a)** jump off; **[mit dem rechten/linken Bein] ~:** take off [on the right/left leg]; **b)** *(herunterspringen)* jump down; **vom Fahrrad/Pferd ~:** jump off one's bicycle/horse; **aus dem Flugzeug ~:** jump out of the aeroplane; **c)** *(abplatzen)* come off; ⟨*paint*⟩ flake off

ab|spritzen *tr. V.* **a)** spray off; **b)** *(reinigen)* spray [down]

Ab·sprung der a) *(das Lospringen)* take-off; **b)** *(das Herunterspringen)* jump; **c)** *(fig.)* break; **den ~ wagen** risk making the break

ab|spulen *tr. V.* unwind; **sich ~:** come unwound

ab|spülen 1. *tr. V.* **a)** wash off ⟨*dirt, dust*⟩; **b)** *(reinigen)* rinse off; **sich** *(Dat.)* **die Hände** *usw.* **~:** rinse one's hands *etc.*; **c)** *(bes. südd.)* **das Geschirr ~:** wash the dishes. **2.** *itr. V. (bes. südd.) s.* **abwaschen 2**

ab|stammen *itr. V.* be descended **(von** from)

Abstammung die; ~, ~en descent

Abstammungs·lehre die theory of evolution

Ab·stand der a) *(Zwischenraum)* distance; **in 20 Meter ~:** at a distance of 20 metres; **im ~ von 10 Metern** 10 metres apart; **~ halten** *(auch fig.)* keep one's distance; **b)** *(Unterschied)* gap; difference; **mit ~:** by far; far and away; **c)** *(Zeitspanne)* interval; *(kürzer)* gap; **in Abständen von 20 Minuten** at 20-minute intervals; **d)** *(geh.: Verzicht)* **von etw. ~ nehmen** refrain from sth.; **e)** *(Entschädigung)* compensation; *(bei Übernahme einer Wohnung) payment for furniture and fittings left by previous tenant*

ab|statten ['ap-ʃtatn̩] *tr. V. (geh.)* **jmdm. einen Besuch ~:** pay sb. a visit

ab|stauben *tr., itr. V.* **a)** dust; **b)** *(ugs.: stehlen)* pinch *(sl.)*; nick *(Brit. sl.)*; lift *(Amer. coll.)*

ab|stechen 1. *unr. tr. V.* **a)** slaughter ⟨*animal*⟩ *(by cutting its throat)*; **b)** *(ablaufen lassen)* tap ⟨*beer, wine*⟩. **2.** *unr. itr. V.* **von etw./jmdm. ~:** contrast with sth./sb.

Abstecher der; ~s, ~: side-trip; *(fig.: Abschweifung)* digression

ab|stecken *tr. V.* **a)** *(abgrenzen)* mark out; *(fig.)* define; **b)** *(Schneiderei)* pin up ‹*hem*›

ab|stehen *unr. itr. V.* **a)** *(nicht anliegen)* stick out; **~de Ohren** protruding ears; **b)** *(entfernt stehen)* **40 cm/zu weit von etw. ~:** be 40 cm. away/too far away from sth.

Ab·steige die; ~, ~n *(ugs. abwertend)* cheap and crummy hotel *(sl.)*

ab|steigen *unr. itr. V.; mit sein* **a)** **[vom Pferd/Fahrrad] ~:** get off [one's horse/bicycle]; **b)** *(abwärts gehen)* go down; descend; **gesellschaftlich ~** *(fig.)* decline in social status; **c)** *(Sport)* be relegated; **d)** *(sich einquartieren)* **in einem Hotel ~:** put up at a hotel

Ab·steiger der; ~s, ~ *(Sport)* *(vor dem Abstieg stehend)* team threatened with *or* facing relegation; *(abgestiegen)* relegated team

ab|stellen *tr. V.* **a)** *(absetzen)* put down; **b)** *(unterbringen, hinstellen)* put; *(parken)* park; **c)** *(ausschalten, abdrehen)* turn *or* switch off; turn off ‹*gas, water*›; **jmdm. das Gas/den Strom ~:** cut sb.'s gas/electricity off; **jmdm. das Telefon ~:** disconnect sb.'s telephone; **d)** *(unterbinden)* put a stop to; **e)** *(sein lassen)* stop; **f)** *(beordern)* assign; detail [off] ‹*soldiers*›

Abstell-: ~gleis das siding; **jmdn. aufs ~ schieben** *(fig. ugs.)* put sb. out of harm's way; **~kammer die** lumber-room; **~raum der** store-room

ab|stempeln *tr. V.* **a)** frank ‹*letter*›; cancel ‹*stamp*›; **b)** *(fig.)* **jmdn. als** *od.* **zum Verbrecher/als geisteskrank ~:** label *or* brand sb. as a criminal/as insane

ab|steppen *tr. V.* back-stitch

ab|sterben *unr. itr. V.; mit sein* **a)** *(eingehen, verfallen)* [gradually] die; **b)** *(gefühllos werden)* go numb; **mir sind die Finger abgestorben** my fingers have gone numb

Ab·stieg der; ~[e]s, ~e a) descent; **b)** *(Niedergang)* decline; **[sozialer** *od.* **gesellschaftlicher] ~:** fall *or* drop in [social] status; **c)** *(Sport)* relegation

ab|stillen 1. *tr. V.* wean. **2.** *itr. V.* stop breast-feeding

ab|stimmen 1. *itr. V.* vote **(über** + *Akk.* on); **über etw.** *(Akk.)* **~ lassen** put sth. to the vote. **2.** *tr. V.* **a) etw. mit jmdm. ~:** discuss and agree sth. with sb.; **b)** *(harmonisieren)* **etw. auf etw.** *(Akk.)* **~:** suit sth. to sth.; **Zeitpläne/Programme aufeinander ~:** coordinate timetables/programmes

Ab·stimmung die a) vote; ballot; **eine geheime ~:** a secret ballot; **bei der ~:** in the vote; *(während der ~)* during the voting; **b)** *(Absprache)* agreement; **c)** *(Harmonisierung)* coordination

Abstimmungs-: ~ergebnis das result of a/the vote; **~nieder·lage die** defeat [in a/the vote]; **~sieg der** victory [in a/the vote]

abstinent [apsti'nɛnt] *Adj.* **a)** teetotal; **~ sein** be a non-drinker *or* a teetotaller; **b) sexuell ~:** sexually abstinent; continent

Abstinenz [apsti'nɛnts] **die; ~ a)** teetotalism; **~ üben** be teetotal; **b) sexuelle ~:** sexual abstinence; continence

Abstinenzler der; ~s, ~: teetotaller; non-drinker

ab|stoppen 1. *tr. V.* **a)** halt; stop; check ‹*advance*›; stop ‹*machine*›; **b)** *(mit der Stoppuhr)* **die Zeit ~:** measure the time with a stopwatch. **2.** *itr. V.* come to a halt; ‹*person*› stop

Ab·stoß der *(Fußball)* goal-kick

ab|stoßen 1. *unr. tr. V.* **a)** *(wegstoßen)* push off *or* away; **das Boot [vom Ufer] ~:** push the boat out [from the bank]; **b)** *(beschädigen)* chip ‹*crockery, paintwork, stucco, plaster*›; batter ‹*furniture*›; *s. auch* **Horn; c)** *(verkaufen)* sell off; **d)** *(Physik)* repel; **e)** *(anwidern)* repel; put off; **sich von jmdm./etw. abgestoßen fühlen** find sb./sth. repulsive. **2.** *unr. itr. V.* **a)** *mit sein od. haben (sich entfernen)* be pushed off; **b)** *(anwidern)* be repulsive. **3.** *refl. V.* **sich [vom Boden] ~:** push oneself off

abstoßend *Adj.* repulsive

ab|stottern *tr. V. (ugs.)* pay for in instalments; pay off ‹*debt*› by instalments

abstrahieren [apstra'hi:rən] *tr., itr. V. (geh.)* abstract **(aus** from)

abstrakt [ap'strakt] **1.** *Adj.* abstract. **2.** *adv.* abstractly; **~ denken** think in the abstract

Abstraktion [apstrak'tsio:n] **die; ~, ~en** *(geh.)* abstraction

ab|streichen *unr. tr. V.* **a)** *(abstreifen)* wipe; *(entfernen)* wipe off; **b)** *(abziehen)* knock off; **davon muß man die Hälfte ~** *(fig.)* you have to take it with a pinch of salt

ab|streifen *tr. V.* **a)** pull off; strip off ‹*berries*›; **sich/jmdm. die Kleidung ~:** take off one's/sb.'s clothes; **die Asche [von der Zigarre] ~:** remove the ash [from one's cigar]; **b)** wipe off; *(säubern)* wipe

Abstreifer der; ~s, ~ *s.* **Fußabstreifer**

ab|streiten *unr. tr. V.* deny; **das kann ihm keiner ~:** you cannot deny him that

Ab·strich der a) *(Med.)* swab; **einen ~ machen** take a swab; **b)** *(Streichung, Kürzung)* cut; **~e [an etw.** *(Dat.)***] machen** make cuts [in sth.]; *(Einschränkungen machen)* make concessions [as regards sth.]

ab|stufen *tr. V.* **a)** *(staffeln)* grade; **b)** *(nuancieren)* differentiate

Ab·stufung die; ~, ~en a) *(Staffelung)* gradation; **b)** *(Nuance)* shade

ab|stumpfen *(fig.)* **1.** *tr. V.* deaden. **2.** *itr. V.; mit sein* **man stumpft ab** one's mind becomes deadened; **gegen etw. ~:** become dead to sth.

Ab·sturz der fall; *(eines Flugzeugs)* crash

ab|stürzen *itr. V.; mit sein* **a)** fall; ‹*aircraft, pilot, passenger*› crash; **b)** *(geh.: abfallen)* ‹*cliff*› plunge

ab|stützen 1. *refl. V.* support oneself **(mit** on, **an** + *Dat.* against). **2.** *tr. V.* support

ab|suchen *tr. V.* search **(nach** for); *(durchkämmen)* comb **(nach** for); drag ‹*pond, river, etc.*› **(nach** for); **den Himmel/Horizont ~:** scan the sky/horizon **(nach** for)

absurd [ap'zʊrt] *Adj.* absurd

Absurdität [apzʊrdi'tɛ:t] **die; ~, ~en** absurdity

Abszeß [aps'tsɛs] **der** *(österr. auch:* **das***);* **Abszesses, Abszesse a)** *(Med.)* abscess; **b)** *(Geschwür)* ulcer

Abszisse [aps'tsɪsə] **die; ~, ~en** *(Math.)* abscissa

Abt [apt] **der; ~[e]s, Äbte** ['ɛptə] abbot

Abt. *Abk.* **Abteilung**

ab|tasten *tr. V.* **etw. ~:** feel sth. all over

ab|tauen 1. *itr. V.; mit sein (wegschmelzen)* melt away; *(eis-/schneefrei werden)* become clear of ice/snow; ‹*refrigerator*› defrost. **2.** *tr. V. (schnee-/eisfrei machen)* melt; thaw; de-ice ‹*vehicle windows*›; defrost ‹*refrigerator*›

Abtei [ap'tai] **die; ~, ~en** abbey

Abteil das; ~[e]s, ~e compartment

ab|teilen *tr. V.* **a)** *(aufteilen)* divide [up]; **b)** *(abtrennen)* divide off

Ab·teilung die a) department; *(einer Behörde)* department; section; **b)** *(Bot.)* division; **c)** *(Milit.)* unit

Abteilungs·leiter der head of department/section; departmental manager

ab|tippen *tr. V. (ugs.)* type out

Äbtissin [ɛp'tɪsɪn] **die; ~, ~nen** abbess

ab|tönen *tr. V.* tint

ab|töten *tr. V.* destroy ‹*parasites, germs*›; deaden ‹*nerve, feeling*›; mortify ‹*desire*›

ab|tragen *unr. tr. V.* **a)** *(abnutzen)* wear out; **b)** *(geh.: abräumen)* clear away; **c)** *(einebnen)* level; *(Geol.)* erode; **d)** *(abbauen)* demolish

abträglich ['aptrɛ:klɪç] *Adj. (geh.)* **einer Sache** *(Dat.)* **~ sein** be detrimental *or* harmful to sth.

ab|trainieren *tr. V.* [sich *(Dat.)*] **Fett/Pfunde ~:** get rid of fat/pounds

Ab·transport der *s.* **abtransportieren:** taking away; removal; dispatch

ab|transportieren *tr. V.* take away; remove ‹*dead, injured*›; *(befördern)* dispatch ‹*goods*›

ab|treiben 1. *unr. tr. V.* **a)** *(wegtreiben)* carry away; **jmdn./ein Schiff vom Kurs ~:** drive *or* carry sb./a ship off course; **b)** abort ‹*foetus*›; **ein Kind ~ lassen** have an abortion. **2.** *unr. itr. V.* **a)** *mit sein (weggetrieben werden)* be carried away; ‹*ship*› be carried off course; **b)** *(einen Abort vornehmen lassen)* have an abortion

Abtreibung die; ~, ~en abortion

ab|trennen *tr. V.* **a)** detach; sever ‹*arm, leg, etc.*›; cut off ‹*button, collar, etc.*›; detach ‹*paper, voucher*›; **b)** *(abteilen)* divide off

ab|treten 1. *unr. tr. V.* **a) sich** *(Dat.)* **die Füße/Schuhe ~:** wipe one's feet; **b) jmdm. etw. ~:** let sb. have sth.; **c)** *(Rechtsw.)* transfer; cede ‹*territory*›; **d)** *(abnutzen)* wear down. **2.** *unr. itr. V.; mit sein* **a)** *(Milit.)* dismiss; **b)** *(Theater, auch fig.)* exit; make one's exit; **von der Bühne ~** *(fig.)* step down; leave the arena; **c)** *(zurücktreten)* step down; ‹*monarch*› abdicate. **3.** *unr. refl. V. (sich abnutzen)* become worn; **sich leicht/schnell ~:** wear [out] easily/quickly

Abtreter der; ~s, ~ *s.* **Fußabtreter**

Ab·tritt der a) *(Theater)* exit; **b)** *(Rücktritt)* resignation; **c)** *(veralt.: Toilette)* privy *(arch.)*

ab|trocknen 1. *tr. V.* dry; **sich** *(Dat.)* **die Hände/das Gesicht/die Tränen ~:** dry one's hands/face/tears. **2.** *itr. V.; mit sein* dry off

ab|tropfen *itr. V.; mit sein* drip off; ‹*lettuce, dishes*› drain; ‹*clothing*› drip-dry; **von etw. ~:** drip off sth.

abtrünnig *Adj. (einer Partei)* renegade; *(einer Religion, Sekte)* apostate; **der Kirche/dem Glauben ~ werden** desert the Church/the faith

Abtrünnige der/die; *adj. Dekl. (einer Partei)* renegade; deserter; *(einer Religion, Sekte)* apostate; turncoat

ab|tun *unr. tr. V.* dismiss; **etw. mit einer Handbewegung ~:** wave sth. aside

ab|tupfen *tr. V.* dab away; **sich** *(Dat.)* **die Stirn ~:** dab one's brow

ab|verlangen *tr. V.* **jmdm. etw. ~:** demand sth. of sb.

ab|wägen *unr. od. regelm. tr. V.* weigh up; **die Vor- und Nachteile gegeneinander ~:** weigh the advantages and disadvantages; *s. auch* **abgewogen**

ab|wählen *tr. V.* vote out; drop ‹*school subject*›

ab|wälzen *tr. V.* pass on (**auf** + *Akk.* to); shift ‹*blame, responsibility*› (**auf** + *Akk.* on to)

ab|wandeln *tr. V.* adapt; modify

ab|wandern *itr. V.; mit sein* **a)** migrate (**aus** from, **in** + *Akk.* to); *(in ein anderes Land)* emigrate (**aus** from, **in** + *Akk.* to); **b)** *(fig.)* move over

Ab·wanderung die a) migration (**aus** from, **in** + *Akk.* to); *(in ein anderes Land)* emigration (**aus** from, **in** + *Akk.* to); **b)** *(fig.)* moving over

Ab·wandlung die adaptation; modification

ab|warten 1. *itr. V.* wait; **sie warteten ab** they awaited events; **warten wir [erst mal] ab** let's wait and see; **sich ~d verhalten** adopt an attitude of 'wait and see'. **2.** *tr. V.* wait for; **etw. ~** *(das Ende von etw. ~)* wait for sth. to end

abwärts ['apvɛrts] *Adv.* downwards; *(bergab)* downhill; **den Fluß ~:** downstream; **der Fahrstuhl fährt ~:** the lift is going down

-abwärts *adv.* **rhein~/fluß~:** down the Rhine/down river *or* downstream

abwärts|gehen *unr. itr. V.; mit sein*; **seit damals ging es eigentlich immer nur ~:** from that time on things really only got worse

[1]Abwasch ['apvaʃ] **der; ~[e]s** washing-up *(Brit.)*; washing dishes *(Amer.)*; **den ~ machen** do the washing-up/wash the dishes

[2]Abwasch die; ~, ~en *(österr.)* sink

abwaschbar *Adj.* washable

ab|waschen 1. *unr. tr. V.* **a)** wash off; **etw. von etw. ~:** wash sth. off sth.; **b)** *(reinigen)* wash [up] ‹*dishes*›; wash down ‹*surface*›. **2.** *unr. itr. V.* wash up, do the washing-up *(Brit.)*; wash the dishes *(Amer.)*

Ab·wasser das; *Pl.* **-wässer** sewage

ab|wechseln *refl., itr. V.* alternate; **wir wechselten uns ab** we took turns; **ich wechsle mich mit ihr beim Geschirrspülen ab** she and I take it in turns to do the dishes; **Regen und Sonne wechselten miteinander ab** it rained and was sunny by turns

abwechselnd *Adv.* alternately

Abwechslung die; ~, ~en variety; *(Wechsel)* change; **etwas/wenig ~:** some/not much variety; **zur ~:** for a change

abwechslungs·reich 1. *Adj.* varied. **2.** *adv.* **der Urlaub verlief sehr ~reich** the holiday *or (Amer.)* vacation was full of variety; **sich ~reich ernähren** eat a varied diet

Ab·weg der: auf ~e kommen *od.* **geraten** go astray; **jmdn. auf ~e führen** lead sb. astray

abwegig *Adj. (irrig)* erroneous; false ‹*suspicion*›; *(falsch)* mistaken; wrong

Ab·wehr die; ~ a) *(Ablehnung)* hostility; **b)** *(Zurückweisung)* repulsion; *(von Schlägen)* fending off; **c)** *(Widerstand)* resistance; **d)** *(Milit.: Geheimdienst)* counter-intelligence; **e)** *(Sport) (Hintermannschaft)* defence; *(~aktion)* clearance; clearing *(Amer.)*

ab|wehren 1. *tr. V.* **a)** repulse; fend off, parry ‹*blow*›; *(Sport)* clear ‹*ball, shot*›; save ‹*match point*›; **b)** *(abwenden)* avert ‹*danger, consequences*›; **c)** *(von sich weisen)* avert ‹*suspicion*›; deny ‹*rumour*›; decline ‹*thanks*›; **d)** *(fernhalten)* deter. **2.** *itr. V.* **a)** *(Sport)* clear; **zur Ecke ~:** clear the ball and give away *or* concede a corner; **b)** *(ablehnend reagieren)* demur

Abwehr-: ~kraft die power of resistance; **~spieler der** *(Sport)* defender

ab|weichen *unr. itr. V.; mit sein* **a)** deviate; **b)** *(sich unterscheiden)* differ

Abweichler ['apvaiçlɐ] **der; ~s, ~** *(Politik)* deviationist

Abweichung die; ~, ~en a) deviation; **b)** *(Unterschied)* difference

ab|weisen *unr. tr. V.* **a)** turn away; turn down ‹*suitor, applicant*›; **b)** *(ablehnen)* reject; dismiss ‹*action, case, complaint*›; disallow ‹*claim*›

abweisend 1. *Adj.* cold ‹*tone of voice; look*›; **in ~em Ton** coldly. **2.** *adv.* coldly

Ab·weisung die; ~, ~en *s.* **abweisen:** turning away; turning down; rejection; dismissal; disallowance

ab|wenden 1. *unr. od. regelm. tr. V.* **a)** *(wegwenden)* turn away; **den Blick ~:** look away; avert one's gaze; **b)** *nur regelm. (verhindern)* avert. **2.** *unr. od. regelm. refl. V.* **a)** turn away; **b)** *(fig.)* **sich von jmdm. ~:** turn one's back on sb.

Ab·wendung die *(Verhinderung)* **zur ~ einer Sache** *(Gen.)* in order to avert sth.

ab|werben *unr. tr. V.* lure away, entice away (*Dat.* from)

ab|werfen *unr. tr. V.* **a)** drop; ‹*tree*› shed ‹*leaves, needles*›; ‹*stag*› shed ‹*antlers*›; throw off ‹*clothing*›; jettison ‹*ballast*›; throw ‹*rider*›; *(Kartenspiel)* discard; *(fig.)* cast *or* throw off ‹*yoke of tyranny etc.*›; **b)** *(herunterstoßen)* knock down; **c)** *(ins Spielfeld werfen)* throw out ‹*ball*›; **d)** *(einbringen)* bring in; **viel ~:** show a big profit. **2.** *unr. itr. V. (Sport)* throw the ball out

ab|werten 1. *tr., itr. V.* devalue. **2.** *tr. V. (fig.: herabwürdigen)* run down; belittle

abwertend 1. *Adj.* derogatory. **2.** *adv.* derogatorily; in a derogatory way

Ab·wertung die a) devaluation; **b)** *(fig.: Herabwürdigung)* reduction in status

abwesend 1. *Adj.* **a)** absent; **b)** *(zerstreut)* absent-minded. **2.** *adv.* absent-mindedly

Abwesende der/die; *adj. Dekl.* absentee

Abwesenheit die; ~ a) absence; **durch ~ glänzen** *(iron.)* be conspicuous by one's absence; **b)** *(fig.: Zerstreutheit)* absent-mindedness

ab|wetzen *tr., refl. V.* wear away

ab|wickeln *tr. V.* **a)** unwind; **b)** *(erledigen)* deal with ‹*case*›; do ‹*business*›; *(im Auftrag)* handle ‹*correspondence*›; conduct, handle ‹*transaction, negotiations*›

Abwicklung die; ~, ~en *s.* **abwickeln b:** dealing (*Gen.* with); doing; handling; conducting

ab|wiegeln 1. *tr. V.* pacify; calm down ‹*crowd*›. **2.** *itr. V. (abwertend)* appease

ab|wiegen *unr. tr. V.* weigh out; weigh ‹*single item*›

Abwieglung die; ~, ~en a) conciliation; **b)** *(abwertend)* appeasement

ab|wimmeln *tr. V. (ugs.)* get rid of ‹*person*›; get out of ‹*duty, responsibility, etc.*›

ab|winken 1. *itr. V.* **uninteressiert ~:** wave it/them aside uninterestedly. **2.** *tr. V. (Motorsport)* **ein Rennen ~:** wave the chequered flag; *(bei einer Unterbrechung)* stop a race

ab|wischen *tr. V.* **a)** *(wegwischen)* wipe away; **sich/jmdm. etw. ~:** wipe sth. off oneself/sb.; **sich/jmdm. die Tränen ~:** dry one's/sb.'s tears; **b)** *(säubern)* wipe; **sich/jmdm. die Nase/die Hände** *usw.* **~:** wipe one's/sb.'s nose/hands *etc.* **(an +** *Dat.* on)

ab|wracken ['apvrakn̩] *tr. V.* scrap

Ab·wurf der a) dropping; *(von Ballast)* jettisoning; **b)** *(Fußball)* **beim ~ stolperte der Torwart** the goalkeeper stumbled as he threw the ball out; **c)** *(Handball, Wasserball)* goal throw

ab|würgen *tr. V. (ugs.)* stifle; choke off; squash ‹*proposal*›; stall ‹*car, engine*›

ab|zahlen *tr. V.* pay off ‹*debt, loan*›

ab|zählen 1. *tr. V.* count; **„bitte das Fahrgeld abgezählt bereithalten!"** 'please tender exact fare'. **2.** *itr. V.* **a)** *(Sport, Milit.)* number off; **zu zweien/vieren ~:** number off in twos/fours; **b)** *(mit Abzählreim)* count out

Abzähl·reim der counting-out rhyme

Ab·zahlung die; ~, ~en paying off; repayment; **etw. auf ~ kaufen/verkaufen** buy/sell sth. on easy terms *or (Brit.)* on HP

ab|zapfen *tr. V.* tap ‹*beer, wine*›; let, draw off ‹*blood*›; draw off ‹*petrol*›; **Strom ~:** tap the electricity supply

Ab·zeichen das a) *(Kennzeichen)* emblem; *(fig.)* badge; **b)** *(Anstecknadel, Plakette)* badge

ab|zeichnen 1. *tr. V.* **a)** *(nachzeichnen, kopieren)* copy; **b)** *(signieren)* initial. **2.** *refl. V.* stand out; *(fig.)* begin to emerge; *(drohend)* loom

Abzieh·bild das transfer

ab|ziehen 1. *unr. tr. V.* **a)** pull off; peel off ‹*skin*›; strip ‹*bed*›; **b)** *(Fot.)* make a print/prints of; **c)** *(Druckw.)* run off; **etw. 50mal ~:** run off 50 copies of sth.; **d)** *(Milit., auch fig.)* withdraw; **e)** *(subtrahieren)* subtract; take away; *(abrechnen)* deduct; **f)** *(schälen)* peel ‹*peach, almond, tomato*›; string ‹*runner bean*›; **g)** *(häuten)* skin; **h) eine Handgranate ~:** pull the pin of a hand-grenade; **i)** *(herausziehen)* take out ‹*key*›. **2.** *unr. itr. V., mit sein* **a)** *(sich verflüchtigen)* escape; **b)** *(Milit.)* withdraw; **c)** *(ugs.: weggehen)* push off *(sl.)*; go away

ab|zielen *itr. V.* **auf etw.** *(Akk.)* **~:** be aimed at *or* directed towards sth.

Ab·zug der a) *(an einer Schußwaffe)* trigger; **b)** *(Fot.)* print; **c)** *(Druckw.)* proof; **d)** *(Verminderung)* deduction; **e)** *o. Pl. (Abmarsch, auch fig.)* withdrawal; **f)** *(Öffnung für Rauch usw.)* vent

abzüglich ['aptsy:klɪç] *Präp. mit Gen. (Kaufmannsspr.)* less; **~ 3% Rabatt** less 3% discount

ab|zweigen 1. *itr. V.; mit sein* branch off. **2.** *tr. V.* **a)** *(bereitstellen)* set *or* put aside; **Geld für einen Plattenspieler ~:** put aside *or* put by money to buy a record-player; **b)** *(verhüll.: sich heimlich aneignen)* appropriate

Abzweigung die; ~, ~en turn-off; *(Gabelung)* fork

Accessoire [aksɛ'so̯a:ɐ̯] **das; ~s, ~s** *(geh.)* accessory

Aceton [atse'to:n] **das; ~s** *(Chemie)* acetone

ach [ax] *Interj.* **a)** *(betroffen, mitleidig)* oh [dear]; **~ Gott** o dear; **b)** *(bedauernd, unwirsch)* oh; **c)** *(klagend)* ah; alas *(dated)*; **d)** *(erstaunt)* oh; **~, ja** *od.* **wirklich?** no, really?; **~, der!** oh, him!; **e)** *in:* **~ so!** oh, I see; **~ nein** no, no; **~ was** *od.* **wo!** of course not

Ach das; ~s: mit ~ und Krach *(ugs.)* by the skin of one's teeth

Achat [a'xa:t] **der; ~[e]s, ~e** *(Min.)* agate

Ach-Laut der velar fricative; ach-laut

Achse ['aksə] **die; ~, ~n a)** *(Rad~)* axle; **auf ~ sein** *(ugs.)* be on the road *or* move; **b)** *(Dreh~, Math., Astron.)* axis; **sich um die** *od.* **seine eigene ~ drehen** turn on one's/its own axis

Achsel ['aksl̩] **die; ~, ~n** *(Schulter)* shoulder; *(~höhle)* armpit; **jmdn. über die ~ ansehen** look down on sb.; look down one's nose at sb.; **die** *od.* **mit den ~n zucken** shrug one's shoulders; **jmdn. unter den ~ packen** seize sb. under the arms

achsel-, Achsel-: ~haare *Pl.* hair *sing.* under one's arms; armpit hair *sing.*; **~höhle die** armpit; **~zucken das** shrug [of the shoulders]; **~zuckend** *Adj.* shrugging; **er ging ~zuckend hinaus** he went out with a shrug [of the shoulders]

-achsig *Adj.* **drei-/sechs~:** three-/six-axle

[1]**acht** [axt] *Kardinalz.* eight; **er ist ~ [Jahre]** he is eight [years old]; **um ~ [Uhr]** at eight [o'clock]; **in ~ Tagen** in a week's time; a week from now; **Freitag/morgen in ~ Tagen** a week on Friday/a week tomorrow; **die Linie ~ [der Straßenbahn]** the number eight [tram]; **es steht ~ zu ~/~ zu 2** *(Sport)* the score is eight all/eight [to] two

[2]**acht: wir waren zu ~:** there were eight of us; **wir rückten ihm zu ~ auf die Bude** *(ugs.)* eight of us dropped in on him

[3]**acht: etw. außer ~ lassen** disregard *or* ignore sth.; **sich in ~ nehmen** take care; be careful; **sich vor jmdm./etw. in ~ nehmen** be wary of sb./sth.

acht... *Ordinalz.* eighth; **der ~e** *od.* **8. September** the eighth of September; *(im Brief auch)* 8 September; **am ~en** *od.* **8. September** on the eighth of September; *(im Brief auch)* 8 September; **München, [den] 8. Mai 1984** Munich, 8 May 1984

[1]**Acht die; ~, ~en a)** *(Zahl)* eight; **eine arabische/römische ~:** an arabic/Roman eight; **b)** *(Figur)* figure eight; **c)** *(ugs.: Verbiegung)* buckle; **das Rad hat eine ~:** the wheel is buckled; **d)** *(Spielkarte)* eight; **e)** *(ugs.: Bus-, Bahnlinie)* [number] eight

[2]**Acht die; ~** *(hist.)* outlawry; **jmdn. in ~ und Bann tun** *(kirchlich)* anathematize sb.; put the ban on sb.; *(fig.)* ostracize sb.

-acht die *(Kartenspiel)* eight of ...

achtbar *Adj. (geh.)* respectable; upright ‹*principles*›; **eine ~e Leistung** a creditable performance

Achte der/die; *adj. Dekl.* eighth; **er war [in der Leistung] der ~:** he came eighth; **der ~ [des Monats]** the eighth [of the month]

acht-, Acht-: ~eck das; ~s, ~e octagon; **~eckig** *Adj.* octagonal; **~einhalb** *Bruchz.* eight and a half

achtel ['axtl̩] *Bruchz.* eighth; **ein ~ Kilo** an eighth of a kilo

Achtel das (*schweiz. meist* **der**); **~s, ~ a)** eighth; **b)** *(ugs.: ~pfund)* eighth of a pound; ≈ two ounces; **c)** *(ugs.: ~liter)* eighth of a litre *(of wine)*; **d)** *(Musik) s.* **Achtelnote**

Achtel-: ~liter der, *auch:* **das** eighth of a litre; **~note die** *(Musik)* quaver; **~pause die** *(Musik)* quaver rest; **~pfund das** eighth of a pound

achte·mal *in* **das ~:** for the eighth time

achten 1. *tr. V.* respect. **2.** *itr. V.* **auf etw.** *(Akk.)* **[nicht] ~** *([nicht] auf etw. aufpassen)* [not] mind *or* look after sth.; *(von etw. [keine] Notiz nehmen)* pay [no] attention *or* heed to sth.; **auf jmdn. ~:** look out for sb.; *(aufpassen)* look after sb.; keep an eye on sb.

ächten ['ɛxtn̩] *tr. V.* **a)** *(hist.)* outlaw; **b)** *(gesellschaftlich)* ostracize; **sich geächtet fühlen** feel like an outcast; **c)** *(verdammen)* ban ‹*war, torture, etc.*›

Acht·ender der; ~s, ~ *(Jägerspr.)* eight-pointer

achten·mal *in* **beim ~:** the eighth time; at the eighth attempt *etc.*; **zum ~:** for the eighth time

achtens ['axtn̩s] *Adv.* eighthly

Achter ['axtɐ] **der; ~s, ~ a)** *(Rudern)* eight; **b)** *s.* [1]**Acht a, b, c**

Achter·bahn die roller-coaster; **~bahn fahren** go *or* ride on the roller-coaster

achterlei *Gattungsz.; indekl.* **a)** *attr.* eight kinds *or* sorts of; eight different ‹*sorts, kinds, sizes*›; **b)** *alleinstehend* eight [different] things

achtern *Adv. (Seemannsspr.)* astern; aft

acht·fach *Vervielfältigungsz.* eightfold; **die ~e Menge** eight times the quantity; **etw. in ~er Ausfertigung schicken** send eight copies of sth.; **~ vergrößert/verkleinert, in ~er Vergrößerung/Verkleinerung** magnified *or* enlarged/reduced eight times

Acht·fache das; *adj. Dekl.* **das ~e von 4 ist 32** eight fours are *or* eight times four makes 32; **um das ~e steigen** increase ninefold *or* nine times

acht|geben *unr. itr. V.* **a) auf jmdn./etw. ~:** take care of *or* mind sb./sth.; **[auf jmds. Worte] ~:** pay attention [to what sb. says]; **~ müssen, daß ...:** have to be careful that ...; **b)** *(vorsichtig sein)* be careful; watch out; **gib acht!** look out!; watch out!; **auf sich** *(Akk.)* **~:** be careful

acht-, Acht-: ~hundert *Kardinalz.* eight hundred; **~jährig** *Adj. (8 Jahre alt)* eight-year-old *attrib.*; eight years old *pred.*; *(8 Jahre dauernd)* eight-year *attrib.*; **~jährige der/die;** *adj. Dekl.* eight-year-old; **~kampf der** *(Turnen)* eight-exercise gymnastic competition; **~kantig 1.** *Adj. (Technik)* eight-sided; **2.** *adv. (salopp)* **~kantig rausfliegen** get kicked *or (sl.)* booted out; **~köpfig** *Adj.* eight-headed ‹*monster*›; ‹*family, committee*› of eight

acht·los 1. *Adj.; nicht präd.* heedless. **2.** *adv.* heedlessly

acht-, Acht-: ~mal *Adv.* eight times; **~mal so groß/soviel/so viele** eight times as big/as much/as many; **~malig** *Adj.; nicht präd.* **nach ~maliger Aufforderung** at the eighth request; after being asked eight times; **~monatig** *Adj. (8 Monate alt)* eight-month-old *attrib.*; eight months old *pred.*; *(8 Monate dauernd)* eight-month *attrib.*; **~monatlich 1.** *Adj.* eight-monthly; **im ~monatlichen Turnus** rotating every eight months; **2.** *adv.* every eight months; **~prozentig** *Adj.* eight per cent

achtsam 1. *Adj. (geh.)* attentive. **2.** *adv. (sorgsam)* carefully; with care; **mit etw. [äußerst] ~ umgehen** handle sth. with [extreme] care

acht-, Acht-: ~seitig *Adj.* eight-page *attrib.* ‹*letter, article*›; **~spaltig** *(Druckw.)* **1.** *Adj.* eight-column *attrib.*; **~spaltig sein** have eight columns; **2.** *adv.* in eight columns; **~spänner der; ~s, ~:** eight-in-hand; **~spurig** *Adj.* eight-lane ‹*road*›; eight-track ‹*tape*›; **~ sein** have eight lanes/tracks; **~stellig** *Adj.* eight-figure *attrib.*; **~stellig sein** have eight figures *or* digits; **~stimmig 1.** *Adj.* eight-part *attrib.*; **2.** *adv.* in eight parts; **~stöckig** *Adj.* eight-storey *attrib.*; **~stöckig sein** have eight storeys *or* floors; **~strophig** [-ʃtroːfɪç] *Adj.* with eight verses *postpos., not pred.*; **~strophig sein** have eight verses; **~stunden·tag** [-'---] **der** eight-hour day; **~stündig** *Adj.* eight-hour *attrib.*; **mit ~stündiger Verspätung** eight hours late; **nach ~stündigem Warten** after waiting for eight hours; **~tägig** *Adj. (8 Tage alt)* eight-day-old *attrib.*; *(8 Tage dauernd)* eight-day[-long] *attrib.*; **mit ~tägiger Verspätung** eight days late; **~tausend** *Kardinalz.* eight thousand; **~tausender der** *mountain over eight thousand metres high*; **~teilig** *Adj.* eight-piece ‹*tea-service, tool-set, etc.*›; eight-part ‹*series, serial*›

Acht·uhr-: eight o'clock ‹*news, train, performance, etc.*›

Achtung die; ~ a) *(Wertschät-*

zung) respect (*Gen.*, **vor** + *Dat.* for); **alle ~!** well done!; **b)** *(Aufmerksamkeit)* attention; **~!** watch out!; **~! Stillgestanden!** *(Milit.)* attention!; **~, ~!** your attention, please!; **„~, Stufe!"** 'mind the step'; **~, fertig, los!** on your marks, get set, go!

Ạchtung die; ~, ~en a) *(hist.)* outlawing; **b)** *(gesellschaftliche ~)* ostracism; **c)** *(Verdammung)* banning

ạcht·zehn *Kardinalz.* eighteen; **mit ~ [Jahren] wird man volljährig** one reaches the age of majority at eighteen; **18 Uhr** 6 p.m.; *(auf der 24-Stunden-Uhr)* eighteen hundred hours; 1800; **18 Uhr 33** 6.33 p.m.; *(auf der 24-Stunden-Uhr)* 1833

ạchtzehn, Ạchtzehn-: ~hundert *Kardinalz.* eighteen hundred; **~jährig** *Adj. (18 Jahre alt)* eighteen-year-old *attrib.*; eighteen years old *pred.*; *(18 Jahre dauernd)* eighteen-year *attrib.*; **~jährige der/die;** *adj. Dekl.* eighteen-year-old

ạchtzehnt... *Ordinalz.* eighteenth; *s. auch* **acht...**

achtzig ['axtsɪç] *Kardinalz.* eighty; **mit ~ [km/h] fahren** drive at *or (coll.)* do eighty [k.p.h.]; **über/etwa ~ [Jahre alt] sein** be over/about eighty [years old]; **mit ~ [Jahren]** at eighty [years of age]; **Mitte [der] Achtzig sein** be in one's mid-eighties; **auf ~ sein** *(fig. ugs.)* be hopping mad *(coll.)*

ạchtziger *indekl. Adj.; nicht präd.* **ein ~ Jahrgang** an '80 vintage; **die ~ Jahre** the eighties

[1]Ạchtziger der; ~s, ~ *(80jähriger)* eighty-year-old [man]; octogenarian

[2]Ạchtziger die; ~, ~ *(ugs.) (Briefmarke)* eighty-pfennig/schilling *etc.* stamp

Ạchtzigerin die; ~, ~nen eighty-year-old [woman]; octogenarian

ạchtzig·jährig *Adj. (80 Jahre alt)* eighty-year-old *attrib.*; eighty years old *pred.*; *(80 Jahre dauernd)* eighty-year *attrib.*

achtzigst... ['axtsɪçst] *Ordinalz.* eightieth

Ạchtzigstel das; ~s, ~: eightieth

Ạcht-: ~zimmer·wohnung die eight-roomed flat; **~zylinder der** *(ugs.)* eight-cylinder [engine/car]

ächzen ['ɛçtsn̩] *itr. V.* **a)** *(schwer stöhnen)* groan; **b)** *(knarren)* creak

Acker ['akɐ] **der; ~s, Äcker** ['ɛkɐ] field; **auf dem ~:** in the field

Ạcker-: ~bau der; *o. Pl.* agriculture *no indef. art.*; farming *no indef. art.*; **~bau treiben** farm; **~bau und Viehzucht** farming and stock-breeding; **~furche die** furrow; **~land das;** *o. Pl.* farmland

ạckern *itr. V. (salopp)* slog one's guts out *(coll.)*

a. D. [a:'de:] *Abk.* **außer Dienst** retd.

A. D. *Abk.* **Anno Domini** AD

ad absurdum [at ap'zʊrdʊm] *in* **etw. ~ ~ führen** demonstrate the absurdity of sth.

ADAC [a:de:a:'tse:] **der; ~** *Abk.* **Allgemeiner Deutscher Automobilclub**

ad acta [at 'akta] *in* **etw. ~ ~ legen** shelve sth.

Adagio [a'da:dʒo] **das; ~s, ~s** *(Musik)* adagio

Adam ['a:dam] **(der)** Adam; **seit ~s Zeiten** since the beginning of time; **bei ~ und Eva anfangen** *(ugs.)* begin from the beginning

A̱dam Ri̱ese: das macht nach ~ ~ 4,50 Mark *(ugs. scherzh.)* my arithmetic makes it 4.50 marks *(coll. joc.)*

A̱dams-: ~apfel der *(ugs.)* Adam's apple; **~kostüm das** *(scherzh.)* **im ~kostüm** in one's birthday suit

adäquat [at|ɛ'kva:t] **1.** *Adj. (passend)* appropriate (*Dat.* to); suitable (*Dat.* for); *(angemessen)* adequate ⟨*reward, payment*⟩; appropriate, suitable ⟨*measures, means*⟩. **2.** *adv. (passend)* suitably; appropriately; *(angemessen)* adequately

addieren [a'di:rən] **1.** *tr. V.* add [up]. **2.** *itr. V.* add

Addition [adi'tsi̯o:n] **die; ~, ~en** addition

ade [a'de:] *Interj. (veralt., landsch.)* farewell; **jmdm. ~ sagen** bid farewell to sb.; take one's leave of sb.

Adel ['a:dl̩] **der; ~s** nobility; **von ~ sein** be of noble blood; **~ verpflichtet** noblesse oblige

a̱delig *s.* **adlig**

A̱delige *s.* **Adlige**

a̱deln *tr. V.* **jmdn. ~:** give sb. a title; *(in den hohen Adel erheben)* raise sb. to the peerage; *(fig.)* ennoble sb.

Adels-: ~familie die, ~geschlecht das noble family; **~prädikat das** title of nobility; **~stand der** nobility; *(hoher Adel)* nobility; peerage; **jmdn. in den ~stand erheben** give sb. a title/raise sb. to the peerage; **~titel der** title

Ader ['a:dɐ] **die; ~, ~n a)** *(Anat., Zool.)* blood-vessel; vein; **b)** *o. Pl. (Anlage, Begabung)* streak; **c)** *(Bot., Geol.)* vein; **d)** *(Elektrot.)* core

Aderlaß ['a:dɐlas] **der; A̱derlasses, A̱derlässe** [-lɛsə] *(Med.)* bleeding

adieu [a'di̯ø:] *Interj. (veralt.)* adieu; farewell; **jmdm. ~ sagen** bid sb. adieu *or* farewell

Adjektiv ['atjɛkti:f] **das; ~s, ~e** *(Sprachw.)* adjective

Adjutant [atju'tant] **der; ~en, ~en** adjutant; aide-de-camp

Adler ['a:dlɐ] **der; ~s, ~:** eagle

A̱dler-: ~auge das *(fig.)* eagle eye; **~horst der** eyrie; **~nase die** aquiline nose

adlig ['a:dlɪç] *Adj.* noble; **~ sein** be a noble [man/woman]

A̱dlige der/die *adj. Dekl.* noble [man/woman]

Administration [atmɪnɪstra'tsi̯o:n] **die; ~, ~en** administration

administrativ [atmɪnɪstra'ti:f] **1.** *Adj.* administrative. **2.** *adv.* administratively

Admiral [atmi'ra:l] **der; ~s, ~e** *od.* **Admiräle** [atmi'rɛ:lə] **a)** admiral; **b)** *(Schmetterling)* red admiral

Admiralität [atmirali'tɛ:t] **die; ~, ~en** admiralty

ADN [a:de:'|ɛn] *Abk. (ehem. DDR)* **Allgemeiner Deutscher Nachrichtendienst** *GDR press agency*

adoptieren [adɔp'ti:rən] *tr. V.* adopt

Adoption [adɔp'tsi̯o:n] **die; ~, ~en** adoption

Adoptiv- [adɔp'ti:f-]: **~eltern** *Pl.* adoptive parents; **~kind das** adoptive *or* adopted child

Adrenalin [adrena'li:n] **das; ~s** *(Physiol., Med.)* adrenalin

Adressat [adrɛ'sa:t] **der; ~en, ~en, Adressa̱tin die; ~, ~nen** addressee

Adrẹß·buch das directory

Adresse [a'drɛsə] **die; ~, ~n a)** address; **unter folgender ~:** at the following address; **bei jmdm. an die falsche ~ kommen** *od.* **geraten** *(fig. ugs.)* come to the wrong address *(fig.)*; **bei jmdm. an der falschen ~ sein** *(fig. ugs.)* have come to the wrong place *(fig.)*; **b)** *(geh.: Botschaft)* message

adressi̱eren *tr. V.* address

adrett [a'drɛt] **1.** *Adj.* smart. **2.** *adv.* smartly

Adria ['a:dri̯a] **die; ~:** Adriatic

adriatisch [adri'a:tɪʃ] *Adj.* Adriatic; **das Adriatische Meer** the Adriatic Sea

A̱-Dur das; ~ *(Musik)* A major

Advent [at'vɛnt] **der; ~s a)** Advent; **b)** *(Adventssonntag)* Sunday in Advent

Advents-: ~kalender der Advent calendar; **~kranz der** *garland of evergreens with four candles for the Sundays in Advent;* **~sonntag der** Sunday in Advent

Adverb [at'vɛrp] **das; ~s, ~ien** *(Sprachw.)* adverb

adverbial [atvɛr'bi̯a:l] *(Sprachw.)* **1.** *Adj.* adverbial. **2.** *adv.* adverbially; as an adverb

Advokat [atvo'ka:t] **der; ~en, ~en** *(österr., schweiz., sonst veralt.)* lawyer; advocate *(arch.); (fig.: Fürsprecher)* advocate

aero-, Aero- [aero- *od.* ɛ:ro-]: **~dynamik die** aerodynamics *sing.;* **~dynamisch** *Adj.* aerodynamic; **~gramm das** air[-mail] letter; **~sol** [~'zo:l] **das; ~s, ~e** aerosol

Affäre die; ~, ~n affair; **sich aus der ~ ziehen** *(ugs.)* get out of it

Affe ['afə] **der; ~n, ~n a)** monkey; *(Menschen~)* ape; **b)** *(dummer Kerl)* oaf; clot *(Brit. sl.);* **c)** *(Geck)* dandy; **ein eingebildeter ~:** a conceited so-and-so *(coll.);* **d)** *(Milit. ugs.)* knapsack

Affekt [a'fɛkt] **der; ~[e]s, ~e** feeling; emotion; affect *(Psych.);* **im ~:** in the heat of the moment

Affekt·handlung die emotive act

affektiert [afɛk'ti:ɐ̯t] *(abwertend)* **1.** *Adj.* affected. **2.** *adv.* affectedly

affen-, Affen-: ~artig *Adj.* apelike; **mit ~artiger Geschwindigkeit** *(ugs.)* like a bat out of hell *(coll.);* **~hitze die** *(salopp)* blazing heat; **gestern war eine ~hitze** yesterday was a real scorcher; **~mensch der** ape-man; **~schande die** *(salopp)* **es ist eine ~schande** it's monstrous; **~tempo das** *(salopp)* **mit einem ~tempo** like mad; like the clappers *(Brit. sl.);* **ein ~tempo anschlagen** move like hell *(coll.);* **~theater das** *(salopp)* farce; **~zahn der** *(salopp) s.* **~tempo**

affig *(ugs. abwertend)* **1.** *Adj.* dandyish; *(lächerlich)* ludicrous; *(affektiert)* affected. **2.** *adv.* in a dandyish/a ludicrous/an affected way

Äffin ['ɛfɪn] **die; ~, ~nen** female ape

Affront [a'frõ:] **der; ~s, ~s** affront

Afghane [af'ga:nə] **der; ~n, ~n a)** Afghan; **b)** *(Hund)* Afghan hound

Afghanin die; ~, ~nen Afghan

afghanisch *Adj.* Afghan

Afghanistan [af'ga:nɪsta:n] **(das); ~s** Afghanistan

Afrika ['a:frika] **(das); ~s** Africa

Afrikaner [afri'ka:nɐ] **der; ~s, ~, Afrikanerin die; ~, ~nen** African

afrikanisch *Adj.* African

Afro·amerikaner der Afro-American

After ['aftɐ] **der; ~s, ~:** anus

AG [a:'ge:] *Abk.* **die; ~, ~s a) Aktiengesellschaft** PLC *(Brit.);* Ltd. *(private company) (Brit.);* Inc. *(Amer.);* **b) Arbeitsgemeinschaft**

Ägäis [ɛ'gɛ:ɪs] **die; ~:** Aegean

ägäisch *Adj.* Aegean

Agave [a'ga:və] **die; ~, ~n** *(Bot.)* agave

Agent [a'gɛnt] **der; ~en, ~en** agent

Agenten-: ~netz das network of agents; **~ring der** spy ring

Agentin die; ~, ~nen [female] agent

Agentur [agɛn'tu:ɐ̯] **die; ~, ~en** agency

Aggregat [agre'ga:t] **das; ~[e]s, ~e** *(Technik)* unit; *(Elektrot.)* set

Aggregat·zustand der *(Chemie, Physik)* state

Aggression [agrɛ'si̯o:n] **die; ~, ~en** aggression

Aggressions·trieb der aggressive drive

aggressiv [agrɛ'si:f] **1.** *Adj.* aggressive. **2.** *adv.* aggressively

Aggressivität die; ~: aggressiveness

Aggressor [a'grɛsɔr] **der; ~s, ~en** [-'so:ren] aggressor

agieren [a'gi:rən] *itr. V.* act

agil [a'gi:l] *Adj. (beweglich)* agile; *(geistig rege)* mentally alert

Agitation [agita'tsi̯o:n] **die; ~** *(Politik)* agitation; **~ betreiben** agitate

Agitator [agi'ta:tɔr] **der; ~s, ~en** [-ta'to:rən] agitator

agitieren 1. *itr. V.* agitate. **2.** *tr. V.* stir up

Agonie [ago'ni:] **die; ~, ~n: in ~ liegen** be in the throes of death

agrarisch *Adj.* agrarian; agricultural

Agrar-: ~land das; *Pl.* **~länder** agrarian country; **~markt der** agrarian *or* agricultural products market; **~politik die** agricultural policy

Agronom [agro'no:m] **der; ~en, ~en** agronomist

Agronomie [agrono'mi:] **die; ~:** agronomy *no art.*

Ägypten [ɛ'gʏptn̩] **(das); ~s** Egypt

Ägypter der; ~s, ~: Egyptian

ägyptisch *Adj.* Egyptian

ah [a:] *Interj. (verwundert)* oh; *(freudig, genießerisch)* ah; *(verstehend)* oh; ah

äh [ɛ(:)] *Interj.* **a)** *(angeekelt)* ugh; **b)** *(stotternd)* er; hum

aha [a'ha(:)] *Interj. (verstehend)* oh[, I see]; *(triumphierend)* aha

Ahn [a:n] **der; ~[e]s,** *od.* **~en, ~en** *(geh.)* forebear; ancestor; *(fig.)* father

ahnden ['a:ndn̩] *tr. V. (geh.)* punish

Ahndung die; ~: punishment

Ahne der; ~n, ~n *s.* **Ahn**

ähneln ['ɛ:nl̩n] *itr. V.* **jmdm. ~:** resemble *or* be like sb.; bear a resemblance to sb.; **jmdm. sehr/wenig ~:** strongly resemble *or* be very like sb./bear little resemblance to sb.; **einer Sache** *(Dat.)* **~:** be similar to sth.; be like sth.; **sich sehr/wenig ~:** resemble each other very strongly *or* be very much alike/bear little resemblance to each other

ahnen ['a:nən] **1.** *tr. V.* **a)** have a presentiment *or* premonition of; **b)** *(vermuten)* suspect; *(erraten)* guess; **wer soll denn ~, daß ...:** who would know that ...; **das konnte ich doch nicht ~!** I had no way of knowing that; **du ahnst es nicht!** *(ugs.)* oh heck *or* Lord! *(coll.);* **c)** *(vage erkennen)* just make out; **die Wagen waren in der Dunkelheit mehr zu ~ als zu sehen** one could sense the cars in the darkness, rather than see them. **2.** *itr. V. (geh.)* **mir ahnt nichts Gutes** I fear the worst; **es ahnte mir, daß ...:** I suspected that ...

Ahnen-: ~forschung die genealogy; **~galerie die** gallery of ancestral portraits; **~tafel die** genealogical table; **~verehrung die** ancestor-worship

ähnlich ['ɛ:nlɪç] **1.** *Adj.* similar; **jmdm. ~ sein** be similar to *or* be like sb.; **jmdm. ~ sehen** resemble sb.; be like sb.; **das Kind ist seinem Vater ~:** the child takes after his father; **[so] ~ wie etw. aussehen/klingen** look/sound like sth.; **das sieht dir/ihm ~!** *(ugs.)* that's you/him all over; that's just like you/him. **2.** *adv.* similarly; ‹*answer, react*› in a similar way *or* manner; **~ dumm/naiv** *usw.* **argumentieren** argue in a similarly stupid/naïve *etc.* way *or* manner; **uns geht es ~:** it is/will be much the same for us; *(wir denken, fühlen ~)* we feel much the same. **3.** *Präp. mit Dat.* like

Ähnlichkeit die; ~, ~en similarity; *(ähnliches Aussehen)* similarity; resemblance; **mit jmdm. ~ haben** be similar to *or* be like sb.; *(im Aussehen)* bear a resemblance to *or* be like sb.; **mit etw. ~ haben** bear a similarity to sth.

Ahnung die; ~, ~en a) presentiment; premonition; **eine ~ haben, daß ...:** have a feeling *or* hunch that ...; **b)** *(Befürchtung)* foreboding; **c)** *(ugs.: Kenntnisse)* knowledge; **von etw. [viel] ~ haben** know [a lot] about sth.; **keine ~!** [I've] no idea; [I] haven't a clue; **haben Sie eine ~, wer/wie ...?** have you any idea who/how ...?

ahnungs·los 1. *Adj.* unsuspecting; *(naiv, unwissend)* naïve; **sich ~ stellen** play the innocent. **2.** *adv.* unsuspectingly; *(naiv, unwissend)* naïvely

ahoi [a'hɔy] *Interj. (Seemannsspr.)* **Boot/Schiff** *usw.* **~!** boat/ship *etc.* ahoy!

Ahorn ['a:hɔrn] **der; ~s, ~e** maple

Ähre ['ɛ:rə] **die; ~, ~n** *(von Getreide)* ear; head; *(von Gräsern)* head

Aids [e:ts] **das; ~:** AIDS

Aids-: ~kranke der/die person suffering from AIDS; **~test der** AIDS test

Akademie [akade'mi:] **die; ~, ~n a)** academy; **b)** *(Bergbau, Forst~, Bau~)* school; college

Akademiker [aka'de:mikɐ] **der; ~s, ~, Akademikerin die; ~, ~nen** [university/college] graduate

akademisch 1. *Adj.* academic. **2.** *adv.* academically

Akazie [a'ka:tsi̯ə] **die; ~, ~n** acacia

Akklamation [aklama'tsi̯o:n] **die; ~, ~en** acclamation

akklimatisieren *refl. V.* become *or* get acclimatized

Akkord [a'kɔrt] **der; ~[e]s, ~e a)** *(Musik)* chord; **b)** *(Wirtsch.)* *(~arbeit)* piece-work; *(~lohn)* piece-work pay *no indef. art., no pl.*; *(~satz)* piece-rate; **im ~ sein** *od.* **arbeiten** be on piece-work

Akkord-: ~arbeit die piece-work; **~arbeiter der** piece-worker

Akkordeon [a'kɔrdeɔn] **das; ~s, ~s** accordion

Akkord·lohn der *(Wirtsch.)* piece-work pay *no indef. art., no pl.*

akkreditieren [akredi'ti:rən] *tr. V. (bes. Dipl.)* accredit (**bei** to)

Akkreditierung die; ~, ~en *(bes. Dipl.)* accreditation (**bei** to)

Akku ['aku] **der; ~s, ~s** *(ugs.) s.* **Akkumulator**

Akkumulator [akumu'la:tɔr] **der; ~s, ~en** [-la'to:rən] *(Technik)* accumulator *(Brit.)*; storage battery *or* cell

akkurat [aku'ra:t] **1.** *Adj.* **a)** *(sorgfältig)* meticulous; *(sauber)* neat; **b)** *(exakt, genau)* precise; exact. **2.** *adv.* *(sorgfältig)* meticulously; *(sauber)* neatly

Akkusativ ['akuzati:f] **der; ~s, ~e** *(Sprachw.)* accusative [case]; **im/mit dem ~ stehen** be in/take the accusative [case]

Akkusativ·objekt das *(Sprachw.)* accusative *or* direct object

Akne ['aknə] **die; ~, ~n** *(Med.)* acne

Akquisiteur [akvizi'tø:ɐ] **der; ~s, ~e** *(Wirtsch.)* canvasser

Akribie [akri'bi:] **die; ~** *(geh.)* meticulousness; meticulous precision

Akrobat [akro'ba:t] **der; ~en, ~en** acrobat

Akrobatik die; ~ a) acrobatic skill; **b)** *(Übungen)* acrobatics *pl.*

Akrobatin die; ~, ~nen acrobat

akrobatisch 1. *Adj.* acrobatic. **2.** *adv.* acrobatically

¹Akt [akt] **der; ~[e]s, ~e a)** act; **b)** *(Zeremonie)* ceremony; ceremonial act; **c)** *(Geschlechtsakt)* sexual act; **d)** *(Aktbild)* nude

²Akt der; ~[e]s, ~en *(bes. südd., österr.) s.* **Akte**

Akt·bild das nude [picture]

Akte die; ~, ~n file; **die ~ Schulze** the Schulze file; **das kommt in die ~n** it goes on file

Akten-: ~deckel der folder; **~koffer der,** *(iron.)*; **~mappe die a)** *(~tasche)* brief-case; **b)** *(~deckel)* folder; **~notiz die** note [for the files]; **~ordner der** file; **~schrank der** filing cabinet; **~tasche die** brief-case; **~zeichen das** reference

Akteur [ak'tø:ɐ] **der; ~s, ~e** person involved; *(Theater)* member of the cast; *(Varieté)* performer

Akt·foto das nude photo

Aktie ['aktsi̯ə] **die; ~, ~n** *(Wirtsch.)* share; **~n** shares *(Brit.)*; stock *(Amer.)*; **die ~n fallen/steigen** share *or* stock prices are falling/rising

Aktien-: ~gesellschaft die joint-stock company; **~index der** share index; **~kurs der** share price; **~mehrheit die** majority shareholding (*Gen.* in); **~paket das** block of shares

Aktion [ak'tsi̯o:n] **die; ~, ~en a)** action *no indef. art.*; *(militärisch)* operation; **politische ~en** political action *sing.*; **b)** *(Kampagne)* campaign; **~ saubere Umwelt** campaign to clean up the environment; **c)** *o. Pl. (das Handeln)* action; **in ~ treten** go into action

Aktionär [aktsi̯o'nɛ:ɐ] **der; ~s, ~e** shareholder

aktions-, Aktions-: ~einheit die united action *no art.* (*Gen.* by); **~fähig** *Adj.* capable of action *postpos.*; **~radius der** radius of action

aktiv [ak'ti:f] **1.** *Adj.* **a)** active; **b)** *(Milit.)* serving *attrib.* ⟨*officer, soldier*⟩; **er ist Soldat im ~en Dienst** he is a serving soldier. **2.** *adv.* actively; **sich ~ verhalten** be active

¹Aktiv ['akti:f] **das; ~s, ~e** *(Sprachw.)* active

²Aktiv [ak'ti:f] **das; ~s, ~e** *od.* **~s** *(bes. ehem. DDR)* committee

Aktiva [ak'ti:va] *Pl. (Wirtsch.)* assets

Aktive [ak'ti:və] **der/die;** *adj. Dekl. (Sport)* participant; *(aktives Mitglied)* active member

aktivieren *tr. V.* mobilize ⟨*party members, group, etc.*⟩; **den Kreislauf ~:** stimulate the circulation

Aktivierung die; ~, ~en *(von Parteimitgliedern, einer Gruppe usw.)* mobilization

aktivisch *(Sprachw.) Adj.* active

Aktivismus der; ~: activism *no art.*

Aktivist der; ~en, ~en, Aktivistin die; ~, ~nen activist

aktivistisch *Adj.* activist

Aktivität [aktivi'tɛ:t] **die; ~, ~en** activity

Aktiv-: ~posten der *(Kaufmannsspr., fig.)* asset; **~seite die** *(Kaufmannsspr.)* assets side

Akt-: ~malerei die nude painting *no art.*; **~modell das** nude model

aktualisieren [aktu̯ali'zi:rən] *tr. V.* update.

Aktualisierung die; ~, ~en updating

Aktualität [aktu̯ali'tɛ:t] **die; ~, ~en a)** *(Gegenwartsbezug)* relevance [to the present]; **b)** *(von Nachrichten usw.)* topicality

aktuell [ak'tu̯ɛl] *Adj.* **a)** *(gegenwartsbezogen)* topical; *(gegenwärtig)* current; *(Mode: modisch)* fashionable; **von ~er Bedeutung** of relevance to the present *or* current situation; **dieses Problem ist nicht mehr ~:** this is no longer a problem; **b)** *(neu)* up-to-the-minute; **das Aktuellste von den Olympischen Spielen** the latest from the Olympics; **eine ~e Sendung** *(Ferns., Rundf.)* a [news and] current affairs programme

Akt-: ~zeichnen das nude drawing *no art.*; **~zeichnung die** nude drawing

akupunktieren *(Med.)* **1.** *tr. V.* perform acupuncture on; **sich ~ lassen** have acupuncture. **2.** *itr. V.* perform acupuncture

Akupunktur [akupʊŋk'tu:ɐ̯] **die; ~, ~en** *(Med.)* acupuncture

Akustik [a'kʊstɪk] **die; ~ a)** *(Lehre vom Schall)* acoustics *sing., no art.;* **b)** *(Schallverhältnisse)* acoustics *pl.*

akustisch 1. *Adj.* acoustic. **2.** *adv.* acoustically; **ich habe Sie ~ nicht verstanden** I didn't hear *or* catch what you said

akut [a'ku:t] **1.** *Adj.* **a)** *(vordringlich)* acute; pressing, urgent ⟨*question, issue*⟩; **b)** *(Med.)* acute. **2.** *adv. (Med.)* in an acute form

AKW [a:ka:'ve:] **das; ~[s], ~s** *Abk.* **Atomkraftwerk**

Akzent [ak'tsɛnt] **der; ~[e]s, ~e a)** *(Sprachw.)* accent; *(Betonung)* accent; stress; **b)** *(Sprachmelodie, Aussprache)* accent; **mit starkem koreanischem ~:** with a strong Korean accent; **c)** *(Nachdruck, Gewicht)* emphasis; stress; **den ~ [besonders] auf etw.** *(Akk.)* **legen** lay *or* put [particular] emphasis *or* stress on sth.; **neue ~e setzen** set new directions

akzent·frei 1. *Adj.* without an *or* any accent *postpos.* **2.** *adv.* without an *or* any accent

akzentuieren [aktsɛntu'i:rən] *tr. V.* **a)** *(deutlich aussprechen)* enunciate; articulate; *(betonen)* accentuate; stress; **b)** *(fig.: hervorheben, auch Mode)* accentuate

akzeptabel [aktsɛp'ta:bl̩] **1.** *Adj.* acceptable. **2.** *adv.* acceptably

akzeptieren *tr. V.* accept

à la [a la] *(Gastr., ugs.)* à la

Alabaster [ala'bastɐ] **der; ~s, ~:** alabaster

à la carte [ala'kart] *(Gastr.)* à la carte

Alarm [a'larm] **der; ~[e]s, ~e a)** alarm; *(Flieger~)* air-raid warning; **~ geben** raise *or* sound *or* give the alarm; **blinder ~:** false alarm; **~ schlagen** *(ugs.)* raise *or* sound the alarm; **b)** *(~zustand)* alert

alarm-, Alarm-: ~anlage die alarm system; **~bereit** *Adj.* on alert *postpos.;* ⟨*fire crew, police*⟩ on stand-by *postpos.*, standing by *pred.;* **~bereitschaft die** *s.* **~bereit:** alert; **in [ständiger] ~bereitschaft** on [permanent] alert/stand-by; **~glocke die** alarm bell

alarmieren *tr. V.* **a)** alarm; **~d** alarming; **b)** *(zu Hilfe rufen)* call [out] ⟨*doctor, police, fire brigade, etc.*⟩

Alarmierung die; ~: bei rechtzeitiger ~ der Bergwacht *usw.* if the mountain rescue service *etc.* is/had been called [out] in time

Alarm-: ~klingel die alarm bell; **~signal das** *(auch fig.)* warning signal; **~sirene die** alarm *or* warning siren; **~stufe die** alert stage; **~stufe eins/zwei/drei** stage one/two/three alert; **~übung die** practice drill; *(Milit.)* practice alert; **~zeichen das** *(fig.)* warning signal; **~zustand der** state of alert; **sich im ~zustand befinden** ⟨*troops*⟩ be on alert; ⟨*fire service, police*⟩ be on stand-by; ⟨*country, province*⟩ be on a state of alert

Alaska [a'laska] **(das); ~s** Alaska

Alaun [a'laun] **der; ~s, ~e** alum

Alaun·stift der styptic pencil

Albaner [al'ba:nɐ] **der; ~s, ~:** Albanian

Albanien [al'ba:ni̯ən] **(das); ~s** Albania

albanisch *Adj.* Albanian; *s. auch* **deutsch; Deutsch, [2]Deutsche**

Albatros ['albatrɔs] **der; ~, ~se** *(Zool.)* albatross

Alben *s.* **Album**

[1]albern *itr. V.* fool about *or* around

[2]albern *Adj.* **a)** silly; foolish; **sich ~ benehmen** act silly; **sich** *(Dat.)* **~ vorkommen** feel silly; feel a fool; **b)** *(ugs.: nebensächlich)* silly; stupid

Albernheit die; ~, ~en a) *o. Pl.* silliness; foolishness; **b)** *(alberne Handlung)* silliness; *(alberne Bemerkung)* silly remark

Albino [al'bi:no] **der; ~s, ~s** albino

Album ['albʊm] **das; ~s, Alben** album

Alchemie [alçe'mi:] *usw. (bes. österr.) s.* **Alchimie** *usw.*

Alchimie [alçi'mi:] **die; ~:** alchemy *no art.*

Alchimist der; ~en, ~en alchemist

alchimistisch *Adj.* alchemical

Aldehyd [alde'hy:t] **der; ~s, ~e** *(Chem.)* aldehyde

Alemanne [alə'manə] **der; ~n, ~n, Alemannin die; ~, ~nen** Alemannian

alemannisch *Adj.* Alemannic

Alge ['algə] **die; ~, ~n** alga

Algebra ['algebra, *österr.:* al'ge:bra] **die; ~,** *(fachspr.)* **Algebren** algebra

algebraisch 1. *Adj.* algebraic. **2.** *adv.* algebraically

Algerien [al'ge:riən] **(das); ~s** Algeria

Algerier der; ~s, ~, Algerierin die; ~, ~nen Algerian

algerisch *Adj.* Algerian

Algier ['alʒi:ɐ̯] **(das); ~s** Algiers

alias ['a:li̯as] *Adv.* alias

Alibi ['a:libi] **das; ~s, ~s a)** *(Rechtsw.)* alibi; **b)** *(Ausrede)* alibi *(coll.);* excuse

Alimente [ali'mɛntə] *Pl. (veralt., noch ugs.)* maintenance *sing. (esp. for illegitimate child)*

Alkali [al'ka:li] **das; ~s, Alkalien** alkali

alkalisch *Adj. (Chemie)* alkaline

Alkohol ['alkoho:l] **der; ~s, ~e** alcohol; **unter ~ stehen** *(ugs.)* be under the influence *(coll.)*

alkohol-, Alkohol-: ~arm *Adj.* low-alcohol *attr.;* low in alcohol *pred.;* **~ausschank der** sale of alcohol[ic drinks]; **~einfluß der** influence of alcohol *or* drink; **unter ~einfluß [stehen]** [be] under the influence of alcohol *or* drink; **~fahne die** smell of alcohol [on one's breath]; **eine ~fahne haben** smell of alcohol; **~frei** *Adj.* **a)** non-alcoholic; **~freie Getränke** soft *or* non-alcoholic drinks; **b)** *(ohne ~ausschank)* dry ⟨*country, state, etc.*⟩; **~gehalt der** alcohol content; **~genuß der;** *o. Pl.* consumption of alcohol; **~haltig** *Adj.* containing alcohol *postpos., not pred.*

Alkoholika [alko'ho:lika] *Pl.* alcoholic drinks

Alkoholiker der; ~s, ~, Alkoholikerin die; ~, ~nen alcoholic

alkoholisch *Adj.* alcoholic

alkoholisiert *Adj.* inebriated; **in ~em Zustand** in a state of inebriation

Alkoholismus der; ~: alcoholism *no art.*

alkohol-, Alkohol-: ~konsum der consumption of alcohol; **~krank** *Adj. (Med.)* alcoholic; **~mißbrauch der** alcohol abuse; **~spiegel der** level of alcohol in one's blood; **~süchtig** *Adj.* addicted to alcohol *postpos.;* alcoholic; **~verbot das** ban on alcohol; **~vergiftung die** alcohol[ic] poisoning

Alkoven [al'ko:vn̩] **der; ~s, ~:** alcove; *(Bettnische)* bed-recess

all [al] *Indefinitpron. u. unbest. Zahlw.* **1.** *attr. (ganz, gesamt...)* all; **in ~er Deutlichkeit** in all clarity; **~e Freude, die sie empfunden hat** all the joy she felt; **~es Geld, das ich noch habe** all the money I have left; **ich kann diese Leute ~e nicht leiden** I can't stand any of these people; **~ unser/mein** *usw.* **...:** all our/my *etc.* ...; **~es andere/Weitere/übrige** everything else; **~es Schöne/Neue/**

Fremde everything *or* all that is beautiful/new/strange; **~e Fenster schließen** close all the windows; **wir/ihr/sie ~e** all of us/you/them; we/you/they all; **~e Beteiligten/Anwesenden** all those involved/present; **~e beide/~e zehn** both of them/all ten of them; **~e Männer/Frauen/Kinder** all men/women/children; **~e Bewohner der Stadt** all the inhabitants of the town; **ohne ~en Anlaß** for no reason [at all]; without any reason [at all]; **~e Jahre wieder** every year; **~e fünf Minuten/Meter** every five minutes/metres; **Bücher ~er Art** books of all kinds; all kinds of books; **in ~er Eile** with all haste; **in ~er Ruhe** in peace and quiet; **trotz ~er Versuche/Anstrengungen** despite all [his/her/their/*etc.*] attempts/efforts. **2.** *alleinstehend* **a) ~e** all; **~e miteinander/auf einmal** all together/at once; **~e, die ...** all those who ...; **b) ~es** *(auf Sachen bezogen)* everything; *(auf Personen bezogen)* everybody; **das ~es** all that; **das ist ~es Unsinn** that is all nonsense; **was gab es ~es zu sehen?** what was there to see?; **was es nicht ~es gibt!** well, would you believe it!; well, I never!; **zu ~em fähig sein** *(fig.)* be capable of anything; **~es in ~em** all in all; **vor ~em** above all; **das ist ~es** that's all *or (coll.)* it; **das sind ~es Gauner** *(ugs.)* they're all scoundrels; **wer ~es war** *od.* **wer war ~es da?** who was there?; **~es mal herhören!** *(ugs.)* listen everybody!; *(stärker befehlend)* everybody listen!; **~es aussteigen!** *(ugs.)* everyone *or* all out!; *(vom Schaffner gesagt)* all change!

All das; ~s space *no art.*; *(Universum)* universe

all·abendlich 1. *Adj.; nicht präd.* regular evening; **2.** *adv.* every evening

alle *Adj.; nicht attr.* **a)** *(ugs.: verbraucht, verkauft usw.)* **~ sein** be all gone; **~ werden** run out; **etw. ~ machen** finish sth. off; **b)** *(salopp: erschöpft)* all in *pred.*

alle·dem *Pron.* in **trotz ~:** in spite of *or* despite all that; **von ~ wußte er nichts** he knew nothing about all that; **bei ~:** for all that

Allee [a'le:] **die; ~, ~n** avenue

Allegorie [alego'ri:] **die; ~, ~n** allegory

allegorisch *Adj.* allegorical

allein [a'lain] **1.** *Adj.; nicht attr.* **a)** *(ohne andere, für sich)* alone; on one's/its own; by oneself/itself; **sie waren ~ im Zimmer** they were alone in the room; **ganz ~:** all on one's/its own; **jmdn. ~ lassen** leave sb. alone *or* on his/her own; **b)** *(einsam)* alone. **2.** *adv.* *(ohne Hilfe)* by oneself/itself; on one's/its own; **sie kann ~ schwimmen** she can swim by herself *or* on her own; **etw. ~ tun** do sth. oneself; **von ~** *(ugs.)* by oneself/itself. **3.** *Adv.* **a)** *(geh.: ausschließlich)* alone; **er ~ trägt die Verantwortung** he alone bears responsibility; **sie denkt ~ an sich** she thinks solely *or* only of herself; **nicht ~ ..., sondern auch ...:** not only ..., but also ...; **b)** *(von allem anderen abgesehen)* **[schon] ~ der Gedanke/[schon] der Gedanke ~:** the mere *or* very thought [of it]; **~ die Nebenkosten** the additional costs alone

alleine *(ugs.) s.* **allein**

allein-, Allein-: ~erbe der sole heir; **~erziehend** *Adj.* single ⟨*mother, father, parent*⟩; **~erziehende der/die;** *adj. Dekl.* single parent; **~gang der** *(fig.)* independent initiative; **etw. im ~gang tun** do sth. off one's own bat; **~herrschaft die;** *o. Pl.* autocratic rule; *(Diktatur)* dictatorship; **~herrscher der** *(auch fig.)* autocrat; *(Diktator, auch fig.)* dictator

alleinig *Adj.; nicht präd.* sole; sole, exclusive ⟨*distribution rights*⟩

allein-, Allein-: ~schuld die; *o. Pl.* sole blame *or* responsibility *no indef. art.*; **~sein das a)** *(das Verlassensein)* loneliness; **b)** *(das Ungestörtsein)* privacy; **~stehend** *Adj.* ⟨*person*⟩ living on his/her own *or* alone; *(ledig)* single ⟨*person*⟩; **ich bin ~stehend** I live on my own *or* alone/am single; **~stehende der/die;** *adj. Dekl.* person living on his/her own *or* alone; *(Ledige[r])* single person; **ich als ~stehender ...:** living on my own I .../as a single person I ...; **~unter·halter der** solo entertainer; **~verdiener der** sole earner

alle·mal *Adv.* *(ugs.)* any time *(coll.)*; **was der kann, das kann ich doch ~:** anything he can do, I can do too

allen·falls *Adv.* **a)** *(höchstens)* at [the] most; at the outside; **~ 40 Leute** 40 people at most *or* at the outside; at most 40 people; **b)** *(bestenfalls)* at best

allenthalben ['alənt'halbn̩] *Adv.* *(geh.)* everywhere

aller-: ~äußerst... *Adj.; nicht präd.* **a)** farthest; **b)** *(schlimmst...)* worst; **im ~äußersten Fall** if the worst comes/came to the worst; **~best... 1.** *Adj.* very best; **der/die/das Allerbeste sein** be the best of all; **es wäre das ~beste, wenn du ihn selbst fragst** the best thing [of all] would be for you to ask him yourself; **am ~besten wäre es, wenn ...:** the best thing [of all] would be if ...; **2.** *adv.* **am ~besten** best of all; **~dings 1.** *Adv.* **a)** *(einschränkend)* though; **es stimmt ~dings, daß ...:** it's true though that ...; **b)** *(zustimmend)* [yes,] certainly; **Habe ich dich geweckt? – Allerdings!** Did I wake you up? – You certainly did!; **2.** *Partikel (anteilnehmend)* to be sure; **das war ~dings Pech** that was bad luck, to be sure; **~erst...** *Adj.; nicht präd.* **a)** very first; **als ~erste[r] etw. tun** be the very first to do sth.; **b)** *(allerbest...)* very best; **~frühestens** *Adv.* at the very earliest

Allergie [alɛr'gi:] **die; ~, ~n** *(Med.)* allergy; **eine ~ gegen etw. haben** have an allergy to sth.

allergisch 1. *Adj. (Med., fig.)* allergic **(gegen** to). **2.** *adv.* **auf etw.** *(Akk.)* **~ reagieren** *(Med.)* have an allergic reaction to sth.; **auf jmdn./etw. ~ reagieren** *(fig.)* be allergic to sb./sth.

aller-, Aller-: ~größt... *Adj.* utmost ⟨*trouble, care, etc.*⟩; biggest *or* largest ⟨*car, house, town, etc.*⟩ of all; tallest ⟨*person*⟩ of all; **am ~größten sein** be [the] biggest *or* largest/tallest of all; **~hand** *indekl. unbest. Gattungsz. (ugs.)* **a)** *attr.* all kinds *or* sorts of; **b)** *alleinstehend* all kinds *or* sorts of things; **das ist ~hand!** *(viel)* that's a lot; *(sehr gut)* that's quite something; **das ist ja** *od.* **doch ~hand!** that's just not on *(Brit. coll.)*; that really is the limit *(coll.)*; **~heiligen das; ~** *(bes. kath. Kirche)* All Saints' Day; All Hallows; **~herzlichst...** *Adj.* warmest ⟨*thanks, greetings, congratulations*⟩; most cordial ⟨*reception, welcome, invitation*⟩; **~höchst... 1.** *Adj.* highest ... of all; **der ~höchste Gipfel** the highest peak of all; **es ist ~höchste Zeit, daß ...:** it really is high time that ...; **die ~höchsten Kreise** the very highest circles; **2.** *adv.* **am ~höchsten** ⟨*fly, jump, etc.*⟩ the highest of all; **~höchstens** *Adv.* at the very most

allerlei *indekl. unbest. Gattungsz.: attr.* all kinds *or* sorts of; *alleinstehend* all kinds *or* sorts of things

Allerlei das; ~s, ~s *(Gemisch)* pot-pourri; *(Durcheinander)* jumble

aller-, Aller-: ~letzt... *Adj.; nicht präd.* **a)** very last; **der/die/das ~letzte** the very last [one]; **b)** *(ugs.: ~schlechtest...)* most dreadful *or* awful *(coll.)*; **das ist [ja** *od.* **wirklich] das Allerletzte** *(ugs.)* that [really] is the absolute limit; **c)** *(~neuest...)* very latest; **~liebst 1.** *Adj.* **a)** most favourite; **es wäre mir das ~liebste** *od.* **am ~liebsten, wenn ...:** I should like it best of all if ...; **ihr Allerliebster/seine Allerliebste** her/his beloved; **b)** *(reizend)* enchanting; delightful; **2.** *adv.* **a) etw. am ~liebsten tun** like doing sth. best of all; **am ~ liebsten trinkt/mag er Wein** he likes wine best of all; **b)** *(reizend)* delightfully; **~meist... 1.** *Indefinitpron. u. unbest. Zahlw.* by far the most *attrib.;* **das ~meiste/am ~meisten** most of all/by far the most; **die ~meiste Zeit** by far the greatest part of the time; **die ~meisten [der Arbeiter** *usw.*] the vast majority [of the workers *etc.*]; **2.** *Adv.* **am ~meisten** most of all; **die am ~meisten befahrene Straße** by far the most travelled road; **~mindest...** *Adj.; nicht präd.* slightest; least; **das ~mindeste** the very least; **nicht das ~mindeste** absolutely nothing; **nicht im ~mindesten** not in the least *or* slightest; **~nächst... 1.** *Adj.* very nearest *attrib.; (räumliche od. zeitliche Reihenfolge ausdrükkend)* very next *attrib.;* very closest ‹*relatives*›; **2.** *adv.* **am ~nächsten** nearest of all; **~neu[e]st...** *Adj.* very latest *attrib.;* **das Allerneu[e]ste** the very latest; **~nötigst..., ~notwendigst... 1.** *Adj.* absolutely necessary; **es am ~nötigsten** *od.* **~notwendigsten haben, etw. zu tun** be most in need of doing sth.; **das Allernötigste** what is/was absolutely necessary; **2.** *adv.* **am ~nötigsten** ‹*need etc.*› most badly; **~schlimmst... 1.** *Adj.* very worst *attrib.;* **der/die/das ~schlimmste** *od.* **am ~schlimmsten sein** be the worst of all; **das Allerschlimmste** the worst of all; **2.** *adv.* **am ~schlimmsten** worst of all; **~schönst... 1.** *Adj.* most beautiful *attrib.;* loveliest *attrib.; (angenehmst...)* very nicest *attrib.;* **das Allerschönste, was ich je gesehen habe** the loveliest thing I have ever seen; **das wäre ja noch das Allerschönste** that would beat everything; **am ~schönsten war, daß ...:** the best thing of all was that ...; **2.** *adv.* **er schreibt/singt** *usw.* **am schönsten** his writing/singing *etc.* is the most beautiful of all; **~seelen das;** ~ *(kath. Kirche)* All Souls' Day; **~seits** *Adv.* **a)** *(alle zusammen)* **guten Morgen ~seits!** good morning everyone *or* everybody; **b)** *(überall)* on all sides; on every side; **~seits sehr geschätzt sein** be highly regarded by everyone; **~spätestens** *Adv.* at the very latest

Allerwelts-: gesicht das nondescript face; **~kerl der** Jack of all trades; **~mittel das** cure-all; **~wort das** hackneyed word

aller~wenigst... 1. *Adj.* least ... of all; *Pl.* fewest ... of all; **die ~wenigsten [Menschen] wissen das** very few [people] know that; **das Allerwenigste, was er hätte tun können** the very least he could have done; **2.** *adv.* **am ~wenigsten abbekommen** get [the] least of all; **das am ~wenigsten!** anything but that!; **~wenigstens** *Adv.* at the very least

alle·samt *indekl. Indefinitpron. u. unbest. Zahlw. (ugs.)* all [of you/us/them]; **wir ~:** all of us; we all

Alles-: ~fresser der omnivore; **~kleber der** all-purpose adhesive *or* glue; **~könner der** all-rounder

all·gemein 1. *Adj.* general; universal ‹*conscription, suffrage*›; universally applicable ‹*law, rule*›; **auf ~en Wunsch** by popular *or* general request; **das ~e Wohl** the common good; **im ~en Interesse** in the common interest; in everybody's interest; **im ~en** in general; generally. **2.** *adv.* **a)** generally; **es ist ~ bekannt, daß ...:** it is common knowledge that ...; **~ zugänglich** open to all *or* everybody; **b)** *(oft abwertend: unverbindlich)* ‹*write, talk, discuss, examine, be worded*› in general terms; **eine ~ gehaltene Einführung** a general introduction

allgemein-, Allgemein-: ~befinden das *(Med.)* general state of health; general condition; **~begriff der** *(Philos., Sprachw.)* general concept; **~besitz der** *(auch fig.)* common property; **~bildend** *Adj.* ‹*school, course, etc.*› providing a general *or* an all-round *or (Amer.)* all-around education; **~bildung die;** *o. Pl.* general *or* all-round *or (Amer.)* all-around education; **~gültig** *Adj.; nicht präd.* universally *or* generally applicable ‹*law, rule*›; universally *or* generally valid ‹*law of nature, definition, thesis*›; **~gültigkeit die** *s.* **~gültig:** universal *or* general applicability/validity; **~gut das** *(fig.)* common knowledge

Allgemeinheit die; ~ **a)** generality; **b)** *(Öffentlichkeit)* general public

allgemein-, Allgemein-: ~medizin die; *o. Pl.* general medicine; **~mediziner der** general practitioner; GP; **~platz der** platitude; commonplace; **~verbindlich 1.** *Adj.* universally binding; **2.** *adv.* in universally binding terms; **~verständlich 1.** *Adj.* comprehensible *or* intelligible to all *postpos.;* **2.** *adv.* in a way comprehensible *or* intelligible to all; **~wissen das** general knowledge; **~wohl das** public welfare *or* good

All·heil·mittel das *(auch fig.)* cure-all; panacea

Alligator [ali'ga:tɔr] **der;** ~s, ~en [...ga'to:rən] alligator

alliiert *Adj.; nicht präd.* allied

Alliierte der; *adj. Dekl.* ally; **die ~n** the Allies

all-, All-: ~jährlich 1. *Adj.* annual; yearly; **2.** *adv.* annually; every year; **~mächtig** *Adj.* omnipotent; all-powerful; **der ~mächtige Gott** Almighty God; **~mächtiger Gott!** *(ugs.)* good God!; heavens above!; **~mächtige der;** *adj. Dekl.* **der ~mächtige** the Almighty; **~mächtiger!** good God!; heavens above!

all·mählich 1. *Adj., nicht präd.* gradual. **2.** *adv.* gradually. **3.** *Adv.* **es wird ~ Zeit** it's about time; **ich werde ~ müde** I'm beginning to get tired; **wir sollten ~ gehen** it's time we got going

all-, All: ~monatlich 1. *Adj.* monthly; **2.** *adv.* monthly; every month; **~morgendlich 1.** *Adj.; nicht präd.* regular morning; **2.** *adv.* every morning; **~parteien·regierung die** all-party government; **~rad·antrieb der** *(Kfz-W.)* all-wheel drive

Allround·man ['ɔ:l'raʊndmən] **der;** ~s, **Allroundmen** all-rounder

all-: ~seitig 1. *Adj.* **a)** general; all-round, *(Amer.)* all-around *attrib.;* **zur ~seitigen Zufriedenheit** to the satisfaction of all *or* everyone; **2.** *adv.* generally; **man war ~seitig einverstanden** there was agreement on all sides *or* general agreement; **~seitig geachtet** highly regarded by everyone;

~seits *Adv.* everywhere; on all sides; *(in jeder Hinsicht)* in all respects; **~seits geschätzt** highly regarded by everyone

All·tag der a) *(Werktag)* weekday; **ein Mantel für den ~:** a coat for everyday wear; **zum ~ gehören** *(fig.)* be part of everyday life; **b)** *o. Pl. (Einerlei)* daily routine; **der graue ~:** the dull routine of everyday life; **der ~ der Ehe** the day-to-day realities of married life

all·täglich *Adj.* ordinary ⟨*face, person, appearance, etc.*⟩; everyday ⟨*topic, event, sight*⟩; **ein nicht ~er Anblick** a sight one doesn't see every day; **etw. Alltägliches sein** be an everyday occurrence

Alltäglichkeit die; ~, ~en a) *o. Pl.* ordinariness; **b)** *(alltäglicher Vorgang)* everyday occurrence

all·tags *Adv.* [on] weekdays

Alltags- everyday *attrib.;* of everyday life *postpos., not pred.;* **~pflicht** daily duty

Alltags-: ~kleidung die everyday *or* workaday clothes *pl.;* **~trott der** *(abwertend)* daily round *or* grind

all-, All-: ~wissend *Adj.* omniscient; **~wissenheit die; ~:** omniscience; **~wöchentlich 1.** *Adj.* weekly; **2.** *adv.* weekly; every week; **~zeit** *Adv. (veralt.)* always; **~zeit bereit!** be prepared!

all·zu *Adv.* all too; **~ viele** far too many; **er war nicht ~ begeistert** he was not too *or* not all that enthusiastic; **nicht ~ viele** not all that many *(coll.);* not too many

allzu-: ~bald *Adv.* all too soon; **~früh** *Adv.* all too early; *(~bald)* all too soon; **nicht ~früh** not too early; **~gern** *Adv.* ⟨*like*⟩ only too much; *(bereitwillig)* only too willingly; **ich esse zwar Fisch, aber nicht ~gern** I'll eat fish but I'm not all that fond *(coll.) or* not overfond of it; **~lang[e]** *Adv.* too long; **~oft** *Adv.* too often; **nicht ~oft** not too often; not all that often *(coll.);* **~sehr** *Adv.* too much; **nicht ~sehr** not too much; not all that much *(coll.);* **~viel** *Adv.* too much

All·zweck- all-purpose

Alm [alm] **die; ~, ~en** mountain pasture; Alpine pasture

Almosen ['almo:zn̩] **das; ~s, ~ a)** *(veralt.: Spende)* alms *pl.;* **von ~ leben** live on charity; **b)** *(abwertend: dürftiges Entgelt)* pittance

Aloe ['a:loe] **die; ~, ~n** aloe

Alp [alp] **die; ~, ~en** *(bes. schweiz.) s.* **Alm**

Alpaka [al'paka] **das; ~s, ~s** alpaca

Alp-: ~druck der; *o. Pl.* nightmare; **~drücken das; ~s** nightmares *pl.*

Alpen *Pl.* **die ~:** the Alps

Alpen- Alpine

Alpen-: ~glühen das; ~s alpenglow; **~land das;** *o. Pl.* Alpine country *or* region; **~rose die** rhododendron; Alpine rose; **~veilchen das** cyclamen

Alpha ['alfa] **das; ~[s], ~[s]** alpha

Alphabet [alfa'be:t] **das; ~[e]s, ~e** alphabet

alphabetisch 1. *Adj.* alphabetical. **2.** *adv.* alphabetically

Alphabetisierung die; ~: teaching literacy skills

alpha·numerisch *(DV)* **1.** *Adj.* alphanumeric. **2.** *adv.* alphanumerically

Alp·horn das alpenhorn

alpin [al'pi:n] *Adj.* Alpine

Alpinist der; ~en, ~en Alpinist

Alp·traum der nightmare

Alraune [al'rau̯nə] **die; ~, ~n** mandrake

als [als] *Konj.* **a)** *Temporalsatz einleitend* when; *(während, indem)* as; **gerade ~:** just as; **gleich ~:** as soon as; **damals, ~:** [in the days] when; **b)** *Kausalsatz einleitend* **um so mehr, ~:** all the more since *or* in that; *s. auch* **insofern, insoweit; c) ..., ~ da sind ...** to wit; namely; **die [drei] Grundfarben, ~ da sind Rot, Blau und Gelb** the [three] primary colours, to wit *or* namely red, blue, and yellow; **d)** *Vergleichspartikel* **größer/älter/mehr** *usw.* **~ ...** bigger/older/more *etc.* than ...; **kein anderer** *od.* **niemand anderes ~ ...** none other than; **niemand** *od.* **keiner ~ ...** nobody but ...; **nirgends anders ~ ...** nowhere but ...; **er ist alles andere ~ schüchtern** he is anything but shy; **du brauchst nichts [anderes] zu tun, ~ abzuwarten** all you need to do is just wait and see; **anders ~ ...** different/differently from ...; **lieber hänge ich mich auf, ~ daß ich ins Gefängnis gehe** I'd rather hang myself than go to prison; **die Kinder sind noch zu klein, ~ daß wir sie allein lassen könnten** the children are still too small for us to be able to leave them on their own; **~ [wenn** *od.* **ob]** (+ *Konjunktiv II*) as if *or* though; **er tut so, ~ ob** *od.* **wenn er nichts wüßte, er tut so, ~ wüßte er nichts** he pretends not to know anything; **~ wenn** *od.* **ob ich das nicht wüßte!** as if I didn't know!; **soviel/soweit ~ möglich** as much/as far as possible; *s. auch* **sowohl; e) ~ Rentner/Arzt** *usw.* as a pensioner/doctor *etc.;* **du ~ Lehrer ...** as a teacher you ...; **jmdn. ~ faul/Dummkopf bezeichnen** call sb. lazy/a fool; **sich ~ wahr/Lüge erweisen** prove to be true/a lie

als·baldig *Adj.; nicht präd. (Papierdt.)* immediate

also ['alzo] **1.** *Adv. (folglich)* so; therefore; **~ kommst du mit?** so you're coming too? **2.** *Partikel* **a)** *(das heißt)* that is; **b)** *(nach Unterbrechung)* well [then]; **~, wie ich schon sagte** well [then], as I was saying; **c)** *(verstärkend)* **~, kommst du jetzt oder nicht?** well, are you coming now or not?; **na ~!** there you are[, you see]; **~ schön** well all right then; **~ so was/nein!** well, I don't know; well, really; **~, gute Nacht** goodnight then; **~ dann** right then

alt [alt]; **älter** ['ɛltɐ], **ältest...** ['ɛltəst...] *Adj.* **a)** old; **~ und jung** old and young; **seine ~en Eltern** his aged parents; **mein älterer/ältester Bruder** my elder/eldest brother; **Cato der Ältere** Cato the Elder; **hier werde ich nicht ~** *(fig. ugs.)* I won't be staying here long; **~ aussehen** *(fig. salopp)* be in the cart *(sl.);* **eine sieben Jahre ~e Tochter** a seven-year-old daughter; **wie ~ bist du?** how old are you?; *s. auch* **Herr a; b)** *(nicht mehr frisch)* old; **~es Brot** stale bread; **c)** *(vom letzten Jahr)* old; **~e Äpfel/Kartoffeln** last year's apples/potatoes; **im ~en Jahr** *(dieses Jahr)* this year; *(letztes Jahr)* last year; **d)** *(seit langem bestehend)* ancient; old; long-standing ⟨*acquaintance*⟩; long-serving ⟨*employee*⟩; **ein ~es Volk/ein ~er Brauch** an ancient people/an ancient *or* old custom; **in ~er Freundschaft, Dein ...** yours, as ever, ...; **am Alten hängen** cling to the past; **e)** *(antik, klassisch)* ancient; **f)** *(vertraut)* old familiar ⟨*streets, sights, etc.*⟩; **ganz der/die ~e sein** be just the same; **es bleibt alles beim ~en** things will stay as they were; **g)** *(ugs.) (vertraulich)* **~er Freund/~es Haus!** old friend/pal *(coll.); (bewundernd)* **ein ~er Fuchs/Gauner** an old fox/rascal; *(verstärkend)* **die ~e Hexe/der ~e Geizkragen** the old witch/skinflint; *s. auch* **Alte**

[1]Alt der; ~s, ~e *(Musik)* alto; *(Frauenstimme)* contralto; alto; *(im Chor)* altos *pl.;* contraltos *pl.*

[2]Alt das; ~[s], ~: *top fermented, dark beer*

Altar [al'taːɐ] **der; ~[e]s,** Altäre [al'tɛːrə] altar

Altar-: ~bild das altar-piece; **~gerät das** altar furniture; **~raum der** chancel

alt-, Alt-: ~backen *Adj.* **a)** stale ‹*bread, roll, etc.*›; **b)** *(abwertend: altmodisch)* outdated ‹*ideas, views, policies*›; **~bau der;** *Pl.* **~bauten** old building; **~bau·wohnung die** flat *(Brit.) or (Amer.)* apartment in an old building; old flat *(Brit.) or (Amer.)* apartment; **~bekannt** *Adj.* well-known; **~bewährt** *Adj.* well-tried; long-standing ‹*tradition, friendship*›; **~bier das** *s.* [2]**Alt; ~bundeskanzler der** former Federal Chancellor; **~bundespräsident der** former Federal President

Alte der/die; *adj. Dekl.* **a)** *(alter Mensch)* old man/woman; *Pl. (alte Menschen)* old people; **b)** *(salopp) (Vater, Ehemann)* old man *(coll.)*; *(Mutter)* old woman *(coll.)*; *(Ehefrau)* missis *(sl.)*; old woman *(coll.)*; *(Chef)* boss *(coll.)*; governor *(sl.)*; *(Chefin)* boss *(coll.)*; **die ~n** *(Eltern)* my/his *etc.* old man and old woman *(coll.)*; **c)** *Pl. (Tiereltern)* parents; **d)** *Pl.* **die ~n** *(geh.: Menschen der Antike)* the ancients

alt: ~ehrwürdig *Adj. (geh.)* venerable; time-honoured ‹*customs*›; **~eingeführt** *Adj.* old-established; **~englisch** *Adj.* Old English

Alten-: ~heim das old people's home; **~tages·stätte die** old people's day centre; **~teil das** *in* **sich aufs ~teil zurückziehen** *(fig.)* retire; **~wohn·heim das** old people's home

Alter das; ~s, ~: age; *(hohes ~)* old age; **im ~:** in one's old age; **mit dem ~:** with age; **er ist in meinem ~:** he is my age; **im ~ von** at the age of; **eine Frau mittleren ~s** a middle-aged woman; **Kinder in diesem ~:** children of this age

älter ['ɛltɐ] **1.** *s.* **alt. 2.** *Adj. (nicht mehr jung)* elderly; **eine Melodie für unsere ~en Hörer** a tune for our older listeners; **Ältere** *(ältere Menschen)* the elderly; *s. auch* **Mitbürger**

Alterchen das; ~s, ~ *(ugs.)* grandad *(coll.)*

altern 1. *itr. V.; mit sein* **a)** age; **b)** *(reifen)* mature. **2.** *tr. V.* age, mature ‹*wine, spirits*›

alternativ [altɛrna'tiːf] **1.** *Adj.* alternative. **2.** *adv.* alternatively; ‹*work, farm*› using alternative methods

Alternativ·bewegung die alternative movement

Alternative die; ~, ~n alternative

alt·erprobt *Adj.* well-tried

alters *in* **seit ~, von ~ her** *(geh.)* from time immemorial

alters-, Alters-: ~erscheinung die sign of old age; **~gemäß 1.** *Adj.* ‹*behaviour, education, etc.*› appropriate to one's/its age; **2.** *adv.* in a manner appropriate to one's/its age **~genosse der, ~genossin die** contemporary; person/child of the same age; **~grenze die** age limit; *(für Rente)* retirement age; **~gründe** *Pl.* reasons of age; **~gruppe die** age-group; **~heim das** old people's home; old-age home *(Amer.)*; **~rente die** old-age pension; **~ruhe·geld das** retirement pension; **~schwach** *Adj.* old and infirm ‹*person*›; old and weak ‹*animal*›; [old and] decrepit ‹*object*›; **eine ~schwache Frau** an infirm *or* frail old woman; **~schwäche die;** *o. Pl. (bei Menschen)* [old] age and infirmity; *(bei Tieren)* [old] age and weakness; *(von Dingen)* [age and] decrepitude; **~sichtig** *Adj.* presbyopic *(Med.)*; **~stufe die** age; **~unterschied der** age difference; **~versorgung die** provision for one's old age; *(System)* pension scheme

Altertum das; ~s, Altertümer [-tyːmɐ] **a)** *o. Pl.* antiquity *no art.*; **b)** *Pl.* antiquities

altertümlich 1. *Adj.* old-fashioned. **2.** in an old-fashioned style

Altertümlichkeit die; ~: old-fashionedness

Altertums·forschung die; *o. Pl.* archaeology

Alterung die; ~, ~en a) *(das Altwerden)* ageing; **b)** *(von Werkstoffen)* ageing; **c)** *(von Wein usw.)* ageing; maturing

Älteste ['ɛltəstə] **der/die;** *adj. Dekl.* **a)** *(Dorf~, Vereins~, Kirchen~ usw.)* elder; **b)** *(Sohn, Tochter)* eldest

alt-, Alt-: ~flöte die *(Querflöte)* alto *or* bass flute; *(Blockflöte)* alto *or* treble recorder; **~fränkisch** *(ugs. scherzh.)* **1.** *Adj.* old-fashioned; **2.** *adv.* in an old-fashioned way; ; **~gedient** *Adj.* long-serving; **~glas·behälter der** bottle bank; **~griechisch** *Adj.* ancient Greek; *(Ling.)* classical *or* ancient Greek; **~griechisch das** classical *or* ancient Greek; **~her·gebracht** *Adj.* traditional; **~hochdeutsch** *Adj.* Old High German; **~hochdeutsch das** Old High German

Altistin die; ~, ~nen *(Musik)* alto; contralto

alt-, Alt-: ~klug; ~kluger, ~klugst... 1. *Adj.* precocious; **2.** *adv.* precociously; **~klugheit die** precociousness; **~last die a)** *(Ökologie)* old, improperly disposed of harmful waste *no indef. art.*; **b)** *(fig.)* inherited problem

ältlich ['ɛltlɪç] *Adj.* rather elderly; oldish

alt-, Alt: ~material das scrap; **~meister der a)** *(Vorbild)* doyen; **b)** *(Sport)* ex-champion; former champion; **~meisterin die a)** *(Vorbild)* doyenne; **b)** *s.* **~meister b; ~metall das** scrap metal; **~modisch 1.** *Adj.* old-fashioned; **2.** *adv.* in an old-fashioned way; **~öl das** used oil; **~papier das** waste paper; **~philologe der** classical scholar; **~philologie die** classical studies *pl., no art.*; **~philologisch** *Adj.* classical; **~rosa** *indekl. Adj.* old rose

altruistisch *(geh.)* **1.** *Adj.* altruistic. **2.** *adv.* altruistically

alt-, Alt-: ~sänger der alto; **~sängerin die** alto; contralto; **~schnee der** old snow; **~stadt die** old [part of the] town; **die Düsseldorfer ~stadt** the old part of Düsseldorf; **~stimme die** alto voice; *(von Frau)* alto *or* contralto voice; **~überliefert** *Adj.* traditional; **~väterisch** [~fɛːtərɪʃ] **1.** *Adj.* old-fashioned; **2.** *adv.* in an old-fashioned way; **~vertraut** *Adj.* old familiar *attrib.*; **~waren·händler der** second-hand dealer

[1]**Alu** ['aːlu] **das; ~s** *(ugs.)* aluminium

[2]**Alu die; ~** *(ugs.) Abk.:* **Arbeitslosenunterstützung** dole [money] *(Brit.)*; unemployment pay

Alu·folie die aluminium foil

Aluminium [alu'miːnium] **das; ~s** aluminium

Aluminium·folie die aluminium foil

am [am] *Präp. + Art.* **a)** = **an dem; b)** *(räumlich)* **am Boden** on the floor; **Frankfurt am Main** Frankfurt on [the] Main; **am Rande** on the edge; **am Marktplatz** on the market square *or* place; **am Meer/Fluß** by the sea/on *or* by the river; **am Anfang/Ende** at the beginning/end; **sich am Kopf stoßen** bang one's head; **c)** *(österr.: auf dem)* on the; **d)** *(zeit-*

lich) on; **am Freitag** on Friday; **am 19. November** on 19 November; **am Anfang/Ende** at the beginning/end; **am letzten Freitag** last Friday; **e)** *(zur Bildung des Superlativs)* **am gescheitesten/schönsten sein** be the cleverest/most beautiful; **am schnellsten laufen** run [the] fastest; **das machen wir am besten nachher** it's best if we do it afterwards; **f)** *(nach bestimmten Verben)* **am Gelingen eines Planes** *usw.* **zweifeln** have doubts about *or* doubt the success of a plan *etc.*; **am Wettbewerb teilnehmen** take part in the contest; **g)** *(zur Bildung der Verlaufsform)* **am Verwelken/Verfallen sein** be wilting/decaying

Amalgam [amal'ga:m] **das; ~s, ~e** *(Chemie, auch fig.)* amalgam

Amaryllis [ama'rʏlɪs] **die; ~, Amaryllen** amaryllis

Amateur [ama'tø:ɐ̯] **der; ~s, ~e** amateur

Amateur- amateur

amateurhaft 1. *Adj.* amateurish. **2.** *adv.* amateurishly

Amazonas [ama'tso:nas] **der; ~:** Amazon

Amazone [ama'tso:nə] **die; ~, ~n** *(Myth.)* Amazon

Amazonien [ama'tso:niən] **(das); ~s** Amazonia

Amboß ['ambɔs] **der; Ambosses, Ambosse** anvil

ambulant [ambu'lant] **1.** *Adj.* **a)** *(Med.)* out-patient *attrib.* ⟨*treatment, therapy, etc.*⟩; **ein ~er Patient** an out-patient; **b)** *(umherziehend)* itinerant. **2.** *adv.* *(Med.)* **jmdn. ~ behandeln** treat sb. as an out-patient *or* give sb. out-patient treatment

Ambulanz [ambu'lants] **die; ~, ~en a)** *(in Kliniken)* out-patient[s'] department; **b)** *(Krankenwagen)* ambulance

Ameise ['a:maizə] **die; ~, ~n** ant

Ameisen·bär der ant-eater

Ameisen-: ~haufen der anthill; **~säure die;** *o. Pl.* formic acid; **~staat der** ant colony

amen ['a:mɛn] *Interj.* *(christl. Rel.)* amen; **zu allem ja und ~ sagen** *(ugs.)* agree to anything

Amen das; ~s, ~: Amen; **das ist so sicher wie das ~ in der Kirche** *(ugs.)* you can bet your bottom dollar on it

Amerika [a'me:rika] **(das); ~s** America

Amerikaner [ameri'ka:nɐ] **der; ~s, ~ a)** American; **b)** *(Gebäck)* *small, flat iced cake*

Amerikanerin die; ~, ~nen American

amerikanisch *Adj.* American *s. auch* **deutsch, Deutsch, ²Deutsche**

amerikanisieren *tr. V.* Americanize

Amethyst [ame'tʏst] **der; ~[e]s, ~e** amethyst

Ami ['ami] **der; ~[s], ~[s]** *(ugs.)* Yank *(coll.)*

Amino·säure [a'mi:no-] **die** *(Chemie)* aminoacid

Ammann ['aman] **der; ~[e]s, Ammänner** *(schweiz.)* *(Gemeinde~, Bezirks~)* ≈ mayor; *(Land~)* cantonal president

Amme ['amə] **die; ~, ~n** wetnurse; *(Tier)* foster-mother

Ammen·märchen das fairy tale *or* story

Ammer ['amɐ] **die; ~, ~n** bunting

Ammoniak [amo'ni̯ak] **das; ~s** *(Chemie)* ammonia

Amnestie [amnɛs'ti:] **die; ~, ~n** amnesty

amnestieren *tr. V.* grant an amnesty to; amnesty

Amöbe [a'mø:bə] **die; ~, ~n** *(Biol.)* amoeba

Amok ['a:mɔk] **der: ~ laufen** run amok; *(ugs.: wütend werden)* go wild *(coll.)*; **~ fahren** go berserk at the wheel

Amok-: ~fahrer der berserk driver; **~lauf der** crazed rampage; **~läufer der** madman; **der ~läufer, der mehrere Menschen erschossen hatte** the man who had gone berserk and shot several people; **~schütze der** crazed gunman

a-Moll das; ~: A minor

Amor ['a:mɔr] **(der)** Cupid

amortisieren *(Wirtsch.)* **1.** *tr. V.* repay ⟨*investment, acquisition costs*⟩. **2.** *refl. V.* pay for itself

Ampel ['ampl̩] **die; ~, ~n a)** *(Verkehrs~)* traffic lights *pl.*; **die ~ sprang auf Rot** the traffic lights turned to red; **an der nächsten ~:** at the next set of traffic lights; **b)** *(Hängelampe)* hanging lamp; **c)** *(für Pflanzen)* hanging flowerpot

Ampel-: ~anlage die set of traffic lights; **~koalition die** *(ugs.)* *coalition between the SPD, FDP, and the Green Party*

Ampere [am'pɛ:ɐ̯] **das; ~[s], ~:** ampere; amp *(coll.)*

Ampere·meter das ammeter

Amphibie [am'fi:bi̯ə] **die; ~, ~n** *(Zool.)* amphibian

Amphibien·fahrzeug das amphibious vehicle

amphibisch *Adj.* amphibious

Amphi·theater [am'fi:-] **das** amphitheatre

Ampulle [am'pʊlə] **die; ~, ~n** *(Med.)* ampoule

Amputation [amputa'tsi̯o:n] **die; ~, ~en** *(Med.)* amputation

amputieren *tr.* *(auch itr.)* *V.* amputate; **jmdm. das Bein/den Arm ~:** amputate sb.'s leg/arm

Amsel ['amzl̩] **die; ~, ~n** blackbird

Amt [amt] **das; ~[e]s, Ämter** ['ɛmtɐ] **a)** *(Stellung)* post; position; *(hohes politisches od. kirchliches ~)* office; **sein ~ antreten** take up one's post/take up office; **im ~ sein** be in office; **für ein ~ kandidieren** be a candidate for a post *or* position/an office; **von ~s wegen** because of one's profession *or* job; **b)** *(Aufgabe)* task; job; *(Obliegenheit)* duty; **seines ~es walten** *(geh.)* discharge the duties of one's office; **c)** *(Behörde)* *(Paß~, Finanz~, ~ für Statistik)* office; *(Sozial~, Fürsorge~, ~ für Denkmalpflege, Vermessungswesen)* department; **von ~s wegen** by order of the authorities; *s. auch* **auswärtig c; d)** *(Gebäude usw.)* office; **e)** *(Fernspr.)* exchange; **das Fräulein vom ~** *(veralt.)* the operator; **vom ~ vermittelt werden** be put through by the operator; **f)** *(kath. Rel.)* [sung] mass

Amt·frau die *s.* **Amtmännin**

amtieren *itr. V.* hold office; **der ~de Generalsekretär** the incumbent Secretary-General

amtlich 1. *Adj.* **a)** *nicht präd.* official; **~es Kennzeichen** *(Kfz-W.)* registration number; **b)** *nicht attr.* *(ugs.: sicher)* definite; certain. **2.** *adv.* officially

Amt·mann der; *Pl.* **Amtmänner** *od.* **Amtleute, Amt·männin** [-mɛnɪn] **die; ~, ~nen** *senior civil servant*

amts-, Amts-: ~anmaßung die *(Rechtsw.)* unauthorized assumption of authority; **~antritt der** assumption of office; **~arzt der** medical officer; **~ärztlich; 1.** *Adj.; nicht präd.* ⟨*examination*⟩ by the medical officer; **2.** *adv.* **sich ~ untersuchen lassen** have an official medical examination; **~blatt das** official gazette; **~deutsch das** *(abwertend)* officialese; **~eid der** oath of office; **~enthebung die** removal *or* dismissal from office; **~führung die;** *o. Pl.* discharge of one's office; **~gericht das a)** *(Instanz)* local *or* district court; **b)** *(Gebäude)* local *or* district court building; **~geschäfte** *Pl.* official duties; **~handlung die** official act *or* duty; **~hilfe die** official assistance *(given by one* *authority to another);* **~kette die** chain of office; **~leitung die** exchange line; **~miene die** *(meist iron.)* official air; **~müde** *Adj.* tired of office *postpos.;* **~nachfolger der** successor in office; **~person die** official; **~schimmel der;** *o. Pl. (scherzh.)* officialism; bureaucracy; **der ~schimmel wiehert** that's bureaucracy for you; **~sprache die a)** *o. Pl. (~deutsch)* official language; officialese *(derog.);* **in der ~sprache** in official language/officialese; **b)** *(eines Landes, einer Organisation)* official language; **~stube die** *(veralt.)* office; **~tracht die** robes *pl.* of office; official dress; **~vorsteher der** head *or* chief [of a/the department]; **~zimmer das** office

Amulett [amu'lɛt] **das; ~[e]s, ~e** amulet; charm

amüsant [amy'zant] **1.** *Adj.* amusing; entertaining. **2.** *adv.* in an amusing *or* entertaining way

amüsieren 1. *refl. V.* **a)** *(sich vergnügen)* enjoy oneself; have a good time; **amüsier dich gut!** enjoy yourself!; have a good time!; **sich mit jmdm. ~:** have fun *or* a good time with sb.; **b)** *(belustigt sein)* be amused; **sich über jmdn./etw. ~:** find sb./sth. funny; *(über jmdn./etw. lachen)* laugh at sb./sth.; *(jmdn. verspotten)* make fun of sb./sth. **2.** *tr. V.* amuse; **amüsiert zusehen** look on with amusement

an [an] **1.** *Präp. mit Dat.* **a)** *(räumlich)* at; *(auf)* on; **an einem Ort** at a place; **an der Wand hängen** be hanging on the wall; **an der Wand stehen** stand by *or* against the wall; **an der Mosel/Donau liegen** be [situated] on the Moselle/Danube; **Frankfurt an der Oder** Frankfurt on [the] Oder; **an etw. lehnen** lean against sth.; **Tür an Tür** next door to one another; b) *(zeitlich)* on; **an jedem Sonntag** every Sunday; **an dem Abend, als er ...:** [on] the evening he ...; **an Ostern** *(bes. südd.)* at Easter; **c)** *(bei bestimmten Substantiven, Adjektiven und Verben)* **arm/reich an Vitaminen** low/rich in vitamins; **jmdn. an etw. erkennen** recognize sb. by sth.; **ein Mangel an etw.** a shortage of sth.; **an etw. leiden** suffer from sth.; **es an der Leber haben** have liver trouble; **an einer Krankheit sterben** die of a disease; **es ist an ihm, das zu tun** it is up to him to do it; **er hat etwas an sich** there ist sth. about him; **d) an [und für] sich** *(eigentlich)* actually. **2.** *Präp. mit Akk.* **a)** to; *(auf, gegen)* on; **etw. an jmdn. schicken** send sth. to sb.; **etw. an etw. hängen** hang sth. on sth.; **b)** *(bei bestimmten Substantiven, Adjektiven und Verben)* **an etw./jmdn. glauben** believe in sth./sb.; **an etw. denken** think of sth.; **sich an etw. erinnern** remember *or* recall sth.; **an die Arbeit gehen** get down to work; **einen Gruß an jmdn. ausrichten lassen** send greetings to sb.; **ich konnte kaum an mich halten vor Lachen/Ärger** I could hardly contain myself for laughing/hardly contain my anger. **3.** *Adv.* **a)** *(Verkehrsw.)* **Köln an: 9.15** arriving Cologne 09.15; **b)** *(ugs.: in Betrieb)* on; **die Waschmaschine/der Fernseher ist an** the washing-machine/television is on; *s. auch* **ansein; c)** *(als Aufforderung)* **Scheinwerfer an!** spotlights on!; **d)** *(ugs.: ungefähr)* around; about; **an [die] 20 000 DM** around *or* about 20,000 DM; *s. auch* **ab 2 d; von 1 a, b**

Anabolikum [ana'bo:likʊm] **das; ~s, Anabolika** *(Med.)* anabolic steroid

Anakonda [ana'kɔnda] **die; ~, ~s** anaconda

Analgetikum [anal'ge:tikʊm] **das; ~s, Analgetika** *(Med.)* analgesic

analog [ana'lo:k] **1.** *Adj.* **a)** analogous (*Dat.*, **zu** to); **b)** *(Technik, DV)* analogue. **2.** *adv.* **a)** analogously; **b)** *(Technik, DV)* ‹*display, reproduce*› in analogue form

Analogie die; ~, ~n *(geh., fachspr.)* analogy; **in ~ zu etw.** in analogy to sth.

Analog-: ~rechner der *(DV)* analogue computer; **~uhr die** analogue clock/watch

An·alphabet der; ~en, ~en illiterate [person]; **~ sein** be illiterate

Analyse [ana'ly:zə] **die; ~, ~n** *(auch: Psycho~)* analysis

analysieren *tr. V.* analyse

analytisch 1. *Adj.* analytical. **2.** *adv.* analytically

Anämie [anɛ'mi:] **die; ~, ~n** *(Med.)* anaemia

Ananas ['ananas] **die; ~, ~** *od.* **~se** pineapple

Anarchie [anar'çi:] **die; ~, ~n** anarchy

anarchisch *Adj.* anarchic

Anarchismus der; ~: anarchism

Anarchist der; ~en, ~en anarchist

anarchistisch *Adj.* anarchistic

Anästhesie [anǀɛste'zi:] **die; ~, ~n** *(Med.)* anaesthesia

Anästhesist der; ~en, ~en, Anästhesistin die; ~, ~nen *(Med.)* anaesthetist

Anatolien [ana'to:li̯ən] **(das); ~s** Anatolia

Anatomie [anato'mi:] **die; ~, ~n a)** anatomy; **b)** *(Institut)* anatomical institute

anatomisch [ana'to:mɪʃ] **1.** *Adj.* anatomical. **2.** *adv.* anatomically

anǀbahnen 1. *tr. V.* initiate ‹*negotiations, talks, process, etc.*›; develop ‹*relationship, connection*›. **2.** *refl. V.* ‹*development*› be in the offing; ‹*friendship, relationship*› start to develop

anǀbandeln ['anbandl̩n] *(südd., österr.),* **anǀbändeln** ['anbɛndl̩n] *itr. V. (ugs.)* **mit jmdm. ~:** get off with sb. *(Brit. coll.);* pick sb. up

An·bau der; *Pl.* **~ten a)** *o. Pl.* building; **die Genehmigung für den ~ einer Garage an ein Haus bekommen** receive permission to build a garage on to a house; **b)** *(Gebäude)* extension; **c)** *o. Pl. (das Anpflanzen)* cultivation; growing

anǀbauen 1. *tr. V.* **a)** build on; add; **eine Garage ans Haus ~:** build a garage on to the house; **b)** *(anpflanzen)* cultivate; grow. **2.** *itr. V. (das Haus vergrößern)* build an extension; *(~ lassen)* have an extension built

Anbau-: ~gebiet das: ~gebiete für Getreide cereal-growing *or* grain-growing areas; **~gebiete für Rotwein** red-wine-growing areas *or* areas for red wine; **~möbel das** unit furniture; **~schrank der** cupboard unit

An·beginn der *(geh.)* beginning; **von ~ [an]** right from the beginning

anǀbehalten *unr. tr. V. (ugs.)* **etw. ~:** keep sth. on

an·bei *Adv. (Amtsspr.)* herewith; **Rückporto ~:** return postage enclosed

anǀbeißen 1. *unr. tr. V.* bite into; take a bite of; **er hat die Banane nur angebissen** he only took one bite of the banana. **2.** *unr. itr. V. (auch fig. ugs.)* bite; **bei ihr hat noch keiner angebissen** *(fig. ugs.)* she hasn't managed to hook anybody yet

anǀbekommen *unr. tr. V. (ugs.)* **a)** *(anziehen können)* **etw. ~:** manage to get sth. on; **b)** *(anzünden od. starten können)* **ein Feuer/Streichholz ~:** manage to get a fire going/a match to light; **den Motor ~:** manage to get the engine going *or* to start

anǀbelangen *tr. V.* **was mich/**

diese Sache *usw.* **anbelangt** as far as I am/this matter is *etc.* concerned

an|bellen *tr. V.* bark at

an|beraumen [-bəraumən] *tr. V.* *(Amtsspr.)* arrange, fix

an|beten *tr. V.* *(auch fig.)* worship

An·betracht der: in ~ einer Sache *(Gen.)* in consideration *or* view of sth.

an|betreffen *unr. tr. V.* **in was mich/diese Sache** *usw.* **anbetrifft** as far as I am/this matter is *etc.* concerned

an|betteln *tr. V.* **jmdn. ~:** beg from sb.; **jmdn. um etw. ~:** beg sb. for sth.

Anbetung die; ~, ~en *(auch fig.)* worship; *(fig.: Verehrung)* adoration

an|biedern [-bi:dɐn] *refl. V.* *(abwertend)* curry favour (**bei** with)

Anbiederung die; ~, ~en *(abwertend)* currying favour (**bei** with)

an|bieten 1. *unr. tr. V.* offer; **jmdm. etw. ~:** offer sb. sth.; **jmdm. ~, etw. zu tun** offer to do sth. for sb.; **Verhandlungen ~:** offer to negotiate. **2.** *unr. refl. V.* **a)** offer one's services (**als** as); **sich ~, etw. zu tun** offer to do sth.; **b)** *(naheliegen)* ‹*opportunity*› present itself; ‹*possibility, solution*› suggest *or* present itself; **es bietet sich an, das zu tun** it would seem to be the thing to do; **c)** *(geeignet sein)* **sich für etw. ~:** be suitable for sth.

an|binden *unr. tr. V.* tie [up] (**an** + *Dat. od. Akk.* to); tie up, moor ‹*boat*› (**an** + *Dat. od. Akk.* to); tether ‹*animal*› (**an** + *Dat. od. Akk.* to); **er läßt sich nicht ~** *(fig.)* he won't be tied down; *s. auch* **angebunden 2; b)** *(verbinden, anschließen)* link (**an** + *Akk.* to)

an|blasen *unr. tr. V.* blow at

an|blecken *tr. V.* bare its/their teeth at

an|bleiben *unr. itr. V.; mit sein* *(ugs.)* stay on

an|blenden *tr. V.* flash [at]

An·blick der sight; **einen erfreulichen/traurigen ~ bieten** be a welcome/sad sight; **beim ~ der Pyramiden** at the sight of the Pyramids

an|blicken *tr. V.* look at

an|blinken *tr. V.* flash [at]

an|blinzeln *tr. V.* **a)** blink at; **b)** *(zuzwinkern)* wink at

an|bohren *tr. V.* **a)** bore into; *(mit der Bohrmaschine)* bore *or* drill into; **b)** *(erschließen)* tap [by drilling]

an|brechen 1. *unr. tr. V.* **a)** crack; **b)** *(öffnen)* open; start; **eine angebrochene Flasche** an opened bottle; **c)** *(zu verbrauchen beginnen)* break into ‹*supplies, reserves*›; **einen Hundertmarkschein ~:** break into *or (Amer.)* break a hundred mark note. **2.** *unr. itr. V.; mit sein (geh.: beginnen)* ‹*dawn*› break; ‹*day*› dawn, break; ‹*darkness, night*› come down, fall; ‹*age, epoch*› dawn

an|brennen 1. *unr. tr. V. (anzünden)* light. **2.** *unr. itr. V.; mit sein* **a)** burn; **ihm ist das Essen angebrannt** he has burnt the food; **nichts ~ lassen** *(fig. ugs.)* not miss out on anything; **b)** *(zu brennen beginnen)* ‹*wood, coal, etc.*› catch

an|bringen *unr. tr. V.* **a)** *(befestigen)* put up ‹*sign, aerial, curtain, plaque*› (**an** + *Dat.* on); fix ‹*lamp, camera*› (**an** + *Dat.* [on] to); **an etw.** *(Dat.)* **angebracht sein** be fixed [on] to sth.; **b)** *(äußern)* make ‹*request, complaint, comment, reference*›; **c)** *(zeigen)* display, demonstrate ‹*knowledge, experience*›; **d)** *(ugs.: herbeibringen)* bring; **e)** *(ugs.: verkaufen)* sell; move

An·bruch der *o. Pl. (geh.)* dawn[ing]; **der ~ des Tages** dawn; daybreak; **vor/nach/bei** *od.* **mit ~ der Nacht** before/after/at nightfall

an|brüllen *tr. V.* **a)** ‹*tiger, lion, etc.*› roar at; ‹*cow, bull, etc.*› bellow at; **b)** *(ugs.: anschreien)* bellow *or* bawl at

an|brummen *tr. V. (auch ugs.: unfreundlich anreden)* growl at

Andacht ['andaxt] **die; ~, ~en a)** *o. Pl. (Sammlung im Gebet)* silent prayer *or* worship; **in tiefer ~:** in deep devotion; **in ~ versunken** sunk in silent prayer *or* worship; sunk in one's devotions; **b)** *o. Pl. (innere Sammlung)* rapt attention; **c)** *(Gottesdienst)* prayers *pl.*; **eine ~ halten** hold a [short] service; **zur ~gehen** go to prayers *or* to the service

andächtig ['andɛçtɪç] **1.** *Adj.* **a)** *(ins Gebet versunken)* devout; reverent; **b)** *(innerlich gesammelt)* rapt; **c)** *nicht präd. (feierlich)* reverent. **2.** *adv.* **a)** *(ins Gebet versunken)* devoutly; reverently; **b)** *(innerlich gesammelt)* raptly

Andalusien [anda'lu:ziən] **(das); ~s** Andalusia

Andante [an'dantə] **das; ~[s], ~s** *(Musik)* andante

an|dauern *itr. V.* ‹*negotiations*› continue, go on; ‹*weather, rain*› last, continue

andauernd 1. *Adj.; nicht präd.* continual; constant. **2.** *adv.* continually; constantly; **warum fragst du denn ~ dasselbe?** why do you keep on asking the same thing?

Anden ['andn̩] *Pl.* **die ~:** the Andes

An·denken das; ~s, ~ a) *o. Pl.* memory; **jmds. ~ bewahren/in Ehren halten** keep/ honour sb.'s memory; **zum ~ an jmdn./etw.** to remind you/us *etc.* of sb./sth.; **das schenke ich dir zum ~:** I'll give you that to remember me/us by; **b)** *(Erinnerungsstück)* memento, souvenir; *(Reise~)* souvenir

ander... ['andɐ...] *Indefinitpron.* **1.** *attr.* **a)** other; **ein ~er Mann/eine ~e Frau/ein ~es Haus** another man/woman/house; **das Kleid gefällt mir nicht, haben Sie noch ~e/ein ~es?** I don't like that dress, do you have any others/ another?; **der/die/das eine oder ~e ...** one or two ...; **b)** *(nächst...)* next; **am/bis zum ~[e]n Tag** [on] the/by the next *or* following day; **c)** *(verschieden)* different; **~er Meinung sein** be of a different opinion; take a different view; **das ~e Geschlecht** the opposite sex; **bei ~er Gelegenheit** another time; **d)** *(neu)* **einen ~en Job finden** find another job; **er ist ein ~er Mensch geworden** he is a changed man. **2.** *alleinstehend* **a)** *(Person)* **jemand ~er** *or* **~es** someone else; *(in Fragen)* anyone else; **ein ~r/eine ~e:** another [one]; **die ~n** the others; **alle ~n** all the others; everyone else; **jeder/jede ~e** anyone *or* anybody else; **kein ~er/keine ~e** nobody *or* no one else; **was ist mit den ~n?** what about the others *or* the rest?; **niemand ~er** *od.* **~es** nobody *or* no one else; **niemand ~er** *od.* **~es als ...:** nobody *or* no one but ...; **einen ~[e]n/eine ~e haben** *(fig. ugs.)* have found somebody *or* someone else; **auf ~e hören** listen to others; **nicht drängeln, einer nach dem ~n** don't push, one after the other; **der eine oder [der] ~e** one or two *or* a few people; *s. auch* **recht c; b)** *(Sache)* **etwas ~es** something else; *(in Fragen)* anything else; **nichts ~es** nothing else; not anything else; **alles ~e** everything else; **ein[e]s nach dem ~[e]n** first things first; **ich will weder das eine noch das ~e** I don't want either; **und ~es/vieles ~e mehr** and more/much more besides; **unter ~[e]m** among[st] other things; **so kam eins zum ~[e]n** what with one thing on top of the other; **das ist etwas [ganz] ~es** that's [something quite] different; **von etwas ~em sprechen** talk about something else; **alles ~e als ...:** anything but ...; **~es zu tun haben** have other things to do

anderen·falls *Adv.* otherwise

anderen·orts *Adv. (geh.)* elsewhere

anderer·seits *Adv.* on the other hand

ander·mal *Adv. in* **ein ~:** another time

andern- *s.* **anderen-**

ändern ['ɛndɐn] **1.** *tr. V.* change; alter; alter ⟨*garment*⟩; change ⟨*person*⟩; amend ⟨*motion*⟩; **daran kann man nichts ~:** nothing can be done *or* there's nothing you/we *etc.* can do about it **2.** *refl. V.* change; alter; ⟨*person, weather*⟩ change

anders ['andɐs] *Adv.* **a)** ⟨*think, act, feel, do*⟩ differently (**als** from *or esp. Brit.* to); ⟨*be, look, sound, taste*⟩ different (**als** from *or esp. Brit.* to); **es war alles ganz ~:** it was all quite different; **wie könnte es ~ sein!** *(iron.)* surprise, surprise! *(iron.)*; **mir wird ganz ~** *(ugs.)* I feel weak at the knees; **es kam ~, als wir dachten** things didn't turn out the way we expected; **ich habe es mit ~ überlegt** I've changed my mind; **ich kann auch ~** *(ugs.)* you'd/he'd *etc.* better watch it *(coll.)*; **so und nicht ~:** this way and no other; exactly like that; **wie nicht ~ zu erwarten [war]** as [was to be] expected; **wenn es nicht ~ geht** if there is no other way; **b)** *(sonst)* else; **irgendwo/nirgendwo ~:** somewhere/nowhere else; **niemand ~:** nobody else; **jemand ~:** someone else; *(verneint, in Fragen)* anyone else; **c)** *(ugs.: andernfalls)* otherwise; or else

anders·artig *Adj.* different

Anders·denkende der/die; *adj. Dekl.* dissident; dissenter

anderseits *Adv. s.* **andererseits**

anders-, Anders-: ~farbig 1. *Adj.* different-coloured *attrib.*; of a different colour *postpos.*; **2.** *adv.* ⟨*decorated*⟩ in a different colour; **~geartet** *Adj.* different; of a different nature *postpos.*; **~gläubig** *Adj.* of a different faith *or* religion *postpos.*; **~gläubige der/die** person of a different faith *or* religion; **die ~gläubigen** those of different faiths *or* religions; **~herum** *Adv.* the other way round *or (Amer.)* around; **etw. ~herum drehen** turn sth. the other way; **~herum gehen/fahren** go/drive round *or (Amer.)* around the other way; **~lautend** [~lautn̩t] *Adj.; nicht präd.* to the contrary *postpos.*; **~rum** *(ugs.)* **1.** *Adv. s.* **~herum; 2.** *Adj.; nicht attr.* **~rum sein** be a poof *(Brit. coll.) or* a fairy *(sl.)*; be queer *(sl.)*; **~wo** *Adv. (ugs.)* elsewhere; *(verneint, in Fragen)* anywhere else; **~woher** *Adv. (ugs.)* from elsewhere; from somewhere else; *(verneint, in Fragen)* from anywhere else; **~wohin** *Adv. (ugs.)* elsewhere; somewhere else; *(verneint, in Fragen)* anywhere else

anderthalb ['andɐt'halp] *Bruchz.* one and a half; **~ Pfund Mehl** a pound and a half of flour; **~ Stunden** an hour and a half

anderthalb·fach *Vervielfältigungsz.* one and a half times; **die ~e Anzahl/Menge** one and a half times the number/amount

anderthalb·mal *Adv.* one and a half times; **~ so groß wie ...:** half as big again as ...

Änderung die; ~, ~en *s.* **ändern 1:** change (*Gen.* in); alteration (*Gen.* to); amendment (*Gen.* to)

Änderungs-: ~schneiderei die tailor's [that does alterations]; **~vor·schlag der** suggestion for a change; *(für Gesetz, Antrag usw.)* suggestion for an amendment; **~wunsch der** request for a change

anderweitig [-vaitɪç] **1.** *Adj.; nicht präd.* other. **2.** *adv.* **a)** *(auf andere Weise)* in another way; **~ beschäftigt sein** be otherwise engaged; **b)** *(an jmd. anderen)* to somebody else

an|deuten 1. *tr. V.* **a)** *(zu verstehen geben)* intimate; hint; **jmdm. etw. ~:** intimate *or* hint sth. to sb.; **b)** *(nicht ausführen)* outline; *(kurz erwähnen)* indicate. **2.** *refl. V. (sich abzeichnen)* be indicated

An·deutung die a) *(Anspielung)* hint; **eine ~ machen** give *or* drop a hint (**über** + *Akk.* about); **b)** *(schwaches Anzeichen)* suggestion; hint

andeutungs·weise *Adv.* in the form of a hint *or* suggestion/hints *or* suggestions; **davon war nur ~ die Rede** it was only hinted at

an|dichten *tr. V.* **jmdm. etw. ~:** impute sth. to sb.

an|dienen 1. *tr. V.* **a) jmdm. etw. ~:** offer sth. to sb.; *(aufdringlich)* press sth. on sb. **2.** *refl. V.* **sich jmdm. ~:** offer oneself *or* one's services to sb.; *(aufdringlich)* press oneself *or* one's services on sb.

an|docken *tr., itr. V. (Raumf.)* dock (**an** + *Dat.* with)

An·drang der; *o. Pl.* crowd; *(Gedränge)* crush; **es herrschte großer ~:** there was a large crowd/great crush

an|drängen *itr. V.; mit sein* surge (**gegen** against); ⟨*crowd*⟩ surge forward; ⟨*army*⟩ push forward

andr... *usw. s.* **ander...** *usw.*

Andreas [an'dre:as] **(der)** Andrew

Andreas·kreuz das a) St Andrew's cross; **b)** *(Verkehrsw.)* diagonal cross

an|drehen *tr. V.* **a)** *(einschalten)* turn on; **b)** *(ugs.: verkaufen)* **jmdm. etw. ~:** palm sb. off with sth.; palm sth. off on sb.; **c)** *(anziehen)* screw ⟨*nut*⟩ on; screw ⟨*screw*⟩ in

andrerseits *s.* **andererseits**

an|drohen *tr. V.* **jmdm. etw. ~:** threaten sb. with sth.

An·drohung die threat; **unter ~ von Gewalt** with the *or* under threat of violence

Android [andro'i:t] **der; ~en, ~en, Androide** [andro'i:də] **der; ~n, ~n** android

an|drücken *tr. V.* press down

an|ecken [-ɛkn̩] *itr. V.; mit sein (fig. ugs.)* **bei jmdm. ~:** rub sb. [up *(Brit.)*] the wrong way

an|eignen *refl. V.* **a)** appropriate; **sich** *(Dat.)* **etw. widerrechtlich ~:** misappropriate sth.; **b)** *(lernen)* acquire; learn

An·eignung die a) appropriation; **widerrechtliche ~:** misappropriation; **b)** *(Lernen)* acquisition; learning

an·einander *Adv.:* **~ denken** think of each other *or* one another; **~ vorbeigehen** pass each other *or* one another; go past each other *or* one another; **sich ~ gewöhnen** get used to each other *or* one another; **~ vorbeireden** talk at cross purposes; **sich ~ festhalten** hold each other *or* one another

aneinander-: ~|binden *unr. tr. V.* tie together; hitch ⟨*horses*⟩ together; **~|drängen 1.** *tr. V.* push *or* press together; **2.** *refl. V.* press together; **~|geraten** *unr. itr. V.; mit sein (sich prügeln)* come to blows (**mit** with); *(sich streiten)* quarrel (**mit** with); **~|grenzen** *itr. V.* ⟨*properties, rooms, etc.*⟩ adjoin [each other *or* one another]; ⟨*countries*⟩ border on each other *or* one another; **~|halten** *unr. tr. V.* hold next to each other *or* one another; ~|**legen** *tr. V.* put *or* place next to each other *or* one another; ~|**liegen** *unr. itr. V.* lie next to each other; ⟨*properties*⟩ adjoin [each other *or* one another]; be adjacent [to each other *or* one another]

Anekdote [anɛk'do:tə] **die; ~, ~n** anecdote

an|ekeln *tr. V.* disgust; nauseate; **du ekelst mich an** you make me sick; **sich angeekelt abwenden** turn away in disgust

Anemone [ane'mo:nə] **die; ~, ~n** anemone

anerkannt *Adj.* recognized; recognized, acknowledged ⟨*authority, expert*⟩

an|erkennen *unr. tr. V.* **a)** recognize ⟨*country, record, verdict, qualification, document*⟩; acknowledge ⟨*debt*⟩; accept ⟨*demand, bill, conditions, rules*⟩; allow ⟨*claim, goal*⟩; **jmdn. als gleichberechtigten Partner ~:** accept sb. as an equal partner; **b)** *(nicht leugnen)* acknowledge; **c)** *(würdigen)* acknowledge, appreciate ⟨*achievement, efforts*⟩; appreciate ⟨*person*⟩; respect ⟨*viewpoint, opinion*⟩; **ein ~der Blick** an appreciative look; **~d nicken** nod appreciatively

anerkennens·wert *Adj.* commendable

Anerkennung die; ~, ~en *s.* **anerkennen: a)** recognition; acknowledgement; acceptance; allowance; **b)** acknowledgement; **c)** acknowledgement; appreciation; respect (*Gen.* for)

an|fachen [-faxn̩] *tr. V.* fan; *(fig.)* arouse ⟨*anger, enthusiasm*⟩; arouse, inflame ⟨*passion*⟩; inspire, stir up ⟨*hatred*⟩; inspire ⟨*hope*⟩; ferment ⟨*discord, war*⟩

an|fahren 1. *unr. tr. V.* **a)** run into; hit; **b)** *(herbeifahren)* deliver; **c)** *(ansteuern)* stop *or* call at ⟨*village etc.*⟩; ⟨*ship*⟩ put in at ⟨*port*⟩; **d)** *(zurechtweisen)* shout at; **e)** *(in Betrieb nehmen)* commission ⟨*power-station, blast furnace*⟩. **2.** *unr. itr. V.; mit sein* **a)** *(starten)* start off; **b) angefahren kommen** come driving/riding along; *(auf einen zu)* come driving/riding up

An·fahrt die a) *(das Anfahren)* journey; **b)** *(Weg)* approach

Anfahrts-: ~weg der journey; **~zeit die** travelling time

An·fall der a) *(Attacke)* attack; *(epileptischer ~, fig.)* fit; **einen ~ bekommen** have an attack/a fit; **in einem ~ von ...** *(fig.)* in a fit of ...; **b)** *o. Pl. (Anfallendes)* amount (**an** + *Dat.* of); *(Ertrag)* yield (**an** + *Dat.* of)

an|fallen 1. *unr. tr. V.* **a)** *(angreifen)* attack; **b)** *(geh.: befallen)* **Zweifel/Angst fiel mich an** I was assailed by doubt/fear. **2.** *unr. itr V.; mit sein* ⟨*costs*⟩ arise, be incurred; ⟨*interest*⟩ accrue; ⟨*work*⟩ come up; ⟨*parcels etc.*⟩ accumulate

an·fällig *Adj.* ⟨*person*⟩ with a delicate constitution; ⟨*machine*⟩ susceptible to faults; **er ist sehr ~:** he has a very delicate constitution; **gegen** *od.* **für etw. ~ sein** be susceptible to sth.; **für eine Krankheit ~ sein** be prone to an illness

Anfälligkeit die *(einer Person)* delicate constitution; *(einer Maschine)* susceptibility to faults

An·fang der beginning; start; *(erster Abschnitt)* beginning; **[ganz] am ~ der Straße** [right] at the start of the street; **am** *od.* **zu ~:** at first; to begin with; **von ~ an** from the beginning *or* outset; **~ 1984/der achtziger Jahre/Mai/der Woche** *usw.* at the beginning of 1984/the eighties/May/the week *etc.*; **von ~ bis Ende** from beginning to end *or* start to finish; **der ~ vom Ende** the beginning of the end; **im ~ war das Wort** *(bibl.)* in the beginning was the Word; **einen ~ machen** make a start; **den ~ machen** make a start; start; *(als erster handeln)* make the first move; **einen neuen ~ machen** make a new *or* fresh start; **aller ~ ist schwer** *(Spr.)* it's always difficult at the beginning; **in den** *od.* **seinen Anfängen stekken** be in its infancy

an|fangen 1. *unr. itr. V.* **a)** begin; start; **das fängt ja gut an!** *(ugs. iron.)* that's a good start! *(iron.)*; **er hat ganz klein/als ganz kleiner Angestellter angefangen** he started small/started [out] as a minor employee; **mit etw. ~:** start [on] sth.; **fang nicht wieder damit an!** don't start [all] that again!; **~, etw. zu tun** start to do sth.; **es fängt an zu schneien** it's starting *or* beginning to snow; **fang doch nicht gleich an zu weinen** don't start crying; **angefangen bei** *od.* **mit** *od.* **von ...:** starting *or* beginning with ...; **Weiß fängt an** white starts; **er hat angefangen** *(mit dem Streit o. ä.)* he started it; **b)** *(zu sprechen ~)* begin; **von etw. ~:** start on about sth.; **c)** *(eine Stelle antreten)* start. **2.** *unr. tr. V.* **a)** begin; start; *(anbrechen)* start; **das Rauchen ~:** start smoking; **b)** *(machen)* do; **damit kann ich nichts/nicht viel ~:** that's no/not much good to me; *(das verstehe ich nicht/kaum)* that doesn't mean anything/much to me; **kannst du noch etwas damit ~?** is it any good *or* use to you?; **nichts mit sich anzufangen wissen** not know what to do with oneself

An·fänger der; ~s, ~, An·fängerin die; ~, ~nen beginner; *(abwertend: Stümper)* amateur

Anfänger·kurs der beginners' course; course for beginners

anfänglich ['anfɛŋlɪç] *Adj.* initial

anfangs *Adv.* at first; initially

Anfangs-: ~buchstabe der initial [letter]; first letter; **~gehalt das** starting salary; **~schwierigkeit die;** *meist Pl.* initial difficulty; **~stadium das** initial stage; **im ~stadium sein** be in its/their initial stages *pl.*; **~zeit die** starting time

an|fassen 1. *tr. V.* **a)** *(fassen, halten)* take hold of; **b)** *(berühren)* touch; **c)** *(bei der Hand nehmen)* **jmdn. ~:** take sb.'s hand; **faßt euch an** take each other's hand; **d)** *(angehen)* approach, tackle ⟨*problem, task, etc.*⟩; **e)** *(behandeln)* treat ⟨*person*⟩. **2.** *itr. V. (mithelfen)* **[mit] ~:** lend a hand

an|fauchen *tr. V.* **a)** ⟨*cat*⟩ spit at; **b)** *(fig.)* snap at

anfechtbar *Adj.* **a)** *(bes. Rechtsw.)* contestable; **b)** *(kritisierbar, bestreitbar)* disputable ⟨*statement, decision*⟩; ⟨*book*⟩ open to criticism

an|fechten *unr. tr. V.* **a)** *(bes. Rechtsw.)* challenge, dispute ⟨*validity, authenticity, statement*⟩; contest ⟨*will*⟩; contest, challenge ⟨*decision*⟩; dispute ⟨*contract*⟩; challenge ⟨*law, opinion*⟩; **b)** *(beunruhigen)* trouble; bother

Anfechtung die; ~, ~en *(bes. Rechtsw.) s.* **anfechten a:** challenging; disputing; contesting

an|feinden [-faindn̩] *tr. V.* treat with hostility

an|fertigen *tr. V.* make; make up ⟨*medicament, preparation*⟩; do ⟨*homework, translation*⟩; prepare, draw up ⟨*report*⟩; cut, make ⟨*key*⟩

An·fertigung die *s.* **anfertigen:** making; doing; making up; preparing; drawing up; cutting

an|feuchten [-fɔyçtn̩] *tr. V.* moisten ⟨*lips, stamp*⟩; dampen, wet ⟨*ironing, cloth, etc.*⟩

an|feuern *tr. V.* spur on; **~de Rufe/Gesten** shouts of encouragement/rousing gestures

An·feuerung die spurring on

an|flehen *tr. V.* beseech; implore; **jmdn. um etw. ~:** beg sb. for sth.

ạn|fliegen 1. *unr. itr. V.; mit sein* ‹*aircraft*› fly in; *(beim Landen)* approach; come in to land; ‹*bird etc.*› fly in; **angeflogen kommen** come flying in; *(auf einen zu)* ‹*bird*› come flying up; **gegen den Wind ~:** fly into the wind. **2.** *unr. tr. V.* **a)** fly to ‹*city, country, airport*›; *(beim Landen)* approach ‹*airport*›; **b)** *(ansteuern)* ‹*aircraft*› approach; ‹*bird*› fly towards, approach

Ạn·flug der a) approach; **die Maschine befindet sich im ~ auf Berlin** the plane is now approaching Berlin; **b)** *(Hauch)* hint; trace; **c)** *(Anwandlung)* fit; **in einem ~ von Großzügigkeit** in a fit of generosity

ạn|flunkern *tr. V. (ugs.)* tell fibs to

ạn|fordern *tr. V.* request, ask for ‹*help*›; ask for ‹*catalogue*›; order ‹*goods, materials*›; send for ‹*ambulance*›

Ạn·forderung die a) *o. Pl. (das Anfordern)* request (*Gen.* for); **b)** *(Anspruch)* demand; **große/hohe ~en an jmdn./etw. stellen** make great demands on sb./sth.; **den ~en nicht gewachsen sein/nicht genügen** not be up to the demands

Ạn·frage die inquiry; *(Parl.)* question; **große/kleine ~** *(Parl.)* oral/written question

ạn|fragen *itr. V.* inquire; ask

ạn|fressen 1. *unr. tr. V.* **a)** nibble [at]; ‹*bird*› peck [at]; **b)** *(zersetzen)* eat away [at]. **2.** *unr. refl. V.* **sich** *(Dat.)* **einen Bauch ~** *(salopp)* develop a paunch

ạn|freunden *refl. V.* make *or* become friends (**mit** with); **sich mit etw. ~** *(fig.)* get to like sth.

ạn|frieren *unr. itr. V.; mit sein* **an etw.** *(Dat.)* **~:** freeze to sth.

ạn|fügen *tr. V.* add

ạn|fühlen *refl. V.* feel

ạn|führen *tr. V.* **a)** lead; lead, head ‹*procession*›; **b)** *(zitieren)* quote; **c)** *(nennen)* quote, give, offer ‹*example*›; give, offer ‹*details, reason, proof*›; **d)** *(ugs.: hereinlegen)* have on *(Brit. coll.)*; dupe

Ạn·führer der a) *(Führer)* leader; **b)** *(Rädelsführer)* ringleader

Ạn·führung die a) leadership; **b)** *(das Zitieren, Zitat)* quotation; **c)** *(Nennung) s.* **anführen c:** quotation; giving; offering

Ạnführungs·zeichen das quotation-mark; inverted comma *(Brit.)*

ạn|füllen *tr. V.* fill [up]; **mit etw. angefüllt sein** be filled *or* full with sth.

Ạn·gabe die a) *(das Mitteilen)* giving; **ohne ~ von Gründen** without giving [any] reasons; **b)** *(Information)* piece of information; **~n** information *sing.*; **c)** *(Anweisung)* instruction; **d)** *o. Pl. (Prahlerei)* boasting; bragging; *(angeberisches Benehmen)* showing-off; **e)** *(Ballspiele)* service; serve; **[eine] ~ machen** serve; **ich habe [die] ~:** it's my serve

ạn|gaffen *tr. V. (abwertend)* gape at

ạn|geben 1. *unr. tr. V.* **a)** give ‹*reason*›; declare ‹*income, dutiable goods*›; name, cite ‹*witness*›; **zur angegebenen Zeit** at the stated time; **wie oben angegeben** as stated *or* mentioned above; **b)** *(bestimmen)* set ‹*course, direction*›; **den Takt ~:** keep time; **c)** *(veralt.: anzeigen, melden)* report ‹*theft etc.*›; give away ‹*accomplice etc.*›. **2.** *unr. itr. V.* **a)** *(prahlen)* boast; brag; *(sich angeberisch benehmen)* show off; **b)** *(Ballspiele)* serve

Ạngeber der; ~s, ~: boaster; braggart

Angeberei die; ~: boasting; bragging; *(angeberisches Benehmen)* showing-off

ạngeberisch *(ugs.)* **1.** *Adj.* boastful ‹*person*›; pretentious, showy ‹*glasses, car, jacket*›. **2.** *adv.* boastfully

Ạngebetete der/die; *adj. Dekl. (meist scherzh.)* beloved; *(Idol)* idol

ạngeblich 1. *Adj.* alleged. **2.** *adv.* supposedly; allegedly; **er ist ~ krank** he is supposed to be ill; *(er sagt, er sei krank)* he says he's ill

ạn·geboren *Adj.* innate ‹*characteristic*›; congenital ‹*disease*›

Ạn·gebot das a) offer; **b)** *(Wirtsch.) (angebotene Menge)* supply; *(Sortiment)* range

ạn·gebracht *Adj.* appropriate

ạn·gebunden *Adj.* **a)** tied down; **b) kurz ~** *(ugs.)* short; abrupt

ạn·gegossen *Adj.* **wie ~ sitzen/passen** *(ugs.)* fit like a glove

angegraut ['angəgraut] *Adj.* greying

ạn·gegriffen *Adj.* weakened ‹*health, stomach*›; strained ‹*nerves, voice*›; *(erschöpft)* exhausted; *(nervlich)* strained

ạn·geheiratet *Adj.* by marriage *postpos.*; **~ sein** be related by marriage

angeheitert ['angəhaitɐt] *Adj.* tipsy; merry *(coll.)*

ạn|gehen 1. *unr. itr. V.; mit sein* **a)** *(sich einschalten, entzünden)* ‹*radio, light, heating*› come on; ‹*fire*› catch, start burning; **b)** *(sich einschalten, entzünden lassen)* ‹*radio, light*› go on; ‹*fire*› light, catch; **c)** *(ugs.: beginnen)* start; **d)** *(anwachsen, wachsen)* ‹*plant*› take root; **e)** *(geschehen dürfen)* **es mag noch ~:** it's [just about] acceptable; **es geht nicht an, daß radikale Elemente die Partei unterwandern** radical elements must not be allowed to infiltrate the party; **f)** *(bes. nordd.: wahr sein)* **das kann doch wohl nicht ~!** that can't be true!; **g) gegen etw./jmdn. ~:** fight sth./sb. **2.** *unr. tr. V.* **a)** *(angreifen)* attack; *(Sport)* tackle; challenge; **b)** *(in Angriff nehmen)* tackle ‹*problem, difficulty*›; take ‹*fence, bend*›; **c)** *(bitten)* ask; **jmdn. um etw. ~:** ask sb. for sth.; **d)** *(betreffen)* concern; **was geht dich das an?** what's it got to do with you?; **das geht dich nichts an** it's none of your business; **was das/mich angeht, [so] ...:** as far as that is/I am concerned ...

ạngehend *Adj.* budding; *(zukünftig)* prospective

ạn|gehören *itr. V.* **jmdm./einer Sache ~:** belong to sb./sth.; **der Regierung/einer Familie ~:** be a member of the government/a family

ạn·gehörig *Adj.* belonging (*Dat.* to)

Ạngehörige der/die; *adj. Dekl.* **a)** *(Verwandte)* relative; relation; **der nächste ~:** the next of kin; **b)** *(Mitglied)* member

Angeklagte ['angəkla:ktə] **der/die;** *adj. Dekl.* accused; defendant

ạngeknackst *Adj. (fig. ugs.)* weakened

Angel ['aŋl̩] **die; ~, ~n a)** fishing-rod; rod and line; **die ~ auswerfen** cast the line; **b)** *(Tür~, Fenster~ usw.)* hinge; **etw. aus den ~n heben** lift sth. off its hinges; *(fig.)* turn sth. upside down

Ạn·gelegenheit die matter; *(Aufgabe, Problem)* affair; concern; **öffentliche/kulturelle ~en** public/cultural affairs; **das ist meine/nicht meine ~:** that is my affair *or* business/not my concern *or* business; **kümmere dich um deine eigenen ~en!** mind your own business; **sich in jmds. ~en mischen** meddle in sb.'s affairs

Ạngel-: ~gerät das *o. Pl.* fishing-tackle; **~haken der** fish-hook

ạngeln 1. *tr. V. (zu fangen suchen)*

fish for; *(fangen)* catch; **sie hat sich** *(Dat.)* **einen reichen Mann geangelt** *(fig.)* she has hooked a rich husband. **2.** *itr. V.* angle; fish; **nach etw. ~** *(fig.)* fish for sth.

Angel-: ~rute die fishing-rod; **~sachse der** Anglo-Saxon; **~sächsisch** *Adj.* Anglo-Saxon; **~schein der** fishing permit *or* licence; **~schnur die** fishing-line

an·gemessen 1. *Adj.* appropriate; reasonable, fair ‹*price, fee*›; adequate ‹*reward*›. **2.** *adv.* ‹*behave*› appropriately; ‹*reward*› adequately; ‹*recompense*› reasonably, fairly

an·genehm 1. *Adj.* pleasant; agreeable; **ist Ihnen die Temperatur/ist es so ~?** is the temperature all right for you/is it all right like that?; **es ist mir gar nicht ~, daß ...:** I don't at all like it that ...; **~e Reise/Ruhe!** [have a] pleasant journey/have a good rest; **[sehr] ~!** delighted to meet you; **das Angenehme mit dem Nützlichen verbinden** combine business with pleasure. **2.** *adv.* pleasantly; agreeably

angepaßt *Adj.* well adjusted

angeregt 1. *Adj.* lively; animated. **2.** *adv.* **sich ~ unterhalten/~ diskutieren** have a lively *or* an animated conversation/discussion

an·geschlagen *Adj.* groggy; poor, weakened ‹*health*›

angeschmutzt ['angəʃmʊtst] *Adj.* slightly soiled

angesehen *Adj.* respected

An·gesicht das; ~[e]s, ~er, *österr. auch* **~e** *(geh.)* **a)** *(Gesicht)* face; **von ~ zu ~:** face to face; **b)** *in* **im ~** *(+ Gen.) s.* **angesichts a**

angesichts *Präp. mit Gen. (geh.)* **a) ~ des Feindes/der Gefahr/des Todes** in the face of the enemy/of danger/of death; **~ der Stadt/der Küste** in sight of the town/coast; **b)** *(fig.: in Anbetracht)* in view of

an·gespannt 1. *Adj.* **a)** *(angestrengt)* close ‹*attention*›; taut ‹*nerves*›; **b)** *(kritisch)* tense ‹*situation*›; tight ‹*market, economic situation*›. **2.** *adv.* ‹*work*› concentratedly; ‹*listen*› with concentrated attention

angestellt *Adj.* **bei jmdm. ~ sein** be employed by sb.; work for sb.; **fest ~ sein** have a permanent position

Angestellte der/die; *adj. Dekl.* [salaried] employee; **Arbeiter und ~:** workers and salaried staff; blue- and white-collar workers; **sie ist ~ bei der Stadt** she works for the town council; *(im Gegensatz zur Beamtin/Arbeiterin)* she has a salaried position with the town council

Angestellten·versicherung die [salaried] employees' insurance

angestrengt 1. *Adj.* close ‹*attention*›; concentrated ‹*work, study, thought*›. **2.** *adv.* ‹*work, think, search*› concentratedly

an·getan *Adj.* **von jmdm./etw. ~ sein** be taken with sb./sth.; **dazu** *od.* **danach ~ sein, etw. zu tun** *(geh.)* be suitable for doing sth.

an·getrunken *Adj.* [slightly] drunk

an·gewandt *Adj.; nicht präd.* applied

an·gewiesen *Adj.* **auf etw.** *(Akk.)* **~ sein** have to rely on sth.; **auf jmdn./jmds. Unterstützung ~ sein** be dependent on *or* have to rely on sb./sb.'s support; **auf sich selbst ~ sein** be thrown back upon one's own resources

an|gewöhnen 1. *tr. V.* **jmdm. etw. ~:** get sb. used to sth.; accustom sb. to sth.; **jmdm. ~, etw. zu tun** get sb. used to *or* accustom sb. to doing sth. **2.** *refl. V.* **sich** *(Dat.)* **etw. ~:** get into the habit of sth.; **sich** *(Dat.)* **schlechte Manieren ~:** become ill-mannered; **[es] sich** *(Dat.)* **~, etw. zu tun** get into the habit of doing sth.; **sich** *(Dat.)* **das Rauchen ~:** take up smoking

An·gewohnheit die habit

Angina [aŋ'gi:na] **die; ~, Anginen** angina

an|gleichen 1. *unr. tr. V.* **etw. einer Sache** *(Dat.) od.* **an etw.** *(Akk.)* **~:** bring sth. into line with sth. **2.** *unr. refl. V.* **sich jmdm./einer Sache** *od.* **an jmdn./etw. ~:** become like sb./sth.

Angler der; ~s, ~: angler

Anglikaner [aŋgli'ka:nɐ] **der; ~s, ~:** Anglican

anglikanisch *Adj.* Anglican

Anglikanismus der; ~: Anglicanism *no art.*

anglisieren *tr. V.* Anglicize

Anglist der; ~en, ~en English specialist *or* scholar; Anglicist; *(Student)* English student

Anglistik die; ~: Anglistics *sing.;* English [language and literature]; English studies *pl., no art.*

Anglistin die; ~, ~nen *s.* **Anglist**

Anglo-: ~amerikaner der Anglo-American; **~-Amerikaner der** Anglo-Saxon; **die ~:** the British and the Americans; **~amerikanerin die** Anglo-American

an|glotzen *tr. V. (ugs.)* gawp at *(coll.)*

Angola [aŋ'go:la] **(das); ~s** Angola

Angolaner der; ~s, ~: Angolan

Angora- [aŋ'go:ra]: **~kaninchen das** angora rabbit; **~katze die** angora cat; **~wolle die** angora [wool]

angreifbar *Adj.* contestable

an|greifen 1. *unr. tr. V.* **a)** *(auch fig.)* attack; **b)** *(schwächen)* weaken, affect ‹*health, heart*›; affect ‹*stomach, intestine, voice*›; weaken ‹*person*›; **c)** *([be]schädigen)* attack ‹*metal*›; harm ‹*hands*›; **d)** *(anbrechen)* break into ‹*supplies, savings, etc.*›. **2.** *unr. itr. V. (auch fig.)* attack

An·greifer der, Angreiferin die; ~, ~nen *(auch fig.)* attacker

an|grenzen *itr. V.* **an etw.** *(Akk.)* **~:** border on *or* adjoin sth.

An·griff der attack; **zum ~ übergehen** go over to the attack; take the offensive; **zum ~ blasen** *(auch fig.)* sound the charge *or* attack; **etw. in ~ nehmen** *(fig.)* set about *or* tackle sth.

angriffs-, Angriffs-: ~krieg der war of aggression; **~lust die** aggression; aggressiveness; **~lustig 1.** *Adj.* aggressive; **2.** *adv.* aggressively; **~punkt der** *(fig.)* target

an|grinsen *tr. V.* grin at

angst [aŋst] *Adj.* **jmdm. ist/wird [es] ~ [und bange]** sb. is/becomes afraid *or* frightened; *(jmd. ist/wird unruhig)* sb. is/becomes very worried *or* anxious; **jmdm. ~ [und bange] machen** frighten *or* scare sb.; *(jmdn. unruhig machen)* make sb. very worried *or* anxious

Angst die; ~, Ängste ['ɛŋstə] **a)** *(Furcht)* fear (**vor** + *Dat.* of); *(Psych.)* anxiety; **~ [vor jmdm./etw.] haben** be afraid *or* frightened [of sb./sth.]; **jmdn. in ~ und Schrecken versetzen** worry and frighten sb.; **~ bekommen** *od. (ugs.)* **kriegen, es mit der ~ [zu tun] bekommen** *od. (ugs.)* **kriegen** become *or* get frightened *or* scared; **jmdm. ~ einflößen/einjagen/machen** frighten *or* scare sb.; **keine ~!** don't be afraid; **sich aus ~ verstecken** hide in fear; **aus ~, sich zu verraten, sagte er kein einziges Wort** he didn't say a word for fear of betraying himself; **b)** *(Sorge)* worry; anxiety; **~ [um jmdn./etw.] haben** be worried *or* anxious [about sb./sth.]; **sie hat ~, ihn zu verletzen** she is worried about hurting him; **keine ~!** don't worry!

ängstigen ['ɛŋstɪgn̩] **1.** *tr. V.* frighten; scare; *(beunruhigen)* worry. **2.** *refl. V.* be frightened *or* afraid; *(sich sorgen)* worry; **sich vor etw.** *(Dat.)*/**um jmdn. ~:** be frightened *or* afraid of sth./worried about sb.

ängstlich ['ɛŋstlɪç] **1.** *Adj.* **a)** *(verängstigt)* anxious; apprehensive; **b)** *(furchtsam, schüchtern)* timorous; timid; **c)** *(besorgt)* worried; anxious. **2.** *adv.* **a)** *(verängstigt)* anxiously; apprehensively; **b)** *(besorgt)* anxiously; **c)** *(übermäßig genau)* meticulously; **~ bemüht** *od.* **darauf bedacht sein, etw. zu tun** be at great pains to do sth.

Ängstlichkeit die; ~ a) *(Furchtsamkeit)* timorousness; timidity; **b)** *(Schüchternheit)* timidity; **c)** *(Besorgnis)* anxiety

angst-, Angst-: ~neurose die anxiety neurosis; **~psychose die** anxiety psychosis; **~röhre die** *(ugs. scherzh.)* topper *(coll.)*; top hat; **~schweiß der** cold sweat; **der ~schweiß brach ihm aus** he broke out in a cold sweat; **~verzerrt** *Adj.* ‹*face*› twisted in fear

an|gucken *tr. V. (ugs.)* look at; **sich** *(Dat.)* **etw./jmdn. ~:** look *or* have a look at sth./ sb.; **guck dir das/den an!** [just] look at that/ him!

an|gurten *tr. V.* strap in; **sich ~:** put on one's seat-belt; *(im Flugzeug)* fasten one's seat-belt

an|haben *unr. tr. V.* **a)** *(ugs.: am Körper tragen)* have on; **b) jmdm./einer Sache etwas ~ können** be able to harm sb./harm *or* damage sth.; **c)** *(ugs.: in Betrieb haben)* have on

An·halt der *s.* **Anhaltspunkt**

an|halten 1. *unr. tr. V.* **a)** stop; **den Atem ~:** hold one's breath; **b)** *(auffordern)* urge; **c) sich** *(Dat.)* **eine Hose/einen Rock** *usw.* **~:** hold a pair of trousers/a skirt *etc.* up against oneself. **2.** *unr. itr. V.* **a)** stop; **b)** *(andauern)* go on; last; **c) [bei jmdm.] um jmdn.** *od.* **jmds. Hand ~** *(veralt.)* ask [sb.] for sb.'s hand [in marriage]

anhaltend 1. *Adj.* constant; continuous. **2.** *adv.* constantly; continuously

An·halter der hitch-hiker; **per ~ fahren** hitch[-hike]

An·halterin die hitch-hiker

Anhalts·punkt der clue (**für** to); *(für eine Vermutung)* grounds *pl.* (**für** for)

an·hand 1. *Präp. mit Gen.* with the help of. **2.** *Adv.* **~ von** with the help of

An·hang der a) *(eines Buches)* appendix; **b)** *(Anhängerschaft)* following; **c)** *(Verwandtschaft)* family

[1]**an|hängen** *(geh.) unr. itr. V.* **a)** *(verbunden sein mit)* be attached *(Dat.* to); **b)** *(glauben an)* subscribe *(Dat.* to) ‹*belief, idea, theory, etc.*›

[2]**an|hängen 1.** *tr. V.* **a)** hang up (**an** + *Akk.* on); **b)** *(ankuppeln)* couple on (**an** + *Akk.* to); hitch up ‹*trailer*› (**an** + *Akk.* to); **c)** *(anfügen)* add (**an** + *Akk.* to); **d)** *(ugs.: zuschreiben, anlasten)* **jmdm. etw. ~:** blame sth. on sb.; **e)** *(ugs.: geben)* **jmdm. etw. ~:** give sb. sth.. **2.** *refl. V.* **a)** hang on (**an** + *Akk.* to); **b)** *(ugs.: sich anschließen)* **sich [an jmdn.** *od.* **bei jmdm.] ~:** tag along [with sb.] *(coll.)*

An·hänger der a) *(Mensch)* supporter; *(einer Sekte)* adherent; follower; **b)** *(Wagen)* trailer; **c)** *(Schmuckstück)* pendant; **d)** *(Schildchen)* label; tag

Anhängerschaft die; ~, ~en supporters *pl.*; *(einer Sekte)* followers *pl.*; adherents *pl.*

anhängig *Adj. (Rechtsw.)* pending ‹*action*›; **etw. ~ machen** start legal proceedings over sth.

anhänglich *Adj.* devoted ‹*dog, friend*›

Anhänglichkeit die; ~: devotion (**an** + *Akk.* to); **aus [alter] ~** *(Nostalgie)* out of old affection

Anhängsel ['anhɛŋzl̩] **das; ~s, ~:** appendage *(Gen.* to)

an|hauchen *tr. V.* breathe on ‹*mirror, glasses*›; blow on ‹*fingers, hands*›

an|hauen *tr. V. (salopp)* accost; **jmdn. um 50 Mark ~:** touch *(sl.)* *or* tap sb. for 50 marks

an|häufen 1. *tr. V.* accumulate; amass. **2.** *refl. V.* accumulate; pile up

An·häufung die a) accumulation; amassing; **b)** *(Haufen)* accumulation

an|heben *unr. tr. V.* **a)** lift [up]; **b)** *(erhöhen)* raise ‹*prices, wages, etc.*›

An·hebung die *(Erhöhung)* increase *(Gen.* in); raising *(Gen.* of)

an|heften *tr. V.* tack [on] ‹*hem, sleeve, etc.*›; attach ‹*label, list*›; put up ‹*sign, notice*›

anheimelnd *Adj.* homely; cosy

anheim-: ~|fallen *unr. itr. V.; mit sein (geh.)* **jmdm./dem Staate ~fallen** ‹*wealth, property*› pass to sb./the state; **der Vergessenheit/ der Zerstörung ~fallen** sink into oblivion/fall prey to destruction

~|stellen *tr. V.* **[es] jmdm. ~stellen, etw. zu tun** *(geh.)* leave it to sb. to do sth.

an|heizen 1. *tr. V.* **a)** fire up ‹*stove, boiler, etc.*›; **b)** *(fig. ugs.)* stimulate ‹*interest*›. **2.** *itr. V.* put the heating on

an|heuern ['anhɔyɐn] **1.** *tr. V.* **a)** *(Seemannsspr.)* sign on; **b)** *(fig. ugs.: einstellen)* sign on *or* up. **2.** *itr. V. (Seemannsspr.)* sign on

An·hieb der *in* **auf [den ersten] ~** *(ugs.)* straight off; first go

an|himmeln [-hɪml̩n] *tr. V. (ugs.)* **a)** *(verehren)* idolize; worship; **b)** *(ansehen)* gaze adoringly at

An·höhe die rise; elevation

an|hören 1. *tr. V.* listen to; **etw. [zufällig] mit ~:** overhear sth.; **sich** *(Dat.)* **jmdn./etw. ~:** listen to sb./sth.; **ich kann das nicht mehr mit ~!** I can't listen to that any longer. **2.** *refl. V.* sound

Anhörung die; ~, ~en hearing

animalisch [ani'ma:lɪʃ] *Adj.* **a)** animal *attrib.*; **b)** *(abwertend: triebhaft)* animal *attrib.*; bestial

Animateur [anima'tø:ɐ̯] **der; ~s, ~e** host

Animier·dame die hostess

animieren [ani'mi:rən] *tr. (auch itr.) V.* encourage; **das soll zum Kaufen ~:** that's to encourage people to buy

An·ion das *(Chemie)* anion

Anis [a'ni:s] **der; ~[es], ~e a)** *(Pflanze)* anise; **b)** *(Gewürz)* aniseed; **c)** *(Branntwein)* aniseed brandy

Ank. *Abk.* **Ankunft** arr.

an|kämpfen *itr. V.* **gegen jmdn./ etw. ~:** fight [against] sb./sth.; **gegen den Strom/Wind ~:** battle against the current/the wind

an|karren *tr. V. (ugs.)* cart along; bring along ‹*supporters, followers*›

An·kauf der purchase; **„Heinrich Meyer, An- und Verkauf"** 'Heinrich Meyer, second-hand dealer'

an|kaufen *tr. V.* purchase; buy

Anker ['aŋkɐ] **der; ~s, ~ a)** anchor; **vor ~ gehen/liegen** drop anchor/lie at anchor; **~ werfen** drop anchor; **b)** *(Elektrot.)* armature

ankern *itr. V.* **a)** *(vor Anker gehen)* anchor; drop anchor; **b)** *(vor Anker liegen)* be anchored; lie at anchor

Anker-: ~platz der anchorage; **~wicklung die** *(Elektrot.)* armature winding; **~winde die** windlass

an|ketten *tr. V.* chain up (**an** + *Akk. od. Dat.* to)

An·klage die a) charge; **der Staatsanwalt hat ~ [wegen Mordes**

gegen ihn] erhoben the public prosecutor brought a charge [of murder against him]; **unter ~ stehen** have been charged (**wegen** with); **b)** *(~vertretung)* prosecution

Anklage-: ~bank die; *Pl.* **~bänke** dock; **auf der ~bank sitzen** *(auch fig.)* be in the dock

an|klagen 1. *tr. V.* **a)** *(Rechtsw.)* charge; accuse; **jmdn. einer Sache** *(Gen.) od.* **wegen etw. ~:** charge sb. with *or* accuse sb. of sth.; **b)** *(geh.: beschuldigen)* accuse **2.** *itr. V.* cry out in accusation; **jmdn. ~d ansehen** look at sb. accusingly

An·kläger der prosecutor

an|klammern 1. *tr. V.* peg *(Brit.)* *or (Amer.)* pin ‹*clothes, washing*› up (**an** + *Akk.* to); clip ‹*copy, sheet, etc.*› (**an** + *Akk.* to); *(mit Heftklammern)* staple ‹*copy, sheet, etc.*› (**an** + *Akk.* on). **2.** *refl. V.* **sich an jmdn./etw. ~:** cling to *or* hang on to sb./sth.

An·klang der *in* **[bei jmdm.] ~ finden** meet with [sb.'s] approval; find favour [with sb.]; **wenig/keinen/großen ~ finden** be poorly/badly/well received (**bei** by)

an|kleben 1. *tr. (auch itr.) V.* stick up ‹*poster, etc.*› (**an** + *Akk.* on). **2.** *itr. V.; mit sein* stick (**an** + *Dat.* to)

Ankleide·kabine die changing cubicle

an|kleiden *tr. V. (geh.)* dress; **sich ~:** get dressed; dress [oneself]

Ankleide·raum der dressing-room

an|klingen *unr. itr. V.; auch mit sein* be discernible; **ein Thema ~ lassen** touch on a theme

an|klopfen *itr. V.* knock (**an** + *Akk. od. Dat.* at *or* on)

an|knabbern *tr. V. (ugs.)* nibble [at]

an|knipsen *tr. V. (ugs.)* switch *or* put on

an|knüpfen 1. *tr. V.* **a)** tie on (**an** + *Akk.* to); **b)** *(beginnen)* start up ‹*conversation*›; establish ‹*relations, business links*›; form ‹*relationship*›; strike up ‹*acquaintance*›. **2.** *itr. V.* **an etw.** *(Akk.)* **~:** take sth. up; **ich knüpfe dort an, wo wir vorige Woche aufgehört haben** I'll pick up where we left off last week

an|kommen *unr. itr. V.; mit sein* **a)** *(eintreffen)* arrive; ‹*letter, parcel*› come, arrive; ‹*bus, train, plane*› arrive, get in; **seid ihr gut angekommen?** did you arrive safely *or* get there all right?; **b)** *(herankommen)* come along; **c)** *(ugs.: Anklang finden)* **[bei jmdm.] [gut] ~:** go down [very] well [with sb.]; **er ist ein Typ, der bei den Frauen ankommt** he is the sort who is a success with women; **d) gegen jmdn./etw. ~:** be able to cope *or* deal with sb./fight sth.; **e)** *unpers.* **es kommt auf jmdn./etw. an** *(jmd./etw. ist ausschlaggebend)* it depends on sb./sth.; **es kommt auf etw.** *(Akk.)* **an** *(etw. ist wichtig)* sth. matters (*Dat.* to); **es kommt [ganz] darauf an, ob ...:** it [all] depends whether ...; **es kommt [ganz] darauf** *od.* **drauf an** *(ugs.)* it [all] depends; **es käme auf einen Versuch an** it's *or* it would be worth a try; **darauf kommt es mir nicht so sehr an** that doesn't matter so much to me; **f) es auf etw.** *(Akk.)* **~ lassen** *(etw. riskieren)* [be prepared to] risk sth.; **es d[a]rauf ~ lassen** *(ugs.)* take a chance; chance it

Ankömmling ['ankœmlɪŋ] **der; ~s, ~e** newcomer

an|koppeln 1. *tr. V.* couple ‹*carriage*› up (**an** + *Akk.* to); hitch ‹*trailer*› up (**an** + *Akk.* to); dock ‹*spacecraft*› (**an** + *Akk.* with). **2.** *itr. V.* ‹*spacecraft*› dock (**an** + *Akk.* with)

an|kreiden *tr. V. (ugs.)* **jmdm. etw. ~:** hold sth. against sb.

an|kreuzen *tr. V.* mark with a cross; put a cross beside

an|kündigen 1. *tr. V.* announce; **ein Gewitter ~:** herald a storm; **eine angekündigte/nicht angekündigte Klassenarbeit** a class test announced in advance/a surprise test. **2.** *refl. V.* ‹*spring, storm*› announce itself; ‹*illness*› show itself

An·kündigung die announcement

Ankunft ['ankʊnft] **die; ~, Ankünfte** arrival; „~" 'arrivals'

Ankunfts·halle die *(Flugw.)* arrival[s] hall

an|kuppeln *tr. V. s.* **ankoppeln 1**

an|kurbeln *tr. V.* **a)** crank [up]; **b)** *(fig.)* boost ‹*economy, production, etc.*›

Ankurb[e]lung die; ~, ~en *(fig.)* boosting; **Maßnahmen zur ~ der Wirtschaft** measures to boost the economy

Anl. *Abk.* **Anlage** encl.

an|lächeln *tr. V.* smile at; **jmdn. freundlich ~:** give a friendly smile to sb.

an|lachen 1. *tr. V.* smile at. **2.** *refl. V.* **sich** *(Dat.)* **jmdn. ~** *(ugs.)* get off with sb. *(Brit. coll.)*; pick sb. up

An·lage die a) *o. Pl. (das Anlegen) (einer Kartei)* establishment; *(eines Parks, Gartens usw.)* laying out; construction; *(eines Parkplatzes, Stausees)* construction; **b)** *(Grün~)* park; *(um ein Schloß, einen Palast usw. herum)* grounds *pl.*; **öffentliche/städtische ~n** public/municipal parks and gardens; **c)** *(Angelegtes, Komplex)* complex; **d)** *(Einrichtung)* facilities *pl.*; **sanitäre/militärische ~n** sanitary facilities/military installations; **die elektrische ~:** the electrical equipment; **e)** *(Werk)* plant; **f)** *(Musik~, Lautsprecher~ usw.)* equipment; system; **g)** *(Geld~)* investment; **h)** *(Konzeption)* conception; *(Struktur)* structure; **i)** *(Veranlagung)* aptitude, gift, talent (**zu** for); *(Neigung)* tendency, predisposition (**zu** to); **j)** *(Beilage zu einem Brief)* enclosure; **als ~ sende ich Ihnen/erhalten Sie ein ärztliches Attest** please find enclosed *or* I enclose a medical certificate

Anlage-: ~berater der investment adviser; **~kapital das** investment capital

an|landen *tr. V.* land

an|langen 1. *itr. V.; mit sein* arrive; **bei/auf/an etw.** *(Dat.)* **~:** arrive at *or* reach sth. **2.** *tr. V.* **a)** *(südd.: anfassen)* touch; **b)** *s.* **anbelangen**

Anlaß ['anlas] **der; Anlasses, Anlässe** ['anlɛsə] **a)** *(Ausgangspunkt, Grund)* cause (**zu** for); **der ~ des Streites** the cause of the dispute; **etw. zum ~ nehmen, etw. zu tun** use *or* take sth. as an opportunity to do sth.; **jmdm. ~ zu Beschwerden geben** give sb. cause for complaint; **~ zur Sorge/Beunruhigung/Klage geben** give cause for concern/unease/complaint; **beim geringsten/kleinsten ~:** for the slightest reason; **aus aktuellem ~:** because of current events; **b)** *(Gelegenheit)* occasion; **bei festlichen Anlässen** on festive occasions

an|lassen 1. *unr. tr. V.* **a)** leave ‹*light, radio, heating, engine, tap, etc.*› on; leave ‹*candle*› burning; **b)** keep ‹*coat, gloves, etc.*› on; **c)** *(in Gang setzen)* start [up]. **2.** *unr. refl. V.* **sich gut/schlecht ~:** make a *or* get off to a good/bad *or* poor start

Anlasser der; ~s, ~ *(Kfz-W.)* starter

an·läßlich *Präp. mit Gen.* on the occasion of

an|lasten *tr. V.* **jmdm. ein Verbrechen ~:** accuse sb. of a crime; **jmdm. die Schuld an etw.** *(Dat.)* **~:** blame sb. for sth.

An·lauf der a) run-up; **[mehr] ~**

nehmen take [more of] a run-up; **mit/ohne** ~: with/without a run-up; **er sprang mit/ohne** ~: he did a running/standing jump; **b)** *(Versuch)* attempt; **beim** *od.* **im ersten/dritten** ~: at the first/third attempt *or (coll.)* go

an|laufen 1. *unr. itr. V.; mit sein* **a) angelaufen kommen** come running along; *(auf einen zu)* come running up; **b) gegen jmdn./etw.** ~: run at sb./sth.; **c)** *(Anlauf nehmen)* take a run-up; **d)** *(zu laufen beginnen)* ‹*engine*› start [up]; *(fig.)* ‹*film*› open; ‹*production, search, campaign*› start; **e)** *(sich färben)* turn; go; **f)** *(beschlagen)* mist *or* steam up. **2.** *unr. tr. V.* put in at ‹*port*›

Anlauf·stelle die place to go

An·laut der *(Sprachw.)* initial sound; **im** ~: in initial position

an|legen 1. *tr. V.* **a)** *(an etw. legen)* put *or* lay ‹*domino, card*› [down] **(an** + *Akk.* next to); place, position ‹*ruler, protractor*› **(an** + *Akk.* on); put ‹*ladder*› up **(an** + *Akk.* against); **einen strengen Maßstab [an etw.** *(Akk.)***]** ~: apply strict standards [to sth.]; **b)** *(an den Körper legen)* **die Flügel/Ohren** ~: close its wings/lay its ears back; **die Arme** ~: put one's arms to one's sides; **c)** *(geh.: anziehen, umlegen)* don; put on; **d)** *(schaffen)* lay out ‹*town, garden, plantation, street*›; start ‹*file, album*›; compile ‹*statistics, index*›; **e)** *(gestalten, entwerfen)* structure ‹*story, novel*›; **f)** *(investieren)* invest; **g)** *(ausgeben)* spend **(für** on); **h) es darauf ~, etw. zu tun** be determined to do sth.; **er legt es auf einen Streit an** he is determined to have a fight. **2.** *itr. V.* **a)** *(landen)* moor; **b)** *(Kartenspiel)* lay a card/cards; **bei jmdm.** ~: lay a card/cards on sb.'s hand; **c)** *(Domino)* play [a domino/dominoes]; **d)** *(zielen)* aim **(auf** + *Akk.* at). **3.** *refl. V.* **sich mit jmdm.** ~: pick an argument *or* quarrel with sb.

Anleger der; ~s, ~ **a)** *(Schiffahrt)* jetty; **b)** *(Investor)* investor

Anlege-: ~steg der landing-stage; jetty; **~stelle die** mooring

an|lehnen 1. *tr. V.* **a)** *(an etw. lehnen)* lean **(an** + *Akk. od. Dat.* against); **b)** leave ‹*door*› slightly open *or* ajar; leave ‹*window*› slightly open; **die Tür war angelehnt** the door was [left] slightly open *or* ajar. **2.** *refl. V.* **sich [an jmdn.** *od.* **jmdm./etw.]** ~: lean [on sb./against sth.]; **sich an ein Vorbild** ~ *(fig.)* follow an example

Anlehnung die; ~, **~en: in** ~ **an jmdn./etw.** in imitation of *or* following sb./sth.

Anleihe die; ~, **~n a)** *(Darlehen)* loan; **b)** *(fig.)* borrowing; **eine** ~ **bei Goethe/Picasso machen** borrow from Goethe/Picasso

an|leimen *tr. V.* stick *or* glue on **(an** + *Akk. od. Dat.* to)

an|leinen [-lainən] *tr. V.* put ‹*dog*› on the lead; **Hunde sind anzuleinen** dogs must be kept on a lead

an|leiten *tr. V.* **a)** instruct; teach; **jmdn. zur Selbständigkeit** ~: teach sb. to be independent

An·leitung die instructions *pl.*

an|lernen *tr. V.* train; **ein angelernter Arbeiter** a semi-skilled worker

an|lesen 1. *unr. tr. V.* begin *or* start reading *or* to read. **2.** *unr. refl. V.* **sich** *(Dat.)* **etw.** ~: learn sth. by reading *or* from books

an|liegen *unr. itr. V.* **a)** *(an etw. liegen)* ‹*pullover etc.*› fit tightly *or* closely; ‹*hair, ears*› lie flat; **ein eng ~der Pullover** a tight- *or* close-fitting pullover; **b)** *(ugs.: vorliegen)* be on; *(zu erledigen sein)* to be done

An·liegen das; ~s, ~ *(Bitte)* request; *(Angelegenheit)* matter; **etw. zu seinem persönlichen** ~ **machen** take a personal interest in sth.

anliegend *Adj.* **a)** *nicht präd. (angrenzend)* adjacent; **b)** *(beiliegend)* enclosed

Anlieger der; ~s, ~: resident; **„~ frei"** 'residents only'

an|locken *tr. V.* attract ‹*customers, tourists, etc.*›; lure ‹*bird, animal*›

an|löten *tr. V.* solder on **(an** + *Akk. od. Dat.* to)

an|lügen *tr. V.* lie to

Anm. *Abk.* **Anmerkung**

an|machen *tr. V.* **a)** *(anschalten, -zünden usw.)* put *or* turn ‹*light, radio, heating*› on; light ‹*fire*›; **b)** *(bereiten)* mix ‹*cement, plaster, paint, etc.*›; dress ‹*salad*›; **c)** *(ugs.: anbringen)* put ‹*curtain, sign*› up; **d)** *(ugs.: ansprechen)* chat up *(Brit. coll.)*; **e)** *(ugs.: begeistern, erregen)* get ‹*audience etc.*› going; **das macht mich ungeheuer/nicht an** it really turns me on *(coll.)*/does nothing for me *(coll.)*

an|mahnen *tr. V.* send a reminder about

an|malen *tr. V.* **a)** *(ugs.: bemalen)* paint; **etw. rot** ~: paint sth. red; **b)** *(ugs.: schminken)* paint; **sich** ~: paint one's face; **c)** *(hinmalen)* paint **(an** + *Akk.* on); *(ugs.: hinzeichnen)* draw **(an** + *Akk.* on); **jmdm./sich einen Bart** ~: paint *or* draw a beard on sb.'s/one's face *or* on sb./oneself

An·marsch der advance; **im** ~ **sein** be advancing; *(ugs. scherzh.: unterwegs sein)* be on one's way

an|marschieren *itr. V.; mit sein* advance; **anmarschiert kommen** *(ugs.)* come marching along; *(auf einen zu)* come marching up

an|maßen [-ma:sn̩] *refl. V.* **sich** *(Dat.)* **etw.** ~: claim sth. [for oneself]; arrogate sth. to oneself; **darüber kannst du dir gar kein Urteil** ~: you have no right *or* it's not your place to pass judgement on that

an·maßend 1. *Adj.* presumptuous; *(arrogant)* arrogant. **2.** *adv.* presumptuously; *(arrogant)* arrogantly

Anmaßung die; ~, **~en** presumptuousness; presumption; *(Arroganz)* arrogance; **es ist eine** ~ **zu behaupten, daß ...:** it is presumptuous to assert that ...

an|melden *tr. V.* **a)** *(als Teilnehmer)* enrol; **jmdn./sich zu einem Kursus/in** *od.* **bei einer Schule** ~: enrol sb./enrol for a course/at a school; **sich schriftlich** ~: register [in writing]; **b)** *(melden, anzeigen)* license, get a licence for ‹*television, radio*›; apply for ‹*patent*›; register ‹*domicile, car, trade mark*›; **sich/seinen neuen Wohnsitz** ~: register one's new address; *s. auch* **Konkurs; c)** *(ankündigen)* announce; **sind Sie angemeldet?** do you have an appointment?; **sich beim Arzt** ~: make an appointment to see the doctor; **d)** *(geltend machen)* express, make known ‹*reservation, doubt, wish*›; put forward ‹*demand*›; **e)** *(Kartenspiele: ansagen)* bid; **f)** *(Fernspr.)* book

An·meldung die a) *(zur Teilnahme)* enrolment; **b)** *s.* **anmelden b:** licensing; registration; **die** ~ **eines Patents** the application for a patent; **c)** *(Ankündigung)* announcement; *(beim Arzt, Anwalt usw.)* making an appointment

an|merken *tr. V.* **a) jmdm. seinen Ärger/seine Verlegenheit** *usw.* ~: notice that sb. is annoyed/embarrassed *etc.*; notice sb.'s annoyance/embarrassment *etc.*; **man merkt ihm [nicht] an, daß er krank ist** you can[not] tell that he is ill; **sich nichts** ~ **lassen** not let it show; **b)** *(geh.: bemerken)* note

Anmerkung die; ~, **~en a)** *(Fußnote)* note; **b)** *(geh.: Bemerkung)* comment; remark

An·mut die; ~ *(geh.)* grace; **mit ~:** gracefully
an|muten [-mu:tn̩] *tr., auch itr. V. (geh.)* **[jmdn.] fremd** *usw.* **~:** seem strange *etc.* [to sb.]
an·mutig *(geh.)* **1.** *Adj.* graceful ‹*girl, gesture, movement, dance*›; charming, delightful ‹*girl, smile, picture, landscape*›. **2.** *adv.* ‹*move, dance*› gracefully; ‹*smile, greet*› charmingly, delightfully
an|nähen *tr. V.* sew on (**an** + *Akk.* to)
an|nähern 1. *refl. V.* **a)** approach; **sich einem Grenzwert ~** *(Math.)* converge towards a limit; **b)** *(fig.: [menschlich] näherkommen)* **sich jmdm. ~:** come *or* get closer to sb.; **c)** *(sich angleichen)* **sich einer Sache** *(Dat.)* **~:** come *or* get closer to sth. **2.** *tr. V. (angleichen)* bring closer (*Dat.* to); **verschiedene Standpunkte einander ~:** bring differing points of view closer together
annähernd 1. *Adv.* almost; nearly; *(ungefähr)* approximately; **nicht ~ so teuer** not nearly as *or* nowhere near as expensive. **2.** *adj.; nicht präd.* approximate; rough
Annäherung die; ~, ~en a) approach (**an** + *Akk.* to); **b)** *(fig.)* **es kam zu einer ~ der beiden Parteien** the two parties came *or* moved closer together; **c)** *(Angleichung)* **eine ~ der gegenseitigen Standpunkte** bringing the points of view on each side closer together
Annäherungs·versuch der advance
Annahme ['anna:mə] **die; ~, ~n a)** acceptance; **die ~ eines Pakets verweigern** refuse to accept [delivery of] a parcel; **b)** *(Vermutung)* assumption; **in der ~, daß ...:** on the assumption that ...; **c)** *s.* **Annahmestelle**
Annahme-: ~schluß der deadline [for acceptance]; **wann ist ~schluß?** when is the deadline [for acceptance]?; **~stelle die** *(für Lotto/Wetten usw.)* place where coupons/bets are accepted; *(für Reparaturen)* repairs counter/department; *(für Lieferungen)* delivery point
Annalen [a'na:lən] *Pl.* annals; **in die ~ der Firma eingehen** go down in the annals of the firm
annehmbar 1. *Adj.* **a)** acceptable; **b)** *(recht gut)* reasonable. **2.** *adv.* reasonably [well]
an|nehmen 1. *unr. tr. V.* **a)** accept; take; accept ‹*alms, invitation, condition, help, fate verdict, punishment*›; take ‹*food, telephone call*›; accept, take [on] ‹*task, job, repairs*›; accept, take up ‹*offer, invitation, challenge*›; **b)** *(Sport)* take; **c)** *(billigen)* approve; approve, adopt ‹*resolution*›; **d)** *(aufnehmen)* take on ‹*worker, patient, pupil*›; **e)** *(adoptieren)* adopt; **jmdn. an Kindes Statt ~** *(veralt.)* adopt sb.; **f)** *(haften lassen)* take ‹*dye, ink*›; **kein Wasser ~:** repel water; be water-repellent; **g)** *(sich aneignen)* adopt ‹*habit, mannerism*›; adopt, assume ‹*name, attitude*›; **h)** *(bekommen)* take on ‹*look, appearance, form, tone, dimension*›; **i)** *(vermuten)* assume; presume; **ich nehme es an/nicht an** I assume *or* presume so/not; **das ist/ist nicht anzunehmen** that can/cannot be assumed; **j)** *(voraussetzen)* assume; **etw. als gegeben** *od.* **Tatsache ~:** take sth. for granted *or* as read; **angenommen, [daß] ...:** assuming [that] ...; **das kannst du ~!** *(ugs.)* you bet! *(coll.)*. **2.** *unr. refl. V. (geh.)* **sich jmds./einer Sache ~:** look after sb./sth.
Annehmlichkeit die; ~, ~en comfort; *(Vorteil)* advantage
annektieren [anɛk'ti:rən] *tr. V.* annex
Annektierung die; ~, ~en, Annexion [anɛ'ksi̯o:n] **die; ~, ~en** annexation
anno ['ano], **Anno** *in* **~ 1910/68** *usw. (veralt.)* in [the year] 1910/'68 *etc.*; **seit ~ 1910** since [the year] 1910; **~ dazumal** *od.* **dunnemals** *od.* **Tobak** *(ugs. scherzh.)* the year dot *(Brit. coll.)*; long ago
Annonce [a'nõ:sə] **die; ~, ~n** advertisement; ad *(coll.)*; advert *(Brit. coll.)*
annoncieren *tr., itr. V.* advertise
annullieren [anʊ'li:rən] *tr. V.* annul
Annullierung die; ~, ~en annulment
Anode [a'no:də] **die; ~, ~n** *(Physik)* anode
an|öden [-ø:dn̩] *tr. V. (ugs.)* bore stiff *(coll.)* *or* to death *(coll.)*
anomal ['anoma:l] **1.** *Adj.* anomalous; abnormal. **2.** *adv.* anomalously; abnormally
Anomalie die; ~, ~n anomaly; abnormality
anonym [ano'ny:m] **1.** *Adj.* anonymous. **2.** *adv.* anonymously
Anonymität [anonymi'tɛ:t] **die; ~:** anonymity
Anorak ['anorak] **der; ~s, ~s** anorak
an|ordnen *tr. V.* **a)** *(arrangieren)* arrange; **b)** *(befehlen)* order
An·ordnung die a) *(Ordnung)* arrangement; **b)** *(Weisung)* order; **auf meine ~/auf ~ des Arztes** on my/doctor's orders *pl.*
an·organisch *Adj.* inorganic
anormal 1. *Adj.* abnormal. **2.** *adv.* abnormally
an|packen 1. *tr. V.* **a)** *(ugs.: anfassen)* grab hold of; **b)** *(angehen)* tackle; **packen wir's an!** let's get down to it; **c)** *(ugs.: behandeln)* treat ‹*person*›. **2.** *itr. V. (ugs.: mithelfen)* **[mit] ~:** lend a hand
an|passen 1. *tr. V.* **a)** *(passend machen)* fit; **b)** *(abstimmen)* suit (*Dat.* to). **2.** *refl. V.* adapt [oneself] (*Dat.* to); ‹*animal*› adapt; *(gesellschaftlich)* conform
Anpassung die; ~, ~en adaptation (**an** + *Akk.* to); *(der Renten, Löhne usw.)* adjustment (**an** + *Akk.* to); *(an die Gesellschaft)* conformity
anpassungs·fähig *Adj.* adaptable
Anpassungs·fähigkeit die; *o. Pl.* adaptability (**an** + *Akk.* to)
an|peilen *tr. V.* **a)** *(Funkw.)* take a bearing on; **b)** *(fig. ugs.)* aim at; **c)** *(anvisieren)* take a sight on
an|pfeifen *(Sport)* **1.** *unr. tr. V.* blow the whistle to start ‹*game, half*›. **2.** *unr. itr. V.* blow the whistle
An·pfiff der a) *(Sport)* whistle for the start of play; **der ~ zur zweiten Halbzeit** the whistle for the start of the second half; **b)** *(salopp: Zurechtweisung)* bawling-out *(coll.)*
an|pflanzen *tr. V.* **a)** plant; **b)** *(anbauen)* grow; cultivate
an|pflaumen [-pflau̯mən] *tr. V. (ugs.)* tease; take the mickey out of *(Brit. sl.)*
an|pflocken [-pflɔkn̩] *tr. V.* tether ‹*animal*›
an|pirschen *refl. V.* creep up (**an** + *Akk.* on)
an|pöbeln *tr. V. (ugs.)* abuse
An·prall der; ~[e]s impact (**auf, an** + *Akk.* with, **gegen** against)
an|prallen *itr. V.; mit sein* crash; **gegen** *od.* **an etw. ~:** crash into sb./against sth.
an|prangern [-praŋɐn] *tr. V.* denounce
an|preisen *unr. tr. V.* extol; **jmdn./etw. jmdm. ~:** extol the virtues of sb./sth. to sb.; recommend sb./sth. highly to sb.
An·probe die fitting
an|probieren *tr. V.* try on
an|pumpen *tr. V. (ugs.)* borrow money from; **jmdn. um 20 Mark ~:** touch *(sl.)* *or* tap sb. for 20 marks

Anrainer ['anrainɐ] **der; ~s, ~ a)** *(Nachbar)* neighbour; **b)** *(bes. österr.: Anlieger)* resident

ạn|rasen *itr. V.; mit sein (ugs.)* **angerast kommen** come racing along; *(auf einen zu)* come racing up

ạn|raten *unr. tr. V.* **jmdm. etw. ~:** recommend sth. to sb.; **auf Anraten des Arztes** on the *or* one's doctor's advice

ạnrechenbar *Adj.* **[auf etw.** *(Akk.)***] ~ sein** count [towards sth.]

ạn|rechnen *tr. V.* **a)** *(gutschreiben, verbuchen)* count; take into account; **er bekam einen Pluspunkt angerechnet** he was given an extra mark; **jmdm. etw. als Verdienst/Fehler ~:** count sth. to sb.'s credit/as sb.'s mistake; **jmdm. etw. hoch ~:** think highly of sb. for sth.; **b)** *(in Rechnung stellen)* **jmdm. etw. ~:** charge sb. for sth.

ạnrechnungs·fähig *Adj. (Papierdt.) s.* **anrechenbar**

Ạn·recht das right; **ein ~ auf etw.** *(Akk.)* **haben** have a right to *or* be entitled to sth.

Ạn·rede die form of address

ạn|reden *tr. V.* address; **jmdn. mit dem Vornamen ~:** address *or* call sb. by his/her Christian name

ạn|regen 1. *tr. V.* **a)** *(ermuntern)* prompt; **jmdn. zum Nachdenken ~:** make sb. think; **b)** *(vorschlagen)* propose; suggest; **~, etw. zu tun** propose *or* suggest doing sth. **2.** *tr. (auch itr.) V.* stimulate ⟨*imagination, digestion*⟩; sharpen, whet, stimulate ⟨*appetite*⟩

ạnregend *Adj.* stimulating; **~ wirken** act as a stimulant

Ạn·regung die a) *s.* **anregen 2:** stimulation; sharpening; whetting; **zur ~ der Verdauung/des Appetits** to stimulate the digestion/whet *etc.* the appetite; **b)** *(Denkanstoß)* stimulus; **c)** *(Vorschlag)* proposal; suggestion

Ạnregungs·mittel das stimulant

ạn|reichern [-raiçɐn] **1.** *tr. V.* **a)** *(auch Kerntechnik)* enrich; **Trinkwasser mit Fluor ~:** add fluoride to drinking water; **b)** *(akkumulieren)* accumulate. **2.** *refl. V.* accumulate

Ạnreicherung die; ~, ~en a) *(auch Kerntechnik)* enrichment; **b)** *(Akkumulation)* accumulation

Ạn·reise die journey [there/here]; **die ~ dauert 10 Stunden** the journey there/here takes ten hours; it takes 10 hours to get there/here

ạn|reisen *itr. V.; mit sein* travel there/here; **mit der Bahn ~:** go/come by train; travel there/here by train; **angereist kommen** come

ạn|reißen *unr. tr. V.* **a)** partly tear; **b)** *(in Gang setzen)* start [up]; **c)** *(anzünden)* strike ⟨*match*⟩; **d)** *(Technik)* mark [out]; **e)** *(kurz ansprechen)* touch on

Ạn·reiz der incentive

ạn|reizen *tr. (auch itr.) V.* **a)** *(anspornen)* stimulate; encourage; **das soll zum Sparen ~:** that is supposed to stimulate *or* act as an incentive to saving; **b)** *(anregen, erregen)* stimulate

ạn|rempeln *tr. V.* barge into; *(absichtlich)* jostle

ạn|rennen 1. *unr. itr. V.; mit sein* **a) angerannt kommen** come running along; *(auf einen zu)* come running up; **b) gegen den Sturm/feindliche Stellungen ~:** run into *or* against the storm/storm enemy positions; **gegen jmdn./etw. ~** *(fig.)* fight against sb./sth. **2.** *unr. refl. V. (ugs.)* **sich** *(Dat.)* **das Knie/den Kopf [an etw.** *(Dat.)***] ~:** bump one's knee/head [on sth.]

Ạnrichte die; ~, ~n sideboard

ạn|richten *tr. V.* **a)** *(auch itr.)* arrange ⟨*food*⟩; *(servieren)* serve; **es ist angerichtet** *(geh.)* dinner is served; **b)** cause ⟨*disaster, confusion, devastation, etc.*⟩

ạn|ritzen *tr. V.* scratch

ạn|rollen *itr. V.; mit sein* **a)** *(zu rollen beginnen)* ⟨*vehicle, column, etc.*⟩ start moving; *(fig.)* ⟨*campaign, search operation*⟩ start; **b)** *(heranrollen)* roll up; ⟨*aircraft*⟩ taxi up; **angerollt kommen** come rolling along; *(auf einen zu)* come rolling up

anrüchig ['anrʏçɪç] *Adj.* **a)** *(verrufen)* disreputable; **b)** *(unanständig)* indecent; *(obszön)* offensive

Ạnrüchigkeit die; ~ *s.* **anrüchig:** disreputableness; indecency

ạn|rücken *itr. V.; mit sein* ⟨*troops*⟩ advance; move forward; ⟨*firemen, police*⟩ move in

Ạn·ruf der a) *(telefonischer ~)* call; **b)** *(Zuruf)* call; *(eines Wachtpostens)* challenge

Ạnruf·beantworter der; ~s, ~: [telephone-]answering machine

ạn|rufen 1. *unr. tr. V.* **a)** call *or* shout to; call ⟨*sleeping person*⟩; hail ⟨*ship*⟩; ⟨*sentry*⟩ challenge; **b)** *(geh.: angehen, bitten)* appeal to ⟨*person, court*⟩ **(um** for); call upon ⟨*God*⟩; **c)** *(telefonisch ~)* ring *(Brit.)*; call. **2.** *unr. itr. V.* ring *(Brit.)*; call; **bei jmdm. ~:** ring *(Brit.) or* call sb.

Ạnrufer der; ~s, ~: caller

Ạnrufung die; ~, ~en a) *(einer Gottheit o. ä.)* invocation; **b)** *(eines Gerichts)* appeal (*Gen.* to)

ạn|rühren *tr. V.* **a)** touch; **b)** *(bereiten)* mix

ans [ans] *Präp. + Art.* **a) = an das; b)** *mit subst. Inf.* **sich ~ Arbeiten machen** set to work

Ạn·sage die a) announcement; **b)** *(Kartenspiel)* bid

ạn|sagen 1. *tr. V. (ankündigen)* announce; *s. auch* **Kampf d; b)** *(Kartenspiel)* bid. **2.** *refl. V.* **sich [bei jmdm.] ~:** say that one is coming [to see sb.]

ạn|sägen *tr. V.* make a saw-cut in; start to saw through

Ạnsager der; ~s, ~, Ạnsagerin die; ~, ~nen a) *(Radio, Fernsehen)* announcer; **b)** *(im Kabarett usw.)* master of ceremonies; *(Brit.)* compère

ạn|sammeln 1. *tr. V.* accumulate; amass ⟨*riches, treasure*⟩. **2.** *refl. V.* **a)** *(zusammenströmen)* gather; **b)** *(sich anhäufen)* accumulate; *(fig.)* ⟨*anger, excitement*⟩ build up

Ạn·sammlung die a) *(von Gegenständen)* collection; *(Haufen)* pile, heap; **b)** *(von Menschen)* crowd

ansässig ['anzɛsɪç] *Adj.* resident

Ạn·satz der a) *(Beginn)* beginnings *pl.*; **die ersten Ansätze** the initial stages; **im ~** *(ansatzweise)* to some extent; **b)** *(eines Körperteils)* base; **c)** *(Math.)* statement

Ạnsatz·punkt der starting-point

ạnsatz·weise *Adv.* to some extent

ạn|saufen *unr. refl. V. (salopp)* **sich** *(Dat.)* **einen [Rausch] ~:** get plastered *(sl.)*

ạn|saugen *tr. V. (geh. auch unr.)* suck in *or* up

ạn|schaffen *tr. V.* **[sich** *(Dat.)***] etw. ~** *(auch fig. ugs.)* get [oneself] sth.; **sich** *(Dat.)* **Kinder ~** *(fig. ugs.)* have children *or (coll.)* kids

Ạn·schaffung die purchase; **sich zur ~ eines Autos entschließen** decide to get *or* buy a car

ạn|schalten *tr. V.* switch on

ạn|schauen *tr. V. (bes. südd., österr., schweiz.) s.* **ansehen**

ạnschaulich 1. *Adj.* *(deutlich)* clear; *(bildhaft, lebendig)*; vivid, graphic ⟨*style, description*⟩; **etw. ~ machen** make sth. vivid; bring sth. to life. **2.** *adv.* *(deutlich)* clearly; *(bildhaft, lebendig)* vividly; ⟨*describe*⟩ vividly, graphically

Ạnschaulichkeit die; ~ *s.* **anschaulich 1:** clarity; vividness; graphicness

Anschauung die; ~, ~**en a)** *(Auffassung)* view; *(Meinung)* opinion; **b)** *(Wahrnehmung)* experience; **aus eigener ~:** from personal *or* one's own experience

Anschauungs-: ~material das illustrative material; *(für den Unterricht)* visual aids *pl.*; **~unterricht der** visual instruction; *(fig.)* object lesson

An·schein der appearance; **allem** *od.* **dem ~ nach** to all appearances; **es hat den ~, als ob ...:** it appears *or* looks as if ...; **sich** *(Dat.)* **den ~ geben, als ob man etw. glaubt** pretend to believe sth.

an·scheinend *Adv.* apparently; seemingly

an|scheißen *(derb) unr. tr. V.* **a)** *(betrügen)* con *(sl.)*; diddle *(sl.)*; **b)** *(zurechtweisen)* **jmdn. ~:** give sb. a bollocking *(Brit. coarse)*; bawl sb. out *(coll.)*

an|schicken *refl. V. (geh.)* **sich ~, etw. zu tun** *(sich bereit machen)* get ready *or* prepare to do sth.; *(anfangen, im Begriff sein)* be about to do sth.; be on the point of doing sth.

an|schieben *unr. tr. V.* push

an|schießen 1. *unr. tr. V.* **a)** *(durch Schuß verletzen)* shoot and wound; **angeschossen** wounded; **b)** *(bes. Fußball)* kick the ball against ‹*player*›; shoot straight at ‹*goalkeeper*›

An·schiß der *(salopp)* bollocking *(Brit. coarse)*; bawling-out *(coll.)*; **einen ~ kriegen** get a bollocking *(Brit. coarse)*; get bawled out *(coll.)*

An·schlag der a) *(Bekanntmachung)* notice; *(Plakat)* poster; **einen ~ machen** put up a notice/poster; **b)** *(Attentat)* assassination attempt; *(auf ein Gebäude, einen Zug o. ä.)* attack; **einen ~ auf jmdn. verüben** make an attempt on sb.'s life; **c)** *(Texterfassung)* keystroke; **50 Anschläge pro Zeile** 50 characters and spaces per line; **d)** *(Musik)* touch; **e)** *(Technik)* stop; **f) mit dem Gewehr im ~:** with rifle/rifles levelled

Anschlag·brett das notice-board *(Brit.)*; bulletin board *(Amer.)*

an|schlagen 1. *unr. tr. V.* **a)** *(aushängen)* put up, post ‹*notice, announcement, message*› **(an + Akk.** on); **b)** *(beschädigen)* chip. **2.** *unr. itr. V.* **a)** *mit sein* **an etw.** *(Akk.)* **~:** knock against sth.; **mit dem Knie/Kopf an etw.** *(Akk.)* **~:** knock one's knee/head on sth. **3.** *unr. refl. V.* **sich** *(Dat.)* **das Knie** *usw.* **~:** knock one's knee *etc.* **(an** + *Dat.* on)

an|schleichen 1. *unr. refl. V.* creep up **(an** + *Akk.* on). **2.** *unr. itr. V.; mit sein* **angeschlichen kommen** come creeping along; *(auf einen zu)* come creeping up

an|schleppen *tr. V.* **a)** *(herbeibringen)* drag along; **b)** *(zum Starten)* tow-start ‹*car etc.*›

an|schließen 1. *unr. tr. V.* **a)** *(mit Schloß)* lock, secure **(an** + *Akk. od. Dat.* to); **b)** *(verbinden)* connect **(an** + *Akk. od. Dat.* to); connect up ‹*electrical device*›; *(mit Stecker)* plug in; **c)** *(anfügen)* add. **2.** *unr. refl. V.* **a) sich jmdm. ~:** join sb.; **b)** *auch itr.* **[sich] an etw.** *(Akk.)* **~** *(zeitlich)* follow sth.; *(räumlich)* adjoin sth.

anschließend 1. *Adv.* afterwards; **~ an etw.** *(Akk.)* after sth. **2.** *adj.* subsequent; **ein Vortrag mit ~er Diskussion** a lecture followed by a discussion

An·schluß der a) connection; *(Kabel)* cable; **~ an etw.** *(Akk.)* **erhalten/haben** be connected [up] to sth.; **b)** *(telefonische Verbindung)* connection; **[keinen] ~ bekommen** [not] get through; **c)** *(Verkehrsw.)* connection; **Sie haben ~ nach ...:** there is a connection to ...; **d)** *(Fernsprecher)* telephone; **kein ~ unter dieser Nummer** number unobtainable; **e)** *o. Pl.* *(Kontakt)* **~ finden** make friends; **~ suchen** want to meet and get to know people; **f) im ~ an etw.** *(Akk.)* following *or* after sth.

Anschluß·zug der connecting train

an|schmiegen 1. *tr. V.* nestle **(an** + *Akk.* against). **2.** *refl. V.* nestle up, snuggle up **(an** + *Akk.* to, against)

an·schmiegsam *Adj.* affectionate ‹*child*›; soft and smooth ‹*material*›

an|schmieren *tr. V. (ugs.: täuschen)* con *(sl.)*; diddle *(sl.)*

an|schnallen *tr. V.* strap on ‹*rucksack*›; put on ‹*skis, skates*›; **sich ~** *(im Auto)* put on one's seat-belt; *(im Flugzeug)* fasten one's seat-belt; **„bitte ~!"** 'fasten your seat-belts, please'

an|schnauzen *tr. V. (ugs.)* shout at

an|schneiden *unr. tr. V.* **a)** cut [the first slice of]; **b)** *(ansprechen)* raise; broach

an|schrauben *tr. V.* screw on **(an** + *Akk. od. Dat.* to)

an|schreiben 1. *unr. tr. V.* **a)** *(hinschreiben)* write up **(an** + *Akk.* on); **b)** *(ugs.: stunden)* **[jmdm.] etw. ~:** chalk sth. up [to sb.'s account]; **bei jmdm. gut/schlecht angeschrieben sein** *(ugs.)* be in sb.'s good/bad books; be on sb.'s good/black list *(Amer.)*; **c)** *(brieflich ansprechen)* write to. **2.** *unr. itr. V. (ugs.: Kredit geben)* give credit; **~ lassen** buy on tick *(coll.)*

an|schreien *unr. tr. V.* shout at

An·schrift die address

an|schuldigen [-ʃʊldɪgn̩] *tr. V.* *(geh.)* accuse (*Gen.*, **wegen** of)

Anschuldigung die; ~, ~**en** accusation

an|schwärzen *tr. V. (ugs.)* **jmdn. ~** *(in Mißkredit bringen)* blacken sb.'s name; *(schlechtmachen)* run sb. down **(bei** to); *(denunzieren)* inform *or* *(Brit. sl.)* grass on sb. **(bei** to)

an|schwellen *unr. itr. V.; mit sein* **a)** *(dicker werden)* swell [up]; **stark angeschwollen** very swollen; **b)** *(lauter werden)* grow louder; ‹*noise*› rise; **c)** *(zunehmen, auch fig.)* swell, grow; ‹*water, river*› rise

an|schwemmen *tr. V.* wash up *or* ashore

an|schwimmen *unr. itr. V.; mit sein* **angeschwommen kommen** come swimming along; *(auf einen zu)* come swimming up

an|schwindeln *tr. V. (ugs.)* **jmdn. ~:** tell sb. fibs

an|sehen *unr. tr. V.* **a)** look at; **jmdn. groß/böse ~:** stare at sb./give sb. an angry look; **hübsch** *usw.* **anzusehen sein** be pretty *etc.* to look at; **sich** *(Dat.)* **etw. ~:** look at sth.; **sich** *(Dat.)* **ein Haus ~:** look at *or* view a house; **sich** *(Dat.)* **ein Fernsehprogramm ~:** watch a television programme; **sich** *(Dat.)* **ein Stück/einen Film ~:** see a play/a film; **sieh [mal] [einer] an!** *(ugs.)* well, I never! *(coll.)*; **das sehe sich einer an!** *(ugs.)* just look at that!; **b)** *(anmerken)* **man sieht ihm sein Alter nicht an** he does not look his age; **man sieht ihr die Strapazen an** she's showing the strain; **man sieht ihr nicht an, daß sie krank ist** there is nothing to show that she is ill; **c)** *(zusehen bei)* **etw. [mit] ~:** watch sth.; **das kann man doch nicht [mit] ~:** I/you can't just stand by and watch that; **ich kann das nicht länger [mit] ~:** I can't stand this any longer; **d)** *(einschätzen)* see; **e)** *(halten für)* regard; consider; **jmdn. als seinen Freund/als Betrüger ~:** regard sb. as a friend/a cheat; consider sb.

[to be] a friend/a cheat; **etw. als/ für seine Pflicht ~:** consider sth. one's duty

Ansehen das; ~s **a)** *(Wertschätzung)* [high] standing *or* reputation; **hohes ~ genießen** enjoy high standing *or* a good reputation; **b)** *(geh.: Aussehen)* appearance; **c) ohne ~ der Person** *(Rechtsw.)* without respect of persons

an·sehnlich *Adj.* **a)** *(beträchtlich)* considerable; **b)** *(gut aussehend, stattlich)* handsome

an|sein *unr. itr. V.; mit sein; nur im Inf. u. Part. zusammengeschrieben (ugs.)* ‹*light, gas, etc.*› be on

an|setzen 1. *tr. V.* **a)** position ‹*ladder, jack, drill, saw*›; put ‹*pen*› to paper; put *or* place ‹*violin bow*› in the bowing position; put ‹*glass, trumpet*› to one's lips; **b)** *(anfügen)* attach, put on **(an** + *Akk. od. Dat.* to); fit **(an** + *Akk. od. Dat.* on to); **c)** *(festlegen)* fix ‹*meeting etc.*›; fix, set ‹*deadline, date, price*›; **d)** *(veranschlagen)* estimate; **die Kosten mit drei Millionen ~:** estimate the cost at three million; **e)** *(anrühren)* mix; prepare; **f)** *(ausbilden)* **Rost/ Grünspan ~:** go rusty/become covered with verdigris; **Fett ~:** put on weight; **Knospen/Früchte ~:** form buds/set fruit. **2.** *itr. V.* **zum Sprechen ~:** open one's mouth to speak; **zur Landung ~:** come in to land; **zum Sprung/ Überholen ~:** get ready *or* prepare to jump/overtake; **b) hier muß die Diskussion/Kritik ~:** this is where the discussion/criticism must start

An·sicht die a) opinion; view; **meiner ~ nach** in my opinion *or* view; **anderer/der gleichen ~ sein** be of a different/the same opinion; **der ~ sein, daß ...:** be of the opinion that ...; **ich bin ganz Ihrer ~:** I entirely agree with you; **da bin ich anderer ~:** I disagree with you there; **die ~en sind geteilt** opinion is divided; **b)** *(Bild)* view; **c) zur ~** *(Kaufmannsspr.)* on approval

Ansichts-: ~karte die picture postcard; **~sache die** *in* **~sache sein** be a matter of opinion

an|siedeln 1. *refl. V.* settle; ‹*industry, bacteria*› become established. **2.** *tr. V.* settle ‹*immigrant, refugee, etc.*›; establish ‹*industry, species, variety, bacteria*›

An·siedlung die a) *s.* **ansiedeln 2:** settlement; establishment; **b)** *(Siedlung)* settlement

An·sinnen das; ~s, ~: [unreasonable] request; **ein freches/ seltsames** *usw.* **~:** an impudent/a strange *etc.* request

ansonsten *Adv. (ugs.)* **a)** *(davon abgesehen)* apart from that; otherwise; **b)** *(andernfalls)* otherwise

an|spannen 1. *tr. V.* **a)** harness, hitch up ‹*horse etc.*› **(an** + *Akk.* to); hitch up, yoke up ‹*oxen*› **(an** + *Akk.* to); hitch up ‹*carriage, cart, etc.*› **(an** + *Akk.* to); **b)** *(anstrengen)* strain. **2.** *itr. V.* hitch up; **~ lassen** have the carriage made ready

An·spannung die strain

an|spielen 1. *itr. V.* **a) auf jmdn./ etw. ~:** allude to sb./sth.; **b)** *(Spiel beginnen)* start; *(Fußball)* kick off; *(Kartenspiele)* lead. **2.** *tr. V.* **a)** *(Sport)* **jmdn. ~:** pass to sb.; **b)** *(Kartenspiele: ins Spiel bringen)* lead

Anspielung die; ~, ~en allusion **(auf** + *Akk.* to); *(verächtlich, böse)* insinuation **(auf** + *Akk.* about)

An·sporn der incentive

an|spornen [-ʃpɔrnən] *tr. V. (fig.)* spur on; encourage

An·sprache die speech; address; **eine ~ halten** make a speech; give an address

ansprechbar *Adj.* **er ist jetzt nicht ~:** you can't talk to him now

an|sprechen 1. *unr. tr. V.* **a)** speak to; *(zudringlich)* accost; **jmdn. mit „Herr Doktor" ~:** address sb. as 'doctor'; **jmdn. mit seinem Vornamen ~:** use sb.'s first name; **jmdn. auf etw./jmdn. ~:** speak to sb. about sth./sb.; **b)** *(gefallen)* appeal to; **c)** *(zur Sprache bringen)* mention; *(kurz, oberflächlich)* touch on. **2.** *unr. itr. V.* **a)** *(reagieren)* ‹*patient, brake, clutch, etc.*› respond **(auf** + *Akk.* to); **b)** *(wirken)* work; **bei jmdm. gut/nicht ~:** have/not have the desired effect on sb.

ansprechend 1. *Adj.* attractive; attractive, appealing ‹*personality*›. **2.** *adv.* attractively

Ansprech·partner der contact

an|springen 1. *unr. itr. V.; mit sein* **a)** ‹*engine, car*› start; **b) angesprungen kommen** come bounding along; *(auf einen zu)* come bounding up; **c)** *(ugs.)* **auf ein Angebot/Geschäft ~:** take up an offer/agree to a deal. **2.** *unr. tr. V.* jump up at

An·spruch der a) claim; *(Forderung)* demand; **hohe Ansprüche [an jmdn.] haben** *od.* **stellen** demand a great deal [of sb.]; **~ auf etw.** *(Akk.)* **erheben** lay claim to sth.; **[keine] Ansprüche stellen** make [no] demands; **in ~ nehmen** take up, take advantage of ‹*offer*›; exercise ‹*right*›; take up ‹*time*›; **jmds. Zeit/Hilfe in ~ nehmen** make demands on sb.'s time/ enlist sb.'s aid; **jmdn. [stark] in ~ nehmen** make [heavy] demands on sb.; **jmdn. völlig in ~ nehmen** take up all [of] sb.'s time; **b)** *(bes. Rechtsspr.: Anrecht)* claim; **[einen] ~/keinen ~ auf etw.** *(Akk.)* **haben** be/not be entitled to sth.; **auf etw.** *(Akk.)* **~ erheben** assert one's entitlement to sth.

an·spruchs-: ~los 1. *Adj.* **a)** undemanding; **b)** *(schlicht)* unpretentious; simple; **2.** *adv.* **a)** undemandingly; ‹*live*› modestly, simply; **b)** *(schlicht)* unpretentiously; simply; **~voll** *Adj. (wählerisch)* demanding, discriminating ‹*gourmet, reader, audience*›; *(schwierig)* demanding; ambitious ‹*subject*›

an|spucken *tr. V.* spit at

an|spülen *tr. V.* wash up *or* ashore

an|stacheln [-ʃtaxl̩n] *tr. V.* spur on **(zu** to)

Anstalt ['anʃtalt] **die;** ~, ~en **a)** institution; **b)** *Pl.* preparations; **[keine] ~ machen** *od. (geh.)* **treffen** make [no] preparations **(für** for); **~ machen/keine ~ machen, etw. zu tun** make a move/make no move to do sth.

An·stand der *o. Pl.* **a)** decency; **keinen ~ haben** have no sense of decency; **b)** *(veralt.: Benehmen)* good manners *pl.*

an·ständig 1. *Adj.* **a)** *(sittlich einwandfrei, rücksichtsvoll)* decent; decent, clean ‹*joke*›; *(ehrbar)* respectable; *(gut angesehen)* decent, respectable ‹*job*›; **b)** *(ugs.: zufriedenstellend)* decent; respectable ‹*result, marks*›; **c)** *(ugs.: beträchtlich)* sizeable ‹*sum, amount, debts*›; **eine ~e Tracht Prügel** a good hiding *(coll.)*. **2.** *adv.* **a)** *(sittlich einwandfrei)* decently; *(ordentlich)* properly; **b)** *(ugs.: zufriedenstellend)* **jmdn. ~ bezahlen** pay sb. pretty well; **ganz ~ abschneiden** do quite well; **~ arbeiten** do good work; **c)** *(ugs.: ziemlich)* **~ ausschlafen** have a decent sleep; **jmdn. ~ eine knallen** really belt sb. one *(coll.)*

anstands·los *Adv.* without [any] objection

an|starren *tr. V.* stare at

an·statt *Konj.* **~ zu arbeiten/~, daß er arbeitet** instead of working

an|stauen 1. *tr. V.* dam up; *(fig.)*

bottle up ⟨*feelings*⟩. **2.** *refl. V.* ⟨*water*⟩ accumulate; *(fig.)* ⟨*feelings*⟩ build up

an|staunen *tr. V.* gaze *or* stare in wonder at; **jmdn./etw. mit offenem Mund ~:** gape at sb./sth. in wonder

an|stechen *unr. tr. V.* **a)** prick; puncture ⟨*tyre*⟩; **b)** *(anzapfen)* tap ⟨*barrel*⟩

an|stecken 1. *tr. V.* **a)** pin on ⟨*badge, brooch*⟩; **jmdm. eine Brosche/einen Ring ~:** pin a brooch on sb./put *or* slip a ring on sb.'s finger; **b)** *(infizieren, auch fig.)* infect **c)** *(bes. nordd., mitteld.) s.* **anzünden. 2.** *itr. V.* be infectious *or* catching; *(durch Berührung)* be contagious; *(fig.)* be infectious *or* contagious

ansteckend *Adj.* infectious; *(durch Berührung)* contagious; *(fig.)* infectious; contagious

Ansteckung die; ~, ~en infection; *(durch Berührung)* contagion

an|stehen *unr. itr. V.* **a)** *(warten)* queue [up], *(Amer.)* stand in line **(nach** for); **b)** *(geh.: sich ziemen)* **jmdm. [wohl/übel] ~:** [well/ill] become sb.

an|steigen *unr. itr. V.; mit sein* **a)** ⟨*hill*⟩ rise; ⟨*road, path*⟩ climb, ascend; ⟨*garden, ground*⟩ slope up, rise; **b)** *(höher werden)* ⟨*water level, temperature, etc.*⟩ rise; ⟨*price, cost, rent, etc.*⟩ rise, go up, increase

an·stelle 1. *Präp. mit Gen.* instead of. **2.** *Adv.* ~ **von** instead of; *s. auch* **Stelle a**

an|stellen 1. *refl. V.* **a)** *(warten)* queue [up], *(Amer.)* stand in line **(nach** for); **b)** *(ugs.: sich verhalten)* act; behave; **sich dumm/ungeschickt ~:** act *or* behave stupidly/be clumsy; **sich dumm/ungeschickt bei etw. ~:** go about sth. stupidly/clumsily; **sich geschickt ~:** go about it well; **stell dich nicht [so] an!** don't make [such] a fuss! **2.** *tr. V.* **a)** *(aufdrehen)* turn on; **b)** *(einschalten)* switch on; turn on, switch on ⟨*radio, television*⟩; start ⟨*engine*⟩; **c)** *(einstellen)* employ **(als** as); **bei jmdm. angestellt sein** be employed by sb.; **d)** *(ugs.: beschäftigen)* **jmdn. zum Kartoffelschälen** *usw.* **~:** get sb. to peel the potatoes *etc.*; **e)** *(anlehnen)* **etw. an etw.** *(Akk.)* **~:** put *or* place sth. against sth.; **f)** *(anrichten)* **etwas/Unfug ~:** get up to something/to mischief; **g)** *(bewerkstelligen)* manage; **h)** *(vornehmen)* do ⟨*calculation*⟩; make ⟨*comparison, assumption*⟩

An·stellung die a) *o. Pl.* employment; **b)** *(Stellung)* job; **ohne ~:** without a job; unemployed

Anstieg der; ~[e]s rise, increase (*Gen.* in)

an|stiften *tr. V.* **a)** *(ins Werk setzen)* instigate; **b)** *(verleiten)* **jmdn. [dazu] ~, etw. zu tun** incite sb. to do sth.; **jmdn. zum Betrug/Mord/zu einem Verbrechen ~:** incite sb. to deception/to murder/to commit a crime

An·stifter der, An·stifterin die instigator

An·stiftung die incitement (zu to)

an|stimmen *tr. V.* start singing ⟨*song*⟩; start playing ⟨*piece of music*⟩; **ein Geschrei ~:** start shouting; **ein Freudengeheul ~:** burst into shouts of joy

An·stoß der a) *(Impuls)* stimulus **(zu** for); **den [ersten] ~ zu etw. geben** initiate sth.; **b) ~ erregen** cause *or* give offence **(bei** to); **[keinen] ~ an etw.** *(Dat.)* **nehmen** [not] object to sth.; *(sich [nicht] beleidigt fühlen)* [not] take offence at sth.; *s. auch* **Stein b; c)** *(Fußball)* kick-off; **den ~ ausführen** kick off

an|stoßen 1. *unr. itr. V.* **a)** *mit sein* **an etw.** *(Akk.)* **~:** bump into sth.; **mit dem Kopf ~:** knock *or* bump one's head; **b)** *(auf etw. trinken)* **[mit den Gläsern] ~:** clink glasses; **auf jmdn./etw. ~:** drink to sb./sth.; **c)** *(Fußball)* kick off. **2.** *unr. tr. V.* **jmdn./etw. ~:** give sb./sth. a push; **jmdn. aus Versehen ~:** knock into sb. inadvertently; **jmdn. mit dem Ellenbogen/Fuß ~** *(als Zeichen)* nudge/kick sb.; **sich** *(Dat.)* **den Kopf/die Zehe ~:** knock *or* bang one's head/stub one's toe

anstößig ['anʃtøːsɪç] **1.** *Adj.* offensive. **2.** *adv.* offensively

an|strahlen *tr. V.* **a)** illuminate; *(mit Scheinwerfer)* floodlight; *(im Theater)* spotlight; **b)** *(anblicken)* beam at

an|streben *tr. V. (geh.)* aspire to; *(mit großer Anstrengung)* strive for

an|streichen *unr. tr. V.* **a)** *(mit Farbe)* paint; *(mit Tünche)* whitewash; **b)** *(markieren)* mark

An·streicher der *(ugs.)* house-painter; painter

an|strengen [-ʃtrɛŋən] **1.** *refl. V.* *(sich einsetzen)* make an effort; exert oneself; *(körperlich)* exert oneself; **sich ~, etw. zu tun** make an effort to do sth.; **sich mehr/sehr ~:** make more of an effort/a great effort. **2.** *tr. V.* **a)** *(anspannen)* strain ⟨*eyes, ears, voice*⟩; **alle seine Kräfte ~:** make every effort; *(körperlich)* use all one's strength; **seine Phantasie ~:** exercise one's imagination; **b)** *(strapazieren)* strain, put a strain on ⟨*eyes*⟩; **jmdn. [zu sehr] ~:** be [too much of] a strain on sb.

anstrengend *Adj.* *(körperlich)* strenuous; *(geistig)* demanding; **~ für die Augen sein** be a strain on the eyes; **es war ~, dem Vortrag zu folgen** following the lecture was a strain

Anstrengung die; ~, ~en a) *(Einsatz)* effort; **~en machen** make an effort; **große ~en machen, etw. zu tun** make every effort to do sth.; **b)** *(Strapaze)* strain

An·strich der *(Farbe)* paint; *(Tünche)* whitewash; **der erste/zweite ~:** the first/second coat

An·sturm der a) *(das Anstürmen)* onslaught; **b)** *(Andrang)* *(auf Kaufhäuser, Schwimmbäder)* rush **(auf +** *Akk.* to); *(auf Banken, Waren)* run **(auf +** *Akk.* on)

an|stürmen *itr. V.; mit sein* **a) gegen etw. ~** ⟨*waves, wind*⟩ pound sth.; *(Milit.)* storm sth.; **b) angestürmt kommen** come charging *or* rushing along; *(auf einen zu)* come charging *or* rushing up

an|tanzen *itr. V.; mit sein (ugs.)* show up *(coll.)*; **angetanzt kommen** turn up

Antarktika [ant'|arktika] **(das); ~s** Antarctica

Antarktis [ant'|arktɪs] **die; ~: die ~:** the Antarctic

antarktisch *Adj.* Antarctic

an|tasten *tr. V.* **a)** *(verbrauchen)* break into ⟨*savings, provisions*⟩; **b)** *(beeinträchtigen)* infringe, encroach on ⟨*right, freedom, privilege*⟩; encroach on ⟨*property, private life*⟩

An·teil der a) *(jmdm. zustehender Teil)* share **(an +** *Dat.* of); **~ an etw.** *(Dat.)* **haben** share in sth.; *(zu etw. beitragen)* play *or* have a part in sth.; **b)** *(Wirtsch.)* share; **c)** *o. Pl.* *(Interesse)* interest **(an +** *Dat.* in); **~ an etw.** *(Dat.)* **nehmen** take an interest in sth.

anteilig 1. *Adj.* proportional; proportionate. **2.** *adv.* proportionally; proportionately

An·teilnahme die a) *(Beteiligung)* participation; **unter reger ~ der Bevölkerung** with the active participation of the public; **b)** *(Interesse)* interest **(an +** *Dat.* in); **c)** *(Mitgefühl)* sympathy **(an +** *Dat.* with); **mit ~ zuhören** listen sympathetically

an|telefonieren *tr. V. (ugs.)* phone *(coll.);* call; ring *(Brit.)*
Antenne die; ~, ~**n** *(Technik)* aerial; antenna *(Amer.);* **eine/keine ~ für etw. haben** *(fig.)* have a/no feeling for sth.
anthrazit [antra'tsi:t] *Adj.; nicht attr.* anthracite[-grey]
[1]**Anthrazit der;** ~s, ~e anthracite
[2]**Anthrazit das;** ~s anthracite grey
anthrazit-: ~**farben,** ~**farbig** *Adj.* anthracite[-coloured]; ~**grau** *Adj.* anthracite-grey
Anthroposoph [antropo'zo:f] **der;** ~**en,** ~**en** anthroposophist
Anthroposophie die; ~: anthroposophy *no art.*
anthroposophisch *Adj.* anthroposophical
anti-, Anti- [anti-]: anti[-]
Anti·alkoholiker der *(Abstinenzler)* teetotaller
anti·autoritär 1. *Adj.* anti-authoritarian; **2.** *adv.* in an anti-authoritarian manner
Antibiotikum [anti'bio:tikʊm] **das;** ~s, **Antibiotika** *(Med.)* antibiotic
Anti·blockier·system das *(Kfz-W.)* anti-lock braking system
antik [an'ti:k] *Adj.* **a)** classical; **b)** *(aus vergangenen Zeiten)* antique ‹*furniture, fittings, etc.*›
Antike [an'ti:kə] **die;** ~, ~**n a)** classical antiquity *no art.;* **b)** *([Kunst]gegenstand)* classical work of art
Antillen [an'tɪlən] *Pl.* **die [Großen/Kleinen]** ~: the [Greater/Lesser] Antilles
Antilope [anti'lo:pə] **die;** ~, ~**n** antelope
Antipathie [antipa'ti:] **die;** ~, ~**n** antipathy (**gegen** to)
an|tippen *tr. V.* give ‹*person, thing*› a [light] tap; touch ‹*accelerator, brake, etc.*›; *(fig.)* touch on ‹*point, question*›
Antiquariat [antikva'ria:t] **das;** ~s, ~**e** *(Laden/Abteilung)* antiquarian bookshop/department; *(mit neueren gebrauchten Büchern)* second-hand bookshop/department; **modernes** ~: *shop/department selling remainders, defective copies, cheap editions, reprints, etc.*
antiquarisch 1. *Adj. (Buchw.)* antiquarian; *(von neueren gebrauchten Büchern)* second-hand. **2.** *adv.* ‹*buy*› second-hand
Antiquität [antikvi'tɛ:t] **die;** ~, ~**en** antique
Antiquitäten-: ~**laden der** antique shop; ~**sammler der** collector of antiques; ~**sammlung die** collection of antiques
anti, Anti: ~**semit der** anti-Semite; ~**semitisch** *Adj.* anti-Semitic; anti-Semite; ~**semitismus der** anti-Semitism; ~**statisch** *Adj. (Physik)* antistatic; ~**these** ['----] **die** antithesis
antizipieren [antitsi'pi:rən] *tr. V. (geh.)* anticipate
Antlitz ['antlɪts] **das;** ~**es,** ~**e** *(dichter., geh.)* countenance *(literary);* face
Antrag ['antra:k] **der;** ~[e]s, **Anträge a)** application, request (**auf** + *Akk.* for); *(Rechtsw.: schriftlich)* petition (**auf** + *Akk.* for); **einen ~ stellen** make an application; *(Rechtsw.: schriftlich)* enter a petition; **b)** *(Formular)* application form; **c)** *(Heirats~)* proposal of marriage; **jmdm. einen ~ machen** propose to sb.
an|tragen *unr. tr. V. (geh.)* offer
Antrags·formular das application form
an|treffen *unr. tr. V.* find; *(zufällig)* come across
an|treiben *unr. tr. V.* **a)** *(vorwärts treiben)* drive ‹*animals, column of prisoners*› on *or* along; *(fig.)* urge; **jmdn. zur Eile/zu immer besseren Leistungen ~** *(fig.)* urge sb. to hurry up/urge *or* drive sb. on to better and better performances; **b)** *(in Bewegung setzen)* drive; *(mit Energie versorgen)* power; **c)** *(veranlassen)* drive; **jmdn. [dazu] ~, etw. zu tun** drive sb. to do sth.
an|treten 1. *unr. itr. V.; mit sein* **a)** *(sich aufstellen)* form up; *(in Linie)* line up; *(Milit.)* fall in; **b)** *(sich stellen)* meet one's opponent; *(als Mannschaft)* line up; **~ gegen** meet; *(als Mannschaft)* line up against; **c)** *(sich einfinden)* report (**bei** to). **2.** *unr. tr. V.* start ‹*job, apprenticeship*›; take up ‹*position, appointment*›; start, set out on ‹*journey*›; begin ‹*prison sentence*›; come into ‹*inheritance*›
An·trieb der a) drive; **ein Fahrzeug mit elektrischem ~:** an electrically powered *or* driven vehicle; **b)** *(Anreiz)* impulse; *(Psych.)* drive; impulse; **jmdm. neuen ~ geben** give sb. fresh impetus; **aus eigenem ~:** of one's own accord; on one's own initiative
Antriebs-: ~**kraft die** *(Technik)* motive *or* driving power; ~**rad das** *(Technik)* drive wheel; ~**welle die** *(Technik)* drive shaft
an|trinken *unr. refl. V.* **sich** *(Dat.)* **einen Rausch/Schwips ~:** get drunk/tipsy; **sich** *(Dat.)* **einen ~** *(ugs.)* get sloshed *(coll.);* **sich** *(Dat.)* **Mut ~:** give oneself Dutch courage
An·tritt der beginning; **vor ~ Ihres Urlaubs** before you go *or* before going on holiday *(Brit.) or (Amer.)* vacation; **vor ~ der Reise** before setting out on the journey; **bei ~ des Erbes/Amtes** on coming into the inheritance/taking up office
an|tuckern *itr. V.; mit sein (ugs.)* **angetuckert kommen** come chugging along; *(auf einen zu)* come chugging up
an|tun *unr. tr. V.* **a)** do (*Dat.* to); **sich** *(Dat.)* **etw. Gutes ~:** give oneself a treat; treat oneself; **jmdm. ein Leid ~:** hurt sb.; **jmdm. etwas Böses/ein Unrecht ~:** do sb. harm/an injustice; **sich** *(Dat.)* **etw. ~** *(ugs. verhüll.)* do away with oneself; **b) das/er** *usw.* **hat es ihr angetan** she was taken with it/him *etc.; s. auch* **angetan**
an|turnen [-tœrnən] *(ugs.) tr. V.* **jmdn. ~** ‹*drugs, music, etc.*› turn sb. on *(coll.)*
Antwerpen [ant'vɛrpn̩] **(das);** ~s Antwerp
Antwort ['antvɔrt] **die;** ~, ~**en a)** answer; reply; **er gab mir keine ~:** he didn't answer [me] *or* reply; he made no answer *or* reply; **er gab mir keine ~ auf meine Frage** he did not reply to *or* answer my question; **keine ~ ist auch eine ~:** your/her *etc.* silence speaks for itself; **b)** *(Reaktion)* response; **als ~ auf etw.** *(Akk.)* in response to sth.
antworten 1. *itr. V.* **a)** answer; reply; **auf etw.** *(Akk.)* **~:** answer sth.; reply to sth.; **jmdm. ~:** answer sb.; reply to sb.; **jmdm. auf seine Frage ~:** reply to *or* answer sb.'s question; **wie soll ich ihm ~?** what answer shall I give him?/what shall I tell him?; **b)** *(reagieren)* respond (**auf** + *Akk.* to). **2.** *tr. V.* answer; **was hat er geantwortet?** what was his answer?
Antwort·schreiben das reply
an|vertrauen 1. *tr. V.* **a) jmdm. etw. ~:** entrust sth. to sb.; entrust sb. with sth.; **b)** *(fig.)* **jmdm./seinem Tagebuch etw. ~:** confide sth. to sb./one's diary. **2.** *refl. V.* **a) sich jmdm./einer Sache ~:** put one's trust in sb./sth.; **b) sich jmdm. ~** *(fig.: sich jmdm. mitteilen)* confide in sb.
an|visieren [-vizi:rən] *tr. V.* align the *or* one's sights on; aim at
an|wachsen *unr. itr. V.; mit sein* **a)** *(festwachsen)* grow on; **wieder**

~ ⟨*finger, toe*⟩ grow back on; **b)** *(Wurzeln schlagen)* take root; **c)** *(zunehmen)* increase; grow

an|wackeln *itr. V.; mit sein (ugs.)* **angewackelt kommen** come waddling along; *(auf einen zu)* come waddling up

an|wählen *tr. V.* dial; **jmdn. ~:** dial sb.'s number

Anwalt ['anvalt] **der; ~[e]s, Anwälte** ['anvɛltə], **Anwältin die; ~, ~nen a)** *(Rechts~)* lawyer; solicitor *(Brit.)*; attorney *(Amer.)*; *(vor Gericht)* barrister *(Brit.)*; attorney[-at-law] *(Amer.)*; advocate *(Scot.)*; **einen ~ nehmen** get a lawyer *or (Amer.)* an attorney; **b)** *(Fürsprecher)* advocate; champion

Anwalts·büro das a) *(Räume)* lawyer's office; solicitor's office *(Brit.)*; **b)** *(Sozietät)* firm of solicitors *(Brit.)*; law firm *(Amer.)*

An·wandlung die mood; **in einer ~ von Großzügigkeit** *usw.* in a fit of generosity *etc.*

an|wärmen *tr. V.* warm up; warm ⟨*hands, feet*⟩

An·wärter der a) candidate **(auf** + *Akk.* for); *(Sport)* contender **(auf** + *Akk.* for); **b)** *(auf den Thron)* claimant; *(Thronerbe)* heir **(auf** + *Akk.* to)

An·wärterin die a) *s.* **Anwärter a; b)** *(auf den Thron)* claimant; *(Thronerbin)* heiress **(auf** + *Akk.* to)

an|weisen *unr. tr. V.* **a)** *(beauftragen)* **jmdn. ~:** give sb. instructions; **jmdn. ~, etw. zu tun** instruct *or* direct sb. to do sth.; **b)** *(zuweisen)* **jmdm. etw. ~:** allocate sth. to sb.

An·weisung die instruction; **~ haben, etw. zu tun** have instructions to do sth.

anwendbar *Adj.* applicable **(auf** + *Akk.* to); **schwer ~:** difficult to apply

an|wenden *unr. (auch regelm.) tr. V.* use, employ ⟨*process, trick, method, violence, force*⟩; use ⟨*medicine, money, time*⟩; apply ⟨*rule, paragraph, proverb, etc.*⟩ **(auf** + *Akk.* to)

An·wendung die *s.* **anwenden:** use; employment; application

an|werben *unr. tr. V.* recruit **(für** to); *(Milit.)* enlist, recruit; **sich ~ lassen** be recruited; *(Milit.)* enlist **(für** in)

An·werbung die recruitment **(für** to)

an|werfen *unr. tr. V. (ugs.: in Gang bringen)* start [up] ⟨*machine, engine, vehicle*⟩

An·wesen das property

anwesend *Adj.* present **(bei** at)

Anwesende der/die; *adj. Dekl.* **die ~n** those present

Anwesenheit die; ~: presence; **in ~ von** in the presence of

Anwesenheits·liste die attendance list

an|wetzen *itr. V.; mit sein (ugs.)* **angewetzt kommen** come rushing *or* tearing along; *(auf einen zu)* come rushing *or* tearing up

an|widern [-vi:dɐn] *tr. V.* nauseate

an|winkeln [-vɪŋkl̩n] *tr. V.* bend

an|winseln *tr. V.* whimper at

Anwohner ['anvo:nɐ] **der; ~s, ~:** resident; **Parken nur für ~:** residents-only parking

An·wurf der [unjustified] reproach; *(Beschuldigung)* [false] accusation

an|wurzeln *itr. V.; mit sein* take root; **wie angewurzelt [da]stehen/stehenbleiben** stand rooted to the spot

An·zahl die; *o. Pl.* number; **eine ganze ~:** a whole lot

an|zahlen *tr. V.* put down ⟨*sum*⟩ as a deposit **(auf** + *Akk.* on); pay a deposit on ⟨*goods*⟩; *(bei Ratenzahlung)* make a down payment on ⟨*goods*⟩; **50 DM ~:** put down 50 marks as a deposit/make a down payment of 50 marks

An·zahlung die deposit; *(bei Ratenzahlung)* down payment; **eine ~ auf etw.** *(Akk.)* **machen** *od.* **leisten** put down *or* pay a deposit on sth./make a down payment on sth.

an|zapfen *tr. V.* tap

An·zeichen das sign; indication; *(Med.)* symptom; **alle ~ deuten darauf hin, daß ...:** all the signs *or* indications are that ...

Anzeige ['antsaigə] **die; ~, ~n a)** *(Straf~)* report; **gegen jmdn. [eine] ~ [wegen etw.] erstatten** report sb. to the police/the authorities [for sth.]; **b)** *(Inserat)* advertisement; **eine ~ aufgeben** place an advertisement; **c)** *(eines Instruments)* display

an|zeigen *tr. V.* **a)** *(Strafanzeige erstatten)* **jmdn./etw. ~:** report sb./sth. to the police/the authorities; **b)** *(zeigen)* show; indicate; show ⟨*time, date*⟩

Anzeigen-: ~blatt das advertiser; **~teil der** advertisement section *or* pages *pl.*

An·zeiger der indicator

Anzeige·tafel die *(Sport)* scoreboard

an|zetteln [-tsɛtl̩n] *tr. V. (abwertend)* hatch ⟨*plot, intrigue*⟩; instigate ⟨*revolt*⟩; foment ⟨*war*⟩

an|ziehen 1. *unr. tr. V.* **a)** *(an sich ziehen)* draw up ⟨*knees, feet, etc.*⟩; **b)** *(anlocken)* attract; draw; **sich von jmdm. angezogen fühlen** feel attracted to sb.; **c)** *(anspannen)* tighten, pull tight ⟨*rope, wire, chain*⟩; **d)** *(festziehen)* tighten ⟨*screw, knot, belt, etc.*⟩; put on, pull on ⟨*handbrake*⟩; **e)** *(ankleiden)* dress; **sich ~:** get dressed; **f)** put on ⟨*clothes*⟩; **sich** *(Dat.)* **etw. ~:** put sth. on; **was soll ich bloß ~?** what shall I wear? **2.** *unr. itr. V.* **a)** ⟨*price, article, share, etc.*⟩ go up; **b)** *unpers.* **es zieht an** *(ugs.)* it's getting colder

anziehend *Adj.* attractive; engaging ⟨*manner, smile*⟩

Anziehungs·kraft die attractive force; *(fig.)* attraction

an|zischen *tr. V.* hiss at

An·zug der a) suit; **b)** *in* **im ~ sein** ⟨*danger*⟩ be imminent; ⟨*storm*⟩ be approaching; ⟨*fever, illness*⟩ be coming on; ⟨*enemy*⟩ be advancing

anzüglich ['antsy:klɪç] **1.** *Adj.* **a)** insinuating ⟨*remark, question*⟩; **b)** *(anstößig)* offensive ⟨*joke, remark*⟩. **2.** *adv.* **a)** in an insinuating way; **b)** *(anstößig)* offensively

Anzüglichkeit die; ~, ~en *s.* **anzüglich 1: a)** *o. Pl.* insinuating nature; offensiveness; **b)** *(Bemerkung)* insinuating remark; offensive remark/joke

an|zünden *tr. V.* light; **ein Gebäude** *usw.* **~:** set fire to a building *etc.*; set a building *etc.* on fire

An·zünder der *(Gas~)* gas lighter; *(Feuer~)* fire-lighter *(Brit.)*

an|zweifeln *tr. V.* doubt; question

an|zwinkern *tr. V.* wink at

AOK [a:|o:'ka:] **die; ~** *Abk.* **Allgemeine Ortskrankenkasse**

Aorta [a'ɔrta] **die; ~, Aorten** *(Med.)* aorta

apart [a'part] **1.** *Adj.* individual *attrib.*; **~ sein** be individual in style. **2.** *adv.* in an individual style

Apartheid [a'pa:ɐthait] **die; ~:** apartheid *no art.*

Apartheid·politik die policy of apartheid

Apartment [a'partmənt] **das; ~s, ~s** studio flat *(Brit.)*; flatlet *(Brit.)*; small flat *(Brit.)*; studio apartment *(Amer.)*

Apathie [apa'ti:] **die; ~, ~n** apathy

apathisch [a'pa:tɪʃ] **1.** *Adj.* apathetic. **2.** *adv.* apathetically

Aperitif [aperi'ti:f] **der; ~s, ~s** aperitif

Apfel ['apf̩l] **der; ~s, Äpfel** ['ɛpf̩l] apple; **[etw.] für einen ~ und ein Ei [kaufen]** [buy sth.] for a song; **in den sauren ~ beißen [und etw. tun]** *(ugs.)* grasp the nettle [and do sth.]
Apfel-: ~baum der apple-tree; **~blüte die a)** apple-blossom; **b)** *(das Blühen)* blossoming of the apple-trees; **während der ~blüte** while the apple-trees are/were in blossom; **~kuchen der** apple-cake; *(mit Äpfeln belegt)* apple flan; **gedeckter ~kuchen** apple pie; **~mus das** apple purée; **~saft der** apple-juice
Apfelsine [apf̩'zi:nə] **die; ~, ~n** orange
Apfel-: ~strudel der apfelstrudel; **~wein der** cider
apolitisch 1. *Adj.* apolitical. **2.** *adv.* apolitically
Apostel [a'pɔstl̩] **der; ~s, ~:** apostle; **die zwölf ~:** the twelve Apostles
Apotheke [apo'te:kə] **die; ~, ~n a)** chemist's [shop] *(Brit.); drugstore (Amer.);* **b)** *(Haus~)* medicine cabinet; *(Reise~, Bord~)* first-aid kit
Apotheker der; ~s, ~, Apothekerin die; ~, ~nen [dispensing] chemist *(Brit.);* druggist
App. *Abk. (Fernspr.)* **Apparat** ext.
Apparat [apa'ra:t] **der; ~[e]s, ~e a)** *(Technik)* apparatus *no pl.; (Haushaltsgerät)* appliance; *(kleiner)* gadget; **b)** *(Radio~)* radio; *(Fernseh~)* television; *(Rasier~)* razor; *(elektrisch)* shaver; *(Foto~)* camera; **c)** *(Telefon)* telephone; *(Nebenstelle)* extension; **am ~!** speaking!; **bleiben Sie am ~!** hold the line; **d)** *(Personen und Hilfsmittel)* organization; *(Verwaltungs~)* system; **e)** *(ugs.: etwas Ausgefallenes, Riesiges)* whopper *(sl.)*
Apparatur [apara'tu:ɐ̯] **die; ~, ~en** apparatus *no pl.;* **komplizierte ~en** complicated equipment
Appartement [apartə'mã:, *schweiz. auch:* -'mɛnt] **das; ~s, ~s** *(schweiz. auch:* **~e) a)** *s.* **Apartment; b)** *(Hotelsuite)* suite
Appell [a'pɛl] **der; ~s, ~e a)** appeal (**zu** for, **an** + *Akk.* to); **einen ~ an jmdn. richten** make an appeal to sb.; appeal to sb.; **b)** *(Milit.)* muster; *(Anwesenheits~)* roll-call; *(Besichtigung)* inspection; **zum ~ antreten** fall in for roll-call inspection
appellieren *itr. V.* appeal (**an** + *Akk.* to)
Appetit [ape'ti:t] **der; ~[e]s, ~e** *(auch fig.)* appetite (**auf** + *Akk.* for); **~ auf etw. haben/bekommen** fancy sth.; **guten ~!** enjoy your meal; **jmdm. den ~ verderben** spoil sb.'s appetite
appetit·anregend *Adj.* **a)** *(appetitlich)* appetizing; **b)** *(den Appetit fördernd)* ⟨*medicine etc.*⟩ that stimulates the appetite
Appetit·happen der canapé
appetitlich 1. *Adj.* **a)** appetizing; **b)** *(sauber, ansprechend)* attractive and hygienic. **2.** *adv.* **a)** appetizingly; **b)** *(sauber, ansprechend)* attractively and hygienically ⟨*packed*⟩
appetit·los 1. *Adj.* without any appetite *postpos.;* **~los sein** have lost one's appetite. **2.** *adv.* without any appetite
Appetit·losigkeit die; ~: lack of appetite
applaudieren [aplau̯'di:rən] *itr. V.* applaud; **jmdm./einer Sache ~:** applaud sb./sth.
Applaus [a'plau̯s] **der; ~es, ~e** applause
Apposition die *(Sprachw.)* apposition
Appretur [apre'tu:ɐ̯] **die; ~, ~en** *(Textilind.)* dressing; finishing
Aprikose [apri'ko:zə] **die; ~, ~n** apricot
April [a'prɪl] **der; ~[s], ~e** April; **der ~:** April; **~, ~!** April fool!; **der 1. ~:** the first of April; *(in bezug auf Aprilscherze)* April Fool's *or* All Fools' Day; **jmdn. in den ~ schicken** make an April fool of sb.
April-: ~scherz der April-fool trick; **das ist doch wohl ein ~scherz!** *(fig.)* you/they *etc.* can't be serious!; **~wetter das;** *o. Pl.* April weather
apropos [apro'po:] *Adv.* apropos; by the way; incidentally
Aquädukt [akvɛ'dʊkt] **der** *od.* **das; ~[e]s, ~e** aqueduct
Aquamarin der; ~s, ~e aquamarine
aquamarin·blau *Adj.* aquamarine
Aquaplaning [akva'pla:nɪŋ] **das; ~[s]** aquaplaning
Aquarell [akva'rɛl] **das; ~s, ~e** water-colour [painting]
Aquarell-: ~farbe die water-colour; **~maler der** water-colour painter; water-colourist; **~malerei die a)** *o. Pl.* water-colour painting; **b)** *(Bild)* water-colour
Aquarium [a'kva:ri̯ʊm] **das; ~s, Aquarien** aquarium
Äquator [ɛ'kva:tɔr] **der; ~s, ~en** equator
Äquator·taufe die crossing-the-line ceremony
Aquavit [akva'vi:t] **der; ~s, ~e** aquavit
äquivalent [ɛkviva'lɛnt] *Adj.* equivalent
Äquivalent das; ~[e]s, ~e equivalent
Äquivalenz [ɛkviva'lɛnt͡s] **die; ~, ~en** equivalence
Ar [a:ɐ̯] **das** *od.* **der; ~s, ~e** are
Ära ['ɛ:ra] **die; ~, Ären** era; **die ~ Kreisky** the Kreisky era
Araber ['a:rabɐ] **der; ~s, ~, Araberin die; ~, ~nen** Arab
Arabien [a'ra:bi̯ən] **(das); ~s** Arabia
arabisch *Adj.* Arabian; Arab; Arabic ⟨*language, numeral, alphabet, literature*⟩; **die Arabische Halbinsel** the Arabian Peninsula; *s. auch* **deutsch, ²Deutsche**
Arabisch das; ~[s] Arabic *s. auch* **Deutsch**
Aralie [a'ra:li̯ə] **die; ~, ~n** *(Bot.)* aralia
Arbeit ['arbai̯t] **die; ~, ~en a)** work *no indef. art.;* **die ~[en] am Staudamm** [the] work on the dam; **an die ~ gehen, sich an die ~ machen** get down to work; **bei der ~ mit Chemikalien** when working with chemicals; **viel ~ haben** have a lot of work [to do]; **[wieder] an die ~!** [back] to work!; **etw. in ~ haben** be working on sth.; **erst die ~, dann das Vergnügen** business before pleasure; **b)** *o. Pl. (Mühe)* trouble; **~ machen** cause bother *or* trouble; **jmdm. ~ machen** make work for sb.; **sich** *(Dat.)* **~ [mit etw.] machen** take trouble [over sth.]; **c)** *o. Pl. (Arbeitssplatz)* work *no indef. art.; (Stellung)* job; **eine ~ suchen/finden** look for/find work *or* a job; **eine ~ als ...:** work *or* a job as ...; **zur** *od. (ugs.)* **auf ~ gehen** go to work; **auf ~ sein** *(ugs.)* be at work; **vor/nach der ~:** before/after work; **d)** *(Aufgabe)* job; **e)** *(Produkt)* work; *(handwerkliche ~)* piece of work; *(kurze schriftliche ~)* article; **f)** *(Schulw.: Klassen~)* test; **eine ~ schreiben/schreiben lassen** do/set a test
arbeiten 1. *itr. V.* **a)** work; **zu ~ haben** have work to do; **an etw.** *(Dat.)* **~:** work on sth.; **bei jmdm./einer Firma** *usw.* **~:** work for sb./a company *etc.;* **seine Frau arbeitet** *(ist berufstätig)* his wife has a job *or* works; **die Zeit arbeitet für/gegen uns** time is on our side/against us; **b)** *(funktionieren)* ⟨*heart, lungs, etc.*⟩ work, function; ⟨*machine*⟩ work, operate; **c)** *(sich verändern)* ⟨*wood*⟩ warp; ⟨*must*⟩ ferment; ⟨*dough*⟩

rise. **2.** *tr. V.* **a)** *(herstellen)* make; *(in Ton, Silber, usw.)* work; make; fashion; **b)** *(tun)* do; **was ~ Sie?** what are you doing?; *(beruflich)* what do you do for a living?; what's your job? **3.** *refl. V.* **a) sich müde/krank ~:** tire oneself out/make oneself ill with work; **sich zu Tode ~:** work oneself to death; **sich** *(Dat.)* **die Hände wund ~:** work one's fingers to the bone; **b)** *(Strecke zurücklegen);* **sich durch etw./in etw.** *(Akk.)* **~:** work one's way through/into sth.; **sich nach oben ~** *(fig.)* work one's way up; **c)** *unpers.* **hier arbeitet es sich gut** this is a good place to work

Arbeiter der; ~s, ~: worker; *(Bau~, Land~)* labourer; *(beim Straßenbau)* workman

Arbeiter-: ~bewegung die *(Politik)* labour movement; **~familie die** working-class family; **~führer der** workers' leader

Arbeiterin die; ~, ~nen *(auch Zool.)* worker

Arbeiter-: ~kind das working-class child; **~klasse die;** *o. Pl.* working class[es *pl.*];**~partei die** workers' party

Arbeiterschaft die; ~: workers *pl.*

Arbeiter-: ~unruhen *Pl.* unrest *sing.* among the workers; **~viertel das** working-class district *or* area; **~wohl·fahrt die;** *o. Pl.* workers' welfare association

Arbeit·geber der employer

Arbeitgeberin die; ~, ~nen [female] employer

Arbeitnehmer der; ~s, ~: employee

Arbeitnehmerin die; ~, ~nen [female] employee

arbeitsam *Adj. (geh. veralt.)* industrious; hard-working

arbeits-, Arbeits-: ~amt das job centre *(Brit.);* employment exchange; labour exchange *(Brit. dated);* **~anfang der** starting-time [at work]; **~anfang ist um 6 Uhr** work starts at 6 a.m.; **~anzug der a)** *(Overall)* overalls *pl.;* **b)** *(Uniform)* fatigue uniform; **~aufwand der: mit großem ~aufwand** with a great deal of work; **~auf·wendig 1.** *Adj.* requiring a great deal of work *postpos., not pred.;* **[sehr] ~aufwendig sein** require a great deal of work; **2.** *adv.* in a way that requires/required a great deal of work; **~ausfall der** loss of working hours; **~bedingungen** *Pl.* working conditions; **~beginn der** *s.* **~anfang; ~belastung die** work-load; **~bereich der** area of work; *(~gebiet)* field of work; **~beschaffung die** job creation; creation of employment; **~beschaffungs·maßnahme die** job-creation measure; **~biene die a)** *(Zool.)* worker bee; **b)** *(ugs.: emsige Frau)* busy bee; **~dienst der a)** *(Arbeit) (low-paid)* community-service work; **b)** *(Organisation)* community service agency; **~eifer der** enthusiasm for one's work; **~ende das** finishing-time [at work]; **nach/bei ~ende** after work/when it's time to go; **um fünf Uhr haben wir** *od.* **ist bei uns ~ende** we finish work at five o'clock; **~erlaubnis die** work permit; **~fähig** *Adj.* fit for work *postpos.; (grundsätzlich)* able to work *postpos.;* viable ‹*government*›; **~fähigkeit die;** *o. Pl.* fitness for work; *(grundsätzlich)* ability to work; **~frei** *Adj.* **zwei ~freie Tage pro Woche** two days off a week; **ein paar Tage/eine Woche ~frei** a few days/a week off; **Montag ist/haben wir ~frei** we've got Monday off; **~gang der a)** *(einzelne Operation)* operation; **b)** *(Ablauf)* process; **~gebiet das** field of work; **~gemeinschaft die** team; *(Hochschulw.)* study group; **~gerät das a)** *(Gegenstand)* tool; **b)** *o. Pl.* *(Gesamtheit)* tools *pl.;* equipment *no indef. art., no pl.;* **~gruppe die** study group; **~intensiv 1.** *Adj.* labour-intensive; **2.** *adv.* labour-intensively; **~kampf der** industrial action; **~kleidung die** work clothes *pl.;* **~kraft die a)** capacity for work; **die menschliche ~kraft wird durch Roboter ersetzt** human labour is being replaced by robots; **b)** *(Mensch)* worker; **~kreis der** study group; **~lager das** labour camp; **~leben das;** *o. Pl.* **a)** *(Berufstätigkeit)* working life; **b)** *(Arbeitswelt)* world of work; working life *no art.;* **~leistung die** rate of output; **~los** *Adj.* unemployed; out of work *postpos.*

Arbeitslose der/die; *adj. Dekl.* unemployed person/man/woman *etc.;* **die ~n** the unemployed *or* jobless

Arbeitslosen-: ~geld das; *o. Pl.* earnings-related unemployment benefit; **~hilfe die;** *o. Pl.* *(reduced-rate)* unemployment benefit; **~unterstützung die** *(volkst.)* unemployment benefit *or* pay; **~versicherung die;** *o. Pl.* unemployment insurance

arbeits-, Arbeits-: ~losigkeit die; ~: unemployment *no indef. art.;* **eine ~losigkeit von 0,5%** a level of unemployment of 0.5%; **~markt der** labour market; **~material das** materials *pl.;* *(einschließlich Werkzeugen)* materials [and equipment *or* tools]; *(für den Unterricht)* teaching aids *pl.;* **~mittel das** tool; **~moral die** morale

Arbeits-: ~pause die break; **~pensum das** work quota; **~pferd das** *(auch fig.)* workhorse; **~platz der a)** work-place; **am ~platz** at one's work-place; **b)** *(Stelle, Job)* job; **~prozeß der** work process; **~raum der a)** workroom; **b)** *s.* **Arbeitszimmer; ~reich** *Adj.* ‹*life, week, etc.*› full of hard work; **~scheu** *Adj.* work-shy; **~schluß der** *s.* **~ende; ~sklave der** slave labourer; **~stelle die** *s.* **Stelle g; ~stil der** style of working; **~suche die** search for a job *or* for work; **auf ~suche sein** be looking for a job; **~tag der** working day; **mein erster ~tag nach dem Urlaub** my first day back at work after the holiday *(Brit.) or (Amer.)* vacation; **~team das** team; **~teilig 1.** *Adj.* ‹*society, mode of production, etc.*› based on the division of labour; **2.** *adv.* **die Produktion ~teilig organisieren** base production on the principle of the division of labour; **~teilung die** division of labour; **~tempo das** rate of work; work rate; **~tier das a)** work animal; **b)** *(Arbeitssüchtiger)* compulsive worker; workaholic *(coll.);* **~tisch der** work-table; *(für Schreibarbeiten)* desk; *(für technische Arbeiten)* [work-] bench; **~überlastung die** overwork

Arbeit·suchende der/die; *adj. Dekl.* person/man/woman looking for work

arbeits-, Arbeits-: ~unfähig *Adj.* unable to work *postpos.;* *(krankheitsbedingt)* unfit for work *postpos.;* **~un·willig** *Adj.* unwilling to work *postpos.;* **~verhältnis das** *contractual relationship between employer and employee;* **ein ~verhältnis eingehen** enter employment; **~vertrag der** contract of employment; **~vor·gang der** work process; **~vor·lage die: etw. als ~vorlage benutzen** work from sth.; **etw. dient jmdm. als ~vorlage** sb. works from sth.; **~weise die a)** way *or* method of working; **b)** *(Funktionsweise)* mode of operation; ~**welt die** world of work; ~**willig** *Adj.* willing to work *postpos.*; ~**wut die** mania for work; ~**zeit die** working time; **die tägliche/wöchentliche ~zeit** the working day/week; **b)** *(der [beruflichen] Arbeit vorbehaltene Zeit)* working hours *pl.*; **während der ~zeit** during working hours; **c)** *(als Ware)* labour time; **2 Stunden ~zeit** two hours' labour; ~**zeit·verkürzung die** reduction in working hours; ~**zeug das**; *o. Pl.* **a)** work things *pl.*; **b)** *(Kleidung)* work[ing] things *pl. or* clothes *pl.*; ~**zimmer das** study

Archäologe [arçɛo'lo:gə] **der; ~n, ~n** archaeologist

Archäologie die; ~: archaeology *no art.*

Archäologin die; ~, ~nen archaeologist

archäologisch 1. *Adj.* archaeological. **2.** *adv.* archaeologically

Arche ['arçə] **die; ~, ~n** ark; **die ~ Noah** Noah's Ark

Archipel [arçi'pe:l] **der; ~s, ~e** archipelago

Architekt [arçi'tɛkt] **der; ~en, ~en, Architektin die; ~, ~nen** architect

Architektur [arçitɛk'tu:ɐ̯] **die; ~:** architecture

Archiv [ar'çi:f] **das; ~s, ~e** archives *pl.*; archive

archivieren *tr. V.* **etw. ~:** archive sth.; put sth. in the archives

ARD [a:|ɛr'de:] **die; ~** *Abk.* **Arbeitsgemeinschaft der öffentlich-rechtlichen Rundfunkanstalten der Bundesrepublik Deutschland** *national radio and television network in the FRG*

Areal [are'a:l] **das; ~s, ~e a)** *(Fläche)* area; **b)** *(Grundstück)* grounds *pl.*

Aren *s.* Ara

Arena [a're:na] **die; ~, Arenen a)** *(hist., Sport, fig.)* arena; **b)** *(für Stierkämpfe, Manege)* ring

arg [ark], **ärger** ['ɛrgɐ], **ärgst...** ['ɛrgst...] **1.** *Adj.* **a)** *(geh., landsch.: schlimm)* bad ⟨*weather, condition, state*⟩; serious ⟨*situation, wound*⟩; hard ⟨*times*⟩; **an nichts Arges denken** be completely unsuspecting; **im ~en liegen** be in a sorry state; **b)** *(geh. veralt.: böse)* wicked; evil; **c)** *(geh., landsch.: unangenehm groß, stark)* severe ⟨*pain, hunger, shock*⟩; severe, bitter ⟨*disappointment*⟩; serious ⟨*dilemma, error*⟩; extreme, *(coll.)* terrible ⟨*embarrassment*⟩; gross ⟨*exaggeration, injustice*⟩; **in ~er Bedrängnis/Not sein** be in desperate straits; **mein ärgster Feind** my worst enemy *or* arch-enemy; **unser ärgster Konkurrent** our most dangerous competitor. **2.** *adv. (geh., landsch.)* extremely, *(coll.)* awfully, *(coll.)* terribly ⟨*painful, cold, steep, expensive, heavy, etc.*⟩; severely, bitterly ⟨*disappointed*⟩; extremely, *(coll.)* terribly ⟨*embarrassed*⟩; ⟨*suffer, weaken*⟩ severely; ⟨*offend*⟩ deeply; ⟨*deceive*⟩ badly; ⟨*rain, pull, punch*⟩ hard; ⟨*hurt*⟩ a great deal; **ihr treibt es gar zu ~!** you're going too far!; **etwas ~ laut** a bit too loud; **es geht ihm ~ schlecht/gut** things are going really badly/well for him

ärger *s.* **arg**

Ärger ['ɛrgɐ] **der; ~s a)** annoyance; *(Zorn)* anger; **jmds. ~ erregen** annoy sb.; **seinem ~ Luft machen** vent one's anger; **seinen ~ an jmdm. auslassen** vent one's anger on sb.; **b)** *(Unannehmlichkeiten)* trouble; **häuslicher/beruflicher ~:** domestic problems *pl.*/problems *pl.* at work; **[jmdm.] ~ machen** cause [sb.] trouble; make trouble [for sb.]; **so ein ~!** how annoying!; **~ bekommen** get into trouble

ärgerlich 1. *Adj.* annoyed; *(zornig)* angry; **ein ~es Gesicht machen** look annoyed/angry; **~ über etw.** *(Akk.)* **sein** be annoyed/angry about sth.; **~ werden** get angry/annoyed; **b)** *(Ärger erregend)* annoying; irritating; **wie ~!** how annoying! **2.** *adv.* **a)** with annoyance; *(zornig)* angrily; **b)** *(Ärger erregend)* annoyingly; irritatingly

ärgern 1. *tr. V.* **a) jmdn. ~:** annoy sb.; *(zornig machen)* make sb. angry; **b)** *(reizen, necken)* tease. **2.** *refl. V.* be annoyed; *(zornig sein)* be angry; *(ärgerlich/zornig werden)* get annoyed/angry; **sich über jmdn./etw. ~:** get annoyed/angry at sb./about sth.; **sich schwarz** *od.* **grün und blau ~:** fret and fume

Ärgernis das; ~ses, ~se a) *o. Pl.* offence; **Erregung öffentlichen ~ses** *(Rechtsspr.)* creating a public nuisance; **b)** *(etw. Ärgerliches)* annoyance; irritation; **c)** *(etw. Anstößiges)* nuisance; *(etw. Skandalöses)* scandal; outrage

arg-, Arg-: ~**list die**; *o. Pl. (geh.) (Hinterlist)* guile; deceit; *(Heimtücke, Rechtsw.)* malice; ~**listig 1.** *Adj. (hinterlistig)* guileful; deceitful; deceitful ⟨*plan*⟩; *(heimtückisch)* malicious; **~listige Täuschung** *(Rechtsspr.)* malicious deception; **2.** *adv. (hinterlistig)* guilefully; deceitfully; *(heimtückisch)* maliciously; ~**listigkeit die; ~** *s.* **~listig 1:** guilefulness; deceitfulness; malice; ~**los 1.** *Adj.* **a)** guileless ⟨*person*⟩; guileless, innocent ⟨*question, remark*⟩; **b)** *(ohne Argwohn)* unsuspecting; **wie kannst du nur so ~los sein?** how can you be so naïve?; **2.** *adv.* **a)** guilelessly; innocently; **b)** *(ohne Argwohn)* unsuspectingly; ~**losigkeit die; ~ a)** *(eines Menschen)* guilelessness; *(einer Äußerung, Absicht)* innocence; **b)** *(Vertrauensseligkeit)* unsuspecting nature

ärgst... *s.* **arg**

Argument [argu'mɛnt] **das; ~[e]s, ~e** argument

Argumentation [argumɛnta'tsio:n] **die; ~, ~en** argumentation

argumentieren *itr. V.* argue; **mit etw. ~:** use sth. as an argument

Argwohn ['arkwo:n] **der; ~[e]s** suspicion; **~ gegen jmdn. hegen** be suspicious of sb.

argwöhnen ['arkwø:nən] *tr. V. (geh.)* suspect

argwöhnisch *(geh.)* **1.** *Adj.* suspicious. **2.** *adv.* suspiciously

Arie ['a:riə] **die; ~, ~n** aria

Aristokrat [arɪsto'kra:t] **der; ~en, ~en** aristocrat

Aristokratie [arɪstokra'ti:] **die; ~, ~n** aristocracy

Aristokratin die; ~, ~nen aristocrat

aristokratisch 1. *Adj.* aristocratic. **2.** *adv.* aristocratically

Aristoteles [arɪs'to:telɛs] **(der)** Aristotle

Arithmetik [arɪt'me:tɪk] **die; ~:** arithmetic *no art.*

arithmetisch 1. *Adj.* arithmetical. **2.** *adv.* arithmetically

Arkade [ar'ka:də] **die; ~, ~n** arcade

Arktis ['arktɪs] **die; ~: die ~:** the Arctic

arktisch 1. *Adj. (auch fig.)* arctic. **2.** *adv.* **das Klima ist ~ beeinflußt** the climate is influenced by the Arctic

Arkus ['arkʊs] **der; ~, ~** ['arku:s] *(Geom.)* arc

arm [arm], **ärmer** ['ɛrmɐ], **ärmst...** ['ɛrmst...] *Adj. (auch fig.)* poor; **um etw. ärmer sein/werden** have lost/lose sth.; **~ an Bodenschätzen/Nährstoffen** poor in mineral resources/nutrients; **das Gebiet ist ~ an Wasser** the area is short of water; **~ an Vitaminen sein** ⟨*food*⟩ be lacking in *or* low in vitamins; **der/die Ärmste** *od.* **Arme**

the poor man/boy/woman/girl; **ach, du Armer** *od.* **Ärmster!** *(meist iron.)* oh, you poor thing!; *s. auch* **dran b**

Arm der; ~[e]s, ~e a) arm; **jmdn. am ~ führen** lead sb. by the arm; **jmds. ~ nehmen** take sb.'s arm; take sb. by the arm; **jmdn. im ~ halten** embrace sb.; **sich** *(Dat.)* **in den ~en liegen** lie in each other's *or* one another's arms; **den längeren ~ haben** *(fig. ugs.)* have more clout *(coll.)*; **jmds. verlängerter ~ sein** *(fig.)* be sb.'s tool *or* instrument; **jmdn. auf den ~ nehmen** *(fig. ugs.)* have sb. on *(Brit. coll.)*; pull sb.'s leg; **jmdm. in die ~e laufen** bump *or* run into sb.; **jmdm. [mit etw.] unter die ~e greifen** help sb. out [with sth.]; *s. auch* **Bein a; b)** *(armartiger Teil)* arm; **c)** *(Ärmel)* arm; sleeve; **ein Hemd/eine Bluse mit halbem ~:** a short-sleeved shirt/blouse

Armada [ar'ma:da] **die; ~, Armaden** *u.* **~s** *(auch fig.)* armada

Armatur [arma'tu:ɐ̯] **die; ~, ~en** *(Technik)* **a)** fitting; **b)** *meist Pl.* *(im Kfz)* instrument

Armaturen·brett das instrument panel; *(im Kfz)* dashboard

Arm-: ~band das bracelet; *(Uhr~)* strap; **~band·uhr die** wrist-watch; **~binde die** armband; **~brust die** crossbow

Ärmchen ['ɛrmçən] **das; ~s, ~:** [little] arm

Arme der/die; *adj. Dekl.* poor man/woman; pauper; **die ~n** the poor *pl; s. auch* **arm**

Armee [ar'me:] **die; ~, ~n a)** *(auch fig.)* army; **b) die ~** *(die Streitkräfte)* the armed forces *pl.*

Armee·fahrzeug das army vehicle

Ärmel ['ɛrməl] **der; ~s, ~:** sleeve; **die ~ hochkrempeln** *(fig. ugs.)* roll up one's sleeves; **[sich** *(Dat.)***] etw. aus dem ~ schütteln** *(ugs.)* produce sth. just like that

Armeleute·essen das poor man's food

Ärmel·kanal der: der ~: the [English] Channel

Armen·haus das *(hist., fig.)* poorhouse

Armenien [ar'me:niən] **(das); ~s** Armenia

Armenier der; ~s, ~, Armenierin, die; ~, ~nen Armenian

armenisch *Adj.* Armenian; *s. auch* **deutsch, Deutsch,** [2]**Deutsche**

ärmer *s.* **arm**

-armig *adj.* -armed

Arm-: ~länge die arm length; *(als Maß)* arm's length; **~lehne die** armrest; **~leuchter der a)** candelabra; **b)** *(ugs.)* berk *(Brit. sl.)*; jerk *(sl.)*

ärmlich ['ɛrmlɪç] **1.** *Adj.* cheap ⟨*clothing*⟩; shabby ⟨*flat, office*⟩; meagre ⟨*meal*⟩; **aus ~en Verhältnissen** from a poor family. **2.** *adv.* cheaply ⟨*dressed, furnished*⟩; **~ leben/wohnen** live in impoverished circumstances

Arm·reif der armlet

arm·selig 1. *Adj.* **a)** *(sehr arm, dürftig, unbefriedigend)* miserable; miserable, wretched ⟨*dwelling*⟩; pathetic ⟨*result, figure*⟩; meagre ⟨*meal, food*⟩; paltry ⟨*return, salary, sum, fee*⟩; **~e 10 Mark** a paltry 10 marks; **b)** *(abwertend: erbärmlich)* miserable, wretched ⟨*swindler, quack*⟩; pathetic, miserable, wretched ⟨*coward*⟩; pathetic, miserable ⟨*amateur, bungler*⟩. **2.** *adv.* **~ leben** lead *or* live a miserable life; **~ eingerichtet** miserably *or* wretchedly furnished

Armseligkeit die; ~ *s.* **armselig 1:** miserableness; wretchedness; patheticness; meagreness; paltriness

Arm·sessel der armchair

ärmst... *s.* **arm**

Arm·stuhl der armchair

Armsünder·miene die *(scherzh.)* expression of misery and remorse

Armut ['armu:t] **die; ~** *(auch fig.)* poverty; **die ~ des Landes an Rohstoffen** *(fig.)* the country's lack of raw materials

Armuts·zeugnis das *in* **ein ~ sein** be a sign of inadequacy

Arnika ['arnika] **die; ~, ~s** arnica

Aroma [a'ro:ma] **das; ~s, Aromen** *(Duft)* aroma; *(Geschmack)* flavour; taste

aromatisch [aro'ma:tɪʃ] *Adj.* **a)** *(duftend)* aromatic; **~ duften** give off an aromatic fragrance; **b)** *(wohlschmeckend)* distinctive ⟨*taste*⟩; **sehr ~ schmecken** have a very distinctive taste

Arrangement [arãʒə'mã:] **das; ~s, ~s** *(geh., Mus.)* arrangement

arrangieren [arã'ʒi:rən] **1.** *tr. V.* *(auch Musik)* arrange. **2.** *itr. V.* *(auch Musik)* **er kann gut ~:** he's a good arranger. **3.** *refl. V.* **sich ~:** adapt, adjust; **sich mit jmdm. ~:** come to an accommodation with sb.; **sich mit etw. ~:** come to terms with sth.

Arrest [a'rɛst] **der; ~[e]s, ~e** *(Milit., Rechtsw., Schule)* detention; **einen Schüler mit ~ bestrafen** *(veralt.)* punish a pupil by putting him in detention

Arrest·zelle die detention cell

arretieren [are'ti:rən] *tr. V.* **a)** *auch itr.* lock; **b)** *(veralt.: festnehmen)* detain; arrest

arrivieren [ari'vi:rən] *itr. V.; mit sein (geh.)* arrive; **zum Superstar/zum Staatsfeind Nummer eins ~:** achieve superstar status/become public enemy number one

Arrivierte der/die; *adj. Dekl.* *(geh.)* man/woman who has/had arrived; *(abwertend)* parvenu

arrogant [aro'gant] *(abwertend)* **1.** *Adj.* arrogant. **2.** *adv.* arrogantly

Arroganz [aro'gants] **die; ~** *(abwertend)* arrogance

Arsch [arʃ] **der; ~[e]s, Ärsche** ['ɛrʃə] *(derb)* **a)** arse *(Brit. coarse)*; bum *(Brit. sl.)*; ass *(Amer. sl.)*; **den ~ voll kriegen** get a bloody good hiding *(Brit. sl.)*; **der ~ der Welt** *(fig.)* the back of beyond; **leck mich am ~!** *(fig.)* piss off *(coarse)*; get stuffed *(Brit. sl.)*; *(verflucht noch mal!; na, so was!)* bugger me! *(Brit. coarse)*; **er kann mich [mal] am ~ lecken** *(fig.)* he can piss off *(coarse)*; he can kiss my arse *(Brit. coarse) or* ass *(Amer. sl.)*; **sich auf den ~ setzen** *(fig.)* *(fleißig arbeiten)* get *or* pull one's finger out *(sl.)*; *(perplex sein)* freak *(sl.)*; **jmdm. in den ~ kriechen** *(fig.)* kiss sb.'s arse *(Brit. coarse) or* ass *(Amer. sl.)*; **jmdm. in den ~ treten** kick sb. *or* give sb. a kick up the arse *(Brit. coarse) or* kick in the ass *(Amer. sl.)*; *(fig.)* give sb. a kick up the backside; **im ~ sein** *(fig.)* be buggered *(coarse)*; **b)** *(widerlicher Mensch)* arse-hole *(Brit. coarse)*; ass-hole *(Amer. sl.)*; **~ mit Ohren** arse-hole *(Brit. coarse)*; ass-hole *(Amer. sl.)*; **c)** *(nichts geltender Mensch)* piece of dirt

Arsch-: ~backe die *(derb)* cheek *(sl.)* [of the/one's arse *(Brit. coarse) or* bum *(Brit. sl.) or* ass *(Amer. sl.)*]; **~kriecher der** *(derb abwertend)* arse-licker *(Brit. coarse)*; ass-licker *(Amer. sl.)*; **~kriecherei die** *(derb abwertend)* arse-licking *(Brit. coarse)*; ass-licking Brit.; **~loch das** *(fig. derb)* arse-hole *(Brit. coarse)*; ass-hole *(Amer. sl.)*

Arsenal [arze'na:l] **das; ~s, ~e** arsenal

Art [a:ɐ̯t] **die; ~, ~en a)** *(Sorte)* kind; sort; *(Biol.: Spezies)* species; **Tische/Bücher aller ~:** tables/books of all kinds *or* sorts; all kinds *or* sorts of tables/books; **einzig in seiner ~:** unique of its kind; **jede ~ von Gewalt ablehnen** reject all forms of violence; **diese**

~ **[von] Menschen** that kind *or* sort of person; people like that; **[so] eine ~ ...:** a sort *or* kind of ...; **aus der ~ schlagen** not be true to type; *(in einer Familie)* be different from all the rest of the family; **b)** *o. Pl. (Wesen)* nature; *(Verhaltensweise)* manner; way; **das entspricht nicht seiner ~:** it's not [in] his nature; that's not his way; **c)** *o. Pl. (gutes Benehmen)* behaviour; **das ist doch keine ~!** that's no way to behave!; **die feine englische ~** *(ugs.)* the proper way to behave; **d)** *(Weise)* way; **auf diese ~:** in this way; **auf grausamste ~:** in the cruellest way; **in einer ~:** in a way; **~ und Weise** way; **nach ~ des Hauses** *(Kochk.)* à la maison; **nach Schweizer** *od.* **auf schweizerische ~** *(Kochk.)* Swiss style

arten·reich *Adj. (Biol.)* species-rich

Arterie [ar'te:ri̯ə] **die; ~, ~n** artery

Arterien·verkalkung die hardening of the arteries; arteriosclerosis *(Med.)*

artig 1. *Adj.* **a)** well-behaved; good; **sei ~:** be good; be a good boy/girl/dog *etc.*; **b)** *(geh. veralt.: höflich)* courteous. **2.** *adv.* **a) sich ~ benehmen** be good; behave well; **b)** *(geh. veralt.: höflich)* courteously

Artikel [ar'ti:kl̩] **der; ~s, ~ a)** article; **b)** *(Ware)* article; item

Artikulation [artikula'tsi̯o:n] **die; ~, ~en** articulation

artikulieren 1. *tr., itr. V.* articulate. **2.** *refl. V. (geh.)* **a)** express oneself; **b)** *(zum Ausdruck kommen)* express itself; be expressed

Artillerie [artɪlə'ri:] **die; ~, ~n** artillery

Artischocke [arti'ʃɔkə] **die; ~, ~n** artichoke

Artist [ar'tɪst] **der; ~en, ~en** [variety/circus] artiste *or* performer

Artistik die; ~ a) circus/variety performance *no art.*; **b)** *(Geschicklichkeit)* skill

Artistin die; ~, ~nen *s.* **Artist**

artistisch 1. *Adj.* **a) eine ~e Glanzleistung** a superb circus/variety performance; **sein ~es Können** his skill as a [circus/variety] artiste; **b)** *(geschickt)* masterly. **2.** *adv.* **a) eine ~ anspruchsvolle Nummer** a circus/variety act of great virtuosity; **b)** *(geschickt)* in a masterly way *or* fashion

Arznei [a:ɐ̯ts'nai̯] **die; ~, ~en** *(veralt.)*, **Arznei·mittel das** medicine; medicament; *(zur äußeren Anwendung)* medicament

Arzt [a:ɐ̯tst] **der; ~es, Ärzte** ['ɛ:ɐ̯tstə] doctor; physician *(arch./formal)*; **zum ~ gehen** go to the doctor['s]; **Sie sollten mal zum ~ gehen** you ought to see a/the doctor; **praktischer ~:** general practitioner; GP

Arzt·helferin die doctor's receptionist

Ärztin ['ɛ:ɐ̯tstɪn] **die; ~, ~nen** doctor; physician *(formal)*

ärztlich ['ɛ:ɐ̯tstlɪç] **1.** *Adj.* medical; **auf ~e Verordnung** on doctor's orders. **2.** *adv.* **sich ~ behandeln lassen** have medical treatment

Arzt·praxis die doctor's surgery *(Brit.) or* practice

as, ¹As [as] **das; ~, ~** *(Musik)* [key of] A flat

²As das; ~ses, ~se ace

Asbest [as'bɛst] **der; ~[e]s, ~e** asbestos

Asche ['aʃə] **die; ~, ~n** ash[es *pl.*]; *(sterbliche Reste)* ashes *pl.*; **in ~ liegen/legen** *(fig. geh.)* lie/lay in ashes; *s. auch* **Friede; Schutt a**

Aschen-: ~bahn die *(Sport)* cinder-track; **~becher der** ashtray; **~brödel** [-brø:dl̩] **das; ~s, ~** *(auch fig.)* Cinderella; **~platz der** *(Tennis)* cinder-court

Ascher der; ~s, ~ *(ugs.)* ashtray

Ascher·mittwoch der Ash Wednesday

Äser *s.* **Aas**

asexuell *Adj.* asexual

Asiat [a'zi̯a:t] **der; ~en, ~en, Asiatin die; ~, ~nen** Asian

asiatisch *Adj.* Asian

Asien ['a:zi̯ən] **(das); ~s** Asia

Askese [as'ke:zə] **die; ~:** asceticism

Asket [as'ke:t] **der; ~en, ~en** ascetic

asketisch 1. *Adj.* ascetic. **2.** *adv.* ascetically

asozial 1. *Adj.* asocial; *(gegen die Gesellschaft gerichtet)* antisocial; **ein ~er Mensch** a social misfit. **2.** *adv.* asocially; antisocially

Asoziale der/die; *adj. Dekl.* social misfit

Aspekt [as'pɛkt] **der; ~[e]s, ~e** aspect

Asphalt [as'falt] **der; ~[e]s, ~e** asphalt

asphaltieren *tr. V.* asphalt

Asphalt·straße die asphalt road

Aspik [as'pi:k] **der** (*österr. auch* **das**); ~s, ~e aspic

Aspirant [aspi'rant] **der; ~en, ~en, Aspirantin die; ~, ~nen** candidate

aß [a:s] *1. u. 3. Pers. Sg. Prät. v.* **essen**

Assel ['asl̩] **die; ~, ~n** *(Keller~, Mauer~)* woodlouse

Assessor [a'sɛsɔr] **der; ~s, ~en** [asɛ'so:rən], **Assessorin die; ~, ~nen** *holder of a higher civil service post, e.g. teacher or lawyer, who has passed the necessary examinations but has not yet completed his/her probationary period*

Assimilation [asimila'tsi̯o:n] **die; ~, ~en** assimilation (**an** + *Akk.* to)

Assistent [asɪs'tɛnt] **der, ~en, ~en, Assistentin die; ~, ~nen** assistant; *s. auch* **wissenschaftlich**

Assistenz·arzt der junior doctor

assistieren *itr. V.* **[jmdm.] ~:** assist [sb.] **(bei** at)

Assoziation [asotsi̯a'tsi̯o:n] **die; ~, ~en** association

assoziieren 1. *tr. V. (bes. Psych., geh.)* associate; **bei einem Namen** *usw.* **etw. ~:** associate sth. with a name *etc.* **2.** *itr. V.* make associations; **frei ~:** free-associate

Ast [ast] **der; ~[e]s, Äste** ['ɛstə] **a)** branch; bough; **den ~ absägen, auf dem man sitzt** *(fig. ugs.)* saw off the branch one is sitting on; **auf dem absteigenden ~ sein** *(fig. ugs.)* be going downhill; **b)** *(in Holz)* knot; **c) sich** *(Dat.)* **einen ~ lachen** *(ugs.)* split one's sides [with laughter]

Aster ['astɐ] **die; ~, ~n** aster; *(Herbst~)* Michaelmas daisy

Ast·gabel die *(zwischen Stamm und Ast)* fork of a/the tree; *(zwischen Ast und Zweig)* fork of a/the branch

Ästhet [ɛs'te:t] **der; ~en, ~en** aesthete

Ästhetik [ɛs'te:tɪk] **die; ~, ~en a)** aesthetics *sing.*; **b)** *(das Ästhetische)* aesthetics *pl.*

ästhetisch 1. *Adj.* aesthetic. **2.** *adv.* aesthetically

ästhetisieren *tr. (auch itr.) V. (geh.)* aestheticize

Asthma ['astma] **das; ~s** asthma

Asthmatiker [ast'ma:tikɐ] **der; ~s, ~:** asthmatic

asthmatisch 1. *Adj.* asthmatic. **2.** *adv.* asthmatically

Ast·loch das knot-hole

ast·rein 1. *Adj.* **a)** *(ugs.)* on the level *(coll.)*; **b)** *(salopp: prima)* fantastic *(coll.)*; great *(coll.)*. **2.** *adv. (salopp)* fantastically *(coll.)*

Astrologe [astro'lo:gə] **der; ~n, ~n** astrologer; *(fig.)* forecaster; pundit

Astrologie die; ~: astrology *no art.*

Astrologin die; ~, ~nen *s.* **Astrologe**

astrologisch 1. *Adj.* astrological. 2. *adv.* astrologically

Astronaut [astro'naut] **der; ~en, ~en, Astronautin die; ~, ~nen** astronaut

Astronom [astro'no:m] **der; ~en, ~en** astronomer

Astronomie die; ~: astronomy *no art.*

astronomisch *Adj.* astronomical

Ast·werk das; ~[e]s branches *pl.*

Asyl [a'zy:l] **das; ~s, ~e a)** [political] asylum; **jmdm. ~ gewähren** grant sb. asylum; **b)** *(Obdachlosen~)* hostel [for the homeless]

Asylant [azy'lant] **der; ~en, ~en, Asylantin die; ~, ~nen** person seeking/granted [political] asylum

asymmetrisch 1. *Adj.* asymmetrical. 2. *adv.* asymmetrically

asynchron 1. *Adj.* asynchronous. 2. *adv.* asynchronously

A. T. *Abk.* **Altes Testament** OT

Atelier [atə'lie:] **das; ~s, ~s** studio

Atem ['a:təm] **der; ~s** breath; **sein ~ wurde schneller** his breathing became faster; **einen langen/den längeren ~ haben** *(fig.)* have great/the greater staying power; **jmdn. in ~ halten** keep sb. in suspense; **den ~ anhalten** hold one's breath; **~ holen** *od. (geh.)* **schöpfen** *(fig.)* get one's breath back; **außer ~ sein/geraten** *od.* **kommen** be/get out of breath; **[wieder] zu ~ kommen** get one's breath back; *s. auch* **ausgehen 1 b**

atem-, Atem-: ~beraubend 1. *Adj.* breath-taking; 2. *adv.* breath-takingly; **~beschwerden** *Pl.* trouble *sing.* with one's breathing; **~los** 1. *Adj.* breathless; 2. *adv.* breathlessly; **~losigkeit die; ~:** breathlessness; **~not die;** *o. Pl.* difficulty in breathing; **~pause die** breathing space; **~übung die** breathing exercise; **~wege** *Pl.* respiratory tract *sing. or* passages; **~zug der** breath; **in einem** *od.* **im selben ~zug** in the same breath

Atheismus [ate'ɪsmʊs] **der; ~:** atheism *no art.*

Atheist der; ~en, ~en atheist

atheistisch 1. *Adj.* atheistic. 2. *adv.* atheistically

Athen [a'te:n] **(das); ~s** Athens

Athener 1. *indekl. Adj.; nicht präd.* Athens *attrib.;* of Athens *postpos.* 2. **der; ~s, ~:** Athenian

Äther ['ɛ:tɐ] **der; ~s, ~** *(Chemie, Physik, geh.)* ether

ätherisch [ɛ'te:rɪʃ] *Adj. (Chemie, dichter.)* ethereal

Äthiopien [ɛ'tio:piən] **(das); ~s** Ethiopia

Athlet [at'le:t] **der; ~en, ~en a)** *(Sportler)* athlete; **b)** *(ugs.: kräftiger Mann)* muscleman

Athletik [at'le:tɪk] **die; ~:** athletics *sing., no art.*

athletisch *Adj.* athletic

Atlanten *s.* **Atlas**

Atlantik [at'lantɪk] **der; -s** Atlantic

atlantisch *(Geogr.) Adj.* Atlantic; **der Atlantische Ozean** the Atlantic Ocean

Atlas ['atlas] **der; ~** *od.* **~ses, Atlanten** *od.* **~se** atlas

atmen ['a:tmən] *itr., tr. V.* breathe

Atmosphäre [atmo'sfɛ:rə] **die; ~, ~n** *(auch fig.)* atmosphere

atmosphärisch 1. *Adj.* atmospheric. 2. *adv.* atmospherically

Atmung die; ~: breathing; respiration *(as tech. term)*

Atom [a'to:m] **das; ~s, ~e** atom

atomar [ato'ma:ɐ̯] 1. *Adj.* atomic; *(Atomwaffen betreffend)* nuclear; nuclear, atomic ⟨*age, weapons*⟩. 2. *adv.* **~ angetrieben** nuclear-powered; atomic-powered; **~ aufrüsten** build up nuclear arms

Atom-: ~bombe die nuclear bomb; atom bomb; **~bombenversuch der** nuclear [weapons] test; **~bunker der** fall-out shelter; **~energie die;** *o. Pl.* nuclear *or* atomic energy *no indef. art.;* **~explosion die** nuclear *or* atomic explosion

atomisieren *tr. V.* **a)** *(zerstören)* **etw. ~:** smash sth. to atoms; **b)** *(zerstäuben)* atomize ⟨*liquid*⟩

atom-, Atom-: ~kern der atomic nucleus; **~kraft die;** *o. Pl.* nuclear *or* atomic power *no indef. art.;* **~kraftwerk das** nuclear *or* atomic power-station; **~krieg der** nuclear war; **~macht die** nuclear power; **~müll der** nuclear *or* atomic waste; **~physik die** nuclear *or* atomic physics *sing., no art.;* **~physiker der** nuclear *or* atomic physicist; **~pilz der** mushroom cloud; **~rakete die** *(Waffe)* nuclear *or* atomic missile; **~reaktor der** nuclear reactor; **~sprengkopf der** nuclear warhead; **~strom der** *(ugs.)* electricity generated by nuclear power; **~-U-Boot das** nuclear[-powered] submarine; **~waffe die** nuclear *or* atomic weapon; **~waffen·frei** *Adj.* nuclear-free; **~waffensperrvertrag der** Nuclear Non-proliferation Treaty; **~zeit·alter das;** *o. Pl.* nuclear *or* atomic age

atonal *(Musik) Adj.* atonal

Atrium ['a:triʊm] **das; ~s, Atrien** atrium

ätsch [ɛ:tʃ] *Interj. (Kinderspr.)* ha ha

Attacke [a'takə] **die; ~, ~n a)** *(auch Med.)* attack **(auf +** *Akk.* on); **b)** *(Reiter~)* [cavalry] charge; **eine ~ [gegen jmdn./etw.] reiten** charge [sb./sth.]; *(fig.)* make an attack [on sb./sth.]

attackieren *tr. V.* **a)** attack; **b)** *(Milit.: zu Pferde)* charge

Attentat ['atnta:t] **das; ~[e]s, ~e** assassination attempt; *(erfolgreiches)* assassination; **ein ~ auf jmdn. verüben** make an attempt on sb.'s life/assassinate sb.

Attentäter ['atntɛ:tɐ] **der; ~s, ~, Attentäterin die; ~, ~nen** would-be assassin; *(bei erfolgreichem Attentat)* assassin

Attest [a'tɛst] **das; ~[e]s, ~e** medical certificate; doctor's certificate

attestieren *tr. V.* certify

Attitüde [ati'ty:də] **die; ~, ~n** *(geh.)* posture

Attraktion [atrak'tsio:n] **die; ~, ~en** attraction

attraktiv [atrak'ti:f] 1. *Adj.* attractive. 2. *adv.* attractively

Attraktivität [atraktivi'tɛ:t] **die; ~:** attractiveness

Attrappe [a'trapə] **die; ~, ~n** dummy

Attribut [atri'bu:t] **das; ~[e]s, ~e** attribute

attributiv [atribu'ti:f] *(Sprachw.)* 1. *Adj.* attributive. 2. *adv.* attributively

atypisch *(geh.)* 1. *Adj.* atypical. 2. *adv.* atypically

ätzen ['ɛtsn̩] 1. *tr. V.* **a)** etch; **b)** *(Med.)* cauterize ⟨*wound*⟩. 2. *itr. V.* corrode

ätzend 1. *Adj.* **a)** corrosive; *(fig.)* caustic ⟨*wit, remark, criticism*⟩; pungent ⟨*smell*⟩; acrid ⟨*smoke*⟩; **b)** *(Jugendspr.)* grotty *(Brit. sl.);* grot *(Brit. sl.).* 2. *adv.* caustically ⟨*ironic, critical*⟩

Ätzung die; ~, ~en a) etching; **b)** *(Med.)* cauterization

au [au] *Interj.* **a)** *(bei Schmerz)* ow; ouch; **b)** *(bei Überraschung, Begeisterung)* oh

Aubergine [obɛr'ʒi:nə] **die; ~, ~n** aubergine *(Brit.);* egg-plant

auch [aux] 1. *Adv.* **a)** *(ebenso, ebenfalls)* as well; too; also; **Klaus war ~ dabei** Klaus was there as well *or* too; Klaus was also there; **Ich gehe jetzt. – Ich ~:** I'm going now – So am I; **Mir ist warm. – Mir ~:** I feel warm – So do I; **... – Ja, das ~:** ... – Yes, that too; ~ **gut!** that's all right too; **das kann ich ~!** I can do that too; **was er verspricht, tut er ~:** what he promises to do, he does; **nicht nur ..., sondern ~ ...:** not only ..., but also ...; **grüß deine Frau und ~ die Kinder** give my regards to your wife and the children too; **sehr gut, aber ~ teuer** very good but expensive too; ~ **das noch!** that's all I/we *etc.* need!; **oder ~:** or; **oder ~ nicht** or not, as the case may be; **das weiß ich ~ nicht** I don't know either; **ich habe ~ keine Lust/kein Geld** I don't feel like it either/don't have any money either; **das hat ~ nichts genützt** that did not help either; *s. auch* **sowohl; b)** *(sogar, selbst)* even; ~ **wenn** even if; **wenn ~:** even if *or* though; **ohne ~ nur zu fragen/eine Sekunde zu zögern** without even asking/hesitating for a second; **c)** *(außerdem, im übrigen)* besides **2.** *Partikel* **a)** *not translated* **etwas anderes habe ich ~ nicht erwartet** I never expected anything else; **so schlimm ist es ~ [wieder] nicht** it's not as bad as all that; **nun hör aber ~ zu!** now listen!; **wozu [denn] ~?** what's the point? why should I/you *etc.?;* **b)** *(zweifelnd)* **bist du dir ~ im klaren, was das bedeutet?** are you sure you understand what that means?; **bist du ~ glücklich?** are you truly happy?; **lügst du ~ nicht?** you're not lying, are you?; **c)** *(mit Interrogativpron.)* **wo .../wer .../wann .../was ...** *usw.* ~ **[immer]** wherever/whoever/whenever/whatever *etc. ...;* **wie dem ~ sei** however that may be; **d)** *(konzessiv)* **mag er ~ noch so klug sein** however clever he may be; no matter how clever he is; **so oft ich ~ anrief** however often I rang; no matter how often I rang; **so sehr er sich ~ bemühte** much as he tried; **wenn ~!** never mind

Audienz [au̯'di̯ɛnts] **die; ~, ~en** audience

Auditorium [au̯di'to:ri̯ʊm] **das; ~s, Auditorien a)** *(Hörsaal)* auditorium; **b)** *(Zuhörerschaft)* audience

Auer·hahn ['au̯ɐ-] **der** [cock] capercaillie

auf [au̯f] **1.** *Präp. mit Dat.* **a)** on; ~ **See** at sea; ~ **dem Baum** in the tree; ~ **der Erde** on earth; ~ **der Welt** in the world; ~ **der Straße** in the street; ~ **dem Platz** in the square; ~ **meinem Konto** in my account; ~ **beiden Augen blind** blind in both eyes; **das Thermometer steht ~ 15°** the thermometer stands at *or* reads 15°; **b)** *(in)* at ‹*post office, town hall, police station*›*;* ~ **seinem Zimmer** *(ugs.)* in his room; ~ **der Schule/Uni** at school/university; **c)** *(bei)* at ‹*party, wedding*›*;* on ‹*course, trip, walk, holiday, tour*›*;* **d) was hat es damit ~ sich?** what's it all about? **2.** *Präp. mit Akk.* **a)** on; on to; **sich ~ einen Stuhl setzen** sit down on a chair; **sich ~ das Bett legen** lie down on the bed; **er nahm den Rucksack ~ den Rücken** he lifted the rucksack up on to his back; ~ **einen Berg steigen** climb up a mountain; **sich** *(Dat.)* **einen Hut ~ den Kopf setzen** put a hat on [one's head]; ~ **den Mond fliegen** fly to the moon; **jmdm. ~ den Fuß treten** step on sb.'s foot; ~ **die Straße gehen** go [out] into the street; **jmdn. ~ den Rücken legen** lay sb. on his/her back; **jmdn. ~ den Rücken drehen** turn sb. on to his/her back; **etw. ~ ein Konto überweisen** transfer sth. to an account; **das Thermometer ist ~ 0° gefallen** the thermometer has fallen to 0°; ~ **ihn!** *(ugs.)* get him!; **b)** *(zu)* to; ~ **die Schule/Uni gehen** go to school/university; ~ **einen Lehrgang gehen** go on a course; **c)** *(bei Entfernungen)* ~ **10 km [Entfernung]** for [a distance of] 10 km; **wir näherten uns der Hütte [bis] ~ 30 m** we approached to within 30 m of the hut; **d)** *(zeitlich)* for; ~ **Jahre [hinaus]** for years [to come]; **etw. ~ nächsten Mittwoch festlegen/verschieben** arrange sth. for/postpone sth. until next Wednesday; **die Nacht von Sonntag ~ Montag** Sunday night; **das fällt ~ einen Montag** it falls on a Monday; **wir verschieben es ~ den 3. Mai** we'll postpone it to 3 May; **e)** *(zur Angabe der Art und Weise)* ~ **diese Art und Weise** in this way; ~ **die Tour erreichst du bei mir nichts** *(ugs.)* you won't get anywhere with me like that; ~ **deutsch** in German; ~ **das sorgfältigste/herzlichste** *(geh.)* most carefully/warmly; ~ **a enden** end in a; **f)** *(auf Grund)* ~ **Wunsch** on request; ~ **vielfachen Wunsch [hin]** in response to numerous requests; ~ **meine Bitte** at my request; ~ **seine Initiative** on his initiative; ~ **Befehl** on command; ~ **meinen Vorschlag [hin]** at my suggestion; **g)** *(sonstige Verwendungen)* **ein Teelöffel ~ einen Liter Wasser** one teaspoon to one litre of water; **das Bier geht ~ mich** *(ugs.)* the beer's on 'me *(coll.);* ~ **wen geht die Cola?** who's paying for the Coke?; **Welle ~ Welle brandete ans Ufer** wave upon wave broke on the shore; **jmdn. ~ Tb untersuchen** examine sb. for TB; **jmdn. ~ seine Eignung prüfen** test sb.'s suitability; ~ **die Sekunde [genau]** [precise] to the second; ~ **ein gutes Gelingen** to our/your success; ~ **deine Gesundheit** your health; ~ **bald/morgen!** *(bes. südd.)* see you soon/tomorrow; ~ **10 zählen** *(bes. südd.)* count [up] to 10; *s. auch* **einmal 1 a; machen 3 f. 3.** *Adv.* **a)** *(Aufforderung, sich zu erheben)* ~**!** up you get!; *(zu einem Hund)* up!; **b) sie waren längst ~ und davon** they had made off long before; **c)** *(bes. südd.: Aufforderung, zu handeln)* ~**!** come on; **d)** *(Aufforderung, sich aufzumachen)* ~ **ins Schwimmbad!** come on, off to the swimming-pool!; **e)** ~ **und ab** up and down; *(hin und her)* up and down; to and fro; **f)** *(Aufforderung, sich etw. aufzusetzen)* **Helm/Hut/Brille ~!** helmet/hat/glasses on!; **g)** *(Aufforderung, etw. zu öffnen)* **Fenster/Türen/Mund ~!** open the window/doors/your mouth! **4.** ~ **daß** *Konj. (veralt.)* so that

auf|arbeiten *tr. V.* **a)** *(erledigen)* catch up with ‹*correspondence etc.*›*;* **b)** *(studieren, analysieren)* review ‹*literature, material*›*;* look back on and reappraise ‹*one's past, childhood*›*;* **c)** *(restaurieren, überholen)* refurbish

auf|atmen *itr. V.* breathe a sigh of relief

auf|backen *regelm. (auch unr.) tr. V.* crisp up ‹*bread, rolls, etc.*›

auf|bahren *tr. V.* lay out ‹*body, corpse*›*;* **jmdn./einen Toten ~:** lay out sb.'s body; **aufgebahrt sein** ‹*king, president, etc.*› lie in state

Auf·bau der; *Pl.* **~ten a)** *o. Pl.* construction; building; *(fig.)* building; **den wirtschaftlichen ~ beschleunigen** speed up economic development; **b)** *o. Pl. (Biol.)* synthesis; **c)** *o. Pl. (Struktur)* structure; **d)** *Pl. (Schiffbau)* superstructure *sing.*

auf|bauen 1. *tr. V.* **a)** *auch itr. (errichten, aufstellen)* erect ‹*hut, kiosk, podium*›*;* set up ‹*equipment, train set*›*;* build ‹*house, bridge*›*;* put up ‹*tent*›*;* **ein Haus neu ~:** rebuild a house; **b)** *(hinstellen, arrangieren)* lay *or* set out ‹*food, presents, etc.*›*;* **c)** *(fig.: schaffen)* build ‹*state, economy, social order, life, political party, etc.*›*;* build up ‹*business, organ-*

ization, army, spy network⟩; **d)** *(fig.: strukturieren)* structure; **e)** *(fig.: fördern)* **jmdn./etw. zu etw. ~:** build sb./sth. up into sth.; **jmdn. als etw. ~:** build sb. up as sth.; **f)** *(gründen)* **etw. auf etw.** *(Dat.)* **~:** base sth. upon sth.; **g)** *(Biol.)* synthesize. **2.** *itr. V.* **auf etw.** *(Dat.)* **~:** be based on sth. **3.** *refl. V.* **a)** *(ugs.: sich hinstellen)* plant oneself; **b)** *(sich zusammensetzen)* be composed **(aus** of)

aufbauend *Adj.* constructive ⟨*criticism, geological process*⟩; restorative ⟨*medicine*⟩; nutrient ⟨*substance*⟩

auf|bäumen ['aufbɔymən] *refl. V.* rear up; **sich gegen jmdn./etw. ~** *(fig.)* rise up against sb./sth.

auf|bauschen *tr. V.* **a)** billow; billow, belly [out] ⟨*sail*⟩; **b)** *(fig.)* blow up *(coll.)*; exaggerate

auf|begehren *itr. V. (geh.)* rebel

auf|behalten *unr. tr. V.* **etw. ~:** keep sth. on

auf|beißen *unr. tr. V.* **etw. ~:** bite sth. open; **sich** *(Dat.)* **die Lippe ~:** bite one's lip [and make it bleed]

auf|bekommen *unr. tr. V.* **a)** *(öffnen können)* **etw. ~:** get sth. open; **b)** *(aufessen können)* manage to eat; **c)** *(aufgegeben bekommen)* be given ⟨*homework*⟩

auf|bessern *tr. V.* improve; increase ⟨*pension, wages, etc.*⟩

Auf·besserung die improvement (*Gen.* in); *(von Renten, Löhnen, Gehältern)* increase (*Gen.* in)

auf|bewahren *tr. V.* keep; store, keep ⟨*medicines, food, provisions*⟩; **etw. kühl ~:** store sth. in a cool place

Auf·bewahrung die *s.* **aufbewahren:** keeping; storage; **jmdm. etw. zur ~ geben/anvertrauen** give sth. to sb. for safe keeping/entrust sb. with the care of sth.

auf|biegen 1. *unr. tr. V.* **etw. ~:** bend sth. open. **2.** *unr. refl. V.* bend open

auf|bieten *unr. tr. V.* **a)** *(aufwenden)* exert ⟨*strength, energy, willpower, authority*⟩; call on ⟨*skill, wit, powers of persuasion*⟩; **b)** *(einsetzen)* call in ⟨*police, troops*⟩

Aufbietung die; ~: unter ~ aller Kräfte/seiner ganzen Überredungskunst summoning up all one's strength/calling on all one's persuasive skills

auf|binden *unr. tr. V.* **a)** *(öffnen, lösen)* untie; undo; **b)** *(hochbinden)* tie *or* put up ⟨*hair*⟩; **c)** *(auf den Rücken binden)* **jmdm./einem Tier etw. ~:** tie sth. on to sb.'s/an animal's back; **d)** *(ugs.: weismachen)* **wer hat dir das aufgebunden?** who spun you that yarn?; **jmdm. ein Märchen/eine Fabel/etwas ~:** spin sb. a yarn; *s.* **auch Bär**

auf|blähen 1. *tr. V.* distend ⟨*body, stomach*⟩; puff out ⟨*cheeks, feathers*⟩; flare ⟨*nostrils*⟩; billow, fill, belly [out] ⟨*sail*⟩; *(fig.: vergrößern)* over-inflate. **2.** *refl. V.* **a)** ⟨*sail*⟩ billow *or* belly out; ⟨*balloon, lungs, chest*⟩ expand; ⟨*stomach*⟩ swell up, become swollen *or* distended; **b)** *(abwertend: sich aufspielen)* puff oneself up

aufblasbar *Adj.* inflatable

auf|blasen 1. *unr. tr. V.* blow up; inflate

auf|bleiben *unr. itr. V.; mit sein* **a)** *(geöffnet bleiben)* stay open; **b)** *(nicht zu Bett gehen)* stay up

auf|blenden 1. *tr. V.* **die Scheinwerfer ~:** switch one's headlights to full beam; **mit aufgeblendeten Scheinwerfern fahren** drive with headlights on full beam. **2.** *itr. V.* switch to full beam

auf|blicken *itr. V.* **a)** look up; *(kurz)* glance up; **von etw. ~:** look/glance up from sth.; **b)** *(verehrend)* **zu jmdm. ~:** look up to sb.

auf|blinken *itr. V.* **a)** ⟨*light*⟩ flash; ⟨*metal*⟩ glint; **b)** *(ugs.: kurz aufblenden)* flash one's headlights

auf|blitzen *itr. V.* flash; ⟨*wave, white-caps*⟩ sparkle

auf|blühen *itr. V.; mit sein* **a)** bloom; come into bloom; ⟨*bud*⟩ open; **b)** *(fig.: aufleben)* blossom [out]; **c)** *(fig.: einen Aufschwung nehmen)* ⟨*trade, business, town, industry*⟩ flourish and expand; ⟨*cultural life, science*⟩ blossom and flourish

auf|bocken *tr. V.* jack up

auf|brauchen *tr. V.* use up

auf|brausen *itr. V.; mit sein (fig.)* flare up; **schnell/leicht ~:** be quick-tempered *or* hot-tempered; have a quick temper

aufbrausend *Adj.* quick-tempered; hot-tempered

auf|brechen 1. *unr. tr. V. (öffnen)* break open ⟨*lock, safe, box, crate, etc.*⟩; break into ⟨*car*⟩; force [open] ⟨*door*⟩. **2.** *unr. itr. V.; mit sein* **a)** *(sich öffnen)* ⟨*bud*⟩ open [up], burst [open]; ⟨*ice [sheet], surface, ground*⟩ break up; ⟨*wound*⟩ open; **b)** *(losgehen, -fahren)* set off; start out

auf|bringen *unr. tr. V.* **a)** *(beschaffen)* find; raise, find ⟨*money*⟩; *(fig.)* find, summon [up] ⟨*strength, energy, courage*⟩; find ⟨*patience*⟩; **b)** *(kreieren)* introduce, start ⟨*fashion, custom*⟩; introduce ⟨*slogan, theory*⟩; start, put about ⟨*rumour*⟩; **c)** *(in Wut bringen)* **jmdn. ~:** make sb. angry; infuriate sb.; **d)** *(aufwiegeln)* **jmdn. gegen jmdn./etw. ~:** set sb. against sb./sth.; **e)** *(Seew.)* seize

Auf·bruch der departure; *(fig. geh.)* awakening; **das Zeichen zum ~ geben** give the signal to set off *or* leave

Aufbruchs·stimmung die; *o. Pl.* **es herrschte allgemeine ~:** everybody was getting ready to go; **bist du schon in ~?** are you all ready to go?

auf|brühen *tr. V.* brew [up]

auf|brüllen *itr. V.* let out *or* give a roar; ⟨*animal*⟩ bellow

auf|brummen *tr. V. (ugs.)* **jmdm. etw. ~:** slap sth. on sb. *(coll.)*

auf|bürden *tr. V. (geh.)* **jmdm./einem Tier etw. ~:** load sth. on to sb./an animal; **jmdm./sich etw. ~** *(fig.)* burden sb./oneself with sth.

auf|decken 1. *tr. V.* **a)** uncover; **das Bett ~:** pull back the covers; **sich im Schlaf ~:** throw off one's covers; **b)** *(Kartenspiele)* show; **die** *od.* **seine Karten ~** *(fig.)* lay one's cards on the table *(fig.)*; **c)** *(enthüllen)* expose ⟨*corruption, error, weakness, crime, plot, abuse, etc.*⟩; *(erkennen und bewußtmachen)* reveal, uncover ⟨*connections, motive, cause, error, weakness, contradiction, etc.*⟩; **d)** *(für eine Mahlzeit)* **etw. ~:** put sth. on the table. **2.** *itr. V.* lay the table

Auf·deckung die *s.* **aufdecken 1c:** exposure; revelation; uncovering

auf|donnern *refl. V. (ugs. abwertend)* tart *(Brit.)* *or* doll oneself up *(coll.)*; get tarted *(Brit.)* *or* dolled up *(coll.)*

auf|drängen 1. *tr. V.* **jmdm. etw. ~:** force sth. on sb. **2.** *refl. V.* **a)** **sich jmdm. ~:** force one's company *or* oneself on sb.; **ich will mich aber nicht ~:** I don't want to impose; **b)** *(fig.: in den Sinn kommen)* **mir drängte sich der Verdacht auf, daß ...:** I couldn't help suspecting that ...; **dieser Gedanke drängt sich [einem] förmlich auf** one simply can't help but think so; the thought is unavoidable

auf|drehen 1. *tr. V.* **a)** *(öffnen)* unscrew ⟨*bottle-cap, nut*⟩; undo ⟨*screw*⟩; turn on ⟨*tap, gas, water*⟩; open ⟨*valve, bottle, vice*⟩; **b)** *(ugs.: laut stellen)* turn up ⟨*radio, record-player, etc.*⟩; **c)** *(ugs.: aufziehen)* wind up ⟨*musical box, watch, toy, etc.*⟩; **sich/jmdm. die Haare**

~: put one's/sb.'s hair in curlers. **2.** *itr. V. (ugs.)* **a)** *(das Tempo steigern)* open up; **b)** *(in Schwung kommen)* get into the mood; get going

auf·dringlich 1. *Adj.* importunate, *(coll.)* pushy ⟨*person*⟩; insistent ⟨*music, advertisement, questioning*⟩; pestering *attrib.* ⟨*journalist*⟩; pungent ⟨*perfume, smell*⟩; loud, gaudy ⟨*colour, wallpaper*⟩; **sei nicht so ~!** don't pester so! **2.** *adv.* ⟨*behave*⟩ importunately, *(coll.)* pushily; ⟨*ask*⟩ insistently

Aufdringlichkeit die; ~, ~en a) *o. Pl. s.* **aufdringlich:** insistent manner; insistence; importunity; pushiness *(coll.)*; pungency; **b) die ~en der Männer** the over-familiarity *sing.* of the men

auf|dröseln ['aufdrø:zln] *tr. V. (ugs., auch fig.)* unravel

Auf·druck der; ~[e]s, ~e imprint

auf|drucken *tr. V.* **etw. auf etw.** *(Akk.)* **~:** print sth. on sth.

auf|drücken *tr. V.* **a)** press **(auf** + *Akk.* on to); **b)** *(aufprägen, -stempeln)* stamp **(auf** + *Akk.* on); **jmdm. einen Kuß ~:** plant a kiss on sb.; *s. auch* **Stempel a; c)** *(öffnen)* push ⟨*door, window, etc.*⟩ open; squeeze ⟨*pimple, boil*⟩

auf·einander *Adv.* **a)** on top of one another *or* each other; **b) ~ warten** wait for each other *or* one another; **~ zufliegen** fly towards one another *or* each other

aufeinander-: ~|beißen *unr. tr. V.* **die Zähne ~beißen** clench one's teeth; **~|drücken** *tr. V.* press together; **~|folgen** *itr. V.; mit sein* follow each other *or* one another; **~folgend** successive; **~|legen 1.** *tr. V.* lay ⟨*planks etc.*⟩ one on top of the other; **2.** *refl. V.* lie on top of each other *or* one another; **~|liegen** *unr. itr. V.* lie on top of each other *or* one another; **~|prallen** *itr. V.; mit sein* crash into each other *or* one another; collide; ⟨*armies*⟩ clash; *(fig.)* ⟨*opinions*⟩ clash; **~|pressen** *tr. V.* press together; **~|schichten** *tr. V.* stack up; **~|schlagen 1.** *unr. tr. V.* strike *or* knock together; bang ⟨*cymbals*⟩ together; **2.** *unr. itr. V.; mit sein* strike *or* knock against each other *or* one another; **~|sitzen** *unr. itr. V.* sit on top of each other *or* one another; **~|stoßen** *unr. itr. V.; mit sein* bump together; ⟨*lines, streets*⟩ meet; *(fig.)* ⟨*opinions*⟩ clash; **~|treffen** *unr. itr. V.; mit sein* hit each other *or* one another; *(fig.)* meet

Aufenthalt ['auf|ɛnthalt] **der; ~[e]s, ~e a)** stay; **der ~ im Depot ist verboten** personnel/the public *etc.* are not permitted to remain within the depot; **b)** *(Fahrtunterbrechung)* stop; *(beim Umsteigen)* wait; **[20 Minuten] ~ haben** stop [for 20 minutes]; *(beim Umsteigen)* have to wait [20 minutes]; **c)** *(geh.: Ort)* residence

Aufenthalts-: ~dauer die length of stay; **~erlaubnis die, ~genehmigung die** residence permit; **~ort der** [place of] residence; **~raum der** *(einer Schule o. ä.)* common-room *(Brit.)*; *(einer Jugendherberge)* day-room; *(eines Betriebs o. ä.)* recreation-room

auf|erlegen *tr. V. (geh.)* **jmdm. etw. ~:** impose sth. on sb.; **du solltest dir etwas Zurückhaltung ~:** you should exercise some restraint

auf|erstehen *unr. itr. V.; mit sein* rise [again]

Auferstehung die; ~, ~en resurrection

auf|essen *unr. tr. (auch itr.) V.* eat up

auf|fädeln *tr. V.* **etw. [auf etw.** *(Akk.)*] **~:** thread sth. on to sth.

auf|fahren 1. *unr. itr. V.; mit sein* **a)** *(aufprallen)* **auf ein anderes Fahrzeug ~:** drive *or* run into the back of another vehicle; **auf etw./jmdn. ~:** drive *or* run into sth./sb.; **b)** *(aufschließen)* **[dem Vordermann] zu dicht ~:** drive too close to the car in front; **c)** *(in Stellung gehen)* move up [into position]; **d)** *(gen Himmel fahren)* ascend; **e)** *(aufschrecken)* start; **aus dem Schlaf ~:** awake with a start; **f)** *(aufbrausen)* flare up. **2.** *unr. tr. V.* **a)** *(in Stellung bringen)* bring *or* move up; **b)** *(ugs.: auftischen)* serve up

Auf·fahrt die a) *(das Hinauffahren)* climb; drive up; **die ~ zum Gipfel** the drive up to the summit; **b)** *(eines Gebäudes)* drive; **c)** *(zur Autobahn)* slip-road *(Brit.)*; access road *(Amer.)*; **d)** *(schweiz.: Himmelfahrt)* Ascension [Day]

auf|fallen *unr. itr. V.; mit sein* **a)** stand out; **diese Fettflecken fallen kaum auf** these grease-marks are hardly noticeable; **mach es so, daß es nicht auffällt** do it so that it doesn't attract attention *or* so that nobody notices; **seine Abwesenheit fiel nicht auf** his absence was not noticed; **um [nicht] aufzufallen** so as [not] to attract attention; **jmdm. fällt etw. auf** sb. notices sth.; sth. strikes sb.; **er ist mir angenehm/unangenehm aufgefallen** he made a good/bad impression on me; **b)** *(auftreffen)* fall **(auf** + *Akk.* on [to]); strike **(auf** + *Akk.* sth.)

auffallend 1. *Adj. (auffällig)* conspicuous; *(eindrucksvoll, bemerkenswert)* striking ⟨*contrast, figure, appearance, beauty, similarity*⟩. **2.** *adv. (auffällig)* conspicuously; *(eindrucksvoll, bemerkenswert)* ⟨*contrast, differ*⟩ strikingly; **stimmt ~!** *(scherzh.)* you're so right!

auf·fällig 1. *Adj.* conspicuous; garish, loud ⟨*colour*⟩; **eine recht ~e Erscheinung sein** have a most striking appearance. **2.** *adv.* conspicuously; **sich ~ kleiden** dress showily

auf|falten *tr. V.* fold open; unfold

auf|fangen *unr. tr. V.* **a)** catch; **b)** *(aufnehmen, sammeln)* collect; collect, catch ⟨*liquid*⟩

auf|fassen *tr. V.* **a)** *(ansehen als)* **etw. als etw. ~:** see *or* regard sth. as sth.; **etw. als Scherz/Kompliment** *usw.* **~:** take sth. as a joke/compliment *etc.*; **etw. persönlich/falsch ~:** take sth. personally/misunderstand sth.; **b)** *(begreifen)* grasp; comprehend

Auf·fassung die view; *(Begriff)* conception; **nach meiner ~:** in my view; **der ~ sein, daß ...:** take the view that ...; be of the opinion that ...

Auffassungs-: ~gabe die, ~vermögen das powers *pl.* of comprehension

auffindbar *Adj.* findable; **es ist nirgends/nicht ~:** it's nowhere to be found/it can't be *or* isn't to be found; **schwer/leicht ~ sein** be hard/easy to find

auf|finden *unr. tr. V.* find

auf|fischen *tr. V. (ugs.)* fish out *(coll.)*

auf|flackern *itr. V.; mit sein* flicker up; *(fig.)* flare up; ⟨*hope*⟩ flicker up

auf|flammen *itr. V.; mit sein (auch fig.)* flare up

auf|fliegen *unr. itr. V.; mit sein* **a)** *(hochfliegen)* fly up; **b)** *(ugs.: scheitern)* ⟨*illegal organization, drug ring*⟩ be *or* get busted *(coll.)*; **einen Schmugglerring ~ lassen** bust a smuggling ring *(coll.)*

auf|fordern *tr. V.* **a) jmdn. ~, etw. zu tun** call upon *or* ask sb. to do sth.; **jmdn. zur Teilnahme/Zahlung ~:** call upon *or* ask sb. to take part/ask sb. for payment; **ich fordere Sie zum letzten Mal auf, ...:** I am asking you for the last time ...; **b)** *(einladen, ermun-*

tern) **jmdn. ~, etw. zu tun** invite *or* ask sb. to do sth.; **jmdn. zu einem Spaziergang/zum Mitspielen ~:** invite sb. for a walk/invite *or* ask sb. to join in; **jmdn. [zum Tanz] ~:** ask sb. to dance

auffordernd **1.** *Adj.* **mit einer ~en Geste** with a gesture of invitation. **2.** *adv.* encouragingly

Auf·forderung die a) request; *(nachdrücklicher)* demand; **nach dreimaliger/mehrmaliger ~:** after three/repeated requests; **b)** *(Einladung, Ermunterung)* invitation

auf|forsten *tr. V.* afforest; **etw. wieder ~:** reforest sth.

auf|fressen 1. *unr. tr. V.* **a)** eat up; *(fig.)* swallow up ‹*small business*›; eat up ‹*savings, money, etc.*›; **er wird dich [deswegen] nicht [gleich] ~** *(ugs.)* he won't *or* isn't going to bite your head off [for that]. **2.** *unr. itr. V.* ‹*animal*› eat [all] its food up; *(salopp)* ‹*person*› eat [everything] up

auf|frischen *tr. V.* freshen up; brighten up ‹*colour, paintwork*›; renovate ‹*polish, furniture*›; *(restaurieren)* restore ‹*tapestry, fresco, etc.*›; *(fig.)* revive ‹*old memories*›; renew ‹*acquaintance, friendship*›; **sein Englisch ~:** brush up one's English

auf|führen 1. *tr. V.* **a)** put on, stage ‹*play, ballet, opera*›; screen, put on ‹*film*›; perform ‹*piece of music*›; **b)** *(nennen)* cite; quote; adduce; *(in Liste)* list. **2.** *refl. V.* behave

Auf·führung die a) performance; **b)** *s.* **aufführen 1 b:** citation; quotation; listing

auf|füllen *tr. V.* **a)** fill up; **b)** *(fig.: ergänzen)* replenish ‹*stocks*›; **c)** *(ugs.: nachfüllen)* **Wasser/Öl ~:** top up *(Brit.)* or *(Amer.)* fill up with water/oil

Auf·gabe die a) *(zu Bewältigendes)* task; **es sich** *(Dat.)* **zur ~ machen, etw. zu tun** make it one's task *or* job to do sth.; **b)** *(Pflicht)* task; responsibility; duty; **c)** *(fig.: Zweck, Funktion)* function; **d)** *(Schulw.) (Übung)* exercise; *(Prüfungs~)* question; **e)** *(Schulw.: Haus~)* piece of homework; **~n** homework *sing.*; **f)** *(Rechen~, Mathematik~)* problem; **g)** *(Beendigung)* abandonment; **h)** *(Kapitulation)* retirement; *(im Schach)* resignation; **jmdn. zur ~ zwingen** force sb. to retire/resign; **i)** *s.* **aufgeben 1 b:** giving up; abandonment; dropping; **j)** *s.* **aufgeben 1 e:** posting *(Brit.)*; mailing *(Amer.)*; handing in; phoning in; placing; checking in

auf|gabeln *tr. V. (salopp)* pick up

Aufgaben·bereich der area of responsibility

Auf·gang der a) *(Mond~, Sonnen~ usw.)* rising; **b)** *(Treppe)* stairs *pl.*; staircase; stairway; *(in einem Bahnhof, zu einer Galerie, einer Tribüne)* steps *pl.*; **c)** *(Weg)* **der ~ zur Burg** the path up to the castle

auf|geben 1. *unr. tr. V.* **a)** give up; give up, stop ‹*smoking, drinking*›; **gib's auf!** *(ugs.)* you might as well give up!; why don't you give up!; **b)** *(sich trennen von)* give up ‹*habit, job, flat, business, practice, etc.*›; give up, abandon, drop ‹*plans, demand*›; give up, abandon ‹*profession, attempt*›; **c)** *(verloren geben)* give up ‹*patient*›; give up hope on *or* with ‹*wayward son, daughter, etc.*›; give up, abandon ‹*chessman*›; **sich selbst ~:** give oneself up for lost; **d)** *(nicht länger zu gewinnen versuchen)* give up ‹*struggle*›; retire from ‹*race, competition*›; **eine Partie ~:** concede a game; **e)** *(übergeben, übermitteln)* post *(Brit.)*, mail ‹*letter, parcel*›; hand in, *(telefonisch)* phone in ‹*telegram*›; place ‹*advertisement, order*›; check ‹*luggage, baggage*› in; **f)** *(Schulw.: als Hausaufgabe)* set *(Brit.)*; assign *(Amer.)*; **g)** *(zur Lösung vorlegen)* **jmdm. ein Rätsel/eine Frage ~:** set *(Brit.)* or *(Amer.)* assign sb. a puzzle/pose sb. a question. **2.** *unr. itr. V.* give up; *(im Sport)* retire; *(Schach)* resign

auf·geblasen *Adj.* puffed up

Auf·gebot das a) *(aufgebotene Menge)* contingent; *(Sport: Mannschaft)* contingent; squad; *(an Arbeitern)* squad; **ein gewaltiges ~ an Polizisten/Fahrzeugen** a huge force of police/array of vehicles; **b)** *(zur Heirat)* notice of an/the intended marriage; *(kirchlich)* banns *pl.*; **das ~ bestellen** give notice of an/the intended marriage; *(kirchlich)* put up the banns

auf·gedreht *Adj. (ugs.)* in high spirits *pred.*

auf·gedunsen *Adj.* bloated

auf|gehen *unr. itr. V.; mit sein* **a)** ‹*sun, moon, etc.*› rise; **b)** *(sich öffnen [lassen])* ‹*door, parachute, wound*› open; ‹*stage curtain*› go up, rise; ‹*knot, button, zip, bandage, shoelace, stitching*› come undone; ‹*boil, pimple, blister*› burst; ‹*flower, bud*› open [up]; **c)** *(keimen)* come up; **d)** *(aufgetrieben werden)* ‹*dough, cake*› rise; **e)** *(Math.)* ‹*calculation*› work out, come out; ‹*equation*› come out; **7 durch 3 geht nicht auf** threes into seven won't go; **seine Rechnung ging nicht auf** *(fig.)* he had miscalculated; **f) etw. geht jmdm. auf** *(etw. wird jmdm. klar)* sb. realizes sth.; **g) in etw.** *(Dat.)* **~:** become absorbed into sth.; ‹*person*› be completely absorbed in sth.; *s. auch* **Flamme a**

auf|geilen *tr. V. (salopp)* **jmdn. [mit/durch etw.] ~:** get sb. randy [with sth.]; **sich [an etw.** *(Dat.)***] ~:** get randy [with sth.]; *(fig.)* get worked up [about sth.]

aufgeklärt *Adj.* enlightened

aufgekratzt *Adj. (ugs.)* in high spirits *pred.*

auf·gelegt *Adj.* **gut/schlecht/heiter** *usw.* **~ sein** be in a good/bad/cheerful *etc.* mood; **zu etw. ~ sein** be in the mood for sth.; **dazu ~ sein, etw. zu tun** be in the mood to do sth.

auf·gelöst *Adj.* distraught; *s. auch* **Träne**

aufgeräumt *Adj.* jovial

aufgeregt 1. *Adj.* excited; *(beunruhigt)* agitated. **2.** *adv.* excitedly; *(beunruhigt)* agitatedly

Aufgeregtheit die; ~: excitement; agitation; *(Nervosität)* agitation

auf·geschlossen *Adj.* open-minded **(gegenüber** as regards, about); *(interessiert, empfänglich)* receptive, open (*Dat.*, **für** to); *(mitteilsam)* communicative; *(zugänglich)* approachable

Auf·geschlossenheit die *s.* **aufgeschlossen:** openmindedness; receptiveness; openness; communicativeness; approachableness

auf·geschmissen *Adj. (ugs.)* in **[ganz schön] ~ sein** be [right] up the creek *(sl.)*; be in a [real] fix

aufgeweckt *Adj.* bright; sharp

Aufgewecktheit die; ~: brightness; sharpness

auf|gießen *unr. tr. V.* make, brew [up] ‹*tea*›; make ‹*coffee*›

auf|gliedern *tr. V.* subdivide, break down, split up (**in** + *Akk.* into)

Auf·gliederung die subdivision; breakdown

auf|greifen *unr. tr. V.* **a)** *(festnehmen)* pick up; **b)** *(sich befassen mit)* take *or* pick up ‹*subject, suggestion*›

auf·grund 1. *Präp. mit Gen.* on the basis *or* strength of. **2.** *Adv.* **~ von** on the basis *or* strength of

Auf·guß der infusion; *(fig.)* rehash

auf|haben *(ugs.)* **1.** *unr. tr. V.* **a)** *(aufgesetzt haben)* have on; wear; **b)** *(geöffnet haben)* have ‹*zip*› undone; have ‹*door, window, jacket, blouse*› open; **die Augen ~:** have one's eyes open; **c)** *(aufbekommen haben)* have got ‹*cupboard, case, safe, etc.*› open; have got ‹*knot, zip*› undone; **d)** *(für die Schule)* **etw. ~:** have sth. as homework; **viel/wenig ~:** have a lot of/not have much homework; **e)** *(aufgegessen haben)* have eaten up *or* finished. **2.** *unr. itr. V.* ‹*shop, office*› be open; **wir haben bis 17.30 auf** we are open until 5.30 p.m.

auf|halsen ['aufhalzn̩] *tr. V.* *(ugs.)* **jmdm./sich etw. ~:** saddle sb./oneself with sth.

auf|halten 1. *unr. tr. V.* **a)** *(anhalten)* halt; halt, check ‹*inflation, advance, rise in unemployment*›; **jmdn. an der Grenze ~:** hold sb. up at the border; **b)** *(stören)* hold up; **c)** *(geöffnet halten)* hold ‹*sack, door, etc.*› open; **die Augen [und Ohren] ~:** keep one's eyes [and ears] open; **die Hand ~** *(auch fig.)* hold out one's hand. **2.** *unr. refl. V.* **a)** *(sich befassen)* **sich mit jmdm./etw. ~:** spend [a long] time on sb./sth.; **sich bei etw. ~:** linger over sth.; **b)** *(sich befinden)* be; *(verweilen)* stay; **tagsüber hielt er sich im Museum auf** he spent the day in the museum

auf|hängen 1. *tr. V.* **a)** hang up; hang ‹*picture, curtains*›; **b)** *(erhängen)* hang **(an +** *Dat.* from). **2.** *refl. V.* hang oneself

Aufhänger der; ~s, ~ a) *(Schlaufe)* loop; **b)** *(fig.: äußerer Anlaß)* peg

auf|häufen 1. *tr. V.* pile up; *(fig.)* amass ‹*treasure, riches*›. **2.** *refl. V.* *(auch fig.)* pile up; accumulate

auf|heben *unr. tr. V.* **a)** pick up; **b)** *(aufbewahren)* keep; preserve; **gut/schlecht aufgehoben sein** be/not be in good hands **(bei** with); **c)** *(abschaffen)* abolish; repeal ‹*law*›; rescind, revoke ‹*order, instruction*›; cancel ‹*contract*›; lift ‹*ban, prohibition, blockade, siege, martial law*›; **d)** *(ausgleichen)* cancel out; neutralize, cancel ‹*effect*›; **sich [gegenseitig] ~:** cancel each other out

Aufheben das; ~s *in* **viel ~[s]/kein ~ von jmdm./etw. machen** make a great fuss/not make any fuss about sb./sth.

Auf·hebung die a) *s.* **aufheben c:** abolition; repeal; rescindment; revocation; cancellation; lifting; **b)** *s.* **aufheben e:** closure; lifting

auf|heitern 1. *tr. V.* cheer up. **2.** *refl. V.* **a)** ‹*mood, face, expression*› brighten; **b)** ‹*weather*› clear *or* brighten up; ‹*sky*› brighten

Aufheiterung die; ~, ~en a) cheering up; **b)** *(des Wetters)* bright period

auf|heizen 1. *tr. V.* heat [up]; *(fig.)* inflame ‹*tensions, conflict*›; fuel ‹*mistrust*›. **2.** *refl. V.* heat up

auf|hellen 1. *tr. V.* **a)** brighten; lighten ‹*hair, shadow, darkness*›; **b)** *(klären)* shed *or* cast *or* throw light on. **2.** *refl. V.* ‹*sky, face, expression*› brighten; ‹*hair*› turn *or* go lighter; ‹*day, weather*› brighten [up]

auf|hetzen *tr. V.* incite; **jmdn. zur Meuterei/zu Gewalttaten ~:** incite sb. to mutiny/violence

auf|holen 1. *tr. V.* make up ‹*time, delay*›. **2.** *itr. V.* catch up; ‹*train*› make up time; ‹*athlete, competitor*› make up ground; *(Zeit ~)* make up time

auf|horchen *itr. V.* prick up one's ears; **jmdn. ~ lassen** *(fig.)* make sb. [sit up and] take notice

auf|hören *itr. V.* stop; ‹*friendship*› end; *(ugs.: das Arbeitsverhältnis aufgeben)* finish; **das muß ~!** this has got to stop!; **da hört [sich] doch alles auf!** *(ugs.)* that really is the limit! *(coll.)*; **es hat aufgehört zu schneien** it's stopped snowing; **[damit] ~, etw. zu tun** stop doing sth.; **nicht [damit] ~, etw. zu tun** keep on doing sth.; **hört mit dem Lärm/Unsinn auf** stop that noise/nonsense

auf|kaufen *tr. V.* buy up

auf|keimen *itr. V.; mit sein* sprout; *(fig.)* ‹*suspicion, doubt, fear, longing, reluctance*› begin to grow; ‹*hope, passion, love, sympathy*› burgeon

auf|klappen *tr. V.* open [up] ‹*suitcase, trunk*›; fold back ‹*car hood*›; open ‹*window, door, book, knife*›

auf|klären 1. *tr. V.* **a)** *(klären)* clear up ‹*matter, mystery, question, misunderstanding, error, confusion*›; solve ‹*crime, problem*›; elucidate, explain ‹*event, incident, cause*›; resolve ‹*contradiction, disagreement*›; **b)** *auch itr. (auch scherzh.: informieren)* enlighten; **jmdn. über jmdn./etw. ~:** enlighten sb. about sb./sth.; **jmdn. [darüber] ~, wie .../ was ...:** enlighten sb. how .../what ...; **c)** *(sexualkundlich)* **ein Kind ~:** tell a child the facts of life; **aufgeklärt sein** know the facts of life. **2.** *refl. V.* **a)** *(sich klären)* ‹*misunderstanding, mystery*› be cleared up; **b)** *(sich aufhellen)* ‹*weather*› clear up; brighten [up]; ‹*sky*› clear, brighten

aufklärerisch 1. *Adj.* ‹*mission, intention*› to instruct and inform. **2.** *adv.* **~ wirken** instruct and inform

Auf·klärung die *o. Pl.* **a)** *s.* **aufklären 1 a:** clearing up; solution; elucidation; explanation; resolution; **b)** *(auch scherzh.: Information)* enlightenment; **c) die ~ der Kinder** *(über Sexualität)*; telling the children the facts of life; **d) die ~** *(hist.)* the Enlightenment

auf|kleben *tr. V.* stick on; *(mit Kleister)* paste on; *(mit Klebstoff, Leim)* stick *or* glue on

Auf·kleber der sticker

auf|knacken *tr. V.* crack [open]

auf|knöpfen *tr. V.* unbutton; undo

auf|knoten *tr. V.* untie, undo

auf|knüpfen *(ugs.)* **1.** *tr. V.* string up *(coll.)*. **2.** *refl. V.* hang oneself

auf|kochen 1. *tr. V.* **a)** bring to the boil. **2.** *itr. V. mit sein* come to the boil; **etw. ~ lassen** bring sth. to the boil

auf|kommen *unr. itr. V.; mit sein* **a)** *(entstehen)* ‹*wind*› spring up; ‹*storm, gale*› blow up; ‹*fog*› come down; ‹*rumour*› start; ‹*suspicion, doubt, feeling*› arise; ‹*fashion, style, invention*› come in; ‹*boredom*› set in; ‹*mood, atmosphere*› develop; **etw. ~ lassen** give rise to sth.; **b) ~ für** *(bezahlen)* bear, pay ‹*costs*›; pay for ‹*damage*›; pay, defray ‹*expenses*›; be liable for ‹*debts*›; stand ‹*loss*›; **für jmdn. ~:** pay for sb.'s upkeep; **c) ~ für** *(verantwortlich sein für)* be responsible for; **d)** *(auftreffen)* land

auf|kratzen *tr. V.* **a)** *(öffnen)* scratch open ‹*wound, sore*›; **b)** *(verletzen)* scratch

auf|krempeln *tr. V.* roll up ‹*sleeves, trousers*›; **jmdm./sich die Ärmel ~:** roll up sb.'s/one's sleeves

auf|kreuzen *itr. V.; mit sein (ugs.: erscheinen)* turn up

auf|kriegen *(ugs.)* *s.* **aufbekommen**

auf|kündigen *tr. V.* terminate ‹*lease, contract*›; cancel ‹*subscription, membership*›; foreclose ‹*mortgage*›; **jmdm. die Freundschaft/den Gehorsam ~** *(geh.)* break off one's friendship with sb./refuse sb. further obedience

Aufl. *Abk.* **Auflage** ed.

auf|lachen *itr. V.* give a laugh; laugh; *(schallend)* burst out laughing

auf|laden 1. *unr. tr. (auch itr.) V.*

a) load (**auf** + *Akk.* on [to]); **jmdm. etw.** ~ *(ugs., fig.)* load sb. with sth.; **b)** charge [up] ⟨*battery*⟩; put ⟨*battery*⟩ on charge; **etw. wieder** ~: recharge sth. **2.** *unr. refl. V.* ⟨*battery*⟩ charge, become charged; **sich wieder** ~: recharge; become recharged

Auf·lage die a) *(Buchw.)* edition; *(gedruckte ~ einer Zeitung)* print run; *(verkaufte ~ einer Zeitung)* circulation; **b)** *(bes. Rechtsw.: Verpflichtung)* condition; **mit der ~, etw. zu tun** with the condition that *or* on condition that one does sth.; **[es] jmdm. zur ~ machen, daß ...:** impose on sb. the condition that ...

Auflagen·höhe die *(Buchw.)* number of copies printed; *(einer Zeitung)* circulation

auf|lassen *unr. tr. V. (ugs.)* **a)** *(offenlassen)* leave open; **b)** *(aufbehalten)* keep on ⟨*hat etc.*⟩

auf|lauern *itr. V.* **jmdm.** ~: lie in wait for sb.; *(um ihn zu überfallen)* waylay sb.

Auf·lauf der a) *(Menschen~)* crowd; **b)** *(Speise)* soufflé

auf|laufen *unr. itr. V.; mit sein* **a)** *(Seemannsspr.)* run aground (**auf** + *Akk.* on); **b)** *(Sport)* **auf jmdn.** ~: run into sb.; **jmdn. ~ lassen** bodycheck sb.; *(fig. ugs.)* put paid to sb.'s [little] plans; **c)** *(sich ansammeln)* accumulate

Auflauf·form die baking-dish; *(für Eierspeisen)* soufflé dish

auf|leben *itr. V.; mit sein* revive; *(fig.: wieder munter werden)* come to life; liven up; **etw. ~ lassen** revive sth.

auf|lecken *tr. V.* lap up

auf|legen 1. *tr. V.* **a)** put on; **noch ein Gedeck ~:** set another place; **den Hörer ~:** put down the receiver; **b)** *(Buchw.)* publish; **ein Buch neu** *od.* **wieder ~:** bring out a new edition of a book; *(nachdrucken)* reprint a book. **2.** *itr. V. (den Hörer ~)* hang up; ring off *(Brit.)*

auf|lehnen *refl. V.* **sich gegen jmdn./etw. ~:** rebel *or* revolt against sb./sth.

Auflehnung die; ~, ~en rebellion; revolt

auf|lesen *unr. tr. V.* pick up; gather [up]; *(fig. ugs.)* pick up, catch ⟨*germ, disease, illness*⟩

auf|leuchten *itr. V.; auch mit sein* light up; *(für kurze Zeit)* flash; ⟨*brake-light*⟩ come on; *(fig.)* ⟨*eyes, face*⟩ light up

auf|liegen *unr. itr. V.* lie; rest

auf|listen *tr. V.* list

auf|lockern 1. *tr. V.* break up, loosen ⟨*soil*⟩; loosen ⟨*stuffing, hair*⟩; *(fig.)* introduce some variety into ⟨*landscape, lesson, lecture*⟩; relieve, break up ⟨*pattern, façade*⟩; make ⟨*mood, atmosphere, evening*⟩ more relaxed. **2.** *refl. V.* ⟨*cloud*⟩ break

Auf·lockerung die *s.* **auflockern 1:** breaking up; loosening; relieving; breaking up; **zur ~ des Vortrags** to introduce some variety into the lecture; **zur ~ der Stimmung/des Abends** to make the mood/evening more relaxed

auf|lodern *itr. V.; mit sein (geh.)* ⟨*fire*⟩ blaze *or* flare up; ⟨*flames*⟩ leap up; *(fig.)* ⟨*jealousy, hatred, anger, passion*⟩ flare up

auflösbar *Adj.* soluble; soluble, solvable ⟨*equation, problem*⟩

auf|lösen 1. *tr. V.* dissolve; resolve ⟨*difficulty, contradiction*⟩; solve ⟨*puzzle, equation*⟩; break off ⟨*engagement*⟩; terminate, cancel ⟨*arrangement, contract, agreement*⟩; dissolve, disband ⟨*organization*⟩; break up ⟨*household*⟩. **2.** *refl. V.* **a)** dissolve (**in** + *Akk.* into); ⟨*parliament*⟩ dissolve itself; ⟨*crowd, demonstration*⟩ break up; ⟨*fog, mist*⟩ disperse, lift; ⟨*cloud*⟩ break up; ⟨*empire, kingdom, social order*⟩ disintegrate; **b)** *(sich aufklären)* ⟨*misunderstanding, difficulty, contradiction*⟩ be resolved; ⟨*puzzle, equation*⟩ be solved

Auf·lösung die a) *s.* **auflösen 1:** dissolving; resolution; solution; breaking off; termination; cancellation; dissolution; disbandment; removal; **b)** *s.* **auflösen 2a:** dissolving; dispersing; lifting; breaking up; disintegration; **c)** *(Verstörtheit)* distraction

auf|machen 1. *tr. V.* **a)** *(öffnen)* open; undo ⟨*button, knot*⟩; open, undo ⟨*parcel, packet*⟩; **b)** *(ugs.: eröffnen)* open [up] ⟨*shop, theatre, business, etc.*⟩; **c)** *(gestalten)* get up; present. **2.** *itr. V.* **a)** *(geöffnet werden)* ⟨*shop, office, etc.*⟩ open; **b)** *(ugs.: die Tür öffnen)* open up; open the door; **jmdm. ~:** open the door to sb.; **mach auf!** open up!; **c)** *(ugs.: eröffnet werden)* ⟨*shop, business*⟩ open [up]. **3.** *refl. V. (aufbrechen)* set out; start [out]

Aufmachung die; ~, ~en presentation; *(Kleidung)* get-up; **ein Buch in ansprechender ~:** an attractively presented book

auf|malen *tr. V.* **etw. [auf etw.** *(Akk.)***] ~:** paint sth. on [sth.]

auf|marschieren *itr. V.; mit sein* draw up; assemble; *(heranmarschieren)* march up; **Truppen sind an der Grenze aufmarschiert** troops were deployed along the border

aufmerksam 1. *Adj.* **a)** attentive ⟨*pupil, reader, observer*⟩; keen, sharp ⟨*eyes*⟩; **jmdn. auf jmdn./etw. ~ machen** draw sb.'s attention to sb./sth.; bring sb./sth. to sb.'s notice; **jmdn. darauf ~ machen, daß ...:** draw sb.'s attention to *or* bring to sb.'s notice the fact that ...; **auf jmdn./etw. ~ werden** become aware of *or* notice sb./sth.; **~ werden** notice; **b)** *(höflich)* attentive; **danke, sehr ~:** thank you, that's very *or* most kind of you. **2.** *adv.* attentively

Aufmerksamkeit die; ~, ~en a) *o. Pl.* attention; **b)** *(Höflichkeit)* attentiveness; **c)** *(Geschenk)* **eine [kleine] ~:** a small gift

auf|möbeln *tr. V. (ugs.)* **a)** *(verbessern)* do up; **b)** *(beleben)* pep *or* buck up *(coll.)*; *(aufmuntern)* buck *(coll.) or* cheer up

auf|motzen *tr. V. (ugs.)* tart up *(Brit. coll.)*; doll up *(coll.)*; soup up *(coll.)* ⟨*car, engine*⟩

auf|mucken, auf|mucksen *itr. V. (ugs.)* kick up *or* make a fuss; **gegen etw. ~:** balk at sth.

auf|muntern *tr. V.* cheer up; *(beleben)* liven up; pep up *(coll.)*

Aufmunterung die; ~, ~en *s.* **aufmuntern:** cheering up; livening up; pepping up *(coll.)*

aufmüpfig ['aufmʏpfɪç] *(ugs.)* **1.** *Adj.* rebellious. **2.** *adv.* rebelliously

Aufmüpfigkeit die; ~: rebelliousness

auf|nähen *tr. V.* sew on; **etw. auf etw.** *(Akk.)* **~:** sew sth. on [to] sth.

Aufnahme die; ~, ~n a) *s.* **aufnehmen b:** opening; starting; establishment; taking up; **b)** *(Empfang)* reception; *(Beherbergung)* accommodation; *(ins Krankenhaus)* admission; **bei jmdm. ~ finden** be taken in [and looked after] by sth.; **c)** *(in einen Verein, eine Schule, Organisation)* admission (**in** + *Akk.* into); **d)** *(Finanzw.: von Geld)* raising; **e)** *(Aufzeichnung)* taking down; *(von Personalien, eines Diktats)* taking [down]; **f)** *(das Fotografieren)* photographing; *(eines Bildes)* taking; *(das Filmen)* shooting; filming; **g)** *(Bild)* picture; shot; photo[graph]; **eine ~ machen** take a picture *or* shot *or* photo[graph]; **h)** *(auf Tonträger)* recording; **i)** *(Anklang)* reception; response (*Gen.* to); **j)** *o. Pl. (Absorption)* absorption; **k)** *(das Einschließen)* inclusion

aufnahme-, Aufnahme-: ~fähig *Adj.* receptive (**für** to); **~fähigkeit die;** *o. Pl.* receptivity (**für** to); **~gebühr die** enrolment fee; **~prüfung die** entrance examination; **~studio das** *(Tonstudio)* recording studio; *(Filmstudio)* film studio

auf|nehmen *unr. tr. V.* **a)** *(hochheben)* pick up; lift up; *(aufsammeln)* pick up; **b)** *(beginnen mit)* open, start ⟨*negotiations, talks*⟩; establish ⟨*relations, contacts*⟩; take up ⟨*studies, activity, fight, idea, occupation*⟩; start ⟨*production, investigation*⟩; **c) es mit jmdm. ~:** take sb. on; **es mit jmdm. ~/nicht ~ können** be a/no match for sb.; **d)** *(empfangen)* receive; *(beherbergen)* take in; **in ein Krankenhaus aufgenommen werden** be admitted to a hospital; **e)** *(beitreten lassen)* admit (**in** + *Akk.* to); **jmdn. als Mitglied in einen Verein ~:** admit sb. as a member of a club; **f)** *(einschließen)* include; **g)** *(fassen)* take; hold; **h)** *(erfassen)* take in, absorb ⟨*impressions, information, etc.*⟩; **i)** *(absorbieren)* absorb; **j)** *(Finanzw.)* raise ⟨*mortgage, money, loan*⟩; **k)** *(reagieren auf)* receive; **etw. positiv/mit Begeisterung ~:** give sth. a positive/an enthusiastic reception; **l)** *(aufschreiben)* take down; take [down] ⟨*dictation, particulars*⟩; **m)** *(fotografieren)* take a photograph of; photograph; take ⟨*picture*⟩; *(filmen)* film; **n)** *(auf Tonträger)* record; **o)** *(Handarbeit)* increase ⟨*stitch*⟩

auf|nötigen *tr. V.* **jmdm. etw. ~:** force sth. on sb.

auf|oktroyieren ['aufɔktroaji:rən] *tr. V.* **jmdm. etw. ~:** impose *or* force sth. on sb.

auf|opfern *refl. V.* devote oneself sacrificingly (**für** to)

aufopfernd 1. *Adj.* self-sacrificing. **2.** *adv.* self-sacrificingly

Auf·opferung die self-sacrifice

aufopferungs·voll *Adj., adv. s.* **aufopfernd**

auf|päppeln *tr. V.* feed up

auf|passen *itr. V.* **a)** look *or* watch out; *(konzentriert sein)* pay attention; **paß mal auf!** *(ugs.) (du wirst sehen)* you just watch!; *(hör mal zu!)* now listen; **aufgepaßt!** *(ugs.)* look *or* watch out!; **kannst du denn nicht ~?** can't you be more careful?; **b) auf jmdn./etw. ~** *(jmdn./etw. beaufsichtigen)* keep an eye on sb./sth.

Aufpasser der; ~s, ~, Aufpasserin die; ~, ~en a) *(abwertend)* spy; **b)** *(Wärter[in], Bewacher[in])* guard

auf|peitschen *tr. V.* whip up ⟨*sea, waves*⟩; inflame ⟨*passions, emotions, senses*⟩; inflame, stir up ⟨*populace, crowd*⟩

auf|pflanzen 1. *tr. V.* **a)** *(aufstellen)* set up; **b)** fix ⟨*bayonet*⟩. **2.** *refl. V. (ugs.)* plant oneself

auf|picken *tr. V.* **a)** *(aufnehmen)* ⟨*bird*⟩ peck up; *(fig. ugs.)* pick up ⟨*expression, idea, piece of information*⟩; **b)** *(öffnen)* peck open

auf|platzen *itr. V.; mit sein* burst open; ⟨*seam, cushion*⟩ split open; ⟨*wound*⟩ open up

auf|plustern ['aufplu:stɐn] **1.** *tr. V.* ruffle [up] ⟨*feathers*⟩; puff up ⟨*cheeks*⟩. **2.** *refl. V.* ⟨*bird*⟩ ruffle [up] its feathers

auf|polieren *tr. V. (auch fig.)* polish up

Auf·prall der; ~[e]s, ~e impact

auf|prallen *itr. V.; mit sein* **auf etw.** *(Akk.)* **~:** strike *or* hit sth.; *(auffahren)* collide with *or* run into sth.

Auf·preis der extra *or* additional charge; **gegen ~:** for an extra *or* additional charge

auf|pumpen *tr. V.* inflate; pump up, inflate ⟨*tyre*⟩; pump up *or* inflate the tyres of *or* on ⟨*bicycle*⟩

auf|putschen *tr. V. (abwertend)* stimulate; arouse ⟨*passions, urge*⟩; **~de Mittel** stimulants; **sich mit Kaffee ~:** drink coffee as a stimulant

Aufputsch·mittel das stimulant

auf|quellen *unr. itr. V.; mit sein* swell up

auf|raffen *refl. V.* **a)** pull oneself up [on to one's feet]; **b)** *(sich überwinden)* pull oneself together; **sich dazu ~, etw. zu tun** bring oneself to do sth.; **sich zu einer Arbeit/Entscheidung ~:** bring oneself to do a piece of work/come to a decision

auf|rappeln *refl. V. (ugs.)* **a)** struggle to one's/its feet; **b)** *(fig.)* recover

auf|rauhen *tr. V.* roughen [up]; nap ⟨*cloth*⟩

auf|räumen 1. *tr. V.* **a)** tidy *or* clear up; *(fig.)* sort out; **b)** *(wegräumen)* clear *or* put away. **2.** *itr. V.* **a)** tidy *or* clear up; *(fig.)* sort things out; **b) mit jmdm./etw. ~** *(jmdn./etw. beseitigen)* eliminate sb./sth.

auf|rechnen *tr. V.* **etw. gegen etw. ~:** set sth. off against sth.

auf·recht 1. *Adj.* **a)** upright ⟨*position*⟩; upright, erect ⟨*posture, bearing*⟩; **etw. ~ hinstellen** place sth. upright *or* in an upright position; **b)** *(redlich)* upright. **2.** *adv.* ⟨*walk, sit, hold oneself*⟩ straight, erect; **sich kaum noch ~ halten können** be hardly able to stand

aufrecht|erhalten *unr. tr. V.* maintain; maintain, keep up ⟨*deception, fiction, contact, custom*⟩

auf|regen 1. *tr. V.* excite; *(ärgerlich machen)* annoy; irritate; *(beunruhigen)* agitate; **du regst mich auf** you're getting on my nerves. **2.** *refl. V.* get worked up (**über** + *Akk.* about)

Auf·regung die excitement *no pl.; (Beunruhigung)* agitation *no pl.;* **jmdn. in ~ versetzen** make sb. excited/agitated; **nur keine ~!** don't get excited!

auf|reiben 1. *unr. tr. V.* **a)** *(zermürben)* wear down; **b)** *(vernichten)* wipe out; **c)** *(wund reiben)* **sich** *(Dat.)* **die Hände/Fersen** *usw.* **~:** rub one's hands/heels *etc.* sore. **2.** *unr. refl. V.* wear oneself out

aufreibend 1. *Adj.* wearing; trying ⟨*day, time*⟩; *(stärker)* gruelling. **2.** *adv.* tryingly; exasperatingly

auf|reißen 1. *unr. tr. V.* **a)** *(öffnen)* tear *or* rip open; tear open ⟨*collar, shirt, etc.*⟩; wrench open ⟨*drawer*⟩; fling open ⟨*window, door*⟩; **die Augen/den Mund ~:** open one's eyes/mouth wide; **b)** *(beschädigen)* tear *or* rip open; rip, tear ⟨*clothes*⟩; break up ⟨*road, soil*⟩; **sich** *(Dat.)* **die Haut/den Ellbogen ~:** gash one's skin/elbow. **2.** *itr. V.; mit sein* ⟨*clothes*⟩ tear, rip; ⟨*seam*⟩ split; ⟨*wound*⟩ open; ⟨*clouds*⟩ break up

auf|reizen *tr. V.* excite ⟨*senses, imagination*⟩; rouse ⟨*passions*⟩; *(provozieren)* provoke

auf·reizend 1. *Adj.* provocative. **2.** *adv.* provocatively

auf|richten 1. *tr. V.* **a)** erect; **den Kopf/Oberkörper ~:** raise one's head/upper body; **jmdn. ~** *(auf die Beine stellen)* help sb. up; **jmdn. im Bett ~:** sit sb. up in bed; **b)** *(trösten)* **jmdn. [wieder] ~:** give fresh heart to sb. **2.** *refl. V.* **a)** stand up [straight]; *(aus gebückter Haltung)* straighten up; *(nach einem Sturz)* get to one's feet; **sich im Bett ~:** sit up in bed; **b)** *(Mut schöpfen)* take heart; **sich an jmdm./etw. [wieder] ~:** take heart from sb./sth.

auf·richtig 1. *Adj.* sincere; honest, sincere ⟨*person, efforts*⟩. **2.** *adv.* sincerely

Auf·richtigkeit die sincerity

auf|ritzen *tr. V.* **a)** *(öffnen)* slit

[open]; **b)** *(verletzen)* scratch; **sich** *(Dat.)* **die Haut/den Arm ~:** scratch oneself/one's arm

auf|rollen *tr. V.* **a)** roll up; **b)** *(auseinanderrollen)* unroll; unfurl ⟨*flag*⟩

auf|rücken *itr. V.; mit sein (auch fig.: befördert werden)* move up

Auf·ruf der a) call; **b)** *(Appell)* appeal (**an** + *Akk.* to)

auf|rufen *unr. tr. V.* **a) jmdn. ~:** call sb.; call sb.'s name; **einen Schüler ~:** call upon a pupil to answer; **b)** *auch itr. (auffordern)* **jmdn. ~, etw. zu tun** call upon sb. to do sth.; **jmdn. zum Widerstand/zum Spenden ~:** call on sb. to resist/for donations; **zum Streik ~:** call a strike

Aufruhr der; ~s, ~e a) *(Rebellion)* revolt; rebellion; **b)** *o. Pl. (Erregung)* turmoil; **jmdn./etw. in ~ versetzen** plunge *or* throw sb./sth. into [a state of] turmoil

auf|rühren *tr. V.* stir up

aufrührerisch 1. *Adj.* **a)** seditious; inflammatory; **b)** rebellious. **2.** *adv. (rebellierend)* seditiously

auf|runden *tr. V.* round off (**auf** + *Akk.* to)

auf|rüsten *tr., itr. V.* arm; **wieder ~:** rearm

Auf·rüstung die armament

auf|rütteln *tr. V. (fig.)* **jmdn. ~:** shake sb. up; **jmds. Gewissen ~:** stir sb.'s conscience; **jmdn. aus seiner Apathie/Lethargie ~:** shake sb. out of his/her apathy/lethargy

aufs *Präp. + Art.* **a) = auf das; b) ~ Klo gehen** *(ugs.)* go to the loo *(Brit. coll.) or (Amer. coll.)* john; **sich ~ Bitten verlegen** resort to appeals

auf|sagen *tr. V.* recite

auf|sammeln *tr. V.* **a)** *(aufheben)* pick *or* gather up; **b)** *(ugs.: aufgreifen)* pick up

aufsässig ['aufzɛsɪç] **1.** *Adj.* **a)** recalcitrant; **b)** *(veralt.: rebellisch)* rebellious. **2.** *adv.* **a)** recalcitrantly; **b)** *(veralt.: rebellisch)* rebelliously

Auf·satz der a) essay; *(in einer Zeitschrift)* article; **b)** *(aufgesetzter Teil)* top *or* upper part

auf|saugen *unr. (auch regelm.) tr. V.* soak up; *(fig.)* absorb

auf|schauen *(südd., österr., schweiz.) s.* **aufblicken**

auf|scheuchen *tr. V.* **a)** put up ⟨*birds, animals*⟩; **b)** *(ugs.: in Unruhe versetzen)* startle

auf|scheuern *tr. V.* chafe; **sich** *(Dat.)* **die Haut/die Fersen ~:** chafe one's skin/heels

auf|schichten *tr. V.* stack up; build [up] ⟨*wall, mound, stack, pile*⟩; pile up ⟨*straw*⟩ [in layers]

auf|schieben *unr. tr. V.* **a)** *(verschieben)* postpone; put off; **aufgeschoben ist nicht aufgehoben** there'll be another opportunity; there is always another time; **b)** slide open ⟨*door, window*⟩; slide *or* draw back ⟨*bolt*⟩

auf|schießen *unr. itr. V.; mit sein* **a)** shoot up; ⟨*flames*⟩ shoot *or* leap up; **b)** *(schnell wachsen)* shoot up; **ein lang aufgeschossener Junge** a tall gangling *or* gangly youth

Auf·schlag der a) *(Aufprall)* impact; **b)** *(Preis~)* extra charge; surcharge; **c)** *(Ärmel~)* cuff; *(Hosen~)* turn-up; *(Revers)* lapel; **d)** *(Tennis usw.)* serve; service; **ich habe ~:** it's my serve

auf|schlagen 1. *unr. itr. V.* **a)** *mit sein* **auf etw.** *(Dat. od. Akk.)* **~:** hit *or* strike sth.; **mit dem Kopf ~:** hit one's head on the ground/pavement *etc.*; **b)** ⟨*price, rent, article*⟩ go up; **c)** *(Tennis usw.)* serve; **Sie schlagen auf!** it's your serve. **2.** *unr. tr. V.* **a)** *(öffnen)* crack ⟨*nut, egg*⟩ [open]; knock a hole in ⟨*ice*⟩; **sich** *(Dat.)* **das Knie/den Kopf ~:** fall and cut one's knee/head; **b)** *(aufblättern)* open ⟨*book, newspaper*⟩; *(zurückschlagen)* turn back ⟨*bedclothes, blanket*⟩; **schlagt Seite 15 auf!** turn to page 15; **c) die Augen ~:** open one's eyes; **d)** *(hoch-, umschlagen)* turn up ⟨*collar, trouser-leg, sleeve*⟩; **e)** *(aufbauen)* set up ⟨*camp*⟩; pitch, put up ⟨*tent*⟩; put up ⟨*bed, hut, scaffolding*⟩; **f) etw. auf einen Betrag/Preis** *usw.* **~:** put sth. on an amount/a price *etc.*

auf|schlecken *tr. V.* lap up

auf|schließen 1. *unr. tr. V.* unlock; **jmdm. die Tür ~:** unlock the door for sb. **2.** *unr. itr. V.* **a)** **[jmdm.] ~:** unlock the door/gate *etc.* [for sb.]; **b)** *(aufrücken)* close up; *(Milit.)* close ranks

auf|schlitzen *tr. V.* slit open; slash open ⟨*stomach, dress*⟩

auf|schluchzen *tr. V.* give a sob

Auf·schluß der information *no pl.*; **über etw.** *(Akk.)* **~ geben** give *or* provide information about sth.; **jmdm. über etw.** *(Akk.)* **~ geben** inform sb. about sth.

auf|schlüsseln *tr. V.* break down (**nach** according to)

aufschluß·reich *Adj.* informative; *(enthüllend)* revealing

auf|schnappen *tr. V. (ugs.)* pick up

auf|schneiden 1. *unr. tr. V.* **a)** cut open; cut ⟨*knot*⟩; lance ⟨*abscess, boil*⟩; **b)** *(zerteilen)* cut; slice. **2.** *unr. itr. V. (ugs. abwertend)* boast, brag (**mit** about)

Auf·schneider der *(ugs. abwertend)* boaster; braggart

auf|schnellen *itr. V.; mit sein* leap up

Auf·schnitt der; *o. Pl.* [assorted] cold meats *pl.*; *(Käse)* [assorted] cheeses *pl.*

auf|schnüren *tr. V.* undo, untie ⟨*knot, parcel, string*⟩; unlace, undo ⟨*shoe, boot, corset*⟩

auf|schrauben *tr. V.* **a)** unscrew; unscrew the top of ⟨*bottle, jar, etc.*⟩; **b)** *(auf etw. schrauben)* screw on (**auf** + *Akk.* to)

auf|schrecken 1. *tr. V.* startle; make ⟨*person*⟩ jump; **jmdn. aus dem Schlaf ~:** startle sb. from his/her sleep. **2.** *regelm., auch unr. itr. V.; mit sein* start [up]; **aus dem Schlaf ~:** awake with a start; start from one's sleep

Auf·schrei der cry; *(stärker)* yell; *(schriller)* scream; **ein ~ [der Empörung** *od.* **Entrüstung]** *(fig.)* an outcry

auf|schreiben *unr. tr. V.* **a)** write down; **[sich** *(Dat.)***] etw. ~** *(etw. notieren)* make a note of sth.; **jmdn. ~** ⟨*policeman*⟩ book sb.; **b)** *(ugs.: verordnen)* prescribe ⟨*medicine*⟩

auf|schreien *unr. itr. V.* cry out; *(stärker)* yell out; *(schrill)* scream

Auf·schrift die inscription

Auf·schub der delay; *(Verschiebung)* postponement; **die Sache duldet keinen ~:** the matter brooks no delay; **jmdm. ~ gewähren** *(Zahlungs~)* allow *or* grant sb. a period of grace

auf|schürfen *tr. V.* **sich** *(Dat.)* **das Knie/die Haut ~:** graze one's knee/oneself

auf|schwatzen, *(bes. südd.)* **auf|schwätzen** *tr. V.* **jmdm. etw. ~:** talk sb. into having sth.; **sich** *(Dat.)* **etw. ~ lassen** be talked into having sth.

auf|schwingen *unr. refl. V.* **sich dazu ~, etw. zu tun** bring oneself to do sth.; **sich zu einem Entschluß ~:** bring oneself to make a decision

Auf·schwung der a) *(Auftrieb)* uplift; **das gab mir neuen ~:** that gave me a lift; **b)** *(gute Entwicklung)* upswing; upturn (*Gen.* in); **c)** *(Turnen)* swing up

auf|sehen *unr. itr. V. s.* **aufblicken**

Aufsehen das; ~s stir; sensation; **~ erregen** cause *or* create a stir *or* sensation; **sich ohne großes**

~ **davonmachen** make off without causing a lot of fuss

aufsehen·erregend *Adj.* sensational

Auf·seher der, Auf·seherin die *(im Gefängnis)* warder *(Brit.)*; [prison] guard *(Amer.)*; *(im Park)* park-keeper; *(im Museum, auf dem Parkplatz)* attendant; *(bei Prüfungen)* invigilator *(Brit.)*; proctor *(Amer.)*; *(auf einem Gut, Sklaven~)* overseer

auf|sein *unr. itr. V.; mit sein; nur im Inf. und Part. zusammengeschrieben (ugs.)* **a)** be open; **b)** *(nicht im Bett sein)* be up

auf|setzen 1. *tr. V.* **a)** put on ⟨*hat, glasses, mask, smile, expression, etc.*⟩; **etw. [auf etw.** *(Akk.)***]** ~: put sth. on [sth.]; **sich** *(Dat.)* **etw.** ~: put sth. on; *s. auch* **Horn a; b)** *(aufs Feuer setzen)* put on; **Wasser [zum Kochen]** ~: put water on [to boil]; **c)** *(verfassen)* draw up ⟨*minutes, contract, will*⟩; **d) jmdn.** ~: sit sb. up; **e)** *(auf eine Unterlage)* set down; put down ⟨*aircraft*⟩; **den Fuß** ~: put one's foot on the ground *or* down. **2.** *itr. V.* ⟨*aircraft*⟩ touch down, land. **3.** *refl. V.* sit up

Auf·sicht die a) *o. Pl.* supervision; *(bei Prüfungen)* invigilation *(Brit.)*; proctoring *(Amer.)*; **[die]** ~ **haben** *od.* **führen** be in charge **(über** + *Akk.* of); *(bei Prüfungen)* invigilate *(Brit.)*; proctor *(Amer.)*; **unter [jmds.]** ~ *(Dat.)* under [sb.'s] supervision; **b)** *(Person)* person in charge; *(Lehrer)* teacher in charge *or* on duty; *(im Museum)* attendant

aufsicht·führend *Adj.; nicht präd.* in charge *postpos.*; supervising ⟨*authority*⟩; ⟨*teacher*⟩ in charge *or* on duty

Aufsichts·rat der *(Wirtsch.)* **a)** *(Gremium)* board of directors; supervisory board; **b)** *(Mitglied)* member of the board [of directors] *or* supervisory board

auf|sitzen *unr. itr. V.; mit sein* **a)** *(auf ein Reittier)* mount; *(auf ein Fahrzeug)* get on; **auf ein Pferd** ~: mount a horse; **b) jmdm./einer Sache** ~: be taken in by sb./sth.

auf|spalten 1. *unr. (auch regelm.) tr. V.* split; *(fig.)* split [up]. **2.** *unr. refl. V.* split

Auf·spaltung die splitting; *(fig.)* splitting [up]

auf|spannen *tr. V.* **a)** open, put up ⟨*umbrella, parasol*⟩; stretch out ⟨*net, jumping-sheet*⟩; put up ⟨*tennis-net, badminton-net, etc.*⟩; **b)** *(befestigen)* stretch, mount ⟨*canvas*⟩ **(auf** + *Akk.* on)

auf|sparen *tr. V. (auch fig.)* save [up]; keep

auf|spielen 1. *refl. V.* **a)** *(ugs. abwertend)* put on airs; **b) sich als Held/Märtyrer** ~: act the hero/martyr. **2.** *itr. V. (musizieren)* play; **zum Tanz** ~: play dance music

auf|spießen *tr. V.* **a)** run ⟨*animal, person*⟩ through; skewer ⟨*piece of meat*⟩; *(mit der Gabel)* take ⟨*piece of food*⟩ on one's fork; *(auf die Hörner nehmen)* gore; **b)** *(befestigen)* pin ⟨*butterfly, insect*⟩

auf|splittern 1. *tr. V.* split up ⟨*party, group, country, etc.*⟩. **2.** *refl. V.* ⟨*party, group, country, etc.*⟩ split up

auf|springen *unr. itr. V.; mit sein* **a)** *(hochspringen)* jump *or* leap up; **b)** *(auf ein Fahrzeug)* jump on; **auf etw.** *(Akk.)* ~: jump on [to] sth.; **c)** *(rissig werden)* crack; ⟨*skin, lips*⟩ crack, chap

auf|sprühen *tr. V.* spray on; **etw. auf etw.** *(Akk.)* ~: spray sth. on [to] sth.

auf|spulen *tr. V.* wind ⟨*cotton, ribbon, fishing-line*⟩ on to a/the reel *or* spool

auf|spüren *tr. V. (auch fig.)* track down

auf|stacheln *tr. V.* incite; **jmdn. zur Revolte/zum Widerstand** ~: incite sb. to revolt/offer resistance

Auf·stand der rebellion; revolt

auf·ständisch *Adj.* rebellious

Aufständische der/die; *adj. Dekl.* rebel

auf|stapeln *tr. V.* stack up

auf|stauen *refl. V.* ⟨*water*⟩ pile up; *(fig.)* ⟨*anger, aggression, bitterness, etc.*⟩ build up

auf|stechen *unr. tr. V.* lance, prick ⟨*boil*⟩; prick ⟨*blister*⟩; lance ⟨*abscess*⟩

auf|stecken 1. *tr. V.* **a) sich** *(Dat.)* **das Haar/die Zöpfe** ~: pin *or* put one's hair/plaits up; **b)** *(ugs.: aufgeben)* **etw.** ~: give sth. up; pack sth. in *(sl.)*. **2.** *itr. V. (ugs.: aufgeben)* retire

auf|stehen *unr. itr. V.* **a)** *mit sein* stand up; *(aus dem Liegen, aus einem Sessel)* get up; **vom Tisch** ~: rise from the table; **b)** *(offenstehen)* ⟨*door, window, etc.*⟩ be open

auf|steigen *unr. itr. V.; mit sein* **a)** *(auf ein Fahrzeug)* get *or* climb on; **auf etw.** *(Akk.)* ~: get *or* climb on [to] sth.; **b)** *(bergan steigen)* climb; **c)** *(hochsteigen)* ⟨*air, smoke, mist, sap, bubble, moon, sun*⟩ rise; **eine ~de Linie** *(fig.)* an ascending line; **d)** *(beruflich, gesellschaftlich)* rise **(zu** to); **zum Direktor** ~: rise to the post of *or* to be manager; **e)** *(hochfliegen)* go up; ⟨*bird*⟩ soar up; **f) in jmdm.** ~ *(geh.)* ⟨*hatred, revulsion, fear, etc.*⟩ rise [up] in sb.; ⟨*memory, thought*⟩ come into sb.'s mind; ⟨*doubt*⟩ arise in sb.'s mind; **g)** *(Sport)* be promoted, go up **(in** + *Akk.* to)

Auf·steiger der a) a social climber; **b)** *(Sport)* promotion side; *(aufgestiegen)* newly promoted side

auf|stellen 1. *tr. V.* **a)** *(hinstellen)* put up **(auf** + *Akk.* on); set up ⟨*skittles*⟩; *(aufrecht stellen)* stand up; **b)** *(postieren)* post; station; **c)** *(bilden)* put together ⟨*team*⟩; raise ⟨*army*⟩; *(Sport)* select, pick ⟨*team*⟩; **d)** *(nominieren)* nominate; put up; *(Sport: auswählen)* select, pick ⟨*player*⟩; **e)** *(errichten)* put up; put up, erect ⟨*scaffolding, monument*⟩; put in, install ⟨*machine*⟩; **f)** *(hochstellen)* erect ⟨*spines*⟩; turn up ⟨*collar*⟩; prick up ⟨*ears*⟩; **g)** *(ausarbeiten)* work out ⟨*programme, budget, plan*⟩; draw up ⟨*statute, balance sheet*⟩; make [out], draw up ⟨*list*⟩; set up ⟨*hypothesis*⟩; establish ⟨*norm*⟩; prepare ⟨*statistics*⟩; devise ⟨*formula*⟩; **h)** *(erzielen)* set up, establish ⟨*record*⟩; **i)** *(formulieren)* put forward ⟨*theory, conjecture, demand*⟩. **2.** *refl. V.* position *or* place oneself; take up position; *(in einer Reihe, zum Tanz)* line up; **sich im Kreis** ~: form a circle

Auf·stellung die a) *s.* **aufstellen 1a:** putting up; setting up; standing up; **b)** *s.* **aufstellen 1e:** putting up; erection; installation; **c)** *s.* **aufstellen 1c:** putting together; raising; selection; picking; **d)** *s.* **aufstellen 1d:** nomination; putting up; selection; picking; **e)** *(Milit.)* ~ **nehmen** *od.* **beziehen** line up; **f)** *s.* **aufstellen 1g:** working out; drawing up; making out; setting up; establishment; preparation; **g)** *(das Erzielen)* setting up; establishment; **h)** *(das Formulieren)* putting forward; **i)** *(Liste)* list; *(Tabelle)* table

Aufstieg der; ~**[e]s,** ~**e a)** climb; ascent; **b)** *(fig.)* rise; **den** ~ **zum Geschäftsleiter/in den Vorstand schaffen** succeed in rising to the position of manager/rising to become a member of the board of directors; **c)** *(Sport)* promotion

Aufstiegs·chance die prospect of promotion

auf|stöbern *tr. V.* **a)** put up ⟨*birds, animals*⟩; **b)** *(entdecken)* track down; run to earth

auf|stocken *tr. V.* **a)** *(auch itr.)* **ein Gebäude ~:** add a storey to a building; **wir haben aufgestockt** we've added another storey; **b)** *(fig.)* increase ‹*capital, budget, funds, pensions*›; build up ‹*supplies*›

auf|stöhnen *itr. V.* groan; **laut/erleichtert ~:** give *or* utter a loud groan/a sigh of relief

auf|stoßen 1. *unr. tr. V.* **a)** *(öffnen)* push open; *(mit einem Fußtritt)* kick open; **b)** *(heftig aufsetzen)* **etw. auf etw.** *(Akk.)* **~:** bang sth. down on sth. **2.** *unr. itr. V.* **a)** belch; burp *(coll.)*; ‹*baby*› bring up wind, *(coll.)* burp; **b)** *auch mit sein* **jmdm. ~:** repeat on sb.; **das könnte dir übel ~** *(fig. ugs.)* you could live to regret that

aufstrebend *Adj.* rising ‹*talent, bourgeoisie, industry*›; ‹*nation, people*› striving for progress; **ein ~er junger Mann** an ambitious and up-and-coming young man

Auf·strich der *(Brot~)* spread

auf|stützen 1. *tr. V.* **die Ellbogen/Arme auf etw.** *(Akk. od. Dat.)* **~:** rest one's elbows/arms on sth.; **mit aufgestütztem Kopf** with one's head resting on one's hands. **2.** *refl. V.* support oneself

auf|suchen *tr. V.* call on, go and see ‹*friends, relatives*›; go to, go and see ‹*doctor*›; **die Toilette ~:** go to the toilet

auf|takeln ['aufta:kln] *refl. V. (ugs. abwertend)* tart *(Brit.) or* doll oneself up *(coll.)*

Auf·takt der a) *(Beginn)* start; **b)** *(Musik)* upbeat; anacrusis

auf|tanken 1. *tr. V.* fill up; refuel ‹*aircraft*›. **2.** *itr. V.* fill up; ‹*aircraft*› refuel

auf|tauchen *itr. V.; mit sein* **a)** *(aus dem Wasser)* surface; ‹*frogman, diver*› surface, come up; **b)** *(sichtbar werden)* appear; *(aus dem Dunkel, dem Nebel)* emerge; appear; **c)** *(kommen, gefunden werden)* turn up; **d)** *(sich ergeben)* ‹*problem, question, difficulties*› crop up, arise

auf|tauen 1. *tr. V.* thaw ‹*ice, frozen food*›; thaw [out] ‹*earth, ground*›. **2.** *itr. V.; mit sein (auch fig.)* thaw; ‹*earth, ground*› thaw [out]

auf|teilen *tr. V.* **a)** *(verteilen)* share out **(unter +** *Akk. od. Dat.* among); **b)** *(aufgliedern)* divide [up] **(in +** *Akk.* into)

Auf·teilung die *s.* **aufteilen a, b:** sharing out **(unter +** *Akk. od. Dat.* among); dividing [up] **(in +** *Akk.* into)

auf|tischen *tr. V.* **a)** serve [up]; **jmdm. etw. ~:** serve sb. with sth.; **b)** *(ugs. abwertend: erzählen)* serve up ‹*excuses, lies, etc.*›; **jmdm. etw. ~:** serve sb. up with sth.

Auftrag der; ~[e]s, Aufträge a) *(Anweisung)* instructions *pl.*; *(Aufgabe)* task; job; **in jmds. ~** *(Dat.) (für jmdn.)* on sb.'s behalf; *(auf jmds. Anweisung)* on sb.'s instructions; **jmdm. den ~ geben** *od.* **erteilen, etw. zu tun** instruct sb. to do sth.; give sb. the job of doing sth.; **einen ~ ausführen** carry out an instruction *or* order; **den ~ haben, etw. zu tun** have been instructed to do sth.; **b)** *(Bestellung)* order; *(bei Künstlern, Architekten usw.)* commission; **ein ~ über etw.** *(Akk.)* an order/a commission for sth.; **etw. in ~ geben** *(Kaufmannsspr.)* order/commission sth. **(bei** from); **c)** *(Mission)* task; mission

auf|tragen 1. *unr. tr. V.* **a) jmdm. ~, etw. zu tun** instruct sb. to do sth.; **er hat mir aufgetragen, dich zu grüßen, er hat mir Grüße aufgetragen** he asked me to pass on his regards; **b)** *(aufstreichen)* apply, put on ‹*paint, make-up, ointment, etc.*›; **etw. auf etw.** *(Akk.)* **~:** apply sth. to sth.; put sth. on sth.; **c)** *(verschleißen)* wear out ‹*clothes*›. **2.** *unr. itr. V.* **a)** ‹*clothes*› be too bulky; **b)** *(ugs.: übertreiben)* **dick ~:** lay it on thick *(sl.)*

Auftrag·geber der client; customer; *(eines Künstlers, Architekten, Schriftstellers usw.)* client

auf|treiben *unr. tr. V.* **a)** *(aufblähen)* bloat; swell; make ‹*dough*› rise; **b)** *(ugs.: ausfindig machen)* get hold of; **ein Quartier ~:** find somewhere to stay

auf|trennen *tr. V.* unpick; undo; unpick ‹*garment*›

auf|treten 1. *unr. itr. V.; mit sein* **a)** tread; **er kann mit dem verletzten Bein nicht ~:** he can't walk on *or* put his weight on his injured leg; **b)** *(sich benehmen)* behave; **forsch/schüchtern ~:** have a forceful/shy manner; **c)** *(fungieren)* appear; **als Zeuge/Kläger ~:** appear as a witness/a plaintiff; **als Vermittler ~:** act as mediator; **d)** *(als Künstler, Sänger usw.)* appear; **sie ist seit Jahren nicht mehr aufgetreten** she hasn't given any public performances for years; **zum ersten Mal ~:** make one's first appearance; **e)** *(die Bühne betreten)* enter; **f)** *(auftauchen)* ‹*problem, question, difficulty*› crop up, arise; ‹*difference of opinion*› arise; **g)** *(vorkommen)* occur; ‹*pest, symptom, danger*› appear. **2.** *unr. tr. V.* kick open ‹*door, gate*›

Auftreten das; ~s a) *(Benehmen)* manner; **b)** *s.* **auftreten 1 g:** occurrence; appearance

Auf·trieb der *o. Pl.* **a)** *(Physik)* buoyancy; *(in der Luft)* lift; **b)** *(Elan, Aufschwung)* impetus; **das hat ihm ~/neuen ~ gegeben** that has given him a lift/given him new impetus

Auf·tritt der a) *s.* **auftreten 1 d:** appearance; **b)** *s.* **auftreten 1 e:** entrance; **c)** *(Szene)* scene

auf|trumpfen *itr. V.* show one's superiority; show how good one is; **„Na siehst du", trumpfte sie auf** there you are, she crowed

auf|tun 1. *unr. refl. V. (geh.: sich öffnen)* open; *(fig.)* open up (*Dat.* before). **2.** *unr. tr. V. (ugs.)* **a)** *(entdecken)* find; **b)** *(servieren)* **jmdm./sich etw. ~:** help sb./oneself to sth. **3.** *unr. itr. V.* **jmdm./sich ~** *(ugs.)* help sb./oneself **(von** to)

auf|türmen 1. *tr. V.* pile up **(zu** into). **2.** *refl. V.* ‹*mountain range*› tower up; *(fig.)* ‹*work, problems, difficulties*› pile up

auf|wachen *itr. V.; mit sein (auch fig.)* wake up, awaken **(aus** from); **aus der Narkose/Ohnmacht ~:** come round from the anaesthetic/faint

auf|wachsen *unr. itr. V.; mit sein* grow up

auf|wallen *itr. V.; mit sein* boil up; **etw. ~ lassen** bring sth. to the boil; **in jmdm. ~** *(fig. geh.)* ‹*joy, tenderness, hatred, passion, etc.*› surge [up] within sb.

Auf·wand der; ~[e]s a) expenditure **(an +** *Dat.* of); *(das Aufgewendete)* cost; expense; **mit einem ~ von 1,5 Mio. Mark** at a cost of 1.5 million marks; **der dazu nötige ~ an Zeit/Kraft** the time/energy needed; **b)** *(Luxus)* extravagance; **[großen] ~ treiben** be [very] extravagant

auf|wärmen 1. *tr. V.* heat *or* warm up ‹*food*›; *(fig. ugs.: wieder erwähnen)* rake *or* drag up. **2.** *refl. V.* warm oneself up

auf|warten *itr. V.* **a)** *(geh.)* **jmdm. mit etw. ~** *(anbieten)* offer sb. sth.; *(vorsetzen)* serve sb. [with] sth.; **b)** *(fig.)* **mit etw. ~:** come up with sth.

aufwärts *Adv.* upwards; *(bergauf)* upwards; uphill; **den Fluß ~:** upstream; **der Fahrstuhl fährt ~:** the lift is going up; **vom Major [an] ~:** from major up

-aufwärts *adv.* **rhein~/fluß~:** up the Rhine/up river *or* upstream
Aufwärts·entwicklung die upward trend
aufwärts|gehen *unr. itr. V.; mit sein (unpers.)* **mit seiner Gesundheit/dem Geschäft geht es ~:** his health is improving/the firm is doing better; **mit ihm geht es ~: he's doing better;** *(gesundheitlich)* he's getting better
Auf·wartung die: jmdm. seine ~ machen *(geh.)* make *or* pay a courtesy call on sb.
auf|wecken *tr. V.* wake [up]; waken; *(fig.)* waken
auf|weichen 1. *tr. V.* soften; **den Boden ~:** make the ground soft *or* sodden. **2.** *itr. V.; mit sein* become soft; soften up
auf|weisen *unr. tr. V.* show; exhibit
auf|wenden *unr. (auch regelm.) tr. V.* use ‹*skill, influence*›; expend ‹*energy, resources*›; spend ‹*money, time*›
auf·wendig 1. *Adj.* lavish; *(kostspielig)* costly; expensive. **2.** *adv.* lavishly; *(kostspielig)* expensively
auf|werfen *unr. tr. V.* **a)** *(aufhäufen)* pile *or* heap up ‹*earth, snow, etc.*›; build, raise ‹*embankment, dam, etc.*›; **b)** *(öffnen)* fling open ‹*door, window*›; **c)** *(ansprechen)* raise ‹*problem, question*›
auf|werten *tr. V.* **a)** *auch itr.* revalue; **b)** *(fig.)* enhance the status of; enhance ‹*standing, reputation, status*›
auf|wickeln *tr. V.* wind up; *(ohne Rolle, Spule)* roll *or* coil up; **jmdm./sich die Haare ~:** put sb.'s/one's hair in curlers
auf|wiegeln ['au̯fvi:gl̩n] *tr. V. (abwertend)* incite; stir up **(gegen** against); **jmdn. zum Aufstand/Streik ~:** incite sb. to rebel/strike
auf|wiegen *unr. tr. V.* make up for; **die Vorteile wiegen die Nachteile auf** the advantages offset the disadvantages
Auf·wind der: im ~ sein *(fig.)* be on the up and up
auf|wirbeln 1. *tr. V.* swirl up; swirl up, raise ‹*dust*›; *s. auch* **Staub. 2.** *itr. V.; mit sein* swirl up
auf|wischen *tr. V.* **a)** wipe *or* mop up; **b)** *auch itr. (säubern)* wipe, *(mit Wasser)* wash ‹*floor*›; **die** *od.* **in der Küche ~:** wipe/wash the kitchen floor
auf|wühlen *tr. V.* churn up ‹*water, sea, mud, soil*›; *(fig.)* stir ‹*person, emotions, passions*› deeply; **ein ~des Erlebnis** *(fig.)* a deeply moving experience
auf|zählen *tr. V.* enumerate; list
Auf·zählung die a) enumeration; listing; **b)** *(Liste)* list
auf|zeichnen *tr. V.* **a)** *(notieren)* record; **b)** *(zeichnen)* draw
Auf·zeichnung die record; *(Magnetband~)* recording; **~en** *(Notizen)* notes
auf|zeigen *tr. V. (nachweisen)* demonstrate; show; *(darlegen)* expound; *(hinweisen auf)* point out; highlight
auf|ziehen 1. *unr. tr. V.* **a)** wind up ‹*clock, toy, etc.*›; **b)** *(öffnen)* pull open ‹*drawer*›; open, draw [back] ‹*curtains*›; undo ‹*zip*›; **c)** *(befestigen)* mount ‹*photograph, print, etc.*› **(auf** + *Akk.* on); stretch ‹*canvas*›; put on ‹*guitar string, violin string, etc.*›; *s. auch* **Saite; d)** *(großziehen)* bring up, raise ‹*children*›; raise, rear ‹*animals*›; raise ‹*plants, vegetables*›; **e)** *(ugs.: gründen)* set up ‹*company, department, business, political party, organization, system*›; **f)** *(ugs.: durchführen)* organize, stage ‹*festival, event, campaign, rally*›; **g)** *(ugs.: necken)* rib *(coll.)*, tease **(mit, wegen** about). **2.** *unr. itr. V.; mit sein* ‹*storm*› gather, come up; ‹*clouds*› gather; ‹*mist, haze*› come up
Auf·zucht die raising; rearing
Auf·zug der a) *(Lift)* lift *(Brit.)*; elevator *(Amer.)*; *(Lasten~, Bau~)* hoist; **b)** *(abwertend: Aufmachung)* get-up; **c)** *(Theater: Akt)* act
auf|zwingen *unr. tr. V.* **jmdm. etw. ~:** force sth. [up]on sb.; **jmdm. seinen Willen ~:** impose one's will [up]on sb.
Aug·apfel der eyeball; **er hütet es wie seinen ~:** it's his most treasured possession
Auge ['au̯gə] **das; ~s, ~n a)** eye; **gute/schlechte ~n haben** have good/poor eyesight; **auf einem ~ blind sein** be blind in one eye; *(fig.)* have a one-sided view; **mit bloßem ~:** with the naked eye; **ihm fallen die ~n zu** his eyelids are drooping; **ganz kleine ~n haben** *(fig.)* be all sleepy; **mit verbundenen ~n** blindfold[ed]; **etw. im ~ haben** have sth. in one's eye; *(fig.: haben wollen)* have one's eye on sth.; **das ~ des Gesetzes** *(fig.: Polizist)* the law *(coll.)*; **ihm/ihr** *usw.* **werden die ~n noch aufgehen** *(fig.)* he/she *etc.* is in for a rude awakening; **[große] ~n machen** *(fig. ugs.)* be wide-eyed; **da wird er ~n machen** *(fig. ugs.)* his eyes will pop out of his head; **ihnen fielen fast die ~n aus dem Kopf** their eyes nearly popped out of their heads; **da blieb kein ~ trocken** *(fig. ugs.)* everyone laughed till they cried; *(es blieb niemand verschont)* no one was safe; **ich traute meinen ~n nicht** *(ugs.)* I couldn't believe my eyes; **ich habe doch hinten keine ~n** *(ugs.)* I haven't got eyes in the back of my head; **ich kann doch meine ~n nicht überall haben!** I can't be looking everywhere at once; **sie hat ihre ~n überall** she doesn't miss a thing; **ein ~** *od.* **beide ~n zudrücken** *(fig.)* turn a blind eye; **ein ~ auf jmdn./etw. geworfen haben** *(fig.)* have taken a liking to sb./have one's eye on sth.; **ein ~ auf jmdn./etw. haben** *(achtgeben)* keep an eye on sb./sth.; **ein ~/ein sicheres ~ für etw. haben** have an eye/a sure eye for sth.; **ich habe ja schließlich ~n im Kopf** *(ugs.)* I'm not blind, you know; **jmdm. die ~n öffnen** *(fig.)* open sb.'s eyes; **jmdn./etw. nicht aus den ~n lassen** not take one's eyes off sb./sth.; not let sb./sth. out of one's sight; **jmdn./etw. aus dem ~** *od.* **den ~n verlieren** lose sight of sb./sth.; *(fig.)* lose contact *or* touch with sb./lose touch with sth.; **aus den ~n, aus dem Sinn!** *(Spr.)* out of sight, out of mind; **jmdn./etw. im ~ behalten** *(fig.)* keep an eye on sb./bear *or* keep sth. in mind; **in jmds. ~n** *(Dat.) (fig.)* to sb.'s mind; in sb.'s opinion; **jmdm. ins ~** *od.* **in die ~n fallen** *od.* **springen** *(fig.)* hit sb. in the eye; **etw. ins ~ fassen** *(fig.)* consider sth.; think about sth.; **einer Sache** *(Dat.)* **ins ~ sehen** *(fig.)* face sth.; **der Wahrheit/Gefahr ins ~ sehen** *(fig.)* face up to the truth/danger; **ins ~ gehen** *(fig. ugs.)* end in disaster; end in failure; **~ um ~, Zahn um Zahn** an eye for an eye, a tooth for a tooth; **unter vier ~n** *(fig.)* in private; **unter jmds. ~n** *(Dat.)* right in front of sb.; right under sb.'s nose; **vor aller ~n** in front of everybody; **jmdm. etw. vor ~n führen** *od.* **halten** *(fig.)* bring sth. home to sb.; **wenn man sich** *(Dat.)* **das mal vor ~n führt** *(fig.)* when you stop and think about it; **b)** *(auf Würfeln, Dominosteinen usw.)* pip; **drei ~n werfen** throw a three; **wie viele ~n hat er geworfen?** how many has he thrown?
äugen ['ɔygn̩] *itr. V.* peer
Augen-: ~arzt der eye specialist; **~aufschlag der** [upward] glance; **~binde die** blindfold
Augen·blick [*auch:* --'-] **der** *s.* [1]**Moment**

augenblicklich [*auch:* --'--] **1.** *Adj.; nicht präd.* **a)** *(unverzüglich)* immediate; **b)** *(gegenwärtig)* present; *(vorübergehend)* temporary. **2.** *adv.* **a)** *(sofort)* immediately; at once; **b)** *(zur Zeit)* at the moment

Augen-: ~braue die eyebrow; **~deckel der** *(ugs.)* eyelid; **~farbe die** colour of one's eyes; **~höhe die** eye-level; **in/auf ~höhe** at/to eye-level; **~klappe die** eye-patch; **~lid das** eyelid; **~maß das;** *o. Pl.* **ein gutes/schlechtes ~maß haben** have a good eye/no eye for distances; **jegliches ~maß verlieren** *(fig.)* lose all sense of proportion; **~merk das: sein ~merk auf jmdn./etw. richten** *od.* **lenken** give one's attention to sb./sth.

Augen·schein der; *o. Pl. (geh.)* **a)** *(Eindruck)* appearance; **dem ~ nach** by all appearances; **dem ersten ~ nach** at first sight; **b)** *(Betrachtung)* inspection; **jmdn./etw. in ~ nehmen** have a close look at sb./sth.; give sb./sth. a close inspection

augen·scheinlich *(geh.)* **1.** *Adj.* evident. **2.** *adv.* evidently

augen-, Augen-: ~weide die; *o. Pl.* feast for the eyes; **~wimper die** eyelash; **~winkel der** corner of one's eye; **~wischerei die** eyewash; **~zeuge der** eyewitness; **~zwinkern das; ~s** wink; **~zwinkernd 1.** *Adj.; nicht präd.* tacit ⟨*agreement*⟩; **2.** *adv.* with a wink

-äugig [-ɔygɪç] *Adj.* -eyed

[1]August [au'gʊst] **der; ~[e]s** *od.* **~, ~e** August; *s. auch* **April**

[2]August ['aʊgʊst] **der; ~** *in* **dummer ~:** clown

Auktion [auk'tsio:n] **die; ~, ~en** auction

Aula ['aula] **die; ~, Aulen** *od.* **~s** *(einer Universität)* [great] hall; *(einer Schule)* [assembly] hall

Au-pair-Mädchen [o'pɛ:r-] **das** au pair [girl]

aus [aus] **1.** *Präp. mit Dat.* **a)** *(räumlich)* out of; **~ dem Bett steigen** get out of bed; **~ der Flasche trinken** drink out of the bottle *or* from the bottle; **b)** *(Herkunft angebend, auch zeitlich)* from; **~ Spanien** from Spain; **er kommt** *od.* **stammt ~ Hamburg** he comes from Hamburg; **jmdm. etw. ~ dem Urlaub mitbringen** bring sth. back from holiday *or (Amer.)* one's vacation for sb.; **~ dem Deutschen ins Englische** from German into English; **c)** *(Veränderung eines Zustandes angebend)* **~ der Mode/Übung sein** be out of fashion/training; **~ tiefem Schlaf erwachen** awake from a deep sleep; **d)** *(Grund, Ursache angebend)* out of; **etw. ~ Erfahrung wissen** know sth. from experience; **~ folgendem Grund** for the following reason; **~ Versehen** inadvertently; by mistake; **~ Furcht vor** for fear of; **~ Spaß/Jux** *(ugs.)* for fun/a laugh; **e)** *(bestehend ~)* of; *(hergestellt ~)* made of; **eine Bank ~ Holz/Stein** a bench made of wood/stone; a wooden/stone bench; **etw. ~ Fertigteilen bauen** build sth. out of prefabricated components; **f)** *(Entwicklung angebend)* **~ ihm ist ein guter Arzt geworden** he made a good doctor; **~ der Sache wird nichts** nothing will come of it; **etwas ~ sich machen** make something of oneself; **~ ihm ist nichts geworden** he never made anything of his life. **2.** *Adv.* **a)** *(ugs.: vorbei, Schluß)* **~ jetzt!** that's enough; **~ und vorbei** over and done with; **b)** *(als Aufforderung: ausschalten)* **„~"** *(an Lichtschaltern)* 'out'; *(an Geräten)* 'off'; **Licht/Radio ~!** lights *pl.* out!/turn the radio off; **c) von ... ~:** from ...; *(fig.)* **von mir ~** *(ugs.)* if you like *or* want; **von sich ~:** of one's own accord; *s. auch* **[2]ein**

Aus das; ~ a) der Ball ging ins ~ *(Tennis)* the ball was out; *(Fußball)* the ball went out of play; **b)** *(fig.)* end

aus|arbeiten *tr. V.* **a)** *(erstellen)* work out, develop ⟨*guidelines, system, method*⟩; prepare, draw up ⟨*agenda, draft, regulations, contract*⟩; prepare ⟨*leaflet*⟩; **b)** *(vollenden)* work out the details of ⟨*plan, proposal, list, lecture, etc.*⟩; elaborate the details of ⟨*picture, drawing*⟩

Ausarbeitung die; ~, ~en a) *s.* **ausarbeiten a:** working out; developing; preparation; drawing up; **b)** *s.* **ausarbeiten b:** working out the details; elaboration of the details

aus|arten *itr. V.; mit sein* degenerate (**in** + *Akk.*, **zu** into)

aus|atmen *itr., tr. V.* breathe out; exhale

aus|baden *tr. V. (ugs.)* carry *or* take the can for *(Brit. sl.)*; take the rap for *(sl.)*

aus|baggern *tr. V.* **a)** excavate ⟨*hole, ditch, etc.*⟩; **b)** *(säubern)* dredge ⟨*channel, river bed, etc.*⟩

aus|balancieren 1. *tr. V. (auch fig.)* balance. **2.** *refl. V.* balance; *(fig.)* balance out

Aus·bau der; ~[e]s a) *(Entfernung)* removal (**aus** from); **b)** *(Erweiterung)* extension; *(einer Straße)* improvement; **ein ~ des Hauses** an extension to the house; **der ~ der Beziehungen zwischen zwei Staaten** the building of closer relations between two states

aus|bauen *tr. V.* **a)** *(entfernen)* remove (**aus** from); **b)** *(erweitern)* extend; improve ⟨*road*⟩; *(fig.)* build up, cultivate ⟨*friendship, relationship*⟩; expand ⟨*theory, knowledge, market*⟩; extend ⟨*one's lead*⟩

aus|bedingen *unr. refl. V.* **sich** *(Dat.)* **etw. ~** *(etw. verlangen)* insist on sth.; *(etw. zur Bedingung machen)* make sth. a condition

aus|bessern *tr. V.* repair; fix *(Amer.)*; mend ⟨*clothes*⟩; touch up ⟨*paintwork*⟩; **einen Schaden an etw.** *(Dat.)* **~:** repair damage to sth.

Aus·besserung die repair

aus|beulen 1. *tr. V.* **a)** make baggy; **ausgebeulte Knie** baggy knees; **b)** *(von Beulen befreien)* remove a dent/the dent/the dents in. **2.** *refl. V.* ⟨*trousers*⟩ go baggy; ⟨*pocket*⟩ bulge

Aus·beute die yield

aus|beuten *tr. V.* exploit

Ausbeuter der; ~s, ~, Ausbeuterin die; ~, ~nen *(abwertend)* exploiter

Ausbeutung die; ~, ~en exploitation

aus|bezahlen *tr. V.* pay [out] ⟨*sum, wages, etc.*⟩; pay off ⟨*employee, worker*⟩; buy out ⟨*partner*⟩; **er bekommt 2000 DM ausbezahlt** his take-home pay is 2,000 marks

aus|bilden 1. *tr. V.* **a)** train; **sich in etw.** *(Dat.)* **~ lassen** take a training in sth.; *(studieren)* study sth.; **sich als** *od.* **zu etw. ~ lassen** train to be sth.; *(studieren)* study to be sth.; **b)** *(fördern)* cultivate, develop ⟨*talent, skill, feeling, etc.*⟩; **c)** *(entwickeln)* develop. **2.** *refl. V.* develop

Ausbilder der; ~s, ~, Ausbilderin die; ~, ~nen instructor

Aus·bildung die a) training; **in der ~ sein** be training; *(an einer Lehranstalt)* be at college; **b)** *(Entwicklung)* development

Ausbildungs-: ~förderung die provision of [education] grants; *(für Berufsschüler, Lehrlinge)* provision of training grants; **~platz der** trainee post; *(für Lehrlinge)* apprenticeship

aus|bitten *unr. refl. V.* **sich** *(Dat.)*

etw. ~: demand sth.; **ich bitte mir Ruhe/mehr Sorgfalt aus** I must insist on silence/that you take more care

aus|blasen *unr. tr. V.* blow out

aus|bleiben *unr. itr. V.; mit sein* **a)** ‹*guests, visitors, customers*› stay away, fail to appear; ‹*order, commission, help, offer, support, rain*› fail to arrive; ‹*effect, disaster, success, reward*› fail to materialize; *(nicht nach Hause kommen)* stay out; **es konnte nicht ~, daß ...:** it was inevitable that ...; **beim Ausbleiben der Regelblutung** if a period is missed; **wenn jahrelang der Regen ausbleibt** if the rains fail year after year; **b)** *(ugs.: ausgeschaltet bleiben)* stay off

aus|blenden **1.** *tr. V. (Rundf., Ferns., Film)* fade out. **2.** *refl. V. (Rundf., Ferns.)* **sich [aus einer Übertragung] ~:** fade oneself out [of a transmission]

Aus·blick der a) view **(auf +** *Akk.* of); **b)** *(Vorausschau)* preview **(auf +** *Akk.* of)

aus|bomben *tr. V.* bomb out

aus|booten *tr. V. (ugs.)* get rid of

aus|borgen *tr. V. (ugs.) s.* **ausleihen**

aus|brechen **1.** *unr. itr. V.; mit sein* **a)** *(entkommen, auch Milit.)* break out **(aus** of); *(fig.)* break free **(aus** from); **b)** *(austreten)* **jmdm. bricht der [kalte] Schweiß aus** sb. breaks into a [cold] sweat; **c)** ‹*volcano*› erupt; **d)** *(beginnen)* break out; ‹*crisis*› break; ‹*misery, despair*› set in; **e) in Gelächter/Weinen ~:** burst out laughing/crying; **in Beifall/Tränen ~:** burst into applause/tears; **in den Ruf ~ „...“** break into the cry, '...'; **in Schweiß ~:** break out into a sweat; **in Zorn/Wut ~:** explode with anger/rage; **f)** *(sich lösen)* ‹*hook, dowel, etc.*› come out. **2.** *unr. tr. V.* **sich** *(Dat.)* **einen Zahn ~:** break a tooth

aus|breiten **1.** *tr. V.* **a)** *(entfalten)* spread [out] ‹*map, cloth, sheet, etc.*›; open out ‹*fan, newspaper*›; *(nebeneinanderlegen)* spread out; **ein Tuch über etw.** *(Akk. od. Dat.)* **~:** spread *or* put a cloth over sth.; **seine Ansichten/sein Leben** *usw.* **vor jmdm. ~** *(fig.)* unfold one's views/life story *etc.* to sb.; **b)** *(ausstrecken)* **die Arme/Flügel ~:** spread one's arms/its wings. **2.** *refl. V.* **a)** spread; **b)** *(ugs.: sich breitmachen)* spread oneself out

aus|brennen **1.** *unr. itr. V.; mit sein* **a)** *(zu Ende brennen)* burn out; **ausgebrannte Kernbrennstäbe** *(fig.)* spent nuclear fuel rods; **b)** *(zerstört werden)* ‹*building, room*› be gutted, be burnt out; ‹*ship, aircraft, vehicle*› be burnt out. **2.** *unr. tr. V.* cauterize ‹*wound*›

Aus·bruch der a) *(Flucht)* escape; *(lit. or fig.)*, break-out *(also Mil.)* **(aus** from); **b)** *(Beginn)* outbreak; **zum ~ kommen** break out; ‹*crisis, storm*› break; **c)** *(Gefühls~)* outburst; *(stärker)* explosion; *(von Wut, Zorn)* eruption; explosion; **d)** *(eines Vulkans)* eruption

aus|brüten *tr. V.* **a)** hatch out; *(im Brutkasten)* incubate; **b)** *(fig. ugs.)* hatch [up] ‹*plot, scheme*›

Ausbuchtung die; ~, ~en bulge

aus|buddeln *tr. V. (ugs.)* dig up

aus|bügeln *tr. V. (fig. ugs.)* iron out ‹*differences, defect*›; make good ‹*mistake*›

aus|buhen *tr. V. (ugs.)* boo

Aus·bund der: ein ~ an *od.* **von Tugend** a paragon *or* model of virtue; **ein ~ an** *od.* **von Bosheit** malice itself *or* personified

Aus·dauer die staying power; stamina; *(Beharrlichkeit)* perseverance; **[beim Lernen] ~/keine ~ haben** have/lack perseverance [when it comes to learning]

aus·dauernd **1.** *Adj.* ‹*runner, swimmer, etc.*› with stamina *or* staying power; *(beharrlich)* persevering; tenacious. **2.** *adv.* perseveringly; tenaciously

aus|dehnen **1.** *tr. V.* **a)** stretch ‹*clothes, piece of elastic*›; *(fig.)* extend ‹*power, borders, trading links*›; expand, increase ‹*capacity*›; **b) etw. auf etw.** *(Akk.)* **~:** extend sth. to sth.; **c)** *(zeitlich)* prolong; **ausgedehnte Ausflüge/Spaziergänge** extended trips/walks. **2.** *refl. V.* **a)** *(räumlich, fig.)* ‹*metal, water, gas, etc.*› expand; ‹*fog, mist, fire, epidemic*› spread; **b)** *(zeitlich)* go on **(bis** until)

Aus·dehnung die a) expansion; *(fig.: der Macht, von Beziehungen, Grenzen)* extension; **b)** *(zeitlich)* prolongation; **c)** *(Ausmaß, Größe)* extent

aus|denken *unr. tr. V.* **sich** *(Dat.)* **etw. ~:** think of sth.; *(erfinden)* think sth. up; *(sich vorstellen)* imagine sth.; **[das ist] nicht auszudenken** it does not bear thinking about

aus|diskutieren *tr. V.* **etw. ~:** discuss sth. fully *or* thoroughly

aus|dörren *tr. V.* dry up; dry up, parch ‹*land, soil*›; parch ‹*throat*›

aus|drehen *tr. V.* turn off

Aus·druck der; ~[e]s, Ausdrücke a) expression; **zum ~ kommen** be expressed; find expression; **etw. zum ~ bringen** express sth.; give expression to sth.; **einer Sache** *(Dat.)* **~ geben** *od.* **verleihen** *(geh.)* express sth.; **b)** *(Wort)* expression; *(Terminus)* term; **du hast dich im ~ vergriffen** your choice of words is most unfortunate; **dumm/ärgerlich** *usw.* **ist gar kein ~:** stupid/angry *etc.* isn't the word for it

aus|drücken **1.** *tr. V.* **a)** squeeze ‹*juice*› out **(aus** of, from); squeeze [out] ‹*lemon, orange, grape, etc.*›; squeeze out ‹*sponge*›; squeeze ‹*boil, pimple*›; **b)** stub out ‹*cigarette*›; pinch out ‹*candle*›; **c)** *(sagen, zum Ausdruck bringen)* express; **anders ausgedrückt** to put it another way; **jmdm. seinen Dank ~:** express one's thanks to sb. **2.** *refl. V.* **a)** *(sich äußern)* express oneself; **b)** *(offenbar werden)* be expressed

ausdrücklich [*od.* -'--] **1.** *Adj.; nicht präd.* express *attrib.* ‹*command, wish, etc.*›; explicit ‹*reservation*›. **2.** *adv.* expressly; ‹*mention*› explicitly

ausdrucks-, Ausdrucks-: **~los** **1.** *Adj.* expressionless‹*face, eyes, etc.*›; unexpressive ‹*style, delivery, etc.*›; **2.** *adv.* ‹*look*› expressionlessly; ‹*write, play*› unexpressively; **~voll** **1.** *Adj.* expressive; **2.** *adv.* expressively; **~weise die** way of expressing oneself

aus|dünsten **1.** *tr. V.* give off. **2.** *itr. V.* transpire

Aus·dünstung die vapour; *(Geruch)* odour

aus·einander *Adv.* **a)** *(voneinander getrennt)* apart; **weit ~ stehende Zähne** widely spaced teeth; **etw. ~ schreiben** write sth. as separate words; **~!** get away from each other!; break it up!; **~ sein** *(ugs.)* ‹*couple*› have separated; have split up; ‹*engagement*› have been broken off, be off; ‹*marriage, relationship, friendship*› have broken up; **b)** *(eines aus dem anderen)* **Behauptungen/Formeln** *usw.* **~ ableiten** deduce propositions/formulae *etc.* one from another

auseinander-, Auseinander-: **~|bekommen** *unr. tr. V.* **etw. ~bekommen** be able to get sth. apart; **~|brechen** **1.** *unr. itr. V.; mit sein (auch fig.)* break up; **2.** *unr. tr. V.* **etw. ~brechen** break sth. up; **~|fallen** *unr. itr. V.; mit sein* fall apart; **~|falten** *tr. V.* unfold; open ‹*newspaper*›; **~|ge-**

hen *unr. itr. V.; mit sein* **a)** *(sich trennen)* part; ⟨*crowd*⟩ disperse; **b)** *(fig.)* ⟨*opinions, views*⟩ differ, diverge; **c)** *(ugs.)* ⟨*relationship, marriage*⟩ break up; **d)** *(ugs.: dick werden)* get round and podgy; *s. auch* **Hefekloß; ~|halten** *unr. tr. V.* keep apart; *(unterscheiden)* distinguish between; **ich kann die beiden Brüder nicht ~halten** I cannot tell the two brothers apart; **~|laufen** *unr. itr. V.; mit sein* run off in different directions; ⟨*crowd*⟩ scatter; **b)** ⟨*paths, roads, etc.*⟩ diverge; **~|leben** *refl. V.* grow apart (**mit** from); **~|nehmen** *unr. tr. V.* **etw. ~nehmen** take sth. apart; dismantle sth.; **~|setzen 1.** *tr. V.* **jmdm. etw. ~setzen** explain sth. to sb.; **2.** *refl. V.* **a) sich mit etw. ~setzen** concern oneself with sth.; **b) sich mit jmdm. ~setzen** have it out with sb.; **~setzung die; ~, ~en a)** *(Beschäftigung)* examination (**mit** of); **b)** *(Streit)* argument; *(zwischen Arbeitgeber und Arbeitnehmer)* dispute; **es kam zu einer ~setzung** an argument/a dispute developed (**wegen** over); **c)** *(Kampfhandlungen, Tätlichkeiten)* clash; **~|treiben** *unr. tr. V.* scatter ⟨*birds, animals*⟩; disperse ⟨*crowd, demonstrators, clouds*⟩

aus|erkiesen *unr. tr. V. (geh., Präsensformen dicht. veralt.)* choose; **zu etw. auserkoren sein** be chosen for sth.

aus·erlesen *Adj. (geh.) s.* **erlesen**

aus|fahren 1. *unr. tr. V.* **a) jmdn. ~** *(im Kinderwagen, Rollstuhl)* take sb. out for a walk; *(im Auto o. ä.)* take sb. out for a drive *or* ride; **b)** *(ausliefern)* deliver ⟨*newspapers, parcels, laundry*⟩; **c)** *(Technik: nach außen bringen)* extend ⟨*aerial, crane, landing-flaps, telescope, etc.*⟩; lower ⟨*undercarriage*⟩; raise ⟨*periscope*⟩; **d)** *(abnutzen)* damage; **ausgefahrene Straßen** rutted and damaged roads; **e)** *(maximal beschleunigen)* drive ⟨*car*⟩ flat out. **2.** *unr. itr. V.; mit sein* **a)** *(spazierenfahren)* go out for a drive; **b)** *(hinausfahren)* ⟨*boat, ship*⟩ put to sea; ⟨*train*⟩ leave, pull out; ⟨*car, lorry*⟩ leave; **aus dem Hafen ~:** leave harbour

Aus·fahrt die a) *(Weg, Straße, Stelle zum Hinausfahren)* exit; *(Autobahn~)* slip road; **b)** *(das Hinausfahren)* departure; **bei der ~ aus dem Hafen tutete das Schiff** as it left [the] harbour, the ship hooted

Aus·fall der a) zum ~ der Zähne führen cause teeth to fall out; **b)** *(das Nichtstattfinden)* cancellation; **c)** *o. Pl. (das Ausscheiden)* retirement; *(vor einem Rennen)* withdrawal; *(Abwesenheit)* absence; **d)** *(Technik) (eines Motors)* failure; *(einer Maschine, eines Autos)* breakdown; *(fig.: eines Organs)* failure; loss of function; **e)** *(Ergebnis)* outcome; result; **f)** *(Einbuße, Verlust)* loss; *(an Einnahmen, Lohn)* drop (*Gen.* in); **g)** *(beleidigende Äußerungen)* attack (**gegen** on)

aus|fallen *unr. itr. V.; mit sein* **a)** *(herausfallen)* fall out; **b)** *(nicht stattfinden)* be cancelled; **etw. ~ lassen** cancel sth.; **c)** *(ausscheiden)* drop out; *(während eines Rennens)* retire; drop out; *(fehlen)* be absent; **d)** *(nicht mehr funktionieren)* ⟨*engine, brakes, signal*⟩ fail; ⟨*machine, car*⟩ break down; **der Strom fiel aus** there was a power failure; **e)** *(geraten)* turn out; **gut/schlecht** *usw.* **~:** turn out well/badly *etc.*; **die Niederlage fiel sehr deutlich aus** the defeat turned out to be *or* was most decisive

ausfallend *Adj.* **[gegen jmdn.] ~ sein/werden** be/become abusive [towards sb.]

Ausfall·straße die *(Verkehrsw.)* main road [leading] out of the/a town/city

aus|fechten *unr. tr. V.* fight out

aus|fegen 1. *tr. V. (bes. nordd.)* sweep out ⟨*room etc.*⟩. **2.** *itr. V.* sweep up

aus|feilen *tr. V.* file down ⟨*key, cogwheel, etc.*⟩; file [out] ⟨*hole*⟩; *(fig.)* polish ⟨*speech, essay, etc.*⟩

aus|fertigen *tr. V. (Amtsspr.)* draw up ⟨*document, agreement, will, etc.*⟩; issue ⟨*passport, certificate*⟩; make out ⟨*bill, receipt*⟩

Aus·fertigung die *(Amtsspr.)* **a)** *s.* **ausfertigen:** drawing up; issuing; making out; **b)** *(Exemplar)* copy; **in doppelter/dreifacher ~:** in duplicate/triplicate; **etw. in vier ~en einreichen** submit four copies of sth.

aus·findig *Adv.* *in* **jmdn./etw. ~ machen** find sb./sth.

aus|fliegen 1. *unr. itr. V.; mit sein* fly out; **die ganze Familie ist ausgeflogen** *(fig. ugs.)* the whole family has gone out [for a walk/drive *etc.*]. **2.** *unr. tr. V.* **jmdn./etw. ~:** fly sb./sth. out

aus|flippen *itr. V.; mit sein (salopp)* freak out *(sl.)*

Aus·flucht die; ~, Ausflüchte [-flʏçtə] excuse; **Ausflüchte machen** make excuses

Aus·flug der outing; *(vom Reisebüro o. ä. organisiert, fig.)* excursion; *(Wanderung)* ramble; walk; **einen ~ machen** go on an outing/on an excursion/for a ramble *or* walk

Ausflügler ['ausfly:klɐ] **der; ~s, ~:** tripper *(Brit.)*; day-tripper; excursionist *(Amer.)*

Ausflugs-: ~dampfer der pleasure steamer; **~lokal das** restaurant/café catering for [day-]trippers; **~ziel das** destination for [day-]trippers

Aus·fluß der a) *o. Pl.* outflow; **b)** *(Med.: Absonderung)* discharge; **c)** *(fig. geh.)* product

aus|formen *tr. V.* shape (**zu** into); give [final] shape to ⟨*text, work of art*⟩

aus|formulieren *tr. V.* formulate ⟨*ideas, questions*⟩; flesh out ⟨*paper*⟩ [from notes]

aus|fragen *tr. V.* **jmdn. ~:** question sb., ask sb. questions; *(verhören)* interrogate sb.

aus|fransen *itr. V.; mit sein* fray

aus|fressen *unr. tr. V.* **[et]was ausgefressen haben** *(ugs.)* have been up to something *(coll.)*

Aus·fuhr die; ~, ~en *s.* **Export a, b**

ausführbar *Adj.* practicable; workable ⟨*plan*⟩

aus|führen *tr. V.* **a)** *(ausgehen mit)* **jmdn. ~:** take sb. out; **b)** *(spazierenführen)* take ⟨*person, animal*⟩ for a walk; **c)** *(exportieren)* export; **d)** *(durchführen)* carry out ⟨*work, repairs, plan, threat*⟩; execute, carry out ⟨*command, order, commission*⟩; execute, perform ⟨*movement, dance-step*⟩; put ⟨*idea, suggestion*⟩ into practice; perform ⟨*operation*⟩; perform, carry out ⟨*experiment, analysis*⟩; **die ~de Gewalt** *(Politik)* the executive power; **e)** *(Fußball, Eishockey usw.)* take ⟨*penalty, free kick, corner*⟩; **f)** *(ausarbeiten)* **etw. ~:** work sth. out in detail *or* fully; **g)** *(erläutern, darlegen)* explain

ausführlich [*auch:* -'--] **1.** *Adj.* detailed, full ⟨*account, description, report, discussion*⟩; thorough, detailed, full ⟨*investigation, debate*⟩; detailed ⟨*introduction, instruction, letter*⟩. **2.** *adv.* in detail; ⟨*investigate*⟩ thoroughly, fully; **etw. ~er/sehr ~ beschreiben** describe sth. in more *or* greater/in great detail

Ausführlichkeit [*auch:* -'---] **die; ~** *s.* **ausführlich 1:** fullness; thoroughness; **mit großer ~:** in great detail

Aus·führung die a) *o. Pl.: s.* **ausführen d:** carrying out; execution; performing; **zur ~ gelangen** *od.* **kommen** *(Papierdt.)* ‹*plan*› be carried out *or* put into effect; **b)** *(Fußball, Eishockey)* taking; **c)** *(Art der Herstellung) (Version)* version; *(Finish)* finish; *(Modell)* model; *(Stil)* style; **d)** *(Darlegung)* explanation; *(Bemerkung)* remark; observation; **e)** *o. Pl. (Ausarbeitung)* **der Entwurf war fertig, jetzt ging es an die ~ des Romans/der Einzelheiten** the draft was ready, and the next task was to work the novel out in detail/to work out the details

aus|füllen *tr. V.* **a)** *(füllen)* fill in ‹*trench, excavation, gravel pit*›; *(zustopfen)* fill ‹*hole, joint*›; **b)** *(beanspruchen, einnehmen)* take up ‹*space, time*›; ‹*person*› fill ‹*chair, doorway, etc.*›; **c)** *(die erforderlichen Angaben eintragen in)* fill in ‹*form, crossword puzzle*›; **d)** *(verbringen)* fill ‹*pause*›; **seine freie Zeit mit etw. ~:** fill [up] one's free time with sth.; **e)** *(innerlich befriedigen)* **jmdn. ~:** fulfil sb.; give sb. fulfilment; **ihr Beruf füllt sie ganz aus** she finds complete fulfilment in her work

Aus·gabe die a) *o. Pl.: s.* **ausgeben a:** distribution; giving out; serving; **die ~ des Essens erfolgt ab ...:** lunch/dinner *etc.* is [served] from ...; **b)** *o. Pl. (das Aushändigen)* issuing; *(von Meldungen, Nachrichten)* release; **c)** *(Geld~)* item of expenditure; expense; **~n** expenditure *sing.* **(für** on); **seine ~n überstiegen seine Einnahmen** his outgoings exceeded his income; **d)** *(Edition, Auflage)* edition

Ausgabe·stelle die *(Schalter)* issuing counter; *(Büro)* issuing office

Aus·gang der a) *o. Pl. (Erlaubnis zum Ausgehen)* time off; *(von Soldaten)* leave; **zwei Tage ~ haben** ‹*servant*› have two days off; ‹*soldier*› have a two-day pass; **bis sechs Uhr ~ haben** ‹*servant*› be free till six; ‹*soldier*› have a pass until six; **b)** *(Tür, Tor)* exit (*Gen.* from); **c)** *(Anat.: Öffnung eines Organs)* outlet; **d)** *(Ende)* end; *(eines Romans, Films usw.)* ending; **e)** *(Ergebnis)* outcome; *(eines Wettbewerbs)* result; **ein Unfall mit tödlichem ~:** an accident with fatal consequences; a fatal accident; **f)** *(Ausgangspunkt)* starting-point; **seinen ~ von etw. nehmen** originate with sth.

Ausgangs-: ~lage die initial position *or* situation; **~position die** initial position; starting position; **~punkt der** starting-point; **~sperre die** curfew; *(für Soldaten)* confinement to barracks; **[eine] ~sperre verhängen** impose a curfew/confine the soldiers/regiment *etc.* to barracks; **~stellung die** starting position

aus|geben *unr. tr. V.* **a)** *(austeilen)* distribute; give out; serve ‹*food, drinks*›; **b)** *(aushändigen, bekanntgeben; Finanzw., Postw.: herausgeben)* issue; **c)** *(verbrauchen)* spend ‹*money*› **(für** on); **d)** *(ugs.: spendieren)* **einen ~:** treat everybody; *(eine Runde geben)* stand a round of drinks *(coll.)*; **ich gebe [dir] einen aus** I'll treat you; **e)** *(fälschlich bezeichnen)* **jmdn./etw. als** *od.* **für jmdn./etw. ~:** pretend sb./sth. is sb./sth.; **sich als jmd./etw.** *od.* **für jmdn./etw. ~:** pretend to be sb./sth.; **f)** *(DV)* output

ausgebucht *Adj.* booked up

ausgebufft ['ausgəbʊft] *Adj. (salopp) (clever)* canny; *(durchtrieben)* crafty

Aus·geburt die *(geh. abwertend)* **a)** *(übles Erzeugnis)* evil product; **eine ~ der Hölle** the spawn of hell; **b)** *(Inbegriff)* epitome

aus·gedient *Adj. (ugs.)* worn out, *(Brit. sl.)* clapped out, *(Amer. sl.)* beat up ‹*vehicle, engine, etc.*›

aus·gefallen *Adj.* unusual

Ausgeflippte der/die; *adj. Dekl. (salopp)* drop-out *(coll.)*

aus·geglichen *Adj.* **a)** *(harmonisch)* balanced, harmonious ‹*structure, façade, etc.*›; well-balanced ‹*person*›; **ein ~es Wesen haben** have an even *or* well-balanced temperament; **b)** *(stabil)* stable; equable ‹*climate*›; **c)** *(Sport)* even

Ausgeglichenheit die; ~ balance; harmony; **die ~ ihres Wesens/ihre ~:** the evenness of her temperament

aus|gehen *unr. itr. V.; mit sein* **a)** go out; **b)** *(fast aufgebraucht sein; auch fig.)* run out; **jmdm. geht etw. aus** sb. is running out of sth.; **ihm geht der Atem** *od.* **die Luft** *od. (ugs.)* **die Puste aus** he is getting short *or* out of breath; he is running out of puff *(Brit. coll.)*; *(fig.: er hat keine Kraft mehr)* he is running out of steam; *(fig.: er ist finanziell am Ende)* he is going broke *(coll.)*; **c)** *(ausfallen)* ‹*hair*› fall out; **d)** *(aufhören zu brennen)* go out; **e)** *(enden)* end; **gut/schlecht ~:** turn out well/badly; ‹*story, film*› end happily/unhappily; **f)** *(herrühren)* **von jmdm./etw. ~:** come from sb./sth.; **g) von etw. ~** *(etw. zugrunde legen)* take sth. as one's starting-point; **du gehst von falschen Voraussetzungen aus** you're starting from false assumptions; **h) auf Abenteuer ~:** look for adventure; **auf Eroberungen ~** *(scherzh.)* set out *or* be aiming to make a few conquests; *s. auch* **leer a; straffrei**

ausgehend *Adj.; nicht präd.* **im ~en Mittelalter** towards the end of the Middle Ages; **das ~e 19. Jahrhundert** the end of *or* closing years of the 19th century

ausgehungert *Adj.* starving; *(abgezehrt)* emaciated

aus·gelassen 1. *Adj.* exuberant ‹*mood, person*›; lively ‹*party, celebration*›; *(wild)* boisterous. **2.** *adv.* exuberantly; *(wild)* boisterously; **es wurde ~ gefeiert** there was a lively party going on

Aus·gelassenheit die exuberance; *(Wildheit)* boisterousness

aus·gemacht 1. *Adj.* **a)** *(beschlossen)* agreed; **b)** *nicht präd. (vollkommen)* complete; complete, utter ‹*nonsense*›. **2.** *adv. (überaus)* extremely; *(ausgesprochen)* decidedly

aus·genommen *Konj.* except [for]; apart from; **alle sind anwesend, ~ er** *od.* **er ~:** everyone is present apart from *or* except [for] him; **er kommt bestimmt, ~ es regnet** he's sure to come, unless it rains

ausgeprägt *Adj.* distinctive ‹*personality, character*›; marked ‹*inclination, tendency, disinclination*›; pronounced ‹*feature, tendency*›

ausgerechnet *Adv.* (*ugs.*) **~ heute/morgen** today/tomorrow of all days; **~ hier** here of all places; **~ du/das** you of all people/that of all things; **~ jetzt kommt er/ muß er kommen** he would have to come [just] now [of all times]

aus·geschlafen *Adj. (ugs.: gewitzt)* wide-awake

aus·geschlossen *Adj.; nicht attr.* **das ist ~:** that is out of the question

aus·geschnitten *Adj.* low-cut ‹*dress, blouse, etc.*›; **ein tief/weit ~es Kleid** a dress with a plunging neckline; a very low-cut dress

aus·gesprochen *Adj.* definite, marked ‹*preference, inclination, resemblance*›; pronounced ‹*dislike*›; marked ‹*contrast*›; **~es Pech/Glück haben** be decidedly unlucky/lucky; **ein ~es Talent für**

etw. a definite talent for sth. **2.** *adv.* decidedly; downright ⟨*stupid, ridiculous, ugly*⟩

aus|gestalten *tr. V.* arrange; *(formulieren)* formulate

aus·gestorben *Adj.* **[wie] ~:** deserted

aus·gewachsen *Adj.* **a)** fully-grown; **b)** *(fig. ugs.)* real ⟨*storm, gale*⟩; full-blown ⟨*scandal*⟩; utter, complete ⟨*nonsense, idiot*⟩

aus·gewogen *Adj.* balanced; [well-]balanced ⟨*personality*⟩

Aus·gewogenheit die; ~: balance

ausgezeichnet [*od.* '--'--] *Adj.* excellent; outstanding ⟨*expert*⟩. **2.** *adv.* excellently; **~ Tennis spielen können** be an excellent tennis player; **sie paßt ~ zu ihm** she suits him very well indeed

ausgiebig ['ausgi:bɪç] **1.** *Adj.* substantial, large ⟨*meal*⟩. **2.** *adv.* ⟨*profit*⟩ handsomely; ⟨*read*⟩ extensively; **von etw. ~ Gebrauch machen** make full use of sth.; **~ frühstücken** eat a substantial breakfast; **etw. ~ betrachten** have a long close look at sth.

aus|gießen *unr. tr. V.* **a)** pour out **(aus** of); **b)** *(leeren)* empty

Ausgleich der; ~[e]s, ~e a) *(von Unregelmäßigkeiten)* evening out; *(von Spannungen)* easing; *(von Differenzen, Gegensätzen)* reconciliation; *(eines Konflikts)* settlement; *(Schadensersatz)* compensation; **einen ~ der verschiedenen Interessen anstreben** strive to reconcile differing interests; **um ~ bemüht sein** be at pains to promote compromise; **als** *od.* **zum ~ für etw.** to make up *or* compensate for sth.; **zum ~ Ihrer Rechnung/Ihres Kontos** in settlement of your invoice/to balance your account; **b)** *o. Pl.* *(Sport)* equalizer; **den ~ erzielen, zum ~ kommen** equalize; score the equalizer

aus|gleichen 1. *unr. tr. V.* even out ⟨*irregularities*⟩; ease ⟨*tensions*⟩; reconcile ⟨*differences of opinion, contradictions*⟩; settle ⟨*conflict*⟩; redress ⟨*injustice*⟩; compensate for ⟨*damage*⟩; equalize, balance ⟨*forces, values*⟩; make up for, compensate for ⟨*misfortune, lack*⟩; **~de Gerechtigkeit** poetic justice. 2. *unr. refl. V. (sich nivellieren)* balance out; *(sich aufheben)* cancel each other out; **das gleicht sich wieder aus** one thing makes up for the other. **3.** *unr. itr. V. (Sport)* equalize; **zum 3:3 ~:** level the score[s] at three all

Ausgleichs·tor das, Ausgleichs·treffer der *(Ballspiele)* equalizer

aus|graben *unr. tr. V.* dig up; *(Archäol.)* excavate; dig out ⟨*trapped person, avalanche victim, etc.*⟩; disinter, exhume ⟨*body, corpse*⟩; *(fig. ugs.)* dig up; dig up, unearth ⟨*old manuscripts, maps, etc.*⟩; **eine alte Geschichte wieder ~** *(fig.)* dig *or* rake up an old story

Aus·grabung die *(Archäol.)* excavation

Ausguck der; ~[e]s, ~e a) *(ugs., Seemannsspr.)* look-out post; **b)** *(Seemannsspr.: Matrose)* look-out

Aus·guß der sink

aus|haben *(ugs.)* **1.** *unr. tr. V.* *(ausgezogen, abgelegt haben)* have taken off. **2.** *unr. itr. V.* *(Schulschluß haben)* have finished school; **wir haben um 12 aus** we finish school at 12

aus|haken 1. *tr. V.* unhook. **2.** *itr. V. (unpers.)* **es hakte bei ihr aus** *(ugs.) (sie begriff es nicht)* she just didn't get it; *(ihre Geduld war zu Ende)* she lost her patience

aus|halten 1. *unr. tr. V.* **a)** stand, bear, endure ⟨*pain, suffering, hunger, blow, noise, misery, heat, etc.*⟩; withstand ⟨*attack, pressure, load, test, wear and tear*⟩; stand up to ⟨*strain, operation*⟩; **er konnte es zu Hause nicht mehr ~:** he couldn't stand it at home any more; **den Vergleich mit jmdm./etw. ~:** stand comparison with sb./sth.; **es läßt sich ~:** it's bearable; I can put up with it; **es ist nicht/nicht mehr zum Aushalten** it is/has become unbearable *or* more than anyone can bear; **b)** *(ugs. abwertend: jmds. Unterhalt bezahlen)* keep; **er läßt sich von seiner Freundin ~:** he gets his girl-friend to keep him. **2.** *unr. itr. V. (durchhalten)* hold out

aus|handeln *tr. V.* negotiate

aus|händigen *tr. V.* **jmdm. etw. ~:** hand sth. over to sb.

Aus·hang der notice; **einen ~ machen** put up a notice

¹aus|hängen *unr. itr. V.* ⟨*notice, timetable, etc.*⟩ have been put up; **am Schwarzen Brett ~:** be up on the notice-board *(Brit.) or (Amer.)* bulletin board

²aus|hängen 1. *tr. V.* **a)** put up ⟨*notice, timetable, etc.*⟩; **b)** take ⟨*door*⟩ off its hinges; take ⟨*window*⟩ out; unhitch ⟨*coupling*⟩. **2.** *refl. V.* ⟨*chain*⟩ come undone *or* unfastened; ⟨*shutter, door, etc.*⟩ come off its hinges

Aushänge·schild das; *Pl.* **~er** [advertising] sign; advertisement *(lit. or fig.)*

aus|harren *itr. V. (geh.)* hold out; **an jmds. Seite** *(Dat.)* **~:** remain at sb.'s side

aus|heben *unr. tr. V.* **a)** dig out ⟨*earth, sand, etc.*⟩; dig ⟨*channel, trench, grave*⟩; **b)** *s.* **²aushängen 1b; c)** *(aus dem Nest nehmen)* steal ⟨*eggs, birds*⟩; *(leeren)* rob ⟨*nest*⟩; *(fig.)* break up ⟨*gang, ring, etc.*⟩; raid ⟨*club, casino, hiding-place*⟩

aus|hecken *tr. V. (ugs.)* hatch ⟨*plan, intrigue*⟩; plan ⟨*attack*⟩

aus|heilen *itr. V.; mit sein* ⟨*injury, organ*⟩ heal [up]; ⟨*patient, illness*⟩ be cured

aus|helfen *unr. itr. V.* help out; **jmdm. ~:** help sb. out

Aus·hilfe die a) *o. Pl.* help; **sie arbeitet in der Kantine zur ~:** she helps out in the canteen; **b)** *s.* **Aushilfskraft**

aushilfs-, Aushilfs-: ~arbeit die temporary work *no pl.*; **~arbeiten** *Pl.* temporary work *sing.*; temporary jobs; **~kraft die** temporary worker; *(in Läden, Gaststätten)* temporary helper *or* assistant; *(Sekretärin)* temporary secretary; temp *(coll.)*; **~weise** *adv.* on a temporary basis

aus|höhlen *tr. V.* hollow out; erode ⟨*rock, cliff, etc.*⟩; *(fig.: untergraben)* undermine

aus|holen *itr. V.* **a) [mit dem Arm] ~:** draw back one's arm; *(zum Schlag)* raise one's arm; **er holte zum Schlag aus** he raised his fist/sword *etc.* to strike; **er holte zum Wurf aus** he drew back his arm ready to throw; **zum Gegenschlag ~** *(fig.)* prepare to counter-attack; **b)** *(fig.: beim Erzählen, Erklären usw.)* go back a long way

aus|horchen *tr. V.* **jmdn. ~:** sound sb. out

aus|kehren *tr., itr. V. (bes. südd.)* *s.* **ausfegen**

aus|kennen *unr. refl. V.* know one's way around *or* about; *(in einer Sache)* know what's what; **sie kennt sich in dieser Stadt aus** she knows her way around the town; **sich [gut] mit/in etw.** *(Dat.)* **~:** know [a lot] about sth.

aus|kippen *tr. V.* **a)** tip out; **b)** *(leeren)* empty

aus|klammern *tr. V.* **a)** *(Math.)* place outside the brackets; **b)** *(beiseite lassen)* leave aside; *(ausschließen)* exclude

Aus·klang der *(geh.)* end; **zum ~ der Saison/des Festes** to end *or* close the season/festival

aus|kleiden 1. *tr. V.* **a)** *(geh.: entkleiden)* undress; **b)** *(innen verkleiden)* line. **2.** *refl. V. (geh.)* undress; disrobe *(formal)*

aus|klingen *unr. itr. V.; mit sein* **a)** ‹*song*› finish; ‹*music, final notes*› die away; **b)** *(fig.)* end

aus|klinken 1. *tr. V.* release. **2.** *refl. V.* release itself/themselves

aus|klopfen *tr. V.* **a)** beat out **(aus +** *Dat.* of); **b)** *(säubern)* beat ‹*carpet*›; knock *or* tap ‹*pipe*› out

aus|klügeln *tr. V.* think out; work out; **ein ausgeklügeltes System** a cleverly devised system

aus|knipsen *tr. V. (ugs.)* switch *or* turn off

aus|knobeln *tr. V. (ugs.)* **a)** *s.* **auswürfeln; b)** *(austüfteln)* work out

aus|kochen *tr. V.* boil; *(keimfrei machen)* sterilize ‹*instruments etc.*› [in boiling water]

aus|kommen *unr. itr. V.; mit sein* **a)** manage, *(coll.)* get by **(mit** on, **ohne** without); **b)** *(sich verstehen)* **mit jmdm. [gut] ~:** get along *or* on [well] with sb.

Auskommen das; ~s livelihood; **sein ~ haben** make a living

aus|kosten *tr. V. (geh.)* **etw. ~:** enjoy sth. to the full

aus|kratzen *tr. V.* scrape out ‹*dirt, remains, etc.*› **(aus** from); scrape [out] ‹*bowl, pan, etc.*›

aus|kugeln *tr. V.* **sich** *(Dat.)* **den Arm** *usw.* **~:** put one's arm *etc.* out [of joint]; dislocate one's arm *etc.*; **jmdm. den Arm ~:** dislocate sb.'s arm

aus|kühlen 1. *tr. V.* chill ‹*person, body*› through. **2.** *itr. V.; mit sein* cool down

aus|kundschaften *tr. V.* find out; find ‹*opportunity*›; track down ‹*refuge, criminal, enemy, etc.*›; spy out ‹*place*›

Auskunft die; ~, Auskünfte a) piece of information; **Auskünfte** information *sing.*; **[jmdm. über etw.** *(Akk.)*] **~ geben** give [sb.] information [about sth.]; **sie gab auf alle Fragen ~:** she answered all the questions; **Auskünfte über jmdn./etw. einholen** obtain information about sb./sth.; **b)** *o. Pl. (Stelle)* information desk/counter/office/centre *etc.*; „~" 'Information'; 'Enquiries' *(Brit.)*; **c)** *(Fernspr.)* directory enquiries *no art. (Brit.)*; directory information *no art. (Amer.)*

Auskunfts-: ~beamte der enquiry office clerk *(Brit.)*; information office clerk *(Amer.)*; **~schalter der** information counter

aus|kuppeln *itr. V.* disengage the clutch; declutch

aus|kurieren *tr. V.* **etw. ~:** heal sth. [completely]

aus|lachen *tr. V.* laugh at

[1]aus|laden *unr. tr. V.* unload

[2]aus|laden *unr. tr. V.* **jmdn. ~:** cancel one's invitation to sb.

Aus·lage die a) *Pl. (Unkosten)* expenses; **unsere ~n für Strom/Heizung** *usw.* our outlay *sing.* on electricity/heating *etc.*; **b)** *(Schaufenster~)* window display; **in der ~:** in the window

Aus·land das; *o. Pl.* foreign countries *pl.*; **im/ins ~:** abroad; **aus dem ~:** from abroad; **die Literatur/Intervention/Hilfe des ~s** foreign literature/intervention/aid; **die Meinung des ~s** opinion abroad; **das ~ hat zurückhaltend reagiert** foreign reaction *or* the reaction of other countries *pl.* was guarded

Ausländer der; ~s, ~: foreigner; alien *(Admin. lang., Law)*

ausländer·feindlich *Adj.* hostile to foreigners *postpos.*

Ausländerin die; ~, ~nen *s.* **Ausländer**

ausländisch *Adj.; nicht präd.* foreign

Auslands-: ~aufenthalt der stay abroad; **~gespräch das** *(Fernspr.)* international call; **~korrespondent der** foreign correspondent; **~reise die** trip abroad

aus|lassen 1. *unr. tr. V.* **a)** *(weglassen)* leave out; leave out, omit ‹*detail, passage, word, etc.*›; **b)** *(versäumen)* miss ‹*chance, opportunity, etc.*›; **c)** *(abreagieren)* vent **(an +** *Dat.* on); **d)** *(ugs.: nicht anziehen, nicht einschalten)* **etw. ~:** leave sth. off. **2.** *unr. refl. V. (abwertend)* talk, speak; *(schriftlich)* write; *(sich verbreiten)* hold forth; **sich im Detail/näher ~:** go into detail/more detail

Auslassung die; ~, ~en a) *(Weglassung)* omission; **b)** *meist Pl. (oft abwertend: Äußerung)* remark

Auslassungs-: ~punkte *Pl.* omission marks; ellipsis *sing.*; **~zeichen das** *(Sprachw.)* apostrophe

aus|lasten *tr. V.* **a)** *(voll ausnutzen)* **etw. ~:** use sth. to full capacity; **ausgelastet sein** ‹*mine, factory, etc.*› be working to full capacity; **b)** *(voll beanspruchen)* fully occupy

aus|latschen *tr. V. (ugs.)* wear ‹*shoes etc.*› out of shape

Auslauf der a) *o. Pl.* **keinen/zuwenig ~ haben** have no/too little chance to run around outside; **der Hund braucht viel ~:** the dog needs plenty of exercise; **b)** *(Raum)* space to run around in; *(für Hühner, Enten usw.)* run; *(für Pferde)* paddock

aus|laufen *unr. itr. V.; mit sein* **a)** run out **(aus** of); ‹*pus*› drain; **b)** *(leer laufen)* empty; ‹*egg*› run out; **c)** *(in See stechen)* sail, set sail **(nach** for); **d)** *(erlöschen)* ‹*contract, agreement, etc.*› run out; **e)** *(nicht fortgesetzt werden)* ‹*model, line*› be dropped *or* discontinued; **etw. ~ lassen** drop *or* discontinue sth.; **f)** *(zum Stillstand kommen)* come *or* roll to a stop

Aus·läufer der a) *(Geogr.)* foothill *usu. in pl.*; **b)** *(Met.) (eines Hochs)* ridge; *(eines Tiefs)* trough

aus|laugen *tr. V.* leach ‹*soil*›; *(fig.)* drain, exhaust, wear out ‹*person*›

aus|leben *tr. V.* give full expression to

aus|lecken *tr. V.* lick out

aus|leeren *tr. V.* empty [out]; empty ‹*ashtray, dustbin, etc.*›

aus|legen *tr. V.* **a)** *(hinlegen)* lay out; display ‹*goods, exhibits*›; lay ‹*bait*›; put down ‹*poison*›; set ‹*trap, net*›; **b)** *(bedecken mit)* **etw. mit Fliesen/Teppichboden ~:** tile/carpet sth.; **einen Schrank [mit Papier] ~:** line a cupboard [with paper]; **c)** *(leihen)* lend; **jmdm. etw.** *od.* **etw. für jmdn. ~:** lend sb. sth.; lend sth. to sb.; **d)** *(interpretieren)* interpret; **etw. falsch ~:** misinterpret sth.; **etw. als Furcht ~:** take sth. to be fear

Auslegung die; ~, ~en interpretation

aus|leiern *(ugs.)* **1.** *itr. V.; mit sein* wear out; ‹*clothes*› go baggy; **ausgeleiert** worn out; baggy ‹*pullover, trousers, etc.*›. **2.** *tr. V.* wear out; make ‹*pullover, trousers, etc.*› go baggy; make ‹*rubber band*› lose its stretch

Ausleihe die; ~, ~n a) *o. Pl. (das Ausleihen)* lending; **b)** *(Stelle)* issue desk

aus|leihen *unr. tr. V.* **a) jmdm.** *od.* **an jmdn. etw. ~:** lend sb. sth.; lend sth. to sb.; **b)** *s.* **leihen b**

aus|lernen *itr. V.* finish one's apprenticeship; **man lernt nie aus** *(Spr.)* you learn something new every day

Aus·lese die *o. Pl.* selection

[1]aus|lesen *unr. tr. V.* pick out **(aus** from)

[2]aus|lesen *unr. tr. V. (ugs.)* finish [reading]

aus|liefern *tr. V.* **a)** *(übergeben)* **jmdm. etw.** *od.* **etw. an jmdn. ~:**

hand sth. over to sb.; **jmdn. an ein Land ~:** extradite sb. to a country; **jmdm./einer Sache ausgeliefert sein** *(fig.)* be at the mercy of sb./sth.; **b)** *auch itr. (Kaufmannsspr.: liefern)* deliver

Aus·lieferung die a) *(Übergabe)* handing over; *(an ein Land)* extradition; **jmds. ~ fordern** demand that sb. be handed over/extradited; **b)** *(Kaufmannsspr.: Lieferung)* delivery

aus|liegen *unr. itr. V.* be displayed; ‹*newspapers, plans, etc.*› be laid out, be available

aus|löffeln *tr. V.* **a)** spoon up [all of] ‹*soup etc.*›; **jetzt muß er die Suppe ~[, die er sich eingebrockt hat]** *(fig.)* he's made his [own] bed and now he must lie in it; **b)** spoon up everything out of ‹*plate, bowl, etc.*›

aus|löschen *tr. V.* **a)** extinguish, put out ‹*fire, lamp*›; snuff, put out, extinguish ‹*candle*›; *(fig.)* extinguish ‹*life*›; **b)** *(beseitigen)* rub out, erase ‹*drawing, writing*›; ‹*wind, rain*› obliterate ‹*tracks, writing*›; *(fig.)* obliterate, wipe out ‹*memory*›; wipe out ‹*people, population*›

aus|losen *tr. V.* **etw. ~:** draw lots for sth.; **es wurde ausgelost, wer beginnt** lots were drawn to decide who would start; **den Gewinner ~:** draw lots to decide the winner

aus|lösen *tr. V.* **a)** trigger ‹*mechanism, device, etc.*›; set off, trigger ‹*alarm*›; release ‹*camera shutter*›; **b)** provoke ‹*discussion, anger, laughter, reaction, outrage, heart attack, sympathy*›; cause ‹*sorrow, horror, surprise, disappointment, panic, war*›; excite, arouse ‹*interest, enthusiasm*›; trigger [off] ‹*crisis, chain of events, rebellion, strike*›

Auslöser der; ~s, ~ a) *(Fot.)* shutter release; **b)** *(fig.)* trigger

Aus·losung die draw

Aus·lösung die a) *(eines Mechanismus)* triggering; *(eines Alarms)* setting off; triggering; **b)** *s.* **auslösen b:** provocation; causing; exciting; arousal; triggering [off]

aus|loten *tr. V. (Seew.)* sound the depth of; sound, plumb ‹*depth*›; *(fig.)* sound out ‹*intentions*›; **ein Problem ~** *(fig.)* try to get to the bottom of a problem

aus|lutschen *tr. V. (ugs.)* suck out ‹*juice*›; suck the juice from ‹*orange, lemon, etc.*›

aus|machen *tr. V.* **a)** *(ugs.)* put out ‹*light, fire, cigarette, candle*›; turn *or* switch off ‹*television, radio, hi-fi*›; turn off ‹*gas*›; **b)** *(vereinbaren)* agree; **c)** *(auszeichnen, kennzeichnen)* make up; constitute; **d)** *(ins Gewicht fallen)* make a difference; **wenig/nichts/viel ~:** make little/no/a great *or* big difference; **e)** *(stören)* **das macht mir nichts aus** I don't mind [that]; **macht es Ihnen etwas aus, wenn ...?** would you mind if ...?; **f)** *(klären)* settle; **etw. mit sich allein/mit seinem Gewissen ~:** sort sth. out for oneself/with one's conscience; **g)** *(erkennen)* make out; **h)** *(betragen)* come to; **der Zeitunterschied/die Entfernung macht ... aus** the time difference/distance is ...

aus|malen 1. *tr. V.* **a)** *(mit Farbe ausfüllen)* colour in; **b)** *(mit Malereien ausschmücken)* **das Innere einer Kirche ~:** decorate the interior of a church with murals/frescoes *etc.*; **c)** *(schildern)* describe. **2.** *refl. V.* **sich** *(Dat.)* **etw. ~:** picture sth. to oneself; imagine sth.

Aus·maß das size; dimensions *pl.*; **gewaltige ~e haben** be of huge *or* vast dimensions; **eine Katastrophe unvorstellbaren ~es** a disaster on an unimaginable scale

aus|mergeln ['ausmɛrgl̩n] *tr. V.* emaciate; **ausgemergelt** gaunt, emaciated ‹*face, body*›

aus|merzen ['ausmɛrtsn̩] *tr. V.* eradicate ‹*pests, insects, weeds, etc.*›; eliminate ‹*errors, slips, offensive passages*›

aus|messen *unr. tr. V.* measure up

aus|misten *tr. (auch itr.) V.* **a)** muck out; **b)** *(fig. ugs.)* clear out

aus|mustern *tr. V.* **a)** *(Milit.)* **jmdn. ~:** reject sb. as unfit [for service]; **b)** *(fig.)* take ‹*vehicle, machine*› out of service

Aus·nahme die; ~, ~n exception; **mit ~ von Peter/des Pfarrers** with the exception of Peter/of the priest; **ohne ~:** without exception; **bei jmdm. eine ~ machen** make an exception in sb.'s case; **~n bestätigen die Regel** the exception proves the rule

Ausnahme·fall der exceptional case

ausnahms-: ~los *Adv.* without exception; **~weise** *Adv.* by way of *or* as an exception; **Dürfen wir mitkommen? – Ausnahmsweise ja** May we come too? – Yes, just this once; **kann ich heute ~weise mal früher weggehen?** can I go earlier today, as a special exception?

aus|nehmen 1. *unr. tr. V.* **a)** gut ‹*fish, rabbit, chicken*›; **b)** *(ausschließen)* exclude; *(gesondert behandeln)* make an exception of; **jeder irrt sich einmal, ich nehme mich nicht aus** everyone makes mistakes once in a while, and I'm no exception; **c)** *(die Eier herausnehmen aus)* rob ‹*nest*›; **d)** *(ugs. abwertend: neppen)* **jmdn. ~:** fleece sb. **2.** *unr. refl. V. (geh.)* look; *(sich anhören)* sound

ausnehmend *(geh.) adv.* exceptionally

aus|nüchtern *tr., itr., refl. V.* sober up

Ausnüchterung die; ~, ~en sobering up; **jmdn. zur ~ auf die Wache bringen** take sb. to the [police] station to sober up

aus|nutzen, *(bes. südd., österr.)* **aus|nützen** *tr. V.* **a)** *(nutzen)* **etw. [voll] ~:** take [full] advantage of sth.; make [full] use of sth.; **b)** *(Vorteil ziehen aus)* take advantage of; *(ausbeuten)* exploit

aus|packen 1. *tr., itr. V.* unpack (**aus** from); unwrap ‹*present*›. **2.** *itr. V. (ugs.)* talk *(coll.)*; squeal *(sl.)*

aus|peitschen *tr. V.* whip; *(auf Grund eines Gerichtsurteils)* flog

aus|pfeifen *unr. tr. V.* **jmdn./etw. ~:** give sb./sth. the bird

aus|pflanzen *tr. V.* plant out

aus|plaudern *tr. V.* let out; blab

aus|plündern *tr. V.* **a)** *(ausrauben)* **jmdn./etw. ~:** rob sb./sth. [of everything]; **b)** *(völlig plündern, auch fig.)* plunder

aus|powern [-po:vɐn] *tr. V. (ugs. abwertend)* bleed ‹*country, nation*› dry *or* white; exploit ‹*workers, masses*›; *(fig.)* impoverish ‹*soil*›

aus|prägen *refl. V.* develop; ‹*peculiarity*› become more pronounced

aus|pressen *tr. V.* press *or* squeeze out ‹*juice*›; squeeze ‹*orange, lemon*›; *(mit einer Presse)* press the juice from ‹*grapes etc.*›; press out ‹*juice, oil*›; *(fig.: ausbeuten)* squeeze ‹*country, population, etc.*› [dry]; *s. auch* **Zitrone**

aus|probieren *tr. V.* try out

Aus·puff der exhaust

aus|pumpen *tr. V.* pump out

aus|pusten *tr. V. (ugs.)* blow out; blow ‹*egg*›

Aus·putzer der; ~s, ~ *(Fußball)* sweeper

aus|quartieren *tr. V.* move out; billet out ‹*troops*›

aus|quatschen *(salopp)* **1.** *tr. V.* let out; blab. **2.** *refl. V.* **sich mit jmdm. ~:** have a really *or (Amer.)* real good chat with sb.; **sich bei**

jmdm. ~: have a heart-to-heart with sb. *(coll.)*

aus|quetschen *tr. V.* **a)** squeeze out ⟨*juice*⟩; squeeze ⟨*orange, lemon, etc.*⟩; **b)** *(ugs.: ausfragen)* grill; *(aus Neugier)* pump; *s. auch* **Zitrone**

aus|radieren *tr. V.* rub out; erase; *(fig.)* annihilate, wipe out ⟨*village, city, etc.*⟩; liquidate ⟨*person*⟩

aus|rangieren *tr. V. (ugs.)* throw out; discard; scrap ⟨*vehicle, machine*⟩

aus|rasieren *tr. V.* shave

aus|rasten *itr. V.; mit sein (Technik)* disengage; **er rastete aus, es rastete bei ihm aus** *(fig. salopp)* something snapped in him

aus|rauben *tr. V.* rob

aus|räuchern *tr. V. (auch fig.)* smoke out; fumigate ⟨*room*⟩

aus|räumen 1. *tr. V.* **a)** clear out; **b)** *(fig.)* clear up; dispel ⟨*prejudice, suspicion, misgivings*⟩. **2.** *itr. V.* clear everything out

aus|rechnen *tr. V.* work out; *(errechnen)* work out; calculate; **das kannst du dir leicht ~** *(fig. ugs.)* you can easily work that out [for yourself]; **sich** *(Dat.)* **Vorteile/gute Chancen ~:** reckon that one has advantages/good prospects

Aus·rede die excuse

aus|reden 1. *itr. V.* finish [speaking]. **2.** *tr. V.* **jmdm. etw. ~:** talk sb. out of sth.

aus|reichen *itr. V.* be enough *or* sufficient (**zu** for); **die Zeit/der Platz reicht [nicht] aus** there's [not] enough *or* sufficient time/space

ausreichend 1. *Adj.* sufficient; enough; *(als Note)* fair. **2.** *adv.* sufficiently

aus|reifen *itr. V.; mit sein* ⟨*fruit, cereal, etc.*⟩ ripen fully; ⟨*cheese, wine, etc.*⟩ mature fully

Aus·reise die: vor/bei der ~: before/when leaving the country; **jmdm. die ~ verweigern** refuse sb. permission to leave [the/a country]

aus|reisen *itr. V.; mit sein* leave [the country]; **nach/aus Italien ~:** go to/leave Italy

aus|reißen 1. *unr. tr. V.* tear out; pull out ⟨*plants, weeds*⟩. **2.** *unr. itr. V.; mit sein* **a)** ⟨*[button] hole etc.*⟩ tear; **b)** *(ugs.: weglaufen)* run away (**von,** *Dat.* from)

Aus·reißer der, Ausreißerin die; ~, ~nen *(ugs.)* runaway

aus|reiten *unr. itr. V.; mit sein* go for a ride; go riding

aus|renken *tr. V.* dislocate; **jmdm./sich den Arm ~:** dislocate sb.'s/one's arm; **sich [nach jmdm.] den Hals ~** *(ugs.)* crane one's neck [to look for sb.]

aus|richten 1. *tr. V.* **a)** *(übermitteln)* **jmdm. etw. ~:** tell sb. sth.; **ich werde es ~:** I'll pass the message on; **kann ich ihm etwas ~?** can I give him a message?; **richte ihr einen Gruß [von mir] aus** give her my regards; **b)** *(einheitlich anordnen)* line up; **c)** *(fig.)* **etw. auf jmdn./etw. ~:** orientate sth. towards sb./sth.; **etw. nach** *od.* **an jmdm./etw. ~:** gear sth. to sb./sth.; **d)** *(erreichen)* accomplish; achieve; **bei jmdm. wenig/nichts ~ können** not be able to get very far/anywhere with sb.; **gegen jmdn./etw. etwas ~ können** be able to do something against sb./sth. **2.** *refl. V.* **a)** *(Milit.)* dress ranks; **sich nach jmdm. ~:** line [oneself] up with sb.; **b) sich an einem Vorbild ~:** follow an example

Aus·ritt der ride [out]

aus|rollen *tr. V.* roll out

aus|rotten *tr. V.* eradicate ⟨*weeds, vermin, etc.*⟩; *(fig.)* wipe out ⟨*family, enemy, species, etc.*⟩; eradicate, stamp out ⟨*superstition, idea, evil, etc.*⟩

aus|rücken *itr. V.; mit sein* **a)** *(bes. Milit.)* move out; ⟨*fire brigade, police*⟩ turn out; **b)** *(ugs.: weglaufen)* make off; **von zu Hause ~:** run away from home

Aus·ruf der cry

aus|rufen *unr. tr. V.* **a)** call out; **„Schön!" rief er aus** 'Lovely', he exclaimed; **jmdn. ~ lassen** have a call put out for sb.; *(im Hotel)* have sb. paged; **die Haltestellen ~:** call out [the names of] the stops; **seine Waren ~:** cry one's wares; **b)** *(offiziell verkünden)* proclaim; declare ⟨*state of emergency*⟩; call ⟨*strike*⟩; **jmdn. zum König/als Präsidenten ~:** proclaim sb. king/president

Ausrufe-: ~satz der *(Sprachw.)* exclamation; exclamatory clause; **~zeichen das** exclamation mark

aus|ruhen 1. *refl., itr. V.* have a rest; **[sich] ein wenig/richtig ~:** rest a little/have a proper *or* good rest; **ausgeruht sein** be rested; *s. auch* **Lorbeer c. 2.** *tr. V. (ruhen lassen)* rest

aus|rüsten *tr. V.* equip; equip, fit out ⟨*ship*⟩; **ein Auto mit Sicherheitsgurten ~:** fit safety belts to a car; fit a car with safety belts

Aus·rüstung die a) *o. Pl. (das Ausrüsten)* equipping; *(von Schiffen)* equipping; fitting out; **die ~ des Autos mit Gurten** *usw.* the fitting of belts *etc.* to the car; **b)** *(Ausrüstungsgegenstände)* equipment *no pl.*; **eine neue ~:** a new set of equipment

aus|rutschen *itr. V.; mit sein* slip; *(fig.)* put one's foot in it

Ausrutscher der; ~s, ~ *(ugs., auch fig.)* slip

Aus·sage die a) statement; stated view; **nach ~ der Experten** according to what the experts say; **b)** *(vor Gericht, bei der Polizei)* statement; **eine ~ machen** make a statement; give evidence; **die ~ verweigern** refuse to make a statement; *(vor Gericht)* refuse to give evidence; **c)** *(geistiger Gehalt)* message

aussage·kräftig *Adj.* meaningful

aus|sagen 1. *tr. V.* **a)** say; **damit wird ausgesagt, daß ...:** this expresses the idea that ...; **b)** *(fig.)* ⟨*picture, novel, etc.*⟩ express; **c)** *(vor Gericht, vor der Polizei)* **~, daß ...:** state that ...; *(unter Eid)* testify that ... **2.** *itr. V.* make a statement; *(unter Eid)* testify

aus|sägen *tr. V.* saw out

Aussage·satz der *(Sprachw.)* affirmative clause

Aus·satz der; *o. Pl. (veralt.)* leprosy

Aussätzige der/die; *adj. Dekl. (veralt., fig.)* leper

aus|saufen *unr. tr. (auch itr.) V.* **a)** ⟨*animal*⟩ drink [up] ⟨*water etc.*⟩; empty ⟨*trough etc.*⟩; **b)** *(derb)* drink

aus|saugen *regelm. (geh. auch unr.) tr. V.* suck out (**aus** of); *(leer saugen)* suck dry; **eine Apfelsine ~:** suck the juice from an orange; **eine Wunde ~:** suck the poison/dirt *etc.* out of a wound

aus|schaben *tr. V.* **a)** *s.* **auskratzen; b)** *(Med.)* curette

Ausschabung die; ~, ~en *(Med.)* curettage

aus|schalten *tr. V.* **a)** *(abstellen)* switch *or* turn off; **b)** *(ausschließen)* eliminate; exclude ⟨*emotion, influence*⟩; dismiss ⟨*doubt, objection*⟩; shut out ⟨*feeling, thought*⟩

Aus·schank der; ~[e]s, Ausschänke ['aʊsʃɛŋkə] **a)** *o. Pl.* serving; **„Kein ~ an Jugendliche unter 16 Jahren"** 'persons under sixteen will not be served with alcoholic drinks'; **b)** *(Schanktisch)* bar; counter

Aus·schau *in* **nach jmdm./etw. ~ halten** look out for *or* keep a look-out for sb./sth.

aus|schauen *itr. V.* **nach jmdm./**

etw. ~: look out for *or* keep a look-out for sb./sth.

aus|scheiden 1. *unr. itr. V.; mit sein* **a) aus etw.** ~: leave sth.; **aus dem Amt** ~: leave office; **b)** *(Sport)* be eliminated; *(aufgeben)* retire; **c)** *(nicht in Betracht kommen)* **diese Möglichkeit/dieser Kandidat scheidet aus** this possibility/candidate has to be ruled out. **2.** *unr. tr. V. (Physiol.)* excrete ‹*waste*›; eliminate, expel ‹*poison*›; exude ‹*sweat*›; *(Chem.)* precipitate

Aus·scheidung die a) *o. Pl. s.* **ausscheiden 2:** excretion; elimination; expulsion; exudation; precipitation; **b)** *Pl. (Physiol.)* excreta; **c)** *(Sport)* qualifier

Ausscheidungs-: **~organ das** *(Physiol.)* excretory organ; **~spiel das** *(Sport)* qualifying game *or* match

aus|schenken *tr. V.* serve

aus|scheren *itr. V.; mit sein* ‹*car, driver*› pull out; ‹*ship*› break out of [the] line; ‹*aircraft*› peel off, break formation; *(fig.)* pull out

aus|schiffen *tr. V.* disembark ‹*passengers*›; unload ‹*cargo*›

aus|schildern *tr. V.* signpost

aus|schimpfen *tr. V.* **jmdn.** ~: give sb. a telling-off; tell sb. off

aus|schlachten *tr. V.* **a)** *(ugs.)* cannibalize ‹*machine, vehicle*›; break ‹*vehicle*› for spares; **b)** *(fig. ugs. abwertend)* exploit; **etw. politisch** ~: make political capital out of sth.

aus|schlafen 1. *unr. itr., refl. V.* have a good *or* proper sleep; **ich war nicht ausgeschlafen** I hadn't had enough sleep. **2.** *unr. tr. V.* **seinen Rausch** ~: sleep off the effects of alcohol

Aus·schlag der a) *(Haut~)* rash; **[einen]** ~ **bekommen** break out *or* come out in a rash; **b)** *(eines Zeigers)* deflection; *(eines Pendels)* swing; **den** ~ **geben** *(fig.)* turn *or* tip the scales *(fig.)*; **das gab den** ~ **für seine Entscheidung** that was the crucial factor in his decision; that decided him

aus|schlagen 1. *unr. tr. V.* **a)** knock out; **jmdm. einen Zahn** ~: knock one of sb.'s teeth out; **b)** *(ablehnen)* turn down; reject; refuse ‹*inheritance*›. **2.** *unr. itr. V.* **a)** ‹*horse*› kick; **b)** *auch mit sein* ‹*needle, pointer*› be deflected, swing; ‹*divining-rod*› dip; ‹*scales*› turn; ‹*pendulum*› swing; **c)** *auch mit sein (sprießen)* come out [in bud]

ausschlag·gebend *Adj.* decisive; **das war** ~ **für seine Entscheidung** that was the crucial factor in his decision; that decided him

aus|schließen *unr. tr. V.* **a)** exclude (**aus** from); **er schließt sich von allem aus** he won't join in anything; **b)** *(ausstoßen)* expel (**aus** from); **c)** *(als nicht möglich, nicht gegeben annehmen, unmöglich machen)* rule out; **einander** ~: be mutually exclusive; **d)** *(aussperren)* lock out; *s. auch* **ausgeschlossen**

aus·schließlich [*od.* '-'--, -'--] **1.** *Adj.; nicht präd.* exclusive; exclusive, sole ‹*concern, right*›. **2.** *Adv. (nur)* exclusively; **das ist** ~ **sein Verdienst** the credit is his alone. **3.** *Präp. mit Gen.* excluding; exclusive of

Ausschließlichkeit [*od.* '-'---] **die;** ~: exclusiveness

aus|schlüpfen *itr. V.; mit sein* hatch [out]; ‹*butterfly*› emerge

aus|schlürfen *tr. V.* sip ‹*drink*› noisily; suck ‹*oyster, egg*›; **sein Glas/seine Tasse** ~: empty one's glass/cup noisily

Aus·schluß der a) exclusion (**von** from); **unter** ~ **der Öffentlichkeit** with the public excluded; *(Rechtsw.)* in camera; **b)** *(Ausstoßung)* expulsion (**aus** from)

aus|schmücken *tr. V.* decorate; deck out; *(fig.)* embellish ‹*story, incident, report, etc.*›

Ausschmückung die; ~, ~**en** *s.* **ausschmücken:** decoration; decking out; embellishment

aus|schneiden *unr. tr. V.* cut out

Aus·schnitt der a) *(Zeitungs~)* cutting, clipping; **b)** *(Hals~)* neck; **ein tiefer** ~: a plunging neck-line; **c)** *(Teil, Auszug)* part; *(eines Textes)* excerpt; *(eines Films)* clip; excerpt; *(Bild~)* detail; **d)** *(Kreis~)* sector; **e)** *(Loch)* [cut-out] opening

aus|schöpfen *tr. V.* **a)** scoop out (**aus** from); *(mit dem Schöpflöffel)* ladle out (**aus** of); **b)** *(leeren)* bale the water *etc.* out of ‹*basin, bath, tank, etc.*›; bale ‹*boat*› out; **c)** *(fig.: ausnutzen)* exhaust

aus|schreiben *unr. tr. V.* **a)** write ‹*word, name, etc.*› out in full; write ‹*number*› out in words; **b)** *(ausstellen)* write *or* make out ‹*cheque, invoice, receipt*›; **c)** *(bekanntgeben)* announce, call ‹*election, meeting*›; advertise ‹*flat, job*›; put ‹*supply order etc.*› out to tender; organize ‹*competition*›

Aus·schreibung die *s.* **ausschreiben c:** announcement; calling; advertisement; invitation to tender; organization

aus|schreiten *(geh.) unr. itr. V.; mit sein* step out

Ausschreitung die; ~, ~**en;** *meist Pl.* act of violence; **es kam zu** ~**en** violence broke out

Aus·schuß der a) *(Kommission)* committee; **b)** *o. Pl. (Aussortiertes)* rejects *pl.*

aus|schütteln *tr. V.* shake ‹*dust, table-cloth, etc.*› out

aus|schütten *tr. V.* **a)** tip out ‹*water, sand, etc.*›; *(leeren)* empty ‹*bucket, bowl, container*›; *(verschütten)* spill; **sich vor Lachen** ~ **[wollen]** *(ugs.)* split one's sides laughing; die laughing *(coll.)*; *s. auch* **Herz b; b)** *(auszahlen)* distribute ‹*dividends, prizes, etc.*›

aus|schwärmen *itr. V.; mit sein (auch fig.)* swarm out; ‹*soldiers*› deploy; *(fächerartig)* fan out

ausschweifend 1. *Adj.* wild ‹*imagination, emotion, hope, desire, orgy*›; extravagant ‹*idea*›; riotous, wild ‹*enjoyment*›; dissolute, dissipated ‹*life*›; dissolute ‹*person*›. **2.** *adv.* ~ **leben** lead a dissolute life

Ausschweifung die; ~, ~**en** dissolution; dissipation

aus|schweigen *unr. refl. V.* remain silent

aus|schwenken *tr. V.* **a)** swing out; **b)** *(ausspülen)* rinse out

aus|schwitzen *tr. V.* sweat out

aus|sehen *unr. itr. V.* look; **gut** ~: look good; ‹*person*› be good-looking; *(gesund ~)* look well; **es sieht nach Regen aus** it looks like rain; **wie sieht ein Okapi aus?** what does an okapi look like?; **ich habe vielleicht ausgesehen!** I looked a real sight!; **es sieht danach** *od.* **so aus, als ob ...:** it looks as if ...; **so siehst du aus!** *(ugs.)* you've got another think coming *(coll.)*; that's what you think!

Aussehen das; ~**s** appearance; **etw. nach dem** ~ **beurteilen** judge sth. by appearances

aus|sein *unr. itr. V.; mit sein; nur im Inf. und Part. zusammengeschrieben* **a)** *(zu Ende sein)* ‹*play, film, war*› be over, have ended; **wann ist die Vorstellung aus?** what time does the performance end?; **die Schule ist aus** school is out *or* has finished; **mit ihm ist es aus** he's had it *(coll.)*; he's finished; **zwischen uns ist es aus** it's [all] over between us; **b)** *(nicht an sein)* ‹*fire, candle, etc.*› be out; ‹*radio, light, etc.*› be off; **c) auf etw.** *(Akk.)* ~: be after *or* interested in sth.; **d)** *(außer Haus sein, Sport: im Aus sein)* be out

außen ['ausn̩] *Adv.* outside; ~ **be-**

malt painted on the outside; ~ **an der Windschutzscheibe** on the outside of the windscreen; **nach ~ hin** on the outside; outwardly; **das Fenster geht nach ~ auf** the window opens outwards; **von dem Skandal darf nichts nach ~ dringen** *(fig.)* nothing must get out about the scandal; **von ~:** from the outside; **Hilfe von ~:** *(fig.)* outside help

Außen-: ~arbeiten *Pl.* outside work *sing.;* **~bahn die** *(Sport)* outside lane

Außenbord·motor der outboard motor

aus|senden *unr. (auch regelm.) tr. V.* send out

außen-, Außen-: ~dienst der: im ~dienst sein *od.* **arbeiten, ~dienst machen** be working out of the office; ⟨*salesman*⟩ be on the road **~handel der;** *o. Pl.* foreign trade *no art.;* **~minister der** Foreign Minister; Foreign Secretary *(Brit.);* Secretary of State *(Amer.);* **~ministerium das** Foreign Ministry; Foreign and Commonwealth Office *(Brit.);* Foreign Office *(Brit. coll.);* State Department *(Amer.);* **~politik die** foreign politics *sing.; (bestimmte)* foreign policy/policies *pl.;* **~politiker der** politician concerned with foreign affairs; **~politisch 1.** *Adj.* foreign-policy *attrib.* ⟨*debate*⟩; ⟨*question*⟩ relating to foreign policy; ⟨*mistake*⟩ in foreign policy; ⟨*reporting*⟩ of foreign affairs; ⟨*experience*⟩ in foreign affairs; ⟨*expert, speaker*⟩ on foreign affairs; **auf ~politischem Gebiet** in foreign affairs; **2.** *adv.* as regards foreign policy; **~politisch gesehen** from the point of view of foreign policy; **~posten der** outpost; **~seite die** outside

Außenseiter der; ~s, ~, Außenseiterin die; ~, ~nen *(Sport, fig.)* outsider

Außen-: ~spiegel der exterior mirror; **~stände** *Pl.* outstanding debts *or* accounts

Außenstehende der/die; *adj. Dekl.* outsider

Außen-: ~stelle die branch; **~stürmer der** *(Ballspiele)* winger; outside forward; **~tasche die** outside pocket; **~temperatur die** outside temperature; **~wand die** external *or* outside wall; **~welt die** outside world; **~winkel der** exterior angle

außer ['ausɐ] **1.** *Präp. mit Dat.* **a)** *auch mit Gen. (außerhalb)* out of; ~ **Atem** out of breath; ~ **Haus sein** be out of the house; ~ **Zweifel stehen** be beyond doubt; ~ **sich sein** be beside oneself (**vor** + *Dat.* with); **b)** *(abgesehen von)* apart from; **alle ~ mir** all except [for] me; **c)** *(zusätzlich zu)* in addition to. **2.** *Präp. mit Akk.* **etw. ~ jeden Zweifel stellen** make sth. very clear *or* clear beyond all doubt; ~ **sich geraten** become beside oneself (**vor** + *Dat.* with). **3.** *Konj. s.* **ausgenommen. 4.** *Adv.* except

äußer... ['ɔysɐ...] *Adj.; nicht präd.* **a)** outer; outer, outside ⟨*wall, door*⟩; external ⟨*diameter, injury, cause, force, form, circumstances*⟩; outside ⟨*pocket*⟩; outlying ⟨*district, area*⟩; outward ⟨*appearance, similarity, effect, etc.*⟩; **b)** *(auswärtig)* foreign

außer-: ~dem [*auch:* --'-] *Adv.* as well; besides; *(im übrigen)* besides; anyway; **er ist Politiker und ~ Schriftsteller** he is a politician and a writer as well; **~dienstlich 1.** *Adj.* private; social, unofficial ⟨*event*⟩; unofficial ⟨*commitment, activity*⟩; **2.** *adv.* out of working hours; **mit jmdm. ~dienstlich verkehren** meet with sb. on a social basis

Äußere das; ~n [outward] appearance; **das ~ täuscht oft** appearances are often deceptive; **der Minister des ~n** the Foreign Minister

außer-: ~ehelich 1. *Adj.* extramarital ⟨*relationship*⟩; illegitimate ⟨*child, birth*⟩; **2.** *adv.* outside marriage; **ein ~ehelich geborenes Kind** a child born out of wedlock; **~fahrplan·mäßig** [--'----] **1.** *Adj.* unscheduled ⟨*train, bus*⟩; **2.** *adv.* **dieser Zug verkehrt ~fahrplanmäßig** this train is not a scheduled one; **~gewöhnlich 1.** *Adj.* **a)** unusual; **b)** *(das Gewohnte übertreffend)* exceptional; **2.** *adv.* **a)** unusually; **b)** *(sehr)* exceptionally; **~halb 1.** *Präp. mit Gen.* outside; **~halb der Sprechstunde** out of *or* outside consulting hours; **2.** *Adv.* **a)** outside; ~ **von** outside; **b)** *(~halb der Stadt)* out of town; **von ~halb** from out of town

äußerlich 1. *Adj.* **a)** external ⟨*use, injury*⟩; **b)** *(nach außen hin)* outward ⟨*appearance, calm, similarity, etc.*⟩. **2.** *adv.* externally; *(nach außen hin)* outwardly; ~ **gesehen** on the face of it

Äußerlichkeit die; ~, ~en a) *(Umgangsform)* formality; **b)** *(Unwesentliches)* minor point

äußern 1. *tr. V.* express, voice ⟨*opinion, view, criticism, reservations, disapproval, doubt*⟩; express ⟨*joy, happiness, wish*⟩; voice ⟨*suspicion*⟩. **2.** *refl. V.* **a)** *(Stellung nehmen)* **sich über etw.** *(Akk.)* **~:** give one's view on sth.; **ich möchte mich dazu jetzt nicht ~:** I don't want to comment on that at present; **b)** *(in Erscheinung treten)* ⟨*illness*⟩ manifest itself; ⟨*emotion*⟩ show itself, be expressed

außer-: ~ordentlich 1. *Adj.* **a)** extraordinary; **b)** *(das Gewohnte übertreffend)* exceptional; *s. auch* **Professor; 2.** *adv. (sehr)* exceptionally; ⟨*value*⟩ highly; extremely ⟨*pleased, relieved*⟩; **~ordentlich viel Mühe** an enormous *or* exceptional amount of trouble; **~plan·mäßig** *Adj.* **a)** unscheduled; unbudgeted ⟨*expenditure*⟩; **b)** *s.* **~fahrplanmäßig**

äußerst *Adv.* extremely; extremely, exceedingly ⟨*important*⟩; ~ **knapp gewinnen/entkommen** *usw.* only just win/escape *etc.*

äußerst... *Adj.; nicht präd.* **a)** farthest; **b)** *(größt...)* extreme; **mit ~er Umsicht** with extreme *or* the utmost circumspection; **aufs ~e erschrocken/angestrengt/verwirrt** frightened in the extreme/strained to the utmost/utterly confused; **von ~er Wichtigkeit sein** be of extreme *or* the utmost importance; **c)** *(letztmöglich)* latest *or* last possible ⟨*date, deadline*⟩; **das Äußerste wagen/versuchen** risk/try everything; **d)** *(schlimmst...)* worst; **im ~en Fall** if the worst comes/came to the worst; **auf das Äußerste gefaßt sein** be prepared for the worst

außerstande *Adv.* ~ **sein, etw. zu tun** *(nicht befähigt)* be unable to do sth.; *(nicht in der Lage)* not be in a position to *or* not be able to do sth.

Äußerung die; ~, ~en comment; remark

aus|setzen 1. *tr. V.* **a)** expose (*Dat.* to); **Belastungen ausgesetzt sein** be subject to strains; **b)** *(sich selbst überlassen)* abandon ⟨*baby, animal*⟩; *(auf einer einsamen Insel)* maroon; **c)** *(hinaussetzen)* release ⟨*animal*⟩ [into the wild]; plant out ⟨*plants, seedlings*⟩; launch, lower ⟨*boat*⟩; **d) an jmdm./etw. nichts auszusetzen haben** have no ovjection to sb./sth.; **an jmdm./etw./allem etwas auszusetzen haben** find fault with sb./sth./everything; **daran war nichts auszusetzen** there was nothing wrong with that; **e)** *(in Aussicht stellen)* offer ‹*reward, prize*›; **eine große Summe für etw. ~:** provide a large sum for sth. **2.** *itr. V.* **a)** *(aufhören)* stop; ‹*engine, machine*› cut out, stop; ‹*heart*› stop [beating]; **b)** *(eine Pause machen)* ‹*player*› miss a turn; **er muß solange ~, bis er eine Sechs würfelt** he must wait until he throws a six; **mit seinem Studium ~:** interrupt one's studies; **mit den Tabletten ~:** stop taking the tablets

Aus·sicht die a) view **(auf +** *Akk.* of); **ein Zimmer mit ~ aufs Meer** a room overlooking the sea; **jmdm. die ~ nehmen/versperren** block *or* obstruct sb.'s view; **b)** *(fig.)* prospect **(auf +** *Akk.* of); **das sind ja vielleicht [heitere] ~en!** *(iron.)* that's a fine prospect! *(iron.)*; **~ auf etw.** *(Akk.)* **haben** have the prospect of sth.; **er hat gute ~en, gewählt zu werden** he stands a good chance of being elected; **etw. in ~ haben** have the prospect of sth.; have sth. in prospect

aussichts-, Aussichts-: ~los 1. *Adj.* hopeless; **2.** *adv.* hopelessly; **~losigkeit die; ~:** hopelessness; **~reich** *Adj.* promising; **~turm der** look-out *or* observation tower

aus|sieben *tr. V.* sift out; screen ‹*coal*›

aus|siedeln *tr. V.* move out and resettle; *(evakuieren)* evacuate

Aus·siedler der; Aus·siedlerin die emigrant

aus|söhnen 1. *refl. V.: s.* **versöhnen 1. 2.** *tr. V.: s.* **versöhnen 2 a**

aus|sondern *tr. V.* **a)** *(ausscheiden)* weed out; **b)** *(auswählen)* sort *or* pick out; select

aus|sortieren *tr. V.* sort out

aus|spannen 1. *tr. V.* **a)** unharness, unhitch ‹*horse, mule*›; unyoke ‹*oxen*›; **b)** *(salopp: wegnehmen)* **jmdm. etw. ~:** get sb. to part with sth.; **jmdm. den Freund/die Freundin ~:** pinch sb.'s boyfriend/girl-friend *(sl.)*. **2.** *itr. V.* *(ausruhen)* take *or* have a break

aus|sparen *tr. V.* leave ‹*line etc.*› blank; *(fig.)* leave out; omit

aus|sperren 1. *tr. V.* lock out; shut ‹*animal*› out. **2.** *itr. V.* organize a lock-out; lock the workforce out

Aus·sperrung die lock-out

aus|spielen 1. *tr. V.* **a)** *(Kartenspiel)* lead; **sein ganzes Wissen ~** *(fig.)* make use of all one's knowledge; **b)** *(manipulieren)* **jmdn./etw. gegen jmdn./etw. ~:** play sb./sth. off against sb./sth. **2.** *itr. V.* *(Kartenspiel)* lead; **wer spielt aus?** whose lead is it?

aus|spionieren *tr. V.* spy out

Aus·sprache die a) pronunciation; *(Artikulation)* articulation; *(Akzent)* accent; **b)** *(Gespräch)* discussion; *(zwangloseres)* talk

Aussprache·wörterbuch das pronouncing dictionary

aussprechbar *Adj.* pronounceable

aus|sprechen 1. *unr. tr. V.* **a)** pronounce; **b)** *(ausdrücken)* express; voice ‹*suspicion, request*›; grant ‹*divorce*›; **der Regierung das Vertrauen ~:** pass a vote of confidence in the government. **2.** *unr. refl. V.* **a)** *(ausgesprochen werden)* be pronounced; **b)** *(sich äußern)* speak; **sich lobend/mißbilligend über jmdn./etw. ~:** speak highly/disapprovingly of sb./sth.; **er hat sich nicht näher darüber ausgesprochen** he did not say anything further about it; **sich für jmdn./etw. ~:** declare *or* pronounce oneself in favour of sb./sth.; **sich gegen jmdn./etw. ~:** declare *or* pronounce oneself against sb./sth.; **c)** *(offen sprechen)* say what's on one's mind; **sich mit** *od.* **bei jmdm. ~:** have a heart-to-heart talk with sb.; **d)** *(Strittiges klären)* have it out, talk things out **(mit** with); **wir haben uns über alles ausgesprochen** we had everything out. **3.** *unr. itr. V.* finish [speaking]

Aus·spruch der remark; *(Sinnspruch)* saying

aus|spucken 1. *itr. V.* spit. **2.** *tr. V.* **a)** spit out; *(fig. ugs.)* spew out ‹*products, data, results, etc.*›; cough up *(sl.)* ‹*money*›; **b)** *(ugs.: erbrechen)* throw up

aus|spülen *tr. V.* **a)** flush *or* wash out; **b)** *(reinigen)* rinse out; *(Med.)* irrigate; wash out; **sich** *(Dat.)* **den Mund ~:** rinse one's mouth out

aus|staffieren ['ausʃta,fi:rən] *tr. V.* kit *or* rig out; fit out, furnish ‹*room etc.*›; *(verkleiden)* dress up

Aus·stand der strike; **im ~ sein** be on strike; **in den ~ treten** go on strike

aus|statten ['ausʃtatn̩] *tr. V.* provide **(mit** with); *(mit Kleidung)* provide; fit out; *(mit Gerät)* equip; *(mit Möbeln, Teppichen usw.)* furnish; **mit Befugnissen ausgestattet sein** be vested with authority *sing.*; **ein prächtig ausgestatteter Band** a splendidly produced volume

Ausstattung die; ~, ~en a) *(das Ausstatten) s.* **ausstatten:** provision; fitting out; equipping; furnishing; vesting; production; **b)** *(Ausrüstung)* equipment; *(Innen~ eines Autos)* trim; **c)** *(Einrichtung)* furnishings *pl.*; **d)** *(Film, Theater)* décor and costumes; **e)** *(Buchw.)* design and layout; *(typographisch)* design

aus|stechen *unr. tr. V.* **a) jmdm. die Augen ~:** put *or* gouge sb.'s eyes out; **b)** *(entfernen)* dig up ‹*plants*›; cut ‹*turf*›; **c)** *(herstellen)* dig [out] ‹*trench, hole, etc.*›; *(Kochk.)* press *or* cut out ‹*biscuits*›; **d)** *(übertreffen)* outdo; **jmdn. bei jmdm. ~:** oust sb. in sb.'s affections/esteem/favour

aus|stehen 1. *unr. itr. V.* **noch ~** ‹*money, amount*› be outstanding; ‹*decision*› be still to be taken, have not yet been taken; ‹*solution*› be still to be found; **ihre Antwort steht noch aus** I am/we are *etc.* still awaiting their reply; **~de Forderungen** outstanding demands. **2.** *unr. tr. V.* endure ‹*pain, suffering*›; suffer ‹*worry, anxiety*›; **ausgestanden sein** be all over; **ich kann ihn/das nicht ~:** I can't stand *or* bear him/it

aus|steigen *unr. itr. V.; mit sein;* **a)** *(aus einem Auto, Boot)* get out **(aus** of); *(aus einem Zug, Bus)* get off; alight *(formal)*; *(Fliegerspr.: abspringen)* bale out; **aus einem Zug/Bus ~:** get off a train/bus; alight from a train/bus *(formal)*; **b)** *(ugs.: sich nicht mehr beteiligen)* **~ aus** opt out of; give up ‹*show business, job*›; leave ‹*project*›; **c)** *(Sport: aus einem Rennen o. ä.)* drop out **(aus** of); retire **(aus** from); **d)** *(ugs.: der Gesellschaft den Rücken kehren)* drop out

Aussteiger der; ~s, ~ *(ugs.)* drop-out *(coll.)*

aus|stellen *tr. V.* **a)** *auch itr.* put on display; display; *(im Museum, auf einer Messe)* exhibit; **ausgestellt sein** ‹*goods*› be on display/be exhibited; ‹*painting*› be exhibited; **b)** *(ausfertigen)* make out, write [out] ‹*cheque, prescription, receipt*›; make out ‹*bill*›; issue ‹*visa, passport, certificate*›; **c)** *(ugs.: ausschalten)* turn *or* switch off

Aussteller der; ~s, ~, Ausstellerin die; ~, ~nen a) *(auf Messen)* exhibitor; **b)** *(eines Dokuments)* issuer; *(Behörde)* issuing authority; *(eines Schecks)* drawer

Aus·stellung die a) exhibiting; **b)** *(das Ausfertigen) s.* **ausstellen b:** making out; writing [out]; issuing; **c)** *(Veranstaltung)* exhibition

Ausstellungs-: ~gelände das exhibition site; **~halle die** exhibition hall; **~raum der** exhibition room; **~stück das** display item; *(in Museen usw.)* exhibit

aus|sterben *unr. itr. V.; mit sein (auch fig.)* die out; ‹*species*› die out, become extinct; **ein ~des Handwerk** *(fig.)* a dying craft

Aus·steuer die trousseau *(consisting mainly of household linen)*

aus|steuern *tr. V. (Elektronik)* modulate ‹*signal, wave*›; *(bei der Aufnahme)* control the recording level of; control the power level of ‹*amplifier*›

Ausstieg der: ~[e]s, ~e a) *(Ausgang)* exit; *(Tür)* door[s]; *(~luke)* hatch; **b)** *o. Pl. (das Aussteigen)* climbing out **(aus** of); **„Kein ~"** 'no exit'

aus|stopfen *tr. V.* stuff

aus|stoßen *unr. tr. V.* **a)** expel; give off, emit ‹*gas, fumes, smoke*›; **b)** give ‹*cry, whistle, laugh, etc.*›; let out ‹*cry, scream, yell*›; heave, give ‹*sigh*›; utter ‹*curse, threat, accusation, etc.*›; **c) jmdm. ein Auge ~:** put sb.'s eye out; **d)** *(ausschließen)* expel **(aus** from); *(aus der Armee)* drum out **(aus** of); **sich ausgestoßen fühlen** feel an outcast

aus|strahlen 1. *tr. V.* **a)** *(auch fig.)* radiate; radiate, give off ‹*heat*›; ‹*lamp*› give out ‹*light*›; **b)** *(Rundf., Ferns.)* broadcast; transmit. **2.** *itr. V.* **a)** radiate; ‹*heat*› radiate, be given off; ‹*light*› be given out; *(fig.)* ‹*pain*› spread, extend; **b)** *(fig.)* **auf jmdn./etw. ~:** communicate itself to sb./influence sth.

Aus·strahlung die a) radiation; *(eines Menschen)* charisma; **b)** *(Rundf., Ferns.)* transmission

aus|strecken 1. *tr. V.* extend, stretch out ‹*arms, legs*›; stretch out ‹*hand*›; put out ‹*feelers*›; stick *or* put out ‹*tongue*›; **mit ausgestreckten Armen** with arms extended; with outstretched arms. **2.** *refl. V.* stretch [oneself] out; **ausgestreckt am Boden liegen** lie stretched out on the floor

aus|streichen *unr. tr. V.* **a)** *(durchstreichen)* cross *or* strike out; delete; **b)** *(verstreichen)* spread; **c)** *(Kochk.)* grease ‹*tin, pan, etc.*›; **d)** *(füllen)* fill, smooth over ‹*cracks*›

aus|strömen 1. *tr. V.* radiate ‹*warmth*›; give off ‹*scent*›; *(fig.)* radiate ‹*optimism, confidence, etc.*›. **2.** *itr. V.; mit sein* stream *or* pour out; ‹*gas, steam*› escape

aus|suchen *tr. V.* choose; pick; **such dir was aus!** choose what you want; take your pick

Aus·tausch der a) exchange; **im ~ gegen** in exchange for; **b)** *(das Ersetzen)* replacement **(gegen** with); *(Sport)* substitution **(gegen** by)

austauschbar *Adj.* interchangeable; *(ersetzbar)* replaceable

aus|tauschen *tr. V.* **a)** exchange **(gegen** for); **b)** *(ersetzen)* replace **(gegen** with); *(Sport)* substitute **(gegen** by)

Austausch-: ~motor der *(Kfz-W.)* replacement engine; **~schüler der** exchange pupil *or* student

aus|teilen *tr. V. (verteilen)* distribute **(an** + *Akk.* to); *(ausgeben)* hand *or* give out ‹*books, post, etc.*› **(an** + *Akk.* to); issue, give ‹*orders*›; deal [out] ‹*cards*›; give out ‹*marks, grades*›; administer ‹*sacrament*›; serve ‹*food etc.*›; give ‹*blessing*›; **Prügel ~** *(fig.)* hand out beatings

Auster ['austɐ] **die; ~, ~n** oyster

aus|tilgen *tr. V.* exterminate ‹*pests, race*›; eradicate ‹*weeds*›; wipe out, eradicate ‹*disease*›

aus|toben *refl. V.* romp about; have a good romp

aus|tragen *unr. tr. V.* **a)** *(zustellen)* deliver ‹*newspapers, post*›; **b)** *(im Mutterleib)* carry ‹*child*› to full term; *(nicht abtreiben)* have ‹*child*›; **c)** *(ausfechten)* settle; settle, resolve ‹*conflict, differences*›; fight out ‹*battle*›; **einen Streit mit jmdm. ~:** have it out with sb.; **d)** *(bes. Sport)* hold ‹*competition, race, event*›

Australien [aus'tra:li̯ən] **(das); ~s** Australia

Australier der; ~s, ~, Australierin die; ~, ~nen Australian

australisch *Adj.* Australian

aus|treiben 1. *unr. tr. V.* **a)** exorcize, cast out ‹*evil spirit, demon*›; **b) jmdm. etw. ~:** cure sb. of sth.; **c)** *(geh.: vertreiben)* drive out **(aus** from). **2.** *unr. itr. V.* ‹*plant*› sprout

aus|treten 1. *unr. tr. V.* **a)** tread out ‹*spark, cigarette-end*›; trample out ‹*fire*›; **b)** *(bahnen)* tread out ‹*path*›; **ausgetretene Pfade** *(fig.)* well-trodden paths; **c)** *(abnutzen)* wear down; **ausgetretene Stufen** worn-down steps; **d)** *(weiten)* wear out ‹*shoes*›; break in ‹*new shoes*›. **2.** *unr. itr. V.; mit sein* **a)** *(ugs.: zur Toilette gehen)* pay a call *(coll.)*; **der Schüler fragte, ob er ~ dürfe** the pupil asked to be excused; **b)** *(ausscheiden)* leave; **aus etw. ~:** leave sth.; **aus einer Vereinigung ~:** resign from a society; **c)** *(nach außen gelangen)* come out; *(entweichen)* escape

aus|tricksen *tr. V. (ugs.)* trick

aus|trinken 1. *tr. V.* finish, drink up ‹*drink*›; finish, drain ‹*glass, cup, etc.*›. **2.** *itr. V.* drink up

Aus·tritt der leaving; *(aus einer Vereinigung)* resignation **(aus** from); **seinen ~ aus der Partei/Kirche erklären** announce that one is leaving the party/church

aus|trocknen 1. *tr. V.* dry out; dry up ‹*river bed, marsh*›; parch ‹*throat*›. **2.** *itr. V.; mit sein* dry out; ‹*river bed, pond, etc.*› dry up; ‹*skin, hair*› become dry; ‹*throat*› become parched

aus|trudeln *tr. V. (ugs.) s.* **auswürfeln**

aus|tüfteln *tr. V. (ugs.)* work out; *(ersinnen)* think up

aus|üben *tr. V.* practise ‹*art, craft*›; follow ‹*profession*›; carry on ‹*trade*›; do ‹*job*›; wield, exercise ‹*power*›; exercise ‹*right, control*›; exert ‹*influence, pressure*›; **welche Tätigkeit üben Sie aus?** what is your occupation?

Aus·übung die *o. Pl.; s.* **ausüben:** practising; following; carrying on; doing; wielding; exercising; exertion

aus|ufern *itr. V.; mit sein* get out of hand

Aus·verkauf der [clearance] sale; *(wegen Geschäftsaufgabe)* closing-down *(Brit.) or (Amer.)* liquidation sale; *(fig.)* sell-out; **etw. im ~ kaufen** buy sth. at the sale[s]

ausverkauft *Adj.* sold out; **Bier ist ~:** there is no beer left [in stock]; **wir sind ~:** we are sold out; **vor ~em Haus spielen** play to a full house

aus|wachsen 1. *unr. refl. V.* **a)** *(verschwinden)* right *or* correct itself; **b)** *(sich entwickeln)* grow **(zu** into). **2.** *unr. itr. V.* **zum Auswachsen sein** *(ugs.)* be enough to drive you up the wall *(coll.)*

Aus·wahl die a) *(das Auswählen)* choice; selection; **Sie haben die [freie] ~:** the choice is yours; you can choose whichever you like; **bei uns stehen Ihnen mehr als 600 Wagen zur ~:** we offer you a choice of over 600 cars; **eine ~ treffen** make a selection; **b)** *(Auslese)* selection; *(von Texten)* anthology; selection; **c)** *(Sortiment)* range; **viel/wenig ~ haben** have a wide/limited selection **(an** + *Dat.*, **von** of); **Spirituosen in reicher ~:** a wide selection of spirits; **d)** *(Sport: Mannschaft)* [selected] team

aus|wählen *tr. V.* choose, select (**aus** from); **sich** *(Dat.)* **etw.** ~: choose *or* select sth. [for oneself]

aus|walzen *tr. V.* roll out; *(fig. ugs.)* drag out ⟨*subject*⟩

Aus·wanderer der emigrant

aus|wandern *itr. V.; mit sein* emigrate

Aus·wanderung die emigration

auswärtig ['ausvɛrtɪç] *Adj.; nicht präd.* **a)** non-local; **b)** *(von auswärts stammend)* ⟨*student, guest, etc.*⟩ from out of town; **c)** *(das Ausland betreffend)* foreign

auswärts *Adv.* **a)** *(nach außen)* outwards; **b)** *(nicht zu Hause)* ⟨*sleep*⟩ away from home; ~ **essen** eat out; **c)** *(nicht am Ort)* in another town; *(Sport)* away; **von/nach** ~: from/to another town

Auswärts- *(Sport)* away ⟨*match, win, etc.*⟩

aus|waschen *unr. tr. V.* **a)** wash out; *(ausspülen)* rinse out; **b)** *(Geol.)* erode

auswechselbar *Adj.* changeable; exchangeable; *(untereinander)* interchangeable; *(ersetzbar)* replaceable

aus|wechseln *tr. V.* **a)** change (**gegen** + *Akk.* for); **b)** *(ersetzen)* replace (**gegen** with); *(Sport)* substitute ⟨*player*⟩; **A gegen B** ~: replace A by B; **sie war wie ausgewechselt** she was a different person

Aus·weg der way out (**aus** of); **der letzte ~ für jmdn. sein** be a last resort for sb.

ausweg·los 1. *Adj.* hopeless. **2.** *adv.* hopelessly

Ausweglosigkeit die; ~: hopelessness

aus|weichen *unr. itr. V.; mit sein* **a)** *(Platz machen)* make way (*Dat.* for); *(wegen Gefahren, Hindernissen)* get out of the way (*Dat.* of); **[nach] rechts/nach der Seite** ~: move to the right/move aside to make way/get out of the way; **einem Schlag/ Angriff** ~: dodge a blow/evade an attack; **dem Feind** ~: avoid [contact with] the enemy; **einer Frage/Entscheidung** ~: evade a question/decision; ~**de Antworten** evasive answers; **b)** *(zurückgreifen)* **auf etw.** *(Akk.)* ~: switch [over] to sth.

Ausweich·manöver das evasive manœuvre; ~ *Pl.* evasive action *sing.*

aus|weinen *refl. V.* have a good cry

Ausweis ['auswais] **der;** ~**es,** ~**e** card; *(Personal~)* identity card; *(Mitglieds~)* membership card

aus|weisen 1. *unr. tr. V.* **a)** *(aus dem Land)* expel (**aus** from); **b)** *(erkennen lassen)* **jmdn. als etw.** ~: show that sb. is/was sth.; **seine Papiere wiesen ihn als ... aus** his papers proved *or* established his identity as ... **2.** *unr. refl. V.* prove *or* establish one's identity [by showing one's papers]; **können Sie sich ~?** do you have any means of identification?

Ausweis·papiere *Pl.* identity papers

Aus·weisung die expulsion (**aus** from)

aus|weiten 1. *tr. V.* stretch. **2.** *refl. V.* **a)** stretch; **b)** *(fig.: sich vergrößern)* expand; **sich zur Krise** ~: develop *or* grow into a crisis

aus·wendig *Adv.* **etw. ~ können/lernen** know/learn sth. [off] by heart; **etw. ~ aufsagen** recite sth. from memory; *s. auch* **inwendig**

Auswendig·lernen das learning by heart

aus|werfen *unr. tr. V.* cast ⟨*net, anchor, rope, line, etc.*⟩

aus|werten *tr. V.* **a)** analyse and evaluate; **b)** *(nutzen)* utilize

Aus·wertung die a) analysis and evaluation; *(Nutzung)* utilization; **b)** *(Ergebnis)* analysis

aus|wickeln *tr. V.* unwrap (**aus** from)

aus|wiegen *unr. tr. V.* weigh

aus|wirken *refl. V.* have an effect (**auf** + *Akk.* on); **sich in etw.** *(Dat.)* ~: result in sth.; **sich günstig/negativ** *usw.* ~: have a favourable/an unfavourable *etc.* effect (**auf** + *Akk.* on); **sich zu jmds. Vorteil** ~: work to sb.'s advantage

Aus·wirkung die effect (**auf** + *Akk.* on); *(Folge)* consequence (**auf** + *Akk.* for)

aus|wischen *tr. V.* **a)** wipe ⟨*dirt etc.*⟩ out (**aus** of); **b)** *(säubern)* wipe [clean]; **sich** *(Dat.)* **die Augen** ~: wipe one's eyes; **c)** *in* **jmdm. eins** ~ *(ugs.)* get one's own back on sb. *(coll.)*

aus|wringen *unr. tr. V. (bes. nordd.)* wring out

Aus·wuchs der a) growth; excrescence *(Med., Bot.)*; **b)** *(fig.)* unhealthy product; *(Folge)* harmful consequence; *(Übersteigerung, Mißstand)* excess

aus|wuchten *tr. V. (Technik)* balance

Aus·wurf der *(Med.)* sputum

aus|würfeln *tr. V.* **eine Runde Bier** *usw.* ~: throw dice to decide who will pay for a round of beer *etc.*; ~, **wer anfangen soll** throw dice to decide who will start

aus|zahlen 1. *tr. V.* pay [out] ⟨*sum, wages, etc.*⟩; pay off ⟨*worker, employee*⟩; buy out ⟨*partner*⟩; **ausgezahlt bekommt er 1650 Mark** his take-home pay is 1,650 marks. **2.** *refl. V.* pay off; ⟨*investment etc.*⟩ pay; **Verbrechen zahlen sich nicht aus** crime doesn't pay

aus|zählen *tr. V.* **a)** count [up] ⟨*votes etc.*⟩; **b)** *(Boxen)* count out

Aus·zahlung die *s.* **auszahlen 1:** paying [out]; paying off; buying out

Aus·zählung die counting [up]; **mit der ~ wurde bereits begonnen** the count has already started

aus|zehren *tr. V. (geh.)* exhaust; **ein ausgezehrtes Gesicht/eine ausgezehrte Gestalt** an emanciated face/figure

aus|zeichnen 1. *tr. V.* **a)** *(mit einem Preisschild)* mark, price; **b)** *(ehren)* honour; **jmdn. mit einem Orden** ~: decorate sb. [with a medal]; **jmdn./etw. mit einem Preis/Titel** ~: award a prize/title to sb./sth.; **c)** *(bevorzugt behandeln)* single out for special favour; *(ehren)* single out for special honour; **d)** *(kennzeichnen)* distinguish (**gegenüber, vor** + *Dat.* from). **2.** *refl. V. (durch eine Eigenschaft)* stand out (**durch** for); *(durch Leistung)* ⟨*person*⟩ distinguish oneself (**durch** by)

Aus·zeichnung die a) *o. Pl. (von Waren)* marking; **b)** *(Ehrung)* honouring; *(mit Orden)* decoration; **c)** *(Orden)* decoration; *(Preis)* award; prize; **d)** *in* **mit** ~: with distinction

ausziehbar *Adj.* extendible; telescopic ⟨*aerial*⟩; extending *attrib.* ⟨*ladder*⟩; sliding-leaf *attrib.* ⟨*table*⟩

aus|ziehen 1. *unr. tr. V.* **a)** *(vergrößern)* pull out ⟨*couch*⟩; extend ⟨*table, tripod, etc.*⟩; **b)** *(ablegen)* take off, remove ⟨*clothes*⟩; **c)** *(entkleiden)* undress; **sich** ~: undress; get undressed; **sich ganz/nackt** ~: strip off *or* undress completely; **d)** *(auszupfen)* pull out ⟨*hair etc.*⟩. **2.** *unr. itr. V.; mit sein* move out (**aus** of)

Auszieh·tisch der extending table; sliding-leaf table

Auszubildende der/die; *adj. Dekl. (bes. Amtsspr.)* trainee; *(im Handwerk)* apprentice

Aus·zug der a) *(Extrakt)* extract; **b)** *(Exzerpt)* extract; excerpt; **c)** *(Konto~)* statement; **d)** *(aus Wohnung)* move

auszugs·weise *Adv.* in extracts *or* excerpts; **etw. ~ lesen** read extracts from sth.

aus|zupfen *tr. V.* pluck out; pull out ⟨*weeds*⟩

autark [au'tark] *Adj. (Wirtsch., fig. geh.)* self-sufficient

Autarkie [autar'ki:] **die; ~, ~n** *(Wirtsch., fig. geh.)* self-sufficiency

authentisch [au'tɛntɪʃ] **1.** *Adj.* authentic. **2.** *adv.* authentically

Auto ['auto] **das; ~s, ~s** car; automobile *(Amer.)*; **~ fahren** drive; *(mitfahren)* go in the car; **mit dem ~ fahren** go by car

Auto-: ~atlas der road atlas; **~bahn die** motorway *(Brit.)*; expressway *(Amer.)*

Autobahn- motorway *(Brit.)*; expressway *(Amer.)* ⟨*exit, intersection, junction, service area, etc.*⟩

Auto·biographie die autobiography

auto·biographisch 1. *Adj.* autobiographical. **2.** *adv.* autobiographically

Auto·bus der *s.* **Bus**

Auto·didakt [-di'dakt] **der; ~en, ~en** autodidact

auto-, Auto-: ~fähre die car ferry; **~fahren das** driving; motoring; **~fahrer der** [car-]driver; **~fahrt die** drive; **~frei** *Adj.* ⟨*place*⟩ where no cars are/were allowed; **X ist ~frei** no cars are allowed in X; **~fried·hof der** *(ugs.)* car dump

auto·gen 1. *Adj.* **a)** *(Technik)* **~es Schneiden/Schweißen** gas *or* oxy-acetylene cutting/welding; **b)** *(Psych.)* **~es Training** autogenic training; autogenics; **2.** *adv. (Technik)* **~ schweißen/schneiden** weld/cut using an oxy-acetylene flame

Auto·gramm das; ~s, ~e autograph

Autogramm·jäger der *(ugs.)* autograph-hunter

Auto-: ~karte die road map; **~kino das** drive-in cinema; **~knacker der** *(ugs.)* car-thief; **~kolonne die** line of cars

Auto·krat [-'kra:t] **der; ~en, ~en** autocrat

auto·kratisch 1. *Adj.* autocratic. **2.** *adv.* autocratically

Auto·mat [-'ma:t] **der; ~en, ~en a)** *(Verkaufs~)* [slot-]machine; vending-machine; **b)** *(in der Produktion, fig.: Mensch)* robot; automaton

Automaten·knacker der *(ugs.)* thief who breaks into slot-machines

Automatik die; ~, ~en automatic control mechanism; *(Getriebe~)* automatic transmission

automatisch 1. *Adj.* automatic. **2.** *adv.* automatically

automatisieren *tr. V.* automate

Automatisierung die; ~, ~en automation

Automatismus der; ~, Automatismen *(Med., Biol., Psych.)* automatism *no art.*

Auto-: ~mechaniker der motor mechanic; **~minute die: zehn ~minuten entfernt sein** be ten minutes [away] by car; be ten minutes' drive [away]

Auto·mobil das; ~s, ~e *(geh.)* motor car; automobile *(Amer.)*

Automobil-: ~aus·stellung die motor show *(Brit.)*; automobile show *(Amer.)*; **~industrie die** motor industry; **~klub der** motoring organization

auto·nom [-'no:m] **1.** *Adj.* autonomous. **2.** *adv.* autonomously

Auto·nomie [-no'mi:] **die; ~, ~n** autonomy

Auto-: ~nummer die [car] registration number; **~papiere** *Pl.* car documents

Autopsie [autɔ'psi:] **die; ~, ~n** *(Med.)* autopsy; post-mortem [examination]

Autor ['autɔr] **der; ~s, ~en** [-'to:rən] author

Auto-: ~radio das car radio; **~reifen der** car tyre; **~reise·zug der** Motorail train *(Brit.)*; auto train *(Amer.)*; **~rennen das** *(Sportart)* motor *(Brit.)* or *(Amer.)* auto racing; *(Veranstaltung)* motor *(Brit.) or (Amer.)* auto race; **~reparatur die** car repair; repair to the/a car

Autorin die; ~, ~nen authoress; author

autoritär [autori'tɛ:ɐ̯] **1.** *Adj.* authoritarian. **2.** *adv.* in an authoritarian manner

Autorität [autori'tɛ:t] **die; ~, ~en** authority

autoritäts·gläubig *Adj.* trusting in authority *pred.*

Auto-: ~schlange die queue *or* line of cars; **~schlüssel der** car key; **~skooter der** dodgem; bumper car; **~stopp der** hitch-hiking; hitching *(coll.)*; **per** *od.* **mit ~stopp fahren, ~stopp machen** hitch-hike; hitch *(coll.)*; **~telefon das** car telephone; **~tür die** car door; **~unfall der** car accident; **~verkehr der** [motor] traffic; **~verleih der, ~vermietung die** car hire *(Brit.)* *or* rental firm *or* service; **~werkstatt die** garage; car repair shop; **~zubehör das** car accessories *pl.*

autsch [autʃ] *Interj.* ouch; ow

Au·wald der riverside forest

auwei[a] [au'vai(a)] *Interj. (ugs.)* oh dear

avancieren [avã'si:rən] *itr. V.; mit sein (geh.)* be promoted *(also iron.)*; rise

Avantgarde [avã'gardə] **die; ~, ~n** avant-garde; *(Politik)* vanguard *(fig.)*

avantgardistisch 1. *Adj.* avant-garde. **2.** *adv.* ⟨*paint etc.*⟩ in an avant-garde style

AvD [a:fau'de:] **der; ~** *Abk.*: **Automobilclub von Deutschland**

Ave-Maria ['a:vema'ri:a] **das; ~[s], ~[s]** *(kath. Kirche)* Ave Maria; Hail Mary

Aversion [avɛr'zio:n] **die; ~, ~en** aversion

Avocado [avo'ka:do] **die; ~, ~s** avocado [pear]

Axel ['aksl̩] **der; ~s, ~** *(Eis-, Rollkunstlauf)* axel

Axiom [a'ksio:m] **das; ~s, ~e** axiom

Axt [akst] **die; ~, Äxte** ['ɛkstə] axe

Azalee [atsa'le:ə] **die; ~, ~n** azalea

Azteke [ats'te:kə] **der; ~n, ~n** Aztec

Azubi ['a:tsubi] **der; ~s, ~s/die; ~, ~s** *(ugs.)* *s.* **Auszubildende**

azur·blau [a'tsu:ɐ̯-] *Adj. (geh.)* azure[-blue]

B

b, B [be:] **das; ~, ~ a)** *(Buchstabe)* b/B; **b)** *(Musik)* [key of] B flat; *s. auch* **a, A**

B *Abk.* **Bundesstraße**

BAB *Abk.* **Bundesautobahn**

Babel ['ba:bl̩] **(das); ~s** Babel; **der Turmbau zu ~:** the building of the Tower of Babel

Baby ['be:bi] **das; ~s, ~s** *od.* **Babies** baby

Baby·ausstattung die layette

babylonisch *Adj.* Babylonian; **ein ~es Sprachengewirr** a babel of languages

baby-, Baby-: ~|sitten [~sɪtn̩] *itr. V.; nur im Inf. (ugs.)* babysit; **~sitter** [~sɪtɐ] **der; ~s, ~:** babysitter; **~sitting** [~sɪtɪŋ] **das; ~s** baby-sitting; **~wäsche die** baby clothes

Bach [bax] **der; ~[e]s, Bäche** ['bɛçə] stream; brook; *(fig.)* stream [of water]; **den ~ runtergehen** *(ugs.)* get pushed into the background

Bache ['baxə] *(Jägerspr.)* **die; ~, ~n** wild sow

Bach·stelze die wagtail

Back·blech das baking-sheet

Back·bord das *(Seew., Luftf.)* port [side]; **über ~:** over the port side

backbord[s] ['bakbɔrt(s)] *Adv.* *(Seew., Luftf.)* on the port side

Bäckchen ['bɛkçən] **das; ~s, ~:** [little] cheek

Backe ['bakə] **die; ~, ~n a)** *(Wange)* cheek; *s. auch* **voll 1a; b)** *(ugs.: Gesäß~)* buttock; cheek *(sl.)*

backen ['bakn̩] **1.** *unr. itr. V.* **a)** bake; **b)** ‹*cake etc.*› bake. **2.** *unr. tr. V.* **a)** bake ‹*cakes, bread, etc.*›; **ich backe vieles selbst** I do a lot of my own baking; *(fig.)* **das frisch gebackene Ehepaar** *(ugs.)* the newly-weds *pl.* *(coll.)*; **ein frisch gebackener Arzt** *(ugs.)* a newly-fledged doctor; **b)** *(bes. südd.: braten)* roast; *(in der Bratpfanne)* fry

Backen-: ~bart der side-whiskers *pl.*; sideboards *pl.* *(sl.)*; **~knochen der** cheek-bone **~zahn der** molar

Bäcker ['bɛkɐ] **der; ~s, ~ a)** baker; **~ lernen** learn the baker's trade; learn to be a baker; **er will ~ werden/ist ~:** he wants to be/is a baker; **b)** *(Geschäft)* baker's [shop]; **zum ~ gehen** go to the baker's; **beim ~:** at the baker's

Bäckerei die; ~, ~en baker's [shop]

Bäckerin die; ~, ~nen baker

Bäcker-: ~laden der baker's shop; **~lehre die** baker's apprenticeship; **~meister der** master baker

Bäckers·frau die baker's wife

Back-: ~fisch der *(bes. südd.)* fried fish *(in breadcrumbs)*; **~form die** baking-tin *(Brit.)*; baking-pan *(Amer.)*

Background ['bɛkgraunt] **der; ~s, ~s** background

Back-: ~hähnchen das *(bes. südd.)*, **~hendl das** *(österr.)* fried chicken *(in breadcrumbs)*; **~obst das** dried fruit; **~ofen der** oven; **~pfeife die** *(bes. nordd.)* slap in the face; **~pulver das** baking-powder; **~röhre die** oven; **~stein der** brick; **~stube die** bakery; bakehouse; **~waren** *Pl.* bread, cakes, and pastries

Bad [ba:t] **das; ~[e]s, Bäder** ['bɛ:dɐ] **a)** *(Wasser)* bath; **[sich** *(Dat.)***] ein ~ einlaufen lassen** run [oneself] a bath; **b)** *(das Baden)* bath; *(das Schwimmen)* swim; *(im Meer o. ä.)* bathe; **ein ~ nehmen** *(geh.)* have *or* take a bath; *(schwimmen)* go for a swim; *(im Meer o. ä.)* bathe; **nach dem ~:** after bathing; **c)** *(Badezimmer)* bathroom; **ein Zimmer mit ~:** a room with [private] bath; **d)** *(Schwimm~)* [swimming-]pool; swimming-bath; **e)** *(Heil~)* spa; *(See~)* [seaside] resort; **f)** *(Technik)* bath

Bade-: ~anstalt die swimming baths *pl.* *(Brit.)*; public pool *(Amer.)*; **~an·zug der** swimming *or* bathing costume; swimsuit; **~gast der** bather; swimmer; **~hose die** swimming *or* bathing trunks *pl*; **~mantel der** dressing-gown; bathrobe; **~meister der** swimming-pool attendant; **~mütze die** swimming *or* bathing cap

baden 1. *itr. V.* **a)** *(in der Wanne)* have *or* take a bath; bath; **warm ~:** have *or* take a hot bath; **b)** *(schwimmen)* bathe; swim; **~ gehen** go for a bathe *or* a swim; go bathing *or* swimming; **[bei** *od.* **mit etw.] ~ gehen** *(fig. ugs.)* come a cropper *(coll.)* [over sth.]. **2.** *tr. V.* bath ‹*person*›; bathe ‹*wound, face, eye, etc.*›; **in Schweiß gebadet** *(fig.)* bathed in sweat

Baden-Württemberg ['ba:dn̩'vʏrtəmbɛrk] **(das); ~s** Baden-Württemberg

baden-württembergisch *Adj.* of Baden-Württemberg *postpos.*; ‹*produce, wine, etc.*› from Baden-Württemberg

Bade-: ~ofen der bath-water heater; **~ort der a)** *(Seebad)* [seaside] resort; **b)** *(Kurort)* spa; **~platz der** bathing-place; **~sachen** *Pl.* bathing *or* swimming things; **~salz das** bath salts *pl.*; **~strand der** bathing-beach; **~tuch das** bath towel; **~wanne die** bath[-tub]; **~wasser das** bath water; **~zeug das** *(ugs.)* bathing *or* swimming things *pl.*; **~zimmer das** bathroom

baff [baf] *in* **~ sein** *(ugs.)* be flabbergasted

Bagage [ba'ga:ʒə] **die; ~, ~n** *(abwertend)* *(Familie)* tribe *(derog.)*; *(Gesindel)* rabble; crowd *(coll.)*; **die ganze ~:** the whole lot of them

Bagatelle [baga'tɛlə] **die; ~, ~n** trifle; bagatelle

Bagger ['bagɐ] **der; ~s, ~:** excavator; digger; *(Schwimm~)* dredger

baggern *tr., itr. V.* excavate; *(mit dem Schwimmbagger)* dredge

Bagger·see der flooded gravel-pit

Baguette [ba'gɛt] **die; ~, ~n** baguette

bah [ba:] *Interj.* ugh

bäh [bɛ:] *Interj.* *(bei Ekel)* ugh; *(schadenfroh)* hee-hee; tee-hee

Bahama·inseln [ba'ha:ma-], **Bahamas** *Pl.* **die ~:** the Bahamas

Bahn [ba:n] **die; ~, ~en a)** *(Weg)* path; way; *(von Wasser)* course; *(fig.)* **einer Sache** *(Dat.)* **~ brechen** pave *or* prepare the way for sth.; **jmdn. aus der ~ werfen** *od.* **bringen** *od.* **schleudern** knock sb. sideways; **auf die schiefe ~ geraten** go astray; **b)** *(Strecke)* path; *(Umlauf~)* orbit; *(einer Rakete)* [flight-] path; *(eines Geschosses)* trajectory; **etw. [wieder] in die richtige ~ lenken** *(fig.)* get sth. [back] on the right track; **c)** *(Sport)* track; *(für Pferderennen)* course *(Brit.)*; track *(Amer.)*; *(für einzelne Teilnehmer)* lane; *(Kegel~)* alley; *(Schlitten~, Bob~)* run; *(Bowling~)* lane; **~ frei!** make way!; get out of the way!; **d)** *(Fahrspur)* lane; **e)** *(Eisen~)* railways *pl.*; railroad *(Amer.)*; *(Zug)* train; **jmdn. zur ~ bringen/an der ~ abholen** take sb. to/pick sb. up from the station; **mit der ~:** by train; **~ fahren** go by train; **f)** *(Straßen~)* tram; streetcar *(Amer.)*; **g)** *(Schienenweg)* railway [track]; **h)** *(Streifen)* *(Stoff~)* length; *(Tapeten~)* strip; length

bahn-, Bahn-: ~beamte der railway *or* *(Amer.)* railroad official; **~brechend** *Adj.* pioneering; **~brechend für etw. sein** pave *or* prepare the way for sth.; **~bus der** railway bus

Bähnchen ['bɛ:nçən] **das; ~s, ~:** little train

Bahn·damm der railway *or* *(Amer.)* railroad embankment

bahnen *tr. V.* clear ‹*way, path*›; **jmdm./einer Sache einen Weg ~:** clear the *or* a way for sb./sth.; *(fig.)* pave *or* prepare the way for sb./sth.; **sich** *(Dat.)* **einen Weg durch etw. ~:** force a *or* one's way through sth.

Bahn-: ~fahrt die train *or* rail journey; **~gleis das** railway *or* *(Amer.)* railroad track *or* line

Bahn·hof der [railway *or* *(Amer.)* railroad] station; **~ Käfertal** Käfertal station; **ich verstehe nur**

~ *(ugs.)* it's [all] double Dutch to me

Bahnhofs-: ~gast·stätte die station restaurant; **~halle die** station concourse; **~hotel das** station hotel; **~mission die** ≈ Travellers' Aid *(charitable organization for helping rail travellers in need of care or assistance)*

Bahn-: ~linie die railway *or (Amer.)* railroad line; **~polizei die** railway *or (Amer.)* railroad police; **~reise die** train *or* rail journey; **~schranke die** level-crossing *(Brit.) or (Amer.)* grade crossing barrier/gate; **~steig** [~ʃtaik] **der; ~[e]s, ~e** [station] platform; **~über·gang der** level-crossing *(Brit.)*; grade *or* railroad crossing *(Amer.)*; **~verbindung die** rail *(Brit.) or* train connection; **~wärter der** level-crossing keeper *(Brit.)*

Bahre ['ba:rə] **die; ~, ~n a)** *(Kranken~)* stretcher; **b)** *(Toten~)* bier

bairisch ['bairɪʃ] *Adj.* Bavarian

Baiser [bɛ'ze:] **das; ~s, ~s** meringue

Bajonett [bajo'nɛt] **das; ~[e]s, ~e** bayonet

Bajuware [baju'va:rə] **der; ~n, ~n** *(scherzh.)* Bavarian

Bakkarat ['bakara(t)] **das; ~s** baccarat

Bakterie [bak'te:riə] **die; ~, ~n** bacterium

bakteriell *Adj.* bacterial

bakteriologisch 1. *Adj.* bacteriological. **2.** *adv.* bacteriologically

Balalaika [bala'laika] **die; ~, ~s** *od.* **Balalaiken** balalaika

Balance [ba'laŋsə] **die; ~, ~n** balance; **die ~ halten/verlieren** keep/lose one's balance

balancieren [balaŋ'si:rən] **1.** *itr. V.; mit sein (auch fig.)* balance **2.** *tr. V.* balance

bald [balt] *Adv.* **a)** soon; **das wird er so ~ nicht vergessen** he won't forget that in a hurry; **wird's ~?** get a move on, will you; **bis ~:** see you soon; **b)** *(ugs.: fast)* almost; nearly; **c)** *(veralt.) in ~ ..., ~ ...:* now ..., now ...; **~ so, ~ so** now this way, now that

Baldachin ['baldaxi:n] **der; ~s, ~e** baldachin; *(über dem Bett)* canopy

baldig *Adj.; nicht präd.* speedy; quick

bald·möglichst *(Papierdt.) Adv.* as soon as possible

Baldrian ['baldria:n] **der; ~s, ~e** valerian

Balearen [bale'a:rən] *Pl.* **die ~:** the Balearic Islands

Balg das; ~[e]s, Bälger ['bɛlgɐ] *od.* **Bälge** ['bɛlgə] *(ugs., oft abwertend)* kid *(sl.)*; brat *(derog.)*

balgen *refl. V. (ugs.)* scrap *(coll.)* **(um** over); *(fig.)* fight **(um** over)

Balgerei die; ~, ~en *(ugs.)* scrap *(coll.)*

Balkan ['balka:n] **der; ~s: der ~:** the Balkans *pl.*; *(Gebirge)* the Balkan Mountains *pl.*; **auf dem ~:** in the Balkans

Balkan- Balkan ⟨*Peninsula, country, etc.*⟩

Balken ['balkn̩] **der; ~s, ~ a)** beam; **lügen, daß sich die ~ biegen** *(ugs.)* tell a [complete] pack of lies; **b)** *(Schwebe~)* beam; **c)** *(Musik)* cross-stroke

Balken·überschrift die banner headline

Balkon [bal'kɔŋ, bal'ko:n] **der; ~s, ~s** [bal'kɔŋs] *od.* **~e** [bal'ko:nə] **a)** balcony; **b)** *(in Theater)* [dress] circle; *(im Kino)* circle

Ball [bal] **der; ~[e]s, Bälle** ['bɛlə] **a)** *(auch fig.: Kugel)* ball; **~ spielen** play ball; **am ~ sein** have the ball; *(fig. ugs.)* be in touch; be on the ball *(coll.)*; **am ~ bleiben** *(fig. ugs.)* stick *(coll.) or* keep at it; **b)** *(Sportjargon: Schuß, Wurf)* ball; *(aufs Tor)* shot; **c)** *(Fest)* ball

ballaballa [bala'bala] *Adj.; nicht attr. (salopp)* crackers *(sl.)*; daft

Ballade [ba'la:də] **die; ~, ~n** ballad

balladenhaft*Adj.* ballad-like

Ballast [ba'last] **der; ~[e]s, ~e** ballast; *(fig.: in Text)* padding; **~ abwerfen** *od.* **über Bord werfen** *(fig.)* rid oneself of unnecessary burdens

Bällchen ['bɛlçən] **das; ~s, ~:** [little] ball

ballen 1. *tr. V.* clench ⟨*fist*⟩; crumple ⟨*paper*⟩ into a ball; press ⟨*snow etc.*⟩ into a ball. **2.** *refl. V.* ⟨*clouds*⟩ gather, build up; ⟨*crowd*⟩ gather; ⟨*fist*⟩ clench; *(fig.)* ⟨*problems, difficulties, etc.*⟩ accumulate, mount up

Ballen der; ~s, ~ a) bale; **b)** *(Hand~, Fuß~)* ball

Ballerei die; ~, ~en *(ugs.)* shoot-out

Baller·mann der; *Pl.* **-männer** *(ugs.)* shooting-iron *(sl.)*; shooter *(coll.)*

ballern *(ugs.)* **1.** *itr. V.* **a)** *(schießen)* fire [away]; bang away; **b)** *(schlagen)* bang, hammer **(gegen** on). **2.** *tr. V.* **a) jmdm. eine ~:** sock sb. one *(sl.)*; **b)** *(Sportjargon)* fire ⟨*ball*⟩

Ballett [ba'lɛt] **das; ~[e]s, ~e** ballet; **beim ~ sein** *(ugs.)* be a ballet-dancer

Ballistik die; ~: ballistics *sing., no art.*

ballistisch *Adj.* ballistic

Ball-: ~junge der ballboy; **~kleid das** ball dress *or* gown

Ballon [ba'lɔŋ] **der; ~s, ~s a)** balloon; **b)** *(salopp: Kopf)* nut *(sl.)*

Ball-: ~saal der ballroom; **~spiel das** ball game; **~spielen das; ~s** playing ball *no art.*; **~spielen verboten** no ball games

Ballungs·gebiet das conurbation

Ball·wechsel der *(Tennis, Tischtennis, Badminton)* rally

Balsam ['balza:m] **der; ~s, ~e** balsam; balm; *(fig.)* balm

Balte ['baltə] **der; ~n, ~n, Baltin die; ~, ~nen** Balt; **er ist Balte** he is from one of the Baltic states

Baltikum ['baltikʊm] **das; ~s: das ~:** the Baltic States *pl.*

baltisch *Adj.* Baltic

Balz [balts] **die; ~, ~en a)** courtship display; **b)** *(Zeit)* mating season

balzen *itr. V.* perform its/their courtship display

Bambus ['bambʊs] **der; ~** *od.* **~ses, ~se** bamboo

Bambus·rohr das bamboo [cane]

Bammel ['baml̩] **der; ~s** *(ugs.)* **~ vor jmdm./etw. haben** be scared stiff of sb./sth. *(coll.)*

banal [ba'na:l] **1.** *Adj.* **a)** *(abwertend: platt)* banal; trite, banal ⟨*speech, reply, etc.*⟩; **b)** *(gewöhnlich)* commonplace; ordinary. **2.** *adv.* **a)** *(abwertend: platt)* banally; tritely; **b)** *(gewöhnlich)* **~ gesagt** to put it plainly and simply

Banalität die; ~, ~en a) *o. Pl. s.* **banal 1:** banality; triteness; commonplaceness; ordinariness; **b)** *(Äußerung)* banality

Banane [ba'na:nə] **die; ~, ~n** banana

Banause [ba'nauzə] **der; ~n, ~n** *(abwertend)* philistine

band [bant] *1. u. 3. Pers. Sg. Prät. v.* **binden**

¹Band das; ~[e]s, Bänder ['bɛndɐ] **a)** *(Schmuck~; auch fig.)* ribbon; *(Haar~, Hut~)* band; *(Schürzen~)* string; **b)** *(Meß~)* measuring-tape; tape-measure; *(Farb~)* ribbon; *(Klebe~, Isolier~)* tape; **c)** *(Ton~)* [magnetic] tape; **etw. auf ~** *(Akk.)* **aufnehmen** tape[-record] sth.; **d)** *s.* **Förderband; e)** *s.* **Fließband; f) am laufenden ~** *(ugs.)* nonstop; continuously; **g)** *(Anat.)* ligament

²Band der; ~[e]s, Bände ['bɛndə] volume; **etw. spricht Bände** *(ugs.)* sth. speaks volumes

[3]**Band** [bɛnt] **die; ~, ~s** band
Bandage [ban'daːʒə] **die; ~, ~n** bandage; **mit harten ~n kämpfen** *(fig.)* fight with the gloves off
bandagieren [banda'ʒiːrən] *tr. V.* bandage
Band-: ~aufnahme die tape recording; **~breite die** *(fig.)* range
[1]**Bändchen** ['bɛntçən] **das; ~s, ~** *(kleines Band)* little ribbon
[2]**Bändchen das; ~s, ~** *(kleiner Band)* little volume
[1]**Bande** ['bandə] **die; ~, ~n a)** gang; **b)** *(ugs.: Gruppe)* mob *(sl.)*; crew
[2]**Bande die; ~, ~n** *(Sport)* [perimeter] barrier; *(mit Reklame)* billboards *pl.*; *(Billard)* cushion; *(der Eisbahn)* boards *pl.*
Bänder-: ~riß der *(Med.)* torn ligament; **~zerrung die** *(Med.)* pulled ligament
-bändig [-bɛndɪç] *adj.* -volume
bändigen ['bɛndɪgn̩] *tr. V.* tame ⟨*animal, sea, river, natural forces*⟩; control ⟨*person, anger, urge*⟩; bring ⟨*fire*⟩ under control
Bändigung die; ~: *s.* **bändigen:** taming; controlling; bringinmg under control
Bandit [ban'diːt] **der; ~en, ~en** bandit; brigand; *(fam. scherzh.)* rascal
Band-: ~säge die band-saw; **~scheibe die** *(Anat.)* [intervertebral] disc; **~wurm der** tapeworm
bang [baŋ] *s.* **bange**
bange; banger, bangst... *od.* **bänger** [bɛŋɐ], **bängst...** [bɛŋst...] **1.** *Adj.* afraid; scared; *(besorgt)* anxious; worried; **mir ist/wurde ~ [zumute]** I am *or* feel/became scared *or* frightened; **jmdm. ~ machen** scare *or* frighten sb. **2.** *adv.* anxiously
Bange die; ~ *(bes. nordd.)* fear; **[nur] keine ~!** don't be afraid; *(sei nicht besorgt)* don't worry
bangen *itr. V.* **um jmdn./etw. ~:** be anxious *or* worried about sb./sth.; worry about sb./sth.
Banjo ['banjo] **das; ~s, ~s** banjo
[1]**Bank** [baŋk] **die; ~, Bänke** ['bɛŋkə] bench; *(mit Lehne)* bench seat; *(Kirchen~)* pew; *(Anklage~)* dock; **etw. auf die lange ~ schieben** *(ugs.)* put sth. off; **durch die ~** *(ugs.)* every single one; the whole lot of them
[2]**Bank die; ~, ~en** bank; **Geld auf der ~ haben** have money in the bank; **ein Konto bei einer ~ eröffnen** open an account with a bank
Bank-: ~angestellte der/die; *adj. Dekl.*, *(veralt.)* **~beamte der** bank employee
Bänkchen ['bɛŋkçən] **das; ~s, ~:** little *or* small bench; *(mit Lehne)* little *or* small seat
Bank·direktor der director of a/the bank
Bänkel- ['bɛŋkl̩-]: **~lied das** street ballad; **~sänger der** singer of street ballads
Bankert ['baŋkɐt] **der; ~s, ~e** *(veralt. abwertend)* bastard
Bankett [baŋ'kɛt] **das; ~[e]s, ~e** banquet
Bankier [baŋ'kieː] **der; ~s, ~s** banker
Bank-: ~konto das bank account; **~leit·zahl die** bank sorting code number; **~note die** banknote; bill *(Amer.)*; **~raub der** bank robbery; **~räuber der** bank robber
bankrott [baŋ'krɔt] *Adj.* bankrupt; **jmdn./etw. ~ machen** bankrupt sb./sth.; **~ gehen** go bankrupt
Bankrott der; ~[e]s, ~e bankruptcy; **seinen ~ erklären** declare oneself bankrupt; **~ machen** go bankrupt
Bank-: ~überfall der bank raid; **~verbindung die** particulars of one's bank account
Bann [ban] **der; ~[e]s a)** *(kath. Kirche)* excommunication; **b)** *(fig. geh.)* spell; **in jmds. ~/im ~ einer Sache** *(Gen.)* **stehen** be under sb.'s spell/under the spell of sth.; **jmdn. in seinen ~ schlagen** cast one's/its spell over sb.
bannen *tr. V.* **a)** *(festhalten)* entrance; captivate; **[wie] gebannt** ⟨*watch, listen, etc.*⟩ spellbound; **b)** *(vertreiben)* exorcize ⟨*spirit*⟩; avert, ward off ⟨*danger*⟩
Banner das; ~s, ~: banner
Bantu ['bantu] **der; ~[s], ~[s]** Bantu
bar [baːɐ̯] **1.** *Adj.* **a)** *nicht präd.* cash; **~es Geld** cash; **in ~:** in cash; **Verkauf nur gegen ~:** cash sales only; *s. auch* **Münze a; b)** *nicht präd.* *(pur)* pure; sheer. **2.** *adv.* in cash; **~ auf die Hand** *(ugs.) od.* *(salopp)* **Kralle** cash on the nail
Bar die; ~, ~s a) *(Nachtlokal)* night-club; bar; **b)** *(Theke)* bar
Bär [bɛːɐ̯] **der; ~en, ~en** bear; **der Große/ Kleine ~** *(Astron.)* the Great/Little Bear; **jmdm. einen ~en aufbinden** have sb. on *(coll.)*; pull sb.'s leg
Baracke [ba'rakə] **die; ~, ~n** hut
Barbar [bar'baːɐ̯] **der; ~en, ~en** barbarian
Barbarei die; ~, ~en a) *(Roheit)* barbarity; **b)** *(Kulturlosigkeit)* barbarism *no indef. art.*
Barbarin die; ~, ~nen *(auch hist.)* barbarian
barbarisch 1. *Adj.* **a)** *(roh)* barbarous; savage; barbarous, brutal ⟨*torture*⟩; **b)** *(unzivilisiert)* barbaric; barbaric, uncivilized ⟨*person*⟩. **2.** *adv.* **a)** *(roh)* barbarously; ⟨*torture*⟩ barbarously, brutally; **b)** *(unzivilisiert)* barbarically; in an uncivilized manner
bar·busig [-buːzɪç] *Adj.* topless
Bar·dame die barmaid
Barde ['bardə] **der; ~n, ~n** bard
bären-, Bären-: ~dienst der *in* **jmdm. einen ~dienst erweisen** do sb. a disservice; **~fell das** bearskin; **~hunger der** *(ugs.)* **einen ~hunger haben/kriegen** be famished *(coll.)* *or* starving *(coll.)*/get famished *(coll.)* *or* ravenous *(coll.)*; **~stark** *Adj.* as strong as an ox *postpos.*
Barett [ba'rɛt] **das; ~[e]s, ~e** *(eines Geistlichen)* biretta; *(eines Richters, Professors)* cap; *(Baskenmütze)* beret
bar-: ~fuß *indekl. Adj.* barefooted; **~fuß herumlaufen/gehen** run about/go barefoot; **~füßig** [~fyːsɪç] *Adj.* *(geh.)* barefooted
barg [bark] *1. u. 3. Pers. Sg. Prät. v.* **bergen**
bar-, Bar-: ~geld das cash; **~geld·los 1.** *Adj.* cashless; **2.** *adv.* without using cash; **~hokker der** bar stool
Bärin ['bɛːrɪn] **die; ~, ~nen** she-bear
Bariton ['ba(ː)ritɔn] **der; ~s, ~e** baritone; *(im Chor)* baritones *pl.*
Bark [bark] **die; ~, ~en** barque
Barkasse [bar'kasə] **die; ~, ~n** launch
Barke ['barkə] **die; ~, ~n** [small] rowing-boat
barmherzig [barm'hɛrtsɪç] *(geh.)* **1.** *Adj.* merciful; compassionate; **~er Gott/Himmel!** merciful God/Heaven!; *s. auch* **Samariter. 2.** *adv.* mercifully; compassionately
Barmherzigkeit die; ~ *(geh.)* mercy; compassion
Bar·mittel *Pl.* cash resources
Bar·mixer der barman; barkeeper *(Amer.)*
barock [ba'rɔk] *Adj.* baroque
Barock das *od.* **der; ~[s]** baroque; *(Zeit)* baroque period *or* age
Barometer [baro'meːtɐ] **das** barometer; **das ~ steht auf Sturm** *(fig.)* the atmosphere is very strained
Baron [ba'roːn] **der; ~s, ~e** baron; *(als Anrede)* **[Herr] ~:** ≈ my lord
Baronin die; ~, ~nen baroness; *(als Anrede)* **[Frau] ~:** ≈ my lady

Barras ['baras] **der;** ~ *(Soldatenspr.)* army; **beim** ~: in the army

Barren ['barən] **der;** ~**s,** ~ **a)** bar; **b)** *(Turngerät)* parallel bars *pl.*

Barriere [ba'ri̯e:rə] **die;** ~, ~**n** *(auch fig.)* barrier

Barrikade [bari'ka:də] **die;** ~, ~**n** barricade; **auf die** ~**n gehen** *od.* **steigen** *(ugs.)* go on the warpath

barsch [barʃ] **1.** *Adj.* curt. **2.** *adv.* curtly; **jmdn.** ~ **anfahren** snap at sb.

Barsch der; ~**[e]s,** ~**e** *(Zool.)* perch

Barschaft die; ~, ~**en** [ready] cash; **seine ganze** ~ **bestand aus 20 Mark** all he had was 20 marks

Bar·scheck der open *or* uncrossed cheque

barst [barst] *1. u. 3. Pers. Sg. Prät. v.* **bersten**

Bart [ba:ɐ̯t] **der;** ~**[e]s, Bärte** ['bɛ:ɐ̯tə] **a)** beard; *(Oberlippen~, Schnurr~)* moustache; *(ugs.: Schnurrhaare)* whiskers *pl.;* **sich** *(Dat.)* **einen** ~ **wachsen** *od.* **stehen lassen** grow a beard; *(fig.)* **der** ~ **ist ab** *(ugs.)* it's all over; **der Witz hat [so] einen** ~ *(ugs.)* that joke is as old as the hills; **etw. in seinen** ~ **brummen** *od.* **murmeln** mumble sth.; **jmdm. um den** ~ **gehen** *(abwertend)* butter sb. up; **b)** *(am Schlüssel)* bit

Bärtchen ['bɛ:ɐ̯tçən] **das;** ~**s,** ~: [small] beard; *(Schnurr~)* [thin] moustache

Bart·haar das hair from sb.'s/one's beard

bärtig ['bɛ:ɐ̯tɪç] *Adj.* bearded

bart-, Bart-: ~**los** *Adj.* beardless; ~**stoppel die** piece of stubble; ~**stoppeln** stubble *sing.;* ~**tracht die** style of beard; ~**träger der** man with a beard; ~**wuchs der** growth of beard

Bar·zahlung die cash payment

Basalt [ba'zalt] **der;** ~**[e]s,** ~**e** basalt

Basar [ba'za:ɐ̯] **der;** ~**s,** ~**e** bazaar

[1]**Base** ['ba:zə] **die;** ~, ~**n a)** *(veralt.: Cousine)* cousin; **b)** *(schweiz.: Tante)* aunt

[2]**Base die;** ~, ~**n** *(Chemie)* base

Baseball ['beɪsbɔ:l] **der;** ~**s** baseball

Basedow-Krankheit ['ba:zədo-] **die, Basedowsche Krankheit die** *(Med.)* exophthalmic goitre; Graves' disease

Basel ['ba:zl̩] **(das);** ~**s** Basle

Basen *s.* **Basis**

basieren *itr. V.* **auf etw.** *(Dat.)* ~: be based on sth.

Basilika [ba'zi:lika] **die;** ~, **Basiliken** *(Kunstwiss.)* basilica

Basilikum [ba'zi:likʊm] **das;** ~**s** basil

Basis ['ba:zɪs] **die;** ~, **Basen a)** *(Grundlage)* basis; **auf einer festen** ~ **ruhen** have a firm basis; **b)** *(Math., Archit., Milit., marx.)* base; **c)** *(Politik)* grass roots *pl.;* **an der** ~: at grass-roots level

basisch *(Chemie)* **1.** *Adj.* basic. **2.** *adv.* ⟨*react*⟩ as a base

Basis·demokratie die *(Politik)* grass-roots democracy

Baske ['baskə] **der;** ~**n,** ~**n** Basque

Basken-: ~**land das** Basque region; ~**mütze die** beret

Basket·ball ['ba(:)skət-] **der** basketball

Baskin die; ~, ~**nen** Basque

baskisch *Adj.* Basque

baß [bas] *Adv. in* ~ **erstaunt sein** *(veralt.)* be quite taken aback

Baß der; Basses, Bässe ['bɛsə] **a)** bass; *(im Chor)* basses *pl.;* **b)** *(Instrument)* double-bass; bass *(coll.)*

Baß·geige die *(volkst.)* double-bass

Bassin [ba'sɛ̃:] **das;** ~**s,** ~**s** *s.* **Becken b**

Bassist der; ~**en,** ~**en** *(Musik)* **a)** *(Sänger)* bass; **b)** *(Instrumentalist)* double-bass player; bassist; *(Gitarrist)* bass guitarist

Baß-: ~**schlüssel der** *(Musik)* bass clef; ~**stimme die** bass voice

Bast [bast] **der;** ~**[e]s,** ~**e** bast; *(Raffia~)* raffia

basta ['basta] *Interj. (ugs.)* that's enough; **und damit** ~! and that's that

Bastard ['bastart] **der;** ~**s,** ~**e a)** *(veralt., salopp)* bastard; **b)** *(Biol.)* hybrid

Bastel·arbeit die piece of handicraft work; ~**en** handicraft work *sing.*

Bastelei die; ~, ~**en a)** handicraft work *no pl.;* **b)** *(Gegenstand)* piece of handicraft work

basteln ['bastl̩n] **1.** *tr. V.* make; make, build ⟨*model, device*⟩. **2.** *itr. V.* make things [with one's hands]; do handicraft work; **an etw.** *(Dat.)* ~: be working on sth.; *(etw. herstellen)* be making sth.; *(etw. laienhaft bearbeiten)* tinker with sth.

Bastion [bas'ti̯o:n] **die;** ~, ~**en** bastion

Bastler ['bastlɐ] **der;** ~**s,** ~: handicraft enthusiast

bat [ba:t] *1. u. 3. Pers. Sg. Prät. v.* **bitten**

Bataillon [batal'jo:n] **das;** ~**s,** ~**e** *(Milit.)* battalion

Batik ['ba:tɪk] **der;** ~**s,** ~**en** *od.* **die;** ~, ~**en** batik

batiken 1. *tr. V.* **etw.** ~: decorate sth. with batik work. **2.** *itr. V.* do batik work

Batist [ba'tɪst] **der;** ~**[e]s,** ~**e** batiste

Batterie [batə'ri:] **die;** ~, ~**n** battery; **eine ganze** ~ **von leeren Flaschen** *(fig. ugs.)* rows of empties *(coll.)*

Batzen ['batsn̩] **der;** ~**s,** ~ **a)** *(ugs.: Klumpen)* lump; **b)** *(ugs.: Menge)* pile *(coll.);* **ein [schöner** *od.* **ganzer]** ~ **Geld** a pile *(coll.)* [of money]

[1]**Bau** [bau̯] **der;** ~**[e]s,** ~**ten a)** *o. Pl.* building; construction; **im** ~ **sein** be under construction; **mit dem** ~ **[von etw.] beginnen** start construction [of sth.]; start building [sth.]; **b)** *(Gebäude)* building; **c)** *o. Pl.* *(~stelle)* building site; **auf dem** ~ **arbeiten** *(Bauarbeiter sein)* be in the building trade; **d)** *o. Pl.* *(Struktur)* structure; **e)** *o. Pl.* *(Körper~)* build; **von schmalem** ~ **sein** be slenderly built; have a slender physique

[2]**Bau** [bau̯] **der;** ~**[e]s,** ~**e** *(Kaninchen~)* burrow; hole; *(Fuchs~)* earth; *(Wolfs~)* lair; *(Dachs~)* sett; earth; **nicht aus dem** ~ **gehen/kommen** *(fig. ugs.)* not stick *or* put one's nose outside the door *(coll.)*

Bau-: ~**arbeiten** *Pl.* building *or* construction work *sing.; (Straßenarbeiten)* road-works; ~**bude die** site hut

Bauch [baux] **der;** ~**[e]s, Bäuche** ['bɔyçə] **a)** stomach; belly; abdomen *(Anat.);* tummy *(coll.); (fig.: von Schiffen, Flugzeugen)* belly; **mir tut der** ~ **weh** I have [a] stomach-ache *or (coll.)* tummy-ache; **sich** *(Dat.)* **den** ~ **vollschlagen** *(ugs.)* stuff oneself *(coll.);* **ich habe nichts im** ~ *(ugs.)* I haven't had anything to eat; **sich** *(Dat.)* **[vor Lachen] den** ~ **halten** *(ugs.)* split one's sides [with laughing]; *(fig.)* **auf den** ~ **fallen** *(ugs.)* come a cropper *(sl.)* **(mit** with); **aus dem hohlen** ~ *(salopp)* off the top of one's head *(sl.);* **b)** *(Wölbung des ~s)* paunch; corporation *(coll.);* **c)** *(Kochk.) (vom Schwein)* belly; *(vom Kalb)* flank

Bauch·binde die a) woollen body-belt; **b)** *(ugs.: bei Zigarren, Büchern)* band

bauchig *Adj.* bulbous

Bauch-: ~**klatscher** [~klatʃɐ] **der;** ~**s,** ~ *(ugs.)* belly-flop *(coll.);* ~**laden der** vendor's tray; ~**landung die** belly-landing

Bäuchlein ['bɔyçlain] **das; ~s, ~:** stomach; tummy *(coll.)*

bauch-, Bauch-: ~muskel der stomach muscle; **~nabel der** *(ugs.)* belly-button *(coll.)*; tummy-button *(coll.)*; **~reden** *itr. V.; nur Inf. gebr.* ventriloquize; **~redner der** ventriloquist; **~schmerz der;** *meist Pl.* stomach pain; **~schmerzen** stomach-ache *sing.;* stomach pains; **~speichel·drüse die** pancreas; **~tanz der** belly-dance; **~tanzen** *itr. V.; nur Inf. gebr.* belly-dance; **~tänzerin die** belly-dancer; **~weh das** *(ugs.)* tummy-ache *(coll.)*; stomach-ache

Bau-: ~denkmal das architectural monument; **~element das** component

bauen 1. *tr. V.* **a)** build; build, construct ‹*house, road, bridge, etc.*›; make ‹*violin, piano, burrow*›; *s. auch* **Bett a; b)** *(ugs.)* **seinen Doktor ~:** do one's Ph.D.; **c)** *(ugs.: verursachen)* **einen Unfall ~:** have an accident. **2.** *itr. V.* **a)** build; **wir wollen ~:** we want to build a house; *(bauen lassen)* we want to have a house built; **an etw.** *(Dat.)* **~:** do building work on sth.; **b)** *fig.)* **auf jmdn./etw. ~:** rely on sb./sth.

[1]**Bauer** ['bauɐ] **der; ~n, ~n a)** farmer; *(mit niedrigem sozialem Status, auch ugs. abwertend)* peasant; **die dümmsten ~n haben die dicksten Kartoffeln** *(abwertend)* fortune favours fools *(prov.)*; **b)** *(Schachfigur)* pawn; **c)** *(Spielkarte)* jack

[2]**Bauer das** *od.* **der; ~s, ~:** bird-cage; cage

Bäuerchen ['bɔyɐçən] **das; ~s, ~: [ein] ~ machen** *(Kinderspr.)* burp

Bäuerin ['bɔyərɪn] **die; ~, ~nen** [lady] farmer; *(Frau eines Bauern)* farmer's wife; *(mit niedrigem sozialem Status)* peasant [woman]

Bäuerlein ['bɔyɐlain] **das; ~s, ~:** [simple] peasant

bäuerlich ['bɔyɐlɪç] **1.** *Adj.* farming *attrib.; (ländlich)* rural. **2.** *adv.* rurally

bauern-, Bauern-: ~fang der *in* **auf ~fang ausgehen** *(ugs. abwertend)* set out to con people out of their money *(coll.)*; **~fänger der** *(ugs. abwertend)* con man *(coll.)*; **~frühstück das** *(Kochk.) fried potatoes mixed with scrambled egg and bacon;* **~haus das** farm-house; **~hoch·zeit die** country wedding; **~hof der** farm; **~krieg der** *(hist.)* **der Große ~krieg, die ~kriege** the Peasant[s'] War; **~regel die** country saying; **~schlau 1.** *Adj.* cunning; sly; crafty; **2.** *adv.* cunningly; slyly; craftily

Bauers-: ~frau die, *s.* **Bäuerin; ~leute** *Pl. (Bauer und Bäuerin)* **die [beiden] ~leute** the farmer and his wife

bau-, Bau-: ~fällig *Adj.* ramshackle; badly dilapidated; unsafe ‹*roof, ceiling*›; **~fälligkeit die;** *o. Pl.* bad state of dilapidation; badly dilapidated state; **~firma die** building *or* construction firm; **~gerüst das** scaffolding; **~hütte die** *(MA.)* stonemasons' lodge; **~ingenieur der** building engineer; **~jahr das** year of construction; *(bei Autos)* year of manufacture; **~kasten der** construction set *or* kit; *(mit Holzklötzen)* box of bricks; **~kasten·system das;** *o. Pl.* unit construction system; **~klotz der** building-brick; **~klötze[r] staunen** *(salopp)* be staggered *(coll.) or* flabbergasted; **~kosten** *Pl.* building *or* construction costs; **~kunst die;** *o. Pl. (geh.)* architecture; **~land das;** *o. Pl.* building land

baulich 1. *Adj.; nicht präd.* structural. **2.** *adv.* **etw. ~ verändern** carry out structural alterations to sth.

Baulichkeit die; ~, ~en building

Baum [baum] **der; ~[e]s, Bäume** ['bɔymə] tree; **er ist stark wie ein ~:** he's as strong as an ox; **Bäume ausreißen können** *(fig. ugs.)* be *or* feel ready to tackle anything

Bau-: ~maschine die piece of construction plant *or* machinery; **~maschinen** construction plant *sing. or* machinery; **~material das** building material

Bäumchen ['bɔymçən] **das; ~s, ~:** small tree; *(junger Baum)* sapling; young tree; **~, wechsle dich** *(Kinderspiel)* puss in the corner

Bau·meister der *(hist.)* [architect and] master builder

baumeln ['baumḷn] *itr. V.* **a)** *(ugs.)* dangle **(an** + *Dat.* from); **die Beine ~ lassen** dangle one's legs; **b)** *(derb: gehängt werden)* swing *(sl.)*

baum-, Baum-: ~grenze die tree-line; timber-line; **~gruppe die** clump of trees; **~krone die** treetop; crown [of the/a tree]; **~lang** *Adj. (ugs.)* tremendously tall *(coll.)*; **~rinde die** bark [of trees]; **~schule die** [tree] nursery; **~stamm der** tree-trunk; **~sterben das** dying-off of trees; **~stumpf der** tree-stump; **~wolle die** cotton; **~wollen** *Adj.; nicht präd.* cotton

Bau-: ~plan der *(Zeichnung)* building plans *pl.; (für eine Maschine)* designs *pl.;* **~planung die** building design; **~platz der** site for building *or* construction

bäurisch ['bɔyrɪʃ] *(abwertend)* **1.** *Adj.* boorish; oafish. **2.** *adv.* boorishly; oafishly

Bau·satz der kit

Bausch [bauʃ] **der; ~[e]s, ~e** *od.* **Bäusche** ['bɔyʃə] **a)** *(am Kleid, Ärmel)* puff; **b) ein ~ Watte** a wad of cotton wool; **c) etw. in ~ und Bogen verwerfen/verdammen** reject/condemn sth. wholesale

bauschen 1. *tr. V.* billow, fill ‹*sail, curtains, etc.*›; **gebauschte Ärmel** puffed *or* puff sleeves. **2.** *refl. V.* ‹*dress, sleeve*› puff out; *(ungewollt)* bunch up; become bunched up; *(im Wind)* ‹*curtain, flag, etc.*› billow [out]

bauschig *Adj.* puffed ‹*dress*›; baggy ‹*trousers*›

bau-, Bau-: ~sparen *itr. V.; nur Inf. gebr.* save with a building society; **~sparer der** building-society investor; **~spar·kasse die** ≈ building society; **~spar·vertrag der** savings contract with a building society *(to save a specified sum which earns interest and is later used to pay for the building of a house)*; **~stein der a)** building stone; **b)** *(Bestandteil)* element; component; *(Elektronik, DV)* module; **die ~steine der Materie** the constituents of matter; **c)** *(~klotz)* building-brick; **~stelle die** building site; *(beim Straßenbau)* road-works *pl.; (bei der Eisenbahn)* site of engineering works; **~stil der** architectural style; **~stoff der** building material; **~teil das** component

Bauten *s.* **Bau**

Bau-: ~tischler der [building] joiner; **~unternehmen das** building firm; **~unternehmer der** building contractor; builder; **~vorhaben das** building project; **~weise die** method of building *or* construction; **~werk das** building; *(Brücke, Staudamm)* structure

Bauxit [bau'ksi:t] **der; ~s, ~e** bauxite

Bau·zaun der site fence

Bayer ['baiɐ] **der; ~n, ~n** Bavarian

bay[e]risch *Adj.* Bavarian

Bayern ['baiɐn] **(das); ~s** Bavaria

Bazille [ba'tsɪlə] **die; ~, ~n** *(ugs.) s.* **Bazillus**

Bazillus [ba'tsɪlʊs] **der; ~, Bazillen a)** bacillus; **b)** *(fig.)* cancer

Bd. *Abk.* **Band** Vol.

B-Dur ['be:-] **das;** ~ *(Musik)* B flat major

beabsichtigen [bə'lapzɪçtɪgn̩] *tr. V.* intend; **~, etw. zu tun** intend *or* mean to do sth.; **die beabsichtigte Wirkung** the intended *or* desired effect

beachten *tr. V.* **a)** observe, follow ⟨*rule, regulations*⟩*;* follow ⟨*instruction*⟩*;* heed, follow ⟨*advice*⟩*;* obey ⟨*traffic signs*⟩*;* observe ⟨*formalities*⟩*;* **b)** *(berücksichtigen)* take account of; *(Aufmerksamkeit schenken)* pay attention to *or* take notice of sth.; **es ist zu ~, daß ...:** please note that ...; **jmdn. nicht ~:** ignore sb.

beachtens·wert *Adj.* remarkable

beachtlich 1. *Adj.* considerable; marked, considerable ⟨*change, increase, improvement, etc.*⟩*;* notable, considerable ⟨*success*⟩*;* **Beachtliches leisten** make one's mark. **2.** *adv.* considerably; ⟨*change, increase, improve, etc.*⟩ markedly, considerably

Beachtung die a) *s.* **beachten a:** observance; following; heeding; obeying; **bei ~ der Regeln** if one observes *or* follows the rules; **b)** *(Berücksichtigung)* consideration; **unter ~ aller Umstände** taking all the circumstances into account; **c)** *(Aufmerksamkeit)* attention; **~/keinerlei ~ finden** receive attention/be ignored completely

Beamte [bə'lamtə] **der;** *adj. Dekl.* official; *(Staats~)* [permanent] civil servant; *(Kommunal~)* [established] local government officer *or* official; *(Polizei~)* [police] officer; *(Zoll~)* [customs] officer *or* official

Beamten-: ~beleidigung die insulting a public servant; **~laufbahn die** career in the civil service *or* as a civil servant

beamtet *Adj.* **~ sein** have permanent civil-servant status

Beamtin die; ~, ~nen *s.* **Beamte**

beängstigend 1. *Adj.* worrying ⟨*feeling*⟩*;* unsettling ⟨*sign*⟩*;* eerie ⟨*silence*⟩*;* **ein ~es Gedränge** a frightening crush of people; **sein Zustand ist ~:** his condition is giving cause for anxiety. **2.** *adv.* alarmingly; **~ schnell** at an alarming speed

beanspruchen *tr. V.* **a)** claim; **etw. ~ können** be entitled to expect sth.; **b)** *(ausnutzen)* make use of ⟨*person, equipment*⟩*;* take advantage of ⟨*hospitality, services*⟩*;* **c)** *(erfordern)* demand ⟨*energy, attention, stamina*⟩*;* take up ⟨*time, space, etc.*⟩

Beanspruchung die; ~, ~en demands *pl.* (*Gen.* on); **die ~ durch den Beruf** the demands of his/her job

beanstanden *tr. V.* object to; take exception to; *(sich beklagen über)* complain about; **an der Arbeit ist nichts/allerlei zu ~:** there ist nothing/there are all sorts of things wrong with the work; **~, daß ...:** complain that ...

Beanstandung die; ~, ~en complaint; **Anlaß zu ~en geben** give cause for complaint *sing.*

beantragen *tr. V.* **a)** apply for; **ich beantrage, mich zu versetzen** I apply to be transferred; **b)** *(fordern)* call for; demand; **c)** *(vorschlagen)* propose

beantworten *tr. V.* answer; reply to, answer ⟨*letter*⟩*;* respond to ⟨*insult*⟩*;* return ⟨*greeting*⟩*;* **jmdm. eine Frage ~:** answer a question for sb.

Beantwortung die; ~, ~en: zur ~ dieser Frage bist du nicht verpflichtet you are not obliged to answer this question; **in ~** *(Amtsspr.)* in reply (*Gen.* to); **zur ~ Ihrer Frage** in order to answer your question

bearbeiten *tr. V.* **a)** deal with; work on, handle ⟨*case*⟩*;* treat ⟨*subject*⟩*;* edit ⟨*book*⟩*;* **ein Buch völlig neu ~:** revise a book completely; **b)** *(adaptieren)* adapt; **ein Stück für Klavier ~:** arrange a piece for the piano; **c)** *(behandeln)* treat; work ⟨*wood, metal, leather, etc.*⟩*;* **etw. mit einer Bürste ~:** work on sth. with a brush; **den Boden ~:** work the soil; **d)** *(ugs.: traktieren)* beat [repeatedly]; hammer away on ⟨*piano, typewriter keys, etc.*⟩*;* **jmdn. mit den Fäusten ~:** pummel sb.; **e)** *(ugs.: überreden)* work on

Bearbeiter der, Bearbeiterin die a) der zuständige Bearbeiter the person who is dealing/who dealt with the matter *etc.;* **b)** *(eines Romans, Schauspiels)* adapter; *(eines Musikstücks)* arranger; **c)** *(eines Buches)* editor

Bearbeitung die; ~, ~en a) die ~ eines Antrags/eines Falles *usw.* dealing with an application/working on *or* handling a case *etc.;* **die ~ eines Themas** the treatment of a subject; **die ~ eines Buches** editing a book; **b)** *(bearbeitete Fassung)* adaptation; *(eines Musikstücks)* arrangement; **c)** *(Behandlung)* treatment; *(von Holz, Metall, Leder usw.)* working; **die ~ des Metalls ist schwer** it is difficult to work the metal; **zur weiteren ~:** in order to be worked further/for further treatment

Bearbeitungs·gebühr die handling charge; *(bei Behörden)* administrative charge

beargwöhnen *tr. V.* **jmdn./etw. ~:** be suspicious of sb./sth.; **beargwöhnt werden** be regarded with suspicion

Beat [bi:t] **der; ~[s]** beat

beatmen *tr. V.* *(Med.)* **jmdn. [künstlich] ~:** administer artificial respiration to sb.; *(während einer Operation)* ventilate sb.

beaufsichtigen *tr. V.* supervise; mind, look after ⟨*child*⟩

Beaufsichtigung die; ~, ~en supervision

beauftragen *tr. V.* **a) jmdn. mit etw. ~:** entrust sb. with sth.; charge sb. with sth.; **jmdn. ~, etw. zu tun** give sb. the job *or* task of doing sth.; **einen Künstler/Architekten ~, etw. zu tun** commission an artist/architect to do sth.; **b)** *(anweisen)* **jmdn. ~, etw. zu tun** order sb. to do sth.

Beauftragte der/die; *adj. Dekl.* representative

beäugen *tr. V.* eye ⟨*person*⟩*;* inspect ⟨*thing*⟩

bebauen *tr. V.* **a)** build on; develop; **ein Gelände mit Häusern ~:** build houses on a site; **b)** *(für den Anbau nutzen)* cultivate

Bebauung die; ~, ~en a) *(mit Gebäuden)* development; **b)** *(Gebäude)* buildings *pl.;* **c)** *(eines Ackers)* cultivation

beben ['be:bn̩] *itr. V.* shake; tremble

Beben das; ~s, ~ a) shaking; trembling; **b)** *(Erd~)* earthquake; quake *(coll.)*

bebildern *tr. V.* illustrate

Becher ['bɛçɐ] **der; ~s, ~** *(Glas~, Porzellan~)* glass; tumbler; *(Plastik~)* beaker; cup; *(Eis~)* *(aus Glas, Metall)* sundae dish; *(aus Pappe)* tub; *(Joghurt~)* carton

bechern *tr., itr. V.* *(ugs. scherzh.)* **[einen] ~:** have a few *(coll.)*

Becken ['bɛkn̩] **das; ~s, ~ a)** *(Wasch~)* basin; *(Abwasch~)* sink; *(Toiletten~)* pan; bowl; **b)** *(Schwimm~)* pool; *(Plansch~)* paddling-pool; *(eines Brunnens)* basin; *(Fisch~)* pond; **c)** *(Anat.)* pelvis; **d)** *Pl.* *(Musik)* cymbals *pl.*

Becken-: ~bruch der *(Med.)* pelvic fracture; **~knochen der** pelvic bone

bedacht 1. *Adj.* **a)** carefully considered; *(umsichtig)* circumspect; **b) auf etw.** *(Akk.)* **~ sein** be intent on sth.; **darauf ~/sehr ~ sein, etw. zu tun** be intent on doing sth./be [most] anxious to do sth. **2.** *adv.* in a carefully considered way; *(umsichtig)* circumspectly

Bedacht der *in* **ohne ~:** rashly; without thinking *or* forethought; **mit ~:** in a carefully considered way; *(umsichtig)* circumspectly

bedächtig [bə'dɛçtɪç] **1.** *Adj.* **a)** deliberate; measured ‹*steps, stride, speech*›; **b)** *(besonnen)* thoughtful; well-considered ‹*words*›; *(vorsichtig)* careful. **2.** *adv.* **a)** deliberately; **~ reden** speak in measured tones; **b)** *(besonnen)* thoughtfully; *(vorsichtig)* carefully

Bedächtigkeit die; ~ a) deliberateness; **b)** *(Besonnenheit)* thoughtfulness; *(Vorsichtigkeit)* carefulness

bedanken *refl. V.* say thank you; express one's thanks; **sich bei jmdm. [für etw.] ~:** thank sb. *or* say thank you to sb. [for sth.]; **sich bei jmdm. ~, daß er etw. getan hat** thank sb. for doing sth.; **dafür kannst du dich bei ihm ~** *(iron. ugs.)* you've got him to thank for that *(iron.)*

Bedarf [bə'darf] **der; ~[e]s a)** need (**an** + *Dat.* of); requirement (**an** + *Dat.* for); *(Bedarfsmenge)* needs *pl.*; requirements *pl.*; **Dinge des täglichen ~s** everyday necessities; **bei ~:** if and when the need arises; if required; **je nach ~:** as required; **kein ~!** *(salopp)* I don't feel like it; **b)** *(Nachfrage)* demand (**an** + *Dat.* for)

Bedarfs-: ~ampel die *(Verkehrsw.)* traffic-controlled lights; *(Fußgängerampel)* pedestrian-controlled *or* -operated lights; **~fall der** *in* **im ~fall[e]** if required; if the need arises/arose; **~güter** *Pl.* consumer goods; **~halte·stelle die** request stop

bedauerlich *Adj.* regrettable; unfortunate

bedauerlicher·weise *Adv.* regrettably; unfortunately

bedauern *tr., itr. V.* **a)** feel sorry for; **sie läßt sich gerne ~:** she likes being pitied; **b)** *(schade finden)* regret; **ich bedaure sehr, daß ...:** I am very sorry that ...; **wir ~, Ihnen mitteilen zu müssen** we regret to [have to] inform you; **bedaure!** sorry!

Bedauern das; ~s a) sympathy; **jmdm. sein ~ ausdrücken** offer one's sympathy to sb.; **b)** *(Betrübnis)* regret; **zu meinem ~:** to my regret; **zu unserem ~ müssen wir Ihnen mitteilen, daß ...:** we regret to [have to] inform you that ...; **mit ~:** with regret

bedauerns·wert *Adj.* unfortunate

bedecken 1. *tr. V.* cover; **von Schlamm/Schmutz bedeckt sein** be covered in mud/dirt. **2.** *refl. V.* cover oneself up

bedeckt *Adj.* **a)** overcast ‹*sky*›; **bei ~em Himmel** when the sky is overcast; **b) sich ~ halten** *(fig.)* keep a low profile

Bedeckung die; ~, ~en a) covering; **b)** *(Schutz)* guard

bedenken *unr. tr. V.* **a)** consider; think about; **wenn ich es recht bedenke/wenn man es recht bedenkt** when I/you stop and think about it; **b)** *(beachten)* take into consideration; **du mußt ~, daß ...:** you should bear in mind that *or* take into consideration the fact that ...; **ich gebe [dir]/er gab [uns] zu ~, daß ...:** I would ask you/he asked us to bear in mind that *or* take into consideration the fact that ...; **c)** *(geh.: beschenken)* **jmdn. reich ~:** shower sb. with gifts; **jmdn. mit etw. ~:** present sb. with sth.; **d)** *(im Testament berücksichtigen)* remember

Bedenken das; ~s, ~: doubt, reservation (**gegen** about); **aber jetzt kommen mir ~:** but now I'm having second thoughts; **ohne ~:** without hesitation

bedenken·los 1. *Adj.* unhesitating; *(skrupellos)* unscrupulous. **2.** *adv.* without hesitation; *(skrupellos)* unscrupulously

bedenkens·wert *Adj.* ‹*argument, suggestion*› worthy of consideration; **~ sein** be worth considering *or* worthy of consideration

bedenklich 1. *Adj.* **a)** *(fragwürdig)* dubious, questionable ‹*methods, transactions, etc.*›; **b)** *(besorgniserregend)* alarming; disturbing; **~ sein/werden** be giving/be starting to give cause for concern; **c)** *(besorgt)* concerned; apprehensive; anxious; **das machte** *od.* **stimmte mich ~:** that gave me cause for concern. **2.** *adv.* **a)** alarmingly; disturbingly; **b)** *(besorgt)* apprehensively; anxiously

Bedenklichkeit die; ~: dubiousness; questionableness

Bedenk·zeit die time for reflection; **ich gebe Ihnen vierundzwanzig Stunden ~:** I'll give you twenty-four hours to think about it

bedeppert [bə'dɛpɐt] *(salopp) Adj.* **a)** *(verwirrt)* confused and embarrassed; **b)** *(dumm)* gormless *(coll.)*

bedeuten *tr. V.* **a)** mean; **was soll das ~?** what does that mean?; **„Ph. D." bedeutet Doktor der Philosophie** 'Ph.D.' stands for Doctor of Philosophy; **b)** *(darstellen)* represent; **das bedeutet ein Wagnis** that is being really daring; **einen Eingriff in die Pressefreiheit ~:** amount to *or* represent an attack on press freedom; **c)** *(hindeuten auf)* mean; **das bedeutet nichts Gutes** that bodes ill; that's a bad sign; **schönes Wetter ~:** mean good weather; **das hat nichts zu ~:** that doesn't mean anything; **d)** *(wert sein)* mean; **Geld bedeutet ihm nichts** money means nothing to him; **e)** *(geh.)* **jmdm. ~, etw. zu tun** intimate *or* indicate to sb. that he/she should do sth.

bedeutend 1. *Adj.* **a)** significant, important ‹*step, event, role, etc.*›; important ‹*city, port, artist, writer, etc.*›; **b)** *(groß)* considerable. **2.** *adv.* considerably

bedeutsam 1. *Adj.* **a)** *s.* **bedeutend 1**; **b)** *(vielsagend)* meaningful; significant. **2.** *adv.* meaningfully; significantly

Bedeutung die; ~, ~en a) *o. Pl.* meaning; significance; **einer Sache** *(Dat.)* **zu große ~ beimessen** attach too much significance to sth.; **b)** *(Wort~)* meaning; **c)** *o. Pl.* *(Wichtigkeit)* importance; *(Tragweite)* significance; importance; **[an] ~ gewinnen** become more significant; **nichts von ~:** nothing important *or* significant; nothing of [any] importance *or* significance

bedeutungs-, Bedeutungs-: ~los *Adj.* insignificant; unimportant; **~losigkeit die; ~:** insignificance; unimportance; **~voll 1.** *Adj.* **a)** significant; **b)** *(vielsagend)* meaningful; meaning ‹*look*›; **2.** *adv.* meaningfully

bedienen 1. *tr. V.* **a)** wait on; ‹*waiter, waitress*› wait on, serve; ‹*sales assistant*› serve; **jmdn. vorn und hinten ~** *(ugs.)* wait on sb. hand and foot; **werden Sie schon bedient?** are you being served?; **b)** operate ‹*machine*›; **c) [mit etw.] gut/schlecht bedient sein** *(ugs.)* be well-served/ill-served [by sth.]; **bedient sein** *(salopp)* have had enough; **d)** *(Kartenspiel)* play; **Kreuz/Trumpf ~:** play a club/trump. **2.** *itr. V.* **a)** serve; **wer bedient hier?** who is serving here?; **b)** *(Kartenspiel)* follow suit. **3.** *refl. V.* **a)** help oneself; **sich selbst ~** *(im Geschäft, Restaurant usw.)* serve oneself; **b) sich einer Sache** *(Gen.)* **~** *(geh.)* make use of sth.; use sth.

Bedienstete der/die; *adj. Dekl.* **a)** *(Amtsspr.)* employee; **b)** *(veralt.: Diener)* servant

Bediente [bə'di:ntə] **der/die;** *adj. Dekl. (veralt.)* servant

Bedienung die; ~, ~en a) *o. Pl. (das Bedienen)* service; **~ inbegriffen** service included; **b)** *o. Pl. (Handhabung)* operation; **c)** *(Person)* waiter/waitress; **hallo, ~!** waiter/waitress!; **d)** *(österr.)* cleaning woman

Bedienungs-: ~an·leitung die operating instructions *pl.; (Heft)* instruction book; **~auf·schlag der, ~geld das, ~zu·schlag der** service charge

bedingen [bə'dɪŋən] *tr. V.* cause; **einander ~:** be interdependent *or* mutually dependent; **psychisch bedingt sein** be psychologically determined

bedingt [bə'dɪŋt] **1.** *Adj.* conditional; qualified ⟨*praise, approval*⟩; *s. auch* **Reflex. 2.** *adv.* partly ⟨*true*⟩; **nur ~ tauglich** fit for certain duties only

-bedingt *Adj.* **krankheitsbedingte Abwesenheit** absence due to illness; **berufsbedingte Krankheiten** occupational illnesses; **witterungsbedingte Schäden** damage caused by the weather

Bedingung die; ~, ~en condition; **etw. zur ~ machen** make sth. a condition; **zu annehmbaren ~en** on acceptable terms; **unter diesen ~en** on these conditions; **unter keiner ~:** under no circumstances; **unter der ~, daß ...:** on condition that ...

bedingungs·los 1. *Adj.* unconditional ⟨*surrender, acceptance, etc.*⟩; absolute, unquestioning ⟨*obedience, loyalty, devotion*⟩. **2.** *adv.* ⟨*surrender, accept, etc.*⟩ unconditionally; ⟨*subordinate oneself*⟩ unquestioningly

Bedingungs·satz der *(Sprachw.)* conditional clause

bedrängen *tr. V.* **a)** besiege ⟨*town, fortress, person*⟩; put ⟨*opposing player*⟩ under pressure; **vom Feind bedrängt sein** be hard pressed by the enemy; **mit Fragen bedrängt werden** be assailed with questions; **in einer bedrängten Lage sein** be hard-pressed *or* in a difficult situation; **b)** *(belästigen)* pester

Bedrängnis die; ~, ~se *(geh.) (innere Not)* distress; *(wirtschaftliche Not)* [great] difficulties *pl.;* **in ~ geraten/sein** get into/be in great difficulties *pl.*

bedrohen *tr. V.* **a)** threaten; **b)** *(gefährden)* threaten; endanger; **den Frieden ~:** be a threat *or* danger to peace; **vom Aussterben bedroht sein** be threatened with extinction

bedrohlich 1. *Adj.* threatening, menacing ⟨*gesture*⟩; *(unheilverkündend)* ominous; *(gefährlich)* dangerous. **2.** *adv.* threateningly; menacingly; *(unheilverkündend)* ominously; *(gefährlich)* dangerously

Bedrohung die threat *(Gen.,* **für** to); **in ständiger ~:** under a constant threat

bedrucken *tr. V.* print; **etw. mit einer Adresse ~:** print an address on sth.

bedrücken *tr. V.* depress; **es bedrückt mich, daß ...:** I feel depressed that ...; **bedrückt dich etwas?** is something weighing on your mind?

bedrückend *Adj.* depressing; oppressive ⟨*atmosphere*⟩

bedruckt *Adj.* printed; print *attrib.* ⟨*dress etc.*⟩

bedrückt *Adj.* depressed

Bedrückung die; ~, ~en depression

Beduine [bedu'i:nə] **der; ~n, ~n** Bed[o]uin

bedürfen *unr. itr. V. (geh.)* **jmds./einer Sache ~:** require *or* need sb./sth.; **es bedarf einiger Mühe** some effort is needed *or* required

Bedürfnis das; ~ses, ~se need **(nach** for); **das ~ haben, etw. zu tun** feel a need to do sth.; **ein ~ nach etw. haben** be in need of sth.; **es war mir ein ~, das zu tun** I felt the need to do it

bedürfnis-, Bedürfnis-: ~anstalt die *(Amtsspr.)* public convenience; **~los** *Adj.* ⟨*person*⟩ with few [material] needs; modest, simple ⟨*life*⟩; **~los sein** have few [material] needs; **~losigkeit die; ~:** lack of [material] needs

bedürftig *Adj.* needy; **die Bedürftigen** the needy; those in need

Bedürftigkeit die; ~: neediness

Beef·steak ['bi:fste:k] **das** [beef]steak; **deutsches ~:** ≈ beefburger

beehren *tr. V.* **jmdn. mit etw. ~** *(geh., auch iron.)* honour sb. with sth.; **~ Sie uns bald wieder** *(gespreizt: besuchen)* we hope to have the pleasure of your custom/company again

beeiden [bə'|aidn̩] *tr. V.* **~, daß ...:** swear [on oath] that ...; **eine Aussage ~:** swear to the truth of a statement

beeilen *refl. V.* hurry [up *(coll.)*]; **beeil dich!** hurry [up]; **sich bei** *od.* **mit etw. ~:** hurry over sth.; **sich ~, etw. zu tun** hasten to do sth.

Beeilung: [los,/ein bißchen] ~! *(ugs.)* get a move on! *(coll.);* hurry up! *(coll.)*

beeindrucken *tr. V.* impress; **sich von etw. ~ lassen** be impressed by sth.; **~d** impressive

beeinflußbar *Adj.* **jmd./etw. ist [nicht] ~:** sb./sth. can[not] be influenced; **leicht/schwer ~ sein** ⟨*person*⟩ be easily influenced/hard *or* difficult to influence

beeinflussen [bə'|ainflʊsn̩] *tr. V.* influence; influence, affect ⟨*result, process, etc.*⟩; **jmdn./etw. positiv ~:** have a positive influence on sb./sth.; **sich leicht ~ lassen** be easily influenced

Beeinflussung die; ~, ~en influencing; **seine ~ durch die Schule** the influence of the school on him

beeinträchtigen [bə'|aintrɛçtɪgn̩] *tr. V.* restrict ⟨*rights, freedom*⟩; detract from, spoil ⟨*pleasure, enjoyment*⟩; spoil ⟨*appetite, good humour*⟩; detract from, diminish ⟨*value*⟩; diminish, impair ⟨*quality*⟩; impair ⟨*efficiency, vision, hearing*⟩; damage, harm ⟨*sales, reputation*⟩; reduce ⟨*production*⟩; **jmdn. in seiner Freiheit ~:** restrict sb.'s freedom; **sich beeinträchtigt fühlen** feel hampered

Beeinträchtigung die; ~, ~en: *s.* **beeinträchtigen:** restriction *(Gen.* on); detraction *(Gen.* from); spoiling; diminution; impairment; damage *(Gen.* to); harm *(Gen.* to); reduction

Beelzebub ['be:ltsə-] **(der)** Beelzebub; **den Teufel mit** *od.* **durch ~ austreiben** *(fig.)* replace one evil by *or* with another

beenden *tr. V.* end; finish ⟨*piece of work*⟩; end, conclude ⟨*negotiations, letter, lecture*⟩; complete, finish ⟨*studies*⟩; end, bring to an end ⟨*meeting, relationship, dispute, strike*⟩; **damit ~ wir unser heutiges Programm** that brings to an end our programmes for today

Beendigung die; ~: zur ~ der Unruhen wurde die Armee eingesetzt the army was called in to put an end to the unrest; **sie wurden zur ~ der Kampfhandlungen aufgefordert** they were called upon to cease *or* stop hostilities

beengen *tr. V.* hinder; restrict; *(fig.)* restrict ‹*freedom [of action]*›; **beengt wohnen** live in cramped surroundings *or* conditions; **sich beengt fühlen** feel cramped; *(fig.)* feel constricted

Beengtheit die; ~: crampedness; **ein Gefühl der ~:** a feeling of being cramped

beerben *tr. V.* **jmdn. ~:** inherit something from sb.; *(allein)* inherit sb's estate

beerdigen [bə'|e:ɐ̯dɪgn̩] *tr. V.* bury; **jmdn. kirchlich ~:** give sb. a Christian burial

Beerdigung die; ~, ~en burial; *(Trauerfeier)* funeral

Beerdigungs·institut das [firm *sing.* of] undertakers *pl. or* funeral directors *pl.*

Beere ['be:rə] **die; ~, ~n** berry

Beet [be:t] **das; ~[e]s, ~e** *(Blumen~)* bed; *(Gemüse~)* plot

befähigen [bə'fɛ:ɪgn̩] *tr. V.* **jmdn. ~, etw. zu tun** enable sb. to do sth.; ‹*qualifications, training, etc.*› qualify sb. to do sth.

befähigt *Adj.* **a)** able; capable **(zu** of); **b)** *(qualifiziert)* qualified

Befähigung die; ~ a) ability; **b)** *(Qualifikation)* qualification; **die ~ zum Internisten/Hochschulstudium/Richteramt** the qualifications *pl.* for becoming an internist/studying at university/being a judge

befahrbar *Adj.* passable; navigable ‹*canal, river*›; **nicht ~:** impassable/unnavigable

befahren *unr. tr. V.* drive on, use ‹*road*›; drive across, use ‹*bridge, pass*›; use ‹*railway line*›; sail ‹*sea*›; navigate ‹*river, canal*›; **die Straße ist nur in einer Richtung zu ~:** traffic can only use the road in one direction; **„Seitenstreifen nicht ~!"** 'keep off verges'; **die Straße ist stark/wenig ~:** the road is heavily/little used; **eine stark ~e Straße/Wasserstraße** a busy road/waterway

Befall der; ~[e]s attack (*Gen.* on; **durch, von, mit** by)

befallen *unr. tr. V.* **a)** overcome; ‹*misfortune*› befall; **Fieber/eine Grippe befiel ihn** *(geh.)* he was stricken by fever/influenza; **von Panik/Angst/Heimweh** *usw.* **~ werden** be seized *or* overcome with *or* by panic/fear/homesickness *etc.*; **b)** ‹*pests*› attack

befangen 1. *Adj.* **a)** self-conscious, awkward; **b)** *(bes. Rechtsw.: voreingenommen)* biased; **einen Richter als ~ ablehnen** challenge a judge on grounds of bias; **c) in einem Glauben/Irrtum ~ sein** *(geh.)* labour under a belief/misapprehension. **2.** *adv.* self-consciously; awkwardly

Befangenheit die; ~ a) self-consciousness; awkwardness; **b)** *(bes. Rechtsw.: Voreingenommenheit)* bias

befassen 1. *refl. V.* **sich mit etw. ~:** occupy oneself with sth.; *(studieren)* study sth.; ‹*article, book*› deal with sth.; **sich mit jmdm./einer Angelegenheit ~:** deal with *or* attend to sb./a matter. **2.** *tr. V. (bes. Amtsspr.)* **jmdn. mit etw. ~:** get *or* instruct sb. to deal with sth.

befehden [bə'fe:dn̩] *tr. V. (hist., fig. geh.)* feud with; **sich ~:** feud

Befehl [bə'fe:l] **der; ~[e]s, ~e a)** order; command; **jmdm. den ~ geben, etw. zu tun** order *or* command sb. to do sth.; **den ~ haben, etw. zu tun** be under orders *or* have been ordered to do sth.; **auf jmds. ~** *(Akk.)* on sb.'s orders; **auf ~** *(Akk.)* **handeln** act under orders; **~ ist ~:** orders are orders; **zu ~!** yes, sir!; aye, aye, sir! *(Navy)*; **dein Wunsch ist mir ~** *(ugs. scherzh.)* your wish is my command; **b)** *(Befehlsgewalt)* command; **den ~ über jmdn./etw. haben** have command of *or* be in command of sb./sth.; **c)** *(DV)* instruction; command

befehlen 1. *unr. tr. V.* **a)** *auch itr.* order; *(Milit.)* order; command; **von Ihnen lasse ich mir nichts ~:** I don't take orders from you; **b)** *(beordern)* order; *(zu sich)* summon; **jmdn. zum Rapport ~:** order/summon sb. to report; **c)** *(geh. veralt.: anvertrauen)* commend. **2.** *unr. itr. V.* **über jmdn./etw. ~:** have command of *or* be in command of sb./sth.

befehligen *tr. V.* have command of; be in command of; **von jmdm. befehligt werden** be commanded by sb.; be under the command of sb.

befehls-, Befehls-: ~empfänger der recipient of an order/orders; **bloße ~empfänger sein** just follow *or* take orders; **~form die** *(Sprachw.)* imperative [form]; **~gewalt die;** *o. Pl.* command **(über +** *Akk.* of); **~haber** [~ha:bɐ] **der; ~s, ~** *(Milit.)* commander; **~ton der;** *o. Pl.* peremptory tone; **~verweigerung die** refusal to obey an order/orders

befestigen *tr. V.* **a)** fix; **etw. mit Stecknadeln/Bindfaden ~:** fasten sth. with pins/string; **etw. mit Schrauben/Leim ~:** fasten *or* fix sth. with screws/fix *or* stick sth. with glue; **etw. an der Wand ~:** fix sth. to the wall; **einen Anhänger an einem Koffer ~:** attach *or* fasten a label to a case; **b)** *(haltbar machen)* stabilize ‹*bank, embankment*›; make up ‹*road, path, etc.*›; **c)** *(sichern)* fortify ‹*town etc.*›; strengthen ‹*border*›

Befestigung die; ~, ~en a) *s.* **befestigen a:** fixing; fastening; attachment; **b)** *s.* **befestigen b:** stabilization; making up; **c)** *(Milit.)* fortification

Befestigungs·anlage die fortifications *pl.*

befeuchten [bə'fɔyçtn̩] *tr. V.* moisten; damp ‹*hair, cloth*›

befeuern *tr. V.* **a)** *(beheizen)* fuel; **b)** *(beschießen)* shoot at; fire on; **c)** *(ugs.: bewerfen)* pelt

befiehlst [bə'fi:lst], **befiehlt** [bə'fi:lt] *2., 3. Pers. Sg. Präsens v.* **befehlen**

befinden 1. *unr. refl. V.* be; **unter ihnen befand sich jemand, der ...:** among them there was somebody who ... **2.** *unr. tr. V. (geh.)* **etw. für** *od.* **als gut/richtig ~:** find *or* consider sth. [to be] good/right; **jmdn. für** *od.* **als schuldig ~:** find sb. guilty. **3.** *unr. itr. V.* **darüber habe ich nicht zu ~:** that's not for me to decide

Befinden das; ~s health; *(eines Patienten)* condition; **sich nach jmds. ~ erkundigen** enquire after *or* about sb.'s health

befindlich *Adj.; nicht präd.* to be found *postpos.*; **das in der Kasse ~e Geld** the money in the till; **die im Bau ~en Häuser** the houses [which are/were] under construction

befingern *tr. V. (salopp)* finger

beflaggen *tr. V.* **etw. ~:** decorate *or* [be]deck sth. with flags; **ein Schiff ~:** dress a ship

beflecken *tr. V.* stain; **sich mit Blut ~** *(verhüll. geh.)* stain one's hands with blood

befleißigen *refl. V. (geh.)* **sich eines klaren Stils/höflicheren Tons** *usw.* **~:** make a great effort to cultivate a clear style/to adopt a more polite tone of voice *etc.*; **sich größter Zurückhaltung ~:** endeavour to exercise the greatest restraint

beflissen [bə'flɪsn̩] *(geh.)* **1.** *Adj.* keen; eager; *(emsig)* assiduous; zealous. **2.** *adv.* keenly; eagerly; *(emsig)* assiduously; zealously

beflügeln *tr. V. (geh.)* **jmdn. ~:** inspire sb.; ‹*success, praise*› spur sb. on, inspire sb.

befohlen [bə'fo:lən] *2. Part. v.* **befehlen**

befolgen *tr. V.* follow, obey ‹*instruction*›; obey, comply with ‹*law, regulation*›; follow, take ‹*advice*›; follow ‹*suggestion*›

Befolgung die; ~: *s.* **befolgen**: following; obedience (*Gen.* to); compliance (*Gen.* with)

befördern *tr. V.* **a)** carry; transport; convey; **jmdn. ins Freie** *od.* **an die Luft** ~ *(ugs.)* chuck *(coll.)* *or* throw sb. out; **b)** *(aufrücken lassen)* promote; **zum Direktor befördert werden** be promoted to director

Beförderung die a) *o. Pl.* carriage; transport; conveyance; **die ~ per Luft/zu Lande** carriage *or* transport by air/road; **b)** *(das Aufrückenlassen)* promotion (**zu** to)

Beförderungs·mittel das means of transport

befrachten *tr. V.* load (**mit** with); **mit Emotionen befrachtet** *(fig.)* ‹*discussion etc.*› charged with emotion

befragen *tr. V.* **a)** question (**über** + *Akk.* about); **einen Zeugen** ~: question *or* examine a witness; **auf Befragen** when questioned; **b)** *(konsultieren)* ask; consult; **jmdn. nach seiner Meinung** ~: ask sb. for his/her opinion; **ein Orakel/die Karten** ~: consult an oracle/the cards

Befragung die; ~, ~**en a)** questioning; *(vor Gericht)* questioning; examination; **b)** *(Konsultation)* consultation; **c)** *(Umfrage)* opinion poll

befreien 1. *tr. V.* **a)** free ‹*prisoner*›; set ‹*animal*› free; liberate ‹*country, people*›; **jmdn. aus den Händen seiner Entführer** ~: rescue sb. from the hands of his/her abductors; **b)** *(freistellen)* exempt; **jmdn. vom Turnunterricht/Wehrdienst/von einer Pflicht** ~: excuse sb. [from] physical education/exempt sb. from military service/release sb. from an obligation; **c)** *(erlösen)* **jmdn. von Schmerzen** ~: free sb. from pain; **von seinen Leiden befreit werden** *(durch den Tod)* be released from one's sufferings; **ein ~des Lachen** a laugh which breaks/broke the tension. **2.** *refl. V.* free oneself (**von** from); **sich von Vorurteilen** ~: rid oneself of prejudice *sing.*

Befreier der, Befreierin die; ~, ~**nen** liberator

befreit *Adj. (erleichtert)* relieved

Befreiung die; ~ **a)** *s.* **befreien 1 a**: freeing; liberation; **die ~ der Frau** the emancipation of women; **b)** *(Erlösung)* **die ~ von Schmerzen** release from pain; **c)** *(Erleichterung)* relief; **d)** *(Freistellung)* exemption; **um ~ vom Sportunterricht/von einer Pflicht bitten** ask to be excused [from] sport/released from an obligation

Befreiungs-: ~bewegung die liberation movement; **~kampf der** liberation struggle; **~krieg der** war of liberation; **die ~kriege** *(hist.)* the Wars of Liberation *(1813–1815)*

befremden [bəˈfrɛmdn̩] **1.** *tr. V.* **jmdn.** ~: put sb. off; *(erstaunen)* take sb. aback; **es befremdete ihn, daß ...**: he was taken aback [to find] that ... **2.** *itr. V.* be disturbing

Befremden das; ~**s** surprise and displeasure

befremdlich *(geh.) Adj.* strange; odd

befreunden [bəˈfrɔyndn̩] *refl. V.* *s.* **anfreunden**

befreundet *Adj.* **[gut** *od.* **eng]** ~ **sein** be [good *or* close] friends (**mit** with); **meine Frau und ich und ein ~es Ehepaar/ein ~er Schauspieler** my wife and I and a couple with whom we are friends/an actor who is a friend of ours; **~e Familien/Kinder** families which are friendly with each other/children who are friends; **das ~e Ausland** friendly [foreign] countries

befrieden *tr. V. (geh.)* bring peace to ‹*country*›

befriedigen [bəˈfriːdɪgn̩] *tr. V.* **a)** *auch itr.* satisfy; satisfy, meet ‹*demand, need*›; satisfy, fulfil ‹*wish*›; satisfy, gratify ‹*lust*›; **seine Gläubiger** ~: satisfy one's creditors; **das Ergebnis befriedigte mich** the result satisfied me *or* was satisfactory to me; **seine Leistung befriedigte [nicht]** his performance was [un]satisfactory; **b)** *auch itr. (ausfüllen)* ‹*job, occupation, etc.*› fulfil; **c)** *(sexuell)* satisfy; **sich [selbst]** ~: masturbate

befriedigend 1. *Adj.* **a)** satisfactory; satisfactory, adequate ‹*reply, performance*›; **nicht ~ sein** be unsatisfactory/inadequate; **b)** fulfilling ‹*job, occupation, etc.*›. **2.** *adv.* satisfactorily; ‹*answer*› satisfactorily, adequately

befriedigt 1. *Adj.* satisfied. **2.** *adv.* with satisfaction

Befriedigung die; ~: *s.* **befriedigen a**: satisfaction; meeting; fulfilment; gratification; **sexuelle** ~: sexual satisfaction; ~ **darin finden, etw. zu tun** get satisfaction from doing sth.

befristen *tr. V.* limit the duration of (**auf** + *Akk.* to)

befristet *Adj.* temporary ‹*visa*›; fixed-term ‹*ban, contract*›; **ein auf zwei Jahre ~er Vertrag** a two-year fixed-term contract; ~ **sein** ‹*visa, permit*› be valid for a limited period [only]; **auf ein Jahr ~ sein** ‹*visa, permit*› be valid for one year

befruchten *tr. V.* **a)** fertilize ‹*egg*›; pollinate ‹*flower*›; impregnate ‹*female*›; **ein Tier künstlich** ~: artificially inseminate an animal; **b)** *(geh.)* **jmdn./etw.** ~, **einen ~den Einfluß auf jmdn./etw. haben** have *or* be a stimulating *or* inspiring influence [up]on sb./sth.

Befruchtung die; ~, ~**en**: *s.* **befruchten a**: fertilization; pollination; impregnation; **künstliche** ~: artificial insemination

befugen *tr. V.* authorize; **[dazu] befugt sein, etw. zu tun** be authorized to do sth.

Befugnis die; ~, ~**se** authority; **seine ~se überschreiten** exceed one's authority *sing.*

befühlen *tr. V.* feel

befummeln *tr. V. (ugs.)* **a)** paw *(coll.)*; **b)** *(sexuell berühren)* grope *(sl.)*; feel up *(sl.)*

Befund der *(bes. Med.)* result[s *pl.*]; **ohne ~ sein** be negative

befürchten *tr. V.* fear; **ich befürchte, daß ...**: I am afraid that ...; **das ist nicht zu** ~: there is no fear of that

Befürchtung die; ~, ~**en** fear; **die ~ haben, daß ...**: be afraid that ...

befürworten [bəˈfyːɐ̯vɔrtn̩] *tr. V.* support

Befürworter der; ~**s**, ~: supporter

Befürwortung die; ~, ~**en** support

begabt [bəˈgaːpt] *Adj.* talented; gifted; **vielseitig ~ sein** be multi-talented; have many talents; **für etw. ~ sein** have a gift *or* talent for sth.

Begabte der/die; *adj. Dekl.* gifted *or* talented person/man/woman *etc.*

Begabung die; ~, ~**en** talent; gift; **eine ~ [für etw.] haben** have a gift *or* talent [for sth.]

begaffen *tr. V. (ugs. abwertend)* gawp at *(coll.)*; stare at

begann [bəˈgan] *1. u. 3. Pers. Sg. Prät. v.* **beginnen**

begatten [bəˈgatn̩] **1.** *tr. V.* mate with; ‹*man*› copulate with; ‹*stallion, bull*› cover. **2.** *refl. V.* mate; ‹*persons*› copulate

Begattung die mating; *(bei Menschen)* copulation

begeben *unr. refl. V. (geh.)* **a)** proceed; make one's way; go; **sich nach Hause ~:** proceed *or* make one's way *or* go home; **sich zu Bett ~:** retire to bed; **sich in ärztliche Behandlung ~:** get medical treatment; go to a doctor for treatment; **sich an die Arbeit ~:** commence work; **b)** *(geschehen)* happen; occur

Begebenheit die; ~, ~en *(geh.)* event; occurrence

begegnen [bə'ge:gnən] *itr. V.; mit sein* **a) jmdm. ~:** meet sb.; **sich** *(Dat.)* **~:** meet [each other]; **ihre Blicke begegneten sich** *(Dat.) (geh.)* their eyes met; **b) etw. begegnet jmdm.** *(jmd. trifft etw. an)* sb. comes across *or* encounters sth.; *(geh.: etw. passiert jmdm.)* sth. happens to sb.; **c) jmdm. freundlich/höflich** *usw.* **~** *(geh.)* behave in a friendly/polite *etc.* way towards sb.; **d)** *(geh.: entgegentreten)* counter ‹*accusation, attack*›; combat ‹*illness, disease; misuse of drugs, alcohol, etc.*›; meet ‹*difficulty, danger*›; deal with ‹*emergency*›

Begegnung die; ~, ~en a) meeting; **eine Stätte internationaler ~en** an international meeting-place; **b)** *(Sport)* match

begehen *unr. tr. V.* **a)** commit ‹*crime, adultery, indiscretion, sin, suicide, faux-pas, etc.*›; make ‹*mistake*›; **eine [furchtbare] Dummheit/Taktlosigkeit ~:** do something [really] stupid/tactless; **b)** *(geh.: feiern)* celebrate; **ein Fest würdig ~:** celebrate an occasion fittingly; **c)** *(abgehen)* inspect [on foot]; **d)** *(betreten)* walk on

begehren *tr. V.* desire; *s. auch* **Herz b**

Begehren das; ~s *(geh.)* desire, wish **(nach** for); *(Bitte)* request

begehrens·wert *Adj.* desirable

begehrlich 1. *Adj.* greedy. **2.** *adv.* greedily

begehrt *Adj.* much sought-after

begeistern 1. *tr. V.* **jmdn. [für etw.] ~:** fill *or* fire sb. with enthusiasm [for sth.]; **das Publikum ~:** fire the audience. **2.** *refl. V.* get enthusiastic; *(begeistert sein)* be enthusiastic **(für** about)

begeistert 1. *Adj.* enthusiastic; **von jmdm./etw. ~ sein** be taken by *or* with sb./be enthusiastic about sth. **2.** *adv.* enthusiastically

Begeisterung die; ~: enthusiasm; **in ~ geraten** become *or* get enthusiastic

begeisterungs-, Begeisterungs-: ~fähig *Adj.* ‹*children, people, etc.*› who are able to get enthusiastic *or* are capable of enthusiasm; **~fähig sein** be able to get enthusiastic; **~fähigkeit die;** *o. Pl.* capacity for enthusiasm; **~sturm der** storm of enthusiastic applause

Begierde [bə'gi:ɐdə] **die; ~, ~n** desire **(nach** for)

begierig 1. *Adj.* eager; *(gierig)* greedy; hungry; **~ sein, etw. zu tun** be [desperately] eager to do sth.; **mit ~en Blicken** with hungry *or* greedy glances. **2.** *adv.* eagerly; *(gierig)* greedily; hungrily

begießen *unr. tr. V.* **a)** water ‹*plants*›; baste ‹*meat*›; **jmdn./etw. mit Wasser ~:** pour water over sb./sth.; **b)** *(ugs.)* **etw. ~:** celebrate sth. with a drink; **das muß begossen werden** that calls for a drink

Beginn [bə'gɪn] **der; ~[e]s** start; beginning; **zu ~:** at the start *or* beginning

beginnen 1. *unr. itr. V.* start; begin; **mit einer Arbeit/dem Studium ~:** start *or* begin a piece of work/one's studies; **mit dem Bau ~:** start *or* begin building; **dort beginnt der Wald** the forest starts there. **2.** *unr. tr. V.* **a)** start; begin; start ‹*argument*›; **b) es ~, etw. zu tun** go *or* set about doing sth.; **was hättet ihr nur ohne mich begonnen?** what would you have done without me?

beginnend *Adj.; nicht präd.* incipient; **mit der ~en Morgendämmerung** as dawn begins/began to break; **im ~en 19. Jahrhundert** at the beginning of the 19th century

beglaubigen [bə'glaubɪgn̩] *tr. V.* certify

Beglaubigung die; ~, ~en certification

Beglaubigungs·schreiben das letter of accreditation

begleichen *unr. tr. V.* settle, pay ‹*bill, debt*›; pay ‹*sum*›; **mit jmdm. eine Rechnung zu ~ haben** *(fig.)* have a score to settle with sb.

Begleit·brief der covering *or* accompanying letter

begleiten *tr. V. (auch Musik, fig.)* accompany; **jmdn. zur Tür ~:** show sb. to the door; **jmdn. nach Hause ~:** see sb. home

Begleiter der; ~s, ~, Begleiterin die; ~, ~nen companion; *(zum Schutz)* escort

Begleit-: ~erscheinung die concomitant; *(einer Krankheit)* accompanying symptom; **~musik die** *(fig.)* accompaniment; **~person die** escort; **~schreiben das** *s.* **~brief**

Begleitung die; ~, ~en a) *o. Pl.* **er bot uns seine ~ an** he offered to accompany us; **in ~ einer Frau/eines Erwachsenen** in the company of *or* accompanied by a woman/an adult; **er ist in ~ hier** he's here with someone; **b)** *(Musik)* accompaniment; **ohne ~:** unaccompanied *or* without accompaniment; **c)** *(Person[en])* companion[s *pl.*]; *(zum Schutz)* escort

beglücken *tr. V. (geh.)* **jmdn. ~:** make sb. happy; delight sb.; **jmdn. mit etw. ~** *(oft iron.)* favour sb. with sth.; **die Frauen/Männer ~:** gratify women/men; **ein ~des Erlebnis** a gladdening experience

beglückt 1. *Adj.* happy; delighted. **2.** *adv.* happily; delightedly

Beglückung die; ~: zur ~ der Menschheit beitragen contribute to the sum of human happiness

beglück·wünschen *tr. V.* congratulate **(zu** on)

begnadet *Adj. (geh.)* divinely gifted

begnadigen *tr. V.* pardon; reprieve

Begnadigung die; ~, ~en pardonning; reprieving; *(Straferlaß)* pardon; reprieve

begnügen [bə'gny:gn̩] *refl. V.* content oneself **(mit** with)

Begonie [be'go:ni̯ə] **die; ~, ~n** begonia

begonnen [bə'gɔnən] *2. Part. v.* **beginnen**

begraben *unr. tr. V.* **a)** bury; **dort möchte ich nicht ~ sein** *(ugs.)* I wouldn't live there if you paid me *(coll.)*; **du kannst dich ~ lassen** *(ugs.)* you may as well give up; **b)** *(fig.)* abandon ‹*hope, plan, etc.*›

Begräbnis [bə'grɛ:pnɪs] **das; ~ses, ~se** burial; *(~feier)* funeral

begradigen [bə'gra:dɪgn̩] *tr. V.* straighten

Begradigung die; ~, ~en straightening

begreifen 1. *unr. tr. V.* **a)** understand; understand, grasp, comprehend ‹*connection, problem, meaning*›; **er konnte nicht ~, was geschehen war** he could not grasp what had happened; **kaum zu ~ sein** be almost incomprehensible; **das begreife, wer will** it's beyond me; **b)** *(geh.: betrachten)* regard, see **(als** as). **2.** *itr. V.* understand; **schnell** *od.* **leicht/langsam** *od.* **schwer ~:** be quick/ slow on the uptake; be quick/slow to grasp things

begreiflich *Adj.* understandable; **das ist mir nicht ~:** I can't understand it; **jmdm. etw. ~ machen** make sb. understand sth.
begreiflicher·weise *Adv.* understandably
begrenzen *tr. V.* **a)** limit, restrict (**auf** + *Akk.* to); **b)** *(die Grenze bilden von)* mark the boundary of; **durch etw. begrenzt sein** be bounded by sth.
begrenzt *Adj.* limited; restricted
Begrenzung die; ~, ~en a) *(Grenze)* boundary; **b)** *(das Begrenzen)* limiting; restriction; *(der Geschwindigkeit)* restriction
Begriff der a) concept; *(Terminus)* term; **b)** *(Auffassung)* idea; **einen/keinen ~ von etw. haben** have an idea/no idea of sth.; **sich** *(Dat.)* **keinen ~ von etw. machen können** not be able to imagine sth.; **für meine ~e** in my estimation; **ein/kein ~ sein** be/not be well known; **c) im ~ sein** *od.* **stehen, etw. zu tun** be about to do sth.; **d) schwer von ~ sein** *(ugs.)* be slow on the uptake
begriffen [bəˈgrɪfn̩] *Adj. in* **im Aufbruch/Fallen** *usw.* **~ sein** be leaving/falling *etc.*
begrifflich 1. *Adj.* conceptual. **2.** *adv.* conceptually
begriffs-, Begriffs-: ~bestimmung die definition [of the/a concept]; **~stutzig** *Adj.* obtuse; slow-witted; gormless *(coll.)*; slow-wittedly; gormlessly *(coll.)*; **~stutzigkeit die; ~:** obtuseness; slow-wittedness; gormlessness *(coll.)*; **~verwirrung die** conceptual confusion
begründen *tr. V.* **a)** substantiate ⟨*statement, charge, claim*⟩; give reasons for ⟨*decision, refusal, opinion*⟩; **b)** *(gründen)* found; establish ⟨*fame, reputation*⟩; **einen Hausstand ~:** set up house
Begründer der founder
begründet *Adj.* well-founded; *(berechtigt)* reasonable; **sachlich ~:** objectively based; **in etw.** *(Dat.)* **~ sein** be the result of sth.
Begründung die; ~, ~en a) reason[s *pl.*]; **mit der ~, daß ...:** on the grounds that ...; **seine ~ war ...:** the reason/reasons he gave was/were ...; **ohne jede ~:** without giving any reasons; **b)** *(Gründung)* founding; establishment; *(eines Hausstands)* setting up
begrüßen *tr. V.* **a)** greet; ⟨*host, hostess*⟩ greet, welcome; **b)** *(gutheißen)* welcome ⟨*suggestion, proposal*⟩; **ich begrüße es, daß ...:** I am glad that ...
begrüßens·wert *Adj.* welcome
Begrüßung die; ~, ~en greeting; *(von Gästen)* welcoming; *(Zeremonie)* welcome (*Gen.* for); **jmdm. zur ~ etw. überreichen** welcome sb. with sth.; **zur ~ die Hand schütteln** shake hands by way of greeting
Begrüßungs-: ~an·sprache die, ~rede die speech of welcome; welcoming speech
begucken *tr. V. (ugs.)* look at; have *or* take a look at; **sich** *(Dat.)* **jmdn./etw. ~:** have *or* take a look at sb./sth.
begünstigen [bəˈgʏnstɪgn̩] *tr. V.* **a)** favour; encourage ⟨*exports, trade, growth*⟩; further ⟨*plan*⟩; **b)** *(bevorzugen)* favour; show favour to; **vom Schicksal begünstigt werden** be blessed by fate
Begünstigung die; ~, ~en a) *s.* **begünstigen a:** favouring; encouragement; furthering; **b)** *(Bevorzugung)* preferential treatment
begutachten *tr. V.* **a)** examine and report on; **b)** *(ugs.: ansehen)* look at; have *or* take a look at; **laß dich mal ~!** let's have *or* take a look at you
begütert [bəˈgy:tɐt] *Adj.* wealthy; affluent
begütigen [bəˈgy:tɪgn̩] *tr. V.* placate; mollify; pacify; **~d auf jmdn. einreden** speak soothingly to sb.
behaart [bəˈha:ɐ̯t] *Adj.* hairy; **schwarz/stark ~ sein** be covered with black hair/covered with hair
Behaarung die; ~, ~en hair *no indef. art.*
behäbig [bəˈhɛ:bɪç] **1.** *Adj.* **a)** stolid and portly; **b)** *(langsam)* slow and ponderous. **2.** *adv.* slowly and ponderously
Behäbigkeit die; ~ a) stolidness and portliness; **b)** *(Langsamkeit)* slowness and ponderousness
behaftet *Adj. (geh.)* **mit einem Makel/Laster ~ sein** be marked with a blemish/tainted with a vice; **mit einem schlechten Ruf/einem Fehler ~ sein** have a bad name/a defect
behagen [bəˈha:gn̩] *itr. V.* **etw. behagt jmdm.** sth. pleases sb.; sb. likes sth.; **er behagt mir gar nicht** I don't like him at all
Behagen das; ~s pleasure; **etw. mit ~ essen** eat sth. with relish
behaglich 1. *Adj.* comfortable; comfortable, cosy ⟨*atmosphere, room, home, etc.*⟩; **es jmdm./sich ~ machen** make sb./oneself comfortable. **2.** *adv.* comfortably, cosily ⟨*warm, furnished*⟩
Behaglichkeit die; ~: *s.* **behaglich 1:** comfortableness; cosiness
behalten *unr. tr. V.* **a)** keep; keep on ⟨*employees*⟩; keep, retain ⟨*value, expressive power, etc.*⟩; **etw. für sich ~:** keep sth. to oneself; **die Nerven/die Ruhe ~:** keep one's nerve/keep calm; **b)** *s.* **zurückbehalten b; c)** *(sich merken)* remember; *s. auch* **Recht d**
Behälter [bəˈhɛltɐ] **der; ~s, ~:** container; *(für Abfälle)* receptacle
Behältnis das; ~ses, ~se *(geh.)* container
behämmert *Adj. (salopp) s.* **bekloppt**
behandeln *tr. V.* **a)** treat ⟨*person*⟩; handle ⟨*matter, thing*⟩; **b)** *(bearbeiten)* treat ⟨*material, wood, etc.*⟩; **c)** *(sich befassen mit)* deal with, treat ⟨*subject, question; theme*⟩; **d)** *(ärztlich)* treat **(auf** + *Akk.*, **wegen** for)
Behandlung die; ~, ~en treatment; **in [ärztlicher] ~ sein** be under medical treatment; **er ist bei Dr. N. in ~:** he is under Dr N.
Behandlungs-: ~methode die method of treatment; **~stuhl der** chair for the patient; *(beim Zahnarzt)* [dentist's] chair
Behang der; ~[e]s, Behänge hanging
behangen *Adj.* **ein mit Äpfeln ~er Baum** a tree laden with apples; **mit Schmuck ~:** festooned with jewellery
behängen *tr. V.* **a) etw. mit etw. ~:** hang *or* decorate sth. with sth.; **b)** *(ugs. abwertend)* **jmdn./sich mit etw. ~:** festoon sb./oneself with sth.
beharren *itr. V.* **auf etw.** *(Dat.)* **~:** persist in sth.; **„...", beharrte er** '...', he insisted
beharrlich 1. *Adj.* dogged; persistent. **2.** *adv.* doggedly; persistently
Beharrlichkeit die; ~: doggedness; persistence
behauchen *tr. V.* breathe on
behauen *unr. tr. V.* hew; **roh ~e Steine** rough-hewn stone blocks
behaupten [bəˈhauptn̩] **1.** *tr. V.* **a)** maintain; assert; **~, jmd. zu sein/etw. zu wissen** claim to be sb./know sth.; **das kann man nicht ~:** you cannot say that; **b)** *(verteidigen)* maintain ⟨*position*⟩; *s. auch* **Feld f. 2.** *refl. V.* **a)** hold one's ground; *(sich durchsetzen)* assert oneself; *(fortbestehen)* survive; **die Kirche/der Dollar konnte sich ~:** the church/the dollar was able to maintain its position; **b)** *(Sport)* win through

Behauptung die; ~, ~en a) claim; assertion; **b)** *(das Sichdurchsetzen)* assertion
Behausung die; ~, ~en *(oft abwertend: Wohnung)* dwelling
beheben *unr. tr. V.* remove; repair ⟨*damage*⟩; remedy ⟨*abuse, defect*⟩
Behebung die; ~, ~en *s.* **beheben:** removal; repair; remedying
beheimatet *Adj.* **an einem Ort/in einem Land ~ sein** ⟨*plant, animal, tribe, race*⟩ be native *or* indigenous to a place/country; ⟨*person*⟩ come from a place/country
beheizbar *Adj.* heatable
beheizen *tr. V.* heat
Behelf [bə'hɛlf] **der; ~[e]s, ~e** stopgap; makeshift
behelfen *unr. refl. V.* get by; manage; **sich mit etw. ~:** make do *or* manage with sth.
Behelfs- temporary ⟨*exit, dwelling, etc.*⟩
behelfs·mäßig 1. *Adj.* makeshift; temporary. **2.** *adv.* in a makeshift way *or* fashion
behelligen [bə'hɛlɪgn̩] *tr. V.* bother
behende [bə'hɛndə] **1.** *Adj.* *(geschickt)* deft; adroit; *(flink)* nimble; agile. **2.** *adv.* *(geschickt)* deftly; adroitly; *(flink)* nimbly; agilely
Behendigkeit die; ~: *s.* **behende 1:** deftness; adroitness; nimbleness; agilety
beherbergen *tr. V.* accommodate, put up ⟨*guest*⟩; *(fig.)* contain
Beherbergung die; ~: accommodation
beherrschen 1. *tr. V.* **a)** rule; **den Markt ~:** dominate *or* control the market; **b)** *(meistern)* control ⟨*vehicle, animal*⟩; be in control of ⟨*situation*⟩; **c)** *(bestimmen, dominieren)* dominate ⟨*townscape, landscape, discussions, relationship*⟩; **d)** *(zügeln)* control ⟨*feelings*⟩; control, curb ⟨*impatience*⟩; **e)** *(gut können)* have mastered ⟨*instrument, trade*⟩; have a good command of ⟨*language*⟩. **2.** *refl. V.* control oneself; **ich kann mich ~** *(iron.)* I can resist the temptation *(iron.)*
Beherrscher der; ~s, ~: ruler
beherrscht 1. *Adj.* self-controlled. **2.** *adv.* with self-control
Beherrschtheit die; ~: self-control
Beherrschung die; ~ a) control; *(eines Volks, Landes usw.)* rule; *(eines Markts)* domination; control; **b)** *(das Meistern)* control; **c)** *(Beherrschtheit)* self-control; **seine** *od.* **die ~ verlieren** lose one's self-control; **d)** *(das Können)* mastery
beherzigen [bə'hɛrtsɪgn̩] *tr. V.* **etw. ~:** take sth. to heart; heed sth.
beherzt 1. *Adj.* spirited; **einige Beherzte** a few brave souls. **2.** *adv.* spiritedly
behexen *tr. V.* bewitch
behilflich [bə'hɪlflɪç] *in* **jmdm. [beim Aufräumen** *usw.*] **~ sein** help sb. [clear up *or* with the clearing-up *etc.*]; **kann ich [Ihnen] ~ sein?** can I help [you]?
behindern *tr. V.* **a)** hinder; hamper, impede ⟨*movement*⟩; hold up ⟨*traffic*⟩; impede ⟨*view*⟩; **b)** *(Sport, Verkehrsw.)* obstruct
behindert *Adj.* handicapped
Behinderte der/die; *adj. Dekl.* handicapped person; **die ~n** the handicapped; **WC für ~:** toilet for disabled persons
Behinderung die; ~, ~en a) hindrance; **b)** *(Sport, Verkehrsw.)* obstruction; **auf der Autobahn A 8 kommt es zu ~en** there are delays on the A 8 motorway; **c)** *(Gebrechen)* handicap
Behörde [bə'hø:ɐ̯də] **die; ~, ~n** authority; *(Amt, Abteilung)* department; **die ~n** the authorities
behördlich 1. *Adj.; nicht präd.* official. **2.** *adv.* officially
behüten *tr. V.* *(bewahren, beschützen)* protect **(vor +** *Dat.* from); *(bewachen)* guard; **jmdn. vor einer Gefahr ~:** keep *or* safeguard sb. from a danger; **[Gott] behüte!** God *or* Heaven forbid!
Behüter der; ~s, ~ *(geh.)* protector
behütet *Adj.* sheltered ⟨*life, upbringing*⟩
behutsam [bə'hu:tza:m] **1.** *Adj.* careful; cautious; *(zartfühlend)* gentle. **2.** *adv.* carefully; cautiously; *(zartfühlend)* gently
Behutsamkeit die; ~: care; caution; *(Zartgefühl)* gentleness
bei [bai] *Präp. mit Dat.* **a)** *(nahe)* near; *(dicht an, neben)* by; **die Schlacht ~ Leipzig** the battle of Leipzig; **~ den Fahrrädern/Kindern bleiben** stay with the bicycles/children; **etw. ~ sich haben** have sth. with *or* on one; **nicht [ganz] ~ sich sein** *(fig.)* be not quite with it; **sich ~ jmdm. entschuldigen/erkundigen** apologize to sb./ask sb.; **wir haben Physik ~ Herrn Meyer** we do physics with Mr Meyer; **b)** *(unter)* among; **war heute ein Brief für mich ~ der Post?** was there a letter for me in the post today?; **c)** *(an)* by; **jmdn. ~ der Hand nehmen** take sb. by the hand; **d)** *(im Wohn-/Lebens-/Arbeitsbereich von)* **~ uns tut man das nicht** we don't do that; **~ mir [zu Hause]** at my house; **~ uns um die Ecke/gegenüber** round the corner from us/opposite us; **~ seinen Eltern leben** live with one's parents; **wir sind ~ ihr eingeladen** we have been invited to her house; **wir treffen uns ~ uns/Peter** we'll meet at our/Peter's place; **~ uns in Österreich** in Austria [where I/we come from/live]; **~ uns in der Firma** in our company; **~ Schmidt** *(auf Briefen)* c/o Schmidt; **~ einer Firma sein** be with a company; **~ jmdm./einem Verlag arbeiten** work for sb./a publishing house; **e)** *(im Bereich eines Vorgangs)* at; **~ einer Hochzeit/einem Empfang** *usw.* at a wedding/reception *etc.*; **~ einem Unfall** in an accident; **f)** *(im Werk von)* **~ Goethe** in Goethe; **~ Schiller heißt es ...:** Schiller says *or* writes that ...; **g)** *(im Falle von)* in the case of; **~ bestimmten Pflanzen** in certain plants; **~ der Hauskatze** in the domestic cat; **wie ~ den Römern** as with the Romans; **hoffentlich geht es nicht wie ~ mir** I hope the same thing doesn't happen as happened in my case; **h)** *(Zeitpunkt)* **~ seiner Ankunft** on his arrival; **~ diesen Worten errötete er** at this he blushed; **~ Sonnenaufgang/-untergang** at sunrise/sunset; **~ unserer Begegnung** at our meeting; **i)** *(modal)* **~ Tag/Nacht** by day/night; **~ Tag und [~] Nacht** day and night; **~ Tageslicht** by daylight; **~ Nebel** in fog; **~ Kälte** when it's cold; **~ offenem Fenster schlafen** sleep with the window open; **j)** *(im Falle des Auftretens von)* **„~ Feuer Scheibe einschlagen"** 'in case of fire, break glass'; **„~ Regen Schleudergefahr"** 'slippery when wet'; **~ hohem Fieber** when sb. has a high temperature; **k)** *(angesichts)* with; **~ dieser Hitze** in this heat; **~ diesem Sturm/Lärm** with this storm blowing/noise going on; **l)** *(trotz)* **~ all seinem Engagement/seinen Bemühungen** in spite of *or* despite *or* for all his commitment/efforts; **~ allem Verständnis, aber ich kann das nicht** much as I sympathize, I cannot do that; **m)** *(in Beteuerungsformeln)* by; **~ Gott!** by God!; **~ meiner Ehre!** *(veralt.)* upon my honour!
bei|behalten *unr. tr. V.* keep; retain; keep up ⟨*custom, habit*⟩;

continue, maintain ⟨*way of life*⟩; keep to ⟨*course, method*⟩; preserve, maintain ⟨*attitude*⟩
Beibehaltung die; ~: *s.* **beibehalten:** keeping; retention; keeping up; continuance; maintenance; preservation
Bei·boot das ship's boat
bei|bringen *unr. tr. V.* **a) jmdm. etw. ~:** teach sb. sth.; **jmdm. Gehorsam ~:** teach sb. obedience; **b)** *(ugs.: mitteilen)* **jmdm. ~, daß ...:** break it to sb. that ...; **c)** *(zufügen)* **jmdm./sich etw. ~:** inflict sth. on sb./oneself; **d)** *(beschaffen)* produce ⟨*witness, evidence*⟩; provide, supply ⟨*reference, proof*⟩; produce, furnish ⟨*money*⟩
Beichte ['baiçtə] **die;** ~, ~**n** confession *no def. art.*; **zur ~ gehen** go to confession; **jmdm. die ~ abnehmen** hear sb.'s confession
beichten 1. *itr. V.* confess; **~ gehen** go to confession. **2.** *tr. V.* *(auch fig.)* confess
Beicht-: ~geheimnis das seal of confession; **~stuhl der** confessional; **~vater der** father confessor
beide ['baidə] *Indefinitpron. u. Zahlw.* **1.** *Pl.* **die ~n** the two; **~:** both; *(der/die/das eine oder der/die/das andere von den ~n)* either *sing;* **die/seine ~n Brüder** the/his two brothers; **die ersten ~n Strophen** the first two verses; **kennst du die ~n?** do you know those two?; **alle ~:** both of us/you/them; **sie sind alle ~ sehr schön** they're both very nice; both of them are very nice; **sie sind ~ nicht hübsch** neither of them is pretty; **ihr/euch ~:** you two; **Ihr/euch ~ nicht** neither of you; **wir/uns ~:** the two of us/both of us; **er hat ~ Eltern verloren** he has lost both [his] parents; **mit ~n Händen** with both hands; **~ Male** both times; **ich habe ~ gekannt** I knew both of them; **einer/eins von ~n** one of the two; **keiner/keins von ~n** neither [of them]. **2.** *Neutr. Sg.* both *pl.*; *(das eine oder das andere)* either; **~s ist möglich** either is possible; **ich glaube ~s/~s nicht** I believe both things/neither thing; **das ist ~s nicht richtig** neither of those is correct; **er hat sich in ~m geirrt** he was wrong on both counts; **er hatte von ~m wenig Ahnung** he had little idea of either
beiderlei ['baidɐ'lai] *Gattungsz., indekl.* **~ Geschlechts** of both sexes; **von ~ Art** of both kinds
beider·seitig *Adj.* mutual ⟨*decision, agreement*⟩; **zur ~en Überraschung** to the surprise of both of us/them; **in ~em Einverständnis** by mutual agreement
beider·seits 1. *Präp. m. Gen.* on both sides of. **2.** *Adv.* on both sides
bei|drehen *itr. V. (Seemannsspr.)* heave to
beid·seitig 1. *Adj.* mutual. **2.** *adv.* ⟨*be printed etc.*⟩ on both sides; **~ gelähmt** paralysed down both sides
bei·einander *Adv.* together; **~ Trost suchen** seek comfort from each other
beieinander-: ~|haben *unr. tr. V.* **etw. ~haben** have got sth. together; **du hast/er hat** *usw.* **[sie] nicht alle ~** *(ugs.)* he's/you're *etc.* not all there *(coll.)*; **~|sein** *unr. itr. V.; mit sein (nur im Inf. und Part. zusammengeschrieben)* **gut/schlecht ~sein** *(ugs.)* be in good/bad shape; **nicht ganz ~sein** *(ugs.)* be not quite all there *(coll.)*
Bei·fahrer der, Bei·fahrerin die a) [front-seat] passenger; *(auf dem Motorrad)* pillion passenger; *(im Beiwagen)* sidecar passenger; **b)** *(berufsmäßig)* co-driver; *(im LKW)* driver's mate
Beifahrer·sitz der passenger seat; *(eines Motorrads)* pillion
Bei·fall der; *o. Pl.* **a)** applause; *(Zurufe)* cheers *pl.*; cheering; **~ klatschen/spenden** applaud; **b)** *(Zustimmung)* approval; **~ finden** meet with approval
bei·fällig 1. *Adj.* approving; favourable ⟨*judgement*⟩. **2.** *adv.* approvingly; **~ nicken** nod approvingly *or* in approval
Beifalls-: ~äußerung die expression of approval; **~bekundung die** demonstration of approval; **~ruf der** shout of approval; cheer; **~sturm der** storm of applause
bei|fügen *tr. V.* **einer Bewerbung etw. ~:** enclose sth. with an application; **einem Paket eine Zollerklärung ~:** attach a customs declaration to a parcel
Bei·gabe die a) *o. Pl.* **unter ~** *(Dat.)* **von etw.** adding sth.; **b)** *(Hinzugefügtes)* addition
beige [be:ʃ] *Adj.* beige
Beige das; ~, ~ *od. (ugs.)* **~s** beige
bei|geben 1. *unr. tr. V.* **a)** add *(Dat.* to). **2.** *unr. itr. V. in* **klein ~** *(ugs.)* give in
Bei·geschmack der; *o. Pl.* **einen bitteren** *usw.* **~ haben** have a slightly bitter *etc.* taste [to it]; taste slightly bitter *etc.*; **dieses Wort hat einen negativen ~** *(fig.)* this word has slightly negative overtones *pl.*
Bei·hilfe die a) [financial] aid *or* assistance; *(Zuschuß)* allowance; *(Subvention)* subsidy; **b)** *o. Pl.* *(Rechtsw.: Mithilfe)* aiding and abetting; **jmdn. wegen ~ zum Mord anklagen** charge sb. with aiding and abetting a murder *or* with acting as accessory to a murder
bei|kommen *unr. itr. V.; mit sein* **jmdm. ~:** get the better of sb.; **den Schwierigkeiten/jmds. Sturheit ~:** overcome the difficulties/cope with sb.'s obstinacy
Beil [bail] **das;** ~[e]s, ~e axe; *(kleiner)* hatchet; *(Fleischer~)* cleaver
Bei·lage die a) *(Zeitungs~)* supplement; **b)** *(zu Speisen)* side-dish; *(Gemüse~)* vegetables *pl.*; **ein Fleischgericht mit diversen ~n** a meat dish with a selection of trimmings
bei·läufig 1. *Adj.* casual; casual, passing ⟨*remark, mention*⟩. **2.** *adv.* casually; **etw. ~ erwähnen** mention sth. casually *or* in passing
Beiläufigkeit die; ~ casualness
bei|legen *tr. V.* **a)** *(dazulegen)* enclose; *(einem Buch, einer Zeitschrift)* insert *(Dat.* in); **einem Brief** *usw.* **etw. ~:** enclose sth. with a letter *etc.*; **b)** *(schlichten)* settle ⟨*dispute, controversy, etc.*⟩; **c)** *s.* **beimessen**
Beilegung die; ~, ~**en** settlement
beileibe [bai'laibə] *Adv.* **~ nicht** certainly not; **er ist ~ kein Genie** he is by no means a genius
Bei·leid das sympathy; **[mein] herzliches** *od.* **aufrichtiges ~!** please accept my sincere condolences; **jmdm. sein [aufrichtiges] ~ [zu etw.] aussprechen** offer one's [sincere] condolences *pl.* to sb. [on sth.]
Beileids-: ~besuch der visit of condolence; **~bezeigung die;** ~, **~en, ~bezeugung die** expression of sympathy
bei|liegen *unr. itr. V.* **einem Brief ~:** be enclosed with a letter; **dem Buch liegt ein Prospekt bei** the book contains a catalogue as an insert
bei·liegend *(Amtsspr.) Adj.* enclosed; **~ senden wir ...:** please find enclosed ...
beim [baim] *Präp. + Art.* **a) = bei dem; b) ~ Bäcker** at the baker's; **jmdn. ~ Arm packen** seize sb. by the arm; **~ Film sein** be in films; **~ Ahorn/Menschen** in the maple/in man; **c)** *(zeitlich)* **er will ~ Arbeiten nicht gestört werden** he

doesn't want to be disturbed when *or* while [he's] working; ~ **Essen spricht man nicht** you shouldn't talk while [you're] eating; ~ **Verlassen des Gebäudes** when *or* on leaving the building; ~ **Fasching** at carnival time; **[gerade]** ~ **Duschen sein** be taking a shower

bei|mengen *tr. V.* add (*Dat.* to)

bei|messen *unr. tr. V.* attach (*Dat.* to)

bei|mischen *tr. V.* add (*Dat.* to)

Bein [bain] **das;** ~**[e]s,** ~**e a)** leg; **jmdm. ~e machen** *(ugs.)* make sb. get a move on *(coll.)*; **er hat sich** *(Dat.)* **kein ~ ausgerissen** *(ugs.)* he didn't over-exert himself; **jmdm. ein ~ stellen** trip sb.; *(fig.)* put *or* throw a spanner *or (Amer.)* a monkey-wrench in sb.'s works; **jmdm. [einen] Knüppel** *od.* **Prügel zwischen die ~e werfen** *(fig.)* put *or* throw a spanner *or (Amer.)* a monkey-wrench in sb.'s works; **das hat ~e gekriegt** *(fig. ugs.)* it seems to have [grown legs and] walked *(coll.)*; **die ~e in die Hand** *od.* **unter die Arme nehmen** *(fig. ugs.)* step on it *(coll.)*; **[wieder] auf die ~e kommen** *(ugs.)* get back on one's/its feet [again]; **jmdn./etw. [wieder] auf die ~e bringen** *(ugs.)* put sb./sth. back on his/her/its feet again; **jmdm. auf die ~e helfen** help sb. to his/her feet; **ich kann mich nicht mehr/kaum noch auf den ~en halten** I can't/can hardly stand up; **auf eigenen ~en stehen** *(fig.)* stand on one's own two feet; support oneself; **mit beiden ~en im Leben** *od.* **[fest] auf der Erde stehen** have both feet [firmly] on the ground; **mit dem linken ~ zuerst aufgestanden sein** *(ugs.)* have got out of bed on the wrong side; **mit einem ~ im Gefängnis/Grab[e] stehen** *(fig.)* stand a good chance of ending up in prison/have one foot in the grave; **von einem ~ aufs andere treten** *(ugs.)* shift from one foot to the other; **b)** *(Hosen~, Tisch~, Stuhl~ usw.)* leg

bei·nah[e] ['baina:(ə)] *Adv.* almost; nearly; **wir wären ~ zu spät gekommen** we were nearly too late

Bei·name der epithet

Bein·bruch der broken leg; **das ist [doch] kein ~bruch!** *(ugs.)* it's not the end of the world *(coll.)*

beinhalten [bə'|ɪnhaltn̩] *tr. V.* *(Papierdt.)* involve

-beinig *adj.* -legged

Bein·schiene die [long] shin pad; *(Cricket, Hockey)* pad

Beipack·zettel der instruction leaflet

bei|pflichten ['baipflɪçtn̩] *itr. V.* **jmdm. [in etw.** *(Dat.)***]** ~: agree with sb. [on sth.]; **einem Vorschlag** *usw.* ~: agree with a proposal *etc.*

Bei·rat der advisory committee *or* board

beirren *tr. V.* **sich durch nichts/von niemandem ~ lassen** not be put off *or* deterred by anything/anybody; not let anything/anybody put one off *or* deter one; **nichts konnte ihn in seinen Ansichten ~**: nothing could shake him in his views

beisammen [bai'zamən] *Adv.* together

beisammen-, Beisammen-: ~**|haben** *unr. tr. V. s.* **beieinanderhaben;** ~**|halten** *unr. tr. V.* keep together; hold on to ⟨*money*⟩; ~**|sein** *unr. itr. V.; mit sein; (nur im Inf. und 2. Part. zusammengeschrieben)* **[gut]** ~**sein** *(ugs.)* be in good health *or* shape; ~**sein das** get-together; ~**|sitzen** *unr. itr. V.* sit together

Bei·schlaf der *(geh., Rechtsw.)* sexual intercourse

Bei·sein das *in* **im ~ von jmdm., in jmds. ~**: in the presence of sb. *or* in sb.'s presence

bei·seite *Adv.* aside; **jmdn. ~ ziehen/schieben** draw/push sb. to one side *or* aside; **etw. ~ bringen** get sth. hidden away; hide sth. away; **etw. ~ lassen** *(fig.)* leave sth. aside; **etw. ~ legen** put *or* lay sth. aside; *(sparen)* put sth. by *or* aside; **jmdn./etw. ~ schaffen** *(ugs.)* get rid of sb./sth.

Beis[e]l ['baizl̩] **das;** ~**s,** ~ *od.* ~**n** *(österr.) s.* **Kneipe**

bei|setzen *tr. V.* bury; inter; lay to rest; inter ⟨*ashes*⟩

Bei·setzung die; ~, ~**en** *(geh.)* funeral; burial

Bei·sitzer der; ~**s,** ~, **Bei·sitzerin die;** ~, ~**nen** assessor; *(bei Ausschüssen)* committee member

Bei·spiel das a) example **(für** of); **zum ~**: for example *or* instance; **wie zum ~**: as for example; such as; **ohne ~ sein** be without parallel; be unparalleled; **b)** *(Vorbild)* example; **ein warnendes ~**: a warning; **jmdm. ein ~ geben** set an example to sb.; **sich** *(Dat.)* **an jmdm./etw. ein ~ nehmen** follow sb.'s example/take sth. as one's example; **mit gutem ~ vorangehen** set a good example

beispielhaft *Adj.* exemplary

beispiel·los 1. *Adj.* unparalleled. **2.** *adv.* incomparably ⟨*well, badly, etc.*⟩; ~ **erfolgreich** with unparalleled success

beispiels-: ~**halber** *Adv.* for example *or* instance; ~**weise** *Adv.* for example *or* instance

bei|springen *unr. itr. V.; mit sein* **jmdm. [in der Not]** ~: leap *or* rush to sb.'s aid *or* assistance [in an emergency]; **jmdm. mit Geld ~**: help sb. out with money

beißen ['baisn̩] **1.** *unr. tr., itr. V.* **a)** bite; *(kauen)* chew; **in etw.** *(Akk.)* ~: bite into sth.; **an den Nägeln ~**: bite one's nails; **ich habe mich** *od.* **mir auf die Zunge/in die Lippe gebissen** I've bitten my tongue/lip; **der Hund hat mir** *od.* **mich ins Bein gebissen** the dog bit me in the leg; **nichts/nicht viel zu ~ haben** *(fig.)* have nothing/not have much to eat; **b)** *(ätzen)* sting; **in die** *od.* **in den Augen ~**: sting one's eyes; make one's eyes sting; **auf der Zunge ~**: burn the tongue. **2.** *unr. refl. V. (ugs.)* ⟨*colours*⟩ clash **(mit** with)

beißend *Adj.; nicht präd.* biting ⟨*cold*⟩; acrid ⟨*smoke, fumes*⟩; sharp ⟨*frost*⟩; pungent, sharp ⟨*smell, taste*⟩; *(fig.)* biting ⟨*ridicule*⟩; cutting ⟨*irony*⟩

Beiß-: ~**ring der** teething-ring; ~**zange die** *s.* **Kneifzange**

Bei·stand der *o. Pl. (geh.)* aid; assistance; help; **jmdm. ~ leisten** give sb. aid *or* assistance; come to sb.'s aid *or* assistance

bei|stehen *unr. itr. V.* **jmdm. ~**: aid *or* assist *or* help sb.; *(zur Seite stehen)* stand by sb.

Beistell-: ~**tisch der,** ~**tischchen das** occasional table; *(im Restaurant)* side-table

bei|steuern *tr. V.* contribute; make ⟨*contribution*⟩

bei|stimmen *itr. V.: s.* **zustimmen**

Bei·strich der *(veralt.)* comma

Beitrag ['baitra:k] **der;** ~**[e]s, Beiträge** ['baitrɛ:gə] contribution; *(Versicherungs~)* premium; *(Mitglieds~)* subscription; **einen ~ zu etw. leisten** make a contribution to sth.

bei|tragen *unr. tr., itr. V.* contribute **(zu** to); **das Seine/viel zu etw. ~**: contribute one's share/a great deal to sth.

beitrags-: ~**frei** *Adj.* non-contributory; ⟨*person*⟩ not liable to pay contributions; ~**pflichtig** *Adj. (Sozialw.)* ⟨*employee*⟩ liable to pay contributions; ⟨*earnings*⟩ on which contributions are payable

bei|treiben *unr. tr. V. (Rechtsw.)* enforce payment of

bei|treten *unr. itr. V.; mit sein*

join; **einem Verein** *usw.* ~: join a club *etc.*; **einem Abkommen/Pakt** ~: accede to an agreement/a pact

Bei·tritt der joining; **seinen ~ erklären** apply for membership

Beitritts·erklärung die application for membership

Bei·wagen der side-car

Bei·werk das; *o. Pl.* accessories *pl.*

beilwohnen *itr. V.* **einer Sache** *(Dat.)* ~: be present at *or* attend sth.

[1]**Beize** ['baitsə] **die;** ~, ~**n** *(Holzbearb.)* [wood-]stain

[2]**Beize die;** ~, ~**n** *(Jagdw.)* hawking

[3]**Beize die;** ~, ~**n** *(schweiz.) s.* **Kneipe**

beizeiten [bai'tsaitn̩] *Adv.* in good time

beizen ['baitsn̩] *tr. V. (Holzbearb.)* stain

Beiz·jagd die *s.* [2]**Beize**

bejahen [bə'ja:ən] *tr. V.* **a) etw.** ~: give an affirmative answer to sth.; answer sth. in the affirmative; **b)** *(gutheißen, befürworten)* approve of; **das Leben** ~: have a positive *or* affirmative attitude to life

bejahend 1. *Adj.* affirmative; affirmative, positive ‹*attitude*›. **2.** *adv.* ‹*answer*› in the affirmative; ‹*nod*› affirmatively

bejahrt [bə'ja:ɐ̯t] *Adj. (geh.)* advanced in years

Bejahung die; ~, ~**en a)** affirmative answer *or* reply; **b)** *(das Gutheißen)* approval

bejammern *tr. V.* lament

bejubeln *tr. V.* cheer; acclaim

bekämpfen *tr. V.* **a)** fight against; **sich [gegenseitig]** ~: fight [one another *or* each other]; **b)** *(fig.)* combat, fight ‹*disease, epidemic, pest*›; combat ‹*unemployment, crime, alcoholism*›; curb ‹*curiosity, prejudice*›

Bekämpfung die; ~ **a)** fight (*Gen.* against); **b)** *s.* **bekämpfen b**: combating; fighting; curbing

bekannt [bə'kant] *Adj.* **a)** well-known; **es wurde ~, daß ...**: it became known that ...; **für etw. ~ sein** be well known for sth.; ~**er sein** be better known; **b) jmd./etw. ist jmdm.** ~: sb. knows sb./sth.; **davon ist mir nichts** ~: I know nothing about that; **mit jmdm. ~ sein/werden** know *or* be acquainted with sb./get to know *or* become acquainted with sb.; **jmdn./sich mit jmdm. ~ machen** introduce sb./oneself to sb.; **Darf ich ~ machen? Meine Eltern** may I introduce my parents?; **jmdn./sich mit etw. ~ machen** acquaint sb./oneself with sth.; **jmdm. ~ vorkommen** seem familiar to sb.; **der Witz kommt mir ~ vor** I think I've heard that joke somewhere before

Bekannte der/die *adj. Dekl.* acquaintance

Bekannten·kreis der circle of acquaintances

bekannter·maßen *Adv. (Papierdt.) s.* **bekanntlich**

Bekannt·gabe die; ~: announcement

bekanntlgeben *unr. tr. V.* announce

Bekanntheit die; ~: **trotz der ~ dieser Tatsache** although this fact is widely known; **wegen Brandts großer** ~: because Brandt is so well known

bekanntlich *Adv.* as is well known; **etw. ist ~ der Fall** sth. is known to be the case; **der Wal ist ~ ein Säugetier** it is well known that the whale is a mammal

bekanntlmachen *tr. V.* announce; *(der Öffentlichkeit)* make public

Bekannt·machung die; ~, ~**en a)** *o. Pl.* announcement; *(Veröffentlichung)* publication; **b)** *(Mitteilung)* announcement; notice

Bekanntschaft die; ~, ~**en a)** *o. Pl.* acquaintance; **bei näherer** ~: on closer acquaintance; **jmds. ~ machen** make sb.'s acquaintance; **b)** *(Bekannter, Bekannte)* acquaintance; *(Bekanntenkreis)* circle of acquaintances

bekanntlwerden *unr. itr. V.; mit sein (nur im Inf. und 2. Part. zusammengeschrieben)* become known; become public knowledge

bekehren 1. *tr. V.* convert (**zu** to). **2.** *refl. V.* become converted (**zu** to)

Bekehrte der/die; *adj. Dekl.* convert

Bekehrung die; ~, ~**en** conversion (**zu** to)

bekennen 1. *unr. tr. V.* **a)** admit ‹*mistake, defeat*›; confess ‹*sin*›; admit, confess ‹*guilt, truth*›; **b)** *(Rel.)* profess; **die Bekennende Kirche** *(hist.)* the Confessional Church. **2.** *refl. V.* **sich zum Islam** *usw.* ~: profess Islam *etc.*; **sich zu Buddha/Mohammed** ~: profess one's faith in Buddha/Muhammad; **seine Freunde bekannten sich zu ihm** his friends stood by him; **sich zu einer Verfehlung** ~: amit to a misdemeanour; **sich zu seiner Vergangenheit** ~: acknowledge one's past; **sich zu seiner Schuld** ~: admit *or* confess one's guilt; **sich schuldig/nicht schuldig** ~: admit *or* confess/not admit *or* not confess one's guilt; *(vor Gericht)* plead guilty/not guilty; **sich zu einem Bombenanschlag** ~: claim responsibility for a bomb attack

Bekenner-: ~brief, der letter claiming responsibility; **~geist der, ~mut der;** *o. Pl.* courage of one's convictions

Bekenntnis das; ~**ses,** ~**se a)** confession; **ein ~ ablegen** make a confession; **b) ein ~ zum Frieden** a declaration for peace; **ein ~ zum Christentum/zur Demokratie ablegen** profess one's faith in Christianity/declare one's belief in democracy; **c)** *(Konfession)* denomination

Bekenntnis-: ~freiheit die; *o. Pl.* religious freedom; freedom of worship; **~schule die** denominational school

bekifft [bə'kɪft] *Adj. (ugs.)* stoned *(sl.)*

beklagen 1. *tr. V. (geh.)* **a)** *(betrauern)* mourn; **Menschenleben waren nicht zu** ~: there were no fatalities; **b)** *(bedauern)* lament; **sein/jmds. Los** ~: lament *or* bewail one's fate/deplore sb.'s fate; **wir haben einen großen Umsatzrückgang zu** ~: we have to note with regret a large drop in sales. **2.** *refl. V.* complain (**bei** to); **ich kann mich nicht** ~: I can't complain

beklagens·wert *Adj.* pitiful ‹*sight, impression*›; pitiable ‹*person*›; lamentable, pitiable, deplorable ‹*condition, state*›; wretched ‹*situation*›

Beklagte der/die; *adj. Dekl.* defendant; *(bei Ehescheidungen)* respondent

beklatschen *tr. V.* clap; applaud

beklauen *tr. V. (salopp)* rob; do *(sl.)*

bekleben *tr. V.* **eine Wand** *usw.* **mit etw.** ~: stick sth. all over a wall *etc.*

bekleckern *(ugs.)* **1.** *tr. V.* **seinen Schlips** *usw.* **mit Soße** *usw.* ~: drop *or* spill sauce *etc.* down one's tie *etc.* **2.** *refl. V.* **sich [mit Soße** *usw.***]** ~: drop *or* spill sauce *etc.* down oneself

bekleiden *tr. V.* **a)** clothe; **mit etw. bekleidet sein** be dressed in *or* be wearing sth.; **b)** *(geh.: innehaben)* occupy, hold ‹*office, position*›

Bekleidung die clothing; clothes *pl.*; garments *pl.*

beklemmend 1. *Adj.* oppressive. **2.** *adv.* oppressively

Beklemmung die; ~, ~**en** oppressive feeling; *(Angst)* [feeling of] unease; *(stärker)* [feeling of] apprehension

beklommen [bə'klɔmən] **1.** *Adj.* uneasy; *(stärker)* apprehensive. **2.** *adv.* uneasily; *(stärker)* apprehensively

Beklommenheit die; ~: uneasiness; *(stärker)* apprehensiveness

bekloppt [bə'klɔpt] *Adj. (salopp)* barmy *(Brit. sl.)*; loony *(sl.)*; **ein Bekloppter** a nut-case *(Brit. sl.)*; a nut *(sl.)*

beknackt *Adj. (salopp)* lousy *(sl.)*; **ein** ~**er Typ** a berk *(Brit. sl.)*

beknien *tr. V. (ugs.)* beg

bekommen 1. *unr. tr. V.* **a)** get; get, receive ‹*money, letter, reply, news, orders*›; *(erlangen)* get; obtain; *(erreichen)* catch ‹*train, bus, flight, etc.*›; **eine Flasche** *usw.* **an den Kopf** ~: get hit on the head with a bottle *etc.*; **was** ~ **Sie?** *(im Geschäft)* can I help you?; *(im Lokal, Restaurant)* what would you like?; **was** ~ **Sie [dafür]?** how much is that?; **noch Geld von jmdm.** ~: be owed money by sb.; **wir** ~ **Regen/besseres Wetter** we're going to get some rain/some better weather; there's rain/better weather on the way; **ich bekomme keine Verbindung** I can't get through; **Besuch** ~: have a visitor/visitors; **sie bekommt ein Kind** she's expecting a baby; **Hunger/Durst** ~: get hungry/thirsty; **einen roten Kopf/eine Glatze** ~: go red/bald; **eine Erkältung** ~: catch a cold; **Krebs** ~: get cancer; **Mut/Angst** ~: take heart/become frightened; **er bekommt einen Bart** he's growing a beard; **sie bekommt eine Brust** her breasts are developing; **Zähne** ~ ‹*baby*› teethe; **wo bekomme ich etwas zu essen/trinken?** where can I get something to eat/drink?; **etw./jmdn. zu fassen** ~: get hold of sth./lay one's hands on sb.; **etw. zu sehen** ~: set eyes on sth.; *s. auch* **hören, spüren; b) etw. durch die Tür/ins Auto** ~: get sth. through the door/into the car; **jmdn. nicht aus dem Bett** ~: be unable to get sb. out of bed *or* up; **jmdn. dazu** ~, **die Wahrheit zu sagen** get sb. to tell the truth; **etw. sauber** ~: get sth. clean; **jmdn. satt** ~: feed sb.; **c) es nicht über sich** *(Akk.)* ~, **etw. zu tun** be unable to bring oneself to do sth. **2.** *unr. V.; in der Funktion eines Hilfsverbs zur Umschreibung des Passivs* get; **etw. geschenkt** ~: get [given] sth. *or* be given sth. as a present; **etw. gestohlen** ~: have sth. stolen; **etw. geliehen** ~: be lent sth.; **einen Zahn gezogen** ~: have a tooth out. **3.** *unr. itr. V.; mit sein* **jmdm. [gut]** ~: do sb. good; be good for sb.; ‹*food, medicine*› agree with sb.; **jmdm. schlecht** *od.* **nicht** ~: not be good for sb.; not do sb. any good; ‹*food, medicine*› not agree with sb.; **wohl bekomm's!** your [very good] health!

bekömmlich [bə'kœmlɪç] *Adj.* easily digestible; **leicht/schwer** ~ **sein** be easily digestible/difficult to digest

Bekömmlichkeit die; ~: easy digestibility

beköstigen [bə'kœstɪgn̩] *tr. V.* cater for; **er wird von seiner Tante beköstigt** he gets his meals provided by his aunt

Beköstigung die; ~: catering *no indef. art.*

bekräftigen *tr. V.* reinforce ‹*statement*›; reaffirm ‹*promise*›

Bekräftigung die: zur ~ **seiner Worte** to reinforce his words; **zur** ~ **seines Versprechens** to reaffirm his promise

bekreuzigen *refl. V.* cross oneself

bekriegen *tr. V.* wage war on; *(fig.)* fight; **sich** ~: be at war; *(fig.)* fight [each other *or* one another]

bekritteln *tr. V. (abwertend)* find fault with *(in a petty way)*

bekritzeln *tr. V.* scribble on; **die Wände waren von oben bis unten bekritzelt** the walls were covered with graffiti

bekümmern *tr. V.* **jmdn.** ~: cause sb. worry

bekümmert [bə'kʏmɐt] *Adj.* worried; troubled; *(stärker)* distressed

bekunden [bə'kʊndn̩] *tr. V. (geh.)* express

Bekunden das: nach eigenem ~: according to his/her *etc.* own statement[s]

Bekundung die; ~, ~**en** expression; *(Aussage)* statement

belächeln *tr. V.* smile [pityingly/tolerantly *etc.*] at; **belächelt werden** meet with a pitying smile

[1]**beladen** *unr. tr. V.* load ‹*ship*›; load [up] ‹*car, wagon*›; load up ‹*horse, donkey, etc.*›; **Be- und Entladen gestattet/verboten** loading and unloading permitted/no loading or unloading

[2]**beladen** *Adj.* loaded, laden (**mit** with); **mit etw.** ~ **sein** be laden with sth.; **sie war schwer mit Paketen** ~: she was loaded *or* laden down with parcels; **mit Sorgen/Schuld** ~ **sein** *(fig.)* ‹*person*› be burdened with cares/guilt

Belag [bə'la:k] **der;** ~**[e]s, Beläge** [bə'lɛ:gə] **a)** coating; film; *(Zahn~)* film; **b)** *(Fußboden~)* covering; *(Straßen~)* surface; *(Brems~)* lining; **c)** *(von Kuchen, Pizza, Scheibe Brot usw.)* topping; *(von Sandwich)* filling

Belagerer [bə'la:gərɐ] **der;** ~**s,** ~: besieger

belagern *tr. V. (Milit.)* besiege; lay siege to; *(fig.)* besiege

Belagerung die; ~, ~**en** *(Milit.)* siege; *(fig.)* besieging

Belang [bə'laŋ] **der;** ~**[e]s,** ~**e a)** *o. Pl. (Bedeutung)* **[für etw.] von/ohne** ~ **sein** be of importance/of no importance [for sth.]; **für jmdn. von/ohne** ~ **sein** be important/not be important to sb.; **b)** *Pl. (Interessen)* interests; **jmds.** ~**e wahrnehmen/vertreten** look after/represent sb.'s interests

belangen *tr. V. (Rechtsw.)* sue; *(strafrechtlich)* prosecute; **jmdn. wegen etw.** ~: sue/prosecute sb. for sth.

belang·los *Adj.* of no importance (**für** for); *(trivial)* trivial

Belang·losigkeit die; ~, ~**en** unimportance; *(Trivialität)* triviality

belassen *unr. tr. V.* leave; ~ **wir es dabei** let's leave it at that

belastbar *Adj.* **a)** tough, resilient ‹*material*›; ‹*material*› able to withstand stress *pred.*; **[nur] mit 3,5 t** ~ **sein** be able to take a load of [only] 3.5 t; **b)** *(beanspruchbar)* tough, resilient ‹*person*›; **seelisch/körperlich** ~ **sein** be emotionally/physically tough *or* resilient; be able to stand emotional/physical stress

Belastbarkeit die; ~, ~**n a)** *(von Material)* ability to withstand stress; *(von Konstruktionen)* load-bearing capacity; **b)** *(von Menschen)* toughness; resilience

belasten *tr. V.* **a) etw.** ~: put sth. under strain; *(durch Gewicht)* put weight on sth.; **b)** *(beeinträchtigen)* pollute ‹*atmosphere*›; put pressure on ‹*environment*›; **c)** *(in Anspruch nehmen)* burden (**mit** with); **d)** *(zu schaffen machen)* **jmdn.** ~ ‹*responsibility, guilt*› weigh upon sb.; ‹*thought*› weigh upon sb.'s mind; **Fett belastet den Magen** fat puts a strain on the stomach; **e)** *(Rechtsw.: schuldig erscheinen lassen)* incriminate; ~**des Material**

incriminating evidence; **f)** *(Geldw.)* **jmds. Konto mit 100 DM ~:** debit sb.'s account with 100 DM; **den Staatshaushalt ~:** place a burden on the national budget; **das Haus ist mit einer Hypothek belastet** the house is encumbered with a mortgage

belästigen [bə'lɛstɪgn̩] *tr. V.* bother; *(sehr aufdringlich)* pester; *(sexuell)* molest; **sich von etw. belästigt fühlen** regard sth. as a nuisance

Belästigung die; ~, ~en: die ~ durch die Reporter/Insekten being pestered by reporters/ bothered by insects; **etw. als ~ empfinden** regard sth. as a nuisance

Belastung [bə'lastʊŋ] **die; ~, ~en a)** strain; *(das Belasten)* straining; *(durch Gewicht)* loading; *(Last)* load; **b) die ~ der Atmosphäre/Umwelt durch Schadstoffe** the pollution of the atmosphere by harmful substances/the pressure on the environment caused by harmful substances; **c)** *(Bürde, Sorge)* burden; **das stellte eine schwere seelische ~ für sie dar** it was causing her great strain and distress; **d)** *(Rechtsw.)* incrimination

Belastungs-: ~material das *(Rechtsw.)* incriminating evidence; **~probe die** *(bei Menschen)* endurance test; *(bei Materialien)* stress test; *(bei Konstruktionen)* load test; **~zeuge der** *(Rechtsw.)* witness for the prosecution

belauern *tr. V.* **jmdn. ~:** eye *or* watch sb. carefully; keep a watchful eye on sb.; *(aus einem Versteck heraus)* watch sb. from hiding

belaufen *unr. refl. V.* **sich auf ...***(Akk.)* **~:** amount *or* come to ...; ‹*rent, price*› come to ..., be ...

belauschen *tr. V.* eavesdrop on

beleben 1. *tr. V.* **a)** enliven; liven up *(coll.)*; ‹*drink*› revive; **neu ~:** put new life into; stimulate ‹*economy*›; **b)** *(lebendig gestalten)* enliven; brighten up; **c)** *(lebendig machen)* give life to. **2.** *refl. V.* **a)** ‹*eyes*› light up; ‹*face*› brighten [up]; ‹*market, economic activity*› revive, pick up; **b)** *(lebendig, bevölkert werden)* come to life

belebend 1. *Adj.* stimulating; invigorating. **2.** *adv.* **~ wirken** have a stimulating *or* invigorating effect

belebt *Adj.* **a)** *(lebhaft, bevölkert)* busy ‹*street, crossing, town, etc.*›; **b)** *(lebendig)* living; **die ~e Natur** the living world

Belebtheit die; ~: bustle; bustling activity

Belebung die; ~: zur ~ ein Glas Sekt trinken have a glass of champagne to revive oneself; **die ~ der Konjunktur** the stimulation of the economy

belecken *tr. V.* lick

Beleg [bə'le:k] **der; ~[e]s, ~e a)** *(Beweisstück)* piece of [supporting] documentary evidence; *(Quittung)* receipt; **als ~ für etw.** as evidence for sth.; **b)** *(Sprachw.: Zitat)* quotation; **für dieses Wort gibt es zwei ~e** there are two instances of this word

belegbar *Adj.* verifiable

belegen *tr. V.* **a)** *(Milit.: beschießen)* bombard; *(mit Bomben)* attack; **b)** *(mit Belag versehen)* cover ‹*floor*› **(mit** with); fill ‹*flan base, sandwich*›; top ‹*open sandwich*›; **eine Scheibe Brot mit Schinken/Käse ~:** put some ham/ cheese on a slice of bread; **c)** *(in Besitz nehmen)* occupy ‹*seat, room, etc.*›; **d)** *(Hochschulw.)* enrol for, register for ‹*seminar, lecture-course*›; **e)** *(Sport)* **den ersten/letzten Platz ~:** come first *or* take first place/come last; **f)** *(nachweisen)* prove; give a reference for ‹*quotation*›; **etw. mit** *od.* **durch Quittungen ~:** support sth. with receipts; **g)** *(versehen)* **jmdn./etw. mit etw. ~:** impose sth. on sb./sth.

Belegschaft die; ~, ~en staff; employees *pl.*

belegt *Adj.* **a) ein ~es Brot** an open *or (Amer.)* openface sandwich; *(zugeklappt)* a sandwich; **ein ~es Brötchen** a roll with topping; an open-face roll *(Amer.)*; *(zugeklappt)* a filled roll; a sandwich roll *(Amer.)*; **b)** *(mit Belag bedeckt)* coated, furred ‹*tongue, tonsils*›; **c)** *(heiser)* husky ‹*voice*›; **d)** *(nicht mehr frei)* ‹*room, flat*› occupied; ‹*[telephone] line, number*› engaged, *(Amer.)* busy; **voll ~** ‹*hotel, hospital*› full

belehren *tr. V.* **a)** teach; instruct; *(aufklären)* enlighten; *(informieren)* inform; advise; **jmdn. über etw.** *(Akk.)* **~:** inform sb. about sth.; **b)** *(von einer irrigen Meinung abbringen)* **sich ~ lassen müssen** learn otherwise; **ich lasse mich gern ~:** I'm quite willing to believe otherwise; *s. auch* **besser**

belehrend *Adj.* didactic

Belehrung die; ~, ~en a) *(das Belehrtwerden)* instruction; **b)** *(Zurechtweisung)* lecture

beleibt [bə'laipt] *Adj. (geh.)* stout; portly; corpulent

Beleibtheit die; ~ *(geh.)* stoutness; portliness; corpulence

beleidigen [bə'laidɪgn̩] *tr. V.* insult; offend; *(fig.)* offend ‹*sb.'s honour, ear, eye*›; **~d** offensive

beleidigt *Adj.* insulted; offended; *(gekränkt)* offended; **er ist schnell ~:** he easily takes offence

Beleidigung die; ~, ~en insult; *(Rechtsw.) (schriftlich)* libel; *(mündlich)* slander; **eine ~ für das Auge/Ohr** *(fig.)* an offence to the eye/ear

beleihen *unr. tr. V.* grant a loan on the security of; grant a mortgage on ‹*home, property*›; raise money on ‹*insurance, policy*›

belemmert *(ugs.) Adj.* miserable; **er stand [wie] ~ da** he stood there miserably

belesen *Adj.* well-read

Belesenheit die; ~: [große] ~: [very] wide reading

beleuchten *tr. V.* **a)** illuminate; light up; light ‹*stairs, room, street, etc.*›; **festlich beleuchtet** festively lit; **b)** *(fig.: untersuchen)* examine ‹*topic, problem*›

Beleuchter der; ~s, ~ *(Theater, Film)* lighting technician

Beleuchtung die; ~, ~en a) *(Licht)* light; **die ~ fiel aus** all the lights *pl.* went out; **b)** *(das Beleuchten)* lighting; *(Anstrahlung)* illumination; **c)** *(fig.: Untersuchung)* examination

beleumdet [bə'lɔymdət], **beleumundet** [bə'lɔymʊndət] *Adj.* **übel/gut ~ sein** have a bad/good reputation

Belgien ['bɛlgi̯ən] **(das); ~s** Belgium

Belgier ['bɛlgi̯ɐ] **der; ~s, ~** Belgian

belgisch ['bɛlgɪʃ] *Adj.* Belgian

belichten 1. *tr. V. (Fot.)* expose; **eine Aufnahme richtig/falsch ~:** give a shot the right/wrong exposure. **2.** *itr. V. (Fot.)* **richtig/ falsch/kurz ~:** use the right/ wrong exposure/a short exposure time

Belichtung die; ~, ~en *(Fot.)* exposure

Belichtungs-: ~messer der *(Fot.)* exposure meter; **~zeit die** *(Fot.)* exposure time

belieben *itr. V. (geh.)* **ihr könnt tun, was euch** *(Dat.)* **beliebt** you can do what you like; *(unpers.)* **[ganz] wie es dir beliebt** [just] as you like

Belieben das; ~s: es steht in deinem ~/es bleibt Ihrem ~ überlassen it is up to you; **nach ~:** just as you/they *etc.* like

beliebig 1. *Adj.* any; **du kannst ein ~es Beispiel wählen** you can choose any example you like; **fünf ~e Personen** any five people; **in ~er Reihenfolge** in any order. **2.** *adv.* as you like/he likes *etc.*; **~ viele** as many as you like/he likes *etc.*; **wähle eine ~ große Zahl** choose any number[, as high as] you like

beliebt *Adj.* popular; favourite *attrib.*; **sich [bei jmdm.] ~ machen** make oneself popular [with sb.]

Beliebtheit die; ~: popularity

beliefern *tr. V.* supply; **jmdn. mit etw. ~:** supply sb. with sth.

Belieferung die supply; **jmds. ~ mit etw.** supplying sb. with sth.

bellen ['bɛlən] **1.** *itr. V.* **a)** ‹*dog, fox*› bark; ‹*hound*› bay; *(fig.)* ‹*cannon*› boom; **b)** *(laut husten)* have a hacking cough. **2.** *tr. V. (abwertend)* bark out ‹*orders*›

Belletristik [bɛle'trɪstɪk] **die; ~:** belles-lettres *pl.*

belletristisch *Adj.* belletristic ‹*literature*›; **ein ~er Verlag** a publishing house specializing in belletristic literature

belobigen [bə'lo:bɪgn̩] *tr. V.* commend

Belobigung die; ~, ~en commendation

belohnen *tr. V.* reward

Belohnung die; ~, ~en a) *(Lohn)* reward; **eine ~ für etw. aussetzen** offer a reward for sth.; **b)** *o. Pl. (das Belohnen)* rewarding

belüften *tr. V.* ventilate

Belüftung die; ~: ventilation

belügen *unr. tr. V.* **jmdn. ~:** lie to sb; tell lies to sb.; **sich selbst ~:** deceive oneself

belustigen *tr. V.* amuse

belustigt 1. *Adj.* amused. **2.** *adv.* in amusement

Belustigung die; ~: amusement; **der allgemeinen ~ dienen** serve to amuse everybody

bemächtigen [bə'mɛçtɪgn̩] *refl. V. (geh.)* **sich jmds./einer Sache ~:** seize sb./sth.; **Angst bemächtigte sich seiner** he was seized by fear

bemäkeln *tr. V. (ugs.)* find fault with

bemalen 1. *tr. V.* paint; *(verzieren)* decorate ‹*porcelain etc.*›; **sich** *(Dat.)* **das Gesicht ~** *(ugs.)* paint one's face. **2.** *refl. V. (ugs.)* paint one's face; put on one's war-paint *(coll.)*

Bemalung die; ~, ~en a) *o. Pl.* painting; *(Verzierung)* decorating; **b)** *(Farbschicht)* painting

bemängeln [bə'mɛŋl̩n] *tr. V.* find fault with; **die Reifen ~:** find the tyres to be faulty; **etw. an jmdm./etw. ~:** criticize sth. about sb./sth.

bemannen *tr. V.* man

bemänteln [bə'mɛntl̩n] *tr. V.* cover up

bemerkbar *Adj.* noticeable; perceptible; **sich ~ machen** *(auf sich aufmerksam machen)* attract attention [to oneself]; *(spürbar werden)* become apparent; ‹*tiredness*› make itself felt

bemerken *tr. V.* **a)** *(wahrnehmen)* notice; **ich wurde nicht bemerkt** I was unobserved; **sie bemerkte zu spät, daß ...:** she realized too late that ...; **b)** *(äußern)* remark; **nebenbei bemerkt** by the way; incidentally

bemerkenswert 1. *Adj.* remarkable; notable. **2.** *adv.* remarkably

Bemerkung die; ~, ~en a) remark; comment; **b)** *(schriftliche Anmerkung)* note

bemessen 1. *unr. tr. V.* **etw. nach etw. ~:** measure sth. according to sth.; **die Zeit ist kurz/sehr knapp ~:** time is short *or* limited/very limited. **2.** *unr. refl. V. (Amtsspr.)* **sich ~ nach** be measured on the basis of

bemitleiden *tr. V.* pity; feel sorry for; **er ist zu ~:** he is to be pitied; **sich selbst ~:** feel sorry for oneself

bemitleidens·wert *Adj.* pitiable

bemogeln *tr. V. (ugs.)* cheat; diddle *(Brit. sl.)*; con *(sl.)*

bemühen 1. *refl. V.* **a)** *(sich anstrengen)* try; make an effort; **sich sehr ~:** try hard; **bemüht sein, etw. zu tun** endeavour to do sth.; **bitte, ~ Sie sich nicht [weiter]!** please do not trouble yourself [any further]; **b)** *(sich kümmern)* **sich um jmdn./etw. ~:** seek to help sb./endeavour *or* strive to achieve sth.; **um das Wohl der Hotelgäste bemüht sein** make every effort to ensure the comfort and enjoyment of the hotel patrons; **c)** *(zu erlangen suchen)* **sich um etw. ~:** try *or* endeavour to obtain sth.; **sich um eine Stelle/Wohnung ~:** try to get a job/a flat *(Brit.) or (Amer.)* apartment; **sich um einen Regisseur/Trainer ~:** try *or* endeavour to obtain the services of a director/manager; **d)** *(geh.: sich begeben)* proceed *(formal)*. **2.** *tr. V. (geh.)* trouble; call in, call upon the services of ‹*lawyer, architect, etc.*›; *(zum Beweis heranziehen)* bring in a quotation/quotations from ‹*author, philosopher, etc.*›

Bemühen das; ~s *(geh.)* effort; endeavour; **trotz jahrelangen ~s** despite years of effort

Bemühung die; ~, ~en effort; endeavour; **alle ~en waren vergeblich** all efforts were in vain; **trotz aller ~en** in spite of *or* despite all our/his *etc.* efforts; **vielen Dank für Ihre ~en** thank you very much for your efforts *or* trouble

bemüßigt [bə'my:sɪçt] *(geh. iron.) in* **sich ~ fühlen, etw. zu tun** feel obliged to do sth.; feel it incumbent on oneself to do sth.

bemuttern *tr. V.* mother

benachbart *Adj.* neighbouring *attrib.*; **~e Fachgebiete** related fields of study

benachrichtigen [bə'na:xrɪçtɪgn̩] *tr. V.* inform, notify (**von** of, about)

Benachrichtigung die; ~, ~en notification; **ich bitte um sofortige ~:** I wish to be informed *or* notified immediately

benachteiligen *tr. V.* put at a disadvantage; *(diskriminieren)* discriminate against; **sich benachteiligt fühlen** feel at a disadvantage/feel discriminated against; **sozial benachteiligt** underprivileged

Benachteiligte der/die; *adj. Dekl.* disadvantaged person; **die sozial ~n** the underprivileged; the socially deprived

Benachteiligung die; ~, ~en *(Vorgang)* discrimination (*Gen.* against); *(Zustand)* disadvantage (*Gen.* to); **der Firma wurde eine ~ der Frauen vorgeworfen** the firm was accused of discriminating against women

benagen *tr. V.* gnaw [at]

benebeln *tr. V.* befuddle

benehmen *unr. refl. V.* behave; *(in bezug auf Umgangsformen)* behave [oneself]; **sich schlecht ~:** behave badly; misbehave

Benehmen das; ~s a) behaviour; **kein ~ haben** have no manners *pl.*; **b)** *(Amtsspr.) in* **sich mit jmdm. ins ~ setzen** make contact with sb.

beneiden *tr. V.* envy; be envious of; **jmdn. um etw. ~:** envy sb. sth.; **du bist [nicht] zu ~:** I [don't] envy you

beneidens·wert 1. *Adj.* enviable. **2.** *adv.* enviably

Benelux·länder ['be:neluks-] *Pl.* Benelux countries

benennen *unr. tr. V.* **a)** name; **etw./jmdn. nach jmdm. ~:** name sth./name *or* call sb. after *or* *(Amer.)* for sb.; **b)** *(namhaft machen)* call ‹*witness*›; **jmdn. als**

Kandidaten ~: nominate sb. as a candidate

Benennung die a) *o. Pl.* naming; **b)** *o. Pl.* **durch ~ zweier weiterer Zeugen** by calling two more witnesses; **c)** *(Name)* name

benetzen *tr. V. (geh.)* moisten; ⟨*dew*⟩ cover

Bengale [bɛŋ'ga:lə] **der; ~n, ~n** Bengali; Bengalese

Bengalen (das); ~s Bengal

bengalisch *Adj.* Bengalese; Bengali, Bengalese ⟨*people, language*⟩; **~e Beleuchtung** Bengal light *or* fire

Bengel ['bɛŋl̩] **der; ~s, ~ a)** *(abwertend: junger Bursche)* young rascal; **b)** *(fam.: kleiner Junge)* little lad *or* boy

Benimm [bə'nɪm] **der; ~s** *(ugs.)* manners *pl.*; **jmdm. ~ beibringen** teach sb. some manners

benommen [be'nɔmən] *Adj.* bemused; dazed; *(durch Fieber, Alkohol)* muzzy **(von** from)

Benommenheit die; ~: bemused *or* dazed state; *(durch Fieber, Alkohol)* muzziness

benoten *tr. V.* mark *(Brit.)*; grade *(Amer.)*; **einen Test mit „gut" ~:** mark a test 'good' *(Brit.)*; assign a grade of 'good' to a test *(Amer.)*

benötigen *tr. V.* need; require; **das benötigte Geld** the necessary money

Benotung die; ~, ~en a) *o. Pl.* marking *(Brit.)*; grading *(Amer.)*; **b)** *(Note)* mark *(Brit.)*; grade *(Amer.)*

benutzbar *Adj.* usable

benutzen *tr. V.* use; take, use ⟨*car, lift*⟩; take ⟨*train, taxi*⟩; use, consult ⟨*reference book*⟩

Benutzer der; ~s, ~: user

Benutzung die; ~: use; **jmdm. etw. zur ~ überlassen** give sb. the use of sth.

Benutzungs·gebühr die charge; *(Maut)* toll

Benzin [bɛn'tsi:n] **das; ~s** petrol *(Brit.)*; gasoline *(Amer.)*; gas *(Amer. coll.)*; *(Wasch~)* benzine

Benzin-: ~feuerzeug das petrol *(Brit.) or (Amer.)* gasoline lighter; **~gut·schein der** petrol (Brit.) or (Amer.) gasoline coupon; **~kanister der** petrol *(Brit.) or (Amer.)* gasoline can; **~motor der** petrol *(Brit.) or (Amer.)* gasoline engine; **~preis der** price of petrol *(Brit.) or (Amer.)* gasoline; **~pumpe die** petrol *(Brit.) or (Amer.)* gasoline pump; **~verbrauch der** fuel consumption

Benzol [bɛn'tso:l] **das; ~s, ~e** *(Chemie)* benzene

beobachten [bə'|o:baxtn̩] *tr. V.* **a)** observe; watch; *(als Zeuge)* see; **er hat beobachtet, wie sie das Radio stahl** he watched her steal the radio; **jmdn. ~ lassen** put sb. under surveillance; have sb. watched; **b)** *(bemerken)* notice; observe; **etw. an jmdm. ~:** notice sth. about sb.

Beobachter der; ~s, ~: observer

Beobachtung die; ~, ~en observation; **zur ~:** for observation; **unter ~ stehen** be kept under surveillance

Beobachtungs-: ~gabe die; *o. Pl.* powers *pl.* of observation; **~posten der** observation post

beordern *tr. V.* order; **jmdn. nach Hause/ins Ausland ~:** order *or* summon sb. home/order sb. [to go] abroad

bepacken *tr. V.* load; **etw./jmdn./sich mit etw. ~:** load sth. up with/sb. with/oneself with sth.

bepflanzen *tr. V.* plant

Bepflanzung die planting

bepinseln *tr. V.* **a)** *(ugs.: einpinseln)* paint ⟨*gums*⟩; brush ⟨*dough, cake-mixture*⟩; **b)** *(ugs. abwertend: anstreichen)* paint; **etw. mit Farbe ~:** paint sth.

bepudern *tr. V.* powder

bequatschen *tr. V. (salopp)* **a)** *(bereden)* have a jaw about *(coll.)*; **b)** *(überreden)* persuade; **jmdn. ~, daß er mitkommt** talk sb. into coming along

bequem [bə'kve:m] **1.** *Adj.* **a)** comfortable; **es sich** *(Dat.)* **~ machen** make oneself comfortable; **machen Sie es sich ~:** make yourself at home; **b)** *(mühelos)* easy; **ein ~es Leben führen** have an easy *or* comfortable life; **c)** *(abwertend: träge)* lazy; idle. **2.** *adv.* **a)** comfortably; **liegen/sitzen Sie ~ so?** are you comfortable like that?; **b)** *(mühelos)* easily; comfortably

bequemen *refl. V. (geh.)* deign; **sich zu einer Antwort ~:** deign to answer; **sich dazu ~, etw. zu tun** deign to do sth.

Bequemlichkeit die; ~, ~en a) comfort; **b)** *o. Pl. (Trägheit)* laziness; idleness; **aus [reiner] ~:** out of [sheer] laziness *or* idleness

berappen [bə'rapn̩] *tr., itr. V. s.* **blechen**

beraten 1. *unr. tr. V.* **a)** advise; **jmdn. gut/schlecht ~:** give sb. good/bad advice; **sich ~ lassen** take *or* get advice **(von** from); **b)** *(besprechen)* discuss ⟨*plan, matter*⟩. **2.** *unr. itr. V.* **über etw.** *(Akk.)* **~:** discuss sth.; **sie berieten lange** they were a long time in discussion. **3.** *unr. refl. V.* **sich mit jmdm. ~, ob ...:** discuss with sb. whether ...; **sich mit seinem Anwalt ~:** consult one's lawyer

beratend *Adj.* advisory, consultative ⟨*function, role, etc.*⟩

Berater der; ~s, ~, Beraterin die; ~, ~nen adviser

beratschlagen [bə'ra:tʃla:gn̩] **1.** *tr. V.* discuss. **2.** *itr. V.* **über etw.** *(Akk.)* **~:** discuss sth.

Beratung die; ~, ~en a) advice *no indef. art.*; *(durch Arzt, Rechtsanwalt)* consultation; **ohne juristische ~:** without [taking] legal advice; **b)** *(Besprechung)* discussion; **Gegenstand der ~ war ...:** the subject under discussion was ...; **sich zur ~ zurückziehen** withdraw for discussions *pl.*

Beratungs·stelle die advice centre *(Brit.)*; counseling center *(Amer.)*

berauben *tr. V. (auch fig.)* rob; **jmdn. einer Sache** *(Gen.)* **~** *(geh.)* rob sb. of sth.; **jmdn. seiner Freiheit/Hoffnungen ~** *(fig.)* deprive sb. of his/her freedom/hopes

Beraubung die; ~: robbing *no indef. art.*

berauschen *(geh.)* **1.** *tr. V. (auch fig.)* intoxicate; ⟨*alcohol*⟩ intoxicate, inebriate; ⟨*drug*⟩ make euphoric; ⟨*speed*⟩ exhilarate; **der Erfolg/die Macht berauschte ihn** he was intoxicated *or* drunk with success/drunk with power. **2.** *refl. V.* become intoxicated; **sich an etw.** *(Dat.)* **~:** become intoxicated with sth.

berauschend 1. *Adj.* intoxicating; heady, intoxicating ⟨*perfume, scent*⟩; **das ist nicht ~** *(ugs.)* it's nothing very special *or (coll.)* nothing to write home about. **2.** *adv.* **~ schön** enchantingly beautiful; **~ wirken** have an intoxicating effect

Berber ['bɛrbɐ] **der; ~s, ~ a)** Berber; **b)** *(Teppich)* Berber carpet/rug

Berber·teppich der Berber carpet/rug

berechenbar [bə'rɛçn̩ba:ɐ̯] *Adj.* calculable; predictable ⟨*behaviour*⟩

berechnen *tr. V.* **a)** calculate; predict ⟨*consequences, behaviour*⟩; **b)** *(anrechnen)* charge; **jmdm. 10 Mark für etw.** *od.* **jmdm. etw. mit 10 Mark ~:** charge sb. 10 marks for sth.; **jmdm. etw. nicht ~:** not charge sb. for sth.; **jmdm. zuviel ~:** overcharge sb.; charge sb. too much; **c)** *(kalkulieren)* calculate; *(vorsehen)* intend; **für sechs Personen berechnet sein** ⟨*recipe, buffet*⟩ be for six people

berechnend *Adj.* calculating

Berechnung die a) calculation; **nach meiner ~, meiner ~ nach** according to my calculations *pl.*; **b)** *o. Pl. (abwertend: Eigennutz)* [calculating] self-interest; **etw. aus ~ tun** do sth. from motives of self-interest; **c)** *o. Pl. (Überlegung)* deliberation; calculation; **mit kühler ~ vorgehen** act with cool deliberation

berechtigen [bəˈrɛçtɪgn̩] **1.** *tr. V.* entitle; **jmdn. ~, etw. zu tun** entitle sb. *or* give sb. the right to do sth.; **das berechtigt ihn zu dieser Kritik** it entitles him *or* gives him the right to criticize [in this way]. **2.** *itr. V.* **die Karte berechtigt zum Eintritt** the ticket entitles the bearer to admission; **das berechtigt zu der Annahme, daß ...:** it justifies the assumption that ...

berechtigt *Adj.* **a)** *(gerechtfertigt)* justified, legitimate; **b)** *(befugt)* authorized

Berechtigung die; ~, ~en a) *(Befugnis)* entitlement; *(Recht)* right; **mit welcher ~ kritisiert er mich?** what right has he to criticize me?; **b)** *(Rechtmäßigkeit)* legitimacy; **seine/ihre ~ haben** be justified *or* legitimate

bereden 1. *tr. V.* **a)** *(besprechen)* talk over; discuss; **b)** *(überreden)* **jmdn. ~, etw. zu tun** talk sb. into doing sth.; **sich ~ lassen, etw. zu tun** let oneself be talked into doing sth. **2.** *refl. V.* **sich [mit jmdm.] über etw.** *(Akk.)* **~:** talk sth. over *or* discuss sth. [with sb.]

beredsam [bəˈreːtzaːm] **1.** *Adj.* eloquent. **2.** *adv.* eloquently

Beredsamkeit die; ~: eloquence

beredt [bəˈreːt] *Adj. (auch fig.)* eloquent

Bereich der; ~[e]s, ~e a) area; **im ~ der Stadt** within the town; **b)** *(fig.)* sphere; area; *(Fachgebiet)* field; area; **in jmds. ~** *(Akk.)* **fallen** be [within] sb.'s province; **im ~ des Möglichen liegen** be within the bounds *pl.* of possibility; **aus dem ~ der Kunst/Politik** from the sphere of art/politics; **im privaten/staatlichen ~:** in the private/public sector

bereichern [bəˈraiçɐn] **1.** *refl. V.* make a profit; **sich an jmdm./etw. ~:** make a great deal of money at sb.'s expense/out of sth. **2.** *tr. V.* enrich

Bereicherung die; ~, ~en enrichment; **eine wertvolle ~ der koreanischen Literatur** a valuable addition to Korean literature

bereifen *tr. V.* put tyres on ⟨*car*⟩; put a tyre on ⟨*wheel*⟩; **neu bereift sein** ⟨*car*⟩ have new tyres

Bereifung die; ~, ~en [set *sing.* of] tyres *pl.*

bereinigen *tr. V.* clear up ⟨*misunderstanding*⟩; settle, resolve ⟨*dispute*⟩; **mit jmdm. etw. zu ~ haben** have sth. to sort out with sb.

Bereinigung die *s.* **bereinigen:** clearing up; settlement; resolution

bereisen *tr. V.* travel around *or* about; *(beruflich)* ⟨*representative etc.*⟩ cover ⟨*area*⟩; **fremde Länder ~:** travel in foreign countries; **ganz Afrika ~:** travel throughout Africa

bereit [bəˈrait] *Adj.* **a)** *(fertig, gerüstet)* **~ sein** be ready; **sich ~ halten** be ready; **etw. ~ haben** have sth. ready; **b)** *(gewillt)* **~ sein, etw. zu tun** be willing *or* ready *or* prepared to do sth.; **sich ~ zeigen/finden, etw. zu tun** show oneself/be willing *or* ready *or* prepared to do sth.; **sich ~ erklären, etw. zu tun** declare oneself willing *or* ready to do sth.; **zu einem Kompromiß ~ sein** be ready to compromise

bereiten *tr. V.* **a)** prepare (*Dat.* for); make ⟨*tea, coffee*⟩ (*Dat.* for); **b)** *(fig.)* **[jmdm.] Schwierigkeiten/Ärger/Kummer ~:** cause [sb.] difficulty/trouble/sorrow; **jmdm. Freude/einen begeisterten Empfang ~:** give sb. great pleasure/an enthusiastic reception; **einer Sache** *(Dat.)* **ein Ende ~:** put an end to sth.

bereit-: ~|halten *unr. tr. V.* have ready; *(für Notfälle)* keep ready; **~|legen** *tr. V.* lay out ready (*Dat.* for); **~|liegen** *unr. itr. V.* be ready; ⟨*surgical instruments, tools, papers*⟩ be laid out ready; **~|machen** *tr. V.* get ready

bereits *Adv. s.* **schon 1 a, d**

Bereitschaft die; ~ a) willingness; readiness; preparedness; **etw. in ~ haben** have sth. ready; **b)** *(ugs.) s.* **Bereitschaftsdienst**

Bereitschafts-: ~arzt der doctor on call; **~dienst der: ~dienst haben** ⟨*doctor, nurse*⟩ be on call; ⟨*policeman, fireman*⟩ be on standby duty; ⟨*chemist's*⟩ be on rota duty *(for dispensing outside normal hours)*

bereit-: ~|stehen *unr. itr. V.* be ready; ⟨*car, train, aircraft*⟩ be waiting; ⟨*troops*⟩ be standing by; **für uns steht ein Auto ~:** a car is/will be waiting for us; **~|stellen** *tr. V.* place ready; get ready ⟨*food, drinks*⟩; provide, make available ⟨*money, funds*⟩

bereit·willig 1. *Adj.* willing. **2.** *adv.* readily

Bereitwilligkeit die; ~: willingness

bereuen 1. *tr. V.* regret; **seine Sünden ~:** repent [of] one's sins. **2.** *itr. V.* be sorry; *(Rel.)* repent

Berg [bɛrk] **der; ~[e]s, ~e a)** hill; *(im Hochgebirge)* mountain; **in die ~e fahren** go up into the mountains; **über ~ und Tal** up hill and down dale; **~ Heil!** *greeting between mountaineers;* **mit etw. hinter dem** *od.* **hinterm ~ halten** *(fig.)* keep sth. to oneself; **über den ~ sein** *(ugs.)* be out of the wood *(Brit.) or (Amer.)* woods; ⟨*patient*⟩ be on the mend, have turned the corner; **[längst] über alle ~e sein** *(ugs.)* be miles away; **b)** *(Haufen)* enormous *or* huge pile; *(von Akten, Abfall auch)* mountain

berg·ab *Adv.* downhill; **einen steilen Weg ~ fahren** go down a steep path; **mit dem Patienten/der Firma geht es ~** *(fig. ugs)* the patient's getting worse/the firm's going downhill

berg·an *Adv. s.* **bergauf**

Berg·arbeiter der miner; mineworker

berg·auf *Adv.* uphill; **es geht ~ mit der Firma** *(fig. ugs.)* things are looking up for the firm; **mit dem Patienten geht es ~:** the patient's on the mend

Berg: ~bahn die mountain railway; *(Seilbahn)* mountain cableway; **~bau der;** *o. Pl.* mining; **~bauer der** mountain farmer

bergen *unr. tr. V.* **a)** *(retten)* rescue, save ⟨*person*⟩; salvage ⟨*ship, wrecked car*⟩; salvage, recover ⟨*cargo, belongings*⟩; **jmdn. tot/lebend ~:** recover sb.'s body/rescue sb. alive; **b)** *(geh.: enthalten)* hold; **Gefahren [in sich** *(Dat.)***] ~** *(fig.)* hold dangers

Berg-: ~fried [-friːt] **der; ~[e]s, ~e** keep; **~führer der** mountain guide; **~gipfel** der mountain peak *or* top; summit; **~hütte die** mountain hut

bergig *Adj.* hilly; *(mit hohen Bergen)* mountainous

berg-, Berg-: ~kessel der corrie; cirque; **~kette die** range *or* chain of mountains; mountain range *or* chain; **~kristall der** rock crystal; **~kuppe die** [rounded] peak *or* mountain-top; **~land das** hilly country *no indef. art.; (mit hohen Bergen)* mountainous country *no indef. art.;* **das spanische ~land** the hill country of Spain; **das Schottische ~land**

the Highlands of Scotland; **~mann der;** *Pl.* **Bergleute** miner; mineworker; **~männisch** [-mɛnɪʃ] *Adj.* miner's *attrib.;* **~not die: in ~not sein/geraten** ⟨*climber*⟩ be/get into difficulties while climbing [in the mountains]; **jmdn. aus ~not retten** rescue sb. who has got into difficulties while climbing [in the mountains]; **~predigt die;** *o. Pl.* Sermon on the Mount; **~rücken der** mountain ridge; **~rutsch der** landslide; landslip; **~sattel der** saddle; col; **~see der** mountain lake; **~spitze die** [mountain] peak; mountain top; **~station die** top station; **~steigen** *unr. itr. V.; mit haben od. sein; nur im Inf. und Part.* go mountaineering *or* mountain-climbing; **~steigen das; ~s** mountaineering *no art.;* mountain-climbing *no art.;* **~steiger der, ~steigerin die; ~, ~nen** mountaineer; mountain-climber; **~tour die** mountain tour; *(kürzere Wanderung)* hike in the mountains; *(Klettertour)* mountain climb; **~-und-Tal-Bahn die** roller-coaster

Bergung die; ~, ~en a) *(von Verunglückten)* rescue; saving; **b)** *(von Schiffen, Gut)* salvaging; salvage

Berg-: ~volk das mountain people; **~wacht die;** *o. Pl.* mountain rescue service; **~wand die** mountain face; **~wanderung die** hike in the mountains; **~werk das** mine; **im ~ arbeiten** work down the mine; **~wiese die** mountain pasture

Bericht [bəˈrɪçt] **der; ~[e]s, ~e** report; **über etw.** *(Akk.)* **geben** give a report on sth.

berichten *tr., itr. V.* report **(von, über** + *Akk.* on); **jmdm. etw. ~:** report sth. to sb.; **es wird soeben berichtet, daß ...:** reports are coming in that ...

Bericht-: ~erstatter [-ɛɐ̯ʃtatɐ] **der; ~s, ~:** reporter; **~erstattung die** reporting *no indef. art.;* **die ~erstattung durch Presse und Rundfunk über diese Ereignisse** press and radio coverage of these events

berichtigen *tr. V.* correct

Berichtigung die; ~, ~en correction

Berichts·heft das *(Schulw.) (apprentice's/trainee's)* record book

beriechen *unr. tr. V.* **a)** smell; sniff [at]; **b)** *(fig. ugs.)* **sich [gegenseitig] ~:** size each other *or* one another up

berieseln *tr. V.* **a)** *(bewässern)* irrigate; **b)** *(ugs. abwertend)* **mit Werbung/Musik berieselt werden** be subjected to a constant [unobtrusive] stream of advertisements/to constant background music; **sich die ganze Zeit [mit Musik] ~ lassen** constantly have music on in the background

Berieselung die; ~ a) *(Bewässerung)* irrigation; **b)** *(ugs. abwertend)* **die ständige ~ mit Musik** subjection to constant background music

Bering·straße [ˈbe:rɪŋ-] **die** *(Geogr.)* Bering Strait

beritten *Adj.* mounted

Berliner [bɛrˈli:nɐ] **1.** *Adj.; nicht präd.* Berlin; **~ Weiße [mit Schuß]** *light, very fizzy beer flavoured with a dash of raspberry juice or woodruff.* **2. der; ~s, ~ a)** Berliner; **b)** *(~ Pfannkuchen)* [jam *(Brit.) or (Amer.)* jelly] doughnut

Berlinerin die; ~, ~nen Berliner

berlinern *itr. V. (ugs.)* speak [in] Berlin dialect

berlinisch *Adj.* Berlin *attrib.*

Bermuda·inseln [bɛrˈmu:da-] *Pl.* Bermuda *sing., no art.;* Bermudas

Bermudas *Pl.* **a)** Bermudas; Bermuda *sing., no art.;* **b)** *s.* **Bermudashorts**

Bermuda·shorts *Pl.* Bermuda shorts

Bern [bɛrn] **(das); ~s** Bern[e]

Berner 1. *Adj.; nicht präd.* Bernese; **eine ~ Zeitung** a Bern[e] newspaper. **2. der; ~s, ~:** Bernese

Bernerin die; ~, ~nen Bernese

Bernhardiner [bɛrnharˈdi:nɐ] **der; ~s, ~:** St. Bernard [dog]

Bern·stein [ˈbɛrn-] **der a)** *o. Pl.* amber; **b)** *(Stück ~)* piece of amber

bernstein·farben *Adj.* amber[-coloured]

Berserker [bɛrˈzɛrkɐ] **der; ~s, ~: wie ein ~ arbeiten** work like mad; **wie ein ~ auf jmdn. einschlagen** go berserk and attack sb.

bersten [ˈbɛrstn̩] *unr. itr. V.; mit sein (geh.)* ⟨*ice*⟩ break *or* crack up; ⟨*glass*⟩ shatter [into pieces]; ⟨*wall*⟩ crack up; **zum Bersten voll sein** be full to bursting-point; **vor Neugier/Ungeduld/Zorn ~** *(fig.)* be bursting with curiosity/impatience/rage

berüchtigt [bəˈrʏçtɪçt] *Adj.* notorious **(wegen** for); *(verrufen)* disreputable

berücksichtigen [bəˈrʏkzɪçtɪgn̩] *tr. V.* **a)** take into account *or* consideration, take account of ⟨*fact*⟩; **b)** consider ⟨*applicant, application, suggestion*⟩

Berücksichtigung die; ~ a) bei ~ aller Umstände taking all the circumstances into account; **unter ~** *(Dat.)* **der Vor- und Nachteile** taking account of all the advantages and disadvantages; **b)** *(Beachtung)* consideration; **eine ~ Ihres Antrags ist nicht möglich** we cannot consider your application

Beruf der; ~[e]s, ~e occupation; *(akademischer, wissenschaftlicher, medizinischer)* profession; *(handwerklicher)* trade; *(Stellung)* job; *(Laufbahn)* career; **was sind Sie von ~?** what do you do for a living?; what is your occupation?; **er ist von ~ Bäcker/Lehrer** he's a baker by trade/a teacher by profession; **den ~ verfehlt haben** *(scherzh.)* have missed one's vocation

[1]**berufen 1.** *unr. tr. V.* **a)** *(einsetzen)* appoint; **jmdn. auf einen Lehrstuhl/in ein Amt ~:** appoint sb. to a chair/an office; **b) berufe es nicht!** *(ugs.)* don't speak too soon! **2.** *unr. refl. V.* **sich auf etw.** *(Akk.)* **~:** refer to sth.; **sich auf jmdn. ~:** quote *or* mention sb.'s name; *(jmdn. zitieren)* quote *or* cite sb.

[2]**berufen** *Adj.* **a)** competent; **aus ~em Munde** from somebody *or* one competent *or* qualified to speak; **b)** *(prädestiniert)* **sich dazu ~ fühlen, etw. zu tun** feel called to do sth.; feel one has a mission to do sth.; **zum Dichter/zu Höherem ~ sein** have a vocation as a poet/be destined for greater things

beruflich 1. *Adj.; nicht präd.* occupational, vocational ⟨*training etc.*⟩; *(bei akademischen Berufen)* professional ⟨*training etc.*⟩; **seine ~e Tätigkeit** his occupation. **2.** *adv.* **~ erfolgreich sein** be successful in one's career; **~ viel unterwegs sein** be away a lot on business; **sich ~ weiterbilden** undertake further job training; **~ verhindert sein** be detained by one's work

berufs-, Berufs-: ~armee die *s.* **~heer; ~aus·bildung die** occupational *or* vocational training; *(als Lehrer, Wissenschaftler, Arzt)* professional training; **~aussichten** *Pl.* job prospects *(in a particular profession etc.);* **~bedingt** *Adj.* occupational ⟨*disease*⟩; ⟨*expenses, difficulties*⟩ connected with one's job; **~berater der** vocational adviser; **~beratung die** vocational guidance; **~bezeichnung die** job title; **~bild das** outline of a/the profession/trade as a career; **~boxer der** professional boxer; **~erfahrung die;** *o. Pl.* [professional] experience; **~fachschule die** vocational college *(providing full-time vocational training);* **~feuerwehr die** [professional] fire service; **~geheimnis das** professional secret; **~gruppe die** occupational group; **~heer das** regular *or* professional army; **~kleidung die** [prescribed] work[ing] clothes *pl.;* **~krankheit die** occupational disease; **~leben das** working life; **im ~leben stehen** be working; **~politiker der** professional politician; **~richter der** full-time salaried judge; **~risiko das** occupational risk; **~schule die** vocational school; **~schüler der** student at a vocational school; **~soldat der** regular *or* professional soldier; **~stand der** profession; **~tätig** *Adj.* working *attrib.;* **~tätig sein** work; **~tätige der/die;** *adj. Dekl.* working person; **~tätige** *Pl.* working people; **~verbrecher der** professional criminal; **~verkehr der** rush-hour traffic; **~wahl die;** *o. Pl.* choice of career

Berufung die; ~, ~en a) *(für ein Amt)* offer of an appointment **(auf, in, an +** *Akk.* to); **b)** *(innerer Auftrag)* vocation; **die ~ zum Künstler in sich** *(Dat.)* **verspüren** feel one has a vocation as an artist; **c)** *(das Sichberufen)* **unter ~** *(Dat.)* **auf jmdn./etw.** referring *or* with reference to sb./sth.; **d)** *(Rechtsw.: Einspruch)* appeal; **~ einlegen** lodge an appeal; appeal; **in die ~ gehen** appeal

Berufungs·verfahren das *(Rechtsw.)* appeal proceedings *pl.*

beruhen *itr. V.* **auf etw.** *(Dat.)* **~:** be based on sth.; **etw. auf sich** *(Dat.)* **~ lassen** let sth. rest

beruhigen [bəˈruːɪgn̩] **1.** *tr. V.* calm [down]; quieten, pacify ⟨*child, baby*⟩; salve, soothe ⟨*conscience*⟩; *(trösten)* soothe; *(die Befürchtung nehmen)* reassure; **die Nerven/den Magen ~:** calm one's nerves/settle the stomach; **beruhigt schlafen/nach Hause gehen können** be able to sleep/go home with one's mind set at ease. **2.** *refl. V.* ⟨*person*⟩ calm down; ⟨*sea*⟩ become calm; ⟨*struggle, traffic*⟩ lessen; ⟨*rush of people*⟩ subside; ⟨*prices, stock exchange, stomach*⟩ settle down

Beruhigung die; ~ a) *s.* **beruhigen 1:** calming [down]; quietening; pacifying; salving; soothing; reassurance; **jmdm. etw. zur ~ geben** give sb. sth. to calm him/her [down]; **b)** *(das Ruhigwerden)* **eine ~ des Wetters ist vorauszusehen** the weather can be expected to become more settled; **zu Ihrer ~ kann ich sagen, ...:** you'll be reassured to know that ...; **eine ~ der politischen Lage ist nicht zu erwarten** we should not expect that the political situation will become more stable

Beruhigungs-: ~mittel das sedative; tranquillizer; **~pille die** sedative [pill]; tranquillizer; **~spritze die** sedative injection; **~zelle die** cooling-off cell

berühmt [bəˈryːmt] *Adj.* famous; **wegen** *od.* **für etw. ~ sein** be famous for sth.

berühmt-berüchtigt *Adj.* notorious

Berühmtheit die; ~, ~en a) *o. Pl.* fame; **~ erlangen/gewinnen** become famous/win fame; **zu trauriger ~ gelangen** become notorious; **b)** *(Mensch)* celebrity

berühren *tr. V.* **a)** touch; *(fig.)* touch on ⟨*topic, issue, question*⟩; **sich** *od.* *(geh.)* **einander ~:** touch; **„Bitte Waren nicht ~!"** 'please do not touch the merchandise'; **b)** *(beeindrucken)* affect; **wir fühlten uns davon unangenehm/peinlich berührt** it made an unpleasant impression on us/made us feel embarrassed; **das berührt mich [überhaupt] nicht** it's a matter of [complete] indifference to me

Berührung die; ~, ~en a) *(das Berühren)* touch; **bei der geringsten ~:** at the slightest touch; **b)** *(Kontakt)* contact; **mit jmdm./etw. in ~** *(Akk.)* **kommen** come into contact with sb./sth.

Berührungs-: ~angst die fear of contact; **~punkt der a)** *(Math.)* point of contact *or* tangency; **b)** *(fig.: Gemeinsamkeit)* point of contact

besagen *tr. V.* say; *(bedeuten)* mean; **das besagt noch gar nichts** that doesn't mean anything

besagt *Adj.; nicht präd. (Amtsspr.)* aforementioned

besamen [bəˈzaːmən] *tr. V.* fertilize; *(künstlich)* inseminate

besänftigen [bəˈzɛnftɪgn̩] *tr. V.* calm [down]; pacify; calm, soothe ⟨*temper*⟩

Besänftigung die; ~: calming [down]; pacifying; *(von jmds. Zorn)* calming; soothing

besät [bəˈzɛːt] *Adj.* sown **(mit** with); *(fig.)* covered **(mit, von** with); **mit Sternen ~:** studded with stars

Besatz der *(Mode: Borte)* trimming *no indef. art.*

Besatzung die a) *(Mannschaft)* crew; **b)** *(Milit.)* occupying troops *pl. or* forces *pl.*

Besatzungs-: ~macht die occupying power; **~zone die** occupied zone

besaufen *unr. refl. V. (salopp)* get boozed up *(Brit. sl.) or* canned *(Brit. sl.) or* bombed *(Amer. sl.)*

Besäufnis [bəˈzɔyfnɪs] **das; ~ses, ~se** *(salopp)* booze-up *(Brit. sl.):* blast *(Amer. sl.)*

beschädigen *tr. V.* damage

Beschädigung die a) *o. Pl. (das Beschädigen)* damaging; **b)** *(Schaden)* damage; **zahlreiche ~en** a lot of damage *sing.*

[1]beschaffen *tr. V.* obtain; get; get ⟨*job*⟩; **ein Quartier ~:** find accommodation; **jmdm. etw. ~:** obtain/get sb. sth. *or* sth. for sb.; **sich** *(Dat.)* **Geld/die Genehmigung ~:** get [hold of] money/get *or* obtain the permit/licence

[2]beschaffen *Adj.* **so ~ sein, daß ...:** be such that ...; ⟨*product*⟩ be made in such a way that ...; **ähnlich ~ wie Leder** similar in nature to leather

Beschaffenheit die; ~: properties *pl.; (Konsistenz)* consistency

Beschaffung die *s.* **beschaffen:** obtaining; getting; finding

beschäftigen [bəˈʃɛftɪgn̩] **1.** *refl. V.* occupy *or* busy oneself; **sich viel mit Musik/den Kindern ~:** devote a great deal of one's time to music/the children; **sich mit den Schriften Hegels ~:** be engaged in a study of the writings of Hegel; **sehr beschäftigt sein** be very busy. **2.** *tr. V.* **a)** occupy; **jmdn. mit etw. ~:** give sb. sth. to occupy him/her; **du mußt die Kinder ~:** you must keep the children occupied; **b)** *(angestellt haben)* employ ⟨*workers, staff*⟩; **bei einer Firma beschäftigt sein** work for a firm; **c) jmdn. ~** *(jmdn. geistig in Anspruch nehmen)* be on sb.'s mind; preoccupy sb.

Beschäftigte der/die; *adj. Dekl.* employee; **die Fabrik/das Kaufhaus hat 500 ~:** the factory has a workforce/the department store has a staff of 500

Beschäftigung die; ~, ~en a) *(Tätigkeit)* activity; occupation; **bei dieser ~ solltest du ihn nicht stören** you shouldn't disturb him while he's occupied with that; **b)** *(Anstellung, Stelle)* job; **ohne ~**

sein not be working; *(unfreiwillig)* be unemployed; **c)** *o. Pl. (geistige Auseinandersetzung)* consideration (**mit** of); *(Studium)* study (**mit** of); **d)** *o. Pl. (von Arbeitskräften)* employment

beschämen *tr. V.* shame; **jmdn. durch seine Großmütigkeit ~:** make sb. ashamed by one's generosity

beschämend 1. *Adj.* **a)** *(schändlich)* shameful; **b)** *(demütigend)* humiliating. **2.** *adv.* shamefully

beschämt *Adj.* ashamed; abashed

Beschämung die; ~: shame

beschatten *tr. V.* **a)** *(geh.)* shade; **b)** *(überwachen)* shadow

beschaulich [bəˈʃaulɪç] **1.** *Adj.* peaceful, tranquil ‹*life, manner, etc.*›; meditative, contemplative ‹*person, character*›. **2.** *adv.* peacefully; tranquilly

Beschaulichkeit die; ~: peacefulness; tranquillity

Bescheid [bəˈʃait] **der; ~[e]s, ~e a)** *(Auskunft)* information; *(Antwort)* answer; reply; **jmdm. ~ geben** *od.* **sagen[, ob ...]** let sb. know or tell sb. [whether ...]; **sage bitte im Restaurant ~, daß ...:** please let the restaurant know *or* let them know in the restaurant that ...; **jmdm. ~ sagen** *(ugs.: sich beschweren)* give sb. a piece of one's mind *(coll.)*; **[über etw.** *(Akk.)***] ~ wissen** know [about sth.]; **b)** *(Entscheidung)* decision; **ein abschlägiger/positiver ~:** a refusal/a positive reply

¹bescheiden 1. *unr. tr. V.* **a)** inform, notify ‹*person*›; **jmdn./etw. abschlägig ~:** turn sb./sth. down; refuse sb./sth.; **b) es war ihm nicht beschieden, ... zu ...** *(geh.)* it was not granted to him to ... **2.** *unr. refl. V. (geh.)* be content

²bescheiden 1. *Adj.* **a)** modest; modest, unassuming ‹*person, behaviour*›; **b)** *(einfach)* modest; simple ‹*meal*›; **in ~en Verhältnissen aufwachsen** grow up in humble circumstances; **c)** *(dürftig)* modest ‹*salary, results, pension, etc.*›; **d)** *(ugs. verhüll.: sehr schlecht)* lousy *(sl.)*; bloody awful *(Brit. coll.)*. **2.** *adv.* modestly

Bescheidenheit die; ~: modesty

bescheinen *unr. tr. V.* shine [up]on; **vom Mond/von der Sonne beschienen** moonlit/sunlit

bescheinigen [bəˈʃainɪgn̩] *tr. V.* **etw. ~:** confirm sth. in writing; **jmdm. den Empfang des Geldes ~:** acknowledge receipt of the money; **sich** *(Dat.)* **~ lassen, daß man arbeitsunfähig ist** get oneself certified as unfit for work

Bescheinigung die; ~, ~en written confirmation *no indef. art.*; *(Schein, Attest)* certificate

bescheißen *unr. tr. V. (derb)* **jmdn. ~:** rip sb. off *(sl.)*; screw sb. *(coarse)*

beschenken *tr. V.* **jmdn. ~:** give sb. a present/presents; **jmdn. reich ~:** shower sb. with presents; **jmdn. mit etw. ~:** give sb. sth. as a present

bescheren 1. *tr. V.* **a) jmdn. [mit etw.] ~** *(zu Weihnachten beschenken)* give sb. [sth. as] a Christmas present/Christmas presents; **b) ich bin gespannt, was uns dieser Tag ~ wird** I wonder what today will bring. **2.** *itr. V.* **nach dem Abendessen wird beschert** the presents are given out after supper

Bescherung die; ~, ~en a) *(zu Weihnachten)* giving out of the Christmas presents; **die Kinder konnten die ~ kaum erwarten** the children could hardly wait for the presents to be given out; **b)** *(ugs. iron.: unangenehme Überraschung)* **das ist ja eine schöne ~:** this is a pretty kettle of fish; **jetzt haben wir die ~:** that's done it, I told you so

bescheuert *Adj. (salopp)* **a)** barmy *(Brit. sl.)*; nuts *(sl.)*; **b)** *(unangenehm)* stupid ‹*task, party, etc.*›; **etw. ~ finden** find sth. a real pain [in the neck] *(coll.)*

beschickert [bəˈʃɪkɐt] *Adj. (ugs.)* tipsy; merry *(Brit. coll.)*

beschießen *unr. tr. V.* fire *or* shoot at; *(mit Artillerie)* bombard

Beschießung die; ~, ~en *s.* **beschießen:** firing (*Gen.* at); shooting (*Gen.* at); bombardement (*Gen.* of)

beschimpfen *tr. V.* abuse; swear at

Beschimpfung die; ~, ~en insult; **~en** abuse *sing.*; insults

Beschiß der; Beschisses *(derb)* rip-off *(sl.)*

beschissen [bəˈʃɪsn̩] *(derb)* **1.** *Adj.* lousy *(sl.)*; shitty *(vulg.)*. **2.** *adv.* ‹*behave*› in a bloody awful manner *(Brit. coll.)*, shittily *(vulg.)*; **ihm geht es ~:** he's having a lousy *or (Brit.)* bloody awful time of it *(sl.)*

Beschlag der a) fitting; **b) jmdn./etw. mit ~ belegen** *od.* **in ~ nehmen** monopolize sb./sth.

¹beschlagen 1. *unr. tr. V.* shoe ‹*horse*›; **Schuhsohlen mit Nägeln ~:** stud the soles of shoes with [hob]nails. **2.** *unr. itr. V.; mit sein* ‹*window*› mist up *(Brit.)*, fog up *(Amer.)*; *(durch Dampf)* steam up; **~e Scheiben** misted-up/fogged-up/steamed-up windows

²beschlagen *Adj.* knowledgeable; **in etw.** *(Dat.)* **[gut] ~ sein** be knowledgeable about sth.

Beschlagenheit die; ~: thorough *or* sound knowledge

Beschlag·nahme [-na:mə] **die; ~, ~n** seizure; confiscation

beschlagnahmen *tr. V.* seize; confiscate

beschleichen *unr. tr. V.* **a)** creep up on *or* to; steal up to; ‹*hunter*› stalk ‹*game, prey*›; **b)** *(geh.: überkommen)* creep over

beschleunigen [bəˈʃlɔynɪgn̩] **1.** *tr. V.* accelerate; quicken ‹*pace, step[s], pulse*›; speed up, expedite ‹*work, delivery*›; hasten ‹*departure, collapse*›; accelerate, speed up, expedite ‹*process*›. **2.** *refl. V.* ‹*speed, heart-rate*› increase; ‹*pulse*› quicken. **3.** *itr. V.* ‹*driver, car, etc.*› accelerate

Beschleunigung die; ~, ~en a) *s.* **beschleunigen 1:** speeding up; quickening; acceleration; expedition; hastening; **b)** *(ugs.: ~svermögen)* acceleration; **eine gute ~ haben** have good acceleration; **c)** *(Physik)* acceleration

beschließen 1. *unr. tr. V.* **a) ~, etw. zu tun** decide *or* resolve to do sth.; ‹*committee, council, etc.*› resolve to do sth.; **den Bau einer Brücke ~:** decide to build a bridge; **das ist beschlossene Sache** it's settled; **b)** *(beenden)* end; end, conclude ‹*lecture*›; end, close ‹*letter*›. **2.** *unr. itr. V.* **über etw.** *(Akk.)* **~:** decide concerning sth.

Beschluß der decision; *(gemeinsam gefaßt)* resolution; **einen ~ fassen** come to a decision/pass a resolution; **gemäß dem ~ des Gerichtes** in accordance with the decision of the court

beschluß·fähig *Adj.* quorate; **~ sein** have a quorum; be quorate

Beschluß·fähigkeit die; *o. Pl.* presence of a quorum

beschmeißen *unr. tr. V. (salopp)* **jmdn. mit etw. ~:** pelt sb. with sth.; **jmdn./etw. mit Dreck ~** *(fig.)* fling mud at sb./sth.

beschmieren *tr. V.* **a) etw./sich ~:** get sth./oneself in a mess; **sich** *(Dat.)* **die Kleidung/Hände mit etw. ~:** smear *or* get sth. [smeared] all over one's clothes/hands; **b)** *(abwertend) (bemalen)* daub paint all over; *(bekritzeln)* scrawl *or* scribble all over; **c)** *(bestreichen)* **sein Brot mit etw. ~:** spread sth. on one's bread; **etw.**

mit Fett/Salbe ~: grease sth./ smear ointment on sth.; **d)** *(abwertend: vollschreiben)* cover ⟨*paper*⟩

beschmutzen *tr. V.* **etw. ~:** make sth. dirty; **ganz beschmutzt sein** be covered in dirt; **jmds. Namen/Gedenken ~** *(fig.)* besmirch sb.'s name/memory

beschneiden *unr. tr. V.* **a)** cut, trim, clip ⟨*hedge*⟩; prune, cut back ⟨*bush*⟩; cut back ⟨*tree*⟩; clip ⟨*bird's wings*⟩; **b)** *(Med., Rel.)* circumcise; **c)** *(fig.)* cut ⟨*salary, income, wages*⟩; restrict ⟨*rights*⟩

Beschneidung die; ~, ~en a) *s.* **beschneiden a:** trimming; cutting; clipping; pruning; cutting back; **b)** *s.* **beschneiden c:** cutting; restriction; **c)** *(Med., Rel.)* circumcision

beschönigen [bəˈʃøːnɪgn̩] *tr. V.* gloss over

Beschönigung die; ~, ~en glossing over; **das wäre eine ~:** that would be to gloss over the true situation

beschränken [bəˈʃrɛŋkn̩] **1.** *tr. V.* restrict; limit; **etw. auf etw.** *(Akk.)* **~:** restrict *or* limit sth. to sth.; **jmdn. in seinen Rechten ~:** restrict sb.'s rights. **2.** *refl. V.* tighten one's belt *(fig.)*; **sich auf etw.** *(Akk.)* **~:** restrict *or* confine oneself to sth.

beschrankt *Adj.* ⟨*level crossing*⟩ with barriers; **~ sein** have barriers

beschränkt 1. *Adj.* **a)** *(abwertend: dumm)* dull-witted; **b)** *(engstirnig)* narrow-minded ⟨*person*⟩; narrow[-minded] ⟨*views, outlook*⟩. **2.** *adv.* narrow-mindedly; in a narrow-minded way

Beschränktheit die; ~: a) *(Dummheit)* lack of intelligence; **b)** *s.* **beschränkt b:** narrow-mindedness; narrowness; **c)** *(das Begrenztsein)* limitedness; restrictedness

Beschränkung die; ~, ~en restriction; **jmdm./einer Sache ~en auferlegen** impose restrictions on sb./sth.

beschreiben *unr. tr. V.* **a)** write on; *(vollschreiben)* write ⟨*page, sheet, etc.*⟩; **eng beschriebene Seiten** closely written pages; **b)** *(darstellen)* describe; **ich kann dir [gar] nicht ~, wie ...:** I [simply] can't tell you how ...; **c) einen Kreis/Bogen** *usw.* **~:** describe a circle/curve *etc.*

Beschreibung die; ~, ~en description

beschreien *unr. tr. V.: s.* **[1]berufen 1 b**

beschreiten *unr. tr. V. (geh.)* walk along ⟨*path etc.*⟩; **neue Wege ~** *(fig.)* tread new paths

beschriften [bəˈʃrɪftn̩] *tr. V.* label; inscribe ⟨*stone*⟩; letter ⟨*sign, label, etc.*⟩; *(mit Adresse)* address

Beschriftung die; ~, ~en a) *o. Pl.* labelling; *(eines Steines)* inscribing; *(eines Etiketts)* lettering; *(mit Adresse)* addressing; **b)** *(Aufschrift)* label; *(eines Steines)* inscription; *(eines Etiketts usw.)* lettering

beschuldigen [bəˈʃʊldɪgn̩] *tr. V.* accuse (*Gen.* of); **jmdn. ~, etw. getan zu haben/etw. zu sein** accuse sb. of doing/being sth.

Beschuldigte der/die; *adj. Dekl.* accused

Beschuldigung die; ~, ~en accusation

beschummeln *tr. V. (ugs.)* cheat; diddle *(Brit. coll.)*; burn *(Amer. sl.)*

Beschuß der fire; *(aus Kanonen)* shelling; *(mit Pfeilen)* shooting; **unter ~ nehmen** fire at/shell/ shoot at; *(fig.: kritisieren)* attack; **unter ~ geraten/liegen** *(auch fig.)* come/be under fire

beschützen *tr. V.* protect (**vor** + *Dat.* from)

Beschützer der; ~s, ~, Beschützerin die; ~, ~nen protector (**vor** from)

beschwatzen *tr. V. (ugs.)* **a) jmdn. ~:** talk sb. round; **jmdn. zu etw. ~:** talk sb. into sth.; **jmdn. ~, etw. zu tun** talk sb. into doing sth.; **b)** *(bereden)* chat about *or* over

Beschwerde [bəˈʃveːɐ̯də] **die; ~, ~n a)** complaint (**gegen, über** + *Akk.* about); **~ führen** *(Amtsspr.)* *od.* **einlegen** *(Rechtsw.)* lodge a complaint; *(gegen einen Entscheid)* lodge an appeal; **b)** *Pl.* *(Schmerz)* pain *sing.*; *(Leiden)* trouble *sing.*

beschweren [bəˈʃveːrən] **1.** *refl. V.* complain (**über** + *Akk.*, **wegen** *about*); **sich bei jmdm. ~:** complain to sb. **2.** *tr. V.* weight; *(durch Auflegen eines schweren Gegenstands)* weight down

beschwerlich *Adj.* arduous; *(ermüdend)* exhausting

beschwichtigen [bəˈʃvɪçtɪgn̩] *tr. V.* pacify; calm ⟨*excitement*⟩; placate, mollify ⟨*anger etc.*⟩

Beschwichtigung die; ~, ~en pacification; *(des Zorns, Hasses)* mollification

beschwindeln *tr. V. (ugs.)* **jmdn. ~:** tell sb. a fib/fibs; *(betrügen)* hoodwink sb.

beschwingt [bəˈʃvɪŋt] *Adj.* elated, lively ⟨*mood*⟩; lively, lilting ⟨*tune, melody*⟩; **~ sein/sich ~ fühlen** ⟨*person*⟩ be/feel elated

beschwipst [bəˈʃvɪpst] *Adj.* *(ugs.)* tipsy

beschwören *unr. tr. V.* **a)** swear to; **~, daß ...:** swear that ...; **eine Aussage ~:** swear a statement on *or* unter oath; **b)** charm ⟨*snake*⟩; **c)** *(erscheinen lassen)* invoke, conjure up ⟨*spirit*⟩; *(fig.)* evoke, conjure up ⟨*pictures, memories, etc.*⟩; **d)** *(bitten)* beg; implore; **in ~dem Ton** in a beseeching *or* imploring tone

Beschwörung die; ~, ~en a) *s.* **beschwören b, c:** charming; invoking; conjuring up; evoking; **b)** *(Zauberformel)* spell; incantation; **c)** *(Bitte)* entreaty

besehen *unr. tr. V.* have a look at; **sich** *(Dat.)* **etw. genau ~:** have a close look at sth.; inspect sth. closely; **er besah sich im Spiegel** he looked at himself in the mirror

beseitigen [bəˈzaitɪgn̩] *tr. V.* **a)** remove; eliminate ⟨*error, difficulty*⟩; dispose of ⟨*rubbish*⟩; eradicate ⟨*injustice, abuse*⟩; **b)** *(verhüll.: ermorden)* dispose of; eliminate

Beseitigung die; ~ a) *s.* **beseitigen a:** removal; elimination; disposal; eradication; **b)** *(verhüll.: Ermordung)* elimination

Besen [ˈbeːzn̩] **der; ~s, ~ a)** broom; *(Reisig~)* besom; *(Hand~)* brush; **ich fress' einen ~, wenn das stimmt** *(salopp)* I'll eat my hat if that's right *(coll.)*; **neue ~ kehren gut** *(Spr.)* a new broom sweeps clean *(prov.)*; **b)** *(salopp abwertend: Frau)* battleaxe *(coll.)*

besen-, Besen-: ~kammer die broom-cupboard; broom-closet *(Amer.)*; **~rein** *Adj.* swept clean *postpos.*; **~schrank der** *s.* **~kammer; ~stiel der** broom-handle; *(eines Reisigbesens)* broomstick

besessen [bəˈzɛsn̩] *Adj.* **a)** possessed; **vom Teufel ~ sein** be possessed by *or (dated)* of the Devil; **wie ~** *od.* **ein Besessener/eine Besessene** like one possessed; **b)** *(fig.)* obsessive ⟨*gambler*⟩; **von einer Idee** *usw.* **~ sein** be obsessed with an idea *etc.*

Besessenheit die; ~ a) possession; **b)** *(fig.)* obsessiveness; **mit wahrer ~:** in a truly obsessive manner

besetzen *tr. V.* **a)** *(mit Pelz, Spitzen)* edge; trim; **mit Perlen/Edelsteinen besetzt** set with pearls/

precious stones; **b)** *(belegen; auch Milit.: erobern)* occupy; *(füllen)* fill (**mit** with); **c)** *(vergeben)* fill ‹*post, position, role, etc.*›

besetzt *Adj.* occupied; ‹*table, seat*› taken *pred.*; *(gefüllt)* full; filled to capacity; **es** *od.* **die Leitung/die Nummer ist ~:** the line/number is engaged *or (Amer.)* busy

Besetzt·zeichen das *(Fernspr.)* engaged tone *(Brit.)*; busy signal *(Amer.)*

Besetzung die; ~, ~en a) *(einer Stellung)* filling; **b)** *(Mitwirkende) (Film, Theater usw.)* cast; **c)** *(Eroberung)* occupation

besichtigen [bəˈzɪçtɪgn̩] *tr. V.* see ‹*sights*›; see the sights of ‹*town*›; look round ‹*building*›; ‹*prospective buyer or tenant*› view ‹*house, flat*›

Besichtigung die; ~, ~en viewing; *(Rundgang, -fahrt)* tour; *(von Truppen)* inspection; **die ~ der Kirche ist zwischen 10 und 16 Uhr möglich** the church is open to visitors between 10 a.m. and 4 p.m.

besiedeln *tr. V.* settle (**mit** with); **ein dicht/dünn besiedeltes Land** a densely/thinly populated country

Besiedlung die settlement

besiegeln *tr. V.* set the seal on

Besieg[e]lung die; ~, ~en sealing; **die ~ von etw. sein** seal sth.; **zur ~ unserer Freundschaft** to seal our friendship

besiegen *tr. V.* **a)** defeat; **b)** *(fig.)* overcome ‹*doubts, curiosity, etc.*›

Besiegte der/die; *adj. Dekl.* loser

besingen *unr. tr. V.* **a)** *(geh.)* celebrate in verse; *(durch ein Lied)* celebrate in song; **b) eine Platte ~:** make a record [of songs]

besinnen *unr. refl. V.* **a)** think it *or* things over; **sich anders/eines Besseren ~:** change one's mind/think better of it; **b)** *(sich erinnern)* **sich [auf jmdn./etw.] ~:** remember *or* recall [sb./sth.]

besinnlich *Adj.* contemplative; thoughtful ‹*person*›; reflective ‹*story*›; **ein ~er Abend** an evening of reflection

Besinnung die; ~ a) consciousness; **die ~ verlieren** lose consciousness; **ohne** *od.* **nicht bei ~:** unconscious; **[wieder] zur ~ kommen** come to; regain consciousness; **b)** *(Nachdenken)* reflection; **zur ~ kommen** stop and think things over; **jmdn. zur ~ bringen** bring sb. to his/her senses

besinnungs·los 1. *Adj.* **a)** unconscious; **b)** *(fig.)* mindless, blind ‹*rage, hatred*›. **2.** *adv.* mindlessly

Besitz der a) property; **nur wenig ~ haben** have only a few possessions *pl.*; **b)** *(das Besitzen)* possession; **sich in jmds. ~** *(Dat.)* **befinden, in jmds. ~** *(Dat.)* **sein** be in sb.'s possession; **sich in privatem ~ befinden** be privately owned; be in private ownership *or* hands; **im ~ einer Sache** *(Gen.)* **sein** be in possession of sth.; possess sth.; **etw. in ~** *(Akk.)* **nehmen, von etw. ~ ergreifen** take possession of sth.; **c)** *(Landgut)* estate

Besitz·anspruch der claim to ownership; **einen ~ auf etw.** *(Akk.)* **anmelden** file a claim to ownership of sth.

besitz·anzeigend *Adj.* *(Sprachw.)* possessive

besitzen *unr. tr. V.* own; have ‹*quality, talent, right, etc.*›; *(nachdrücklicher)* possess; **keinen Pfennig ~** *(ugs.)* not have a penny to one's name; **er besaß die Frechheit, zu ...:** he had the cheek *or* nerve to ...

Besitzer der; ~s, ~: owner; *(eines Betriebs usw.)* proprietor *(formal)*; **den ~ wechseln** change hands *pl.*

Besitzerin die; ~, ~nen *s.* **Besitzer**

Besitzer·stolz der pride of ownership

besitz·los *Adj.* destitute

Besitz·stand der standard of living; **den ~stand wahren** maintain living standards

Besitztum das; ~s, Besitztümer [-ty:mɐ] possession

besoffen [bəˈzɔfn̩] **1.** *2. Part. v.* **besaufen. 2.** *Adj.* *(salopp)* boozed [up] *(sl.)*; plastered *(sl.)*; pissed *pred.* *(sl.)*; **völlig ~:** completely stoned *(sl.)*; blind drunk

Besoffene der/die; *adj. Dekl.* *(salopp)* drunk

besohlen *tr. V.* sole; **neu ~:** resole

Besoldung [bəˈzɔldʊŋ] **die; ~, ~en** pay

besonder... [bəˈzɔndɐ...] *Adj.*; *nicht präd.* special; *(größer als gewohnt)* particular ‹*pleasure, enthusiasm, effort, etc.*›; *(hervorragend)* exceptional ‹*quality, beauty, etc.*›; **im ~en** in particular; **ein ~es Ereignis** an unusual *or* a special event; **keine ~en Vorkommnisse wurden gemeldet** no incidents of any particular note were reported; **~e Merkmale** *(im Paß usw.)* distinguishing marks; **keine ~e Leistung** no great achievement

Besondere das; *adj. Dekl.* **etwas [ganz] ~s** something [really] special; **nichts ~s** nothing special; **das ist doch nichts ~s** there's nothing special *or* unusual about that

Besonderheit die; ~, ~en special *or* distinctive feature; *(Eigenart)* peculiarity

besonders 1. *Adv.* **a)** particularly; **~ du solltest das wissen** you of all people should know that; **~ bei schönem Wetter** especially in fine weather; **b)** *nur verneint (ugs.: besonders gut)* particularly well; **es geht ihm nicht ~:** he doesn't feel too well. **2.** *Adj.*; *nicht attr.*; *nur verneint (ugs.)* **nicht ~ sein** be nothing special; be nothing to write home about

besonnen [bəˈzɔnən] **1.** *Adj.* prudent; *(umsichtig)* circumspect; **ruhig und ~:** calm and collected. **2.** *adv.* prudently; *(umsichtig)* circumspectly

Besonnenheit die; ~: prudence *(Umsichtigkeit)*; circumspection

besorgen *tr. V.* **a)** get; *(kaufen)* buy; **jmdm. etw. ~:** get/buy sb. sth. *or* sth. for sb.; **sich** *(Dat.)* **etw. ~:** get/buy sth.; *(ugs. verhüll.: stehlen)* help oneself to sth.; **b)** *(erledigen)* take care of; deal with; **jmdm. den Haushalt/die Wäsche ~:** keep house/do the washing for sb.

Besorgnis die; ~, ~se concern; **jmds. ~ erregen** cause sb. concern

besorgnis·erregend *Adj.* serious; **~ sein** give cause for concern

besorgt 1. *Adj.* worried (**über +** *Akk.*, **um** about); concerned *usu. pred.* (**über +** *Akk.*, **um** about); **sie war rührend um das Wohl ihrer Gäste ~:** she showed a touching concern for the well-being of her guests. **2.** *adv.* with concern; *(ängstlich)* anxiously

Besorgung die; ~, ~en a) purchase; **[einige] ~en machen** do some shopping; **b)** *o. Pl.* *(das Beschaffen)* getting; *(das Kaufen)* buying

bespannen *tr. V.* cover ‹*wall, chair, car, etc.*›; string ‹*racket, instrument*›

Bespannung die; ~, ~en covering; *(eines Schlägers, eines Instruments)* stringing

bespielbar *Adj.* *(Sport)* playable ‹*ground, tennis-court*›

bespielen *tr. V.* record on ‹*tape, cassette*›; **ein Band mit etw. ~:** record sth. on a tape; **bespielt** used ‹*tape, cassette*›; *(vom Hersteller)* prerecorded ‹*cassette*›; **ist**

dieses Band bespielt$ is there anything on this tape?

bespitzeln *tr. V.* spy on

Bespitz[e]lung die; ~, ~**en** spying

besprechen 1. *unr. tr. V.* **a)** discuss; talk over; **b)** *(rezensieren)* review; **gut/schlecht besprochen werden** get a good/bad review; *(mehrfach)* get good/bad reviews; **c) eine Kassette** ~: make a [voice] recording on a cassette; *(statt eines Briefes)* record a message on a cassette; **d) etw.** ~ *(beschwören)* utter a magic incantation *or* spell over sth. **2.** *unr. refl. V.* confer (**über** + *Akk.* about); **sich mit jmdm.** ~: have a talk with sb.

Besprechung die; ~, ~**en a)** discussion; *(Konferenz)* meeting; **in einer** ~ **sein, [gerade] eine** ~ **haben** be in a meeting; **b)** *(Rezension)* review (*Gen.*, **von** of)

bespritzen *tr. V.* **a)** splash; *(mit einem Wasserstrahl)* spray; **b)** *(beschmutzen)* bespatter

besprühen *tr. V.* spray

bespucken *tr. V.* **jmdn. [mit etw.]** ~: spit [sth.] at sb.

besser ['bɛsɐ] **1.** *Adj.* **a)** better; ~ **werden** get better; ⟨*work etc.*⟩ improve; **um so** ~: so much the better; all the better; that wasn't the best of it *(iron.)*; **ich habe Besseres zu tun** I've got better things to do; **jmdn. eines Besseren belehren** *(geh.)* put sb. right; *s. auch* **besinnen a; b)** *(sozial höher gestellt)* superior; upper-class; ~**e** *od.* **die** ~**en Kreise** more elevated circles; **eine** ~**e Gegend/Adresse** a smart[er] *or* [more] respectable area/address; **c)** *(abwertend)* glorified; **wir arbeiten in einer** ~**en Baracke** we work in a glorified hut. **2.** *adv.* better; **[immer] alles** ~ **wissen** always know better; **es** ~ **haben** be better off; *(es leichter haben)* have an easier time of it; **es kommt noch** ~ *(iron.)* it gets even better *(iron.)*; ~ **gesagt** to be [more] precise; **er täte** ~ **daran, zu ...**: he would do better to ... **3.** *Adv. (lieber)* **das läßt du** ~ **sein** *od. (ugs.)* **bleiben** you'd better not do that

besser|gehen *unr. itr. V.; mit sein* **jmdm. geht es besser** sb. feels better

bessern 1. *refl. V.* improve; ⟨*person*⟩ mend one's ways. **2.** *tr. V.* improve; reform ⟨*criminal*⟩

Besserung die; ~: **a)** *(Genesung)* recovery; **[ich wünsche dir] gute** ~: [I hope you] get well soon; **sich auf dem Wege der** ~ **befinden** be on the road to recovery *or* on the mend; **b)** *(Verbesserung)* improvement (*Gen.* in); *(eines Kriminellen)* reform; ~ **geloben** promise to mend one's ways

Besser-: ~wisser der; ~**s,** ~ *(abwertend)* know-all; smart aleck; **~wisserei** [----'-] **die;** ~ *(abwertend)* superior attitude

best... ['bɛst...] **1.** *Adj.* **a)** *attr.* best; **bei** ~**er Gesundheit/Laune sein** be in the best of health/spirits *pl.*; **im** ~**en Falle** at best; **in den** ~**en Jahren, im** ~**en Alter** in one's prime; ~**e** *od.* **die** ~**en Grüße an ...** *(Akk.)* best wishes to ...; **mit den** ~**en Grüßen** *od.* **Wünschen** with best wishes; *(als Briefschluß)* ≈ yours sincerely; ~**en Dank** many thanks *pl.*; **der/die/das nächste** ~**e ...**: the first ... one comes across; **es steht nicht zum** ~**en mit etw.** things are not going too well for sth.; **eine Geschichte/einen Witz zum** ~**en geben** entertain [those present] with a story/a joke; **jmdn. zum** ~**en halten** *od.* **haben** pull sb.'s leg; **das Beste vom Besten** the very best; **sein Bestes tun** do one's best; **das Beste aus etw. machen** make the best of sth.; **das Beste hoffen** hope for the best; **ich will nur dein Bestes** I am doing this for your own good; **zu deinem Besten** for your benefit; in your best interests *pl.*; **b) es ist** *od.* **wäre das** ~**e** *od.* **am** ~**en, wenn ...**: it would be best if ...; **es wäre das** ~**e, ... zu ...**: it would be best to ... **2.** *adv.* **a) am** ~**en** the best; **b) am** ~**en fährst du mit dem Zug** you'd best go by train

Bestand der a) *o. Pl.* existence; *(Fort~)* continued existence; survival; **keinen** ~ **haben, nicht von** ~ **sein** not last; not last long; **b)** *(Vorrat)* stock (**an** + *Dat.* of)

bestanden [bə'ʃtandn̩] *Adj.* **von** *od.* **mit etw.** ~ **sein** have sth. growing on it; **mit Tannen** ~**e Hügel** fir-covered hills

beständig 1. *Adj.* **a)** *nicht präd. (dauernd)* constant; **b)** *(gleichbleibend)* constant; steadfast ⟨*person*⟩; settled ⟨*weather*⟩; *(Chemie)* stable ⟨*compound*⟩; *(zuverlässig)* reliable; **c)** *(widerstandsfähig)* resistant (**gegen, gegenüber** to). **2.** *adv.* **a)** *(dauernd)* constantly; **b)** *(gleichbleibend)* consistently

-beständig *adj.* **hitze~/wetter~/säure~**: heat-/weather-/acid-resistant

Beständigkeit die; ~ **a)** constancy; steadfastness; *(bei der Arbeit)* consistency; *(Zuverlässigkeit)* reliability; **b)** *(Widerstandsfähigkeit)* resistance (**gegen, gegenüber** to)

Bestands·aufnahme die stocktaking; **[eine]** ~ **machen** do a stock-taking; take inventory *(Amer.)*; *(fig.)* take stock

Bestand·teil der component; **sich in seine** ~**e auflösen** fall apart; fall to pieces; **etw. in seine [sämtlichen]** ~**e zerlegen** dismantle sth. [completely]

bestärken *tr. V.* confirm; **jmdn. in seinem Plan** *od.* **Vorsatz** *od.* **darin** ~, **etw. zu tun** strengthen sb.'s resolve *or* confirm sb. in his/her resolve to do sth.

bestätigen [bə'ʃtɛ:tɪgn̩] **1.** *tr. V.* confirm; endorse ⟨*document*⟩; acknowledge ⟨*receipt of letter, money, goods, etc.*⟩; **ein Urteil** ~ *(Rechtsw.)* uphold a judgement; **jmdn. [im Amt]** ~: confirm sb.'s appointment. **2.** *refl. V.* be confirmed; ⟨*rumour*⟩ prove to be true

Bestätigung die; ~, ~**en** confirmation; *(des Empfangs)* acknowledgement; *(schriftlich)* letter of confirmation; **die** ~ **in seinem Amt** the confirmation of his appointment

bestatten [bə'ʃtatn̩] *tr. V. (geh.)* inter *(formal)*; bury; **bestattet werden** be laid to rest

Bestattung die; ~, ~**en** *(geh.)* interment *(formal)*; burial; *(Feierlichkeit)* funeral

Bestattungs·unternehmen das [firm of] undertakers *pl. or* funeral directors *pl.*; funeral parlor *(Amer.)*

bestäuben [bə'ʃtɔybn̩] *tr. V.* **a)** dust; **b)** *(Biol.)* pollinate

Bestäubung die; ~, ~**en** *(Biol.)* pollination

bestaunen *tr. V.* marvel at; *(bewundernd anstarren)* gaze in wonder at

bestechen 1. *unr. tr. V.* bribe. **2.** *unr. itr. V.* be attractive (**durch** on account of)

bestechend *Adj.* attractive; captivating, winning ⟨*smile, charm*⟩; persuasive ⟨*argument, logic*⟩; tempting ⟨*offer*⟩

bestechlich *Adj.* corruptible; open to bribery *postpos.*

Bestechlichkeit die; ~: corruptibility

Bestechung die; ~, ~**en** bribery *no indef. art.*; **eine** ~: a case of bribery; **aktive** ~ *(Rechtsw.)* giving bribes; **passive** ~ *(Rechtsw.)* accepting bribes

Bestechungs-: ~geld das bribe; **~versuch der** attempted bribery

Besteck [bəˈʃtɛk] **das; ~[e]s, ~e a)** cutlery setting; *(ugs.: Gesamtheit der Bestecke)* cutlery; **b)** *(Med.)* [set *sing.* of] instruments *pl.*

Besteck-: ~kasten der cutlery-box; *(größer)* canteen; **~schublade die** cutlery-drawer

bestehen 1. *unr. itr. V.* **a)** exist; **die Schule besteht noch nicht sehr lange** the school has not been in existence *or* has not been going for very long; **es besteht [die] Aussicht/Gefahr, daß ...:** there is a prospect/danger that ...; **noch besteht die Hoffnung, daß ...:** there is still hope that ...; **b)** *(fortdauern)* survive; last; *(standhalten)* hold one's own; **in einer Gefahr** *usw.* **~:** prove oneself in a dangerous situation *etc.;* **c) aus etw. ~:** consist of sth.; *(aus einem Material)* be made of sth.; **d) ihre Aufgabe besteht in der Aufstellung der Liste** her task is to draw up the list; **der Unterschied besteht darin, daß ...:** the difference is that ...; **eine Möglichkeit besteht darin, zu beweisen ...:** one possibility would be to prove ...; **e) auf etw.** *(Dat.)* **~:** insist on sth.; **er bestand darauf, den Chef zu sprechen** he insisted on seeing the boss; **f)** *(die Prüfung ~)* pass [the examination]. **2.** *unr. tr. V.* pass ‹*test, examination*›; **nach bestandener Prüfung** after passing one's examination

Bestehen das; ~s existence; **die Firma feiert ihr 10jähriges ~:** the firm is celebrating its tenth anniversary; **seit ~ der Bundesrepublik** since the Federal Republic came into existence

bestehen|bleiben *unr. itr. V.; mit sein* remain; ‹*doubt*› persist; ‹*regulation*› remain in force

bestehend *Adj.* existing; current ‹*conditions*›

bestehlen *unr. tr. V.* rob

besteigen *unr. tr. V.* **a)** climb; mount ‹*horse, bicycle*›*;* ascend ‹*throne*›*;* **b)** *(betreten)* board ‹*ship, aircraft*›*;* get on ‹*bus, train*›

Besteigung die ascent

bestellen 1. *tr. V.* **a)** order **(bei** from); **sich** *(Dat.)* **etw. ~:** order sth. [for oneself]; **würden Sie mir bitte ein Taxi ~?** would you order me a taxi?; **b)** *(reservieren lassen)* reserve ‹*table, tickets*›*;* **c)** *(kommen lassen)* **jmdn. [für 10 Uhr] zu sich ~:** ask sb. to go/come to see one [at 10 o'clock]; **beim** *od.* **zum Arzt bestellt sein** have an appointment with the doctor; **d)** *(ausrichten)* **jmdm. etw. ~:** pass on sth. to sb.; tell sb. sth.; **bestell ihm schöne Grüße von mir** give him my regards; **er läßt dir ~, daß ...:** he left a message [for you] that ...; **nichts/nicht viel zu ~ haben** have no say/little *or* not much say; **e)** *(ernennen)* appoint **(zu, als** as); **f)** *(bearbeiten)* cultivate, till ‹*field*›*;* **g) es ist um jmdn./etw.** *od.* **mit jmdm./etw. schlecht bestellt** sb./sth. is in a bad way; **mit seiner Gesundheit ist es schlecht bestellt** he is in poor health. **2.** *itr. V.* order

Bestell-: ~nummer die order number; **~schein der** order form

Bestellung die a) order **(über +** *Akk.* for); *(das Bestellen)* ordering *no indef. art.;* **auf ~:** to order; **b)** *(Reservierung)* reservation; **c)** *(das Ernennen)* appointment; **d)** *(das Bearbeiten)* cultivation; tilling

Bestell·zettel der order-form

besten·falls *Adv.* at best

bestens *Adv.* excellently; extremely well; **sich ~ verstehen** get on splendidly; **jmdn. ~ grüßen** give sb. one's best wishes

besteuern *tr. V.* tax

Besteuerung die taxation

Best·form die; *o. Pl. (Sport)* best form; **in ~form** in top form

best·gehaßt *Adj.; nicht präd. (ugs. iron.)* most heartily disliked

bestialisch [bɛsˈti̯a:lɪʃ] **1.** *Adj.* **a)** bestial; **b)** *nicht präd. (ugs.: schrecklich)* ghastly *(coll.);* awful *(coll.).* **2.** *adv.* **a)** in a bestial manner; **b)** *(ugs.: schrecklich)* awfully *(coll.);* unbearably

Bestialität [bɛsti̯aliˈtɛ:t] **die; ~, ~en a)** *o. Pl.* bestiality; **ein Verbrechen von solcher ~:** a crime of such a bestial nature *or* of such brutality; **b)** *(Tat)* brutality; atrocity

besticken *tr. V.* embroider

Bestie [ˈbɛsti̯ə] **die; ~, ~n** *(auch fig. abwertend)* beast

bestimmbar *Adj.* ascertainable; *(identifizierbar)* identifiable; **nicht [genau] ~ sein** be impossible to ascertain/identify [precisely]

bestimmen 1. *tr. V.* **a)** *(festsetzen)* decide on; fix ‹*price, time, etc.*›*;* **jmdn. zum** *od.* **als Nachfolger ~:** decide on sb. as one's successor; *(nennen)* name sb. as one's successor; **b)** *(vorsehen)* destine; intend; set aside ‹*money*›*;* **das ist für dich bestimmt** that is meant for you; **er ist zu Höherem bestimmt** he is destined for higher things; **c)** *(ermitteln, definieren)* identify ‹*part of speech, find, plant, etc.*›*;* determine ‹*age, position*›*;* define ‹*meaning*›*;* **d)** *(prägen)* determine the character of; give ‹*landscape, townscape*› its character. **2.** *itr. V.* **a)** make the decisions; **hier bestimme ich** I'm in charge *or* the boss here; my word goes around here; **b)** *(verfügen)* **über jmdn. ~:** tell sb. what to do; **[frei] über etw.** *(Akk.)* **~:** do as one wishes with sth.

bestimmend 1. *Adj.* decisive; determining. **2.** *adv.* decisively

bestimmt 1. *Adj.* **a)** *nicht präd. (speziell)* particular; *(gewiß)* certain; *(genau)* definite; **ich habe nichts Bestimmtes vor** I am not doing anything in particular; **b)** *(festgelegt)* fixed; given ‹*quantity*›*;* **c)** *(Sprachw.)* definite ‹*article etc.*›*;* **d)** *(entschieden)* firm. **2.** *adv. (entschieden)* firmly. **3.** *Adv.* for certain; **du weißt es doch [ganz] ~ noch** I'm sure you must remember it; **ganz ~, ich komme** I'll definitely come; yes, certainly, I'll come; **ich habe das ~ liegengelassen** I must have left it behind

Bestimmtheit die; ~ a) *(Entschiedenheit)* firmness; *(im Auftreten)* decisiveness; **etw. mit aller ~ sagen/ablehnen** say sth. very firmly/reject sth. categorically; **b)** *(Gewißheit)* **mit ~:** for certain

Bestimmung die a) *o. Pl. (das Festsetzen)* fixing; **b)** *(Vorschrift)* regulation; **gesetzliche ~en** legal requirements; **c)** *o. Pl. (Zweck)* purpose; **eine Brücke** *usw.* **ihrer ~ übergeben** [officially] open a bridge *etc.;* **d)** *(das Ermitteln)* identification; *(eines Begriffs, der Bedeutung)* definition; *(des Alters, der Position)* determination; **e)** *(Sprachw.)* **adverbiale ~:** adverbial qualification

Best·leistung die *(Sport)* best performance; **persönliche ~:** personal best

best·möglich *Adj.* best possible; **das Bestmögliche tun** do the best one can

bestrafen *tr. V.* punish **(für, wegen** for); **es wird mit Gefängnis bestraft** it is punishable by imprisonment

Bestrafung die; ~, ~en punishment; *(Rechtsw.)* penalty

bestrahlen *tr. V.* **a)** illuminate; floodlight ‹*building*›*; (scheinen auf)* ‹*sun etc.*› shine on; **b)** *(Med.)* treat ‹*tumour, part of body*› using radiotherapy; *(mit Höhensonne)* use sun-lamp treatment on ‹*part of body*›

Bestrahlung die; ~, ~en *(Med.)*

radiation [treatment] *no indef. art.*; *(mit Röntgenstrahlen)* radiotherapy *no art.*; *(mit Höhensonne)* sun-lamp treatment

Bestreben das endeavour[s *pl.*]

bestrebt *Adj.* **~ sein, etw. zu tun** endeavour to do sth.

Bestrebung die; ~, ~en effort; *(Versuch)* attempt

bestreichen *unr. tr. V.* **A mit B ~**: spread B on A; **sein Brot mit Butter ~**: spread butter on one's bread; butter one's bread

bestreiken *tr. V.* take strike action against; **diese Firma wird bestreikt** there is a strike [on] at this firm

bestreitbar *Adj.* disputable; **es ist nicht ~[, daß ...]**: it is indisputable *or* cannot be denied [that ...]

bestreiten *unr. tr. V.* **a)** dispute; contest; *(leugnen)* deny; **er bestreitet, daß ...**: he denies that ...; **es läßt sich nicht ~, daß ...**: it cannot be denied *or* there is no disputing that ...; **jmdm. das Recht auf etw.** *(Akk.)* **~**: dispute *or* challenge sb.'s right to sth.; **b)** *(finanzieren)* finance ‹*studies*›; pay for ‹*studies, sb.'s keep, etc.*›; meet ‹*costs, expenses*›; **c)** *(gestalten)* carry ‹*programme, conversation, etc.*›

bestreuen *tr. V.* **etw. mit Zucker ~**: sprinkle sth. with sugar; **einen Weg mit Sand/Salz ~**: scatter sand on a path/salt a path

Bestseller ['bɛstzɛlɐ] **der; ~s, ~**: best seller

bestürmen *tr. V.* **a)** storm; **b)** *(bedrängen)* besiege (**mit** with)

bestürzen *tr. V.* dismay; *(erschüttern)* shake

bestürzend *Adj.* disturbing

bestürzt 1. *Adj.* dismayed (**über** + *Akk.* about). **2.** *adv.* with dismay *or* consternation; **jmdn. [sehr] ~ ansehen** look at sb. in *or* with [great] consternation

Bestürzung die; ~: dismay; consternation; **mit ~ feststellen, daß ...**: find to one's consternation that ...

Best·zeit die *(Sport)* best time; **persönliche ~zeit** personal best [time]

Besuch [bə'zu:x] **der; ~[e]s, ~e a)** visit; **ein ~ bei jmdm.** a visit to sb.; *(kurz)* a call on sb.; **~ eines Museums** *usw.* visit to a museum *etc.*; **bei seinem letzten ~**: on his last visit; **~ von jmdm. bekommen** receive a visit from sb.; **ich bekomme gleich ~**: I've got visitors/a visitor coming any minute; **auf** *od.* **zu ~ kommen** come for a visit; *(für länger)* come to stay; **b)** *(das Besuchen)* visiting; *(Teilnahme)* attendance (*Gen.* at); **c)** *(Gast)* visitor; *(Gäste)* visitors *pl.*

besuchen *tr. V.* **a)** visit ‹*person*›; *(weniger formell)* go to see, call on ‹*person*›; **b)** visit ‹*place*›; go to ‹*exhibition, theatre, museum, etc.*›; *(zur Besichtigung)* go to see ‹*church, exhibition, etc.*›; **die Schule/Universität ~**: go to *or* *(formal)* attend school/university

Besucher der; ~s, ~, Besucherin die; ~, ~nen visitor (*Gen* to); **die Besucher der Vorstellung** those attending the performance

Besuchs-: ~erlaubnis die visiting permit; **~tag der** visiting day; **~zeit die** visiting time *or* hours *pl.*

besucht *Adj.* **gut/schlecht ~**: well/poorly attended ‹*lecture, performance, etc.*›; much/little frequented ‹*restaurant etc.*›

besudeln *tr. V. (geh. abwertend)* besmirch; **jmds. Andenken/Namen ~** *(fig.)* cast a slur on sb.'s memory/name

Beta ['be:ta] **das; ~[s], ~s** beta

betagt [bə'ta:kt] *Adj. (geh.)* elderly; *(scherzh.)* ancient ‹*car etc.*›

betasten *tr. V.* feel [with one's fingers]

betätigen 1. *refl. V.* busy *or* occupy oneself; **sich politisch/literarisch/körperlich ~**: engage in political/literary/physical activity; **sich als etw. ~**: act as sth. **2.** *tr. V.* operate ‹*lever, switch, flush, etc.*›; apply ‹*brake*›

Betätigung die; ~, ~en a) activity; **b)** *o. Pl. (das Bedienen)* operation; *(einer Bremse)* application

betäuben [bə'tɔybn̩] *tr. V.* **a)** *(Med.)* anaesthetize; make numb, deaden ‹*nerve*›; **jmdn. örtlich ~**: give sb. a local anaesthetic; **b)** *(unterdrücken)* ease, deaden ‹*pain*›; quell, still ‹*unease, fear*›; **seinen Kummer mit Alkohol ~** *(fig.)* drown one's sorrows [in drink]; **c)** *(benommen machen)* daze; *(mit einem Schlag)* stun; **ein ~der Duft** a heady *or* intoxicating scent

Betäubung die; ~, ~en a) *(Med.)* anaesthetization; *(Narkose)* anaesthesia; **b)** *(Benommenheit)* daze

Betäubungs·mittel das narcotic; *(Med.)* anaesthetic

Bete ['be:tə] **die; ~, ~n** *in* **rote ~**: beetroot *(Brit.)*; [red] beet *(Amer.)*

beteiligen 1. *refl. V.* **sich an etw.** *(Dat.)* **~**: participate *or* take part in sth.; **er hat sich kaum an der Diskussion beteiligt** he took hardly any part in the discussion; **sich an einem Geschäft ~**: take a share in *or* come in on a deal. **2.** *tr. V.* **jmdn. [mit 10%] an etw.** *(Dat.)* **~**: give sb. a [10%] share of sth.

beteiligt *Adj.* **a)** involved (**an** + *Dat.* in); **b)** *(finanziell)* **an einem Unternehmen/am Gewinn ~ sein** have a share in a business/in the profit; **er ist mit 20000 DM ~**: he has a 20,000 mark share

Beteiligte der/die; *adj. Dekl.* **a)** person involved (**an** + *Dat.* in); **b)** *s.* **Teilnehmer**

Beteiligung die; ~, ~en a) participation (**an** + *Dat.* in); *(an einem Verbrechen)* involvement (**an** + *Dat.* in); **unter ~ von** with the participation of; **b)** *(Anteil)* share (**an** + *Dat.* in)

beten ['be:tn̩] **1.** *itr. V.* pray (**für, um** for). **2.** *tr. V.* say ‹*prayer*›

beteuern [bə'tɔyɐn] *tr. V.* affirm; assert, protest ‹*one's innocence*›

Beteuerung die; ~, ~en *s.* **beteuern**: affirmation; assertion; protestation

betiteln [bə'ti:tl̩n] *tr. V.* **a)** give ‹*book etc.*› a title; **b)** *s.* **titulieren**

Beton [be'tɔŋ, *bes. österr.:* be'to:n] **der; ~s, ~s** [be'tɔŋs] *od.* **~e** [be'to:nə] concrete

Beton-: ~bau der; *Pl.* **~bauten** concrete building; **~bunker der** *(abwertend: ~bau)* concrete box

betonen [bə'to:nən] *tr. V.* **a)** stress ‹*word, syllable*›; accent ‹*syllable, beat*›; **ein Wort falsch ~**: put the wrong stress on a word; **b)** *(hervorheben)* emphasize; **die Taille ~**: accentuate the waist

betonieren [beto'ni:rən] *tr. V.* concrete; surface ‹*road etc.*› with concrete

Beton-: ~klotz der *(abwertend: massiver ~bau)* concrete monolith; **~kopf der** *(abwertend)* hardliner

betont [bə'to:nt] **1.** *Adj.* **a)** stressed; accented; **b)** *(bewußt)* pointed, studied; deliberate, studied ‹*simplicity, elegance*›. **2.** *adv.* pointedly; deliberately; **sich ~ sportlich kleiden** wear clothes with a strong *or* pronounced sporting character; **sich ~ zurückhaltend verhalten** behave with studied reserve

Betonung die; ~, ~en a) *o. Pl.* stressing; accenting; **b)** *(Akzent)* stress; accent *(esp. Mus.)*; *(Intonation)* intonation; **c)** *o. Pl.: s.* **betonen b**; emphasis (*Gen.* on); accentuation

betören [bə'tø:rən] *tr. V. (geh.)* captivate; bewitch

betr. *Abk.* **betreffs, betrifft** re

Betr. *Abk.* **Betreff** re

Betracht [bə'traxt] *in* **jmdn./etw. in ~ ziehen** consider sb./sth.; **jmdn./etw. außer ~ lassen** discount *or* disregard sb./sth.; **sie kommt/kommt nicht in ~:** she can/cannot be considered

betrachten *tr. V.* **a)** look at; **sich** *(Dat.)* **etw. [genau] ~:** take a [close] look at sth.; watch *or* observe sth. [closely]; **sich im Spiegel ~:** look at oneself in the mirror; *(längere Zeit)* contemplate oneself in the mirror; **genau/bei Licht betrachtet** *(fig.)* upon closer consideration/seen in the light of day; **objektiv betrachtet** viewed objectively; from an objective point of view; **so betrachtet** seen in this light *or* from this point of view; **b) jmdn./etw. als etw. ~:** regard sb./sth. as sth.

Betrachter der; ~s, ~: observer

beträchtlich [bə'trɛçtlɪç] **1.** *Adj.* considerable; **um ein ~es** to a considerable degree. **2.** *adv.* considerably

Betrachtung die; ~, ~en **a)** *o. Pl.* contemplation; *(Untersuchung)* examination; **bei genauer[er] ~:** upon close[r] examination; *(fig.)* upon close[r] consideration; **b)** *(Überlegung)* reflection; **~en über etw.** *(Akk.)* **anstellen** reflect on sth.

Betrachtungs·weise die way of looking at things

Betrag [be'tra:k] **der;** ~[e]s, **Beträge** [bə'trɛ:gə] sum; amount; **~ dankend erhalten** *(auf Quittungen)* received *or* paid with thanks

betragen 1. *unr. itr. V.* be; *(bei Geldsummen)* come to; amount to. **2.** *unr. refl. V. s.* **benehmen**

Betragen das; ~s behaviour; *(in der Schule)* conduct

betrauen *tr. V.* **jmdn. mit etw. ~:** entrust sb. with sth.; **jmdn. damit ~, etw. zu tun** entrust sb. with the task of doing sth.

betrauern *tr. V.* mourn ‹*death, loss*›; mourn for ‹*person*›

Betreff [bə'trɛf] **der;** ~[e]s, ~e *(Amtsspr., Kaufmannsspr.)* subject; matter; *(~zeile)* heading; reference line; **~: Ihr Schreiben vom 26. d. M.** *(im Brief)* re: your letter of the 26th inst.

betreffen *unr. tr. V.* concern; ‹*new rule, change, etc.*› affect; **was mich betrifft, ...:** as far as I'm concerned ...; **was das betrifft, ...:** as regards that; as far as that goes

betreffend *Adj.* concerning; **der ~e Sachbearbeiter** the person concerned with *or* dealing with this matter; **in dem ~en Fall** in the case concerned *or* in question

Betreffende der/die; *adj. Dekl.* person concerned; **die ~n** the people concerned

betreffs *Präp. mit Gen. (Amtsspr., Kaufmannsspr.)* concerning

betreiben *unr. tr. V.* **a)** tackle ‹*task*›; proceed with, *(energisch)* press ahead with ‹*task, case, etc.*›; pursue ‹*policy, studies*›; carry on ‹*trade*›; **auf jmds./sein Betreiben** *(Akk.)* **[hin]** at the instigation of sb./at his instigation; **b)** *(führen)* run ‹*business, shop*›; **Radsport ~:** go in for cycling as a sport; **c)** *(in Betrieb halten)* operate **(mit** by); **die Kühlbox kannst du auch mit Gas ~:** the fridge runs on *or* you can run the fridge on gas

¹betreten *unr. tr. V. (eintreten in)* enter; *(treten auf)* walk *or* step on to; *(begehen)* walk on ‹*carpet, grass, etc.*›; **er hat das Haus nie wieder ~:** he never set foot in the house again; **„Betreten verboten"** 'Keep off'; *(kein Eintritt)* 'Keep out'; **„Betreten der Baustelle verboten"** 'Building site. No entry *or* Keep out'; **den Rasen nicht ~:** keep off the grass; **ein Grundstück unerlaubt ~:** trespass on sb.'s property

²betreten 1. *Adj.* embarrassed; **ein ~es Gesicht machen** look embarrassed. **2.** *adv.* with embarrassment

Betretenheit die; ~: embarrassment

betreuen [bə'trɔyən] *tr. V.* look after; care for ‹*invalid*›; supervise ‹*youth group*›; see to the needs of ‹*tourists, sportsmen*›

Betreuer der; ~s, ~, **Betreuerin die;** ~, ~nen *s.* **betreuen:** *person who looks after/cares for/etc. others; (einer Jugendgruppe)* supervisor

Betreuung die; ~: care *no indef. art.;* **zwei Reiseleiter waren zu unserer ~ vorhanden** there were two couriers *or* travel guides to see to our needs

Betrieb der; ~[e]s, ~e **a)** business; *(Firma)* firm; **ein landwirtschaftlicher ~:** an agricultural holding; **im ~** *(am Arbeitsplatz)* at work; **b)** *o. Pl. (das In-Funktion-Sein)* operation; **in ~ sein** be running; be in operation; **außer ~ sein** not operate; *(wegen Störung)* be out of order; **in/außer ~ setzen** start up/stop ‹*machine etc.*›; **in ~ nehmen** put into operation; put ‹*bus, train*› into service; **den ~ einstellen** close down *or* cease operations; *(in einer Fabrik)* stop work; **den [ganzen] ~ aufhalten** *(ugs.)* hold everybody up; **c)** *o. Pl. (ugs.: Treiben)* bustle; commotion; *(Verkehr)* traffic; **es herrscht großer ~, es ist viel ~:** it's very busy

betrieblich *Adj.; nicht präd.* firm's; company

betriebsam 1. *Adj.* busy; *(ständig ~)* constantly on the go *postpos.* **2.** *adv.* busily

Betriebsamkeit die; ~: [bustling] activity; **eine hektische ~ an den Tag legen** become frantically busy

betriebs-, Betriebs-: ~angehörige der/die employee; **~an·leitung die** operating instructions *pl.; (Heft)* instruction manual; **~aus·flug der** staff outing; **~blind** *Adj.* inured to the shortcomings of working methods *postpos.;* professionally blinkered; **~blind werden** get into a rut *or* become blinkered in one's work; **~ferien** *Pl.* firm's annual close-down *sing.;* **„Wegen ~ferien geschlossen"** 'closed for annual holidays'; **~fest das** firm's party; **~geheimnis das** company secret; trade secret *(also fig.);* **~klima das** working atmosphere; **~leiter der** manager; *(einer Fabrik)* works manager; **~leitung die** management [of the firm]; **~rat der;** *Pl.:* **~räte a)** works committee; **b)** *(Person)* member of a/the works committee; **~ruhe die: ~ruhe haben** ‹*business, factory*› be closed; **~stillegung die** closure [of a/the firm]; *(eines Werks)* works closure; **~unfall der** *(veralt.)* industrial accident; **~verfassungs·gesetz das** industrial relations law *(for the private sector);* **~versammlung die** meeting of the work-force; **~wirt der** graduate in business management; **~wirtschaft die;** *o. Pl.* business management; **~wirtschaftlich** *Adj.* business management *attrib.;* **~wirtschafts·lehre die;** *o. Pl.* [theory of] business management; *(Fach)* management studies *sing., no art.*

betrinken *unr. refl. V.* get drunk; **sich fürchterlich/sinnlos ~:** get terribly/blind drunk

betroffen [bə'trɔfn̩] **1.** *Adj.* upset; *(bestürzt)* dismayed. **2.** *adv.* in dismay *or* consternation; **~ schweigen** be too upset/dismayed to say anything

Betroffene der/die; *adj. Dekl.* person affected; **die von ... ~n** those affected by ...

Betroffenheit die; ~: dismay; consternation
betrüben *tr. V.* sadden
betrüblich *Adj.* gloomy; *(deprimierend)* depressing
betrübt [bə'try:pt] **1.** *Adj.* sad **(über +** *Akk.* about); *(deprimiert)* dismayed, depressed **(über +** *Akk.* about); gloomy ‹*face etc.*›. **2.** sadly; *(schwermütig)* gloomily
Betrug der; ~[e]s deception; *(Mogelei)* cheating *no indef. art.; (Rechtsw.)* fraud; **das ist [glatter] ~:** that's [plain] fraud/cheating
betrügen 1. *unr. tr. V.* deceive; be unfaithful to ‹*husband, wife*›; *(Rechtsw.)* defraud; *(beim Spielen)* cheat; **sich selbst ~:** deceive oneself; **jmdn. um 100 DM ~:** cheat *or (coll.)* do sb. out of 100 marks; *(arglistig)* swindle sb. out of 100 marks. **2.** *unr. itr. V.* cheat; *(bei Geschäften)* swindle people
Betrüger der; ~s, ~: swindler; *(Hochstapler)* con man *(coll.); (beim Spielen)* cheat
Betrügerei die; ~, ~en deception; *(beim Spielen usw.)* cheating; *(bei Geschäften)* swindling
Betrügerin die; ~, ~nen swindler; *(beim Spielen)* cheat
betrügerisch *Adj.* deceitful; *(Rechtsw.)* fraudulent
betrunken [bə'trʊŋkn̩] *Adj.* drunken *attrib.;* drunk *pred.*
Betrunkene der/die; *adj. Dekl.* drunk; **eine ~:** a drunken woman
Bett [bɛt] **das; ~[e]s, ~en a)** bed; **die ~en machen** *od. (ugs. scherzh.)* **bauen** make the beds; **jmdm. das Frühstück ans ~ bringen** bring sb. breakfast in bed; **jmdn. aus dem ~ holen** *(ugs.)* get sb. out of bed; **er kommt nur schwer aus dem ~:** he doesn't like getting up; **im ~:** in bed; **ins** *od.* **zu ~ gehen, sich ins** *od.* **zu ~ legen** go to bed; **ins ~ fallen** *(ugs.)* fall into bed; **die Kinder ins ~ bringen** put the children to bed; **das ~ hüten [müssen]** *(fig.)* [have to] stay in bed; **mit jmdm. ins ~ gehen** *od.* **steigen** *(fig. ugs.)* go to bed with sb.; **b)** *(Feder~)* duvet; **c)** *(Fluß~)* bed
Bet·tag der *s.* **Buß- und Bettag**
Bett-: ~bezug der duvet cover; **~couch die** bed-settee; studio couch; **~decke die** blanket; *(gesteppt)* [continental] quilt; *(Federbett)* duvet
Bettel ['bɛtl̩] **der; ~s** *(ugs.)* junk *(coll.)*
bettel·arm *Adj.* destitute; penniless
Bettelei die; ~, ~en begging *no art.*
Bettel·mönch der mendicant friar
betteln ['bɛtl̩n] *itr. V.* beg **(um** for); **„Betteln verboten!"** 'No begging'; **bei jmdm. um etw. ~:** beg sb. for sth.
Bettel-: ~orden der mendicant order; **~stab der** *in* **jmdn. an den ~stab bringen** reduce sb. to penry
betten *(geh.)* **1.** *tr. V.* lay; **jmdn. flach ~:** lay sb. [down] flat. **2.** *refl. V. (fig.)* **wie man sich bettet, so liegt man** as you make your bed, so you must lie on it
bett-, Bett-: ~hupferl [~hʊpfɐl] **das; ~s, ~:** bedtime treat; **~kante die** edge of the bed; **~kasten der** bedding box *(under a bed);* **~lägerig** [~lɛ:gərɪç] *Adj.* bedridden; **~laken das** sheet; **~lektüre die** bedtime reading *no indef. art.*
Bettler ['bɛtlɐ] **der; ~s, ~:** beggar
Bettlerin die; ~, ~nen beggar [woman]
Bett-: ~nässer der; ~s, ~: bedwetter; **~pfanne die** bedpan; **~ruhe die** bed-rest; **~tuch das** *s.* **Bettuch**
Bettuch das sheet
Bett-: ~vorleger der bedside rug; **~wäsche die** bed-linen; **~zeug das;** *o. Pl. (ugs.)* bedclothes *pl.*
betucht [bə'tu:xt] *Adj. (ugs.)* **[gut] ~:** well-heeled *(coll.);* well-off
betulich [bə'tu:lɪç] **1.** *Adj.* **a)** fussy; **b)** *(gemächlich)* leisurely; unhurried. **2.** *adv.* **a)** fussily; **b)** *(gemächlich)* in a calm unhurried way
Betulichkeit die; ~ a) fussiness; *(Besorgtheit)* agitation; **b)** *(Gemächlichkeit)* calm unhurried manner
betupfen *tr. V.* dab
Beuge ['bɔygə] **die; ~, ~n** *(einer Gliedmaße)* crook; **in der ~ des linken Arms** in the crook of his/her *etc.* left arm
beugen 1. *tr. V.* **a)** bend; bow ‹*head*›; **den Rumpf ~:** bend from the waist; **gebeugt gehen** walk with a stoop; **vom Alter/vom Kummer gebeugt** *(geh.)* bent *or* bowed with age *postpos.*/bowed down with grief *postpos.;* **b)** *(geh.: brechen)* **jmdn. ~:** break sb.'s resistance; **jmds. Starrsinn/Stolz ~:** break sb.'s stubborn/proud nature; **c)** *(Sprachw.) s.* **flektieren 1; d)** *(Rechtsw.)* bend ‹*law*›; **das Recht ~:** pervert the course of justice. **2.** *refl. V.* **a)** bend over; *(sich bücken)* stoop; **sich nach vorn/hinten ~:** bend forwards/bend over backwards; **sich aus dem Fenster ~:** lean out of the window; **b)** *(sich fügen)* give way; give in; **sich der Mehrheit ~:** bow to the will of the majority
Beugung die; ~, ~en a) bending; **b)** *(Sprachw.) s.* Flexion
Beule ['bɔylə] **die; ~, ~n** bump; *(Vertiefung)* dent; *(eiternd)* boil
beulen *itr. V.* bulge
beunruhigen [bə'|ʊnru:ɪgn̩] **1.** *tr. V.* worry; **es beunruhigte ihn sehr** it made him very worried; **über etw.** *(Akk.)* **beunruhigt sein** be worried about sth. **2.** *refl. V.* worry **(um, wegen** about)
Beunruhigung die; ~, ~en worry; concern
beurkunden [bə'|u:ɐ̯kʊndn̩] *tr. V.* record; *(belegen)* document, provide a record of
beurlauben [bə'|u:ɐ̯laubn̩] *tr. V.* **a) jmdn. [für zwei Tage] ~:** give sb. [two days'] leave of absence; **sich ~ lassen** obtain leave of absence; **b)** *(suspendieren)* suspend
Beurlaubung die; ~, ~en a) leave of absence *no indef. art.;* **b)** *(Suspendierung)* suspension
beurteilen *tr. V.* judge; assess ‹*situation etc.*›; **etw. falsch ~:** misjudge sth./assess sth. wrongly
Beurteilung die; ~, ~en a) judgement; *(einer Lage usw.)* assessment; **b)** *(Gutachten)* assessment
Beute ['bɔytə] **die; ~, ~n a)** *(Gestohlenes)* haul; loot *no indef. art.; (Kriegs~)* booty; spoils *pl.;* **fette ~ machen** get rich pickings *pl.;* **b)** *(eines Tiers)* prey; *(eines Jägers)* bag; **[seine] ~ schlagen** catch its prey; **c)** *(geh.: Opfer)* prey **(+** *Gen.* to); **eine ~ der Flammen werden** be consumed by the flames
Beutel ['bɔytl̩] **der; ~s, ~ a)** bag; *(kleiner, für Tabak usw.)* pouch; **b)** *(ugs.: Geld~)* purse; **c)** *(Zool.)* pouch
beuteln *tr. V.* **a)** *(südd., österr.: schütteln)* shake; **b) das Leben hat ihn gebeutelt** *(fig.)* life has given him some hard knocks
Beutel-: ~ratte die opossum; **~schneider der** *(veralt., geh.: Nepper)* shark; racketeer; **~tier das** marsupial
bevölkern [bə'fœlkɐn] **1.** *tr. V.* populate; inhabit; *(fig.)* fill; invade; **ein stark/dünn** *od.* **wenig bevölkertes Land** a densely/thinly *or* sparsely populated country. **2.** *refl. V.* become populated; ‹*bar, restaurant, etc.*› fill up
Bevölkerung die; ~, ~en population; *(Volk)* people

Bevölkerungs-: ~dichte die population density; **~explosion die** population explosion
bevollmächtigen [bəˈfɔlmɛçtɪgn̩] *tr. V.* **jmdn. [dazu] ~, etw. zu tun** authorize sb. to do sth.; *(in Rechtshandlungen)* give sb. power of attorney to do sth.
Bevollmächtigte der/die; *adj. Dekl.* authorized representative
Bevollmächtigung die; ~, ~en authorization; *(Rechtsw.)* power of attorney
bevor [bəˈfoːɐ̯] *Konj.* before; **~ du nicht unterschrieben hast** until you sign/have signed
bevormunden *tr. V.* **jmdn. ~:** impose one's will on sb.; **bevormundet werden** be dictated to
Bevormundung die; ~, ~en imposing one's will (+ *Gen.* on); **wie kann sie sich diese ~ durch ihre Eltern gefallen lassen?** how can she put up with her parents telling her what to do?
bevor|stehen *unr. itr. V.* be near; be about to happen; *(unmittelbar)* be imminent; **mir steht etwas Unangenehmes bevor** there's something unpleasant in store for me
bevorstehend *Adj.* forthcoming; *(unmittelbar)* imminent
bevorzugen [bəˈfoːɐ̯tsuːgn̩] *tr. V.* **a)** prefer (**vor** + *Dat.* to); **b)** *(begünstigen)* favour; give preference *or* preferential treatment to (**vor** + *Dat.* over)
bevorzugt 1. *Adj.* favoured; preferential ‹*treatment*›; *(privilegiert)* privileged. **2.** *adv.* **jmdn. ~ behandeln** give sb. preferential treatment; **jmdn. ~ abfertigen** give sb. priority *or* precedence
Bevorzugung die; ~, ~en preferential treatment; preference (+ *Gen.*, **von** for)
bewachen *tr. V.* guard; *(Ballspiele)* mark; **bewachter Parkplatz** car park with an attendant
Bewacher der; ~s, ~: guard; *(Ballspiele)* marker
bewachsen *unr. tr. V.* grow over; cover; **mit Efeu ~:** overgrown with ivy *postpos.*; ivy-covered
Bewachung die; ~: guarding; *(Ballspiele)* marking; **unter scharfer ~:** closely guarded; **jmdn. unter ~ stellen** put sb. under guard
bewaffnen [bəˈvafnən] **1.** *tr. V.* arm; **ein Heer [neu] ~:** supply an army with [new] weapons. **2.** *refl. V. (auch fig.)* arm oneself
bewaffnet *Adj. (auch fig.)* armed
Bewaffnete der/die; *adj. Dekl.* armed man/woman/person
Bewaffnung die; ~, ~en a) arming; **b)** *(Waffen)* weapons *pl.*
bewahren *tr. V.* **a) jmdn. vor etw.** *(Dat.)* **~:** protect *or* preserve sb. from sth.; **[Gott** *od.* **i] bewahre!** good Lord, no!; *(Gott behüte)* God forbid!; **b)** *(erhalten)* **seine Fassung** *od.* **Haltung ~:** keep *or* retain one's composure; **Stillschweigen/Treue ~:** remain silent/faithful; **sich** *(Dat.)* **etw. ~:** retain *or* preserve sth.; **etw. im Gedächtnis ~** *(fig. geh.)* preserve the memory of sth.
bewähren *refl. V.* prove oneself/itself; prove one's/its worth; **sich als [guter] Freund ~:** prove to be a [good] friend; **sich im Leben ~:** make a success of one's life; **sich gut/schlecht ~:** prove/not prove to be worthwhile *or* a success
bewahrheiten [bəˈvaːɐ̯haitn̩] *refl. V.* prove to be true
bewährt *Adj.* proven ‹*method, design, etc.*›; well-tried, tried and tested ‹*recipe, cure*›; reliable ‹*worker*›
Bewährung die; ~, ~en *(Rechtsw.)* probation; **3 Monate Gefängnis mit ~:** three months suspended sentence [with probation]; **eine Strafe zur ~ aussetzen** [conditionally] suspend a sentence on probation
Bewährungs-: ~frist die period of probation; **~helfer der** probation officer
bewaldet [bəˈvaldət] *Adj.* wooded
Bewaldung die; ~, ~en tree cover; *(Wälder)* woodlands *pl.*
bewältigen [bəˈvɛltɪgn̩] *tr. V.* deal with; cope with; overcome ‹*difficulty, problem*›; cover ‹*distance*›; *(innerlich verarbeiten)* get over ‹*experience*›
Bewältigung die; ~, ~en *s.* **bewältigen:** coping with; overcoming; covering; getting over; **zur ~ der Arbeit** *usw.* to deal *or* cope with the work *etc.*
bewandert [bəˈvandɐt] *Adj.* well-versed; knowledgeable; **auf einem Gebiet/in etw.** *(Dat.)* **~ sein** be well-versed *or* well up in a subject/in sth.
Bewandtnis [bəˈvantnɪs] **die; ~, ~se: mit etw. hat es [s]eine eigene/besondere ~:** there's a particular explanation for sth. *or* a [special] story behind sth.; **damit hat es folgende ~:** the story behind *or* reason for it is this
bewässern *tr. V.* irrigate
Bewässerung die; ~, ~en irrigation
bewegbar *Adj.* movable
[1]bewegen [bəˈveːgn̩] **1.** *tr. V.* **a)** move; **etw. von der Stelle ~:** move *or* shift sth. [from the spot]; **b)** *(ergreifen)* move; **eine ~de Rede** a moving speech; **c)** *(innerlich beschäftigen)* preoccupy; **das bewegt mich schon lange** I have been preoccupied with this *or* this has exercised my mind for a long time. **2.** *refl. V.* **a)** move; **b)** *(ugs.: sich Bewegung verschaffen)* **ich muß mich ein bißchen ~:** I must get some exercise; **du solltest/mußt dich mehr ~:** you ought to/must take more exercise; **c) seine Ausführungen ~ sich in der gleichen Richtung** *(fig.)* his comments have the same drift *or* are on the same lines; **d)** *(sich verhalten)* behave
[2]bewegen *unr. tr. V.* **jmdn. dazu ~, etw. zu tun** ‹*thing*› make sb. do sth., induce sb. to do sth.; ‹*person*› prevail upon *or* persuade sb. to do sth.; **jmdn. zur Teilnahme ~** ‹*person*› talk sb. into taking part; ‹*thing*› make sb. take part; induce sb. to take part
Beweg·grund der motive
beweglich *Adj.* **a)** movable; moving ‹*target*›; **seine ~e Habe** one's goods and chattels *pl.*; one's personal effects *pl.*; **~e Feste** movable feasts; **etw. ist leicht/schwer ~:** sth. is easy/difficult to move; **b)** *(rege)* agile, active ‹*mind*›; **geistig ~ sein** be nimble-minded; have an agile mind
Beweglichkeit die; ~ a) mobility; **b)** *(Regheit)* agility
bewegt [bəˈveːkt] *Adj.* **a)** eventful; *(unruhig)* turbulent; **ein ~es Leben** an eventful/turbulent life; **b)** *(gerührt)* moved *pred.*; emotional ‹*words, voice*›; **mit tief ~en Worten/~er Stimme** in words/a voice heavy with emotion; **c)** *(unruhig)* **leicht/stark ~** ‹*sea*› slightly choppy/very rough
Bewegung die; ~, ~en a) movement; *(bes. Technik, Physik)* motion; **in ~ sein** ‹*person*› be on the move; ‹*thing*› be in motion; **eine Maschine in ~ setzen** start [up] a machine; **sich in ~ setzen** ‹*train etc.*› start to move; ‹*procession*› move off; ‹*person*› get moving; **b)** *(körperliche ~)* exercise; **c)** *(Ergriffenheit)* emotion; **d)** *(Bestreben, Gruppe)* movement
Bewegungs·freiheit die; *o. Pl.* freedom of movement
bewegungs·los 1. *Adj.* motionless; **vor Schreck ~los** paralysed with fright. **2.** *adv.* without moving; **~los liegen/sitzen/stehen** lie/sit/stand motionless

beweih·räuchern [bə'vairɔy-çan] *tr. V.* surround with incense; *(fig. abwertend)* idolize; **sich selbst ~:** sing one's own praises

Beweis [bə'vais] **der; ~es, ~e** proof (*Gen.*, **für** of); *(Zeugnis)* evidence; **einen ~/~e für etw. haben** have proof/evidence of sth.; **als** *od.* **zum ~ seiner Aussage/ Theorie** to substantiate *or* in support of his statement/theory; **aus Mangel an ~en** owing to lack of evidence; **jmdm. einen ~ seines Vertrauens/seiner Hochachtung geben** give sb. a token of one's trust/esteem

beweisbar *Adj.* provable; susceptible of proof *postpos.*

beweisen 1. *unr. tr. V.* **a)** prove; **b)** *(zeigen)* show. **2.** *unr. refl. V.* prove oneself *or* one's worth (**vor** + *Dat.* to)

Beweis-: ~führung die a) *(Rechtsw.)* presentation of the evidence *or* case; **b)** *(Argumentation)* reasoning; argumentation; **~material das** evidence

bewenden *unr. V.* **es bei** *od.* **mit etw. ~ lassen** content oneself with sth.

bewerben *unr. refl. V.* apply (**um** for); **sich bei einer Firma** *usw.* **~:** apply to a company *etc.* [for a job]; **sich als Buchhalter** *usw.* **~:** apply for a job as a bookkeeper *etc.*

Bewerber der; ~s, ~, Bewerberin die; ~, ~nen applicant

Bewerbung die application (**um** for)

Bewerbungs-: ~schreiben das letter of application; **~unterlagen** *Pl.* documents in support of an/the application

bewerfen *unr. tr. V.* **jmdn./etw. mit etw. ~:** throw sth. at sb./sth.; **jmdn. mit [faulen] Eiern ~:** pelt sb. with [rotten] eggs

bewerkstelligen [bə'vɛrkʃtɛlɪgn̩] *tr. V.* manage

bewerten *tr. V.* assess; rate; *(dem Geldwert nach)* value (**mit** at); *(Schulw., Sport)* mark; grade *(Amer.)*; **einen Aufsatz mit [der Note] „gut" ~:** mark *or (Amer.)* grade an essay 'good'

Bewertung die a) *s.* **bewerten:** assessment; valuation; marking; grading *(Amer.)*; **b)** *(Note)* mark; grade *(Amer.)*

Bewertungs·maß·stab der criterion of assessment

bewilligen [bə'vɪlɪgn̩] *tr. V.* grant; award ‹*salary, grant*›; *(im Parlament usw.)* approve ‹*sum, tax increase, etc.*›

Bewilligung die; ~, ~en granting; *(Zustimmung)* approval; *(eines Gehalts, Stipendiums)* award

bewirken *tr. V.* bring about; cause; **~, daß etw. geschieht** cause sth. to happen; **das bewirkt bei ihm nichts/das Gegenteil** it has no effect/the opposite effect on him

bewirten [bə'vɪrtn̩] *tr. V.* feed; **jmdn. mit etw. ~:** serve sth. to sb.; serve sb. sth.

bewirtschaften *tr. V.* **a)** run, manage ‹*estate, farm, restaurant, business, etc.*›; **b)** *(bestellen)* farm ‹*fields, land*›; cultivate ‹*field*›

Bewirtung die; ~, ~en provision of food and drink; **die ~ der Gäste** catering for the guests

bewog [bə'vo:k] *1. u. 3. Pers. Sg. Prät. v.* **²bewegen**

bewohnbar *Adj.* habitable

bewohnen *tr. V.* inhabit, live in ‹*house, area*›; live in ‹*room, flat*›; live on ‹*floor, storey*›; ‹*animal, plant*› be found in

Bewohner der; ~s, ~, Bewohnerin die; ~, ~nen *(eines Hauses, einer Wohnung)* occupant; *(einer Stadt, eines Gebietes)* inhabitant; **ein ~ des Waldes** a forest-dweller; *(Tier)* a woodland creature

bewohnt *Adj.* occupied ‹*house etc.*›; inhabited ‹*area*›; **ist das Haus noch ~?** is the house still lived in *or* occupied?

bewölken [bə'vœlkn̩] *refl. V.* cloud over; become overcast

bewölkt *Adj.* cloudy; overcast; **dicht** *od.* **stark ~:** heavily overcast; **der Himmel ist nur leicht ~:** there is only a light cloud cover

Bewölkung die; ~, ~en a) *o. Pl.* clouding over; **b)** *(Wolkendecke)* cloud [cover]; **wechselnde ~:** variable amounts *pl.* of cloud

Bewunderer der; ~s, ~, Bewunderin die; ~, ~nen admirer

bewundern *tr. V.* admire (**wegen, für** for)

bewunderns·wert 1. *Adj.* admirable; worthy of admiration *postpos.* **2.** *adv.* admirably; in an admirable fashion

Bewunderung die; ~: admiration

bewunderungs·würdig *s.* **bewundernswert**

bewußt [bə'vʊst] **1.** *Adj.* **a)** conscious ‹*reaction, behaviour, etc.*›; *(absichtlich)* deliberate ‹*lie, deception, attack, etc.*›; **etw. ist/wird jmdm. ~:** sb. is/becomes aware of sth.; sb. realizes sth.; **b)** *(denkend)* **ein ~er Mensch** a thinking person; **sich** *(Dat.)* **einer Sache** *(Gen.)* **~ sein/werden** be/become aware *or* conscious of something; **c)** *nicht präd.* *(bekannt)* particular; *(fraglich)* in question *postpos.* **2.** *adv.* consciously; *(absichtlich)* deliberately; **~er leben** live with greater awareness

bewußt·los *Adj.* unconscious

Bewußtlosigkeit die; ~: unconsciousness; **bis zur ~** *(ugs.)* ad nauseam

bewußt|machen *tr. V.* **jmdm./ sich etw. ~:** make sb. realize/realize sth.

Bewußt·sein das a) *(deutliches Wissen)* awareness; **etw. mit ~ erleben** be fully aware of sth. [one is experiencing]; **jetzt erst kam ihr zu ~, daß ...:** only now did she realize that ...; **b)** *(geistige Klarheit)* consciousness; **das ~ verlieren** lose consciousness; **wieder zu ~ kommen, das ~ wiedererlangen** regain consciousness; **bei [vollem] ~ sein** be [fully] conscious

Bewußtseins·spaltung die *(Med., Psych.)* split consciousness; schizophrenia

bez. *Abk.* **bezahlt** pd.

bezahlbar *Adj.* affordable

bezahlen 1. *tr. V.* pay ‹*person, bill, taxes, rent, amount*›; pay for ‹*goods etc.*›; **jmdm. etw. ~:** pay for sth. for sb.; **bekommst du das Essen bezahlt?** do you get your meals paid for?; **bezahlter Urlaub** paid leave; holiday[s] with pay; **er mußte seinen Leichtsinn teuer ~** *(fig.)* he had to pay dearly for his carelessness; **das macht sich bezahlt** it pays off. **2.** *itr. V.* pay; **Herr Ober, ich möchte ~** *od.* **bitte ~:** waiter, the bill *or (Amer.)* check please

Bezahlung die payment; *(Lohn, Gehalt)* pay; **die ~ der Waren** the payment for the goods; **gegen ~** ‹*work*› for payment *or* money

bezähmen 1. *tr. V.* contain, control ‹*wrath, curiosity, impatience*›; restrain ‹*desire*›. **2.** *refl. V.* restrain oneself

bezaubern *tr. V.* enchant; **von etw. bezaubert** enchanted with *or* by sth.

bezaubernd 1. *Adj.* enchanting. **2.** *adv.* enchantingly

bezeichnen *tr. V.* **a) jmdn./sich/ etw. als etw. ~:** call sb./oneself/ sth. sth.; describe sb./oneself/sth. as sth.; **wie bezeichnet man das?** what is it called?; **mit dem Wort bezeichnet man eine Art Jacke** this word is used to denote *or* describe a kind of jacket; **b)** *(Name sein für)* denote; **c)** *(markieren)* mark; *(durch Zeichen angeben)* indicate

bezeichnend *Adj.* characteristic, typical (**für** of); *(bedeutsam)* significant

bezeichnender·weise *Adv.* characteristically; typically

Bezeichnung die a) *o. Pl.* marking; *(Angabe durch Zeichen)* indication; **b)** *(Name)* name; **mir fällt die richtige ~ dafür nicht ein** I can't think of the right word for it/them

bezeugen *tr. V.* testify to; **~, daß ...:** testify that ...

bezichtigen [bə'tsɪçtɪgn̩] *tr. V.* accuse; **jmdn. des Verrats ~:** accuse sb. of treachery; **jmdn. ~, etw. getan zu haben** accuse sb. of having done sth.

Bezichtigung die; ~, ~en accusation

beziehbar *Adj.* **a)** ⟨*flat, house, etc.*⟩ ready for occupation; **b) auf jmdn./etw. ~:** applicable to sb./sth. *postpos.*

beziehen 1. *unr. tr. V.* **a)** cover, put a cover/covers on ⟨*seat, cushion, umbrella, etc.*⟩; **die Betten frisch ~:** put clean sheets on the beds; **das Sofa ist mit Leder bezogen** the sofa is upholstered in leather; **b)** *(einziehen in)* move into ⟨*house, office*⟩; **c)** *(Milit.)* take up ⟨*position, post*⟩; **einen klaren Standpunkt ~** *(fig.)* adopt a clear position; take a definite stand; **d)** *(erhalten)* receive, obtain [one's supply of] ⟨*goods*⟩; take ⟨*newspaper*⟩; draw, receive ⟨*pension, salary*⟩; **Prügel ~** *(ugs.)* get a hiding *(coll.)*; **e)** *(in Beziehung setzen)* apply (**auf** + *Akk.* to); **etw. auf sich** *(Akk.)* **~:** take sth. personally; **bezogen auf jmdn./etw.** [seen] in relation to sb./sth. **2.** *unr. refl. V.* **a) es/der Himmel bezieht sich** it/the sky is clouding over *or* becoming overcast; **b) sich auf jmdn./etw. ~** ⟨*person, letter, etc.*⟩ refer to sb./sth.; ⟨*question, statement, etc.*⟩ relate to sb./sth.; **wir ~ uns auf Ihr Schreiben vom 28. 8., und ...:** with reference to your letter of 28 August, we ...

Bezieher der; ~s, ~, Bezieherin die; ~, ~nen *(einer Zeitung)* subscriber (*Gen.*, **von** to); *(einer Rente, eines Gehalts)* recipient

Beziehung die a) relation; *(Zusammenhang)* connection (**zu** with); **gute ~en** *od.* **eine gute ~ zu jmdm. haben** have good relations with sb.; be on good terms with sb.; **~en haben** *(gewisse Leute kennen)* have [got] connections; **etw. durch ~en bekommen** get sth. through connections; **seine ~en spielen lassen** pull some strings; **zu jmdm. keine ~ haben** be unable to relate to sb.; **er hat keine ~ zur Kunst** he has a blind spot where the arts are concerned; the arts are a closed book to him; **zwischen A und B besteht keine/eine ~:** there is no/a connection between A and B; **A zu B in ~** *(Akk.)* **setzen** relate A to B; see A in relation to B; **A und B in ~** *(Akk.)* **zueinander setzen** relate A and B to each other; connect *or* link A and B; **b)** *(Freundschaft, Liebes~)* relationsship; **c)** *(Hinsicht)* respect; **in mancher ~:** in many respects

beziehungs-: ~los 1. *Adj.* unconnected; unrelated; **2.** *adv.* without any connection; **~reich** *Adj.* evocative; rich in associations *postpos.*; **~weise** *Konj.* **a)** and ... respectively; *(oder)* or; **die beiden Münzen waren aus Kupfer ~weise aus Nickel** the two coins were made of copper and of nickel respectively; **b)** *(ugs.: oder vielmehr)* that is; or to be precise

beziffern [bə'tsɪfɐn] **1.** *tr. V.* **a)** *(numerieren)* number; **b)** *(angeben)* estimate (**auf** + *Akk.* at). **2.** *refl. V.* **sich auf 10 Millionen** *(Akk.)* **DM ~:** come *or* amount to 10 million marks

Bezirk [bə'tsɪrk] **der; ~[e]s, ~e a)** district; **b)** *(Verwaltungs~)* [administrative] district

bezug [bə'tsu:k] *in* **in ~ auf jmdn./etw.** concerning *or* regarding sb./sth.

Bezug der a) *(für Kissen usw.)* cover; *(für Polstermöbel)* loose cover; slip-cover *(Amer.)*; *(für Betten)* duvet cover; *(für Kopfkissen)* pillowcase; **b)** *o. Pl. (Erwerb)* obtaining; *(Kauf)* purchase; **~ einer Zeitung** taking a newspaper; **c)** *Pl.* salary *sing.*; **die Bezüge der Beamten** the salaries of the civil servants; **d)** *(Papierdt.) in* **mit** *od.* **unter ~ auf etw.** *(Akk.)* with reference to sth.; **auf etw.** *(Akk.)* **~ nehmen** refer to sth.; **~ nehmend auf unser Telex** with reference to our telex; **e)** *(Verbindung)* connection; link

bezüglich [bə'tsy:klɪç] **1.** *Präp. mit Gen.* concerning; regarding. **2.** *Adj.* **auf etw.** *(Akk.)* **~:** relating to sth.; **die darauf ~en Paragraphen** the relevant paragraphs

Bezugnahme [bə'tsu:kna:mə] **die; ~, ~n** *(Papierdt.)* reference; **unter ~ auf etw.** *(Akk.)* with reference to sth.

bezuschussen [bə'tsu:ʃʊsn̩] *tr. V. (Papierdt.)* subsidize

bezwecken [bə'tsvɛkn̩] *tr. V.* aim to achieve; aim at; **was willst du damit ~?** what do you expect to achieve by [doing] that?

bezweifeln *tr. V.* doubt; question; **ich bezweifle nicht, daß ...:** I do not doubt that ...; **das ist nicht zu ~:** there is no doubt about that

bezwingen *unr. tr. V.* conquer ⟨*enemy, mountain, pain, etc.*⟩; defeat ⟨*opponent*⟩; take, capture ⟨*fortress*⟩; master ⟨*pain, hunger*⟩; **seinen Zorn/seine Neugier ~:** keep one's anger/curiosity under control

Bezwinger der, Bezwingerin die; ~, ~nen conqueror

BGB [be:ge:'be:] *Abk.:* **Bürgerliches Gesetzbuch**

BH [be:'ha:] **der; ~[s], ~[s]** *Abk.:* **Büstenhalter** bra

Biathlon ['bi:atlɔn] **das; ~s, ~s** *(Sport)* biathlon

bibbern ['bɪbɐn] *itr. V. (ugs.) (vor Kälte)* shiver (**vor** with); *(vor Angst)* shake, tremble (**vor** with); **um jmdn./etw. ~:** fear *or* tremble for sb./sth.

Bibel ['bi:bl̩] **die; ~, ~n** *(auch fig.)* Bible

Bibel-: ~spruch der biblical saying; **~vers der** verse from the Bible

[1]**Biber** ['bi:bɐ] **der; ~s, ~:** beaver

[2]**Biber der** *od.* **das; ~s** *(Stoff)* flannelette

Bibliographie die; ~, ~n bibliography

Bibliothek [biblio'te:k] **die; ~, ~en** library

Bibliothekar [bibliote'ka:ɐ̯] **der; ~s, ~e, Bibliothekarin die; ~, ~nen** librarian

biblisch ['bi:blɪʃ] *Adj.* biblical; **ein ~es Alter** a grand old age

Bidet [bi'de:] **das; ~s, ~s** bidet

bieder ['bi:dɐ] **1.** *Adj.* **a)** unsophisticated; *(langweilig)* stolid; **b)** *(veralt.: rechtschaffen)* upright. **2.** *adv.* in an unsophisticated manner

Bieder·mann der; *Pl.* **Biedermänner a)** *(veralt.)* man of integrity *or* probity; **b)** *(Spießer)* petty bourgeois

Biedermeier das; ~s Biedermeier [period/style]

biegen ['bi:gn̩] **1.** *unr. tr. V.* bend. **2.** *unr. refl. V.* bend; *(nachgeben)* give; sag; **der Tisch bog sich unter der Last der Speisen** the table sagged *or* groaned under the weight of the food. **3.** *unr. itr. V.; mit sein* turn; **um die Ecke ~:** turn the corner; ⟨*car*⟩ take the corner. **4.** *in* **auf Biegen oder Brechen** *(ugs.)* at all costs; by hook or by crook; **es geht auf Biegen oder**

Brechen *(ugs.)* it has come to the crunch *or (Amer.)* showdown
biegsam *Adj.* flexible; pliable ‹*material*›; *(gelenkig)* supple
Biegsamkeit die; ~ *s.* **biegsam:** flexibility; pliability; suppleness
Biegung die; ~, ~**en** bend; **eine [scharfe]** ~ **nach rechts machen** bend [sharply] to the right
Biene ['bi:nə] **die;** ~, ~**n a)** bee; **b)** *(ugs. veralt.: Mädchen)* bird *(Brit. + l.)*; dame *(Amer. sl.)*
Bienen-: ~**fleiß der** unflagging industry; ~**honig der** bees' honey; ~**königin die** queen bee; ~**korb der** straw hive; ~**schwarm der** swarm of bees; ~**stich der a)** bee-sting; **b)** *(Kuchen) cake with a topping of sugar and almonds (and sometimes a cream filling)*; ~**stock der** bee-hive
Bier [bi:ɐ̯] **das;** ~**[e]s,** ~**e** beer; **ein kleines/großes** ~**:** a small/large [glass of] beer; **zwei** ~**:** two beers; two glasses of beer; **das ist [nicht] mein** ~ *(ugs.)* that is [not] my affair *or* business
Bier·bauch der *(ugs. spött.)* beer belly *(coll.)*
Bierchen das; ~s, ~ *(ugs.)* little [glass of] beer
bier-, Bier-: ~**deckel der** beer-mat; ~**dose die** beer can; ~**ernst** *(ugs.)* **1.** *Adj.* deadly serious; solemn; **2.** *adv.* solemnly; ~**faß das** beer-barrel; ~**filz der** beer-mat; ~**flasche die** beer-bottle; ~**garten der** beer garden; ~**glas das** beer-glass; ~**kasten der** beer-crate; ~**keller der** beer cellar; ~**krug der** beer-mug; *(aus Glas, Zinn)* tankard; ~**laune die** *(ugs.)* **in einer** ~**laune, aus einer** ~**laune heraus** in an exuberant mood; ~**schinken der** *slicing sausage containing pieces of ham*; ~**selig** *(scherzh.)* **1.** *Adj.* beery ‹*mood*›; ~**selig, wie er war** in his beerily happy state; **2.** *adv.* in a beerily happy state; ‹*laugh*› in beery merriment; ~**tisch der: am** ~**tisch** over a glass of beer; in the pub *(Brit. coll.) or (Amer.)* bar; ~**trinker der,** ~**trinkerin die** beer-drinker; ~**zeitung die** joke newspaper *(made up for a closed group)*; ~**zelt das** beer tent
Biest [bi:st] **das;** ~**[e]s,** ~**er** *(ugs. abwertend)* **a)** *(Tier, Gegenstand)* wretched thing; *(Bestie)* creature; **b)** *(Mensch)* beast *(derog.)*; wretch; **das freche** ~**:** the cheeky devil *(coll.)*
bieten ['bi:tn̩] **1.** *unr. tr. V.* **a)** offer; put on ‹*programme etc.*›; provide ‹*shelter, guarantee, etc.*›; *(bei Auktionen, Kartenspielen)* bid **(für, auf** + *Akk.* for); **jmdm. Geld/eine Chance** ~**:** offer sb. money/a chance; **jmdm. den Arm** ~ *(geh.)* offer sb. one's arm; **eine hervorragende Leistung** ~**:** put up an outstanding performance; **das bietet keine Schwierigkeiten** that presents no difficulties; **das Stadion bietet 40 000 Personen Platz** the stadium has room for *or* can hold 40,000 people; **b) ein schreckliches/gespenstisches** *usw.* **Bild** ~**:** present a terrible/eerie *etc.* picture; be a terrible/eerie *etc.* sight; **einen prächtigen Anblick** ~**:** look splendid; be a splendid sight; **c)** *(zumuten)* **das lasse ich mir nicht** ~**:** I won't put up with *or* stand for that. **2.** *unr. refl. V.* **sich jmdm.** ~**:** present itself to sb.; **hier bietet sich dir eine Chance** this is an opportunity for you; this offers you an opportunity; **ihnen bot sich ein Bild des Grauens** a horrific sight confronted them. **3.** *unr. itr. V.* bid **(auf** + *Akk.* for)
Bigamie [biga'mi:] **die;** ~, ~**n** bigamy *no def. art.*
Big Band ['bɪg 'bænd] **die;** ~ ~, ~ ~**s** big band
bigott [bi'gɔt] *(abwertend) Adj.* **a)** religiose; over-devout; **b)** *(scheinheilig)* sanctimonious; holier-than-thou
Bigotterie [bigɔtə'ri:] **die;** ~ *(abwertend)* religious bigotry; religiosity; *(Scheinheiligkeit)* sanctimoniousness
Bikini [bi'ki:ni] **der;** ~s, ~s bikini
Bilanz [bi'lants] **die;** ~, ~**en a)** *(Kaufmannsspr., Wirtsch.)* balance sheet; **eine** ~ **aufstellen** make up the accounts *pl.*; draw up a balance sheet; **b)** *(Ergebnis)* outcome; *(Endeffekt)* net result; ~ **ziehen** take stock; sum things up; **die** ~ **aus etw. ziehen** draw conclusions *pl.* about sth.; *(rückblickend)* take stock of sth.
bi·lateral *(Politik)* **1.** *Adj.* bilateral. **2.** *adv.* bilaterally
Bild [bɪlt] **das;** ~**[e]s,** ~**er a)** picture; *(in einem Buch usw.)* illustration; *(Spielkarte)* picture *or* court card; **ein** ~ **[von jmdm./etw.] machen** take a picture [of sb./sth.]; **ein** ~ **von einem Mann/einer Frau sein** be a fine specimen of a man/woman; be a fine-looking man/woman; **b)** *(Aussehen)* appearance; *(Anblick)* sight; **ein** ~ **des Jammers** a pathetic sight; **ein** ~ **für [die] Götter** *(scherzh.)* a sight for sore eyes; **c)** *(Metapher)* image; metaphor; **d)** *(Abbild)* image; *(Spiegel~)* reflection; **e)** *(Vorstellung)* image; **ein falsches/merkwürdiges** ~ **von etw. haben** have a wrong impression/curious idea of sth.; **sich** *(Dat.)* **ein** ~ **von jmdm./etw. machen** form an impression of sb./sth.; **f)** *in* **jmdn. [über etw.** *(Akk.)***] ins** ~ **setzen** put sb. in the picture [about sth.]; **[über etw.** *(Akk.)***] im** ~**e sein** be in the picture [about sth.]; **g)** *(Theater)* scene
Bild·band der copiously illustrated book
bildbar *Adj.* formable **(aus** from); malleable ‹*personality, mind*›
Bild-: ~**bei·lage die** pictorial *or* illustrated supplement; ~**beschreibung die** picture description
bilden 1. *tr. V.* **a)** form **(aus** from); *(modellieren)* mould **(aus** from); **den Charakter** ~**:** form *or* mould sb.'s/one's personality; **eine Gasse** ~**:** make a path *or* passage; **sich** *(Dat.)* **ein Urteil [über jmdn./etw.]** ~**:** form an opinion [of sb./sth.]; **b)** *(ansammeln)* build up ‹*fund, capital*›; **c)** *(darstellen)* be, represent ‹*exception etc.*›; constitute ‹*rule etc.*›; **den Höhepunkt des Abends bildete sein Auftritt** his appearance was the high spot of the evening; **d)** *(erziehen)* educate; *itr.* **Reisen bildet** travel broadens the mind. **2.** *refl. V.* **a)** *(entstehen)* form; **b)** *(lernen)* educate oneself
bildend *Adj.* **a) die** ~**e Kunst** the plastic arts *pl. (including painting and architecture)*; **b)** *(belehrend)* educational
Bilder·buch das picture-book *(for children)*; **aussehen wie im** *od.* **aus dem** ~**:** look a picture
Bilderbuch- perfect ‹*landing, weather*›; picture-book ‹*weather, village*›; story-book ‹*marriage, career*›
Bilder-: ~**geschichte die** picture story; *(Comic)* strip cartoon; ~**rahmen der** picture-frame; ~**rätsel das** picture puzzle; *(Rebus)* rebus
bild-, Bild-: ~**fläche die: auf der** ~**fläche erscheinen** *(ugs.)* appear on the scene; *(auftauchen)* turn up; **von der** ~**fläche verschwinden** *(ugs.) (rasch weggehen)* make oneself scarce *(coll.)*; *(aus der Öffentlichkeit verschwinden)* disappear from the scene; ~**geschichte die** *s.* **Bildergeschichte;** ~**haft 1.** *Adj.* graphic; pictorial, illustrative ‹*language, sense, etc.*›; vivid ‹*imagination, clarity,*

etc.⟩; **2.** *adv.* graphically; *(lebhaft)* vividly; **~hauer der** sculptor; **~hauerei** [---'-] **die** sculpture *no def. art.*; **~hauerin die; ~, ~nen** sculptress; **~hübsch** *Adj.* really lovely; stunningly beautiful ⟨*girl*⟩

bildlich 1. *Adj.* **a)** pictorial; **b)** *(übertragen)* figurative; **~er Ausdruck, ~e Wendung** figure of speech; image. **2.** *adv.* **a)** pictorially; **sich etw. ~ vorstellen** picture sth. to oneself; **b)** *(übertragen)* figuratively; **~ gesprochen** metaphorically speaking

Bild·material das pictures *pl.*; *(Fotos/Film)* photographic/film material

bildnerisch *Adj.* artistic; creative ⟨*abilities*⟩

Bildnis ['bıltnıs] **das; ~ses, ~se** portrait; *(Plastik)* sculpture

Bild-: ~reportage die photoreportage; **~röhre die** *(Ferns.)* picture tube; **~schirm der** screen; **am ~ arbeiten** work at *or* with a VDU

Bildschirm-: ~arbeit die VDU work *no art., no pl.*; **~gerät das** VDU; visual display unit; **~zeitung die** teletext

bild·schön *Adj.* really lovely; stunningly beautiful ⟨*girl, woman*⟩

Bild·störung die interference *no def. art.* on vision

Bildung die; ~, ~en a) *(Erziehung)* education; *(Kultur)* culture; **[keine] ~ haben** be [un]educated; *([un]kultiviert sein)* be [un]cultivated *or* [un]cultured; **b)** *(Schaffung)* formation; **die ~ einer Kommission** setting up a committee

bildungs-, Bildungs-: ~chancen *Pl.* educational opportunities **~hunger der** thirst for education; **~hungrig** *Adj.* eager to be educated *postpos.*; **~lücke die** gap in one's education; **das ist eine ~lücke!** that's culpable ignorance!; **~politik die** educational policy; **~urlaub der** educational leave

Bild-: ~unter·schrift die caption; **~wörter·buch das** pictorial dictionary

Billard ['bıljart, *österr.:* bi'jaːɐ̯] **das; ~s, ~e** billiards

Billard-: ~kugel die billiard-ball; **~stock der** billiard-cue; **~tisch der** billiard-table

Billett [bıl'jɛt] **das; ~[e]s, ~e** *od.* **~s** *(schweiz., veralt.)* ticket

Billiarde [bı'li̯ardə] **die; ~, ~n** thousand million million; quadrillion *(Amer.)*

billig ['bılıç] **1.** *Adj.* **a)** cheap; **b)** *(abwertend: primitiv)* shabby, cheap ⟨*trick*⟩; feeble ⟨*excuse*⟩; **ist dir das nicht zu ~?** isn't that beneath you? **2.** *adv.* cheaply; **~ einkaufen** shop cheaply; **~ abzugeben** *(in Anzeigen)* for sale cheap

billigen *tr. V.* approve; **~, daß jmd. etw. tut** approve of sb.'s doing sth.; **etw. stillschweigend ~:** give sth. one's tacit approval

Billig·flug der cheap flight

Billigung die; ~: approval; **jmds. ~ finden** meet with *or* receive sb.'s approval

Billig·ware die cheap goods *pl.*

Billion [bı'li̯oːn] **die; ~, ~en** million million; trillion *(Amer.)*

bim [bım] *Interj.* ding; **~, bam** ding dong

Bimbam *in* **[ach du] heiliger ~!** *(ugs.)* [oh] my sainted aunt! *(sl.)*; glory be! *(sl.)*

Bimmel ['bıml̩] **die; ~, ~n** *(ugs.)* [ting-a-ling] bell

Bimmel·bahn die *(ugs. scherzh.)* narrow-gauge railway *(with a warning bell)*

Bimmelei die; ~ *(ugs. abwertend)* constant ringing

bimmeln *itr. V. (ugs.)* ring

Bims·stein ['bıms-] **der** pumice-stone

bin [bın] *1. Pers. Sg. Präsens v.* **[1]sein**

Binde ['bındə] **die; ~, ~n a)** *(Verband)* bandage; *(Augen~)* blindfold; **b)** *(Arm~)* armband; **c)** *(veralt.: Krawatte)* tie; **sich** *(Dat.)* **einen hinter die ~ gießen** *od.* **kippen** *(ugs.)* have a drink or two

Binde-: ~gewebe das *(Anat.)* connective tissue; **~glied das** [connecting] link; **~haut die** *(Anat.)* conjunctiva; **~mittel das** binder

binden 1. *unr. tr. V.* **a)** *(bündeln)* tie; **etw. zu etw. ~:** tie sth. into sth.; **b)** *(herstellen)* make up ⟨*wreath, bouquet*⟩; make ⟨*broom*⟩; **c)** *(fesseln)* bind; **d)** *(verpflichten)* bind; **e)** *(befestigen, auch fig.)* tie **(an + *Dat.*** to); **nicht an einen Ort gebunden sein** *(fig.)* not be tied to one place; **jmdn. an sich** *(Akk.)* **~** *(fig.)* make sb. dependent on one; **f)** *(knüpfen)* tie ⟨*knot, bow, etc.*⟩; knot ⟨*tie*⟩; **g)** *(festhalten)* bind ⟨*soil, mixture, etc.*⟩; thicken ⟨*sauce*⟩; **h)** *(Buchw.)* bind. **2.** *unr. itr. V. (als Bindemittel wirken)* bind. **3.** *unr. refl. V.* tie oneself down; **ich bin zu jung, um mich schon zu ~:** I am too young to be tied down

bindend *Adj.* binding **(für** on); definite ⟨*answer*⟩

Binder der; ~s, ~ a) *(Krawatte)* tie; **b)** *(Bindemittel)* binder

Binde-: ~strich der hyphen; **~wort das** *(Sprachw.)* conjunction

Bind·faden der string; **ein [Stück] ~:** a piece of string; **es regnet Bindfäden** *(ugs.)* it's raining cats and dogs *(coll.)*

Bindung die; ~, ~en a) *(Beziehung)* relationship **(an + *Akk.*** to); **b)** *(Verbundenheit)* attachment **(an + *Akk.*** to); **c)** *(Ski~)* binding; **d)** *(Chemie)* bond

binnen ['bınən] *Präp. mit Dat. od. (geh.) Gen.* within; **~ Jahresfrist** within a year

Binnen-: ~gewässer das inland water; **~hafen der** inland port; **~land das** *o. Pl.* interior; **~meer das** inland sea; **~see der** lake

Binom [bi'noːm] **das; ~s, ~e** *(Math.)* binomial

binomisch *Adj. (Math.)* binomial

Binse ['bınzə] **die; ~, ~n** *(Bot.)* rush; **in die ~n gehen** *(ugs.)* fall through

Binsen·weisheit die truism

Bio ['biːo] *o. Art. (Schülerspr.)* biol *(school sl.)*; biology

bio-, Bio-: ~chemie die biochemistry; **~chemisch** *Adj.* biochemical; **~gas** ['---] **das** biogas; **~graph der; ~en, ~en** biographer; **~graphie die; ~, ~n** biography; **~graphisch** *Adj.* biographical; **~loge** [-'loːgə] **der; ~n, ~n** biologist; **~logie die; ~:** biology *no art.*; **~login die; ~, ~nen** biologist; **~logisch** *Adj.* **a)** biological; **b)** *(natürlich)* natural ⟨*medicine, cosmetic, etc.*⟩; **~masse** ['----] **die** biomass

Biotop [bio'toːp] **der** *od.* **das; ~s, ~e** *(Biol.)* biotope

Bio·wissenschaften *Pl.* life sciences

Birke ['bırkə] **die; ~, ~n a)** birch[-tree]; **b)** *o. Pl. (Holz)* birch[wood]

Birk-: ~hahn der blackcock; **~huhn das** black grouse

Birma ['bırma] **(das); ~s** Burma

Birn·baum der pear-tree

Birne ['bırnə] **die; ~, ~n a)** pear; **b)** *(Glüh~)* [light-]bulb; **c)** *(salopp: Kopf)* nut *(sl.)*

bis [bıs] **1.** *Präp. mit Akk.* **a)** *(zeitlich)* until; till; *(die ganze Zeit über und bis zu einem bestimmten Zeitpunkt)* up to; up until; up till; *(nicht später als)* by; **ich muß ~ fünf Uhr warten** I have to wait until *or* till five o'clock; **~ gestern glaubte ich ...:** [up] until yesterday I had thought ...; **von Dienstag ~ Donnerstag** from Tuesday to Thursday; Tuesday through Thursday *(Amer.)*; **von sechs ~ sieben [Uhr]** from six until *or* till seven [o'clock]; **~ Ende März ist er zurück/verreist** he'll be back by/away until the end of March; **~ wann dauert das Konzert?** till *or* until when does the concert go on?; **~ dann/gleich/später/morgen/nachher!** see you then/in a while/later/tomorrow/later!; **b)** *(räumlich, fig.)* to; **dieser Zug fährt nur ~ Offenburg** this train only goes to *or* as far as Offenburg; **~ wohin fährt der Bus?** how far does the bus go?; **nur ~ Seite 100** only up to *or* as far as page 100; **~ 5000 Mark** up to 5,000 marks; **Kinder ~ 6 Jahre** children up to the age of six *or* up to six years of age. **2.** *Adv.* **a) Städte ~ zu 50 000 Einwohnern** towns of up to 50,000 inhabitants; **~ zu 6 Personen** up to six people; **~ nach Köln** to Cologne; **~ an die Decke** up to the ceiling; **b) ~ auf** *(einschließlich)* down to; *(mit Ausnahme von)* except for. **3.** *Konj.* **a)** *(nebenordnend)* to; **vier ~ fünf** four to five; **b)** *(unterordnend)* until; till; *(österr.: sobald)* when

Bisam·ratte die musk-rat

Bischof ['bɪʃɔf] **der; ~s, Bischöfe** ['bɪʃœfə] bishop

bischöflich *Adj.* episcopal

Bischofs-: ~mütze die [bishop's] mitre; **~sitz der** seat of a/the bishopric; **~stab der** [bishop's] crosier *or* crook

Bi·sexualität die bisexuality

bi·sexuell 1. *Adj.* bisexual. **2.** *adv.* bisexually

bis·her *Adv.* up to now; *(aber jetzt nicht mehr)* until now; till now; **er hat sich ~ nicht gemeldet** he hasn't been in touch up to now *or* as yet

bisherig *Adj.; nicht präd. (vorherig)* previous; *(momentan)* present; **sie ziehen um, ihre ~e Wohnung wird zu klein** they are moving – their present flat is getting too small; **sie sind umgezogen, ihre ~e Wohnung wurde zu klein** they have moved – their previous flat became too small

Biskaya [bɪs'ka:ja] **die; ~: die ~/der Golf von ~:** the Bay of Biscay

Biskuit [bɪs'kvi:t] **das** *od.* **der; ~[e]s, ~s** *od.* **~e a)** sponge biscuit; **b)** *(~teig)* sponge

bis·lang *Adv.: s.* **bisher**

Bismarck·hering ['bɪsmark-] **der** Bismarck herring

Bison ['bi:zɔn] **der; ~s, ~s** bison

Biß [bɪs] **der; Bisses, Bisse** bite

bißchen *indekl. Indefinitpron.* **1.** *adj.* **ein ~ Geld/Brot/Milch/Wasser** a bit of *or* a little money/bread/a drop of *or* a little milk/water; **ich würde ihm kein ~ Geld mehr leihen** I wouldn't lend him any more money at all; **ein/kein ~ Angst haben** be a bit/not a bit frightened. **2.** *adv.* **ein/kein ~:** a bit *or* a little/not a *or* one bit; **ich werde mich ein ~ aufs Ohr legen** I'm going to lie down for a bit; **ein klein ~:** a little bit; **ein ~ zuviel/mehr** a bit too much/a bit more. **3.** *subst.* **ein ~:** a bit; a little; *(bei Flüssigkeiten)* a drop; a little; **von dem ~ werde ich nicht satt** that little bit/drop won't fill me up; **das/kein ~:** the little [bit]/not a *or* one bit

Bissen der; ~s, ~: mouthful; **sie bekam keinen ~ herunter** she couldn't eat a thing; **ihm blieb der ~ im Hals[e] stecken** *(ugs.)* the food stuck in his throat; **sich** *(Dat.)* **den letzten ~ vom Munde absparen** scrimp [and save]

bissig 1. *Adj.* **a) ~ sein** ⟨dog⟩ bite; **ein ~er Hund** a dog that bites; **„Vorsicht, ~er Hund"** 'beware of the dog'; **b)** *(fig.)* cutting, caustic ⟨*remark, tone, etc.*⟩. **2.** *adv. (fig.)* ⟨*say*⟩ cuttingly, caustically

Biß·wunde die bite

bist [bɪst] *2. Pers. Sg. Präsens v.* **sein**

biste *(ugs.)* = **bist du;** *s. auch* **haste**

Bistum ['bɪstu:m] **das; ~s, Bistümer** ['bɪsty:mɐ] bishopric; diocese

bis·weilen *Adv. (geh.)* from time to time; now and then

Bitt·brief der letter of request; *(Bittgesuch)* petition

bitte ['bɪtə] **1.** *Adv.* please; **können Sie mir ~ sagen ...?** could you please tell me ...?; **~ nicht!** no, please don't!; **~ nach Ihnen** after you. **2.** *Interj.* **a)** *(Bitte, Aufforderung)* please; **~[, nehmen sie doch Platz]!** do take a seat; **~[, treten sie ein]!** come in!; **zwei Tassen Tee, ~:** two cups of tea, please; **Noch eine Tasse Tee? – [Ja] ~!** Another cup of tea? – Yes, please; **b)** *(Aufforderung etw. entgegenzunehmen)* **~ [schön** *od.* **sehr]!** there you are!; **na ~!** *(da siehst du es!)* there you are!; **c)** *(Ausdruck des Einverständnisses)* **~ [gern]!** certainly; of course; **aber ~!** yes do; **~, macht doch, was ihr wollt** just [go ahead and] do what you want; **Entschuldigung! – Bitte!** [I'm] sorry! – That's all right!; **d)** *(Aufforderung, sich zu äußern)* **~ [schön** *od.* **sehr]!** *(im Laden, Lokal)* yes, please?; **ja, ~?** *(am Telefon)* hello?; yes?; **e)** *(Nachfrage)* **[wie] ~?** sorry; *(überrascht, empört)* what?; **f)** *(Erwiderung einer Dankesformel)* **~ [schön** *od.* **sehr]** not at all; you're welcome

Bitte die; ~, ~n request; *(inständig)* plea; **eine große ~ [an jmdn.]/nur die eine ~ haben** have a [great] favour to ask [of sb.]/have [just] one request *or* just one thing to ask

bitten 1. *unr. itr. V.* **a) um etw. ~:** ask for *or* request sth.; *(inständig)* beg for sth.; **der Blinde bat um eine milde Gabe** the blind man begged for alms; **ich bitte einen Moment um Geduld/Ihre Aufmerksamkeit** I must ask you to be patient for a moment/may I ask for your attention for a moment; **b)** *(einladen)* ask; **ich lasse ~:** [please] ask him/her/them to come in. **2.** *unr. tr. V.* **a) jmdn. um etw. ~:** ask sb. for sth.; **darf ich Sie um Feuer/ein Glas Wasser ~?** could I ask you for a light/a glass of water, please?; **darf ich die Herrschaften um Geduld/Ruhe ~?** could I ask you to be patient/silent?; **[aber] ich bitte dich/Sie!** [please] don't mention it; **b)** *(einladen)* ask, invite; **jmdn. zum Tee [zu sich] ~:** ask *or* invite sb. to tea; **jmdn. ins Haus/Zimmer ~:** ask *or* invite sb. [to come] in

bitter 1. *Adj.* **a)** bitter; plain ⟨*chocolate*⟩; **b)** *(schmerzlich)* bitter ⟨*experience, disappointment, etc.*⟩; painful, hard ⟨*loss*⟩; painful, bitter, hard ⟨*truth*⟩; hard ⟨*time, fate, etc.*⟩; **eine ~e Lehre** a hard lesson; **c)** *(beißend)* bitter ⟨*irony, sarcasm*⟩; **d)** *(verbittert)* bitter; **ein ~es Gefühl** a feeling of bitterness; **e)** *(groß, schwer)* bitter ⟨*cold, tears, grief, remorse, regret*⟩; dire ⟨*need*⟩; desperate ⟨*poverty*⟩; grievous ⟨*injustice, harm*⟩. **2.** *adv.* **a)** *(verbittert)* bitterly; **b)** *(sehr stark)* desperately; ⟨*regret*⟩ bitterly

bitter-: ~böse 1. *Adj.* furious; **2.** *adv.* furiously; **~ernst 1.** *Adj.* deadly serious; **damit ist es mir ~ernst!** I am deadly serious; **2.** *adv.* **ich meine das ~ernst** I mean it deadly seriously; **~kalt** *Adj.; präd. getrennt geschr.* bitterly cold

Bitterkeit die; ~ *(auch fig.)* bitterness

bitterlich 1. *Adj.* slightly bitter ⟨*taste*⟩. **2.** *adv. (heftig)* ⟨*cry, complain, etc.*⟩ bitterly

Bitter·mandel die bitter almond

bitter·süß *Adj. (auch fig.)* bittersweet

Bitt-: ~gang der: einen ~ zu jmdm. machen go to sb. with a request; **~gesuch das** petition; **~schrift die** petition; **~steller** [~ʃtɛlɐ] **der; ~s, ~:** petitioner

Biwak ['bi:vak] **das; ~s, ~s** *(bes. Milit., Bergsteigen)* bivouac

bizarr [bi'tsar] **1.** *Adj.* bizarre; *(phantastisch)* fantastic. **2.** *adv.* bizarrely

Bizeps ['bi:tsɛps] **der; ~[es], ~e** biceps

BKA [be:ka:'|a:] **das; ~[s]** *Abk.:* **Bundeskriminalamt**

Blackout ['blɛkaut] **das** *od.* **der; ~[s], ~s** black-out

blaffen ['blafn̩], **bläffen** ['blɛfn̩] *itr. V.* **a)** bark; give a short bark; *(kläffen)* yap; **b)** *(schimpfen)* snap

blähen ['blɛ:ən] **1.** *tr. V.* **a)** billow, fill, belly [out] ‹*sail*›; billow ‹*sheet, curtain, clothing*›; **b)** *(aufblasen)* flare ‹*nostrils*›. **2.** *refl. V.* ‹*sail*› billow *or* belly out; ‹*nostrils*› dilate. **3.** *itr. V. (Blähungen verursachen)* cause flatulence *or* wind; **~de Speisen** flatulent foods

Blähung die; ~, ~en flatulence *no art., no pl.;* wind *no art., no pl.;* **~en** flatulence *sing.;* wind *sing.*

blamabel [bla'ma:bl̩] **1.** *Adj.* shameful, disgraceful ‹*behaviour etc.*›. **2.** *adv.* shamefully; disgracefully

Blamage [bla'ma:ʒə] **die; ~, ~n** disgrace

blamieren [bla'mi:rən] **1.** *tr. V.* disgrace. **2.** *refl. V.* disgrace oneself; *(sich lächerlich machen)* make a fool of oneself

blanchieren [blã'ʃi:rən] *tr. V. (Kochk.)* blanch

blank [blaŋk] *Adj.* **a)** *(glänzend)* shiny; **etw. ~ reiben/polieren** rub/polish sth. till it shines; **b)** *(unbekleidet)* bare; naked; **c)** *(ugs.: mittellos)* **~ sein** be broke *(coll.);* **d)** *(bloß)* bare ‹*wood, plaster, earth, etc.*›; **e)** *(rein)* pure; sheer; utter ‹*mockery*›

Blanko- ['blaŋko-]: **~scheck der** *(Wirtsch., fig.)* blank cheque; **~vollmacht die** *(Wirtsch., fig.)* carte blanche

blank·poliert *Adj.; präd. getrennt geschr.* brightly polished

Blank·vers der blank verse

Bläschen ['blɛ:sçən] **das; ~s, ~ a)** [small] bubble; **b)** *(in der Haut)* [small] blister

Blase ['bla:zə] **die; ~, ~n a)** bubble; *(in einem Anstrich)* blister; **~n werfen** *od.* **ziehen** ‹*paint*› blister; ‹*wallpaper*› bubble; **b)** *(in der Haut)* blister; **sich** *(Dat.)* **~n laufen** get blisters [from walking/running]; **c)** *(Harn~)* bladder; **d)** *(salopp: Leute)* mob *(sl.)*

Blase·balg der bellows *pl.;* pair of bellows

blasen 1. *unr. itr. V.* **a)** blow; **b) auf dem Kamm ~:** play the comb; **c) zum Angriff/Rückzug/Aufbruch ~:** sound the charge/retreat/departure; **d)** *(wehen)* ‹*wind*› blow; *unpers.:* **es bläst** it's windy *or* blowy. **2.** *unr. tr. V.* **a)** blow; **b)** *(spielen)* play ‹*musical instrument, tune, melody, etc.*›; **c)** *(wehen)* ‹*wind*› blow

Blasen-: ~bildung die blistering **~katarrh der** *(Med.)* cystitis *no indef. art.*

Bläser ['blɛ:zɐ] **der; ~s, ~** *(Musik)* wind player

blasiert [bla'zi:ɐt] *(abwertend)* **1.** *Adj.* blasé. **2.** *adv.* in a blasé way

Blas-: ~instrument das wind instrument; **~kapelle die** brass band; **~musik die** brass-band music; *(~kapelle)* brass band; **~orchester das** brass band

Blasphemie [blasfe'mi:] **die; ~, ~n** blasphemy

blasphemisch *Adj.* blasphemous

Blas·rohr das blowpipe

blaß [blas] **1.** *Adj.* **a)** pale; *(fig.)* colourless ‹*account, portrayal, etc.*›; **~ werden** turn *or* go pale; **Rot macht dich ~:** red makes you look pale [in the face]; **~ vor Neid sein/werden** *(fig.)* be/turn *or* go green with envy; **b)** *(schwach)* faint ‹*recollection, hope*›. **2.** *adv.* palely

Blässe ['blɛsə] **die; ~:** paleness

Bläß·huhn ['blɛs-] **das** coot

bläßlich 1. *Adj.* **a)** rather pale; palish; **b)** *(fig.)* colourless ‹*person, account, portrayal, etc.*›. **2.** *adv. (fig.)* colourlessly

Blatt [blat] **das; ~[e]s, Blätter** ['blɛtɐ] **a)** *(von Pflanzen)* leaf; **kein ~ vor den Mund nehmen** not mince one's words; **b)** *(Papier)* sheet; **ein ~ Papier** a sheet of paper; **[noch] ein unbeschriebenes ~ sein** *(ugs.) (unerfahren sein)* be inexperienced; *(unbekannt sein)* be an unknown quantity; **c)** *(Buchseite usw.)* page; leaf; **etw. vom ~ spielen** sight-read sth.; **auf einem anderen ~ stehen** *(fig.)* be [quite] another *or* a different matter; **d)** *(Zeitung)* paper; **e)** *(Spielkarten)* hand; **f)** *(am Werkzeug, Ruder)* blade; **g)** *(Graphik)* print

Blättchen ['blɛtçən] **das; ~s, ~ a)** *(von Pflanzen)* [small] leaf; **b)** *(Papier)* [small] sheet; **c)** *(abwertend: Zeitung)* rag *(derog.)*

Blattern *Pl.* smallpox *sing.*

blättern ['blɛtɐn] **1.** *itr. V.* **in einem Buch ~:** leaf through a book. **2.** *tr. V.* put down [one by one]; **er blätterte mir 50 Mark auf den Tisch** he counted me out fifty marks in notes on the table

Blätter·teig der puff pastry

Blatt-: ~gold das; *o. Pl.* gold leaf; **~laus die** aphid; greenfly; **~pflanze die** foliage plant; **~säge die** wide-bladed [hand]-saw; **~salat der** green salad

blau [blau] *Adj.* blue; **ein ~es Auge** *(ugs.)* a black eye; **mit einem ~en Auge davonkommen** *(fig. ugs.)* get off fairly lightly; **ein ~er Fleck** a bruise; **ein ~er Brief** *(ugs.) (Kündigung)* one's cards *pl.; (Schulw.) letter informing parents that their child is in danger of having to repeat a year;* **einen ~en Montag einlegen** *od.* **machen** *(ugs.)* skip work on Monday; **sein ~es Wunder erleben** *(ugs.)* get a nasty surprise; **~ sein** *(fig. ugs.)* be tight *(coll.) or* canned *(sl.)*

Blau das; ~s, ~ *od. (ugs.:)* **~s** blue

blau-, Blau-: ~alge die blue-green alga; **~äugig** *Adj.* **a)** blue-eyed; **b)** *(naiv)* naive; **~beere die** bilberry; whortleberry; **~blütig** *Adj. (meist iron.)* blue-blooded

[1]**Blaue** ['blauə] **das; ~n** blue; **das ~ vom Himmel [herunter]lügen** *(ugs.)* lie like anything; tell a pack of lies; **wir wollen einfach ins ~ fahren** we'll just set off and see where we end up; *s. auch* **Fahrt c**

[2]**Blaue der; ~n, ~n** *(ugs.)* hundred-mark note

Bläue ['blɔyə] **die; ~** *(geh.)* blue; blueness; *(des Himmels)* blue

blau-, Blau-: ~filter der *od.* **das** *(Fot.)* blue filter; **~grau** *Adj.* blue-grey; bluish grey; **~grün** *Adj.* blue-green; bluish green; **~kraut das** *(südd., österr.) s.* **Rotkohl**

bläulich *Adj.* bluish

blau-, Blau-: ~licht das flashing blue light; **ein Krankenwagen raste mit ~ vorbei** an ambulance raced past with [its] blue light flashing; **~|machen** *itr. V. (ugs.)* skip work; **~mann der;** *Pl.* **~männer** *(ugs.)* boiler suit; **~meise die** blue tit; **~papier das** [blue] carbon paper; **~pause die** blueprint; **~rot** *Adj.* purple; **~säure die;** *o. Pl. (Chemie)* prussic acid; hydrocyanic acid; **~stichig** [~ʃtɪçɪç] *Adj. (Fot.)* with a blue cast *postpos., not pred.;* **~sti-**

chig sein have a blue cast; **~strumpf der** *(abwertend)* bluestocking; **~tanne die** blue spruce; Colorado spruce; **~wal der** blue whale

Blazer ['ble:zɐ] **der; ~s, ~:** blazer

Blech [blɛç] **das; ~[e]s, ~e a)** sheet metal; *(Stück Blech)* metal sheet; **b)** *(Back~)* [baking] tray; **c)** *o. Pl. (ugs.: Unsinn)* rubbish; nonsense; tripe *(sl.)*

Blech-: ~blas·instrument das brass instrument; **~büchse die, ~dose die** can; tin *(Brit.)*

blechen *tr., itr. V. (ugs.)* cough up *(sl.)*; fork out *(sl.)*

blechern ['blɛçɐn] *Adj.* **1. a)** *nicht präd. (aus Blech)* metal; **b)** *(metallisch klingend)* tinny ⟨*sound, voice*⟩. **2.** *adv. (metallisch)* tinnily

Blech-: ~kiste die *(ugs. abwertend)* crate *(sl.)*; **~musik die** *(abwertend)* brass-band music; **~napf der** metal bowl

Blechner der; ~s, ~ *(südd.) s.* **Klempner**

Blech-: ~schaden der *(Kfz-W.)* damage *no indef. art.* to the bodywork; **~trommel die** tin drum

blecken ['blɛkn̩] *tr. V.* **die Zähne ~:** bare one's/its teeth

[1]Blei [blai̯] **das; ~[e]s, ~e** lead

[2]Blei der *od.* **das; ~[e]s, ~e** *(ugs.: ~stift)* pencil

Bleibe die; ~, ~n place to stay; **keine ~ haben** have nowhere to stay

bleiben ['blai̯bn̩] *unr. itr. V.; mit sein* **a)** stay; remain; **~ Sie bitte am Apparat** hold the line please; **wo bleibt er so lange?** where has he got to?; **wo bleibst du denn so lange?** where have you been *or* what's been keeping you all this time?; **zum Abendessen ~:** stay for supper; **auf dem Weg ~:** keep to *or* stay on the path; **bei etw. ~** *(fig.: an etw. festhalten)* keep *or* stick to sth.; **jmdm. in Erinnerung** *od.* **im Gedächtnis ~:** stay in sb.'s mind *or* memory; **das bleibt unter uns** *(Dat.)* that's [just] between ourselves; **zusehen können, wo man bleibt** *(ugs.)* have to fend for oneself; **jmdn. zum Bleiben auffordern** ask sb. to stay; **im Feld/im Krieg/auf See ~** *(verhüll. geh.)* die *or* fall in action/die in the war/die at sea; **der Kuchen bleibt mehrere Tage frisch** the cake will keep for several days; **bleib ruhig!** keep calm!; **das Geschäft bleibt heute geschlossen** the shop is closed today; **unbestraft/unbemerkt ~:** go unpunished/go unnoticed *or* escape notice; **sitzen ~:** stay *or* remain sitting down *or* seated; **dabei bleibt es!** *(ugs.: daran wird nichts mehr geändert)* that's that; that's the end of it; **b) das bleibt abzuwarten** that remains to be seen; **es bleibt zu hoffen, daß ...:** we can only hope that ...; **c)** *(übrigbleiben)* be left; remain; **uns** *(Dat.)* **bleibt noch Zeit** we still have time; **es blieb ihm keine Hoffnung mehr** he had no hope left

bleibend *Adj.* lasting; permanent ⟨*damage*⟩

bleiben|lassen *unr. tr. V.* **a)** *(nicht tun)* **etw. ~:** give sth. a miss; forget sth.; **b)** *(aufhören)* **das Rauchen ~:** give up *or* stop smoking

bleich [blai̯ç] *Adj.* pale; **~ werden** turn *or* go pale; *(vor Angst, Schreck)* pale; turn *or* go pale

[1]bleichen *tr. V.* bleach

[2]bleichen *regelm. (veralt. auch unr.) itr. V.* become bleached; bleach; **in der Sonne ~:** be bleached by the sun

Bleich-: ~gesicht das *Pl.* **~gesichter** *(scherzh.: Weißer)* paleface; **~mittel das** bleach; bleaching agent

bleiern ['blai̯ɐn] **1.** *Adj.* **a)** *nicht präd.* lead; **b)** *(geh.: bleifarben)* leaden ⟨*sky, grey*⟩; **c)** *(schwer)* heavy ⟨*sleep, tiredness, etc.*⟩; leaden ⟨*heaviness*⟩

blei-, Blei-: ~frei *Adj.* unleaded ⟨*fuel*⟩; **~frei das; ~s** unleaded; **~gießen das** *pouring lead into cold water to tell one's fortune for the coming year;* **~kristall das** lead crystal; **~kugel die** lead ball; *(Geschoß)* lead bullet; **~schwer 1.** *Adj.* heavy as lead *postpos.*; **2.** *adv.* heavily; like a heavy *or* lead weight; **~soldat der** lead soldier

Blei·stift der pencil; **mit ~:** in pencil

Bleistift-: ~absatz der stiletto heel; **~mine die** [pencil] lead; **~spitzer der** pencil-sharpener

Blende die; ~, ~n a) *(Lichtschutz)* shade; *(im Auto)* [sun-]visor; **b)** *(Optik, Film, Fot.)* diaphragm; **die ~ öffnen/schließen** open up the aperture/stop down; **c)** *(Film, Fot.: Blendenzahl)* aperture setting; f-number; **mit** *od.* **bei ~ 8** at [an aperture setting of] f/8

blenden 1. *tr. V.* **a)** *(auch fig.)* dazzle; **b)** *(blind machen)* blind. **2.** *itr. V.* ⟨*light*⟩ be dazzling

blendend 1. *Adj.* splendid; brilliant ⟨*musician, dancer, speech, achievement, etc.*⟩; **es geht mir ~:** I feel wonderfully well *or* wonderful. **2.** *adv.* **wir haben uns ~ amüsiert** we had a wonderful *or* marvellous time

Blendung die; ~, ~en a) dazzling; **b)** *(das Blindmachen)* blinding

Blesse ['blɛsə] **die; ~, ~n** blaze

blich [blɪç] *1. u. 3. Pers. Sg. Prät. v.* **[2]bleichen**

Blick [blɪk] **der; ~[e]s, ~e a)** look; *(flüchtig)* glance; **jmdm. einen ~/sich ~e zuwerfen** give sb. a look/exchange glances; **einen kurzen ~ auf etw.** *(Akk.)* **werfen** take a quick look at *or* glance [briefly] at sth.; **auf den ersten ~:** at first glance; **auf den zweiten ~:** looking at it again *or* a second time; **mein ~ fiel auf den Brief** my eye fell on the letter; the letter caught my eye; **b)** *o. Pl. (Ausdruck)* look in one's eyes; **mit mißtrauischem ~:** with a suspicious look in one's eye; **c)** *(Aussicht)* view; **ein Zimmer mit ~ aufs Meer** a room with a sea view; **jmdn./etw. aus dem ~ verlieren** lose sight of sb./sth.; **etw. im ~ haben** be able to see sth.; **d)** *o. Pl. (Urteil[skraft])* eye; **einen sicheren/geschulten ~ für etw. haben** have a sure/trained eye for sth.; **keinen ~ für etw. haben** have no eye for sth.

blicken 1. *itr. V.* look; *(flüchtig)* glance; **jmdm. gerade in die Augen ~:** look sb. straight in the eye. **2.** *tr. V.*: *in* **sich ~ lassen** put in an appearance; **laß dich mal wieder ~:** come again some time

Blick-: ~fang der eye-catcher; **~feld das** field of vision *or* view; **~punkt der** view; field of vision; **jmdn. in den ~punkt rücken** *(fig.)* single sb. out; **~richtung die** line of sight *or* vision; **~winkel der a)** angle of vision; **b)** *(fig.)* point of view; viewpoint; perspective

blieb [bli:p] *1. u. 3. Pers. Sg. Prät. v.* **bleiben**

blies [bli:s] *1. u. 3. Pers. Sg. Prät. v.* **blasen**

blind [blɪnt] **1.** *Adj.* **a)** blind; **~ werden** go blind; **auf einem Auge ~ sein** be blind in one eye; **~ für etw. sein** be blind to sth.; **b)** *(maßlos)* blind ⟨*rage, hatred, fear, etc.*⟩; indiscriminate ⟨*violence*⟩; **c)** *(kritiklos)* blind ⟨*obedience, enthusiasm, belief, etc.* ⟩; **d)** *(trübe)* clouded ⟨*glass*⟩; dull, tarnished ⟨*metal*⟩; **e)** *(verdeckt)* concealed; invisible ⟨*seam*⟩; **ein ~er Passagier** a stowaway; **f) ~er Alarm** a false alarm; **g) der ~e Zufall** pure *or* sheer chance. **2.** *adv.* **a)** *(ohne* *hinzusehen)* without looking; *(wahllos)* blindly; wildly; **b)** *(unkritisch)* ‹*trust*› implicitly; ‹*obey*› blindly

Blind·darm der a) *(Anat.)* caecum; **b)** *(volkst.: Wurmfortsatz)* appendix

Blind·darm-: ~entzündung die *(volkst.)* appendicitis; **~operation die** *(volkst.)* appendix operation

Blinde der/die; *adj. Dekl.* blind person; blind man/woman; **die ~n** the blind; **das sieht doch ein ~r [mit dem Krückstock]** *(ugs.)* anyone *or* any fool can see that

Blinde·kuh *o. Art.* blind man's buff

Blinden-: ~hund der guide-dog; **~schrift die** Braille

Blind·gänger der a) *(Geschoß)* unexploded shell; dud *(sl.)*; **b)** *(salopp: Versager)* dead loss *(coll.)*

Blindheit die; ~ *(auch fig.)* blindness

blindlings ['blɪntlɪŋs] *Adv.* blindly; ‹*trust*› implicitly

Blind·schleiche [~ʃlaiçə] **die;** ~, **~n** slowworm; blindworm

blind·wütig 1. *Adj.* raging ‹*anger, hatred, fury, etc.*›; wild ‹*rage*›. **2.** *adv.* in a blind rage *or* fury

blinken ['blɪŋkn̩] **1.** *itr. V.* **a)** ‹*light, glass, crystal*› flash; ‹*star*› twinkle; ‹*metal, fish*› gleam; **b)** *(Verkehrsw.)* indicate. **2.** *tr. V.* flash; **SOS ~:** flash an SOS [signal]

Blinker der; **~s,** ~ **a)** *(am Auto)* indicator [light]; winker; **b)** *(Angeln)* spoon[-bait]

Blink-: ~feuer das *(Seew.)* flashing light; **~licht das** *(Verkehrsw.)* flashing light; *(Blinker)* indicator light; **~licht·anlage die** *(Verkehrsw.)* flashing lights *pl.* **~zeichen das** flashlight signal

blinzeln ['blɪntsln̩] *itr. V.* blink; *(mit einem Auge, um ein Zeichen zu geben)* wink

Blitz [blɪts] **der; ~es, ~e a)** lightning *no indef. art.*; **ein ~:** a flash of lightning; **der ~ hat eingeschlagen** lightning has struck; **[schnell] wie der ~:** like lightning; as fast as lightning; **wie ein geölter ~** *(ugs.)* like greased lightning; **wie ein ~ aus heiterem Himmel** like a bolt from the blue; **b)** *(Fot.)* flash

blitz-, Blitz-: ~ab·leiter der lightning-conductor; **~aktion die** lightning operation; **~angriff der** *(Milit.)* lightning attack; **~artig 1.** *Adj.; nicht präd.* lightning; **2.** *adv.* like lightning; ‹*disappear*› in a flash; **~blank** *Adj.* *(ugs.)* **~blank [geputzt]** sparkling clean; brightly polished ‹*shoes*›

blitzeblank *s.* **blitzblank**

blitzen 1. *itr. V.* **a)** *unpers.* **es blitzte** *(einmal)* there was a flash of lightning; *(mehrmals)* there was lightning; there were flashes of lightning; **b)** *(glänzen)* ‹*light, glass, crystal*› flash; ‹*metal*› gleam; **das Haus blitzte vor Sauberkeit** the house was sparkling clean; **c)** *(ugs.: mit Blitzlicht)* use [a] flash. **2.** *tr. V.* *(ugs.)* take a flash photo of

blitz-, Blitz-: ~gerät das *(Fot.)* flash [unit]; flash-gun; **~krieg der** *(Milit.)* blitzkrieg; **~licht das;** *Pl.* **~lichter** flash[light]; **~sauber** *Adj.* sparkling clean; **~schlag der** flash of lightning; **von einem ~schlag getroffen werden** be struck *or* hit by lightning; **~schnell 1.** *Adj.* lightning *attrib.*; **~schnell sein** be like lightning; **2.** *adv.* like lightning; ‹*disappear*› in a flash; **~sieg der** *(Milit.)* lightning victory; **~start der** lightning start; **~würfel der** *(Fot.)* flash-cube

Block [blɔk] **der; ~[e]s, Blöcke** ['blœkə] *od.* **~s a)** *Pl.* **Blöcke** *(Brocken)* block; *(Fels~)* boulder; **b)** *(Wohn~)* block; **c)** *Pl.* **Blöcke** *(Gruppierung von politischen Kräften, Staaten)* bloc; **d)** *(Schreib~)* pad

Blockade [blɔ'ka:də] **die; ~, ~n** blockade

block-, Block-: ~buchstabe der block capital *or* letter; **~flöte die** recorder; **~frei** *Adj.* non-aligned ‹*country, state*›; **die Blockfreien** the non-aligned countries *or* states; **~haus das, ~hütte die** log cabin

blockieren 1. *tr. V.* block; jam ‹*telephone line*›; stop, halt ‹*traffic*›; lock ‹*wheel, machine, etc.*›. **2.** *itr. V.* ‹*wheels*› lock; ‹*gears*› jam

Block-: ~schokolade die cooking chocolate; **~schrift die** block capitals *pl. or* letters *pl.*; **~stunde die** *(Schulw.)* double period

blöd[e] ['blø:t, 'blø:də] **1.** *Adj.* **a)** *(schwachsinnig)* mentally deficient; imbecilic; **b)** *(ugs.: dumm)* stupid; idiotic *(coll.)*; **c)** *(ugs.: unangenehm)* stupid; **das Blöde ist nur, daß ...:** the stupid thing is that ... **2.** *adv.* **a)** *(schwachsinnig)* imbecilically; **b)** *(ugs.: dumm)* stupidly; idiotically *(coll.)*; **frag doch nicht so ~:** don't ask such stupid *or (coll.)* idiotic questions; **c)** *(ärgerlich)* stupidly

Blödel ['blø:dl̩] **der; ~s,** ~ *(ugs. abwertend) s.* **Blödian**

Blödelei die; ~, ~en a) *o. Pl.* messing *or* fooling about *no indef. art.*; **b)** *(Äußerung)* silly joke

blödeln *itr. V.* **a)** mess *or* fool about; **b)** *(alberne Witze machen)* make silly jokes

blöder·weise *Adv.* *(ugs.)* stupidly

Blöd·hammel der *(salopp abwertend)* stupid fool *or (coll.)* idiot *or (Brit. sl.)* twit *or (Amer. sl.)* jerk

Blödheit die; ~, ~en a) *o. Pl.* *(Dummsein)* stupidity; **b)** *(dumme Äußerung)* stupid remark; *(dumme Tat)* stupidity; **c)** *o. Pl.* *(Schwachsinnigkeit)* mental deficiency; imbecility

Blödian ['blø:dia:n] **der; ~s, ~e** *(ugs. abwertend)* idiot *(coll.)*; fool

blöd-, Blöd-: ~mann der; *Pl.* **~männer** *(salopp)* stupid idiot *(coll.) or* fool; **~sinn der;** *o. Pl.* *(ugs. abwertend)* nonsense; **jetzt habe ich ~sinn gemacht** now I've [gone and] messed it up; **mach doch keinen ~sinn!** don't be stupid; **~sinnig 1.** *Adj.* **a)** *(ugs.)* stupid; idiotic *(coll.)*; **b)** *(schwachsinnig)* mentally deficient; imbecilic; **2.** *adv.* *(ugs.)* stupidly; idiotically *(coll.)*

blöken ['blø:kn̩] *itr. V.* ‹*sheep*› bleat; ‹*cattle*› low

blond [blɔnt] *Adj.* fair-haired, blond ‹*man, race*›; blonde, fair-haired ‹*woman*›; blond/blonde, fair ‹*hair*›

Blond das; ~s blond; *(von Frauenhaar)* blonde

blond·gelockt *Adj.; präd. getrennt geschr.* fair curly *attrib.* ‹*hair*›; ‹*girl, child, etc.*› with fair curly hair

blondieren *tr. V.* bleach; *(mit Färbemittel)* dye blond/blonde; **sich ~ lassen** have one's hair bleached/dyed blond/blonde

Blondine [blɔn'di:nə] **die; ~, ~n** blonde

bloß [blo:s] **1.** *Adj.* **a)** *(nackt)* bare; naked; **mit ~em Oberkörper** stripped to the waist; **mit ~em Kopf** bare-headed; **mit ~en Händen** with one's bare hands; **b)** *(nichts als)* mere ‹*words, promises, triviality, suspicion, etc.*›; **der ~e Gedanke daran** the mere *or* very thought of it; **ein ~er Zufall** mere *or* pure chance; **~es Gerede** mere gossip. **2.** *Adv.* *(ugs.: nur)* only. **3.** *Partikel* **was hast du dir ~ dabei gedacht?** what on earth *or* whatever were you thinking of?; **wie konnte das ~ geschehen?** how on earth did it happen?; **wenn ich**

das ~ wüßte! if only I knew!; **sei ~ pünktlich!** just make sure you're on time

Blöße ['blø:sə] **die; ~, ~n a)** *(geh.: Nacktheit)* nakedness; **b)** *in* **sich** *(Dat.)* **eine/keine ~ geben** show a/not show any weakness

bloß-: ~|legen *tr. V.* uncover; expose; *(fig.)* expose; reveal ‹*error, defect, etc.*›; **~|liegen** *unr. itr. V.; mit sein* be uncovered *or* exposed; **~|stellen** *tr. V.* show up; **~|strampeln** *refl. V.* kick the *or* one's covers off

Blouson [blu'zõ:] **das** *od.* **der; ~|s|, ~s** blouson; bomber jacket

blubbern ['blʊbɐn] *itr. V. (ugs.)* bubble

Blücher ['blʏçɐ] *in* **er/sie geht ran wie ~** *(ugs.)* he/she really goes hard at it

Bluejeans, Blue jeans ['blu:-dʒi:ns] *Pl. od.* **die; ~, ~:** [blue] jeans *pl.*; denims *pl.*

Blues [blu:s] **der; ~, ~** *(Musik, Tanz)* blues; **der ~:** the blues *sing. or pl.*

Bluff [blʊf] **der; ~s, ~s** bluff

bluffen *tr., itr. V.* bluff

blühen ['bly:ən] *itr. V.* **a)** ‹*plant*› flower, bloom, be in flower *or* bloom; ‹*flower*› bloom, be in bloom, be out; ‹*tree*› be in blossom; **blau ~:** have blue flowers; **~de Gärten** gardens full of flowers; **es blüht** there are flowers in bloom; **b)** *(florieren)* flourish; thrive; **c)** *(ugs.: bevorstehen)* **jmdm. ~:** be in store for sb.; **das kann dir auch noch ~:** the same may *or* could happen to you; **sonst blüht dir was!** otherwise you'll catch it!

blühend *Adj.* **a)** *(frisch, gesund)* glowing ‹*colour, complexion, etc.*›; radiant ‹*health*›; **sie starb im ~en Alter von 20 Jahren** she died at 20, in the full bloom of youth; **b)** *(übertrieben)* vivid, lively ‹*imagination*›

Blümchen ['bly:mçən] **das; ~s, ~:** [little] flower

Blume ['blu:mə] **die; ~, ~n a)** flower; **etw. durch die ~ sagen** say sth. in a roundabout way; **b)** *(des Weines)* bouquet; **c)** *(des Biers)* head

blumen-, Blumen-: ~beet das flower-bed; **~erde die** potting compost; **~frau die** flower-woman; **~garten der** flower-garden; **~geschäft das** florist's; flower-shop; **~geschmückt** *Adj.* flower-bedecked; adorned with flowers *postpos.*; **~kasten der** flower-box; *(vor einem Fenster)* window box; **~kohl der** cauliflower; **~laden der** *s.* **~geschäft; ~mädchen das** flower-girl; **~muster das** floral pattern; **~rabatte die** flower-border; herbaceous border; **~stock der** [flowering] pot plant; **~strauß der** bunch of flowers; *(Bukett)* bouquet of flowers; **~topf der** flowerpot; **~vase die** [flower] vase; **~zwiebel die** bulb

blümerant [blymə'rant] *Adj.* queasy

Bluse ['blu:zə] **die; ~, ~n** blouse

Blut [blu:t] **das; ~|e|s** blood; **gleich ins ~ gehen** pass straight into the bloodstream; **es wurde viel ~ vergossen** there was a great deal of bloodshed; **den Zuschauern gefror** *od.* **stockte** *od.* **gerann das ~ in den Adern** *(fig.)* the spectators' blood ran cold; **an jmds. Händen klebt ~** *(fig. geh.)* there is blood on sb.'s hands *(fig.)*; **blaues ~ in den Adern haben** *(fig.)* have blue blood in one's veins *(fig.)*; **böses ~ machen** *od.* **schaffen** *(fig.)* cause *or* create bad blood; **~ und Wasser schwitzen** *(fig. ugs.)* sweat blood *(fig. coll.)*; **[nur/immer] ruhig ~!** *(ugs.)* keep your hair on! *(Brit. sl.)*; keep your cool! *(coll.)*; **jmdn. bis aufs ~ quälen** *od.* **peinigen** *(fig.)* torment sb. mercilessly; **jmdm. im ~ liegen** *(fig.)* be in sb.'s blood *(fig.)*

blut-, Blut-: ~alkohol der blood alcohol level; **~apfelsine die** *s.* **~orange; ~arm** *Adj. (Med.)* anaemic; **~armut die** *(Med.)* anaemia; **~bad das** blood-bath; **~bahn die** bloodstream; **~befleckt** *Adj.* bloodstained; **seine Hände sind ~befleckt** *(fig.)* he has blood on his hands; **~beschmiert** *Adj.* smeared with blood *postpos.*; **~bild das** *(Med.)* blood picture; **~buche die** copper beech; **~druck der;** *o. Pl.* blood pressure

Blüte ['bly:tə] **die; ~, ~n a)** flower; bloom; *(eines Baums)* blossom; **~n treiben** flower; bloom; ‹*tree*› blossom; **b)** *(das Blühen)* flowering; blooming; *(Baum~)* blossoming; **in [voller] ~ stehen** be in [full] flower *or* bloom/blossom; **c)** *(fig. geh.)* **seine ~ erreichen** ‹*culture*› reach its full flowering; **die Renaissance war für die Kunst eine Zeit der ~:** art flourished during the Renaissance; **d)** *(ugs.: falsche Banknote)* dud note *(sl.)*

Blut·egel der leech

bluten *itr. V.* **a)** bleed (**aus** from); **b)** *(ugs.: viel bezahlen)* **[ganz schön] ~:** cough up *(sl.) or* fork out *(sl.)* a[n awful] lot of money

blüten-, Blüten-: ~blatt das petal; **~honig der** blossom honey; **~kelch der** *(Bot.)* calyx; **~knospe die** flower-bud; **~pflanze die** *(Bot.)* flowering plant; **~stand der** *(Bot.)* inflorescence; **~staub der** *(Bot.)* pollen; **~weiß** *Adj.* sparkling white

Bluter ['blu:tɐ] **der; ~s, ~** *(Med.)* haemophiliac

Blut·erguß der haematoma; *(blauer Fleck)* bruise

Bluter·krankheit die haemophilia *no art.*

Blüte·zeit die a) die ~ der Geranien ist von Mai bis Oktober geraniums flower *or* are in flower from May to October; **während der ~ der Obstbäume** when the fruit-trees are/were in blossom; **b)** *(fig.)* heyday; **seine ~ erleben** ‹*culture, empire*› be in its heyday

Blut-: ~fleck[en] der bloodstain; **~gefäß das** *(Anat.)* blood-vessel; **~gerinnsel das** blood clot; **~gruppe die** *(Med.)* blood group; blood type; **jmds. ~gruppe bestimmen** blood-type sb.; type sb.'s blood; **er hat ~gruppe 0** he is blood group 0; **~hochdruck der** *(Med.)* high blood pressure; **~hund der** bloodhound

blutig a) bloody; **jmdn. ~ schlagen** beat sb. to a pulp; **~ geschlagen werden** be left battered and bleeding; **b)** *nicht präd. (fig. ugs.)* absolute, complete ‹*beginner, layman, etc.*›

blut-, Blut-: ~jung *Adj.* very young; **~konserve die** *(Med.)* container of stored blood; **~konserven** stored blood; **~körperchen das** *(Anat.)* blood corpuscle; **rote/weiße ~körperchen** red/white corpuscles; **~krebs der** *(Med.)* leukaemia; **~kreislauf der** *(Physiol.)* blood circulation; **~lache die** pool of blood; **~leer** *Adj. (auch fig.)* anaemic; **~orange die** blood orange; **~plasma das** *(Physiol.)* blood plasma; **~probe die** *(Med.)* **a)** blood sample; **b)** *(~untersuchung)* blood test; **~rache die** blood revenge; blood vengeance; **~rausch der** *(geh.)* murderous frenzy; **~rot** *Adj.* blood-red; **~rünstig** [-rʏnstɪç] **1.** *Adj.* bloodthirsty; **2.** *adv.* bloodthirstily; **~sauger der** *(auch fig.)* bloodsucker

Bluts·brüderschaft die blood brotherhood

blut-, Blut-: ~schande die incest; **~serum das** *(Physiol.)* blood serum; **~spende die** *(das Spenden)* giving *no indef. art.* of blood; donation of blood; *(~menge)* blood-donation; **~spender der** blood donor; **~spur die a)** trail of blood; **b)** *Pl. (auf Kleidung o. ä.)* traces of blood; **~stillend** *Adj.* styptic; **~stillende Mittel** styptics

bluts-, Bluts-: ~tropfen der drop of blood; **~verwandt** *Adj.* related by blood *postpos.*; **sie sind nicht ~verwandt** they are not blood relations; **~verwandtschaft die** blood relationship

blut-, Blut-: ~tat die *(geh.)* bloody deed; **~transfusion die** blood-transfusion; **~triefend** *Adj.; nicht präd.* dripping with blood *pred.*; **~überströmt** *Adj.* streaming with blood *pred.*; covered in blood *pred.*

Blutung die; ~, ~en a) bleeding *no indef. art., no pl.*; haemorrhage; **innere/äußere ~en** internal/external bleeding *sing.*; **b)** *(Regel~)* period

blut-, Blut-: ~unterlaufen *Adj.* suffused with blood *postpos.*; bloodshot ⟨*eyes*⟩; **~untersuchung die** *(Med.)* blood test; **~vergießen das; ~s** bloodshed; **~vergiftung die** blood-poisoning *no indef. art., no pl.*; **~verlust der** loss of blood; **~verschmiert** *Adj.* blood-stained, smeared with blood *pred.*; **~wurst die** black pudding; **~zucker der** *(Physiol.)* blood sugar

b-Moll ['be:mɔl] **das; ~:** B flat minor

Bö [bø:] **die; ~, ~en** gust [of wind]

Bob [bɔp] **der; ~, ~s** bob[-sleigh]

Bob-: ~bahn die bob[-sleigh] run; **~fahrer der** bobber; **~rennen das** bob[-sleigh] racing; *(Veranstaltung)* bob [-sleigh] race

[1]**Bock** [bɔk] **der; ~[e]s, Böcke** ['bœkə] **a)** *(Reh~, Kaninchen~)* buck; *(Ziegen~)* billy-goat; he-goat; *(Schafs~)* ram; **stur wie ein ~ sein** *(ugs.)* be as stubborn as a mule; **einen ~ schießen** *(fig. ugs.)* boob *(Brit. sl.)*; make a boo-boo *(Amer. coll.)*; *(einen Fauxpas begehen)* drop a clanger *(sl.)*; **den ~ zum Gärtner machen** *(ugs.)* be asking for trouble; **einen/keinen ~ auf etw.** *(Akk.)* **haben** *(ugs.)* fancy/not fancy sth.; **b)** *(ugs.: Schimpfwort)* **der geile alte ~:** the randy old goat; **sturer ~!** you stubborn git *(sl. derog.)*; **c)** *(Gestell)* trestle; **d)** *(Turnen)* buck; **e)** *(Kutsch~)* box

[2]**Bock das; ~s, Bock·bier das** bock [beer]

bocken *itr. V.* **a)** *(nicht weitergehen)* refuse to go on; *(vor einer Hürde)* refuse; *(sich aufbäumen)* buck; rear; **b)** *(fam.: trotzig sein)* be stubborn and awkward; play up *(coll.)*; *(fig. ugs.)* ⟨*car*⟩ play up *(coll.)*

bockig 1. *Adj.* stubborn and awkward; contrary *(coll.)*. **2.** *adv.* stubbornly [and awkwardly]; contrarily *(coll.)*

Bock·mist der *(salopp)* bilge *no indef. art. (sl.)*; bullshit *no indef. art. (coarse)*

Bocks-: ~beutel der a) bocksbeutel; *wide, bulbous bottle for fine Franconian wines;* **b)** *(Wein) o. Pl.* bocksbeutel wine; **~horn das** *in* **sich ins ~horn jagen lassen** *(ugs.)* let oneself be browbeaten

Bock-: ~springen das *(Turnen)* vaulting [over the buck]; *(ohne Gerät)* leapfrog; **~sprung der** *(ungelenker Sprung)* [ungainly] jump *or* leap; **~sprünge machen** jump *or* leap about; **~wurst die** bockwurst

Boden ['bo:dn̩] **der; ~s, Böden** ['bø:dn̩] **a)** *(Erde)* ground; soil; **etw. [nicht] aus dem ~ stampfen können** [not] be able to conjure sth. up [out of thin air]; **b)** *(Fuß~)* floor; **zu ~ fallen/sich zu ~ fallen lassen** fall/drop to the ground; **der Boxer ging zu ~:** the boxer went down; **jmdn. zu ~ schlagen** *od. (geh.)* **strecken** knock sb. down; floor sb.; *(fig.)* **sich auf unsicherem ~ bewegen** be on shaky ground *(fig.)*; **am ~ liegen** be bankrupt; **am ~ zerstört [sein]** *(ugs.)* [be] shattered *(coll.)*; **bleiben wir doch auf dem ~ der Tatsachen!** let's stick to the facts; **c)** *o. Pl. (Terrain)* **heiliger ~:** holy ground; **feindlicher ~:** enemy territory; **auf französischem ~:** on French soil; **[an] ~ gewinnen/verlieren** gain/lose ground; **d)** *(unterste Fläche)* bottom; *(Hosen~)* seat; *(Torten~)* base; *s. auch* **doppelt**; **e)** *(Dach~)* loft; **auf dem ~:** in the loft

boden-, Boden-: ~belag der *(Teppich, Linoleum)* floor-covering; *(Fliesen, Parkett)* flooring; **~erosion die** soil erosion; **~frost der** ground frost; **~kammer die** attic; **~los** *Adj.* **a)** *(tief)* bottomless; **ins Bodenlose fallen** fall into a bottomless abyss; **b)** *(ugs.: unerhört)* incredible, unbelievable ⟨*foolishness, meanness, etc.*⟩; **~nebel der** ground mist; *(dichter)* ground fog; **~personal das** *(Flugw.)* ground staff; **~reform die** land reform; **~satz der** sediment; **~schätze** *Pl.* mineral resources

Boden·see der; *o. Pl.* Lake Constance

boden-, Boden-: ~ständig *Adj.* indigenous, native ⟨*culture, population, etc.*⟩; local ⟨*custom, craft, cuisine, tradition*⟩; ⟨*novel*⟩ rooted in the soil; **~station die** *(Raumf.)* ground station; **~streitkräfte** *Pl.*, **~truppen** *Pl.* ground forces *or* troops; **~turnen das** floor exercises *pl.*; **~vase die** large vase *(standing on the floor)*; **~welle die** bump

Bodybuilding [bɔdibɪldɪŋ] **das; ~s** body-building *no art.*

Bodycheck ['bɔditʃɛk] **der; ~s, ~s** *(Eishockey)* body-check

Böe ['bø:ə] **die; ~, ~n** *s.* **Bö**

Bofist ['bo:fɪst] **der; ~[e]s, ~e** puff-ball

bog [bo:k] *1. u. 3. Pers. Sg. Prät. v.* **biegen**

Bogen ['bo:gn̩] **der; ~s, ~,** *(südd., österr.:)* **Bögen** ['bø:gn̩] **a)** *(gebogene Linie)* curve; *(Math.)* arc; *(Skifahren)* turn; *(Schlittschuhlaufen)* curve; **einen ~ schlagen** move in a curve; **der Weg macht/beschreibt einen ~:** the path bends/the path describes a curve; **immer, wenn ich sie auf der Straße sehe, mache ich einen großen ~** *(fig. ugs.)* whenever I see her in the street I make a detour [round her]; **einen großen ~ um jmdn./etw. machen** *(fig. ugs.)* give sb./sth. a wide berth; **in hohem ~ hinausfliegen** *(fig. ugs.)* be chucked out *(sl.)*; **b)** *(Archit.)* arch; **c)** *(Waffe)* bow; **den ~ überspannen** *(fig.)* go too far; **d)** *(Musik: Geigen~ usw.)* bow; **e)** *(Papier~)* sheet; **ein ~ Packpapier** a sheet of wrapping-paper; **ein A4-~:** a sheet of A4 paper; **f)** *(Musik: Zeichen)* slur; *(bei gleicher Notenhöhe)* tie

bogen-, Bogen-: ~fenster das arched window; **~förmig** *Adj.* arched; **~gang der** arcade; **~schießen das** *(Sport)* archery *no art.*; **~schütze der** *(Sport)* archer

Boheme [bo'e:m] **die; ~:** bohemian world *or* society

Bohemien [boe'mi̯ɛ̃:] **der; ~s, ~s** bohemian

Bohle ['bo:lə] **die; ~, ~n** [thick] plank

Böhme ['bø:mə] **der; ~n, ~n** Bohemian

Böhmen ['bø:mən] *(das);* **~s** Bohemia

Böhmer·wald der Bohemian Forest
Böhmin die; ~, **~nen** Bohemian
böhmisch *Adj.* Bohemian
Böhnchen ['bø:nçən] **das; ~s, ~:** [small] bean
Bohne ['bo:nə] **die;** ~, **~n** bean; **grüne ~n** green beans; French beans *(Brit.)*; **dicke/weiße ~n** broad/haricot beans; **blaue ~n** *(scherzh.)* bullets; **nicht die ~** *(ugs.)* not one little bit
Bohnen-: ~ein·topf der bean stew; **~kaffee der** real coffee; **~kraut das** savory; **~salat der** bean salad; **~stange die** *(auch ugs.: Mensch)* beanpole; **~stroh** *in* **dumm wie ~stroh** *(ugs.)* as thick as two short planks *(coll.)*; **~suppe die** bean soup
bohnern *tr., itr. V.* polish; **„Vorsicht, frisch gebohnert!“** 'freshly polished floor/stairs *etc.*'
Bohner·wachs das floor-polish
bohren ['bo:rən] **1.** *tr. V.* **a)** bore; *(mit Bohrer, Bohrmaschine)* drill, bore ‹*hole*›; sink ‹*well, shaft*›; bore, drive ‹*tunnel*›; sink ‹*pole, post etc.*› **(in** + *Akk.* into); **b)** *(bearbeiten)* drill ‹*wood, concrete, etc.*›; **c)** *(drücken in)* poke **(in** + *Akk.* in[to]). **2.** *itr. V.* **a)** *(eine Bohrung vornehmen)* drill; **in der Nase ~:** pick one's nose; **nach Öl/Wasser** *usw.* **~:** drill for oil/water *etc.*; **b)** *(ugs.: drängen, fragen)* keep on; **jetzt hört auf zu ~:** now, don't keep on. **3.** *refl. V.* **sich in/durch etw. ~:** bore its way into/through sth.
bohrend *Adj.* **a)** gnawing ‹*pain, hunger, remorse*›; **b)** *(hartnäckig)* piercing ‹*look etc.*›; probing ‹*question*›
Bohrer der; ~s, ~: drill; *(zum Vorbohren)* gimlet
Bohr-: ~insel die drilling rig; **~maschine die** drill; **~meißel der** bit; **~turm der** derrick
Bohrung die; ~, ~en a) *s.* **bohren 1a:** boring; drilling; sinking; driving; **b)** *(Loch)* drill-hole
böig *Adj.* gusty
Boiler ['bɔylɐ] **der; ~s, ~:** water-heater
Boje ['bo:jə] **die;** ~, **~n** buoy
Bolero [bo'le:ro] **der; ~s, ~s** bolero
Bolivianer [boli'vi̯a:nɐ] **der; ~s, ~, Bolivianerin die;** ~, **~nen** Bolivian
bolivianisch *Adj.* Bolivian
Bolivien [bo'li:vi̯ən] **(das); ~s** Bolivia
Böller·schuß ['bœlɐ-] **der** gun salute; **der Admiral wurde mit fünf Böllerschüssen begrüßt** the admiral was greeted with a five-gun salute
Boll·werk das bulwark; *(fig.)* bulwark; bastion; stronghold
Bolschewik [bɔlʃe'vɪk] **der; ~en, ~i,** *(abwertend:)* **~en** Bolshevik
Bolschewismus [bɔlʃe'vɪsmʊs] **der; ~:** Bolshevism *no art.*
Bolschewist der; ~en, ~en Bolshevist
bolschewistisch *Adj.* Bolshevik
bolzen ['bɔltsn̩] *(ugs.) itr. V.* kick the ball about
Bolzen der; ~s, ~ a) pin; bolt; *(mit Gewinde)* bolt; **b)** *(Geschoß)* bolt
Bolzerei die; ~, **~en** *(ugs.)* [aimless] kick-about
Bolz·platz der [children's] football area
bombardieren [bɔmbar'di:rən] *tr. V.* **a)** *(Milit)* bomb; **b)** *(fig. ugs.)* bombard
Bombardierung die; ~, **~en a)** *(Milit.)* bombing; **b)** *(fig. ugs.)* bombardment
bombastisch *(abwertend)* **1.** *Adj.* bombastic ‹*speech, language, style, etc.*›; ostentatious ‹*architecture, theatrical, production*›. **2.** *adv.* ‹*speak, write*› bombastically; ostentatiously ‹*dressed*›
Bombe ['bɔmbə] **die;** ~, **~n a)** bomb; **die Nachricht schlug ein wie eine ~:** the news came as a bombshell; **die ~ ist geplatzt** *(fig. ugs.)* the balloon has gone up *(fig.)*; **b)** *(Sportjargon: Schuß)* thunderbolt; tremendous shot *(coll.)*
bomben-, Bomben-: ~angriff der bomb attack; bombing raid; **~an·schlag der, ~attentat das** bomb attack; **~drohung die** bomb threat; **~erfolg der** *(ugs.)* smash hit *(sl.)*; **~form die** *(ugs.)* top form; **~gehalt das** *(ugs.)* tremendous salary *(coll.)*; **~geschäft das** *(ugs.)* **ein ~geschäft machen** do a roaring trade; **~krater der** bomb crater; **~sicher** *Adj.* **a)** bomb-proof; **b)** ['--'--] *(ugs.: gewiß)* dead certain; **~stimmung die** *(ugs.)* tremendous *or* fantastic atmosphere *(coll.)*
Bomber der; ~s, ~ *(ugs.)* bomber
bombig *(ugs.) Adj.* super *(coll.)*; smashing *(coll.)*; terrific *(coll.)*; fantastic *(coll.)*
Bommel ['bɔml̩] **die;** ~, **~n** *od.* **der; ~s, ~** *(bes. nordd.)* bobble; pompom
Bon [bɔŋ] **der; ~s, ~s a)** voucher; coupon; **b)** *(Kassen~)* sales slip
Bonbon [bɔŋ'bɔŋ] **der** *od. (österr. nur)* **das; ~s, ~s** sweet; candy *(Amer.)*; *(fig.)* treat
bonbon-: ~farben, ~farbig *Adj. (abwertend)* candy-coloured
bongen ['bɔŋən] *(ugs.) tr. V.* ring up; **gebongt sein** *(fig.)* be fine
Bongo ['bɔŋgo] **das; ~[s], ~s** *od.* **die;** ~, **~s** bongo [drum]
Bonmot [bõ'mo:] **das; ~s, ~s** bon mot
Bonsai der; ~s, ~s bonsai [tree]
Bonus ['bo:nʊs] **der;** ~ *od.* **Bonusses,** ~ *od.* **Bonusse** *(Kaufmannsspr.) (Rabatt)* discount; *(Dividende)* extra dividend; *(Versicherungsw.)* bonus; **b)** *(Punktvorteil)* bonus points *pl.*
Bonze ['bɔntsə] **der; ~n, ~n** *(abwertend: Funktionär)* bigwig *(coll.)*; big noise *(sl.)*; big wheel *(Amer. sl.)*
Boogie-Woogie ['bʊgi'vʊgi] **der; ~[s], ~s** boogie-woogie
Boom [bu:m] **der; ~s, ~s** boom
Boot [bo:t] **das; ~[e]s, ~e** boat; **wir sitzen alle in einem** *od.* **im selben ~** *(fig. ugs.)* we're all in the same boat
Boots-: ~fahrt die boat trip; **~haus das** boathouse; **~länge die** [boat's] length; **~mann der** *Pl.* **~leute a)** ≈ boatswain, bosun; **b)** *(Milit.: Rang)* ≈ petty officer; **~steg der** landing-stage; **~verleih der** boat-hire [business]
Bor [bo:ɐ] **das; ~s** *(Chemie)* boron
[1]**Bord** [bɔrt] **das; ~[e]s, ~e** shelf
[2]**Bord der; ~[e]s, ~e: an ~:** on board; **an ~ eines Schiffes/der „Baltic“** on board *or* aboard a ship/the 'Baltic'; **alle Mann an ~!** all aboard!; **über ~:** overboard; **etw. über ~ werfen** *(auch fig.)* throw sth. overboard; **von ~ gehen** leave the ship/aircraft
Bord-: ~buch das log[-book]; **~computer der** on-board computer
Bordeaux [bɔr'do:] **der;** ~, ~ [bɔr'do:s] *s.* **Bordeauxwein**
bordeaux·rot *Adj.* bordeaux-red; claret
Bordeaux·wein der Bordeaux [wine]; **roter ~:** claret
Bordell [bɔr'dɛl] **das; ~s, ~e** brothel
Bord-: ~funk der [ship's/aircraft] radio; **~funker der** radio operator **~personal das** *(Flugw.)* cabin crew; **~stein der** kerb; **~stein·kante die** [edge of the] kerb
Bordüre [bɔr'dy:rə] **die;** ~, **~n** edging
borgen ['bɔrgn̩] *tr. V.: s.* **leihen**

Borke ['bɔrkə] **die**; ~, ~**n** bark
Borken·käfer der bark beetle
borkig *Adj.* cracked ⟨*earth*⟩; chapped, cracked ⟨*skin*⟩
borniert [bɔr'ni:ɐt] *(abwertend)* **1.** *Adj.* narrow-minded; bigoted. **2.** *adv.* in a narrow-minded *or* bigoted way
Borniertheit die; ~: narrow-mindedness; bigotry
Borretsch ['bɔrɛtʃ] **der**; ~**[e]s** borage
Bor·salbe die boric acid ointment
Börse ['bœrzə] **die**; ~, ~**n a)** *(Aktien~)* stock market; **an der** ~: on the stock market; **b)** *(Gebäude)* stock exchange; **c)** *(geh. veralt.: Geld~)* purse
Börsen-: ~krach der stock-market crash; collapse of the [stock] market; **~kurs der** [stock-]market price; **~makler der** stockbroker
Borste ['bɔrstə] **die**; ~, ~**n** bristle
borstig *Adj.* bristly
Borte ['bɔrtə] **die**; ~, ~**n** braiding *no indef. art.*; trimming *no indef. art.*; edging *no indef. art.*
Bor·wasser das boric acid lotion
bös [bø:s] *s.* **böse 1 c, d, 2**
bös·artig 1. *Adj.* **a)** malicious ⟨*person, remark, etc.*⟩; vicious ⟨*animal*⟩; **b)** *(Med.)* malignant. **2.** *adv.* maliciously
Bös·artigkeit die a) maliciousness; *(von Tieren)* viciousness; **b)** *(Med.)* malignancy
Böschung ['bœʃʊŋ] **die**; ~, ~**en** *(an der Straße)* bank; embankment; *(am Bahndamm)* embankment; *(am Fluß)* bank
böse ['bø:zə] **1.** *Adj.* **a)** wicked; evil; **jmdm. Böses tun** *(geh.)* do sb. harm; **ich will dir doch nichts Böses** I don't mean you any harm; **b)** *nicht präd. (schlimm, übel)* bad ⟨*times, illness, dream, etc.*⟩; nasty ⟨*experience, affair, situation, trick, surprise, etc.*⟩; **ein ~s Ende nehmen** end in disaster; **eine ~ Geschichte** a bad *or* nasty business; **nichts Böses ahnend** unsuspectingly; **c)** *(ugs.) (wütend)* mad *(coll.)*; *(verärgert)* cross *(coll.)*; ~ **auf jmdn.** *od.* **mit jmdm. sein** be mad at/cross with sb. *(coll.)*; ~ **über etw.** *(Akk.)* **sein** be mad at/cross about sth. *(coll.)*; **d)** *(fam.: ungezogen)* naughty; **e)** *nicht präd. (ugs.: arg)* terrible *(coll.)* ⟨*pain, fall, shock, disappointment, storm, etc.*⟩. **2.** *adv.* **a)** *(schlimm, übel)* ⟨*end*⟩ badly; **mit ihm wird es noch ~ enden** he'll come to a bad end; **das wird ~ enden** it is bound to end in disaster; **es war nicht ~ gemeint** I didn't mean it nastily; **b)** *(ugs.) (wütend)* angrily; *(verärgert)* crossly *(coll.)*; **c)** *(ugs.: sehr)* terribly *(coll.)*; ⟨*hurt*⟩ badly
Böse·wicht der; ~**[e]s**, ~**er a)** *(ugs. scherzh.: Schlingel)* rascal; **b)** *(veralt., fig.: Schuft)* villain
boshaft ['bo:shaft] **1.** *Adj.* malicious. **2.** *adv.* maliciously
Boshaftigkeit die; ~, ~**en a)** *o. Pl.* maliciousness; **b)** *(Bemerkung)* malicious remark; *(Handlung)* piece of maliciousness
Bosheit die; ~, ~**en a)** *o. Pl. (Art)* malice; **b)** *(Bemerkung)* malicious remark; *(Handlung)* piece of maliciousness
Boskop ['bɔskɔp], **Boskoop** ['bɔsko:p] **der**; ~**s**, ~: russet
Boß [bɔs] **der**; **Bosses, Bosse** *(ugs.)* boss *(coll.)*
Bossa Nova ['bɔsa'no:va] **der**; ~~, ~~**s** bossa nova
bosseln ['bɔsl̩n] *tr., itr. V. (ugs.)* **etw./an etw.** *(Dat.)* ~: beaver away *(Brit.) or* slave away making sth.; **er braucht immer was zu ~**: he always needs to be working on *or* making something
bös·willig 1. *Adj.* malicious; wilful ⟨*desertion*⟩. **2.** *adv.* maliciously; wilfully ⟨*desert*⟩
Böswilligkeit die; ~: malice; maliciousness
bot [bo:t] *1. u. 3. Pers. Sg. Prät. v.* **bieten**
Botanik [bo'ta:nɪk] **die**; ~: botany *no art.*
Botaniker der; ~**s**, ~: botanist
botanisch 1. *Adj.* botanical. **2.** *adv.* botanically
Bötchen ['bø:tçən] **das**; ~**s**, ~: little boat
Bote ['bo:tə] **der**; ~**n**, ~**n a)** messenger; *(fig.)* herald; harbinger; **b)** *(Laufbursche)* errand-boy; messenger[-boy]
Boten-: ~dienst der job as a messenger/errand-boy; **~gang der** errand; **~gänge erledigen** run errands
bot·mäßig *(geh. veralt.)* **1.** *Adj. (gehorsam)* obedient; *(untertänig)* submissive. **2.** *adv. (gehorsam)* obediently; *(untertänig)* submissively
Botschaft die; ~, ~**en a)** message; **die Frohe ~** *(das Evangelium)* the Gospel; **b)** *(diplomatische Vertretung)* embassy
Botschafter der; ~**s**, ~, **Botschafterin die**; ~, ~**nen** ambassador; **der irische Botschafter in Japan** the Irish ambassador to Japan
Böttcher ['bœtçɐ] **der**; ~**s**, ~: cooper
Böttcherei die; ~, ~**en a)** *o. Pl. (Handwerk)* cooper's trade; cooperage *no art.*; **b)** *(Werkstatt)* cooper's workshop; cooperage
Bottich ['bɔtɪç] **der**; ~**s**, ~**e** tub
Bouillon [bul'jɔŋ] **die**; ~, ~**s** bouillon; consommé
Bouillon·würfel der bouillon cube
Boulevard [bulə'va:ɐ] **der**; ~**s**, ~**s** boulevard
Boulevard-: ~blatt das *s.* **~zeitung**; **~presse die** *(abwertend)* popular press; **~theater das** light theatre; **~zeitung die** *(abwertend)* popular rag *(derog.)*; tabloid
bourgeois [bʊr'ʒoa] *Adj. (abwertend, Soziol.)* bourgeois
Bourgeois der; ~, ~ *(abwertend, Soziol.)* bourgeois
Bourgeoisie [bʊrʒoa'zi:] **die**; ~, ~**n** *(abwertend, Soziol.)* bourgeoisie
Boutique [bu'ti:k] **die**; ~, ~**s** *od.* ~**n** boutique
Bowle ['bo:lə] **die**; ~, ~**n** punch *(made of wine, champagne, sugar, and fruit or spices)*
Bowling ['boʊlɪŋ] **das**; ~**s**, ~**s** [ten-pin] bowling
Bowling·bahn die [ten-pin] bowling-alley
Box [bɔks] **die**; ~, ~**en a)** box; **b)** *(Lautsprecher~)* speaker; **c)** *(Pferde~)* [loose] box; **d)** *(Motorsport)* pit; **an den ~en** in the pits
boxen 1. *itr. V.* box; **gegen jmdn. ~**: fight sb.; box [against] sb.; **jmdm. in den Magen ~**: punch sb. in the stomach. **2.** *tr. V.* **a)** *(ugs.)* punch; **b)** *(Sportjargon: boxen gegen)* fight. **3.** *refl. V.* **a) sich ins Freie/durch die Menge** *usw.* ~ *(ugs.)* fight one's way outside/through the crowd *etc.*; **b)** *(ugs.: sich prügeln)* have a punch-up *(coll.) or* fight
Boxer der; ~**s**, ~ *(Sportler, Hund)* boxer
Boxer-: ~motor der *(Technik)* horizontally opposed engine; **~nase die** boxer's nose
Box·hand·schuh der boxing-glove
Box·kalf [-kalf] **das**; ~**s**, ~**s** box-calf
Box-: ~kampf der boxing match; *(Prügelei)* fist-fight; **~ring der** boxing ring; **~sport der**; *o. Pl.* boxing *no art.*
Boy [bɔy] **der**; ~**s**, ~**s** servant; *(im Hotel)* page-boy
Boykott [bɔy'kɔt] **der**; ~**[e]s**, ~**s** boycott

boykottieren *tr. V.* boycott
brabbeln ['brabl̩n] *tr., itr. V. (ugs.)* mutter; mumble; ⟨*baby*⟩ babble
brach [bra:x] *1. u. 3. Pers. Sg. Prät. v.* **brechen**
brachial [bra'xi̯a:l] *Adj.* violent; **~e Gewalt** brute force
Brachial·gewalt die; *o. Pl.* brute force
Brach·land das fallow [land]; *(auf Dauer)* uncultivated *or* waste land
brach|liegen *unr. itr. V. (auch fig.)* lie fallow; *(auf Dauer)* lie waste
brachte ['braxtə] *1. u. 3. Pers. Sg. Prät. v.* **bringen**
Brach·vogel der curlew
Brahmane [bra'ma:nə] **der; ~n, ~n** Brahmin
Branche ['brã:ʃə] **die; ~, ~n** [branch of] industry; **er kennt sich in der ~ am besten aus** he has the most knowledge of the industry
Branchen·verzeichnis das classified directory; *(Telefonbuch)* yellow pages *pl.*
Brand [brant] **der; ~[e]s, Brände** ['brɛndə] **a)** fire; **b)** *(Brennen)* **beim ~ der Scheune** when the barn caught fire; **in ~ geraten** catch fire; **etw. in ~ setzen** *od.* **stecken** set fire to sth.; set sth. on fire; **c)** *(ugs.: Durst)* raging thirst
brand-, Brand-: ~aktuell *Adj.* up-to-the-minute ⟨*report*⟩; red-hot ⟨*news item, issue*⟩; highly topical ⟨*book*⟩; **~binde die** dressing [for burns]; **~blase die** [burn] blister; **~eilig** *Adj. (ugs.)* extremely urgent
branden *itr. V. (geh.)* break
brand-, Brand-: ~fleck der burn mark; **~gefahr die** danger of fire; **~geruch der** smell of burning; **~herd der** source of the fire; **~mal das** *(geh.)* burn mark; **~marken** *tr. V.* brand ⟨*person*⟩; denounce ⟨*thing*⟩; **jmdn. als Verräter ~marken** brand sb. as a traitor; **~neu** *Adj. (ugs.)* brand-new; **~rede die** fiery tirade; **~salbe die** ointment for burns; **~schaden der** fire damage *no pl., no indef. art.*; **~schatzen** *tr. V. (hist.)* pillage and threaten to burn; **~sohle die** insole; **~stelle die** *(verbrannte Stelle)* burn; *(größer)* burnt patch; **~stifter der** arsonist; **~stiftung die** arson *no pl., no indef. art.*
Brandung die; ~, ~en surf; breakers *pl.*
Brandungs·welle die breaker
Brand·wunde die burn; *(Verbrühung)* scald
brannte ['brantə] *1. u. 3. Pers. Sg. Prät. v.* **brennen**
Brannt·wein der a) spirit; **b)** *o. Pl. (Spirituosen)* spirits *pl.*
Brasil die; ~, ~[s] Brazil cigar
Brasilianer [brazi'li̯a:nɐ] **der; ~s, ~, Brasilianerin die; ~, ~nen** Brazilian
brasilianisch *Adj.* Brazilian
Brasilien [bra'zi:li̯ən] **(das); ~s** Brazil
brät [brɛ:t] *3. Pers. Sg. Präsens v.* **braten**
Brat·apfel der baked apple
braten ['bra:tn̩] *unr. tr., itr. V.* fry; *(im Backofen, am Spieß)* roast
Braten der; ~s, ~ a) joint; **b)** *o. Pl.* roast [meat] *no indef. art.*; **kalter ~:** cold meat; **c) den ~ riechen** *(fig. ugs.)* get wind of what's going on; *(merken, daß etwas nicht stimmt)* smell a rat
Braten-: ~saft der meat juice[s *pl.*]; **~soße die** gravy
Brat-: ~fett das [cooking] fat; **~fisch der** fried fish; **~hähnchen das,** *(südd., österr.)* **~hendl das a)** roast chicken; *(gegrillt)* broiled chicken; **b)** *(Hähnchen zum Braten)* roasting chicken; *(zum Grillen)* broiling chicken; **~hering der** fried herring; **~huhn das, ~hühnchen das** *s.* **~hähnchen; ~kartoffeln** *Pl.* fried potatoes; home fries *(Amer.)*; **~ofen der** oven; **~pfanne die** frying-pan; **~röhre die** *s.* **~ofen; ~rost der** grill
Bratsche ['bra:tʃə] **die; ~, ~n** *(Musik)* viola
Bratschist der; ~en, ~en violist; viola-player
Brat-: ~spieß der spit; **~wurst die a)** [fried/grilled] sausage; **b)** *(Wurst zum Braten)* sausage [for frying/grilling]
Brauch [braux] **der; ~[e]s, Bräuche** ['brɔyçə] custom; **das ist bei ihnen so ~:** that's their custom; **nach altem ~:** in accordance with an old custom
brauchbar 1. *Adj.* useful; *(benutzbar)* usable; wearable ⟨*clothes*⟩. **2.** *adv.* **er schreibt/arbeitet ganz ~:** he's a useful writer/he does useful work
brauchen 1. *tr. V.* **a)** *(benötigen)* need; **alles, was man zum Leben braucht** everything one needs in order to live reasonably; **b)** *(aufwenden müssen)* **mit dem Auto braucht er nur zehn Minuten** it only takes him ten minutes by car; **er hat für die Arbeit Jahre gebraucht** the work took him years; **wie lange hast du dafür gebraucht?** how long did it take you?; **c)** *(benutzen, gebrauchen)* use; **ich könnte es gut ~:** I could do with it. **2.** *mod. V.; 2. Part.* **~:** need; **du brauchst nicht zu helfen** there is no need [for you] to help; you don't need to help; **du brauchst doch nicht gleich zu weinen** there's no need to start crying; **das hättest du nicht zu tun ~:** there was no need to do it; you needn't have done that; **du brauchst es [mir] nur zu sagen** you only have to tell me
Brauchtum das; ~s, Brauchtümer [-ty:mɐ] custom
Braue ['brau̯ə] **die; ~, ~n** [eye]brow
brauen *tr. V.* **a)** brew; **b)** *(ugs.: aufbrühen, zubereiten)* brew [up] ⟨*tea, coffee*⟩; concoct ⟨*potion etc.*⟩
Brauerei die; ~, ~en a) *o. Pl.* brewing; **b)** *(Betrieb)* brewery
braun [brau̯n] *Adj.* **a)** brown; **~ werden** *(sonnengebräunt)* get brown; get a tan; **b)** *(abwertend: nationalsozialistisch)* Nazi; **~ sein** ⟨*person*⟩ be a Nazi; **die Zeitung ist ziemlich ~:** the paper has definite Nazi tendencies
Braun das; ~s, ~, *(ugs.)* **~s** brown
braun·äugig *Adj.* brown-eyed; **~ sein** have brown eyes
Braun·bär der brown bear
Bräune ['brɔynə] **die; ~:** [sun-]tan
bräunen 1. *tr. V.* **a)** tan ⟨*skin, body, etc.*⟩; **sich ~:** get a tan; **b)** *(Kochk.)* brown. **2.** *itr. V.* **die Sonne bräunt stark** the sun gives you a good tan. **3.** *refl. V.* go brown; ⟨*skin*⟩ tan
braun·gebrannt *Adj.* [sun-]tanned
Braun·kohle die brown coal; lignite
bräunlich *Adj.* brownish
Bräunung die; ~, ~en browning
Braus [brau̯s] *s.* **Saus**
Brause ['brau̯zə] **die; ~, ~n a)** fizzy drink; *(~pulver)* sherbet; **b)** *(veralt.: Dusche)* shower
brausen 1. *itr. V.* **a)** ⟨*wind, water, etc.*⟩ roar; *(fig.)* ⟨*organ, applause, etc.*⟩ thunder; **b)** *(sich schnell bewegen)* race; **c)** *auch refl.: s.* **duschen 1. 2.** *tr. V. s.* **duschen 2**
Brausen das; ~s roar
Brause-: ~pulver das sherbet; **~tablette die** effervescent tablet
Braut [brau̯t] **die; ~, Bräute** ['brɔytə] **a)** bride; **b)** *(Verlobte)* fiancée; bride-to-be; **c)** *(ugs.: Freundin)* girl[-friend]
Braut·eltern *Pl.* bride's parents
Bräutigam ['brɔytɪgam] **der; ~s, ~e a)** [bride]groom; **b)** *(veralt.: Verlobter)* fiancé; husband-to-be
Braut-: ~jungfer die brides-

maid; ~**kleid das** wedding dress; ~**kranz der** bridal wreath; ~**mutter die** bride's mother; ~**paar das** bridal couple; bride and groom; ~**schleier der** bridal veil; ~**vater der** bride's father

brav [bra:f] **1.** *Adj.* **a)** *(artig)* good; **sei [schön] ~:** be good; **b)** *(redlich)* honest; upright; **c)** *(hausbacken)* plain and conservative ‹*clothes*›. **2.** *adv.* **nun iß schön ~ deine Suppe** be a good boy/girl and eat up your soup; eat up your soup like a good boy/girl

bravo ['bra:vo] *Interj.* bravo

Bravo das; ~s, ~s cheer; **ein ~ für ...:** three cheers for ...

Bravo·ruf der cheer

Bravour [bra'vu:ɐ̯] **die; ~:** stylishness; **mit ~:** with style and élan

Bravour·leistung die brilliant performance

bravourös [bravu'rø:s] **1.** *Adj.* brilliant. **2.** *adv.* brilliantly

Bravour·stück das piece of bravura; brilliant performance

BRD [be:|ɛr'de:] **die; ~** *Abk.* **Bundesrepublik Deutschland** FRG

Breakdance ['breɪkdæns] **der; ~[s]** breakdancing

Brech-: ~bohne die French bean *(Brit.);* green bean; ~**eisen das** crowbar

brechen ['brɛçn̩] **1.** *unr. tr. V.* **a)** break; **sich** *(Dat.)* **den Arm/das Genick ~:** break one's arm/neck; **b)** *(abbauen)* cut ‹*marble, slate, etc.*›; **c)** *(ablenken)* break the force of ‹*waves*›; refract ‹*light*›; **d)** *(bezwingen)* overcome ‹*resistance*›; break ‹*will, silence, record, blockade, etc.*›; **e)** *(nicht einhalten)* break ‹*agreement, contract, promise, the law, etc.*›; **f)** *(ugs.: erbrechen)* bring up. **2.** *unr. itr. V.* **a)** *mit sein* break; **mir bricht das Herz** *(fig.)* it breaks my heart; **brechend voll sein** be full to bursting; **b) mit jmdm. ~:** break with sb.; **c)** *mit sein* **durch etw. ~:** break through sth.; **d)** *(ugs.: sich erbrechen)* throw up. **3.** *unr. refl. V.* ‹*waves etc.*› break; ‹*rays etc.*› be refracted

Brecher der; ~s, ~: breaker

Brech-: ~mittel das emetic; ~**reiz der** nausea; ~**stange die** crowbar; **mit der ~stange vorgehen** *(fig.)* go about it with a sledgehammer

Brechung die; ~, ~en *(Physik)* refraction

Brechungs·winkel der *(Physik)* angle of refraction

Brei [brai̯] **der; ~[e]s, ~e** *(Hafer~)* porridge *(Brit.),* oatmeal *(Amer.) no indef. art.; (Reis~)* rice pudding; *(Grieß~)* semolina *no indef. art.;* **etw. zu einem ~ verrühren** make sth. into a mash *or* purée; **um den heißen ~ herumreden** *(fig. ugs.)* beat about the bush

breiig *Adj.* mushy

breit [brai̯t] **1.** *Adj.* **a)** wide; broad, wide ‹*hips, shoulders, forehead, etc.*›; **etw. ~er machen** widen sth.; **die Beine ~ machen** open one's legs; **ein ~es Lachen** a guffaw; **ein 5 cm ~er Saum** a hem 5 cm wide; **b)** *(groß)* **die ~e Masse** the general public; most people *pl.;* **die ~e Öffentlichkeit** the general public; **ein ~es Interesse finden** arouse a great deal of interest. **2.** *adv.* **~ gebaut** sturdily *or* well built; **~ lachen** guffaw; **etw. ~ darstellen** *(fig.)* describe sth. in great detail

breit·beinig *Adj.; nicht attr.* with one's legs apart; **er stand ~ vor uns** he stood squarely in front of us

Breite die; ~, ~n a) *s.* **breit 1a:** width; breadth; **in die ~ gehen** *(ugs.)* put on weight; **b)** *(Geogr.)* latitude; **auf/unter 50° nördlicher ~:** at/below latitude 50° north; **in diesen ~n** in these latitudes

breiten *(geh.) tr., refl. V.* spread

Breiten-: ~grad der degree of latitude; **der 30. ~grad** the 30th parallel; ~**kreis der** [line of] latitude; parallel; ~**sport der** popular sport

breit-, Breit-: ~|machen *refl. V. (ugs.)* **a)** take up room; **b)** *(sich ausbreiten)* be spreading; ~**|schlagen** *unr. tr. V. (ugs.)* **sich zu etw. ~schlagen lassen** let oneself be talked into sth.; **er ließ sich ~schlagen** he let himself be persuaded; ~**schult[e]rig** *Adj.* broad-shouldered; ~**seite die** long side; ~**|treten** *unr. tr. V. (fig. ugs. abwertend)* go on about; ~**wand die** *(Kino)* wide *or* big screen; ~**wand·film der** wide-screen *or* big-screen film

Brems-: ~backe die brake-shoe; ~**belag der** brake lining

¹Bremse ['brɛmzə] **die; ~, ~n** brake; **auf die ~ treten** put on the brakes

²Bremse die; ~, ~n *(Insekt)* horse-fly

bremsen 1. *itr. V.* brake. **2.** *tr. V.* **a)** brake; *(um zu halten)* stop; **b)** *(fig.)* slow down ‹*rate, development, production, etc.*›; restrict ‹*imports etc.*›; **jmdn. ~** *(ugs.)* stop sb. **3.** *refl. V. (ugs.)* stop oneself; hold oneself back

Brems-: ~flüssigkeit die brake fluid; ~**hebel der** brake arm; ~**klotz der** brake pad; ~**licht das;** *Pl.* ~**lichter** brake-light; ~**pedal das** brake-pedal; ~**scheibe die** brake-disc; ~**spur die** skid-mark; ~**trommel die** brake-drum

Bremsung die; ~, ~en braking

Brems·weg der braking distance

brenn·bar *Adj.* [in]flammable; combustible; **leicht ~:** highly [in]flammable *or* combustible

brennen ['brɛnən] **1.** *unr. itr. V.* **a)** burn; ‹*house etc.*› burn, be on fire; **schnell/leicht ~:** catch fire quickly/easily; **es brennt!** fire!; **wo brennt's denn?** *(fig. ugs.)* what's the panic?; **b)** *(glühen)* be alight; **c)** *(leuchten)* be on; **in ihrem Zimmer brennt Licht** there is a light on in her room; **das Licht ~ lassen** leave the light on; **d)** *(scheinen)* **die Sonne brannte** the sun was burning down; **e)** *(schmerzen)* ‹*wound etc.*› burn, sting; ‹*feet etc.*› hurt, be sore; **mir ~ die Augen** my eyes are stinging *or* smarting; **f) darauf ~, etw. zu tun** be dying *or* longing to do sth. **2.** *unr. tr. V.* **a)** burn ‹*hole, pattern, etc.*›; **einem Tier ein Zeichen ins Fell ~:** brand an animal; **b)** *(mit Hitze behandeln)* fire ‹*porcelain etc.*›; distil ‹*spirits*›; **gebrannter Kalk** quicklime; **c)** *(rösten)* roast ‹*coffee-beans, almonds, etc.*›

brennend 1. *Adj. (auch fig.)* burning; lighted ‹*cigarette*›; raging ‹*thirst*›; urgent ‹*topic, subject*›. **2.** *adv.* **es scheint dich ja ~ zu interessieren, was besprochen wurde** you seem to be dying to know what was discussed

Brennerei die; ~, ~en a) *o. Pl.* distilling; **b)** *(Betrieb)* distillery

Brennessel ['brɛnnɛsl̩] **die; ~, ~n** stinging nettle

Brenn-: ~glas das burning-glass; ~**holz das;** *o. Pl.* firewood; ~**material das** fuel; ~**nessel die** *s.* **Brennessel;** ~**punkt der** *(Math., Optik, fig.)* focus; **im ~punkt des Interesses stehen** be the focus of attention *or* interest; ~**schere die** curling-tongs *pl. (Brit.);* curling iron *(Amer.);* ~**spiritus der** methylated spirits *pl.;* ~**stab der** *(Kerntechnik)* fuel rod; ~**stoff der** fuel; ~**weite die** *(Optik)* focal length

brenzlig ['brɛnt͜slɪç] *Adj.* **a)** ‹*smell, taste, etc.*› of burning *not pred.;* **~ riechen/schmecken** smell of burning/taste burnt; **b)** *(ugs.: bedenklich)* dicey *(sl.);* **mir wird die Sa-**

che zu ~: things are getting too hot for me

Bresche ['brɛʃə] **die; ~, ~n** gap; breach; *(fig.)* **[für jmdn.] in die ~ springen** stand in [for sb.]; **für jmdn./etw. eine ~ schlagen** give one's backing to sb./sth.

Bretagne [bre'tanje] **die; ~**: Brittany

Brett [brɛt] **das; ~[e]s, ~er a)** board; *(lang und dick)* plank; **Schwarzes ~**: notice-board; **ein ~ vor dem Kopf haben** *(fig. ugs.)* be thick; **b)** *(für Spiele)* board; **c)** *Pl. (Ski)* skis; **d)** *Pl. (Bühne)* stage *sing.;* boards; **die ~er, die die Welt bedeuten** the stage *sing.;* the boards; **e)** *Pl. (Boxen)* floor *sing.;* canvas *sing.*

Brettchen das; ~s, ~ a) *wooden board used for breakfast;* **b)** *(zum Schneiden)* board

Bretter-: ~boden der wooden floor; **~bude die** [wooden] hut, shack; **~verschlag der** [wooden] shed; **~wand die** wooden wall *or* partition; **~zaun der** wooden fence

Brett·spiel das board game

Brezel ['bre:tsl̩] **die; ~, ~n,** *(österr.)* **Brezen** ['bre:tsn̩] **der; ~s, ~** *od.* **die; ~, ~**: pretzel

Bridge [brɪtʃ] **das; ~**: bridge

Brief [bri:f] **der; ~[e]s, ~e** letter; **jmdm. ~ und Siegel [auf etw.** *(Akk.)***] geben** *(fig.)* promise sb. faithfully *or* give sb. one's word [on sth.]

Brief-: ~beschwerer der; ~s, ~: paperweight; **~block der;** *Pl.* **~blöcke** *od.* **~blocks** writing-pad; letter-pad; **~bogen der** sheet of writing-paper *or* note-paper

Briefchen das; ~s, ~ a) ein ~ Streichhölzer a book of matches; **ein ~ Nähnadeln** a packet of needles; **b)** *(kurzer Brief)* note

Brief-: ~druck·sache die *(Postw.)* printed paper *(sent as a letter);* **~freund der, ~freundin die** pen-friend; pen-pal *(coll.);* **~geheimnis das** privacy of the post; secrecy of correspondence; **~karte die** correspondence card; **~kasten der a)** post-box; **b)** *(privat)* letter-box

Briefkasten-: ~firma die accommodation address; **~tante die** *(ugs. scherzh.)* agony aunt *(coll.)*

Brief-: ~kopf der a) letter-heading; **b)** *(aufgedruckt)* letter-head; **~kuvert das** *(veralt.)* s. **~umschlag**

brieflich 1. *Adj.; nicht präd.* written. **2.** *adv.* by letter

Brief·marke die [postage] stamp

Briefmarken-: ~album das stamp-album; **~sammler der** stamp-collector; philatelist; **~sammlung die** stamp-collection

Brief-: ~öffner der letter-opener; **~papier das** writing-paper; notepaper; **~partner der, ~partnerin die** pen-friend; **~schreiber der** [letter-]writer; **~tasche die** wallet; **~taube die** carrier pigeon; **~träger der** postman; letter-carrier *(Amer.);* **~trägerin die** postwoman; **~um·schlag der** envelope; **~waage die** letter-scales *pl.;* **~wahl die** postal vote; **~wechsel der a)** correspondence; **einen ~wechsel führen** have a *or* be in correspondence; **b)** *(gesammelte Briefe)* correspondence

Bries [bri:s] **das; ~es, ~e** *(Kochk.)* sweetbreads *pl.*

briet [bri:t] *1. u. 3. Pers. Sg. Prät. v.* **braten**

Brigade [bri'ga:də] **die; ~, ~n** *(Milit.)* brigade

Brikett [bri'kɛt] **das; ~s, ~s** briquette

brillant [brɪl'jant] **1.** *Adj.* brilliant. **2.** *adv.* brilliantly

Brillant [brɪl'jant] **der; ~en, ~en** brilliant

Brillant-: ~ring der *(brilliant-cut)* diamond ring; **~schmuck der;** *o. Pl. (brilliant-cut)* diamond jewellery

Brillanz [brɪl'jants] **die; ~**: brilliance

Brille ['brɪlə] **die; ~, ~n a)** glasses *pl.;* spectacles *pl.;* specs *(coll.) pl.;* **eine ~**: a pair of glasses *or* spectacles; **eine ~ tragen** wear glasses *or* spectacles; **etw. durch eine rosa[rote] ~ sehen** *od.* **betrachten** *(fig.)* see sth. through rose-coloured *or* rose-tinted spectacles; **b)** *(ugs.: Klosett~)* [lavatory *or* toilet] seat

Brillen-: ~etui das, ~futteral das glasses-case; spectacle-case; **~glas das** [spectacle-]lens; **~schlange die** spectacled cobra; **~träger der** spectacle-user *or* -wearer; person who wears glasses; **~ sein** wear glasses

brillieren [brɪl'ji:rən] *itr. V. (geh.)* be brilliant

Brimborium [brɪm'bo:ri̯ʊm] **das; ~s** *(ugs. abwertend)* hoo-ha *(coll.)*

bringen ['brɪŋən] *unr. tr. V.* **a)** *(her~)* bring; *(hin~)* take; **sie brachte mir/ich brachte ihr ein Geschenk** she brought me/I took her a present; **Unglück/Unheil [über jmdn.] ~**: bring misfortune/disaster [upon sb.]; **jmdm. Glück/Unglück ~**: bring sb. [good] luck/bad luck; **jmdm. eine Nachricht ~**: bring sb. news; **b)** *(begleiten)* take; **jmdn. nach Hause/zum Bahnhof ~**: take sb. home/to the station; **die Kinder ins Bett** *od.* **zu Bett ~**: put the children to bed; **c) es zu etwas/nichts ~**: get somewhere/get nowhere *or* not get anywhere; **es bis zum Direktor ~**: make it to director; **es weit ~**: get on *or* do very well; **es im Leben weit ~**: go far in life; **d) jmdn. ins Gefängnis ~** ⟨*crime, misdeed*⟩ land sb. in prison *or* gaol; **eine Sache vor Gericht ~**: take a matter to court; **das Gespräch auf etw./ein anderes Thema ~**: bring the conversation round to sth./change the topic of conversation; **jmdn. wieder auf den rechten Weg ~** *(fig.)* get sb. back on the straight and narrow; **jmdn. zum Lachen/zur Verzweiflung ~**: make sb. laugh/drive sb. to despair; **jmdn. dazu ~, etw. zu tun** get sb. to do sth.; **du hast mich auf eine gute Idee gebracht** you have given me a good idea; **etw. hinter sich ~** *(ugs.)* get sth. over and done with; **es nicht über sich** *(Akk.)* **~ [können], etw. zu tun** not be able to bring oneself to do sth.; **etw. an sich** *(Akk.)* **~** *(ugs.)* collar sth. *(sl.);* **e) jmdn. um seinen Besitz ~**: do sb. out of his property; **jmdn. um den Schlaf/Verstand ~**: rob sb. of his/her sleep/drive sb. mad; **f)** *(veröffentlichen)* publish; **alle Zeitungen brachten Berichte über das Massaker** all the papers carried reports of the massacre; **g)** *(senden)* broadcast; **das Fernsehen bringt eine Sondersendung** there is a special programme on television; **das Fernsehen hat nichts darüber gebracht** there was nothing about it on television; **h)** *(darbringen)* **das/ein Opfer ~**: make the/a sacrifice; **eine Nummer/ein Ständchen ~**: perform a number/a serenade; **das kannst du nicht ~** *(ugs.)* you can't do that; **i)** *(erbringen)* **einen großen Gewinn/hohe Zinsen ~**: make a large profit/earn high interest; **das Gemälde brachte 50 000 DM** the painting fetched 50,000 marks; **das bringt nichts** *od.* **bringt's nicht** *(ugs.)* it's pointless; **j) das bringt es mit sich, daß ...**: that means that ...; **k)** *(verursachen)* cause ⟨*trouble, confusion*⟩; **es kann dir doch nur Vorteile ~**: it can only be to your advantage; **l)** *(salopp: schaffen, er-*

reichen) **das bringst du doch nicht** you'll never do it; **m)** *(bes. südd.)* *s.* **bekommen 1 b**

brisant [bri'zant] *Adj.* explosive

Brisanz [bri'zants] **die;** ~explosiveness; explosive nature

Brise ['bri:zə] **die;** ~, ~**n** breeze

Britannien [bri'taniən] **(das);** ~s Britain; *(hist.)* Britannia

Brite ['brɪtə] **der;** ~**n**, ~**n** Briton; **die** ~**n** the British; **er ist [kein]** ~: he is [not] British

Britin die; ~, ~**nen** Briton; British girl/woman; **die** ~**nen** the British [girls/women]; **sie ist [keine]** ~: she is [not] British

britisch *Adj.* British; **die Britischen Inseln** the British Isles

bröckelig *Adj.* crumbly

bröckeln ['brœkl̩n] **1.** *itr. V.* **a)** crumble; **b)** *mit sein* **von der Decke/Wand** ~: crumble away from the ceiling/wall. **2.** *tr. V.* crumble

Brocken ['brɔkn̩] **der;** ~s, ~ **a)** *(von Brot)* hunk, chunk; *(von Fleisch)* chunk; *(von Lehm, Kohle, Erde)* lump; **b)** *(fig.)* **ein paar** ~ **Englisch** a smattering of English; **ein harter** ~ *(ugs.)* a tough *or* hard nut to crack; **c)** *(ugs.: dicke Person)* lump

brocken·weise *Adv. (auch fig.)* bit by bit

bröcklig *s.* **bröckelig**

brodeln ['bro:dl̩n] *itr. V.* bubble

Broiler ['brɔylɐ] **der;** ~s, ~ *(regional) s.* **Brathähnchen**

Brokat [bro'ka:t] **der;** ~**[e]s**, ~**e** brocade

Brokkoli ['brɔkoli] *Pl.* broccoli *sing.*

Brom [bro:m] **das;** ~**s** *(Chemie)* bromine

Brom·beere ['brɔm-] **die** blackberry

Bronchial-: ~**katarrh der** *(Med.) s.* **Bronchitis;** ~**tee der** bronchial tea

Bronchie ['brɔnçiə] **die;** ~, ~**n** *(Med.)* bronchial tube; bronchus

Bronchitis ['brɔnçi:tɪs] **die;** ~, **Bronchitiden** *(Med.)* bronchitis

Bronze ['brõ:sə] **die;** ~: bronze

Bronze·medaille die bronze medal

Bronze·zeit die Bronze Age

Brosche ['brɔʃə] **die;** ~, ~**n** brooch

broschiert [brɔ'ʃi:rt] *Adj.* paperback; **eine** ~**e Ausgabe** a paperback *or* soft-cover edition

Broschüre [brɔ'ʃy:rə] **die;** ~, ~**n** booklet; pamphlet

Brösel ['brø:zl̩] **der;** ~s, ~: breadcrumb

bröselig *Adj.* crumbly

bröseln *itr., tr. V.* crumble

Brot [bro:t] **das;** ~**[e]s**, ~**e a)** bread *no pl., no indef. art.; (Laib* ~*)* loaf [of bread]; *(Scheibe* ~*)* slice [of bread]; **b)** *(Lebensunterhalt)* daily bread *(fig.);* **das ist ein hartes** ~: it's a hard way to earn a *or* your living

Brot-: ~**aufstrich der** spread; ~**belag der** topping; *(im zusammengeklappten Brot)* filling

Brötchen ['brø:tçən] **das;** ~s, ~: roll; **kleinere** ~ **backen [müssen]** *(fig. ugs.)* [have to] lower one's sights; **seine/die** ~ **verdienen** *(ugs.)* earn one's/the daily bread

Brötchen·geber der; ~s, ~ *(scherzh.)* employer

brot-, Brot-: ~**erwerb der** way to earn a living; ~**fabrik die** bakery *(producing bread on a large scale);* ~**kasten der** bread-bin; ~**korb der** bread-basket; **jmdm. den** ~**korb höher hängen** *(fig. ugs.)* put sb. on short rations; ~**krume die,** ~**krümel der** breadcrumb; ~**kruste die** [bread] crust; ~**laib der** loaf [of bread]; ~**los** *Adj.* unemployed; ~**maschine die** bread-slicer; ~**messer das** bread-knife; ~**rinde die** [bread] crust; ~**teig der** bread dough; ~**zeit die** *(südd.)* **a)** *(Pause)* [tea-/coffee-/lunch-]break; **b)** *o. Pl. (Vesper)* snack; *(Vesperbrot)* sandwiches *pl.*

BRT *Abk.* **Bruttoregistertonne** grt

Bruch [brʊx] **der;** ~**[e]s, Brüche** ['brʏçə] **a)** *(auch fig.)* break; *(eines Versprechens)* breaking; **der** ~ **des Deiches/Dammes** the breaching *(Brit.) or (Amer.)* breaking of the dike/dam; ~ **machen** *(ugs.)* break things; **in die Brüche gehen** *(zerbrechen)* break; get broken; *(enden)* break up; **zu** ~ **gehen** break; get broken; **etw. zu** ~ **fahren** smash sth. up; **b)** *(*~*stelle)* break; **die Brüche im Deich** the breaches *(Brit.) or (Amer.)* breaks in the dike; **c)** *(Med.: Knochen*~*)* fracture; break; **d)** *(Med.: Eingeweide*~*)* hernia; rupture; **sich** *(Dat.)* **einen** ~ **heben** rupture oneself *or* give oneself a hernia [by lifting sth.]; **e)** *(Math.)* fraction

Bruch·bude die *(ugs. abwertend)* hovel; dump *(coll.)*

brüchig ['brʏçɪç] *Adj.* **a)** brittle, crumbly ⟨*rock, brickwork*⟩; **der Stoff ist ziemlich** ~: the material is splitting quite a bit; **b)** *(fig.)* crumbling ⟨*relationship, marriage, etc.*⟩

bruch-, Bruch-: ~**landung die** crash-landing; ~**rechnen** *itr. V.; nur im Inf.* do fractions; ~**rechnen das** fractions *pl.;* **beim** ~**rechnen ...:** when doing fractions ...; ~**rechnung die** fractions *pl.;* ~**strich der** fraction line; ~**stück das** fragment; ~**stückhaft 1.** *Adj.* fragmentary; **2.** *adv.* in a fragmentary way; ~**teil der** fraction; **im** ~**teil einer Sekunde** in a fraction of a second; in a split second

Brücke ['brʏkə] **die;** ~, ~**n a)** *(auch: Schiffs*~*, Zahnmed., Turnen, Ringen)* bridge; **die** *od.* **alle** ~**n hinter sich** *(Dat.)* **abbrechen** *(fig.);* burn one's bridges *(fig.);* **jmdm. eine [goldene]** ~ *od.* **[goldene]** ~**n bauen** *(fig.)* make things easier for sb.; **b)** *(Teppich)* rug

Brücken-: ~**bogen der** arch [of a/the bridge]; ~**geländer das** parapet; railing; ~**kopf der** *(Milit., fig.)* bridgehead; ~**pfeiler der** pier [of a/the bridge]

Bruder ['bru:dɐ] **der;** ~s, **Brüder** ['bry:dɐ] **a)** *(auch fig.)* brother; **die Brüder Müller** the Müller brothers; the brothers Müller; **der große** ~ *(fig.)* Big Brother; **unter Brüdern** *(fig. ugs. scherzh.)* between *or* amongst friends; **b)** *(ugs. abwertend: Mann)* guy *(sl.)*

Brüderchen ['bry:dɐçən] **das;** ~s, ~: little brother

Bruder-: ~**krieg der** fratricidal war; ~**kuß der** brotherly kiss

brüderlich 1. *Adj.* brotherly; *(im politischen Bereich)* fraternal. **2.** *adv.* in a brotherly way; *(im politischen Bereich)* fraternally; **etw.** ~ **[mit jmdm.] teilen** share sth. [with sb.] in a fair and generous way

Brüderlichkeit die; ~: brotherliness; *(im politischen Bereich)* fraternity

Brüderschaft die; ~: **[mit jmdm.]** ~ **trinken** drink to close friendship [with sb.] *(agreeing to use the familiar 'du' form)*

Brühe ['bry:ə] **die;** ~, ~**n a)** stock; *(als Suppe)* clear soup; broth; **b)** *(ugs. abwertend) (Getränk)* muck; *(verschmutztes Wasser)* dirty *or* filthy water

brühen *tr. V.* **a)** blanch; **b)** *(auf*~*)* brew, make ⟨*tea*⟩; make ⟨*coffee*⟩

brüh-, Brüh-: ~**warm** *Adj.: in* **etw.** ~**warm weitererzählen** *(ugs.)* pass sth. on *or* spread sth. around straight away; ~**würfel der** stock cube; ~**wurst die** sausage *(which is heated in boiling water)*

brüllen ['brʏlən] **1.** *itr. V.* **a)** ⟨*bull, cow, etc.*⟩ bellow; ⟨*lion, tiger, etc.*⟩ roar; ⟨*elephant*⟩ trumpet; **b)** *(ugs.: schreien)* roar; shout; **vor Schmerzen/Lachen** ~: roar with pain/laughter; **nach jmdm. ~:** shout to *or* for sb.; **das ist [ja] zum Brüllen** *(ugs.)* it's a [real] scream; what a scream; **c)** *(ugs.: weinen)* howl; bawl; **er brüllte wie am Spieß** he bawled his head off. **2.** *tr. V.* yell; shout

Brumm-: ~bär der *(ugs.)* grouch *(coll.)*; **~baß der** *(ugs.)* deep *or* bass voice

brummeln ['brʊml̩n] *tr., itr. V. (ugs.)* mumble; mutter

brummen ['brʊmən] *tr., itr. V.* **a)** ‹*insect*› buzz; ‹*bear*› growl; ‹*engine etc.*› drone; **mir brummt der Schädel** *od.* **Kopf** *(ugs.)* my head is buzzing; **b)** *(unmelodisch singen)* drone; **c)** *(mürrisch sprechen)* mumble; mutter

Brummer der; ~s, ~ *(ugs.)* **a)** *(Fliege)* bluebottle; **b)** *(Lkw)* heavy lorry *(Brit.) or* truck

Brummi der; ~s, ~s *(ugs.)* lorry *(Brit.)*; truck

brummig *(ugs.)* **1.** *Adj.* grumpy. **2.** *adv.* grumpily

Brumm-: ~kreisel der humming top; **~schädel der** *(ugs.)* thick head

brünett [bry'nɛt] *Adj.* dark-haired ‹*person*›; dark ‹*hair*›; **sie ist ~:** she's [a] brunette

Brünette die; ~, ~n brunette

Brunft [brʊnft] **die; ~, Brünfte** ['brʏnftə] *(Jägerspr.) s.* **Brunst**

Brunnen ['brʊnən] **der; ~s, ~ a)** well; **b)** *(Spring~)* fountain; **c)** *(Heilwasser)* spring water

Brunnen-: ~kresse die watercress; **~vergifter der; ~s, ~:** water-poisoner; *(fig. abwertend)* trouble-maker; **~vergiftung die** water-poisoning; *(fig. abwertend)* trouble-making

Brunst [brʊnst] **die; ~, Brünste** ['brʏnstə] *(von männlichen Tieren)* rut; *(von weiblichen Tieren)* heat; **Männchen/Weibchen in der ~:** rutting males/females in *or* on heat

Brunst·zeit die *(bei männlichen Tieren)* rut; rutting season; *(bei weiblichen Tieren)* [season of] heat

brüsk [brʏsk] **1.** *Adj.* brusque; abrupt. **2.** *adv.* brusquely; abruptly

brüskieren *tr. V.* offend; *(stärker)* insult; *(schneiden)* snub

Brüssel ['brʏsl̩] **(das); ~s** Brussels

Brüsseler 1. *indekl. Adj.; nicht präd.* Brussels; **~ Spitzen** Brussels lace *sing.* **2. der; ~s, ~:** inhabitant of Brussels; *(von Geburt)* native of Brussels; *s. auch* **Kölner**

Brust [brʊst] **die; ~, Brüste** ['brʏstə] **a)** chest; *(fig. geh.)* breast; heart; **sich in die ~ werfen** puff oneself up; **b)** *(der Frau)* breast; **einem Kind die ~ geben** breast-feed a baby; **c)** *(Hähnchen~)* breast; *(Rinder~)* brisket; **d)** *o. Pl.: s.* **Brustschwimmen**

Brust-: ~bein das breastbone; **~beutel der** purse *(worn around the neck)*

brüsten ['brʏstn̩] *refl. V. (abwertend)* **sich mit etw. ~:** boast *or* brag about sth.

Brust-: ~flosse die *(Zool.)* pectoral fin; **~kasten** *(ugs.)* chest; **~korb der** *(Anat.)* thorax *(Anat.)*; **~krebs der** breast cancer; cancer of the breast; **~schwimmen** *itr. unr. V.; nur im Inf.* do [the] breast-stroke; **~schwimmen das** breast-stroke; **~stück das** *(Kochk.)* breast; *(vom Rind)* brisket; **~tasche die** breast pocket; *(Innentasche)* inside breast pocket; **~ton der** *in* **im ~ton der Überzeugung** *(fig.)* with utter conviction; **~umfang der** chest measurement; *(bei Frauen)* bust measurement

Brüstung die; ~, ~en parapet; *(Balkon~)* balustrade

Brust-: ~warze die nipple; **~wickel der** *(Med.)* chest compress; **~wirbel der** *(Anat.)* thoracic vertebra

Brut [bru:t] **die; ~, ~en a)** *(das Brüten)* brooding; **b)** *(Jungtiere, auch fig. scherzh.: Kinder)* brood

brutal [bru'ta:l] **1.** *Adj.* brutal; violent ‹*attack, film, etc.*›; brute ‹*force*›. **2.** *adv.* brutally

brutalisieren *tr. V.* brutalize

Brutalität [brutali'tɛ:t] **die; ~, ~en a)** *o. Pl.* brutality; **b)** *(Handlung)* act of brutality *or* violence

brüten ['bry:tn̩] *itr. V.* **a)** brood; **~de Hitze** *(fig.)* stifling heat; **b)** *(grübeln)* ponder **(über** + *Dat.* over); **über einem Plan ~:** work on a plan

brütend·heiß *Adj.; nicht präd. (ugs.)* boiling *or* stifling hot

Brüter der; ~s, ~ *(Kerntechnik)* breeder

Brut-: ~kasten der incubator; **~stätte die** breeding-ground; *(fig.)* breeding-ground *(Gen.,* **für** for); hotbed *(Gen.,* **für** for)

brutto ['brʊto] *Adv.* gross; **~ 4 000 DM, 4 000 DM ~:** 4,000 marks gross; **~ 800 kg** 800 kilos gross

Brutto- gross ‹*income, weight, etc.*›; full ‹*price*›

brutzeln ['brʊtsl̩n] **1.** *itr. V.* sizzle. **2.** *tr. V. (ugs.)* fry [up]

Btx [be:te:'ɪks] *Abk.* **Bildschirmtext**

Bub [bu:p] **der; ~en, ~en** *(südd., österr., schweiz.)* boy; lad

Bube ['bu:bə] **der; ~n, ~n** *(Kartenspiele)* jack; knave

Buben·streich der childish prank

Bubi ['bu:bi] **der; ~s, ~s a)** [little] boy *or* lad *or* fellow; **b)** *(salopp abwertend: Schnösel)* young lad

Bubi·kopf der bobbed hair[cut]; bob; **sich** *(Dat.)* **einen ~kopf schneiden lassen** have one's hair bobbed

Buch [bu:x] **das; ~[e]s, Bücher** ['by:çɐ] **a)** book; **das ~ der Bücher** the Book of Books; **wie ein ~ reden** *(ugs.)* talk nineteen to the dozen; **ein Detektiv/ein Faulpelz, wie er im ~e steht** a classic [example of a] detective/a complete lazybones; **ein ~ mit sieben Siegeln** a closed book; a complete mystery; **ein schlaues ~** *(ugs.)* a reference book/textbook; **b)** *(Dreh~)* script; **c)** *(Geschäfts~)* book; **über etw.** *(Akk.)* **~ führen/genau ~ führen** keep a record/an exact record of sth.; **zu ~[e] schlagen** be reflected in the budget; *(fig.)* have a big influence; **mit 200 DM zu ~[e] schlagen** make a difference of 200 marks

Buch-: ~besprechung die book-review; **~binder der** bookbinder; **~deckel der** [book] cover *(front or back)*; **~druck der;** *o. Pl.* letterpress printing; **im ~druck** in letterpress; **~drucker der** printer; **~druckerei die a)** *o. Pl.* letterpress printing; **b)** *(Betrieb)* printing works; **~drukker·kunst die;** *o. Pl.* art of printing

Buche die; ~, ~n a) beech[-tree]; **b)** *o. Pl. (Holz)* beech[wood]

Buch·ecker die beech-nut

buchen *tr. V.* **a)** enter; **etw. auf ein Konto ~:** enter sth. into an account; **etw. als Erfolg ~** *(fig.)* count sth. as a success; **b)** *(vorbestellen)* book

Buchen-: ~holz das beechwood; **~wald der** beech-wood

Bücher-: ~bord das a) *s.* **~brett; b)** *s.* **~regal; ~brett das** bookshelf

Bücherei die; ~, ~en library

Bücher-: ~regal das bookshelves *pl.*; **~schrank der** bookcase; **~stütze die** bookend; **~verbrennung die** burning of books; **~wand die a)** *(Möbel)* bookshelf unit; **b)** *(Wand mit ~regal)* wall of bookshelves; **~wurm der** *(scherzh.)* bookworm

Buch-: ~fink der chaffinch;

~führung die bookkeeping; **~halter der, ~halterin die** bookkeeper; **~haltung die a)** accountancy; **b)** *(Abteilung)* accounts department; **~handel der;** *o. Pl.* book trade; **im ~handel erhältlich** available from bookshops; **~händler der** bookseller; **~handlung die** bookshop; **~klub der** book club; **~laden der** bookschop; **~messe die** book fair; **~prüfer der** auditor; **~rücken der** spine

Buchs·baum ['bʊks-] **der** box-tree; box

Buchse ['bʊksə] **die; ~, ~n a)** *(Elektrot.)* socket; **b)** *(Technik)* bush; liner

Büchse ['bʏksə] **die; ~, ~n a)** can; tin *(Brit.);* **b)** *(ugs.: Sammel~)* box; **c)** *(Gewehr)* rifle; *(Schrot~)* shotgun

Büchsen-: ~fleisch das tinned *(Brit.) or (Amer.)* canned meat; **~milch die** tinned *(Brit.) or (Amer.)* canned milk; **~öffner der** tin-opener *(Brit.);* can opener *(Amer.)*

Buchstabe ['bu:xʃta:bə] **der; ~ns, ~n** letter; *(Druckw.)* character; **ein großer/kleiner ~:** a capital [letter]/small letter; **sich auf seine vier ~n setzen** *(ugs. scherzh.)* sit [oneself] down

buchstaben·getreu 1. *Adj.* literal. **2.** *adv.* to the letter

buchstabieren *tr. V.* **a)** spell; **b)** *(mühsam lesen)* spell out

buchstäblich ['bu:xʃtɛ:plɪç] *Adv.* literally

Buch·stütze die *s.* **Bücherstütze**

Bucht [bʊxt] **die; ~, ~en** bay

Buch·titel der title

Buchung die; ~, ~en a) entry; **b)** *(Vorbestellung)* booking

Buch-: ~weizen der buckwheat; **~wissen das** book-learning; **~zeichen das** bookmark[er]

Buckel ['bʊkl̩] **der; ~s, ~ a)** *(ugs.: Rücken)* back; **einen ~ machen** ‹*cat*› arch its back; ‹*person*› hunch one's shoulders; **rutsch mir den ~ runter!** *(fig. salopp)* get lost! *(sl.);* **den ~ hinhalten** *(fig.)* take the blame; carry the can *(sl.);* **einen krummen ~ machen***(fig.)* bow and scrape; kowtow; **schon 40 Jahre auf dem ~ haben** be 40 already; **b)** *(Rückenverkrümmung)* hunchback; hump; **c)** *(ugs.: Hügel)* hillock; **d)** *(ugs.: gewölbte Stelle)* bump

buckeln *itr. V. (ugs.) (abwertend)* bow and scrape; kowtow; **vor jmdm. ~:** kowtow to sb.; **nach oben ~ und nach unten treten** bow to superiors and kick underlings

bücken ['bʏkn̩] *refl. V.* bend down; **sich nach etw. ~:** bend down to pick sth. up

bucklig *Adj.* **a)** hunchbacked; humpbacked; **b)** *(ugs.: uneben)* bumpy

Bucklige der/die; *adj. Dekl.* hunchback; humpback

[1]Bückling ['bʏklɪŋ] **der; ~s, ~e** *(ugs. scherzh.: Verbeugung)* bow

[2]Bückling der; ~s, ~e *(Hering)* bloater

Buddel ['bʊdl̩] **die; ~, ~n** *(nordd.)* bottle

buddeln *itr., tr. V. (ugs.)* dig; **die Kinder ~ im Sand** the children are digging about in the sand

Buddha ['bʊda] **der; ~s, ~s** Buddha

Buddhismus der; ~: Buddhism *no art.*

Buddhist der; ~en, ~en Buddhist

buddhistisch *Adj.* Buddhist *attrib.*

Bude ['bu:də] **die; ~, ~n a)** kiosk; *(Markt~)* stall; *(Jahrmarkts~)* booth; **b)** *(Bau~)* hut; **c)** *(ugs.: Haus)* dump *(coll.);* **d)** *(ugs.: Zimmer)* room; digs *pl. (Brit. coll.);* **Leben in die ~ bringen** liven the place up; **e)** *(ugs. abwertend: Laden, Lokal)* outfit *(coll.)*

Budget [bʏ'dʒe:] **das; ~s, ~s** budget

Büfett [by'fɛt] **das; ~[e]s, ~s** *od.* **~e a)** sideboard; **b)** *(Schanktisch)* bar; **c)** *(Verkaufstisch)* counter; **d) kaltes ~:** cold buffet

Büffel ['bʏfl̩] **der; ~s, ~:** buffalo

Büffelei die; ~ *(ugs.)* swotting *no pl. (Brit. sl.)*

büffeln *(ugs.)* **1.** *itr. V.* swot *(Brit. sl.);* cram. **2.** *tr. V.* swot up *(Brit. sl.);* cram

Buffet [by'fe:] **das; ~s, ~s** *s.* **Büfett**

Bug [bʊ:k] **der; ~[e]s, ~e** *u.* **Büge** ['by:gə] **a)** *(Schiffs~)* bow; *(Flugzeug~)* nose; **b)** *(Schulterstück)* shoulder

Bügel ['by:gl̩] **der; ~s, ~ a)** *(Kleider ~)* hanger; **b)** *(Brillen~)* earpiece; **c)** *(an einer Tasche, Geldbörse)* frame

bügel-, Bügel-: ~brett das ironing-board; **~eisen das** iron; **~falte die** [trouser] crease; **~frei** *Adj.* non-iron

bügeln *tr., itr. V.* iron; *s. auch* **gebügelt**

Buggy ['bagi] **der; ~s, ~s** buggy

Bügler der; ~s, ~, Büglerin die; ~, ~nen ironer

bugsieren [bʊ'ksi:rən] *tr. V. (ugs.)* shift; manœuvre; steer ‹*person*›

Bug·welle die bow wave

buh [bu:] *Interj.* boo; **~ rufen** boo

Buh das; ~s, ~s *(ugs.)* boo

buhen *itr. V. (ugs.)* boo

buhlen ['bu:lən] *itr. V. (abwertend)* **um jmds. Gunst ~:** court sb.'s favour; **um jmds. Anerkennung ~:** strive for recognition by sb.

Buh·mann der; *Pl.* **Buhmänner** *(ugs.)* **a)** whipping-boy; scapegoat; **b)** *(Schreckgestalt)* bogyman

Bühne ['by:nə] **die; ~, ~n a)** stage; **ein Stück auf die ~ bringen** put on *or* stage a play; **auf der politischen ~** *(fig.)* on the political scene; **über die ~ bringen** *(ugs.)* finish ‹*process*›*;* get ‹*event*› over; **über die ~ gehen** *(ugs.)* go off; **b)** *(Theater)* theatre; **die Städtischen ~n Köln** the Cologne municipal theatres; **zur ~ gehen** go on the stage *or* into the theatre

bühnen-, Bühnen-: ~anweisung die stage direction; **~arbeiter der** stage-hand; **~autor der** playwright; **~bearbeitung die** stage adaptation; **~bildner der; ~s, ~, ~bildnerin die; ~, ~nen** stage *or* set designer; **~reif** *Adj.* ‹*play etc.*› ready for the stage; ‹*imitation etc.*› worthy of the stage; dramatic ‹*entrance etc.*›*;* **~stück das** stage play

Buh·ruf der boo

buk [bu:k] *1. u. 3. Pers. Sg. Prät. v.* **backen**

Bukett [bu'kɛt] **das; ~s, ~s** *od.* **~e** *(geh.)* bouquet

Bulette [bu'lɛtə] **die; ~, ~n** *(bes. berl.)* rissole

Bulgare [bʊl'ga:rə] **der; ~n, ~n** Bulgarian

Bulgarien [bʊl'ga:ri̯ən] **(das); ~s** Bulgaria

bulgarisch *Adj.* Bulgarian

Bull- [bʊl-]: **~auge das** circular porthole; **~dogge die** bulldog

Bulldozer ['bʊldo:zɐ] **der; ~s, ~:** bulldozer

[1]Bulle ['bʊlə] **der; ~n, ~n a)** bull; **b)** *(ugs. abwertend: Mann)* great ox; big bull; **c)** *(salopp abwertend: Polizist)* cop *(sl.)*

[2]Bulle die; ~, ~n *(päpstlicher Erlaß)* bull

Bullen·hitze die *(ugs.)* sweltering *or* boiling heat

Bulletin [byl'tɛ̃:] **das; ~s, ~s** bulletin

bullig 1. *Adj.* **a)** beefy, stocky ‹*person, appearance, etc.*›*;* chunky, hefty ‹*car*›*;* **b)** *(drückend)* sweltering, boiling ‹*heat*›. **2.** *adv.* **~ heiß** boiling hot

Bull·terrier der bull-terrier

bum [bʊm] *Interj.* bang

Bumerang ['bʊmərаŋ] **der; ~s, ~e** *od.* ~s boomerang; **es erwies sich als ~** *(fig.)* it boomeranged [on him/her/them]

Bummel ['bʊml̩] **der; ~s, ~:** stroll **(durch** around); *(durch Lokale)* pub-crawl *(coll.)*

Bummelant [bʊmə'lant] **der; ~en, ~en** *(ugs.)* **a)** slowcoach *(Brit.)*; slowpoke *(Amer.)*; dawdler; **b)** *(Faulenzer)* idler; loafer

Bummelei die; ~, ~en *(ugs.)* **a)** dawdling; **b)** *(Faulenzerei)* idling *or* loafing about

bummelig *(ugs. abwertend)* **1.** *Adj.* slow. **2.** *adv.* slowly

bummeln *itr. V.* **a)** *mit sein (ugs.)* stroll **(durch** around); **~ gehen** go for *or* take a stroll; **durch die Kneipen ~:** go round the pubs *(Brit. coll.)*; go on a pub-crawl *(Brit. coll.)*; **b)** *(ugs.: trödeln)* dawdle; **bei den Schulaufgaben ~:** dawdle over one's homework; **c)** *(ugs.: faulenzen)* laze about; do nothing

Bummel-: ~streik der go-slow; *(bei Beamten usw.)* work to rule; **in einen ~streik treten** go on a go-slow; **~zug der** *(ugs.)* slow *or* stopping train

bums [bʊms] *Interj.* bang!; **es machte laut ~:** there was a loud bang *or* thud

Bums der; ~es, ~e *(ugs.)* bang; *(dumpfer)* thud; thump

bumsen 1. *itr. V. (ugs.)* **a)** *unpers.* **es bumste ganz furchtbar** there was a terrible bang/thud *or* thump; **an dieser Kreuzung bumst es mindestens einmal am Tag** *(fig.)* there's at least one smash *or* crash a day at this junction; **b)** *(schlagen)* bang; *(dumpfer)* thump; **gegen die Tür ~:** bang/thump on the door; **c)** *mit sein (stoßen)* bang; bash; **er ist mit dem Kopf gegen die Wand gebumst** he banged *or* bashed his head on the wall; **d)** *(salopp: koitieren)* have it off *(sl.)*; screw *(vulg.)*. **2.** *tr. V. (salopp: koitieren mit)* have it off with *(sl.)*; screw *(vulg.)*

Bums-: ~lokal das *(ugs. abwertend)* dive *(coll.)*; **~musik die** *(ugs. abwertend)* oompah music *(coll.)*

[1]Bund [bʊnt] **der; ~[e]s, Bünde** ['bʏndə] **a)** *(Verband, Vereinigung)* association; society; *(Bündnis, Pakt)* alliance; **der Dritte im ~e** *(fig.)* the third in the trio; **den ~ der Ehe eingehen, den ~ fürs Leben schließen** *(geh.)* enter into the bond of marriage; **b)** *(föderativer Staat)* federation; **c)** *(ugs.: Bundeswehr)* forces *pl;* **beim ~:** in the forces *pl.;* **d)** *(an Röcken od. Hosen)* waistband

[2]Bund das; ~[e]s, ~e bunch; **ein ~ Petersilie** a bunch of parsley

Bündchen ['bʏntçən] **das; ~s, ~:** band

Bündel ['bʏndl̩] **das; ~s, ~:** bundle; **ein ~ von Fragen** *(fig.)* a set *or* cluster of questions; **sein ~ packen** *od.* **schnüren** pack one's bags *pl.*

bündeln *tr. V.* bundle up ⟨*newspapers, old clothes, rags, etc.*⟩; tie ⟨*banknotes etc.*⟩ into bundles/a bundle; tie ⟨*flowers, carrots, etc.*⟩ into bunches/a bunch; sheave ⟨*straw, hay, etc.*⟩

bündel·weise *Adv.* by the bundle; in bundles; *(bei Blumen, Möhren usw.)* by the bunch; in bunches

Bundes- federal ⟨*motorway, civil servant, territory, capital, state, authority, etc.*⟩; *(in Namen, Titeln)* Federal ⟨*Railway, Government, Republic, Chancellor, etc.*⟩

bundes-, Bundes-: ~bürger der *(Bürger der alten BRD)* West German citizen; **~deutsch** *Adj. (auf die alte BRD bezogen)* West German; **~ebene die: auf ~ebene** at federal *or* national level; **~genosse der** ally; **~grenzschutz der** Federal Border Police; **~haus das** Federal Parliament building; **~kabinett das** Federal Cabinet; **~kanzler der** Federal Chancellor; **~kanzleramt das** Federal Chancellery; **~lade die** *(jüd. Rel.)* Ark of the Covenant; **~land das** [federal] state; *(österr.)* province; **~liga die** national *or* federal division; **~ligist** [~ligıst] **der; ~en, ~en** team in the national *or* federal division; **~präsident der a)** [Federal] President; **b)** *(schweiz.)* President of the Confederation; **~rat der a)** Bundesrat; **b)** *(österr., schweiz.)* Federal Council; **~republik die** federal republic; **~republik Deutschland** Federal Republic of Germany; **~staat der a)** federal state; **b)** *(Gliedstaat)* state; **~straße die** federal highway; ≈ A road *(Brit.)*; **~tag der** Bundestag

Bundestags-: ~abgeordnete der/die member of parliament; member of the Bundestag; **~wahl die** parliamentary *or* general election

bundes-, Bundes-: ~trainer der national team manager; national coach; **~wehr die** [Federal] Armed Forces *pl.*; **~weit 1.** *Adj.; nicht präd.* nation-wide; national; **2.** *adv.* nation-wide; nationally

Bund-: ~falten *Pl.* pleats; **~hose die** knee-breeches *pl.*

bündig ['bʏndıç] **1.** *Adj.* **a)** concise; succinct; **b)** *(schlüssig)* conclusive. **2.** *adv.* **a)** concisely; succinctly; **b)** *(schlüssig)* conclusively

Bündnis ['bʏntnıs] **das; ~ses, ~se** alliance

Bund·weite die waist; *(Maß)* waist measurement

Bungalow ['bʊŋgalo] **der; ~s, ~s** bungalow

Bunker ['bʊŋkɐ] **der; ~s, ~ a)** *(auch Behälter)* bunker; **b)** *(Luftschutz~)* air-raid shelter; **c)** *(salopp: Gefängnis)* clink *(sl.)*

Bunsen·brenner ['bʊnzn̩-] **der** Bunsen burner

bunt [bʊnt] **1.** *Adj.* **a)** colourful; *(farbig)* coloured; **~e Farben/Kleidung** bright colours/brightly coloured *or* colourful clothes; **b)** *(fig.)* colourful ⟨*sight*⟩; varied ⟨*programme etc.*⟩; **ein ~er Abend** a social [evening]; *s. auch* **Hund a; c)** *(ungeordnet)* confused ⟨*muddle etc.*⟩; **ein ~es Treiben** a real hustle and bustle; **jetzt wird es mir zu ~** *(ugs.)* that's *or* it's too much. **2.** *adv.* **a)** colourfully; **die Vorhänge waren ~ geblümt** the curtains had a colourful floral pattern; **etw. ~ bemalen** paint sth. in bright colours; **~ gekleidet sein** be colourfully dressed; have colourful clothes; **b) ein ~ gemischtes Programm** a varied programme; **c) ~ durcheinander liegen** be in a complete muddle; **es zu ~ treiben** *(ugs.)* go too far; overdo it

bunt-, Bunt-: ~bemalt *Adj.; präd. getrennt geschrieben* brightly *or* colourfully painted; **~papier das** coloured paper; **~sand·stein der** red sandstone; **~scheckig** *Adj.* spotted; **~specht der** spotted woodpecker; **~stift der** coloured pencil/crayon; **~wäsche die** coloureds *pl.*

Bürde ['bʏrdə] **die; ~, ~n** *(geh.)* weight; load; *(fig.)* burden

Burg die; ~, ~en a) castle; **b)** *(Strand~)* wall of sand

Bürge ['bʏrgə] **der; ~n, ~n** guarantor

bürgen *itr. V.* **a) für jmdn./etw. ~:** vouch for *or* act as guarantor for sb./vouch for *or* guarantee sth.; **b)** *(fig.)* guarantee; **der Name bürgt für Qualität** the name is a guarantee of quality

Bürger der; ~s, ~, **Bürgerin die**; ~, ~nen citizen

Bürger-: **~initiative die** citizens' action group; **~krieg der** civil war

bürgerlich 1. *Adj.* **a)** *nicht präd.* *(staats~)* civil ‹*rights, marriage, etc.*›; civic ‹*duties*›; **das Bürgerliche Gesetzbuch** the [German] Civil Code; **sein ~er Name** his real name; **b)** *(dem Bürgertum zugehörig)* middle-class; **die ~e Küche** good plain cooking; good home cooking; **c)** *(Polit.)* non-socialist; *(nicht marxistisch)* non-Marxist; **d)** *(abwertend: spießerhaft)* bourgeois. **2.** *adv.* **a)** ‹*think, etc.*› in a middle-class way; **~ leben** live a middle-class life; **gut ~ essen** have a good plain meal; *(gewohnheitsmäßig)* eat good plain food; **b)** *(abwertend: spießerhaft)* in a bourgeois way

Bürgerliche der/die; *adj. Dekl.* **a)** *(Nichtadlige)* commoner; **b)** *(Polit.)* non-socialist

bürger-, Bürger-: **~meister der** mayor; **~meisterin die** mayor; **~nah** *Adj.* which/who reflects the general public's interests *postpos., not pred.*; **~pflicht die** civic duty; duty as a citizen; **~recht das** one of the civil rights; **~rechte** civil rights; **~rechtler der**; ~s, ~: civil-rights campaigner

Bürgerschaft die; ~, ~en **a)** citizens *pl.*; **b)** *(Stadtparlament)* city parliament

Bürger·schreck der bogey of the middle classes

Bürger·steig der pavement *(Brit.)*; sidewalk *(Amer.)*

Bürgertum das; ~s **a)** middle class; **b)** *(Groß~)* bourgeoisie

Bürger·wehr die vigilante group

Burg-: **~friede der** truce; **~graben der** [castle] moat

Bürgin ['bʏrgɪn] **die**; ~, ~nen *s.* **Bürge**

Bürgschaft die; ~, ~en **a)** *(Rechtsw.)* guarantee; security; **die ~ für jmdn./etw. übernehmen** agree to act as sb.'s guarantor/to guarantee sth.; **b)** *(Garantie)* guarantee; **c)** *(Betrag)* penalty

Burgund [bʊr'gʊnt] **(das)**; ~s Burgundy

Burgunder der; ~s, ~ *(Wein)* burgundy

burlesk [bʊr'lɛsk] *Adj.* burlesque

Burma ['bʊrma] **(das)**; ~s Burma

Büro [bʏ'ro] **das**; ~s, ~s office

Büro-: **~angestellte der/die** office-worker; **~bedarf der** office supplies *pl.*; **~haus das** office-block; **~klammer die** paper-clip; **~kraft die** clerical worker

Bürokrat [bʏro'kra:t] **der**; ~en, ~en *(abwertend)* bureaucrat

Bürokratie [byrokra'ti:] **die**; ~, ~n bureaucracy

bürokratisch 1. *Adj.* bureaucratic. **2.** *adv.* bureaucratically

Büro-: **~maschine die** office machine; **~zeit die** office hours *pl.*; **während der ~zeit** during office hours

Bürschchen ['bʏrʃçən] **das**; ~, ~: little fellow; little chap; **ein freches ~**: a cheeky little devil

Bursche ['bʊrʃə] **der**; ~n, ~n **a)** boy; lad; **b)** *(veralt.: junger Mann)* young man; **die jungen ~n aus dem Dorf** the village youths; **er ist ein toller ~** *(ugs.)* he's a reckless devil; **c)** *(abwertend: Kerl)* guy *(sl.)*; **d)** *(ugs.: Prachtexemplar)* specimen; **e)** *(Milit. hist.)* batman; orderly

Burschenschaft die; ~, ~en students' duelling society

burschikos [bʊrʃi'ko:s] **1.** *Adj.* **a)** sporty ‹*clothes, look*›; [tom]boyish ‹*behaviour, girl, haircut*›; **b)** *(ungezwungen)* casual ‹*comment, behaviour, etc.*›. **2.** *adv.* **a)** [tom]boyishly; **b)** *(ungezwungen)* ‹*express oneself*› in a colloquial way

Bürste ['bʏrstə] **die**; ~, ~n **a)** brush; **b)** *(Haarschnitt)* crew cut

bürsten *tr. V.* brush

Bürsten·schnitt der crew cut

Bus [bʊs] **der**; ~ses, ~se bus; *(Privat- und Reisebus)* coach; bus

Bus·bahn·hof der bus station; *(für Reisebusse)* coach station; bus station

Busch [bʊʃ] **der**; ~[e]s, **Büsche** ['bʏʃə] **a)** bush; *(fig.)* **auf den ~ klopfen** *(ugs.)* sound things out; **bei jmdm. auf den ~ klopfen** *(ugs.)* sound sb. out; **es ist etw. im ~** *(ugs.)* something's up; **b)** *(Geogr.)* bush; **c)** *(ugs.: Urwald)* jungle

Büschel ['bʏʃl̩] **das**; ~s, ~ *(von Haaren, Federn, Gras usw.)* tuft; *(von Heu, Stroh)* handful

buschig *Adj.* bushy

Busch-: **~mann der** Bushman; **~messer das** machete; **~wind·röschen das** wood anemone

Busen ['bu:zn̩] **der**; ~s, ~: bust; **sie hat wenig ~** *(ugs.)* she has very little bosom

busen-, Busen-: **~frei** *Adj.* topless; **~freund der**, **~freundin die** *(oft iron.)* bosom friend

Bus-: **~fahrer der** bus-/coach-driver; **~halte·stelle die** bus-/coach-stop; **~linie die** bus-/coach-route

Bussard ['bʊsart] **der**; ~s, ~e *(Zool.)* buzzard

Buße ['bu:sə] **die**; ~, ~n **a)** *(Rel.)* penance *no art.*; **b)** *(Rechtsw.)* damages *pl.*

büßen ['bʏ:sn̩] **1.** *tr. V.* **a)** *(Rel.: sühnen)* atone for; expiate; **b)** *(bestraft werden für)* atone for; **c)** *(fig.)* pay for; **das sollst du mir ~**: you'll pay for that. **2.** *itr. V.* **a)** *(Rel.)* **für etw. ~**: atone for *or* expiate sth.; **b)** *(bestraft werden)* suffer; **c)** *(fig.)* pay

Büßer der; ~s, ~ *(Rel.)* penitent

Buß-: **~geld das** fine; **~geld·bescheid der** official demand for payment of a fine; **~prediger der** repentance-preacher

Büsten·halter der bra; brassière *(formal)*

Bus-: **~verbindung die a)** *(Linie)* bus service; **b)** *(Anschluß)* bus/coach connection; **~verkehr der** bus/coach service

Butan·gas das butane gas

Butt [bʊt] **der**; ~[e]s, ~e flounder; butt

Büttel ['bʏtl̩] **der**; ~s, ~ *(abwertend)* lackey

Bütten das; ~s *s.* **~papier**

Bütten-: **~papier das** handmade paper *(with deckle-edge)*; **~rede die** carnival speech

Butter ['bʊtɐ] **die**; ~: butter; *(fig.)* **es ist alles in ~** *(ugs.)* everything's fine; **sie läßt sich** *(Dat.)* **nicht die ~ vom Brot nehmen** *(ugs.)* she doesn't let anyone put one over on her

Butter-: **~blume die** *(Sumpfdotterblume)* marsh marigold; *(Hahnenfuß)* buttercup; **~brot das** piece *or* slice of bread and butter; *(zugeklappt)* sandwich; **ein ~brot mit Schinken** a slice of bread and butter with ham on it/a ham sandwich; **~brot·papier das** grease-proof paper; **~creme die** butter-cream; **~creme·torte die** butter-cream cake; **~dose die** butter-dish

Butterfly ['bʌtəflai] *(Schwimmen)* butterfly [stroke]

Butter-: **~käse der** rich creamy cheese; **~keks der** butter biscuit; **~milch die** buttermilk

buttern 1. *itr. V.* make butter. **2.** *tr. V.* **a)** butter; grease ‹*baking tray*› with butter

butter·weich *Adj.* **a)** beautifully soft; **b)** *(fig.)* vague ‹*agreement, promise*›

Button [bʌtn] **der**; ~s, ~s badge

Butzen·scheibe die bull's-eye pane

b.w. *Abk.* **bitte wenden** p.t.o.
bzw. *Abk.* **beziehungsweise**

C

c, C [tse:] **das; ~, ~: a)** *(Buchstabe)* c/C; **b)** *(Musik)* [key of] C; *s. auch* **a, A**
C *Abk.* **Celsius** C
ca. *Abk.* **cirka** c.
Café [ka'fe:] **das; ~s, ~s** café
Cafeteria [kafetə'ri:a] **die; ~, ~s** cafeteria
cal *Abk.* **[Gramm]kalorie** cal.
Callgirl ['kɔ:lgə:l] **das; ~s, ~s** call-girl
Calypso [ka'lɪpso] **der; ~[s], ~s** calypso
Camembert ['kaməmbe:ɐ] **der; ~s, ~s** Camembert
Camp [kɛmp] **das; ~s, ~s** camp
campen ['kɛmpn̩] *itr. V.* camp
Camper der; ~s, ~, Camperin die; ~, ~nen camper
Camping ['kɛmpɪŋ] **das; ~s** camping; **zum ~ [nach X] fahren** go camping [in X]
Camping-: ~bus der motor caravan; camper; **~platz der** campsite; campground *(Amer.)*
Campus ['kampʊs] **der; ~** *(Hochschulw.)* campus
Canasta [ka'nasta] **das; ~s** canasta
Cape [ke:p] **das; ~s, ~s** cape
Caravan ['ka(:)ravan] **der; ~s, ~s a)** *(Kombi)* estate car; station wagon *(Amer.)*; **b)** *(Wohnwagen)* caravan; trailer *(Amer.)*
Caritas ['ka:ritas] **die; ~:** Caritas *(Catholic welfare organization)*
Cartoon [kar'tu:n] **der** *od.* **das; ~[s], ~s** cartoon
Casanova [kaza'no:va] **der; ~[s], ~s** Casanova
Cäsar ['tsɛ:zar] **(der)** Caesar
Cassata [ka'sa:ta] **die** *od.* **das; ~, ~s** cassata
catchen ['kɛtʃn̩] *itr. V.* do all-in wrestling
Catcher ['kɛtʃɐ] **der; ~s, ~:** all-in wrestler
Cayenne·pfeffer [ka'jɛn-] **der** cayenne [pepper]
CB-Funk [tse:'be:-] **der; ~s** CB radio
ccm *Abk.* **Kubikzentimeter** c.c.
CD [tse:'de:] **die; ~, ~s** CD
CD-Spieler der CD-player
CDU [tse:de:'|u:] **die; ~** *Abk.* **Christlich-Demokratische Union [Deutschlands]** [German] Christian Democratic Party
C-Dur ['tse:-] **das; ~:** C major; *s. auch* **A-Dur**
Cedille [se'di:j(ə)] **die; ~, ~n** *(Sprachw.)* cedilla
Cellist [tʃɛ'lɪst] **der; ~en, ~en** cellist
Cello ['tʃɛlo] **das; ~s, ~s** *od.* **Celli** cello
Celsius ['tsɛlziʊs] *o. Art.* **1 Grad/20 Grad ~:** 1 degree/20 degrees Celsius *or* centigrade
Cembalo ['tʃɛmbalo] **das; ~s, ~s** *od.* **Cembali** harpsichord
Ceylon ['tsailɔn] **(das); ~s** *(hist.)* Ceylon *(Hist.)*
C-Flöte ['tse:-] **die** soprano recorder
Chamäleon [ka'mɛ:leɔn] **das; ~s, ~s** *(auch fig.)* chameleon
Champagner [ʃam'panjɐ] **der; ~s, ~:** champagne
Champignon ['ʃampɪnjɔn] **der; ~s, ~s** mushroom
Champion ['tʃɛmpiən] **der; ~s, ~s** *(Sport)* champion
Chance ['ʃã:sə] **die; ~, ~n a)** *(Gelegenheit)* chance; **die ~n [zu gewinnen] stehen eins zu hundert** the chances [of winning] are one in a hundred; *(bes. beim Wetten)* the odds [against winning] are 100:1 *or* a hundred to one; **b)** *Pl. (Aussichten)* prospects; **[bei jmdm] ~n haben** stand a chance [with sb.]
Chancen·gleichheit die; *o. Pl. (Päd., Soziol.)* equality of opportunity *no art.*
changieren [ʃã'ʒi:rən] *itr. V.* shimmer *(in different colours)*; iridesce
Chanson [ʃã'sõ:] **das; ~s, ~s** chanson; cabaret-style song
Chaos ['ka:ɔs] **das; ~:** chaos *no art.*
Charakter [ka'raktɐ] **der; ~s, ~e** [...'te:rə] **a)** character; *(eines Menschen)* character; personality; **Geld verdirbt den ~:** money spoils people; **b)** *o. Pl. (~stärke)* [strength of] character; **keinen ~ haben** lack [strength of] character; be spineless
charakter-, Charakter-: ~darsteller der actor of complex parts; **~darstellerin die** actress of complex parts; **~eigenschaft die** characteristic; trait; **~fest** *Adj.* steadfast
charakterisieren *tr. V.* characterize
charakteristisch *Adj.* characteristic, typical **(für** of)
Charakter·kopf der striking head
charakterlich 1. *Adj.* character *attrib.* ‹*defect, development, training*›; personal ‹*qualities*›. **2.** *adv.* in [respect of] character
charakter-, Charakter-: ~los 1. *Adj.* unprincipled; *(niederträchtig)* despicable; *(labil)* spineless; **2.** *adv. (niederträchtig)* despicably; *(labil)* spinelessly; **~losigkeit die; ~:** lack of principle; *(Niederträchtigkeit)* despicableness; *(Labilität)* weakness of character; spinelessness; **~schwäche die** weakness of character; spinelessness *no pl.*; **~schwein das** *(salopp abwertend)* unprincipled bastard *(coll.)*; **~stärke die;** *o. Pl.* strength of character; **~voll** *Adj.* **a)** *(~fest)* steadfast; showing strength of character *postpos., not pred.*; **b)** *(ausdrucksvoll)* distinctive; ‹*house etc.*› of character; strongly characterized, individual ‹*features*›; **~zug der** characteristic
Charge ['ʃarʒə] **die; ~, ~n** rank; **die unteren ~n** the lower ranks *(Mil.)*/orders; **die oberen ~n** the upper ranks *(Mil.)*/echelons
charmant [ʃar'mant] **1.** *Adj.* charming. **2.** *adv.* charmingly; with much charm
Charme [ʃarm] **der; ~s** charm; **seinen ganzen ~ aufwenden** use all one's charms
Charta ['karta] **die; ~, ~s** *(Politik)* charter
Charter- [tʃartɐ-]: **~flug der** charter flight; **~maschine die** chartered aircraft
chartern *tr. V.* charter ‹*aircraft, boat*›; hire [the services of] ‹*guide, firm*›
Chassis [ʃa'si:] **das; ~** [ʃa'si:(s)], **~** [ʃa'si:s] *(Kfz-W., Elektrot.)* chassis
Chauffeur [ʃɔ'fø:ɐ] **der; ~s, ~e** driver; *(privat angestellt)* chauffeur
chauffieren *(veralt.) tr., itr. V.* drive
Chaussee [ʃo'se:] **die; ~, ~n** *(veralt.) (surfaced)* [high] road; highway *(Amer.)*
Chauvi ['ʃo:vi] **der; ~s, ~s** *(ugs. abwertend)* male chauvinist *(coll. derog.)*
Chauvinismus [ʃovi'nɪsmʊs] **der; ~** *(auch fig. abwertend)* chauvinism

Chauvinist der; ~en, ~en *(auch fig. abwertend)* chauvinist
chauvinistisch *(auch fig. abwertend) Adj.* chauvinistic; *(männlich-~)* male chauvinist
Check *(schweiz.) s.* **Scheck**
checken ['tʃɛkn̩] *tr. V.* **a)** *(bes. Technik: kontrollieren)* check; examine; **b)** *(salopp: begreifen)* twig *(coll.)*; *(bemerken)* spot
Check·liste ['tʃɛk-] **die** check-list
Chef [ʃɛf] **der; ~s, ~s a)** *(Leiter) (einer Firma, Abteilung, Regierung)* head; *(der Polizei, des Generalstabs)* chief; *(einer Partei, Bande)* leader; *(Vorgesetzter)* superior; boss *(coll.)*; **wer ist denn hier der ~?** who's in charge here?; **b)** *(salopp: Anrede)* **hallo, ~:** hey, chief *or* squire *(Brit. coll.)*; hey mister *(Amer. coll.)*
Chef- chief ‹*editor, ideologist, etc.*›
Chef-: ~arzt der *head of one or more specialist departments in a hospital; (Direktor)* superintendent *(of small hospital)*; **~etage die** management floor
Chefin die; ~, ~nen a) *(Leiterin) (einer Firma, Abteilung, Regierung)* head; *(einer Partei, Bande)* leader; *(Vorgesetzte)* superior; boss *(coll.)*; **b)** *(ugs.: Frau des Chefs)* boss's wife *(coll.)*; **c)** *(salopp: Anrede)* missis *(sl.)*; ma'am *(Amer.)*
Chef-: ~koch der chef; head cook; **~redakteur der** chief editor; **~sekretärin die** director's secretary
Chemie [çe'mi:] **die; ~ a)** chemistry *no art.*; **b)** *(ugs.: Chemikalien)* chemicals *pl.*
Chemie-: ~arbeiter der chemical worker; **~betrieb der** chemical firm; **~faser die** synthetic *or* man-made fibre; **~laborant der** chemical laboratory assistant
Chemikalie [çemi'ka:li̯ə] **die; ~, ~n** chemical
Chemiker ['çe:mikɐ] **der; ~s, ~, Chemikerin die; ~, ~nen** *(graduate)* chemist
chemisch 1. *Adj.* chemical; **~er Versuch** chemistry experiment. **2.** *adv.* chemically
Chester·käse ['tʃɛstɐ-] **der** *(usu. processed)* Cheddar cheese
Chicorée ['ʃikore] **der; ~s** *od.* **die; ~:** chicory
Chiffon ['ʃɪfõ] **der; ~s, ~s** chiffon
Chiffre ['ʃɪfrə] **die; ~, ~n a)** *(Zeichen)* symbol; **b)** *(Geheimzeichen)* cipher; **~n** cipher *sing.*; **c)** *(in Annoncen)* box number; **Zuschriften unter ~ ...:** reply quoting box no. ...
Chile ['tʃi:le, 'çi:lə] **(das); ~s** Chile
Chilene [tʃi'le:nə, çi'le:nə] **der; ~n, ~n, Chilenin die; ~, ~nen** Chilean
chilenisch *Adj.* Chilean
Chili ['tʃi:li] **der; ~s, ~es a)** *Pl. (Schoten)* chillies; **b)** *o. Pl. (Gewürz)* chilli [powder]
Chimäre [çi:mɛ:rə] **die; ~, ~n** *s.* **Schimäre**
China ['çi:na] **(das); ~s** China
Chinese [çi'ne:zə] **der; ~n, ~n, Chinesin die; ~, ~nen** Chinese
chinesisch *Adj.* Chinese; **die Chinesische Mauer** the Great Wall of China
Chinin [çi'ni:n] **das; ~s** quinine
Chip [tʃɪp] **der; ~s, ~s a)** *(Spielmarke)* chip; **b)** *(Kartoffel~)* [potato] crisp *(Brit.) or (Amer.)* chip; **c)** *(Elektronik)* [micro]chip
Chiropraktiker [çiro...] **der; ~s, ~** *(Med.)* chiropractor
Chirurg [çi'rʊrk] **der; ~en, ~en** surgeon
Chirurgie [çirʊr'gi] **die; ~, ~n a)** *o. Pl. (Disziplin)* surgery *no art.*; **b)** *(Abteilung)* surgical department; *(Station)* surgical ward
Chirurgin die; ~, ~nen surgeon
chirurgisch 1. *Adj.; nicht präd.* surgical. **2.** *adv. (operativ)* surgically; by surgery
Chitin [çi'ti:n] **das; ~s** chitin
Chitin·panzer der *(Zool.)* chitinous exoskeleton
Chlor [klo:ɐ̯] **das; ~s** chlorine
chloren *tr. V.* chlorinate
Chloroform [kloro'fɔrm] **das; ~s** chloroform
Chlorophyll [kloro'fʏl] **das; ~s** *(Bot.)* chlorophyll
Choke [tʃoʊk] **der; ~s, ~s** *(Kfz-W.)* [manual] choke
Cholera ['ko:lera] **die; ~** *(Med.)* cholera
Choleriker [ko'le:rikɐ] **der; ~s, ~ a)** choleric type; **b)** *(ugs.: jähzorniger Mensch)* irascible person; **ein ~ sein** have a short fuse
cholerisch 1. *Adj.* irascible; choleric ‹*temperament*›. **2.** *adv.* irascibly
Cholesterin [çolɛste'ri:n] **das; ~s** *(Med.)* cholesterol
Chor [ko:ɐ̯] **der; ~[e]s, Chöre** ['kø:rə] *(auch Archit.)* choir; *(in Oper, Sinfonie, Theater; Komposition)* chorus; **im ~ rufen** shout in chorus
Choral [ko'ra:l] **der; ~s, Choräle** [ko'rɛ:lə] **a)** chorale; **b)** *(Gregorianischer ~)* [Gregorian] chant
Choreograph [koreo'gra:f] **der; ~en, ~en** choreographer
Choreographie die; ~, ~n choreography
Chor-: ~knabe der choirboy; chorister; **~leiter der** chorus-master; *(eines Kirchenchors)* choirmaster; **~musik die** choral music; **~sänger der, ~sängerin die** member of the chorus
Chose ['ʃo:zə] **die; ~, ~n** *(ugs.)* **a)** *(Angelegenheit)* business *(derog.)*; **b)** *(Gegenstände)* stuff; **die ganze ~:** the whole lot *(coll.) or (sl.)* shoot *or (sl.)* caboodle
Chow-Chow [tʃau̯ 'tʃau̯] **der; ~s, ~s** chow
Christ [krɪst] **der; ~en, ~en** Christian
christ-, Christ-: ~baum der *(bes. südd.)* Christmas tree; **~demokrat der** *(Politik)* Christian Democrat; **~demokratisch** *(Politik) Adj.* Christian-Democrat
Christen·gemeinde die Christian community
Christenheit die; ~: Christendom *no art.*
Christentum das; ~s Christianity *no art.*; *(Glaube)* Christian faith
Christen·verfolgung die persecution of Christians
Christ·fest das *(veralt., noch südd., österr.) s.* **Weihnachtsfest**
Christin die; ~, ~nen Christian
Christ·kind das; *o. Pl.* Christchild *(as bringer of Christmas gifts)*
christlich 1. *Adj.* Christian. **2.** *adv.* in a [truly] Christian spirit; **Kinder ~ erziehen** give children a Christian upbringing
Christ-: ~messe die *(kath. Rel.)* Christmas Mass; **~mette die** *(kath. Rel.)* Christmas Mass; *(ev. Rel.)* midnight service [on Christmas Eve]; **~rose die** Christmas rose; **~stollen der** [German] Christmas loaf *(with candied fruit, almonds, etc.)*
Christus ['krɪstʊs] **(der); ~** *od.* **Christi** Christ
Christ·vesper die *(christl. Rel.)* Christmas Eve vespers *(with music)*
Chrom [kro:m] **das; ~s** chromium
Chromatik [kro'ma:tɪk] **die; ~** *(Musik)* chromaticism
chromatisch *Adj.* chromatic
Chromosom [kromo'zo:m] **das; ~s, ~en** *(Biol.)* chromosome
Chronik ['kro:nɪk] **die; ~, ~en** chronicle
chronisch *Adj.* chronic
Chronist [kro'nɪst] **der; ~en, ~en** chronicler
Chronologie die; ~: chronology
chronologisch 1. *Adj.* chronological. **2.** *adv.* chronologically; in chronological order

Chrysantheme [krʏzan'te:mə] **die; ~, ~n** chrysanthemum
CIA ['si:aɪ'eɪ] **der** *od.* **die; ~:** CIA
circa *s.* **zirka**
cis, Cis [tsɪs] **das; ~, ~** *(Musik)* C sharp
City ['sɪti] **die; ~, ~s** city centre
clever ['klɛvɐ] **1.** *Adj. (raffiniert)* shrewd; *(intelligent, geschickt)* clever; smart. **2.** *adv.: s. Adj.:* shrewdly; cleverly; smartly
Clinch [klɪntʃ] **der; ~[e]s a)** *(Boxen)* clinch; **b)** *(ugs.: Auseinandersetzung)* conflict; **mit jmdm. im ~ liegen** be locked in dispute with sb.
Clique ['klɪkə] **die; ~, ~n a)** *(abwertend: Interessengemeinschaft)* clique; **b)** *(Freundeskreis)* set; lot *(coll.); (größere Gruppe)* crowd *(coll.); (Jugendliche)* gang *(coll.)*
Clou [klu:] **der; ~s, ~s** *(ugs.)* main point; **der besondere ~:** the really special thing [about it]
Clown [klaun] **der; ~s, ~s** clown
Club *s.* **Klub**
cm *Abk.:* **Zentimeter** cm.
Co. *Abk.:* **Compagnie** Co.
Coach [koʊtʃ] **der; ~[s], ~s** *(Sport)* coach; *(bes. Fußball: Trainer)* manager
Cocker·spaniel ['kɔkɐ-] **der; ~s, ~s** cocker spaniel
Cockpit ['kɔkpɪt] **das; ~s, ~s** cockpit
Cocktail ['kɔkteɪl] **der; ~s, ~s** cocktail
Cocktail-: ~kleid das cocktail dress; **~party die** cocktail party
Cognac der; ~s, ~s Cognac
Collage [kɔ'la:ʒə] **die; ~, ~n** collage
Color- [ko'lɔ:ɐ-] *(Fot.)* colour ⟨*film, slide, etc.*⟩
Colt ® [kɔlt] **der; ~s, ~s** Colt **(P)** [revolver]
Combo ['kɔmbo] **die; ~, ~s** small (jazz *or* dance) band; combo *(sl.)*
Comeback [kam'bɛk] **das; ~[s], ~s** come-back; **ein ~ feiern** stage a come-back
Comic ['kɔmɪk] **der; ~s, ~s** comic strip; *(Heft)* comic
Comic·heft das comic
Computer [kɔm'pju:tɐ] **der; ~s, ~:** computer; **auf ~** *(Akk.)* **umstellen** computerize
Conférencier [kõferã'sie:] **der; ~s, ~s** compère *(Brit.);* master of ceremonies
Container [kɔn'te:nɐ] **der; ~s, ~:** container; *(für Müll)* [refuse] skip
Contergan·kind das thalidomide child
cool [ku:l] *(ugs.)* **1.** *Adj.* cool; **~ bleiben** keep one's cool *(sl.).* **2.** *adv.* coolly
Copyright ['kɔpirait] **das; ~s, ~s** copyright
Cord [kɔrt] **der; ~[e]s, ~e** *od.* **~s** cord; *(~samt)* corduroy
Cord-: ~anzug der cord/corduroy suit; **~hose die** [pair *sing.* of] corduroy trousers *pl. or* cords *pl.;* **~jeans** *Pl.* corduroy jeans; cords
Cordon bleu [kɔrdõ'blø] **das; ~ ~, ~s ~s** [kɔrdõ'blø] *(Kochk.)* veal escalope cordon bleu
Cord·samt der corduroy
Corned beef ['kɔ:nd 'bi:f] **das; ~ ~:** corned beef
Corn-flakes ['kɔ:nfleɪks] *Pl.* cornflakes
Couch [kautʃ] **die,** *(schweiz. auch:)* **der; ~, ~es** sofa
Couch-: ~garnitur die three-piece suite; **~tisch der** coffee-table
Couleur [ku'lø:ɐ] **die; ~, ~s** *o. Pl.* shade [of opinion]; persuasion
Countdown ['kaunt'daun] **der** *od.* **das; ~[s], ~s** *(Raumf., auch fig.)* countdown
Country-music ['kʌntrɪmju:zɪk] **die; ~:** country music
Coup [ku:] **der; ~s, ~s** coup; **einen ~ landen** *(ugs.)* pull off a coup
Coupé [ku'pe:] **das; ~s, ~s** *(Auto)* coupé
Coupon [ku'põ:] **der; ~s, ~s** coupon; voucher; **auf** *od.* **für** *od.* **gegen diesen ~ bekommen Sie ...:** for this voucher you will receive ...
Courage [ku'ra:ʒə] **die; ~** *(ugs.)* courage
couragiert [kura'ʒi:ɐt] **1.** *Adj. (mutig)* courageous; *(beherzt)* spirited. **2.** *adv. s.* **1:** courageously; spiritedly
Cousin [ku'zɛ̃:] **der; ~s, ~s** *(male)* cousin
Cousine [ku'zi:nə] **die; ~, ~n** *(female)* cousin
Cover ['kavɐ] **das; ~s, ~s a)** *(von Illustrierten)* cover; **b)** *(von Schallplatten)* sleeve
Cowboy ['kaubɔy] **der; ~s, ~s** cowboy
Cox Orange ['kɔks|orã:ʒə] **der; ~ ~, ~ ~:** Cox's orange pippin
Cracker ['krɛkɐ] **der; ~s, ~[s]** cracker
Credo *s.* **Kredo**
Creme [kre:m] **die; ~, ~s,** *(schweiz.:)* **~n a)** cream; **b)** *o. Pl. (oft iron.: Oberschicht)* cream; top people
creme·farben *Adj.* cream[-coloured]
Creme·torte die cream cake *or* gateau
cremig *Adj.* creamy; **etw. ~ schlagen** beat sth. into a cream
Crew [kru:] **die; ~, ~s** team; *(eines Schiffs/Flugzeugs)* crew
C-Schlüssel ['tse:-] **der** *(Musik)* C clef
ČSFR [tʃe:|ɛs|ɛf'|ɛr] **die; ~: die ~:** Czechoslovakia
ČSSR [tʃe:|ɛs|ɛs'|ɛr] **die; ~** *(1960–1990)* **die ~:** Czechoslovakia
CSU [tse:|ɛs'|u:] **die; ~** *Abk.:* **Christlich-Soziale Union** CSU
Cup [kap] **der; ~s, ~s** *(Sport)* cup
Curriculum [kʊ'ri:kulʊm] **das; ~s, Curricula** *(Päd.)* curriculum
Curry ['kœri] **das; ~s, ~s** curry-powder
Curry-: ~sauce, ~soße die curry sauce; **~wurst die** *sliced fried sausage sprinkled with curry powder and served with ketchup*
Cutter ['katɐ] **der; ~s, ~, Cutterin die; ~, ~nen** *(Film, Ferns., Rundf.)* editor
CVJM [tse:faujɔt'|ɛm] **der; ~** *Abk.:* **a) Christlicher Verein Junger Männer** YMCA; **b) Christlicher Verein Junger Menschen** *combined form of YMCA and YWCA*

D

d, D [de:] **das; ~, ~ a)** *(Buchstabe)* d/D; **b)** *(Musik)* [key of] D
D *Abk.* **Damen**
da [da:] **1.** *Adv.* **a)** *(dort)* there; **da draußen/drüben/unten** out/over/down there; **da hinten/vorn[e]** [there] at the back/front; **he, Sie da!** hey, you there!; **der Kerl da** that fellow [over there]; **halt, wer da?** *(Milit.)* halt, who goes there?; **da bist du ja!** there you are [at last]!; **da, ein Reh!** look, [there's] a deer!; **da, wo die Straße nach X abzweigt** where the road to X turns off; at the turning for X; **da und da** at such-and-such a place; **da und dort** here and there; *(manchmal)* now and again *or* then; **b)** *(hier)* here; **da hast du das Buch** here's the book; **da, nimm schon!** here [you are], take it!; *s. auch* **dasein, dahaben; c)** *(zeitlich)* then; *(in dem Augen-*

blick) at that moment; **von da an** from then on; **in meiner Jugend, da war alles besser** back in my young days, everything was better [then]; **d)** *(deshalb)* **der Zug war schon weg, da habe ich den Bus genommen** the train had already gone, so I took the bus; **e)** *(ugs.: in diesem Fall)* **da kann man nichts machen** there's nothing one can do about it *or* that; **da kann ich [ja] nur lachen!** that's plain ridiculous!; **was tut man da?** what does one do in a case like this?; **f)** *(altertümelnd: nach Relativpronomen; wird nicht übersetzt)* **..., der da sagt ...,** who says; **g)** *(hervorhebend; wird meist nicht übersetzt)* **ich habe da einen Kollegen, der ...:** I have a colleague who ...; **da fällt mir noch was ein** [oh yes] another thought strikes me. **2.** *Konj. (weil)* as; since

d. Ä. *Abk.:* **der Ältere**

da|behalten *unr. tr. V.* keep [there]; *(hierbehalten)* keep here

da·bei [*(hinweisend:)* '--] *Adv.* **a)** with it/him/her/them; **eine Tankstelle mit einer Werkstatt ~:** a filling station with its own workshop [attached]; **nahe ~:** near it; close by; **b)** *(währenddessen)* at the same time; *(bei diesem Anlaß)* then; on that occasion; **die ~ entstehenden Kosten** the expense involved; **er ist ~ gesehen worden, wie er das Geld nahm** he was seen [in the act of] taking the money; **ein Unfall – ~ gab es zwei Tote** an accident – two people were killed [in it]; **er suchte nach dem Brief, ~ hatte er ihn in der Hand** he was looking for the letter and all the time he had it in his hand; **c)** *(außerdem)* **~ [auch]** what is more; **er ist sehr beschäftigt, aber ~** *(dennoch)* **immer freundlich** he is very busy but even so always friendly; **d)** *(hinsichtlich dessen)* **ich fühle mich gar nicht wohl ~:** I'm not at all happy about it; **was hast du dir denn ~ gedacht?** what were you thinking of?; what came over you?; **er hat sich nichts ~ gedacht** he saw no harm in it; **e)** **da ist doch nichts ~!** there's really no harm in it!; *(es ist nicht schwierig)* there's nothing to it!; *s. auch* **bleiben a**

dabei-: ~|bleiben *unr. itr. V.; mit sein (dort)* stay there; be there; *(bei einer Tätigkeit)* stick to it; **~|haben** *unr. tr. V.* have with one; **ich habe kein Geld ~:** I haven't got any money with me *or* on me; **~|sein** *unr. itr. V.; mit sein (Zusschr. nur im Inf. u. 2. Part.)* **a)** *(anwesend sein)* be there; be present (**bei** at); *(teilnehmen)* take part (**bei** in); **Dabeisein ist alles!** it's taking part that counts; **b) [gerade] ~sein, etw. zu tun** be just doing sth.; **~|sitzen** *unr. itr. V.* sit there; **~|stehen** *unr. itr. V.* stand by; stand there

da|bleiben *unr. itr. V.; mit sein* stay there; *(hierbleiben)* stay here; **[noch] ~:** stay on

Dach [dax] **das; ~[e]s, Dächer** ['dɛçɐ] **a)** roof; **[ganz oben] unterm ~** [right up] in the attic; **ein/kein ~ über dem Kopf haben** *(ugs.)* have a/no roof over one's head; **etw. unter ~ und Fach bringen** get sth. [safely] under cover; bring in sth.; *(fig.: erfolgreich beenden)* get sth. all wrapped up; **b)** *(fig. ugs.)* **jmdm. aufs ~ steigen** give sb. a piece of one's mind; **jmdm. eins aufs ~ geben** bash sb. over the head; *(tadeln)* give sb. a dressing down; tear a strip off sb. *(sl.);* **eins aufs ~ kriegen** get a bash on the head; *(eine Rüge erhalten)* get it in the neck *(coll.)*

Dach-: ~boden der loft; **auf dem ~boden** in the loft; **~decker** [~dɛkɐ] **der; ~s, ~:** roofer **~fenster das** skylight; *(Dachgaube)* dormer window **~garten der** roof-garden; **~gepäckträger der** *(Kfz-W.)* roof-rack; **~geschoß das** attic [storey]; **~kammer die** attic [room]; *(ärmlich)* garret; **~lawine die** *mass of snow sliding from a roof;* **~luke die** skylight; **~rinne die** gutter

Dachs [daks] **der; ~es, ~e** badger

Dachs·bau der; *Pl.* **~e** badger's earth *or* set

Dach-: ~schaden der a) *o. Pl. (ugs.)* **einen ~schaden haben** be not quite right in the head; be slightly screwy *(sl.);* **b)** *(Schaden am Dach)* roof-damage; **~stube die** *(veralt.) s.* **~kammer; ~stuhl der** roof-truss

dachte ['daxtə] *1. u. 3. Pers. Sg. Prät. v.* **denken**

Dach-: ~terrasse die roof-terrace; **~wohnung die** attic flat *(Brit.) or (Amer.)* apartment; **~ziegel der** roof-tile; **~zimmer das** attic room

Dackel ['dakl̩] **der; ~s, ~:** dachshund

Dackel·beine *Pl. (ugs. scherzh.)* [stumpy] bow legs

da·durch [*(hinweisend:)* '--] *Adv.* **a)** through it/them; **b)** *(durch diesen Umstand)* as a result; *(durch dieses Mittel)* in this way; by this [means]; **ich nehme den D-Zug, ~ bin ich zwanzig Minuten eher da** I'll take the express, that way I'll get there twenty minutes earlier; **~, daß er älter ist, hat er einige Vorteile** he has several advantages by virtue of being older *or* because he is older; **~ gekennzeichnet sein, daß ...** be characterized by the fact that ...

da·für [*(hinweisend:)* '--] *Adv.* **a)** for it/them; **~, daß ...** *(damit)* so that ...; **~ sorgen [, daß ...]** see to it [that ...]; **der Grund ~, daß ...** the reason why ...; **~ sein** be in favour [of it]; **ich bin ganz ~:** I'm all for it; **das ist ein Beweis ~, daß ...:** this is proof that ...; **ein Beispiel ~ ist ...:** an example of this is ...; **alles spricht ~, daß ...:** all the evidence *or* everything suggests that ...; **b)** *(als Gegenleistung)* in return [for it]; *(beim Tausch)* in exchange; *(statt dessen)* instead; **heute hat er keine Zeit, ~ will er morgen kommen** he has no time today, so he wants to come tomorrow instead; **c) er ist schon 60, aber ~ hält ihn niemand** he is 60 but nobody would think so; **d)** *(wenn man das berücksichtigt)* **~ ist sein Französisch nicht sehr gut** his French is not very good, considering; **~ daß ...** considering that ...

dafür·können *unr. tr. V.* **etwas/nichts ~** be/not be responsible; **dafür kann er nichts[, daß ...]:** it's not his fault [that ...]; he can't help it [that ...]

dagegen [*(hinweisend:)* '---] *Adv.* **a)** against it/them; **er stieß aus Versehen ~:** he knocked into it by mistake; **ich protestiere energisch ~, daß Sie mich verleumden** I must protest strongly against this slander; **ich habe nichts ~:** I've no objection; I don't mind; **was hat er ~, daß wir Freunde sind?** why does he object to our being friends?; **~ sein** be opposed to it *or* against it; **~ sein, etw. zu tun** be opposed to doing sth.; **was spricht ~?** what is the objection?; **~ kann man nichts machen** there is nothing one can do about it; **b)** *(im Vergleich dazu)* by *or* in comparison; compared with that; *(jedoch)* on the other hand; **c)** *(als Gegenwert)* in exchange

dagegen-: ~|halten *unr. tr. V.* **a)** *(entgegnen)* counter; *(einwenden)* object; **b)** *(ugs.: vergleichen)* hold it/them against; compare it/them with; **~|stellen** *refl. V.* oppose it

da|haben *unr. tr. V. (Zusschr. nur im Inf. u. 2. Part.) (ugs.)* have

[here]; *(im Hause)* have in the house; **mal sehen, ob ich noch eins da habe** I'll see whether I've got one left
da·heim *Adv. (bes. südd., österr., schweiz.)* **a)** *(zu Hause)* at home; *(nach Präp.)* home; ~ **anrufen** phone *or* ring home; **bei mir ~:** at my place; **b)** *(in der Heimat)* [back] home; **bei uns ~:** back home where I/we come from
da·her *Adv.* **a)** from there; ~ **habe ich meine neuen Stiefel** that's where I got my new boots from; ~ **weht also der Wind!** *(ugs.)* so 'that's the way the wind blows! *(fig.)*; **b)** *(durch diesen Umstand)* hence; ~ **kommt seine gute Laune** that's why he's in a good mood; ~ **wußte er das** *od.* **hat er das** that's how he knew; that's where he got it from; **c)** *(deshalb)* therefore; so
daher-: ~gelaufen *Adj.; nicht präd. (abwertend)* that nobody's heard of *postpos.*; **jeder ~gelaufene Kerl** any guy who comes along; any Tom, Dick, or Harry; **~|kommen** *unr. itr. V.* come along; **~reden** *(abwertend)* **1.** *itr. V.* talk off the cuff; **[so] dumm ~reden** talk [such] rubbish; **2.** *tr. V.* say off the cuff
da·hin *Adv.* **a)** there; **b)** *(fig.)* ~ **mußte es kommen** it had to come to that; **du wirst es ~ bringen, daß ...:** you'll carry things *or* matters so far that ...; **c)** *in* **bis ~:** to there; *(zeitlich)* until then; **bis ~ sind es 75 km** it's 75 km from here; **es steht mir bis ~** *(ugs.)* I am sick and tired of it *or* fed up to the back teeth with it *(coll.)*; **d)** [-'-] *(verloren, vorbei)* ~ **sein** be *or* have gone; **e)** *(in diesem Sinne)* ~ **[gehend], daß ...:** to the effect that ...; **man kann dieses Schreiben auch ~ [gehend] auslegen, daß ...:** one can also interpret this letter as meaning that ...
da-: ~hinab *Adv.* down there; down that way; **~hinauf** *Adv.* up there; up that way; **~hinaus** *Adv.* out there; *(in die Richtung)* out that way
dahin-: ~|dämmern *itr. V.; mit sein* be semi-conscious; **~|eilen** *itr. V.; mit sein (geh.)* hurry along *or* on one's way; ‹*time*› fly [past]
da·hinein *Adv.* in there; *(hier hinein)* in here
dahin-: ~|gehen *unr. itr. V.; mit sein (geh.: vergehen)* pass; ‹*years*› go by; **~gestellt** *in* **es ist** *od.* **bleibt ~gestellt** it remains to be seen; **etw. ~gestellt sein lassen** leave sth. open [for the moment]; **~|jagen** *itr. V.; mit sein (geh.)* tear *or* race along; **~|sagen** *tr. V.* say without thinking; **das war nur so ~gesagt** that was just a casual *or* off-the-cuff remark
da·hinten *Adv.* over there
da·hinter [*(hinweisend:)* '---] *Adv.* behind it/them; *(folgend)* after it/them; **ein Haus mit einem Garten ~:** a house with a garden behind *or* at the back
dahinter-: ~|klemmen *refl. V.(ugs.)* buckle down to it; pull one's finger out *(sl.)*; **~|kommen** *unr. itr. V.; mit sein (ugs.)* find out; **~|stecken** *itr. V. (ugs.)* **a)** *(als Grund, Urheber)* be behind it/them; **b)** *(Sinn haben)* **es steckt nichts/nicht viel ~:** there is nothing/not much to it/them; **~|stehen** *unr. itr. V. (fig.)* be behind it/them
dahin|ziehen 1. *unr. itr. V.; mit sein* go *or* move on one's/its way; ‹*clouds*› drift by; **2.** *unr. refl. V.* ‹*path*› pass along
Dahlie ['da:liə] **die; ~, ~n** dahlia
da-: ~|lassen *unr. tr. V. (ugs.)* leave there; *(hierlassen)* have [here]; **~|liegen** *unr. itr. V.* lie there
dalli ['dali] *Adv. (ugs.)* **aber [ein bißchen] ~!** and make it snappy *(coll.)*; **[~] ~!** get a move on!
damalig ['da:ma:lıç] *Adj.; nicht präd.* at that *or* the time *postpos.*; **der ~e Bundeskanzler** the then Federal Chancellor; the Federal Chancellor at that *or* the time; **die ~e Regierung** the government of the day; **im ~en Gallien** in what was then Gaul
damals ['da:ma:ls] *Adv.* then; at that time; **~, als ...:** at the time *or* in the days when ...; **von ~:** of that time *or* those days; *(aus dieser Zeit)* from that time *or* those days; **seit ~:** since then
Damast [da'mast] **der; ~[e]s, ~e** damask
Dame ['da:mə] **die; ~, ~n a)** lady; **sehr verehrte** *od.* **meine ~n und Herren!** ladies and gentlemen; **die Abfahrt/die 200 Meter der ~n** *(Sport)* the women's downhill/200 metres; **b)** *(Schach, Kartenspiele)* queen; **c)** *o. Pl. (Spiel)* draughts *(Brit.)*; checkers *(Amer.)*; **d)** *(Doppelstein im Damespiel)* king
Damen-: ~binde die sanitary towel *(Brit.) or (Amer.)* napkin; **~fahr·rad das** lady's bicycle; **~friseur der** ladies' hairdresser
damenhaft 1. *Adj.* ladylike. **2.** *adv.* like a lady; in a ladylike manner
Damen-: ~mannschaft die women's team; **~rad das** lady's bicycle; **~salon der** ladies' hairdressing salon *(Brit.)*; beauty salon *(Amer.)*; **~sitz der** *(Reiten)* **im ~sitz reiten** ride side-saddle; **~toilette die** ladies' toilet; **~wahl die;** *o. Pl.* ladies' choice
Dame·spiel das draughts *(Brit.)*; checkers *(Amer.)*
Dam·hirsch ['dam-] **der** fallow deer; *(männliches Tier)* fallow buck
da·mit [*(hinweisend:)* '--] **1.** *Adv.* **a)** *(mit dieser Sache)* with it/them; **ich bin gleich ~ fertig** I'll be finished in a moment; **er hatte nicht ~ gerechnet** he had not expected that *or* reckoned with that; **was ist denn ~?** what's the matter with it/them?; what about it/them?; **wie wäre es ~?** how about it?; **b)** *(gleichzeitig)* with that; thereupon; **c)** *(daher)* thus; as a result. **2.** *Konj.* so that
dämlich ['dɛ:mlıç] *(ugs. abwertend)* **1.** *Adj.* stupid. **2.** *adv.* stupidly; **~fragen** ask stupid questions
Dämlichkeit die; ~ *(ugs. abwertend)* stupidity
Damm [dam] **der; ~[e]s, Dämme** ['dɛmə] **a)** embankment; levee *(Amer.)*; *(Deich)* dike; *(Stau~)* dam; *(fig.)* bulwark; **b)** *(Straßen~, Bahn~)* embankment
Dämmer·licht das; *o. Pl.* twilight; *(trübes Licht)* dim light
dämmern ['dɛmɐn] *itr. V.* **a) es dämmert** *(morgens)* it is getting light; *(abends)* it is getting dark; **der Morgen dämmert** the day is dawning *or* breaking; **der Abend dämmert** dusk is falling; **b)** *(ugs.: klarwerden)* **jmdm. ~:** dawn upon sb.; **mir dämmert da etwas** the penny is beginning to drop; *(ich habe einen Verdacht)* I am beginning to smell a rat; **c)** *(halb schlafen)* doze
Dämmerung die; ~, ~en *(Abend~)* twilight; dusk; *(Morgen~)* dawn; daybreak
dämmrig *Adj.* **a) es ist/wird schon ~** *(morgens)* it is beginning to get light; day is breaking; *(abends)* it is beginning to get dark; night is falling; **b)** *(halbdunkel)* gloomy; dim ‹*light*›
Dämon ['dɛ:mɔn] **der; ~s, ~en** [dɛ'mo:nən] demon
dämonisch 1. *Adj.* daemonic. **2.** *adv.* daemonically
Dampf [dampf] **der; ~[e]s, Dämpfe** ['dɛmpfə] steam *no pl., no indef. art.*; *(Physik)* [water] vapour *as tech. term, no pl., no indef. art.*; ~

dahinter/hinter etw. *(Akk.)* **machen** *(ugs.) (sich beeilen)* get a move on/get a move on with sth.; *(andere zur Eile treiben)* get things *pl.*/sth. moving

Dampf·bügeleisen das steam iron

dampfen *itr. V.* steam

dämpfen ['dɛmpfn̩] *tr. V.* **a)** *(mit Dampf garen)* steam ⟨*fish, vegetables, potatoes*⟩; **b)** *(mildern)* muffle, deaden ⟨*sound*⟩; lower ⟨*voice*⟩; dim, turn down ⟨*lights*⟩; cushion, absorb ⟨*blow, impact, shock*⟩; *(fig.)* temper, diminish ⟨*joy*⟩; dampen ⟨*enthusiasm*⟩; assuage ⟨*sb.'s wrath*⟩; calm ⟨*anger, excitement*⟩

Dampfer der; ~s, ~: steamer; **auf dem falschen ~ sein** *(fig. ugs.)* be barking up the wrong tree; have got it wrong

Dämpfer der; ~s, ~: a) *(beim Klavier)* damper; *(bei Streich- u. Blasinstrumenten)* mute; **b)** *(fig.)* **einen ~ bekommen** *(ugs.)* have one's enthusiasm dampened; *(gerügt werden)* be taken down a peg or two

Dampf-: ~kessel der boiler; **~kochtopf der** pressure-cooker; **~lok[omotive] die** steam locomotive *or* engine; **~maschine die** steam engine; **~nudel die** *(südd., Kochk.)* steamed yeast dumpling; **~schiff das** steamer

Dämpfung die; ~, ~en a) *(der Stimme)* lowering; *(von Licht)* dimming; **b)** *(Stoß~)* cushioning; absorption; *(von Schwingungen)* damping; *(fig.) (von Freude, Leidenschaft)* tempering; diminishing; *(von Begeisterung)* dampening; *(von Wut, Aufregung)* calming

Dampf·walze die steamroller

da·nach [*(hinweisend:)* '--] *Adv.* **a)** *(zeitlich)* after it/that; then; **noch tagelang ~:** for days after[wards]; **eine Stunde ~:** an hour later; **b)** *(räumlich: dahinter)* after it/them; **voran gingen die Eltern, ~ kamen die Kinder** the parents went in front, the children following after *or* behind; **c)** *(ein Ziel angebend)* towards it/them; **er griff ~:** he made a grab for it/them; **~ laßt uns alle streben** let us all strive for that; **~ fragen** ask about it/them; **d)** *(entsprechend)* in accordance with it/them; **ein Brief ist gekommen; ~ ist sie schon unterwegs** a letter has arrived, according to which she is already on her way; **ihr kennt die Regeln, nun richtet euch ~!** you know the rules, so stick to *or* abide by them

Däne ['dɛ:nə] **der; ~n, ~n** Dane; **er ist ~:** he is Danish *or* a Dane

da·neben [*(hinweisend:)* '---] *Adv.* **a)** next to *or* beside him/her/it/them *etc.*; **b)** *(im Vergleich dazu)* in comparison; **c)** *(außerdem)* in addition [to that]; besides [that]

daneben-: ~|benehmen *unr. refl. V. (ugs.)* blot one's copybook *(coll.)*; spoil one's record; *(sich aufführen)* make an exhibition of oneself; **~|gehen** *unr. itr. V.; mit sein* **a)** *(das Ziel verfehlen)* miss [the target]; **b)** *(ugs.: fehlschlagen)* misfire; be a flop *(sl.)*; **~|schießen** *unr. itr. V.* miss [the target]; **mit Absicht ~ schießen** shoot to miss; **~|tippen** *itr. V. (ugs.)* guess wrong

Dänemark ['dɛ:nəmark] **(das); ~s** Denmark

dang [daŋ] *1. u. 3. Pers. Sg. Prät. v.* **dingen**

danieder|liegen *unr. itr. V. (geh.)* **a)** *(krank sein)* be laid low; **schwer [krank]/sterbend ~:** lie seriously ill/dying; **b)** *(fig.)* ⟨*trade, economy*⟩ be depressed

Dänin die; ~, ~nen Dane; Danish woman/girl

dänisch ['dɛ:nɪʃ] *Adj.* Danish

dank [daŋk] *Präp. mit Dat. u. Gen.* thanks to

Dank der; ~[e]s a) thanks *pl.*; **jmdm. seinen ~ abstatten** offer one's thanks to sb.; **jmdm. [großen] ~ schulden** *od.* **schuldig sein** *(geh.)*, **jmdm. zu [großem] ~ verpflichtet sein** owe sb. a [great] debt of gratitude; **und das ist nun der ~ dafür** *(iron.)* so that's all the thanks I get!; **mit vielem** *od.* **bestem ~ zurück** thanks for the loan; *(bes. geschrieben)* returned with thanks!; **b)** *(in Dankesformeln)* **vielen/besten/herzlichen ~!** thank you very much; many thanks; **vielen ~, daß du mir geholfen hast** thank you very much for helping me; **tausend ~!** *(ugs.)* very many thanks [indeed]

dankbar 1. *Adj.* **a)** grateful; *(anerkennend)* appreciative ⟨*child, audience, etc.*⟩; **sich ~ zeigen** show one's gratitude *or* appreciation; **für eine baldige Antwort wären wir ~:** we should be grateful for an early reply; **b)** *(lohnend)* rewarding ⟨*job, part, task, etc.*⟩. **2.** *adv.* gratefully; **jmdn. ~ anblicken** give sb. a look of gratitude

Dankbarkeit die; ~: gratitude

danke ['daŋkə] *Interj.* thank you; *(ablehnend)* no, thank you; **ja ~[, gern]** yes, please; **nein ~:** no, thank you; **~ schön/sehr/vielmals** thank you very much; **~ schön sagen** say 'thank you'; **sonst geht's dir [wohl] ~!** *(ugs.)* what do you think you're doing?; have you taken leave of your senses?

danken 1. *itr. V.* thank; **ich danke Ihnen vielmals** thank you very much; **Betrag ~d erhalten** [payment] received with thanks; **na, ich danke!** *(ugs.)* no, 'thank you!. **2.** *tr. V.* **[aber bitte,] nichts zu ~:** don't mention it; not at all; **sie hat ihm seine Hilfe schlecht gedankt** she gave him a poor reward for his help

dankens·wert *Adj.* commendable ⟨*effort etc.*⟩; **es ist ~, daß er uns hilft** it is kind *or* very good of him to help [us]

Danke·schön das; ~s thank-you; **ein [herzliches] ~ sagen** express one's [sincere] thanks

dann [dan] *Adv.* **a)** then; **was ~?** what happens then?; **noch drei Tage, ~ ist Ostern** another three days and it will be Easter; **bis ~:** see you then; **~ und wann** now and then; **er ist der Klassenbeste, ~ kommt sein Bruder** he is top of the class, followed by his brother *or* then comes his brother; **b)** *(unter diesen Umständen)* then; in that case **[na,] ~ eben nicht!** in that case, forget it!; **~ bis morgen** see you tomorrow, then; **nur ~, wenn ...:** only if ...; **c)** *(außerdem)* **~ noch ...:** then ... as well; **zuletzt fiel ~ noch der Strom aus** finally to top it all there was a power failure

dannen ['danən] *Adv. in* **von ~** *(veralt.)* from thence *(arch./literary)*

daran [da'ran, *(hinweisend:)* '--] *Adv.* **a)** on it/them; **es hängt etwas ~:** something is hanging from it/them; **er klammert sich ~** *(auch fig.)* he clings to it; **~ riechen** take a sniff at it/them; **dicht ~:** close to it/them; **nahe ~ sein, etw. zu tun** be on the point of doing sth.; **b) ~ ist nichts zu machen** there's nothing one can do about it; **~ wird sich nichts ändern** nothing will alter this fact; **kein Wort ~ ist wahr** not a word of it is true; **~ arbeiten** work on it/them; **wir haben keinen Bedarf mehr ~:** we no longer have any need of it/them; **mir liegt viel ~:** it means a lot to me; **c) ich wäre beinahe ~ erstickt** I almost choked on it; it almost made me choke; **er ist ~ gestorben** he died of it; **d) im Anschluß**

~ **fand eine Diskussion statt** after that there was a discussion

daran-: ~|**gehen** *unr. itr. V.; mit sein* set about it; ~**gehen, etw. zu tun** set about doing sth.; ~|**machen** *refl. V. (ugs.)* set about it; *(energisch)* get down to it; ~|**setzen** *tr. V.* devote ‹*energy etc.*› to it; summon up ‹*ambition*› for it; *(aufs Spiel setzen)* risk ‹*one's life, one's honour*› for it

darauf [da'rauf, *(hinweisend:)* '--] *Adv.* **a)** on it/them; *(oben ~)* on top of it/them; **b) er hat ~ geschossen** he shot at it/them; ~ **müßt ihr zugehen** that's what you must head towards *or* make for; **er ist ganz versessen ~:** he is mad [keen] on it *(sl.)*; **also darauf willst du hinaus** so 'that's what you're getting at; **c) wie kommst du nur ~?** what makes you think that?; **d)** *(danach)* after that; **ein Jahr ~ / kurz ~ starb er** he died a year later/shortly afterwards; **zuerst kamen die Kinder, ~ folgten die Festwagen** first came the children, then followed *or* followed by the floats; **e)** *(infolgedessen, daraufhin)* because of that; as a result

darauf-: ~**folgend** *Adj.; nicht präd.* following; **am ~folgenden Tag** the following day; next day; ~**hin** [--'-] *Adv.* **a)** *(infolgedessen)* as a result [of this/that]; consequently; *(danach)* thereupon; **b)** *(unter diesem Gesichtspunkt)* with a view to this/that; **etw. ~hin prüfen, ob es geeignet ist** examine sth. to see whether it is suitable

daraus [da'raus, *(hinweisend:)* '--] *Adv.* **a)** from it/them; out of it/them; **b) mach dir nichts ~** don't worry about it; ~ **ist eine große Firma geworden** it has become *or* turned into a large business; **was ist ~ geworden?** what has become of it?; ~ **wird nichts** nothing will come of it

darben ['darbn̩] *itr. V. (geh.)* live in want; *(hungern)* go hungry

dar|bieten *(geh.)* **1.** *unr. tr. V.* perform. **2.** *unr. refl. V.* **sich jmds. Blicken ~:** expose oneself to sb.'s gaze

Darbietung die; ~, ~en *(geh.)* **a)** presentation; **b)** *(Aufführung)* performance; *(beim Varieté usw.)* act

darf [darf] *1. u. 3. Pers. Sg. Präsens v.* **dürfen**

darfst [darfst] *2. Pers. Sg. Präsens v.* **dürfen**

darin [da'rɪn, *(hinweisend:)* '--] *Adv.* **a)** in it/them; *(drinnen)* inside [it/them]; **b)** *(in dieser Hinsicht)* in that respect; ~ **stimme ich völlig mit Ihnen überein** I entirely agree with you there

dar|legen *tr. V.* explain *(Dat.* to); set forth ‹*reasons, facts*›; expound ‹*theory*›

Darlegung die; ~, ~en explanation

Darlehen ['da:ɐ̯le:ən] **das; ~s, ~:** loan; **ein ~ aufnehmen** get *or* raise a loan

Darm [darm] **der; ~[e]s, Därme** ['dɛrmə] **a)** intestines *pl.*; bowels *pl.*; **b)** *(Wursthaut)* skin; **c)** *o. Pl. (Material)* gut

Darm-: ~**grippe die** gastric influenza; ~**trägheit die** *(Med.)* constipation

dar|stellen 1. *tr. V.* **a)** depict; portray; **etw. graphisch ~:** present sth. graphically; **b)** *(verkörpern)* play; act; **etwas/nichts ~:** make [a bit of] an impression/not make any sort of an impression; ‹*gift etc.*› look good/not look anything special; **c)** *(schildern)* describe ‹*person, incident, etc.*›; present ‹*matter, argument*›; **d)** *(sein, bedeuten)* represent; constitute. **2.** *refl. V.* **a)** *(sich erweisen, sich zeigen)* prove [to be]; turn out to be; **sich jmdm. als ... ~:** appear to sb. as ...; **b)** *(sich selbst schildern)* portray oneself

Darsteller der; ~s, ~ actor

Darstellerin die; ~, ~nen actress

darstellerisch *Adj.; nicht präd.* acting *attrib.*; **ihre ~en Fähigkeiten** her abilities as an actress

Darstellung die a) representation; *(Schilderung)* portrayal ; *(Bild)* picture; **graphische/schematische ~:** diagram; *(Graph)* graph; **b)** *(Beschreibung, Bericht)* description; account; **c)** *(Theater)* interpretation; performance

darüber [da'ry:bɐ, *(hinweisend:)* '---] *Adv.* **a)** *(über diesem/diesen)* over *or* above it/them; *(über dies/diese)* over it/them; **wir wohnen im zweiten Stock und er ~:** we live on the second floor and he lives above us; **b) ~ hinaus** in additon [to that]; *(noch obendrein)* what is more; **c)** *(über dieser/diese Angelegenheit)* about it/them; ~ **wollen wir hinwegsehen** we will overlook it; **d)** *(über diese Grenze, dieses Maß hinaus)* above [that]; over [that]; **Ist es schon 12 Uhr? – Aber ja, es ist schon 10 Minuten ~:** Is it twelve o'clock yet? – Oh yes, it's already ten past; **e)** *(währenddessen)* meanwhile; **f)** *(währenddessen und deshalb)* because of it/them; as a result

darüber-: ~|**fahren** *unr. itr. V.; mit sein* run over it/them; ~|**liegen** *unr. itr. V.* be higher ~|**stehen** *unr. itr. V. (fig.)* be above such things; ~|**steigen** *unr. itr. V.; mit sein* climb over it/ them

darum [da'rʊm, *(hinweisend:)* '--] *Adv.* **a)** [a]round it/them; **b) ich werde mich ~ bemühen** I will try to deal with it; *(versuchen, es zu bekommen)* I'll try to get it; **sie wird nicht ~ herumkommen, es zu tun** she won't get out of *or* avoid doing it; **es geht mir ~, eine Einigung zu erzielen** my concern *or* aim is to reach an agreement; **c)** ['--] *(deswegen)* because of that; for that reason; **ach, ~ ist er so schlecht gelaunt!** so that's why he's in such a bad mood!; **Warum weinst du? – Darum!** Why are you crying? – Because!

darum|legen *tr. V.* put around it/them

darunter [da'rʊntɐ, *(hinweisend:)* '---] *Adv.* **a)** under *or* beneath it/them; **wir wohnen im 2. Stock und er ~:** we live on the second floor and he lives under us *or* on the floor below; **b) 10° oder etwas ~:** 10° or a bit less; **Bewerber im Alter von 40 Jahren und ~:** applicants aged 40 and under; **c) was verstehen Sie ~?** what do you understand by that?; **sie hat sehr ~ gelitten** she suffered a great deal from *or* because of it/that; **d)** *(dabei, dazwischen)* amongst them; **in vielen Ländern, ~ der Schweiz** in many countries, including Switzerland

darunter-: ~|**bleiben** *unr. itr. V.; mit sein (fig.)* keep below this; ~|**fallen** *unr. itr. V.; mit sein (fig.)* be included; be amongst them; *(in diese Kategorie)* come under it; ~|**liegen** *unr. itr. V. (fig.)* be lower; ~|**mischen 1.** *tr. V.* mix in; mix with it; **2.** *refl. V.* mingle with it/them; ~|**schreiben** *unr. tr. V.* write underneath *or* at the bottom; ~|**setzen** *tr. V.* put ‹*signature, name*› to it

das [das] **1.** *best. Art. Nom. u. Akk.* the; **das Leben im Dschungel** life in the jungle; **das Weihnachtsfest** Christmas; **das Laufen fällt ihm schwer** walking is difficult for him. **2.** *Demonstrativpron.* **a)** *attr.* **das Kind war es** it was 'that child; **b)** *alleinstehend* **das [da]** that one; **das [hier]** this one [here]; **das mit dem blonden Haar** the one with the fair hair. **3.** *Relativpron. (Mensch)* who; that; *(Sache, Tier)* which; that; **das Mädchen, das da drüben entlanggeht** the girl walking along over there

da|sein *unr. itr. V.; mit sein; Zusschr. nur im Inf. u. Part.* **a)** be there; *(hiersein)* be here; **noch ~** *(übrig sein)* be left; **ist Herr X da?** is Mr X about *or* available?; **er ist schon da** he has already arrived; **ich bin gleich wieder da** I'll be right *or* straight back; **dafür** *od.* **dazu ist es ja da!** that's what it's [there] for!; **b)** *(fig.)* ⟨*case*⟩ occur; ⟨*moment*⟩ have arrived; ⟨*situation*⟩ have arisen; **c)** *(existieren, leben)* be left; be still alive; **da warst du noch gar nicht da** *(ugs.)* you weren't around then; **d)** *(ugs.: klar bei Bewußtsein sein)* **ganz** *od.* **voll ~:** be completely with it *(coll.)*

Da·sein das existence

Daseins·berechtigung die right to exist

da|sitzen *unr. itr. V.* **a)** sit there; **b)** *(ugs.: in Schwierigkeiten sein)* be left [there]; **ich saß ohne Geld da** I was stuck there without any money

dasjenige *s.* **derjenige**

daß [das] *Konj.* **a)** that; **entschuldigen Sie bitte, ~ ich mich verspätet habe** please forgive me for being late; please forgive my being late; **ich weiß, ~ du recht hast** I know [that] you are right; **ich verstehe nicht, ~ sie ihn geheiratet hat** I don't understand why she married him; **es ist schon 3 Jahre her, ~ wir zum letzten Mal im Theater waren** it is three years since *or* it was three years ago when we last went to the theatre; **b)** *(nach Pronominaladverbien o. ä.)* [the fact] that; **Wissen erwirbt man dadurch, ~ man viel liest** one acquires knowledge by reading a great deal; **das liegt daran, ~ du nicht aufgepaßt hast** that is due to the fact that you did not pay attention; that comes from your not paying attention; **ich bin dagegen, ~ er geht** I am against his going; **c)** *(im Konsekutivsatz)* that; **er lachte so [sehr], ~ ihm die Tränen in die Augen traten** he laughed so much that he almost cried; **d)** *(im Finalsatz)* so that; **e)** *(im Wunschsatz)* if only; **~ mir das nicht noch einmal passiert!** see that it doesn't happen again!; **f)** *(im Ausruf)* **~ er so jung sterben mußte!** how terrible *or* it's so sad that he had to die so young!; **~ mir das passieren mußte!** why did it have to [go and] happen to me!; *s. auch* **als, [an]statt, auf, außer, nur, ohne, kaum**

dasselbe *s.* **derselbe**

da|stehen *unr. itr. V.* **a)** *([untätig] stehen)* [just] stand there; **krumm ~:** slouch; **~ wie der Ochs vorm Berg** *(salopp)* be completely baffled; **b)** *(in einer bestimmten Lage sein)* find oneself; **gut ~:** be in a good position; **[ganz] allein ~:** be [all] alone in the world; **mit leeren Händen/als Lügner** *usw.* **~:** be left empty-handed/looking like a liar *etc.*

Daten ['da:tn̩] **1.** *s.* **Datum. 2.** *Pl.* data; **die technischen ~ eines Typs** the technical specification *sing.* of a model

Daten-: ~bank die; *Pl.* **~banken** data bank; **~erfassung die;** *o. Pl.* data collection *or* capture; **~schutz der** data protection; **~schutzbeauftragte der/die** data protection officer; **~verarbeitung die** data processing *no def. art.*

datieren [da'ti:rən] *tr. V.* date; **vom 1. Mai datiert** dated 1 May

Dativ ['da:ti:f] **der; ~s, ~e** *(Sprachw.)* dative [case]

Dativ·objekt das *(Sprachw.)* indirect object

Dattel ['datl̩] **die; ~, ~n** date

Dattel·palme die date-palm

Datum ['da:tʊm] **das; ~s, Daten** ['da:tn̩] date; **welches ~ haben wir heute?** what is the date today?

Dauer ['daʊɐ] **die; ~: a)** length; duration; **die ~ eines Vertrags** the term of a contract; **von kurzer** *od.* **nicht von [langer] ~ sein** not last long; be short-lived; **für die ~ eines Jahres** *od.* **von einem Jahr** for a period of one year; **b) von ~ sein** last [long]; **auf die ~:** in the long run; **auf ~:** permanently; for good; **er hat die Stelle jetzt auf ~:** his job is now permanent

dauer-, Dauer-: ~auftrag der *(Finanzw.)* standing order; **~frost der** long period of frost; **~gast der** *(im Hotel usw.)* long-stay guest *or* resident; **~haft 1.** *Adj.* **a)** [long-]lasting, enduring ⟨*peace, friendship, etc.*⟩; **b)** *(haltbar)* durable; hard-wearing; **2.** *adv.* lastingly; with long-lasting effect; **~karte die** season ticket; **~lauf der** jogging *no art.;* **einen ~lauf machen** go for a jog; go jogging; **im ~lauf** at a jog; **~lutscher der** large lollipop; all-day sucker *(Amer.)*

dauern *itr. V.* last; ⟨*job etc.*⟩ take; **der Film dauert zwei Stunden** the film lasts [for] *or* goes on for two hours; **bei ihm dauert alles furchtbar lange** everything takes him a terribly long time; **einen Moment, es dauert nicht lange** just a minute, it won't take long; **das dauert** *(ugs.)* that will take [some] time

dauernd 1. *Adj.; nicht präd.* constant, perpetual ⟨*noise, interruptions, etc.*⟩; permanent ⟨*institution*⟩. **2.** *adv.* constantly; *(immer)* always; the whole time; **er kommt ~ zu spät** he is for ever *or* keeps on arriving late

Dauer-: ~regen der continuous rain; **~stellung die** permanent position; **~welle die** perm; permanent wave; **~wurst die** smoked sausage **~zustand der** permanent state [of affairs]

Däumchen ['dɔymçən] **das; ~s, ~:** little thumb; **~ drehen** *(ugs.)* twiddle one's thumbs

Daumen ['daʊmən] **der; ~s, ~:** thumb; **am ~ lutschen** suck one's thumb; **jmdm. den** *od.* **die ~ drükken** *od.* **halten** keep one's fingers crossed for sb.; **auf etw.** *(Dat.)* **den ~ haben, auf etw.** *(Akk.)* **den ~ halten** *(ugs.)* keep a careful eye *or* check on sth.; **[etw.] über den ~ peilen** *(ugs.)* make a guesstimate [of sth.] *(coll.)*

daumen·breit *Adj.* as wide as your thumb *postpos.;* ≈ an inch across *postpos.*

Daumen·nagel der thumb-nail

Daune ['daʊnə] **die; ~, ~n** down [feather]; **~n** down *sing.*

Daunen·bett das down-filled quilt

da·von [*(hinweisend:)* '--] *Adv.* **a)** from it/them; *(von dort)* from there; *(mit Entfernungsangabe)* away [from it/them]; **wir sind noch weit ~ entfernt** *(fig.)* we are still a long way from that; **b) dies ist die Hauptstraße, und ~ zweigen einige Nebenstraßen ab** this is the main road and a few side-roads branch off it; **c)** *(darüber)* about it/them; **d)** *(dadurch)* by it/them; thereby; **~ wirst du krank** it will make you ill; **~ kriegt man Durchfall** you get diarrhoea from [eating] that/those; **das kommt ~!** *(ugs.)* [there you are,] that's what happens; **e) das Gegenteil ~ ist wahr** the opposite [of this] is true; **geben Sie mir vier ~:** give me four of them; **f)** *(aus diesem Material, auf dieser Grundlage)* from *or* out of it/them; **~ kann man nicht leben** you can't live on that

davon-: ~|fahren *unr. itr. V.; mit sein* leave; *(mit dem Auto)* drive away *or* off; *(mit dem Fahrrad, Motorrad)* ride away *or* off; **jmdm. ~fahren** leave sb. behind; **~|kommen** *unr. itr. V.; mit sein* get away; escape; **mit dem**

Schrecken/einer Geldstrafe ~kommen get off with a fright/a fine; ~|**laufen** *unr. itr. V.; mit sein* **a)** run away; **er ist mir ~gelaufen** he's made off; **es ist zum Davonlaufen** *(ugs.)* it really turns you off *(coll.);* it makes you want to run a mile; **b)** *(ugs.: überraschend verlassen)* **jmdm. ~laufen** walk out on sb.; ~|**machen** *refl. V.* make off (**mit** with); ~|**stehlen** *unr. refl. V. (geh.)* steal away; ~|**tragen** *unr. tr. V.* **a)** carry away; **b)** *(geh.: erringen)* win, gain ⟨*a victory, fame*⟩; **c)** *(geh.: sich zuziehen)* receive, suffer ⟨*injuries*⟩

da·vor [*(hinweisend:)* '--] *Adv.* **a)** in front of it/them; **b)** *(zeitlich)* before [it/them]; **c) jmdn. ~ warnen** warn sb. of *or* about it/them.; **er hat Angst ~, erwischt zu werden** he is afraid of being caught; **wir sind ~ geschützt** we are protected from it/them

davor-: ~|**legen 1.** *tr. V.* put in front of it/them; **2.** *refl. V.* lie down in front of it/them; ~|**liegen** *unr. itr. V.* lie in front of it/them; ~|**schieben a)** *unr. tr. V.* push in front of it/them; **b)** *unr. refl. V.* move in front of it/them; ~|**stehen** *unr. itr. V.* **a)** stand in front of it/them; *(vor einem Haus usw.)* stand outside; **b)** *(fig.)* **kurz ~stehen** *(vor diesem Ereignis usw.)* be close to it; *(vor dieser Tat)* be about to do it; ~|**stellen 1.** *tr. V.* put in front of it/them; **2.** *refl. V.* plant oneself in front of it/them

da·zu [*(hinweisend:)* '--] *Adv.* **a)** with it/them; *(gleichzeitig)* at the same time; *(außerdem)* what is more; **~ reicht man am besten Salat** it's/they're best served with lettuce/salad; **b)** *(darüber)* about *or* on it/them; **c)** *(zu diesem Zweck)* for it; *(es zu tun)* to do it; **~ reicht das Geld nicht** we haven't enough money for that; **d) im Widerspruch** *od.* **Gegensatz ~:** contrary to this/that; **~ war sie nicht in der Lage** she was not in a position to do it *or* do so; **er hatte ~ keine Lust** he didn't want to *or* didn't feel like it; **wie komme ich ~?** *(ugs.)* it would never occur to me; why on earth should I?

dazu-: ~|**geben** *unr. tr. V.* **a)** *(beisteuern)* give towards it; **b)** *(zusätzlich geben)* add; give as well; ~|**gehören** *tr. V.* **a)** belong to it/them; *(als Zusatz)* go with it/them; **b)** *(erforderlich sein);* **es gehört Mut/schon einiges ~:** it takes courage/quite something; **~gehörig** *Adj.; nicht präd.* appropriate; which goes/go with it/them *postpos.;* ~|**kommen** *unr. itr. V.; mit sein* **a)** *(hinkommen)* arrive [on the scene]; turn up; **b)** *(hinzukommen)* **kommt noch etwas ~?** *(fig.)* is there anything else [you would like]?; **~ kommt, daß ...** *(fig.)* what's more, ...; on top of that, ...; ~|**lernen** *tr., itr. V.* **[etwas] ~ lernen** learn [something new]; ~|**rechnen** *tr. V.* add on; ~|**setzen** *refl. V.* sit down next to him/her/you/them; ~|**tun** *unr. tr. V. (ugs.)* add; **das Seine ~tun** do one's bit; **ohne jmds. Dazutun** without sb.'s help; ~|**verdienen** *tr., itr. V.* earn ⟨*sth.*⟩ extra; *(durch Nebenbeschäftigung)* earn ⟨*sth.*⟩ on the side

da·zwischen [*(hinweisend:)* '---] *Adv.* in between; between them; *(darunter)* among them

dazwischen-: ~|**fahren** *unr. itr. V.; mit sein (eingreifen)* step in [and sort things out]; ~|**funken** *itr. V. (ugs.)* put a spanner in the works; *(sich einmischen)* put one's oar in ~|**kommen** *unr. itr. V.; mit sein* **a) mit dem Finger ~kommen** get one's finger caught [in it]; **b)** *(als Hindernis auftreten)* **mir ist etwas ~gekommen** I had problems; ~|**liegen** *unr. itr. V.;* lie in between; **Jahre lagen ~:** years had passed; **die ~liegende Zeit/Strecke** the intervening period/distance; ~|**reden** *itr. V.* interrupt; ~|**rufen 1.** *unr. itr. V.* interrupt [by shouting]; **2.** *unr. tr. V.* interrupt [loudly] with; interject

DDR [de:de:'|ɛr] **die; ~** *Abk. (1949–1990)* **Deutsche Demokratische Republik** GDR; East Germany *(in popular use)*

dealen ['di:lən] *itr. V. (ugs.)* push drugs

Dealer der; ~s, ~ *(ugs.)* pusher

Debatte [de'batə] **die; ~, ~n** debate (**über** + *Akk.* on); **etw. in die ~ werfen** introduce *or* bring sth. into the debate; **[nicht] zur ~ stehen** [not] be under discussion

debattieren *tr., itr. V.* debate; **[mit jmdm.] über etw. ~:** discuss sth. [with sb.]

Debüt [de'by:] **das; ~s, ~s** debut; **sein ~ geben** make one's debut

Debütant [deby'tant] **der; ~en, ~en, Debütantin die; ~, ~nen** newcomer [making his debut]

dechiffrieren *tr. V.* decipher ⟨*code, message*⟩

Deck [dɛk] **das; ~[e]s, ~s a)** deck; **alle Mann an ~!** all hands on deck!; **b)** *(Park~)* storey; level

Deck-: **~adresse die** accommodation *or (Amer.)* cover address; **~anstrich der** top coat; **~bett das** *s.* **Oberbett**

Decke ['dɛkə] **die; ~, ~n a)** *(Tisch~)* tablecloth; **b)** *(Woll~, Pferde~, auch fig.)* blanket; *(Reise~)* rug; *(Deckbett, Stepp~)* quilt; **mit jmdm. unter einer ~ stecken** *(ugs.)* be hand in glove with sb.; be in cahoots with sb. *(sl.);* **c)** *(Zimmer~)* ceiling; **mir fällt die ~ auf den Kopf** *(ugs.)* I get sick of [the sight of] these four walls; **an die ~ gehen** *(ugs.)* hit the roof *(coll.)*

Deckel ['dɛkl̩] **der; ~s, ~ a)** lid; *(von Flaschen, Gläsern usw.)* top; *(Schacht~, Uhr~, Buch~ usw.)* cover; **b)** *(Bier~)* beer-mat; **c) jmdm. eins auf den ~ geben** *(ugs.)* haul sb. over the coals; take sb. to task

decken 1. *tr. V.* **a) etw. über etw.** *(Akk.)* **~:** spread sth. over sth.; **b)** roof ⟨*house*⟩; cover ⟨*roof*⟩; **ein Dach/Haus mit Ziegeln/Stroh ~:** tile/thatch a roof/house; **c) den Tisch ~:** lay *or* set the table; **d)** *(schützen)* cover; *(bes. Fußball)* mark ⟨*player*⟩; *(vor Gericht usw.)* cover up for ⟨*accomplice, crime, etc.*⟩; **e)** *(befriedigen)* satisfy, meet ⟨*need, demand*⟩; **mein Bedarf ist gedeckt** *(ugs.)* I've had enough; **f)** *(Finanzw., Versicherungsw.)* cover; **g)** *(begatten)* cover; ⟨*stallion*⟩ serve ⟨*mare*⟩. **2.** *itr. V.* **a)** *(Fußball)* mark; *(Boxen)* keep up one's guard; **b)** *(den Tisch ~)* lay *or* set the table; **c)** ⟨*colour*⟩ cover. **3.** *refl. V.* coincide; tally

Decken-: **~beleuchtung die** ceiling light; **~gemälde das** ceiling painting; **~malerei die** ceiling painting

Deck-: **~farbe die** paint *(which covers well);* body-colour; **~hengst der** stud-horse; breeding stallion; **~mantel der;** *o. Pl. (abwertend)* cover; **unter dem ~mantel der Entwicklungshilfe** using development aid as a blind *or* cover; under the guise of development aid; **~name der** alias; assumed name

Deckung die; ~, ~en a) *(Schutz; auch fig.)* cover *(esp. Mil.); (Schach)* defence; *(Boxen)* guard; *(bes. Fußball: die deckenden Spieler)* defence; **~ nehmen, in ~ gehen** take cover; **b)** *(Finanzw.: das Begleichen) o. Pl. (von Schulden)* meeting; *(von Schecks)* cover[ing]; **als ~ für seine Schulden** as security for his debts; **c)**

(Befriedigung) satisfaction; **d)** *(Übereinstimmung)* **Pläne** *usw.* **zur ~ bringen** make plans *etc.* agree; bring plans *etc.* into line

deckungs·gleich *Adj. (Geom.)* congruent

Deck·weiß das opaque white

de facto [de: 'fakto] *Adv.* de facto *(esp. Polit., Law)*; in reality

defekt [de'fɛkt] *Adj.* defective; faulty; **~ sein** have a defect; be faulty; *(nicht funktionieren)* not be working

Defekt der; ~[e]s, ~e defect, fault (**an** + *Dat.* in)

defensiv [defɛn'zi:f] **1.** *Adj.* defensive. **2.** *adv.* defensively

Defensive die; ~, ~n defensive; **in der ~:** on the defensive

Defensiv·krieg der defensive war

definierbar *Adj.* definable

definieren [defi'ni:rən] *tr. V.* define

Definition [defini'tsi̯o:n] **die; ~, ~en** definition

definitiv [defini'ti:f] **1.** *Adj.* definitive; final ‹*answer, decision*›. **2.** *adv.* finally

Defizit ['de:fitsɪt] **das; ~s, ~e a)** deficit; **b)** *(Mangel)* deficiency

Deformation die deformation; *(Mißbildung)* deformity

deformieren *tr. V.* **a)** distort; put out of shape; **b)** *(entstellen)* deform *(also fig.)*; *(verunstalten)* disfigure ‹*face etc.*›; *(verstümmeln)* mutilate

deftig ['dɛftɪç] *(ugs.) Adj.* **a)** [good] solid *attrib.*, good and solid *pred.* ‹*meal etc.*›; [nice] big, [nice] fat ‹*sausage etc.*›; **b)** *(derb)* crude, coarse ‹*joke, speech, etc.*›

Degen ['de:gn̩] **der; ~s, ~ a)** *(Waffe)* [light] sword; **b)** *(Sportgerät)* épée

degenerieren [degene'ri:rən] *itr. V.; mit sein* degenerate (**zu** into)

degeneriert *Adj.* degenerate

degradieren [degra'di:rən] *tr. V.* demote; **jmdn./etw. zu etw. ~** *(fig.)* reduce sb./sth. to [the level of] sth.

Degradierung die; ~, ~en demotion; *(fig.)* degradation; reduction (**zu** to the level of)

dehnbar *Adj.* **a)** elastic ‹*waistband etc.*›; stretch ‹*fabric*›; **etw. ist ~:** sth. can be stretched; **b)** *(fig.: vage)* elastic; **das ist ein ~er Begriff** it's a loose concept

Dehnbarkeit die; ~ *(auch fig.)* elasticity

dehnen ['de:nən] **1.** *tr. V.* **a)** stretch; **b)** lengthen, draw out ‹*vowel, word*›. **2.** *refl. V.* stretch

Dehnung die; ~, ~en stretching

Deich [daiç] **der; ~[e]s, ~e** dike

Deichsel ['daiksl̩] **die; ~, ~n** shaft; *(in der Mitte)* pole; *(aus zwei Stangen)* shafts *pl.*

deichseln *tr. V. (ugs.)* fix; *(durch eine List)* wangle *(sl.)*

dein, *(in Briefen)* **Dein** [dain] *Possessivpron.* your; *(Rel., auch altertümelnd)* thy; **viele Grüße von Deinem Emil** with best wishes, yours Emil; **das Buch dort, ist das ~[e]s?** that book over there, is it yours?; **du und die Deinen** *(geh.)* you and yours *or* your family; **der/die Deine** *(geh.)* your husband/wife; **das Deine** *(geh.)* your possessions *pl. or* property; **du mußt das Deine tun** you must do your bit *or* share

deiner *Gen. von* **du** *(geh.)* of you; **ich gedenke ~ auf ewig** I will always remember you

deiner·seits ['dainɐ'zaits] *Adv.* *(von deiner Seite)* on your part; *(auf deiner Seite)* for your part

deines·gleichen *indekl. Pron.* people *pl.* like you; *(abwertend)* the likes *pl.* of you; your sort *or* kind; **unter ~:** amongst your own sort *or* kind

deinet·wegen *Adv.* **a)** because of you; on your account; *(für dich)* on your behalf; *(dir zuliebe)* for your sake; **b)** *(von dir aus)* **du hast gesagt, ~ könnten wir gehen** you said we could go as far as you were concerned

deinet·willen *Adv. in* **um ~:** for your sake

de jure [de: 'ju:rə] *Adv.* de jure; legally

Dekade [de'ka:də] **die; ~, ~n** decade

dekadent [deka'dɛnt] *Adj.* decadent

Dekadenz [deka'dɛnts] **die; ~:** decadence

Dekan [de'ka:n] **der; ~s, ~e** dean

Dekanat [deka'na:t] **das; ~s, ~e** dean's office

deklamieren [dekla'mi:rən] *tr., itr. V.* recite

Deklaration [deklara'tsi̯o:n] **die; ~, ~en** declaration

deklarieren [dekla'ri:rən] *tr. V.* declare; **etw. als etw. ~:** declare sth. to be sth.

deklassieren *tr. V.* **a)** *(herabsetzen)* reduce; downgrade; **b)** *(Sport)* outclass; *(beim Rennen)* leave standing

Deklination [deklina'tsi̯o:n] **die; ~, ~en** *(Sprachw.)* declension

deklinierbar *Adj. (Sprachw.)* declinable

deklinieren [dekli'ni:rən] *tr. V.* *(Sprachw.)* decline; **ein Wort schwach/stark ~:** decline a word as weak/strong

Dekolleté [dekɔl'te:] **das; ~s, ~s** low[-cut] neckline; décolletage

Dekor [de'ko:ɐ̯] **das; ~s, ~s** *od.* **~e** decoration; *(Muster)* pattern

Dekorateur [dekora'tø:ɐ̯] **der; ~s, ~e, Dekorateurin die; ~, ~nen** *(Schaufenster~)* window-dresser; *(von Innenräumen)* interior decorator *or* designer

Dekoration [dekora'tsi̯o:n] **die; ~, ~en a)** *o. Pl.* decoration; *(von Schaufenstern)* window-dressing; **b)** *(Schmuck, Ausstattung)* decorations *pl.*; *(Schaufenster~)* window display; *(Theater, Film)* set; scenery *no pl.*

dekorativ [dekora'ti:f] **1.** *Adj.* decorative. **2.** *adv.* decoratively

dekorieren [deko'ri:rən] *tr. V.* **a)** decorate ‹*room etc.*›; dress ‹*shop-window*›; **b)** *(mit Orden auszeichnen)* decorate (**mit** with)

Dekostoff ['de:ko-] **der** furnishing fabric

Dekret [de'kre:t] **das; ~[e]s, ~e** decree

dekretieren *tr. V.* decree

Delegation [delega'tsi̯o:n] **die; ~, ~en** delegation

delegieren [dele'gi:rən] *tr. V.* **a)** send as a delegate/as delegates (**zu** to); **jmdn. ins Komitee ~:** select sb. as one's representative on the committee; **b)** delegate ‹*task etc.*› (**an** + *Akk.* to)

Delegierte der/die; *adj. Dekl.* delegate

Delegierten·konferenz die delegates' *or* delegate conference

delikat [deli'ka:t] *Adj.* **a)** *(wohlschmeckend)* delicious; *(fein)* subtle, delicate ‹*bouquet, aroma*›; **b)** *(heikel)* delicate

Delikatesse [delika'tɛsə] **die; ~, ~n** delicacy; *(fig.)* treat

Delikatessen·geschäft, Delikateß·geschäft das delicatessen

Delikt [de'lɪkt] **das; ~[e]s, ~e** offence

Delinquent der; ~en, ~en offender

delirieren [deli'ri:rən] *itr. V.* be delirious

Delirium [de'li:ri̯ʊm] **das; ~s, Delirien** delirium

Delle ['dɛlə] **die; ~, ~n** *(ugs.)* dent

[1]**Delphin** [dɛl'fi:n] **der; ~s, ~e** dolphin

[2]**Delphin das; ~s** *(Schwimmen)* butterfly [stroke]

Delphin·schwimmen das butterfly

[1]**Delta** ['dɛlta] **das; ~[s], ~[s]** *(Buchstabe)* delta

²Delta das; ~s, ~s *od.* **Delten** *(Fluß~)* delta

Delta·mündung die delta

dem [de:m] **1.** *best. Art., Dat. Sg. v.* **¹der 1** *u.* **das 1:** the; **ich gab dem Mann das Buch** I gave the man the book; I gave the book to the man; **er hat sich dem Okkultismus zugewandt** he turned to occultism; **aus dem Libanon** from Lebanon. **2.** *Demonstrativpron., Dat. Sg. v.* **¹der 2** *u.* **das 2: a)** *attr.* that; **gib es dem Mann** give it to 'that man; **b)** *alleinstehend* **gib es nicht dem, sondern dem da!** don't give it to him, give it to that man/child *etc.;* **Zwiebeln schneide ich nicht mit dem [hier], sondern mit dem da** I chop onions with 'that knife, not with this one. **3.** *Relativpron., Dat. Sg. v.* **¹der 3** *u.* **das 3** *(Person)* that/whom; *(Sache)* that/which; **der Mann, dem ich das Geld gab** the man to whom I gave the money *or (coll.)* [that] I gave the money to; **der Mann, dem ich geholfen habe** the man [whom *or* that] I helped

Demagoge [dema'go:gə] **der; ~n, ~n** *(abwertend)* demagogue

demagogisch *(abwertend)* **1.** *Adj.* demagogic. **2.** *adv.* by demagogic means

Demarkations·linie die demarcation line

demaskieren 1. *refl. V. (fig.)* reveal oneself. **2.** *tr. V. (fig.)* unmask; expose

Dementi [de'mɛnti] **das; ~s, ~s** denial

dementieren 1. *tr. V.* deny. **2.** *itr. V.* deny it

dem-: ~entsprechend 1. *Adj.* appropriate; **das Wetter war schlecht und die Stimmung ~ entsprechend** the weather was bad and the general mood was correspondingly bad *or* bad too. **2.** *adv.* accordingly; *(vor Adjektiven)* correspondingly; **~gegenüber** *Adv.* in contrast; *(jedoch)* on the other hand; **~gemäß** *Adv.* **a)** *(infolgedessen)* consequently; **b)** *(entsprechend)* accordingly; **~jenigen** *s.* **derjenige**; **~nach** *Adv.* therefore; **~nächst** *Adv.* in the near future; shortly

Demo ['dɛmo] **die; ~, ~s** *(ugs.)* demo

Demo·graphie die demography *no art.*

demo·graphisch 1. *Adj.* demographic. **2.** *adv.* demographically

Demokrat [demo'kra:t] **der; ~en, ~en** democrat

Demokratie [demokra'ti:] **die; ~, ~n a)** *o. Pl. (Prinzip)* democracy *no art.;* **b)** *(Staat)* democracy

demokratisch 1. *Adj.* **a)** democratic. **2.** *adv.* democratically; **es wurde ~ gewählt** democratic elections were held; **bei uns geht es ~ zu** we run things on democratic lines

demokratisieren *tr. V.* democratize

Demokratisierung die; ~ democratization

demolieren [demo'li:rən] *tr. V.* **a)** wreck; smash up ‹*furniture*›; **b)** *(österr.: abreißen)* demolish

Demonstrant [demɔn'strant] **der; ~en, ~en, Demonstrantin die; ~, ~nen** demonstrator

Demonstration [demɔnstra'tsio:n] **die; ~, ~en** demonstration **(für** in support of, **gegen** against)

Demonstrations-: ~recht das right to demonstrate; **~verbot das** ban on demonstrations; **~zug der** column *or* procession of demonstrators

demonstrativ [demɔnstra'ti:f] **1.** *Adj.* **a)** demonstrative; pointed; **ein ~es Nein** an emphatic no; **b)** *(Sprachw.)* demonstrative. **2.** *adv.* pointedly; **ich sah ~ weg** I intentionally looked the other way

Demonstrativ·pronomen das *(Sprachw.)* demonstrative pronoun

demonstrieren [demɔn'stri:rən] **1.** *itr. V.* demonstrate **(für** in support of, **gegen** against). **2.** *tr. V.* demonstrate

demontieren *tr. V.* **a)** dismantle; break up ‹*ship, aircraft*›; **b)** *(abmontieren)* take off

demoralisieren *tr. V.* demoralize

Demoralisierung die; ~, ~en demoralization

Demoskop [demo'sko:p] **der; ~en, ~en** opinion pollster

Demoskopie [demosko'pi:] **die; ~:** [public] opinion research *no art.*

demoskopisch 1. *Adj.; nicht präd.* opinion research ‹*institute, methods, data, etc.*›; ‹*data etc.*› from opinion polls *or* opinion research; **~e Umfrage** [public] opinion poll. **2.** *adv.* through opinion polls *or* research

dem·selben *s.* **derselbe**

Demut ['de:mu:t] **die; ~:** humility

demütig ['de:my:tɪç] **1.** *Adj.* humble. **2.** *adv.* humbly

demütigen 1. *tr. V.* humiliate; humble ‹*sb.'s pride*›. **2.** *refl. V.* humble oneself

Demütigung die; ~, ~en humiliation

Demuts·gebärde die *(Verhaltensf.)* attitude of submission

dem·zufolge *Adv.* therefore; consequently

¹den [de:n] **1.** *best. Art., Akk. Sg. v.* **¹der 1:** the; **ich sah den Mann** I saw the man; **wir haben den „Faust" gelesen** we read 'Faust'; **in den Libanon reisen** travel to Lebanon; **den Sozialismus ablehnen** reject socialism. **2.** *Demonstrativpron., Akk. Sg. v.* **¹der 2: a)** *attr.* 'that; **ich meine den Mann, nicht den anderen** I mean 'that man, not the other; **b)** *alleinstehend* **ich meine den [da]** I mean 'that one. **3.** *Relativpron., Akk. Sg. v.* **¹der 3:** *(Person)* that/whom; *(Sache)* that/which; **der Mann, den ich gesehen habe** the man [that] I saw

²den 1. *best. Art., Dat. Pl. v.* **¹der 1, ¹die 1, das 1:** the; **ich gab es den Männern** I gave it to the men. **2.** *Demonstrativpron., Dat. Pl. v.* **¹der 2 a; ¹die 2 a, das 2 a** those

denen ['de:nən] **1.** *Demonstrativpron., Dat. Pl. v.* **¹der 2 b, ¹die 2 b, das 2 b:** them; **gib es ~, nicht den anderen** give it to 'them, not to the others. **2.** *Relativpron., Dat. Pl. v.* **¹der 3, ¹die 3, das 3:** *(Person)* that/whom; *(Sache)* that/which; **die Menschen, ~ wir Geld gegeben haben** the people to whom we gave money; **die Tiere, ~ er geholfen hat** the animals that he helped

Den Haag [de:n 'ha:k] **(das); ~ ~s** The Hague

denjenigen *s.* derjenige

denkbar 1. *Adj.* conceivable; **in einem Zustand, wie er schlimmer nicht ~ ist** in the worst state imaginable. **2.** *adv. (äußerst)* extremely; **die Lösung ist ~ leicht** the solution could not be easier

denken ['dɛŋkn̩] **1.** *unr. itr. V.* think **(an** *od. [südd., österr.]* **auf** + *Akk.* of, **über** + *Akk.* about); **liberal ~:** be liberal-minded; **wie denkst du darüber?** what do you think about it?; what's your opinion of it?; **erst ~, dann handeln** think before you act; **Denken ist Glückssache** you/he/she *etc.* thought wrong; **jmdm. zu ~ geben** make sb. think; *(stutzig machen)* make sb. suspicious; **denk daran, daß .../zu ...:** don't forget that .../to ...; **ich denke nicht daran!** no way!; not on your life!; **ich denke nicht daran, das zu tun** I've no intention *or* I wouldn't dream of doing that. **2.** *unr. tr. V.* think; **er dachte den gleichen Gedanken** the same thought occurred to him; **ich denke es** I think so; **denkste!** *(ugs.)* how wrong can one be!; *(da irrst du dich)* that's what 'you think!; **eine gedachte Linie** an imaginary line. **3.** *unr. refl. V.* **a)** *(sich vorstellen)* imagine; **das kann ich mir ~/nicht ~:** I can well believe/cannot believe that; **das hast du dir so gedacht!** that's what you thought; **du hättest dir doch ~ können, daß ...:** you should have realized that ...; **das habe ich mir [gleich] gedacht** that's [just] what I thought; *(bei Verdacht)* I thought *or* suspected as much; **ich denke mir mein[en] Teil** I can put two and two together *or* work things out for myself; **b) sich** *(Dat.)* **bei etw. etwas ~** *(etw. ganz bewußt tun)* mean something by sth.; **ich habe mir nichts [Böses] dabei gedacht** I didn't mean any harm [by it]

Denken das; ~s thinking; *(Denkweise)* thought

Denker der; ~s, ~: thinker

denk- **Denk-:** **~faul** *Adj.* mentally lazy; **sei nicht so ~faul** use your brains; **~fehler der** flaw in one's reasoning; **~mal das; ~s, Denkmäler** *od.* **Denkmale a)** monument; memorial; **b)** *(historisches Zeugnis)* monument; **~pause die** pause for thought; **~sport·auf·gabe die** brainteaser; **~vermögen das: [kreatives] ~vermögen** ability to think [creatively]; **~weise die** way of thinking; [mental] attitude; **~würdig** *Adj.* memorable; **~zettel der** warning; lesson; **jmdm. einen ~zettel verpassen** teach sb. a lesson

denn [dɛn] **1.** *Konj.* **a)** *(kausal)* for; because; **b)** *(geh.: als)* than; **schöner ~ je [zuvor]** more beautiful than ever. **2.** *Adv.: in* **es sei ~, [daß] ...:** unless ...; *s. auch* **geschweige. 3.** *Partikel* **a)** *(in Fragesätzen: oft nicht übersetzt);* **was ist ~ da los?** what 'is going on there?; **wie geht es dir ~?** tell me, how are you?; **ist das ~ so wichtig?** is that really so important?; **was muß ich ~ machen?** what am I to do, then?; **wie heißt du ~?** tell me your name; **warum ~ nicht?** why ever not?; **was soll das ~?** what's all this about?; **was ~ [sonst]?** well, what [else] then?; **b)** *(verstärkend)* **das ist ~ doch die Höhe!** that really is the limit!

dennoch ['dɛnɔx] *Adv.* nevertheless; even so; **ein höfliches und ~ eisiges Lächeln** a polite yet frosty smile

denselben *s.* **derselbe**

Denunziant [denʊn'tsi̯ant] **der; ~en, ~en** *(abwertend)* informer; grass *(sl.)*

Denunziation [denʊntsi̯a'tsi̯o:n] **die; ~, ~en** *(abwertend)* denunciation

denunzieren [denʊn'tsi:rən] *tr. V. (abwertend) (anzeigen)* denounce; *(bei der Polizei)* inform against; grass on *(sl.)* (**bei** to)

Deo ['de:o] **das; ~s, ~s, Deodorant** [de|odo'rant] **das; ~s, ~s** *(auch:)* **~e** deodorant

Deo·spray das deodorant spray

deplaciert, deplaziert [depla'tsi:ɐ̯t] *Adj.* out of place *pred.*; misplaced ⟨*remark etc.*⟩

Deponie [depo'ni:] **die; ~, ~n** tip *(Brit.);* dump

deponieren *tr. V.* put *(im Safe o. ä.)* deposit (**bei** with)

Deportation [depɔrta'tsi̯o:n] **die; ~, ~en** transportation (**in** + *Akk.*, **nach** to); *(ins Ausland)* deportation (**in** + *Akk.*, **nach** to)

deportieren [depɔr'ti:rən] *tr. V.* transport (**in** + *Akk.*, **nach** to); *(ins Ausland)* deport (**in** + *Akk.*, **nach** to)

Deportierte der/die; *adj. Dekl.* transportee; *(ins Ausland)* deportee

Depot [de'po:] **das; ~s, ~s a)** depot; *(Lagerhaus)* warehouse; *(für Möbel usw.)* depository; *(im Freien, für Munition o. ä.)* dump; *(in einer Bank)* strong-room; safe deposit; **b)** *(hinterlegte Wertgegenstände)* deposits *pl.*

Depp [dɛp] **der; ~en ~s, ~en ~e** *(bes. südd., österr., schweiz. abwertend) s.* **Dummkopf**

Depression [deprɛ'si̯o:n] **die; ~, ~en** depression

depressiv [deprɛ'si:f] **1.** *Adj.* depressive. **2.** *adv.* **~ veranlagt sein** have a tendency towards depression

deprimieren [depri:mi:rən] *tr. V.* depress

deprimierend *Adj.* depressing

deprimiert *Adj.* depressed

Deputierte der/die; *adj. Dekl. (Abgeordnete[r])* deputy

[1]**der** [de:ɐ̯] **1.** *best. Art. Nom.* the; **der Kleine** the little boy; **der Tod** death; **der April/Winter** April/winter; **der „Faust"** 'Faust'; **der Dieter** *(ugs.)* Dieter; **der Kapitalismus/Islam** capitalism/Islam; **der Bodensee/Mount Everest** Lake Constance/Mount Everest; **der Iran** Iran. **2.** *Demonstrativpron.* **a)** *attr.* that; **der Mann war es** it was 'that man; **b)** *alleinstehend* he; **der war es** it was 'him; **der und arbeiten!** *(ugs.)* [what,] him work! *(coll.);* **der [da]** *(Mann)* that man; *(Gegenstand, Tier)* that one; **der [hier]** *(Mann)* this man; *(Gegenstand, Tier)* this one. **3.** *Relativpron. (Mensch)* who/that; *(Sache)* which/that; **der Mann, der da drüben entlanggeht** the man walking along over there. **4.** *Relativ- u. Demonstrativpron.* the one who

[2]**der 1.** *best. Art.* **a)** *Gen. Sg. v.* [1]**die 1: der Hut der Frau** the woman's hat; **der Henkel der Tasse** the handle of the cup; **b)** *Dat. Sg. v.* [1]**die 1:** to the; *(nach Präp.)* the; **in der Türkei** in Turkey; **c)** *Gen. Pl. v.* [1]**der 1,** [1]**die 1,** [1]**das 1: das Haus der Freunde** our/their *etc.* friends' house; **das Bellen der Hunde** the barking of the dogs. **2.** *Demonstrativpron.* **a)** *Gen. Sg. v.* [1]**die 2a:** of the; of that; **b)** *Dat. Sg. v.* [1]**die 2** *attr.* **der Frau [da/hier] gehört es** it belongs to that woman there/this woman here; *alleinstehend* **gib es der da!** *(ugs.)* give it to 'her; **c)** *Gen. Pl. v.* [1]**der 2a,** [1]**die 2a, das 2a:** of those. **3.** *Relativpron.; Dat. Sg. v.* [1]**die 3** *(Person)* whom; **die Frau, der ich es gegeben habe** the woman to whom I gave it; the woman I gave it to; *(Sache)* that/which; **die Katze, der er einen Tritt gab** the cat [that] he kicked

der·art *Adv.* **jmdn. ~ schlecht/unfreundlich behandeln, daß ...;** treat sb. so badly/in such an unfriendly way that ...; **es hat lange nicht mehr ~ geregnet** it hasn't rained as hard as that for a long time; **sie hat ~ geschrien, daß ...:** she screamed so much that ...

der·artig 1. *Adj.; nicht präd.* such; **etwas Derartiges** a thing like that; such a thing. **2.** *adv. s.* **derart**

derb [dɛrp] **1.** *Adj.* **a)** strong, tough ⟨*material*⟩; stout, strong, sturdy ⟨*shoes*⟩; **b)** *(kraftvoll, deftig)* earthy ⟨*scenes, humour*⟩; **c)** *(unverblümt)* crude, coarse ⟨*expression, language*⟩. **2.** *adv.* **a)** strongly ⟨*made, woven, etc.*⟩; **b)** *(kraftvoll, deftig)* earthily; **c)** *(unverblümt)* crudely; coarsely

Derbheit die; ~: *s.* **derb 1b, c:** earthiness; crudity; coarseness

Derby ['dɛrbi] **das; ~s, ~s a)** *(Pferdesport)* Derby; **b)** *(Fußball)* derby

deren ['de:rən] **1.** *Relativpron.* **a)** *Gen. Sg. v.* [1]**die 3** *(Menschen)* whose; *(Sachen)* of which; **die Katastrophe, ~ Folgen furchtbar waren** the disaster, the consequences of which were frightful; **die Großmutter, ~ wir uns gerne erinnern** our grandmother, of whom we have fond memories; **b)** *Gen., Pl. v.* [1]**der 3,** [1]**die 3, das 3** *(Menschen)* whose; *(Sachen)* **Maßnahmen, ~ Folgen wir noch nicht absehen können** measures, the consequences of which we cannot yet foresee. **2.** *Demonstrativpron.* **a)** *Gen. Sg. v.* [1]**die 2 meine Tante, ihre Freundin und ~ Hund** my aunt, her friend and 'her dog; **b)** *Gen. Pl. v.* [1]**der 2b,** [1]**die 2b, das 2b: meine Verwandten und ~ Kinder** my relatives and their children; **Bücher? Deren hat er genug** *(geh.)* Books? He's got enough of those

derent-: ~wegen *Adv.* **1.** *relativ* on whose account; on account of whom; because of whom; *(von Sachen)* on account of which; because of which; **2.** *demonstrativ* because of them; **~willen** *Adv.* **1.** *relativ* **um ~willen** for whose sake; for the sake of whom; *(von Sachen)* for the sake of which; **die Erbstücke, um ~willen sich die Kinder zerstritten** the heirlooms over which the children fell out; **2.** *demonstrativ* **um ~willen** for her/their sake

derer ['de:rɐ] *Demonstrativpron.; Gen. Pl. v.* [1]**der 2b,** [1]**die 2b, das 2b** of those; **die Zahl ~, die das glauben, nimmt ab** the number of people who believe that is declining

der·gestalt *Adv. (geh.)* **~, daß ...:** in such a way that ...

der·gleichen *indekl. Demonstrativpron.* **a)** *attr.* such; like that *postpos., not pred.;* **b)** *alleinstehend* that sort of thing; such things *pl.;* things *pl.* like that; **nichts ~:** nothing of the sort; **und ~ [mehr]** and suchlike

der·jenige [-je:nɪgə], **die·jenige, das·jenige** *Demonstrativpron.* **a)** *attr.* that; *Pl.* those; **diejenige Person, die ...:** the *or* that person who ...; **b)** *alleinstehend* that one; *Pl.* those; **derjenige, der .../diejenige, die ...:** the person who ...; **diejenigen, die ...:** those [poeple] who ...; **dasjenige, was ...:** that which ...

derlei ['de:ɐ̯lai̯] *indekl. Demonstrativpron.* **a)** *attr.* such; like that *postpos., not pred.;* **b)** *alleinstehend* that sort of thing; such things *pl.;* things *pl.* like that

der·maßen *Adv.* **~ schön** *usw.*, **daß ...:** so beautiful *etc.* that ...; **ein ~ intelligenter Mensch** such an intelligent person

derselbe [de:ɐ̯'zɛlbə], **dieselbe, dasselbe,** *Pl.* **dieselben** *Demonstrativpron.* **a)** *attr.* the same; **b)** *alleinstehend* the same one; **sie ist immer noch [ganz] dieselbe** she is still [exactly] the same; **es sind immer dieselben, die ...:** it's always the same people *or* ones who ...; **noch einmal dasselbe, bitte** *(ugs.)* [the] same again please; **er sagt immer dasselbe** he always says the same thing

der·weil[en] *(veralt.)* **1.** *Adv. s.* **inzwischen c. 2.** *Konj.* while

der·zeit *Adv.* at present; at the moment

der·zeitig *Adj.; nicht präd.* present; current

[1]**des** [dɛs] **1.** *best. Art.; Gen. Sg. v.* [1]**der 1, das 1: die Mütze des Jungen** the boy's cap; **das Klingeln des Telefons** the ringing of the telephone. **2.** *Demonstrativpron.; Gen. Sg. v.* [1]**der 2a, das 2a: er ist der Sohn des Mannes, der ...** he's the son of the man who ...

[3]**des, Des das; ~, ~** *(Musik)* D flat

Desaster [de'zastɐ] **das; ~s, ~:** disaster

Deserteur [dezɛr'tø:ɐ̯] **der; ~s, ~e** *(Milit.)* deserter

desertieren *itr. V.; mit sein (Milit., fig.)* desert

des·gleichen *Adv.* likewise; **er ist Arzt, ~ sein Sohn** he is a doctor, as is *or* and so is his son; **es fehlt an Papier, ~ an Bleistiften** there's a shortage of paper and also [of] pencils

des·halb *Adv.* for that reason; because of that; **~ bin ich zu dir gekommen** that is why I came to you; **aber ~ ist sie nicht dumm** but that doesn't mean she is stupid

Design [di'zai̯n] **das; ~s, ~s** design

Designer [di'zai̯nɐ] **der; ~s, ~, Designerin die; ~, ~nen** designer

desillusionieren [dɛs|iluzi̯o'ni:rən] *tr. V.* disillusion

Des·infektion [dɛs|-] **die** disinfection

Desinfektions·mittel das disinfectant

des·infizieren *tr. V.* disinfect

Des·information die disinformation *no indef. art.*

Des·interesse das lack of interest **(an +** *Dat.* in)

des·interessiert 1. *Adj.* uninterested. **2.** *adv.* uninterestedly

deskriptiv [deskrɪp'ti:f] **1.** *Adj.* descriptive. **2.** *adv.* descriptively

desodorierend [dɛs|odo'ri:rənt] *Adj.* deodorant

desolat [dezo'la:t] *Adj. (geh.)* wretched

Desperado [dɛspe'ra:do] **der; ~s, ~s** desperado

Despot [dɛs'po:t] **der; ~en, ~en** despot; *(fig. abwertend)* tyrant

Despotie [dɛspo'ti:] **die, ~, ~n** despotism

despotisch 1. *Adj.* despotic. **2.** *adv.* despotically

des·selben *s.* **derselbe**

dessen ['dɛsn̩] **1.** *Relativpron.; Gen. Sg. v.* [1]**der 3, das 3** *(Mensch)* whose; *(Sache, Tier)* of which; **der Großvater, ~ wir uns gern erinnern** our grandfather, of whom we have fond memories. **2.** *Demonstrativpron.; Gen. Sg. v.* [1]**der 2b, das 2b: mein Onkel, sein Sohn und ~ Hund** my uncle, his son, and 'his dog; **das Waldsterben und ~ Folgen** the death of the forests and its consequences; **Onkel August? Dessen erinnere ich mich noch sehr gut** Uncle August? I remember 'him well

dessent-: ~wegen *Adv.* **1.** *relativ* on whose account; on account of whom; because of whom; *(von Sachen)* on account of which; because of which; **das Verbrechen, ~wegen er verurteilt wurde** the crime of which he was convicted; **2.** *demonstrativ* because of him; *(von Sachen)* because of this; **~willen** *Adv.* **1.** *relativ* **um ~willen** for whose sake; for the sake of whom; *(von Sachen)* for the sake of which; **2.** *demonstrativ* **um ~willen** for his sake

dessen·ungeachtet *Adv.* nevertheless; notwithstanding [this]

Dessert [dɛ'se:ɐ̯] **das; ~s, ~s** dessert

Dessin [dɛ'sɛ̃:] **das; ~s, ~s** design; pattern

destillieren *tr. V. (Chemie)* distil; **destilliertes Wasser** distilled water

desto *Konj., nur vor Komp.* **je eher, ~ besser** the sooner the better; **~ ängstlicher** the more anxious/anxiously; **ich schätzte ihn ~ mehr** I appreciated him all the more

Destruktion [destrʊk'tsi̯o:n] **die; ~, ~en** destruction

destruktiv [destrʊk'ti:f] **1.** *Adj.* destructive. **2.** *adv.* destructively

des·wegen *Adv. s.* **deshalb**

Detail [de'tai̯] **das; ~s, ~s** detail; **ins ~ gehen** go into detail; **in allen ~s** in the fullest detail

detailliert [deta'ji:ɐt] **1.** *Adj.* detailed. **2.** *adv.* in detail; **sehr ~:** in great detail

Detektei [detɛk'tai] **die; ~, ~en** [private] detective agency

Detektiv [detɛk'ti:f] **der; ~s, ~e, Detektivin die; ~, ~nen** [private] detective

Detektiv·roman der detective novel

Detonation [detona'tsio:n] **die; ~, ~en** detonation; explosion; **etw. zur ~ bringen** detonate sth.

detonieren [deto'ni:rən] *itr. V.; mit sein* detonate; explode

Deut [dɔyt] *in* **kein[en] ~ besser** not one bit *or* whit better

deutbar *Adj.* interpretable

deuteln ['dɔytl̩n] *itr. V.* quibble **(an** + *Dat.* about); **daran gibt es nichts zu ~:** there are no ifs and buts about it

deuten ['dɔytn̩] **1.** *itr. V.* point; **[mit dem Finger] auf jmdn./etw. ~:** point [one's finger] at sb./sth. **2.** *tr. V.* interpret

deutlich 1. *Adj.* **a)** clear; **daraus wird ~, daß/wie ...:** this makes it clear that/how ...; **b)** *(eindeutig)* clear, distinct ⟨*recollection, feeling*⟩; **~ werden** make oneself plain *or* clear. **2.** *adv.* **a)** clearly; **b)** *(eindeutig)* clearly; plainly; **jmdm. etw. ~ zu verstehen geben** make sth. clear *or* plain to sb.

Deutlichkeit die; ~: a) clarity; **b)** *(Eindeutigkeit)* clearness; distinctness; **in** *od.* **mit aller ~ sagen, daß ...:** make it perfectly clear *or* plain that ...

deutsch [dɔytʃ] **1.** *Adj.* **a)** German; **Deutsche Mark** Deutschmark; German mark; **Deutsche Demokratische Republik** *(1949 bis 1990)* German Democratic Republic; **das Deutsche Reich** *(hist.)* the German Reich *or* Empire; **alles Deutsche** all things *pl.* or everything German; **das typisch Deutsche daran** what is/was typically German about it; **b)** *(die Sprache betreffend)* German; **etw. auf ~ sagen** say sth. in German; **was heißt das Wort auf ~?** what is the word in German?; what is the German for that word?; **auf [gut] ~** *(ugs.)* in plain English; **die ~e Schweiz** German-speaking Switzerland. **2.** *adv.; ~* **sprechen/schreiben** speak/write German; **~ geschrieben sein** be written in German

Deutsch das; ~[s] a) German; **gutes/fließend ~ sprechen** speak good/fluent German; **kein ~ [mehr] verstehen** *(ugs.)* not understand plain English; **b)** *o. Art. (Unterrichtsfach)* German *no art.*; **er ist gut in ~:** he's good at German

Deutsch·amerikaner der German-American

deutsch-amerikanisch *Adj.; nicht präd.* German-American

deutsch-deutsch *Adj.; nicht präd.* intra-German

[1]**Deutsche** ['dɔytʃə] **der/die;** *adj. Dekl.* German; **~[r] sein** be German

[2]**Deutsche das;** *adj. Dekl.* **das ~:** German; **aus dem ~n/ins ~ übersetzen** translate from/into German

deutsch-französisch *Adj.* Franco-German ⟨*relations, border, etc.*⟩; German-French ⟨*dictionary, anthology, etc.*⟩

Deutschland (das); ~s Germany

Deutschland·lied das *the song 'Deutschland, Deutschland über alles'*; **~politik die** *(innerdeutsche Politik)* intra-German policy; *(gegenüber ~)* policy towards Germany

deutsch-, Deutsch-: ~lehrer der German teacher; **~sprachig** *Adj.* **a)** German-speaking; **Deutschsprachige** *Pl.* German speakers; **b)** *(in deutscher Sprache)* German-language *attrib.* ⟨*newspaper, edition, broadcast*⟩; ⟨*teaching*⟩ in German; German ⟨*literature*⟩; **~stämmig** *Adj.* of German origin *postpos.*

Deutschtum das; ~s Germanness

Deutsch·unterricht der German teaching; *(Unterrichtsstunde)* German lesson; **~ erteilen** *or* **geben** teach German

Deutung die; ~, ~en interpretation

Devise [de'vi:zə] **die; ~, ~n** motto

Devisen *Pl.* foreign exchange *sing.*; *(Sorten)* foreign currency *sing. or* exchange *sing.*

Devisen-: ~börse die foreign exchange market; **~geschäft das** foreign exchange business *or* dealings *pl.*; *(einzelne Transaktion)* foreign exchange transaction; **~schmuggel der** [foreign] currency smuggling

devot [de'vo:t] *(geh. abwertend)* **1.** *Adj.* obsequious. **2.** *adv.* obsequiously

Dextrose [dɛks'tro:zə] **die; ~:** dextrose

Dezember [de'tsɛmbɐ] **der; ~s, ~:** December

dezent [de'tsɛnt] **1.** *Adj.* quiet ⟨*colour, pattern, suit*⟩; subdued ⟨*lighting, music*⟩; discreet ⟨*smile, behaviour*⟩. **2.** *adv.* discreetly; ⟨*dress*⟩ unostentatiously

dezentralisieren *tr. V.* decentralize

Dezentralisierung die; ~, ~en decentralization

Dezernat [detsɛr'na:t] **das; ~[e]s, ~e** department

Dezernent [detsɛr'nɛnt] **der; ~en, ~en** head of department

Dezi- ['de:tsi-]: deci⟨*litre, metre, etc.*⟩

Dezibel [detsi'bɛl] **das; ~s, ~:** decibel

dezimal [detsi'ma:l] *Adj.* decimal

Dezimal-: ~rechnung die decimal arithmetic *no art.*; **~stelle die** decimal place; **~system das** decimal system; **~zahl die** decimal [number]

dezimieren [detsi'mi:rən] *tr. V.* decimate

Dezimierung die; ~, ~en decimation

DFB [de:|ɛf'be:] **der; ~** *Abk.* **Deutscher Fußball-Bund**

DGB [de:ge:'be:] **der; ~** *Abk.* **Deutscher Gewerkschaftsbund** West German Trade Union Federation

dgl. *Abk.* **dergleichen, desgleichen**

d. Gr. *Abk.* **der/die Große**

d. h. *Abk.* **das heißt** i. e.

Di. *Abk.* **Dienstag** Tue[s].

Dia ['di:a] **das; ~s, ~s** slide

Diabetes [dia'be:tɛs] **der; ~:** diabetes

Diabetiker [dia'be:tikɐ] **der; ~s, ~, Diabetikerin, die; ~, ~nen** diabetic

diabolisch *(geh.)* **1.** *Adj.* diabolic. **2.** *adv.* with diabolic malevolence

Diadem [dia'de:m] **das; ~s, ~e** diadem

Diagnose [dia'gno:zə] **die; ~, ~n** diagnosis; **eine ~ stellen** make a diagnosis

diagnostisch 1. *Adj.* diagnostic. **2.** *adv.* diagnostically

diagnostizieren [diagnɔsti'tsi:rən] *tr. V.* diagnose

diagonal [diago'na:l] **1.** *Adj.* diagonal. **2.** *adv.* diagonally; **etw. ~ lesen** *(ugs.)* skim through sth.

Diagonale die; ~, ~n diagonal

Diagramm das graph; *(schematische Darstellung)* diagram

Diakon [dia'ko:n] **der; ~s** *od.* **~en, ~e[n]** *(christl. Kirche)* deacon

Diakonisse [diako'nɪsə] **die; ~, ~n** *(ev. Kirche)* deaconess

Dialekt [dia'lɛkt] **der; ~[e]s, ~e** dialect

Dialekt·ausdruck der; *Pl.* **~ausdrücke** dialect expression

dialekt·frei 1. *Adj.* **~es Deutsch sprechen** speak German without a trace of [any] dialect. **2.** *adv.* ⟨*speak*⟩ without a trace of [any] dialect

Dialog [dia'lo:k] **der;** ~[e]s, ~e dialogue
Dialyse [dia'ly:zə] **die;** ~, ~n *(fachspr.)* dialysis
Diamant [dia'mant] **der;** ~en, ~en diamond
diametral [diame'tra:l] *(fig. geh.)* 1. *Adj.* diametrical ‹*opposition*›. 2. *adv.* diametrically
Dia-: ~**positiv das** slide; ~**projektor der** slide projector
diät [di'ɛ:t] *adv.* ~ **kochen** cook according to a/one's diet; ~ **essen** be on a diet
Diät die; ~, ~en diet; **eine ~ einhalten** keep to a diet
Diäten *Pl.* [parliamentary] allowance *sing.*
dich [dɪç], *(in Briefen)* **Dich** 1. *Akk. von* **du** you. 2. *Akk. des Reflexivpron. der 2. Pers. Sg.* yourself; **wäschst du dich?** are you washing [yourself]?; **entschuldige dich!** apologize!
dicht [dɪçt] 1. *Adj.* **a)** thick ‹*hair, fur, plumage, moss*›; thick, dense ‹*foliage, fog, cloud*›; dense ‹*forest, thicket, hedge, crowd*›; heavy, dense ‹*traffic*›; densely ranked, close-ranked ‹*rows of houses*›; heavy ‹*snowstorm, traffic*›; *(fig.)* full, packed ‹*programme*›; **in ~er Folge** in rapid *or* quick succession; **b)** *(undurchlässig) (für Luft)* airtight; *(für Wasser)* watertight ‹*shoes*›; *(für Licht)* heavy ‹*curtains, shutters*›; ~ **machen** seal ‹*crack*›; seal the crack[s]/leak[s] in ‹*roof, window, etc.*›; waterproof ‹*material, umbrella, etc.*›; **nicht ganz ~ sein** *(salopp)* have a screw loose *(coll.)*; **c)** *(ugs.: geschlossen)* shut; closed. 2. *adv.* **a)** densely ‹*populated*›; tightly ‹*packed*›; thickly, densely ‹*wooded*›; heavily ‹*built up*›; ~ **verschneit** thick with snow; ~ **besetzt** full; packed; ~ **behaart** [very] hairy; ~ **an** ~ *od.* ~ **gedrängt stehen/sitzen** stand/sit close together; **b)** *(undurchlässig)* tightly; **c)** *mit Präp. (nahe)* ~ **neben** right next to; ~ **daran** hard by; ~ **beieinander** close together; ~ **vor/hinter ihm** right *or* just in front of/behind him; **die Polizei ist ihm ~ auf den Fersen** the police are hard *or* close on his heels; **d)** *(zeitlich: unmittelbar)* **ich war ~ daran, es zu tun** I was just about to do it; ~ **bevorstehen** be imminent
dicht-: ~**bebaut** *Adj. (präd. getrennt geschrieben)* heavily built-up; ~**behaart** *Adj. (präd. getrennt geschrieben)* [very] hairy; ~**besiedelt** *Adj. (präd. getrennt geschrieben)* densely populated; ~**bewachsen** *Adj. (präd. getrennt geschrieben)* covered with dense vegetation *postpos.*
Dichte ['dɪçtə] **die;** ~ *(Physik, fig.)* density
[1]**dichten** ['dɪçtn̩] 1. *itr. V.* **[gut]** ~: make a good seal. 2. *s.* **abdichten**
[2]**dichten** 1. *itr. V.* write poetry. 2. *tr. V.* write; compose
Dichter der; ~s, ~: poet; *(Schriftsteller)* writer; author
Dichterin die; ~, ~nen poet[ess]; *(Schriftstellerin)* writer; author[ess]
dichterisch 1. *Adj.* poetic; *(schriftstellerisch)* literary. 2. *adv.*; *s.* 1: poetically; literarily
dicht- ~**gedrängt** *Adj. (präd. getrennt geschrieben)* tightly *or* closely packed; ~|**halten** *unr. itr. V. (ugs.)* keep one's mouth shut *(coll.)*
dicht|machen *tr., itr. V. (ugs.)* shut; close; *(endgültig)* shut *or* close down
[1]**Dichtung die;** ~, ~en **a)** *o. Pl.* sealing; **b)** *(dichtendes Teil)* seal; *(am Hahn usw.)* washer; *(am Vergaser, Zylinder usw.)* gasket
[2]**Dichtung die;** ~, ~en **a)** literary work; work of literature; *(in Versform)* poetic work; poem; *(fig. ugs.)* fiction; ~ **und Wahrheit** fact and fiction; truth and fantasy; **b)** *o. Pl. (Dichtkunst)* literature; *(in Versform)* poetry
dick [dɪk] 1. *Adj.* **a)** thick; thick, chunky ‹*pullover*›; stout ‹*tree*›; fat ‹*person, arms, legs, behind, etc.*›; big ‹*bust*›; ~ **und rund** *od.* **fett sein** *(ugs.)* be round and fat; ~ **machen** ‹*drink, food*› be fattening; **das Kleid macht ~:** the dress makes you look fat; **im ~sten Verkehr** *(fig. ugs.)* in the heaviest traffic; **mit jmdm. durch ~ und dünn gehen** stay *or* stick with sb. through thick and thin; **b)** *(ugs.: angeschwollen)* swollen ‹*cheek, ankle, tonsils, etc.*›; **c)** *(ugs.: groß)* big ‹*mistake, order*›; hefty, *(coll.)* fat ‹*fee, premium, salary*›; **ein ~es Auto** *(ugs.)* a great big car *(coll.)*; **jmdm. ein ~es Lob aussprechen** give sb. a great deal of prise *or* high praise; **das ~e Ende kommt noch** *(ugs.)* the worst is yet to come; **d)** *(ugs.: eng)* close ‹*friends, friendship, etc.*›. 2. *adv.* **a)** thickly; **etw. 5 cm ~ schneiden/auftragen** *usw.* cut/apply sth. 5 cm. thick; **etw. ~ unterstreichen** underline sth. heavily; **sich ~ anziehen** wrap up warm[ly]; ~ **geschminkt** heavily made up; ~ **auftragen** *(ugs. abwertend)* lay it on thick *(sl.)*; **b)** ~ **geschwollen** *(ugs.)* badly swollen; **c)** ~ **befreundet sein** *(ugs.)* be close friends
Dick-: ~**bauch der** *(scherzh.)* fatty; *(mit Spitzbauch)* pot-belly; ~**darm der** *(Anat.)* large intestine
dicke *Adv. (ugs.)* easily; **jmdn./etw. ~ haben** *(salopp)* have had a bellyful of sb./sth.
[1]**Dicke die;** ~: thickness; *(von Menschen, Körperteilen)* fatness
[2]**Dicke der/die;** *adj. Dekl. (ugs.)* fatty *(coll.)*; fat man/woman; **die ~n** fatties *(coll.)*; fat people
Dickerchen das; ~s, ~ *(ugs. scherzh.)* podge *(coll.)*
dick-, Dick-: ~**fellig** *(ugs. abwertend)* 1. *Adj.* thick-skinned. 2. *adv.* in a thick-skinned way; ~**felligkeit die;** ~ *(ugs. abwertend)* insensitivity; ~**flüssig** *Adj.* thick; ~**häuter der;** ~s, ~: pachyderm
Dickicht ['dɪkɪçt] **das;** ~[e]s, ~e thicket; *(fig.)* jungle
Dick·kopf der *(ugs.)* mule *(coll.)*; **du bist ein ~** you're as stubborn as a mule
dick·köpfig *(ugs.)* 1. *Adj.* stubborn; pigheaded; 2. *adv.* stubbornly; pigheadedly
dicklich *Adj.* plumpish; chubby
Dick-: ~**milch die** sour milk; ~**schädel der** *s.* ~**kopf**
Didaktik [di'daktɪk] **die;** ~, ~en **a)** *o. Pl.* didactics *sing., no art.*; **b)** *(Unterrichtsmethode)* teaching method
didaktisch 1. *Adj.* didactic. 2. *adv.* didactically
[1]**die** 1. *best. Art. Nom.* the; **die Kleine** the little girl; **die Liebe/Freundschaft** love/friendship; **die „Iphigenie"/***(ugs.)* **Helga** 'Iphigenia'/Helga; **die Demokratie** democracy; **die Marktstraße** Market Street; **die Schweiz** Switzerland; **die Frau/Menschheit** women *pl.*/mankind; **die „Concorde"/„Klaus Störtebeker"** 'Concorde'/the 'Klaus Störtebeker'; **die Kunst/Oper** art/opera. 2. *Demonstrativpron.* **a)** *attr.* **die Frau war es** it was 'that woman; **b)** *alleinstehend* she; **die war es** it was 'her; **die und arbeiten!** *(ugs.)* [what,] her work!; **die mit dem Hund** *(ugs.)* her with the dog; **die [da]** *(Frau, Mädchen)* that woman/girl; *(Gegenstand, Tier)* that one; **die blöde Kuh, die!** *(fig. salopp)* what a silly cow! *(sl.)*. 3. *Relativpron. Nom. (Mensch)* who; that; *(Sache, Tier)* which; that; **die Frau, die da drüben entlanggeht** the woman walking along

over there. **4.** *Relativ- u. Demonstrativpron.* the one who; **die das getan hat** the woman *etc.* who did it

²die 1. *best. Art.* **a)** *Akk. Sg. v.* **¹die 1:** the; **hast du die Ute gesehen?** *(ugs.)* have you seen Ute?; **b)** *Nom. u. Akk. Pl. v.* **¹der 1, ¹die 1, das 1:** the. **2.** *Demonstrativpron. Nom. u. Akk. Pl. v.* **¹der 1, ¹die 1, das 1:** *attr.* **ich meine die Männer, die gestern hier waren** I mean those men who were here yesterday; *alleinstehend* **ich meine die [da]** I mean 'them. **3.** *Relativpron.* **a)** *Akk. Sg. v.* **¹die 3:** *(bei Menschen)* who; that; *(bei Sachen, Tieren)* which; that; **b)** *Nom. u. Akk. Pl. v.* **¹der 3, ¹die 3, das 3:** *(bei Menschen)* whom; **die Männer, die ich gesehen habe** the men I saw; *(bei Sachen, Tieren)* which; **die Bücher, die da liegen** the books lying there

Dieb [di:p] **der; ~[e]s, ~e** thief; **haltet den ~!** stop thief!

Diebes-: ~bande die *(abwertend)* gang of thieves; **~gut das** stolen goods *pl. or* property

Diebin die; ~, ~nen [woman] thief

diebisch 1. *Adj.* **a)** thieving; **b)** *(verstohlen)* mischievous. **2.** *adv.* mischievously; **sich ~ über etw.** *(Akk.)* **freuen** take a mischievous pleasure in sth.

Diebstahl ['di:p-ʃta:l] **der; ~[e]s, Diebstähle** ['di:p-ʃtɛ:lə] theft

die·jenige, diejenigen *s.* **derjenige**

Diele ['di:lə] **die; ~, ~n a)** hall[way]; **b)** *(Fußbodenbrett)* floor-board

dienen ['di:nən] *itr. V.* **a)** be in service; **jmdm. als Magd ~:** serve sb. as a maid; **b)** *(veralt.: Militärdienst tun)* do military service; **beim Heer ~:** serve in the army; **c)** *(dienlich sein)* serve; **das dient einer guten Sache** it is in a good cause; **d)** *(helfen)* help (**in** + *Dat.* in); **womit kann ich ~?** what can I do for you?; can I help you?; **mit 20 DM wäre mir schon gedient** 20 marks would do; **e)** *(verwendet werden)* serve; **als Museum ~:** serve *or* be used as a museum; **das soll dir als Warnung ~:** let that serve as *or* be a warning to you

Diener der; ~s, ~ servant; **einen ~ machen** *(ugs.)* bow; make a bow

Dienerin ['di:nərın] **die; ~, ~nen** maid; servant

dienern *itr. V. (abwertend)* bow; *(fig.)* bow and scrape

Dienerschaft die; ~: servants *pl.;* domestic staff

dienlich *Adj.* helpful; useful; **jmdm./einer Sache ~ sein** be helpful *or* of help to sb./sth.; **kann ich Ihnen mit etwas ~ sein?** *(geh.)* can I be of any assistance to you?

Dienst [di:nst] **der; ~[e]s, ~e a)** *o. Pl. (Tätigkeit)* work; *(von Soldaten, Polizeibeamten, Krankenhauspersonal usw.)* duty; **seinen ~ antreten** start work/go on duty; **~ haben** be at work/on duty; ⟨*chemist*⟩ be open; **außerhalb des ~es** outside work/when off duty; **seinen ~ tun** ⟨*machine, appliance*⟩ serve its purpose; **~ ist ~, und Schnaps ist Schnaps** *(ugs.)* you shouldn't mix business and pleasure; **b)** *(Arbeitsverhältnis)* post; **den** *od.* **seinen ~ quittieren** resign one's post; *(Milit.)* leave the service; ⟨*officer*⟩ resign one's commission; **Major** *usw.* **außer ~:** retired major *etc.;* **in ~ stellen** put sth. into service *or* commission; **c)** *o. Pl. (Tätigkeitsbereich)* service; **der höhere ~ der Beamtenlaufbahn** the senior civil service; **d)** *(Hilfe)* service; **~ am Kunden** *(ugs.)* customer service; **jmdm. mit etw. einen schlechten ~ erweisen** do sb. a disservice *or* a bad turn with sth.; **zu jmds. ~en** *od.* **jmdm. zu ~en sein** *od.* **stehen** *(geh.)* be at sb.'s disposal *or* service; **e)** *(Hilfs~)* service; *(Nachrichten~, Spionage~)* [intelligence] service

Diens·tag [di:ns-] **der** Tuesday; **am ~:** on Tuesday; **~, der 1. Juni** Tuesday the first of June; Tuesday, 1 June; **er kommt ~:** he is coming on Tuesday; **eines ~s** one Tuesday; **den ganzen ~ über** all day Tuesday; the whole of Tuesday; **ab nächsten** *od.* **nächstem ~:** from next Tuesday [onwards]; **die Nacht von ~ auf** *od.* **zum Mittwoch** Tuesday night; **~ in einer Woche** *od.* **in acht Tagen** Tuesday week; a week on Tuesday; **~ vor einer Woche** a week last Tuesday

Dienstag·abend der Tuesday evening *or (coll.)* night

dienstäglich 1. *Adj.; nicht präd.* [regular] Tuesday. **2.** *adv.* on Tuesday

Dienstag-: ~mittag der Tuesday lunchtime; **~morgen der** Tuesday morning; **~nachmittag der** Tuesday afternoon; **~nacht die** Tuesday night

diens·tags *Adv.* on Tuesday[s]; **~ abends/morgens** on Tuesday evening[s]/morning[s]; on a Tuesday evening/morning

Dienstag·vormittag der Tuesday morning

dienst-, Dienst-: ~alter das length of service; **~an·tritt der** commencement of one's duties; **~auf·fassung die** conception of duty; **~ausweis der** [official] identity card; **~beflissen 1.** *Adj.* zealous; eager; **2.** *adv.* zealously; eagerly; **~beginn der** start of work; **~bereit** *Adj.* ⟨*chemist*⟩ open *pred.;* ⟨*doctor*⟩ on call *or* duty; ⟨*dentist*⟩ on duty; **~bezüge** *Pl.* salary *sing.;* **~bote der** servant; **~boten·eingang der** tradesmen's entrance; **~eid der** official oath; **~eifer der** zeal; eagerness; **~eifrig 1.** *Adj.* zealous; eager; **2.** *adv.* zealously; eagerly; **~fähig** *Adj.* fit for work *postpos.; (Milit.)* fit for service *postpos.;* **~fahrt die** *s.* **~reise; ~frei** *Adj.* free ⟨*time*⟩; **an ~freien Tagen** on days off; **~frei haben/bekommen** have/get time off; **~geheimnis das a)** professional secret; *(im Staatsdienst)* official secret; **b)** *o. Pl.* professional secrecy; *(im Staatsdienst)* official secrecy; **unter das ~geheimnis fallen** be a professional/official secret; **~grad der** *(Milit.)* rank; **~habend** *Adj.; nicht präd.* duty ⟨*officer*⟩; ⟨*official, doctor*⟩ on duty; **~hund der** dog used for police/security work; **~jahr das** year of service; **~kleidung die** uniform; **~leistung die** *(auch Wirtsch.)* service

Dienstleistungs-: ~abend der late opening evening; **~betrieb der** *(Wirtsch.)* business in the service sector

dienstlich 1. *Adj.* **a)** business ⟨*call*⟩; *(im Staatsdienst)* official ⟨*letter, call, etc.*⟩; **b)** *(offiziell)* official; **~ werden** *(ugs.)* get businesslike and formal. **2.** *adv.* on business; *(im Staatsdienst)* on official business

dienst-, Dienst-: ~mädchen das *(veralt.)* maid; **~marke die** [police] identification badge; ≈ warrant card *(Brit.) or (Amer.)* ID card; **~ordnung die** official regulations *pl.;* **~pflicht die a)** *o. Pl.* compulsory service; **b)** *(bei Beamten)* duty; **~pistole die** service pistol; **~reise die** business trip; **~schluß der;** *o. Pl* end of work; **um 17 Uhr ist ~schluß** work finishes at 5 o'clock; **nach ~schluß** after work; **~stelle die** office; *(Abteilung)* department; **~stunden** *Pl.* **a)** working hours; **b)** *(Öffnungszeiten)* **~stunden haben** be open;

~tuend *Adj.; nicht präd. s.* **~habend; ~unfähig** *Adj.* unfit for work *postpos.; (Milit.)* unfit for service *postpos.*; **~vergehen das** offence against [official] regulations; **~vorschrift die** regulations *pl.; (Milit.)* service regulations; **~wagen der** official car; *(Geschäftswagen)* company car; **~weg der** proper *or* official channels *pl.*; **den ~weg gehen** *od.* **einhalten** go through the proper *or* official channels; **~wohnung die** *(von Firmen)* company flat *(Brit.) or (Amer.)* apartment; *(von staatlichen Stellen)* government flat *(Brit.) or (Amer.)* apartment; *(vom Militär)* army/navy/air force flat *(Brit.) or (Amer.)* apartment; **~zeit die a)** period of service; **b)** *(tägliche Arbeitszeit)* working hours *pl.*; **~zimmer das** office

dies [di:s] *s.* **dieser**

dies·bezüglich 1. *Adj.; nicht präd.* relating to *or* regarding this *postpos., not pred.* **2.** *adv.* regarding this; on this matter

diese ['di:zə] *s.* **dieser**

Diesel ['di:zl̩] **der; ~[s], ~** diesel

die·selbe *s.* **derselbe**

Diesel-: ~lokomotive die diesel locomotive; **~motor der** diesel engine

dieser ['di:zɐ], **diese, dieses, dies** *Demonstrativpron.* **a)** *attr.* this; *Pl.* these; **dieses Buch/diese Bücher [da]** that book/those books [there]; **in dieser Nacht wird es noch schneien/begann es zu schneien** it will snow tonight/it started to snow that night; **er hat dieser Tage Geburtstag** it's his birthday within the next few days; **ich habe ihn dieser Tage noch gesehen** I saw him the other day; **diese Inge ist doch ein Goldschatz** that Inge is a treasure, isn't she?; **b)** *alleinstehend* **diese[r] [hier/da]** this one [here]/that one [there]; **diese** *Pl.* **[hier/da]** these [here]/those [there]; **dies alles** all this; **diese ..., jene ...** *(geh.)* the latter ..., the former ...; **dies und das,** *(geh.)* **dieses und jenes** this and that; **dieser und jener** *(geh.) (einige)* some [people] *pl.*; *(ein paar)* a few [people] *pl.*

dieser·art *(geh.) indekl. Demonstrativpron.* of this/that kind *postpos.*

dieses *s.* **dieser**

diesig *Adj.* hazy

dies-: ~jährig *Adj.; nicht präd.* this year's; **unser ~jähriges Treffen** our meeting this year; **~mal** *Adv.* this time; **~seitig** *Adj.; nicht präd.* **das ~seitige Rheinufer** this side of the Rhine; **~seits 1.** *Präp. mit Gen.* on this side of; **2.** *Adv.* **~seits von** on this side of

Diesseits das; ~: das ~: this world

Dietrich ['di:trɪç] **der; ~s, ~e** picklock

die·weil *(veralt.)* **1.** *Konj.* **a)** *(zeitlich)* while; **b)** *(kausal)* because. **2.** *adv.* in the mean time *or* the mean while

diffamieren [dɪfa'mi:rən] *tr. V.* defame; **~de Äußerungen** defamatory utterances

Diffamierung die; ~, ~en defamation; *(Bemerkung)* defamatory statement

Differential [dɪfərɛn'tsia:l] **das; ~s, ~e a)** *(Math.)* differential; **b)** *(Technik)* differential [gear]

Differential- differential ⟨*gear, equation, calculus, etc.*⟩

Differenz [dɪfə'rɛnts] **die; ~, ~en** difference; *(Meinungsverschiedenheit)* difference [of opinion]

Differenz·betrag der difference

differenzieren 1. *tr. V. (Math.)* differentiate. **2.** *itr. V.* differentiate; make a distinction/distinctions **(zwischen** between); *(bei einem Urteil, einer Behauptung)* be discriminating

differenziert 1. *Adj.* subtly differentiated ⟨*methods, colours*⟩; complex ⟨*life, language, person, emotional life*⟩; sophisticated ⟨*taste*⟩; diverse ⟨*range*⟩. **2.** *adv.* **~ urteilen** be discriminating in one's judgement

differieren [dɪfə'ri:rən] *itr. V. (geh.)* differ **(um** by)

diffizil [dɪfi'tsi:l] *Adj. (geh.)* difficult

diffus [dɪ'fu:s] **1.** *Adj.* **a)** *(Physik, Chemie)* diffuse; **b)** *(geh.)* vague; vague and confused ⟨*idea, statement, etc.*⟩. **2.** *adv.* in a vague and confused way

digital [digi'ta:l] *(fachspr.)* **1.** *Adj.* digital. **2.** *adv.* digitally

Digital- digital ⟨*clock, display, recording, etc.*⟩

digitalisieren *tr. V. (DV)* digitalize

Diktat [dɪk'ta:t] **das; ~[e]s, ~e a)** dictation; **nach ~ schreiben** take dictation; **b)** *(geh.: Befehl)* dictate; *(Politik)* diktat

Diktator [dɪk'ta:tɔr] **der; ~s, ~en** [-'to:rən] *(auch fig.)* dictator

diktatorisch *(auch fig.)* **1.** *Adj.* dictatorial. **2.** *adv.* dictatorially

Diktatur [dɪkta'tu:ɐ̯] **die; ~, ~en** *(auch fig.)* dictatorship

diktieren [dɪk'tieren] *tr. V.* dictate

Diktier·gerät das dictating machine

Dilemma [di'lɛma] **das; ~s, ~s** dilemma

Dilettant [dilɛ'tant] **der; ~en, ~en, Dilettantin die; ~, ~nen** *(auch abwertend)* dilettante

dilettantisch *(abwertend)* **1.** *Adj.* dilettante; amateurish. **2.** *adv.* amateurishly

Dill [dɪl] **der; ~[e]s, ~e** dill

Dimension [dimɛn'zi̯o:n] **die; ~, ~en** *(Physik, fig.)* dimension

-dimensional [dimɛnzi̯o'na:l] *Adj.* -dimensional

DIN [di:n] *Abk.* **Deutsche Industrie-Norm[en]** *German Industrial Standard[s]*; DIN; **DIN-Format** DIN size; **DIN-A4-Format** A4

[1]**Ding** [dɪŋ] **das; ~[e]s, ~e a)** thing; **jedes ~ hat zwei Seiten** *(fig.)* there are two sides to everything; **b)** *meist Pl.* **nach Lage der ~e** the way things are; **über den ~en stehen** be above such things; **persönliche/private ~e** personal/private matters; **ein ~ der Unmöglichkeit sein** be quite impossible; **das geht nicht mit rechten ~en zu** there's something funny about it; **vor allen ~en** above all; **c) guter ~e sein** *(geh.)* be in good spirits

[2]**Ding das; [e]s, ~er** *(ugs.)* **a)** thing; **das ist ja ein ~!** that's really something; **ein ~ drehen** ⟨*criminal*⟩ pull a job *(sl.)*; **mach keine ~er!** stop having me on *(Brit. coll.)*; stop putting me on *(Amer. coll.)*; **b)** *(Mädchen)* thing; creature

dingen *unr. tr. V. (geh.)* hire

ding·fest *in* **jmdn. ~ machen** arrest *or* apprehend sb.

[1]**Dings** [dɪŋs] **(der/die); ~** *(ugs.: für einen Personennamen)* thingamy *(coll.)*; thingumajig *(coll.)*; what's-his-name/-her-name

[2]**Dings** *(ugs.)* **1. das; ~** *(Gegenstand)* thingamy *(coll.)*; thingumajig *(coll.)* what-d'you-call-it. **2. (das); ~** *(für einen Ortsnamen)* what's-its-name; what's-it-called

Dings·bums *s.* [1]**Dings,** [2]**Dings**

Dinosaurier [dino'zaurjɐ] dinosaur

Diode [di'o:də] **die; ~, ~n** *(Elektrot.)* diode

Dioxyd ['di:lɔksy:t] **das; ~s, ~e** *(Chemie)* dioxide

Dioxin ['di:lɔksi:n] **das; ~s** *(Chemie)* dioxin

Diözese [diø'tse:zə] **die; ~, ~n** diocese

Diphtherie [dɪfte'ri:] **die; ~, ~n** *(Med.)* diphtheria

Dipl.-Ing. *Abk.* **Diplomingenieur** *academically qualified engineer*

Diplom [di'plo:m] **das; ~s, ~e a)** ≈ [first] degree *(in a scientific or technical subject); (für einen Handwerksberuf)* diploma; **b)** *(Urkunde)* ≈ degree certificate *(in a scientific or technical subject); (für einen Handwerksberuf)* diploma

Diplom-: qualified

Diplom·arbeit die ≈ degree dissertation *(for a first degree in a scientific or technical subject); (für einen Handwerksberuf)* dissertation [submitted for a/the diploma]

Diplomat [diplo'ma:t] **der; ~en, ~en** *(auch fig.)* diplomat

Diplomaten-: ~koffer der attaché case; executive case; **~viertel das** embassy district

Diplomatie [diploma'ti:] **die; ~** diplomacy

Diplomatin die; ~, ~nen *(auch fig.)* diplomat

diplomatisch *(auch fig.)* **1.** *Adj.* diplomatic. **2.** *adv.* diplomatically

diplomiert *Adj.* qualified

Diplom·prüfung die ≈ degree examination *(in a scientific or technical subject); (für einen Handwerksberuf)* diploma examination

dir [di:ɐ̯], *(in Briefen)* **Dir 1.** *Dat. von* **du** to you; *(nach Präp.)* you; **ich gab ~ das Buch** I gave you the book; **Freunde von ~:** friends of yours; **gehen wir zu ~:** let's go to your place. **2.** *Dat. des Reflexivpron. der 2. Pers. Sg.* yourself; **hast du ~ gedacht, daß ...:** did you think that ...; **nimm ~ noch von dem Braten** help yourself to some more roast

direkt [di'rɛkt] **1.** *Adj.* direct. **2.** *adv.* **a)** *(geradewegs, sofort)* straight; directly; ⟨*broadcast sth.*⟩ live; **b)** *(nahe)* directly; **~ am Marktplatz** right by the market square; **c)** *(unmittelbar)* direct; **sich ~ mit jmdm. verbinden lassen** get a direct line to sb.; **d)** *(unverblümt)* directly; **e)** *(ugs.: geradezu)* really, positively ⟨*dangerous, witty*⟩

Direkt·flug der direct flight

Direktheit die; ~: directness

Direktion [dirɛk'tsi̯o:n] **die; ~, ~en a)** *o. Pl.* management; *(von gemeinnützigen, staatlichen Einrichtungen)* administration; **b)** *(die Geschäftsleiter)* management; **c)** *(Büroräume)* managers' offices *pl.*

Direktor [di'rɛktɔr] **der; ~s, ~en** [...'to:rən], **Direktorin** [dirɛk'to:rɪn] **die; ~, ~nen** director; *(einer Schule)* headmaster/headmistress; *(einer Fachschule)* principal; *(einer Strafanstalt)* governor; *(einer Abteilung)* manager

Direkt-: ~sendung die, ~übertragung die live transmission *or* broadcast; **~verbindung die a)** *(Eisenb.)* direct connection; through train; *(Flugw.)* direct flight; **b)** *(Fernspr.)* direct [telephone] connection; **~wahl die a)** *(Polit.)* direct election; **b)** *o. Pl. (Fernspr.)* direct dialling

Direx ['di:rɛks] **der; ~, ~e,/die; ~, ~en** *(Schülerspr.)* head

Dirigent [diri'gɛnt] **der; ~en, ~en** conductor

Dirigenten-: ~pult das conductor's rostrum; **~stab der, ~stock der** [conductor's] baton

dirigieren [diri'gi:rən] *tr. V.* **a)** *(Musik) auch itr.* conduct; **b)** *(führen)* steer ⟨*vehicle, person*⟩; **jmdn. an einen Ort ~:** send sb. to a place

dirigistisch *(Wirtsch.)* **1.** *Adj.* dirigiste. **2.** *adv.* in a dirigiste manner

Dirndl ['dɪrndl̩] **das; ~s, ~, Dirndl·kleid das** dirndl

Dis [dɪs] **das; ~, ~** *(Musik)* D sharp

Disco ['dɪsko:] **die; ~, ~s** disco

Discount- [dɪs'kaunt-] discount ⟨*shop, price, etc.*⟩

Diskette [dɪs'kɛtə] **die; ~, ~n** *(DV)* floppy disc

Disketten·lauf·werk das *(DV)* [floppy-] disc drive

Disk·jockey ['dɪskdʒɔke] **der** disc jockey

Diskont [dɪs'kɔnt] **der; ~s, ~e** *(Finanzw.)* discount

Diskont·satz der *(Finanzw.)* discount rate

Diskothek [dɪsko'te:k] **die; ~, ~en a)** *(Tanzlokal)* discothèque; **b)** *(Schallplatten)* record collection

diskreditieren [dɪskredi'ti:rən] *tr. V.* discredit

Diskrepanz [dɪskre'pants] **die; ~, ~en** discrepancy

diskret [dɪs'kre:t] **1.** *Adj.* **a)** *(vertraulich)* confidential ⟨*discussion, report*⟩; *(unauffällig)* discreet ⟨*action*⟩; **b)** *(taktvoll)* discreet; tactful ⟨*behaviour, reserve*⟩. **2.** *adv.* **a)** *(vertraulich)* confidentially; **etw. ~ behandeln** treat sth. in confidence; **b)** *(taktvoll)* discreetly; tactfully

Diskretion [dɪskre'tsi̯o:n] **die; ~ a)** *(Verschwiegenheit, Takt)* discretion; **~ [ist] Ehrensache** you can rely on my discretion; **b)** *(Unaufdringlichkeit)* discreetness

diskriminieren [dɪskrimi'ni:rən] *tr. V.* **a)** discriminate against; **b)** *(herabwürdigen)* disparage

diskriminierend *Adj.* disparaging

Diskriminierung die; ~, ~en discrimination (*Gen.* against)

Diskus ['dɪskʊs] **der; ~** *od.* **~ses, Disken** *od.* **~se** *(Leichtathletik)* discus

Diskussion [dɪskʊ'si̯o:n] **die; ~, ~en** discussion; **etw. zur ~ stellen** put sth. up for discussion; **[nicht] zur ~ stehen** [not] be under discussion

Diskussions-: ~beitrag der contribution to a/the discussion; **~leiter der** chairman [of the discussion]; **~teilnehmer der** participant [in a/the discussion]

Diskus·werfen das; ~s *(Leichtathletik)* [throwing the] discus

diskutieren [dɪsku'ti:rən] **1.** *itr. V.* **a) über etw.** *(Akk.)* **~:** discuss sth.; **wir haben stundenlang diskutiert** our discussion went on for hours. **2.** *tr. V.* discuss

disponieren *itr. V.* make plans; *(vorausplanen)* plan ahead; **anders ~:** make other plans

Dispositions·kredit der *(Finanzw.)* overdraft facility

Disput [dɪs'pu:t] **der; ~[e]s, ~e** *(geh.)* dispute, argument

Disqualifikation die *(auch Sport)* disqualification

disqualifizieren *tr. V.* *(auch Sport)* disqualify

Dissertation [dɪsɛrta'tsi̯o:n] **die; ~, ~en** [doctoral] dissertation *or* thesis

Dissonanz [dɪso'nants] **die; ~, ~en** *(Musik, fig.)* dissonance

Distanz [dɪs'tants] **die; ~, ~en** *(auch Sport, fig.)* distance; **~ zu etw. gewinnen** *(fig.)* distance oneself from sth.; **auf ~** *(Akk.)* **gehen, ~ wahren** *od.* **halten** *(fig.)* keep one's distance

distanzieren *refl. V.* dissociate (**von** from) oneself

distanziert 1. *Adj.* distant; reserved. **2.** *adv.* in a distant *or* reserved manner; with reserve

Distel ['dɪstl̩] **die; ~, ~n** thistle

distinguiert [dɪstɪŋ'gi:ɐ̯t] *(geh.)* **1.** *Adj.* distinguished. **2.** *adv.* in a distinguished manner

Disziplin [dɪstsi'pli:n] **die; ~, ~en a)** *o. Pl.* discipline; *(Selbstbeherrschung)* [self-]discipline; **~ halten** keep discipline; *(sich diszipliniert verhalten)* behave in a disciplined way; **b)** *(Wissenschaftszweig, Sportart)* discipline

disziplinarisch 1. *Adj.* disciplinary. **2.** *adv.* **gegen jmdn. ~ vorge-**

hen take disciplinary action against sb.
Disziplinar- disciplinary ‹*measure, proceedings, etc.*›
disziplinieren **1.** *tr. V.* discipline. **2.** *refl. V.* discipline oneself
diszipliniert **1.** *Adj.* **a)** well-disciplined; **b)** *(beherrscht)* disciplined. **2.** *adv.* **a)** in a well-disciplined way; **b)** *(beherrscht)* in a disciplined way
disziplin-, Disziplin-: **~los** **1.** *Adj.* undisciplined; **2.** *adv.* in an undisciplined way; **~losigkeit** **die; ~:** lack of discipline; **~schwierigkeiten** *Pl.* discipline problems; problems in maintaining discipline
dito ['di:to] *Adv. (Kaufmannsspr., auch ugs.)* ditto
Diva ['di:va] **die; ~, ~s** *u.* **Diven** **a)** prima donna; diva; *(Film~)* great [film] star; **b)** *(eingebildeter Mensch)* prima donna
divers... [di'vɛrs...] *Adj.; nicht präd.* various; *(mehrer...)* several
Dividende [divi'dɛndə] **die; ~, ~n** *(Börsenw., Wirtsch.)* dividend
dividieren [divi'di:rən] *tr. V. (Math.)* divide
Division [divi'zi̯o:n] **die; ~, ~en** *(Math., Milit.)* division
Diwan ['di:va:n] **der; ~s, ~e** *(veralt.)* divan
d. J. *Abk.* **a) dieses Jahres; b) der/die Jüngere**
dm *Abk.* **Dezimeter** dm
DM *Abk.* **Deutsche Mark** DM
Do. *Abk.* **Donnerstag** Thur[s].
Dobermann ['do:bɐman] **der; ~s, Dobermänner** Dobermann [pinscher]
doch [dɔx] **1.** *Konj.* but. **2.** *Adv.* **a)** *(jedoch)* but; **b)** *(dennoch)* all the same; still; *(wider Erwarten)* after all; **und ~:** and yet; **c)** *(geh.: nämlich)* **wußte er ~, daß ...:** because he knew that ...; **d)** *(als Antwort)* [oh] yes; **Hast du keinen Hunger? – Doch!** Arent't you hungry? – Yes [I am]!; **e)** *(trotz allem, was dagegen sprechen/gesprochen haben mag)* **er war also ~ der Mörder!** so he 'was the murderer!; **sie hat es also ~ gesagt** so she 'did say it; **f)** *(ohnehin)* in any case; **du kannst mir ~ nicht helfen** there's nothing you can do to help me. **3.** *Interj.* **a)** *(widersprechende Antwort auf eine verneinte Aussage)* **Das stimmt nicht. – Doch!** That's not right. – [Oh] yes it is!; **b)** *(negative Antwort auf eine verneinte Frage)* **Hast du keinen Hunger? – Doch!** Aren't you hungry? – Yes [I am]! **4.** *Partikel* **a)** *(auffordernd, Ungeduld, Empörung ausdrückend)* **das hättest du ~ wissen müssen** you [really] should have known that; **du hast ~ selbst gesagt, daß ...** *(rechtfertigend)* you did say yourself that ...; **gib mir ~ bitte mal die Zeitung** pass me the paper, please; **reg dich ~ nicht so auf!** don't get so worked up!; **paß ~ auf!** [oh,] do be careful!; **das ist ~ nicht zu glauben** that's just incredible; **b)** *(Zweifel ausdrükkend)* **du hast ~ meinen Brief erhalten?** you did get my letter, didn't you?; **c)** *(Überraschung ausdrückend)* **das ist ~ Karl!** there's Karl! **d)** *(an Bekanntes erinnernd)* **er ist ~ nicht mehr der jüngste** he's not as young as he used to be[, you know]; **e)** *(nach Vergessenem fragend)* **wie war ~ sein Name?** now what was his name?; **f)** *(verstärkt Bejahung/Verneinung ausdrückend)* **gewiß/sicher ~:** [why] certainly; of course; **ja ~:** [yes,] all right *or (coll.)* OK; **nicht ~!** *(abwehrend)* [no,] don't!; **g)** *(Wunsch verstärkend)* **wäre es ~ ...:** if only it were ...
Docht [dɔxt] **der; ~[e]s, ~** wick
Dock [dɔk] **das; ~s, ~s** dock
docken **1.** *itr. V. (Seew., Raumf.)* dock. **2.** *tr. V. (Seew.)* dock ‹*ship*›; put ‹*ship*› in dock
Dogge ['dɔgə] **die; ~, ~n** **a) deutsche ~:** Great Dane; **b) englische ~:** mastiff
Dogma ['dɔgma] **das; ~s, Dogmen** *(auch fig.)* dogma
dogmatisch *(auch fig.)* **1.** *Adj.* dogmatic. **2.** *adv.* dogmatically
Dohle ['do:lə] **die; ~, ~n** jackdaw
Doktor ['dɔktɔr] **der; ~s, ~en** [-'to:rən] **a)** *o. Pl. (Titel)* doctorate; doctor's degree; **den/seinen ~ machen** do a/one's doctorate; **b)** *(Träger)* doctor; **Herr ~ Krause** Doctor Krause; **c)** *(ugs.: Arzt)* doctor; **der Onkel ~** *(Kinderspr.)* the nice doctor
Doktorand [dɔkto'rant] **der; ~en, ~en, Doktorandin die; ~, ~nen** student taking his/her doctorate
Doktor·arbeit **die** doctoral thesis *or* dissertation
Doktorin die; ~, ~nen doctor
Doktor-: **~titel der** title of doctor; **~vater der** *(ugs.)* [thesis] supervisor
Doktrin [dɔk'tri:n] **die; ~, ~en** doctrine
Dokument [doku'mɛnt] **das; ~[e]s, ~e** document
Dokumentar·film **der** documentary [film]
dokumentarisch **1.** *Adj.* documentary. **2.** *adv.* **etw. ~ belegen** provide documentary evidence of *or* for sth.; **etw. ~ festhalten** make a documentary record of sth.
Dokumentation [dɔkumɛnta'tsi̯o:n] **die; ~, ~en** **a)** *o. Pl.* documentation; *(fig.)* demonstration; **b)** *(Material)* documentary account; *(Bericht)* documentary report
dokumentieren *tr. V.* **a)** *(belegen)* document; *(fig.)* demonstrate; **b)** *(festhalten, darstellen)* record ‹*behaviour, event*›
Dolch [dɔlç] **der; ~[e]s, ~e** dagger
Dolde ['dɔldə] **die; ~, ~n** *(Bot.)* umbel
doll [dɔl] *(bes. nordd., salopp)* **1.** *Adj.* **a)** *(ungewöhnlich)* incredible; amazing; **b)** *(großartig)* fantastic *(coll.)*; great *(coll.)*. **2.** *adv.* **a)** *(großartig)* fantastically [well] *(coll.)*; **b)** *(sehr)* ‹*hurt*› dreadfully *(coll.)*; ‹*shake, rain*› good and hard *(coll.)*
Dollar ['dɔla:ɐ̯] **der; ~[s], ~s** dollar
Dollar·kurs der dollar rate
dolmetschen *itr. V.* act as interpreter (**bei** at)
Dolmetscher der; ~s, ~, Dolmetscherin die; ~, ~nen interpreter
Dom [do:m] **der; ~[e]s, ~e** cathedral; *(fig.)* dome; **der Kölner ~, der ~ zu Köln** Cologne Cathedral
Domäne [do'mɛ:nə] **die; ~, ~n** *(fig.)* domain
domestizieren [domɛsti'tsi:rən] *tr. V.* domesticate
dominant [domi'nant] *Adj. (auch Biol.)* dominant
Dominante die; ~, ~n *(Musik)* *(Quint)* dominant; *(Akkord)* dominant chord
Dominanz [domi'nants] **die; ~, ~en** *(auch Biol.)* dominance
dominieren [domi'ni:rən] *itr. V.* dominate; **~d** dominant
Dominikaner [domini'ka:nɐ] **der; ~s, ~, Dominikanerin die; ~, ~nen** *(Mönch/Nonne, Einwohner/Einwohnerin der Dominikanischen Republik)* Dominican
Dominikaner·orden der *o. Pl.* Dominican order
dominikanisch *Adj.* Dominican; **die Dominikanische Republik** the Dominican Republic
Domino das; ~s, ~s dominoes *sing.*
Domino·stein der **a)** domino; **b)** *(Gebäck)* *small chocolate-covered cake with layers of marzipan, jam, and gingerbread*
Domizil [domi'tsi:l] **das; ~s, ~e** *(geh.)* domicile; residence

Dompteur [dɔmp'tøːɐ] **der; ~s, ~e, Dompteuse** [dɔmp'tøːzə] **die; ~, ~n** tamer

Donau ['doːnau̯] **die; ~:** Danube

Donner ['dɔnɐ] **der; ~s, ~** *(auch fig.)* thunder; **wie vom ~ gerührt dastehen** *od.* **sein** be thunderstruck

donnern 1. *itr. V.* **a)** *(unpers.)* thunder; **es hat gedonnert und geblitzt** there was thunder and lightning; **b)** *(fig.)* ⟨*gun*⟩ thunder, boom [out]; ⟨*engine*⟩ roar; ⟨*hooves*⟩ thunder; **~der Applaus** thunderous applause; **c)** *mit sein (sich laut fortbewegen)* ⟨*train, avalanche, etc.*⟩ thunder; **d)** *(ugs.: schlagen)* thump, hammer **(an +** *Akk.*, **gegen** on); **e)** *mit sein (ugs.: prallen)* **gegen etw. ~:** smash into sth. **2.** *tr. V.* **a)** *(ugs.: schleudern)* sling *(coll.)*; hurl

Donner·schlag der clap *or* peal of thunder; **die Nachricht traf uns wie ein ~:** the news completely stunned us

Donners·tag der Thursday; *s. auch* **Dienstag**

donnerstags *Adv.* on Thursday[s]; *s. auch* **dienstags**

Donner·wetter das *(ugs.)* **a)** *(Krach)* row; **b)** ['--'--] **zum ~ [noch einmal]!** *(Ausruf der Verärgerung)* damn it!; **~!** *(Ausruf der Bewunderung)* my word; wow

doof [doːf] *(ugs. abwertend)* **1.** *Adj.* **a)** *(einfältig)* stupid; dumb *(coll.)*; dopey *(sl.)*; **b)** *(langweilig)* boring; **c)** *nicht präd. (ärgerlich)* stupid. **2.** *adv.* stupidly

Doofheit die; *~(ugs. abwertend)* stupidity; dumbness *(coll.)*

Doofmann der; ~[e]s, Doofmänner *(ugs. abwertend)* dope *(coll.)*; dummy; [stupid] twit *(sl.)*

dopen ['dɔpn̩] *tr. V.* dope ⟨*horse etc.*⟩; **jmdn. ~:** give sb. drugs; **gedopt sein** ⟨*athlete*⟩ have taken drugs

Doping ['dɔpɪŋ] **das; ~s, ~s a)** *(bei Sportlern)* taking drugs; **b)** *(von Pferden usw.)* doping

Doppel ['dɔpl̩] **das, ~s, ~ a)** *(Kopie)* duplicate; copy; **b)** *(Sport)* doubles *sing. or pl.*; **ein ~:** a game of doubles; *(im Turnier)* a doubles match

doppel-, Doppel-: ~album das double album *or* LP; **~bett das** double bed; **~bock das** extra-strong bock beer; **~decker der; ~s, ~ a)** *(Flugzeug)* biplane; **b)** *(Omnibus)* double-decker [bus]; **~deutig** [~dɔy̯tɪç] **1.** *Adj.* ambiguous; **2.** *adv.* ambiguously; **~deutigkeit die; ~, ~en** ambiguity; **~fenster das** double-glazed window; **~gänger der; ~s, ~, ~gängerin die; ~, ~nen** double; **~kinn das** double chin; **~kopf der;** *o. Pl. (Kartenspiel)* Doppelkopf; **~leben das** double life; **~moral die** double standards *pl.*; **~mord der** double murder; **~name der** double-barrelled name *(Brit.)*; hyphenated name; **~paß der** *(Fußball)* one-two; **~punkt der** colon; **~rolle die** dual role; **~seitig** *Adj. (Med.)* double ⟨*pleurisy, pneumonia*⟩; bilateral ⟨*paralysis*⟩; **~sinnig 1.** *Adj.* ambiguous; **2.** *adv.* ambiguously; **~stecker der** *(Elektrot.)* two-way plug *or* adapter; **~stunde die** *(Schulw.)* double period

doppelt 1. *Adj.* **a)** *(zweifach)* double; dual ⟨*nationality*⟩; **die ~e Länge** double *or* twice the length; **ein ~er Klarer** *(ugs.)* a double schnapps; **ein ~er Boden** a false bottom; **b)** *(besonders groß, stark)* redoubled ⟨*enthusiasm, attention*⟩; **mit ~er Kraft arbeiten** work with twice as much energy; **2.** *adv.* **a)** *(zweimal)* twice; **~ genäht hält besser** *(Spr.)* it's better to be on the safe side; better safe than sorry; **das ist ~ gemoppelt** *(ugs.)* that's just saying the same thing twice over; **~ soviel** twice as much; **das habe ich ~:** I have two of them; **~ sehen** see double; **b)** *(ganz besonders, noch mehr)* **~ einsam** twice as lonely; **sich ~ anstrengen** try twice as hard

¹Doppelte das; *adj. Dekl.* **das ~ bezahlen** pay twice as much; pay double; **auf das ~ steigen** double

²Doppelte der; *adj. Dekl. (ugs.)* double

doppel-, Doppel-: ~zentner der 100 kilograms; quintal; **~zimmer das** double room; **~züngig** [-tsʏnɪç] *(abwertend)* **1.** *Adj.* two-faced; **2.** *adv.* **~züngig reden** be two-faced

Dorf [dɔrf] **das; ~[e]s, Dörfer** ['dœrfɐ] village; **auf dem ~:** in the country; **vom ~** from the country; **über die Dörfer** from village to village; **das sind für mich böhmische Dörfer** *(ugs.)* it's all Greek to me

Dorf-: ~bewohner der villager; **~jugend die** young people *pl.* of the village; village youth

dörflich *Adj.* village *attrib.* ⟨*life, traditions, etc.*⟩; *(ländlich)* rural ⟨*character*⟩

Dorn [dɔrn] **der; ~[e]s, ~en** thorn; **jmdm. ein ~ im Auge sein** annoy sb. intensely

Dornen-: ~krone die crown of thorns; **~strauch der** thornbush

dornig *Adj.* thorny

Dorn·röschen (das) the Sleeping Beauty

dorren ['dɔrən] *itr. V.; mit sein (geh.)* dry up

dörren ['dœrən] **1.** *tr. V.* dry. **2.** *itr. V.; mit sein* dry

Dörr-: ~fleisch das *(südd.)* ≈ streaky bacon; **~obst das** dried fruit; **~pflaume die** prune

Dorsch [dɔrʃ] **der; ~[e]s, ~e** cod; *(junger Kabeljau)* codling

dort [dɔrt] *Adv.* there

dort-: ~|behalten *unr. tr. V.* keep there; **~|bleiben** *unr. itr. V.; mit sein* stay there; **~her** *Adv.* **[von] ~her** from there; **~hin** *Adv.* there; **bis ~hin** as far as there; **~hin, wo ...** to where ...; **~hinab** *Adv.* down there; down that way; **~hinauf** *Adv.* up there; up that way; **~hinaus** *Adv.* **a)** out there; *(in diese Richtung)* out that way; **b)** ['---] **frech bis ~hinaus** *(ugs.)* [as] cheeky as anything; **~hinein** *Adv.* in there

dortig *Adj.; nicht präd.* there *postpos.*

Dose ['doːzə] **die; ~, ~n a)** *(Blech~)* tin; *(Pillen~)* box; *(Zukker~)* bowl; **b)** *(Konserven~)* can; tin *(Brit.)*; *(Bier~)* can

dösen ['døːzn̩] *itr. V. (ugs.)* doze

Dosen-: ~bier das canned beer; **~milch die** canned *or (Brit.)* tinned milk; **~öffner der** can opener; tin-opener *(Brit.)*

dosieren *tr. V.* **etw. ~:** measure out the required dose of sth.; **sorgfältig dosierte Mengen** carefully measured doses

Dosierung die; ~, ~en a) *o. Pl.* measuring out; **b)** *s.* **Dosis**

dösig ['døːzɪç] *(ugs.)* **1.** *Adj.* drowsy; dozy. **2.** *adv.* drowsily

Dosis ['doːzɪs] **die; ~, Dosen** *(auch fig.)* dose; **die tägliche ~:** the daily dosage

Döskopp ['døːskɔp] **der; ~s, Dösköppe** ['døːskœpə] *(salopp)* dozy twit *(Brit. sl.)*; dim-wit

Dossier [dɔ'si̯eː] **das ~s, ~s** dossier

dotieren [do'tiːrən] *tr. V.* **eine Position gut/mit 5 000 DM ~:** offer a good salary/a salary of 5,000 marks with a position; **eine gut dotierte Stellung** a well-paid position

Dotter ['dɔtɐ] **der** *od.* **das; ~s, ~:** yolk

Double ['duːbl̩] **das; ~s, ~s a)** *(Ersatzdarsteller[in])* stand-in; **b)** *(Doppelgänger)* double

Dozent [do'tsɛnt] **der; ~en, ~en,**

Dozentin die; ~, ~nen lecturer **(für in)**

dozieren [do'tsi:rən] *itr. V. (auch fig.)* lecture

dpa [de:pe:'|a:] **die;** ~ *Abk.* **Deutsche Presse-Agentur** West German Press Agency

Dr. *Abk.:* **Doktor** Dr

Drache ['draxə] **der; ~n, ~n** *(Myth.)* dragon

Drachen der; ~s, ~ a) *(Papier~)* kite; **einen ~ steigen lassen** fly a kite; **b)** *(salopp: zänkische Frau)* dragon

Drachen·fliegen das; ~s *(Sport)* hang-gliding

Dragée, Dragee [dra'ʒe:] **das; ~s, ~s** dragée

Draht [dra:t] **der; ~[e]s, Drähte** ['drɛ:tə] **a)** wire; **b)** *(Leitung)* wire; cable; *(Telefonleitung)* line; wire; **heißer ~:** hot line; **c) auf ~ sein** *(ugs.)* be on the ball *(coll.)*

Draht·bürste die wire brush

Draht-: ~esel der *(ugs. scherzh.)* bike *(coll.)*; **~geflecht das** wire mesh

drahtig *Adj.* wiry ‹*person, hair*›

draht-, Draht-: ~los *(Nachrichtenw.)* **1.** *Adj.* wireless; **2.** *adv.* **etw. ~los telegrafieren/übermitteln** radio sth.; **~schere die** wire-cutters *pl.*; **~seil das** [steel] cable; **~seil·bahn die** cable railway; **~zieher** [-tsi:ɐ] **der** *(fig.)* wire puller

drakonisch [dra'ko:nɪʃ] **1.** *Adj.* Draconian. **2.** *adv.* in a Draconian way

drall [dral] *Adj.* strapping ‹*girl*›; full, rounded ‹*cheeks, bottom*›

Drall der; ~[e]s, ~e spin

Drama ['dra:ma] **das; ~s, Dramen** drama; *(fig. ugs.: Katastrophe)* disaster

Dramatik [dra'ma:tɪk] **die; ~:** drama

Dramatiker der; ~s, ~: dramatist

dramatisch 1. *Adj.* dramatic. **2.** *adv.* dramatically

dramatisieren *tr. V.* dramatize

Dramaturg [drama'tʊrk] **der; ~en, ~en** *(Theater)* literary and artistic director; *(Rundf., Ferns.)* script editor

Dramaturgie [dramatʊr'gi:] **die; ~, ~n a)** dramaturgy; **b)** *(Abteilung) (Theater)* literary and artistic director's department; *(Rundf., Ferns.)* script department

dramaturgisch 1. *Adj.* dramaturgical. **2.** *adv.* **~ wirkungsvoll in Szene gesetzt** staged effectively

dran [dran] *Adv. (ugs.)* **a) das Schild bleibt ~:** the sign stays up; **häng das Schild ~!** put the sign up!; **ich komme/kann nicht ~:** I can't reach; **b) arm ~ sein** be in a bad way; **gut/schlecht ~ sein** be well off/badly off; **früh/spät ~ sein** be early/late; **an dem Gerücht ist was ~:** there is something in the rumour; **ich bin ~** *od. (scherzh.)* **am ~sten** *(ich bin an der Reihe)* it's my turn; I'm next; *(ich werde zur Verantwortung gezogen)* I'll be for the high jump *or (sl.)* for it *(Brit.)*; I'll be under the gun *(Amer.)*; **nicht wissen wo man ~ ist** not know where one stands; *s. auch:* **daran; dranbleiben; dranhängen** *usw.;* **glauben**

dran|bleiben *unr. itr. V.; mit sein (ugs.) (am Telefon)* hold *or (coll.)* hang on; *(an der Arbeit)* stick at it *(coll.)*

drang [draŋ] *1. u. 3. Pers. Sg. Prät. v.* **dringen**

Drang der; ~[e]s, Dränge ['drɛŋə] urge; **ein ~ nach Bewegung/Freiheit** an urge to move/be free

dränge ['drɛŋə] *1. u. 3. Pers. Sg. Konjunktiv II v.* **dringen**

dran-: ~|geben *unr. tr. V.* give up ‹*time*›; give, sacrifice ‹*one's life*›; **~|gehen** *unr. itr. V.; mit sein (ugs.)* **a)** *(berühren)* touch; **b)** *(in Angriff nehmen)* **~gehen, etw. zu tun** get down to doing sth.

Drängelei die; ~, ~en *(abwertend)* **a)** pushing [and shoving]; **b)** *(mit Wünschen, Bitten)* pestering

drängeln ['drɛŋl̩n] *(ugs.)* **1.** *itr. V.* **a)** push [and shove]; **b)** *(auf jmdn. einreden)* go on *(coll.)*; **zum Aufbruch ~:** go on about it being time to leave *(coll.)*. **2.** *tr. V.* **a)** push; shove; **b)** *(einreden auf)* pester; go on at *(coll.)*. **3.** *refl. V.* **sich nach vorn ~:** push one's way to the front

drängen ['drɛŋən] **1.** *itr. V.* **a)** *(schieben)* push; **die Menge drängte zum Ausgang** the crowd pressed towards the exit; **b) auf etw.** *(Akk.)* **~:** press for sth.; **zum Aufbruch ~:** insist that it is/was time to leave; **zur Eile ~:** hurry us/them *etc.* up; **die Zeit drängt** time is pressing. **2.** *tr. V.* **a)** push; **b)** *(antreiben)* press; urge. **3.** *refl. V.* ‹*visitors, spectators, etc.*› crowd, throng; ‹*crowd*› throng; **sich nach vorn ~:** push one's way to the front; **sich in den Vordergrund ~** *(fig.)* make oneself the centre of attention

Drangsal ['draŋza:l] **die; ~, ~e** *(geh.) (Not)* hardship; *(Qual)* suffering

drangsalieren *tr. V. (abwertend) (quälen)* torment; *(plagen)* plague

dran-: ~|halten *unr. refl. V. (ugs.)* get a move on *(coll.)*; **~|hängen** *(ugs.)* **1.** *tr. V.* **a)** *(aufwenden)* invest; **viel Zeit/Geld ~hängen** put a lot of time/money into it; **b)** *(anschließen)* add **(an + Akk.** to). **2.** *refl. V. (verfolgen)* stay *or (coll.)* stick close behind; **~|kommen** *unr. itr. V.; mit sein (ugs.)* have one's turn; **ich kam als erste/erster ~:** it was my turn first; *(beim Arzt, Zahnarzt usw.)* I was the first one; **wer kommt jetzt ~?** who's next?; **ich bin heute in Latein ~gekommen** *(aufgerufen worden)* I got picked on to answer in Latin today *(coll.)*; **~|kriegen** *tr. V. (ugs.)* **jmdn. ~kriegen** get sb.; *(zum Arbeiten bringen)* get sb. at it *(coll.)*; **~|machen** *refl. V. (ugs.)* **sich ~machen, etw. zu tun** get down to doing sth.; **wenn sich die Kinder ~machen, ist der Kuchen gleich weg** once the children get started on it the cake won't last long; **~|nehmen** *unr. tr. V. (ugs.) (beim Friseur usw.)* see to; *(beim Arzt)* see; *(in der Schule)* pick on; **~|setzen** *(ugs.)* **1.** *tr. V. (einsetzen)* **seine ganze Kraft ~setzen, etw. zu erreichen** put all one's energy into achieving sth.; **2.** *refl. V. (beginnen)* get down to it

drastisch 1. *Adj.* **a)** *(grob)* crudely explicit ‹*joke, story, etc.*›; graphic ‹*report, account*›; **b)** *(empfindlich spürbar)* drastic ‹*measure, means*›. **2.** *adv.* **a)** *(grob)* with crude explicitness; *(deutlich)* graphically; **b)** *(einschneidend)* drastically

drauf [drauf] *Adv. (ugs.)* on it; **den Deckel ~ machen** put the lid on; **die dollsten Sprüche ~ haben** *(fig.)* have the most amazing patter; **90 Sachen ~ haben** *(fig.)* be doing 90; **~ und dran sein, etw. zu tun** be just about to do *or* be on the verge of doing sth.

drauf-, Drauf-: ~|bekommen *unr. tr. V. (ugs.)* **in eins ~bekommen** *(gescholten werden)* get it in the neck *(coll.)*; *(geschlagen werden)* get a smack; **~gänger der** daredevil; **~gängerisch** *Adj.* daring; audacious; **~|geben** *unr. tr. V. (ugs.)* **a)** *(dazugeben)* add; **b) jmdm. eins ~geben** *(schlagen)* give sb. a smack; *(zurechtweisen)* put sb. in his/her place; **~|gehen** *unr. itr. V.; mit sein (ugs.)* **a)** *(umkommen)* kick the bucket *(sl.)*; **b)** *(verbraucht werden)* go; **für etw. ~gehen** ‹*money*› go on sth.; **c)** *(entzweigehen)* get busted *(coll.) or* broken; **~|legen**

(ugs.) **1.** *tr. V.* **150 DM/noch etwas ~legen** fork out *(sl.)* an extra 150 marks/a bit more; **2.** *itr. V.* lay out *(sl.)*; **ich lege dabei noch ~:** it's costing me money

drauf·los *Adv.* **nichts wie ~!** go on!

drauflos-: ~|arbeiten *itr. V.* work away; *(anfangen zu arbeiten)* get straight down to work; **~|gehen** *unr. itr. V.; mit sein (ugs.)* get going; **~|reden** *itr. V.* talk away; *(anfangen zu reden)* start talking away

drauf-: ~|machen *in* **einen ~machen** *(ugs.)* paint the town red; **~|stehen** *unr. itr. V. (ugs.)* be on it; **~|zahlen** *(ugs.)* **1.** *tr. V.* **noch etwas/1 250 DM ~zahlen** fork out *(sl.) or* pay a bit more/an extra 1,250 marks; **2.** *itr. V. (Unkosten haben)* **ich zahle dabei noch ~:** it's costing me money

draus [draus] *Adv. (ugs.) s.* **daraus**

draußen ['drausn̩] *Adv.* outside; **hier/da ~:** out here/there; **~ vor der Tür** at the door; **nach/von ~:** outside/from outside; **~ in der Welt** *(fig.)* in the world outside

drechseln ['drɛksl̩n] *tr. V.* turn

Dreck [drɛk] **der; ~[e]s a)** *(ugs.)* dirt; *(sehr viel/ekelerregend)* filth; *(Schlamm)* mud; **vor ~ starren** be covered in dirt; be filthy [dirty]; **~ machen** make a mess; **b)** *(salopp abwertend: Angelegenheit)* **bei/wegen jedem ~ regt er sich auf** he gets worked up about every piddling little thing *(coll.)*; **mach deinen ~ allein** do it yourself; **kümmere dich um deinen eigenen ~:** mind your own damn business; **das geht dich einen ~ an** *(salopp)* none of your damned business *(sl.)*; **jmdn. wie [den letzten] ~ behandeln** *(ugs.)* treat sb. like dirt; **c)** *(salopp: Zeug)* rubbish *no indef. art.;* junk *no indef. art.*

Dreck-: ~arbeit die *(salopp)* **a)** dirty *or* messy work *no indef. art., no pl.*/job; **b)** *(fig.)* dirty *or* menial work *no indef. art., no pl.*/job; **~fink der** *(ugs.) s.* **Schmutzfink**

dreckig 1. *Adj.* **a)** *(ugs.: schmutzig, ungepflegt, auch fig.)* dirty; *(sehr/ekelerregend schmutzig, auch fig.)* filthy; **mach dich nicht ~:** don't get yourself dirty; **b)** *(salopp abwertend: unverschämt)* cheeky; **c)** *nicht präd. (salopp abwertend: gemein)* dirty, filthy ⟨*swine etc.*⟩. **2.** *adv.* **a) es geht ihm ~** *(ugs.)* he's in a bad way; **b)** *(salopp abwertend: unverschämt)* cheekily; **~ grinsen** have a cheeky grin on one's face

Dreck-: ~loch *(salopp abwertend)* dump *(coll.)*; **~sack der** *(derb)* bastard *(coll.)*; **~sau die** *(derb)* dirty *or* filthy swine; **~schleuder die a)** *(derb abwertend) (Mundwerk)* foul mouth; *(Mensch)* foul-mouth; **b)** *(ugs. abwertend: Quelle schädlicher Emissionen)* factory/power-station pumping out clouds of pollutants; *(Fahrzeug)* car/lorry *etc.* belching clouds of exhaust fumes; **~schwein das** *s.* **~sau**

Drecks·kerl der *(derb abwertend)* dirty *or* filthy swine

Dreck·spatz der *(fam.: Kind)* grubby little so-and-so *(coll.)*; *(Kind, das etw. schmutzig macht)* mucky pup *(Brit. coll.)*

Dreh [dre:] **der; ~s, ~s** *(ugs.)* **a) den ~ heraushaben** have [got] the knack; **b) [so] um den ~** *(so ungefähr)* about that

Dreh-: ~arbeiten *Pl. (Film)* shooting *sing.*; **~bank die** lathe

drehbar 1. *Adj.* revolving *attrib.* ⟨*stand, stage*⟩; swivel *attrib.* ⟨*chair*⟩; **~ sein** revolve/swivel. **2.** *adv.* **~ gelagert** pivoted

Dreh-: ~bewegung die rotary motion; rotation; **~bleistift der** propelling pencil *(Brit.)*; mechanical pencil *(Amer.)*; **~buch das** screenplay; [film] script

drehen 1. *tr. V.* **a)** turn; **b)** *(ugs.: einstellen)* **das Radio laut/leise ~:** turn the radio up/down; **die Flamme klein/die Heizung auf klein ~:** turn the heat/heating down; **c)** *(formen)* twist ⟨*rope, thread*⟩; roll ⟨*cigarette*⟩; **d)** *(Film)* shoot; film ⟨*report*⟩; **e)** *(ugs. abwertend: beeinflussen)* **es so ~, daß ...:** work it so that ... *(sl.)*. **2.** *itr. V.* **a)** ⟨*car, driver*⟩ turn; ⟨*wind*⟩ change, shift; **b) an etw.** *(Dat.)* **~:** turn sth.; **da muß einer dran gedreht haben** *(salopp)* somebody must have fiddled about *or* messed around with it; **c)** *(Film)* shoot [a/the film]; film. **3.** *refl. V.* turn; ⟨*wind*⟩ change, shift; *(um eine Achse)* turn; rotate; revolve; *(um einen Mittelpunkt)* revolve (**um** around); *(sehr schnell)* spin; **mir dreht sich alles** *(ugs.)* everything's going round and round; **sich auf den Bauch ~:** turn over on to one's stomach; **b) sich um etw. ~** *(fig. ugs.)* be about sth.

Dreher der; ~s, ~ lathe-operator

Dreh-: ~orgel die barrel-organ; **~punkt der** pivot; **der ~- und Angelpunkt einer Sache** *(fig.)* the key element in sth.; **~scheibe die** *(fig.)* hub; **~stuhl der** swivel chair; **~tür die** revolving door

Drehung die; ~, ~en a) *(um eine Achse)* turn; rotation; revolution; *(um einen Mittelpunkt)* revolution; *(sehr schnell)* spin; **eine halbe/ganze ~:** a half/complete turn; **eine ~ um 180° [machen]** [do] a 180° turn; *(fig.)* [do] a complete about-face; **b)** *(das Drehen)* turning; *(sehr schnell)* spinning

Dreh-: ~wurm der *in* **einen** *od.* **den ~wurm kriegen/haben** *(salopp)* get/feel giddy; **~zahl die** number of revolutions *or (coll.)* revs [per minute]

drei *Kardinalz.* three; **aller guten Dinge sind ~!** all good things come in threes; *(nach zwei mißglückten Versuchen)* third time lucky!; **nicht bis ~ zählen können** *(ugs.)* be as thick as two [short] planks *(Brit. coll.)*; be dead from the neck up *(coll.)*; *s. auch* [1]**acht**

Drei die; ~, ~en three; **eine ~ schreiben/bekommen** *(Schulw.)* get a C; *s. auch* [1]**Acht a, b, d, e; Zwei b**

drei-, Drei-: ~achtel·takt der *(Musik)* three-eight time; **~akter der** three-act play; **~dimensional 1.** *Adj.* three-dimensional; **2.** *adv.* three-dimensionally; in three dimensions; **~eck das; ~s, ~e** triangle; **~eckig** *Adj.* triangular; three-cornered; **~ecks·verhältnis das** eternal triangle; **~ein·halb** *Bruchz.* three and a half; **~einigkeit die** *(christl. Rel.)* trinity

Dreier der; ~s, ~ a) *(ugs.) s.* **Drei; b)** *(ugs.: im Lotto)* three winning numbers; **c)** *(ugs.: Sprungbrett)* three-metre board

dreierlei *Gattungsz.; indekl.* **a)** *attr.* three kinds *or* sorts of; three different ⟨*sorts, kinds, sizes, possibilities*⟩; **b)** *subst.* three [different] things

drei-, Drei-: ~fach *Vervielfältigungsz.* triple; **die ~fache Menge** three times *or* triple the amount; three times as much; *s. auch* **achtfach; ~fache das;** *adj. Dekl.* **das ~fache von 3 ist 9** three times three is nine; **auf ein ~faches** *od.* **auf das ~fache steigen** treble; triple; **~gang·schaltung die** three-speed gearbox *or* gears *pl. or (Amer.)* gear-shift; **~hundert** *Kardinalz.* three hundred; **~jährig** *Adj. (3 Jahre alt)* three-year-old *attrib.*; *(3 Jahre dauernd)* three-year *attrib.*; **~kampf der** *(Sport)* triathlon; **~käse·hoch** [-'---] **der; ~s, ~s** *(ugs. scherzh.)* [little] nipper *(Brit. sl.)*; little kid *(sl.)*; **~klang der** triad; **~könige** [-'---] *Pl.; o. Art.* Epiphany

sing.; **~köpfig** *Adj.* ⟨*family, crew*⟩ of three; **~ländereck** [-'---] **das**; **~s, ~e** *region where three countries meet*; **~mal** *Adv.* three times; **~malig** *Adj.*; *nicht präd.* **eine ~malige Warnung** three warnings; **~meilen·zone** [-'----] **die** three-mile zone; **~meter·brett** [-'---] **das** three-metre board
drein *(ugs.) s.* **darein**
drein-: **~|blicken, ~|schauen** *itr. V.* look
drei-, Drei-: **~rad das** tricycle; **~räd[e]rig** *Adj.* three-wheeled; **~satz der** *(Math.)* rule of three; **~seitig** *Adj.* three-sided ⟨*figure*⟩; three-page ⟨*letter, leaflet, etc.*⟩; **~silbig** *Adj.* trisyllabic; three-syllable *attrib.*; **~spaltig** *(Druckw.) Adj.* three-column *attrib.*; **~sprachig** *Adj.* trilingual; **~sprung der** triple jump
dreißig ['draisɪç] *Kardinalz.* thirty
dreißiger *indekl. Adj.*; *nicht präd.* **die ~ Jahre** the thirties
[1]**Dreißiger der**; **~s, ~** *(30jähriger)* thirty-year-old
[2]**Dreißiger die**; **~, ~** *(ugs.) (Briefmarke)* thirty-pfennig/schilling *etc.* stamp
dreißigjährig *Adj.* *(30 Jahre alt)* thirty-year-old *attrib.*; *(30 Jahre dauernd)* thirty-year *attrib.*
dreißigst... ['draisɪçst...] *Ordinalz.* thirtieth
Dreißigstel das; **~s, ~**: thirtieth
dreist [draist] **1.** *Adj.* brazen; barefaced ⟨*lie*⟩. **2.** *adv.* brazenly
drei·stellig *Adj.* three-figure *attrib.* ⟨*number, sum*⟩
Dreistigkeit die; **~, ~en a)** *o. Pl.* *(Art)* brazenness; **b)** *(Handlung)* brazen act; *(Bemerkung)* brazen remark
drei-, Drei-: **~stimmig 1.** *Adj.* ⟨*song*⟩ for three voices; three-voice ⟨*choir*⟩; three-part ⟨*singing*⟩; **2.** *adv.* ⟨*sing*⟩ in three voices; ⟨*play*⟩ in three parts; **~stöckig 1.** *Adj.* three-storey *attrib.*; **2.** *adv.* ⟨*build*⟩ three storeys high; **~stündig** *Adj.* three-hour *attrib.*; *s. auch* **achtstündig**; **~tägig** *Adj.* *(3 Tage alt)* three-day-old *attrib.*; *(3 Tage dauernd)* three-day *attrib.*; *s. auch* **achttägig**; **~tausend** *Kardinalz.* three thousand; **~tausender der** mountain more than three thousand metres high; **~teilig** *Adj.* three-part *attrib.* ⟨*documentary, novel, etc.*⟩; three-piece *attrib.* ⟨*suit*⟩; **~viertel** *Bruchz.* three-quarters; **~viertel Liter** three-quarters of a litre; **~viertel·liter·flasche** [---'----] **die** three-quarter-litre bottle; **~viertel·mehrheit** [-'----] **die** three-quarters majority; **~viertel·stunde** [---'--] **die** three-quarters of an hour; **~viertel·takt** [-'---] **der** three-four time; **~wege·katalysator der** *(Kfz-W.)* three-way catalytic converter; **~wertig** *Adj.* *(Chemie)* trivalent; **~zehn** *Kardinalz.* thirteen; **jetzt schlägt's aber ~zehn!** *(ugs.)* that's going too far; *s. auch* **achtzehn**
Dresche ['drɛʃə] **die**; **~** *(salopp)* walloping *(sl.)*; thrashing
dreschen 1. *unr. tr. V.* **a)** thresh; **b)** *(salopp: prügeln)* wallop *(sl.)*; thrash; **c)** *(salopp: schießen)* wallop *(sl.)* ⟨*ball*⟩. **2.** *unr. itr. V.* **a)** thresh; **b)** *(salopp: schlagen)* thump; bang
Dresch-: **~flegel der** flail; **~maschine die** threshing-machine
Dreß [drɛs] **der**; **~** *od.* **Dresses, Dresse**; *(österr. auch* **die**; **~, Dressen)** *(Sportkleidung)* kit *(Brit.)*
dressieren *tr. V.* train ⟨*animal*⟩
Dressman ['drɛsmən] **der**; **~s, Dressmen** male model
Dressur [drɛ'su:ɐ̯] **die**; **~, ~en a)** training; **b)** *(Kunststück)* trick; **c)** *(Dressurreiten)* dressage
Dressur·pferd das dressage horse
dribbeln ['drɪbl̩n] *itr. V.* *(Ballspiele)* dribble [the ball]
Dribbling ['drɪblɪŋ] **das**; **~s, ~s** *(Ballspiele)* piece of dribbling
Drill [drɪl] **der**; **~[e]s** drilling; *(Milit.)* drill
drillen *tr. V.* *(auch Milit.)* drill
Drillich ['drɪlɪç] **der**; **~s, ~e** drill
Drilling ['drɪlɪŋ] **der**; **~s, ~e** triplet
drin [drɪn] *Adv.* **a)** *(ugs.: darin)* in it; **mehr als 2 000 DM ist nicht ~**: any more than 2,000 marks is not on *(Brit. coll.) or (Amer. coll.)* is no go; **es ist noch alles ~** *(bei einem Fußballspiel usw.)* there's still everything to play for; **nach drei Tagen ist man wieder ~** *(wieder eingearbeitet)* after three days you're back in the swing of things; **b)** *(ugs.: drinnen)* inside; **hier/da ~**: in here/there
dringen ['drɪŋən] *unr. itr. V.* **a)** *mit sein (gelangen)* ⟨*water, smell, etc.*⟩ penetrate, come through; ⟨*news*⟩ get through; **in etw.** *(Akk.)* **~**: get into *or* penetrate sth.; **durch etw. ~**: come through *or* penetrate sth.; ⟨*person*⟩ push one's way through sth.; **b)** *mit sein* **in jmdn. ~** *(geh.)* press *or* urge sb.; **c) auf etw.** *(Akk.)* **~**: insist upon sth.
dringend 1. *Adj.* **a)** *(eilig)* urgent; **b)** *(eindringlich, stark)* urgent ⟨*appeal*⟩; strong ⟨*suspicion, advice*⟩; compelling ⟨*need*⟩. **2.** *adv.* **a)** *(sofort)* urgently; **b)** *(zwingend)* ⟨*recommend, advise, suspect*⟩ strongly; **~ erforderlich sein** be imperative *or* essential
dringlich ['drɪŋlɪç] **1.** *Adj.* urgent. **2.** *adv.* urgently; **jmdn. ~ bitten, etw. zu tun** plead hard with sb. to do sth.
Dringlichkeit die; **~**: urgency
Drink [drɪŋk] **der**; **~[s], ~s** drink
drinnen ['drɪnən] *Adv.* inside; *(im Haus)* indoors; inside; **nach ~ gehen** go in[side]/indoors; **hier/da ~**: in here/there
drin-: **~|sitzen** *unr. itr. V.* *(ugs.)* be right in it *(coll.)*; **~|stecken** *itr. V.* *(ugs.)* **a) [bis über beide Ohren] in etw.** *(Dat.)* **~stecken** be up to one's ears in sth. *(coll.)*; **b) ich bin überzeugt, daß viel in ihm ~steckt** I am convinced he has a lot in him; **da steckt viel Arbeit ~**: there's a lot of work in that; **c) da steckt man nicht ~** *(das kann man nicht wissen)* there's no [way of] telling
dritt [drɪt] **wir waren zu ~**: there were three of us; *s. auch* [2]**acht**
dritt... *Ordinalz.* third; **in Gegenwart Dritter** in the presence of other people; **der lachende Dritte** the one to benefit *(from a dispute between two others)*; *s. auch* **acht...**
dritt·best... *Adj.* third-best
Drittel das, *(schweiz. meist* **der**); **~s, ~**: third
dritteln *tr. V.* split *or* divide ⟨*cost, profit*⟩ three ways; divide ⟨*number*⟩ by three
drittens *Adv.* thirdly
Dritt·kläßler der; **~s ~**: third-former
Drive [draif] **der**; **~s, ~s** *(auch Jazz, Golf, Tennis)* drive
Dr. jur. *Abk.* **doctor juris** LL D
DRK [de:|ɛr'ka:] **das**; **~** *Abk.* **Deutsches Rotes Kreuz** German Red Cross
Dr. med. *Abk.* **doctor medicinae** MD
droben ['dro:bn̩] *Adv.* *(südd., österr., sonst geh.)* up there
Droge ['dro:gə] **die**; **~, ~n** drug; **unter ~n stehen** be on drugs
drogen-, Drogen-: **~abhängig** *Adj.* addicted to drugs *postpos.*; **~abhängige der/die**; *adj. Dekl.* drug addict; **~süchtig** *Adj. s.* **~abhängig**; **~szene die** drug scene

Drogerie [drogə'ri:] **die; ~, ~n** chemist's [shop] *(Brit.);* drugstore *(Amer.)*
Drogist der; ~en, ~en, Drogistin die; ~, ~nen chemist *(Brit.);* druggist *(Amer.)*
Droh·brief der threatening letter
drohen ['dro:ən] **1.** *itr. V.* **a)** threaten; **er drohte mit [seiner] Kündigung** he threatened to give notice; **er drohte ihm mit erhobenem Zeigefinger** he raised a warning finger to him; **b)** *(bevorstehen)* be threatening; **jmdm. droht etw.** sb. is threatened with sth. **2.** *mod. V.* **etw. zu tun ~:** threaten to do sth.
drohend *Adj.* **a)** threatening; **b)** *(bevorstehend)* impending ‹*danger, strike, disaster*›
Drohne die; ~, ~n drone
dröhnen ['drø:nən] *itr. V.* **a)** ‹*voice, music*› boom; ‹*machine*› roar; ‹*room etc.*› resound **(von** with); **~er Applaus** thunderous applause
Drohung die; ~, ~en threat; **eine ~ wahr machen** carry out a threat
drollig ['drɔlɪç] **1.** *Adj.* **a)** *(spaßig)* funny; comical; *(niedlich)* sweet; cute *(Amer.);* **b)** *(seltsam)* odd; peculiar. **2.** *adv.* **a)** *(spaßig)* comically; *(niedlich)* sweetly; cutely *(Amer.);* **b)** *(seltsam)* oddly; peculiarly
Dromedar ['dro:meda:ɐ̯] **das; ~s, ~e** dromedary
Drops [drɔps] **der** *od.* **das; ~, ~:** fruit *or (Brit.)* acid drop; **saurer** *od.* **saures ~:** acid drop *(Brit.);* sour ball *(Amer.)*
drosch [drɔʃ] *1. u. 3. Pers. Sg. Prät. v.* **dreschen**
Droschke ['drɔʃkə] **die; ~, ~n a)** hackney carriage; **b)** *(veralt.: Taxi)* [taxi-]cab
Drossel ['drɔsl̩] **die; ~, ~n** thrush
drosseln *tr. V.* **a)** turn down ‹*heating, air-conditioning*›; throttle back ‹*engine*›; reduce *or* restrict the flow of ‹*steam, air*›; check ‹*flow*›; **b)** *(herabsetzen)* reduce; cut back *or* down
Dr. phil. *Abk.* **doctor philosophiae** Dr; **~ ~ Hans Schulz** Dr Hans Schulz; Hans Schulz, Ph. D.
drüben ['dry:bn̩] *Adv.* **a) dort** *od.* **da ~:** over there; **~ auf der anderen Seite** over on the other side; **von ~ kommen** come from across the border/sea *etc.*; **b)** *(veralt.) (in der DDR)* in the East; *(in der BRD, in West-Berlin)* in the West
drüber ['dry:bɐ] *(ugs.) s.* **darüber**
drüber- *(ugs.) s.* **darüber-**
¹Druck [drʊk] **der; ~[e]s, Drücke** ['drʏkə] **a)** *(Physik)* pressure; **einen ~ im Kopf haben** *(fig.)* have a feeling of pressure in one's head; **b)** *o. Pl. (das Drücken)* **ein ~ auf den Knopf** a touch of *or* on the button; **c)** *o. Pl. (Zwang)* pressure; **auf jmdn. ~ ausüben** put pressure on sb.; **unter ~ stehen** be under pressure; **jmdn. unter ~ setzen** put pressure on sb.; **~ dahinter machen** *(ugs.)* put some pressure on
²Druck der; ~[e]s, ~e a) *o. Pl. (das Drucken)* printing; *(Art des Drucks)* print; **in ~ gehen** go to press; **im ~ sein** be being printed; **b)** *(Bild, Graphik usw.)* print
Druck-: ~abfall der *(Physik)* drop *or* fall in pressure; **~buchstabe der** printed letter
Drückeberger ['drʏkəbɛrgɐ] **der; ~s, ~** *(ugs.)* shirker
druck·empfindlich *Adj.* pressure-sensitive ‹*material*›; easily bruised ‹*fruit*›
drucken *tr., itr. V.* print
drücken ['drʏkn̩] **1.** *tr. V.* **a)** press; press, push ‹*button*›; squeeze ‹*juice, pus*› **(aus** out of); **jmdm. die Hand ~:** squeeze sb.'s hand; **jmdn. an die Wand ~:** push sb. against the wall; **jmdn. ans Herz** *od.* **an sich** *(Akk.)* **~:** clasp sb. to one's breast; **jmdm. etw. in die Hand ~:** press sth. into sb.'s hand; **b)** *(liebkosen)* **jmdn. ~:** hug [and squeeze] sb.; **c)** *(Druck verursachen, quetschen)* ‹*shoe, corset, bandage, etc.*› pinch; **d)** *(geh.: be~)* ‹*consience*› weigh heavily [up]on sb.; **e)** *(herabsetzen)* push *or* force down ‹*price, rate*›; depress ‹*sales*›; bring down ‹*standard*›; **f)** *(Gewichtheben)* press. **2.** *itr. V.* **a)** press; **auf den Knopf ~:** press *or* push the button; **„[bitte] ~“:** 'push'; **das drückte auf die Stimmung/unsere gute Laune** *(fig.)* it spoilt the atmosphere/dampened our spirits; **b)** *(Druck verursachen)* ‹*shoe, corset, bandage*› pinch; **mein Rucksack drückt** my rucksack is pressing *or* digging into me; **c) auf etw.** *(Akk.)* **~** *(fig.: etw. sinken lassen)* push *or* force sth. down. **3.** *refl. V.* **a) sich in die Ecke ~:** squeeze [oneself] into the corner; **b)** *(ugs.)* shirk; **sich vor etw.** *(Dat.)* **~:** get out of *or* dodge sth.
drückend *Adj.* **a)** burdensome ‹*responsibility*›; grinding ‹*poverty*›; heavy ‹*debt, taxes*›; serious ‹*worries*›; **b)** *(schwül)* oppressive
Drucker der; ~s, ~: printer
Drücker der; ~s, ~ a) *(Tür~)* handle; **auf den letzten ~** *(ugs.)* at the very last minute; **b)** *(Knopf)* [push-]button; **am ~ sitzen** *od.* **sein** *(fig. ugs.)* be in charge
Druckerei die; ~, ~en printing-works; *(Firma)* printing-house; printer's
Drucker·schwärze die printing *or* printer's ink
druck-, Druck-: ~farbe die printer's *or* printing ink; **~fehler der** misprint; printer's error; **~fest** *Adj.* pressure-resistant; **~knopf der a)** press-stud *(Brit.);* snap-fastener; **b)** *(an Geräten)* push-button; **~luft die** *(Physik)* compressed air; **~maschine die** printing-press; **~mittel das** means of bringing pressure to bear **(gegenüber** on); **~reif 1.** *Adj.* ready for publication; *(fig.)* polished, perfectly formulated ‹*phrase, reply*›; **2.** *adv.* ‹*speak*› in a polished manner; **~sache die a)** *(Postw.)* printed matter; **b)** *(Druckw.)* printed stationery; **~schrift die a)** printed writing; **b)** *(Schriftart)* type[-face]; **c)** *(Schriftwerk)* pamphlet
drucksen ['drʊksn̩] *itr. V. (ugs.)* hum and haw *(coll.)*
druck-, Druck-: ~stelle die mark *(where pressure has been applied); (an Obst)* bruise; **~taste die** push-button; **~verband der** pressure bandage; **~verfahren das** printing process; **~welle die** *(Physik)* shock wave
drum [drʊm] *Adv. (ugs.)* **a)** *s.* **darum; b)** [a]round; **um etw. ~ herum** [all] [a]round sth.; **~ rumreden** beat about *or (Amer.)* around the bush; **sei's ~:** never mind; [that's] too bad; **alles** *od.* **das [ganze] Drum und Dran** *(bei einer Mahlzeit)* all the trimmings; *(bei einer Feierlichkeit)* all the palaver that goes with it *(coll.)*
Drum·herum das; ~s everything that goes/went with it
Drummer ['dramɐ] **der; ~s, ~** *(Musik)* drummer
drunter ['drʊntɐ] *Adv. (ugs.)* underneath; **es** *od.* **alles geht ~ und drüber** everything is topsy-turvy; things are completely chaotic
Drüse ['dry:zə] **die; ~, ~n** gland
Dschungel ['dʒʊnl̩] **der; ~s, ~** *(auch fig.)* jungle
dt. *Abk.* **deutsch** G.
Dtzd. *Abk.* **Dutzend** doz.
du [du:] *Personalpron.; 2. Pers. Sg. Nom.* you; thou *(arch.); (in Briefen)* **Du** you; **mit jmdm. auf du und du stehen** be on familiar terms with sb.; **du bist es** it's 'you; **mach du das doch** 'you do it; *s.*

auch (Gen.) **deiner,** *(Dat.)* **dir,** *(Akk.)* **dich**
Du das; ~[s], ~[s] 'du' *no art.;* the familiar form 'du'; **jmdm. das ~ anbieten** suggest to sb. that he/she use [the familiar form] 'du' *or* the familiar form of address
Dübel ['dy:bl̩] **der; ~s, ~:** plug
dübeln *tr. V.* **etw. ~:** fix sth. using a plug/plugs
dubios [du'bio:s] *Adj. (geh.)* dubious
Dublette [du'blɛtə] **die; ~, ~n** duplicate
ducken ['dʊkn̩] **1.** *refl. V.* duck; *(vor Angst)* cower. **2.** *tr. V. (abwertend) (einschüchtern)* intimidate; *(demütigen)* humiliate. **3.** *itr. V.* humble oneself
Duckmäuser ['dʊkmɔyzɐ] **der; ~s, ~** *(abwertend)* moral coward
Dudel·kasten der *(salopp abwertend) (Radio)* radio; *(Plattenspieler)* record-player
dudeln ['du:dl̩n] **1.** *tr. V. (auf Blasinstrument)* tootle; *(singen)* sing tunelessly. **2.** *itr. V.* ‹*radio, television, etc.*› drone on; ‹*barrel organ*› grind away
Dudel·sack der bagpipes *pl.*
Duell [du'ɛl] **das; ~s, ~e a)** duel; **b)** *(Sport)* contest
duellieren *refl. V.* fight a duel
Duett [du'ɛt] **das; ~[e]s, ~e** *(Musik)* duet; **im ~ singen** sing a duet
Dufflecoat ['dʌfəlkoʊt] **der; ~s, ~s** duffle-coat
Duft [dʊft] **der; ~[e]s, Düfte** ['dʏftə] pleasant smell; scent; *(Zool.)* scent; *(von Parfüm, Blumen)* scent; fragrance; *(von Kaffee, frischem Brot, Tabak)* aroma; *(iron.)* beautiful smell *(iron.);* **den ~ der großen, weiten Welt schnuppern** *(fig.)* get a taste of the big, wide world
dufte ['dʊftə] *(ugs.)* **1.** *Adj.* great *(coll.).* **2.** *adv.* ‹*dressed, behave*› smashingly *(coll.);* ‹*taste*› great *(coll.)*
duften ['dʊftn̩] *itr. V.* smell **(nach** of); **die Rosen ~ gut** the roses smell lovely *or* have a lovely scent
duftend *Adj.* sweet-smelling; fragrant
Duft-: ~stoff der a) aromatic substance; **b)** *(Biol.)* scent; **~wasser das;** *Pl.* **~wässer** *(scherzh.: Parfüm)* perfume; scent; **~wolke die** cloud of perfume
dulden ['dʊldn̩] **1.** *tr. V.* **a)** tolerate; put up with; **die Arbeit duldet keinen Aufschub** the work will admit no delay; **b)** *(Aufenthalt gestatten)* **jmdn. ~:** tolerate *or* put up with sb.'s presence. **2.** *itr. V. (geh.)* suffer
duldsam ['dʊltza:m] **1.** *Adj.* tolerant **(gegen** towards). **2.** *adv.* tolerantly
Duldsamkeit die; ~: tolerance
Duldung die; ~: toleration
dumm [dʊm], **dümmer** ['dʏmɐ], **dümmst...** ['dʏmst...] **1.** *Adj.* **a)** stupid; stupid, thick, dense ‹*person*›; **sich ~ stellen** act stupid; **sich nicht für ~ verkaufen lassen** *(ugs.)* not be taken in; **sich ~ und dämlich** *od.* **dusselig reden/verdienen** *(ugs.)* talk till one is blue in the face/earn a fortune; **b)** *(unvernünftig)* foolish; stupid; daft; **so etwas Dummes!** how stupid!; **c)** *(ugs.: töricht, albern)* idiotic; silly; stupid; **das ist mir [einfach] zu ~** *(ugs.)* I've had enough of it; **d)** *(ugs.: unangenehm)* nasty ‹*feeling, suspicion*›; annoying ‹*habit*›; awful *(coll.)* ‹*coincidence*›; **so etwas Dummes!** how annoying! **2.** *adv.* **a)** *(ugs.: töricht)* foolishly; stupidly; **frag nicht so ~:** don't ask such silly *or* stupid questions; **b)** *(ugs.: unangenehm)* ‹*end*› badly *or* unpleasantly; **jmdm. ~ kommen** be cheeky *or* insolent to sb.
Dumme der/die; *adj. Dekl.* fool; **einen ~n finden, der etw. macht** find somebody stupid enough to do sth.; **der ~ sein** *(ugs.)* be the loser
Dumme·jungen·streich der *(ugs.)* silly prank
dummer·weise *Adv.* **a)** *(leider)* unfortunately; *(ärgerlicherweise)* annoyingly; irritatingly; **b)** *(törichterweise)* foolishly; like a fool; stupidly
Dummheit die; ~, ~en a) *o. Pl.* stupidity; **b)** *(unkluge Handlung)* stupid *or* foolish thing; **[mach] keine ~en!** don't do anything stupid *or* foolish; **lauter** *od.* **nur ~en im Kopf haben** have a head full of silly ideas
Dumm·kopf der *(ugs.)* [silly] fool *or* idiot
dümmlich ['dʏmlɪç] **1.** *Adj.* simple-minded. **2.** *adv.* ‹*grin, smile*› [rather] foolishly *or* stupidly
dumpf [dʊmpf] **1.** *Adj.* **a)** dull ‹*thud, rumble of thunder*›; muffled ‹*sound, thump*›; **b)** *(muffig)* musty; **c)** *(stumpfsinnig)* dull; numb ‹*indifference*›; **d)** *(undeutlich)* dull ‹*pain, anger*›. **2.** *adv.* **a)** ‹*echo*› hollowly; **~ auf etw.** *(Akk.)* **aufschlagen** land with a dull thud on sth.; **b)** *(stumpfsinnig)* apathetically; numbly
Dumpfheit die; ~ *(Stumpfsinn)* torpor; apathy
Dumping·preis der dumping price
Düne ['dy:nə] **die; ~, ~n** dune
Dung [dʊŋ] **der; ~[e]s** dung; manure
Dünge·mittel das fertilizer
düngen ['dʏŋən] **1.** *tr. V.* fertilize ‹*soil, lawn, etc.*›; spread fertilizer on ‹*field*›; scatter fertilizer around ‹*plants*›. **2.** *itr. V.* ‹*person*› put on fertilizer; **gut ~** ‹*substance*› be a good fertilizer
Dünger der; ~s, ~: fertilizer
Dung·haufen der dunghill; dung *or* manure heap
Düngung die; ~, ~en use of fertilizers
dunkel ['dʊŋkl̩] **1.** *Adj.* **a)** dark; **es wird um 22 h ~:** it gets dark about 10 o'clock; **im Dunkeln** in the dark; **im ~n bleiben** *(fig.)* remain a mystery; remain unidentified; **im ~n tappen** *(fig.)* grope around *or* about in the dark; **b)** *(unerfreulich)* dark ‹*chapter in one's life*›; black ‹*day*›; **c)** *(fast schwarz)* dark; **dunkles Brot** brown bread; **d)** *(tief)* deep ‹*voice, sound*›; **e)** *(unbestimmt)* vague; dim, faint, vague ‹*recollection*›; dark ‹*hint, foreboding, suspicion*›; **f)** *(abwertend: zweifelhaft)* dubious; shady. **2.** *adv.* **a)** ‹*dress, paint sth., etc.*› in a dark colour/in dark colours; **b)** *(unbestimmt)* vaguely
Dunkel das; ~s *(geh.)* darkness; **in ~ gehüllt sein** *(fig.)* be shrouded in mystery
Dünkel ['dʏŋkl̩] **der; ~s** *(geh. abwertend) (Überheblichkeit)* arrogance; haughtiness; *(Einbildung)* conceit[edness]
dunkel- dark ‹*blue, grey, etc.*›
dunkel-: ~blond *Adj.* light brown ‹*hair*›; ‹*person*› with light brown hair; **~haarig** *Adj.* dark-haired; **~häutig** *Adj.* dark-skinned
Dunkelheit die; ~: darkness; **bei ~:** during the hours of darkness; **bei Einbruch der ~:** at nightfall
Dunkel-: ~kammer die darkroom; **~mann der;** *Pl.* **~männer** *(abwertend)* shady character
dunkeln *itr. V.* **a)** *unpers.* **es dunkelt** *(geh.)* it is growing dark; **b)** *mit sein* grow *or* go darker; darken
Dunkel·ziffer die number of unrecorded cases
dünken ['dʏŋkn̩] *(geh. veralt.)* **1.** *tr. V.* **jmdn. gut/schlecht/gerecht** *usw.* **~:** strike sb. as good/bad/just *etc.;* **mich dünkt, er hat recht** me thinks he is right *(arch.).* **2.**

refl. V. **er dünkt sich etwas Besseres/ein Held** he thinks of himself as superior/a hero; **ich dünkte mich sicher** *od.* **in Sicherheit** I imagined I was safe

dünn [dʏn] **1.** *Adj.* **a)** thin ‹*slice, layer, etc.*›; slim ‹*book*›; **b)** *(mager)* thin ‹*person*›; **sich ~ machen** *(scherzh.)* squash *or (Amer.)* scrunch up [a bit]; **c)** *(leicht)* thin, light ‹*clothing, fabric*›; fine ‹*stocking*›; *(fig.)* thin, rarefied ‹*air*›; fine ‹*rain*›; **d)** *(spärlich)* thin ‹*hair*›; sparse ‹*tree, cover, vegetation*›; **e)** *(wenig gehaltvoll)* thin ‹*soup*›; weak, watery ‹*coffee, tea*›; watery ‹*beer*›; **f)** *(~flüssig)* thin ‹*paint, lubricating oil*›; runny ‹*batter*›; **g)** *(schwach)* thin ‹*voice*›; weak, faint ‹*smile*›; faint ‹*scent*›. **2.** *adv.* **a)** thinly **b)** *(leicht)* lightly ‹*dressed*›; **c)** *(spärlich)* thinly, sparsely ‹*populated*›; **d)** *(schwach)* ‹*smile*› weakly, faintly

dünn-, Dünn-: ~besiedelt *Adj. (präd. getrennt geschrieben)* thinly *or* sparsely populated *or* inhabited; **~bier das** *(veralt.)* small beer; **~darm der** *(Anat.)* small intestine

Dünne der/die; *adj. Dekl. (ugs.)* thin man/woman

dünn-: ~flüssig *Adj.* thin; runny ‹*batter etc.*›; **~gesät** *Adj. (präd. getrennt geschrieben) (ugs.)* rare; **~|machen** *refl. V. (ugs.)* make oneself scarce *(coll.)*

Dunst [dʊnst] **der; ~[e]s, Dünste** ['dʏnstə] **a)** *o. Pl.* haze; *(Nebel)* mist; **b)** *(Geruch)* smell; *(Ausdünstung)* fumes *pl.*; *(stickige, dumpfe Luft)* fug *(coll.)*; **c) keinen [blassen] ~ von etw. haben** *(ugs.)* have not the foggiest *or* faintest idea about sth.

Dunst·abzugs·haube die extractor hood

dünsten ['dʏnstn̩] *tr. V.* steam ‹*fish, vegetables*›; braise ‹*meat*›; stew ‹*fruit*›

Dunst·glocke die pall of haze

dunstig *Adj.* **a)** hazy; *(neblig)* misty; **b)** *(verräuchert)* smoky

Dunst·kreis der *(fig.)* orbit

Dunst·schleier der veil of haze; *(Nebelschleier)* veil of mist

Duo ['du:o] **das; ~s, ~s** *(Musik)* **a)** duet; **b)** *(fig. scherzh.)* duo; pair

Duplikat [dupli'ka:t] **das; ~[e]s, ~e** duplicate

Dur [du:ɐ̯] **das; ~** *(Musik)* major [key]; **C-~:** C major

durch [dʊrç] **1.** *Präp. mit Akk.* **a)** *(räumlich)* through; **~ ganz Europa reisen** travel all over *or* throughout Europe; **~ einen Fluß waten** wade across a river; **b)** *(modal)* by; **etw. ~ die Post schikken** send sth. by post *(Brit.) or* mail; **etw. ~ das Fernsehen bekanntgeben** announce sth. on television; **etw. ~ jmdn. bekommen** get *or* obtain sth. through sb.; **zehn [geteilt] ~ zwei** ten divided by two. **2.** *Adv.* **a)** *(hin~)* **das ganze Jahr ~:** throughout the whole year; all year; **die ganze Zeit ~:** the whole time; **b)** *(ugs.: vorbei)* **es war 3 Uhr ~:** it was past *or* gone 3 o'clock; **c) ~ und ~ naß** wet through [and through]; **er ist ein Lügner ~ und ~:** he's an out and out liar; *s. auch* **durchsein**

durch|arbeiten 1. *tr. V.* **a)** work *or* go through ‹*book, article*›; **b)** *(durchkneten)* work *or* knead thoroughly ‹*dough*›; massage *or* knead thoroughly ‹*muscles*›. **2.** *itr. V.* work through; **die Nacht/Pause ~:** work through the night/break. **3.** *refl. V. (auch fig,)* work one's way through

durch|atmen *itr. V.* breathe deeply

durch·aus *Adv.* **a)** *(ganz und gar)* absolutely; perfectly, quite ‹*correct, possible, understandable*›; **das ist ~ richtig** that is entirely right; **ich bin ~ Ihrer Meinung** I am entirely of your opinion; **das hat ~ nichts damit zu tun** that's got nothing at all *or* whatsoever to do with it; **es ist ~ nicht so einfach wie ...:** it is by no means as easy as ...; **b)** *(unbedingt)* **~ mitkommen wollen** [absolutely] insist on coming too; **~ nicht ins Wasser wollen** absolutely refuse to go into the water

durch|beißen 1. *unr. tr. V.* bite through. **2.** *unr. refl. V. (ugs.)* [manage to] struggle through

durch|bekommen *unr. tr. V.* **a)** *(hindurchbekommen)* **etw. ~:** get sth. through; **b)** *(zerteilen)* get *or* cut through ‹*rope, brauch, etc.*›; **c)** *(durchlesen)* get through; finish

durch|biegen *unr. refl. V.* sag

durch|blasen 1. *unr. tr. V.* **a)** *(reinigen)* **etw. ~:** clear sth. by blowing through it; **b)** *(treiben)* **etw. durch etw. ~:** blow sth. through sth. **2.** *unr. itr. V.* **durch etw. ~** ‹*wind*› blow through sth.

durch|blättern *tr. V.* leaf through ‹*book, file, etc.*›

Durch·blick der *(ugs.)* **den [absoluten] ~ haben** know [exactly] what's going on; **den ~ verlieren** no longer know what's going on

durch|blicken *itr. V.* **a)** look through; **durch etw. ~:** look through sth.; **b)** *(ugs.)* **ich blicke da nicht durch** I can't make head or tail of it; **c) etw. ~ lassen** hint at sth.

durch·bluten *tr. V.* supply ‹*body, limb, etc.*› with blood; **seine Beine sind schlecht durchblutet** the circulation in his legs is poor

Durch·blutung die; *o. Pl.* flow *or* supply of blood *(Gen.* to); [blood-]circulation

[1]**durch|bohren 1.** *tr. V.* drill *or* bore through ‹*wall, plank*›. **2.** *itr. V.* **durch etw. ~:** drill *or* bore through sth.

[2]**durch·bohren** *tr. V.* pierce; **jmdn. mit Blicken ~** *(fig.)* look piercingly *or* penetratingly at sb.

durch|boxen *(ugs.)* **1.** *refl. V.* fight one's way through; *(fig.)* battle through. **2.** *tr. V.* force *or* push through ‹*law, measure, bill, etc.*›

durch|braten *unr. tr. V.* **etw. ~:** cook *or* roast sth. till it is well done; **ich möchte mein Steak durchgebraten** I'd like my steak well done

[1]**durch|brechen 1.** *unr. tr. V.* break in two. **2.** *unr. itr. V.; mit sein* **a)** break in two; **der Blinddarm/das Magengeschwür ist durchgebrochen** *(Med.)* the appendix has burst/the gastric ulcer has perforated; **b)** *(hervorkommen)* ‹*sun*› break through; **c)** *(einbrechen)* fall through; **durch etw. ~:** fall through sth.

[2]**durch·brechen** *unr. tr. V.* break through ‹*sound barrier*›; break *or* burst through ‹*crowd barrier*›; ‹*car*› crash through ‹*railings etc.*›

durch|brennen *unr. itr. V.; mit sein* **a)** ‹*heating coil, light bulb*› burn out; ‹*fuse*› blow; **b)** *(ugs.: weglaufen) (von zu Hause)* run away; *(mit der Kasse)* run off; abscond; *(mit dem Geliebten/der Geliebten)* run off

durch|bringen *unr. tr. V.* **a)** *s.* **durchbekommen; b)** *(durch eine Kontrolle)* **etw. ~:** get sth. through; **c)** *(bei Wahlen)* **jmdn. ~:** get sb. elected; **d)** *(durchsetzen)* get ‹*bill*› through; get ‹*motion*› passed; get ‹*proposal*› accepted; **e)** *(versorgen)* **seine Familie ~:** support one's family; **f)** *(verschwenden)* get through

Durch·bruch der *(fig.)* breakthrough; **einer Idee** *(Dat.)* **zum ~ verhelfen** get an idea generally accepted

durch|bürsten *tr. V.* brush ‹*hair*› thoroughly

durch|checken *tr. V.* check ‹*list, documents*› thoroughly; check ‹*car*› over thoroughly

durch·dacht *Adj.* **wenig/gut ~:** badly/well thought-out

durch·denken *unr. tr. V.* think over *or* through

durch|drängen *refl. V.* **sich [durch etw.] ~:** push *or* force one's way through [sth.]

durch|drehen 1. *tr. V.* put ⟨*meat etc.*⟩ through the mincer *or (Amer.)* grinder; chop ⟨*nuts etc.*⟩ in the blender. **2.** *itr. V.* **a)** *auch mit sein (ugs.)* crack up *(coll.)*; go to pieces; **b)** ⟨*wheels*⟩ spin

[1]**durch|dringen** *unr. itr. V.; mit sein* ⟨*rain, sun*⟩ come through; **durch etw. ~:** penetrate sth.; come through sth.; **der Redner drang mit seiner Stimme nicht durch** the speaker couldn't make himself heard

[2]**durch·dringen** *unr. tr. V.* penetrate

durch·dringend 1. *Adj.* **a)** *(intensiv)* piercing, penetrating ⟨*voice, look, scream, sound*⟩; **b)** *(penetrant)* pungent, penetrating ⟨*smell*⟩. **2.** *adv.* piercingly; penetratingly

durch|drücken *tr. V.* **a) etw. [durch etw.] ~:** press sth. through [sth.]; **b)** *(strecken)* straighten ⟨*limb, back*⟩; **c)** *(ugs.: durchsetzen)* manage to get ⟨*extra holiday etc.*⟩; manage to force ⟨*application*⟩ through

durch|dürfen *unr. itr. V. (ugs.)*; **darf ich mal [hier] durch?** can I get through here?

durch·einander *Adv.* **~ sein** ⟨*papers, desk, etc.*⟩ be in a mess *or* a muddle; *(verwirrt sein)* be confused *or* in a state of confusion; *(aufgeregt sein)* be flustered *or (coll.)* in a state

Durcheinander das; ~s a) muddle; mess; **b)** *(Wirrwarr)* confusion

durcheinander-: ~|bringen *unr. tr. V.* **a)** get ⟨*room, flat*⟩ into a mess; get ⟨*papers, file*⟩ into a muddle; muddle up ⟨*papers, file*⟩; **b)** *(verwirren)* confuse; **c)** *(verwechseln)* confuse ⟨*names etc.*⟩; get ⟨*names etc.*⟩ mixed up *or* muddled; **~|geraten** *unr. itr. V.; mit sein* ⟨*collection, letters*⟩ get in a muddle; **~|kommen** *unr. itr. V.; mit sein* get into a muddle; **~|laufen** *unr. itr. V.; mit sein* run [around] in all directions; **~|reden** *itr. V.* all talk at once *or* at the same time

durch|exerzieren *tr. V. (ugs.)* rehearse ⟨*situation etc.*⟩

[1]**durch|fahren** *unr. itr. V.; mit sein* **a) [durch etw.] ~:** drive through [sth.]; **b)** *(nicht anhalten)* go straight through; *(mit dem Auto)* drive straight through; *(fahren, ohne umsteigen zu müssen)* travel direct; go straight through; **der Zug fährt [in H.] durch** the train doesn't stop [at H.]; **der Zug fährt bis München durch** the train is non-stop to Munich

[2]**durch·fahren** *unr. tr. V.* **a)** travel through; ⟨*train*⟩ pass through; *(mit dem Auto)* drive through; **b)** *(zurücklegen)* cover ⟨*distance*⟩; complete ⟨*course, lap*⟩; **c) plötzlich durchfuhr ihn ein Schreck** he was seized with sudden fright

Durch·fahrt die a) *o. Pl. (das Durchfahren)* passage; **„~ verboten"** 'no entry except for access' **die ~ freigeben** allow vehicles through; **b)** *o. Pl. (Durchreise)* **auf der ~ sein** be passing through; be on the way through; **c)** *(Weg)* thoroughfare; **„bitte [die] ~ freihalten"** 'please do not obstruct'

Durch·fall der diarrhoea *no art.*

durch|fallen *unr. itr. V.; mit sein* **a)** fall through; **durch etw. ~:** fall through sth.; **b)** *(ugs.: nicht bestehen)* fail; flunk *(Amer. coll.)*; **bei etw./in etw.** *(Dat.)***/durch etw. ~:** fail *or* flunk sth.; **c)** *(ugs.: erfolglos sein)* ⟨*play, performance*⟩ flop *(sl.)*; be a flop *(sl.) or* failure; **d)** *(ugs.: die Wahl verlieren)* lose the election

durch|finden *unr. refl. V.* **sich [durch etw.] ~:** find one's way through [sth.]

durch|fliegen *unr. itr. V.; mit sein* **a) [durch etw.] ~:** fly through [sth.]; **unter der Brücke ~:** fly under the bridge; **b)** *(nicht zwischenlanden)* fly non-stop

durch|fließen *unr. itr. V.; mit sein* **[durch etw.] ~:** flow through [sth.]

durch·fluten *tr. V. (geh.)* ⟨*warmth, pleasant feeling*⟩ flood through ⟨*person*⟩; ⟨*light*⟩ flood ⟨*room*⟩

durch·forsten *tr. V. (fig.)* sift through

durch|fragen *refl. V.* **sich [zum Museum] ~:** find one's way [to the museum] by asking

durch|fressen 1. *unr. tr. V.* eat through; ⟨*moths*⟩ eat holes in ⟨*pullover etc.*⟩. **2.** *unr. refl. V.* ⟨*maggot, woodworm*⟩ eat [its way] through; ⟨*rust*⟩ eat through

durch|frieren *unr. itr. V.; mit sein* **a) durchgefroren sein** ⟨*person*⟩ be frozen stiff *or* chilled to the bone; **b)** ⟨*water, lake*⟩ freeze solid

durchführbar *Adj.* practicable; feasible; workable; **ein leicht ~er Plan** a plan that is easy to carry out

Durchführbarkeit die; ~: practicability; feasibility; workability

durch|führen 1. *tr. V.* carry out; carry out, put into effect, implement ⟨*plan*⟩; perform, carry out ⟨*operation*⟩; take ⟨*measurement*⟩; make ⟨*charity collection*⟩; hold ⟨*meeting, election, examination*⟩. **2.** *itr. V.* **durch etw./unter etw.** *(Dat.)* **~** ⟨*track, road*⟩ go *or* run *or* pass through/under sth.

Durch·führung die carrying out; *(einer Operation)* performing; *(einer Messung)* taking; *(eines Kongresses usw.)* holding

durch|füttern *tr. V. (ugs.)* feed; support; **sich von jmdm. ~ lassen** live off sb.

Durch·gang der a) „kein ~", „~ verboten" 'no thoroughfare'; **b)** *(Weg)* passage[way]; **c)** *(Phase)* stage; *(einer Versuchsreihe)* run; *(Sport, bei Wahlen, Wettbewerb)* round

durch·gängig 1. *Adj.* general; *(universell)* universal; constant ⟨*feature*⟩. **2.** *adv.* generally, universally ⟨*accepted*⟩

Durchgangs-: ~lager das transit camp; **~straße die** through road; thoroughfare; **~verkehr der** through traffic

durch|geben *unr. tr. V.* announce ⟨*news*⟩; give ⟨*results, winning numbers, weather report*⟩; make ⟨*announcement*⟩

durch|gehen 1. *unr. itr. V.; mit sein* **a) [durch etw.] ~:** go *or* walk through [sth.]; **„bitte ~!"** 'pass *or* move right down, please'; **b) [durch etw.] ~:** ⟨*rain, water*⟩ come through [sth.]; ⟨*wind*⟩ go through [sth.]; **c)** ⟨*train, bus, flight*⟩ go [right] through (**bis** to); go direct; **d)** ⟨*path etc.*⟩ go *or* run through (**bis zu** to); ⟨*stripe*⟩ go *or* run right through; **e)** *(angenommen werden)* ⟨*application, claim*⟩ be accepted; ⟨*law*⟩ be passed; ⟨*motion*⟩ be carried; ⟨*bill*⟩ be passed, get through; **f)** *(hingenommen werden)* ⟨*discrepancy*⟩ be tolerated; ⟨*mistake, discourtesy*⟩ be allowed to *or* let pass, be overlooked; **[jmdm.] etw. ~ lassen** let sb. get away with sth.; **g)** *(davonstürmen)* ⟨*horse*⟩ bolt; **h)** *(ugs.: davonlaufen)* run off; **i)** *(außer Kontrolle geraten)* **die Nerven gehen mit ihm durch** he loses his temper; **ihr Temperament/ihre Begeisterung geht mit ihr durch** her temperament/enthusiasm gets the better of her; **j)** *(ugs.: durchgebracht* *werden können)* **[durch etw.]** ~: go through [sth.]; **k)** *(gehalten werden für)* **für neu/30 Jahre** *usw.* ~: be taken to be *or* pass for new/thirty *etc.*. **2.** *unr. tr. V.; mit sein* go through ⟨*newspaper, text*⟩

durch·gehend **1.** *Adj.* **a)** continuous ⟨*line, pattern, etc.*⟩; constantly recurring ⟨*motif*⟩; **b)** *(direkt)* through *attrib.* ⟨*train, carriage*⟩; direct ⟨*flight, connection*⟩. **2.** *adv.* ~ **geöffnet haben** be open all day

durch·geschwitzt *Adj.* ⟨*person*⟩ soaked *or* bathed in sweat; ⟨*clothes*⟩ soaked with sweat; sweat-soaked *attrib.* ⟨*clothes*⟩

durch|gießen *unr. tr. V.;* **etw. [durch etw.]** ~: pour sth. through [sth.]

durch|greifen *unr. itr. V.* **a)** **[hart]** ~: take drastic measures *or* steps; **b)** **[durch etw.]** ~: reach through [sth.]

durch|gucken *itr. V. (ugs.)* **[durch etw.]** ~: peep *or* look through [sth.]

durch|haben *unr. tr. V. (ugs.)* have finished with ⟨*book, newspaper*⟩

durch|hacken *tr. V.* hack *or* chop through

durch|halten **1.** *unr. itr. V. (bei einem Kampf)* hold out; *(bei einer schwierigen Aufgabe)* see it through; *(beim Rennen)* stay the course. **2.** *unr. tr. V.* stand

durch|hängen *unr. itr. V.* sag;

durch|hauen **1.** *regelm. (auch unr.) tr. V.: s.* **durchschlagen 1 a.** **2.** *tr. V. (ugs.)* **jmdn.** ~: give sb. a good hiding *(coll.) or (sl.)* walloping

durch|heizen **1.** *tr. V.* heat ⟨*house, offices, etc.*⟩ through. **2.** *itr. V.* have *or* keep the heating on

[1]**durch|kämmen** *tr. V.* comb ⟨*hair*⟩ through

[2]**durch·kämmen** *tr. V.* comb ⟨*area etc.*⟩

durch|kämpfen **1.** *tr. V.* **a)** fight ⟨*case*⟩ [right] to the end; fight one's way through ⟨*adversity*⟩; **b)** *(durchsetzen)* force through. **2.** *refl. V.* **sich [durch etw.]** ~: fight *or* battle one's way through [sth.]

durch|kauen *tr. V.* **a)** **etw. [gut]** ~: chew sth. thoroughly *or* well; **b)** *(ugs.: besprechen)* go over and over

durch|kneten *tr. V.* knead ⟨*dough etc.*⟩ thoroughly

durch|kommen *unr. itr. V.; mit sein* **a)** come through; *(mit Mühe hindurchgelangen)* get through; **es gab kein Durchkommen** there was no way through; **b)** *(ugs.: beim Telefonieren)* get through; **c)** *(durchgehen, -fahren usw.)* **durch etw.** ~: come *or* pass through sth.; **d)** *(sich zeigen)* ⟨*sun*⟩ come out; ⟨*character trait, upbringing*⟩ come through, become apparent; **e)** *(erfolgreich sein)* **damit kommst du bei mir nicht durch** you won't get anywhere with me like that; **f)** *(ugs.: überleben)* pull through; **g)** *(ugs.: durchdringen)* **[durch etw.]** ~ ⟨*water, sand, etc.*⟩ come through [sth.]; **h)** *(bestehen)* get through; pass; **i)** *(auskommen)* manage; get by

durch|können *unr. itr. V. (ugs.)* **[durch etw.]** ~: be able to go/come through [sth.]; **kann ich bitte mal durch?** can I get by, please?; excuse me, please

durch·kreuzen *tr. V.* thwart, frustrate ⟨*plan, policy*⟩

durch|kriechen *unr. itr. V.; mit sein* **[durch etw.]** ~: crawl through [sth.]; **unter etw.** *(Dat.)* ~: crawl [through] under sth.

durch|kriegen *tr. V. (ugs.) s.* **durchbekommen**

durch|laden **1.** *unr. tr. V.* cock ⟨*pistol etc.*⟩ and rotate the cylinder. **2.** *unr. itr. V.* cock the trigger and rotate the cylinder

Durchlaß ['dʊrçlas] **der; Durchlasses, Durchlässe** ['dʊrçlɛsə] *(Öffnung)* gap; opening

durch|lassen *unr. tr. V.* **a)** **jmdn. [durch etw.]** ~: let *or* allow sb. through [sth.]; **den Ball** ~ *(Sport)* ⟨*goalkeeper*⟩ let a goal in; **b)** let ⟨*light, water, etc.*⟩ through

durchlässig ['dʊrçlɛsɪç] *Adj.* permeable; **die Grenzen müssen durchlässiger werden** *(fig.)* the borders must be opened up further

Durchlässigkeit **die;** ~ permeability

Durch·lauf **der** **a)** *(Sport, DV)* run; **b)** *(von Wasser)* flow

[1]**durch|laufen** **1.** *unr. itr. V.; mit sein* **a)** **[durch etw.]** ~: run through [sth.]; **b)** *(durchrinnen)* **[durch etw.]** ~: trickle through [sth.]; **der Kaffee ist durchgelaufen** the coffee is filtered; **c)** *(ohne Pause laufen)* run without stopping. **2.** *unr. tr. V.* go through ⟨*socks, soles of shoes*⟩

[2]**durch·laufen** *unr. tr. V.* go *or* pass through ⟨*phase, stage*⟩

durchlaufend **1.** *Adj.* continuous. **2.** *adv.* ⟨*numbered, marked*⟩ in sequence

Durchlauf·erhitzer **der;** ~s, ~: geyser; instantaneous water-heater

durch·leben *tr. V.* live through; experience; experience ⟨*moments of bliss, terror, fright*⟩

durch|lesen *unr. tr. V.* **etw. [ganz]** ~: read sth. [all the way] through; **sich** *(Dat.)* **etw.** ~: read sth. through

durch·leuchten *tr. V.* **a)** x-ray ⟨*patient, part of body*⟩; **sich** ~ **lassen** have an x-ray; **b)** *(fig.: analysieren)* investigate ⟨*case, matter, problem, sb.'s past, etc.*⟩ thoroughly; vet ⟨*applicant*⟩

Durchleuchtung **die;** ~, ~**en** **a)** x-ray examination; **b)** *(fig.: Analyse)* [thorough] investigation; *(von Bewerbern usw.)* vetting

durch·löchern *tr. V.* **a)** make holes in; **völlig durchlöchert sein** be full of holes; **b)** *(fig.: schwächen)* undermine ⟨*system*⟩ completely; render ⟨*principle*⟩ meaningless

durch|lüften **1.** *tr. V.* air ⟨*room, flat, etc.*⟩ thoroughly. **2.** *itr. V.* air the place

durch|machen *(ugs.)* **1.** *tr. V.* **a)** undergo ⟨*change*⟩; go through ⟨*stage, phase*⟩; **b)** *(erleiden)* go through; suffer ⟨*illness*⟩. **2.** *itr. V. (durcharbeiten)* work [right] through; *(durchfeiern)* celebrate all night/day *etc.*; keep going all night/day *etc.*

durch|marschieren *itr. V.; mit sein* **[durch etw.]** ~: march through [sth.]

durch·messen *unr. tr. V. (geh.)* cross ⟨*room*⟩

Durchmesser **der;** ~s, ~: diameter

durch|mischen *tr. V.* mix ⟨*ingredients etc.*⟩ thoroughly

durch|mogeln *refl. V. (ugs. abwertend)* cheat one's way through; **sich bei einer Prüfung** *usw.* ~: get through an examination *etc.* by cheating

durch|müssen *unr. itr. V. (ugs.)* **[durch etw.]** ~: have to go through [sth.]; **da werden wir** ~ *(fig.)* we'll have to see it *or* the thing through

durch|nagen *tr. V.* gnaw through

durch·nässen *tr. V.* soak; drench; **[völlig] durchnäßt sein** be soaking wet *or* wet through

durch|nehmen *unr. tr. V. (Schulw.: behandeln)* deal with; do

durch|numerieren *tr. V.* number ⟨*pages, seats, etc.*⟩ consecutively from beginning to end

durch|pausen *tr. V.* trace

durch|peitschen *tr. V. (ugs. abwertend)* railroad ⟨*law, application, etc.*⟩ through

durch|probieren *tr. V.* taste *or* try ⟨*wines, cakes, etc.*⟩ one after another; try on ⟨*dresses, suits, etc.*⟩ one after another

durch|prügeln *tr. V. (ugs.)* give ⟨*child*⟩ a good hiding *or (sl.)* walloping

durch·queren *tr. V.* cross; travel across ⟨*country*⟩

durch|rechnen *tr. V.* calculate

durch|regnen *itr. V. (unpers.)* **in der Küche** *usw.* **regnet es durch** the rain is coming through in the kitchen *etc.*; **die ganze Nacht ~:** rain all [through the] night

Durchreiche die; ~, ~n [serving]hatch

durch|reichen *tr. V.* **etw. [durch etw.] ~:** pass *or* hand sth. through [sth.]

Durch·reise die journey through; **auf der ~ sein** be on the way through *or* passing through

durch|reisen *itr. V.; mit sein* travel *or* pass through

Durch·reisende der/die person travelling through

Durchreise·visum das transit visa

durch|reißen 1. *unr. tr. V.* **etw. ~:** tear sth. in two *or* in half. **2.** *unr. itr. V.; mit sein* ⟨*fabric, garment*⟩ rip, tear; ⟨*thread, rope*⟩ snap *or* break [in two]

durch|reiten *unr. itr. V.; mit sein* **[durch etw.] ~:** ride through [sth.]

durch|rennen *unr. itr. V.; mit sein* **[durch etw.] ~:** run through [sth.]

durch|ringen *unr. refl. V.* **sie hat sich endlich [zu einem Entschluß] durchgerungen** finally she managed to come to a decision; **wann wirst du dich dazu ~, es zu tun?** when are you going to bring yourself to do it?

durch|rosten *itr. V.; mit sein* rust through

durch|rühren *tr. V.* **etw. [gut] ~:** stir sth. [well]

durch|rutschen *itr. V.; mit sein* **[durch etw.] ~** slip through [sth.]

durchs [dʊrçs] *Präp. + Art.* = **durch das**

Durch·sage die announcement

durch|sagen *tr. V.* announce

durch|sägen *tr. V.* saw through

durchschaubar *Adj.* transparent; **leicht ~** easy to see through

durch·schauen *tr. V.* see through ⟨*lie, plan, intention, person, etc.*⟩; see ⟨*situation*⟩ clearly; **du bist durchschaut** I've/we've seen through you; I/we know what you're up to

durch|scheinen *unr. itr. V.* **[durch etw.] ~** ⟨*sun, light*⟩ shine through [sth.]; ⟨*colour, pattern*⟩ show through [sth.]

durchscheinend *Adj.* translucent

durch|scheuern 1. *tr. V.* wear through; **ein durchgescheuertes Kabel** a worn cable. **2.** *refl. V.* wear through

durch|schimmern *itr. V.* **a) [durch etw.] ~** ⟨*light*⟩ shimmer through [sth.]; ⟨*colour*⟩ gleam through [sth.]; **b)** *(fig.)* ⟨*qualities, emotions*⟩ show through

durch|schlafen *unr. itr. V.* sleep [right] through; **die ganze Nacht ~:** sleep all night [without waking]

Durch·schlag der carbon [copy]

¹durch|schlagen 1. *unr. tr. V.* **a) etw. ~:** chop *or* split sth. in two; **b)** *(schlagen)* **einen Nagel [durch etw.] ~:** knock *or* drive a nail through [sth.]. **2.** *unr. itr. V. mit sein* **[durch etw.] ~** ⟨*dampness, water*⟩ come through [sth.]; **das schlägt auf die Preise durch** *(fig.)* it has an effect on prices. **3.** *refl. V.* **a)** struggle along; **b)** *(ein Ziel erreichen) (mit Gewalt)* fight one's way through; *(mit List)* make one's way through

²durch·schlagen *unr. tr. V.* smash

durchschlagend *Adj.* resounding ⟨*success*⟩; decisive ⟨*effect*⟩

Durchschlag·papier das copy paper

Durchschlags·kraft die *(fig.: Wirkung)* power; force

durch|schlängeln *refl. V.* **sich [durch etw.] ~** *(auch fig.)* thread one's way through [sth.]

durch|schleusen *tr. V. (ugs.)* **jmdn./etw. [durch etw.] ~:** guide sb./sth. through [sth.]; *(durchschmuggeln)* get sb./sth. through [sth.]

Durchschlupf [ˈdʊrçʃlʊpf] **der; ~[e]s, ~e** gap; *(Loch)* hole

durch|schlüpfen *itr. V.; mit sein* **[durch etw.] ~:** slip through [sth.]

durch|schmuggeln *tr. V.* **etw. [durch etw.] ~:** smuggle sth. through [sth.]

durch|schneiden *unr. tr. V.* cut through ⟨*thread, cable*⟩; cut ⟨*ribbon, sheet of paper*⟩ in two; cut ⟨*throat, umbilical cord*⟩; **etw. in der Mitte ~:** cut sth. in half

Durch·schnitt der average; **im ~:** on average; **im ~ 110 km/h fahren** average 110 k.p.h.; do 110 k.p.h. on average; **über/unter dem ~ liegen** be above/below average

durchschnittlich 1. *Adj.* **a)** *nicht präd.* average ⟨*growth, performance, output*⟩; **b)** *(ugs.: nicht außergewöhnlich)* ordinary ⟨*life, person, etc.*⟩; **c)** *(mittelmäßig)* modest ⟨*intelligence, talent, performance, achievements*⟩; ordinary ⟨*appearance*⟩. **2.** *adv.* ⟨*produce, spend, earn, etc.*⟩ on [an] average; **~ groß sein** be of average height; **~ begabt sein** be moderately talented

Durchschnitts- average ⟨*age, speed, person, etc.*⟩

durch·schreiten *unr. tr. V. (geh.)* stride across ⟨*room*⟩; stride through ⟨*door, hall*⟩

Durch·schrift die carbon [copy]

Durch·schuß der bullet *or* gunshot wound *(where the bullet has passed right through)*

durch|schütteln *tr. V.* **jmdn. ~:** give sb. a good shaking; **wir wurden im Bus tüchtig durchgeschüttelt** we were shaken about all over the place in the bus

¹durch|schwimmen *unr. itr. V.; mit sein* **[durch etw.] ~:** swim through [sth.]

²durch·schwimmen *unr. tr. V.* swim ⟨*the Channel, course, etc.*⟩

durch|schwitzen *tr. V.* **ich habe mein Hemd** *usw.* **durchgeschwitzt** my shirt *etc.* is soaked in sweat

durch|sehen 1. *unr. itr. V.* **a) [durch etw.] ~:** look through [sth.]; **b)** *s.* **durchblicken b. 2.** *unr. tr. V.* look through ⟨*essay, homework, newspaper, etc.*⟩; **etw. auf Fehler ~:** look *or* check through sth. for mistakes

durch|seihen *tr. V. (Kochk.)* strain; pass ⟨*sauce, gravy*⟩ through a sieve

durch|sein *unr. itr. V., mit sein; nur im Inf. u. Part. zusammengeschrieben (ugs.)* **a) [durch etw.] ~:** be through *or* have got through [sth.]; **ist die Post/der Briefträger schon durch?** has the postman *(Brit.) or (Amer.)* mailman been?; **b)** *(vorbeigefahren sein)* ⟨*train, cyclist*⟩ have gone through; *(abgefahren sein)* ⟨*train, bus, etc.*⟩ have gone; **c)** *(fertig sein)* have finished; **durch/mit etw. ~:** have got through sth.; **d)** *(durchgescheuert sein)* have worn through; **e)** *(reif sein)* ⟨*cheese*⟩ be ripe; **f)** *(durchgebraten sein)* ⟨*meat*⟩ be well done; **g)** *(angenommen sein)* ⟨*law, regulation*⟩ have gone through; ⟨*35-hour week etc.*⟩ have been adopted; **h)** *(gerettet sein)* ⟨*sick or injured person*⟩ be out of danger; **i) bei jmdm. unten ~:** be in sb.'s bad books

durchsetzbar *Adj.* enforceable ⟨*demand, claim*⟩

[1]**durch|setzen** 1. *tr. V.* carry *or* put through ⟨*programme, reform*⟩; carry through ⟨*intention, plan*⟩; accomplish, achieve ⟨*objective*⟩; enforce ⟨*demand, claim*⟩; get ⟨*resolution*⟩ accepted; **seinen Willen ~**: have one's [own] way. **2.** *refl. V.* assert oneself (**gegen** against); ⟨*idea*⟩ find *or* gain acceptance, become generally accepted *or* established; ⟨*fashion*⟩ catch on *(coll.)*, find *or* gain acceptance

[2]**durch·setzen** *tr. V.* **ein Land mit Spionen ~**: infiltrate spies into a country; **mit Nadelbäumen durchsetzt sein** be interspersed with conifers

Durchsetzung die; ~ *s.* [1]**durchsetzen**: carrying through; putting through; accomplishment; achievement; enforcement

Durchsetzungs·kraft die, ~vermögen das ability to assert oneself

Durch·sicht die: nach [einer] ~ der Unterlagen after looking *or* checking through the documents; **jmdm. etw. zur ~ geben** give sb. sth. to look *or* check through

durchsichtig *Adj. (auch fig.)* transparent; see-through, transparent ⟨*night-dress, blouse*⟩

durch|sickern *itr. V.; mit sein* **a)** seep through; **b)** *(bekannt werden)* ⟨*news*⟩ leak out; **es ist durchgesickert, daß ...**: news has leaked out that ...

[1]**durch|sieben** *tr. V.* sift, sieve ⟨*flour etc.*⟩; strain ⟨*tea etc.*⟩

[2]**durch·sieben** *tr. V.* ⟨*bullets*⟩ riddle

durch|spielen *tr. V.* **a)** act ⟨*scene*⟩ through; play ⟨*piece of music*⟩ through; **b)** *(fig.)* go through ⟨*alternatives, options*⟩

durch|sprechen *unr. tr. V.* talk ⟨*matter etc.*⟩ over; discuss ⟨*matter etc.*⟩ thoroughly

durch|springen *unr. itr. V.; mit sein* **[durch etw.] ~**: jump *or* leap through [sth.]

durch|spülen *tr. V.* **etw. [gut/gründlich] ~**: rinse sth. thoroughly

durch|starten *itr. V.; mit sein* **a)** *(Flugw.)* begin climbing again; **b)** *(Kfz-W.)* accelerate away again

durch|stechen *unr. tr. V.* pierce

durch|stecken *tr. V.* **etw. [durch etw.] ~**: put *or (coll.)* stick sth. through [sth.]

durch|stehen *unr. tr. V.* stand ⟨*pace, boring job*⟩; come through ⟨*adventure, difficult situation*⟩; pass ⟨*test*⟩; get over ⟨*illness*⟩

durch|steigen *unr. itr. V.; mit sein* **a) [durch etw.] ~**: climb through [sth.]; **b)** *(salopp: verstehen)* get it *(coll.)*

durch|stellen *tr. V.* put ⟨*call*⟩ through

[1]**durch|stöbern** *tr. V. (ugs.)* search all through ⟨*house*⟩; rummage through ⟨*cupboard, case, etc.*⟩; scour ⟨*wood, area*⟩

[2]**durch·stöbern** *tr. V. (ugs.)* **a)** *s.* [1]**durch|stöbern; b)** rummage through (**nach** in search of); rummage around ⟨*shop*⟩ (**nach** in search of)

[1]**durch|stoßen** *unr. itr. V.* **a) durch etw. ~**: knock a hole through sth.; break through sth.; **b)** *mit sein (Milit.)* break through (**bis zu** to)

[2]**durch·stoßen** *unr. tr. V.* break through

durch|streichen *unr. tr. V.* cross through *or* out; *(in Formularen)* delete

durch·streifen *tr. V. (geh.)* roam, wander through ⟨*fields, countryside*⟩

durch·strömen *tr. V.* flow through

[1]**durch|suchen** *tr. V.* search through

[2]**durch·suchen** *tr. V.* search (**nach** for)

Durchsuchung die; ~, ~**en** search

Durchsuchungs·befehl der search warrant

durch|trainieren *tr. V.* get ⟨*athlete, team, body*⟩ into condition; **ein gut durchtrainierter Körper** a body in peak condition

[1]**durch|trennen,** [2]**durch·trennen** *tr. V.* cut [through] ⟨*wire, rope*⟩; sever ⟨*nerve etc.*⟩

durch|treten 1. *unr. tr. V.* press ⟨*clutch-, brake-pedal*⟩ right down; depress ⟨*clutch-, brake-pedal*⟩ completely. **2.** *unr. itr. V. mit sein* **[durch etw.] ~** ⟨*liquid, gas*⟩ come through [sth.]

durchtrieben *(abwertend)* **1.** *Adj.* crafty; sly. **2.** *adv.* craftily; slyly

Durchtriebenheit die; ~: craftiness; slyness

durch·wachen *tr. V.* **die Nacht ~**: stay awake all night

[1]**durch|wachsen** *unr. itr. V.; mit sein* **[durch etw.] ~** ⟨*plant*⟩ grow through sth.

[2]**durch·wachsen 1.** *Adj.* **a) ~er Speck** streaky bacon; **b)** *nicht attr. (ugs. scherzh.)* so-so. **2.** *adv.* **ihr geht es ~**: she has her ups and downs

Durchwahl die; *o. Pl.* **a)** direct dialling; **b) mein Apparat hat keine ~**: I don't have an outside line; **c)** *s.* **Durchwahlnummer**

durch|wählen *itr. V.* **a)** dial direct; **direkt nach Nairobi ~**: dial Nairobi direct; **b)** *(bei Nebenstellenanlagen)* dial straight through

Durchwahl·nummer die number of the/one's direct line

[1]**durch|wandern** *itr. V.; mit sein* walk *or* hike without a break

[2]**durch·wandern** *tr. V.* walk *or* hike through

durch|waschen *unr. tr. V. (ugs.)* **etw. ~**: wash sth. through

[1]**durch|waten** *itr. V.; mit sein* **[durch etw.] ~**: wade through [sth.]

[2]**durch·waten** *tr. V.* wade across

durchweg ['dʊrçvɛk] *Adv.* without exception

[1]**durch|weichen** *itr. V.; mit sein* ⟨*cardboard, paper*⟩ become *or* go [soft and] soggy

[2]**durch·weichen** *tr. V.* make ⟨*earth, path, etc.*⟩ sodden

durch|werfen *unr. tr. V.* **etw. [durch etw.] ~**: throw sth. through [sth.]

durch|wetzen wear through ⟨*sleeves, knees, elbows, etc.*⟩

durch|wollen *unr. itr. V. (ugs.)* **[durch etw.] ~**: want to go/come/get through [sth.]

[1]**durch|wühlen 1.** *tr. V.* rummage through, ransack ⟨*drawers, cupboard, case*⟩ (**nach** in search of, looking for); turn ⟨*room, house*⟩ upside down (**nach** in search of, looking for). **2.** *refl. V. (ugs.)* **sich durch die Erde ~** ⟨*mole*⟩ burrow through the earth; **sich durch einen Aktenstoß ~** *(fig.)* plough through a pile of documents

[2]**durch·wühlen** *tr. V. s.* [1]**durchwühlen 1**

durch|zählen *tr. V.* count; count up ⟨*money, people*⟩

[1]**durch|ziehen 1.** *unr. tr. V.* **a) jmdn./etw. [durch etw.] ~**: pull sb./sth. through [sth.]; **ein Gummiband [durch etw.] ~**: draw an elastic through [sth.]; **b)** *(ugs.: durchführen)* get through ⟨*syllabus, programme*⟩; **wir müssen die Sache ~**: we must see the matter through; **c)** *(salopp: rauchen)* smoke. **2.** *unr. itr. V.; mit sein* **a) [durch ein Gebiet** *usw.*] **~**: pass through [an area *etc.*]; ⟨*soldiers*⟩ march through [an area *etc.*]; **b)** *(Kochk.)* ⟨*fruit, meat, etc.*⟩ soak

[2]**durch·ziehen** *unr. tr. V.* ⟨*river, road, ravine*⟩ run through, traverse ⟨*landscape*⟩; ⟨*theme, motif, etc.*⟩ run through ⟨*book etc.*⟩

durch·zucken *tr. V.* ⟨*lightning, beam of light*⟩ flash across; **jmdn. ~** *(fig.)*⟨*thought*⟩ flash through *or* cross sb.'s mind

Durch·zug der a) *o. Pl.* draught; **~ machen** create a draught; **die Ohren auf ~ stellen** *(ugs.)* let it go in one ear and out the other; **b)** *(das Durchziehen)* passage through; *(von Truppen)* march through

dürfen ['dʏrfn̩] **1.** *unr. Modalverb; 2. Part.* **~ a)** *(Erlaubnis haben zu)* **etw. tun ~:** be allowed *or* permitted to do sth.; **darf ich [das tun]?** may I [do that]?; **das darf man nicht tun** *(ist einem nicht erlaubt)* that is not allowed *or* permitted; *(sollte man nicht)* one shouldn't do that; **er hat es nicht tun ~:** he was not allowed *or* permitted to do it; **nein, das darfst du nicht** no, you may not; **ich darf morgen nicht verschlafen** I mustn't oversleep tomorrow; **du darfst nicht lügen/jetzt nicht aufgeben!** you mustn't tell lies/give up now!; *(solltest nicht)* you shouldn't tell lies/give up now!; **ihm darf nichts geschehen** nothing must happen to him; **das darf nicht wahr sein** *(ugs.)* that's incredible; **hier darf man nicht rauchen** smoking is prohibited here; **b)** *(in Höflichkeitsformeln)* **darf ich rauchen?** may I smoke?; **darf ich Sie bitten, das zu tun?** could I ask you to do that?; **darf** *od.* **dürfte ich mal Ihre Papiere sehen?** may I see your papers?; **darf ich um diesen Tanz bitten?** may I have [the pleasure of] this dance?; **was darf es sein?** can I help you?; **was möchten Sie trinken, was darf es sein?** what can I get you to drink?; **darf ich bitten?** *(um einen Tanz)* may I have the pleasure?; *(einzutreten)* won't you come in?; **Ruhe, wenn ich bitten darf!** will you please be quiet!; **c)** *(Grund haben zu)* **ich darf Ihnen mitteilen, daß ...:** I am able to inform you that ...; **darf ich annehmen, daß ...?** can I assume that ...?; **sie darf sich nicht beklagen** she can't complain; she has no reason to complain; **das darfst du mir glauben** you can take my word for it; **d)** *Konjunktiv II + Inf.* **das dürfte der Grund sein** that is probably the reason; *(ich nehme an, daß das der Grund ist)* that must be the reason; **das dürfte reichen** that should be enough. **2.** *unr. tr., itr. V.* **er hat nicht gedurft** he was not allowed *or* permitted to; **darf ich ins Theater?** may I go to the theatre?; **darfst du das?** are you allowed to?

durfte ['dʊrftə], *1. u. 3. Pers. Sg. Prät. v.* **dürfen**

dürfte ['dʏrftə] *1. u. 3. Pers. Sg. Konjunktiv II v.* **dürfen**

dürftig ['dʏrftɪç] **1.** *Adj.* **a)** *(ärmlich)* poor; scanty, meagre ⟨*meal*⟩; scanty, poor ⟨*clothing*⟩; **b)** *(abwertend: unzulänglich)* poor ⟨*substitute, performance, light*⟩; feeble, poor ⟨*explanation*⟩; lame, feeble ⟨*excuse*⟩; scanty ⟨*knowledge, evidence, results*⟩; sparse ⟨*growth of hair*⟩; paltry, meagre ⟨*income*⟩. **2.** *adv.* **a)** ⟨*live*⟩ poorly; scantily ⟨*dressed*⟩; **b)** *(abwertend: unzulänglich)* skimpily, scantily ⟨*furnished*⟩; poorly ⟨*attended*⟩; ⟨*report, formulate*⟩ sketchily; thinly ⟨*concealed*⟩

dürr [dʏr] *Adj.* **a)** withered ⟨*branch*⟩; dry, dried up, withered ⟨*grass, leaves*⟩; arid, barren ⟨*ground, earth*⟩; **b)** *(mager)* skinny, scraggy, scrawny ⟨*legs, arms, body, person*⟩; **c)** *(unergiebig)* lean ⟨*years*⟩; bare ⟨*words, description*⟩

Dürre die; ~, ~n drought

Dürre·periode die period of drought

Durst [dʊrst] **der; ~[e]s** thirst; **~ haben** be thirsty; **~ bekommen** get *or* become thirsty; **seinen ~ löschen** *od.* **stillen** quench *or* slake one's thirst; **ich habe ~ auf ein Bier** I could just drink a beer; **~ nach Wissen** *(fig. geh.)* a thirst for knowledge; **ein Glas** *od.* **einen über den ~ trinken** *(ugs. scherzh.)* have one too many

dursten *(geh.) itr. V.* thirst; **~ müssen** have to go thirsty

dürsten ['dʏrstn̩] *tr. V. (unpers.)* **mich dürstet** *(geh.)* I am thirsty

durstig *Adj.* thirsty

durst-, Durst-: ~löschend, ~stillend *Adj.* thirst-quenching; **~strecke die** *(fig.)* lean period *or* time

Dusch·bad das shower[-bath]

Dusche ['dʊʃə] **die; ~, ~n** shower; **unter die ~ gehen** take *or* have a shower; **unter der ~ sein** be in the shower; **eine kalte ~ [für jmdn.] sein** *(fig. ugs.)* be like a cold douche *or* a douche of cold water [on sb.]

duschen 1. *itr., refl. V.* take *or* have a shower; **kalt ~:** take *or* have a cold shower. **2.** *tr. V.* **jmdn. ~:** give sb. a shower

Düse ['dy:zə] **die; ~, ~n** *(Technik)* nozzle; *(eines Vergasers)* jet

Dusel ['du:zl̩] **der; ~s** *(ugs.)* luck; **~ haben** be jammy *(Brit. coll.) or* lucky

düsen *itr. V.; mit sein (ugs.)* dash

Düsen-: ~an·trieb der jet propulsion; **~flugzeug das** jet aeroplane *or* aircraft *or* plane; **~trieb·werk das** jet power plant; jet engine

Dussel ['dʊsl̩] **der; ~s, ~** *(ugs.)* dope *(coll.)*; idiot; clot *(Brit. sl.)*

dusselig, dußlig *(ugs.)* **1.** *Adj.* gormless *(Brit. coll.)*; stupid; idiotic. **2.** *adv.* gormlessly *(Brit. coll.)*; stupidly

düster ['dy:stɐ] **1.** *Adj.* **a)** dark; gloomy; dim ⟨*light*⟩; **b)** *(fig.)* gloomy; sombre ⟨*colour, music*⟩; dark ⟨*foreboding*⟩. **2.** *adv. (fig.)* gloomily

Düsterheit, Düsterkeit die; ~a) *s.* **düster a:** darkness; gloom; dimness; **b)** *s.* **düster b:** gloominess; sombreness; darkness

Dutt [dʊt] **der; ~[e]s, ~e** *od.* **~s** bun

Dutzend ['dʊtsn̩t] **das; ~s, ~e** dozen; **zwei ~:** two dozen; **ein ~ Eier** a dozen eggs; **das ~ Schnecken kostet** *od.* **kosten 16 Mark** snails cost 16 marks a dozen; **sie kamen zu ~en** they came in [their] dozens *(coll.)*

dutzend-, Dutzend-: ~fach, ~mal *Adv.* a dozen times; dozens of times; **~ware die** *(abwertend)* cheap mass-produced item **~weise** *Adv.* ⟨*arrive, leave*⟩ in [their] dozens *(coll.)*; **etw. ~weise verkaufen** sell sth. by the dozen

duzen ['du:tsn̩] *tr. V.* call ⟨*sb*⟩ 'du' *(the familiar form of address)*; **sich ~:** call each other 'du'; **sich mit jmdm. ~:** call sb. 'du'

Duz·freund der good friend *(whom one addresses with 'du')*

Dynamik [dy'na:mɪk] **die; ~ a)** *(Physik)* dynamics *sing., no art.*; **b)** *(Triebkraft)* dynamism; **c)** *(Musik)* dynamics *pl.*

dynamisch [dy'na:mɪʃ] **1.** *Adj.* dynamic; **~e Renten** ≈ index-linked pensions *(linked to changes in the national product)*. **2.** *adv.* dynamically

Dynamit [dyna'mi:t] **das; ~s** dynamite

Dynamo [dy'na:mo] **der; ~s, ~s** dynamo

Dynastie die; ~, ~n dynasty

D-Zug ['de:-] **der** fast *or* express train; **ein alter Mann/eine alte Frau ist doch kein ~!** *(salopp)* I'm too old to hurry

D-Zug-Zuschlag der fast train supplement

E

e, E [e:] **das; ~, ~** *(Buchstabe)* e/E

E *Abk.* **Europastraße**

Ebbe ['ɛbə] **die; ~, ~n a)** *(Bewegung)* ebb tide; **es ist ~:** the tide is going out; **bei ~:** at ebb tide; when the tide is going out; **~ und Flut** ebb and flow; **b)** *(Zustand)* low tide; **es ist ~:** the tide is out; **bei ~:** at low tide; when the tide is/was out; **es herrschte ~ in seinem Portemonnaie** *(fig. ugs.)* he was short of cash *(coll.)*

eben ['e:bn̩] **1.** *Adj.* **a)** *(flach)* flat; **b)** *(glatt)* level ⟨*ground, path, stretch*⟩. **2.** *Adv.* **a)** *(gerade jetzt)* just; **b)** *(kurz)* [for] a moment; **c)** *(gerade noch)* just [about]; **etw. ~ noch schaffen** only just manage sth.; **d)** *(genau)* precisely; **ja, ~:** yes, exactly *or* precisely; **ja, ~ das meine ich auch** yes, that's just *or* exactly what I think. **3.** *Partikel* **a) nicht ~:** not exactly; **b)** *(nun einmal)* simply; **das ist ~ so** that's just the way it is

eben-, Eben-: ~bild das image; **ganz jmds. ~bild sein** be the spitting image of sb.; **~bürtig** [~bʏrtɪç] *Adj.* equal (*Dat.* to); **~da** *Adv.* there; *(bei Literaturangaben)* ibid *abbr.*; ibidem; **~der, ~die, ~das** *Demonstrativpron.* **~das meine ich** that's exactly what I mean; **~die, von der wir sprachen** the very one we were talking about; **~der war krank** he was the very one who was ill; **~derselbe, ~dieselbe, ~dasselbe** *Demonstrativpron.; attr.* the very same ⟨*person, thing*⟩; *alleinstehend* **~dieselbe meine ich** she's just the one I mean; **~dieser, ~diese, ~dieses** *Demonstrativpron.; attr.* **~dieses Thema wurde behandelt** this very topic was discussed; *alleinstehend* **~dieser wurde genannt** he was the very one who was mentioned

Ebene die, ~, ~n a) *(flaches Land)* plain; **in der ~:** on the plain; **b)** *(Geom., Physik)* plane; **c)** *(fig.)* level

eben-: ~erdig 1. *Adj.* ground-level; **2.** *adv.* at ground level; **~falls** *Adv.* likewise; as well; **danke, ~falls** thank you, [and] [the] same to you

Eben·holz das ebony

Eben·maß das; *o. Pl.* *(der Gesichtszüge)* regularity; *(des Körperbaus)* symmetry; even proportions *pl.*; *(von Versen)* regularity; harmony

eben·mäßig *Adj.* regular ⟨*features*⟩; well-proportioned ⟨*figure*⟩; regular, harmonious ⟨*verse*⟩; even ⟨*proportions*⟩

eben·so *Adv.* **a)** *mit Adjektiven* just as; **~ groß wie ... sein** be just as big as ...; **ein ~ frecher wie dummer Kerl** a fellow who is/was as impudent as he is/was stupid; **b)** *mit Verben* in exactly the same way; *(in demselben Maße)* just as much; **mir geht es ~:** its just the same for me

ebenso-: ~gern *Adv.* **~gern mag ich Erdbeeren [wie ...]** I like strawberries just as much [as ...]; **~gern würde ich an den Strand gehen** I would just as soon go to the beach; **~gut** *Adv.* just as well; **ich kann ~gut ein Taxi nehmen** I can just as easily take a taxi; **~oft** *Adv.* just as often; just as frequently

eben·solch... *Demonstrativpron.* the same; **ich habe ebensolche Angst wie du** I am just as afraid as you are

ebenso-: ~sehr *Adv.* just as much; **~viel** *Indefinitpron., Adv.* just as much; **~wenig** *Indefinitpron., Adv.* just as little; **man kann dieses ~wenig wie jenes tun** one cannot do this, any more than that

Eber ['e:bɐ] **der; ~s, ~:** boar

ebnen *tr. V.* level ⟨*ground*⟩; **jmdm. den Weg** *od.* **die Bahn ~** *(fig.)* smooth the way for sb.

Echo ['ɛço] **das; ~s, ~s** echo; *(fig.)* response **(auf** + *Akk.* to); **das ~ in der Presse** *(fig.)* the press reaction; the reaction in the press

Echo·lot das echo sounder

Echse ['ɛksə] **die; ~, ~n** *(Zool.)* **a)** saurian; **b)** *(Eid~)* lizard

echt [ɛçt] **1.** *Adj.* **a)** genuine; authentic, genuine ⟨*signature, document*⟩; **b)** *(wahr)* true, real ⟨*love, friendship*⟩; real, genuine ⟨*concern, sorrow, emergency*⟩; **c)** *nicht präd. (typisch)* real, typical ⟨*Bavarian, American, etc.*⟩; **d)** *(Math.)* proper ⟨*fraction*⟩. **2.** *adv.* **a) ~ golden/italienisch** *usw.* real gold/real *or* genuine Italian *etc.*; **b)** *(ugs.: wirklich)* really; **das ist ~ wahr/blöd** that's absolutely true/stupid; **c)** *(typisch)* typically

echt- real ⟨*silver, silk, leather, etc.*⟩

Echtheit die; ~ genuineness; *(einer Unterschrift, eines Dokuments)* authenticity

Eck [ɛk] **das; ~s, ~e** *(bes. südd., österr.)* corner; **über(s) ~:** diagonally

Eck-: ~ball der *(Sport)* corner [-kick/-hit/-throw]; **einen ~ball treten** take a corner; **~bank die** corner seat

Ecke ['ɛkə] **die; ~, ~n a)** corner; **Nietzschestr., ~ Goethestr.** on the corner of Nietzschestrasse and Goethestrasse; **um die ~ biegen** turn the corner; go/come round the corner; **die lange/kurze ~** *(Ballspiele)* the far/near corner; **jmdn. um die ~ bringen** *(fig. salopp)* bump sb. off *(sl.)*; **mit jmdm. um** *od.* **über sieben ~n verwandt sein** *(fig. ugs.)* be distantly related to sb.; **an allen ~n [und Enden** *od.* **Kanten]** *(ugs.)* everywhere; **b)** *(Ballspiele)* corner; **eine ~ treten** take a corner; **c)** *(ugs.: Gegend)* corner; **eine schöne ~:** a lovely spot; **e)** *(ugs., bes. nordd.: Strecke)* **bis dahin ist es noch eine ganze ~:** it's still quite some way there

Ecker ['ɛkɐ] **die; ~, ~n** beech-nut

Eck·haus das corner house; house on the/a corner; *(einer Häuserreihe)* end house

eckig 1. *Adj.* **a)** square; angular ⟨*features*⟩; **b)** *(ruckartig)* jerky ⟨*movement, walk, gait*⟩. **2.** *adv.* jerkily

Eck-: ~kneipe die *small friendly pub on a street-corner;* **~pfeiler der** corner pillar; *(fig.)* cornerstone; **~schrank der** corner cupboard; **~stoß der** *(Fußball)* corner-kick; **~zahn der** canine tooth; **~zimmer das** corner room

Economy·klasse [ɪ'kɔnəmɪ-] **die** economy class; tourist class

edel ['e:dl̩] **1.** *Adj.* **a)** *nicht präd.* thoroughbred ⟨*horse*⟩; **b)** *(großmütig)* noble[-minded], high-minded ⟨*person*⟩; noble ⟨*thought, gesture, deed*⟩; honourable ⟨*motive*⟩; **edle Gesinnung** nobility of mind; noble-mindedness; **c)** *(geh.: wohlgeformt)* finely-shaped; **von edlem Wuchs** of noble stature; **d)** *(geh.: vortrefflich)* fine ⟨*wine*⟩; high-grade ⟨*wood, timber*⟩; **e)** *nicht präd. (veralt.: adlig)* noble. **2.** *adv.* nobly

Edel-: ~holz das high-grade wood; high-grade timber; **~kitsch der** grandly pretentious kitsch; **~mann der;** *Pl.* **~leute** *od.* **~männer** *(hist.)* nobleman; noble; **~metall das** noble metal; **~mut der** *(geh.)* nobility of mind; noble-mindedness; magnanimity; **~schnulze die** *(abwertend)* example of pretentious schmaltz; **~stahl der** stainless steel; **~stein der** precious stone; gem[stone]; **~tanne die** silver fir; **~weiß das; ~[es], ~e** edelweiss

Eden ['e:dn̩] *in* **der Garten ~** *(bibl.)* the Garden of Eden

Edikt [e'dɪkt] **das; ~[e]s, ~e** *(hist.)* edict

Edition [edi'tsi̯o:n] **die; ~, ~en** edition

EDV *Abk.* **elektronische Datenverarbeitung** EDP

EEG [e:e:'ge:] **das; ~[s], ~[s]** *Abk.* **Elektroenzephalogramm** EEG

Efeu ['e:fɔy] **der; ~s** ivy

efeu·bewachsen *Adj.* ivy-covered; ivy-clad

Effeff [ɛf'|ɛf] *in* **etw. aus dem ~ beherrschen** know sth. inside out

Effekt [ɛ'fɛkt] **der; ~[e]s, ~e** effect

Effekten [ɛ'fɛktn̩] *Pl. (Finanzw.)* securities

Effekt·hascherei [-haʃə'rai̯] **die; ~, ~en** *(abwertend)* straining for effect; showiness

effektiv [ɛfɛk'ti:f] **1.** *Adj.* effective. **2.** *adv.* effectively

Effektivität [ɛfɛktivi'tɛ:t] **die; ~:** effectiveness

Effektiv·lohn der real wage[s]

effekt·voll 1. *Adj.* effective ⟨*speech, poem, contrast, pattern*⟩; dramatic ⟨*pause, gesture, entrance*⟩. **2.** *adv.* effectively

Effet [ɛ'fe:] **der; ~s, ~s** spin; *(Billard)* side; **den Ball mit ~ schlagen** put spin/side on the ball

effizient [ɛfi'tsi̯ɛnt] **1.** *Adj. (geh.)* efficient. **2.** *adv.* efficiently

EG ['e:'ge:] *Abk.* **a) die; ~: Europäische Gemeinschaft** EC; **b) Erdgeschoß**

egal [e'ga:l] *Adj.* **a)** *nicht attr. (ugs.: einerlei)* **es ist jmdm. ~:** it's all the same to sb.; **das ist ~:** that doesn't make any difference; **[ganz] ~, wie/wer/ob** *usw.* **...:** no matter how/who/whether *etc.* ...; **b)** *(ugs.: gleich[artig])* identical

egalisieren *tr. V. (Sport)* equal ⟨*record*⟩

Egel ['e:gl̩] **der; ~s, ~:** leech

Egge ['ɛgə] **die; ~, ~n** *(Landw.)* harrow

eggen *tr. V. (Landw.)* harrow

Ego ['e:go] **das; ~, ~s** *(Psych.)* ego

Egoismus [ego'ɪsmʊs] **der; ~:** egoism

Egoist [ego'ɪst] **der; ~en, ~en, Egoistin die; ~, ~nen** egoist

egoistisch 1. *Adj.* egoistic[al]. **2.** *adv.* egoistically

Egozentriker der; ~s, ~, Egozentrikerin die; ~, ~nen egocentric

egozentrisch *Adj.* egocentric

[1]**eh** [e:] *Interj. (ugs.)* **a)** hey; **b) das hast du nicht erwartet, ~?** you didn't expect that, did you [,eh]?

[2]**eh** *Adv.* **a)** *(bes. südd., österr.: sowieso)* anyway; in any case; **es ist ~ alles zu spät** it's too late anyway *or* in any case; **b) seit ~ und je** for as long as anyone can remember; for donkey's years *(coll.)*; **wie ~ und je** just as before

ehe ['e:ə] *Konj. s.* **bevor**

Ehe ['e:ə] **die; ~, ~n** marriage; **eine glückliche ~ führen** be happily married; lead a happy married life; **die ~ brechen** commit adultery *(geh. veralt.)*; **jmdm. die ~ versprechen** promise to marry sb.; **aus erster ~:** from his/her first marriage

ehe-, Ehe-: ~beratung die marriage guidance *(Brit.)*; marriage counselling; **~bett das** marriage-bed; *(Doppelbett)* double bed; **~brecher der; ~s, ~:** adulterer; **~brecherin die; ~, ~nen** adulteress; **~brecherisch** *Adj.* adulterous; **~bruch der** adultery

ehe·dem *Adv. (geh.)* formerly; in former times

Ehe-: ~frau die wife; *(verheiratete Frau)* married woman; **~gatte der** *(geh.)* husband; *(~mann od. ~frau)* spouse; **beide ~gatten** both husband and wife; **~glück das** wedded *or* married bliss; **~hälfte die** *(scherzh.)* better half *(joc.)*; **~krach der** *(ugs.)* row; quarrel; **~leben das** married life; **~leute** *Pl.* married couple; **die beiden ~leute** the husband and wife

ehelich *Adj.* marital; matrimonial; conjugal ⟨*rights, duties*⟩; legitimate ⟨*child*⟩; **~e Gemeinschaft** marriage partnership

ehemalig ['e:əmalɪç] *Adj.* former; **seine ~e Frau** his ex-wife; **seine Ehemalige/ihr Ehemaliger** *(ugs.)* his/her ex *(coll.)*

ehe-, Ehe-: ~mann der; *Pl.* **~männer** husband; *(verheirateter Mann)* a married man; **~müde** *Adj.* tired of married life *postpos.*; **~paar das** married couple

eher ['e:ɐ] *Adv.* **a)** *(früher)* earlier; sooner; **je ~, desto lieber** *od.* **besser** the sooner the better; **b)** *(lieber)* rather; sooner; **alles ~ als das** anything but that; **c)** *(wahrscheinlicher)* more likely; *(leichter)* more easily; **das ist schon ~ möglich** that's more likely; **d)** *(mehr)* **er ist ~ faul als dumm** he is lazy rather than stupid; he's more lazy than stupid *(coll.)*; **alles ~ sein als ...:** be anything but ...

Ehe-: ~ring der wedding-ring; **~scheidung die** divorce

ehest... ['e:əst] **1.** *Adj.; nicht präd.* earliest; **zum ~en Termin** at the earliest possible date. **2.** *adv.* **am ~en** *(am liebsten)* best of all; *(am wahrscheinlichsten)* most likely

Ehe·stand der; *o. Pl.* marriage *no art.*; matrimony *no art.*

ehestens ['e:əstn̩s] *Adv. s.* **frühestens**

Ehe-: ~streit der marital *or* matrimonial dispute; **~vermittlungs·institut das** marriage bureau

ehrbar 1. *Adj. (geh.)* respectable, worthy ⟨*person, occupation*⟩; honourable ⟨*intentions*⟩. **2.** *adv.* respectably

Ehre ['e:rə] **die; ~, ~n a)** honour; **es ist mit eine ~, ... zu ...:** it is an honour for me to ...; **die ~ haben, etw. zu tun** have the ~ of doing sth.; **jmdm./einer Sache [alle] ~ machen** do sb./sth. [great] credit; **jmds. Andenken** *(Akk.)* **in ~n halten** honour sb.'s memory; **auf ~ und Gewissen** in all truthfulness *or* honesty; **jmdm./einer Sache zuviel ~ antun** *(fig.: jmdn./etw. überschätzen)* overvalue sb./sth.; **jmdm. zur ~ gereichen** *(geh.)* bring honour to sb.; **~, wem ~ gebührt** [give] credit where credit is due; **jmdm. die letzte ~ erweisen** pay one's last respects to sb.; **um der Wahrheit die ~ zu geben** *(fig.)* to tell the truth; to be [perfectly] honest; **zu ~n des Königs, dem König zu ~n** in honour of the king; **wieder zu ~n kommen** *(fig.)* come back into favour; **damit kannst du keine ~ einlegen** that does you no credit; **b)** *o. Pl. (Ehrgefühl)* sense of honour; **er hat keine ~ im Leib[e]** he doesn't have an ounce of integrity in him

ehren *tr. V.* **a)** *(Ehre erweisen)* honour; **jmdn. mit einem Orden ~:** award sb. a medal; **sehr geehrter Herr Müller/sehr geehrte Frau Müller** *usw.* Dear Herr Müller/ Dear Frau Müller *etc.*; **b)** *(Ehre machen)* **deine Hilfsbereitschaft ehrt dich** your willingness to help does you credit; **sein Vertrauen**

ehrt mich I'm honoured by his confidence in me

ehren-, Ehren-: ~abzeichen das medal; **~amt das** honorary position *or* post; **~amtlich 1.** *Adj.* honorary ⟨*position, membership*⟩; voluntary ⟨*help, worker*⟩; **2.** *adv.* in an honorary capacity; *(freiwillig)* on a voluntary basis

Ehren-: ~bürger der, ~bürgerin die honorary citizen; **jmdn. zum ~bürger der Stadt ernennen** give sb. the freedom *or* make sb. a freeman of the town/city; **~doktor der a)** honorary doctor; **b)** *(Titel)* honorary doctorate; **~gast der** guest of honour; **~geleit das** official escort

ehrenhaft 1. *Adj.* honourable ⟨*intentions, person*⟩. **2.** *adv.* ⟨*act*⟩ honourably

ehren-, Ehren-: ~halber *Adv.* **jmdm. den Doktortitel ~halber verleihen** confer an honorary doctorate on sb.; **Doktor ~halber** honorary doctor; **~kodex der** code of honour; **~mal das** monument; **~mann der;** *Pl.* **~männer** man of honour; **~mitglied das** honorary member; **~platz der** place of honour; **~preis der** special prize; **~rechte** *Pl.* **die bürgerlichen ~rechte** civil rights *or* liberties; **~rettung die: zu ihrer ~rettung muß ich sagen, daß ...:** it must be said in her defence that ...; **~rührig** *Adj.* defamatory ⟨*allegations*⟩; insulting ⟨*behaviour*⟩; **~runde die** lap of honour; **~sache die: das ist ~sache** that is a point of honour; **Verschwiegenheit ist ~sache** I/we feel honour bound to stay silent; **~sache!** you can count on me!; **~tag der** *(geh.)* special day; **~tor das, ~treffer der** *(Sport)* consolation goal; **~tribüne die** VIP stand; **~urkunde die** certificate; **~voll 1.** *Adj.* honourable ⟨*peace, death, compromise, occupation*⟩; creditable, gallant ⟨*attempt, conduct*⟩; **2.** *adv.* ⟨*act*⟩ honourably; **~vorsitzender der/die** honorary chairman; **~wert** *Adj. (geh.)* worthy, honourable ⟨*person, occupation*⟩; **die Ehrenwerte Gesellschaft** the Mafia; **~wort das;** *pl.* **~worte: ~wort [!/?]** word of honour [!/?]; **sein ~wort brechen** break one's word

ehrerbietig ['e:ɐ̯|ɛɐ̯bi:tɪç] **1.** *Adj.* *(geh.)* respectful. **2.** *adv.* ⟨*greet*⟩ respectfully

Ehrerbietung die; ~ *(geh.)* respect

Ehr·furcht die reverence **(vor +** *Dat.* for); **[große] ~ vor jmdm./etw. haben** have [a great] respect for sb./sth.; **jmdm. ~ einflößen** fill sb. with awe

ehrfürchtig 1. *Adj.* reverent. **2.** *adv.* reverently

ehr-, Ehr-: ~gefühl das; *o. Pl.* sense of honour; **~geiz der** ambition; **~geizig 1.** *Adj.* ambitious; **2.** *adv.* ambitiously

ehrlich 1. *Adj.* honest ⟨*person, face, answer, deal*⟩; genuine ⟨*concern, desire, admiration*⟩; upright ⟨*character*⟩; honourable ⟨*intentions*⟩; *(wahrheitsgetreu)* truthful ⟨*answer, statement*⟩; **der ~e Finder gab die Brieftasche ab** the person who found the wallet handed *(Brit.) or (Amer.)* turned it in; **~ währt am längsten** *(Spr.)* honesty is the best policy *(prov.)*. **2.** *adv.* honestly; **etw. ~ teilen** share sth.; **~ spielen** play fairly; **es ~ mit jmdm. meinen** play straight with sb.; **~ gesagt** quite honestly; to be honest

Ehrlichkeit die; ~ *s.* **ehrlich 1:** honesty; genuineness; uprightness; honourableness; truthfulness

ehr·los 1. *Adj.* dishonourable; **2.** *adv.* dishonourably

Ehr·losigkeit die; ~: dishonourableness

Ehrung die; ~, **~en a) die ~ der Preisträger** the prize-giving *(Brit.) or (Amer.)* awards ceremony; **bei der ~ der Sieger** when the winners were awarded their medals/trophies; **b)** *(etw. Ehrendes)* honour

ehr·würdig *Adj.* **a)** venerable ⟨*person*⟩; **ein ~es Alter haben** ⟨*person*⟩ have reached a grand old age; ⟨*building*⟩ be of great age; **b)** *(kath. Kirche)* **~er Vater/~e Mutter** Reverend Father/Mother

ei [ai] *Interj.* hey; *(abschätzig)* oho

Ei [ai] **das; ~[e]s, ~er a)** egg; *(Physiol., Zool.)* ovum; **aus dem ~ schlüpfen** hatch [out]; **verlorene** *od.* **pochierte ~er** poached eggs; **russische ~er** egg mayonnaise; Russian eggs; **sie geht wie auf [rohen] ~ern** *(fig.)* she is walking very carefully; **ach, du dickes ~!** *(ugs.)* dash it! *(Brit. coll.)*; darn it! *(Amer. coll.)*; **das ~ des Kolumbus** *(fig.)* an inspired discovery; **wie aus dem ~ gepellt sein** *(fig.)* be dressed to the nines; **sich gleichen wie ein ~ dem anderen** be as like as two peas in a pod; *s. auch* **Apfel; b)** *(derb: Hoden) meist Pl.* **~er** balls *(coarse)*; nuts *(Amer. coarse)*

Eibe ['aibə] **die;** ~, **~n** yew[-tree]

Eiche ['aiçə] **die;** ~, **~n** oak[-tree]; *(Holz)* oak[-wood]

Eichel ['aiçl̩] **die;** ~, **~n a)** *(Frucht)* acorn; **b)** *(Anat.)* glans

Eichel·häher der jay

eichen *tr. V.* calibrate ⟨*measuring instrument, thermometer*⟩; standardize ⟨*weights, measures*⟩

Eichen-: ~holz das oak[-wood]; **~wald der** oak-wood; *(größer)* oak forest

Eich·hörnchen das, *(landsch.)* **Eich·kätzchen das** squirrel

Eid [ait] **der; ~[e]s, ~e** oath; **einen ~ leisten** *od.* **ablegen** swear *or* take an oath; **einen ~ auf die Verfassung schwören** solemnly swear to preserve, protect, and defend the constitution

Eidechse ['aidɛksə] **die;** ~, **~n** lizard

Eides·formel die *(jur.)* wording of the oath

eides·stattlich *(Rechtsw.)* **1.** *Adj.* **~e Erklärung** statutory declaration. **2.** *adv.* **~ erklären** *od.* **versichern, daß ...:** attest in a statutory declaration that ...

eid-, Eid-: ~genosse der Swiss; **~genossenschaft die;** *o. Pl.* **die Schweizerische ~genossenschaft** the Swiss Confederation; **~genössisch** *Adj.* Swiss

Ei·dotter der *od.* **das** egg yolk

Eier-: ~becher der egg-cup; **~farbe die** *paint for decorating eggs as Easter gifts;* **~kuchen der** pancake; *(Omelett)* omelette; **~laufen das** egg-and-spoon race; **~likör der** egg-liqueur; **~löffel der** egg-spoon

eiern *itr. V. (ugs.)* wobble

eier-, Eier-: ~nudel die egg noodle; **~pfann·kuchen der** *s.* **~kuchen; ~schale die** eggshell; **~schalen·farben** *Adj.* off-white; **~speise die a)** egg dish; **b)** *(österr.)* scrambled egg; **~stock der** *(Anat.)* ovary; **~uhr die** egg-timer

Eifer ['aifɐ] **der;** **~s** eagerness; *(Emsigkeit)* zeal; *(Begeisterung)* enthusiasm; **im ~ des Gefechts** in the *or* with all the excitement

Eiferer der; ~s, ~ *(geh.)* zealot

eifern *itr. V.* **für/gegen etw. ~:** agitate for/against sth.

Eifer·sucht die jealousy **(auf +** *Akk.* of)

eifer·süchtig 1. *Adj.* jealous **(auf +** *Akk.* of). **2.** *adv.* jealously

ei·förmig *Adj.* egg-shaped

eifrig 1. *Adj.* eager; enthusiastic ⟨*supporter, collector*⟩; *(fleißig)* assiduous; **~ bei etw. sein** show keen interest in doing sth. **2.** *adv.*

eagerly; ~ **dabei sein, etw. zu tun** be busy doing sth.; ~ **bemüht sein, etw. zu tun** be eager to do sth.

Ei·gelb das; ~**[e]s,** ~**e** egg yolk; **drei** ~: the yolks of three eggs

eigen ['aign̩] *Adj.* **a)** *nicht präd.* own; **eine** ~**e Wohnung haben** have one's own flat *(Brit.) or (Amer.)* apartment; **ein Zimmer mit** ~**em Eingang** a room with a separate entrance; **auf** ~**en Füßen** *od.* **Beinen stehen** stand on one's own two feet; **sich** *(Dat.)* **etw. zu** ~ **machen** adopt sth.; **b)** *(kennzeichnend)* characteristic; **c)** *(landsch.: penibel)* particular (**mit** about)

eigen-, Eigen-: ~**art die** peculiarity; **eine** ~**art dieser Stadt** one of the characteristic features of this city; ~**artig** *Adj.* peculiar; strange; odd; ~**artigerweise** *Adv.* strangely [enough]; oddly [enough]; ~**brötler** [~brø:tlɐ] **der;** ~**s,** ~: loner; lone wolf; ~**brötlerisch 1.** *Adj.* solitary; **2.** *adv.* **sich** ~**brötlerisch verhalten** behave like a loner *or* a lone wolf; ~**gewicht das** own weight; ~**händig 1.** *Adj.* personal ‹*signature*›; personally inscribed ‹*dedication*›; holographic ‹*will, document*›; **2.** *adv.* personally; ~**heim das** house of one's own

Eigenheit die; ~, ~**en** peculiarity

eigen-, Eigen-: ~**initiative die** initiative of one's own; ~**interesse das** personal interest; ~**leben das;** *o. Pl.* life of one's own; ~**liebe die** amour propre; ~**lob das** self-praise; ~**lob stinkt!** *(ugs.)* self-praise is no recommendation; ~**mächtig 1.** *Adj.* unauthorized ‹*decision*›; *(selbstherrlich)* high-handed; **2.** *adv.* ~**mächtig handeln** act on one's own authority; *(selbstherrlich)* act high-handedly; **etw.** ~**mächtig tun** do sth. without asking; ~**mächtigkeit die;** ~, ~**en a)** *o. Pl.* high-handedness; **b)** *(Handlung)* unauthorized action; ~**name der** proper name; *(Ling.)* proper noun; ~**nutz der;** ~**es** self-interest; ~**nützig** [~nʏtsɪç] **1.** *Adj.* self-interested, self-seeking ‹*person*›; selfish ‹*motive*›; **2.** *adv.* selfishly

eigens *Adv.* specially; ~ **für diesen Zweck** specifically for this purpose

Eigenschaft die; ~, ~**en** *(von Lebewesen)* quality; characteristic; *(von Sachen, Stoffen)* property; **in seiner** ~ **als Mann/Vorsitzender** as a man/in his capacity as chairman

Eigenschafts·wort das adjective

eigen-, Eigen-: ~**sinn der;** *o. Pl.* obstinacy; stubbornness; ~**sinnig 1.** *Adj.* obstinate; stubborn; **2.** *adv.* obstinately; stubbornly; ~**sinnigkeit die** *s.* ~**sinn;** ~**ständig 1.** *Adj.* independent; **2.** *adv.* independently; ~**ständigkeit die;** ~: independence; ~**süchtig 1.** *Adj.* selfish; **2.** *adv.* selfishly

eigentlich ['aign̩tlɪç] **1.** *Adj.; nicht präd. (wirklich)* actual; real; *(wahr)* true; *(ursprünglich)* original; **das Eigentliche** the essential thing. **2.** *Adv. (tatsächlich, genaugenommen)* actually; really; ~ **müßte ich ja jetzt gehen, aber ...:** really, I ought to go now, but ...; **es ist** ~ **schade, daß ...:** actually, it's a pity that... **3.** *Partikel* **wann erscheint** ~ **der letzte Band?** tell me, when will the last volume come out; **sind sie** ~ **verheiratet?** are they in fact married?; **wer sind Sie** ~**?** who do you think you are?; **was willst du** ~**?** what exactly do you want?

Eigen·tor das *(Ballspiele, fig.)* own goal

Eigentum das; ~**s a)** property; *(einschließlich Geld usw.)* assets *pl.*; **geistiges** ~: [one's own] intellectual creation; **b)** *(Recht des Eigentümers)* ownership (**an** + *Dat.* of)

Eigentümer ['aign̩ty:mɐ] **der;** ~**s,** ~: owner; *(Hotel~, Geschäfts~)* proprietor; owner

Eigentümerin die owner; *(Hotel~, Geschäfts~)* proprietress; proprietor; owner

eigentümlich ['aign̩ty:mlɪç] **1.** *Adj.* **a)** *(typisch)* peculiar; characteristic; **b)** *(eigenartig)* peculiar; strange; odd. **2.** *adv.* peculiarly; strangely; oddly

Eigentümlichkeit die; ~, ~**en a)** *o. Pl. (Eigenartigkeit)* peculiarity; strangeness; **b)** *(typischer Zug)* peculiarity

Eigentums-: ~**delikt das** offence against property; ~**wohnung die** owner-occupied flat *(Brit.)*; condominium *or* co-op apartment *(Amer.)*; **eine** ~**wohnung kaufen** buy a flat *(Brit.) or (Amer.)* an apartment

eigen-, Eigen-: ~**verantwortlich 1.** *Adj.* responsible; **2.** *adv.* ~**verantwortlich handeln** act on one's own authority; **etw.** ~**verantwortlich bestimmen/entscheiden** decide sth. on one's own responsibility; ~**wert der** intrinsic value; ~**willig** *Adj.* **a)** self-willed ‹*person*›; individual ‹*style, idea*›; **b)** *(~sinnig)* obstinate; stubborn; ~**willigkeit die;** ~, ~**en a)** *o. Pl.* individualism; independence of mind; **b)** *(Handlung)* display of self-will

eignen *refl. V.* be suitable; **sich als** *od.* **zum Lehrer** ~: be suitable as a teacher; **das Buch eignet sich gut als Geschenk** this book makes a good present; **für solche Arbeiten eignet er sich besonders** he is particularly well suited for that kind of work; *s. auch* **geeignet**

Eigner der; ~**s,** ~: owner

Eignung die; ~: suitability; aptitude; **seine** ~ **zum Fliegen** his aptitude for flying

Eignungs-: ~**prüfung die,** ~**test der** aptitude test

Ei·klar, das; ~**s,** ~ *(österr.) s.* **Eiweiß a**

Ei·land das; ~**[e]s,** ~**e** *(veralt., dichter.)* isle *(poet.)*

Eil- [ail-]: ~**bote der** special messenger; **„durch** *od.* **per** ~**boten"** *(veralt.)* 'express'; ~**brief der** express letter

Eile ['ailə] **die;** ~: hurry; **ich habe keine** *od.* **bin nicht in** ~: I'm not in a *or* any hurry; **die Sache hat keine** ~: there's no hurry; it's not urgent; **in aller** ~: in great haste; **jmdn. zur** ~ **antreiben** hurry sb. up

Ei·leiter der *(Anat.)* Fallopian tube

eilen 1. *itr. V.* **a)** *mit sein* hurry; hasten; *(besonders schnell)* rush; **nach Hause** ~: hurry/rush home; **jmdm. zu Hilfe** ~: rush to sb.'s aid; **b)** *(dringend sein)* ‹*matter*› be urgent; **„eilt!"** 'urgent'; **„eilt sehr!"** 'immediate'. **2.** *refl. V.* hurry; make haste

eilends *Adv. (geh.)* hastily

Eil·gut das fast freight; express goods *pl.*

eilig 1. *Adj.* **a)** *(schnell)* hurried; **mit** ~**en Schritten** hurriedly; **es** ~ **haben** be in a hurry; **b)** *(dringend)* urgent ‹*news*›; **es [sehr]** ~ **mit etw. haben** be in a [great] hurry about sth.; **nichts Eiligeres zu tun haben, als ...** *(iron)* have nothing better to do than.... **2.** *adv.* hurriedly

Eil-: ~**schritt der: im** ~**schritt laufen** walk with short, quick steps; ~**sendung die** express consignment; ~**tempo das** *(ugs.)* **im** ~**tempo** in a rush; ~**zug der** semi-fast train; stopping train *(Brit.)*

Eimer ['aimɐ] **der;** ~**s,** ~ bucket; *(Abfall~)* bin; **ein** ~ **[voll] Wasser**

a bucket of water; **es gießt wie aus ~n** *(ugs.)* it's raining cats and dogs *(coll.)*; it's coming down in buckets *(coll.)*; **im ~ sein** *(salopp)* be up the spout *(sl.)*

eimer·weise *Adv.* by the bucketful; in bucketfuls

[1]**ein** [ain] **1.** *Kardinalz.* **ich will dir noch ~[e]s sagen** there's one more thing I'd like to tell you; **~er von beiden** one of the two; one or the other; **~er für alle, alle für ~en** one for all and all for one; **~ für allemal** once and for all; **~ und derselbe** one and the same; **~er Meinung sein** be of the same opinion. **2.** *unbest. Art.* a/an; **~ Kleid/Apfel** a dress/an apple; **~ bißchen** *od.* **wenig** a little [bit]; **~ anderer** somebody else; **~ jeder** *(geh.)* each and every one; **~e Kälte ist das hier!** it's freezing here! **3.** *Indefinitpron.: s.* **irgendein a**; *s. auch* **einer**

[2]**ein** *(elliptisch)* **~ – aus** *(an Schaltern)* on – off; **~ und aus gehen** go in and out; **bei jmdm. ~ und aus gehen** be a regular visitor at sb.'s house; **ich wußte nicht ~ noch aus** I didn't know where to turn *or* what to do

Einakter ['ainlaktɐ] **der; ~s, ~:** one-act play

einander [ai'nandɐ] *reziprokes Pron.; Dat u. Akk. (geh.)* each other; one another

ein|arbeiten *tr. V.* **a)** *(ausbilden)* train ‹*employee*›; **sich in etw.** *(Akk.)* **~:** become familiar *or* familiarize oneself with sth.; **b)** *(einfügen)* incorporate ‹*quotation etc.*› **(in +** *Akk.* into)

Einarbeitung die; ~, ~en training

ein·armig *Adj.* one-armed; **ein Einarmiger** a one-armed man

ein|äschern ['ainlɛʃɐn] *tr. V.* **a)** reduce ‹*building etc.*› to ashes; **b)** cremate ‹*corpse*›

Einäscherung die; ~, ~en a) *(das Niederbrennen)* burning down; **die ~ der Stadt** the destruction of the town by fire; **b)** *(Leichenverbrennung)* cremation

ein|atmen *tr., itr. V.* breathe in

ein·äugig *Adj.* one-eyed; single-lens ‹*camera*›

Ein·bahn·straße die one-way street

ein|balsamieren *tr. V.* embalm

Ein·band der; *Pl.* **-bände** binding; [book-] cover

ein·bändig *Adj.* one-volume

Ein·bau der; *o. Pl.* **a)** fitting; *(eines Motors)* installation; **b)** *s.* **einbauen b:** insertion; incorporation

ein|bauen *tr. V.* **a)** build in, fit ‹*cupboard, kitchen*›; install ‹*engine, motor*›; **b)** *(einfügen)* insert, incorporate ‹*chapter*›

Einbau-: ~küche die fitted kitchen; **~möbel** *Pl.* built-in furniture *sing.*; *(Regale)* fitted shelves; **~schrank der** built-in cupboard; *(für Kleidung)* built-in wardrobe

ein|behalten *unr. tr. V.* withhold

ein·beinig *Adj.* one-legged

ein|berufen *unr. tr. V.* **a)** summon; call; **den Bundestag ~:** summon the Bundestag; **b)** *(zur Wehrpflicht)* call up; conscript; draft *(Amer.)*

Ein·berufung die a) calling; **die ~ des Parlaments** the summoning of Parliament; **b)** *(zur Wehrpflicht)* call-up; conscription; draft *(Amer.)*

Einberufungs-: ~befehl der, ~bescheid der call-up papers *pl.*; draft card *(Amer.)*

ein|betonieren *tr. V.* concrete in

ein|betten *tr. V.* embed **(in +** *Akk.* in)

Einbett-: ~kabine die single-berth cabin; **~zimmer das** single room

ein|beulen *tr. V.* **etw. ~:** dent sth.; make a dent in sth.

ein|beziehen *unr. tr. V.* include **(in +** *Akk.* in)

Ein·beziehung die *o. Pl.* inclusion **(in +** *Akk.* in)

ein|biegen 1. *unr. itr. V.; mit sein* turn **(in +** *Akk.* into); **[nach] links/rechts ~:** turn left/right. **2.** *unr. tr. V.* bend

ein|bilden *refl. V.* **a) sich** *(Dat.)* **etw. ~:** imagine sth.; **eine eingebildete Krankheit** an imaginary illness; **was bildest du dir eigentlich ein?** *(ugs.)* what do you think you are doing?; **b)** *(ugs.)* **sich etwas ~:** be conceited **(auf +** *Akk.* about); **er bildet sich** *(Dat.)* **ganz schön viel ein** he thinks no end of himself *(coll.)*; **darauf brauchst du dir nichts einzubilden** there's no need to be stuck-up about it

Ein·bildung die; ~, ~en a) *o. Pl. (Phantasie)* imagination; **b)** *(falsche Vorstellung)* fantasy; **das ist alles nur ~:** it's all in the mind; **c)** *o. Pl. (Hochmut)* conceitedness

Einbildungs-: ~kraft die, ~vermögen das [powers *pl.* of] imagination; imaginative powers *pl.*

ein|binden *unr. tr. V.* **a)** bind ‹*book*›; **etw. neu ~:** rebind sth.; **b)** *(fig.: integrieren)* link **(in +** *Akk.* into); **in ein System eingebunden bleiben** remain part of a system

ein|blenden *(Rundf., Ferns., Film)* **1.** *tr. V.* insert; **eine Nachricht in eine Sendung ~:** interrupt a programme with a news flash; **Musik nachträglich ~:** dub in music. **2.** *refl. V.* **sich in ein Fußballspiel ~:** go over to a football match

Ein·blendung die *(Rundf., Ferns., Film)* insertion

ein|bleuen *tr. V.* **jmdm. etw. ~:** drum *or* hammer sth. into sb.

Ein·blick der a) view **(in +** *Akk.* into); **~ in etw.** *(Akk.)* **haben** be able to see into sth.; **b)** *s.* **Einsicht b; c)** *(fig.: Kenntnis)* insight **(in +** *Akk.* into)

ein|brechen *unr. itr. V.* **a)** *mit haben od. sein* break in; **in eine Bank ~:** break into a bank; **bei jmdm. ~:** burgle sb.; **b)** *mit sein (einstürzen)* ‹*roof, ceiling*› fall in, cave in; **c)** *mit sein (durchbrechen)* fall through; **d)** *mit sein (eindringen)* **in ein Land ~:** invade a country; **e)** *mit sein (geh.: beginnen)* ‹*night, darkness*› fall; ‹*winter*› set in

Einbrecher der; ~s, ~: burglar

ein|bringen 1. *unr. tr. V.* **a)** bring *or* gather in ‹*harvest*›; **b)** *(verschaffen)* yield ‹*profit*› bring in ‹*interest, money*›; bring ‹*fame, honour*›; **das bringt nichts ein** it isn't worth it; **c)** *(Parl.: vorlegen)* introduce ‹*bill*›; **d)** *(in eine Gemeinschaft, Gesellschaft usw.)* invest ‹*capital, money*›; **etw. in eine Ehe ~:** bring sth. into a marriage. **2.** *unr. refl. V.* **sich in eine Beziehung ~:** make one's own contribution to a relationship

ein|brocken *tr. V. (ugs.)* **sich/jmdm. etwas [Schönes] ~, sich/jmdm. eine schöne Suppe ~:** land oneself/sb. in the soup *or* in it *(coll.)*; **das hast du dir selbst eingebrockt** you've only yourself to thank for that *(coll.)*

Ein·bruch der a) burglary; break-in **(in +** *Akk.* at); **b)** *(das Einstürzen)* collapse; **ein ~ der Börsenkurse** *(fig.)* a slump in stock-market prices; **c)** *s.* **einbrechen d:** invasion **(in +** *Akk.* of); **d)** *(Beginn)* **vor ~ der Dunkelheit** before it gets dark; **der ~ des Winters** the onset of winter; **bei ~ der Nacht** at nightfall; when night closes/closed in

einbruch[s]·sicher *Adj.* burglar-proof

ein|buchten *tr. V. (salopp)* **jmdn. ~:** lock sb. up *(coll.)*; put sb. away *(coll.)*

Einbuchtung die; ~, ~en a) *(Bucht)* bay; inlet; **b)** *(Delle)* dent

einbürgern 1. *tr. V.* naturalize

⟨*person, plant, animal*⟩; introduce ⟨*custom, practice*⟩. **2.** *refl. V.* ⟨*custom, practice*⟩ become established; ⟨*person, plant, animal*⟩ become naturalized

Einbürgerung die; ~, ~en naturalization

Ein·buße die loss (**an** + *Dat.* of)

ein|büßen 1. *tr. V.* lose; *(durch eigene Schuld)* forfeit. **2.** *itr. V.* **sie büßte an Ansehen ein** her reputation suffered

ein|checken *tr., itr. V. (Flugw.)* check in

ein|cremen *tr. V.* put cream on ⟨*hands, back*⟩; **sich ~:** put cream on

ein|dämmen *tr. V. (fig.)* check; stem

Eindämmung die; ~, ~en *s.* **eindämmen:** checking, stemming

ein|decken 1. *refl. V.* stock up. **2.** *tr. V. (ugs.: überhäufen)* swamp

ein|dellen *tr. V. (ugs.)* dent [in]

eindeutig ['aindɔytɪç] **1.** *Adj.* **a)** *(klar)* clear; clear, definite ⟨*proof*⟩; **b)** *(nicht mehrdeutig)* unambiguous. **2.** *adv. s. Adj.* clearly; unambiguously

Eindeutigkeit die; ~: *s.* **eindeutig:** clarity; unambiguity

ein|deutschen *tr. V.* Germanize

Eindeutschung die; ~, ~en a) *o. Pl.* Germanization; **b)** *(Wort)* Germanized word

ein·dimensional *Adj.* one-dimensional

ein|dösen *itr. V. (ugs.) mit sein* doze off

ein|drängen *itr. V.; mit sein* **auf jmdn. ~:** crowd around sb.; **Eindrücke/Erinnerungen drängten auf ihn ein** *(fig.)* impressions/memories crowded in [up]on him

ein|drehen *tr. V.* **a)** screw in ⟨*light bulb*⟩ (**in** + *Akk.* into); **b) sich** *(Dat.)* **die Haare ~:** put one's hair in curlers *or* rollers

ein|dringen *unr. itr. V.; mit sein* **a) in etw.** *(Akk.)* **~:** penetrate into sth.; ⟨*vermin*⟩ get into sth.; ⟨*bullet*⟩ pierce sth.; *(allmählich)* ⟨*water, sand, etc.*⟩ seep into sth.; **b)** *(einbrechen)* **in ein Gebäude ~:** force an entry *or* one's way into a building; **Feinde sind in das Land eingedrungen** *(geh.)* enemies invaded the country; **c) ~ auf** (+ *Akk.*) set upon, attack ⟨*person*⟩; **mit Fragen auf jmdn. ~:** besiege *or* ply sb. with questions

ein·dringlich 1. *Adj.* urgent ⟨*warning, entreaty*⟩; impressive ⟨*voice*⟩; forceful, powerful ⟨*speech, words*⟩. **2.** *adv.* ⟨*urge*⟩ strongly; ⟨*talk*⟩ insistently

Ein·dringlichkeit die; ~ *s.* **eindringlich:** urgency; impressiveness; forcefulness

Eindringling ['aindrɪŋlɪŋ] **der; ~s, ~e** intruder

Ein·druck der; ~[e]s, Eindrücke *(Druckstelle, fig.)* impression; **~ auf jmdn. machen** make an impression on sb.; **er tat es nur, um [bei ihr] ~ zu schinden** *(ugs.)* he only did it to impress [her]

ein|drücken *tr. V.* **a)** smash in ⟨*mudguard, bumper*⟩; stave in ⟨*side of ship*⟩; smash ⟨*pier, support*⟩; break ⟨*window*⟩; crush ⟨*ribs*⟩; flatten ⟨*nose*⟩; **b)** *(hineindrücken)* **etw. [in etw.** *(Akk.)*] **~:** press *or* push sth. in[to sth.]

eindrucks·voll 1. *Adj.* impressive; **2.** *adv.* impressively

eine *s.* **einer**

ein|ebnen *tr. V.* level *(fig.)* eliminate ⟨*difference*⟩

Einebnung die; ~, ~en levelling; *(fig.)* elimination

eineiig ['ainlaiɪç] *Adj.* identical ⟨*twins*⟩

ein·ein·halb *Bruchz.* one and a half; **~ Stunden** an hour and a half; one and a half hours; **~ Jahre** eighteen months

ein·ein·halb·fach *Vervielfältigungsz.* one and a half times; **die ~e Anzahl** one and a half times the number

ein|engen *tr. V.* **a) jmdn. ~:** restrict sb.'s movement[s]; **sich eingeengt fühlen** feel hemmed in *or* shut in; **b)** *(fig.: einschränken)* restrict; **jmdn. in seiner Freiheit ~:** restrict *or* curb sb.'s freedom

einer, eine, eines, eins *Indefinitpron. (man)* one; *(jemand)* someone; somebody; *(fragend, verneint)* anyone; anybody; **das mach mal einem verständlich** try explaining that to anybody; **eine/einer/ein[e]s der besten** one of the best [people/things]; **kaum einer** hardly anybody; **einer nach dem anderen** one after the other; one by one; **die einen ..., die anderen ...:** some ..., the others ...; **er trinkt ganz gerne einen** *(ugs.)* he likes [to have] a drink; **ein[e]s ist sicher** one thing is for sure

Einer der; ~s, ~ a) *(Math.)* unit; **b)** *(Sport)* single sculler; **im ~:** in the single sculls

einerlei ['ainɐlai] *Adj.; nicht attr. (unwichtig)* **~, ob/wo/wer** *usw.* no matter whether/where/who *etc.*; **es ist ~:** it makes no difference; **es ist ihm ~:** it is all the same *or* all one to him

Einerlei das; ~s monotony

einerseits ['ainɐ'zaits] *Adv.* on the one hand

Einer·stelle die *(Math.)* units place

eines *s.* **einer**

ein·fach 1. *Adj.* **a)** simple; simple, easy ⟨*task*⟩; plain, simple ⟨*food*⟩; **b)** *(nicht mehrfach)* single ⟨*knot, ticket, journey*⟩. **2.** *adv.* **a)** simply; **b)** *(nicht mehrfach)* **etw. ~ falten** fold sth. once; **zweimal ~ [nach Köln]** two singles [to Cologne]. **3.** *Partikel* simply; just

Einfachheit die; ~ simplicity; *(der Nahrung)* plainness; simplicity; **der ~ halber** for the sake of simplicity; for simplicity's sake

ein|fädeln 1. *tr. V.* **a)** thread ⟨*needle, film, tape*⟩ (**in** + *Akk.* into); thread up ⟨*sewing-machine*⟩; **einen [neuen] Faden ~:** [re]thread the needle; **b)** *(ugs.: geschickt einleiten)* engineer ⟨*scheme, plot*⟩; **das hat sie fein/schlau eingefädelt** she worked that nicely/craftily *(coll.)*. **2.** *refl. V. (Verkehrsw.)* filter in; **sich in den fließenden Verkehr ~:** filter into the flow of traffic

ein|fahren 1. *unr. itr. V.; mit sein* come in; ⟨*train*⟩ come *or* pull in; **in den Bahnhof ~:** come *or* pull into the station; **der Zug nach Hamburg ist soeben auf Gleis 5 eingefahren** the Hamburg train has just arrived at platform 5. **2.** *unr. tr. V.* **a)** bring in ⟨*harvest*⟩; **b)** *(beschädigen)* knock down ⟨*wall*⟩; smash in ⟨*mudguard*⟩; **c)** *(Kfz-W.)* run in ⟨*car*⟩; **d)** *(Technik)* retract ⟨*undercarriage, antenna, aerial, etc.*⟩; *s. auch* **eingefahren**

Ein·fahrt die a) *(Weg, Straße, Stelle zum Hineinfahren)* entrance; *(Autobahn~)* slip road; **b)** *(das Hineinfahren)* entry; **Vorsicht bei der ~ des Zuges!** stand clear [of the edge of the platform], the train is approaching

Ein·fall der a) *(Idee)* idea; **b)** *o. Pl. (Licht~)* incidence *(Optics)*; **c)** *(in ein Land usw.)* invasion (**in** + *Akk.* of)

ein|fallen *unr. itr. V.; mit sein* **a) jmdm. fällt etw. ein** sb. thinks of sth.; sth. occurs to sb.; **ihm fallen immer wieder neue Ausreden ein** he can always think of *or (coll.)* come up with new excuses; **was fällt dir denn ein!** what do you think you're doing?; how dare you?; **b)** *(in Erinnerung kommen)* **ihr Name fällt mir nicht ein** I cannot think of her name; **es wird dir schon [wieder] ~:** it will come [back] to you; **plötzlich fiel ihr ein, daß ...** suddenly she remembered that ...; **c)** *(von Licht)* come in; **d)** *(gewaltsam eindringen)* **in ein Land ~:** invade a country; **e)** *(einstimmen, mitreden usw.)* join in

einfalls-, Einfalls-: ~los 1. *Adj.* unimaginative; lacking in ideas; **2.** *adv.* unimaginatively; without imagination; **~losigkeit die; ~:** unimaginativeness; lack of ideas; **~reich 1.** *Adj.* imaginative; full of ideas; **2.** *adv.* imaginatively; with imagination; **~reichtum der;** *o. Pl.* imaginativeness; wealth of ideas

Einfalt ['ainfalt] **die; ~:** simpleness; simple-mindedness

einfältig ['ainfɛltɪç] *Adj.* **a)** *(arglos)* simple; naïve; artless; naïve ‹*remarks*›; **sei nicht so ~!** don't be so naïve!; **b)** *(beschränkt)* simple; simple-minded

Einfältigkeit die; ~ a) *(Arglosigkeit)* simplicity; naïvety; **b)** *(Beschränktheit)* simpleness; simple-mindedness

Einfalts·pinsel der *(ugs. abwertend)* nincompoop

Ein·familien·haus das [detached] house

ein|fangen 1. *unr. tr. V.* **a)** catch, capture ‹*fugitive, animal*›; **b)** *(fig. geh.)* capture ‹*atmosphere, aura, etc.*›. **2.** *unr. refl. V. (ugs.: bekommen)* **sich** *(Dat.)* **eine Erkältung** *usw.* **~:** catch *or* get a cold *etc.*

ein|färben *tr. V.* dye

ein·farbig *Adj.* single-colour; of one colour *postpos.*; *(ohne Muster)* plain

ein|fassen *tr. V.* border, hem, edge ‹*material, dress, tablecloth*›; set ‹*gem*›; edge ‹*lawn, flower-bed, grave*›; curb ‹*source, spring*›

Ein·fassung die *s.* **einfassen:** border; hem; edging; setting; *(von Brunnen, Quelle)* enclosure

ein|fetten *tr. V.* grease; dubbin ‹*leather*›; **sich** *(Dat.)* **die Haut/Hände ~:** rub cream into one's skin/hands

ein|finden *unr. refl. V.* arrive; ‹*crowd*› gather

ein|flechten *unr. tr. V.* **Episoden in einen Roman ~:** weave episodes into a novel; **wenn ich das kurz ~ darf** if I could turn to this for a moment

ein|fliegen 1. *unr. tr. V.* fly in ‹*supplies, troops*›. **2.** *unr. itr. V.; mit sein* fly in; **in etw. ~:** fly into sth.

ein|fließen *unr. itr. V.; mit sein* flow in; **von Norden fließt Kaltluft nach Westeuropa ein** *(fig.)* a cold northerly airstream is moving into Western Europe; **etw. in ein Gespräch ~ lassen** *(fig.)* slip sth. into a conversation

ein|flößen *tr. V.* **a) jmdm. Tee/Medizin ~:** pour tea/medicine into sb.'s mouth; **b)** *(fig.)* **jmdm. Angst ~:** put fear into sb.; arouse fear in sb.

Einflug·schneise die *(Flugw.)* approach path

Ein·fluß der influence **(auf** + *Akk.* on); **unter jmds. ~** *(Dat.)* **stehen** be under sb.'s influence

Einfluß·bereich der sphere of influence

einfluß·reich *Adj.* influential

ein·förmig 1. *Adj.* monotonous. **2.** *adv.* monotonously

Ein·förmigkeit die; ~, ~en monotony

ein|fressen *unr. refl. V.* **sich in etw.** *(Akk.)* **~:** eat into sth.

ein|frieren 1. *unr. itr. V.; mit sein* ‹*water*› freeze, turn to ice; ‹*pond*› freeze over; ‹*pipes*› freeze up; ‹*ship*› be frozen in. **2.** *unr. tr. V.* **a)** deep-freeze ‹*food*›; **b)** *(fig.)* freeze

ein|fügen 1. *tr. V.* fit in; **etw. in etw.** *(Akk.)* **~:** fit sth. into sth.; **etw. in einen Text ~:** insert sth. into a text. **2.** *refl. V.* adapt; **sich in etw.** *(Akk.)* **~:** adapt oneself to sth.; **sich überall gut ~:** fit in well anywhere

ein|fühlen *refl. V.* **sich in jmdn. ~:** empathize with sb.; **ich kann mich gut in deine Lage ~:** I know exactly how you feel; **er kann sich gut in eine Rolle ~:** he is good at getting into a part

einfühlsam *Adj.* understanding; sensitive ‹*interpretation, performance*›

Einfühlungs·vermögen das ability to empathize

Ein·fuhr die; ~, ~en import

ein|führen *tr. V.* **a)** *(importieren)* import; **b)** *(als Neuerung)* introduce ‹*method, technology*›; **c)** *(ein-, unterweisen)* introduce **(in** + *Akk.* to); **jmdn. in sein Amt ~:** install sb. in office; **d)** *(hineinschieben)* introduce, insert ‹*catheter etc.*› **(in** + *Akk.* into)

Einfuhr-: ~sperre die, ~stopp der embargo *or* ban on imports

Ein·führung die a) introduction; **die ~ in sein Amt** his installation in office; **b)** *(Einarbeitung)* introduction; initiation; induction; **c)** *(das Hineinschieben)* introduction; insertion

Einführungs-: ~kurs[us] der *(Schulw.)* introductory course; **~preis der** *(Kaufmannsspr.)* introductory price

ein|füllen *tr. V.* **etw. in etwas** *(Akk.)* **~:** pour *or* put sth. into something

Ein·gabe die *(Gesuch)* petition; *(Beschwerde)* complaint

Ein·gang der a) entrance; **„kein ~"** 'no entry'; **in etw.** *(Akk.)* **~ finden** *(fig.)* become established in sth.; **b)** *o. Pl. (von Post, Geld)* receipt

ein·gängig *Adj.* catchy ‹*song, melody*›

eingangs *Adv.* at the beginning; at the start

Eingangs-: ~datum das *(Büro.)* date of receipt; **~halle die** entrance hall; *(eines Hotels, Theaters)* foyer; **~tür die** [entrance] door; *(von Wohnung, Haus usw.)* front door

ein|geben *unr. tr. V. (DV)* feed in; **etw. in den Computer ~:** feed sth. into the computer

ein·gebildet *Adj.* **a)** *(imaginär)* imaginary ‹*illness*›; **ein ~er Kranker** a malade imaginaire; **~e Schwangerschaft** false pregnancy; **b)** *(arrogant)* conceited

ein·geboren *Adj.* native ‹*population etc.*›

Eingeborene der/die; *adj. Dekl. (veralt.)* native

Eingebung die; ~, ~en inspiration; **einer ~ folgend** acting on a sudden impulse

eingedenk ['aingədɛŋk] *Adj.; nicht attr.* **einer Sache** *(Gen.)* **~ sein** *(geh.)* be mindful of sth.; **~ der Tatsache, daß ...:** bearing in mind that ...

ein·gefahren *Adj.* long-established; **sich auf** *od.* **in ~en Bahnen** *od.* **Gleisen bewegen** go on in the same old way

ein·gefallen *Adj.* gaunt ‹*face*›; sunken, hollow ‹*cheeks*›

eingefleischt ['aingəflaiʃt] *Adj., nicht präd.* confirmed ‹*bachelor*›; inveterate ‹*smoker*›

ein|gehen 1. *unr. itr. V.; mit sein* **a)** *(eintreffen)* arrive; be received; **b)** *(fig.)* **in die Geschichte ~:** go down in history; **in die Weltliteratur ~:** find one's/its place in world literature; **c)** *(schrumpfen)* shrink; **d) auf eine Frage/ein Problem ~/nicht ~:** go into *or* deal with/ignore a question/problem; **auf jmdn. ~:** be responsive to sb.; **auf jmdn. nicht ~:** ignore sb.'s wishes; **auf ein Angebot ~/nicht ~:** accept/reject an offer; **e)** *(sterben)* die; **f)** *(bankrott gehen)* close down. **2.** *unr. tr. V.* enter into ‹*contract, matrimony*›; take ‹*risk*›; accept ‹*obligation*›; **darauf gehe ich jede Wette ein** *(ugs.)* I'll bet you anything on that *(coll.)*

eingehend 1. *Adj.* detailed. **2.** *adv.* in detail

ein·gekeilt *Adj. (von beiden Seiten)* wedged in (**in, zwischen** + *Dat.* between); *(von allen Seiten)* hemmed in (**in** among)

Ein·gemachte das; ~**n** preserved fruit/vegetables; *(fig.: Substanz)* **ans ~ gehen** *(ugs.)* draw on one's reserves

ein|gemeinden *tr. V.* incorporate ⟨*village*⟩ (**in** + *Akk.*, **nach** into)

ein·genommen *(Adj.)* **von sich ~ sein** be conceited; **von etw. ~ sein** be conceited about sth.

ein·geschnappt *Adj. (ugs.: beleidigt)* huffy

ein·geschossig *Adj.* single-storey; one-storey

ein·geschränkt *Adj.* reduced; **~es Haltverbot** prohibition of stopping except for certain purposes

ein·geschrieben *Adj.* registered ⟨*letter, member*⟩; enrolled ⟨*student*⟩

ein·geschworen *Adj.* dedicated (**auf** + *Akk.* to)

ein·gespannt *Adj.* **stark ~:** very busy

ein·gespielt *Adj.* in practice; **aufeinander ~:** playing well together

Ein·geständnis das confession; admission

ein|gestehen *unr. tr. V.* admit, confess ⟨*guilt*⟩; admit, confess to ⟨*mistake, theft*⟩; **[sich] ~, daß ...:** admit [to oneself] that ...

ein·gestellt *Adj.* **fortschrittlich ~:** progressively minded; **wie ist er [politisch] ~?** what are his [political] views?

Eingeweide ['aingəvaidə] **das; ~s, ~;** *meist Pl.* entrails *pl.*; innards *pl.*

ein|gewöhnen 1. *refl. V.* get used *or* accustomed to one's new surroundings; accustom oneself to one's new surroundings; **sich an seinem neuen Arbeitsplatz/in eine neue Tätigkeit ~:** settle in at one's new place of work/get used to a new job. **2.** *tr. V.* **jmdn. in etw.** *(Akk.)* **~:** get sb. used *or* accustomed to sth.

Ein·gewöhnung die; *o. Pl.* settling in *no art.*; **die ~ in seiner neuen Umgebung/an seinem neuen Arbeitsplatz fiel ihm schwer** he found it difficult to get used to his new surroundings/job

ein|gießen *unr. tr. (auch itr.) V.* pour in; **etw. in etw.** *(Akk.)* **~:** pour sth. into sth.; **den Kaffee ~:** pour [out] the coffee

ein|gipsen *tr. V.* **a)** fix ⟨*nail, hook, etc.*⟩ in with plaster; **b)** put *or* set ⟨*arm, leg, etc.*⟩ in plaster

eingleisig ['ainglaizıç] *Adj.* single-track ⟨*railway line*⟩

ein|gliedern 1. *tr. V.* integrate (**in** + *Akk.* into); incorporate ⟨*village, company*⟩ (**in** + *Akk.* into); *(einordnen)* include (**in** + *Akk.* in). **2.** *refl. V.* **sich in etw.** *(Akk.)* **~:** fit into sth.

Ein·gliederung die *s.* **eingliedern 1:** integration; incorporation; inclusion

ein|graben *unr. tr. V.* bury (**in** + *Akk.* in); sink ⟨*pile, pipe*⟩ (**in** + *Akk.* into)

ein|gravieren *tr. V.* engrave (**in** + *Akk.* on)

ein|greifen *unr. itr. V.* intervene (**in** + *Akk.* in)

ein|grenzen *tr. V.* **a)** enclose; **b)** *(fig.: beschränken)* limit; restrict (**auf** + *Akk.* to)

Ein·griff der a) intervention (**in** + *Akk.* in); **ein ~ in jmds. Rechte** an infringement of sb.'s rights; **b)** *(Med.)* operation

ein|hacken *itr. V.* **auf jmdn./etw. ~:** peck at sb./sth.

ein|haken 1. *tr. V.* **a)** *(mit Haken befestigen)* fasten; **b)** *reziprok* **sich ~:** link arms; **sie gingen eingehakt** they walked arm in arm. **2.** *refl. V.* link arms (**bei** with). **3.** *itr. V. (fig. ugs.)* butt in

Ein·halt der: jmdm./einer Sache ~ gebieten *od.* **tun** *(geh.)* stop *or* halt sb./sth.

ein|halten 1. *unr. tr. V.* keep ⟨*appointment*⟩; meet ⟨*deadline, commitments*⟩; keep to ⟨*diet, speed-limit, agreement*⟩; observe ⟨*regulation*⟩; obey ⟨*laws*⟩. **2.** *unr. itr. V. (geh.)* stop

Ein·haltung die; *o. Pl. (einer Verabredung)* keeping; *(einer Vorschrift)* observance; **die ~ einer Frist** meeting a deadline

ein|hämmern 1. *itr. V.* **auf etw.** *(Akk.)* **~:** hammer on sth. **2.** *tr. V.* **jmdm. etw. ~:** hammer *or* drum sth. into sb. *or* sb.'s head

ein|handeln *refl. V.* **sich** *(Dat.)* **etw. ~** *(fig. ugs.)* let oneself in for sth. *(coll.)*

einhändig ['ainhɛndıç] **1.** *Adj.* one-handed. **2.** *adv.* with [only] one hand

ein|hängen 1. *tr. V.* hang ⟨*door*⟩; fit ⟨*window*⟩; put down ⟨*receiver*⟩. **2.** *itr. V.* hang up. **3.** *refl. V.* **sich bei jmdm. ~:** take sb.'s arm

ein|hauen 1. *unr. tr. V. s.* **einschlagen 1a, b. 2.** *unr. itr. V. s.* **einschlagen 2b**

ein|heften *tr. V.* file

ein·heimisch *Adj.* native; indigenous ⟨*population, plant*⟩

Einheimische der/die; *adj. Dekl.* local

ein|heimsen ['ainhaimzn̩] *tr. V. (ugs.)* rake in *(coll.)* ⟨*profits*⟩

ein|heiraten *itr. V.* **in eine Familie ~:** marry into a family

Einheit die; ~, ~en a) unity; **b)** *(Maß~, Milit.)* unit

einheitlich 1. *Adj.* **a)** *(in sich geschlossen)* unified; integrated; **b)** *(unterschiedslos)* uniform ⟨*dress*⟩; standardized ⟨*education*⟩; standard ⟨*procedure, practice*⟩. **2.** *adv.* **~ gekleidet sein** be dressed the same; **~ gestaltet sein** be designed along the same lines

Einheits-: ~format das standard size; **~front die** united front; **~gewerkschaft die** general trade union; **~staat der** centralized state

ein|heizen 1. *tr. V.* put on ⟨*stove, boiler*⟩; heat ⟨*room*⟩. **2.** *itr. V. (ugs.: bedrängen)* **jmdm. ~:** give sb. a kick up the backside *(coll.)*

einhellig ['ainhɛlıç] **1.** *Adj.* unanimous. **2.** *adv.* unanimously

Einhelligkeit die; ~: unanimity

ein·her|gehen *unr. itr. V.; mit sein (fig.: begleitet sein)* **mit etw. ~:** be accompanied by sth.

ein|holen 1. *tr. V.* **a)** *(erreichen)* **jmdn./ein Fahrzeug ~:** catch up with sb./a vehicle; **b)** *(ausgleichen)* make up ⟨*arrears, time*⟩; **c)** *(einziehen)* haul in, pull in ⟨*nets*⟩; lower ⟨*flag*⟩; **d)** *(ugs.: einkaufen)* buy, get ⟨*groceries*⟩; **e)** *(erbitten)* ask for, seek ⟨*reference, advice*⟩; make ⟨*enquiries*⟩. **2.** *itr. V. (ugs.)* **~ gehen** go shopping

Ein·horn das unicorn

ein|hüllen *tr. V.* **sich/jmdn. in etw.** *(Akk.)* **~:** wrap oneself/sb. up in sth.

ein·hundert *Kardinalz.* a *or* one hundred; *s. auch* **hundert**

einig ['ainıç] *Adj.* **sich** *(Dat.)* **~ sein** be agreed *or* in agreement; **sich** *(Dat.)* **~ werden** reach agreement; **mit jmdm. über etw.** *(Akk.)* **~ sein** be in agreement *or* agree with sb. about *or* on sth.

einig... ['ainıg...] *Indefinitpron. u. unbest. Zahlwort* some ⟨*effort, hope, courage*⟩; **in ~er Entfernung** some distance away; **~e wenige** a few; **~e hundert** several hundred; **~er Ärger** *(viel Ärger)* quite a bit *or* quite a lot of trouble

einigen 1. *tr. V.* unite. **2.** *refl. V.* come to *or* reach an agreement (**mit** with, **über** + *Akk.* about); **sich auf jmdn./etw. ~:** agree on sb./sth.

einigermaßen *Adv.* rather; somewhat; ~ **zufrieden** fairly *or* reasonably satisfied; **Wie geht's dir? – Einigermaßen** How are you? – Not too bad

Einigkeit die; ~ **a)** unity; **b)** *(Übereinstimmung)* agreement

Einigung die; ~, ~**en a)** *(Übereinkunft)* agreement; **b)** *(Vereinigung)* unification

Einigungs·vertrag der *(Politik)* unification treaty

ein|impfen *tr. V. (ugs.)* **jmdm. etw.** ~: drum sth. into sb.

ein|jagen *tr. V.* **jmdm. Angst/einen Schrecken** ~: give sb. a fright

ein·jährig *Adj.* **a)** *(ein Jahr alt)* one-year-old *attrib.*; one year old *pred.*; *(ein Jahr dauernd)* **one-year** *attrib.*; **eine** ~**e Abwesenheit** a year's absence; **b)** *(Bot.)* annual

Einjährige der/die; *adj. Dekl.* one-year-old

ein|kalkulieren *tr. V.* take into account

ein|kassieren *tr. V.* **a)** collect; **b)** *(ugs.: entwenden)* pinch *(sl.)*; nick *(Brit. sl.)*; **c)** *(salopp: festnehmen)* pinch *(sl.)*; nab

Ein·kauf der a) buying; *(für eine Firma)* buying; purchasing; **[einige] Einkäufe machen** do some shopping; **b)** *(eingekaufte Ware)* purchase; **ein guter/schlechter** ~: a good/bad buy; **c)** *o. Pl. (Kaufmannsspr.)* buying *or* purchasing department

ein|kaufen 1. *itr. V.* shop; ~ **gehen** go shopping; **beim Bäcker/im Supermarkt** ~: shop at the baker's/the supermarket. **2.** *tr. V.* buy; purchase; buy in ‹*stores, provisions*›

Ein·käufer der, Ein·käuferin die *(Berufsbez.)* buyer; purchaser

Einkaufs-: ~**abteilung die** *s.* Einkauf c; ~**bummel der: einen** ~**bummel machen** go on a shopping expedition; ~**korb der** shopping basket; *(im Geschäft)* [wire-]basket; ~**netz das** string bag; ~**tasche die** shopping bag; ~**wagen der** [shopping] trolley *(Brit.) or (Amer.)* cart; ~**zentrum das** shopping centre; ~**zettel der** shopping-list

Einkehr ['ainke:ɐ̯] **die;** ~ *(geh.: Selbstbesinnung)* ~ **halten** take stock of oneself and one's attitudes

ein|kehren *itr. V.; mit sein* stop; **in einem Wirtshaus** ~: stop at an inn

ein|kellern *tr. V.* store in the/a cellar

ein|kerben *tr. V.* cut *or* carve a notch/notches in; notch

Einkerbung die; ~, ~**en** notch

ein|klagen *tr. V.* sue for ‹*damages, compensation, etc.*›; sue for the recovery of ‹*debts*›

ein|klammern *tr. V.* **etw.** ~: put sth. in brackets; bracket sth.

Ein·klang der harmony; **im** ~ **mit jmdm. sein** be in accord *or* agreement with sb.; **im** *od.* **in** ~ **mit etw. stehen** accord with sth.; **zwei Dinge in** ~ **bringen** harmonize two things

ein·klassig *Adj. (Schulw.)* one-room ‹*school*›

ein|kleben *tr. V.* stick in; **etw. in etw.** *(Akk.)* ~: stick sth. into sth.

ein|kleiden *tr. V.* **a) sich/jmdn.** ~: clothe oneself/sb.; **sich/jmdn. neu** ~: fit oneself/sb. out with a new set of clothes; **b)** *(Milit.)* kit out ‹*soldier*›

ein|klemmen *tr. V.* **a)** *(quetschen)* catch; **jmdm./sich die Hand [in etw.** *(Dat.)***]** ~: catch *or* trap sb.'s/one's hand [in sth.]; **b)** *(fest einfügen)* clamp

ein|klopfen *tr. V.* knock in ‹*nail*›

ein|knicken 1. *tr. V.* bend; *(brechen)* snap. **2.** *itr. V.; mit sein* bend; *(brechen)* snap; **sie knickte beim Gehen ein** she went over on her ankle while walking along

ein|kochen 1. *tr. V.* preserve ‹*fruit, vegetables*›. **2.** *itr. V.* thicken

Einkommen das; ~**s,** ~: income

Einkommen·steuer die income tax

ein|kreisen *tr. V.* **a)** *(markieren)* **etw.** ~: put a circle round sth.; **b)** *(umzingeln)* surround; **c)** *(fig.: eingrenzen)* circumscribe ‹*problem*›

ein|kriegen *(ugs.)* **1.** *tr. V. s.* **einholen 1 a. 2.** *refl. V.* control oneself; **sie konnte sich vor Lachen nicht** ~: she couldn't stop laughing

Einkünfte ['ainkʏnftə] *Pl.* income *sing.*

[1]**ein|laden** *unr. tr. V.* load (**in** + *Akk.* into) ‹*goods*›

[2]**ein|laden 1.** *unr. tr. V.* **a)** invite; **jmdn. zum Essen** ~: invite sb. for a meal; *(im Restaurant)* invite sb. out for a meal; **sich einladen** *(scherzh.)* invite oneself; **jmdn. zu sich nach Hause** ~: invite sb. over; **b)** *(freihalten)* treat (**zu** to); **ich lade euch ein** this is on me. **2.** *unr. itr. V.* **a) die Feuerwehr lädt zu einem Tag der offenen Tür ein** the fire station is having an open day; **der Direktor des Goethe-Instituts lädt zu einem Empfang ...:** the Director of the Goethe-Institut requests the pleasure of your company at a reception ...; **b)** *(fig.)* **das Meer lädt zum Baden ein** the sea looks inviting; **das lädt zum Diebstahl geradezu ein** that's inviting theft

einladend 1. *Adj.* inviting; tempting, appetizing ‹*meal*›. **2.** *adv.* invitingly

Ein·ladung die invitation

Ein·lage die a) *(in einem Brief)* enclosure; **b)** *(Kochk.) vegetables, meat balls, dumplings, etc. added to a clear soup;* **eine Brühe mit** ~: a clear soup with meat balls/dumpling *etc.*; **c)** *(Schuh~)* arch-support; **d)** *(Darbietung)* **eine witzige** ~: a witty *or* humorous aside; **eine musikalische** ~: a musical interlude; **e)** *(Finanzw.) (Guthaben)* deposit; *(Beteiligung)* investment

ein|lagern 1. *tr. V.* store; lay in ‹*stores*›. **2.** *refl. V.* **sich [in etw.** *(Akk.)***]** ~: be deposited [in sth.]

Ein·lagerung die storage

Einlaß ['ainlas] **der; Einlasses, Einlässe** ['ainlɛsə] admission, admittance (**in** + *Akk.* to); **„~ ab 20 Uhr"** 'doors open 8 p.m.'

ein|lassen 1. *unr. tr. V.* **a)** admit; let in; **b)** *(einfüllen)* run ‹*water*›; **c)** *(einpassen)* **etw. in etw.** *(Akk.)* ~: set sth. into sth. **2.** *unr. refl. V.* **a)** *(meist abwertend)* **sich mit jmdm.** ~: get mixed up *or* involved with sb.; **b) sich auf etw.** *(Akk.)* ~: get involved in sth.

Ein·lauf der a) *(Med.)* enema; **jmdm. einen** ~ **machen** give sb. an enema; **b)** *o. Pl. (Sport)* **beim** ~ **in die Gerade/das Stadion** entering the straight/the stadium

ein|laufen 1. *unr. itr. V.; mit sein* **a)** *(Sport)* **ins Stadion** ~: run into *or* enter the stadium; **in die letzte Runde** ~: start the last lap; **b)** *(ankommen)* **das Schiff läuft ein** the ship is coming in; **in den Hafen** ~: come into *or* enter port; **c)** *(kleiner werden)* ‹*clothes*› shrink; **d)** *(hineinfließen)* run in; **e)** *(eingehen)* ‹*news, information*› come in. **2.** *unr. tr. V.* **a)** wear in ‹*shoes*›; **b)** *s.* **einrennen 1 b. 3.** *unr. refl. V. (Sport)* warm up

ein|läuten *tr. V.* ring in ‹*Sunday, New Year*›

ein|leben *refl. V.* settle down; *(in einem Haus)* settle in

Einlege·arbeit die *(Kunsthandwerk)* inlaid work; *(Gegenstand)* piece of inlaid work

ein|legen *tr. V.* **a)** put in; **etw. in etw.** *(Akk.)* ~: put sth. in sth.; **den ersten Gang** ~: engage first gear; **b) jmdm./sich das Haar** ~: set sb.'s/one's hair; **c)** *(Kunsthand-*

werk) inlay; **eingelegte Muster** inlaid patterns; **d)** *(Kochk.)* pickle; **e)** *(fig.: einschieben)* **eine Rast ~:** stop for a rest; **einen Spurt ~:** put on a spurt; **eine Pause ~:** take a break; **f)** *(geltend machen)* lodge ⟨*protest, appeal*⟩; **Widerspruch ~:** protest; *s. auch* **Ehre a; Veto; Wort b**

Einlege·sohle die insole

ein|leiten *tr. V.* **a)** introduce; institute, start ⟨*search*⟩; open ⟨*negotiations, investigation*⟩; launch, open ⟨*campaign*⟩; induce ⟨*birth*⟩; **einige ~de Worte** a few introductory remarks; **b) etw. in etw.** *(Akk.)* **~:** lead sth. into sth.

Ein·leitung die: *s.* **einleiten a:** introduction; institution; opening; launching; induction

ein|lenken *itr. V.* give way; make concessions

ein|lesen 1. *unr. refl. V.* **sich in ein Buch ~:** get into a book. **2.** *unr. tr. V. (DV)* read in; **etw. in den Speicher ~:** read sth. into the memory

ein|leuchten *itr. V.* **jmdm. ~:** be clear to sb.; **es leuchtet ihr nicht ein, daß sie es machen soll** she doesn't see why she should do it

ein·leuchtend 1. *Adj.* plausible. **2.** *adv.* plausibly

ein|liefern *tr. V.* **a)** post *(Brit.)*, mail ⟨*letter, parcel*⟩; **b) jmdn. ins Krankenhaus/Gefängnis ~:** take sb. to hospital/jail

Ein·lieferung die a) *s.* **einliefern a:** posting *(Brit.)*; mailing; **b) die ~ eines Verurteilten [ins Gefängnis]** taking a convicted prisoner to jail

Einlieferungs·schein der *(Postw.)* certificate of posting

ein|lochen *tr. V. (salopp)* **jmdn. ~:** put sb. away *(coll.)*

ein|lösen *tr. V.* **a)** cash ⟨*cheque*⟩; cash [in] ⟨*token, voucher, bill of exchange*⟩; redeem ⟨*pawned article*⟩; **b)** *(geh.: erfüllen)* redeem ⟨*pledge*⟩; **sein Wort ~:** keep one's word

ein|machen *tr. V.* preserve ⟨*fruit, vegetables*⟩; *(in Gläser)* bottle

Einmach·glas das preserving jar

einmal 1. *Adv.* **a)** *(ein Mal)* once; **noch ~ so groß [wie]** twice as big [as]; **etw. noch ~ tun** do sth. again; **~ sagt er dies, ein andermal das** first he says one thing, then another; **~ ist keinmal** *(Spr.)* it won't matter just this once; **auf ~:** all at once; suddenly; *(zugleich)* at once; **b)** ['-'-] *(später)* some day; one day; *(früher)* once; **es war ~ ein König, der ...:** once upon a time there was a king who ... **2.** *Partikel* **a) daran ist nun ~ nichts mehr zu ändern** there's nothing more that can be done about it; **nicht ~:** not even; **wieder ~:** yet again; **b) alle ~ zuhören!** listen everybody!

Einmal·eins das; ~: [multiplication] tables *pl.*; **das kleine/große ~:** tables from 1 to 10/11 to 20; *(fig. Anfangsgründe)* fundamentals *pl.*

Einmal·hand·tuch das disposable towel

einmalig 1. *Adj.* **a)** unique ⟨*opportunity, chance*⟩; one-off, single ⟨*payment, purchase*⟩; **b)** *(hervorragend)* superb ⟨*film, book, play, etc.*⟩; *(ugs.)* fantastic *(coll.)* ⟨*girl, woman*⟩. **2.** *adv. (ugs.)* really fantastic *or* superb *(coll.)*

Ein·mann·betrieb der a) *(Firma)* one-man business; **b)** *(Arbeitsweise)* one-man operation

Ein·mark·stück das one-mark piece

Ein·marsch der a) entry; **der ~ ins Stadion** the march into the stadium; **b)** *(Besetzung)* invasion **(in +** *Akk.* of)

ein|marschieren *itr. V.; mit sein* **a)** march in; **in etw.** *(Akk.)* **~:** march into sth.; **b) in ein Land ~** *(Milit.)* march into *or* invade a country

ein|mauern *tr. V.* **a)** immure ⟨*prisoner, traitor*⟩; wall in ⟨*relic, treasure*⟩; **b)** *(ins Mauerwerk einfügen)* **etw. in die Wand** *usw.* **~:** set sth. into the wall *etc.*

Ein·meter·brett das one-metre board

ein|mieten *refl. V.* **sich in einer Villa/Pension ~:** rent a villa/a room in a boarding-house

ein|mischen *refl. V.* interfere **(in +** *Akk.* in)

Ein·mischung die interference **(in +** *Akk.* in)

einmonatig *Adj.; nicht präd.* **a)** *(einen Monat alt)* one-month-old *attrib.*; **b)** *(einen Monat dauernd)* one-month *attrib.*; *s. auch* **achtmonatig**

ein·monatlich 1. *Adj.* monthly; *s. auch* **achtmonatlich 1. 2.** *adv.* monthly; once a month

ein·motorig *Adj.* single-engined

ein|motten *tr. V. (fig.)* mothball

ein|mumme[l]n *tr. V. (ugs.)* wrap up; **sich ~:** wrap [oneself] up

ein|münden *itr. V.; auch mit sein* flow in; enter; **in etw. ~:** flow into *or* enter sth.

Ein·mündung die *(von Straßen)* junction; **die ~ der Straße in die Hauptstraße** the junction of the street and the main road

einmütig ['ainmy:tɪç] **1.** *Adj.* unanimous. **2.** *adv.* unanimously

Einmütigkeit die; ~: unanimity

ein|nähen *tr. V.* sew in; **etw. in etw.** *(Akk.)* **~:** sew sth. into sth.

Einnahme die; ~, ~en a) *meist Pl.* income; *(Staats~)* revenue; *(Kassen~)* takings *pl.*; **b)** *o. Pl. (von Arzneimitteln, einer Mahlzeit)* taking; **c)** *o. Pl. (einer Stadt, Burg)* capture; taking

Einnahme·quelle die source of income; *(des Staates)* source of revenue

ein|nehmen *unr. tr. V.* **a)** *(kassieren)* take; **b)** *(zu sich nehmen)* take ⟨*medicine, tablets, meal*⟩; **c)** *(besetzen)* capture, take ⟨*town, fortress*⟩; **d) seinen Platz ~:** take one's place; *(sich setzen)* take one's seat *or* place; **einen Standpunkt ~** *(fig.)* take up *or* adopt a position; **eine wichtige Stellung ~** *(fig.)* occupy an important place; **e)** *(ausfüllen)* take up ⟨*amount of room*⟩; **f)** *(beeinflussen)* **jmdn. für sich ~:** win sb. over; **gegen jmdn. eingenommen sein** be prejudiced against sb.; **von sich eingenommen sein** think a lot of oneself *(coll.)*

einnehmend *Adj.* winning ⟨*manner*⟩

ein|nisten *refl. V.* **a) sich bei jmdm. ~** *(fig. abwertend)* park oneself on sb. *(coll.)*; **b)** *(ein Nest bauen)* build a nest/their nests; nest

Ein·öde die barren *or* featureless waste; *(Einsamkeit)* isolation

ein|ölen *tr. V.* **a)** oil; **b) sich/jmdn. ~:** put *or* rub oil on oneself/sb.

ein|ordnen 1. *tr. V.* **a)** arrange; put in order; **b)** *(klassifizieren)* classify; categorize, classify ⟨*writer, thinker, artist*⟩. **2.** *refl. V.* **a)** *(Verkehrsw.)* get into the correct lane; **sich rechts/links ~:** get into the right-hand/left-hand lane; **„bitte ~"** 'get in lane'; **b) sich [in die Gemeinschaft] ~:** fit in[to the community]

ein|packen 1. *tr. V.* **a)** pack **(in +** *Akk.* in); *(einwickeln)* wrap [up]; **b)** *(ugs.: warm anziehen)* wrap up. **2.** *itr. V. (ugs.)* **er kann ~:** he's had it *(coll.)*; **pack ein!** pack it in! *(coll.)*; give it a rest! *(coll.)*

ein|parken *tr., itr. V.* park

ein|pauken *tr. V.* **etw. ~:** mug up *(Brit.) or (Amer.)* bone up on sth. *(coll.)*; **jmdm. etw. ~:** drum *or* hammer sth. into sb.

ein|pendeln *refl. V.* settle down

Ein·pfennig·stück das one-pfennig piece

ein|pferchen *tr. V.* **eingepfercht**

stehen/sein stand/be crammed *or* crushed together

ein|pflanzen *tr. V.* **a)** plant ⟨*flowers, shrubs, etc.*⟩; **b)** *(Med.)* implant; **jmdm. ein Organ ~:** implant an organ in[to] sb.

ein|planen *tr. V.* **etw. ~:** include sth. in one's plans; **das war nicht eingeplant** we/they *etc.* didn't plan on that

ein|pökeln *tr. V. (Kochk.)* salt

ein·polig *Adj. (Physik, Elektrot.)* single-pole

ein|prägen 1. *tr. V.* **a)** stamp **(in** + *Akk.* into, on); **b)** *(fig.)* **sich** *(Dat.)* **etw. ~:** memorize sth.; commit sth. to memory. **2.** *refl. V.* **das prägte sich ihm [für immer] ein** it made an [indelible] impression on him

einprägsam 1. *Adj.* easily remembered. **2.** *adv.* **er hat das sehr ~ dargelegt** he expounded it in a way that made it easy to remember

ein|programmieren *tr. V. (DV)* programme in

ein|prügeln 1. *itr. V.* **auf jmdn. ~:** beat sb. **2.** *tr. V.* **jmdm. etw. ~** *(fig.)* drub *or* beat sth. into sb.

ein|pudern *tr. V.* powder

ein|quartieren 1. *tr. V.* quarter, billet ⟨*troops*⟩; **die Opfer wurden vorläufig in Hotels einquartiert** the victims were given temporary accommodation in hotels. **2.** *refl. V.* **sich bei jmdm. ~** *(Milit.)* be quartered with *or* billeted on sb.

ein|rahmen *tr. V.* frame

ein|rammen *tr. V.* ram in

ein|rasten *itr. V.; mit sein* engage

ein|räuchern *tr. V.* envelope in smoke; fill ⟨*room*⟩ *with smoke*

ein|räumen *tr. V.* **a)** put away; **etw. in etw.** *(Akk.)* **~:** put sth. away in sth.; **Bücher wieder [ins Regal] ~:** put books back [on the shelf]; **b) er mußte seinen Schrank ~:** he had to put his things away in his cupboard; **das Zimmer wieder ~:** put everything *or* all the furniture back into the room; **c)** *(zugestehen)* admit; concede; **jmdm. ein Recht/einen Kredit ~:** give *or* grant sb. a right/loan

ein|rechnen *tr. V.* include, take account of ⟨*costs etc.*⟩

ein|reden 1. *tr. V.* **jmdm. etw. ~:** talk sb. into believing sth.; **sich** *(Dat.)* **~, daß ...:** persuade oneself that...; **das redest du dir bloß ein** you're just imagining it. **2.** *itr. V.* **auf jmdn. ~:** talk insistently to sb.

ein|regnen *refl. V. (unpers.)* **es hat sich eingeregnet** it's begun to rain steadily

ein|reiben *unr. tr. V.* **Salbe [in die Haut] ~:** rub ointment in[to one's skin]; **jmdm. den Rücken ~:** rub lotion/ointment *etc.* into sb.'s back; **sich** *(Dat.)* **das Gesicht mit etw. ~:** rub sth. into one's face

ein|reichen *tr. V.* **a)** submit ⟨*application*⟩; hand in, submit ⟨*piece of work, dissertation, thesis*⟩; lodge, make ⟨*complaint*⟩; tender ⟨*resignation*⟩; **b)** *(jur.)* file ⟨*suit, petition for divorce*⟩

ein|reihen 1. *refl. V.* **sich in etw.** *(Akk.)* **~:** join sth. **2.** *tr. V.* place **(in** + *Akk.* in)

Einreiher der; ~s, ~: single-breasted suit/jacket/coat

einreihig ['ainraiıç] *Adj.* single-breasted ⟨*suit, jacket, coat*⟩

Ein·reise die entry; **bei der ~ nach Frankreich** on entry into France

ein|reisen *itr. V.; mit sein* enter; **nach Schweden ~:** enter Sweden

Einreise-: ~verbot das: jmdm. ~verbot erteilen refuse sb. entry; **~visum das** entry visa

ein|reißen 1. *unr. tr. V.* **a)** tear; rip; **b)** *s.* **abreißen 1 b. 2.** *unr. itr. V.; mit sein* **a)** tear; rip; **b)** *(ugs.: zur Gewohnheit werden)* become a habit

ein|reiten 1. *unr. itr. V.; mit sein* ride in. **2.** *unr. tr. V.* break in ⟨*horse*⟩

ein|renken 1. *tr. V.* **a)** *(Med.)* set; reduce *(Med.)*; **jmdm. den Fuß/Arm [wieder] ~:** [re]set sb.'s foot/arm; **b)** *(ugs.: bereinigen)* **etw. ~:** sort *or* straighten sth. out. **2.** *refl. V.* **das renkt sich ein** that will sort *or* straighten itself out

ein|rennen 1. *unr. tr. V.* **a)** break down ⟨*door*⟩; *s. auch* **offen 1 a; b) jmdm. das Haus** *od.* **die Tür** *od.* **die Bude ~** *(ugs.)* pester sb. all the time. **2.** *unr. refl. V. (ugs.: sich verletzen)* **sich** *(Dat.)* **den Kopf an etw.** *(Dat.)* **~:** bash *or* bang one's head on *or* against sth.

ein|richten 1. *refl. V.* **a) sich gemütlich/schön ~:** furnish one's home comfortably/beautifully; **sich häuslich ~:** make oneself at home; **b)** *(auskommen)* **sich [mit seinem Gehalt] ~:** get by *or* make ends meet [on one's salary]; **c)** *(sich vorbereiten)* **sich auf jmdn./etw. ~:** prepare for sb./sth.. **2.** *tr. V.* **a)** furnish ⟨*flat, house*⟩; fit out ⟨*shop, restaurant*⟩; equip ⟨*laboratory*⟩; **b)** *(ermöglichen)* arrange; **das läßt sich ~:** that can be arranged; **c)** *(eröffnen)* open ⟨*branch, shop*⟩; set up ⟨*advisory centre*⟩; start, set up ⟨*business*⟩

Ein·richtung die a) *o. Pl. (das Einrichten)* furnishing; **b)** *(Mobiliar)* furnishings *pl.*; **c)** *(Geräte)* **~en** *(Geschäfts~)* fittings; *(Labor~)* equipment *sing.*; **sanitäre ~en** sanitation *sing.*; **d)** *(Institution, Gewohnheit)* institution

Einrichtungs·gegen·stand der piece of furniture

ein|ritzen *tr. V.* carve

ein|rollen *tr. V.* roll up ⟨*carpet etc.*⟩; **sich/jmdm. die Haare ~:** put one's/sb.'s hair in curlers *or* rollers; **sich ~:** ⟨*hedgehog, cat*⟩ curl up

ein|rosten *itr. V.; mit sein* go rusty; rust up

ein|rücken 1. *itr. V.; mit sein* *(Milit.: einmarschieren)* move in; **in ein Land ~:** march into a country. **2.** *tr. V.* indent ⟨*line, heading, etc.*⟩

ein|rühren *tr. V.* stir in

eins [ains] **1.** *Kardinalz.* one; **es ist ~:** it is one o'clock; **~ zu null für dich!** *(ugs.)* that's one up to you!; **die Nummer ~ sein** *(fig.)* be number one; **„~, zwei, drei!"** 'ready, steady, go'; *s. auch* [1]**acht. 2.** *Adj.; nicht attr.* **mir ist alles ~:** it's all the same *or* all one to me; **mit jmdm. über etw.** *(Akk.)* **~ sein/werden** be in/reach agreement with sb. about *or* on sth. **3.** *Indefinitpron.: s.* **einer**

Eins die; ~, ~en a) one; **wie eine ~ stehen** *(ugs.)* stand as straight as a ramrod; *s. auch* [1]**Acht a, e; b)** *(Schulnote)* one; A

[1]**ein|sacken** *tr. V.* **a)** *(in Säcke füllen)* **etw. ~:** put sth. into sacks; **b)** *(ugs.: einstecken)* grab; pocket ⟨*money*⟩

[2]**ein|sacken** *itr. V.; mit sein* sink in; ⟨*building, pavement*⟩ subside

einsam *Adj.* **a)** *(verlassen)* lonely ⟨*person, decision*⟩; **~ leben** live a lonely *or* solitary life; **b)** *(einzeln)* solitary ⟨*rock, tree, wanderer*⟩; **c)** *(abgelegen)* isolated; **~ liegen** be situated miles from anywhere; **d)** *(menschenleer)* empty; deserted

Einsamkeit die; ~, ~en a) *(Verlassenheit)* loneliness; **b)** *(Alleinsein)* solitude; **c)** *(Abgeschiedenheit)* isolation

ein|sammeln *tr. V.* **a)** *(auflesen)* pick up; gather up; **b)** *(sich aushändigen lassen)* collect in; collect ⟨*tickets*⟩

Ein·satz der a) *(eingesetztes Teil) (in Tischdecke, Kopfkissen usw.)* inset; *(in Kochtopf, Nähkasten usw.)* compartment; **b)** *(eingesetzter Betrag)* stake; **den ~ erhöhen** raise the stakes *pl.*; **c)** *(das Einsetzen) (von Maschinen, Waffen usw.)* use; *(von Truppen)* deployment; **unter ~ seines Lebens** at the risk of his life; **zum ~ kommen** *od.* **gelangen** *(Papierdt.)* ‹*machine*› come into operation; ‹*police, troops*› be used; **jmdn./etw. zum ~ bringen** use sb./sth.; **d)** *(Engagement)* commitment; dedication; **e)** *(Milit.)* **im ~ sein/fallen** be in action *or* on active service/ die in action; **einen ~ fliegen** fly a mission; **f)** *(Musik)* **der ~ der Instrumente** the entry of the instruments

einsatz-, Einsatz-: ~befehl der order to go into action; **~bereit** *Adj.* **a)** ready for use; **b)** *(Milit.)* combat-ready *attrib.*; ready for action *postpos.*; **~gruppe die, ~kommando das** task force; **~leiter der** head of operations; **~plan der** plan of action; **~wagen der** *(der Polizei)* police car; *(der Feuerwehr)* fire-engine; *(Notarztwagen)* ambulance; *(der Straßenbahn)* relief; **~zentrale die** operations centre

ein|saugen *unr. (auch regelm.) tr. V.* suck in ‹*air, liquid*›

ein|schalten 1. *tr. V.* **a)** switch on; turn on; **einen anderen Sender ~:** switch to another station; **b)** *(fig.: beteiligen)* call in ‹*press, police, expert, etc.*›; **jmdn. in die Verhandlungen ~:** bring sb. into the negotiations. **2.** *refl. V.* **a)** switch [itself] on; come on; **b)** *(eingreifen)* intervene **(in** + *Akk.* in)

Einschalt·quote die *(Rundf.)* listening figures *pl.*; *(Ferns.)* viewing figures *pl.*

ein|schärfen *tr. V.* **jmdm. etw. ~:** impress sth. [up]on sb.

ein|schätzen *tr. V.* judge ‹*person*›; assess ‹*situation*›; *(schätzen)* estimate; **jmdn./eine Situation falsch ~:** misjudge sb./a situation

Ein·schätzung die *s.* **einschätzen:** judging; assessment; estimation; **nach meiner ~:** in my estimation *or* judgement

ein|schäumen *tr. V.* lather

ein|schenken *tr., itr. V.* **a)** pour [out]; **jmdm. etw. ~:** pour out sth. for sb.; **b)** *(füllen)* fill [up] ‹*glass, cup*›

ein|scheren *itr. V.; mit sein (Verkehrsw.)* **in** *od.* **auf eine Fahrspur ~:** get *or* move into a lane

ein|schicken *tr. V.* send in

ein|schieben *unr. tr. V.* **a)** *(hineinschieben)* push in; **b)** *(einfügen)* put in; insert; put on ‹*trains, buses*›; fit in ‹*client, patient*›; **etw. in etw.** *(Akk.)* **~:** put *or* insert sth. into sth.

ein|schießen 1. *unr. tr. V.* **a)** *(zerstören)* demolish ‹*wall, building*› by gunfire; **das Fenster [mit einem Ball] ~** *(fig.)* smash the window [with a ball]; **b)** *(treffsicher machen)* try out, test ‹*gun etc.*›; **c)** *(Sport)* kick in ‹*ball*›; **den Ball zum 1:1 ~:** shoot a goal to make it *or* the score 1-1. **2.** *unr. refl. V. (auch Sport)* find *or* get the range **(auf** + *Akk.* of)

ein|schiffen 1. *tr. V.* embark ‹*passengers*›; load ‹*cargo*›. **2.** *refl. V.* embark **(nach** for)

Einschiffung die; ~, ~en *s.* **einschiffen:** embarkation; loading

ein|schlafen *unr. itr. V.; mit sein* **a)** fall asleep; go to sleep; **ich kann nicht ~:** I can't get to sleep; **b)** *(verhüll.: sterben)* pass away *(euphem.)*; **c)** *(gefühllos werden)* ‹*arm, leg*› go to sleep; **d)** *(aufhören)* peter out

ein|schläfern *tr. V.* **a)** *(in Schlaf versetzen)* **jmdn. ~:** send sb. to sleep; **b)** *(betäuben)* **jmdn. ~:** put sb. to sleep; **c)** *(schmerzlos töten)* **ein Tier ~:** put an animal to sleep; **d)** *(beruhigen)* soothe, salve ‹*conscience*›; dull ‹*critical faculties*›

einschläfernd 1. *Adj.* soporific. **2.** *adv.* **~ wirken** have a soporific effect

Ein·schlag der a) *(Einschlagen)* **wir sahen den ~ des Blitzes/der Bomben** we saw the lightning strike/the bombs land; **b)** *(Stelle)* **wir sahen die Einschläge der Kugeln/der Bomben** we saw the bullet-holes/where the bombs had fallen *or* landed; **c)** *(Anteil)* element; **eine Familie mit südländischem ~:** a family with southern blood in it; **d)** *(Kfz-W.) (des Lenkrads)* turning; *(der Räder)* lock

ein|schlagen 1. *unr. tr. V.* **a)** *(hineinschlagen)* knock in; hammer in; **etw. in etw.** *(Akk.)* **~:** knock *or* hammer sth. into sth.; **b)** *(zertrümmern)* smash [in]; **c)** *(einwickeln)* wrap up ‹*present*›; cover ‹*book*›; **d)** *(wählen)* take ‹*route, direction*›; take up ‹*career*›; adopt ‹*policy*›; **einen Kurs ~** *(auch fig.)* follow a course; **einen anderen Kurs ~** *(auch fig.)* change *or* alter course; **e)** *(Kfz-W.)* turn ‹*[steering-]wheel*›. **2.** *unr. itr. V.* **a)** *(auftreffen)* ‹*bomb*› land; ‹*lightning*› strike; **bei uns hat es eingeschlagen** our house was struck by lightning; **b)** *(einprügeln)* **auf jmdn./etw. ~:** rain blows on *or* beat sb./sth.; **c)** *(durch Händedruck)* shake [hands] on it; *(fig.)* accept; **schlag ein!** shake on it!; **d)** *(Kfz-W.)* **nach links/rechts ~:** steer to the left/ right

einschlägig ['ainʃlɛ:gɪç] **1.** *Adj.* relevant. **2.** *adv.* **er ist ~ vorbestraft** he has previous convictions for a similar offence/similar offences

ein|schleichen *unr. refl. V.* steal *or* sneak *or* creep in; *(fig.)* creep in; **sich in etw.** *(Akk.)* **~:** steal *or* sneak *or* creep into sth.

ein|schleifen *unr. tr. V.* cut in

ein|schleppen *tr. V.* bring in, introduce ‹*disease, pest*›

ein|schleusen *tr. V.* smuggle in; infiltrate ‹agents› **(in** + *Akk.* into)

ein|schließen *unr. tr. V.* **a) etw. in etw.** *(Dat.)* **~:** lock sth. up [in sth.]; **jmdn./sich ~:** lock sb./oneself in; **b)** *(umgeben)* ‹*wall*› surround, enclose; ‹*people*› surround, encircle; **c)** *(einbeziehen)* **etw. in etw.** *(Akk.)* **~:** include sth. in sth.

einschließlich 1. *Präp. mit Gen.* including; inclusive of; **~ der Unkosten** including expenses. **2.** *adv.* **bis ~ 30. Juni/Montag** up to and including 30 June/Monday

Ein·schluß der a) *(Einbeziehung)* inclusion; **unter** *od.* **mit ~ von** including; **b)** *(Geol.)* inclusion

ein|schmeicheln *refl. V.* ingratiate oneself **(bei** with)

einschmeichelnd *Adj.* beguiling ‹*music, voice*›; ingratiating ‹*manner*›

ein|schmelzen *unr. tr. V.* melt down

ein|schmieren *tr. V. (ugs.) (mit Creme)* cream ‹*face, hands, etc.*›; *(mit Fett)* grease; *(mit Öl)* oil

ein|schmuggeln *tr. V.* **a)** smuggle in; **b) sich in etw.** *(Akk.)* **~** *(ugs.)* sneak into sth.

ein|schnappen *itr. V.; mit sein* **a)** ‹*door, lock*› click to; **b)** *(ugs.: schmollen)* go into a huff

ein|schneiden 1. *unr. tr. V.* make a cut in; cut. **2.** *unr. itr. V.* **das Kleid schneidet an den Schultern ein** the dress cuts into my shoulders

einschneidend *Adj.* drastic, radical ‹*measure, change*›; drastic, far-reaching ‹*effect*›

ein|schneien *itr. V.; mit sein* get snowed in; **eingeschneit sein** be snowed in

Ein·schnitt der a) cut; *(Med.)* incision; *(im Gebirge)* cleft; **b)** *(Zäsur)* break; **c)** *(einschneidendes Ereignis)* [decisive] turning-point; decisive event

ein|schnüren *tr. V.* **a) sich/**

jmdm. die Taille ~: lace one's/sb.'s waist; **b)** *(einengen)* ⟨*belt, elastic*⟩ cut in

ein|schränken 1. *tr. V.* **a)** reduce, curb ⟨*expenditure, consumption, power*⟩; **das Trinken/Rauchen ~:** cut down on the amount one drinks/smokes; **b)** *(einengen)* limit; restrict; **jmdn. in seinen Rechten/seiner Bewegungsfreiheit ~:** limit *or* restrict sb.'s rights/freedom of movement; **c)** *(relativieren)* qualify, modify ⟨*remark*⟩. **2.** *refl. V.* economize; cut back on spending; **sich finanziell ~ müssen** have to cut back on one's spending; **sich im Rauchen/Trinken ~:** cut down on the amount one smokes/drinks

Einschränkung die; ~, ~en a) restriction; limitation; **jmdm. ~en auferlegen** impose restrictions on sb.; **b)** *(Vorbehalt)* reservation; **nur mit ~[en]** only with reservations *pl.*; **ohne ~[en]** without reservation; **mit der ~, daß...:** with the [one] reservation that ...

ein|schrauben *tr. V.* screw in

Einschreibe-: ~brief der registered letter; **~gebühr die** *(Postw., Hochschulw.)* registration fee

ein|schreiben *unr. tr. V.* **a)** *(hineinschreiben)* write up; **b)** *(Postw.)* register ⟨*letter*⟩; **c)** *(eintragen)* **sich/jmdn. [in eine Liste] ~:** enter sb.'s/one's name [on a list]; **sich an einer Universität ~:** register at a university; **sich für einen Abendkurs ~:** enrol for an evening class

Ein·schreiben das *(Postw.)* registered letter; **per ~:** by registered mail

Ein·schreibung die *(Hochschulw.)* registration; *(für einen Abendkurs)* enrolment

ein|schreiten *unr. itr. V.* intervene; **gegen jmdn./etw. ~:** take action against sb./sth.

ein|schrumpfen *itr. V.; mit sein* shrivel up

Ein·schub der insertion

ein|schüchtern *tr. V.* intimidate

ein|schulen *tr. V.* **eingeschult werden** start school

Ein·schuß der bullet wound; wound at point of entry

ein|schütten *tr. V.* pour in

ein|schweißen *tr. V.* **a)** weld in; **b)** *(in Klarsichtfolie)* **etw. ~:** seal sth. in transparent film

ein|schwenken *itr. V.; mit sein* **a)** turn in; **in die Toreinfahrt ~:** turn into the gateway; **nach links ~:** wheel left; **b)** *(fig.)* fall into line

ein|sehen *unr. tr. V.* **a)** see into ⟨*building, garden, etc.*⟩; **b)** *(prüfend lesen)* look at, see ⟨*files*⟩; **c)** *(erkennen)* see; realize; **d)** *(begreifen)* understand; see

Einsehen das; ~s: ein ~ haben show [some] understanding

ein|seifen *tr. V.* **a)** lather; **jmdn. mit Schnee ~:** rub snow in sb.'s face; **b)** *(ugs.: betrügen)* con *(coll.)*

ein·seitig 1. *Adj.* **a)** on one side *postpos.*; unrequited ⟨*love*⟩; one-sided ⟨*friendship*⟩; **er hat eine ~e Lähmung** he's paralysed down one side; **b)** one-sided, biased ⟨*view, statement, etc.*⟩; one-sided ⟨*person*⟩; **c)** unbalanced ⟨*diet*⟩; one-sided ⟨*education*⟩. **2.** *adv.* **a) etw. ~ bedrucken** print sth. on one side; **b)** *s.* **einseitig 1b:** one-sidedly; **c) sich ~ ernähren** have an unbalanced diet

Einseitigkeit die; ~, ~en *s.* **einseitig 1b:** one-sidedness; bias

ein|senden *unr. (auch regelm.) tr. V.* send [in]; **etw. einem Verlag** *od.* **an einen Verlag ~:** send sth. to a publisher

Ein·sender der sender; *(bei einem Preisausschreiben)* entrant

Einsende·schluß der closing date

Ein·sendung die letter/card/contribution/article *etc.*; *(bei einem Preisausschreiben)* entry

Einser der; ~s, ~ *(ugs.: Schulnote)* one; A

ein|setzen 1. *tr. V.* **a)** *(hineinsetzen)* put in; put in, fit ⟨*window*⟩; insert, put in ⟨*tooth, piece of fabric, value, word*⟩; **etw. in etw.** *(Akk.)* **~:** put/fit/insert sth. into sth.; **b)** *(Verkehrsw.)* put on ⟨*special train etc.*⟩; **c)** *(ernennen, in eine Position setzen)* appoint; **jmdn. in ein Amt ~:** appoint sb. to an office; **d)** *(in Aktion treten lassen)* use ⟨*weapon, machine, strength*⟩; bring into action, use ⟨*troops, police*⟩; bring on, use ⟨*reserve player*⟩; **e)** *(aufs Spiel setzen)* stake ⟨*money*⟩; **f)** *(riskieren)* risk ⟨*life, reputation*⟩. **2.** *itr. V.* start; begin; ⟨*storm*⟩ break; **mit etw. ~:** start *or* begin sth.. **3.** *refl. V.* **a)** *(sich engagieren)* **ich werde mich dafür ~, daß ...:** I shall do what I can to see that ...; **der Schüler/Minister setzt sich nicht genug ein** the pupil is lacking application/the minister is lacking in commitment; **b)** *(Fürsprache einlegen)* **sich für jmdn. ~:** support sb.'s cause

Einsetzung die, ~, ~en appointment (**in** + *Akk.* to)

Ein·sicht die a) *(das Einsehen)* view (**in** + *Akk.* into); **b)** *o. Pl.* *(Einblick)* **~ in die Akten nehmen** take *or* have a look at the files; **jmdm. ~ in etw.** *(Akk.)* **gewähren** allow sb. to look at *or* see sth.; **c)** *(Erkenntnis)* insight; **zu der ~ kommen, daß ...:** come to realize that ...; **d)** *o. Pl.* *(Vernunft)* sense; reason; *(Verständnis)* understanding; **zur ~ kommen** come to one's senses

einsichtig 1. *Adj.* **a)** *(verständnisvoll)* understanding; **b)** *(verständlich)* comprehensible, understandable, clear; **ihm war nicht ~, warum ...:** he was not clear why ... **2.** *adv.* **sehr ~ vorgehen** show a great deal of understanding

ein|sickern *itr. V.; mit sein* seep in

Einsiedelei [ainzi:də'lai] **die; ~, ~en** hermitage

Ein·siedler der hermit; *(fig.)* recluse

ein·silbig *Adj.* **a)** monosyllabic ⟨*word*⟩; **b)** *(fig.)* taciturn ⟨*person*⟩; monosyllabic ⟨*answer*⟩

ein|sinken *unr. itr. V.* sink in; **in etw.** *(Dat.)* **~:** sink into sth.; **eingesunkene Wangen** sunken cheeks

ein|sortieren *tr. V.* sort ⟨*books, papers, etc.*⟩ and put them away; **Karteikarten ~:** file cards; **Briefe in Fächer ~:** sort letters into pigeon-holes

ein·spaltig *(Druckw.)* **1.** *Adj.* single-column *attrib.* **2.** *adv.* ⟨*print, set*⟩ in one column

ein|spannen *tr. V.* **a)** harness ⟨*horse*⟩; **b)** *(in etw. spannen)*; **den Bogen [in die Schreibmaschine] ~:** put the sheet of paper in[to the typewriter]; **das Werkstück [in den Schraubstock] ~:** clamp the work [in the vice]; **c)** *(ugs.: heranziehen)* rope in *(coll.)*

ein|sparen *tr. V.* save, cut down on ⟨*costs, expenditure*⟩; save ⟨*time*⟩; save, economize on ⟨*energy, materials*⟩; **Stellen/Arbeitsplätze ~:** cut down on the number of posts/cut down on staff

Einsparung die; ~, ~en saving; **~en an Kosten/Energie** savings *or* economies in costs/energy; **durch ~ von Material** by economizing on *or* saving materials

ein|speichern *tr. V.* *(DV)* feed in; input

ein|sperren *tr. V.* lock ⟨*sb.*⟩ up

ein|spielen 1. *refl. V.* **a)** ⟨*musician, athlete, team, etc.*⟩ warm up; **sich aufeinander ~** *(fig.)* get used to each other's ways *or* one another; **b)** *(fig.)* get going [properly]. **2.** *tr. V.* **a)** *(einbringen)* make; bring in; **b)** play *or* break in ‹*musical instrument*›; **c)** *(aufnehmen)* record

einsprachig ['ainʃpra:xɪç] **1.** *Adj.* monolingual. **2.** *adv.* ~ **aufwachsen** grow up speaking only one language

ein|springen *unr. itr. V.; mit sein (als Stellvertreter)* stand in; *(fig.: aushelfen)* step in and help out

ein|spritzen *tr. V.* inject; **jmdm. etw.** ~: inject sb. with sth.

Einspritz-: **~motor der** fuel-injection engine; **~pumpe die** injection pump

Ein·spruch der *(bes. Rechtsw.)* objection; *(gegen Urteil, Entscheidung)* appeal; **[gegen etw.]** ~ **einlegen/erheben** raise an objection [to sth.]; *(gegen Urteil, Entscheidung)* lodge an appeal [against sth.]

ein|sprühen *tr. V.* **x mit y** ~: spray y on [to] x; **sich** *(Dat.)* **das Haar** ~: put hair-spray on one's hair

einspurig ['ainʃpu:rɪç] **1.** *Adj.* single-track ‹*road*›. **2.** *adv.* **die Straße ist nur ~ befahrbar** only one lane of the road is open

einst [ainst] *Adv. (geh.)* **a)** *(früher)* once; **b)** *(der~)* some *or* one day

ein|stampfen *tr. V.* pulp ‹*books*›

Ein·stand der a) seinen ~ geben/feiern celebrate starting a new job; **b)** *o. Pl. (Sport: erstes Spiel)* début; **c)** *o. Pl. (Tennis)* deuce

ein|stanzen *tr. V.* stamp in

ein|stauben *itr. V.; mit sein* get dusty; get covered in dust

ein|stechen 1. *unr. itr. V.* **auf jmdn.** ~: stab sb. **2.** *unr. tr. V.* prick

ein|stecken *tr. V.* **a)** put in; **das Bügeleisen** ~: plug in the iron; **er steckte die Pistole/das Messer wieder ein** he put the pistol back in the holster/the knife back in the sheath; **b)** *(mitnehmen)* **[sich** *(Dat.)***] etw.** ~: take sth. with one; **c)** mail ‹*letter*›; **d)** *(abwertend: für sich behalten)* pocket ‹*money, profits*›; **e)** *(hinnehmen)* take ‹*criticism, defeat, etc.*›; take, swallow ‹*insult*›

ein|stehen *unr. itr. V.* **für jmdn.** ~: vouch for sb.; **für etw.** ~: take responsibility for *or* assume liability for sth.

ein|steigen *unr. itr. V.; mit sein* **a)** *(in ein Fahrzeug)* get in; **in ein Auto** ~: get into a car; **in den Bus** ~: get on the bus; **vorn/hinten** ~ *(ins Auto)* get into the front/back; *(in den Bus)* get on at the front/back; **b)** *(eindringen)* **durch ein Fenster/über den Balkon** ~: climb in *or* get in through a window/over the balcony; **c)** *(ugs.: sich engagieren)* **in ein Geschäft/die Politik** ~: go into a business/into politics

einstellbar *Adj.* adjustable

ein|stellen 1. *tr. V.* **a)** *(einordnen)* put away ‹*books etc.*›; **b)** *(unterstellen)* put in ‹*car, bicycle*›; **c)** *(auch itr.) (beschäftigen)* take on, employ ‹*workers*›; **d)** *(regulieren)* adjust; set; focus ‹*camera, telescope, binoculars*›; adjust ‹*headlights*›; **e)** *(beenden)* stop; call off ‹*search, strike*›; **das Feuer** ~: cease fire; **ein Gerichtsverfahren** ~: abandon court proceedings; **die Arbeit** ~ ‹*factory*› close; ‹*workers*› stop work; **f)** *(Sport)* equal ‹*record*›. **2.** *refl. V.* **a)** *(ankommen, auch fig.)* arrive; **b)** *(eintreten)* ‹*pain*› begin; ‹*success*› come; ‹*symptoms, consequences*› appear; **c) sich auf jmdn./etw.** ~: adapt to sb./prepare oneself *or* get ready for sth.; **sich schnell auf neue Situationen** ~: adjust quickly to new situations

ein·stellig *Adj.* single-figure *attrib.* ‹*number*›

Einstell·platz der parking space; *(auf eigenem Grundstück)* carport

Ein·stellung die a) *(von Arbeitskräften)* employment; taking on; **b)** *(Regulierung)* adjustment; setting; *(eines Fernglases, einer Kamera)* focusing; **c)** *(Beendigung)* stopping; *(einer Suchaktion, eines Streiks)* calling off; **d)** *(Sport)* **die ~ eines Rekordes** the equalling of a record; **e)** *(Ansicht)* attitude; **ihre politische/religiöse** ~: her political/religious views *pl.*; **f)** *(Film)* take

Einstellungs-: **~gespräch das** interview; **~stopp der** freeze on recruitment

Ein·stich der puncture; prick

Ein·stieg der; **~[e]s, ~e a)** *(Eingang)* entrance; *(Tür)* door/doors; **b)** *o. Pl. (das Einsteigen)* entry; **„kein ~"** 'exit only'; **c)** *(fig.)* **der ~ in diese Problematik ist schwierig** these are difficult problems to approach

einstig *Adj.; nicht präd.* former

ein|stimmen 1. *itr. V.* join in; **in den Gesang** ~: join in the singing. **2.** *tr. V.* **jmdn. auf etw.** *(Akk.)* ~: get sb. in the [right] mood for sth.

einstimmig 1. *Adj.* **a)** *(Musik)* **ein ~es Lied** a song for one voice; **b)** *(einmütig)* unanimous ‹*decision, vote*›. **2.** *adv.* **a)** *(Musik)* ~ **singen** sing in unison; **b)** *(einmütig)* unanimously

Einstimmigkeit die; ~: unanimity; ~ **erzielen** achieve unanimity

Ein·stimmung die: zur ~: to get in the [right] mood **(auf** + *Akk.* for)

ein·stöckig *Adj.* single-storey *attrib.*; one-storey *attrib.*; ~ **sein** have one storey

ein|stöpseln *tr. V.* plug in ‹*telephone, electrical device*›

ein|stoßen *unr. tr. V.* break down ‹*door, wall*›; smash [in] ‹*window*›

ein|streichen *unr. tr. V. (ugs.)* pocket ‹*money, winnings, etc.*›; *(abwertend)* rake in *(coll.)* ‹*money, profits, etc.*›

ein|streuen *tr. V.* **a) etw. mit Sand** ~: strew *or* scatter sand on sth.; **b)** *(einfügen)* **er streute witzige Bemerkungen in seinen Vortrag ein** he sprinkled his lecture with witty remarks

ein|strömen *itr. V.* ‹*water*› pour *or* flood *or* stream in; ‹*air, light*› stream in; *(fig.)* ‹*crowd, supporters*› stream *or* pour in

ein|studieren *tr. V.* rehearse

Einstudierung die; **~, ~en a)** *o. Pl.* rehearsal; **b)** *(Inszenierung)* production

ein|stufen *tr. V.* classify; categorize; **jmdn. in eine Kategorie** ~: put sb. in a category

Einstufung ['ainʃtu:fʊŋ] **die,** **~, ~en** classification; categorization

ein·stündig *Adj.* one-hour *attrib.* ‹*wait, delay*›; *s. auch* **achtstündig**

ein|stürmen *itr. V.* **mit Fragen/Bitten auf jmdn.** ~: besiege sb. with questions/requests

Ein·sturz der collapse

ein|stürzen *itr. V.; mit sein* **a)** collapse; **eine Welt stürzte für sie ein** *(fig.)* her whole world collapsed *or* fell apart; **b)** *(fig.)* **auf jmdn.** ~ ‹*worries, problems*› crowd in [up]on sb.

Einsturz·gefahr die; *o. Pl.* danger of collapse; **„Achtung, ~!"** 'danger – building unsafe'

einst·weilen *Adv.* for the time being; temporarily

ein·tägig *Adj.* one-day *attrib.*; *s. auch* **achttägig**

Eintags·fliege die *(Zool.)* mayfly; *(fig. ugs.)* seven-day wonder

ein|tauchen 1. *tr. V.* dip; *(untertauchen)* immerse. **2.** *itr. V.; mit sein* dive in

ein|tauschen *tr. V.* exchange **(gegen** for)

ein·tausend *Kardinalz.* a *or* one thousand; *s. auch* **¹acht**

ein|teilen *tr. V.* **a)** divide up; classify ⟨*plants, species*⟩; **b)** *(disponieren, verplanen)* organize; plan [out] ⟨*work, time*⟩; **sein Geld [besser] ~:** plan *or* organize one's finances [better]; **c)** *(delegieren, abkommandieren)* **jmdn. für** *od.* **zu etw. ~:** assign sb. to sth.

einteilig ['aintailɪç] *Adj.* one-piece ⟨*dress, bathing-suit*⟩

Ein·teilung die a) *(Gliederung)* division; dividing up; *(Biol.)* classification; **b)** *(planvolles Disponieren)* organization; planning; **c)** *(Delegierung, Abkommandierung)* assignment

Eintel ['aintl̩] **das** (*schweiz. meist* **der**); **~s, ~:** whole

ein|tippen *tr. V. (in die Kasse)* register; *(in einen Rechner)* key in

eintönig ['aintø:nɪç] **1.** *Adj.* monotonous. **2.** *adv.* monotonously

Eintönigkeit die; ~: monotony

Ein·topf der, Eintopf·gericht das *(Kochk.)* stew

Ein·tracht die; *o. Pl.* harmony; concord

ein·trächtig 1. *Adj.* harmonious. **2.** *adv.* harmoniously; **~ zusammenleben** live together in harmony

Eintrag ['aintra:k] **der; ~[e]s, Einträge** ['aintrɛ:gə] entry

ein|tragen *unr. tr. V.* **a)** *(einschreiben)* enter; copy out ⟨*essay*⟩; *(einzeichnen)* mark in; enter; **seinen Namen** *od.* **sich [in eine Liste] ~:** enter one's name [on a list]; **b)** *(Amtsspr.)* register; **sich ~ lassen** register; **etw. auf seinen Namen ~ lassen** have sth. registered in one's name; **ein eingetragenes Warenzeichen** a registered trade mark; **c)** *(einbringen)* bring in ⟨*money*⟩; bring ⟨*criticism*⟩; win ⟨*goodwill*⟩; **das Geschäft trägt [einen] Gewinn ein** the business makes a profit

einträglich ['aintrɛ:klɪç] *Adj.* profitable, lucrative ⟨*business, sideline*⟩; lucrative ⟨*work, job*⟩

Eintragung die; ~, ~en a) *(das Eintragen)* entering; **b)** *(Eingetragenes)* entry

ein|treffen *unr. itr. V.; mit sein* **a)** arrive; **b)** ⟨*prophecy*⟩ come true

Ein·treffen das; *o. Pl.* arrival

ein|treiben *unr. tr. V.* collect ⟨*taxes, debts*⟩; *(durch Gerichtsverfahren)* recover ⟨*debts, money*⟩

Eintreibung die; ~, ~en collection; *(durch Gerichtsverfahren)* recovery

ein|treten 1. *unr. itr. V.; mit sein* **a)** *(auch fig.)* enter; **bitte, treten Sie ein!** please come in; **in Verhandlungen ~:** enter into negotiations; **b)** *(Mitglied werden)* **in einen Verein/einen Orden ~:** join a club/enter a religious order; **c)** *(sich ereignen)* occur; **d)** *(sich einsetzen)* **für jmdn./etw. ~:** stand up for sb./sth.; *(vor Gericht)* speak in sb.'s defence. **2.** *unr. tr. V.* kick in ⟨*door, window, etc.*⟩

ein|trichtern *tr. V.* **jmdm. etw. ~** *(salopp)* drum sth. into sb.

Ein·tritt der a) entry; entrance; **sich** *(Dat.)* **[in etw.** *(Akk.)***] ~ verschaffen** gain entry [to sth.]; **vor dem ~ in die Verhandlungen** *(fig.)* before entering into negotiations; **b)** *(Beitritt)* **der ~ in einen Verein/Orden** joining a club/entering a religious order; **c)** *(Zugang, Eintrittsgeld)* admission; **[der] ~ [ist] frei** admission [is] free; **d)** *(Beginn)* onset; **e)** *(eines Ereignisses)* occurrence; **bei ~ des Todes** when death occurs

Eintritts-: ~geld das admission charge *or* fee; entrance charge *or* fee; **~karte die** admission *or* entrance ticket; **~preis der** admission *or* entrance charge

ein|trocknen *itr. V.; mit sein* **a)** ⟨*paint, blood*⟩ dry; ⟨*water, toothpaste*⟩ dry up; **b)** *(verdorren)* ⟨*leather*⟩ dry out; ⟨*berry, fruit*⟩ shrivel

ein|trudeln *itr. V.; mit sein (ugs.)* drift in *(coll.)*

ein|tüten *tr. V.* bag

ein|üben *tr. V.* practise

ein·und·ein·halb *s.* **anderthalb**

ein|verleiben [-fɛɐ̯laibn̩] **1.** *tr. V.* annex ⟨*land, country*⟩. **2.** *refl. V.* assimilate, absorb ⟨*knowledge, experience*⟩; *(scherzh.: zu sich nehmen)* put away *(coll.)*

Ein·vernehmen das; ~s harmony; *(Übereinstimmung)* agreement

einverstanden *Adj.; nicht attr.* **~ sein** agree; **mit jmdm./etw. ~ sein** approve of sb./sth.; **~!** *(ugs.)* okay! *(coll.)*; agreed!

Ein·verständnis das a) *(Billigung)* consent, approval **(zu** of); **b)** *(Übereinstimmung)* agreement

Einwand der; ~[e]s, Einwände ['ainvɛndə] objection **(gegen** to)

Ein·wanderer der immigrant

ein|wandern *itr. V.; mit sein* immigrate **(in +** *Akk.* into)

Ein·wanderung die immigration

einwand·frei 1. *Adj.* flawless; perfect; impeccable ⟨*behaviour*⟩; indisputable, definite ⟨*proof*⟩; watertight ⟨*alibi*⟩. **2.** *adv.* perfectly; flawlessly; ⟨*behave*⟩ impeccably; **es ist ~ erwiesen, daß ...:** it has been proved beyond question *or* doubt that ...

einwärts ['ainvɛrts] *Adv.* inwards

ein|wechseln *tr. V.* **a)** change ⟨*money*⟩; **b)** *(Sport)* substitute ⟨*player*⟩

ein|wecken *tr. V.* preserve; preserve, bottle ⟨*fruit, vegetables*⟩

Einweck·glas das preserving-jar

Ein·weg·flasche die non-returnable bottle

ein|weichen *tr. V.* soak

ein|weihen *tr. V.* **a)** open [officially] ⟨*bridge, road*⟩; consecrate ⟨*church*⟩; dedicate ⟨*monument*⟩; **b)** *(ugs. scherzh.: zum erstenmal benutzen)* christen *(coll.)*; **c)** *(vertraut machen)* **jmdn. in etw.** *(Akk.)* **~:** let sb. in on sth.

Einweihung die; ~, ~en [official] opening

ein|weisen *unr. tr. V.* **a) jmdn. in ein Krankenhaus ~:** have sb. admitted to hospital; **b)** *(in eine Tätigkeit)* **jmdn. [in eine/die Arbeit] ~:** show sb. what a/the job involves; **c)** *(in ein Amt)* install; **jmdn. in sein Amt ~:** install sb.; **d)** *(Verkehrsw.)* direct

Ein·weisung die a) ~ in ein Krankenhaus admission to a hospital; **b)** *(in eine Tätigkeit)* introduction

ein|wenden *unr. (auch regelm.) tr. V.* **gegen etw. nichts einzuwenden haben** have no objection to sth.; **dagegen läßt sich manches ~:** there are a number of things to be said against that; **„...", wandte er ein** '...,' he objected

Ein·wendung die objection **[gegen** to]

ein|werfen 1. *unr. tr. V.* **a)** put in, insert ⟨*coin*⟩; mail ⟨*letter, mail*⟩; **b)** *(zertrümmern)* smash, break ⟨*window*⟩; **c)** *(Ballspiele)* throw in ⟨*ball*⟩; **d)** *(bemerken, sagen)* throw in ⟨*remark*⟩; **„...", warf sie ein** '...,' she interjected. **2.** *unr. itr. V. (Ballspiele) (vom Rand)* take the throw-in; *(ins Tor)* score

ein|wickeln *tr. V.* **a)** wrap [up]; **b) jmdn. ~** *(ugs.)* take sb. in

ein|willigen *itr. V.* agree, consent **(in +** *Akk.* to)

Einwilligung die; ~, ~en agreement; consent; **seine ~ zu etw. geben** give one's consent to sth.

ein|winken *tr. V. (Verkehrsw.)* guide in ⟨*aircraft*⟩; guide *or* direct in ⟨*car*⟩

ein|wirken a) auf jmdn. ~: influence sb.; **beruhigend auf jmdn. ~:** exert a soothing *or* calming influence on sb.; **b)** *(eine Wirkung ausüben)* have an effect **(auf +**

Akk. on); **die Creme ~ lassen** let the cream work in

Ein·wirkung die *(Einfluß)* influence; *(Wirkung)* effect; **unter ~ von Drogen stehen** be under the influence of drugs

ein·wöchig *Adj.* one-week *attrib.*; week-old ‹*baby*›; week-long ‹*conference*›

Einwohner der; ~s, ~, Einwohnerin die; ~, ~nen inhabitant; **die Stadt hat 3 Millionen ~:** the town has 3 million inhabitants *or* a population of 3 million

Einwohner·meldeamt das *local government office for registration of residents*

Einwohnerschaft die; ~: population; inhabitants *pl.*

Ein·wurf der a) *(Einwerfen)* insertion; *(von Briefen)* mailing; **b)** *(Ballspiele)* throw-in; **c)** *(Öffnung) (eines Briefkastens)* slit; *(einer Tür)* letter-box; **d)** *(Zwischenbemerkung)* interjection; *(kritisch)* objection

Ein·zahl die; *o. Pl. (Sprachw.)* singular

ein|zahlen *tr. V.* pay in; deposit; **Geld auf sein Konto ~:** pay *or* deposit money into one's account

Ein·zahlung die payment; deposit; *(Überweisung)* payment

Einzahlungs·beleg der counterfoil

ein|zäunen *tr. V.* fence in; enclose

Einzäunung die; ~, ~en a) *(das Einzäunen)* fencing-in; enclosure; **b)** *(Zaun)* fence; enclosure

ein|zeichnen *tr. V.* draw *or* mark in; **etw. in eine Karte ~:** draw *or* mark sth. in on a map

ein·zeilig *Adj.* one-line *attrib.*

Einzel ['aintsḷ] **das; ~s, ~** *(Sport)* singles *pl.*; **~ spielen** to play a singles match

Einzel-: ~aktion die independent action; **~anfertigung die** custom-made article; **~ausgabe die** separate edition; **~band der** individual *or* single volume; **~bett das** single bed; **~erscheinung die** isolated occurrence; **~fahrschein der** single; **~fall der a)** particular case; **im ~fall** in particular cases; **b)** *(Ausnahme)* isolated case; exception; **~frage die** individual question

Einzelgänger [-gɛŋɐ] **der; ~s, ~ a)** solitary person; loner; **b)** *(Tier)* lone animal

Einzel·haft die solitary confinement

Einzel·handel der retail trade; **etw. im ~ kaufen** buy sth. retail

Einzelhandels-: ~geschäft das retail shop; retail store *(Amer.)*; **~preis der** retail price

Einzel·händler der retailer; retail trader

Einzelheit die; ~, ~en a) detail; **b)** *(einzelner Umstand)* particular

Einzel·kind das only child

einzeln *Adj.* **a)** individual; **die ~en Bände eines Werkes** the individual *or* separate volumes of a work; **jede ~e Insel** each individual island; **ein ~er Schuh/Handschuh** an odd shoe/glove; **schon ein ~es von diesen Gläsern** just one of these glasses on its own; **b)** solitary ‹*building, tree*›; **eine ~e Dame/ein ~er Herr** a single lady/gentleman; **c) ~e** *(wenige)* a few; *(einige)* some; **~e Regenschauer** scattered *or* isolated showers; **d)** *substantivisch (~er Mensch)* **der/jeder ~e** the/each individual; **als ~er** as an individual; **jeder ~e der Betroffenen** every [single] one of those concerned; **ein ~er** one individual; **e)** *substantivisch* **~es** *(manches)* some things *pl.*; **etw. im ~en besprechen** discuss sth. in detail; **ins ~e gehen** go into detail[s *pl.*]; **bis ins ~e** right down to the last detail

Einzel-: ~stück das individual piece *or* item; **~teil das** [individual *or* separate] part; **etw. in [seine] ~teile zerlegen** take sth. to pieces; **~unterricht der** individual tuition; **~zelle die** single cell; **~zimmer das** single room

Einzieh·decke die duvet *(Brit.)*; continental quilt *(Brit.)*; stuffed quilt *(Amer.)*

ein|ziehen 1. *unr. tr. V.* **a)** put in ‹*duvet*›; thread in ‹*tape, elastic*›; **b)** *(einbauen)* put in ‹*wall, ceiling*›; **c)** *(einholen)* haul in, pull in ‹*net*›; retract, draw in ‹*feelers, claws*›; **den Kopf ~:** duck; **der Hund zog den Schwanz ein** the dog put its tail between its legs; **d)** *(einatmen)* breathe in ‹*scent, fresh air*›; inhale ‹*smoke*›; **e)** *(einberufen)* call up, conscript ‹*recruits*›; **f)** *(beitreiben)* collect; **er läßt die Miete vom Konto ~:** he pays his rent by direct debit; **g)** *(beschlagnahmen)* confiscate; seize; **h)** *(aus dem Verkehr ziehen)* withdraw, call in ‹*coins, banknotes*›; **i)** *(Papierdt.: einholen)* **Informationen/Erkundigungen ~:** gather information/make enquiries. **2.** *unr. itr. V.; mit sein* **a)** ‹*liquid*› soak in; **b)** *(einkehren)* enter; **der Frühling zieht ein** *(geh.)* spring comes *or* arrives; **c)** *(in eine Wohnung)* move in

Ein·ziehung die a) *(Einberufung)* call-up; conscription; drafting *(Amer.)*; **b)** *(Beitreibung)* collection; **c)** *(von Eigentum)* confiscation, seizure; *(von Münzen, Banknoten usw.)* withdrawal

einzig ['aintsɪç] **1.** *Adj.* **a)** only; **nur ein ~er** only one; **kein** *od.* **nicht ein ~es Stück** not one single piece; **ihre ~e Freude** her one and only joy; **das ~e** the only thing; **b)** *nicht präd. (völlig)* complete; absolute; **eine ~e Qual** one long torment; **c)** *nicht attr. (geh.: unvergleichlich)* unique; unparalleled. **2.** *adv.* **a)** *(intensivierend bei Adj.)* singularly; extraordinarily; **b)** *(ausschließlich)* only; **das ~ Wahre** the only thing; **das ~ Vernünftige/Richtige** the only sensible/right thing [to do]; **~ und allein** nobody/nothing but; solely

einzig·artig 1. *Adj.* unique. **2.** *adv.* uniquely; **~ schön** extraordinarily beautiful

Einzigartigkeit die uniqueness

Ein·zimmer-: ~appartement das, ~wohnung die one-room flat *or (Amer.)* apartment

Ein·zug der a) entry (**in** + *Akk.* into); **der ~ des Winters** *(geh.)* the advent of winter; **[seinen] ~ halten** make one's entrance; **b)** *(in eine Wohnung)* move

Einzugs-: ~bereich der, ~gebiet das catchment area

ein|zwängen *tr. V.* squeeze *or* hem in; ‹*corset*› constrict

Eis [ais] **das; ~es a)** ice; **ein Whisky mit ~:** a whisky with ice *or* on the rocks; **etw. auf ~ legen** *(auch fig. ugs.)* put sth. on ice; **b)** *(Speise~)* ice-cream; **ein ~ am Stiel** an ice-lolly *(Brit.) or (Amer.)* ice pop

Eis-: ~bahn die ice-rink; **~bär der** polar bear; **~becher der a)** *(~portion)* ice-cream sundae; **b)** *(Gefäß)* [ice-cream] sundae dish; **~bein das** *(Kochk.)* knuckle of pork; **~berg der** iceberg; **die Spitze des ~bergs** *(fig.)* the tip of the iceberg; **~beutel der** ice-bag; ice-pack; **~blume die** frost flower; **~bombe die** *(Gastr.)* bombe glacée; **~brecher der** ice-breaker; **~café das** ice-cream parlour

Ei·schnee der stiffly beaten egg-white

Eisen ['aizn̩] **das; ~s, ~: a)** *o. Pl.* iron; **aus ~ sein** be made of iron; **b)** *(Werkzeug, Golf~)* iron; *(fig.)* **das ist ein heißes ~:** that is a hot potato; **noch ein ~ im Feuer haben** have another iron in the fire; **zum alten ~ gehören** belong on the scrap heap

Eisen·bahn die a) railway; railroad *(Amer.)*; **mit der ~ fahren** go *or* travel by train *or* rail; **es ist [aller]höchste ~** *(ugs.)* it's high time; its' getting late; **b)** *(Bahnstrecke)* railway line; railroad track *(Amer.)*; **c)** *(Spielbahn)* train *or* railway set

Eisenbahn·abteil das railway *or (Amer.)* railroad compartment

Eisenbahner der; ~s, ~: railwayman; railroader *(Amer.)*

Eisenbahn-: ~fähre die train ferry; **~knotenpunkt der** railway *or (Amer.)* railroad junction; **~netz das** railway *or (Amer.)* railroad network; **~schaffner der** railway guard; railroad conductor *(Amer.)*; **~tunnel der** railway *or (Amer.)* railroad tunnel; **~unglück das** train crash; **~wagen der** railway carriage; railroad car *(Amer.)*; *(Güterwagen)* railway wagon; railroad car *(Amer.)*

eisen-, Eisen-: ~berg·werk das iron mine; **~erz das** iron ore; **~haltig** *Adj.* iron-bearing ⟨*stone*⟩; ⟨*food*⟩ containing iron; **~hütte die** ironworks *sing. or pl.*; iron foundry; **~mangel der** *(Med.)* iron deficiency; **~nagel der** iron nail; **~säge die** hacksaw; **~stange die** iron bar; **~teil das** iron part; **~träger der** iron girder; **~verarbeitend** *Adj.; nicht präd.* iron-processing ⟨*industry, firm, etc.*⟩; **~waren** *Pl.* ironmongery *sing.*; **~waren·händler der** ironmonger; **~zeit die** Iron Age

eisern ['aizɐn] **1.** *Adj.* **a)** *nicht präd. (aus Eisen)* iron; **~e Lunge** *(Med.)* iron lung; **der Eiserne Vorhang** *(Politik)* the Iron Curtain; **b)** *(unerschütterlich)* iron ⟨*discipline*⟩; unflagging ⟨*energy*⟩; **mit ~em Willen** with a will of iron; **c)** *(unerbittlich)* iron; unyielding; iron ⟨*discipline*⟩; **d)** *(bleibend)* **~er Bestand/~e Reserve** emergency stock/reserves *pl.*; **die ~e Ration** the iron rations *pl.*; *(fig.)* one's last reserves *pl. or* standby. **2.** *adv.* **a)** *(unerschütterlich)* resolutely; **~ bei etw. bleiben** stick tenaciously to sth.; **sich ~ an etw.** *(Akk.)* **halten** keep resolutely to sth.; **~ sparen/trainieren** save/train with iron determination; **b)** *(unerbittlich)* **~ durchgreifen** take drastic measures *or* action

eis-, Eis-: ~fach das freezing compartment; **~fläche die** sheet *or* surface of ice; **~frei** *Adj.* ice-free; free of ice *postpos.*; **~gekühlt** *Adj.* iced ⟨*drink*⟩; **~glatt** *Adj.* icy ⟨*road*⟩; **~glätte die** black ice; **~hockey das** ice hockey

eisig 1. *Adj.* **a)** icy ⟨*wind, cold*⟩; icy [cold] ⟨*water*⟩; **b)** *(fig.)* frosty, icy ⟨*atmosphere*⟩; frosty ⟨*smile*⟩. **2.** *adv.* **a) ~ kalt sein** be icy cold; **b)** *(fig.)* ⟨*smile*⟩ frostily; **~ schweigen** maintain an icy silence

eis-, Eis-: ~kaffee der iced coffee; **~kalt 1.** *Adj.* **a)** ice-cold ⟨*drink*⟩; freezing cold ⟨*weather*⟩; **b)** *(fig.)* icy; ice-cold ⟨*technocrat, businessman*⟩; **ein ~kalter Blick** a cold look. **2.** *adv. (fig.)* **a) es lief mir ~kalt über den Rücken** a cold shiver went down my spine; **b) etw. ~kalt tun** *(kaltblütig)* do sth. in cold blood; *(lässig)* do sth. without turning a hair; **~kristall das** ice crystal; **~kübel der** ice bucket

Eis·kunst-: ~lauf der figure skating; **~laufen das** figure skating; **~läufer der** figure skater

eis-, Eis-: ~lauf der ice-skating; **~|laufen** *unr. itr. V.; mit sein* ice-skate; **~laufen das** ice-skating; **~läufer der** ice-skater; **~mann der;** *Pl.* **~männer** *(ugs.)* ice-cream man; **~maschine die** ice-cream maker; freezer *(Amer.)*; **~meer das: das Nördliche/Südliche ~meer** the Arctic/Antarctic Ocean; **~pickel der** *(Bergsteigen)* ice-pick

Ei·sprung der *(Physiol.)* ovulation

eis-, Eis-: ~regen der sleet; **~revue die** ice show; **~schicht die** layer of ice; **~scholle die** ice-floe; **~schrank der** *(ugs.)* refrigerator; **~sport der** ice sports *pl.*; **~stadion das** ice rink; **~stock·schießen das** *(Sport)* ice-stick shooting; **~tanz der** *(Sport)* ice-dancing; **~wasser das;** *o. Pl.* iced water; **~wein der** *wine made from grapes frozen on the vine*; **~würfel der** ice cube; **~zapfen der** icicle; **~zeit die** ice age; **~zeitlich** *Adj.* ice-age *attrib.*, of the ice age *postpos.*

eitel ['aitl̩] *Adj.* **a)** *(abwertend)* vain; **b)** *(veralt.: nichtig)* vain ⟨*hope*⟩; futile, vain ⟨*endeavour*⟩; **c)** *indekl., nicht präd. (veralt.: rein)* pure

Eitelkeit die; ~, ~en vanity

Eiter ['aitɐ] **der; ~s** pus

Eiter-: ~beule die boil; abscess; **~pickel der** spot; pimple

eitern *itr. V.* suppurate

eitrig *Adj.* suppurating; festering

Ei·weiß das a) egg-white; **b)** *(Chemie, Biol.)* protein

eiweiß-, Eiweiß-: ~arm *Adj.* low-protein *attrib.*; low in protein *postpos.*; **~haltig** *Adj.* ⟨*food*⟩ containing protein; **~mangel der** protein deficiency; **~reich** *Adj.* high-protein *attrib.*; rich in protein *postpos.*

Ejakulation [ejakula'tsio:n] **die; ~, ~en** ejaculation

EKD [e:ka:'de:] **die; ~** *Abk.* **Evangelische Kirche in Deutschland**

[1]Ekel der; ~s disgust; loathing; revulsion; **[einen] ~ vor etw.** *(Dat.)* **haben** have a loathing *or* revulsion for sth.

[2]Ekel das; ~s, ~ *(ugs. abwertend)* horror; **er ist ein [altes] ~:** he is a perfect horror *or* quite obnoxious

ekel·erregend *Adj.* disgusting; nauseating; revolting

ekelhaft *s.* **eklig 1 a, 2**

ekeln ['e:kl̩n] **1.** *refl. V.* be *or* feel disgusted *or* sickened; **sie ekelt sich vor Spinnen** she finds spiders repulsive; **sich vor jmdm./etw. ~:** find sb./sth. disgusting *or* revolting. **2.** *tr., itr. V.; unpers.* **mich** *od.* **mir ekelt davor** I find it disgusting *or* revolting. **3.** *tr. V.* **a) Hunde ~ ihn** he finds dogs repulsive; **b)** *(vertreiben)* **jmdn. aus dem Haus ~:** hound sb. out of the house

ekelig *s.* **eklig**

EKG [e:ka:'ge:] **das; ~[s], ~[s]** *Abk.* **Elektrokardiogramm** ECG

Eklat [e'kla(:)] **der; ~s, ~s** *(geh.) (Aufsehen, Skandal)* sensation; stir; *(Konfrontation)* row; altercation

eklatant [ekla'tant] *Adj.* *(geh.)* striking ⟨*difference*⟩; flagrant, scandalous ⟨*offence*⟩

eklig ['e:klıç] **1.** *Adj.* **a)** disgusting, revolting, nauseating ⟨*sight*⟩; nasty *(coll.)*, horrible ⟨*weather, person*⟩; **~ riechen/schmecken** smell/taste disgusting *or* revolting; **b)** *(ugs.: gemein)* mean; nasty; **sich ~ benehmen** be mean *or* nasty. **2.** *adv.* **a)** in a disgusting *or* revolting *or* nauseating way; **b)** *(ugs.: sehr)* terribly *(coll.)*, dreadfully *(coll.)* ⟨*hot, cold*⟩

Eklipse [e'klıpsə] **die; ~, ~n** *(Astron.)* eclipse

Ekstase [ɛk'sta:zə] **die; ~, ~n** ecstasy; **in ~ geraten** go into ecstasies; **jmdn. in ~ versetzen** send sb. into ecstasies

ekstatisch [ɛk'sta:tıʃ] **1.** *Adj.* ecstatic. **2.** *adv.* ecstatically

Ekzem das; ~s, ~e *(Med.)* eczema

Elan [e'la:n] **der; ~s** zest; vigour

Elaste [e'lastə] *Pl.* *(Chemie)* elastomers

elastisch *Adj.* **a)** elastic; *(Textilw.: Gummifäden o. ä. enthal-*

tend) elasticated ⟨*fabric*⟩; *(federnd)* springy, resilient; **b)** *(geschmeidig)* supple, lithe ⟨*person, body*⟩

Elastizität [elastitsi'tɛ:t] **die; ~: a)** elasticity; *(Federkraft)* springiness; **b)** *(Geschmeidigkeit)* suppleness

Elch [ɛlç] **der; ~[e]s, ~e** elk; *(in Nordamerika)* moose

Eldorado [ɛldo'ra:do] **das; ~s, ~s** eldorado; **ein ~ der** *od.* **für Taucher** *(fig.)* a divers' paradise

Elefant [ele'fant] **der; ~en, ~en** elephant; **wie ein ~ im Porzellanladen** *(ugs.)* like a bull in a china shop; *s. auch* **Mücke**

Elefanten-: ~bulle der bull elephant; **~herde die** elephant herd; **~kuh die** cow elephant

elegant [ele'gant] **1.** *Adj.* elegant, stylish ⟨*dress, appearance*⟩; elegant ⟨*society*⟩; elegant, graceful ⟨*movement*⟩; neat ⟨*solution*⟩; elegant, civilized ⟨*taste*⟩; elegant ⟨*style*⟩; civilized ⟨*manner*⟩. **2.** *adv.* elegantly, stylishly ⟨*dressed*⟩

Eleganz [ele'gants] **die; ~** elegance

Elegie [ele'gi:] **die; ~, ~n** elegy

elektrifizieren [elɛktrifi'tsi:rən] *tr. V.* electrify

Elektrifizierung die; ~, ~en electrification

Elektrik [e'lɛktrɪk] **die; ~, ~en** electrics *pl.*

Elektriker der; ~s, ~: electrician

elektrisch 1. *Adj.* electric ⟨*current, light, heating, shock*⟩; electrical ⟨*resistance, wiring, system*⟩; **der ~e Stuhl** the electric chair. **2.** *adv.* **~ kochen** cook with electricity; **~ geladen sein** be charged with electricity; **sich ~ rasieren** use an electric shaver

elektrisieren 1. *tr. V. (fig.)* electrify. **2.** *refl. V.* give oneself *or* get an electric shock

Elektrizität [elɛktritsi'tɛ:t] **die; ~** *(Physik)* electricity; *(elektrische Energie)* electricity; [electric] power

Elektrizitäts-: ~versorgung die [electric] power supply; **~werk das** power station

Elektro-: ~antrieb der electric drive; **~auto das** electric car

Elektrode [elɛk'tro:də] **die; ~, ~n** electrode

elektro-, Elektro-: ~fahrzeug das electric vehicle; **~gerät das** electrical appliance; **~geschäft das** electrical shop *or (Amer.)* store; **~herd der** electric cooker; **~industrie die** electrical goods industry; **~ingenieur der** electrical engineer; **~installateur der** electrical fitter; electrician; **~konzern der** electrical company; **~magnet der** electromagnet; **~magnetisch** *Adj.* electromagnetic; **~mobil das** electric car; **~motor der** electric motor

Elektron [e'lɛktrɔn] **das; ~s, ~en** [-'tro:nən] *(Kernphysik)* electron

Elektronen-: ~blitz der electronic flash; **~[ge]hirn das** *(ugs.)* electronic brain *(coll.)*; **~mikroskop das** electron microscope; **~röhre die** electron tube *or* valve

Elektronik [elɛk'tro:nɪk] **die; ~ a)** *o. Pl.* electronics *sing., no art.*; **b)** *(elektronisches System)* electronics *pl.*

Elektroniker der; ~s, ~: electronics engineer

elektronisch 1. *Adj.* electronic. **2.** *adv.* electronically

elektro-, Elektro-: ~ofen der *(Technik)* electric furnace; **~rasierer der** electric shaver *or* razor; **~schock der** *(Med.)* electric shock; **~technik die** electrical engineering *no art.*

Element [ele'mɛnt] **das; ~[e]s, ~e a)** element; **er war/fühlte sich in seinem ~:** he was/felt in his element; **zwielichtige/kriminelle ~e** shady/criminal elements; **b)** *(Bauteil)* element; *(einer Schrankwand)* unit; **c)** *(Elektrot.)* cell

elementar [elemɛn'ta:ɐ̯] **1.** *Adj.* **a)** *(grundlegend)* fundamental ⟨*requirement, right, condition, etc.*⟩; **b)** *(einfach)* elementary, rudimentary ⟨*knowledge*⟩; **c)** *(naturhaft)* elemental ⟨*force, forces*⟩. **2.** *adv.* with elemental force

Elementar-: ~kenntnisse *Pl.* elementary *or* rudimentary knowledge *sing.*; **~stufe die** *(Schulw.)* pre-school level; **~teilchen das** *(Physik)* elementary particle; **~unterricht a)** elementary instruction; **b)** *(Unterricht in der ~stufe)* pre-school teaching

elend ['e:lɛnt] **1.** *Adj.* **a)** wretched, miserable ⟨*existence, life conditions, environment*⟩; **b)** *(krank)* **sich ~ fühlen** feel wretched *or (coll.)* awful; **mir ist/wird ~:** I feel/I am beginning to feel awful *or* terrible *(coll.)*; **c)** *(gemein)* despicable ⟨*person, coward, allegation*⟩; **d)** *nicht präd. (ugs.: besonders groß)* dreadful *(coll.)* ⟨*hunger, pain*⟩. **2.** *adv.* **a)** *(jämmerlich)* wretchedly; miserably; **~ zugrunde gehen** come to a miserable *or* wretched end; **b)** *(ugs.: intensivierend)* dreadfully *(coll.)*

Elend das; ~s a) *(Leid)* misery; wretchedness; *s. auch* **Häufchen; b)** *(Armut)* misery; destitution

elendig, elendiglich *Adv. (geh.)* miserably; wretchedly

Elends-: ~quartier das slum [dwelling]; **~viertel das** slum area

Eleve [e'le:və] **der; ~n, ~n, Elevin die; ~, ~nen** *(Theater, Ballett)* student

elf [ɛlf] *Kardinalz.* eleven; *s. auch* [1]**acht**

[1]**Elf die; ~, ~en a)** eleven; *s. auch* [1]**Acht a, e; b)** *(Sport)* team; side

[2]**Elf der; ~en, ~en** elf

Elfe ['ɛlfə] **die; ~, ~n** fairy

Elfen·bein das ivory

elfenbein-, Elfenbein-: ~farben *Adj.* ivory-coloured; **~küste die** Ivory Coast; **~schnitzerei die a)** *o. Pl.* ivory-carving; **b)** *(Gegenstand)* ivory carving; **~turm der** *(fig.)* ivory tower

Elfer der; ~s, ~ a) *(Fußballjargon)* penalty; **b)** *(landsch.: Zahl Elf)* eleven; **c)** *(Buslinie)* number eleven

elf-: ~fach *Vervielfältigungsz.* elevenfold; *s. auch* **achtfach; ~mal** *Wiederholungsz.* eleven times; *s. auch* **achtmal**

Elf·meter der *(Fußball)* penalty; **einen ~ schießen** take a penalty

Elfmeter-: ~punkt der *(Fußball)* penalty spot; **~schießen das** *(Fußball)* **durch ~schießen** by *or* on penalties

elft: wir waren zu ~: there were eleven of us; *s. auch* [2]**acht**

elft... *Ordinalz.* eleventh; *s. auch* **acht...**

elf·tausend *Kardinalz.* eleven thousand

Elftel ['ɛlftl̩] **das; ~s, ~:** eleventh

elftens *Adv.* eleventh

Elimination [elimina'tsio:n] **die; ~, ~en** elimination

eliminieren [elimi'ni:rən] *tr. V.* eliminate

Eliminierung die; ~, ~en elimination

Elisabeth [e'li:zabɛt] **(die)** Elizabeth

elisabethanisch *Adj.* Elizabethan

elitär [eli'tɛ:ɐ̯] **1.** *Adj.* élitist. **2.** *adv.* in an élitist fashion

Elite [e'li:tə] **die; ~, ~n** élite

Elite-: ~denken das élitist thinking; élitism; **~truppe die** *(Milit.)* élite *or* crack force

Elixier [elɪ'ksi:ɐ̯] **das; ~s, ~e** elixir

Ell·bogen der; ~s, ~: elbow; **er/sie hat keine ~** *(fig. ugs.)* he/she isn't pushy enough *(coll.)*

Ellbogen·freiheit die elbowroom

Elle ['ɛlə] **die; ~, ~n a)** *(Anat.)* ulna; **b)** *(frühere Längeneinheit)* cubit; **c)** *(veralt.: Maßstock)* ≈ yardstick; **alles mit einer ~ messen** *(fig.)* measure everything by the same yardstick

Ellen·bogen *s.* **Ellbogen**

ellen·lang *Adj. (ugs.)* ⟨*list*⟩ as long as your arm; interminable ⟨*lecture, sermon*⟩; terribly long *(coll.)* ⟨*letter*⟩

Ellipse [ɛ'lɪpsə] **die; ~, ~n** ellipse; *(Sprachw., Rhet.)* ellipsis

elliptisch *Adj.* elliptical

eloquent [elo'kvɛnt] *(geh.) Adj.* eloquent

Elsaß ['ɛlzas] **das;** ~ *od.* **Elsasses: das ~:** Alsace; **im/aus dem ~:** in/from Alsace

Elsässer ['ɛlzɛsɐ] **1.** *indekl. Adj.; nicht präd.* Alsatian. **2. der; ~s, ~:** Alsatian

Elster ['ɛlstɐ] **die; ~, ~n** *(Zool.)* magpie; **wie eine ~ stehlen** be light-fingered; **eine diebische ~** *(fig.)* a pilferer

Elter ['ɛltɐ] **das** *od.* **der; ~s, ~n** *(Biol.)* parent

elterlich *Adj.; nicht präd.* parental

Eltern *Pl.* parents

eltern-, Eltern-: ~abend der parents' evening; **~beirat der** *(Schulw.)* parents' association; **~haus das** parental home; **aus einem armen/katholischen ~haus kommen** come from a poor/Catholic home; **~liebe die** parental love; **~los 1.** *Adj.* parentless; orphaned; **ein ~loses Kind** a child without parents; an orphan; **2.** *adv.* **~los aufwachsen** grow up an orphan *or* without parents; **~sprech·tag** parents' day; **~teil der** parent; **~versammlung die** parents' meeting

Email [e'mai] **das; ~s, ~s, Emaille** [e'maljə] **die; ~, ~n** enamel

emaillieren *tr. V.* enamel

Emanze [e'mantsə] **die; ~, ~n** *(ugs. abwertend)* women's libber *(coll.)*

Emanzipation [emantsipa'tsio:n] **die; ~, ~en** emancipation

Emanzipations·bewegung die liberation movement

emanzipieren [emantsi'pi:rən] *refl. V.* emancipate oneself **(von** from)

emanzipiert *Adj.* emancipated; emancipated, liberated ⟨*woman*⟩

Embargo [ɛm'bargo] **das; ~s, ~s** embargo

Emblem [ɛm'ble:m] **das; ~s, ~e** emblem

Embolie [ɛmbo'li:] **die; ~, ~n** *(Med.)* embolism

Embryo ['ɛmbryo] **der; ~s, ~nen** [-y'o:nən] *od.* **~s** embryo

embryonal *Adj.; nicht präd. (auch fig.)* embryonic

emeritieren [emeri'ti:rən] *tr. V.* confer emeritus status on; **emeritierter Professor** emeritus professor; professor emeritus

Emigrant [emi'grant] **der; ~en, ~en** emigrant; *(politischer Flüchtling)* emigré

Emigration [emigra'tsio:n] **die; ~, ~en** emigration; **in der ~ leben** live in exile

emigrieren [emi'gri:rən] *itr. V.; mit sein* emigrate

eminent [emi'nɛnt] **1.** *(geh.) Adj.* eminent; **von ~er Bedeutung sein** be of the utmost significance. **2.** *adv.* eminently

Eminenz [emi'nɛnts] **die; ~, ~en** *(kath. Kirche)* eminence; **Eure/Seine ~:** Your/His Eminence

Emir ['e:mɪr] **der; ~s, ~e** emir

Emirat das; ~[e]s, ~e emirate

Emission [emɪ'sio:n] *(fachspr.)* **a)** emission; **b)** *(Ausgabe [von Briefmarken, Wertpapieren])* issue

emittieren *tr. V. (fachspr.)* **a)** emit; **b)** issue ⟨*stamps, shares*⟩

e-Moll das E minor

Emotion [emo'tsio:n] **die; ~, ~en** emotion

emotional 1. *Adj.* emotional. **2.** *adv.* emotionally

Emotionalität die; ~: emotionalism

empfahl [ɛm'pfa:l] *1. u. 3. Pers. Sg. Prät. v.* **empfehlen**

empfand [ɛm'pfant] *1. u. 3. Pers. Sg. Prät. v.* **empfinden**

Empfang [ɛm'pfaŋ] **der; ~[e]s, Empfänge a)** *(auch Funkw., Rundf., Ferns.)* reception; **b)** *(Entgegennahme)* receipt; **bei ~:** on receipt; **etw. in ~ nehmen** accept sth.

empfangen *unr. tr. V.* **a)** *(auch Funkw., Rundf., Ferns.)* receive; receive, greet ⟨*person*⟩; **b)** *(geh.: erhalten)* conceive ⟨*idea*⟩; **c)** *auch itr. (geh.)* **[ein Kind] ~:** conceive [a child]

Empfänger [ɛm'pfɛŋɐ] **der; ~s, ~ a)** recipient; *(eines Briefs)* addressee; **~ unbekannt** not known at this address; **b)** *(Empfangsgerät)* receiver

Empfängerin die; ~, ~nen *s.* **Empfänger a**

empfänglich *Adj.* **a)** receptive **(für** to); **b)** *(anfällig, auch fig.)* susceptible **(für** to)

Empfänglichkeit die; ~ a) *(Zugänglichkeit)* receptivity, receptiveness **(für** to); **b)** *(Auffälligkeit, auch fig.)* susceptibility **(für** to)

Empfängnis die; ~: conception

empfängnis·verhütend *Adj.* **ein ~es Mittel** a contraceptive

Empfängnis·verhütung die contraception

Empfängnisverhütungs·mittel das contraceptive

Empfangs-: ~antenne die [receiving] aerial *(Brit.) or (Amer.)* antenna; **~bestätigung die** receipt; **~chef der** head receptionist; **~dame die** receptionist; **~gerät das** receiver; **~halle die** reception lobby

empfehlen [ɛm'pfe:lən] **1.** *unr. tr. V.* **jmdm. etw./jmdn. ~:** recommend sth./sb. to sb.; **etw. ist sehr zu ~:** sth. is to be highly recommended. **2.** *unr. refl. V.* **a)** *(geh.: sich verabschieden und gehen)* take one's leave; **darf ich mich ~?** may I take my leave?; **b)** *(ratsam sein)* be advisable; **es empfiehlt sich, ... zu ...:** it is advisable to ...; **c)** *(geh.: sich als geeignet ausweisen)* **sich [durch/wegen etw.] ~:** commend oneself/itself [because of sth.]

empfehlens·wert *Adj.* **a)** to be recommended *postpos.;* recommendable; **b)** *(ratsam)* advisable

Empfehlung die; ~, ~en a) recommendation; **b)** *(Empfehlungsschreiben)* letter of recommendation; **c)** *(höflicher Gruß)* **„mit freundlicher ~"** 'with kindest regards'

Empfehlungs·schreiben das letter of recommendation

empfiehl [ɛm'pfi:l] *Imperativ Sg. v.* **empfehlen**

empfiehlst *2. Pers. Sg. Präsens v.* **empfehlen**

empfiehlt *3. Pers. Sg. Präsens v.* **empfehlen**

empfinden [ɛm'pfɪndn̩] *unr. tr. V.* feel ⟨*pain, pleasure, bitterness, etc.*⟩; **etwas/nichts für jmdn. ~:** feel something/nothing for sb.; **etw. als Beleidigung ~:** feel sth. to be an insult

Empfinden das; ~s feeling; **für mein** *od.* **nach meinem ~:** to my mind

empfindlich 1. *Adj.* **a)** *(sensibel, feinfühlig, auch fig.)* sensitive; fast ⟨*film*⟩; **eine ~e Stelle** a tender spot; **b)** *(leicht beleidigt)* sensitive, touchy ⟨*person*⟩; **c)** *(anfällig)* delicate; **~ gegen Viruserkrankungen** prone to virus infections; **d)** *(spürbar)* severe ⟨*punishment, shortage*⟩; harsh ⟨*punishment, measure*⟩; sharp ⟨*increase*⟩. **2.** *adv.* **a) ~ auf etw.** *(Akk.)* **reagieren** *(sensibel)* be susceptible to sth.; *(beleidigt)* react oversensitively to sth.; **b)** *(spürbar)* ⟨*punish*⟩ severely, harshly; ⟨*increase*⟩ sharply; **c)** *(intensivierend)* ⟨*hurt*⟩ badly; bitterly ⟨*cold*⟩

Empfịndlichkeit die; ~, ~en *s.* **empfindlich:** sensitivity; touchiness; severity; harshness; *(eines Films)* speed

empfịndsam 1. *Adj.* sensitive ⟨*nature*⟩; *(gefühlvoll)* sentimental. **2.** *adv.* sensitively; *(gefühlvoll)* sentimentally

Empfịndsamkeit die; ~: sensitivity; *(Literaturw.)* sentimentality

Empfịndung die; ~, ~en a) *(sinnliche Wahrnehmung)* sensation; sensory perception; **b)** *(Gefühl)* feeling; emotion

empfịndungs-, Empfịndungs-: ~los *Adj.* **a)** *(körperlich)* numb; **b)** *(seelisch)* insensitive; unfeeling; **~losigkeit die; ~ a)** *(körperlich)* numbness; **b)** *(seelisch)* insensitivity

empfing [ɛm'pfɪŋ] *1. u. 3. Pers. Sg. Prät. v.* **empfangen**

empfohlen [ɛm'pfo:lən] *Adj.* recommended

empịrisch [ɛm'pi:rɪʃ] **1.** *Adj.* empirical. **2.** *adv.* empirically

empor [ɛm'po:ɐ̯] *Adv. (geh.)* upwards; up

empọr-: ~|arbeiten *refl. V. (geh.)* work one's way up; **~|blicken** *itr. V. (geh.)* look upwards *or (literary)* heavenwards

Empọre die; ~, ~n gallery

empö̤ren [ɛm'pø:rən] **1.** *tr. V.* fill with indignation; incense; outrage. **2.** *refl. V.* become indignant *or* incensed *or* outraged (**über** + *Akk.* about)

empö̤rend *Adj.* outrageous

empọr-: ~|heben *unr. tr. V. (geh.)* raise; **~|kommen** *unr. itr. V.; mit sein (geh.)* **a)** *(nach oben kommen)* come up; **b)** *(fig.: aufsteigen)* rise

Empọrkömmling [-kœmlɪŋ] **der; ~s, ~e** *(abwertend)* upstart; parvenu

empọr-: ~|ragen *itr. V. (geh.)* rise [up]; **~|schauen** *itr. V. (geh.)* raise one's eyes; **~|schwingen** *unr. refl. V. (geh.)* swing oneself aloft; **~|steigen** *unr. itr. V.; mit sein (geh.)* **a)** climb up; **b)** ⟨*balloon, kite*⟩ rise aloft

empö̤rt 1. *Adj.* outraged ⟨*letter, look*⟩; **über jmdn./etw. ~ sein** be outraged at sth./about sb. **2.** *adv.* **jmdn./etw. ~ zurückweisen** reject sb./sth. indignantly *or* angrily

Empö̤rung die; ~ outrage

emsig ['ɛmzɪç] **1.** *Adj. (fleißig)* industrious, busy ⟨*person*⟩; *(geschäftig)* bustling ⟨*activity*⟩; *(übereifrig)* sedulous; **ein ~es Treiben** bustling activity; a hustle and bustle. **2.** *adv. (fleißig)* industriously; busily; *(übereifrig)* sedulously

Ẹmsigkeit die; ~ *(Fleiß)* industriousness; business; *(Übereifer)* sedulousness

Emu ['e:mu] **der; ~s, ~s** *(Zool.)* emu

Emulsion [emʊl'zi̯o:n] **die; ~, ~en** *(fachspr.)* emulsion

E̤-Musik die; ~: serious music

Ẹnd-: ~abnehmer der *(Wirtsch.)* ultimate buyer; **~abrechnung die** final account; **~bahnhof der** terminus

Ende ['ɛndə] **das; ~s, ~n a)** end; **am ~:** at the end; *(schließlich)* in the end; **am ~ der Welt** *(scherzh.)* at the back of beyond; **am/bis/gegen ~ des Monats/der Woche** at/by/towards the end of the month/week; **~ April** at the end of April; **bis ~ der Woche** by the end of the week; **zu ~ sein** ⟨*patience, hostility, war*⟩ be at an end; **die Schule/das Kino/das Spiel ist zu ~:** school is over/the film/game has finished; **zu ~ gehen** ⟨*period of time*⟩ come to an end; ⟨*supplies, savings*⟩ run out; ⟨*contract*⟩ expire; **etw. zu ~ führen** *od.* **bringen** finish sth.; **ein Buch zu ~ lesen** read a book to the end; **~ gut, alles gut** all's well that ends well *(prov.)*; **ein/kein ~ nehmen** come to an end/never come to an end; **einer Sache/seinem Leben ein ~ machen** *od.* **setzen** *(geh.)* put an end to sth./take one's life; **am ~ sein** *(ugs.)* be at the end of one's tether; **ich bin mit meiner Geduld am ~:** my patience is at an end; **b)** *(ugs.: kleines Stück)* bit; piece; **c)** *(ugs.: Strecke)* **ein ganzes ~:** a pretty long way; **d)** *(Jägerspr.)* point

Ẹnd·effekt der: im ~: in the end; in the final analysis

ẹnden *itr. V.* **a)** end; ⟨*programme*⟩ end, finish; **der Zug endet hier** this train terminates here; **gut ~:** end well; **nicht ~ wollender Beifall** unending applause; **b)** *(sterben) mit sein* **in der Gosse/im Gefängnis ~:** die in the gutter/end one's days in prison

Ẹnd·ergebnis das final result

ẹnd·gültig 1. *Adj.* final ⟨*consent, answer, decision*⟩; conclusive ⟨*evidence*⟩; **etwas/nichts Endgültiges sagen/hören** say/hear something/nothing definite. **2.** *adv.* **das ist ~ vorbei** that's all over and done with; **sich ~ trennen** separate for good

Endivie [ɛn'di:vi̯ə] **die; ~, ~n** endive

Endịvien·salat der a) endive; **b)** *(Speise)* endive salad

Ẹnd-: ~kampf der *(Sport)* final; *(Milit.)* final battle; **~lagerung die** permanent disposal *(of nuclear waste)*; **~lauf der** final

ẹndlich 1. *Adv.* **a)** *(nach langer Zeit)* at last; **na ~ [kommst du]!** [so you've arrived] at [long] last; **b)** *(schließlich)* in the end; eventually. **2.** *Adj.* finite ⟨*size, number*⟩

ẹnd·los 1. *Adj.* **a)** *(ohne Ende)* infinite; *(ringförmig)* endless, continuous ⟨*belt, chain*⟩; **b)** *(nicht enden wollend)* endless ⟨*road, desert, expanse, etc.*⟩; interminable ⟨*speech*⟩. **2.** *adv.* **~ lange dauern** be interminably long

Ẹnd-: ~phase die final stages *pl.*; **~produkt das** final *or* end product; **~punkt der** end; **~resultat das** final result; **~runde die** *(Sport)* final; **~runden·teilnehmer der** *(Sport)* finalist; **~sieg der** *(bes. ns.)* final *or* ultimate victory; **~spiel das a)** *(Sport)* final; **b)** *(Schach)* endgame; **~spurt der** *(bes. Leichtathletik)* final spurt; **einen guten ~spurt haben** have a good finish; **~stadium das** final stage; *(Med.)* terminal stage; **Krebs im ~stadium** terminal cancer; **~stand der** *(Sport)* final result; **~station die** terminus; **~stück das** end; *(eines Brotes)* crust

Ẹndung die; ~, ~en *(Sprachw.)* ending

Ẹnd-: ~ziel das *(einer Reise)* final destination; *(Zweck)* ultimate aim *or* goal; **~zweck der** ultimate purpose *or* object

Energie [enɛr'gi:] **die; ~, ~n a)** *(Physik)* energy; **b)** *o. Pl. (Tatkraft)* energy; vigour

energie̲-, Energie̲-: ~bedarf der energy requirement; **~geladen** *Adj.* energetic, dynamic ⟨*person*⟩; **~gewinnung die** energy production; **~haushalt der** *(Physiol.)* energy balance; **~krise die** energy crisis; **~los** *Adj.* lacking [in] energy *postpos.*; sluggish; **~politik die** energy policy; **~quelle die** energy source; source of energy; **~sparend** *Adj.* energy-saving; **~verbrauch der** energy consumption; **~verschwendung die** wasting of energy; **~versorgung die** energy supply; **~wirtschaft die** energy sector

energisch [e'nɛrgɪʃ] **1.** *Adj.* **a)**

(tatkräftig) energetic, vigorous ‹*person*›; firm ‹*action*›; ~ **werden** put one's foot down; **b)** *(von starkem Willen zeugend)* determined; forceful; strong ‹*chin*›; **c)** *(entschlossen)* forceful, firm ‹*voice, words*›. **2.** *adv.* **a)** *(tatkräftig)* ~ **durchgreifen** take drastic action; **b)** *(entschlossen)* ‹*reject, say*› forcefully, firmly; ‹*stress*› emphatically; ‹*deny*› strenuously

eng [ɛŋ] **1.** *Adj.* **a)** *(schmal)* narrow ‹*valley, road, bed*›; **einen ~en Horizont haben** *(fig.)* have a narrow *or* limited outlook; **b)** *(fest anliegend)* close-fitting, tight; **der Anzug/Rock ist zu ~:** the suit/skirt is too tight; **c)** *(beschränkt)* narrow, restricted ‹*interpretation, concept*›; cramped, constricted ‹*room, space*›; **d)** *im Komp. u. Sup. (begrenzt)* **in die ~ere Wahl kommen** be short-listed *(Brit.)*; **im ~eren Sinne** in the stricter sense; **e)** *(nahe)* close ‹*friend*›; **im ~sten Freundeskreis** among close friends; **die ~ere Verwandtschaft** one's immediate relatives. **2.** *adv.* **a)** *(dicht)* ~ **schreiben** write closely together; ~ **[zusammen] sitzen/stehen** sit/stand close together; **b)** *(fest anliegend)* ~ **anliegen/sitzen** fit closely; **c)** *(beschränkt)* **etw. zu ~ auslegen** interpret sth. too narrowly; **d)** *(nahe)* closely; **mit jmdm. ~ befreundet sein** be a close friend of sb.

Engadin ['ɛŋgadi:n] **das; ~s** Engadine

Engagement [ãgaʒə'mã:] **das; ~s, ~s a)** *o. Pl. (Einsatz)* involvement; **sein ~ für etw.** his commitment to sth.; **sein ~ gegen etw.** his committed stand against sth.; **b)** *(eines Künstlers)* engagement

engagieren [ãga'ʒi:rən] **1.** *refl. V.* commit oneself, become committed **(für** to); **sich politisch ~:** become politically involved. **2.** *tr. V.* engage ‹*artist, actor, etc.*›

engagiert *Adj.* committed ‹*literature, film, director*›; **politisch/sozial ~ sein** be politically/socially committed *or* involved

eng-: ~anliegend *Adj. (präd. getrennt geschrieben)* tight-fitting; close-fitting; **~bedruckt** *Adj. (präd. getrennt geschrieben)* closely-printed ‹*page*›

Enge ['ɛŋə] **die; ~, ~n a)** *o. Pl.* confinement; restriction; **b) jmdn. in die ~ treiben** *(fig.)* drive sb. into a corner

Engel ['ɛŋl̩] **der; ~s, ~:** angel; **sie ist mein guter/ein rettender ~:** she is my good/a guardian angel

Engel-: ~macher der, ~macherin die; ~, ~nen *(ugs. verhüll.)* backstreet abortionist; **~schar die** heavenly host; host of angels

Engels-: ~geduld die patience of a saint; **~miene die** innocent look; **~zungen** *Pl.*: **mit ~zungen auf jmdn. einreden** use all one's powers of persuasion on sb.

eng·herzig 1. *Adj.* petty. **2.** *adv.* in a petty way

Eng·herzigkeit die; ~: pettiness

England (das); ~s a) England; **b)** *(ugs.: Großbritannien)* Britain

Engländer ['ɛŋlɛndɐ] **der; ~s, ~ a)** Englishman/English boy; **er ist ~:** he is English *or* an Englishman; **die ~:** the English; **b)** *(ugs.: Brite)* British person/man; Britisher *(Amer.)*; **die ~:** the British; **c)** *(Schraubenschlüssel)* monkey wrench

Engländerin die; ~, ~nen a) Englishwoman/English girl; **sie ist ~:** she is English *or* an Englishwoman; **b)** *(ugs.: Britin)* British person/woman; **die ~nen sind ...:** British women are ...

englisch 1. *Adj.* English; **~-deutsch** Anglo-German; English-German ‹*dictionary*›; **die ~e Sprache/Literatur** the English language/English literature; **die ~e Krankheit** *(veralt.)* rickets. **2.** *adv.* ~ **[gebraten]** rare; underdone; **ein ~ abgefaßter Artikel** an article in English; *s. auch* **deutsch, ²Deutsche**

Englisch das; ~[s] English; **ein gutes/fehlerfreies ~ sprechen** speak good/perfect English; *s. auch* **Deutsch**

englisch-, Englisch-: ~horn das *(Musik)* cor anglais; **~lehrer der** 'English teacher; **~sprachig** *Adj.* **a)** *(in ~er Sprache)* English-language ‹*book, magazine*›; **die ~sprachige Literatur** English literature; **b)** *(~ sprechend)* English-speaking ‹*population, country*›; **~unterricht der** English teaching; *(Unterrichtsstunde)* English lesson; **er gibt ~unterricht** he teaches English

eng·maschig *Adj.* close-meshed ‹*fabric*›

Eng·paß der a) [narrow] pass; defile; **b)** *(fig.: in der Versorgung usw.)* bottle-neck

en gros [ã'gro] *(Kaufmannsspr.)* wholesale

eng·stirnig 1. *Adj. (abwertend)* narrow-minded ‹*person*›. **2.** *adv.* **~stirnig handeln** be narrow-minded in the way one acts

Eng·stirnigkeit die; ~: narrow-mindedness

Enkel der; ~s, ~ a) grandson; **b)** *(Nachfahr)* grandchild

Enkelin die; ~, ~nen granddaughter

Enkel-: ~kind das grandchild; **~sohn der** grandson; **~tochter die** granddaughter

Enklave [ɛn'kla:və] **die; ~, ~en** enclave

en masse [ã'mas] *(ugs.)* en masse

enorm [e'nɔrm] **1.** *Adj.* enormous ‹*sum, costs*›; tremendous *(coll.)* ‹*effort*›; immense ‹*strain*›; vast ‹*knowledge, sum*›. **2.** *adv.* tremendously *(coll.)* ‹*expensive, practical*›; ~ **viel/viele** a tremendous *(coll.) or* an enormous amount/number

en passant [ãpa'sã] en passant; in passing

Ensemble [ã'sã:bl̩] **das a)** *(auch fig. geh.)* ensemble; *(von Schauspielern)* company; **b)** *(Kleidungsstücke)* outfit

entarten *itr. V.; mit sein* degenerate; **entartet** degenerate; **zu** *od.* **in** *(Akk.)* **etw. ~:** degenerate into sth.

Entartung die; ~, ~en degeneration

entäußern *refl. V. (geh.)* **sich einer Sache** *(Gen.)* ~ *(entsagen)* renounce sth.; *(weggeben)* relinquish *or* give up sth.

entbehren [ɛnt'be:rən] **1.** *tr. V.* **a)** *(geh.: vermissen)* miss ‹*person*›; **b)** *(verzichten)* do without; spare; **etw./jmdn. nicht ~ können** not be able to do without sth./sb.; **viel[es] ~ müssen** have to go without [a lot of things]. **2.** *itr. V. (geh.: ermangeln)* **einer Sache** *(Gen.)* **~:** lack *or* be without sth.

entbehrlich *Adj.* dispensable

Entbehrung die; ~, ~en privation; **große ~en auf sich** *(Akk.)* **nehmen** make great sacrifices

entbehrungs-: ~reich, ~voll *Adj.* ‹*life, years*› of privation

entbinden 1. *unr. tr. V.* **a)** *(befreien)* **jmdn. von einem Versprechen ~:** release sb. from a promise; **seines Amtes** *od.* **von seinem Amt entbunden werden** be relieved of [one's] office; **b)** *(Geburtshilfe leisten)* **jmdn. ~:** deliver sb.; deliver sb.'s baby; **von einem Jungen/Mädchen entbunden werden** give birth to a boy/girl. **2.** *unr. itr. V. (gebären)* give birth; **zu Hause ~:** have one's baby at home

Entbindung die a) *(das Gebären)* birth; delivery; **bei der ~ anwesend sein** be present at the birth; **b)** *(Befreiung)* release

Entbindungs·station die maternity ward

entblättern *refl. V.* **a)** ‹*trees, shrubs*› shed its/their leaves; **b)** *(scherzh.: sich ausziehen)* strip; take one's clothes off

entblößen **1.** *refl. V.* take one's clothes off; ‹*exhibitionist*› expose oneself. **2.** *tr. V.* **a)** uncover; **entblößt** bare; **b)** *(fig.)* reveal ‹*feelings, thoughts*›

entbrennen *unr. itr. V.; mit sein (geh.)* **a)** *(beginnen)* ‹*battle*› break out; ‹*quarrel*› flare up; **b) in Liebe entbrannt sein** be passionately in love

Entchen ['ɛntçən] **das; ~s, ~:** duckling

entdecken *tr. V.* **a)** *(finden)* discover; **b)** *(ausfindig machen)* **jmdn. ~:** find *or* spot sb.; **etw. ~:** find *or* discover sth.; **c)** *(überraschend bemerken)* discover ‹*theft*›; come across ‹*acquaintance*›

Entdecker der; ~s, ~: discoverer; *(Forschungsreisender)* explorer

Entdeckung die; ~, ~en discovery

Entdeckungs·reise die voyage of discovery; *(zu Lande)* expedition

Ente ['ɛntə] **die; ~, ~n a)** duck; **eine lahme ~** *(ugs.)* a slow-coach *(coll.)*; **b)** *(ugs.: Falschmeldung)* canard; spoof *(coll.)*; **c) kalte ~:** [cold] punch; **d)** *(ugs.: Auto)* Citroën 2CV car; **e)** *(ugs.: Uringefäß)* [bed-]bottle

enteignen *tr. V.* expropriate

Enteignung die expropriation

enteisen *tr. V.* de-ice

Enten-: ~braten der roast duck; **~ei das** duck's egg; **~küken das** duckling

Entente [ã'tã:t(ə)] **die; ~, ~n** *(Politik)* entente

Enten·teich der duck pond

enterben *tr. V.* disinherit

Enter·haken der grapnel; grappling iron

Enterich ['ɛntərıç] **der; ~s, ~e** drake

entern ['ɛntɐn] *tr. V.* board ‹*ship*›

entfachen *tr. V. (geh.)* **a)** kindle, light ‹*fire*›; **einen Brand ~:** start a fire; **b)** *(fig.)* provoke, start ‹*quarrel, argument*›; arouse ‹*passion, enthusiasm*›

entfahren *unr. itr. V.; mit sein* **ihm entfuhr ein Fluch/ein Seufzer** he swore inadvertently/he let out a sigh

entfallen *unr. itr. V.; mit sein* **a) der Name/das Wort ist mir ~:** the name/word escapes me *or* has slipped my mind; **b) auf jmdn./etw. ~:** be allotted to sb./sth.; **auf jeden Erben/Miteigentümer entfielen 10 000 Mark** each heir received/each of the joint owners had to pay 10,000 marks; **c)** *(wegfallen)* lapse; **für Kinder ~ diese Gebühren** these charges do not apply to children

entfalten **1.** *tr. V.* **a)** open [up]; unfold, spread out ‹*map etc.*›; **b)** *(fig.)* show, display ‹*ability, talent*›; **c)** *(fig.)* expound ‹*ideas, thoughts*›; present ‹*plan*›. **2.** *refl. V.* **a)** ‹*flower, parachute*› open [up]; **b)** *(fig.)* ‹*personality, talent*› develop; **sich frei ~:** develop one's own personality to the full

Entfaltung die; ~ *(fig.)* **a)** *(Entwicklung)* development; **zur ~ kommen** develop; **b)** *s.* **entfalten 1b:** display; **c)** *s.* **entfalten 1c:** exposition; presentation

entfärben **1.** *tr. V.* take the colour out of ‹*material, clothing*›. **2.** *refl. V.* ‹*material, clothing, etc.*› fade

Entfärber der colour *or* dye remover

entfernen **1.** *tr. V.* **a)** remove ‹*stain, wart, etc.*›; take out ‹*tonsils etc.*›; **jmdn. aus seinem Amt ~:** dismiss sb. from office; **b)** *(geh.: fortbringen)* remove. **2.** *refl. V.* go away; **sich vom Weg ~:** go off *or* leave the path; **sich unerlaubt von der Truppe ~:** go absent without leave

entfernt **1.** *Adj.* **a)** away **(von** from); **voneinander ~:** apart; **das ist** *od.* **liegt weit ~ von der Stadt** it is a long way from the town *or* out of town; **er ist weit davon ~, das zu tun** *(fig.)* he does not have the slightest intention of doing that; **b)** *(fern, entlegen)* remote; **c)** *nicht präd. (weitläufig)* slight ‹*acquaintance*›; distant ‹*relation*›; **d)** *(schwach)* slight, vague ‹*resemblance*›. **2.** *adv.* **a)** *(fern)* remotely; **nicht im ~esten** not in the slightest *or* in the least; **b)** *(weitläufig)* slightly ‹*acquainted*›; distantly ‹*related*›; **c)** *(schwach)* slightly, vaguely

Entfernung die; ~, ~en a) *(Abstand)* distance; *(beim Schießen)* range; **in einer ~ von 100 m** at a distance/range of 100 m.; **auf eine ~ von 100 m** from a distance of 100 m.; **aus der ~:** from a distance; **b)** *(das Beseitigen)* removal

entfesseln *tr. V.* unleash ‹*war, riot, etc.*›; **entfesselt** raging ‹*elements*›

entfetten *tr. V.* skim ‹*milk*›

Entfettungs·kur die diet to remove one's excess fat

entflammbar *Adj.* inflammable

entflammen **1.** *tr. V.* arouse ‹*enthusiasm etc*›; **jmdn. für etw. ~:** arouse sb.'s enthusiasm for sth. **2.** *itr. V.; mit sein* ‹*hatred etc.*› flare up; ‹*battle, strike*› break out; **für jmdn. entflammt sein** be passionately in love with sb.

entflechten *unr. (auch regelm.) tr. V.* **a)** *(entwirren)* disentangle; **b)** *(Wirtsch.)* break up ‹*cartel etc.*›

Entflechtung die; ~, ~en *(Wirtsch.)* breaking-up; break-up

entfliegen *unr. itr. V.; mit sein* fly away

entfliehen *unr. itr. V.; mit sein* **jmdm./einer Sache ~:** escape sb./sth.; **dem Alltag ~** *(geh.)* escape from the daily routine

entfremden **1.** *tr. V.* **jmdn. einer Sache** *(Dat.)* **~:** alienate *or* estrange sb. from sth.; **etw. seinem Zweck ~:** use sth. for a different purpose. **2.** *refl. V.* **sich jmdm./einer Sache ~:** become estranged from sb./unfamiliar with sth.

entfremdet *Adj. (Philos., Soziol.)* alienated

Entfremdung die; ~, ~en a) alienation; estrangement; **b)** *(Philos., Soziol.)* alienation

entfrosten *tr. V.* defrost

Entfroster der; ~s, ~: defroster

entführen *tr. V.* **a)** kidnap, abduct ‹*child etc.*›; hijack ‹*plane, lorry, etc.*›; **b)** *(scherzh.: mitnehmen)* steal; make off with

Entführer der *s.* **entführen a:** kidnapper; abducter; hijacker

Entführung die *s.* **entführen a:** kidnap; kidnapping; abduction; hijack; hijacking

entgegen **1.** *Adv.* towards; **der Sonne ~!** on towards the sun! **2.** *Präp. mit Dat.* **~ meinem Wunsch** against my wishes; **~ dem Befehl** contrary to orders

entgegen-, Entgegen-: ~|bringen *unr. tr. V. (fig.)* **jmdm. Liebe/Verständnis ~bringen** show sb. love/understanding; **~|gehen** *unr. itr. V.; mit sein* **a) jmdm. [ein Stück] ~gehen** go [a little way] to meet sb.; **b)** *(fig.)* **einer Katastrophe/schweren Zeiten ~gehen** be heading for *or* towards a catastrophe/hard times; **der Vollendung/dem Ende ~gehen** be approaching completion/its end; **~gesetzt** **1.** *Adj.* **a)** opposite ‹*end, direction*›; **b)** *(gegensätzlich)* opposing; **~gesetzter Meinung sein** hold opposing views; **das Entgegengesetzte tun** do the opposite; **2.** *adv.* **genau ~gesetzt handeln/denken** do/think exactly the opposite; **~|halten** *unr. tr. V.*

a) jmdm. etw. ~halten offer sth. to sb.; **b)** *(fig.: einwenden)* **einem Argument ein anderes ~halten** counter an argument with another; **~|kommen** *unr. itr. V.; mit sein* **a) jmdm. ~kommen** come to meet sb.; **der ~kommende Verkehr** oncoming traffic; **b)** *(fig.)* **jmdm. ~kommen** be accommodating towards sb.; *(in Verhandlungen)* make concessions; **sie/das kam unseren Wünschen ~:** she complied with our wishes/it was what we wanted; **~kommen das a)** *(Konzilianz)* co-operation; **b)** *(Zugeständnis)* concession; **~kommend** *Adj.* obliging; **~kommenderweise** *Adv.* obligingly; **~nahme die** *(Amtsdt.)* receipt; **~|nehmen** *unr. itr. V.* receive; accept ⟨*parcel*⟩; **~|sehen** *unr. itr. V.* **einer Sache** *(Dat.)* **[freudig] ~sehen** look forward [eagerly] to sth.; **~|setzen** *tr. V.* **a) einer Sache** *(Dat.)* **etw. ~setzen** oppose sth. with sth.; **einer Sache** *(Dat.)* **Widerstand ~setzen** resist sth.; **b) einer Behauptung/einem Argument etw. ~setzen** counter a claim/an argument with sth.; **~|stellen** *tr. V. s.* **~setzen b; ~|treten** *unr. itr. V.; mit sein* go/come up to; *(fig.)* **Schwierigkeiten** *(Dat.)* **~treten** stand up to difficulties; **einem Angriff ~treten** answer an attack; **Vorwürfen/Anschuldigungen ~treten** answer reproaches/accusations; **~|wirken** *itr. V.* **einer Sache** *(Dat.)* **~wirken** [actively] oppose sth.

entgegnen [ɛnt'ge:gnən] *tr. V.* retort; reply; **auf etw.** *(Dat.)* *od.* **einer Sache** *(Dat.)* **etw. ~:** say sth. in reply to sth.; **jmdm. ~, daß ...:** reply that ...

Entgegnung die; ~, ~en retort; reply; **als ~ darauf** in reply

entgehen *unr. itr. V.; mit sein* **a) einer Gefahr/Strafe** *(Dat.)* **~:** escape *or* avoid danger/punishment; **das darf man sich** *(Dat.)* **nicht ~ lassen** *(fig.)* that is not to be missed; **b) jmdm. entgeht etw.** sb. misses sth.; **ihm ist nicht entgangen, daß ...:** it has not escaped his notice that ...

entgeistert [ɛnt'gaistɐt] *Adj.* dumbfounded; **jmdn. ~ anstarren** stare at sb. in amazement *or* astonishment

Entgelt [ɛnt'gɛlt] **das; ~[e]s, ~e** payment; fee; **gegen** *od.* **für ein geringes ~:** for a small fee

entgelten *unr. tr. V. (geh.)* pay for *(also fig.)*; **jmdm. eine Arbeit ~:** pay sb. a job

entgleisen *itr. V.; mit sein* **a)** be derailed; **der Zug ist entgleist** the train was derailed; **b)** *(fig.: aus der Rolle fallen)* make *or* commit a/some faux pas

Entgleisung die; ~, ~en *s.* **entgleisen: a)** derailment; **b)** *(fig.)* faux pas

entgleiten *unr. itr. V.; mit sein (geh.)* **a)** slip; **jmds. Händen ~:** slip from sb.'s hands; **b)** *(fig.)* **jmdm. entgleitet etw.** sb. loses his/her grip on sth.

entgräten *tr. V.* fillet; bone; **entgräteter Fisch** filleted fish

enthaaren *tr. V.* remove hair from; depilate *(formal)*

Enthaarungs·mittel das hair remover; depilatory

[1]**enthalten 1.** *unr. tr. V.* contain. **2.** *unr. refl. V.* **sich einer Sache** *(Gen.)* **~:** abstain from sth.; **sich der Stimme ~:** abstain; **sich jeder Meinung/Äußerung ~:** refrain from giving any opinion/making any comment

[2]**enthalten** *Adj.* **in etw.** *(Dat.)* **~ sein** be contained in sth.; **das ist im Preis ~:** that is included in the price

enthaltsam 1. *Adj.* abstemious; *(sexuell)* abstinent. **2.** *adv.* **~ leben** live in abstinence

Enthaltsamkeit die; ~: abstinence

Enthaltung die abstention

enthärten *tr. V.* soften ⟨*water*⟩

enthaupten *tr. V. (geh.)* behead

Enthauptung die; ~, ~en *(geh.)* beheading

enthäuten *tr. V.* skin

entheben *unr. tr. V. (geh.)* relieve; **jmdn. seines Amtes ~:** relieve sb. of his/her office

enthemmen *tr. V.* **jmdn. ~:** make sb. lose his/her inhibitions

enthemmend 1. *Adj.* disinhibitory ⟨*effect, etc.*⟩. **2.** *adv.* **~ wirken** take away sb.'s inhibitions

enthemmt *Adj.* uninhibited

Enthemmung die loss of inhibition[s]; disinhibition *(Psych.)*

enthüllen 1. *tr. V.* **a)** unveil ⟨*monument etc.*⟩; reveal ⟨*face, etc.*⟩; **b)** *(offenbaren)* reveal ⟨*truth, secret*⟩; disclose ⟨*secret*⟩; *(Zeitungsw.)* expose ⟨*scandal*⟩. **2.** *refl. V.* **sich [jmdm.] ~:** be revealed [to sb.]

Enthüllung die; ~, ~en *s.* **enthüllen 1:** unveiling; revelation; disclosure; exposé

enthülsen *tr. V.* shell; pod

Enthusiasmus [ɛntu'zi̯asmʊs] **der; ~:** enthusiasm

Enthusiast der; ~en, ~en enthusiast

enthusiastisch 1. *Adj.* enthusiastic. **2.** *adv.* enthusiastically

entjungfern *tr. V.* deflower

entkalken *tr. V.* decalcify

entkernen *tr. V.* core ⟨*apple etc.*⟩; stone, remove stone from ⟨*plum etc.*⟩; remove pips from ⟨*grape etc.*⟩

entkleiden *tr. V.* **jmdn./sich ~** *(geh.)* undress sb./undress

entkoffeiniert [ɛntkɔfei'ni:ɐ̯t] *Adj.* decaffeinated

entkommen *unr. itr. V.; mit sein* escape; **jmdm./einer Sache ~:** escape *or* get away from sb./sth.; **es gibt kein Entkommen** there is no escape

entkorken *tr. V.* uncork ⟨*bottle*⟩

entkräften [ɛnt'krɛftn̩] *tr. V.* **a)** weaken; **völlig ~:** exhaust; **[völlig] entkräftet sein** be [utterly] exhausted; **b)** *(fig.)* refute, invalidate ⟨*argument etc.*⟩; remove ⟨*suspicion etc.*⟩

Entkräftung die; ~, ~en a) debility; **völlige ~:** exhaustion; **b)** *(fig.)* refutation; invalidation

entkrampfen 1. *tr. V.* **a)** relax; **b)** *(fig.)* ease ⟨*situation, tension*⟩. **2.** *refl. V.* **a)** relax; **b)** *(fig.)* ⟨*atmosphere etc.*⟩ become relaxed

Entkrampfung die; ~, ~en *s.* **entkrampfen:** relaxation; easing

entladen 1. *unr. tr. V.* unload; discharge ⟨*battery*⟩. **2.** *unr. refl. V.* **a)** ⟨*storm*⟩ break; **b)** *(fig.: hervorbrechen)* ⟨*anger etc.*⟩ erupt; ⟨*aggression etc.*⟩ be released; **c)** *(Elektrot.)* ⟨*battery*⟩ run down

Entladung die *s.* **entladen 1, 2b:** unloading; discharge; eruption; release

entlang 1. *Präp. mit Akk. u. Dat.* along; **den Weg ~, ~ dem Weg** along the path. **2.** *Adv.* along; **dort ~, bitte!** that way please!

entlang-: ~|fahren *unr. itr. V.; mit sein* **a)** drive along; **die Straße/den** *od.* **am Fluß ~fahren** drive *or* go down the street/along the river; **b)** *(streichen)* go along; **~|führen 1.** *tr. V.* lead along; **jmdn. die Straße ~führen** lead sb. along *or* down the street; **2.** *itr. V.* *(verlaufen)* run *or* go along; **~|gehen** *unr. itr. V.; mit sein* ⟨*person*⟩ go *or* walk along; **bitte gehen Sie hier ~:** [go] this way please; **~|laufen** *unr. itr. V.; mit sein* **a)** go *or* walk/run along; **b)** *(verlaufen)* go *or* run along

entlarven *tr. V.* expose

Entlarvung die; ~, ~en exposure

entlassen *unr. tr. V.* **a)** *(aus dem Gefängnis)* release; *(aus dem Krankenhaus, der Armee)* discharge; **jmd. wird aus der Schule**

~: sb. leaves school; **b)** *(aus einem Arbeitsverhältnis)* dismiss; *(wegen Arbeitsmangels)* make redundant *(Brit.);* lay off; **bei einer Firma ~ werden** be dismissed from/be made redundant *(Brit.)* *or* laid off by a company; **c)** *(geh.: gehen lassen)* release

Entlassung die; ~, ~en a) *(aus dem Gefängnis)* release; *(aus dem Krankenhaus, der Armee)* discharge; *(aus der Schule)* leaving; **b)** *(aus einem Arbeitsverhältnis)* dismissal; *(wegen Arbeitsmangels)* redundancy *(Brit.);* laying off; **c)** *s.* **Entlassungsschreiben**

Entlassungs-: ~feier die *(Schulw.)* school-leaving *or* *(Amer.)* graduation ceremony; **~schreiben das** *(Arbeitsw.)* notice of dismissal; *(wegen Arbeitsmangels)* redundancy notice *(Brit.);* pink slip *(Amer.)*

entlasten *tr. V.* **a)** relieve; **jmdn. ~:** relieve *or* take the load off sb.; **den Kreislauf ~:** relieve the strain on the circulation; **sein Gewissen ~:** ease *or* relieve one's conscience; **b)** *(Rechtsspr.)* exonerate ‹*defendant*›

Entlastung die; ~, ~en a) *(Rechtsw.)* exoneration; defence; **zu jmds. ~:** in sb.'s defence; **b)** *(Minderung der Belastung)* relief; **die ~ eines Menschen/des Körpers/der Straßen** relief of the burden on a person/the body/the roads; **c)** *(Erleichterung)* easing; relief

Entlastungs-: ~material das *(Rechtsw.)* evidence for the defence; **~zeuge der** *(Rechtsw.)* witness for the defence; defence witness; **~zug der** *(Eisenb.)* relief train

entlauben *tr. V.* strip ‹*branch*›; defoliate ‹*forest, area*›

entlaufen *unr. itr. V.; mit sein* run away; **jmdm. ~:** run away from sb.; **ein ~er Sträfling/Sklave** an escaped convict/a runaway slave

entledigen *refl. V. (geh.)* **a) sich jmds./einer Sache** *(Gen.)* **~:** dispose of *or* rid oneself of sb./sth.; **b) sich eines Kleidungsstücks ~:** remove an item of clothing; **c)** *(erledigen)* **sich einer Aufgabe/einer Schuld/seiner Pflichten ~:** carry out a task/discharge a debt/one's duty

entleeren 1. *tr. V.* empty ‹*ashtray etc.*›; evacuate ‹*bowels, bladder*›. **2.** *refl. V.* empty; become empty

Entleerung die emptying

entlegen *Adj. (entfernt)* remote, out-of-the-way ‹*place*›

entlehnen *tr. V. (Sprachw.)* borrow (*Dat.*, **aus** from)

entleihen *tr. V.* borrow

Entleiher der; ~s, ~: borrower

Entlein ['ɛntlain] **das; ~s, ~:** duckling; **ein häßliches ~** *(ugs. scherzh.)* an ugly duckling

entloben *refl. V.* break off one's *or* the engagement

entlocken *tr. V. (geh.)* **jmdm. etw. ~:** elicit sth. from sb.; **jmdm. ein Geheimnis ~:** worm a secret out of sb.

entlohnen, *(bes. schweiz.)* **entlöhnen** *tr. V.* pay

Entlohnung die; ~, ~en payment; *(Lohn)* pay

entlüften *tr. V.* **a)** ventilate; **b)** *(Technik)* bleed ‹*brakes, radiator, etc.*›

Entlüftung die a) ventilation; *(Anlage)* ventilation [system]; **b)** *(Technik)* bleeding

entmachten *tr. V.* deprive of power

Entmachtung die; ~, ~en deprivation of power

entmannen *tr. V.* castrate

entmilitarisieren *tr. V.* demilitarize

Entmilitarisierung die demilitarization

entmündigen *tr. V. (Rechtsw.)* incapacitate; *(fig.)* deprive of the right of decision

Entmündigung die; ~, ~en *(Rechtsw.)* incapacitation; *(fig.)* deprivation of the right of decision

entmutigen *tr. V.* discourage; dishearten; **laß dich nicht ~:** don't be discouraged

Entmutigung die; ~, ~en discouragement

Entnahme die; ~, ~n *(von Wasser)* drawing; *(von Geld, Blutprobe)* taking; *(von Blut)* extraction; *(von Organen)* removal

entnehmen *unr. tr. V.* **a) etw. [einer Sache** *(Dat.)***] ~:** take sth. [from sth.]; **der Kasse Geld ~:** take money out of the till; **jmdm. Blut/eine Blutprobe ~:** take a blood sample from sb.; **Organe ~:** remove organs; **b)** *(ersehen aus)* gather (*Dat.* from)

entnerven *tr. V.* **jmdn. ~:** be nerve-racking for sb.

entnervend *Adj.* nerve-racking

entpuppen *refl. V.* **sich als etw. ~:** turn out to be sth.

entrahmen *tr. V.* skim ‹*milk*›

enträtseln *tr. V.* decipher ‹*code etc.*›; understand, fathom ‹*behaviour etc.*›

entrechten *tr. V.* **jmdn. ~:** deprive sb. of his/her rights

entreißen *unr. tr. V.* **jmdm. etw. ~:** snatch sth. from sb.; **jmdn. dem Tod ~** *(fig.)* save sb. from imminent death

entrichten *tr. V. (Amtsspr.)* pay ‹*fee*›

entriegeln *tr. V.* unbolt

entrinnen *unr. itr. V.; mit sein* *(geh.)* **einer Sache** *(Dat.)* **~:** escape sth.

entrosten *tr. V.* derust

entrückt *Adj. (geh.)* carried away; *(gedankenverloren)* lost in reverie

entrümpeln [ɛnt'rʏmpl̩n] *tr. V.* clear out

Entrümpelung die; ~, ~en clear-out; clearing out

entrußen *tr. V.* clear of soot

entrüsten 1. *refl. V.* **sich [über etw.** *(Akk.)***] ~:** be indignant [at *or* about sth.]. **2.** *tr. V.* **jmdn. ~:** make sb. indignant; **über etw.** *(Akk.)* **entrüstet/aufs höchste entrüstet sein** be indignant/outraged at sth.

Entrüstung die indignation (**über** + *Akk.* at, about)

entsaften *tr. V.* extract the juice from

Entsafter der; ~s, ~: juice-extractor

entsagen *itr. V. (geh.)* **einem Genuß ~:** renounce *or* forgo a pleasure

Entsagung die; ~, ~en *(geh.)* renunciation

entsagungs·voll *Adj.* **a)** full of self-denial *postpos.*; **b)** *(Entsagungen verlangend)* full of privation *postpos.*

entsalzen *tr. V.* desalinate

entschädigen *tr. V.* compensate (**für** for); **jmdn. für etw. ~** *(fig.)* make up for sth.

Entschädigung die compensation *no indef. art.*

Entschädigungs·summe die compensation *no indef. art.*

entschärfen *tr. V.* **a)** defuse, deactivate ‹*bomb etc.*›; **b)** *(fig.)* defuse ‹*situation*›; tone down ‹*discussion, criticism*›

Entschärfung die; ~, ~en a) *(von Bomben usw.)* defusing; deactivation; **b)** *(fig.)* defusing; toning down

Entscheid [ɛnt'ʃait] **der; ~[e]s, ~e** decision

entscheiden 1. *unr. refl. V.* **a)** decide; **sich für/gegen jmdn./etw. ~:** decide on *or* in favour of/against sb./sth.; **sich nicht ~ können** be unable to make up one's mind; **b)** *(entschieden werden)* be decided; **morgen entscheidet es sich, ob ...:** I/we/you will know tomorrow whether ... **2.** *unr. itr. V.* **über etw.** *(Akk.)* **~:** decide on *or* settle sth. **3.** *unr. tr. V.* **a)** *(bestimmen)* decide on ‹*dispute*›*;* **der Richter entschied, daß ...:** the judge decided *or* ruled that ...; **b)** *(den Ausschlag geben für)* decide ‹*outcome, result*›

entscheidend 1. *Adj.* crucial ‹*problem, question, significance*›*;* decisive ‹*action*›*;* **die ~e Stimme** the deciding vote. **2.** *adv.* **jmdn./etw. ~ beeinflussen** have a crucial *or* decisive influence on sb./sth.

Entscheidung die decision; *(Gerichts~)* ruling; *(Schwurgerichts~)* verdict; **etw. steht vor der ~:** sth. is just about to be decided

Entscheidungs-: ~befugnis die decision-making powers *pl.;* **~kampf der** decisive struggle; **~schlacht die** decisive battle; **~spiel das** deciding match; *(bei gleichem Rang)* play-off

entschieden 1. *Adj.* **a)** *(entschlossen)* determined; resolute; **b)** *(eindeutig)* definite. **2.** *adv.* resolutely; **etw. ~/auf das ~ste ablehnen** reject sth. emphatically *or* categorically; **das geht ~ zu weit** that is going much too far

Entschiedenheit die; ~: decisiveness; **etw. mit ~ fordern** demand sth. emphatically

entschlacken *tr. V.* cleanse

entschlafen *unr. itr. V.; mit sein (verhüll.: sterben)* pass away; fall asleep *(euphem.)*

entschleiern *tr. V. (geh.)* **a)** *(fig.)* reveal; uncover; **b)** unveil ‹*face*›

entschließen *unr. refl. V.* decide; make up one's mind; **sich ~, etw. zu tun** decide *or* resolve to do sth.; **sich dazu ~:** decide to do it

Entschließung die resolution

entschlossen 1. *Adj.* determined, resolute ‹*person*›*;* determined ‹*look etc.*›*;* **fest ~ [sein], etw. zu tun** [be] absolutely determined to do sth. **2.** *adv.* **~ handeln** act resolutely *or* with determination; **kurz ~:** on the spur of the moment; *(als Reaktion)* immediately

Entschlossenheit die; ~: determination; resolution

entschlummern *itr. V.; mit sein (dichter.: einschlafen)* fall asleep

entschlüpfen *itr. V.; mit sein* **a)** escape; slip away; **b)** ‹*remarks, words*› slip out

Entschluß der decision; **seinen ~ ändern** change one's mind; **aus eigenem ~:** of one's own volition

entschlüsseln *tr. V.* decipher; decode

Entschlüsselung die; ~, ~en deciphering; decoding

Entschluß·kraft die decisiveness

entschuldbar *Adj.* excusable; pardonable

entschuldigen 1. *refl. V.* apologize; **sich bei jmdm. wegen** *od.* **für etw. ~:** apologize to sb. for sth. **2.** *tr., auch itr. V.* excuse ‹*person*›*;* **sich ~ lassen** ask to be excused; **~ Sie [bitte]!** *(bei Fragen, Bitten)* excuse me; *(bedauernd)* excuse me; I'm sorry

Entschuldigung die; ~, ~en a) *(Rechtfertigung)* excuse; **etw. zu seiner ~ sagen/anführen** say sth. in one's defence; **b)** *(schriftliche Mitteilung)* [excuse] note; letter of excuse; **c) jmdn. für** *od.* **wegen etw. um ~ bitten** apologize to sb. for sth.; **~!** *(bei Fragen, Bitten)* excuse me; *(bedauernd)* excuse me; [I'm] sorry; **d)** *(entschuldigende Äußerung)* apology

Entschuldigungs-: ~grund der excuse; **~schreiben das** letter of apology

entschwinden *unr. itr. V.; mit sein (geh.)* disappear; vanish

entsenden *unr., auch regelm. tr. V.* dispatch

entsetzen 1. *refl. V.* be horrified; **sich vor** *od.* **bei dem Anblick von etw. ~:** be horrified at the sight of sth. **2.** *tr. V.* **a)** *(erschrecken)* horrify; **über etw.** *(Akk.)* **entsetzt sein** be horrified by sth.; **b)** *(Milit.)* relieve

Entsetzen das; ~s horror; **er bemerkte mit ~, daß ...:** he noticed to his horror that ...

entsetzlich 1. *Adj.* **a)** horrible, dreadful ‹*accident, crime, etc.*›*;* **b)** *nicht präd. (ugs.: stark)* terrible ‹*thirst, hunger*›. **2.** *adv.* terribly *(coll.);* awfully

entseuchen *tr. V.* decontaminate

entsichern *tr. V.* release the safety catch of ‹*pistol etc.*›

entsinnen *unr. refl. V.* **sich jmds./einer Sache ~, sich an jmdn./etw. ~:** remember sb./sth.

entsorgen *tr. V. (Amtsspr., Wirtsch.)* dispose of ‹*waste etc.*›*;* **eine Stadt/ein Kernkraftwerk ~:** dispose of a town's/a nuclear power station's waste

Entsorgung die; ~, ~en *(Amtsspr., Wirtsch.)* waste disposal

entspannen 1. *tr. V.* relax ‹*body etc.*›*;* relax, loosen ‹*muscles*›. **2.** *refl. V.* **a)** ‹*person*› relax; **b)** *(fig.)* ‹*situation, tension*› ease

Entspannung die; *o. Pl.* **a)** relaxation; **b)** *(politisch)* easing of tension; détente

Entspannungs-: ~politik die policy of détente; **~übung die** relaxation exercise

entsprechen *unr. itr. V.* **einer Sache** *(Dat.)* **~:** correspond to sth.; **der Wahrheit/den Tatsachen ~:** be in accordance with the truth/the facts; **den Erwartungen ~:** live up to one's expectations; **sich** *(Dat.) od. (geh.)* **einander ~:** correspond; **einem Wunsch/einer Bitte ~:** comply with a wish/request; **den Anforderungen ~:** meet the requirements; **dem Anlaß ~:** be appropriate for the occasion; **dem Zweck ~:** suit the purpose

entsprechend 1. *Adj.* **a)** corresponding; *(angemessen)* appropriate ‹*payment, reply, etc.*›*;* **b)** *nicht attr. (dem~)* in accordance *postpos.;* **das Wetter war schlecht und die Stimmung ~:** the weather was bad and the mood was the same; **c)** *nicht präd. (betreffend, zuständig)* relevant ‹*department etc.*›*;* ‹*person*› concerned. **2.** *adv.* **a)** *(angemessen)* appropriately; **b)** *(dem~)* accordingly. **3.** *Präp. mit Dativ* in accordance with; **es geht ihm den Umständen ~:** he is as well as can be expected [in the circumstances]

Entsprechung die; ~, ~en a) correspondence; **b)** *(Analogie)* parallel

entspringen *unr. itr. V.; mit sein* **a)** ‹*river*› rise, have its source; **b)** *(entstehen aus)* **einer Sache** *(Dat.)* **~:** spring from sth.; **c)** *(entweichen aus)* escape

entstammen *itr. V.; mit sein* **einer Sache** *(Dat.)* **~:** come from sth.; *(von etw. herrühren)* derive from sth.

entstehen *unr. itr. V.; mit sein* **a)** originate; ‹*quarrel, friendship, etc.*› arise; ‹*work of art*› be created; ‹*building, town, etc.*› be built; ‹*industry*› emerge; ‹*novel etc.*› be written; **b)** *(gebildet werden)* be formed **(aus** from, **durch** by); **c)** *(sich ergeben)* occur; *(als Folge)* result; **jmdm. ~ Kosten** sb. incurs costs; **hoffentlich ist nicht der Eindruck entstanden, daß ...:** I/we hope I/we have not given the impression that ...

Entstehung die; ~: origin; **die ~ dieser Industrie** the emergence of this industry

Entstehungs-: ~geschichte die history of the origin[s]; **~ort der** place of origin; **~zeit die** time of origin

entsteinen *tr. V.* stone
entstellen *tr. V.* **a)** disfigure ‹*person*›; distort ‹*face*›; **b)** *(fig.)* distort ‹*text, facts*›
Entstellung die a) disfigurement; **b)** *(fig.)* distortion
entstören *tr. V. (Elektrot.)* suppress ‹*engine, distributor, electrical appliance*›
enttarnen *tr. V.* uncover; *(fig.)* discover; **etw. als etw. ~:** reveal sth. as sth.
Enttarnung die uncovering
enttäuschen 1. *tr. V.* disappoint; **unsere Hoffnungen wurden enttäuscht** our hopes were dashed. **2.** *itr. V.* be a disappointment
enttäuscht *Adj.* disappointed; dashed ‹*hopes*›; **von jmdm. ~ sein** be disappointed in sb.; **von** *od.* **über etw. ~ sein** be disappointed by *or* at sth.
Enttäuschung die disappointment (**für** to); **jmdm. eine ~ bereiten** be a disappointment to sb.
entthronen *tr. V. (geh.)* dethrone
entvölkern [ɛnt'fœlkɐn] **1.** *tr. V.* depopulate. **2.** *refl. V.* become depopulated *or* deserted
ent·wachsen *unr. itr. V.; mit sein* **einer Sache** *(Dat.)* **~:** grow out of *or* outgrow sth.
entwaffnen *tr. V. (auch fig.)* disarm
entwaffnend 1. *Adj.* disarming. **2.** *adv.* disarmingly
Entwaffnung die; ~: disarming
entwarnen *itr. V.* sound *or* give the all-clear
Entwarnung die [sounding of the] all-clear
entwässern *tr. V.* drain ‹*area, meadow*›
Entwässerung die; ~, ~en drainage
entweder *Konj.*: **~ ... oder ...:** either ... or ...
entweichen *unr. itr. V.; mit sein* escape
entweihen *tr. V.* desecrate; profane
entwenden *tr. V. (geh.)* purloin (*Dat.* **from**)
entwerfen *unr. tr. V.* design ‹*furniture, dress*›; draft ‹*novel, text, etc.*›; draw up ‹*plans etc.*›
entwerten *tr. V.* **a)** cancel ‹*ticket, postage stamp*›; **b)** *(Finanzw.)* devalue ‹*currency*›
Entwerter der; ~s, ~: ticket-cancelling machine
Entwertung die *s.* **entwerten:** cancellation; cancelling; devaluation
entwickeln 1. *refl. V.* develop (**aus** from, **zu** into). **2.** *tr. V.* **a)** *(auch Fot.)* develop; **b)** *(hervorbringen)* give off, produce ‹*vapour, smell*›; show, display ‹*ability, characteristic*›; elaborate ‹*theory, ideas*›
Entwickler der; ~s, ~ *(Fot.)* developer
Entwicklung die; ~, ~en a) *(auch Fot.)* development; **in der ~ sein** ‹*young person*› be adolescent *or* in one's adolescence; **in seiner [körperlichen] ~ zurückbleiben** be physically underdeveloped; **etw. befindet sich in der ~:** sth. is [still] in the development stage; **b)** *(einer Theorie usw.)* elaboration
entwicklungs-, Entwicklungs-: ~dienst der development aid service; **~fähig** *Adj.* capable of development; **~geschichte die** history of the development; **die ~geschichte der Menschheit/der Meerestiere** the evolution of man/of marine animals; **~helfer der** development aid worker; **~hilfe die** [development] aid; **~land das;** *Pl.* **~länder** developing country; **~politik die** development aid policy; **~störung die** developmental disturbance; **~zeit die** period of development
entwirren *tr. V.* **a)** unravel, disentangle ‹*wool etc.*›; **b)** *(fig.)* unravel, sort out ‹*situation etc.*›
entwischen *itr. V.; mit sein (ugs.)* get away; **jmdm. ~:** give sb. the slip *(coll.)*
entwöhnen [ɛnt'vø:nən] *tr. V.* **a)** wean ‹*baby*›; **b)** *(geh.)* **jmdn. einer Sache** *(Dat.)* **~:** break sb. of the habit of [doing] sth.; **jmdn. [von einer Sucht] ~:** cure sb. [of an addiction]
entwürdigen *tr. V.* degrade
entwürdigend 1. *Adj.* degrading. **2.** *adv.* ‹*treat sb.*› in a degrading manner
Entwurf der a) design; **b)** *(Konzept)* draft; **der ~ zu einem Roman** the outline *or* draft of a novel
entwurzeln *tr. V. (auch fig.)* uproot
entzerren *tr. V.* **a)** *(Technik)* correct; rectify; **b)** *(Fot.)* rectify
entziehen 1. *unr. tr. V.* **a)** take away; **etw. jmdm./einer Sache ~:** take sth. away from sb./sth.; **jmdm. den Führerschein ~:** take sb.'s driving licence away; **jmdm. das Wort ~:** ask sb. to stop [speaking]; **jmdm. das Vertrauen/seine Unterstützung ~:** withdraw one's confidence in sb./one's support from sb.; **b) etw. einer Sache** *(Dat.)* **~** *(entfernen von, aus)* remove sth. from sth.; *(herausziehen aus)* extract sth. from sth.. **2.** *unr. refl. V.* **sich der Gesellschaft** *(Dat.)* **~** *(geh.)* withdraw from society; **sich seinen Pflichten** *(Dat.)* **~:** shirk *or* evade one's duty; **das entzieht sich meiner Kontrolle/Kenntnis** that is beyond my control/knowledge
Entziehung die a) withdrawal; **b)** *(Entziehungskur)* withdrawal treatment *no indef. art.*
Entziehungs·kur die course of withdrawal treatment; withdrawal programme
entzifferbar *Adj.* decipherable
entziffern *tr. V.* decipher
entzücken *tr. V.* delight
entzückend 1. *Adj.* delightful; **das ist ja ~!** *(iron.)* [that's] charming! **2.** *adv.* delightfully
entzückt *Adj.* delighted; **von/über etw.** *(Akk.)* **~ sein** be delighted by/at sth.
Entzug der; ~[e]s a) withdrawal; *(das Herausziehen)* extraction; **b)** *s.* **Entziehung b**
Entzugs·erscheinung die withdrawal symptom
entzündbar *Adj.* [in]flammable
entzünden 1. *tr. V.* **a)** *(geh.: anzünden)* light ‹*fire*›; strike, light ‹*match*›; **b)** *(geh.: erregen)* kindle, arouse ‹*passion*›; arouse ‹*hatred*›. **2.** *refl. V.* **a)** catch fire; ignite; **b)** *(anschwellen)* become inflamed; **entzündet** inflamed; **c)** *(geh.: entstehen)* **sich an etw.** *(Dat.)* **~:** ‹*quarrel*› be sparked off by sth.; ‹*temper*› flare at sth.
entzündlich *Adj.* **a)** [in]flammable ‹*substance*›; **b)** *(Med.)* inflammatory
Entzündung die; ~, ~en inflammation
entzwei *Adj.; nicht attr. (geh.)* in pieces
entzweien 1. *refl. V.* **sich [mit jmdm.] ~:** fall out [with sb.]. **2.** *tr. V.* cause ‹*persons*› to fall out
entzwei·gehen *(geh.) unr. itr. V.; mit sein* break; ‹*machine*› break down; ‹*shoes, clothes*› fall to pieces
Enzian ['ɛntsia:n] **der; ~s, ~e a)** *(Bot.)* gentian; **b)** *(Schnaps)* enzian liqueur
Enzyklika [ɛn'tsy:klika] **die; ~, Enzykliken** encyclical
Enzyklopädie [ɛntsyklopɛ'di:] **die; ~, ~n** encyclopaedia
enzyklopädisch *Adj.* encyclopaedic
Enzym [ɛn'tsy:m] **das; ~s, ~e** *(Chemie)* enzyme
Epen *s.* **Epos**
Epidemie [epide'mi:] **die; ~, ~n** *(auch fig.)* epidemic

epidemisch *Adj.* epidemic
Epik ['e:pɪk] **die; ~** *(Literaturw.)* epic poetry
Epiker der; ~s, ~: epic poet
Epilepsie [epilɛ'psi:] **die; ~, ~n** *(Med.)* epilepsy *no art.*
Epileptiker [epi'lɛptikɐ] **der; ~s, ~:** epileptic
epileptisch *Adj.* epileptic
Epilog [epi'lo:k] **der; ~s, ~e** epilogue
episch ['e:pɪʃ] *Adj.* epic
Episkopat [epɪsko'pa:t] **das** *od.* **der; ~[e]s, ~e** episcopate
Episode [epi'zo:də] **die; ~, ~n** episode
episodenhaft *Adj.* episodic
Epistel [e'pɪstl̩] **die; ~, ~n a)** *(bibl.)* epistle; **b)** *(kath. Kirche)* epistle; lesson;
Epitaph [epi'ta:f] **das; ~s, ~e** *(geh.)* epitaph
Epi·zentrum das *(Geol.)* epicentre
epochal [epɔ'xa:l] *Adj.* epochal; epoch-making ⟨*invention*⟩; *(fig. iron.)* world-shattering; monumental
Epoche [e'pɔxə] **die; ~, ~n** epoch
epoche·machend *Adj.* epoch-making
Epos ['e:pɔs] **das; ~, Epen** epic [poem]; epos
Equipe [e'kɪp] **die; ~, ~n** team
er [e:ɐ̯] *Personalpron.; 3. Pers. Sg. Nom. Mask.* he; *(betont)* him; *(bei Dingen/Tieren)* it; *(bei männlichen Tieren)* he/him; it; **„Er"** *(auf Handtüchern, an Türen)* 'His'; **bring Er den Wein!** *(veralt.)* fetch the wine!; *s. auch* **ihm; ihn; seiner**
Er der; ~, ~s *(ugs.)* he; **ist es ein Er oder eine Sie?** is it a he or a she?
erachten *tr. V. (geh.)* consider; **etw. als** *od.* **für seine Pflicht ~:** consider sth. [to be] one's duty; **etw. als** *od.* **für notwendig ~:** consider *or* think sth. necessary
Erachten das: meines ~s in my opinion
erahnen *tr. V.* imagine; guess
erarbeiten *tr. V.* **a)** *(erwerben)* work for; **[sich** *(Dat.)***] ein Vermögen ~:** make [oneself] a fortune; **b)** *(zu eigen machen)* work on; study; **c)** *(erstellen)* work out ⟨*plan, programme, etc.*⟩
Erb- ['ɛrp-]: **~adel der** hereditary nobility; **~anlage die** *(Biol.)* hereditary disposition
erbarmen [ɛɐ̯'barmən] *(geh.)* **1.** *refl. V.* **sich jmds./einer Sache ~:** take pity on sb./sth.; **Herr, erbarme dich unser!** Lord, have mercy upon us. **2.** *tr. V.* **jmdn. ~:** arouse sb.'s pity; move sb. to pity
Erbarmen das; ~s pity; **mit jmdm. ~ haben** take pity on *or* feel pity for sb.; **er kennt kein ~:** he knows no pity *or* mercy; **zum ~ sein** be pitiful *or* pathetic
erbärmlich [ɛɐ̯'bɛrmlɪç] **1.** *Adj.* **a)** *(elend)* wretched; **b)** *(unzulänglich)* pathetic; **c)** *(abwertend: gemein)* mean; wretched; **d)** *nicht präd. (sehr groß)* terrible ⟨*hunger, thirst, fear, etc.*⟩. **2.** *adv. (intensivierend)* terribly ⟨*cold, thirsty, etc.*⟩
Erbärmlichkeit die; ~: a) *(Elend)* wretchedness; **b)** *(abwertend: Gemeinheit)* meanness; wretchedness
erbarmungs·los *Adj.* merciless
erbauen 1. *tr. V.* **a)** build; **b)** *(geh.: erheben)* uplift; edify; **wir waren von seinen Plänen wenig erbaut** we were not exactly delighted about his plans. **2.** *refl. V.* **sich an etw.** *(Dat.)* **~** *(geh.)* be uplifted *or* edified by sth.
Erbauer der; ~s, ~: architect
erbaulich *Adj.* edifying
Erbauung die; ~ *(fig. geh.)* edification
erb·berechtigt *Adj.* entitled to inherit; entitled to an/the inheritance
[1]**Erbe** ['ɛrbə] **das; ~s a)** *(Vermögen)* inheritance; **das väterliche/mütterliche ~:** patrimony/maternal inheritance; **sein ~ antreten** come into one's inheritance; **b)** *(Vermächtnis)* heritage; legacy
[2]**Erbe der; ~n ~n** heir; **jmdn. zum** *od.* **als ~n einsetzen** appoint sb. as one's heir; **die lachenden ~n** *(ugs.)* my/his *etc.* heirs and successors
erbeben *itr. V.; mit sein (geh.)* **a)** shake; tremble; **b)** *(fig.: erregt werden)* shake; quiver
Erb·eigenschaft die *(Biol.)* hereditary characteristic
erben *tr., auch itr. V.* inherit; **bei mir ist nichts zu ~** *(ugs.)* you won't get anything out of me
erbetteln *tr. V.* get by begging
erbeuten [ɛɐ̯'bɔytn̩] *tr. V.* carry off, get away with ⟨*valuables, prey, etc.*⟩; *(Milit.)* capture
Erb-: ~feind der traditional enemy; **~folge die** succession; **die gesetzliche ~:** intestate succession; **~forschung die** genetics *sing., no art.;* **~gut das** *(Biol.)* genotype; genetic make-up; **~hof der** ancestral estate
erbieten *unr. refl. V. (geh.)* **sich ~, etw. zu tun** offer to do sth.
Erbin die; ~, ~nen heiress
erbitten *unr. tr. V. (geh.)* request
erbittern *tr. V.* enrage; incense
erbittert 1. *Adj.* bitter ⟨*resistance, struggle*⟩. **2.** *adv.* **~ kämpfen** wage a bitter struggle
Erbitterung die; ~: bitterness
Erb·krankheit die hereditary disease
erblassen [ɛɐ̯'blasn̩] *itr. V.; mit sein (geh.) s.* **erbleichen**
Erblasser ['ɛrplasɐ] **der; ~s, ~** *(Rechtsw.)* testator
erbleichen *itr. V.; mit sein (geh.)* go *or* turn pale; blanch *(literary)*
erblich 1. *Adj.* hereditary ⟨*title, disease*⟩. **2.** *adv.* **er ist ~ belastet** he suffers from a hereditary condition; *(scherzh.)* it runs in his family
erblicken *tr. V. (geh.)* **a)** catch sight of; see; **b)** *(fig.)* see
erblinden *itr. V.; mit sein* go blind; lose one's sight
Erblindung die; ~: loss of sight
erblühen *itr. V.; mit sein (geh.)* bloom; blossom
Erb·masse die a) *(Biol.)* genotype; genetic make-up; **b)** *(Rechtsspr.)* estate
erbosen [ɛɐ̯'bo:zn̩] *tr. V. (geh.)* infuriate
erbost *Adj.* angry, furious **(über** + *Akk.* at)
Erb·pacht die *(Rechtsw.)* hereditary lease
erbrechen 1. *unr. tr. V.* bring up ⟨*food*⟩. **2.** *unr. itr., refl. V.* vomit; be sick
Erbrechen das; ~s vomiting; **bis zum ~** *(ugs.)* ad nauseam
Erb·recht das *o. Pl.* law of inheritance
erbringen *unr. tr. V.* **a)** produce ⟨*proof, evidence*⟩; **b)** *(liefern)* produce ⟨*result etc.*⟩; yield ⟨*amount*⟩; result in ⟨*savings etc.*⟩; **die vorgesehene Leistung ~:** do the required work
Erb·schaden der *(Genetik)* hereditary defect
Erbschaft die; ~, ~en inheritance; **eine ~ machen** come into an inheritance
Erbschaft[s]·steuer die estate *or* death duties *pl.*
Erb·schleicher der; ~s, ~ *(abwertend)* legacy-hunter
Erbse ['ɛrpsə] **die; ~, ~n** pea
erbsen·groß *Adj.* pea-size; the size of a pea *postpos.*
Erbsen·suppe die a) pea soup; **b)** *(ugs.: Nebel)* pea-souper
Erb-: ~stück das heirloom; **~sünde die** original sin; **~teil das** share of an/the inheritance
Erd-: ~achse die earth's axis; **~anziehung die** earth's gravitational pull; **~apfel der** *(bes. österr.)* potato; **~atmosphäre**

die earth's atmosphere; **~ball der** *(geh.)* globe; earth; **~beben das** earthquake

erdbeben·sicher *Adj.* earthquake-proof ‹*building, construction*›; ‹*region etc.*› free from earthquakes

Erd·beere die strawberry

Erd-: ~bevölkerung die earth's population; **~bewohner der** inhabitant of the earth; **~boden der** ground; earth; **etw. dem ~boden gleichmachen** raze sth. to the ground; **vom ~boden verschwinden** disappear from *or* off the face of the earth

Erde ['e:ɐ̯də] **die; ~, ~n a)** *(Erdreich)* soil; earth; **ein Klumpen ~:** a lump of earth; **etw. in die ~ rammen** ram sth. into the ground; **b)** *o. Pl. (fester Boden)* ground; **etw. auf die ~ legen/stellen** put sth. down [on the ground]; **zu ebener ~:** on the ground floor *or (Amer.)* the first floor; **auf der ~ bleiben** *(fig.)* keep one's feet on the ground *(fig.)*; **unter der ~ liegen** *(geh. verhüll.)* be in one's grave; **jmdn. unter die ~ bringen** *(ugs.)* bury sb.; *(fig.: töten)* be the death of sb. *(coll.)*; **c)** *o. Pl. (Welt)* earth; world; **auf ~n** *(bibl.)*, **auf der ~:** on earth; **auf der ganzen ~:** throughout the world; **ein ruhiges/idyllisches Fleckchen ~:** a peaceful/idyllic spot; **d)** *o. Pl. (Planet)* Earth; **e)** *(Elektrot.)* earth

erden *tr. V. (Elektrot.)* earth

Erden·bürger der earth-dweller

erdenken *unr. tr. V.* think *or* make up

erdenklich *Adj.* conceivable; imaginable; **sich** *(Dat.)* **alle** *od.* **jede ~e Mühe geben** take the greatest possible trouble

Erd-: ~gas das natural gas; **~geist der** earth spirit; **~geschichte die;** *o. Pl.* history of the earth; **~geschoß das** ground floor; first floor *(Amer.)*; **~hörnchen das** *(Zool.)* chipmunk; ground-squirrel

erdig *Adj.* **a)** earthy ‹*mass, smell, taste*›; **b)** *(geh.: mit Erde beschmutzt)* muddy

erd-, Erd-: ~innere das interior of the earth; **~kabel das** underground cable; **~kruste die** earth's crust; **~kugel die** terrestrial globe; earth; **~kunde die** geography; **~magnetismus der** terrestrial magnetism; **~mittelpunkt der** centre of the earth; **~nuß die** peanut; ground-nut; **~nuß·butter die** peanut butter; **~nuß·öl das** ground-nut oil; **~ober·fläche die** earth's surface; **~öl das** oil; petroleum *(as tech. term)*

erdolchen *tr. V. (geh.)* stab to death

erdöl-, Erdöl-: ~exportierend *Adj.* oil-exporting ‹*country*›; **~feld das** oilfield; **~gewinnung die** oil production; **~leitung die** oil pipeline; **~produzent der** oil-producing country; **~raffinerie die** oil refinery

Erd·reich das soil

erdreisten *refl. V.* **sich ~, etw. zu tun** have the audacity to do sth.

erdrosseln *tr. V.* strangle

erdrücken *tr. V.* **a)** crush; **b)** *(fig.)* overwhelm

erdrückend *Adj.* overwhelming ‹*evidence, superiority*›

Erd-: ~rutsch der landslide; landslip; **ein politischer ~rutsch** a political landslide; **~schicht die a)** layer of earth; **b)** *(Geol.)* stratum; **~stoß der** earth tremor; **~teil der** continent

erdulden *tr. V.* endure ‹*sorrow, misfortune*›; tolerate ‹*insults*›; *(über sich ergehen lassen)* undergo

Erd-: ~um·drehung die rotation of the earth; **~um·fang der** circumference of the earth; **~umlaufbahn die** orbit [of the earth]; **in die ~umlaufbahn eintreten** enter into orbit

Erdung die; ~, ~en *(Elektrot.)* **a)** earthing; **b)** *(Leitung)* earth [connection]

Erd-: ~wall der wall of earth; *(Milit., Straßenbau)* earthwork; **~zeit·alter das** geological era

ereifern *refl. V.* get excited (**über** + *Akk.* about)

ereignen *refl. V.* happen; ‹*accident, mishap*› occur

Ereignis [ɛɐ̯'|aiknɪs] **das; ~ses, ~se** event; occurrence; **die ~se überstürzten sich** everything seemed to happen at once; **ein freudiges ~:** a happy event

ereignis-: los *Adj.* uneventful; **~reich** *Adj.* eventful

ereilen *tr. V. (geh.)* **der Tod ereilte ihn** he died [suddenly]; **das gleiche Schicksal ereilte ihn** he met the same fate

Erektion [erɛk'tsio:n] **die; ~, ~en** erection

Eremit [ere'mi:t] **der; ~en, ~en** hermit

ererbt *Adj.* inherited

[1]**erfahren** *unr. tr. V.* **a)** find out; learn; *(hören)* hear; **etw. von jmdm. ~:** find sth. out from sb.; **etw. über jmdn./etw. ~:** find out *or* hear sth. about sb./sth.; **etw. von etw. ~:** find out *or* learn/hear sth. about sth.; **etw. durch jmdn./etw. ~:** learn of sth. from sb./sth.; **b)** *(geh.: erleben)* experience; **viel Leid/Kummer ~:** suffer much sorrow/anxiety; **c)** *(mitmachen)* undergo ‹*change, development, etc.*›; suffer ‹*set-back*›

[2]**erfahren** *Adj.* experienced

Erfahrung die; ~, ~en a) experience; **über reiche/langjährige ~en verfügen** have extensive/years of experience; **~en sammeln** gain experience *sing.*; **die ~ machen, daß ...:** learn by experience that ...; **wir haben schlechte ~en mit ihm/damit gemacht** our experience of him/it has not been very good; **b) etw. in ~ bringen** discover sth.

erfahrungs-, Erfahrungs-: ~austausch der exchange of experiences **~gemäß** *Adv.* in our/my experience; **~gemäß ist es so, daß ...:** experience shows that ...

erfassen *tr. V.* **a)** *(mitreißen)* catch; **b)** *(begreifen)* grasp ‹*situation, implications, etc.*›; **c)** *(registrieren)* register; record; **d)** *(einbeziehen)* cover; **e)** *(packen)* seize; **Angst/Freude erfaßte ihn** he was seized by fear/overcome with joy

Erfassung die registration

erfinden *unr. tr. V.* **a)** invent; **b)** *(ausdenken)* make up ‹*story, words*›; make up, invent ‹*excuse*›; **eine erfundene Geschichte** a fictional story; **das ist alles erfunden** it is pure fabrication; *s. auch* **Pulver b**

Erfinder der; ~s, ~ a) inventor; **b)** *(Urheber)* creator; **das ist nicht im Sinne des ~s** *(ugs.)* that's not what it was meant for

erfinderisch *Adj.* inventive; *s. auch* **Not b**

Erfindung die; ~, ~en a) invention; **eine ~ machen** invent something; **er hat viele ~en gemacht** he has many inventions to his credit; **b)** *(Ausgedachtes)* invention; fabrication

erfindungs-, Erfindungs-: ~gabe die inventiveness; **~reich** *Adj.* imaginative

erflehen *tr. V. (geh.)* beg; **jmds. Hilfe/Hilfe von jmdm. ~:** beg sb.'s help/beg help from sb.

Erfolg [ɛɐ̯'fɔlk] **der; ~[e]s, ~e** success; **viel/keinen ~ haben** be very successful/be unsuccessful; **viel ~!** good luck!; **etw. mit/ohne ~ tun** do something successfully/without success; **der ~ blieb aus** success was not forthcoming; **der**

~ war, daß ... *(ugs.)* the upshot was that ...
erfolgen *itr. V.; mit sein* take place; occur; **auf seine Beschwerden erfolgte keine Reaktion** there was no reaction to his complaints
erfolg-, Erfolg-: ~los 1. *Adj.* unsuccessful; **2.** *adv.* unsuccessfully; **~losigkeit die;** ~: lack of success; **~reich 1.** *Adj.* successful; **2.** *adv.* successfully
Erfolgs-: ~aus·sicht die; *meist Pl.* prospect of success; **~autor der** successful author; **~erlebnis das** feeling of achievement; **~mensch der** successful individual; **~prämie die** *(eines Vertreters)* commission; *(eines Arbeiters)* bonus; **~quote die** success rate; *(bei Prüfungen)* pass rate; **~rezept das** recipe for success; **~roman der** successful novel; **~zwang der** pressure to succeed
erfolg·versprechend *Adj.* promising
erforderlich *Adj.* necessary; required;
erfordern *tr. V.* require; demand
Erfordernis das; ~ses, ~se requirement
erforschen *tr. V.* discover ‹*facts, causes, etc.*›; explore ‹*country*›; find out ‹*truth*›; **sein Gewissen ~:** search one's conscience
Erforschung die research (*Gen.* into); *(der Erde, des Weltalls usw.)* exploration
erfragen *tr. V.* ascertain [by asking]
erfreuen 1. *tr. V.* please; **sehr erfreut!** pleased to meet you. **2.** *refl. V.* **a) sich an etw.** *(Dat.)* **~:** take pleasure in sth.; **b) sich einer Sache** *(Gen.)* **~** *(geh.)* enjoy sth.
erfreulich *Adj.* pleasant; **eine ~e Mitteilung** a piece of good news
erfreulicherweise *Adv.* happily
erfrieren *unr. itr. V.; mit sein* **a)** ‹*person, animal*› freeze to death; ‹*plant, harvest, etc.*› be damaged by frost; suffer frost-damage; **ihm sind die Zehen erfroren** he got frostbite in his toes; **er ist ganz erfroren** *(ugs.)* he's absolutely frozen; **b)** *(fig.: erstarren)* freeze
Erfrierung die; ~, ~en frostbite *no pl.*; **~en an den Händen/Füßen** frostbitten hands/feet
erfrischen 1. *tr., auch itr. V.* refresh; **ein Spaziergang erfrischt sehr** a walk is very refreshing. **2.** *refl. V.* freshen oneself up
erfrischend *(auch fig.)* **1.** *Adj.* refreshing. **2.** *adv.* refreshingly
Erfrischung die; ~, ~en *(auch fig.)* refreshment
Erfrischungs-: ~getränk das soft drink; **~raum der** refreshment room; **~tuch das;** *Pl.* **~tücher** tissue wipe; towelette
erfüllbar *Adj.* ‹*wish*› which can be granted; ‹*condition*› which can be met
erfüllen 1. *tr. V.* **a)** grant ‹*wish, request*›; fulfil ‹*contract*›; carry out ‹*duty*›; meet ‹*condition*›; serve ‹*purpose*›; **b)** *(füllen)* fill; *(fig. geh.)* **ein erfülltes Leben** a full life; **eine Sehnsucht erfüllte sein Herz** a longing came over him; **jmdn. mit etw. ~:** fill sb. with sth.. **2.** *refl. V.* come true
Erfüllung die *(einer Pflicht)* performance; *(eines Wunsches)* fulfilment; **in ~ gehen** come true
ergänzen [ɛɐ̯'gɛntsn̩] *tr. V.* **a)** *(vervollständigen)* complete; *(erweitern)* add to; replenish ‹*supply*›; amplify ‹*remark, statement, etc.*›; amend ‹*statute*›; **b)** *(hinzufügen)* add ‹*remark*›; **c)** *(hinzukommen zu)* complement; **d) sich** *od. (geh.)* **einander ~:** complement each other
Ergänzung die; ~, ~en a) *(Vervollständigung)* completion; *(Erweiterung)* enlargement; *(von Vorräten)* replenishment; **zur ~ des Gesagten/einer Sammlung** to amplify what has been said/in order to enlarge a collection; **b)** *(Zusatz)* addition; *(zu einem Gesetz)* amendment; **c)** *(zusätzliche Bemerkung)* further remark; **d)** *(Sprachw.: Objekt)* object
Ergänzungs·band der; *Pl.* **~bände** supplementary volume; supplement
ergattern *tr. V. (ugs.)* manage to grab
ergaunern *tr. V.* get by underhand means
[1]**ergeben 1.** *unr. refl. V.* **a)** *(sich fügen)* **sich in etw.** *(Akk.)* **~:** submit to sth.; **sich in sein Schicksal ~:** resign oneself *or* become resigned to one's fate; **b)** *(kapitulieren)* surrender (*Dat.* to); **c)** *(entstehen)* ‹*opportunity, difficulty, problem*› arise (**aus** from); **bald ergab sich ein angeregtes Gespräch** soon a lively discussion was taking place; **d) sich dem Alkohol/***(ugs.)* **Suff ~** *(fig.)* take to alcohol/drink *or* the bottle. **2.** *unr. tr. V.* result in; **die Ernte ergab rund 400 Zentner Kartoffeln** the harvest produced about 400 hundredweight of potatoes; **eins und eins ergibt zwei** one and one makes two
[2]**ergeben 1.** *Adj.* devoted; **Ihr sehr ~er ...** *(geh.)* yours most obediently, ... **2.** *adv.* devotedly
Ergebenheit die; ~: devotion
Ergebnis das; ~ses, ~se result; *(von Verhandlungen, Überlegungen usw.)* conclusion; **zu einem ~ führen** produce a result
ergebnis·los 1. *Adj.* fruitless ‹*discussion*›; **die Verhandlungen blieben ~/wurden ~ abgebrochen** negotiations remained inconclusive/were broken off without a conclusion having been reached. **2.** *adv.* fruitlessly
ergehen 1. *unr. refl. V.* **a) sich in etw.** *(Dat.)* **~:** indulge in sth.; **b)** *(geh.: lustwandeln)* take a turn. **2.** *unr. itr. V.; mit sein* **a)** *(geh.: erlassen werden)* ‹*law*› be enacted; **die Einladungen ergingen an alle Mitglieder** the invitations went to all members; **b)** *unpers.* **jmdm. ist es gut/schlecht** *usw.* **ergangen** things went well/badly *etc.* for someone; **c) etw. über sich** *(Akk.)* **~ lassen** let sth. wash over one
ergiebig [ɛɐ̯'gi:bɪç] *Adj.* rich ‹*deposits, resources*›; productive ‹*mine*›; fertile ‹*fisheries, topic*›
Ergiebigkeit die; ~: *s.* **ergiebig:** richness; productivity; fertility
ergießen *unr. refl. V.* pour
ergo ['ɛrgo] *Adv.* ergo
ergötzen *(geh.)* **1.** *tr. V.* enthrall; captivate. **2.** *refl. V.* **sich an etw.** *(Dat.)* **~:** be delighted by sth.
Ergötzen das; ~s *(geh.)* delight
ergötzlich *(geh.) Adj.* delightful
ergrauen *itr. V.; mit sein* go *or* turn grey
ergreifen *unr. tr. V.* **a)** grab; **jmds. Hand ~:** grasp sb.'s hand; **die Macht ~** *(fig.)* seize power; **b)** *(festnehmen)* catch ‹*thief etc.*›; **c)** *(fig.: erfassen)* seize; **von blindem Zorn ergriffen** *(geh.)* in the grip of blind anger; **d)** *(fig.: aufnehmen)* **einen Beruf ~:** take up a career; **die Initiative/eine Gelegenheit ~:** take the initiative/an opportunity; **e)** *(fig.: bewegen)* move
ergreifend 1. *Adj.* moving. **2.** *adv.* movingly
Ergreifung die; ~ a) *(Festnahme)* capture; **b)** *(der Macht)* seizure
ergriffen *Adj.* moved
Ergriffenheit die; ~: vor ~ schweigen be too moved to speak; **vor ~ weinen** be moved to tears
ergründen *tr. V.* ascertain; discover ‹*cause*›; fathom ‹*mystery*›
Ergründung die *s.* **ergründen:** ascertainment; discovery; fathoming
Erguß der a) *(Med.)* *(Blut~)* bruise; contusion; *(Samen~)* ejaculation; **b)** *(geh. abwertend)* outburst

erhaben *Adj.* **a)** solemn ⟨*moment*⟩; awe-inspiring ⟨*sight*⟩; sublime ⟨*beauty*⟩; **b) über etw.** *(Akk.)* ~ **sein** be above sth.; **über jeden Zweifel** ~: beyond all criticism

Erhabenheit die; ~: grandeur

Erhalt der; ~**[e]s** *(Amtsdt.)* **a)** receipt; **bei** ~ **zahlen** pay on receipt; **b)** *s.* **Erhaltung**

erhalten 1. *unr. tr. V.* **a)** *(bekommen)* receive ⟨*letter, news, gift*⟩; be given ⟨*order*⟩; get ⟨*good mark, impression*⟩; **eine hohe Geldstrafe** ~: be fined heavily; **er erhielt 3 Jahre Gefängnis** he was sentenced to 3 years in prison; **b)** *(bewahren)* preserve ⟨*town, building*⟩; conserve ⟨*energy*⟩; **gut** ~ **sein** ⟨*clothes etc.*⟩ be in good condition; **jmdn. am Leben** ~: keep sb. alive. **2.** *unr. refl. V.* survive

erhältlich [ɛɐ̯'hɛltlɪç] *Adj.* obtainable

Erhaltung die; ~ *(des Friedens)* maintenance; *(der Arten, von Kunstschätzen)* preservation; *(der Energie)* conservation

erhängen *tr. V.* **jmdn./sich** ~: hang sb./oneself; **Tod durch Erhängen** death by hanging

erhärten *tr. V.* strengthen ⟨*suspicion, assumption*⟩; substantiate ⟨*claim*⟩

erheben 1. *unr. tr. V.* **a)** *(emporheben)* raise ⟨*one's arm/hand/glass*⟩; **erhobenen Hauptes** with head held high; **die Stimme** ~: raise one's voice; **b)** levy ⟨*tax*⟩; charge ⟨*fee*⟩; **c) jmdn. in den Adelsstand** ~: elevate sb. to the nobility; **d)** gather, collect ⟨*data, material*⟩; **e) Anklage** ~: bring *or* prefer charges. **2.** *unr. refl. V.* **a)** rise; **b)** *(rebellieren)* rise up **(gegen** against)

erhebend *Adj.* uplifting

erheblich [ɛɐ̯'he:plɪç] **1.** *Adj.* considerable. **2.** *adv.* considerably

Erhebung die; ~, ~**en a)** *(Anhöhe)* elevation; **b)** *(Aufstand)* uprising; **c)** *(Umfrage)* survey; **d)** *(von Steuern)* levying; *(von Gebühren)* charging

erheitern *tr. V.* **jmdn.** ~: cheer sb. up

Erheiterung die; ~, ~**en** amusement

erhellen 1. *tr. V.* **a)** light up, illuminate ⟨*room, sky*⟩; **b)** *(erklären)* shed light on, illuminate ⟨*reason, relationship*⟩. **2.** *refl. V.* *(geh.)* ⟨*eyes, face*⟩ brighten

Erhellung die; ~ *(Erklärung)* illumination

erhitzen 1. *tr. V.* **a)** heat ⟨*liquid*⟩; **jmdn.** ~: make sb. hot; **b)** *(fig.: erregen)* **die Gemüter** ~: make feelings run high. **2.** *refl. V.* **a)** heat up; ⟨*person*⟩ become hot; **b)** *(fig.: sich erregen)* ⟨*feelings*⟩ become heated

Erhitzung die; ~, ~**en** heating; *(Hitze)* heat

erhoffen *tr. V.* **sich** *(Dat.)* **viel/wenig von etw.** ~: expect a lot/little from sth.; **die erhoffte Änderung/Lohnerhöhung** the change/pay rise we/they had expected

erhöhen 1. *tr. V.* increase, raise ⟨*prices, productivity, etc.*⟩; increase ⟨*dose*⟩; **erhöhte Temperatur haben** have a temperature; **erhöhter Blutdruck** somewhat high blood pressure; **erhöhte Vorsicht** extra care. **2.** *refl. V.* ⟨*rent, prices*⟩ rise

Erhöhung die; ~, ~**en: eine** ~ **der Preise/Steuern** an increase in prices/taxes; **eine Erhöhung des Blutdrucks** a rise in blood pressure; **die** ~ **einer Dosis** the increasing of a dose

erholen *refl. V.* **a)** recover **(von** from); *(nach Krankheit)* recuperate; *(sich ausruhen)* rest; have a rest; *(sich entspannen, ausspannen)* relax; **b)** *(fig.)* recover

erholsam *Adj.* restful ⟨*weekend, holiday*⟩; **wandern ist sehr** ~: walking is very refreshing

Erholung die; ~ *s.* **erholen:** recovery; recuperation; rest; relaxation; ~ **brauchen** need a rest; **nach der langen Krankheit hat er** ~ **nötig** he needs to recuperate after his long illness; **zur** ~ **fahren** go on holiday to rest/relax; *(nach Krankheit)* go on holiday to convalesce; **eine** ~ **sein** be relaxing; *(fig.)* be a refreshing change

erholungs-, Erholungs-: ~**bedürftig** *Adj.* in need of a rest *postpos.*; ~**bedürftig sein** need a rest; ~**gebiet das** holiday area; ~**heim das** holiday home; ~**ort der;** *Pl.* ~**e** resort; ~**pause die** break

erhören *tr. V.* *(geh.)* hear ⟨*plea, prayer*⟩; **jmdn.** ~ *(veralt.)* yield to sb.

erigieren [eri'gi:rən] *itr. V.; mit sein* become erect; **erigiert** erect

Erika ['e:rika] **die;** ~, ~**s** *od.* **Eriken** [-kən] *(Bot.)* erica

erinnern [ɛɐ̯'ɪnɐn] **1.** *refl. V.* **sich an jmdn./etw. [gut/genau]** ~: remember sb./sth. [well/clearly]; **sich [daran]** ~, **daß ...:** remember *or* recall that ...; **wenn ich mich recht erinnere** if I remember rightly. **2.** *tr. V.* **a) jmdn. an etw./jmdn.** ~: remind sb. of sth./sb.; **jmdn. daran** ~, **etw. zu tun** remind sb. to do sth.; **b)** *(bes. nordd.: sich erinnern an)* remember. **3.** *itr. V.* **a) jmd./etw. erinnert an jmdn./etw.** sb./sth. reminds one of sb./sth.; **b)** *(zu bedenken geben)* **an etw.** *(Akk.)* ~: remind sb. of sth.; **ich möchte daran** ~, **daß ...:** let us not forget *or* overlook that ...

Erinnerung die; ~, ~**en a)** memory **(an** + *Akk.* of); **etw. [noch gut] in** ~ **haben** [still] remember sth. [well]; **wenn mich die** ~ **nicht täuscht** if my memory does not deceive me; **nach meiner** ~, **meiner** ~ **nach** as far as I remember; **jmdn./etw. in guter** ~ **behalten** have pleasant memories of sb./sth.; **zur** ~ **an jmdn./etw.** in memory of sb./sth.; **b)** *(Erinnerungsstück)* remembrance; souvenir; **c)** *Pl.* *(Memoiren)* memoirs

Erinnerungs-: ~**lücke die** gap in one's memory; ~**stück das** keepsake; *(von einer Reise)* souvenir; ~**wert der** sentimental value

erkalten *tr. V.; mit sein* cool; ⟨*limbs*⟩ grow cold; *(fig.)* ⟨*passion, feeling*⟩ cool

erkälten *refl. V.* catch cold

Erkältung die; ~, ~**en** cold; **sich** *(Dat.)* **eine** ~ **zuziehen** *od.* *(ugs.)* **holen** catch a cold

Erkältungs·krankheit die cold

erkämpfen *tr. V.* win; **sich** *(Dat.)* **etw.** ~ **müssen** have to fight for sth.

erkaufen *tr. V.* **a)** buy; **b)** *(fig.)* win; **etw. teuer** ~: win something at great cost

erkennbar *Adj.* recognizable; *(sichtbar)* visible; *(schwach sichtbar)* discernible

erkennen 1. *unr. tr. V.* **a)** *(deutlich sehen)* make out; **deutlich zu** ~ **sein** be clearly visible; **b)** *(identifizieren)* recognize **(an** + *Dat.* by); **sich zu** ~ **geben** reveal one's identity; **c)** *(fig.)* recognize; realize. **2.** *unr. itr. V.* **a)** *(Rechtsspr.)* **auf Freispruch** ~: grant an acquittal; **b)** *(Sport)* **auf Elfmeter/Freistoß** ~: award a penalty/free kick

erkenntlich *Adj.* **sich [für etw.]** ~ **zeigen** show one's appreciation [for sth.]

Erkenntnis die; ~, ~**se a)** discovery; **wissenschaftliche/gesicherte** ~**se** scientific findings/firm insights; **zu der** ~ **kommen, daß ...:** come to the realization that ...; **b)** *o. Pl.* *(das Erkennen)* cognition

erkennungs-, Erkennungs-: ~**dienst der** police records department; ~**dienstlich 1.** *Adj.* ~**dienstliche Behandlung** fingerprinting and photographing; **2.** *adv.* **jmdn. ~dienstlich behandeln** take sb.'s fingerprints and photograph; **~melodie die** *(einer Sendung)* theme music; *(eines Senders)* signature tune; **~zeichen das** sign [to recognize sb. by]

Erker ['ɛrkɐ] **der; ~s, ~:** bay window

Erker-: ~fenster das bay window; **~zimmer das** room with a bay window

erklärbar *Adj.* explicable; **etw. ist ~:** sth. can be explained

erklären 1. *tr. V.* **a)** explain (*Dat.* to, **durch** by); **b)** *(mitteilen)* state; declare; announce ‹*one's resignation*›; **jmdm. den Krieg ~:** declare war on sb.; **c)** *(bezeichnen)* **jmdn. für tot ~:** pronounce someone dead; **etw. für ungültig/verbindlich ~:** declare something to be invalid/binding; **jmdn. zu etw. ~:** name sb. as sth. **2.** *refl. V.* **a) sich einverstanden/bereit ~:** declare oneself [to be] in agreement/willing; **sich für/gegen jmdn./etw. ~** *(geh.)* declare one's support for/opposition to sb./sth.; **b)** *(seine Begründung finden)* be explained; **das erklärt sich einfach/von selbst** that is easily explained/self-evident

erklärend *Adj.* explanatory; **mit einigen ~en Worten** with a few words of explanation

erklärlich *Adj.* understandable; **es ist mir einfach nicht ~, wie ...:** I just can't understand how ...

erklärt *Adj.; nicht präd.* declared ‹*opponent, intention*›

Erklärung die; ~, ~en a) explanation; **b)** *(Mitteilung)* statement

Erklärungs·versuch der attempt at an explanation

erklecklich [ɛɐ̯'klɛklɪç] *Adj.* considerable ‹*sum, profit*›

erklettern *tr. V.* climb to the top of ‹*rock, wall, mountain*›; climb to ‹*summit*›

erklimmen *unr. tr. V. (geh.)* climb ‹*wall, tree*›

erklingen *unr. itr. V.; mit sein* ring out

erkranken *itr. V.; mit sein* become ill (**an** + *Dat.* with); **er ist an einer Lungenentzündung erkrankt** he's got pneumonia; **schwer erkrankt sein** be seriously ill; **ein erkrankter Kollege** a sick colleague

Erkrankung die; ~, ~en *(eines Menschen, Tieres)* illness; *(eines Körperteils)* disease

Erkrankungs·fall der: im ~: in event of illness

erkunden *tr. V.* reconnoitre ‹*terrain*›; **die Situation ~:** find out what the situation is

erkundigen *refl. V.* **sich nach jmdm./etw. ~:** ask after sb./enquire about sth.; **sich ~, wann ...:** enquire when ...

Erkundigung die; ~, ~en enquiry; **~en einholen** *od.* **einziehen** make enquiries

Erkundung die; ~, ~en *(meist Milit.)* reconnaissance; **auf ~ gehen** go out on reconnaissance

erlahmen *itr. V.; mit sein* tire; become tired; ‹*strength*› flag; ‹*enthusiasm etc.*› wane

erlangen *tr. V.* gain; obtain ‹*credit, visa*›; reach ‹*age*›

Erlaß [ɛɐ̯'las] **der; Erlasses, Erlasse a)** *(Anordnung)* decree (*Gen.* by); **b)** *(Straf~, Schulden~ usw.)* remission; **c)** *o. Pl. (eines Gesetzes, einer Bestimmung)* enactment; *(eines Dekrets)* issue; *(eines Verbots)* imposition

erlassen *unr. tr. V.* **a)** *(verkünden)* enact ‹*law*›; declare ‹*amnesty*›; issue ‹*warrant*›; **b)** remit ‹*sentence*›

erlauben 1. *tr. V.* **a)** allow; **jmdm. ~, etw. zu tun** allow sb. to do sth.; **~ Sie mir, das Fenster zu öffnen?** *(geh.)* would you mind if I opened the window?; **[na], ~ Sie mal!** *(ugs.)* do you mind! *(coll.)*; **b)** *(ermöglichen)* permit; **meine Zeit erlaubt es mir nicht** time does not allow. **2.** *refl. V.* **a)** *(sich die Freiheit nehmen)* **sich** *(Dat.)* **etw. ~:** permit oneself sth.; **sich** *(Dat.)* **Freiheiten ~:** take liberties; **sich** *(Dat.)* **über jmdn./etw. kein Urteil ~ können** not feel free to comment on sb./sth.; **sich** *(Dat.)* **einen Scherz [mit jmdm.] ~:** play a trick [on someone]; **b)** *(sich leisten)* **sich** *(Dat.)* **etw. ~:** treat oneself to sth.

Erlaubnis die; ~, ~se permission; *(Schriftstück)* permit; **jmdn. um ~ bitten, etw. zu tun** ask sb.'s permission to do sth.; **jmdm. die ~ erteilen/verweigern, etw. zu tun** give/refuse sb. permission to do sth.

erläutern *tr. V.* explain; comment on ‹*picture etc.*›; annotate ‹*text*›; **näher ~:** clarify; **~de Anmerkungen** explanatory notes

Erläuterung die explanation; *(zu einem Bild usw.)* commentary; *(zu einem Text)* [explanatory] note

Erle ['ɛrlə] **die; ~, ~n** alder

erleben *tr. V.* experience; **etwas Schönes/Schreckliches ~:** have a pleasant/terrible experience; **das habe ich noch nie erlebt!** I've never heard of such a thing!; **große Abenteuer ~:** have great adventures; **so ängstlich hatte er sie noch nie erlebt** he had never seen her so afraid before; **etw. bewußt/intensiv ~:** be fully aware of sth./experience sth. to the full; **sie wünschte sich nur, die Hochzeit ihrer Tochter noch zu ~:** her only remaining wish was to be at her daughter's wedding; **er wird das nächste Jahr nicht mehr ~:** he won't see next year; **du kannst was ~!** *(ugs.)* you won't know what's hit you!

Erlebnis das; ~ses, ~se experience; **das war ein ~:** what an experience!

erledigen 1. *tr. V.* **a) einen Auftrag ~:** deal with a task; **ich muß noch einige Dinge erledigen** I must see to a few things; **die Angelegenheit ist erledigt** the matter is settled; **sie hat alles pünktlich erledigt** she got everything done on time; **schon erledigt!** that's already done; **b)** *(erschöpfen)* finish *(coll.)* ‹*person*›; *(ugs.: töten)* knock off *(sl.)*; *(fig.: zerstören)* destroy. **2.** *refl. V.* ‹*matter, problem*› resolve itself; **damit hat sich die Sache erledigt** that's that; **sich von selbst ~:** sort it'self out

erledigt *Adj.* **a)** closed ‹*case*›; **b)** *(ugs.)* worn out ‹*person*›

Erledigung die; ~, ~en a) *o. Pl.* carrying out; *(Beendigung)* completion; *(einer Angelegenheit)* settling; **um baldige ~ wird gebeten** please give this matter your prompt attention; **b)** *(Besorgung)* **er hat einige ~en zu machen** he's got one or two things to see to

erlegen *tr. V.* **a)** shoot ‹*animal*›; **b)** *(österr.: entrichten)* pay ‹*fee, charge*›

erleichtern 1. *tr. V.* **a)** *(einfacher machen)* make easier; **jmdm./sich die Arbeit ~:** make sb.'s/one's work easier; **b)** *(befreien)* relieve; **das hat ihn erleichtert** that came as a relief to him; **erleichtert aufatmen** breathe a sigh of relief; **c)** *(Gewicht verringern, fig.)* lighten; **sein Herz/sein Gewissen ~:** open one's heart/unburden one's conscience; **jmdn. um etw. ~** *(ugs. scherzh.)* relieve sb. of sth. **2.** *refl. V. (verhüll.: seine Notdurft verrichten)* relieve oneself

Erleichterung die; ~, ~en a) *o. Pl. (Vereinfachung)* **zur ~ der Arbeit** to make the work easier; **b)** *o. Pl. (Befreiung)* relief; **~ empfinden** feel relieved; **c)** *(Verbesserung, Milderung)* alleviation

erleiden *unr. tr. V.* suffer

erlernen *tr. V.* learn

erlesen *Adj.* superior ‹*wine*›; choice ‹*dish*›; select ‹*circle*›

erleuchten *tr. V.* **a)** light; **Blitze erleuchteten den Himmel** the sky was lit up by flashes of lightning; **hell erleuchtet** brightly lit; **b)** *(geh.: inspirieren)* inspire

Erleuchtung die; ~, ~en inspiration

erliegen *unr. itr. V.; mit sein* **a)** succumb (*Dat.* to); **einem Irrtum ~:** be misled; **b)** *(zum Opfer fallen)* **einer Krankheit** *(Dat.)* **~:** die from an illness; **c) zum Erliegen kommen** come to a standstill

erlisch [ɛɐ̯'lɪʃ], **erlischst, erlischt** *Imperativ, 2. u. 3. Pers. Sg. Präsens v.* **erlöschen**

erlogen *Adj.* made up; untruthful ‹*story*›

Erlös [ɛɐ̯'lø:s] **der; ~es, ~e** proceeds *pl.*

erlosch [ɛɐ̯'lɔʃ] *1. u. 3. Pers. Sg. Präteritum v.* **erlöschen**

erloschen *2. Partizip v.* **erlöschen**

erlöschen *unr. itr. V.; mit sein* **a)** ‹*fire*› go out; **ein erloschener Vulkan** an extinct volcano; **die Lichter waren schon erloschen** the lights were already out; **b)** *(fig.)* ‹*hope, feelings*› wane; ‹*family, clan*› die out; ‹*claim, obligation*› cease; ‹*firm, membership*› cease to exist

erlösen *tr. V.* save, rescue (**von** from); **jmdn. von seinen Schmerzen ~:** release sb. from pain; **und erlöse uns von dem Übel** *od.* **Bösen** *(bibl.)* and deliver us from evil

erlösend *Adj.* **das ~e Wort sprechen** say the magic word

Erlöser der; ~s, ~ a) saviour; **b)** *(christl. Rel.)* redeemer

Erlösung die release (**von** from); *(christl. Rel.)* redemption

ermächtigen *tr. V.* authorize

Ermächtigung die; ~, ~en authorization

ermahnen *tr. V.* admonish; tell *(coll.)*; *(warnen)* warn

Ermahnung die admonition; *(Warnung)* warning

Ermang[e]lung die; ~: in ~ einer Sache (*Gen.*) *(geh.)* in the absence of sth.; **in ~ eines Besseren** for lack of anything better

ermannen *refl. V. (geh.)* **sich ~, etw. zu tun** pluck up courage to do sth.

ermäßigen 1. *tr. V.* reduce. **2.** *refl. V.* be reduced

Ermäßigung die reduction

ermatten *(geh.)* **1.** *itr. V.; mit sein* ‹*person*› become exhausted; *(fig.)* ‹*enthusiasm*› wane. **2.** *tr. V.* exhaust, tire ‹*person*›

ermessen *unr. tr. V.* estimate, gauge ‹*consequences*›

Ermessen das; ~s estimation; **nach eigenem ~:** in one's own estimation; **in jmds. ~** *(Dat.)* **liegen** be at sb.'s discretion; **nach menschlichem ~:** as far as anyone can judge

Ermessens-: ~frage die matter of discretion; **~spielraum der** powers *pl.* of discretion

ermitteln 1. *tr. V.* **a)** ascertain, determine ‹*facts*›; discover ‹*culprit, hideout, address*›; establish, determine ‹*identity, origin*›; decide ‹*winner*›; **b)** *(errechnen)* calculate ‹*quota, rates, data*›. **2.** *itr. V.* investigate; **gegen jmdn. ~:** investigate sb.; **in einer Sache ~:** investigate sth.

Ermittlung die; ~, ~en a) *s.* **ermitteln a:** ascertainment; determination; discovery; establishment; **die ~ eines Gewinners** deciding a winner; **b)** *meist Pl. (der Polizei, Staatsanwaltschaft)* investigation

Ermittlungs-: ~arbeit die investigatory work; **~beamte der** investigating officer; **~verfahren das** *(Rechtsw.)* preliminary inquiry

ermöglichen *tr. V.* enable; **jmdm. etw. ~:** make sth. possible for sb.; **es jmdm. ~, etw. zu tun** enable sb. to do sth.

ermorden *tr. V.* murder; *(aus politischen Gründen)* assassinate

Ermordung die; ~, ~en *s.* **ermorden:** murder; assassination

ermüden 1. *itr. V.; mit sein* tire; become tired. **2.** *tr. V.* tire; make tired

ermüdend *Adj.* tiring

Ermüdung die; ~, ~en tiredness

ermuntern *tr. V.* encourage; **jmdn. zu etw. ~, jmdn. [dazu] ~, etw. zu tun** encourage sb. to do sth.

ermunternd *Adj.* encouraging

Ermunterung die; ~, ~en a) encouragement; **zur ~:** to encourage; **b)** *(ermunternde Worte)* words *pl.* of encouragement

ermutigen *tr. V. s.* **ermuntern a**

ermutigend *Adj.* encouraging

Ermutigung die; ~, ~en *s.* **Ermunterung**

ernähren 1. *tr. V.* **a)** feed ‹*young, child*›; **mit der Flasche ernährt werden** be bottle-fed; **b)** *(unterhalten)* keep ‹*family, wife*›. **2.** *refl. V.* feed oneself; **sich von etw. ~:** live on sth.; ‹*animal*› feed on sth.

Ernährer der; ~s, ~, Ernährerin die; ~, ~nen breadwinner; provider

Ernährung die; ~: a) feeding; **b)** *(Ernährungsweise)* diet; **gesunde/ungesunde ~:** a healthy/an unhealthy diet

Ernährungs-: ~weise die diet; **~wissenschaft die** dietetics *sing., no art.*

ernennen *unr. tr. V.* appoint ‹*deputy, ambassador, etc.*›; **jmdn. zu etw. ~:** make sb. sth.

Ernennung die appointment (**zu** as)

erneuern 1. *tr. V.* **a)** *(auswechseln)* replace; **b)** *(wiederherstellen)* renovate ‹*roof, building*›; *(fig.)* thoroughly reform ‹*system*›; **c)** *(verlängern lassen)* extend, renew ‹*permit, licence, contract*›. **2.** *refl. V.* ‹*nature, growth*› renew itself

Erneuerung die a) *(Auswechslung)* replacement; **b)** *(Wiederherstellung)* renovation; *(fig.)* thorough reform; **demokratische/religiöse ~:** democratic/religious revival; **c)** *(Verlängerung eines Vertrages usw.)* renewal; extension

erneut 1. *Adj.; nicht präd.* renewed. **2.** *adv.* once again

erniedrigen *tr. V.* humiliate; **sich [selbst] ~:** lower oneself

erniedrigend *Adj.* humiliating

Erniedrigung die; ~, ~en humiliation

ernst [ɛrnst] **1.** *Adj.* **a)** serious ‹*face, expression, music, doubts*›; **b)** *(aufrichtig)* genuine ‹*intention, offer*›; **c)** *(schlimm)* serious ‹*injury*›; grave ‹*situation*›. **2.** *adv.* seriously; **jmdn./etw. ~ nehmen** take sb./sth. seriously

Ernst der; ~[e]s a) seriousness; **das ist mein [voller] ~:** I mean that [quite] seriously; **es ist mir [bitterer] ~ damit** I'm [deadly] serious about it; **allen ~es** in all seriousness; **b) daraus wurde [blutiger/bitterer] ~:** it became [deadly] serious; **der ~ des Lebens** the serious side of life; **der ~ der Lage** the seriousness of the situation; **er wird mit seiner Drohung ~ machen** he will carry out his threat; **er wird ~ machen** he will carry it out; **c)** *(gemessene Haltung)* gravity

Ernst·fall der: im ~: when the real thing happens

ernst·gemeint *Adj. (präd. getrennt geschrieben)* serious ‹*offer, reply*›; sincere ‹*wish*›

ernsthaft 1. *Adj.* serious; **etwas/nichts Ernsthaftes** something/nothing serious. **2.** *adv.* seriously

Ernsthaftigkeit die; ~: seriousness

ernstlich 1. *Adj.* serious; genuine ⟨*wish*⟩. **2.** *adv.* seriously; genuinely ⟨*sorry, repentant*⟩
Ernte ['ɛrntə] **die; ~, ~n a)** harvest; **bei der ~ sein** be bringing in the harvest; **während der ~:** at harvest time; **b)** *(Ertrag)* crop; **die ~ einbringen** bring in the harvest
Ernte-: ~aus·fall der crop failure; **~dank·fest das** harvest festival; **~ertrag der** yield; **~maschine die** harvester
ernten *tr. V.* harvest ⟨*cereal, fruit*⟩; *(fig.)* get ⟨*mockery, ingratitude*⟩; win ⟨*fame, praise*⟩
ernüchtern *tr. V.* **a)** sober up; **b)** *(fig.)* **jmdn. [völlig] ~:** bring sb. down to earth [with a bang]; **~d** sobering
Ernüchterung die; ~, ~en *(fig.)* disillusionment
Eroberer [ɛɐ̯'lo:bərɐ] **der; ~s, ~, Eroberin die; ~, ~nen** conqueror
erobern *tr. V.* **a)** conquer ⟨*country*⟩; take ⟨*town, fortress*⟩; **b)** *(fig.)* conquer ⟨*woman, market*⟩; seize ⟨*power*⟩; **[sich** *(Dat.)*] **die Herzen ~:** win hearts; **eine Stadt/ein Land ~** *(scherzh.)* take a town/ country by storm
Eroberung die; ~, ~en *(auch fig. scherzh.)* conquest; *(einer Stadt, Festung)* taking; *(der Macht)* seizing; **~en machen** make conquests
eröffnen 1. *tr. V.* **a)** open ⟨*shop, gallery, account*⟩; start ⟨*business, practice*⟩; **b)** *(beginnen)* open ⟨*meeting, conference*⟩; begin ⟨*event*⟩; **das Feuer ~:** open fire; **c) jmdm. etw. ~** *(mitteilen)* reveal sth. to sb.; **d) ein Testament ~:** read a will; **e)** *(Rechtsw., Wirtsch.)* **den Konkurs ~:** institute bankruptcy proceedings; **das Verfahren ~:** begin proceedings; **f) jmdm. neue Möglichkeiten ~:** open up new possibilities to sb.. **2.** *refl. V. (sich bieten)* **sich jmdm. ~** ⟨*opportunity, possibility*⟩ present itself
Eröffnung die a) opening; *(einer Sitzung)* start; *(einer Schachpartie)* opening [move]; **b)** *(Mitteilung)* revelation; **c)** *(Testaments~)* reading; **d)** *(Wirtsch.)* **die ~ des Konkurses** the institution of bankruptcy proceedings
Eröffnungs-: ~an·sprache die opening speech; **~feier die** opening ceremony
erogen [ero'ge:n] *Adj.* erogenous
erörtern [ɛɐ̯'lœrtɐn] *tr. V.* discuss
Erörterung die; ~, ~en discussion
Erosion [ero'zi̯o:n] **die; ~, ~en** erosion
Erotik [e'ro:tɪk] **die; ~:** eroticism
erotisch 1. *Adj.* erotic. **2.** *adv.* erotically
Erpel ['ɛrpl̩] **der; ~s, ~:** drake
erpicht [ɛɐ̯'pɪçt] *Adj.:* **auf etw.** *(Akk.)* **~ sein** be keen on sth.
erpressen *tr. V.* **a)** blackmail ⟨*person*⟩; **b)** extort ⟨*money, confession*⟩ **(von** from)
Erpresser der; ~s, ~, Erpresserin die; ~, ~nen blackmailer
erpresserisch *Adj.* blackmailing *attrib.*; **in ~er Absicht** for the purpose of blackmail
Erpressung die blackmail *no indef. art.*; *(von Geld, Geständnis)* extortion
erproben *tr. V.* test ⟨*medicine*⟩ **(an +** *Akk.* on); put ⟨*reliability etc.*⟩ to the test; **ein erprobter Soldat** an experienced soldier
Erprobung die; ~, ~en testing
erquicken [ɛɐ̯'kvɪkn̩] *tr. V. (geh.)* refresh
erquickend *Adj. (geh.)* refreshing
erraten *unr. tr. V.* guess
errechenbar *Adj.* calculable
errechnen *tr., auch itr. V.* calculate
erregbar *Adj.* excitable
erregen 1. *tr. V.* **a)** annoy; **b)** *(sexuell)* arouse; **c)** *(verursachen)* arouse; **Ärgernis/Aufsehen ~:** cause annoyance/ a stir. **2.** *refl. V.* **sich über etw.** *(Akk.)* **~:** get excited about sth.
erregend *Adj.* exciting; *(sexuell)* arousing
Erreger der; ~s, ~ *(Med.)* pathogen
erregt *Adj.* excited; hot ⟨*temper*⟩ *(sexuell)* aroused
Erregung die a) excitement; *(sexuell)* arousal; **in ~ geraten** become excited; **b) ~ öffentlichen Ärgernisses** *(Rechtsspr.)* causing a public nuisance
erreichbar *Adj.* **a)** within reach *postpos.*; reachable; **der Ort ist mit dem Auto/Zug ~:** the place can be reached by car/train; **leicht ~ sein** be easy to reach; **b) er ist [telefonisch] ~:** he can be contacted [by telephone]
erreichen *tr. V.* **a)** reach; **den Zug ~:** catch the train; **etw. ist zu Fuß zu ~:** sth. can be reached on foot; **b) er ist telefonisch zu ~:** he can be contacted by telephone; **c)** achieve ⟨*goal, aim*⟩; **[bei jmdm.] etwas/nichts ~:** get somewhere/ not get anywhere [with sb.]
erretten *tr. V. (geh.)* save
Erretter der *(geh.)* saviour
errichten *tr. V.* **a)** build ⟨*house, bridge, etc.*⟩; **b)** erect, put up ⟨*rostrum, barrier, etc.*⟩; **c)** found ⟨*company*⟩; set up ⟨*fund*⟩
erringen *unr. tr. V.* gain ⟨*victory*⟩; reach ⟨*first etc. place*⟩; win ⟨*majority*⟩; gain, win ⟨*sb.'s trust*⟩
erröten *itr. V.; mit sein* blush **(vor** with)
Errungenschaft [ɛɐ̯'rʊŋənʃaft] **die; ~, ~en** achievement
Ersatz der; ~es a) replacement; *(nicht gleichartig)* substitute; **als ~ für jmdn.** in place of sb.; **b)** *(Entschädigung)* compensation
ersatz-, Ersatz-: ~befriedigung die *(Psych.)* vicarious satisfaction; **~dienst der** *community service as an alternative to military service*; **~kasse die** private health insurance company; **~los 1.** *Adj.* without replacement *postpos.*; **2.** *adv.* **etw. ~los streichen** cancel sth.; **~mann der;** *Pl.* **~männer, ~leute** replacement; *(Sport)* substitute; **~spieler der** *(Sport)* substitute [player]; **~teil das** spare part; spare *(Brit.)*; **~weise** *Adv.* as an alternative
ersaufen *unr. itr. V.; mit sein (salopp)* drown
ersäufen [ɛɐ̯'zɔyfn̩] *tr. V.* drown; **seinen Kummer [im Alkohol] ~** *(fig.)* drown one's sorrows [in drink]
erschaffen *unr. tr. V.* create
Erschaffung die creation
erschallen *unr. od. regelm. itr. V.; mit sein* ⟨*song, call*⟩ ring out; ⟨*music*⟩ sound
erschaudern *itr. V.; mit sein (geh.)* shudder **(bei** at)
erschauern *itr. V.; mit sein (geh.)* tremble **(vor +** *Dat.* with)
erscheinen *unr. itr. V.; mit sein* **a)** appear; **jmdm. ~:** appear to sb.; **vor Gericht ~:** appear in court; **um rechtzeitiges/zahlreiches Erscheinen wird gebeten** a punctual arrival/a full turn-out is requested; **b)** ⟨*newspaper, periodical*⟩ appear; ⟨*book*⟩ be published; **c)** *(zu sein scheinen)* seem *(Dat.* to)
Erscheinung die; ~, ~en a) *(Phänomen)* phenomenon; **in ~ treten** become evident; **b)** *(äußere Gestalt)* appearance; **eine stattliche/elegante ~ sein** be an imposing/elegant figure; **c)** *(Vision)* apparition; **eine ~/~en haben** see a vision/visions
Erscheinungs-: ~bild das appearance; **~form die** manifestation; **~jahr das** year of publication
erschießen *unr. tr. V.* shoot dead; **Tod durch Erschießen** death by firing squad; **erschossen**

sein *(fig. ugs.)* be completely whacked *(Brit. coll.)*
Erschießung die; ~, ~en shooting
Erschießungs·kommando das firing squad
erschlaffen *itr. V.; mit sein* **a)** ⟨*muscle, limb*⟩ become limp; *(fig.)* ⟨*resistance, will*⟩ weaken; **b)** ⟨*skin*⟩ grow slack
[1]**erschlagen** *unr. tr. V.* strike dead; kill; **jmdn. mit Argumenten ~** *(fig.)* defeat sb. with arguments
[2]**erschlagen** *Adj. (ugs.)* **a)** *(erschöpft)* worn out; **b)** *(verblüfft)* **wie ~ sein** be flabbergasted *(coll.)* *or* thunderstruck
erschleichen *unr. refl. V. (abwertend)* **sich** *(Dat.)* **etw. ~:** get sth. by devious means
erschließen 1. *unr. tr. V.* **a)** *(zugänglich machen)* develop ⟨*area, building land*⟩; open up ⟨*market*⟩; **jmdm. etw. ~** *(fig.)* make sth. accessible to sb.; **b)** *(nutzbar machen)* tap ⟨*resources, energy sources*⟩; **c)** *(ermitteln)* deduce ⟨*meaning, wording*⟩. **2.** *unr. refl. V.* **sich jmdm. ~:** become accessible to sb.
Erschließung die a) *(eines Gebiets, von Bauland)* development; *(von Märkten)* opening up; **b)** *(von Rohstoffen)* tapping
erschöpfen 1. *tr. V. (auch fig.)* exhaust. **2.** *refl. V.* **darin ~ sich ihre Kenntnisse** her knowledge does not go beyond that
erschöpfend 1. *Adj.* exhaustive. **2.** *adv.* exhaustively
erschöpft *Adj.* exhausted
Erschöpfung die exhaustion; **bis zur ~:** to the point of exhaustion
[1]**erschrecken** *unr. itr. V.; mit sein* be startled; **vor etw.** *(Dat.)* *od.* **über etw.** *(Akk.)* **~:** be startled by sth.
[2]**erschrecken** *tr. V.* frighten; scare; **du hast mich erschreckt!** you gave me a scare
[3]**erschrecken** *unr. od. regelm. refl. V.* get a fright; **erschrick dich nicht!** don't be frightened
erschreckend *Adj.* **1.** alarming. **2.** *adv.* alarmingly
erschrocken *Adj.* frightened; **sie wandte sich ~ ab** she turned away in fright
erschüttern *tr. V. (auch fig.)* shake; **über etw.** *(Akk.)* **erschüttert sein** be shaken by sth.; **das kann mich nicht ~** *(ugs.)* that doesn't worry me
erschütternd *Adj.* deeply distressing ⟨*account, picture, news*⟩; deeply shocking ⟨*conditions*⟩
Erschütterung die; ~, ~en a) vibration; *(der Erde)* tremor; **wirtschaftliche ~en** *(fig.)* economic upheavals; **b)** *(Ergriffenheit)* shock; *(Trauer)* distress
erschweren *tr. V.* **etw. ~:** make sth. more difficult; **etw. durch etw. ~:** impede *or* hinder sth. by sth.
erschwerend 1. *Adj.* complicating ⟨*factor*⟩. **2.** *adv.* **es kommt ~ hinzu, daß er ...:** to make matters worse he ...; **das kommt ~ hinzu** that is an added problem
Erschwerung die; ~, ~en impediment *(für* to); **das ist eine ~ seiner Arbeit** that makes his job more difficult
erschwindeln *refl. V.* get by swindling; **sich** *(Dat.)* **etw. von jmdm. ~:** swindle sb. out of sth.
erschwinglich *Adj.* reasonable ⟨*price*⟩; affordable ⟨*rent*⟩; **für jmdn. nicht ~ sein** not be within sb.'s reach
ersehen *unr. tr. V.* see; **aus etw. [klar] zu ~ sein** be evident from sth.
ersehnen *tr. V. (geh.)* long for
ersetzbar *Adj.* replaceable
ersetzen *tr. V.* **a)** replace; **etw./jmdn. durch etw./jmdn. ~:** replace sth./sb. by sth./sb.; **Talent durch Fleiß ~:** substitute hard work for talent; **b)** *(erstatten)* reimburse ⟨*expenses etc.*⟩; **jmdm. einen Schaden ~:** compensate sb. for damages
Ersetzung die; ~, ~en *(Erstattung)* reimbursement; **die ~ von Schäden** compensation for damage
ersichtlich *Adj.* apparent
ersinnen *unr. tr. V. (geh.)* devise
ersparen *tr. V.* **a)** save ⟨*money*⟩; **mein erspartes Geld** *od.* **Erspartes** my savings; **b) jmdm./sich etw. ~:** save *or* spare sb./oneself sth.; **das würde mir viel Arbeit ~:** that would save me a lot of work; **es bleibt einem nichts erspart** *(ugs.)* at least I/you *etc.* could have been spared that
Ersparnis die; ~, ~se a) *(österr. auch* **das; ~ses, ~se)** *(ersparte Summe)* savings *pl.*; **b)** *(Einsparung)* saving
erst [eːɐ̯st] **1.** *Adv.* **a)** *(zu~)* first; **~ einmal** first [of all]; **~ noch** first; **eine solche Frau muß ~ noch geboren werden** such a woman has not yet been born; **b)** *(nicht eher als)* **eben ~:** only just; **er will ~ in drei Tagen/einer Stunde zurückkommen** he won't be back for three days/an hour; **~ nächste Woche/um 12 Uhr** not until next week/12 o'clock; **er war ~ zufrieden, als ...:** he was not satisfied until ...; **c)** *(noch nicht mehr als)* only; **~ eine Stunde/halb soviel** only an hour/half as much; **sie ist mit ihrer Arbeit ~ am Anfang** she is only just beginning her work. **2.** *Partikel* **so was lese ich gar nicht ~:** I don't even start reading that sort of stuff; **jetzt tue ich es ~ recht!** that makes me even more determined to do it
erst... *Ordinalz.* **a)** first; **der ~e Stock** the first *or (Amer.)* second floor; **etw. das ~e Mal tun** do sth. for the first time; **am Ersten [des Monats]** on the first [of the month]; **als erstes** first of all; **als ~er/~e etw. tun** be the first to do sth.; **Karl der Erste** Charles the First; **fürs ~e** for the moment; **sie kam als ~e ins Ziel** she was the first to reach the finish; **b)** *(best...)* **das ~e Hotel** the best hotel
erstarren *itr. V.; mit sein* **a)** ⟨*jelly, plaster*⟩ set; **b)** ⟨*limbs, fingers*⟩ grow stiff; **c) vor Schreck/Entsetzen ~:** be paralysed by fear/with horror
erstatten *tr. V.* **a)** reimburse ⟨*expenses*⟩; **b) Anzeige gegen jmdn. ~:** report sb. [to the police]; **jmdm. Bericht über etw.** *(Akk.)* **~:** report on sth. to sb.
Erstattung die; ~, ~en a) *(von Kosten)* reimbursement; **b) die ~ einer Anzeige** the reporting of sth. [to the police]
Erst-: ~aufführung die première; **~auflage die** first impression
erstaunen *tr. V.* astonish; amaze; **es erstaunte ihn nicht** he wasn't surprised
Erstaunen das; ~s astonishment; amazement; **jmdn. in ~ versetzen** astonish *or* amaze sb.
erstaunlich 1. *Adj.* astonishing, amazing ⟨*achievement, number, amount*⟩. **2.** *adv.* astonishingly; amazingly
erstaunlicher·weise *Adv.* astonishingly *or* amazingly [enough]
erstaunt *Adj.* astonished; amazed
erst-, Erst-: ~aus·gabe die first edition; **~best...** *Adj.* **der/die/das ~beste** the first suitable; **der ~beste Wagen, der ihr angeboten wurde** the first car she was offered
erstechen *unr. tr. V.* stab [to death]
erstehen 1. *unr. tr. V. (geh.: kaufen)* purchase. **2.** *unr. itr. V.; mit sein (geh.)* **a)** *(entstehen)* ⟨*diffi-*

culties, problems⟩ arise; **b)** *(auferstehen)* rise
Erste-Hilfe-Ausrüstung die first-aid kit
ersteigen *unr. tr. V.* climb
ersteigern *tr. V.* buy [at an auction]
Ersteigung die ascent
erstellen *tr. V. (Papierdt.)* **a)** *(bauen)* build; **b)** *(anfertigen)* make ⟨*assessment*⟩; draw up ⟨*plan, report, list*⟩
erste·mal: das ~: for the first time
ersten·mal: zum ~: for the first time; **beim** ~: the first time
erstens ['e:ɐ̯stn̩s] *Adv.* firstly; in the first place
erster... ['e:ɐ̯stɐ...] *Adj.* **~er/der ~e** the former
Erste[r]-Klasse-Abteil das first-class compartment
erst-: ~geboren *Adj.; nicht präd.* first-born; **der/die Erstgeborene** the first-born child; **~genannt** *Adj.; nicht präd.* mentioned first *postpos.*
ersticken 1. *itr. V.; mit sein* suffocate; *(sich verschlucken)* choke; **an einem Knochen ~:** choke on a bone; **vor Lachen ~** *(ugs.)* choke with laughter; **zum Ersticken sein** ⟨*heat*⟩ be stifling; **in Arbeit ~** *(ugs.)* be swamped with work. **2.** *tr. V.* **a)** suffocate; **der Widerstand wurde erstickt** *(fig.)* resistance was suppressed; **etw. sofort** *od.* **im Keim ~** *(fig.)* nip sth. in the bud; **b)** *(löschen)* smother ⟨*flames*⟩
Erstickung die; ~: suffocation; asphyxiation
Erstickungs-: ~gefahr die danger of suffocation; **~tod der** death from suffocation
erst-, Erst-: ~klassig 1. *Adj.* first-class; **~klassige Bedingungen** excellent conditions; **2.** *adv.* superbly; **da kann man ~klassig essen** you can get a first-class meal there; **~kläßler der; ~s, ~** *(südd., schweiz.) pupil in first class of primary school;* first-year pupil; **~kommunion die** *(kath. Rel.)* first communion
Erstling der; ~s, ~e first work
Erstlings-: ~film der first film; **~roman der** first novel; **~werk das** first work
erstmalig 1. *Adj.* first. **2.** *adv.* for the first time
erstmals *Adv.* for the first time
erstrahlen *itr. V.; mit sein* shine
erstrangig *Adj.* **a)** *(vordringlich)* of top priority *postpos.;* **von ~er Bedeutung** of the utmost importance; **b)** *s.* **erstklassig 1**
erstreben *tr. V.* strive for
erstrebens·wert *Adj.* ⟨*ideals etc.*⟩ worth striving for; desirable ⟨*situation*⟩
erstrecken *refl. V.* **a)** stretch; **sich bis an etw.** *(Akk.)* **~:** extend as far as sth.; **sich über ein Gebiet ~:** extend over *or* cover an area; **b)** *(dauern)* **sich über 10 Jahre ~:** carry on for 10 years; **c)** *(betreffen)* **sich auf jmdn./etw. ~:** affect sb./sth.; ⟨*laws, regulations*⟩ apply to sb./sth.
Erst-: ~schlag der first strike; **~schlag·waffe die** first-strike weapon; **~stimme die** first vote
erstunken [ɛɐ̯'ʃtʊŋkn̩]: **~ und erlogen sein** *(salopp)* be a pack of lies
erstürmen *tr. V.* take ⟨*fortress, town*⟩ by storm
Erst·wähler der first-time voter
ersuchen *tr. V. (geh.)* ask; **jmdn. um etw. ~:** request sth. of sb.; **jmdn. ~, etw. zu tun** request sb. to do sth.
Ersuchen das; ~s, ~: request **(an** + *Akk.* to); **auf ~ von .../des ...:** at the request of ...
ertappen *tr. V.* catch ⟨*thief, burglar*⟩; **jmdn. dabei ~, wie er etw. tut** catch sb. in the act of doing sth.; **sich bei etw. ~:** catch oneself doing sth.; *s. auch* **frisch 1 a**
erteilen *tr. V.* give ⟨*advice, information*⟩; give, grant ⟨*permission*⟩; **Unterricht ~:** teach; **Deutschunterricht ~:** give German lessons
Erteilung die giving; *(einer Genehmigung)* granting
ertönen *itr. V.; mit sein* sound
Ertrag [ɛɐ̯'tra:k] **der; ~[e]s, Erträge** [ɛɐ̯'trɛ:gə] yield
ertragen *unr. tr. V.* bear ⟨*pain, shame, uncertainty*⟩; **es ist nicht mehr zu ~:** I can't stand it any longer
erträglich [ɛɐ̯'trɛ:klɪç] **1.** *Adj.* **a)** bearable ⟨*pain*⟩; tolerable ⟨*conditions, climate*⟩; **die Grenze des Erträglichen erreichen** be as much as one can endure; **b)** *(ugs.: annehmbar)* tolerable. **2.** *adv. (ugs.: annehmbar)* tolerably
ertrag·reich *Adj.* lucrative ⟨*business*⟩; productive ⟨*land, soil*⟩
ertränken *tr. V.* drown; **seinen Kummer [im Alkohol] ~** *(fig.)* drown one's sorrows [in drink]
erträumen *refl. V.* dream of
ertrinken *unr. itr. V.; mit sein* be drowned; drown; *(fig.)* be inundated
Ertrinkende der/die; *adj. Dekl.* drowning person
Ertrunkene der/die; *adj. Dekl.* drowned person
ertüchtigen 1. *tr. V.* toughen up ⟨*body*⟩. **2.** *refl. V.* **sich körperlich ~:** get/keep oneself fit
Ertüchtigung die; ~, ~en getting/keeping fit
erübrigen 1. *tr. V.* spare ⟨*money, time*⟩; **etw. Geld/Zeit ~ können** have some money/time to spare. **2.** *refl. V.* be unnecessary; **es erübrigt sich, etw. zu tun** there's no point in doing sth.
eruieren [eru'i:rən] *tr. V. (geh.)* find out
Eruption [erʊp'tsi̯o:n] **die; ~, ~en** *(Geol., Med.)* eruption
erwachen *itr. V.; mit sein (geh.)* awake; wake up; *(fig.)* awake; **aus tiefem Schlaf ~:** awake from a deep sleep; **aus der Narkose ~:** come round; **ein neuer Tag erwacht** *(geh.)* a new day dawns
Erwachen das; ~s *(auch fig.)* awakening; **es wird ein böses ~ [für ihn] geben** *(fig.)* it'll be a rude awakening [for him]
[1]**erwachsen** *unr. itr. V.; mit sein* **a)** grow **(aus** out of); **b)** *(sich ergeben)* ⟨*difficulties, tasks*⟩ arise
[2]**erwachsen 1.** *Adj.* grown-up *attrib.;* **~ sein** be grown up. **2.** *adv.* ⟨*behave*⟩ in an adult way
Erwachsene der/die; *adj. Dekl.* adult; grown-up
Erwachsenen·bildung die; *o. Pl.* adult education *no art.*
Erwachsen·sein das being an adult/adults *no art.*
erwägen *unr. tr. V.* consider
Erwägung die; ~, ~en consideration; **etw. in ~ ziehen** consider sth.; take sth. into consideration
erwählen *tr. V. (geh.)* choose
Erwählte der/die; *adj. Dekl.* *(Freund[in])* sweetheart; *(Bevorrechtigte)* **er gehört zu den wenigen ~n** he belongs to the select few
erwähnen *tr. V.* mention; **etw. mit keinem Wort ~:** make no mention of sth.; **jmdn. lobend ~:** speak in praise of sb.
erwähnens·wert *Adj.* worth mentioning *postpos.*
Erwähnung die; ~, ~en mention; **~ verdienen** be worth mentioning
erwandern *tr., refl. V.* **er hat [sich** *(Dat.)***] ganz Frankreich erwandert** he's walked all round France
erwärmen 1. *tr. V.* **a)** heat; **b) jmdn. für etw. ~** *(fig.)* win sb. over to sth. **2.** *refl. V.* **a)** warm up; **b) sich für jmdn./etw. ~** *(fig.)* warm to sb./sth.
erwarten *tr. V.* **a)** expect ⟨*guests, phone call, post*⟩; **etw. ungeduldig/sehnlich ~:** wait impatiently/eagerly for sth; **jmdn. am Bahn-**

hof ~: wait for sb. at the station; **wir ~ ihn um 7 Uhr** we are expecting him at 7 o'clock; **ein Kind ~:** be expecting a baby; be expecting *(coll.)*; **b)** *(rechnen mit)* **etw. von jmdm. ~:** expect sth. of sb.; **von jmdm. ~, daß er etw. tut** expect sb. to do sth.; **es ist** *od. (geh.)* **steht zu ~, daß ...:** it is to be expected that ...; **wider Erwarten** contrary to expectation; **[sich** *(Dat.)*] **von etw. viel/wenig/nichts ~:** expect a lot/little/nothing from sth.

Erwartung die; ~, ~en expectation; **~en in etw.** *(Akk.)* **setzen** have expectations of sth.; **in freudiger ~:** in joyful anticipation; **die ~en [nicht] erfüllen** [not] come up to one's expectations

erwartungs-: ~gemäß *Adv.* as expected; **~voll 1.** *Adj.* expectant; **2.** *adv.* expectantly

erwecken *tr. V.* **a)** resurrect; **b)** *(fig.)* arouse ‹*longing, mistrust, pity*›; **den Eindruck ~, als ...:** give the impression that ...

Erweckung die; ~, ~en a) resurrection; **b)** *(fig.)* arousal

erwehren *refl. V. (geh.)* **sich jmds./einer Sache ~:** fend *or* ward sb./sth. off; **sie konnte sich des Gefühls/des Eindrucks nicht ~, daß ...:** she could not help feeling/thinking that ...

erweichen *tr. V.* soften; **jmdn./jmds. Herz ~** *(fig.)* soften sb.'s heart

Erweichung die; ~, ~en softening

erweisen 1. *unr. tr. V.* **a)** prove; **b) jmdm. Achtung ~:** show respect to sb.; **jmdm. einen Gefallen ~:** do sb. a favour. **2.** *unr. refl. V.* **sich als etw. ~:** prove to be sth.; **sich als falsch ~:** prove false

erweitern 1. *tr. V.* widen ‹*river, road*›; expand ‹*library, business*›; enlarge ‹*collection*›; dilate ‹*pupil, blood vessel*›; extend ‹*power*›; broaden ‹*horizons, knowledge*›; **eine erweiterte Neuauflage** a new, expanded edition; **erweiterte Oberschule** *(ehem. DDR) (Stufe)* ≈ sixth form; *(Schule)* ≈ sixth-form college. **2.** *refl. V.* ‹*road, river*› widen; ‹*pupil, blood vessel*› dilate; **sich zu etw. ~:** widen into sth.

Erweiterung die; ~, ~en *s.* **erweitern:** widening; expansion; enlargement; dilation; extension; broadening

Erwerb [ɛɐ̯'vɛrp] **der; ~[e]s a) der ~ des Lebensunterhaltes** earning a living; **b)** *(Arbeit)* occupation; **ohne ~ sein** be unemployed; **c)** *(Aneignung)* acquisition; **d)** *(Kauf)* purchase

erwerben *unr. tr. V.* **a)** *(verdienen)* earn; *(fig.)* win ‹*fame*›; **sich** *(Dat.)* **großen Ruhm ~:** win great fame; **b)** *(sich aneignen)* gain ‹*experience, influence*›; acquire, gain ‹*knowledge*›; **c)** acquire ‹*property, works of art, etc.*›; **etw. käuflich ~** *(Papierdt.)* purchase sth.; **d)** *(Biol., Psych.)* acquire

erwerbs-, Erwerbs-: ~fähig *Adj.* capable of gainful employment *postpos.*; able to work *postpos.*; **~fähigkeit die;** *o. Pl.* ability to work; **~leben das** working life; **~los** *Adj.*: *s.* **arbeitslos; ~lose der/die;** *adj. Dekl.*: *s.* **Arbeitslose; ~tätig** *Adj.* gainfully employed; **~tätige der/die;** *adj. Dekl.* person in work; **die ~tätigen** those in work; **~unfähig** *Adj.* incapable of gainful employment *postpos.*; unable to work *postpos.*; **~unfähigkeit die** inability to work

Erwerbung die acquisition

erwidern [ɛɐ̯'vi:dɐn] *tr. V.* **a)** reply; **etw. auf etw.** *(Akk.)* **~:** say sth. in reply to sth.; **b)** *(reagieren auf)* return ‹*greeting, visit*›; reciprocate ‹*sb.'s feelings*›

Erwiderung die; ~, ~en a) *(Antwort)* reply **(auf +** *Akk.* to); **b)** *s.* **erwidern b:** return; reciprocation

erwiesen *Adj.* proved; **eine ~e Tatsache** a proven fact

erwiesener·maßen *Adv.* as has been proved; **er hat ~ gelogen** it has been proved that he lied

erwirtschaften *tr. V.* **etw. ~:** obtain sth. by careful management

erwischen *tr. V. (ugs.)* **a)** *(fassen, ertappen, erreichen)* catch ‹*culprit, train, bus*›; **jmdn. beim Abschreiben ~:** catch sb. copying; **b)** *(greifen)* grab; **jmdn. am Ärmel ~:** grab sb. by the sleeve; **c)** *(bekommen)* manage to catch *or* get; **d)** *unpers.* **es hat ihn erwischt** *(ugs.) (er ist tot)* he has bought it *(sl.)*; *(er ist krank)* he has got it; *(er ist verletzt)* he's been hurt; *(scherzh.: er ist verliebt)* he's got it bad *(coll.)*

erwünscht [ɛɐ̯'vʏnʃt] *Adj.* wanted; desired ‹*result*›

erwürgen *tr. V.* strangle

Erz [ɛrts *od.* e:ɐ̯ts] **das; ~es, ~e** ore

erzählen *tr., auch itr. V.* tell ‹*joke, story*›; recount ‹*dream, experience*›; **jmdm. etw. ~:** tell sb. sth.; **erzähl keine Märchen!** *(ugs.)* don't tell stories!; **jmdm. von etw. ~:** tell sb. about sth.; **von etw. ~:** talk about sth.; **etw. über jmdn. ~:** tell sth. about sb.

Erzähler der a) story-teller; **der ~ eines Romans** the narrator of a novel; **b)** *(Autor)* writer [of stories]; narrative writer

erzählerisch *Adj.* narrative *attrib.*

Erzählung die; ~, ~en a) narration; *(Bericht)* account; **b)** *(Literaturw.)* story; *(märchenhafte Geschichte)* tale

Erz-: ~bergbau der ore-mining *no art.*; **~bergwerk das** ore mine; **~bischof der** archbishop; **~bistum das, ~diözese die** archbishopric; archdiocese; **~engel der** archangel

erzeugen *tr. V.* **a)** produce; generate ‹*electricity*›; **b)** *(österr.: anfertigen)* manufacture

Erzeuger der; ~s, ~ a) *(Vater)* father; **b)** *(Produzent)* producer; **c)** *(österr.: Hersteller)* manufacturer

Erzeuger-: ~land das country of origin; **~preis der** manufacturer's price

Erzeugnis das *(auch fig.)* product; **landwirtschaftliche ~se** agricultural products *or* produce

Erzeugung die a) *(von Lebensmitteln usw.)* production; *(von Industriewaren)* manufacture; *(von Strom)* generation; **b)** *(österr.: Herstellung)* manufacture

erz-, Erz-: ~feind der arch enemy; **~gang der** lode of ore; **~grube die** ore mine; **~haltig** *Adj.* ore-bearing; **~herzog der** archduke; **~herzogin die** archduchess; **~hütte die** ore-smelting works *sing.*

erziehbar *Adj.* educable; **der Junge ist sehr schwer ~:** the boy is a very difficult child

erziehen *unr. tr. V.* **a)** bring up; *(in der Schule)* educate; **ein Kind streng/sehr frei ~:** give a child a strict/very liberal upbringing/education; **gut/schlecht erzogen sein** have been brought up/not have been brought up properly; **b) jmdn. zum Verbrecher ~:** bring sb. up to criminal ways; **ein Kind zur Ordnung ~:** bring a child up to be tidy; **jmdn./sich dazu ~, etw. zu tun** train sb./oneself to do sth.

Erzieher der; ~s, ~, Erzieherin die; ~, ~nen educator; *(Lehrer[in])* teacher; *(Kindergärtner[in]* nursery-school teacher

erzieherisch *s.* **pädagogisch 1, 2**

Erziehung die; *o. Pl.* **a)** upbringing; *(Schul~)* education; **b)** *(Manieren)* upbringing; breeding; **seine gute ~ vergessen** *(fig.)* forget oneself

erziehungs-, Erziehungs-: ~anstalt die *(veralt.)* approved school; Borstal *(Brit.)*; **~beratung die a)** child guidance; **b)** *(Beratungsstelle)* child guidance clinic; **~berechtigt** *Adj.* having parental authority *postpos., not pred.*; **~berechtigte der/die;** *adj. Dekl.* parent or [legal] guardian; **~heim das** community home; **~methode die** educational method; teaching method; **~wesen das** educational system; education; **~wissenschaft die** education

erzielen *tr. V.* reach ‹*agreement, compromise, speed*›; achieve ‹*result, effect*›; make ‹*profit*›; obtain ‹*price*›; score ‹*goal*›

erzittern *itr. V.; mit sein* [begin to] shake *or* tremble; **etw. ~ lassen** shake sth.

erz·konservativ *Adj.* ultra-conservative

erzürnen *(geh.)* **1.** *tr. V.* anger; *(stärker)* incense; **erzürne ihn nicht** don't make him angry. **2.** *refl. V.* **sich über jmdn./etw. ~:** become *or* grow angry with sb./about sth.

erzwingen *unr. tr. V.* force; **sich** *(Dat.)* **den Zutritt ~:** force an entry

[1]**es** [ɛs] *Personalpron.; 3. Pers. Sg. Nom. u. Akk. Neutr.* **a)** (*s. auch Gen.* **seiner;** *Dat.* **ihm)** *(bei Dingen)* it; *(bei weiblichen Personen)* she/her; *(bei männlichen Personen)* he/him; **b)** *ohne Bezug auf ein bestimmtes Subst., mit unpers. konstruierten Verben, als formales Satzglied* it; **keiner will es gewesen sein** no one will admit to it; **ich bin es** it's me; it is I *(formal)*; **wir sind traurig, ihr seid es auch** we are sad, and you are too *or* so are you; **er hatte es nicht anders erwartet** he hadn't expected anything else; **es war einmal ein König** once upon a time there was a king; **es gibt keinen anderen Weg** there is no other way; **es wundert mich, daß ...:** I'm surprised that ...; **es sei denn, [daß] ...:** unless ...; **es ist genug!** that's enough; **wir schaffen es** we'll manage it; **es regnet/schneit/donnert** it rains/snows/thunders; *(jetzt)* it is raining/snowing/thundering; **es hat geklopft** there was a knock; **es klingelte** there was a ring; **es klingelt** someone is ringing; **es friert mich** I am cold; **es ist 9 Uhr/spät/Nacht** it is 9 o'clock/late/night-time; **es wird schöner** the weather is improving; **es wird kälter** it's getting colder; **es wird Frühling** spring is on the way; **es geht ihm gut/schlecht** he is well/unwell; **es wird gelacht** there is laughter; **es wird um 6 Uhr angefangen** we/they *etc.* start at 6 o'clock; **es läßt sich aushalten** it is bearable; **es lebt sich gut hier** it's a good life here; **er hat es gut** he has it good; it's all right for him; **er meinte es gut** he meant well; **sie hat es mit dem Herzen** *(ugs.)* she has got heart trouble *or* something wrong with her heart; *s. auch* **haben 1 n**

[2]**es, Es das; ~, ~** *(Musik)* E flat

E-Saite die E-string

Esche ['ɛʃə] **die; ~, ~n** *(Bot.)* ash

Esel ['e:zl̩] **der; ~s, ~ a)** donkey; ass; **b)** *(ugs.: Dummkopf)* ass *(coll.)*; idiot *(coll.)*; **so ein alter ~:** what a stupid ass *or* idiot *(coll.)*; **du ~:** you ass!

Eselin die; ~, ~nen she-donkey; jenny-ass

Esels-: ~brücke die *(ugs.)* mnemonic; **~ohr das** *(ugs.)* **a) ~ohren haben** *(fig.)* have donkey's ears; **b)** *(umgeknickte Ecke)* dog-ear; **ein Buch voller ~ohren** a dog-eared book

Esel·treiber der donkey-driver

Eskalation [ɛskala'tsi̯o:n] **die; ~, ~en** escalation

eskalieren *tr., itr. V.* escalate

Eskapade [ɛska'pa:də] **die; ~, ~n** escapade; *(Seitensprung)* amorous adventure

Eskimo ['ɛskimo] **der; ~[s], ~[s]** Eskimo

Eskorte [ɛs'kɔrtə] **die; ~, ~n** escort; *(fig.)* entourage

eskortieren *tr. V.* escort

esoterisch [ezo'te:rɪʃ] *Adj.* esoteric

Espe ['ɛspə] **die; ~, ~n** aspen

Espen·laub das: wie ~ zittern shake like a leaf

Esperanto [ɛspe'ranto] **das; ~[s]** Esperanto

Esplanade [ɛspla'na:də] **die; ~, ~n** esplanade

Espresso [ɛs'prɛso] **der; ~[s], ~s** espresso [coffee]; **zwei ~, bitte** two espressos, please

Esprit [ɛs'pri:] **der; ~s** esprit

Essay ['ɛse] **der** *od.* **das; ~s, ~s** essay

eßbar *Adj.* edible; **ist etwas Eßbares im Haus?** *(ugs.)* is there anything to eat in the house?; **nicht ~ sein** be inedible

Eß·besteck das knife, fork, and spoon

Eß·ecke die dining area

essen ['ɛsn̩] *unr. tr., itr. V.* eat; eat, drink ‹*soup*›; **etw. gern ~:** like sth.; **möchten Sie ein Stück Kuchen ~?** would you like a piece of cake?; **was gibt es zu ~?** what's for lunch/dinner/supper?; **von etw. ~:** eat some of sth.; **jmdm. etwas zu ~ machen** get sb. something to eat; **sich satt ~:** eat one's fill; **den Teller leer ~:** clear one's plate; **gut ~:** have a good meal; *(immer)* eat well; **warm/kalt ~:** have a hot/cold meal; **das Kind ißt schlecht** the child doesn't eat very much *or* has a poor appetite; **~ gehen** got out for a meal; **er ißt bei seiner Tante** he has his meals with his aunt; **es wird nichts so heiß gegessen, wie es gekocht wird** *(Spr.)* nothing is ever as bad as it seems; **selber ~ macht fett** *(ugs.)* I'm all right, Jack *(coll.)*; *s. auch* **Abend; Mittag**

Essen das; ~s, ~ a) *o. Pl.* **beim ~ sein** be having lunch/dinner/supper; **zum ~ gehen** go to lunch; **jmdn. zum ~ einladen** invite sb. for a meal; **b)** *(Mahlzeit)* meal; *(Fest~)* banquet; **c)** *(Speise)* food; **[das] ~ machen/kochen** get/cook the meal; **das ~ warm stellen** keep the lunch/dinner/supper hot; **~ auf Rädern** meals on wheels; **d)** *(Verpflegung) o. Pl.* food; **~ und Trinken** food and drink

Essen[s]-: ~ausgabe die a) *(das Ausgeben)* serving of meals; **die ~ausgabe ist um 12 Uhr** meals are *or* lunch is served at 12 [o'clock]; **b)** *(Stelle)* serving-hatch; **~marke die** meal-ticket; **~zeit die** mealtime

essentiell [ɛsɛn'tsi̯ɛl] *Adj. (geh., fachspr.)* essential

Essenz [ɛ'sɛnts] **die; ~, ~en** essence

Esser der; ~s, ~: er ist ein guter/schlechter ~: he has a healthy/poor appetite

Eß·geschirr das a) pots and pans; **b)** *(Milit.)* mess-kit

Essig ['ɛsɪç] **der; ~s, ~e** vinegar; **~ und Öl** oil and vinegar; **es ist mit etw. ~** *(ugs.)* sth. has fallen through completely *(coll.)*

Essig-: ~essenz die vinegar essence; **~gurke die** pickled gherkin; **~sauer** *Adj.* acetic; **~saure Tonerde** basic aluminium acetate; **~säure die** *(Chemie)* acetic acid

eß-, Eß-: ~kastanie die sweet chestnut; **~löffel der** soup-spoon; **~lokal das** restaurant; **~stäbchen das** chopstick; **~teller der** dinner plate; **~tisch der** dining-table; **~waren** *Pl.* food *sing.*; **~zimmer das** dining-room; *(Möbel)* dining-room suite

Establishment [ɪs'tɛblɪʃmənt] **das; ~s, ~s** Establishment

Este ['e:stə] **der; ~n, ~n, Estin die; ~, ~nen** Estonian
Est·land (das); ~s Estonia
estnisch *Adj.* Estonian
Estragon ['ɛstragɔn] **der; ~s** tarragon
Estrich ['ɛstrɪç] **der; ~s, ~e** composition *or* jointless floor
Eszett [ɛs'tsɛt] **das; ~, ~:** *(the letter)* ß
etablieren [eta'bli:rən] **1.** *tr. V.* establish; set up. **2.** *refl. V.* **a)** *(sich niederlassen)* ‹*shop*› open up; ‹*chain store*› open up *or* set up branches; **b)** *(sich einrichten)* settle in; **c)** *(gesellschaftlich)* become established
etabliert *Adj.* established
Etablissement [etablɪs(ə)'mã:] **das; ~s, ~s** establishment
Etage [e'ta:ʒə] **die; ~, ~n** floor; storey; **in** *od.* **auf der dritten ~ wohnen** live on the third *or* *(Amer.)* fourth floor
Etagen-: ~bett das bunk-bed; **~wohnung die** *flat (Brit.) or (Amer.) apartment occupying an entire floor*
Etappe [e'tapə] **die; ~, ~n a)** *(Teilstrecke)* stage; leg; *(Rennsport)* stage; **b)** *(Stadium)* stage; **c)** *(Milit.)* back area; base
Etappen-: ~sieg der *(Rennsport)* stage-win; **~wertung die** *(Rennsport)* daily points classification
Etat [e'ta:] **der; ~s, ~s** budget
Etat·kürzung die cut in the budget
etc. *Abk.* **et cetera** etc.
et cetera [ɛt'tse:tera] et cetera
etepetete [e:təpe'te:tə] *Adj.; nicht attrib. (ugs.)* fussy; finicky; pernickety *(coll.)*
Eternit Ⓦⓩ [etɛr'ni:t] **das** *od.* **der; ~s** asbestos cement
Ethik ['e:tɪk] **die; ~, ~en a)** *(Sittenlehre)* ethics *sing.;* **b)** *o. Pl. (sittliche Normen)* ethics *pl.;* **c)** *(Werk über Ethik)* ethical work
ethisch *Adj.* ethical
ethnisch ['ɛtnɪʃ] *Adj.* ethnic
Ethnologe [ɛtno'lo:gə] **der; ~n, ~n** ethnologist
Ethnologie die; ~, ~n ethnology *no art.*
ethnologisch *Adj.* ethnological
Ethologie [etolo'gi:] **die; ~, ~n** ethology *no art.*
Ethos ['e:tɔs] **das; ~:** ethos
Etikett [eti'kɛt] **das; ~[e]s, ~en** *od.* **~e** *od.* **~s** label; **jmdn./etw. mit einem ~ versehen** *(fig.)* pin a label on sb./sth.
Etikette die; ~, ~n etiquette; **die ~ wahren** observe the proprieties
etikettieren [etikɛ'ti:rən] *tr. V.* label

etlich... ['ɛtlɪç...] *Indefinitpron. u. unbest. Zahlwort (ugs.) Sg.* quite a lot of; *Pl.* quite a few; a number of; **vor ~en Wochen** several *or* some weeks ago
Etrusker [e'trʊskɐ] **der; ~s, ~:** Etruscan
etruskisch *Adj.* Etruscan
Etüde [e'ty:də] **die; ~, ~n** *(Musik)* étude
Etui [ɛt'vi:] **das; ~s, ~s** case
etwa ['ɛtva] **1.** *Adv.* **a)** *(ungefähr)* about; approximately; **~ so groß wie ...:** about as large as ...; **~ so** roughly like this; **in ~:** to some *or* a certain extent *or* degree; **b)** *(beispielsweise)* for example; for instance; **vergleicht man ~ ...:** for example, if one compares ...; **wie ~ ...:** as, for example .... **2.** *Part. (womöglich)* **hast du das ~ vergessen?** you haven't forgotten that, have you?; **störe ich ~?** am I disturbing you at all?
etwaig... ['ɛtva(:)ɪg...] *(Papierdt.) Adj.* possible ‹*delays*›; **~e Mängel/Beschwerden** any faults/complaints [which might arise]
etwas ['ɛtvas] *Indefinitpron.* **a)** something; *(fragend, verneint)* anything; **irgend ~:** something; **erzähl ihm einfach irgend ~:** just tell him anything!; **~ gegen jmdn. haben** have something against sb.; **sie haben ~ miteinander** *(ugs.)* there is something going on between them; **~ für sich haben** *(ugs.)* have sth. in it; **so ~:** a thing like that; **so ~ habe ich noch nie gesehen** I've never seen anything like it; **so ~ Schönes habe ich noch nie gesehen** I've never seen anything so beautiful before; **~ anderes** something else; *(fragend, verneinend)* anything else; **b)** *(Bedeutsames)* **aus ihm wird ~:** he'll make something of himself *or* his life; **es zu ~ bringen** get somewhere; **das will ~ heißen** that really is something; **c)** *(ein Teil)* some; *(fragend, verneinend)* any; **~ von dem Geld/davon** some of the money/it; **d)** *(ein wenig)* a little; **noch ~ Milch** a little more *or* some more milk; **kannst du mir ~ Geld leihen?** can you lend me some money?; **[noch] ~ spielen/lesen** play/read for a little while [longer]; **~ Englisch** a little *or* some English
Etwas das; ~, ~: something; **das gewisse ~:** that certain something
Etymologe [etymo'lo:gə] **der; ~n, ~n** etymologist
Etymologie die; ~, ~n etymology
etymologisch *(Sprachw.)* **1.** *Adj.* etymological. **2.** *adv.* etymologically

euch, *(in Briefen)* **Euch** [ɔyç] **1.** *Dat. u. Akk. von* **ihr, Ihr** you; **ich gebe ~ das** I'll give you it; I'll give it to you. **2.** *Dat. u. Akk. Pl. des Reflexivpron. der 2. Pers. Pl.* **a)** *refl.* yourselves; **b)** *reziprok* one another; *s. auch* **uns**
Eucharistie [ɔyçarɪs'ti:] **die; ~, ~n** *(kath. Rel.)* Eucharist
eucharistisch *Adj. (kath. Rel.)* Eucharistic
[1]**euer,** *(in Briefen)* **Euer** ['ɔyɐ] *Possessivpron. der 2. Pers. Pl.* your; **Grüße von Eu[e]rer Helga/Eu[e]rem Hans** Best wishes, Yours, Helga/Hans; **Eu[e]re** *od.* **Euer Exzellenz** Your Excellency; **ist das/sind das eure?** is that/are they yours?; *s. auch* [1]**unser**
[2]**euer,** *(in Briefen)* **Euer** *Gen. von* **ihr** *(geh.)* of you; **wir werden ~ gedenken** we will remember you
Eukalyptus [ɔyka'lʏptʊs] **der; ~, Eukalypten** *od.* **~:** eucalyptus
Eule ['ɔylə] **die; ~, ~n** owl; **~n nach Athen tragen** carry coals to Newcastle
Eulen-: ~spiegel der joker; *s. auch* **Till; ~spiegelei die; ~, ~en** caper
Eumel ['ɔyml̩] **der; ~s, ~** *(Jugendspr.)* twerp *(sl. derog.)*
Eunuch [ɔy'nu:x] **der; ~en, ~en** eunuch
Euphorie [ɔyfo'ri:] **die; ~, ~n** *(geh., fachspr.)* euphoria
euphorisch *(geh., fachspr.)* **1.** *Adj.* euphoric. **2.** *adv.* euphorically
Eurasien [ɔy'ra:zi̯ən] **(das); ~s** Eurasia
eurasisch *Adj.* Eurasian
eure, Eure ['ɔyrə] *s.* [1]**euer**
eurer·seits *Adv. (von eurer Seite)* on your part; *(auf eurer Seite)* for your part
eures·gleichen *indekl. Pron.* people *pl.* like you; *(abwertend)* the likes *pl.* of you; your sort *or* kind; *s. auch* **deinesgleichen**
euret-: ~halben [-halbn̩] *(veralt.),* **~wegen** *Adv. (wegen euch)* because of you; on your account; *(für euch)* on your behalf; *(euch zuliebe)* for your sake; **ich mache mir ~wegen keine Sorgen** I don't worry about you; **~willen** *Adv.* **um ~willen** for your sake
Eurocheque ['ɔyroʃɛk] **der; ~s, ~s** Eurocheque
Euro~: ~dollar der *(Wirtsch.)* Eurodollar; **~krat** [~'kra:t] **der; ~en, ~en** Eurocrat
Europa [ɔy'ro:pa] **(das); ~s** Europe

Europa·cup der *(Sport)* European cup
Europäer [ɔyro'pɛ:ɐ] **der; ~s, ~, Europäerin die; ~, ~nen** European
Europa·flagge die flag of the Council of Europe
europäisch [ɔyro'pɛ:ɪʃ] *Adj.* European; **die Europäische Gemeinschaft** the European Community; **Europäische Wirtschaftsgemeinschaft** European Economic Community
europäisieren *tr. V.* Europeanize
Europa-: ~meister der *(Sport)* European champion; **~meisterschaft die** *(Sport)* **a)** *(Wettbewerb)* European Championship; **b)** *(Titel)* championship of Europe, European title; **~minister der** minister for Europe; **~parlament das;** *o. Pl.* European Parliament *or* Assembly; **~pokal der** *(Sport)* European cup; **~politik die** policy towards the EC; **~rat der;** *o. Pl.* Council of Europe; **~rekord der** *(Sport)* European record; **~straße die** European long-distance road; **~wahlen** *Pl.* European elections
Euro·scheck der Eurocheque
Euro·vision die Eurovision
Euter ['ɔytɐ] **das** *od.* **der; ~s, ~:** udder
Euthanasie [ɔytana'zi:] **die; ~:** euthanasia *no art.*
e.V., E.V. *Abk.* **eingetragener Verein**
ev. *Abk.* **evangelisch** ev.
Eva ['e:fa *od.* 'e:va] **(die)** Eve; *s. auch* **Adam**
evakuieren [evaku'i:rən] *tr. V.* evacuate
Evakuierte der/die; *adj. Dekl.* evacuee
Evakuierung die; ~, ~en evacuation
evangelisch [evaŋ'ge:lɪʃ] *Adj.* Protestant; **die ~e Kirche** the Protestant Church
evangelisch-lutherisch *Adj.* Lutheran
evangelisch-reformiert *Adj.* Reformed
Evangelist der; ~en, ~en evangelist
Evangelium [evaŋ'ge:liʊm] **das; ~s, Evangelien a)** *o. Pl. (auch fig.)* gospel; **b)** *(christl. Rel.)* Gospel; **das ~ des Johannes** St. John's Gospel
Evas- ['e:fas- *od.* 'e:vas-]: **~kostüm das: im ~kostüm** *(ugs. scherzh.)* in her birthday suit/their birthday suits *(coll. joc.)*; **~tochter die** *(scherzh.)* **eine echte ~tochter** a real little Eve
Eventual·fall [evɛn'tua:l-] **der** eventuality; contingency; **für den ~fall** should the eventuality arise
Eventualität [evɛntuali'tɛ:t] **die; ~, ~n** eventuality; contingency
eventuell [evɛn'tuɛl] **1.** *Adj.; nicht präd.* possible ⟨*objections, difficulties, applicants*⟩; ⟨*objections, difficulties*⟩ which might occur; **bei ~en Schäden** in the event *or* case of damage. **2.** *adv.* possibly; perhaps
Evergreen ['ɛvəgri:n] **der; ~s, ~s** old favourite
Evolution [evolu'tsio:n] **die; ~, ~en** evolution
evolutionär [evolutsio'nɛ:ɐ] **1.** *Adj.* evolutionary. **2.** *adv.* by evolution
Evolutions·theorie die theory of evolution
evtl. *Abk.* **eventuell**
E-Werk das power station
EWG [e:ve:'ge:] **die; ~:** EEC
ewig ['e:vɪç] **1.** *Adj.* eternal, everlasting ⟨*life, peace*⟩; eternal, undying ⟨*love*⟩; *(abwertend)* never-ending; **die Ewige Stadt** the Eternal City; **ein ~er Student** *(scherzh.)* an eternal student; **seit ~en Zeiten** for ages *(coll.)*; for donkey's years *(coll.)*; **das Ewige Licht** *(kath. Rel.)* the Sanctuary Lamp. **2.** *adv.* eternally; for ever; **~ warten** wait for ever; **~ dauern** take ages; **~ halten** last for ever *or* indefinitely; **auf ~:** for ever
Ewigkeit die; ~, ~en a) eternity; **in ~:** for ever and ever; **in die ~ eingehen** *(geh. verhüll.)* find eternal rest; **b) eine [halbe] ~, ~en** *(ugs.)* ages; **es dauert eine ~:** it takes ages *(coll.)*; **in alle ~:** for ever
ex [ɛks] *Adv. (ugs.)* **etw. ex trinken** down sth. in one *(coll.)*; **ex!** down in one! *(sl.)*
Ex- *(vor Personenbez.: vormalig)* ex-⟨*wife, husband, president, etc.*⟩
exakt [ɛ'ksakt] **1.** *Adj.* exact; precise. **2.** *adv.* ⟨*work etc.*⟩ accurately; **~ um 12 Uhr** at 12 o'clock precisely
Exaktheit die; ~: precision; exactness
exaltiert 1. *Adj. (hysterisch)* over-excited; *(überspannt)* exaggerated ⟨*behaviour, gestures*⟩; *(überschwenglich)* effusive. **2.** *adv. (hysterisch)* over-excitedly; *(überschwenglich)* effusively
Examen [ɛ'ksa:mən] **das; ~s, ~** *od.* **Examina** [ɛ'ksa:mina] examination; exam *(coll.)*; **ein ~ machen** *od.* **ablegen** sit *or* take an examination; **~ haben** *(ugs.)* have examinations
Examens-: ~angst die examination nerves *pl.*; **~arbeit die** *written work presented for an examination;* **~kandidat der** examination candidate
examinieren *tr. V.* **a)** examine; **b)** *(ausfragen)* question
Exegese [ɛkse'ge:zə] **die; ~, ~n** *(Theol.)* exegesis
exekutieren [ɛkseku'ti:rən] *tr. V.* execute
Exekution [ɛkseku'tsio:n] **die; ~, ~en** execution
Exekutions·kommando das firing squad
exekutiv [ɛkseku'ti:f] *Adj. (bes. Politik, Rechtsw.)* executive
Exekutive [ɛkseku'ti:və] **die; ~, ~n** *(Rechtsw., Politik)* executive
Exempel [ɛ'ksɛmpl̩] **das; ~s, ~:** example; **ein ~ [an jmdm.] statuieren** make an example [of sb.]
Exemplar [ɛksɛm'pla:ɐ] **das; ~s, ~e** specimen; *(Buch, Zeitung, Zeitschrift)* copy
exemplarisch [ɛksɛm'pla:rɪʃ] *Adj.* exemplary
exerzieren [ɛksɛr'tsi:rən] *(Milit.)* **1.** *itr. V.* drill. **2.** *tr. V.* drill ⟨*soldiers*⟩
Exerzier-: ~munition die *(Milit.)* dummy ammunition; **~platz der** *(Milit.)* parade ground
Exerzitien [ɛksɛr'tsi:tsiən] *Pl. (kath. Rel.)* religious *or* spiritual exercises
Exhibitionismus der; ~ *(Psych., fig.)* exhibitionism
Exhibitionist der; ~en, ~en exhibitionist
exhibitionistisch *Adj.* exhibitionist
exhumieren [ɛkshu'mi:rən] *tr. V.* exhume
Exhumierung die; ~, ~en exhumation
Exil [ɛ'ksi:l] **das; ~s, ~e** exile; **ins ~ gehen** go into exile
Exilant [ɛksi'lant] **der; ~en, ~en** exile
exiliert [ɛksi'li:ɐt] *Adj.* exiled
Exilierte der/die; *adj. Dekl.* exile
Exil-: ~literatur die literature written in exile; **~regierung die** government in exile
existent [ɛksɪs'tɛnt] *Adj.* existing; existent
Existentialismus [ɛksɪstɛntsia'lɪsmʊs] **der; ~:** existentialism *no art.*
Existentialist der; ~en, ~en existentialist
existentialistisch *Adj.* existentialist
existentiell [ɛksɪstɛn'tsiɛl] *Adj.*

existential; **in etw.** *(Dat.)* **eine ~e Bedrohung sehen** see in sth. a threat to one's existence
Existenz [ɛksɪs'tɛnts] **die; ~, ~en a)** existence; **b)** *(Lebensgrundlage)* livelihood; **sich** *(Dat.)* **eine ~ aufbauen** build a life for oneself; **c)** *(Mensch)* **zweifelhafte ~en** dubious characters
existenz-, Existenz-: ~berechtigung die right to exist; **~fähig** *Adj.* able to exist *or* to survive *postpos.;* **~grund·lage die** basis of one's livelihood; **~kampf der** struggle for existence; **~minimum das** subsistence level; **am Rande des ~minimums leben** live at subsistence level; **~philosophie die** existential philosophy *no art.*
existieren [ɛksɪs'ti:rən] *itr. V.* exist
Exitus ['ɛksitʊs] **der; ~** *(Med.)* death
exkl. *Abk.* **exklusiv[e]** excl.
exklusiv [ɛksklu'zi:f] **1.** *Adj.* exclusive. **2.** *adv.* exclusively
Exklusiv·bericht der exclusive [report]
exklusive [ɛksklu'zi:və] *Präp.* + *Gen.* *(Kaufmannsspr.)* exclusive of; excluding
Exklusivität [ɛkskluzivi'tɛ:t] **die; ~:** exclusiveness; exclusivity
Ex·kommunikation die *(kath. Kirche)* excommunication
ex·kommunizieren *tr. V.* *(kath. Kirche)* excommunicate
Exkrement [ɛkskre'mɛnt] **das; ~[e]s, ~e** *(fachspr., geh.)* excrement
Exkurs [ɛks'kʊrs] **der; ~es, ~e** digression; *(in einem Buch)* excursus
Exkursion [ɛkskʊr'zi̯o:n] **die; ~, ~en** study trip *or* tour
Exmatrikulation [ɛksmatrikula'tsi̯o:n] **die; ~, ~en** *(Hochschulw.)* *removal of a student's name from the register on leaving a university*
exmatrikulieren [ɛksmatriku'li:rən] *tr. V.* *(Hochschulw.)* **jmdn./sich ~:** remove sb.'s name/have one's name removed from the university register
Exodus ['ɛksodʊs] **der; ~, ~se** *(geh.)* exodus
Exorzismus [ɛksɔr'tsɪmʊs] **der; ~, Exorzismen** *(Rel.)* exorcism
Exot [ɛ'kso:t] **der; ~en, ~en, Exotin die; ~, ~nen** strange foreigner
exotisch 1. *Adj.* exotic. **2.** *adv.* exotically
Expander [ɛks'pandɐ] **der; ~s, ~** *(Sport)* chest-expander
expandieren [ɛkspan'di:rən] *tr., itr. V.* expand
Expansion [ɛkspan'zi̯o:n] **die; ~, ~en** expansion
expansiv [ɛkspan'zi:f] *Adj.* **a)** *(Politik)* expansionist; **b)** *(Wirtsch.)* expansionary
Expedition [ɛkspedi'tsi̯o:n] **die; ~, ~en** expedition
Experiment [ɛksperi'mɛnt] **das; ~[e]s, ~e** experiment
experimentell [ɛksperimɛn'tɛl] **1.** *Adj.; nicht präd.* experimental. **2.** *adv.* experimentally
experimentieren *itr. V.* experiment
Experte [ɛks'pɛrtə] **der; ~n, ~n, Expertin die; ~, ~nen** expert **(für** in)
Expertise [ɛkspɛr'ti:zə] **die; ~, ~n** expert's report
explizit [ɛkspli'tsi:t] *(geh.)* **1.** *Adj.* explicit. **2.** *adv.* explicitly
explodieren [ɛksplo'di:rən] *itr. V.; mit sein (auch fig.)* explode; ‹*costs*› rocket
Explosion [ɛksplo'zi̯o:n] **die; ~, ~en** explosion; **etw. zur ~ bringen** detonate sth.
explosions-, Explosions-: ~artig 1. *Adj.* explosive, astronomical ‹*growth, increase*›; **2.** *adv.* ‹*rise*› astronomically; **~gefahr die;** *o. Pl.* danger of explosion; **„~gefahr!"** '[Danger,] Explosives!'; **~welle die** shock wave
explosiv [ɛksplo'zi:f] **1.** *Adj. (auch fig.)* explosive. **2.** *adv.* explosively; **~ reagieren** *(fig.)* react violently
Explosivität [ɛksplozivi'tɛ:t] **die** explosiveness
Exponat [ɛkspo'na:t] **das; ~[e]s, ~e** exhibit
Exponent [ɛkspo'nɛnt] **der; ~en, ~en** *(Math.)* exponent; *(fig.)* leading exponent
Exponential- [ɛksponɛn'tsi̯a:l-] *(Math.)* exponential ‹*function, equation, curve*›
exponieren [ɛkspo'ni:rən] *tr. V.* **jmdn./sich ~** *(geh.)* *(der Aufmerksamkeit aussetzen)* draw attention to sb./oneself; *(der Gefahr aussetzen)* lay sb./oneself open to attack
exponiert *Adj.* exposed
Export [ɛks'pɔrt] **der; ~[e]s, ~e a)** *o. Pl. (das Exportieren)* export; exporting; **der ~ nach Afrika** exports to Africa; **b)** *(das Exportierte)* export; **c)** *(~bier)* export; **zwei ~:** two export
Export-: ~artikel der export; **~bier das** export beer
Exporteur [ɛkspɔr'tø:ɐ̯] **der; ~s, ~e** *(Wirtsch.)* exporter
Export-: ~firma die exporter; **~geschäft das a)** export business; **b)** *(geschäftlicher Abschluß)* export deal; **~handel der** export trade; **~händler der** exporter
exportieren *tr., itr. V.* export
Exposé [ɛkspo'ze:] **das; ~s, ~s a)** exposé; report; **b)** *(eines Drehbuchs, Romans usw.)* outline
Exposition [ɛkspozi'tsi̯o:n] **die; ~, ~en** exposition
expreß [ɛks'prɛs] *Adv.* *(veralt.)* express
Expreß der; Expresses *(bes. österr.)* express [train]
Expreß·gut das express freight; express goods *pl.;* **etw. als ~gut schicken** send sth. by express goods
Expressionismus [ɛksprɛsi̯o'nɪsmʊs] **der** expressionism *no art.*
Expressionist der; ~en, ~en expressionist
expressionistisch 1. *Adj.* expressionist. **2.** *adv.* expressionistically; ‹*influenced*› by expressionism
exquisit [ɛkskvi'zi:t] **1.** *Adj.* exquisite. **2.** *adv.* exquisitely
extensiv [ɛkstɛn'zi:f] **1.** *Adj. (auch Landw.)* extensive. **2.** *adv. (auch Landw.)* extensively
extern [ɛks'tɛrn] *(Schulw.)* **1.** *Adj.* external; **ein ~er Schüler** a day boy/girl. **2.** *adv.* **eine Prüfung ~ ablegen** take an examination as an external candidate
exterritorial [ɛkstɛrito'ri̯a:l] *Adj.* *(Völkerr.)* extraterritorial
extra ['ɛkstra] *Adv.* **a)** *(gesondert)* ‹*pay*› separately; **Getränke werden ~ berechnet** drinks are extra; **b)** *(zusätzlich, besonders)* extra; **dafür brauche ich aber noch 10 DM ~:** but I need another 10 marks for that; **c)** *(eigens)* especially; **etw. ~ für jmdn. tun** do sth. especially *or* just for sb.; **~ deinetwegen** just because of you; **d)** *(ugs.: absichtlich)* **etw. ~ tun** do sth. on purpose
Extra das; ~s, ~s; *meist Pl.* extra
Extra-: ~ausgabe die a) *(Zeitung)* special edition; extra; **b)** *(Geldausgabe)* extra *or* additional expense; **~blatt das** special edition; extra; **~fahrt die** *(bes. schweiz.)* special excursion
Extrakt [ɛks'trakt] **der; ~[e]s, ~e a)** *fachspr. auch* **das** extract; **b)** *(Zusammenfassung)* summary; synopsis
Extra·ration die extra ration
extraterrestrisch [-tɛ'rɛstrɪʃ] *Adj. (Astron.)* extraterrestrial
extravagant [-va'gant] **1.** *Adj.* flamboyant. **2.** *adv.* flamboyantly

Extravaganz [-va'gants] **die;** ~, ~**en a)** *o. Pl.* flamboyance; **b)** *(Sache)* ~**en** flamboyance *sing.*
extravertiert [-vɛr'ti:ɐ̯t] *Adj. (Psych.)* extrovert[ed]
Extra·wurst die *(fig. ugs.)* **eine ~ bekommen** get special treatment *or* special favours
extrem [ɛks'tre:m] **1.** *Adj.* extreme. **2.** *adv.* extremely; ~ **reagieren** react in an extreme manner
Extrem das; ~s, ~e extreme; **von einem ~ ins andere fallen** go from one extreme to another
Extrem·fall der extreme case
Extremismus der; ~, **Extremismen** extremism
Extremist der; ~en, ~en, **Extremistin die;** ~, ~**nen** extremist
extremistisch *Adj.* extremist
Extremität [ɛkstremi'tɛ:t] **die;** ~, ~**en** extremity
Extrem-: ~punkt der, ~wert der *(Math.)* extremum
extrovertiert [ɛkstrovɛr'ti:ɐ̯t] *s.* **extravertiert**
exzellent [ɛkstsɛ'lɛnt] *(geh.)* **1.** *Adj.* excellent. **2.** *adv.* excellently
Exzellenz [ɛkstsɛ'lɛnts] **die;** ~, ~**en** Excellency; **Eure/Seine ~:** Your/His Excellency
Exzentriker [ɛks'tsɛntrikɐ] **der;** ~s, ~: eccentric
exzentrisch 1. *Adj.* eccentric. **2.** *adv.* eccentrically
Exzentrizität [ɛkstsɛntritsi'tɛ:t] **die;** ~, ~**en** eccentricity
Exzeß [ɛks'tsɛs] **der; Exzesses, Exzesse** excess; **etw. bis zum ~ treiben** carry sth. to excess
exzessiv [ɛkstsɛ'si:f] *Adj.* excessive
E-Zug der *s.* **Eilzug**

F

f, F [ɛf] **das;** ~, ~ **a)** *(Buchstabe)* f/F; **nach Schema F** according to a set pattern *or* routine; **b)** *(Musik)* [key of] F; *s. auch* **a/A**
f. *Abk.* **folgend** f.
F *Abk.* **Fahrenheit** F
Fa. *Abk.* **Firma**
Fabel ['fa:bl̩] **die;** ~, ~**n a)** *(Literaturw.) (Gattung)* fable; *(Kern einer Handlung)* plot; **b)** *(Erfundenes)* story; tale; fable; **ins Reich der ~ gehören** belong in the realm of fantasy
fabelhaft 1. *Adj.* **a)** *(ugs.: großartig)* fantastic *(coll.)*; **b)** *nicht präd. (unglaublich)* fabulous ‹*riches*›. **2.** *adv. (ugs.)* fantastically *(coll.)*; fabulously *(coll.)*
Fabel·tier das mythological *or* fabulous creature
Fabrik [fa'bri:k] **die;** ~, ~**en** factory; *(Papier~, Baumwollspinnerei)* mill; **eine chemische ~:** a chemical works
Fabrik·anlage die factory; *(Maschinen)* factory plant
Fabrikant [fabri'kant] **der;** ~en, ~**en** manufacturer
Fabrik·arbeiter der factory-worker
Fabrikat [fabri'ka:t] **das;** ~[e]s, ~e product; *(Marke)* make
Fabrikation [fabrika'tsio:n] **die;** ~: production
Fabrikations·fehler der manufacturing fault
Fabrik-: ~besitzer der factory-owner; **~direktor der** works *or* production manager; **~gebäude das** factory building; **~gelände das** factory site
fabrizieren [fabri'tsi:rən] *tr. V.* **a)** *(ugs. abwertend)* knock together *(coll.)*; **b)** *(veralt.: herstellen)* manufacture; produce
fabulieren [fabu'li:rən] *itr. V.* invent stories; spin yarns
Facette [fa'sɛtə] **die;** ~, ~**n** facet
Fach [fax] **das;** ~[e]s, **Fächer** ['fɛçɐ] **a)** compartment; *(für Post)* pigeon-hole; *(im Schrank)* shelf; **b)** *(Studienrichtung, Unterrichts~)* subject; *(Wissensgebiet)* field; *(Berufszweig)* trade; **ein Meister seines ~es** a master of his trade; **vom ~ sein** be an expert; **ein Mann vom ~:** an expert
fach-, Fach-: ~arbeiter der skilled worker; craftsman; **~arzt der** specialist **(für** in); **~ausdruck der** technical *or* specialist term; **~bezogen** *Adj.* specialized ‹*training*›; **~buch das** specialist book
fächeln ['fɛçl̩n] *tr. V.* fan
Fächer ['fɛçɐ] **der;** ~s, ~: fan
fächer·artig *Adj.* fan-like
fach-, Fach-: ~frau die expert; **~gebiet das** field; **~gelehrte der/die** specialist **(für** in); **~gerecht 1.** *Adj.* correct; **2.** *adv.* correctly; **~geschäft das** specialist shop; **ein ~geschäft für Sportartikel/Eisenwaren** a specialist sports shop/ironmonger's; **~hochschule die** college *(offering courses in a special subject)*; **~idiot der** *(abwertend) person who has no interests outside his/her subject*; **~jargon der** *(abwertend)* technical jargon; **~kenntnis die** specialized *or* specialist knowledge; **~kraft die** skilled worker; **~kreise** *Pl.* **in ~kreisen** in specialist circles; **~kundig 1.** *Adj.* knowledgeable; **2.** *adv.* **jmdn. ~kundig beraten** give sb. informed *or* expert advice
fachlich 1. *Adj.* specialist ‹*knowledge, work*›; technical ‹*problem, explanation, experience*›; **~e Ausbildung/Qualifikation** training/qualification in the subject. **2.** *adv.* **etw. ~ beurteilen** give a professional opinion on sth.; **~ qualifiziert** qualified in the subject
fach-, Fach-: ~literatur die specialist literature; *(bes. naturwissenschaftlich auch)* technical literature; **in der medizinischen ~literatur** in the specialist medical literature; **~mann der** expert; **~männisch 1.** *Adj.* expert; **2.** *adv.* **jmdn. ~männisch beraten** give sb. expert advice
Fachschaft die; ~, ~**en a)** *(einer Berufsgruppe)* professional association; **b)** *(von Studenten)* student body of the/a faculty
fach-, Fach-: ~schule die technical college; **~simpelei** [~zɪmpə'lai] **die;** ~, ~**en** *(ugs.)* shop-talk; **~simpeln** [~zɪmpl̩n] *itr. V. (ugs.)* talk shop; **~sprache die** technical terminology *or* language; **~sprachlich** *Adj.* technical; **~übergreifend** *Adj.* inter-disciplinary; **~welt die;** *o. Pl.* experts *pl.*; **in der ~welt** among experts
Fach·werk das *o. Pl. (Bauweise)* half-timbered construction
Fachwerk·haus das half-timbered house
Fach-: ~wissenschaftler der specialist; **~wort das;** *Pl.* **~wörter** technical *or* specialist term; **~zeitschrift die** *s.* **~literatur:** specialist/technical journal
Fackel ['fakl̩] **die;** ~, ~**n** torch
fackeln *itr. V. (ugs.)* shilly-shally *(coll.)*; dither; **nicht lange gefakkelt!** no shilly-shallying! *(coll.)*; don't dither about!
Fackel-: ~schein der torchlight; **~zug der** torchlight procession
fade ['fa:də] *Adj.* **a)** *(schal)* insipid; **ein ~r Beigeschmack** *(fig.)* a flat after-taste; **b)** *(bes. südd., österr.: langweilig)* dull

¹Faden ['fa:dn̩] **der; ~s, Fäden** ['fɛ:dn̩] **a)** *(Garn)* thread; **ein ~:** a piece of thread; **der rote ~** *(fig.)* the central theme; **den ~ verlieren** *(fig.)* lose the thread; **er hat** *od.* **hält alle Fäden in der Hand** *(fig.)* he holds the reins; **an einem dünnen** *od.* **seidenen ~ hängen** *(fig.)* hang by a single thread; **Fäden ziehen** ⟨*cheese etc.*⟩ be soft and stringy; **b)** *(Med.)* suture; **die Fäden ziehen** remove the stitches

²Faden der; ~s, ~ *(Seemannsspr.)* fathom

faden-, Faden-: ~kreuz das cross-hairs *pl.*; **~scheinig** [~ʃainɪç] *Adj.* **a)** *(fig. abwertend)* threadbare ⟨*morality*⟩; flimsy ⟨*argument, reason, excuse*⟩; **b)** *(abgewetzt)* threadbare ⟨*clothes*⟩; **~wurm der** *(Zool.)* threadworm

Fagott [fa'gɔt] **das; ~[e]s, ~e** bassoon

Fagottist der; ~en, ~en, bassoonist

fähig ['fɛ:ɪç] *Adj.* **a)** *(begabt)* able; capable; **ein ~er Kopf sein** have an able mind; **b)** *(bereit, in der Lage)* **zu etw. ~ sein** be capable of sth.; **~ sein, etw. zu tun** be capable of doing sth.

Fähigkeit die; ~, ~en a) *meist Pl.* *(Tüchtigkeit)* ability; capability; **geistige ~en** intellectual faculties *or* abilities; **praktische ~en** practical skills; **jmds. ~en wecken** awaken sb.'s talents; **b)** *o. Pl.* *(Imstandesein)* ability **(zu** to)

fahl [fa:l] *Adj.* pale; pallid; wan ⟨*light, smile*⟩

Fähnchen ['fɛ:nçən] **das; ~s, ~ a)** little flag; **b)** *(ugs. abwertend: Kleid)* **ein billiges ~:** a cheap frock *(Brit.) or* dress

fahnden ['fa:ndn̩] *itr. V.* search **(nach** for)

Fahndung die; ~, ~en search

Fahne ['fa:nə] **die; ~, ~n a)** flag; **b)** *(fig.)* **etw. auf seine ~n schreiben** espouse the cause of sth.; **seine ~ nach dem Wind[e] hängen** trim one's sails to the wind; **mit fliegenden ~n zu jmdm./etw. überlaufen** openly and suddenly turn one's coat; **c)** *o. Pl.* *(ugs.: Alkoholgeruch)* **eine ~ haben** reek of alcohol

fahnen-, Fahnen-: ~eid der oath of allegiance; **~flucht die** desertion; **~flucht begehen** desert; **~flüchtig** *Adj.* **~flüchtig werden/sein** desert/be a deserter; **~mast der, ~stange die** flagpole; **~träger der** standard-bearer

Fähnrich ['fɛ:nrɪç] **der; ~s, ~e** *(Milit.)* ensign; **~ zur See** ensign

Fahr·ausweis der a) *(Amtsspr.: Fahrschein)* ticket; **b)** *(schweiz.: Führerschein)* driving licence

Fahr·bahn die carriageway; **beim Überqueren der ~:** when crossing the road

Fahrbahn·markierung die road-marking

fahrbar *Adj.* ⟨*table, bed*⟩ on castors; mobile ⟨*crane, kitchen, etc.*⟩; **ein ~er Untersatz** *(ugs.)* wheels *pl.* *(joc.)*

Fähre ['fɛ:rə] **die; ~, ~n** ferry

fahren ['fa:rən] **1.** *unr. itr. V.; mit sein* **a)** *(als Fahrzeuglenker)* drive; *(mit dem Fahrrad, Motorrad usw.)* ride; **mit dem Auto ~:** drive; *(her~ auch)* come by car; *(hin~ auch)* go by car; **mit dem Fahrrad/Motorrad ~:** cycle/motor cycle; come/go by bicycle/motor cycle; **mit 80 km/h ~:** drive/ride at 80 k.p.h.; **links/rechts ~:** drive on the left/right; *(abbiegen)* bear *or* turn left/right; **langsam ~:** drive/ride slowly; **gegen etw. ~:** go into sth.; **b)** *(mit dem Auto usw. als Mitfahrer; mit öffentlichen Verkehrsmitteln usw./als Fahrgast)* go **(mit** by); *(mit dem Aufzug/der Rolltreppe/der Seilbahn/dem Skilift)* take the lift *(Brit.) or (Amer.)* elevator/escalator/cable-car/ski-lift; *(mit der Achterbahn, dem Karussell usw.)* ride **(auf** + *Dat.* on); *(per Anhalter)* hitch-hike; **erster/zweiter Klasse/zum halben Preis ~:** travel *or* go first/second class/at half-price; **ich fahre nicht gern [im] Auto/Bus** I don't like travelling in cars/buses; **fährst du mit mir?** are you coming with me?; **c)** *(reisen)* go; **in Urlaub ~:** go on holiday; **d)** *(los~)* go; leave; **e)** ⟨*motor vehicle, train, lift, cable-car*⟩ go; ⟨*ship*⟩ sail; **mein Auto fährt nicht** my car won't go; **der Aufzug fährt heute nicht** the lift *(Brit.) or (Amer.)* elevator is out of service today; **f)** *(verkehren)* run; **der Bus fährt alle fünf Minuten/bis Goetheplatz** the bus runs *or* goes every five minutes/goes to Goetheplatz; **von München nach Passau fährt ein D-Zug** there's a fast train from Munich to Passau; **g)** *(betrieben werden)* **mit Diesel/Benzin ~:** run on diesel/petrol *(Brit.) or (Amer.)* gasoline; **mit Dampf/Atomkraft ~:** be steam-powered/atomic-powered; **h)** *(schnelle Bewegungen ausführen)* **in die Kleider ~:** leap into one's clothes; **in die Höhe ~:** jump up [with a start]; **der Blitz ist in einen Baum ge~:** the lightning struck a tree; **jmdm. an die Kehle ~:** leap at sb.'s throat; **sich** *(Dat.)* **mit der Hand durchs Haar ~:** run one's fingers through one's hair; **was ist denn in dich ge~?** *(fig.)* what's got into you?; **der Schreck fuhr ihm in die Glieder** *(fig.)* the shock went right through him; **jmdm. über den Mund ~** *(fig.)* shut sb. up; **aus der Haut ~** *(ugs.)* blow one's top *(coll.)*; **i)** *(Erfahrungen machen)* **gut/schlecht mit jmdm./einer Sache ~:** get on well/badly with sb./sth. **2.** *unr. tr. V.* **a)** *(fortbewegen)* drive ⟨*car, lorry, train, etc.*⟩; ride ⟨*bicycle, motor cycle*⟩; **ein Boot ~:** sail a boat; **Auto/Motorrad/Roller ~:** drive [a car]/ride a motor cycle/scooter; **Bahn/Bus** *usw.* **~:** **go by train/bus** *etc.*; **Kahn** *od.* **Boot/Kanu ~:** go boating/canoeing; **Ski ~:** ski; **Schlitten ~:** toboggan; **Rollschuh ~:** [roller-]skate; **Schlittschuh ~:** [ice-]skate; **Aufzug/Rolltreppe ~:** take the lift *(Brit.) or (Amer.)* elevator/use the escalator; **Sessellift ~:** ride in a/the chair-lift; **U-Bahn ~:** ride on the underground *(Brit.) or (Amer.)* subway; **Karusell ~:** ride on the merry-go-round; **b)** *mit sein ([als Strecke] zurücklegen)* drive; *(mit dem Motorrad, Fahrrad)* ride; take ⟨*curve*⟩; **einen Umweg/eine Umleitung ~:** make a detour/follow a diversion; **der Zug fährt jetzt eine andere Strecke** the train takes a different route now; **c)** *(befördern)* drive, take ⟨*person*⟩; take ⟨*thing*⟩; ⟨*vehicle*⟩ take; ⟨*ship, lorry, etc.*⟩ carry ⟨*goods*⟩; *(zum Sprecher)* drive, bring ⟨*person*⟩; bring ⟨*thing*⟩; ⟨*vehicle*⟩ bring; **d)** *mit sein* **80 km/h ~:** do 80 k.p.h.; **hier muß man 50 km/h ~:** you've got to keep to 50 k.p.h. here; **e)** *meist mit sein* **ein Rennen ~:** take part in a race; **f)** *meist mit sein* **einen Rekord ~:** set a record; **1:23:45/eine gute Zeit ~:** do *or* clock 1.23.45/a good time; **g) ein Auto schrottreif ~:** write off a car; *(durch lange Beanspruchung)* run *or* drive a car into the ground; **eine Beule in den Kotflügel ~:** dent the wing; **i)** *(als Treibstoff benutzen)* use ⟨*diesel, regular*⟩. **3.** *unr. refl. V.* **a) sich gut ~** ⟨*car*⟩ handle well, be easy to drive; **b)** *unpers.* **in dem Wagen/mit dem Zug fährt es sich bequem** the car gives a comfortable ride/it is comfortable travelling by train

fahrend *Adj.* itinerant; **~es Volk** travelling people *pl.*

fahren|lassen *unr. tr. V.* **a)** *(loslassen)* let go; **b)** *(aufgeben)* abandon ‹*hope*›

Fahrer der; ~s, ~: driver

Fahrer·flucht die: ~flucht begehen fail to stop after [being involved in] an accident

Fahrerin die; ~, ~nen driver

Fahr-: ~erlaubnis die *(Amtsspr.)* driving licence; **jmdm. die ~ entziehen** disqualify sb. from driving; **~gast der** passenger; **~geld das** fare; **~gemeinschaft die** car pool; **~gestell das** *(Kfz-W.)* chassis

fahrig ['fa:rıç] *Adj.* nervous, agitated ‹*movements*›

Fahr·karte die ticket

Fahrkarten-: ~ausgabe die ticket office; **~automat der** ticket machine; **~schalter der** ticket window

fahr-, Fahr-: ~lässig 1. *Adj.* negligent ‹*behaviour*›; **~e Tötung/Körperverletzung** *(Rechtsw.)* causing death/injury through *or* by [culpable] negligence; **2.** *adv.* negligently; **~lässigkeit die** negligence; **~lehrer der** driving instructor

Fähr·mann der ferryman

Fahr·plan der a) timetable; schedule *(Amer.)*; **den ~ einhalten** run to schedule *or* on time; **b)** *(ugs.: Programm)* plans *pl.*

fahrplan·mäßig 1. *Adj.* scheduled ‹*departure, arrival*›. **2.** *adv.* ‹*depart, arrive*› according to schedule, on time

Fahr-: ~praxis die driving experience; **~preis der** fare; **~prüfung die** driving test

Fahr·rad das bicycle; cycle; **mit dem ~ fahren** cycle; ride a bicycle

Fahrrad-: ~fahrer der cyclist; **~geschäft das** bicycle shop; **~händler der** bicycle dealer; **etw. beim ~händler kaufen** buy sth. from a/the bicycle shop; **~ständer der** bicycle rack *or* stand

Fahr·rinne die shipping channel; fairway

Fahr·schein der ticket

Fahrschein-: ~automat der ticket machine; **~entwerter der** ticket cancelling machine; **~heft das** book of tickets

Fahr-: ~schule die a) *(Unternehmen)* driving school; **b)** *(ugs.: Unterricht)* driving lessons *pl;* **~schüler der a)** learner driver; **b)** *pupil who must use transport to get to school;* **~spur die** traffic-lane; **die ~spur wechseln/beibehalten** change lanes/stay in one's lane

fährst [fɛ:ɐst] *2. Pers. Sg. Präsens v.* **fahren**

Fahr-: ~stil der style of driving; *(mit dem Rad)* style of riding; **~streifen der** *s.* **~spur; ~stuhl der** lift *(Brit.)*; elevator *(Amer.)*; *(für Lasten)* hoist; **mit dem ~ fahren** take the lift/elevator; **~stunde die** driving lesson

Fahrt [fa:ɐt] **die; ~, ~en a)** *o. Pl. (das Fahren)* journey; **freie ~ haben** have a clear run; *(fig.)* have been given the green light; **b)** *(Reise)* journey; *(Schiffsreise)* voyage; **auf der ~:** on the journey; **c)** *(kurze Reise, Ausflug)* trip; *(Wanderung)* hike; **eine ~ [nach/zu X] machen** go on *or* take a trip [to X]; **eine ~ ins Blaue machen** *(mit dem Auto)* go for a drive; *(Veranstaltung)* go on a mystery tour; **auf ~ gehen** *(veralt.)* go hiking; **d)** *o. Pl. (Bewegung)* **~ machen** *(Seemannsspr.)* make way; **in voller ~:** at full speed; **die ~ verlangsamen** slow down; decelerate; **die ~ beschleunigen** speed up; accelerate; **~ aufnehmen** gather speed; pick up speed; **in ~ kommen** *od.* **geraten** *(ugs.)* get going; *(böse werden)* get worked up

fährt [fɛ:ɐt] *3. Pers. Sg. Präsens v.* **fahren**

fahr·tauglich *Adj.* fit to drive *postpos.*

Fahr·tauglichkeit die fitness to drive

Fährte ['fɛ:ɐtə] **die** tracks *pl.;* trail; **jmds. ~ verfolgen** track sb.; **die richtige ~ finden** *(fig.)* get on the right track; **die falsche ~ verfolgen** *(fig.)* be on the wrong track

Fahrten·messer das sheath-knife

Fahrt·kosten *Pl. (für öffentliche Verkehrsmittel)* fare/fares; *(für Autoreisen)* travel costs; **die ~ erstatten** pay travelling expenses

Fahr·treppe die escalator

Fahrt·richtung die direction; **in ~ parken** park in the direction of the traffic; **gegen die ~ sitzen** *(im Zug)* sit with one's back to the engine; *(im Bus)* sit facing backwards; **in ~ sitzen** *(im Zug)* sit facing the engine; *(im Bus)* sit facing forwards

Fahrtrichtungs·anzeiger der *(Kfz-W.)* [direction] indicator

fahr·tüchtig *Adj.* ‹*driver*› fit to drive; ‹*vehicle*› roadworthy

Fahr·tüchtigkeit die *(des Fahrers)* fitness to drive; *(des Fahrzeugs)* roadworthiness

Fahrt-: ~unterbrechung die break [in the journey]; stop; **~wind der** airflow; **~ziel das** destination

fahr-, Fahr-: ~untüchtig *Adj.* ‹*driver*› unfit to drive; ‹*vehicle*› unroadworthy; **~verbot das** disqualification from driving; driving ban; **jmdm. [ein] ~verbot erteilen** ban *or* disqualify sb. from driving; **~wasser das** shipping channel; fairway; **in ein gefährliches ~wasser geraten** *(fig.)* get on to dangerous ground; **in jmds. ~wasser schwimmen** *od.* **segeln** *(fig.)* follow [along] in sb.'s wake; **~werk das** *(Flugw.)* undercarriage; **~zeit die** travelling time

Fahr·zeug das vehicle; *(Luft~)* aircraft; *(Wasser~)* vessel

Fahrzeug-: ~bau der motor manufacturing industry; **~führer der** driver of a/the motor vehicle; **~halter der** registered keeper [of a/the vehicle]; **~papiere** *Pl.* vehicle documents *pl.*

Faible ['fɛ:bl̩] **das; ~s, ~s** liking; *(Schwäche)* weakness **(für** for)

fair [fɛ:ɐ] **1.** *Adj.* fair **(gegen** to). **2.** *adv.* fairly; **~ spielen** play fairly *or (coll.)* fair

Fairneß ['fɛ:ɐnɛs] **die; ~:** fairness

Fakir ['fa:ki:ɐ, *österr.:* fa'ki:ɐ] **der; ~s, ~e** fakir

Faksimile [fak'zi:mile] **das; ~s, ~s** facsimile

Fakten *s.* **Faktum**

faktisch 1. *Adj.; nicht präd.* real; actual; practical ‹*disadvantage, usefulness*›. **2.** *adv.* **a) das bedeutet ~ ...:** it means in effect ...; **es ist ~ möglich/unmöglich** it is in actual fact possible/impossible; **b)** *(bes. österr. ugs.: praktisch, eigentlich)* more or less; virtually

Faktor ['faktɔr] **der; ~s, ~en** [-'to:rən] factor

Faktotum [fak'to:tʊm] **das; ~s, ~s** *od.* **Faktoten** *(scherzh.)* factotum

Faktum ['faktʊm] **das; ~s, Fakten** fact

Fakultät [fakʊl'tɛ:t] **die; ~, ~en** *(Hochschulw.)* faculty

fakultativ [fakʊlta'ti:f] *Adj.* optional ‹*subject, participation*›

Falke ['falkə] **der; ~n, ~n** *(auch Politik fig.)* hawk

Falkland·inseln *Pl.* Falkland Islands; Falklands

Falkner der; ~s, ~: falconer

Fall [fal] **der; ~[e]s, Fälle** ['fɛlə] **a)** *(Sturz)* fall; **zu ~ kommen** have a fall; *(fig.)* come to grief; **jmdn. zu ~ bringen** *(fig.)* bring about sb.'s downfall; **etw. zu ~ bringen** *(fig.)* stop sth.; **der ~ einer Stadt** *(fig.)* the fall of a town; **b)** *(das Fallen)*

descent; **der freie ~:** free fall; **c)** *(Ereignis, Vorkommnis)* case; *(zu erwartender Umstand)* eventuality; **für den äußersten** *od.* **schlimmsten ~, im schlimmsten ~:** if the worst comes to the worst; **im besten ~:** at best; **es ist [nicht] der ~:** it is [not] the case; **gesetzt den ~:** assuming; supposing; **auf jeden ~, in jedem Fall, auf alle Fälle** in any case; **auf keinen ~:** on no account; **das ist doch ein ganz klarer ~:** it's perfectly clear; **nicht jmds. ~ sein** *(fig. ugs.)* not be sb.'s cup of tea; **d)** *(Rechtsw., Med., Grammatik)* case; **der 1./2./3./4. ~** *(Grammatik)* the nominative/genitive/dative/accusative case

Fall·beil das guillotine

Falle ['falə] **die; ~, ~n a)** *(auch fig.)* trap; **in die ~ gehen** walk into the trap; **jmdm. eine ~ stellen** *(fig.)* set a trap for sb.; **jmdm. in die ~ gehen** *(fig.)* fall into sb.'s trap; **b)** *(salopp: Bett)* **in die ~ gehen** turn in *(coll.)*

fallen *unr. itr. V.; mit sein* **a)** fall; **etw. ~ lassen** drop sth.; **sich ins Gras/Bett/Heu ~ lassen** fall on to the grass/into bed/into the hay; *(fig.)* **in Trümmer ~:** collapse in ruins; **in Schwermut ~:** be overcome by melancholy; **b)** *(hin~, stürzen)* fall [over]; **auf die Knie/in den Schmutz ~:** fall to one's knees/in the dirt; **über einen Stein ~:** trip over a stone; **c)** *(sinken)* ‹*prices*› fall; ‹*temperature, water level*› fall, drop; ‹*fever*› subside; **im Preis ~:** go down *or* fall in price; **d)** *(an einen bestimmten Ort gelangen)* ‹*light, shadow, glance, choice, suspicion*› fall; **die Wahl fiel auf ihn** the choice fell on him; **e)** *(abgegeben werden)* ‹*shot*› be fired; *(Sport: erzielt werden)* ‹*goal*› be scored; *(geäußert werden)* ‹*word*› be spoken; ‹*remark*› be made; *(getroffen werden)* ‹*decision*› be taken *or* made; **f)** *(nach unten hängen)* ‹*hair*› fall; **die Haare ~ ihr ins Gesicht/auf die Schulter** her hair falls over her face/to her shoulders; **g)** *(im Kampf sterben)* die; fall *(literary)*; **im Krieg ~:** die in the war; **h)** *(aufgehoben, beseitigt werden)* ‹*ban*› be lifted; ‹*tax*› be abolished; ‹*obstacle*› be removed; ‹*limitation*› be overcome; **i)** *(zu einer bestimmten Zeit stattfinden)* **in eine Zeit ~:** occur at a time; **mein Geburtstag fällt auf einen Samstag** my birthday falls on a Saturday; **j)** *(zu einem Bereich gehören)* **in/unter eine Kategorie ~:** fall into *or* within a category; **unter ein Gesetz/eine Bestimmung ~:** come under a law/a regulation; **k)** *(zu~, zuteil werden)* ‹*inheritance, territory*› fall **(an** + *Akk.* to); **jmdm. in die Hände ~:** fall into the hands of sb.

fällen ['fɛlən] *tr. V.* **a)** fell ‹*tree, timber*›; **b) ein Urteil ~** ‹*judge*› pass sentence; ‹*jury*› return a verdict; **einen Schiedsspruch ~:** make a ruling

fallen|lassen *unr. tr. V. (fig.)* **a)** abandon ‹*plan, aim, project*›; **b)** drop ‹*friend, colleague*›; **c)** let fall ‹*remark*›; drop ‹*hint*›

fällig ['fɛlıç] *Adj.* **a)** due; **eine ~e Reform** an overdue reform; **b)** ‹*sum of money*› payable, due; **ein ~er Wechsel/~e Zinsen** a bill to mature/interest payable

Fall·obst das windfalls *pl.*

Fall·rückzieher der *(Fußball)* bicycle kick

falls [fals] *Konj.* if; *(für den Fall, daß)* in case; **~ es regnen sollte** in case it should rain

Fall·schirm der parachute; **mit dem ~ abspringen** *(im Notfall)* parachute out; *(als Sport)* make a [parachute] jump

Fallschirm-: ~springen das parachuting *no art.*; **~springer der** parachutist

Fall-: ~strick der trap; snare; **jmdm. ~stricke legen** *(fig.)* set traps for sb.; **~studie die** case-study; **~tür die** trapdoor

falsch [falʃ] **1.** *Adj.* **a)** *(unecht, imitiert)* false ‹*teeth, plait*›; imitation ‹*jewellery*›; **~er Hase** *(Kochk.)* meat loaf; **b)** *(gefälscht)* counterfeit, forged ‹*banknote*›; false, forged ‹*passport*›; assumed ‹*name*›; **c)** *(irrig, fehlerhaft)* wrong ‹*impression, track, pronunciation*›; wrong, incorrect ‹*answer*›; **logisch ~ sein** be logically false; **an den Falschen geraten** come to the wrong man; **etw. in die ~e Kehle** *od.* **den ~en Hals bekommen** *(fig. ugs.)* take sth. the wrong way; **d)** *(unangebracht)* false ‹*shame, modesty*›; **e)** *(irreführend)* false ‹*statement, promise*›; **unter Vorspiegelung ~er Tatsachen** under false pretences; **f)** *(abwertend: hinterhältig)* false ‹*friend*›; **ein ~er Hund** *(salopp)* a two-faced so-and-so *(sl.)*; **eine ~e Schlange** *(fig.)* a snake in the grass; **ein ~es Spiel [mit jmdm.] treiben** play false with sb.. **2.** *adv.* **a)** *(fehlerhaft)* wrongly; incorrectly; **~ singen** sing wrongly; **~ gehen/fahren** go the wrong way; **etw. ~ verstehen** misunderstand sth.; **die Uhr geht ~:** the clock is wrong; **~ informiert** *od.* **unterrichtet sein** be misinformed; **~ herum** *(verkehrt)* back to front; the wrong way round; *(auf dem Kopf)* upside down; *(links)* inside-out; **b)** *(irreführend)* **~ schwören** lie on oath

Falsch·aus·sage die *(Rechtsspr.)* **[eidliche] ~aussage** false testimony *or* evidence; **uneidliche ~aussage** false statement [not on oath]

fälschen ['fɛlʃn̩] *tr. V.* forge, fake ‹*signature, document, passport*›; forge, counterfeit ‹*coin, banknote*›

Fälscher der; ~s, ~ *s.* **fälschen:** forger; counterfeiter

Falsch·geld das counterfeit money

Falschheit die; ~: duplicity; deceitfulness

fälschlich 1. *Adj.; nicht präd.* false ‹*claim, accusation*›; *(irrtümlich)* mistaken, false ‹*assumption, suspicion*›. **2.** *adv.* falsely, wrongly ‹*claim, accuse*›; mistakenly, falsely ‹*assume, suspect*›

fälschlicher·weise *Adv.* by mistake; mistakenly

falsch-, Falsch-: ~|liegen *unr. itr. V. (ugs.)* be mistaken; **~meldung die** false report; **~münzer** ['~mʏntsɐ] **der** forger; counterfeiter

Fälschung die; ~, ~en a) fake; counterfeit; **b)** *(das Fälschen) s.* **fälschen:** forging; counterfeiting

Falsett [fal'zɛt] **das; ~[e]s, ~e** *(Musik)* falsetto [voice]

Falt-: ~blatt das leaflet; *(in Zeitungen, Zeitschriften, Büchern)* insert; **~boot das** collapsible boat

Falte ['faltə] **die; ~, ~n a)** crease; **~n schlagen** crease; **b)** *(im Stoff)* fold; *(mit scharfer Kante)* pleat; **c)** *(Haut~)* wrinkle; line; **die Stirn in ~n legen** *od.* **ziehen** *(nachdenklich)* knit one's brow; *(verärgert)* frown; **d)** *(Geol.)* fold

fälteln ['fɛltl̩n] *tr. V.* pleat

falten 1. *tr. V.* fold; **die Hände ~:** fold one's hands. **2.** *refl. V. (auch Geol.)* fold; ‹*skin*› wrinkle, become wrinkled

falten-, Falten-: ~bildung die folding; *(der Haut)* wrinkling; **~gebirge das** [range *sing.* of] fold mountains *pl.*; **~los** *Adj.* uncreased ‹*garment*›; unwrinkled ‹*skin*›; **~rock der** pleated skirt

Falter der; ~s, ~ *(Nacht~)* moth; *(Tag~)* butterfly

faltig a) *Adj.* ‹*clothes*› gathered [in

folds]; wrinkled ⟨*skin, hands*⟩; **b)** *(zerknittert)* creased

-fältig [-fɛltɪç] *Adj., adv.* -fold

Falz [falts] **der; ~es, ~e** *(Buchbinderei) (scharfe Faltlinie)* fold; *(Übergang zwischen Buchdeckel und -rücken)* groove

falzen *tr. V. (Buchbinderei)* fold; *(Technik)* seam

familiär [fami'liɛ:ɐ] **1.** *Adj.* **a)** family ⟨*problems, worries*⟩; **aus ~en Gründen** for family reasons; **b)** *(zwanglos)* familiar; informal; informal ⟨*tone, relationship*⟩. **2.** *adv. (zwanglos)* **sich ~ ausdrücken** to talk in a familiar way

Familie [fa'mi:li̯ə] **die; ~, ~n a)** family; **~ Meyer** the Meyer family; **~ haben** *(ugs.)* have a family; **eine ~ gründen** *(heiraten)* marry; *(Kinder bekommen)* start a family; **das bleibt in der ~:** it will stay in the family; **das kommt in den besten ~n vor** it happens in the best families; **das liegt in der ~:** it runs in the family; **b)** *(Biol.)* family

familien-, Familien-: ~angehörige der/die; *adj. Dekl.* member of the family; **~angelegenheit die** family affair *or* matter; **~anschluß der** personal contact [with a/the family]; **~anzeigen** *Pl.* births, deaths, and marriages; **~besitz der** family property; **im ~besitz** in the family's possession; **~betrieb der** family business *or* firm; **~feier die** family party; **~krach der** *(ugs.)* family row; **~kreis der** family circle; **im engsten ~kreis** in the immediate family; **~leben das;** *o. Pl.* family life; **~mit·glied das** member of the family; **~name der** surname; family name; **~ober·haupt das** head of the family; **~planung die;** *o. Pl.* family planning *no art.*; **~politik die;** *o. Pl.* policy/policies relating to the family; **~stand der** marital status; **~vater der: ~vater sein** be the father of a family; **ein guter ~vater** a good husband and father; **~verhältnisse** *Pl.* family circumstances; family background

famos [fa'mo:s] *(veralt.)* **1.** *Adj.* splendid. **2.** *adv.* splendidly

Fan [fɛn] **der; ~s, ~s** fan

Fanal [fa'na:l] **das; ~s, ~e** *(geh.)* torch

Fanatiker [fa'na:tikɐ] **der; ~s, ~:** fanatic; *(religiös)* fanatic; zealot

fanatisch 1. *Adj.* fanatical. **2.** *adv.* fanatically

Fanatismus der; ~: fanaticism

fand [fant] *1. u. 3. Pers. Sg. Prät. v.* **finden**

Fanfare [fan'fa:rə] **die; ~, ~n a)** herald's trumpet; **b)** *(Signal)* fanfare; flourish

Fang [faŋ] **der; ~[e]s, Fänge** ['fɛŋə] **a)** *o. Pl. (Tier~)* trapping; *(von Fischen)* catching; **b)** *o. Pl. (Beute)* bag; *(von Fischen)* catch; haul; **einen guten ~ machen** *od.* **tun** *(fig.)* make a good catch

Fang·arm der *(Zool.)* tentacle

fangen 1. *unr. tr. V.* **a)** *(ergreifen, fassen)* catch, trap ⟨*bird, animal*⟩; catch ⟨*fish*⟩; **die Katze fängt eine Maus** the cat catches a mouse; **eine ~** *(südd., österr. ugs.)* get a clip round the ear *(coll.)*; **b)** *(gefangennehmen)* catch, capture ⟨*fugitive etc.*⟩; **gefangene Soldaten** captured soldiers; **von etw. [ganz] gefangen sein** *(fig.)* be [quite] enthralled by sth.; **c)** *auch itr. (auffangen)* catch ⟨*ball*⟩; **er kann gut/nicht ~:** he's good/not good at catching. **2.** *unr. refl. V.* **a)** *(in eine Falle geraten, nicht mehr frei kommen)* get *or* be caught; **b)** *(wieder in die normale Lage kommen)* **sich [gerade] noch ~:** [just] manage to steady oneself; **sich wieder ~** *(fig.)* recover

Fangen das; ~s: ~ spielen play tag *or* catch

Fänger ['fɛŋɐ] **der; ~s, ~:** catcher; *(von Großwild)* hunter

Fang-: ~frage die catch question; trick question; **~netz das** *(Fischereiw.)* [fishing] net; **~schuß der** *(Jagdw.)* coup de grâce

Fan- [fɛn-]: **~klub der** fan club; **~post die** fan mail

Farb-: ~band das [typewriter] ribbon; **~bild das a)** *(Foto)* colour photo; **b)** *(Illustration)* colour picture; **~dia das** colour slide; colour transparency

Farbe ['farbə] **die; ~, ~n a)** colour; **~ bekommen/verlieren** get some colour/lose one's colour; **an ~ gewinnen/verlieren** *(fig.)* become more/less colourful; **b)** *(Substanz) (zum Malen, Anstreichen)* paint; *(zum Färben)* dye; **~n mischen/auftragen** mix/apply paint; **c)** *o. Pl. (Farbigkeit)* colour; **der Film ist in ~:** the film is in colour; **d)** *(Kartenspiel)* suit; **eine ~ bedienen** follow suit; **~ bekennen** *(fig. ugs.)* come clean *(coll.)*

farb·echt *Adj.* colour-fast

Färbe·mittel das dye

färben ['fɛrbn̩] **1.** *tr. V.* **a)** dye ⟨*wool, material, hair*⟩; **etw. grün ~:** dye sth. green; **b) eine politisch gefärbte Rede** a speech with a political slant. **2.** *refl. V.* **sich schwarz/rot** *usw.* **~:** turn black/red *etc.* **3.** *itr. V. (ugs.: ab~)* ⟨*material, blouse etc.*⟩ run

-farben *Adj., adv.* -coloured; **creme~ angestrichen** painted cream

farben-, Farben-: ~blind *Adj.* colour-blind; **~freudig** *Adj.*, **~froh** *Adj.* colourful; **~pracht die** colourful splendour; **~prächtig** *Adj.* vibrant with colour *postpos.*

Färber der; ~s, ~: dyer

Färberei die; ~, ~en dye-works *sing.*

Färberin die; ~, ~nen dyer

Farb-: ~fernsehen das colour television; **~fernseher der** *(ugs.)* colour telly *(coll.) or* television; **~fernsehgerät das** colour television [set]; **~film der** colour film; **~foto das** colour photo; **~fotografie die** *o. Pl.* colour photography

farbig 1. *Adj.* **a)** coloured; **b)** *(bunt, fig.)* colourful ⟨*dress, picture, description, tale*⟩; **~e [Kirchen]fenster** stained-glass [church-]windows. **2.** *adv.* colourfully

-farbig *Adj., adv. s.* **-farben**

Farbige der/die; *adj. Dekl.* coloured man/woman; coloured; **die ~n** the coloured people; *(in Südafrika)* the Coloureds

farblich 1. *Adj.* in colour *postpos.*; as regards colour *postpos.* **2.** *adv.* **etw. ~ aufeinander abstimmen** match sth. in colour

farb-, Farb-: ~los *Adj. (auch fig.)* colourless; clear ⟨*varnish*⟩; neutral ⟨*shoe polish*⟩; **~schicht die** layer of paint; *(Anstrich)* coat of paint; **~stift der a)** *(Buntstift)* coloured pencil; **b)** *(Filzstift)* coloured felt-tip *or* pen; **~stoff der a)** *(Med., Biol.)* pigment; **b)** *(für Lebensmittel)* colouring; **~ton der** shade

Färbung die; ~, ~en a) *(Farbgebung)* colouring; colour; **b)** *(das Färben)* dyeing; **c)** *(fig.: Tendenz)* slant

Farce ['farsə] **die; ~, ~n** farce

Farm [farm] **die; ~, ~en** farm

Farmer der; ~s, ~: farmer

Farn [farn] **der; ~[e]s, ~e** fern

Farn·kraut das fern

Fasan [fa'za:n] **der; ~[e]s, ~e[n]** pheasant

Faschierte das; *adj. Dekl. (österr.)* minced meat; mince

Fasching ['faʃɪŋ] **der; ~s, ~e** *od.* **~s** [pre-Lent] carnival

Faschismus [fa'ʃɪsmʊs] **der; ~:** fascism *no art.*

Faschist der; ~en, ~en fascist

faschistisch *Adj.* fascist

faseln ['fa:zl̩n] *itr. V. (ugs. abwertend)* drivel; blather

Faser ['fa:zɐ] **die; ~, ~n** fibre

fa̲serig *Adj.* fibrous ⟨*paper*⟩; stringy ⟨*meat*⟩

fa̲sern *itr. V.* fray

Fa̲ser·pflanze die fibre-plant

Fas·nacht ['fas-] **die;** *o. Pl. (bes. südd.) s.* **Fastnacht**

Faß [fas] **das; Fa̩sses, Fässer** ['fɛsɐ] barrel; *(Öl~, Benzin~ usw.)* drum; *(kleines Bier~)* keg; *(kleines Sherry~, Portwein~ usw.)* cask; **Bier vom ~:** draught beer; **das schlägt dem ~ den Boden aus** *(ugs.)* that takes the biscuit *(Brit. coll.) or (coll.)* cake; **ein ~ ohne Boden sein** be an endless drain on sb.'s resources; **ein ~ aufmachen** *(ugs.)* paint the town red

Fassade [fa'sa:də] **die; ~, ~n a)** façade; frontage; **b)** *(abwertend: äußere Erscheinung)* façade; front

fa̩ßbar *Adj.* **a)** *(greifbar, konkret)* tangible, concrete ⟨*results*⟩; **b)** *(verständlich)* comprehensible

Fa̩ß·bier das beer on draught; *(Bier vom Faß)* draught beer

Fäßchen ['fɛsçən] **das; ~s, ~:** small barrel; [small] cask

fassen ['fasn̩] **1.** *tr. V.* **a)** *(greifen)* grasp; take hold of; **jmdn. am Arm ~:** take hold of sb.'s arm; **jmdn. bei der Hand ~:** take sb. by the hand; **etw. zu ~ bekommen** get a hold on sth.; **b)** *(festnehmen)* catch ⟨*thief, culprit*⟩; **c)** *(aufnehmen können)* ⟨*hall, tank*⟩ hold; **d)** *(begreifen)* **ich kann es nicht ~:** I cannot take it in; **das ist [doch] nicht zu fassen!** it's incredible; **e)** *(in verblaßter Bedeutung)* make, take ⟨*decision*⟩; **Vertrauen** *od.* **Zutrauen zu jmdm. ~:** begin to feel confidence in *or* to trust sb.; **Mut ~:** take courage; **f)** *(in eine Fassung bringen)* set, mount ⟨*jewel*⟩; curb ⟨*spring, well*⟩; **g)** *(formulieren, gestalten)* **etw. in Worte/Verse ~:** put sth. into words/verse; **einen Begriff eng/weit ~:** define a concept narrowly/widely; **h)** *(fachspr.: aufnehmen)* take on ⟨*load, goods*⟩; **i)** *(Soldatenspr.)* draw ⟨*rations, supplies, ammunition*⟩. **2.** *itr. V.* **a)** *(greifen)* **nach etw. ~:** reach for sth.; **in etw.** *(Akk.)* **~:** put one's hand in sth.; **an etw.** *(Akk.)* **~:** touch sth.; **ins Leere ~:** grasp thin air; **b)** *(einrasten)* ⟨*screw*⟩ bite; ⟨*cog*⟩ mesh. **3.** *refl. V.* **a)** pull oneself together; recover [oneself]; **b) sich kurz ~:** be brief

Fasson [fa'sõ:] **die; ~, ~s** style; shape; **jeder muß nach seiner [eigenen] *od.* auf seine [eigene] ~ selig werden** everyone has to work out his own salvation

Fasso̩n·schnitt der short back and sides

Fa̩ssung die; ~, ~en a) *(Version)* version; **b)** *o. Pl. (Selbstbeherrschung, Haltung)* composure; self-control; **die ~ bewahren** keep one's composure; **die ~ verlieren** lose one's self-control; **jmdn. aus der ~ bringen** upset *or* ruffle sb.; **etw. mit ~ tragen** bear sth. calmly; **c)** *(für Glühlampen)* holder; **d)** *(von Juwelen)* setting; *(Bilder~, Brillen~)* frame

fa̩ssungs-, Fa̩ssungs-: ~los *Adj.* stunned; **jmdn. ~los anstarren** gaze at sb. in bewilderment; **~losigkeit die** state of bewilderment; **~vermögen das;** *o. Pl.* capacity

fast [fast] *Adv.* almost; nearly; **~ nie** almost never; hardly ever; **~ nirgends** hardly anywhere

fa̩sten *itr. V.* fast

Fa̩sten-: ~kur die drastic reducing diet; **~zeit die a)** *(Rel.)* time of fasting; **b)** *(kath. Rel.)* Lent

Fa̩st·nacht die a) *(Faschingsdienstag)* Shrove Tuesday; **b)** *(Karneval)* carnival; Shrovetide

Fa̩stnachts-: ~dienstag der Shrove Tuesday; **~zug der** carnival procession

Faszination [fastsina'tsio:n] **die; ~:** fascination; **eine ~ auf jmdn. ausüben** fascinate sb.

faszinieren [fastsi'ni:rən] *tr. V.* fascinate

fatal [fa'ta:l] *Adj.* **a)** fatal; **b)** *(peinlich, mißlich)* awkward; embarrassing; **~e Folgen haben** have unfortunate consequences

Fatali̩smus der; ~: fatalism

Fatali̩st der; ~en, ~en fatalist

fatali̩stisch *Adj.* fatalistic

Fata Morgana ['fa:ta mɔr'ga:na] **die; ~ ~, ~ Morga̲nen** *od.* **~ ~s** fata morgana; mirage; *(fig.)* illusion

Fatzke ['fatskə] **der; ~n** *od.* **~s, ~n** *od.* **~s** *(ugs. abwertend)* twit *(Brit. sl.)*; jerk *(sl.)*

fauchen ['fauxn̩] *itr. V.* **a)** ⟨*cat*⟩ hiss; ⟨*tiger*⟩ snarl; *(fig.)* ⟨*engine*⟩ hiss; **b)** *(sich gereizt äußern)* snarl

faul [faul] **1.** *Adj.* **a)** *(verdorben)* rotten, bad ⟨*food*⟩; bad ⟨*tooth*⟩; rotten ⟨*wood*⟩; foul, stale ⟨*air*⟩; foul ⟨*water*⟩; **b)** *(träge)* lazy; idle; **zu ~ zu etw. sein/zu ~ sein, etw. zu tun** be too lazy *or* idle for sth./to do sth.; **auf der ~en Haut liegen/sich auf die ~e Haut legen** take it easy; **c)** *(ugs.: nicht einwandfrei)* bad ⟨*joke*⟩; dud ⟨*cheque*⟩; false ⟨*peace*⟩; lame ⟨*excuse*⟩; shabby ⟨*compromise*⟩; shady ⟨*business, customer*⟩; **das ist doch [alles] ~er Zauber** it's [all] quite bogus; **etwas ist ~ im Staate Dänemark** something is rotten in the state of Denmark; **d)** *(säumig)* bad ⟨*debtor*⟩. **2.** *adv. (träge)* lazily; idly

fa̲ulen *itr. V.; meist mit sein* rot; ⟨*water*⟩ go foul, stagnate; ⟨*meat*⟩ go off, putrefy; ⟨*fish*⟩ go off, go bad

faulenzen ['faulɛntsn̩] *itr. V.* laze about; loaf about *(derog.)*

Fa̲ulenzer der; ~s, ~: idler; lazybones *sing. (coll.)*

Faulenzere̲i die; ~, ~en *(abwertend)* idleness; laziness

Fa̲ulheit die; ~: laziness; idleness

fa̲ulig *Adj.* stagnating ⟨*water*⟩; putrefying ⟨*meat*⟩; ⟨*meat*⟩ which is going bad; rotting ⟨*vegetables, fruit*⟩; foul, putrid ⟨*smell*⟩; **~ schmecken/riechen** taste/smell bad *or* off

Fäulnis ['fɔylnɪs] **die; ~:** rottenness; **in ~ übergehen** begin to rot

Fa̲ul-: ~pelz der *(fam.)* lazybones *sing. (coll.)*; **~tier das a)** *(Zool.)* sloth; **b)** *(ugs.: Faulenzer) s.* **~pelz**

Fauna ['fauna] **die; ~, Fa̲unen** *(Zool.)* fauna

Faust [faust] **die; ~, Fäuste** ['fɔystə] fist; **eine ~ machen, die Hand zur ~ ballen** clench one's fist; **die ~ ballen/öffnen** clench/unclench one's fist; **jmdm. mit der ~ ins Gesicht schlagen** punch sb. in the face; **das paßt wie die ~ aufs Auge** *(ugs.) (paßt nicht)* that clashes horribly; *(paßt)* that matches perfectly; **die ~/Fäuste in der Tasche ballen** *(fig.)* be seething inwardly; **auf eigene ~:** on one's own initiative; off one's own bat *(coll.)*; **mit der ~ auf den Tisch schlagen** *od.* **hauen** *(fig.)* put one's foot down

Fa̲ust·ball der faustball

Fäustchen ['fɔystçən] **das; ~s, ~:** fist; **sich** *(Dat.)* **ins ~ lachen** laugh up one's sleeve; *(aus finanziellen Gründen)* laugh all the way to the bank

fa̲ust·di̩ck 1. *Adj.* as thick as a man's fist *postpos.*; **eine ~e Lüge** *(fig.)* a bare-faced lie. **2.** *adv.* **er hat es ~ hinter den Ohren** *(ugs.)* he's a crafty *or* sly one

fa̲usten *tr. V.* fist, punch ⟨*ball*⟩

fa̲ust-, Fa̲ust-: ~groß *Adj.* as big as a fist *postpos.*; **~handschuh der** mitten; **~kampf der** *(geh.)* pugilism; boxing; *(Wett-*

kampf) boxing contest; **~kämpfer der** *(geh.)* pugilist; boxer

Fäustling ['fɔystlɪŋ] **der; ~s, ~e** mitten

Faust-: ~pfand das security; **~recht das;** *o. Pl.* rule of force; **~regel die** rule of thumb; **~schlag der** punch

Fauxpas [fo'pa] **der; ~, ~:** faux pas

Favorit [favo'ri:t] **der; ~en, ~en** favourite

Fax [faks] **das; ~, ~[e]** fax

faxen *tr. V.* fax

Faxen *Pl. (ugs.)* **a)** *(dumme Späße)* fooling around; **laß die ~!** stop fooling around *or* playing the fool!; **b)** *(Grimassen)* **~ machen** *od.* **schneiden** make *or* pull faces

Fazit ['fa:tsɪt] **das; ~s, ~s** *od.* **~e** result; **das ~ [aus etw.] ziehen** sum [sth.] up

FDJ [ɛfde:'jɔt] **die; ~** *Abk. (ehem. DDR)* **Freie Deutsche Jugend** Free German Youth

F.D.P. ['ɛfde:pe:] **die; ~** *Abk.* **Freie Demokratische Partei**

F-Dur ['ɛf-] **das; ~** *(Musik)* [key of] F major

Feature ['fi:tʃɐ] **das; ~s, ~s** *od.* **die; ~, ~s** *(Rundf., Ferns., Zeitungsw.)* feature

Februar ['fe:brua:ɐ̯] **der; ~[s], ~e** February; *s. auch* **April**

fechten ['fɛçtn̩] *unr. itr., tr. V.* fence; *(fig. geh.)* fight

Fechter der; ~s, ~, Fechterin die; ~, ~nen fencer

Fecht·kampf der rapier fight; *(Sport)* fencing bout

Feder ['fe:dɐ] **die; ~, ~n a)** *(Vogel~)* feather; *(Gänse~)* quill; *(lange Hut~)* plume; **[noch] in den ~n liegen** *(ugs.)* [still] be in one's bed; **sich mit fremden ~n schmükken** strut in borrowed plumes; **b)** *(zum Schreiben)* nib; *(mit Halter)* pen; *(Gänse~)* quill[-pen]; **eine spitze ~ führen** *(geh.)* wield a sharp pen; **zur ~ greifen** *(geh.)* take up one's pen; **c)** *(Technik)* spring

feder-, Feder-: ~ball der a) *o. Pl. (Spiel)* badminton; **b)** *(Ball)* shuttlecock; **~bett das** duvet *(Brit.)*; continental quilt *(Brit.)*; stuffed quilt *(Amer.)*; **~fuchser** [~fʊksɐ] **der; ~s, ~** *(abwertend)* pen-pusher; **~führend** *Adj.* in charge *postpos.*; **~führung die: unter der ~führung des Ministers** under the overall control of the minister; **~gewicht** *(Schwerathletik)* featherweight; *s. auch* **Fliegengewicht a; ~halter der** fountain-pen; **~kiel der** quill; **~kissen das** feather cushion; *(im Bett)* feather pillow; **~kraft die** tension [of a/the spring]; **~leicht 1.** *Adj.* ⟨*person*⟩ as light as a feather; featherweight ⟨*object*⟩; **2.** *adv.* as lightly as a feather; **~lesen das: nicht viel ~lesen[s] mit jmdm./etw. machen** make short work of sb./sth.; **ohne viel ~lesen[s], ohne langes ~lesen** without much ado; **viel zu viel ~lesen[s] machen** make far too much fuss; **~mappe die** pen and pencil case

federn 1. *itr. V.* ⟨*springboard, floor, etc.*⟩ be springy; **in den Knien ~:** bend at the knees. **2.** *tr. V.* **a)** spring; **das Auto ist gut/schlecht gefedert** the car has good/poor suspension; **das Bett ist gut gefedert** the bed is well-sprung; **b)** *s. auch* **teeren**

Feder·strich der stroke of the pen

Federung die; ~, ~en *(in Möbeln)* springs *pl.*; *(Kfz-W.)* suspension

Feder-: ~vieh das *(ugs.)* poultry; **~zeichnung die** pen-and-ink drawing

Fee [fe:] **die; ~, ~n** fairy

Fege·feuer das purgatory

fegen ['fe:gn̩] **1.** *tr. V.* **a)** *(bes. nordd.: säubern)* sweep; *(mit einem Handfeger)* brush; **b)** *(schnell entfernen)* brush; **etwas vom Tisch ~:** brush sth. off the table; *(fig.)* brush sth. aside; **c)** *(schnell treiben)* sweep; drive. **2.** *itr. V.* **a)** sweep; do the sweeping; **b)** *mit sein (rasen, stürmen)* sweep; tear *(coll.)*

Fehde ['fe:də] **die; ~, ~n** feud; **mit jmdm. in ~ liegen** be at feud with sb.

fehl [fe:l] *Adv.* **~ am Platz[e] sein** be out of place

Fehl·anzeige die a) ~! *(ugs.)* no chance *(coll.)*; **b)** *(Milit.)* nil return

fehlbar *Adj.* fallible

Fehl-: ~besetzung die: [als Ophelia] eine ~besetzung sein be miscast [in the role of Ophelia]; **~betrag der** *(bes. Kaufmannsspr.)* deficit; **~diagnose die** incorrect diagnosis; **~einschätzung die** false assessment; *(einer Entwicklung)* misjudgement

fehlen *itr. V.* **a)** *(nicht vorhanden sein)* be lacking; **ihm fehlt der Vater/das Geld** he has no father/no money; **ihr fehlt der Sinn dafür** she lacks a *or* has no feeling for it; **b)** *(ausbleiben)* be missing; be absent; **[un]entschuldigt ~:** be absent with[out] permission; **du darfst bei dieser Party nicht ~:** you mustn't miss this party; **Knoblauch darf bei dieser Soße nicht ~:** garlic is a must in this sauce; **c)** *(verschwunden sein)* be missing; be gone; **in der Kasse fehlt Geld** money is missing *or* has gone from the till; **d)** *(vermißt werden)* **er/das wird mir ~:** I shall miss him/that; **e)** *(erforderlich sein)* be needed; **ihm ~ noch zwei Punkte zum Sieg** he needs only two points to win; **es fehlte nicht viel, und ich wäre eingeschlafen** I all but fell asleep; **das fehlte mir gerade noch [zu meinem Glück], das hat mir gerade noch gefehlt** *(ugs.)* that's all I needed; **f)** *unpers. (mangeln)* **es fehlt an Lehrern** there is a lack of teachers; **bei ihnen fehlt es am Nötigsten** they lack what is most needed; **es an nichts ~ lassen** provide everything that is needed; **es fehlt an allen Ecken und Enden** *od.* **Kanten [bei jmdm.]** sb. is short of everything; **g) was fehlt Ihnen?** what seems to be the matter?; **fehlt dir etwas?** is there something wrong?; are you all right?; **h) weit gefehlt!** *(geh.)* far from it!

Fehl·entscheidung die wrong decision

Fehler der; ~s, ~ a) mistake; error; *(falsches Verhalten, Sport)* fault; **b)** *(schlechte Eigenschaft)* fault; shortcoming; *(Gebrechen)* [physical] defect; **c)** *(schadhafte Stelle)* flaw; blemish; **Porzellan mit kleinen ~n** porcelain with small flaws *or* imperfections

fehler·frei 1. *Adj.* faultless, perfect ⟨*piece of work, dictation, etc.*⟩; correct ⟨*measurement*⟩; **ein ~es Deutsch sprechen/schreiben** speak/write faultless *or* perfect German; *(Reiten)* **ein ~er Durchgang** a clear round. **2.** *adv.* without any mistakes; *(Reiten)* without any faults

fehlerhaft *Adj.* faulty; defective; incorrect ⟨*measurement*⟩; **eine ~e Stelle im Material** a defect in the material

fehler-, Fehler-: ~los 1. *Adj.* flawless; **2.** *adv.* flawlessly; without a mistake; **~quelle die** source of error; **~quote die** *(Statistik, Schulw.)* error rate; **~zahl die** number of mistakes *or* errors

fehl-, Fehl-: ~farbe die *(Kartenspiel) (Farbe, die einem Spieler fehlt)* void suit; *(Farbe, die nicht Trumpf ist)* plain suit; **~geburt die** miscarriage; **~|gehen** *unr. itr. V.; mit sein (geh.)* **a)** *(sich ir-*

ren) go *or* be wrong; **b)** *(sich verlaufen)* lose one's way; **Sie können nicht ~gehen** you cannot go [far] wrong; **~griff der** mistake; wrong choice; **~information die** piece of wrong information; **auf einer ~information beruhen** be based on [a piece of] incorrect information; **~interpretation die** misinterpretation; **~investition die** bad investment; **~konstruktion die: eine ~konstruktion sein** be badly designed; **~paß der** *(Ballspiele)* bad pass; **~planung die** [piece of] bad planning *no art.*; **~schlag der** failure; **~|schlagen** *unr. itr. V.; mit sein* fail; **~start der a)** *(Leichtathletik)* false start; **b)** *(Flugw.)* faulty start; **c)** *(Raumf.)* abortive launch; **~tritt der a)** false step; **b)** *(geh.: Verfehlung)* slip; indiscretion; **~urteil das a)** *(Rechtsw.)* **ein ~urteil fällen** ⟨*jury*⟩ return a wrong verdict; ⟨*judge*⟩ pass a wrong judgement; **b)** *(falsche Beurteilung)* error of judgement; **~zündung die** *(Technik)* misfire

feien ['faiən] *tr. V. (geh.)* protect **(gegen** against)

Feier ['faiɐ] **die; ~, ~n a)** *(Veranstaltung)* party; *(aus festlichem Anlaß)* celebration; **eine ~ in kleinem Rahmen/im Familienkreis** a small/family celebration/party; **b)** *(Zeremonie)* ceremony; **zur ~ des Tages** *(oft scherzh.)* in honour of the occasion

Feier·abend der a) *(Zeit nach der Arbeit)* evening; **schönen ~!** have a nice evening; **b)** *(Arbeitsschluß)* finishing time; **nach ~:** after work; **~ machen** finish work; knock off *(coll.)*; **für mich ist ~, dann ist** *od.* **mache ich ~** *(fig. ugs.)* I'm finished; I've had enough *(coll.)*

feierlich 1. *Adj.* **a)** ceremonial; solemn; **eine ~e Handlung** a ceremonial act; **das ist ja [schon] nicht mehr ~** *(ugs.)* it's got beyond a joke; **b)** *(emphatisch)* solemn ⟨*declaration*⟩. **2.** *adv.* **a)** solemnly; ceremoniously; **~ verabschiedet werden** be given a ceremonious farewell; **b)** *(emphatisch)* solemnly ⟨*declare, swear, etc.*⟩

Feierlichkeit die; ~, ~en a) *o. Pl.* solemnity; **b)** *meist Pl. (Veranstaltung)* celebration; festivity

feiern 1. *tr. V.* **a)** *(festlich begehen)* celebrate ⟨*birthday, wedding, etc.*⟩; **man muß die Feste ~, wie sie fallen** you have to enjoy yourself while you can; **b)** *(ehren, umjubeln)* acclaim ⟨*artist, sportsman, etc.*⟩. **2.** *itr. V.* celebrate; have a party

Feier-: ~schicht die *(Arbeitswelt)* cancelled shift; **eine ~schicht einlegen müssen** have one's shift cancelled; **~tag der** holiday; **ein gesetzlicher/kirchlicher ~tag** a public holiday/religious festival; **an Sonn- und ~tagen** on Sundays and public holidays

feig, feige [faik, 'faigə] **1.** *Adj.* cowardly. **2.** *adv.* like a coward/like cowards; in a cowardly way

Feige die; ~, ~n fig

Feigen-: ~baum der fig tree; **~blatt das a)** fig-leaf; **b)** *(fig.)* front; cover

Feigheit die; ~: cowardice; cowardliness

Feigling der; ~s, ~e coward

feil [fail] *Adj. (veralt.)* for sale *postpos.*; *(fig.)* venal

feil|bieten *unr. tr. V. (geh.)* offer ⟨*goods*⟩ for sale

Feile die; ~, ~n file

feilen *tr., itr. V.* file

feilschen ['failʃn̩] *itr. V.* haggle **(um** over)

fein [fain] **1.** *Adj.* **a)** *(zart)* fine ⟨*material, line, etc.*⟩; **b)** *(~körnig)* fine ⟨*sand, powder*⟩; finely-ground ⟨*flour*⟩; finely-granulated ⟨*sugar*⟩; **etw. ~ mahlen** grind sth. fine; **c)** *(hochwertig)* high-quality ⟨*fruit, soap, etc.*⟩; fine ⟨*silver, gold, etc.*⟩; fancy ⟨*cakes, pastries, etc.*⟩; **nur das Feinste vom Feinen kaufen** buy only the best; **d)** *(ugs.: erfreulich)* great *(coll.)*; marvellous; **e)** *(~geschnitten)* finely shaped, delicate ⟨*hands, features, etc.*⟩; **f)** *(scharf, exakt)* keen, sensitive ⟨*hearing*⟩; keen ⟨*sense of smell*⟩; **g)** *(ugs.: anständig, nett)* great *(coll.)*, splendid ⟨*person*⟩; **eine ~e Verwandtschaft/Gesellschaft** *(iron.)* a fine *or* nice family/crowd; **h)** *(einfühlsam)* delicate ⟨*sense of humour*⟩; keen ⟨*sense, understanding*⟩; **i)** *(gediegen, vornehm)* refined ⟨*gentleman, lady*⟩. **2.** *adv.* **a) ~ [he]raussein** *(ugs.)* be sitting pretty *(coll.)*; **Unterschiede ~ herausarbeiten** bring out subtle differences; **b)** *(ugs.: bekräftigend)* **etw. ~ säuberlich aufschreiben** write sth. down nice and neatly

Fein·arbeit die detailed work; *(Technik)* precision work

Feind der; ~[e]s, ~e a) enemy; **sich** *(Dat.)* **~e machen** make enemies; **sich** *(Dat.)* **jmdn. zum ~ machen** make an enemy of sb.; **b) der ~** *(Milit.)* the enemy *constr. as pl.*

Feind·berührung die *(Milit.)* contact with the enemy

Feindin die; ~, ~nen *s.* **Feind a**

feindlich 1. *Adj.* **a)** hostile; **b)** *nicht präd. (Milit.)* enemy ⟨*attack, broadcast, activity*⟩. **2.** *adv.* in a hostile manner; with hostility

Feindschaft die; ~, ~en enmity; **sich** *(Dat.)* **jmds. ~ zuziehen** make an enemy of sb.

feind·selig 1. *Adj.* hostile. **2.** *adv.* **sich ~ ansehen** look at each other in a hostile manner *or* with hostility

Feind·seligkeit die; ~, ~en hostility; **~en** *(Milit.)* hostilities

fein-, Fein-: ~frost der; *o. Pl. (regional)* deep-frozen foods *pl.*; **~fühlig 1.** *Adj.* sensitive; **2.** *adv.* sensitively; **~fühligkeit die; ~:** sensitivity; **~gebäck das** [fancy] cakes and pastries *pl.*; **~gefühl das** sensitivity; **~gold das** fine gold

Feinheit die; ~, ~en a) *o. Pl.* fineness; delicacy; **b)** *(Nuance)* subtlety; **die stilistischen ~en** the stylistic subtleties *or* nuances

fein-, Fein-: ~körnig *Adj.* **a)** fine-grained, fine ⟨*sand, gravel, etc.*⟩; finely-granulated ⟨*sugar*⟩; **b)** *(Fot.)* fine-grain ⟨*film*⟩; **~kost die** delicatessen *pl.*; **~kost·geschäft das** delicatessen; **~|machen** *refl. V. (ugs.)* dress up; **~maschig** *Adj.* finely meshed, fine-mesh *attrib.* ⟨*net etc.*⟩; **~mechaniker der** precision engineer; **~schmecker der; ~s, ~:** gourmet; **~schmecker·lokal das** gourmet restaurant; **~schnitt der** *(Tabak)* fine cut; **~sinnig 1.** *Adj.* sensitive and subtle; **2.** *adv.* in a sensitive and subtle manner; **~strumpfhose die** sheer tights *pl. or* pantihose; **~wäsche die** delicates *pl.*; **~waschmittel das** mild detergent

feist *Adj. (meist abwertend)* fat ⟨*face, fingers, etc.*⟩

feixen ['faiksn̩] *itr. V. (ugs.)* smirk

Feld [fɛlt] **das; ~[e]s, ~er a)** *o. Pl. (geh.: unbebaute Bodenfläche)* country[side]; **freies ~:** open country[side]; **b)** *(bebaute Bodenfläche)* field; **auf dem ~ arbeiten** work in the field; **das ~ bestellen** till the field; **c)** *(Sport: Spiel~)* pitch; field [of play]; **d)** *(auf Formularen)* box; space; *(auf Brettspielen)* space; *(auf dem Schachbrett)* square; **e)** *o. Pl. (Tätigkeitsbereich)* field; sphere; **das ~ der Wissenschaften** the field of science; **ein weites ~ [sein]** *(fig.)* [be] a wide sphere; **f)** *o. Pl. (ver-*

alt.: Schlacht~) field [of battle]; **gegen/für jmdn./etw. ins ~ ziehen** *(fig.)* crusade against/for sb./sth.; **das ~ räumen** leave; get out; **jmdn. aus dem ~[e] schlagen** eliminate sb.; get rid of sb.; **g)** *(Sport: geschlossene Gruppe)* field

feld-, Feld-: ~arbeit die a) work in the field; **b)** *(Wissensch.)* fieldwork; **~blume die** field flower; wild flower; **~flasche die** *(Milit.)* canteen; water-bottle; **~frucht die** arable crop; **~gottesdienst der** field-service; **~hase der** common hare; European hare; **~herr der** *(veralt.)* commander; **~jäger der** *(Polizist)* military policeman; *(Polizei)* military police; **~küche die** *(bes. Milit.)* field kitchen; **~marschall der** Field Marshal; **~maus die** [European] common vole; **~post die** forces' *(Brit.)* or *(Amer.)* military postal service; **~salat der** corn salad; lamb's lettuce; **~spat** [~ʃpa:t] **der; ~[e]s, ~späte** [~ʃpɛ:tə] *od.* **~spate** feldspar; **~spieler der** player *(excluding goalkeeper)*; **~stecher der** binoculars *pl.*; field glasses *pl.*; **~verweis der** *(Sport)* sending-off

Feld-Wald-und-Wiesen- *(ugs.)* run-of-the-mill; common-or-garden

Feld-: ~webel [-ve:bl̩] **der; ~s, ~** *(Milit.)* sergeant; **~weg der** path; track; **~zug der** *(Milit., fig.)* campaign

Felge ['fɛlgə] **die; ~, ~n a)** [wheel] rim; **b)** *(Turnen)* circle

Fell [fɛl] **das; ~[e]s, ~e a)** *(Haarkleid)* fur; *(Pferde~, Hunde~, Katzen~)* coat; *(Schaf~)* fleece; skin; **einem Tier das ~ abziehen** skin an animal; **jmdm. das ~ über die Ohren ziehen** *(fig. salopp)* take sb. for a ride *(sl.)*; **b)** *o. Pl. (Material)* fur; furskin; **c)** *(abgezogene behaarte Haut)* skin; hide; **d)** *(salopp: Haut des Menschen)* skin; *(fig.)* **ihm** *od.* **ihn juckt das ~** *(ugs.)* he is asking for a good hiding *(coll.)*; **ein dickes ~ haben** *(ugs.)* be thick-skinned *or* have a thick skin

Fell-: ~jacke die fur jacket; **~mütze die** fur cap

Fels [fɛls] **der; ~en, ~en a)** *o. Pl.* rock; **b)** *(geh.: Felsen)* rock; **wie ein ~ in der Brandung stehen** stand as firm as a rock

Fels·block der; *Pl.* **~blöcke** rock; boulder

Felsen ['fɛlzn̩] **der; ~s, ~:** rock; *(an der Steilküste)* cliff

felsen-, Felsen-: ~fest 1. *Adj.* firm; unshakeable ⟨*opinion, belief*⟩; **2.** *adv.* ⟨*believe, be convinced*⟩ firmly; **~klippe die** rocky cliff; **~riff das** rocky reef

felsig *Adj.* rocky

Fels-: ~spalte die crevice [in the rock]; **~vorsprung der** ledge; **~wand die** rock face

Feme ['fe:mə] **die; ~, ~n a)** *(hist.)* vehmgericht; **b)** *(Geheimgericht)* kangaroo court

Feme·mord der lynching

feminin [femi'ni:n] *Adj.* **a)** *(geh.: weiblich)* feminine ⟨*characteristic, behaviour*⟩; **b)** *(abwertend: unmännlich)* effeminate ⟨*man, type*⟩; **c)** *(Sprachw.)* feminine

Femininum ['fe:mini:nʊm] **das; ~s, Feminina** feminine noun

Feminismus der; ~, Feminismen feminism *no art.*

Feminist der; ~en, ~en, Feministin die; ~, ~nen feminist

feministisch *Adj.* feminist

Fenchel ['fɛnçl̩] **der; ~s** fennel

Fenn [fɛn] **das; ~[e]s, ~e** *(bes. nordd.)* fen

Fenster ['fɛnstɐ] **das; ~s, ~** *(auch DV)* window; **[sein] Geld zum ~ hinauswerfen** *(fig.)* throw [one's] money down the drain; **weg vom ~ sein** *(ugs.)* be right out of it

Fenster-: ~bank die, ~brett das window-sill; window-ledge; **~glas das a)** *o. Pl.* window glass; **b)** *Pl.* **~gläser** *(ungeschliffenes Glas)* plain glass; **~kreuz das** mullion and transom; **~laden der** [window] shutter; **~leder das** wash-leather

fensterln ['fɛnstɐln] *itr. V. (bes. südd., österr.)* climb through one's sweetheart's window

fenster-, Fenster-: ~los *Adj.* windowless; **~putzer der** window-cleaner; **~rahmen der** window-frame; **~scheibe die** window-pane

Ferien ['fe:ri̯ən] *Pl.* **a)** holiday *(Brit.)*; vacation *(Amer.)*; *(Werks~)* shut-down; holiday *(Brit.)*; *(Parlaments~)* recess; *(Hochschul~)* vacation; **~ haben** have a *or* be on holiday/vacation; **b)** *(Urlaub)* holiday[s *pl.*] *(Brit.)*; vacation *(Amer.)*; **in die ~ fahren** go on holiday/vacation

Ferien-: holiday... *(Brit.)*; vacation... *(Amer.)* ⟨*house, camp, resort, trip, etc.*⟩; *s. auch* **Urlaubs-**

Ferien-: ~arbeit die vacation work; **eine ~arbeit** a vacation job; **~kolonie die** [children's] holiday/vacation camp

Ferkel ['fɛrkl̩] **das; ~s, ~ a)** piglet; **b)** *(ugs. abwertend)* pig

Ferkelei die; ~, ~en *(ugs. abwertend)* *(Benehmen)* filthy behaviour; *(Bemerkung)* dirty remark

ferkeln *itr. V.* farrow

Ferment [fɛr'mɛnt] **das; ~[e]s, ~e** *(veralt.)* ferment *(arch.)*; enzyme

fern [fɛrn] **1.** *Adj.* **a)** *(räumlich)* distant, far-off, faraway ⟨*country, region, etc.*⟩; **b)** *(zeitlich)* distant ⟨*past, future*⟩; **in [nicht allzu] ~er Zukunft** in the [not too] distant future; **der Tag ist nicht mehr ~:** the day is not far off. **2.** *adv.* **~ von der Heimat [sein/leben]** [be/live] far from home; **etw. von ~ betrachten** look at sth. from a distance; *s. auch* **Osten c; nahe 2a. 3.** *Präp. mit Dat. (geh.)* far [away] from; a long way from; **~ der Heimat [leben]** [live] far from home *or* a long way from home

fern-, Fern-: ~ab [-'-] *(geh.)* **1.** *Adv.* far away; **2.** *Präp. mit Dat.* **~ab aller Zivilisation** far [away] from all civilization; **~amt das** *(veralt.)* telephone exchange; **~bedienung die** remote control; **~|bleiben** *unr. itr. V.; mit sein (geh.)* stay away (*Dat.* from)

ferne: von ~ *(geh.)* from far off *or* away

Ferne die; ~, ~n distance; **etw. in weiter ~ erblicken** see sth. in the far distance; **b) das liegt noch/schon in weiter ~** *(zeitlich)* that is still a long time away/that was a long time ago

ferner *Adv.* **a)** in addition; furthermore; **er rangiert unter „~ liefen"** *(fig.)* he is an also-ran; **b)** *(geh.: künftig)* in [the] future

fern-, Fern-: ~fahrer der long-distance lorry-driver *(Brit.)* or *(Amer.)* trucker; **~gelenkt** *Adj.* remote-controlled; *(fig.: durch Geheimdienste usw.)* controlled; **~gespräch das** long-distance call; trunk call; **ein ~gespräch mit jmdm./London führen** speak to *or* with sb./London long-distance; **~glas das** binoculars *pl.*; **~|halten 1.** *unr. tr. V.* **jmdn./etw. von jmdm./etw. ~halten** keep sb./sth. away from sb./sth.; **2.** *unr. refl. V.* **sich von jmdm./etw. ~halten** keep away from sb./sth.; **~heizung die** district heating system; **~kopierer der** fax machine; **~kurs[us] der** correspondence course; **~laster der** *(ugs.)* long-distance lorry *(Brit.)* or *(Amer.)* truck; **~last·zug der** [long-distance] articulated lorry; **~leitung die a)** *(Postw.)* long-distance line; **b)** *(Energiewirtsch.)* long-distance cable; **~|lenken** *tr. V.* operate by remote control; **~lenkung die** remote control;

~licht das *(Kfz-W.)* full beam; **das ~licht anhaben** drive on full beam; **~|liegen** *unr. itr. V.* **das liegt mir ~:** that is the last thing I want to do

Fern·melde-: ~amt das local telephone headquarters; **~gebühren** *Pl.* telephone charges; **~technik die;** *o. Pl.* telecommunications *sing., no art.*

fern-, Fern-: ~ost *o. Art.* Far East; **in/nach ~ost** in/to the Far East; **~östlich** *Adj.; nicht präd.* Far Eastern; **~rohr das** telescope; **~ruf der** telephone number; **~schreiben das** telex [message]; **~schreiber der** telex [machine]; teleprinter

Fernseh-: ~ansager der, ~ansagerin die television announcer; **~antenne die** television aerial *(Brit.) or (Amer.)* antenna; **~apparat der** television [set]

fern|sehen *unr. itr. V.* watch television

Fern·sehen das; ~s television; **im ~:** on television; **vom** *od.* **im ~ übertragen werden** be televised; be shown on television

Fern·seher der; ~s, ~ *(ugs.)* **a)** *(Gerät)* telly *(Brit. coll.);* TV; television; **b)** *(Zuschauer)* [television] viewer

Fernseh-: ~film der television film; **~gebühren** *Pl.* television licence fee; **~gerät das** television [set]; **~kamera die** television camera; **~programm das a)** *(Sendungen)* television programmes *pl.;* **b)** *(Kanal)* television channel; **c)** *(Blatt, Programmheft)* television [programme] guide; **~sendung die** television programme; **~spiel das** television play; **~studio das** television studio; **~turm der** television tower; **~übertragung die** television broadcast; **~zuschauer der** television viewer

Fern·sicht die *(Aussicht)* view; *(gute Sicht)* visibility

fern·sichtig *Adj. s.* **weitsichtig**

Fern·sprech- *(bes. Amtsspr.) s.* **Telefon-**

Fernsprech-: ~anschluß der telephone; line; **~auskunft die** directory enquiries *sing., no art.*

Fern·sprecher der telephone

Fernsprech-: ~gebühren *Pl.* telephone charges; **~teilnehmer der** telephone subscriber; telephone customer *(Amer.)*

fern-, Fern-: ~|steuern *tr. V. s.* **~lenken; ~steuerung die** *(Technik)* remote control; *(fig.: durch Geheimdienste usw.)* control; **~straße die** [principal] trunk road; major road; **~studium das** correspondence course; **~unterricht der** correspondence courses *pl.;* **~verkehr der** long-distance traffic; **~wärme die** district heating; **~weh das** *(geh.)* wanderlust; **~ziel das a)** *(zeitlich)* long-term aim; **b)** *(räumlich)* distant destination

Ferse ['fɛrzə] **die; ~, ~n** heel; *(fig.)* **sich an jmds. ~n** *(Akk.)***/sich jmdm. an die ~n heften** stick [hard] on sb.'s heels; **jmdm. [dicht] auf den ~n sitzen** *od.* **sein** *(ugs.)* be [hard *or* close] on sb.'s heels

Fersen·geld das: ~geld geben *(ugs. scherzh.)* take to one's heels

fertig ['fɛrtɪç] *Adj.* **a)** *(völlig hergestellt)* finished ‹*manuscript, picture, etc.*›; **das Essen ist ~:** lunch/dinner *etc.* is ready; **und ~ ist der Lack** *od.* **die Laube** *(ugs.)* and there you are; and bob's your uncle *(Brit. coll.);* **b)** *nicht attr. (zu Ende)* finished; **[mit etw.] ~ sein/werden** have finished/finish [sth.]; **bist du ~?** have you finished?; **mit jmdm. ~ sein** *(ugs.)* be finished *or* through with sb.; **mit etw. ~ werden** *(fig.)* cope with sth.; **c)** *nicht attr. (bereit, verfügbar)* ready **(zu, für** for); **zum Abmarsch/Start ~ sein** be ready to march/ready for take-off; **auf die Plätze – ~ – los!** on your marks, get set, go! *(Sport); (bei Kindern auch:)* ready, steady, go!; **d)** *nicht attr. (ugs.: erschöpft)* shattered *(coll.);* **mit den Nerven ~ sein** be at the end of one's tether; **e)** *(reif)* mature ‹*person, artist, etc.*›

fertig-, Fertig-: ~bau der; *Pl.* **~ten** prefabricated building; **~bauweise die** prefabricated construction; prefabrication; **~|bekommen** *unr. tr. V.,* **~|bringen** *unr. tr. V.* **a)** manage; **ich brächte es nicht ~, das zu tun** I couldn't bring myself to do that; **der bringt das ~!** *(iron.)* I wouldn't put it past him; **b)** *(zu Ende bringen)* finish

fertigen *tr. V.* make; **von Hand/maschinell gefertigt** hand-made/machine-produced

Fertig-: ~gericht das ready-to-serve meal; **~haus das** prefabricated house; prefab *(coll.)*

Fertigkeit die; ~, ~en skill

fertig-, Fertig-: ~|machen *tr. V.* **a)** *(ugs.: beenden)* finish ‹*task, job, etc.*›; **b)** *(ugs.: bereitmachen)* get ‹*meals, beds*› ready; **sich für etw. ~machen** get ready for sth.; **c) jmdn. ~machen** *(erschöpfen)* wear sb. out; *(durch Schikanen)* wear sb. down; *(deprimieren)* get sb. down; *(salopp: zusammenschlagen, töten)* do sb. in *(sl.); (ugs.: zurechtweisen)* tear sb. off a strip *(sl.);* **~|stellen** *tr. V.* complete; finish; **~stellung die** completion

Fertigung die; ~: production; manufacture

Fes [fe:s] **der; ~[es], ~[e]** fez

fesch [fɛʃ] *Adj. (bes. österr.: hübsch)* smart ‹*woman, suit, etc.*›

¹Fessel ['fɛsl̩] **die; ~, ~n** *meist Pl. (auch fig.)* fetter; shackle; *(Kette)* chain; **jmdm. ~n anlegen** fetter sb./put sb. in chains

²Fessel die; ~, ~n *(Anat.)* **a)** *(bei Huftieren)* pastern; **b)** *(bei Menschen)* ankle

fesseln *tr. V.* **a)** tie up; *(mit Ketten)* chain up; **jmdn. an Händen und Füßen ~:** tie sb. hand and foot; **jmdm. die Hände auf den Rücken ~:** tie sb.'s hands behind his/her back; **ans Bett/Haus/an den Rollstuhl gefesselt sein** *(fig.)* be confined to [one's] bed/tied to the house/confined to a wheelchair; **b)** *(faszinieren)* ‹*book*› grip; ‹*work, person*› fascinate; ‹*personality*› captivate; ‹*idea*› possess; **das Buch hat mich so gefesselt** I was so gripped by the book

fest [fɛst] **1.** *Adj.* **a)** *(nicht flüssig od. gasförmig)* solid; **~e Nahrung** solid food; **~e Gestalt** *od.* **Form[en] annehmen** take on a definite shape; **b)** *(straff)* firm, tight ‹*bandage*›; **c)** *(kräftig)* firm ‹*handshake*›; *(tief)* sound ‹*sleep*›; **d)** *(haltbar, solide)* sturdy ‹*shoes*›; tough, strong ‹*fabric*›; solid ‹*house, shell*›; **e)** *(energisch)* firm ‹*tread*›; steady ‹*voice*›; **eine ~e Hand brauchen** *(fig.)* need a firm hand; **f)** *(unbeirrbar)* **der ~en Überzeugung/Meinung sein, daß ...:** be firmly convinced/of the firm opinion that ...; **g)** *(endgültig)* firm ‹*appointment, date*›; firm, definite ‹*commitment*›; **h)** *nicht präd. (konstant)* fixed, permanent ‹*address*›; fixed ‹*income*›; **einen ~en Platz in etw.** *(Dat.)* **haben** *(fig.)* be firmly established in sth.. **2.** *adv.* **a)** *(straff)* ‹*tie, grip*› tight[ly]; **b)** (*ugs. auch* **~e)** *(tüchtig)* ‹*work*› with a will; ‹*eat*› heartily; ‹*sleep*› soundly; **~ zuschlagen** plant a solid punch; **er schläft ~:** he is fast asleep; **c)** *(unbeirrbar)* ‹*believe, be convinced*› firmly; **sich auf jmdn./etw. ~ verlassen** rely one hundred per cent on sb./sth.; **d)** *(endgültig)* firmly; definitely; **etw. ~ verein-**

baren come to a firm *or* definite arrangement about sth.; **e)** *(auf Dauer)* permanently; ~ **angestellt sein** be permanently employed; ~ **befreundet sein** be close friends; *(als Paar)* be going steady

Fest das; ~[e]s, ~e **a)** *(Veranstaltung)* celebration; *(Party)* party; **b)** *(Feiertag)* festival; *(Kirchen~)* feast; festival; **frohes ~!** happy Christmas/Easter!

fest-, Fest-: ~**akt der** ceremony; ~**ansprache die** address; ~|**beißen** *unr. refl. V.* **sich in etw.** *(Dat.)* ~**beißen** ‹*dog etc.*› sink its teeth firmly into sth.; ~**beleuchtung die** festive lighting; **in ~beleuchtung erstrahlen** be ablaze with festive illuminations; ~|**binden** *unr. tr. V.* tie [up] (**an** + *Dat.* to)

feste *Adv. (ugs.) s.* **fest 2 b**

fest-, Fest-: ~**essen das** banquet; ~|**fahren** *unr. itr., refl. V. (itr. mit sein)* get stuck; *(fig.)* get bogged down; ~**halle die** festival hall; ~|**halten 1.** *unr. tr. V.* **a)** *(halten, packen)* hold on to; **jmdn. am Arm ~halten** hold on to sb.'s arm; **etw. mit den Händen ~halten** hold sth. in one's hands; **b)** *(nicht weiterleiten)* withhold ‹*letter, parcel, etc.*›; **c)** *(verhaftet haben)* hold, detain ‹*suspect*›; **d)** *(aufzeichnen, fixieren)* record; capture; **etw. mit der Kamera ~halten** capture sth. with the camera; **e)** *(konstatieren)* record; **2.** *unr. refl. V. (sich anklammern)* **sich an jmdm./etw. ~halten** hold on to sb./sth.; **halt dich ~!** hold tight!; *(fig. ugs.)* brace yourself!; **3.** *unr. itr. V.* **an jmdm./etw. ~halten** stand by sb./sth.

festigen ['fɛstɪgn̩] **1.** *tr. V.* strengthen ‹*friendship, alliance, marriage, etc.*›; consolidate ‹*position*›; **in sich** *(Dat.)* **gefestigt sein** be strong. **2.** *refl. V.* ‹*friendship, ties*› become stronger

Festigkeit die; ~ **a)** *(Entschlossenheit)* firmness; **b)** *(Standhaftigkeit)* steadfastness; resolution; **c)** *(von Stoffen)* strength

Festigung die; ~: strengthening; *(einer Stellung)* consolidation

Festival ['fɛstivəl] **das;** ~s, ~s festival

Festivität [fɛstivi'tɛ:t] **die;** ~, ~en *(veralt., scherzh.)* festivity; celebration

fest-, Fest-: ~|**klammern** *refl. V.* **sich an jmdm./etw. ~klammern** cling [on] to sb./sth.; ~|**kleben 1.** *itr. V.; mit sein* stick (**an** + *Dat.* to); **2.** *tr. V.* stick; **etw. an etw.** *(Dat.)* ~**kleben** stick sth. to sth.; ~|**klemmen 1.** *itr. V.; mit sein* ~**geklemmt sein** be stuck *or* jammed; **2.** *tr. V.* wedge; jam; ~**körper der** *(Physik)* solid; ~|**krallen** *refl. V.* **sich in etw.** *(Dat.)* ~**krallen** ‹*cat etc.*› dig its claws into sth.; **sich an jmdm. ~krallen** ‹*cat etc.*› cling to sb. with its claws; ‹*person*› cling [on] to sb.; ~**land das;** *o. Pl.* mainland; **das europäische ~land** the continent of Europe/the European mainland; ~**ländisch** *Adj.; nicht präd.* **a)** mainland *attrib.*; **b)** *(kontinental)* continental ‹*climate, shelf, etc.*›; ~**land·sockel der** *(Geogr.)* continental shelf; ~|**legen** *tr. V.* **a)** *(verbindlich regeln)* fix ‹*time, deadline, price*›; arrange ‹*programme*›; **etw. gesetzlich ~legen** prescribe sth. by law; **b)** *(verpflichten)* **sich [auf etw.** *(Akk.)*] ~**legen [lassen]** commit oneself [to sth.]; **jmdn. [auf etw.** *(Akk.)*] ~**legen** tie sb. down [to sth.]; **c)** *(Bankw.)* tie up ‹*money*›; ~**legung die;** ~, ~en *s.* ~**legen: a)** fixing; arrangement; **b)** commitment

festlich 1. *Adj.* **a)** festive ‹*atmosphere*›; **b)** *(einem Fest gemäß)* formal ‹*dress*›. **2.** *adv.* **a)** festively; **b)** *(einem Fest gemäß)* formally; **etw. ~ begehen** celebrate sth.

Festlichkeit die; ~, ~en **a)** *(Feier)* celebration; **b)** *(der Stimmung, Atmosphäre)* festiveness; *(Feierlichkeit, Würde)* solemnity

fest-, Fest-: ~|**liegen** *unr. itr. V.* **a)** *(nicht weiterkommen)* be stuck; **b)** *(~stehen)* have been fixed; **c)** *(Bankw.)* ‹*money*› be tied up; ~|**machen 1.** *tr. V.* **a)** *(befestigen)* fix; **b)** *(fest vereinbaren)* arrange ‹*meeting etc.*›; **c)** *(Seemannsspr.)* moor ‹*boat*›; **2.** *itr. V. (Seemannsspr.)* moor; ~|**nageln** *tr. V.* **a)** *(befestigen)* nail (**an** + *Dat.* to); **b)** *(ugs.: festlegen)* **jmdn. [auf etw.** *(Akk.)*] ~**nageln** tie sb. down [to sth.]; **sich auf etw.** *(Akk.)* ~**nageln lassen** let oneself be tied [down] to sth.; ~**nahme die;** ~, ~n arrest; **bei seiner ~nahme** when he was/is arrested; ~|**nehmen** *unr. tr. V.* arrest; **jmdn. vorläufig ~nehmen** take sb. into custody; ~**platte die** *(DV)* hard disc; ~**platz der** fairground; ~**rede die** speech; ~**redner der** speaker; ~**saal der** banqueting hall; *(Ballsaal)* ballroom; ~|**saugen** *regelm. (auch unr.) refl. V.* attach itself (**an** + *Dat.* to); ~|**schrauben** *tr. V.* screw [up] tight; ~|**schreiben** *unr. tr. V.* establish; ~|**schrift die** commemorative volume; ~|**setzen 1.** *tr. V.* **a)** *(~legen)* fix ‹*time, deadline, price*›; lay down ‹*duties*›; **b)** *(in Haft nehmen)* detain; **2.** *refl. V.* ‹*dust*› collect, settle; *(fig.)* ‹*idea*› take root; ~**setzung die;** ~, ~en *s.* ~**setzen 1:** fixing; laying down; ~|**sitzen** *unr. itr. V.* be stuck; ~**spiel das a)** *Pl.* festival *sing.*; **b)** *(Bühnenstück)* festival production; ~|**stehen** *unr. itr. V.* **a)** *(~gelegt sein)* ‹*order, appointment, etc.*› have been fixed; **b)** *(unumstößlich sein)* ‹*decision*› be definite; ‹*fact*› be certain; ~ **steht** *od.* **es steht ~, daß ...:** it is certain *or* definite that ...; ~**stellbar** *Adj.* **a)** *(zu ermitteln)* ascertainable; **b)** *(wahrnehmbar)* detectable; diagnosable ‹*illness*›; ~|**stellen** *tr. V.* **a)** *(ermitteln)* establish ‹*identity, age, facts*›; **b)** *(wahrnehmen)* detect; diagnose ‹*illness*›; **er stellte ~, daß er sich geirrt hatte** he realized that he was wrong; **die Ärzte konnten nur noch den Tod ~stellen** all the doctors could do was [to] confirm that the patient/victim *etc.* was dead; **c)** *(aussprechen)* state ‹*fact*›; **ich muß ~stellen, daß ...:** I must *or* am bound to say that ...

Fest·stellung die a) *(Ermittlung)* establishment; **b)** *(Wahrnehmung)* realization; **die ~ machen, daß ...:** realize that ...; **c)** *(Erklärung)* statement; **die ~ treffen, daß ...:** observe that ...

Fest·tag der holiday; *(Kirchenfest)* [religious] feast-day; *(Ehrentag)* special day

Festung die; ~, ~en fortress

Festungs-: ~**anlage die** fortification; ~**mauer die** wall of a/the fortress

fest-, Fest-: ~**verzinslich** *Adj. (Bankw.)* fixed-interest *attrib.*; fixed-income *attrib.*; ~**vortrag der** lecture; ~**wiese die** festival site; ~**zelt das** marquee; ~|**ziehen** *unr. tr. V.* pull tight; ~**zug der** procession

Fete ['fe:tə] **die;** ~, ~n *(ugs.)* party; **eine ~ geben** *od.* **feiern** have *or* throw a party

Fetisch ['fe:tɪʃ] **der;** ~s, ~e *(Völkerk., fig.)* fetish

Fetischismus der; ~: fetishism *no art.*

fett [fɛt] **1.** *Adj.* **a)** fatty ‹*food*›; ~**er Speck** fat bacon; **b)** *(sehr dick)* fat; **c)** *(ugs.: üppig, reich)* fat ‹*inheritance, wallet*›; ~**e Jahre/Zeiten** rich years/good times; ~**e Beute machen** make a rich haul; **d)** *(ertragreich)* rich ‹*soil*›; luxuriant ⟨*vegetation*⟩; **e)** *(Druckw.)* bold; *(breiter, größer)* extra bold; **etw. ~ drucken** print sth. in bold/ extra bold [type]. **2.** *adv.* **~ essen** eat fatty foods; **~ kochen** use a lot of fat [in cooking]

Fett das; ~[e]s, ~e a) fat; **sein ~ [ab]bekommen** *od.* **[ab]kriegen** *(ugs.)* get one's come-uppance *(Amer.)*; **sein ~ [weg]haben** *(ugs.)* have been put in one's place *or* taught a lesson; **b)** *o. Pl. (~gewebe)* fat; **~ ansetzen** ⟨*animal*⟩ fatten up; ⟨*person*⟩ put on weight; **~ schwimmt oben** *(Spr.)* fat people never drown!; *(fig.)* the rich never suffer; **c)** *(Schmiermittel, Pflegemittel)* grease

fett-, Fett-: ~arm 1. *Adj.* low-fat ⟨*food*⟩; low in fat *pred.*; **2.** *adv.* **~arm essen** eat low-fat foods; **~auge das** speck of fat; **~creme die** enriched [skim] cream; **~druck der** bold type; **in ~druck** in bold [type]

fetten 1. *tr. V. (mit Fett einreiben)* grease. **2.** *itr. V. (Fett absondern)* be greasy

fett-, Fett-: ~fleck[en] der grease mark *or* spot; **~frei** *Adj.* fat-free; grease-free ⟨*surface*⟩; **~ sein** be fat-free/be free of grease; **~gedruckt** *Adj. (präd. getrennt geschrieben)* bold; **~haltig** *Adj.* fatty; **[sehr] ~haltig sein** contain [a lot of] fat

fettig *Adj.* greasy; oily; greasy ⟨*skin, saucepan, etc.*⟩

fett-, Fett-: ~kloß der *(ugs. abwertend)* fatty; fatso *(sl.)*; **~leibig** *Adj.* obese; **~leibigkeit die; ~:** obesity; **~näpfchen das: [bei jmdm.] ins ~näpfchen treten** *(scherzh.)* put one's foot in it [with sb.]; **~reich 1.** *Adj.* high-fat ⟨*food*⟩; **2.** *adv.* **~reich essen** eat high-fat foods; **~sack der** *(salopp abwertend)* fatso *(sl.)*; **~schicht die** layer of fat; **~stift der a)** *(Schreibgerät)* grease pencil; lithographic pencil; **b)** *(Lippenstift)* lip salve; **~sucht die** *(Med.)* obesity; **~wanst der** *(salopp abwertend)* fatso *(sl.)*

Fetus ['fe:tʊs] **der; ~** *od.* **~ses, ~se** *od.* **Feten** *(Med.)* foetus

Fetzen der; ~s, ~ a) scrap; **etw. in ~ [zer]reißen** tear sth. to pieces *or* shreds; **in ~ gehen** *(ugs.)* fall apart *or* to pieces; **daß die ~ fliegen** *(ugs.)* like mad; **b)** *(abwertend: Kleid)* **ein billiger ~:** cheap rags *pl.*

feucht [fɔyçt] *Adj.* damp ⟨*cloth, wall, hair*⟩; tacky ⟨*paint*⟩; humid ⟨*climate*⟩; sweaty, clammy ⟨*hands*⟩; moist ⟨*lips*⟩; **eine ~e Aussprache haben** *(scherzh.)* spit when one speaks; **~e Augen bekommen** be close to tears

feucht-, Feucht-: ~fröhlich *Adj. (ugs. scherzh.)* merry ⟨*company*⟩; boozy *(coll.)* ⟨*evening*⟩; **~gebiet das** wet area; **~heiß** *Adj.* hot and humid

Feuchtigkeit die a) *(leichte Nässe)* moisture; **b)** *(das Feuchtsein)* dampness; *(des Bodens)* wetness; (Luft~) humidity

Feuchtigkeits·creme die *(Kosmetik)* moisturizing cream; moisturizer

feucht-: ~kalt *Adj.* cold and damp; **~warm** *Adj.* muggy; humid

feudal [fɔy'da:l] **1.** *Adj.* **a)** feudal ⟨*system*⟩; **b)** *(aristokratisch)* aristocratic ⟨*regiment etc.*⟩; **c)** *(ugs.: vornehm)* plush ⟨*hotel etc.*⟩. **2.** *adv. (ugs.: vornehm)* **~ essen** have a slap-up meal *(coll.)*

Feudalismus der; ~: feudalism *no art.*

Feuer ['fɔyɐ] **das; ~s, ~ a)** fire; **[ein Gegensatz] wie ~ und Wasser sein** be as different as chalk and cheese; **das Essen aufs ~ stellen/vom ~ nehmen** put the food on to cook/take the food off the heat; **jmdn. um ~ bitten** ask sb. for a light; **jmdm. ~ geben** give sb. a light; **mit dem ~ spielen** play with fire; **er ist absolut ehrlich, für ihn** *od.* **dafür lege ich die Hand ins ~:** he is totally honest, I'd swear to it; **[für etw.] ~ und Flamme sein** be full of enthusiasm [for sth.]; **~ fangen** catch fire; *(fig.: sich verlieben)* be smitten; *(fig.: sich schnell begeistern)* be fired with enthusiasm; **für jmdn. durchs ~ gehen** go through hell and high water for sb.; **b)** *(Brand)* fire; blaze; **~!** fire!; **c)** *o. Pl. (Milit.)* fire; **unter feindliches ~ geraten** come under enemy fire; **das ~ einstellen** cease fire; **jmdn./etw. unter ~ nehmen** fire on sb./sth.; **d)** *o. Pl. (Leuchten, Funkeln)* sparkle; blaze; **ihre Augen sprühten ~:** her eyes blazed [with fire]; **e)** *o. Pl. (innerer Schwung)* fire; passion

feuer-, Feuer-: ~alarm der fire alarm; **~bekämpfung die** fire-fighting; **~beständig** *Adj.* fire-resistant; **~bestattung die** cremation; **~eifer der** enthusiasm; zest; **~fest** *Adj.* heat-resistant ⟨*dish, plate*⟩; fire-proof ⟨*material*⟩; **~gefahr die** fire hazard *or* risk; **bei ~gefahr** when there is a risk of fire; **~gefährlich** *Adj.* [in]flammable; **~haken der** poker

Feuer·land (das); ~s Tierra del Fuego

Feuer-: ~leiter die *(bei Häusern)* fire escape; *(beim ~wehrauto)* [fireman's] ladder; *(fahrbar)* turntable ladder; **~löscher der; ~s, ~:** fire extinguisher; **~melder der** fire alarm

feuern 1. *tr. V.* **a)** *(ugs.: entlassen)* fire *(coll.)*; sack *(coll.)*; **b)** *(ugs.: schleudern, werfen)* fling; **jmdm. eine ~** *(salopp)* belt sb. one; **c)** *(heizen)* fire ⟨*stove*⟩; **mit Holz ~:** have wood fires. **2.** *itr. V. (Milit.)* fire **(auf** + *Akk.* at)

feuer-, Feuer-: ~polizei die *authorities responsible for fire precautions and fire-fighting;* **~probe die** *(fig.)* test; **die ~probe bestehen** pass the [acid] test; **~rot** *Adj.* fiery red; flaming red

Feuers·brunst die *(geh.)* great fire; conflagration

feuer-, Feuer-: ~schein der glow of the/a fire; **~schiff das** lightship; **~schlucker der** fire-eater; **~schutz der a)** *(Brandschutz)* fire prevention *or* protection; **b)** *(Milit.)* covering fire; **jmdm. ~schutz geben** cover sb.; **~speiend** *Adj.; nicht präd.* fire-breathing ⟨*dragon*⟩; ⟨*volcano*⟩ spewing fire; **~spritze die** fire hose; **~stein der** flint; **~stelle die** [camp]fire; **~stuhl der** *(ugs. scherzh.)* [motor]bike *(coll.)*; machine; **~taufe die** baptism of fire; **~tod der** *(geh.)* [death at] the stake; **~treppe die** fire escape

Feuerung die; ~, ~en a) *(Vorrichtung)* firing [system]; **b)** *o. Pl. (das Heizen)* heating

Feuer-: ~versicherung die fire insurance; **~wache die** fire station; **~waffe die** firearm

Feuer·wehr die; ~, ~en fire service

Feuerwehr-: ~auto das fire engine; **~mann der;** *Pl.* **~männer** *od.* **~leute** fireman

Feuer-: ~werk das firework display; *(~werkskörper)* fireworks *pl.*; *(fig.)* barrage; **~werks·körper der** firework; **~zangen·bowle die** *burnt rum and red wine punch;* **~zeug das** lighter

Feuilleton [fœjə'tõ:] **das; ~s, ~s a)** arts section; **b)** *(literarischer Beitrag)* [literary] article

feurig *Adj.* fiery ⟨*horse, spice, wine*⟩; passionate ⟨*speech*⟩

Fez [fe:ts] **der; ~es** *(ugs.)* lark *(coll.)*; **~ machen** lark about *(coll.)*; **hört mit dem ~ auf!** stop larking about *(coll.)*

ff [ɛf'ɛf] *Abk.* **sehr fein** superior-quality ‹*sweets, pastries, etc.*›
ff. *Abk.* **folgende [Seiten]** ff.
Fiaker ['fi̯akɐ] **der; ~s, ~** *(österr.)* hackney carriage; cab
Fiasko ['fi̯asko] **das; ~s, ~s** fiasco; **unser Urlaub war ein einziges ~:** our holiday was a total disaster *(coll.)*
Fibel ['fi:bl̩] **die; ~, ~n a)** *(Lesebuch)* reader; primer; **b)** *(Lehrbuch)* handbook; guide
Fiber ['fi:bɐ] **die; ~, ~n** fibre
ficht [fɪçt] *Imperativ Sg. u. 3. Pers. Sg. Präsens v.* **fechten**
Fichte die; ~, ~n a) spruce; **b)** *(Rottanne)* Norway spruce
Fichten-: ~holz das spruce [wood]; **~nadel die** spruce needle; **~wald der** spruce forest
Fick [fɪk] **der; ~s, ~s** *(vulg.)* fuck *(coarse)*
ficken *tr., itr. V. (vulg.)* fuck *(coarse);* **mit jmdm. ~:** fuck sb.
fidel [fi'de:l] *Adj. (ugs.)* jolly, merry ‹*company, person*›
Fidschi·inseln ['fɪdʒi-] *Pl.* **die ~:** Fiji; the Fiji Islands
Fieber ['fi:bɐ] **das; ~s a)** [high] temperature; *(über 38 °C)* fever; **~ haben** have a [high] temperature/a fever; **~ messen/bei jmdm. ~ messen** take one's/sb.'s temperature; **im ~:** in one's fever; **b)** *(geh.: Besessenheit)* fever
fieber-, Fieber-: ~anfall der attack *or* bout of fever; **~frei** *Adj.* ‹*person*› free from fever; **er ist wieder ~frei** his temperature is back to normal; **~haft 1.** *Adj.* **a)** feverish, febrile ‹*infection, state, condition*›; **b)** *(fig.)* feverish ‹*activity*›. **2.** *adv. (fig.)* feverishly
fieberig *Adj. s.* **fiebrig**
Fieber·kurve die temperature chart
fiebern *itr. V.* **a)** have *or* run a temperature; **b)** *(fig.)* **vor Aufregung/Erwartung** *(Dat.)* **~:** be in a fever of excitement/anticipation; **nach etw. ~:** long desperately for sth.
fieber-, Fieber-: ~senkend *Adj.* antipyretic; **~senkende Mittel** antipyretics; **~thermometer das** [clinical] thermometer
fiebrig *Adj. (auch fig.)* feverish
Fiedel ['fi:dl̩] **die; ~, ~n** *(veralt., scherzh.)* fiddle
fiedeln *tr., itr. V. (scherzh., abwertend)* fiddle
fiel [fi:l] *1. u. 3. Pers. Sg. Prät. v.* **fallen**
fiepen ['fi:pn̩] *itr. V.* ‹*dog*› whimper; ‹*bird*› cheep
fies [fi:s] **1.** *Adj. (ugs.)* **a)** *(charakterlich)* nasty ‹*person, character*›; **das finde ich ~:** I think that's mean; **b)** *(geschmacklich)* horrid *(coll.);* awful *(coll.).* **2.** *adv.* in a nasty way
Fifa, FIFA ['fi:fa] **die; ~:** FIFA; International Football Federation
Figur [fi'gu:ɐ̯] **die; ~, ~en a)** *(Wuchs, Gestalt) (einer Frau)* figure; *(eines Mannes)* physique; **eine gute/schlechte ~ machen** cut a good/poor *or* sorry figure; **b)** *(Bildwerk)* figure; **c)** *(geometrisches Gebilde)* shape; **d)** *(Spielstein)* piece; **e)** *(Persönlichkeit)* figure; **f)** *(literarische Gestalt)* character; **die komische ~** *(Theater)* the comic character *or* figure; **g)** *(Tanzen, Eissport usw.)* figure; **~en laufen** skate figures
figurativ [figura'ti:f] *(Sprachw., Kunstw.)* **1.** *Adj.* figurative. **2.** *adv.* figuratively
figürlich [fi'gy:ɐ̯lɪç] **1.** *Adj. (Kunstwiss.)* figured. **2.** *adv. (in bezug auf die Figur)* as far as her figure/his physique is concerned
Fiktion [fɪk'tsi̯o:n] **die; ~, ~en** fiction
fiktiv [fɪk'ti:f] *Adj. (geh.)* fictitious
[1]**Filet** [fi'le:] **das; ~s, ~s** *(Textilw.)* filet; netting
[2]**Filet das; ~s, ~s** fillet; *(Rinder~, Schweine~)* fillet; filet
Filet·steak das fillet steak
Filiale [fi'li̯a:lə] **die; ~, ~n** branch
Filial·leiter der branch manager
Filius ['fi:li̯ʊs] **der; ~, ~se** *(scherzh.)* son
Film [fɪlm] **der; ~[e]s, ~e a)** *(Fot.)* film; **b)** *(Kino~)* film; movie *(Amer. coll.);* **da ist bei ihm der ~ gerissen** *(fig. ugs.)* he's had a mental blackout; **c)** *o. Pl. (~branche)* films *pl.;* **beim ~ sein** be in films; **d)** *(dünne Schicht)* film
Film·atelier das film studio
Filme·macher der; ~s, ~, Filme·macherin die; ~, ~nen film-maker
filmen 1. *tr. V.* **a)** film; **b)** *(ugs.: hereinlegen)* **jmdn. ~:** take sb. for a ride *(sl.).* **2.** *itr. V.* film; make a film/films
Film·festspiele *Pl.* film festival *sing.*
filmisch 1. *Adj.* cinematic ‹*art etc.*›. **2.** *adv.* cinematically
Film-: ~kamera die film camera; *(Schmalfilmkamera)* cine-camera; **~kritik die a)** *(Besprechung)* film review; **b)** *(~kritiker)* film critics *pl.;* **~musik die** film music; *(eines einzelnen Films)* theme music; **~preis der** film award; **~produzent der** film producer; **~regisseur der** film director; **~rolle die a)** *(schauspielerische Rolle)* film part *or* role; **b)** *(Spule)* reel of film; **~schauspieler der** film actor; **~schauspielerin die** film actress; **~star der** film star; **~verleih der** film distributor[s]; **~vorstellung die** film show
Filou [fi'lu:] **der** *(abwertend)* **a)** *(Spitzbube)* dog *(derog.);* rogue; **b)** *(Verführer)* devil *(derog.)*
Filter ['fɪltɐ] **der,** *(fachspr. meist)* **das; ~s, ~:** filter; **Zigarette ohne/mit ~:** plain/[filter-]tipped cigarette
filter·fein *Adj.* finely-ground *attrib.*, filter-fine *attrib.* ‹*coffee*›
Filter·kaffee der filter coffee
filtern *tr. V.* filter
Filter-: ~papier das filter paper; **~tüte die** filter; **~zigarette die** [filter-]tipped cigarette
Filtration [fɪltra'tsi̯o:n] **die; ~, ~en** *(Technik)* filtration
filtrieren *tr. V.* filter
Filz [fɪlts] **der; ~es, ~e a)** felt; **b)** *(filzartig Verschlungenes)* mass; mat; **c)** *(Bierdeckel)* beer-mat
filzen 1. *itr. V.* felt. **2.** *tr. V. (ugs.: durchsuchen)* search ‹*room, car, etc.*›*;* frisk ‹*person*›
Filz·hut der felt hat
filzig *Adj.* felted ‹*wool*›*;* matted ‹*hair*›
Filzokratie [fɪltsokra'ti:] **die; ~, ~n** *(abwertend)* corruption; graft *(coll.)*
Filz-: ~pantoffel der slipper; **~schreiber der, ~stift der** felt-tip pen
Fimmel ['fɪml̩] **der; ~s, ~** *(ugs. abwertend)* **einen ~ für etw. haben** have a thing about sth. *(coll.);* **du hast wohl einen ~!** there must be something the matter with you; you must be dotty *(Brit.)*
Finale [fi'na:lə] **das; ~s, ~[s] a)** *(Sport)* final; **b)** *(Musik, fig.)* finale
Final·satz der *(Sprachw.)* final clause
Finanz [fi'nants] **die; ~ a)** *(Geldwesen)* finance *no art.;* **b)** *(~leute)* financial world
Finanz-: ~amt das a) *(Behörde)* ≈ Inland Revenue; **das ~amt** *(ugs.: die Steuerbehörden)* the taxman; **b)** *(Gebäude)* tax office; **~beamte der** tax officer
Finanzen *Pl.* **a)** finance *sing.;* **b)** *(ugs.: finanzielle Verhältnisse)* finances; **c)** *(Einkünfte des Staates)* [government] finances
Finanz·hoheit die fiscal prerogative
finanziell [finan'tsi̯ɛl] **1.** *Adj.* financial. **2.** *adv.* financially

finanzieren *tr. V.* finance; *(fig.: bezahlen)* pay for; **frei/staatlich finanziert sein** be privately financed/financed by the state
Finanzierung die; ~, ~en financing
finanz-, Finanz-: ~kraft die; *o. Pl.* financial strength; **~kräftig** *Adj.* financially powerful; **~minister der** minister of finance; ≈ Chancellor of the Exchequer *(Brit.)*; ≈ Secretary of the Treasury *(Amer.)*; **~ministerium das** Ministry of Finance; *(in GB u. USA)* ≈ Treasury; **~politik die** politics of finance; **eine neue ~politik** a new financial policy; **~politisch 1.** *Adj.* ‹*questions etc.*› relating to financial policy; **2.** *adv.* from the point of view of financial policy; **~stark** *Adj.* financially strong; **~wesen das** *o. Pl.* system of public finances
Findel·kind ['fɪndl̩-] **das** foundling
finden ['fɪndn̩] **1.** *unr. tr. V.* **a)** *(entdecken)* find; **eine Spur von jmdm. ~:** get a lead on sb.; **keine Spur von jmdm. ~:** find no trace of sb.; **b)** *(erlangen, erwerben)* find ‹*work, flat, wife, etc.*›; **Freunde ~:** make friends; **c)** *(heraus~)* find ‹*solution, mistake, pretext, excuse, answer*›; **d)** *(einschätzen, beurteilen)* **etw. gut ~:** think sth. is good; **nichts bei etw. ~:** not mind sth.; **ich finde nichts dabei** I don't mind; **e)** *(erhalten)* **Hilfe [bei jmdm.] ~:** get help [from sb.]. **2.** *unr. refl. V.* **sich ~:** turn up; **es fand sich niemand/jemand, der das tun wollte** nobody wanted to do that/ there was somebody who wanted to do that; **das/es wird sich alles ~** it will all work out all right. **3.** *unr. itr. V.* **zu jmdm. ~:** find sb.; **nach Hause ~:** find the way home; **zu sich selbst ~** *(fig.)* come to terms with oneself
Finder der; ~s, ~, Finderin die; ~, ~nen finder
Finder·lohn der reward [for finding sth.]
findig *Adj.* resourceful
Finesse [fi'nɛsə] **die; ~, ~n a)** *meist Pl. (Kunstgriff)* trick; **b)** *meist Pl. (in der Ausstattung)* refinement; **mit allen ~n** with every refinement; **c)** *(Schlauheit)* flair
fing [fɪŋ] *1. u. 3. Pers. Sg. Prät. v.* **fangen**
Finger ['fɪŋɐ] **der; ~s, ~ a)** finger; **mit dem ~ auf jmdn./etw. zeigen** *(auch fig.)* point one's finger at sb./sth.; **b)** *(fig.)* **wenn man ihm den kleinen ~ reicht, nimmt er gleich die ganze Hand** if you give him an inch he takes a mile; **die ~ davonlassen/von etw. lassen** *(ugs.)* steer clear of it/of sth.; **sie macht keinen ~ krumm** *(ugs.)* she never lifts a finger; **er rührte keinen ~:** he wouldn't lift a finger; **lange ~ machen** *(ugs.)* get itchy fingers; **ich würde mir alle [zehn] ~ danach lecken** *(ugs.)* I'd give my eyeteeth for it; **die ~ in etw.** *(Dat.)***/im Spiel haben** *(ugs.)* have a hand in sth./have one's finger in the pie; **sich** *(Dat.)* **die ~ verbrennen** *(ugs.)* burn one's fingers *(fig.)*; **sich** *(Dat.)* **die ~ schmutzig machen** get one's hands dirty; **sich** *(Dat.)* **etw. an den [fünf** *od.* **zehn] ~n abzählen können** be able to see sth. straight away; **jmdm. auf die ~ klopfen** *(ugs.)* rap sb. across the knuckles; **sich** *(Dat.)* **etw. aus den ~n saugen** *(ugs.)* make sth. up; **ihm** *od.* **ihn juckt es in den ~n [, etw. zu tun]** *(ugs.)* he is itching [to do sth.]; **wenn ich den in die ~ kriege!** *(ugs.)* wait till I get my hands on him *(coll.)*; **jmdn. um den [kleinen] ~ wickeln** *(ugs.)* wrap sb. round one's little finger
finger-, Finger-: ~abdruck der fingerprint; **~breit der; ~, ~** *(fig.)* inch; **~fertigkeit die;** *o. Pl.* dexterity; **~hakeln** [~ha:kl̩n] **das; ~s** finger-wrestling; **~handschuh der** glove [with fingers]; **~hut der a)** thimble; **b)** *(Bot.)* foxglove; **~knöchel der** knuckle; **~kuppe die** fingertip
fingern *itr. V.* fiddle; **an etw.** *(Dat.)* **~:** fiddle with sth.; **nach etw. ~:** fumble [around] for sth.
Finger-: ~nagel der fingernail; **~spitze die** fingertip; **das muß man in den ~n haben** *(fig.)* you have to have a feel for it; **~spitzen·gefühl das;** *o. Pl.* feeling; **~zeig** [~tsaik] **der; ~s, ~e** hint; *(an die Polizei)* tip-off
fingieren [fɪn'gi:rən] *tr. V.* fake
Fink [fɪŋk] **der; ~en, ~en** finch
Finne ['fɪnə] **der; ~n, ~n, Finnin die; ~, ~nen** Finn
finnisch *Adj.* Finnish
Finnland ['fɪnlant] **(das); ~s** Finland
finster ['fɪnstɐ] **1.** *Adj.* **a)** dark; **im Finstern** in the dark; **b)** *(düster)* dark ‹*house, forest, alleyway*›; dimly-lit ‹*pub, district*›; **c)** *(dubios)* shady ‹*plan, affair*›; sinister ‹*figure*›; **d)** *(verdüstert, feindselig)* **eine ~e Miene** a black expression; **e)** *(fig.)* **im ~n tappen** be groping in the dark. **2.** *adv.* **jmdn. ~ ansehen** give sb. a black look
Finsternis die; ~, ~se a) darkness; *(bibl., fig.)* dark; **b)** *(Astron.)* eclipse
Finte ['fɪntə] **die; ~, ~n a)** *(List)* trick; **b)** *(Fechten)* feint
Firlefanz ['fɪrləfants] **der; ~es** *(ugs. abwertend)* **a)** *(Tand, Flitter)* frippery; trumpery; **b)** *(Unsinn)* nonsense; **~ machen** fool around
firm [fɪrm] *Adj.* **in etw.** *(Dat.)* **~ sein** be well up in sth.; know sth. thoroughly
Firma ['fɪrma] **die; ~, Firmen** firm; company
Firmament [fɪrma'mɛnt] **das; ~[e]s** *(dichter.)* firmament
firmen *tr. V. (kath. Rel.)* confirm
Firmen-: ~inhaber der owner of the/a company; **~name der** name of a/the company *or* firm; **~schild das** company's name plate; **~wagen der** company car; **~zeichen das** trademark
firmieren *itr. V.* trade
Firmung die; ~, ~en confirmation; **jmdm. die ~ erteilen** confirm sb.
Firn der; ~[e]s firn
Firnis ['fɪrnɪs] **der; ~ses, ~se** varnish
First [fɪrst] **der; ~[e]s, ~e** ridge
Fis [fɪs] **das; ~, ~** *(Musik)* F sharp
Fisch [fɪʃ] **der; ~[e]s, ~e a)** fish; **[fünf] ~e fangen** catch [five] fish **gesund und munter wie ein ~ im Wasser** as fit as a fiddle; **stumm wie ein ~ sein** keep a stony silence; *(fig.)* **kleine ~e** *(ugs.)* small fry; **b)** *o. Pl. (Nahrungsmittel)* fish; **das ist weder ~ noch Fleisch** *(fig.)* that's neither fish nor fowl; **c)** *(Astrol.)* **die ~e** Pisces; the Fishes; **er ist [ein] ~:** he is a Piscean; **im Zeichen der ~e geboren sein** be born under [the sign of] Pisces
Fisch-: ~becken das fish-pond; **~dampfer der** steam trawler
fischen 1. *tr. V.* **a)** fish for; **b)** *(ugs.)* **etw. aus etw. ~:** fish sth. out of sth. **2.** *itr. V.* fish **(nach** for); **~ gehen** go fishing; *s. auch* **trüb 1 a**
Fischer der; ~s, ~ fisherman
Fischer-: ~boot das fishing boat; **~dorf das** fishing village
Fischerei die; ~: fishing
Fisch·fang der; *o. Pl.* **vom ~ leben** make a/one's living by fishing; **auf ~ gehen** go fishing
Fisch-: ~filet das fish fillet; **~geruch der** smell of fish; **~geschäft das** fishmonger's [shop] *(Brit.)*; fish store *(Amer.)*; **~gräte[en]·muster das** *(Textilw.)* herringbone pattern; **~gründe** *Pl.* fishing grounds; **~händler der** fishmonger

(Brit.); fish dealer *(Amer.);* **~konserve die** canned fish; **~kutter der** fishing trawler; **~laden der** *s.* **~geschäft; ~markt der** fish market; **~mehl das** fish-meal; **~otter der** otter; **~schuppe die** fish scale; **~stäbchen das** *(Kochk.)* fish finger; **~sterben das** death of the fish; **~zucht die** fish farming; **~zug der** *(ugs.)* killing

Fịs-Dur [auch: '-'-] **das; ~** *(Musik)* F sharp major

Fisimatenten [fizima'tɛntn̩] *Pl.* *(ugs.)* messing about *sing.;* **mach keine ~!** stop messing about

fiskalisch [fɪs'ka:lɪʃ] *Adj.* fiscal

Fiskus ['fɪskʊs] **der; ~, Fịsken** *od.* **~se** Government *(as managing the State finances)*

fịs-Moll [auch: '-'-] **das; ~** *(Musik)* F sharp minor

Fistel ['fɪstl̩] **die; ~, ~n** *(Med.)* fistula

Fịstel·stimme die thin high-pitched voice

fit [fɪt] *Adj.; nicht attr.* fit; **sich ~ halten** keep fit; **das hält ~:** it keeps you fit

Fitness, Fitneß ['fɪtnɛs] **die; ~:** fitness

Fittich ['fɪtɪç] **der; ~[e]s, ~e** *(dichter.)* wing; pinion; **jmdn. unter seine ~e nehmen** *(ugs. scherzh.)* take sb. under one's wing

Fitzelchen ['fɪtsl̩çən] **das; ~s, ~** *(ugs.)* scrap

fix [fɪks] **1.** *Adj.* **a)** *(ugs.: flink, wendig)* quick; **ein ~er Bursche** a bright lad; **b)** *(ugs.)* **~ und fertig** *(fertig vorbereitet)* quite finished; *(völlig erschöpft)* completely shattered *(coll.);* **c)** *(festgelegt)* fixed ‹*cost, salary*›; **eine ~e Idee** an idée fixe. **2.** *adv. (ugs.)* quickly; **das geht ganz ~:** it won't take a jiffy *(coll.);* **[mach] ~!** hurry up!

fixen ['fɪksn̩] *itr. V. (Drogenjargon)* fix *(sl.)*

Fịxer der; ~s, ~, Fịxerin, die ~, ~nen *(Drogenjargon)* fixer

Fixier·bad das *(Fot.)* fixer

fixieren [fɪ'ksi:rən] *tr. V.* **a)** *(scharf ansehen)* fix one's gaze on; **b)** *(geh.: schriftlich niederlegen)* take down; **c)** *(Fot.)* fix

Fixierung die; ~, ~en a) *(starres Festlegen, -halten)* **die ~ auf seine Mutter** his mother-fixation; **b)** *(Festlegung)* determination

Fịx·stern der *(Astron.)* fixed star

Fixum ['fɪksʊm] **das; ~s, Fịxa** basic salary

Fjord [fjɔrt] **der; ~[e]s, ~e** fiord

FKK [ɛf ka: 'ka:] *Abk.* **Freikörperkultur** nudism *no art.;* naturism *no art.*

FKK-: ~-Anhänger der nudist; naturist; **~-Strand der** nudist beach

flach [flax] *Adj.* **a)** flat ‹*countryside, region, roof*›; **~ liegen** lie flat; **die ~e Hand** the flat of one's hand; **b)** *(niedrig)* low ‹*heels, building*›; flat ‹*shoe*›; **c)** *(nicht tief)* shallow ‹*water, river, dish*›; **d)** *(fig. abwertend)* shallow

flạch·brüstig *Adj.* flat-chested

Flạch·dach das flat roof

Fläche ['flɛçə] **die; ~, ~n a)** *(ebenes Gebiet)* area; **b)** *(Ober~, Außenseite)* surface; **c)** *(Math.)* area; *(einer dreidimensionalen Figur)* side; face; **d)** *(weite Land~, Wasser~)* expanse

Flạ̈chen-: ~brand der extensive blaze; **~inhalt der** *(Math.)* area; **~maß das** *(Math.)* unit of square measure

flạch-, Flạch-: ~|fallen *itr. V.; mit sein (ugs.)* ‹*trip*› fall through; ‹*event*› be cancelled; **~hang der** slip-off slope; **~land das;** *o. Pl.* lowland; **~|legen 1.** *refl. V. (ugs.)* lie down; **2.** *tr. V. (zu Boden strecken)* floor ‹*opponent*›; **~|liegen** *unr. itr. V. (ugs.)* be flat on one's back; **~mann der;** *Pl.* **~männer** *(ugs. scherzh.)* hip-flask

Flachs [flaks] **der; ~es a)** flax; **b)** *(ugs.: Ulk)* **das war doch nur ~:** I/he *etc.* was just having you on *(Brit. coll.) or (Amer. coll.)* putting you on; **ganz ohne ~:** no kidding *(coll.)*

flạchs·blond *Adj.* flaxen ‹*hair*›

flachsen ['flaksn̩] *itr. V. (ugs.)* **mit jmdm. ~:** joke with sb.; **gerne ~:** like a joke

Flạch·zange die flat tongs *pl.*

flackern ['flakɐn] *itr. V.* flicker

Fladen ['fla:dn̩] **der; ~s, ~ a)** *flat, round unleavened cake made with oat or barley flour;* ≈ [large] oatcake *(Scot.);* **b)** *(Kuh~)* cowpat

Fladen·brot das unleavened bread

Flagge ['flagə] **die; ~, ~n** flag; **die ~ streichen** *(fig.)* strike the flag *(fig.);* **~ zeigen** *(fig.)* show one's colours

flạggen *itr. V.* put out the flags

Flạggen-: ~alphabet das international code of signals; **~gruß der** flag salute; **~mast der** flagstaff

Flạgg-: ~leine die halyard; **~offizier der** flag officer; **~schiff das** flagship

Flair [flɛ:ɐ̯] **das** *od.* **der; ~s a)** *(Fluidum, Aura)* air; **b)** *(Talent)* flair

Flak [flak] **die; ~, ~** *(Milit.)* anti-aircraft gun; AA gun

Flakon [fla'kõ:] **das** *od.* **der; ~s, ~s** bottle

flambieren [flam'bi:rən] *tr. V.* flambé

Flame ['fla:mə] **der; ~n, ~n** Fleming

Flamenco [fla'mɛŋko] **der; ~[s], ~s** flamenco

Flämin ['flɛ:mɪn] **die; ~, ~nen** Fleming

Flamingo [fla'mɪŋgo] **der; ~s, ~s** flamingo

flämisch *Adj.* Flemish

Flamme ['flamə] **die; ~, ~n a)** flame; **in ~n stehen/aufgehen** be/go up in flames; **b)** *(Brennstelle)* burner; **c)** *(ugs. veralt.: Freundin)* flame

flammen ['flamən] *itr. V. (geh.)* blaze

flạmmend *Adj.* **a)** flaming; **~es Haar** flaming red hair; **b)** *(fig.)* fiery ‹*speech*›

Flạmmen-: ~meer das *(geh.)* sea of flame[s]; **~tod der** *(geh.)* death by burning; **~werfer der** *(Milit.)* flame-thrower

Flandern ['flandɐn] **(das); ~s** Flanders

Flanell [fla'nɛl] **der; ~s, ~e** flannel

Flanẹll·anzug der flannel suit

Flaneur [fla'nø:ɐ̯] **der; ~s, ~e** *(geh.)* flâneur

flanieren [fla'ni:rən] *itr. V.; mit Richtungsangabe mit sein* stroll

Flanke ['flaŋkə] **die; ~, ~n a)** *(auch Milit.)* flank; **b)** *(Ballspiele) (Flankenball)* centre; *(Teil des Spielfeldes)* wing; **c)** *(Turnen)* flank vault

flạnken *itr. V.* **a)** *(Ballspiele)* **[in die Mitte] ~:** centre the ball; **b)** *(Turnen)* flank vault

Flạnken-: ~ball der *(Ballspiele)* centre; **~deckung die** *(Milit.)* flank defence

flankieren *tr. V.* flank; **~de Maßnahmen** *(fig.)* additional measures

Flansch [flanʃ] **der; ~[e]s, ~e** *(Technik)* flange

flapsig ['flapsɪç] *(ugs.)* **1.** *Adj.* rude. **2.** *adv.* rudely

Flasche ['flaʃə] **die; ~, ~n a)** bottle; **ein Tier mit der ~ großziehen** rear an animal by bottle; **ich muß dem Kind noch die ~ geben** I must just feed the baby; **zur ~ greifen** *(fig.)* take to the bottle; **b)** *(ugs. abwertend) (Feigling)* wet *(sl.);* *(unfähiger Mensch)* **eine ~ sein** be useless; **du ~!** you useless item! *(coll.)*

Flạschen-: ~bier das bottled beer; **~hals der** *(auch fig.)* bottleneck; **~kind das** bottle-fed baby; ~**milch die** bottled milk; ~**öffner der** bottle-opener; ~**pfand das** deposit [on a/the bottle]; ~**post die** message in a/the bottle; ~**wein der** wine by the bottle; ~**zug der** block and tackle

Flaschner ['flaʃnɐ] **der;** ~**s,** ~ *(südd., schweiz.)* plumber

flatterhaft *Adj.* fickle

Flatterhaftigkeit die; ~: fickleness

Flatter·mann der; *Pl.* ~**männer** *(salopp)* **a)** *o. Pl. (nervöse Unruhe)* jitters *pl. (coll.);* **b)** *(scherzh.: Brathuhn)* roast chicken

flattern *itr. V.* **a)** *mit Richtungsangabe mit sein* flutter; **b)** *(zittern)* ‹*hands*› shake; ‹*eyelids*› flutter; **seine Nerven flatterten** *(fig.)* he got in a flap *(coll.)*

flau [flau] **1.** *Adj.* **a)** *(schwach, matt)* slack ‹*breeze*›; flat ‹*atmosphere*›; **b)** *(leicht übel)* queasy ‹*feeling*›; **mir ist** ~: I feel queasy. **2.** *adv. (Kaufmannsspr.)* **das Geschäft geht** ~: business is slack

Flaum [flaum] **der;** ~**[e]s a)** fuzz; **b)** *(~federn)* down

Flaum·bart der downy beard

Flaum·feder die down feather

flaumig *Adj.* downy

Flausch [flauʃ] **der;** ~**[e]s,** ~**e** brushed wool

flauschig *Adj.* fluffy

Flause ['flauzə] **die;** ~, ~**n;** *meist Pl. (ugs.)* **er hat nur** ~**n im Kopf** he can never think of anything sensible; **jmdm. die** ~**n austreiben** knock some sense into sb.

Flaute ['flautə] **die;** ~, ~**n a)** *(Seemannsspr.)* calm; **b)** *(Kaufmannsspr.)* fall[-off] in trade; **in der** ~: in the doldrums

Flecht·arbeit die piece of wickerwork; ~**en** wickerwork *sing.*

Flechte ['flɛçtə] **die;** ~, ~**n a)** *(Bot.)* lichen; **b)** *(Med.)* eczema

flechten *unr. tr. V.* plait ‹*hair*›; weave ‹*basket, mat*›

Flechter der; ~**s,** ~, **Flechterin die;** ~, ~**nen** basket-weaver

Flecht·werk das a) *(Geflecht)* wickerwork; **b)** *(Archit.)* wattle and daub

Fleck [flɛk] **der;** ~**[e]s,** ~**e a)** *(verschmutzte Stelle)* stain; ~**e machen** leave stains; **einen** ~ **auf der [weißen] Weste haben** *(fig. ugs.)* have blotted one's copybook; **b)** *(andersfarbige Stelle)* patch; *s. auch* **blau; c)** *(Stelle, Punkt)* spot; **sich nicht vom** ~ **rühren** not to move an inch; **auf demselben** ~: in the same place; **wir kriegten den Stein nicht vom** ~: we couldn't budge the stone; **ich bin nicht vom** ~ **gekommen** *(fig.)* I didn't get anywhere; **vom** ~ **weg** *(fig.)* on the spot; *s. auch* **Herz a**

Fleckchen das; ~**s,** ~: spot; **ein schönes** ~ **Erde** a lovely little spot

flecken *itr. V.* stain

Flecken der; ~**s,** ~ **a)** *s.* **Fleck a, b; b)** *(Ortschaft)* little place

Fleck·entferner der, Flekken·wasser das stain *or* spot remover

fleckig ['flɛkɪç] *Adj.* **a)** *(verschmutzt)* stained; **b)** *(gepunktet)* speckled ‹*apple*›; blotchy ‹*face, skin*›

fleddern ['flɛdɐn] *tr. V.* plunder, rob ‹*person*›

Fleder·maus ['fle:dɐ-] **die** bat

Fleder·wisch der feather duster

Flegel ['fle:gl̩] **der;** ~**s,** ~ *(abwertend)* lout

Flegel·alter das *s.* **Flegeljahre**

flegelhaft *Adj. (abwertend)* loutish; boorish ‹*tone of voice*›

Flegel·jahre *Pl.* uncouth adolescence *sing.;* **in die** ~ **kommen/aus den** ~**n heraussein** reach/be past the awkward age *sing.*

flegeln *refl. V. (abwertend)* **sich auf ein Sofa** ~: flop on to a sofa

flehen ['fle:ən] *itr. V.* plead; **[bei jmdm.] um etw.** ~: plead [with sb.] for sth.; **zu Gott** ~: beg God

flehentlich *Adv. (geh.)* pleadingly

Fleisch [flaiʃ] **das;** ~**[e]s a)** *(Muskelgewebe)* flesh; **das nackte/rohe** ~: one's bare/raw flesh; *(fig.)* **sein eigen[es]** ~ **und Blut** *(geh.)* his own flesh and blood; **jmdm. in** ~ **und Blut übergehen** become second nature to sb.; **sich** *(Dat.)* **ins eigene** ~ **schneiden** cut off one's nose to spite one's face; **vom** ~ **fallen** *(ugs.)* waste away; **b)** *(Nahrungsmittel)* meat; **c)** *(Frucht~)* flesh

fleisch-, Fleisch-: ~**arm** *Adj.* ‹*diet*› low in meat; ~**beschau die** meat inspection; ~**brocken der** chunk of meat; ~**brühe die** bouillon; consommé

Fleischer ['flaiʃɐ] **der;** ~**s,** ~: butcher; *s. auch* **Bäcker**

Fleischerei die; ~, ~**en** butcher's shop; **in der** ~: at the butcher's; *s. auch* **Bäckerei**

Fleischer-: ~**haken der** meat hook; ~**meister der** master butcher; ~**messer das** butcher's knife

fleisch-, Fleisch-: ~**farben,** ~**farbig** *Adj.* flesh-coloured; ~**fondue das** *(Kochk.)* meat fondue; ~**fressend** *Adj.; nicht präd. (Biol.)* carnivorous; ~**fresser der;** ~**s,** ~ *(Biol.)* carnivore; ~**gang der** *(Gastr.)* meat course; ~**gericht das** meat dish

fleischig *Adj.* plump ‹*hands, face*›; fleshy ‹*leaf, fruit*›

fleisch-, Fleisch-: ~**käse der** meat loaf; ~**kloß der a)** *(Kochk.)* meat ball; **b)** *s.* ~**klumpen;** ~**klößchen das** small meat ball; ~**klumpen der** *(ugs.)* chunk of meat; ~**konserve die** tin of meat *(Brit.);* can of meat *(Amer.);* ~**los** *Adj.* **a)** without meat ‹*meal*›; **b)** *(hager, mager)* bony ‹*hands, face*›; ~**pastete die** *(Kochk.)* pâté; ~**salat der** *(Kochk.)* meat salad; ~**vergiftung die** food poisoning [from meat]; ~**waren** *Pl.* meat products; ~**wolf der** mincer; ~**wunde die** flesh-wound; ~**wurst die** pork sausage

Fleiß [flais] **der;** ~**es a)** *(eifriges Streben)* hard work; *(Eigenschaft)* diligence; **viel** ~ **auf etw.** *(Akk.)* **verwenden** put a lot of effort into sth.; **durch** ~ **etw. erreichen** achieve sth. by hard work; **ohne** ~ **kein Preis** *(Spr.)* success never comes easily; **b)** *(veralt., südd.: Absicht)* **mit** ~: on purpose

fleißig ['flaisɪç] **1.** *Adj.* **a)** *(arbeitsam)* hard-working; willing ‹*hands*›; **b)** *nicht präd. (von Fleiß zeugend)* diligent ‹*piece of work*›; **c)** *(regelmäßig, häufig)* frequent ‹*visitor*›; **d)** *(unermüdlich)* indefatigable ‹*collector*›; great ‹*walker*›. **2.** *adv.* **a)** ‹*work, study*› hard; **b)** *(unermüdlich)* ‹*drink, spend*› steadily; ‹*collect*› regularly; **c)** *(regelmäßig)* frequently

flektieren [flɛk'ti:rən] *(Sprachw.)* **1.** *tr. V.* inflect. **2.** *itr. V.* be inflected

flennen ['flɛnən] *itr. V. (ugs. abwertend)* blubber

fletschen ['flɛtʃn̩] *tr. V.* **die Zähne** ~: bare one's/its teeth

Fleurop Ⓦ ['flɔyrɔp] **die** Interflora **(P)**

flexibel [flɛ'ksi:bl̩] **1.** *Adj.* flexible. **2.** *adv.* flexibly

Flexibilität [flɛksibili'tɛ:t] **die;** ~: flexibility

Flexion [flɛ'ksio:n] **die;** ~, ~**en** *(Sprachw.)* inflexion

Flexions·endung die; ~, ~**en** *(Sprachw.)* inflectional suffix *or* ending

flicht *Imperativ Sg. u. 3. Pers. Sg. Präsens v.* **flechten**

flicken ['flɪkn̩] *tr. V.* mend ‹*trousers, dress*›; repair ‹*engine, cable*›; mend, repair ‹*wall, roof*›

Flicken der; ~**s,** ~: patch

Flicken·decke die patchwork quilt

Flickflack ['flıkflak] **der; ~s, ~s** *(Turnen)* flik-flak

Flick-: ~werk das; *o. Pl. (abwertend)* botched-up job; **~zeug das** repair kit

Flieder ['fli:dɐ] **der; ~s, ~:** lilac

flieder·farben, flieder·farbig *Adj.* lilac

Fliege ['fli:gə] **die; ~, ~n a)** fly; **sie starben wie die ~n** they were dying like flies; **er tut keiner ~ etwas zuleide/könnte keiner ~ etwas zuleide tun** he wouldn't/couldn't hurt a fly; *(fig.)* **ihn stört die ~ an der Wand** the least little thing annoys him; **zwei ~n mit einer Klappe schlagen** kill two birds with one stone; **die** *od.* **'ne ~ machen** *(salopp)* beat it *(sl.)*; **b)** *(Schleife)* bow-tie; **c)** *(Bärtchen)* shadow

fliegen 1. *unr. itr. V.* **a)** *mit sein* fly; **das ~de Personal** the air-crew; **in die Luft ~** *(durch Explosion)* blow up; **b)** *mit sein (ugs.: geworfen werden)* **aus der Kurve ~:** skid off a/the bend; **vom Pferd ~:** fall off a/the horse; **c)** *mit sein (ugs.: entlassen werden)* be sacked *(coll.)*; get the sack *(coll.)*; **auf die Straße/aus einer Stellung ~:** get the sack *(coll.)*; **von der Schule ~:** be chucked out [of the school] *(coll.)*; **d)** *mit sein (ugs.: hinfallen, stürzen)* fall; **über etw.** *(Akk.)* **~:** trip over sth.; **durch das Examen ~** *(fig.)* fail the exam; **e)** *mit sein* **auf jmdn./etw. ~** *(ugs.)* go for sb./sth.. **2.** *unr. tr. V.* **a)** *(steuern, fliegend befördern)* fly ⟨*aircraft, passengers, goods*⟩; **b)** *auch mit sein (fliegend ausführen)* **einen Einsatz ~:** fly a mission; **einen Umweg ~:** make a detour

fliegend *Adj.; nicht präd.* flying; **ein ~er Händler** a pedlar

Fliegen-: ~fänger der fly-paper; **~fenster das** wire-mesh window; **~gewicht das** *(Schwerathletik)* flyweight; **im ~gewicht starten** compete at flyweight; **~klatsche die** fly swat; **~pilz der** fly agaric

Flieger der; ~s, ~ a) pilot; **er ist bei den ~n** *(Milit.)* he's in the air force; **b)** *(Radsport)* sprinter

Flieger·alarm der air-raid warning

Fliegerei die; ~: flying *no art.*

Fliegerin die; ~, ~nen [woman] pilot

fliehen ['fli:ən] **1.** *unr. itr. V.; mit sein (flüchten)* flee **(vor** + *Dat.* from); *(entkommen)* escape **(aus** from); **ins Ausland/über die Grenze ~:** flee the country/escape over the border. **2.** *unr. tr. V. (geh.: meiden)* shun

fliehend *Adj.; nicht präd.* sloping ⟨*forehead*⟩; receding ⟨*chin*⟩

Flieh·kraft die *(Physik)* centrifugal force

Fliese ['fli:zə] **die; ~, ~n** tile; **etw. mit ~n auslegen** tile sth.

Fliesenleger [-le:gɐ] **der; ~s, ~:** tiler

Fließ-: ~band das conveyor belt; **am ~band arbeiten** work on the assembly line; **~band·arbeit die** assembly-line work

fließen ['fli:sn̩] *unr. itr. V.; mit sein* flow; **es floß Blut** blood was shed; **die Gaben flossen reichlich** donations were pouring in

fließend 1. *Adj.* running ⟨*water*⟩; moving ⟨*traffic*⟩; fluid ⟨*transition*⟩; fluent ⟨*English, French, etc.*⟩; **die Grenzen sind ~:** the dividing-line ist blurred. **2.** *adv.* ⟨*speak a language*⟩ fluently

Fließ·heck das *(Kfz-W.)* fastback

Flimmer-: ~kasten der, ~kiste die *(ugs.)* telly *(coll.)*; box *(coll.)*

flimmern ['flımɐn] *itr. V.* ⟨*water, air, surface*⟩ shimmer; ⟨*film*⟩ flicker; **ihm flimmerte es vor den Augen** everything was swimming in front of his eyes

flink [flıŋk] **1.** *Adj.* nimble ⟨*fingers*⟩; sharp ⟨*eyes*⟩; quick ⟨*hands*⟩; **~ wie ein Wiesel** as quick as a flash. **2.** *adv.* quickly

Flinkheit die; ~ *s.* **flink 1:** nimbleness; sharpness; quickness

Flinte ['flıntə] **die; ~, ~n** shotgun; **der soll mir nur vor die ~ kommen!** *(fig. salopp)* if I can just get my hands on him; **die ~ ins Korn werfen** *(fig.)* throw in the towel

Flinten-: ~kugel die shotgun pellet; **~lauf der** shotgun barrel

Flip [flıp] **der; ~s, ~s** flip

Flipper ['flıpɐ] **der; ~s, ~, Flipper·automat der** pinball machine

flippern ['flıpɐn] *itr. V. (ugs.)* play pinball

Flirt [flırt] **der; ~s, ~s** flirtation; **einen ~ mit jmdm. haben** flirt with sb.

flirten *itr. V.* flirt

Flittchen ['flıtçən] **das; ~s, ~** *(ugs. abwertend)* floozie

Flitter ['flıtɐ] **der; ~s, ~ a)** *o. Pl.* frippery; trumpery; **b)** *(Metallplättchen)* sequin

Flitter·kram der *(ugs. abwertend)* frippery; trumpery

Flitter·wochen *Pl.* honeymoon *sing.*; **in die ~wochen fahren** go on one's honeymoon

Flitz[e]·bogen ['flıts(ə)-] **der** bow; *(fig.)* **gespannt sein wie ein ~:** be on tenterhooks

flitzen ['flıtsn̩] *itr. V.; mit sein (ugs.)* shoot; dart; **ich flitze mal eben zum Fleischer** I'll just dash to the butcher's

Flitzer der; ~s, ~ *(ugs.: kleines, schnelles Auto)* sporty job *(coll.)*

floaten ['floʊtn̩] *tr., itr. V. (Wirtsch.)* float

flocht [flɔxt] *1. u. 3. Pers. Sg. Prät. v.* **flechten**

Flocke ['flɔkə] **die; ~, ~n a)** flake; **eine ~ Watte/Wolle** a bit of cottonwool/tuft of wool; **b)** *(Staub~)* piece of fluff

flockig *Adj.* fluffy

flog [flo:k] *1. u. 3. Pers. Sg. Prät. v.* **fliegen**

floh [flo:] *1. u. 3. Pers. Sg. Prät. v.* **fliehen**

Floh [flo:] **der; ~[e]s, Flöhe** ['flø:ə] **a)** flea; **jmdm. einen ~ ins Ohr setzen** *(ugs.)* put an idea into sb.'s head; **b)** *Pl. (salopp: Geld)* dough *sing. (sl.)*; bread *sing. (sl.)*

Floh-: ~biß der flea-bite; **~kino das** *(ugs.)* flea-pit *(sl.)*; **~markt der** flea market; **~zirkus der** flea-circus

[1]**Flor** [flo:ɐ̯] **der; ~s, ~e** *(geh.)* **a)** *(Blütenpracht)* **im ~ stehen** be in full bloom; **b)** *(Blumenfülle)* display

[2]**Flor der; ~s, ~e a)** *(zartes Gewebe)* gauze; **b)** *(Faserenden)* pile; **c)** *s.* **Trauerflor**

Flora ['flo:ra] **die; ~, Floren** flora

[1]**Florentiner** [florɛn'ti:nɐ] **der; ~s, ~:** Florentine

[2]**Florentiner der; ~s, ~ a)** *(Hut)* picture hat; **b)** *(Gebäck)* florentine

Florentinerin die; ~, ~nen Florentine

Florenz [flo'rɛnts] **(das); Florenz'** Florence

Florett [flo'rɛt] **das; ~[e]s, ~e a)** foil; **b)** *o. Pl. (~fechten)* foils *sing.*; foil fencing *no art.*

florieren [flo'ri:rən] *itr. V.* ⟨*business*⟩ flourish

Florist [flo'rıst] **der; ~en, ~en, Floristin die; ~, ~nen a)** *(Blumenbinder)* flower-arranger; **b)** *(Blumenhändler)* florist

Floskel ['flɔskl̩] **die; ~, ~n** cliché

floskelhaft *Adj.* cliché-ridden; clichéd

floß [flɔs] *1. u. 3. Pers. Sg. Prät. v.* **fließen**

Floß [flo:s] **das; ~es, Flöße** ['flø:sə] raft

Flosse ['flɔsə] **die; ~, ~n a)** *(Zool., Flugw.)* fin; **b)** *(zum Tauchen)* flipper; **c)** *(ugs. scherzh. od. abwertend: Hand)* paw

flößen ['flø:sn̩] *tr., itr. V.* float; **Baumstämme ~:** raft tree trunks

Flößer ['flø:sɐ] **der; ~s, ~:** raftsman
Flößerei die; ~: rafting
Flöte ['flø:tə] **die; ~, ~n a)** *(Musik)* flute; *(Block~)* recorder; **b)** *(Skat)* **die [ganze] ~ herunterspielen** play a [straight] flush
flöten ['flø:tn̩] **1.** *itr. V.* **a)** play the flute; *(Blockflöte spielen)* play the recorder; ‹*bird*› flute; **b)** *(ugs.: pfeifen)* whistle; **c)** *(ugs.: affektiert sprechen)* pipe. **2.** *tr. V.* **a)** play ‹*tune etc.*› on the flute/recorder; **b)** *(ugs.: pfeifen)* whistle ‹*tune etc.*›
flöten-, Flöten-: ~|gehen *unr. itr. V.; mit sein (ugs.)* ‹*money*› go down the drain; ‹*time*› be wasted; **seine Illusionen gingen ~:** his illusions went for a burton *(Brit. sl.)*; **~konzert das a)** *(Musikstück)* flute concerto; **b)** *(Veranstaltung)* flute concert; **~spiel das** flute-playing; **~spieler der** flute-player; **~ton der: jmdm. die ~töne beibringen** *(fig. ugs.)* teach sb. a thing or two *(coll.)*
Flötist [flø'tɪst] **der; ~en, ~en, Flötistin die; ~, ~nen** flautist
flott [flɔt] **1.** *Adj.* **a)** *(ugs.: schwungvoll)* lively ‹*music, dance, pace, style*›; snappy ‹*dialogue*›; **b)** *(ugs.: schick, modisch)* smart ‹*hat, suit, car*›; **c)** *(munter, hübsch)* stylish; smart; **~ aussehen** look attractive; **d)** *(leichtlebig)* **ein ~es Leben führen** be fast-living; **e)** *nicht attr. (fahrbereit, wiederhergestellt)* seaworthy ‹*vessel*›; *(ugs.)* roadworthy ‹*vehicle*›; airworthy ‹*aircraft*›; **mein Auto ist wieder ~:** my car ist back on the road again. **2.** *adv.* ‹*work*› quickly; ‹*dance, write*› in a lively manner; ‹*be dressed*› smartly
Flotte ['flɔtə] **die; ~, ~n** fleet
Flotten-: ~stützpunkt der naval base; **~verband der** naval unit
flott-: ~|kriegen *tr. V.* get ‹*boat*› afloat; get ‹*car*› going; **~|machen** *tr. V.* refloat ‹*ship*›; get ‹*car*› back on the road
Flöz [flø:ts] **das; ~es, ~e** *(Bergbau)* seam
Fluch [flu:x] **der; ~[e]s, Flüche** ['fly:çə] **a)** *(Kraftwort)* curse; oath; **b)** *(Verwünschung)* curse; **c)** *o. Pl. (Unheil, Verderben)* curse; **ein ~ liegt über/lastet auf jmdm.** there's a curse on sb.
fluchen *itr. V.* curse; swear; **auf/über jmdn./etw. ~:** swear at *or* curse sb./sth.
[1]**Flucht** [flʊxt] **die; ~ a)** flight; **auf/während der ~:** while fleeing; *(von Gefangenen)* on the run; **jmdn. auf der ~ erschießen** shoot sb. while he/she is trying to escape; **die ~ ergreifen** ‹*prisoner*› make a dash for freedom; *(fig.: weglaufen)* make a dash for it; **jmdn. in die ~ schlagen** put sb. to flight; **b)** *(fig.)* refuge; **die ~ in die Anonymität** taking refuge in anonymity; **die ~ aus der Wirklichkeit** escape from reality; **die ~ nach vorn antreten** take the bull by the horns
[2]**Flucht die; ~, ~en a)** *(Bauw.: Häuser~, Arkaden~)* row; **die ~ der Fenster** the line of the windows; **b)** *(Zimmer~)* suite
flucht·artig 1. *Adj.* hurried; hasty; **2.** *adv.* hurriedly; hastily
Flucht·auto das getaway car
flüchten ['flʏçtn̩] **1.** *itr. V.; mit sein* **vor jmdm./etw. ~:** flee from sb./sth.; **vor der Polizei ~:** run away from the police; *(mit Erfolg)* escape from the police; **zu jmdm. ~:** take refuge with sb.; **ins Ausland ~:** escape abroad. **2.** *refl. V.* **sich in ein Bauernhaus ~:** take refuge in a farmhouse; **sich aufs Dach ~:** escape on to the roof
Flucht-: ~fahrzeug das getaway vehicle; **~gefahr die** risk of an escape attempt; **~helfer der** person who aids/aided an/the escape
flüchtig ['flʏçtɪç] **1.** *Adj.* **a)** *(flüchtend)* fugitive; wanted ‹*thief, criminal*› **~ sein** be at large; **b)** *(oberflächlich)* cursory; superficial ‹*insight*›; hurried ‹*piece of work*›; **c)** *(eilig, schnell)* quick; short ‹*visit, greeting*›; fleeting ‹*glance*›; **d)** *(vergänglich)* fleeting ‹*moment*›. **2.** *adv* **a)** *(oberflächlich)* cursorily; **b)** *(eilig)* hurriedly
Flüchtigkeit die; ~, ~en a) *(Oberflächlichkeit)* cursoriness; **b)** *s.* **Flüchtigkeitsfehler; c)** *(Vergänglichkeit)* fleetingness
Flüchtigkeits·fehler der slip; *(tadelnswert)* careless mistake
Flüchtling ['flʏçtlɪŋ] **der; ~s, ~e** refugee
Flüchtlings-: ~lager das refugee camp; **~treck der** long stream of refugees
Flucht-: ~linie die vanishing-line; **~plan der** escape plan; **~versuch der** escape attempt; **~weg der** escape route
Flug [flu:k] **der; ~[e]s, Flüge** ['fly:gə] **a)** *o. Pl.* flight; **im ~:** in flight; **etw. vergeht [wie] im ~e** sth. flows by; **b)** *(Flugreise)* flight
Flug-: ~abwehr die *(Milit.)* anti-aircraft defence; **~angst die** fear of flying; **~bahn die** trajectory; **~ball der** *(Tennis)* volley; **~begleiter der** steward; **~begleiterin die** stewardess; **~blatt das** pamphlet; leaflet; **~boot das** flying boat
Flügel ['fly:gl̩] **der; ~s, ~ a)** wing; **die ~ hängen lassen** *(fig. ugs.)* become disheartened; **jmdm. die ~ stutzen** *(fig.)* clip sb.'s wings; **b)** *(Altar~)* wing; *(Fenster~)* casement; *(Nasen~)* nostril; **c)** *(Klavier)* grand piano; **d)** *(Milit., Ballspiele)* wing
flügel-, Flügel-: ~horn das *(Musik)* flugelhorn; **~lahm** *Adj.* **a)** ‹*bird*› with an injured wing; **b)** *(fig.: mutlos, kraftlos)* lacking energy *postpos.*; limping ‹*organization*›; **~spannweite die** *(Flugw., Zool.)* wing span; **~stürmer der** *(Ballspiele)* wing forward; winger; **~tür die** double door
Flug·gast der [air] passenger
flügge ['flʏgə] *Adj.* fully-fledged; *(fig.: selbständig)* independent
Flug-: ~geschwindigkeit die *(eines Flugzeugs)* flying speed; *(eines Vogels)* speed of flight; **~gesellschaft die** airline; **~hafen der** airport; **~höhe die** altitude; **~kapitän der** captain; **~lärm der** aircraft noise; **~lehrer der** flying instructor; **~linie die a)** *(Strecke)* air route; **b)** *(Gesellschaft)* airline; **~lotse der** air traffic controller; **~objekt das** flying object; **ein unbekanntes ~objekt** an unidentified flying object; **~personal das** flight personnel; **~plan der** flight schedule; **~platz der** airfield; aerodrome; **~preis der** air fare; **~reise die** air journey
flugs [flʊks] *Adv. (veralt.)* swiftly
flug-, Flug-: ~sand der wind-borne sand; **~schein der** air ticket; **~schreiber der** flight-recorder; **~schrift die** pamphlet; **~sicherung die** air traffic control; **~steig der** pier; *(Ausgang)* **~steig 5** gate 5; **~ticket das** air ticket; **~verkehr der** air traffic; **~zeit die** flight time
Flug·zeug das; ~[e]s, ~e aeroplane *(Brit.)*; airplane *(Amer.)*; aircraft; **mit dem ~ reisen** travel by plane *or* air
Flugzeug-: ~absturz der plane crash; **~bau der;** *o. Pl.* aircraft construction; **~besatzung die** crew; **~entführer der** [aircraft] hijacker; **~entführung die** [aircraft] hijack[ing]; **~modell das** model aeroplane; **~träger der** aircraft carrier; **~unglück das** plane crash; **~wrack das** wreckage of the/a plane

Fluidum ['flu:idʊm] **das; ~s, Fluida** aura; atmosphere

Fluktuation [flʊktua'tsio:n] **die; ~, ~en** *(bes. Wirtsch., Soziol.)* fluctuation (*Gen.* in)

fluktuieren [flʊktu'i:rən] *itr. V. (bes. Wirtsch., Soziol.)* fluctuate

Flunder ['flʊndɐ] **die; ~, ~n** flounder

Flunkerei [flʊŋkə'rai] **die; ~, ~en** *(ugs.)* **a)** *o. Pl.* story-telling; **b)** *(Lügengeschichte)* tall story

flunkern ['flʊŋkɐn] *itr. V.* tell stories

Flunsch [flʊnʃ] **der; ~[e]s, ~e** *od.* **die; ~, ~en** *(ugs.)* pout; **eine[n] ~ ziehen** *od.* **machen** pout

Fluor ['flu:ɔr] **das; ~s** *(Chemie)* fluorine

Fluoreszenz [fluorɛs'tsɛnts] **die; ~:** fluorescence

fluoreszieren *itr. V.* fluoresce; be fluorescent

[1]**Flur** [flu:ɐ] **der; ~[e]s, ~e** corridor; *(Diele)* [entrance] hall

[2]**Flur die; ~, ~en a)** *(landwirtschaftliche Nutzfläche)* farmland *no indef. art.;* **die ~en** the fields; **b)** *(geh.: offenes Kulturland)* fields *pl.;* **allein auf weiter ~ sein** *od.* **stehen** *(fig.)* be all alone in the world

Flur-: ~bereinigung die reallocation of land; **~garderobe die** hall-stand; **~schaden der** damage *no pl., no indef. art.* to farmland; **~tür die** front door

Fluß [flʊs] **der; Flusses, Flüsse** ['flʏsɐ] **a)** river; **b)** *o. Pl. (fließende Bewegung)* flow; **die Dinge sind im ~:** things are in a state of flux; **etw. in ~ bringen** get sth. going

fluß-, Fluß-: ~aal der freshwater eel; **~ab[wärts]** *Adv.* downstream; **~arm der** river branch; river arm; **~auf[wärts]** *Adv.* upstream; **~bett das** river bed

Flüßchen ['flʏsçən] **das; ~s, ~:** small river

flüssig ['flʏsɪç] **1.** *Adj.* **a)** liquid ⟨*nourishment, fuel*⟩; molten ⟨*ore, glass*⟩; melted ⟨*butter*⟩; runny ⟨*honey*⟩; **etw. ~ machen** melt sth.; **b)** *(fließend, geläufig)* fluent; free-flowing ⟨*traffic*⟩; **c)** *(verfügbar, solvent)* ready ⟨*capital, money*⟩; liquid ⟨*assets*⟩; **wieder ~ sein** *(ugs.)* have got some cash to play with again *(coll.);* **nicht ~ sein** *(ugs.)* be skint *(Brit. coll.) or (coll.)* [flat]broke. **2.** *adv.* ⟨*write, speak*⟩ fluently

Flüssiggas das liquid gas

Flüssigkeit die; ~, ~en a) liquid; *(Körper~, Brems~ usw.)* fluid; **b)** *(Geläufigkeit)* fluency

Flüssigkeits·maß das liquid measure

Flüssig·kristall·anzeige die *(Technik)* liquid crystal display

flüssig|machen *tr. V.* make available ⟨*money, funds*⟩

Flüssigseife die liquid soap

Fluß-: ~krebs, der *(Zool.)* crayfish; **~landschaft die** *(Geogr.)* fluvial topography; **~mündung die** river mouth **~pferd das** hippopotamus; **~tal das** river valley; **~ufer das** river bank; **das diesseitige/jenseitige ~ufer** the near/opposite bank [of the river]

flüstern ['flʏstɐn] **1.** *itr. V.* whisper; **sich ~d unterhalten** speak in whispers. **2.** *tr. V.* whisper; **jmdm. [et]was ~** *(ugs.)* give sb. something to think about; **das kann ich dir ~** *(ugs.)* I can promise you that

Flüster-: ~propaganda die underground propaganda; **~ton der** whisper; **im ~ton sprechen** speak in whispers; **~tüte die** *(ugs.)* megaphone; **~witz der** underground joke

Flut [flu:t] **die; ~, ~en a)** *o. Pl.* tide; **es ist ~:** the tide is coming in; **b)** *meist Pl. (geh.: Wassermasse)* flood; **schmutzige ~en** dirty waters; **eine ~ von Protesten** *(fig.)* a flood of protests

fluten 1. *itr. V.; mit sein (geh.)* flood; **in etw.** *(Akk.)* **~:** flood sth.. **2.** *tr. V. (Seemannsspr.: unter Wasser setzen)* flood

Flut·licht das; *o. Pl.* floodlight

flutschen ['flʊtʃn̩] *itr. V. (ugs., bes. nordd.)* **a)** *mit sein (gleiten)* slip; **b)** *(glatt vonstatten gehen)* go smoothly; **es flutscht nur so** it's going extremely well

Flut·welle die tidal wave

fluvial [flu'via:l] *Adj. (Geol.)* fluvial

f-Moll ['ɛf-] F minor

focht [fɔxt] *1. u. 3. Pers. Sg. Prät. v.* **fechten**

Fock [fɔk] **die; ~, ~en** *(Seew.)* foresail; *(auf einer Jacht)* jib

Föderalismus der; ~: federalism *no art.*

föderalistisch *Adj.* federalist

Föderation [fødera'tsio:n] **die; ~, ~en** federation

föderativ [fødera'ti:f] *Adj.* federal

Fohlen das; ~s, ~: foal

Föhn [fø:n] **der; ~[e]s, ~e** föhn

Föhre ['fø:rə] **die; ~, ~n** *(landsch.) s.* [2]**Kiefer**

Fokus ['fo:kʊs] **der; ~, ~se** *(Optik, Med.)* focus

Folge ['fɔlgə] **die; ~, ~n a)** consequence; *(Ergebnis)* result; **an den ~n eines Unfalls sterben** die as a result of an accident; **etw. zur ~ haben** result in sth.; lead to sth.; **b)** *(Aufeinander~)* succession; *(zusammengehörend)* sequence; **in rascher ~:** in quick succession; **c)** *(Fortsetzung) (einer Sendung)* episode; *(eines Romans)* instalment; **d) einem Aufruf/einem Befehl/einer Einladung ~ leisten** *(Amtsspr.)* respond to an appeal/obey *or* follow an order/accept an invitation

Folge·erscheinung die consequence

folgen ['fɔlgn̩] *itr. V.* **a)** *mit sein* follow; **jmdm. im Amt ~:** succeed sb. in office; **auf etw.** *(Akk.)* **~:** follow sth.; come after sth.; **kannst du mir ~?** *(oft scherzh.)* do you follow me?; **daraus folgt, daß ...:** it follows from this that ...; **b)** *auch mit sein* **jmds. Anordnungen/Befehlen ~:** follow *or* obey sb.'s orders; **seiner inneren Stimme/seinem Gefühl ~:** listen to one's inner voice/be ruled by one's feelings

folgend *Adj.* **der/die/das ~e** the next in order; **er sagte ~es** *od.* **das ~e ...:** he said this ...; **Folgendes, das Folgende** the following [passage *etc.*]; **im ~en, in ~em** in [the course of] the following passage *etc.*

folgendermaßen *Adv.* as follows; *(in folgender Weise)* in the following way

folgen-: ~los *Adj.* without consequences *postpos.;* **das ist nicht ~los geblieben** that hasn't been without its consequences; **~reich** *Adj.* ⟨*decision, event*⟩ fraught with consequences; **~schwer** Adj. with serious consequences *postpos.*

folge·richtig 1. *Adj.* logical ⟨*decision, conclusion*⟩; consistent ⟨*behaviour, action*⟩. **2.** *adv.* ⟨*think, develop, conclude*⟩ logically; ⟨*act, behave*⟩ consistently

Folge·richtigkeit die *s.* **folgerichtig:** logicality; consistency

folgern ['fɔlgɐn] **1.** *tr. V.* deduce **(aus** from); **~, daß ...:** conclude that ... **2.** *itr. V.* **richtig ~:** draw a/the correct conclusion

Folgerung die; ~, ~en conclusion

Folge·schaden der *(Versicherungsw.)* consequential damage

folglich ['fɔlklɪç] *Adv.* consequently; as a result; *(ugs.: deshalb)* consequently; therefore

folgsam 1. *Adj.* obedient. **2.** *adv.* obediently

Folie ['fo:liə] **die; ~, ~n** *(Metall~)* foil; *(Plastik~)* film

Folklore [fɔlk'lo:rə] **die;** ~ **a)** folklore; **b)** *(Musik)* folk-music
folkloristisch 1. *Adj.* folkloric. **2.** *adv.* in a folkloric way
Follikel [fɔ'li:kl̩] **der; ~s,** ~ *(Med., Bot.)* follicle
Follikel·sprung der *(Med.)* ovulation
Folter ['fɔltɐ] **die;** ~, **~n** torture; **jmdn. auf die ~ spannen** *(fig.)* keep sb. in an agony of suspense
Folterer der; ~s, ~: torturer
Folter·kammer die torture-chamber
foltern 1. *tr. V.* torture. **2.** *itr. V.* use torture
Folterung die; ~, **~en** torture
Fön ® [fø:n] **der; ~[e]s, ~e** hair-drier
[1]**Fond** [fõ:] **der; ~s, ~s** *(geh.)* rear compartment; back
[2]**Fond der; ~s, ~s** *(Kochk.)* juices *pl.*
Fonds [fõ:] **der;** ~ [fõ:(s)], ~ [fõ:s] fund
Fondue [fõ'dy:] **die;** ~, **~s** *od.* **das; ~s, ~s** *(Kochk.)* fondue
fönen ['fø:nən] *tr. V.* blow-dry ⟨*hair*⟩
Fontäne [fɔn'tɛ:nə] **die;** ~, **~n** jet; *(Springbrunnen)* fountain
foppen ['fɔpn̩] *tr. V.* **jmdn.** ~ *(ugs.)* pull sb.'s leg *(coll.)*; put sb. on *(Amer. coll.)*
forcieren [fɔr'si:rən] *tr. V.* step up ⟨*production*⟩; redouble, intensify ⟨*efforts*⟩; speed up, push forward ⟨*developments*⟩; **das Tempo** ~ *(Sport)* force the pace
Forcierung die; ~, **~en** *s.* **forcieren:** stepping up; redoubling; intensification; speeding up; pushing forward; forcing
Förde ['fø:ɐ̯də] **die;** ~, **~n** long narrow inlet
Förder-: ~anlage die *(Technik)* conveyor; **~band das** *(Technik)* conveyor belt
Förderer ['fœrdərɐ] **der; ~s,** ~ *(Gönner)* patron
Förder·korb der *(Bergbau)* cage
förderlich *Adj.* beneficial (*Dat.* to); **guten Beziehungen ~ sein** be conducive to *or* promote good relations
fordern ['fɔrdɐn] *tr. V.* **a)** demand; **Rechenschaft von jmdm.** ~: call sb. to account; **b)** *(fig.: kosten)* claim ⟨*lives*⟩; **c)** *(in Anspruch nehmen)* make demands on; **von etw. gefordert werden** be stretched by sth.; **d) jmdn. [zum Duell]** ~: challenge sb. [to a duel]
fördern ['fœrdɐn] *tr. V.* **a)** promote ⟨*trade, plan, project, good relations*⟩; patronize, support ⟨*artist, art*⟩; further ⟨*investigation*⟩; foster ⟨*talent, tendency, new generation*⟩; improve ⟨*appetite*⟩; aid ⟨*digestion, sleep*⟩; **b)** *(Bergbau)* mine ⟨*coal, ore*⟩; extract ⟨*oil*⟩
Förder·schacht der *(Bergbau)* winding shaft
Forderung die; ~, **~en a)** *(Anspruch)* demand; *(in bestimmter Höhe)* claim; **b)** *(Kaufmannsspr.)* claim (**an** + *Akk.* against); **eine ~ einklagen** sue for payment of a debt; **c)** *(zum Duell)* challenge
Förderung die; ~ **a)** *s.* **fördern a:** promotion; patronage; support; furthering; fostering; improvement; aiding; **b)** *(Bergbau) s.* **fördern b:** mining; extraction; **c)** *(Bergbau: geförderte Menge)* output
Förder·wagen der *(Bergbau)* mine car
Forelle [fo'rɛlə] **die;** ~, **~n** trout
forensisch [fo'rɛnzɪʃ] *Adj.* forensic
Forke ['fɔrkə] **die;** ~, **~n** *(bes. nordd.)* fork
Form [fɔrm] **die;** ~, **~en a)** *(Gestalt)* shape; **die Demonstration nahm häßliche ~en an** the demonstration began to look ugly; **in ~ von Tabletten** in the form of tablets; **b)** *(bes. Sport: Verfassung)* form; **in ~ sein/sich in ~ bringen** be/get on form; **in guter/schlechter ~ sein** be in good/off form; **c)** *(vorgeformtes Modell)* mould; *(Back~)* baking tin; **d)** *(Gestaltungsweise, Erscheinungs~)* form; **e)** *(Umgangs~)* form; **die ~[en] wahren** observe the proprieties; **in aller ~:** formally
formal [fɔr'ma:l] **1.** *Adj.* formal; **ein ~er Fehler** a technical error; *(Rechtsw.)* procedural error. **2.** *adv.* formally; **~ im Recht sein** be technically in the right
Formalin ® [fɔrma'li:n] **das; ~s** formalin
formalisieren *tr. V.* formalize
Formalisierung die; ~, **~en** formalization
Formalismus der; ~, **Formalismen** formalism
Formalist der; ~en, ~en formalist
Formalität [fɔrmali'tɛ:t] **die;** ~, **~en** formality
formal·juristisch 1. *Adj.* technical; **ein rein ~er Standpunkt** a narrowly legalistic view. **2.** *adv.* technically
Format [fɔr'ma:t] **das; ~[e]s, ~e a)** size; *(Buch~, Papier~, Bild~)* format; **b)** *o. Pl.* *(Persönlichkeit)* stature; **c)** *o. Pl.* *(besonderes Niveau)* quality; **etw. hat/ist ohne ~:** sth. has/lacks class
Formation [fɔrma'ts̮io:n] **die;** ~, **~en a)** *(Herausbildung, Anordnung)* formation; *(einer Generation, Gesellschaft)* development; **b)** *(Gruppe)* group; **c)** *(Milit.)* *(von Flugzeugen)* formation; *(von Soldaten)* unit
Formations·flug der formation flying
formbar *Adj.* malleable; soft ⟨*bone*⟩; *(fig.)* malleable, pliable ⟨*character, person*⟩
Formbarkeit die; ~: malleability; *(fig.)* malleability; pliability
Form·blatt das form
Formel ['fɔrml̩] **die;** ~, **~n** formula; **~ 1** *(Motorsport)* Formula One
formelhaft *Adj.* stereotyped ⟨*style, mode of expression, phrase*⟩
formell [fɔr'mɛl] **1.** *Adj.* formal. **2.** *adv.* formally; **die Einladung wurde rein ~ ausgesprochen** the invitation was made only as a matter of form; **er ist nur ~ im Recht** he's only technically in the right
Formel·zeichen das symbol
formen 1. *tr. V.* form; shape; mould, form ⟨*character, personality*⟩; mould ⟨*person*⟩. **2.** *refl. V.* take on a shape; *(fig.)* form; take shape
formen-, Formen-: ~lehre die *(Sprachw., Biol.)* morphology; **~reich** *Adj.* with its/their great variety of forms *postpos.*
form-, Form-: ~fehler der a) *(in einem Verfahren, Dokument)* irregularity; **b)** *(Taktlosigkeit)* faux pas; breach of etiquette; **~frage die** formality; **~gebung die;** ~, **~en** design; **~gerecht 1.** *Adj.* correct; proper; **2.** *adv.* correctly; properly
formieren 1. *tr. V.* form ⟨*team, party, organization*⟩. **2.** *refl. V.* **a)** form; **b)** *(sich zusammenschließen)* be formed
Formierung die; ~, **~en** formation; *(von Truppen)* drawing up
-förmig [-fœrmɪç] -shaped
förmlich ['fœrmlɪç] **1.** *Adj.* **a)** formal; **b)** *nicht präd.* *(regelrecht)* positive; **ein ~er Schreck** a real fright. **2.** *adv.* **a)** *(steif, unpersönlich, offiziell)* formally; **b)** *(geradezu)* positively; **sich ~ fürchten** be really afraid
Förmlichkeit die; ~, **~en** formality
form-, Form-: ~los 1. *Adj.* **a)** informal; **einen ~losen Antrag stellen** make an application without the official form[s]; apply informally; **b)** *(gestaltlos)* shapeless; **2.** *adv.* informally; **~losigkeit**

die; ~ **a)** informality; **b)** *(Gestaltlosigkeit)* shapelessness; ~**sache die** formality; ~**schön** *Adj.* elegant

Formular [fɔrmuˈla:ɐ̯] **das;** ~**s,** ~**e** form

formulieren [fɔrmuˈli:rən] *tr. V.* formulate

Formulierung die; ~, ~**en a)** *o. Pl. (das Formulieren)* formulation; *(eines Entwurfes, Gesetzes)* drafting; **b)** *(formulierter Text)* formulation

Formung die; ~, ~**en a)** design; **die strenge** ~ **des Sonetts** strict sonnet form; **b)** *o. Pl. (Bildung, Erziehung)* moulding

form·vollendet 1. *Adj.* perfectly executed ⟨*pirouette, bow, etc.*⟩; perfect in form ⟨*poem*⟩; **2.** *adv.* **etw.** ~**vollendet tun** do sth. faultlessly

forsch [fɔrʃ] **1.** *Adj.* self-assertive; forceful. **2.** *adv.* self-assertively; forcefully

forschen *itr. V.* **a) nach jmdm./etw.** ~: search *or* look for sb./sth.; **jmdn.** ~**d** *od.* **mit** ~**dem Blick betrachten** look at sb. searchingly; give sb. a searching look; **b)** *(als Wissenschaftler)* research; do research

Forscher der; ~**s,** ~, **Forscherin die;** ~, ~**nen a)** researcher; research scientist; **b)** *(Forschungsreisender)* explorer

Forscher-: ~**drang der a)** *(Wissensdurst)* thirst for new knowledge; **b)** *(Entdeckerfreude)* urge to explore; ~**team das** research team

Forschheit die; ~: self-assertiveness; forcefulness

Forschung die; ~, ~**en** research; ~**en [auf einem Gebiet] betreiben** do research [in a field]; ~ **und Lehre** teaching and research

Forschungs-: ~**auftrag der** research assignment; ~**bericht der** research report; ~**gebiet das** field of research; ~**programm das** research programme; ~**reise die** expedition; ~**reisende der/die** explorer; ~**satellit der** research satellite; ~**tätigkeit die** research work; ~**zweck der** purpose of the research; **für** ~**zwecke** for research purposes

Forst [fɔrst] **der;** ~**[e]s,** ~**e[n]** forest

Forst-: ~**amt das** forestry office; ~**beamte der** forestry official

Förster [ˈfœrstɐ] **der;** ~**s,** ~: forest warden; forester; ranger *(Amer.)*

Forst-: ~**frevel der** offence against the forest law; ~**frevel begehen** break the forest law; ~**haus das** forester's house; ~**revier das** forest district; ~**wirtschaft die** forestry

Forsythie [fɔrˈzy:tsi̯ə] **die;** ~, ~**n** forsythia

fort [fɔrt] *Adv.* **a)** *s.* **weg a; b)** *(weiter)* **nur immer so** ~: just carry on as you are *or* like that; **und so** ~: and so on; and so forth; **in einem** ~: continuously

Fort [fo:ɐ̯] **das;** ~**s,** ~**s** fort

fort-, Fort-: ~**an** [-ˈ-] *Adv.* from now/then on; ~**bestand der;** *o. Pl.* continuation; *(eines Staates)* continued existence; ~|**bestehen** *unr. itr. V.* remain; continue; ⟨*nation*⟩ remain in existence; ~|**bewegen 1.** *tr. V.* move; shift; **2.** *refl. V.* move [along]; ~**bewegung die;** *o. Pl.* locomotion; ~|**bilden** *tr. V.* **jmdn./sich** ~**bilden** continue sb.'s/one's education; ~**bildung die;** *o. Pl.* further education; *(beruflich)* further training; ~**bildungs·kurs der** further education course; *(beruflich)* training course; ~|**bleiben** *unr. itr. V.; mit sein s.* **wegbleiben;** ~|**bringen** *unr. tr. V.: s.* **wegbringen;** ~**dauer die** continuation; ~|**dauern** *itr. V.* continue

forte [ˈfɔrtə] *Adv. (Musik, Pharm.)* forte

fort-, Fort-: ~|**eilen** *itr. V.; mit sein (geh.)* hurry off *or* away; hasten away; ~|**entwickeln 1.** *tr. V.* **etw.** ~**entwickeln** develop sth. further; **2.** *refl. V.* develop; ~|**fahren 1.** *unr. itr. V.* **a)** *mit sein s.* **wegfahren 1; b)** *auch mit sein (weitermachen)* continue, go on **(mit** with); ~**fahren, etw. zu tun** continue *or* go on doing sth. **2.** *unr. tr. V. s.* **wegfahren 2;** ~|**fallen** *unr. itr. V.; mit sein s.* **wegfallen;** ~|**fliegen** *unr. itr. V.; mit sein: s.* **wegfliegen;** ~|**führen** *tr. V.* **a)** lead away; **b)** *(fortsetzen)* continue, keep up ⟨*tradition, business*⟩; continue, carry on ⟨*another's work*⟩; ~**führung die** *s.* ~**führen 1b;** continuation; keeping up; carrying on; ~**gang der;** *o. Pl.* **a)** *s.* **Weggang; b)** *(Weiterentwicklung)* progress; ~|**geben** *unr. tr. V. s.* **weggeben;** ~|**gehen** *unr. itr. V.; mit sein* **a)** *s.* **weggehen; b)** *(andauern, verlaufen)* continue; go on; ~**geschritten** *Adj.* advanced ⟨*age, stage of illness*⟩; **zu** ~**geschrittener Tageszeit** at a late hour; ~**geschrittene der/die;** *adj. Dekl.* advanced student/player; ~**geschrittenen·kurs[us] der** advanced course; ~**gesetzt 1.** *Adj.; nicht präd.* continual; constant; ~**gesetzter Betrug** repeated fraud; **2.** *adv.* continually; constantly

fortissimo [fɔrˈtɪsimo] *Adv. (Musik)* fortissimo

fort-, Fort-: ~|**jagen** *tr. V.: s.* **wegjagen;** ~|**kommen** *unr. itr. V.; mit sein* **a)** *s.* **wegkommen a, b, d, f; b)** *(Erfolg haben)* get on; do well; ~**kommen das;** ~**s** progress; ~|**können** *unr. itr. V.: s.* **wegkönnen;** ~|**lassen** *unr. tr. V.: s.* **weglassen;** ~|**laufen** *unr. itr. V.; mit sein* **a)** *s.* **weglaufen; b)** *(sich ~setzen)* continue; ~**laufend 1.** *Adj.* continuous; ongoing ⟨*plot of a series*⟩; consecutive ⟨*numbers, issues*⟩. **2.** *adv.* continuously; consecutively ⟨*numbered*⟩; ~|**leben** *itr. V.: s.* **weiterleben c;** ~|**legen** *tr. V. s.* **weglegen;** ~|**machen** *(ugs.) refl. V.* get away; ~|**müssen** *unr. itr. V.: s.* **wegmüssen;** ~|**nehmen** *unr. tr. V.* take away; ~|**pflanzen** *refl. V.* **a)** *(sich vermehren)* reproduce [oneself/itself]; **b)** *(sich verbreiten)* ⟨*idea, mood*⟩ spread; ⟨*sound, light*⟩ travel, propagate; ~**pflanzung die a)** *(Vermehrung)* reproduction; **b)** *(Verbreitung)* transmission; *(von Schall, Licht)* propagation; *(von Ideen)* spread; ~**pflanzungs·fähig** *Adj.* capable of reproduction *postpos.;* ~|**reißen** *unr. tr. V.* **a)** tear away; ⟨*floods*⟩ sweep away; **b)** *(fig.)* **jmdn.** ~**reißen** carry *or* sweep sb. along; ~|**rennen** *unr. itr. V.; mit sein (ugs.)* run off *or* away; ~|**schaffen** *tr. V.* take *or* carry away; ~|**scheren** *refl. V. (ugs.)* clear off *(coll.);* ~|**scheuchen** *tr. V.* shoo *or* chase away; ~|**schicken** *tr. V.: s.* **wegschikken;** ~|**schleichen** *unr. itr., refl. V.: s.* **wegschleichen;** ~|**schleppen** *refl., tr. V.: s.* **wegschleppen;** ~|**schleudern** *tr. V.* fling away; ~|**schreiben** *unr. tr. V.* update; *(in die Zukunft)* project forward; ~**schreibung die** updating; *(in die Zukunft)* forward projection; ~|**schreiten** *unr. itr. V.; mit sein* ⟨*process*⟩ progress, continue; ⟨*time*⟩ move on; **der Sommer ist [weit]** ~**geschritten** we are well into summer; ~**schreitend** *Adj.* progressive; advancing ⟨*age*⟩; **mit** ~**schreitender Jahreszeit** as the year goes/went on; ~**schritt der** progress; ~**schritte** progress *sing.;* **ein** ~**schritt** a step forward; ~**schrittlich 1.** *Adj.* progressive; **2.** *adv.* progressively; ~**schritts·feindlich** *Adj.* anti-progressive; ~**schritts·gläu-**

big *Adj.* **~schrittsgläubig sein** put one's faith in progress; **~|setzen 1.** *tr. V.* continue; carry on; **2.** *refl. V.* continue; **~setzung die; ~, ~en a)** *(das ~setzen)* continuation; **[s]eine ~setzung finden** resume; **b)** *(anschließender Teil)* instalment; **~setzung folgt** to be continued; **~setzungs·roman der** serial; serialized novel; **~|spülen** *tr. V.: s.* **wegspülen; ~|stehlen** *unr. refl. V.* steal *or* sneak away; **~|stoßen** *unr. tr. V.: s.* **wegstoßen; ~|tragen** *unr. tr. V.: s.* **wegtragen; ~|treiben 1.** *unr. tr. V.* drive off *or* away; **2.** *itr. V.; mit sein* float away

Fortuna [fɔr'tu:na] **(die)** Fortune

fort-: ~|währen *itr. V. (geh.)* continue; **~|während 1.** *Adj.; nicht präd.* continual; incessant; **2.** *adv.* continually; incessantly; **~|werfen** *unr. tr. V.: s.* **wegwerfen; ~|wollen** *unr. itr. V. s.* **wegwollen; ~|ziehen** *unr. tr., itr. V.: s.* **wegziehen**

Forum ['fo:rʊm] **das; ~s, Foren** forum; *(Diskussionsveranstaltung)* forum discussion

fossil [fɔ'si:l] *Adj.* fossilized; fossil *attrib.*

Fossil das; ~s, ~ien fossil

Foto ['fo:to] **das; ~s, ~s** photo; **~s machen** *od. (ugs.:)* **schießen** take photos; **auf einem ~:** in a photo

Foto-: ~album das photo album; **~apparat der** camera; **~atelier das** photographic studio; **~ecke die** [mounting] corner

fotogen [foto'ge:n] *Adj.* photogenic

Foto·geschäft das photographic shop

Fotograf der; ~en, ~en photographer

Fotografie die; ~, ~n a) *o. Pl.* photography *no art.;* **b)** *(Lichtbild)* photograph

fotografieren 1. *tr. V.* photograph; take a photograph/photographs of **2.** *itr. V.* take photographs

Fotografin die; ~, ~nen photographer

fotografisch 1. *Adj.* photographic. **2.** *adv.* photographically

foto-, Foto-: ~kopie die photocopy; **~kopieren** *tr., itr. V.* photocopy; **~kopierer der, ~kopier·gerät das** photocopier; photocopying machine

Foto-: ~labor das photographic laboratory; **~modell das** photographic model; **~montage die** photomontage; **~reporter der** press photographer; newspaper photographer; **~wettbewerb der** photographic competition; **~zeitschrift die** photographic magazine

Fotze ['fɔtsə] **die; ~, ~n** *(vulg.)* cunt *(coarse)*

Foul das; ~s, ~s *(Sport)* foul **(an + *Dat.*** on)

foulen ['faulən] *(Sport)* **1.** *tr. V.* foul. **2.** *itr. V.* commit a foul

Fox [fɔks] **der; ~[es], ~e a)** *s.* **Foxterrier; b)** *s.* **Foxtrott**

Fox·terrier der fox-terrier

Fox·trott [-trɔt] **der; ~s, ~e** *od.* **~s** foxtrot

Foyer [foa'je:] **das; ~s, ~s** foyer

FPÖ *Abk.* **Freiheitliche Partei Österreichs**

[1]**Fr.** *Abk.* **Franken** SFr.

[2]**Fr.** *Abk.* **Frau**

[3]**Fr.** *Abk.* **Freitag** Fri.

Fracht [fraxt] **die; ~, ~en a)** *(Schiffs~, Luft~)* cargo; freight; *(Bahn~, LKW~)* goods *pl.;* freight; **b)** *(~kosten) (Schiffs~, Luft~)* freight; freightage; *(Bahn~, LKW~)* carriage

Fracht·brief der consignment note; waybill

Frachter der; ~s, ~: freighter

Fracht-: ~geld das *s.* **Fracht b; ~gut das** slow freight; slow goods *pl.;* **~raum der;** *o. Pl.* [cargo] hold; *(Platz)* [cargo] space; **~schiff das** cargo ship

Frack [frak] **der; ~[e]s, Fräcke** ['frɛkə] tails *pl.;* evening dress; **im ~:** in tails *or* evening dress

Frack-: ~hemd das dress shirt; **~sausen das** *in* **~sausen haben** *(ugs.)* get the wind up *(sl.);* **~schoß der;** *meist Pl.* coat-tail; **~weste die** waistcoat *(worn with evening dress)*

Frage ['fra:gə] **die; ~, ~n a)** question; **jmdm.** *od.* **an jmdn. eine ~ stellen** put a question to sb.; **jmdm. eine ~ beantworten/auf jmds. ~** *(Akk.)* **antworten** reply to *or* answer sb.'s question; **eine ~ verneinen/bejahen** give a negative/positive answer to a question; **b)** *(Problem)* question; *(Angelegenheit)* issue; **das ist [nur] eine ~ der Zeit** that is [only] a question *or* matter of time; **c)** *in* **das ist noch sehr die ~:** that is still very much the question; **das ist die große ~:** that is the big question; **das ist gar keine ~:** there's no doubt *or* question about it; **etw. in ~ stellen** call sth. into question; question sth.; **in ~ kommen** be possible; **für ein Stipendium kommen nur gute Schüler in ~:** only good pupils can be considered for a grant; **dieses Kleid kommt für mich nicht in ~:** I couldn't possibly wear this dress; **das kommt nicht in ~** *(ugs.)* that is out of the question; **ohne ~:** without question

Frage·bogen der questionnaire; *(Formular)* form

fragen 1. *tr., itr. V.* **a)** ask; **er fragt immer so klug** he always asks *or* puts such astute questions; **frag nicht so dumm!** *(ugs.)* don't ask such silly questions; **das fragst du noch?** *(ugs.)* need you ask?; **da fragst du mich zuviel** that I don't know; I really can't say; **jmdn. ~d ansehen** look at sb. inquiringly; give sb. a questioning look; **b)** *(sich erkundigen)* **nach etw. ~:** ask *or* inquire about sth.; **jmdn. nach/wegen etw. ~:** ask sb. about sth.; **nach dem Weg/jmds. Meinung ~:** ask the way/[for] sb.'s opinion; **nach jmdm. ~** *(jmdn. suchen)* ask for sb.; *(nach jmds. Befinden ~)* ask after *or* about sb.; **c)** *(nachfragen)* ask for; **jmdn. um Rat ~:** ask sb. for advice; **d)** *(verneint: sich nicht kümmern)* **nach jmdm./etw. nicht ~:** not care about sb./sth.. **2.** *refl. V.* **sich ~, ob ...:** wonder whether ...; **das frage ich mich auch** I was wondering that, too

Frage-: ~satz der interrogative sentence/clause; **ein direkter/indirekter ~satz** a direct/an indirect question; **~stellung die a)** *(Formulierung)* formulation of a/the question; **durch eine geschickte ~stellung** by skilled questioning; **b)** *(Problem)* problem; **~stunde die** *(Parl.)* question time; **~-und-Antwort-Spiel das** question-and-answer game; **~zeichen das** question mark

fragil [fra'gi:l] *Adj. (geh.)* fragile

fraglich ['fra:klɪç] *Adj.* **a)** doubtful; **b)** *nicht präd. (betreffend)* in question *postpos.;* relevant

frag·los *Adv.* without question; unquestionably

Fragment [fra'gmɛnt] **das; ~[e]s, ~e** fragment

fragmentarisch [fragmɛn'ta:rɪʃ] *Adj.* fragmentary

frag·würdig *Adj.* **a)** questionable; **b)** *(zwielichtig)* dubious

Fragwürdigkeit die; ~, ~en a) questionableness; **b)** *(Zwielichtigkeit)* dubiousness

Fraktion [frak'tsio:n] **die; ~, ~en a)** *(Parl.)* parliamentary party; *(mit zwei Parteien)* parliamentary coalition; **b)** *(Sondergruppe)* faction

fraktionell [fraktsio'nɛl] *Adj.* within a/the party/group *postpos.;* internal ‹*conflict, agreement*›

fraktions-, Fraktions- *(Parl.):* **~beschluß der** party/coalition decision; **~los** *Adj.* independent; **~sitzung die** meeting of the parliamentary party/coalition; **~vorsitzende der/die** leader of the parliamentary party/coalition; **~zwang der** obligation to vote in accordance with party policy; **den ~zwang aufheben** allow a free vote

Fraktur [frak'tu:ɐ] **die; ~, ~en a)** *(Med.)* fracture; **b)** *(Schriftart)* Fraktur; **mit jmdm. ~ reden** *(ugs.)* talk straight with sb.

Franc [frã:] **der; ~, ~s** franc

frank [fraŋk] *Adv.* **~ und frei** frankly and openly; openly and honestly

Franke der; ~n, ~n a) Franconian; **b)** *(hist.)* Frank

[1]**Franken (das); ~s** Franconia

[2]**Franken der; ~s, ~:** [Swiss] franc

[1]**Frankfurter** ['fraŋkfʊrtɐ] **die; ~, ~** *(Wurst)* frankfurter

[2]**Frankfurter 1.** *indekl. Adj.; nicht präd.* Frankfurt. **2. der; ~s, ~:** Frankfurter

frankieren [fraŋ'ki:rən] *tr. V.* frank

fränkisch ['frɛŋkɪʃ] *Adj.* **a)** Franconian; **b)** *(hist.)* Frankish

franko ['fraŋko] *Adv.* *(Kaufmannsspr. veralt.)* carriage paid; *(mit der Post)* post-free

Frank·reich (das); ~s France

Franse ['franzə] **die; ~, ~n** strand [of a/the fringe]; **die ~n des Teppichs** the fringe of the carpet

Franz [frants] **(der)** Francis

Franz·branntwein der; *o. Pl.* *(veralt.)* alcoholic liniment

Franziskaner [frantsɪs'ka:nɐ] **der; ~s, ~:** Franciscan

Franzose [fran'tso:zə] **der; ~n, ~n a)** Frenchman; **die ~n** the French; **b)** *(ugs.: Schraubenschlüssel)* screw wrench

Französin [fran'tsø:zɪn] **die; ~, ~nen** Frenchwoman

französisch 1. *Adj.* French; **ein ~es Bett** a double bed; **die Französische Schweiz** French-speaking Switzerland. **2.** *adv.* **sich [auf] ~ empfehlen** *od.* **verabschieden** *(ugs.)* take French leave

Französisch das; ~[s] French

frappant [fra'pant] *Adj.* striking ‹*similarity*›; remarkable ‹*success, discovery*›

frappieren [fra'pi:rən] *tr. V.* *(geh.)* astonish; astound

frappierend *Adj.* astonishing; remarkable

Fräse ['frɛ:zə] **die; ~, ~n a)** *(für Holz)* moulding machine; *(für Metall)* milling machine; **b)** *(Boden~)* rotary cultivator

fräsen *tr. V.* shape ‹*wood*›; mill ‹*metal*›; form ‹*groove, thread*›

Fräser der; ~s, ~ a) *(Werkzeug)* cutter; **b)** *(Metallbearb.)* milling-machine operator; *(Holzverarb.)* moulding-machine operator

fraß [fra:s] *1. u. 3. Pers. Sg. Prät. v.* **fressen**

Fraß der; ~es a) food; **einem Tier etw. zum ~ vorwerfen** feed an animal with sth.; **jmdm. etw. zum ~ vorwerfen** *(fig. abwertend)* let sb. have sth.; **b)** *(derb: schlechtes Essen)* muck; swill

Fratz [frats] **der; ~es, ~e,** *(österr.:)* **~en, ~en a)** *(ugs.: niedliches Kind)* [little] rascal; **b)** *(bes. südd., österr.: ungezogenes Kind)* brat

Fratze ['fratsə] **die; ~, ~n a)** *(häßliches Gesicht)* hideous features *pl.;* *(abwertend)* mug *(sl.);* **b)** *(ugs.: Grimasse)* grimace; **jmdm. ~n schneiden** pull faces at sb.

Frau die; ~, ~en a) woman; **b)** *(Ehe~)* wife; **c)** *(Titel, Anrede)* **~ Schulze** Mrs Schulze; **~ Professor/Dr. Schulze** Professor/Dr. Schulze; **~ Ministerin/Direktorin/Studienrätin Schulze** Mrs/Miss/Ms Schulze; **~ Ministerin/Professor/Doktor** Minister/Professor/doctor; **~ Vorsitzende/Präsidentin** Madam Chairman/President; *(in Briefen)* **Sehr geehrte ~ Schulze** Dear Madam; *(bei persönlicher Bekanntschaft)* Dear Mrs/Miss/Ms Schulze; **[Sehr verehrte] gnädige ~:** [Dear] Madam; **Ihre ~ Mutter** your mother; **d)** *(Herrin)* lady; mistress; **die ~ des Hauses** the lady of the house

Frauchen ['frauçən] **das; ~s, ~ a)** *(ugs.: Ehefrau)* wifie; **b)** *(Herrin eines Hundes)* mistress

frauen-, Frauen-: ~arbeit die a) *o. Pl.* *(Erwerbstätigkeit)* women's employment; **b)** *(für ~ geeignete Arbeit)* women's work; **~arzt der, ~ärztin die** gynaecologist; **~beruf der** women's occupation; **~bewegung die;** *o. Pl.* women's movement; **~emanzipation die** female emancipation; women's emancipation; **~feind der** misogynist; **~feindlich** *Adj.* anti-women; **~haus das** battered wives' refuge; **~held der** lady-killer; **~mörder der** killer of women; **~rechtlerin** [-rɛçtlərɪn] **die; ~, ~nen** feminist; Women's Libber *(coll.)*

Frauens·person die *(veralt.)* female

Frauen-: ~sport der women's sport; **~station** *die* women's ward; **~stimme die** woman's voice

Frauen-: ~wahl·recht das women's franchise; women's right to vote; **~zeitschrift die** women's magazine; **~zimmer das** *(abwertend)* female

Fräulein ['frɔylain] **das; ~s, ~** (*ugs.* **~s) a)** *(junges ~)* young lady; *(ältliches ~)* spinster; **b)** *(Titel, Anrede)* **~ Mayer** Miss Mayer; **[sehr verehrtes] gnädiges ~ [X]** Dear Miss X; **Ihr ~ Tochter** your daughter; **c)** *(Kellnerin)* waitress; **~, wir möchten zahlen** [Miss,] could we have the bill *(Brit.) or (Amer.)* check, please?; **d) das ~ vom Amt** *(veralt.)* the operator

fraulich 1. *Adj.* feminine; *(reif)* womanly. **2.** *adv.* in a feminine/womanly way

Fraulichkeit die; ~: femininity; *(reifes Wesen)* womanliness

frech [frɛç] **1.** *Adj.* **a)** *(respektlos, unverschämt)* impertinent; impudent; cheeky; bare-faced ‹*lie*›; **etw. mit ~er Stirn behaupten** *(fig.)* have the bare-faced cheek to say sth.; **b)** *(keck, keß)* saucy. **2.** *adv.* *(respektlos, unverschämt)* impertinently; impudently; cheekily; **jmdn. ~ anlügen** tell sb. bare-faced lies

Frech·dachs der *(ugs., meist scherzh.)* cheeky little thing

Frechheit die; ~, ~en a) *o. Pl.* *(Benehmen)* impertinence; impudence; cheek; **die ~ haben, etw. zu tun** have the impertinence *etc.* to do sth.; **b)** *(Äußerung)* impertinent *or* impudent *or* cheeky remark; **sich** *(Dat.)* **~en erlauben** be impertinent

Freesie ['fre:ziə] **die; ~, ~n** freesia

Fregatte [fre'gatə] **die; ~, ~n** frigate

Fregatten·kapitän der commander

frei [frai] **1.** *Adj.* **a)** free ‹*man, will, life, people, decision, etc.*›; **b)** *(nicht angestellt)* free-lance ‹*writer, worker, etc.*›; **die ~en Berufe** the independent professions; **c)** *(ungezwungen)* free and easy; lax *(derog.);* **d)** *(nicht in Haft)* free; at liberty *pred.;* **e)** *(offen)* open; **unter ~em Himmel** in the open [air]; outdoors; **auf ~er Strecke** *(Straße)* on the open road; *(Eisenbahn)* between stations; **ins Freie gehen** walk out into the open; **f)** *(unbesetzt)* vacant; unoccupied; free; **ein ~er Stuhl/Platz** a vacant *or* free chair/seat; **Entschuldigung, ist hier noch ~?** excuse me, is this anyone's seat *etc.?*; **eine ~e Stelle** a vacancy; **ein Bett ist [noch] ~:** one bed is [still] free *or* not taken; **ist der Tisch ~?** is this table free?; **einige Seiten ~ lassen** leave some pages blank; **g)** *(kostenlos)* free ⟨*food, admission*⟩; **20 kg Gepäck ~ haben** have *or* be allowed a 20 kilogram baggage allowance; **Lieferung ~ Haus** carriage free; **h)** *(ungenau)* **eine ~e Übersetzung** a free *or* loose translation; **i)** *(ohne Vorlage)* improvised; **j)** *(uneingeschränkt)* free; **der Zug hat ~e Fahrt** the train can proceed; **der ~e Fall** *(Physik)* free fall; **k)** *(nicht beeinträchtigt)* free; **~ von Schmerzen sein** be free of pain; **~ von Fehlern** without faults; **l)** *(verfügbar)* spare; free; **ich habe heute ~/meinen ~en Abend** I've got today off/this is my evening off; **sich** *(Dat.)* **~ nehmen** *(ugs.)* take some time off; **er ist noch/nicht mehr ~:** he is still/no longer unattached; **m)** *(ohne Hilfsmittel)* **eine ~e Rede** an extempore speech; **n)** *(unbekleidet)* bare; **den Oberkörper ~ machen** strip to the waist; **o)** *(bes. Fußball)* unmarked; **p)** *(Chemie, Physik)* free; **~ werden** *(bei einer Reaktion)* be given off; **q)** *(in festen Wendungen)* **~e Hand haben/jmdm. ~e Hand lassen** have/give sb. a free hand; **aus ~en Stücken** *(ugs.)* of one's own accord; voluntarily; **jmdn. auf ~en Fuß setzen** set sb. free; **auf ~em Fuß** *(von Verbrechern etc.)* at large. **2.** *adv.* ⟨*act, speak, choose*⟩ freely; ⟨*translate*⟩ freely, loosely; **etw. ~ heraus sagen** say sth. freely; **~ herumlaufen** run around scot-free; **eine Rede ~ halten** make a speech without notes; **~ stehen** ⟨*player*⟩ be unmarked

frei-, Frei-: ~bad das open-air *or* outdoor swimming-pool; **~|bekommen 1.** *unr. itr. V. (ugs.)* get time off; **2.** *unr. tr. V.* **jmdn./etw. ~bekommen** get sb./sth. released; **~beruflich 1.** *Adj.* self-employed; freelance ⟨*journalist, editor, architect, etc.*⟩; ⟨*doctor, lawyer*⟩ in private practice; **2.** *adv.* **~beruflich tätig sein/arbeiten** work freelance/practise privately; **~betrag der** *(Steuerw.)* [tax] allowance; **~beuter** [-bɔytɐ] **der; ~s, ~** *(hist.: Pirat)* freebooter; **~bier das** free beer; **~brief der: kein ~brief für etw. sein** be no excuse for sth.; **jmdm. einen ~brief für etw. ausstellen** give sb. a licence for sth.

freien 1. *tr. V. (veralt.)* marry; wed. **2.** *itr. V.* **um ein Mädchen ~:** court *or* woo a girl

Freier der; ~s, ~ a) *(veralt.)* suitor; **b)** *(salopp: Kunde einer Dirne)* punter *(sl.)*

Freiers·füße *Pl. in* **auf ~n gehen** *(scherzh.)* be courting

frei-, Frei-: ~exemplar das *(Buch)* free copy; *(Zeitung)* free issue; **~frau die** baroness; **~gabe die a)** release; *(der Wechselkurse)* floating; **b)** *(einer Straße, Brücke usw.)* opening **(für** to); *(eines Films)* passing; **~|geben 1.** *unr. tr. V.* **a)** release ⟨*prisoner, footballer*⟩; float ⟨*exchange rates*⟩; **jmdm. den Weg ~geben** let sb. through; **b)** open ⟨*road, bridge, etc.*⟩ **(für** to); pass ⟨*film*⟩; **der Film ist ab 18 freigegeben** the film has been passed 18; **2.** *unr. tr., itr. V.* **jmdm. ~geben** give sb. time off; **~gebig** [-geːbɪç] *Adj.* generous; open-handed; **~gebigkeit die** generosity; open-handedness; **~gehege das** outdoor *or* open-air enclosure; **~|haben** *unr. tr., itr. V. (ugs.)* **ich habe [am** *od.* **den] Montag ~:** I've got Monday off; **~hafen der** free port; **~|halten** *unr. tr. V.* **a)** treat; **er hielt das ganze Lokal ~:** he stood drinks for everyone in the pub *(Brit.) or (Amer.)* bar; **b)** *(offenhalten)* keep ⟨*entrance, roadway*⟩ clear; **Einfahrt ~halten!** no parking in front of entrance; keep clear; **c)** *(reservieren)* **jmdm.** *od.* **für jmdn. einen Platz ~halten** keep a place for sb.; **~handel der** free trade; **~handels·zone die** free-trade zone; **~händig** [-hɛndɪç] **1.** *Adj.* free-hand ⟨*drawing*⟩; **2.** *adv.* ⟨*cycle*⟩ without holding on; ⟨*draw*⟩ free-hand

Freiheit die; ~, ~en a) freedom; **die persönliche ~:** personal freedom *or* liberty; **jmdm. völlige ~ lassen** give sb. a completely free hand; **b)** *(Vorrecht)* freedom; privilege; **sich** *(Dat.)* **~en herausnehmen** take liberties **(gegen** with); **die dichterische ~:** poetic licence

freiheitlich 1. *Adj.* liberal ⟨*philosophy, conscience*⟩; **~ und demokratisch** free and democratic. **2.** *adv.* liberally

Freiheits-: ~beraubung die *(jur.)* wrongful detention; **~bewegung die** liberation movement; **~entzug der** imprisonment; **~kampf der** struggle for freedom; **~kämpfer der** freedom fighter; **~liebe die;** *o. Pl.* love of freedom *or* liberty; **~liebend** *Adj.* freedom-loving; **~rechte** *Pl.* civil rights; **~statue die** Statue of Liberty; **~strafe die** *(Rechtsw.)* term of imprisonment; prison sentence

frei·heraus *Adv.* frankly; openly

frei-, Frei-: ~herr der baron; **~kämpfen** *tr. V.* liberate; **sich ~kämpfen** fight one's way out; **~karte die** complimentary *or* free ticket; **~|kaufen** *tr. V.* ransom ⟨*hostage*⟩; buy the freedom of ⟨*slave*⟩; **sich von der Verantwortung ~kaufen** *(fig.)* buy off one's responsibility; **~|kommen** *unr. itr. V.; mit sein* **aus dem Gefängnis ~kommen** be released from prison; leave prison; **aus jmds. Fängen ~kommen** escape from sb.'s clutches; **~körper·kultur die;** *o. Pl.* nudism *no art.*; naturism *no art.*; **~land·gemüse das** outdoor vegetables *pl.*; **~|lassen** *unr. tr. V.* set free; release; **~lassung die** release; **~lauf der** *(Technik)* free-wheel; **im ~lauf fahren** free-wheel; **~lebend** *Adj.* living in the wild *postpos.*; **~|legen** *tr. V.* uncover

freilich *Adv.* **a)** *(einschränkend)* **er arbeitet schnell, ~ nicht sehr gründlich** he works quickly, though admittedly he's not very thorough; **sie hat sehr viel Talent, ~ fehlt es ihr an Ausdauer** she has a great deal of talent, but she does lack staying power; **b)** *(einräumend)* **man muß ~ bedenken, daß ...:** one must of course bear in mind that ...; **~ scheinen die Tatsachen gegen meine Überlegungen zu sprechen ...:** admittedly the facts seem to contradict my ideas, but ...; **c)** *(bes. südd.: selbstverständlich)* of course; **ja ~:** [why] yes; of course

Frei·licht-: ~bühne die, ~theater das open-air *or* outdoor theatre

frei-, Frei-: ~|machen 1. *refl. V. (ugs.: frei nehmen)* **sich ~machen** take time off; **2.** *tr. V. (Postw.)* frank; **~maurer der** Freemason; **~maurer·loge die** Freemasons' lodge; **~mut der** candidness; frankness; **~mütig 1.** *Adj.* candid; frank; **2.** *adv.* candidly; frankly; **~mütigkeit die; ~:** candidness; frankness; **~raum der** *(Psych., Soziol.)* space *no indef. art.* to be oneself; **~schaf-**

fend *Adj.* free-lance; **~schärler** [-ʃɛːɐ̯lɐ] **der; ~s, ~:** irregular [soldier]; **~|schwimmen** *unr. refl. V.* pass the 15-minute swimming test; **~|setzen** *tr. V.* **a)** *(Physik, Chemie)* release ‹*energy*›; emit ‹*rays, electrons, neutrons*›; release, give off ‹*gas*›; **b)** *(Wirtsch.)* release ‹*staff*›; **~spiel das** free turn; **~|spielen** *tr. V.* *(Ballspiele)* **jmdn./sich ~spielen** create space for sb./oneself; **~|sprechen** *unr. tr. V.* **a)** *(Rechtsw.)* acquit; **jmdn. von einer Anklage ~sprechen** acquit sb. of a charge; **b)** *(fig.)* exonerate **(von** from); **~spruch der** *(Rechtsw.)* acquittal; **~|stehen** *unr. itr. V.* **a) es steht jmdm. ~, etw. zu tun** sb. is free to do sth.; **b)** ‹*flat, house*› be empty *or* vacant; ‹*storeroom etc.*› be empty; **~stehend** *Adj.* detached ‹*house*›; **~|stellen** *tr. V.* **a) jmdm. etw. ~stellen** leave sth. up to sb.; let sb. decide sth.; **b)** *(befreien)* release ‹*person*›; **jmdn. vom Wehrdienst ~stellen** exempt sb. from military service; **~stellung die** release; *(befristet)* leave

Frei·stil der; *o. Pl.* *(Sport)* **a)** *s.* **~ringen; b)** *s.* **~schwimmen**

Freistil-: ~ringen das free-style wrestling; **~schwimmen das** free-style swimming

Frei-: ~stoß der *(Fußball)* free kick; **~stunde die** *(Schulw.)* free period

Frei·tag der Friday; *s. auch* **Dienstag, Dienstag-**

freitags *Adv.* on Friday[s]; *s. auch* **dienstags**

frei-, Frei-: ~tod der *(verhüll.)* suicide *no art.*; **den ~tod wählen** choose to take one's own life; **~tragend** *Adj.* *(Bauw.)* suspended ‹*floor*›; cantilever ‹*bridge*›; **~treppe die** [flight of] steps; **~übung die;** *meist Pl.* *(Sport)* keep-fit exercise; **~umschlag der** stamped addressed envelope; s.a.e.; **~weg** *Adv.* *(ugs.)* openly; freely ‹*talk*› **sag es ~weg** say it straight out; **~wild das** fair game; **~willig 1.** *Adj.* voluntary ‹*decision*›; optional ‹*subject*›; **2.** *adv.* voluntarily; of one's own accord; **sich ~willig melden** volunteer; **~willige der/die;** *adj. Dekl.* volunteer; **~wurf der** free throw; **~zeichen das** ringing tone

Frei·zeit die; *o. Pl.* **a)** spare time; leisure time; **b)** *(Zusammenkunft)* [holiday/weekend] course; *(der Kirche)* retreat

Freizeit-: ~anzug der leisure suit; **~beschäftigung die** leisure pursuit *or* activity; **~wert der: eine Stadt mit hohem ~wert** a town with many leisure amenities

frei-, Frei-: ~zügig 1. *Adj.* **a)** *(großzügig)* generous, liberal ‹*dosage, spending*›; liberal, flexible ‹*interpretation of rule etc.*›; **b)** *(gewagt, unmoralisch)* risqué, daring ‹*remark, film, dress*›; permissive ‹*attitude*›; **2.** *adv.* **Geld ~zügig ausgeben** be generous with one's money; **ein Gesetz ~zügig auslegen** interpret a law flexibly; **~zügigkeit die; ~ a)** *(Großzügigkeit)* liberalness; *(in Geldsachen)* generosity; *(von Interpretation)* flexibility; **b)** *(von Einstellung)* permissiveness; **c)** *(freie Wahl des Wohnsitzes)* freedom of domicile

fremd [frɛmt] *Adj.* **a)** strange; foreign ‹*country, government, customs, language*›; **b)** *nicht präd.* *(nicht eigen)* other people's; of others *postpos.*; **ohne ~e Hilfe** without anyone else's help; **c)** *(unbekannt)* strange; strange, unknown ‹*surroundings*›; **sich sehr ~ fühlen** feel very much a stranger; **einander ~ werden** become estranged; grow apart

fremd·artig *Adj.* strange

Fremd·artigkeit die strangeness

[1]Fremde [ˈfrɛmdə] **der/die;** *adj. Dekl.* **a)** stranger; **b)** *(Ausländer)* foreigner; alien *(Admin. lang.)*; **c)** *(Tourist)* visitor

[2]Fremde die; ~ *(geh.)* foreign parts *pl.*; **in die ~ ziehen** go off to foreign parts; go abroad

fremden-, Fremden-: ~feindlich *Adj.* hostile to strangers/foreigners *postpos.*; **~feindlichkeit die; ~:** xenophobia; **~heim das** guest-house; boarding-house; **~legion die;** *o. Pl.* foreign legion; **~paß der** alien's passport; **~verkehr der** tourism *no art.*; **~zimmer das** room; **~zimmer frei** vacancies *(Brit.)*; vacancy *(Amer.)*

fremd|gehen *unr. itr. V.*; *mit sein* *(ugs.)* be unfaithful

Fremdheit die; ~: strangeness

fremd-, Fremd-: ~herrschaft die foreign domination *no art. or* rule *no art.*; **~körper der** *(Med., Biol.)* foreign body; **ein ~körper sein** *(fig.)* be out of place; **~ländisch** [-lɛndɪʃ] *Adj.* foreign; *(exotisch)* exotic

Fremdling der; ~s, ~e *(veralt.)* stranger

Fremd·sprache die foreign language

Fremdsprachen-: ~korrespondentin die bilingual/multilingual secretary; **~unterricht der** teaching of foreign languages

fremd-, Fremd-: ~sprachig *Adj.* foreign ‹*literature*›; foreign-language ‹*edition, teaching*›; **~sprachlich** *Adj.*; *nicht präd.* foreign-language ‹*teaching*›; foreign ‹*word*›; **~wort das;** *Pl.* **~wörter** foreign word; **Liebe ist für ihn ein ~wort** *(fig.)* he doesn't know the meaning of the word love; **~wörter·buch das** dictionary of foreign words

frenetisch [freˈneːtɪʃ] **1.** *Adj.* frenetic. **2.** *adv.* frenetically

frequentieren *tr. V.* frequent ‹*pub, café*›; use ‹*library*›

Frequenz [freˈkvɛnts] **die; ~, ~en** *(fachspr.)* frequency; *(Med.: Puls~)* rate

Fresko [ˈfrɛsko] **das; ~s, Fresken** *(Kunstwiss.)* fresco

Fressalien [frɛˈsaːli̯ən] *Pl.* *(ugs. scherzh.)* grub *(sl.)*

Fresse [ˈfrɛsə] **die; ~, ~n** *(derb)* **a)** *(Mund)* gob *(sl.)*; trap *(sl.)*; **eine große ~ haben** *(fig.)* have a big mouth *(coll.)*; **[ach] du meine ~!** bloody hell! *(sl.)*; **die ~ halten** keep one's trap *or* gob shut *(sl.)*; **b)** *(Gesicht)* mug *(sl.)*; **jmdm. die ~ polieren** smash sb.'s face in *(sl.)*

fressen 1. *unr. tr. V.* **a)** ‹*animal*› eat; *(sich ernähren von)* feed on; **sich satt ~:** eat its/her/his fill; **b)** *(ugs.: verschlingen)* swallow up ‹*money, time, distance*›; drink ‹*petrol*›; **c)** *(zerstören)* eat away; **d)** *(derb: von Menschen)* guzzle; *(fig.)* **er wird dich schon nicht ~** *(salopp)* he won't eat you *(coll.)*; **etw. ge~ haben** *(ugs.)* have understood sth.; **jmdn. ge~ haben** *(ugs.)* hate sb.'s guts *(coll.)*; **jmdn. zum Fressen gern haben** like sb. so much one could eat him/her. **2.** *unr. itr. V.* **a)** *(von Tieren)* feed; **einem Tier zu ~ geben** feed an animal; **b)** *(zerstören)* **an etw.** *(Dat.)* **~:** ‹*rust*› eat away at sth.; ‹*fire*› begin to consume sth.; **c)** *(derb: von Menschen)* stuff oneself *or* one's face *(sl.)*. **3.** *unr. refl. V.* **sich durch/in etw.** *(Akk.)* **~:** eat its way through/into sth.

Fressen das; ~s a) *(für Hunde, Katzen usw.)* food; *(für Vieh)* feed; **b)** *(derb: Essen)* grub *(sl.)*; **das ist ein gefundenes ~ für sie** *(fig.)* that's just what she needed; that's a real gift for her

Fresserei die; ~, ~en *(derb abwertend)* guzzling; stuffing; **eine große ~:** a big blow-out *(sl.)*

Freß-: ~korb der *(ugs.)* **a)** *(Ver-*

pflegungskorb) picnic basket; **b)** *(Geschenkkorb)* hamper; **~napf der** feeding-bowl; **~paket das** *(ugs.)* food parcel; **~sack der** *(derb)* greedy pig *(sl.)*

Frettchen ['frɛtçən] **das; ~s, ~** ferret

Freude ['frɔydə] **die; ~, ~n** joy; *(Vergnügen)* pleasure; *(Wonne)* delight; **~ an etw.** *(Dat.)* **haben** take pleasure in sth.; **~ am Leben haben** enjoy life; **das war eine große ~ für uns** that was a great pleasure for us; **jmdm. eine ~ machen** *od.* **bereiten** make sb. happy; **zu unserer ~:** to our delight; **mit ~n** with pleasure; **die ~n des Alltags/der Liebe** the pleasures of everyday life/the joys of love

Freuden-: ~fest das celebration; **~feuer das** bonfire; **~haus das** house of pleasure; **~mädchen das** *(verhüll.)* woman of easy virtue; **~schrei der** cry *or* shout of joy; **~tanz der** *in* **einen [wilden** *od.* **wahren] ~tanz aufführen** *od.* **vollführen** dance for joy; **~taumel der** transport of delight *or* joy; **~tränen** *Pl.* tears of happiness *or* joy

freude·strahlend *Adj.* beaming *or* radiant with joy

freudig 1. *Adj.* **a)** joyful, happy ⟨*face, feeling, greeting*⟩; joyous ⟨*heart*⟩; **in ~er Erwartung** in joyful anticipation; **b)** *(erfreulich)* delightful ⟨*surprise*⟩; **ein ~es Ereignis** *(verhüll.)* a happy event. **2.** *adv.* **~ erregt** happy and excited; **von etw. ~ überrascht sein** be surprised and delighted about sth.; **etw. ~ erwarten** look forward to sth. with pleasure

freud·los 1. *Adj.* joyless ⟨*days, existence*⟩; cheerless ⟨*surroundings*⟩. **2.** *adv.* joylessly

freuen ['frɔyən] **1.** *refl. V.* be pleased *or* glad **(über +** *Akk.* about); *(froh sein)* be happy; **sich zu früh ~:** get carried away *or* rejoice too soon; **sich auf etw.** *(Akk.)***/jmdn. ~:** look forward to sth./to seeing sb.; **sich mit jmdm. ~:** rejoice with sb. **2.** *tr. V.* please; **es freut mich, daß ...:** I am pleased *or* glad that ...; **freut mich!** pleased to meet you

Freund der; ~es, ~e a) friend; **alter ~!** *(ugs. scherzh. drohend)* mate!; **b)** *(Geliebter)* boy-friend; *(älter)* gentleman-friend; **c)** *(Anhänger, Liebhaber)* lover; **ich bin kein ~ von großen Worten** *(fig.)* I am not one for fine words

Freundchen das; ~s, ~ *(Anrede; scherzh. drohend)* mate

Freundes·kreis der circle of friends; **im engen ~kreis** among close friends

Freundin die; ~, ~nen a) friend; **b)** *(Geliebte)* girl-friend; *(älter)* lady-friend; **c)** *s.* **Freund c**

freundlich 1. *Adj.* **a)** kind ⟨*face*⟩; kind, friendly ⟨*reception*⟩; friendly ⟨*smile*⟩; fond ⟨*farewell*⟩; **zu jmdm. ~ sein** be kind to sb.; **er war so ~, mir zu helfen** he was kind *or* good enough to help me; **würden Sie bitte so ~ sein und das Fenster schließen?** would you be so kind *or* good as to close the window?; **b)** *(angenehm)* pleasant ⟨*weather, surroundings*⟩; pleasant, congenial ⟨*atmosphere*⟩; pleasant, mild ⟨*climate*⟩; **c)** *(freundschaftlich)* friendly, amiable ⟨*person, manner*⟩; friendly ⟨*disposition, attitude, warning*⟩. **2.** *adv.* **jmdn. ~ begrüßen** greet sb. amiably; **jmdm. ~ danken** thank sb. kindly; **jmdm. ~ gesinnt sein** be well-disposed towards sb.

freundlicher·weise *Adv.* kindly

Freundlichkeit die; ~, ~en a) kindness; **jmdm. ein paar ~en sagen** make a few kind remarks to sb.; **b)** *o. Pl. (angenehme Art)* pleasantness; friendliness; *(eines Zimmers, Hauses)* cheerfulness

Freundschaft die; ~, ~en friendship; **mit jmdm. ~ schließen** make *or* become friends with sb.; **jmdm. etw. in aller ~ sagen** tell sb. sth. as a friend

freundschaftlich 1. *Adj.* friendly; amicable. **2.** *adv.* in a friendly way; amicably

Freundschafts-: ~besuch der *(bes. Politik)* goodwill visit; **~dienst der** service rendered out of friendship; **~spiel das** *(Sport)* friendly match *or* game; friendly *(coll.)*; **~vertrag der** *(Politik)* treaty of friendship

Frevel ['fre:fl̩] **der; ~s, ~** *(geh., veralt.)* crime; outrage; **~ gegen Gott** sacrilege

frevelhaft *(geh.)* **1.** *Adj.* wicked ⟨*deed, rebellion, person*⟩; criminal ⟨*stupidity*⟩. **2.** *adv.* wickedly

freveln *itr. V. (geh.)* **an jmdm./gegen etw. ~:** commit a crime against sb./sth.

Friede ['fri:də] **der; ~ns, ~n a)** *(älter, geh.) s.* **Frieden; b)** *(geh.)* **~ seiner Asche** *(Dat.)* God rest his soul; **~ auf Erden** peace on earth

Frieden der; ~s, ~ peace; **~ schließen/stiften** make peace; **mit jmdm. ~ schließen** make one's peace with sb.; **mitten im ~:** in the middle of peace-time; **um des lieben ~s willen** for the sake of peace and quiet; **laß mich in ~!** *(ugs.)* leave me in peace!; leave me alone!; **ich traue dem ~ nicht** *(ugs.)* it's too good to be true

friedens-, Friedens-: ~bedingungen *Pl.* peace terms; terms for peace; **~bewegung die** peace movement; **~bruch der** violation of the peace; **~diktat das** dictated peace terms *pl.*; **~forschung die** peace studies *pl., no art.*; **~göttin die** goddess of peace; **~konferenz die** peace conference; **~liebe die** love of peace; **~nobelpreis der** Nobel Peace Prize; **~pfeife die** pipe of peace; **~politik die** policy of peace; **~richter der** *lay magistrate dealing with minor offences;* ≈ Justice of the Peace; **~schluß der** peace settlement; **~sicherung die** peace-keeping; **~stifter der** peacemaker; **~taube die** dove of peace; **~truppe die** peace-keeping force; **~verhandlungen** *Pl.* peace negotiations; peace talks; **~vertrag der** peace treaty; **~zeiten** *Pl.* peacetime *sing.*

fried·fertig *Adj.* peaceable ⟨*person, character*⟩; peaceful ⟨*intentions*⟩

Fried·fertigkeit die; ~: peaceableness

Fried·hof der cemetery; *(Kirchhof)* graveyard; churchyard

Friedhofs-: ~gärtner der cemetery gardener; **~kapelle die** cemetery chapel

friedlich ['fri:tlɪç] **1.** *Adj.* **a)** peaceful; **auf ~em Wege** by peaceful means; **b)** *(ruhig, verträglich)* peaceable, peaceful ⟨*character, person*⟩; peaceful, tranquil ⟨*life, atmosphere, valley*⟩; **sei ~!** *(ugs.)* be quiet!. **2.** *adv.* ⟨*live, sleep*⟩ peacefully

fried·liebend *Adj.* peace-loving

Friedrich Wilhelm der; ~ ~s, ~ ~s *(ugs. scherzh.: Unterschrift)* monicker *(coll. joc.)*

frieren ['fri:rən] *unr. itr. V.* **a)** be *or* feel cold; **erbärmlich/sehr ~:** be freezing/terribly cold; **er fror an den Händen** he had [freezing] cold hands; *unpers.:* **jmdn. friert [es]** sb. is cold; **b)** *mit sein (ge~)* freeze; **das Wasser ist gefroren** the water is *or* has frozen; **steif gefroren sein** be frozen stiff; **blau gefroren sein** be blue with cold; **c)** *(unpers.)* **es friert** it is freezing; *s. auch* **Stein**

frigid[e] [fri'gi:d(ə)] *Adj.* frigid

Frikadelle [frika'dɛlə] **die; ~, ~n** rissole

Frikassee [frika'se:] **das; ~s, ~s** *(Kochk.)* fricassee

frisch [frɪʃ] **1.** *Adj.* **a)** fresh; new-laid ‹*egg*›; fresh, clean ‹*linen*›; clean ‹*underwear*›; wet ‹*paint*›; **mit ~en Kräften** with renewed strength; **sich ~ machen** freshen oneself up; **jmdn. auf ~er Tat ertappen** catch sb. red-handed; **b)** *(munter)* fresh; **~ und munter sein** *(ugs.)* be bright and cheerful. **2.** *adv.* freshly; **~ gewaschen sein** ‹*person*› have just had a wash; ‹*garment*› have just been washed; **~ von der Universität kommen** have come straight from the university; **~ gestrichene Bänke** newly painted seats; **„Vorsicht, ~ gestrichen!"** 'wet paint'; **die Betten ~ beziehen** put fresh *or* clean sheets on the beds; **~ gewagt ist halb gewonnen** *(Spr.)* nothing ventured, nothing gained *(prov.)*

Frische die; ~: freshness; **geistige ~:** mental alertness; **körperliche ~:** physical fitness; vigour

frisch-, Frisch-: ~fisch der fresh fish; **~fleisch das** fresh meat; **~gebacken** *Adj.; nicht präd. (ugs.)* **ein ~gebackenes Ehepaar** a newly-wed couple; newly-weds *pl.*; **ein ~gebackener Doktor** a newly-qualified doctor; **~gemüse das** fresh vegetables *pl.*; **~halte·beutel der** airtight bag; **~halte·packung die** airtight pack; **~käse der** curd cheese

Frischling der; ~s, ~e a) *(Jägerspr.)* young boar; **b)** *(scherzh.)* new boy *or* girl

Frisch-: ~luft die fresh air **~milch die** fresh milk; **~obst das** fresh fruit; **~zelle die** *(Med.)* living cell

Friseur [fri'zø:ɐ̯] **der; ~s, ~e** hairdresser; *(Herren~)* hairdresser; barber; *s. auch* **Bäcker**

Friseurin [fri'zø:rɪn], **die; ~, ~nen** hairdresser

Friseur·salon der hairdressing *or* hairdresser's salon *(Brit.)*; beauty salon *(Amer.)*; *(für Herren)* barber shop *(Amer.)*

Friseuse [fri'zø:zə] **die; ~, ~n** hairdresser

Frisier·creme die hair cream

frisieren [fri'zi:rən] *tr. V.* **a) jmdn./sich ~:** do sb.'s/one's hair; **sich ~ lassen** have one's hair done; **b)** *(ugs.: verfälschen)* doctor ‹*reports, statistics*›; fiddle *(coll.)* ‹*accounts*›; **c)** *(Kfz-W.)* soup up *(coll.)* ‹*engine, vehicle*›

Frisör *s.* **Friseur**

friß [frɪs] *Imperativ Sg. v.* **fressen**

Frist [frɪst] **die; ~, ~en** time; period; **[sich** *(Dat.)***] eine ~ von 3 Wochen setzen** set [oneself] a time limit of 3 weeks; **die ~ verlängern** extend the deadline; **in kürzester ~:** within a very short time; **jmdm. eine ~ von drei Tagen geben** give sb. three days' time; **eine letzte ~** *(Aufschub)* a final extension

fristen *tr. V.* **ein kümmerliches Dasein** *od.* **Leben ~:** eke out a wretched existence; barely manage to survive

frist-: ~gemäß, ~gerecht *Adj., adv.* within the specified time *postpos.*; *(bei Anmeldung usw.)* before the closing date *postpos.*; **~los 1.** *Adj.* instant ‹*dismissal*›; **2.** *adv.* without notice; **jmdm. ~ kündigen** dismiss sb. without notice; **jmdm. ~los die Wohnung kündigen** ask sb. to quit without notice

Frisur [fri'zu:ɐ̯] **die; ~, ~en** hairstyle; hair-do *(coll.)*

Friteuse [fri'tø:zə] **die; ~, ~n** deep fryer

fritieren [fri'ti:rən] *tr. V.* deep-fry

Fritte ['frɪtə] **die; ~, ~n** *(ugs.)* chip

frivol [fri'vo:l] *Adj.* **a)** *(schamlos)* suggestive ‹*picture, etc.*›; risqué ‹*remark, joke*›; earthy ‹*man*›; flighty ‹*woman*›; **b)** *(leichtfertig)* frivolous; irresponsible

Frivolität [frivoli'tɛ:t] **die; ~, ~en a)** *o. Pl. s.* **frivol a:** suggestiveness; risqué nature; earthiness; flightiness; **b)** *(frivole Bemerkung)* risqué remark

froh [fro:] *Adj.* **a)** *(glücklich)* happy; cheerful ‹*person, mood*›; **jmdn. ~ machen** make sb. happy; cheer sb. up; **b)** *(ugs.: erleichtert)* pleased, glad **(über** + *Akk.* about); **du kannst ~ sein, daß ...:** you can be thankful *or* glad that ...; **da bin ich aber ~ [, daß...]** I am glad [that ...]; **seines Lebens nicht mehr ~ werden** not enjoy life any more; **c)** *nicht präd. (erfreulich)* good ‹*news*›; happy ‹*event*›

froh-: ~gelaunt *Adj.* cheerful; **~gemut 1.** *Adj.* happy; **2.** *adv.* happily; in good spirits

fröhlich ['frø:lɪç] **1.** *Adj.* cheerful; happy; **~es Treiben** merry-making. **2.** *adv.* *(unbekümmert)* blithely; cheerfully

Fröhlichkeit die; ~: cheerfulness; *(eines Festes, einer Feier)* gaiety

froh·locken *itr. V. (geh.)* rejoice; exult; **frohlocket dem Herrn** sing joyfully unto the Lord

Froh-: ~natur die *(Mensch)* cheerful person; **~sinn der;** *o. Pl.* cheerfulness; gaiety

fromm [frɔm]; **~er** *od.* **frömmer** ['frœmɐ], **~st...** *od.* **frömmst... 1.** *Adj.* **a)** pious, devout ‹*person*›; devout ‹*life, Christian*›; **b)** *(scheinheilig)* **~es Getue** pious affectation; **c)** *(wohlgemeint)* **eine ~e Lüge** a white lie; **ein ~er Wunsch** a pious hope. **2.** *adv.* piously

Frömmelei die; ~*(abwertend)* affected piety

Frömmigkeit ['frœmɪçkait] **die; ~:** piety; devoutness

Frömmler ['frœmlɐ] **der; ~s, ~** *(abwertend)* [pious] hypocrite

Fron [fro:n] **die; ~, ~en a)** *(hist.)* corvée; **b)** *(geh.: aufgezwungene Mühsal)* drudgery

frönen ['frø:nən] *itr. V. (geh.)* **einer Neigung/einem Laster ~:** indulge an inclination/in a vice

Fron·leichnam *o. Art.* [the feast of] Corpus Christi

Front [frɔnt] **die; ~, ~en a)** *(Gebäude~)* front; façade; **b)** *(Kampfgebiet)* front [line]; **an die ~ gehen/an der ~ sein** go to the front/fight at the front; **c)** *(Milit.: vorderste Linie)* front line; **in vorderster ~ kämpfen** fight at the very front; **die ~en haben sich verhärtet** *(fig.)* attitudes have hardened; **an zwei ~en kämpfen** *(fig.)* fight on two fronts; **d)** *(Milit.: einer Truppe)* **die ~ abnehmen/abschreiten** inspect the troops/guard of honour *etc.*; **gegen jmdn./etw. ~ machen** *(fig.)* make a stand against sb./sth.; **e)** *(Sport)* **in ~ liegen/gehen** be in front *or* in the lead/go in front; **f)** *(Met.)* front

frontal [frɔn'ta:l] **1.** *Adj.; nicht präd.* **a)** *(von vorn)* head-on ‹*collision*›; **b)** *(nach vorn)* frontal ‹*attack*›. **2.** *adv.* ‹*collide*› head-on; ‹*attack*› from the front

Frontal-: ~angriff der frontal attack; **~zusammenstoß der** head-on collision

Front-: ~antrieb der *(Kfz-W.)* front-wheel drive; **~scheibe die** windscreen (Brit.); windshield *(Amer.)*; **~urlaub der** *(Milit.)* leave from the front; **~wechsel der** *(fig.)* U-turn; volte-face

fror [fro:ɐ̯] *1. u. 3. Pers. Sg. Prät. v.* **frieren**

Frosch [frɔʃ] **der; ~[e]s, Frösche** ['frœʃə] **a)** frog; **sei kein ~** *(ugs.)* don't be a spoilsport; **b)** *(Musik)* nut

Frosch-: ~könig der Frog Prince; **~mann der;** *Pl.* **~männer** frogman; **~perspektive die** worm's-eye view; **~schenkel der.** frog's leg

Frost [frɔst] **der; ~[e]s, Fröste** ['frœstə] frost; **es herrscht [strenger] ~:** there is a [severe] frost; it is [very] frosty

frost·beständig *Adj.* frost-resistant

Frost·beule die chilblain

frösteln ['frœstl̩n] *itr. V.* feel chilly; **vor Kälte ~:** shiver with cold; *unpers.:* **es fröstelt ihn, ihn fröstelt** he feels chilly

Froster der; ~s, ~: freezing compartment

frostig ['frɔstɪç] **1.** *Adj. (auch fig.)* frosty. **2.** *adv.* frostily

Frost-: ~schaden der frost damage; **~schutz der** frost protection; protection from frost; **~schutz·mittel das a)** frost protection agent; **b)** *(Kfz-W.)* anti-freeze; **~warnung die** *(Met.)* frost warning

Frottee [frɔ'te:] **das** u. **der; ~s, ~s** terry towelling

Frottee-: ~handtuch das terry towel; **~kleid das** towelling dress

frottieren [frɔ'ti:rən] *tr. V.* rub; towel; **sich ~:** rub oneself down

Frotzelei die; ~, ~en *(ugs.)* **a)** *o. Pl.* teasing; **b)** *(Bemerkung)* teasing remark

frotzeln ['frɔts̮l̩n] **1.** *tr. V.* tease. **2.** *itr. V.* **über jmdn./etw. ~:** make fun of sb./sth.

Frucht [frʊxt] **die; ~, Früchte** ['frʏçtə] **a)** *(auch fig. geh.)* fruit; **Früchte tragen** *(auch fig.)* bear fruit; **b)** *o. Pl. (landsch.: Getreide)* corn; crops *pl.*

frucht·bar *Adj.* fertile ⟨*soil, field, man, woman*⟩; prolific ⟨*breed*⟩; fruitful ⟨*work, idea*⟩; fruitful, rewarding ⟨*conversation*⟩; **eine Idee** *usw.* **für etw. ~ machen** allow sth. to benefit from an idea *etc.*

Fruchtbarkeit die; ~ *s.* **fruchtbar:** fertility; prolificness; fruitfulness

Frucht-: ~becher der a) *(Eisbecher)* fruit sundae; **b)** *(Bot.)* cupule; **~blase die** *(Anat.)* amniotic sac; **~bonbon das** *od.* **der** fruit drop

Früchtchen ['frʏçtçən] **das; ~s, ~** *(ugs. abwertend: Tunichtgut)* good-for-nothing

Frucht·eis das fruit ice-cream

fruchten *itr. V.* **nichts ~:** be [of] no use; be of no avail; **[bei jmdm.] nicht[s] ~:** have no effect [on sb.]

Frucht·fleisch das flesh; pulp

fruchtig *Adj.* fruity

frucht-, Frucht-: ~joghurt der *od.* **das** fruit yoghurt; **~los** *Adj.* fruitless, vain ⟨*efforts*⟩; **~saft der** fruit juice; **~wasser das** *(Anat.)* amniotic fluid; waters *pl. (coll.)*; **~zucker der** fruit sugar; fructose

früh [fry:] **1.** *Adj.* **a)** early; **am ~en Morgen** early in the morning; **b)** *(vorzeitig)* premature; **ein ~es Ende finden** come to an untimely end; **einen ~en Tod sterben** die an untimely *or* premature death. **2.** *adv.* **a)** early; **~ am Tage** early in the day; **~ genug kommen** arrive in [good] time; **~er oder später** sooner or later; **b)** *(morgens)* in the morning; **heute/morgen/gestern ~:** this/tomorrow/yesterday morning; **von ~ bis spät** from morning till night; from dawn to dusk; *s. auch* **früher**

früh-, Früh-: ~auf *in* **von ~auf** from early childhood on[wards]; **~aufsteher der; ~s, ~:** early riser; early bird *(coll.)*; **~dienst der** early duty; *(im Betrieb)* early shift

Frühe die; ~: in der ~: *(geh.)* in the early morning; **in aller ~:** at the crack of dawn

früher *Adv.* formerly; **~ war er ganz anders** he used to be quite different at one time; **meine Bekannten von ~:** my former acquaintances; **ich kenne ihn [noch] von ~ [her]** I know him from some time ago; **an ~ denken** think back

früher... *Adj., nicht präd.* **a)** *(vergangen)* earlier; former; **in ~en Zeiten** in the past; in former times; **aus ~en Jahrhunderten** from past centuries; **b)** *(ehemalig)* former ⟨*owner, occupant, friend*⟩

Früh·erkennung die *(Med.)* early recognition *or* diagnosis

frühestens ['fry:əstn̩s] *Adv.* at the earliest

frühest·möglich ['fry:əst-'mø:klɪç] *Adj.; nicht präd.* earliest possible

Früh-: ~geburt die a) premature birth; **b)** *(Kind)* premature baby; **~invalide der/die** premature invalid

Früh·jahr das spring

Frühjahrs-: ~müdigkeit die springtime tiredness; **~putz der** spring-cleaning

Früh-: ~kapitalismus der early capitalism *no art.*; **~kartoffel die** early potato

Frühling ['fry:lɪŋ] **der; ~s, ~e** spring; **im ~:** in [the] spring; **der ~ kommt** spring is coming; **im ~ des Lebens** *(geh.)* in the springtime of one's life; **seinen zweiten ~ erleben** *(fig. iron.)* relive one's youth

frühlings-, Frühlings-: ~anfang der first day of spring; **~haft** *Adj.* springlike; **~tag der** spring day

früh-, Früh-: ~messe, ~mette die *(kath. Kirche)* early [morning] mass; **~morgens** [-'--] *Adv.* early in the morning; **~nebel der** early morning fog/mist; **~reif** *Adj.* precocious ⟨*child*⟩; **~rentner der** *person who has retired early;* **~rentner werden/sein** retire/have retired early; **~schicht die** early shift; **~schoppen der** morning drink; *(um Mittag)* lunchtime drink; **~sport der** early-morning exercise; **~stadium das** early stage; **~start der** *(Sport)* false start

Früh·stück das; ~s, ~e breakfast; **zweites ~:** mid-morning snack

frühstücken 1. *itr. V.* breakfast; have breakfast; **ausgiebig ~:** have a hearty breakfast. **2.** *tr. V.* **etw. ~:** breakfast on sth.; have sth. for breakfast

Frühstücks-: ~fernsehen das breakfast television; **~fleisch das** luncheon meat; **~pause die** morning break; coffee break

früh-, Früh-: ~verstorben *Adj. (präd. getrennt geschrieben)* **seine ~verstorbene Mutter** his mother, who died young; **~werk das** early work; *(gesamtes)* early works *pl.*; **~zeit die** early period; **~zeitig 1.** *Adj.* early; *(vorzeitig)* premature; untimely ⟨*death*⟩; **2.** *adv.* early; *(im Leben, in der Entwicklung)* at an early stage; *(vorzeitig)* prematurely; **jmdn. ~zeitig benachrichtigen** let someone know in good time; **~zug der** early [morning] train

Frust [frʊst] **der; ~[e]s** *(ugs.)* frustration; **ihre Arbeit war der absolute ~:** her work was a real drag *(coll.)*; **der große ~ überkam ihn** he began to feel really browned off *(coll.)*

Frustration [frʊstra'ts̮io:n] **die; ~, ~en** *(Psych.)* frustration

frustrieren [frʊs'tri:rən] *tr. V.* frustrate

FU[B] ['ɛf|'u:('be:)] **die; ~** *Abk.* **Freie Universität [Berlin]**

Fuchs [fʊks] **der; ~es, Füchse** ['fʏksə] **a)** *(auch Pelz)* fox; **dort sagen sich ~ und Hase** *od.* **die Füchse gute Nacht** *(scherzh.)* it's in the middle of nowhere *or* at the back of beyond; **b)** *(ugs.: schlauer Mensch)* **ein [schlauer] ~:** a sly *or* cunning devil; **c)** *(Pferd)* chestnut; *(heller)* sorrel

Fuchs·bau der; *Pl.* **~baue** foxden

fuchsen 1. *tr. V.* annoy; vex. **2.** *refl. V.* **sich [über etw.** *(Akk.)***] ~:** be annoyed [about sth.]

Fuchsie ['fʊksi̯ə] **die; ~, ~n** *(Bot.)* fuchsia

Füchsin ['fʏksɪn] **die; ~, ~nen** vixen

fuchs-, Fuchs-: ~jagd die fox-hunt; *(Schleppjagd)* drag-hunt; **~pelz der** fox-fur; **~rot** *Adj.* ginger; **~schwanz der a)** [fox's] brush; foxtail; **b)** *(Bot.)* amaranth; love-lies-bleeding; **c)** *(Werkzeug)* handsaw; **~teu·fels·wild** *Adj. (ugs.)* livid *(coll.);* hopping mad *(coll.)*

Fuchtel ['fʊxtl̩] **die; ~, ~n** *o. Pl. (ugs.: strenge Zucht)* **jmdn. unter der/seiner ~ haben/halten** have/keep sb. under one's thumb

fuchteln *itr. V.* **mit etw. ~** *(ugs.)* wave sth. about

Fuder ['fu:dɐ] **das; ~s, ~ a)** *(Wagenladung)* cart-load; **b)** *(ugs.: große Menge)* load *(coll.)*

Fuffziger der; ~s, ~ *(ugs.)* fifty-pfennig piece; **ein falscher ~** *(salopp)* a real crook

Fug [fu:k] **der** *in* **mit ~ [und Recht]** rightly; justifiably

¹Fuge ['fu:gə] **die; ~, ~n** joint; *(Zwischenraum)* gap; **der Tisch kracht in allen ~n** *(ugs.)* every joint in the table creaks; **aus den ~n gehen** *od.* **geraten/sein** *(fig.)* be turned completely upside down *(fig.)*

²Fuge die; ~, ~n *(Musik)* fugue

fügen ['fy:gn̩] **1.** *tr. V.* **a)** *(hinzu~)* place; set; **Wort an Wort ~:** string words together; **b)** *(geh.: zusammen~)* put together; **lose gefügte Bretter** loosely jointed boards; **c)** *(geh.: bewirken)* ⟨*fate*⟩ ordain, decree; ⟨*person*⟩ arrange. **2.** *refl. V.* **a)** *(sich ein~)* **sich in etw.** *(Akk.)* **~:** fit into sth.; **b)** *(gehorchen)* **sich ~:** fall into line; **sich jmdm./einer Sache** *(Dat.)* **~:** fall into line with sb./sth.; **er muß lernen, sich zu ~:** he must learn to toe the line; **sich in sein Schicksal ~:** submit to *or* accept one's fate; **c)** *(geh.: geschehen)* **es fügt sich gut, daß ...:** it is fortunate that ...

fügsam 1. *Adj.* obedient. **2.** *adv.* obediently

Fügsamkeit die; ~: obedience

Fügung die; ~, ~en a) eine ~ Gottes divine providence; **eine ~ des Schicksals** a stroke of fate; **b)** *(Sprachw.)* construction

fühlbar 1. *Adj.* **a)** noticeable; **b)** *(wahrnehmbar)* perceptible. **2.** *adv.* **a)** noticeably; **b)** *(wahrnehmbar)* perceptibly

fühlen ['fy:lən] **1.** *tr. V.* feel. **2.** *refl. V.* **sich krank/bedroht/schuldig ~:** feel sick/threatend/guilty; **sich zu etw. berufen ~:** feel called to be sth.; **sich als Künstler ~:** feel oneself to be an artist; feel one is an artist. **3.** *itr. V.* feel; **nach etw. ~:** feel for sth.

Fühler der; ~s, ~ feeler; antenna; **seine/die ~ ausstrecken** *(fig.)* put out feelers

Fühlung die; ~: contact; **mit jmdm. ~ bekommen/[auf]nehmen** get into contact with sb.

Fühlungnahme die; ~: initial contact

fuhr [fu:ɐ̯] *1. u. 3. Pers. Sg. Prät. v.* **fahren**

Fuhre ['fu:rə] **die; ~, ~n a)** *(Wagenladung)* load; **b)** *(Transport)* trip; journey; *(mit Taxi)* fare

führen ['fy:rən] **1.** *tr. V.* **a)** lead; **ein Tier an der Leine ~:** walk an animal on a lead; **jmdn. durch eine Stadt ~:** show sb. around a town; **durch das Programm führt [Sie] Klaus Frank** Klaus Frank will present the programme; **jmdn. auf die richtige Spur ~:** put sb. on the right track; **b)** *(Kaufmannsspr.)* stock, sell ⟨*goods*⟩; **c)** *(durch~)* **Gespräche ~:** hold conversations; **ein Orts-/Ferngespräch ~:** make a local/long-distance call; **ein unruhiges Leben ~:** lead a turbulent life; **eine glückliche Ehe ~:** be happily married; **einen Prozeß [gegen jmdn.] ~:** take legal action [against sb.]; **d)** *(verantwortlich leiten)* manage, run ⟨*company, business, pub, etc.*⟩; lead ⟨*party, country*⟩; command ⟨*regiment*⟩; chair ⟨*committee*⟩; **eine Reisegruppe ~:** be courier to a group of tourists; **e)** *(gelangen lassen)* ⟨*journey, road*⟩ take; **was führt Sie zu mir?** what brings you to me?; **f)** *(Amtsspr.)* drive ⟨*train, motor, vehicle*⟩; navigate ⟨*ship*⟩; fly ⟨*aircraft*⟩; **g)** *(verlaufen lassen)* take ⟨*road, cable, etc.*⟩; **h)** *(als Kennzeichnung, Bezeichnung haben)* bear; **einen Titel/Künstlernamen ~:** have a title/use a stage name; **den Titel „Professor" ~:** use the title of professor; **i)** *(angelegt haben)* keep ⟨*diary, list, file*⟩; **j)** *(befördern)* carry; **der Zug führt einen Speisewagen** the train has a dining-car; **der Fluß führt Hochwasser** the river is in flood; **k)** *(registriert haben)* **jmdn. in einer Kartei ~:** have sb. on file; **l)** *(tragen)* **etw. bei** *od.* **mit sich ~:** have sth. on one; **eine Waffe bei sich ~:** carry a weapon. **2.** *itr. V.* **a)** lead; **die Straße führt nach .../durch .../über ...:** the road leads *or* goes to .../goes through .../goes over ...; **das würde zu weit ~** *(fig.)* that would be taking things too far; **b)** *(an der Spitze liegen)* lead; be ahead; **nach Punkten ~:** be ahead on points; **in der Tabelle ~:** be the league leaders; be at the top of the league; **c) zu etw. ~** *(etw. bewirken)* lead to sth.; **zum Ziel ~:** bring the desired result; **das führt zu nichts** *(ugs.)* that won't get you/us *etc.* anywhere *(coll.).* **3.** *refl. V.* **sich gut/schlecht ~:** conduct oneself *or* behave well/badly

führend *Adj.* leading ⟨*politician, figure, role*⟩; prominent ⟨*position*⟩; **auf einem Gebiet ~ sein** be a leader in a field

Führer der; ~s, ~ a) leader; **der ~** *(ns.)* the Führer; **b)** *(Fremden~)* guide; **c)** *(Handbuch)* guide, guidebook](**durch** to)

Führer·haus das driver's cab

Führerin die; ~, ~nen a) leader; **b)** *(Fremden~)* guide

führer·los 1. *Adj.* **a)** leaderless; **b)** *(ohne Lenker)* driverless ⟨*car*⟩; pilotless ⟨*aircraft*⟩; **2.** *adv.* without a leader; **b)** *(ohne Lenker) s.* **1 b):** without a driver; without a pilot

Führer·schein der driving licence *(Brit.);* driver's license *(Amer.);* **den ~ machen** *(ugs.)* learn to drive

Fuhr·park der transport fleet

Führung die; ~, ~en a) *o. Pl. s.* **führen 1 d:** management; running; leadership; command; chairmanship; **b)** *(Fremden~)* guided tour; **c)** *o. Pl. (führende Position)* lead; **in etw.** *(Dat.)* **die ~ haben** be leading *or* the leader/leaders in sth.; **in ~ liegen/gehen** *(Sport)* be in/go into the lead; **d)** *o. Pl. (Erziehung)* guidance; **eine feste ~:** a firm hand; firm guidance; **e)** *o. Pl. (leitende Gruppe)* leaders *pl.; (einer Partei)* leadership; *(einer Firma)* directors *pl.; (eines Regiments)* commanders *pl.;* **f)** *o. Pl. (Betragen)* conduct; **g)** *o. Pl. (eines Registers, Protokolls usw.)* keeping

Führungs-: ~anspruch der claim to leadership; **einen ~anspruch erheben** lay claim to the leadership; **~aufgabe die** *(im Betrieb)* management function; *(Politik)* leadership function; **~kraft die** manager; **~spitze die** *(Politik)* top leadership; *(im Betrieb)* top management; **~zeugnis das** *document issued* *by police certifying that holder has no criminal record*

Fuhr-: ~unternehmen das haulage business; **~unternehmer der** haulage contractor; **~werk das** cart *(drawn by horse[s], ox[en], etc.)*

Fülle ['fʏlə] **die; ~ a)** wealth; abundance; **eine ~ von Arbeit** an enormous amount of work; **in ~:** in plenty; in abundance; **b)** *(Körper~)* corpulence

füllen 1. *tr. V.* **a)** fill; **bis zum Rand gefüllt sein** be full to the brim; **der Saal ist bis auf den letzten Platz gefüllt** the hall is completely full; *s. auch* **gefüllt; b)** *(fig.)* fill in ‹*gap, time*›; **c)** *(mit einer Füllung versehen)* stuff ‹*fowl, tomato, apple, mattress, toy*›; fill ‹*tooth*›; *s. auch* **gefüllt; d)** *(schütten)* pour; **etw. in Flaschen/Säcke ~:** bottle sth./put sth. into sacks; **e)** *(einnehmen)* fill ‹*space etc.*›. **2.** *refl. V.* fill [up]; **sich mit etw. ~:** fill up with sth.

Füller der; ~s, ~, Füll·federhalter der; ~s, ~ [fountain-]pen

Füll·gewicht das net weight

füllig *Adj.* corpulent, portly ‹*person*›; ample, portly ‹*figure*›; full ‹*face*›; ample ‹*bosom*›

Füllung die; ~, ~en a) *(in Geflügel, Paprika usw., in Kissen, Matratzen)* stuffing; *(in Pasteten, Kuchen)* filling; *(in Schokolade, Pralinen)* centre; **b)** *(Zahnmed.)* filling; **c)** *(Teil der Tür)* panel

Füll·wort das; *Pl.* **-wörter** filler; *(Sprachw., Literaturw.)* expletive

Fummel [fʊml̩] **der; ~s, ~** *(salopp)* rags *pl.*

Fummelei die; ~, ~en *(ugs.)* **a)** twiddling; **das ist eine furchtbare ~:** it's terribly fiddly; **b)** *(Petting)* petting; groping *(coll.)*

fummeln *itr. V.* **a)** *(ugs.: fingern)* fiddle; **an etw.** *(Dat.)* **~:** fiddle [around] with sth.; **nach etw. ~:** grope for *or* feel for sth.; **b)** *(ugs.: erotisch)* pet

Fund [fʊnt] **der; ~[e]s, ~e** find

Fundament [fʊnda'mɛnt] **das; ~[e]s, ~e a)** *(Bauw.)* foundations *pl.*; **das ~ legen** *od.* **mauern** lay the foundations; **etw. in seinen ~en erschüttern** *(fig.)* strike at the very foundations of sth.; **b)** *(Basis)* base; basis

fundamental [fʊndamɛn'ta:l] *Adj.* fundamental

Fund-: ~büro das lost property office *(Brit.)*; lost and found office *(Amer.)*; **~grube die** treasure-house

fundieren [fʊn'di:rən] *tr. V.* underpin; **ein wissenschaftlich fundierter Vortrag** a scientifically sound lecture

fündig ['fʏndɪç] *Adj.* **~ sein** yield something; **~ werden** make a find; *(bei Bohrungen)* make a strike

Fund-: ~ort der, ~stelle die place *or* site where sth. is/was found

Fundus ['fʊndʊs] **der; ~, ~ a)** *(Requisition)* equipment store; **b)** *(Bestand)* **einen ~ von/an etw.** *(Dat.)* **haben** have a fund of sth.

fünf [fʏnf] *Kardinalz.* five; **~[e] gerade sein lassen** *(fig. ugs.)* let sth. pass; **man muß manchmal ~[e] gerade sein lassen** *(ugs.)* one has to turn a blind eye sometimes; **[um] ~ Minuten vor zwölf** *(fig.)* at the eleventh hour; at the last minute; *s. auch* [1]**acht; Sinn a**

Fünf die; ~, ~en five; **eine ~ schreiben/bekommen** *(Schulw.)* get an E; *s. auch* [1]**Acht a, d, e**

fünf-, Fünf- (*s. auch* **acht-, Acht-**): **~eck das; ~s, ~e** pentagon; **~eckig** *Adj.* pentagonal; five-cornered

Fünfer der; ~s, ~ *(ugs.)* **a)** *(Geldschein, Münze)* five; **b)** *(ugs.: Ziffer)* five; **c)** *(Lottogewinn)* five out of six; **d)** *(ugs.: Sprungturm)* five-metre platform

fünf-, Fünf-: ~fach *Vervielfältigungsz.* fivefold; quintuple; *s. auch* **achtfach; ~fache das;** *adj. Dekl.* five times as much; quintuple; *s. auch* **Achtfache; ~hundert** *Kardinalz.* five hundred; **~hundert·jahr·feier die** quincentenary; **~jahr[es]·plan der** five-year plan; **~jährig** *Adj.* *(~ Jahre alt)* five-year-old; *(~ Jahre dauernd)* five-year; **~kampf der** *(Sport)* pentathlon; **~kämpfer der, ~kämpferin die** *(Sport)* pentathlete; **~köpfig** *Adj.* ‹*family, crew*› of five; five-headed ‹*monster*›;

Fünfling ['fʏnflɪŋ] **der; ~s, ~e** quintuplet; quin *(coll.)*

fünf-, Fünf-: ~mal *Adv.* five times; *s. auch* **achtmal; ~mark·stück das** five-mark piece; **~meter·raum der** *(Fußball)* goal area; **~pfennig·stück das** five-pfennig piece; **~prozentig** *Adj.* five per cent; **~stellig** *Adj.* five-figure ‹*number, sum*›; *s. auch* **achtstellig; ~stöckig** *Adj.* five-storey *attrib.*; *s. auch* **achtstöckig**

fünft [fʏnft] *in* **wir/sie waren zu ~:** there were five of us/them; *s. auch* [2]**acht**

fünft... *Ordinalz.* fifth; *s. auch* **acht...**

fünf-, Fünf-: ~tage·woche die five-day [working] week; **~tägig** *Adj.* five-day; *s. auch* **achttägig; ~tausend** *Kardinalz.* five thousand

fünfteilig *Adj.* five-part; *s. auch* **achtteilig**

fünftel ['fʏnftl̩] *Bruchz.* fifth; *s. auch* **achtel**

Fünftel das (*schweiz. meist* **der**); **~s, ~:** fifth

fünftens ['fʏnftn̩s] *Adv.* fifthly; in the fifth place

fünf-: ~zehn *Kardinalz.* fifteen; *s. auch* **achtzehn; ~zehn·jährig** *Adj.* *(15 Jahre alt)* fifteen-year-old *attrib.*; *(15 Jahre dauernd)* fifteen-year *attrib.*

fünfzig ['fʏnftsɪç] *Kardinalz.* fifty; *s. auch* **achtzig**

fünfziger *indekl. Adj.*; *nicht präd.* **die ~ Jahre** the fifties; *s. auch* **achtziger**

Fünfziger der; ~s, ~ a) fifty-year-old; **b)** *(Münze, Geldschein)* fifty

fünfzig·jährig *Adj.* *(50 Jahre alt)* fifty-year-old *attrib.*; *(50 Jahre dauernd)* fifty-year *attrib.*

fünfzigst... ['fʏnftsɪçst...] *Ordinalz.* fiftieth; *s. auch* **acht...**

fungieren [fʊŋ'gi:rən] *itr. V.* **als etw. ~** ‹*person*› act as sth.; ‹*word etc.*› function as sth.

Funk [fʊŋk] **der; ~s a)** *(drahtlose Übermittlung)* radio; **über ~:** by radio; **b)** *(Rund~)* radio; **beim ~ sein** *(ugs.)* *od.* **arbeiten** be *(coll.)* *or* work in radio

Funk·amateur der radio ham

Fünkchen ['fʏŋkçən] **das; ~s, ~:** *s.* **Funke b**

Funke ['fʊŋkə] **der; ~ns, ~n a)** spark; **~n sprühen** send out a shower of sparks; *(fig.)* ‹*eyes*› flash; **b)** *(fig.)* **der ~ der Begeisterung** the spark of enthusiasm; **kein ~** *od.* **Fünkchen [von] Verstand/Ehrgefühl/Mitleid** not a glimmer of understanding/shred of honour/scrap of sympathy

funkeln ['fʊŋkl̩n] *itr. V.* ‹*light, star*› twinkle, sparkle; ‹*gold, diamonds*› glitter, sparkle; ‹*eyes*› blaze

funkel·nagel·neu *Adj.* *(ugs.)* brand new; spanking new *(coll.)*

funken ['fʊŋkn̩] **1.** *tr. V.* radio; ‹*transmitter*› broadcast; **SOS ~:** send out an SOS. **2.** *itr. V.*; *unpers.* *(fig. ugs.)* **es hat gefunkt** *(es hat Streit gegeben)* the sparks flew; *(man hat sich verliebt)* something clicked between them/us *(coll.)*; **es hat bei ihm gefunkt** the penny's dropped [with him] *(coll.)*

Funker der; ~s, ~: radio operator

Funk-: ~gerät das radio set; *(tragbar)* walkie-talkie; **~haus das** broadcasting centre; **~kolleg das** radio-based [adult education] course; **~sprech·gerät das** radiophone; *(tragbar)* walkie-talkie; **~sprech·verkehr der** radio telephony; **~spruch der** radio signal; *(Nachricht)* radio message; **~stille die** radio silence; **bei ihm herrscht ~stille** *(fig.)* he's keeping quiet; **~streife die** [police] radio patrol; **~taxi das** radio taxi; **~technik die** radio technology

Funktion [fʊnk'tsi̯o:n] **die; ~, ~en a)** function; **b)** *o. Pl. (Tätigkeit, Arbeiten)* functioning, working; **in ~ sein/in ~** *(Akk.)* **treten** be in operation/come into operation; **jmdn./etw. außer ~ setzen** put sb./sth. out of operation

funktional [fʊŋktsi̯o'na:l] **1.** *Adj.* functional. **2.** *adv.* functionally

Funktionär [fʊŋktsi̯o'nɛ:ɐ̯] **der; ~s, ~e, Funktionärin die; ~, ~nen** official; functionary

funktionieren *itr. V.* work; function

funktions-, Funktions-: ~fähig *Adj.* able to function *or* work *pred.*; **~störung die** *(Med.)* functional disorder; dysfunction; **~tüchtig** *Adj.* working ⟨*equipment, part*⟩; sound ⟨*organ*⟩

Funk-: ~turm der radio tower; **~verbindung die** radio contact; **~verkehr der** radio communication

Funzel ['fʊntsl̩] **die; ~, ~n** *(ugs., abwertend)* useless lamp *or* light; **bei dieser ~:** in this gloomy light

für [fy:ɐ̯] **1.** *Präp. mit Akk.* **a)** for; **~ jmdn. bestimmt sein** be meant for sb.; **das ist nichts ~ mich** that's not for me; **Lehrer ~ etw. sein** be a teacher of sth.; **~ sich** by oneself; on one's own; **sich ~ jmdn. freuen** be pleased for sb.; **~ immer** for ever; for good; **~ gewöhnlich** usually; **~ nichts und wieder nichts** for absolutely nothing; **b)** *(zugunsten)* for; **~ jmdn./etw. sein** be for *or* in favour of sb./sth.; **das hat etwas ~ sich** it has something to be said for it; **das Für und Wider** the pros and cons *pl.*; **c)** *(als)* **etw. ~ zulässig erklären** declare sth. admissible; **jmdn. ~ tot erklären** declare sb. dead; **d)** *(an Stelle)* for; **~ jmdn. einspringen** take sb.'s place; **~ zwei arbeiten** do the work of two people; **e)** *(als Stellvertreter)* for; on behalf of; **f)** *(um)* **Jahr ~ Jahr** year after year; **Wort ~ Wort** word for word; **Schritt ~ Schritt** step by step *s. auch* **was 1**

Für·bitte die intercession; **[bei jmdm.] für jmdn. ~ einlegen** intercede [with sb.] for sb.

Furche ['fʊrçə] **die; ~, ~n a)** furrow; **~n auf der Stirn haben** have a furrowed brow; **b)** *(Rille)* groove

furchen *tr. V. (geh.)* furrow

Furcht [fʊrçt] **die; ~:** fear; **~ vor jmdm./etw. haben** fear sb./sth.; **jmdm. ~ einflößen** frighten sb.; **aus ~ vor jmdm./etw.** for fear of sb./sth.; **jmdn. in ~ und Schrecken versetzen** fill sb. with terror; terrify sb.

furchtbar 1. *Adj.* **a)** awful; frightful; dreadful; **es war mir ~, das tun zu müssen** it was awful [for me] to have to do it; **b)** *(ugs.: unangenehm)* awful *(coll.)*; terrible *(coll.)*; **ein ~er Angeber** an awful *or* frightful show-off *(coll.)*. **2.** *adv. (ugs.)* awfully *(coll.)*; terribly *(coll.)*; **~ lachen [müssen]** laugh oneself silly *(coll.)*; **es dauerte ~ lange** it took an awfully long time

furcht·einflößend *Adj.* fearsome; frightening

fürchten 1. *refl. V.* **sich [vor jmdm./etw.] ~:** be afraid *or* frightened [of sb./sth.]; **es ist zum Fürchten** it is quite frightening. **2.** *tr. V.* fear; be afraid of; **ein gefürchteter Kritiker** a feared critic; **ich fürchte, [daß] ...:** I'm afraid [that] ... **3.** *itr. V.* **für** *od.* **um jmdn./etw. ~:** fear for sb./sth.

fürchterlich *Adj. s.* **furchtbar**

furcht·erregend *Adj.* frightening

furcht·los 1. *Adj.* fearless. **2.** *adv.* fearlessly

Furcht·losigkeit die; ~: fearlessness

furchtsam ['fʊrçtza:m] **1.** *Adj.* timid; fearful. **2.** *adv.* timidly; fearfully

Furchtsamkeit die; ~, ~en timidity; fearfulness

für·einander *Adv.* for one another; for each other

Furie ['fu:ri̯ə] **die; ~, ~n** Fury; **sie wurde zur ~** *(fig.)* she started acting like a woman possessed

Furnier [fʊr'ni:ɐ̯] **das; ~s, ~e** veneer

Furore [fu'ro:rə] *in* **~ machen** cause a sensation *or* stir

fürs [fy:ɐ̯s] *Präp. + Art.* **a)** = **für das; b) ~ erste** for the time being

Für·sorge die; ~ a) care; **b)** *(veralt.: Sozialhilfe)* welfare; **c)** *(veralt.: Sozialamt)* social services *pl.*; **d)** *(ugs.: Unterstützungsgeld)* social security *(Brit.)*; welfare *(Amer.)*

für·sorgend 1. *Adj.* caring; thoughtful. **2.** *adv.* caringly; thoughtfully

für·sorglich 1. *Adj.* considerate; thoughtful. **2.** *adv.* considerately; thoughtfully

Fürsorglichkeit die; ~: considerateness; thoughtfulness

Für·sprache die support; **bei jmdm. für jmdn. ~ einlegen** put in a good word for sb. with sb.

Für·sprecher der, Für·sprecherin die advocate

Fürst [fʏrst] **der; ~en, ~en** prince

Fürsten-: ~geschlecht das, ~haus das royal house

Fürstentum das; ~s, Fürstentümer [-ty:mɐ] principality

Fürstin die; ~, ~nen princess

fürstlich 1. *Adj.* **a)** *nicht präd.* royal; **b)** *(fig.: üppig)* handsome; lavish. **2.** *adv. (fig.)* handsomely; lavishly

Furt [fʊrt] **die; ~, ~en** ford

Furunkel [fu'rʊŋkl̩] **der** *od.* **das; ~s, ~:** boil; furuncle

Für·wort das; *Pl.* **-wörter** pronoun

Furz [fʊrts] **der; ~es, Fürze** ['fʏrtsə] *(derb)* fart *(coarse)*

furzen *(derb) itr. V.* fart *(coarse)*

Fusel ['fu:zl̩] **der; ~s, ~** *(ugs. abwertend)* rotgut *(coll. derog.)*

Fusion [fu'zi̯o:n] **die; ~, ~en a)** amalgamation; *(von Konzernen)* merger; **b)** *(Naturw.)* fusion

fusionieren *itr. V.* merge

Fuß [fu:s] **der; ~es, Füße** ['fy:sə] **a)** foot; **sich** *(Dat.)* **den ~ verstauchen/brechen** sprain one's ankle/break a bone in one's foot; **mit bloßen Füßen** barefoot; with bare feet; **jmdm. auf den ~ treten** tread on sb.'s foot; **zu ~ gehen** go on foot; walk; **gut/schlecht zu ~ sein** be a good/bad walker; **jmdm. auf dem ~e folgen** follow at sb.'s heels; **bei ~!** heel!; **nimm die Füße weg!** *(ugs.)* move your feet!; **b)** *(fig.)* **stehenden ~es** *(veralt., geh.)* without delay; instanter *(arch.)*; **sich die Füße nach etw. ablaufen** *od.* **wund laufen** chase round everywhere for sth.; **[festen] ~ fassen** find one's feet; **kalte Füße kriegen** *(ugs.)* get cold feet *(coll.)* **auf freiem ~ sein** be at large; **jmdn. auf freien ~ setzen** set sb. free; **auf großem ~ leben** live in great style; **jmdm. auf die Füße treten** *(ugs.)* give sb. a good talking-to; **jmdn./etw. mit Füßen treten** trample on sb./sth.; **jmdm. etw. vor die Füße werfen** throw sth. in sb.'s face; **jmdm. zu Füßen liegen** *(geh.: bewundern)* adore *or* worship sb.; **c)** *(tragender Teil) (einer Lampe)* base; *(eines Weinglases)* foot; *(eines Schranks, Sessels, Klaviers)* leg; **auf tönernen Füßen stehen** *(fig.)* be unsoundly based; **d)** *o. Pl. (eines Berges)* foot; *(einer Säule)* base; **e)** *Pl.: ~ (Längenmaß)* foot; **zwei/drei ~:** two/three feet *or* foot; **f)** *(Teil des Strumpfes)* foot

fuß-, Fuß-: ~abdruck der footprint; **~abstreifer, ~abtreter der** shoe scraper; **~angel die** mantrap; *(fig.)* trap; **~bad das** foot-bath

Fuß·ball der a) *o. Pl. (Ballspiel)* [Association] football; soccer *(coll.)*; **b)** *(Ball)* football; soccer ball *(coll.)*

Fuß·ballen der the ball of the/one's foot

Fußballer der; ~s, ~, Fußballerin die; ~, ~nen: footballer; soccer player *(coll.)*

Fußball-: ~mannschaft die football team; **~meisterschaft die** football championship; **~platz der** football ground; *(Spielfeld)* football pitch; **~schuh der** football boot; **~spiel das a)** football match; **b)** *o. Pl. (Sportart)* football *no art.*; **~spieler der** football player; **~toto das** *od.* **der** football pools *pl.*; **~verein der** football club

Fuß-: ~bank die foot-stool; **~boden der** floor

Fußboden-: ~belag der floor covering; **~heizung die** underfloor heating

Fuß·breit der; ~: foot

Füßchen ['fy:sçən] **das; ~s, ~:** [little] foot

Fussel ['fʊsl̩] **die; ~, ~n** *od.* **der; ~s, ~[n]** fluff; **ein[e] ~:** a piece of fluff; some fluff

fusselig *Adj.* covered in fluff *postpos.*; *(ausgefranst)* frayed; **sich** *(Dat.)* **den Mund ~ reden** *(salopp)* talk till one is blue in the face *(coll.)*

fusseln *itr. V.* make fluff

fußen *itr. V.* **auf etw.** *(Dat.)* **~:** be based on sth.

Fuß·ende das foot

Fußgänger [-gɛŋɐ] **der; ~s, ~, Fußgängerin, die; ~, ~nen:** pedestrian

Fußgänger-: ~brücke die footbridge; **~übergang der, ~überweg der** pedestrian crossing; **~unterführung die** pedestrian subway; **~zone die** pedestrian precinct

fuß-, Fuß-: ~gelenk das ankle; **~hebel der** foot pedal; **~kalt** *Adj.* **das Zimmer ist ~kalt** the room has a cold floor; **~kettchen das** anklet; **~leiste die** skirting-board *(Brit.)*; baseboard *(Amer.)*; **~marsch der** march; **~matte die** doormat; **~nagel der** toe-nail; **~note die** footnote; **~pflege die** foot treatment; *(beruflich)* chiropody; **~pfleger der, ~pflegerin die** chiropodist; **~pilz der** athlete's foot; **~schweiß der** foot perspiration; **~sohle die** sole [of the/one's foot]; **~spitze die: auf den ~spitzen gehen** walk on tiptoe; **~spur die** footprint; *(Fährte)* line of footprints; tracks *pl.*; **~stapfen der; ~s, ~:** footprint; **in jmds. ~stapfen** *(Akk.)* **treten** *(fig.)* follow in sb.'s footsteps; **~tritt der** kick; **jmdm./einer Sache einen ~tritt geben** *od.* **versetzen** *(fig.)* give sb./sth. a kick; **einen ~tritt bekommen** *(fig.)* get a kick in the teeth *(coll.)*; **~volk das a)** *(hist.)* footmen *pl.*; **b)** *(abwertend: Untergeordnete)* lower ranks *pl.*; dogsbodies *pl. (coll.)*; **~wanderung die** ramble; **~weg der a)** *(Gehweg, Bürgersteig)* footpath; **b)** *(Gehen zu ~)* walk; **eine Stunde/zwei Stunden ~weg** one hour's/two hours' walk

futsch [fʊtʃ] *Adj.; nicht attr. (salopp)* **~ sein** have gone for a burton *(Brit. sl.)*

[1]Futter ['fʊtɐ] **das; ~s** *(Tiernahrung)* feed; *(für Pferde, Kühe)* fodder; **dem Vieh ~ geben** feed the cattle; **gut im ~ sein** *od.* **stehen** *(ugs.)* be well-fed

[2]Futter das; ~s *(von Kleidungsstücken)* lining

Futteral [fʊtə'ra:l] **das; ~s, ~e** case

Futter·krippe die manger; *(fig.)* **an der ~krippe sitzen** *(ugs.)* be in clover

futtern *(ugs.)* **1.** *tr. V.* eat. **2.** *itr. V.* feed

[1]füttern ['fʏtɐn] *tr. V.* feed

[2]füttern *tr. V. (mit [2]Futter ausstatten)* line

Futter-: ~napf der bowl; **~neid der a)** *(Verhaltensf.)* jealousy [as regards food]; **b)** *(fig. ugs.: Neid)* jealousy; envy; **~pflanze die** fodder plant; forage plant; **~suche die** search for food; **auf ~suche/bei der ~suche** searching for food; **~trog der** feeding trough

Fütterung die; ~, ~en feeding

Futur [fu'tu:ɐ̯] **das; ~s, ~e** *(Sprachw.)* future [tense]; **das erste/zweite ~:** future/future perfect [tense]

futuristisch a) futuristic; **b)** *(den Futurismus betreffend)* Futurist

G

g, G [ge:] **das; ~, ~ a)** *(Buchstabe)* g/G; **b)** *(Musik)* [key of] G; *s. auch* **a, A**

g *Abk.* **Gramm** g

gab [ga:p] *1. u. 3. Pers. Sg. Prät. v.* **geben**

Gabardine ['gabardi:n] **der; ~s** *od.* **die; ~:** gabardine

Gabe ['ga:bə] **die; ~, ~n a)** *(geh.: Geschenk)* gift; present; **eine ~ Gottes** a gift of God; **b)** *(Almosen, Spende)* alms *pl.*; *(an eine Sammlung)* donation; **c)** *(geh.: Begabung, Talent)* gift; **die ~ haben, etw. zu tun** have the gift *or (iron.)* knack of doing sth.

gäbe ['gɛ:bə] *1. u. 3. Pers. Sg. Konjunktiv II v.* **geben**; *s. auch* **gang**

Gabel ['ga:bl̩] **die; ~, ~n a)** fork; **b)** *(Heu~, Mist~)* pitchfork; **c)** *(Telefon~)* rest; cradle; **d)** *(Fahrrad~)* fork; **e)** *(Ast~)* fork

gabel·förmig *Adj.* forked

Gabel·frühstück das cold buffet; fork lunch

gabeln *refl. V.* fork; *(fig.: sich teilen)* divide

Gabel·stapler [-ʃta:plɐ] **der; ~s, ~:** fork-lift truck

Gabelung die; ~, ~en fork

Gaben·tisch der gift table *(at Christmas and on birthdays)*

gackern ['gakɐn] *itr. V.* **a)** cluck; **b)** *(ugs.: kichern, lachen)* cackle

gaffen ['gafn̩] *itr. V. (abwertend)* gape; gawp *(coll.)*

Gaffer der; ~s, ~: gaper; starer

Gag [gɛk] **der; ~s, ~s a)** *(Theater, Film)* gag; **b)** *(Besonderheit)* gimmick

Gage ['ga:ʒə] **die; ~, ~n** salary; *(für einzelnen Auftritt)* fee

gähnen ['gɛ:nən] *itr. V.* **a)** yawn; **im Saal herrschte ~de Leere** the hall was totally empty; **b)** *(geh.: sich auftun)* ⟨*chasm, abyss*⟩ yawn; ⟨*hole*⟩ gape

Gala ['ga:la, *auch* 'gala] **die; ~ a)**

(Festkleidung) formal *or* gala dress; **b)** *(Veranstaltung)* gala

Gala-: ~abend der [evening] gala; **~diner das** formal dinner; banquet; **~empfang der** gala *or* formal reception

galaktisch [ga'laktɪʃ] *Adj.* galactic; **~er Nebel** [galactic] nebula

galant [ga'lant] **1.** *Adj.* **a)** *(veralt.)* gallant; **b)** *(amourös)* amorous ⟨*adventure*⟩. **2.** *adv.* gallantly

Gala·vorstellung die gala performance

Galaxie [gala'ksi:] **die; ~, ~n** *(Astron.)* galaxy

Galeere [ga'le:rə] **die; ~, ~n** galley

Galerie [galə'ri:] **die; ~, ~n a)** gallery; **b)** *(bes. österr., schweiz.: Tunnel)* tunnel

Galerist [galə'rɪst] **der; ~en, ~en** gallery-owner

Galgen ['galgn̩] **der** gallows *sing.*; gibbet; **jmdn. an den ~ bringen** *(ugs.)* bring sb. to the gallows

Galgen-: ~frist die reprieve; **~humor der** gallows humour; **~strick der, ~vogel der** *(ugs. abwertend)* rogue

Galions·figur [ga'li̯o:ns-] **die** figurehead

Galle ['galə] **die; ~, ~n a)** *(Gallenblase)* gall[-bladder]; **b)** *(Sekret) (bei Tieren)* gall; *(bei Menschen)* bile; **bitter wie ~:** extremely bitter; **mir lief die ~ über** *od.* **kam die ~ hoch** *(fig.)* my blood boiled

Gallen-: ~blase die gall-bladder; **~kolik die** biliary colic; **~leiden das** gall-bladder complaint; **~stein der** gallstone

Gallert ['galɐt] **das; ~[e]s** jelly

Galopp [ga'lɔp] **der; ~s, ~s** *od.* **~e** gallop; **im/in gestrecktem ~:** at a/at full gallop; **in ~ fallen** break into a gallop; **etw. im ~ machen** *(fig. ugs.)* race through sth.

Galopp·bahn die *(Pferdesport)* race-track; racecourse

Galopper der; ~s, ~ *(Pferd)* racehorse; *(Reiter)* jockey

galoppieren *itr. V.; meist mit sein* gallop; **die ~de Inflation** galloping inflation

Galopp·rennen das *(Pferdesport)* race

galt [galt] *1. u. 3. Pers. Sg. Prät. v.* **gelten**

galvanisch [gal'va:nɪʃ] *Adj.* galvanic

galvanisieren *tr. V.* electroplate

Gamasche [ga'maʃə] **die; ~, ~n** gaiter; *(bis zum Knöchel)* spat

Gambe ['gambə] **die; ~, ~n** *(Musik)* viola da gamba

Gamma ['gama] **das; ~[s], ~s** gamma

Gamma·strahlen *Pl. (Physik, Med.)* gamma rays

Gammel ['gaml̩] **der; ~s** *(ugs.)* junk *(coll.)*

gammelig ['gam(ə)lɪç] *Adj. (ugs.)* **a)** bad; rotten; **b)** *(unordentlich)* scruffy

gammeln ['gamln̩] *itr. V. (ugs.)* **a)** go bad; go off; **b)** *(nichts tun)* loaf around; bum around *(Amer. coll.)*

Gammler ['gamlɐ] **der; ~s, ~, Gammlerin die; ~, ~nen** *(ugs.)* drop-out *(coll.)*

gang [gaŋ] *in* **~ und gäbe sein** be quite usual; be the usual *or* accepted thing

¹Gang [gaŋ] **der; ~[e]s, Gänge** ['gɛŋə] **a)** *(Gehweise)* walk; gait; **jmdn. am ~ erkennen** recognise sb. by the way he/she walks; **b)** *(zu einem Ort)* **einen ~ in die Stadt machen** go to town; **einen schweren ~ tun** *od.* **gehen [müssen]** *(fig.)* [have to] do a difficult thing; **c)** *(Besorgung)* errand; **d)** *o. Pl. (Bewegung)* running; **etw. in ~ bringen** *od.* **setzen/halten** get/keep sth. going; **in ~ sein** be going; *(Maschine)* be running; **in ~ kommen** get going; get off the ground; **e)** *o. Pl. (Verlauf)* course; **seinen [gewohnten] ~ gehen** go on as usual; **im ~[e] sein** be in progress; **f)** *(Technik)* gear; **den ersten ~ einlegen** engage first gear; **in den ersten ~ [zurück]schalten** change [down] into first gear; **einen ~ zulegen** *(fig. ugs.)* get a move on *(coll.)*; **g)** *(Flur) (in Zügen, Gebäuden usw.)* corridor; *(Verbindungs~)* passage[-way]; *(im Theater, Kino, Flugzeug)* aisle; **h)** *(unterirdisch)* tunnel; passage[way]; *(im Bergwerk)* gallery; *(eines Tierbaus)* tunnel; **i)** *(Kochk.)* course

²Gang [gɛŋ] **die; ~, ~s** *(Bande)* gang

Gang·art die walk; way of walking; gait; *(eines Pferdes)* gait; **eine schnellere ~ anschlagen** step up the pace

gangbar *Adj.* passable; **ein ~er Weg** *(fig.)* a feasible *or* practicable way

Gängel·band ['gɛŋəl-] **das** *in* **jmdn. am ~ führen** keep sb. in leading-reins

gängeln ['gɛŋln̩] *tr. V.* **jmdn. ~** *(ugs.)* boss sb. around; tell sb. what to do

gängig ['gɛŋɪç] *Adj.* **a)** *(üblich)* common; **b)** *(leicht verkäuflich)* popular; in demand *postpos.*

Gang·schaltung die gear-change; **ein Fahrrad mit ~:** a bicycle with gears

Gangster ['gɛŋstɐ] **der; ~s, ~** *(abwertend)* gangster

Gangster-: ~bande die gang [of criminals]; **~boß der** *(ugs.)* gang boss

Gangway ['gæŋweɪ] **die; ~, ~s** gangway

Ganove [ga'no:və] **der; ~n, ~n** *(ugs. abwertend)* crook *(coll.)*

Gans [gans] **die; ~, Gänse** ['gɛnzə] **a)** goose; **b)** *(abwertend: weibliche Person)* **eine [dumme/alberne/blöde] ~:** a silly goose

Gänse-: ~blümchen das daisy; **~braten der** roast goose; **~feder die** goose-feather; goose-quill; **~füßchen das;** *meist Pl. (ugs.) s.* **Anführungszeichen; ~haut die** *(fig.)* goose-flesh; goose pimples *pl.*; **~leber·pastete die** pâté de foie gras; **~marsch** *in* **im ~marsch** in single *or* Indian file

Gänserich ['gɛnzərɪç] **der; ~s, ~e** gander

Gänse·schmalz das goose dripping

ganz [gants] **1.** *Adj.* **a)** *nicht präd. (gesamt)* whole; entire; **den ~en Tag** all day; **die ~e Welt** the whole world; **die ~e Straße** *(alle Bewohner)* everybody in the street; **~ Europa** the whole of Europe; **wir fuhren durch ~ Frankreich** we travelled all over France; **~e Arbeit leisten** do a complete *or* proper job; **die ~e Geschichte** *od.* **Sache** *(ugs.)* the whole story *or* business; **b)** *nicht präd. (ugs.: sämtlich)* **die ~e Milch** all the milk; **die ~en Leute** *usw.* all the people *etc.*; **c)** *nicht präd. (vollständig)* whole ⟨*number, truth*⟩; **eine ~e Note** *(Musik)* a semibreve *(Brit.)*; a whole note *(Amer.)*; **im ~en sechs Tage** six days in all *or* altogether; **im [großen und] ~en** on the whole; all in all; **d)** *nicht präd. (ugs.: ziemlich groß)* **eine ~e Menge/ein ~er Haufen** quite a lot/quite a pile; **e)** *(ugs.: unversehrt)* intact; **etw. wieder ~ machen** mend sth.; **f)** *nicht präd. (ugs.: nur)* all of; **~e 14 Jahre alt** all of fourteen [years old]. **2.** *adv.* **a)** *(vollkommen)* quite; **das ist mir ~ egal** it's all the same to me; I don't care; **etw. ~ vergessen** completely *or* quite forget sth.; **etwas ~ anderes** something quite different; **etw. ~ allein tun** *od.* **machen** do sth. entirely on one's own; **nicht ~:** not quite; **~ besonders** especially; **sie ist ~ die Mutter** she's the image of *or* just like her mother; **~ und gar** totally; utterly; **b)** *(sehr, ziemlich)* quite; **es**

ist mir ~ recht it's quite all right with me

Ganze das; *adj. Dekl.* **a)** *(Einheit)* whole; **b)** *(alles)* **das ~:** the whole thing; **aufs ~ gehen** *(ugs.)* go the whole hog *(coll.)*

Gänze ['gɛntsə] *in* **in seiner/ihrer ~** *(geh.)* in its/their entirety

ganz·jährig 1. *Adj.; nicht präd.* **die ~e Trockenperiode** the dry period lasting all year. **2.** *adv.* **~ geöffnet** open throughout the year *or* all the year round

gänzlich ['gɛntslɪç] **1.** *Adv.* completely; entirely. **2.** *Adj.* complete; total

ganz·tägig 1. *Adj.; nicht präd.* all-day; **eine ~tägige Arbeit** a full-time job; **2.** *adv.* all day

ganz·tags *Adv.* **~ arbeiten** work full-time

Ganztags-: ~beschäftigung die; *o. Pl.* full-time job; **~schule die** all-day school; *(System)* all-day schooling *no art.*

¹gar [ga:ɐ̯] *Adj.* cooked; done *pred.;* **etw. ~ kochen** cook sth. [until it is done]

²gar *Partikel* **a)** *(überhaupt)* **~ nicht [wahr]** not [true] at all; **~ nichts** nothing at all *or* whatsoever; **~ niemand** *od.* **keiner** nobody at all *or* whatsoever; **~ keines** not a single one; **b)** *(südd., österr., schweiz.: verstärkend)* **~ zu** only too; **er wäre ~ zu gern gekommen** he would so much have liked to come; **c)** *(geh.: sogar)* even; **d)** *(veralt.: sehr)* very

Garage [ga'ra:ʒə] **die; ~, ~n** garage

Garantie [garan'ti:] **die; ~, ~n a)** *(Gewähr)* guarantee **(für** of); **b)** *(Kaufmannsspr.)* guarantee; warranty; **eine ~ auf etw.** *(Akk.)* **geben** guarantee sth.; **für** *od.* **auf etw.** *(Akk.)* **ein Jahr ~ erhalten** get a one year guarantee on sth.; **c)** *(Sicherheit)* guarantee; surety

Garantie·frist die guarantee period

garantieren 1. *tr. V.* guarantee; **jmdm. etw. ~:** guarantee sb. sth. **2.** *itr. V.* **für etw. ~:** guarantee sth.

garantiert *Adv. (ugs.)* **wir kommen ~ zu spät** we're dead certain to arrive late *(coll.)*

Garantie·schein der guarantee [certificate]

Garaus ['ga:ɐ̯laus] *in* **jmdm. den ~ machen** do sb. in *(coll.);* **dem Unkraut den ~ machen** get rid of the weeds

Garbe ['garbə] **die; ~, ~n a)** *(Getreide~)* sheaf; **b)** *(Geschoß~)* burst [of fire]

Garde ['gardə] **die; ~, ~n a)** *(Mil., Leib~)* guard; **b)** *(Gruppe)* team

Garderobe [gardə'ro:bə] **die; ~, ~n a)** *o. Pl.* *(Oberbekleidung)* wardrobe; clothes *pl.;* **die passende ~:** suitable clothes; **für ~ wird nicht gehaftet!** clothes are left at the owner's risk; **b)** *(Flur~)* coat-rack; **c)** *(im Theater o. ä.)* cloakroom; checkroom *(Amer.);* **d)** *(Ankleideraum)* dressing-room

Garderoben-: ~frau die cloakroom *or (Amer.)* checkroom attendant; **~marke die** cloakroom *or (Amer.)* checkroom ticket; **~ständer der** coat-stand

Gardine [gar'di:nə] **die; ~, ~n a)** net curtain; **b)** *(landsch., veralt.)* curtain; *s. auch* **schwedisch**

Gardinen-: ~leiste die curtain rail; **~predigt die** *(ugs.)* telling-off *(coll.); (einer Ehefrau zu ihrem Mann)* curtain lecture; **~stange die** curtain rail

garen ['ga:rən] *tr., itr. V.* cook

gären ['gɛ:rən] *regelm. (auch unr.) itr. V.* ferment; *(fig.)* seethe

Garn [garn] **das; ~[e]s, ~e a)** thread; *(Näh~)* cotton; **b)** *(fig.)* **[s]ein ~ spinnen** spin a yarn; **jmdm. ins ~ gehen** fall *or* walk into sb.'s trap

Garnele [gar'ne:lə] **die; ~, ~n** shrimp

garnieren [gar'ni:rən] *tr. V.* **a)** *(schmücken)* decorate (*mit* with); **b)** *(Gastr.)* garnish

Garnierung die; ~, ~en a) garnish; **b)** *(Vorgang)* garnishing

Garnison [garni'zo:n] **die; ~, ~en** garrison

Garnison·stadt die garrison town

Garnitur [garni'tu:ɐ̯] **die; ~, ~en a)** set; *(Wäsche)* set of [matching] underwear; *(Möbel)* suite; **eine zweiteilige ~:** a two-piece suite; **b)** *(ugs.)* **die erste/zweite ~:** the first/second-rate people *pl.;* **zur ersten/zweiten ~ gehören** be first-/second-rate; **c)** *(Gastr.)* garnishing; garniture

Garn-: ~knäuel das *od.* **der** ball of thread; **~rolle die** reel; bobbin; *(von Nähgarn)* cotton reel

garstig ['garstɪç] *Adj.* **a)** *(boshaft)* nasty **(zu** to); bad ⟨*behaviour*⟩; nasty, naughty, *(coll.)* horrid ⟨*child*⟩; **b)** *(abscheulich)* horrible; nasty

Garten ['gartn̩] **der; ~s, Gärten** ['gɛrtn̩] garden; *s. auch* **zoologisch**

Garten-: ~arbeit die gardening; **~bau der** *o. Pl.* horticulture; **~erde die** garden mould; **~fest das** garden party; **~gerät das** garden tool; **~haus das, ~laube die** summer-house; garden house; **~lokal das** beer garden; *(Restaurant)* open-air café; **~party die** *s.* **~fest; ~schlauch der** garden hose; **~stuhl der** garden chair; **~zaun der** garden fence; **~zwerg der** garden gnome

Gärtner ['gɛrtnɐ] **der; ~s, ~:** gardener

Gärtnerei die; ~, ~en nursery

Gärtnerin die; ~, ~nen gardener

Gärung die; ~, ~en a) fermentation; **b)** *(fig.: Unruhe)* ferment

Gas [ga:s] **das; ~es, ~e a)** gas; **b)** *(Kfz.-W.)* **~ wegnehmen** decelerate; take one's foot off the accelerator; **~ geben** accelerate; put one's foot down *(coll.);* **c)** *(ugs.) s.* **Gaspedal**

gas-, Gas-: ~anzünder der gas-lighter; **~explosion die** gas explosion; **~feuerzeug das** gas lighter; **~flamme die** gas flame; **~flasche die** gas-cylinder; *(für einen Herd, Ofen)* gas bottle; gas container; **~förmig** *Adj.* gaseous; **~hahn der** gas tap; **den ~hahn aufdrehen** *(ugs. verhüll.)* end it all *(coll. euphem.);* **~heizung die** gas heating; **~herd der** gas cooker; **~kammer die** gas chamber; **~kocher der** camping stove; **~laterne die** gas lamp; **~leitung die** gas pipe; *(Hauptrohr)* gas main; **~licht das** gaslight; **~mann der** *(ugs.)* gas man; **~maske die** gas mask; **~ofen der** gas heater; **~pedal das** accelerator [pedal]; gas pedal *(Amer.);* **~rechnung die** gas bill

Gasse ['gasə] **die; ~, ~n a)** lane; narrow street; *(österr.)* street; **[für jmdn.] eine ~ bilden** *(fig.)* make way *or* clear a path [for sb.]; **b)** *(Fußball)* opening

Gassen-: ~hauer der *(ugs.)* popular song; **~junge der** *(abwertend)* street urchin

Gassi ['gasi] *in* **~ gehen** *(ugs.)* go walkies *(Brit. coll.)*

Gast [gast] **der; ~[e]s, Gäste** ['gɛstə] **a)** guest; **ungebetene Gäste** *(auch fig.)* uninvited guests; **bei jmdm. zu ~ sein** be sb.'s guest/guests; **jmdn. zu ~ haben** have sb. as one's guest/guests; **b)** *(Besucher eines Lokals)* patron; **c)** *(Besucher)* visitor

Gast·arbeiter der immigrant *or* foreign *or* guest worker

Gäste-: ~buch das guest book; **~handtuch das** guest-towel; **~zimmer das** *(privat)* guest room; spare room; *(im Hotel)* room

gast-, Gast-: ~freundlich *Adj.*

hospitable; **~freundlichkeit die, ~freundschaft die** hospitality; **~geber der** host; **~geberin die; ~, ~nen** hostess; **~haus das, ~hof der** inn; **~hörer der** auditor *(Amer.)*

gastieren *itr. V.* give a guest performance

gastlich 1. *Adj.* hospitable. **2.** *adv.* hospitably

Gastlichkeit die; ~: hospitality

Gast-: ~mahl das *(geh.)* banquet; **~mannschaft die** *(Sport)* visiting team; **~recht das** right to hospitality; **~recht genießen** enjoy the privileges of a guest; **das ~recht mißbrauchen** abuse one's position as a guest

Gastritis [gas'tri:tɪs] **die; ~, Gastritiden** *(Med.)* gastritis

gastro-, Gastro- [gastro:-]: **~nom** [~'no:m] **der; ~en, ~en** restaurateur; **~nomie** [~no'mi:] **die; ~ a)** restaurant trade; *(Versorgung, Service)* catering *no art.;* **b)** *(Kochk.)* gastronomy; **~nomisch** [~'no:mɪʃ] *Adj.* gastronomic

Gast·spiel das guest performance; **ein [kurzes] ~ geben** *(fig. scherzh.)* stay for a short time

Gastspiel·reise die tour; **eine ~ durch Japan** a tour of Japan

Gast-: ~stätte die public house; *(Speiselokal)* restaurant; **~stube die** bar; *(in einem Speiselokal)* restaurant; **~vorlesung die** guest lecture; **~wirt der** publican; landlord; *(eines Restaurants)* [restaurant] proprietor *or* owner; *(Pächter)* restaurant manager; **~wirtschaft die** *s.* **~stätte**

Gas-: ~uhr die gas meter; **~verbrauch der** gas consumption; **~vergiftung die** gas-poisoning *no indef. art.;* **~versorgung die** gas supply; **~werk das** gasworks *sing.;* **~zähler der** gas meter

Gatte ['gatə] **der; ~n, ~n** *(geh.)* husband

Gatter ['gatɐ] **das; ~s, ~ a)** *(Zaun)* fence; *(Lattenzaun)* fence; paling; **b)** *(Tor)* gate

Gattin ['gatɪn] **die; ~, ~nen** *(geh.)* wife

Gattung ['gatʊŋ] **die; ~, ~en a)** kind; sort; *(Kunst~)* genre; form; **b)** *(Biol.)* genus; **c)** *(Milit.)* service

GAU [gau] **der; ~s, ~s** *(Kerntechnik)* MCA

Gaucho ['gautʃo] **der; ~[s], ~s** gaucho

Gaudi ['gaudi] **das; ~s** *(bayr., österr.:* **die; ~***) (ugs.)* bit of fun; **ein[e] ~ sein** be great fun

Gaukelei die; ~, ~en *(geh.)* **a)** *(Vorspiegelung)* trickery *no indef. art., no pl.;* **b)** *(Possenspiel)* trick

Gaukler der; ~s, ~ a) *(veralt.: Taschenspieler)* itinerant entertainer; **b)** *(geh.: Betrüger)* charlatan; mountebank; trickster

Gaul [gaul] **der; ~[e]s, Gäule** ['gɔylə] **a)** *(abwertend)* nag *(derog.);* hack *(derog.);* **b)** *(veralt.)* horse; **einem geschenkten ~ schaut man nicht ins Maul** *(Spr.)* never look a gift-horse in the mouth

Gaumen ['gaumən] **der; ~s, ~:** palate; roof of the mouth; **das ist etwas für einen verwöhnten ~:** this is something for the real gourmet

Gaumen-: ~freude die; *meist Pl. (geh.),* **~kitzel der** *(geh.)* delicacy

Gauner ['gaunɐ] **der; ~s, ~ a)** *(abwertend)* crook *(coll.);* rogue; **ein kleiner ~:** a small-time crook *(coll.);* **b)** *(ugs.: schlauer Mensch)* cunning devil *(coll.);* sly customer *(coll.)*

Gaunerei die; ~, ~en swindle; *(das Gaunern)* swindling

Gauner-: ~sprache die thieves' cant *or* Latin; **~streich der** swindle

Gaze ['ga:zə] **die; ~, ~n** gauze; *(Draht~)* gauze; [wire-]mesh

Gazelle [ga'tsɛlə] **die; ~, ~n** gazelle

Gazette [ga'tsɛtə] **die; ~, ~n** newspaper; rag *(coll. derog.)*

G-Dur ['ge:-] **das; ~** *(Musik)* G major

geachtet *Adj.* respected; **bei jmdm. ~ sein** be respected *or* held in esteem by sb.

Geächtete der/die; *adj. Dekl.* outlaw

geadert, geädert *Adj.* veined

geartet *Adj.* **kein wie auch immer ~er Reiz** no stimulus of any kind; **besonders ~:** special; **sie ist so ~, daß ...:** her nature is such that ...; **sie ist ganz anders ~:** she is quite different; she has quite a different nature

Geäst [gə'lɛst] **das; ~[e]s** branches *pl.;* boughs *pl.*

geb. *Abk.* **a) geboren; b) geborene**

Gebäck [gə'bɛk] **das; ~[e]s, ~e** cakes and pastries *pl.; (Kekse)* biscuits *pl.; (Törtchen)* tarts *pl.*

gebacken *2. Part. v.* **backen**

Gebälk [gə'bɛlk] **das; ~[e]s, ~e a)** beams *pl.; (Dach~)* rafters *pl.;* **es knistert** *od.* **kracht im ~** *(fig.)* there are signs that things are beginning to fall apart *(fig.)*

Geballere das; ~s *(ugs. abwertend)* banging

geballt [gə'balt] *Adj.; nicht präd.* concentrated; **jmdm. eine ~e Ladung Sand ins Gesicht werfen** *(ugs.)* chuck a load of sand in sb.'s face *(coll.)*

gebar *1. u. 3. Pers. Sg. Prät. v.* **gebären**

Gebärde [gə'bɛ:ɐ̯də] **die; ~, ~n** gesture; **mit vielen ~n** with much gesticulation

gebärden *refl. V.* **sich seltsam/wie ein Rasender/wie toll ~:** behave *or* act oddly/like a madman/as if one were mad

Gebärden·sprache die sign language; *(Taubstummensprache)* deaf-and-dumb language

Gebaren das; ~s *(oft abwertend)* conduct; behaviour

gebären [gə'bɛ:rən] **1.** *unr. tr. V.* bear; give birth to; **jmdm. ein Kind ~** *(geh.)* bear sb. a child; **wo bist du geboren?** where were you born?; **er ist blind/taub geboren** he was born blind/deaf; *s. auch* **geboren. 2.** *unr. itr. V.* give birth

Gebär·mutter die; ~, -mütter womb

gebauchpinselt [gə'bauxpɪnzl̩t] *in* **sich ~ fühlen** *(ugs. scherzh.)* feel flattered

Gebäude [gə'bɔydə] **das; ~s, ~ a)** building; **b)** *(fig.)* structure; **ein ~ von Lügen** a tissue of lies

Gebäude·komplex der complex of buildings

gebaut *Adj.* **gut ~ sein** have a good figure; **so wie du ~ bist ...** *(ugs.)* with a figure like yours ...; *(fig.)* you being what you are ... *(coll.)*

Gebein das; ~s, ~e a) *(geh.: Skelett)* bones *pl.;* **b)** *Pl. (sterbliche Reste)* [mortal] remains

Gebell das; ~[e]s barking; *(der Jagdhunde)* baying; *(fig.: von Geschützen)* booming

geben ['ge:bn̩] **1.** *unr. tr. V.* **a)** give; *(reichen)* give; hand; pass; **jmdm. zu essen ~:** give sb. sth. to eat; **~ Sie mir bitte Herrn N.** please put me through to Mr N.; **ich gäbe viel darum, wenn ich das machen könnte** I'd give a lot to be able to do that; **jmdm. etw. in die Hand ~:** give sb. sth.; **etw. [nicht] aus der Hand ~:** [not] let go of sth.; **~ Sie mir bitte ein Bier** I'll have a beer, please; **Geben ist seliger denn Nehmen** *(Spr.)* it is more blessed to give than to receive *(prov.);* **b)** *(über~)* **jmdn. zu jmdm. in die Lehre ~:** apprentice sb. to sb.; **etw. in Druck** *(Akk.) od.* **zum Druck ~:** send sth. to press *or* to be printed; *s. auch* **Pflege; c)** *(gewähren)* give; **einen Elfmeter ~** *(Sport)* award a penalty; **d)**

(bieten) give; **jmdm. ein gutes Beispiel ~:** set sb. a good example; **e)** *(versetzen)* give ⟨*slap, kick, etc.*⟩; **es jmdm. ~** *(ugs.: jmdm. die Meinung sagen)* give sb. what for *(coll.)*; *(jmdn. verprügeln)* let sb. have it; **gib [es] ihm!** *(ugs.)* let him have it!; **f)** *(erteilen)* give; **Unterricht ~:** teach; **Französisch ~:** teach French; **g)** *(hervorbringen)* give ⟨*milk, shade, light*⟩; **h)** *(veranstalten)* give, throw ⟨*party*⟩; give, lay on ⟨*banquet*⟩; give ⟨*dinner-party, ball*⟩; **i)** *(aufführen)* give ⟨*concert, performance*⟩; **das Theater gibt den „Faust"** the theatre is putting on 'Faust'; **was wird heute ge~?** what's on today?; **sein Debüt ~:** make one's debut; **j)** *(er~)* **drei mal drei gibt neun** three threes are nine; three times three is *or* makes nine; **eins plus eins gibt zwei** one and one is *or* makes two; **das gibt [k]einen Sinn** that makes [no] sense; **ein Wort gab das andere** one word led to another; **k)** *in* **etw. ist jmdm. nicht ge~:** sb. just hasn't got sth.; **l)** *(äußern)* **etw. von sich ~:** utter sth.; **Unsinn/dummes Zeug von sich ~** *(abwertend)* talk nonsense/rubbish; **keinen Laut/Ton von sich ~:** not make a sound; **m)** *in* **viel/wenig auf etw.** *(Akk.)* **~:** set great/little store by sth.; **n)** *(hinzu~)* add; put in; **etw. an das Essen ~:** add sth. to *or* put sth. into the food; **o)** *(ugs.: erbrechen)* **alles wieder von sich ~:** bring *or (coll.)* sick everything up again. **2.** *unr. tr. V.; unpers.* **a)** *(vorhanden sein)* **es gibt** there is/are; **das gibt es wohl häufiger** it happens all the time; **daß es so etwas heutzutage überhaupt noch gibt!** I'm surprised that such things still go on nowadays; **zu meiner Zeit gab es das nicht** it wasn't like that in my day; **das gibt es ja gar nicht** I don't believe it; you're joking *(coll.)*; **Ein Hund mit fünf Beinen? Das gibt es ja gar nicht** A dog with five legs? There's no such thing!; **Kommen Sie herein. Was gibt es?** Come in. What's the matter *or (coll.)* what's up?; **was gibt's denn da?** what's going on over there?; **was es nicht alles gibt!** *(ugs.)* what will they think of next?; **da gibt's nichts** *(ugs.)* there's no denying it *or* no doubt about it; **da gibt's nichts, da würde ich sofort protestieren** there's nothing else for it, I'd protest immediately in that case; **b)** *(angeboten werden)* **was gibt es zu essen/trinken?** what is there to eat/drink?; **was gibt es denn zum Mittagessen?** what's for lunch?; **heute gibt's Schweinefleisch** we're having pork today; **c)** *(kommen zu)* **morgen gibt es Schnee/Sturm** it'll snow tomorrow/there'll be a storm tomorrow; **gleich/sonst gibt's was** *(ugs.)* there'll be trouble in a minute/otherwise. **3.** *unr. itr. V.* **a)** *(Karten austeilen)* deal; **wer gibt?** whose deal is it?; **b)** *(Sport: aufschlagen)* serve. **4.** *unr. refl. V.* **a) sich [natürlich] ~:** act *or* behave [naturally]; **sich nach außen hin gelassen geben** give the appearance of being relaxed; **deine Art, dich zu ~:** the way you behave; **b)** *(nachlassen)* **das Fieber wird sich ~:** his/her *etc.* temperature will drop; **sein Eifer wird sich bald ~:** his enthusiasm will soon wear off *or* cool; **das gibt sich/wird sich noch ~:** it will get better

Geber der; ~s, ~, Geberin die; ~, ~nen a) *(veralt.)* giver; donor

Gebet [gə'be:t] **das; ~[e]s, ~e** prayer; **sein ~ verrichten** say one's prayers *pl.*; **jmdn. ins ~ nehmen** *(ugs.)* give sb. a dressing down; take sb. to task

Gebet·buch das prayer-book

gebeten *2. Part. v.* **bitten**

gebierst [gə'bi:ɐ̯st], **gebiert** [gə'bi:ɐ̯t] *2., 3. Pers. Sg. Präsens v.* **gebären**

Gebiet [gə'bi:t] **das; ~[e]s, ~e a)** region; area; **b)** *(Staats~)* territory; **c)** *(Bereich)* field; sphere; **d)** *(Fach)* field; **auf einem ~:** in a field

gebieten *(geh.)* **1.** *unr. tr. V.* **a)** command; order; **jmdm. ~, etw. zu tun** command *or* order sb. to do sth.; **eine Respekt ~de Persönlichkeit** a figure who commands/commanded respect; **b)** *(erfordern)* demand; bid; *s. auch* **Einhalt. 2.** *unr. itr. V.* **a) über etw.** *(Akk.)* **~:** command sth.; have command over sth.; **über ein Land ~:** hold sway over a country; **b)** *(verfügen)* **über Geld ~:** have money at one's disposal

Gebieter der; ~s, ~ *(veralt.)* master

Gebieterin die; ~, ~nen *(veralt.)* mistress

gebieterisch *(geh.)* **1.** *Adj.* imperious; *(herrisch)* domineering; overbearing; peremptory ⟨*tone*⟩. **2.** *adv.* imperiously

Gebiets·anspruch der territorial claim

Gebilde [gə'bɪldə] **das; ~s, ~** object; *(Bauwerk)* construction; structure; **diese Dinge sind ~ seiner Phantasie** *(fig.)* these things are products of his imagination

gebildet *Adj.* educated; *(kultiviert)* cultured

Gebimmel das; ~s *(ugs.)* ringing; *(von kleinen Glocken)* tinkling

Gebinde das; ~s, ~ *(Blumenarrangement)* arrangement; *(Bund, Strauß)* bunch; *(von kleinen Blumen)* posy

Gebirge [gə'bɪrgə] **das; ~s, ~ a)** mountain range; range of mountains; **ein ~ von Schutt** *(fig.)* a mountain of rubble *(fig.)*; **b)** *(Gebirgsgegend)* mountains *pl.*

gebirgig *Adj.* mountainous

Gebirgs-: ~ausläufer der foothill; **~bach der** mountain stream; **~kette die** mountain chain *or* range; **~landschaft die** mountainous region; *(Ausblick)* mountain scenery; *(Gemälde)* mountain landscape; **~massiv das** massif; **~paß der** mountain pass; **~zug der** mountain range

Gebiß das; Gebisses, Gebisse a) set of teeth; teeth *pl.*; **b)** *(Zahnersatz)* denture; plate *(coll.)*; *(für beide Kiefer)* dentures *pl.*; set of false teeth; false teeth *pl.*

gebissen [gə'bɪsn̩] *2. Part. v.* **beißen**

Gebläse [gə'blɛ:zə] **das; ~s, ~** *(Technik)* fan

geblasen *2. Part. v.* **blasen**

geblichen [gə'blɪçn̩] *2. Part. v.* **bleichen**

Geblödel das; ~s *(ugs.)* silly chatter; twaddle *(coll.)*

geblümt [gə'bly:mt] *Adj.* flowered

Geblüt [gə'bly:t] **das; ~[e]s** *(geh.)* blood; **von königlichem ~ sein** be of royal blood; **eine Prinzessin von ~:** a princess of the blood

gebogen 1. *2. Part. v.* **biegen. 2.** *Adj.* bent; **eine aufwärts ~e Nase** an upturned nose

geboren [gə'bo:rən] **1.** *2. Part. v.* **gebären. 2.** *Adj.* **Frau Anna Schmitz ~e Meyer** Mrs Anna Schmitz née Meyer; **sie ist eine ~e von Schiller** she is a von Schiller by birth; **der ~e Schauspieler** *usw.* **sein** be a born actor *etc.*

geborgen 1. *2. Part. v.* **bergen. 2.** *Adj.* safe; secure; **sich bei jmdm. ~ fühlen** feel safe and secure with sb.

Geborgenheit die; ~: security

geborsten [gə'bɔrstn̩] *2. Part. v.* **bersten**

gebot *1. u. 3. Pers. Sg. Prät. v.* **gebieten**

Gebot das; ~[e]s, ~e a) precept;

die Zehn ~e *(Rel.)* the Ten Commandments; **b)** *(Vorschrift)* regulation; **c)** *(geh.: Befehl)* command; *(Verordnung)* decree; **auf jmds. ~** *(Akk.)* **[hin]** at sb.'s command; **d)** *in* **jmdm. zu ~[e] stehen** *(geh.)* be at sb.'s command/disposal; **e)** *(Erfordernis)* **ein ~ der Klugheit** a dictate of good sense; **f)** *(Kaufmannsspr.)* bid; **verkaufe X gegen ~:** offers [are] invited for X

geboten 1. *2. Part. v.* **bieten, gebieten. 2.** *Adj. (ratsam)* advisable; *(notwendig)* necessary; *(unbedingt ~)* imperative; **mit der ~en Sorgfalt/mit dem ~en Respekt** with all due care/respect

Gebots·schild das *(Verkehrsw.)* regulatory sign

Gebr. *Abk.* **Gebrüder** Bros.

gebracht [gə'braxt] *2. Part. v.* **bringen**

gebrannt [gə'brant] *2. Part. v.* **brennen**

gebraten *2. Part. v.* **braten**

Gebräu [gə'brɔy] **das; ~[e]s, ~e** *(meist abwertend)* brew; concoction *(derog.)*

Gebrauch der a) *o. Pl.* use; **vor ~ gut schütteln** shake well before use; **von etw. ~ machen** make use of sth.; **von seinem Recht ~ machen** avail oneself of *or* exercise one's rights *pl.*; **außer ~ kommen** fall into disuse; **etw. in ~** *(Akk.)* **nehmen/in** *od.* **im ~ haben** start/be using sth.; **in** *od.* **im ~ sein** be in use; **b)** *meist Pl. (Brauch)* custom

gebrauchen *tr. V.* use; **das kann ich gut ~:** I can make good use of that; I can just do with that *(coll.)*; **er ist zu nichts zu ~** *(ugs.)* he is useless; **den Verstand ~:** use one's common sense; **er könnte einen neuen Mantel ~** *(ugs.)* he could do with *or (coll.)* use a new coat; **ich kann jetzt keine Störung ~** *(ugs.)* I don't want to be disturbed just now

gebräuchlich [gə'brɔyçlɪç] *Adj.* **a)** *(üblich)* normal; usual; customary; **b)** *(häufig)* common

gebrauchs-, Gebrauchs-: ~anleitung die, ~anweisung die instructions *pl. or* directions *pl.* [for use]; **~fähig** *Adj.* usable; in working order *pred.*; **~fertig** *Adj.* ready for use *pred.*; **~gegen·stand der** item of practical use

gebraucht *Adj.* second-hand ‹*bicycle, clothes, etc.*›; used, second-hand ‹*car*›; used ‹*handkerchief*›; **etw. ~ kaufen** buy sth. second-hand

Gebraucht-: ~wagen der used *or* second-hand car; **~wagen·händler der** used-car dealer; second-hand car dealer; **~waren** *Pl.* second-hand goods

Gebrechen das; ~s, ~ *(geh.)* affliction

gebrechlich *Adj.* infirm; frail

Gebrechlichkeit die; ~: infirmity; frailty

gebrochen [gə'brɔxn̩] **1.** *2. Part. v.* **brechen. 2.** *Adj.* **a)** *(fehlerhaft)* **~es Englisch** broken English; **b)** *(niedergedrückt)* broken; **c)** *(gestört)* **ein ~es Verhältnis zu jmdm./etw. haben** have a disturbed relationship to sb./sth. **3.** *adv.* **~ Deutsch sprechen** speak broken German

Gebrüder *Pl.* **a)** *(Kaufmannsspr.)* **die ~ Meyer** Meyer Brothers; **b)** *(veralt.)* **die ~ Schulze** the brothers Schulze

Gebrüll das; ~[e]s a) roaring; *(von Rindern)* bellowing; **b)** *(ugs.) (lautes Schreien)* bellowing; yelling; *(einer Menschenmenge)* roaring; **auf sie mit ~!** *(scherzh.)* go for *or* get them!; **c)** *(ugs.: lautes Weinen)* bawling

gebückt *Adj.* **in ~er Haltung** bending forward; **~ gehen** walk with a stoop

Gebühr [gə'by:ɐ̯] **die; ~, ~en a)** charge; *(Maut)* toll; *(Anwalts~)* fee; *(Fernseh~)* licence fee; *(Vermittlungs~)* commission *no pl.*; fee; *(Post~)* postage *no pl.*; **~ bezahlt Empfänger** postage will be paid by addressee; **b) über ~** *(Akk.)* unduly; excessively

gebühren *(geh.)* **1.** *itr. V.* **jmdm. gebührt Achtung** *usw.* **[für etw.]** sb. deserves respect *etc.* [for sth.]; respect *etc.* is due to sb. [for sth.]. **2.** *refl. V.* **wie es sich gebührt** as is fitting *or* proper

gebührend 1. *Adj.* fitting; proper; *(angemessen)* fitting; suitable; **mit ~er Sorgfalt** with due care. **2.** *adv.* fittingly; in a fitting manner

gebühren-, Gebühren-: ~einheit die *(Fernspr.)* [tariff] unit; **~erhöhung die** increase in charges; *(der Anwaltsgebühren)* increase in fees; *(der Fernsehgebühren)* licence fee increase; **~ermäßigung die** reduction of charges; *(der Anwaltsgebühren)* reduction of fees; **~frei 1.** *Adj.* free of charge *pred.*; post-free ‹*letter, packet, etc.*›; **2.** *adv.* free of charge; *(portofrei)* post-free

gebunden 1. *2. Part. v.* **binden. 2.** *Adj.* **a)** *(verpflichtet)* bound; **an ein Versprechen/das Haus ~ sein** be bound by a promise/tied to one's home; **sich [an etw.** *(Akk.)***] ~ fühlen** feel bound by sth.; **b)** *(verlobt)* engaged; *(verheiratet)* married

Geburt [gə'bu:ɐ̯t] **die; ~, ~en** birth; **von ~ an** from birth; **vor/nach Christi ~:** before/after the birth of Christ; **das war eine schwere ~** *(fig. ugs.)* it wasn't easy; it took some doing *(coll.)*

Geburten·kontrolle die; *o. Pl.* birth control

gebürtig [gə'bʏrtɪç] *Adj.* **ein ~er Schwabe** a Swabian by birth; **aus Ungarn/Paris ~ sein** be Hungarian/Parisian by birth

Geburts-: ~datum das date of birth; **~haus das: das ~haus Beethovens** the house where Beethoven was born; Beethoven's birthplace; **~helfer der** *(Arzt)* obstetrician; *(Laie)* assistant [at a/the birth]; **~helferin die** obstetrician; *(Hebamme)* midwife; **~jahr das** year of birth; **~ort der** place of birth; birthplace; **~stunde die** hour of birth

Geburts·tag der a) birthday; **jmdm. zum ~ gratulieren** wish sb. [a] happy birthday *or* many happy returns of the day; **er hat morgen ~:** it's his birthday tomorrow; **b)** *(Geburtsdatum)* date of birth

Geburtstags-: ~feier die birthday party; **~geschenk das** birthday present; **~kind das** *(scherzh.)* birthday boy/girl; **~überraschung die** birthday surprise

Geburts·ur·kunde die birth certificate

Gebüsch [gə'bʏʃ] **das; ~[e]s, ~e** bushes *pl.*; clump of bushes; **ein niedriges ~:** a clump of low bushes; some low bushes *pl.*

Geck [gɛk] **der; ~en, ~en** *(abwertend)* dandy; fop

geckenhaft 1. *Adj.* dandyish; foppish. **2.** *adv.* **er kleidet sich ~:** he dresses like a dandy

gedacht [gə'daxt] **1.** *2. Part. v.* **denken, gedenken. 2.** *Adj.* **für jmdn./etw. ~ sein** be meant *or* intended for sb./sth.; **so war das nicht ~:** that wasn't what I intended

Gedächtnis [gə'dɛçtnɪs] **das; ~ses, ~se a)** memory; **sich** *(Dat.)* **etw. ins ~ [zurück]rufen** recall sth.; **aus dem ~:** from memory; **ein ~ wie ein Sieb** *(ugs.)* a memory like a sieve *(coll.)*; **b)** *(Andenken)* memory; remembrance; **zum ~ an jmdn.** in memory *or* remembrance of sb.

Gedächtnis-: ~**lücke die** gap in one's memory; ~**schwund der** loss of memory; amnesia; ~**stütze die** memory aid; mnemonic

gedämpft *Adj.* subdued ⟨*mood*⟩; subdued, soft ⟨*light*⟩; subdued, muted ⟨*colour*⟩; muffled ⟨*sound*⟩; low, hushed ⟨*voice*⟩

Gedanke der; ~**ns,** ~**n a)** thought; **seinen** ~**n nachhängen** abandon oneself to one's thoughts; **jmdn. auf andere** ~**n bringen** take sb.'s mind off things; **in** ~**n verloren** *od.* **versunken [sein]** [be] lost *or* deep in thought; **mit seinen** ~**n nicht bei der Sache sein** have one's mind on something else; **sich mit einem** ~**n vertraut machen/einen** ~**n aufgreifen** get used to/take up an idea; ~**n lesen können** be able to read people's thoughts *or* to mind-read; **sich** *(Dat.)* **[um jmdn./etw.** *od.* **wegen jmds./etw.]** ~**n machen** be worried [about sb./sth.]; **sich über etw.** *(Akk.)* ~**n machen** *(länger nachdenken)* think about *or* ponder sth.; **b)** *o. Pl.* **der** ~ **an etw.** *(Akk.)* the thought of sth.; **bei dem** ~**n, hingehen zu müssen** at the thought of having to go; **kein** ~ **[daran]!** *(ugs.)* out of the question!; no way! *(coll.)*; **c)** *Pl. (Meinung)* ideas; **seine** ~**n [über etw.** *(Akk.)***] austauschen** exchange views [about sth.]; **d)** *(Einfall)* idea; **das bringt mich auf einen** ~**n** that gives me an idea; **mir kam der** ~**, wir könnten ...:** it occurred to me that we could ...; **auf dumme** ~**n kommen** *(ugs.)* get silly ideas *(coll.)*; **mit dem** ~**n spielen[, etw. zu tun]** be toying with the idea [of doing sth.]; **e)** *(Idee)* idea; **der** ~ **des Friedens** the idea of peace

gedanken-, Gedanken-: ~**austausch der** exchange of ideas; ~**blitz der** *(ugs. scherzh.)* brainwave *(coll.)*; ~**freiheit die;** *o. Pl.* freedom of thought; ~**gang der** train of thought; ~**gut das** thought; ~**lesen das** mind-reading; ~**los 1.** *Adj.* unconsidered; thoughtless; **2.** *adv.* without thinking; thoughtlessly; ~**losigkeit die** lack of thought; thoughtlessness; ~**strich der** dash; ~**übertragung die** telepathy *no indef. art.*; thought-transference *no indef. art.*; ~**verloren** *Adv.* lost in thought; ~**voll 1.** *Adj.* pensive; thoughtful; **2.** *adv.* pensively; thoughtfully

gedanklich [gəˈdaŋklɪç] **1.** *Adj.; nicht präd.* intellectual. **2.** *adv.* intellectually

Gedärm [gəˈdɛrm] **das;** ~**[e]s,** ~**e** intestines *pl.*; bowels *pl.*, *(eines Tieres)* entrails *pl.*

Gedeck das; ~**[e]s,** ~**e a)** place setting; cover; **ein** ~ **auflegen** lay *or* set a place; **b)** *(Menü)* set meal; **c)** *(Getränk)* drink [with a cover charge]

gedeckt *Adj.* subdued, muted ⟨*colour*⟩

Gedeih *in* **auf** ~ **und Verderb** for good or ill; for better or [for] worse; **jmdm. auf** ~ **und Verderb ausgeliefert sein** be entirely at sb.'s mercy

gedeihen [gəˈdaiən] *unr. itr. V.; mit sein* **a)** thrive; *(wirtschaftlich)* flourish; prosper; **b)** *(fortschreiten)* progress

gedeihlich *Adj. (geh.)* thriving, flourishing, successful ⟨*business*⟩; successful ⟨*development, co-operation*⟩; beneficial ⟨*effect etc.*⟩

gedenken *unr. itr. V.* **a) jmds./einer Sache** ~ *(geh.)* remember sb./sth.; *(erwähnen)* recall sb./sth.; *(in einer Feier)* commemorate sb./sth.; **b) etw. zu tun** ~**:** intend to do *or* doing sth.

Gedenken das; ~**s** *(geh.)* remembrance; memory; **zum** ~ **an jmdn./etw.** in memory *or* remembrance of sb./sth.

Gedenk-: ~**feier die** commemoration; commemorative ceremony; ~**minute die** minute's silence; **eine** ~**minute einlegen** observe a minute's silence; ~**stätte die** memorial; ~**stein der** memorial *or* commemorative stone; ~**tag der** day of remembrance; commemoration day

Gedicht das; ~**[e]s,** ~**e** poem; **Goethes** ~**e** Goethe's poetry *sing. or* poems; **das Steak/Kleid ist ein** ~ *(fig. ugs.)* the steak is just superb/the dress is just heavenly

Gedicht·sammlung die collection of poems; *(von mehreren Dichtern)* anthology of poetry *or* verse; poetry anthology

gediegen [gəˈdi:gn̩] **1.** *Adj.* **a)** *(solide)* solid, solidly-made ⟨*furniture*⟩; sound, solid ⟨*piece of work*⟩; well-made ⟨*clothing*⟩; sound ⟨*knowledge*⟩; **b)** *(rein)* pure ⟨*gold, silver, etc.*⟩. **2.** *adv.* ~ **gebaut/verarbeitet** solidly built/made

gedieh [gəˈdi:] *1. u. 3. Pers. Sg. Prät. v.* **gedeihen**

gediehen *2. Part. v.* **gedeihen**

gedient [gəˈdi:nt] *Adj.* **ein** ~**er Soldat** a former soldier

Gedränge das; ~**s a)** pushing and shoving; *(Menschenmenge)* crush; crowd; **b) ins** ~ **kommen** *od.* **geraten** *(fig. ugs.)* get into difficulties

gedrängt *Adj.* compressed, condensed ⟨*account*⟩; terse, succinct ⟨*style, description*⟩; crowded ⟨*timetable, agenda*⟩

gedroschen [gəˈdrɔʃn̩] *2. Part. v.* **dreschen**

gedrückt *Adj.* dejected, depressed ⟨*mood*⟩

gedrungen [gəˈdrʊŋən] **1.** *2. Part. v.* **dringen. 2.** *Adj.* stocky; thickset ⟨*build*⟩

Gedudel das; ~**s** *(ugs. abwertend)* tootling; *(im Radio)* noise

Geduld [gəˈdʊlt] **die;** ~**:** patience; **keine** ~ **[zu etw.] haben** have no patience [with sth.]; **mit jmdm.** ~ **haben** be patient with sb.

gedulden *refl. V.* be patient; ~ **Sie sich bitte ein paar Minuten** please be so good as to wait a few minutes

geduldig 1. *Adj.* patient. **2.** *adv.* patiently

Geduld-s-: ~**faden der** *in* **mir/ihm** *etc.* **reißt der** ~**faden** *(ugs.)* my/his *etc.* patience is wearing thin; ~**probe die** trial of one's patience; **auf eine harte** ~**probe gestellt werden** have one's patience sorely tried; ~**spiel das** puzzle; *(fig.)* Chinese puzzle

gedungen [gəˈdʊŋən] *2. Part. v.* **dingen**

gedunsen [gəˈdʊnzn̩] *Adj. s.* **aufgedunsen**

gedurft [gəˈdʊrft] *2. Part. v.* **dürfen**

geeignet *Adj.* suitable; *(richtig)* right

Gefahr die; ~**,** ~**en** danger; *(Bedrohung)* danger; threat; *(Risiko)* risk; **die** ~**en meines Berufs** the hazards of my job; **die** ~**en des Dschungels** the perils of the jungle; **eine** ~ **für jmdn./etw.** a danger to sb./sth.; **in** ~ **kommen/geraten** get into danger; **jmdn./etw. in** ~ **bringen** put sb./sth. in danger; **sich in** ~ **begeben** put oneself in danger; expose oneself to danger; **in** ~ **sein/schweben** be in danger; ⟨*rights, plans*⟩ be in jeopardy *or* peril; **außer** ~ **sein** be out of danger; **bei** ~**:** in case of emergency; **jmdn./sich einer** ~ **aussetzen** run *or* take a risk; **es besteht die Gefahr, daß ...:** there is a danger *or* risk that ...; **auf die** ~ **hin, daß das passiert** at the risk of that happening; ~ **laufen, etw. zu tun** risk *or* run the risk of doing sth.; **auf eigene** ~**:** at one's own risk; **wer sich in ~ begibt, kommt darin um** if you keep on taking risks, you'll come to grief eventually

gefährden [gəˈfɛːɐ̯dn̩] *tr. V.* endanger; jeopardize ⟨*enterprise, success, position, etc.*⟩; *(aufs Spiel setzen)* put at risk

gefährdet *Adj.* ⟨*people, adolescents, etc.*⟩ at risk *postpos.*

Gefährdung die; ~, ~**en a)** *o. Pl.* endangering; *(eines Unternehmens, einer Position usw.)* jeopardizing; **b)** *(Gefahr)* threat (*Gen.* to)

gefahren *2. Part. v.* **fahren**

Gefahren-: ~**bereich der** danger area *or* zone; ~**herd der,** ~**quelle die** source of danger; ~**zone die** *s.* ~**bereich;** ~**zulage die** danger money *no indef. art.*

gefährlich [gəˈfɛːɐ̯lɪç] **1.** *Adj.* dangerous; *(gewagt)* risky; **[für jmdn./etw.]** ~ **sein** be dangerous [for sb./sth.]; **er könnte mir ~ werden** he could be a threat *or* a danger to me; *(fig.)* I could fall for him [in a big way]. **2.** *adv.* dangerously

Gefährlichkeit die; ~: dangerousness; *(Gewagtheit)* riskiness

gefahr·los 1. *Adj.* safe. **2.** *adv.* safely

Gefährt das; ~**[e]s,** ~**e** *(geh.)* vehicle

Gefährte der; ~**n,** ~**n, Gefährtin die;** ~, ~**en** *(geh.)* companion; *(Lebens~)* partner in life

Gefälle [gəˈfɛlə] **das;** ~**s,** ~ slope; incline; *(eines Flusses)* drop; *(einer Straße)* gradient

[1]**gefallen** *unr. itr. V.* **a) das gefällt mir [gut]/[gar] nicht** I like it [very much *or (coll.)* a lot]/ don't like it [at all]; **es gefiel ihr, wie er sich bewegte** she liked the way he moved; **weißt du, was mir an dir/ dem Bild so gut gefällt?** do you know what I like so much about you/the picture?; **mir gefällt es hier** I like it here; **er gefällt mir [ganz und gar] nicht** *(ugs.: sieht krank aus)* he looks in a bad way to me *(coll.)*; **die Sache gefällt mir nicht** *(ugs.)* I don't like [the look of] it *(coll.)*; **b) sich** *(Dat.)* **etw. ~ lassen** put up with sth.; **c)** *(abwertend)* **sich** *(Dat.)* **in einer Rolle ~:** enjoy *or* like playing a role; fancy oneself in a role *(coll.)*; **er gefällt sich in Übertreibungen** he likes to exaggerate

[2]**gefallen 1.** *2. Part. v.* **fallen, gefallen. 2.** *Adj.* fallen ⟨*angel etc.*⟩; **ein ~es Mädchen** *(veralt.)* a fallen woman

[1]**Gefallen der;** ~**s,** ~: favour; **jmdm. einen ~ tun** *od.* **erweisen** do sb. a favour; **tu mir den** *od.* **einen ~, und ...!** *(ugs.)* do me a favour and ...; **jmdn. um einen ~ bitten** ask a favour of sb.

[2]**Gefallen das;** ~s pleasure; ~ **an jmdm./aneinander finden** like sb./ each other; **an etw.** *(Dat.)* ~ **finden** get *or* derive pleasure from sth.; enjoy sth.; **etw. jmdm. zu ~ tun** do sth. to please sb.

Gefallene der; *adj. Dekl.* soldier killed in action; **die ~n** the fallen; those killed *or* those who fell in action

ge·fällig 1. *Adj.* **a)** *(hilfsbereit)* obliging; helpful; **jmdm. ~ sein** oblige *or* help sb.; **b)** *(anziehend)* pleasing; agreeable; pleasant, agreeable ⟨*programme, behaviour*⟩; **c)** *nicht attr.* **noch ein Kaffee ~?** would you like *or* care for another coffee?. **2.** *adv.* pleasingly; agreeably

Gefälligkeit die; ~, ~**en a)** *(Hilfeleistung)* favour; **jmdm. eine ~ erweisen** do sb. a favour; **b)** *o. Pl.* *(Hilfsbereitschaft)* obligingness; helpfulness; **etw. aus reiner ~ tun** do sth. just to be obliging

gefälligst [gəˈfɛlɪçst] *Adv. (ugs.)* kindly; **laß das ~!** kindly stop that

gefangen *2. Part. v.* **fangen**

Gefangene der/die; *adj. Dekl.* **a)** prisoner; captive; ~ **machen** take prisoners; **b)** *(Häftling, Kriegs~)* prisoner

Gefangenen·lager das prisoner of war camp; prison camp

gefangen-, Gefangen-: ~**|halten** *unr. tr. V.* **a) jmdn./ein Tier ~halten** hold sb. prisoner *or* captive/keep an animal in captivity; **b)** *(fig. geh.: fesseln)* **jmdn. ~halten** hold sb. enthralled; ~**nahme die;** ~: capture; ~**|nehmen** *unr. tr. V.* **a) jmdn. ~nehmen** capture sb.; take sb. prisoner; **b)** *(fig. fesseln)* captivate; enthral

Gefangenschaft die; ~, ~**en** captivity; **in ~ sein/geraten** be a prisoner/be taken prisoner

Gefängnis [gəˈfɛŋnɪs] **das;** ~**ses,** ~**se a)** prison; gaol; **jmdn. ins ~ bringen/werfen** put/throw sb. in[to] prison; **im ~ sein** *od.* **sitzen** be in prison; **ins ~ kommen** be sent to prison; **b)** *(Strafe)* imprisonment; **darauf steht ~:** that is punishable by imprisonment *or* a prison sentence; **jmdn. zu zwei Jahren ~ verurteilen** sentence sb. to two years' imprisonment *or* two years in prison

Gefängnis-: ~**direktor der** prison governor; ~**kleidung die** prison uniform; ~**mauer die** prison wall; ~**strafe die** prison sentence; **eine ~strafe verbüßen** serve a prison sentence; ~**wärter der** prison officer; [prison] warder; ~**zelle die** prison cell

Gefasel das; ~s *(ugs. abwertend)* twaddle *(coll.)*; drivel *(derog.)*

Gefäß [gəˈfɛːs] **das;** ~**es,** ~**e a)** *(Behälter)* vessel; container; **b)** *(Med.)* vessel

gefaßt [gəˈfast] **1.** *Adj.* **a)** calm; composed; **mit ~er Haltung** with composure; **b)** *in* **auf etw.** *(Akk.)* **[nicht] ~ sein** [not] be prepared for sth.; **sich auf etw.** *(Akk.)* ~ **machen** prepare oneself for sth.; **der kann sich auf was ~ machen** *(ugs.)* he'll catch it *or* be for it *(coll.)*. **2.** *adv.* calmly; with composure

Gefecht das; ~**[e]s,** ~**e a)** battle; engagement *(Milit.)*; **ein schweres/kurzes ~:** fierce fighting/a skirmish; **sich** *(Dat.)***/dem Feind ein ~ liefern** engage each other/ the enemy in battle; **jmdn./etw. außer ~ setzen** put sb./sth. out of action; **b)** *(Fechten)* bout; *s. auch* **Eifer**

gefechts-, Gefechts-: ~**bereit,** ~**klar** *Adj. (Milit.)* ready for action *or* battle *postpos.*; combat-ready; ~**stand der** *(Milit.)* battle headquarters *pl.*; command post; *(Luftw.)* operations room

gefehlt *Adj.* **weit ~!** wide of the mark!

gefestigt *Adj.* assured ⟨*beliefs*⟩; secure ⟨*person*⟩; established ⟨*tradition*⟩

Gefieder [gəˈfiːdɐ] **das;** ~**s,** ~: plumage; feathers *pl.*

gefiedert *Adj.* **a)** feathered; **b)** *(Bot.)* pinnate

Gefilde [gəˈfɪldə] **das;** ~**s,** ~ *(geh.)* **sonnige ~:** sunny climes *(literary)*; **wieder in heimatlichen ~n sein** *(scherzh.)* be back under one's native skies

Geflecht das; ~**[e]s,** ~**e a)** wickerwork *no art.*; **b)** *(fig.)* tangle; **ein wirres/dichtes ~ von Zweigen** a tangled/dense network of twigs

gefleckt *Adj.* spotty, blotchy ⟨*skin, face*⟩; spotted ⟨*leopard skin*⟩

geflissentlich [gəˈflɪsn̩tlɪç] **1.** *Adj.; nicht präd.* deliberate. **2.** *adv.* deliberately

geflochten [gəˈflɔxtn̩] *2. Part. v.* **flechten**

geflogen [gəˈfloːgn̩] *2. Part. v.* **fliegen**

geflohen [gəˈfloːən] *2. Part. v.* **fliehen**

geflossen [gəˈflɔsn̩] *2. Part. v.* **fließen**

Geflügel das; ~s poultry

Geflügel-: ~farm die poultry farm; **~schere die** poultry shears *pl.*

geflügelt *Adj.* winged ⟨*insect, seed*⟩; **ein ~es Wort** *(fig.)* a standard *or* familiar quotation

Geflüster das; ~s whispering

gefochten [gə'fɔxtn̩] *2. Part. v.* **fechten**

Gefolge das; ~s, ~ a) *(Begleitung)* entourage; retinue; **b)** *(Trauergeleit)* cortège

Gefolgschaft die; ~: allegiance; **jmdm. ~ leisten** give one's allegiance to sb.

gefragt *Adj.* ⟨*artist, craftsman, product*⟩ in great demand; sought-after ⟨*artist, craftsman, product*⟩

gefräßig [gə'frɛ:sɪç] *Adj. (abwertend)* greedy; gluttonous; voracious ⟨*animal, insect*⟩

Gefräßigkeit die; ~ *(abwertend)* greediness; gluttony; *(von Tieren)* voracity

Gefreite [gə'fraitə] **der;** *adj. Dekl. (Milit.)* lance-corporal *(Brit.);* private first class *(Amer.); (Marine)* able seaman; *(Luftw.)* aircraftman first class *(Brit.);* airman third class *(Amer.)*

gefressen *2. Part. v.* **fressen**

gefrieren 1. *unr. itr. V.; mit sein* freeze. **2.** *unr. tr. V.* [deep-]freeze ⟨*food*⟩

gefrier-, Gefrier-: ~fach das freezing compartment; **~punkt der** freezing-point; **Temperaturen über/unter dem ~punkt** temperatures above/below freezing; **~schrank der** [upright] freezer; **~|trocknen** *tr. V.; meist im Inf. u. 2. Part.* freeze-dry; **~truhe die** [chest] freezer

gefroren [gə'fro:rən] *2. Part. v.* **frieren, gefrieren**

Gefüge das; ~s, ~: structure; **das soziale ~:** the social fabric

gefügig *Adj.* submissive; compliant; docile ⟨*animal*⟩; **ein ~es Werkzeug** *(fig.)* a willing tool; **sich** *(Dat.)* **jmdn. ~ machen** make sb. submit to one's will

Gefühl das; ~s, ~e a) sensation; feeling; **ein ~ des Schmerzes** a sensation of pain; **b)** *(Gemütsregung)* feeling; **ein ~ der Einsamkeit** a sense *or* feeling of loneliness; **kein ~ haben** have no feelings; **das ist das höchste der ~e** *(ugs.)* that's the absolute limit; **c)** *(Ahnung)* feeling; **etw. im ~ haben** have a feeling *or* a premonition of sth.; **d)** *(Verständnis, Gespür)* sense; instinct; **sich auf sein ~ verlassen** trust one's feelings *or* instinct; **etw. nach ~ tun** do sth. by instinct

gefühl·los *Adj.* **a)** numb; **b)** *(herzlos)* unfeeling; callous

Gefühllosigkeit die; ~ a) numbness; lack of sensation; **b)** *(Herzlosigkeit)* unfeelingness; callousness

gefühls-, Gefühls-: ~arm *Adj.* lacking in feeling; **~ausbruch der** outburst [of emotion]; **~betont 1.** *Adj.* emotional ⟨*speech, argument*⟩; **2.** *adv.* **~betont handeln** be guided by one's emotions; **~duselei** [-du:zə'lai] **die;** ~ *(ugs. abwertend)* mawkishness; mawkish sentimentality; **~leben das** emotional life; **~mäßig 1.** *Adj.* emotional ⟨*reaction*⟩; ⟨*action*⟩ based on emotion; **2.** *adv.* **rein ~mäßig würde ich sagen, daß ...:** my own, purely instinctive, feeling would be to say that ...; **~regung die** emotion

gefühl·voll 1. *Adj.* **a)** *(empfindsam)* sensitive; **b)** *(ausdrucksvoll)* expressive. **2.** *adv.* sensitively; expressively; with feeling

gefüllt *Adj.* full ⟨*wallet*⟩; double ⟨*lilac, geraniums*⟩; stuffed ⟨*fowl, tomato etc.*⟩; **~e Bonbons** sweets *(Brit.) or (Amer.)* candies with centres

gefunden [gə'fʊndn̩] *2. Part. v.* **finden;** *s. auch* **Fressen b**

gefurcht *Adj.* lined; wrinkled

gefürchtet *Adj.* dreaded; feared ⟨*despot, opponent*⟩

gegangen *2. Part. v.* **gehen**

gegeben 1. *2. Part. v.* **geben. 2.** *Adj.* **a)** given; **etw. als ~ voraussetzen/hinnehmen** take sth. for granted; **aus ~em Anlaß** for certain reasons *(specified or not);* **unter den ~en Umständen** in these circumstances; **b)** *(passend)* right; proper; **das ist das gegebene** that's the best thing; **zu ~er Zeit** in due course; at the appropriate time

gegebenen·falls *Adv.* should the occasion arise

Gegebenheit die; ~, ~en; *meist Pl.* condition; fact

gegen ['ge:gn̩] **1.** *Präp. mit Akk.* **a)** towards; *(an)* against; **das Dia ~ das Licht halten** hold the slide up to *or* against the light; **~ die Tür schlagen** bang on the door; **~ etw. stoßen** knock into *or* against sth.; **ein Mittel ~ Husten/Krebs** a cough medicine/a cure for cancer; **~ die Abmachung** contrary to *or* against the agreement; **~ alle Vernunft/bessere Einsicht** against all reason/one's better judgement; **b)** *(ungefähr um)* around ⟨*midnight, 4 o'clock, etc.*⟩; **~ Abend/Morgen** towards evening/dawn; **c)** *(im Vergleich zu)* compared with; in comparison with; **ich wette hundert ~ eins, daß er ...:** I'll bet you a hundred to one he ...; **d)** *(im Ausgleich für)* for; **etw. ~ bar verkaufen** sell sth. for cash; **etw. ~ Quittung erhalten** receive sth. against a receipt; **e)** *(veralt.: gegenüber)* to; towards; **~ jmdn./sich streng sein** be strict with sb./oneself. **2.** *Adv. (ungefähr)* about; around

Gegen-: ~angriff der counter-attack; **~argument das** counter-argument; **~beispiel das** example to the contrary; counter-example; **~besuch der** return visit; **~bewegung die** counter-movement; **~beweis der** evidence to the contrary, counter-evidence *no indef. art., no pl.;* **den ~beweis antreten** *od.* **führen** produce evidence to the contrary *or* counter-evidence

Gegend ['ge:gn̩t] **die; ~, ~en a)** *(Landschaft)* landscape; *(geographisches Gebiet)* region; **durch die ~ latschen/kurven** *(salopp)* traipse around *(coll.)*/drive around; **b)** *(Umgebung)* area; neighbourhood; **in der ~ von/um Hamburg** in the Hamburg area; **in der ~ des Parks** in the neighbourhood of the park; **c) in der ~ des Magens** in the region of the stomach

Gegen·darstellung die: eine ~darstellung [der Sache] an account [of the matter] from an opposing point of view

gegen·einander *Adv.* **a)** against each other *or* one another; *(im Austausch)* **man tauschte die Geiseln ~ aus** the hostages were exchanged; **zwei Begriffe/Epochen ~ abgrenzen** distinguish two concepts/divide two periods from each other; **b)** *(zueinander)* to[wards] each other *or* one another

gegeneinander-: ~|halten *unr. tr. V.* **a) zwei Dinge ~halten** hold two things up together *or* side by side; **b)** *(fig.: vergleichen)* compare; put side by side; **~|prallen** *itr. V.; mit sein* collide

gegen-, Gegen-: ~entwurf der alternative draft; **~fahrbahn die** opposite carriageway; **~frage die** question in return; counter-question; **~gewicht das** counterweight; **ein ~gewicht zu** *od.* **gegen etw. bilden** *(fig.)* counterbalance sth.; **~gift das** antidote; **~grund der** *s.* **Grund**

d; ~kandidat der opposing candidate; rival candidate; **~leistung die** service in return; consideration; **als ~leistung für etw.** in return for sth.; **~|lenken** *itr. V.* turn the wheel to correct the line

Gegen·licht das; *o. Pl. (bes. Fot.)* back-lighting

Gegenlicht·aufnahme die *(Fot.)* photograph taken against the light; contre-jour photograph

gegen-, Gegen-: ~liebe die *in* **[bei jmdm.] ~liebe finden** *od.* **auf ~liebe stoßen** find favour [with sb.]; **~maßnahme die** countermeasure; **~mittel das** *(gegen Gift)* antidote; **~partei die** opposing side; other side; **~pol der** *(auch fig.)* opposite pole; *(Math.)* antipole; **~probe die** cross-check; **die ~probe machen** *(bei einer Behauptung, These)* carry out a cross-check; *(bei einer Rechnung)* work the sum the other way round; *(bei Abstimmungen) carry out a recount in which the opposite motion is put;* **~rede die a)** *(geh.: Erwiderung)* reply; rejoinder; **b)** *(Widerrede)* contradiction; *(Einspruch)* objection; **~richtung die** opposite direction; **~satz der a)** opposite; **einen schroffen/diametralen ~satz zu etw./jmdm. bilden** contrast sharply with/be diametrically opposed to sth./sb.; **im ~satz zu** in contrast to *or* with; unlike; **b)** *(Widerspruch)* conflict; **im krassen/scharfen ~satz zu etw. stehen** be in stark/sharp conflict with sth.; **c)** *Pl. (Meinungsverschiedenheiten)* differences *Pl.*; **~sätzlich 1.** *Adj.* conflicting ‹*views, opinions, etc.*›; opposing ‹*alignments*›; **2.** *adv.* **etw. ~sätzlich beurteilen** judge sth. completely differently; **~schlag der** counterstroke; **zum ~schlag ausholen** prepare to counter-attack *or* strike back; **~seite die a)** other side; far side; **b)** *s.* **~partei; ~seitig 1.** *Adj.* mutual ‹*aid, consideration, love, consent, services*›; reciprocal ‹*aid, obligation, services*›; **in ~seitiger Abhängigkeit stehen** be mutually dependent; be dependent on each other *or* one another; **in ~seitigem Einvernehmen** by mutual agreement; **2.** *adv.* **sich ~seitig helfen** help each other *or* one another; **~seitigkeit die** reciprocity; **auf ~seitigkeit** *(Dat.)* **beruhen** be mutual; **~spieler der a)** *(Widersacher)* opponent; **b)** *(Sport)* opposite number; **c)** *(Theater)* antagonist; **~sprechanlage die** intercom [system]; *(Fernspr.)* duplex system

Gegen·stand der a) object; **Gegenstände des täglichen Bedarfs** objects *or* articles of everyday use; **b)** *o. Pl. (Thema)* subject; topic; **etw. zum ~ haben** deal with sth.; be concerned with sth.; **c)** *(Ziel) (der Zuneigung, des Hasses)* object; *(der Kritik)* target; butt

gegenständlich ['ge:gn̩ʃtɛntlɪç] *Adj. (Kunst)* representational

gegenstands·los *Adj.* **a)** *(hinfällig)* invalid; **b)** *(grundlos)* unsubstantiated, unfounded ‹*accusation, complaint*›; baseless ‹*fear*›; unfounded ‹*jealousy*›

gegen-, Gegen-: ~stimme die a) vote against; **ohne ~stimme** unanimously; **b)** *(gegenteilige Meinung)* dissenting voice; **~stück das** companion piece; *(fig.)* counterpart; **~teil das** opposite; **im ~teil** on the contrary; **~teilig** *Adj.* opposite; contrary; **~teiliger Meinung sein** be of the opposite opinion; **~tor das, ~treffer der** *(Ballspiele)* goal for the other side

gegen·über 1. *Präp. mit Dat.* **a)** opposite; **~ dem Bahnhof, dem Bahnhof ~:** opposite the station; **b)** *(in bezug auf)* **~ jmdm.** *od.* **jmdm. ~ freundlich/streng sein** be kind to/strict with sb.; **~ einer Sache** *od.* **einer Sache ~ skeptisch sein** be sceptical about sth.; **c)** *(im Vergleich zu)* compared with; in comparison with; **~ jmdm. im Vorteil sein** have an advantage over sb. **2.** *Adv.* opposite

Gegen·über das; ~s, ~ person [sitting/standing] opposite

gegenüber-, Gegenüber-: ~|liegen *unr. itr. V.* **sich** *(Dat.) od.* **einander ~liegen** face each other *or* one another; **auf der ~liegenden Seite** on the opposite side; **~|sitzen** *unr. itr. V.* **jmdm./sich ~sitzen** sit opposite *or* facing sb./each other; **~|stehen** *unr. itr. V.* **a) jmdm./einer Sache ~stehen** stand facing sb./sth.; **Schwierigkeiten ~stehen** *(fig.)* be faced *or* confronted with difficulties; **b) jmdm./einer Sache feindlich/wohlwollend ~stehen** *(fig.)* be ill/well disposed towards sb./sth.; *s. auch* **ablehnend 2; c) sich ~stehen** *(Sport)* face each other *or* one another; meet; **d) sich** *(Dat.)* **~stehen** *(fig.: im Widerstreit stehen)* stand directly opposed to each other *or* one another; **~|stellen** *tr. V.* **a)** confront; **jmdn. einem Zeugen ~stellen** to confront sb. with a witness; **b)** *(vergleichen)* compare; **~stellung die a)** confrontation; **b)** *(Vergleich)* comparison; **~|treten** *unr. itr. V.; mit sein (auch fig.)* face ‹*person, difficulties*›

Gegen-: ~verkehr der oncoming traffic; **~vorschlag der** counter-proposal

Gegenwart [-vart] **die; ~ a)** present; *(heutige Zeit)* present [time *or* day]; **die Musik der ~:** contemporary music; **b)** *(Anwesenheit)* presence; **in ~ von anderen** in the presence of others; **c)** *(Grammatik)* present [tense]

gegenwärtig [-vɛrtɪç] **1.** *Adj.* **a)** *nicht präd.* present; *(heutig)* present[-day]; current; **b)** *(veralt.: anwesend, zugegen)* present; **bei etw. ~ sein** be present at sth.. **2.** *adv.* at present; at the moment; *(heute)* at present; currently

gegenwarts-: ~bezogen, ~nah[e] 1. *Adj.* relevant to the present day *or* to today *postpos.*; *(aktuell)* topical; **2.** *adv.* **~nah** *od.* **~bezogen unterrichten** teach in accordance with contemporary ideas

Gegen-: ~wehr die; *o. Pl.* resistance; **[keine] ~wehr leisten** put up [no] resistance; **~wert der** equivalent; **~wind der** head wind; **~zug der a)** *(Brettspiele, fig.)* countermove; *(Politik)* reciprocal gesture; **b)** *(entgegenkommender Zug)* train in the opposite direction

gegessen [gə'gɛsn̩] *2. Part. v.* **essen**

geglichen [gə'glɪçn̩] *2. Part. v.* **gleichen**

geglitten [gə'glɪtn̩] *2. Part. v.* **gleiten**

geglommen [gə'glɔmən] *2. Part. v.* **glimmen**

Gegner ['ge:gnɐ] **der; ~s, ~ a)** adversary; opponent; *(Rivale)* rival; **ein ~ einer Sache** *(Gen.)* **sein** oppose sth.; be an opponent of sth.; **b)** *(Sport)* opponent; *(Mannschaft)* opposing team; **c)** *(Milit.)* enemy

Gegnerin die; ~, ~nen a) *s.* **Gegner a; b)** *(Sport)* opponent

gegnerisch *Adj.; nicht präd.* **a)** opposing; **b)** *(Sport)* opposing ‹*team, player, etc.*›; opponents' ‹*goal*›; **c)** *(Milit.)* enemy

Gegnerschaft die; ~ hostility; antagonism

gegolten [gə'gɔltn̩] *2. Part. v.* **gelten**

gegoren [gə'go:rən] *2. Part. v.* **gären**

gegossen [gə'gɔsn̩] *2. Part. v.* **gießen**

gegriffen [gə'grɪfn̩] *2. Part. v.* **greifen**

Gegröle das; ~s *(ugs. abwertend)* [raucous] bawling and shouting; *(Gesang)* raucous singing

Gehabe das; ~s *(abwertend)* affected behaviour

gehaben *refl. V. (veralt., noch scherzh.) in* **gehab dich wohl!/gehabt euch wohl!** farewell!

gehabt 1. *2. Part. v.* **haben. 2.** *Adj.; nicht präd. (ugs.: schon dagewesen)* same old *(coll.);* usual; **wie ~:** as before

Gehackte [gə'haktə] **das;** *adj. Dekl.* mince[meat]; **~s vom Rind/Schwein** minced beef/pork

[1]**Gehalt der; ~[e]s, ~e a)** *(gedanklicher Inhalt)* meaning; **religiöser ~:** religious content; **b)** *(Anteil)* content; **ein hoher ~ an Gold** a high gold content

[2]**Gehalt das,** *österr. auch:* **der; ~[e]s, Gehälter** [gə'hɛltɐ] salary; **1 000 DM ~, ein ~ von 1 000 DM beziehen** draw a salary of 1,000 marks

gehalten 1. *2. Part. v.* **halten. 2.** *Adj. (geh.)* **~ sein, etw. zu tun** be obliged *or* required to do sth.

gehalt·los *Adj.* unnutritious ‹*food*›; ‹*wine*› lacking in body; *(fig.)* vacuous; empty; lacking in substance *postpos., not pred.*

Gehalts-: ~anspruch der salary claim; pay claim; **~aufbesserung die** increase in salary; **~empfänger der** salary earner; **~erhöhung die** salary increase; rise [in salary]; **~liste die** payroll; **auf jmds. ~liste** *(Dat.)* **stehen** *(fig.)* be on sb.'s payroll; be in sb.'s pocket; **~vorschuß der** advance [on one's salary]

gehalt·voll *Adj.* nutritious, nourishing ‹*food*›; full-bodied ‹*wine*›; ‹*novel, speech*› rich in substance *postpos.*

Gehänge das; ~s, ~ *(Girlande)* festoon; *(Kranz)* garland

geharnischt [gə'harnɪʃt] *Adj.* **a)** *(scharf, energisch)* strongly-worded; **b)** *(hist.: gepanzert)* **ein ~er Ritter** a knight in armour

gehässig [gə'hɛsɪç] *Adj. (abwertend)* spiteful; **~ von jmdm. reden** be spiteful about sb.

Gehässigkeit die; ~, ~en a) *o. Pl. (Wesen)* spitefulness; **b)** *meist Pl. (Äußerung)* spiteful remark

gehauen *2. Part. v.* **hauen**

gehäuft 1. *Adj.* heaped ‹*spoon*›. **2.** *adv.* in large numbers

Gehäuse [gə'hɔyzə] **das; ~s, ~ a)** *(einer Maschine, Welle)* casing; housing; *(einer Kamera, Uhr)* case; casing; *(einer Lampe)* housing; **b)** *(Schnecken~ usw.)* shell; **c)** *(Kern~)* core; **d)** *(Sportjargon: Tor)* goal

geh·behindert *Adj.* able to walk only with difficulty *postpos.;* disabled; **sie ist stark ~behindert** she can walk only with great difficulty

Gehege das; ~s, ~ a) *(Jägerspr.: Revier)* preserve; **jmdm. ins ~ kommen** *(fig.)* poach on sb.'s preserve; **sich** *(Dat.)* **[gegenseitig] ins ~ kommen** *(fig.)* encroach on each other's territory; **b)** *(im Zoo)* enclosure

geheim 1. *Adj.* **a)** secret; **streng ~:** top *or* highly secret; **b)** *(mysteriös)* mysterious. **2.** *adv.* **~ abstimmen** vote by secret ballot

Geheim- secret ‹*agent, society, code, service, etc.*›

geheim|halten *unr. tr. V.* keep secret; **etw. ~halten** keep sth. secret

Geheim·haltung die; *o. Pl.* observance of secrecy

Geheimnis das; ~ses, ~se a) secret; **vor jmdm. [keine] ~se haben** have [no] secrets from sb.; **das ist das ganze ~:** that's all there is to it; **b)** *(Unerforschtes)* mystery; secret; **die ~se der Natur** the mysteries *or* secrets of nature

geheimnis-, Geheimnis-: ~krämer der *(ugs.)* mystery-monger; **~krämerei die; ~, ~tuerei die; ~** *(ugs. abwertend)* secretiveness; mystery-mongering; **~voll 1.** *Adj.* mysterious; **auf ~volle Weise** in a mysterious way; mysteriously; **2.** *adv.* mysteriously; **~voll tun** be mysterious; act mysteriously

Geheim·polizei die secret police

Geheimrats·ecken *Pl. (ugs. scherzh.)* receding hairline *sing.*

Geheim-: ~rezept das secret recipe; **~tip der** inside tip; **~tür die** secret door; **~waffe die** *(Milit.)* secret weapon

Geheiß das; ~es *(geh.)* behest *(literary);* command; **auf jmds. ~:** at sb.'s behest *or* command

gehemmt *Adj.* inhibited

gehen ['ge:ən] *unr. itr. V.; mit sein* **a)** *(sich zu Fuß fortbewegen)* walk; go; **auf und ab ~:** walk up and down; **über die Straße ~:** cross the street; **wo er geht und steht** wherever he goes *or* is; no matter where he goes *or* is; **etw. geht durch die Presse** *(fig.)* sth. is in the papers; **b)** *(sich irgendwohin begeben)* go; **tanzen ~:** go dancing; **schlafen ~:** go to bed; **zu jmdm. ~:** go to see sb.; go and see sb. *(coll.);* **zum Arzt ~:** go to the doctor; **nach London ~:** move to London; **an die Arbeit ~** *(fig.)* get down to work; **in sich** *(Akk.)* **~** *(fig.)* take stock of oneself; **c)** *(regelmäßig besuchen)* attend; **in die** *od.* **zur Schule ~:** be at *or* attend school; **d)** *(weg~)* go; leave; **Sie können ~:** you may go; **der Minister mußte ~:** the Minister had to resign; **er ist von uns gegangen** *(verhüll.)* he has passed away *or* passed over *(euphem.);* **e)** *(ugs.: [ab]fahren)* ‹*train*› leave; **f)** *(in Funktion sein)* work; **meine Uhr geht falsch/richtig** my watch is wrong/right; **das Telefon geht ununterbrochen** the telephone never stops ringing; **g)** *(möglich sein)* **ja, das geht** yes, I/we can manage that; **das geht nicht** that can't be done; that's impossible; *(ist nicht zulässig)* that's not on *(Brit. coll.);* no way *(coll.);* **Donnerstag geht auch** Thursday's a possibility *or* all right too; **es geht einfach nicht, daß du so spät nach Hause kommst** it simply won't do for you to come home so late; **es geht leider nicht anders** unfortunately there's nothing else for it; **das wird schwer/schlecht ~:** that will be difficult; **auf diese Weise geht es nicht/sicher** it won't/is bound to work this way; **h)** *(ugs.: angehen)* **es geht so** it could be worse; **das Essen ging ja noch, aber der Wein war ungenießbar** the food was passable, but the wine was undrinkable; **Hast du gut geschlafen? – Es geht** Did you sleep well? – Not too bad *or* So-so; **i)** *(ablaufen)* **das Geschäft geht gut/gar nicht** business is doing well/not doing well at all; **es geht alles nach Plan** everything is going according to plan; **alles geht drunter und drüber** *(ugs.)* everything's at sixes and sevens; **wie geht die Melodie?** *(fig.)* how does the tune go?; what's the tune?; **vor sich ~:** go on; happen; **j)** *(reichen)* **das Wasser geht mir bis an die Knie** the water comes up to *or* reaches my knees; **ich gehe ihm bis zu den Schultern** I come up to his shoulders; **in die Hunderte ~:** run into [the] hundreds; **das geht über mein Vermögen/meinen Horizont** *(fig.)* that is beyond me; **es geht [doch] nichts über ...** (+ *Akk.*) *(fig.)* there is nothing like *or* nothing to beat ...; nothing beats ...; **das geht zu weit** *(fig.)* that's going too far; **k)** *unpers.* **jmdm. geht es gut/schlecht** *(gesundheitlich)* sb. is well/not well; *(geschäftlich)* sb. is doing well/badly; **wie geht es dir/Ihnen?** how are you?; **wie geht's, wie steht's?** *(ugs.)* how are things?; **l)** *unpers. (sich um etw. handeln)* **es geht um mehr als ...:** there is more at stake than ...; **jmdm. geht es um etw.** sth. matters to sb.; **worum geht es hier?** what is this all about?; **m)** *(tätig werden)* **in den Staatsdienst/in die Politik ~:** join the Civil Service/go into politics; **zum Film/Theater ~:** go into films/on the stage; **n)** *(ugs.: sich kleiden)* **in Hosen ~:** wear trousers; **in kurz/lang ~:** wear a short/long dress/skirt; **als Zigeuner ~:** go as a gypsy; **o)** *(ugs.: sich zu schaffen machen an)* **du sollst nicht an meine Sachen ~:** you must not mess around with my things; *(benutzen)* you must not take my things; **die Kinder sind an den Kuchen gegangen** the children have been at the cake *(coll.)*; **p) mit jmdm. ~:** go out with sb.; **q)** *(absetzbar sein)* **[gut/schlecht] ~:** sell [well/slowly]; **r)** *(passen)* go; **s)** *(verlaufen)* go; **die Straße geht geradeaus/nach links** the road goes *or* runs straight ahead/turns to the left; **wohin geht diese Straße?** where does this road go *or* lead to?; **t)** *(gerichtet sein auf)* **nach der Straße/nach Süden ~** ⟨*room, window, etc.*⟩ face the road/face south; **gegen jmdn./etw. ~** *(fig.)* be aimed *or* directed at sb./sth.; **das geht gegen meine Überzeugung** that goes against my convictions; **u)** *(als Maßstab nehmen)* **nach jmdm./etw. ~:** go by sb./sth.; **v) in Stücke/Scherben ~:** get smashed; **w) etw. geht auf jmdn./jmds. Rechnung** sb. is paying for sth.; **x)** *(bestimmt sein)* **an jmdn. ~:** go to sb.; **die Briefe ~ nach Oxford** the letters are going to Oxford. **2.** *unr. tr. V. (zurücklegen)* **eine Strecke ~:** cover *or* do a distance; **einen Umweg ~:** make a detour; **einen Weg in 30 Minuten ~:** do a walk in 30 minutes; **seine eigenen Wege ~** *(fig.)* go one's own way

Gehen das; ~s a) walking; **er hat Schmerzen beim ~:** it hurts him to walk; **b)** *(Leichtathletik)* walking; **der Sieger im 50-km-~:** the winner of the 50 km walk

gehen|lassen 1. *unr. refl. V.* lose control of oneself; *(sich vernachlässigen)* let oneself go. **2.** *unr. tr. V. (ugs.)* leave alone

Geher ['ge:ɐ] **der; ~s, ~** *(Leichtathletik)* walker

geheuer [gə'hɔyɐ] *Adj.* **a) in diesem Gebäude ist es nicht ~:** this building is eerie; this building feels as if it's haunted *(coll.)*; **b) ihr war doch nicht [ganz] ~:** she felt [a little] uneasy; **c) die Sache ist [mir] nicht ganz ~:** [I feel] there's something odd *or* suspicious about this business

Geheul das; ~[e]s a) *(auch fig.)* howling; **b)** *(ugs. abwertend: Weinen)* bawling; wailing

Geheule das; ~s *s.* **Geheul b**

Gehilfe [gə'hɪlfə] **der; ~n, ~n, Gehilfin die; ~, ~nen a)** qualified assistant; **b)** *(veralt.: Helfer/Helferin)* helper; assistant

Gehirn das; ~[e]s, ~e a) brain; **b)** *(ugs.: Verstand)* mind; **sein ~ anstrengen** *od.* **sich** *(Dat.)* **das ~ zermartern** rack one's brain[s]

Gehirn-: ~erschütterung die *(Med.)* concussion; **~schlag der** *(Med.)* stroke; [cerebral] apoplexy *no art. (Med.)*; **~wäsche die** brainwashing *no indef. art.*; **jmdn. einer ~wäsche unterziehen** brainwash sb.; **~zelle die** brain cell

gehoben [gə'ho:bn̩] **1.** *2. Part. v.* **heben. 2.** *Adj.* **a)** higher ⟨*income*⟩; senior ⟨*position*⟩; **der ~e Dienst** the higher [levels of the] Civil Service; **der ~e Mittelstand** the upper middle class; **b)** *(anspruchsvoll)* **Kleidung für den ~en Geschmack** clothes for those with discerning taste; **Artikel für den ~en Bedarf** luxury goods; **c)** *(gewählt)* elevated, refined ⟨*language, expression*⟩; **d)** *(feierlich)* festive ⟨*mood*⟩. **3.** *adv.* **sich ~ ausdrücken** use elevated *or* refined language

Gehöft [gə'hœft, -'hø:ft] **das; ~[e]s, ~e** farm[stead]

geholfen [gə'hɔlfn̩] *2. Part. v.* **helfen**

Gehölz [gə'hœlts] **das; ~es, ~e a)** copse; spinney *(Brit.)*; **b)** *meist Pl. (Holzgewächs)* woody plant

Gehör [gə'hø:ɐ] **das; ~[e]s** [sense of] hearing; **ein gutes ~ haben** have good hearing; **[etw.] nach dem ~ singen/spielen** sing/play [sth.] by ear; **das absolute ~ haben** *(Musik)* have absolute pitch; **~/kein ~ finden** meet with *or* get a/no response; **jmdm./einer Sache [kein] ~ schenken** [not] listen to sb./sth.

gehorchen [gə'hɔrçn̩] *itr. V.* **jmdm. ~:** obey sb.; **einer Sache** *(Dat.)* **~:** respond to sth.; **einer Laune** *(Dat.)* **~:** yield to a caprice

gehören 1. *itr. V.* **a)** *(Eigentum sein)* **jmdm. ~:** belong to sb.; **das Haus gehört uns nicht** the house doesn't belong to us; we don't own the house; **wem gehört das Buch?** whose book is it?; who does the book belong to?; **b)** *(Teil eines Ganzen sein)* **zu jmds. Freunden ~:** be one of sb.'s friends; **zu jmds. Aufgaben ~:** be part of sb.'s duties; **c)** *(passend sein)* **dein Roller gehört doch nicht in die Küche!** your scooter does not belong in the kitchen!; **das gehört nicht/durchaus zur Sache** that is not to the point/is very much to the point; **d)** *(sein sollen)* **das gehört verboten** that should be forbidden; that shouldn't be allowed; **du gehörst ins Bett** you should be in bed; **d)** *(nötig sein)* **es hat viel Fleiß dazu gehört** it took *or* called for a lot of hard work; **dazu gehört sehr viel/einiges** that takes a lot/something; **dazu gehört nicht viel** that doesn't take much; **e)** *(bes. südd.)* **er gehört geohrfeigt** he deserves *or* *(coll.)* needs a box round the ears. **2.** *refl. V. (sich schicken)* be fitting; **es gehört sich [nicht], ... zu ...:** it is [not] good manners to ...; **wie es sich gehört** comme il faut

gehörig 1. *Adj.* **a)** *nicht präd. (gebührend)* proper; **jmdm. den ~en Respekt/die ~e Achtung erweisen** show sb. proper *or* due respect; **b)** *nicht präd. (ugs.: beträchtlich)* **ein ~er Schrecken/eine ~e Portion Mut** a good fright/a good deal of courage. **2.** *adv.* **a)** *(gebührend)* properly; **b)** *(ugs.: beträchtlich)* **~ essen** eat properly *or* heartily; **er hat ~ geschimpft** he didn't half grumble *(coll.)*

gehör·los *Adj.* deaf

gehörnt *Adj.* **a)** horned; *(mit einem Geweih)* antlered; **b)** *(scherzh. verhüll.: betrogen)* cuckolded; **ein ~er Ehemann** a cuckold

gehorsam [gə'ho:ɐza:m] *Adj.* obedient

Gehorsam der; ~s obedience; **jmdm. ~ leisten/den ~ verweigern** obey/refuse to obey sb.

Gehorsams·verweigerung die; *o. Pl. (Milit.)* insubordination; refusal to obey orders

Geh-: ~steig der pavement *(Brit.)*; sidewalk *(Amer.)*; **~versuch der;** *meist Pl.* attempt at walking; *(nach einem Unfall)* attempt at walking again

Geier ['gaiɐ] **der; ~s, ~:** vulture; **hol's der ~** *(ugs.)* to hell with it *(coll.)*; **weiß der ~** *(salopp)* God only knows *(sl.)*; Christ knows *(sl.)*

Geifer ['gaifɐ] **der; ~s a)** slaver; slobber; **b)** *(geh. abwertend: Gehässigkeit)* venom; vituperation

geifern *itr. V.* **a)** slaver; slobber; **b)** *(abwertend: gehässig reden)* **gegen jmdn./über etw.** *(Akk.)* **~:** discharge one's venom at sb./sth.

Geige ['gaigə] **die; ~, ~n** violin; **die erste ~ spielen** *(fig. ugs.)* play first fiddle; call the tune

geigen 1. *itr. V.* **a)** *(ugs.: Geige spielen)* play the fiddle *(coll.) or* the violin; **b)** *(ugs.: von Insekten)* chirp; chirr. **2.** *tr. V.* **a)** *(ugs.: auf der Geige spielen)* **einen Walzer ~:** play a waltz on the fiddle *(coll.) or* violin

Geigen-: ~bauer der violin-maker; **~bogen der** violin bow; **~kasten der** violin case

Geiger der; ~s, ~, Geigerin die; ~, ~nen: violin-player; violinist

Geiger·zähler der *(Physik)* Geiger counter

geil [gail] **1.** *Adj. (oft abwertend) (sexuell erregt)* randy; horny *(sl.); (lüstern)* lecherous; **auf jmdn. ~ sein** lust for *or* after sb.. **2.** *adv. (oft abwertend)* lecherously

Geisel ['gaizl̩] **die; ~, ~n** hostage; **jmdn. als** *od.* **zur ~ nehmen** take sb. hostage

Geisel-: ~nahme die taking of hostages; **~nehmer der** terrorist/guerrilla *etc.* holding the hostages

Geiß [gais] **die; ~, ~en a)** *(südd., österr., schweiz.: Ziege)* [nanny-]goat; **b)** *(Jägerspr.)* doe

Geißel ['gaisl̩] **die; ~, ~n** *(auch fig.)* scourge

geißeln *tr. V.* **a)** *(tadeln)* castigate; **b)** *(züchtigen)* scourge

Geißelung die; ~, ~en a) *(Tadelung)* castigation; **b)** *(Züchtigung)* scourging

Geist [gaist] **der; ~[e]s, ~er a)** *o. Pl. (Verstand)* mind; **jmds. ~ ist verwirrt/gestört** sb. is mentally deranged/disturbed; **jmdm. mit etw. auf den ~ gehen** *(salopp)* get on sb.'s nerves with sth.; **den ~ aufgeben** *(geh./ugs. scherzh., auch fig.)* give up the ghost; **im ~[e]** in my/his *etc.* mind's eye; **b)** *o. Pl. (Scharfsinn)* wit; **c)** *o. Pl. (innere Einstellung)* spirit; **d)** *(denkender Mensch)* mind; intellect; **ein großer/kleiner ~:** a great mind/a person of limited intellect; **hier** *od.* **da scheiden sich die ~er** this is where opinions differ; **e)** *(überirdisches Wesen)* spirit; **der Heilige ~** *(christl. Rel.)* the Holy Ghost *or* Spirit; **der böse ~:** the evil spirit; **von allen guten ~ern verlassen sein** have taken leave of one's senses; be out of one's mind; **f)** *(Gespenst)* ghost; **~er gehen im Schloß um/spuken im Schloß** the castle is haunted

Geister-: ~bahn die ghost train; **~fahrer der** *(ugs.)* ghost-driver *(Amer.); person driving on the wrong side of the road or the wrong carriageway;* **~geschichte die** ghost story

geisterhaft *Adj.* ghostly; spectral; eerie ‹*atmosphere*›

Geister·hand die *in* **wie von ~:** as if by an invisible hand

geistern ['gaistɐn] *itr. V.; mit sein* ‹*ghost*› wander; *(fig.)* wander like a ghost; **diese Idee geisterte immer noch durch seinen Kopf** he still had this idea in his head

Geister-: ~stadt die ghost town; **~stunde die** witching hour

geistes-, Geistes-: ~abwesend 1. *Adj.* absent-minded; **2.** *adv.* absent-mindedly; **~abwesenheit die** absent-mindedness; **~blitz der** *(ugs.)* brainwave; flash of inspiration; **~gaben** *Pl.* intellectual gifts; **~gegenwart die** presence of mind; **~gegenwärtig 1.** *Adj.* quick-witted; **2.** *adv.* with great presence of mind; **~geschichte die** history of ideas; intellectual history; **~gestört** mentally disturbed; **~haltung die** attitude [of mind]; **~krank** *Adj.* mentally ill; [mentally] deranged; **~kranke der/die** mentally ill person; *(im Krankenhaus)* mental patient; **~krankheit die** mental illness; **~schwäche die;** *o. Pl.* feeble-mindedness; mental deficiency; **~wissenschaften** *Pl.* arts; humanities; **~wissenschaftler der** arts scholar; scholar in the humanities; **~wissenschaftlich** *Adj.* **~wissenschaftliche Fächer** arts subjects; **~zustand der** mental condition; mental state

geistig 1. *Adj.; nicht präd.* **a)** intellectual; spiritual ‹*legacy, father, author*›; *(Psych.)* mental; **~e Arbeit** brain-work; **~er Diebstahl** plagiarism; **b)** *(alkoholisch)* **~e Getränke** alcoholic drinks *or* beverages. **2.** *adv.* intellectually ‹*superior*›; mentally ‹*lazy, active, retarded, disabled*›; **~ weggetreten sein** *(ugs.)* be miles away *(coll.)*

geistlich *Adj.; nicht präd.* sacred ‹*song, music*›; religious ‹*order*›; religious, devotional ‹*book, writings*›; spiritual ‹*matter, support*›; spiritual, religious ‹*leader*›; ecclesiastical ‹*office, dignitary*›; **der ~e Stand** the clergy

Geistliche der; *adj. Dekl.* clergyman; priest; *(einer Freikirche)* minister; *(Militär~, Gefängnis~)* chaplain

geist-, Geist-: ~los *Adj.* dim-witted; witless; *(trivial)* trivial; **~losigkeit die; ~:** dim-wittedness; witlessness; *(Trivialität)* triviality; **~reich 1.** *Adj.* witty; *(klug)* clever; **2.** *adv.* wittily; *(klug)* cleverly; **~tötend** *Adj.* soul-destroying ‹*work, job*›; stupefyingly boring ‹*chatter, drivel*›; **~voll** *Adj.* brilliantly witty ‹*joke, satire*›; brilliant ‹*idea*›; intellectually stimulating ‹*conversation, book*›

Geiz [gaits] **der; ~es** meanness; *(Knauserigkeit)* miserliness

geizen *itr. V.* be mean; **mit etw. ~:** be mean *or* stingy with sth.; **mit Lob ~** *(fig.)* be sparing with one's praise

Geiz·hals der *(abwertend)* skinflint

geizig *Adj.* mean; *(knauserig)* miserly

Geiz·kragen der *(ugs. abwertend)* skinflint

gekannt [gə'kant] *2. Part. v.* **kennen**

Gekicher das; ~s giggling

Geklimper das; ~s *(abwertend)* plunking

geklungen [gə'klʊŋən] *2. Part. v.* **klingen**

geknickt *Adj. (ugs.)* dejected; downcast

gekniffen *2. Part. v.* **kneifen**

Geknister das; ~s rustling; rustle; *(von Holz, Feuer)* crackling; crackle

gekommen *2. Part. v.* **kommen**

gekonnt [gə'kɔnt] **1.** *2. Part. v.* **können. 2.** *Adj.* accomplished; *(hervorragend ausgeführt)* masterly. **3.** *adv.* in an accomplished manner; *(hervorragend)* in masterly fashion

gekoren [gə'ko:rən] *2. Part. v.* **küren, kiesen**

Gekreisch[e] das; ~s *(von Vögeln)* screeching; *(von Menschen)* shrieking; squealing; *(von Rädern, Bremsen)* squealing

Gekritzel[e] das; ~s *(abwertend)* scribble; scrawl

gekrochen *2. Part. v.* **kriechen**

gekünstelt [gə'kʏnstl̩t] **1.** *Adj.* artificial; forced ‹*smile*›; affected ‹*behaviour*›. **2.** *adv.* **~ lächeln** give a forced smile; **~ sprechen** talk affectedly

Gel [ge:l] **das; ~s, ~e** *(Chemie)* gel

Gelaber[e] das; ~s *(ugs. abwertend)* rabbiting *(coll.) or* babbling on

Gelächter [gəˈlɛçtɐ] **das; ~s, ~:** laughter; **in ~ ausbrechen** burst out laughing

gelackmeiert [gəˈlakmaiɐt] *Adj.; nicht attr. (salopp scherzh.)* had *(sl.)*; conned *(coll.)*; **der Gelackmeierte sein** be the one who's been had *(sl.)*

geladen 1. *2. Part. v.* **laden. 2.** *in* **~ sein** be furious *or (Brit. coll.)* livid

Gelage das; ~s, ~: feast; banquet; *(abwertend)* orgy of eating and drinking

Gelähmte der/die; *adj. Dekl.* paralytic

Gelände [gəˈlɛndə] **das; ~s, ~ a)** ground; terrain; **das ~ steigt an/fällt ab** the ground rises/falls; **b)** *(Grundstück)* site; *(von Schule, Krankenhaus usw.)* grounds *pl.*

gelände-, Gelände-: ~fahrt die cross-country drive; *(das Fahren)* cross-country driving; **~fahrzeug das** cross-country vehicle; **~gängig** *Adj.* cross-country *attrib.* ⟨*vehicle*⟩; ⟨*vehicle*⟩ suitable for cross-country driving

Geländer [gəˈlɛndɐ] **das; ~s, ~:** banisters *pl.*; handrail; *(am Balkon, an einer Brücke)* railing[s *pl.*]; *(aus Stein)* balustrades; parapet

Gelände·wagen der *s.* **Geländefahrzeug**

gelang *3. Pers. Sg. Prät. v.* **gelingen**

gelangen *itr. V.; mit sein* **a) an etw.** *(Akk.)***/zu etw. ~:** arrive at *or* reach sth.; **an die Öffentlichkeit ~:** reach the public; leak out; **in jmds. Besitz ~:** come into sb.'s possession; **b)** *(fig.)* **zu Ansehen ~:** gain esteem *or* standing; **zu Ruhm ~:** achieve fame; **zu der Erkenntnis ~, daß ...:** come to the realization that ...; realize that ...; **c)** *als Funktionsverb* **zur Aufführung ~:** be presented *or* performed; **zur Auszahlung ~:** be paid [out]

gelassen 1. *2. Part. v.* **lassen. 2.** *Adj.* calm; *(gefaßt)* composed; **~ bleiben** keep calm *or* cool. **3.** *adv.* calmly

Gelassenheit die; ~: calmness; *(Gefaßtheit)* composure

Gelatine [ʒelaˈtiːnə] **die; ~:** gelatine

gelaufen *2. Part. v.* **laufen**

geläufig *Adj.* **a)** *(vertraut)* familiar, common ⟨*expression, concept*⟩; **etw. ist jmdm. ~:** sb. is familiar with sth.; **b)** *(fließend, perfekt)* fluent

gelaunt [gəˈlaunt] **gut/schlecht ~ sein** be in a good/bad mood; **wie ist sie ~?** what sort of mood is she in?

gelb [gɛlp] *Adj.* yellow; **vor Neid ~ werden** turn green with envy; **das ist nicht das Gelbe vom Ei** *(fig. ugs.)* that's no great shakes *(sl.)*

Gelb das; ~s, ~ *od. (ugs.)* **~s** yellow; **bei ~ über die Ampel fahren** go through *or* crash the lights on amber

gelb-, Gelb-: ~braun *Adj.* yellowish-brown; **~fieber das** *(Med.)* yellow fever; **~grün** *Adj.* yellowish-green

gelblich *Adj.* yellowish; yellowed ⟨*paper*⟩; sallow ⟨*skin*⟩

Gelb·sucht die; *o. Pl. (Med.)* jaundice; icterus *(Med.)*

Geld [gɛlt] **das; ~es, ~er** money; **großes ~:** large denominations *pl.*; **kleines/bares ~:** change/cash; **es ist für ~ nicht zu haben** money cannot buy it; **das ist hinausgeworfenes ~:** that is a waste of money *or (coll.)* money down the drain; **ins ~ gehen** *(ugs.)* run away with the money *(coll.)*; **~ stinkt nicht** *(Spr.)* money has no smell; **~ regiert die Welt** *(Spr.)* money makes the world go round; **~ allein macht nicht glücklich** [*(scherzh.)*, **aber es hilft**] *(Spr.)* money isn't everything[, but it helps]; **das große ~ machen** make a lot of money; **~ wie Heu haben, im ~ schwimmen** be rolling in money *or* in it *(coll.)*; **nicht für ~ und gute Worte** *(ugs.)* not for love or money; **zu ~ kommen** get hold of [some] money; **etw. zu ~ machen** turn sth. into money *or* cash; **öffentliche ~er** public money *sing. or* funds

geld-, Geld-: ~angelegenheit die; *meist Pl.* money *or* financial matter; **~anlage die** investment; **~automat der** cash dispenser; **~betrag der** sum *or* amount [of money]; **~beutel der** *(bes. südd.)*, **~börse die** purse; **~geber der** financial backer; *(für Forschungen usw.)* sponsor; **~gier die** *(abwertend)* greed; avarice; **~gierig** *Adj. (abwertend)* greedy; avaricious; **~hahn der** *in* **[jmdm.] den ~hahn zudrehen** *(ugs.)* cut off sb.'s supply of money; **~institut das** financial institution

geldlich *Adj.; nicht präd.* financial

Geld-: ~mittel *Pl.* financial resources; funds; **~prämie die** cash bonus; *(~preis)* cash prize; **~preis der** cash prize; *(bei einem Turnier)* prize money; **~quelle die** source of income; *(für den Staat)* source of revenue; **~schein der** banknote; bill *(Amer.)*; **~schrank der** safe; **~schrank·knacker der** *(ugs.)* safe-breaker; safe-cracker; **~schwierigkeiten** *Pl.* financial difficulties *or* straits; **~sorgen** *Pl.* money troubles; financial worries; **~strafe die** fine; **jmdn. zu einer ~strafe verurteilen** fine sb.; **~stück das** coin; **~verschwendung die** waste of money; **~waschanlage die** *(ugs.)* money-laundering scheme; **~wechsel der** exchanging of money; **„~wechsel"** 'bureau de change'; 'change'

geleckt *Adj. in* **wie ~ aussehen** *(ugs.)* look all spruced up

Gelee [ʒeˈleː] **der** *od.* **das; ~s, ~s** jelly; **Aale in ~:** jellied eels

gelegen 1. *2. Part. v.* **liegen. 2.** *Adj.* **a)** *(passend)* convenient; **das kommt mir ~:** that comes just at the right time for me; **b)** *(liegend)* situated

Gelegenheit die; ~, ~en a) opportunity; **die ~ nutzen** make the most of the opportunity; **bei nächster ~:** at the next opportunity; **bei ~:** some time; **die ~ beim Schopf[e] fassen** *od.* **ergreifen** grab *or* seize the opportunity with both hands; **b)** *(Anlaß)* occasion

Gelegenheits-: ~arbeit die casual work; **~arbeiter der** casual worker; **~kauf der** bargain

gelegentlich 1. *Adj.; nicht präd.* occasional. **2.** *adv.* **a)** *(manchmal)* occasionally; **b)** *(bei Gelegenheit)* some time

gelehrig [gəˈleːrɪç] *Adj.* ⟨*child*⟩ who is quick to learn *or* quick at picking things up; ⟨*animal*⟩ that is quick to learn

gelehrt *Adj.* **a)** *(kenntnisreich)* learned; erudite; **b)** *(wissenschaftlich)* scholarly

Gelehrte der/die; *adj. Dekl.* scholar; **darüber streiten sich die ~n** the experts disagree on that; *(fig.)* that's a moot point

Geleit das; ~[e]s, ~e *(geh.)* **jmdm. sein ~ anbieten** offer to accompany *or* escort sb.; **freies** *od.* **sicheres ~** *(Rechtsw.)* safe-conduct; **jmdm. das letzte ~ geben** *(geh. verhüll.)* attend sb.'s funeral

geleiten *tr. V. (geh.)* escort; *(begleiten)* accompany; escort; **jmdn. zur Tür ~:** see sb. to the door; show sb. out

Geleit·schutz der *(Milit.)* escort; **jmdm. ~schutz geben** provide an escort for sb.

Gelenk [gəˈlɛŋk] **das; ~[e]s, ~e** *(Anat., Technik)* joint; *(Scharnier)* hinge

gelenkig 1. *Adj.* agile ⟨*person*⟩; *(geschmeidig)* supple ⟨*limb*⟩. **2.** *adv.* agilely

Gelenkigkeit die; ~: agility; *(von Gliedmaßen)* suppleness

gelernt *Adj.; nicht präd.* qualified

gelesen *2. Part. v.* **lesen**

Geliebte [gəˈli:ptə] **der/die;** *adj. Dekl.* **a)** lover/mistress; **b)** *(geh. veralt.)* beloved

geliefert 1. *2. Part. v.* **liefern. 2.** *Adj. in* **~ sein** *(salopp)* be sunk *(coll.);* have had it *(coll.)*

geliehen [gəˈli:ən] *2. Part. v.* **leihen**

gelieren [ʒeˈli:rən] *itr. V.* set

gelind[e] 1. *Adj.* **a)** *(schonend)* mild; **b)** *(geh. veralt.: mild, sanft)* mild ⟨*climate*⟩; light ⟨*punishment*⟩; slight ⟨*pain*⟩. **2.** *adv.* mildly; **~e gesagt** to put it mildly

gelingen [gəˈlɪŋən] *unr. itr. V.; mit sein* succeed; **es gelang ihr, es zu tun** she succeeded in doing it; **es gelang ihr nicht, es zu tun** she did not succeed in doing it; she failed to do it; **eine gelungene Arbeit** a successful piece of work; *s. auch* **gelungen 2**

Gelingen das; ~s success; **auf ein gutes ~ hoffen** hope for success; **jmdm. gutes ~ wünschen** wish sb. every success; **gutes ~!** the best of luck!

gelitten [gəˈlɪtn̩] *2. Part. v.* **leiden**

gell[e] [ˈgɛl(ə)] *Interj. (südd.) s.* **gelt**

gellen [ˈgɛlən] *itr. V.* ring out; **ein Schrei gellte durch die Nacht** a scream *or* shriek pierced the night; **jmdm. in den Ohren ~:** make sb.'s ears ring; **~des Gelächter** shrill peals of laughter

geloben *tr. V. (geh.)* vow; **Besserung ~:** promise solemnly to improve; **jmdm. Treue ~:** vow to be faithful to sb.; **sich** *(Dat.)* **~, etw. zu tun** vow to oneself *or* make a solemn resolve to do sth.; **das Gelobte Land** the Promised Land

Gelöbnis [gəˈlø:pnɪs] **das; ~ses, ~se** *(geh.)* vow; **ein ~ ablegen** *od.* **leisten** make *or* take a vow

gelockt [gəˈlɔkt] *Adj.* curly

gelogen *2. Part. v.* **lügen**

gelöst [gəˈlø:st] *Adj.* relaxed

gelt [gɛlt] *Interj. (südd., österr. ugs.)* **~, du bist mir doch nicht böse?** you're not angry with me, are you?; **er kommt doch morgen zurück, ~?** he'll be coming back tomorrow, won't he *or (coll.)* right?

gelten [ˈgɛltn̩] **1.** *unr. itr. V.* **a)** *(gültig sein)* be valid; ⟨*banknote, coin*⟩ be legal tender; ⟨*law, regulation, agreement*⟩ be in force; ⟨*price*⟩ be effective; **etw. gilt für jmdn.** sth. applies to sb.; **das gilt auch für dich/Sie!** *(ugs.)* that includes you!; that goes for you too!; **das gilt nicht!** that doesn't count!; **nach ~dem Recht** in accordance with the law as it [now] stands; **die ~de Meinung** the generally accepted opinion; **etw. [nicht] ~ lassen** [not] accept sth.; **b)** *(angesehen werden)* **als etw. ~:** be regarded as sth.; be considered [to be] sth.; **c)** (+ *Dat.*) *(bestimmt sein für)* be directed at; **die Bemerkung gilt dir** the remark is aimed at you; **der Beifall galt auch dem Regisseur** the applause was also for the director. **2.** *unr. tr. V.* **a)** *(wert sein)* **sein Wort gilt viel/wenig** his word carries a lot of/little weight; **was gilt die Wette?** what do you bet?; **etw. gilt jmdm. mehr als ...:** sth. is worth *or* means more to sb. than ...; **b)** *unpers. (darauf ankommen, daß)* **es gilt, rasch zu handeln** it is essential to act swiftly; **c)** *unpers. (geh.: auf dem Spiel stehen)* **es gilt dein Leben** *od.* **deinen Kopf** your life is at stake

geltend *in* **etw. ~ machen** assert sth.; **einige Bedenken/einen Einwand ~ machen** express some doubts/raise an objection; *s. auch* **gelten 1 a**

Geltung die; ~ a) *(Gültigkeit)* validity; **~ haben** ⟨*banknote, coin*⟩ be legal tender; ⟨*law, regulation, agreement*⟩ be in force; ⟨*price*⟩ be effective; **für jmdn. ~ haben** apply to sb.; **b)** *(Wirkung)* recognition; **jmdm./sich/einer Sache ~ verschaffen** gain *or* win recognition for sb./oneself/sth.; **an ~ verlieren** ⟨*value, principle, etc.*⟩ lose its importance, become less important; **etw. zur ~ bringen** show sth. to its best advantage; **zur ~ kommen** show to [its best] advantage

Geltungs-: ~bedürfnis das need for recognition; **~bereich der** scope; **~dauer die** period of validity; **~drang der** *s.* **~bedürfnis; ~sucht die** [pathological] craving for recognition

gelungen [gəˈlʊŋən] **1.** *2. Part. v.* **gelingen. 2.** *Adj.* **a)** *(ugs.: spaßig)* priceless; **das finde ich ~:** what a laugh!; **b)** *(ansprechend)* inspired

Gelüst das; ~[e]s, ~e, Gelüste das; ~s, ~ *(geh.)* longing; strong desire; *(zwingend, krankhaft)* craving; **ein ~ nach** *od.* **auf etw.** *(Akk.)* **haben** have a longing *or* a strong desire/a craving for sth.

gelüsten *tr. V.; unpers.* **es gelüstet ihn nach ...:** he has a longing for ...; *(zwingend, krankhaft)* he has a craving for ...

Gemach [gəˈma(:)x] **das; ~[e]s, Gemächer** [gəˈmɛ(:)çɐ] *(veralt. geh.)* apartment

gemächlich [gəˈmɛ(:)çlɪç] **1.** *Adj.* leisurely; **ein ~es Leben führen** take life easily. **2.** *adv.* in a leisurely manner

gemacht *Adj. in* **ein ~er Mann sein** *(ugs.)* be a made man

Gemahl der; ~s, ~e *(geh.)* consort; husband; **bitte grüßen Sie Ihren Herrn ~:** please give my regards to your husband

Gemahlin die; ~, ~nen *(geh.)* consort; wife; **eine Empfehlung an Ihre Frau ~:** my compliments to your wife

gemahnen *tr., itr. V. (geh.)* **jmdn. an etw.** *(Akk.)* **~:** remind sb. of sth.

Gemälde [gəˈmɛ:ldə] **das; ~s, ~:** painting

Gemälde-: ~ausstellung die exhibition of paintings; **~galerie die** picture gallery

gemäß [gəˈmɛ:s] **1.** *Präp. + Dat.* in accordance with. **2.** *Adj.* **jmdm./einer Sache ~ sein** be appropriate for sb./to sth.

gemäßigt *Adj.* moderate; temperate ⟨*climate*⟩

Gemäuer [gəˈmɔyɐ] **das; ~s, ~:** walls *pl.; (Ruine)* ruin

Gemecker[e] das; ~s a) *(von Schafen, Ziegen)* bleating; **b)** *(ugs.: Nörgelei)* griping *(coll.);* grousing *(sl.);* moaning

gemein 1. *Adj.* **a)** *(abstoßend)* coarse, vulgar ⟨*joke, expression*⟩; nasty ⟨*person*⟩; **b)** *(niederträchtig)* mean; base, dirty ⟨*lie*⟩; mean, dirty ⟨*trick*⟩; **du bist ~!/das ist ~ [von dir]!** you're mean *or* nasty!/that's mean *or* nasty [of you]!; **c)** *(ärgerlich)* infuriating; damned annoying *(coll.);* **d)** *nicht präd. (Bot., Zool., sonst veralt.: allgemein vorkommend)* common; **der ~e Mann** the ordinary man; the man in the street; **e)** *(veralt.: allgemein)* general; **etw. mit jmdm./etw. ~ haben** have sth. in common with sb./sth.. **2.** *adv.* **a) jmdn. ~ behandeln** treat sb. in a mean *or* nasty way; **b) es hat ganz ~ weh getan** *(ugs.)* it hurt like hell *(coll.)*

Gemeinde [gəˈmaində] **die; ~, ~n a)** *(staatliche Verwaltungseinheit)* municipality; *(ugs.: ~amt)* local authority; **die ~ X** the municipality of X; **b)** *(Seelsorgebezirk) (christlich)* parish; *(nichtchristlich)* community; **c)** *(Einwohnerschaft)* community; local population; **d)** *(Gottesdienstteilnehmer)* congregation; **e)** *(Anhängerschaft)* body of followers; **die ~ seiner Anhänger** his following

Gemeinde-: ~haus das parish hall; **~mitglied das** parishioner; **~rat der a)** *(Gremium)* local council; **b)** *(Mitglied)* local councillor; **~schwester die** district nurse; **~verwaltung die** local administration; **~zentrum das** community centre

gemein·gefährlich *Adj.* dangerous to the public; dangerous ‹*criminal*›; **~ sein** be a danger to the public

Gemein·gut das; *o. Pl. (geh.)* common property

Gemein·heit die; ~, ~en a) *o. Pl.* meanness; nastiness; **b)** *(gemeine Handlung)* mean *or* nasty *or* dirty trick; **das war eine ~:** that was a mean *or* nasty thing to do/say

gemein-, Gemein-: ~hin *Adv.* commonly; generally; **~nutz der; ~es** public good; **~nützig** [-nʏtsɪç] *Adj.* serving the public good *postpos., not pred.; (wohltätig)* charitable; **eine ~nützige Institution** a charitable *or* non-profit-making institution; **~platz der** platitude; commonplace

gemeinsam 1. *Adj.* **a)** common ‹*interests, characteristics*›; mutual ‹*acquaintance, friend*›; joint ‹*property, account*›; shared ‹*experience*›; **der Gemeinsame Markt** the Common Market; **~e Interessen haben** have interests in common; **~e Kasse machen** pool funds *or* resources; **b)** *nicht präd. (miteinander unternommen)* joint ‹*undertaking, consultations*›; joint, concerted ‹*efforts, action, measures*›; **[mit jmdm.] ~e Sache machen** join forces *or* up [with sb.]; **c)** *nicht attr.* **viel Gemeinsames haben, viel[es] ~ haben** have a lot in common; **das ist ihnen ~:** that is something they have in common. **2.** *adv.* together; **es gehört ihnen ~:** it is owned by them jointly

Gemeinsamkeit die; ~, ~en a) *(gemeinsames Merkmal)* common feature; point in common; **b)** *o. Pl. (Verbundenheit)* community of interest; **ein Gefühl der ~:** a sense of community

Gemeinschaft die; ~, ~en a) community; **die Europäische ~:** the European Community; **b)** *o. Pl. (Verbundenheit)* coexistence; **in unserer Klasse herrscht keine echte ~:** there is no real sense of community in our class; **in ~ mit jmdm.** together with sb.

gemeinschaftlich 1. *Adj.; nicht präd.* common ‹*interests, characteristics*›; joint ‹*property, undertaking*›; joint, concerted ‹*efforts, action*›. **2.** *adv.* together; **wir führen die Firma ~:** we run the firm jointly *or* together

Gemeinschafts-: ~antenne die community aerial *(Brit.) or (Amer.)* antenna; **~arbeit die a)** *o. Pl.* joint work; **b)** *(Ergebnis)* joint product *or* effort; **~gefühl das;** *o. Pl.* community spirit; **~kunde die;** *o. Pl.* social studies *sing.*; **~raum der** common-room *(Brit.)*

gemein-, Gemein-: ~sinn der; *o. Pl.* public spirit; **~verständlich 1.** *Adj.* generally comprehensible *or* intelligible; **2.** *adv.* **sich ~verständlich ausdrükken** make oneself generally comprehensible *or* intelligible; **~wesen das** community; *(staatlich)* political unit; polity; **~wohl das** public *or* common good; **etw./jmd. dient dem ~wohl** sth. is/sb. acts in the public interest

Gemenge das; ~s, ~ mixture

gemessen 1. *2. Part. v.* **messen. 2.** *Adj.* measured ‹*steps, tones, language*›; deliberate ‹*words, manner of speaking*›; **~en Schrittes** with measured tread *or* steps *pl.*

Gemetzel das; ~s, ~ *(abwertend)* blood-bath; massacre

gemieden [gə'mi:dn̩] *2. Part. v.* **meiden**

Gemisch das; ~[e]s, ~e mixture **(aus, von** of**)**; mix *(coll.)*

gemischt *Adj.* **a)** mixed; **~e Kost** a varied diet; **eine ~e Klasse** a mixed *or* coeducational *or (coll.)* coed class; **b)** *(abwertend: anrüchig)* disreputable ‹*crowd*›

gemocht [gə'mɔxt] *2. Part. v.* **mögen**

gemolken *2. Part. v.* **melken**

gemoppelt [gə'mɔpl̩t] *s.* **doppelt 2a**

Gemotze das; ~s *(salopp)* grouching *(coll.)*; belly-aching *(sl.)*

Gemse ['gɛmzə] **die; ~, ~n** chamois

Gemurmel das; ~s murmuring

Gemüse [gə'my:zə] **das; ~s, ~:** vegetables *pl.*; **ein ~:** a vegetable; **junges ~** *(fig. ugs.)* youngsters *pl.*

Gemüse-: ~beet das vegetable patch *or* plot; **~beilage die** vegetables *pl.*; **~eintopf der** vegetable stew; **~frau die** vegetable seller; **~garten der** vegetable *or* kitchen garden; *(Teil eines Gartens)* vegetable patch *or* plot; **~händler der** greengrocer; **~laden der** greengrocer's [shop]; **~saft der** vegetable juice; **~suppe die** vegetable soup

gemußt [gə'mʊst] *2. Part. v.* **müssen**

Gemüt [gə'my:t] **das; ~[e]s, ~er a)** nature; disposition; **ein sonniges/kindliches ~ haben** *(iron.)* be [really] naive; **b)** *(Empfindungsvermögen)* heart; soul; **das rührt ans** *od.* **ist etw. fürs ~:** that touches the heart *or* tears at one's heart-strings; **jmdm. aufs ~ schlagen** *od.* **gehen** make sb. depressed; **c)** *(Mensch)* soul; **etw. erhitzt/erregt die ~er** sth. makes feelings run high

gemütlich 1. *Adj.* **a)** *(behaglich)* snug; cosy; gemütlich *(literary)*; *(bequem)* comfortable; **mach es dir ~!** make yourself comfortable *or* at home!; **b)** *(ungezwungen)* informal ‹*get-together*›; **c)** *(umgänglich)* sociable; friendly; **d)** *(gemächlich)* leisurely; comfortable ‹*pace*›. **2.** *adv.* **a)** *(behaglich)* cosily; *(bequem)* comfortably; **b)** *(ungezwungen)* **~ beisammensitzen** sit pleasantly together; **sich ~ unterhalten** have a pleasant chat; **c)** *(gemächlich)* at a leisurely *or* comfortable pace; unhurriedly

Gemütlichkeit die; ~ a) *(Behaglichkeit)* snugness; **b)** *(Ungezwungenheit)* informality; **die ~ stören** disturb the atmosphere *or* mood of informality; **c)** *(Gemächlichkeit)* **in aller ~:** quite unhurriedly; **d) da hört die ~ auf** *(fig. ugs.)* that's going too far

gemüts-, Gemüts-: ~bewegung die emotion; **~krank** *Adj.* emotionally disturbed; **~mensch der** *(ugs.)* good-natured *or* even-tempered person; **~regung die** emotion; **~ruhe die** peace of mind; **in aller ~ruhe** *(ugs.) (ohne Sorge)* completely unconcerned; *(ohne Hast)* as if there were all the time in the world

gemüt·voll *Adj.* warm-hearted; *(empfindsam)* sentimental

gen [gɛn] *Präp. + Akk. (veralt., bibl., dichter.)* toward[s]; **~ Süden** southwards

Gen [ge:n] **das; ~s, ~e** *(Biol.)* gene

genannt [gə'nant] *2. Part. v.* **nennen**

genas [gə'na:s] *1. u. 3. Pers. Sg. Prät. v.* **genesen**

genau [gə'nau̯] **1.** *Adj.* **a)** exact; precise; accurate ⟨*scales*⟩; exact, right ⟨*time*⟩; **Genaues/Genaueres wissen** know the/more exact *or* precise details; **ich weiß nichts Genaues/Genaueres** I don't know anything definite/more definite; **b)** *(sorgfältig, gründlich)* meticulous, painstaking ⟨*person*⟩; careful ⟨*study*⟩; precise ⟨*use of language*⟩; detailed, thorough ⟨*knowledge*⟩. **2.** *adv.* **a)** exactly; precisely; ~ **um 8⁰⁰** at 8 o'clock precisely; at exactly 8 o'clock; **die Uhr geht [auf die Minute]** ~: the watch/clock keeps perfect time; **b)** *(gerade, eben)* just; ~ **reichen** be just enough; **c)** *(als Verstärkung)* just; exactly; precisely; **d)** *(als Zustimmung)* exactly; precisely; quite [so]; **e)** *(sorgfältig)* **etw.** ~ **durchdenken** think sth. out carefully *or* meticulously; **jmdn.** ~ **kennen** know exactly what sb. is like; **etw.** ~ **beachten** observe sth. meticulously *or* painstakingly; **es mit etw. [nicht so]** ~ **nehmen** be [not too] particular about sth.

genau·genommen *Adv.* strictly speaking

Genauigkeit die; ~ **a)** exactness; exactitude; precision; *(einer Waage)* accuracy; **b)** *(Sorgfalt)* meticulousness

genau·so *Adv. s.* **ebenso**

Gendarm [ʒan'darm] **der;** ~**en,** ~**en** *(österr., sonst veralt.)* village *or* local policeman *or* constable

Gendarmerie [ʒandarmə'ri:] **die;** ~, ~**n** *(österr., sonst veralt.)* village *or* local constabulary

genehm [gə'ne:m] *Adj.* **jmdm.** ~ **sein** *(geh.)* be convenient to *or* suit sb.; *(für jmdn. annehmbar sein)* be acceptable to sb.

genehmigen *tr. V.* approve ⟨*plan, alterations*⟩; grant, approve ⟨*application*⟩; authorize ⟨*stay*⟩; grant, agree to ⟨*request*⟩; give permission for ⟨*demonstration*⟩; **sich** *(Dat.)* **etw.** ~ *(ugs.)* treat oneself to sth.; **sich** *(Dat.)* **einen** ~ *(ugs.)* have a drink

Genehmigung die; ~, ~**en a)** *(eines Plans, Antrags, einer Veränderung)* approval; *(eines Aufenthalts)* authorization; *(einer Bitte)* granting; *(einer Demonstration)* permission (*Gen.* for); **b)** *(Schriftstück)* permit; *(Lizenz)* licence

geneigt [gə'naikt] *Adj.* ~ **sein** *od.* **sich** ~ **zeigen, etw. zu tun** be inclined to do sth.; *(bereit sein)* be ready *or* willing to do sth.

Genera *s.* **Genus**

General [genə'ra:l] **der;** ~**s,** ~**e** *od.* **Generäle** [genə'rɛ:lə] general; **Herr** ~: General

general-, General-: ~**amnestie die** general amnesty; ~**baß der** *(Musik)* [basso] continuo; thorough-bass; ~**direktor der** chairman; president *(Amer.)*; ~**probe die a)** *(auch fig.)* dress *or* final rehearsal; **b)** *(Sport: letztes Testspiel)* final trial; ~**sekretär der** Secretary General; *(einer Partei)* general secretary; ~**stab der** *(Milit.)* general staff; ~**streik der** general strike; ~**überholen** *tr. V.; nur im Inf. und 2. Part. gebr. (bes. Technik)* **etw.** ~**überholen** give sth. a general overhaul; ~**versammlung die** general meeting; **die** ~**versammlung der Vereinten Nationen** the General Assembly of the United Nations; ~**vertreter der** general representative; ~**vollmacht die** *(Rechtsw.)* full *or* unlimited power of attorney

Generation [genəra'tsi̯o:n] **die;** ~, ~**en** generation

Generations-: ~**konflikt der** generation gap; ~**problem das** generation problem; ~**wechsel der a)** new generation; **b)** *(Biol.)* alternation of generations

Generator [genə'ra:tɔr] **der;** ~**s,** ~**en** [---'--] generator

generell [genə'rɛl] **1.** *Adj.* general. **2.** *adv.* generally; **man kann ganz** ~ **sagen, daß ...**: generally speaking *or* in general, it can be said that ...; **es sollte sonnabends** ~ **schulfrei sein** all schools should close on Saturdays

genesen [gə'ne:zn̩] *unr. itr. V.; mit sein (geh.)* recover; recuperate; *(fig.)* recover

Genesende der/die; *adj. Dekl.* convalescent

Genesung die; ~, ~**en** *(geh.)* recovery

Genetik [ge'ne:tɪk] **die;** ~ *(Biol.)* genetics *sing., no art.*

genetisch *(Biol.)* **1.** *Adj.* genetic. **2.** *adv.* genetically

Genf [gɛnf] **(das);** ~**s** Geneva

Genfer 1. der; ~**s,** ~: Genevese. **2.** *Adj.* Genevese; **der** ~ **See** Lake Geneva

Genferin die; ~, ~**nen** Genevese

genial [ge'ni̯a:l] **1.** *Adj.* brillant ⟨*idea, invention, solution, etc.*⟩; **ein** ~**er Musiker** an inspired musician; a musician of genius. **2.** *adv.* brilliantly

Genialität [geniali'tɛ:t] **die;** ~: genius

Genick [gə'nɪk] **das;** ~**[e]s,** ~**e** back *or* nape of the neck; **sich das** ~ **brechen** *(auch fig.)* break one's neck; **am** ~: by the scruff of the neck; **jmdm./einer Sache das** ~ **brechen** *(ugs.)* ruin sb./sth.

Genick·starre die stiffness of the neck

Genie [ʒe'ni:] **das;** ~**s,** ~**s** genius; **sie ist ein** ~ **im Kochen** she is a brilliant cook

genieren [ʒe'ni:rən] *refl. V.* be *or* feel embarrassed (**wegen** about); **sich vor jmdm.** ~: be *or* feel embarrassed *or* shy in sb.'s presence

genießbar *Adj. (eßbar)* edible; *(trinkbar)* drinkable; **er ist heute nicht** ~ *(fig. ugs.)* he is unbearable today

genießen [gə'ni:sn̩] *unr. tr. V.* **a)** enjoy; **er hat eine gute Ausbildung genossen** he had [the benefit of] a good education; **b)** *(geh.: essen/trinken)* eat/drink; **nicht mehr zu** ~ **sein** be no longer edible/drinkable

Genießer der; ~**s,** ~: **er ist ein richtiger** ~: he is a regular 'bon viveur'; he really knows how to enjoy life [to the full]; **er ist ein stiller** ~: he enjoys life [to the full] in his own quiet way

genießerisch 1. *Adj.* appreciative. **2.** *adv.* appreciatively; ⟨*drink, eat*⟩ with relish

Genitale [geni'ta:lə] **das;** ~**s, Genitalien** [geni'ta:li̯ən] genital organ

Genitiv ['ge:niti:f] **der;** ~**s,** ~**e** *(Sprachw.)* genitive [case]

Genius ['ge:ni̯ʊs] **der;** ~, **Genien** ['ge:ni̯ən] *(geh.)* genius

genommen [gə'nɔmən] *2. Part. v.* **nehmen**

genoppt [gə'nɔpt] *Adj.* knop ⟨*yarn, wool*⟩; pimpled ⟨*rubber*⟩; made of knop yarn ⟨*suit*⟩

genoß [gə'nɔs] *1. u. 3. Pers. Sg. Prät. v.* **genießen**

Genosse [gə'nɔsə] **der;** ~**n,** ~**n a)** comrade; *(als Titel, Anrede)* Comrade; **b)** *(veralt.: Kamerad)* comrade; companion

genossen [gə'nɔsn̩] *2. Part. v.* **genießen**

Genossenschaft die; ~, ~**en** co-operative

genossenschaftlich 1. *Adj.* co-operative; collective ⟨*ownership*⟩; jointly owned ⟨*property*⟩. **2.** *adv.* on a co-operative basis

Genossin die; ~, ~**nen a)** *s.* **Genosse a; b)** *(veralt.: Kameradin)* companion

Genre ['ʒã:rə] **das;** ~**s,** ~**s** genre

Gen-: ~**technik die,** ~**technologie die** genetic engineering *no art.*

genug [gə'nu:k] *Adv.* enough; ~ **Geld/Geld ~ haben:** have enough *or* sufficient money; **das ist ~:** that's enough *or* sufficient; ~ **gearbeitet haben** have done enough work; **ich habe jetzt ~ [davon]** now I've had enough [of it] **nicht ~ damit, daß er faul ist, er ist auch frech** not only is he lazy, he is cheeky as well; **das ist ihm nicht gut ~:** that is not good enough for him; **sich** *(Dat.)* **selbst ~ sein** be quite happy in one's own company; **er kann nie ~ kriegen** *(ugs.)* he is very greedy; **davon kann er nicht ~ kriegen** *(ugs.)* he can't get enough of it *(fig. coll.)*

Genüge [gə'ny:gə] *(geh.)* **jmdm. ~ tun** *od.* **leisten** satisfy sb.; **einer Anordnung/einer Pflicht ~ tun** *od.* **leisten** comply with an order; **zur ~** *(ausreichend)* enough; sufficiently; *(im Übermaß)* quite enough; **etw. zur ~ kennen** know sth. only too well; be only too familiar with sth.

genügen *itr. V.* **a)** be enough *or* sufficient; **das genügt mir** that is enough *or* sufficient [for me]; that will do [for me]; **b)** *(erfüllen)* satisfy; **den Bestimmungen ~:** comply with the regulations

genügend 1. *Adj.* **a)** enough; sufficient; **b)** *(befriedigend)* satisfactory. **2.** *adv.* enough; sufficiently; **~ Geld haben** have enough *or* sufficient money

genügsam [gə'ny:kza:m] **1.** *Adj.* modest ⟨*life*⟩; **ein ~er Mensch** a person who lives modestly; **Schafe sind sehr ~e Tiere** sheep can live *or* subsist on very little. **2.** *adv.* **~ leben** live modestly

Genügsamkeit die; ~: wegen ihrer ~ sind Schafe ...: as they can live *or* subsist on very little, sheep are ...

Genugtuung [-tu:ʊŋ] **die; ~, ~en** satisfaction; **es ist mir eine ~, das zu hören** it gives me satisfaction to hear that

Genus ['gɛnʊs] **das; ~, Genera** ['gɛnera] *(Sprachw.)* gender

Genuß [gə'nʊs] **der; Genusses, Genüsse** [gə'nʏsə] **a)** *o. Pl.* consumption; **b)** *(Wohlbehagen)* **etw. mit/ohne ~ essen/trinken** eat/drink sth. with/without relish; **etw. mit ~ lesen** enjoy reading sth.; **das Konzert/der Kuchen ist ein ~:** the concert is thoroughly enjoyable/the cake is delicious; **in den ~ von etw. kommen** enjoy sth.

genüßlich [gə'nʏslɪç] **1.** *Adj.* appreciative; comfortable ⟨*feeling*⟩; *(schadenfroh)* gleeful. **2.** *adv.* appreciatively; ⟨*eat, drink*⟩ with relish; *(schadenfroh)* ⟨*smile*⟩ gleefully; **sich ~ im Sessel zurücklehnen** lie back luxuriously in the armchair

genuß-, Genuß-: ~mittel das *tea, coffee, alcoholic drinks, tobacco, etc.*; **~sucht die;** *o. Pl. (oft abwertend)* craving for pleasure; **~süchtig** *Adj. (oft abwertend)* pleasure-seeking

Geograph [-'gra:f] **der; ~en, ~en** geographer

Geographie die; ~: geography *no art.*

geographisch 1. *Adj.* geographic[al]. **2.** *adv.* geographically

Geologe [-'lo:gə] **der; ~n, ~n** geologist

Geologie die; ~: geology *no art.*

geologisch 1. *Adj.; nicht präd.* geological; **2.** *adv.* geologically

Geometrie die; ~: geometry *no art.*

geometrisch 1. *Adj.* geometric[al]. **2.** *adv.* geometrically

geordnet *Adj.* **in ~en Verhältnissen leben, ein geordnetes Leben führen** live a settled life

Gepäck [gə'pɛk] **das; ~[e]s** luggage *(Brit.)*; baggage *(Amer.)*

Gepäck-: ~abfertigung die a) *o. Pl.* checking in the luggage/baggage; **b)** *(Schalter) (am Bahnhof)* luggage office *(Brit.)*; baggage office *(Amer.)*; *(am Flughafen)* baggage check-in; **~ablage die** luggage rack *(Brit.)*; baggage rack *(Amer.)*; **~annahme die a)** *o. Pl.* checking in the luggage/baggage; **b)** *(Schalter)* [in-counter of the] luggage office *(Brit.) or (Amer.)* baggage office; *(zur Aufbewahrung)* [in-counter of the] left-luggage office *(Brit.) or (Amer.)* checkroom; *(am Flughafen)* baggage check-in; **~aufbewahrung die** left-luggage office *(Brit.)*; checkroom *(Amer.)*; **~ausgabe die** [out-counter of the] luggage office *(Brit.) or (Amer.)* baggage office; *(zur Aufbewahrung)* [out-counter of the] left-luggage office *(Brit.) or (Amer.)* checkroom; *(am Flughafen)* baggage reclaim; **~kontrolle die** baggage check; **~netz das** *s.* **~ablage; ~raum der** luggage/baggage compartment; **~schein der** luggage ticket *(Brit.)*; baggage check *(Amer.)*; **~stück das** piece *or* item of luggage/baggage; **~träger der a)** porter; **b)** *(am Fahrrad)* carrier; rack; **~wagen der** luggage van *(Brit.)*; baggage car *(Amer.)*

Gepard ['ge:part] **der; ~s, ~e** cheetah; hunting leopard

gepfeffert *Adj. (ugs.)* steep *(coll.)* ⟨*price, rent, etc.*⟩

gepfiffen [gə'pfɪfn̩] *2. Part. v.* **pfeifen**

gepflegt *Adj.* **a)** well-groomed, spruce ⟨*appearance*⟩; neat ⟨*clothing*⟩; cultured ⟨*conversation*⟩; cultured, sophisticated ⟨*atmosphere, environment*⟩; stylish ⟨*living*⟩; well-kept, well-tended ⟨*garden, park*⟩; well cared-for ⟨*hands, house*⟩; **b)** *(hochwertig)* choice ⟨*food, drink*⟩. **3.** *adv.* **~ essen** dine in style; **sich ~ ausdrücken** express oneself in a cultured manner

Gepflogenheit die; ~, ~en *(geh.) (Sitte, Brauch)* custom; tradition; *(Gewohnheit)* habit; *(Verfahrensweise)* practice

Geplänkel [gə'plɛŋkl̩] **das; ~s, ~ a)** *(Wort~)* banter *no indef. art.*; **b)** *(Milit. veralt.)* skirmish

Geplapper das; ~s *(ugs., oft abwertend)* prattling; **das ~ des Babys** the baby's babbling

Gepolter das; ~s a) clatter; **mit ~ die Treppe hinunterrennen** clatter down the stairs; **b)** *(Schimpfen)* grumbling; moaning

Gepräge das; ~s, ~ *(fig. geh.)* [special] character; **einer Sache** *(Dat.)* **ihr ~ geben** give sth. its character

gepriesen *2. Part. v.* **preisen**

gepunktet *Adj.* spotted ⟨*tie, blouse, etc.*⟩; *(regelmäßig)* polka-dot; dotted ⟨*line*⟩

gequält *Adj.* forced ⟨*smile, gaiety*⟩; pained ⟨*expression*⟩

Gequassel, Gequatsche das; ~s *(ugs. abwertend)* jabbering

gequollen *2. Part. v.* **quellen**

gerade [gə'ra:də], *(ugs.)* **grade** ['gra:də] **1.** *Adj.* **a)** straight; **den ~n Weg verfolgen** *(fig.)* keep to the straight and narrow; **b)** *(nicht schief)* upright; **~ gewachsen sein** ⟨*plant*⟩ have grown straight; ⟨*person*⟩ have grown up straight; **c)** *(aufrichtig)* forthright; direct; **d)** *nicht präd. (genau)* **das ~ Gegenteil** the direct *or* exact opposite; **e)** *(Math.)* even ⟨*number*⟩. **2.** *Adv.* **a)** just; **haben Sie ~ Zeit?** do you have time just now?; **~ erst** only just; **ich frage ihn ~ [mal]** *(bes. südd.: mal eben)* I'll just ask him; **b)** *(direkt)* right; **~ gegenüber/um die Ecke** right opposite/just round the corner; **c)** *(knapp)* just; **~ noch** only just; **er hat das Examen ~ so bestanden** he just scraped through the examination; **~ so viel, daß ...:** just enough to ...; ~ **noch rechtzeitig** only just in time; **d)** *(genau)* ~ **diese Angelegenheit** precisely *or* just this matter; **e)** *(ausgerechnet)* ~ **du/dieser Idiot** you/this idiot, of all people; **warum ~ ich/heute?** why me of all people/today of all days?. **3.** *Partikel* **a)** *(besonders)* particularly; **nicht ~:** not exactly; **b)** *(ugs.: erst recht)* **jetzt [tue ich es] ~:** [you] just watch me; [you] just try and stop me now

Gerade die; ~n, ~n a) *(Geom.)* straight line; **b)** *(Leichtathletik)* straight; **c)** *(Boxen)* straight-arm punch; **linke/rechte ~:** straight left/right

gerade·aus *Adv.* straight ahead; ‹*walk, drive*› straight on, straight ahead; **immer ~ gehen/fahren** carry straight on

gerade-: ~|biegen *unr. tr. V.* **a)** bend straight; straighten [out]; **b)** *(ugs.: bereinigen)* straighten out; put right; **~|halten 1.** *unr. tr. V.* **etw. ~halten** hold sth. straight; **den Kopf ~halten** hold one's head up; **2.** *unr. refl. V.* hold oneself [up] straight; **~heraus** [----'-] *(ugs.)* **1.** *Adv.* **etw. ~heraus sagen** say sth. straight out; **jmdm. ~heraus sagen/jmdn. ~heraus fragen** tell/ask sb. straight; **2.** *adj.; nicht präd.* straightforward; direct; **~|machen** *tr. V. (ugs.)* straighten [out]; **~|richten** *tr. V.* straighten [out]; put *or* set straight

gerädert *Adj. (ugs.)* whacked *(coll.)*; tired out

gerade-: ~|sitzen *unr. itr. V.* sit up straight; **~so** *Adv.* **etw. ~ machen wie jmd. anderes** do sth. just like sb. else; **~so groß/lang wie ...:** just as big/long as ...; **~sogut** *Adv.* just as well; equally well; **~sogut wie ...:** just as well as ...; **~|stehen** *unr. itr. V.* **a)** stand up straight; **b)** *(fig.: einstehen)* **für etw. ~stehen** accept responsibility for sth.; **für jmdn. ~stehen** answer for sb.; **~wegs** *Adv.* **a)** straight; **b)** *(ohne Umschweife)* straight away; directly; **~wegs zum Thema kommen** come straight to the point; **~zu** *Adv.* really; perfectly; **das ist ~zu lächerlich** that is downright ridiculous; **ein ~zu ideales Beispiel** an absolutely perfect example

gerad·linig [~li:nɪç] **1.** *Adj.* **a)** straight; direct, lineal ‹*descent, descendant*›; **b)** *(fig.)* straightforward; **2.** *adv.* **a) ~linig verlaufen** run in a straight line; **b)** *(fig.)* **~linig handeln/denken** be straightforward

gerammelt *Adv. in* **~ voll** *(ugs.)* [jam-]packed *(coll.)*; packed out *(coll.)*

Gerangel [gə'raŋl̩] **das; ~s** *(ugs.)* **a)** scrapping *(coll.)*; **b)** *(abwertend: Kampf)* free-for-all; scramble

Geranie [ge'ra:ni̯ə] **die; ~, ~n** geranium

gerann [gə'ran] *3. Pers. Sg. Prät. v.* **gerinnen**

gerannt [gə'rant] *2. Part. v.* **rennen**

Geraschel das; ~s *(ugs.)* rustling

Gerät [gə'rɛ:t] **das; ~[e]s, ~e a)** piece of equipment; *(Fernseher, Radio)* set; *(Garten~)* tool; *(Küchen~)* utensil; *(Meß~)* instrument; **elektrische ~e** electrical appliances; **b)** *(Turnen)* piece of apparatus; **an den ~en turnen** do gymnastics on the apparatus; **c)** *o. Pl. (Ausrüstung)* equipment *no pl.*

¹geraten *unr. itr. V.; mit sein* **a)** get; **in ein Unwetter ~:** be caught in a storm; **an jmdn. ~:** meet sb.; **an den Richtigen/Falschen ~:** come to the right/wrong person; **in Panik ~:** panic *or* get into a panic; **b)** *(gelingen)* turn out well; **sie ist zu kurz/lang ~** *(scherzh.)* she has turned out on the short/tall side; **c)** *(ähneln)* **nach jmdm. ~:** take after sb.

²geraten *Adj.; nicht attr.* advisable; **es scheint mir ~, ...:** I think it advisable ...

Geräte-: ~schuppen der tool shed; **~turnen das** apparatus gymnastics *sing.*

Geratewohl *in* **wir fuhren aufs ~ los** *(ugs.)* we went for a drive just to see where we ended up; **er hat sich aufs ~ einige Firmen ausgewählt** *(ugs.)* he selected a few firms at random

Geratter das; ~s *(ugs.)* clatter

geraum *Adj.; nicht präd. (geh.)* considerable; **nach ~er Zeit** after some [considerable] time

geräumig [gə'rɔymɪç] *Adj.* roomy; spacious ‹*room*›

Geräusch [gə'rɔyʃ] **das; ~[e]s, ~e** sound; *(unerwünscht)* noise

geräusch-, Geräusch-: ~empfindlich *Adj.* sensitive to noise *pred.*; **~empfindliche Menschen** people who are sensitive to noise; **~los 1.** *Adj.* silent; noiseless; **2.** *adv.* **a)** silently; without a sound; noiselessly; **b)** *(fig. ugs.: ohne Aufsehen)* without [any] fuss; quietly; **~pegel der** noise level; **~voll 1.** *Adj.* noisy; **2.** *adv.* noisily

gerben ['gɛrbn̩] *tr. V.* tan ‹*hides, skins*›; **von Wind und Wetter gegerbte Haut** *(fig.)* skin tanned by wind and sun

Gerberei die; ~, ~en tannery

gerecht 1. *Adj.* just ‹*teacher, cause, verdict, punishment*›; *(unparteiisch)* just; fair; impartial ‹*judge*›; righteous ‹*anger*›; **der ~e Gott** *(bibl.)* our righteous Lord; **~ gegen jmdn. sein** be fair *or* just to sb.; **jmdm./einer Sache ~ werden** do justice to sb./sth.; **einer Aufgabe ~ werden** cope with a task. **2.** *adv.* justly; ‹*judge, treat*› fairly

gerechtfertigt *Adj.* justified

Gerechtigkeit die; ~ a) justice; **die ~ Gottes** *(christl. Rel.)* the righteousness of God; **~ üben** *(geh.)* act justly; be just; **jmdm. ~ widerfahren lassen** *(geh.)* treat sb. justly; **b) die ~ nimmt ihren Lauf** the law takes its course

gerechtigkeits-, Gerechtigkeits-: ~gefühl das sense of justice; **~liebend** *Adj.* **~liebend sein** have a love of justice; **~sinn der** *s.* **~gefühl**

Gerede das; ~s *(abwertend)* **a)** *(ugs.)* talk; **b)** *(Klatsch)* gossip; **ins ~ kommen/jmdn. ins ~ bringen** get into/bring sb. into disrepute

geregelt *Adj.; nicht präd.* regular, steady ‹*job*›; orderly, well-ordered ‹*life*›; computer-controlled ‹*catalytic converter*›

gereichen *itr. V. (geh.)* **jmdm. zur Ehre/zum Vorteil ~:** redound to sb.'s honour *or* credit/advantage

gereift *Adj.; nicht präd.* mature

gereizt *Adj.* irritable; touchy

¹Gericht [gə'rɪçt] **das; ~[e]s, ~e a)** *(Institution)* court; **jmdn. vor ~ laden** summon sb. to appear in court; **vor ~ erscheinen/aussagen** appear/testify in court; **vor ~ stehen** be on *or* stand trial; **b)** *(Richter)* bench; **Hohes ~!** Your Honour!; **c)** *(Gebäude)* court[-house]; **d)** *in* **das Jüngste** *od.* **Letzte ~** *(Rel.)* the Last Judgement; **mit jmdm. [hart** *od.* **scharf] ins ~ gehen** take sb. [severely] to task

²Gericht das; ~[e]s, ~e dish

gerichtlich 1. *Adj.; nicht präd.* judicial; forensic ‹*psychology, medicine*›; legal ‹*proceedings*›; court ‹*order*›; **ein ~es Nachspiel haben** have legal consequences. **2.** *adv.* **jmdn. ~ verfolgen** prosecute sb.; take sb. to court; **gegen jmdn. ~ vorgehen** take legal action against sb.; take sb. to court

gerichts-, Gerichts-: ~beschluß der decision of the/a court; the/a court's decision; **~hof der** Court of Justice; **~ko-**

sten *Pl.* legal costs; costs of the case; **~notorisch** *(Rechtsspr.) Adj.* ‹*person, event, fact*› known to the court; **~saal der** courtroom; **~stand der** *(Rechtsspr.)* place of jurisdiction; **~urteil das** judgement [of the court]; **~verfahren das** legal proceedings *pl.;* **ein ~verfahren einleiten** institute legal *or* court proceedings; **~verhandlung die** *(strafrechtlich)* trial; *(zivil)* hearing; **~vollzieher der; ~s, ~:** bailiff

gerieben 1. *2. Part. v.* **reiben. 2.** *Adj. (ugs.)* artful

geriffelt *Adj.* corrugated ‹*surface, sheet metal*›; fluted ‹*column*›; ribbed ‹*glass*›

gering [gə'rɪŋ] *Adj.* **a)** low ‹*temperature, pressure, price*›; low, small ‹*income, fee*›; little ‹*value*›; small ‹*quantity, amount*›; short ‹*distance, time*›; **in ~er Höhe** low down; **b)** *(unbedeutend)* slight; minor ‹*role*›; **meine ~ste Sorge** the least of my worries; **nicht das ~ste** nothing at all; **nicht im ~sten** not in the slightest *or* least; **c)** *(veralt.: niedrigstehend)* humble ‹*origin, person*›; **kein Geringerer als ...:** no less a person than ...; **d)** *(geh.: schlecht)* poor, low, inferior ‹*quality, opinion*›

gering|achten *tr. V.: s.* **geringschätzen**

geringelt *Adj.* curly; ‹*hair*› in ringlets; ‹*pattern, socks, jumper*› with horizontal stripes

gering·fügig [-fy:gɪç] **1.** *Adj.* slight ‹*difference, deviation, improvement*›; slight, minor ‹*alteration, injury*›; small, trivial ‹*amount*›; minor, trivial ‹*detail*›. **2.** *adv.* slightly

Geringfügigkeit die; ~, ~en a) *o. Pl.* triviality; insignificance;; **b)** *(Kleinigkeit)* triviality; trifle

gering|schätzen *tr. V.* have a low opinion of, think very little of ‹*person, achievement*›; set little store by ‹*success, riches*›; make light of ‹*danger*›; **sein eigenes Leben ~:** have scant regard for one's own life

gering·schätzig [-ʃɛtsɪç] **1.** *Adj.* disdainful; contemptuous; disparaging ‹*remark*›. **2.** *adv.* disdainfully; contemptuously; **von jmdm. ~ sprechen** speak disparagingly of sb.

Gering·schätzung die; *o. Pl.* disdain; contempt

gerinnen *unr. itr. V.; mit sein* coagulate; ‹*blood*› coagulate, clot; ‹*milk*› curdle; *s. auch* **Blut**

Gerinnsel [gə'rɪnzḷ] **das; ~s, ~** *(Blut)* clot

Gerinnung die; ~, ~en coagulating; *(von Blut auch)* clotting; *(von Milch)* curdling

Gerippe das; ~s, ~ a) skeleton; **sie ist bis zum ~ abgemagert** *(fig.)* she has lost so much weight that she is only skin and bones; **b)** *(fig.)* framework; *(von Schiffen, Gebäuden)* skeleton

gerippt [gə'rɪpt] *Adj.* ribbed ‹*fabric, garment*›; fluted ‹*glass, column*›; laid ‹*paper*›

gerissen [gə'rɪsn̩] **1.** *2. Part. v.* **reißen. 2.** *Adj. (ugs.)* crafty

geritten *2. Part. v.* **reiten**

geritzt *Adj. (salopp) in* **etw. ist ~:** sth. is [all] settled; **ist ~!** will do! *(coll.)*

Germane [gɛr'ma:nə] **der; ~n, ~n, Germanin die; ~, ~nen** *(hist.)* ancient German; Teuton

germanisch *Adj.* Germanic; Teutonic

Germanist der; ~en, ~en Germanist; German scholar

Germanistik die; ~: German studies *pl., no art.*

Germanistin die; ~, ~nen *s.* **Germanist**

gern[e] ['gɛrn(ə)]; **lieber** ['li:bɐ], **am liebsten** [-'li:pstn̩] *Adv.* **a)** *(mit Vergnügen)* **etw. ~ tun** like *or* enjoy *or* be fond of doing sth.; **er spielt lieber Tennis als Golf** he prefers playing tennis to golf; **etw. ~ essen/trinken** like sth.; **am liebsten trinkt er Wein** he likes wine best; **ja, ~/aber ~:** yes, of course; certainly!; **Kommst du mit? – Ja, ~!** Are you coming too? – Yes I'd like to!; **[das ist] ~ geschehen** it is *or* was a pleasure; **jmdn./etw. ~ haben** like *or* be fond of sb./sth.; **jmdn./etw. am liebsten haben** *od.* **mögen** like sb./sth. best; **~ gesehen sein** be welcome; **der kann mich ~ haben!** *(ugs.)* he can go to hell! *(coll.);* he can get stuffed *(sl.);* **b)** *(drückt Billigung aus: durchaus)* **das glaube ich ~:** I can quite *or* well believe that; **das kannst du ~ tun** you are welcome to do that; **c)** *(drückt Wunsch aus)* **ich hätte ~ einen Apfel** I would like an apple; **er wäre ~ mitgekommen** he would have liked to come along; **das hättest du lieber nicht tun sollen** it would have been better if you had not done that; **laß das lieber** better not do that; **noch ein Stück Kuchen? – Lieber nicht** Another piece of cake? – I'd better not; **ich bleibe heute lieber im Bett** I'd better stay in bed today; **d) etw. ~ tun** *(etw. oft tun)* usually do sth.

Gerne·groß der; ~, ~e *(ugs. scherzh.)* **er ist ein [kleiner] ~:** he likes to act big *(coll.)*

gerochen *2. Part. v.* **riechen**

Geröll [gə'rœl] **das; ~s, ~e** detritus; debris; *(größer)* boulders *pl.; (im Gebirge auch)* scree

geronnen [gə'rɔnən] *2. Part. v.* **rinnen, gerinnen**

Gerste ['gɛrstə] **die; ~:** barley

Gersten·korn das a) *(Frucht)* barleycorn; **b)** *(Augenentzündung)* sty

Gerte ['gɛrtə] **die; ~, ~n** switch

Geruch [gə'rʊx] **der; ~[e]s, Gerüche** [gə'rʏçə] smell; odour; *(von Blumen)* scent; fragrance; *(von Brot, Kuchen)* smell; aroma; **einen unangenehmen ~ verbreiten** give off an unpleasant smell *or* odour *or* a stench

geruch·los *Adj.* odourless; *(ohne Duft)* unscented, scentless ‹*flower etc.*›

Geruchs-: ~organ das olfactory organ; **~sinn der;** *o. Pl.* sense of smell; olfactory sense

Gerücht [gə'rʏçt] **das; ~[e]s, ~e** rumour; **ein ~ in die Welt** *od.* **in Umlauf setzen** start a rumour; **das halte ich für ein ~!** *(ugs.)* I can't believe that!

gerücht·weise *Adv.* **ich habe ~ vernommen** *od.* **gehört, daß ...:** I've heard it rumoured that ...

gerufen *2. Part. v.* **rufen**

gerührt *Adj.* touched *(also iron.);* moved

geruhsam 1. *Adj.* peaceful; quiet; leisurely ‹*stroll*›. **2.** *adv.* leisurely; *(ungestört)* quietly

Gerümpel [gə'rʏmpḷ] **das; ~s** *(abwertend)* junk; [useless] rubbish

Gerundium [ge'rʊndi̯ʊm] **das; ~s, Gerundien** *(Sprachw.)* gerund

gerungen [gə'rʊŋən] *2. Part. v.* **ringen**

Gerüst [gə'rʏst] **das; ~[e]s, ~e** scaffolding *no pl., no indef. art.; (fig.: eines Romans usw.)* framework

gerüttelt *Adj.: in* **ein ~ Maß** *(veralt.)* a good measure

ges, Ges [gɛs] **das; ~, ~** *(Musik)* [key of] G flat; *s. auch* **a, A**

gesalzen *Adj. (salopp)* steep *(coll.)* ‹*price, bill*›

gesammelt *Adj.* concentrated ‹*attention, energy*›; **~e Werke** collected works

gesamt *Adj.; nicht präd.* whole; entire; **das ~e Vermögen** the entire *or* total wealth

gesamt-, Gesamt-: ~auflage die *(Druckw.)* total edition; *(einer Zeitung)* total circulation; **~ausgabe die** *(Buchw.)* complete edition; ~**betrag der** total amount; ~**eindruck der** general *or* overall impression; ~**ergebnis das** overall result; ~**gewicht das** total weight; **das zulässige ~gewicht** the permissible maximum weight

Gesamtheit die; ~ die ~ der Beamten all civil servants; **die ~ der Bevölkerung** the whole of the *or* the entire population; **die Lehrer in ihrer ~:** teachers as a whole

gesamt-, Gesamt-: ~hoch·schule die *(Hochschulw.) institution with colleges teaching at various levels, so that students can more readily extend their courses;* ~**note die** overall mark; ~**schule die** comprehensive [school]; ~**sieger der, ~siegerin die** *(Sport)* overall winner; ~**summe die** *s.* ~**betrag;** ~**werk das** œuvre; *(Bücher)* complete works *pl.;* ~**zahl die** total number

gesandt [gə'zant] *2. Part. v.* **senden**

Gesandte der/die; *adj. Dekl.* envoy; **der päpstliche ~:** the papal legate *or* nuncio

Gesandtschaft die; ~, ~en legation

Gesang [gə'zaŋ] **der; ~[e]s, Gesänge** [gə'zɛŋə] **a)** *o. Pl.* singing; **b)** *(Lied)* song; **c)** *(Literaturw.)* canto

Gesang·buch das hymn-book

Gesang[s]-: ~stunde die singing-lesson; ~**unterricht der** singing instruction; ~**unterricht nehmen/geben** take/give singing-lessons *pl.*

Gesang·verein der choral society

Gesäß [gə'zɛ:s] **das; ~es, ~e** backside; buttocks *pl.*

Gesäß·tasche die back pocket

gesättigt *Adj. (Chemie)* saturated

Geschädigte [gə'ʃɛ:dıçtə] **der/die;** *adj. Dekl.* injured party

geschaffen *2. Part. v.* **schaffen 1**

geschafft 1. *2. Part. v.* **schaffen 2, 3. 2.** *Adj.; nicht attr. (ugs.)* all in *(coll.)*

Geschäft [gə'ʃɛft] **das; ~[e]s, ~e a)** business; *(Abmachung)* [business] deal *or* transaction; **mit jmdm. ~e/ein ~ machen** do business/strike a bargain *or* do a deal with sb.; **mit etw. ein gutes/schlechtes ~ machen** make a good/poor profit on sth.; **b)** *o. Pl. (Absatz)* business *no art.;* **das ~ blüht** business *or* trade is booming; **c)** *(Firma)* business; **im ~** *(südd.)* at work; **ein ~ führen** run *or* manage a business; **e)** *(Laden)* shop; store *(Amer.); (Kaufhaus)* store; **f)** *(Aufgabe)* task; duty; **seinen ~en nachgehen** go about one's business; **g) sein großes/kleines ~ erledigen** *od.* **machen** *(ugs. verhüll.)* do big jobs *or* number two/small jobs *or* number one *(child language)*

Geschäfte·macher der *(abwertend)* profit-seeker

geschäftig 1. *Adj.* bustling; **ein ~es Treiben** bustling activity; hustle and bustle. **2.** *adv.* **~ hin und her laufen** bustle about

Geschäftigkeit die; ~: bustle

geschäftlich 1. *Adj.* **a)** business *attrib.* ⟨*conference, appointment*⟩; **b)** *(sachlich, kühl)* business-like. **2.** *adv.* **a)** on business; **er hat dort ~ zu tun** he has [some] business to do there; **b)** *(sachlich, kühl)* in a business-like way *or* manner

geschäfts-, Geschäfts-: ~abschluß der conclusion of the/a business transaction *or* deal; ~**aufgabe die** closure of the/a business; ~**bedingungen** *Pl.* terms [and conditions] of trade; ~**bericht der** company report; *(jährlich)* annual report; ~**beziehungen** *Pl.* business dealings; ~**brief der** business letter; ~**bücher** *Pl.* books; accounts; ~**eröffnung die** opening of a/the shop *or (Amer.)* store; ~**frau die** businesswoman; ~**freund der** business associate; ~**führend** *Adj.; nicht präd.* managing ⟨*director*⟩; executive ⟨*chairman*⟩; ~**führer der a)** manager; **b)** *(Vereinswesen)* secretary; ~**führung die** management; ~**haus das** office-block *(with or without shops);* ~**interesse das;** *meist Pl.* **das ~interesse/die ~interessen** the interests *pl.* of the business; ~**kosten** *Pl.: in* **auf ~kosten** on expenses; ~**lage die a) die ~lage der Firma** the [business] position of the firm; **die allgemeine ~lage** the general business situation; **b)** *(Ort)* **in guter ~lage** well situated [for business]; ~**leben das** business [life]; ~**leitung die** *s.* ~**führung;** ~**mann der;** *Pl.* ~**leute** businessman; ~**ordnung die** standing orders *pl.; (im Parlament)* [rules *pl.* of] procedure; **Fragen zur ~ordnung** questions on points of order; ~**partner der** business partner; ~**räume** *Pl.* business premises; *(Büroräume)* offices; ~**reise die** business trip; ~**schädigend** *Adj.* bad for business; ⟨*conduct*⟩ damaging to the interests of the company; ~**schluß der** closing-time; **nach ~schluß** after business hours; *(im Büro)* after office hours; ~**sinn der;** *o. Pl.* business sense *or* acumen; ~**stelle die** *(einer Bank, Firma)* branch; *(einer Partei, eines Vereins)* office; ~**straße die** shopping-street; ~**stunden** *Pl.* business hours; *(im Büro)* office hours; ~**tüchtig** *Adj.* able, capable, efficient ⟨*businessman, landlord, etc.*⟩; ~**wagen der** company car; ~**zeit die** *s.* ~**stunden**

gescheckt [gə'ʃɛkt] *Adj.* spotted ⟨*cow, bull, rabbit, etc.*⟩; skewbald ⟨*horse*⟩

geschehen [gə'ʃe:ən] *unr. itr. V.; mit sein* **a)** *(passieren)* happen; occur; **er ließ es ~:** he let it happen; **~ ist ~:** what's done is done; **b)** *(ausgeführt werden)* ⟨*deed*⟩ be done; **der Mord geschah aus Eifersucht** the murder was committed out of jealousy; **es muß etwas ~:** something must be done; **was geschieht damit?** what's to be done with it?; **c)** *(widerfahren)* **jmdm. geschieht etw.** sth. happens to sb.; **es geschieht dir nichts** nothing will happen to you; **das geschieht ihm recht** it serves him right; **e) es ist um ihn ~:** it's all up with him; **es ist um seine Gesundheit/Stellung ~:** his health is ruined/he has lost his job

Geschehen das; ~s, ~ *(geh.)* **a)** *(Ereignisse)* events *pl.;* happenings *pl.;* **b)** *(Vorgang)* action

gescheit [gə'ʃait] **1.** *Adj.* **a)** *(intelligent)* clever; **daraus werde ich nicht ~:** I can't make head or tail of it; **b)** *(ugs.: vernünftig)* sensible; **sei doch ~:** be sensible; **du bist wohl nicht ganz** *od.* **nicht recht ~:** you can't be quite right in the head; **c)** *(ugs.: ordentlich, gut)* decent. **2.** *adv.* cleverly

Geschenk [gə'ʃɛŋk] **das; ~[e]s, ~e** present; gift; **jmdm. ein ~ machen** give sb. a present; **ein ~ des Himmels** a godsend

Geschenk-: ~artikel der gift; ~**packung die** gift pack; ~**papier das** gift wrapping-paper

Geschichte [gə'ʃıçtə] **die; ~, ~n a)** *o. Pl.* history; **~ machen** make history; **in die ~ eingehen** *(geh.)* go down in history; **b)** *(Erzählung)* story; *(Fabel, Märchen)* story; tale; **c)** *(ugs.: Sache)* **das sind alte ~n** that's old hat *(coll.);* **das ist [wieder] die alte ~:** it's the [same] old story [all over again]; **das sind ja schöne ~n!** *(iron.)* that's a fine thing *or* state of affairs! *(iron.);* **die ganze ~:** the whole business *or* thing; **mach keine ~n!** don't do anything silly; **mach keine langen ~n** don't make a [great] fuss

geschichtlich 1. *Adj.* **a)** historical; **b)** *(bedeutungsvoll)* historic. **2.** *adv.* historically

Geschichts-: ~atlas der historical atlas; **~bewußtsein das** awareness of history; historical awareness; **~buch das** history book; **~lehrer der** history teacher; **~schreibung die** historiography; **~unterricht der** history teaching; *(Unterrichtsstunde)* history lesson; **im ~unterricht** in history; **~wissenschaftler der** [academic] historian

[1]**Geschick** [gə'ʃɪk] **das; ~[e]s, ~e a)** *(geh.)* fate; **ihn ereilte sein ~:** he met his fate; **ein glückliches ~:** a kindly Providence; **b)** *Pl. (Lebensumstände)* destiny *sing.*

[2]**Geschick das; ~[e]s** skill; **ein ~ für etw. haben** be skilled at sth.

Geschicklichkeit die; ~: skilfulness; skill

Geschicklichkeits·spiel das game of skill

geschickt 1. *Adj.* **a)** skilful; *(fingerfertig)* skilful; dexterous; *(beweglich)* agile ⟨*climber*⟩; **b)** *(klug)* clever; adroit. **3.** *adv.* **a)** skilfully; *(fingerfertig)* skilfully; dexterously; **b)** *(klug)* cleverly; adroitly

geschieden *2. Part. v.* **scheiden**

Geschiedene der/die; *adj. Dekl.* divorcee; **seine ~:** his ex-wife

geschienen *2. Part. v.* **scheinen**

Geschirr [gə'ʃɪr] **das; ~[e]s, ~e a)** *(Riemenzeug)* harness; **dem Pferd das ~ anlegen** harness the horse; **sich ins ~ legen** *(kräftig ziehen)* pull hard; *(angestrengt arbeiten)* work like a slave; **b)** *(Teller, Tassen usw.)* crockery; *(benutzt)* dishes *pl.*; *(zusammenpassend)* [dinner/tea] service; *(Küchen~)* pots and pans *pl.*; kitchenware; **das gute ~:** the good china; **das ~ abwaschen** wash up *or* do the dishes

Geschirr-: ~schrank der china cupboard; **~spülen das; ~s** washing-up; **~spüler der, ~spül·maschine die** dishwashing machine; dishwasher; **~tuch das;** *Pl.* **-tücher** tea-towel; drying-up cloth *(Brit.)*; dish towel *(Amer.)*

geschissen [gə'ʃɪsn̩] *2. Part. v.* **scheißen**

geschlafen *2. Part. v.* **schlafen**

geschlagen *2. Part. v.* **schlagen**

Geschlecht das; ~[e]s, ~er a) sex; **männlichen/weiblichen ~s sein** be male/female; **das starke ~** *(ugs. scherzh.)* the stronger sex; **das schwache/schöne/zarte ~** *(ugs. scherzh.)* the weaker/fair/gentle sex; **b)** *(Sippe)* family; **von altem ~:** of ancient lineage; **das ~ der Habsburger** the house of Habsburg; **c)** *(Sprachw.)* gender; **d)** *o. Pl. (Geschlechtsteil)* sex

geschlechtlich 1. *Adj.* sexual. **2.** *adv.* **mit jmdm. ~ verkehren** have sexual intercourse with sb.

geschlechts-, Geschlechts-: ~akt der sex[ual] act; **~hormon das** sex hormone; **~krank** *Adj.* ⟨*person*⟩ suffering from VD *or* a venereal disease; **~krank sein** have VD; be suffering from a venereal disease; **~krankheit die** venereal disease; **~leben das** sex life; **~los** *Adj. (Biol.)* asexual; *(fig.)* sexless; **~merkmal das** sex[ual] characteristic; **~organ das** sex[ual] organ; genital organ; **~reif** *Adj.* sexually mature; **~reife die** sexual maturity; **~spezifisch** *Adj. (Soziol.)* sex-specific; **~teil das: die ~teile/das ~teil** the genitals *pl.*; **~trieb der** sex[ual] drive *or* urge; **~verkehr der** sexual intercourse; **~wort das** *(Sprachw.)* article

geschlichen *2. Part. v.* **schleichen**

geschliffen 1. *2. Part. v.* **schleifen. 2.** *Adj.* polished, refined ⟨*style, manners, etc.*⟩; polished ⟨*sentence*⟩. **3.** *adv.* in a polished manner

geschlossen 1. *2. Part. v.* **schließen. 2.** *Adj.* **a)** *(gemeinsam)* united ⟨*action, front*⟩; unified ⟨*procedure*⟩; *s. auch* **Gesellschaft c; b)** *(zusammenhängend)* **eine ~e Ortschaft** a built-up area; **c)** *(abgerundet)* full; complete ⟨*picture, impression*⟩. **3.** *adv.* **~ für etw. stimmen/sein** vote/be unanimously in favour of sth.; **wir verließen ~ unser Büro** we walked out in a body *or* en masse; **~ gegen etw. vorgehen** take concerted action against sth.; **~ hinter jmdm. stehen** be solidly behind sb.

Geschlossenheit die; ~ a) *(Gemeinschaft)* unity; **b)** *(Einheitlichkeit)* unity; uniformity

geschlungen [gə'ʃlʊŋən] *2. Part. v.* **schlingen**

Geschmack [gə'ʃmak] **der; ~[e]s, Geschmäcke** [gə'ʃmɛkə] *od. ugs. scherzh.:* **Geschmäcker** [gə'ʃmɛkɐ]) **a)** taste; **einen guten/schlechten ~ haben** have good/bad taste; **das ist [nicht] mein** *od.* **nach meinem ~:** that is [not] to my taste; **das verstößt gegen den guten ~:** that offends against good taste; **im ~ jener Zeit** in the style of that period; **über ~ läßt sich nicht streiten** there's no accounting for taste[s]; **an etw.** *(Dat.)* **~ finden** *od.* **gewinnen** acquire a taste for sth.; take a liking to sth.; **sie kann solchen Bildern keinen ~ abgewinnen** she cannot appreciate such pictures; **auf den ~ kommen** acquire the taste for it; get to like it; **b)** *o. Pl. (Geschmackssinn)* sense of taste

geschmacklos 1. *Adj.* **a)** tasteless; insipid; **b)** *(fig.)* tasteless; **~ sein** be in bad taste; ⟨*person*⟩ be lacking in taste. **2.** *adv.* tastelessly

Geschmacklosigkeit die; ~, ~en a) lack of [good] taste; bad taste; **b)** *o. Pl. (fig.)* tastelessness; bad taste; **c)** *(fig.) (Äußerung)* tasteless remark; *(Handlung)* tasteless behaviour *sing., no indef. art.*

geschmacks-, Geschmacks-: ~frage die question *or* matter of taste; **~neutral** *Adj.* tasteless; flavourless; **~richtung die a)** flavour; **b)** *(fig.)* taste

Geschmack[s]·sache die *in* **das ist ~:** that is a question *or* matter of taste

Geschmacks-: ~sinn der; *o. Pl.* sense of taste; **~verirrung die** *(abwertend)* lapse of taste; **an** *od.* **unter ~verirrung** *(Dat.)* **leiden** *(ugs.)* suffer from a lapse in taste

geschmack·voll 1. *Adj.* tasteful; **die Bemerkung war nicht sehr ~:** the remark was not in very good taste. **2.** *adv.* tastefully

Geschmeide [gə'ʃmaidə] **das; ~s, ~** *(geh.)* jewellery *no pl.*

geschmeidig 1. *Adj.* **a)** sleek ⟨*hair, fur*⟩; supple, soft ⟨*leather, boots, skin*⟩; smooth ⟨*dough*⟩; **b)** *(gelenkig)* supple ⟨*fingers*⟩; supple, lithe ⟨*body, movement, person*⟩; **c)** *(fig.: anpassungsfähig)* adaptable. **2.** *adv.* **a)** *(gelenkig)* agilely; **b)** *(fig.)* adaptably

Geschmeidigkeit die; ~: *s.* **geschmeidig 1:** sleekness; suppleness; softness; smoothness; litheness

Geschmiere das; ~s *(ugs. abwertend)* **a)** [filthy] mess; **b)** *(Geschriebenes)* scribble; scrawl; **c)** *(Machwerk)* rubbish; bilge *(sl.)*

geschmissen [gə'ʃmɪsn̩] *2. Part. v.* **schmeißen**

geschmolzen *2. Part. v.* **schmelzen**

Geschnatter das; ~s *(ugs.)* **a)** *(das Schnattern)* cackling;

cackle; **b)** *(abwertend: das Sprechen)* chatter[ing]; nattering *(coll.)*
geschniegelt *Adj. (ugs. abwertend)* nattily dressed; ~ **und gebügelt** *od.* **gestriegelt** all spruced up
geschnitten [gəˈʃnɪtn̩] *2. Part. v.* **schneiden**
geschnoben *2. Part. v.* **schnauben**
geschoben [gəˈʃoːbn̩] *2. Part. v.* **schieben**
geschollen *2. Part. v.* **schallen**
gescholten [gəˈʃɔltn̩] *2. Part. v.* **schelten**
Geschöpf [gəˈʃœpf] **das; ~[e]s, ~e a)** creature; **b)** *(erfundene Gestalt)* creation
geschoren *2. Part. v.* **scheren**
[1]**Geschoß das; Geschosses, Geschosse** projectile; *(Kugel)* bullet; *(Rakete)* rocket; missile; *(Granate)* shell; grenade
[2]**Geschoß das; Geschosses, Geschosse** *(Etage)* floor; *(Stockwerk)* storey
geschossen [gəˈʃɔsn̩] *2. Part. v.* **schießen**
-geschossig 1. *Adj.* -storey; **ein~/mehr~:** single-storey/multistorey; **zwei~ sein:** have two storeys. **2.** *adv.* **drei~ bauen** build three storeys high
geschraubt *Adj. (ugs. abwertend)* stilted ⟨*language*⟩; *(schwülstig)* affected, pretentious ⟨*way of speaking, style*⟩
Geschrei das; ~s a) shouting; shouts *pl.; (durchdringend)* yelling; yells *pl.; (schrill)* shrieking; shrieks *pl.; (von Verletzten, Tieren)* screaming; screams *pl.;* **b)** *(ugs.: das Lamentieren)* fuss; to-do; **ein großes ~ wegen etw. machen** make *or* kick up a great fuss about sth.; make a great to-do about sth.
geschrieben *2. Part. v.* **schreiben**
geschrie[e]n [gəˈʃriː(ə)n] *2. Part. v.* **schreien**
geschritten *2. Part. v.* **schreiten**
geschunden *2. Part. v.* **schinden**
Geschütz das; ~es, ~e [big] gun; piece of artillery; **die ~e** the artillery *sing.;* the [big] guns; **schweres ~ auffahren** *(fig. ugs.)* bring up the big guns *or* heavy artillery *(fig.)*
Geschütz·feuer das artillery-fire; shell-fire
geschützt *Adj.* **a)** sheltered; **b)** *(unter Naturschutz)* protected
Geschwader [gəˈʃvaːdɐ] **das; ~s, ~** *(Marine)* squadron; *(Luftwaffe)* wing *(Brit.);* group *(Amer.)*
Geschwafel das; ~s *(ugs. abwertend)* waffle
Geschwätz das; ~es *(ugs. abwertend)* **a)** prattle; prattling; **b)** *(Klatsch)* gossip; tittle-tattle
geschwätzig *Adj. (abwertend)* talkative
Geschwätzigkeit die; ~ *(abwertend)* talkativeness
geschweift *Adj.* curved; **~e Klammern** *(Druckw.)* braces
geschweige *Konj.* ~ **[denn]** let alone; never mind
geschwiegen *2. Part. v.* **schweigen**
geschwind [gəˈʃvɪnt] *(bes. südd.)* **1.** *Adj.* swift; quick. **2.** *adv.* swiftly; quickly; **ich laufe ~ zum Kaufmann** I'm just dashing to the grocer's
Geschwindigkeit die; ~, ~en speed; **mit großer ~:** at great speed; **mit einer ~ von 50 km/h** at a speed of 50 km/h
Geschwindigkeits-: ~begrenzung die, ~beschränkung die speed limit; **die ~beschränkung nicht beachten** exceed the speed limit; **~kontrolle die** speed check
Geschwister [gəˈʃvɪstɐ] **das; ~s, ~ a)** *Pl.* brothers and sisters; **Hans und Maria sind ~:** Hans and Maria are brother and sister; **b)** *(bes. Biol., Psych.)* sibling
geschwisterlich *Adj.* brotherly/sisterly ⟨*affection, love*⟩
Geschwister·paar das brother and sister
geschwollen 1. *2. Part. v.* **schwellen. 2.** *Adj. (abwertend)* pompous; bombastic. **3.** *adv. (abwertend)* pompously; bombastically
geschwommen [gəˈʃvɔmən] *2. Part. v.* **schwimmen**
geschworen 1. *2. Part. v.* **schwören. 2.** *Adj.; nicht präd.* sworn ⟨*enemy*⟩
Geschworene der/die; *adj. Dekl.* juror; **die ~n** the jury
Geschworenen-: ~bank jury-box; *(fig.)* jury; **~gericht das** *s.* **Schwurgericht**
Geschwulst [gəˈʃvʊlst] **die; ~, Geschwülste** [gəˈʃvʏlstə] tumour
geschwunden *2. Part. v.* **schwinden**
geschwungen 1. *2. Part. v.* **schwingen. 2.** *Adj.* curved
Geschwür [gəˈʃvyːɐ̯] **das; ~s, ~e** ulcer; *(Furunkel)* boil; *(fig.)* running sore
Geselle [gəˈzɛlə] **der; ~n, ~n a)** journeyman; **b)** *(Kerl)* fellow
gesellen *refl. V.* **sich zu jmdm. ~:** join sb.
Gesellen·brief der journeyman's diploma *or* certificate
gesellig 1. *Adj.* **a)** sociable; gregarious; **ein ~er Abend/~es Beisammensein** a convivial *or* sociable evening/a friendly get-together; **b)** *(Biol.)* gregarious. **2.** *adv.* ~ **leben** live gregariously; be gregarious; ~ **zusammensitzen** sit [together] and chat [sociably]
Geselligkeit die; ~: die ~ lieben enjoy [good] company
Gesellschaft die; ~, ~en a) society; **jmdn. in die ~ einführen** introduce sb. into society; **b)** *(Anwesenheit anderer)* company; ~ **bekommen** get company; **in schlechte ~ geraten** get into bad company; **jmdm. ~ leisten** keep sb. company; **c)** *(Veranstaltung)* party; **eine ~ geben** give a party; **eine geschlossene ~:** a private function *or* party; **d)** *(Kreis von Menschen)* group of people; crowd; *(abwertend)* crew; lot *(coll.);* **e)** *(Wirtschaft)* company
Gesellschafter der; ~s, ~ a) ein guter ~ sein be good company; **b)** *(Wirtsch.)* partner; *(Teilhaber)* shareholder; **stiller ~:** sleeping partner; silent partner *(Amer.)*
Gesellschafterin die; ~, ~nen a) [lady] companion; **b)** *(Wirtsch.)* partner; *(Teilhaber)* shareholder
gesellschaftlich 1. *Adj.; nicht präd.* **a)** social; **b)** *(Soziol.)* society; **die ~e Produktion** production by society; **~es Eigentum an etw.** *(Dat.)* social ownership of sth. **2.** *adv.* socially
gesellschafts-, Gesellschafts-: ~anzug der dress-suit; **~fähig** *Adj. (auch fig.)* socially acceptable; **~form die** form of society; social system; **~klasse die** social class; **~kritik die** social criticism; **~kritisch** *Adj.* critical of society *postpos.;* **~ordnung die** social order; **~politik die** social policy; **~politisch** *Adj.* socio-political; **~reise die** group tour; **~schicht die** stratum of society; **~spiel das** parlour *or* party game; **~system das** social system; **~tanz der** ballroom dance; *(das Tanzen)* ballroom dancing; **~wissenschaften** *Pl.* social sciences; **~wissenschaftlich** *Adj.* sociological ⟨*studies, analyses*⟩
gesessen [gəˈzɛsn̩] *2. Part. v.* **sitzen**
Gesetz [gəˈzɛts] **das; ~es, ~e a)** law; *(geschrieben)* statute; **ein ~ verabschieden/einbringen** pass/introduce a bill; **das ~ des Handelns an sich reißen** seize the ini-

tiative; **etw. hat seine eigenen ~e** *(fig.)* sth. is a law unto itself; **b)** *(Regel)* rule; law

Gesetz-: ~blatt das law gazette; **~buch das** statute-book; **das Bürgerliche ~buch** the Civil Code; **~entwurf der** bill

gesetzes-, Gesetzes-: ~brecher der law-breaker; **~hüter der** *(iron.)* guardian of the law; **~novelle die** amendment; **~text der** wording of the/a law; **~treu 1.** *Adj.* law-abiding; **2.** *adv.* in accordance with the law; **~übertretung die** violation of the law; **~vorlage die** bill

gesetz-, Gesetz-: ~gebend *Adj.* legislative; **die ~gebende Gewalt** the legislative power; **~geber der** legislator; law-maker; *(Organ)* legislature; **~gebung die; ~:** legislation; law-making

gesetzlich 1. *Adj.* legal ⟨*requirement, definition, respresentative, interest*⟩; legal, statutory ⟨*obligation*⟩; statutory ⟨*period of notice, holiday*⟩; lawful, legitimate ⟨*heir, claim*⟩. **2.** *adv.* legally; **~ verankert sein** be established in law; **~ geschützt** registered ⟨*patent, design*⟩; ⟨*symbol*⟩ registered as a trade mark

gesetz-, Gesetz-: ~los *Adj.* lawless; **~mäßig 1.** *Adj.* **a)** law-governed ⟨*development, process*⟩; **~mäßig sein** be governed by *or* obey a [natural] law/ [natural] laws; **b)** *(gesetzlich)* legal; *(rechtmäßig)* lawful; legitimate; **2.** *adv.* in accordance with a [natural] law/[natural] laws; **~mäßigkeit die a)** conformity to a [natural] law/[natural] laws; **b)** *(Gesetzlichkeit)* legality; *(Rechtmäßigkeit)* lawfulness; legitimacy

gesetzt *Adj.* staid; **eine Dame ~en Alters** a woman of mature years

gesetz·widrig 1. *Adj.* illegal; unlawful. **2.** *adv.* illegally; unlawfully

Gesicht [gəˈzɪçt] **das; ~[e]s, ~er a)** face; **ein fröhliches ~ machen** look pleasant *or* cheerful; **über das ganze ~ strahlen** *(ugs.)* beam all over one's face; (fig.) **sein wahres ~ zeigen** show oneself in one's true colours; show one's true character; **jmdm. wie aus dem ~ geschnitten sein** be the [very *or* dead] spit [and image] of sb.; **das ist ein Schlag ins ~:** that is a slap in the face; **jmdm. ins ~ lachen/lügen** laugh in/lie to sb.'s face; **jmdm. etw. ins ~ sagen** say sth. to sb.'s face; **jmdm. ins ~ sehen** to look sb. in the face; **den Tatsachen ins ~ sehen** face the facts; **jmdm. [nicht] zu ~[e] stehen** [not] become sb.; **ein anderes ~ aufsetzen** *od.* **machen** put on a different expression; **das ~ verlieren** lose face; **ein ~ machen wie drei** *od.* **acht** *od.* **vierzehn Tage Regenwetter** look as miserable as sin; **ein langes ~/lange ~er machen** pull a long face; **~er schneiden** pull *or* make faces; **das stand ihm im ~ geschrieben** it was written all over his face; **b)** *(fig.: Aussehen)* **das ~ einer Stadt** the appearance of a town; **die vielen ~er Chinas** the many faces of China; **ein anderes ~ bekommen** take on a different complexion; **c)** *o. Pl. (geh., veralt.: Sehvermögen)* sight; **das Zweite ~ [haben]** [have] second sight; **jmdn./etw. zu ~ bekommen** set eyes on *or* see sb./sth.

Gesichts-: ~ausdruck der expression; look; **~creme die** face-cream; **~farbe die** complexion; **~kreis der** *(veralt.)* field of view; field *or* range of vision; *(fig.)* horizon; outlook; **~lotion die** face lotion; **~maske die a)** *(Larve)* mask; **b)** *(Kosmetik)* face-mask; face-pack; **~puder der** face-powder; **~punkt der** point of view; **~verlust der** loss of face; **~wasser das** face-lotion; **~winkel der a)** angle of vision; visual angle; **b)** *s.* **~punkt; ~züge** *Pl.* features

Gesindel [gəˈzɪndl̩] **das; ~s** *(abwertend)* rabble; riff-raff *pl.*

gesinnt [gəˈzɪnt] *Adj.* **christlich/sozial ~ [sein]** [be] Christian-minded/public-spirited; **jmdm. freundlich/übel ~ sein** be well-disposed/ill-disposed towards sb.

Gesinnung die; ~, ~en [basic] convictions *pl.;* [fundamental] beliefs *pl.;* **eine niedrige ~:** a low cast of mind

gesinnungs-, Gesinnungs-: ~genosse der like-minded person; **~los** *(abwertend)* **1.** *Adj.* unprincipled; **2.** *adv.* in an unprincipled manner; **~losigkeit die; ~:** lack of principle; **~wandel der** change *or* shift of attitude *or* views

gesittet [gəˈzɪtət] **1.** *Adj.* **a)** well-behaved; well-mannered ⟨*behaviour*⟩; **b)** *(zivilisiert)* civilized. **2.** *adv.* **a) sich ~ benehmen** be well-behaved; **b)** *(zivilisiert)* in a civilized manner

Gesöff [gəˈzœf] **das; ~[e]s, ~e** *(salopp abwertend)* muck *(coll.);* awful stuff *(coll.)*

gesogen *2. Part. v.* **saugen**

gesondert [gəˈzɔndɐt] **1.** *Adj.* separate. **2.** *adv.* separately

gesonnen 1. *2. Part. v.* **sinnen. 2.** *Adj.* **~ sein, etw. zu tun** feel disposed to do sth.

gesotten *2. Part. v.* **sieden**

Gespann [gəˈʃpan] **das; ~[e]s, ~e a)** *(Zugtiere)* team; **b)** *(Wagen)* horse and carriage; *(zur Güterbeförderung)* horse and cart; **c)** *(Menschen)* couple; pair

gespannt *Adj.* **a)** eager; expectant; rapt ⟨*attention*⟩; **ich bin ~, ob ...:** I'm keen *or* eager to know/see whether ...; **b)** *(konfliktbeladen)* tense ⟨*situation, atmosphere*⟩; strained ⟨*relations, relationships*⟩. **3.** *adv.* eagerly; expectantly; **einer Geschichte ~ zuhören** listen with rapt attention to a story

Gespenst [gəˈʃpɛnst] **das; ~[e]s, ~er a)** ghost; **~er sehen** *(fig.)* be imagining things; **b)** *(geh.: Gefahr)* spectre

Gespenster-: ~geschichte die ghost story; **~stunde die** witching hour

gespenstisch *Adj.* ghostly; ghostly, eerie ⟨*appearance*⟩; eerie ⟨*building, atmosphere*⟩

gespie[e]n [gəˈʃpiː(ə)n] *2. Part. v.* **speien**

Gespiele der; ~n, ~n, Gespielin die; ~, ~nen *(geh. veralt.)* playmate

Gespinst [gəˈʃpɪnst] **das; ~[e]s, ~e** gossamer-like material; **das ~ der Seidenraupe** the cocoon of the silkworm; **ein ~ von Lügen** *(fig.)* a tissue of lies

gesponnen [gəˈʃpɔnən] *2. Part. v.* **spinnen**

gespornt *Adj. s.* **gestiefelt**

Gespött [gəˈʃpœt] **das; ~[e]s** mockery; ridicule; **jmdn./sich zum ~ machen** make sb./oneself a laughing-stock

Gespräch [gəˈʃprɛːç] **das; ~[e]s, ~e a)** conversation; *(Diskussion)* discussion; **der Gegenstand des ~[e]s** the subject *or* topic under discussion; **ein ~ mit jmdm. führen** have a conversation *or* talk with sb.; **jmdn. in ein ~ verwickeln** engage sb. in conversation; **mit jmdm. ins ~ kommen** get into *or* engage in conversation with sb.; *(fig.: sich annähern)* enter into a dialogue with sb.; **im ~ sein** be under discussion;; **b)** *(Telefonanruf)* call

gesprächig *Adj.* talkative; **der Alkohol machte ihn ~:** the alcohol loosened his tongue

gesprächs-, Gesprächs-:

~bereit *Adj.* ready to talk *postpos.; (zu Verhandlungen bereit auch)* ready for discussions *postpos.;* **~bereitschaft die** readiness for discussions; **~fetzen der** fragment *or* snatch of conversation; **~gegenstand der** topic of conversation; **~kreis der** discussion group **~partner der: mein heutiger ~partner wird der Innenminister sein** today I shall be talking to the Minister of the Interior; **~pause die** break in the discussions *or* talks; **~stoff der** subjects *pl. or* topics *pl.* of conversation; **~teilnehmer der** participant in the discussion; **~thema das** topic of conversation; **~zeit die** *(Fernspr.)* call-time

gespreizt *Adj.* *(abwertend)* stilted; affected

gesprenkelt *Adj.* mottled; speckled ‹*egg*›

Gespritzte der; *adj. Dekl. (südd.)* wine with soda water

gesprochen [gəˈʃprɔxn̩] *2. Part. v.* **sprechen**

gesprossen [gəˈʃprɔsn̩] *2. Part. v.* **sprießen**

gesprungen *2. Part. v.* **springen**

Gespür [gəˈʃpy:ɐ̯] **das; ~s** feel

Gestade [gəˈʃta:də] **das; ~s, ~** *(dichter.)* shore(s)

Gestalt [gəˈʃtalt] **die; ~, ~en a)** build; **b)** *(Mensch, Persönlichkeit)* figure; **eine zwielichtige ~:** a shady character; **c)** *(in der Dichtung)* character; **d)** *(Form)* form; **~ annehmen** *od.* **gewinnen** take shape; **in ~ von etw.** *od.* **einer Sache** *(Gen.)* in the form of sth.

gestalten 1. *tr. V.* fashion, shape, form ‹*vase, figure, etc.*›; design ‹*furnishings, stage-set, etc.*›; lay out ‹*public gardens*›; dress ‹*shop-window*›; mould, shape ‹*character, personality*›; arrange ‹*party, conference, etc.*›; frame ‹*sentence, reply, etc.*›. **2.** *refl. V.* turn out; **sich schwieriger ~ als erwartet** turn out *or* prove to be more difficult than had been expected

gestalt·los *Adj.* shapeless; formless

Gestaltung die; ~, ~en *s.* **gestalten:** fashioning; shaping; forming; designing; laying out; dressing; moulding; shaping; arranging; framing

Gestaltungs·prinzip das formal principle

Gestammel das; ~s stammering; stuttering

gestand *1. u. 3. Pers. Sg. Prät. v.* **gestehen**

gestanden 1. *2. Part. v.* **stehen, gestehen. 2.** *Adj.* **ein ~er Mann** a grown man

geständig *Adj.* **~ sein** have confessed

Geständnis [gəˈʃtɛntnɪs] **das; ~ses, ~se** confession

Gestänge [gəˈʃtɛŋə] **das; ~s, ~ a)** *(Stangen)* struts *pl.;* **b)** *(Technik)* linkage; *(des Kolbens)* connecting rod

Gestank [gəˈʃtaŋk] **der; ~[e]s** *(abwertend)* stench; stink

gestatten [gəˈʃtatn̩] **1.** *tr., itr. V.* permit; allow; **„Rauchen nicht gestattet!"** 'no smoking'; **~ Sie, daß ich ...:** may I ...?; **wenn Sie ~:** if I may. **2.** *refl. V. (geh.)* **sich** *(Dat.)* **etw. ~:** allow oneself sth.

Geste [ˈgɛstə, ˈge:stə] **die; ~, ~n** *(auch fig.)* gesture

Gesteck [gəˈʃtɛk] **das; ~[e]s, ~e** flower arrangement

gestehen *tr., itr. V.* confess; **die Tat** *usw.* **~:** confess to the deed *etc.;* **jmdm. seine Gefühle ~:** confess one's feelings to sb.; **offen gestanden ...:** frankly *or* to be honest ...

Gestein das; ~[e]s, ~e rock

Gesteins·kunde die petrology

Gestell [gəˈʃtɛl] **das; ~[e]s, ~e a)** *(für Weinflaschen)* rack; *(zum Wäschetrocknen)* horse; *(für Pflanzen)* planter; **b)** *(Unterbau)* frame; *(eines Wagens)* chassis

gestelzt *Adj.* stilted; affected

gestern [ˈgɛstɐn] *Adv.* **a)** yesterday; **~ morgen/abend/mittag** yesterday morning/evening/[at] midday yesterday; **~ vor einer Woche** a week ago yesterday; **die Zeitung von ~:** yesterday's [news]-paper; **von ~ sein** be outdated *or* outmoded; **sie ist nicht von ~** *(ugs.)* she wasn't born yesterday *(coll.)*

gestiefelt *Adj.* booted; **der gestiefelte Kater** Puss in Boots; **~ und gespornt** *(ugs. scherzh.)* ready and waiting

gestiegen *2. Part. v.* **steigen**

Gestik [ˈgɛstɪk] **die; ~:** gestures *pl.*

Gestikulation [gɛstikulaˈtsi̯o:n] **die; ~, ~en** gesticulation

gestikulieren [gɛstikuˈli:rən] *itr. V.* gesticulate

gestimmt *Adj.* **freudig/heiter ~:** in a joyful/cheerful mood *pred.*

Gestirn das; ~[e]s, ~e heavenly body; *(Stern)* star

gestoben [gəˈʃto:bn̩] *2. Part. v.* **stieben**

gestochen [gəˈʃtɔxn̩] **1.** *2. Part. v.* **stechen. 2.** *Adj.* **eine ~e Handschrift** extremely neat *or* careful handwriting. **3.** *adv.* **~ scharfe Bilder** crystal-clear photographs

gestohlen [gəˈʃto:lən] **1.** *2. Part. v.* **stehlen. 2.** *Adj.* **der/das kann mir ~ bleiben** *(ugs.)* he can get lost *(sl.)*/you can keep it *(coll.)*

gestorben [gəˈʃtɔrbn̩] *2. Part. v.* **sterben**

gestört [gəˈʃtø:ɐ̯t] *Adj.* disturbed

gestoßen *2. Part. v.* **stoßen**

Gestotter [gəˈʃtɔtɐ] **das; ~s** *(ugs., meist abwertend)* stuttering

Gesträuch [gəˈʃtrɔy̯ç] **das; ~[e]s, ~e** shrubbery; bushes *pl.*

gestreift *Adj.* striped

gestrichen 1. *2. Part. v.* **streichen. 2.** *Adj.* level ‹*measure*›; **ein ~er Teelöffel [Zucker** *usw.*] a level teaspoon[ful] [of sugar *etc.*]

gestrig [ˈgɛstrɪç] *Adj.; nicht präd.* yesterday's; **der ~e Abend** yesterday evening; *(spät)* last night; **der ~e Tag** yesterday

gestritten [gəˈʃtrɪtn̩] *2. Part. v.* **streiten**

Gestrüpp [gəˈʃtrʏp] **das; ~[e]s, ~e** undergrowth

Gestühl [gəˈʃty:l] **das; ~[e]s, ~e** seats *pl.; (Kirchen~)* pews *pl.*

gestunken [gəˈʃtʊŋkn̩] *2. Part. v.* **stinken**

Gestüt [gəˈʃty:t] **das; ~[e]s, ~e** stud[-farm]

Gesuch [gəˈzu:x] **das; ~[e]s, ~e** request (**um** for); *(Antrag)* application (**um** for); **ein ~ einreichen/zurückziehen** submit/withdraw a request/an application

gesucht *Adj.* **a)** *(begehrt)* [much] sought-after; **b)** *(gekünstelt)* affected ‹*style*›; laboured ‹*expression*›; far-fetched ‹*comparison*›

Gesumm [gəˈzʊm] **das; ~[e]s** buzzing; humming

gesund [gəˈzʊnt]; **gesünder** [gəˈzʏndɐ], *seltener:* **gesunder, gesündest...** [gəˈzʏndəst...], *seltener:* **gesundest...** *Adj.* **a)** healthy; *(fig.)* viable, financially sound ‹*company, business*›; **wieder ~ werden** get better; recover; **~ sein** ‹*person*› be healthy; *(im Augenblick)* be in good health; **jmdn. ~ pflegen** nurse sb. back to health; **jmdn. ~ schreiben** pass sb. fit; **~ und munter** hale and hearty; **bleib ~!** look after yourself!; **b)** *(natürlich, normal)* healthy ‹*mistrust, ambition, etc.*›; sound ‹*construction*›; healthy, sound ‹*attitude, approach*›; **der ~e Menschenverstand** common sense

gesund|beten *tr. V.* **jmdn. ~beten** heal sb. *or* restore sb. to health by prayer

gesunden *itr. V.; mit sein* ‹*person*› recover, get well, regain one's health; ‹*tissue*› heal; *(fig.)* ‹*economy etc.*› recover

Gesundheit die; ~: health; **bei bester ~ sein** be in the best of health; **~!** *(ugs.: Zuruf beim Niesen)* bless you!

gesundheitlich 1. *Adj.; nicht präd.* **aus ~en Gründen** for reasons of health. **2.** *adv.* **~ geht es ihm nicht sehr gut** he is not in very good health

gesundheits-, Gesundheits-: ~amt das [local] public health department; **~attest das** health certificate; **~gefährdung die** risk to health; **~schaden der** damage *no pl., no indef. art.* to [one's] health; **das kann ~schäden bewirken** that can damage one's health; **~schädlich** *Adj.* detrimental to [one's] health *postpos.;* unhealthy; **Rauchen ist ~schädlich** smoking can damage your health; **~wesen das** [public] health service; **~zeugnis das** certificate of health; health certificate; **~zustand der** state of health

gesund-: ~|schrumpfen *itr. (auch refl.) V. (ugs.)* ‹*industry, firm*› be slimmed down; **~|stoßen** *unr. refl. V. (salopp)* grow fat *(coll.)*

Gesundung die; ~ *(geh., auch fig.)* recovery

gesungen [gə'zʊŋən] *2. Part. v.* **singen**

gesunken [gə'zʊŋkn̩] *2. Part. v.* **sinken**

getäfelt *2. Part. v.* **täfeln**

getan [gə'ta:n] *2. Part. v.* **tun**

Getier das; ~[e]s *(geh.)* animals *pl.;* wildlife

getigert [gə'ti:gɐt] *Adj.* **a)** *(mit ungleichen Flecken)* patterned like a tiger *postpos.;* **b)** *(mit Querstreifen)* striped

Getöse das; ~s [thunderous] roar; *(von vielen Menschen)* din

getragen 1. *2. Part. v.* **tragen. 2.** *Adj.* solemn ‹*music, voice, etc.*›. **3.** *adv.* solemnly

Getrampel das; ~s *(ugs.)* tramping

Getränk [gə'trɛŋk] **das; ~[e]s, ~e** drink; beverage *(formal)*

Getränke-: ~automat der drinks machine *or* dispenser; **~karte die** list of beverages; *(in einem Restaurant)* wine list

Getratsch[e] das; ~s *(ugs. abwertend)* gossip; gossiping

getrauen *refl. V. s.* **trauen 2**

Getreide [gə'traidə] **das; ~s** grain; corn

Getreide-: ~anbau der growing of cereals *or* grain; **~art die** kind of grain *or* cereal; **~ernte die** grain harvest; **~feld das** cornfield; **~handel der** corn-trade; **~speicher der** grain silo

getrennt 1. *Adj.* separate. **2.** *adv.* ‹*pay*› separately; ‹*sleep*› in separate rooms; **[von jmdm.] ~ leben** live apart [from sb.]

Getrennt·schreibung die *writing a lexical item as two or more separate words*

getreten *2. Part. v.* **treten**

getreu 1. *Adj. (geh.)* **a)** *(genau entsprechend)* exact ‹*wording*›; true, faithful ‹*image*›; **b)** *(treu)* faithful, loyal ‹*friend, servant*›. **2.** *adv. (geh.)* **a)** *(genau entsprechend)* ‹*report, describe*› faithfully, accurately; **b)** *(treu)* faithfully, loyally. **3.** *präpositional (geh.)* **~ einer Abmachung handeln** act in accordance with an agreement

getreulich *Adv. s.* **getreu 2**

Getriebe das; ~s, ~ a) gears *pl.; (in einer Maschine)* gear system; *(~kasten)* gearbox; **b)** *(Betriebsamkeit)* hustle and bustle

getrieben *2. Part. v.* **treiben**

Getriebe·schaden der gearbox damage

getroffen *2. Part. v.* **treffen, triefen**

getrogen [gə'tro:gn̩] *2. Part. v.* **trügen**

getrost 1. *Adj.* confident. **2.** *adv.* **a)** *(zuversichtlich)* confidently; **b)** *(ruhig)* **du kannst das Kind ~ allein lassen** you need have no qualms about leaving the child on its own

getrunken *2. Part. v.* **trinken**

Getto ['gɛto] **das; ~s, ~s** ghetto

Getue [gə'tu:ə] **das; ~s** *(ugs. abwertend)* fuss **(um** about)

Getümmel [gə'tʏml̩] **das; ~s** tumult; **mitten im dichtesten** *od.* **dicksten ~:** in the thick of it

getupft *Adj.* speckled ‹*garment, fabric, etc.*›

Getuschel das; ~s *(ugs.)* whispering

geübt [gə'|y:pt] *Adj.* experienced, accomplished, proficient ‹*horseman, speaker, etc.*›; trained, practised ‹*eye, ear*›; **in etw.** *(Dat.)* **~ sein** be proficient at sth.

Gewächs [gə'vɛks] **das; ~es, ~e a)** *(Pflanze)* plant; **b)** *(Weinsorte)* wine; *(Weinjahrgang)* vintage; **c)** *(Med.: Geschwulst)* growth

gewachsen 1. *2. Part. v.* **wachsen. 2.** *Adj.* **jmdm./einer Sache ~ sein** be a match for sb./be equal to sth.

Gewächs·haus das greenhouse

gewagt *Adj.* **a)** *(kühn)* daring; *(gefährlich)* risky; **b)** *(fast anstößig)* risqué ‹*joke, song, etc.*›; daring ‹*neckline etc.*›

gewählt 1. *Adj.* refined, elegant. **2.** *adv.* in a refined manner; elegantly

gewahr [gə'va:ɐ̯] *in* **jmdn./etw.** *od. (geh.)* **jmds./einer Sache ~ werden** catch sight of sb./sth.; **etw.** *(Akk.) od. (geh.)* **einer Sache** *(Gen.)* **~ werden** *(etw. erkennen, feststellen)* become aware of sth.

Gewähr [gə'vɛ:ɐ̯] **die; ~:** guarantee; **keine ~ übernehmen** be unable to guarantee sth.; **die Angaben erfolgen ohne ~:** no responsibility is accepted for the accuracy of this information; **ohne ~** *(auf Fahrplänen usw.)* subject to change

gewahren *tr. V. (geh.)* become aware of

gewähren 1. *tr. V.* **a)** *(zugestehen)* give; grant, give ‹*asylum, credit, loan*›; **jmdm. einen Aufschub ~:** grant *or* allow sb. a period of grace; **b)** *(erfüllen)* grant. **2.** *itr. V. in* **jmdn. ~ lassen** let sb. do as he/she likes

gewähr·leisten *tr. V.* guarantee

Gewähr·leistung die guarantee; *(von Sicherheit)* ensuring

Gewahrsam [gə'va:ɐ̯za:m] **der; ~s a)** *(Obhut)* safe-keeping; **etw. in ~ nehmen/behalten** take sth. into safe-keeping/keep sth. safe; **b)** *(Haft)* custody

Gewährs·mann der; *Pl.* **...männer** *od.* **...leute** source

Gewährung die; ~: *s.* **gewähren 1:** granting; giving; offering

Gewalt [gə'valt] **die; ~, ~en a)** *(Macht, Befugnis)* power; **jmdn./ein Land in seine ~ bekommen/bringen** catch sb./bring a country under one's control; **die ~ über sein Fahrzeug verlieren** *(fig.)* lose control of one's vehicle; **sich/seine Beine in der ~ haben** have oneself under control/have control over one's legs; **b)** *o. Pl. (Willkür)* force; **er versuchte mit aller ~, seinen Ehrgeiz zu befriedigen** he did everything he could to achieve his ambition; **c)** *o. Pl. (körperliche Kraft)* force; violence; **~ anwenden** use force *or* violence; **etw. mit ~ öffnen** force sth. open; **d)** *(geh.: elementare Kraft)* force; **höhere ~ [sein]** [be] an act of God

Gewalt-: ~akt der act of violence; **~anwendung die** use of force *or* violence

Gewalten·teilung die separation of powers

Gewạlt·herrschaft die; *o. Pl.* tyranny; despotism

gewạltig 1. *Adj.* **a)** *(immens)* enormous, huge ‹*sum, amount, difference, loss*›; tremendous ‹*progress*›; **b)** *(imponierend)* mighty, huge, massive ‹*wall, pillar, building, rock*›; monumental ‹*literary work etc.*›; mighty ‹*spectacle of nature*›; **c)** *(mächtig; auch fig.)* powerful. **2.** *adv.* *(ugs.: sehr, überaus)* **sich ~ irren/täuschen** be very much mistaken

gewạlt-, Gewạlt-: ~los 1. *Adj.* non-violent; **2.** *adv.* without violence; **~losigkeit die;** ~: non-violence; **~marsch der** forced march

gewạltsam 1. *Adj.* forcible ‹*expulsion*›; enforced ‹*separation*›; violent ‹*death*›. **2.** *adv.* forcibly; **~ die Tür öffnen** open the door by force

gewạlt-, Gewạlt-: ~tat die *s.* **~verbrechen; ~tätig** *Adj.* violent; **~tätigkeit die a)** *o. Pl.* *(gewalttätige Art)* violence; **b)** *s.* **~akt; ~verbrechen das** crime of violence; **~verbrecher der** violent criminal; **~verzicht der** renunciation of the use of force; **~verzichts·abkommen das** non-aggression treaty

Gewạnd das; ~[e]s, Gewänder [gəˈvɛndɐ] *(geh.)* robe; gown; *(Abendkleid)* gown; **im neuen ~** *(fig.)* dressed up as new

gewandt [gəˈvant] **1.** *2. Part. v.* **wenden. 2.** *Adj.* skilful; *(körperlich)* agile; expert ‹*skier*›; **~e Umgangsformen** easy social manners. **3.** *adv.* skilfully; *(körperlich)* agilely

Gewạndtheit die; ~: *s.* **gewandt 2:** skill; skilfulness; agility; expertness; easiness

gewạnn [gəˈvan] *1. u. 3. Pers. Sg. Prät. v.* **gewinnen**

gewärtig [gəˈvɛrtɪç] *Adj.* **einer Sache** *(Gen.)* **~ sein** *(geh.)* be prepared for sth.

Gewäsch [gəˈvɛʃ] **das; ~[e]s** *(ugs. abwertend)* twaddle; garbage *(Amer. coll.)*

gewạschen *2. Part. v.* **waschen**

Gewässer [gəˈvɛsɐ] **das; ~s, ~:** stretch of water; **sich in arktische ~ wagen** venture into Arctic waters

Gewẹbe das; ~s, ~ a) *(Stoff)* fabric; **b)** *(Med., Biol.)* tissue

Gewehr [gəˈveːɐ̯] **das; ~[e]s, ~e** rifle; *(Schrot~)* shotgun; **mit dem ~ auf jmdn./etw. zielen** aim [one's rifle/shotgun] at sb./sth.

Gewe̱hr-: ~feuer das; *o. Pl.* rifle fire; **~kolben der** rifle/shotgun butt; **~kugel die** rifle bullet; **~lauf der** rifle/shotgun barrel; **~schuß der** rifle shot

Geweih [gəˈvai̯] **das; ~[e]s, ~e** antlers *pl.*; **ein ~:** a set of antlers

Gewe̱ih·stange die *(Jägerspr.)* beam; main trunk

Gewẹrbe das; ~s, ~ a) business; *(Handel, Handwerk)* trade; **b)** *o. Pl.* *(kleine Betriebe)* [small and medium-sized] businesses and industries

Gewẹrbe·aufsicht die *enforcement of laws governing health and safety and conditions of work*

Gewẹrbe-: ~schein der licence to carry on a business *or* trade; **~steuer die** trade tax; **~treibende der/die;** *adj. Dekl.* tradesman/tradeswoman

gewẹrblich 1. *Adj.; nicht präd.* commercial; business *attrib.*; *(industriell)* industrial; trade *attrib.* ‹*union, apprentice*›; **~e Nutzung** use for commercial *or* business/industrial purposes. **2.** *adv.* **~ tätig sein** work; **etw. ~ nutzen** use sth. for commercial *or* business/industrial purposes

gewẹrbs·mäßig 1. *Adj.; nicht präd.* professional. **2.** *adv.* **etw. ~ betreiben** do sth. professionally *or* for gain

Gewerkschaft [gəˈvɛrkʃaft] **die;** ~, **~en** trade union

Gewẹrkschaft[l]er der; ~s, ~, Gewẹrkschaft[l]erin die; ~, ~nen trade unionist

gewẹrkschaftlich 1. *Adj.* [trade] union *attrib.*; ‹*rights, duties*› as a [trade] union member; **~er Vertrauensmann/~e Vertrauensfrau** shop steward. **2.** *adv.* **~ organisiert sein** belong to a [trade] union; **sich ~ engagieren** devote oneself to trade union work

Gewẹrkschafts-: ~bewegung die; *o. Pl.* [trade] union movement; **~bund der** federation of trade unions; ≈ Trades Union Congress *(Brit.)*; ≈ AFL-CIO *(Amer.)* **~führer der** [trade] union leader; **~funktionär der** [trade] union official; **~kongreß der** *s.* **~tag; ~mitglied das** member of a [trade] union; **~tag der** [trade] union conference

gewe̱sen *2. Part. v.* [1]**sein**

gewịchen *2. Part. v.* **weichen**

Gewicht [gəˈvɪçt] **das; ~[e]s, ~e** *(auch Physik, auch fig.)* weight; **ein ~ von 75 kg/ein großes ~ haben** weigh 75 kg/be very heavy; **das spezifische ~** *(Physik)* the specific gravity; **sein ~ halten** stay the same weight; **einer Sache** *(Dat.)* **[kein] ~ beimessen** *od.* **beilegen** attach [no] importance to sth.; **[nicht] ins ~ fallen** be of [no] consequence

Gewịcht-: ~heben das; ~s weight-lifting; **~heber der; ~s, ~:** weight-lifter

gewịchtig *Adj.* **a)** *(veralt.: schwer)* heavy; weighty; **b)** *(bedeutungsvoll)* weighty, important ‹*reason, question, decision, etc.*›

Gewịchts-: ~klasse die a) *(Sport)* weight [division *or* class]; **b)** *(Kaufmannsspr.)* weight class; **~verlagerung die** shift *or* transfer of weight; *(fig.)* shift in *or* of emphasis; **~verlust der** loss of weight

gewieft [gəˈviːft] *Adj.* *(ugs.)* cunning; wily

gewie̱sen *2. Part. v.* **weisen**

gewillt [gəˈvɪlt] *Adj.* *in* **~/nicht ~ sein, etw. zu tun** be willing/unwilling to do sth.

Gewịmmel das; ~s throng; milling crowd; *(von Insekten)* teeming mass

Gewịmmer das; ~s whimpering

Gewịnde das; ~s, ~ *(Technik)* thread

Gewịnde-: ~bohrer der [screw] tap; **~stift der** grub-screw

Gewinn [gəˈvɪn] **der; ~[e]s, ~e a)** *(Reinertrag)* profit; **etw. mit ~ verkaufen** sell sth. at a profit; **b)** *(Preis einer Lotterie)* prize; *(beim Wetten, Kartenspiel usw.)* winnings *pl.*; **die ~e auslosen** draw the winners *or* winning numbers; **c)** *(Nutzen)* gain; profit; **d)** *(Sieg)* win

gewịnn-, Gewịnn-: ~beteiligung die *(Wirtsch.)* profit-sharing; *(Betrag)* profit-sharing bonus; **~bringend** *Adj.* profitable; lucrative; **~chance die** chance of winning

gewinnen [gəˈvɪnən] **1.** *unr. tr. V.* **a)** *(siegen in)* win ‹*contest, race, etc.*›; *s. auch* **Spiel b; b)** *(erringen, erreichen, erhalten)* gain, win ‹*respect, sympathy, etc.*›; gain ‹*time, lead, influence, validity, confidence*›; win ‹*prize*›; **wie gewonnen, so zerronnen** *(Spr.)* easy come, easy go; *s. auch* **Oberhand; c)** *(Unterstützung erlangen)* **jmdn. für etw. ~:** win sb. over [to sth.]; **d)** *(abbauen, fördern)* mine, extract ‹*coal, ore, metal*›; recover ‹*oil*›; **e)** *(erzeugen)* produce **(aus** from); *(durch Recycling)* reclaim; recover. **2.** *unr. itr. V.* **a)** win **(bei** at); **jedes zweite Los gewinnt!** every other ticket [is] a winner!; **b)** *(sich vorteilhaft verändern)* improve; **c)** *(zunehmen)* **an Höhe/**

Fahrt ~: gain height/gain *or* pick up speed; **an Bedeutung** ~: gain in importance

gewinnend 1. *Adj.* winning, engaging, winsome ⟨*manner, smile, way*⟩. 2. *adv.* ⟨*smile*⟩ winningly, engagingly, winsomely

Gewinner der; ~s, ~ , **Gewinnerin die;** ~, ~nen winner

gewinn-, Gewinn-: ~**spanne die** profit margin; ~**streben das** pursuit of profit; ~**sucht die;** *o. Pl.* greed for profit; ~**süchtig** *Adj.* greedy for profits *pred.;* ~**trächtig** *Adj.* profitable; lucrative

Gewinnung die; ~ **a)** *(von Kohle, Erz usw.)* mining; extraction; *(von Öl)* recovery; *(von Metall aus Erz)* extraction; **b)** *(Erzeugung)* production

Gewinn·zahl die winning number

Gewinsel das; ~s whimpering; whining

Gewirr das; ~[e]s tangle; **ein** ~ **von Paragraphen** a maze *or* jungle of regulations; **ein** ~ **von Stimmen** a [confused] babble of voices

Gewisper das; ~s whispering

gewiß [gə'vɪs] 1. *Adj.* **a)** *nicht präd. (nicht sehr viel/groß)* certain; **in gewisser Beziehung** in some respects; *s. auch* **Etwas;** [1]**Maß d; b)** *(sicher)* certain (*Gen.* of); **etw. ist jmdm.** ~: sb. is certain *or* sure of sth. 2. *adv.* certainly; **ja** *od.* **aber** ~ **[doch]!** but of course!

Gewissen das; ~s, ~: conscience; **ruhigen** ~**s etw. tun** do sth. with a clear conscience; **mit gutem** ~: with a clear conscience; **etw./jmdn. auf dem** ~ **haben** have sth./sb. on one's conscience; **jmdm. ins** ~ **reden [, etw. zu tun]** have a serious talk with sb. [and persuade him/her to do sth.]

gewissenhaft 1. *Adj.* conscientious. 2. *adv.* conscientiously

Gewissenhaftigkeit die; ~: conscientiousness

gewissen·los 1. *Adj.* conscienceless; unscrupulous. 2. *adv.* ~ **handeln** act with a complete lack of conscience

Gewissenlosigkeit die; ~ lack of conscience

Gewissens-: ~**bisse** *Pl.* pangs of conscience; ~**frage die** question *or* matter of conscience; matter for one's conscience; ~**freiheit die;** *o. Pl.* freedom of conscience; ~**gründe** *Pl.* reasons of conscience; ~**konflikt der** moral conflict

gewissermaßen *Adv.* *(sozusagen)* as it were; *(in gewissem Sinne)* to a certain extent

Gewißheit die; ~, ~en certainty; **sich** *(Dat.)* ~ **verschaffen** find out for certain

Gewitter [gə'vɪtɐ] **das;** ~s, ~: thunderstorm; *(fig.)* storm

Gewitter·front die storm front

gewittern *itr. V. (unpers.)* **es gewitterte/wird bald** ~: there was/will soon be thunder and lightning

Gewitter-: ~**neigung die;** *o. Pl.* likelihood of thunderstorms; ~**regen der** thundery shower; ~**wolke die** thunder-cloud

gewittrig [gə'vɪtrɪç] *Adj.* thundery; ~**e Schwüle** sultry heat

gewitzt [gə'vɪtst] *Adj.* shrewd; **ein** ~**er Junge** a smart lad

gewoben [gə'vo:bn̩] *2. Part. v.* **weben**

gewogen 1. *2. Part. v.* **wiegen.** 2. *Adj. (geh.)* well disposed, favourably inclined (*Dat.* towards)

gewöhnen [gə'vø:nən] 1. *tr. V.* **jmdn. an jmdn./etw.** ~: get sb. used *or* accustomed to sb./sth.; accustom sb. to sb./sth.; **an jmdn./etw. gewöhnt sein** be used *or* accustomed to sb./sth.. 2. *refl. V.* **sich an jmdn./etw.** ~: get used *or* get *or* become accustomed to sb./sth.; accustom oneself to sb./sth.

Gewohnheit [gə'vo:nhait] **die;** ~, ~en habit; **die** ~ **haben, etw. zu tun** be in the habit of doing sth.; **sich** *(Dat.)* **etw. zur** ~ **machen** make a habit of sth.; **nach alter** ~: from long-established habit

gewohnheits-, Gewohnheits-: ~**mäßig** 1. *Adj.* habitual ⟨*drinker etc.*⟩; automatic ⟨*reaction etc.*⟩; 2. *adv.* **a)** *(regelmäßig)* habitually; **b)** *(einer Gewohnheit folgend)* as is/was my/his *etc.* habit; ~**recht das** *(Rechtsw.)* **a)** *o. Pl. (System)* common law; **b)** *(einzelnes Recht)* established right; ~**tier das** *(scherzh.)* creature of habit; ~**trinker der** habitual drinker; ~**verbrecher der** *(Rechtsw.)* habitual criminal

gewöhnlich [gə'vø:nlɪç] 1. *Adj.* **a)** *nicht präd. (alltäglich)* normal; ordinary; **b)** *nicht präd. (gewohnt, üblich)* usual; normal; customary; **c)** *(abwertend: ordinär)* common. 2. *adv.* **a) [für]** ~: usually; normally; **wie** ~: as usual; **b)** *(abwertend: ordinär)* in a common way

gewohnt *Adj.* **a)** *nicht präd. (vertraut)* usual; **b) es** ~ **sein, etw. zu tun** be used *or* accustomed to doing sth.

Gewöhnung die; ~ **a)** habituation (**an** + *Akk.* to); **b)** *(Sucht)* habit; addiction

Gewölbe [gə'vœlbə] **das;** ~s, ~: vault

gewonnen [gə'vɔnən] *2. Part. v.* **gewinnen**

geworben [gə'vɔrbn̩] *2. Part. v.* **werben**

geworfen [gə'vɔrfn̩] *2. Part. v.* **werfen**

gewrungen [gə'vrʊŋən] *2. Part. v.* **wringen**

Gewühl das; ~[e]s **a)** milling crowd; **b)** *(das Wühlen)* rooting about

gewunden *2. Part. v.* **winden**

gewunken [gə'vʊŋkn̩] *2. Part. v.* **winken**

gewürfelt *Adj. (kariert)* check; checked

Gewürz das; ~es, ~e spice; *(würzende Zutat)* seasoning; condiment; *(Kraut)* herb

Gewürz-: ~**gurke die** pickled gherkin; ~**mischung die** mixed spices *pl.*/herbs *pl.;* ~**nelke die** clove

gewußt *2. Part. v.* **wissen**

Geysir ['gaizɪr] **der;** ~s, ~e geyser

gez. *Abk.* **gezeichnet** sgd.

Gezänk [gə'tsɛŋk] **das;** ~[e]s, **Gezanke** [gə'tsaŋkə] **das;** ~s *(abwertend)* quarrelling

Gezappel das; ~s *(ugs., oft abwertend)* wriggling

Gezeiten *Pl.* tides

Gezeiten·kraft·werk das tidal powerstation

Gezeter das; ~s *(abwertend)* scolding; nagging

geziehen [gə'tsi:ən] *2. Part. v.* **zeihen**

gezielt 1. *Adj.* specific ⟨*questions, measures, etc.*⟩; deliberate ⟨*insult, indiscretion*⟩; well-directed ⟨*advertising campaign*⟩. 2. *adv.* ⟨*proceed, act*⟩ purposefully, in a purposeful manner; ~ **nach etw. forschen** search specifically for sth.

geziemen *(geh. veralt.) refl. V.* be proper *or* right; **sich für jmdn.** ~: befit sb.

geziemend *(geh.)* 1. *Adj.* fitting; proper, due ⟨*respect*⟩. 2. *adv.* in a fitting manner

geziert 1. *Adj.* affected. 2. *adv.* affectedly

Geziertheit die; ~ *(abwertend)* affectedness

Gezirp[e] das; ~s *(oft abwertend)* chirping; chirruping

gezogen [gə'tso:gn̩] *2. Part. v.* **ziehen**

Gezwitscher das; ~s twittering; chirping; chirruping

gezwungen [gə'tsvʊŋən] 1. *2.*

Part. v. **zwingen. 2.** *Adj.* forced ⟨*laugh, smile, etc.*⟩; stiff ⟨*behaviour*⟩. **3.** *adv.* ⟨*laugh*⟩ in a forced way *or* manner; ⟨*behave*⟩ stiffly

gezwungenermaßen *Adv.* of necessity; **etw. ~ machen** be forced to do sth.

GG *Abk.* **Grundgesetz**

ggf. *Abk.* **gegebenenfalls**

gib [gi:p] *Imperativ Sg. Präsens v.* **geben**

Gibbon ['gɪbɔn] **der; ~s, ~s** *(Zool.)* gibbon

gibst [gi:pst] *2. Pers. Sg. Präsens v.* **geben**

gibt [gi:pt] *3. Pers. Sg. Präsens v.* **geben**

Gicht die; ~: gout

gichtig, gichtisch *Adj.* gouty

Giebel ['gi:bl̩] **der; ~s, ~ a)** gable; **b)** *(von Portalen)* pediment

Giebel-: ~dach das gable roof; **~fenster das** gable-window

Gier [gi:ɐ̯] **die; ~** greed (**nach** for); **mit solcher ~:** so greedily; **~ nach Macht/Ruhm** lust *or* craving for power/craving for fame

gierig 1. *Adj.* greedy; avid ⟨*desire, reader*⟩; **nach etw. ~ sein** be greedy for sth. **2.** *adv.* greedily

gießen ['gi:sn̩] **1.** *unr. tr. V.* **a)** *(rinnen lassen/schütten)* pour (**in** + *Akk.* into, **über** + *Akk.* over); **b)** *(verschütten)* spill (**über** + *Akk.* over); **c)** *(begießen)* water ⟨*plants, flowers, garden*⟩; **d)** cast ⟨*machine part, statue, candles, etc.*⟩; cast, found ⟨*metal*⟩; found ⟨*glass*⟩. **2.** *unr. itr. V. (unpers., ugs.)* pour [with rain]; **es gießt in Strömen** it is coming down in buckets; it's raining cats and dogs

Gießer der; ~s, ~: caster; founder

Gießerei die; ~, ~en a) *(Betrieb)* foundry; **b)** *o. Pl. (Zweig der Metallindustrie)* casting; founding

Gießkannen·prinzip das; *o. Pl. (scherzh.)* principle of 'equal shares for all'

Gift [gɪft] **das; ~[e]s, ~e a)** poison; *(Schlangen~)* venom; **b)** *(fig.)* **~ für jmdn./etw. sein** be extremely bad for sb./sth.; **~ und Galle speien** *od.* **spucken** *(sehr wütend sein)* be in a terrible rage; *(gehässig reagieren)* give vent to one's spleen

gift-, Gift-: ~frei *Adj.* non-toxic; non-poisonous; **~gas das** poison gas; **~grün** *Adj.* garish green

giftig 1. *Adj.* **a)** poisonous; venomous, poisonous ⟨*snake*⟩; toxic, poisonous ⟨*substance, gas, chemical*⟩; **b)** *(ugs.: bösartig)* venomous, spiteful ⟨*remark, person, words, etc.*⟩; venomous ⟨*look*⟩; **~ werden** turn nasty; **c)** *(grell, schreiend)* garish, loud ⟨*colour*⟩. **2.** *adv.* venomously

Gift-: ~mischer der *(ugs.)* **a)** maker of poisons; **b)** *(scherzh.: Apotheker)* chemist; **~mord der** [murder by] poisoning; **~mörder der** poisoner; **~müll der** toxic waste; **~müll·deponie die** toxic [waste] tip *or* dump; **~pfeil der** poisoned arrow; **~pilz der** poisonous mushroom; [poisonous] toadstool; **~schlange die** poisonous *or* venomous snake; **~stachel der** poisonous sting; **~stoff der** poisonous *or* toxic substance; **~zahn der** poison fang; **~zwerg der** *(ugs. abwertend)* [nasty] spiteful little man

Giga- [giga-] giga⟨*hertz etc.*⟩

Gigant [gi'gant] **der; ~en, ~en** giant

gigantisch *Adj.* gigantic; huge ⟨*success*⟩

Gigolo ['ʒi:golo] **der; ~s, ~s** gigolo

gilt [gɪlt] *3. Pers. Sg. Präsens v.* **gelten**

Gimpel ['gɪmpl̩] **der; ~s, ~ a)** *(Vogel)* bullfinch; **b)** *(ugs. abwertend: einfältiger Mensch)* ninny

ging [gɪŋ] *1. u. 3. Pers. Sg. Prät. v.* **gehen**

Ginster ['gɪnstɐ] **der; ~s, ~:** broom; *(Stech~)* gorse; furze

Gipfel ['gɪpfl̩] **der; ~s, ~ a)** peak; *(höchster Punkt des Berges)* summit; **b)** *(Höhepunkt)* height; *(von Begeisterung, Glück, Ruhm, Macht auch)* peak; **auf dem ~ der Macht/des Ruhmes** at the height of one's power/fame; **das ist [doch] der ~!** *(ugs.)* that's the limit!; **c)** *(~konferenz)* summit

Gipfel-: ~konferenz die summit conference; **~kreuz das** cross on the summit of a/the mountain

gipfeln *itr. V.* **in etw.** *(Dat.)* **~:** culminate in sth.

Gipfel·punkt der highest point; top; *(fig.)* high point; **der ~ seines künstlerischen Schaffens** the peak of his artistic powers

Gips [gɪps] **der; ~es, ~e** plaster; gypsum *(Chem.)*; *(zum Modellieren)* plaster of Paris

Gips-: ~abdruck der, ~abguß der plaster cast; **~bein das** *(ugs.)* **ich komme mit meinem ~bein nicht mit** I can't keep up, with this plaster on my leg

gipsen *tr. V.* **a)** plaster ⟨*wall, ceiling*⟩; put ⟨*leg, arm, etc.*⟩ in plaster; **b)** *(ausbessern)* repair with plaster

Gipser der; ~s, ~: plasterer

Gips-: ~figur die plaster [of Paris] figure; **~verband der** plaster cast

Giraffe [gi'rafə] **die; ~, ~n** giraffe

Girlande [gɪr'landə] **die; ~, ~n** festoon

Giro ['ʒi:ro] **das; ~s, ~s,** *österr. auch* **Giri** *(Finanzw.)* giro

Giro·konto das *(Finanzw.)* current account

girren ['gɪrən] *itr. V. (auch fig.)* coo

Gis, gis [gɪs] **das; ~, ~** *(Musik)* G sharp

Gischt [gɪʃt] **der; ~[e]s, ~e** *od.* **die; ~, ~en a)** *(Schaumkronen)* foam; surf; **b)** *(Sprühwasser)* spray

Gis-Dur [*auch:* '-'-] **das; ~** *(Musik)* G sharp major

gis-Moll [*auch:* '-'-] **das; ~** *(Musik)* G sharp minor

Gitarre [gi'tarə] **die; ~, ~n** guitar

Gitarrist der; ~en, ~en guitarist

Gitter ['gɪtɐ] **das; ~s, ~ a)** *(parallele Stäbe)* bars *pl.*; *(Drahtgeflecht)* grille; *(in der Straßendecke, im Fußboden)* grating; *(Geländer)* railing[s *pl.*]; *(Spalier)* trellis; *(feines Draht~)* mesh; **hinter ~n** *(ugs.)* behind bars; **b)** *(Physik, Chemie)* lattice

Glace ['glasə] **die; ~, ~n** *(schweiz.)* ice cream

Glacé·handschuh [gla'se:...] **der** kid glove; **jmdn./etw. mit ~en anfassen** *(ugs.)* handle sb./sth. with kid gloves

Gladiator [gla'di̯a:tɔr] **der; ~s, ~en** [-'to:rən] gladiator

Gladiole [gla'di̯o:lə] **die; ~, ~n** gladiolus

Glanz [glants] **der; ~es a)** *(von Licht, Sternen)* brightness; brilliance; *(von Haar, Metall, Perlen, Leder usw.)* shine; lustre; sheen; *(von Augen)* shine; brightness; lustre; **den ~ verlieren** ⟨*diamonds, eyes*⟩ lose their sparkle; ⟨*metal, leather*⟩ lose its shine; **b) mit ~ und Gloria** *(ugs. iron.)* in grand style

glänzen ['glɛntsn̩] *itr. V.* **a)** *(Glanz ausstrahlen)* shine; ⟨*car, hair, metal, paintwork, etc.*⟩ gleam; ⟨*elbows, trousers, etc.*⟩ be shiny; **vor Sauberkeit ~:** be so clean [that] it shines; **b)** *(Bewunderung erregen)* shine (**bei** at); **durch Abwesenheit ~** *(iron.)* be conspicuous by one's absence

glänzend *(ugs.)* **1.** *Adj.* **a)** shining; gleaming ⟨*car, hair, metal, paintwork, etc.*⟩; shiny ⟨*elbows,* *trousers, etc.*⟩; **b)** *(bewundernswert)* brilliant ⟨*idea, career, victory, pupil, prospects, etc.*⟩; splendid, excellent, outstanding ⟨*references, marks, results, etc.*⟩. **2.** *adv.* ~ **mit jmdm. auskommen** get on very well with sb.; **es geht mir/uns ~:** I am/we are very well; *(finanziell)* I am/we are doing very well *or* very nicely; **eine Aufgabe ~ lösen** solve a problem brilliantly

glanz-, Glanz-: ~leistung die *(auch iron.)* brilliant performance; **~licht das** *(bild. Kunst)* highlight; **einer Sache** *(Dat.)* **[noch einige] ~lichter aufsetzen** give sth. [more] sparkle; **~los** *Adj.* dull; lacklustre; **~nummer die** star turn; **~papier das** glossy paper; **~rolle die** star role; **~stück das a)** *(Meisterwerk)* pièce de résistance; **b)** *(der kostbarste Gegenstand)* show-piece; **~voll 1.** *Adj.* **a)** *(ausgezeichnet)* brilliant; sparkling ⟨*variety number*⟩; **b)** *(prachtvoll)* magnificent; **2.** *adv.* **a)** *(ausgezeichnet)* brilliantly; **b)** *(prachtvoll)* **Louis XIV pflegte ~voll Hof zu halten** Louis XIV used to hold court in glittering style; **~zeit die** heyday

[1]Glas ['gla:s] **das; ~es, Gläser** ['glɛ:zɐ] **a)** *o. Pl.* glass; **unter ~:** behind glass; ⟨*plants*⟩ under glass; **b)** *(Trinkgefäß)* glass; **ein ~ über den Durst trinken, zu tief ins ~ gucken** *(ugs. scherzh.)* have one too many *or* one over the eight; **c)** *(Behälter aus ~)* jar

[2]Glas das; ~es, ~en *(Seemannsspr.)* bell; **es schlug acht ~en** it struck eight bells

Glas-: ~auge das glass eye; **~baustein der** glass brick *or* block; **~bläser der** glass-blower

Gläschen ['glɛ:sçən] **das; ~s, ~:** [little] glass

Glaser der; ~s, ~: glazier

Glaserei die; ~, ~en glazing business

gläsern ['glɛ:zɐn] *Adj.; nicht präd.* *(aus Glas)* glass

glas-, Glas-: ~fabrik die glassworks *sing. or pl.*; **~faser die;** *meist Pl.* glass fibre; **~fenster das** [glass] window; **bemalte ~fenster** stained glass windows; **~fiber die** *s.* **~faser; ~fiber·stab der** *(Leichtathletik)* glass-fibre pole; **~flasche die** glass bottle; **~haus das** greenhouse; glasshouse; **wer [selbst] im ~haus sitzt, soll nicht mit Steinen werfen** *(Spr.)* those who live in glass houses shouldn't throw stones *(prov.)*

glasieren *tr. V.* **a)** *(glätten und haltbar machen)* glaze; **b)** *(Kochk.)* ice ⟨*cake etc.*⟩; glaze ⟨*meat*⟩

glasig *Adj.* **a)** *(starr)* glassy ⟨*stare, eyes, etc.*⟩; **b)** *(Kochk.: durchsichtig)* transparent

Glas-: ~kasten der glass case; *(kleiner)* glass box; **~kugel die** glass ball; *(einer Wahrsagerin)* crystal ball; *(Murmel)* marble; **~malerei die** stained glass; *(Verfahren)* glass-staining; **~perle die** glass bead; **~platte die** glass plate; *(eines Tisches)* glass top; *(im Fenster)* pane of glass; **~scheibe die** sheet of glass; *(im Fenster)* pane of glass; **~scherbe die** piece of broken glass; **~schneider der** glass-cutter; **~splitter der** splinter of glass; **~tür die** glass door

Glasur [gla'zu:ɐ̯] **die; ~, ~en a)** *(Schmelz)* glaze; **b)** *(Kochk.)* *(auf Kuchen)* icing; *(auf Fleisch)* glaze

Glas·wolle die glass wool

glatt [glat] **1.** *Adj.* **a)** smooth; straight ⟨*hair*⟩; **eine ~e Eins/Fünf** a clear A/E; **b)** *(rutschig)* slippery; **c)** *nicht präd.* *(komplikationslos)* smooth ⟨*landing, journey*⟩; clean, straightforward ⟨*fracture*⟩; **d)** *nicht präd.* *(ugs.: offensichtlich)* downright, outright ⟨*lie*⟩; outright ⟨*deception, fraud*⟩; sheer, utter ⟨*nonsense, madness, etc.*⟩; pure, sheer ⟨*invention*⟩, flat ⟨*refusal*⟩; complete ⟨*failure*⟩; **e)** *(allzu gewandt)* smooth. **2.** *adv.* **a)** **die Rechnung geht ~ auf** the calculation works out exactly; **b)** *(komplikationslos)* smoothly; **c)** *(ugs.: rückhaltlos)* **jmdm. etw. ~ ins Gesicht sagen** tell sb. sth. straight to his/her face

glatt|bügeln *tr. V.* iron smooth

Glätte ['glɛtə] **die; ~ a)** smoothness; **b)** *(Rutschigkeit)* slipperiness

Glatt·eis das glaze; ice; *(auf der Straße)* black ice; **jmdn. aufs ~ führen** *(fig.)* catch sb. out

Glatteis·gefahr die; *o. Pl.* danger of black ice

glätten 1. *tr. V.* smooth out ⟨*piece of paper, banknote, etc.*⟩; smooth [down] ⟨*feathers, fur, etc.*⟩; plane ⟨*wood etc.*⟩. **2.** *refl. V.* ⟨*waves*⟩ subside; ⟨*sea*⟩ become calm *or* smooth; *(fig.)* subside; die down

glatt-: ~|gehen *unr. itr. V.; mit sein* *(ugs.)* go smoothly; **~hobeln** *tr. V.* plane smooth; **~machen** *tr. V.* *(ebnen, glätten)* smooth out; level ⟨*ground*⟩; **~rasiert** *Adj.* clean-shaven; **~weg** *Adv.* *(ugs.)* **etw. ~weg ablehnen/ignorieren** turn sth. down flat/just *or* simply ignore sth.; **~|ziehen** *unr. tr. V.* pull straight

Glatze ['glatsə] **die; ~, ~n** bald head; **eine ~ haben/bekommen** be/go bald

Glatz·kopf der a) *(Kopf)* bald head; **b)** *(ugs.: Person)* baldhead

glatz·köpfig *Adj.* bald[-headed]

Glaube ['glaubə] **der; ~ns** faith **(an + *Akk.*** in); *(Überzeugung, Meinung)* belief **(an + *Akk.*** in); **jmdm./jmds. Worten ~n schenken** believe sb./what sb. says; **[bei jmdm.] ~n finden** be believed [by sb.]; **in dem ~n leben, daß ...:** live in the belief that ...; **laß ihn in seinem ~n** don't disillusion him; **[der] ~ versetzt Berge** faith can move mountains

glauben 1. *tr. V.* **a)** *(annehmen, meinen)* think; believe; **ich glaube, ja** I think *or* believe so; **b)** *(für wahr halten)* believe; **das glaubst du doch selbst nicht!** [surely] you can't be serious; **ob du es glaubst oder nicht ...:** believe it or not ...; **wer hätte das [je] geglaubt?** who would [ever] have thought it?; **du glaubst [gar] nicht, wie ...:** you have no idea how ...; **wer's glaubt, wird selig** *(ugs. scherzh.)* if you believe that, you'll believe anything; **das ist doch kaum zu ~** *(ugs.)* it's incredible. **2.** *itr. V.* **a)** *(vertrauen)* **an jmdn./etw./sich [selbst] ~:** believe in *or* have faith in sb./sth./oneself; **b)** *(gläubig sein)* hold religious beliefs; believe; **fest/unbeirrbar ~:** have a strong/unshakeable religious belief; **c)** *(von der Existenz von etw. überzeugt sein)* believe **(an + *Akk.*** in); **d)** **dran ~ müssen** *(salopp: getötet werden)* buy it *(sl.)*; *(salopp: sterben)* peg out *(sl.)*; kick the bucket *(sl.)*

Glaubens-: ~bekenntnis das *o. Pl.* *(auch fig.: Überzeugung)* creed; **~frage die** question of faith *or* belief; **~freiheit die;** *o. Pl.* religious freedom; freedom of worship; **~gemeinschaft die** religious sect; denomination; **~streit der** religious dispute

Glauber·salz ['glaubɐ-] **das; ~es** *(Chemie)* Glauber's salt

glaubhaft 1. *Adj.* credible; believable. **2.** *adv.* convincingly

Glaubhaftigkeit die; ~: credibility

gläubig ['glɔybɪç] **1.** *Adj.* **a)** *(religiös)* devout; **sehr/zutiefst ~ sein** be very/deeply religious; **b)** *(vertrauensvoll)* trusting. **2.** *adv.* **a)** *(religiös)* devoutly; **b)** *(vertrauensvoll)* trustingly

Gläubige der/die; *adj. Dekl.* believer; **die ~n** the faithful

Gläubiger der; ~s, ~, Gläubigerin die; ~, ~nen creditor

Gläubigkeit die; ~ a) *(religiöse Überzeugung)* religious faith; **b)** *(Vertrauen)* trustfulness

glaub·würdig 1. *Adj.* credible; believable. **2.** *adv.* convincingly

Glaubwürdigkeit die credibility

Glaukom [glau'ko:m] **das; ~s, ~e** *(Med.)* glaucoma

glazial [gla'tsia:l] *Adj.* *(Geol.)* glacial

gleich [glaiç] **1.** *Adj.* **a)** *(identisch, von derselben Art)* same; *(~berechtigt, ~wertig, Math.)* equal; **~er Lohn für ~e Arbeit** equal pay for equal work; **~es Recht für alle** equal rights for all; **dreimal zwei [ist] ~ sechs** three times two equals *or* is six; **das ~e wollen/beabsichtigen** have the same objective[s *pl.*]/intentions *pl.*; **das kommt auf das ~e** *od.* **aufs ~e heraus** it amounts *or* comes to the same thing; **~es mit ~em vergelten** pay sb. back in his/her own coin *or* in kind; **~ und ~ gesellt sich gern** *(Spr.)* birds of a feather flock together *(prov.)*; **b)** *(ugs.: gleichgültig)*; **ganz ~, wer anruft, ...:** no matter who calls, ... **2.** *adv.* **a)** *(übereinstimmend)* **~ groß/alt** *usw.* **sein** be the same height/age *etc.*; **~ gut/schlecht** *usw.* equally good/bad *etc.*; **b)** *(in derselben Weise)* **~ aufgebaut/gekleidet** having the same structure/wearing identical clothes; **alle Menschen ~ behandeln** treat everyone alike; **c)** *(sofort)* at once; right *or* straight away; *(bald)* in a moment *or* minute; **ich komme ~:** I'm just coming; **es muß nicht ~ sein** there's no immediate hurry; **es ist ~ zehn Uhr** it is almost *or* nearly ten o'clock; **das habe ich [euch] ~ gesagt** I told you so; what did I tell you?; **bis ~!** see you later!; **d)** *(räumlich)* right; immediately; just; **~ rechts/links** just *or* immediately on the right/left. **3.** *Präp.* + *Dat.* *(geh.)* like. **4.** *Partikel* **a) nun wein' nicht ~/sei nicht ~ böse** don't start crying/don't get cross; **b)** *(in Fragesätzen)* **wie hieß er ~?** what was his name [again]?

gleich-, Gleich-: ~alt[e]rig [~alt[ə]rıç] *Adj.* of the same age (**mit** as); **die beiden sind ~alt[e]rig** they are both the same age; **~artig 1.** *Adj.* of the same kind *postpos.* (*Dat.* as); *(sehr ähnlich)* very similar (*Dat.* to); **2.** *adv.* in the same way; **~bedeutend** *Adj.* **~bedeutend mit** synonymous with; ⟨*action*⟩ tantamount to; **~berechtigt** *Adj.* having *or* enjoying *or* with equal rights *postpos.*; **~berechtigte Partner/Mitglieder** equal partners/members; **~berechtigt sein** have *or* enjoy equal rights; **~berechtigung die** equal rights *pl.*; **~|bleiben** *unr. itr. V.; mit sein* remain *or* stay the same; ⟨*speed, temperature, etc.*⟩ remain *or* stay constant *or* steady; **sich** *(Dat.)* **~bleiben** remain the same; **das bleibt sich [doch] gleich** *(ugs.)* it makes no difference; **~bleibend** *Adj.* constant, steady ⟨*temperature, speed, etc.*⟩

gleichen *unr. itr. V.* **jmdm./einer Sache ~:** be like *or* resemble sb./sth.; *(sehr ähnlich aussehen)* closely resemble sb./sth.; **sich** *(Dat.)* **~:** be alike; *(sehr ähnlich aussehen)* closely resemble each other

gleichen·orts *Adv.* *(schweiz.)* in the same place

gleichermaßen *Adv.* equally

gleich-, Gleich-: ~falls *Adv.* *(auch)* also; *(ebenfalls)* likewise; **danke ~falls!** thank you, [and] the same to you; **~förmig 1.** *Adj.* **a)** *(einheitlich)* uniform; steady ⟨*development*⟩; **b)** *(langweilig, monoton)* monotonous; **2.** *adv.* **a)** *(einheitlich)* uniformly; **b)** *(langweilig, monoton)* monotonously; **~förmigkeit die; ~ a)** *(Einheitlichkeit)* uniformity; **b)** *(Monotonie)* monotony; **~geschlechtlich** *Adj.* homosexual; **~gesinnt** *Adj.; nicht präd.* like-minded

Gleich·gewicht das; *o. Pl.* **a)** balance; **das ~ halten/verlieren** keep/lose one's balance; **im ~ sein** be in equilibrium; **b)** *(Ausgewogenheit)* balance; **das europäische ~:** the balance of power in Europe; **das ~ der Kräfte** the balance of power; **c)** *(innere Ausgeglichenheit)* equilibrium; **jmdn. aus dem ~ bringen** throw sb. off balance

Gleichgewichts-: ~organ das *(Anat.)* organ of equilibrium; **~sinn der** sense of balance; **~störung die** disturbance of one's sense of balance

gleich·gültig 1. *Adj.* **a)** indifferent (**gegenüber** towards); **b)** *(egal)* **sie war ihm [nicht] ~** *(verhüll.)* he was [by no means] indifferent to her; **das ist mir [vollkommen] ~:** it's a matter of [complete] indifference to me; **es ist ~, ob ...:** it does not matter whether ... **2.** *adv.* indifferently; ⟨*look on*⟩ with indifference

Gleich·gültigkeit die indifference (**gegenüber** towards)

Gleichheit die; ~, ~en a) *(Identität)* identity; *(Ähnlichkeit)* similarity; **b)** *o. Pl.* *(gleiche Rechte)* equality

Gleichheits-: ~[grund]satz der principle of equality before the law; **~zeichen das** equals sign

gleich-, Gleich-: ~klang der harmony; **~|kommen** *unr. itr. V.; mit sein* amount to; be tantamount to; **jmdm./einer Sache [an etw.** *(Dat.)***] ~kommen** equal sb./sth. [in sth.]; **~laufend** *Adj.* parallel (**mit** with); **~lautend** *Adj.* identical homonymous ⟨*words*⟩ *(Ling.)*; **~|machen** *tr. V.* make equal; *s. auch* **Erdboden; ~macherei die; ~, ~en** *(abwertend)* levelling down *(derog.)*; egalitarianism; **~macherisch** *Adj.* *(abwertend)* egalitarian; **~maß das;** *o. Pl.* **a)** *(Ebenmaß) (von Bewegung, Strophen)* regularity; *(von Zügen, Proportionen)* symmetry; **b)** *(Ausgeglichenheit)* equilibrium; **~mäßig 1.** *Adj.* regular ⟨*interval, rhythm*⟩; uniform ⟨*acceleration, distribution*⟩; even ⟨*heat*⟩; **2.** *adv.* ⟨*breathe*⟩ regularly; **etw. ~mäßig verteilen/auftragen** distribute sth. equally/apply sth. evenly; **~mäßig hohe Temperaturen** constantly high temperatures; **~mäßigkeit die** *s.* **~mäßig:** regularity; uniformity; evenness; **~mut der** equanimity; calmness; composure; **~mütig 1.** *Adj.* calm; composed; unruffled ⟨*calm*⟩; **2.** *adv.* calmly; **~namig** [-na:mıç] *Adj.* **a)** of the same name *postpos.*; **b)** *(Math.)* **~namige Brüche** fractions with a common denominator; **Brüche ~namig machen** reduce fractions to a common denominator

Gleichnis das; ~ses, ~se *(Allegorie)* allegory; *(Parabel)* parable

gleich·rangig [-raŋıç] **1.** *Adj.* ⟨*principle, problem, etc.*⟩ of equal importance *or* status; equally important ⟨*principle, problem, etc.*⟩; ⟨*official, job*⟩ of equal rank. **2.** *adv.* **alle Punkte ~ behandeln** give all points equal treatment

Gleich·richter der *(Elektrot.)* rectifier

gleichsam *Adv.* *(geh.)* as it were; so to speak; **~ als [ob] ...:** just as if ...

gleich-, Gleich-: ~|schalten *tr. V.* *(abwertend)* force *or* bring into line; ~**schenk[e]lig** *Adj. (Math.)* isosceles; ~**schritt der**; *o. Pl.* marching in step; ~**seitig** *Adj. (Math.)* equilateral; ~**setzen** *tr. V.* **zwei Dinge ~setzen** equate two things; **etw. einer Sache** *(Dat.) od.* **mit etw. ~setzen** equate sth. with sth.; ~**stand der**; *o. Pl.* **a)** *(Sport: gleicher Spielstand)* **den ~stand herstellen/erzielen** level the score; ~|**stellen** *tr. V.* **etw. einer Sache** *(Dat.) od.* **mit etw. ~stellen** equate sth. with sth.; ~**stellung die: die rechtliche ~stellung unehelicher Kinder** giving equal rights to illegitimate children; **soziale ~stellung** social equality; ~**strom der** *(Elektrot.)* direct current; ~|**tun** *unr. tr. V.* **es jmdm. ~tun** match *or* equal sb.; *(nachahmen)* copy sb.

Gleichung die; ~, ~**en** equation

gleich-, Gleich-: ~**viel** [-'- *od.* '--] *Adv.* no matter; ~**wertig** *Adj.* **a)** of equal *or* the same value *postpos.*; **b)** *(Sport: gleich stark)* evenly matched ⟨*opponents, teams*⟩; ~**wink[e]lig** *Adj.* equiangular; ~**wohl** [-'- *od.* '--] *Adv.* nevertheless; ~**zeitig 1.** *Adj.; nicht präd.* simultaneous; **2.** *adv.* **a)** simultaneously; at the same time; **b)** *(auch noch)* at the same time; ~**zeitigkeit die** simultaneity; simultaneousness; ~|**ziehen** *unr. itr. V.* catch up; draw level

Gleis [glais] **das**; ~**es**, ~**e a)** *(Fahrspur)* track; line; rails *pl.*; permanent way *as Brit. tech. term*; *(Bahnsteig)* platform; *(einzelne Schiene)* rail; **auf ~ 5 einlaufen** ⟨*train*⟩ arrive at platform 5; **jmdn. aufs tote ~ schieben** put sb. out of harm's way *(fig.)*; **jmdn. aus dem ~ bringen** *od.* **werfen** put sb. off [his/her stroke]; *(von jmdm. psychisch nicht bewältigt werden)* upset *or* affect sb. deeply

Gleis-: ~**an·lage die** [railway] lines *pl. or* tracks *pl.*; ~**anschluß der** siding

gleißen ['glaisn̩] *itr. V. (dichter.)* blaze

gleiten ['glaitn̩] *unr. itr. V.; mit sein* **a)** glide; ⟨*hand*⟩ slide; **aus dem Sattel/ins Wasser ~:** slide out of the saddle/slide *or* slip into the water; **jmdm. aus den Händen ~:** slip from sb.'s hands; **b)** *(ugs.: in bezug auf Arbeitszeit)* work flexitime

gleitend *Adj.; nicht präd.* ~**e Arbeitszeit** flexitime; flexible working hours *pl.*; ~**e Lohnskala** index-linked wage scale

Gleit-: ~**flug der** glide; **im ~flug landen** glide-land; ~**zeit die** flexible working hours *pl.*

Gletscher ['glɛtʃɐ] **der**; ~**s**, ~: glacier

Gletscher-: ~**bach der** glacial stream; ~**eis das** glacial ice; ~**spalte die** crevasse

glich [glɪç] *1. u. 3. Pers. Sg. Prät. v.* **gleichen**

Glied [gli:t] **das**; ~**[e]s**, ~**er a)** *(Körperteil)* limb; *(Finger~, Zehen~)* joint; **der Schreck sitzt** *od.* **steckt ihm noch in den ~ern** he is [still] shaking with the shock; **der Schreck fuhr ihr in die** *od.* **durch alle ~er** the shock made her shake all over; **b)** *(Ketten~, auch fig.)* link; **c)** *(Teil eines Ganzen)* section; part; *(Mitglied)* member; *(eines Satzes)* part; *(einer Gleichung)* term; **d)** *(Penis)* penis; **e)** *(Mannschaftsreihe)* rank

Glieder·füßer der *(Zool.)* arthropod

gliedern ['gli:dɐn] **1.** *tr. V.* structure; organize ⟨*thoughts*⟩; **nach Eigenschaften ~:** classify according to properties. **2.** *refl. V.* **sich in Gruppen/Abschnitte** *usw.* ~: divide *or* be divided into groups/sections *etc.*

Glieder-: ~**puppe die** jointed doll; ~**schmerz der** rheumatic pains *pl.*

Gliederung die; ~, ~**en a)** *(Aufbau, Einteilung)* structure; **b)** *(das Gliedern)* structuring; *(von Gedanken)* organization; *(nach Eigenschaften)* classification; *(in Teile)* arrangement

Glied-: ~**maße** [-ma:sə] **die**; ~, ~**n** limb; ~**satz der** *(Sprachw.)* subordinate clause

glimmen ['glɪmən] *unr. od. regelm. itr. V.* glow

Glimmer der; ~**s**, ~: mica

glimmern *itr. V.* glimmer; ⟨*lake etc.*⟩ glisten

Glimm·stengel der *(ugs. scherzh.)* fag *(sl.)*; ciggy *(coll.)*

glimpflich ['glɪmpflɪç] **1.** *Adj.* **a) der Unfall nahm ein ~es Ende** the accident turned out not to be too serious; **b)** *(mild)* lenient ⟨*sentence, punishment*⟩. **2.** *adv.* **a)** *(ohne Schaden)* ~ **davonkommen** get off lightly; **es ist ~ abgegangen** it turned out not to be too bad; **b)** *(mild)* mildly; leniently

glitschen ['glɪtʃn̩] *itr. V.; mit sein (ugs.)* slip

glitschig ['glɪtʃɪç] *Adj. (ugs.)* slippery

glitt [glɪt] *1. u. 3. Pers. Sg. Prät. v.* **gleiten**

glitzern ['glɪtsɐn] *itr. V.* ⟨*star*⟩ twinkle; ⟨*diamond, decorations*⟩ sparkle, glitter; ⟨*snow, eyes, tears*⟩ glisten

global [glo'ba:l] **1.** *Adj.* **a)** *(weltweit)* global; world-wide; **b)** *(umfassend)* general, all-round ⟨*education*⟩; overall ⟨*control, planning, etc.*⟩; **c)** *(allgemein)* general. **2.** *adv.* **a)** *(weltweit)* world-wide; globally; **b)** *(umfassend)* in overall terms; **c)** *(allgemein)* in general terms

Globen *s.* **Globus**

Globetrotter ['glo:bətrɔtɐ] **der**; ~**s**, ~: globetrotter

Globus ['glo:bʊs] **der**; ~ *od.* ~**ses, Globen** ['glo:bn̩] globe

Glöckchen ['glœkçən] **das**; ~**s**, ~: [little] bell

Glocke ['glɔkə] **die**; ~, ~**n a)** *(auch: Tür~, Taucher~, Blüte)* bell; **etw. an die große ~ hängen** *(ugs.)* tell the whole world about sth.; **b)** *(Hut)* cloche; **c)** *(Käse~, Butter~, Kuchen~)* cover; bell

glocken-, Glocken-: ~**blume die** *(Bot.)* bell-flower; campanula; ~**förmig** *Adj.* bell-shaped; widely flared ⟨*skirt etc.*⟩; ~**hell** *Adj.* bell-like; **eine ~helle Stimme** a high, clear voice; ~**läuten das** pealing *or* ringing of bells; ~**rock der** widely flared skirt; ~**schlag der** stroke; **mit dem** *od.* **auf den ~schlag** *(ugs.)* on the dot *(coll.)*; ~**spiel das a)** carillon; *(mit einer Uhr gekoppelt auch)* chimes *pl.*; **b)** *(Musikinstrument)* glockenspiel; ~**turm der** bell tower; belfry

glockig ['glɔkɪç] *s.* **glockenförmig**

Glöckner ['glœknɐ] **der**; ~**s**, ~ *(veralt.)* bellringer; **der ~ von Notre Dame** the Hunchback of Notre Dame

glomm [glɔm] *1. u. 3. Pers. Sg. Prät. v.* **glimmen**

glorifizieren [glorifi'tsi:rən] *tr. V.* glorify

Glorifizierung die; ~, ~**en** glorification

Gloriole [glo'rio:lə] **die**; ~, ~**n a)** *(auch fig.)* glory; **b)** *(um den Kopf)* halo; aura

glor·reich ['glo:ɐ̯-] **1.** *Adj.* glorious. **2.** *adv.* gloriously

Glossar [glɔ'sa:ɐ̯] **das**; ~**s**, ~**e** glossary

Glosse ['glɔsə] **die**; ~, ~**n a)** *(in den Medien)* commentary; **b)** *(spöttische Bemerkung)* sneering *or (coll.)* snide comment

glossieren *tr. V.* **a)** commentate on; **b)** *(bespötteln)* sneer at

Glotz·augen *Pl. (salopp abwertend)* goggle eyes; ~ **machen/kriegen** go goggle-eyed; goggle

Glotze ['glɔtsə] **die; ~, ~n** *(salopp)* box *(coll.)*; goggle-box *(Brit. sl.)*

glotzen *itr. V. (abwertend)* goggle; gawk, gawp *(coll.)*

Gloxinie [glɔ'ksi:ni̯ə] **die; ~, ~n** *(Bot.)* gloxinia

Glück [glʏk] **das; ~[e]s a)** luck; **ein großes/unverdientes ~:** a great/an undeserved stroke of luck; **[es ist/war] ein ~, daß ...:** it's/it was lucky that ...; **er hat [kein] ~ gehabt** he was [un]lucky; **~ bei Frauen haben** be successful with women; **jmdm. ~ wünschen** wish sb. [good] luck; **viel ~!** [the] best of luck!; good luck!; **~ bringen** bring [good] luck; **mehr ~ als Verstand haben** have more luck than judgement; **sein ~ versuchen** *od.* **probieren** try one's luck; **auf gut ~:** trusting to luck; **zum ~** *od.* **zu meinem/seinem** *usw.* **~:** luckily *or* fortunately [for me/him *etc.*]; **b)** *(Hochstimmung)* happiness; **das häusliche ~:** domestic bliss; **jmdn. zu seinem ~ zwingen** make sb. do what is good for him/her; **jeder ist seines ~es Schmied** *(Spr.)* life is what you make it; **c)** *(Fortuna)* fortune; luck

glück·bringend *Adj.* lucky

Glucke ['glʊkə] **die; ~, ~n** brood-hen; mother hen

glucken *itr. V.* **a)** *(brüten)* brood; **b)** *(ugs.: herumsitzen)* sit around; **c)** *(Laut hervorbringen)* cluck

glücken *tr. V.; mit sein* succeed; be successful; **etw. glückt jmdm.** sb. is successful with sth.; **ein geglückter Versuch** a successful attempt; **die Flucht ist nicht geglückt** the escape[-attempt] failed

gluckern ['glʊkɐn] *itr. V.* gurgle; glug

glücklich 1. *Adj.* **a)** *(von Glück erfüllt)* happy **(über +** *Akk.* about); **b)** *(erfolgreich)* lucky ‹*winner*›; successful ‹*outcome*›; safe ‹*journey*›; happy ‹*ending*›; **c)** *(vorteilhaft)* fortunate; **ein ~er Zufall** a happy coincidence; a lucky chance; *s. auch* **Hand f. 2.** *adv.* **a)** *(erfolgreich)* successfully; **b)** *(vorteilhaft, zufrieden)* happily ‹*chosen, married*›; **c)** *(endlich)* at last; eventually

glücklicher·weise *Adv.* fortunately; luckily

glück·los *Adj.* luckless ‹*enterprise*›; unhappy ‹*existence etc.*›

Glücks·bringer der lucky *or* good-luck charm; [lucky] mascot

glück·selig *Adj.* blissfully happy; blissful

Glück·seligkeit die; ~: bliss; blissful happiness

glucksen ['glʊksn̩] *itr. V.* **a)** *s.* **gluckern; b)** *(lachen)* chuckle

Glücks-: ~fall der piece *or* stroke of luck; **~göttin die** goddess of fortune; Fortune *no art.*; **~käfer der** *s.* **Marienkäfer; ~kind das** lucky person; **~klee der** four-leaf *or* four-leaved clover; **~pfennig der** lucky penny; **~pilz der** *(ugs.)* lucky devil *(coll.) or* beggar *(coll.)*; **~sache die: das ist ~sache** it's a matter of luck

Glücks-: ~spiel das a) game of chance; **dem ~spiel verfallen sein** be addicted to gambling; **b)** *(fig.)* matter of luck; lottery; **~spieler der** gambler; **~stern der** lucky star; **~strähne die** lucky streak; **eine ~strähne haben** have hit a lucky streak; have a run of good luck; **~tag der** lucky day

glück·strahlend *Adj.* radiant; radiantly happy; **sie verkündete uns ~, daß sie heiraten werde** she was radiant with happiness *or* radiantly happy as she told us she was going to get married

Glücks-: ~treffer der a) *(Gewinn)* bit *or* piece of luck; **b)** *(beim Schießen)* lucky hit; fluke; **~zahl die** lucky number

Glück·wunsch der congratulations *pl.*; **herzlichen ~ zum Geburtstag!** happy birthday!; many happy returns of the day!

Glückwunsch-: ~karte die congratulations card; *(zum Geburtstag)* greetings card; **~telegramm das** telegram of congratulations; congratulatory telegram; *(zum Geburtstag)* greetings telegram

Glucose [glu'ko:ze] **die; ~** *(Chemie)* glucose

Glüh·birne die light-bulb

glühen ['gly:ən] **1.** *itr. V.* **a)** *(leuchten)* glow; *(fig.)* ‹*eyes, cheeks, etc.*› be aglow, glow; **b)** *(geh.: erregt sein)* burn. **2.** *tr. V. (zum Leuchten bringen)* heat until red-hot

glühend 1. *Adj.* **a)** *(heiß)* red-hot ‹*metal etc.*›; *(fig.)* blazing ‹*heat*›; burning ‹*hatred*›; flushed, burning ‹*cheeks*›; **b)** *(begeistert)* ardent ‹*admirer etc.*›; passionate ‹*words, letter, etc.*›. **2.** *adv.* ‹*love*› passionately; ‹*admire*› ardently

glühend-: ~heiß *Adj. (präd. getrennt geschrieben)* scorching *or* blazing hot; **~rot** *Adj. (präd. getrennt geschrieben)* red-hot

Glüh-: ~faden der filament; **~lampe die** light bulb; **~wein der** mulled wine; glühwein; **~würmchen das** *(ugs.) (weiblich)* glow-worm; *(männlich)* firefly

Glukose *s.* **Glucose**

Glupsch·augen ['glʊpʃ-] *Pl. (nordd.)* goggle-eyes

Glut [glu:t] **die; ~, ~en a)** embers *pl.*; *(fig.)* [blazing] heat; **b)** *(geh.: Leidenschaft)* passion

Glutamat [gluta'ma:t] **das; ~[e]s, ~e** *(Chemie)* glutamate

Glut·hitze die blazing *or* sweltering heat

Glycerin *(fachspr.)*, **Glyzerin** [glytse'ri:n] **das; ~s** glycerine

GmbH *Abk.* **Gesellschaft mit beschränkter Haftung** ≈ p.l.c.

g-Moll ['ge:mɔl] **das; ~** *(Musik)* G minor

Gnade ['gna:də] **die; ~, ~n a)** *(Gewogenheit)* favour; **vor jmdm.** *od.* **vor jmds. Augen ~ finden** find favour with sb. *or* in sb.'s eyes; **jmdm. auf ~ und** *od.* **oder Ungnade ausgeliefert sein** be [completely] at sb.'s mercy; **in ~n wieder aufgenommen werden** be restored to favour; **b)** *(Rel.: Güte)* grace; **c)** *(Milde)* mercy; **~ vor** *od.* **für Recht ergehen lassen** temper justice with mercy; **d)** *(veraltete Anrede)* **Euer** *od.* **Ihro** *od.* **Ihre ~n** Your Grace

gnaden *itr. V. in* **gnade mir/dir Gott!** God *or* Heaven help me/you!

gnaden-, Gnaden-: ~akt der act of mercy; **~brot das: jmdm./einem Tier das ~brot geben** keep sb./an animal in his/ her/its old age; **~frist die** reprieve; **~gesuch das** plea for clemency; **~los** *(auch fig.)* **1.** *Adj.* merciless; **2.** *adv.* mercilessly; **~losigkeit die; ~:** mercilessness; **~schuß der** coup de grâce *(by shooting)*; **~stoß der** coup de grâce *(with sword etc.)*; **~tod der** euthanasia; mercy killing

gnädig ['gnɛ:dɪç] **1.** *Adj.* **a)** *(oft iron.)* gracious; **~er Herr** *(veralt.)* sir; **die ~e Frau/das ~e Fräulein/der ~e Herr** *(veralt.)* madam/the young lady/the master; **b)** *(glimpflich)* lenient, light ‹*sentence etc.*›; **c)** *(Rel.)* gracious ‹*God*›; **Gott sei uns ~:** [may] the good Lord preserve us. **2.** *adv.* **a)** *(oft iron.)* graciously; **b)** *(glimpflich)* **das ist ~ abgegangen** it turned out not to be too bad

Gneis [gnai̯s] **der; ~es, ~e** *(Geol.)* gneiss

Gnom [gno:m] **der; ~en, ~en** gnome; *(fig.: ugs.)* little twerp *(sl.)*

Gnu [gnu:] **das; ~s, ~s** gnu

Gobelin [gobəˈlɛ̃:] **der; ~s, ~s** Gobelin [tapestry]

Gockel [ˈgɔkl̩] **der; ~s, ~** *(bes. südd., sonst ugs. scherzh.)* cock

Go-go-Girl [ˈgo:gogø:ɐ̯l] **das** go-go girl *or* dancer

Go-Kart [ˈgo:kart] **der; ~[s], ~s** go-kart *(Brit.)*; kart

Golan·höhen [goˈla:n-] *Pl.* Golan Heights

Gold das; ~[e]s gold; **das schwarze ~** *(fig.)* black gold *(fig.)*; **es ist nicht alles ~, was glänzt** *(Spr.)* all that glitters *or* glistens is not gold *(prov.)*; **~ in der Kehle haben** *(fig.)* have a golden voice; **olympisches ~**: Olympic gold

gold-, Gold-: ~ader die vein of gold; **~barren der** gold bar *or* ingot; **~bestickt** *Adj.* embroidered with gold [thread] *postpos.*; **~dublee das** rolled gold

golden 1. *Adj.* **a)** gold ⟨*bracelet, watch, etc.*⟩; **der Tanz ums Goldene Kalb** the worship of the golden calf *or* Mammon; **eine ~e Schallplatte** a gold disc; **das Goldene Vlies** *(Myth.)* the Golden Fleece; **b)** *(dichter.: goldfarben)* golden; **c)** *(herrlich)* golden ⟨*memories, days, etc.*⟩; blissful ⟨*freedom etc.*⟩; **die ~e Mitte** *od.* **den ~en Mittelweg finden/wählen** find/strike a happy medium; **die goldenen zwanziger Jahre** the roaring twenties; **der Goldene Schnitt** *(Math.)* the golden section. **2.** *adv.* like gold

gold-, Gold-: ~farben, ~farbig *Adj.* gold-coloured; golden; **~füllung die** gold filling; **~fund der** gold find *or* strike; **~gehalt der** gold content; **~gelb** *Adj.* golden yellow; **~glänzend** *Adj.* shining gold; **~gräber der** gold-digger; **~grube die** *(auch fig.)* gold-mine; **~haltig** *Adj.* gold-bearing; auriferous; **~hamster der** golden hamster

goldig 1. *Adj.* *(niedlich, landsch.: nett)* sweet. **2.** *adv.* sweetly

Gold-: ~junge der *(Kosewort)* good [little] boy; **~kette die** gold chain; **~klumpen der** gold nugget; **~krone die** *(Zahnmed.)* gold crown; **~kurs der** *(Börsenw.)* price of gold; gold price; **~küste die; ~** *(Geogr.)* Gold Coast; **~lack der a)** gold lacquer; **b)** *(Bot.)* wallflower; **~macher der** alchemist

Gold·medaille die gold medal

Goldmedaillen-: ~gewinner der, ~gewinnerin die gold medallist; gold-medal winner

gold-, Gold-: ~mine die gold mine; **~münze die** gold coin; **~rausch der** gold fever; **~regen der a)** *(Bot.)* laburnum; golden rain; **b)** *(Feuerwerk)* golden rain; **c)** *(Reichtum)* riches *pl.*; wealth; **~reserve die** gold reserve; **~richtig** *(ugs.)* **1.** *Adj.* absolutely *or* dead right; **2.** *adv.* absolutely right; **~schatz der** gold treasure; *(verborgen auch)* hoard of gold

Gold·schmied der goldsmith

Goldschmiede·kunst die; *o. Pl.* goldsmith's art; goldwork *no art.*

Gold-: ~schmuck der gold jewelry *or* *(Brit.)* jewellery; **~schnitt der** gilt edging; **~staub der** gold dust; **~stück das** *(hist.)* gold piece; **sie ist ein ~stück** *(fig.)* she is a [real] treasure; **~sucher der** gold prospector; **~vorkommen das** gold deposit; **~waage die** gold balance; **alles** *od.* **jedes Wort auf die ~waage legen** *(wörtlich nehmen)* take everything *or* every word [too] literally; *(vorsichtig äußern)* weigh one's words very carefully; **~währung die** *(Wirtsch.)* currency tied to the gold standard; **~zahn der** *(ugs.)* gold tooth

¹Golf [gɔlf] **der; ~[e]s, ~e** gulf; **der ~ von Neapel** the Bay of Naples

²Golf das; ~s *(Sport)* golf

Golf·ball der golf ball

Golfer der; ~s, ~, Golferin die; ~, ~nen golfer

Golf-: ~krieg der Gulf War; **~mütze die** golf[ing] cap; **~platz der** golf-course; **~schläger der** golf club; **~spieler der, ~spielerin die** golfer; **~staat der** Gulf State; **~strom der** Gulf Stream; **~turnier das** golf tournament

Gomorrha [goˈmɔra] *s.* **Sodom**

Gondel [ˈgɔndl̩] **die; ~, ~n** gondola

gondeln *itr. V.; mit sein (ugs.)* **durch die Stadt/die Ägäis ~**: cruise around town/the Aegean; **durch die Gegend ~**: cruise around

Gong [gɔŋ] **der; ~s, ~s** gong

Gong·schlag der stroke of the/a gong; **beim ~**: when the gong sounds/sounded

gönnen [ˈgœnən] *tr. V.* **a) jmdm. etw. ~**: not begrudge sb. sth.; **ich gönne ihm diesen Erfolg von ganzem Herzen** I'm delighted *or* very pleased for him that he has had this success; **b)** *(zukommen lassen)* **sich/jmdm. etw. ~**: give *or* allow oneself/sb. sth.; **sie gönnte ihm keinen Blick** she didn't spare him a single glance

Gönner der; ~s, ~: patron

gönnerhaft *(abwertend)* **1.** *Adj.* patronizing. **2.** *adv.* patronizingly; in a patronizing manner

Gönnerin die; ~, ~nen patroness

Gönner·miene die *(abwertend)* patronizing expression

Gonorrhö[e] [gonɔˈrø:] **die; ~, Gonorrhöen** *(Med.)* gonorrhoea

gor [go:ɐ̯] *3. Pers. Sg. Prät. v.* **gären**

gordisch [ˈgɔrdɪʃ] *Adj.* **der Gordische Knoten** the Gordian knot

Göre [ˈgø:rə] **die; ~, ~n** *(nordd., oft abwertend)* **a)** *(Kind)* child; kid *(coll.)*; brat *(coll. derog.)*; **b)** *(freches Mädchen)* [cheeky *or* saucy] little madam *(coll.)*

Gorilla [goˈrɪla] **der; ~s, ~s a)** gorilla; **b)** *(ugs.: Leibwächter)* heavy *(coll.)*

Gosch[e] [ˈgɔʃ(ə)], **Goschen** [ˈgɔʃn̩] **die; ~, Goschen** *(südd., österr. meist abwertend)* mouth

Gospel [ˈgɔspl̩] **das** *od.* **der; ~s, ~s, Gospel·song der; ~s, ~s** gospel song

goß [gɔs] *1. u. 3. Pers. Sg. Prät. v.* **gießen**

Gosse [ˈgɔsə] **die; ~, ~n** gutter; *(fig. abwertend)* **in der ~ enden** end up in the gutter

Gote [ˈgo:tə] **der; ~n, ~n** Goth

Gotik [ˈgo:tɪk] **die; ~** *(Stil)* Gothic [style]; *(Epoche)* Gothic period

gotisch *Adj.* Gothic; **die ~e Schrift** Gothic [script]

Gott [gɔt] **der; ~es, Götter** [ˈgœtɐ] **a)** *o. Pl.; o. Art.* God; **~ Vater** God the Father; **vergelt's ~!** *(landsch.)* thank you! God bless you!; **großer** *od.* **mein ~!** good God!; **o** *od.* **ach [du lieber] ~!** goodness me!; **~ behüte** God *or* Heaven forbid; **~ und die Welt** all the world and his wife; **über ~ und die Welt quatschen** *(ugs.)* talk about everything under the sun *(coll.)*; **~ sei Dank!** *(ugs.)* thank God!; **um ~es Willen** *(bei Erschrecken)* for God's sake; *(bei einer Bitte)* for heaven's *or* goodness' sake; **tue es in ~es Namen** *(ugs.)* do it and have done with it; **wie ~ in Frankreich leben** *(ugs.)* live in the lap of luxury; **den lieben ~ einen guten Mann sein lassen** *(ugs.)* take things as they come; **b)** *(übermenschliches Wesen)* god; **wie ein junger ~ spielen/tanzen** play/dance divinely; **das wissen die Götter** *(ugs.)* God *or* heaven only knows

gott·ähnlich *Adj.* godlike

Gott·erbarmen das *in* **zum ~ sein** *(mitleiderregend)* be pitiful; *(schlecht)* be pathetic

gott·ergeben 1. *Adj.* meek. 2. *adv.* meekly

Götter·speise die a) *o. Pl. (Myth.)* food of the gods; **b)** *(Kochk.)* jelly

gottes-, Gottes-: ~dienst der service; **den ~dienst besuchen** go to church; **~furcht die** fear of God; **~haus das** *(geh.)* house of God; **~lästerer der** blasphemer; **~lästerlich** 1. *Adj.* blasphemous; 2. *adv.* blasphemously; **~lästerung die** blasphemy; **~mutter die;** *o. Pl.* Mother of God; **~urteil das** *(hist.)* trial by ordeal

gott-: ~gefällig *Adj. (geh.)* pleasing to God *postpos.*; **~gewollt** *Adj.* ordained by God *postpos.*

Gottheit die; ~, ~en a) *(Gott, Göttin)* deity; **b)** *o. Pl. (geh.: Gottsein)* divinity

Göttin ['gœtɪn] **die; ~, ~nen** goddess

göttlich ['gœtlɪç] 1. *Adj.* **a)** *(Gott eigen od. ähnlich; herrlich)* divine ‹*grace, beauty, etc.*›; **b)** *(einem Gott zukommend)* god-like ‹*status etc.*›. 2. *adv. (herrlich)* divinely

gott-: ~lob *adv.* thank goodness; **~los** 1. *Adj.* **a)** *(verwerflich)* ungodly, wicked ‹*life etc.*›; impious ‹*words, speech, etc.*›; *(pietätlos)* irreverent; **b)** *(Gott leugnend)* godless ‹*theory etc.*›; 2. *adv. (verwerflich)* irreverently

gott-, Gott-: ~vater der; *o. Pl.* God the Father; **~verdammt** *Adj.; nicht präd. (salopp)*, **~verflucht** *Adj.; nicht präd. (salopp)* goddamn[ed] *(sl.)*; **~verlassen** *Adj.* **a)** *(ugs.: abseits)* godforsaken; **b)** *(von Gott verlassen)* forsaken by God *postpos.*; **~vertrauen das** trust in God

Götze ['gœtsə] **der; ~n, ~n** *(auch fig.)* idol

Götzen-: ~bild das idol; graven image *(bibl.)*; *(fig.)* idol; **~diener der** idolater; *(fig.)* worshipper

Götz·zitat ['gœts-] **das** *the insulting remark 'du kannst mich am Arsch lecken' or the like, frequently used in altercations; a verbal equivalent of the V-sign*

Goulasch *s.* **Gulasch**

Gouvernante [guvɛr'nantə] **die; ~, ~n** governess

Gouverneur [guvɛr'nøːɐ̯] **der; ~s, ~e** governor

Grab [graːp] **das; ~[e]s, Gräber** ['grɛːbɐ] grave; **er würde sich im ~[e] herumdrehen** *(fig. ugs.)* he would turn in his grave; **das ~ des Unbekannten Soldaten** the tomb of the Unknown Soldier *or* Warrior; **verschwiegen wie ein** *od.* **das ~ sein** *(ugs.)* keep absolutely mum *(coll.)*; **sich** *(Dat.)* **selbst sein ~ schaufeln** *(fig.)* dig one's own grave *(fig.)*; **mit einem Fuß** *od.* **Bein im ~[e] stehen** *(fig.)* have one foot in the grave *(fig.)*; **jmdn. ins ~ bringen** be the death of sb.; **etw. mit ins ~ nehmen** *(geh.)* take sth. with one to the grave; **jmdn. zu ~e tragen** *(geh.)* bury sb.; **seine Hoffnungen zu ~e tragen** *(fig. geh.)* abandon one's hopes; **jmdn. an den Rand des ~es bringen** *(fig. geh.)* drive sb. to distraction

graben 1. *unr. tr. V.* dig. 2. *unr. itr. V.* dig (**nach** for). 3. *unr. refl. V. (geh.)* **sich in etw.** *(Akk.)* **~:** dig into sth.

Graben der; ~s, Gräben ['grɛːbn̩] **a)** ditch; **b)** *(Schützengraben)* trench; **c)** *(Festungsgraben)* moat

Gräber-: ~feld das [large] cemetery; **~fund der** grave find

Grabes-: ~kälte die *(geh.)* deathly cold; **~stille die** deathly silence *or* hush; **~stimme die** *(ugs.)* sepulchral voice

Grab-: ~hügel der grave mound; **~inschrift die** inscription [on a/the gravestone]; epitaph; **~kammer die** burial chamber; **~kreuz das** cross [on the/a grave]; **~mal das;** *Pl.* **~mäler,** *geh.* **~male** monument; *(~stein)* gravestone; **das ~mal des Unbekannten Soldaten** the tomb of the Unknown Soldier *or* Warrior; **~platte die** memorial slab; *(aus Metall)* memorial plate; **~rede die** funeral oration *or* speech; **~schändung die** desecration of a/the grave/of [the] graves

grabschen ['grapʃn̩] 1. *tr. V.* grab; snatch. 2. *itr. V.* **nach etw. ~:** grab at sth.

Grab-: ~stätte die tomb; grave; **~stein der** gravestone; tombstone; **~stelle die** burial plot

gräbst [grɛːpst] *2. Pers. Sg. Präsens v.* **graben**

gräbt *3. Pers. Sg. Präsens v.* **graben**

Grabung die; ~, ~en *(bes. Archäol.)* excavation

Grab·urne die funeral urn

Gracht [graxt] **die; ~, ~en** canal

Grad [graːt] **der; ~[e]s, ~e a)** degree; **Verbrennungen ersten/zweiten ~es** first-/second-degree burns; **ein Verwandter ersten/zweiten ~es** an immediate relation/a relation once removed; **in hohem ~e** to a great *or* large extent; **b)** *(akademischer ~)* degree; *(Milit.)* rank; **c)** *(Maßeinheit, Math., Geogr.)* degree; **10 ~ Wärme/Kälte** 10 degrees above zero/below [zero]; **39 ~ Fieber haben** have a temperature of 39 degrees; **minus 5 ~/5 ~ minus** minus 5 degrees; **null ~:** zero; **Gleichungen zweiten ~es** equations of the second degree; quadratic equations; **sich um hundertachtzig ~ drehen** *(fig.)* completely change [one's views]; **der 50. ~ nördlicher Breite** [latitude] 50 degrees North

grad-, ¹Grad- *s.* **gerad[e]-, Gerad[e]-**

²Grad-: ~einteilung die graduation; **~messer der** gauge, yardstick (**für** of)

graduell [gra'du̯ɛl] 1. *Adj.* gradual ‹*development etc.*›; slight ‹*difference etc.*›. 2. *adv.* gradually; by degrees; ‹*different*› in degree

graduiert *Adj.* graduate; **ein ~er Ingenieur/eine ~e Ingenieurin** an engineering graduate

Graf [graːf] **der; ~en, ~en a)** count; *(britischer ~)* earl; **b)** *o. Pl. (Titel)* Count; *(britischer ~)* Earl; **~ Koks [von der Gasanstalt]** *(salopp)* Lord Muck *(Brit. joc.)*

Grafik usw. *s.* **Graphik** usw.

Gräfin ['grɛːfɪn] **die; ~, ~nen** countess; *(Titel)* Countess

gräflich ['grɛːflɪç] *Adj.* count's *attrib.*; of the count *postpos., not pred.*; *(in Großbritannien)* earl's; of the earl

Grafschaft die; ~, ~en a) *(Amtsbezirk des Grafen)* count's land; *(in Großbritannien)* earldom; **b)** *(Verwaltungsbezirk)* county

Graham·brot ['graːham-] **das** wholemeal *(Brit.) or (Amer.)* wheatmeal bread

Gral [graːl] **der; ~[e]s: der [Heilige] ~:** the [Holy] Grail

Grals-: ~ritter der knight of the [Holy] Grail

gram [graːm] *in* **jmdm. ~ sein** be aggrieved at sb.

Gram der; ~[e]s *(geh.)* grief; sorrow; **aus ~ um** *od.* **über etw.** *(Akk.)* out of grief *or* sorrow at sth.

grämen ['grɛːmən] 1. *tr. V.* grieve. 2. *refl. V.* grieve (**über** + *Akk.*, **um** over)

gram·gebeugt *Adj.* bowed down with grief *or* sorrow *postpos.*

grämlich ['grɛːmlɪç] 1. *Adj.* morose; sullen; morose ‹*thought*›. 2. *adv.* morosely; sullenly

Gramm [gram] **das; ~s, ~e** gram; **250 ~ Käse** 250 grams of cheese

Grammatik [gra'matɪk] **die; ~, ~en a)** grammar; **b)** *(Lehrbuch)* grammar [book]

grammatikalisch [gramati'ka:lɪʃ], **grammatisch 1.** *Adj.* grammatical. **2.** *adv.* grammatically

Grammophon Ⓦz [gramo'fo:n] **das; ~s, ~e** gramophone

Granat [gra'na:t] **der; ~[e]s, ~e a)** *(Schmuckstein)* garnet; **b)** *(Garnele)* [common] shrimp

Granat·apfel der pomegranate

Granate [gra'na:tə] **die; ~, ~n** shell; *(Hand~)* grenade

Granat-: ~feuer das shell-fire *no pl., no indef. art.*; **~splitter der** shell splinter; **~werfer der** *(Milit.)* mortar

Grand [grã: *od.* graŋ] **der; ~s, ~s** *(Skat)* grand

Grand·hotel ['grã:-] **das** luxury *or* five-star hotel

grandios [gran'dio:s] **1.** *Adj.* magnificent. **2.** *adv.* magnificently

Grand Prix [grã'pri:] **der; ~ ~** [- pri:(s)], **~ ~** [- pri:s] Grand Prix

Granit [gra'ni:t] **der; ~s, ~e** granite; **auf ~** *(Akk.)* **beißen** *(fig.)* bang one's head against a brick wall *(fig.)*; **bei jmdm. auf ~** *(Akk.)* **beißen** *(fig.)* get nowhere with sb. *(fig.)*

Granit·block der; *Pl.* **-blöcke** block of granite; granite block

Granne ['granə] **die; ~, ~n** awn; beard

grantig ['grantɪç] *(südd., österr. ugs.)* **1.** *Adj.* bad-tempered; grumpy. **2.** *adv.* bad-temperedly; grumpily

Granulat [granu'la:t] **das; ~[e]s, ~e** *(bes. Chemie)* granules *pl.*

granulieren *itr., tr. V. (bes. Chemie)* granulate

Grapefruit ['gre:pfru:t] **die; ~, ~s** grapefruit

Graphik ['gra:fɪk] **die; ~, ~en a)** *o. Pl.* graphic art[s *pl.*]; **b)** *(Kunstwerk)* graphic; *(Druck)* print; **c)** *(Illustration)* diagram

Graphiker der; ~s, ~, Graphikerin die; ~, ~nen [graphic] designer; *(Künstler[in])* graphic artist

graphisch 1. *Adj.* **a)** graphic; **das ~e Gewerbe** *(veralt.)* the printing trade; **b)** *(schematisch)* graphic; diagrammatic; **eine ~e Darstellung** a diagram. **2.** *adv.* graphically

Graphit [gra'fi:t] **der; ~s, ~e** graphite

Graphologe [grafo'lo:gə] **der; ~n, ~n** graphologist

Graphologie die; ~: graphology *no art.*

graphologisch 1. *Adj.* graphological. **2.** *adv.* graphologically

grapschen ['grapʃn̩] *s.* **grabschen**

Gras [gra:s] **das; ~es, Gräser** ['grɛ:zɐ] grass; **das ~ wachsen hören** *(ugs. spött.)* read too much into things; **über etw.** *(Akk.)* **~ wachsen lassen** *(ugs.)* let the dust settle on sth.; **ins ~ beißen [müssen]** *(salopp)* bite the dust *(coll.)*

gras-: ~bedeckt, ~bewachsen *Adj.* grass-covered; grassy

Gras·büschel das tuft of grass

grasen *itr. V.* graze

gras-, Gras-: ~fläche die area of grass; *(Rasen)* lawn; **~fleck der a)** patch of grass; **b)** *(auf der Kleidung)* grass stain; **~grün** *Adj.* grass-green; **~halm der** blade of grass; **~hüpfer der** *(ugs.)* grasshopper; **~land das;** *o. Pl.* grassland; **~mücke die** warbler; **~narbe die** turf

grassieren [gra'si:rən] *itr. V.* ‹*disease etc.*› rage, be rampant; ‹*craze etc.*› be [all] the rage; ‹*rumour*› be rife

gräßlich ['grɛslɪç] **1.** *Adj.* **a)** *(abscheulich)* horrible; terrible ‹*accident*›; **b)** *(ugs.: unangenehm)* dreadful *(coll.)*; awful; **c)** *(ugs.: sehr stark)* terrible *(coll.)*; awful. **2.** *adv.* **a)** *(abscheulich)* horribly; terribly; **b)** *(ugs.: unangenehm)* terribly *(coll.)*; **c)** *(ugs.: sehr)* terribly *(coll.)*; dreadfully *(coll.)*

Gräßlichkeit die; ~, ~en a) *o. Pl.* *(Abscheulichkeit)* horribleness; *(eines Unfalls)* terribleness; **b)** *o. Pl.* *(unangenehme Art)* dreadfulness *(coll.)*; awfulness

Gras-: ~steppe die *(Geogr.)* [grassy] steppe; **~streifen der** strip of grass; *(längs einer Straße)* grass verge

Grat [gra:t] **der; ~[e]s, ~e a)** *(Bergrücken)* ridge; **b)** *(Archit.)* hip; **c)** *(Technik)* burr

Gräte ['grɛ:tə] **die; ~, ~n** bone *(of fish)*

gräten·los *Adj.* boneless

Gräten·muster das herring-bone [pattern]

Gratifikation [gratifika'tsi̯o:n] **die; ~, ~en** bonus

gratis ['gra:tɪs] *Adv.* free [of charge]; gratis

Gratis-: ~muster das, ~probe die free sample; **~vorstellung die** free performance

Grätsche ['grɛ:tʃə] **die; ~, ~n** *(Turnen)* straddle; *(Sprung)* straddle-vault; **in die ~ gehen** go into the straddle position

grätschen 1. *tr. V.* **die Beine ~:** straddle one's legs. **2.** *itr. V.; mit sein* straddle; do *or* perform a straddle; **über etw.** *(Akk.)* **~:** do a straddle-vault over sth.

Gratulant [gratu'lant] **der; ~en, ~en, Gratulantin die; ~, ~nen** well-wisher

Gratulation [gratula'tsi̯o:n] **die; ~, ~en** congratulations *pl.*

Gratulations·schreiben das congratulatory letter

gratulieren *itr. V.* **jmdm. ~:** congratulate sb.; **jmdm. zum Geburtstag ~:** wish sb. many happy returns [of the day]; **[ich] gratuliere!** congratulations!

Grat·wanderung die ridge walk; *(fig.)* balancing act

grau [grau̯] *Adj.* **a)** grey; **~ werden** go grey; **~ in ~:** grey and drab; **b)** *(trostlos)* dreary; drab; depressing; **der ~e Alltag** the dull routine *or* monotony of daily life; **c)** *(zwischen legal und illegal)* grey; **d)** *(unbestimmt)* vague; **in ~er Vorzeit** in the dim and distant past

Grau das; ~s, ~ a) grey; **b)** *o. Pl.* *(Trostlosigkeit)* dreariness; drabness

grau·blau *Adj.* grey-blue

Grau·brot das *bread made with rye- and wheat-flour*

Grau·bünden [-'bʏndn̩] **(das); ~s** the Grisons

[1]**grauen** *itr. V. (geh.)* **der Morgen/der Tag graut** morning is breaking; day is dawning *or* breaking

[2]**grauen** *itr. V. (unpers.)* **ihm graut [es] davor/vor ihr** he dreads [the thought of] it/he's terrified of her; **mir graut es, wenn ich nur daran denke** I dread the [mere] thought of it

Grauen das; ~s, ~ a) *o. Pl.* horror **(vor + *Dat.*** of**)**; **ein Bild des ~s** a scene of horror; **b)** *(Schreckbild)* horror

grauen·haft, grauen·voll 1. *Adj.* **a)** horrifying; **b)** *(ugs.: sehr unangenehm)* terrible *(coll.)*; dreadful *(coll.)*. **2.** *adv.* **a)** horrifyingly; **b)** *(ugs.: sehr unangenehm)* terribly *(coll.)*; dreadfully *(coll.)*

grau-, Grau-: ~gans die grey goose; greylag [goose]; **~grün** *Adj.* grey-green; **~haarig** *Adj.* grey-haired

gräulich ['grɔ̯ylɪç] *Adj.* greyish

grau·meliert *Adj. (präd. getrennt geschrieben)* greying ‹*hair*›

Graupe ['grau̯pə] **die; ~, ~n** grain of pearl barley; **~n** pearl barley *sing.*

Graupel ['grau̯pl̩] **die; ~, ~n** soft hail pellet; **~n** soft hail; graupel

Graupel·schauer der shower of soft hail

Graupen·suppe die barley soup *or* broth

Graus [graus] **der; ~es: es ist ein ~:** it's terrible; **o ~!** *(ugs. scherzh.)* oh horror! *(joc.)*

grausam 1. *Adj.* **a)** cruel; **~ gegen jmdn. sein** be cruel to sb.; **b)** *(furchtbar)* terrible; dreadful; **c)** *(ugs.: sehr schlimm)* terrible *(coll.)*; dreadful *(coll.)*. **2.** *adv.* **a)** cruelly; **sich ~ für etw. rächen** take cruel revenge for sth.; **b)** *(furchtbar)* terribly, dreadfully; **c)** *(ugs.: sehr stark)* terribly *(coll.)*; dreadfully *(coll.)*

Grausamkeit die; ~, ~en a) *o. Pl.* cruelty; **b)** *(Handlung)* act of cruelty; *(Greueltat)* atrocity

Grau·schimmel der a) *(Pferd)* grey [horse]; **b)** *(Pilz)* grey mould

grau·schwarz *Adj.* grey-black

grausen *s.* **²grauen**

Grausen das; ~s horror; **das kalte ~ kriegen** *(ugs.)* be scared stiff *or* to death *(coll.)*

grausig *s.* **grauenhaft**

grauslich *(bes. bayr., österr.) s.* **gräßlich**

grau-, Grau-: ~tier das *(ugs. scherzh.)* *(Esel)* ass; donkey; *(Maultier)* mule; **~weiß** *Adj.* greyish white; **~zone die** grey area *(fig.)*

Graveur [gra'vø:ɐ̯] **der; ~s, ~e, Graveurin die; ~, ~nen** engraver

gravieren [gra'vi:rən] *tr. V.* engrave

gravierend *Adj.* serious, grave ‹*matter, accusation, error, etc.*›; important ‹*difference, decision*›

Gravierung die; ~, ~en engraving

Gravitation [gravita'tsio:n] **die; ~** *(Physik, Astron.)* gravitation

Gravitations-: ~feld das *(Physik, Astron.)* gravitational field; **~gesetz das** *(Physik, Astron.)* law of gravitation

gravitätisch 1. *Adj.* grave; solemn. **2.** *adv.* gravely; solemnly

Gravur [gra'vu:ɐ̯] **die; ~, ~en, Gravüre** [gra'vy:rə] **die; ~, ~n** engraving

Grazie ['gra:tsiə] **die; ~, ~n a)** *o. Pl. (Anmut)* grace; gracefulness; **b)** *Pl. (Myth.)* Graces; **c)** *Pl. (fig. scherzh.)* beauties

grazil [gra'tsi:l] *Adj. (auch fig.)* delicate

graziös [gra'tsiø:s] **1.** *Adj.* graceful. **2.** *adv.* gracefully

Gregor ['gre:gɔr] **(der)** Gregory

Gregorianisch [grego'ria:nɪʃ] *Adj.* Gregorian; **~er Gesang** Gregorian chant

Greif [graif] **der; ~[e]s** *od.* **~en, ~en a)** *(Wappentier)* griffin; gryphon; **b)** *s.* **Greifvogel**

greif·bar 1. *Adj.* **a) etw. ~ haben** have sth. to hand; **~ sein** be within reach; **in ~er Nähe** *(fig.)* within reach; **b)** *(deutlich)* tangible; concrete; **c)** *nicht attr. (ugs.: verfügbar)* available. **2.** *adv.* **~ nahe** *(fig.)* within reach

greifen 1. *unr. tr. V.* **a)** *(er~)* take hold of; grasp; *(rasch ~)* grab; seize; **sich** *(Dat.)* **etw. ~:** help oneself to sth.; **von hier scheint der See zum Greifen nah[e]** from here the lake seems close enough to reach out and touch; **zum Greifen nahe sein** ‹*end, liberation*› be imminent; ‹*goal, success*› be within sb.'s grasp; **b)** *(fangen)* catch; **c) einen Akkord ~** *(auf dem Klavier usw.)* play a chord; *(auf der Gitarre usw.)* finger a chord; **d)** *(schätzen)* **tausend ist zu hoch/niedrig gegriffen** one thousand is an overestimate/underestimate. **2.** *unr. itr. V.* **a) in/unter/hinter etw./sich** *(Akk.)* **~:** reach into/under/behind sth./one; **nach etw. ~:** reach for sth.; *(hastig)* make a grab for sth.; **zu Drogen/zur Zigarette ~:** turn to drugs/reach for a cigarette; **nach der Macht ~** *(fig.)* try to seize power; **etw. greift um sich** sth. is spreading; **b)** *(Technik)* grip; **c)** *(ugs.: spielen)* **in die Tasten/Saiten ~:** sweep one's hand over the keys/across the strings

Greifer der; ~s, ~ *(Technik)* grab[-bucket]

Greif-: ~vogel der *(Zool.)* diurnal bird of prey; **~zange die** tongs *pl.*

greis [grais] *Adj. (geh.)* aged; white ‹*hair, head*›

Greis der; ~es, ~e old man

Greisen·alter das old age

greisen·haft *Adj.* old man's/woman's *attrib.*; aged; *(von jüngerem Menschen)* ‹*face etc.*› like that of an old man/woman

Greisin die; ~, ~nen old woman *or* lady

grell [grɛl] **1.** *Adj.* **a)** *(hell)* glaring, dazzling ‹*light, sun, etc.*›; **b)** *(auffallend)* garish, gaudy ‹*colour etc.*›; flashy, loud ‹*dress, pattern, etc.*›; **c)** *(schrill)* shrill, piercing ‹*cry, voice, etc.*›. **2.** *adv.* **a)** *(hell)* with glaring *or* dazzling brightness; **b)** *(schrill)* shrilly; piercingly

grell-: ~beleuchtet *Adj. (präd. getrennt geschrieben)* dazzlingly lit; **~bunt** *Adj.* gaudily coloured; **~rot** *Adj.* garish *or* bright red

Gremium ['gre:miʊm] **das; ~s, Gremien** committee

Grenadier [grena'di:ɐ̯] **der; ~s, ~e** *(Milit.)* **a)** *(Infanterist)* infantryman; **b)** *(hist.)* grenadier

Grenz-: ~abfertigung die *(Zollw.)* passport control and customs clearance [at the/a border]; **~beamte der** border official; **~befestigung die** *(Milit.)* border fortification; **~bereich der a)** *o. Pl.* border *or* frontier zone *or* area; **b)** *(äußerster Bereich)* limit[s *pl.*]; **~bezirk der** border *or* frontier district

Grenze ['grɛntsə] **die; ~, ~n a)** *(zwischen Staaten)* border; frontier; **die ~ zu Italien** the border with Italy; **an der ~ wohnen** live on the border *or* frontier; **b)** *(zwischen Gebieten)* boundary; **c)** *(gedachte Trennungslinie)* borderline; dividing line; **d)** *(Schranke)* limit; **jmdm. [keine] ~n setzen** impose [no] limits on sb.; **an seine ~n stoßen** reach its limit[s]; **sich in ~n halten** *(begrenzt sein)* keep *or* stay within limits; **seine Leistungen hielten sich in ~n** his achievements were not [all that *(coll.)*] outstanding

grenzen *itr. V.* **an etw.** *(Akk.)* **~:** border [on] sth.; *(fig.)* verge on sth.

grenzen·los 1. *Adj.* boundless; endless; *(fig.)* boundless, unbounded ‹*joy, wonder, jealousy, grief, etc.*›; unlimited ‹*wealth, power*›; limitless ‹*patience, ambition*›; extreme ‹*tiredness, anger, foolishness*›. **2.** *adv.* endlessly; *(fig.)* beyond all measure

Grenzen·losigkeit die; ~: boundlessness; immensity

grenz-, Grenz-: ~fall der *(nicht eindeutiger Fall)* borderline case; *(Sonderfall)* limiting case; **~formalitäten** *Pl.* passport and customs formalities [at the/a border]; **~gänger der; ~s, ~:** [regular] commuter across the border *or* frontier; **~gebiet das a)** border *or* frontier area *or* zone; **b)** *(Sachgebiet zwischen Disziplinen)* adjacent field; **~konflikt der** border *or* frontier conflict; **~kontrolle die** border *or* frontier check; **~linie die** border; **~nah** *Adj.; nicht präd.* close to the border *or* frontier *postpos.*; **~posten der** border *or* frontier guard; **~schutz der a)** border *or* frontier protection; **b)** *(ugs.: Bundesgrenzschutz)* border *or* frontier police; **~soldat der** border *or* frontier guard; **~stadt die** border *or* frontier town; **~stein der** boundary stone; **~streitigkeit die** boundary dispute; *(we-*

gen einer Staatsgrenze) border *or* frontier dispute; **~übergang der a)** border crossing-point; frontier crossing-point; [border] check-point; **b)** *(das Passieren der Grenze)* crossing of the border *or* frontier; **~verkehr der** [cross-]-border traffic; frontier traffic; **~verletzung die** border *or* frontier violation; **~wall der** border *or* frontier rampart; **~wert der** *(Math.)* limit; **~zwischenfall der** border incident

Gretchen·frage die; ~: crucial question; sixty-four-thousand-dollar question *(coll.)*

Greuel ['grɔyəl] **der; ~s, ~** *(geh.)* **a)** *o. Pl. (Abscheu)* horror; **er/sie/es ist mir ein ~:** I loathe *or* detest him/her/it; **b)** *meist Pl. (~tat)* atrocity

Greuel-: ~märchen das horror story; **~propaganda die** atrocity propaganda; stories *pl.* of atrocities; **~tat die** atrocity

greulich *s.* **gräßlich**

Griebe ['gri:bə] **die; ~, ~n** crackling *no indef. art.*; greaves *pl.*

Grieben·schmalz das dripping with crackling *or* greaves

Grieche ['gri:çə] **der; ~n, ~n** Greek

Griechen·land (das); ~s Greece

Griechin die; ~, ~nen Greek

griechisch 1. *Adj.* Greek ‹*language, mythology, island, etc.*›; Grecian, Greek ‹*vase, style, etc.*›; **die ~e Tragödie** Greek tragedy. **2.** *adv.* **~ sprechen/schreiben** speak/write in Greek; *s. auch* **deutsch**

Griechisch das; ~[s] Greek *no art.*; *s. auch* **Deutsch**

griechisch-: ~-orthodox *Adj.* Greek Orthodox; **~-römisch** *Adj. (Ringen)* Graeco-Roman

Griesgram ['gri:sgra:m] **der; ~[e]s, ~e** grouch *(coll.)*

griesgrämig ['gri:sgrɛ:mɪç] **1.** *Adj.* grouchy *(coll.)*; grumpy. **2.** *adv.* in a grouchy *(coll.) or* grumpy manner

Grieß [gri:s] **der; ~es, ~e** semolina

Grieß·brei der semolina

griff [grɪf] *1. u. 3. Pers. Sg. Prät. v.* **greifen**

Griff der; ~[e]s, ~e a) grip; grasp; **mit eisernem/festem ~:** with a grip of iron/a firm grip; **der ~ nach etw./in etw.** *(Akk.)***/an etw.** *(Akk.)* reaching for sth./dipping into sth./taking hold of *or* grasping sth.; **[mit jmdm./etw.] einen guten/glücklichen ~ tun** make a good choice [with sb./sth.]; **b)** *(beim Ringen, Bergsteigen)* hold; *(beim Turnen)* grip; **etw. im ~ haben** *(etw. routinemäßig beherrschen)* have the hang of sth. *(coll.)*; *(etw. unter Kontrolle haben)* have sth. under control; **c)** *(Knauf, Henkel)* handle; *(eines Gewehrs, einer Pistole)* butt; *(eines Schwerts)* hilt; **d)** *(Musik)* finger-placing

griff·bereit *Adj.* ready to hand *postpos.*

Griff·brett das *(Musik)* fingerboard

Griffel ['grɪfl̩] **der; ~s, ~ a)** *(Schreibgerät)* slate-pencil; **b)** *(Bot.)* style

griffig *Adj.* **a)** *(handlich)* handy; ‹*tool etc.*› that is easy to handle; **b)** *(gut greifend)* that grips well *postpos., not pred.*; non-slip ‹*surface, floor*›

Grill [grɪl] **der; ~s, ~s** grill; *(Rost)* barbecue

Grille ['grɪlə] **die; ~, ~n a)** *(Insekt)* cricket; **b)** *(sonderbarer Einfall)* whim; fancy

grillen 1. *tr. V.* grill. **2.** *itr. V.* **im Garten ~:** have a barbecue in the garden

Grill·platz der barbecue area

Grimasse [gri'masə] **die; ~, ~n** grimace; **eine ~ schneiden** *od.* **machen** grimace; pull a face

Grimm [grɪm] **der; ~[e]s** *(geh.)* fury

grimmig 1. *Adj.* **a)** *(zornig)* furious ‹*person*›; grim ‹*face; expression*›; fierce, ferocious ‹*enemy, lion, etc.*›; **b)** *(heftig)* fierce, severe ‹*cold, hunger, pain, etc.*›. **2.** *adv.* **a)** *(wütend)* furiously; **~ lachen** laugh grimly; **b)** *(heftig)* fiercely

Grind [grɪnt] **der; ~[e]s, ~e** *(Wundschorf)* scab

grindig *Adj.* scabby

grinsen ['grɪnzn̩] *itr. V.* grin; *(höhnisch)* smirk

Grinsen das; ~s: ein fröhliches/unverschämtes ~: a happy grin/an insolent smirk

Grippe ['grɪpə] **die; ~, ~n a)** influenza; flu *(coll.)*; **b)** *(volkst.: Erkältung)* cold

Grippe·welle die wave of influenza *or (coll.)* flu

Grips [grɪps] **der; ~es** brains *pl.*; nous *(coll.)*; **streng deinen ~ an** use your brains *or* nous

Grisly·bär, Grizzly·bär ['grɪsli-] **der** grizzly bear

grob [gro:p] **1.** *Adj.* **a)** coarse ‹*sand, gravel, paper, sieve, etc.*›; thick ‹*wire*›; rough, dirty ‹*work*›; **b)** *(ungefähr)* rough; **in ~en Umrissen** in rough outline; **c)** *(schwerwiegend)* gross; flagrant ‹*lie*›; **ein ~er Fehler/Irrtum** a bad mistake *or* gross error; **aus dem Gröbsten heraussein** *(ugs.)* be over the worst; **d)** *(barsch)* coarse; rude; **~ werden** become abusive *or* rude; **e)** *(nicht sanft)* rough; **~ [zu jmdm.] sein** be rough [with sb.]. **2.** *adv.* **a)** coarsely; **b)** *(ungefähr)* roughly; **~ geschätzt** at a rough estimate; **c)** *(schwerwiegend)* grossly; **d)** *(barsch)* coarsely; rudely; **e)** *(nicht sanft)* roughly

grob·gemahlen *Adj. (präd. getrennt geschrieben)* coarsely ground; coarse-ground

Grobheit die; ~, ~en a) *o. Pl.* rudeness; coarseness; **b)** *(Äußerung)* rude remark

Grobian ['gro:bia:n] **der; ~[e]s, ~e** boor; lout

grob-: ~körnig *Adj.* coarse ‹*sand, flour, etc.*›; *(Fot.)* coarse-grained ‹*film*›; **~maschig** *Adj.* wide-meshed ‹*sieve, net, etc.*›; loose-knit ‹*pullover etc.*›; **~schlächtig** [~ʃlɛçtɪç] *Adj.* heavily built

Grog [grɔk] **der; ~s, ~s** grog

groggy ['grɔgi] *Adj.; nicht attr.* **a)** *(Boxen)* groggy; **b)** *(ugs.: erschöpft)* whacked [out] *(coll.)*; all in *(coll.)*

grölen ['grø:lən] **1.** *tr. V. (ugs. abwertend)* bawl [out]; roar, howl ‹*approval*›. **2.** *itr. V.* bawl

Groll [grɔl] **der; ~[e]s** *(geh.)* rancour; resentment; **einen ~ gegen jmdn./etw. hegen** harbour resentment *or* a grudge against sb./sth.

grollen *itr. V. (geh.)* **a)** *(verstimmt sein)* be sullen; **[mit] jmdm. ~:** bear a grudge against sb.; bear sb. a grudge; **b)** *(dröhnen)* rumble; ‹*thunder*› roll, rumble

Grön·land ['grø:n-] **(das); ~s** Greenland

Gros [gro:] **das; ~** [gro:(s)], **~** [gro:s] bulk; main body

Groschen ['grɔʃn̩] **der; ~s, ~ a)** *(österreichische Münze)* groschen; **b)** *(ugs.: Zehnpfennigstück)* ten-pfennig piece; *(fig.)* penny; cent *(Amer.)*; **[sich** *(Dat.)***] ein paar ~ verdienen** *(ugs.)* earn [oneself] a few pennies *or* pence; **der ~ ist [bei ihm] gefallen** *(fig.)* the penny has dropped

Groschen·roman der *(abwertend)* cheap novel; dime novel *(Amer.)*

groß [gro:s] **größer** ['grø:sɐ], **größt...** ['grø:st...] **1.** *Adj.* **a)** big; big, large ‹*house, window, area, room, etc.*›; large ‹*pack, size, can, etc.*›; great ‹*length, width,* *height*›; tall ‹*person*›; **~e Eier/Kartoffeln** large eggs/potatoes; **eine ~e Terz/Sekunde** *(Musik)* a major third/second; **ein ~es Bier, bitte** a pint, please; **b)** *(eine bestimmte Größe aufweisend)* **1 m²/2 ha ~:** 1 m²/2 ha in area; **sie ist 1,75 m ~:** she is 1.75 m tall; **doppelt/dreimal so ~ wie ...:** twice/three times the size of...; **c)** *(älter)* big ‹*brother, sister*›; **seine größere Schwester** his elder sister; **unsere Große/unser Großer** our eldest *or* oldest daughter/son; **d)** *(erwachsen)* grown-up ‹*children, son, daughter*›; **[mit etw.] ~ werden** grow up [with sth.]; **die Großen** *(Erwachsene)* the grown-ups; *(ältere Kinder)* the older children; **~ und klein** old and young [alike]; **e)** *(lange dauernd)* long, lengthy ‹*delay, talk, explanation, pause*›; **die ~en Ferien** *(Schulw.)* the summer holidays *or (Amer.)* long vacation *sing.*; **die ~e Pause** *(Schulw.)* [mid-morning] break; **f)** *(beträchtlich)* **~e Summen/Kosten** large sums/heavy costs; **eine ~e Auswahl** a wide selection *or* range; **g)** *(außerordentlich)* great ‹*pleasure, pain, hunger, anxiety, hurry, progress, difficulty, mistake, importance*›; intense ‹*heat, cold*›; high ‹*speed*›; **mit dem größten Vergnügen** with the greatest of pleasure; **~en Hunger haben** be very hungry; **ihre/seine ~e Liebe** her/his great love; **h)** *(gewichtig)* great; major ‹*producer, exporter*›; great, major ‹*event*›; **ein ~er Augenblick/Tag** a great moment/day; **~e Worte** grand *or* fine words; **[k]eine ~e Rolle spielen** [not] play a great *or* an important part; **die Großen [der Welt]** the great figures [of our world]; **i)** *nicht präd. (glanzvoll)* grand ‹*celebration, ball, etc.*›; **die ~e Dame/den ~en Herrn spielen** *(iron.)* play the fine lady/gentleman; **j)** *(bedeutend)* great, major ‹*artist, painter, work*›; **Katharina die Große** Catherine the Great; *s. auch* **Karl; k)** *(wesentlich)* **die ~e Linie/der ~e Zusammenhang** the basic line/the overall context; **in ~en Zügen** *od.* **Umrissen** in broad outline; **im ~en [und] ganzen** by and large; on the whole; **l)** *(geh.: selbstlos)* noble ‹*deed etc.*›; **ein ~es Herz haben** be great-hearted; **m)** *(ugs.: ~spurig)* **~e Reden schwingen** *od. (salopp)* **Töne spukken** talk big *(coll.)*. **2.** *adv.* **a) ein Wort ~ schreiben** write a word with a capital [initial] letter *or* a capital; **jmdn. ~ ansehen** stare hard at sb.; **~ machen** *(Kindersp r.)* do number two *(child lang.)*; **~ und breit** at great length; **b)** *(ugs.: aufwendig)* **~ ausgehen** go out for a big celebration; **etw. ~ feiern** celebrate sth. in a big way; **c)** *(ugs.: besonders)* greatly; particularly; **d)** *(ugs.: ~artig)* **sie steht ganz ~ da** she has made it big *(coll.) or* made the big time *(coll.)*

groß-, Groß-: ~abnehmer der bulk buyer *or* purchaser; **~angelegt** *Adj.; nicht präd.* large-scale ‹*project, plan, etc.*›; full-scale ‹*attack, investigation*›; **~artig 1.** *Adj.* magnificent; splendid; wonderful ‹*person*›; **2.** *adv.* magnificently; splendidly; **~aufnahme die** *(Film)* close-up

Groß·britannien (das) the United Kingdom; [Great] Britain

Groß·buchstabe der capital [letter]; upper-case letter *(Printing)*

Größe ['grø:sə] **die; ~, ~n a)** size; *(Kleider~)* **in ~ 38** in size 38; **b)** *(Höhe, Körper~)* height; **der ~ nach** by height; **c)** *(Bedeutsamkeit, sittlicher Wert)* greatness; **d)** *(Genie)* outstanding *or* important figure; **e)** *(Math., Physik)* quantity

Groß-: ~ein·kauf der bulk purchase; **~eltern** *Pl.* grandparents; **~enkel der** great-grandchild; *(Junge)* great-grandson; **~enkelin die** great-granddaughter

größen-, Größen-: ~ordnung die order [of magnitude]; **~verhältnis das a)** *(Maßstab)* scale; **b)** *(Proportion)* proportions *pl.*; **~wahn der** *(abwertend)* megalomania; delusions *pl.* of grandeur; **~wahnsinnig** *Adj.* megalomaniacal

größer *s.* **groß**

groß-, Groß-: ~fahndung die large-scale search *or* manhunt; **~grund·besitzer der** big landowner; **~handel der** wholesale trade; **~händler der** wholesaler; **~handlung die** wholesale business; **~herzig** *(geh.)* **1.** *Adj.* magnanimous; **2.** *adv.* magnanimously; **~herzog der** Grand Duke; **~herzogin die** Grand Duchess; **~herzogtum das** grand duchy; **~herzogtum Luxemburg** Grand Duchy of Luxembourg; **~hirn das** *(Anat.)* cerebrum; **~industrielle der/die;** *adj. Dekl.* big industrialist

Grossist der; ~en, ~en *(Kaufmannsspr.)* wholesaler

groß-, Groß-: ~kapital das *(Wirtsch.)* big business *or* capital; **~konzern der** big *or* large combine; **~kotzig** [~kɔtsɪç] **1.** *Adj. (salopp abwertend)* pretentious ‹*style*›; swanky *(coll.)* ‹*present etc.*›; **2.** *adv.* boastfully; **~kundgebung die** mass rally *or* meeting; **~macht die** great power; **~mama die** *(ugs.)* grandma *(coll./child lang.)*; granny *(coll./child lang.)*; **~markt der** central market; **~mast der** *(Seemannsspr.)* mainmast; **~maul das** *(ugs. abwertend)* big-mouth *(coll.)*; braggart; **~mäulig** [~mɔylɪç] *Adj. (ugs. abwertend)* big-mouthed *(coll.)*; **~meister der** Grand Master; **~mut die; ~:** magnanimity; generosity; **~mütig** [~my:tɪç] **1.** *Adj.* magnanimous; generous; **2.** *adv.* magnanimously; generously; **~mutter die a)** grandmother; **das kannst du deiner ~mutter erzählen** *(ugs.)* tell that to the marines; **b)** *(ugs.: alte Frau)* old lady; **~neffe der** great-nephew; grandnephew; **~nichte die** great-niece; grandniece; **~onkel der** great-uncle; granduncle; **~papa der** *(ugs.)* grandpa *(coll./child lang.)*; granddad *(coll./child lang.)*; **~raum der** area; **im ~raum Hamburg** in the [Greater] Hamburg area; **~raum·büro das** open-plan office

großräumig [-rɔymɪç] **1.** *Adj.* extensive; over a wide *or* large area *postpos., not pred.*; *(viel Platz bietend)* spacious, roomy ‹*office, house, etc.*›. **2.** *adv.* over a wide *or* large area

groß-, Groß-: ~reinemachen das *(ugs.)* thorough cleaning; spring-clean; **~|schreiben** *unr. tr. V.* **~geschrieben werden** *(fig. ugs.)* be stressed *or* emphasized; *s. auch* **groß 2a; ~segel das** *(Seemannsspr.)* mainsail; **~sprecherisch** *Adj. (abwertend)* boastful; **~spurig** *(abwertend)* **1.** *Adj.* boastful; *(hochtrabend)* pretentious ‹*word, language*›; grandiose ‹*plan*›; **2.** *adv.* boastfully; *(hochtrabend)* pretentiously; **~stadt die** city; large town; **~städter der** urbanite; city-dweller; **~städtisch** *Adj.* [big-]city *attrib.* ‹*life*›

größt... *s.* **groß**

Groß-: ~tante die great-aunt; grandaunt; **~tat die** *(geh.)* great feat; **~teil der** large part; *(Hauptteil)* major part; **zum ~teil** mostly; for the most part

größten·teils *Adv.* for the most part

größt·möglich *Adj.; nicht präd.* greatest possible

groß-, Groß-: ~|tun *unr. itr. V.* boast; brag; **~unternehmen das** *(Wirtsch.)* large-scale enterprise; big concern; **~vater der** grandfather; **~verdiener der** big earner; **~wild das** big game; **~wild·jagd die** big-game hunting *no art.*; **~|ziehen** *unr. tr. V.* bring up; raise; rear ⟨*animal*⟩; **~zügig 1.** *Adj.* generous; generous, handsome ⟨*tip*⟩; **b)** *(in großem Stil)* grand and spacious ⟨*building, gardens, etc.*⟩; generous, liberal ⟨*working conditions*⟩; large-scale ⟨*measures*⟩; **2.** *adv.* **a)** generously; **sich ~zügig über etw.** *(Akk.)* **hinwegsetzen** be broad-minded enough to disregard sth.; **b)** *(in großem Stil)* **ein ~zügig eingerichtetes Büro** a handsomely equipped office; **~zügigkeit die a)** generosity; **b)** *(großes Ausmaß)* grand scale

grotesk [gro'tɛsk] **1.** *Adj.* grotesque. **2.** *adv.* grotesquely

Groteske die; ~, ~n a) *(Ornamentik)* grotesque; **b)** *(Literaturwiss.)* grotesque tale

Grotte ['grɔtə] **die; ~, ~n** grotto

grub [gru:p] *1. u. 3. Pers. Sg. Prät. v.* **graben**

Grübchen ['gry:pçən] **das; ~s, ~:** dimple

Grube die; ~, ~n a) pit; hole; **wer andern eine ~ gräbt, fällt selbst hinein** *(Spr.)* take care that you are not hoist with your own petard; **b)** *(Bergbau)* mine; pit; **c)** *(veralt.: offenes Grab)* grave; **in die ~ fahren** *(veralt.)* yield up the ghost *(arch.)*

Grübelei die; ~, ~en pondering

grübeln ['gry:bl̩n] *itr. V.* ponder **(über** + *Dat.* on, over)

Gruben-: ~arbeiter der miner; mineworker; **~unglück das** pit *or* mine disaster

Grübler der; ~s, ~meditative person

grüezi ['gry:ɛtsi] *Adv. (schweiz.)* hallo

Gruft [grʊft] **die; ~, Grüfte** ['grʏftə] vault; *(in einer Kirche)* crypt

grummeln ['grʊml̩n] *itr. V.* **a)** *(dröhnen)* rumble; **b)** *(murmeln)* mumble

grün [gry:n] *Adj.* **a)** green; **~er Salat** lettuce; **die Ampel ist ~** *(ugs.)* the lights are green; **die Grüne Insel** the Emerald Isle; **~e Bohnen/Erbsen** French beans/green peas; **~e Heringe** fresh herrings; **ein ~er Junge** *(abwertend)* a greenhorn; **~es Licht geben** give the go-ahead; **jmdn. ~ und blau** *od.* **gelb schlagen** *(ugs.)* beat sb. black and blue; **sich ~ und blau** *od.* **gelb ärgern** *(ugs.)* be livid *(coll.) or* furious; **b)** *(ugs.: wohlgesinnt)* **ich bin ihr nicht ~:** she's not someone I care for; **c)** *(Politik)* Green; *s. auch* [2]**Grüne**

Grün das; ~s, ~ *od. (ugs.)* **~s a)** green; **die Ampel zeigt ~:** the lights *pl.* are at green; **das ist dasselbe in ~** *(ugs.)* it makes *or* there is no real difference; **b)** *o. Pl. (Pflanzen)* greenery; **c)** *(Golf)* green

Grün·anlage die green space; *(Park)* park

grün·blau *Adj.* greenish blue

Grund [grʊnt] **der; ~[e]s, Gründe** ['grʏndə] **a)** *(Erdoberfläche)* ground; **etw. bis auf den ~ abreißen** raze sth. to the ground; **sich in ~ und Boden schämen** be utterly ashamed; **b)** *o. Pl. (eines Gewässers, geh.: eines Gefäßes)* bottom; **auf ~ laufen** run aground; **im ~e seines Herzens/seiner Seele** *(fig. geh.)* at heart *or* deep down/in his innermost soul; **der Sache** *(Dat.)* **auf den ~ gehen/kommen** get to the bottom *or* root of the matter; **im ~e [genommen]** basically; **c)** *(Ursache, Veranlassung)* reason; *(Beweg~)* grounds *pl.*; reason; **[k]einen ~ zum Feiern/Klagen haben** have [no] cause for [a] celebration/to complain *or* for complaint; **aus dem einfachen ~, weil ...** *(ugs.)* for the simple reason that ...; **ohne ersichtlichen ~:** for no obvious *or* apparent reason; **d) Gründe und Gegengründe** pros and cons; arguments for and against; **e)** *(Land)* land; **~ und Boden** land; **f) auf ~** *s.* **aufgrund**

grund-, Grund-: ~ausbildung die *(Milit.)* basic training; **~ausstattung die** basic equipment; **~besitz der a)** *(Eigentum an Land)* ownership of land; **b)** *(Land)* land; landed property; **~besitzer der** landowner; **~bestandteil der** [basic] element; **~buch das** land register; **~ehrlich** *Adj.* thoroughly honest; **~einheit die** fundamental unit

gründen ['grʏndn̩] **1.** *tr. V.* **a)** *(neu schaffen)* found, set up, establish ⟨*organization, party, etc.*⟩; set up, establish ⟨*business*⟩; start [up] ⟨*club*⟩; **eine Familie/ein Heim ~:** start a family/set up home; **b)** *(aufbauen)* base ⟨*plan, theory, etc.*⟩ **(auf** + *Akk.* on). **2.** *refl. V.* **sich auf etw.** *(Akk.)* **~:** be based on sth.

Gründer der; ~s, ~, Gründerin die; ~, ~nen: founder

grund-, Grund-: ~falsch *Adj.* utterly wrong; **~festen** *Pl.* **in etw. in seinen** *od.* **bis in seine ~festen erschüttern** shake sth. to its [very] foundations; **~fläche die a)** *(eines Zimmers)* [floor] area; **b)** *(Math.)* base; **~form die a)** *(Hauptform)* basic form; **b)** *(Urform)* original form; **c)** *(Sprachw.)* infinitive; **~gebühr die** basic *or* standing charge; **~gedanke der** basic idea; **~gesetz das a)** *(Verfassung)* Basic Law; **b)** *(wichtiges Gesetz)* fundamental *or* basic law

grundieren [grʊn'di:rən] *tr. V.* prime; *(Ölmalerei)* ground; apply the ground to

Grundierung die; ~, ~en *(erster Anstrich)* priming coat; *(Ölmalerei)* ground coat

Grund-: ~kapital das *(Wirtsch.)* equity *or* share capital; **~kenntnis die;** *meist Pl.* basic knowledge *no pl.* **(in** + *Dat.* of); **~kurs der** basic course

Grund·lage die basis; foundation; **auf der ~:** on the basis; **jeder ~ entbehren** be completely unfounded *or* without any foundation

grund·legend 1. *Adj.* fundamental, basic **(für** to); seminal ⟨*idea, work*⟩. **2.** *adv.* fundamentally

gründlich ['grʏntlɪç] **1.** *Adj.* thorough. **2.** *adv.* **a)** *(gewissenhaft)* thoroughly; **b)** *(ugs.: gehörig)* **sich ~ täuschen** be sadly *or* greatly mistaken; **~ mit etw. aufräumen** do away completely with sth.

Gründlichkeit die; ~: thoroughness

grund-, Grund-: ~linie die a) *(Math.)* base; **b)** *(Sport)* baseline; **c)** *(Hauptzug)* main *or* principal feature *or* characteristic; **~los 1.** *Adj.* **a)** *(unbegründet)* groundless; unfounded; **b)** *(ohne festen Boden)* bottomless ⟨*sea, depths, etc.*⟩; **2.** *adv.* **~los lachen** laugh for no reason [at all]; **jmdn. ~los verdächtigen** be suspicious of sb. without reason; **~mauer die** foundation wall; **das Haus war bis auf die ~mauern abgebrannt** the house had burnt to the ground; **~nahrungsmittel das** basic food[stuff]

Grün·donnerstag der Maundy Thursday

Grund-: ~prinzip das fundamental *or* basic principle; **~recht das** basic *or* fundamental *or* constitutional right; **~regel die** fundamental *or* basic rule; **~riß der a)** *(Bauw.)* [ground-] plan; **b)** *(Leitfaden)* outline; **~satz der** principle; **aus ~satz** on principle

grund·sätzlich 1. *Adj.* **a)** fundamental ‹*difference, question, etc.*›; **b)** *(aus Prinzip)* ‹*rejection, opponent, etc.*› on principle; **c)** *(allgemein)* ‹*agreement, readiness, etc.*› in principle. **2.** *adv.* **a)** fundamentally; **zu etw. ~ Stellung nehmen** make a statement of principle on sth.; **b)** *(aus Prinzip)* as a matter of principle; on principle; **c)** *(allgemein)* in principle

Grund-: ~schule die primary school; **~schüler der** primary school pupil; **~schul·lehrer der** primary-school teacher; **~stein der** foundation-stone; **~stein·legung die; ~, ~en** laying of the foundation-stone; **~stellung die** *(Sport)* basic position; **~steuer die** *(Steuerw.)* property tax [under German law]; **~stock der** basis; foundation; **~stoff der a)** *(Chemie: Element)* element; **b)** *(Rohstoff)* [basic] raw material

Grund·stück das piece of land; *(Bau~)* plot of land; *(größer)* site; **jmds. ~ betreten** enter sb.'s property

Grundstücks·makler der estate agent

Grund-: ~studium das basic course; **~tendenz die** basic trend; **~ton der a)** *(Farbton)* basic colour; **b)** *(~stimmung)* basic *or* prevailing tone *or* mood; **c)** *(Musik)* fundamental [tone]; root; **~übel das** basic evil

Gründung die; ~, ~en *(Partei~, Vereins~)* foundation; establishment; setting up; *(Geschäfts~)* setting up; establishing; *(Klub~)* starting [up]

grund-, Grund-: ~verkehrt *Adj.* completely *or* entirely wrong; **~verschieden** *Adj.* totally *or* completely different; **~wasser das** *(Geol.)* ground water; **~wasser·spiegel der** water table; ground-water level; **~wehr·dienst der** basic military service; national service; **~zug der** essential feature

[1]**Grüne das;** *adj. Dekl.* **a)** green; **b) im ~n/ins ~:** [out] in/into the country; **c)** *(ugs.) s.* **Grün b**

[2]**Grüne der/die;** *adj. Dekl. (Politik)* member of the Green Party; **die ~n** the Greens

grünen *itr. V. (geh.)* be green; *(grün werden)* turn green

grün-, Grün-: ~fink der greenfinch; **~fläche die** green space; *(im Park)* lawn; **~gelb** *Adj.* greenish yellow; **~gürtel der** green belt; **~kohl der** curly kale

grünlich *Adj.* greenish

Grün-: ~schnabel der *(abwertend)* [young] whippersnapper; *(Neuling)* greenhorn; **~span der** verdigris; **~streifen der** central reservation; centre strip *(grassed and often with trees and bushes)*

grunzen ['grʊntsn̩] *tr., itr. V.* grunt

Grün·zeug das *(ugs.) s.* [1]**Grüne c**

Gruppe ['grʊpə] **die; ~, ~n a)** *(auch fachspr.)* group; **b)** *(Kategorie, Klasse)* class; category

Gruppen-: ~arbeit die; *o. Pl.* group work; **~bild das** group photograph; **~reise die** *(Touristik)* group travel *no pl., no art.*; **eine ~reise nach London machen** travel to London with a group; **~sex der** group sex; **~sieg der** *(Sport)* top place in the group; **den ~sieg erreichen** win the group; **~sieger der** *(Sport)* winner of the/a group; **~therapie die** *(Psych.)* group therapy

gruppieren 1. *tr. V.* arrange. **2.** *refl. V.* form a group/groups

Gruppierung die; ~, ~en a) *(Personengruppe)* grouping; group; *(Politik)* faction; **b)** *(Anordnung)* arrangement; grouping

Grusel·geschichte die horror story

gruselig ['gru:zəlɪç] *Adj.* eerie; creepy; blood-curdling ‹*apparition, scream*›; spine-chilling ‹*story, film*›

gruseln 1. *tr., itr. V. (unpers.)* **es gruselt jmdn.** *od.* **jmdm.** sb.'s flesh creeps. **2.** *refl. V.* be frightened; get the creeps *(coll.)*

Gruß [gru:s] **der; ~es, Grüße** ['gry:sə] **a)** greeting; *(Milit.)* salute; **bestell Barbara bitte viele Grüße von mir** please give Barbara my regards; please remember me to Barbara; **einen [schönen] ~ an jmdn./von jmdm.** [best] regards *pl.* to/from sb.; **b)** *(im Brief)* **mit herzlichen Grüßen** [with] best wishes; **viele liebe Grüße Euer Hans** love, Hans

grüßen ['gry:sn̩] **1.** *tr. V.* **a)** greet; *(Milit.)* salute; **grüß [dich] Gott!** *(südd.)* hello; **b)** *(Grüße senden)* **grüße deine Eltern [ganz herzlich] von mir** please give your parents my [kindest] regards; **jmdn. ~ lassen** send one's regards to sb.. **2.** *itr. V.* say hello; *(Milit.)* salute

gruß·los *Adv.* without a word of greeting/farewell

Grütze ['grʏtsə] **die; ~, ~n a)** groats *pl.*; **rote ~:** red fruit pudding *(made with fruit juice, fruit and cornflour, etc.)*; **b)** *o. Pl. (ugs.: Verstand)* brains *pl.*; nous *(coll.)*

G-Schlüssel ['ge:-] **der** *(Musik) s.* **Violinschlüssel**

Guatemala [guate'ma:la] **(das); ~s** Guatemala

Guatemalteke [guatemal'te:kə] **der; ~n, ~n** Guatemalan

gucken ['gʊkn̩] **1.** *itr. V. (ugs.)* **a)** look; *(heimlich)* peep; **jmdm. über die Schulter ~:** look *or* peer over sb.'s shoulder; **laß [mich] mal ~!** let's have a look! *(coll.)*; **b)** *(hervorsehen)* stick out; **c)** *(dreinschauen)* look; **finster/freundlich ~:** look grim/affable. **2.** *tr. V. (ugs.)* **Fernsehen ~:** watch TV *or (coll.)* telly *or (coll.)* the box

Guck·loch das spy-hole; peephole

Guerilla [ge'rɪlja] **die; ~, ~s a)** *(Krieg)* guerrilla war; **b)** *(Einheit)* guerrilla unit

Guerilla-: ~kämpfer der guerrilla; **~krieg der** guerrilla war

Gugel·hupf [-hʊpf] **der; ~[e]s, ~e** *(südd., österr.)* gugelhupf

Guillotine [gijo'ti:nə] **die; ~, ~n** guillotine

Guinea [gi'ne:a] **(das); ~s** Guinea

Gulasch ['gʊlaʃ, 'gu:laʃ] **das** *od.* **der; ~[e]s, ~e** *od.* **~s** goulash

Gulasch·suppe die goulash soup

Gulden ['gʊldn̩] **der; ~s, ~:** guilder; florin

Gully ['gʊli] **der; ~s, ~s** drain

gültig ['gʏltɪç] *Adj.* valid; current ‹*note, coin*›; **diese Münze/dieser Geldschein ist nicht mehr ~:** this coin/note is no longer legal tender; **der Fahrplan ist ab 1. Oktober ~:** the timetable comes into operation on 1 October

Gültigkeit die; ~: validity; **~ haben/erlangen** be/become valid; **die ~ verlieren** become invalid

[1]**Gummi** ['gʊmi] **der** *od.* **das; ~s, ~[s] a)** [india] rubber; **b)** *(~ring)* rubber *or* elastic band

[2]**Gummi der; ~s, ~s a)** *(Radier~)* rubber; eraserrubber *(sl.)*; **b)** *(salopp: Präservativ)* rubber *(sl.)*

[3]**Gummi das; ~s, ~s** *(~band)* elastic *no indef. art.*

gummi-, Gummi-: ~artig *Adj.* rubbery; rubber-like ‹*material*›; **~ball der** rubber ball; **~band das;** *Pl.* **~bänder a)** rubber *or* elastic band; **b)** *(in Kleidung)* elastic *no indef. art.*; **~bär der, ~bärchen das** jelly baby; **~baum der** rubber plant; **~bonbon das** gumdrop

gummieren *tr. V.* gum
Gummierung die; ~, ~en gum
Gummi-: ~hand·schuh der rubber glove; **~knüppel der** [rubber] truncheon; **~paragraph der** *(ugs.)* paragraph *or* section with an elastic interpretation; **~reifen der** rubber tyre; **~ring der a)** rubber band; **b)** *(Spielzeug)* rubber ring; quoit; **c)** *(Weckglasring)* rubber seal; **~sohle die** rubber sole; **~stiefel der** rubber boot; *(für Regenwetter)* wellington [boot] *(Brit.)*; **~zelle die** padded cell
Gunst [gʊnst] **die; ~:** favour; goodwill; **die ~ der Stunde nutzen** *(fig.)* take advantage of the favourable *or* propitious moment; **zu jmds. ~en** in sb.'s favour
günstig ['gʏnstɪç] **1.** *Adj.* **a)** *(vorteilhaft)* favourable; propitious ⟨*sign*⟩; auspicious ⟨*moment*⟩; beneficial ⟨*influence*⟩; good, reasonable ⟨*price*⟩; **bei ~em Wetter** if the weather is favourable; weather permitting; **b)** *(wohlwollend)* well-disposed; favourably disposed. **2.** *adv.* **a)** *(vorteilhaft)* favourably; **etw. ~ beeinflussen** have *or* exert a beneficial influence on sth.; **etw. ~ kaufen/verkaufen** buy/sell sth. at a good price; **b)** *(wohlwollend)* **jmdn./etw. ~ aufnehmen** receive sb./sth. well *or* favourably
günstig[st]en·falls *Adv.* at best
Günstling ['gʏnstlɪŋ] **der; ~s, ~e** favourite
Guppy ['gʊpi] **der; ~s, ~s** *(Zool.)* guppy
Gurgel ['gʊrgl̩] **die; ~, ~n** throat; **jmdm. die ~ zudrücken** strangle *or* throttle sb.; **jmdm. die ~ durchschneiden** cut sb.'s throat; **jmdm. an die ~ wollen** fly at sb.
gurgeln *itr. V.* **a)** *(spülen)* gargle; **b)** *(blubbern)* gurgle
Gürkchen ['gʏrkçən] **das; ~s, ~:** [cocktail] gherkin
Gurke ['gʊrkə] **die; ~, ~n a)** cucumber; *(eingelegt)* gherkin; **saure ~n** pickled gherkins; **b)** *(salopp: Nase)* hooter *(sl.)*; snout *(coll.)*
Gurken-: ~hobel der cucumber-slicer; **~salat der** cucumber salad; **~truppe die** *(salopp)* useless *or* feeble bunch *(coll.)*
gurren ['gʊrən] *itr. V.* coo
Gurt [gʊrt] **der; ~[e]s, ~e** strap; *(im Auto, Flugzeug)* [seat-]belt
Gürtel ['gʏrtl̩] **der; ~s, ~:** belt; **den ~ enger schnallen** *(fig. ugs.)* tighten one's belt *(fig.)*
Gürtel-: ~linie die waist[line]; **ein Schlag unter die ~linie** *(Boxen)* a punch *or* blow below the belt; **~reifen der** radial[-ply] tyre; **~schnalle die** belt buckle; **~tier das** armadillo
gürten ['gʏrtn̩] *(geh. veralt.)* **1.** *tr. V.* gird *(arch./literary)*. **2.** *refl. V.* **sich [zum Kampf] ~:** gird oneself
Guru ['gʊru] **der; ~s, ~s** guru
Guß [gʊs] **der; Gusses, Güsse** ['gʏsə] **a)** *(das Gießen)* casting; founding; **[wie] aus einem ~:** forming a unified *or* an integrated whole; fully co-ordinated ⟨*plan*⟩; **b)** *(ugs.: Regenschauer)* downpour; **c)** *(gegossenes Erzeugnis)* casting; cast; **d)** *(das Begießen)* stream
guß-, Guß-: ~eisen das cast iron; **~eisern** *Adj.* cast-iron; **~form die** casting mould
Gusto ['gʊsto] **der; ~s, ~s** taste; liking
gut [gu:t]; **besser** ['bɛsɐ], **best...** ['bɛst...] **1.** *Adj.* **a)** good; fine ⟨*wine*⟩; **in Französisch ~ sein** be good at French; **ist der Kuchen ~ geworden?** did the cake turn out all right?; **es wäre ~, wenn ...:** it would be as well if ...; **also ~:** very well; all right; **schon ~:** [it's] all right *or (coll.)* OK; **das ist ja alles ~ und schön** that's all very well *or* all well and good; **etwas Gutes zu essen/trinken** something good to eat/drink; **es ~ sein lassen** *(ugs.)* leave it at that; **das ist ~ gegen** *od.* **für Kopfschmerzen** it's good for headaches; **~en Tag!** good morning/afternoon!; **~en Morgen!** good morning!; **~en Abend!** good evening!; **~e Nacht!** good night!; **ein ~es neues Jahr** a happy new year; **er hat es doch ~ bei uns** he's well enough off with us; **mir ist nicht ~:** I'm not feeling well; I don't feel well; **alles Gute!** all the best!; **~en Appetit!** enjoy your lunch/dinner *etc.!*; **es dürfte eine ~e Stunde [von hier] sein** it must be a good hour [from here]; **sich** *(Dat.)* **zu ~ für etw. sein** consider sth. beneath one *or* beneath one's dignity; **du bist ~!** *(iron.)* you're joking!; you must be joking!; **sei [bitte] so ~ und reich mir das Buch** would you be good *or* kind enough to pass me the book?; **im ~en auseinandergehen** part amicably *or* on amicable terms; **b)** *(besonderen Anlässen vorbehalten)* best; **sein ~er Anzug** his best suit. **2.** *adv.* **a)** well; **~ reiten/schwimmen** be a good rider/swimmer; **etw. ~ können** be good at sth.; **seine Sache ~ machen** do well; **~ hören/sehen** [be able to] hear/see well *or* clearly; **[das hast du] ~ gemacht!** well done!; **der Laden/das Geschäft geht ~:** the shop/business is doing well; **~ zwei Pfund wiegen** weigh a good two pounds; **~ und gern** *(ugs.)* easily; at least *(ugs.)*; **so ~ wie nichts** next to nothing; **so ~ ich kann** as best I can; **jmdm. ~ zureden** coax sb. [gently]; **es ~ meinen** mean well; **b)** *(mühelos)* easily; **~ zu Fuß sein** *(ugs.)* be a strong walker; **hinterher hat** *od.* **kann man ~ reden** it's easy to be wise after the event; **du hast ~ lachen** it's all right for you to laugh; **es kann ~ sein, daß ...:** it may well be that ...; *s. auch* **besser, best...**
Gut das; ~[e]s, Güter ['gy:tɐ] **a)** *(Eigentum)* property; *(Besitztum, auch fig.)* possession; **irdische Güter** earthly goods *or* possessions; **unrecht ~ gedeihet nicht** *(Spr.)* ill-gotten goods *or* gains never *or* seldom prosper; **b)** *(landwirtschaftlicher Grundbesitz)* estate; **c)** *(Fracht~, Ware)* item; **Güter** goods; *(Fracht~)* freight *sing.*; goods *(Brit.)*; **d)** *(das Gute)* **~ und Böse** good and evil; **jenseits von ~ und Böse sein** *(iron.)* be past it *(coll.)*
gut-, Gut-: ~achten das; ~s, ~: [expert's] report; **~achter der; ~s, ~:** expert; *(in einem Prozeß)* expert witness; **~artig** *Adj.* **a)** good-natured; **b)** *(nicht gefährlich)* benign; **~aussehend** *Adj.; nicht präd.* good-looking; **~bezahlt** *Adj.; nicht präd.* well-paid; **~bürgerlich** *Adj.* good middle-class; **~bürgerliche Küche** good plain cooking; **~dünken das; ~s** discretion; judgement; **nach [eigenem] ~dünken** at one's own discretion
Güte ['gy:tə] **die; ~ a)** goodness; kindness; *(~ Gottes)* loving-kindness; goodness; **ein Vorschlag zur ~:** a suggestion for an amicable agreement; **[ach] du meine** *od.* **liebe ~!** *(ugs.)* my goodness!; goodness me; **b)** *(Qualität)* quality
Gute·nacht·kuß der good-night kiss
Güter-: ~abfertigung die a) dispatch of freight *or (Brit.)* goods; **b)** *(Annahmestelle)* freight *or (Brit.)* goods office; **~bahnhof der** freight depot; goods station *(Brit.)*; **~trennung die** *(Rechtsw.)* separation of property; **~wagen der** goods wagon *(Brit.)*; freight car *(Amer.)*; **~zug der** goods train *(Brit.)*; freight train *(Amer.)*

gut-, Gut-: ~|**gehen** *unr. itr. V.; mit sein* **a)** *(unpers.)* **es geht jmdm. ~** *(gesundheitlich)* sb. is well *or (coll.)* fine; *(geschäftlich, beruflich)* sb. is doing well; **b)** *(~ ausgehen)* turn out well; **es ist noch einmal ~gegangen** it worked out all right again this time; **~gehend** *Adj.; nicht präd.* flourishing; thriving; **~gelaunt** *Adj. (präd. getrennt geschrieben)* good-humoured; cheerful; **~gemeint** *Adj. (präd. getrennt geschrieben)* well-meant; **~gläubig** *Adj.* innocently trusting; **~gläubigkeit die** innocent trust; ~|**haben** *unr. tr. V.* **etw. bei jmdm. ~haben** be owed sth. by sb.; **~haben das; ~s, ~:** credit balance; **Sie haben ein ~haben von 450 DM auf Ihrem Konto** your account is 450 marks in credit; ~|**heißen** *unr. tr. V.* approve of; **~herzig** *Adj.* kind-hearted; good-hearted

gütig ['gy:tɪç] **1.** *Adj.* kindly; kind ⟨*heart*⟩. **2.** *adv.* **~ lächeln/nicken** give a kindly smile/nod

gütlich ['gy:tlɪç] **1.** *Adj.; nicht präd.* amicable. **2.** *adv.* amicably; **sich ~ an etw.** *(Dat.)* **tun** regale oneself with sth.

gut-, Gut-: ~|**machen** *tr. V.* **a)** *(in Ordnung bringen)* make good ⟨*damage*⟩; put right, correct ⟨*omission, mistake, etc.*⟩; **b)** *(Überschuß erzielen)* make [a profit of] **(bei** on); **~mütig 1.** *Adj.* good-natured; **2.** *adv.* good-naturedly; **~mütigkeit die; ~:** good nature; goodnaturedness; **~nachbarlich 1.** *Adj.* good-neigbourly ⟨*relations etc.*⟩; **2.** *adv.* as good neighbours

Guts·besitzer der owner of a/the estate; landowner

gut-, Gut-: ~schein der voucher, coupon **(für, auf +** *Akk.* for); ~|**schreiben** *unr. tr. V.* credit; **etw. jmdm./jmds. Konto ~schreiben** credit sb./sb.'s account with sth.; **~schrift die** credit

Guts-: ~herr der lord of the manor; **~hof der** estate; manor; **~verwalter der** steward; bailiff

gut-, Gut-: ~|**tun** *unr. itr. V.* do good; **~willig 1.** *Adj.* willing; *(entgegenkommend)* obliging; **sich ~willig zeigen** be obliging; show willing *(coll.)*; **2.** *adv.* **etw. ~willig herausgeben** hand sth. over voluntarily; **~willigkeit die** willingness; *(Entgegenkommen)* obligingness

Gymnasiast [gʏmna'zi̯ast] **der; ~en, ~en, Gymnasiastin die; ~, ~nen** ≈ grammar-school pupil

Gymnasium [gʏm'na:zi̯ʊm] **das; ~s, Gymnasien** ≈ grammar school; **aufs ~ gehen** ≈ be at *or* attend grammar school

Gymnastik [gʏm'nastɪk] **die; ~:** physical exercises *pl.; (Turnen)* gymnastics *sing.*

gymnastisch *Adj.* gymnastic

Gynäkologe [gʏnɛko'lo:gə] **der; ~n, ~n** gynaecologist

Gynäkologie die; ~: gynaecology *no art.*

Gynäkologin die; ~, ~nen gynaecologist

gynäkologisch *Adj.; nicht präd.* gynaecological

H

h, H [ha:] **das; ~, ~ a)** *(Buchstabe)* h/H; **b)** *(Musik)* [key of] B; *s. auch* **a, A**

h *Abk.* **a) Uhr** hrs; **b) Stunde** hr[s]

H *Abk.* **a) Herren; b) Haltestelle**

[1]**ha** [ha(:)] *Interj.* ha!; oh!; ah!; *(Triumph)* aha!

[2]**ha** *Abk.* **Hektar** ha

Haag [ha:k] **(das)** *od.* **der; ~s** The Hague

Haar [ha:ɐ̯] **das; ~[e]s, ~e a)** *(auch Zool., Bot.)* hair; **blonde ~e** *od.* **blondes ~ haben** have fair hair; **[sich** *(Dat.)***] das ~** *od.* **die ~e waschen** wash one's hair; **sich** *(Dat.)* **das ~** *od.* **die ~e schneiden lassen** have *or* get one's hair cut; **ihm geht das ~ aus** he's losing his hair; **sich** *(Dat.)* **die ~e [aus]raufen** *(ugs.)* tear one's hair [out]; **b)** *(fig.)* **ihr stehen die ~e zu Berge** *od.* **sträuben sich die ~e** *(ugs.)* her hair stands on end; **ein ~ in der Suppe finden** *(ugs.)* find something to quibble about *or* find fault with; **kein gutes ~ an jmdm./etw. lassen** *(ugs.)* pull sb./sth. to pieces *(fig. coll.)*; **~e auf den Zähnen haben** *(ugs. scherzh.)* be a tough customer; **sich** *(Dat.)* **über** *od.* **wegen etw. keine grauen ~e wachsen lassen** not lose any sleep over sth.; not worry one's head about sth.; **er wird dir kein ~ krümmen** *(ugs.)* he won't harm a hair of your head; **das ist an den ~en herbeigezogen** *(ugs.)* that's far-fetched; **jmdm. aufs ~ gleichen** be the spitting image of sb.; **sich in die ~e kriegen** *(ugs.)* quarrel, squabble **(wegen** over); **sich** *(Dat.)* **in den ~en liegen** *(ugs.)* be at loggerheads; **um ein ~** *(ugs.)* very nearly

Haar-: ~ausfall der loss of hair; hair loss; **~bürste die** hairbrush; **~büschel das** tuft of hair

haaren *itr. V.* moult; lose *or* shed its hair

Haares·breite die *in* **um ~:** by a hair's breadth

haar-, Haar-: ~farbe die hair colour; **~fein** *Adj.* fine as a hair *postpos.*; **~festiger der** setting lotion; **~genau** *(ugs.)* **1.** *Adj.* exact; **2.** *adv.* exactly; **das stimmt ~genau** that is absolutely right

haarig *Adj.* **a)** *(behaart)* hairy; **b)** *(ugs.: heikel)* tricky

haar-, Haar-: ~klammer die hair-grip; **~klein 1.** *Adj.* minute; **2.** *adv.* in minute detail; **~klemme die** hair-grip; **~kranz der a)** fringe *or* circle of hair; **b)** *(Frisur)* chaplet [of plaited hair]; **~los** *Adj.* hairless; **~mode die** hairstyle; **~nadel die** hairpin; **~nadel·kurve die** hairpin bend; **~netz das** hair-net; **~scharf** *Adv.* **a)** *(sehr nah)* **die Kugel flog ~scharf an ihm vorbei** the bullet missed him by a hair's breadth; **b)** *(sehr genau)* with great precision; **~schleife die** bow; hair-ribbon; **~schnitt der** haircut; *(modisch)* hair-style; **~schopf der** mop *or* shock of hair; **~spalter der; ~s, ~** *(abwertend)* hair-splitter; **~spalterei die; ~** *(abwertend)* hair-splitting; **das ist doch ~spalterei** that's splitting hairs; **~spange die** hair-slide; **~spray der** *od.* **das** hair spray; **~sträubend** *Adj.* **a)** *(grauenhaft)* hair-raising; horrifying; **b)** *(empörend)* outrageous; shocking; **~teil das** hair-piece; **~wasch·mittel das** shampoo; **~wasser das;** *Pl.* **~wässer** hair lotion; **~wuchs der** hair growth; growth of hair; **einen spärlichen/starken ~wuchs haben** have little/a lot of hair; **~wuchs·mittel das** hair-restorer

Hab [ha:p] *in* **~ und Gut** *(geh.)* possessions *pl.*; belongings *pl.*

Habe ['ha:bə] **die; ~** *(geh.)* possessions *pl.*; belongings *pl.*

haben 1. *unr. tr. V.* **a)** have; have

got; **er hat nichts** *(ugs.)* he has nothing; **da hast du das Geld** there's the money; **ich habe Zeit/keine Zeit** I have [got] [the] time/I have [got] no time *or* I haven't [got] any time; **die Sache hat Zeit** it's not urgent; it can wait *(coll.)*; **heute ~ wir schönes Wetter/30°** the weather is fine/it's 30° today; **wann hast du Urlaub?** when is your holiday?; *s. auch* **Schuld b**; **b)** *(empfinden)* **Hunger/Durst ~:** be hungry/thirsty; **Sehnsucht nach etw. ~:** long for sth.; **Heimweh/Furcht ~:** be homesick/afraid; **Husten/Fieber/Schmerzen ~:** have [got] a cough/a temperature/have pain; **was hast du denn?** *(ugs.)* what's the matter?; what's wrong?; **ich kann das nicht ~** *(ugs.)* I can't stand it; **c)** *mit Adj. u. „es"* **es gut/schlecht/schwer/eilig ~:** have it good *(coll.)*/have a bad time [of it]/have a difficult *or* tough time/be in a hurry; **d)** *mit „zu" u. Inf.* **nichts zu essen/trinken ~:** have nothing to eat/drink; *(müssen)* **du hast zu gehorchen** you must obey; **etw. zu tun/erledigen ~:** have [got] sth. to do *or* that one must do; **er hat zu tun** he's busy; *(dürfen)* **er hat mir nichts zu befehlen** he has [got] no right to order me about; **e)** *(sich zusammensetzen aus)* **das Jahr hat 12 Monate** there are 12 months in a year; **ein Kilometer hat 1000 Meter** there are 1,000 metres in a kilometre; **diese Stadt hat 10000 Einwohner** this town has 10,000 inhabitants; **f)** *(bekommen)* have; **zu ~ sein** *(ugs.)* be unattached; **dafür ist er immer zu ~:** he's always game for that; **da hast du's** *(ugs.)* there you are; **g)** *(ugs.: in der Schule)* **morgen ~ wir Geschichte** we've got history tomorrow; **h)** *(ugs.: gebrauchen)* **man hat das nicht mehr** it is no longer in use/in fashion; **i)** *(ugs.: gefaßt ~)* have ‹*thief etc.*›; **jetzt hab' ich dich** now I've got you; **j)** *(bekommen ~)* **Nachricht von jmdm. ~:** have heard from sb.; **was** *(ugs.)*/**welche Note hast du diesmal in Physik?** what did you get in *or* for physics this time?; **k)** *(gefunden ~)* **ich hab's!** *(ugs.)* I've got it!; **das werden wir gleich ~** *(ugs.)* we'll soon find out; **l)** *(ugs.: repariert, beendet ~)* **das werden wir gleich ~:** we'll soon fix that; **m)** *mit Präp.* **wir ~ viele Bilder an der Wand [hängen]** we have quite a lot of pictures up; **etwas/nichts gegen jmdn.** *od.* **etw. ~:** have something/nothing against sb. *or* sth.; **etwas mit jmdm. ~** *(ugs.)* have a thing *or* something going with sb. *(coll.)*; **viel/wenig von jmdm. ~:** see a lot/little of sb.; **etw. von etw. ~:** get sth. out of sth.; **n)** *unpers. (bes. österr., südd.: vorhanden sein)* **es hat ...:** there is/are ... **2.** *refl. V.* **a)** *(ugs.)* **hab dich nicht so!** don't make *or* stop making such a fuss!; **b)** *(ugs.: sich erledigt ~)* **und damit hat es sich** *od.* **hat sich die Sache** then that's that; **hat sich was!** far from it! **3.** *Hilfsverb* have; **ich habe/hatte ihn eben gesehen** I have *or* I've/I had *or* I'd just seen him; **sie ~ gelacht** they laughed; **er hat das gewußt** he knew it; **das hättest du früher machen können** you could have done that earlier

Haben das; ~s, ~ *(Kaufmannsspr.)* credit; *s. auch* **Soll a**

Habe·nichts der; ~, ~e pauper

Haben-: ~seite die *(Kaufmannsspr.)* credit side; **~zinsen** *Pl.* interest *sing.* on deposits

Haber der; ~s *(südd., österr., schweiz.) s.* **Hafer**

Hab·gier die *(abwertend)* greed

hab·gierig 1. *Adj. (abwertend)* greedy. **2.** *adv.* greedily

habhaft *in* **jmds./einer Sache ~ werden** catch *or* apprehend sb./get hold of sth.

Habicht ['ha:bɪçt] **der; ~s, ~e** hawk

Habilitation [habilita'tsi̯o:n] **die; ~, ~en** habilitation *(qualification as a university lecturer)*

habilitieren [habili'ti:rən] *refl. V.* habilitate *(qualify as a university lecturer)*

Habsburger ['ha:psbʊrgɐ] **der; ~s, ~** *(hist.)* Habsburg

hab-, Hab-: ~seligkeiten *Pl.* [meagre] possessions *or* belongings; **~sucht die; ~** *(abwertend)* greed; avarice; **~süchtig** *(abwertend)* **1.** *Adj.* greedy; avaricious; **2.** *adv.* greedily; avariciously

hach [hax] *Interj.* oh!

Hachse ['haksə] **die; ~, ~n** *(südd.)* **a)** knuckle; **b)** *(ugs. scherzh.)* leg

Hack [hak] **das; ~s** *(ugs., bes. nordd.)* mince; minced meat

Hack-: ~beil das chopper; cleaver; **~braten der** *(Kochk.)* meat loaf

[1]**Hacke die; ~, ~n** hoe; *(Pickel)* pick[axe]

[2]**Hacke die; ~, ~n** *(bes. nordd. u. md.)* heel; **sich** *(Dat.)* **die ~n nach etw. ablaufen** wear oneself out running around looking for sth.

hacken 1. *itr. V.* **a)** *(mit der Hacke arbeiten)* hoe; **b) sich** *(Dat.)* **ins Bein ~:** cut one's leg [with an axe *etc.*]; **c)** *(picken)* peck. **2.** *tr. V.* **a)** *(mit der Hacke bearbeiten)* hoe ‹*garden, flower-bed, etc.*›; **b)** *(zerkleinern)* chop ‹*wood etc.*›; chop [up] ‹*meat, vegetables, etc.*›; **etw. in Stücke ~:** chop sth. up; **c) ein Loch in etw.** *(Akk.)* **~:** chop *or* hack a hole in sth.

Hacker der; ~s ~ *(DV-Jargon)* hacker

Hack-: ~fleisch das minced meat; mince; **aus jmdm. ~fleisch machen** *(fig. ugs.)* make mincemeat of sb.; **~klotz der** chopping-block

Häcksel ['hɛksl̩] **der** *od.* **das; ~s** *(Landw.)* chaff

Hader ['ha:dɐ] **der; ~s** *(geh.)* discord

hadern *itr. V. (geh.)* **mit etw. ~:** be at odds with sth.; **er haderte mit seinem Schicksal** he railed against his fate

Hafen ['ha:fn̩] **der; ~s, Häfen** harbour; port; **der Hamburger ~:** the port of Hamburg; **ein Schiff läuft den ~ an/aus dem ~ aus/in den ~ ein** a ship is putting into/leaving/entering port *or* harbour

Hafen-: ~anlagen *Pl.* docks; **~arbeiter der** dock-worker; docker; **~kneipe die** dockland pub *(Brit. coll.) or (Amer.)* bar; **~polizei die** port *or* dock police; **~rund·fahrt die** trip round the harbour; **~stadt die** port; **~viertel das** dock area; dockland *no art.*

Hafer ['ha:fɐ] **der; ~s** oats *pl.*; **jmdn. sticht der ~** *(ugs.)* sb. is feeling his oats

Hafer-: ~brei der porridge; **~flocken** *Pl.* rolled oats; porridge oats; **~grütze die a)** oat groats; **b)** *(Brei)* porridge; **~schleim der** gruel

Haff [haf] **das; ~[e]s, ~s** *od.* **~e** lagoon

Haft [haft] **die; ~:** custody; *(aus politischen Gründen)* detention; **jmdn. aus der ~ entlassen** release sb. from custody/detention; **jmdn. zu zwei Jahren ~ verurteilen** sentence sb. to two years in prison *or* two years' imprisonment

-haft *Adj., adv.* -like

Haft·anstalt die prison

haftbar *Adj.* [legally] liable; **jmdn. für etw. ~ machen** make *or* hold sb. [legally] liable for sth.

Haft-: ~befehl der *(Rechtsw.)* warrant [of arrest]; **~creme die** *(Pharm.)* fixative cream

[1]**haften** *itr. V.* **a)** *(festkleben)* stick; **an/auf etw.** *(Dat.)* **~:** stick to sth.; **b)** *(sich festsetzen)* ‹*smell, dirt, etc.*› cling (**an** + *Dat.* to); **an ihm haftet ein Makel** *(fig.)* he carries a stigma

[2]**haften** *itr. V. (Rechtsw., Wirtsch.)* be liable

haften|bleiben *unr. itr. V.; mit sein* stick (**an/auf** + *Dat.* to)

Häftling ['hɛftlɪŋ] **der; ~s, ~e** prisoner

Häftlings·kleidung die prison clothing

Haft·pflicht die a) liability (**für** for); **b)** *s.* **Haftpflichtversicherung**

Haftpflicht·versicherung die personal liability insurance; *(für Autofahrer)* third party insurance

haft-, Haft-: ~reibung die *(Physik)* static friction; **~richter der** *(Rechtsw.)* magistrate; **~schale die;** *meist Pl.* contact lens; **~strafe die** *(Rechtsspr. veralt.)* prison sentence; **~unfähig** *Adj.* unfit to be kept in prison *postpos.*

[1]**Haftung die; ~:** adhesion; *(von Reifen)* grip

[2]**Haftung die; ~, ~en a)** *(Verantwortlichkeit)* liability; responsibility; *s. auch* **Garderobe; b)** *(Rechtsw., Wirtsch.)* liability; **Gesellschaft mit [un]beschränkter ~:** [un]limited [liability] company

Hagebutte ['ha:gəbʊtə] **die; ~, ~n a)** *(Frucht)* rose-hip; **b)** *(ugs.: Heckenrose)* dog-rose

Hagel ['ha:gl̩] **der; ~s, ~** *(auch fig.)* hail

Hagel·korn das hailstone

hageln *itr., tr. V. (unpers.)* hail; **es hagelt** it is hailing; **es hagelte Steine und leere Bierdosen** *(fig.)* there was a hail of stones and empty beer-cans

Hagel-: ~schauer der [short] hailstorm; **~schlag der** hail

hager ['ha:gɐ] *Adj.* gaunt ‹*person, figure, face*›; thin ‹*neck, arm, fingers*›

Häher ['hɛ:ɐ] **der; ~s, ~:** jay

[1]**Hahn** [ha:n] **der; ~[e]s, Hähne** ['hɛ:nə] **a)** cock; **~ im Korb sein** *(ugs.)* be cock of the walk; **nach ihr/danach kräht kein ~** *(ugs.)* no one could care less about her/it; **b)** *(Wetter~)* weathercock

[2]**Hahn der; ~[e]s, Hähne,** *fachspr.:* **~en a)** tap; faucet *(Amer.)*; **b)** *(bei Waffen)* hammer; **den ~ spannen** cock a/the gun

Hähnchen ['hɛ:nçən] **das; ~s, ~:** chicken

Hahnen-: ~fuß der buttercup; **~kampf der** cock-fighting; *(einzelner Wettkampf)* cock-fight; **~schrei der** cock-crow; **beim ersten ~schrei** at cock-crow; **~tritt·muster das** dog-tooth *or* dog's tooth check

Hai [hai] **der; ~s, ~e** shark

Hai·fisch der shark

Haifisch·flossen·suppe die *(Kochk.)* shark-fin soup

Hain [hain] **der; ~[e]s, ~e** *(dichter. veralt.)* grove

Hain·buche die hornbeam

Haiti [ha'i:ti] **(das); ~s** Haiti

haitianisch *Adj.* Haitian

Häkchen ['hɛ:kçən] **das; ~s, ~ a)** [small] hook; **b)** *(Zeichen)* mark; *(beim Abhaken)* tick

Häkel·garn das crochet thread *or* yarn

häkeln ['hɛ:kl̩n] *tr., itr. V.* crochet

Häkel·nadel die crochet-hook

haken ['ha:kn̩] **1.** *tr. V.* hook (**an** + *Akk.* on to). **2.** *itr. V. (klemmen)* be stuck

Haken der; ~s, ~ a) hook; **~ und Öse** hook and eye; **einen ~ schlagen** dart sideways; **b)** *(Zeichen)* tick; **c)** *(ugs.: Schwierigkeit)* catch; snag; **der ~ an etw.** *(Dat.)* the catch in sth.; **d)** *(Boxen)* hook

haken-, Haken-: ~förmig *Adj.* hooked; hook-shaped; **~kreuz das** swastika; **~nase die** hooked nose; hook-nose

halb [halp] **1.** *Adj. u. Bruchz.* **a)** half; **eine ~e Stunde/ein ~er Meter/ein ~es Glas** half an hour/a metre/a glass; **zum ~en Preis** [at] half price; **~ Europa/die ~e Welt** half of Europe/half the world; **es ist ~ eins** it's half past twelve; **5 Minuten vor/nach ~:** 25 [minutes] past/to; *s. auch* **Weg d; b)** *(unvollständig, vermindert)* **die ~e Wahrheit** half [of] *or* part of the truth; **nichts Halbes und nichts Ganzes [sein]** [be] neither one thing nor the other ; **c)** *(fast)* **[noch] ein ~es Kind sein** be hardly *or* scarcely more than a child; **die ~e Stadt** half the town. **2.** *adv.* **a) ~ voll/leer** half-full/-empty; **~ lachend, ~ weinend** half laughing, half crying; **b)** *(unvollständig)* **~ gar/angezogen** half-done *or* -cooked/half dressed; **c)** *(fast)* **~ blind/verhungert/tot** half blind/starved/dead; **~ und ~** *(ugs.)* more or less

halb-, Halb-: ~amtlich *Adj.* semi-official; **~bildung die** *(abwertend)* superficial education; **~bitter** *Adj.* plain ‹*chocolate*›; **~blut das a)** *(bei Pferden)* cross-breed; **b)** *(Mischling)* half-caste; half-breed; **~bruder der** half-brother; **~dunkel das** semi-darkness

Halbe die *od.* **das;** *adj. Dekl. (ugs.)* half litre *(of beer etc.)*

Halb·edelstein der *(veralt.)* semi-precious stone

halbe-halbe *in* **[mit jmdm.] ~ machen** *(ugs.)* go halves [with sb.]

halber ['halbɐ] *Präp. mit Gen. (wegen)* on account of; *(um ... willen)* for the sake of; **der Ordnung ~:** as a matter of form

halb-, Halb-: ~fertig *Adj. (präd. getrennt geschrieben)* half-finished; **~fett 1.** *Adj. (Druckw.)* bold ‹*type*›; *(schmaler, kleiner)* semibold; **2.** *adv.* **etw. ~fett drukken** print sth. in bold/semibold [type]; **~finale das** *(Sport)* semifinal; **~gar** *Adj.* half-cooked; half-done; **~gebildet** *Adj. (abwertend)* half-educated; **~gefror[e]ne das;** *adj. Dekl.* soft ice cream; **~gott der** *(Myth., fig. iron.)* demigod

Halbheit die; ~, ~en *(abwertend)* half-measure

halb-: ~herzig 1. *Adj.* half-hearted; **2.** *adv.* half-heartedly; **~hoch** *Adj.* calf-length ‹*boot*›

halbieren *tr. V.* cut/tear ‹*object*› in half; halve ‹*amount, number*›; *(Math.)* bisect

halb-, Halb-: ~insel die peninsula; **~jahr das** six months *pl.*; half year; **im ersten/zweiten ~jahr** in the first/last six months [of the year]; **~jährig** *Adj.; nicht präd.* **a)** *(ein halbes Jahr alt)* six-months-old ‹*baby, pony, etc.*›; **b)** *(ein halbes Jahr dauernd)* six-month ‹*contract, course, etc.*›; **~jährlich 1.** *Adj.* half-yearly; six-monthly; **2.** *adv.* every six months; twice a year; **~kreis der** semicircle; **sich im ~kreis aufstellen** form a semicircle; **~kreis·förmig 1.** *Adj.* semicircular; **2.** *adv.* in a semicircle; **~kugel die** hemisphere; **~kugel·förmig** *Adj.* hemispherical; **~lang** *Adj.* mid-length ‹*hair*›; mid-calf length ‹*coat, dress, etc.*›; **~laut 1.** *Adj.* low; quiet; **2.** *adv.* in a low voice; in an undertone; **~leder das** *(Buchw.)* half-leather; **~leinen das a)** *(Gewebe)* fifty-percent linen material; **b)** *(Buchw.)* half-cloth; **~leiter der** *(Elektronik)* semiconductor; **~links** [-'-] *Adv. (Fußball)* ‹*play*› [at] inside left; **~mast** *Adv.* at half-mast; **~mast flaggen** fly a flag/the flags at half-mast; **~messer der** *(Math.)* radius; **~monatlich** *Adj.* fortnightly; twice-monthly; **~mond der a)** *(Mond)* half-moon; **b)** *(Figur)* crescent; **~nackt** *Adj. (präd. getrennt ge-*

schrieben) half-naked; **~offen** *Adj. (präd. getrennt geschrieben)* half-open ⟨*door etc.*⟩; **~part** *Adv. in* **[mit jmdm.] ~part machen** *(ugs.)* go halves [with sb.]; **~pension die;** *o. Pl., meist o. Art.* half-board; **~rechts** [-'-] *Adv. (Fußball)* ⟨*play*⟩ [at] inside right; **~roh** *Adj. (präd. getrennt geschrieben)* half-cooked; half-done; **~rund** *Adj. (präd. getrennt geschrieben)* semicircular; **~rund das** semicircle; **~schatten der** half shadow **~schlaf der** light sleep; **im ~schlaf liegen** be half asleep; doze; **~schuh der** shoe; **~schwer·gewicht das** *(Schwerathletik)* light-heavyweight; **~schwergewichtler** [-gəvıçtlɐ] **der; ~s, ~** *(Schwerathletik)* light-heavyweight; **~schwester die** half-sister; **~seiden** *Adj.* **a)** fifty-per-cent silk; **b)** *(ugs. abwertend: unmännlich)* poofy *(coll.)*; pansyish *(coll.)*; **c)** *(ugs. abwertend: anrüchig)* dubious ⟨*business practice etc.*⟩; fast ⟨*woman*⟩; **~seitig 1.** *Adv.* **a)** half-page ⟨*advertisement, article, etc.*⟩; **b)** *(Med.: einseitig)* of one side of the body *postpos.*; **2.** *adv.* **a) ~ annoncieren** place a half-page advert; **b)** *(Med.: einseitig)* **~ gelähmt** paralysed down one side; **~starke der;** *adj. Dekl. (ugs. abwertend)* young rowdy; [young] hooligan; **~stiefel der** half-boot; ankle boot; **~stündig** *Adj.; nicht präd.* half-hour; lasting half an hour *postpos., not pred.*; **~stündlich 1.** *Adj.; nicht präd.* half-hourly; **2.** *adv.* half-hourly; every half an hour; **~stürmer der** *(bes. Fußball)* midfield player

halb·tags *Adv.* ⟨*work*⟩ part-time; *(morgens/nachmittags)* ⟨*work*⟩ [in the] mornings/afternoons

Halbtags-: ~arbeit die; *o. Pl.* part-time job; *(morgens/nachmittags)* morning/afternoon job; **~kraft die** part-time worker; part-timer

halb-, Halb-: ~ton der; *Pl.* **~töne a)** *(Musik)* semitone; halftone *(Amer.)*; **b)** *(Malerei)* half-tone; **~verhungert** *Adj. (präd. getrennt geschrieben)* half-starved; **~voll** *Adj. (präd. getrennt geschrieben)* half-full; half-filled; **~wach** *Adj. (präd. getrennt geschrieben)* half-awake; **~wahrheit die** half-truth; **~wegs** ['~'ve:ks] *Adv.* to some extent; reasonably ⟨*good, clear, etc.*⟩; **~welt die;** *o. Pl.* demi-monde; **~wüchsig** [~vy:ksıç] *Adj.* adolescent; teenage; **~wüchsige der/die;** *adj. Dekl.* adolescent; teenager; **~zeit die** *(bes. Fußball)* **a)** half; **die erste/zweite ~zeit** the first/second half; **b)** *(Pause)* half-time

Halde ['haldə] **die; ~, ~n a)** *(Bergbau)* slag-heap; *(von Vorräten)* pile; *(fig.)* mountain; pile; **b)** *(geh.: Hang)* slope

half [half] *1. u. 3. Pers. Sg. Prät. v.* **helfen**

Hälfte ['hɛlftə] **die; ~, ~n a)** half; **die ~ einer Sache** *(Gen.) od.* **von etw.** half [of] sth.; **Studenten bezahlen die ~ des Preises** students pay *or (coll.)* are half-price; **er füllte sein Glas nur bis zur ~:** he only half-filled his glass; **über die ~:** more than *or* over half; **um die ~ größer/kleiner** half as big/small again; **etw. zur ~ zahlen** pay half of sth.; **die gegnerische ~** *(Sport)* the opponents' half; **ich habe die ~ vergessen** I've forgotten half of it; **meine bessere ~** *(ugs. scherzh.)* my better half *(coll. joc.)*; **b)** *(ugs.: Teil)* part

[1]**Halfter** ['halftɐ] **der** *od.* **das; ~s, ~;** *veralt. auch* **die; ~, ~n** halter

[2]**Halfter die; ~, ~n;** *auch* **das; ~s, ~:** holster

Hall [hal] **der; ~[e]s, ~e a)** *(geh.)* reverberation; **b)** *(Echo)* echo

Halle ['halə] **die; ~, ~n** *(Saal, Gebäude)* hall; *(Fabrik~)* shed; *(Hotel~, Theater~)* lobby; foyer; *(Sport~)* [sports] hall

halleluja [hale'lu:ja] *Interj.* hallelujah!; *(scherzh.: hurra)* hurrah!

Halleluja das; ~s, ~s hallelujah

hallen *itr. V.* **a)** reverberate; ring; ⟨*shot, bell, cry*⟩ ring out; **b)** *(widerhallen)* echo

Hallen- indoor ⟨*swimming-pool, handball, football, hockey, tennis, etc.*⟩

Hallig ['halıç] **die; ~, ~en** small low island *(off the North Sea coast of Schleswig-Holstein)*

hallo *Interj.* **a)** *meist* ['halo] *(am Telefon)* hello; **~, warte doch mal auf mich!** hey! wait for me!; **~, gehört Ihnen diese Tasche?** excuse me! is this your bag?; **b)** *meist* [ha'lo:] *(überrascht)* hello

Hallo [ha'lo:] **das; ~s, ~s** cheering; cheers *pl.*; **mit großem ~:** with loud cheering *or* cheers

Halluzination [halutsina'tsio:n] **die; ~, ~en** hallucination

Halm [halm] **der; ~[e]s, ~e** stalk; stem

Halma ['halma] **das; ~s** halma

Halogen [halo'ge:n] **das; ~s, ~e** *(Chemie)* halogen

Halogen- halogen ⟨*lamp, headlamp*⟩

Hals [hals] **der; ~es, Hälse** ['hɛlzə] **a)** neck; **sich** *(Dat.)* **den ~ brechen** break one's neck; **jmdm. um den ~ fallen** throw *or* fling one's arms around sb.['s neck]; **~ über Kopf** *(ugs.)* in a rush *or* hurry: **sich ~ über Kopf verlieben** fall head over heels in love; **einen langen ~ machen** *(ugs.)* crane one's neck; **jmdm. den ~ brechen** *(ugs.)* drive sb. to the wall; **das kostete ihn** *od.* **ihm den ~** *(ugs.)* that did for him *(coll.)*; **sich jmdm. an den ~ werfen** *(ugs.)* throw oneself at sb.; **jmdm. jmdn. auf den ~ schicken** *od.* **hetzen** *(ugs.)* get *or* put sb. on [to] sb.; **sich** *(Dat.)* **jmdn./etw. auf den ~ laden** *(ugs.)* lumber *or* saddle oneself with sb./sth. *(coll.)*; **jmdm. steht das Wasser bis zum ~** *(ugs.: jmd. hat Schulden)* sb. is up to his/her eyes in debt; *(ugs.: jmd. hat Schwierigkeiten)* sb. is up to his/her neck in it; **b)** *(Kehle)* throat; **aus vollem ~[e]** at the top of one's voice; **er hat es in den falschen ~ bekommen** *(ugs.: falsch verstanden)* he took it the wrong way; *(ugs.: sich verschluckt)* it went down [his throat] the wrong way; **er kann den ~ nicht voll [genug] kriegen** *(ugs.)* he can't get enough; he's insatiable; **das hängt/wächst mir zum ~[e] heraus** *(ugs.)* I'm sick and tired of it *(coll.)*; **c)** *(einer Flasche)* neck; **d)** *(Musik) (einer Note)* stem; *(eines Saiteninstruments)* neck

hals-, Hals-: ~abschneider der *(ugs. abwertend)* shark; **~ausschnitt der** neckline; **~band das;** *Pl.* **~bänder a)** *(für Tiere)* collar; **b)** *(Samtband)* choker; neck-band; **~brecherisch** [~brɛçərıʃ] *Adj.* dangerous, risky ⟨*climb, action, etc.*⟩; hazardous ⟨*road*⟩; breakneck *attrib.* ⟨*speed*⟩; **~entzündung die** inflammation of the throat; **~kette die** necklace; **~krause die** ruff; **~-Nasen-Ohren-Arzt der** ear, nose, and throat specialist; **~schlagader die** carotid [artery]; **~schmerzen** *Pl.* sore throat *sing.*; **[starke] ~schmerzen haben** have a[n extremely] sore throat; **~starrig** [~ʃtarıç] *(abwertend)* **1.** *Adj.* stubborn; obstinate; **2.** *adv.* stubbornly; obstinately; **~starrigkeit die** *(abwertend)* stubbornness; obstinacy; **~tuch das** cravat; *(des Cowboys)* neckerchief; **~- und Beinbruch** *Interj. (scherzh.)* good luck!; best of luck!; **~weh das;** *(ugs.) s.*

~schmerzen; ~wirbel der *(Anat.)* cervical vertebra

[1]halt [halt] *Partikel (südd., österr., schweiz.) s.* **eben 3 b**

[2]halt *Interj.* stop; *(Milit.)* halt

Halt der; ~[e]s, ~e a) *o. Pl. (Stütze)* hold; **seine Füße/Hände fanden keinen ~:** he couldn't find *or* get a foothold/handhold; **den ~ verlieren** lose one's hold; **b)** *(Anhalten)* stop; **ohne ~:** non-stop; without stopping

haltbar *Adj.* **a)** *(nicht verderblich)* **~ sein** ⟨*food*⟩ keep [well]; **etw. ~ machen** preserve sth.; **~ bis 5. 3.** use by 5 March; **b)** *(nicht verschleißend)* hard-wearing, durable ⟨*material, clothes*⟩; **c)** *(aufrechtzuerhalten)* tenable ⟨*hypothesis etc.*⟩; **d)** *(Ballspiele)* stoppable, savable ⟨*shot*⟩

Haltbarkeit die; ~ a) Lebensmittel von beschränkter ~: perishable foods; **b)** *(Strapazierfähigkeit)* durability; **c)** *s.* **haltbar c:** tenability

Halte-: ~griff der a) [grab] handle; *(Riemen)* [grab] strap; **b)** *(Budo, Ringen)* pinning hold; **~linie die** *(Verkehrsw.)* stop line

halten 1. *unr. tr. V.* **a)** *(auch Milit.)* hold; **etw. an einem Ende ~:** hold one end of sth.; **sich** *(Dat.)* **den Kopf/den Bauch ~:** hold one's head/stomach; **jmdn. an** *od.* **bei der Hand ~:** hold sb.'s hand; hold sb. by the hand; **die Hand vor den Mund ~:** put one's hand in front of one's mouth; **etw. ins Licht/gegen das Licht ~:** hold sth. to/up to the light; **b)** *(Ballspiele)* save ⟨*shot, penalty, etc.*⟩; **c)** *(bewahren)* keep; *(beibehalten, aufrechterhalten)* keep up ⟨*speed etc.*⟩; maintain ⟨*temperature, equilibrium*⟩; **einen Ton ~:** stay in tune; *(lange an~)* sustain a note; **den Takt ~:** keep time; **Diät ~:** keep to a diet; **den Kurs ~:** stay on course; **diese Behauptung läßt sich nicht ~:** this statement does not hold up; **Ordnung/Frieden ~:** keep order/the peace; **d)** *(erfüllen)* keep; **sein Wort/ein Versprechen ~:** keep one's word/a promise; **e)** *(besitzen, beschäftigen, beziehen)* keep ⟨*chickens etc.*⟩; take ⟨*newspaper, magazine, etc.*⟩; **ein Auto ~:** run a car; **f)** *(einschätzen)* **jmdn. für reich/ehrlich ~:** think sb. is *or* consider sb. to be rich/honest; **ich halte es für das beste/möglich/meine Pflicht** I think it best/possible/my duty; **viel/nichts/wenig von jmdm./etw. ~:** think a lot/nothing/not think much of sb./sth.; **g)** *(ab~, veranstalten)* give, make ⟨*speech*⟩; give, hold ⟨*lecture*⟩; **Unterricht ~:** give lessons; teach; **seinen Mittagsschlaf ~:** have one's *or* an afternoon nap; **h)** *(Halt geben)* hold up, support ⟨*bridge etc.*⟩; hold back ⟨*curtain, hair*⟩; fasten ⟨*dress*⟩; **i)** *(zurück~)* keep; **ihn hält hier nichts** there's nothing to keep him here; **es hält dich niemand** nobody's stopping you; **j)** *(bei sich be~)* **das Wasser ~:** hold one's water; **k)** *(nicht aufgeben)* **ein Geschäft** *usw.* **~:** keep a business *etc.* going; **l)** *(behandeln)* treat; **jmdn. streng ~:** be strict with sb.; **m)** *(vorziehen)* **es mehr** *od.* **lieber mit jmdm./etw. ~:** prefer sb./sth.; **n)** *(verfahren)* **es mit einer Sache so/anders ~:** deal with *or* handle sth. like this/differently; **o)** *(gestalten)* **das Badezimmer ist in Grün ge~:** the bathroom is decorated in green; **die Rede war sehr allgemein ge~:** the speech was very general. **2.** *unr. itr. V.* **a)** *(stehenbleiben)* stop; **b)** *(unverändert, an seinem Platz bleiben)* last; **der Nagel/das Seil hält nicht mehr länger** the nail/rope won't hold much longer; **diese Freundschaft hält nicht [lange]** *(fig.)* this friendship won't last [long]; **c)** *(Sport)* save; **er hat gut ge~:** he made some good saves; **d)** *(beistehen)* **zu jmdm. ~:** stand *or* stick by sb.; **e)** *(zielen)* aim **(auf** + *Akk.* at); **f)** *(Seemannsspr.)* head; **auf etw.** *(Akk.)* **~:** head for *or* towards sth.; **g)** *(sich beherrschen)* **an sich** *(Akk.)* **~:** control oneself; **h)** *(achten)* **auf Ordnung ~:** attach importance to tidiness. **3.** *unr. refl. V.* **a)** *(sich durchsetzen, behaupten)* **wir werden uns nicht länger ~ können** we won't be able to hold out much longer; **das Geschäft wird sich nicht ~ können** the shop won't keep going [for long]; **b)** *(sich bewähren)* **sich gut ~:** do well; make a good showing; **halte dich tapfer** be brave; **c)** *(unverändert bleiben)* ⟨*weather, flowers, etc.*⟩ last; ⟨*milk, meat, etc.*⟩ keep; **d)** *(Körperhaltung haben)* **sich schlecht/gerade/aufrecht ~:** hold *or* carry oneself badly/straight/erect; **e)** *(bleiben)* **sich auf den Beinen/im Sattel ~:** stay on one's feet/in the saddle; **f)** *(gehen, bleiben)* **sich links/rechts ~:** keep [to the] left/right; **sich an jmds. Seite** *(Dat.)***/hinter jmdm. ~:** stay *or* keep next to/behind sb.; **g)** *(befolgen)* **sich an etw.** *(Akk.)* **~:** keep to *or* follow sth.; **h)** *(sich wenden)* **sich an jmdn. ~:** ask sb.; **i)** *(ugs.: jung, gesund bleiben)* **sie hat sich gut ge~:** she is well preserved for her age *(coll.)*

Halter der; ~s, ~ a) *(Fahrzeug~)* keeper; **b)** *(Tier~)* owner; **c)** *(Vorrichtung)* holder; **d)** *(ugs.: Feder~)* pen

Halterin die; ~, ~nen *s.* **Halter a, b**

Halterung die; ~, ~en support

Halte-: ~signal das stop signal; **~stelle die** stop; **~verbot das a) „~verbot"** 'no stopping'; **„absolutes/eingeschränktes ~verbot"** 'no stopping/no waiting'; **b)** *(Stelle)* no-stopping zone; **~verbots·schild das** no-stopping sign

-haltig [-haltɪç], *(österr.)* **-hältig** [-hɛltɪç] **vitamin~/silber~** *usw.* containing vitamins/silver *etc. postpos., not pred.;* **vitamin~ sein** contain vitamins

halt-, Halt-: ~los *Adj.* **a)** *(labil)* **ein ~loser Mensch** a weak character; **b)** *(unbegründet)* unfounded; **~losigkeit die; ~ a)** *(Labilität)* weakness of character; **b)** *(mangelnde Begründung)* unfoundedness; **~|machen** *itr. V.* stop; **vor jmdm./etw. nicht ~machen** not spare sb./sth.

Haltung die; ~, ~en a) *(Körper~)* posture; *(Sport)* stance; *(in der Bewegung)* style; **~ annehmen** *(Milit.)* stand to attention; **b)** *(Pose)* manner; **c)** *(Einstellung)* attitude; **d)** *o. Pl. (Fassung)* composure; **~ zeigen/bewahren** keep one's composure; **e)** *(Tier~)* keeping

Haltungs·fehler der a) *(Med.)* bad posture; **b)** *(Sport)* style fault

Halunke [ha'lʊnkə] **der; ~n, ~n a)** scoundrel; villain; **b)** *(scherzh.: Lausbub)* rascal; scamp

Hamburg ['hambʊrk] **(das); ~s** Hamburg

[1]Hamburger 1. der; ~s, ~: native of Hamburg; *(Einwohner)* inhabitant of Hamburg; **Schmidt ist ~:** Schmidt comes from Hamburg. **2.** *indekl. Adj.* Hamburg; **der ~ Hafen** the harbour at Hamburg; Hamburg harbour

[2]Hamburger der; ~s, ~ *od.* **~s** *(Frikadelle)* hamburger

hämisch ['hɛːmɪʃ] **1.** *Adj.* malicious. **2.** *adv.* maliciously

Hammel ['haml̩] **der; ~s, ~ a)** wether; **b)** *(Fleisch)* mutton; **c)** *(salopp abwertend)* oaf; dolt

Hammel-: ~bein das: jmdm. die ~beine langziehen *(ugs.)* give sb. a good telling-off; **~fleisch**

das mutton; **~herde die** *(salopp abwertend)* flock of sheep; **~keule die** leg of mutton; **~sprung der** *(Parl.)* division

Hammer ['hamɐ] **der; ~s, Hämmer** ['hɛmɐ] **a)** hammer; *(Holz~)* mallet; **~ und Sichel** hammer and sickle; **unter den ~ kommen** come under the hammer; **b)** *(Technik)* tup; ram; **c)** *(Musik)* hammer; **d)** *(Leichtathletik)* hammer; **e)** *(ugs.: Fehler)* bad mistake; *(in einer Aufgabe)* howler *(coll.)*; **ein dikker ~:** an awful blunder

Hämmerchen ['hɛmɐçən] **das; ~s, ~:** [small] hammer

hämmern ['hɛmɐn] **1.** *itr. V.* **a)** hammer; **es hämmert** sb. is hammering; **b)** *(schlagen)* hammer; *(mit der Faust)* hammer; pound; **gegen die Wand/die Tür ~:** hammer/pound on the wall/door; **c)** *(klopfen)* pound; ⟨*pulse*⟩ race. **2.** *tr. V.* **a)** hammer; beat, hammer ⟨*tin, silver, etc.*⟩; beat ⟨*jewellery*⟩; **b)** *(ugs.)* hammer *or* pound out ⟨*melody etc.*⟩; **c)** *(ugs.: einprägen)* **jmdm. etw. in den Schädel ~:** hammer *or* knock sth. into sb.'s head *(coll.)*

Hammer-: ~werfen das *(Leichtathletik)* throwing the hammer; **er ist Weltmeister im ~werfen** he's world champion in the hammer; **~werfer der; ~s, ~** *(Leichtathletik)* hammer-thrower; **~wurf der** *(Leichtathletik) s.* **~werfen**

Hammond·orgel ['hæmənd-] **die** Hammond organ

Hämorrhoiden [hɛmɔro'i:dn̩] *Pl. (Med.)* haemorrhoids; piles

Hampel·mann ['hampl̩-] **der; ~[e]s, Hampelmänner a)** jumping jack; **b)** *(ugs. abwertend)* puppet

hampeln *itr. V. (ugs.)* jump about

Hamster ['hamstɐ] **der; ~s, ~:** hamster

Hamsterer der; ~s, ~, Hamsterin die; ~, ~nen *(ugs.)* hoarder

Hamster-: ~fahrt die foraging trip; **auf ~fahrt gehen** go foraging; **~kauf der** panic-buying *no pl.*; **~käufe machen** panic-buy

hamstern *tr., itr. V. (horten)* hoard; *(Hamsterkäufe machen)* panic-buy

Hand [hant] **die; ~, Hände** ['hɛndə] **a)** hand; **mit der rechten/linken ~:** with one's right/left hand; **jmdm. die ~ geben** *od. (geh.)* **reichen** shake sb.'s hand; shake sb. by the hand; **jmdm. die ~ drücken/schütteln** press/shake sb.'s hand; **eine ~ frei haben** have a free hand; **Hände hoch!** hands up!; **jmdn. an die** *od. (geh.)* **bei der ~ nehmen** take sb. by the hand; **jmdm. etw. aus der ~ nehmen** take sth. out of sb.'s hand/hands; **etw. aus der ~ legen** put sth. down; **jmdm. aus der ~ lesen** read sb.'s hand *or* palm; **etw. in die/zur ~ nehmen** pick sth. up; **etw. in der ~/den Händen haben** *od. (geh.)* **halten** have got *or* hold sth. in one's hand/hands; **in die Hände klatschen** clap one's hands; **mit Händen und Füßen reden** use gestures to make oneself understood; **etw. mit der ~ schreiben/nähen** write/sew sth. by hand; **von ~:** by hand; **~ in ~ gehen** go *or* walk hand-in-hand; **jmdm. etw. in die ~ versprechen** promise sb. sth. faithfully; **b)** *o. Pl. (Fußball)* handball; **c)** *(in Wendungen)* **was hältst du davon – ~ aufs Herz!** what do you think? – be honest; **eine ~ wäscht die andere** you scratch my back and I'll scratch yours; **jmdm. sind die Hände gebunden** sb.'s hands are tied; **~ und Fuß/weder ~ noch Fuß haben** *(ugs.)* make sense/no sense; **[bei etw. selbst mit] ~ anlegen** lend a hand [with sth.]; **die** *od.* **seine ~ aufhalten** *(ugs.)* hold out one's hand; **letzte ~ an etw.** *(Akk.)* **legen** put the finishing touches *pl.* to sth.; **sich** *(Dat.) od. (geh.)* **alle** *od.* **beide Hände damit voll haben, etw. zu tun** *(ugs.)* have one's hands full doing sth.; **die Hände in den Schoß legen** sit back and do nothing; **bei etw. die** *od.* **seine Hände [mit] im Spiel haben** have a hand in sth.; **die Hände über dem Kopf zusammenschlagen** *(ugs.)* throw up one's hands in horror; **zwei linke Hände haben** *(ugs.)* have two left hands *(coll.)*; **eine lockere** *od.* **lose ~ haben** *(ugs.)* hit out at the slightest provocation; **eine glückliche ~ bei etw. haben** have a feel for the right choice in sth.; **linker/rechter ~:** on *or* to the left/right; **[klar] auf der ~ liegen** *(ugs.)* be obvious; **jmdn. auf Händen tragen** lavish every kind of care and attention on sb.; **ein Auto/Möbel aus erster ~:** a car/furniture which has/had had one [previous] owner; **etw. aus erster ~ wissen** know sth. at first hand; have first-hand knowledge of sth.; **Kleidung aus zweiter ~:** second-hand clothes *pl.*; **jmdm. aus der ~ fressen** eat out of sb.'s hand *(fig.)*; **etw. aus der ~ geben** *(weggeben)* let sth. out of one's hands; *(aufgeben)* give sth. up; **jmdm. etw. aus der ~ nehmen** relieve sb. of sth.; **etw. bei der ~ haben** *(greifbar haben)* have sth. handy; *(parat haben)* have sth. ready; **mit etw. schnell** *od.* **rasch bei der ~ sein** *(ugs.)* be ready [with sth.]; **~ in ~ arbeiten** work hand in hand; **mit etw. ~ in ~ gehen** go hand in hand with sth.; **hinter vorgehaltener ~:** off the record; **in die Hände spucken** spit on one's hands; *(fig. ugs.)* roll up one's sleeves *(fig.)*; **jmdn./etw. in die ~** *od.* **Hände bekommen** lay *or* get one's hands on sb./get one's hands on sth.; **jmdm. in die Hände fallen** fall into sb.'s hands; **jmdn. in der ~ haben** have *or* hold sb. in the palm of one's hand; **etw. in die ~ nehmen** take sth. in hand; **in jmds. ~** *(Dat.)* **sein** *od. (geh.)* **liegen** be in sb.'s hands; **in sicheren** *od.* **guten Händen sein** be in safe *or* good hands; **sich mit Händen und Füßen gegen etw. sträuben** *od.* **wehren** *(ugs.)* fight tooth and nail against sth.; **mit leeren Händen** emptyhanded; **das Geld mit vollen Händen ausgeben** spend money like water; **um jmds. ~ anhalten** *od.* **bitten** *(geh. veralt.)* ask for sb.'s hand [in marriage]; **das geht ihm gut/leicht von der ~:** he finds that no trouble; **etw. von langer ~ vorbereiten** plan sth. well in advance; **die Nachteile/seine Argumente sind nicht von der ~ zu weisen** the disadvantages cannot be denied/his arguments cannot [simply] be dismissed; **von der ~ in den Mund leben** live from hand to mouth; **jmdm. zur ~ gehen** lend sb. a hand; **zu Händen [von] Herrn Müller** for the attention of Herr Müller; attention Herr Müller; *s. auch* **öffentlich 1; d) an ~** *s.* anhand

Hand·arbeit die a) *o. Pl.* handicraft; craft work; **etw. in ~ herstellen** make sth. by hand; **b)** *(Gegenstand)* handmade article; **c)** *(Arbeit aus Stoff, Wolle usw.)* **sie macht gerne ~en** she likes doing needlework/knitting/crocheting; **d)** *o. Pl. (ugs.: ~arbeitsunterricht)* needlework

hand·arbeiten *itr. V.* do needlework

Handarbeits-: ~geschäft das wool and needlework shop; **~korb der** workbasket; **~lehrerin die** needlework teacher

hand-, Hand-: ~auflegen das; ~s *(bes. Rel.)* laying on *or* imposition of hands; **~ball der** handball; **~ballen der** ball of the thumb; **~bedienung die;** *o. Pl. s.* **~betrieb; ~besen der** brush; **~betrieb der;** *o. Pl.* manual operation; **mit ~betrieb** manually operated *or* hand-operated; **~bewegung die a)** movement of the hand; **b)** *(Geste)* gesture; **~bibliothek die a)** reference library; **b)** *(~apparat)* set of reference books; reference collection; **~bohrer der** *(mit Kurbel)* hand-drill; *(zum Vorbohren)* gimlet; **~bohr·maschine die** hand-drill; *(elektrisch)* drill; **~brause die** shower handset; **~breit** *Adj.* ‹*seam etc.*› a few inches wide; **~breit die; ~, ~: eine/zwei ~breit** a few/several inches; **~bremse die** hand-brake; **~buch das** handbook; *(technisches ~buch)* manual

Händchen ['hɛntçən] **das; ~s, ~:** [little] hand; **~ halten** *(ugs. scherzh.)* hold hands

Hand·creme die hand cream

Hände *s.* **Hand**

Hände-: ~druck der; *Pl.* **~drükke** handshake; **~klatschen das; ~s** clapping; applause

[1]**Handel** ['handl̩] **der; ~s a)** *(Wirtschaft)* trade; commerce; **b)** *(Handeln)* trade; **der ~ mit Waffen/Drogen** the traffic in arms/drugs; **c)** *(Geschäftsverkehr)* trade; **das ist [nicht mehr] im ~:** it is [no longer] on the market; **d)** *(Vereinbarung)* deal

[2]**Handel der; ~s, Händel** ['hɛndl̩]; *meist Pl. (geh.)* **Händel suchen** [try to] pick a quarrel

Hand·elfmeter der *(Fußball)* penalty for handball

handeln 1. *itr. V.* **a)** trade; deal; **mit** *od.* **in Gemüse/Gebrauchtwagen ~:** deal in vegetables/second-hand cars; **b)** *(feilschen)* haggle; bargain; **um den Preis ~:** haggle over the price; **mit ihm läßt sich [nicht] ~:** he is [not] open to negotiation; **c)** *(eingreifen)* act; **auf Befehl/aus Überzeugung ~:** act on orders/out of conviction; **im Affekt/in Notwehr ~:** act in the heat of the moment/in self-defence; **d)** *(verfahren)* act; **eigenmächtig/richtig/fahrlässig ~:** act on one's own authority/correctly/carelessly; **e)** *(sich verhalten)* behave; **f) von etw.** *od.* **über etw.** *(Akk.)* **~** ‹*book, film, etc.*› be about *or* deal with sth. **2.** *refl. V. (unpers.)* **bei dem Besucher handelte es sich um einen entfernten Verwandten** the visitor was a distant relative; **es handelt sich um ...:** it is a matter of ...; *(es dreht sich um)* it's about *or* it concerns ... **3.** *tr. V.* sell **(für** at, for); **diese Papiere werden nicht an der Börse gehandelt** these securities are not traded on the stock exchange

Handeln das; ~s a) *(das Feilschen)* haggling; bargaining; **b)** *(das Eingreifen)* action; **c)** *(Verhalten)* action[s *pl.*]

handels-, Handels-: ~abkommen das trade agreement; **~bank die** merchant bank; **~beziehungen** *Pl.* trade relations; **~bilanz die a)** *(eines Betriebes)* balance-sheet; **b)** *(eines Staates)* balance of trade; **eine aktive/passive ~bilanz** a balance of trade surplus/deficit; **~boykott der** trade boycott; **~einig, ~eins** *in* **mit jmdm. ~einig** *od.* **~eins werden/sein** agree/have agreed terms *or* come/have come to an agreement with sb.; **~firma die** [business *or* commercial] firm; business concern; **~flagge die** merchant flag; **~flotte die** merchant fleet; **~gesellschaft die** company; **offene ~gesellschaft** general partnership; **~gesetz·buch das;** *o. Pl.* commercial code; **~hafen der** commercial *or* trading port; **~kammer die** *s.* **Industrie- und ~kammer; ~klasse die** grade; **~macht die** trading power; **~marine die** merchant navy; **~marke die** trade mark; **~minister der** minister of trade; *(in UK)* Secretary of State for Trade; Trade Secretary *(coll.)*; **~ministerium das** ministry of trade; *(in UK)* Department of Trade; **~mission die** trade mission; **~name der** trade *or* business name; **~niederlassung die** branch; **~organisation die a)** trading organization; **b)** *(ehem. DDR) [state-owned] commercial concern running shops, hotels, etc.*; **~partner der** trading partner; **~politik die** trade *or* commercial policy; **~recht das;** *o. Pl.* commercial law; **~rechtlich 1.** *Adj.; nicht präd.* relating to commercial law *postpos.*; ‹*offence*› against commercial law; **2.** *adv.* from the point of view of commercial law; **~register das** register of companies; **~reisende der/die;** *adj. Dekl. s.* **~vertreter; ~schiff das** merchant ship; trading vessel; **~schiffahrt die** merchant shipping; *(Schiffsverkehr)* movement of merchant shipping; **~schranke die;** *meist Pl.* trade barrier; **~schule die** commercial college; **~spanne die** *(Kaufmannsspr.)* margin; **~straße die** *(hist.)* trade route; **~üblich** *Adj.* **~übliche Größen** standard [commercial] sizes; **~unternehmen das** trading concern; **~verbindung die;** *meist Pl.* trade link; **~vertreter der** [sales] representative; travelling salesmann/saleswoman; commercial traveller; **~vertretung die** *s.* **~mission; ~volumen das** *(Wirtsch.)* volume of trade; **~ware die** commodity; **„keine ~ware“** *(Postw.)* 'no commercial value'; **~zentrum das** trading *or* commercial centre

handel·treibend *Adj.; nicht präd.* trading ‹*nation*›

hände-, Hände-: ~ringend *Adv.* **a)** wringing one's hands; **b)** *(ugs.: dringend)* **~ringend nach jmdm./etw. suchen** search desperately for sb./sth.; **~schütteln das; ~s** hand-shaking *no pl.*; **~waschen das; ~s** washing *no art.* one's hands

hand-, Hand-: ~feger der brush; **~fest** *Adj.* **a)** *(kräftig)* robust; sturdy; **b)** *(deftig)* substantial ‹*meal etc.*›; **etwas Handfestes** something substantial; **c)** *(gewichtig)* solid, tangible ‹*proof*›; concrete ‹*suggestion*›; full-blooded, violent ‹*row*›; complete ‹*lie*›; well-founded ‹*argument*›; real, thorough ‹*beating*›; **~fläche die** palm [of one's/the hand]; flat of one's/the hand; **~gearbeitet** *Adj.* hand-made ‹*furniture, jewellery, etc.*›; **~geld das** lump sum [payment]; **~gelenk das** wrist; **ein loses** *od.* **lockeres ~gelenk haben** *(ugs.)* lash out at the slightest provocation; **etw. aus dem ~gelenk schütteln** *(ugs.)* do sth. just like that *(coll.)*; **~gemacht** *Adj.* handmade; **~gemenge das** fight; **~gepäck das** hand-baggage; **~geschöpft** *Adj.* handmade; **~geschrieben** *Adj.* handwritten; **~gesteuert** *Adj.* manually operated; manually controlled ‹*vehicle*›; **~gewebt** *Adj.* hand-woven; **~granate die** hand-grenade; **~greiflich** *Adj.* **a)** *(tätlich)* **eine ~greifliche Auseinandersetzung** a scuffle; **~greiflich werden** start using one's fists; **b)** tangible ‹*success, advantage, proof, etc.*›; palpable ‹*contradiction, error*›; obvious ‹*fact*›; **~greiflichkeit die; ~, ~en es kam zu ~greiflichkeiten** a fight broke out; **~griff der a) ein falscher ~griff** a false move; **mit einem ~griff/wenigen ~griffen** in one movement/without much trouble; *(schnell)* in no time at all/next to no time; **jeder ~griff muß sitzen** every movement must be exactly right; **b)** *(am Koffer, an einem Werkzeug)* handle; **~habe die; ~, ~n: eine [rechtliche] ~habe [gegen jmdn.]** a legal handle [against sb.]; **~haben** *tr. V.* **a)** handle; operate ⟨*machine; device*⟩; **b)** *(praktizieren)* implement ⟨*law etc.*⟩; **~habung die; ~, ~en a)** handling; *(eines Gerätes, einer Maschine)* operation; **b)** *(Durchführung)* implementation

Handikap ['hɛndikɛp] **das; ~s, ~s** *(auch Sport)* handicap

hand-, Hand-: ~kante die edge of the/one's hand; **Handkanten·schlag der** chop; **~käse der** *(landsch.) small, hand-formed curd cheese;* **~käse mit Musik** *(landsch.)* marinaded hand-formed curd cheese; **~koffer der** [small] suitcase; **~koloriert** *Adj.* hand-coloured; **~kuß der** kiss on sb.'s hand; **etw. mit ~kuß tun** *(fig. ugs.)* do sth. with [the greatest of] pleasure; **~langer der; ~s, ~ a)** *(ungelernter Arbeiter)* labourer; *(abwertend)* lackey; general dogsbody; **b)** *(abwertend: Büttel)* henchman; **~lauf der** handrail

Händler ['hɛndlɐ] **der; ~s, ~, Händlerin die; ~, ~nen** trader; tradesman/tradeswoman; **ein fliegender ~:** a hawker *or* street-trader

handlich ['hantlɪç] **1.** *Adj.* handy; easily carried ⟨*parcel, suitcase*⟩. **2.** *adv.* **~ verpackt** wrapped as a manageable parcel

Handlichkeit die; ~: handiness; *(eines Buches)* handy size

Handlung die; ~, ~en a) *(Vorgehen)* action; *(Tat)* act; **eine symbolische/feierliche ~:** a symbolic/ceremonial act; **b)** *(Fabel)* plot; **Einheit der ~:** unity of action

handlungs-, Handlungs-: ~fähig *Adj.* able to act *pred.;* working *attrib.* ⟨*majority*⟩; **~fähigkeit die;** *o. Pl.* ability to act; **~freiheit die;** *o. Pl.* freedom of action *or* to act; **~spiel·raum der** scope for action; **~unfähig** *Adj.* **a)** unable to act *pred.;* **b)** *(Rechtsw.)* unable to act on one's own account *pred.;* **~unfähigkeit die** inability to act; **~weise die** behaviour; conduct

hand-, Hand-: ~presse die hand-press; **~puppe die** glove *or* hand puppet; **~puppen·spiel das** glove puppet *or* hand puppet show; **~rücken der** the back of the/one's hand; **~säge die** hand-saw; **~schelle die** handcuff; **jmdm. ~schellen anlegen** handcuff sb.; put handcuffs on sb.; **~schlag der a)** handshake; **etw. durch einen ~schlag besiegeln** shake hands on sth.; **b)** *in* **er tat keinen ~schlag** *(ugs.)* he did not lift a finger; **~schrift die a)** handwriting; **b)** *(Ausdrucksweise)* personal style; **c)** *(Text)* manuscript; **~schriftlich 1.** *Adj.* hand-written; **~schriftliche Quellen** manuscript sources; **2.** *adv.* by hand; **~schuh der** glove; **~schuh·fach das** glove compartment *or* box; **~signiert** *Adj.* signed; **~spiegel der** hand-mirror; **~spiel das** *(Fußball)* handball; **~stand der** *(Turnen)* handstand; **einen ~stand machen** do a handstand; **~stand·über·schlag der** *(Turnen)* handspring; **~steuerung die a)** *o. Pl.* manual operation *or* control; **b)** *(Apparatur)* manual control; **~streich der** *(bes. Milit.)* lightning *or* surprise attack; **~tasche die** handbag; **~teller der** palm [of the/one's hand]; **~tuch das;** *Pl.* **-tücher** towel; **das ~tuch werfen** *(Boxen, fig.)* throw in the towel; **~tuch·halter der** towel-rail; **~umdrehen: im ~umdrehen** in no time at all; **~voll die; ~** *(auch fig.)* handful; **~waffe die** hand weapon; **~wagen der** handcart; **~warm 1.** *Adj.* hand-hot; **2.** *adv.* **etw. ~warm waschen** wash sth. in hand-hot water

Handwerk das a) craft; *(als Beruf)* trade; **ein ~ ausüben/betreiben** carry on/ply a trade; **b)** *(Beruf)* **sein ~ verstehen/beherrschen** know one's job; ⟨*tradesman*⟩ know/be master of one's trade; **jmdm. das ~ legen** put a stop to sb.'s activities; **jmdm. ins ~ pfuschen** try to do sb.'s job for him/her; **c)** *o. Pl. (Berufsstand)* craft professions *pl.*

Handwerker der; ~s, ~: tradesman; craftsman; **die ~ im Haus haben** have the workmen in

handwerklich *Adj.; nicht präd.* **a)** ⟨*training, skill, ability*⟩ as a craftsman; **ein ~er Beruf** a [skilled] trade; **b)** *(fig.)* technical

Handwerks-: ~betrieb der workshop; **~bursche der** *(veralt.)* travelling journeyman *(arch.);* **~kammer die** Chamber of Crafts; **~zeug das** tools *pl.; (fig.)* tools *pl.* of the trade

Hand-: ~wörter·buch das concise dictionary; **~zeichen das a)** sign [with one's hand]; *(eines Autofahrers)* hand signal; **b)** *(Abstimmung)* show of hands; **durch ~zeichen** by a show of hands; **~zettel der** handbill; leaflet

hanebüchen ['ha:nəby:çn̩] *Adj.* outrageous

Hanf [hanf] **der; ~[e]s a)** hemp; **b)** *(Samen)* hempseed

Hänfling ['hɛnflɪŋ] **der; ~s, ~e a)** *(Vogel)* linnet; **b)** *(abwertend)* weakling

Hang [haŋ] **der; ~[e]s, Hänge** ['hɛŋə] **a)** *(Berg~)* slope; hillside/mountainside; *(Ski~)* slope; **das Haus am ~:** the house on the hillside; **b)** *(Neigung)* tendency; **einen ~ zum Träumen/Lügen** *usw.* **haben** have a tendency to dream/lie *etc.;* **c)** *(Turnen)* hang

Hangar ['haŋga:ɐ̯] **der; ~s, ~s** hangar

Hänge-: ~backe die flabby cheek; **~bauch der** paunch; **~brücke die** suspension bridge; **~brust die, ~busen der** sagging breasts *pl.;* **~lampe die** pendant-light; drop-light

hangeln ['haŋl̩n] *itr., refl. V.; meist mit sein* make one's way hand over hand; **[sich] an einem Seil über die Schlucht ~:** make one's way hand over hand along a rope over the ravine

Hänge·matte die hammock

[1]hängen ['hɛŋən] *unr. itr. V.; südd., österr., schweiz. mit sein* **a)** hang; **die Bilder ~ [schon]** the pictures are [already] up; **der Schrank hängt voller Kleider** the wardrobe is full of clothes; **der Weihnachtsbaum hängt voller Süßigkeiten** the Christmas tree is laden with sweets; **an einem Faden ~:** be hanging by a thread; **b)** *(sich festhalten)* hang, dangle **(an** + *Dat.* from); **jmdm. am Hals ~:** hang round sb.'s neck; *s. auch* **Rockzipfel; c)** *(erhängt werden)* hang; be hanged; **d)** *(an einem Fahrzeug)* be hitched *or* attached **(an** + *Dat.* to); **e)** *(herab~)* hang down; **bis auf den Boden ~:** hang down to the ground; **die Beine ins Wasser ~ lassen** let one's legs dangle in the water; **f)** *(unordentlich sitzen)* **im Sessel ~** *(erschöpft, betrunken)* be *or* sit slumped in one's/the chair; *(flegelhaft)* lounge in one's/the chair; **g)** *(geh.: schweben, auch fig.)* hang **(über** + *Dat.* over); **h)** *(haften)* cling, stick **(an** + *Dat.* to); **i)** *(fest~)* **sie hing mit dem Rock am Zaun/in der Fahrradkette** her skirt was caught on the fence/in the bicycle chain; **j)** *(ugs.: sich aufhalten, sein)* hang around *(coll.);* **[schon wieder] am Telefon/vorm Fernseher ~:** be on the telephone [again]/be in front of the television [again]; **k)** *(sich nicht* *trennen wollen)* **an jmdm./etw.** ~: be very attached to sb./sth.; **l)** *(sich neigen)* lean; **m)** *(ugs.: angeschlossen sein)* **an etw.** *(Dat.)* ~: be on sth.; **n)** *(ugs.: nicht weiterkommen)* be stuck; **o)** *(ugs.: zurück sein)* be behind; **p)** *(entschieden werden)* **an/bei jmdm./etw.** ~: depend on sb./sth.

²**hängen 1.** *tr. V.* **a) etw. in/über etw.** *(Akk.)* ~: hang sth. in/over sth.; **etw. an/auf etw.** *(Akk.)* ~: hang sth. on sth.; **b)** *(befestigen)* hitch up (**an** + *Akk.* to); couple on ‹*railway carriage, trailer, etc.*› (**an** + *Akk.* to); **c)** *(~ lassen)* hang; **die Beine ins Wasser** ~: let one's legs dangle in the water; **d)** *(er~)* hang; **Tod durch Hängen** death by hanging; **mit Hängen und Würgen** by the skin of one's teeth; **e)** *(ugs.: aufwenden)* **an/in etw.** *(Akk.)* ~: put ‹*work, time, money*› into sth.; spend ‹*time, money*› on sth.; **f)** *(ugs.: anschließen)* **jmdn./etw. an etw.** *(Akk.)* ~: put sb./sth. on sth.; *s. auch* **Glokke a; Nagel b. 2.** *refl. V.* **a)** *(ergreifen)* **sich an etw.** *(Akk.)* ~: hang on to sth.; **sich jmdm. an den Hals** ~: cling to sb.'s neck; **sich ans Telefon** ~ *(fig. ugs.)* get on the telephone; **b)** *(sich festsetzen)* ‹*smell*› cling (**an** + *Akk.* to); ‹*burr, hairs, etc.*› cling, stick (**an** + *Akk.* to); **c)** *(anschließen)* **sich an jmdn.** ~: attach oneself to sb.; latch on to sb. *(coll.)*; **d)** *(verfolgen)* **sich an jmdn./ein Auto** ~: follow *or* *(coll.)* tail sb./a car

hängen|bleiben *unr. itr. V.; mit sein (ugs.)* **a)** *(festgehalten werden)* **[mit dem Ärmel** *usw.***] an/in etw.** *(Dat.)* ~: get one's sleeve *etc.* caught on/in sth.; **b)** *(verweilen)* get stuck *(coll.)*; **c)** *(haften)* **an/auf etw.** *(Dat.)* ~: stick to sth.; **von dem Vortrag blieb [bei ihm] nicht viel hängen** *(fig.)* not much of the lecture stuck *(coll.)*; **ein Verdacht bleibt an ihr hängen** *(fig.)* suspicion rests on her; **d)** *(ugs.: sitzenbleiben)* stay down; have to repeat a year

hängend *Adj.* hanging; **mit ~em Kopf** with head hanging

hängen|lassen 1. *unr. tr. V.* **a)** *(vergessen)* **etw.** ~: leave sth. behind; **b)** *(ugs.: nicht helfen)* **jmdn.** ~: let sb. down. **2.** *unr. refl. V.* let oneself go; **laß dich nicht so hängen!** [you must] pull yourself together!

Hänge-: **~ohr das** lop ear; **~partie die** *(Schach)* adjourned game; **~schrank der** wall-cupboard

Hannover [ha'no:fɐ] **(das);** ~s Hanover

Hannoveraner 1. der; ~s, ~ Hanoverian. **2.** *indekl. Adj.* Hanover

Hans [hans] **der;** ~, **Hänse** ['hɛnzə] ~ **im Glück** lucky devil; *(Märchenfigur)* Hans in Luck

Hansaplast Ⓦ [hanza'plast] **das;** ~[e]s sticking plaster; Elastoplast (P)

Hans·dampf der; ~[e]s, ~e: ~ **[in allen Gassen]** Jack of all trades

Hanse ['hanzə] **die;** ~ *(hist.)* Hanse; Hanseatic league

Hanseat [hanze'a:t] **der;** ~en, ~en **a)** citizen of a Hanseatic city; **b)** *(hist.)* member of the Hanseatic League

hanseatisch *Adj.* Hanseatic

Hänselei die; ~: teasing

hänseln ['hɛnzl̩n] *tr. V.* tease

Hanse·stadt die Hanseatic city

Hans·wurst der; ~[e]s, ~e **a)** *(dummer Mensch)* clown; **b)** *(Theater)* fool; hanswurst

Hantel ['hantl̩] **die;** ~, ~n *(Sport) (kurz)* dumb-bell; *(lang)* barbell

hantieren [han'ti:rən] *itr. V.* be busy

hapern ['ha:pɐn] *itr. V. (unpers.)* **a)** *(fehlen)* **es hapert bei jmdm. an etw.** sb. is short of sth.; **b)** *(nicht klappen)* **es hapert mit etw.** there's a problem with sth.

Häppchen ['hɛpçən] **das;** ~s, ~ **a)** [small] morsel; **b)** *(Appetithappen)* canapé

Happen ['hapn̩] **der;** ~s, ~: morsel; **einen** ~ **essen** have a bite to eat; **ein fetter** ~ *(fig.)* a real plum

happig ['hapɪç] *Adj. (ugs.)* **~e Preise** fancy prices *(coll.)*

Happy-End ['hɛpi'|ɛnt] **das;** ~[s], ~s happy ending

Härchen ['hɛ:ɐ̯çən] **das;** ~s, ~: little *or* tiny hair

Harem ['ha:rɛm] **der;** ~s, ~s *(auch ugs. scherzh.)* harem

Harems-: **~dame die** lady of the harem; **~wächter der** guardian of the harem

Häretiker [hɛ're:tikɐ] **der;** ~s, ~: heretic

Harfe ['harfə] **die;** ~, ~n harp

Harke ['harkə] **die;** ~, ~n rake; **jmdm. zeigen, was eine ~ ist** *(fig. salopp)* give sb. what for *(coll.)*

harken *tr. V.* rake

Harlekin ['harleki:n] **der;** ~s, ~e harlequin

härmen ['hɛrmən] *refl. V. (geh.)* grieve (**um** over)

harm·los 1. *Adj.* **a)** harmless; slight ‹*injury, cold, etc.*›; mild ‹*illness*›; safe ‹*medicine, bend, road, etc.*›; **eine ~e Grippe** a mild bout of flu; **b)** *(arglos)* innocent; harmless ‹*fun, pastime, etc.*›. **2.** *adv.* **a)** harmlessly; **b)** *(arglos)* innocently; **ganz ~ tun** act innocent

Harmlosigkeit die; ~ **a)** harmlessness; *(einer Krankheit)* mildness; *(eines Medikamentes)* safety; **b)** *(Arglosigkeit)* innocence

Harmonie [harmo'ni:] **die;** ~, ~n *(auch fig.)* harmony

Harmonie·lehre die; *o. Pl.* theory of harmony

harmonieren *itr. V.* harmonize; go together; match; **mit etw.** ~: harmonize *or* go together with sth.

Harmonik [har'mo:nɪk] **die;** ~: harmony

Harmonika [har'mo:nika] **die;** ~, ~s *od.* **Harmoniken** harmonica

harmonisch 1. *Adj.* **a)** *(Musik)* harmonic ‹*tone, minor*›; **b)** *(wohlklingend, übereinstimmend)* harmonious; **c)** *(Math.)* **~e Teilung** harmonic division. **2.** *adv.* **a)** *(Musik)* harmonically; **b)** *(wohlklingend, übereinstimmend)* harmoniously; ~ **zusammenleben** live together in harmony

harmonisieren *tr. V.* **a)** *(Musik)* harmonize; **b)** *(in Einklang bringen)* co-ordinate

Harmonium [har'mo:ni̯ʊm] **das;** ~s, **Harmonien** harmonium

Harn [harn] **der;** ~[e]s, ~e *(Med.)* urine; ~ **lassen** *(ugs.)* pass water; urinate

Harn-: **~blase die** bladder; **~drang der** desire to urinate *or* pass water

Harnisch ['harnɪʃ] **der;** ~s, ~e **a)** armour; **b) jmdn. in ~ bringen** get sb.'s hackles up; make sb. see red

harn·treibend *Adj.* diuretic

Harpune [har'pu:nə] **die;** ~, ~n harpoon

Harpunier [harpu'ni:ɐ̯] **der;** ~s, ~e harpooner

harpunieren 1. *tr. V.* harpoon. **2.** *itr. V.* throw/fire the harpoon

harren ['harən] *itr. V. (geh.)* **jmds./einer Sache** ~: wait for *or* await sb./sth.; *(fig.)* await sb./sth.; **der Dinge ~, die da kommen sollen** wait and see what happens

harsch [harʃ] **1.** *Adj.* **a)** *(vereist)* crusted ‹*snow*›; **b)** *(barsch)* harsh. **2.** *adv.* harshly

Harsch der; ~[e]s crusted *or* hard snow

hart [hart]; **härter** ['hɛrtɐ], **härtest...** ['hɛrtəst...] **1.** *Adj.* **a)** hard; **~e Eier** hard-boiled eggs; **Eier ~ kochen** hard-boil eggs; ~ **gefroren** frozen solid; *s. auch* **Nuß a; b)** *(abgehärtet)* tough; ~ **im Neh-**

men sein *(Schläge ertragen können)* be able to take a punch; *(Enttäuschungen ertragen können)* be able to take the rough with the smooth; **c)** *(schwer erträglich)* hard ⟨*work, life, fate, lot, times*⟩; tough ⟨*childhood, situation, job*⟩; harsh ⟨*reality, truth*⟩; **ein ~er Schlag für jmdn. sein** be a heavy *or* severe blow for sb.; **d)** *(streng)* severe, harsh ⟨*penalty, punishment, judgement*⟩; tough ⟨*measure, law, course*⟩; harsh ⟨*treatment*⟩; severe, hard ⟨*features*⟩; **e)** *(heftig)* hard, violent ⟨*impact, jolt*⟩; heavy ⟨*fall*⟩; **f)** *(rauh, scharf)* rough ⟨*game, opponent*⟩; hard, severe ⟨*winter, frost*⟩; harsh ⟨*accent, contrast*⟩. **2.** *adv.* **a)** *(mühevoll)* ⟨*work*⟩ hard; **es kommt mich ~ an** it is hard for me; **b)** *(streng)* severely; harshly; **~ durchgreifen** take tough measures; **jmdn. ~ anfassen** be tough with sb.; **c)** *(heftig)* **jmdm. ~ zusetzen, jmdn. ~ bedrängen** press sb. hard; **es geht ~ auf ~:** the chips are down; **d)** *(nahe)* close **(an +** *Dat.* to); **~ am Wind segeln** *(Seemannsspr.)* sail near *or* close to the wind

Härte ['hɛrtə] **die; ~, ~n a)** *(auch Physik)* hardness; **b)** *o. Pl. (Widerstandsfähigkeit)* toughness; **c)** *(schwere Belastung)* hardship; **eine soziale ~:** a case of social hardship; **d)** *o. Pl. (Strenge)* severity; **e)** *o. Pl. (Heftigkeit) (eines Aufpralls usw.)* force; *(eines Streits)* violence; **f)** *(Rauheit)* roughness; **g)** *o. Pl. (Stabilität)* hardness; **h)** *(von Wasser)* hardness; **i)** *(von Licht, Farbe)* harshness; *(von Frost)* hardness

Härte-: ~fall der a) case of hardship; **b)** *(ugs.: Person)* hardship case; **~grad der** degree of hardness

härten 1. *tr. V.* harden; harden, temper ⟨*steel*⟩. **2.** *itr. V.* harden

härter, härtest... *s.* **hart**

hart-, Hart-: ~faser·platte die hardboard; **~gekocht** *Adj.* **a)** hard-boiled ⟨*egg*⟩; **b)** *s.* **~gesotten; ~geld das** coins *pl.;* small change; **~gesotten** *Adj.* **a)** *(gefühllos)* hard-bitten; hard-boiled; **b)** *(unbelehrbar)* hardened; **~gummi das** hard rubber; **~herzig 1.** *Adj.* hard-hearted; **2.** *adv.* hard-heartedly; **~herzigkeit die; ~:** hard-heartedness; **~holz das** hardwood; **~metall das** hard metal; **~näckig** [~nɛkɪç] **1.** *Adj.* **a)** obstinate; stubborn; **b)** *(ausdauernd)* persistent; dogged; inveterate ⟨*liar*⟩; stubborn, dogged ⟨*resistance*⟩; persistent ⟨*questioning, questioner*⟩; **2.** *adv.* **a)** obstinately; stubbornly; **b)** *(ausdauernd)* persistently; doggedly; **~näckigkeit die; ~ a)** obstinacy; stubbornness; **b)** *(Ausdauer)* persistence; doggedness; **~platz der** *(Sport) (Tennis)* hard court; *(Fußball)* asphalt pitch; **~schalig** [~ʃa:lɪç] *Adj.* hardshell; hard-shelled; thick-skinned ⟨*apple, pear, etc.*⟩

Härtung die; ~, ~en hardening; *(von Stahl auch)* tempering

Hart·wurst die dry sausage

Harz [ha:ɐ̯ts] **das; ~es, ~e** resin

Harzer Käse der; ~ ~s, ~ ~: Harz [Mountain] cheese

harzig *Adj.* resinous

Hasch [haʃ] **das; ~s** *(ugs.)* hash *(coll.)*

Haschee [ha'ʃe:] *(Kochk.)* **das; ~s, ~s** hash

¹haschen *(veralt.)* **1.** *tr. V.* catch. **2.** *itr. V.* **nach etw. ~:** make a grab for sth.

²haschen *itr. V. (ugs.)* smoke [hash] *(coll.)*

Häschen ['hɛ:sçən] **das; ~, ~:** bunny

Häscher ['hɛʃɐ] **der; ~s, ~** *(geh. veralt.)* pursuer

Haschisch ['haʃɪʃ] **das** *od.* **der; ~[s]** hashish

Hase ['ha:zə] **der; ~n, ~n** hare; **ein alter ~ sein** *(ugs.)* be an old hand; **falscher ~** *(Kochk.)* meat loaf; **da liegt der ~ im Pfeffer** *(ugs.)* that's the real trouble; **sehen/wissen wie der ~ läuft** *(ugs.)* see/know which way the wind blows; **mein Name ist ~** *(ugs. scherzh.)* I'm not saying anything

Hasel-: ~kätzchen das hazel catkin; **~nuß die a)** hazel-nut; **b)** hazel [tree]; **~[nuß]·strauch der** hazel [tree]

hasen-, Hasen-: ~fuß der *(spöttisch abwertend)* coward; chicken *(sl.);* **~jagd die** hare shoot; **~panier** *in* **das ~panier ergreifen** take to one's heels; **~pfeffer der** *(Kochk.)* marinaded and stewed trimmings *pl.* of hare; **~rein** *Adj.* **er/das ist nicht ganz ~rein** *(fig.)* there's something fishy *(coll.)* about him/it; **~scharte die** *(Med.)* harelip

Haspel ['haspl̩] **die; ~, ~n** *(Technik)* **a)** *(für Garn)* reel; bobbin; *(für ein Seil, Kabel)* drum; **b)** *(Seilwinde)* windlass

Haß [has] **der; Hasses** hate; hatred **(auf +** *Akk.,* **gegen** of, for); **sein [ganzer] ~:** [all] his hatred

hassen *tr., itr. V.* hate; *s. auch* **Pest**

haß·erfüllt 1. *Adj.* filled with hatred *or* hate *postpos.* **2.** *adv.* **jmdn. ~ ansehen** look at sb. with [one's] eyes full of hatred *or* hate

häßlich ['hɛslɪç] **1.** *Adj.* **a)** ugly; **~ wie die Nacht** as ugly as sin *(coll.);* **b)** *(gemein)* nasty; hateful; **c)** *(unangenehm)* terrible *(coll.),* awful ⟨*weather, cold, situation, etc.*⟩. **2.** *adv.* **a)** ⟨*dress*⟩ unattractively; **b)** *(gemein)* nastily; hatefully

Häßlichkeit die; ~, ~en a) *o. Pl. (Aussehen)* ugliness; **b)** *o. Pl. (Gesinnung)* meanness; nastiness; hatefulness

Haß·liebe die love-hate relationship

hast [hast] *2. Pers. Sg. Präsens v.* **haben**

Hast die; ~: haste; **etw. in** *od.* **mit größter ~ tun** do sth. in great haste; **ohne ~:** unhurriedly; without hurrying *or* haste

haste ['hastə] *(ugs.)* = **hast du; [was] ~ was kannste** as fast as he/you/they *etc.* can/could; **~ was, biste was** money talks

hasten *itr. V.; mit sein* hurry; hasten

hastig 1. *Adj.* hasty; hurried. **2.** *adv.* hastily; hurriedly; **nur nicht so ~!** not so fast!

hat [hat] *3. Pers. Sg. Präsens v.* **haben**

Hätschel·kind das pampered child; *(fig.)* darling

hätscheln ['hɛ:tʃl̩n] *tr. V.* **a)** *(liebkosen)* fondle; caress; **b)** *(verwöhnen)* pamper; *(fig.)* lionize

hatschi [ha'tʃi:] *Interj.* atishoo; atchoo

hatte ['hatə] *1. u. 3. Pers. Sg. Prät. v.* **haben**

hätte ['hɛtə] *1. u. 3. Pers. Sg. Konjunktiv II v.* **haben**

Hatz [hats] **die; ~, ~en a)** *(Hetzjagd, auch fig. ugs.)* hunt; **b)** *(ugs., bes. bayr.: Eile)* mad rush

Haube ['haubə] **die; ~, ~n a)** bonnet; *(einer Krankenschwester)* cap; **unter die ~ kommen** *(ugs. scherzh.)* get hitched *(coll.);* **b)** *(Kfz-W.)* bonnet *(Brit.);* hood *(Amer.);* **c)** *(Zool.)* crest; **d)** *(Bedeckung)* cover; *(über Teekanne, Kaffeekanne, Ei)* cosy

Hauben·taucher der great crested grebe

Haubitze [hau'bɪtsə] **die; ~, ~n** *(Milit.)* howitzer

Hauch [haux] **der; ~[e]s, ~e** *(geh.)* **a)** *(Atem, auch fig.)* breath; **b)** *(Luftzug)* breath of wind; breeze; **c)** *(leichter Duft)* delicate smell; waft; **d)** *(dünne Schicht)* [gossamer-] thin layer

hauch·dünn 1. *Adj.* gossamer-thin ‹*material, dress*›; wafer-thin, paper-thin ‹*layer, slice, majority*›. **2.** *adv.* **etw. ~ auftragen** apply sth. very sparingly; **etw. ~ schneiden** cut sth. wafer-thin *or* into wafer-thin slices

hauchen 1. *itr. V.* breathe **(gegen, auf** + *Akk.* on). **2.** *tr. V.* *(auch fig.: flüstern)* breathe; **jmdm. etw. ins Ohr ~:** breathe sth. in sb.'s ear

hauch-: ~fein *Adj.* extremely fine; **~zart** *Adj.* extremely delicate; gossamer-thin

Hau·degen der: [alter] ~: old soldier *or* warhorse

Haue ['hauə] **die; ~, ~n a)** *(südd., österr.: Hacke)* hoe; **b)** *o. Pl. (ugs.: Prügel)* a hiding *(coll.)*

hauen 1. *tr. V.* **a)** *(ugs.: schlagen)* belt; clobber *(coll.)*; beat; **jmdn. windelweich/grün und blau ~:** beat sb. black and blue; **b)** *(ugs.: auf einen Körperteil)* belt *(coll.)*; hit; *(mit der Faust auch)* smash *(sl.)*; punch; *(mit offener Hand auch)* slap; smack; **c)** *(ugs.: hineinschlagen)* knock; **d)** *(herstellen)* carve ‹*figure, statue, etc.*› **(in** + *Akk.* in); cut, chop ‹*hole*›; **Stufen in den Fels ~:** cut steps in the rock; **e)** *(mit einer Waffe schlagen)* **jmdn. aus dem Sattel/vom Pferd ~:** knock sb. out of the saddle/off his/her horse; **f)** *(salopp: schleudern)* sling *(coll.)*; fling; **g)** *(landsch.: fällen)* fell; cut down; **h)** *(Bergbau)* cut ‹*coal, ore*›. **2.** *itr. V.* **a) jmdm. auf die Schulter ~:** slap *or* clap sb. on the shoulder; **jmdm. ins Gesicht ~:** belt/slap sb. in the face; **mit der Faust auf den Tisch ~:** thump the table [with one's fist]; **b)** *mit sein (ugs.: stoßen)* bump; **mit dem Kopf/Bein gegen etw. ~:** bang *or* hit *or* bump one's head/leg against sth. **3.** *refl. V.* **a)** *(ugs.: sich prügeln)* have a punch-up *(coll.) or* a fight; fight; **b)** *(salopp: sich setzen, legen)* fling *or* throw oneself; **sich ins Bett ~:** hit the sack *(sl.)*

Hauer der; ~s, ~ a) *(Bergmannsspr.)* face-worker; **b)** *(Jägerspr.)* tusk; *(fig.)* fang

Häufchen ['hɔyfçən] **das; ~s, ~:** [small *or* little] pile *or* heap; **nur noch ein ~ Unglück** *od.* **Elend sein** *(ugs.)* be nothing but a small bundle of misery

Haufen ['haufn̩] **der; ~s, ~ a)** heap; pile; **etw. zu ~ aufschichten** stack sth. up in piles; **alles auf einen ~ werfen** throw everything in a heap; **der Hund hat da einen ~ gemacht** *(ugs.)* the dog has done his business there *(coll.)*; **etw. über den ~ werfen** *(ugs.) (aufgeben)* chuck sth. in *(coll.)*; *(zunichte machen)* mess sth. up; **jmdn. über den ~ fahren/rennen** *(ugs.)* knock sb. down; run sb. over; **jmdn. über den ~ schießen** *od.* **knallen** *(ugs.)* gun *or* shoot sb. down *(coll.)*; **b)** *(ugs.: große Menge)* heap *(coll.)*; pile *(coll.)*; load *(coll.)*; **ein ~ Arbeit/Bücher** a load *or* heap *or* pile of work/books *(coll.)*; loads *or* heaps *or* piles of work/books *(coll.)*; **ein ~ Geld** loads of money *(coll.)*; **c)** *(Ansammlung von Menschen)* crowd; **so viele Idioten auf einem ~** *(ugs.)* so many idiots in one place

häufen ['hɔyfn̩] **1.** *tr. V.* heap, pile **(auf** + *Akk.* on to). **2.** *refl. V. (sich mehren)* pile up

haufen·weise *Adv. (ugs.)* **~ Geld ausgeben/Eis essen** spend loads of money/eat heaps *or* loads of ice cream *(coll.)*

häufig ['hɔyfıç] **1.** *Adj.* frequent. **2.** *adv.* frequently; often

Häufigkeit die; ~, ~en frequency

Häufung die; ~, ~en increasing frequency

Haupt [haupt] **das; ~[e]s, Häupter** ['hɔyptɐ] **a)** *(geh.: Kopf)* head; **erhobenen ~es** with one's head [held] high; **gesenkten ~es** with head bowed; **gekrönte Häupter** crowned heads; **b)** *(geh.: wichtigste Person)* head

haupt-, Haupt-: ~aktionär der principal shareholder; **~akzent der** *(Phon.)* main *or* primary stress; *(fig.)* main emphasis; **~amtlich 1.** *Adj.* full-time; **2.** *adv.* **~amtlich tätig sein** work full-time *or* on a full-time basis; **~arbeit die** main part of the work; **~bahnhof der** main station; **Amsterdam ~bahnhof** Amsterdam Central; **~beruflich 1.** *Adj.* **seine ~berufliche Tätigkeit** his main occupation; **2.** *adv.* **er ist ~beruflich als Elektriker tätig** his main occupation is that of electrician; **~beschäftigung die** main occupation; **~buch das** *(Kaufmannsspr.)* ledger; **~darsteller der** *(Theater, Film)* leading man; male lead; **~darstellerin die** *(Theater, Film)* leading lady; female lead; **~eingang der** main entrance; **~einnahme·quelle die** main *or* principal source of income; *(eines Staates)* main *or* principal source of revenue; **~fach das a)** *(Universität)* main subject; major; **etw. im ~fach studieren** study sth. as one's main subject; **b)** *(Schule)* main subject; **~fehler der** main *or* principal *or* chief mistake/*(im Charakter)* fault/*(in einer Theorie, einem Argument)* flaw; **~feld das** *(Sport)* [main] bunch; **~feldwebel der** *(Milit.)* ≈ staff sergeant *(Brit.)*; ≈ sergeant first class *(Amer.)*; **~figur die** main *or* principal character; **~film der** main feature *or* film; **~gang der a)** main corridor; **b)** *s.* **~gericht; ~gebäude das** main building; **~gericht das** main course; **~geschäft das a)** *(Laden)* main branch; **b)** *(größter Umsatz)* peak sales *pl.*; *(wichtigster Geschäftszweig)* main line; **~geschäfts·straße die** main shopping street; **~gewicht das** main emphasis; **~gewinn der** first *or* top prize; **~grund der** main *or* principal *or* chief reason; **~hahn der** mains stopcock; **~interesse das** main interest; **~last die** main burden; **~leitung die** *(Gas-, Wasserleitung)* main; *(Stromleitung)* main[s *pl.*]

Häuptling ['hɔyptlıŋ] **der; ~s, ~e** chief[tain]; *(iron.)* bigwig *(coll.)*

haupt-, Haupt-: ~mahlzeit die main meal; **~mann der;** *Pl.* **~leute a)** *(Milit.)* captain; **b)** *(hist.)* leader; **~merkmal das** main *or* principal *or* chief characteristic; **~motiv das a)** *(Gegenstand)* main *or* principal motif; **b)** *(Beweggrund)* main *or* principal *or* chief motive; **~person die** central figure; **sie will immer und überall die ~person sein** *(fig.)* she always wants to be the centre of everything *or* of attention; **~post die, ~post·amt das** main post office; **~problem das** main *or* chief problem; **~quartier das** *(Milit., auch fig.)* headquarters *sing. or pl.*; **~redner der** main *or* principal speaker; **~reise·zeit die** high season; peak [holiday] season; **~rolle die** leading *or* main role; lead; **die ~rolle spielen** play the leading role *or* the lead **(in** + *Dat.* in); **die ~rolle [in** *od.* **bei etw.] spielen** *(fig.)* play the leading role [in sth.]; **~sache die** main *or* most important thing; **in der ~sache** mainly; in the main; **~sächlich 1.** *Adv.* mainly; principally; chiefly; **2.** *Adj.; nicht präd.* main; principal; chief; **~saison die** high season; **~satz der** *(Sprachw.)* main clause; *(alleinstehend)* sentence; **~schalter der** *(Elektrot.)* mains switch;

~schlagader die aorta; **~schlüssel der** master key; pass key; **~schul·abschluß der** ≈ secondary school leaving certificate; **~schuld die** main share of the blame; **~schuldige der/die** person mainly to blame; *(an einem Verbrechen)* main *or* chief offender; **~schule die** ≈ secondary modern school; **~schul·lehrer der** ≈ secondary modern school teacher; **~sicherung die** *(Elektrot.)* mains fuse; **~sitz der** head office; headquarters *pl.*; **~stadt die** capital [city]; **~städtisch** *Adj.* metropolitan; **~straße die a)** *(wichtigste Geschäftsstraße)* high *or* main street; **b)** *(Durchgangsstraße)* main road; **~strecke die** *(Eisenb.)* main line; **~teil der** major part; **~treffer der** *s.* **~gewinn; ~tribüne die** *(Sport)* main stand; **~unterschied der** main *or* principal *or* chief difference; **~ursache die** main *or* principal *or* chief cause; **~verantwortliche der/die** person mainly responsible; **~verhandlung die** *(Rechtsw.)* main hearing

Hauptverkehrs-: ~straße die main road; **~zeit die** rush hour

Haupt-: ~versammlung die *(Wirtsch.)* shareholders' meeting; **~verwaltung die** head office; **~wohn·sitz der** main place of residence; **~wort das** *(Sprachw.)* noun

hau ruck ['hau'rʊk] *Interj.* heave[-ho]

Haus [haus] **das; ~es, Häuser** ['hɔyzɐ] **a)** house; *(Firmengebäude)* building; **er ist gerade aus dem ~ gegangen** he has just gone out; **im ~ spielen** play indoors; **kommt ins ~, es regnet** come inside, it's raining; **~ und Hof** *(fig.)* house and home; **jmdm. ins ~ stehen** *(fig. ugs.)* be in store for sb.; **b)** *(Heim)* home; **jmdm. das ~ verbieten** not allow sb. in one's *or* the house; **etw. ins ~/frei ~ liefern** deliver sth. to sb.'s door/free of charge; **das ~ auf den Kopf stellen** *(ugs.)* turn the place upside down; **außer ~[e] sein/essen** be/eat out; **ist Ihre Frau im ~[e]?** is your wife at home?; **nach ~e** home; **zu ~e** at home; **fühlt euch wie zu ~e** make yourselves at home; **das ~ hüten** stay at home *or* indoors; **jmdm. das ~ einrennen** *(ugs.)* be constantly on sb.'s doorstep; **auf einem Gebiet/in etw.** *(Dat.)* **zu ~e sein** *(ugs.)* be at home in a field/in sth.; **c)** *(Theater)* theatre; *(Publikum)* house; **das große/kleine ~:** the large/small theatre; **vor vollen/ausverkauften Häusern spielen** play to full *or* packed houses; **d)** *(Gasthof, Geschäft)* **das erste ~ am Platze** the best shop of its kind/hotel in the town/village *etc.*; **eine Spezialität des ~es** a speciality of the house; **e)** *(Firma)* firm; business house; **das ~ Meyer** the firm of Meyer; **f)** *(geh.: Parlament)* **das Hohe ~:** the House; **g)** *(geh.: Familie)* household; **der Herr/die Dame des ~es** the master/lady of the house; **aus gutem ~e kommen** come from a *or* be of good family; **der Herr im eigenen ~ sein** be master in one's own house; **von ~[e] aus** *(von der Familie her)* by birth; *(eigentlich)* really; actually; **h)** *(~halt)* household; **jmdm. das ~ führen** keep house for sb.; **i)** *(Dynastie)* **das ~ Tudor/[der] Hohenzollern** the House of Tudor/Hohenzollern; **j) ein gelehrtes/lustiges** *usw.* ~ *(ugs. scherzh.)* a scholarly/amusing *etc.* sort *(coll.)*; **k)** *(Schnecken~)* shell

haus-, Haus-: ~angestellte der/die domestic servant; **~apotheke die** medicine cabinet; **~arbeit die a)** housework; **b)** *(Schulw.)* item of homework; **~arrest der a)** house arrest; **b)** *(in der Familie)* **mein Bruder hat ~arrest** my brother is being kept in; **~arzt der** family doctor; **~aufgabe die** piece of homework; **~aufgaben aufhaben** *(ugs.)* have homework *sing.*; **~aufsatz der** homework essay; **~backen 1.** *Adj.* plain; unadventurous, boring ⟨*clothes*⟩; **2.** *adv.* ⟨*dress*⟩ unadventurously; **~bau der** house-building; **beim ~bau** when building a/one's house; **~besetzer der** squatter; **~besetzung die** *(Vorgang)* squatting; *(Ergebnis)* squat; **~besitzer der** house-owner; *(Vermieter)* landlord; **~besitzerin die** house-owner; *(Vermieterin)* landlady; **~besuch der** house-call; **~bewohner der** occupant [of the house]; **~boot das** houseboat

Häuschen ['hɔysçən] **das; ~s, ~: a)** little *or* small house; **b) [ganz** *od.* **rein] aus dem ~ sein** *(ugs.)* be [completely] over the moon *(coll.)*; **c)** *(ugs.: Toilette)* privy

haus-, Haus-: ~dame die housekeeper; **~detektiv der** house detective; **~diener der** domestic servant; **~drachen der** *(ugs. abwertend)* dragon *(coll.)*; **~ecke die** corner of the house; **~eigen** *Adj.* **der ~eigene Kindergarten** the company's/hotel's *etc.* own kindergarten; **das Hotel hat einen ~eigenen Swimming-pool/Strand** the hotel has its own swimming-pool/[private] beach; **~eigentümer der** *s.* **~besitzer; ~eingang der** entrance [to the house]

hausen *itr. V. (ugs.)* **a)** *(wohnen)* live; **b)** *(Verwüstungen anrichten)* **[furchtbar] ~:** cause *or* wreak havoc

Häuser·block der block [of houses]

Haus-: ~flur der hall[way]; entrance-hall; *(im Obergeschoß)* landing; **~frau die** housewife

hausfraulich *Adj.* housewifely; **ihre ~en Fähigkeiten** her abilities as a housewife

Haus·freund der a) friend of the family; family friend; **b)** *(verhüll.: Liebhaber)* man-friend *(euphem.)*

Hausfriedens·bruch der *(Rechtsw.)* trespass

haus-, Haus-: ~gebrauch der domestic use; **das reicht für den ~gebrauch** *(ugs.)* it's good enough to get by *(coll.)*; **~gehilfin die** [home] help; **~gemacht** *Adj.* home-made; **~gemeinschaft die a)** *(gemeinsamer ~halt)* household; **b)** *(Bewohner eines Hauses)* occupants *pl.* of the block

Haus·halt der a) household; **einen ~ gründen/auflösen** set up home/break up a household; **b)** *(Arbeit im ~)* housekeeping; **jmdm. den ~ führen** keep house for sb.; **im ~ helfen** help with the housework; **c)** *(Politik)* budget

haus|halten *unr. itr. V.* be economical **(mit** with)

Haushälterin die; ~, ~nen housekeeper

Haushalts-: ~artikel der household article; **~auflösung die** house clearance; **~buch das** housekeeping book; **~debatte die** *(Politik)* budget debate; **~defizit das** budgetary deficit; **~führung die** housekeeping; **~geld das;** *o. Pl.* housekeeping money; **~gerät das** household appliance; **~hilfe die** home help; **~jahr das** *(Rechnungsjahr)* financial year; **~kasse die** housekeeping money; **die ~kasse war leer** there was no housekeeping money left; **~plan der** budget; **~politik die** budgetary policy; **~waren** *Pl.* household goods

haus-, Haus-: ~haltung die a)

s. **Haushalt a; b)** *(Haushaltsführung)* housekeeping; **~herr der a)** *(Familienoberhaupt)* head of the household; **b)** *(als Gastgeber)* host; **c)** *(Rechtsspr.) (Eigentümer)* owner; *(Mieter)* occupier; **d)** *(südd., österr.) s.* **~besitzer; ~herrin die a)** *(Familienoberhaupt)* lady of the house; **b)** *(als Gastgeberin)* hostess; **c)** *(südd., österr.) s.* **~besitzerin; ~hoch 1.** *Adj.* ‹*flames/waves etc.*› as high as a house; *(fig.)* overwhelming ‹*superiority etc.*›; **die ~hohe Favoritin** the hot favourite; **2.** *adv.* **~hoch gewinnen/jmdn. ~hoch schlagen** win hands down/beat sb. hands down; **jmdm. ~hoch überlegen sein** be vastly superior to sb.

hausieren *itr. V.* **[mit etw.] ~:** hawk [sth.]; peddle [sth.]; **„Hausieren verboten"** 'no hawkers'

Hausierer der; ~s, ~: pedlar; hawker

haus-, Haus-: ~intern 1. *Adj.* internal ‹*regulations, purposes, information*›; ‹*agreement, custom*› within the company; **2.** *adv.* internally; within the company; **~katze die** domestic cat; **~kleid das** house dress; **~lehrer der** private tutor

häuslich ['hɔyslɪç] **1.** *Adj.* **a)** *nicht präd.* domestic ‹*bliss, peace, affairs, duties, etc.*›; **am ~en Kaminfeuer** at one's own fireside; *s. auch* **Herd a; b)** *(das Zuhause liebend)* home-loving. **2.** *adv.* **sich [bei jmdm./irgendwo] ~ niederlassen** *(ugs.)* make oneself at home [in sb.'s house/somewhere]

Hausmacher-: ~art die *in* **nach ~art** home-made-style *attrib.*; **~wurst die** home-made sausage

Haus-: ~macht die *(hist.)* allodium; *(fig.)* power base; **~mädchen das** [home] help

Haus·mann der *man who stays at home and does the housework; (Ehemann)* househusband

Hausmanns·kost die plain cooking

Haus-: ~marke die a) *(Wein, Sekt)* house wine; **b)** *(ugs.: bevorzugtes Getränk)* usual *or* favourite tipple *(coll.)*; **~meister der, ~meisterin die** caretaker; **~mittel das** household remedy; **~musik die** music at home; **~mütterchen das** *(ugs. scherzh.)* little housewife; **~nummer die** house number; **ihre ~nummer** the number of her house; **~ordnung die** house rules *pl.*; **~putz der** spring-clean; *(regelmäßig)* clean-out; **~putz halten** *od.* **machen** spring-clean the house

Haus·rat der household goods *pl.*

Hausrat·versicherung die [household *or* home] contents insurance

Haus-: ~recht das *(Rechtsw.) right of a householder or owner of a property to forbid sb. entrance or order sb. to leave;* **~schlachtung die** home slaughtering; **~schlüssel der** front-door key; house-key; **~schuh der** slipper

Hausse ['ho:s(ə)] **die; ~, ~n** *(Börsenw.)* rise [in prices]; *(fig.)* boom

Haus-: ~segen der: bei ihnen hängt der ~segen schief *(ugs. scherzh.)* they've been having a row; **~stand der** household; **einen [eigenen] ~stand gründen** set up home [on their own]; **~suchung die; ~, ~en** house search; **~suchungs·befehl der** search warrant; **~tier das a)** pet; **b)** *(Nutztier)* domestic animal; **~tür die** front door; **etw. direkt vor der ~tür haben** *(ugs. fig.)* have sth. on one's doorstep; **~tyrann der** *(ugs.)* tyrant [in one's own home]; **~verbot das** ban on entering the house/pub/restaurant *etc.*; **jmdm. ~verbot erteilen** ban sb. [from the house/pub/restaurant *etc.*]; **~verwalter der** manager [of the block]; **~verwaltung die** management [of the block]; **~wand die** [house] wall; **~wirt der** landlord; **~wirtin die** landlady

Haus·wirtschaft die; *o. Pl.* domestic science and home economics

hauswirtschaftlich *Adj.; nicht präd.* domestic

Hauswirtschafts-: ~lehrerin die home economics and domestic science teacher; **~schule die** college of domestic science and home economics

Haut [haut] **die; ~, Häute** ['hɔytə] **a)** skin; **sich** *(Dat.)* **die ~ abschürfen** graze oneself; **viel ~ zeigen** *(ugs. scherzh.)* show a lot of bare flesh *(coll.)*; **naß bis auf die ~:** soaked to the skin; wet through; **nur noch ~ und Knochen sein** *(ugs.)* be nothing but skin and bone; **seine eigene ~ retten** save one's own skin; **seine ~ so teuer wie möglich verkaufen** *(ugs.)* sell oneself as dearly as possible; **sich seiner ~** *(Gen.)* **wehren** *(ugs.)* stand up for oneself; **aus der ~ fahren** *(ugs.)* go up the wall *(coll.)*; **er/sie kann nicht aus seiner/ihrer ~ heraus** *(ugs.)* a leopard cannot change its spots *(prov.)*; **sich in seiner ~ nicht wohl fühlen** *(ugs.)* feel uneasy; *(unzufrieden sein)* feel discontented [with one's lot]; **ich möchte nicht in deiner ~ stecken** *(ugs.)* I shouldn't like to be in your shoes *(coll.)*; **mit heiler ~ davonkommen** *(ugs.)* get away with it; **b)** *(Fell)* skin; *(von größerem Tier auch)* hide; **auf der faulen ~ liegen** *(ugs.)* sit around and do nothing; **c)** *(Schale, dünne Schicht, Bespannung)* skin; **d)** *(ugs.)* **eine gute/ehrliche ~:** a good/honest sort *(coll.)*

Haut-: ~abschürfung die graze; **~arzt der** skin specialist; dermatologist; **~ausschlag der** [skin-]rash; **~creme die** skin cream

häuten ['hɔytn̩] **1.** *tr. V.* skin, flay ‹*animal*›; skin ‹*tomato, almond, etc.*›. **2.** *refl. V.* shed its skin/their skins; ‹*snake*› shed *or* slough its skin

haut-, Haut-: ~eng *Adj.* skin-tight; **~farbe die** [skin] colour; **wegen seiner ~farbe** because of the colour of his skin; **~freundlich** *Adj.* kind to the/one's skin *pred.*; **~krankheit die** skin disease; **~nah 1.** *Adj.* **a)** *(unmittelbar)* immediate ‹*contact*›; eyeball-to-eyeball ‹*confrontation*›; **b)** *(ugs.: packend, anschaulich)* realistic and gripping ‹*description*›; **2.** *adv. (unmittelbar)* **mit etw. ~nah in Berührung/Kontakt kommen** come into very close contact with sth.; **~pflege die** skin care; **~transplantation die** *(Med.)* skin graft

Häutung die; ~, ~en a) *s.* **häuten 1:** skinning; flaying; **b)** *(das Sichhäuten)* **Schlangen machen viele ~en durch** snakes shed *or* slough their skin many times

Havarie [hava'ri:] **die; ~, ~n** *(Seew., Flugw., österr. auch: ~ eines Autos)* accident; *(Schaden)* damage *no indef. art.*

havarieren *itr. V. (Seew., Flugw.)* ‹*aircraft*› crash; ‹*ship*› have an accident; **ein havariertes Schiff** a damaged ship

Hawaii [ha'vai] **(das); ~s** Hawaii

Haxe die; ~, ~n *s.* **Hachse**

H-Bombe ['ha:-] **die** H-bomb

H-Dur ['ha:-] **das; ~** *(Musik)* B major

he [he:] *Interj. (ugs.)* **a)** *(Zuruf, Ausruf)* hey; **~ [du], komm mal her!** hey [you], come here!; **b)** *(zur Verstärkung einer Frage)* eh

Heb·amme die midwife

Hebe-: ~balken der lever;

~bühne die hydraulic lift; **~figur die** *(Eis-, Rollkunstlauf)* lift
Hebel ['he:bl̩] **der; ~s, ~** *(auch Griff, Physik)* lever; **den ~ ansetzen** position the lever; **alle ~ in Bewegung setzen** *(ugs.)* move heaven and earth; **am längeren ~ sitzen** *(ugs.)* have the whip hand
Hebel-: ~gesetz das *(Physik)* principle of the lever; **~kraft, die, ~wirkung die** leverage
heben ['he:bn̩] **1.** *unr. tr. V.* **a)** *(nach oben bewegen)* lift; raise; raise ‹*baton, camera, glass*›; **eine Last ~:** lift a load; **die Hand/den Arm ~:** raise one's hand/arm; **schlurft nicht, hebt die Füße!** pick your feet up!; **die Stimme ~** *(geh.)* raise one's voice; **einen ~** *(ugs.)* have a drink; **b)** *(an eine andere Stelle bringen)* lift; **jmdn. auf die Schulter/von der Mauer ~:** lift sb. [up] on to one's shoulders/[down] from the wall; **c)** *(heraufholen)* dig up ‹*treasure etc.*›; raise ‹*wreck*›; **d)** *(verbessern)* raise, improve ‹*standard, level*›; increase ‹*turnover, self-confidence*›; improve ‹*mood*›; enhance ‹*standing*›; boost ‹*morale*›; **e)** *(unpers.)* **es hebt jmdm. den Magen** sb.'s stomach heaves. **2.** *unr. refl. V.* **a)** *(geh.: sich recken, sich er~)* rise; **b)** *(hochgehen, hochsteigen)* rise; ‹*curtain*› rise, go up; ‹*mist, fog*› lift; **sich ~ und senken** rise and fall; ‹*sea, chest*› rise and fall, heave; **c)** *(sich verbessern)* ‹*mood*› improve; ‹*trade*› pick up; ‹*standard, level*› rise, improve, go up
Heber der; ~s, ~ a) *(Technik)* jack; **b)** *(Chemie)* pipette; **c)** *(Sport: Gewicht~)* weight-lifter
hebräisch [he'brεɪʃ] Hebrew
Hebung die; ~, ~en a) die ~ eines Schiffes the raising of a ship; **bei der ~ des Schatzes ...:** when the treasure is/was dug up ...; **b)** *o. Pl. (Verbesserung)* raising; improvement; **zur ~ des Selbstvertrauens/der Moral** to improve sb.'s self-confidence/morale; **c)** *(Geol.)* uplift; **d)** *(Verslehre)* stressed syllable
hecheln *itr. V.* pant [for breath]
Hecht [hεçt] **der; ~[e]s, ~e a)** pike; **der ~ im Karpfenteich sein** *(ugs.)* be the kingpin; **b)** *(ugs.: Bursche)* **ein toller ~:** an incredible fellow; **c)** *(Tabaksqualm)* fug *(coll).*
hechten *itr. V.; mit sein* dive headlong; make a headlong dive; *(schräg nach oben)* throw oneself sideways; *(vom Sprungturm)* perform *or* do a pike-dive; *(Turnen)* do a long-fly
Hecht-: ~sprung der a) *(Turnen)* Hecht vault; **b)** *(Schwimmen)* racing dive; *(vom Sprungturm)* pike-dive; **~suppe die** *in* **es zieht wie ~suppe** *(ugs.)* there's a terrible draught *(coll.)*
Heck [hεk] **das; ~[e]s, ~e** *od.* **~s a)** *(Schiffs~)* stern; **b)** *(Flugzeug~)* tail; **im ~ der Maschine** at the rear of the plane; **c)** *(Auto~)* rear; back
Heck·antrieb der *(Kfz-W.)* rear-wheel drive
Hecke die; ~, ~n a) hedge; **b)** *(wildwachsend)* thicket
Hecken-: ~rose die dogrose; **~schere die** hedge shears *pl.; (elektrisch)* hedge trimmer; **~schütze der** sniper
Heck·fenster das rear *or* back window
Heckmeck ['hεkmεk] **der; ~s** *(ugs. abwertend)* **a)** *(Getue)* fuss; **b)** *(Unsinn)* rubbish
Heck-: ~motor der rear engine; **~scheibe die** rear *or* back window
Heer [he:ɐ̯] **das; ~[e]s, ~e a)** *(Gesamtheit der Streitkräfte)* armed forces *pl.;* **das stehende ~:** the standing army; **b)** *(für den Landkrieg, auch fig.)* army
Heeres·leitung die *(Milit.)* army command staff; **die oberste ~** the high command
Heer·schar die *(veralt., noch fig.)* host *(arch.); s. auch* **himmlisch 1 a**
Hefe ['he:fə] **die; ~, ~n** yeast; *(fig.)* driving force
Hefe-: ~gebäck das pastry *(made with yeast dough);* **~kloß der** dumpling made with yeast dough; **aufgehen** *od.* **auseinandergehen wie ein ~kloß** *(ugs. scherzh.)* blow up like a balloon; **~kuchen der** yeast cake; **~teig der** yeast dough; **~zopf der** plaited bun
[1]**Heft** [hεft] **das; ~[e]s, ~e** *(geh.) (am Dolch, Messer)* haft; handle; *(am Schwert)* hilt; **das ~ in der Hand haben/behalten** *(geh.)* be in/keep control
[2]**Heft das; ~[e]s, ~e a)** exercise-book; **b)** *(Nummer einer Zeitschrift)* issue; **Jahrgang 10, Heft 12** Volume 10, No. 12; **c)** *(kleines Buch) (small stapled)* book
Heftchen das; ~s, ~ a) *(Comic)* comic; *(Groschenroman)* novelette; **b)** *(Block)* book [of tickets/stamps *etc.*]
heften 1. *tr. V.* **a)** *(mit einer Nadel)* pin; fix; *(mit einer Klammer)* clip; fix; *(mit Klebstoff)* stick; **etw. an/in etw.** *(Akk.)* **~:** pin/stick/clip sth. to/into sth.; **b)** *(richten)* **den Blick auf jmdn./etw. ~:** fasten one's gaze on sb./sth.; **c)** *(Schneiderei)* tack; baste; **d)** *(Buchbinderei)* stitch; *(mit Klammern)* staple. **2.** *refl. V.* **sich an jmds. Fersen** *(Akk.)* **~:** stick hard on sb.'s heels
Hefter der; ~s, ~: [loose-leaf] file
Heft·garn das tacking-thread; basting-thread
heftig 1. *Adj.* violent; heavy ‹*rain, shower, blow*›; intense, burning ‹*hatred, desire*›; fierce ‹*controversy, criticism, competition*›; severe ‹*pain, cold*›; loud ‹*bang*›; rapid ‹*breathing*›; bitter ‹*weeping*›; heated, vehement ‹*tone, words*›; **~ werden** fly into a temper. **2.** *adv.* ‹*rain, snow, breathe*› heavily; ‹*hit*› hard; ‹*hurt*› a great deal; ‹*answer*› angrily, heatedly; ‹*react*› angrily, violently
Heftigkeit die; ~ a) *s.* **heftig 1:** violence; heaviness; intensity; fierceness; severity; loudness; rapidity; bitterness; **b)** *(Unbeherrschtheit)* vehemence
Heft-: ~klammer staple; **~maschine die** stapler; *(Buchbinderei)* stitcher; **~pflaster das** sticking plaster; **~zwecke die** *s.* **Reißzwecke**
Hege ['he:gə] **die; ~** *(Forstw., Jagdw.)* care and protection; *(fig.)* care
hegen *tr. V.* **a)** *(bes. Forstw., Jagdw.)* look after, tend ‹*plants, animals*›; **b)** *(geh.: umsorgen)* look after; take care of; preserve ‹*old customs*›; **jmdn./etw. ~ und pflegen** lavish care and attention on sb./sth.; **c)** *(in sich tragen)* feel ‹*contempt, hatred, mistrust*›; cherish ‹*hope, wish, desire*›; harbour, nurse ‹*grudge, suspicion*›; **eine Abneigung gegen jmdn. ~:** have a dislike for sb.; **ich hege den Verdacht, daß ...:** I have a suspicion that ...
Hehl [he:l] *in* **kein[en] ~ aus etw. machen** make no secret of sth.
Hehler der; ~s, ~: fence; receiver [of stolen goods]
Hehlerei die; ~, ~en *(Rechtsw.)* receiving [stolen goods] *no art.*
hehr [he:ɐ̯] *Adj. (geh.)* majestic ‹*sight*›; glorious ‹*moment*›; noble ‹*ideal*›
heia ['haia] *(Kinderspr.) in* **~ machen** go bye-byes *or* beddy-byes *(child lang.)*
Heia die; ~, ~[s], Heia·bett das *(Kinderspr.)* bye-byes, beddy-byes *(child lang.);* **ab in die ~:** off to bye-byes *or* beddy-byes

[1]**Heide** ['haidə] **der; ~n, ~n** heathen; pagan

[2]**Heide die; ~, ~n a)** moor; heath; *(~landschaft)* moorland; heathland; **die Lüneburger ~:** the Luneburg Heath; **b)** *s.* **~kraut**

Heide-: ~kraut das; *o. Pl.* heather; ling; **~land das** moorland; heathland

Heidel·beere ['haidl̩-] **die** bilberry; blueberry; whortleberry

Heiden-: ~angst die; *o. Pl. (ugs.)* **eine ~angst vor etw.** *(Dat.)* **haben** be scared stiff of sth. *(coll.)*; **~arbeit die;** *o. Pl. (ugs.)* a heck of a lot of work *(coll.)*; **~krach der;** *o. Pl. (ugs.)* **a)** *s.* **~lärm; b)** *(Streit)* flaming row *(coll.)*; **~lärm der** *(ugs.)* unholy *or* dreadful din *or* row *(coll.)*; dreadful racket *(coll.)*; **~respekt der** *(ugs.)* healthy respect **(vor +** *Dat.* for); **~spaß der;** *o. Pl. (ugs.)* terrific fun *(coll.)*; **es macht einen ~spaß** it's terrific fun *(coll.)*; **~spektakel der** *(ugs.) (Lärm)* unholy *or* dreadful din *or* row *(coll.)*; *(Aufregung)* great *or (coll.)* dreadful commotion

Heidentum das; ~s heathenism; paganism

Heidin die; ~, ~nen heathen; pagan

heidnisch *Adj.* heathen; pagan

Heid·schnucke die; ~, ~n German Heath [sheep]

heikel ['haikl̩] *Adj.* **a)** *(schwierig)* delicate, ticklish ⟨*matter, subject*⟩; ticklish, awkward, tricky ⟨*problem, question, situation*⟩; **b)** *(wählerisch, empfindlich)* finicky, fussy, fastidious **(in bezug auf +** *Akk.* about)

heil [hail] *Adj.* **a)** *(unverletzt)* unhurt, unharmed ⟨*person*⟩; **~ ankommen** arrive safely *or* safe and sound; **etw. ~ überstehen** survive sth. unscathed; *s. auch* **Haut a; b)** *nicht attr. (wieder gesund)* **~ werden/wieder ~ sein** ⟨*injured part*⟩ heal [up]/have healed [up]; **c)** *(nicht entzwei)* intact; in one piece; **eine ~e Welt** *(fig.)* an ideal *or* a perfect world

Heil das; ~s a) *(Wohlergehen)* benefit; **bei jmdm./irgendwo sein ~ versuchen** try one's luck with sb./somewhere; **sein ~ in der Flucht suchen** seek refuge in flight; **b)** *(Rel.)* salvation

Heiland ['hailant] **der; ~[e]s, ~e a)** *(Christus)* Saviour; Redeemer; **b)** *(geh.: Retter)* saviour

Heil-: ~anstalt die a) *(Anstalt für Kranke, Süchtige)* sanatorium; **b)** *(psychiatrische Klinik)* mental hospital *or* home; **~bad das a)** *(Kurort)* spa; watering-place; **b)** *(medizinisches Bad)* medicinal bath

heilbar *Adj.* curable

Heil·butt der halibut

heilen 1. *tr. V.* **a)** cure ⟨*disease*⟩; heal ⟨*wound*⟩; **jmdn. ~:** cure sb.; restore sb. to health; **b)** *(befreien)* **jmdn. von etw. ~:** cure sb. of sth.; **davon/von ihm bin ich geheilt** *(ugs.)* I've been cured of it/my attachment to him. **2.** *itr. V.; mit sein* ⟨*wound*⟩ heal [up]; ⟨*infection*⟩ clear up; ⟨*fracture*⟩ mend

heil·froh *Adj.; nicht attr.* very *or (Brit. coll.)* jolly glad

heilig *Adj.* **a)** holy; **der Heilige Vater** the Holy Father; **die Heilige Jungfrau** the Blessed Virgin; **die ~e Barbara/der ~e Augustinus** Saint Barbara/Saint Augustine; **die Heilige Familie/Dreifaltigkeit** the Holy Family/Trinity; **der Heilige Geist** the Holy Spirit; **die Heiligen Drei Könige** the Three Kings *or* Wise Men; the Magi; **die Heilige Schrift** the Holy Scriptures *pl.*; **das Heilige Römische Reich** *(hist.)* the Holy Roman Empire; **b)** *(besonders geweiht)* holy; sacred; **~e Stätten** holy *or* sacred places; **der Heilige Abend/die Heilige Nacht** Christmas Eve/Night; **das Heilige Land** the Holy Land; **c)** *(geh.: unantastbar)* sacred ⟨*right, tradition, cause, etc.*⟩; sacred, solemn ⟨*duty*⟩; gospel ⟨*truth*⟩; solemn ⟨*conviction, oath*⟩; righteous ⟨*anger, zeal*⟩; awed ⟨*silence*⟩; **etw. ist jmdm. ~:** sth. is sacred to sb.; **bei allem, was mir ~ ist** by all that I hold sacred; *s. auch* **hoch 2d; d)** *(ugs.: groß)* incredible *(coll.)*; healthy ⟨*respect*⟩

Heilig·abend der Christmas Eve

Heilige der/die; *adj. Dekl.* saint; **ein sonderbarer** *od.* **komischer ~r** *(ugs. iron.)* a queer fish *(coll.)*

heiligen *tr. V.* keep, observe ⟨*tradition, Sabbath, etc.*⟩; **der Zweck heiligt die Mittel** the end justifies the means

Heiligen-: ~bild das picture of a saint; **~legende die** life of a saint; **~schein der** gloriole; aureole; *(um den Kopf)* halo; **jmdn. mit einem ~schein umgeben** *(fig.)* be unable to see sb.'s faults

Heiligkeit die; ~ a) holiness; **Seine/Euere ~** *(Anrede)* His/Your Holiness; **b)** *(der Ehe, Taufe usw.)* sanctity; sacredness

heilig|sprechen *unr. tr. V. (kath. Kirche)* canonize

Heilig·sprechung die; ~, ~en *(kath. Kirche)* canonization

Heiligtum das; ~s, Heiligtümer shrine; **sein Arbeitszimmer ist sein ~** *(fig.)* his study is his sanctuary *or* sanctum

heil-, Heil-: ~kraft die healing *or* curative power; **~kräftig** *Adj.* medicinal ⟨*herb, plant, etc.*⟩; curative ⟨*effect*⟩; **~kraut das** medicinal *or* officinal herb; **~kunde die** medicine; **~kundig** *Adj.* skilled in medicine *or* the art of healing *postpos.*; **~los 1.** *Adj.* hopeless, awful ⟨*mess, muddle*⟩; utter, *(coll.)* terrible ⟨*confusion*⟩; **2.** *adv.* hopelessly; **~methode die** method of treatment; **~mittel das** *(auch fig.)* remedy **(gegen** for); *(Medikament)* medicament; **~pflanze die** medicinal *or* officinal plant *or* herb; **~praktiker der** non-medical practitioner; **~quelle die** mineral spring

heilsam *Adj.* salutary ⟨*lesson, effect, experience, etc.*⟩

Heils-: ~armee die Salvation Army; **~botschaft die** message of salvation; **~lehre die** *(auch fig.)* doctrine of salvation

Heilung die; ~, ~en healing; *(von Krankheit, Kranken)* curing; **wenig Hoffnung auf ~ haben** have little hope of being cured; **~ suchen** seek a cure; **diese Salbe wird die ~ der Wunde beschleunigen** this ointment will help the wound to heal faster

Heilungs·prozeß der healing process

heim [haim] *Adv.* home

Heim das; ~[e]s, ~e a) *(Zuhause)* home; **ein eigenes ~:** a home of his/their *etc.* own; **b)** *(Anstalt, Alters~)* home; *(für Obdachlose)* hostel; *(für Studenten)* hall of residence; hostel

Heim-: ~arbeit die outwork; **etw. in ~arbeit herstellen lassen** have sth. produced by homeworkers; **~arbeiter der, ~arbeiterin die** home-worker; outworker

Heimat ['haima:t] **die; ~, ~en** home; homeland

Heimat-: ~dichter der regional writer; **~erde die** native soil; **~film der** [sentimental] film in a[n idealized] regional setting; **~hafen der** home port; **~kunde die** local history, geography, and natural history; **~land das** homeland; native land; *(fig.)* home

heimatlich *Adj.* native ⟨*dialect*⟩; **die ~en Berge** the mountains of [one's] home; **~e Klänge** sounds which evoke memories of home

heimat-, Heimat-: ~los *Adj.* homeless; **durch den Krieg ~los werden** be displaced by the war; **~lose der/die;** *adj. Dekl.* homeless person; **die ~losen** the homeless; **~museum das** museum of local history; **~stadt die** home town; **~vertriebene der/die;** *adj. Dekl.* expellee [from his/her homeland]

heim-: ~|begleiten *tr. V.* **jmdn. ~begleiten** take *or* see sb. home; **~|bringen** *unr. tr. V.* **a)** *s.* **~begleiten; b)** bring home

Heimchen das; ~s, ~ a) *(ugs. abwertend: Frau)* **~ [am Herd]** little hausfrau *or* housewife; **b)** *(Grille)* house cricket

Heim·computer der home computer

heim|dürfen *unr. itr. V.* be allowed [to go] home; **darf ich heim?** may I go home?

heimelig ['haimәlıç] *Adj.* cosy

heim-, Heim-: ~erzieher der *counsellor in a home for children or young people;* **~|fahren 1.** *unr. itr. V.; mit sein* go home; **2.** *unr. tr. V. (mit dem Auto)* drive ⟨*person*⟩ home; **~fahrt die** way home; *(~reise)* journey home; **~|finden** *unr. itr. V.* find one's way home; **~|führen** *tr. V.* **a)** *(geleiten)* take home; **b)** *(geh. veralt.: heiraten)* **er führte sie ~:** he took her to wife *(arch.)*; **~|gehen** *unr. itr. V.; mit sein* **a)** go home; **b)** *(geh. verhüll.: sterben)* pass away; **~|holen** *tr. V.* **a)** fetch home; **b)** *(geh. verhüll.)* **Gott hat ihn [zu sich] ~geholt** he has been called to his Maker

heimisch *Adj.* **a)** *(einheimisch)* indigenous, native ⟨*plants, animals, etc.*⟩ **(in** + *Dat.* to); domestic, home ⟨*industry*⟩*;* **die ~en Flüsse und Seen** the rivers and lakes of his/her *etc.* native land; **vor ~em Publikum** *(Sport)* in front of a home crowd; **b)** *nicht präd. (zum Heim gehörend)* **an den ~en Herd zurückkehren** go back home; **c) ~ sein/sich ~ fühlen** be/feel at home

heim-, Heim-: ~kehr die; ~: return home; homecoming; **~|kehren** *itr. V.; mit sein* return home **(aus** from); **~kehrer der** home-comer; **die ~kehrer aus dem Urlaub/Krieg** the holiday-makers returning home/the soldiers returning from the war; **~kind das** child brought up in a home; **~|kommen** *unr. itr. V.; mit sein* come *or* return home; **~|laufen** *unr. itr. V.; mit sein* run [back] home; **schnell ~laufen** dash home; **~leiter der** warden; *(eines Kinderheims/Jugendheims)* superintendent; *(eines Pflegeheims)* director; **~leiterin die** warden; *(eines Kinderheims/Jugendheims)* superintendent; *(eines Pflegeheims)* matron

heimlich 1. *Adj.* secret. **2.** *adv.* secretly; ⟨*meet*⟩ secretly, in secret; **er ist ~ weggelaufen** he slipped *or* stole away; **~, still und leise** *(ugs.)* on the quiet; quietly

Heimlichkeit die; ~, ~en; *meist Pl.* secret; **in aller ~:** in secret; secretly

Heimlichtuer [-tu:ɐ] **der; ~s, ~** *(abwertend)* secretive person

heim-, Heim-: ~mannschaft die *(Sport)* home team *or* side; **~|müssen** *unr. itr. V.* have to go home; **~niederlage die** *(Sport)* home defeat; **~reise die** journey home; **~|schicken** *tr. V.* send home; **~sieg der** *(Sport)* home win; **~spiel das** *(Sport)* home match *or* game; **~|suchen** *tr. V.* **a)** ⟨*storm, earthquake, epidemic*⟩ strike; ⟨*disease*⟩ afflict; ⟨*nightmares, doubts*⟩ plague; ⟨*catastrophe, fate*⟩ overtake; **von Dürre ~gesucht** drought-ridden; **b)** *(aufsuchen)* ⟨*visitor, salesman, etc.*⟩ descend [up]on; **~suchung die; ~, ~en** affliction; visitation

Heim·tücke die; ~ *(Bösartigkeit)* [concealed] malice; *(Hinterlistigkeit, fig.: einer Krankheit)* insidiousness

heim·tückisch 1. *Adj. (bösartig)* malicious; *(fig.)* insidious ⟨*disease*⟩; *(hinterlistig)* insidious. **2.** *adv.* maliciously

heim-, Heim-: ~vorteil der *(Sport)* advantage of playing at home; home advantage; **~wärts** [~vɛrts] *(nach Hause zu)* home; *(in Richtung Heimat)* homeward[s]; **~weg der** way home; **sich auf den ~weg machen** set off [for] home

Heim·weh das homesickness; **nach einem Ort ~ haben** be homesick for a place

heimweh·krank *Adj.* homesick

Heim·werker der handyman; do-it-yourselfer

heim|zahlen *tr. V.* **jmdm. etw. ~** pay sb. back *or* get even with sb. for sth.; **es jmdm. in gleicher Münze ~** pay sb. back in the same coin

Heini ['haini] **der; ~s, ~s** *(ugs. Schimpfwort)* idiot; clot *(sl.)*

Heinzel·männchen ['haintsl̩-] **das; ~s, ~:** brownie

Heirat ['haira:t] **die; ~, ~en** marriage

heiraten 1. *itr. V.* marry; get married; **~ müssen** *(verhüll.)* have to get married. **2.** *tr. V.* marry

heirats-, Heirats-: ~absichten *Pl.* marriage plans; **~annonce die** advertisement for a marriage partner; **~antrag der** proposal *or* offer of marriage; **jmdm. einen ~antrag machen** propose to sb.; **~anzeige die a)** *(Anzeige, daß jemand heiratet)* announcement of a/the forthcoming marriage; **b)** *s.* **~annonce; ~fähig** *Adj.* ⟨*person*⟩ of marriageable age; **~schwindler der** *person who makes a spurious offer of marriage for purposes of fraud;* **~urkunde die** marriage certificate; **~vermittler der** marriage broker

heischen ['haiʃn̩] *tr. V. (geh.)* demand

heiser ['haizɐ] **1.** *Adj.* hoarse. **2.** *adv.* hoarsely; in a hoarse voice

Heiserkeit die; ~: hoarseness

heiß [hais] **1.** *Adj.* **a)** hot; hot, torrid ⟨*zone*⟩*;* **brennend/glühend ~:** burning/scorching hot; **jmdm. ist ~:** sb. feels hot; **es überläuft mich ~ und kalt** I feel hot and cold all over; **sie haben sich die Köpfe ~ geredet** the conversation/debate became heated; **b)** *(heftig)* heated ⟨*debate, argument*⟩*;* impassioned ⟨*anger*⟩*;* burning, fervent ⟨*desire*⟩*;* fierce ⟨*fight, battle*⟩*;* **c)** *(innig)* ardent, passionate ⟨*wish, love*⟩*;* **~e Tränen weinen** weep bitterly; cry one's heart out; **~en Dank** *(ugs.)* thanks a lot! *(coll.);* **d)** *(aufreizend)* hot ⟨*rhythm etc.*⟩*;* sexy ⟨*blouse, dress, etc.*⟩*;* **was für'n ~er Typ!** *(salopp)* what a guy! *(coll.);* **e)** *(ugs.: gefährlich)* hot *(coll.)*⟨*goods, money*⟩*;* **ein ~es Thema** a controversial subject; *s. auch* **Eisen b; f)** *nicht präd. (ugs.: Aussichten habend)* hot ⟨*favourite, tip, contender, etc.*⟩*;* **auf einer ~en Spur sein** be hot on the scent; **g)** *nicht präd. (ugs.: schnell)* hot; *s. auch* **Ofen e; h)** *(ugs.: brünstig)* on heat; **i)** *(salopp: aufgereizt)* **jmdn. ~ machen** turn sb. on *(coll.).* **2.** *adv.* **a)** *(heftig)* ⟨*fight*⟩ fiercely; **es ging ~ her** things got heated; sparks flew *(coll.); (auf einer Party usw.)* things got wild; **b)** *(innig)* **jmdn. ~ und innig lieben** love sb. dearly *or* with all one's heart

heiß·blütig *Adj.* hot-blooded; ardent, passionate ⟨*lover*⟩*; (leicht erregbar)* hot-tempered

[1]**heißen 1.** *unr. itr. V.* **a)** *(den Namen tragen)* be called; **ich heiße Hans** I am called Hans; my name

is Hans; **er heißt mit Nachnamen Müller** his surname is Müller; **so wahr ich ... heiße** *(ugs.)* as sure as I'm standing here; **dann will ich Emil ~** *(ugs.)* then I'm a Dutchman *(coll.)*; **b)** *(bedeuten)* mean; **was heißt „danke" auf Französisch?** what's the French for 'thanks'?; **das will viel/nicht viel ~:** that means a lot/doesn't mean much; **was soll das denn ~?** what's that supposed to mean?; **was heißt hier: morgen?** what do you mean, tomorrow?; **das heißt** that is [to say]; **c)** *(lauten)* ‹*saying*› go; **der Titel/sein Motto heißt ...:** the title/his motto is ...; **d)** *unpers.* **es heißt, daß ...:** they say *or* it is said that ...; **es heißt, daß sie unheilbar krank ist** she is said to be incurably ill; **es soll nicht ~, daß ...:** never let it be said that ...; **e)** *unpers.* **in dem Gedicht/Roman/Artikel heißt es ...:** in the poem/novel/article it says that ...; **f)** *unpers.* **jetzt heißt es aufgepaßt!** *(geh.)* you'd better watch out now! **2.** *unr. tr. V.* **a)** *(geh.: auffordern)* tell; bid; **jmdn. etw. tun ~:** tell sb. to do sth.; bid sb. do sth.; **b)** *(geh.: bezeichnen als)* call; **jmdn. einen Lügner ~:** call sb. a liar; **jmdn. willkommen ~:** bid sb. welcome; **c)** *(veralt.: nennen)* name; call

²heißen *tr. V. s.* **hissen**

heiß-, Heiß-: ~ersehnt *Adj. (präd. getrennt geschrieben)* **das ~ersehnte Fahrrad** the bicycle he/she has/had longed for so fervently; **~geliebt** *Adj. (präd. getrennt geschrieben)* dearly beloved ‹*husband, son, etc.*›; beloved ‹*doll, car, etc.*›; **~hunger der: einen ~hunger auf etw.** *(Akk.) od.* **nach etw. [haben]** [have] a craving for sth.; **etw. mit [wahrem] ~hunger verschlingen** devour sth. ravenously; [absolutely *(coll.)*] wolf sth. down; **~hungrig 1.** *Adj.* ravenous; **2.** *adv.* ravenously; voraciously; **~|laufen 1.** *unr. itr. V.; mit sein* run hot; ‹*engine*› run hot, overheat; **sie hat soviel telefoniert, daß die Drähte heißliefen** she made so many telephone calls that the wires were buzzing; **2.** *unr. refl. V.* run hot; ‹*engine*› run hot, overheat

Heiß·luft die hot air

Heißluft·ballon der hot-air balloon

heiß-, Heiß-: ~mangel die rotary ironer; **~sporn der** hothead; **~umkämpft** *Adj. (präd. getrennt geschrieben)* fiercely contested *or* disputed; **~umstritten** *Adj. (präd. getrennt geschrieben)* hotly debated ‹*matter, subject, etc.*›; highly controversial ‹*figure, director, etc*›

Heißwasser·bereiter der water heater

heiter ['haitɐ] *Adj.* **a)** *(fröhlich)* cheerful, happy ‹*person, nature*›; happy, merry ‹*laughter*›; **b)** *(froh stimmend)* cheerful ‹*music etc.*›; *(amüsant)* funny, amusing ‹*story etc.*›; **einer Sache** *(Dat.)* **die ~e Seite abgewinnen** look on the bright side of sth.; **das kann ja ~ werden!** *(ugs. iron.)* that'll be fun *(iron.)*; **c)** *(sonnig)* fine ‹*weather*›

Heiterkeit die; ~ a) *(Frohsinn)* cheerfulness; **b)** *(Belustigung)* merriment; **allgemeine ~ erregen** provoke *or* cause general merriment

heizbar *Adj.* heated ‹*windscreen, room, etc.*›; **das Zimmer ist nicht/schwer ~:** the room has no heating/is difficult to heat

Heiz·decke die electric blanket

heizen ['haitsn̩] **1.** *itr. V.* have the heating on; **der Ofen heizt gut** the stove gives off *or* throws out a good heat; **mit Kohle ~:** use coal for heating. **2.** *tr. V.* **a)** heat ‹*room etc.*›; **b)** stoke ‹*furnace, boiler, etc.*›; **sie ~ ihre Öfen mit Öl** their boilers are oil-fired

Heizer der; ~s, ~ *(einer Lokomotive)* fireman; stoker; *(eines Schiffes)* stoker

Heiz-: ~gerät das heater; **~kessel der** boiler; **~kissen das** heating pad; **~körper der** radiator; **~kosten** *Pl.* heating costs; **~lüfter der** fan heater; **~ofen der** stove; heater; **ein elektrischer ~ofen** an electric heater; **~öl das** heating oil; fuel oil; **~periode die** heating period; **~strahler der** radiant heater

Heizung die; ~, ~en a) [central] heating *no pl., no indef. art.*; **b)** *(ugs.: Heizkörper)* radiator

Heizungs-: ~anlage die heating system; **~keller der** boiler-room *(in the basement)*; **~monteur der** heating engineer

Hektar ['hɛkta:ɐ̯] **das** *od.* **der; ~s, ~e** hectare

Hektik ['hɛktɪk] **die; ~:** hectic rush; *(des Lebens)* hectic pace; **nur keine ~!** *(ugs.)* take it easy!

hektisch 1. *Adj.* hectic; **nun mal nicht so ~!** take it easy!. **2.** *adv.* ‹*work, run to and fro*› frantically; **~ zugehen** be hectic; **~ leben** lead a hectic life

Hekto·liter der *od.* **das** hectolitre

helau [he'lau̯] *Interj.: cheer or greeting used at Carnival time*

Held [hɛlt] **der; ~en, ~en** hero; **du bist mir ein schöner ~** *(scherzh.)* a fine one you are!

Helden-: ~dichtung die *(Literaturw.)* epic *or* heroic poetry; **~epos das** *(Literaturw.)* heroic epic

heldenhaft 1. *Adj.* heroic. **2.** *adv.* heroically

helden-, Helden-: ~mut der heroism; **~mütig 1.** *Adj.* heroic; **2.** *adv.* heroically; **~sage die** *(Literaturw.)* heroic legend; *(aus Norwegen, Island)* heroic saga; **~tat die** heroic feat *or* deed; **das war keine ~tat** *(spött.)* that was nothing to be proud of

Heldentum das; ~s heroism

Heldin die; ~, ~nen heroine

helfen ['hɛlfn̩] *unr. itr. V.* **a)** help; **jmdm. ~ [etw. zu tun]** help *or* assist sb. [to do sth.]; lend *or* give sb. a hand [in doing sth.]; **jmdm. bei etw. ~:** help *or* assist sb. with sth.; **jmdm. in den/aus dem Mantel ~:** help sb. into *or* on with/out of *or* off with his/her coat; **jmdm. über die Straße/in den Bus ~:** help sb. across the road/on to the bus; **dem Kranken war nicht mehr zu ~:** the patient was beyond [all] help; **dir ist nicht zu ~** *(ugs.)* you're a hopeless case; **sich** *(Dat.)* **nicht mehr zu ~ wissen** be at one's wits' end; **dem werde ich ~, einfach die Schule zu schwänzen!** *(ugs.)* I'll teach him to play truant; **ich kann mir nicht ~, aber ...:** I'm sorry, but [I have to say that] ...; **hilf dir selbst, so hilft dir Gott** *(Spr.)* God helps those who help themselves; **b)** *(hilfreich sein, nützen)* help; **das hilft gegen** *od.* **bei Kopfschmerzen** it is good for *or* helps to relieve headaches; **da hilft alles nichts** there's nothing *or* no help for it; **es hilft nichts** it's no use *or* good; **was hilft's?** what's the use *or* good?; **damit ist uns nicht geholfen** that is no help to us; that doesn't help us

Helfer der; ~s, ~, Helferin die; ~, ~nen helper; *(Mitarbeiter)* assistant; *(bei einem Verbrechen)* accomplice; **ein ~ in der Not** a friend in need

Helfers·helfer der; ~s, ~ *(abwertend)* accomplice

Helikopter [heli'kɔptɐ] **der; ~s, ~:** helicopter

Helium ['he:li̯ʊm] **das; ~s** helium

hell [hɛl] **1.** *Adj.* **a)** *(von Licht erfüllt)* light ‹*room etc.*›; well-lit ‹*stairs*›; **es wird ~:** it's getting light; **es war schon ~er Morgen/Tag** it was already broad daylight; **am ~en Tag** *(ugs.)* in broad daylight; **in ~en Flammen stehen** be in flames *or* ablaze; **b)** *(klar, viel Licht spendend)* bright; **c)** *(blaß)* light ‹*colour*›; fair ‹*skin, hair*›; light-coloured ‹*clothes*›; **~es Bier** ≈ lager; **d)** *(akustisch)* **ein ~er Ton/Klang** a high, clear sound; **eine ~e Stimme** a high, clear voice; **ein ~es Lachen** a ringing laugh; **e) ein ~er Kopf sein** be bright; **f)** *(voll bewußt)* lucid ‹*moment, interval*›; **g)** *nicht präd. (ugs.: absolut)* sheer, utter ‹*madness, foolishness, despair, nonsense*›; unbounded, boundless ‹*enthusiasm*›; unrestrained ‹*jubilation*›; **in ~e Wut geraten** fly into a blind rage; **er hat seine ~e Freude an ihr/daran** she/it is his great joy. **2.** *adv.* **a)** brightly ‹*lit*›; ‹*shine, blaze*› brightly; **b)** *(in hoher Tonlage)* **~ läuteten die Glokken** the bells rang out high and clear; **~ lachen** give a ringing laugh; **c)** *(sehr)* highly ‹*enthusiastic, delighted, indignant, etc.*›

hell-: ~auf *Adv.* highly ‹*enthusiastic, indignant, etc.*›; **~auf lachen** laugh out loud; **~blau** *Adj.* light blue; **~blond** *Adj.* very fair; light blonde; **~braun** *Adj.* light brown

helle *Adj.; nicht attr. (landsch.)* bright; intelligent

Helle das; *adj. Dekl.* ≈ lager

Hellebarde [hɛlə'bardə] **die; ~, ~n** *(hist.)* halberd

Heller der; ~s, ~: heller; **bis auf den letzten ~/bis auf ~ und Pfennig** *(ugs.)* down to the last penny *or (Amer.)* cent

hell-: ~erleuchtet *Adj. (präd. getrennt geschrieben)* brightly-lit; **~gelb** *Adj.* light yellow; **~grau** *Adj.* light grey; **~grün** *Adj.* light green; **~haarig** *Adj.* fair[-haired]; **~häutig** *Adj.* fair[-skinned]; fair-skinned, pale-skinned ‹*race*›; **~hörig** *Adj.* **a)** *(aufmerksam)* **~hörig werden** sit up and take notice *(coll.)*; **jmdn. ~hörig machen** make sb. sit up and take notice *(coll.)*; **b)** *(schalldurchlässig)* badly *or* poorly sound-proofed

hellicht ['hɛl'lɪçt] *Adj.* **es ist ~er Tag** it's broad daylight; **am ~en Tag** in broad daylight

Helligkeit die; ~, ~en *(auch Physik)* brightness

hellodernd ['hɛlo:dɐnt] *Adj.; nicht präd.* blazing; **~e Flammen** raging flames

hell-, Hell-: ~rot *Adj.* light red; **~sehen** *unr. itr. V.; nur im Inf.* **~sehen können** have second sight; be clairvoyant; **~seher der, ~seherin die** clairvoyant; **~seherisch** *Adj.* clairvoyant; **~sichtig** *Adj.* **a)** *(durchschauend)* perceptive; **b)** *(weitblickend)* far-sighted; **~wach** *Adj.* **a)** *(ganz wach)* wide awake; **b)** *(ugs.: klug)* bright

Helm [hɛlm] **der; ~[e]s, ~e** helmet; **~ ab zum Gebet!** *(Milit.)* helmets off for prayers!

Helm·busch der plume; crest

Hemd [hɛmt] **das; ~[e]s, ~en a)** *(Oberhemd)* shirt; **b)** *(Unterhemd)* [under]vest; undershirt; **c)** *in* **etw. wechseln wie sein ~** *(ugs. abwertend)* change sth. as often as one changes one's clothes; **das ~ ist mir näher als der Rock** for me charity begins at home; **mach dir nicht ins ~** *(salopp)* don't get [all] uptight *(coll.)*; **für sie gibt er sein letztes** *od.* **das letzte ~ her** *(ugs.)* he'd sell the shirt off his back to help her; **jmdn. bis aufs ~ ausziehen** *(ugs.)* have the shirt off sb.'s back *(coll.)*

Hemd-: ~bluse die shirt; **~blusen·kleid das** shirt-waist dress

Hemds·ärmel der shirt-sleeve; **in ~n** in [one's] shirt-sleeves

hemdsärmelig [~ɛrməlɪç] *Adj.* **a)** *(im Hemd)* shirt-sleeved *attrib.*; in [one's] shirt-sleeves *postpos.*; **b)** *(ugs.: leger)* casual ‹*manner*›; informal ‹*style*›

Hemisphäre [hemi'sfɛ:rə] **die; ~, ~n** hemisphere

hemmen ['hɛmən] *tr. V.* **a)** *(verlangsamen)* slow [down]; retard; **b)** *(aufhalten)* check; stem ‹*flow*›; **c)** *(beeinträchtigen)* hinder; hamper

Hemmnis das; ~ses, ~se obstacle, hindrance **(für** to)

Hemm·schuh der a) *(Hemmnis)* obstacle, hindrance **(für** to); **b)** *(Eisenb.)* slipper [brake]

Hemmung die; ~, ~en a) inhibition; **~en haben** have inhibitions; be inhibited; **b)** *(Skrupel)* scruple

hemmungs·los 1. *Adj.* unrestrained; unrestrained, unbridled ‹*passion*›; *(skrupellos)* unscrupulous. **2.** *adv.* unrestrainedly; without restraint; ‹*cry, laugh, scream*› uncontrollably; *(skrupellos)* unscrupulously

Hemmungslosigkeit die; ~: lack of restraint; *(Skrupellosigkeit)* unscrupulousness

Hendl ['hɛndl̩] **das; ~s, ~[n]** *(bayr., österr.)* chicken

Hengst [hɛŋst] **der; ~[e]s, ~e** *(Pferd)* stallion; *(Kamel)* male; *(Esel)* male; jackass

Hengst-: ~fohlen das, ~füllen das colt; [male] foal

Henkel ['hɛnkl̩] **der; ~s, ~:** handle

Henkel-: ~kanne die jug; *(größer)* pitcher; **~mann der** *(ugs.) portable set of stacked containers for taking a hot meal to one's work*

henken ['hɛŋkn̩] *tr. V. (veralt.)* hang

Henker der; ~s, ~ hangman; *(Scharfrichter, auch fig.)* executioner; **hol's der ~!** damn [it]! *(coll.)*; the devil take it! *(coll.)*; **weiß der ~!** the devil only knows *(coll.)*

Henkers-: ~knecht der hangman's assistant; *(eines Scharfrichters)* executioner's assistant; *(fig.)* henchman; **~mahlzeit die** last meal *(before execution)*

Henna ['hɛna] **die; ~** *od.* **das; ~[s]** henna

Henne ['hɛnə] **die; ~, ~n** hen

her [he:ɐ̯] *Adv.* **a) ~ damit** give it to me; give it here *(coll.)*; **~ mit dem Geld** hand over *or* give me the money; **vom Fenster ~:** from the window; **von weit ~:** from far away *or* a long way off; **b)** *(zeitlich)* **jmdn. von früher/von der Schulzeit ~ kennen** know sb. from earlier times/from one's schooldays; *s. auch* **hersein; c) von der Konzeption ~:** as far as the basic design is concerned; **das ist von der Sache ~ nicht vertretbar** it is unjustifiable in the nature of the matter

herab [hɛ'rap] *Adv.* **bis ~ auf etw.** *(Akk.)* down to sth.; **die Treppe/den Berg ~:** down the stairs/the mountain; **von oben ~** *(fig.)* condescendingly

herab-, Herab- (*s. auch* **herunter-**): **~|fließen** *unr. itr. V.; mit sein (geh.)* flow down; **~|hängen** *unr. itr. V.* **a)** *(nach unten hängen)* hang [down] **(von** from); *(fig.)* ‹*clouds*› hang low [in the sky]; **b)** *(schlaff hängen)* ‹*hair, arms, etc.*› hang down; **~|lassen 1.** *unr. tr. V.* let down; lower; **2.** *unr. refl. V.* **sich ~lassen, etw. zu tun** condescend *or* deign to do sth.; **~lassend 1.** *Adj.* condescending; patronizing **(zu** towards); **2.** *adv.* condescendingly; patronizingly; in a condescending *or* patronizing manner; **~|mindern** *tr. V.* **a)** reduce; **b)** *(schlechtmachen)* belittle, disparage ‹*achievement, qualities, etc.*›; **~|regnen** *itr. V.; mit sein* ‹*drops of rain*› fall; *(fig.)* rain down;

~**|sehen** *unr. itr. V.* **a)** *(nach unten sehen)* look down (**auf** + *Akk.* on); **b)** *(geringschätzig betrachten)* **auf jmdn. ~sehen** look down on sb.; ~**|senken** *refl. V. (geh.)* ⟨*night, evening*⟩ fall; ⟨*mist, fog*⟩ settle, descend (**auf** + *Akk.* on, over); ~**|setzen** *tr. V.* **a)** *(reduzieren)* reduce, cut ⟨*cost, price, working hours, etc.*⟩; reduce ⟨*speed*⟩; **zu ~gesetzten Preisen** at reduced prices; **b)** *(abwerten)* belittle; disparage; ~**setzung die;** ~: *s.* **~setzen: a)** reduction, cut (*Gen.* in); **b)** belittling; disparagement; ~**|sinken** *unr. itr. V.; mit sein* sink [down]; *(fig.)* ⟨*night*⟩ fall; descend; ⟨*mist, fog*⟩ settle, descend (**auf** + *Akk.* on, over); ~**|steigen** *unr. itr. V.; mit sein (geh.)* descend; climb down; *(vom Pferd)* dismount; ~**|stürzen 1.** *itr. V.; mit sein* plummet down; **er stürzte vom Gerüst ~:** he fell from *or* off the scaffolding; **~stürzende Felsbrocken** falling rocks; **2.** *refl. V.* throw oneself (**von** from *or* off); ~**|würdigen** *tr. V.* belittle; disparage

heran [hɛ'ran] *Adv.* **an etw.** *(Akk.)* ~: close to *or* right up to sth.; **nur ~ zu mir!, immer ~!** come closer!

herạn-, Herạn-: ~**|bilden 1.** *tr. V.* train [up]; *(auf der Schule, Universität)* educate; **2.** *refl. V. (sich entwickeln)* develop; ~**|bringen** *unr. tr. V.* **a)** *(zu jmdm. bringen)* bring [up] (**an** + *Akk.*, **zu** to); **b)** *(vertraut machen)* **jmdn. an etw.** *(Akk.)* **~bringen** introduce sb. to sth.; ~**|führen 1.** *tr. V.* **a)** *(in die Nähe führen)* lead up; bring up ⟨*troops*⟩; **b)** *(nahe bringen)* bring up (**an** + *Akk.* to); **c)** *(vertraut machen)* **jmdn. an etw.** *(Akk.)* **~führen** introduce sb. to sth.; **2.** *itr. V.* **an etw.** *(Akk.)* **~führen** lead to sth.; ~**|gehen** *unr. itr. V.; mit sein* **a)** go up (**an** + *Akk.* to); **näher ~gehen** go [up] closer; **b)** *(anpacken)* **an ein Problem/eine Aufgabe/die Arbeit** *usw.* **~gehen** tackle a problem/a task/the work *etc.*; ~**|kommen** *unr. itr. V.; mit sein* **a) an etw.** *(Akk.)* **~kommen** come *or* draw near to sth.; approach sth.; **ganz nahe an etw.** *(Akk.)* **~kommen** come right up to sth.; **b)** *(zeitlich)* **der große Tag war ~gekommen** the big day had arrived; **c) an etw.** *(Akk.)* **~kommen** *(erreichen)* reach sth.; *(erwerben)* obtain sth.; get hold of sth.; **an jmdn. ~kommen** *(fig.)* get hold of sb.; **an jmds. Erfolg/Rekord ~kommen** *(fig.)* equal sb.'s success/record; ~**|machen** *refl. V. (ugs.)* **a)** *(beginnen)* **sich an etw.** *(Akk.)* **~machen** get down to *or (coll.)* get going on sth.; **b)** *(nähern)* **sich an jmdn. ~machen** chat sb. up *(coll.)*; ~**|nahen** *itr. V.; mit sein (geh.)* approach; draw near; ~**|reichen** *itr. V.* **a) [an etw.** *(Akk.)***] ~ reichen** reach [sth.]; **b) an jmdn./etw. ~reichen** *(fig.)* come *or* measure up to the standard of sb./sth.; ~**|reifen** *itr. V.; mit sein* ⟨*fruit, crops*⟩ ripen; *(fig.)* ⟨*plan*⟩ mature; **zur Frau/zum Mann/zu einer großen Malerin ~reifen** mature into a woman/man/great painter; ~**|rücken 1.** *tr. V.* pull up ⟨*table*⟩; draw *or* pull *or* bring up ⟨*chair*⟩; **2.** *itr. V.; mit sein* move *or* come closer *or* nearer; ⟨*troops*⟩ advance (**an** + *Akk.* towards); **dicht** *od.* **nah ~rücken** move up close (**an** + *Akk.* to); **mit seinem Stuhl ~rükken** draw *or* pull *or* bring one's chair up closer; ~**|schaffen** *tr. V.* bring; *(liefern)* supply; ~**|tasten** *refl. V.* **sich [an etw.** *Akk.* **] ~tasten** grope *or* feel one's way [over to sth.]; *(fig.)* feel one's way [towards sth.]; ~**|tragen** *unr. tr. V.* **a)** bring [over]; **b) eine Bitte/Beschwerde an jmdn. ~tragen** *(jmdm. vortragen)* go/come to sb. with a request/complaint; ~**|treten** *unr. itr. V.; mit sein* **a)** *(an eine Stelle treten)* come/go up (**an** + *Akk.* to); **b)** *(sich wenden)* **an jmdn. ~treten** approach sb.; ~**|wachsen** *unr. itr. V.; mit sein* grow up; *(fig.)* develop; **zum Mann/zur Frau ~wachsen** grow up into *or* to be a man/woman; **die ~wachsende Generation** the rising *or* up-and-coming generation; **~wachsende der/die;** *adj. Dekl.* **a)** young person; **b)** *(Rechtsw.)* adolescent; ~**|wagen** *refl. V.* venture near; dare to go near; **sich an etw.** *(Akk.)* **~wagen** venture near sth.; dare to go near sth.; *(fig.)* venture *or* dare to tackle *or* attempt sth.; ~**|ziehen 1.** *unr. tr. V.* **a)** *(an eine Stelle ziehen)* pull *or* draw over; pull *or* draw up ⟨*chair*⟩; **etw. zu sich ~ziehen** pull *or* draw sth. towards one; **b)** *(fig.: beauftragen)* call *or* bring in; **weitere Arbeitskräfte ~ziehen** bring in more labour; **c)** *(fig.) (in Betracht ziehen)* refer to; *(geltend machen)* invoke; quote; **2.** *unr. itr. V.; mit sein (auch fig.)* approach; *(Milit.)* advance

herauf [hɛ'rau̯f] *Adv.* up; **vom Tal ~:** up from the valley

herauf-: ~**|arbeiten** *refl. V.* **a)** work one's/its way up; **b)** *(hocharbeiten)* work one's way up; ~**|beschwören** *tr. V.* **a)** *(verursachen)* cause, bring about ⟨*disaster, war, crisis*⟩; cause, provoke ⟨*dispute, argument*⟩; give rise to ⟨*criticism*⟩; **b)** *(erinnern)* evoke ⟨*memories etc.*⟩; ~**|bringen** *unr. tr. V.* bring up; ~**|führen 1.** *itr. V.* **es führen zwei Wege ~:** there are two paths up; **2.** *tr. V.* show ⟨*person*⟩ up; ~**|kommen** *unr. itr. V.; mit sein* **a)** *(nach oben kommen)* come up; **auf den Baum/die Mauer ~kommen** climb *or* get up the tree/up on the wall; **b)** *(aufsteigen)* rise; come up; **c)** *(bevorstehen)* ⟨*storm*⟩ be approaching *or* gathering *or* brewing; ~**|setzen** *tr. V.* increase, raise, put up ⟨*prices, rents, interest rates, etc.*⟩; ~**|ziehen 1.** *unr. tr. V.* pull up; **2.** *unr. itr. V.; mit sein* ⟨*storm*⟩ be approaching *or* gathering *or* brewing; ⟨*disaster*⟩ be approaching

heraus [hɛ'rau̯s] *Adv.* **~ aus dem Bett!** out of bed!; **~ mit dir!** get out of here!; **~ damit!** *(gib her!)* hand it over!; *(weg damit!)* get rid of it!; **~ mit der Sprache!** out with it!; **aus einem Gefühl der Einsamkeit ~:** out of a feeling of loneliness

heraus-, Heraus-: ~**|arbeiten 1.** *tr. V. (aus Stein, Holz)* fashion, carve (**aus** out of); **b)** *(hervorheben)* bring out ⟨*difference, aspect, point of view, etc.*⟩; develop ⟨*observation, remark*⟩; ~**|bekommen 1.** *unr. tr. V.* **a)** *(entfernen)* get out (**aus** of); **b)** *(ugs.: lösen)* work out ⟨*problem, answer, etc.*⟩; solve ⟨*puzzle*⟩; **c)** *(ermitteln)* find out; **etw. aus jmdm. ~bekommen** get sth. out of sb.; **d)** *(als Wechselgeld bekommen)* **5 DM ~bekommen** get back 5 marks change; **e)** *(von sich geben)* utter; say; **2.** *unr. itr. V. (Wechselgeld bekommen)* **richtig/falsch ~bekommen** *(ugs.)* get the right/wrong change; ~**|bilden** *refl. V.* develop; ~**|bitten** *unr. tr. V.* **jmdn. ~bitten** ask sb. to come out[side]; ~**|brechen 1.** *unr. tr. V.* knock out; *(mit brutaler Gewalt)* wrench out; pull up ⟨*paving-stone*⟩; **2.** *unr. itr. V.; mit sein* ⟨*anger, hatred*⟩ burst forth, erupt; ~**|bringen** *unr. tr. V.* **a)** *(nach außen bringen)* bring out (**aus** of); **b)** *(nach draußen begleiten)* show out; **c)** *(veröffentlichen)* bring out; publish; *(aufführen)* put on, stage ⟨*play*⟩; screen ⟨*film*⟩; **d)** *(auf den Markt bringen)* bring out; launch; **e)** *(populär machen)* make widely known;

jmdn./etw. ganz groß ~bringen launch sb./sth. in a big way; **f)** *(ugs.: ermitteln) s.* **~bekommen 1 c; g)** *(ugs.: lösen) s.* **~bekommen 1 b; h)** *(von sich geben)* utter; say; **~|drehen** *tr. V.* unscrew; **~|drücken** *tr. V.* **a) etw. ~drükken** squeeze sth. out **(aus** of); squeeze *or* press ⟨*juice, oil*⟩ out **(aus** of); **b)** *(vorwölben)* stick out ⟨*chest etc.*⟩; **~|fahren 1.** *unr. itr. V.; mit sein* **a)** *(nach außen fahren)* **aus etw. ~fahren** drive out of sth.; *(mit dem Rad, Motorrad)* ride out of sth.; **der Zug fuhr aus dem Bahnhof ~:** the train pulled out of the station; **b)** *(fahrend ~kommen)* come out; **c)** *(ugs.: entschlüpfen)* ⟨*word, remark, etc.*⟩ slip out; **2.** *unr. tr. V.* **a) den Wagen/das Fahrrad [aus dem Hof] ~fahren** drive the car/ride the bicycle out [of the yard]; **b)** *(Sport)* **eine gute Zeit/einen Sieg ~fahren** record a good *or* fast time/a victory; **~|finden 1.** *unr. tr. V.* **a)** *(entdecken)* find out; trace ⟨*fault*⟩; **man fand ~, daß ...:** it was found *or* discovered that ...; **b)** *(aus einer Menge)* pick out **(aus** from [among]); find **(aus** among); **2.** *unr. itr. V.* find one's way out **(aus** of); **~|fliegen 1.** *unr. itr. V.; mit sein* **a)** fly out **(aus** of); **b)** *(aus etw. fallen)* be thrown out **(aus** of); **c)** *(ugs.: entlassen werden)* be fired *or (coll.)* sacked **(bei** from); **2.** *unr. tr. V.* fly out **(aus** of); **~forderer der; ~s, ~** *(auch Sport)* challenger; **~|fordern 1.** *tr. V.* **a)** *(auch Sport)* challenge; **b)** *(heraufbeschwören)* provoke ⟨*person, resistance, etc.*⟩; invite ⟨*criticism*⟩; court ⟨*danger*⟩; **sein Schicksal ~fordern** tempt fate *or* providence; **2.** *itr. V. (provozieren)* **zu etw. ~fordern** provoke sth.; **~fordernd 1.** *Adj.* provocative; *(Streit suchend)* challenging, defiant ⟨*words, speech, look*⟩; **2.** *adv.: s. Adj.:* provocatively; challengingly; defiantly

Heraus·forderung die *(auch Sport)* challenge; *(Provokation)* provocation

heraus-, Heraus-: ~|führen *tr., itr. V.* lead out **(aus** of); **~gabe die;** *o. Pl.* **a)** *(von Eigentum, Personen, Geiseln usw.)* handing over; *(Rückgabe)* return; **b)** *(das Veröffentlichen)* publication; *(Redaktion)* editing; **~|geben 1.** *unr. tr. V.* **a)** *(nach außen geben)* hand *or* pass out; **b)** *(aushändigen)* hand over ⟨*property, person, hostage, etc.*⟩; *(zurückgeben)* return; give back; **c)** *(als Wechselgeld zurückgeben)* **5 DM/zuviel ~geben** give 5 marks/too much change; **d)** *(veröffentlichen)* publish; *(für die Veröffentlichung bearbeiten)* edit [for publication]; **e)** issue ⟨*stamp, coin, etc.*⟩; **f)** *(erlassen)* issue; **2.** *itr. V.* give change; **können Sie [auf 100 DM] ~geben?** do you have *or* can you give me change [for 100 marks]?; **~geber der, ~geberin die** publisher; *(Redakteur[in])* editor; **~|gehen** *unr. itr. V.; mit sein* **a)** *(nach außen)* go out; leave; **aus sich ~gehen** come out of one's shell; **b)** *(sich entfernen lassen)* ⟨*stain, cork, nail, etc.*⟩ come out; **~|greifen** *unr. tr. V.* pick out; select; **sich** *(Dat.)* **jmdn. ~greifen** pick *or* single sb. out **(aus** from); *(fig.)* take ⟨*example, aspect, etc.*⟩ **(aus** from); **~|halten 1.** *unr. tr. V.* **a)** *(nach außen halten)* put *or* stick out **(aus** of); **b)** *(ugs.: fernhalten, nicht verwickeln)* keep out **(aus** of); **2.** *unr. refl. V.* keep *or* stay out; [1]**~|hängen** *unr. itr. V.* hang out **(aus** of); [2]**~|hängen** *tr. V.* hang out **(aus** of); **~|hauen** *unr. tr. V.* **a)** *(durch Hauen fertigen)* carve ⟨*figure, letters, relief, etc.*⟩ **(aus** from, out of); **b)** *(ugs.: befreien)* get out; *(aus Schwierigkeiten)* bail out; **~|heben 1.** *unr. tr. V.* **a)** *(nach außen heben)* lift out **(aus** of); **b)** *(hervorheben)* bring out; **es ist diese Eigenschaft, die ihn aus der Masse ~hebt** it is this quality that raises him above *or* sets him apart from the rest; **2.** *unr. refl. V.* stand out **(aus** from); **~|holen** *tr. V.* **a)** *(nach außen holen)* bring out; **b)** *(ugs.: abgewinnen)* get out; gain, win ⟨*victory, points*⟩; **er holte das Letzte aus sich ~:** he made an all-out *or* supreme effort; **c)** *(ugs.: erwirken)* gain, win ⟨*wage increase, advantage, etc.*⟩; get, achieve ⟨*result*⟩; **d)** *(ugs.: durch Fragen)* get out; **e)** *(ugs.: ~arbeiten)* bring out ⟨*difference, aspect, point of view*⟩; **~|hören** *tr. V.* **a)** hear; **b)** *(erkennen)* detect, sense **(aus** in); **~|kehren** *tr. V.* parade; **den Vorgesetzten ~kehren** parade the fact that one is in charge; **~|kommen** *unr. itr. V.; mit sein* **a)** *(nach außen kommen)* come out **(aus** of); **b)** *(ein Gebiet verlassen)* **er ist nie aus seiner Heimatstadt ~gekommen** he's never been out of *or* never left his home town; **wir kamen aus dem Staunen/Lachen nicht ~** *(fig.)* we couldn't get over our surprise/stop laughing; **c)** *(ugs.: einen Ausweg finden)* get out **(aus** of); **d)** *(ugs.: auf den Markt kommen)* come out; **mit einem Produkt ~kommen** bring out *or* launch a product; **e)** *(erscheinen)* ⟨*book, timetable, etc.*⟩ come out, be published, appear; ⟨*postage stamp, coin*⟩ be issued; ⟨*play*⟩ be staged; **f)** *(ugs.: bekannt werden)* come out; **g)** *(ugs.: zur Sprache kommen)* **mit etw. ~kommen** come out with sth.; **h)** *(ugs.: sich erfolgreich produzieren)* **ganz groß ~kommen** make a big splash; **i)** *(deutlich werden)* come out; ⟨*colour*⟩ show up; **j)** *(ugs.: sich als Resultat ergeben)* **bei etw. ~kommen** come out of *or* emerge from sth.; **auf dasselbe ~kommen** amount to the same thing; **was kommt bei der Aufgabe ~?** what is the answer to the question?; **dabei kommt nichts ~:** nothing will come of it; **was soll dabei ~kommen?** what's that supposed to achieve?; **k)** *(ugs.: ausspielen)* lead; **wer kommt ~?** whose lead is it?; **mit etw. ~kommen** lead sth.; **~|kriegen** *tr. V. (ugs.) s.* **~bekommen; ~|kristallisieren 1.** *tr. V.* **a)** *(Chemie)* crystallize [out]; **b)** *(zusammenfassen)* extract; **2.** *refl. V.* **a)** *(Chemie)* crystallize [out]; ⟨*crystal*⟩ form; **b)** *(entwickeln)* crystallize **(aus** out of); **~|lesen** *unr. tr. V.* **a)** *(entnehmen)* tell **(aus** from); **b)** *(interpretieren)* **etw. aus etw. ~lesen** read sth. into sth.; **c)** *(auswählen)* pick out **(aus** from); **~|locken** *tr. V.* entice out **(aus** of); lure ⟨*enemy, victim, etc.*⟩ out **(aus** of); **jmdn. aus seiner Reserve ~locken** draw sb. out of his/her shell; **~|machen** *(ugs.)* **1.** *tr. V.* take out; get out ⟨*stain*⟩; **2.** *refl. V.* come on well; **~|müssen** *unr. itr. V. (ugs.)* **a) aus etw. ~müssen** have to leave sth.; **dieser Zahn muß ~:** this tooth has to come out; **b)** *(aufstehen müssen)* have to get up; **c)** *(gesagt werden müssen)* **das mußte einfach ~!** I simply had to get that off my chest *(coll.)*; **~nehmbar** *Adj.* removable; detachable ⟨*lining*⟩; **~|nehmen** *unr. tr. V.* **a)** take out **(aus** of); **den Gang ~nehmen** *(fig.)* put the car into neutral; **b)** *(ugs.: entfernen)* take out, remove ⟨*appendix, tonsils, tooth, etc.*⟩; **sich** *(Dat.)* **die Mandeln ~nehmen lassen** have one's tonsils out; **c)** *(ugs.)* **sich** *(Dat.)* **Freiheiten ~nehmen** take liberties; **sich** *(Dat.)* **zuviel ~nehmen** go too far; **~|picken** *tr. V. (fig.)* pick out **(aus** of); **~|platzen** *itr. V.; mit sein (ugs.)* **a)** *(~lachen)* burst out laughing; **b)** *(spontan äußern)* **mit etw. ~platzen** blurt sth. out; **~|putzen** *tr. V.* **a)** *(festlich kleiden)* dress up; **sich ~putzen** get dressed up; **b)** *(festlich schmükken)* deck out; **sich ~putzen** be decked out; **~|ragen** *itr. V.* **a)** jut out, project (**aus** from); *(sich erheben über)* **aus etw. ~ragen** rise above sth.; **b)** *(hervortreten)* stand out (**aus** from); **~ragend** *Adj.* outstanding; **~|reden** *refl. V. (ugs.)* talk one's way out (**aus** of); **~|reißen** *unr. tr. V.* **a)** tear *or* rip out (**aus** of); pull up *or* out ‹*plant*›; pull out ‹*hair*›; pull up ‹*floor*›; rip out ‹*tiles*›; **b)** *(aus der Umgebung, der Arbeit)* tear away (**aus** from); **jmdn. aus einem Gespräch/seiner Lethargie ~reißen** drag sb. away from a conversation/jolt *or* shake sb. out of his/her lethargy; **c)** *(ugs.: befreien)* save; **~|rücken 1.** *tr. V.* **a)** *(nach außen rücken)* move out (**aus** of); **b)** *(ugs.: hergeben)* hand over; cough up *(coll.)* ‹*money*›; **2.** *itr. V.; mit sein* **mit etw./der Sprache ~rücken** come out with sth./it; **~|rufen 1.** *unr. itr. V.* call *or* shout out (**aus** of); **2.** *unr. tr. V.* call out; **jmdn. aus einer Sitzung ~rufen** call sb. out of a meeting; **~|rutschen** *itr. V.; mit sein* **a)** slip out (**aus** of); **b)** *(ugs.: entschlüpfen)* ‹*remark etc.*› slip out; **die Bemerkung war ihr nur so ~gerutscht** the remark just slipped out somehow; **~|schlagen 1.** *unr. tr. V.* **a)** knock out; **b)** *(ugs.: gewinnen)* get ‹*discount, advantage, etc.*›; make ‹*money, profit*›; **2.** *unr. itr. V.; mit sein* ‹*flames*› leap out (**aus** of); **~|schleudern** *tr. V.* hurl *or* fling out (**aus** of); **~|schlüpfen** *itr. V.; mit sein* slip out (**aus** of); **~|schmuggeln** *tr. V.* smuggle out (**aus** of); **~|schneiden** *unr. tr. V.* cut out (**aus** of); **~|schrauben** *tr. V.* unscrew; **~|schreien** *unr. tr. V.* **seine Wut/seinen Zorn/seinen Haß ~schreien** vent *or* give vent to one's anger/rage/hatred in a loud outburst; **~|sein** *unr. itr. V. mit sein (nur im Inf. u. Part. zusammengeschrieben) (ugs.)* **a)** *(draußen sein)* be out; **b)** *(entfernt sein)* ‹*tooth, appendix, nail, etc.*› be out; **c)** *(hervorgekommen sein)* ‹*flowers, stars, etc.*› be out; **d)** *(hinter sich gelassen haben)* **aus etw. ~sein** be out of sth.; **e)** *(überstanden haben)* **aus dem Gröbsten ~sein** be over the worst; **f)** *(ugs.)* **fein ~sein** be sitting pretty *(coll.)*; **~|springen** *unr. itr. V.; mit sein* **a)** jump *or* leap out (**aus** of); **b)** *(sich lösen)* come out; **c)** *(ugs.: zu erwarten sein)* **dabei springt nicht viel für ihn ~:** there's not much in it for him; **~|stehen** *unr. itr. V.* protrude; stick out; **~|stellen 1.** *tr. V.* **a)** put out[side]; **einen Spieler ~stellen** *(Sport)* send a player off; **b)** *(hervorheben)* emphasize; bring out; present, set out ‹*principles etc.*›; **2.** *refl. V.* **es stellte sich ~, daß ...:** it turned out *or* emerged that ...; **wie sich später ~stellte, hatte er ...:** it turned out later that he had ...; **sich als falsch/wahr** *usw.* **~stellen** turn out *or* prove to be wrong/true *etc.*; **~|strecken** *tr. V.* stick out (**aus** of); **jmdm. die Zunge ~strekken** stick *or* put one's tongue out at sb.; **seinen Arm/Kopf zum Fenster ~strecken** stick *or* put one's arm/head out of the window; **~|streichen** *unr. tr. V.* **a)** *(ausstreichen)* cross out; delete (**aus** from); **b)** *(hervorheben)* point out; **~|stürzen** *itr. V.; mit sein* **a)** *(~fallen)* fall out (**aus** of); **b)** *(eilen)* rush *or* dash out (**aus** of); **~|suchen** *tr. V.* pick out; look out ‹*file*›; **~|treten** *unr. itr. V.; mit sein* **a)** come out (**aus** of); **auf den Balkon ~treten** come *or* step out onto the balcony; **b)** *(sich abzeichnen)* ‹*veins etc.*› stand out; **~|werfen** *unr. tr. V.* **a)** throw out (**aus** of); **b)** *s.* **hinauswerfen; ~|winden** *unr. refl. V.* wriggle out (**aus** of); **~|wirtschaften** *tr. V.* make ‹*profit etc.*› (**aus** out of); **~|ziehen 1.** *unr. tr. V.* **a)** pull out (**aus** of); **b)** *(wegbringen)* pull out, withdraw ‹*troops etc.*›; **2.** *unr. itr. V.; mit sein* move out (**aus** of)

herb [hɛrp] *Adj.* **a)** [slightly] sharp *or* astringent ‹*taste*›; dry ‹*wine*›; [slightly] sharp *or* tangy ‹*smell, perfume*›; **b)** bitter ‹*disappointment, loss*›; severe ‹*face, features*›; austere ‹*beauty*›; **c)** *(unfreundlich)* harsh ‹*words, criticism*›

herbei [hɛɐ̯'bai̯] *Adv.* **~ [zu mir]!** come [over] here!

herbei-: ~|bringen *unr. tr. V.* bring [over]; **~|eilen** *itr. V.; mit sein* hurry over; come hurrying up; **~|führen** *tr. V.* produce, bring about ‹*decision*›; bring about, cause ‹*downfall*›; cause ‹*accident, death*›; **~|holen** *tr. V.* fetch; **~|kommen** *unr. itr. V.; mit sein* come up *or* along; **~|laufen** *unr. itr. V.; mit sein* come running up; **~|reden** *tr. V.* talk ‹*crisis, problem, etc.*› into existence; **~|rufen** *unr. tr. V.* call over; **Hilfe/einen Arzt ~rufen** summon help/call a doctor; **~|schaffen** *tr. V.* bring; *(besorgen)* get; **~|sehnen** *tr. V.* long for; **~|strömen** *itr. V.; mit sein* come in crowds; come flocking

her|bemühen *(geh.)* **1.** *tr. V.* **jmdn. ~:** trouble sb. to come. **2.** *refl. V.* take the trouble to come

Herberge ['hɛrbɛrgə] **die; ~, ~n a)** *(veralt.: Gasthaus)* inn; **b)** *(Jugend~)* [youth] hostel

Herbergs-: ~mutter die, ~vater der warden [of the/a youth hostel]

her-: ~|bestellen *tr. V.* **jmdn. ~bestellen** ask sb. to come; *(~beordern)* summon sb.; **~bitten** *unr. tr. V.* **jmdn. ~bitten** ask sb. to come; **~bringen** *unr. tr. V.* **etw. ~bringen** bring sth. [here]

Herbst [hɛrpst] **der; ~[e]s, ~e** autumn; fall *(Amer.)*; *s. auch* **Frühling**

Herbst-: ~anfang der beginning of autumn; **~blume die** autumn flower; **~ferien** *Pl.* autumn half-term holiday *sing.*

herbstlich 1. *Adj.* autumn *attrib.*; autumnal; **es wird ~:** autumn is coming. **2.** *adv.* **sich ~ färben** take on the colours of autumn

Herbst-: ~tag der autumn day; **~zeit·lose die; ~, ~n** *(Bot.)* meadow saffron

Herd [he:ɐ̯t] **der; ~[e]s, ~e a)** *(Kochstelle)* cooker; stove; **das Essen auf dem ~ haben** *(ugs.)* be cooking something; **eigener ~ ist Goldes wert** there's no place like home *(prov.)*; **b)** *(Ausgangspunkt)* centre *(of disturbance/rebellion)*; **c)** *(Med.)* focus; seat

Herde ['he:ɐ̯də] **die; ~, ~n** herd; **eine ~ Schafe** a flock of sheep

Herden-: ~tier das a) gregarious animal; **b)** *(abwertend: Mensch)* sheep; **~trieb der** *(auch fig. abwertend)* herd instinct

Herd·platte die hotplate; *(eines Kohlenherds)* top

herein [hɛ'rai̯n] *Adv.* **~!** come in!; **[immer] nur ~ mit dir!** come on in!

herein-: ~|bekommen *unr. tr. V. (ugs.)* get in ‹*fresh stocks*›; pick up ‹*radio station*›; recover ‹*investment*›; **~|bitten** *unr. tr. V.* **jmdn. ~bitten** ask *or* invite sb. in; **~|brechen** *unr. itr. V.; mit sein* **a)** *(geh.: hart treffen)* **über jmdn./etw. ~brechen** ‹*fate, disaster, misfortune, etc.*› befall *or* overtake sb./sth.; **b)** *(geh.: beginnen)* ‹*night, evening, dusk*› fall; ‹*winter*› set in; ‹*storm*› strike, break; **~|bringen** *unr. tr. V.* **a)** bring in;

b) *(wettmachen)* make up ⟨*loss*⟩; make up for ⟨*delay*⟩; recoup ⟨*costs*⟩; ~|**fallen** *unr. itr. V.; mit sein* **a)** ⟨*light*⟩ shine in; **b)** *(ugs.: betrogen werden)* be taken for a ride *(coll.);* be done *(coll.);* **bei/mit etw. ~fallen** be taken for a ride with sth.; **auf jmdn./etw. ~fallen** be taken in by sb./sth.; ~|**führen** *tr. V.* **jmdn. ~führen** show sb. in; ~|**holen** *tr. V.* **a)** bring in; **b)** *(ugs.: verdienen)* make *(coll.);* ~|**kommen** *unr. itr. V.; mit sein* come in; **in das Haus/zur Tür ~kommen** come into the house/in through the door; ~|**kriegen** *tr. V. (ugs.) s.* **~bekommen;** ~|**lassen** *unr. tr. V.* let *or* allow in; ~|**legen** *tr. V. (ugs.)* **jmdn. ~legen** take sb. for a ride *(coll.)* **(mit, bei** with); ~|**platzen** *itr. V.; mit sein (ugs.)* burst in; come bursting in; ~|**regnen** *itr. V. (unpers.)* **es regnet ~:** the rain's coming in; ~|**rufen** *unr. tr. V.* **jmdn. ~rufen** call sb. in; ~|**schneien** *unr. itr. V.* **a)** *mit sein (ugs.)* turn up out of the blue *(coll.);* **b)** *(unpers.)* **es schneit ~:** the snow's coming in; ~|**sehen** *unr. itr. V.* **a)** see in; *(hereinblicken)* look in; **b)** *(kurz besuchen)* look *or* drop in **(bei** on); ~|**spazieren** *itr. V.; mit sein (ugs.)* walk in; stroll in; **nur ~spaziert!** come right in!; ~|**stecken** *tr. V.* **den Kopf zur Tür ~stecken** put one's head round the door; ~|**stürmen** *itr. V.; mit sein* rush *or* dash in; come rushing *or* dashing in; *(wütend)* storm in; come storming in

her-, Her-: ~|**fahren 1.** *unr. itr. V.; mit sein* come here; *(mit einem Auto)* drive *or* come here; *(mit einem [Motor]rad)* ride *or* come here; **hinter/vor jmdm./etw. ~fahren** drive/ride along behind/in front of sb./sth.; **2.** *unr. tr. V.* **jmdn. ~fahren** drive sb. here; **~fahrt die** journey here; ~|**fallen** *unr. itr. V.; mit sein* **a) über jmdn. ~fallen** set upon *or* attack sb.; ⟨*animal*⟩ attack sb.; **b)** *(gierig zu essen beginnen)* **über etw.** *(Akk.)* **~fallen** fall upon sth.; ~|**finden** *unr. itr. V.* find one's way here; ~|**führen** *tr. V.* **jmdn. ~führen** bring sb. here; **~gang der: der ~gang der Ereignisse** the sequence of events; **schildern Sie den ~gang des Überfalls** describe what happened during the attack; ~|**geben** *unr. tr. V.* **a)** hand over; *(weggeben)* give away; **sein Geld für etw. ~geben** put one's money into sth.; **er hat sein letztes ~gegeben** he gave everything he had; **dazu gebe ich mich nicht ~:** I won't have anything to do with it; **b)** *(reichen)* give; **gib es ~!** hand it over!; **c)** *(erbringen)* ; **was seine Beine ~gaben** as fast as his legs could carry him; **~gebracht** *Adj.; nicht präd.* time-honoured; ~|**gehen** *unr. itr. V.; mit sein* **a)** *(begleiten)* **neben/vor/hinter jmdm. ~gehen** walk along beside/in front of/behind sb.; **b)** *(ugs.)* **~gehen und etw. tun** just [go and] do sth.; **c)** *(unpers.) (ugs.)* **bei der Debatte ging es heiß ~:** the sparks really flew in the debate; **~gelaufen** *Adj.; nicht präd.* **dieser ~gelaufene Strolch** this good-for-nothing rascal from Heaven knows where; ~|**haben** *unr. tr. V. (ugs.)* **wo hat er/sie das ~?** where did he/she get that from?; ~|**halten 1.** *unr. itr. V.* **~halten müssen [für jmdn./etw.]** be the one to suffer [for sb./sth.]; **2.** *unr. tr. V.* hold out; ~|**holen** *tr. V.* fetch; **etw. von weit ~holen** get sth. from a long way away; **weit ~geholt** far-fetched; ~|**hören** *itr. V.* listen; **alle mal ~hören!** listen everybody

Hering ['he:rɪŋ] **der; ~s, ~e a)** herring; **wie die ~e** *(fig.)* packed together like sardines; **b)** *(Zeltpflock)* peg

Herings·salat der herring salad

herinnen [hɛ'rɪnən] *Adv. (südd., österr.)* in here

her-: ~|**kommen** *unr. itr. V.; mit sein* **a)** come here; **komm [mal] ~!** come here!; **b)** *(stammen)* come; **~kömmlich** [~kœmlɪç] *Adj.* conventional; traditional ⟨*custom*⟩

Herkunft ['he:ɐ̯kʊnft] **die; ~, Herkünfte** ['he:ɐ̯kʏnftə] origin[s *pl.*]; **einfacher** *(Gen.) od.* **von einfacher ~ sein** be of humble origin *or* stock

her-: ~|**laufen** *unr. itr. V.; mit sein* **a) vor/hinter/neben jmdm. ~laufen** run [along] in front of/behind/alongside sb.; **b)** *(nachlaufen)* **hinter jmdm. ~laufen** run after sb.; *(fig.)* chase sb. up; **c)** *(zum Sprechenden laufen)* come on foot; *(schneller)* come running up; ~|**leiten 1.** *tr. V.* derive **(aus, von** from); **etw. von jmdm. ~leiten** derive sth. from sb.; **2.** *refl. V.* **sich von/aus etw. ~leiten** derive *or* be derived from sth.; ~|**machen** *(ugs.)* **1.** *refl. V.* **a) sich über etw.** *(Akk.)* **~machen** get stuck into sth. *(coll.);* **sich über das Essen ~machen** fall upon the food; **b)** *(~fallen)* **sich über jmdn. ~machen** set on *or* attack sb.; **2.** *tr. V.* **wenig ~machen** not look much *(coll.);* **viel ~machen** look great *(coll.)*

Hermelin [hɛrmə'li:n] **das; ~s, ~e** ermine; *(im Sommerfell)* stoat

hermetisch [hɛr'me:tɪʃ] **1.** *Adj.* hermetic. **2.** *adv.* hermetically; **ein Dorf** *usw.* **~ abriegeln** seal a village *etc.* off completely

her·nach *Adv. (veralt.)* after that

her|nehmen *unr. tr. V.* **wo soll ich das Geld ~?** where am I supposed to get the money from *or* find the money?

heroben [hɛ'ro:bn̩] *Adv. (südd., österr.)* up here

Heroin [hero'i:n] **das; ~s** heroin

heroin·süchtig *Adj.* addicted to heroin *postpos.*

heroisch *Adj.* heroic

Herold ['he:rɔlt] **der; ~[e]s, ~e** herald

Herr [hɛr] **der; ~n** *(selten:* **~en), ~en a)** *(Mann)* gentleman; **ein feiner ~:** a refined gentleman; **das Kugelstoßen der ~en** *(Sport)* the men's shot-put; **mein Alter ~** *(ugs. scherzh.: Vater)* my old man *(coll.);* **Alter ~** *(Studentenspr.)* former member; **b)** *(Titel, Anrede)* **~ Schulze** Mr Schulze; **~ Professor/Dr. Schulze** Professor/Dr Schulze; **~ Minister/Direktor/Studienrat Schulze** Mr Schulze; **~ Minister/Professor/Doktor** Minister/Professor/doctor; **~ Vorsitzender/Präsident** Mr Chairman/President; **Sehr geehrter ~ Schulze!** Dear Sir; *(bei persönlicher Bekanntschaft)* Dear Mr Schulze; **Sehr geehrte ~en!** Dear Sirs; **~ Ober!** waiter!; **mein ~:** sir; **meine ~en** gentlemen; **bitte sehr, der ~!** there you are, sir; **Ihr ~ Vater/Sohn** your father/son; **c)** *(Gebieter)* master; **mein ~ und Gebieter** *(scherzh.)* my lord and master *(joc.);* **die ~en der Schöpfung** *(ugs. scherzh.)* their lordships *(coll. joc.);* **sein eigener ~ sein** be one's own master; **~ der Lage sein/bleiben** be/remain master of the situation; **nicht mehr ~ seiner Sinne sein** be no longer in control of oneself; **aus aller ~en Länder[n]** *(geh.)* from the four corners of the earth; from all over the world; **d)** *(Besitzer)* master **(über +** *Akk.* of); **e)** *(christl. Rel.: Gott)* Lord; **Gott der ~:** Lord God

Herrchen das; ~s, ~: master

herren-, Herren-: **~abend der** stag evening; **~ausstatter der** [gentle]men's outfitter; **~bekanntschaft die** gentleman acquaintance; ~**besuch der** gentleman visitor/visitors; ~**besuch haben** have a gentleman visitor/gentlemen visitors; ~**friseur der** men's hairdresser; ~**los** *Adj.* abandoned ⟨*car, luggage*⟩; stray ⟨*dog, cat*⟩; ~**mode die** men's fashion; ~**schuh der** man's shoe; ~**schuhe** men's shoes; ~**toilette die** [gentle]men's toilet

Herr·gott der; ~s a) *(ugs.: Gott)* **der [liebe]/unser ~:** the Lord [God]; God; ~ **noch mal!** for Heaven's sake!; for God's sake!; **b)** *(südd., österr.: Kruzifix)* crucifix

Herrgotts·frühe die *in* **in aller ~:** at the crack of dawn

her·richten 1. *tr. V.* **a)** *(bereitmachen)* get ⟨*room, refreshments, etc.*⟩ ready; dress ⟨*shop-window*⟩; arrange ⟨*table*⟩; **b)** *(in Ordnung bringen)* renovate; do up *(coll.)*. **2.** *refl. V.* get ready

Herrin die; ~, ~en mistress; *(als Anrede)* my lady

herrisch 1. *Adj.* overbearing; peremptory; imperious. **2.** *adv.* peremptorily; imperiously

herr·je, herrjemine [hɛr'je:mine] *Interj. (ugs.)* goodness gracious [me]; heavens [above]

herrlich 1. *Adj.* marvellous; marvellous, glorious ⟨*weather*⟩; magnificent, splendid ⟨*view*⟩; magnificent, gorgeous ⟨*clothes*⟩; ⟨*sth. tastes, looks, sounds*⟩ wonderful, marvellous. **2.** *adv.* marvellously; ~ **und in Freuden leben** live in clover

Herrlichkeit die; ~, ~en a) *o. Pl. (Schönheit)* magnificence; splendour; **die ~ Gottes** the glory of God; **b)** *meist Pl. (herrliche Sache)* marvellous *or* wonderful thing

Herrschaft die; ~, ~en a) *o. Pl.* rule; *(Macht)* power; **die ~ an sich reißen/erringen** seize/gain power; **die ~ über sich/das Auto verlieren** *(fig.)* lose control of oneself/the car; **b)** *Pl. (Damen u. Herren)* ladies and gentlemen; **meine ~en!** ladies and gentlemen!

herrschaftlich *Adj.* **a)** *(zu einer Herrschaft gehörend)* master's/mistress's ⟨*coach etc.*⟩; **b)** *(einer Herrschaft gemäß)* grand

Herrschafts·form die system of government

herrschen ['hɛrʃn̩] *itr. V.* **a)** *(regieren)* rule; ⟨*monarch*⟩ reign, rule; **b)** *(vorhanden sein)* **draußen ~ 30° Kälte** it's 30° below outside; **überall herrschte große Freude/Trauer** there was great joy/sorrow everywhere; **jetzt herrscht hier wieder Ordnung** order has been restored here; **c)** *(unpers.)* prevail; **es herrscht jetzt Einigkeit** there is now agreement

herrschend *Adj., nicht präd.* **a)** ruling ⟨*power, party, etc.*⟩; reigning ⟨*monarch*⟩; **die Herrschenden** the rulers; those in power; **b)** *(vorhanden)* prevailing ⟨*opinion, view, conditions, etc.*⟩

Herrscher der; ~s, ~: ruler; ~ **über ein Volk sein** be [the] ruler of a people

Herrscher-: ~geschlecht das ruling dynasty; ~**haus das** ruling house

Herrscherin die; ~, ~nen *s.* **Herrscher**

Herrsch·sucht die thirst for power; *(herrisches Wesen)* domineering nature

herrsch·süchtig *Adj.* domineering

her-: ~|rufen *unr. tr. V.* call ⟨*dog*⟩; **jmdn. ~rufen** call sb. [over]; **etw. hinter jmdm. ~rufen** call sth. after sb.; ~|**rühren** *itr. V.* **von jmdm./etw. ~rühren** come from sb./stem from sth.; ~|**schieben** *unr. tr. V.* **etw. ~schieben** push sth. here; **etw. vor sich** *(Dat.)* **~schieben** push sth. along in front of one; *(fig.)* put sth. off; ~|**sehen** *unr. itr. V.* look [over] here *or* this way; **seht mal alle ~!** look here *or* this way, everyone!; ~|**sein** *unr. itr. V.; mit sein* **a) einen Monat/einige Zeit/lange ~sein** be a month/some time/a long time ago; **es ist lange ~, daß wir ...:** it is a long time since we ...; **es muß 5 Jahre ~sein, daß wir ...:** it must be five years since we ...; **b)** *(stammen)* **von Köln ~sein** be *or* come from Cologne; **c)** *in* **es ist nicht weit ~ mit jmdm./etw.** *(ugs.)* sb./sth. isn't all that hot *(coll.)*; **hinter jmdm.** *(ugs.)*/**etw. ~sein** be after sb./sth.; ~|**stellen 1.** *tr. V.* **a)** *(anfertigen)* produce; manufacture; make; **in Deutschland ~gestellt** made in Germany; **etw. serienmäßig ~stellen** mass-produce sth.; **b)** *(zustande bringen)* establish ⟨*contact, relationship, etc.*⟩; bring about ⟨*peace, order, etc.*⟩; **c)** *(gesund machen)* **sie** *od.* **ihre Gesundheit ist [ganz] ~gestellt** she has [quite] recovered; **d)** *(zum Sprechenden stellen)* **etw. ~stellen** put sth. [over] here; **2.** *refl. V.* **stell dich ~ [zu mir]** [come and] stand over here [next to *or* by me]

Her·steller der; ~s, ~ producer; manufacturer

Her·stellung die a) *(Anfertigung)* production; manufacture; **b)** *s.* **herstellen b:** establishment; bringing about

her|tragen *unr. tr. V.* **a)** *(zum Sprecher)* **etw. ~:** carry sth. here; **b)** *(begleiten und tragen)* **etw. hinter/vor jmdm. ~:** carry sth. along behind/in front of sb.;

herüben [hɛ'ry:bn̩] *Adv. (südd., österr.)* over here

herüber [hɛ'ry:bɐ] *Adv.* over

herüber-: ~|bringen *unr. tr. V.* **jmdn./etw. ~bringen** bring sb./sth. over; ~|**fahren 1.** *unr. itr. V.; mit sein* drive *or* come over; *(mit dem Motorrad, Rad)* come *or* ride over; **2.** *unr. tr. V.* **jmdn./etw. ~fahren** drive sb./sth. over; ~|**kommen** *unr. itr. V.; mit sein* come over; **über den Zaun/Fluß ~kommen** get over the fence/across the river; **kommt doch ~!** come over!; ~|**schicken** *tr. V.* **jmdn./etw. ~schicken** send sb./sth. over

herum [hɛ'rʊm] *Adv.* **a)** *(Richtung)* round; **im Kreis ~:** round in a circle; **verkehrt/richtig ~:** the wrong/right way round; *(mit Ober- und Unterseite)* upside down/the right way up; **etw. falsch** *od.* **verkehrt ~ anziehen** put sth. on back-to-front/*(Innenseite nach außen)* inside-out; **b)** *(Anordnung)* **um jmdn./etw. ~:** around sb./sth.; **c)** *(in enger Umgebung)* **um jmdn. ~:** around sb.; **um München ~:** around Munich; **d)** *(ugs.: ungefähr)* **um Weihnachten/Ostern ~:** around Christmas/Easter; **um das Jahr 1050 ~:** around *or* about the year 1050

herum-, Herum-: ~|albern *itr. V. (ugs.)* fool around *or* about; ~|**ärgern** *refl. V. (ugs.)* **sich mit jmdm./etw. ~ärgern** keep getting annoyed with sb./sth.; ~|**blättern** *itr. V.* **in etw.** *(Dat.)* **~blättern** keep leafing through sth.; ~|**brüllen** *itr. V. (ugs.)* go on shouting one's head off *(coll.)*; ~|**bummeln** *itr. V. (ugs.)* **a)** *mit sein (spazieren)* stroll *or* wander around; **b)** *(trödeln)* **[mit etw.] ~bummeln** dawdle [over sth.]; ~|**drehen 1.** *tr. V. (ugs.)* turn ⟨*key*⟩; turn over ⟨*coin, mattress, hand, etc.*⟩; **den Kopf ~drehen** turn one's head; **2.** *refl. V.* turn [a]round; **sich [auf die andere Seite] ~drehen** turn over [on to one's other side]; **3.** *itr. V. (ugs.)* **an etw.** *(Dat.)* **~drehen** fiddle [around *or* about] with sth.; ~|**drücken** *refl. V.* **a)** *(ugs.: vermeiden)* **sich um etw. ~drücken** get out of *or (coll.)*

dodge sth.; **b)** *(ugs.: sich aufhalten)* hang around; **wo hast du dich ~gedrückt?** where have you been?; **~|drucksen** *itr. V. (ugs.)* hum and haw *(coll.)*; **~|erzählen** *tr. V. (ugs.)* **etw. ~erzählen** spread sth. around; **er erzählte überall ~, daß ...:** he went around telling everyone that ...; **~|fahren** *(ugs.)* **1.** *unr. itr. V.; mit sein* **a) um etw. ~fahren** drive *or* go round sth.; *(mit einem Motorrad, Rad)* ride *or* go round sth.; *(mit einem Schiff)* sail round sth.; **b)** *(irgendwohin fahren)* drive/ride/sail around; **c)** *(sich plötzlich herumdrehen)* spin round; **2.** *unr. tr. V.* **jmdn. [in der Stadt] ~fahren** drive sb. around the town; **~|fragen** *itr. V. (ugs.)* ask around (**bei** among); **~|fuchteln** *itr. V. (ugs.)* **mit den Armen/einem Messer** *usw.* **~fuchteln** wave one's arms/a knife *etc.* around *or* about; **~|führen 1.** *tr. V.* **a) jmdn. [in der Stadt] ~führen** show sb. around the town; *s. auch* **Nase; b)** *(rund um etw. führen)* **jmdn. um etw. ~führen** lead *or* take sb. round sth.; **c)** *(um etw. bauen)* **die Straße um die Stadt ~führen** take the road round the town; **2.** *itr. V.* **um etw. ~führen** ⟨*road etc.*⟩ go round sth.; **~|fuhr·werken** *itr. V. (ugs.)* mess about; **~|fummeln** *itr. V. (ugs.)* **a) an etw.** *(Dat.)* **~fummeln** fiddle about with sth.; **b)** *(sich handwerklich beschäftigen)* fiddle *or* mess around with something; **c)** *(betasten)* **an jmdm. ~fummeln** touch sb. up *(sl.)*; **~|gehen** *unr. itr. V.; mit sein* **a) um etw. ~gehen** go *or* walk round sth.; **b)** *(ziellos gehen)* walk around; **im Garten ~gehen** walk around the garden; **c)** *(die Runde machen)* go around; *(~gereicht werden)* be passed *or* handed around; **etw. ~gehen lassen** circulate sth.; **d)** *(vergehen)* pass; go by; **~|geistern** *itr. V.; mit sein (ugs.)* wander around *or* about; *(fig.)* ⟨*idea, rumour, etc.*⟩ go round; **~|hacken** *itr. V. (ugs.)* **auf jmdm. ~hacken** keep getting at sb. *(coll.)*; **~|hängen** *unr. itr. V. (ugs.)* **a)** *(aufgehängt sein)* **überall ~hängen** be hung up all over the place; **b)** *s.* **rumhängen a; ~|horchen** *itr. V. (ugs.)* keep one's ears open; **~|irren** *itr. V.; mit sein* wander around *or* about; **im Wald ~irren** wander about the wood; **~|kommandieren** *(ugs.) tr. V.* **jmdn. ~kommandieren** boss *(coll.) or* order sb. around *or* about; **~|kommen** *unr. itr. V.; mit sein (ugs.)* **a)** *(vorbeikommen können)* get round; **b)** *(sich herumbewegen)* come round; **um die Ecke ~kommen** come round the corner; **c)** *(vermeiden können)* **um etw. [nicht] ~kommen** [not] be able to get out of sth.; **d)** *(viel reisen)* get around *or* about; **in der Welt ~kommen** see a lot of the world; **viel ~kommen** get around *or* about a lot *or* a great deal; **~|kramen** *itr. V. (ugs.)* keep rummaging around *or* about; **~|kriegen** *tr. V. (salopp)* **jmdn. ~kriegen** talk sb. into it; *(verführen)* get sb. into bed *(coll.)*; **~|laufen** *unr. itr. V.; mit sein* **a)** walk/*(schneller)* run around *or* about; **in der Stadt ~laufen** walk/*(schneller)* run around the town; **b)** *(umrunden)* **um etw. ~laufen** walk *or* go round sth.; **c)** *(gekleidet sein)* **wie läufst du wieder ~!** what do you look like!; **~|liegen** *unr. itr. V. (ugs.)* lie around *or* about; **~|lungern** *itr. V. (salopp)* loaf around; **~|machen** *itr. V. (ugs.)* be busy; *(abwertend)* mess about *or* around; **~|nörgeln** *itr. V. (ugs. abwertend)* moan; grumble; **an jmdm./etw. ~nörgeln** moan *or* grumble about sb./sth.; **~|quälen** *refl. V. (ugs.)* **sich [mit einem Problem] ~quälen** struggle [with a problem]; **~|reden** *itr. V. (ugs.)* **um etw. ~reden** talk round sth.; **red nicht lange um die Sache ~!** don't beat about the bush!; **~|reichen** *(ugs.) tr. V.* **etw. ~reichen** pass sth. round; **~|reißen** *unr. tr. V.* **den Wagen/das Pferd ~reißen** swing the car/horse round; *s. auch* **[1]Steuer; ~|reiten** *unr. itr. V.; mit sein* **a) in der Gegend ~reiten** ride around the area; **b)** *(salopp; auf dasselbe zurückkommen)* **auf etw.** *(Dat.)* **~reiten** go on about sth. *(coll.)*; harp on sth.; **~|rennen** *unr. itr. V.; mit sein (ugs.)* **a)** *(ziellos rennen)* run around *or* about; **b)** *(im Bogen rennen)* **um etw. ~rennen** run round sth.; **im Kreis ~rennen** run round in a circle; **~|schlagen** *unr. refl. V. (ugs.)* **a)** *(sich schlagen)* **sich mit jmdm. ~schlagen** keep fighting *or* getting into fights with sb.; **b)** *(sich auseinandersetzen)* **sich mit Problemen/Einwänden ~schlagen** grapple with problems/battle against objections; **~|schleichen** *unr. itr. V.; mit sein (ugs.)* creep around *or* about; **um etw. ~schleichen** creep round sth.; **~|schleppen** *tr. V. (ugs.)* **etw. um etw. ~schleppen** lug sth. round sth.; **eine Erkältung/ein Problem mit sich** *(Dat.)* **~schleppen** *(fig.)* go around with a cold/be worried by a problem; **~|schnüffeln** *itr. V. (ugs. abwertend)* nose *or* snoop around *or* about *(coll.)*; **~|sein** *unr. itr. V.; mit sein; Zusammenschreibung nur im Inf. und Part. (ugs.)* **a)** have passed; have gone by; **seine Probezeit ist noch nicht ~:** his probationary period is not yet over; **b) immer um jmdn. ~sein** be always around sb; **~|sitzen** *unr. itr. V. (ugs.)* sit around *or* about; **um etw. ~sitzen** sit round sth.; **tatenlos ~sitzen** sit around *or* about doing nothing; **~|sprechen** *unr. refl. V.* get around *or* about; **schnell hatte sich ~gesprochen, daß ...:** it had quickly got around that ...; **~|stehen** *unr. itr. V. (ugs.)* stand around *or* about; **~|stöbern** *itr. V. (ugs.) (in einem Schreibtisch usw.)* keep rummaging around *or* about (**in** + *Dat.* in); **~|stochern** *itr. V.* poke around *or* about; **im Essen ~stochern** pick at one's food; **~|stoßen** *unr. tr. V. (ugs.)* **jmdn. ~stoßen** push sb. around; **~|tanzen** *itr. V.; mit sein (ugs.)* dance around *or* about; *s. auch* **Nase b; ~|tollen** *itr. V.; mit sein* romp around *or* about; **auf dem Hof ~tollen** romp around the yard; **~|tragen** *unr. tr. V. (ugs.)* **a)** *(überallhin tragen)* **jmdn./etw. ~tragen** carry sb./sth. around *or* about; **b) eine Idee/einen Plan mit sich ~tragen** nurse an idea/a plan; **c)** *(abwertend: weitererzählen)* **etw. ~tragen** spread sth. around; **~|trampeln** *itr. V.; mit haben od. sein* **auf etw.** *(Dat.)* **~trampeln** trample [around] on sth.; trample all over sth.; **auf jmds. Nerven/Gefühlen ~trampeln** *(fig.)* really get on sb.'s nerves/trample on sb.'s feelings; **~|treiben** *unr. refl. V. (ugs. abwertend)* **sich auf den Straßen/in Spelunken ~treiben** hang around the streets/*(coll.)* in dives; **sich mit Männern ~treiben** hang around with men; **wo hast du dich nur ~getrieben?** where have you been?; **~treiber der** *(ugs.)* layabout; *(Streuner)* vagabond; **~|trödeln** *itr. V. (ugs.)* dawdle around *or* about (**mit** over); **~|wälzen** *(ugs.)* **1.** *tr. V.* **etw. ~wälzen** roll sth. over; **2.** *refl. V.* roll around *or* about; **sich im Bett ~wälzen** toss and turn in bed; **~|werfen 1.** *unr. tr. V.* **a)** *(ugs.: umherwerfen)* **etw. ~werfen** chuck *(coll.) or* throw sth. around *or* about; **b)** *(in eine andere Richtung drehen)* throw ⟨*helm, steering-wheel, etc.*⟩ [hard] over; **den Kopf ~werfen** turn one's head quickly; **2.** *unr. refl. V.* **sich im Bett ~werfen** toss and turn in bed; **~|zeigen** *tr. V. (ugs.)* **etw. ~zeigen** show sth. round; **~|ziehen 1.** *unr. itr. V.* move around *or* about; **im Land ~ziehen** move around *or* about the country; **2.** *unr. tr. V. (ugs.: mit sich ziehen)* **jmdn./etw. ~ziehen** drag sb./sth. round *(coll.)*

herunten [hɛˈrʊntn̩] *Adv. (südd., österr.)* down here

herunter [hɛˈrʊntɐ] *Adv.* **a)** *(nach unten)* down; **von Kiel nach München ~** *(fig.)* from Kiel down to Munich; **b)** *(fort)* off; **~ vom Sofa!** [get] off the sofa!

herunter-: ~|bekommen *unr. tr. V. (ugs.)* **a)** *(essen können)* be able to eat; *(~schlucken)* swallow; **b)** *(entfernen können)* **etw. [von etw.] ~bekommen** be able to get sth. off [sth.]; **~|beten** *tr. V. (abwertend)* **etw. ~beten** recite sth. mechanically; **~|bringen** *unr. tr. V.* **a)** *(nach unten bringen)* bring down; **b)** *(zugrunde richten)* ruin; **c)** *(ugs.: herunterschlucken) s.* **~bekommen a; ~|drücken** *tr. V.* **a)** *(nach unten drücken)* **etw. ~drücken** press sth. down; **b)** *(auf ein niedriges Niveau bringen, verringern)* force down ⟨*prices, wages, etc.*⟩; bring down ⟨*temperature*⟩; reduce ⟨*marks*⟩; **~|fahren 1.** *unr. itr. V.; mit sein* drive *or* come down; ⟨*skier*⟩ ski down; *(mit einem Motorrad, Rad)* ride down; **2.** *unr. tr. V.* **jmdn./etw. ~fahren** drive *or* bring sb. down/bring sth. down; **~|fallen** *unr. itr. V.; mit sein* fall down; **vom Tisch/Stuhl ~fallen** fall off the table/chair; **die Treppe ~fallen** fall down the stairs; **jmdm. fällt etw. herunter** sb. drops sth.; **~|gehen** *unr. itr. V.; mit sein* **a)** *(nach unten gehen)* come down; **b)** *(niedriger werden)* ⟨*temperature*⟩ go down, drop, fall; ⟨*prices*⟩ come down, fall; **c)** *(die Höhe senken)* **auf eine Flughöhe von 2000 m ~gehen** descend to 6,000 ft.; **mit den Preisen ~gehen** reduce one's/its prices; **d) von etw. ~gehen** *(ugs.: räumen)* get off sth.; **e)** *(ugs.: sich lösen)* come off; **~gekommen 1.** *2. Part. v.* **~kommen; 2.** *Adj.* poor ⟨*health*⟩; dilapidated, run-down ⟨*building*⟩; run-down ⟨*area*⟩; down and out ⟨*person*⟩; **~|handeln** *tr. V. (ugs.)* **einen Preis ~handeln** beat down a price; **100 DM vom Kaufpreis ~handeln** get 100 marks knocked off the price *(coll.)*; **~|hängen** *unr. itr. V.* hang down; **~|hauen** *unr. tr. V. (ugs.)* **a)** *(ohrfeigen)* **jmdm. eine ~hauen** give sb. a clout round the ear *(coll.)*; **b)** *(schlecht ausführen)* dash off; **~|holen** *tr. V.; mit sein* **jmdn./etw. ~holen** fetch sb./sth. down; **~|kommen** *unr. itr. V.; mit sein* **a)** *(kommen)* come down; *(nach unten kommen können)* manage to come down; **b)** *(ugs.: verfallen)* go to the dogs *(coll.)*; **er ist so weit ~gekommen, daß ...**: he has sunk so low that ...; **c)** *(ugs.: wegkommen)* **von Drogen/vom Alkohol ~kommen** come off drugs/alcohol; kick the habit *(sl.)*; **~|können** *unr. itr. V. (ugs.)* be able to get down; **~|lassen** *unr. tr. V. (schließen)* let down, lower ⟨*blind, shutter*⟩; lower ⟨*barrier*⟩; shut ⟨*window*⟩; *(nach unten gleiten lassen)* wind down ⟨*car window*⟩; **jmdn./etw. an etw.** *(Dat.)* **~lassen** lower sb./sth. by sth.; **die Hose ~lassen** take one's trousers down; **~|leiern** *tr. V. (salopp)* **a)** *(abwertend)* drone out *(coll.)*; **b)** wind down ⟨*car window*⟩; **~machen** *tr. V. (salopp)* **a)** *(zurechtweisen)* **jmdn. ~machen** give sb. a rocket *(sl.)*; tear sb. off a strip *(coll.)*; **b)** *(herabsetzen)* slate *(coll.)*; run down *(coll.)*; **~|nehmen** *unr. tr. V.* take down; **etw. von etw. ~nehmen** take sth. off sth.; **~|putzen** *tr. V. (salopp) s.* **~machen a; ~|reißen** *unr. tr. V. (ugs.)* **a)** *(nach unten reißen)* pull down; **b)** *(abreißen)* pull off ⟨*plaster, wallpaper*⟩; tear down ⟨*poster*⟩; **c)** *(salopp: ableisten)* get through; **~|rutschen** *itr. V.; mit sein (ugs.)* slide down; ⟨*trousers, socks*⟩ slip down; **~|schalten** *itr. V. (Kfz-Jargon)* change down; **~|schlucken** *tr. V.* swallow; **~|schrauben** *tr. V.* turn down ⟨*wick etc.*⟩; **seine Ansprüche/Erwartungen ~schrauben** *(fig.)* reduce one's requirements/lower one's expectations; **~|sein** *unr. itr. V.; mit sein; Zusammenschreibung nur im Inf. und Part. (ugs.)* **a)** *(unten sein)* be down; **b)** *(am Ende der Kräfte sein)* **[körperlich] ~sein** be in poor health; **~|spielen** *tr. V. (ugs.)* **a)** *(als unbedeutend darstellen)* play down *(coll.)*; **b)** *(ausdruckslos spielen)* **etw. ~spielen** play sth. through mechanically; **~|steigen** *unr. itr. V.; mit sein* climb down; **~|stürzen 1.** *itr. V.; mit sein* fall down; *(steil herabfallen)* ⟨*aircraft, person, etc.*⟩ plunge down; *(~eilen)* rush down; **vom Dach ~stürzen** fall off the roof; **2.** *tr. V.* **a)** *(schnell trinken)* gulp down; **b) jmdn. ~stürzen** throw sb. down; **~|tragen** *unr. tr. V.* **etw. ~tragen** carry sth. down; **~|werfen** *unr. tr. V.* **a)** *(nach unten werfen)* **etw. ~werfen** throw sth. down; **b)** *(ugs.: herunterfallen lassen)* drop; **~|ziehen 1.** *unr. tr. V.* pull down; **2.** *unr. itr. V.; mit sein* go *or* move down

her·vor *Adv.* **aus etw. ~**: out of sth.; **aus der Ecke ~ kam ...**: from out of the corner came ...

hervor-, Hervor-: ~|bringen *unr. tr. V.* **a)** *(zum Vorschein bringen)* bring out **(aus** of); produce **(aus** from); **b)** *(wachsen, entstehen lassen; auch fig.)* produce; **c)** *(von sich geben)* say; produce ⟨*sound*⟩; **~|gehen** *unr. itr. V.; mit sein (geh.)* **a)** *(seinen Ursprung haben)* **viele große Musiker gingen aus dieser Stadt ~**: this city produced many great musicians; **drei Kinder gingen aus der Ehe ~**: the marriage produced three children; there were three children of the marriage; **b)** *(herauskommen, sich ergeben)* emerge **(aus** from); **aus seinem Brief geht klar ~, daß ...**: it is clear from his letter that ...; **c)** *(zu folgern sein)* follow; **daraus geht ~, daß ...**: from this it follows that ...; **~|heben** *unr. tr. V.* emphasize; stress; **~|holen** *tr. V.* take out **(aus** of); **~|kommen** *unr. itr. V.; mit sein* come out **(aus** of, **unter** + *Dat.* from under); **~|locken** *tr. V.* lure *or* entice ⟨*person, animal*⟩ out **(aus** of); **~|ragen** *itr. V.* **a)** *(aus etw. ragen)* project; jut out; ⟨*cheekbones*⟩ stand out; **b)** *(sich auszeichnen)* stand out; **~ragend 1.** *Adj.* outstanding[ly good]; **2.** *adv.* **~ragend geschult** outstandingly well trained; **~ragend spielen/arbeiten** play/work outstandingly well *or* excellently; **~ruf der** curtain-call; **~|rufen** *unr. tr. V.* **a)** *(nach vorn rufen)* **jmdn. ~rufen** call for sb. to come out; *(Theater usw.)* call sb. back; **b)** *(verursachen)* elicit, provoke ⟨*response*⟩; arouse ⟨*admiration*⟩; cause ⟨*unease, disquiet, confusion, merriment, disease*⟩; provoke ⟨*protest, displeasure*⟩; **~|stechen** *unr. itr. V.* stick out **(aus** of); **~stechend** *Adj.* outstanding; striking; **~|stehen** *unr. itr. V.* protrude; stick out;

⟨*cheekbones*⟩ stand out; ~|**treten** *unr. itr. V.; mit sein* emerge, step out (**hinter** + *Dat.* from behind); ⟨*veins, ribs, etc.*⟩ stand out; ⟨*similarity etc.*⟩ become apparent *or* evident; ⟨*eyes*⟩ bulge, protrude; ~|**tun** *unr. refl. V.* **a)** *(Besonderes leisten)* distinguish oneself; **sich mit/als etw. ~tun** make one's mark with/as sth.; **b)** *(wichtig tun)* show off; ~|**wagen** *refl. V.* dare to come out (**aus** of); **du kannst dich wieder ~wagen** you can come out again

Her·weg der: auf dem ~weg on the way here

Herz [hɛrts] **das; ~ens, ~en a)** *(auch: herzförmiger Gegenstand, zentraler Teil)* heart; **sie hat es am ~en** *(ugs.)* she has a bad heart; *(fig.)* **komm an mein ~, Geliebter** come into my arms, my darling; **mir blutet das ~** *(auch iron.)* my heart bleeds; **ihm rutschte** *od.* **fiel das ~ in die Hose[n]** *(ugs., oft scherzh.)* his heart sank into his boots; **jmds. ~ höher schlagen lassen** make sb.'s heart beat faster; **jmdm. das ~ brechen** *(geh.)* break sb.'s heart; **das ~ auf dem rechten Fleck haben** have one's heart in the right place; **jmdn./etw. auf ~ und Nieren prüfen** *(ugs.)* grill sb./go over sth. with a fine toothcomb; **b)** *(meist geh.: Gemüt)* heart; **die ~en bewegen/rühren** touch people's hearts; **von ~en kommen** come from the heart; **im Grunde seines Herzens** in his heart of hearts; **ein ~ und eine Seele sein** be bosom friends; **jmds. ~ hängt an etw.** *(Dat.) (jmd. möchte etw. sehr gerne behalten)* sb. is attached to sth.; *(jmd. möchte etw. sehr gerne haben)* sb.'s heart is set on sth.; **ihm war/wurde das ~ schwer** his heart was/grew heavy; **alles, was das ~ begehrt** everything one's heart desires; **sich** *(Dat.)* **ein ~ fassen** pluck up one's courage; take one's courage in both hands; **sein ~ für etw. entdecken** *(geh.)* discover a passion for sth.; **ein ~ für Kinder/die Kunst haben** have a love of children/art; **jmdm. sein ~ ausschütten** pour out one's heart to sb.; **das ~ auf der Zunge tragen** wear one's heart on one's sleeve; **seinem ~en einen Stoß geben** [suddenly] pluck up courage; **seinem ~en Luft machen** *(ugs.)* give vent to one's feelings; **leichten ~ens** easily; happily; **schweren ~ens** with a heavy heart; **jmd./etw. liegt jmdm. am ~en** sb. has the interests of sb./sth. at heart; **jmdm. etw. ans ~ legen** entrust sb. with sth.; **jmd./etw. ist jmdm. ans ~ gewachsen** sb. has grown very fond of sb./sth.; **etw. auf dem ~en haben** have sth. on one's mind; **jmdn. ins** *od.* **in sein ~ schließen** take to sb.; **mit halbem ~en** *(geh.)* half-heartedly; **es nicht übers ~ bringen, etw. zu tun** not have the heart to do sth.; **von ~en gern** [most] gladly; **von ganzem ~en** *(aufrichtig)* with all one's heart; *(aus voller Überzeugung)* wholeheartedly; **sich** *(Dat.)* **etw. zu ~en nehmen** take sth. to heart; **mit ganzem ~en** *(geh.)* wholeheartedly; **jmdm. aus dem ~en sprechen** express just what sb. is/was thinking; *s. auch* **Luft c; Stein b; Stich e; c)** *(Kartenspiel)* hearts *pl.; (Karte)* heart; *s. auch* [2]**Pik; d)** *(Kosewort)* **mein ~:** my dear

herz-, Herz-: ~an·fall der heart attack; **~as das** ace of hearts; **~beklemmend** *Adj.* oppressive; **~beschwerden** *Pl.* heart trouble *sing.;* **~bube der** jack of hearts

Herzchen das; ~s, ~ a) *(abwertend: naive/unzuverlässige Person)* simpleton/unreliable person; **b)** *(Kosewort)* darling; sweetheart; **c)** *(kleines Herz)* little heart

Herz·dame die queen of hearts

her|zeigen *tr. V. (ugs.)* show; **zeig [es] mal her!** let me see [it]!

herzen *tr. V. (veralt.)* hug

herzens-, Herzens-: ~angelegenheit die *(Liebesangelegenheit)* affair of the heart; *(Leidenschaft)* passion; **~bedürfnis das** *in* **jmdm. ein ~bedürfnis sein** *(geh.)* be very important to sb.; **~brecher der** lady-killer; **~gut** ['--'-] *Adj.* kind-hearted; good-hearted; **~lust die: etw. nach ~lust tun** do sth. to one's heart's content; **~wunsch der** dearest *or* fondest wish

herz-, Herz-: ~erfrischend 1. *Adj.* refreshing; **2.** *adv.* refreshingly; **~fehler der** heart defect; **~förmig** *Adj.* heart-shaped

herzhaft 1. *Adj.* **a)** *(kräftig)* hearty; **b)** *(nahrhaft)* hearty, substantial ⟨*meal*⟩; *(von kräftigem Geschmack)* tasty; **ein ~er Eintopf** a substantial/tasty stew. **2.** *adv. (kräftig)* heartily; **~ gähnen** give a wide yawn

her|ziehen 1. *unr. itr. V.* **a)** *mit sein od. haben (ugs.: abfällig reden)* **über jmdn./etw. ~:** run sb./sth. down; pull sb./sth. to pieces; **b)** *mit sein (mitgehen)* **vor/hinter/neben jmdm./etw. ~:** march along in front of/behind/beside sb./sth.; **c)** *(umziehen) mit sein* move here. **2.** *unr. tr. V.* **a)** *(ugs.: zum Sprechenden bewegen)* **etw. ~:** pull sth. over [here]; **b)** *(mit sich führen)* **jmdn./etw. hinter sich** *(Dat.)* **~:** pull sb./sth. along behind one

herzig 1. *Adj.* sweet; dear; delightful. **2.** *adv.* sweetly; delightfully

herz-, Herz-: ~infarkt der heart attack; cardiac infarction *(Med.);* **~insuffizienz die** *(Med.)* cardiac insufficiency; **~kammer die** *(Anat.)* ventricle; **~klappe die** *(Anat.)* heart-valve; **~klopfen das; ~s: jmd. bekommt ~klopfen** sb.'s heart starts to pound; **mit ~klopfen** with a pounding heart; **~könig der** king of hearts; **~krank** *Adj.* ⟨*person*⟩ with *or* suffering from a heart condition; **~kranke der/die** person with *or* suffering from a heart condition; *(Patient)* cardiac patient; **~kranz·gefäß das;** *meist Pl.* coronary vessel

herzlich 1. *Adj.* warm ⟨*smile, reception*⟩; kind ⟨*words*⟩; **~e Grüße/~en Dank** kind regards/many thanks; **sein ~es Beileid zum Ausdruck bringen** express one's sincere condolences *pl.; s. auch* **Glückwunsch. 2.** *adv.* warmly; ⟨*congratulate*⟩ heartily; **es grüßt euch ~ Eure Viktoria** *(als Briefschluß)* kind regards, Victoria; **c)** *(sehr)* **~ wenig** very *or (coll.)* precious little; **~ gern!** gladly

Herzlichkeit die *s.* **herzlich:** warmth; kindness

herz-, Herz-: ~los 1. *Adj.* heartless; callous; **2.** *adv.* heartlessly; callously; **~losigkeit die; ~, ~en a)** *o. Pl.* heartlessness; callousness; **b)** *(herzlose Tat/Bemerkung)* heartless act/remark; **~massage die** cardiac massage; heart massage; **~mittel das** *(ugs.)* heart pills *pl.;* **~muskel der** *(Anat.)* heart muscle; cardiac muscle

Herzog ['hɛrtso:k] **der; ~s, Herzöge** ['hɛrtsø:gə] duke; **[Herr] Friedrich ~ von Meiningen** Frederick, Duke of Meiningen

Herzogin die; ~, ~nen duchess

herzoglich *Adj.; nicht präd.* ducal; of the duke *postpos., not pred.;* **die ~e Familie** the family of the duke

Herzogtum das; ~s, Herzogtümer duchy

herz-, Herz-: ~schlag der a) heartbeat; **einen ~schlag lang**

(geh.) for a *or* one fleeting moment; **b)** *o. Pl. (Abfolge der Herzschläge, auch fig. geh.)* pulse; **c)** *(Herzversagen)* heart failure; **an einem ~schlag sterben** die of heart failure; **~schrittmacher der** *(Anat., Med.)* [cardiac] pacemaker; **~spezialist der** heart specialist; **~stärkend** *Adj.* **ein ~stärkendes Mittel** a cardiac tonic; **~still·stand der** *(Med.)* cardiac arrest; **~stück das** *(geh.)* heart; **~transplantation die** *(Med.)* heart transplantation; **~versagen das; ~s** heart failure; **~zerreißend 1.** *Adj.* heart-rending; **2.** *adv.* heart-rendingly

Hesse ['hɛsə] **der; ~n, ~n** Hessian

Hessen (das) Hesse

hessisch *Adj.* Hessian

hetero-, Hetero- [hetero-] hetero-

heterogen [-'ge:n] *Adj.* heterogeneous

hetero·sexuell *Adj.* heterosexual

Hetz [hɛts] **die; ~, ~en** *(österr. ugs.)* **das war eine ~!** that was a [good] laugh; **aus ~:** for fun

Hetz·blatt das *(abwertend)* political smear-sheet

Hetze ['hɛtsə] **die; ~ a)** *(große Hast)* [mad] rush; **in großer ~:** in a mad rush *or* hurry; **b)** *o. Pl. (abwertend: Aufhetzung)* smear campaign; *(gegen eine Minderheit)* hate campaign

hetzen 1. *tr. V.* **a)** hunt; **die Hunde/die Polizei auf jmdn. ~:** set the dogs on [to] sb./get the police on to sb.; **b)** *(antreiben)* rush; hurry. **2.** *itr. V.* **a)** *(in großer Eile sein)* rush; **den ganzen Tag ~:** be in a rush all day long; **b)** *mit sein (hasten)* rush; hurry; *(rennen)* dash; race; **c)** *(abwertend: Haß entfachen)* stir up hatred; *(schmähen)* say malicious things; **gegen jmdn./etw. ~:** smear sb./agitate against sth.

Hetzer der; ~s, ~, Hetzerin die; ~, ~nen malicious agitator

Hetz-: ~jagd die a) *(Jagdw.)* hunting *(with hounds)*; *(einzelne Jagd)* hunt *(with hounds)*; **b)** *(Hast)* [mad] rush; **~kampagne die** *(abwertend)* smear campaign

Heu [hɔy] **das; ~[e]s** hay; **~ machen** make hay

Heu·boden der hayloft

Heuchelei die; ~ *(abwertend)* hypocrisy

heucheln ['hɔyçl̩n] **1.** *itr. V.* be a hypocrite. **2.** *tr. V.* feign

Heuchler der; ~s, ~, Heuchlerin die; ~, ~nen hypocrite

heuchlerisch 1. *Adj.* **a)** *(unaufrichtig)* hypocritical; **b)** *(geheuchelt)* feigned ⟨*interest, sympathy, etc.*⟩. **2.** *adv.* hypocritically

heuer ['hɔyɐ] *Adv. (südd., österr., schweiz.)* this year

Heuer die; ~, ~n *(Seemannsspr.)* **a)** *(Lohn)* pay; wages *pl.*; **b)** *(Anstellung)* **auf einem Frachter ~ nehmen** ship on board a freighter; **eine ~ bekommen** get hired

Heu-: ~ernte die a) hay harvest; haymaking; **b)** *(Ertrag)* hay crop; **~gabel die** hay-fork; **~haufen der** haystack; hayrick

Heul·boje die *(Seew.)* whistling-buoy

heulen ['hɔylən] *itr. V.* **a)** ⟨*wolf, dog, jackal, etc.*⟩ howl; *(fig.)* ⟨*wind, gale*⟩ howl; ⟨*storm*⟩ roar; **b)** ⟨*siren, buoy, etc.*⟩ wail; **c)** *(ugs.: weinen)* howl; bawl; **vor Wut/Schmerz/Freude ~:** howl and weep with rage/pain/howl with delight; **das ist zum Heulen** *(ugs.)* it's enough to make you weep

Heulsuse ['hɔylzu:zə] **die; ~, ~n** *(ugs. abwertend)* cry-baby

Heu·pferd das grasshopper

Heurige der; *adj. Dekl. (bes. österr.)* new wine; **sie saßen beim ~n** they sat drinking the new wine

Heu-: ~schnupfen der hay fever; **~schober der** *(südd., österr.)* haystack; hayrick; **~schrecke die** grasshopper; *(in Afrika, Asien)* locust; grasshopper

heut [hɔyt] *(ugs.)*, **heute** ['hɔytə] *Adv.* today; **~ früh** early this morning; **~ morgen/abend** this morning/evening; **~ mittag** [at] midday today; **~ nacht** tonight; *(letzte Nacht)* last night; **~ in einer Woche** a week [from] today; today week; **~ vor einer Woche** a week ago today; **seit ~:** from today; **ab ~, von ~ an** from today [on]; **bis ~:** until today; **bis ~ nicht** *(erst ~)* not until today; *(überhaupt noch nicht)* not to this day; *(bis jetzt noch nicht)* not as yet; **für ~:** for today; **lieber ~ als morgen** *(ugs.)* the sooner, the better; **von ~ auf morgen** from one day to the next; **der/die Frau von ~:** the woman of today

heutig *Adj.; nicht präd.* **a)** *(von diesem Tag)* today's; **der ~e Tag/am ~en Tage** today; **bis zum ~en Tag** until the present day *or* today; **b)** *(gegenwärtig)* today's; of today *postpos.*; **die ~e Jugend/Generation** today's youth/generation; the youth/generation of today; **in der ~en Zeit** today; nowadays

heut·zu·tage *Adv.* nowadays

Hexe ['hɛksə] **die; ~, ~n a)** witch; **b) diese kleine ~** *(abwertend)* this little minx

hexen *itr. V.* work magic; **ich kann doch nicht ~** *(ugs.)* I'm not a magician *(coll.)*

Hexen-: ~jagd die *(auch fig.)* witch-hunt; **~kessel der: ein [wahrer] ~kessel sein** be [absolute] bedlam; **das Fußballstadion glich einem ~kessel** there was pandemonium *or* bedlam in the football-ground; **~meister der** sorcerer; **~schuß der;** *o. Pl.* lumbago *no indef. art.*; **~verbrennung die** *(hist.)* burning of a witch/of witches; **~verfolgung die** *(hist.)* witch-hunt

Hexer der; ~s, ~: sorcerer

Hexerei die; ~, ~en sorcery; witchcraft; *(Zauberkunststücke)* magic; **das ist doch keine ~:** there's no magic about it

Hickhack ['hɪkhak] **das** *od.* **der; ~s, ~s** *(ugs.)* squabbling; bickering

hie [hi:] *Adv.* **~ und da** *(an manchen Stellen)* here and there; *(manchmal)* [every] now and then

hieb [hi:p] *1. u. 3. Pers. Sg. Prät. v.* **hauen**

Hieb der; ~[e]s, ~e a) *(Schlag)* blow; *(mit der Peitsche)* lash; *(im Fechten)* cut; *(fig.)* dig (**gegen** at); **jmdm. einen ~ mit der Faust versetzen** punch sb.; **b)** *Pl. (ugs.: Prügel)* **~e bekommen/kriegen** get a hiding *or* beating *or (sl.)* walloping; **es gibt/setzt ~e!** you'll get a hiding *or* beating *or (sl.)* walloping

hieb·fest *Adj. in* **hieb- und stichfest** watertight; cast-iron

hielt [hi:lt] *1. u. 3. Pers. Sg. Prät. v.* **halten**

hier [hi:ɐ̯] *Adv.* **a)** here; **[von] ~ oben/unten** [from] up/down here; **~ vorn** here in front; **~ draußen/drinnen** out/in here; **~ entlang** along here; **von ~ [aus]** from here; **er ist nicht von ~:** he's not from this area *or* around here; **~ spricht Hans Schulze** this is Hans Schulze [speaking]; **~ und da** *od.* **dort** *(an manchen Stellen)* here and there; *(manchmal)* [every] now and then; **~ und jetzt** *od.* **heute** *(geh.)* here and now; **b)** *(zu diesem Zeitpunkt)* now; **von ~ an** from now on

hieran ['hi:'ran] *Adv.* **a)** *(an dieser/diese Stelle)* here; **b)** *(fig.)* **im Anschluß ~:** immediately after this

Hierarchie [hierar'çi:] **die; ~, ~n** hierarchy

hierarchisch 1. *Adj.* hierarchical. **2.** *adv.* hierarchically

hierauf ['hi:'rauf] *Adv.* **a)** *(auf diese/diese Stelle)* on here; **b)** *(darauf)* on this; **wir werden ~ zurückkommen** we'll come back to this; **c)** *(danach)* after that; then; **d)** *(infolgedessen)* whereupon

hierauf·hin *Adv.* hereupon

hieraus ['hi:'raus] *Adv.* **a)** *(aus dem eben Erwähnten)* out of *or* from here; **b)** *(aus dieser Tatsache, Quelle)* from this; **c)** *(aus diesem Material)* out of this

hier|behalten *unr. tr. V.* **jmdn./etw. ~:** keep sb./sth. here

hier·bei *Adv.* **a)** *(bei dieser Gelegenheit)* **Diese Übung ist sehr schwierig. Man kann sich ~ leicht verletzen.** This exercise is very difficult. You can easily injure yourself doing it; **Ich habe ihn gestern getroffen. Hierbei habe ich gleich ...:** I met him yesterday, and straightaway I ...; **b)** *(bei der erwähnten Sache)* here

hier|bleiben *unr. itr. V.; mit sein* stay here

hier·durch *Adv.* **a)** *(hier hindurch)* through here; **b)** *(auf Grund dieser Sache)* because of this; as a result of this

hier·für *Adv.* for this

hier·her *Adv.* here; **ich gehe bis ~ und nicht weiter** I'm going this far and no further; **bis ~ und nicht weiter** *(als Warnung)* so far and no further

hierher-: ~|gehören *itr. V.* **a)** *(an diesen Ort gehören)* belong here; **b)** *(hierfür wichtig sein)* be relevant [here]; **~|kommen** *unr. itr. V.; mit sein* come here; **wie bist du ~gekommen?** how did you get here?

hier·hin *Adv.* here; **sie blickte bald ~, bald dorthin** she looked this way and that; **bis ~:** up to here *or* this point

hierin ['hi:'rɪn] *Adv.* in this

hier|lassen *unr. tr. V.* **etw. ~:** leave sth. here

hier·mit *Adv.* with this/these; **~ ist der Fall erledigt** that puts an end to the matter; **~ erkläre ich, daß ...** *(Amtsspr.)* I hereby declare that ...

hier·nach *Adv.* **a)** *(zeitlich, räumlich)* after this *or* that; **b)** *(diesem entsprechend)* in accordance with this/these; **c)** *(demnach)* according to this/these

Hieroglyphe [hiero'gly:fə] **die; ~, ~n** hieroglyph; **~n** hieroglyphics

hier|sein *unr. itr. V.; mit sein; Zusammenschreibung nur im Inf. u. Part.* be here

hierüber ['hi:'ry:bɐ] *Adv.* **a)** *(über dem Erwähnten)* above here; **b)** *(über das Erwähnte)* over here; **c)** *(das Erwähnte betreffend)* about this/these; **d)** *(geh.: währenddessen)* **er war ~ eingeschlafen** he had fallen asleep while doing so

hierum ['hi:'rʊm] *Adv.* about this; **~ geht es gar nicht** that's not the point; it's not a question of that

hierunter ['hi:'rʊntɐ] *Adv.* **a)** *(unter diese[r] Stelle)* under here; **b)** **~ leiden** suffer from this; **etw. ~ verstehen** *od.* **sich** *(Dat.)* **etw. ~ vorstellen** understand sth. by this; **c)** *(unter die genannte/der genannten Gruppe)* among these

hier·von *Adv.* **a)** *(von dieser Sache)* of this; **~ zeugen** bear witness to this; **b)** *(dadurch)* because of this; **c)** *(aus dieser Menge)* of this/these; **d)** *(aus diesem Material)* out of this

hier·vor *Adv.* **a)** *(vor dieser/diese Stelle)* in front of this *or* here; **b)** **Respekt ~ haben** have respect for this; **Angst ~ haben** be afraid of this

hier·zu *Adv.* **a)** *(zu dieser Sache)* with this; **vgl. ~:** cf.; **b)** *(zu dieser Gruppe)* **~ gehört/gehören ...:** this includes/these include; **c)** **ich kann dir ~ nur raten** I can only recommend you to do this/buy this/go *etc.*; **ich wünsche dir ~ viel Erfolg** I wish you every success with this; **~ reicht mein Geld nicht** I haven't got enough money for that; **d)** *(hinsichtlich dieser Sache)* about this

hierzu·lande *Adv.* *(in diesem Land)* [here] in this country

hiesig ['hi:zɪç] *Adj.; nicht präd.* local

hieß [hi:s] *1. u. 3. Pers. Sg. Prät. v.* **heißen**

hieven ['hi:vn̩] *tr. V.* heave

Hi-Fi ['haifai, 'haifi-] hi-fi ‹*system, unit, etc.*›

high [haɪ] *Adj.; nur präd. (ugs.)* high *(coll.)*

Highlife ['hailaif] **das; ~[s]** *(ugs.)* high life; **~ machen** live it up

High-Society [-sə'saɪətɪ] **die; ~:** high society

High-Tech- ['hai'tɛk-] high-tech ‹*equipment, device, etc.*›

hihi [hi'hi:], **hihihi** [hihi'hi:] *Interj.* he-he[-he]

Hilfe ['hɪlfə] **die; ~, ~n a)** help; *(für Notleidende)* aid; relief; **wirtschaftliche/finanzielle ~:** economic aid/financial assistance; **jmdm. ~ leisten** help sb.; **mit ~** (+ *Gen.*) with the help *or* aid of; **jmdn. um ~ bitten** ask sb. for help *or* assistance; **um ~ rufen** shout for help; **jmdn. zu ~ rufen** call on sb. for help; **jmdm. zu ~ kommen/eilen** come/hurry to sb.'s aid *or* assistance; **[zu] ~!** help!; **Erste ~:** first aid; **b)** *(Hilfskraft)* help; *(im Geschäft)* assistant

hilfe-, Hilfe-: ~leistung die help; assistance; **unterlassene ~leistung** *(Rechtsspr.)* failure to render assistance in an emergency; **~ruf der** cry for help; *(Notsignal)* distress signal; **~stellung die** *(Turnen)* **jmdm. ~stellung geben** act as spotter for sb.; **~suchend 1.** *Adj.* **sein ~suchender Blick ging zum Fenster** he looked towards the window, seeking help; **2.** *adv.* **sich ~suchend umschauen** look round for help

hilf-, Hilf-: ~los 1. *Adj.* helpless; **2.** *adv.* helplessly; **~losigkeit die; ~:** helplessness; **~reich** *(geh.) Adj.* helpful

hilfs-, Hilfs-: ~aktion die relief programme; **~arbeiter der** labourer; *(in einer Fabrik)* unskilled worker; **~bedürftig** *Adj.* **a)** *(schwach)* in need of help *postpos.*; **b)** *(notleidend)* in need; needy; **~bedürftigkeit die** need; neediness; **~bereit** *Adj.* helpful; **~bereitschaft die** helpfulness; readiness *or* willingness to help; **~dienst der** *(Organisation)* emergency service; *(bei Katastrophen)* [emergency] relief service; *(für Autofahrer)* [emergency] breakdown service; **~kraft die** assistant; **~maßnahme die** aid *or* relief measure; **~mittel das** aid; **~organisation die** aid *or* relief organization; **~schule die** *(veralt., ugs.)* special school; **~schüler der** *(veralt., ugs.)* pupil at a special school; **~verb das, ~zeitwort das** *(Sprachw.)* auxiliary [verb]

Himalaja [hi'ma:laja] **der; ~[s]: der/im ~:** the/in the Himalayas *pl.*

Him·beere ['hɪm-] **die** raspberry

Himbeer-: ~eis das raspberry ice [cream]; **~strauch der** raspberry bush

Himmel ['hɪml̩] **der; ~s, ~ a)** sky; **am ~:** in the sky; **unter freiem ~:** in the open [air]; outdoors; **aus heiterem ~** *(ugs.)* out of the blue; **b)** *(Aufenthalt Gottes)* heaven; **in den ~ kommen** go to heaven; **im ~ sein** *(verhüll.)* be in heaven; **gen ~ fahren** *(geh.)* ascend into heaven; **~ und Hölle in Bewegung setzen**

(ugs.) move heaven and earth; **im sieb[en]ten ~ sein/sich [wie] im sieb[en]ten ~ fühlen** *(ugs.)* be in the seventh heaven; **zum ~ schreien** be scandalous *or* a scandal; **zum ~ stinken** *(salopp)* stink to high heaven; **c)** *(verhüll.: Schicksal)* Heaven; **gerechter/gütiger/[ach] du lieber ~!** good Heavens!; Heavens above!; **dem ~ sei Dank** thank Heaven[s]; **weiß der ~!** *(ugs.)* Heaven knows; **um [des] ~s willen!** *(Ausruf des Schreckens)* good Heavens!; good God!; *(inständige Bitte)* for Heaven's sake; **~ noch [ein]mal!** for Heaven's *or* goodness' sake!; **~, Arsch und Zwirn!** *(derb)* bloody hell! *(Brit. sl.)*; **d)** *(Baldachin)* canopy; **e)** *(im Auto)* roof lining

himmel-, Himmel-: ~angst *Adj.* **in mir ist/wird ~angst** I am scared to death; **~bett das** four-poster bed; **~blau** *Adj.* sky-blue; azure; clear blue ⟨*eyes*⟩

Himmel·fahrt die *(Rel.)* **a) Christi ~:** the Ascension of Christ; **die ~ Marias** the Assumption of the Virgin Mary; **b)** *(Festtag)* **[Christi] ~:** Ascension Day *no art.*; **Mariä ~:** the [feast of the] Assumption

Himmelfahrts·kommando das suicide mission *or* operation

Himmel·herrgott *in* **~herrgott noch [ein]mal!** *(salopp)* hell's bells! *(sl.)*

himmel-, Himmel-: ~hoch 1. *Adj.* soaring; towering; **2.** *adv.* ⟨*rise up etc.*⟩ high into the sky; **~hoch jauchzend, zu Tode betrübt** up one minute, down the next; on top of the world one minute, down in the dumps the next; **~reich das** *(christl. Rel.)* kingdom of heaven; **~schreiend** *Adj.* scandalous; outrageous; scandalous, appalling ⟨*conditions, disgrace*⟩; arrant *attrib.* ⟨*nonsense*⟩; **eine ~e Ungerechtigkeit** an injustice that cries out to heaven

Himmels-: ~gabe die *(geh.)* gift from heaven; **~karte die** *(Astron.)* star map; **~körper der** celestial body; **~kunde die** astronomy; **~macht die** *(geh.)* heavenly power; **~richtung die** point of the compass; cardinal point; **aus allen ~richtungen** from all directions; **in alle ~richtungen verstreut sein** be scattered to all four corners of the earth; **~schlüssel der, ~schlüsselchen das** cowslip; *(Waldschlüsselblume)* oxlip; **~stürmer der** *(geh.)* unshakeable idealist

himmel·stürmend *Adj.; nicht präd. (geh.)* boundless ⟨*enthusiasm*⟩; wildly ambitious ⟨*plan*⟩

Himmels·zelt das *(dichter.)* firmament

himmel·weit *Adj.* enormous, vast ⟨*difference*⟩

himmlisch 1. *Adj.* **a)** *nicht präd.* heavenly; **die ~en Heerscharen** the heavenly host[s]; **der ~e Vater** our Heavenly Father; **eine ~e Fügung** divine providence; **b)** *(herrlich)* heavenly; divine; wonderful ⟨*weather, day, view*⟩. **2.** *adv.* divinely; wonderfully, gloriously ⟨*comfortable, warm*⟩

hin [hɪn] *Adv.* **a)** *(räumlich)* **zur Straße ~ liegen** face the road; **nach Frankfurt ~:** in the direction of Frankfurt; **bis zu dieser Stelle ~:** [up] to this point; as far as here; **b)** *(zeitlich)* **gegen Mittag ~:** towards midday; **zum Herbst ~:** towards the autumn; as autumn approaches/approached; **c)** *(in Verbindungen)* **nach außen ~:** outwardly; **auf meinen Rat ~:** on my advice; **auf seine Bitte ~:** at his request; **selbst/auch auf die Gefahr ~, einen Fehler zu begehen** even at the risk of making a mistake; **d) ~ und zurück** there and back; **einmal Köln ~ und zurück** a return [ticket] to Cologne; **Hin und zurück? – Nein, nur ~:** Return? – No, just a single; **~ und her** to and fro; back and forth; **~ und her beraten/reden** go backwards and forwards over the same old ground; **das Hin und Her** the toing and froing; **nach langem Hin und Her** after a great deal of argument; **~ und wieder** [every] now and then; **e)** *(elliptisch)* **nichts wie ~!** what are we waiting for?; **~ zu ihm!** [hurry up,] to him!; *s. auch* **hinsein**

hinab [hɪ'nap] *Adv. s.* **hinunter**

hinab|- *s.* **hinunter|-**

hinan [hɪ'nan] *Adv. (geh.) s.* **hinauf**

hin|arbeiten *itr. V.* **auf etw.** *(Akk.)* **~:** work towards sth.

hinauf [hɪ'nauf] *Adv.* up; **den Hügel ~:** up the hill

hinauf|- ⟨*go, walk, take, throw, look, let, etc.*⟩ up; *(in ein weiter oben gelegenes Stockwerk)* ⟨*go, take, etc.*⟩ upstairs; **auf einen Berg ~steigen** climb up a mountain

hinauf-: ~|arbeiten *refl. V. s.* **hocharbeiten; ~|fallen** *unr. itr. V.; mit sein s.* **Treppe a; ~|gehen** *unr. itr. V.; mit sein* **a)** go up; *(in ein höheres Stockwerk)* go upstairs; **die Treppe ~gehen** go up the stairs; **b)** *(nach oben führen)* ⟨*path, road, etc.*⟩ lead up; **es geht steil ~:** the road/path climbs steeply; **c)** *(ugs.: steigen)* ⟨*prices, taxes, etc.*⟩ go up; rise; **d) mit dem Preis/der Miete ~gehen** *(ugs.)* put the price/rent up; **~|setzen** *tr. V.* *(erhöhen)* raise; increase; put up; **die Preise ~setzen** increase *or* raise prices

hinaus [hɪ'naus] *Adv.* **a)** *(räumlich)* out; **~ [mit dir]!** out you go!; out with you!; **zum Fenster ~:** out of the window; **hier/dort ~:** this/that way out; **nach hinten/vorne ~ liegen** ⟨*room*⟩ be situated at the back/front; **b)** *(zeitlich)* **auf Jahre ~:** for years to come; **c) über etw.** *(Akk.)* **~:** beyond sth.; *(zusätzlich zu etw.)* over and above *or* in addition to sth.; *s. auch* **darüber**

hinaus|- ⟨*go, look, drive, ride, let, carry, etc.*⟩ out; *(ins Freie)* ⟨*go, look, carry, etc.*⟩ outside; **aus etw. ~gehen/~sehen** *usw.* go/look etc. out of sth.; **etw. aus etw. ~bringen/~werfen** *usw.* take/throw *etc.* sth. out of sth.; **zur Tür ~laufen** run out of the door; **etw. zum Fenster ~werfen** throw sth. out of the window; **über etw.** *(Akk.)* **~gehen/~kommen** go/get beyond sth.

hinaus-: ~|ekeln *tr. V. (ugs.)* **jmdn. ~ekeln** drive sb. out; **~|fliegen 1.** *unr. itr. V.; mit sein* **a)** *s.* **hinaus|-; b)** *(fig. ugs.: ~geworfen werden)* be chucked out *(coll.)*; *(als Arbeitnehmer)* get the sack *(coll.)*; be fired *(coll.)*; *(als Mieter)* be thrown out *(coll.)*; **2.** *unr. tr. V. s.* **hinaus|-; ~|gehen** *unr. itr. V.; mit sein* **a)** *s.* **hinaus|-; b) das Zimmer geht zum Garten/nach Westen ~:** the room looks out on to *or* faces the garden/faces west; **die Tür geht auf den Hof ~:** the door leads *or* opens into the yard; **die Schlafzimmer gehen nach hinten ~:** the bedrooms are at the back; **c)** *unpers.* **wo geht es ~?** which is the way out?; **hier/da geht es ~:** this/that is the way out; **~|kommen** *unr. itr. V.; mit sein* **a)** *s.* **hinaus|-; b) ich bin schon seit zwei Tagen nicht mehr ~gekommen** I've not got *or* been out of the house for two days; **er ist nie aus dem Dorf ~gekommen** he has never been out of *or* outside his village; **~|komplimentieren** *tr. V.* **jmdn. ~komplimentieren** usher sb. out; **~|laufen** *unr. itr. V.; mit sein* **a)** *s.* **hinaus|-; b) auf etw.** *(Akk.)* **~laufen** lead to sth.; **das läuft auf dasselbe ~:** it comes to the same thing; ~|**ragen** *itr. V.* **über etw.** *(Akk.)* ~ **ragen** rise up above sth.; *(horizontal)* jut out *or* project beyond sth.; ~|**schieben 1.** *unr. tr. V.* **a)** *s.* **hinaus|-; b)** *(aufschieben)* put off; postpone; **eine Entscheidung [um einen Tag] ~schieben** put off *or* postpone *or* defer a decision [by one day]; **2.** *unr. refl. V.: s.* **hinaus|-**; ~|**sein** *unr. itr. V.; mit sein; nur im Inf. u. Part. zusammengeschrieben* **über etw.** *(Akk.)* **~sein** be past *or* beyond sth.; ~|**setzen 1.** *tr. V. s.* **hinaus|-; 2.** *refl. V.* go and sit outside; ~|**stellen 1.** *tr. V.* **a)** *s.* **hinaus|-; 2.** *refl. V.* go and stand outside; ~|**wachsen** *unr. itr. V.; mit sein* **a) über etw.** *(Akk.)* **~wachsen** grow taller than *or* up above sth.; **b)** *(fig.)* **über etw.** *(Akk.)* **~wachsen** outgrow sth.; **über jmdn./sich [selbst] ~wachsen** surpass sb./rise above oneself; ~|**werfen** *unr. tr. V.* **a)** *s.* **hinaus|-; b)** *(ugs.: ausschließen, die Wohnung kündigen)* **jmdn. ~werfen** throw sb. out (**aus** of); *(ugs.: entlassen)* sack sb. *(coll.)*; ~|**wollen** *unr. itr. V.* **a)** want to get *or* go out (**aus** of); **b) [zu] hoch ~wollen** *(fig.)* aim [too] high; set one's sights [too] high; **c) worauf willst du ~?** *(fig.)* what are you getting *or* driving at?; **auf etwas Bestimmtes ~wollen** *(fig.)* have something particular in mind

Hinaus·wurf der *(ugs.)* throwing out; *(eines Angestellten)* sacking *(coll.)*

hinaus-: ~|**ziehen 1.** *unr. tr. V.* **a)** *s.* **hinaus|-; b)** *(verzögern)* put off; delay; **c)** *unpers.* **es zog sie in die Natur ~:** she felt the urge to get out into the countryside; **2.** *unr. itr. V.; mit sein s.* **hinaus|-; 3.** *unr. refl. V. (sich verzögern)* be delayed; ~|**zögern 1.** *tr. V.* delay; put off; **2.** *refl. V.* be delayed; be put off

hin-, Hin-: ~|**bauen** *tr. V.* build; put up; ~|**begeben** *unr. refl. V. s.* **hingehen a**; ~|**begleiten** *tr. V.* **jmdn. ~begleiten** accompany sb. [there]; ~|**bekommen** *unr. tr. V. (ugs.) s.* **hinkriegen**; ~|**bestellen** *tr. V.* **jmdn. ~bestellen** tell sb. to be there; ~|**blättern** *tr. V. (ugs.)* fork *or* shell out *(sl.)*, pay out ‹*sum of money*›; **~blick der** *in* **im** *od.* **in ~blick auf etw.** *(Akk.) (wegen)* in view of; *(hinsichtlich)* with regard to; ~|**blicken** *itr. V.* look; **zu jmdm. ~blicken** look [across] at sb.; ~|**bringen** *unr. tr. V.* **jmdn./etw. ~bringen** take sb./sth. [there]

hin|denken *unr. itr. V.* **wo denkst du hin?** *(ugs.)* whatever are you thinking of?; what an idea!

hinderlich *Adj.* ~ **sein** get in the way; **jmdm.** ~ **sein** get in sb.'s way; **einer Sache** *(Dat.)* **sein** be an obstacle to sth.

hindern ['hındɐn] *tr. V.* **jmdn. ~:** stop *or* prevent sb.; **jmdn. [daran] ~, etw. zu tun** prevent *or* stop sb. [from] doing sth.; **jmdn. am Sprechen ~:** prevent *or* stop sb. [from] speaking; **ich werde dich nicht ~** *(iron.)* I'm not stopping you

Hindernis das; ~ses, ~se obstacle; *(Springreiten)* jump; obstacle; *(Pferderennen)* fence

Hindernis-: ~lauf der, ~laufen das *(Leichtathletik)* steeplechase; **~rennen das** *(Pferdesport)* steeplechase

Hinderung die; ~, ~en hindrance

Hinderungs·grund der: das ist kein ~ für mich it does not prevent *or* stop me

hin|deuten *itr. V.* **a) auf jmdn./etw.** *od.* **zu jmdm./etw. ~:** point to sb./sth.; **b) auf etw.** *(Akk.)* ~ *(fig.)* suggest sth.; point to sth.; **alles deutet darauf hin, daß ...:** everything suggests that ...

hin|drängen 1. *tr. V.* **jmdn. zu etw. ~:** force sb. towards sth. **2.** *itr. V.; mit sein; refl. V.* **[sich] zu jmdm./etw. ~:** push [one's way] towards sb./sth.

Hindu ['hındu] **der; ~[s], ~[s]** Hindu

Hinduismus der; ~: Hinduism *no art.*

hinduistisch *Adj.* Hindu

hin·durch *Adv.* **a)** *(räumlich)* **durch den Wald ~:** through the wood; **mitten/quer durch etw. ~:** straight through sth.; **b)** *(zeitlich)* **das ganze Jahr ~:** throughout the year; **die ganze Nacht ~:** all night [long]; throughout the night; all through the night

hindurch|- ‹*go, run, look, throw, etc.*› through; **durch etw. ~laufen** run through sth.; **etw. durch etw. ~werfen** throw sth. through sth.; **sich durch etw. ~finden** find one's way through sth.; **unter etw.** *(Dat.)* **~gehen** walk under sth.; **durch etw. ~müssen** have to go through sth.

hin-: ~|**dürfen** *unr. itr. V. (ugs.)* be allowed to go (**zu** to); **ihr dürft nicht mehr ~:** you're not to go there any more; ~|**eilen** *itr. V.; mit sein* **a)** hurry (**zu** to); **alle eilten ~:** everyone hurried there

hinein [hı'naın] *Adv.* **a)** *(räumlich)* in; ~ **mit euch!** in you go!; in with you!; **in etw.** *(Akk.)* **~:** into sth.; **nur ~!** go *or* walk right in!; **b)** *(zeitlich)* **bis in den Morgen/tief in die Nacht ~:** till morning/far into the night

hinein|- ‹*go, drive, ride, run, look, throw, etc.*› in; **in etw.** *(Akk.)* **~gehen/~sehen** go/look into sth.; **etw. in etw.** *(Akk.)* **~legen/~bekommen** put/get sth. in[to] sth.; **sich in etw.** *(Akk.)* **~bohren** bore one's/its way into sth.; **in etw.** *(Akk.)* **~dürfen** be allowed into sth.

hinein-: ~|**denken** *unr. refl. V.* **sich in jmdn./in jmds. Lage ~denken** put oneself in sb.'s position; ~|**finden** *unr. refl. V.* **sich in etw.** *(Akk.)* **~finden** get used to sth.; ~|**fressen 1.** *unr. tr. V.* **etw. in sich ~fressen** ‹*animal, (derb) person*› gobble sth. down *or* up, wolf sth. down; **seine Sorgen/seinen Ärger in sich ~fressen** *(fig.)* bottle up one's worries/anger; **2.** *unr. refl. V.* **sich in etw.** *(Akk.)* **~fressen** eat into sth.; ~|**gebären** *unr. tr. V.* **in eine Zeit/Umwelt** *usw.* **~geboren werden/sein** be/have been born into an age/environment *etc.*; ~|**gehen** *unr. itr. V.; mit sein* **a)** go in; **in etw.** *(Akk.)* **~gehen** go into sth.; **b)** *(~passen)* **in den Eimer gehen drei Liter ~:** the bucket holds three litres; ~|**helfen** *unr. itr. V.* **jmdm. in den Mantel ~helfen** help sb. on with his/her coat; **jmdm. in den Bus ~helfen** help sb. on to the bus; ~|**knien** *refl. V. (ugs.)* **sich in etw.** *(Akk.)* **~knien** get one's teeth into sth.; ~|**laufen** *unr. itr. V.; mit sein* **a)** run in; *(gehen)* walk in; **in etw.** *(Akk.)* **~laufen** run/walk into sth.; **in sein Verderben ~laufen** *(fig.)* be heading [straight] for disaster; **in ein Fahrzeug ~laufen** run under a vehicle; **b)** *(fließen)* **in etw.** *(Akk.)* **~laufen** run into sth.; ~|**reden** *itr. V. (abwertend: sich einmischen)* **jmdm. in seine Angelegenheiten/Entscheidungen** *usw.* **~reden** meddle *or* interfere in sb.'s affairs/decisions *etc.*; ~|**rennen** *unr. itr. V.; mit sein (ugs.)* run in; race in; **in etw.** *(Akk.)* **~rennen** run *or* race into sth.; **in sein Verderben ~rennen** *(fig.)* be heading [straight] for disaster; ~|**schauen** *itr. V.* **a)** *(bes. südd., österr.) s.* **~sehen; b)** *(bes. südd., österr.: kurz besuchen)* **bei jmdm. ~schauen** look in on sb.; ~|**steigern** *refl. V.* **sich in große Erregung/seine Wut ~steigern** work oneself up into a state of great excitement/into a rage;

~|**versetzen** *refl. V.* **sich in jmdn.** *od.* **jmds. Lage ~versetzen** put oneself in sb.'s position; ~|**wachsen** *unr. itr. V.; mit sein* **a) in etw.** *(Akk.)* **~wachsen** grow into sth.; **b)** *(ugs.)* **in ein Kleid** *usw.* **~wachsen** grow into a dress *etc.*; **c)** *(fig.)* **in eine Aufgabe/Rolle ~wachsen** get to know a job/get into *or* inside a part; ~|**wollen** *unr. itr. V. (ugs.: ~gelangen wollen)* want to get *or* go in; **in etw.** *(Akk.)* **~wollen** want to go/get into sth.; ~|**ziehen 1.** *unr. tr. V.* **a)** *s.* **hinein|-**; **b)** *(fig.)* **jmdn. in eine Angelegenheit/einen Streit/Skandal ~ziehen** drag sb. into an affair/a dispute/scandal; **2.** *unr. itr. V.; mit sein: s.* **hinein|-**

hin-, Hin-: ~|**fahren 1.** *unr. itr. V.; mit sein* go/drive/ride there; **wo ist er ~gefahren?** where has he gone?; **2.** *unr. tr. V.* **jmdn. ~fahren** drive *or* take sb. there; **jmdn. zum Bahnhof ~fahren** drive *or* take sb. to the station; **~fahrt die** journey there; *(Seereise)* voyage out; **auf der ~fahrt** on the way *or* journey there/the voyage out; ~|**fallen** *unr. itr. V.; mit sein* **a)** fall down *or* over; **lang ~fallen** fall flat [on one's face/back]; **b) jmdm. fällt etw. ~:** sb. drops sth.; **etw. ~fallen lassen** drop sth.; **~fällig** *Adj.* **a)** *(schwächlich)* infirm; frail; **b)** *(ungültig)* invalid; **~fälligkeit die; ~ a)** *(Schwäche)* infirmity; frailty; **b)** *(Ungültigkeit)* invalidity; ~|**finden** *unr. itr. V.* find one's way there; **zu jmdm./zu einem Ort ~finden** find one's way to sb./a place; ~|**flegeln** *refl. V. (ugs. abwertend)* loll around *or* about; ~|**fliegen 1.** *unr. itr. V.; mit sein* **a)** fly there; **er fliegt heute ~:** he's flying [out] there today; **b)** *(ugs.: fallen)* come a cropper *(coll.)*; fall over; **2.** *unr. tr. V.* **jmdn./etw. ~fliegen** fly sb./sth. [out] there; **~flug der** outward flight; ~|**führen 1.** *tr. V.* **jmdn. ~:** lead *or* take sb. there; **jmdn. zu etw. ~:** lead sb. to sth. **2.** *itr. V.* **zu etw. ~:** lead to sth.

hing [hɪŋ] *1. u. 3. Pers. Sg. Prät. v.* **hängen**

Hi̱n·gabe die; ~ a) devotion; *(Eifer)* dedication; **etw. mit ~ tun** do sth. with dedication; **b)** *(geh.: das Opfern)* **unter ~ des Lebens** at the cost of one's life

hi̱n|geben 1. *unr. tr. V. (geh.)* give; sacrifice; **sein Leben ~:** lay down *or* sacrifice one's life. **2.** *unr. refl. V.* **a) sich einer Illusion/einem Genuß ~:** entertain an illusion/abandon oneself to a pleasure; **b)** *(verhüll.)* **sich einem Mann ~:** give oneself to a man

hi̱ngebungs·voll 1. *Adj.* devoted. **2.** *adv.* devotedly; with devotion; ‹*listen*› raptly, with rapt attention; ‹*dance, play*› with abandon

hi̱n·ge̱gen *Adv.* however; *(andererseits)* on the other hand

hi̱n-: **~gegossen** *Adj. (ugs. scherzh.)* **wie ~gegossen auf der Couch liegen/sitzen** have draped oneself over the couch; ~|**gehen** *unr. itr. V.; mit sein* **a)** go [there]; **zu jmdm./etw. ~gehen** go to sb./sth.; **wo gehst du ~?** where are you going?; **b)** *(verstreichen)* ‹*years, time*› pass, go by; **darüber gingen Jahre ~:** it took years; **c)** *(~genommen werden)* pass; **diesmal mag das noch ~gehen** I'll/we'll *etc.* let it pass this time; ~|**gehören** *itr. V. (ugs.)* go; belong; ‹*person*› belong; **wo gehört das ~?** where does this go *or* belong *or (coll.)* live?; ~|**gelangen** *itr. V.; mit sein* get there; **zu jmdm./etw. ~gelangen** get to sb./sth.; ~|**geraten** *unr. itr. V.; mit sein* get there; **wo ist er/der Brief ~geraten?** where has he/the letter got to?; **~gerissen** *Adj.; nicht attr.* carried away; spellbound; **~gerissen der Musik lauschen** listen spellbound to the music; ~|**halten** *unr. tr. V.* **a)** hold out (*Dat.* to); **b)** *(warten lassen)* **jmdn. ~halten** put sb. off; keep sb. waiting

hi̱n-: ~|**hängen** *tr. V. (ugs.)* hang up; ~|**hauen 1.** *unr. tr. V.* **a)** *(salopp) s.* **~schmeißen b**; **b)** *(salopp abwertend: flüchtig anfertigen)* knock off *(coll.)*; dash off; **c)** *unpers. (salopp)* **es hat mich ~gehauen** I came a cropper *(coll.)*; **2.** *unr. itr. V.* **a)** *(ugs.: schlagen)* take a swipe *(coll.)*; **b)** *mit sein (hinfallen)* fall [down] heavily; **c)** *(salopp: gutgehen)* ‹*plan*› work [all right]; **es wird schon ~hauen** it'll work out *or* be all right *or (coll.)* OK; **d)** *(salopp: richtig sein)* ‹*calculation*› be right; **3.** *unr. refl. V. (salopp)* lie down and have a kip *(coll.)*; ~|**hören** *itr. V.* listen

hinken ['hɪŋkn̩] *itr. V.* **a)** limp; walk with a limp; **auf** *od.* **mit dem rechten Bein ~:** have a limp in one's right leg; **b)** *mit sein* limp; hobble; **c)** *(fig.)* ‹*comparison*› be poor *or* feeble

hi̱n-, Hi̱n-: ~|**knien** *refl. V.* kneel [down]; ~|**kommen** *unr. itr. V.; mit sein* **a)** get there; **wie kommt man zu ihm ~?** how do you get to his place?; **b)** *(an einen Ort gehören)* go; belong; **wo kommen die Gläser ~?** where do the glasses go *or* belong?; **c)** *(~geraten)* **wo ist meine Uhr ~gekommen?** where has my watch got to *or* gone?; **wo kommen** *od.* **kämen wir ~, wenn ...** *(fig.)* where would we be if ...; **d)** *(ugs.: auskommen)* manage; **e)** *(ugs.: in Ordnung kommen)* work out *or* turn out all right *or (coll.)* OK; **f)** *(ugs.: stimmen)* be right; ~|**kriegen** *tr. V. (ugs.)* **a)** *(fertigbringen)* **das hat sie toll ~gekriegt** she made a great job of that *(coll.)*; **das wird er schon ~kriegen** he'll manage it all right *or (coll.)* OK; **b) etw. wieder ~ kriegen** fix sth.; **jmdn. wieder ~kriegen** put sb. right; ~|**langen** *itr. V.* **a)** *(ugs.: fassen)* **er langte ~ und steckte einige Uhren in seine Tasche** he reached over and stuck some watches in his pocket *(coll.)*; **b)** *(salopp: zuschlagen)* **[kräftig] ~langen** take a [hefty] swipe *(coll.)*; **c)** *(salopp: sich bedienen)* help oneself; **schön/ordentlich ~langen** help oneself in a big way *(coll.)*; **~länglich 1.** *Adj.* sufficient; *(angemessen)* adequate; **2.** *adv.* sufficiently; *(angemessen)* adequately; ~|**lassen** *unr. tr. V. (ugs.)* **jmdn. ~lassen** allow sb. to go there; let sb. go there; **jmdn. zu etw. ~lassen** allow sb. to go to sth.; let sb. go to sth.; ~|**laufen** *unr. itr. V.; mit sein* **a)** run there; **zu jmdm./zu einer Stelle ~laufen** run to sb./a place; **b)** *(zu Fuß gehen)* walk [there]; ~|**legen 1.** *tr. V.* **a)** put; **sie legte den Kindern frische Wäsche ~:** she put out clean underwear for the children; **b)** *(weglegen)* put down; **c)** *(zu Bett bringen)* **jmdn. ~legen** lay sb. down; **d)** *(ugs.: zahlen)* pay *or (sl.)* shell out; **e)** *(salopp: ausführen)* **eine hervorragende Rede ~legen** do a brilliant speech; **eine gekonnte Übung auf dem Trampolin ~legen** turn in a splendid performance on the trampoline; **2.** *refl. V.* **a)** lie down; **da legst du dich [lang] ~** *(ugs.)* you won't believe your ears; **b)** *(sich schlafen legen)* lie down; **c)** *(ugs.: hinfallen)* come a cropper *(coll.)*; fall [down *or* over]; ~|**leiten** *tr. V.* lead there; **etw. zu etw. ~leiten** lead sth. to sth.; ~|**lenken** *tr. V.* **a) etw. zu etw. ~lenken** steer sth. to sth.; **b)** *(fig.)* steer ‹*conversation*› **(auf +** *Akk.* round to); direct ‹*attention*› **(auf +** *Akk.* towards); turn ‹*gaze*› **(auf +** *Akk.* towards); ~|**machen** *(ugs.)* **1.** *tr. V.* put up

⟨*curtain, picture, fence, etc.*⟩; put on ⟨*paint, oil, cream*⟩; put in ⟨*comma etc.*⟩; put ⟨*cross, ring, etc.*⟩; make ⟨*dirty mark etc.*⟩; **2.** *itr. V. (seine Notdurft verrichten)* do one's/its business *(coll.)*; **~|nehmen** *unr. tr. V.* **a)** *(annehmen)* accept; take; put up with, swallow, accept ⟨*insult*⟩; **b)** *(ugs.: mitnehmen)* **kannst du das Buch mit ~nehmen** can you take the book with you?; **kannst du mich mit ~nehmen?** can you take me there?; **~|neigen 1.** *tr. V.* incline; **den Kopf zu jmdm. ~neigen** incline *or* bend one's head towards sb.; **2.** *refl. V.* lean [over]; **~|passen** *itr. V. (ugs.)* **a)** fit *or* go in; **b)** *(harmonieren)* fit in; go; **~|reichen 1.** *tr. V.* hand; pass; **jmdm. etw. ~reichen** hand *or* pass sth. to sb.; **2.** *itr. V.* **a)** reach; **bis zu etw. ~reichen** reach to *or* as far as sth.; **b)** *(ausreichen)* be enough *or* sufficient; **~reichend 1.** *Adj.* sufficient; *(angemessen)* adequate; **2.** *adv.* sufficiently; *(angemessen)* adequately; **~reise die** journey there; outward journey; *(mit dem Schiff)* voyage out; outward voyage; **[die] Hin- und Rückreise** the journey there and back; **~|reisen** *itr. V.; mit sein* travel there; **~|reißen** *unr. tr. V.* **a) jmdn. zu sich ~reißen** pull sb. to one; **b)** *(begeistern)* enrapture; **das Publikum zu Beifallsstürmen ~reißen** elicit thunderous *or* rapturous applause from the audience; **c) jmdn. zu etw. ~reißen** drive sb. to sth.; **sich dazu ~reißen lassen, etw. zu tun** let oneself get *or* be carried away and do sth.; **er ließ sich zu einer Beleidigung ~reißen** he let himself be carried away and insulted him/her *etc.*; **~reißend 1.** *Adj.* enchanting ⟨*person, picture, view*⟩; captivating ⟨*speaker, play*⟩; **2.** *adv.* enchantingly; **~|richten** *tr. V.* execute; **~richtung die** execution; **~|rücken 1.** *tr. V.* **etw. ~rücken** move *or* push sth. over; **2.** *itr. V.; mit sein* move over; **~|sagen** *tr. V.* say without thinking; *(nur beiläufig sagen)* say casually; **das hat er nur so ~gesagt** he just said it without thinking; **~|schauen** *itr. V. (bes. südd., österr.) s.* **~sehen; ~|scheiden das; ~s** *(geh. verhüll.)* decease; demise; **~|scheißen** *unr. itr. V. (derb)* crap *(vulg.)*; **~|schicken** *tr. V.* send; **~|schieben** *unr. tr. V.* **jmdm. etw. ~schieben** push sth. over to sb.; **~|schielen** *itr. V.* steal a glance/glances (**zu** at); **~|schlagen** *unr. itr. V.* **a)** strike; hit; **b)** *mit sein (ugs.: fallen)* **[der Länge nach** *od.* **lang] ~schlagen** fall flat on one's face/back; **~|schleichen** *unr. itr. V.; mit sein; unr. refl. V.* creep *or* steal over; **~|schleppen 1.** *refl. V.* drag oneself along; **sich zu etw. ~schleppen** drag oneself to sth.; **2.** *tr. V.* **etw. ~schleppen** drag sth. there; **etw. zu etw. ~schleppen** drag sth. to sth.; **~|schmeißen** *unr. tr. V. (salopp)* **a)** chuck down *(coll.)*; **b)** *(aufgeben)* chuck in *(coll.)*; **c)** *(fallen lassen)* drop; **~|schreiben 1.** *unr. tr. V.* write down; **2.** *unr. itr. V.* write; **~|sehen** *unr. itr. V.* look; **ich kann nicht ~sehen** I can't [bear to] look; **bei genauerem Hinsehen** on closer inspection; **~|sein** *unr. itr. V.; mit sein (nur im Inf. u. Part. zusammengeschrieben) (ugs.)* **a)** *(nicht mehr brauchbar sein)* have had it *(coll.)*; ⟨*car*⟩ be a write-off; **b)** *(salopp: tot sein)* have snuffed it *(sl.)*; have pegged out *(sl.)*; **wenn er richtig zuschlägt, bist du ~:** if he really hits you you've had it *(coll.)*; **c)** *(ugs.: hingerissen sein)* **von jmdm./etw. ganz ~sein** be mad about sb./bowled over by sth.; **d)** *(~gegangen/~gefahren sein)* have gone; **e)** *(in der Zukunft liegen)* **das ist noch lange ~:** that's not for a long time yet; **bis zu dem Termin ist es noch einige Zeit ~:** there's some time to go before the deadline; **~|setzen 1.** *tr. V.* put; seat, put ⟨*person*⟩; **das Kind ~setzen** sit the child/baby down; **2.** *refl. V.* **a)** sit down; **setz dich doch ~!** do sit down!; **sich gerade ~setzen** sit up straight; **sich ~setzen und etw. tun** *(fig.)* sit down and do sth.; get down to doing sth.; **b)** *(ugs.: fallen)* land on one's backside; **c)** *(salopp: überrascht sein)* **er wird sich ~setzen** he won't believe his ears; **~sicht die;** *o. Pl.* respect; **in finanzieller ~sicht** financially; **~sichtlich** *Präp. mit Gen. (Amtsspr.)* with regard to; **~|sollen** *unr. itr. V. (ugs.)* **wo sollen die Sachen ~?** where do these things go?; where do you want these things [to go]?; **sie weiß nicht, wo sie ~soll** she doesn't know where to go; **~spiel das** *(Sport)* first leg; **~|starren** *itr. V.* stare (**zu, nach** at); **~|stellen 1.** *tr. V.* **a)** put; put up ⟨*building*⟩; put, park ⟨*car*⟩; **b)** *(auf den Boden stellen)* put down; **c)** *(darstellen)* **etw. als falsch ~stellen** make sth. out to be *or* represent sth. as false; **jmdn. als Lügner ~stellen** make sb. out to be *or* represent sb. as a liar; **jmdn. als Vorbild ~stellen** hold sb. up as an example; **sich als Opfer/als unschuldig ~stellen** make out that one is a victim/is innocent; **2.** *refl. V.* stand; ⟨*driver*⟩ park; *(aufstehen)* stand up; **sich gerade ~stellen** stand up straight; **~|strecken 1.** *tr. V.* **a)** stretch out; hold out; **jmdm. die Hand ~strecken** hold out one's hand to sb.; **b)** *(geh. veralt.: töten)* fell; slay *(liter.)*; **2.** *refl. V.* stretch [oneself] out; lie down full length; **~|strömen** *itr. V.; mit sein* ⟨*people*⟩ flock there; **zu etw. ~strömen** flock to sth.; **~|stürzen** *itr. V.; mit sein* **a)** fall down [heavily]; **b)** *(hineilen)* rush *or* dash there; **zum Ausgang ~stürzen** rush *or* dash towards the exit

hinten ['hɪntn̩] *Adv.* at the back; in *or* at the rear; **~ im Bus** in the back of the bus; **~ einsteigen** get on at the back; **sich ~ anstellen** join the back of the queue *(Brit.)* *or (Amer.)* line; **~ im Buch** at the back *or* end of the book; **weiter ~:** further back; *(in einem Buch)* further on; **von ~ nach vorne** backwards; *(in einem Buch)* from back to front; **nach ~ gehen** go *or* walk to the back/into the room behind; **die Adresse steht ~ auf dem Brief** the address is on the back of the envelope; **~ am Haus** at the back *or* rear of the house; **von ~ kommen/jmdn. von ~ erstechen** come from behind/stab sb. from behind; **von ~ sah sie jünger aus** she looked younger from the back; **jmdm. ~ drauffahren** *(ugs.)* run into the back of sb.; **die anderen sind ganz weit ~:** the others are a long way back *or* behind; **~ in Sibirien** far away in Siberia; **~ und vorn[e] bedient werden** *(ugs.)* be waited on hand and foot; **jmdn. am liebsten von ~ sehen** *(ugs.)* be glad to see the back of sb.

hinten-: ~drauf *Adv. (ugs.)* on the back; **jmdm. eins** *od.* **ein paar ~drauf geben** *(ugs.)* smack sb.'s bottom; **~herum** ['----], **~rum** ['---] *Adv. (ugs.)* **a)** round the back; **mir ist ~herum kalt** my back's cold; **b) etw. ~herum erfahren** hear sth. indirectly

hinten·über *Adv.* backwards

hintenüber-: ~|fallen *unr. itr. V.; mit sein* fall [over] backwards; **~|kippen** *itr. V.; mit sein* tip [over] backwards; **~|stürzen** *itr. V.; mit sein: s.* **~fallen**

hinter ['hɪntɐ] **1.** *Präp. mit Dat.* **a)** behind; ~ **dem Haus** behind *or* at the back of the house; ~ **jmdm. zurückbleiben** lag behind sb.; **eine große Strecke ~ sich haben** have put a good distance behind one; ~ **jmdm. stehen** *(fig.)* be behind sb.; back *or* support sb.; ~ **etw.** *(Dat.)* **stehen** *(fig.)* support sth.; **jmdn. ~ sich haben** *(fig.)* have sb.'s backing; **3 km ~ der Grenze** 3 km beyond the frontier; **die nächste Station ~ Mannheim** the next stop after Mannheim; ~ **der Entwicklung/der Zeit zurückbleiben** lag behind in development/be behind in times; **er ist ~ unseren Erwartungen zurückgeblieben** he has fallen short of our expectations; **b) eine Prüfung/Aufgabe ~ sich haben** *(fig.)* have got an examination/a job over [and done] with; **viele Enttäuschungen/eine Krankheit ~ sich haben** have experienced many disappointments/have got over an illness; **wenn er das Studium ~ sich hat** when he's finished his studies. **2.** *Präp. mit Akk.* **a)** behind; ~ **das Haus gehen** go behind the house; **b) etw. ~ sich bringen** get sth. over [and done] with; **c)** *(fig.)* ~ **ein Geheimnis/die Wahrheit/seine Geschichte kommen** find out a secret/get to the truth/get to the bottom of his story

hinter... *Adj.; nicht präd.* back; **das ~e Ende des Ganges/des Zimmers** the far end of the corridor/the far end *or* the back of the room; **das ~e Ende des Zuges** the back *or* rear [end] of the train; **die ~ste Reihe** the back row

Hinter-: ~achse die rear *or* back axle; **~an·sicht die** rear *or* back view; **~ausgang der** rear *or* back exit; **~bänkler** [~bɛŋklɐ] **der; ~s, ~** *(ugs.)* inconspicuous back-bencher; **~bein das** hind leg; **sich auf die ~beine stellen** *(ugs.)* put up a fight

Hinterbliebene [-'bli:bənə] **der/die;** *adj. Dekl.* **a)** *Pl.* **die ~n** the bereaved [family]; **b)** *(Rechtsspr.)* surviving dependant

hinter·bringen *unr. tr. V.* **jmdm. etw. ~bringen** inform sb. [confidentially] of sth.

hinter·einander *Adv.* **a)** *(räumlich)* one behind the other; **sie liefen dicht ~:** they were running close behind one another; **b)** *(zeitlich)* one after another *or* the other; **an drei Tagen ~:** for three days running *or* in succession

hinter-, Hinter-: ~ein·gang der rear *or* back entrance; **~fragen** [--'--] *tr. V.* examine; analyse; **~fuß der** hind foot; **~gebäude das** *s.* **~haus; ~gedanke der** ulterior motive; **einen ~gedanken bei etw. haben** have an ulterior motive for sth.; **~gehen** [--'--] *unr. tr. V.* deceive

Hinter·grund der *(auch fig.)* background; **der akustische/musikalische ~:** the background sounds/music; **etw. im ~ haben** have sth. up one's sleeve

hinter·gründig 1. *Adj.* enigmatic; cryptic. **2.** *adv.* enigmatically; cryptically

Hintergründigkeit die; ~ enigmaticness; crypticness

Hintergrund-: ~information die item *or* piece of background information; **~informationen** [items *or* pieces of] background information *sing.;* **~musik die** background music

hinter-, Hinter-: ~halt der ambush; **in einen ~halt geraten** be ambushed; **jmdn. aus dem ~halt überfallen** ambush sb.; **im ~halt lauern** lie in ambush; **~hältig 1.** *Adj.* underhand; **2.** *adv.* in an underhand fashion *or* manner; **~hältigkeit die; ~, ~en a)** *o. Pl.* underhandedness; **b)** *(Handlung)* underhand act; **~hand die a)** *(bei Tieren)* hindquarters *pl.;* **b) etw. in der ~hand haben** have sth. up one's sleeve *or* in reserve; **~haus das** *dwelling situated at or forming the rear of a house [and accessible only from a courtyard]*

hinter·her *Adv.* **a)** *(räumlich)* behind; **nichts wie ihm ~!** quick, after him!; **b)** *(zeitlich)* afterwards; **es ~ besser wissen** be wise after the event

hinterher-: ~|fahren *unr. itr. V.; mit sein* **a)** go/drive/ride [along] behind; **jmdm. ~fahren** drive/ride [along] behind sb.; **b)** *s.* **nachfahren; ~|gehen** *unr. itr. V.; mit sein* **a)** walk [along] behind; **jmdm. ~gehen** walk [along] behind sb.; **b)** *(nachgehen)* **jmdm. ~gehen** follow sb.; **~|hinken** *itr. V.; mit sein* **a)** limp *or* hobble [along] behind; **jmdm. ~hinken** limp *or* hobble [along] behind sb.; **b)** *(fig.)* **einer Sache** *(Dat.)* **~hinken** lag behind sth.; **mit etw. ~hinken** be behind with sth.; **~|kommen** *unr. itr. V.; mit sein* follow behind; **~|laufen** *unr. itr. V.; mit sein* **a)** run [along] behind; **jmdm. ~laufen** run[along] behind sb.; **b)** *(nachlaufen)* **jmdm. ~laufen** follow sb.; **c)** *s.* **~gehen; d)** *(fig. ugs.) s.* **nachlaufen b; ~|schicken** *tr. V.* **jmdm. jmdn./etw. ~schicken** send sb. after sb./send sth. on to sb.

hinter-, Hinter-: ~hof der courtyard; **~kopf der** back of the/one's head; **etw. im ~kopf haben/behalten** *(ugs.)* have/keep sth. at the back of one's mind; **~land das** hinterland; *(Milit.)* back area; **~lassen** [--'--] *unr. tr. V.* leave; *(testamentarisch)* leave; bequeath; **die ~lassenen Schriften** the posthumous works; **keine Spuren ~lassen** leave no trace[s] [behind]; **~lassenschaft** [--'---] **die; ~, ~en** estate; **jmds. ~lassenschaft antreten** inherit sb.'s estate; **~legen** [--'--] *tr. V.* deposit (**bei** with); *(als Pfand)* deposit, leave (**bei** with)

Hinterlegung die; ~, ~en *s.* **hinterlegen:** depositing; leaving; **jmdn. gegen ~ einer Kaution freilassen** release sb. on bail

Hinter·list die *o. Pl.* guile; deceit

hinter·listig *Adj.* deceitful

hinterm ['hɪntɐm] *(ugs.) Präp. + Art.* = **hinter dem**

Hinter-: ~mann der; *Pl.* **~männer a)** person behind; **sein ~mann** the person behind [him]; **b)** *(jmd., der aus dem Hintergrund lenkt)* **der ~mann/die ~männer** the brains behind the operation; **~mannschaft die** *(Sport)* defence

hintern *(ugs.) Präp. + Art.* = **hinter den**

Hintern ['hɪntɐn] **der; ~s, ~** *(ugs.)* behind; backside; bottom; **jmdm. den ~ verhauen** *od.* **versohlen** tan sb.'s hide; **jmdn.** *od.* **jmdm. in den ~ treten** kick sb. in the pants *(coll.) or* up the backside; *(fig.)* kick sb. in the teeth *(fig.);* **sich [vor Wut** *od.* **Ärger] in den ~ beißen** *(salopp)* kick oneself; **jmdm. in den ~ kriechen** *(derb)* lick sb.'s arse *(coarse);* suck up to sb. *(sl.);* **sich auf den ~ setzen** *(salopp) (sich anstrengen)* get *or* knuckle down to it; *(aufs Gesäß fallen)* fall on one's behind; *(überrascht sein)* be flabbergasted

Hinter-: ~pfote die hind paw; **~rad das** back *or* rear wheel; **~rad·antrieb der** rear-wheel drive

hinter·rücks ['hɪntɐrʏks] *Adv.* from behind

hinters ['hɪntɐs] *(ugs.) Präp. + Art.* = **hinter das**

hinter-, Hinter-: ~seite die *s.* **Rückseite; ~sinn der** deeper meaning; **~sinnig** *Adj.* ‹*remark, story, etc.*› with a deeper meaning; subtle ‹*sense of hu-*

mour›; **~teil das** *(ugs.)* backside; behind; **~treffen das** *(ugs.) in* **ins ~treffen geraten** *od.* **kommen** fall behind; **~treiben** [--'--] *unr. tr. V.* foil, thwart, frustrate ‹*plan*›; prevent ‹*marriage, promotion*›; block ‹*law, investigation, reform*›; **~treppe die** back stairs *pl.*; **~tupfingen** [~'tʊpfɪŋən] **(das); ~s** *(ugs. spött.)* the back of beyond; **~tür die** back door; **durch die ~tür** *(auch fig.)* by the back door; **sich** *(Dat.)* **eine ~tür offenhalten** *(fig.)* leave oneself a way out *(fig.)*; **~wäldler** [~vɛltlɐ] **der; ~s, ~** *(spött.)* backwoodsman; **~wäldlerisch** *Adj. (spött.)* backwoods *attrib.* ‹*views, attitudes, manners, etc.*›; **~ziehen** [--'--] *unr. tr. V.* misappropriate ‹*materials, goods*›; **Steuern ~ziehen** evade [payment of] tax; **~ziehung die** *s.* **~ziehen:** misappropriation; evasion; **~zimmer das** back room

hịn-: ~|tragen *unr. tr. V.* **jmdn./etw. ~tragen** carry sb./take *or* carry sth. there; **etw. zu jmdm.** *od.* **jmdm. etw. ~tragen** take sth. to sb.; **~|treiben 1.** *unr. tr. V.* **a)** drive ‹*animals*› there; **die Strömung/der Wind trieb das Boot zum Ufer ~:** the current carried/the wind blew the boat to the shore; **b)** *unpers.* **es trieb ihn immer wieder zu ihr ~:** something always drove him back to her; **2.** *unr. itr. V.; mit sein* drift *or* float there; **~|treten** *unr. itr. V.; mit sein* **vor jmdn. ~treten** go up to sb.

hịn|tun *unr. tr. V. (ugs.)* put; **wo soll ich ihn bloß ~?** *(fig.)* I can't place him

hinüber [hɪ'ny:bɐ] *Adv.* over; across; **bis zur anderen Seite ~:** over *or* across to the other side; **~ und herüber** back and forth

hinüber-: ~|blicken *itr. V.* look across; **zu** *od.* **nach jmdm. ~blicken** look across at sb.; **~|bringen** *unr. tr. V.* **jmdn./etw. ~bringen** take sb./sth. across *or* over; **~|fahren 1.** *unr. itr. V.; mit sein (mit dem Auto/Fahrrad)* drive/ride *or* go over *or* across; **über den Fluß ~fahren** cross the river; **2.** *unr. tr. V.* **jmdn./ein Auto ~fahren** drive *or* take sb./drive a car over *or* across; **~|führen 1.** *tr. V.* **jmdn. über die Straße/in den Saal ~führen** take sb. across the road/take *or* show sb. across to the hall; **2.** *itr. V.* ‹*street, path, etc.*› lead *or* go over *or* across; **über etw.** *(Akk.)* **~führen** lead *or* go over sth.; **~|gehen** *unr. itr. V.; mit sein* walk *or* go over *or* across; **~|helfen** *unr. itr. V.* **jmdm. [über etw.** *(Akk.)***] ~helfen** help sb. over *or* across [sth.]; **~|kommen** *unr. itr. V.; mit sein* **a)** come over *or* across; *(~kommen können)* get across; **über etw.** *(Akk.)* **~kommen** get across sth.; **b)** *(ugs.: Besuch machen)* come over; pop over *(coll.)*; **~|lassen** *unr. tr. V.* **jmdn. ~lassen** allow *or* let sb. over *or* across; **~|reichen** *tr. V.* **[jmdm.] etw. ~reichen** pass *or* hand sth. across [to sb.]; **~|retten** *tr. V. (fig.)* preserve ‹*tradition etc.*›; **~|rufen 1.** *unr. tr. V.* call over; call out ‹*greeting, order, etc.*›; **2.** *unr. itr. V.* call over; **~|schauen** *itr. V. s.* **~blicken; ~|schicken** *tr. V.* **jmdn./etw. ~schicken** send sb./sth. over; **~|schwimmen** *unr. itr. V.; mit sein* swim over *or* across; **~|sehen** *unr. itr. V. s.* **~blicken; ~|sein** *unr. itr. V.; mit sein (nur im Inf. u. 2. Part. zusammengeschrieben) (ugs.)* **a)** *(tot, unbrauchbar sein)* have had it *(coll.)*; **er ist ~:** he's had it *(coll.)*; **b)** *(verdorben sein)* be off; have gone off; **c)** *(eingeschlafen sein)* have dropped off; *(bewußtlos sein)* be out for the count *(coll.)*; **d)** *(betrunken sein)* be well away *(coll.) or* plastered *(sl.)*; **~|spielen** *tr. V. (Sport)* cross ‹*ball*›; **~|springen** *unr. itr. V.; mit sein* jump over; **über etw.** *(Akk.)* **~springen** jump over sth.; **~|steigen** *unr. itr. V.; mit sein* climb over; **über etw.** *(Akk.)* **~steigen** climb over sth.; **~|wechseln** *itr. V.; mit haben od. sein* cross over; **zu einer anderen Partei ~wechseln** go over *or* switch to another party; **~|werfen** *unr. tr. V.* throw ‹*sth.*› over *or* across; **einen Blick ~werfen** *(fig.)* glance over *or* across; **~|ziehen 1.** *unr. tr. V.* draw *or* pull ‹*sb./sth.*› over *or* across; **2.** *unr. itr. V.; mit sein* move across

hịn- und hẹr|- ‹*move, travel, go, walk, etc.*› to and fro, back and forth

Hịn- und Rụck-: ~fahrt die journey there and back; round trip *(Amer.)*; **~flug der** outward and return flight; **~reise die, ~weg der** journey there and back

hinunter [hɪ'nʊntɐ] *Adv.* down; **den Hang ~:** down the slope

hinụnter|- ‹*go, walk, fall, take, throw, let, etc.*› down; *(in ein weiter unten gelegenes Stockwerk)* ‹*go, take, etc.*› downstairs

hinụnter-: ~|blicken *itr. V.* look down; **auf jmdn. ~blicken** *(fig.)* look down on sb.; **~|fallen** *unr. itr. V.; mit sein* fall down; **etw. ~fallen lassen** drop sth.; **mir ist die Vase ~gefallen** I dropped the vase; **~|lassen** *unr. tr. V.* **a)** *(mit einem Seil usw.)* **jmdn./etw. ~lassen** lower sb./sth.; let sb./sth. down; **b)** *(~gehen lassen)* **jmdn. ~lassen** let sb. [go] down; **~|schlucken** *tr. V.* **a)** swallow; **b)** *(hinnehmen)* swallow ‹*insult etc.*›; **c)** *(unterdrücken)* bite back ‹*remark, oath, etc.*›; choke back ‹*tears, anger*›; **~|spülen** *tr. V.* **a)** **etw. [den Ausguß] ~spülen** swill sth. down [the sink]; **etw. [die Toilette] ~spülen** flush sth. down [the toilet]; **b)** *(ugs.: hinunterschlukken)* wash down ‹*tablets etc.*›; **seinen Kummer [mit Alkohol] ~spülen** *(fig.)* drown one's sorrows [in drink]

hịn|wagen *refl. V.* dare [to] go there; venture there

hin·wẹg *Adv.* **a)** *(geh.)* **~ mit diesem Unrat!** away with this rubbish!; **~ mit dir!** away with you!; **b) über etw. ~:** over sth.; **über den Brillenrand ~:** over [the top of] his/her spectacles; **über jmdn. ~** *(fig.)* over sb.'s head; **über Jahre/lange Zeit ~:** for many years/a long time

Hịn·weg der way there

hinwẹg-: ~|brausen *itr. V.; mit sein* **über etw.** *(Akk.)* **~brausen** roar over sth.; **~|gehen** *unr. itr. V.; mit sein* **über etw.** *(Akk.)* **~gehen** *(auch fig.)* pass over sth.; **~|helfen** *unr. itr. V.* **jmdm. über etw.** *(Akk.)* **~helfen** help sb. [to] get over sth.; **~|kommen** *unr. itr. V.; mit sein* **über etw.** *(Akk.)* **~kommen** get over sth.; **~|lesen** *unr. itr. V.* **über etw.** *(Akk.)* **~lesen** read past sth. without noticing it; **~|sehen** *unr. itr. V.* **a) über jmdn./etw. ~sehen** see over sb. *or* sb.'s head/sth.; **b) über etw.** *(Akk.)* **~sehen** *(fig.)* overlook sth.; **~|setzen 1.** *itr. V.; auch mit sein* **über etw.** *(Akk.)* **~setzen** leap *or* jump over sth.; **2.** *refl. V.* **sich über etw.** *(Akk.)* **~setzen** ignore *or* disregard sth.; **~|täuschen** *tr., auch itr. V.* **jmdn. über etw.** *(Akk.)* **~täuschen** blind sb. to sth.; deceive *or* mislead sb. about sth.; **darüber ~täuschen, daß ...:** hide *or* obscure the fact that ...; **~|trösten** *tr. V.* **jmdn. über etw.** *(Akk.)* **~trösten** console sb. for sth.

Hinweis ['hɪnvais] **der; ~es, ~e a)** *(Wink)* hint; tip; **jmdm. einen ~ geben** give sb. a hint; **wenn ich mir den ~ erlauben darf** if I may [just] point something out *or* draw your attention to something; **b) unter ~ auf** (+ *Akk.*) with reference to; **c)** *(Anzeichen)* hint; indication

hin-: ~|weisen 1. *unr. itr. V.* **a)** *s.* **hindeuten a; b)** *s.* **hindeuten b; c) auf etw. ~weisen** *(fig.: aufmerksam machen)* point sth. out; refer to sth.; **darauf ~weisen, daß ...:** point out that; **2.** *unr. tr. V.* **jmdn. auf etw.** *(Akk.)* **~weisen** point sth. out to sb.; draw sb.'s attention to sth.; **~weisend** *Adj. (Sprachw.)* demonstrative ‹*pronoun, adjective, etc.*›

Hinweis-: ~schild das sign; *(Straßenschild)* [road] sign; **~tafel die** information board

hin-: ~|wenden 1. *unr. tr. V.* turn (**zu** towards); **2.** *unr. refl. V.* turn (**zu** to, towards); **~|werfen 1.** *unr. tr. V.* **a)** throw down; **b)** *(ugs.: aufgeben)* chuck in *(coll.)*; **c)** *(flüchtig schreiben)* jot down; *(flüchtig zeichnen)* dash off; **d)** *(beiläufig äußern)* drop [casually] ‹*remark*›; ask casually ‹*question*›; say casually ‹*words*›; **e)** *(ugs.: fallen lassen)* drop; **2.** *unr. refl. V.* **sich [vor jmdm.] ~werfen** throw oneself down [before sb.]

hin|wirken *itr. V.* **auf etw.** *(Akk.)* **~:** work towards sth.

Hinz [hɪnts] *in* **~ und Kunz** *(ugs. abwertend)* every Tom, Dick and Harry

hin-: ~|zaubern *tr. V. (ugs.)* **etw. ~zaubern** produce sth. as if by magic; **~|zeigen** *itr. V.* point (**zu** to, towards); **~|ziehen 1.** *unr. tr. V.* **a)** pull, draw (**zu** to, towards); **sich zu jmdm./etw. ~gezogen fühlen** be *or* feel attracted to sb./sth.; **b)** *(in die Länge ziehen)* draw out; protract; **2.** *unr. itr. V.; mit sein* move there; **wo ist sie ~gezogen?** where did she move to?; **3.** *unr. refl. V.* drag on (**über** + *Akk.* for); **~|zielen** *itr. V.* **auf etw.** *(Akk.)* **~zielen** aim at sth.; ‹*policies, efforts, etc.*› be aimed at sth.

hinzu-, Hinzu-: ~|bekommen *unr. tr. V.* get in addition; **~|denken** *unr. refl. V.* **sich** *(Dat.)* **etw. ~denken** add sth. in one's imagination; **~|fügen** *tr. V.* add; **~fügung die** addition; **unter ~fügung** *(Dat.)* **einer Sache** *(Gen.)* *od.* **von etw.** with the addition of sth.; **~|geben** *unr. tr. V.* add; **~|gesellen** *refl. V.* **sich [zu] jmdm./etw. ~gesellen** join sb./sth.; **~|kommen** *unr. itr. V.; mit sein* **a)** *s.* **dazukommen a; b)** *(hinzugefügt werden)* **zu etw. ~kommen** be added to sth.; **es kommt noch ~, daß ...** *(fig.)* there is also the fact that ...; **~|nehmen** *unr. tr. V.* add; **~|setzen 1.** *refl. V. s.* **dazusetzen; 2.** *tr. V.* add; **~|tun** *unr. tr. V. (ugs.)* add; **~|verdienen** *tr. V.* earn ‹*sth.*› extra; **~|zählen** *tr. V.* add [on]; **~|ziehen** *unr. tr. V.* consult; call in; **~ziehung die; ~:** consultation; **unter ~ziehung einschlägiger Literatur** by consulting the relevant literature

Hiob ['hi:ɔp] **(der)** Job

Hiobs·botschaft die bad news

Hippe ['hɪpə] **die; ~, ~n** pruning-knife; *(des Todes)* scythe

hipp, hipp, hurra ['hɪp'hɪphʊ'ra:] *Interj.* hip, hip, hooray *or* hurrah

Hippie ['hɪpi] **der; ~s, ~s** hippie *(coll.)*

Hirn [hɪrn] **das; ~[e]s, ~e a)** brain; **b)** *(Speise; ugs.: Verstand)* brains *pl.*; **sich** *(Dat.)* **das ~ zermartern** rack one's brains

hirn-, Hirn-: ~gespinst das *(abwertend)* fantasy; **~hautentzündung die** *(Med.)* meningitis; **~los** *(abwertend)* **1.** *Adj.* brainless; **2.** *adv.* brainlessly; **~rissig** *Adj. (salopp abwertend) s.* **~verbrannt; ~tumor der** *(Med.)* brain tumour; **~verbrannt** *Adj. (ugs. abwertend)* crazy; crack-brained *(coll.)*

Hirsch [hɪrʃ] **der; ~[e]s, ~e a)** deer; **b)** *(Rothirsch)* red deer; **c)** *(männlicher Hirsch)* stag; hart *(literary, in pub names etc.)*; **d)** *(Speise)* venison; **e)** *(Schimpfwort)* bastard *(coll.)*

Hirsch-: ~brunft, ~brunst die rut [of the stags]; **während der ~brunft** *od.* **~brunst** while the stags are in rut; **~geweih das** [stag's] antlers *pl.*; **ein ~geweih** a set of antlers; **~horn das** staghorn; **~kalb das** [male] deer calf; [male] fawn; **~kuh die** hind; **~leder das** buckskin

Hirse ['hɪrzə] **die; ~, ~n** millet

Hirse·brei der millet gruel

Hirt [hɪrt] **der; ~en, ~en** *s.* **Hirte**

Hirte der; ~n, ~n herdsman; *(Schaf~)* shepherd; **der Gute ~:** the Good Shepherd

Hirten-: ~amt das *(kath. Rel.)* pastorate; pastoral office; **~brief der** *(kath. Rel.)* pastoral letter; **~hund der** sheep-dog; **~stab der a)** *(geh.)* shepherd's crook; **b)** *(kath. Rel.)* pastoral staff; crosier

his, His [hɪs] **das; ~, ~** *(Musik)* B sharp; *s. auch* **a, A**

hissen ['hɪsn̩] *tr. V.* hoist, run up ‹*flag*›

Historiker [hɪs'to:rikɐ] **der; ~s, ~:** historian

historisch 1. *Adj.* **a)** historical; **b)** *(geschichtlich bedeutungsvoll)* historic. **2.** *adv.* historically

Hit [hɪt] **der; ~[s], ~s** *(ugs.)* hit

Hitler- ['hɪtlɐ-]: **~gruß der** Nazi salute; **~jugend die** Hitler Youth; **~junge der** member of the Hitler Youth

Hit-: ~liste die top ten/twenty *etc.*; **~parade die** hit parade

Hitze ['hɪtsə] **die; ~** heat; **bei dieser ~:** in this heat; **etw. bei mittlerer/mäßiger ~ backen** bake sth. in a medium/moderate oven; **die fliegende ~:** the hot flushes; **in der ~ des Gefechts** in the heat of the moment

hitze-, Hitze-: ~beständig *Adj.* heat-resistant, heat-resisting ‹*metal etc.*›; heat-proof, heat-resistant ‹*glass etc.*›; **~bläschen das** heat spot; **~empfindlich** *Adj.* sensitive to heat *postpos.*; heat-sensitive ‹*material*›; **~frei** *Adj.* **~frei haben/bekommen** have/be given the rest of the day off [school/work] because of excessively hot weather; **~periode die** hot spell; spell *or* period of hot weather; **~welle die** heat wave

hitzig *Adj.* **a)** *(heftig)* hot-tempered; quick-tempered; **~ werden** flare up; fly into a temper; **b)** *(leidenschaftlich)* hot-blooded ‹*person, race, etc.*›; hot ‹*blood*›; **c)** *(erregt)* heated ‹*discussion, argument, words, etc.*›

Hitzigkeit die; ~ hot *or* quick temper

hitz-, Hitz-: ~kopf der hothead; **~köpfig** *Adj.* hot-headed; **~schlag der** heat-stroke

Hiwi ['hi:vi] **der; ~s, ~s** laboratory *or (coll.)* lab/departmental/library assistant

HJ die; ~ *Abk. (ns.)* **Hitlerjugend**

hl. *Abk.* **heilig** St.

hm [hm̩] *Interj.* h'm; hem

H-Milch ['ha:-] **die; ~:** long-life *or* UHT milk

h-Moll ['ha:mɔl] **das; ~** *(Musik)* B minor; *s. auch* **a-Moll**

HNO-Arzt [ha:|en'|o:-] **der** ENT specialist

hob [ho:p] *1. u. 3. Pers. Sg. Prät. v.* **heben**

Hobby ['hɔbi] **das; ~s, ~s** hobby

Hobby- amateur ‹*gardener, archaeologist, astronomer, etc.*›

Hobby·raum der hobby room

Hobel ['ho:bl̩] **der; ~s, ~ a)** plane; **b)** *(Gemüse~)* [vegetable] slicer

Hobel·bank die carpenter's *or* woodworker's bench

hobeln *tr., itr. V.* **a)** plane; **an etw.** *(Dat.)* **~:** plane sth.; **b)** slice 〈*vegetables etc.*〉

Hobel·span der shaving

hoch [ho:x]; **höher** ['hø:ɐ], **höchst...** ['hø:çst...] **1.** *Adj.* **a)** high; high, tall 〈*building*〉; tall 〈*tree, mast*〉; long 〈*grass*〉; deep 〈*snow, water*〉; long, tall 〈*ladder*〉; high-ceilinged 〈*room*〉; **10 m ~:** 10 m high; **eine hohe Stirn** a high forehead; **von hoher Gestalt** *(geh.)* tall in stature; of tall stature; **hohe Absätze** high heels; **hohe Schuhe** *(mit hohem Schaft)* high boots; *(mit hohen Absätzen)* high-heeled shoes; **der hohe Norden** *(fig.)* the far North; **b)** *(mengenmäßig groß)* high 〈*price, wage, rent, speed, pressure, temperature, sensitivity*〉; heavy 〈*fine*〉; great 〈*weight*〉; large 〈*sum, amount*〉; high, large, big 〈*profit*〉; severe, extensive 〈*damage*〉; **c)** *(zeitlich fortgeschritten)* great 〈*age*〉; **ein hohes Alter erreichen** live to *or* reach a ripe old age; **es ist höchste Zeit, daß ...:** it is high time that ...; **d)** *(oben in einer Rangordnung)* high 〈*birth, office*〉; high-ranking 〈*officer, civil servant*〉; senior 〈*official, officer, post*〉; high-level 〈*diplomacy, politics*〉; important 〈*guest, festival*〉; great 〈*honour, discretion, urgency*〉; **Verhandlungen auf höchster Ebene** top-level negotiations; **der hohe Adel** the higher ranks of the nobility; **höchste Gefahr** extreme danger; **im höchsten Fall[e]** at the most; **e)** *(qualitativ ~stehend)* high 〈*standard, opinion*〉; great 〈*responsibility, concentration, talent, happiness, good, importance*〉; **die Hohe Schule** *(Reiten)* haute école; **f)** *(Musik)* high 〈*voice, note*〉; **das hohe C** top C; **g)** *(Math.)* **vier ~ fünf** four to the power [of] five; **h)** *(auf dem Höhepunkt)* **das hohe Mittelalter** the High Middle Ages; **i)** *in* **das ist mir zu ~** *(ugs.)* that's beyond me; that went over my head. **2.** *adv.* **a)** high; **~ oben am Himmel** high up in the sky; **~ über uns** high above us; **die Sonne steht ~:** the sun ist high in the sky; **~ zu Roß** *(geh.)* on horseback; **er wohnt drei Treppen ~:** he lives on the third *(Brit.) or (Amer.)* fourth floor; **~ auf etw.** *(Dat.)* **sitzen** sit high up on sth.; **b)** *(nach oben)* up; **Kopf ~!** chin up! **die Flammen loderten ~:** the flames leapt up high; **ein ~ aufgeschossener Junge** a very tall lad; **einen Ball ~ in die Luft werfen** throw a ball high in the air; **die Nase ~ tragen** walk around with one's nose in the air; **c)** *(zahlenmäßig viel)* highly 〈*taxed, paid*〉; **~ verschuldet/versichert** heavily in debt/insured for a large sum [of money]; **~ gewinnen/verlieren** *(Sport)* win/lose by a large margin; **wenn es ~ kommt** at [the] most; **d)** *(sehr)* highly 〈*gifted, delighted, satisfied*〉; most 〈*welcome*〉; highly, greatly 〈*esteemed*〉; **jmdm. etw. ~ anrechnen** consider sth. [to be] greatly to sb.'s credit; **jmdn. ~ verehren** esteem sb. highly *or* greatly; **e)** *(zeitlich fortgeschritten)* **~ in den Siebzigern** well into his/her seventies; **f)** *(Musik)* high; **g)** *(in Wendungen)* **etw. ~ und heilig versprechen** promise sth. faithfully; **das Herz höher schlagen lassen** make sb.'s heart beat faster; **es ging ~ her** things were pretty lively; **sie kamen drei Mann ~:** three of them came; there were three of them

Hoch das; **~s, ~s a)** *(Hochruf)* **ein [dreifaches] ~ auf jmdn. ausbringen** give three cheers for sb.; **ein ~ dem Gastgeber!** three cheers for the host!; **b)** *(Met.)* high

hoch-: ~|achten *tr. V. (geh.)* **jmdn./etw. ~achten** respect sb./sth. greatly; have a high regard for sb./sth.

Hoch·achtung die great respect; high esteem; **~ vor jmdm. haben** have a great respect for sb.; hold sb. in high esteem; **meine ~!** may I congratulate you

hochachtungs·voll *Adv. (Briefschluß)* yours faithfully

hoch-, Hoch-: ~adel der higher ranks *pl.* of the nobility; **~aktuell** *Adj.* highly topical; **~alpin** *Adj.* high alpine *attrib.* 〈*landscape, flora, fauna, etc.*〉; **~altar der** high altar; **~amt das** *(kath. Rel.)* high mass; **~angesehen** *Adj. (präd. getrennt geschrieben)* highly respected *or* regarded; **~anständig 1.** *Adj.* very decent; **2.** *adv.* very *or* most decently; **~|arbeiten** *refl. V.* work one's way up; **~bahn die** overhead railway; elevated railroad *(Amer.)*; **~barren der** *(Sport)* parallel bars *pl. (set at international height of 180 cm.)*; **~bau der;** *o. Pl.* [building] construction *no art.*; **~- und Tiefbau** [building] construction and civil engineering *no art.*; **~befriedigt** *Adj. (präd. getrennt geschrieben)* highly satisfied; **~begabt** *Adj. (präd. getrennt geschrieben)* highly gifted *or* talented; **~beinig** *Adj.* long-legged 〈*person, animal*〉; 〈*table, sofa, etc.*〉 with long legs; **~beladen** *Adj. (präd. getrennt geschrieben)* heavily laden; **~berühmt** *Adj.* very famous; **~betagt** *Adj.* aged; 〈*person*〉 advanced in years *postpos.*; **~betrieb der;** *o. Pl. (ugs.)* **es herrschte ~betrieb im Geschäft** the shop was at its busiest; **~blüte die** golden age; **~|bringen** *unr. tr. V.* **a)** bring up; *(ugs.: in die Wohnung bringen)* bring in[to the flat *(Brit.) or (Amer.)* apartment]; **b)** *(gesund machen)* **jmdn. ~bringen** put sb. on his/her feet; **c)** *(ugs.: ärgern)* **jmdn. ~bringen** put sb.'s back up; **~burg die** stronghold; **~deutsch** *Adj.* standard *or* High German; **~deutsch das, ~deutsche das** standard *or* High German; **~|dienen** *refl. V.* work one's way up; **~dotiert** *Adj. (präd. getrennt geschrieben)* highly paid; **~|drehen** *tr. V.* **a)** wind up 〈*window, barrier, etc.*〉; **b)** rev [up] *(coll.)* 〈*engine*〉

[1]Hoch·druck der a) *(Technik, Met.)* high pressure; **b)** *(Geschäftigkeit)* **mit** *od.* **unter ~ arbeiten** *(ugs.)* work flat out *or* at full stretch; **c)** *(Med.)* high blood pressure; hypertension *(Med.)*

[2]Hoch·druck der *(Druckw.)* **a)** *(Verfahren)* relief *or* letterpress printing; **etw. im ~ herstellen/drucken** produce/print sth. by letterpress; **b)** *(Erzeugnis)* piece of letterpress work

Hochdruck·gebiet das high-pressure area

hoch-, Hoch-: ~ebene die plateau; tableland; **~empfindlich** *Adj.* highly sensitive 〈*instrument, device, material, etc.*〉; high-speed, fast 〈*film*〉; extremely delicate 〈*fabric*〉; **~entwickelt** *Adj. (präd. getrennt geschrieben)* highly developed 〈*country etc.*〉; [highly] sophisticated 〈*method, device, etc.*〉; **~erhoben** *Adj. (präd. getrennt geschrieben)* **mit ~erhobenen Armen** with arms raised *or* held high; **~erhobenen Hauptes** with head held high; **~explosiv** *Adj. (auch fig.)* highly explosive; **~|fahren 1.** *unr. itr. V.; mit sein* **a)** *(ugs.)* go/drive/ride up; **b)** *(auffahren)* start up; **aus dem Sessel ~fahren** start [up] from one's chair; **aus dem Schlaf ~fahren** wake up with a start; **c)** *(aufbrausen)* flare up; **2.** *unr. tr. V. (ugs.)* drive 〈*sb./sth.*〉 up; **~fahrend** *Adj.* arrogant; supercilious; **~finanz die** high finance; ~|**fliegen** *unr. itr. V.; mit sein* fly up [into the air]; ~**fliegend** *Adj.* ambitious ‹*plan, idea, etc.*›; ~**form die** peak *or* top form; ~**format das** upright format; **in ~format** with an upright format; ~**frequenz die** *(Physik)* high frequency; ~**frisur die** upswept hair-style; ~**geachtet** *Adj. (präd. getrennt geschrieben)* highly respected *or* regarded; ~**gebildet** *Adj. (präd. getrennt geschrieben)* highly cultured; ~**gebirge das** [high] mountains *pl.*; ~**geehrt** *Adj.; nicht präd.* highly honoured; ~**gefühl das** [feeling of] elation; **im ~gefühl des Erfolges/Sieges** in his/her *etc.* elation at success/victory; ~|**gehen** *unr. itr. V.; mit sein (ugs.)* **a)** *s.* **hinaufgehen; b)** *(zornig werden)* blow one's top *(coll.)*; explode; **c)** *(explodieren)* ‹*bomb, mine*› go off; ‹*bridge, building, etc.*› go up; **etw. ~gehen lassen** *(salopp)* blow sth. up; **d)** *(aufgedeckt werden)* get caught *or (sl.)* nabbed; **jmdn. ~gehen lassen** ‹*informer*› grass *or* squeal on sb. *(sl.)*; ~**geistig** *Adj.* highly intellectual; ~**gelegen** *Adj. (präd. getrennt geschrieben)* high-lying; ~**gelehrt** *Adj.* extremely *or* very learned *or* erudite; ~**genuß der** *in* **ein ~genuß sein** be a real delight; ‹*meal, concert, etc.*› be a real treat; ~**geschätzt** *Adj. (präd. getrennt geschrieben)* highly esteemed *or* respected; ~**geschlossen** *Adj.* high-necked ‹*dress*›; ~**gespannt** *Adj.* great, high ‹*expectations*›; ~**gestellt** *Adj.; nicht präd.* ‹*person*› in a high position; important ‹*person*›; ~**gestochen** *(ugs. abwertend)* **1.** *Adj.* highbrow *(coll.)*; **2.** *adv.* in a highbrow way *(coll.)*; ~**gewachsen** *Adj.* tall; ~**gezüchtet** *Adj.* highly-bred ‹*animal*›; highly sophisticated ‹*engine, system*›

Hoch·glanz der: ein Foto in ~ *(Dat.)* a high-gloss print; **etw. auf ~** *(Akk.)* **polieren** polish sth. until it shines *or* gleams; **etw. auf ~** *(Akk.)* **bringen** give sth. a high polish; *(fig.)* make sth. spick and span

hoch-, Hoch-: ~**gradig 1.** *Adj.; nicht präd.* extreme; **2.** *adv.* extremely; ~**hackig** *Adj.* high-heeled ‹*shoe*›; ~|**halten** *unr. tr. V.* **a)** hold up ‹*arms*›; **b)** *(fig. geh.)* uphold ‹*truth, tradition, etc.*›; ~**haus das** high-rise-building; ~|**heben** *unr. tr. V.* lift up; raise ‹*arm, leg, etc.*›; raise, hold up ‹*hand*›; ~**herrschaftlich** *Adj.* palatial ‹*house, apartment*›; ~**herzig** *Adj. (geh.)* magnanimous; generous; ~**intelligent** *Adj.* highly intelligent; ~**interessant** *Adj.* extremely *or* most interesting; fascinating; ~|**jagen** *tr. V.* scare up ‹*birds*›; forcibly rouse ‹*sleeper*›; ~|**jubeln** *tr. V. (ugs.)* **jmdn./etw. ~jubeln** build sb. up as a star/sth. up as a hit; ~**kant** *Adv.* **a)** on end; **b)** *(ugs.) in* **jmdn. ~kant rauswerfen** chuck sb. out *(sl.)*; throw sb. out on his/her ear *(coll.)*; **~kant rausfliegen** be chucked out *(sl.)*; be thrown out on one's ear *(coll.)*; ~**kantig** *Adv. (ugs.) s.* **~kant b**; ~**karätig** [~ka'rɛ:tɪç] *Adj.* **a)** high-carat ‹*gold, diamond*›; **b)** *(fig.)* top-flight *(coll.)*; ~|**klappen 1.** *tr. V.* fold up ‹*chair, table*›; raise, lift up ‹*lid, car-bonnet*›; turn up ‹*collar*›; **2.** *itr. V.; mit sein* fold up; ~|**klettern** *itr. V.; mit sein (ugs.)* climb up; **den Baum ~klettern** climb [up] the tree; ~|**kommen** *unr. itr. V.; mit sein (ugs.)* **a)** come up; **b)** *(fig.: vorwärtskommen)* get on; **c)** *(aus dem Magen)* **ihr kam das Essen ~:** she threw up *(coll.) or* brought up her meal; **es kommt einem ~, wenn ...** *(fig.)* it makes you sick when ...; **d)** *(sich erheben)* get up; *(sich erheben können)* be able to get up; ~**konjunktur die** *(Wirtsch.)* boom; **auf dem Automarkt herrscht ~konjunktur** the car market is booming; ~|**können** *unr. V. (ugs.)* be able to get up; ~|**krempeln** *tr. V.* roll up ‹*sleeve, trouserleg*›; ~**kultur die** advanced civilization *or* culture; ~**land das;** *Pl.* **~länder** highlands *pl.*; ~|**leben** *itr. V. in* **jmdn./etw. ~leben lassen** cheer sb./sth.; **er lebe ~!** three cheers for him!; **der König lebe ~!** long live the king!; ~|**legen** *tr. V.* **ein gebrochenes Bein ~legen** support a broken leg in a raised position; **die Beine ~legen** put one's feet up

Hoch·leistung die outstanding performance

Hochleistungs·sport der top-level sport

hoch-, Hoch-: ~**mittel·alter das** High Middle Ages; ~**modern** ultra-modern; ~**moor das** *(Geogr.)* high-moor bog; ~**motiviert** *Adj.* highly motivated; ~**mut der** arrogance; **~mut kommt vor dem Fall** *(Spr.)* pride goes before a fall *(prov.)*; ~**mütig** *Adj.* arrogant; ~**näsig** [~nɛ:zɪç] *Adj. (abwertend)* stuck-up; conceited; ~**näsigkeit die;** ~ *(abwertend)* conceitedness; ~|**nehmen** *unr. tr. V.* **a)** lift *or* pick up; **b)** *(ugs.: verspotten)* **jmdn. ~nehmen** pull sb.'s leg; **c)** *(ugs.: nach oben nehmen)* **jmdn./etw. mit ~nehmen** take sb./sth. up with one; **d)** *(salopp: verhaften)* run in; ~**ofen der** blast furnace; ~**parterre das** upper ground floor; ~**politisch** *Adj.* highly political; ~**prozentig** *Adj.* high-proof ‹*spirits*›; ~**qualifiziert** *Adj. (präd. getrennt geschrieben)* highly qualified; ~|**ragen** *itr. V.* rise *or* tower up; ~|**rappeln** *refl. V. s.* **aufrappeln;** ~**rechnung die** *(Statistik)* projection; ~|**reißen** *unr. tr. V.* whip up; pull up ‹*aircraft*›; **die Arme ~reißen** throw one's arms up; ~**rot** *Adj.* bright red; **~rot im Gesicht werden** *(aus Verlegenheit)* go as red as a beetroot; ~**ruf der** cheer; ~**saison die** high season; ~|**schaukeln** *(ugs.) tr. V.* blow up ‹*problem, incident, etc.*›; ~|**schieben** *unr. tr. V. (ugs.)* push up; ~|**schießen 1.** *unr. tr. V.* send up, launch ‹*rocket, space probe, etc.*›; **2.** *unr. itr. V.; mit sein (auch fig.)* shoot up; ~|**schlagen 1.** *unr. tr. V.* turn up ‹*collar, brim*›; **2.** *unr. itr. V.; mit sein* ‹*waves*› surge up; ‹*flames*› leap up; ~|**schnellen** *itr. V.; mit sein* leap up; ~|**schrauben 1.** *tr. V.* **a)** raise ‹*seat*› *(by screwing)*; **b)** *(fig.)* force up ‹*prices*›; step up, increase ‹*demands*›; raise ‹*expectations*›; **2.** *refl. V.* circle up[wards]

Hoch-: ~**schule die** college; *(Universität)* university; ~**schüler der** college/university student

Hochschul-: ~**lehrer der** college/university lecturer *or* teacher; ~**studium das** college/university studies *pl., no art.*

hoch·schwanger *Adj.* in an advanced stage of pregnancy *postpos.*; very pregnant *(coll.)*

Hoch·see die; *o. Pl.* open sea

Hochsee-: ~**fischerei die** deep-sea fishing *no art.*; ~**flotte die** deep sea fleet; ~**jacht die** ocean-going yacht

Hoch·seil das high wire

Hochseil·akrobat der performer on the high wire

hoch-, Hoch-: ~**sitz der** *(Jagdw.)* raised hide; ~**sommer der** high summer; midsummer; ~**sommerlich** *Adj.* very summery ‹*weather etc.*›

Hoch·spannung die a) *(Elektrot.)* high voltage *or* tension; **Vorsicht, ~spannung!** danger –

high voltage; **b)** *o. Pl. (gespannte Stimmung)* high tension
Hochspannungs-: ~leitung die high voltage *or* high tension [transmission] line; power line; **~mast der** electricity pylon
hoch-, Hoch-: ~|spielen *tr. V.* blow up ⟨*incident, affair, etc.*⟩; **~sprache die** standard language; **~|springen** *unr. itr. V.; mit sein* jump *or* leap up; **an jmdm. ~springen** ⟨*dog etc.*⟩ jump up at sb.; **~springer der** *(Sport)* high jumper; **~sprung der** *(Sport)* high jump
höchst [hø:çst] *Adv.* extremely; most
höchst... *s.* **hoch**
hoch-, Hoch-: ~stand der *(Jagdw.)* raised stand; **~stapelei** [~ʃta:pəˈlai] **die; ~, ~en a)** fraud; **eine ~stapelei** a confidence trick; **b)** *(Aufschneiderei)* empty boasting; **~|stapeln** *itr. V.* **a)** perpetrate a fraud/frauds; **b)** *(aufschneiden)* make empty boasts; **~stapler** [~ʃta:plɐ] **der; ~s, ~ a)** confidence trickster; con-man *(coll.)*; **b)** *(Aufschneider)* fraud
Höchst-: ~betrag der maximum amount; **~bietende der/die;** *adj. Dekl.* highest bidder
hoch-: ~stehend *Adj.* ⟨*person*⟩ of high standing; **geistig ~stehend** intellectually distinguished; ⟨*person*⟩ of high intellect; **sittlich ~stehend** high-minded; **~|steigen** *unr. itr. V.; mit sein* **a)** climb; **die Treppe/Stufen ~steigen** climb the stairs/steps; **b)** ⟨*bubbles, smoke, etc.*⟩ rise; ⟨*rocket*⟩ go up; **c)** *(langsam entstehen)* rise up; ⟨*tears*⟩ well up
hoch|stellen *tr. V.* **a)** put up; **b)** *(hochklappen)* turn up ⟨*collar*⟩
höchsten·falls [ˈhø:çstn̩-] *Adv.* at [the] most *or* the outside; at the very most
höchstens *Adv.* at most; *(bestenfalls)* at best
Höchst-: ~fall der *in* **im ~fall[e]** at [the] most *or* the outside; at the very most; **~form die** *(bes. Sport)* peak *or* top form; **~gebot das** highest bid *or* offer; **~geschwindigkeit die** top *or* maximum speed
hoch|stilisieren *tr. V. (abwertend)* build up **(zu** into)
Hoch·stimmung die festive mood; high spirits *pl.*; **in ~ sein** be in a festive mood
höchst-, Höchst-: ~leistung die supreme performance; *(Ergebnis)* supreme achievement; *(Technik)* maximum performance; **~maß das: ein ~maß an etw.** *(Dat.)* a very high degree of sth.; **ein ~maß von etw.** *(Dat.)* a maximum [amount] of sth.; **~möglich** *Adj.; nicht präd.* highest possible; **~persönlich 1.** *Adj.; nicht präd.* personal; **2.** *adv.* in person; **~preis der** *(höchstmöglicher Preis)* highest price; *(höchstzulässiger Preis)* maximum price
Hoch·straße die elevated road; flyover *(Brit.)*; overpass
hoch|streifen *tr. V.* pull up
höchst-, Höchst-: ~satz der maximum *or* top rate; **~stand der** highest level; **~strafe die** maximum penalty; **~wahrscheinlich** *Adv.* very probably; **~wert der** maximum value; **~zulässig** [auch: ˈ-ˈ---] *Adj.; nicht präd.* maximum [permissible] ⟨*weight, speed, etc.*⟩
hoch-, Hoch-: ~tour die *in* **auf ~touren laufen** run at top *or* full speed; *(intensiv betrieben werden)* be in full swing; **~tourig** [~tu:rɪç] *(Technik)* **1.** *Adj.* fast-revving *(coll.)* ⟨*engine*⟩; **2.** *adv.* **~tourig fahren** drive at high revs *(coll.)*; **~trabend** *(abwertend)* **1.** *Adj.* pretentious; high-flown; **2.** *adv.* pretentiously; in a high-flown manner; **~|tragen** *unr. tr. V.* carry up; **~|treiben** *unr. tr. V.* **a)** *(ugs.: hinauftreiben)* drive up; **b)** *(fig.)* force *or* push up ⟨*prices etc.*⟩; **~verdient** *Adj. (präd. getrennt geschrieben)* **a)** ⟨*scientist etc.*⟩ of outstanding merit; **b)** richly deserved ⟨*victory, success, etc.*⟩; **~verehrt** *Adj.; nicht präd.* highly respected *or* esteemed; *(als Anrede)* **meine ~verehrten Damen und Herren!** ladies and gentlemen!; **~verrat der** high treason; **~verschuldet** *Adj. (präd. getrennt geschrieben)* heavily *or* deep in debt *postpos.*
Hoch·wasser das *(Flut)* high tide *or* water; *(Überschwemmung)* flood; **der Fluß hat** *od.* **führt ~:** the river is in flood; **er hat ~** *(ugs. scherzh.)* his trousers are at half-mast *(coll.)*
hoch-, Hoch-: ~|werfen *unr. tr. V.* **etw. ~werfen** throw sth. up; **~wertig** *Adj.* high-quality ⟨*goods*⟩; highly nutritious ⟨*food*⟩; **~willkommen** *Adj. (präd. getrennt geschrieben)* very *or* most welcome; **~|winden 1.** *unr. tr. V.* wind up; weigh ⟨*anchor*⟩; **2.** *unr. refl. V.* wind its way up; **~wirksam** *Adj. (präd. getrennt geschrieben)* highly *or* extremely effective; **~wohlgeboren** *Adj. (veralt.)* high-born; **Euer Hochwohlgeboren** Your Honour; **~würden** *o. Art.;* **~[s]** *(veralt.)* Reverend Father; **~zahl die** *(Math.)* exponent
[1]**Hoch·zeit die** *(geh.)* Golden Age
[2]**Hochzeit** [ˈhɔxtsait] **die; ~, ~en** wedding; **silberne/goldene ~:** silver/golden wedding [anniversary]; **man kann nicht auf zwei ~en tanzen** *(fig. ugs.)* you can't be in two places at once
Hochzeiter der; ~s, ~ *(landsch.)* [bride] groom; **die ~:** the bride and groom
Hochzeiterin die; ~, ~nen *(landsch.)* bride
Hochzeits-: ~feier die wedding; **~geschenk das** wedding gift *or* present; **~kleid das** *(Zool.)* nuptial coloration; *(von Vögeln)* nuptial plumage; **~nacht die** wedding night; **~reise die** honeymoon [trip]; **wir haben unsere ~reise nach Berlin gemacht** we went to Berlin for our honeymoon; **~tag der a)** wedding day; **b)** *(Jahrestag)* wedding anniversary
hoch-: ~|ziehen 1. *unr. tr. V.* **a)** pull up; pull up, raise ⟨*shutters, blind*⟩; hoist, raise, run up ⟨*flag*⟩; hoist ⟨*sail*⟩; **die Schultern/Brauen ~ziehen** hunch one's shoulders/raise one's eyebrows; **die Nase ~ziehen** sniff [loudly]; **b) ein Flugzeug ~ziehen** put an aircraft into a steep climb; **c)** *(bauen)* put up, build ⟨*wall, building*⟩; **2.** *unr. refl. V.* **sich [an etw.** *(Dat.)*] **~ziehen** pull oneself up [by hanging on to sth.]; **sich an etw.** *(Dat.)* **~ziehen** *(fig.)* latch on to sth.
Hocke [ˈhɔkə] **die; ~, ~n a)** squat; crouch; **in der ~ sitzen** squat; crouch; **in die ~ gehen** squat [down]; crouch down; **b)** *(Turnen)* squat vault
hocken 1. *itr. V.* **a)** *mit haben od. (südd.) sein* squat; crouch; **b)** *mit sein (südd.: sitzen)* sit; **d)** *mit sein (Turnen)* perform *or* do a squat vault **(über +** *Akk.* over). **2.** *refl. V.* **a)** crouch down; squat [down]; **b)** *(südd.: sich setzen)* sit down
Hocker der; ~s, ~: stool
Höcker [ˈhœkɐ] **der; ~s, ~:** hump; *(auf der Nase)* bump; *(auf dem Schnabel)* knob
Hockey [ˈhɔki] **das; ~s** hockey
Hockey-: ~schläger der hockey stick; **~spieler der** hockey player
Hoden [ˈho:dn̩] **der; ~s, ~:** testicle
Hoden·sack der scrotum

Hof [ho:f] **der; ~[e]s, Höfe** ['hø:fə] **a)** courtyard; *(Schul~)* playground; *(Gefängnis~)* [prison] yard; **b)** *(Bauern~)* farm; **c)** *(eines Herrschers)* court; **am ~[e]** at court; **d) jmdm. den ~ machen** *(veralt.)* pay court to sb.; **e)** *(Aureole)* corona; aureole

hof-, Hof-: ~amt das *(hist.)* [hereditary] office at court; **~dame die** lady of the court; *(Begleiterin der Königin)* lady-in-waiting; **~fähig** *Adj.* presentable at court *pred.; (fig.)* [socially] acceptable

Hoffart ['hɔfart] **die; ~** *(veralt.)* overweening pride; haughtiness

hoffärtig ['hɔfɛrtɪç] *(veralt. abwertend)* **1.** *Adj.* haughty. **2.** *adv.* haughtily

hoffen ['hɔfn̩] **1.** *tr. V.* hope; **ich hoffe es/will es ~:** I hope so/can only hope so; **ich will es nicht ~, ich hoffe es nicht** I hope not; **es bleibt zu ~, daß ...:** let us hope that ...; **~ wir das Beste** let's hope for the best. **2.** *itr. V.* **a)** hope; **auf etw.** *(Akk.)* **~:** hope for sth.; **b)** *(Vertrauen setzen auf)* **auf jmdn./etw. ~:** put one's trust *or* faith in sb./sth.

hoffentlich ['hɔfn̩tlɪç] *Adv.* hopefully; **~!** let's hope so; **~ ist ihr nichts passiert** I do hope nothing's happened to her; **es ist dir doch ~ recht** I hope it's all right with you

Hoffnung ['hɔfnʊŋ] **die; ~, ~en** hope; **seine ~ auf jmdn./etw. setzen** pin one's hopes *pl.* on sb./sth.; **keine ~ mehr haben** have given up [all] hope; **sich** *(Dat.)* **[falsche] ~en machen** have [false] hopes; **jmdm. ~en machen** raise sb.'s hopes

hoffnungs-, Hoffnungs-: ~los 1. *Adj.* hopeless; despairing ‹*person*›; **2.** *adv.* hopelessly; **~losigkeit die; ~:** hopelessness; *(Verzweiflung)* despair; **~schimmer der** *(geh.)* glimmer of hope; **~voll 1.** *Adj.* **a)** hopeful; full of hope *pred.;* **jmdn. ~voll stimmen** give sb. cause to hope *or* make sb. hopeful; **b)** *(erfolgversprechend)* promising; **2.** *adv.* **a)** full of hope; **b)** *(erfolgversprechend)* promisingly

hof|halten *unr. itr. V.* hold court

Hof·hund der watchdog

hofieren [ho'fi:rən] *tr. V. (geh.)* pay court to

höfisch ['hø:fɪʃ] *Adj.* courtly

höflich ['hø:flɪç] **1.** *Adj.* polite; courteous; **etw. in ~em Ton fragen/sagen** ask/say sth. politely. **2.** *adv.* politely; courteously

Höflichkeit die; ~, ~en a) *o. Pl.* politeness; courteousness; **etw. [nur] aus ~ tun/sagen** do/say sth. [only] to be polite *or* out of politeness; **b)** *meist Pl. (höfliche Redensart)* civility; courtesy

Höflichkeits-: ~besuch der courtesy visit; **~floskel die** polite phrase

Höfling ['hø:flɪŋ] **der; ~s, ~e** courtier

Hof-: ~marschall der major-domo; **~narr der** *(hist.)* court jester; **~rat der** *(veralt., noch österr.) honorary title conferred on senior civil servant;* **~schranze die** *od.* **der** *(veralt. abwertend)* fawning courtier; **~staat der;** *o. Pl.* court; **~tor das** courtyard gate

hoh... ['ho:...] *s.* **hoch**

Höhe ['hø:ə] **die; ~, ~n a)** height; *(Entfernung nach oben)* height; altitude; **in einer ~ von 4000 m fliegen** fly at a height *or* altitude of 4,000 m.; **an ~ gewinnen/verlieren** gain/lose height *or* altitude; **auf halber ~:** at mid-altitude; **b)** *(Richtung)* **etw. in die ~ heben** lift sth. up; **in die ~ [auf]steigen** rise up[wards]; **c)** *(Gipfelpunkt)* height; **auf der ~ seines Ruhmes/Könnens/Erfolges sein** be at the height of one's fame/ability/success; **auf der ~ sein** *(fig. ugs.) (gesund sein)* be fit; *(sich wohl fühlen)* feel fine; **nicht [ganz] auf der ~ sein** *(fig. ugs.)* be/feel a bit under the weather *(coll.);* not be/feel quite oneself; **das ist ja die ~!** *(fig. ugs.)* that's the limit; **d)** *(meßbare Größe)* level; *(von Einkommen)* size; level; **die ~ der Geschwindigkeit/Temperatur** the speed/temperature level; **Unkosten in ~ von 5000 DM** expenses of 5,000 DM; **e)** *(Linie)* **auf gleicher ~ sein/fahren** be in line abreast *or* be level/travel in line abreast; **auf ~ des Leuchtturms/von Hull sein** *(Seemannsspr.)* be level with *or* abreast of the lighthouse/be off Hull; **f)** *(Anhöhe)* hill; **die ~n und Tiefen des Lebens** *(fig.)* the ups and downs of life; **g)** *(Math., Astron.)* altitude; **h)** *Pl. (Akustik)* treble *sing.*

Hoheit ['ho:hait] **die; ~, ~en a)** *o. Pl. (Souveränität)* sovereignty **(über + *Akk.*** over); **unter der ~ eines Staates stehen** be under the sovereignty of a state; **b) Seine/Ihre ~:** His/Your Highness

Hoheits-: ~ab·zeichen das national emblem; **~gebiet das** [sovereign] territory; **~gewässer das;** *meist Pl.* territorial waters *pl.;* **~recht das;** *meist Pl.* right of the state; **~zeichen das** national emblem

Hohe·lied das; Hohenlied[e]s a) *(bibl.)* **das ~:** The Song of Songs; **b)** *(fig. geh.)* song of praise

Höhen-: ~angst die; *o. Pl.* fear of heights; **~flug der** *(Flugw.)* high-altitude flight; *(fig.)* flight; **~krankheit die** altitude sickness; **~lage die** altitude; **in ~lage** at high altitude; **~luft die;** *o. Pl.* mountain air; air at high altitude; **~messer der** altimeter; **~ruder das** *(Flugw.)* elevator; **~sonne die a)** *(Gerät)* sun lamp; **b)** *(Bestrahlung)* sun lamp treatment; **~unterschied der** altitude difference; difference in altitude; **~zug der** *(Geogr.)* range of hills

Hohe·priester der; Hohenpriesters, Hohenpriester *(bibl.)* high priest

Höhepunkt der high point; *(einer Veranstaltung)* high spot; highlight; *(einer Laufbahn, des Ruhms)* peak; pinnacle; *(einer Krankheit)* crisis; critical point; *(einer Krise)* turning-point; *(der Macht)* summit; pinnacle; *(des Glücks)* height; *(Orgasmus; eines Stückes)* climax; **auf dem ~ seiner Laufbahn stehen** be at the peak of one's career

höher ['hø:ɐ] *s.* **hoch**

höher-: ~gestellt *Adj. (präd. getrennt geschrieben)* senior ‹*official, civil servant*›; **~|schrauben** *tr. V.* force *or* push up ‹*prices*›

hohl [ho:l] **1.** *Adj.* **a)** hollow; **sich innerlich ~ fühlen** *(fig.)* feel empty inside; **b)** *(eingewölbt)* cupped ‹*hand*›*;* sunken, hollow ‹*cheeks, eyes*›*;* **ein ~es Kreuz** a hollow back; **c)** *(dumpf)* hollow ‹*sound, voice, etc.*›*;* **d)** *(abwertend: geistlos)* hollow, empty ‹*phrases, slogans*›*;* empty ‹*talk, chatter*›. **2.** *adv.* **a)** *(dumpf)* hollowly; **b)** *(abwertend: geistlos)* inanely

hohl·äugig *Adj.* hollow-eyed; sunken-eyed

Höhle ['hø:lə] **die; ~, ~n a)** cave; *(größer)* cavern; **b)** *(Tierbau)* den; lair; **sich in die ~ des Löwen begeben** enter the lion's den; **c)** *(abwertend: Wohnung)* hole; **d)** *meist Pl. (Augen~)* socket

höhlen ['hø:lən] *tr. V.* hollow out; **steter Tropfen höhlt den Stein** *(Spr.)* these things take their toll eventually

Höhlen-: ~forscher der speleologist; *(Sportler)* caver; **~malerei die** cave-painting; **~mensch der** cave-dweller; cave-man

hohl-, Hohl-: ~kopf der *(abwertend)* idiot *(coll.)*; dimwit; **~köpfig** *Adj. (abwertend)* idiotic *(coll.)*; blockheaded; **~körper der** hollow body; **~kreuz das** hollow back; lordosis *(Med.)*; **~kugel die** hollow sphere; **~maß das** measure of capacity; **~raum der** cavity; [hollow] space; **~saum der** *(Handarb.)* hem-stitch; **~spiegel der** concave mirror

Höhlung die; ~, ~en hollow

hohl·wangig *Adj.* hollow-cheeked; sunken-cheeked

Hohl·weg der defile

Hohn [ho:n] **der; ~[e]s** scorn; derision; **jmdn. mit ~ und Spott überschütten** pour *or* heap scorn on sb.

höhnen ['hø:nən] *(geh.) itr. V.* jeer; sneer

Hohn·gelächter das derisive *or* scornful laughter

höhnisch ['hø:nɪʃ] **1.** *Adj.* scornful; derisive. **2.** *adv.* scornfully; derisively

hohn-: ~|lachen *itr. V. (Präs. u. Prät. auch fest; ich hohnlache/hohnlachte usw.)* laugh scornfully *or* derisively; **ein Hohnlachen** a scornful *or* derisive laugh; **~|sprechen** *unr. itr. V.* **einer Sache** *(Dat.)* **~sprechen** fly in the face of sth.

Hokuspokus [ho:kʊs'po:kʊs] **der; ~:** hocus-pocus; *(abwertend: Getue)* fuss

hold [hɔlt] **1.** *Adj.* **a)** *(dichter. veralt.)* fair; lovely; lovely ‹*sight*›; sweet, lovely ‹*smile*›; **die ~e Weiblichkeit** *(scherzh.)* the fair sex; **das Glück war uns** *(Dat.)* **~** *(geh.)* fortune smiled upon us. **2.** *adv.* sweetly

Holder der; ~s, ~ *(bes. südd.) s.* **Holunder**

holen ['ho:lən] **1.** *tr. V.* **a)** fetch; get; **sich** *(Dat.)* **Hilfe/Rat** *usw.* **~:** get [some] help/advice *etc.*; **jmdn. aus dem Bett ~:** get *or (coll.)* drag sb. out of bed; **da/bei ihr ist nichts zu ~** *(fig.)* you won't get anything there/out of her; **b)** *(ab~)* fetch; pick up; collect; *(ugs. verhüll.: verhaften)* take away; **c)** *(ugs.: erlangen)* get, win ‹*prize*›; get, carry off, win ‹*medal, trophy, etc.*›; get, score ‹*points*›; **sich** *(Dat.)* **die Meisterschaft/den Preis** *usw.* **~:** win *or* take the championship/prize *etc.*; **d)** *(ugs. landsch.: kaufen)* get; **sich** *(Dat.)* **etw. ~:** get [oneself] sth. **2.** *refl. V. (ugs.: sich zuziehen)* catch; **sich** *(Dat.)* **[beim Baden** *usw.*] **einen Schnupfen/die Grippe ~:** catch a cold/the flu [swimming *etc.*]; **sich** *(Dat.)* **den Tod ~** *(fig.)* catch one's death [of cold]

Holland ['hɔlant] **(das); ~s** Holland

Holländer ['hɔlɛndɐ] **der; ~s, ~ a)** Dutchman; **er ist ~:** he is Dutch *or* a Dutchman; **die ~:** the Dutch; **b)** *(Käse)* Dutch cheese

Holländerin die; ~, ~nen Dutchwoman/Dutch girl

holländisch *Adj.* Dutch

Hölle ['hœlə] **die; ~, ~n a)** hell *no art.*; **in die ~ kommen** go to hell; **zur ~ fahren** *(geh.)* descend into hell; **jmdn. zur ~ wünschen** *(geh.)* wish sb. to hell; **zur ~ mit ihm/damit!** to hell with him/it *(coll.)*; **b)** *(fig.)* **die ~ ist los** *(ugs.)* all hell has broken loose *(coll.)*; **die ~ auf Erden haben** suffer hell on earth; **jmdm. das Leben zur ~ machen** make sb.'s life hell *(coll.)*; **jmdm. die ~ heiß machen** give sb. hell *(coll.)*

Höllen-: ~angst die *(salopp)* **eine ~angst vor etw.** *(Dat.)* **haben** be scared to death of sth. *(coll.)*; be terrified of sth.; **~lärm der** *(ugs.)* diabolical noise *or* row *(coll.)*; **~maschine die** infernal machine *(arch.)*; time bomb; **~pein, ~qual die** agony; **~qualen erleiden** suffer the torments of hell *(fig.)*; suffer terrible agony *sing.*; **~spektakel das** *(ugs.) s.* **~lärm**; **~tempo das** *(ugs.)* breakneck speed

Holler ['hɔlɐ] **der; ~s, ~** *(bes. südd., österr.) s.* **Holunder**

höllisch ['hœlɪʃ] **1.** *Adj.* **a)** *nicht präd.* infernal; ‹*spirits, torments*› of hell; **b)** *(schrecklich)* terrible ‹*war, situation*›; fiendish, diabolical ‹*invention, laughter*›; **~e Schmerzen** terrible agony *sing.*; **c)** *(ugs.: sehr groß)* tremendous *(coll.)* ‹*noise, shock, respect*› *(coll.)*; enormous *(coll.)* ‹*pleasure*›; **~e Angst vor etw.** *(Dat.)* **haben** be scared stiff of sth. *(coll.)*. **2.** *adv. (ugs.: sehr)* terribly, hellishly *(coll.)* ‹*cold, difficult*›; **sich ~ zusammennehmen** make a tremendous effort to control oneself *(coll.)*; **es tut ~ weh** it hurts like hell *(coll.)*

Hollywood·schaukel ['hɔlɪwʊd-] **die** swinging garden hammock

Holm [hɔlm] **der; ~[e]s, ~e a)** *(Turnen)* bar; **b)** *(Leiter~)* upright; side-piece

Holocaust [holo'kaust] **der; ~[s], ~s** Holocaust

holperig *s.* **holprig**

holpern ['hɔlpɐn] *itr. V.* **a)** jolt; **b)** *mit sein (holpernd fahren)* jolt; bump

holprig ['hɔlprɪç] **1.** *Adj.* **a)** *(uneben)* bumpy; uneven; rough; **b)** *(stockend)* stumbling, halting ‹*speech*›; clumsy ‹*verses, translation, language, style, wording, etc.*›. **2.** *adv.* haltingly; **~ lesen** stumble over one's words when reading

Holster ['hɔlstɐ] **das; ~s, ~:** holster

holterdiepolter [hɔltɐdi'pɔltɐ] *Adv. (ugs.)* helter-skelter; **alles ging ~:** there was a mad rush

Holunder [ho'lʊndɐ] **der; ~s, ~ a)** *(Strauch)* elder; **b)** *o. Pl. (Früchte)* elderberries *pl.*

Holunder-: ~beere die elderberry; **~strauch der** elder[berry] bush

Holz [hɔlts] **das; ~es, Hölzer** ['hœltsɐ] **a)** wood; *(Bau~, Tischler~)* timber; wood; **bearbeitetes ~:** timber *(Brit.)*; lumber *(Amer.)*; **ein Festmeter ~:** a cubic metre of timber; **[viel] ~ vor der Hütte** *od.* **Tür haben** *(fig. ugs. scherzh.)* be well stacked *(coll.)* *or* well endowed; **aus dem ~ sein, aus dem man Helden macht** be of the stuff heroes are made of; **aus dem gleichen ~ [geschnitzt] sein** *(fig.)* be cast in the same mould; **b)** *(Streich~)* match; *(zweckentfremdet)* matchstick

Holz-: ~bein das wooden leg; **~bläser der** woodwind player; **~blas·instrument das** woodwind instrument; **~block der;** *Pl.* **~blöcke** block of wood; **~bock der a)** *(Gestell)* wooden stand *or* trestle; **b)** *(Zecke)* castor-bean tick

Hölzchen ['hœltsçən] **das; ~s, ~ a)** small piece of wood; *(Stöckchen)* stick; **b)** *s.* **Holz b**

holzen *itr. V. (Fußballjargon)* play dirty *(coll.)*

hölzern ['hœltsɐn] *Adj.*; *nicht präd. (auch fig.)* wooden

holz-, Holz-: ~fäller der woodcutter; lumberjack *(Amer.)*; **~feuer das** wood fire; **~frei** *Adj.* wood-free ‹*paper*›; **~gas das** wood-gas; **~hacker der a)** *(bes. österr.) s.* **~fäller**; **b)** *(Fußballjargon)* dirty player; **~haltig** *Adj.* woody ‹*paper*›; ‹*paper*› containing mechanical wood pulp; **~hammer der** [wooden] mallet

Holzhammer·methode die *(ugs.)* sledge-hammer method

Holz·haus das timber *or* wooden house

holzig *Adj.* woody

holz-, Holz-: ~kitt der plastic

wood; **~klotz der** block of wood; *(als Spielzeug)* wooden block; **~kohle die** charcoal; **~kopf der** *(salopp abwertend)* blockhead; numskull; **~kreuz das** wooden cross; **~leim der** wood-glue; **~leiste die** batten; **~nagel der** wooden nail; **~pantine die** *(landsch.)*, **~pantoffel der** clog; **~pflock der** wooden stake; **~scheit das** piece of wood; *(Brenn~)* piece of firewood; **~schnitt der a)** *o. Pl.* woodcutting *no art.*; **b)** *(Bild)* woodcut; **~schnitt·artig** *Adj. (fig.)* simplistic; **~schnitzer der** wood-carver; **~schnitzerei die** wood-carving; **~schuh der** clog; **~span der a)** *(zum Feueranzünden)* stick of firewood; *(zum Rühren usw.)* small stick [of wood]; **b)** *meist Pl. (Hobelspan)* [wood] shaving; **~splitter der** splinter of wood; **~stab der** wooden rod; **~stoß der** pile of wood; **~verarbeitend** *Adj.; nicht präd.* timber processing ‹*industry etc.*›; **~weg der** *(fig.) in* **auf dem ~weg sein** *od.* **sich auf dem ~weg befinden** be on the wrong track *(fig.)*; be barking up the wrong tree *(fig.)*; **~wolle die;** *o. Pl.* woodwool; **~wurm der** woodworm

Homburg ['hɔmburk] **der; ~s, ~s** Homburg

Homo ['ho:mo] **der; ~s, ~s** *(ugs.)* queer *(coll.)*, homo *(coll.)*

homo-, Homo-: ~gen [~'ge:n] *Adj.* homogeneous; **~genisieren** *tr. V. (fachspr.)* homogenize; **~genität** [~geni'tɛ:t] **die;** ~ *(geh.)* homogeneity

homöo-, Homöo- [homøo-]: **~path** [~'pa:t] **der; ~en, ~en** homoeopath; **~pathie die; ~:** homoeopathy *no art.*; **~pathisch** *Adj.* homoeopathic

homo-, Homo-: ~sexualität die; ~: homosexuality; **~sexuell 1.** *Adj.* homosexual; **2.** *adv.* **~sexuell veranlagt sein** have homosexual tendencies; **~sexuelle der/die;** *adj. Dekl.* homosexual

Honduras [hɔn'du:ras] **(das); Honduras'** Honduras

Hongkong ['hɔŋkɔŋ] **(das); ~s** Hong Kong

Honig ['ho:nɪç] **der; ~s, ~e** honey; **jmdm. ~ um den Bart** *(ugs.) od. (salopp)* **ums Maul schmieren** *(fig.)* butter sb. up

honig-, Honig-: ~biene die honey-bee; **~brot das** bread and honey; **~kuchen der** honey cake; **~kuchen·pferd das** *in* **grinsen wie ein ~kuchenpferd** *(ugs. scherzh.)* grin like a Cheshire cat; **~lecken das** *in* **das ist kein ~lecken** *(ugs.)* it is not a bed of roses; **~melone die** honeydew melon; **~schlecken das** *s.* **~lecken; ~süß 1.** *Adj.* ‹*grapes, taste, etc.*› as sweet as honey; *(fig.)* honey-sweet ‹*voice*›; **ein ~süßes Lächeln** *(fig.)* the sweetest of smiles; **2.** *adv. (fig.)* **~süß lächeln/antworten** smile a honey-sweet smile/answer in honeyed tones

Honorar [hono'ra:ɐ̯] **das; ~s, ~e** fee; *(Autoren~)* royalty

Honoratioren [honora'tsi̯o:rən] *Pl.* notabilities

honorieren *tr. V. (würdigen)* appreciate; *(belohnen)* reward

honorig *Adj.* honourable; respectable

Hopfen der; ~s, ~: hop; *(Blüten)* hops *pl.*; **bei ihm ist ~ und Malz verloren** *(ugs.)* he's a hopeless case

hopp [hɔp] *Interj.* quick; look sharp

hoppeln ['hɔpl̩n] *itr. V.; mit sein* hop **(über +** *Akk.* across, over); *(fig.)* bump, jolt **(über +** *Akk.* across, over)

hopp·hopp 1. *Interj. s.* **hopp. 2.** *Adv.* in double-quick time

hoppla ['hɔpla] *Interj.* oops; whoops

hopp|nehmen *unr. tr. V. (salopp)* nab *(sl.)*; nick *(sl.)*

hops [hɔps] *Interj.* up; jump

hopsen *itr. V.; mit sein (ugs.) s.* **hüpfen**

Hopser der; ~s, ~ *(ugs.)* [little] jump

hops-: ~|gehen *unr. itr. V.; mit sein (salopp)* **a)** *(umkommen)* buy it *(sl.)*; **b)** *(entzweigehen)* get broken; *(abhanden kommen)* go missing; **~|nehmen** *unr. tr. V. s.* **hoppnehmen**

hörbar 1. *Adj.* audible. **2.** *adv.* audibly; *(geräuschvoll)* noisily

horchen ['hɔrçn̩] *itr. V.* listen **(auf +** *Akk.* to); *(heimlich zuhören)* eavesdrop; listen; **an der Tür/Wand ~:** listen at the door/through the wall

Horcher der; ~s, ~: *s.* **Lauscher a**

[1]**Horde** ['hɔrdə] **die; ~, ~n** *(auch Völkerk.)* horde; *(von Halbstarken)* mob; crowd

[2]**Horde die; ~, ~n** *(Gestell)* rack

hören ['hø:rən] **1.** *tr. V.* **a)** hear; **jmdn. kommen/sprechen ~:** hear sb. coming/speaking; **ich höre nichts** I can't hear anything; *s. auch* **Gras; b)** *(anhören)* listen to, hear ‹*programme, broadcast, performance, etc.*›; hear ‹*singer, musician*›; **Radio ~:** listen to the radio; **den Angeklagten/Zeugen ~:** hear the accused/witness; **das läßt sich ~:** that's good news; **c)** *(erfahren)* hear; **ich habe gehört, daß** I hear that; **etw. von jmdm. ~:** hear sth. from sb.; **er läßt nichts von sich ~:** I/we *etc.* haven't heard from him; **laß mal etwas von dir ~!** keep in touch; **von jmdm. etwas zu ~ bekommen** *od. (ugs.)* **kriegen** get a good talking-to from sb. *(coll.)*; **d)** *(erkennen)* **an etw.** *(Dat.)* **~, daß ...:** hear *or* tell by sth. that ... **2.** *itr. V.* **a)** hear; **gut ~:** have good hearing; **schlecht ~:** have bad hearing; be hard of hearing; **nur auf einem Ohr ~:** be deaf in one ear; **b)** *(aufmerksam verfolgen)* **auf etw.** *(Akk.)* **~:** listen to sth.; **c)** *(zuhören)* listen; **ich höre** I'm listening; **hörst du!** listen [here]!; **hörst du?** are you listening?; **man höre und staune** would you believe it!; wonders will never cease *(iron.)*; **hör mal!/~ Sie mal!** listen [here]!; **d) auf jmdn./jmds. Rat ~:** listen to *or* heed sb./sb.'s advice; **auf den Namen Monika ~:** answer to the name [of] Monika; **alles hört auf mein Kommando!** *(Milit.)* I'm taking command; *(scherzh.)* everyone do as I say; **e)** *(Kenntnis erhalten)* **von jmdm./etw. ~:** hear of sb./sth.; **von jmdm. ~** *(Nachricht bekommen)* hear from sb.; **Sie hören noch von mir** you'll be hearing from me again; you haven't heard the last of this; **f)** *(ugs.: gehorchen)* do as one is told; **wer nicht ~ will, muß fühlen** *(Spr.)* if you don't do as you're told, you'll suffer for it

Hören·sagen das; ~s *in* **vom ~:** by *or* from hearsay

Hörer der; ~s, ~ a) listener; **b)** *(Telefon~)* receiver

Hörer·brief der listener's letter

Hörerin die; ~, ~nen listener

Hör-: ~fehler der a) das war ein ~fehler he/she *etc.* misheard; **b)** *(Schwerhörigkeit)* hearing defect; **~folge die** radio series; *(in Fortsetzungen)* radio serial; **~funk der** radio; **im ~funk** on the radio; **~gerät das** hearing-aid

hörig *Adj.* **a)** *in* **jmdm. ~ sein** be submissively dependent on sb.; *(sexuell)* be sexually dependent on *or* enslaved to sb.; be sb.'s sexual slave; **b)** *(hist.)* **die ~en Bauern** the serfs; **~ sein** be in bondage

Hörige der/die; *adj. Dekl. (hist.)* serf; bondsman/bondswoman

Hörigkeit die; ~ **a)** enslavement; *(sexuell)* sexual dependence; **b)** *(hist.)* bondage; serfdom
Horizont [hori'tsɔnt] **der; ~[e]s, ~e** *(auch fig.)* horizon; **am ~:** on the horizon; **seinen ~ erweitern** *(fig.)* widen *or* expand one's horizons *pl.*; **hinter dem ~:** below the horizon; **über jmds. ~** *(Akk.)* **gehen** *(fig.)* be beyond sb.; go over sb.'s head
horizontal [horitsɔn'ta:l] **1.** *Adj.* horizontal. **2.** *adv.* horizontally
Horizontale die; ~, ~n a) *(Linie)* horizontal line; **b)** *o. Pl.* *(Lage)* **die ~:** the horizontal; **sich in die ~ begeben** *(scherzh.)* lie down
Hormon [hɔr'mo:n] **das; ~s, ~e** hormone
hormonal [hɔrmo'na:l] **1.** *Adj.* hormonal. **2.** *adv.* hormonally
Hör·muschel die ear-piece
Horn [hɔrn] **das; ~[e]s, Hörner** ['hœrnɐ] **a)** horn; **jmdm. Hörner aufsetzen** *(fig. ugs.)* cuckold sb.; **sich** *(Dat.)* **die Hörner abstoßen** *(fig.)* sow one's wild oats; **b)** *(Blasinstrument)* horn; *(Milit.)* bugle; **ins gleiche ~ stoßen** *(fig.)* take the same line; **c)** *o. Pl.* *(Substanz)* horn; **d)** *(Signal~)* *(eines Autos usw.)* horn; hooter *(Brit.)*; *(eines Zuges)* horn
Hornberger ['hɔrnbɛrgɐ] *in* **wie das ~ Schießen ausgehen** all come to nothing
Horn·brille die horn-rimmed spectacles *pl. or* glasses *pl.*
Hörnchen ['hœrnçən] **das; ~s, ~ a)** small *or* little horn; **b)** *(Gebäck)* croissant
Horn·haut die a) callus; hard *or* callused skin *no indef. art.*; **b)** *(am Auge)* cornea
Hornisse [hɔr'nɪsə] **die; ~, ~n** hornet
Hornist der; ~en, ~en a) horn player; **b)** *(Milit.)* bugler
Horn-: ~kamm der horn comb; **~ochse der** *(ugs.)* stupid ass
Horoskop [horo'sko:p] **das; ~s, ~e** horoscope
horrend [hɔ'rɛnt] *Adj.* shocking *(coll.)*, horrendous *(coll.)* ⟨*price*⟩; colossal *(coll.)* ⟨*sum, amount, rent*⟩
Hör·rohr das a) *(Stethoskop)* stethoscope; **b)** *(Hörgerät)* ear-trumpet
Horror ['hɔrɔr] **der; ~s** horror; **einen ~ vor jmdm./etw. haben** loathe and fear sb./have a horror of sth.
Horror·film der horror film
Hör·saal der a) lecture theatre *or* hall *or* room; **b)** *o. Pl.* *(Zuhörerschaft)* audience
Hör·spiel das radio play
Horst [hɔrst] **der; ~[e]s, ~e** *(Nest)* eyrie
Hort [hɔrt] **der; ~[e]s, ~e a)** *(dichter.: Goldschatz)* hoard [of gold]; **b)** *(geh.: sicherer Ort)* refuge; sanctuary; **ein ~ der Freiheit** a stronghold *or* bulwark of liberty; **c)** *s.* **Kinderhort**
horten *tr. V.* hoard; stockpile ⟨*raw materials*⟩
Hortensie [hɔr'tɛnziə] **die; ~, ~n** hydrangea
Hör-: ~test der hearing test; **~vermögen das;** *o. Pl.* hearing; **~weite die** hearing range; **in/außer ~weite** in/out of hearing range *or* of earshot
Höschen ['hø:sçən] **das; ~s, ~ a)** trousers *pl.*; pair of trousers; *(kurze Hose)* short trousers *pl.*; shorts *pl.*; pair of shorts; **b)** *(Slip)* panties *pl.*; pair of panties
Hose ['ho:zə] **die; ~, ~n a)** trousers *pl.*; pants *pl.* *(Amer.)*; *(Unter~)* pants *pl.*; *(Freizeit~)* slacks *pl.*; *(Bund~)* breeches *pl.*; *(Reit~)* jodhpurs *pl.*; riding breeches *pl.*; **eine ~:** a pair of trousers/pants/slacks *etc.*; **eine kurze/lange ~:** [a pair of] short trousers *or* shorts/long trousers; **ein/zwei Paar ~n** one/two pairs of trousers; **das Kind hat in die ~[n] gemacht** the child has made a mess in its pants; **b)** *(fig.)* **die ~n anhaben** *(ugs.)* wear the trousers; **jmdm. die ~n strammziehen** *(ugs.)* give sb. a good hiding *(coll.)*; **sich [vor Angst] in die ~[n] machen** *(salopp)* shit oneself *(coarse)*; get into a blue funk *(sl.)*; **es ist tote ~** *(Jugendspr.)* there's nothing doing *(coll.)*
Hosen-: ~an·zug der trouser suit *(Brit.)*; pant suit; **~aufschlag der** [trouser *or* *(Amer.)* pants] turn-up; **~bein das** trouser-leg; pants leg *(Amer.)*; **~boden der** seat of the/one's/sb.'s trousers *or* *(Amer.)* pants; **sich auf den ~boden setzen** *(fig.)* knuckle down to it; **jmdm. den ~boden strammziehen** *(fig. ugs.)* give sb. a good hiding *(coll.)*; **~bügel der** trouser-hanger; **~bund der** waistband; **~klammer die** bicycle-clip; **~knopf der** trouser-button; pants button *(Amer.)*; **~matz der** *(ugs. scherzh.)* toddler; [tiny] tot; **~rock der** culottes *pl.*; divided skirt; **~schlitz der** fly; flies *pl.*; **~tasche die** trouser-pocket; pants pocket *(Amer.)*; **etw. wie seine ~tasche kennen** *(fig. ugs.)* know sth. like the back of one's hand; **~träger** *Pl.* braces; suspenders *(Amer.)*; pair of braces/suspenders
hosianna [ho'ziana] *Interj.* *(christl. Rel.)* hosanna
Hospital [hɔspi'ta:l] **das; ~s, ~e** *od.* **Hospitäler** [hɔspi'tɛ:lɐ] hospital
hospitieren *itr. V.* **bei jmdm. ~:** sit in on sb.'s lectures/seminars; **in einem Seminar/einer Vorlesung ~:** sit in on a seminar/lecture
Hostess, Hosteß [hɔs'tɛs] **die; ~, Hostessen** hostess
Hostie ['hɔstiə] **die; ~, ~n** *(christl. Rel.)* host
Hotel [ho'tɛl] **das; ~s, ~s** hotel
Hotel-: ~boy der page[-boy]; bellboy *(Amer.)*; **~führer der** hotel guide
Hotel garni [~ gar'ni:] **das; ~ ~, ~s ~s** bed-and-breakfast hotel
Hotel·halle die hotel lobby
Hotelier [hotɛ'lie:] **der; ~s, ~s** hotelier
Hotel-: ~page der *s.* **~boy; ~portier der** [hotel] commissionaire; **~zimmer das** hotel room
hott [hɔt] *Interj.* gee[-up]; *s. auch* **hü**
hrsg. *Abk.* **herausgegeben** ed.
Hrsg. *Abk.* **Herausgeber** ed.
hu [hu:] *Interj.* **a)** ugh; **b)** *(bei Kälte)* brrr; **c)** *(zum Erschrecken)* boo
hü [hy:] *Interj.* **a)** *(vorwärts)* giddap; gee[-up]; **b)** *(halt)* whoa; **einmal sagt sie ~ und einmal hott** *(fig. ugs.)* first she says one thing, then another
Hub [hu:p] **der; ~[e]s, Hübe** ['hy:bə] *(Technik: Weg des Kolbens)* stroke
Hubbel ['hʊbl̩] **der; ~s, ~** *(bes. südd.)* bump
hubbelig *Adj.* *(bes. südd.)* bumpy
Hub·brücke die lift bridge
hüben ['hy:bn̩] *Adv.* on this side; over here; **~ und drüben** on both sides
Hub·raum der *(Technik)* cubic capacity
hübsch [hʏpʃ] **1.** *Adj.* **a)** pretty; nice-looking ⟨*boy, person*⟩; nice, pleasant ⟨*area, flat, voice, tune, etc.*⟩; nice ⟨*phrase, idea, present*⟩; **sich ~ machen** make oneself look nice; **b)** *nicht präd.* *(ugs.: ziemlich groß)* **eine ~e Stange Geld kosten** cost a pretty penny; **ein ~es Sümmchen** a tidy sum *(coll.)*; a nice little sum; **ein ~es Stück Arbeit** a fair amount *or* quite a lot of work; **c)** *(ugs. iron.: unangenehm)* **das ist eine ~e Geschichte/hier herrschen ~e Zustände** this is a fine *or* pretty kettle of fish *(coll.)*

or a fine state of affairs. **2.** *adv.* **a)** prettily; **sich ~ anziehen** dress nicely; wear nice clothes; **~ eingerichtet/gekleidet** nicely *or* attractively furnished/dressed; **b)** *(ugs.: sehr)* **[ganz] ~ kalt** perishing cold; **c)** *(ugs.: ordentlich)* **immer ~ der Reihe nach** everybody must take his turn; **sei ~ brav** be a good boy/girl

Hub·schrauber der; ~s, ~: helicopter

Hubschrauber·landeplatz der heliport; *(kleiner)* helicopter pad; landing pad

huch [hʊx] *Interj.* ugh; *(bei Kälte)* brrr

Hucke ['hʊkə] **die; ~, ~n** pannier; **jmdm. die ~ voll hauen** *(fig. ugs.)* give sb. a good hiding *(coll.)*; *(bei einer Prügelei)* beat hell out of sb. *(coll.)*; **jmdm. die ~ voll lügen** *(fig. ugs.)* tell sb. a pack of lies; **die ~ voll kriegen** *(fig. ugs.)* get a good hiding *(coll.)*; *(bei einer Prügelei)* get a proper beating *(coll.)*

huckepack ['hʊkəpak] *Adv. (ugs.) in* **jmdn. ~ tragen** carry sb. piggyback; give sb. a piggyback; **etw. ~ tragen** carry sth. piggyback; **jmdn./etw. ~ nehmen** take sth. up on one's back

hudeln ['hu:dl̩n] *itr. V. (bes. südd., österr.)* work sloppily; be sloppy *or* slapdash (**bei** in); **nur nicht ~!** don't be in such a hurry!; take it easy!

Huf [hu:f] **der; ~[e]s, ~e** hoof; **einem Pferd die ~e beschlagen** shoe a horse

huf-, Huf-: ~eisen das horseshoe; **~eisenförmig** *Adj.* horseshoe-shaped; **~lattich der** coltsfoot; **~nagel der** horseshoe nail; **~schmied der** farrier; blacksmith

Hüft·bein das *(Anat.)* hip bone; innominate bone *(Anat.)*

Hüfte ['hʏftə] **die; ~, ~n** hip; **aus der ~ schießen** shoot from the hip

hüft-, Hüft-: ~gelenk das *(Anat.)* hip-joint; **~gürtel der, ~halter der** girdle; **~hoch** *Adj.* ≈ waist-high; **~hoch sein** be almost waist-high

Huf·tier das hoofed animal; ungulate *(Zool.)*

Hüft·weite die hip size

Hügel ['hy:gl̩] **der; ~s, ~:** hill; *(fig.)* heap; pile

Hügel·grab das *(Archäol.)* barrow; tumulus

hügelig *Adj.* hilly

Hügel·kette die chain *or* range of hills

Hugenotte [hugə'nɔtə] **der; ~n, ~n** Huguenot

Huhn [hu:n] **das; ~[e]s, Hühner** ['hy:nɐ] **a)** chicken; [domestic] fowl; *(Henne)* chicken; hen; **gebratenes ~:** roast chicken; **herumlaufen wie ein aufgescheuchtes ~** *(ugs.)* run about in a great panic *(coll.)*; **da lachen [ja] die Hühner** *(ugs.)* you/he *etc.* must be joking *(coll.)*; **ein blindes ~ findet auch mal ein Korn** *(Spr.)* anyone can have a stroke of luck once in a while; **mit den Hühnern zu Bett gehen/aufstehen** *(scherzh.)* go to bed early/get up with the lark; **b)** *(ugs.: Mensch)* **ein dummes/fideles ~:** a stupid twit *(Brit. sl.) or* an idiot/a cheerful sort *(coll.)*

Hühnchen ['hy:nçən] **das; ~s, ~:** little *or* small chicken; **mit jmdm. [noch] ein ~ zu rupfen haben** *(ugs.)* [still] have a bone to pick with sb.

Hühner-: ~auge das *(am Fuß)* corn; **jmdm. auf die ~augen treten** *(fig. ugs.)* tread on sb.'s corns *or* toes; **~brühe die** chicken broth; **~brust die a)** *(Med.)* chicken-breast; pigeon-breast; **b)** *(ugs.: flacher Brustkorb)* scrawny chest; **~dieb der** chicken-thief; **~ei das** hen's egg; **~farm die** chicken farm; **~frikassee das** chicken fricassee; fricassee of chicken; **~futter das** chicken-feed; **~habicht der** [northern] goshawk; **~hof der** chicken-run; **~klein das; ~s** trimmings *pl.* of chicken *(in stew etc.)*; **~leiter die** chicken-ladder; **~stall der** chicken-coop; hen-coop; **~suppe die** chicken soup

hui [hu̯i] *Interj.* whoosh; **außen ~ und innen pfui** *(von Dingen)* the outside's fine but inside it's a different story; *(von Personen)* he/she seems very nice on the surface, but underneath it's a different story

huldigen ['hʊldɪgn̩] *itr. V.* **a)** **jmdm. ~:** pay tribute to *or* honour sb.; **b)** *(geh.: anhängen)* **einem Grundsatz/einer Ansicht/Mode ~:** hold [devotedly] to a principle/a point of view/follow a fashion

Huldigung die; ~, ~en tribute; homage

Hülle ['hʏlə] **die; ~, ~n a)** cover; *(für Ausweis, Zeitkarte)* cover; holder; *(Schallplatten~)* cover; sleeve; **die sterbliche ~** *(geh. verhüll.)* the mortal remains *pl.*; **b)** *(ugs. scherzh.: Kleidung)* **seine** *od.* **die ~ fallen lassen** strip off [one's clothes]; **c) in ~ und Fülle** in abundance; in plenty

hüllen *tr. V. (geh.)* wrap; **jmdn./sich in etw.** *(Akk.)* **~:** wrap sb./oneself in sth.; **in Wolken** *(Akk.)* **gehüllt** *(fig.)* enveloped in clouds

hüllenlos *Adj.* **a)** *nicht präd. (unverhüllt)* plain; clear; **b)** *(scherzh.: nackt)* naked

Hülse ['hʏlzə] **die; ~, ~n a)** *(für Füllhalter, Thermometer, Patrone)* case; *(für Film)* [cassette] container; **b)** *(Bot.)* pod; hull

Hülsen·frucht die; *meist Pl.* **a)** fruit of a leguminous plant; **Hülsenfrüchte** pulse *sing.*; **b)** *(Pflanze)* legume; leguminous plant

human [hu'ma:n] **1.** *Adj.* **a)** humane; **b)** *(nachsichtig)* considerate; **c)** *(Med.)* human. **2.** *adv.* **a)** humanly; **b)** *(nachsichtig)* considerately

humanisieren *tr. V.* humanize

Humanisierung die; ~: humanization

Humanismus der; ~: humanism; *(Epoche)* Humanism *no art.*

Humanist der; ~en, ~en a) humanist; *(hist.)* Humanist; **b)** *(Altsprachler)* classical scholar

humanistisch *Adj.* **a)** humanist[ic]; *(hist.)* Humanist; **b)** *(altsprachlich)* classical; **ein ~es Gymnasium** *secondary school emphasizing classical languages*

humanitär [humani'tɛ:ɐ̯] *Adj.* humanitarian

Humanität [humani'tɛ:t] **die; ~:** respect for humanity

Human·medizin die human medicine *no art.*

Humbug ['hʊmbʊk] **der; ~s** *(ugs. abwertend)* humbug

Hummel ['hʊml̩] **die; ~, ~n** bumble-bee; humble-bee; **eine wilde ~** *(scherzh.)* a proper tomboy

Hummer ['hʊmɐ] **der; ~s, ~:** lobster

Hummer·krabbe die king prawn

Humor [hu'mo:ɐ̯] **der; ~s** humour; *(Sinn für ~)* sense of humour; **etw. mit ~ tragen/nehmen** bear/take sth. with a sense of humour *or* cheerfully; **den ~ nicht verlieren** remain good-humoured

humorig *Adj.* humorous

Humorist der; ~en, ~en a) humorist; **b)** *(Komiker)* comedian

humoristisch 1. *Adj.* humorous. **2.** *adv.* with humour

humor-, Humor-: ~los 1. *Adj.* humourless; **2.** *adv.* without humour; **~losigkeit die; ~:** humourlessness; lack of humour; **~voll 1.** *Adj.* humorous; **2.** *adv.* humorously; in a humorous way

humpeln ['hʊmpl̩n] *itr. V.* **a)** *auch mit sein* walk with *or* have a limp; **b)** *mit sein (sich ~d fortbewegen)* hobble; limp

Humpen ['hʊmpn̩] **der; ~s, ~:** tankard; [beer-]mug; *(aus Ton auch)* stein

Humus ['hu:mʊs] **der; ~:** humus

Hund [hʊnt] **der; ~es, ~e a)** dog; *(Jagd~)* hound; dog; **bekannt sein wie ein bunter ~** *(ugs.)* be a well-known figure; **bei diesem Wetter würde man keinen ~ vor die Tür schicken** I wouldn't turn a dog out in weather like this; **da liegt der ~ begraben** *(fig. ugs.) (Ursache)* that's what's causing it; *(Grund)* that's the real reason; **da wird der ~ in der Pfanne verrückt** *(salopp)* it's quite incredible; **~e, die bellen, beißen nicht** *(Spr.)* barking dogs seldom bite; **den letzten beißen die ~e** *(fig.)* late-comers must expect to be unlucky; **ein dicker ~** *(ugs.: grober Fehler)* a real bloomer *(Brit. sl.) or (sl.)* goof; **das ist ein dicker ~** *(ugs.: Frechheit)* that's a bit thick *(coll.)*; **wie ~ und Katze leben** *(ugs.)* lead a cat-and-dog life; **damit kannst du keinen ~ hinter dem Ofen hervorlocken** that won't tempt anybody; **auf den ~ kommen** *(ugs.)* go to the dogs *(coll.)*; **vor die ~e gehen** *(ugs.)* go to the dogs *(coll.)*; *(sterben)* die; kick the bucket *(sl.)*; **b)** *(salopp: Mann)* bloke *(Brit. coll.)*; *(abwertend)* bastard *(coll.)*; **so ein blöder ~!** [what a] stupid bastard!

Hündchen ['hʏntçən] **das; ~s, ~** *(kleiner Hund)* little dog; *(Koseform)* doggie *(coll.)*; *(junger Hund)* puppy; pup

hunde-, Hunde-: ~dreck der *(ugs.)* dog's mess *or* muck; **~elend** *Adj.; nicht attr.* [really] wretched *or* awful; **~futter das** dog food; **~halsband das** dog-collar; **~hütte die** *(auch fig. abwertend)* [dog-]kennel; **~kalt** *Adj.; nicht attr. (ugs.)* freezing cold; **~kälte die** *(ugs.)* freezing cold; **~kot der** *(geh.)* dog-dirt; **~kuchen der** dog-biscuit; **~leben das** *(ugs.)* dog's life; **~marke die** dog-licence disc; dog-tag; **~müde** *Adj.; nicht attr. (ugs.)* dog-tired; **~rasse die** breed of dog; **~rennen das** dog-racing; greyhound-racing

hundert ['hʊndɐt] *Kardinalz.* **a)** a *or* one hundred; **mehrere/einige ~ Menschen** several/a few hundred people; **b)** *(ugs.: viele)* hundreds of

Hundert das; ~s, ~e *od. (nach unbest. Zahlwörtern)* **~ a)** hundred; **ein halbes ~:** fifty; **fünf vom ~:** five per cent; **b)** *Pl. (große Anzahl)* **~e von Menschen** hundreds of people; **in die ~e gehen** *(ugs.)* run into hundreds

hundert·ein[s] *Kardinalz.* a *or* one hundred and one

Hunderter der; ~s, ~ a) *(ugs.)* hundred-mark/-dollar *etc.* note; **b)** *(Math.)* hundred

hunderterlei *Gattungsz.; indekl. (ugs.)* **a)** *attr.* a hundred and one different ⟨*answers, kinds, etc.*⟩; **b)** *attr. (viele)* a hundred and one; **c)** *subst.* a hundred and one different things; **d)** *subst. (vieles)* a hundred and one things

hundert·fach *Vervielfältigungsz.* hundredfold; **die ~e Menge/der ~e Preis** a hundred times the amount/price; **das Hundertfache** a hundred times as much; *s. auch* **achtfach**

Hundert·jahr·feier die centenary; centennial

hundert·jährig *Adj.* **a)** *(100 Jahre alt)* [one-]hundred-year-old; **b)** *(100 Jahre dauernd)* **nach ~em Kampf** after a hundred years of war; **der Hundertjährige Krieg** *(hist.)* the Hundred Years' War

hundert·mal *Adv.* a hundred times; **auch wenn du dich ~ beschwerst** *(ugs.)* however much *or* no matter how much you complain; *s. auch* **achtmal**

Hundert-: ~mark·schein der hundred-mark note; **~meter·hürden·lauf der** *(Leichtathletik)* hundred-metres hurdles *sing.*; **~meter·lauf der** *(Leichtathletik)* hundred metres *sing.*

hundert·prozentig 1. *Adj.* **a)** [one-]hundred per cent *attrib.*; **b)** *(ugs.: völlig)* a hundred per cent, complete, absolute ⟨*certainty, agreement, etc.*⟩. **2.** *adv. (ugs.)* **ich bin nicht ~ sicher** I'm not a hundred per cent sure; **etw. ~ wissen** know sth. for sure

Hundertschaft die; ~, ~en group of a hundred; **einige ~en Polizei** several hundred police

hundertst... ['hʊndɐtst...] *Ordinalz.* hundredth; **zum ~en Mal fragen** *(ugs.)* ask for the hundredth time; **vom Hundertsten ins Tausendste kommen** get carried away so that one subject just leads another

hundertstel ['hʊndɐtstl̩] *Bruchz.* hundredth; *s. auch* **achtel**

Hundertstel das (*schweiz. meist* **der**); **~s, ~:** hundredth

Hundertstel·sekunde die hundredth of a second

hundert·tausend *Kardinalz.* a *or* one hundred thousand

hundert·und·ein[s] *Kardinalz.* a *or* one hundred and one

hundert·zehn *Kardinalz.* a *or* one hundred and ten

Hunde-: ~scheiße die *(derb)* dog-shit *(coarse)*; **~schlitten der** dog-sledge; dog-sled *(Amer.)*; **~steuer die** dog-licence fee; **~wetter das** *(ugs.)* filthy *or (sl.)* lousy weather; **~zwinger der** dog run

Hündin ['hʏndɪn] **die; ~, ~nen** bitch

hündisch ['hʏndɪʃ] **1.** *Adj.* **a)** *(würdelos)* doglike, servile ⟨*obedience*⟩; doglike ⟨*devotion*⟩; fawning, abject ⟨*submissiveness*⟩; **b)** *(gemein)* mean; nasty. **2.** *adv.* **jmdm. ~ ergeben sein** have a doglike devotion to sb.

hunds-, Hunds-: ~erbärmlich *(ugs.)* **1.** *Adj.* **a)** [really] dreadful *(coll.)*; **b)** *(verabscheuenswürdig)* dirty *attrib.* ⟨*lie, coward*⟩; **2.** *adv.* **a)** *(sehr)* terribly *(coll.)*, dreadfully *(coll.)* ⟨*cold*⟩; **b)** *(sehr schlecht)* [really] abysmally *(coll.) or* dreadfully *(coll.)*; **~gemein** *(ugs.)* **1.** *Adj.* **a)** *(abwertend: überaus gemein)* really mean *or* shabby; dirty ⟨*liar*⟩; **b)** *(sehr stark)* terrible *(coll.)*, dreadful *(coll.)* ⟨*cold, weather, pain, etc.*⟩; **2.** *adv.* **a)** *(gemein)* ⟨*deceive, behave*⟩ really meanly *or* shabbily; **b)** *(sehr stark)* **das tut ~gemein weh** it hurts like hell *(coll.) or* terribly *(coll.)*; **~miserabel** *(salopp abwertend)* **1.** *Adj.* [really] lousy *(sl.) or* dreadful *(coll.)*; **2.** *adv.* ⟨*behave*⟩ [really] appallingly *(coll.) or* dreadfully *(coll.)*; **~tage** *Pl.* dog-days

Hüne ['hy:nə] **der; ~n, ~n** giant

Hünen·grab das megalithic tomb; *(Hügelgrab)* barrow; tumulus

hünenhaft *Adj.* gigantic ⟨*build, stature*⟩

Hunger ['hʊŋɐ] **der; ~s a) ~ bekommen/haben** get/be hungry; **ich habe ~ wie ein Bär** *od.* **Wolf** I'm so hungry I could eat a horse; **sein ~ war groß** he was very hungry; **~ auf etw.** *(Akk.)* **haben** fancy sth., feel like sth. *(coll.)*; **~ leiden** go hungry; starve; **vor ~ sterben** die of starvation *or* hunger; starve to death; **b)** *(Hungersnot)* famine; **c)** *(geh.: Verlangen)* hunger; *(nach Ruhm, Macht)* craving; thirst

Hunger-: ~gefühl das feeling of hunger; **~kur die** starvation diet; **~leider der** *(ugs. abwertend)* starving pauper; **~lohn der** *(abwertend)* starvation wage[s *pl.*]

hungern ['hʊŋɐn] **1.** *itr. V.* **a)** go hungry; starve; **b)** *(verlangen)* **nach etw.** ~: hunger *or* be hungry for sth.; *(nach Macht, Ruhm)* crave sth.; thirst for sth. **2.** *refl. V.* **sich zu Tode** ~: starve oneself to death; **sich schlank** ~: slim by going on a starvation diet

Hungers·not die famine

Hunger-: **~streik der** hunger-strike; **~tod der** death from starvation; **~tuch das** *in* **am ~tuch nagen** *(ugs. scherzh.)* be on the breadline

hungrig *Adj.* hungry; **das macht [einen]** ~: it makes you hungry *or* gives you an appetite; ~ **nach etw. sein** fancy sth.; feel like sth. *(coll.)*

Hunne ['hʊnə] **der; ~n, ~n** *(hist.)* Hun

Hupe ['hu:pə] **die; ~, ~n** horn

hupen *itr. V.* sound the *or* one's horn; **dreimal** ~: hoot three times; give three toots on the horn

hupfen ['hʊpfn̩] *itr. V.; mit sein (südd., österr.)* hop; **das ist gehupft wie gesprungen** *(ugs.)* it doesn't make any difference; it doesn't matter either way

hüpfen ['hʏpfn̩] *itr. V.; mit sein* hop; ⟨*child*⟩ skip; ⟨*ball*⟩ bounce; **Hüpfen spielen** play [at] hopscotch; **mein Herz hüpfte vor Freude** my heart leapt for joy; **das ist gehüpft wie gesprungen** *(ugs.)* *s.* **hupfen**

Hüpfer der; ~s, ~: skip; *(auf einem Bein)* hop

Hup·konzert das *(ugs. scherzh.)* chorus of hooting

Hürde ['hʏrdə] **die; ~, ~n a)** *(Leichtathletik, fig.)* hurdle; **eine ~ nehmen** clear a hurdle; *(fig.)* get over a hurdle

Hürden-: **~lauf der** *(Leichtathletik)* hurdling; *(Wettbewerb)* hurdles *pl.;* hurdle race; **~läufer der** *(Leichtathletik)* hurdler

Hure ['hu:rə] **die; ~, ~n** *(abwertend)* whore

huren *itr. V. (abwertend)* whore; fornicate

Huren·sohn der *(abwertend)* bastard *(coll.);* son of a bitch *(derog.)*

hurra [hʊ'ra:] *Interj.* hurray; hurrah; ~ **schreien** cheer; *s. auch* **hipp, hipp, hurra**

Hurra das; ~s, ~s cheer; **jmdn. mit ~ begrüßen** greet sb. with cheering *or* cheers *pl.*

Hurra·ruf der cheering; cheers *pl.*

Hurrikan ['hʌrɪkən] **der; ~s, ~s** hurricane

hurtig ['hʊrtɪç] **1.** *Adj.; nicht präd.* rapid. **2.** *adv.* quickly; ⟨*work*⟩ fast, quickly

husch [hʊʃ] *Interj.* quick; quickly; ~, ~! away with you!; be off with you!; *(zu einem Tier)* shoo!

huschen *itr. V.; mit sein* ⟨*person*⟩ flit, dart; ⟨*mouse, lizard, etc.*⟩ dart; ⟨*smile*⟩ flit; ⟨*light*⟩ flash; ⟨*shadow*⟩ slide *or* glide quickly

hüsteln ['hy:stl̩n] *itr. V.* cough slightly; give a slight cough

husten ['hu:stn̩] **1.** *itr. V.* **a)** cough; **b)** *(Husten haben)* have a cough; be coughing. **2.** *tr. V.* cough up ⟨*blood, phlegm*⟩; **jmdm. etwas ~** *(salopp spött.)* tell sb. where he/she can get off *(coll.)*

Husten der; ~s, ~: cough; ~ **haben** have a cough

Husten-: **~an·fall der** coughing-fit; fit of coughing; **~bonbon das** cough sweet *(Brit.);* cough-drop; **~reiz der** tickling in the throat; **~saft der** cough-syrup; cough mixture

[1]**Hut** [hu:t] **der; ~es, Hüte** ['hy:tə] **a)** hat; **den ~ ziehen** raise one's hat; **in ~ und Mantel** wearing one's hat and coat; with one's hat and coat on; **b)** *(fig.)* **da geht einem/mir der ~ hoch** *(ugs.)* it makes you/me mad *or* wild *(coll.);* **~ ab!** *(ugs.)* hats off to him/her *etc.;* I take my hat off to him/her *etc.;* **ein alter ~ sein** *(ugs.)* be old hat; **seinen ~ nehmen** *(ugs.)* pack one's bags and go; **vor jmdm./etw. den ~ ziehen** *(ugs.)* take off one's hat to sb./sth.; **das kann er sich** *(Dat.)* **an den ~ stecken** *(ugs. abwertend)* he can keep it *(coll.) or (sl.)* stick it; **mit etw. nichts am ~ haben** *(ugs.)* have nothing to do with sth.; **jmdm. eins auf den ~ geben** *(ugs.)* give sb. a dressing down *or (Brit. sl.)* rocket; **eins auf den ~ kriegen** *(ugs.)* get a dressing down *or (Brit. sl.)* rocket; **verschiedene Interessen/Personen unter einen ~ bringen** *(ugs.)* reconcile different interests/the interests of different people; **c)** *(Bot.: eines Pilzes)* cap

[2]**Hut die; ~:** *in* **auf der ~ sein** *(geh.)* be on one's guard

Hut-: **~ab·lage die** hat rack; **~band das** hat-band; *(eines Damenhutes)* hat-ribbon

hüten ['hy:tn̩] **1.** *tr. V.* look after; take care of; tend, keep watch over ⟨*sheep, cattle, etc.*⟩; **ein Geheimnis ~** *(fig.)* keep *or* guard a secret; *s. auch* **Bett a. 2.** *refl. V.* *(vorsehen)* be on one's guard; **sich vor jmdm./etw. ~:** be on one's guard against sb./sth.; **sich ~, etw. zu tun** take [good] care not to do sth.; **ich werde mich ~!** *(ugs.)* no fear! *(coll.);* not likely! *(coll.)*

Hüter der; ~s, ~, Hüterin die; ~, ~nen guardian; custodian

Hut-: **~feder die** hat feather; **~geschäft das** hat shop; hatter's [shop]; *(für Damen)* hat shop; milliner's (shop); **~größe die** hat size; size of hat; **~krempe die** [hat] brim; **~macher der, ~macherin die; ~, ~nen** hatter; hat maker; *(für Damen)* milliner; **~nadel die** hat-pin; **~schachtel die** hat-box; **~schnur die** *in* **das geht mir über die ~schnur** *(ugs.)* that's going too far

Hütte ['hʏtə] **die; ~, ~n a)** hut; *(Holz~)* cabin; hut; *(ärmliches Haus)* shack; hut; **b)** *(Eisen~)* iron [and steel] works *sing. or pl.;* *(Glas~)* glassworks *sing. or pl.;* *(Blei~)* lead works *sing. or pl.;* **c)** *(Jagd~)* [hunting-]lodge

Hütten-: **~käse der** cottage cheese; **~schuh der** slipper-sock

hutzelig *Adj. (ugs.)* wizened ⟨*person, face*⟩; shrivelled, dried-up ⟨*fruit*⟩

Hyäne ['hy̆ɛ:nə] **die; ~, ~n** hyena

Hyazinthe [hy̆a'tsɪntə] **die; ~, ~n** hyacinth

Hydrant [hy'drant] **der; ~en, ~en** hydrant

Hydrat [hy'dra:t] **das; ~[e]s, ~e** *(Chemie)* hydrate

Hydraulik [hy'draulɪk] **die; ~** *(Technik)* **a)** hydraulics *sing., no art.;* **b)** *(System)* hydraulics *pl.;* hydraulic system

hydraulisch *(Technik)* **1.** *Adj.* hydraulic. **2.** *adv.* hydraulically

Hydrid [hy'dri:t] **das; ~[e]s, ~e** *(Chemie)* hydride

hydrieren [hy'dri:rən] *tr. V. (Chemie)* hydrogenate

hydro-, Hydro- [hydro-]: **~kultur die; ~, ~en** *(Gartenbau)* hydroponics *sing.;* **~lyse** [~'ly:zə] **die; ~, ~n** *(Chemie)* hydrolysis; **~pneumatisch** *Adj. (Technik)* hydropneumatic

Hygiene [hy'gi̯e:nə] **die; ~:** hygiene

hygienisch **1.** *Adj.* hygienic. **2.** *adv.* hygienically

Hymen ['hy:mən] **das** *od.* **der; ~s, ~** *(Anat.)* hymen

Hymne ['hʏmnə] **die; ~, ~n a)** hymn; **b)** *(Nationalhymne)* national anthem

hymnisch *Adj.* hymnic

Hyperbel [hy'pɛrbl̩] **die; ~, ~n** *(Geom.)* hyperbola

hyper-[hypɐ-]: **~korrekt** *(ugs. abwertend, Sprachw.)* **1.** *Adj.* hypercorrect; **2.** *adv.* in a hyper-

correct way; ~**modern 1.** *Adj.* ultra-modern; ultra-fashionable ⟨*clothes*⟩; **2.** *adv.* ultra-modernly; ⟨*dress*⟩ ultra-fashionably; ~**sensibel** *Adj.* hypersensitive
Hypnose [hyp'no:zə] **die; ~, ~n** hypnosis; **jmdn. in ~ versetzen** put sb. under hypnosis; **unter ~ stehen** be under hypnosis
hypnotisch *Adj.* hypnotic; hypnotic, soporific ⟨*drug*⟩
Hypnotiseur [hypnoti'zø:ɐ̯] **der; ~s, ~e** hypnotist
hypnotisieren *tr. V.* hypnotize
Hypochonder [hypo'xɔndɐ] **der; ~s, ~:** hypochondriac
Hypochondrie die; ~, ~n *(Med.)* hypochondria *no art.*
hypochondrisch *Adj.* hypochondriac
Hypo- [hypo-]: ~**physe** [~'fy:zə] **die; ~, ~n** *(Anat.)* hypophysis; ~**tenuse** [~te'nu:zə] **die; ~, ~n** *(Math.)* hypotenuse
Hypothek [-'te:k] **die; ~, ~en a)** *(Bankw.)* mortgage; **eine ~ aufnehmen** take out a mortgage; **etw. mit einer ~ belasten** encumber sth. with a mortgage; mortgage sth.; **b)** *(fig.: Bürde)* burden
Hypotheken·zins der; *meist Pl.* mortgage interest
Hypo·these [-'te:zə] **die; ~, ~n** hypothesis
hypo·thetisch 1. *Adj.* hypothetical; **2.** *adv.* hypothetically
Hysterie [hyste'ri:] **die; ~, ~n** [-i:ən] hysteria
Hysteriker [hys'te:rikɐ] **der; ~s, ~, Hysterikerin die; ~, ~nen** hysterical person; hysteric
hysterisch 1. *Adj.* hysterical. **2.** *adv.* hysterically
Hz *Abk.* **Hertz** Hz

I

i, I [i:] **das; ~, ~:** i/I; **das Tüpfelchen auf dem i** *(fig.)* the final touch; *s. auch* **a, A**
i *Interj.* ugh; **i bewahre, i wo** *(ugs.)* [good] heavens, no!
i. A. *Abk.* **im Auftrag[e]** p.p.
iah ['i:'a:] *Interj.* hee-haw
iahen ['i:'a:ən] *itr. V.* hee-haw; bray
iberisch [i'be:rɪʃ] *Adj.* Iberian; **Iberische Halbinsel** Iberian Peninsula
Ibis ['i:bɪs] **der; ~ses, ~se** *(Zool.)* ibis
IC *Abk.* **Intercity** IC
ich [ɪç] *Personalpron.; 1. Pers. Sg. Nom.* I; **Wer ist da? – Ich bin's!** Who's there? – It's me!; **Wer hat das gemacht? – Ich war's** Who did that? – I did *or* It was me; **Hat sie mich gerufen? – Nein, ~:** Was it she who called me? – No, I did; **und ~ Esel/Idiot habe es gemacht** and I, silly ass/idiot that I am, did it; and, like a fool, I did it; **immer ~** *(ugs.)* [it's] always me; **~ selbst** I myself; **~ nicht** not me; **Menschen wie du und ~:** people like you and I *or* me; *s. auch (Gen.)* **meiner,** *(Dat.)* **mir,** *(Akk.)* **mich**
Ich das; ~[s], ~[s] a) self; **das eigene ~:** one's own self; **b)** *(Psych.)* ego
ich-, Ich-: ~**bezogen 1.** *Adj.* egocentric; **2.** *adv.* **~bezogen denken** think in an egocentric way; ~**bezogenheit die; ~:** egocentricity; ~**-Form die;** *o. Pl.* first person; ~**-Laut der** *(Sprachw.)* palatal fricative; ich-laut
Ichthyo·saurier [ɪçtyo-] **der** ichthyosaurus
ideal [ide'a:l] **1.** *Adj.* ideal. **2.** *adv.* ideally
Ideal das; ~s, ~e ideal
Ideal-: ~**bild das** ideal; ~**fall der** ideal case; **im ~fall** in ideal circumstances *pl.;* ~**figur die** ideal figure; ~**gewicht das** ideal weight
idealisieren *tr. V.* idealize
Idealismus der; ~: idealism
Idealist der; ~en, ~en idealist
idealistisch 1. *Adj.* idealistic. **2.** *adv.* idealistically
Ideal·vorstellung die ideal
Idee [i'de:] **die; ~, ~n a)** idea; **du hast [vielleicht] ~n** *(iron.)* you do get some ideas, don't you!; **auf eine ~ kommen** hit [up]on an idea; **jmdn. auf eine ~ bringen** give sb. an idea; **eine fixe ~:** an obsession; an idée fixe; **b)** *(ein bißchen)* **eine ~:** a shade *or* trifle; **eine ~ Salz** a touch of salt
ideell [ide'ɛl] **1.** *Adj.* non-material; *(geistig-seelisch)* spiritual. **2.** *adv.* **etw. ~ unterstützen** support sth. in non-material ways
ideen·reich *Adj.* full of ideas *postpos.;* inventive
Identifikation [idɛntifika'tsi̯o:n] **die; ~, ~en** identification
identifizierbar *Adj.* identifiable
identifizieren [idɛntifi'tsi:rən] **1.** *tr. V.* identify. **2.** *refl. V.* **sich mit jmdm./etw. ~:** identify with sb./sth.
Identifizierung die; ~, ~en identification
identisch [i'dɛntɪʃ] *Adj.* identical
Identität [idɛnti'tɛ:t] **die; ~:** identity
Identitäts-: ~**krise die** identity crisis; ~**verlust der** loss of identity
Ideologe [ideo'lo:gə] **der; ~n, ~n** ideologue
Ideologie die; ~, ~n [-i:ən] ideology
ideologisch 1. *Adj.* ideological. **2.** *adv.* ideologically; **jmdn. ~ schulen** give sb. ideological instruction
Idiom [i'di̯o:m] **das; ~s, ~e** *(Sprachw.)* idiom
idiomatisch 1. *Adj.* idiomatic. **2.** *adv.* idiomatically
Idiot [i'di̯o:t] **der; ~en, ~en a)** idiot; **b)** *(ugs. abwertend)* fool; *(stärker)* idiot *(coll.)*
Idioten-: ~**hang der,** ~**hügel der** *(ugs. scherzh.)* nursery slope
idioten·sicher *Adj. (ugs. scherzh.)* foolproof
Idiotie [idi̯o'ti:] **die; ~, ~n** [-i:ən] **a)** idiocy; **b)** *(ugs. abwertend: Dummheit)* lunacy; madness
Idiotin die; ~, ~nen a) idiot; **b)** *(ugs. abwertend)* fool; *(stärker)* idiot
idiotisch 1. *Adj.* **a)** severely subnormal; idiotic *(as tech. term);* **b)** *(ugs. abwertend: unsinnig)* stupid; *(stärker)* idiotic. **2.** *adv.* **a)** idiotically; **b)** *(ugs. abwertend: unsinnig)* stupidly; *(stärker)* idiotically
Idol [i'do:l] **das; ~s, ~e** idol
Idyll [i'dyl] **das; ~s, ~e** idyll
Idylle die; ~, ~n idyll
idyllisch 1. *Adj.* idyllic. **2.** *adv.* **~ gelegen** in an idyllic spot
IG *Abk.* **Industriegewerkschaft**
Igel ['i:gl̩] **der; ~s, ~:** hedgehog
igitt[igitt] [i'gɪt(i'gɪt)] *Interj.* ugh
Iglu ['i:glu] **der** *od.* **das; ~s, ~s** igloo
Ignorant der; ~en, ~en *(abwertend)* ignoramus
Ignoranz [ɪgno'rants] **die; ~** *(abwertend)* ignorance
ignorieren *tr. V.* ignore
ihm [i:m] *Dat. von* **er, es** *(bei Personen)* him; *(bei Dingen, Tieren)* it; *(bei männlichen Tieren)* him; it; **gib es ~:** give it to him; give him it; *(dem Tier)* give it to it/him; **~ geht es gut** he's well; **sie sah ~ ins Gesicht** she looked him in the

face; **Freunde von ~:** friends of his

ihn [i:n] *Akk. von* **er** *(bei Personen)* him; *(bei Dingen, Tieren)* it; *(bei männlichen Tieren)* him; it

ihnen ['i:nən] *Dat. von* **sie,** *Pl.* them; **gib es ~:** give it to them; give them it; **Freunde von ~:** friends of theirs; *s. auch* **ihm**

Ihnen *Dat. von* **Sie** *(Anrede)* you; **ich habe es ~ gegeben** I gave it to you; I gave you it; **geht es ~ gut?** are you well?; **Freunde von ~:** friends of yours; *s. auch* **ihm**

[1]**ihr** [i:ɐ̯] *Dat. von* **sie,** *Sg. (bei Personen)* her; *(bei Dingen, Tieren)* it; *(bei weiblichen Tieren)* her; it; *s. auch* **ihm**

[2]**ihr,** *(in Briefen)* **Ihr** *Personalpron.; 2. Pers. Pl. Nom. (Anrede an vertraute Personen)* you; **Ihr Lieben** *(im Brief)* dear all; *s. auch (Gen.)* **euer/Euer,** *(Dat., Akk.)* **euch/Euch**

[3]**ihr** *Possessivpron.* **a)** *(einer Person)* her; **Ihre Majestät** Her Majesty; **das Buch dort, ist das ~[e]s?** that book there, is it hers?; is that book hers?; **das ist nicht mein Mann, sondern ~er** that is not my husband, but hers; **der/die/das ~e** hers; **die ~en** hers; **die Ihren** her family; **b)** *(eines Tieres, einer Sache)* its; *(eines weiblichen Tieres)* her; its; **die Lok fährt glatt ~e 200 Sachen** *(ugs.)* the locomotive does a good 200 kilometres an hour; **c)** *(mehrerer Personen, Tiere, Sachen)* their; **der/die/das ~e** theirs; **die ~en** theirs; **die Ihren** their family; **sie haben das Ihre getan** they did their bit

Ihr *Possessivpron. (Anrede)* your; **~ Hans Meier** *(Briefschluß)* yours, Hans Meier; **das Buch dort, ist das ~[e]s?** that book there, is it yours?; **welcher Mantel ist ~er?** which coat is yours?; **der/die/das ~e** yours; **die ~en** yours; *s. auch* [3]**ihr**

ihrer ['i:rɐ] **a)** *Gen. von* **sie,** *Sg. (geh.)* **wir gedachten ~:** we remembered her; **b)** *Gen. von* **sie,** *Pl. (geh.)* **wir werden ~ gedenken** we will remember them; **es waren ~ zwölf** there were twelve of them

Ihrer *Gen. von* **Sie** *(Anrede) (geh.)* **wir werden ~ gedenken** we will remember you

ihrerseits [-zaits] *Adv.* **a)** *Sg. (von ihrer Seite)* on her part; *(auf ihrer Seite)* for her part; **b)** *Pl. (von ihrer Seite)* on their part; *(auf ihrer Seite)* for their part

Ihrerseits *Adv. (von Ihrer Seite)* on your part; *(auf Ihrer Seite)* for your part

ihres·gleichen *indekl. Pron.* **a)** *Sg.* people *pl.* like her; *(abwertend)* the likes of her; her sort *or* kind; **sie fühlt sich nur unter ~ wohl** she only feels at home among people like herself *or* her own kind; **b)** *Pl.* people like them; *(abwertend)* the likes of them; their sort *or* kind; **sie sollten unter ~ bleiben** they should stay among their own kind

Ihresgleichen *indekl. Pron.* people *pl.* like you; *(abwertend)* the likes of you; your sort *or* kind; **Sie sollten besser unter ~ bleiben** you should stay among your own kind

ihret·halben *(veralt.),* **ihret·wegen** *Adv.* **a)** *Sg. (wegen ihr)* because of her; on her account; *(für sie)* on her behalf; *(ihr zuliebe)* for her sake; **b)** *Pl. (wegen ihnen)* because of them; on their account; *(für sie)* on their behalf; *(ihnen zuliebe)* for their sake[s]

Ihrethalben *(veralt.),* **Ihretwegen** *Adv. (wegen Ihnen)* because of you; on your account; *(für Sie)* on your behalf; *(Ihnen zuliebe)* for your sake; *Pl.* for your sake[s]

ihret·willen *Adv. in* **um ~** *(Sg.)* for her sake; *(Pl.)* for their sake[s]

Ihret·willen *Adv. in* **um ~** *(Sg.)* for your sake; *(Pl.)* for your sake[s]

ihrige ['i:rɪgə] *Possessivpron. (geh. veralt.)* **a)** *Sg.* **der/die/das ~:** hers; **b)** *Pl.* **der/die/das ~:** theirs

Ihrige *Possessivpron. (Anrede) (geh. veralt.)* **der/die/das ~:** yours

Ikone [i'ko:nə] **die; ~, ~n** icon

illegal ['ɪlega:l] **1.** *Adj.* illegal. **2.** *adv.* illegally

Illegalität [ɪlegali'tɛ:t] **die; ~, ~en** illegality

Illumination [ɪlumina'tsio:n] **die; ~, ~en** illumination

illuminieren *tr. V.* illuminate

Illuminierung die; ~, ~en illumination

Illusion [ɪlu'zio:n] **die; ~, ~en** illusion; **sich** *(Dat.)* **~en machen** delude oneself

illusionär [ɪluzio'nɛ:ɐ̯] *Adj. (geh.)* illusory ⟨*conception, expectation, thing*⟩

Illusionist der; ~en, ~en a) *(geh.)* dreamer; **b)** *(Zauberkünstler)* illusionist

illusionistisch *Adj. (Kunstw.)* illusionistic

illusions·los *Adj.* [sober and] realistic; **~ sein** have no illusions

illusorisch [ɪlu'zo:rɪʃ] *Adj.* illusory

Illustration [ɪlʊstra'tsio:n] **die; ~, ~en** illustration

Illustrator [ɪlʊs'tra:tɔr] **der; ~s, ~en** [-'to:rən] illustrator

illustrieren *tr. V. (auch fig.)* illustrate; **eine illustrierte Zeitschrift** a magazine

Illustrierte die; *adj. Dekl.* magazine

Iltis ['ɪltɪs] **der; ~ses, ~se** polecat; *(Pelz)* fitch

im [ɪm] *Präp. + Art.* **a) = in dem; b)** *(räumlich)* in the; **im vierten Stock** on the fourth floor; **im Theater** at the theatre; **im Fernsehen** on television; **im Bett** in bed; **c)** *(zeitlich)* **im Mai/Januar** in May/January; **im Jahre 1648** in [the year] 1648; **im letzten Jahr** last year; **im Alter von 50 Jahren** at the age of 50; **d)** *(Verlauf)* **etw. im Sitzen tun** do sth. [while] sitting down; **im Kommen sein** be coming

Image ['ɪmɪtʃ] **das; ~[s], ~s** image

imaginär [imagi'nɛ:ɐ̯] *Adj. (geh., Math.)* imaginary

Imagination [imagina'tsio:n] **die; ~, ~en** *(geh.)* imagination

Imam [i'ma:m] **der; ~s, ~s** *od.* **~e** imam

Imbiß ['ɪmbɪs] **der; Imbisses, Imbisse a)** snack; **b)** *s.* **Imbißlokal**

Imbiß-: ~bude die *(ugs.)* ≈ hot-dog stall *or* stand; **~lokal das** café; **~stand der** *s.* **~bude; ~stube die** café

Imitation [imita'tsio:n] **die; ~, ~en** imitation

Imitator [imi'ta:tɔr] **der; ~s, ~en** [-ta'to:rən] imitator; mimic; *(im Kabarett usw.)* impressionist

imitieren *tr. V.* imitate

Imker ['ɪmkɐ] **der; ~s, ~:** beekeeper

Imkerei die; ~, ~en a) *o. Pl.* beekeeping *no art.;* **b)** *(Betrieb)* apiary

immanent [ɪma'nɛnt] *Adj. (geh.)* inherent; **einer Sache** *(Dat.)* **~ sein** be inherent in sth.

Immatrikulation [ɪmatrikula'tsio:n] **die; ~, ~en a)** *(an der Hochschule)* registration; **b)** *(schweiz.: eines Fahrzeugs)* registration

immatrikulieren 1. *tr. V.* **a)** *(an der Hochschule)* register; **b)** *(schweiz.)* register ⟨*vehicle*⟩. **2.** *refl. V. (an der Hochschule)* register

immens [ɪ'mɛns] **1.** *Adj.* immense. **2.** *adv.* immensely

immer ['ɪmɐ] *Adv.* **a)** always; **wie ~:** as always; as usual; **~ dieser Streit** you're/they're *etc.* always arguing; **~ diese Kinder!** these wretched children!; **schon ~:** always; **~ und ewig** for ever; *(jedes-*

mal) always; **auf** *od.* **für ~ [und ewig]** for ever [and ever]; **~ wieder** again and again; time and time again; **~ wieder von vorne anfangen** keep on starting from the beginning again; **~, wenn** every time that; whenever; **b) ~ dunkler/häufiger** darker and darker/more and more often; **~ mehr** more and more; **c)** *(ugs.: jeweils)* **es durften ~ zwei auf einmal eintreten** we/they were allowed in two at a time; **d) wo/wer/wann/wie [auch] ~:** wherever/whoever/whenever/however; **e)** *(verstärkend)* **~ noch, noch ~:** still; **f)** *(ugs.: bei Aufforderung)* **~ langsam!/mit der Ruhe!** take it easy!; **~ der Nase nach!** keep following your nose!

immer-, Immer-: ~fort *Adv.* all the time; constantly; **~grün** *Adj.* evergreen; **~grün das** periwinkle; **~hin** *Adv.* **a)** *(wenigstens)* at any rate; anyhow; at least; **er ist zwar nicht reich, aber ~hin!** he's not rich, it's true, but still; **b)** *(trotz allem)* nevertheless; all the same; **c)** *(schließlich)* after all; **~während 1.** *Adj.; nicht präd.* perpetual; **2.** *adv.* perpetually; **~zu** *Adv. (ugs.)* the whole time; all the time; constantly

Immigrant [ɪmi'grant] **der; ~en, ~en, Immigrantin die; ~, ~nen** immigrant

Immigration [ɪmigra'tsi̯o:n] **die; ~, ~en** immigration

immigrieren *itr. V.; mit sein* immigrate

Immobilie [ɪmo'bi:li̯ə] **die; ~, ~n** [piece of] property; **mit ~n handeln** deal in real estate *sing.*

Immobilien-: ~geschäft das real-estate business; **~makler der** estate agent *(Brit.)*; realtor *(Amer.)*; **~markt der** property market

Immortelle [ɪmɔr'tɛlə] **die; ~, ~n** everlasting [flower]; immortelle

immun [ɪ'mu:n] **a)** *(Med., fig.)* immune **(gegen** to); **b)** *(Rechtsspr.)* **~ sein** have *or* enjoy immunity

immunisieren *tr. V.* immunize **(gegen** against)

Immunisierung die; ~, ~en immunization **(gegen** against)

Immunität [ɪmuni'tɛ:t] **die; ~, ~en a)** *(Med.)* immunity **(gegen** to); **b)** *(Rechtsspr.)* immunity

Imperativ ['ɪmperati:f] **der; ~s, ~e** *(Sprachw.)* imperative

imperativisch *Adj. (Sprachw.)* imperative

Imperator [ɪmpe'ra:tɔr] **der; ~s, ~en** [-'to:rən] *(hist.)* **a)** *(Feldherr)* imperator; **b)** *(Kaiser)* emperor

Imperfekt ['ɪmpɛrfɛkt] **das; ~s, ~e** *(Sprachw.)* imperfect [tense]

Imperialismus [ɪmperi̯a'lɪsmus] **der; ~:** imperialism *no art.*

Imperialist der; ~en, ~en imperialist

imperialistisch *Adj.* imperialistic

Imperium [ɪm'pe:ri̯ʊm] **das; ~s, Imperien** *(auch fig.)* empire

impertinent [ɪmpɛrti'nɛnt] **1.** *Adj.* impertinent; impudent. **2.** *adv.* impertinently; impudently

Impertinenz [ɪmpɛrti'nɛnts] **die; ~, ~en** impertinence; impudence

Impetus ['ɪmpetʊs] **der; ~:** impetus; *(Schwung)* verve; zest

Impf·ausweis der vaccination certificate

impfen ['ɪmpfn̩] *tr. V.* vaccinate, inoculate; **sich ~ lassen** be vaccinated *or* inoculated

Impf-: ~paß der vaccination certificate; **~schutz der** protection given by vaccination; **~stoff der** vaccine

Impfung die; ~, ~en vaccination; inoculation

Impf·zeugnis das *s.* **~ausweis**

Implantat [ɪmplan'ta:t] **das; ~[e]s, ~e** *(Med.)* implant

implantieren *tr. V. (Med.)* implant; **jmdm. etw. implantieren** implant sth. in sb.

Implikation [ɪmplika'tsi̯o:n] **die; ~, ~en** *(geh.)* implication

implizieren [ɪmpli'tsi:rən] *tr. V. (geh.)* imply

implizit [ɪmpli'tsi:t] *(geh.)* **1.** *Adj.* implicit. **2.** *adv.* implicitly

imponieren [ɪmpo'ni:rən] *itr. V.* impress; **jmdm. durch etw./mit etw. ~:** impress sb. by sth.

imponierend 1. *Adj.* impressive. **2.** *adv.* impressively

Imponier·gehabe[n] das *(Verhaltensf.)* display

Import [ɪm'pɔrt] **der; ~[e]s, ~e** import; **den ~ erhöhen** increase imports; **eine Firma für ~ und Export** an import/export firm

Importeur [ɪmpɔr'tø:ɐ̯] **der; ~s, ~e** importer

importieren *tr., itr. V.* import

imposant [ɪmpo'zant] **1.** *Adj.* imposing; impressive ⟨*achievement*⟩. **2.** *adv.* imposingly

impotent ['ɪmpotɛnt] *Adj.* impotent

Impotenz ['ɪmpotɛnts] **die; ~:** impotence

imprägnieren [ɪmprɛ'gni:rən] *tr. V.* impregnate; *(wasserdicht machen)* waterproof

Imprägnierung die; ~, ~en *s.* **imprägnieren a, b:** impregnation; waterproofing

Impresario [ɪmpre'za:ri̯o] **der; ~s, ~s** *od.* **Impresari** *(veralt.)* impresario

Impressen *s.* **Impressum**

Impression [ɪmprɛ'si̯o:n] **die; ~, ~en** impression

Impressionismus der; ~: impressionism *no art.*

impressionistisch *Adj.* impressionistic

Impressum [ɪm'prɛsum] **das; ~s, Impressen** imprint

Improvisation [ɪmproviza'tsi̯o:n] **die; ~, ~en** improvisation

Improvisations·gabe die gift *or* talent for improvisation

improvisieren *tr., itr. V.* improvise; **über ein Thema ~** *(Musik)* improvise on a theme

Impuls [ɪm'pʊls] **der; ~es, ~e a)** *(Anstoß)* stimulus; **einer Sache** *(Dat.)* **neue ~e geben** give sth. fresh stimulus *sing. or* impetus *sing.*; **b)** *(innere Regung)* impulse; **einem ~ folgen** act on [an] impulse; **etw. aus einem ~ heraus tun** do sth. on impulse; **c)** *(Physik)* impulse; **d)** *(Elektrot.)* pulse

impulsiv [ɪmpʊl'zi:f] **1.** *Adj.* impulsive. **2.** *adv.* impulsively

Impulsivität [ɪmpʊlzivi'tɛ:t] **die; ~:** impulsiveness

imstande [ɪm'ʃtandə] *Adv.* **~ sein, etw. zu tun** *(fähig sein)* be able to do sth.; be capable of doing sth.; *(die Möglichkeit haben)* be in a position to do sth.; **zu etw. ~ sein** be capable of sth.; **er ist ~ und schiebt mir die Schuld in die Schuhe** he's [quite] capable of putting the blame on to me

¹in [ɪn] **1.** *Präp. mit Dat.* **a)** *(räumlich, fig.)* in; **er hat ~ Tübingen studiert** he studied at Tübingen; **~ Deutschland/der Schweiz** in Germany/Switzerland; **sind Sie schon mal ~ China gewesen?** have you ever been to China?; **~ der Schule/Kirche** at school/church; **~ der Schule/Kirche steht noch eine alte Orgel** there's still an old organ in the school/church; **~ einer Partei** in a party; **b)** *(zeitlich)* in; **~ zwei Tagen/einer Woche** in two days/a week; **~ diesem Sommer** this summer; **[gerade] ~ dem Moment, als er kam** the [very] moment he came; **~ diesem Jahr/Monat** this/that year/month; **c)** *(modal)* in; **~ Farbe/Schwarzweiß** in colour/black and white; **~ deutsch/englisch** in German/English; **~ Mathematik/Englisch** in mathematics/English; **sich ~ jmdm. täuschen** be wrong about sb.; **d)** *in* **er hat es ~ sich** *(ugs.)* he's got what it takes *(coll.)*; **der**

Schnaps/diese Übersetzung hat es ~ sich *(ugs.)* this schnapps packs a punch *(coll.)*/this translation is a tough one; **e)** *(Kaufmannsspr.)* **~ etw. handeln** deal in sth.; *s. auch* **im. 2.** *Präp. mit Akk.* **a)** *(räumlich, fig.)* into; **~ die Stadt/das Dorf** into town/the village; **~ die Schweiz** to Switzerland; **~ die Kirche/Schule gehen** go to church/school; **~ eine Partei eintreten** join a party; **b)** *(zeitlich)* into; **bis ~ den Herbst** into the autumn; **c)** *(fig.)* **~ die Millionen gehen** run into millions; **sich ~ jmdn. verlieben** fall in love with sb.; **~ etw. einwilligen** agree *or* consent to sth.; *s. auch* **ins**

²in *Adj. (ugs.)* **~ sein** be in

in·adäquat 1. *Adj. (geh.)* inadequate. **2.** *adv.* inadequately

in·akzeptabel *Adj. (geh.)* unacceptable

In·angriffnahme [-na:mə] **die; ~, ~n** *(Papierdt.)* commencement; *(eines Problems)* tackling

In·anspruchnahme die; ~, ~n a) *(Papierdt.)* use; **bei häufiger ~ der Versicherung** if frequent [insurance] claims are made; **b)** *(starke Belastung)* demands *pl.*; **die starke berufliche ~:** the heavy demands made on him/her by his/her job; **c)** *(von Maschinen, Material)* use; *(von Einrichtungen)* utilization

In·begriff der quintessence; **der ~ des Spießers** the epitome of the petit bourgeois; the quintessential petit bourgeois

inbegriffen *Adj.* included

In·besitznahme die; ~, ~n *(Papierdt.)* appropriation

In·betriebnahme die; ~, ~n, In·betriebsetzung [-zɛtsʊŋ] **die; ~, ~en** *(Papierdt.)* **a)** *(von [öffentlichen] Einrichtungen)* opening; **b)** *(von Maschinen)* bringing into service; **vor ~ der Maschine** before bringing the machine into service; **c)** *(eines Kraftwerks)* commissioning

In·brunst die; ~ *(geh.)* fervour; *(der Liebe)* ardour; **jmdn. mit ~ lieben** love sb. ardently

in·brünstig *(geh.)* **1.** *Adj.* fervent; ardent ⟨*love*⟩. **2.** *adv.* fervently; ⟨*love*⟩ ardently

in·dem *Konj.* **a)** *(während)* while; *(gerade als)* as; **b)** *(dadurch, daß)* **~ man etw. tut** by doing sth.

Inder ['ɪndɐ] **der; ~s, ~, Inderin die; ~, ~nen** Indian

in·des *(veralt.)*, **in·dessen** *Adv.* **a)** *(inzwischen)* meanwhile; in the mean time; **b)** *(jedoch)* however

Index ['ɪndɛks] **der; ~** *od.* **~es, ~e** *od.* **Indizes** ['ɪnditse:s] index; **der ~** *(kath. Kirche)* the Index

Indianer [ɪn'dia:nɐ] **der; ~s, ~:** [American] Indian

Indianer·häuptling der Indian chief

Indianerin die; ~, ~nen [American] Indian

Indianer·krapfen der *(österr.) s.* **Mohrenkopf b**

indianisch *Adj.* Indian

Indien ['ɪndiən] **(das); ~s** India

in·different *(geh.)* **1.** *Adj.* indifferent. **2.** *adv.* indifferently

indigniert [ɪndɪ'gni:ɐt] *Adj. (geh)* indignant

Indigo ['ɪndigo] **der** *od.* **das; ~s, ~s** indigo

Indikation [ɪndika'tsio:n] **die; ~, ~en a)** *(Med.)* indication; **b)** *(Rechtsw.)* **[medizinische/soziale/ethische] ~:** [medical/social/ethical] grounds *pl.* for abortion

Indikativ ['ɪndikati:f] **der; ~s, ~e** [-i:və] *(Sprachw.)* indicative [mood]

indikativisch *Adj. (Sprachw.)* indicative

Indikator [ɪndi'ka:tɔr] **der; ~s, ~en** [-ka'to:rən] indicator

Indio ['ɪndio] **der; ~s, ~s** *(Central/South American)* Indian

in·direkt 1. *Adj.* indirect; **~e Rede** *(Sprachw.)* indirect *or* reported speech. **2.** *adv.* indirectly; **einen Freistoß ~ ausführen** *(Sport)* take an indirect free kick

indisch ['ɪndɪʃ] *Adj.* Indian

in·diskret *Adj.* indiscreet

In·diskretion die; ~, ~en indiscretion

in·disponiert *Adj.* indisposed

Individualismus [ɪndividua'lɪsmʊs] **der; ~** individualism

Individualist der; ~en, ~en individualist

individualistisch *Adj.* individualistic

Individualität [ɪndividuali'tɛ:t] **die; ~** *(geh.)* individuality

individuell [ɪndivi'duɛl] **1.** *Adj.* **a)** individual; **b)** *nicht präd. (privat)* private ⟨*property, vehicle, etc.*⟩. **2.** *adv.* individually; **etw. ~ gestalten** give sth. one's own personal touch

Individuum [ɪndi'vi:duʊm] **das; ~s, Individuen** *(auch fachspr.)* individual

Indiz [ɪn'di:ts] **das; ~es, ~ien a)** *(Rechtsw.)* piece of circumstantial evidence; **~ien** circumstantial evidence *sing.*; **b)** *(geh.: Anzeichen)* sign **(für** of)

Indizien-: ~beweis der *(Rechtsw.)* circumstantial evidence *no pl., no art.*; **~prozeß der** *(Rechtsw.)* trial based on circumstantial evidence

indizieren [ɪndi'tsi:rən] *tr. V.* **a)** *(Med.)* indicate; **b)** *(kath. Kirche)* **ein Buch ~:** place a book on the Index

indo-, Indo- [ɪndo-]: **~china (das)** Indo-China; **~europäer** *Pl. s.* **~germanen; ~europäisch** *Adj. s.* **~germanisch; ~germanen** *Pl.* Indo-Europeans; **~germanisch** *Adj.* Indo-European; Indo-Germanic

Indoktrination [ɪndɔktrina'tsio:n] **die; ~, ~en** *(geh.)* indoctrination

indoktrinieren *tr. V. (geh.)* indoctrinate

Indonesien [ɪndo'ne:ziən] **(das); ~s** Indonesia

Indonesier der; ~s, ~, Indonesierin die; ~, ~nen Indonesian

indonesisch *Adj.* Indonesian

Induktion [ɪndʊk'tsio:n] **die; ~, ~en** *(fachspr.)* induction

Induktions-: ~spule die *(Elektrot.)* induction coil; **~strom der** *(Elektrot.)* induced current

industrialisieren [ɪndʊstriali'zi:rən] *tr. V.* industrialize

Industrialisierung die; ~: industrialization

Industrie [ɪndʊs'tri:] **die; ~, ~n** industry

Industrie- industrial ⟨*plant, product, area, etc.*⟩; ⟨*branch*⟩ of industry

industriell [ɪndʊstri'ɛl] **1.** *Adj.*; *nicht präd.* industrial; **die ~e Revolution** *(hist.)* the Industrial Revolution. **2.** *adv.* industrially

Industrielle der/die; *adj. Dekl.* industrialist

Industrie- und Handels·kammer die Chamber of Industry and Commerce

induzieren [ɪndu'tsi:rən] *tr. V.* *(fachspr.)* induce

in·einander *Adv.* **~ verliebt sein** be in love with each other *or* one another; **~ verschlungene Ornamente** intertwined decorations

ineinander-: ~|fließen *unr. itr. V.; mit sein* flow together; ⟨*dyes, colours*⟩ run into each other *or* one another; **~|fügen 1.** *tr. V.* fit into each other *or* one another; fit together; **2.** *refl. V.* fit into each other *or* one another; fit together; **~|greifen** *unr. itr. V.* mesh *or* engage [with each other *or* one another]; mesh together *(lit. or fig.)*; **~|passen** *itr. V.* fit into each other *or* one another; fit together; **~|schieben** *unr. tr., refl. V.* telescope

infam [ɪn'fa:m] *(geh.)* **1.** *Adj.* disgraceful. **2.** *adv.* disgracefully
Infamie [ɪnfa'mi:] **die; ~, ~n** *(geh.)* **a)** *o. Pl.* disgracefulness; **b)** *(Äußerung)* disgraceful remark; *(Handlung)* disgraceful action
Infanterie ['ɪnfant(ə)ri:] **die; ~, ~n** *(Milit.)* infantry
Infanterist der; ~en, ~en *(Milit.)* infantryman
infantil [ɪnfan'ti:l] *(Psych., Med., sonst abwertend)* **1.** *Adj.* infantile. **2.** *adv.* in an infantile way
Infarkt [ɪn'farkt] **der; ~[e]s, ~e** *(Med.)* infarct; infarction
Infekt [ɪn'fɛkt] **der; ~[e]s, ~e** *(Med.)* infection
Infektion [ɪnfɛk'tsi̯o:n] **die; ~, ~en** *(Med.)* infection
Infektions-: ~gefahr die *(Med.)* danger *or* risk of infection; **~herd der** *(Med.)* seat of the/an infection; **~krankheit die** *(Med.)* infectious disease
infektiös [ɪnfɛk'tsi̯ø:s] *Adj.* *(Med.)* infectious
infernalisch [ɪnfɛr'na:lɪʃ] *(geh.)* **1.** *Adj.* infernal. **2.** *adv.* infernally; **~ stinken** stink dreadfully
Inferno [ɪn'fɛrno] **das; ~s** *(geh.)* inferno
Infiltration [ɪnfɪltra'tsi̯o:n] **die; ~, ~en** infiltration
infiltrieren *tr. V.* infiltrate
in·finit *Adj. (Sprachw.)* infinite
Infinitesimal·rechnung [ɪnfinitezi'ma:l-] **die** *(Math.)* infinitesimal calculus
Infinitv ['ɪnfiniti:f] **der; ~s, ~e** [-ti:və] *(Sprachw.)* infinitive
infizieren [ɪnfi'tsi:rən] **1.** *tr. V.* *(auch fig.)* infect. **2.** *refl. V.* become *or* get infected; **sich bei jmdm. ~:** be infected by sb.; catch an infection from sb.
in flagranti [ɪn fla'granti] *Adv.* *(geh.)* in flagrante [delicto]
Inflation [ɪnfla'tsi̯o:n] **die; ~, ~en** *(Wirtsch.)* inflation; *(Zeit der ~)* period of inflation; **eine schleichende ~:** creeping inflation
inflationär [ɪnflatsi̯o'nɛ:ɐ̯] *Adj.* inflationary
[1]**Info** ['ɪnfo] **das; ~s, ~s** *(ugs.)* hand-out
[2]**Info die; ~, ~s** *(ugs.)* info *no pl., no indef. art. (coll.)*; **eine ~:** a piece of info; **weitere ~s** more info
in·folge 1. *Präp. + Gen.* as a result of; owing to. **2.** *Adv.* **~ von etw.** *(Dat.)* as a result of *or* owing to sth.
infolge·dessen *Adv.* consequently; as a result of this
Informant [ɪnfɔr'mant] **der; ~en, ~en** informant
Informatik [ɪnfɔr'ma:tɪk] **die; ~:** computer science *no art.*
Informatiker der; ~s, ~: computer scientist
Information [ɪnfɔrma'tsi̯o:n] **die; ~, ~en a)** information *no pl., no indef. art.* **(über** + *Akk.* about, on); **eine ~:** [a piece of] information; **zu Ihrer ~:** for your information; **nach neuesten ~en** according to the latest information; **nähere ~en erhalten Sie ...:** you can obtain more information ...; **b)** *(Büro)* information bureau; *(Stand)* information desk
Informations-: ~aus·tausch der exchange of information; **~büro das** information bureau *or* office; **~material das** informational literature; **~quelle die** source of information; **~stand der a)** information stand; **b)** *(Grad der Informiertheit)* **bei meinem jetzigen ~stand** with the information I have at present
informativ [ɪnfɔrma'ti:f] **1.** *Adj.* informative. **2.** *adv.* informatively
in·formell *Adj.* informal
informieren 1. *tr. V.* inform **(über** + *Akk.* about); **falsch/einseitig informiert sein** be misinformed/have biased information; **aus gut informierten Kreisen** from well-informed circles. **2.** *refl. V.* obtain information; **sich über etw.** *(Akk.)* **~:** inform oneself *or* find out about sth.
infra·rot ['ɪnfra-] *Adj.* *(Physik)* infra-red
Infra·rot das; ~s *(Physik)* infrared radiation
Infra·struktur die infrastructure
Infusion [ɪnfu'zi̯o:n] **die; ~, ~en** *(Med.)* infusion
Ing. *Abk.* **Ingenieur**
Ingenieur [ɪnʒe'ni̯ø:ɐ̯] **der; ~s, ~e** [qualified] engineer
Ingenieur·büro das firm of consulting engineers
Ingenieurin die; ~, ~nen [qualified] engineer
Ingredienz [ɪngre'di̯ɛnts] **die; ~, ~en;** *meist Pl.* ingredient
In·grimm der *(geh.)* inward rage *or* wrath
Ingwer ['ɪŋvɐ] **der; ~s, ~:** ginger
Inhaber ['ɪnha:bɐ] **der; ~s, ~, Inhaberin die; ~, ~nen a)** holder; **b)** *(Besitzer)* owner
inhaftieren [ɪnhaf'ti:rən] *tr. V.* take into custody; detain
Inhaftierte der/die; *adj. Dekl.* prisoner
Inhaftierung die; ~, ~en detention
Inhalation [ɪnhala'tsi̯o:n] **die; ~, ~en** *(Med.)* inhalation
inhalieren *tr. V.* inhale
In·halt der; ~[e]s, ~e a) contents *pl.*; **b)** *(das Dargestellte/geistiger Gehalt)* content; **ein Buch politischen ~s** a political book; **c)** *(Flächen~)* area; *(Raum~)* volume
inhaltlich 1. *Adj.* **die ~e Struktur des Dramas** the plot-structure of the drama; **an ~en Gesichtspunkten gemessen** from the point of view of content. **2.** *adv.* **~ ist der Aufsatz gut** the essay is good as regards content; **~ übereinstimmen** be the same in content
inhalts-, Inhalts-: ~angabe die summary [of contents]; synopsis; *(eines Films, Dramas)* [plot] summary; synopsis; **~los** *Adj.* lacking in content *postpos.*; meaningless ⟨*word, phrase*⟩; empty ⟨*life*⟩; **~verzeichnis das** table of contents; *(auf einem Paket)* list of contents; *(als Überschrift)* [table of] contents
in·human *Adj.* **a)** inhuman; **b)** *(rücksichtslos)* inhumane
In·humanität die; ~: inhumanity
Initiale die; ~, ~n initial [letter]
Initial·zündung die detonation
initiativ [initsi̯a'ti:f] *Adj.* **~ werden** take the initiative
Initiative die; ~, ~n a) initiative; **die ~ ergreifen** take the initiative; **auf jmds. ~** *(Akk.)* **[hin]** on sb.'s initiative; **~ entwickeln/entfalten** develop initiative; **nur der ~** *(Dat.)* **der Opposition ist es zu verdanken, daß ...:** it is only thanks to the Opposition that ...; **b)** *s.* **Bürgerinitiative**
Initiator [ini'tsi̯a:tɔr] **der; ~s, ~en, Initiatorin die; ~, ~nen** initiator; *(einer Organisation)* founder
initiieren [initsi'i:rən] *tr. V. (geh.)* initiate
Injektion [ɪnjɛk'tsi̯o:n] **die; ~, ~en** *(Med.)* injection
Injektions-: ~nadel die hypodermic needle; **~spritze die** hypodermic syringe
injizieren [ɪnji'tsi:rən] *tr. V.* *(Med.)* inject; **jmdm. etw. ~:** inject sb. with sth.
Inkarnation [ɪnkarna'tsi̯o:n] **die; ~, ~en** incarnation
inkl. *Abk.* **inklusive** incl.
inklusive [ɪnklu'zi:və] **1.** *Präp.* + *Gen. (bes. Kaufmannsspr.)* inclusive of; including; **der Preis versteht sich ~ der Verpackung** the price includes *or* is inclusive of packing; **wir bezahlten ~ Frühstück 40 DM** we paid 40 DM,

breakfast included *or* including breakfast. **2.** *Adv.* inclusive

inkognito [ɪn'kɔgnito] *Adv. (geh.)* incognito

Inkognito das; ~s, ~s incognito

in·kompetent *Adj.* incompetent

In·kompetenz die incompetence

in·konsequent 1. *Adj.* inconsistent. **2.** *adv.* inconsistently

In·konsequenz die inconsistency

in·korrekt 1. *Adj.* incorrect. **2.** *adv.* incorrectly

Inkorrektheit die; ~, ~en a) *o. Pl.* incorrectness; **b)** *(Fehler)* mistake

In·kraft·treten das; ~s: mit dem ~ des Gesetzes when the law comes/came into effect *or* force

Inkubation [ɪnkuba'tsi̯o:n] **die; ~, ~en** *(Med.)* incubation

Inkubations·zeit die *(Med.)* incubation period

In·land das; ~[e]s a) im ~: at home; **im ~ hergestellte Waren, Produktionen des ~es** home-produced goods; **im In- und Ausland** at home and abroad

Inländer ['ɪnlɛndɐ] **der; ~s, ~, Inländerin die; ~, ~nen** native citizen

inländisch *Adj.* domestic; internal, domestic ‹*trade, traffic*›; home, domestic ‹*market*›; home-produced, domestic ‹*goods*›

Inlands-: ~markt der home *or* domestic market; **~porto das** inland postage

Inlett ['ɪnlɛt] **das; ~[e]s, ~e** *od.* **~s** *(Stoff)* tick; ticking; *(Hülle)* tick

in·mitten 1. *Präp. + Gen. (geh.)* in the midst of; surrounded by. **2.** *Adv.* **~ von** in the midst of; surrounded by

inne- ['ɪnə-]: **~|haben** *unr. tr. V.* hold, occupy ‹*position*›; hold ‹*office*›; **die Führung/Leitung ~haben** be in charge; **~|halten** *unr. itr. V.* pause; **in** *od.* **mit etw. ~halten** stop sth. for a moment

innen ['ɪnən] *Adv.* **a)** inside; *(auf/an der Innenseite)* on the inside; **etw. von ~ nach außen kehren** turn sth. inside out; **die Leitung verlief von ~ nach außen** the cable ran from the inside to the outside; **~ und außen** inside and out[side]; **nach ~ aufgehen** open inwards; **etw. von ~ besichtigen/ansehen** look round/look at the inside of sth.; **von ~ heraus** from within; **b)** *(österr.: drinnen)* inside; *(im Haus)* indoors

innen-, Innen-: ~arbeiten *Pl.* interior work *sing.*; **~architekt der** interior designer; **~aufnahme die** *(Fot.)* indoor photo[graph]; *(Film)* indoor *or* interior shot; **~ausstattung die: [eine] ~ausstattung** decoration and furnishings; *(eines Autos)* [an] interior trim; **~bahn die** *(Sport)* inside lane; **~dienst der: ~dienst haben** be working in the office; ‹*policeman*› be on station duty; **im ~dienst tätig sein** work in the office; ‹*policeman*› do station duty; **~einrichtung die** furnishings *pl.*; **~hof der** inner courtyard; **~leben das;** *o. Pl.* **a)** [inner] thoughts and feelings *pl.*; **b)** *(oft scherzh.: Ausstattung)* inside; *(eines Hauses)* interior; *(eines Autos, Fernsehers usw.)* inner workings *pl.*; **~minister der** Minister of the Interior; ≈ Home Secretary *(Brit.)*; ≈ Secretary of the Interior *(Amer.)*; **~ministerium das** Ministry of the Interior; ≈ Home Office *(Brit.)*; ≈ Department of the Interior *(Amer.)*; **~politik die** *domestic politics sing.*; *(bestimmte)* domestic policy/policies *pl.*; **~politiker der** politician concerned with home affairs; **~politisch 1.** *Adj.* ‹*question*› relating to domestic policy; ‹*mistake*› in domestic policy; ‹*experience*› in home affairs; **eine ~politische Debatte** a debate on domestic policy; *s. auch* **außenpolitisch 1; 2.** *adv.* as regards domestic policy; **~politisch betrachtet** from the point of view of domestic policy; **~raum der a)** inner room; **b)** *o. Pl. (Platz im Innern)* room inside; **ein Auto/Haus mit großem ~raum** a car/house with a spacious interior; **~seite die** inside; **~spiegel der** rear-view mirror; **~stadt die** town centre; downtown *(Amer.)*; *(einer Großstadt)* city centre; **~tasche die** inside pocket; **~temperatur die** inside temperature; **~wand die** interior wall; **~winkel der** interior angle

inner... ['ɪnɐ...] *Adj.; nicht präd.* **a)** inner; internal ‹*organs, structure, stability, etc.*›; inside ‹*pocket, lane*›; **b)** *(inländisch)* internal

inner-: ~betrieblich 1. *Adj.* internal ‹*problem, question, regulation, agreement*›; **2.** *adv.* internally; **~deutsch** *Adj. (hist.)* ‹*trade, relations, border*› between the two German states

Innere ['ɪnərə] **das;** *adj. Dekl.; o. Pl.* **a)** inside; *(eines Gebäudes, Wagens, Schiffes)* interior; inside; *(eines Landes)* interior; **der Minister des Innern** the Minister of the Interior; **b)** *(Empfindung)* inner being; **in seinem tiefsten ~n** in his heart of hearts; deep [down] inside; **c)** *(Kern)* heart

Innereien [ɪnə'rai̯ən] *Pl.* entrails; *(Kochk.)* offal *sing.*

inner·halb 1. *Präp. + Gen.* **a)** *(räumlich; fig.)* within, inside; **~ der Familie/Partei** *(fig.)* within the family/party; **b)** *(zeitlich)* within; **~ einer Woche** within a week; **~ der Arbeitszeit** during *or* in working hours. **2.** *Adv.* **a)** *(räumlich; fig.)* **~ von** within; inside; **b)** *(zeitlich)* **~ von** within

innerlich 1. *Adj.* **a)** inner; *(von außen nicht erkennbar)* inward; **b)** *(im Körper)* internal ‹*use, effect*›. **2.** *adv.* **a)** inwardly; **~ lachen** laugh inwardly *or* to oneself; **b)** *(im Körper)* internally

inner-: ~parteilich *Adj.* **~parteiliche Auseinandersetzungen** internal [party] disputes; disputes within the party; **~parteiliche Diskussionen** discussions within the party; **~staatlich** *Adj.* internal; domestic

innerst... *Adj.; nicht präd.* inmost; innermost; **ihre ~e Überzeugung** her deepest *or* most profound conviction

Innerste das; *adj. Dekl.; o. Pl.* innermost being; **in meinem ~n** in my heart of hearts; deep [down] inside

innert ['ɪnɐt] *Präp. + Gen. od. Dat. (schweiz., österr.)* within

inne|wohnen *itr. V. (geh.)* **etw. wohnt jmdm./einer Sache ~:** sb./sth. possesses sth.

innig ['ɪnɪç] **1.** *Adj.* **a)** heartfelt, deep ‹*affection, sympathy*›; heartfelt, fervent ‹*wish*›; intimate ‹*relation, relationship, friendship*›; **b)** *(Chemie)* intimate. **2.** *adv.* ‹*hope*› fervently; ‹*love*› deeply, with all one's heart

Innigkeit die; ~: depth; *(einer Beziehung)* intimacy

inniglich ['ɪnɪklɪç] *Adj., adv. (geh.) s.* **innig 1 a, 2**

Innovation [ɪnova'tsi̯o:n] **die; ~, ~en** innovation

innovativ *Adj.* innovative

Innung ['ɪnʊŋ] **die; ~, ~en** [trade] guild; **die ganze ~ blamieren** *(ugs. scherzh.)* let the side down

in·offiziell 1. *Adj.* unofficial. **2.** *adv.* unofficially

in petto [ɪn 'pɛto] *in* **etw. ~ ~ haben** *(ugs.)* have sth. up one's sleeve

in puncto [ɪn 'pʊŋkto] **~ ~ Pünktlichkeit** *usw.* as regards punctuality *etc.*; where punctuality *etc.* is concerned

Input ['ɪnpʊt] **der** *od.* **das; ~s, ~s** *(fachspr.)* input

Inquisition [ɪnkvizi'tsi̯o:n] **die;** *~(hist.)* Inquisition

Inquisitor [ɪnkvi'zi:tɔr] **der; ~s, ~en** [-zi'to:rən] *(hist.)* inquisitor

ins [ɪns] *Präp.* + *Art.* **a)** = **in das; b) ~ Bett/Theater gehen** go to bed/the theatre; **er kam ~ Stottern** he began to stutter

Insasse ['ɪnzasə] **der; ~n, ~n, Insassin die; ~, ~nen a)** *(Fahrgast)* passenger; **die ~n eines Autos/Flugzeuges** the passengers in a car/an aircraft; **b)** *(Bewohner)* inmate

ins·besond[e]re *Adv.* *(Papierdt.)* *s.* **besonders 1**

In·schrift die inscription

Insekt [ɪn'zɛkt] **das; ~s, ~en** insect

insekten-, Insekten-: ~fressend *Adj.; nicht präd.* insectivorous; insect-eating; **~plage die** plague of insects; **~stich der** *(einer Wespe, Biene)* insect-sting; *(einer Mücke)* insect-bite

Insektizid [ɪnzɛkti'tsi:t] **das; ~s, ~e** *(fachspr.)* insecticide

Insel ['ɪnzl̩] **die; ~, ~n** *(auch fig.)* island; **die ~ Helgoland** the island of Heligoland; **die ~ Man** the Isle of Man

Insel-: ~gruppe die group of islands **~volk das** island race *or* people; **~welt die** islands *pl.*

Inserat [ɪnze'ra:t] **das; ~[e]s, ~e** advertisement *(in a newspaper);* **sich auf ein ~ melden** reply to an advertisement; **ein ~ aufgeben** put in an advertisement

Inserent [ɪnze'rɛnt] **der; ~en, ~en** advertiser

inserieren *itr., tr. V.* advertise; **[wegen etw.] in einer Zeitung ~:** advertise [sth.] in a newspaper

ins·geheim *Adv.* secretly

ins·gesamt *Adv.* **a)** in all; altogether; **es waren ~ 500** there were 500 in all *or* altogether; **b)** *(alles in allem)* all in all; **~ gesehen** all in all

Insider ['ɪnsaɪdə] **der; ~s, ~:** insider

Insigne [ɪn'zɪgnə] **das; ~s, Insignien;** *meist Pl.* insignia

insistieren [ɪnzɪs'ti:rən] *itr. V.* *(geh.)* insist **(auf** + *Dat.* on)

insofern 1. *Adv.* [ɪn'zo:fɛrn] *(in dieser Hinsicht)* in this respect; to this extent; **~, als** in so far as. **2.** *Konj.* [ɪnzo'fɛrn] *(falls)* provided [that]; so *or* as long as

insoweit [ɪn'zo:vai̯t/ɪnzo'vai̯t] *Adv./Konj. s.* **insofern**

in spe [ɪn 'spe:] **mein Schwiegersohn ~ ~:** my future son-in-law

Inspekteur [ɪnspɛk'tø:ɐ̯] **der; ~s, ~e** *(Milit.)* Chief of Staff

Inspektion [ɪnspɛk'tsi̯o:n] **die; ~, ~en a)** inspection; **b)** *(Kfz-W.)* service; **das Auto zur ~ bringen** take the car in for a service

Inspektor [ɪns'pɛktɔr] **der; ~s, ~en** [-'to:rən], **Inspektorin die; ~, ~nen** inspector; *(als Titel)* Inspector

Inspiration [ɪnspira'tsi̯o:n] **die; ~, ~en** inspiration

inspirieren *tr. V.* inspire; **das inspirierte ihn zu einem Roman** it inspired him to write a novel; **sich von jmdm./etw. ~ lassen** be inspired by sb./sth.

Inspizient [ɪnspi'tsi̯ɛnt] **der; ~en, ~en** *(Theater)* stage-manager; *(Ferns., Rundf.)* studio manager

inspizieren *tr. V.* inspect

Inspizierung die; ~, ~en inspection

in·stabil *Adj. (geh., fachspr.)* unstable

Installateur [ɪnstala'tø:ɐ̯] **der; ~s, ~e a)** *(Klempner)* plumber; **b)** *(Gas~)* [gas-]fitter; **c)** *(Heizungs~)* heating engineer; **d)** *(Elektro~)* electrician

Installation [ɪnstala'tsi̯o:n] **die; ~, ~en a)** *o. Pl.* installation; **b)** *(Anlage)* installation; **~en** *(installierte Rohre)* plumbing *no pl.*

installieren 1. *tr. V.* **a)** install; **b)** *(einrichten)* set up. **2.** *refl. V.* settle in

in·stand *Adv.* **etw. ist gut/ schlecht ~:** sth. is in good/poor condition; **etw. ~ halten** keep sth. in good condition *or* repair; *(funktionsfähig halten)* keep sth. in working order; **etw. ~ setzen/ bringen** repair sth.; *(funktionsfähig machen)* get sth. into working order

instand-, Instand-: ~besetzen *tr. V.* **ein Haus ~besetzen** occupy and renovate a house *(illegally, to prove that its demolition is not desirable);* **~besetzung die** *(illegal)* occupation and renovation; **~haltung die** maintenance; upkeep

in·ständig 1. *Adj.* urgent. **2.** *adv.* urgently; **~ um etw. bitten** beg for sth.; **jmdn. ~ bitten, etw. zu tun** beg *or* implore *or* beseech sb. to do sth.; **~ auf etw.** *(Akk.)* **hoffen** hope fervently for sth.

Instandsetzung die; ~, ~en *(Papierdt.)* repair

Instanz [ɪn'stants] **die; ~, ~en a)** authority; **durch alle ~en gehen** *od.* **alle ~en durchlaufen** go *or* pass through all the official channels; **b)** *(Rechtsw.)* **[die] erste/ zweite/letzte ~:** the court of first instance *or* of original jurisdiction/the appeal court/the court of final appeal; **durch alle ~en gehen** go through all the courts

Instinkt [ɪn'stɪŋkt] **der; ~[e]s, ~e** instinct; **einen ~ für etw. haben** have a flair for sth.; **seinem ~ folgen** follow one's instincts *pl.*

Instinkt·handlung die instinctive action

instinktiv [ɪnstɪŋk'ti:f] **1.** *Adj.* instinctive. **2.** *adv.* instinctively

instinkt·los 1. *Adj.* insensitive. **2.** *adv.* insensitively

Instinktlosigkeit die; ~, ~en insensitivity

Institut [ɪnsti'tu:t] **das; ~[e]s, ~e** institute

Institution [ɪnstitu'tsi̯o:n] **die; ~, ~en** *(auch fig.)* institution

institutionalisieren [ɪnstitutsi̯onali'zi:rən] *tr. V. (geh.)* institutionalize

institutionell [ɪnstitutsi̯o'nɛl] *Adj.* institutional

instruieren [ɪnstru'i:rən] *tr. V.* **a)** inform; **b)** *(anweisen)* instruct

Instruktion [ɪnstrʊk'tsi̯o:n] **die; ~, ~en** instruction

instruktiv [ɪnstrʊk'ti:f] **1.** *Adj.* instructive; informative. **2.** *adv.* instructively; informatively

Instrument [ɪnstru'mɛnt] **das; ~[e]s, ~e** instrument

instrumental [ɪnstrumɛn'ta:l] *(Musik)* **1.** *Adj.* instrumental. **2.** *adv.* instrumentally

Instrumental- instrumental ⟨*music, accompaniment, etc.*⟩

Instrumentarium [ɪnstrumɛn'ta:ri̯ʊm] **das; ~s, Instrumentarien a)** *(Technik)* equipment; instruments *pl.;* **b)** *(Musik)* instruments *pl.;* **c)** *(geh.: Gesamtheit der Mittel)* apparatus

Instrumenten-: ~brett das instrument panel; **~flug der** *(Flugw.)* flying on instruments; instrument-flying

instrumentieren *tr. V. (Musik)* instrument

Insuffizienz ['ɪnzʊfitsi̯ɛnts] **die; ~, ~en** *(Med.)* insufficiency

Insulin [ɪnzu'li:n] **das; ~s** insulin

inszenieren [ɪnstse'ni:rən] *tr. V.* **a)** stage, put on ⟨*play, opera*⟩; *(Regie führen bei)* direct; *(Ferns.)* direct; produce; **b)** *(oft abwertend) (einfädeln)* engineer; *(organisieren)* stage

Inszenierung die; ~, ~en a) staging; *(Regie)* direction; **b)** *(Produktion)* production; **c)** *(oft abwertend) (das Einfädeln)* engineering; *(das Organisieren)* staging

intakt [ɪn'takt] *Adj.* **a)** intact; **b)** *(funktionsfähig)* in [proper] working order *postpos.*

Intarsie [ɪn'tarziə] **die;** ~, ~**n** intarsia

integer [ɪn'te:gɐ] *Adj.* **eine integre Persönlichkeit** a person of integrity; ~ **sein** be a person of integrity

integral [ɪnte'gra:l] *Adj. (geh.)* integral

Integral das; ~**s,** ~**s** *(Math.)* integral

Integral-: ~**helm der** integral helmet; ~**rechnung die a)** integral calculus; **b)** *(einzelne Rechnung)* problem in integral calculus

Integration [ɪntegra'tsio:n] **die;** ~, ~**en** *(auch Math.)* integration

integrieren *tr. V. (auch Math.)* integrate

Integrierung die; ~, ~**en** *(auch Math.)* integration

Integrität [ɪntegri'tɛ:t] **die;** ~: integrity

Intellekt [ɪntɛ'lɛkt] **der;** ~**[e]s** intellect

intellektuell [ɪntɛlɛk'tuɛl] *Adj.* intellectual

Intellektuelle der/die; *adj. Dekl.* intellectual

intelligent [ɪntɛli'gɛnt] **1.** *Adj.* intelligent. **2.** *adv.* intelligently

Intelligenz [ɪntɛli'gɛnts] **die;** ~ **a)** intelligence; **b)** *(Gesamtheit der Intellektuellen)* intelligentsia

Intelligenz·bestie die *(ugs.)* egghead *(coll.)*; brain *(coll.)*

Intelligenz-: ~**quotient der** intelligence quotient; ~**test der** intelligence test

Intendant [ɪntɛn'dant] **der;** ~**en,** ~**en** manager and artistic director; *(Fernseh~, Rundfunk~)* director-general

Intendanz [ɪntɛn'dants] **die;** ~, ~**en a)** *(Amt)* management and artistic directorship; *(Ferns., Rundf.)* director-generalship; **b)** *(Büro)* office of the manager and artistic director; *(Ferns., Rundf.)* director-general's office

intendieren [ɪntɛn'di:rən] *tr. V. (geh.)* intend

Intensität [ɪntɛnzi'tɛ:t] **die;** ~: intensity

intensiv [ɪntɛn'zi:f] **1.** *Adj.* intensive ‹*research, efforts, cultivation, etc.*›; *(kräftig)* intense; strong ‹*smell, taste*›. **2.** *adv.* intensively; ‹*think*› hard; *(kräftig)* intensely; ‹*smell, taste*› strongly; **sich ~ mit etw. beschäftigen** be deeply involved with sth.

-intensiv ‹*time, labour, etc.*›-intensive

intensivieren *tr. V.* intensify

Intensivierung die; ~, ~**en** intensification

Intensiv-: ~**kurs der** intensive course; ~**station die** intensive-care unit

Intention [ɪntɛn'tsio:n] **die;** ~, ~**en** *(geh.)* intention

Inter·city der; ~**s,** ~**s** inter-city [train]

inter·disziplinär 1. *Adj.* interdisciplinary. **2.** *adv.* ~ **forschen** do interdisciplinary research

interessant [ɪntərɛ'sant] **1.** *Adj.* interesting; **sich ~ machen** attract attention to oneself. **2.** *adv.* ~ **schreiben** write in an interesting way

interessanterweise *Adv.* interestingly enough

Interesse [ɪntə'rɛsə] **das;** ~**s,** ~**n** interest; **[großes] ~ an jmdm./etw. haben** be [very] interested in sb./sth.; ~ **für jmdn./etw. haben/zeigen** have/show an interest in sb./sth.; **gemeinsame ~n haben** have interests in common; **im eigenen ~ handeln** act in one's own interest; **jmds. ~n wahrnehmen** look after *or* represent sb.'s interests; **in jmds. ~** *(Dat.)* **liegen** be in sb.'s interest

interesse-, Interesse-: ~**halber** [~halbɐ] *Adv.* out of interest; ~**los 1.** *Adj.* uninterested; **2.** *adv.* without interest; uninterestedly; ~**losigkeit die;** ~: lack of interest

Interessen-: ~**ausgleich der** reconciliation of [conflicting] interests; ~**gegensatz der** *s.* ~**konflikt;** ~**gruppe die** interest group; ~**konflikt der** conflict of interests; ~**lage die** interests *pl.*

Interessent [ɪntərɛ'sɛnt] **der;** ~**en,** ~**en, Interessentin die;** ~, ~**nen a)** interested person; **wenn es genug Interessenten gibt** if enough people are interested; **b)** *(möglicher Käufer)* potential buyer

Interessen·vertretung die a) representation; **b)** *(Gremium, Organisation)* representative body

interessieren 1. *refl. V.* **sich für jmdn./etw.** ~: be interested in sb./sth. **2.** *tr. V.* interest; **jmdn. für etw.** ~: interest sb. in sth.; **interessiert dich denn nicht, was passiert ist?** aren't you interested to know what happened?; **das interessiert mich nicht** I'm not interested [in it]; it doesn't interest me

interessiert *Adj.* interested; **an jmdm./etw. ~ sein** be interested in sb./sth.; **er ist daran ~, daß sie nichts davon erfahren** he doesn't want them to find out anything about it; **vielseitig ~ sein** have a wide range of interests; ~ **zuhören** listen with interest

Inter·feron [-fe'ro:n] **das;** ~**s,** ~**e** *(Med.)* interferon

Interim ['ɪnterɪm] **das;** ~**s,** ~**s** *(geh.)* interim measure

Interims·lösung die *(geh.)* interim solution

Interjektion [ɪntɐjɛk'tsio:n] **die;** ~, ~**en** *(Sprachw.)* interjection

inter-, Inter-: ~**kontinental** *Adj. (geh.)* intercontinental; ~**kontinental·rakete die** *(Milit.)* intercontinental ballistic missile; ~**mezzo** [~'mɛtso] **das;** ~**mezzos,** ~**mezzos** *od.* ~**mezzi** *(Theat., Musik)* intermezzo; *(fig.)* interlude; intermezzo

intern [ɪn'tɛrn] **1.** *Adj.* **a)** internal; **b)** *(im Internat wohnend)* **ein ~er Schüler** a boarder. **2.** *adv.* internally; **wir haben das Jubiläum nur ~ gefeiert** we only celebrated the anniversary among ourselves

internalisieren [ɪntɐnali'zi:rən] *tr. V. (Soziol., Psych.)* internalize

Internalisierung die; ~, ~**en** *(Soziol., Psych.)* internalization

Internat [ɪntə'na:t] **das;** ~**[e]s,** ~**e a)** boarding-school; **b)** *(einer Schule angeschlossenes Heim)* dormitory block

inter-, Inter-: ~**national 1.** *Adj.* international; **2.** *adv.* internationally; ~**nationale die;** ~, ~**n a)** *(Internationale Arbeiterassoziation)* International; Internationale; **b)** *(Lied)* Internationale; ~**nationalismus der** *o. Pl. (Politik)* internationalism

Internats-: ~**schule die** boarding-school; ~**schüler der,** ~**schülerin die** boarding-school pupil; boarder

internieren *tr. V. (Milit.)* intern

Internierte der/die; *adj. Dekl. (Milit.)* internee

Internierung die; ~, ~**en** internment

Internierungs·lager das internment camp

Internist der; ~**en,** ~**en** *(Med.)* internist

Interpol ['ɪntɐpo:l] **(die);** ~ Interpol

Interpolation [ɪntɐpola'tsio:n] **die;** ~, ~**en** *(fachspr.)* interpolation

interpolieren *itr., tr. V. (fachspr.)* interpolate

Interpret [ɪntɐ'prɛ:t] **der;** ~**en,** ~**en** interpreter

Interpretation [ɪntɐpreta'tsio:n] **die;** ~, ~**en** interpretation

interpretieren *tr. V.* interpret;

etw. falsch ~: misinterpret sth.; interpret sth. wrongly
Interpretin die; ~, ~nen *s.* **Interpret**
Interpunktion [ɪntɐpʊnk'tsi̯o:n] **die;** ~ *(Sprachw.)* punctuation
Interrail·karte ['ɪntəreɪl-] **die** *(Eisenbahnw.)* Interrail card
interrogativ [ɪntɐroga'ti:f] *Adj. (Sprachw.)* interrogative
Interrogativ·pronomen das interrogative pronoun
Intervall [ɪntɐ'val] **das; ~s, ~e** interval
intervenieren [ɪntɐve'ni:rən] *itr. V. (geh., Politik)* intervene; *(protestieren)* make representations **(bei** to)
Intervention [ɪntɐvɛn'tsi̯o:n] **die; ~, ~en** *(geh., Politik)* intervention; *(Protest)* representations *pl.*
Interventions·krieg der war of intervention
Interview [ɪntɐ'vju:] **das; ~s, ~s** interview
interviewen [ɪntɐ'vju:ən] *tr. V.* interview
Interviewer [ɪntɐ'vju:ɐ] **der; ~s, ~, Interviewerin die; ~, ~nen** interviewer
Inthronisation [ɪntroniza'tsi̯o:n] **die; ~, ~en** enthronement
inthronisieren *tr. V.* enthrone
intim [ɪn'ti:m] **1.** *Adj.* intimate; **mit jmdm. ~ sein/werden** *(verhüll.)* be/become intimate with sb. *(euphem.).* **2.** *adv.* **mit jmdm. ~ verkehren** *(verhüll.)* have intimate relations with sb. *(euphem.)*
Intim-: ~bereich der a) *s.* **~sphäre; b)** *(Genitalbereich)* genital area; **~hygiene die** intimate personal hygiene
Intimität [ɪntimi'tɛ:t] **die; ~, ~en** intimacy; **es ist zu ~en gekommen** *(verhüll.)* intimacy took place *(euphem.)*; **~en austauschen** *(verhüll.)* be engaged in intimacy *(euphem.)*
Intim-: ~leben das *(verhüll.)* intimate life; **~pflege die** *s.* **~hygiene; ~sphäre die** private life; **jmds. ~sphäre verletzen** invade sb.'s privacy; **~spray der** *od.* **das** intimate deodorant
Intimus ['ɪntimʊs] **der; ~, Intimi** intimate friend; *(Vertrauter)* confidant
Intim·verkehr der *(verhüll.)* intimate relations *pl. (euphem.)*
in·tolerant 1. *Adj.* intolerant **(gegenüber** of). **2.** *adv.* intolerantly
In·toleranz die; ~, ~en intolerance **(gegenüber** of)
Intonation [ɪntona'tsi̯o:n] **die; ~, ~en** *(Sprachw., Musik)* intonation
intonieren *tr. V.* **a)** *(Musik) (anstimmen)* **etw. ~:** sing/play the first few bars of sth.; start to sing/play sth.; **b)** *(Sprachw.)* **etw. richtig/anders ~:** say sth. with the right/a different intonation
intransitiv 1. *Adj. (Sprachw.)* intransitive. **2.** *adv.* intransitively
Intrigant der; ~en, ~en, Intrigantin die; ~, ~nen schemer; intriguer
Intrige [ɪn'tri:gə] **die; ~, ~n** intrigue
intrigieren *itr. V.* intrigue; scheme; **gegen jmdn. ~:** intrigue *or* scheme against sb.
introvertiert [ɪntrovɛr'ti:ɐ̯t] *Adj. (Psych.)* introverted
Intuition [ɪntui̯'tsi̯o:n] **die; ~, ~en** intuition
intuitiv [ɪntui̯'ti:f] **1.** *Adj.* intuitive. **2.** *adv.* intuitively
intus ['ɪntʊs] *in* **etw. ~ haben** *(ugs.) (begriffen haben)* have got sth. into one's head; *(gegessen od. getrunken haben)* have put sth. away *(coll.)*
invalid [ɪnva'li:t], **invalide** [ɪnva'li:də] *Adj.* invalid *attrib.*; **~ sein** be an invalid
Invalide der; *adj. Dekl.* invalid
Invalidität [ɪnvalidi'tɛ:t] **die; ~:** invalidity
in·variabel *Adj.* invariable
Invasion [ɪnva'zi̯o:n] **die; ~, ~en** *(auch fig. scherzh.)* invasion
Inventar [ɪnvɛn'ta:ɐ̯] **das; ~s, ~e a) [totes] ~** *(einer Firma)* fittings and equipment *pl.*; *(eines Hauses, Büros)* furnishings and fittings *pl.*; *(eines Hofes)* machinery and equipment; **lebendes ~:** livestock; **zum ~ gehören** *(fig.)* ⟨*person*⟩ be part of the scenery; **b)** *(Verzeichnis)* inventory
Inventur [ɪnvɛn'tu:ɐ̯] **die; ~, ~en** stock-taking; **~ machen** carry out a stock-taking
Inversion [ɪnvɛr'zi̯o:n] **die; ~, ~en** *(fachspr.)* inversion
investieren *tr., itr. V. (auch fig.)* invest **(in +** *Akk.* in)
Investition [ɪnvɛsti'tsi̯o:n] **die; ~, ~en** investment; **die privaten ~en sind zurückgegangen** private investment has fallen
Investitions-: ~güter *Pl. (Wirtsch.)* capital goods; **~lenkung die** investment control
Investment [ɪn'vɛstmɛnt] **das; ~s, ~s** *(Finanzw.) s.* **Investition**
Investment·fonds der investment fund
Investor [ɪn'vɛstɔr] **der; ~s, ~en** [-'to:rən] *(Wirtsch.)* investor
in·wendig 1. *Adj.* inside ⟨*pocket*⟩; inner ⟨*part*⟩; *(fig.)* inner, inward ⟨*happiness, strength*⟩. **2.** *adv.* [on the] inside; *(fig.)* inwardly; deep down [inside]; **etw./jmdn. in- und auswendig kennen** *(ugs.)* know sb./sth. inside out
in·wie·fern *Adv. (in welcher Hinsicht)* in what way; *(bis zu welchem Grade)* to what extent; how far
in·wie·weit *Adv.* to what extent; how far
In·zahlungnahme die; ~, ~n part-exchange; trade in *(Amer.)*
Inzest [ɪn'tsɛst] **der; ~[e]s, ~e** incest
In·zucht die; ~: inbreeding
in·zwischen *Adv.* **a)** *(seither)* in the meantime; since [then]; **es hatte sich ~ nichts geändert** nothing had changed in the meantime *or* since; **b)** *(bis zu einem Zeitpunkt) (in der Gegenwart)* by now; *(in der Vergangenheit)* by then; *(in der Zukunft)* by then; by that time; **c)** *(währenddessen)* meanwhile; in the meantime
IOK [i:o:'ka:] **das; ~[s] Internationales Olympisches Komitee** IOC
Ion [i̯o:n] **das; ~s, ~en** *(Physik, Chemie)* ion
Ionen-: ~austauscher der *(Physik, Chemie)* ion exchanger; **~gitter das** *(Chemie)* ionic lattice
Ionisation [i̯oniza'tsi̯o:n] **die; ~, ~en** *(Physik, Chemie)* ionization
Iono·sphäre [i̯ono-] **die** ionosphere
I-Punkt ['i:-] **der** dot over *or* on the i; **bis auf den ~** *(fig.)* down to the last detail
IQ [i:'ku: *od.* aɪ'kju:] **der; ~[s], ~[s]** IQ
i. R. [i:'ɛr] *Abk.* retd.
IRA [i:|ɛr'|a:] **die; ~:** IRA
Irak [i'ra:k] **(das); ~s** *od.* **der; ~[s]** Iraq; **in/nach/aus** *od.* **im/in den/aus dem ~:** in/to/from Iraq
Iraker der; ~s, ~, Irakerin die; ~, ~nen Iraqi
irakisch Iraqi
Iran [i'ra:n] **(das); ~s** *od.* **der; ~[s]** Iran; *s. auch* **Irak**
Iraner der; ~s, ~, Iranerin die; ~, ~nen Iranian
iranisch *Adj.* Iranian
irden ['ɪrdn̩] *Adj.* earthen[ware] ⟨*bowl, pot, jug*⟩; **~es Geschirr** earthenware
irdisch *Adj.* **a)** earthly ⟨*joys, paradise, love*⟩; mortal, earthly ⟨*creature, being*⟩; temporal ⟨*power, justice*⟩; worldly ⟨*goods, pleasures, possessions*⟩; **den Weg alles Irdischen gehen** go the way of all flesh; ⟨*object*⟩ go the way of all

things; **b)** *(zur Erde gehörig)* terrestrial; **das ~e Leben** life on earth

Ire ['i:rə] **der; ~n, ~n** Irishman; **die ~n** the Irish; **er ist ~:** he is Irish *or* an Irishman

irgend ['ɪrgn̩t] *Adv.* **a)** ~ **jemand** someone; somebody; somebody or other *(coll.)*; *(fragend, verneint)* anyone; anybody; ~ **etwas** something; *(fragend, verneint)* anything; ~ **so ein Politiker** *(ugs.)* some politician [or other]; ~ **so etwas** something like that; something of the sort *or* kind; **b)** *(irgendwie)* **wenn ~ möglich** if at all possible

irgend-: ~ein *Indefinitpron.* **a)** *attr.* some; *(fragend, verneint)* any; **~ein Idiot** some idiot [or other]; **in ~einer Zeitung habe ich neulich gelesen, daß ...:** I read in one of the papers recently that ...; **Welche Zeitung soll es sein? – Irgendeine** What newspaper do you want? – Just any; **~ein anderer/~eine andere** someone *or* somebody else; *(fragend, verneint)* anyone *or* anybody else; **mehr als ~ein anderer** more than anyone *or* anybody else; **b)** *(alleinstehend)* **~einer/~eine** someone; somebody; *(fragend, verneint)* anyone; anybody; **~eines** *od. (ugs.)* **~eins** any one; **~einer muß es machen** someone *or* somebody [or other] must do it; **nicht ~einer** not just anyone; **~wann** *Adv.* [at] some time [or other]; somewhen; *(zu jeder beliebigen Zeit)* [at] any time; **~wann einmal** [at] some time [or other]; **~was** *Indefinitpron. (ugs.)* something [or other]; *(fragend, verneint)* anything; **[nimm] ~was** [take] anything [you like]; **ist ~was?** is [there] something wrong *or* the matter?; **~welch...** *Indefinitpron.* some; *(fragend, verneint)* any; **~wer** *Indefinitpron. (ugs.)* somebody or other *(coll.)*; someone; somebody; *(fragend, verneint)* anyone; anybody; **~wie** *Adv.* somehow; somehow or other *(coll.)*; **kann man das ~wie anders/besser machen?** is there some other/better way of doing this?; **er tut mir ~wie leid, aber ...:** I feel sorry for him in a way, but ...; **~wo** *Adv.* **a)** somewhere; some place [or other] *(coll.)*; *(fragend, verneint)* anywhere; **ist hier ~wo ein Lokal?** is there a pub anywhere around here?; **~wo anders** somewhere/anywhere else; **b)** *(ugs.: irgendwie)* **er tut mir ~wo leid, aber ...:** I feel sorry for him in a way, but ...; **~woher** *Adv.* from somewhere; from some place; from somewhere or other *(coll.)*; *(fragend, verneint)* from anywhere; from any place; **~wohin** *Adv.* somewhere; somewhere or other *(coll.)*; *(fragend, verneint)* anywhere

Irin die; ~, ~nen Irishwoman; **sie ist ~:** she is Irish *or* an Irishwoman

Iris ['i:rɪs] **die; ~, ~** *(Bot., Anat.)* iris

irisch *Adj.* Irish; **Irisch/das Irische** Irish

irisieren *itr. V.* iridesce; be iridescent; **~d** iridescent

Irland ['ɪrlant] **(das); ~s** Ireland; *(die Republik)* Ireland; Eire

Ironie [iro'ni:] **die; ~, ~n** irony; **es war eine ~ des Schicksals, daß ...:** it was one of the ironies of fate *or* an irony of fate that ...

ironisch 1. *Adj.* ironic; ironical. **2.** *adv.* ironically

ironischer·weise *Adv.* ironically

ironisieren *tr. V.* ironize

Ironisierung die; ~, ~en ironizing

irr [ɪr] *Adj. s.* **irre 1 a**

irrational ['ɪratsiona:l] **1.** *Adj.* irrational. **2.** *adv.* irrationally

Irrationalismus der; ~, Irrationalismen irrationalism

Irrationalität die; ~: irrationality

irre ['ɪrə] **1.** *Adj.* **a)** *(geistesgestört)* mad, insane ‹*person*›; insane ‹*laughter*›; demented ‹*grin, look*›; insane, crazy ‹*idea, thought, suggestion*›; **davon kann man ja ~ werden** it's enough to drive you mad *or* crazy; **b)** *nicht präd. (salopp: stark)* terrific *(coll.)*; terrible *(coll.)*; **eine ~ Arbeit** a hell of a job *(coll.)*; **c)** *(salopp: faszinierend)* amazing *(coll.)*; **d)** *(geh.)* **an jmdm. ~ werden** lose faith in sb.; **an sich** *(Dat.)* **selbst ~ werden** doubt oneself. **2.** *adv. (salopp)* terrifically *(coll.)*; terribly *(coll.)*; **sich ~ freuen** be thrilled to bits *(coll.)*

[1]Irre ['ɪrə] **der/die;** *adj. Dekl.* madman/madwoman; lunatic; *(fig.)* fool; idiot; lunatic; **er fährt wie ein ~r** he drives like a maniac *or* lunatic; **er schreit/arbeitet wie ein ~r** he shouts/works like mad

[2]Irre die *(geh.)* **in die ~ gehen** go astray; *(fig.: sich irren)* make a mistake; **jmdn. in die ~ führen** mislead sb.; *(täuschen)* deceive sb.

irreal ['ɪrea:l] *Adj.* unreal

irre-, Irre-: ~|führen *tr. V.* mislead; *(täuschen)* deceive; **~führend** *Adj.* misleading; *(täuschend)* deceptive; **~führung die: das war eine bewußte ~führung** that was a deliberate attempt to mislead; **~führung der Öffentlichkeit** misleading the public; **~|gehen** *unr. V.; mit sein (geh.)* be mistaken; **~|leiten** *tr. V. (geh.)* lead astray; **irregeleitete Emotionen/Jugend** misguided emotions/youth

irrelevant ['ɪrelevant] *Adj.* irrelevant **(für** to)

Irrelevanz die irrelevance **(für** to)

irre|machen *tr. V.* **a)** *(verwirren)* disconcert; put off; **laß dich durch ihn nicht ~:** don't be disconcerted *or (coll.)* put off by him; **b)** *(zweifeln lassen)* **jmdn. in seinem Glauben ~:** shake sb.'s faith; **sie ließ sich in ihrer Hoffnung/ihrem Plan nicht ~:** she would not let anything confound her hopes/her plan

irren ['ɪrən] **1.** *refl. V.* be mistaken; **man kann sich auch mal ~:** everybody makes *or* we all make mistakes [sometimes]; **Sie ~ sich, wenn ...:** you are making a mistake if ...; **er hat sich in einigen Punkten geirrt** he got a few things wrong; **Sie haben sich in der Person/Hausnummer geirrt** you've got the wrong person/number; **sich um 1 DM ~:** be out by 1 DM; be 1 DM out. **2.** *itr. V.* **a)** *(sich irren)* **da ~ Sie** you are mistaken *or* wrong there; **Irren ist menschlich** to err is human *(prov.)*; **b)** *mit sein* **durch die Straßen/den Park ~:** wander the streets/about in the park

Irren-: ~anstalt die *(veralt.)* mental home; madhouse *(derog.)*; **~arzt der** *(veralt.)* maddoctor *(arch.)*; **~haus das** *(abwertend)* [lunatic] asylum; madhouse *(derog.)*; **das war das reinste ~haus** *(ugs.)* it was bedlam *or* an absolute madhouse; **er ist reif fürs ~haus** *(ugs.)* he'll crack up soon *(coll.)*

irreparabel [ɪrepa'ra:bl̩] *Adj.* irreparable; beyond repair *pred.*

Irr-: ~fahrt die wandering; **meine Reise wurde zu einer endlosen ~fahrt** my journey turned into an endless series of wanderings; **~glaube[n] der** *(Irrtum)* misconception

irrig ['ɪrɪç] *Adj.* erroneous

irrigerweise *Adv.* mistakenly; erroneously

Irritation [ɪrita'tsio:n] **die; ~, ~en** *(Med., geh.)* irritation

irritieren *tr., itr. V.* **a)** *(verwirren)* bother; put off; **das irritiert** it's off-putting; **laß dich dadurch nicht ~:** don't be put off by it; **b)** *(stören)* disturb; **c)** *(befremden)* annoy; irritate

irr-, Irr-: ~licht das will o' the wisp; jack o' lantern; **~lichter entstehen durch ...:** will o' the wisp *or* jack o' lantern is caused by ...; **~sinn der;** *o. Pl.* **a)** insanity; madness; **b)** *(ugs. abwertend)* madness; lunacy; **so ein ~sinn!** what lunacy!; **~sinnig 1.** *Adj.* **a)** insane; mad; *(absurd)* idiotic; **bist du ~sinnig?** are you mad?; **wie ~sinnig schreien/rasen** scream/rush like mad; **b)** *(ugs.: extrem)* terrible *(coll.)*, horrific *(coll.)* ‹*pain, screams, prices, etc.*›; terrific *(coll.)* ‹*speed, heat, cold*›; **2.** *adv.* *(ugs.)* terribly *(coll.)*; frightfully *(coll.)*; **~sinnig schuften** slog away like mad *or* *(coll.)* crazy; **~sinnige der/die;** *adj. Dekl.* madman/madwoman; lunatic

Irrtum der; ~s, Irrtümer ['ɪrty:mɐ] a) fallacy; misconception; **im ~ sein, sich im ~ befinden** be wrong *or* mistaken; **b)** *(Fehler)* mistake; error

irrtümlich ['ɪrty:mlɪç] **1.** *Adj.; nicht präd.* incorrect; wrong. **2.** *adv.* by mistake; **wie man oft ~ meint** as is often erroneously *or* mistakenly thought

Irrung die; ~, ~en *(geh.)* **die ~en und Wirrungen seiner verfehlten Jugend** the vagaries of his misspent youth

irr-, Irr-: ~weg der error; **diese Methode hat sich als ~weg erwiesen** this method has proved to be wrong; **~witzig** *Adj. (geh.)* mad

Ischias ['ɪʃi̯as] **der** *od.* **das** *od. Med.* **die; ~:** sciatica

Ischias·nerv der sciatic nerve

Isegrim ['i:zəgrɪm] **der; ~s, ~e** *(Myth.)* **[Meister] ~:** Isegrim; Isgrin

Islam [ɪs'la:m *od.* 'ɪslam] **der; ~[s]: der ~:** Islam; **die Welt des ~[s]** the Islamic world; the world of Islam

islamisch *Adj.* Islamic; Islamitic

Is·land ['i:s-] **(das); ~s** Iceland

Isländer ['i:slɛndɐ] **der; ~s, ~, Isländerin die; ~, ~nen** Icelander

isländisch ['i:slɛndɪʃ] *Adj.* Icelandic; **Isländisch/das Isländische** Icelandic

Iso·bare die; ~, ~n *(Met.)* isobar

Isolation [izola'tsi̯o:n] **die; ~, ~en** *s.* **Isolierung**

Isolations·haft die solitary confinement

Isolator [izo'la:tɔr] **der; ~s, ~en** [-'to:ren] insulator

Isolier·band das; *Pl.* **~bänder** insulating tape

isolieren *tr. V.* **a)** isolate ‹*prisoner, patient, bacterium, element*›; **von der Umwelt isoliert** cut off from the outside world; **etw. isoliert betrachten** look at sth. out of context; **b)** *(Technik)* insulate ‹*wiring, wall, etc.*›; lag ‹*boilers, pipes, etc.*›; *(gegen Schall)* sound-proof; insulate ‹*room, door, window, etc.*›

Isolier-: ~kanne die Thermos jug (P); vacuum jug; **~schicht die** insulating layer; **~station die** *(Med.)* isolation ward

Isolierung die; ~, ~en a) *(auch fig.)* isolation; **in der ~:** in isolation; **in die ~ geraten** *(fig.)* become isolated *or* detached; **b)** *(Technik)* insulation; insulating; *(von Kesseln, Röhren)* lagging; *(gegen Schall)* soundproofing; **c)** *(Isoliermaterial)* insulation; *(für Kessel, Röhren)* lagging

Isotherme [izo'tɛrmə] **die; ~, ~n** *(Met.)* isotherm

Isotop [izo'to:p] **das; ~s, ~e** isotope

Israel ['ɪsrae:l] **(das); ~s** Israel; **das Volk ~** *(bibl.)* the Israelites; the people of Israel; **die Kinder ~[s]** *(bibl.)* the Children of Israel

Israeli der; ~[s], ~[s]/die; ~, ~[s] Israeli

israelisch *Adj.* Israeli

Israelit der; ~en, ~en, Israelitin die; ~, ~nen Israelite

israelitisch *Adj.* Israelite

iß [ɪs] *Imperativ Sg. v.* **essen**

ißt [ɪst] *2. u. 3. Pers. Sg. Präsens v.* **essen**

ist [ɪst] *3. Pers. Sg. Präsens v.* **[1]sein**

Ist-: ~-Bestand der *(Kaufmannsspr.)* actual stocks *pl.*; **~-Stärke die** *(Milit.)* actual strength

Isthmus ['ɪstmʊs] **der; ~, Isthmen** isthmus

Italien [i'ta:li̯ən] **(das); ~s** Italy

Italiener [ita'li̯e:nɐ] **der; ~s, ~, Italienerin die; ~, ~nen** Italian

italienisch *Adj.* Italian; **Italienisch/das Italienische** Italian

Italo·western ['i:talo-] **der** Italian-made Western; spaghetti western *(derog.)*

I-Tüpfel[chen] das; ~s, ~, *(österr.)* **I-Tüpferl** ['i:typfɐl] **das; ~s, ~n** final *or* finishing touch; **bis aufs [letzte] ~:** down to the last *or* smallest detail

i. V. [i:'fau̯] *Abk.* **in Vertretung**

i. w. S. *Abk.* **im weiteren Sinne**

J

j, J [jɔt, *österr.:* je:] **das; ~, ~:** j/J; *s. auch* **a, A**

ja [ja:] **1.** *Interj.* **a)** yes; **Wohnen Sie hier? – Ja** Do you live here? – Yes[, I do]; **Hast du ihm Bescheid gesagt? – Ja** Have you told him? – Yes[, I have]; **b)** *(Bitte um Bestätigung)* **du bleibst doch noch ein bißchen, ja?** but you'll stay on a bit, won't you *or* surely?; **Sie kommen. – Ja?** They're coming. – Are they?; *(ungläubig)* **Der König ist tot. – Ja?** The King is dead. – [Is he] really? **c)** *(Antwort auf Anrede, Anruf usw.)* yes; **ja [bitte]?** *(am Telefon)* yes? **2.** *Partikel* **a)** *(beschwichtigend)* **ich komme ja schon** I'm [just] coming; **b)** *(Bekanntheit unterstellend)* **die Delphine, die ja Säugetiere sind** dolphins, which are known to be mammals; **ich habe es ja gleich gesagt** I said that in the first place, didn't I?; **Sie wissen ja, daß ...:** you know, of course, that ...; **du kennst ihn ja** you know what he's like; you know him; **c)** *(Überraschung ausdrückend)* **es schneit ja!** it's [actually] snowing!; **da seid ihr ja!** there you are!; **d)** *(einräumend)* **er mag ja recht haben** he may [well] be right. **3.** *Adv.* **a)** *(unbedingt)* **laß ja die Finger davon!** [just you] leave it alone!; **sag das ja nicht weiter!** don't [you dare] pass it on, whatever you do!; **damit er ja alles mitbekommt** to make sure he knows all *or* everything that's going on; **damit wir ja nicht zu spät kommen** so that there's no risk of us being late; **b)** *konjunktional (sogar)* indeed; even; **ich schätze, ja bewundere ihn** I like him, indeed admire him *or* admire him even

Ja das; ~[s], ~[s] yes; **mit ~ stimmen** vote yes

Jacht [jaxt] **die; ~, ~en** yacht

Jacht·hafen der yacht harbour; marina

Jäckchen ['jɛkçən] **das; ~s, ~:** jacket; *(gestrickt)* cardigan

Jacke ['jakə] **die; ~, ~n** jacket; *(gestrickt)* cardigan; **das ist ~ wie Hose** *(ugs.)* it makes no odds *(coll.)*

Jacken·tasche die jacket pocket

Jacket·krone ['dʒɛkɪt-] **die** *(Zahnmed.)* jacket crown

Jackett [ʒa'kɛt] **das; ~s, ~s** jacket

Jade ['ja:də] **der; ~[s]** *od.* **die; ~:** jade

Jagd [ja:kt] **die; ~, ~en a)** *o. Pl. (Weidwerk)* **die ~:** shooting; hunting; **die ~ auf Hasen** hare-hunting; **~ auf Fasanen/Wildschweine machen** shoot pheasant/hunt wild boar; **auf der ~ sein** be hunting/shooting; **auf die ~ gehen** go hunting/shooting; **b)** *(Veranstaltung)* shoot; *(Hetzjagd)* hunt; **c)** *(Revier)* preserve; shoot; **eine ~ pachten** rent a hunting-preserve *or* shoot; **d)** *(Verfolgung)* hunt; *(Verfolgungsjagd)* chase; **auf jmdn./etw. ~ machen** hunt for sb./sth.; **die ~ nach Geld/Besitz** *(fig.)* the constant pursuit of money/possessions

jagdbar *Adj.* **~e Tiere** animals that can be hunted/shot

Jagd-: ~beute die bag; kill; **~bomber der** *(Luftwaffe)* fighter-bomber; **~falke der** falcon; **~fieber das** hunting-fever; **vom ~fieber gepackt** in the fever of the hunt; **~flugzeug das** *(Luftwaffe)* fighter aircraft; **~frevel der** poaching; **~gesellschaft die** shooting-party; hunting-party; **~gewehr das** sporting gun; **~glück das: ~glück/kein ~glück haben** be lucky/unlucky [in the hunt]; **~grund der;** *meist Pl.* hunting-ground; **in die ewigen Jagdgründe eingehen** *(verhüll.)* go to the happy hunting-grounds; **~haus das** hunting *or* shooting lodge; hunting *or* shooting box; **~horn das** hunting-horn; **~hund der** gun-dog; *(bei Hetzjagden)* hunting-dog; hound; **~hütte die** hunting *or* shooting box; **~messer das** hunting-knife; **~revier das** preserve; shoot; *(fig.)* hunting-ground; **~schein der** game licence; **~wurst die** chasseur sausage; **~zeit die** open *or* hunting *or* shooting season

jagen ['ja:gn̩] **1.** *tr. V.* **a)** hunt ⟨*game, fugitive, criminal, etc.*⟩; shoot ⟨*game, game birds*⟩; *(hetzen)* chase, pursue ⟨*fugitive, criminal, etc.*⟩; *(wegscheuchen)* chase; run after; **von Todesangst gejagt** stricken by the fear of death; **ein Gedanke jagte den anderen** thoughts raced through his/her *etc.* mind; **b)** *(treiben)* drive; **jmdn. aus dem Haus ~:** throw sb. out of the house; **jmdn. aus dem Bett ~:** turn sb. out of bed; **jmdn. in die Flucht ~:** put sb. to flight; **c)** *(ugs.)* **sich/jmdm. eine Spritze in den Arm ~:** jab *or* stick a needle in one's/sb.'s arm; **sich/jmdm. eine Kugel durch den Kopf ~:** blow one's/sb.'s brains out. **2.** *itr. V.* **a)** *(die Jagd ausüben)* go shooting *or* hunting; *(auf Hetzjagd gehen)* go hunting; **b)** *mit sein (eilen)* race; rush; **Wolken ~ am Himmel** *(fig.)* clouds race *or* scud across the sky; **mit ~dem Puls** *(fig.)* with his/her *etc.* pulse racing

Jäger ['jɛ:gɐ] **der; ~s, ~ a)** hunter; *(bei Hetzjagden)* huntsman; **b)** *(Jagdflugzeug)* fighter

Jäger·art die *(Kochk.)* **Schnitzel nach ~:** escalope chasseur

Jägerin die; ~, ~nen huntress; huntswoman

Jäger-: ~latein das *(scherzh.)* [hunter's *or* huntsman's] tall story/stories; **~schnitzel das** *(Kochk.)* escalope chasseur; **~sprache die** hunting language

Jaguar ['ja:gu̯a:ɐ̯] **der; ~s, ~e** jaguar

jäh [jɛ:] **1.** *Adj. (geh.)* **a)** sudden; sudden, abrupt ⟨*change, movement, stop*⟩; sudden, sharp ⟨*pain*⟩; **er fand einen ~en Tod** he met his death suddenly; **ein ~es Erwachen** *(fig.)* a rude awakening; **b)** *(steil)* steep; precipitous ⟨*slope, ravine, ridge*⟩. **2.** *adv.* **a) die Stimmung schlug ~ um** the mood changed suddenly *or* abruptly; **b)** *(steil)* ⟨*fall, drop*⟩ steeply, abruptly

Jahr [ja:ɐ̯] **das; ~[e]s, ~e a)** year; **ein halbes ~:** six months; **anderthalb ~e** eighteen months; a year and a half; **im ~[e] 1908** in [the year] 1908; **jedes ~:** every year; **jedes zweite ~:** [once] every two years; **alle ~e** every year; **lange ~e [hindurch]** for many years; **~ für** *od.* **um ~:** year after year; **von ~ zu ~:** from one year to the next; from year to year; **zwischen den ~en** between Christmas and the New Year; **auf ~ und Tag** to the exact day; **nach ~ und Tag** after many years; **vor ~ und Tag** *(mit Präteritum)* many years ago; *(mit Plusquamperfekt)* many years before; **b)** *(Lebens~)* year; **er ist zwanzig ~e [alt]** he is twenty years old *or* of age; **Kinder bis zu zwölf ~en** children up to the age of twelve *or* up to twelve years of age; **Kinder über 14 ~e** children over the age of 14 *or* over 14 years of age; **Kinder ab zwei ~en** children of two years and over; **alle Männer zwischen 18 und 45 ~en** all men between the ages of 18 and 45; **mit 65 ~en** *od.* **im Alter von 65 ~en** at the age of 65; **das hat er schon in jungen ~en gelernt** he learned that at an early age *or* while he was still young; **mit den ~en** as he/she *etc.* grows/grew older; **er ist um ~e gealtert** he's put on years; **in die ~e kommen** reach middle age

jahr·aus *Adv.* **~, jahrein** year in, year out

Jahr·buch das year-book

Jährchen ['jɛ:ɐ̯çən] **das; ~s, ~** *(scherzh.)* year; **die paar ~, die ich noch zu leben habe!** the few short years I have left to live; **einige ~ auf dem Buckel haben** be knocking on a bit *(coll.)*

jahr·ein *Adv.; s.* **jahraus**

jahre·lang 1. *Adj.; nicht präd.* [many] years of ⟨*practice, imprisonment, experience, etc.*⟩; long-standing ⟨*feud, friendship*⟩; **mit ~er Verspätung** years late; **nach ~er Abwesenheit** after being away for years. **2.** *adv.* for [many] years

jähren ['jɛ:rən] *refl. V.* **heute jährt sich [zum fünften Male] sein Todestag** today is the [fifth] anniversary of his death; **heute jährt [es] sich zum zehntenmal, daß ...:** it is ten years ago today that ...; it is ten years since ...

jahres-, Jahres-: ~anfang der beginning *or* start of the year; **~ausgleich der** *(Steuerw.)* end-of-year adjustment; **den ~ausgleich beantragen** send in one's tax return; **~beginn der** beginning *or* start of the new year; **~beitrag der** annual *or* yearly subscription; **~bilanz die** *(Wirtsch., Kaufmannsspr.)* annual balance [of accounts]; *(Dokument)* annual balance sheet; **~einkommen das** annual income; **~ende das** end of the year; **~frist die** *in* **in** *od.* **innerhalb** *od.* **binnen ~frist** within [a period of] a *or* one year; **vor ~frist** in less than a year; *(vor einem Jahr)* a year ago; **nach ~frist** after [a period of] a *or* one year; **~gehalt das** annual salary; **zwei ~gehälter** two years' salary; **~hälfte die: die erste/zweite ~hälfte** the first/second half *or* six months of the year; **~haupt·versammlung die**

(Wirtsch.) annual general meeting; **~karte die** yearly season ticket; **~miete die** annual *or* yearly rent; **~mittel das** annual mean; **im ~mittel fallen 3,2 cm Niederschlag** the mean annual precipitation is 3.2 cm; **~ring der;** *meist Pl. (Bot.)* annual ring; **~tag der** anniversary; **~umsatz der** annual turnover; **~urlaub der** annual holiday *or (formal)* leave *or (Amer.)* vacation; **~versammlung die** annual [general] meeting; **~wechsel der** turn of the year; **zum ~wechsel die besten Wünsche** best wishes for the New Year; **~wende die** turn of the year; **~zahl die** date; **~zeit die** season; **für die ~zeit ist es kalt** it's cold for the time of the year; **trotz der vorgerückten ~zeit** although it is/was late in the year; **~zeitlich 1.** *Adj.; nicht präd.* seasonal; **2.** *adv.* **~zeitlich schwanken** vary with the seasons *or* according to the time of year; **~zeitlich bedingt sein** be governed by seasonal factors

Jahr·gang der a) year; **der ~ 1900** those *pl.* born in 1900; **sie ist ~ 1943** she was born in 1943; **er ist mein ~:** he was born in the same year as I was; **b)** *(eines Weines)* vintage; **der 81er soll ein guter ~ werden** 81 should be a vintage year; **ein Edelzwicker ~ 1978** a 1978 Edelzwicker; **c)** *(einer Zeitschrift)* set [of issues] for a/the year; **die beiden letzten Jahrgänge** the sets of back numbers for the past two years

Jahr·hundert das century; **im 19. und 20. ~:** in the 19th and 20th centuries; **durch die ~e** over *or* through the centuries; **im ersten ~ vor/nach Christi Geburt** in the first century BC/AD; **die Literatur des 19. ~s** 19th-century literature; the literature of the 19th century

Jahrhundert·wende die turn of the century; **aus der Zeit um die ~wende** from the turn of the century

-jährig ['jɛ:rɪç] **a)** *(... Jahre alt)* **ein elfjähriges/halbjähriges Kind** an eleven-year-old/a six-month-old child; **kaum acht~:** hardly eight years old; **b)** *(... Jahre dauernd)* ... year's/years'; -year; **nach vierjähriger/halbjähriger Vorbereitung** after four years'/six months' preparation; **mit dreijähriger/halbjähriger Verspätung** three years/six months late

jährlich ['jɛ:ɐlɪç] **1.** *Adj.; nicht präd.* annual; yearly. **2.** *adv.* annually; yearly; **einmal/zweimal ~:** once/twice a *or* per year; **ein Umsatz von 5 Millionen ~:** a turnover of five million per annum

Jahr·markt der fair; fun-fair

Jahr·millionen *Pl.* millions of years

Jahr·tausend das thousand years; millennium; **vor ~en** thousands of years ago; **das dritte ~ nach Christi Geburt** the third millennium AD

Jahr·zehnt das decade

jahrzehnte·lang 1. *Adj.; nicht präd.* decades of ⟨*practice, experience, etc.*⟩; **mit ~er Verspätung** decades late; **nach ~er Abwesenheit** after being away for decades. **2.** *adv.* for decades

Jäh·zorn der violent anger; **er neigt zum ~:** he tends towards violent fits *or* outbursts of temper *or* anger; **in wildem ~:** in blind anger *or* a blind rage

jäh·zornig 1. *Adj.* violent-tempered; **ein ~es Temperament** a violent temper. **2.** *adv.* in blind anger; in a blind rage

ja·ja *Partikel (ugs.)* **a)** *(seufzend)* **~[, so ist das Leben]** o well[, that's life]; **b)** *(ungeduldig)* **~[, ich komme schon]!** OK, OK *or* all right, all right[, I'm coming]!

Jakob ['ja:kɔp] **(der)** James; *(in der Bibel)* Jacob; **ich weiß ja nicht, ob das der wahre ~ ist** *(ugs.)* I don't know if that is really quite the thing

Jakobiner [jako'bi:nɐ] **der; ~s, ~** *(hist.)* Jacobin

Jalousette [[ʒalu'zɛtə] **die; ~, ~n, Jalousie** [ʒalu'zi:] **die; ~, ~n** Venetian blind

Jamaika [ja'maika] **(das); ~s** Jamaica

Jamaikaner der; ~s, ~, Jamaikanerin die; ~, ~nen Jamaican

jamaikanisch *Adj.* Jamaican

Jamaika·rum der Jamaica rum

jambisch *Adj. (Verslehre)* iambic

Jambus ['jambʊs] **der; ~, Jamben** *(Verslehre)* iambus; iamb; **ein Drama in Jamben** a drama in iambic verse *or* in iambics

Jammer ['jamɐ] **der; ~s a)** *(Wehklagen)* [mournful] wailing; **b)** *(Elend)* misery; **ein Bild des ~s** a picture of misery; **es ist ein ~, daß ...** *(ugs.)* it's a crying shame that ...

Jammer-: ~bild das miserable sight; **~gestalt die a)** pitiful creature; **b)** *(ugs. abwertend)* miserable wretch; **~lappen der** *(ugs. abwertend) (Feigling)* coward; *(Schwächling)* sniveller

jämmerlich ['jɛmɐlɪç] **1.** *Adj.* **a)** *(Jammer ausdrückend)* pathetic; pitiful; **b)** *(beklagenswert)* miserable ⟨*existence, conditions, etc.*⟩; wretched ⟨*appearance, existence, etc.*⟩; **c)** *(ärmlich)* pathetic; pitiful ⟨*conditions, clothing, housing*⟩; paltry, meagre ⟨*quantity*⟩; pitiful, sorry ⟨*state*⟩; **d)** *(abwertend: minderwertig)* contemptible ⟨*person*⟩; pathetic, paltry ⟨*wages, sum*⟩; pathetic, useless ⟨*piece of work etc.*⟩. **2.** *adv.* **a)** *(Jammer ausdrükkend)* pathetically; pitifully; **b)** *(beklagenswert)* miserably; hopelessly; pitifully; **~ versagen** fail miserably *or* hopelessly; **c)** *(ärmlich)* pitifully; miserably; **d)** *(abwertend: schlecht)* pathetically; hopelessly; **e)** *(sehr, stark)* terribly *(coll.)*; **~ frieren** be frozen stiff

jammern *itr. V.* **a)** wail; **ohne zu ~:** without so much as a groan; **b)** *(sich beklagen)* moan; grumble; **über sein Schicksal ~:** bemoan one's fate; **c)** *(verlangen)* cry [out]; **die Kinder jammerten nach einem Stück Brot** the children were crying out for *or* crying after a piece of bread

jammer-, Jammer-: ~schade *Adj.; nicht attr. (ugs.)* **es ist ~schade, daß ...:** it's a crying shame that ...; **es ist ~schade um ihn** it's a great pity about him; **~tal das;** *o. Pl. (geh.)* vale of tears; **~voll 1.** *Adj.* **a)** pathetic; pitiful ⟨*cry etc.*⟩; **b)** *(beklagenswert)* miserable; **2.** *adv.* **a)** pathetically; pitifully; **b)** *(beklagenswert)* miserably; wretchedly

Janker ['jaŋkɐ] **der; ~s, ~** *(südd., österr.)* Alpine jacket

Jänner ['jɛnɐ] **der; ~s, ~** *(österr.)*, **Januar** ['janu̯a:ɐ̯] **der; ~[s], ~e** January; *s. auch* **April**

Japan ['ja:pan] **(das); ~s** Japan

Japaner der; ~s, ~, Japanerin die; ~, ~nen Japanese

japanisch *Adj.* Japanese; **Japanisch/das Japanische** Japanese; *s. auch* **Deutsch**

Japs [japs] **der; ~es, ~e** *(ugs. abwertend)* Jap *(derog.)*

japsen ['japsn̩] *itr. V. (ugs.)* pant; **ich kann kaum noch ~:** I'm gasping for breath

Japser der; ~s, ~ *(ugs.)* gasp of breath

Jargon [jar'gõ:] **der; ~s, ~s a)** jargon; **der ~ der Juristen/Mediziner** legal/medical jargon; **der Berliner ~:** Berlin slang; **im „Spiegel"-~:** in the jargon of the 'Spiegel'; **b)** *(abwertend)* language; **er redet in einem ganz or-**

dinären ~: he uses very vulgar language

Ja·sager [-zagɐ] **der;** ~s, ~ *(abwertend)* yes-man

Jasmin [jas'mi:n] **der;** ~s, ~e **a)** jasmine; **b)** *(Falscher ~)* mock orange

Jasmin·tee der jasmine-tea

Jaß [jas] **der; Jasses** *(schweiz.)* jass

jassen *itr. V. (schweiz.)* play jass

Ja·stimme die yes-vote; **die ~n** the votes in favour; the ayes *(Brit. Parl.)*

jäten ['jɛ:tn̩] **1.** *tr. V.* weed [out]; ⟨*dandelions, thistles, etc.*⟩; weed ⟨*flower-bed*⟩; **Unkraut ~:** weed. **2.** *itr. V.* weed

Jauche ['jauxə] **die;** ~, ~n **a)** liquid manure; **b)** *(ugs. abwertend)* muck

Jauche·grube die liquid-manure reservoir

jauchzen ['jauxtsn] *itr. V.* **a)** cheer; **vor Freude ~:** shout for joy; **das Publikum jauchzte** the audience was in raptures; **b)** *(veralt.)* rejoice; **jauchzet dem Herrn** rejoice in the Lord

Jauchzer der; ~s, ~: cry of delight

jaulen ['jaulən] *itr. V.* ⟨*dog, cat, etc.*⟩ howl, yowl; ⟨*wind*⟩ howl; ⟨*engine*⟩ scream

Jause ['jauzə] **die;** ~, ~n *(österr.)* **a)** snack; **eine ~ machen** have a snack; **b)** *(Nachmittagskaffee)* [afternoon] tea

Jausen·station die *(österr.)* café

jausnen *itr. V. (österr.)* have a snack

ja·wohl 1. *Interj.* certainly; ~, **Herr Oberst!** yes, sir! **2.** *Partikel* **Kant, ~ Kant** Kant, no less

jawoll [ja'vɔl] *(ugs.) s.* **jawohl**

Ja·wort das consent; **jmdm. das ~ geben** consent to marry sb.; **sich** *(Dat.)* **das ~ geben** accept each other in marriage

Jazz [dʒæz *od.* dʒɛs *od.* jats] **der;** ~: jazz

Jazz-: ~band die, ~kapelle die jazz band; **~keller der** jazz cellar; **~musik die** jazz music

[1]**je** [je:] **1.** *Adv.* **a)** *(jemals)* ever; **mehr/besser denn je** more/better than ever; **seit** *od.* **von je** always; for as long as anyone can remember; *s. auch* [2]**eh b; b)** *(jeweils)* **je zehn Personen** ten people at a time; **die Kinder stellen sich je zwei und zwei auf** the children arrange themselves in twos *or* in pairs; **sie kosten je 30 DM** they cost 30 DM each; **er gab den Mädchen je eine Birne** he gave each of the girls a pear; **in Schachteln mit** *od.* **zu je 10 Stück verpackt** packed in boxes of ten; **c) je nach Gewicht/Geschmack** according to weight/taste. **2.** *Präp. mit Akk.* per; for each; **je angebrochene Stunde** for each *or* per hour or part of an hour. **3.** *Konj.* **a) je länger, je lieber** the longer the better; **je früher du kommst, desto** *od.* **um so mehr Zeit haben wir** the earlier you come, the more time we'll have; **b) je nachdem** it all depends; **wir gehen hin, je nachdem [ob] wir Zeit haben oder nicht** we'll go, depending on whether we have the time or not

[2]**je** *Interj.* **ach je, wie schade!** oh dear *or* dear me, what a shame!

Jeans [dʒi:nz] *Pl. od.* **die;** ~, ~ jeans *pl.;* denims *pl.*

Jeans-: ~hose die [pair of] jeans; **~jacke die** denim jacket; **~stoff der** denim; jean[s] material

jeck [jɛk] *Adj. (rhein., meist abwertend) (leicht verrückt)* stupid; daft; *(wahnsinnig)* crazy

Jeck der; ~en, ~en *(rhein.)* **a)** *(abwertend: Verrückter)* idiot; **b)** *(Fastnachter)* carnival clown

jede *s.* **jeder**

jeden·falls *Adv.* **a)** in any case; at any rate; anyway; **das steht ~ fest** that much is certain, in any case *or* at any rate *or* anyway; **b)** *(zumindest)* at any rate; **ich ~ habe keine Lust mehr** I at any rate *or* for one have had enough

jeder ['je:dɐ], **jede, jedes** *Indefinitpron. u. unbest. Zahlwort* **1.** *attr.* **a)** *(alle)* every; **jeder einzelne Schüler** every single pupil; **jeder zweite Bürger** one out of *or* in every two citizens; **der Zug fährt jeden Tag/viermal jeden Tag** the train runs every day/four times a day; **das kann Ihnen jedes Kind sagen** any child could tell you that; **ohne jeden Zweifel** without any doubt; **ohne jeden Grund** without any reason whatever; for no reason whatever; **b)** *(alle einzeln)* each; **c)** *(jeglicher)* all; **jede Hilfe kam zu spät** all help came too late; **hier wurde jedes Maß überschritten** that went beyond all bounds; **Menschen jeden** *od.* **jedes Alters** people of all ages. **2.** *alleinstehend* **a)** *(alle)* everyone; everybody; **jeder** *od. (geh.)* **ein jeder darf mitkommen** everyone *or* everybody can come; **hier kennt jeder jeden** everybody knows everybody else here; *(verstärkend)* **jeder, der Lust hat, ist willkommen** anyone who wants to come is welcome; **das kann ja jeder** anyone can do that; **b)** *(alle einzeln)* **jedes der Kinder** every one *or* each of the children; **jeder von uns kann helfen** each *or* every one of us can help; **jedem nach seinem Verdienst** to each according to his merits

jeder-: ~art *unbest. Gattungsz.; indekl.; nicht präd.* any kind *or* sort *or* type of; **~lei** *unbest. Gattungsz.; indekl.; nicht präd. (geh.)* all kinds *or* sorts of; **~mann** *Indefinitpron.* everyone; everybody; **hier kann ~mann mitmachen** everyone *or* everybody *or* anyone *or* anybody can come along and join in; **Schnecken sind nicht ~manns Sache/Geschmack** snails are not to everyone's *or* everybody's taste; **~zeit** *Adv.* [at] any time

jedes *s.* **jeder**

jedes·mal *Adv.* every time; ~, **wenn das Telefon klingelt, ...:** every time the telephone rings ...

je·doch *Konj., Adv.* however; **es war ~ zu spät** it was too late, however; it was, however, too late

Jeep ⓦ [dʒi:p] **der;** ~s, ~s jeep (P)

jeglicher ['je:klɪçɐ], **jegliche, jegliches** *Indefinitpron. u. unbest. Zahlw. s.* **jeder 1 c, 2 b**

je·her [od. '-'-] *Adv.* **seit** *od.* **von ~:** always; since *or* from time immemorial

Jehova [je'ho:va] *s.* **Zeuge**

jemals ['je:ma:ls] *Adv.* ever

jemand ['je:mant] *Indefinitpron.* someone; somebody; *(fragend, verneint)* anyone; anybody; **ich kenne ~[en], der ...:** I know someone *or* somebody who ...; **sich mit ~[em] treffen** to meet someone *or* somebody; **ist da ~?** is anybody there?; **ich glaube nicht, daß da ~ ist** I don't think there's anybody there; **~ anders/Fremdes** someone *or* somebody else/strange; **kaum ~:** hardly *or* scarcely anyone *or* anybody; *s. auch* **irgend a**

Jemen ['je:mən] **(das);** ~s *od.* **der;** ~[s] Yemen; *s. auch* **Irak**

jener ['je:nɐ], **jene, jenes** *Demonstrativpron. (geh.)* **1.** *attr.* that; *Pl.* those; **in jenem Haus dort** in that house [over] there; **zu jenem Zeitpunkt** at that time; **in jenen Tagen** in those days. **2.** *alleinstehend* that one; *Pl.* those; **jene, die ...:** those who ...

jenseitig ['je:n- *od.* 'jɛn-] *Adj.; nicht präd.* opposite; far, opposite ⟨*bank, shore*⟩

jenseits ['je:n-] **1.** *Präp. mit Gen.* on the other side of; *(in größerer Entfernung)* beyond; **~ des Flusses** on the other *or* far *or* opposite side of the river. **2.** *Adv.* beyond; on the other side; ~ **von** on the other side of; beyond; **eine Welt ~ von Haß und Gewalt** a world free from hatred and violence

Jenseits das; ~: hereafter; beyond; **jmdn. ins ~ befördern** *(salopp)* bump sb. off *(sl.)*

[1]**Jersey** ['dʒøːɐzi] **der; ~[s], ~s** jersey

[2]**Jersey das; ~s, ~s** jersey

Jersey·kleid das jersey dress

Jesuit [je'zuiːt] **der; ~en, ~en** *(Rel., auch fig. abwertend)* Jesuit

Jesus ['jeːzʊs] **(der); Jesu** ['jeːzu] Jesus; ~ **Christus** Jesus Christ

Jesus·kind das: das ~: the Infant Jesus; baby Jesus *(child lang.)*

Jet [dʒɛt] **der; ~[s], ~s** jet

Jet-set ['dʒɛtsɛt] **der; ~[s], ~s** jet set

jetten ['dʒɛtn̩] *itr. V.; mit sein (ugs.)* jet

jetzig ['jɛtsɪç] *Adj.; nicht präd.* present; current

jetzt [jɛtst] *Adv.* **a)** at the moment; just now; **bis ~:** up to now; **bis ~ noch nicht** not yet; not so far; **von ~ an** *od.* **ab** from now on[wards]; ~ **noch** still; **was, ~ [so spät] noch?** what, now?; ~ **oder nie!** it's now or never; ~ **ist aber Schluß!** that's [quite] enough!; ~ **ist es aus mit uns** we've had it now; ~ **endlich** [now,] at last; **erst ~** *od.* **~ erst** only just; **schon ~:** already; **er ist ~ schon drei Wochen krank** he has been ill for three weeks now; **b)** *(heutzutage)* now; these days; nowadays

Jetzt das; ~ *(geh.)* **das ~:** the present

jeweilig ['jeːvailɪç] *Adj.; nicht präd.* **a)** *(in einem bestimmten Fall)* particular; **b)** *(zu einer bestimmten Zeit)* current; of the time *postpos., not pred.;* **c)** *(zugehörig, zugewiesen)* respective

jeweils ['jeːvails] *Adv.* **a)** *(jedesmal)* ~ **am ersten/letzten Mittwoch des Monats** on the first/last Wednesday of each month; **b)** *(zur Zeit)* currently; at the time

Jh. *Abk.* **Jahrhundert** c.

jiddisch ['jidɪʃ] *Adj.* Yiddish; **Jiddisch/das Jiddische** Yiddish; *s. auch* **Deutsch**

Jiu-Jitsu [dʒiːu'dʒɪtsu] **das; ~[s]** j[i]u-jitsu

Job [dʒɔp] **der; ~s, ~s** *(ugs.; auch DV)* job

jobben *itr. V. (ugs.)* do a job/jobs; **als Taxifahrer ~:** do [some] taxi-driving

Jobber der; ~s, ~ *(Börsenw.)* [stock] jobber

Joch [jɔx] **das; ~[e]s, ~e a)** *(auch fig.)* yoke; **Ochsen ins/unters ~ spannen** yoke oxen; **b)** *(Geogr.)* col; saddle

Joch·bein das *(Anat.)* zygomatic bone; malar bone

Jockei, Jockey ['dʒɔke *od.* 'dʒɔki] **der; ~s, ~s** jockey

Jod [joːt] **das; ~[e]s** iodine

Jodel·lied das yodelling song

jodeln ['joːdl̩n] *itr., tr. V.* yodel

Jodler der; ~s, ~ a) *(Person)* yodeller; **b)** *(kurzes Jodeln)* yodel

Jodlerin die; ~, ~nen yodeller

Jod·tinktur die tincture of iodine; iodine tincture

Joga ['joːga] **der** *od.* **das; ~[s]** yoga

Joga·übung die yoga exercise

joggen ['dʒɔgn̩] *itr. V.; mit Richtungsangabe mit sein* jog; **[zwei Kilometer] ~:** go jogging [for two km]

Jogging ['dʒɔgɪŋ] **das; ~s** jogging *no art.*

Joghurt ['joːgʊrt] **der** *od.* **das; ~[s], ~[s]** yoghurt

Joghurt·becher der yoghurt pot *(Brit.) or (Amer.)* container

Jogi ['joːgi] **der; ~s, ~s** yogi

Johann ['joːhan] **(der)** John

Johanna [jo'hana] **(die)** Joan; ~ **von Orléans** Joan of Arc

Johannes [jo'hanəs] **(der); Johannes'** John; ~ **der Täufer** John the Baptist

Johanni [jo'hani] **(das);** *indekl. s.* **Johannistag**

Johannis·beere die currant; **rote/weiße/schwarze ~n** redcurrants/white currants/blackcurrants

Johannisbeer-: ~saft der currant juice; **~strauch der** currant bush

Johannis-: ~brot das *(Bot.)* Saint-John's-bread; carob [bean]; **~feuer das** Saint John's fire; **~käfer der** *(südd.) s.* **Leuchtkäfer; ~tag der** Saint John the Baptist's day

Johanniter·orden der Order of [the Hospital of] St. John of Jerusalem

johlen ['joːlən] *itr. V.* yell; *(vor Wut)* howl

Joint [dʒɔɪnt] **der; ~s, ~s** *(ugs.)* joint *(sl.)*

Jo-Jo [jo(ː)'joː] **das; ~s, ~s** yo-yo

Joker ['joːkɐ *od.* dʒoːkɐ] **der; ~s, ~** *(Kartensp.)* joker

Jolle ['jɔlə] **die; ~, ~n** keel-centreboard yawl

Jollen·kreuzer der dinghy cruiser

Jongleur [ʒɔŋ'løːɐ̯] **der; ~s, ~e** juggler

jonglieren *tr., itr. V.* juggle; **Bälle** *od.* **mit Bällen ~:** juggle with balls; **mit Zahlen ~** *(fig.)* juggle [about] with figures

Jordan ['jɔrdan] **der; ~[s]** Jordan; **über den ~ gehen** *(verhüll.)* go the way of all flesh

Jordanien [jɔr'daːni̯ən] **(das); ~s** Jordan

Jordanier der; ~s, ~, Jordanierin die; ~, ~nen Jordanian

jordanisch *Adj.* Jordanian

Jot [jɔt] **das; ~, ~:** j, J; *s. auch* **a, A**

Jota ['joːta] **das; ~[s], ~s** iota; **kein/nicht ein/um kein ~** *(geh.)* not an iota; not one jot

Joule [dʒuːl *od.* dʒau̯l] **das; ~[s], ~** *(Physik)* joule

Journal [ʒʊr'naːl] **das; ~s, ~e a)** *(veralt.: Tageszeitung)* journal *(dated);* newspaper; **b)** *(geh.: Zeitschrift)* journal; periodical; **c)** *(veralt.: Tagebuch)* journal *(dated);* diary

Journal·beamte der *(österr.)* official *or* officer [on duty]

Journalismus der; ~: journalism *no art.*

Journalist der; ~en, ~en, Journalistin die; ~, ~nen journalist

journalistisch 1. *Adj.; nicht präd.* journalistic; **eine ~e Ausbildung** a training in journalism. **2.** *adv.* journalistically; ~ **tätig sein** work as *or* be a journalist

jovial [jo'vi̯aːl] *Adj.* jovial

Jovialität [jovi̯ali'tɛːt] **die; ~:** joviality

jr. *Abk.* **junior** Jr.

Jubel ['juːbl̩] **der; ~s** rejoicing; jubilation; *(laut)* cheering; **großer ~ brach aus** a loud cheer went up; **unter dem ~ der Zuschauer** amid the cheering *or* cheers of the spectators

Jubel-: ~feier die jubilee; anniversary; *(Feierlichkeiten)* jubilee *or* anniversary celebrations *pl.* **~jahr das** jubilee; **alle ~jahre [einmal]** once in a blue moon

jubeln *itr. V.* cheer; **über etw.** *(Akk.)* **~:** rejoice over sth.

Jubel-: ~paar das couple celebrating their wedding anniversary; **~ruf der** cheer; joyful shout

Jubilar [jubi'laːɐ̯] **der; ~s, ~e** man celebrating his anniversary/birthday

Jubilarin die; ~, ~nen woman celebrating her anniversary/birthday

Jubiläum [jubi'lɛːʊm] **das; ~s, Jubiläen** anniversary; *(eines Monarchen)* jubilee; **fünfundzwanzigjähriges/fünfzigjähriges ~:** twenty-fifth/fiftieth anniversary/jubilee; **hundertjähriges ~:**

hundredth anniversary; centenary

Jubiläums-: ~aus·gabe die jubilee edition; **~aus·stellung die** jubilee exhibition

jubilieren *itr. V. (geh. veralt.)* jubilate *(literary);* rejoice

juchhe [jʊx'he:], **juchheißa** [jʊx'haisa] *(veralt.) Interj.* hurrah

Juchten ['jʊxtn̩] **der** *od.* **das; ~s a)** *(Leder)* Russia [leather]; **b)** *(Duftstoff)* Russian leather

juchzen ['jʊxtsn̩] *itr. V. (ugs.)* shout with glee

Juchzer der; ~s, ~ *(ugs.)* shout of glee; **einen ~ ausstoßen** shout with glee

jucken ['jʊkn̩] **1.** *tr., itr. V.* **a) mir juckt die Haut** I itch; **es juckt mir** *od.* **mich auf dem Kopf** my head itches; **es juckt mich hier** I've got an itch here; **b)** *(Juckreiz verursachen)* irritate; **die Wolle juckt ihn** *od.* **ihm auf der Haut** the wool makes him itch; the wool irritates his skin; **ein ~der Hautausschlag** an itching rash. **2.** *tr. V. (ugs.)* **a)** *(reizen)* **es juckt mich, das zu tun** I am itching *or* dying to do it; **b)** *(stören)* **das juckt mich nicht** I couldn't care less *(coll.).* **3.** *refl. V. (ugs.: sich kratzen)* scratch

Jucken das; ~s itching; **ein ~ verspüren** feel an itch

Juck-: ~pulver das itching powder; **~reiz der** itch

Judaist der; ~en, ~en, Judaistin die; ~, ~nen specialist in Jewish studies

Judaistik die; ~: Jewish studies *pl., no art.*

Judas ['ju:das] **1. (der); Judas'** Judas; **~ Ischariot** Judas Iscariot. **2. der; ~, ~se** *(fig.)* Judas

Jude ['ju:də] **der; ~n, ~n** Jew; **er ist ~:** he is a Jew; he is Jewish

Juden-: ~haß der anti-Semitism; hatred of [the] Jews; **~hetze die** Jew-baiting; **~pogrom der** *od.* **das** pogrom against the Jews; **~stern der** *(ns.)* Star of David

Judentum das; ~s a) *(Volk)* Jewry; Jews *pl.;* **das gesamte ~:** the whole of Jewry; **b)** *(Kultur u. Religion)* Judaism

Juden-: ~verfolgung die persecution of [the] Jews; **~viertel das** Jewish quarter; *(hist.)* Jewry

Judikative [judika'ti:və] **die; ~, ~n** *(Rechtsw., Politik)* judiciary

Jüdin ['jy:dɪn] **die; ~, ~nen** Jewess; **sie ist ~:** she is Jewish *or* a Jewess

jüdisch *Adj.* Jewish

Judo ['ju:do] **das; ~[s]** judo *no art.*

Judo·griff der judo throw

Judoka [ju'do:ka] **der; ~[s], ~[s]** judoka; judoist

Jugend ['ju:gn̩t] **die; ~ a)** youth; **in ihrer ~:** in her youth; when she was young; **schon in früher ~:** at an early age; **schon von ~ auf** from an early age; from his/her *etc.* youth; **b)** *(Jugendliche)* young people; **die weibliche/ männliche ~:** girls *pl.*/boys *pl.*

jugend-, Jugend-: ~alter das adolescence; **~amt das** youth office *(agency responsible for education and welfare of young people);* **~arrest der** detention in a community home; **vier Wochen ~arrest** four weeks in a community home; **~buch das** book for young people; **~frei** *Adj.* ⟨*film, book, etc.*⟩ suitable for persons under 18; **nicht ~frei** ⟨*film*⟩ not U-certificate *pred.; (scherzh.)* ⟨*joke, story, etc.*⟩ not for young ears *pred.;* **~freund der** friend of [the days of] one's youth; **er ist ein ~freund von ihr** he used to be a friend of hers when she was young; **~gefährdend** *Adj.* liable to have an undesirable influence on the moral development of young people *postpos.;* **~gericht das** juvenile court; **~gruppe die** youth group; **~heim das** youth centre; **~herberge die** youth hostel; **~klub der** youth club; **~kriminalität die** juvenile delinquency

jugendlich ['ju:gn̩tlɪç] **1.** *Adj.* **a)** *nicht präd.* young ⟨*offender, customer, etc.*⟩; **noch in ~em Alter sein** still be a youngster; still be young; **b)** *(jung, für Jugendliche charakteristisch)* youthful; **in ~er Begeisterung** fired by the spirit of youth *or* by youthful enthusiasm; **sie wirkt noch sehr ~:** she still looks very young; **c)** *(bes. Werbespr.)* young ⟨*fashions, dress, hairstyle, etc.*⟩. **2.** *adv.* **sich ~ kleiden** dress young

Jugendliche ['ju:gn̩tlɪçə] **der/ die;** *adj. Dekl.* **a)** young person; **für ~:** for young people; **b)** *(Rechtsspr.)* juvenile; young person; **zwei ~:** two juveniles; two young persons; **ein 16jähriger ~r/eine 16jährige ~:** a 16-year-old youth/girl

Jugendlichkeit die youth; *(jugendliche Wirkung)* youthfulness

Jugend-: ~liebe die love *or* sweetheart of one's youth; **~mannschaft die** *(Sport)* youth team *or* side; **~meister der** youth champion; *(Mannschaft)* youth champions; **~psychologie die** psychology of adolescence; adolescent psychology *no art.;* **~schutz der** protection of young people; **~sprache die** young people's language *no art.;* **~stil der** art nouveau; *(in Deutschland)* Jugendstil; **~straf·anstalt die** detention centre; **~strafe die** youth custody sentence; **sechs Monate ~strafe** get six months in a detention centre; **~sünde die, ~torheit die** youthful folly; **~weihe die** *(ehem. DDR) ceremony in which fourteen-year-olds are given adult social status;* **~werk das** early *or* youthful work; *(gesamtes)* early *or* youthful works *pl.;* juvenilia *pl.;* **~zeit die** youth; younger days *pl.;* **~zentrum das** youth centre

Jugo·slawe [jugo-] **der** Yugoslav

Jugo·slawien (das); ~s Yugoslavia

Jugo·slawin die Yugoslav

jugo·slawisch *Adj.* Yugoslav[ian]

Juice [dʒu:s] **der** *od.* **das; ~, ~s** ['dʒu:sɪs] *(bes. österr.)* [fruit] juice

Julei [ju:'lai] **der; ~[s], ~s** *(ugs.; bes. zur Verdeutlichung) s.* **Juli**

Juli ['ju:li] **der; ~[s], ~s** July; *s. auch* **April**

Jumper ['dʒampɐ] **der; ~s, ~:** jumper *(Brit.);* pullover

jun. *Abk.* **junior** Jr.

jung [jʊŋ] *Adj.;* **jünger** ['jʏŋɐ], **jüngst...** ['jʏŋst...] young; new ⟨*project, undertaking, sport, marriage, etc.*⟩; **er ist ~ gestorben** he died young; **~ an Jahren** young in years; **Cato der Jüngere** Cato the Younger; **[ganze] 30 Jahre ~** *(ugs. scherzh.)* 30 years young; **Sport hält ~:** sport keeps you young; **die Nacht ist noch ~:** the night is young; **der ~e Tag** *(geh.)* the new day; **in jüngster Zeit** recently; lately; **ein Ereignis der jüngeren/jüngsten Geschichte** an event in recent/very recent history; **die jüngsten Geschehnisse** the latest *or* [most] recent happenings; **der Jüngste Tag** *(Rel.)* doomsday

Jung-: ~brunnen der Fountain of Youth; **das ist ein wahrer ~brunnen** *(fig.)* that's a real tonic; **~bürger der** *(bes. österr.)* first-time voter; new voter

Jungchen das; ~s, ~ *(bes. ostd.)* little boy; little lad; **mein ~:** my boy *or* lad

[1]**Junge** ['jʊŋə] **der; ~n, ~n** *od.* *(ugs.)* **Jung[en]s** boy; **Tag, alter ~!** *(ugs.)* hello, old pal! *(coll.);* **jmdn. wie einen dummen ~n behandeln**

(ugs.) treat sb. like a child; ~, ~! *(ugs.)* [boy], oh boy!; *s. auch* **schwer 1 e**

[2]**Junge das;** *adj. Dekl.* **ein ~s** one of the young; **~ kriegen** give birth to young; **eine Löwin und ihr ~s** a lioness and her cub

Jüngelchen ['jʏŋl̩çən] **das; ~s, ~** *(ugs. abwertend)* young puppy *or* cub

jungenhaft *Adj.* boyish

Jungen-: ~klasse die boys' class; **wir waren eine reine ~klasse** our class was all boys; **~schule die** boys' school; school for boys; **~streich der** boyish prank

jünger ['jʏŋɐ] *Adj.* youngish; **sie ist noch ~:** she is still quite young; **Jüngere** *(jüngere Menschen)* [the] younger people; **die Jüngeren unter Ihnen** the younger ones amongst you; *s. auch* **jung**

Jünger der; ~s, ~: follower; disciple; *(der Kunst, Literatur)* devotee; **Jesus und seine ~:** Jesus and his disciples

Jungfer ['jʊŋfɐ] **die; ~, ~n a)** *(veralt.)* young lady; **b)** *(abwertend: ältere ledige Frau)* spinster; **eine alte ~:** an old maid

Jungfern-: ~fahrt die maiden voyage; **~flug der** maiden flight; **~häutchen das** hymen

Jung·frau die a) virgin; **sie ist noch ~:** she is still a virgin; **die ~ Maria** the Virgin Mary; **b)** *(Astrol.)* Virgo; **c)** *(veralt.: junges Mädchen)* young maid *or* maiden *(arch.)*

jung·fräulich [-frɔylɪç] *Adj.* *(geh., auch fig.)* virgin; **~ in die Ehe gehen** be a virgin bride

Jungfräulichkeit die *(geh.)* virginity; *(fig.)* virgin state

Jung·geselle der bachelor

Jung·gesellen-: ~bude die *(ugs.)* bachelor pad *(coll.)*; **~leben das** bachelor[’s] life; **~wohnung die** bachelor flat; **~zeit die** bachelor days *pl.*; bachelorhood

Jung·gesellin die bachelor girl; **sie ist ~gesellin geblieben** she never married

Jüngling ['jʏŋlɪŋ] **der; ~s, ~e** *(geh./spött.)* youth; boy

jüngst ['jʏŋst] *Adv. (geh.)* recently

jüngst... *s.* **jung**

Jüngste der/die; *adj. Dekl.* *(Sohn, Tochter)* youngest [one]

Jung-: ~steinzeit die Neolithic period; New Stone Age; **~tier das** young animal; **~verheiratete der/die;** *adj. Dekl.*, *(geh.)* **~vermählte der/die;** *adj. Dekl.* young married man/woman; **die ~verheirateten** the newly-weds; **~volk das;** *o. Pl.* *(veralt., scherzh.)* young folk; **~wähler der** first-time voter; new voter

Juni ['ju:ni] **der; ~[s], ~s** June; *s. auch* **April**

junior ['ju:ni̯ɔr] *indekl. Adj.; nach Personennamen* junior

Junior der; ~s, ~en [-'ni̯o:rən] **a)** *(oft scherzh.)* junior *(joc.)*; **mit seinem ~:** with junior; **b)** *(Kaufmannsspr.)* junior partner

Junior·chef der owner's *or* *(coll.)* boss's son

Junioren-: ~mannschaft die youth team; **~meister der** junior champion; *(Mannschaft)* junior champions

Junior·partner der junior partner

Junker ['jʊŋkɐ] **der; ~s, ~** *(hist., oft abwertend)* junker; squire

Junkie [dʒʌŋki] **der; ~s, ~s** *(Drogenjargon)* junkie *(sl.)*

Junktim das; ~s, ~s package [deal]; **zwischen den beiden Abkommen besteht ein ~:** the two agreements form one package

Juno [ju'no:] **der; ~[s], ~s** *(ugs.; bes. zur Verdeutlichung)* June

Junta ['xʊnta] **die; ~, Junten** junta

[1]**Jupiter** ['ju:pitɐ] **der; ~s** *(Astron.)* Jupiter

[1]**Jura** ['ju:ra] *o. Art., o. Pl.* law; **~ studieren** read *or* study Law

[2]**Jura der; ~s** *(Geol.)* Jurassic [period/system]

juridisch [ju'ri:dɪʃ] *Adj. (österr., veralt.) s.* **juristisch**

Jurisdiktion [jurɪsdɪk'tsi̯o:n] **die; ~, ~en** *(geh.)* jurisdiction

Jurisprudenz [jurɪspru'dɛnts] **die; ~:** jurisprudence *no art.*

Jurist der; ~en, ~en lawyer; jurist

Juristerei die; ~ *(oft scherzh.)* law *no art.*

Juristin die; ~, ~nen *s.* **Jurist**

juristisch 1. *Adj.* legal ⟨*wrangle, term, training, career*⟩; law ⟨*examination*⟩; **die Juristische Fakultät** the Law Faculty. **2.** *adv.* **~ denken** think in legal terms; **~ argumentieren** use legal arguments

Juror ['ju:rɔr] **der; ~s, ~en** [-'ro:ren], **Jurorin die; ~, ~nen** judge

Jury [ʒy'ri:] **die; ~, ~s** panel [of judges]; jury

Jus [ju:s] **das; ~** *(österr., schweiz.) s.* [1]**Jura**

just [jʊst] *Adv. (veralt., scherzh.)* just; **~ an jenem Tag** on that very day

justieren *tr. V.* adjust

Justitiar [jʊstɪtsi̯a:ɐ̯] **der; ~s, ~e** company lawyer

Justiz [jʊs'ti:ts] **die; ~:** justice; *(Behörden)* judiciary; **ein Vertreter der ~:** a representative of justice *or* of the law

Justiz-: ~beamte der court official; **~behörde die** judicial authority; **~irrtum der** miscarriage of justice; **~minister der** Minister of Justice; **~ministerium das** Ministry of Justice; **~mord der** judicial murder; **~vollzugs·anstalt die** *(Amtsspr.)* penal institution *(formal)*

Jute ['ju:tə] **die; ~:** jute

Jute·sack der jute *or* gunny sack

Jüt·land ['jy:t-] **(das); ~s** Jutland

[1]**Juwel** [ju've:l] **das** *od.* **der; ~s, ~en** piece *or* item of jewellery; *(Edelstein)* jewel; gem

[2]**Juwel das; ~s, ~e** *(fig.)* gem; **ein ~ gotischer Baukunst** a gem *or* jewel of Gothic architecture

Juwelen·raub der jewel robbery

Juwelier [juvə'li:ɐ̯] **der; ~s** jeweller; *s. auch* **Bäcker**

Juwelier·geschäft das jeweller's shop

Jux [jʊks] **der; ~es, ~e** *(ugs.)* joke; **aus ~:** as a joke; for fun; **sie machten sich** *(Dat.)* **einen ~ daraus, das zu tun** they did it as a joke *or* for a lark

jwd [jɔtve:'de:] *Adv. (ugs. scherzh.)* in *or* at the back of beyond; miles out

K

k, K [ka:] **das; ~, ~:** k/K; *s. auch* **a, A**

Kabarett [kaba'rɛt] **das; ~s, ~s** *od.* **~e a)** satirical cabaret [show]; satirical revue; **ein politisches ~:** a satirical political revue; **b)** *(Ensemble)* cabaret act

Kabarettist der; ~en, ~en, Kabarettistin die; ~, ~nen revue performer

kabarettistisch *Adj.* [satirical] revue *attrib.*; **~e Szenen** scenes in the style of a [satirical] revue

Kabäuschen [ka'bɔysçən] **das; ~s, ~** *(ugs.) (Zimmer)* cubbyhole; *(Häuschen)* little hut

Kabbelei die; ~, ~**en** squabble
kabbeln ['kabln] *refl. V. (ugs.)* squabble, bicker (**mit** with)
Kabel ['ka:bl] **das;** ~**s,** ~ *(auch veralt.: Telegramm)* cable; *(für kleineres Gerät)* flex
Kabel·fernsehen das cable television
Kabeljau ['ka:bljau] **der;** ~**s,** ~**e** *od.* ~**s** cod
kabeln *tr., itr. V. (veralt.)* cable
Kabine [ka'bi:nə] **die;** ~, ~**n a)** cabin; **b)** *(Umkleideraum, abgeteilter Raum)* cubicle; **in die** ~**n gehen** *(Fußball)* go back into the dressing-rooms; **c)** *(einer Seilbahn)* [cable-]car
Kabinen-: ~**bahn die** cableway; ~**roller der** bubble car
Kabinett [kabi'nɛt] **das;** ~**s,** ~**e a)** Cabinet; **b)** *(österr.: kleines Zimmer)* small room with one window; box-room *(Brit.)*; **c)** *o. Art.; o. Pl. (Weinprädikat)* Kabinett
Kabinetts-: ~**beschluß der** Cabinet decision; ~**bildung die** formation of a/the Cabinet; ~**sitzung die** Cabinet meeting
Kabinett·stück[chen] das tour de force
Kabrio ['ka:brio] **das;** ~**s,** ~**s, Kabriolett** [kabrio'lɛt] **das;** ~**s,** ~**s** convertible
Kabuff [ka'bʊf] **das;** ~**s,** ~**s** *(ugs., oft abwertend)* [poky little] cubbyhole
Kachel ['kaxl] **die;** ~, ~**n** [glazed] tile; **etw. mit** ~**n auslegen** tile sth.
kacheln *tr. V.* tile; **eine grün gekachelte Wand** a wall covered with green tiles
Kachel·ofen der tiled stove
Kacke ['kakə] **die;** ~ *(derb; auch fig.)* shit *(coarse)*; crap *(coarse)*; **so eine** ~! shit! *(coarse)*
kacken ['kakn] *itr. V. (derb)* shit *(coarse)*; crap *(coarse)*
Kadaver [ka'da:vɐ] **der;** ~**s,** ~ *(auch fig., abwertend)* carcass
Kadaver·gehorsam der *(abwertend)* blind obedience
Kadenz [ka'dɛnts] **die;** ~, ~**en** *(Musik)* cadence; *(solistische Paraphrasierung)* cadenza
Kader ['ka:dɐ] **der** *od. (schweiz.)* **das;** ~**s,** ~ **a)** cadre; **b)** *(Sport)* squad
Kader-: ~**abteilung die** *(ehem. DDR)* personnel department; ~**akte die** *(ehem. DDR)* personal file; ~**arbeit die** *(ehem. DDR)* cadre work; ~**leiter der** *(ehem. DDR)* [chief] personnel officer
Kadi ['ka:di] **der;** ~**s,** ~**s** cadi; **jmdn. vor den** ~ **schleppen** *(ugs.)* haul sb. up before a judge *or (Brit. sl.)* the beak
Käfer ['kɛ:fɐ] **der;** ~**s,** ~: beetle
Kaff [kaf] **das;** ~**s,** ~**s** *od.* **Käffer** ['kɛfɐ] *(ugs. abwertend)* dump *(coll.)*; hole *(coll.)*
Kaffee ['kafe *od. (österr.)* ka'fe:] **der;** ~**s,** ~**s a)** coffee; ~ **kochen** make coffee; ~ **mit Milch** white coffee *(Brit.)*; coffee with milk; **dir haben sie wohl was in den** ~ **getan?** *(ugs.)* have you gone soft in the head? *(coll.)*; **das ist kalter** ~ *(ugs.) (ist längst bekannt)* that's old hat *(coll.)*; *(ist Unsinn)* that's a load of old rubbish *(coll.)*; **b)** *(Nachmittags~)* afternoon coffee; ~ **trinken** have afternoon coffee
kaffee-, Kaffee-: ~**bohne die** coffee-bean; ~**braun** *Adj.* coffee-coloured; ~**filter der** coffee filter; *(Filtertüte)* filter [paper]
Kaffee·haus das *(bes. österr.)* coffee-house
Kaffee-: ~**kanne die** coffee-pot; ~**klatsch der** *(ugs. scherzh.)* get-together and a chat over coffee; coffeeklatsch *(Amer.)*; ~**kränzchen das** *(veralt.)* **a)** *(Zusammentreffen)* coffee afternoon; **b)** *(Gruppe)* coffee circle; ~**löffel der** coffee-spoon; ~**maschine die** coffee-maker; ~**mühle die** coffee-grinder; ~**pulver das** coffee powder; ~**satz der** coffee-grounds *pl.*; ~**service das** coffee-service *or* -set; ~**sieb das** coffee-strainer; ~**tante die** *(ugs. scherzh.)* coffee addict; ~**tasse die** coffee-cup; ~**tisch der: sie saßen gerade am** ~**tisch** they were [sitting] having coffee and cakes; ~**wärmer der** coffee-pot cosy *or* cover; ~**wasser das:** ~**wasser/das** ~**wasser aufsetzen** put on some water for coffee/the water for the coffee
Kaffer ['kafɐ] **der;** ~**s,** ~**n a)** Xhosa; **b)** *(Schimpfwort)* blockhead; thickhead
Käfig ['kɛ:fɪç] **der;** ~**s,** ~**e** cage; **in einem goldenen** ~ **sitzen** *(fig.)* be a bird in a gilded cage
Kaftan ['kaftan] **der;** ~**s,** ~**e** caftan
kahl [ka:l] *Adj.* **a)** *(ohne Haare, Federn)* bald; ~ **werden** go bald; **b)** *(ohne Grün, schmucklos)* bare
kahl|fressen *unr. tr. V.* **etw.** ~**fressen** strip sth. bare
Kahlheit die; ~: *s.* **kahl a, b:** baldness; bareness
kahl-, Kahl-: ~**kopf der a)** bald head; **b)** *(ugs.: Person)* baldhead; ~**köpfig** *Adj.* bald[-headed]; ~|**scheren** *unr. tr. V.* **jmdn.** ~**scheren** shave sb.'s hair off; shave sb.'s head; **sein** ~**geschorener Kopf** his shaven head; ~**schlag der a)** clear-felling *no indef. art.*; clear-cutting *no indef. art.*; **b)** *(Waldfläche)* clear-felled area; **c)** *(fig.)* clearance
Kahn [ka:n] **der;** ~**[e]s, Kähne** ['kɛ:nə] **a)** *(Ruder~)* rowing-boat; *(Stech~)* punt; ~ **fahren** go rowing/punting; **b)** *(Lastschiff)* barge; **c)** *(ugs.: Schiff)* tub
Kahn·fahrt die; ~, ~**en** trip in a rowing-boat/punt
Kai [kai] **der;** ~**s,** ~**s** quay
Kai·anlage die quays *pl.*
Kaiman ['kaiman] **der;** ~**s,** ~**e** *(Zool.)* cayman
Kai·mauer die quay wall
Kairo ['kairo] **(das);** ~**s** Cairo
Kaiser ['kaizɐ] **der;** ~**s,** ~: emperor; **sich um des** ~**s Bart streiten** engage in pointless argument
Kaiserin die; ~, ~**nen** empress
Kaiser·krone die a) imperial crown; **b)** *(Zierpflanze)* crown imperial
kaiserlich *Adj.* imperial
kaiserlich-königlich *Adj.; nicht präd.* imperial and royal
Kaiser-: ~**reich das** empire; ~**schmarren der** *(österr., südd.) pancake pulled to pieces and sprinkled with powdered sugar and raisins;* ~**schnitt der** Caesarean section; ~**wetter das** *(scherzh.)* glorious, sunny weather *(for an event)*
Kajak ['ka:jak] **der;** ~**s,** ~**s** kayak
Kajüte [ka'jy:tə] **die;** ~, ~**n** *(Seemannsspr.)* cabin
Kakadu ['kakadu] **der;** ~**s,** ~**s** cockatoo
Kakao [ka'kau] **der;** ~**s,** ~**s** cocoa; **jmdn./etw. durch den** ~ **ziehen** *(ugs.)* make fun of sb./sth.; take the mickey out of sb./sth. *(Brit. sl.)*
Kakao·pulver das cocoa powder
Kakerlak ['ka:kɐlak] **der;** ~**s** *od.* ~**en,** ~**en** cockroach; black-beetle
Kaktee [kak'te:ə] **die;** ~, ~**n, Kaktus** ['kaktʊs] **der;** ~, **Kakteen** cactus
Kalauer ['ka:lauɐ] **der;** ~**s,** ~: laboured *or (coll.)* corny joke; *(Wortspiel)* atrocious *or (coll.)* corny pun
kalauern *itr. V.* tell laboured *or (coll.)* corny jokes; *(mit Wortspielen)* make atrocious *or (coll.)* corny puns
Kalb [kalp] **das;** ~**[e]s, Kälber** ['kɛlbɐ] **a)** calf; *(Hirsch~)* fawn; **b)** *(~fleisch)* veal
Kälbchen ['kɛlpçən] **das;** ~**s,** ~: little calf

kalben *itr. V. (auch Geogr.)* calve
kalbern ['kalbɐn] *itr. V. (ugs.)* mess *or* fool about *or* around
Kalb·fleisch das veal
Kalbs-: ~braten der roast veal *no indef. art.; (Gericht)* roast of veal; **~brust die** breast of veal; **~frikassee das** fricassee of veal; **~hachse,** *(südd.:)* **~haxe die** knuckle of veal; **~leder das** calfskin; calf-leather; **~schnitzel das** veal cutlet
Kaldaune [kal'daunə] **die; ~, ~n** entrails *pl.*
Kaleidoskop [kalaido'sko:p] **das; ~s, ~e** *(auch fig.)* kaleidoscope
Kalender [ka'lɛndɐ] **der; ~s, ~:** calendar; *(Taschen~)* diary; **sich** *(Dat.)* **etw./einen Tag im ~ [rot] anstreichen** *(oft iron.)* mark sth. in red on the calendar/mark a day as a red-letter day
Kalender-: ~blatt das calendar sheet; **~jahr das** calendar year
Kali ['ka:li] **das; ~s, ~s** potash
Kaliber [ka'li:bɐ] **das; ~s, ~ a)** *(Technik, Waffenkunde)* calibre; **b)** *(ugs., oft abwertend)* sort; kind
Kali·dünger der potash fertilizer
Kalif [ka'li:f] **der; ~en, ~en** *(hist.)* caliph
Kalifornien [kali'fɔrni̯ən] **(das); ~s** California
kalifornisch *Adj.* Californian
Kalium ['ka:li̯ʊm] *(Chemie)* **das; ~s** potassium
Kalk [kalk] **der; ~[e]s, ~e a)** *(Kalziumkarbonat)* calcium carbonate; **b)** *(Baustoff)* lime; quicklime; burnt lime; **bei ihm rieselt schon der ~** *(salopp)* he's going a bit senile
Kalk-: ~ablagerung die deposit of calcium carbonate; **~boden der** limy soil; lime soil
kalken ['kalkn̩] *tr. V. (tünchen)* whitewash
kalk-, Kalk-: ~erde die a) *(gebrannter Kalk)* lime; quicklime; burnt lime; **b)** *s.* **~boden; ~haltig** *Adj. (bes. Geol., Mineral.)* limy ‹*soil*›; calcareous ‹*soil, rock*› *(Geol., Min.);* ‹*water*› containing calcium carbonate; **das Wasser ist sehr ~haltig** the water is high in calcium carbonate; **~mangel der;** *o. Pl.* **a)** *(Mangel an Kalzium)* calcium deficiency; **b)** *(Mangel an Kalk)* deficiency of lime; **~stein der** limestone
Kalkül [kal'ky:l] **das** *od.* **der; ~s, ~e** *(geh.)* calculation
Kalkulation [kalkula'tsi̯o:n] **die; ~, ~en** *(auch Wirtsch.)* calculation; **nach meiner ~:** according to my calculations *pl.*
kalkulieren 1. *tr. V.* **a)** *(Kaufmannsspr.)* calculate ‹*cost, price*›; cost ‹*product, article*›; **b)** *(abschätzen)* calculate. **2.** *itr. V.* calculate; **falsch ~:** miscalculate
Kalkutta [kal'kʊta] **(das); ~s** Calcutta
kalk·weiß *Adj.* **a)** chalk-white; **b)** *(sehr bleich)* deathly pale; chalky white; **~ sein** be as white as a sheet
Kalorie [kalo'ri:] **die; ~, ~n** calorie
kalorien-, Kalorien-: ~arm 1. *Adj.* low-calorie *attrib.;* **~arm sein** be low in calories; **2.** *adv.* **~arm kochen/essen** cook low-calorie meals/eat low-calorie foods; **~gehalt der** calorie content; **~reich 1.** *Adj.* high-calorie *attrib.;* **~reich sein** be high in calories; **2.** *adv.* **~reich kochen/essen** cook high-calorie meals/eat high-calorie foods
kalt [kalt]; **kälter** ['kɛltɐ], **kältest...** ['kɛltəst...] **1.** *Adj.* cold; chilly, frosty ‹*atmosphere, smile*›; **ein ~es Buffet** a cold buffet; **mir ist/wird ~:** I am/am getting cold; **das Essen wird ~:** the food is getting cold; **im Kalten sitzen** sit in the cold; **~ und berechnend sein** be cold and calculating; **es packte uns das ~e Grausen/Entsetzen** our blood ran cold; **jmdm. die ~e Schulter zeigen** give sb. the cold shoulder; cold-shoulder sb. **2.** *adv.* **a) ~ duschen** have *or* take a cold shower; **~ schlafen** sleep in a cold room; **Getränke/Sekt ~ stellen** cool drinks/chill champagne; **jmdn. ~ erwischen** *(bes. Sportjargon)* catch sb. on the hop; **b)** *(nüchtern)* coldly; **c)** *(abweisend, unfreundlich)* coldly; frostily; **jmdn. ~ anblicken** look at sb. coldly; **~ lächeln** smile coldly *or* frostily; **d) mich überlief** *od.* **durchrieselte es ~:** cold shivers ran down my spine
kalt-, Kalt-: ~|bleiben *unr. itr. V.; mit sein* remain unmoved; **~blüter** [-bly:tɐ] **der** *(Zool.)* cold-blooded animal; **~blütig 1.** *Adj.* **a)** *(beherrscht)* cool-headed; **b)** *(abwertend: skrupellos)* cold-blooded; **c)** *(Zool.)* cold-blooded; **2.** *adv.* **a)** *(beherrscht)* coolly; calmly; **b)** *(abwertend: skrupellos)* cold-bloodedly; **~blütigkeit die; ~** *s.* **~blütig a, b:** cool-headedness; cold-bloodedness
Kälte ['kɛltə] **die; ~ a)** cold; **10 Grad ~:** 10 degrees of frost; 10 degrees below freezing; **vor ~ zittern** shiver with cold; **bei dieser ~:** in this cold; when it's as cold as this; **b)** *(fig.)* coldness
kälte-, Kälte-: ~beständig *Adj.* cold-resistant; **~beständig sein** be resistant to cold; **~einbruch der** *(Met.)* sudden onset of cold weather; **~empfindlich** *Adj.* sensitive to cold *pred.;* **~empfindliche Pflanzen** plants which are sensitive to cold; **~grad der** degree of frost
kälter, kältest... *s.* **kalt**
Kälte-: ~technik die refrigeration engineering *no art.;* **~tod der: den ~tod erleiden** freeze to death; die of cold; **~welle die** cold wave *or* spell
kalt-, Kalt-: ~front die *(Met.)* cold front; **~herzig** *Adj.* cold-hearted; **~lächelnd** *Adv. (ugs. abwertend)* **etw. ~lächelnd tun** take callous pleasure in doing sth.; **~|lassen** *unr. tr. V. (ugs.)* **jmdn. ~lassen** leave sb. unmoved; *(nicht interessieren)* leave sb. cold *(coll.);* **~luft die;** *o. Pl.* cold air; **~|machen** *tr. V. (salopp)* **jmdn. ~machen** do sb. in *(sl.);* **~mamsell die** *girl/woman who prepares and serves cold dishes in a restaurant, hotel, etc.;* **~miete die** rent exclusive of heating; **~schale die** *cold sweet soup made with fruit, beer, wine, or milk;* **~schnäuzig** [~ʃnɔy̆tsɪç] *(ugs.)* **1.** *Adj.* cold and insensitive; **2.** *adv.* coldly and insensitively; **~schnäuzigkeit die; ~** *(ugs.)* coldness and insensitivity; *(Frechheit)* insolence; **~|stellen** *tr. V. (ugs.)* **jmdn. ~stellen** put sb. out of the way *(coll. joc.);* **den Mittelstürmer ~stellen** cut the centre-forward out of the game
Kalvinismus [kalvi'nɪsmʊs] **der; ~:** Calvinism *no art.*
Kalvinist der; ~en, ~en, Kalvinistin die; ~, ~nen Calvinist
kalvinistisch *Adj.* Calvinist
Kalzium ['kaltsi̯ʊm] **das; ~s** calcium
kam [ka:m] *1. u. 3. Pers. Prät. v.* **kommen**
Kambodscha [kam'bɔdʒa] **(das); ~s** Cambodia
Kambodschaner [kambo'dʒa:nɐ] **der; ~s, ~:** Cambodian
käme ['kɛ:mə] *1. u. 3. Pers. Konjunktiv II v.* **kommen**
Kamel [ka'me:l] **das; ~s, ~e a)** camel; **b)** *(salopp)* clot *(Brit. sl.);* twit *(Brit. sl.);* fathead
Kamel-: ~haar das camel-hair; **~haar·mantel der** camel-hair coat
Kamelie [ka'me:li̯ə] **die; ~, ~n** camellia

Kamellen [ka'mɛlən] *Pl. (ugs.) in* **alte** *od.* **olle ~:** old hat *sing. (coll.)*

Kamera ['kaməra] **die; ~, ~s** camera

Kamerad [kamə'ra:t] **der; ~en, ~en** *(Gefährte)* companion; *(Freund)* friend; *(Mitschüler)* mate; friend; *(Soldat)* comrade; *(Sport)* team-mate

Kameraderie [kamәradә'ri:] **die; ~** *(meist abwertend)* loyalty to a/the clique

Kameradin die; ~, ~nen *s.* **Kamerad**

Kameradschaft die; ~: comradeship; **die ~ zwischen ihnen** the sense of comradeship between them

kameradschaftlich 1. *Adj.* comradely. **2.** *adv.* in a comradely way

Kameradschaftlichkeit die; ~: comradeliness

Kamera-: ~führung die *(Film)* camerawork *no indef. art.*; **~mann der** *Pl.* **~männer** *od.* **~leute** cameraman

Kamerun ['kaməru:n] **(das); ~s** Cameroon; the Cameroons *pl.*

Kameruner der; ~s, ~: Cameroonian

Kamille [ka'mılə] **die; ~, ~n** camomile

Kamillen·tee der camomile tea

Kamin [ka'mi:n] **der,** *schweiz.:* **das; ~s, ~e a)** fireplace; **sie saßen am ~:** they sat by the hearth *or* the fireside; **b)** *(bes. südd.: Schornstein; Felsspalt)* chimney

Kamin-: ~feger der *(bes. südd.) s.* **Schornsteinfeger; ~feuer das** [open] fire; **~kehrer der** *(bes. südd.) s.* **Schornsteinfeger; ~sims der** *od.* **das** mantelpiece; mantelshelf

Kamm [kam] **der; ~[e]s, Kämme** ['kɛmə] **a)** comb; **alle/alles über einen ~ scheren** lump everyone/everything together; **b)** *(bei Hühnern usw.)* comb; *(bei Reptilien, Amphibien)* crest; **ihm schwillt der ~** *(ugs.)* he gets cocky and big-headed *(coll.)*; **c)** *(Gebirgs~)* ridge; crest; **d)** *(Wellen~)* crest; **e)** *(Rinder~)* neck; *(Schweine~)* spare rib

kämmen ['kɛmən] *tr. V.* comb; **jmdm./sich die Haare ~, jmdn./sich ~:** comb sb.'s/one's hair; **jmdm. einen Scheitel/Pony ~:** put a parting in sb.'s hair/comb sb.'s hair into a fringe

Kammer ['kamɐ] **die; ~, ~n a)** store-room; *(veralt.: Schlafraum)* chamber; **b)** *(Biol., Med., Technik)* chamber; **c)** *(Parl.)* chamber; House; **die erste/zweite ~:** the upper/lower chamber *or* House; **d)** *(Rechtsw.)* court *(dealing with a particular branch of judicial business)*; **e)** *(gewerbliche Vereinigung)* professional association; **f)** *(Milit.)* stores *pl.*

Kammer·chor der chamber choir

Kämmerchen ['kɛmɐçən] **das; ~s, ~:** small room; *(Abstellkammer)* [small] store-room

Kammer·diener der *(veralt.)* valet

Kämmerer ['kɛmɐrɐ] **der; ~s, ~** *(veralt.)* [town/city] treasurer

Kammer-: ~jäger der pest controller; **~konzert das** chamber concert

Kämmerlein ['kɛmɐlain] **das; ~s, ~** *(oft scherzh.) in* **im stillen ~ über etw.** *(Akk.)* **nachdenken** think about sth. in peace and quiet

Kammer-: ~musik die; *o. Pl.* chamber music; **~orchester das** chamber orchestra; **~ton der** *(Musik)* standard pitch; **~zofe die** *(veralt.)* lady's maid

Kamm·garn das worsted

Kampagne [kam'panjə] **die; ~, ~n** campaign (**für** for, on behalf of; **gegen** against)

Kämpe ['kɛmpə] **der; ~n, ~n** *(veralt.)* [brave] warrior *or* fighter; **ein alter ~** *(scherzh.)* an old campaigner; a seasoned veteran

Kampf [kampf] **der; ~[e]s, Kämpfe a)** *(militärisch)* battle (**um** for); **nach wochenlangen erbitterten Kämpfen** after weeks of bitter fighting; **er ist im ~ gefallen** he fell *or* was killed in action *or* combat; **b)** *(zwischen persönlichen Gegnern)* fight; *(fig.)* struggle; **ein ~ aller gegen alle** a free-for-all; **ein ~ Mann gegen Mann** a hand-to-hand fight; **ein ~ auf Leben und Tod** a fight to the death; **c)** *(Wett~)* contest; *(Boxen)* contest; fight; bout; **sich einen spannenden ~ liefern** produce an exciting contest; **d)** *(fig.)* struggle, fight (**um, für** for; **gegen** against); **der ~ ums Dasein** the struggle for existence; **jmdm./einer Sache den ~ ansagen** declare war on sb./sth.; **der ~ zwischen den Geschlechtern** the battle of the sexes

kampf-, Kampf-: ~ab·stimmung die *(Politik)* crucial vote; **~an·sage die** declaration of war; **~bahn die** *(für Gladiatoren)* arena; *(für Stiere)* ring; **~bereit** *Adj.* **a)** *(vorbereitet)* ready to fight *postpos.*; ⟨*army*⟩ ready for battle; ⟨*troops*⟩ ready for battle *or* action; **b)** *(willens)* willing to fight *postpos.*; **~bereitschaft die** *s.* **~bereit a, b:** readiness to fight; readiness for battle; readiness for battle *or* action; willingness to fight

kämpfen ['kɛmpfn̩] **1.** *itr. V.* **a)** fight (**um, für** for); **mit jmdm. ~:** fight [with] sb.; **gegen jmdn. ~:** fight [against] sb.; **mit den Tränen ~** *(fig.)* fight back one's tears; **mit dem Schlaf ~** *(fig.)* struggle to keep awake; **mit dem Tod ~** *(fig.)* fight for one's life *or* to stay alive; **mit etw. zu ~ haben** *(fig.)* have to contend with sth.; **[lange] mit sich** *(Dat.)* **~:** have a [long] struggle with oneself; **b)** *(Sport: sich messen)* ⟨*team*⟩ play; ⟨*wrestler, boxer*⟩ fight; **gegen jmdn. ~:** play/fight sb. **2.** *refl. V. (auch fig.)* fight one's way. **3.** *tr. V.* **einen Kampf ~** *(auch fig.)* fight a battle

Kampfer ['kampfɐ] **der; ~s** camphor

Kämpfer ['kɛmpfɐ] **der; ~s, ~, Kämpferin die; ~, ~nen** fighter

kämpferisch 1. *Adj.* **a)** fighting ⟨*spirit, mood*⟩; ⟨*person*⟩ full of fighting spirit; **eine ~e Natur sein** be full of fighting spirit; **b)** *(Sport)* spirited. **2.** *adv.* **a)** in a fighting spirit; **b)** *(Sport)* spiritedly

Kämpfer·natur die fighter

kampf·erprobt *Adj.* battle-tried; battle-tested ⟨*equipment*⟩

kampf-, Kampf-: ~flugzeug das bomber; **~gas das** war gas; **~gebiet das** battle area; combat zone; **~geist der;** *o. Pl.* fighting spirit; **~gericht das** *(Sport)* [panel of] judges *pl.*; **~hahn der a)** fighting cock; **b)** *(fig. ugs.)* fighter; brawler; **~handlungen** *Pl.* fighting *sing.*; **die ~handlungen einstellen** cease hostilities *or* fighting; **~kraft die** fighting power *or* strength; **~lied das** battle song; *(einer Bewegung)* battle anthem; **~los 1.** *Adj.* **an eine ~lose Übergabe der Stadt war nicht zu denken** to hand over the town without a fight was unthinkable; **2.** *adv.* without a fight; **~lustig 1.** *Adj.* belligerent; **2.** *adv.* belligerently; **~maßnahme die;** *meist Pl.* active measure; **~maßnahmen [beschließen]** [decide to take] action *sing.*; **~platz der** battlefield; **~richter der** *(Sport)* judge; **~stark** *Adj.* powerful ⟨*army*⟩; efficient ⟨*troops*⟩; strong, powerful ⟨*team*⟩; **~stärke die** *(eines Heeres)* fighting strength *or* power; *(einer Mannschaft)* strength; **~stoff der;** *meist Pl.* warfare agent; **~unfähig** *Adj.*

‹*troops*› unfit for action *or* battle; ‹*boxer etc.*› unfit to fight; **jmdn./etw. ~unfähig machen** put sb./sth. out of action; **~zone die** *(Milit.)* battle zone; combat zone

kampieren [kam'pi:rən] *itr. V.* camp; *(ugs.: wohnen)* camp down *or* out; *(ugs.: übernachten)* bed *or (Brit. sl.)* doss down

Kanada ['kanada] **(das); ~s** Canada

Kanadier [ka'na:di̯ɐ] **der; ~s, ~ a)** *(Einwohner Kanadas)* Canadian; **b)** *(Boot)* Canadian canoe

kanadisch [ka'na:dɪʃ] *Adj.* Canadian

Kanaille [ka'naljə] **die; ~, ~n** *(abwertend)* scoundrel; villain

Kanal [ka'na:l] **der; ~s, Kanäle** [ka'nɛ:lə] **a)** canal; **b)** *(Geogr.)* **der ~:** the [English] Channel; **c)** *(für Abwässer)* sewer; **d)** *(zur Entwässerung, Bewässerung)* channel; *(Graben)* ditch; **e)** *(Rundf., Ferns., Weg der Information)* channel; **f)** *(salopp)* **den ~ voll haben** *(betrunken sein)* be canned *or* plastered *(sl.)*; *(überdrüssig sein)* have had a bellyful *or* as much as one can take

Kanal-: ~arbeiter der sewerage worker; **~deckel der** manhole cover; **~inseln** *Pl.* **die ~inseln** the Channel Islands

Kanalisation [kanaliza'tsi̯o:n] **die; ~, ~en a)** *(System der Abwasserkanäle)* sewerage system; sewers *pl.*; **b)** *(Ausbau eines Flusses)* canalization

kanalisieren *tr. V.* **a)** *(lenken)* channel ‹*energies, goods, etc.*›; **b)** *(schiffbar machen)* canalize

Kanalisierung die; ~, ~en a) *(Lenkung)* channelling; **b)** *(Schiffbarmachen)* canalization

Kanal·tunnel der Channel Tunnel

Kanapee ['kanape] **das; ~s, ~s a)** *(veralt., scherzh.: Sofa)* sofa; settee; **b)** *(belegtes Weißbrotschnittchen)* canapé

Kanaren [ka'na:rən] *Pl.* **die ~:** the Canaries

Kanarien·vogel [ka'na:ri̯ən-] **der** canary

Kanarische Inseln *Pl.* **die ~n, ~:** the Canary Islands

Kandare [kan'da:rə] **die; ~, ~n** curb bit; **jmdn. an die ~ nehmen** *(fig.)* take sb. in hand

Kandelaber [kande'la:bɐ] **der; ~s, ~:** candelabrum

Kandidat [kandi'da:t] **der; ~en, ~en, Kandidatin die; ~, ~nen a)** candidate; **b)** *(beim Quiz usw.)* contestant

Kandidatur [kandida'tu:ɐ̯] **die; ~, ~en** candidature **(auf +** *Akk.* for)

kandidieren *itr. V.* stand [as a candidate] **(für** for)

kandieren [kan'di:rən] *tr. V.* candy; **kandiert** crystallized ‹*orange, petal*›; glacé ‹*cherry, pear*›; candied ‹*peel*›

Kandis ['kandɪs] **der; ~, Kandis·zucker der** rock candy

Känguruh ['kɛŋguru] **das; ~s, ~s** kangaroo

Kanin [ka'ni:n] **das; ~s, ~e** *(fachspr.)* rabbit [fur]

Kaninchen [ka'ni:nçən] **das; ~s, ~:** rabbit

Kaninchen-: ~bau der; *Pl.* **~baue** rabbit-burrow; rabbit-hole; **~fell das** rabbit fur; **~stall der** rabbit-hutch

Kanister [ka'nɪstɐ] **der; ~s, ~:** can; [metal/plastic] container

kann [kan] *1. u. 3. Pers. Sg. Präsens v.* **können**

Kännchen ['kɛnçən] **das; ~s, ~:** [small] pot; *(für Milch)* [small] jug; **ein ~ Kaffee/Milch** a [small] pot of coffee/jug of milk

Kanne ['kanə] **die; ~, ~n a)** *(Krug) (Tee~, Kaffee~)* pot; *(Milch~, Wasser~)* jug; **b)** *(Henkel~)* can; *(große Milch~)* churn; *(Gieß~)* watering-can

kannen·weise *Adv.* by the jugful

Kannibale [kani'ba:lə] **der; ~n, ~n** cannibal

kannibalisch *Adj.* cannibalistic

Kannibalismus der; ~ *(auch Zool.)* cannibalism *no art.*

kannst [kanst] *2. Pers. Sg. Präsens v.* **können**

kannste ['kanstə] *(ugs.)* = **kannst du;** *s. auch* **haste**

kannte ['kantə] *1. u. 3. Pers. Sg. Prät. v.* **kennen**

Kanon ['ka:nɔn] **der; ~s, ~s** *(Musik, Lit., Theol., geh.)* canon

Kanonade [kano'na:də] **die; ~, ~n** *(Milit.)* cannonade; *(fig. ugs.)* barrage

Kanone [ka'no:nə] **die; ~, ~n a)** cannon; big gun; **mit ~n auf Spatzen** *(Akk.)* **schießen** *(fig.)* take a sledgehammer to crack a nut; **das ist unter aller ~** *(ugs.)* it's appallingly bad *or* indescribably dreadful *(coll.)*; **b)** *(ugs.: Könner)* ace; **c)** *(salopp: Revolver)* shooting-iron *(sl.)*; rod *(Amer. sl.)*

Kanonen·boot das gunboat

Kanonen-: ~donner der [rumble of] gunfire; **~futter das** *(ugs.)* cannon-fodder; **~kugel die** cannon-ball; **~ofen der** cylindrical [iron] stove; **~rohr das** gun-barrel

Kanonier [kano'ni:ɐ̯] **der; ~s, ~e** *(Milit.)* gunner; artilleryman

kanonisch *Adj. (kath. Kirche)* canonical; **~es Recht** canon law

kanonisieren *tr. V.* canonize

Kanossa·gang [ka'nɔsa-] **der** *(geh.)* humiliation; **einen ~ antreten/machen** eat humble pie; go to Canossa *(literary)*

Kantate [kan'ta:tə] **die; ~, ~n** *(Musik)* cantata

Kante ['kantə] **die; ~, ~n** edge; *(bei Stoffen)* selvedge; **etw. auf die hohe ~ legen** *(ugs.)* put sth. away *or* by; **etw. auf der hohen ~ haben** *(ugs.)* have sth. put away *or* by; *s. auch* **Ecke a**

kantig *Adj.* square-cut ‹*timber, stone*›; rough-edged ‹*rock*›; angular ‹*face, figure, etc.*›; sharp ‹*nose*›; square ‹*chin*›; jerky, awkward ‹*movement*›

Kantine [kan'ti:nə] **die; ~, ~n** canteen

Kanton [kan'to:n] **der; ~s, ~e** canton

kantonal [kanto'na:l] **1.** *Adj.* cantonal. **2.** *adv.* on a cantonal basis

Kantons-: ~rat der *(schweiz.)* cantonal great council; **~regierung die** *(schweiz.)* cantonal government

Kantor ['kantɔr] **der; ~s, ~en** [-'to:rən] choirmaster and organist

Kant·stein der kerb

Kanu ['ka:nu] **das; ~s, ~s** canoe

Kanüle [ka'ny:lə] **die; ~, ~n** *(Med.)* cannula; *(einer Injektionsspritze)* [hypodermic] needle

Kanute [ka'nu:tə] **der; ~n, ~n** *(Sport)* canoeist

Kanzel ['kants̮l] **die; ~, ~n a)** pulpit; **auf der ~:** in the pulpit; **b)** *(Flugw.)* cockpit

kanzerogen [kantsero'ge:n] *Adj. (Med.)* carcinogenic

Kanzlei [kants'lai̯] **die; ~, ~en a)** *(veralt.: Büro)* office; **b)** *(Anwalts~)* chambers *pl.* *(of barrister)*; office *(of lawyer)*

Kanzlei·sprache die; *o. Pl.* language of officialdom; officialese

Kanzler ['kantslɐ] **der; ~s, ~ a)** chancellor; **b)** *(an Hochschulen)* vice-chancellor

Kanzler-: ~amt das Chancellery; **~kandidat der** candidate for the chancellorship

Kap [kap] **das; ~s, ~s** cape; **das ~ der Guten Hoffnung** Cape of Good Hope; **~ Hoorn** Cape Horn

Kapazität [kapatsi'tɛ:t] **die; ~, ~en a)** *(auch Wirtsch.)* capacity; **ungenutzte ~en** *(Wirtsch.)* unused capacity; **b)** *(Experte)* expert

Kapee [ka'pe] *in* **schwer von ~ sein** *(salopp)* be slow on the uptake

Kapelle [ka'pɛlə] **die; ~, ~n a)** *(Archit.)* chapel; **b)** *(Musik)* band; [light] orchestra

Kapẹll·meister der bandleader; bandmaster; *(im Orchester)* conductor; *(im Theater usw.)* musical director

Kaper ['ka:pɐ] **die; ~, ~n** caper *usu. in pl.*

ka̲pern *tr. V. (hist.)* capture, seize ‹*ship*›

Ka̲pern·soße die caper sauce

Ka̲per·schiff das *(hist.)* privateer

kapieren [ka'pi:rən] *(ugs.)* **1.** *tr. V. (ugs.)* get *(coll.)*; understand; **kapier das endlich!** get that into your thick skull! *(coll.)*. **2.** *itr. V.* **schnell ~:** be quick to catch on *(coll.)*; **kapiert?** got it? *(coll.)*

kapital [kapi'ta:l] *Adj.; nicht präd.* major ‹*error, blunder, etc.*›; *(Jägerspr.)* large and powerful; royal ‹*stag*›

Kapita̲l das; ~s, ~e *od.* **~ien a)** capital; **b)** *(fig.)* asset; **~ aus etw. schlagen** make capital out of sth.; capitalize on sth.

Kapita̲l-: ~anlage die *(Wirtsch.)* capital investment; **~flucht die** flight of capital

Kapitalịsmus der; ~ capitalism *no art.*

Kapitalịst der; ~en, ~en capitalist

kapitalịstisch 1. *Adj.* capitalistic. **2.** *adv.* capitalistically

Kapita̲l-: ~markt der *(Wirtsch.)* capital market; **~verbrechen das** serious offence *or* crime; *(mit Todesstrafe bedroht)* capital offence *or* crime; **~verbrecher der** serious/capital offender; **~verflechtung die** interlacing of capital interests

Kapitän [kapi'tɛ:n] **der; ~s, ~e** *(Seew., Flugw., Sport)* captain; **~ der Landstraße** *(ugs.)* knight of the road

Kapitä̲ns·patent das master's certificate

Kapitel [ka'pɪtl̩] **das; ~s, ~** *(auch fig.)* chapter; **das ist ein anderes ~** *(fig.)* that's another story; **das ist ein ~ für sich** *(fig.)* that's a complicated subject

Kapitell [kapi'tɛl] **das; ~s, ~e** capital

Kapitulation [kapitula'tsi̯o:n] **die; ~, ~en a)** *(Milit.)* surrender; capitulation; *(Vertrag)* surrender *or* capitulation document; **seine ~ erklären** admit defeat; **b)** *(fig.)* giving up

kapituli̲eren *itr. V.* **a)** *(Milit.)* surrender; capitulate; **vor dem Feind ~:** surrender to the enemy; **b)** *(fig.)* give up; **vor etw.** *(Dat.)* **~:** give up in the face of sth.

Kaplan [ka'pla:n] **der; ~s, Kapläne** *(kath. Kirche)* **a)** *(Hilfsgeistlicher)* curate; **b)** *(Geistlicher mit besonderen Aufgaben)* chaplain

Kappe ['kapə] **die; ~, ~n** cap; **etw. auf seine [eigene] ~ nehmen** *(ugs.)* take the responsibility for sth.

kạppen *tr. V.* **a)** *(Seemannsspr.)* cut; **b)** *(beschneiden)* cut back ‹*hedge etc.*›; *(fig.)* cut; **c)** *(abschneiden)* cut off ‹*branches, shoots, crown, etc.*›

Kappes ['kapəs] **der; ~** *(bes. westd.)* **a)** *(Weißkohl)* cabbage; **b)** *(ugs.: Unsinn)* rubbish; nonsense

Käppi ['kɛpi] **das; ~s, ~s** overseas cap; garrison cap

Kaprice [ka'pri:sə] **die; ~, ~n** *(geh.)* caprice; whim

Kapriole [kapri'o:lə] **die; ~, ~n a)** caper; capriole; **~n schlagen** cut capers; **b)** *(Streich)* trick

kapriziös [kapri'tsi̯ø:s] **1.** *Adj.* capricious. **2.** *adv.* capriciously

Kapsel ['kapsl̩] **die; ~, ~n** capsule

kaputt [ka'pʊt] *Adj.* **a)** *(entzwei)* broken ‹*toy, cup, plate, arm, leg, etc.*›; **die Maschine/das Auto ist ~:** the machine/car has broken down; *(ganz und gar)* the machine/car has had it *(coll.)*; **irgend etwas ist am Auto ~:** there's something wrong with the car; **diese Jacke ist ~:** this jacket needs mending; *(ist zerrissen)* this jacket's torn; **die Birne ist ~:** the bulb has gone; *(ist zerbrochen)* the bulb is smashed; **das Telefon ist ~:** the phone is not working *or* is out of order; **der Fernseher ist ~:** the television has gone wrong; **sein Leben ist ~:** his life is in ruins; **ein ~er Typ** *(fig. ugs.)* a down-and-out; **eine ~e Lunge/ein ~es Herz haben** *(ugs.)* have bad lungs/a bad heart; **die Ehe ist ~:** the marriage has failed *or (coll.)* is on the rocks; **was ist denn jetzt ~?** *(ugs.)* what's wrong *or* the matter now?; **b)** *(ugs.: erschöpft)* shattered *(coll.)*; whacked *(Brit. coll.)*; pooped *(coll.)*; **c)** *(salopp: abartig)* sick

kapụtt-: ~|arbeiten *refl. V. (ugs.)* work oneself into the ground *(coll.)*; **~|fahren** *unr. tr. V.* **a)** *(ugs.)* smash up ‹*car etc.*›; **b)** *(salopp) s.* **totfahren; ~|gehen** *unr. itr. V.; mit sein (ugs.)* **a)** *(entzweigehen)* break; ‹*machine*› break down, *(sl.)* pack up; ‹*clothes, shoes*› fall to pieces; ‹*light-bulb*› go; *(zerbrechen)* be smashed; *(eingehen)* ‹*plant*› die; *(verderben)* ‹*fish, fruit, etc.*› go off; *(fig.)* ‹*marriage*› fail; ‹*community, relationship, etc.*› break up; **b)** *(zugrunde gehen)* ‹*firm*› go bust *(coll.)*; ‹*person*› go to pieces; **~|kriegen** *tr. V. (ugs.)* break; **wie hast du das ~gekriegt?** how did you [manage to] break it?; **~|lachen** *refl. V. (ugs.)* kill oneself [laughing] *(coll.)*; **das ist ja zum Kaputtlachen!** that's a laugh!; **~|machen** (ugs.) **1.** *tr. V.* break ‹*watch, spectacles, plate, etc.*›; spoil ‹*sth. made with effort*›; ruin ‹*clothes, furniture, etc.*›; burst ‹*balloon*›; drive ‹*business, company*› to the wall; destroy ‹*political party*›; finish ‹*person*› off; **2.** *refl. V.* wear oneself out; **~|schlagen** *unr. tr. V. (ugs.)* smash

Kapuze [ka'pu:tsə] **die; ~, ~n** hood; *(bei Mönchen)* cowl; hood

Kapuziner [kapu'tsi:nɐ] **der; ~s, ~:** Capuchin [friar]

Karabiner [kara'bi:nɐ] **der; ~s, ~ a)** *(Gewehr)* carbine; **b)** *(österr.) s.* **Karabinerhaken**

Karabi̲ner·haken der snap hook; spring hook; *(Bergsteigen)* karabiner

Karacho [ka'raxo] **das; ~s** *in* **mit ~** *od.* **in vollem ~** *(ugs.)* hell for leather *(coll.)*

Karaffe [ka'rafə] **die; ~, ~n** carafe; *(mit Glasstöpsel)* decanter

Karambolage [karambo'la:ʒə] **die; ~, ~n** *(ugs.)* crash; collision

Karamel [kara'mɛl] **der** (*schweiz.:* **das**); **~s** caramel

Karamẹl-: ~bonbon der *od.* **das** caramel [toffee]; **~creme die** crème caramel

Karamelle [kara'mɛlə] **die; ~, ~n** caramel [toffee]

Karat [ka'ra:t] **das; ~[e]s, ~e** carat; **ein Diamant von 5 ~:** a 5-carat diamond; **reines Gold hat 24 ~:** pure gold is 24 carats

Karate [ka'ra:tə] **das; ~[s]** karate

-karäter [-karɛ:tɐ] **der Zehnkaräter/Fünfkaräter** ten-carat/five-carat diamond/stone

Kara̲te·schlag der karate chop

-karätig [-karɛ:tɪç] **zehnkarätig/fünfkarätig** ten-carat/five-carat

Karavelle [kara'vɛlə] **die; ~, ~n** *(hist.)* caravel

Karawane [kara'va:nə] **die; ~, ~n** caravan

Karawa̲nen·straße die caravan route

Karbid [kar'bi:t] **das; ~[e]s, ~e** *(Chem.)* carbide

Karbol [kar'bo:l] **das; ~s** carbolic acid

Karbunkel [kar'bʊŋkl̩] **der; ~s, ~** *(Med.)* carbuncle

Kardan- [kar'da:n-] *(Technik)* cardan ‹*drive, shaft, tunnel, etc.*›

Kardinal [kardi'na:l] **der; ~s, Kardinäle** [kardi'nɛ:lə] cardinal

Kardinal-: ~fehler der cardinal error; **~tugend die;** *meist Pl.* cardinal virtue; **~zahl die** cardinal [number]

Kardiogramm [kardio'gram] **das; ~s, ~e** *(Med.)* cardiogram

Karenz [ka'rɛnts] **die; ~, ~en, Karenz·zeit die** waiting period

Kar·freitag [ka:ɐ̯-] **der** Good Friday

Karfunkel [kar'fʊŋkl̩] **der; ~s, ~** *(Edelstein; volkst.: Geschwür)* carbuncle

karg [kark] **1.** *Adj.* meagre ‹*wages, pay, etc.*›; frugal ‹*meal etc.*›; poor ‹*light, accommodation*›; scanty ‹*supply*›; meagre, scant ‹*applause*›; sparse ‹*furnishings*›; barren, poor ‹*soil*›. **2.** *adv.* **~ bemessen sein** ‹*helping*› be mingy *(Brit. coll.)*; ‹*supply*› be scanty; **~ leben** live frugally; **~ möbliert** sparsely furnished; **~ ausgestattet** scantily equipped

Kargheit die; ~: *s.* **karg 1:** meagreness; frugality; poorness; scantiness; sparseness; barrenness

kärglich ['kɛrklɪç] **1.** *Adj.* meagre, poor ‹*wages, pension, etc.*›; poor ‹*light*›; frugal ‹*meal*›; scanty ‹*supply*›; meagre ‹*existence*›; meagre, scant ‹*applause*›; sparse ‹*furnishing*›. **2.** *adv.* sparsely ‹*furnished*›; poorly ‹*lit, paid, rewarded*›

Karibik [ka'ri:bɪk] **die; ~: die ~:** the Caribbean; **in die ~:** to the Caribbean

karibisch *Adj.* Caribbean

kariert [ka'ri:ɐ̯t] **1.** *Adj.* check, checked ‹*material, pattern*›; check ‹*jacket etc.*›; squared ‹*paper*›. **2.** *adv. (ugs.)* **~ reden** *od.* **quatschen** talk rubbish

Karies ['ka:riɛs] **die; ~** *(Zahnmed.)* caries

Karikatur [karika'tu:ɐ̯] **die; ~, ~en a)** cartoon; *(Porträt)* caricature; **b)** *(abwertend: Zerrbild)* caricature

Karikaturist der; ~en, ~en cartoonist; *(Porträtist)* caricaturist

karikaturistisch 1. *Adj.* caricatural; **eine ~e Darstellung** a caricature. **2.** *adv.* **etw. ~ überzeichnen** caricature sth.

karikieren [kari'ki:rən] *tr. V.* caricature

kariös [ka'riø:s] *Adj. (Zahnmed.)* carious

karitativ [karita'ti:f] **1.** *Adj.* charitable. **2.** *adv.* **sich ~ betätigen** do work for charity

Karl [karl] **(der)** Charles; **~ der Große** Charlemagne

Karmeliter [karme'li:tɐ] **der; ~s, ~:** Carmelite [friar]

Karmelit[er]in die; ~, ~nen Carmelite [nun]

Karmin [kar'mi:n] **das; ~s** carmine

karmin·rot *Adj.* carmine

Karneval ['karnəval] **der; ~s, ~e** *od.* **~s** carnival; **im ~:** at carnival time; **~ feiern** join in the carnival festivities

Karnevalist der; ~en, ~en carnival reveller; *(Vortragender)* carnival performer

karnevalistisch *Adj.* carnival *attrib.* ‹*festivities etc.*›

Karnevals- carnival ‹*costume, society, procession, etc.*›

Karnickel [kar'nɪkl̩] **das; ~s, ~** *(landsch.)* rabbit

Kärnten ['kɛrntn̩] **(das); ~s** Carinthia

Kärnt[e]ner der; ~s, ~, Kärntnerin die; ~, ~nen Carinthian

Karo ['ka:ro] **das; ~s, ~s a)** square; *(Raute)* diamond; **b)** *o. Pl. (~muster)* check; **c)** *(Kartenspiel) (Farbe)* diamonds *pl.*; *(Karte)* diamond; *s. auch* **²Pik**

Karo-: ~as das ace of diamonds; **~bube der** jack of diamonds; **~dame die** queen of diamonds; **~könig der** king of diamonds; **~muster das** check; check[ed] pattern

Karolinger ['ka:rolɪŋɐ] **der; ~s, ~** *(hist.)* Carolingian

Karosse [ka'rɔsə] **die; ~, ~n a)** *(Prunkwagen)* [state-]coach; **b)** *(scherzh. iron.: Auto)* limousine

Karosserie [karɔsə'ri:] **die; ~, ~n** bodywork; coachwork

Karotin [karo'ti:n] **das; ~s** carotene

Karotte [ka'rɔtə] **die; ~, ~n** small carrot

Karpaten [kar'pa:tn̩] *Pl.* **die ~:** the Carpathians; the Carpathian Mountains

Karpfen ['karpfn̩] **der; ~s, ~:** carp

Karpfen·teich der carp pond; *s. auch* **Hecht a**

Karre ['karə] **die; ~, ~n** *(bes. nordd.)* **a)** *s.* **Karren; b)** *(abwertend: Fahrzeug)* [old] heap *(coll.)*

Karree [ka're:] **das; ~s, ~s a)** *(Rechteck)* rectangle; *(Quadrat; auch Milit.: Formation)* square; **b)** *(Häuserblock)* block

karren *tr. V.* **a)** cart; **b)** *(salopp: mit einem Auto)* run *(coll.)*

Karren der; ~s, ~ *(bes. südd., österr.)* cart; *(zweirädrig)* barrow; *(Schubkarren)* [wheel]barrow; *(für Gepäck usw.)* trolley; **ein ~ voll Sand** a cartload/barrowload of sand; *(fig.)* **den ~ in den Dreck fahren** *(ugs.)* get things into a mess; mess things up; **den ~ [für jmdn.] aus dem Dreck ziehen** *(ugs.)* sort out the mess [for sb.]; **jmdm. an den ~ fahren** *(ugs.)* tell sb. where he/she gets off *(coll.)*

Karriere [ka'rie:rə] **die; ~, ~n** career; **~ machen** make a [successful] career for oneself

Karriere-: ~frau die career woman/girl; **~macher der** *(abwertend)* careerist

Karrierist der; ~en, ~en *(abwertend)* careerist

Kar·samstag [ka:ɐ̯-] **der** Easter Saturday; Holy Saturday

Karst der; ~[e]s, ~e *(Geol.)* karst

karstig *Adj.* karstic

Karst·landschaft die karst landscape

Kartäuser [kar'tɔyzɐ] **der; ~s, ~ a)** *(Mönch)* Carthusian [monk]; **b)** *(Likör)* chartreuse

Karte ['kartə] **die; ~, ~n a)** card; **die gelbe/rote ~** *(Fußball)* the yellow/red card; **b)** *(Speise~)* menu; *(Wein~)* wine-list; **nach der ~ essen** eat à la carte; **c)** *(Fahr~, Flug~, Eintritts~)* ticket; **d)** *(Lebensmittel~)* ration-card; **auf ~n** on coupons; **e)** *(Land~)* map; *(See~)* chart; **~n lesen** map-read; **f)** *(Spiel~)* card; **jmdm. die ~n legen** read sb.'s fortune from the cards; **die** *od.* **seine ~n aufdecken** *od.* **[offen] auf den Tisch legen** *od.* **offenlegen** put one's cards on the table; **alles auf eine ~ setzen** stake everything on one chance; **auf die falsche ~ setzen** back the wrong horse; **jmdm. in die ~n sehen** *od. (ugs.)* **gucken** find out *or* see what sb. is up to; **sich** *(Dat.)* **nicht in die ~n sehen** *od. (ugs.)* **gucken lassen** play one's cards close to one's chest; not show one's hand; **mit offenen/verdeckten ~n spielen** put one's cards on the table/play one's cards close to one's chest; **g)** *(Anzahl von Spielkarten)* hand; **eine schlechte ~ haben** have a poor hand

Kartei [kar'tai] **die; ~, ~en** card file *or* index

Kartei-: ~karte die file *or* index card; **~kasten der** file-card *or* index-card box; **~leiche die** *(ugs. scherzh.: passives Mitglied)* inactive member

Kartell [kar'tɛl] **das; ~s, ~e** *(Wirtsch., Politik)* cartel

Kartell-: ~amt das, ~behörde die *government body concerned with the control and supervision of cartels;* ≈ Monopolies and Mergers Commission *(Brit.)*; **~recht das** law relating to cartels; ≈ monopolies law *(Brit.)*

Karten-: ~gruß der greeting *or* short message on a [post-]card; **an jmdn. einen ~gruß schicken** send sb. a card; **~haus das** house of cards; **~legen das; ~s** reading the cards *no art.*; cartomancy; **~leger der, ~legerin die** fortune-teller *(who tells fortunes by reading the cards)*; **~lesen das** map-reading; **~spiel das a)** card-game; **b)** *(Satz Spielkarten)* pack *or (Amer.)* deck [of cards]; **c)** *(das Kartenspielen)* card-playing *no art.*; **~spieler der** card-player; **~vorverkauf der;** *o. Pl.* advance booking

Karthager [kar'ta:gɐ] **der; ~s, ~:** Carthaginian

Karthago [kar'ta:go] **(das); ~s** Carthage

Kartoffel [kar'tɔfl̩] **die; ~, ~n** potato

Kartoffel-: ~acker der potato-field; **~brei der** mashed *or* creamed potatoes *pl.*; mash *(coll.)*; **~chips** *Pl.* [potato] crisps *(Brit.) or (Amer.)* chips; **~käfer der** Colorado beetle; potato-beetle; **~kloß der, ~knödel der** *(südd.)* potato dumpling; **~puffer der** potato pancake *(made from grated raw potatoes)*; **~püree das** *s.* **~brei; ~salat der** potato salad; **~suppe die** potato soup

Kartograph [karto'gra:f] **der; ~en, ~en** cartographer

Kartographie die; ~: cartography *no art.*

kartographisch *Adj.* cartographic

Karton [kar'tɔŋ] **der; ~s, ~s a)** *(Pappe)* card[board]; **b)** *(Behälter)* cardboard box; *(kleiner und dünner)* carton; **zwei ~[s] Seife** two boxes *or* packs of soap

kartonieren [karto'ni:rən] *tr. V. (Buchw.)* bind in [paper] boards

Karussell [karʊ'sɛl] **das; ~s, ~s** *od.* **~e** merry-go-round; carousel *(Amer.)*; *(kleineres)* roundabout; **~ fahren** have a ride on *or* go on the merry-go-round/roundabout

Kar·woche ['ka:ɐ̯vɔxə] **die** Holy Week; Passion Week

Karzer ['kartsɐ] **der; ~s, ~** *(hist.) (Raum)* detention room *(in university, school)*; *(Strafe)* detention *(often lasting several days)*

karzinogen [kartsino'ge:n] *Adj. (Med.)* carcinogenic

Karzinom [kartsi'no:m] **das; ~s, ~e** *(Med.)* carcinoma

Kaschemme [ka'ʃɛmə] **die; ~, ~n** *(abwertend)* [low] dive *(coll.)*

kaschen ['kaʃn̩] *tr. V. (salopp)* nab *(sl.)*; nick *(sl.)*

kaschieren [ka'ʃi:rən] *tr. V.* **a)** *(geh.)* conceal; hide; disguise ‹*fault*›; **b)** *(Buchw.)* laminate ‹*jacket etc.*›; line ‹*cover etc.*› [with paper]

[1]Kaschmir ['kaʃmi:ɐ̯] **(das); ~s** Kashmir

[2]Kaschmir der; ~s, ~e *(Textilw.)* cashmere

Käse ['kɛ:zə] **der; ~s, ~ a)** cheese; **b)** *(ugs. abwertend: Unsinn)* rubbish; nonsense; codswallop *(Brit. sl.)*

Käse-: ~blatt das *(salopp abwertend)* rag; **~brot das** slice of bread and cheese; *(zugeklappt)* cheese sandwich; **~fondue das** cheese fondue; **~gebäck das** cheese savouries *pl.*; **~glocke die** cheese dome; **~kuchen der** cheesecake

Kasematte [kazə'matə] **die; ~, ~n** *(Milit., Marine)* casemate

Käse·platte die *(Gericht)* [selection of] assorted cheeses *pl.*

Käserei [kɛ:zə'rai] **die; ~, ~en a)** *o. Pl. (Herstellung von Käse)* cheese-making *no art.* **b)** *(Betrieb)* cheese-factory

Kaserne [ka'zɛrnə] **die; ~, ~n** barracks *sing. or pl.*

Kasernen·hof der barrack square

kasernieren *tr. V.* quarter in barracks

Kasernierung die; ~, ~en quartering in barracks *no art.*

käse-, Käse-: ~stange die cheese straw; **~torte die** cheesecake; **~weiß** *Adj. (ugs.)* [as] white as a sheet

käsig *Adj.* **a)** *(ugs.: bleich)* pasty; pale; *(vor Schreck)* as white as a sheet; **b)** *(wie Käse)* cheesy; cheeselike

Kasino [ka'zi:no] **das; ~s, ~s a)** *(Spiel~)* casino; **b)** *(Offiziers~)* [officers'] mess

Kaskade [kas'ka:də] **die; ~, ~n** *(auch fig.)* cascade; **eine ~ von Verwünschungen/Flüchen** *(fig.)* a barrage of curses

Kasko·versicherung die *(Voll~)* comprehensive insurance; *(Teil~) insurance against theft, fire, or act of God*

Kasper ['kaspɐ] **der; ~s, ~ a)** ≈ Punch; **b)** *(ugs.: alberner Mensch)* clown; fool

Kasperl ['kaspɐl] **das; ~s, ~[n]** *(österr.)*, **Kasperle** ['kaspɐlə] **das** *od.* **der; ~s, ~** *s.* **Kasper**

Kasper-: ~puppe die ≈ Punch and Judy puppet; **~theater das** ≈ Punch and Judy show; *(Puppenbühne)* ≈ Punch and Judy theatre

Kaspische Meer ['kaspɪʃə -] **das** Caspian Sea

Kasse ['kasə] **die; ~, ~n a)** *(Kassette)* cash-box; *(Registrier~)* till; cash register; **in die ~ greifen** *od.* **einen Griff in die ~ tun** *(ugs.; auch fig.)* help oneself from the till; **er wurde beim Griff in die ~ ertappt** *(auch fig.)* he was caught with his fingers in the till; **b)** *(Ort zum Bezahlen)* cash *or* pay desk; *(im Supermarkt)* check-out; *(in einer Bank)* counter; **~ machen** *(Kaufmannsspr.)* cash up; **jmdn. zur ~ bitten** *(ugs.)* ask sb. to pay up; **c)** *(Geld)* cash; **gemeinsame ~ führen** *od.* **machen** share expenses; **getrennte ~ haben** pay separately; **gut/knapp bei ~ sein** be well-off *or* flush/be short of cash *or* money; **etw. reißt ein Loch in die ~** *(ugs.)* sth. makes a hole in sb.'s pocket *or* a dent in sb.'s finances; **die ~ führen** be in charge of the money *or* finances *pl.*; **d)** *(Kassenraum)* cashier's office; **e)** *(Theater~, Kino~, Stadion~)* box-office; **f)** *s.* **Krankenkasse**

Kasseler ['kasəlɐ] **das; ~s** smoked loin of pork

kassen-, Kassen-: ~arzt der *doctor who treats members of health insurance schemes;* **~bon der** sales slip; receipt; **~buch das** cash-book; **~erfolg der** box-office success; **~magnet der** *(ugs.)* box-office draw; **~patient der** *patient who is a member of a health insurance scheme;* **~raum der** counter hall; **~schlager der** *(ugs.)* **a)** *(Film, Theater)* box-office hit; **b)** *(von Waren)* top seller; **~stunden** *Pl.* hours of business, business hours *(of bank, cashier's office, etc.)*; **~sturz der** *(ugs.)* **~sturz machen** check up on one's ready cash; **~wart der** treasurer; **~zettel der** receipt; *(~bon)* sales slip

Kasserolle [kasə'rɔlə] **die; ~, ~n** saucepan

Kassette [ka'sɛtə] **die; ~, ~n a)** *(für Geld u. Wertsachen)* box; case; **b)** *(mit Büchern, Schallplatten)* boxed set; **c)** *(Tonband~, Film~)* cassette; **etw. auf ~ aufnehmen** record *or* tape sth. on cassette

Kassetten·recorder **der** cassette recorder

[1]**kassieren** **1.** *tr. V.* **a)** *(einziehen)* collect ‹*rent etc.*›; **b)** *(ugs.: einnehmen)* collect ‹*money, fee, etc.*›; *(fig.)* receive, get ‹*recognition, praise, etc.*›; **bei der Transaktion hat er 100 000 DM kassiert** he made 100,000 marks on the deal; **c)** *(ugs.: hinnehmen müssen)* receive, get ‹*penalty points, scorn, ingratitude, etc.*›; **d)** *(ugs.: wegnehmen)* confiscate; take away ‹*driving licence*›; **e)** *(ugs.: verhaften/gefangennehmen)* pick up; nab *(sl.)*; nick *(Brit. sl.)*. **2.** *itr. V.* **a) bei jmdm.** ~ ‹*waiter*› give sb. his/her bill *or (Amer.)* check; *(ohne Rechnung)* ‹*waiter*› settle up with sb.; **darf ich bei Ihnen ~?** would you like your bill?/can I settle up with you?; **b)** *(ugs.: Geld einnehmen)* collect the money; **[bei einem Geschäft] ganz schön ~:** make a packet *(coll.) or (sl.)* a bomb [on a deal]

[2]**kassieren** *tr. V. (Rechtsw.)* quash ‹*judgement etc.*›

Kassierer der; ~s, ~, Kassiererin die; ~, ~nen cashier; teller

Kastagnette [kastan'jɛtə] **die; ~, ~n** castanet

Kastanie [kas'ta:ni̯ə] **die; ~, ~n** chestnut; **[für jmdn.] die ~n aus dem Feuer holen** *(ugs.)* pull the chestnuts out of the fire [for sb.]

Kastanien·baum der chestnut-tree

kastanien·braun *Adj.* chestnut

Kästchen ['kɛstçən] **das; ~s, ~ a)** small box; **b)** *(vorgedrucktes Quadrat)* square; *(auf Fragebögen)* box

Kaste ['kastə] **die; ~, ~n** caste

kasteien [kas'tai̯ən] *refl. V.* chastise oneself; *(fig.)* deny oneself

Kasteiung die; ~, ~en self-chastisement; *(fig.)* self-denial

Kastell [kas'tɛl] **das; ~s, ~e a)** *(hist.: röm. Lager)* fort; **b)** *(Burg)* castle

Kasten ['kastn̩] **der; ~s, Kästen** ['kɛstn̩] **a)** box; **b)** *(für Flaschen)* crate; **c)** *(ugs.: Briefkasten)* postbox; **d)** *(ugs. abwertend: Gebäude)* barracks *sing. or pl.*; **das ist ja ein furchtbarer alter ~:** that's a terrible old barracks of a place; **e)** *(ugs. abwertend: Fernseher, Radio)* box *(coll.)*; **f)** *(ugs.: Kamera)* **ein Bild im ~ haben** have got a picture; **eine Szene im ~ haben** have a picture in the can; **g) etw. auf dem ~ haben** *(ugs.)* have got it up top *(coll.)*; have plenty of grey matter; **h)** *(Schaukasten)* showcase; display case; **i)** *(Turnen)* box; **j)** *(Ballspiele Jargon)* goal; **k)** *(bes. nordd.: Schublade)* drawer; **l)** *(südd., österr., schweiz.: Schrank)* cupboard

Kasten-: ~brot das tin[-loaf]; **~form die** *(Backform)* [rectangular] tin; **~wagen der** van; **~wesen das;** *o. Pl.* caste system

Kastrat [kas'tra:t] **der; ~en, ~en a)** *(Eunuch)* eunuch; **b)** *(Musik hist.)* castrato

Kastraten·stimme die a) *(Musik)* castrato voice; **b)** *(abwertend)* falsetto voice

Kastration [kastra'tsi̯o:n] **die; ~, ~en** castration

kastrieren *tr. V.* castrate

Kasus ['ka:zʊs] **der; ~, ~** ['ka:zu:s] *(Sprachw.)* case

Katafalk [kata'falk] **der; ~s, ~e** catafalque

Katakombe [kata'kɔmbə] **die; ~, ~n;** *meist Pl.* catacomb

Katalog [kata'lo:k] **der; ~[e]s, ~e** *(auch fig.)* catalogue

katalogisieren *tr. V.* catalogue

Katalysator [kataly'za:tɔr] **der; ~s, ~en** [-za'to:rən] **a)** *(Chemie, fig. geh.)* catalyst; **b)** *(Kfz-W.)* catalytic converter; *s. auch* **geregelt 2**

Katamaran [katama'ra:n] **der** *od.* **das; ~s, ~e** catamaran

Katapult [kata'pʊlt] **das** *od.* **der; ~[e]s, ~e** catapult

katapultieren *tr. V. (auch fig.)* catapult; eject ‹*pilot*›

Katarakt [kata'rakt] **der; ~[e]s, ~e** *(Stromschnelle)* rapids *pl.*; *(Wasserfall)* cataract

Katarrh [ka'tar] **der; ~s, ~e** *(Med.)* catarrh; **einen ~ haben** have catarrh

Kataster [ka'tastɐ] **der** *od.* **das; ~s, ~:** land register

Kataster·amt das land registry

katastrophal [katastro'fa:l] **1.** *Adj.* disastrous; *(stärker)* catastrophic. **2.** *adv.* disastrously; *(stärker)* catastrophically; **sich ~ auswirken** have a disastrous/catastrophic effect; **~ enden** end in disaster/catastrophe

Katastrophe [katas'tro:fə] **die; ~, ~n** disaster; *(stärker)* catastrophe; **jmd. ist eine ~** *(ugs.)* sb. is a disaster

Katastrophen-: ~alarm der emergency *or* disaster alert; **~dienst der** emergency services *pl.*; **~einsatz der: den ~einsatz üben** practise procedures in case of a disaster; **~fall der** disaster [situation]; **~gebiet das** disaster area; **~schutz der a)** *(Organisation)* emergency services *pl.*; **b)** *(Maßnahmen)* disaster procedures *pl.*; **dem ~schutz dienen** be useful in the event of a disaster

Kate ['ka:tə] **die; ~, ~n** *(bes. nordd.)* small cottage

Katechismus [katɛ'çɪsmʊs] **der; ~, Katechismen** *(christl. Kirche)* catechism

Kategorie [katego'ri:] **die; ~, ~n** [-i:ən] category

kategorisch 1. *Adj.* categorical. **2.** *adv.* categorically

kategorisieren *tr. V.* categorize

Kater ['ka:tɐ] **der; ~s, ~ a)** tom-cat; **wie ein verliebter ~:** like an amorous tom-cat; **b)** *(ugs.: schlechte Verfassung)* hangover; **einen ~ haben** have a hangover; be hung-over

Kater-: ~frühstück das *breakfast, usually of pickled herrings and gherkins, supposed to cure a hangover;* **~stimmung die** morning-after feeling

Katheder [ka'te:dɐ] **das** *od.* **der; ~s, ~:** lectern; *(Pult des Lehrers)* teacher's desk

Kathedrale [kate'dra:lə] **die; ~, ~n** cathedral

Kathete [ka'te:tə] **die; ~, ~n** *(Math.)* leg *(of a right-angled triangle)*

Katheter [ka'te:tɐ] **der; ~s, ~** *(Med.)* catheter

Kathode [ka'to:də] **die; ~, ~n** *(Physik)* cathode

Katholik [kato'li:k] **der; ~en, ~en, Katholikin die; ~, ~nen** [Roman] Catholic

katholisch *Adj.* [Roman] Catholic

Katholizismus [katoli'tsɪsmʊs] **der; ~:** [Roman] Catholicism *no art.*

Kattun [ka'tu:n] **der; ~s, ~e** calico

Katz [kats] **die** *in* **~ und Maus [mit jmdm.] spielen** *(ugs.)* play cat and mouse [with sb.]; **für die ~ sein** *(salopp)* be a waste of time

katzbuckeln *itr. V. (abwertend)* bow and scrape (**vor** + *Dat.* to)

Kätzchen ['kɛtsçən] **das; ~s, ~ a)** little cat; *(liebkosend)* pussy; *(junge Katze)* kitten; **b)** *(Blüte der Birke, Erle u. a.)* catkin

Katze ['katsə] **die; ~, ~n a)** cat; **die ~ läßt das Mausen nicht** *(Spr.)* a leopard cannot change its spots *(prov.)*; **bei Nacht sind alle ~n grau** it's impossible to see any details in the dark; **wenn die ~ aus dem Haus ist, tanzen die Mäuse [auf dem Tisch]** *(Spr.)* when the cat's away the mice will play *(prov.)*; **die ~ aus dem Sack lassen** *(ugs.)* let the cat out of the bag; **die ~ im Sack kaufen** *(ugs.)* buy a pig in a poke; **um etw. herumgehen wie die ~ um den heißen Brei** *(ugs.)* beat about the bush; *s. auch* **Katz**

katzen-, Katzen-: ~auge das a) *(ugs.: Rückstrahler)* reflector; **b)** *(Mineral.)* cat's-eye; **~buckel der** hunched back; **einen ~buckel machen** hunch one's back; **~fell das** cat's skin; **~haft** *Adj.* catlike; **~jammer der** *(fig.)* mood of depression; **~klo das** *(ugs.)* cat's [litter] tray; **~kopf der** *(ugs.)* cobble[-stone]; **~musik die** *(ugs. abwertend)* terrible row *(coll.)*; cacophony; **~sprung der** *(fig.)* stone's throw; **bis zum Strand ist es nur ein ~sprung** the beach is only a stone's throw away; **~tisch der** *(ugs. scherzh.)* children's table; **~wäsche die** *(ugs.)* lick and a promise *(coll.)*; catlick *(coll.)*

Kauderwelsch ['kaudɐvɛlʃ] **das; ~[s]** gibberish *no indef. art.*; double Dutch *no indef. art.*; **juristisches ~:** legal jargon; **ein ~ aus Deutsch, Englisch und Französisch** an incomprehensible hotchpotch of German, English, and French

kauen ['kauən] **1.** *tr. V.* chew; **Nägel ~:** bite *or* chew one's nails. **2.** *itr. V.* **a)** chew; **an etw.** *(Dat.)* **~:** chew [on] sth.; **mit vollen Backen ~** *(ugs.)* chew with one's mouth [stuffed] full; **b)** *(nagen, knabbern)* chew; bite; **an einem Bleistift/den Fingernägeln ~:** chew a pencil/bite *or* chew one's nails

kauern ['kauɐn] *itr., refl. V.* crouch [down]; *(ängstlich)* cower

Kauf [kauf] **der; ~[e]s, Käufe** ['kɔyfə] **a)** *(das Kaufen)* buying; purchasing *(formal)*; **einen ~ abschließen/tätigen** complete/make a purchase; **jmdn. zum ~ ermuntern** encourage sb. to buy; **jmdm. etw. zum ~ anbieten** offer sb. sth. for sale; **etw. in ~ nehmen** *(fig.)* accept sth.; **b)** *(Gekauftes)* purchase

kaufen 1. *tr. V.* **a)** buy; purchase; **etw. billig/zu teuer ~:** buy sth. cheaply/pay too much for sth.; **sich/jmdm. etw. ~:** buy sth. for oneself/sb.; buy oneself/sb. sth.; **etw. auf Raten** *od.* **Abzahlung ~:** buy sth. on hire-purchase *(Brit.)* *or (Amer.)* the installment plan; **etw. für viel** *od.* **teures Geld ~:** pay a lot of money for sth.; **das wird viel** *od.* **gern gekauft** it sells well; **sich** *(Dat.)* **jmdn. ~** *(ugs.)* give sb. what for *(sl.)*; let sb. have *or* give sb. a piece of one's mind; **b)** *(ugs.: bestechen)* buy. **2.** *itr. V.* *(einkaufen)* shop; **in diesem Laden kaufe ich nicht mehr** I'm not getting anything in that shop again

Käufer ['kɔyfɐ] **der; ~s, ~, Käuferin die; ~, ~nen** buyer; purchaser; *(Kunde/Kundin)* customer

Kauf-: ~frau die businesswoman; *(Händlerin)* trader; merchant; **~haus das** department store; **~haus·detektiv der** store detective; **~kraft die** *(Wirtsch.)* **a)** *(des Geldes)* purchasing power; **b)** *(von Personen)* spending power; **~laden der** *(Kinderspielzeug)* toy shop

käuflich ['kɔyflɪç] **1.** *Adj.* **a)** for sale *postpos.*; **ein ~es Mädchen** *(fig.)* a woman/girl of easy virtue; **~e Liebe** prostitution *no art.*; **b)** *(bestechlich)* venal; **~ sein** be easily bought. **2.** *adv.* **etw. ~ erwerben/erstehen** buy *or* purchase sth.; **~ zu erwerben sein** be for sale

kauf·lustig *Adj.* eager to buy *pred.*; **die Kauflustigen** the eager shoppers

Kauf·mann der; *Pl.* **Kaufleute a)** *(Geschäftsmann)* businessman; *(Händler)* trader; merchant; **gelernter ~:** *person who has completed a course of training in some branch of business;* **b)** *(veralt.: Lebensmittelhändler)* grocer; **zum ~ gehen** go to the grocer's

kaufmännisch 1. *Adj.* commercial; business *attrib.*; commercial ‹*bookkeeping*›; **~er Angestellter** clerk; employee in business; **einen ~en Beruf ergreifen/erlernen** go into business/receive a business training; **~es Geschick/~e Erfahrung haben** possess business skill/experience. **2.** *adv.* **~ tätig sein** be in business; **~ denken** think along commercial lines

Kaufmanns·sprache die business parlance

Kauf-: ~preis der purchase price; **~rausch der** frantic urge to spend; **~vertrag der** contract of sale; *(beim Hauskauf)* title-deed; **~zwang der** obligation to buy *or* purchase

Kau·gummi der *od.* **das; ~s, ~s** chewing gum

Kaukasus ['kaukazʊs] **der; ~: der ~:** the Caucasus

Kaulquappe ['kaulkvapə] **die; ~, ~n** tadpole

kaum [kaum] *Adv.* **a)** *(fast gar nicht)* hardly; scarcely; **~ jemand/etwas** hardly anybody *or* anyone/anything; **~ älter/größer/besser** hardly *or* scarcely any older/bigger/better; **ich kann es ~ glauben/erwarten** I can hardly believe it/wait; **ich konnte ~ rechtzeitig damit fertig werden** I could hardly *or* barely finish it in time; **diese Schrift ist ~ zu entziffern** this writing is barely decipherable; **b)** *(vermutlich nicht)* hardly; scarcely; **er wird [wohl] ~ zustimmen** he is hardly likely to agree; **ich glaube ~:** I hardly *or* scarcely think so; **c)** *(eben erst)* **~ hatte er Platz genommen, als ...:** no sooner had he sat down than ...; **d) ~ daß** almost as soon as; **~ daß er aus dem Gefängnis gekommen war ...:** hardly *or* scarcely had he left prison when ...

kausal [kau'za:l] *Adj.* *(geh., Sprachw.)* causal

Kausalität [kauzali'tɛ:t] **die; ~, ~en** causality

Kausal- causal ‹*clause, connection, etc.*›

Kau·tabak der chewing tobacco

Kaution [kau'tsio:n] **die; ~, ~en a)** *(für Gefangenen)* bail; **eine ~ stellen** stand bail *or* surety; **gegen ~:** on bail; **jmdn. gegen ~ freibekommen** bail sb. out; **b)** *(für Wohnung)* deposit

Kautschuk ['kautʃʊk] **der; ~s, ~e** [india] rubber

Kauz [kauts] **der; ~es, Käuze** ['kɔytsə] **a)** owl; *(Stein~)* little owl; **b)** *(Sonderling)* odd *or* strange fellow; oddball *(coll.)*; **ein komischer ~:** an odd *or* a queer bird *(coll.)*

Käuzchen ['kɔytsçən] **das; ~s, ~:** *s.* **Kauz a**

kauzig *Adj.* odd; queer; funny *(coll.)*

Kavalier [kava'li:ɐ] **der; ~s, ~e** gentleman

Kavaliers·delikt das trifling offence; peccadillo

Kavalier[s]·start der racing start

Kavallerie [kavalə'ri:] **die; ~, ~n** *(Milit. hist.)* cavalry

Kaviar ['ka:viar] **der; ~s, ~e** caviare

kcal *Abk.* **Kilo[gramm]kalorie** kcal

keck [kɛk] **1.** *Adj.* **a)** impertinent; cheeky; saucy *(Brit.)*; **b)** *(flott)* jaunty, pert ‹*hat etc.*›. **2.** *adv.* **a)** impertinently; cheekily; saucily *(Brit.)*; **b)** *(flott)* jauntily

Keckheit die; ~, ~en impertinence; cheek; sauce *(Brit.)*

Kefir ['ke:fɪr] **der; ~s** kefir

Kegel ['ke:gl̩] **der; ~s, ~ a)** cone; **b)** *(Spielfigur)* skittle; *(beim Bowling)* pin; **~ schieben** *s.* **kegelschieben; c)** *(Licht~)* beam

kegel-, Kegel-: ~bahn die skittle alley; **~förmig** *Adj.* conical; cone-shaped; **~klub der** skittle club

kegeln 1. *itr. V.* play skittles *or* ninepins. **2.** *tr. V.* **eine Partie ~:** play a game of skittles *or* ninepins; **eine Neun ~:** score a nine

kegel-, Kegel-: ~|schieben *unr. itr. V.; Zusammenschreibung nur im Inf.* play skittles *or* ninepins; **~schnitt der** *(Geom.)* conic section; **~stumpf der** truncated cone; frustum of a cone

Kegler der; ~s, ~: skittle player

Kehle ['ke:lə] **die; ~, ~n** throat; **sich** *(Dat.)* **die ~ anfeuchten** wet one's whistle *(coll.)*; **sich** *(Dat.)* **die ~ aus dem Hals schreien** *(ugs.)* shout *or* yell one's head off; **aus voller ~:** at the top of one's voice; **sein ganzes Geld durch die ~ jagen** pour all one's money down one's throat; **jmdm. in der ~ stekkenbleiben** stick in sb.'s throat *or* gullet; **etw. in die falsche ~ bekommen** *(ugs.)* *(fig.: etw. mißverstehen)* take sth. the wrong way; *(sich an etw. verschlucken)* have sth. go down the wrong way

kehlig 1. *Adj.* guttural ⟨*speech, sound, etc.*⟩; throaty, guttural ⟨*voice, laugh, etc.*⟩. **2.** *adv.* throatily; gutturally; in a throaty *or* guttural voice

Kehl·kopf der *(Anat.)* larynx

Kehlkopf·krebs der *(Med.)* cancer of the larynx

Kehl·laut der a) guttural sound; **b)** *(Sprachw.)* guttural

Kehr-: ~aus der; ~: last dance; **~blech das** *(landsch.)* dustpan

Kehre ['ke:rə] **die; ~, ~n a)** sharp bend *or* turn; *(Haarnadelkurve)* hairpin bend; **b)** *(Turnen)* back *or* rear vault

¹kehren 1. *tr. V.* turn; **die Innenseite von etw. nach außen ~:** turn sth. inside out; **jmdm. den Rücken ~:** turn one's back on sb. **2.** *refl. V.* **a)** turn; **sich gegen jmdn./etw. ~:** turn against sb./sth.; **b) sich an etw.** *(Dat.)* **nicht ~:** pay no attention to *or* not care about sth. **3.** *itr. V.* **in sich** *(Akk.)* **gekehrt** lost in thought; in a brown study

²kehren *(bes. südd.)* **1.** *itr. V.* sweep; do the sweeping. **2.** *tr. V.* sweep; *(mit einem Handfeger)* brush

Kehricht ['ke:rɪçt] **der** *od.* **das; ~s a)** *(geh.)* rubbish; **das geht dich einen feuchten ~ an!** *(salopp)* mind your own damned business!; **b)** *(schweiz.: Müll)* refuse; garbage *(Amer.)*

Kehricht-: ~eimer der dustbin; garbage can *(Amer.)*; **~haufen der** pile *or* heap of rubbish

Kehr-: ~maschine die [mechanical] road sweeper; **~reim der** refrain; **~schaufel die** dustpan; **~seite die a)** back; *(einer Münze, Medaille)* reverse; **die ~seite der Medaille** *(fig.)* the other side of the coin; **b)** *(scherzh.: Gesäß)* backside; **c)** *(nachteiliger Aspekt)* drawback; disadvantage

kehrt *Interj. (Milit.)* about turn; about face *(Amer.)*

kehrt|machen *itr. V. (ugs.)* turn [round and go] back; *(plötzlich)* turn in one's tracks; **auf dem Absatz ~:** turn on one's heel

Kehrt·wendung die *(bes. Milit.; fig.)* about-turn; about-face *(Amer.)*

keifen ['kaifn̩] *itr. V. (abwertend)* nag; scold

Keil [kail] **der; ~[e]s, ~e** wedge; **einen ~ zwischen die beiden Freunde treiben** *(fig.)* drive a wedge between the two friends

Keil·absatz der wedge [heel]

Keile ['kailə] **die; ~** *(nordd.)* walloping *(sl.)*; thrashing; **kriegen** get a walloping *(sl.) or* thrashing

keilen 1. *refl. V. (ugs.: sich prügeln)* fight; scrap; **sich um etw. ~:** fight over sth. **2.** *tr. V. (ugs.: anwerben)* rope in *(coll.)*; recruit

Keiler der; ~s, ~ *(Jägerspr.)* wild boar

Keilerei die; ~, ~en *(ugs.)* punch-up *(coll.)*; brawl; fight

keil-, Keil-: ~förmig wedge-shaped; **~hose die** tapering trousers; **~kissen das** wedge-shaped bolster; **~riemen der** *(Technik)* V-belt; *(Kfz-W.: zum Antrieb des Kühlergebläses)* fan belt; **~schrift die** cuneiform script

Keim [kaim] **der; ~[e]s, ~e a)** *(Bot.: erster Trieb)* shoot; **b)** *(Biol.: befruchtete Eizelle)* embryo; **c)** *(fig.: Ursprung)* seed[s *pl.*]; **etw. im ~ ersticken** nip sth. in the bud; **d)** *(Biol., Med.: Krankheitserreger)* germ

Keim-: ~blatt das cotyledon; seed leaf; **~drüse die** *(Zool., Med.)* gonad

keimen *itr. V.* **a)** germinate; sprout; **b)** *(fig.)* ⟨*hope*⟩ stir; ⟨*thought, belief, decision*⟩ form; ⟨*love, yearning*⟩ awaken

keim-, Keim-: ~fähig *Adj.* viable; capable of germinating *postpos.*; **~fähigkeit die;** *o. Pl.* viability; ability to germinate; **~frei** *Adj.* germ-free; sterile; **etw. ~frei machen** sterilize sth.

Keimling ['kaimlɪŋ] **der; ~s, ~e** *(Bot.)* embryo

Keim·zelle die a) *(fig.)* nucleus; **b)** *(Bot.)* germ-cell

kein [kain] *Indefinitpron.; attr.* **a)** no; **ich habe ~ Geld/~e Zeit** I have no money/time; I don't have any money/time; **er hat ~ Wort gesagt** he didn't say a word; he said not a word; **er konnte ~e Arbeit finden** he could find no work; he could not find any work; **~ Mensch/~ einziger** nobody *or* no one/not a single one; **in ~er Weise/unter ~en Umständen** in no way/in *or* under no circumstances; **das ist ~ schlechter Vorschlag** that's not a bad suggestion; **~ anderer als er kann es gewesen sein** it can't have been anybody else but him; **b)** *(ugs.: weniger als)* less than; **es ist ~e drei Tage her, daß ich zuletzt dort war** it's not *or* it's less than three days since I was last there; **es dauert ~e fünf Minuten** it won't take five minutes; **c)** *nachgestellt (ugs.)* **Lust habe ich ~e** I don't feel like it; **Kinder waren ~e da** there weren't any children there; *s. auch* **kein...**

kein... *Indefinitpron. (niemand, nichts)* **~er/~e** nobody; no one; **~er von uns** not one of us; none of us; **ich kenne ~en, der dir helfen kann** I don't know anyone who can help you; **keins von beiden** neither [of them]; **ich wollte ~es von beiden** I didn't want either of them; **mir kann ~er!** *(salopp)* I can look after myself!

keinerlei *indekl. unbest. Gattungsz.* no ... at all; no ... what[so]ever

keines·falls *Adv.* on no account; **die Aufgabe ist schwer, aber ~ unlösbar** the problem is difficult but by no means insoluble

keines·wegs *Adv.* by no means; not by any means; not at all; **sein Einfluß darf ~ unterschätzt werden** his influence must in no way be underestimated

kein·mal *Adv.* not [even] once; *s. auch* **einmal a**

Keks [ke:ks] **der; ~** *od.* **~es, ~** *od.* **~e** biscuit *(Brit.)*; cookie *(Amer.)*

Kelch [kɛlç] **der; ~[e]s, ~e a)** goblet; **der ~ ist an ihm vorübergegangen** *(geh.)* he was spared that ordeal; **b)** *(Rel.)* chalice; communion cup; **c)** *(Bot.)* calyx

Kelch·blatt das *(Bot.)* sepal

kelch·förmig *Adj.* goblet-shaped

Kelle ['kɛlə] **die; ~, ~n a)** *(Schöpf-*

löffel) ladle; **b)** *(Signalstab)* signalling disc; **c)** *(Maurer~)* trowel

Keller ['kɛlɐ] **der; ~s, ~ a)** cellar; *(~geschoß)* basement; **der Dollar[kurs] ist in den ~ gegangen** *(fig.)* the dollar has gone through the floor *(fig.)*; **im ~ sein** *(Skat Jargon)* have a minus score *or* minus points; **b)** *(Luftschutz~)* [air-raid] shelter; **c)** *s.* **Kellerlokal**

Keller·assel die wood-louse

Kellerei die; ~, ~en winery; wine producer's

Keller-: ~fenster das cellar window; *(von ~geschoß)* basement window; **~geschoß das** basement; **~gewölbe das** underground vault; **~lokal das** cellar bar/disco/restaurant *etc.*; **~treppe die** cellar stairs *pl.*; **~wohnung die** basement flat *(Brit.) or (Amer.)* apartment

Kellner ['kɛlnɐ] **der; ~s, ~:** waiter

Kellnerin die; ~, ~nen waitress

kellnern *itr. V. (ugs.)* work as a waiter/waitress

Kelte ['kɛltə] **der; ~n, ~n** Celt

Kelter ['kɛltɐ] **die; ~, ~n** fruit-press; *(für Trauben)* winepress

keltern *tr. V.* press ‹*grapes etc.*›

keltisch *Adj.* Celtic

Keltisch das; ~[s] Celtic

Kenia ['ke:nia] **(das); ~s** Kenya

Kenianer [ke'nia:nɐ] **der; ~s, ~, Kenianerin die; ~, ~nen** Kenyan

kennen ['kɛnən] *unr. tr. V.* **a)** know; **das Leben ~:** know about life; know the ways of the world; **das ~ wir gar nicht anders** it's always been like that; **kennst du ihn?** do you know who he is?; *(bist du mit ihm bekannt)* are you acquainted with him?; **kennst du den?** *(diesen Witz)* have you heard this one?; **jmds. Bücher/Werk ~:** know *or* be acquainted with sb.'s books/work; **da kennst du mich aber schlecht** *(ugs.)* that just shows you don't know me very well; **das kennen wir [schon]** *(ugs. abwertend)* we've heard all that before; **sich nicht mehr ~ [vor ...]** be beside oneself [with ...]; **da kenne ich/da kennt er nichts** *(ugs.)* and to hell with everything else *(coll.)*; **b)** *(bekannt sein mit)* know; be acquainted with; **jmdn. flüchtig/persönlich ~:** know sb. slightly/personally; **die beiden ~ sich nicht mehr** *(fig.)* the two are no longer on speaking terms; **ich glaube, wir beide ~ uns noch nicht** I don't think we've been introduced; **c)** *(haben)* have; **keinen Winter/Sommer ~:** have no winter/summer; **er kennt keine Kopfschmerzen** he never gets a headache; **kein Mitleid ~:** know *or* have no pity; **d)** *(wiedererkennen)* know; recognize; **na, kennst du mich noch?** well, do you remember me?

kennen|lernen *tr. V.* get to know; **jmdn./etw. [näher] ~:** get to know sb./sth. [better]; become [better] acquainted with sb./sth.; **jmdn. ~** *(jmdm. erstmals begegnen)* meet sb.; **jmdn. von einer bestimmten Seite ~:** see a particular side of sb.; **jmdn. als einen bescheidenen Menschen** *usw.* **~:** come to know sb. as a modest person *etc.*; **du wirst mich noch ~!** you'll find out I don't stand for any nonsense; **[es] freut mich, Sie kennenzulernen** pleased to meet you; pleased to make your acquaintance *(formal)*

Kenner der; ~s, ~ a) expert, authority (+ *Gen.* on); **b)** *(von Wein, Speisen)* connoisseur

Kennerblick der expert eye; **mit ~:** with an expert eye

Kenn-: ~karte die identity card; **~marke die** [police] identification badge; ≈ [police] warrant card *or (Amer.)* ID card; **~melodie die** *(Rundf.)* signature tune; **~nummer die** *s.* **Kennummer**

kenntlich ['kɛntlɪç] *Adj. in* **~ sein** be recognizable *or* distinguishable **(an** by); **jmdn./etw. ~ machen** mark sth./make sb. [easily] identifiable; **etw. als Gift ~ machen** mark *or* label sth. as a poison

Kenntnis ['kɛntnɪs] **die; ~, ~se a)** *o. Pl.* knowledge; **von etw. ~ haben/erhalten** be informed on sth. *or* have knowledge of sth./learn *or* hear about sth.; **das entzieht sich meiner ~** *(geh.)* I have no knowledge of that; **von etw. ~ nehmen, etw. zur ~ nehmen** take note of sth.; **jmdn. von etw. in ~ setzen** inform *or* notify sb. of sth.; **jmdn. zur ~ nehmen** take notice of sb.; **b)** *Pl.* knowledge *sing.*; **oberflächliche/gründliche ~se in etw.** *(Dat.)* **haben** have a superficial/thorough knowledge of sth.

Kenntnisnahme die; ~ *(Papierdt.)* **jmdm. etw. zur ~ vorlegen** submit sth. to sb. for his/her attention; **nach ~ der Akten** after giving the documents my/his *etc.* attention

kenntnis·reich *Adj.* well-informed; knowledgeable

Kennummer die; ~, ~n reference number; code number

kenn-, Kenn-: ~wort das; *Pl.* **~wörter a)** *(Erkennungszeichen)* code-word; reference; **b)** *(Parole)* password; code-word; **~zahl die a)** *s.* **~ziffer; b)** *(Fernspr.)* code; **~zeichen das a)** *(Merkmal)* sign; mark; **ein ~zeichen eines Genies** a [hall]mark of a genius; **besondere ~zeichen** distinguishing marks; **b)** *(Erkennungszeichen)* badge; *(auf einem Behälter, einer Ware usw.)* label; **c)** *(Kfz-W.)* registration number; **~zeichnen** *tr. V.* **a)** mark; label ‹*container, goods, etc.*›; mark, signpost ‹*way*›; tag ‹*bird, animal*›; **etw. als ... ~zeichnen** mark *or* identify sth. as ...; **b)** *(charakterisieren)* characterize; **jmdn. als ... ~zeichnen** characterize sb. as ...; **c)** *(in seiner Eigenart erkennen lassen)* typify; **jmdn. als ... ~zeichnen** mark sb. out as ...; **~zeichnend** *Adj.* typical, characteristic **(für** of); **~zeichnung die a)** marking; *(von Behältern, Waren)* labelling; *(von Vögeln, Tieren)* tagging; **b)** *(Charakterisierung)* characterization; **c)** *(Kennzeichen)* mark; **~ziffer die** reference number; *(bei einem Zeitungsinserat)* box number

kentern ['kɛntɐn] *itr. V.; mit sein* capsize

Keramik [ke'ra:mɪk] **die; ~, ~en a)** *o. Pl. (gebrannter Ton)* ceramics *pl.*; pottery; **b)** *(~gegenstand)* ceramic; piece of pottery; **c)** *(Material)* fired clay; **d)** *o. Pl. (Technik)* ceramics *sing.*; pottery

keramisch *Adj.; nicht präd.* ceramic

Kerbe ['kɛrbə] **die; ~, ~n** notch; **in dieselbe** *od.* **die gleiche ~ hauen** *(ugs.)* take the same line

Kerbel ['kɛrbl̩] **der; ~s** chervil

kerben *tr. V.* notch; **etw. in etw.** *(Akk.)* **~:** carve sth. into sth.

Kerb-: ~holz das *in* **etwas/einiges auf dem ~holz haben** *(ugs.)* have done a job/a job or two *(sl.)*; **~tier das** insect

Kerker ['kɛrkɐ] **der; ~s, ~ a)** *(hist.) (Gefängnis)* dungeons *pl.*; *(Zelle)* dungeon; **b)** *(österr., hist.: Strafe)* imprisonment

Kerl [kɛrl] **der; ~s, ~e** (*nordd., md. auch:* **~s) a)** *(ugs.: männliche Person)* fellow *(coll.)*; chap *(coll.)*; bloke *(Brit. sl.)*; **ein ganzer** *od.* **richtiger ~:** a splendid fellow *(coll.) or* chap *(coll.)*; **ein gemeiner/frecher ~** *(abwertend)* a nasty so-and-so *(coll.)*/an impudent fellow *(coll.)*; **b)** *(ugs.: sympathischer Mensch)* **er ist ein feiner ~:** he's a fine chap *(coll.) or (sl.)* a good bloke; **sie ist ein netter/feiner ~:** she's a nice/fine woman

Kern [kɛrn] **der;** ~**[e]s,** ~**e a)** *(Fruchtsamen)* pip; *(von Steinobst)* stone; *(von Nüssen, Mandeln usw.)* kernel; **der** ~ **eines Problems/Vorschlags** *(fig.)* the crux *or* gist of a problem/gist of a suggestion; **er hat einen guten** *od.* **in ihm steckt ein guter** ~ *(fig.)* he is good at heart; **zum** ~ **einer Sache** *(Gen.)* **kommen** *(fig.)* get to the heart of a matter; **b)** *(wichtigster Teil einer Gruppe)* core; nucleus; **der harte** ~**:** the hard core; **c)** *(Physik: Atom~)* nucleus; **d)** *(einer elektrischen Spule, eines Reaktors)* core

kern-, Kern-: ~**brenn·stoff der** nuclear fuel; ~**energie die** nuclear energy *no art.;* ~**explosion die** nuclear explosion; ~**fach das** *(Schulw.)* core subject; ~**forschung die** nuclear research; ~**frage die** central question; ~**gehäuse das** core; ~**gesund** *Adj.* fit as a fiddle *pred.;* sound as a bell *pred.*

kernig 1. *Adj.* **a)** *(urwüchsig, markig)* robust, earthy ⟨*language*⟩; down-to-earth ⟨*remarks*⟩; *(kraftvoll)* powerful, forceful ⟨*speech*⟩; pithy ⟨*saying*⟩; **ein** ~**er Mann/Typ** *(ugs.)* a robust and athletic man/type; **b)** *(gehaltvoll, kräftig)* full-bodied ⟨*wine*⟩. **2.** *adv. (urwüchsig, markig)* robustly; *(kraftvoll)* forcefully

Kern·kraft die a) nuclear power *no art.;* **b)** *Pl. (Physik)* nuclear forces

Kernkraft-: ~**gegner der** opponent of nuclear power; ~**werk das** nuclear power station *or* plant

kern-, Kern-: ~**los** *Adj.* seedless; ~**obst das** pomaceous fruit; pomes *pl.;* ~**pflicht·fach das** *(Schulw.)* core-curriculum subject; ~**physik die** nuclear physics *sing., no art.;* ~**physiker der** nuclear physicist; ~**punkt der** central point; ~**reaktor der** nuclear reactor; ~**satz der** key sentence *or* statement; ~**seife die** washing soap; hard soap; ~**spaltung die** *(Physik)* nuclear fission *no art.;* ~**stück das** centre-piece; ~**technik die** nuclear engineering *no art.;* ~**waffe die;** *meist Pl.* nuclear weapon; ~**waffen·frei** *Adj.* nuclear-free; ~**zeit die** core time

Kerosin [kero'zi:n] **das;** ~**s** kerosene

Kerze ['kɛrtsə] **die;** ~, ~**n a)** candle; **elektrische** ~**:** candle bulb; **b)** *(Zünd~)* spark-plug; sparking-plug; **c)** *(Turnen)* shoulder stand

kerzen-, Kerzen-: ~**beleuchtung die** candle-light *no indef. art.;* ~**docht der** [candle] wick; ~**gerade,** *(ugs.)* ~**grade 1.** *Adj.* dead straight ⟨*tree, post, etc.*⟩; very stiff ⟨*bow*⟩; **2.** *adv.* bolt upright; ⟨*rise*⟩ straight upwards; ~**halter der** candle-holder; ~**leuchter der** candlestick; *(für mehrere Kerzen)* candelabrum; ~**licht das;** *o. Pl.* the light of a candle/of candles; **bei** ~**licht** by candle-light; ~**schein der;** *o. Pl.* candle-light *no pl.;* ~**stummel der,** ~**stumpf der** stump of a/the candle

keß [kɛs] **1.** *Adj.* **a)** *(flott)* pert; pert, jaunty ⟨*hat, dress, etc.*⟩; **b)** *(frech)* cheeky. **2.** *adv.* **a)** *(flott)* jauntily; **b)** *(frech)* cheekily

Kessel ['kɛsl̩] **der;** ~**s,** ~ **a)** *(Tee~)* kettle; **b)** *(zum Kochen)* pot; *(für offenes Feuer)* cauldron; *(in einer Brauerei)* vat; *(Wasch~)* copper; wash-boiler; **c)** *(Berg~)* basin-shaped valley; **d)** *(Milit.)* encircled area; *(kleiner)* pocket; **e)** *(Dampf~, Heiz~)* boiler

Kessel-: ~**fleisch das** *s.* **Wellfleisch;** ~**pauke die** kettledrum; ~**schlacht die** battle of encirclement; ~**stein der;** *o. Pl.* fur; scale; ~**treiben das a)** *(Jägerspr.)* battue *(using a circle of hunters and beaters);* **b)** *(fig.)* witch-hunt

Ketchup ['kɛtʃap] **der** *od.* **das;** ~**[s],** ~**s** ketchup

Kettchen ['kɛtçən] **das;** ~**s,** ~**:** [neck-]chain *(with cross etc. attached); (Fuß~)* anklet; *(Arm~)* bracelet

Kette ['kɛtə] **die;** ~, ~**n a)** chain; *(von Kettenfahrzeugen)* track; **die** ~ **[an der Tür] vorlegen** put the chain across [the door]; **an der** ~ **liegen** ⟨*dog*⟩ be chained up; **jmdn. in** ~**n legen** put sb. in chains; **die** ~**n abwerfen/zerreißen** *(fig. geh.)* cast off *or* throw off/break one's chains *or* shackles; **jmdn. an die** ~ **legen** *(fig.)* keep sb. on a [tight *or* short] leash; **b)** *(Halsschmuck)* necklace; *(eines Bürgermeisters usw.)* chain; **c)** *(fig.: Reihe)* chain; *(von Ereignissen)* string; series; ~ **rauchen** *(ugs.)* chain-smoke; **d)** *(Weberei)* warp

ketten *tr. V.* **a)** chain (**an** + *Akk.* to); **b)** *(fig.)* bind; **jmdn. an sich** *(Akk.)* ~**:** bind sb. to oneself; **sich an jmdn.** ~**:** tie oneself to sb.

Ketten-: ~**glied das** [chain-]link; ~**hemd das** *(hist.)* coat of chain mail; ~**hund der** watch-dog *or* guard-dog *(kept on a chain);* ~**laden der** chain store; ~**panzer der** chain mail; chain-armour; ~**rauchen das;** ~**s** chain-smoking *no art.;* ~**raucher der** chain-smoker; ~**reaktion die** chain reaction; **eine** ~**reaktion auslösen** trigger a chain reaction; ~**schutz der** chain guard

Ketzer ['kɛtsɐ] **der;** ~**s,** ~ *(auch fig.)* heretic

Ketzerei die; ~, ~**en** *(auch fig.)* heresy

Ketzerin die; ~, ~**nen** *(auch fig.)* heretic

ketzerisch *(auch fig.)* **1.** *Adj.* heretical. **2.** *adv.* heretically

keuchen ['kɔyçn̩] *itr. V.* **a)** pant; gasp for breath; **b)** *mit sein (sich keuchend fortbewegen)* puff *or* pant one's way; come/go puffing *or* panting along

Keuch·husten der whooping cough *no art.*

Keule ['kɔylə] **die;** ~, ~**n a)** club; cudgel; **chemische** ~**:** Chemical Mace **(P); b)** *(Gymnastik)* [Indian] club; **c)** *(Kochk.)* leg; *(Reh~, Hasen~)* haunch; *(Gänse~, Hühner~)* drumstick; leg

Keulen-: ~**hieb der,** ~**schlag der** blow with a club *or* cudgel; ~**schwingen das** *(Gymnastik)* club swinging; swinging [Indian] clubs

keusch [kɔyʃ] **1.** *Adj.* **a)** chaste; pure; **b)** *(geh. veralt.) (sittsam)* modest; demure; *(sittlich rein)* pure. **2.** *adv.* **a)** ~ **leben** lead a chaste life; **b)** *(sittsam)* modestly; demurely; *(sittlich rein)* in a pure manner

Keuschheit die; ~ **a)** chastity; **b)** *(geh. veralt.) (Sittsamkeit)* modesty; *(sittliche Reinheit)* purity

Keuschheits-: ~**gelübde das** vow of chastity; ~**gürtel der** chastity belt

Kfz [ka:|ɛf'tsɛt] *Abk.* **Kraftfahrzeug**

kg *Abk.* **Kilogramm** kg

KG *Abk.* **Kommanditgesellschaft**

khaki·farben *Adj.* khaki[-coloured]

Kibbuz [kɪ'bu:ts] **der;** ~, ~**im** [kɪbu'tsi:m] *od.* ~**e** kibbutz

Kicher·erbse die chick-pea

kichern ['kɪçɐn] *itr. V.* giggle; **vor sich hin** ~**:** giggle to oneself

kicken *(ugs.)* **1.** *itr. V.* play football. **2.** *tr. V.* kick

Kicker der; ~**s,** ~**[s]** *(ugs.)* footballer; [football-]player

kidnappen ['kɪtnɛpn̩] *tr. V.* kidnap

Kidnapper der; ~**s,** ~**:** kidnapper

Kidnapping ['kɪtnɛpɪŋ] **das; ~s, ~s** kidnapping
kiebig ['ki:bɪç] *Adj. (bes. nordd.) (frech)* cheeky; impertinent; *(gereizt)* touchy
Kiebitz ['ki:bɪts] **der; ~es, ~e a)** lapwing; peewit; **b)** *(ugs.: Zuschauer beim Spiel)* kibitzer *(coll.)*
kiebitzen *itr. V. (ugs. scherzh.)* kibitz *(coll.)*
[1]**Kiefer** ['ki:fɐ] **der; ~s, ~:** jaw; *(~knochen)* jaw-bone
[2]**Kiefer die; ~, ~n a)** pine[tree]; **b)** *o. Pl. (Holz)* pine[-wood]
Kiefer-: ~höhle die *(Anat.)* maxillary sinus; **~knochen der** jawbone
Kiefern-: ~holz das pine[-wood]; **~nadel die** pine-needle; **~zapfen der** pine-cone
Kiefer·orthopädie die orthodontics *sing., no art.*
kieken ['ki:kn̩] *itr. V. (nordd.)* look
Kieker ['ki:kɐ] **der; ~s, ~** *in* **jmdn. auf dem ~ haben** *(ugs.)* have it in for sb. *(coll.)*
[1]**Kiel** [ki:l] **der; ~[e]s, ~e** keel; **ein Schiff auf ~ legen** lay down a ship; lay the keel of a ship
[2]**Kiel der; ~[e]s, ~e** *(Feder~)* quill
kiel-, Kiel-: ~holen *tr. V. (Seemannsspr.)* keel-haul ⟨*person*⟩; **~oben** [-'--] *Adv.* bottom up; **~raum der** bilge; **~wasser das** wake; **in jmds. ~wasser segeln** *(fig.)* follow in sb.'s wake
Kieme ['ki:mə] **die; ~, ~n;** *meist Pl.* gill
Kien [ki:n] **der; ~[e]s** resinous wood; *(Kiefernholz)* resinous pine-wood
Kien-: ~fackel die pine[-wood] torch; **~span der** pine-wood chip; *(zum Anzünden)* pine-wood spill
Kiepe ['ki:pə] **die; ~, ~n** *(nordd., md.)* dosser; pannier
Kies [ki:s] **der; ~es, ~e a)** gravel; *(auf dem Strand)* shingle; **b)** *(Mineral.)* pyrites *sing.;* **c)** *(salopp: Geld)* dough *(sl.);* bread *(sl.)*
Kiesel ['ki:zl̩] **der; ~s, ~:** pebble
Kiesel-: ~erde die siliceous earth; **~säure die** *(Chemie)* silicic acid; **~stein der** pebble
kiesen ['ki:zn̩] *unr., auch regelm. tr. V. (dichter.)* choose; select
Kies·grube die gravel pit
Kiez [ki:ts] **der; ~es, ~e** *(bes. berlin.)* neighbourhood
kiffen ['kɪfn̩] *itr. V. (ugs.)* smoke pot *(sl.) or* grass *(sl.)*
Kiffer der; ~s, ~ *(ugs.)* pot-head *(sl.)*
kikeriki [kikəri'ki:] *Interj. (Kinderspr.)* cock-a-doodle-doo
killen ['kɪlən] *tr. V. (salopp)* do in *(sl.);* bump off *(sl.)*
Killer der; ~s, ~ *(salopp)* killer; *(gegen Bezahlung)* hit man *(sl.)*
Kilo ['ki:lo] **das; ~s, ~[s]** kilo
Kilo-: ~gramm das kilogram; **~hertz das; ~, ~** *(Physik)* kilohertz; **~meter der; ~s, ~:** kilometre
kilometer-, Kilometer-: ~fresser der *(ugs.)* **er ist ein ~fresser** he really burns up the miles *(coll.);* **~geld das** mileage allowance; **~lang 1.** *Adj.* miles long *pred.;* **eine ~lange Autoschlange** a traffic jam stretching [back] for miles; **2.** *adv.* for miles [and miles]; **~stand der** mileage reading; **~weit 1.** *Adj.; nicht präd.* **in ~weiter Entfernung** miles away in the distance; **2.** *adv.* for miles [and miles]; **~ entfernt** miles away
Kilowatt·stunde die *(Physik; bes. Elektrot.)* kilowatt-hour
Kimme ['kɪmə] **die; ~, ~n** sighting notch
Kimono [ki'mo:no] **der; ~s, ~s** kimono
Kind [kɪnt] **das; ~[e]s, ~er a)** child; kid *(coll.); (Kleinkind)* child; infant; *(Baby)* child; baby; **jmdm. ein ~ machen** *(ugs.)* put sb. in the family way *(coll.) or* in the club *(sl.);* **ein ~ erwarten/bekommen** *od. (ugs.)* **kriegen** be expecting/have a baby; **ein ~ zur Welt bringen** *(geh.)* give birth to a child; **ein ~/~er in die Welt setzen** bring a child/children into the world; **wir werden das ~ schon [richtig] schaukeln** *(ugs.)* we'll soon sort things out *or* have things sorted out; **das ~ mit dem Bade ausschütten** *(fig.)* throw the baby out with the bath-water; **das ~ beim Namen nennen** *(fig.)* call a spade a spade; **jmdn. wie ein [kleines] ~ behandeln** treat sb. like a [small] child; **das weiß/kann doch jedes ~:** any child *or* five-year old knows/can do that; **von ~ an** *od.* **auf** from childhood; **sich wie ein ~ freuen** be [as] pleased as Punch; **dann kommt bei ihm das ~ im Manne durch** *(scherzh.)* then he shows that he is [still] a child at heart; **sich bei jmdm. lieb ~ machen** *(ugs.)* get on the right side of sb.; **einziges ~ sein** be an only child; **armer/reicher Leute ~ sein** be the child of poor/wealthy parents; come from a poor/wealthy family; **ein ~ der Liebe** *(geh. verhüll.)* a love-child; **er ist/du bist** *usw.* **kein ~ von Traurigkeit** *(ugs.)* he knows/you know *etc.* how to enjoy himself/yourself *etc.;* **jmdn. an ~es Statt annehmen** *(veralt.)* adopt sb.; *s. auch* **totgeboren; b)** *(ugs.: als Anrede)* **mein [liebes] ~:** my [dear] child; **~er, hört mal alle her!** listen to this, all of you *(coll.);* **[~er,] ~er!** my goodness!
Kindchen ['kɪntçən] **a)** [small *or* little] child; **b)** *(Anrede)* dear child
kinder-, Kinder-: ~arbeit die; *o. Pl.* child labour; **~arzt der** paediatrician; **~bett das** cot; *(für größeres Kind)* child's bed **~bild das** *(Foto)* photograph of a child; *(Malerei usw.)* portrait of a child; **ein ~bild von jmdm.** a photograph/portrait of sb. as a child; **~buch das** children's book; **~chor der** children's choir; **~dorf das** children's village
Kinderei die; ~, ~en childishness *no indef. art., no pl.;* **eine ~:** a childish prank; **~en** childishness *sing.;* childish behaviour *sing.*
kinder-, Kinder-: ~erziehung die bringing up of children; **~fahrrad das** child's bicycle; **~feindlich 1.** *Adj.* hostile to children *pred.;* anti-children *pred.;* **2.** *adv.* **sich ~feindlich verhalten** act in a manner hostile to children; **~feindlichkeit die;** *o. Pl.* hostility to children; *(von Planung, Politik)* failure to cater for children; **~fest das** children's party; children's fête; **~film der** children's film; **~freund der: ein [großer] ~freund/[große] ~freunde sein** be [very] fond of children; **~freundlich 1.** *Adj.* fond of children *pred.;* ⟨*planning, policy*⟩ which caters for the needs of children; **2.** *adv.* **sich ~freundlich verhalten** act in a manner friendly to children; **~freundlichkeit die** fondness for children; **~funk der** children's programmes *pl.;* **~garten der** kindergarten; nursery school; **~gärtnerin die** kindergarten teacher; nursery-school teacher; **~geld das** child benefit; **~gesicht das** child's face; *(eines Erwachsenen)* child-like face; baby-face; **~glaube der** childlike belief *or* faith; *(abwertend)* childish belief *or* faith; **~gottes·dienst der** children's service; **~heim das** children's home; **~hort der** day-home for schoolchildren; **~karussell das** children's roundabout; **~kleidung die** children's clothes *pl.;* children's wear; **~krankheit die a)** children's disease *or* illness;

welche ~krankheiten hatten Sie? what childhood diseases have you had?; **b)** *Pl. (fig.)* teething troubles; **~kriegen das; ~s** *(ugs.)* having children; **~krippe die** crèche; day nursey; **~lähmung die** poliomyelitis; infantile paralysis *no art.*; **~leicht** *(ugs.)* **1.** *(Adj.)* childishly simple *or* easy; dead easy; **das ist ~leicht** it's child's play *or (coll.)* kid's stuff; **2.** *adv.* **es ist ~leicht zu bedienen** it's childishly simple to use; **~lieb** *Adj.* fond of children *pred.*; **~liebe die** love of children; **~lied das** nursery rhyme; **~los** *Adj.* childless; **~losigkeit die** childlessness; **~mädchen das** nursemaid; nanny; **~märchen das** [children's] fairy-tale; **~mord der** child-murder; **~mund der** child's mouth; **~mund tut Wahrheit kund** *(Prov.)* it takes a child to point out the truth; **~narr der: er ist ein ~narr** he adores children; **~popo der** *(ugs.)* [baby's] bottom; **glatt wie ein ~popo** [as] smooth as a baby's bottom; **~reich** *Adj.* with many children *postpos., not pred.*; **eine ~reiche Familie** a large family; **~reichtum der;** *o. Pl.* large number of children; **~reim der** nursery rhyme; **~schreck der;** *o. Pl.* bogyman; **~schuh der** child's shoe; **ich bin/du bist den ~schuhen entwachsen** *(fig.)* I'm/you're not a child any more; **noch in den ~schuhen stecken** ⟨*process, technique, etc.*⟩ be still in its infancy; **~schwester die** children's nurse; **~segen der;** *o. Pl. (oft scherzh.)* **eine Familie mit reichem ~segen** a family blessed with a large number of children; **~sitz der** child's seat; *(an einem Fahrrad)* child-carrier [seat]; *(im Auto)* child's safety seat; **~spiel das** children's game; **[für jmdn.] ein ~spiel sein** be child's play [to sb.]; **~spielplatz der** [children's] playground; **~spielzeug das** [children's] toys *pl. or* playthings *pl.*; *(einzeln)* [child's] toy *or* plaything; **~sterblichkeit die** child mortality; **~stimme die** child's voice; **~stube die;** *o. Pl.* **eine gute/schlechte ~stube gehabt** *od.* **genossen haben** have been well/badly brought up; **hast du gar keine ~stube?** didn't you ever learn any manners?; **~teller der a)** child's plate; **b)** *(Gericht)* children's menu; **~wagen der** pram *(Brit.)*; baby carriage *(Amer.)*; *(Sportwagen)* push-chair *(Brit.)*; stroller *(Amer.)*; **~zimmer das** children's room; *(für Kleinkinder)* nursery

Kịndes-: ~alter das; *o. Pl.* childhood; **im ~alter** at an early age; **~beine** *Pl.* *in* **von ~beinen an** from *or* since childhood; from an early age; **~entführung die** kidnapping [of a child]; child abduction; **~kind das** *(veralt.)* grandchild; **Kinder und ~kinder** children and children's children; **~mißhandlung die** *(Rechtsw.)* child abuse; **~mord der** child murder; *(Mord am eigenen Kind)* infanticide; **~mörderin die** infanticide

kịnd·gemäß *Adj.* suitable for children *postpos.*

Kịndheit die; ~: childhood; **seit frühester ~:** from earliest childhood; from infancy

Kịndheits·erinnerung die childhood memory

kịndisch 1. *Adj.* childish, infantile ⟨*behaviour, enjoyment*⟩; naïve ⟨*ideas*⟩; **~ werden** become childish; **werd nicht ~!** do behave sensibly. **2.** *adv.* childishly; **sich ~ an etw.** *(Dat.)* **freuen** take childish pleasure in sth.

kịndlich 1. *Adj.* childlike. **2.** *adv.* ⟨*behave*⟩ in a childlike way *or* manner; **sich ~ über etw.** *(Akk.)* **freuen** take a childlike pleasure in sth.

Kịndlichkeit die; ~: childlike quality

Kịnds·kopf der overgrown child; **sei doch kein ~!** don't be so childish!; act your age!

Kịnd·taufe die christening

Kinematographie [kınematogra'fi:] **die; ~:** cinematography *no art.*

Kinetik [ki'ne:tık] **die; ~** *(Physik)* kinetics *sing., no art.*

Kinkerlitzchen ['kıŋkɐlıts̮çən] *Pl. (ugs.)* trifles

Kinn [kın] **das; ~[e]s, ~e** chin

Kịnn-: ~backe die, ~backen der *(südd.)* cheek; **~bart der** chin-beard; chin-tuft; **~haken der** hook to the chin; **~lade die** jaw

Kino ['ki:no] **das; ~s, ~s a)** cinema *(Brit.)*; movie theatre *or* house *(Amer.)*; **in die [deutschen] ~s kommen** go on general release [in Germany]; **b)** *(Vorstellung)* film; movie *(Amer.)*; **ins ~ gehen** go to the cinema *(Brit.) or* pictures *(Brit.) or (Amer.)* movies *pl.*; **c)** *o. Pl. (Film als Medium)* cinema

Kino-: ~karte die cinema ticket *(Brit.)*; movie ticket *(Amer.)*; **~kasse die** cinema *(Brit.) or (Amer.)* movie box-office; **~programm das** cinema guide

Kintopp ['ki:ntɔp] **der** *od.* **das; ~s, ~s** *od.* **Kintöppe** ['ki:ntœpə] *(ugs.)* cinema

Kiosk [ki̯ɔsk] **der; ~[e]s, ~e** kiosk

Kipfel ['kıpf̩l] **das; ~s, ~, Kipferl** ['kıpfɐl] **das; ~s, ~n** *(bayr., österr.) s.* **Hörnchen b**

[1]**Kippe** ['kıpə] **die; ~, ~n** *(ugs.)* cigarette end; fag-end *(sl.)*; dog-end *(sl.)*

[2]**Kịppe die; ~, ~n a)** *(Müll~)* tip; dump; **b)** *in* **auf der ~ stehen** *(ugs.)* be balanced precariously; **etw. steht auf der ~** *(fig.) (etw. befindet sich in einer kritischen Lage)* it's touch and go with sth.; *(etw. ist noch nicht entschieden)* sth. hangs in the balance

kippelig ['kıpəlıç] *Adj. (ugs.)* wobbly; rickety, wobbly ⟨*chair, table*⟩

kippeln ['kıpl̩n] *itr. V. (ugs.)* wobble; ⟨*chair, table*⟩ wobble; be wobbly *or* rickety; **[mit seinem Stuhl] ~:** rock one's chair backwards and forwards

kịppen 1. *tr. V.* **a)** *(neigen)* tip [up]; tilt; **b)** *(ausschütten)* tip; **c)** *(ugs.: trinken)* knock back *(sl.)*; **einen ~:** have a quick one *(coll.) or* a drink; **d)** *(ugs.: abbrechen)* give ⟨*project, series*⟩ the chop *(coll.)*. **2.** *itr. V.*; *mit sein* tip over; ⟨*top-heavy object*⟩ topple over; ⟨*person*⟩ fall, topple; ⟨*boat*⟩ overturn; ⟨*car*⟩ roll over; **von etw. ~:** topple *or* fall off sth.

Kịpper der; ~s, ~: tipper lorry *or* truck; dump truck; *(Eisenb.)* tipper *or* tipping wagon; dump car *(Amer.)*

Kịpp-: ~fenster das horizontally pivoted window; **~lore die** tipper *or* tipping wagon; **~schalter der** tumbler *or* toggle switch; **~wagen der** *s.* **~lore**

Kirche ['kırçə] **die; ~, ~n a)** church; *(fig.)* **die ~ im Dorf lassen** keep a sense of proportion; **mit der ~ ums Dorf gehen/fahren** do things in a roundabout way; **b)** *o. Pl. (Gottesdienst)* church *no art.*; **in der ~ sein** be at church; **in die ~ gehen** go to church; **c)** *(Institution)* Church; **aus der ~ austreten** secede from *or* leave the Church

kịrchen-, Kịrchen-: ~älteste der *(ev. Kirche)* [church-]elder; **~bank die** [church-]pew; **~chor der** church choir; **~feindlich** *Adj.* hostile to the Church *postpos.*; **~fenster das** church window; **~fürst der** *(geh.)* high ecclesiastical dignitary; high dignitary of the Church; *(kath. Kirche: Kardinal)* Prince of the Church; **~gemeinde die** parish; **~glocke die** church bell; **~jahr das** ecclesiastical year; Church year; **~lied das** hymn; **~maus die** *in* **arm sein wie eine ~maus** *(ugs. scherzh.)* be as poor as a church mouse; **~musik die** church music; sacred music; **~portal das** portal *or* main door of the/a church; **~schiff das** *(Archit.)* nave; **~steuer die** church tax; **~tag der** Church congress; **~tür die** church door

Kirch-: ~gang der: der sonntägliche ~gang going to church on Sunday; **~gänger der** churchgoer; **~hof der** *(veralt.)* churchyard; graveyard

kirchlich 1. *Adj.* ecclesiastical; Church *attrib.;* ecclesiastical ⟨*law, building*⟩*;* religious, church ⟨*festival*⟩*;* church *attrib.* ⟨*wedding, funeral*⟩. **2.** *adv.* **~ getraut/begraben werden** have a church wedding *or* be married in church/have a church funeral

Kirch·turm der church tower; *(mit Spitze)* [church] steeple

Kirchturm-: ~spitze die church spire; **~uhr die** church clock

Kirch-: ~weih die; ~, ~en fair; **~weihe die** consecration of a/the church

Kirmes ['kɪrməs] **die; ~, Kirmessen** ['kɪrmɛsn̩] *(bes. md., niederd.)* *s.* **Kirchweih**

kirre ['kɪrə] *Adj.; nicht attr.* **jmdn. ~ machen** *(ugs.)* bring sb. to heel

Kirsch [kɪrʃ] **der; ~[e]s, ~:** *s.* **Kirschwasser**

Kirsch-: ~baum der a) cherry[-tree]; **b)** *(Holz) o. Pl.* cherry[-wood]; **~blüte die a)** *(Blüte des ~baums)* cherry blossom; **b)** *(Zeit der ~blüte)* cherry blossom time

Kirsche ['kɪrʃə] **die; ~, ~n** cherry; **mit ihm ist nicht gut ~n essen** *(ugs.)* it's best not to tangle with him

kirsch-, Kirsch-: ~kern der cherry stone; **~likör der** cherry liqueur; *(Weinbrand)* cherry brandy; **~saft der** cherry juice; **~stein der** *s.* **~kern; ~torte die** cherry gateau; *(mit Tortenboden)* cherry flan; **Schwarzwälder ~torte** Black Forest gateau; **~wasser das** kirsch

Kissen ['kɪsn̩] **das; ~s, ~:** cushion; *(Kopf~)* pillow

Kissen-: ~bezug der cushion cover; *(für Kopfkissen)* pillowcase; pillow-slip; **~schlacht die** *(ugs.)* pillow-fight

Kiste ['kɪstə] **die; ~, ~n a)** box; *(Truhe)* chest; *(Latten~)* crate; *(für Obst)* case; box; *(für Wein)* case; **b)** *(salopp) (Flugzeug, Auto)* bus *(coll.); (Fernseher)* box *(coll.);* **c)** *(ugs., bes. berlin.: Sache, Angelegenheit)* affair; business

Kitsch [kɪtʃ] **der; ~[e]s** kitsch

kitschig *Adj.* kitschy

Kitt [kɪt] **der; ~[e]s, ~e** *(Fenster~)* putty; *(für Porzellan, Kacheln usw.)* cement; *(Füllmasse)* filler

Kittchen das; ~s, ~ *(ugs.)* clink *(sl.);* jug *(sl.);* jail; **im ~ sitzen** be inside *(coll.);* be in clink *or* jug *(sl.)*

Kittel ['kɪtl̩] **der; ~s, ~ a)** overall; *(eines Arztes, Laboranten usw.)* white coat; **b)** *(hemdartige Bluse)* smock

kitten *tr. V.* cement [together]; stick [together] with cement; *(fig.)* mend ⟨*breach*⟩*;* patch up ⟨*broken marriage, friendship*⟩

Kitz [kɪts] **das; ~es, ~e** *(Reh~)* fawn; *(Ziegen~, Gemsen~)* kid

Kitzel ['kɪtsl̩] **der; ~s, ~** *(Reiz, Antrieb)* itch; urge; *(freudige Erregung)* thrill

kitzelig *s.* **kitzlig**

kitzeln 1. *tr. V.* tickle; **es kitzelt mich in der Nase** my nose tickles. **2.** *itr. V.* tickle; **auf der Haut ~:** tickle [the skin]

Kitzler der; ~s, ~ *(Anat.)* clitoris

kitzlig *Adj.* **a)** ticklish; **b)** *(schwierig, heikel)* ticklish

Kiwi ['ki:vi] **die; ~, ~s** kiwi [fruit]

KKW *Abk.* **Kernkraftwerk**

Klacks [klaks] **der; ~es, ~e** *(ugs.)* *(~ Schlagsahne, Kartoffelbrei)* dollop *(coll.); (~ Senf)* blob; dab; **etw. ist ein ~ [für jmdn.]** *(fig.)* sth. is no trouble at all [for sb.]

Kladde ['kladə] **die; ~, ~n** rough book; **etw. in ~ schreiben** write sth. in rough

klaffen ['klafn̩] *itr. V.* gape; yawn; ⟨*hole, wound*⟩ gape; ⟨*gap*⟩ yawn; **in der Mauer klaffte ein großes Loch** there was a gaping hole in the wall

kläffen ['klɛfn̩] *itr. V. (abwertend)* yap

klaffend *Adj.* gaping; yawning; gaping ⟨*hole, wound*⟩*;* yawning ⟨*gap*⟩

Kläffer der; ~s, ~ *(ugs. abwertend)* yapping dog; yapper

Klage ['kla:gə] **die; ~, ~n a)** *(aus Trauer)* lamentation; lament; *(wegen Schmerzen)* complaint; **b)** *(Beschwerde)* complaint; **keinen Grund zur ~ geben/haben** give/have no grounds *pl. or* reason for complaint; **bei jmdm. über jmdn./etw. ~ führen** make a complaint to sb. *or* lodge a complaint with sb. about sb./sth.; **c)** *(Rechtsw.) (im Zivilrecht)* action; suit; *(im Strafrecht)* charge; **[öffentliche] ~ gegen jmdn. einreichen/erheben** bring an action against sb.; institute [criminal] proceedings against sb.

Klage-: ~laut der plaintive cry; *(von Schmerzen verursacht)* cry of pain; *(stöhnend)* moan; **~lied das** lament; **~mauer die** Wailing Wall

klagen 1. *itr. V.* **a)** *(geh.: jammern)* wail; *(stöhnend)* moan; ⟨*animal*⟩ cry plaintively; **b)** *(sich beschweren)* complain; **über etw.** *(Akk.)* **~:** complain about sth.; **über Rückenschmerzen ~:** complain of backache *sing.;* **[ich] kann nicht ~:** [I] can't complain; [I] mustn't grumble; **c)** *(geh.)* **um jmdn./jmds. Tod ~:** mourn sb./sb.'s death; **über den Verlust seines Vermögens ~:** lament *or* bewail the loss of one's fortune; **d)** *(bei Gericht)* sue; take legal action; **auf Schadenersatz ~:** sue for damages; bring an action for damages; **gegen jmdn. ~:** sue sb.; take legal action against sb. **2.** *tr. V.* **jmdm. sein Leid ~:** pour out one's sorrows *pl.* to sb.

Kläger ['klɛ:gɐ] **der; ~s, ~, Klägerin die; ~, ~nen** *(im Zivilrecht)* plaintiff; *(im Strafrecht)* prosecuting party; *(bei einer Scheidung)* petitioner

Klage·schrift die *(Rechtsw.) (im Zivilrecht)* statement of claim; *(im Strafrecht)* charge/list of charges; *(bei einer Scheidung)* petition

kläglich ['klɛ:klɪç] *Adj.* **a)** *(mitleiderregend)* pitiful ⟨*expression, voice, cry*⟩*;* pitiful, wretched ⟨*condition, appearance*⟩*;* **b)** *(minderwertig)* pathetic ⟨*achievement, result, etc.*⟩*;* **c)** *(erbärmlich)* despicable, wretched ⟨*behaviour, role, compromise*⟩*;* pathetic ⟨*result, defeat*⟩

klaglos 1. *Adj.* uncomplaining. **2.** *adv.* uncomplainingly; without complaint

Klamauk [kla'mauk] **der; ~s** *(ugs. abwertend)* fuss; to-do; *(Lärm, Krach)* row *(coll.);* racket; *(im Theater)* slapstick

klamm [klam] *Adj.* **a)** *(feucht)* cold and damp; **b)** *(steif)* numb

Klammer die; ~, ~n a) *(Wäsche~)* peg; **b)** *(Haar~)* [hair-]grip; **c)** *(Zahn~)* brace; **d)** *(Wund~)* clip; **e)** *(Büro~)* paperclip; *(Heft~)* staple; **f)** *(Schriftzeichen)* bracket; **~ auf/zu** open/close brackets

klammern 1. *refl. V.* **sich an jmdn./etw.** ~ *(auch fig.)* cling to sb./sth. **2.** *tr. V.* **a) eine Wunde ~:** close a wound with a clip/clips; **b)** *(mit einer Büroklammer)* clip; *(mit einer Heftmaschine)* staple; *(mit einer Wäscheklammer)* peg. **3.** *itr. V. (Boxen)* clinch

klamm·heimlich 1. *Adj.; nicht präd. (ugs.)* on the quiet *postpos.* **2.** *adv.* on the quiet

Klamotte [kla'mɔtə] **die; ~, ~n a)** *Pl. (salopp: Kleidung)* clobber *sing. (sl.);* gear *sing. (sl.);* **b)** *Pl. (salopp: Kram)* junk *sing.;* stuff *sing.;* **c)** *(ugs. abwertend: Schwank)* rubbishy play/film *etc.*

klang [klaŋ] *1. u. 3. Pers. Sg. Prät. v.* **klingen**

Klang der; ~[e]s, Klänge ['klɛŋə] **a)** *(Ton)* sound; **b)** *(~farbe)* tone; **c)** *Pl. (Melodie)* **alte, wohlbekannte Klänge** old familiar tunes; **nach den Klängen eines Walzers tanzen** dance to the strains of a waltz

Klang·farbe die tone colour *or* quality

klanglich 1. *Adj.; nicht präd.* tonal ‹*beauty, quality, etc.*›; tonal, tone *attr.* ‹*characteristics*›. **2.** *adv.* tonally

klanglos 1. *Adj.* toneless ‹*voice*›. **2.** *adv.* tonelessly; *s. auch* **sanglos**

klang·voll *Adj.* sonorous ‹*voice, language*›; *(fig.)* illustrious ‹*name, title*›

Klapp·bett das folding bed

Klappe die; ~, ~n a) [hinged] lid; *(am Briefkasten)* flap; **b)** *(am LKW)* tail-board; tail-gate; *(seitlich)* side-gate; *(am Kombiwagen)* back; **c)** *(an Kleidertaschen)* flap; **d)** *(am Ofen)* [drop-]door; **e)** *(an Musikinstrumenten)* key; *(an einer Trompete)* valve; **f)** *(Herz~)* valve; **g)** *(Augen~)* [eye-] patch; **h)** *(Achselstück)* shoulder-strap; **i)** *(Filmjargon)* clapper-board; **j)** *(salopp: Mund)* trap *(sl.);* **die** *od.* **seine ~ halten** shut one's trap *(sl.);* **eine große ~ haben** *(abwertend)* have a big mouth; **k)** *(ugs.: Bett) s.* **Falle b**

klappen 1. *tr. V.* **nach oben/unten ~:** turn up/down ‹*collar, hat-brim*›; lift up/put down *or* lower ‹*lid*›; **nach vorne/hinten ~:** tilt forward/back ‹*seat*›. **2.** *itr. V.* **a)** ‹*door, shutter*› bang; **b)** *(stoßen)* bang; **c)** *(ugs.: gelingen)* work out all right; ‹*rehearsal, performance, etc.*› go [off] all right; **hat es mit den Karten geklappt?** did you get the tickets all right?

Klappen·text der *(Buchw.)* blurb

Klapper die; ~, ~n rattle

Klapper·kiste die *(ugs.)* rattle-trap

klappern *itr. V.* **a)** rattle; **b)** *(ein Klappern erzeugen)* make a clatter; **vor Kälte klapperte er mit den Zähnen** his teeth were chattering with cold; **mit den Augen ~** *(ugs.)* keep blinking; *(kokettieren)* flutter one's eyelashes

Klapper-: ~schlange die rattlesnake; **~storch der** *(Kinderspr.)* stork

Klapp-: ~fenster das top-hung window; **~messer das** clasp-knife; **~rad das** folding bicycle

klapprig *Adj.* **a)** *(alt)* rickety; ramshackle; **b)** *(wenig stabil)* rickety; wobbly; **c)** *(ugs.: hinfällig)* decrepit

Klapp-: ~sitz der folding seat; tip-up seat; **~stuhl der** folding chair; **~tisch der** folding table

Klaps [klaps] **der; ~es, ~e a)** *(ugs.: leichter Schlag)* smack; slap; **b)** *(salopp)* **einen ~ haben** have a screw loose *(coll.);* be a bit bonkers *(sl.)*

Klaps·mühle die *(salopp)* loony-bin *(sl.);* nut-house *(sl.)*

klar [kla:ɐ̯] **1.** *Adj.* **a)** clear; **bei ~er Sicht** when it's clear; on a clear day; **ein ~er Verstand** clear judgement; **~ [im Kopf] sein** have a clear head; be able to think clearly *or* straight; **er ist nicht ganz ~ im Kopf** *(salopp)* he's not quite right in the head *(sl.);* **b)** *(eindeutig)* clear ‹*decision, aim, objective*›; straight ‹*question, answer*›; **~e Verhältnisse schaffen** set things straight; **[ist] alles ~?** [is] everything clear?; **jetzt ist mir alles ~:** now I understand; **na ~!** *(ugs.),* **aber ~!** *(ugs.)* of course!; **ist dir ~, daß ...?** are you aware that ...?; **sich** *(Dat.)* **über etw.** *(Akk.)* **im ~en sein** realize *or* be aware of sth.; **c)** *nicht attr. (fertig)* ready. **2.** *adv.* clearly; **etw. ~ und deutlich sagen** say sth. clearly and unambiguously

Klär·anlage die sewage treatment plant; *(einer Fabrik)* waste-water treatment plant

Klare der; ~n, ~n schnapps

klären ['klɛ:rən] **1.** *tr. V.* **a)** *(aufklären)* settle, resolve ‹*question, issue, matter*›; clarify ‹*situation*›; clear up ‹*case, affair, misunderstanding*›; **b)** *(reinigen)* purify; treat ‹*effluent, sewage*›; clear ‹*beer, wine*›. **2.** *refl. V.* **a)** *(klar werden)* ‹*situation*› become clear; ‹*question, issue, matter*› be settled *or* resolved; **b)** *(rein werden)* ‹*liquid, sky*› clear. **3.** *itr. V. (Ballspiele)* clear [the ball]

klar|gehen *unr. itr. V.; mit sein (ugs.)* go OK *(coll.);* **es wird schon ~:** it'll be OK *(coll.)*

Klarheit die; ~, ~en a) *o. Pl.* clarity; *(von Ausführungen, Rede usw.)* clarity; lucidity; **b)** *o. Pl. (Gewißheit)* **sich** *(Dat.)* **über etw.** *(Akk.)* **~ verschaffen** clarify sth.; **c)** *(ugs. scherzh.)* **jetzt sind alle ~en beseitigt** now I'm/everyone's *etc.* totally confused

Klarinette [klari'nɛtə] **die; ~, ~n** clarinet

Klarinettist der; ~en, ~en, Klarinettistin die; ~, ~nen clarinettist

klar-, Klar-: ~|kommen *unr. itr. V.; mit sein (ugs.)* manage; cope; **mit jmdm. ~kommen** get on with sb.; **~|machen** *tr. V.* **a)** *(ugs.)* make clear; **jmdm./sich etw. ~machen** make sth. clear to sb./realize sth.; **b)** *(Seemannsspr.)* get ready; prepare; **~schriftleser der** *(DV)* optical character reader; **~|sehen** *unr. itr. V.* understand the matter

Klarsicht-: ~folie die transparent film; **~packung die** transparent pack

klar-, Klar-: ~|spülen *itr. V.* rinse; **~|stellen** *tr. V.* clear up; clarify; **ich möchte ~stellen, daß ...:** I should like to make it clear that ...; **~stellung die** clarification; **~text der** *(auch DV)* clear *or* plain text; **im ~text** *(fig.)* in plain language

Klärung die; ~, ~en a) clarification; **b)** *(Reinigung)* purification; *(von Abwässern)* treatment

klar|werden *unr. V.; mit sein; nur im Inf. und Part. zusammengeschrieben* **1.** *refl. V.* **sich** *(Dat.)* **über etw.** *(Akk.)* **~:** realize *or* grasp sth.. **2.** *itr. V.* **jmdm. ~:** become clear to sb.

Klär·werk das sewage works *sing. or pl.; (einer Fabrik)* waste-water treatment works *sing. or pl.*

klasse ['klasə] *(ugs.)* **1.** *indekl. Adj.* great *(coll.);* marvellous. **2.** *adv.* marvellously

Klasse die; ~, ~n a) *(Schul~)* class; form *(esp. Brit.); (Raum)* class-room; *(Stufe)* year; grade *(Amer.);* **b)** *(Bevölkerungsgruppe)* class; **die ~ der Werktätigen** the working class; **c)** *(Sport)* league; *(Boxen)* division; class; **d)** *(Kategorie)* class; **eine Fahrkarte erster ~:** a first-class ticket; **zweiter ~ liegen** occupy a second-class hospital-bed; **er ist ein Künstler erster ~** *(ugs.)* he is a first-class *or* first-rate artist; **das ist [einsame** *od.* **ganz große] ~!** *(ugs.)* that's

[just] great *(coll.)* *or* marvellous!; **g)** *(Biol.)* class

Klasse-: ~frau die *(ugs.)* stunner *(coll.)*; smasher *(coll.)*; **~mann der** *(ugs.)* marvellous man; fantastic guy *(sl.)*

klassen-, Klassen-: ~arbeit die [written] class test; **~ausflug der** class outing; **~beste der/die;** *adj. Dekl.* top pupil in the class; **~bewußtsein das** *(Soziol.)* classconsciousness; **~buch das** *(Schulw.)* ≈ [class-]register; **~fahrt die** *s.* **~ausflug; ~feind der** *(marx.)* class enemy; **~gegensatz der** class difference; **~gesellschaft die** *(Soziol.)* class society; **~justiz die** *(Soziol.)* legal system with a built-in class bias; **~kamerad der, ~kameradin die** class-fellow; class-mate; **~kampf der** *(marx.)* class struggle; **~lehrer der, ~lehrerin die** class *or* form teacher; form master/mistress; **~los** *Adj.* *(Soziol.)* classless; **~sprecher der, ~sprecherin die** class spokesman; ≈ form leader *or* captain; **~treffen das** class reunion; **~unterschied der** *(Soziol.)* class difference; **~ziel das** *(Schulw.)* required standard *(for pupils in a particular class)*; **das ~ziel erreichen** reach the required standard; *(fig.)* make the grade; come up to scratch; **~zimmer das** classroom

Klassifikation [klasifika'tsi̯o:n] **die; ~, ~en** classification

klassifizieren [klasifi'tsi:rən] *tr. V.* classify (**als** as)

Klassifizierung die; ~, ~en classification

Klassik ['klasɪk] **die; ~ a)** *(Antike)* classical antiquity *no art.*; **b)** *(Zeit kultureller Höchstleistung)* classical period *or* age

Klassiker der; ~s, ~ classic; *(Schriftsteller)* classic; classical writer; *(Komponist)* classic; classical composer

klassisch *Adj.* **a)** classical; **b)** *(vollendet, zeitlos; auch iron.)* classic

Klassizismus [klasi'tsɪsmʊs] **der; ~:** classicism

klassizistisch *Adj.* classical

klatsch *Interj.* smack

Klatsch [klatʃ] **der; ~[e]s, ~e a)** *o. Pl.* *(ugs. abwertend: Gerede)* gossip; tittle-tattle; **b)** *(Geräusch)* smack

Klatsch·base die *(ugs. abwertend)* gossip

klatschen 1. *itr. V.* **a)** *auch mit sein* ‹*waves, wet sails*› slap (**gegen** against); **der Regen klatscht gegen die Scheiben** the rain beats against the windows; **jmdm. eine Ohrfeige geben, daß es nur so klatscht** give sb. a resounding smack *or* slap round the face; **b)** *(mit den Händen; applaudieren)* clap; **in die Hände ~:** clap one's hands; **lautes Klatschen** loud applause; **c)** *(schlagen)* slap; **sich** *(Dat.)* **auf die Schenkel ~:** slap one's thighs; **d)** *(ugs. abwertend: reden)* gossip (**über** + *Akk.* about). **2.** *tr. V.* **a)** *(ugs.: werfen)* slap; chuck *(coll.)* ‹*book etc.*›; **b) den Takt ~** clap time; **jmdm. Beifall ~:** clap *or* applaud sb.; **c)** *(ugs.: schlagen)* **jmdm. eine ~:** slap sb. *or* give sb. a slap across the face

Klatscherei die; ~, ~en *(ugs. abwertend)* gossiping

klatsch-, Klatsch-: ~mohn der corn-poppy; field poppy; **~naß** *Adj.* *(ugs.)* soaking *or* sopping wet ‹*clothes*›; dripping wet ‹*hair*›; **~naß werden** get soaked [to the skin] *or* drenched; **~spalte die** *(ugs. abwertend)* gossip column; **~süchtig** *Adj.* extremely gossipy; **~süchtig sein** be a compulsive gossip/compulsive gossips; **~tante die** *(ugs. abwertend)*, **~weib das** *(ugs. abwertend)* gossip

Klaue ['klau̯ə] **die; ~, ~n a)** claw; *(von Raubvögeln)* talon; *(fig. geh.)* **in den ~n eines Erpressers** in the clutches of a blackmailer; **b)** *(Huf)* hoof; **c)** *(salopp: Hand)* mitt *(sl.)*; paw *(coll.)*; **d)** *o. Pl.* *(salopp abwertend: Handschrift)* scrawl

klauen *(ugs.)* **1.** *tr. V.* pinch *(sl.)*; nick *(Brit. sl.)*; *(fig.)* pinch *(sl.)*, nick *(Brit. sl.)*, crib ‹*idea*›; **jmdm. etw. ~:** pinch *or* *(Brit.)* nick/crib sth. from sb. **2.** *itr. V.* pinch *(sl.)* *or* nick *(Brit. sl.)* things

Klauen·seuche die *s.* **Maul- und Klauenseuche**

Klausel ['klau̯zl̩] **die; ~, ~n** clause; *(Bedingung)* stipulation; condition; *(Vorbehalt)* proviso

Klausur [klau̯'zu:ɐ̯] **die; ~, ~en a) in ~:** ‹*meet*› in private; **b)** *(Klausurarbeit)* [examination] paper; **eine ~ schreiben** take a[n examination] paper

Klausur·tagung die private meeting

Klaviatur [klavi̯a'tu:ɐ̯] **die; ~, ~en** keyboard

Klavier [kla'vi:ɐ̯] **das; ~s, ~e** piano

Klavier-: ~begleitung die piano accompaniment; **~hocker der** piano-stool; **~konzert das a)** *(Komposition)* piano concerto; **b)** *(Veranstaltung)* piano recital; **~lehrer der, ~lehrerin die** piano teacher; **~sonate die** piano sonata; **~spiel das** piano-playing; **~spieler der, ~spielerin die** pianist; piano-player; **~stunde die** piano-lesson; **~unterricht der** piano-lessons *pl.*

Klebe-: ~band das adhesive *or* sticky tape; **~folie die** adhesive film

kleben 1. *itr. V.* **a)** stick (**an** + *Dat.* to); **das Hemd klebte ihm am Körper** his shirt stuck *or* clung to his body; **an seinen Händen klebt Blut** *(fig.)* he has blood on his hands *(fig.)*; his hands are stained with blood *(fig.)*; **b)** *(ugs.: klebrig sein)* be sticky (**von, vor** + *Dat.* with); **c)** *(ugs.: sich klammern)* **an seinem Stuhl/an der Theke ~:** stay put in one's chair *(coll.)*/prop the bar up *(coll.)*. **2.** *tr. V.* **a)** *(befestigen)* stick; *(mit Klebstoff)* stick; glue; *(mit Leim)* stick; paste; **jmdm. eine ~** *(salopp)* belt sb. one *(coll.)*; **b)** *(mit Klebstoff reparieren)* stick *or* glue ‹*vase etc.*› back together

kleben|bleiben *unr. itr. V.; mit sein* **a)** *(haftenbleiben)* stick; remain stuck; **b)** *(salopp)* *s.* **sitzenbleiben a**

Kleber der; ~s, ~: adhesive; glue

Kleb·pflaster das adhesive plaster; sticking-plaster

klebrig *Adj.* sticky; *(von Schweiß)* clammy ‹*hands etc.*›

Kleb-: ~stelle die join; *(eines Films, Tonbandes)* splice; **~stoff der** adhesive; glue; **~streifen der** adhesive *or* sticky tape

kleckern ['klɛkɐn] *(ugs.)* **1.** *itr. V.* **a)** *(Flecken machen)* make a mess; **oje, jetzt habe ich gekleckert** oh dear, now I've gone and spilled something *(coll.)*; **b)** *mit sein (heruntertropfen)* drip; spill; **c) nicht ~, sondern klotzen** *(ugs.)* not mess about with half-measures, but do the thing properly. **2.** *tr. V.* spill; splash ‹*paint*›

kleckerweise *Adv.* *(ugs.)* in dribs and drabs

Klecks [klɛks] **der; ~es, ~e a)** stain; *(nicht aufgesogen)* blob; *(Tintenfleck)* [ink-]blot; **b)** *(ugs.: kleine Menge)* spot; *(von Senf, Mayonnaise)* dab

klecksen 1. *itr. V.* **a)** make a stain/stains; *(mit Tinte)* make a blot/blots; ‹*pen*› blot; **b)** *(ugs. abwertend: schlecht malen)* daub. **2.** *tr. V.* *(ugs.)* daub ‹*paint*›

Klee [kle:] **der;** ~s clover; **jmdn./etw. über den grünen ~ loben** *(ugs.)* praise sb./sth. to the skies

Klee·blatt das a) clover-leaf; **ein vierblättriges ~:** a four-leaf *or* four-leaved clover; **b)** *(ugs.: drei Personen)* trio; threesome

Kleid [klait] **das;** ~es, ~er **a)** dress; **ein zweiteiliges ~:** a two-piece [suit]; **b)** *Pl. (Kleidung)* clothes; **~er machen Leute** *(Spr.)* clothes make the man; the apparel oft proclaims the man *(literary)*

kleiden 1. *refl. V.* dress. **2.** *tr. V.* **a)** dress; **b)** suit; look well on; **die Farbe kleidet dich gut** the colour suits you *or* looks well on you; **c) etw. in Worte ~:** express sth. in words; put sth. into words

Kleider-: ~ablage die a) *(Ablage)* coat rack; **b)** *(Raum)* cloakroom; checkroom *(Amer.)*; **~bügel der** clothes-hanger; coat-hanger; **~bürste die** clothes-brush; **~größe die** size; **~haken der** coat-hook; **~schrank der** wardrobe; **~ständer der** coat-stand; **~stange die** clothes-rail

kleidsam *Adj.* becoming

Kleidung die; ~: clothes *pl.*; clothing

Kleidungs·stück das garment; article of clothing; **~e** clothes

Kleie ['klaiə] **die;** ~: bran

klein [klain] **1.** *Adj.* **a)** little; small ⟨*format, letter*⟩; little ⟨*finger, toe*⟩; small, short ⟨*steps*⟩; **das Kleid ist mir zu ~:** the dress is too small for me; **ein ~es Bier** a small beer; ≈ a half[-pint]; **sich ~ machen** make oneself small; **auf ~stem Raum** in the minimum of space; **sie ist ~ [von Gestalt/für ihr Alter]** she is small [in stature/for her age]; **er ist [einen Kopf] ~er als ich** he is [a head] shorter than me *or* shorter than I am [by a head]; **im ~en** in miniature; on a small scale; **~, aber oho** he/she may be small, but he/she certainly makes up for it; **~, aber fein** little, but very nice; **b)** *(jung)* little ⟨*brother, sister*⟩; **als ich [noch] ~ war** when I was small *or* little; **für die Kleinen** for the little ones; **von ~ auf** from an early age; *s. auch* **1,2,3Kleine; c)** *(von kurzer Dauer)* little, short ⟨*while*⟩; short ⟨*walk, break*⟩; short, brief ⟨*delay, introduction*⟩; brief ⟨*moment*⟩; **d)** *(von geringer Menge)* small ⟨*family, amount, audience, staff*⟩; small, low ⟨*salary*⟩; low ⟨*price*⟩; **~es Geld haben** have some [small] change; **haben Sie es ~?** *(ugs.)* do you have the right money?; **~er habe ich es nicht** I don't have anything smaller; **e)** *(von geringem Ausmaß)* light ⟨*refreshment*⟩; small ⟨*party, gift*⟩; scant, little ⟨*attention*⟩; slight ⟨*cold, indisposition*⟩; slight, small ⟨*mistake, irregularity*⟩; minor ⟨*event, error*⟩; **die ~en Dinge des Alltags** the little everyday things; **das ~ere Übel** the lesser evil; the lesser of the two evils; **ein ~[es] bißchen** a little *or* tiny bit; **ein ~ wenig** a little bit; **im Kleinen wie im Großen** in little things as well as in big ones; **bis ins Kleinste** down to the smallest *or* tiniest detail; **f)** *(unbedeutend)* lowly ⟨*employee, sales assistant*⟩; minor ⟨*official*⟩; **der ~e Mann** the ordinary citizen; the man in the street; **die ~en Leute** ordinary people; the man *sing.* in the street; **~ anfangen** *(ugs.)* start off in a small way; **g) ganz ~ [und häßlich] werden** become meek and subdued; **h) ein ~er Geist** *(engstirnig)* a narrow-minded person; *(beschränkt)* a person of limited intellect. **2.** *adv.* **die Heizung ~/~er einstellen** turn the heating down low/lower; **ein Wort ~ schreiben** write a word with a small initial letter; **~ machen** *(Kinderspr.)* do number one *(child lang.)*; *s. auch* **beigeben 2**

klein-, Klein-: ~anzeige die *(Zeitungsw.)* small *or* classified advertisement *or (coll.)* ad; **~arbeit die;** *o. Pl.* painstaking and detailed work; **~asien (das)** Asia Minor; **~bauer der** small farmer; smallholder; **~|bekommen** *unr. tr. V.: s.* **~kriegen; ~betrieb der a)** *(Industrie)* small business; **b)** *(Landw.)* small farm; smallholding; **~bild·kamera die** *(Fot.)* miniature camera; 35 mm camera; **~buchstabe der** small letter; lower-case letter *(Printing)*; **~bürger der** lower middle-class person; *(abwertend: Spießbürger)* petit bourgeois; **~bürgerlich 1.** *Adj.* **a)** lower middle-class; **b)** *(abwertend: spießbürgerlich)* petit bourgeois; **2.** *adv. (abwertend: spießbürgerlich)* **~bürgerlich denken** have a petit-bourgeois way of thinking; **~bürgertum das** lower middle class; petite bourgeoisie; **~bus der** minibus

1Kleine der; *adj. Dekl.* **a)** *(kleiner Junge)* little boy; **b)** *(ugs. Anrede)* little man

2Kleine die; *adj. Dekl.* **a)** *(kleines Mädchen)* little girl; **b)** *(ugs. Anrede)* love; *(abwertend)* little madam; **c)** *(ugs.: Freundin)* girl[-friend]

3Kleine das; *adj. Dekl.* **a)** *(ugs. scherzh.)* little boy/girl *(joc.)*; **b)** *(von Tieren)* baby; little one

klein-, Klein-: ~familie die *(Soziol.)* nuclear family; **~format das** small size; *(bei Büchern)* small format *or* size; **~garten der** ≈ allotment; **~gärtner der** ≈ allotment-holder; **~gedruckt** *Adj. (präd. getrennt geschrieben)* in small print *postpos.*; **~gedruckte das;** *adj. Dekl.* small print; **~geist der** *(abwertend)* small-minded person; **~geld das;** *o. Pl.* [small] change; **~gemustert** *Adj. (präd. getrennt geschrieben)* small-patterned; **~gläubig** *Adj.* of little faith *postpos.*; sceptical; *(ängstlich, zweifelnd)* faint-hearted; **~gläubigkeit die** *s.* **~gläubig:** lack of faith; scepticism; faint-heartedness; **~|hacken** *tr. V.* chop up

Kleinheit die; ~ smallness; small size

Klein·holz das; *o. Pl.* chopped wood; **~ aus etw./jmdm. machen, etw./jmdn. zu ~ machen** *(ugs.)* smash sth. to pieces/make mincemeat of sb.

Kleinigkeit die; ~, ~en **a)** *(kleine Sache)* small thing; *(kleines Geschenk)* small *or* litte gift *or* present *(Einzelheit)* [small] detail; minor point; **ich habe noch eine ~ zu erledigen** I still have a small matter to attend to; **eine ~ essen** have a [small] bite to eat; **das kostet eine ~** *(ugs. iron.)* that costs a bob *or* two *(Brit. coll.) or* a tidy sum *(coll.)*; **die ~ von 50 000 DM** *(ugs. iron.)* the small *or* little matter of 50,000 marks; **sich nicht mit ~en abgeben** not concern oneself with details *or* trifles; **b)** *(leichte Aufgabe)* **eine ~ für jmdn. sein** be no trouble for sb.; be a simple matter for sb.; **das war eine ~:** it was nothing; **c) eine ~** *(ugs.: ein bißchen)* a little bit

Kleinigkeits·krämer der *(abwertend)* pettifogger

klein-, Klein-: ~kaliber·gewehr das small-bore rifle; **~kariert 1.** *Adj.* **a)** ⟨*skirt, shirt, etc.*⟩ with a small check *or* with a small-checked pattern; **b)** *(ugs. abwertend: engstirnig)* narrow-minded; **2.** *adv. (ugs. abwertend)* narrow-mindedly; in a narrow-minded way; **~kind das** small child; **~kram der** *(ugs.)* **a)** *(kleine Dinge)* odds and ends *pl.*; **b)** *(unbedeutende Dinge)* trivial matters *pl.*; *(Einzelheiten)* trivial details; ~**kredit der** *(Bankw.)* personal loan; ~**krieg der** *(fig.)* running battle; ~|**kriegen** *tr. V. (ugs.)* **a)** *(zerkleinern)* crush [to pieces]; *(zerkauen)* get one's teeth through ⟨*tough meat*⟩; **b)** *(zerstören)* smash; break; **nicht ~zukriegen sein** be indestructible; **c)** *(aufbrauchen)* get through, *(sl.)* blow ⟨*money*⟩; get through, *(joc.)* demolish ⟨*sweets, cakes, etc.*⟩; **d) jmdn. ~kriegen** get sb. down *(coll.)*; ~**kunst die**; *o. Pl.* cabaret; ~**laut 1.** *Adj.* subdued; **2.** *adv.* in a subdued fashion

kleinlich *(abwertend)* **1.** *Adj.* pernickety; *(ohne Großzügigkeit)* mean; *(engstirnig)* small-minded; petty; *(in bezug auf Sauberkeit und Ordnung)* pernickety; fussy; petty ⟨*regulations*⟩. **2.** *adv.* meticulously; punctiliously

Kleinlichkeit die; ~ *(abwertend) s.* **kleinlich**: pernicketiness; meanness; small-mindedness; pettiness; fussiness

klein-, Klein-: ~|**machen** *tr. V.* **a)** *(zerkleinern)* cut up small; chop up ⟨*wood*⟩; **b)** *(ugs.: aufbrauchen)* get through, *(sl.)* blow ⟨*money*⟩; **c)** *(ugs.: wechseln)* change; **kann mir jemand ein 5-Mark-Stück ~machen?** can anyone give me change for a five-mark piece?; ~**mut der** *(geh.)* faint-heartedness; timidity; ~**mütig** *Adj. (geh.)* faint-hearted; timid

Kleinod ['klaino:t] **das**; ~[e]s, ~e *od.* ~**ien** [-'no:diən] *(geh.)* **a)** *(Schmuckstück)* piece of jewellery; *(Edelstein)* jewel; **b)** *(fig.: Kostbarkeit)* gem

klein-, Klein-: ~|**schneiden** *unr. tr. V.* cut up small; cut into small pieces; chop up ⟨*onion*⟩ [small]; ~**staat der** small state; ~**stadt die** small town; ~**städter der** small-town dweller; ~**städtisch** *Adj.* small-town *attrib.*; ~|**stellen** *tr. V.* turn down [low]

Kleinst·lebewesen das micro-organism

kleinst·möglich *Adj.*; *nicht präd.* smallest possible

Klein·tier das pet; *(Nutztier)* small domestic animal

Kleintier·zucht die [professional] breeding of small animals

Klein·vieh das small farm *or* domestic animals *pl.*; small livestock; ~ **macht auch Mist** *(ugs.)* many a mickle makes a muckle *(prov.)*; every little helps

Klein·wagen der small car

klein·wüchsig [-vy:ksıç] *Adj.* ⟨*person*⟩ of small stature; small, short ⟨*person, race*⟩; small ⟨*variety, species*⟩

Kleister ['klaistɐ] **der**; ~s, ~: paste

kleistern *tr. V. (ugs.)* **a)** *(kleben)* paste, stick **(an** + *Akk.* on); **b)** *(reparieren)* stick; **c)** *(dick auftragen)* plaster **(auf** + *Akk.* on)

Klementine [klemɛn'ti:nə] **die**; ~, ~n clementine

Klemme ['klɛmə] **die**; ~, ~**n a)** *(Haar~)* [hair-]clip; *(Med.)* clip; **b)** *(ugs.: schwierige Lage)* **in der ~ sein** *od.* **sitzen** be in a fix *or* jam *(coll.)*; **jmdm. aus der ~ helfen** help sb. out of a fix *or* jam *(coll.)*

klemmen 1. *tr. V.* **a)** *(befestigen)* tuck; stick *(coll.)*; **etw. unter den Arm ~**: tuck *or (coll.)* stick sth. under one's arm; **b)** *(quetschen)* **sich** *(Dat.)* **die Hand ~**: get one's hand caught *or* trapped; catch *or* trap one's hand. **2.** *refl. V.* **sich hinter etw.** *(Akk.)* **~**: wedge oneself behind sth.; *(fig. ugs.: sich einsetzen)* put some hard work into sth.; **sich hinter jmdn. ~**: *(fig. ugs.)* get to work on sb. *(coll.)*. **3.** *itr. V.* ⟨*door, drawer, etc.*⟩ stick

Klempner ['klɛmpnɐ] **der**; ~s, ~: tinsmith; *(Installateur)* plumber

Kleptomane [klɛpto'ma:nə] **der**; ~n, ~n *(Psych.)* kleptomaniac

Kleptomanie [klɛptoma'ni:] **die**; ~ *(Psych.)* kleptomania *no art.*

Kleptomanin die; ~, ~**nen** kleptomaniac

klerikal [kleri'ka:l] *Adj.* clerical

Kleriker ['kle:rikɐ] **der**; ~s, ~: cleric

Klerus ['kle:rʊs] **der**; ~: clergy

Klette ['klɛtə] **die**; ~, ~**n** bur; *(Pflanze)* burdock; **sich wie eine ~ an jmdn. hängen** *(ugs.)* stick like a bur to sb.

Kletterer der; ~s, ~: climber

Kletter-: ~**gerüst das** climbing-frame; ~**maxe der** *(ugs. scherzh.)* climbing-mad child

klettern ['klɛtɐn] *itr. V.*; *mit sein (auch fig.)* climb; *(mit Mühe)* clamber; **auf einen Baum ~**: climb a tree

Kletter-: ~**partie die a)** *(Bergsteigen)* climb; **b)** *(ugs.: anstrengende Wanderung)* climbing expedition; ~**pflanze die** creeper; *(Bot.)* climbing plant; climber; ~**tour die** *s.* ~**partie**

klick [klık] *Interj.* click; ~ **machen** click; go click

klicken *itr. V.* click; **es klickte** there was a click

Klient [kli'ɛnt] **der**; ~**en**, ~**en**, **Klientin die**; ~, ~**nen** client

Kliff [klıf] **das**; ~[e]s, ~e cliff

Klima ['kli:ma] **das**; ~s, ~s *od.* **Klimate** [kli'ma:tə] climate; **das politische** ~ *(fig.)* the political climate; **im Büro herrscht ein angenehmes** ~ *(fig.)* there's a pleasant atmosphere in the office

Klima·anlage die air-conditioning *no indef. art.*; air-conditioning system; **mit ~**: air-conditioned

klimatisch [kli'ma:tıʃ] **1.** *Adj.*; *nicht präd.* climatic. **2.** *adv.* climatically

Klimbim [klım'bım] **der**; ~s *(ugs.)* **a)** *(Kram)* junk; odds and ends *pl.*; **b)** *(Wirbel)* fuss; ~ **um etw. machen** make a fuss about sth.

Klimm·zug der *(Turnen)* pull-up

klimpern ['klımpɐn] **1.** *itr. V.* jingle; tinkle; ⟨*coins, keys*⟩ jingle; **mit den Schlüsseln ~**: jingle the keys; **mit den Wimpern** ~ *(scherzh.)* flutter one's eyelashes [seductively]; **auf dem Klavier** ~ *(ugs.)* plunk away on the piano. **2.** *tr. V. (ugs. abwertend)* plunk out ⟨*tune etc.*⟩

Klinge die; ~, ~**n a)** blade; **b) jmdn. über die ~ springen lassen** *(fig.) (ugs.: ruinieren)* ruin sb.; *(beruflich)* put paid to sb.'s career *(coll.)*

Klingel ['klıŋḷ] **die**; ~, ~**n** bell

Klingel-: ~**beutel der** offertory-bag; collection-bag; ~**knopf der** bell-button; bell-push

klingeln *itr. V.* **a)** ring; ⟨*alarm clock*⟩ go off; ring; **es klingelt** somebody is ringing the doorbell; there is a ring at the door; **es klingelte zur Pause** the bell went for the break; **es hat bei ihm/ihr** *usw.* **geklingelt** *(ugs.)* the penny's dropped *(coll.)*; **b)** *(die Klingel betätigen)* ring [the bell]; **nach jmdm. ~**: ring for sb.; **c)** *(Kfz-W.)* ⟨*engine*⟩ pink

klingen *unr. itr. V.* **a)** ⟨*bell*⟩ ring; ⟨*glass*⟩ clink; **aus dem Haus klangen fröhliche Stimmen** the sound of merry voices came from the house; **die Gläser ~ lassen** clink glasses [in a toast]; **b)** *(einen bestimmten Klang haben)* sound; **es klang, als ob ...**: it sounded as if ...

klingend *Adj.* **~e Münze** [hard] cash

Klinik ['kli:nık] **die**; ~, ~**en** hospital; *(spezialisiert)* clinic

Klinikum ['kli:nikʊm] **das**; ~s, **Klinika** *od.* **Kliniken** hospital complex

klinisch 1. *Adj.*; *nicht präd.* *(Med.)* clinical. **2.** *adv.* ~ **tot** clinically dead

Klinke ['klıŋkə] **die**; ~, ~**n** door-handle; **sich** *(Dat.)* **die ~ in die Hand geben** *(ugs.)* come and go in a continuous stream

Klinken·putzer der *(ugs. abwertend)* door-to-door salesman

Klinker der; ~s, ~: [Dutch] clinker

klipp [klɪp] *in* **~ und klar** *(ugs.)* quite plainly *or* clearly

Klippe die; ~, ~n rock; **alle ~n umschiffen** *(fig.)* negotiate every obstacle [successfully]

klirren ['klɪrən] *itr. V.* ⟨*glasses, ice-cubes*⟩ clink; ⟨*weapons in fight*⟩ clash; ⟨*window-pane*⟩ rattle; ⟨*chains, spurs*⟩ clank, rattle; ⟨*harness*⟩ jingle; **~der Frost** *(fig.)* sharp frost

Klischee [kli'ʃe:] **das; ~s, ~s a)** cliché; **das ~ vom braven Hausmütterchen** the conventional picture *or* stereotype of the good little housewife; **b)** *(Druckw.)* block; plate

Klitoris ['kli:tɔrɪs] **die; ~, ~** *od.* **Klitorides** [kli'to:ride:s] *(Anat.)* clitoris

Klitsche die; ~, ~n *(ugs.)* **a)** *(kleiner Betrieb)* little shoestring outfit *(coll.)*; **b)** *(Schmierentheater)* third-rate little theatre

klitsch·naß *Adj. (ugs.)* soaking *or* sopping wet; *(tropfnaß)* dripping wet; **wir sind ~ geworden** we got soaked [to the skin] *or* drenched

klitze·klein ['klɪtsə-] *Adj. (ugs.)* teeny[-weeny] *(coll.)*

Klo [klo:] **das; ~s, ~s** *(ugs.)* loo *(Brit. coll.)*; john *(Amer. coll.)*; **aufs ~ müssen** have to go to the loo; **etw. ins ~ schütten** tip sth. down the loo

Kloake [klo'a:kə] **die; ~, ~n** cesspit; *(Kanal)* sewer

Kloben ['klo:bn̩] **der; ~s, ~** log

klobig *Adj.* **a)** *(kantig)* heavy and clumsy[-looking] ⟨*shoes, furniture*⟩; heavily-built, bulky ⟨*figure*⟩; **b)** *(plump)* clumsy; boorish

Klo-: ~bürste die *(ugs.)* loo-brush *(Brit. coll.)*; toilet-brush; **~frau die** *(ugs.)* loo-attendant *(Brit. coll.)*; bathroom attendant *(Amer.)*

klomm [klɔm] *1. u. 3. Pers. Sg. Prät. v.* **klimmen**

Klo·papier das *(ugs.)* loo-paper *(Brit. coll.)*; toilet-paper

klopfen ['klɔpfn̩] **1.** *itr. V.* **a)** *(schlagen)* knock; **an die Tür ~:** knock at the door; **es hat geklopft** there's somebody knocking at the door; **jmdm.** *od.* **jmdn. auf die Schulter ~:** slap sb. on the shoulder; **„bitte ~!"** 'please knock'; **b)** *(pulsieren)* ⟨*heart*⟩ beat; ⟨*pulse*⟩ throb; **mit ~dem Herzen** with pounding *or* beating heart; **ein ~der Schmerz** a throbbing pain; **c)** *(Kfz-W.)* ⟨*engine*⟩ knock. **2.** *tr. V.* beat ⟨*carpet*⟩; **den Takt [zur Musik] ~:** beat time [to the music]; **Staub vom Mantel ~:** beat dust from one's coat; **einen Nagel in die Wand ~:** knock *or* hammer a nail into the wall

Klopfer der; ~s, ~ a) *(Teppich~)* carpet-beater; **b)** *(Tür~)* [door-] knocker; **c)** *(Fleisch~)* meat mallet *or* tenderizer

Klopf·zeichen das knock; *(leiser)* tap

Klöppel ['klœpl̩] **der; ~s, ~:** clapper

Klöppel·arbeit die piece of pillow-lace *or* bobbin-lace

klöppeln *tr., itr. V.* **[etw.] ~:** make *or* work [sth. in] pillow-lace *or* bobbin-lace

Klöppel·spitze die pillow-lace; bobbin-lace

kloppen *(nordd., md.)* **1.** *tr. V.* hit. **2.** *refl. V.* fight; scrap *(coll.)*

Klöpplerin die; ~, ~nen pillow-lace *or* bobbin-lace maker

Klops [klɔps] **der; ~es, ~e** *(nordostd.)* meat ball

Klosett [klo'zɛt] **das; ~s, ~s** *od.* **~e** lavatory; **etw. ins ~ schütten** tip sth. down the lavatory

Klosett-: ~brille die *(ugs.)* loo-seat *(Brit. coll.)*; toilet-seat; **~bürste die** lavatory-brush; toilet-brush; **~deckel der** toilet-lid; **~papier das** toilet-paper; lavatory-paper

Kloß [klo:s] **der; ~es, Klöße** ['klø:sə] dumpling; *(Fleisch~)* meat ball; **einen ~ im Hals haben** *(ugs.)* have a lump in one's throat

Kloster ['klo:stɐ] **das; ~s, Klöster** ['klø:stɐ] *(Mönchs~)* monastery; *(Nonnen~)* convent; nunnery

Kloster·kirche die monastery/ convent church

klösterlich *Adj.* monastic; monastic/convent ⟨*life*⟩

Kloster-: ~regel die rules *pl.* of the monastery/convent; **~schule die** monastery-school/ convent-school; **~schüler der** monastery-school/convent-school pupil

Klotz [klɔts] **der; ~es, Klötze** ['klœtsə] **a)** block [of wood]; *(Stück eines Baumstamms)* log; **[jmdm.] ein ~ am Bein sein** *(ugs.)* be a millstone round sb.'s neck; **sich** *(Dat.)* **einen ~ ans Bein binden** *(ugs.)* tie a millstone round one's neck; **b)** *(salopp abwertend)* *(ungehobelter Mensch)* clod; oaf; *(roher Mensch)* lout

Klötzchen ['klœtsçən] **das; ~s, ~:** small block of wood

klotzen *(ugs.) itr. V. (ugs.: großzügig vorgehen)* lash out in a big way *(coll.)*; *s. auch* **kleckern 1c**

klotzig *Adj. (abwertend)* large and ugly[-looking] ⟨*building*⟩; large and clumsy[-looking] ⟨*furniture*⟩

Klub [klʊp] **der; ~s, ~s** club

Klub-: ~haus das club-house; **~jacke die** blazer; **~mitglied das** club-member; **~sessel der** club chair

[1]**Kluft** [klʊft] **die; ~, ~en** *(ugs.)* rig-out *(coll.)*; gear *(coll.)*; *(Uniform)* uniform; garb

[2]**Kluft die; ~, Klüfte** ['klʏftə] **a)** *(veralt.) (Spalte)* cleft; fissure; *(im Gletscher)* crevasse; *(Abgrund)* chasm; **b)** *(Gegensatz)* gulf

klug [klu:k]; **klüger** ['kly:gɐ], **klügst...** ['kly:kst...] **1.** *Adj.* **a)** clever; intelligent; clever, bright ⟨*child, pupil*⟩; intelligent ⟨*eyes*⟩; **er ist ein ~er Kopf** he's clever *or* bright; he's got brains; **b)** *(gelehrt, weise)* wise; **so ~ wie vorher** *od.* **zuvor sein** be none the wiser; **hinterher ist man immer klüger** it's easy to be wise after the event; **daraus werde ich nicht ~, daraus soll ein Mensch ~ werden** I can't make head or tail of it; **aus jmdm. nicht ~ werden** not know what to make of sb.; **c)** *(vernünftig)* wise; wise, sound ⟨*advice*⟩; wise, prudent ⟨*remark, course of action*⟩; *(geschickt)* clever, shrewd ⟨*politician, negotiator, question*⟩; shrewd, astute ⟨*businessman*⟩; great ⟨*foresight*⟩; **der Klügere gibt nach** *(Spr.)* discretion is the better part of valour *(prov.)*. **2.** *adv.* **a)** cleverly; intelligently; **~ daherreden** talk as if one knows it all; **b)** *(vernünftig)* wisely; *(geschickt)* cleverly; shrewdly

klüger *s.* **klug**

klugerweise *Adv.* wisely

Klugheit die; ~ *s.* **klug a, b, c:** cleverness; intelligence; brightness; wisdom; soundness; prudence; shrewdness; astuteness

Klug·scheißer der *(salopp abwertend)* know-it-all *(coll.)*; smart aleck *(coll.)*

klügst... *s.* **klug**

klumpen ['klʊmpn̩] *itr. V.* go lumpy

Klumpen der; ~s, ~ lump; **ein ~ Erde** a lump *or* clod of earth; **ein ~ Gold** a gold nugget

Klump·fuß der club foot

klumpig *Adj.* lumpy

Klüngel ['klʏŋl̩] **der; ~s, ~** *(abwertend)* clique

Klunker ['klʊŋkɐ] **die; ~, ~n** *od.* **der; ~s, ~** *(ugs.)* rock *(sl.)*

km *Abk.* **Kilometer** km.

knabbern ['knabɐn] **1.** *tr. V.* nibble. **2.** *itr. V.* **an etw.** *(Dat.)* **~:** nibble *or* gnaw [at] sth.; **an etw.** *(Dat.)* **[noch lange] zu ~ haben** *(ugs.) (sich anstrengen müssen)* have sth. to think about *or* chew on; *(leiden müssen)* take a long time to get over sth.

Knabe ['kna:bə] **der; ~n, ~n a)** *(geh. veralt./ südd., österr., schweiz.)* boy; **b)** *(ugs.: Bursche)* chap *(coll.)*

Knaben·chor der boys' choir

knabenhaft 1. *Adj.* boyish. **2.** *adv.* boyishly

Knäcke·brot ['knɛkə-] **das** crispbread; *(Scheibe)* slice of crispbread

knacken 1. *itr. V.* ⟨*bed, floor, etc.*⟩ creak; **es knackte im Gebälk** the beams creaked. **2.** *tr. V.* **a)** crack ⟨*nut, shell*⟩; *(salopp)* squash ⟨*louse, bug*⟩; **b)** *(aufbrechen)* crack ⟨*safe*⟩ [open]; break into ⟨*car, bank, etc.*⟩; crack, break ⟨*code*⟩

Knacker der; ~s, ~: alter ~ *(salopp)* old fogey

knackig *Adj.* **a)** *(knusprig)* crisp; crisp, crunchy ⟨*apple*⟩; **b)** *(ugs.: attraktiv)* luscious, delectable ⟨*girl*⟩

knacks *Interj.* crack

Knacks der; ~es, ~e *(ugs.)* **a)** *(Ton)* crack; **b)** *(Sprung)* crack; **c)** *(fig.: Defekt)* **einen ~ bekommen** ⟨*person*⟩ have *or* suffer a breakdown; ⟨*health*⟩ suffer; **die Ehe hatte einen ~:** the marriage was in difficulties

Knall [knal] **der; ~[e]s, ~e** bang; *(fig.)* big row; **einen ~ haben** *(salopp)* be barmy *(sl.) or* off one's rocker *(sl.)*; **auf ~ und Fall, ~ auf Fall** *(ugs.)* without warning

Knall·effekt der *(ugs.)* astonishing part

knallen 1. *itr. V.* **a)** ⟨*shot*⟩ ring out; ⟨*firework*⟩ go bang; ⟨*cork*⟩ pop; ⟨*door*⟩ bang, slam; ⟨*whip, rifle*⟩ crack; **an der Kreuzung hat es geknallt** *(ugs.)* there was a crash at the crossroads; **b)** *(ugs.: schießen)* shoot, fire (**auf +** *Akk.* at); *(mehrere Male)* blaze *or (coll.)* bang away (**auf +** *Akk.* at); **c)** *mit sein (ugs.: prallen)* **die Tür knallte ins Schloß** the door slammed *or* banged shut; **sie knallte mit dem Fahrrad gegen einen Laternenpfahl** she crashed into a lamp-post on her bicycle; **der Ball knallte gegen die Latte** the ball slammed against the crossbar; **d)** *(ugs.: scheinen)* ⟨*sun*⟩ blaze *or* beat down. **2.** *tr. V.* **a)** *(ugs.) (hart aufsetzen)* slam *or* bang down; *(werfen)* sling *(coll.)*; **b)** *(ugs.: schlagen)* **jmdm. eine ~** *(salopp)* belt *or* clout sb. one *(coll.)*; **c)** *(Ballspiele ugs.)* belt ⟨*ball*⟩

Knaller der; ~s, ~: banger

knall-, Knall-: ~erbse die ≈ cap-bomb; **~frosch der** jumping jack; **~gelb** *Adj. (ugs.)* bright *or* vivid yellow; **~hart** *(ugs.)* **1.** *Adj.* **a)** very tough ⟨*job, demands, action, measures, etc.*⟩; ⟨*person*⟩ as hard as nails; **b)** *(kraftvoll)* crashing ⟨*blow*⟩; **2.** *adv.* **a)** *(rücksichtslos, brutal)* brutally; **gegen etw. ~hart vorgehen** take very tough action against sth.; **b)** *(kraftvoll)* ⟨*hit*⟩ really hard

knallig *Adj. (ugs.)* loud; gaudy

knall-, Knall-: ~kopf der, ~kopp der *(salopp)* [stupid] berk *(Brit. sl.) or (Amer. sl.)* jerk; **~rot** *Adj.* bright *or* vivid red; **sie bekam einen ~roten Kopf** she *or* her face turned [bright] scarlet *or* as red as a beetroot

knapp [knap] **1.** *Adj.* **a)** meagre, low ⟨*pension, wage, salary*⟩; meagre ⟨*pocket-money*⟩; **~ sein** be scarce *or* in short supply; **~ werden** ⟨*supplies*⟩ run short; ⟨*money*⟩ get tight; **~ mit etw. sein** be short of sth.; **..., und nicht zu ~!** ... and how!; **b)** narrow ⟨*victory, lead*⟩; narrow, bare ⟨*majority*⟩; close ⟨*result*⟩; **c) vor einer ~en Stunde** just under an hour ago; **d)** *(eng)* tight-fitting ⟨*garment*⟩; **zu ~ sein** ⟨*garment*⟩ be too tight; **e)** *(kurz)* terse ⟨*reply, greeting*⟩; concise, succinct ⟨*description, account, report*⟩. **2.** *adv.* **a) ~ bemessen sein** be meagre; ⟨*time*⟩ be limited; **~ gerechnet** at the lowest estimate; **b) ~ gewinnen** win narrowly *or* by a narrow margin; **eine Prüfung ~ bestehen** just pass an examination; **c)** *(sehr nahe)* just; **~ über dem Knie enden** come to just above the knee; **d)** *(nicht ganz)* just under; **vor ~ einer Stunde** just under an hour ago; **e)** *(eng)* **~ sitzen** fit tightly; **f)** *(kurz)* ⟨*reply*⟩ tersely; ⟨*describe*⟩ concisely, succinctly

knapp|halten *unr. tr. V. (ugs.)* **jmdn. [mit Geld] ~:** keep sb. short [of money]

Knappheit die; ~ a) *(Mangel)* shortage, scarcity (**an +** *Dat.* of); **b)** *(Kürze) (einer Antwort)* terseness; *(einer Beschreibung)* conciseness, succinctness

Knarre ['knarə] **die; ~, ~n a)** *(Rassel)* rattle; **b)** *(salopp: Gewehr)* shooting-iron *(sl.)*

knarren *itr. V.* creak; **mit ~der Stimme** in a rasping *or* grating voice

Knast [knast] **der; ~[e]s, Knäste** ['knɛstə] *od.* **~e** *(ugs.)* **a)** *o. Pl. (Strafe)* bird *(sl.)*; time; **er hat zwei Jahre ~ gekriegt** he got two years' bird *(sl.)*; **b)** *(Gefängnis)* clink *(sl.)*; jug *(sl.)*; prison; **im ~:** in clink *or* jug *(sl.)*

Knatsch [kna:tʃ] **der; ~[e]s** *(ugs.: Ärger)* trouble; **die beiden haben ~:** the two of them are rowing

knattern ['knatɐn] *itr. V.* **a)** ⟨*machine-gun*⟩ rattle, clatter; ⟨*sail*⟩ flap; ⟨*motor vehicle, engine*⟩ clatter; **b)** *mit sein (~d fahren)* clatter

Knäuel ['knɔyəl] **der** *od.* **das; ~s, ~ a)** ball; *(wirres ~)* tangle; **b)** *(fig.) (von Menschen)* knot

Knauf [knauf] **der; ~[e]s, Knäufe** ['knɔyfə] knob; *(eines Schwertes, Dolches)* pommel

Knauserei die; ~ *(ugs. abwertend)* stinginess; penny-pinching; miserliness

knaus[e]rig *Adj. (ugs. abwertend)* stingy; tight-fisted; close-fisted

knausern ['knauzɐn] *itr. V. (ugs. abwertend)* be stingy; scrimp

knautschen ['knautʃn̩] *(ugs.)* **1.** *tr. V.* crumple; crumple, crease ⟨*dress*⟩. **2.** *itr. V.* ⟨*dress, material*⟩ crease, get creased

Knautsch·zone die *(Kfz-W.)* crumple zone

Knebel ['kne:bl̩] **der; ~s, ~ a)** gag; **b)** *(Griff)* toggle

knebeln *tr. V.* gag; *(fig.)* gag, muzzle ⟨*the press, a people*⟩

Knecht [knɛçt] **der; ~[e]s, ~e** farm-labourer; farm-hand; *(fig.)* slave; vassal

knechten *tr. V. (geh.)* reduce to servitude *or* slavery; enslave; *(unterdrücken)* oppress ⟨*people*⟩

Knechtschaft die; ~, ~en *(geh.)* bondage; servitude; slavery

kneifen ['knaifn̩] **1.** *unr. tr., itr. V.* pinch; **jmdm.** *od.* **jmdn. in den Arm ~:** pinch sb.'s arm. **2.** *unr. itr. V.* **a)** *(drücken)* ⟨*clothes*⟩ be too tight; **b)** *(ugs. abwertend: sich drücken)* chicken *(sl.) or* back out (**vor +** *Dat.* of); **vor einer Prüfung/Verantwortung ~:** funk an examination *(sl.)*/*(coll.)* duck [out of] a responsibility

Kneifer der; ~s, ~: pince-nez

Kneif·zange die pincers *pl.*; **eine ~:** a pair of pincers

Kneipe ['knaipə] **die; ~, ~n** *(ugs.)* pub *(Brit. coll.)*; bar *(Amer.)*

Kneipp·kur die Kneipp cure

Knete die; ~ **a)** *(ugs.) s.* **Knetmasse; b)** *(salopp: Geld)* dough *(sl.)*

kneten ['kne:tn̩] *tr. V.* **a)** knead ‹*dough, muscles*›; work ‹*clay*›; **b)** *(formen)* model ‹*figure*›

Knet·masse die Plasticine (P); plastic modelling-material

Knick [knɪk] **der;** ~**[e]s,** ~**e a)** *(Biegung)* sharp bend; *(in einem Draht)* kink; **b)** *(Falz)* crease

knicken 1. *tr. V.* **a)** *(brechen)* snap; **b)** *(falten)* crease ‹*page, paper, etc.*›; **„Bitte nicht** ~**!"** 'please do not bend'; *(bitte nicht falten)* 'please do not fold'. **2.** *itr. V.; mit sein* snap

knick[e]rig *Adj. (ugs. abwertend)* stingy; tight-fisted

Knicks [knɪks] **der;** ~**es,** ~**e** curtsy; **einen** ~ **machen** make *or* drop a curtsy (**vor** + *Dat.* to)

Knie [kni:] **das;** ~**s,** ~ ['kni:(ə)] **a)** knee; **jmdm. auf [den]** ~**n danken** go down on one's knees and thank sb.; **jmdn. auf** ~**n bitten** beg sb. on bended knees; **vor jmdm. auf die** ~ **fallen** go down on one's knees before sb.; **er hatte/bekam weiche** ~ *(ugs.)* his knees trembled/started to tremble; **jmdn. auf** *od.* **in die** ~ **zwingen** *(geh.)* force sb. to his knees; **in die** ~ **gehen** sink to one's knees; *(fig.)* submit, bow (**vor** + *Dat.* to); **jmdn. übers** ~ **legen** *(ugs.)* put sb. across one's knee; **etw. übers** ~ **brechen** *(ugs.)* rush sth.; **b)** *(Biegung)* sharp bend; *(eines Rohres)* elbow

knie-, Knie-: ~**beuge die** knee-bend; ~**bund·hose die** knee-breeches *pl.;* **eine** ~: a pair of knee-breeches; ~**fall der: einen** ~**fall tun** *od.* **machen** *(auch fig.)* go down on one's knees (**vor** + *Dat.* before); ~**gelenk das** knee-joint; ~**hoch** *Adj.* knee-high; knee-length ‹*boots*›; ~**kehle die** hollow of the knee; ~**lang** *Adj.* knee-length

knien ['kni:(ə)n] **1.** *itr. V.* kneel; ~**d, im Knien** kneeling; on one's knees. **2.** *refl. V.* kneel [down]; get down on one's knees; **sich in die Arbeit** ~ *(fig. ugs.)* get stuck into one's work *(sl.)*

knie-, Knie-: ~**scheibe die** kneecap; ~**schützer der** *(Sport)* knee-pad; ~**strumpf der** knee-length sock; knee-sock; ~**tief** *Adj.* knee-deep

Kniff [knɪf] **der;** ~**[e]s,** ~**e a)** *(Kunstgriff)* trick; dodge; **den** ~ **[bei etw.] heraushaben** have got the knack [of sth.]; **b)** *(Falte)* crease; *(in Papier)* crease; fold

kniff[e]lig *Adj.* tricky

knipsen ['knɪpsn̩] **1.** *tr. V.* **a)** *(entwerten)* clip; punch; **b)** *(fotografieren)* snap; take a snap[shot] of. **2.** *itr. V.* **a)** *(fotografieren)* take snapshots

Knirps [knɪrps] **der;** ~**es,** ~**e a)** (Ⓦ *Taschenschirm*) telescopic umbrella; **b)** *(ugs.: Junge)* nipper *(coll.);* **c)** *(ugs. abwertend: kleiner Mann)* [little] squirt *(coll.)*

knirschen ['knɪrʃn̩] *itr. V.* **a)** crunch; **b) mit den Zähnen** ~: grind one's teeth

knistern ['knɪstɐn] *itr. V.* rustle; ‹*wood, fire*› crackle; **mit etw.** ~: rustle sth.; **eine** ~**de Atmosphäre** *(fig.)* a tense *or* charged atmosphere; *s. auch* **Gebälk**

knitter·frei *Adj.* non-crease

knittern *tr., itr. V.* crease; crumple

knobeln *itr. V.* **a)** *(mit Würfeln)* play dice; *(mit Streichhölzern)* play spoof; *(mit Handzeichen)* play scissors, paper, stone; **b)** *(ugs.: nachdenken)* puzzle (**an** + *Dat.* over)

Knob·lauch ['kno:p-] **der** garlic

Knoblauch-: ~**butter die** garlic butter; ~**zehe die** clove of garlic

Knöchel ['knœçl̩] **der;** ~**s,** ~ **a)** *(am Fuß)* ankle; **bis an/über die** ~: up to the *or* one's ankles/to above ankle level; **b)** *(am Finger)* knuckle

knöchel-: ~**lang** *Adj.* ankle-length; ~**tief** *Adj.* ankle-deep

Knochen ['knɔxn̩] **der;** ~**s,** ~ bone; **Fleisch mit/ohne** ~: meat on/off the bone; **mir tun sämtliche** ~ **weh** *(ugs.)* every bone in my body aches; **der Schreck fuhr ihm in die** ~ *(ugs.)* he was shaken to the core; **keinen Mumm in den** ~ **haben** *(ugs.)* be a weed; **naß/abgemagert bis auf die** ~ **sein** be soaked to the skin/just skin and bones *(coll.);* **jmdn. bis auf die** ~ **blamieren** *(ugs.)* make a complete fool of sb. *(coll.);* **seine** ~ **für etw. hinhalten [müssen]** *(ugs.)* [have to] risk one's neck fighting for sth.

knochen-, Knochen-: ~**arbeit die** *(ugs.)* back-breaking work; ~**bruch der** fracture; ~**hart** *Adj. (ugs.)* rock-hard; ~**mark das** bone marrow; ~**mehl das** bone-meal; ~**trocken** *Adj. (ugs.)* bone-dry

knochig 1. *Adj.* bony. **2.** *adv.* **sehr** ~ **gebaut sein** be very bony

Knockout [nɔk|'aut] **der;** ~**[s],** ~**s** *(Boxen)* knock-out

Knödel ['knø:dl̩] **der;** ~**s,** ~ *(bes. südd., österr.)* dumpling

Knolle ['knɔlə] **die;** ~, ~**n a)** *(einer Pflanze)* tuber; **b)** *(ugs.) (Auswuchs)* large round lump; *(Nase)* big fat conk *(sl.) or (Amer.)* schnozzle

Knollen der; ~**s,** ~ *(ugs.: Strafzettel)* [parking-]ticket

Knollen-: ~**blätterpilz der** amanita; ~**nase die** large bulbous nose

Knopf [knɔpf] **der;** ~**[e]s, Knöpfe** ['knœpfə] **a)** button; **b)** *(Knauf)* knob; **c)** *(ugs.: Kind)* little thing *(coll.)*

knöpfen ['knœpfn̩] *tr. V.* button [up]; **hinten/vorn geknöpft werden** ‹*dress etc.*› button up at the back/in front

Knopf·loch das buttonhole

Knorpel ['knɔrpl̩] **der;** ~**s,** ~ **a)** *(Anat.)* cartilage; **b)** *(im Steak o. ä.)* gristle

knorp[e]lig *Adj.* **a)** *(Anat.)* cartilaginous; **b)** gristly ‹*meat*›

knorrig *Adj.* **a)** gnarled ‹*tree, branch*›; **b)** gruff ‹*person*›

Knospe ['knɔspə] **die;** ~, ~**n** bud; ~**n ansetzen** put forth buds; bud

knospen *itr. V.* bud

knoten ['kno:tn̩] *tr. V.* knot; tie a knot in; do *or* tie up ‹*shoelace*›; **etw. um/an etw.** *(Akk.)* ~: tie sth. round/to sth.

Knoten der; ~**s,** ~ **a)** knot; **b)** *(Haartracht)* bun; knot; **c)** *(Maßeinheit)* knot; **d)** *(Bot.)* node; **e)** *(Med.)* lump

Knoten·punkt der *(Verkehrs~)* junction; intersection

knotig *Adj.* **a)** knobby; knobbly; gnarled; knobbly ‹*fabric*›; **b)** *(knotenförmig)* nodular

Know-how [noʊ'hau] **das;** ~**[s]** know-how

Knuff [knʊf] **der;** ~**[e]s, Knüffe** ['knʏfə] *(ugs.)* poke

knuffen ['knʊfn̩] *tr. V.* poke

knüllen *tr. V.* crumple [up]

Knüller der; ~**s,** ~ *(ugs.)* sensation; *(Film, Buch usw.)* sensation; sensational success; *(Angebot, Verkaufsartikel)* sensational offer

knüpfen ['knʏpfn̩] **1.** *tr. V.* **a)** tie (**an** + *Akk.* to); **b)** *(durch Knoten herstellen)* knot; make ‹*net*›; **c)** *(fig.)* **große Erwartungen an etw.** *(Akk.)* ~: have great expectations of sth.; **Bedingungen an etw.** *(Akk.)* ~: attach conditions to sth. **2.** *refl. V.* **sich an etw.** *(Akk.)* ~: be connected with sth.

Knüppel ['knʏpl̩] **der;** ~**s,** ~ **a)** cudgel; club; *(Polizei~)* truncheon; *s. auch* **Bein; b)** *s.* **Steuerknüppel; c)** *s.* **Schaltknüppel**

knüppel·dick *Adv.* **es kam** ~ *(ugs.)* it was one disaster after the other

knüppeln 1. *tr. V.* cudgel; club; beat with a cudgel *or* club/*(Polizeiknüppel)* truncheon. 2. *itr. V.* use a/one's cudgel *or* club/truncheon

knurren ['knʊrən] 1. *itr. V.* **a)** ‹*animal*› growl; *(wütend)* snarl; *(fig.)* ‹*stomach*› rumble; **b)** *(murren)* grumble (**über** + *Akk.* about); **c)** *(verärgert reden)* growl. 2. *tr. V. (verärgert sagen)* growl

knuspern ['knʊspɐn] *tr., itr. V.* nibble; *(geräuschvoll)* crunch; **an etw.** *(Dat.)* ~: nibble [at] sth.

knusprig 1. *Adj.* **a)** crisp; crisp, crusty ‹*roll*›; crusty ‹*bread*›; crunchy ‹*nuts, crisps*›; **etw. ~ braten** roast/fry sth. crisp and brown; **b)** *(ugs.: frisch u. adrett)* delightfully fresh and attractive. 2. *adv.* **~-frisch** crunchy fresh ‹*crisps, nuts*›; crispy fresh ‹*rolls*›

Knute ['knu:tə] **die; ~, ~n** knout; **unter jmds. ~ [stehen]** *(fig.)* [be] under sb.'s heel

knutschen ['knu:tʃn̩] *(ugs.)* 1. *tr. V.* smooch with *(coll.)*; neck with *(sl.)*; *(sexuell berühren)* pet. 2. *itr. V.* smooch *(coll.)*, neck *(sl.)* (**mit** with); *(sich sexuell berühren)* pet

Knutscherei die; ~, ~en *(ugs.)* smooching *(coll.)*; necking *(sl.)*; *(sexuelle Berührung)* petting

Knutsch·fleck der *(ugs.)* love bite

k. o. [ka:'|o:] *Adj.; nicht attr.* **a)** *(Boxen)* **jmdn. ~ schlagen** knock sb. out; **b)** *(ugs.: übermüdet)* all in *(coll.)*; whacked *(coll.)*

K. o. der; ~, ~ *(Boxen)* knock-out

Koala [ko'a:la] **der; ~s, ~s** koala [bear]

koalieren [ko|a'li:rən] *itr. V. (Politik)* form a coalition (**mit** with)

Koalition [ko|ali'tsi̯o:n] **die; ~, ~en** coalition

Koalitions·regierung die coalition government

Kobalt ['ko:balt] **das; ~s** *(Chemie)* cobalt

Kobold ['ko:bɔlt] **der; ~[e]s, ~e** goblin; kobold; *(fig.)* imp

Kobra ['ko:bra] **die; ~, ~s** cobra

Koch [kɔx] **der; ~[e]s, Köche** ['kœçə] cook; *(als Beruf, Küchenchef)* chef; **viele Köche verderben den Brei** *(Spr.)* too many cooks spoil the broth *(prov.)*

Koch·buch das cookery book; cookbook

kochen 1. *tr. V.* **a)** boil; *(zubereiten)* cook ‹*meal*›; make ‹*purée, jam*›; **Tee ~/sich** *(Dat.)* **einen Tee ~**: make some tea; **die Eier hart/weich ~**: hard-/soft-boil the eggs; **etw. weich/gar ~**: cook sth. until it is soft/[properly] done; **b)** *(waschen)* boil ‹*washing*›; **c)** *(verflüssigen)* heat ‹*tar, glue, etc.*›. 2. *itr. V.* **a)** *(Speisen zubereiten)* cook; *(das Kochen übernehmen)* do the cooking; **gerne/gut ~**: like cooking/be a good cook; **fett/fettarm ~**: use a lot of fat/little fat in cooking; **b)** *(sieden)* ‹*water, milk, etc.*› boil; *(fig.)* ‹*sea*› boil, seethe; **etw. zum Kochen bringen** bring sth. to the boil; **c)** *(gekocht werden)* ‹*meat, vegetables, washing, etc.*› be boiled; **d)** *(ugs.: wütend sein)* **vor Wut/innerlich ~**: be boiling *or* seething with rage/inwardly

kochend·heiß *Adj.; nicht präd.* boiling hot; piping hot ‹*soup etc.*›

Kocher der; ~s, ~ [small] stove

Köcher ['kœçɐ] **der; ~s, ~ a)** *(für Pfeile)* quiver; **b)** *(für Fernglas o. ä.)* case

koch·fertig *Adj.* ready-to-cook *attrib.*; ready to cook *pred.*

Koch·gelegenheit die cooking facilities *pl.*

Köchin ['kœçɪn] **die; ~, ~nen** cook

Koch-: ~kunst die a) culinary art; **b)** *(ugs.: Fertigkeit im Kochen)* culinary skill[s *pl.*]; **~kurs[us] der** cookery course; **~löffel der** wooden spoon; **~nische die** kitchenette; **~rezept das** recipe; **~salz das** common salt; sodium chloride *(Chem.)*; **~topf der** [cooking] pot; **~wäsche die;** *o. Pl.* washing that is to be boiled

Kode [ko:t] **der; ~s, ~s** code

Köder ['kø:dɐ] **der; ~s, ~**: bait; *(fig.)* bait; lure; **einen/mehrere ~ auslegen** put out bait/a number of baits

ködern *tr. V.* lure; **sich von jmdm./etw. nicht ~ lassen** *(fig. ugs.)* not be tempted by sb.'s offer/by sth.

kodieren *tr. V.* code; encode

Koeffizient [ko|ɛfi'tsi̯ɛnt] **der; ~en, ~en** *(bes. Math.)* coefficient

Koexistenz die; ~: coexistence

Koffein [kɔfe'i:n] **das; ~s** caffeine

koffein·frei *Adj.* decaffeinated

Koffer ['kɔfɐ] **der; ~s, ~** [suit]case; **die ~ packen** pack one's bags [and leave]

Koffer-: ~anhänger der luggage tag *or* label; **~kuli der** luggage trolley; **~radio das** portable radio; **~raum der** boot *(Brit.)*; trunk *(Amer.)*

Kognak ['kɔnjak] **der; ~s, ~s** brandy; *s. auch* **Cognac**

Kohl [ko:l] **der; ~[e]s a)** cabbage; **das macht den ~ [auch] nicht fett** *(ugs.)* that doesn't help a lot; **b)** *(ugs. abwertend: Unsinn)* rubbish; rot *(sl.)*; **red keinen ~!** don't talk rot! *(sl.)*

Kohl·dampf der: ~ haben/schieben *(salopp)* be ravenously hungry/go hungry

Kohle ['ko:lə] **die; ~, ~n a)** coal; **wir haben keine ~n mehr** we have run out of coal; **[wie] auf [glühenden] ~n sitzen** be fidgeting on one's seat *(fig.)*; **b)** *(salopp: Geld)* dough *(sl.)*; **Hauptsache, die ~n stimmen!** as long as the money's right; **c)** *(Zeichen~)* charcoal

Kohle-: ~hydrat *s.* **Kohlenhydrat; ~kraft·werk das** coal-fired power station

kohlen-, Kohlen-: ~berg·werk das coal-mine; colliery; **~dioxid das** *(fachspr.)*, **~dioxyd** [--'---] **das** *(Chemie)* carbon dioxide; **~grube die** coal-mine; [coal-]pit; **~halde die** coal heap; **~hydrat das** *(Chemie)* carbohydrate; **~keller der** coal-cellar; **~monoxid** *(fachspr.)*, **~monoxyd** [--'---] **das** *(Chemie)* carbon monoxide; **~ofen der** coal-burning stove; **~säure die** carbonic acid; **~schaufel die** coal-shovel; **~stoff der;** *o. Pl.* carbon; **~wasserstoff** [--'---] **der** *(Chemie)* hydrocarbon

Kohle-: ~ofen *s.* **Kohlenofen; ~papier das** carbon paper; **~tablette die** charcoal tablet; **~zeichnung die** charcoal drawing

kohl-, Kohl-: ~kopf der [head of] cabbage; **~meise die** great tit; **~rabenschwarz** *Adj. s.* **rabenschwarz; ~rabi** [~'ra:bi] **der; ~[s], ~[s]** kohlrabi; **~roulade die** *(Kochk.)* stuffed cabbage; **~rübe die** swede; **~weißling der** cabbage white; cabbage butterfly

Koitus ['ko:itʊs] **der; ~, Koitus** *(geh.)* sexual intercourse; coitus *(formal)*

Koje ['ko:jə] **die; ~, ~n a)** *(Seemannsspr.)* bunk; berth; **b)** *(ugs. scherzh.: Bett)* bed

Kokain [koka'i:n] **das; ~s** cocaine

kokain·süchtig *Adj.* addicted to cocaine *postpos.*

kokett [ko'kɛt] 1. *Adj.* coquettish. 2. *adv.* coquettishly

Koketterie [kokɛtə'ri:] **die; ~**: coquetry; coquettishness

kokettieren *itr. V.* play the coquette; flirt; **mit etw. ~**: make much play with sth.

Kokolores [koko'lo:rɛs] **der; ~** *(ugs.)* rubbish; nonsense; rot *(sl.)*

Kokos- ['ko:kɔs-]: **~flocken** *Pl.* coconut ice *sing.*; *(als Füllung)* desiccated coconut *sing.*;

~milch die coconut milk; **~nuß die** coconut; **~palme die** coconut palm; coconut tree

[1]**Koks** [ko:ks] **der; ~es** coke

[2]**Koks der; ~es** *(Drogenjargon: Kokain)* coke *(sl.)*; snow *(sl.)*

Kolben ['kɔlbn̩] **der; ~s, ~ a)** *(Technik)* piston; **b)** *(Chemie: Glasgefäß)* flask; **c)** *(Teil des Gewehrs)* butt; **d)** *(Bot.)* spadix; *(Mais~)* cob; **e)** *(salopp: dicke Nase)* hooter *(Brit. sl.)*; conk *(sl.)*

Kolchose [kɔl'ço:zə] **die; ~, ~n** kolkhoz; Soviet collective farm

Kolibri ['ko:libri] **der; ~s, ~s** humming-bird

Kolik ['ko:lɪk] **die; ~, ~en** colic

kollabieren [kɔla'bi:rən] *itr. V.; mit sein (Med., fig.)* collapse

Kollaborateur [kɔlabora'tø:ɐ̯] **der; ~s, ~e** collaborator

Kollaboration [kɔlabora'tsi̯o:n] **die; ~:** collaboration

kollaborieren *itr. V.* collaborate **(mit** with)

Kollaps ['kɔlaps] **der; ~es, ~e** *(Med., fig.)* collapse; **einen ~ erleiden** collapse

Kolleg [kɔ'le:k] **das; ~s, ~s** *(Vorlesung)* lecture; *(Vorlesungsreihe)* course of lectures

Kollege [kɔ'le:gə] **der; ~n, ~n** colleague; *(Arbeiter)* workmate; **Herr ~!** Mr. Smith/Jones *etc.!*; **Herr ~ [Müller** *usw.*] *(Abgeordneter)* ≈ the Honourable Gentleman

kollegial [kɔle'gi̯a:l] **1.** *Adj.* helpful and considerate. **2.** *adv.* ‹*act etc.*› like a good colleague/good colleagues

Kollegialität [kɔlegi̯ali'tɛ:t] **die; ~:** helpfulness and consideration

Kollegin die; ~, ~nen *s.* **Kollege**

Kollegium [kɔ'le:gi̯ʊm] **das; ~s, Kollegien** *(Lehrkörper)* [teaching] staff

Kollekte [kɔ'lɛktə] **die; ~, ~n** collection

Kollektion [kɔlɛk'tsi̯o:n] **die; ~, ~en a)** *(auch Mode)* collection; **b)** *(Sortiment)* range

kollektiv [kɔlɛk'ti:f] **1.** *Adj.* collective; joint ‹*collaboration*›. **2.** *adv.* collectively

Kollektiv das; ~s, ~e *od.* **~s a)** group; **b)** *(bes. ehem. DDR: Arbeitsgruppe)* collective

Kollektiv·schuld die; *o. Pl.* collective guilt

Koller ['kɔlɐ] **der; ~s, ~** *(ugs.)* rage; **einen ~ haben/bekommen** be in/fly *or* get into a rage

kollidieren [kɔli'di:rən] *itr. V.* **a)** *mit sein* collide, be in collision **(mit** with); **b)** *(fig.)* clash, conflict **(mit** with)

Kollier [kɔ'li̯e:] **das; ~s, ~s** necklace

Kollision [kɔli'zi̯o:n] **die; ~, ~en a)** collision; **b)** *(fig.)* conflict, clash (*Gen.* between)

Kollisions·kurs der; *o. Pl.* collision course; **auf ~ gehen** *(fig.)* be heading for a confrontation

Kolloquium [kɔ'lo̞kvi̯ʊm] **das; ~s, Kolloquien** colloquium

Köln [kœln] **(das); ~s** Cologne

Kölner 1. *indekl. Adj.; nicht präd.* Cologne *attrib.; (in Köln)* in Cologne *postpos., not pred;* ‹*suburb, archbishop, mayor, speciality*› of Cologne; ‹*car factory, river bank*› at Cologne; **der ~ Dom/Karneval** Cologne Cathedral/the Cologne carnival. **2. der; ~s, ~:** inhabitant of Cologne; *(von Geburt)* native of Cologne; **er ist ~:** he comes from Cologne; **die ~:** the people of Cologne

Kölnerin die; ~, ~nen *s.* **Kölner 2**

kölnisch *Adj.* Cologne *attrib.*; of Cologne *postpos., not pred.;* **Kölnisch Wasser** eau-de-Cologne

kolonial [kolo'ni̯a:l] *Adj.; nicht präd.* colonial

Kolonialismus der; ~: colonialism *no art.*

Kolonial-: ~macht die colonial power; **~zeit die;** *o. Pl.* colonial era *or* period

Kolonie [kolo'ni:] **die; ~, ~n a)** *(auch Biol.)* colony; **b)** *(Siedlung)* colony; settlement

Kolonne [ko'lɔnə] **die; ~, ~n a)** *(Truppe, Gruppe von Menschen, Zahlenreihe)* column; **b)** *(Fahrzeuge)* column; *(Konvoi)* convoy; **~ fahren** drive in a [long] line of traffic; **c)** *(Arbeits~)* gang

kolorieren [kolo'ri:rən] *tr. V.* colour

Kolorit [kolo'ri:t] **das; ~[e]s, ~e** *od.* **~s** *(geh.)* colour

Koloß [ko'lɔs] **der; Kolosses, Kolosse** *(fig.: riesiges Gebilde, ugs. scherzh.: große Person)* colossus; giant

kolossal [kolɔ'sa:l] **1.** *Adj.* **a)** *(riesenhaft)* colossal; gigantic; enormous; **b)** *(ugs.: sehr groß)* tremendous *(coll.)*; incredible *(coll.)* ‹*rubbish, nonsense*›; **~es Glück haben** be incredibly lucky *(coll.)*. **2.** *adv. (ugs.)* tremendously *(coll.)*; **~ viel Geld** a tremendous *or* vast amount of money *(coll.)*

Kolossal·schinken der *(salopp abwertend)* **a)** *(Film)* massive great epic *(coll.)*; **b)** *(Gemälde)* whacking great painting *(sl.)*

kolportieren [kɔlpɔr'ti:rən] *tr. V. (geh.)* spread; circulate

Kolumbianer [kolʊm'bi̯a:nɐ] **der; ~s, ~, Kolumbianerin die; ~, ~nen** Colombian

kolumbianisch *Adj.* Colombian

Kolumbien [ko'lʊmbi̯ən] **(das); ~s** Colombia

Kolumbus [ko'lʊmbʊs] **(der)** Columbus; *s. auch* **Ei a**

Kolumne [ko'lʊmnə] **die; ~, ~n** *(Druckw., Zeitungsw.)* column

Koma ['ko:ma] **das; ~s, ~s** *od.* **~ta** *(Med.)* coma

Kombi ['kɔmbi] **der; ~[s], ~s** *s.* **Kombiwagen**

Kombinat [kɔmbi'na:t] **das; ~[e]s, ~e** *(Wirtsch., bes. ehem. DDR)* combine

Kombination [kɔmbina'tsi̯o:n] **die; ~, ~en a)** *(auch Schach)* combination; **b)** *(gedankliche Verknüpfung)* deduction; piece of reasoning; **c)** *(Kleidungsstücke)* ensemble; suit; *(Herren~)* suit; *(Flieger~)* flying-suit; **d)** *(Ballspiele)* combined move

Kombinations·gabe die; *o. Pl.* powers *pl.* of reasoning *or* deduction

kombinieren [kɔmbi'ni:rən] **1.** *tr. V.* combine **(zu** into). **2.** *itr. V.* **a)** *(Zusammenhänge herstellen)* deduce; reason; **b)** *(Ballspiele)* combine

Kombi-: ~wagen der estate [car]; station wagon *(Amer.)*; **~zange die** combination pliers *pl.*; **eine ~zange** a pair of combination pliers

Komet [ko'me:t] **der; ~en, ~en** comet

kometen·haft *Adj.* meteoric ‹*rise, career*›

Komfort [kɔm'fo:ɐ̯] **der; ~s** comfort; **mit allem ~** ‹*flat, house*› with all modern conveniences *pl.*; ‹*car*› with all the latest luxury features *pl.*

komfortabel [kɔmfɔr'ta:bl̩] **1.** *Adj.* comfortable. **2.** *adv.* comfortably

Komik ['ko:mɪk] **die; ~:** comic effect; *(komisches Element)* comic element *or* aspect; **Sinn für ~ haben** have a sense of the comic

Komiker der; ~s, ~ comedian; comic *(coll.)*

komisch ['ko:mɪʃ] *Adj.* **a)** *(lustig)* comical; funny; **ich finde das gar nicht ~** *(ugs.)* I don't think that's at all funny; **b)** *(seltsam)* funny; strange; odd; **~ [zu jmdm.] sein** act *or* behave strangely [towards sb.]; **mir ist/wird so ~:** I'm feeling funny *or* peculiar; **c)** *(Literaturw., Theater)* comic ‹*part*›

komischer·weise *Adv. (ugs.)* strangely enough

Komitee [komi'te:] **das; ~s, ~s** committee

Komma ['kɔma] **das; ~s, ~s** *od.* **~ta a)** *(Satzzeichen)* comma; **b)** *(Math.)* decimal point; **zwei ~ acht** two point eight; **zwei Stellen hinter dem ~:** two decimal places

Kommandant [kɔman'dant] **der; ~en, ~en** *(einer Stadt, Festung)* commandant; *(eines Panzers, Raumschiffs)* commander; *(einer Militäreinheit)* commander; commanding officer; *(eines Flugzeugs, Schiffs)* captain

Kommandeur [kɔman'dø:ɐ̯] **der; ~s, ~e** *(Milit.)* commander; commanding officer

kommandieren 1. *tr. V.* **a)** *(befehligen)* command; be in command of; **b)** *(abkommandieren)* **jmdn. an die Front ~:** order sb. to the front; **c)** *(ugs.: herumkommandieren)* **jmdn. ~:** order *or (sl.)* boss sb. about. **2.** *itr. V. (ugs.)* order *or (sl.)* boss people about

Kommandit·gesellschaft [kɔman'di:t-] **die** *(Wirtsch.)* limited partnership

Kommando [kɔ'mando] **das; ~s, ~s,** *österr. auch:* **Kommanden a)** *(Befehl)* command; **das ~ zum Schießen geben** give the command *or* order to shoot; **wie auf ~:** as if by command; **b)** *o. Pl. (Befehlsgewalt)* command; **das ~ haben** *od.* **führen/übernehmen** be in/assume *or* take command; **c)** *(Milit.) (Einheit)* detachment; *(Stoßtrupp)* commando

Kommando-: ~brücke die bridge; **~sache die** *in* **geheime ~sache** *(Milit.)* military secret

Kommata *s.* **Komma**

kommen ['kɔmən] *unr. itr. V.; mit sein* **a)** come; *(eintreffen)* come; arrive; **ich komme schon!** I'm coming!; **der Kellner kommt sofort** the waiter will be with you directly; **angelaufen/angebraust** *usw.* **~:** come running/roaring *etc.* along; *(auf jmdn. zu)* come running/roaring *etc.* up; **angekrochen ~** *(fig.)* come crawling up; **zu spät ~:** be late; **durch eine Gegend ~:** pass through a region; **nach Hause ~:** come *or* get home; **zu jmdm. ~** *(jmdn. besuchen)* come and see sb.; **ist für mich keine Post ge~?** is/was there no post for me?; **etw. ~ lassen** *(etw. bestellen)* order sth.; **jmdn. ~ lassen** send for *or* call sb.; **da könnte ja jeder ~!** *(ugs.)* who do you think you are?/who does he think he is? *etc.;* **komm mir bloß nicht damit!** *(ugs.)* don't give me that!; **[bitte] ~!** *(im Funkverkehr)* come in[, please]; **b)** *(gelangen)* get; **ans Ufer/Ziel ~:** reach the bank/finishing-line; **wie komme ich nach Paris?** how do I get to Paris?; *(fig.)* **auf etw.** *(Akk.)* **zu sprechen ~:** turn to the discussion of sth.; **jmdm. auf die Spur/Schliche ~:** get on sb.'s trail/get wise to sb.'s tricks; **wie kommst du darauf?** what gives you that idea?; **dazu ~, etw. zu tun** get round to doing sth.; **zum Einkaufen/Waschen ~:** get round to doing the shopping/washing; **c)** *(auftauchen)* ‹*seeds, plants*› come up; ‹*buds, flowers*› come out; ‹*peas, beans*› form; ‹*teeth*› come through; **zur Welt ~:** be born; **ihr ist ein Gedanke/eine Idee ge~:** she had a thought/an idea; a thought/an idea came to her; **jmdm. ~ die Tränen** tears come to sb.'s eyes; **d)** *(aufgenommen werden)* **zur Schule ~:** go to *or* start school; **ins Krankenhaus/Gefängnis ~:** go into hospital/to prison; **in den Himmel/in die Hölle ~** *(fig.)* go to heaven/hell; **e)** *(gehören)* go; belong; **in die Schublade/ins Regal ~:** go *or* belong in the drawer/on the shelf; **f)** *(gebracht, befördert werden)* go; **in den Müll ~:** be thrown out; **g)** *(geraten)* get; **in Gefahr/Not/Verlegenheit ~:** get into danger/serious difficulties/get *or* become embarrassed; **unter ein Auto/zu Tode ~:** be knocked down by a car/be *or* get killed; **ins Schleudern ~:** go into a skid; **neben jmdn. zu sitzen ~:** get to sit next to sb.; *s. auch* **Schwung; Stimmung; h)** *(nahen)* **ein Gewitter/die Flut kommt** a storm is approaching/the tide's coming in; **der Tag/die Nacht kommt** *(geh.)* day is breaking/night is falling; **dieses Unglück habe ich schon lange ~ sehen** I saw this disaster coming a long time ago; **im Kommen sein** ‹*fashion etc.*› be coming in; ‹*person*› be on the way up; **i)** *(sich ereignen)* come about; happen; **was auch immer ~ mag** come what may; **das durfte [jetzt] nicht ~** *(ugs. spött.)* that's hardly the thing to say now; **gelegen/ungelegen ~** ‹*offer, opportunity*› come/not come at the right moment; ‹*visit*› be/not be convenient; **überraschend [für jmdn.] ~:** come as a surprise [to sb.]; **daher kommt es, daß ...:** that's [the reason] why ...; **das kommt davon, daß ...:** that's because ...; **vom vielen Rauchen/vom Vitaminmangel ~:** be due to smoking/vitamin deficiency; **wie kommt es, daß ...:** how is it that you/he *etc.* ...; how come that ... (coll.); **das kommt davon!** see what happens!; **j)** *unpers.* **es kam zum Streit/Kampf** there was a quarrel/fight; **es kam alles ganz anders** it all *or* everything turned out quite differently; **so weit kommt es noch [, daß ich euern Dreck wieder wegräume]!** *(ugs. iron.)* that really is the limit[, expecting me to clear up your rubbish after you]!; **k)** *(ugs.: erreicht werden)* **wann kommt der nächste Bahnhof?** when do we get to the next station? *(coll.);* **jetzt kommt gleich Mannheim** we'll be at Mannheim any moment; **da vorn kommt eine Tankstelle** there's a petrol station coming up *(coll.);* **l) zu Geld ~:** become wealthy; **zu Erfolg/Ruhm** *usw.* **~:** gain success/fame *etc.;* **nie zu etwas ~** *(ugs.)* never get anywhere; **wieder zu Kräften ~:** regain one's strength; **[wieder] zu sich ~:** regain consciousness; come round; **um etw. ~:** lose sth.; **ums Leben ~:** lose one's life; **m)** *(an der Reihe sein; folgen)* **zuerst/zuletzt kam ...:** first/last came ...; **als erster/letzter ~:** come first/last; **jetzt komme ich [an die Reihe]** it is my turn now; **n)** *(ugs.: sich verhalten)* **jmdm. frech/unverschämt/grob ~:** be cheeky/impertinent/rude to sb.; **so lasse ich mir nicht ~!** I don't stand for that sort of thing!; **o) ich lasse auf ihn** *usw.* **nichts ~:** I won't hear anything said against him *etc.;* **über jmdn. ~** *(jmdn. erfassen)* ‹*feeling*› come over sb.; **p)** *(entfallen)* **auf hundert Berufstätige ~ vier Arbeitslose** for every hundred people in employment, there are four people unemployed; **q) woher ~ diese Sachen?** where do these things come from?; **seine Eltern ~ aus Sachsen** his parents come *or* are from Saxony; **r)** *(ugs.: kosten)* **auf 100 Mark ~:** cost 100 marks; **alles zusammen kam auf ...:** altogether it came to ...; **wie teuer kommt der Stoff?** how much *or* dear is that material?; **etw. kommt [jmdn.] teuer** sth. comes expensive [for sb.]; **s)** *(ugs.: anspringen)* ‹*engine*› start; **t)** *(salopp: Orgasmus haben)* come *(sl.);* **u)** *(ugs.: als Aufforderung, Ermahnung)* **komm/kommt/kommen Sie** come on, now; **komm, komm** oh, come on; **v)** *(Sportjargon: gelingen)* **[gut] ~/nicht ~** ‹*serve, backhand, forehand, etc.*›

be going/not be going well; **w)** *in festen Wendungen: s.* **Ausbruch b; Einsatz c; Entfaltung a; Fall**

kommend *Adj.; nicht präd.* **a)** *(nächst...)* next; **das ~e Wochenende/am ~en Sonntag** next weekend/Sunday; **b)** *(zukünftig)* **~e Generationen** generations to come; future generations; **c)** *(mit großer Zukunft)* **der ~e Mann** the coming man

Kommentar [kɔmɛnˈtaːɐ̯] **der; ~s, ~e a)** *(Erläuterung)* commentary; **b)** *(Stellungnahme)* commentary; comment; **kein ~!** no comment!; **c)** *(oft abwertend: Anmerkung)* comment

Kommentator [kɔmɛnˈtaːtɔr] **der; ~s, ~en** [-taˈtoːrən], **Kommentatorin die; ~, ~nen** commentator

kommentieren *tr. V.* comment on

Kommerz [kɔˈmɛrts] **der; ~es** *(abwertend)* business interests *pl.*

kommerzialisieren [kɔmɛrtsi̯aliˈziːrən] *tr. V.* commercialize

kommerziell [kɔmɛrˈtsi̯ɛl] **1.** *Adj.* commercial. **2.** *adv.* commercially

Kommilitone [kɔmiliˈtoːnə] **der; ~n, ~n, Kommilitonin die; ~, ~nen** *(Studentenspr.)* fellow student

Kommiß [kɔˈmɪs] **der; Kommisses** *(Soldatenspr.)* army

Kommissar [kɔmɪˈsaːɐ̯] **der; ~s, ~e a)** commissioner; **b)** *(Polizist)* detective superintendent

Kommissariat [kɔmɪsaˈri̯aːt] **das; ~s, ~e a)** *s.* **Kommissar:** commissioner's office; detective superintendent's office; **b)** *(österr.)* police station

kommissarisch 1. *Adj.; nicht präd.* acting. **2.** *adv.* in an acting capacity

Kommission [kɔmɪˈsi̯oːn] **die; ~, ~en a)** *(Gremium)* committee; *(Prüfungs~)* commission; **b) etw. in ~ nehmen/haben/geben** *(Wirtsch.)* take/have sth. on commission/give sth. to a dealer for sale on commission

Kommode [kɔˈmoːdə] **die; ~, ~n** chest of drawers

kommunal [kɔmuˈnaːl] **1.** *Adj.* local; *(städtisch)* municipal; local. **2.** *adv.* **etw. wird ~ verwaltet** sth. comes under local government

Kommunal-: ~politik die local politics *sing.*; **~wahl die** local [government] elections *pl.*

Kommune [kɔˈmuːnə] **die; ~, ~n** local authority; *(städtische Gemeinde)* municipality

Kommunikation [kɔmunikaˈtsi̯oːn] **die; ~, ~en** communication

Kommunion [kɔmuˈni̯oːn] **die; ~, ~en** *(kath. Kirche)* [Holy] Communion

Kommuniqué [kɔmyniˈkeː] **das; ~s, ~s** communiqué

Kommunismus der; ~: communism

Kommunist der; ~en, ~en, Kommunistin die; ~, ~nen communist

kommunistisch 1. *Adj.* communist; *(die ~e Partei betreffend)* Communist. **2.** *adv.* Communist‑ ⟨*influenced, led, ruled, etc.*⟩

kommunizieren [kɔmuniˈtsiːrən] *itr. V.* **a)** *(geh.)* communicate; **b)** *(kath. Kirche)* receive [Holy] Communion

Komödiant [komøˈdi̯ant] **der; ~en, ~en, Komödiantin die; ~, ~nen a)** *(veralt.)* actor/actress; player; **b)** *(abwertend: Heuchler/~in)* play-actor

Komödie [koˈmøːdi̯ə] **die; ~, ~n a)** comedy; *(fig.)* farce; **b)** *(Theater)* comedy theatre; **c)** *(Heuchelei)* play-acting

Kompagnon [kɔmpanˈjõː] **der; ~s, ~s** *(Wirtsch.)* partner; associate

kompakt [kɔmˈpakt] *Adj.* **a)** *(massiv)* solid; **b)** *(ugs.: gedrungen)* stocky

Kompanie [kɔmpaˈniː] **die; ~, ~n** *(Milit.)* company

Komparativ [ˈkɔmparatiːf] **der; ~s, ~e** *(Sprachw.)* comparative

Komparse [kɔmˈparzə] **der; ~n, ~n** *(Theater)* supernumerary; super *(coll.)*; *(Film)* extra

Kompaß [ˈkɔmpas] **der; Kompasses, Kompasse** compass

Kompaß·nadel die compass needle

Kompensation [kɔmpɛnzaˈtsi̯oːn] **die; ~, ~en** *(fachspr., geh.)* compensation

kompensieren *tr. V.* **etw. mit etw.** *od.* **durch etw. ~:** compensate for *or* make up for sth. by sth.

kompetent [kɔmpeˈtɛnt] *Adj.* **a)** *(sachverständig)* competent; **b)** *(bes. Rechtsw.: zuständig)* competent, responsible ⟨*authority*⟩

Kompetenz [kɔmpeˈtɛnts] **die; ~, ~en a)** competence; **b)** *(bes. Rechtsw.: Zuständigkeit)* authority; powers *pl.*; **in jmds. ~** *(Dat.)* **liegen/in jmds. ~** *(Akk.)* **fallen** be/come within sb.'s authority *or* powers; **das liegt außerhalb meiner ~:** that doesn't lie within my authority *or* powers

komplementär [kɔmplemɛnˈtɛːɐ̯] *Adj.* complementary

Komplementär·farbe die *(Optik)* complementary colour

komplett [kɔmˈplɛt] **1.** *Adj.* **a)** complete; **es kostet ~ 1 500 Mark** it costs 1,500 marks complete; **heute sind wir ~** *(ugs.)* today we are all here; **b)** *nicht präd. (ugs.: ganz und gar)* complete; utter; **c)** *(österr.: voll)* full ⟨*hotel, tram, etc.*⟩. **2.** *adv.* **a)** fully ⟨*furnished, equipped*⟩; **b)** *(ugs.: ganz und gar)* completely; totally

komplettieren *tr. V.* complete

komplex [kɔmˈplɛks] *Adj. (geh.)* complex

Komplex der; ~es, ~e *(auch Psych.)* complex

Komplexität [kɔmplɛksiˈtɛːt] **die; ~:** complexity

Komplikation [kɔmplikaˈtsi̯oːn] **die; ~, ~en** complication

Kompliment [kɔmpliˈmɛnt] **das; ~[e]s, ~e** compliment; **jmdm. ein ~ machen** pay sb. a compliment (**über** + *Akk.* on); **nicht gerade ein ~ für jmdn. sein** *(fig.)* not exactly do sb. credit

komplimentieren *tr. V. (geh.)* **jmdn. ins Haus ~:** usher *or* show sb. into the house; **jmdn. aus dem Zimmer ~** *(verhüll.)* usher sb. out of the room

Komplize [kɔmˈpliːtsə] **der; ~n, ~n** *(abwertend)* accomplice

komplizieren *tr. V.* complicate

kompliziert 1. *Adj.* complicated; complicated, intricate ⟨*device, piece of apparatus*⟩; complicated, involved ⟨*problem, procedure*⟩; *(Med.)* compound ⟨*fracture*⟩. **2.** *adv.* **sich ~ ausdrücken** express oneself in a complicated *or* an involved way *or* manner

Komplizin die; ~, ~nen *(abwertend)* accomplice

Komplott [kɔmˈplɔt] **das; ~[e]s, ~e** plot; conspiracy; **ein ~ schmieden** hatch a plot

Komponente [kɔmpoˈnɛntə] **die; ~, ~n** component

komponieren *tr., itr. V. (auch fig. geh.)* compose

Komponist der; ~en, ~en, Komponistin die; ~, ~nen composer

Komposition [kɔmpoziˈtsi̯oːn] **die; ~, ~en** *(auch fig. geh.)* composition

Kompositum [kɔmˈpoːzitʊm] **das; ~s, Komposita** *(Sprachw.)* compound [word]

Kompost [kɔmˈpɔst] **der; ~[e]s, ~e** compost

Kompost·haufen der compost heap

Kompott [kɔm'pɔt] **das; ~[e]s, ~e** stewed fruit; compote
Kompresse [kɔm'prɛsə] **die; ~, ~n** *(Med.)* **a)** *(Umschlag)* [wet] compress; **b)** *(Mull)* [gauze] pad
Kompression [kɔmprɛ'sio:n] **die; ~, ~en** *(fachspr.)* compression
Kompressor [kɔm'prɛsɔr] **der; ~s, ~en** [-'so:rən] *(Technik)* compressor
komprimieren [kɔmpri'mi:rən] *tr. V. (fachspr., geh.)* compress
komprimiert *Adj. (fig. geh.)* condensed ⟨*account, form, etc.*⟩
Kompromiß [kɔmpro'mɪs] **der; Kompromisses, Kompromisse** compromise; **einen ~ schließen** make a compromise; compromise; **ein fauler ~** *(ugs.)* a poor sort of compromise *(coll.)*
kompromiß-, Kompromiß-: ~bereit *Adj.* ready *or* willing to compromise *pred.;* **~bereitschaft die;** *o. Pl.* readiness *or* willingness to compromise; **~los 1.** *Adj.* uncompromising; **2.** *adv.* uncompromisingly; **~lösung die** compromise solution; **~vorschlag der** compromise proposal *or* suggestion
kompromittieren [kɔmpromɪ'ti:rən] *tr. V.* compromise; **sich ~:** compromise oneself
Kondensation [kɔndɛnza'tsio:n] **die; ~, ~en** *(Physik, Chemie)* condensation
Kondensator [kɔndɛn'za:tɔr] **der; ~s, ~en** [-za'to:rən] **a)** *(Elektrot.)* capacitor; condenser; **b)** *(Technik)* condenser
kondensieren *tr., itr. V. (itr. auch mit sein) (Physik, Chemie)* condense
Kondens-: ~milch die condensed milk; **~streifen der** condensation trail; vapour trail; **~wasser das** condensation
Kondition [kɔndi'tsio:n] **die; ~, ~en a)** *meist Pl. (bes. Kaufmannsspr., Finanzw.)* condition; **zu günstigen ~en** on favourable terms *or* conditions; **b)** *o. Pl. (körperlich-seelische Verfassung)* condition; **eine gute/schlechte ~ haben** be/not be in good condition *or* shape; **keine ~ haben** be out of condition; *(fig.)* have no stamina
Konditions-: ~schwäche die lack of condition *or* fitness; **~training das** fitness training
Konditor [kɔn'di:tɔr] **der; ~s, ~en** [-di'to:rən] confectioner; pastry-cook; **beim ~:** at the cake-shop
Konditorei die; ~, ~en cake-shop; *(Lokal)* café
kondolieren [kɔndo'li:rən] *itr. V.* offer one's condolences; **jmdm. [zu jmds. Tod] ~:** offer one's condolences to sb. *or* condole with sb. [on sb.'s death]
Kondom [kɔn'do:m] **das** *od.* **der; ~s, ~e** condom; [contraceptive] sheath
Kondukteur [kɔndʊk'tø:ɐ̯] **der; ~s, ~e** *(schweiz.) s.* **Schaffner**
Konfekt [kɔn'fɛkt] **das; ~[e]s a)** confectionery; sweets *pl. (Brit.);* candies *pl. (Amer.);* **b)** *(bes. südd., österr., schweiz.: Teegebäck)* [small] fancy biscuits *pl. (Brit.) or (Amer.)* cookies *pl.*
Konfektion [kɔnfɛk'tsio:n] **die; ~, ~en** ready-made *or* off-the-peg *(Brit.) or (Amer.)* off-the-rack clothes *pl. or* garments *pl.*
Konfektions-: ~geschäft das [ready-made *or* off-the-peg *(Brit.) or (Amer.)* off-the-rack] clothes shop; **~ware die** *s.* **Konfektion**
Konferenz [kɔnfe'rɛnts] **die; ~, ~en** conference; *(Besprechung)* meeting
konferieren [kɔnfe'ri:rən] *itr. V.* confer **(über +** *Akk.* on, about)
Konfession [kɔnfɛ'sio:n] **die; ~, ~en** denomination; religion; **welche ~ haben Sie?** what denomination *or* religion are you?
konfessionell [kɔnfɛsio'nɛl] **1.** *Adj.; nicht präd.* denominational. **2.** *adv.* as regards denomination
konfessions·los *Adj.* not belonging to any denomination *or* religion *postpos., not pred.*
Konfetti [kɔn'fɛti] **das; ~[s]** confetti
Konfirmand [kɔnfɪr'mant] **der; ~en, ~en, Konfirmandin die; ~, ~nen** *(ev. Rel.)* confirmand
Konfirmation [kɔnfɪrma'tsio:n] **die; ~, ~en** *(ev. Rel.)* confirmation
konfiszieren [kɔnfɪs'tsi:rən] *tr. V. (bes. Rechtsw.)* confiscate
Konfitüre [kɔnfi'ty:rə] **die; ~, ~n** jam
Konflikt [kɔn'flɪkt] **der; ~[e]s, ~e** conflict; **mit etw. in ~ geraten** come into conflict with sth.
konflikt-, Konflikt-: ~frei *Adj.* conflict-free; **~situation die** conflict situation; **~stoff der** cause for conflict *or* dispute
Konföderation [kɔnfødera'tsio:n] **die; ~, ~en** confederation
konform [kɔn'fɔrm] *Adj.* concurring *attrib.* ⟨*views*⟩; **mit jmdm./etw. ~ gehen** be in agreement with sb./sth.
Konformist der; ~en, ~en, Konformistin, die; ~, ~nen conformist
konformistisch 1. *Adj.* conformist. **2.** *adv.* in a conformist way
Konfrontation [kɔnfrɔnta'tsio:n] **die; ~, ~en** confrontation
konfrontieren *tr. V.* confront
konfus [kɔn'fu:s] **1.** *Adj.* confused; muddled; **jmdn. ~ machen** confuse *or* muddle sb. **2.** *adv.* in a confused *or* muddled fashion; confusedly
Konglomerat [kɔnglome'ra:t] **das; ~[e]s, ~e** *(geh.)* conglomeration
Kongreß [kɔn'grɛs] **der; Kongresses, Kongresse a)** congress; conference; **b) der ~** *(USA)* Congress
Kongreß-: ~halle die conference hall; **~mitglied das** *(USA)* Congressman/Congresswoman; **~teilnehmer der** congress *or* conference participant
kongruent [kɔngru'ɛnt] *Adj.* *(Math.)* congruent
Kongruenz [kɔngru'ɛnts] **die; ~, ~en** *(Math.)* congruence
König ['kø:nɪç] **der; ~s, ~e** *(auch Schach, Kartenspiele, fig.)* king; **der Kunde ist ~:** the customer is always right
Königin die; ~, ~nen *(auch Bienen~)* queen
Königin·mutter die; *Pl.* **Königinmütter** queen mother
königlich 1. *Adj.* **a)** *nicht präd.* royal; **b)** *(vornehm)* regal; **c)** *(reichlich)* princely ⟨*gift, salary, wage*⟩; lavish ⟨*hospitality*⟩; **d)** *(ugs.: außerordentlich)* tremendous *(coll.)*⟨*fun*⟩. **2.** *adv.* **a)** *(reichlich)* ⟨*entertain*⟩ lavishly; ⟨*pay*⟩ handsomely; **~ beschenkt werden** be showered with lavish presents; **b)** *(ugs.: außerordentlich)* ⟨*enjoy oneself*⟩ immensely *(coll.)*
König·reich das kingdom
königs-, Königs-: ~blau *Adj.* royal blue; **~haus das** royal house; **~hof der** royal court; king's court; **~paar das** royal couple; **~sohn der** prince; king's son; **~tochter die** princess; king's daughter; **~treu** *Adj.* loyal to the king *postpos.; (der Monarchie treu)* royalist
konisch ['ko:nɪʃ] **1.** *Adj.* conical
Konjugation [kɔnjuga'tsio:n] **die; ~, ~en** *(Sprachw.)* conjugation
konjugieren *tr. V. (Sprachw.)* conjugate
Konjunktion [kɔnjʊŋk'tsio:n] **die; ~, ~en** *(Sprachw.)* conjunction
Konjunktiv ['kɔnjʊŋkti:f] **der; ~s, ~e** *(Sprachw.)* subjunctive
Konjunktur [kɔnjʊŋk'tu:ɐ̯] **die; ~, ~en** *(Wirtsch.)* **a)** *(wirtschaftli-*

che Lage) [level of] economic activity; economy; *(Tendenz)* economic trend; **die ~ beleben/bremsen** stimulate/slow down the economy; **b)** *(Hoch~)* boom; *(Aufschwung)* upturn [in the economy]; **~ haben** *(fig.)* be in great demand

konjunktur·abhängig *Adj.* *(Wirtsch.)* dependent on economic trends *postpos.*

konjunkturell [kɔnjʊŋktu'rɛl] **1.** *Adj.; nicht präd.* economic; **die ~e Entwicklung** the development of the economy. **2.** *adv.* **~ bedingt** due to economic trends *postpos.*

Konjunktur- *(Wirtsch.)*: **~politik die** stabilization policy; measures *pl.* aimed at avoiding violent fluctuations in the economy; **~schwankung die** fluctuation in the level of economic activity

konkav [kɔn'ka:f] *(Optik)* **1.** *Adj.* concave. **2.** *adv.* concavely

konkret [kɔn'kre:t] **1.** *Adj.* concrete. **2.** *adv.* in concrete terms; **kannst du mal ~ sagen, was du damit meinst?** could you tell me exactly what you mean by that?

konkretisieren *tr. V.* **etw. ~:** put sth. in concrete terms

Konkurrent [kɔnkʊ'rɛnt] **der; ~en, ~en, Konkurrentin die; ~, ~nen** rival; *(Sport, Wirtsch.)* competitor

Konkurrenz [kɔnkʊ'rɛnts] **die; ~, ~en a)** *o. Pl. (Rivalität)* rivalry *no indef. art.; (Sport, Wirtsch.)* competition *no indef. art.*; **jmdm. ~ machen** compete with sb.; **mit jmdm. in ~ treten/stehen** enter into/be in competition with sb.; **außer ~ starten/teilnehmen** *(bes. Sport)* take part as an unofficial competitor; **b)** *o. Pl. (die Konkurrenten)* competition

konkurrenz-, Konkurrenz-: ~druck der; *o. Pl.* pressure of competition; **~fähig** *Adj.* competitive; **~kampf der** competition; *(zwischen zwei Menschen)* rivalry; **~los** *Adj.* ‹*product, firm, etc.*› that has no competition *or* competitors; *(unvergleichlich)* unrivalled

konkurrieren *itr. V.* compete; **mit jmdm./etw. [um etw.] ~:** compete with sb./sth. [for sth.]

Konkurs [kɔn'kʊrs] **der; ~es, ~e a)** bankruptcy; **~ machen** *od.* **in ~ gehen** go bankrupt; **[den] ~ anmelden** file for bankruptcy; have oneself declared bankrupt; **b)** *(~verfahren)* bankruptcy proceedings *pl.*

Konkurs- *(Wirtsch.)*: **~masse die** bankrupt's assets *pl.*; **~verfahren das** bankruptcy proceedings *pl.*

können ['kœnən] **1.** *unr. Modalverb; 2. Part.* **~ a)** be able to; **er hat/hätte es machen ~:** he was able to *or* he could do it/he could have done it; **er kann es machen/nicht machen** he can do it *or* is able to do it/cannot *or (coll.)* can't do it *or* is unable to do it; **er kann gut reden/tanzen** he can talk/dance well; he is a good talker/dancer; **ich kann nicht schlafen** I cannot *or (coll.)* can't sleep; **ich kann das nicht mehr hören/sehen** I can't stand *or* bear to hear it/can't stand *or* bear the sight of it any longer *(coll.)*; **ich kann dir sagen!** *(ugs.)* I can tell you; **kann das explodieren?** could it explode?; **er kann jeden Moment kommen** he could *or* might come at any moment; **wer kann es sein/gewesen sein?** who can it be/could it have been?; **man kann nie wissen** you never know; one never knows; **es kann sein, daß ...:** it could *or* might be that ...; **das könnte [gut] sein** that could [well] be the case; **das kann nicht sein** that's not possible; **kann ich Ihnen helfen?** can I help you?; **können Sie mir sagen, ...?** can you tell me ...?; **kannst du nicht aufpassen?** can't you be more careful?; **kann sein** *(ugs.)* could be *(coll.)*; **Kommst du morgen? – Kann sein** Are you coming tomorrow? – Might do; **b)** *(Grund haben)* **du kannst ganz ruhig sein** you don't have to worry; **wir ~ uns/er kann sich freuen, daß ...:** we can/he should be glad that ...; **er kann einem leid tun** *(ugs.)* you have to feel sorry for him; **das kann man wohl sagen!** you could well say that; **c)** *(dürfen)* **kann ich gehen?** can I go?; **~ wir mit[kommen]?** can we come too?; **du kannst mich [mal]!** *(salopp verhüll.)* you can get stuffed *(sl.)*; you know what you can do *(coll.)*. **2.** *unr. tr. V.* **a)** *(beherrschen)* know ‹*language*›; be able to play, know how to play ‹*game*›; **sie kann das [gut]** she can do that [well]; **sie kann/kann keine Mathe** she can/can't do maths; **er kann etwas auf seinem Gebiet** he has quite a lot of know-how in his field; **hast du die Hausaufgabe gekonnt?** could you do the homework?; **b) etwas/nichts für etw. ~:** be/not be responsible for sth. **3.** *unr. itr. V.* **a)** *(fähig sein)* **er kann nicht anders** there's nothing else he can do; *(es ist seine Art)* he can't help it *(coll.)*; **b)** *(Zeit haben)* **ich kann heute nicht** I can't today *(coll.)*; **c)** *(ugs.: Kraft haben)* **kannst du noch?** can you go on?; **d)** *(ugs.: essen ~)* **ich kann nicht mehr** I couldn't manage any more; **e)** *(ugs.: umgehen ~)* **[gut] mit jmdm. ~:** get on *or* along [well] with sb.

Können das; ~s ability; *(Kunstfertigkeit)* skill

Könner der; ~s, ~ expert

konnte ['kɔntə] *1. u. 3. Pers. Sg. Prät. v.* **können**

könnte ['kœntə] *1. u. 3. Pers. Sg. Konjunktiv II v.* **können**

konsekutiv [kɔnzeku'ti:f] *Adj.* *(Sprachw.)* consecutive

konsequent [kɔnze'kvɛnt] **1.** *Adj.* **a)** *(folgerichtig)* logical; **b)** *(unbeirrbar)* consistent. **2.** *adv.* **a)** *(folgerichtig)* logically; **b)** *(unbeirrbar)* consistently; **ein Ziel ~ verfolgen** resolutely and single-mindedly pursue a goal; **~ durchgreifen** take rigorous action

Konsequenz [kɔnze'kvɛnts] **die; ~, ~en a)** *(Folge)* consequence; **die ~en tragen** take the consequences; **[aus etw.] die ~en ziehen** draw the obvious conclusion [from sth.]; *(gezwungenermaßen)* accept the obvious consequences [of sth.]; **b)** *o. Pl. (Unbeirrbarkeit)* resolution; determination; **c)** *o. Pl. (Folgerichtigkeit)* logicality; *(eines Gedankenganges)* logical consistency; logicality

konservativ [kɔnzɛrva'ti:f] **1.** *Adj.* conservative. **2.** *adv.* conservatively

Konservative der/die; *adj. Dekl.* conservative

Konservatorium [kɔnzɛrva'to:riʊm] **das; ~s, Konservatorien** conservatoire; conservatory *(Amer.)*

Konserve [kɔn'zɛrvə] **die; ~, ~n** preserved food; *(in Dosen)* canned *or (Brit.)* tinned food; *(ugs.: Dose)* can; tin *(Brit.)*; **von ~n leben** eat out of cans *or (Brit.)* tins; live on canned *or (Brit.)* tinned food; **Musik aus der ~** *(fig. ugs.)* canned music *(coll.)*

Konserven-: ~büchse die, ~dose die can; tin *(Brit.)*

konservieren *tr. V.* preserve; conserve, preserve ‹*building, work of art*›

Konservierung die; ~, ~en preservation

Konservierungs·mittel das preservative

konsistent [kɔnzɪs'tɛnt] *Adj.* consistent

Konsistenz [kɔnzɪs'tɛnts] **die**; ~, ~**en** consistency

konsolidieren [kɔnzoli'di:rən] **1.** *tr. V.* consolidate. **2.** *refl. V.* become consolidated

Konsolidierung die; ~, ~**en** consolidation

Konsonant [kɔnzo'nant] **der**; ~**en**, ~**en** consonant

Konsorten [kɔn'zɔrtn̩] *Pl. (abwertend)* **Meier und** ~: Meier and his lot *or* crowd *(coll.)*; Meier and Co. *(coll.)*

Konsortium [kɔn'zɔrtsi̯ʊm] **das**; ~**s**, **Konsortien** *(Wirtsch.)* consortium

Konspiration [kɔnspira'tsi̯o:n] **die**; ~, ~**en** conspiracy

konspirativ [kɔnspira'ti:f] *Adj.* conspiratorial

konspirieren *itr. V.* conspire, plot (**gegen** against)

konstant [kɔn'stant] **1.** *Adj.* **a)** constant; **eine** ~**e Leistung zeigen** maintain a consistent standard; **b)** *(beharrlich)* consistent; persistent. **2.** *adv.* **a)** constantly; **b)** *(beharrlich)* consistently; persistently

Konstante die; ~, ~**n** *od. adj. Dekl. (Math., Physik)* constant; *(fig.)* constant factor (+ *Gen.* in)

konstatieren [kɔnsta'ti:rən] *tr. V.* **a)** establish ⟨*facts*⟩; detect ⟨*changes etc.*⟩; **b)** *(erklären)* state

Konstellation [kɔnstɛla'tsi̯o:n] **die**; ~, ~**en a)** *(von Parteien usw.)* grouping; *(von Umständen)* combination; **b)** *(Astron., Astrol.)* constellation

konsterniert *Adj.* filled with consternation *pred.*

konstituieren [kɔnstitu'i:rən] **1.** *tr. V.* constitute; set up; **die** ~**de Versammlung** the constituent assembly. **2.** *refl. V.* be constituted

Konstitution [kɔnstitu'tsi̯o:n] **die**; ~, ~**en** constitution

konstitutionell [kɔnstitutsi̯o'nɛl] *Adj.* constitutional

konstruieren [kɔnstru'i:rən] *tr. V.* **a)** *(entwerfen)* design; *(entwerfen und zusammenbauen)* design and construct; **b)** *(aufbauen, Geom., Sprachw.)* construct; **dieses Verb wird mit dem Dativ konstruiert** this verb takes the dative *or* is construed with the dative; **c)** *(künstlich aufbauen)* fabricate; **ein konstruierter Fall** a hypothetical *or* fictitious case; **die Handlung wirkt sehr konstruiert** the plot seems very contrived

Konstrukteur [kɔnstrʊk'tø:ɐ̯] **der**; ~**s**, ~**e**, **Konstrukteurin die**; ~, ~**nen** designer; design engineer

Konstruktion [kɔnstrʊk'tsi̯o:n] **die**; ~, ~**en a)** *(Aufbau, Geom., Sprachw.)* construction; *(das Entwerfen)* designing; *(das Entwerfen und Zusammenbauen)* designing and construction; **b)** *(Entwurf)* design; *(Bau)* construction; structure

Konstruktions·fehler der design fault

konstruktiv [kɔnstrʊk'ti:f] **1.** *Adj.* **a)** constructive; **b)** *(Technik)* constructional. **2.** *adv.* **a)** constructively; **b)** *(Technik)* with regard to construction

Konsul ['kɔnzʊl] **der**; ~**s**, ~**n** *(Dipl., hist.)* consul

Konsulat [kɔnzu'la:t] **das**; ~**[e]s**, ~**e** *(Dipl., hist.)* consulate

Konsultation [kɔnzʊlta'tsi̯o:n] **die**; ~, ~**en** consultation

konsultieren *tr. V. (auch fig.)* consult

Konsum [kɔn'zu:m] **der**; ~**s** consumption (**an** + *Dat.* of)

Konsum·artikel der *(Wirtsch.)* consumer item *or* article; ~ *Pl.* consumer goods

Konsument [kɔnzu'mɛnt] **der**; ~**en**, ~**en**, **Konsumentin die**; ~, ~**nen** consumer

Konsum·gesellschaft die consumer society

Konsumgüter·industrie die *(Wirtsch.)* consumer goods industry

konsumieren *tr. V.* consume; *(fig.)* devour ⟨*book*⟩

Konsum-: ~**terror der** *(abwertend)*, ~**zwang**, **der**; *o. Pl.* pressure to buy

Kontakt [kɔn'takt] **der**; ~**[e]s**, ~**e** *(auch fachspr.)* contact; **mit** *od.* **zu jmdm.** ~ **haben, in** ~ **mit jmdm. stehen** be in contact *or* touch with sb.; **[den]** ~ **mit jmdm./etw. finden/suchen** establish/try to establish contact with sb./sth.; **den** ~ **zu jmdm. abbrechen/verlieren** break off contact/lose contact *or* touch with sb.; **mit jmdm.** ~ **aufnehmen** get into contact with sb.; contact sb.

kontakt-, Kontakt-: ~**anzeige die** contact advertisement; ~**freudig** *Adj.* sociable; ~**freudig sein** make friends easily; ~**linse die** contact lens; ~**mann der**; *Pl.*: ~**männer** *od.* ~**leute** *(Agent)* contact; ~**schale die** *s.* ~**linse**; ~**schwierigkeiten** *Pl.* problems in mixing with others

Kontamination [kɔntamina'tsi̯o:n] **die**; ~, ~**en** *(fachspr.)* contamination

kontaminieren *tr. V. (fachspr.)* contaminate

Konter ['kɔntɐ] **der**; ~**s**, ~ **a)** *(Boxen)* counter; **b)** *(Ballspiele, fig.)* counter-attack

Konterfei ['kɔntɐfai̯] **das**; ~**s**, ~**s** *od.* ~**e** *(veralt., scherzh.)* likeness

kontern *tr., itr. V. (Boxen)* counter; *(Ballspiele)* counter-attack; *(fig.)* counter (**mit** with)

Konter-: ~**revolution die** counter-revolution; ~**schlag der** *s.* **Konter**

Kontext [kɔn'tɛkst] **der**; ~**[e]s**, ~**e** context

Kontinent [kɔnti'nɛnt] **der**; ~**[e]s**, ~**e** continent

kontinental [kɔntinɛn'ta:l] *Adj.* continental

Kontinental-: ~**klima das** *(Geogr.)* continental climate; ~**verschiebung die** *(Geol.)* continental drift

Kontingent [kɔntɪŋ'gɛnt] **das**; ~**[e]s**, ~**e** contingent; *(begrenzte Menge)* quota

kontinuierlich [kɔntinu'i:ɐ̯lɪç] **1.** *Adj.* steady; continuous. **2.** *adv.* steadily

Kontinuität [kɔntinui̯'tɛ:t] **die**; ~: continuity

Konto ['kɔnto] **das**; ~**s**, **Konten** *od.* **Konti** account; **ein laufendes** ~: a current account; **etw. geht auf jmds.** ~ *(ugs.: jmd. ist schuld an etw.)* sb. is to blame *or* is responsible for sth.

Konto-: ~**auszug der** *(Bankw.)* [bank] statement; statement of account; ~**führungsgebühr die** *(Bankw.)* bank charges *pl.*; ~**nummer die** account number

Kontor [kɔn'to:ɐ̯] **das**; ~**s**, ~**e** *(veralt.)* office; *s. auch* **Schlag a**

Konto·stand der *(Bankw.)* balance; state of an/one's account

kontra ['kɔntra] **1.** *Präp. mit Akk. (Rechtsspr., auch fig.)* versus. **2.** *Adv.* against

Kontra das; ~**s**, ~**s** *(Kartenspiele)* double; ~ **sagen** *od.* **geben** double; **jmdm.** ~ **geben** *(fig. ugs.)* flatly contradict sb.

Kontra·baß der double-bass

Kontrahent [kɔntra'hɛnt] **der**; ~**en**, ~**en** adversary; opponent

Kontrakt [kɔn'trakt] **der**; ~**[e]s**, ~**e** contract

Kontra·punkt der *(Musik, fig.)* counterpoint

konträr [kɔn'trɛ:ɐ̯] contrary; opposite

Kontrast [kɔn'trast] **der**; ~**[e]s**, ~**e** contrast; **etw. steht im/in** ~ **zu etw. anderem** sth. is in contrast with sth. else

kontrastieren *tr., itr. V.* contrast

Kontrast·mittel das *(Med.)* contrast medium

kontrast·reich *Adj.* rich in *or* full of contrasts *pred.*
Kontroll·abschnitt der stub
Kontrollampe die pilot-light; indicator light; *(Warnleuchte)* warning light
Kontrolle [kɔn'trɔlə] **die; ~, ~n a)** *(Überwachung)* surveillance; **unter ~ stehen** be under surveillance; **b)** *(Überprüfung)* check; *(bei Waren)* check; inspection; *(bei Lebensmitteln)* inspection; **jmdn./etw. einer ~ unterziehen** check sb./sth.; **in eine ~ kommen** be stopped at a police check; **zur ~:** as a check; **c)** *(Herrschaft)* control; **die ~ über etw./sich** *(Akk.)* **verlieren** lose control of sth./oneself; **außer ~ geraten** get out of control; **etw. unter ~** *(Akk.)* **bringen/halten** get *or* bring/keep sth. under control
Kontrolleur [kɔntrɔ'lø:ɐ̯] **der; ~s, ~e** inspector
Kontroll·gang der tour of inspection; *(eines Nachtwächters)* round; *(eines Polizisten)* patrol
kontrollieren 1. *tr. V.* **a)** *(überwachen)* check; monitor; **die Regierung ~:** scrutinize the actions of the government; **die Lebensmittelproduktion wird streng kontrolliert** strict checks are kept *or* made on the production of food; **b)** *(überprüfen)* check; check, inspect ‹*goods*›; inspect ‹*food*›; **jmdn./etw. auf etw.** *(Akk.)* **[hin] ~:** check sb./check *or* inspect sth. for sth.; **c)** *(beherrschen)* control. **2.** *itr. V.* carry out a check/checks
Kontroll-: ~punkt der checkpoint; *(bei einer Rallye)* control [point]; **~turm der** control tower; **~uhr die** time clock; *(für Wächter)* tell-tale clock
kontrovers [kɔntro'vɛrs] *Adj.* conflicting; *(strittig)* controversial
Kontroverse die; ~, ~n controversy **(um, über +** *Akk.* about)
Kontur [kɔn'tu:ɐ̯] **die; ~, ~en;** *meist Pl.* contour; outline; **~ gewinnen/an ~ verlieren** *(fig.)* become clearer/fade
Konvention [kɔnvɛn'tsi̯o:n] **die; ~, ~en** convention
Konventional·strafe die *(Rechtsw.)* liquidated damages *pl.*
konventionell 1. *Adj.* **a)** conventional; **b)** *(förmlich)* formal. **2.** *adv.* **a)** conventionally; in a conventional way; **b)** *(förmlich)* formally; **hier geht es sehr ~ zu** things are very formal here
Konversation [kɔnvɛrza'tsi̯o:n] **die; ~, ~en** conversation; **~ machen** make conversation
Konversations·lexikon das encyclopaedia
Konversion die; ~, ~en *(fachspr.)* conversion
konvertierbar *Adj.* *(Wirtsch.)* convertible ‹*currency*›
konvertieren *itr. V.; auch mit sein (Rel.)* be converted
konvex [kɔn'vɛks] *(Optik)* **1.** *Adj.* convex. **2.** *adv.* convexly
Konvoi [kɔn'vɔy] **der; ~s, ~s** *(bes. Milit.)* convoy; **im ~ fahren** travel in convoy
Konzentrat [kɔntsɛn'tra:t] **das; ~[e]s, ~e** concentrate
Konzentration [kɔntsɛntra'tsi̯o:n] **die; ~, ~en** concentration
Konzentrations-: ~fähigkeit die; *o. Pl.* ability to concentrate; powers *pl.* of concentration; **~lager das** *(bes. ns.)* concentration camp; **~schwäche die** poor powers *pl.* of concentration
konzentrieren 1. *refl. V.* **a)** concentrate; **sich auf etw.** *(Akk.)* **~:** concentrate on sth.; **b)** *(gerichtet sein)* be concentrated. **2.** *tr. V.* concentrate
konzentriert 1. *Adj.* concentrated. **2.** *adv.* with concentration; **sehr ~ arbeiten** work with great concentration
Konzept [kɔn'tsɛpt] **das; ~[e]s, ~e a)** *(Rohfassung)* [rough] draft; **b) aus dem ~ kommen** *od.* **geraten** lose one's thread; **jmdn. aus dem ~ bringen** put sb. off his/her stroke; **c)** *(Programm)* programme; *(Plan)* plan; **jmdm. das ~ verderben** *(ugs.)* ruin sb.'s plans; **jmdm. nicht ins ~ passen** *(ugs.)* not suit sb.'s plans
Konzeption [kɔntsɛp'tsi̯o:n] **die; ~, ~en** central idea; *(Entwurf)* conception
konzeptionslos 1. *Adj.* haphazard. **2.** *adv.* haphazardly; with no clear plan
Konzept·papier das; *o. Pl.* rough paper
Konzern [kɔn'tsɛrn] **der; ~[e]s, ~e** *(Wirtsch.)* group [of companies]
Konzert [kɔn'tsɛrt] **das; ~[e]s, ~e a)** *(Komposition)* concerto; **b)** *(Veranstaltung)* concert; **ins ~ gehen** go to a concert
Konzert-: ~abend der concert evening; **~agentur die** concert artists' agency; **~flügel der** concert grand; **~meister der, ~meisterin die** leader [of a/the orchestra]; concert-master; **~pianist der, ~pianistin die;** concert pianist; **~saal der** concert-hall
Konzession [kɔntsɛ'si̯o:n] **die; ~, ~en a)** *(Amtsspr.)* licence; **b)** *(Zugeständnis)* concession; **~en [an jmdn./etw.] machen** make concessions [to sb./sth.]
konzessions·bereit *Adj.* ready *or* willing *or* prepared to make concessions *pred.*
konzessiv [kɔntsɛ'si:f] *Adj.* *(Sprachw.)* concessive
Konzil [kɔn'tsi:l] **das; ~s, ~e** *od.* **~ien** *(kath. Kirche)* council
konzipieren [kɔntsi'pi:rən] *tr. V.* draft ‹*speech, essay*›; draw up, draft ‹*plan, policy, etc.*›; design ‹*device, car, etc.*›
Kooperation die; ~, ~en co-operation *no indef. art.*
kooperations·bereit *Adj.* ready *or* willing *or* prepared to co-operate *pred.*
kooperativ 1. *Adj.* co-operative. **2.** *adv.* co-operatively
kooperieren *tr. V.* co-operate
Koordinate die; ~, ~n coordinate
Koordinaten-: ~achse die coordinate axis; **~kreuz das** coordinate axes *pl.*; **~system das** system of coordinates
Koordination die; ~, ~en coordination
koordinieren *tr. V.* co-ordinate
Koordinierung die; ~, ~en coordination
Kopf [kɔpf] **der; ~[e]s, Köpfe** ['kœpfə] **a)** head; **jmdm. den ~ waschen** wash sb.'s hair; *(fig. ugs.: jmdn. zurechtweisen)* give sb. a good talking-to *(coll.)*; give sb. what for *(sl.)*; **[um] einen ganzen/halben ~ größer sein** be a good head/a few inches taller; **die Köpfe zusammenstecken** go into a huddle; **sie haben sich die Köpfe heiß geredet** the conversation/debate became heated; **~ an ~** *(im Wettlauf)* neck and neck; **den ~ einziehen** duck; *(fig.: sich einschüchtern lassen)* be intimidated; **und wenn du dich auf den ~ stellst** you can talk until you're blue in the face; **ich werde/er wird dir nicht gleich den ~ abreißen** *(ugs.)* I'm/he's not going to bite your head off; **jmdm. schwirrt/raucht der ~:** sb.'s head is spinning; **nicht wissen, wo einem der ~ steht** not know whether one is coming or going; **einen dicken ~ haben** *(vom Alkohol)* have a thick head *(coll.)* *or* a hangover; **jmdm.** *od.* **jmdn. den ~ kosten** cost sb. dearly; *(jmdn. das Leben kosten)* cost sb. his/her life; **~ hoch!** chin up!; **den ~ hängen lassen** become disheartened; **~ und Kragen riskieren** risk one's neck; **den ~ hinhalten [müssen]** *(ugs.)* [have to]

face the music; [have to] take the blame *or (sl.)* rap; **den ~ aus der Schlinge ziehen** avoid any adverse consequences *or (sl.)* the rap; **den ~ in den Sand stecken** bury one's head in the sand; **den ~ hoch tragen** hold one's head high; **jmdm. den ~ zurechtrücken** *(ugs.)* bring sb. to his/her senses; **sich [gegenseitig] die Köpfe einschlagen** be at each other's throats; **sich** *(Dat.)* **an den ~ fassen** *od.* **greifen** *(ugs.)* throw up one's hands in despair; **jmdm. Beleidigungen an den ~ werfen** hurl insults at sb.; **sein Geld auf den ~ hauen** *(ugs.)* blow one's money *(sl.)*; **etw. auf den ~ stellen** *(ugs.)* turn sth. upside down; **auf dem ~ stehen** *(ugs.)* be upside down; **den Ablauf der Ereignisse auf den ~ stellen** get the order of events completely *or* entirely wrong; **jmdm. auf dem ~ herumtanzen** *(ugs.)* treat sb. just as one likes; do what one likes with sb.; **jmdm. auf den ~ spucken können** *(salopp scherzh.)* be head and shoulders taller than sb.; **er ist nicht auf den ~ gefallen** *(ugs.)* there are no flies on him *(coll.)*; **jmdm. etw. auf den ~ zusagen** say sth. to sb.'s face; **jmdm. in den** *od.* **zu ~ steigen** go to sb.'s head; **mit dem ~ durch die Wand wollen** *(ugs.)*/**sich** *(Dat.)* **den ~ einrennen** beat *or* run one's head against a brick wall; **etw. über jmds. ~ [hin]weg entscheiden/über jmds. ~ hinwegreden** decide sth./talk over sb.'s head; **jmdm. über den ~ wachsen** *(ugs.)* outgrow sb.; *(jmdn. überfordern)* become too much for sb.; **bis über den ~ in etw. stecken** *(ugs.)* be up to one's ears in sth.; **es geht um ~ und Kragen** *(ugs.)* it's a matter of life and death; **sich um ~ und Kragen reden** *(ugs.)* risk one's neck with careless talk; **von ~ bis Fuß** from head to toe *or* foot; **jmdn. vor den ~ stoßen** *(ugs.)* offend sb.; *s. auch* **Hand c; b)** *(Person)* person; **ein kluger/fähiger ~ sein** be a clever/able man/woman; **pro ~:** per head *or* person; **c)** *(geistige Leitung)* **er ist der ~ der Firma** he's the brains of the firm; **die führenden Köpfe der Wirtschaft** the leading minds in the field of economics; **d)** *(Wille)* **seinen ~ durchsetzen** make sb. do what one wants; **muß es immer nach deinem ~ gehen?** why must 'you always decide?; **e)** *(Verstand)* mind; head; **er hat die Zahlen im ~** *(ugs.)* he has the figures in his head; **er hat nur Autos im ~** *(ugs.)* all he ever thinks about is cars; **was wohl in ihrem ~ vorgeht?** what's going on in her mind?; **sie ist nicht ganz richtig im ~** *(ugs.)* she's not quite right in the head; **einen klaren/kühlen ~ bewahren** *od.* **behalten** keep a cool head; keep one's head; **ich habe den ~ voll mit anderen Dingen** I've got a lot of other things on my mind; **den ~ verlieren** lose one's head; **jmdm. den ~ verdrehen** *(ugs.)* steal sb.'s heart [away]; **sich** *(Dat.)* **den ~ zerbrechen** *(ugs.)* rack one's brains (**über** + *Akk.* over); *(sich Sorgen machen)* worry (**über** + *Akk.* about); **aus dem ~** *(aus dem Gedächtnis)* off the top of one's head; **das geht** *od.* **will ihm nicht aus dem ~:** he can't get it out of his mind; **sich** *(Dat.)* **etw. aus dem ~ schlagen** put sth. out of one's head; **sich** *(Dat.)* **etw. durch den ~ gehen lassen** think sth. over; **jmdm. im ~ herumgehen** *(ugs.)* go round and round in sb.'s mind; **jmdm./sich etw. in den ~ setzen** put sth. into sb.'s head/get sth. into one's head; **etw. im ~ [aus]rechnen** work sth. out in one's head; **was man nicht im ~ hat, muß man in den Beinen haben** a short memory makes work for the legs; **jmdm. geht** *od.* **will etw. nicht in den ~ [hinein]** *(ugs.)* sb. can't get sth. into his/her head; **f)** *(von Nadeln, Nägeln, Blumen)* head; *(von Pfeifen)* bowl; **g) ein ~ Salat/Blumenkohl/Rotkohl** a lettuce/cauliflower/red cabbage; **h)** *(oberer Teil)* head; **i)** *(auf Münzen)* **~ [oder Zahl?]** heads [or tails?]

kopf-, Kopf-: ~-an-Kopf-Rennen das *(Sport, auch fig.)* neck-and-neck race (*Gen.* between); **~arbeit die** brain-work; intellectual work; **~bahnhof der** terminal station; **~ball der** *(Fußball)* header; **~ball·tor das** *(Fußball)* headed goal; **~bedeckung die** head-covering; **ohne ~bedeckung** without anything on one's head; without a hat

Köpfchen ['kœpfçən] **das; ~s, ~ a)** little head; **b)** *(Findigkeit)* brains *pl.*; **~ muß man haben** you've got to have it up here *(coll.)*; **~, ~!** clever, eh? *(coll.)*

köpfen ['kœpfn̩] *tr. V.* **a)** decapitate; *(hinrichten)* behead; *(fig.)* break *or* crack open ⟨*bottle*⟩; slice the top off ⟨*egg*⟩; **b)** *(Fußball)* head; **das 2:0 ~:** head [in] the goal to make it 2–0

kopf-, Kopf-: ~ende das head end; **~form die** head shape; shape of the head; **~geld das** reward; bounty; **~haut die** [skin of the] scalp; **~höhe die: in ~höhe** at head height; **~hörer der** headphones *pl.*

-köpfig *Adj.* **a)** -headed; **b) eine dreiköpfige/fünfköpfige Familie** a family of three/five

kopf-, Kopf-: ~jäger der headhunter; **~kissen das** pillow; **~kissen·bezug der** pillowcase; **~länge die: mit einer ~länge Vorsprung** by a head; **~lastig** *Adj.* down by the head *pred.*; *(fig.)* top-heavy; **~los 1.** *Adj.* **a)** rash; *(in Panik)* panic-stricken; **b)** *(ohne Kopf)* headless; **2.** *adv.* rashly; **~los davonrennen/umherrennen** flee in panic/run round in a panic; **~nicken das; ~s** nod [of the head]; **~nuß die** *(ugs.)* rap on the head with one's *or* the knuckles; **~rechnen das** mental arithmetic; **~salat der** cabbage *or* head lettuce; **~schmerz der;** *meist Pl.* headache; **~schmerzen haben** have a headache *sing.*; **sich** *(Dat.)* **über etw.** *(Akk.) od.* **wegen etw. keine ~schmerzen machen** *(ugs.)* not worry about *or* concern oneself about sth.; **etw. bereitet** *od.* **macht jmdm. ~schmerzen** *(ugs.)* sth. weighs on sb.'s mind; **~schmuck der** head-dress; **~schuppen** *Pl.* dandruff *sing.*; **~schuß der** bullet wound in the head; **durch einen ~schuß getötet werden** be killed by a bullet in the head; **~schütteln das; ~s** shake of the head; **ein allgemeines ~schütteln auslösen** cause everyone to shake their heads; **~schutz der** *(Sport)* protective headgear; **~sprung der** header; **einen ~sprung machen** dive head first; **~stand der** headstand; **~|stehen** *unr. itr. V.* **a)** stand on one's head; **b)** *(ugs.: überrascht sein)* be bowled over; **~steinpflaster das** cobblestones *pl.*; **~stütze die** head-rest; **~tuch das** headscarf; **~über** [-'--] *Adv.* head first; *(fig.: ohne Zögern)* headlong; **~verband der** head bandage; **~verletzung die** head injury; **~wäsche die** hair-wash; shampoo; **~weh das;** *o. Pl.* *(ugs.)* headache; **~weh haben** have a headache; **~zerbrechen das; ~s: etw. bereitet** *od.* **macht jmdm. ~zerbrechen** sb. has to rack his/her brains about sth.; *(etw. macht jmdm. Sorgen)* sth. is a worry to sb.; **sich** *(Dat.)* **über etw.** *(Akk.)* **[kein] ~zerbrechen machen** [not] worry about sth.

Kopie [ko'pi:] **die; ~, ~n a)** copy; *(Imitation)* imitation; **b)** *(Durchschrift)* carbon copy; **c)** *(Fotokopie)* photocopy; **d)** *(Fot., Film)* print

kopieren *tr. V.* **a)** copy; *(imitieren)* imitate; **b)** *(fotokopieren)* photocopy; **c)** *(Fot., Film)* print

Kopierer der; ~s, ~ *(ugs.)* [photo]copier

Kopier-: ~gerät das photocopier; photocopying machine; **~stift der** indelible pencil

Kopilot der; ~en, ~en, Kopilotin die; ~, ~nen *(Flugw.)* copilot; *(Motorsport)* co-driver

Koppel die; ~, ~n a) *(Weide)* paddock; **auf** *od.* **in der ~:** in the paddock; **b)** *(Hunde~)* pack

koppeln *tr. V.* **a)** dock ⟨*spacecraft*⟩; couple [up] ⟨*railway carriage, trailers, etc.*⟩ **(an** + *Akk.* to); **b)** *(verbinden)* link; couple ⟨*circuits, systems, etc.*⟩; **etw. an etw.** *(Akk.)* **~:** link sth. to sth.; **mit etw. gekoppelt sein** be associated with sth.

Kopplung die; ~, ~en; *s.* **koppeln:** docking; coupling [up]; linking; coupling

Koproduktion die; ~, ~en coproduction; joint production

Koproduzent der; ~en, ~en coproducer

Kopulation [kopula'tsio:n] **die; ~, ~en** copulation

kopulieren 1. *itr. V.* copulate

kor [ko:ɐ̯] *s.* **küren; kiesen**

Koralle [ko'ralə] **die; ~, ~n** coral

korallen-, Korallen-: ~bank die; *Pl.* **~bänke** coral reef; **~fischer der** coral fisherman; **~insel die** coral island; **~riff das** coral reef; **~rot** *Adj.* coral-red

Koran [ko'ra:n] **der; ~s, ~e** Koran

Korb [kɔrp] **der; ~es, Körbe** ['kœrbə] **a)** basket; *(Last~ auf einem Tier)* pannier; *(Bienen~)* hive; *(Förder~)* cage; **ein ~ Kartoffeln** a basket[ful] of potatoes; **b)** *(Gondel)* basket; **c)** *(Korbball)* net; *(Basketball)* basket; *(Treffer)* goal; **d)** *o. Pl.* *(Flechtwerk)* wicker[work]; **e) jmdm. einen ~ geben** turn sb. down; **einen ~ bekommen** be turned down

Korb·ball der; *o. Pl.* netball

Körbchen ['kœrpçən] **das; ~s, ~ a)** [little] basket; **husch, husch ins ~** *(fam.)* time for bye-bye[s] *or* beddy-byes *(child lang.)*; **b)** *(des Büstenhalters)* cup

Korb-: ~flasche die wicker bottle; **~macher der** basketmaker; **~möbel das** piece of wicker[work] furniture; **~möbel** *Pl.* wickerwork furniture *sing.*

Kord [kɔrt] **der; ~[e]s a)** corduroy; cord; **b)** *s.* **Kordsamt**

Kordel ['kɔrdl̩] **die; ~, ~n a)** cord; **b)** *(landsch.: Bindfaden)* string

Kord·hose die corduroy *or* cord trousers *pl.*

Kordon [kɔr'dõ, *österr.:* -'do:n] **der; ~s, ~s** *od. österr.:* **~e** cordon

Kord·samt der cord velvet

Korea [ko're:a] **(das); ~s** Korea

Koreaner [kore'a:nɐ] **der; ~s, ~, Koreanerin die; ~, ~nen** Korean

koreanisch *Adj.* Korean

Koriander [ko'riandɐ] **der; ~s, ~:** coriander

Korinthe [ko'rɪntə] **die; ~, ~n** currant

Kork [kɔrk] **der; ~s, ~e** cork

Kork·eiche die cork-oak

Korken der; ~s, ~: cork

Korken·zieher der corkscrew

Kormoran [kɔrmo'ra:n] **der; ~s, ~e** cormorant

[1]Korn [kɔrn] **das; ~[e]s, Körner** ['kœrnɐ] **a)** *(Frucht)* seed; grain; *(Getreide~)* grain [of corn]; *(Pfeffer~)* corn; **b)** *o. Pl.* *(Getreide)* corn; grain; **das ~ steht gut** the grain harvest looks promising; **c)** *(Salz~, Sand~)* grain; *(Hagel~)* stone; **d)** *Pl.* **~e** *(an Handfeuerwaffen)* front sight; foresight; **etw./jmdn. aufs ~ nehmen** take aim at *or* draw a bead on sth./sb.; *(fig. ugs.)* attack sth./start to keep close tabs on sb. *(coll.)*

[2]Korn der; ~[e]s, ~ *(ugs.)* corn schnapps; corn liquor *(Amer.)*

korn-, Korn-: ~ähre die ear of corn; **~blume die** cornflower; **~blumen·blau** *Adj.* cornflower [blue]

Körnchen ['kœrnçən] **das; ~s, ~** *(Frucht)* tiny seed *or* grain; *(von Sand usw.)* [tiny] grain; granule; **ein ~ Wahrheit** *(fig.)* a grain of truth

körnen ['kœrnən] *tr. V.* **a)** granulate; **gekörnte Brühe** stock granules *pl. (for soup)*; **b)** *(Handw.: markieren)* punch

Körner *s.* **Korn**

Korn·feld das cornfield

körnig ['kœrnɪç] *Adj.* granular

Korn·kammer die granary

Korona [ko'ro:na] **die; ~, Koronen a)** *(Astron.)* corona; **b)** *(fig.)* crowd *(coll.)*

Körper ['kœrpɐ] **der; ~s, ~ a)** body; **~ und Geist** body and mind; **am ganzen ~ frieren/zittern** be [freezing] cold/shake all over; **b)** *(Rumpf)* trunk; body; **c)** *(Physik, Chemie)* body; **d)** *(Geom.)* solid body; solid

körper-, Körper-: ~bau der; *o. Pl.* physique; **~behaarung die** body hair *no indef. art.*; **~beherrschung die** body control; **~behindert** *Adj.* physically handicapped *or* disabled; **~behinderte der/die** physically handicapped *or* disabled person; **~fülle die** corpulence; **~geruch der** body odour; BO *(coll.)*; **~gewicht das** body weight; **~größe die** height; **~haltung die** posture; **~hygiene die** *s.* **~pflege**; **~kontakt der** *(Psych.)* physical contact; **~kraft die** physical strength

körperlich 1. *Adj.* physical. **2.** *adv.* physically; **~ [hart] arbeiten** do [hard] physical work

Körper-: ~maße *Pl.* measurements; **~pflege die** body care *no art.*; *(Reinigung)* personal hygiene

Körperschaft die; ~, ~en *(Rechtsw.)* corporation; corporate body; **~ des öffentlichen Rechts** public corporation

Körper-: ~spray der *od.* **das** aerosol deodorant; deodorant spray; **~teil der** part of the/one's body; **~temperatur die** body temperature; **~verletzung die** *(Rechtsw.)* bodily harm *no indef. art.*; **schwere/leichte ~verletzung** grievous/actual bodily harm; **~wärme die** body heat

Korpora *s.* **[2]Korpus**

Korps [ko:ɐ̯] **das; ~** [ko:ɐ̯(s)], **~** [ko:ɐ̯s] *(Milit.)* corps

korpulent [kɔrpu'lɛnt] *Adj.* corpulent

Korpulenz [kɔrpu'lɛnts] **die; ~:** corpulence

[1]Korpus ['kɔrpʊs] **der; ~, ~se** *(usg. scherzh.)* body

[2]Korpus das; ~, Korpora ['kɔrpora] **a)** *(Sprachw.)* corpus; **b)** *o. Pl.* *(Musik)* body

korrekt [kɔ'rɛkt] **1.** *Adj.* correct; **es wäre ~ gewesen, ...:** the correct thing would have been .... **2.** *adv.* correctly

korrekter·weise *Adv.* to be [strictly] correct

Korrektheit die; ~: correctness

Korrektor [kɔ'rɛktɔr] **der; ~s, ~en** [-'to:rən] proof-reader

Korrektur [kɔrɛk'tu:ɐ̯] **die; ~, ~en a)** correction; *(von Ansichten usw.)* revision; **b)** *(Druckw.)* proof-reading; *(Verbesserung)* proof-correction; **~ lesen** read/correct the proofs

Korrektur-: ~abzug der, ~fahne die galley [proof]; **~zeichen das** proof-correction mark

Korrespondent [kɔrɛspɔn'dɛnt] **der; ~en, ~en, Korrespondenten-**

tin **die; ~, ~nen a)** *(Zeitungsw.)* correspondent; **b)** *(Wirtsch.)* correspondence clerk

Korrespondenz [kɔrɛspɔn'dɛnts] **die; ~, ~en** correspondence; **die ~ erledigen** deal with the correspondence; **in ~ mit jmdm. stehen** correspond with sb.

korrespondieren *itr. V.* **a)** correspond **(mit** with); **b)** *(fig. geh.)* correspond **(mit** to, with)

Korridor ['kɔrido:ɐ̯] **der; ~s, ~e** corridor

korrigierbar *Adj.* correctable

korrigieren [kɔri'gi:rən] *tr. V.* correct; revise ⟨*opinion, view*⟩

Korrosion [kɔro'zi̯o:n] **die; ~, ~en** *(fachspr.)* corrosion

korrumpieren [kɔrʊm'pi:rən] *tr. V.* corrupt

korrupt [kɔ'rʊpt] *Adj.* corrupt

Korruption [kɔrʊp'tsi̯o:n] **die; ~, ~en** corruption

Korsett [kɔr'zɛt] **das; ~s, ~s** *od.* **~e** corset; *(fig.)* strait-jacket

Korso ['kɔrzo] **der; ~s, ~s** procession

Kortison [kɔrti'zo:n] **das; ~s** *(Med.)* cortisone

Koryphäe [kory'fɛ:ə] **die; ~, ~n** eminent authority; distinguished expert

koscher ['ko:ʃɐ] *Adj. (auch fig. ugs.)* kosher

K.-o.-Schlag der *(Boxen)* knock-out punch

Kose·form die familiar form

kosen ['ko:zn̩] *(dichter. veralt.)* **1.** *tr. V.* caress. **2.** *itr. V.* **mit jmdm. ~:** caress sb.

Kose·name der pet name

K.-o.-Sieg der *(Boxen)* knock-out victory; victory by a knock-out

Kosinus ['ko:zinʊs] **der; ~, ~** *od.* **~se** *(Math.)* cosine

Kosmetik [kɔs'me:tɪk] **die; ~ a)** beauty culture *no art.;* **b)** *(fig.)* cosmetic procedures *pl.*

Kosmetika *s.* **Kosmetikum**

Kosmetiker der; ~s, ~, Kosmetikerin die; ~, ~nen cosmetician; beautician

Kosmetik·salon der beauty salon

Kosmetikum [kɔs'me:tikʊm] **das; ~s, Kosmetika** cosmetic

kosmetisch *Adj. (auch fig.)* cosmetic

kosmisch ['kɔsmɪʃ] *Adj.* cosmic

Kosmologie [kɔsmolo'gi:] **die; ~, ~n** cosmology

Kosmonaut [kɔsmo'naut] **der; ~en, ~en, Kosmonautin die; ~, ~nen** cosmonaut

Kosmopolit [kɔsmopo'li:t] **der; ~en, ~en** *(geh.)* cosmopolitan

kosmopolitisch *Adj.* cosmopolitan

Kosmos ['kɔsmɔs] **der; ~** cosmos

Kost [kɔst] **die; ~** food; **geistige ~** *(fig.)* intellectual nourishment; **leichte/schwere ~** *(fig.)* easy/heavy going; **~ und Logis** board and lodging

kostbar 1. *Adj.* valuable; precious ⟨*time*⟩. **2.** *adv.* expensively ⟨*dressed*⟩; luxuriously ⟨*decorated*⟩

Kostbarkeit die; ~, ~en a) *(Sache)* treasure; precious object; **b)** *o. Pl. (Eigenschaft)* value

[1]kosten 1. *tr. V.* **a)** *(probieren)* taste; try; sample; **b)** *(geh.: empfinden)* taste; *(fig. iron.)* have a taste of. **2.** *itr. V. (probieren)* have a taste; **von etw. ~:** have a taste of *or* taste sth.

[2]kosten *tr. V.* **a)** cost; **wieviel kostet .../was kostet ...?** how much/what does ... cost?; how much is ...?; **koste es** *od.* **es koste, was es wolle** whatever the cost; **sich etw. etwas ~ lassen** *(ugs.)* spend a fair bit of money on sth.; **b)** *(erfordern)* take; cost ⟨*lives*⟩; **viel Arbeit ~:** take a great deal of work; **c)** *(Verlust nach sich ziehen)* **jmdn.** *od.* **jmdm. etw. ~:** cost sb. sth.

Kosten *Pl.* cost *sing.;* costs; *(Auslagen)* expenses; *(Rechtsw.)* costs; **die ~ tragen, für die ~ aufkommen** bear the cost[s]; **auf seine ~ kommen** cover one's costs; *(fig.)* get one's money's worth; **auf jmds. ~:** at sb.'s expense; **auf ~ einer Sache** *(Gen.)* at the expense of sth.

kosten-, Kosten-: ~aufwand der expense; cost; **mit einem ~aufwand von ...:** at a cost of ...; **~deckend** *Adj.* that covers/cover [one's] costs *postpos., not pred.;* **~ersparnis die** cost saving; **~erstattung die** reimbursement of costs; **~frage die** question of cost; **~los 1.** *Adj.* free; **2.** *adv.* free of charge; **~pflichtig 1.** *Adj.* with costs *postpos.;* **2.** *adv.* **eine Klage ~pflichtig abweisen** dismiss a case with costs; **ein Auto ~pflichtig abschleppen** tow a car away at the owner's expense; **~punkt der: ~punkt?** *(ugs.)* how much is it/are they?; **~sparend** *Adj. (Wirtsch.)* cost-saving; **~voranschlag der** estimate

köstlich ['kœstlɪç] **1.** *Adj.* **a)** delicious; **b)** *(unterhaltsam)* delightful. **2.** *adv.* **a)** ⟨*taste*⟩ delicious; **b) sich ~ amüsieren/unterhalten** enjoy oneself enormously *(coll.)*

Köstlichkeit die; ~, ~en a) *(Sache)* delicacy; **eine literarische ~:** a literary gem; **b)** *o. Pl. (geh.: Eigenschaft)* deliciousness

Kost·probe die; ~, ~n taste; *(fig.)* sample

kost·spielig [-ʃpi:lɪç] **1.** *Adj.* expensive; costly. **2.** *adv.* expensively

Kostüm [kɔs'ty:m] **das; ~s, ~e a)** costume; **b)** *(Mode)* suit

Kostüm-: ~ball der fancy-dress ball; **~bildner der, ~bildnerin die** *(Theater, Film)* costume-designer

kostümieren *tr. V.* **jmdn./sich ~:** dress sb. up/dress [oneself] up; **wie hatte er sich kostümiert?** what was he dressed [up] as?

Kostüm-: ~probe die *(Theater)* dress rehearsal; **~verleih der** [theatrical] costume agency

Kost·verächter der *in* **kein ~ sein** *(scherzh.)* be fond of one's food; *(fig.)* be one for the ladies

K.-o.-System das *(Sport)* knock-out system

Kot [ko:t] **der; ~[e]s, ~e** excrement

Kotangens ['ko:taŋgɛns] **der; ~, ~** *(Math.)* cotangent

Kotelett [kɔt'lɛt] **das; ~s, ~s** chop; *(vom Nacken)* cutlet

Koteletten *Pl.* side-whiskers

Köter ['kø:tɐ] **der; ~s, ~** *(abwertend)* cur; tyke

Kot·flügel der *(Kfz-W.)* wing

Kotze ['kɔtsə] **die; ~** *(derb)* vomit; puke *(coarse)*

kotzen *itr. V. (derb)* puke *(coarse);* throw up *(coll.);* **das ist/ich finde ihn zum Kotzen** it/he makes me sick; it/he makes me want to puke *(coarse)*

kotzübel *Adj.* **mir ist ~** *(derb)* I feel as if I'm going to throw up *(coll.) or (coarse)* puke

KP *Abk.* **Kommunistische Partei** CP

Krabbe ['krabə] **die; ~, ~n a)** *(Zool.)* crab; **b)** *(Garnele)* shrimp; *(größer)* prawn

Krabbel·alter das *(ugs.)* crawling stage

krabbeln ['krabl̩n] *itr. V.; mit sein* crawl

Krach [krax] **der; ~[e]s, Kräche** ['krɛçə] **a)** *o. Pl. (Lärm)* noise; row; **~ machen** make a noise *or (coll.)* a row; be noisy; **b)** *(lautes Geräusch)* crash; bang; **c)** *(ugs.: Streit)* row; **mit jmdm. ~ anfangen/kriegen** start/have a row with sb. *(coll.);* **~ machen** *od.* **schlagen** *(ugs.)* kick up *or* make a fuss; **d)** *(ugs.: Börsen~)* crash

krachen 1. *itr. V.* **a)** ⟨*thunder*⟩ crash; ⟨*shot*⟩ ring out; ⟨*floorboard*⟩ creak; **~de Kälte/~der**

Frost *(fig.)* bitter cold/heavy frost; **b)** *mit sein (ugs.: bersten)* ‹*ice*› crack; ‹*bed*› collapse; ‹*trousers, dress, etc.*› split; **c)** *mit sein (ugs.: krachend auftreffen)* crash; **die Tür krachte ins Schloß** the door banged *or* slammed shut; **d)** *(ugs.: Bankrott machen)* crash; **e)** *(unpers.)* **an der Kreuzung kracht es dauernd** there are frequent crashes at that junction; **sonst kracht's!** *(fig. ugs.)* or there'll be trouble. **2.** *refl. V. (ugs.)* row *(coll.)*; have a row *(coll.)*

Kracher der; ~s, ~ *(ugs.: Knallkörper)* banger

krächzen ['krɛçtsn̩] *itr. V.* ‹*raven, crow*› caw; ‹*parrot*› squawk; ‹*person*› croak; *(fig.)* ‹*loudspeaker etc.*› crackle and splutter

Kräcker ['krɛkɐ] **der; ~s, ~:** *s.* **Cracker**

kraft [kraft] *Präp. + Gen. (Amtsspr.)* **~ [meines] Amtes** by virtue of my office; **~ Gesetzes** by law; **~ [des] Gesetzes hat der Richter ihn zum Tode verurteilt** as empowered by the law, the judge sentenced him to death

Kraft die; ~, Kräfte ['krɛftə] **a)** strength; **geistige/schöpferische Kräfte** mental/creative powers; **unter Aufbietung aller Kräfte** applying all one's energies; **jmds. Kräfte übersteigen** be too much for sb.; **wieder bei Kräften sein** have [got] one's strength back; **bei Kräften bleiben** keep one's strength up; **mit letzter ~:** with one's last ounce of strength; **mit frischer ~:** with renewed energy; **aus eigener ~:** by oneself *or* one's own efforts; **ich werde tun, was in meinen Kräften steht** I shall do everything [with]in my power; **mit vereinten Kräften sollte es gelingen** if we join forces *or* combine our efforts we should succeed; **nach [besten] Kräften** to the best of one's ability; **b)** *(Wirksamkeit)* power; **c)** *(Arbeits~)* employee; *(in einer Fabrik)* employee; worker; **Kräfte** employees/workers; personnel *pl.; (Angestellte auch)* staff *pl.*; **d)** *Pl. (Gruppe)* forces; **e)** *(Physik)* force; **die treibende ~** *(fig.)* the driving force; **f)** *(Seemannsspr.)* **volle/halbe ~ voraus!** full/half speed ahead!; **g)** *in* **außer ~ setzen** repeal ‹*law*›; countermand ‹*order*›; **außer ~ sein/treten** no longer be/cease to be in force; **in ~ treten/sein/bleiben** come into/be in/remain in force

Kraft-: ~akt der feat of strength; *(im Zirkus usw.)* strong-man act; **~aufwand der** effort; **~ausdruck der** swear-word; **~brühe die** strong meat broth

Kräfte·verhältnis das *(bes. Politik)* balance of power

Kraft·fahrer der *(bes. Amtsspr.)* driver; motorist

Kraft·fahrzeug das; ~[e]s, ~e *(bes. Amtsspr.)* motor vehicle

Kraftfahrzeug-: ~brief der vehicle registration document; log-book *(Brit.)*; **~mechaniker der** motor mechanic; **~schein der** vehicle registration document; **~steuer die** vehicle *or* road tax

Kraft·feld das *(Physik)* force field

kräftig ['krɛftɪç] **1.** *Adj.* **a)** strong ‹*person*›; strong, powerful ‹*arms, voice*›; vigorous ‹*plant, shoot*›; **b)** *(fest)* powerful, hefty, hard ‹*blow, kick, etc.*›; firm ‹*handshake*›; **c)** *(ausgeprägt)* strong ‹*breeze, high-pressure area*›; considerable ‹*increase*›; **einen ~en Schluck nehmen** take a deep drink *or (coll.)* good swig; **eine ~e Tracht [Prügel]** a good hiding *(coll.)*; a sound beating; **d)** *(intensiv)* strong, powerful ‹*smell, taste, etc.*›; bold ‹*pattern*›; strong ‹*colour*›; **e)** *(gehaltvoll)* nourishing ‹*soup, bread, meal, etc.*›; **etw. Kräftiges essen** eat a good nourishing meal; **f)** *(grob)* strong ‹*language*›; coarse ‹*expression, oath, etc.*›. **2.** *adv.* **a)** strongly, powerfully ‹*built*›; ‹*hit, kick, press, push*› hard; ‹*sneeze*› loudly; **b)** *(tüchtig)* ‹*rain, snow*› heavily; ‹*eat*› heartily; ‹*sing*› lustily; **etw. ~ schütteln** shake sth. vigorously; give sth. a good shake; **die Preise sind ~ gestiegen** prices have risen steeply; **dem Alkohol ~ zusprechen** hit the bottle in a big way *(coll.)*; **jmdm. ~ die Meinung sagen** give sb. a piece of one's mind

kräftigen *tr. V.* ‹*holiday, air, etc.*› invigorate; ‹*food etc.*› fortify; **sich ~:** build up one's strength

Kräftigung die; ~, ~en strengthening

Kräftigungs·mittel das tonic

kraft-, Kraft-: ~los *Adj.* weak; feeble; *(fig.)* weak ‹*sun*›; **~losigkeit die; ~:** weakness; feebleness; **~meier der; ~s, ~** *(ugs.: abwertend)* muscleman; **~meierei die; ~** *(ugs. abwertend)* playing the muscleman; **~probe die** trial of strength; **~protz der** *(abwertend) s.* **~meier; ~rad das** *(Amtsspr.)* motorcycle; **~reserven** *Pl.* reserves of strength; **~stoff der** *(Kfz-W.)* fuel; **~stoff·verbrauch der** fuel consumption; **~strotzend** *Adj.* vigorous; bursting with vigour *postpos.*; **~voll 1.** *Adj.* powerful; **2.** *adv.* powerfully; **~wagen der** motor vehicle; **~werk das** power station

Kragen ['kra:gn̩] **der; ~s, ~,** *südd., österr. u. schweiz. auch:* **Krägen** ['krɛ:gn̩] **a)** collar; **jmdn. am** *od.* **beim ~ packen** *od.* **nehmen** *(ugs.)* collar sb.; **b)** *(fig.)* **ihm platzte der ~** *(salopp)* he blew his top *(coll.)*; **jetzt platzt mir aber der ~!** *(salopp)* that's the last straw!; **es geht ihm an den ~** *(ugs.)* he's in for it now; **jmdm. an den ~ wollen** *(ugs.)* get at *or* be after sb.; *(jmdn. verantwortlich machen)* try to hang sth. on sb. *(coll.)*

Kragen·weite die collar size; **[nicht] jmds. ~ sein** *(salopp)* [not] be sb.'s cup of tea *(coll.)*

Krähe ['krɛ:ə] **die; ~, ~n** crow; **eine ~ hackt der anderen kein Auge aus** *(Spr.)* dog does not eat dog *(prov.)*

krähen *itr. V. (auch fig.)* crow; *s. auch* **Hahn**

Krähen-: ~füße *Pl. (ugs.) (Hautfalten)* crow's feet; **~nest das** *(auch Seemannsspr.)* crow's nest

Krake ['kra:kə] **der; ~n, ~n a)** *(Tintenfisch)* octopus; **b)** *(Meeresungeheuer)* kraken

krakeelen 1. *itr. V. (ugs. abwertend)* kick up a row *(coll.)*. **2.** *tr. V.* scream

Krakel ['kra:kl̩] **der; ~s, ~** *(ugs. abwertend)* scrawl; scribble

krakeln *tr., itr. V. (ugs. abwertend)* scrawl; scribble

krak[e]lig *Adj. (ugs. abwertend)* scrawly

Kralle ['kralə] **die; ~, ~n** claw

krallen 1. *refl. V.* **sich an etw.** *(Akk.)* **~** ‹*cat*› dig its claws into sth.; ‹*bird*› dig its claws *or* talons into sth.; ‹*person*› clutch sth. [tightly]; **sich in/um etw.** *(Akk.)* **~:** dig into/clutch sth. **2.** *tr. V.* **a)** *(fest greifen)* **die Finger in/um etw.** *(Akk.)* **~:** dig one's fingers into sth./clutch sth. [tightly] with one's fingers; **b)** *(salopp: stehlen)* pinch *(sl.)*; nick *(Brit. sl.)*; **c)** *(salopp: ergreifen)* collar; *(verhaften)* nab *(sl.)*

Kram [kra:m] **der; ~[e]s** *(ugs.)* **a)** stuff; *(Gerümpel)* junk; **den ganzen ~ hinschmeißen** *(fig. ugs.)* chuck the whole thing in *(coll.)*; **b)** *(Angelegenheit)* business; affair; **mach deinen ~ alleine!** do it yourself!; **jmdm. [genau] in den ~ passen** suit sb. [down to the ground *(coll.)*]

kramen 1. *itr. V. (ugs.: herumwühlen)* **in etw.** *(Dat.)* **~**: rummage about in *or* rummage through sth.; **nach etw. ~**: rummage about looking for sth. 2. *tr. V. (ugs.)* **etw. aus etw. ~**: fish *(coll.) or* get sth. out of sth.

Krämer ['krɛ:mɐ] **der; ~s, ~ a)** *(veralt.)* grocer; **b)** *(geiziger Mensch)* skinflint; stingy person; *(engstirniger Mensch)* petty-minded *or* small-minded person

Krämer·seele die *(abwertend) s.* **Krämer b**

Kram·laden der *(ugs. abwertend)* junk shop

Krampe ['krampə] **die; ~, ~n** staple

Krampf [krampf] **der; ~[e]s, Krämpfe** ['krɛmpfə] **a)** cramp; *(Zuckung)* spasm; *(bei Anfällen)* convulsion; **einen ~ bekommen** *od. (ugs.)* **kriegen** get cramp; **b)** *(ugs.) o. Pl. (gequältes Tun)* painful strain; *(sinnloses Tun)* senseless waste of effort

Krampf·ader die varicose vein

krampf·artig 1. *Adj.* convulsive. 2. *adv.* convulsively

krampfen 1. *itr. V.* be affected with cramp; *(bei Anfällen)* be convulsed. 2. *refl. V.* be affected with cramp; *(bei Anfällen)* be convulsed. 3. *tr. V.* **die Fäuste/Finger um/in etw.** *(Akk.)* **~**: clench sth./dig one's hands/fingers into sth.

krampfhaft 1. *Adj.* **a)** convulsive; **b)** *(verbissen)* desperate; forced ⟨*cheerfulness*⟩. 2. *adv.* **a)** convulsively; **b)** *(verbissen)* desperately

Kran [kra:n] **der; ~[e]s, Kräne** ['krɛ:nə] **a)** crane; **b)** *Pl.*: **Kräne** *od.* **~en** *(südwestd.: Wasserhahn)* tap; faucet *(Amer.)*

Kran·führer der crane-operator; *(~fahrer)* crane-driver

Kranich ['kra:nɪç] **der; ~s, ~e** crane

krank [kraŋk]; **kränker** ['krɛŋkɐ], **kränkst...** ['krɛŋkst...] *Adj.* **a)** ill *usu. pred.*; sick; bad ⟨*leg, tooth*⟩; diseased ⟨*plant, organ*⟩; *(fig.)* sick, ailing ⟨*economy, business*⟩; **ein ~es Herz/eine ~e Leber haben** have a bad heart/a liver complaint; **[schwer] ~ werden** be taken *or* fall [seriously *or* very] ill; **er wurde immer kränker** he got steadily worse; **sie liegt ~ zu/im Bett** she is ill in bed; **jmdn. ~ machen** make sb. ill; *(fig.)* get on sb.'s nerves; **vor Heimweh/Liebe ~ sein** be homesick/lovesick; **sich ~ melden** let the office/boss *etc.* know that one is off sick; **jmdn. ~ schreiben** give sb. a medical certificate; **b)** *(Jägerspr.: angeschossen)* wounded

Kranke ['kraŋkə] **der/die;** *adj. Dekl.* sick man/woman; *(Patient)* patient; **die ~n** the sick/the patients

kränkeln ['krɛŋkl̩n] *itr. V.* be in poor health; not be well; *(fig.)* be in poor shape; **er kränkelt leicht** he is always ailing

kranken *itr. V.* **an etw.** *(Dat.)* **~** ⟨*firm, project, etc.*⟩ suffer from sth.

kränken ['krɛŋkn̩] *tr. V.* **jmdn. ~**: hurt *or* wound sb. *or* sb.'s feelings; **jmdn. in seiner Ehre/seinem Stolz/seiner Eitelkeit ~**: wound sb.'s honour/injure *or* wound sb.'s pride/vanity; **~d sein** be hurtful; **tief/schwer gekränkt sein** be deeply hurt

Kranken-: **~besuch der** visit to a sick person; **~geld das** sickness benefit; **~geschichte die** case history; **~gymnastik die** remedial *or* medical gymnastics *sing.*; physiotherapy; **~gymnastin die** remedial gymnast; physiotherapist; **~haus das** hospital; **jmdn. ins ~haus einliefern/aus dem ~haus entlassen** take sb. to hospital/discharge sb. from hospital; **im ~haus liegen** be in hospital; **ins ~haus müssen** have to go [in]to hospital; **~haus·aufenthalt der** stay in hospital; **~kasse die** health insurance institution; *(privat)* health insurance company; **~pflege die** nursing; **~pfleger der** male nurse; **~schein der** health insurance certificate; **~schwester die** nurse; **~versicherung die** health insurance; **~wagen der** ambulance

krank|feiern *itr. V. (ugs.)* skive off work *(sl.)* [pretending to be ill]

kränker *s.* **krank**

krankhaft 1. *Adj.* **a)** pathological ⟨*change etc.*⟩; morbid ⟨*growth, state, swelling, etc.*⟩; **b)** *(abnorm gesteigert)* pathological; pathological, morbid ⟨*fear, obsession*⟩. 2. *adv.* **a)** pathologically; morbidly ⟨*swollen, grown*⟩; **b)** *(abnorm gesteigert)* pathologically; pathologically, morbidly ⟨*obsessed, sensitive*⟩

Krankheit die; ~, ~en a) illness; *(bestimmte Art, von Pflanzen, Organen)* disease; **sich** *(Dat.)* **eine ~ zuziehen** contract *or* catch an illness/a disease; **an einer ~ leiden/sterben** suffer from/die of an illness/a disease; **das ist doch kein Auto, das ist eine ~** *(fig. ugs.)* that's just an apology for a car; **b)** *o. Pl. (Zeit des Krankseins)* illness; **nach langer/schwerer ~**: after a long/serious illness

Krankheits-: **~erreger der** pathogen; disease-causing agent; **~fall der** case of illness; **im ~fall** in the event of illness

kränklich ['krɛŋklɪç] *Adj.* sickly; ailing

Krank·meldung die notification of absence through illness

kränkst... *s.* **krank**

Kränkung die; ~, ~en: eine ~: an injury to one's/sb.'s feelings; **etw. als ~ empfinden** be hurt by sth.; take offence at sth.

Kranz [krants] **der; ~es, Kränze** ['krɛntsə] **a)** wreath; garland; *(auf einem Grab, Sarg, an einem Denkmal)* wreath; **b)** *(Haar~)* chaplet *(of plaited hair)*; **c)** *(Kuchen)* ring cake

Kränzchen ['krɛntsçən] **das; ~s, ~ a)** *(Kaffee~)* coffee circle; coffee klatch *(Amer.)*; **b)** *(kleiner Kranz)* small wreath *or* garland

Kranz-: **~gefäß das** *s.* **Herzkranzgefäß; ~kuchen der** *s.* **Kranz c; ~niederlegung die** laying of a wreath

Krapfen ['krapfn̩] **der; ~s, ~** doughnut

kraß [kras] 1. *Adj.* blatant ⟨*case*⟩; gross, flagrant ⟨*injustice*⟩; rank, complete ⟨*outsider*⟩; glaring, stark ⟨*contrast*⟩; complete ⟨*contradiction*⟩; sharp ⟨*difference*⟩; gross ⟨*discrepancy, imbalance*⟩; out-and-out ⟨*egoist*⟩. 2. *adv.* **sich ~ ausdrücken** put sth. bluntly; **sich von etw. ~ unterscheiden** be in stark contrast to sth.

Krater ['kra:tɐ] **der; ~s, ~**: crater

Kratz·bürste die *(ugs. scherzh.)* stroppy *(Brit. sl.) or* prickly so-and-so

kratzbürstig *Adj. (ugs. scherzh.)* stroppy *(Brit. sl.)*; prickly

Krätze ['krɛtsə] **die; ~**: scabies *sing.*

kratzen ['kratsn̩] 1. *tr. V.* **a)** scratch; **jmdm./sich den Arm blutig ~**: scratch sb.'s/one's arm and make it bleed; **seinen Namen in die Wand ~**: scratch one's name on the wall; **b) etw. aus/von etw. ~**: scrape sth. out of/off sth.; **c)** *(ugs.: stören)* bother; **jmdn. wenig ~**: not bother sb. all that much. 2. *itr. V.* **a)** scratch; **das Kratzen** scratching; **b)** *(jucken)* itch; be scratchy *or* itchy; **c)** *(brennen)* **im Hals ~** ⟨*wine*⟩ taste rough; ⟨*tobacco*⟩ be rough on the throat; ⟨*smoke*⟩ irritate the throat. 3. *refl. V.* scratch [oneself]; **sich hinter**

dem Ohr/am Kopf ~: scratch oneself behind the ear/scratch one's head

Kratzer der; ~s, ~ a) *(ugs.)* scratch; b) *(Schaber)* scraper

kratzig *Adj.* scratchy, itchy ⟨*material, pullover, etc.*⟩; scratchy, rough ⟨*voice*⟩

Kraul [kraul] das; ~s *(Sport)* crawl

[1]**kraulen** 1. *itr. V.; mit Richtungsangabe mit sein* do *or* swim the crawl; **über den See/ans Ufer ~:** swim across the lake/to the bank using the crawl. 2. *tr. V.; auch mit sein* **eine Strecke ~:** cover a distance using the crawl

[2]**kraulen** *tr. V.* **jmdm. das Kinn ~:** tickle sb. under the chin; **jmdn. in den Haaren ~:** run one's fingers through sb.'s hair

kraus [kraus] *Adj.* a) frizzy ⟨*hair, beard*⟩; creased ⟨*skirt etc.*⟩; wavy ⟨*sea*⟩; wrinkled ⟨*brow*⟩; **die Stirn ~ ziehen** wrinkle one's brow; *(unmutig)* frown; b) *(abwertend: verworren)* muddled; confused

Krause die; ~, ~n a) *(Kragen)* ruff; *(am Ärmel)* ruffle; frill; b) *(im Haar)* frizziness

kräuseln ['krɔyzl̩n] 1. *tr. V.* ruffle ⟨*water, surface*⟩; gather ⟨*material etc.*⟩; frizz ⟨*hair*⟩; pucker [up] ⟨*lips*⟩. 2. *refl. V.* ⟨*hair*⟩ go frizzy; ⟨*water*⟩ ripple; ⟨*smoke*⟩ curl up; ⟨*material*⟩ pucker up

krausen 1. *tr. V.* gather ⟨*material etc.*⟩; frizz ⟨*hair*⟩; wrinkle [up] ⟨*forehead, nose*⟩. 2. *itr. V.* ⟨*material, clothes*⟩ crease

kraus·haarig *Adj.* frizzy-haired ⟨*person*⟩; curly-coated ⟨*dog etc.*⟩

Kraus·kopf der frizzy hair; **einen ~ haben** have frizzy hair; be frizzy-haired

Kraut [kraut] das; ~[e]s, Kräuter ['krɔytɐ] a) herb; **dagegen ist kein ~ gewachsen** *(ugs.)* there's nothing anyone can do about it; b) *o. Pl. (Blätter)* foliage; stems and leaves *pl.*; *(von Kartoffeln, Bohnen usw.)* haulm; **ins ~ schießen** put on too much foliage; bolt; *(fig.)* run wild; **wie ~ und Rüben** *(ugs.)* all over the place; in a complete muddle; c) *o. Pl. (bes. südd., österr.: Kohl)* cabbage; d) *(ugs. abwertend: Tabak)* tobacco

Kräuter-: ~**butter die** herb butter; ~**essig der** herb vinegar; ~**likör der** herb liqueur; ~**tee der** herb tea

Kraut·salat der coleslaw

Krawall [kra'val] der; ~s, ~e a) *(Tumult)* riot; b) *o. Pl. (ugs.: Lärm)* row *(coll.)*; racket; **~ machen** kick up *or* make a row *(coll.) or* racket

Krawall·macher der rowdy

Krawatte [kra'vatə] die; ~, ~n tie

Krawatten-: ~**nadel die** tie-pin; ~**zwang der: hier herrscht [kein] ~zwang** you [do not] have to wear a tie here

kraxeln ['kraksl̩n] *itr. V.; mit sein (bes. südd., österr. ugs.)* climb; *(mit Mühe)* clamber; **auf etw. (Akk.) ~:** climb [up] sth.; *(mit Mühe)* clamber up sth.

Kreation [krea'tsio:n] die; ~, ~en *(bes. Mode)* creation

kreativ [krea'ti:f] 1. *Adj.* creative. 2. *adv.* **~ veranlagt sein** have a creative bent

Kreativität [kreativi'tɛ:t] die; ~: creativity

Kreativ·urlaub der [arts and crafts] activity holiday

Kreatur [krea'tu:ɐ̯] die; ~, ~en a) *(Geschöpf)* creature; **Gott schuf alle ~:** God made all creatures *pl.*; b) *(willenloser Mensch)* minion; creature

Krebs [kre:ps] der; ~es, ~e a) crustacean; *(Fluß~)* crayfish; *(Krabbe)* crab; **rot wie ein ~:** as red as a lobster; b) *(Krankheit)* cancer; c) *(Astrol.)* Cancer; the Crab

krebs·artig 1. *Adj.* cancerous. 2. *adv.* cancerously; in the manner of a cancer

krebsen *itr. V. (ugs.: sich abmühen)* **mit etw. zu ~ haben** find sth. a real *or* uphill struggle

krebs-, Krebs-: ~**erregend,** ~**erzeugend** *Adj.* carcinogenic; cancer-producing *usu. attrib.*; ~**forschung die** cancer research; ~**gang der** *o. Pl.* retrogression; **im ~gang gehen** go backwards; ~**geschwulst die** cancerous growth *or* tumour; ~**geschwür das** *(volkst.)* cancerous ulcer; *(fig. geh.)* cancer; ~**krank** *Adj.* cancer *attrib.* ⟨*patient etc.*⟩; **~krank sein** suffer from *or* have cancer; ~**kranke der/die** person suffering from cancer; *(Patient)* cancer patient; ~**rot** *Adj.* as red as a lobster *postpos.*; *(aus Verlegenheit)* as red as a beetroot *postpos.*

Kredit [kre'di:t] der; ~[e]s, ~e a) *(Darlehen)* loan; credit; **jmdm. einen ~ gewähren** *od.* **einräumen** give *or* grant sb. a loan *or* a credit; b) *o. Pl. (Zahlungsaufschub)* credit; **er hat bei uns ~:** his credit is good with us; **auf ~:** on credit; c) *o. Pl. (Kaufmannsspr.: Vertrauenswürdigkeit)* good reputation *or* name

kredit-, Kredit-: ~**abteilung die** credit department; ~**anstalt die** credit institution; ~**geber der** lender; ~**hai der** *(ugs. abwertend)* loan shark *(coll.)*; ~**institut das** credit institution; ~**karte die** credit card; ~**nehmer der** borrower; ~**würdig** *Adj. (Finanzw.)* credit-worthy

Kredo ['kre:do] das; ~s, ~s a) *(kath. Kirche)* creed; credo; b) *(fig. geh.)* credo

Kreide ['kraidə] die; ~, ~n a) *o. Pl. (Kalkstein)* chalk; b) *(zum Schreiben)* chalk; **mit ~ zeichnen/schreiben** draw/write in *or* with chalk; **bei jmdm. [tief] in der ~ stehen** be [deep] in debt to sb.; owe sb. [a lot of] money; c) *o. Pl. (Geol.)* Cretaceous [period]

kreide-, Kreide-: ~**bleich** *Adj.* as white as a sheet *postpos.*; ~**felsen der** chalk cliff; ~**weiß** *Adj.: s.* **~bleich**; ~**zeit die** *s.* **Kreide c**

kreieren [kre'i:rən] *tr. V.* create

Kreis [krais] der; ~es, ~e a) circle; **einen ~ schlagen** *od.* **beschreiben** describe a circle; **einen ~ bilden** *od.* **schließen** form *or* make a circle; **in einem** *od.* **im ~ sitzen** sit in a circle; **sich im ~ drehen** *od.* **bewegen** go *or* turn round in a circle; *(fig.)* go round in circles; **~e ziehen** *(fig.)* ⟨*court case*⟩ have [wide] repercussions; ⟨*movement*⟩ grow in size and influence; b) *(Gruppe)* circle; **im ~e der Freunde/Familie** among *or* with friends/within the family; **im kleinen** *od.* **engsten ~:** with a few close friends [and relatives]; **der ~ seiner Leser/Anhänger** his readers *pl.*/followers *pl.*; **in seinen ~en** in the circles in which he moves/moved; **in weiten** *od.* **breiten ~en der Bevölkerung** amongst wide sections of the population; **die besseren/besten ~e** the best circles; c) *(von Problemen, Lösungen usw.)* range; d) *(Verwaltungsbezirk)* district; *(Wahl~)* ward; **der ~ Heidelberg** the Heidelberg district *or* district of Heidelberg; e) *(Elektrot.)* circuit

Kreis-: ~**bahn die** orbit; ~**bogen der** *(Geom.)* arc [of a/the circle]

kreischen ['kraiʃn̩] *itr. V.* ⟨*person*⟩ screech, shriek; ⟨*bird*⟩ screech; ⟨*brakes*⟩ squeal, screech; ⟨*door*⟩ creak; ⟨*saw*⟩ screech; **mit ~den Bremsen** with a squeal *or* screech of brakes

Kreisel ['kraizl̩] der; ~s, ~ a) *(Technik)* gyroscope; b) *(Kinderspielzeug)* top; c) *(ugs.: Kreisverkehr)* roundabout

kreiseln *itr. V.* a) *auch mit sein*

(sich drehen) spin [round]; gyrate; **b)** *(mit einem Kreisel spielen)* play with a top; spin a top

kreisen *itr. V.* **a)** *auch mit sein* ⟨*planet*⟩ revolve (**um** around); ⟨*satellite etc.*⟩ orbit; ⟨*aircraft, bird*⟩ circle; **die Flasche ~ lassen** *(fig.)* pass the bottle round; **seine Gedanken kreisten immer um dasselbe Thema** *(fig.)* his thoughts always revolved around the same subject; **b)** *(Sport)* **die Arme ~ lassen** swing one's arms round [in a circle]

kreis-, Kreis-: ~förmig *Adj.* circular; **~lauf der a)** *(der Natur, der Wirtschaft, des Lebens usw.)* cycle; *(des Geldes; Technik)* circulation; **b)** *(Physiol.)* circulation; **~lauf·kollaps der** *(Med.)* circulatory collapse; **~lauf·mittel das** circulatory preparation; **~lauf·störungen** *Pl.* circulatory trouble *sing.*; **~rund** *Adj.* [perfectly] circular *or* round; **~säge die** circular saw

Kreiß·saal der *(Med.)* delivery room

Kreis-: ~stadt die chief town of a/the district; **~tag der** district assembly; **~verkehr der** roundabout

Krematorium [krema'to:ri̯ʊm] **das; ~s, Krematorien** crematorium

Kreml ['krɛml̩] **der; ~s** Kremlin

Krempe ['krɛmpə] **die; ~, ~n** brim

Krempel ['krɛmpl̩] **der; ~s** *(ugs. abwertend)* stuff; *(Gerümpel)* junk; **den ganzen ~ hinwerfen** *(fig.)* chuck the whole thing in *(coll.)*

Kren [kre:n] **der; ~[e]s** *(südd., bes. österr.) s.* **Meerrettich**

krepieren [kre'pi:rən] *itr. V.; mit sein* **a)** *(zerplatzen)* explode; go off; **b)** *(salopp: sterben)* ⟨*animal*⟩ die; ⟨*person*⟩ snuff it *(sl.)*

Krepp [krɛp] **der; ~s, ~s** *od.* **~e** crêpe

Kreppapier das crêpe paper

Krepp-: ~papier *s.* **Kreppapier; ~sohle die** crêpe sole

Kresse ['krɛsə] **die; ~, ~n** *(Bot.)* cress

Kretin [kre'tɛ̃:] **der; ~s, ~s a)** *(Med.)* cretin; **b)** *(fig. abwertend)* imbecile

Kreuz [krɔyts] **das; ~es, ~e a)** cross; *(Symbol)* cross; crucifix; **etw. über ~ legen/falten** lay sth. down/fold sth. crosswise; **zu ~e kriechen** humble oneself; **jmdn. ans ~ schlagen** *od.* **nageln** nail sb. to the cross; **das/ein ~ schlagen** make the sign of the cross; *(sich bekreuzigen)* cross oneself; **drei ~e machen** *(ugs.)* heave a sigh of relief; **b)** *o. Pl. (Leid)* cross; **sein ~ auf sich nehmen/tragen** take up/bear one's cross; **es ist ein ~ mit jmdm./etw.** *(ugs.)* sb. is a real strain *or* is really trying/sth. is a real problem; **c)** *(Teil des Rückens)* small of the back; **ein steifes ~ haben** have a stiff back; **Schmerzen im ~:** pain in the small of the back; **ich hab's im ~** *(ugs.)* I've got back trouble *or* a bad back; **jmdn. aufs ~ legen** *(salopp)* take sb. for a ride *(sl.)*; **d)** *(Kartenspiel)* clubs *pl.*; *(Karte)* club; *s. auch* **²Pik; e)** *(Kreuzung)* interchange; **f)** *(Musik)* sharp

kreuz-, Kreuz-: ~as das ace of clubs; **~brav** *Adj.* thoroughly good and honest ⟨*person*⟩; very good *or* well-behaved ⟨*child*⟩; **~bube der** jack of clubs; **~dame die** queen of clubs

kreuzen 1. *tr. V. (auch Biol.)* cross; **die Arme/Beine ~:** cross *or* fold one's arms/cross one's legs. **2.** *refl. V.* **a)** cross; intersect; **b)** *(zuwiderlaufen)* clash (**mit** with). **3.** *itr. V.* **a)** *mit haben od. sein (hin und her fahren)* cruise; **b)** *(Seemannsspr.)* tack

Kreuzer der; ~s, ~ a) *(Milit.: Kriegsschiff)* cruiser; **b)** *(Segelsport)* cruising yacht; cruiser

kreuz-, Kreuz-: ~fahrer der *(hist.)* crusader; **~fahrt die** *(Seereise)* cruise; **eine ~fahrt machen** go on a cruise; **~feuer das** *(Milit., auch fig.)* cross-fire; **im ~feuer stehen/ins ~ geraten** *(fig.)* be/come under fire from all sides; **~fidel** *Adj. (ugs.) (sehr gut gelaunt)* very cheerful; *(sehr lustig)* very jolly; **~förmig 1.** *Adj.* cross-shaped; cruciform; **2.** *adv.* ⟨*built, arranged, etc.*⟩ in the shape of a cross; **~gang der** cloister

kreuzigen ['krɔytsɪgn̩] *tr. V.* crucify; **der Gekreuzigte** Christ crucified

Kreuzigung die; ~, ~en crucifixion

Kreuz-: ~könig der king of clubs; **~otter die** adder; [common] viper; **~schlitzschraube die** Phillips screw (P); **~schlüssel der** four-way wheel-brace; **~schmerzen** *Pl.* pain *sing.* in the small of the back; **~spinne die** cross *or* garden spider

Kreuzung die; ~, ~en a) junction; crossroads *sing.*; **b)** *(Biol.)* crossing; cross-breeding; *(Ergebnis)* cross; cross-breed; **eine ~ aus ... und ...:** a cross between ... and ...

kreuz·unglücklich *Adj. (ugs.)* terribly miserable *(coll.)*

kreuz-, Kreuz-: ~verhör das cross-examination; **jmdn. ins ~verhör nehmen** *(fig.)* cross-examine sb.; **~weise** *Adv.* crosswise; crossways; **du kannst mich mal ~weise!** *(derb)* [you can] get stuffed! *(sl.)*; **~wort·rätsel das** crossword [puzzle]; **~zeichen das** *(bes. kath. Kirche)* sign of the cross; **~zug der** *(hist., fig.)* crusade

kribbelig *Adj. (ugs.)* fidgety; *(nervös)* edgy

kribbeln ['krɪbl̩n] *itr. V.* tickle; *(prickeln)* tingle; **es kribbelt mir** *od.* **mich in der Nase/in den Füßen/unter der Haut** I've got a tickle in my nose/my feet are tingling *or* I've got pins and needles in my feet/my skin is itching *or* prickling; **es kribbelt mir in den Fingern, es zu tun** *(fig.)* I'm just itching to do it

Kricket ['krɪkət] **das; ~s** cricket

kriechen ['kri:çn̩] *unr. itr. V.* **a)** *mit sein* ⟨*insect, baby*⟩ crawl; ⟨*plant*⟩ creep; ⟨*person, animal*⟩ creep, crawl; ⟨*car, train, etc.*⟩ crawl *or* creep [along]; **aus dem Ei/der Puppe ~:** hatch [out]/emerge from the chrysalis; **auf allen vieren/auf dem Bauch ~:** crawl on all fours/crawl [along] on one's stomach; **die Zeit kriecht** *(fig.)* time creeps by; **kaum noch ~ können** hardly be able to get about *or* walk; *s. auch* **Kreuz a; b)** *auch mit sein (abwertend: sich unterwürfig verhalten)* crawl, grovel (**vor** + *Dat.* to)

Kriecher der; ~s, ~ *(abwertend)* crawler; groveller

kriecherisch *Adj. (abwertend)* crawling; grovelling

Kriech-: ~spur die *(Verkehrsw.)* crawler lane; **~tier das** *(Zool.)* reptile

Krieg [kri:k] **der; ~[e]s, ~e** war; *(~sführung)* warfare; **~ führen** wage war (**gegen** on); **den ~ erklären** declare war (**Dat.** on); **in den ~ ziehen** go to war; **der kalte ~:** the cold war

kriegen *(ugs.)* **1.** *tr. V.* **a)** *s.* **bekommen 1 a; am Ende des Films ~ sie sich** at the end of the film boy gets girl; *s. auch* **genug; Motte a; zuviel 1 a; b)** *s.* **bekommen 1 b; c)** *s.* **bekommen 1 c; d)** *(fangen, festnehmen)* catch; **e)** *(bewältigen)* **wir werden das schon ~:** we'll soon sort it out. **2.** *zur Umschreibung des Passivs: s.* **bekommen 2**

Krieger der; ~s, ~: warrior; *(indianischer ~)* brave; **kalter ~**

(Politik) cold warrior; **ein müder ~** *(fig.)* a tired old thing

kriegerisch *Adj.* **a)** *(kampflustig)* warlike; **b)** *nicht präd.* *(militärisch)* military; **eine ~e Auseinandersetzung** an armed conflict

krieg·führend *Adj.; nicht präd.* warring; belligerent

kriegs-, Kriegs-: ~aus·bruch der outbreak of war; **~beil das** tomahawk; **das ~beil ausgraben/begraben** *(fig. scherzh.)* start fighting/bury the hatchet; **~bemalung die** *(Völkerk., fig. scherzh.)* war-paint; **~beschädigt** *Adj.* war-disabled; **~beschädigte der/die** war-disabled person; war invalid; **~dienst der a)** *(im Krieg)* active service; **b)** *(Wehrdienst)* military service; **~dienst·verweigerer der** conscientious objector; **~ende das** end of the war; **~erklärung die** declaration of war; **~fall der: im ~fall[e]** in the event of war; **~fuß der** *in* **mit jmdm. auf [dem] ~fuß stehen** *od.* **leben** *(scherzh.)* be at loggerheads with sb.; **mit etw. auf [dem] ~fuß stehen** *(scherzh.)* be totally lost when it comes to sth.; **~gefangene der** prisoner of war; POW; **~gefangenschaft die** captivity; **in ~gefangenschaft sein/geraten** be a prisoner of war/be taken prisoner; **~gegner der a)** *(Gegner im Krieg)* enemy; **b)** *(Gegner des Krieges)* opponent of the/a war; *(Pazifist)* opponent of war; **~gericht das** court martial; **jmdn. vor ein ~gericht stellen** court-martial sb.; **~hetze die;** *o. Pl.* *(abwertend)* warmongering; **~hetze betreiben** stir up war; **~invalide der/die** *s.* **~beschädigte; ~kamerad der** wartime comrade; **~list die** military stratagem; *(fig. scherzh.)* ruse; **~marine die** navy; **~maschinerie die** *(abwertend)* machinery of war; **~opfer das** war victim; **~pfad der** *in* **auf dem ~pfad** *(auch fig.)* on the war-path; **~rat der** *in* **~rat [ab]halten** *(scherzh.)* have a pow-wow; **~recht das;** *o. Pl.* martial law; **~schau·platz der** theatre of war; **~schiff das** warship; **~verbrechen das** war crime; **~verbrecher der** war criminal; **~verletzung die** war wound *or* injury; **~zeit die** wartime; **in ~zeiten** *Pl.* in wartime

Krim [krɪm] **die; ~: die ~:** the Crimea

Krimi ['kri:mi] **der; ~[s], ~[s]** *(ugs.)* *(Film, Stück, Roman)* crime thriller; whodunit *(coll.)*; *(Roman mit Detektiv als Held)* detective story

Kriminal-: ~beamte der [plain-clothes] detective; **~fall der** criminal case; **~film der** crime film *or* thriller

kriminalisieren *tr. V.* **a) jmdn./etw. ~:** make sb. into a criminal/make sth. a criminal offence; **jmdn. ~** *(zu Straftaten treiben)* make sb. turn to crime; **b)** *(als kriminell hinstellen)* **jmdn./etw. ~:** present sb. as a criminal/sth. as [being] criminal *or* a criminal act

Kriminalistik die; ~: criminalistics *sing., no art.*

kriminalistisch 1. *Adj.* ‹*methods, practice*› of criminalistics; ‹*abilities*› in the field of criminalistics. **2.** *adv.* ‹*proceed etc.*› using the methods of criminalistics

Kriminalität [kriminali'tɛ:t] **die; ~ a)** crime *no art.*; **b)** *(Straffälligkeit)* criminality

Kriminal-: ~kommissar der ≈ detective superintendent; **~polizei die** criminal investigation department; **~roman der** crime novel *or* thriller; *(mit Detektiv als Held)* detective novel

kriminell [krimi'nɛl] **1.** *Adj.* *(auch ugs.: rücksichtslos)* criminal; **~ werden/sein** become a criminal *or* turn to crime/be a criminal. **2.** *adv.* **a) ~ veranlagt sein** have criminal tendencies; **~ handeln** act illegally; break the law; **b)** *(ugs.: rücksichtslos)* criminally; ‹*drive*› with criminal recklessness

Kriminelle der/die; *adj. Dekl.* criminal

Krimskrams ['krɪmskrams] **der; ~[es]** *(ugs.)* stuff

Kringel ['krɪŋl̩] **der; ~s, ~ a)** *(Kreis)* [small] ring; *(Kritzelei)* round squiggle; **b)** *(Gebäck)* [ring-shaped] biscuit; ring

kringelig *Adj.* crinkly ‹*hair*›; squiggly ‹*shape, line, etc.*›; **sich ~ lachen** *(ugs.)* laugh one's head off; kill oneself [laughing] *(coll.)*

kringeln 1. *tr. V.* curl [up] ‹*tail*›. **2.** *refl. V.* curl [up]; ‹*hair*› go curly; **sich [vor Lachen] ~** *(ugs.)* laugh one's head off; kill oneself [laughing] *(coll.)*

Kripo ['kri:po] **die; ~** *(ugs.)* **die ~:** ≈ the CID

Krippe ['krɪpə] **die; ~, ~n a)** *(Futtertrog)* manger; crib; **b)** *(Weihnachts~)* model of a nativity scene; crib; **c)** *(Kinder~)* crèche; day nursery

Krippen·spiel das nativity play

Krise ['kri:zə] **die; ~, ~n** crisis; **in eine ~ geraten** enter a state of crisis

kriseln ['kri:zl̩n] *itr. V.; unpers.* **es kriselt in ihrer Ehe/in der Partei** *(eine Krise droht)* their marriage is running into trouble/there is a crisis looming in the party; *(eine Krise ist vorhanden)* their marriage is in trouble/the party is in a state of crisis

krisen-, Krisen-: ~fest *Adj.* that is/are unaffected by crises *postpos., not pred.*; **~gebiet das** crisis area; **~herd der** trouble spot

[1]**Kristall** [krɪs'tal] **der; ~s, ~e** crystal

[2]**Kristall das; ~s** crystal *no indef. art.*

Kristall- crystal ‹*glass, chandelier, etc.*›

Kristallisation [krɪstaliza'tsi̯o:n] **die; ~, ~en** crystallization

kristallisieren *itr., refl. V.* *(auch fig.)* crystallize

kristall·klar *Adj.* *(auch fig.)* crystal-clear

Kriterium [kri'te:ri̯ʊm] **das; ~s, Kriterien** criterion

Kritik [kri'ti:k] **die; ~, ~en a)** criticism *no indef. art.* **(an + *Dat.*** of); **an jmdm./etw. ~ üben** criticize sb./sth.; **unter aller ~ sein** *(ugs.)* be absolutely hopeless; **b)** *(Besprechung)* review; notice; **eine gute/schlechte ~** *od.* **gute/schlechte ~en bekommen** get good/bad reviews *or* notices; **c) die ~:** the critics *pl. or* reviewers *pl.*

Kritiker ['kri:tikɐ] **der; ~s, ~, Kritikerin die; ~, ~nen** critic

kritik·los 1. *Adj.* uncritical; **2.** *adv.* uncritically; **etw. ~ hinnehmen** accept sth. without criticism

kritisch ['kri:tɪʃ] **1.** *Adj.* critical. **2.** *adv.* critically; **sich mit etw. ~ auseinandersetzen** make a critical study of sth.

kritisieren [kriti'zi:rən] *tr. V.* criticize; review ‹*book, play, etc.*›

Krittelei die; ~, ~en *(abwertend)* fault-finding; carping

kritteln ['krɪtl̩n] *itr. V.* *(abwertend)* find fault **(an + *Dat.*, über + *Akk.*** with); carp **(an + *Dat.*, über + *Akk.*** at)

Kritzelei die; ~, ~en a) *o. Pl.* *(das Schreiben)* scribbling; *(das Zeichnen)* doodling; **b)** *(Geschriebenes)* scribble; *(Zeichnung)* doodle; *(an Wänden)* graffiti *sing. or pl.*

kritzeln ['krɪtsl̩n] **1.** *itr. V.* *(schreiben)* scribble; *(zeichnen)* doodle. **2.** *tr. V.* scribble

kroch *1. u. 3. Pers. Sg. Prät. v.* **kriechen**

Krokant [kro'kant] **der; ~s** praline

Krokette [kro'kɛtə] **die; ~, ~n** *(Kochk.)* croquette

Kroko ['kro:ko] **das; ~[s]** crocodile [leather]
Krokodil [kroko'di:l] **das; ~s, ~e** crocodile
Krokodils·tränen *Pl. (ugs.)* crocodile tears
Kroko·tasche die crocodile[-skin] [hand]bag
Krokus ['kro:kʊs] **der; ~, ~** *od.* **~se** crocus
Krone ['kro:nə] **die; ~, ~n a)** crown; *(kleinere, eines Herzogs, eines Grafen)* coronet; **die ~** *(fig.: Herrscherhaus)* the Crown; **einer Sache** *(Dat.)* **die ~ aufsetzen** cap sth.; **einen in der ~ haben** *(ugs.)* have had a drop too much *(coll.)*; *o. Pl. (das Beste)* **die ~ der Schöpfung/meiner Sammlung** *(fig.)* the pride of creation/my collection; **b)** *(eines Baumes)* top; crown; *(einer Welle)* crest; **c)** *(Zahnmed.)* crown
krönen ['krø:nən] *tr. V. (auch fig.)* crown; **jmdn. zum König ~:** crown sb. king; **von Erfolg gekrönt sein** *od.* **werden** *(fig.)* be crowned with success; **der ~de Abschluß** the culmination
Kronen·korken der crown cap *or* cork
Kron-: ~juwelen *Pl.* crown jewels; **~kolonie die** crown colony; **~leuchter der** chandelier; **~prinz der** crown prince; **~prinzessin die** crown princess
Krönung die; ~, ~en a) coronation; **b)** *(Höhepunkt)* culmination
Kron·zeuge der *(Rechtsw.)* person who turns Queen's/King's evidence; **als ~ auftreten** turn Queen's/King's evidence
Kropf [krɔpf] **der; ~[e]s, Kröpfe** ['krœpfə] **a)** crop; **b)** *(Med.)* goitre
kroß [krɔs] *(nordd.) Adj., adv.: s.* **knusprig**
Krösus ['krø:zʊs] **der; ~** *od.* **~ses, ~se** *(oft scherzh.)* Croesus; **ich bin doch kein ~:** I'm not made of money
Kröte ['krø:tə] **die; ~, ~n a)** toad; **b)** *Pl. (salopp: Geld)* **ein paar/eine ganze Menge ~n verdienen** earn a few bob *(Brit. sl.)*/a fair old whack *(sl.)*; **meine letzten paar ~n** my last few bob *(Brit. sl.)*/bucks *(Amer. sl.)*; **c)** *(ugs. abwertend: Mensch)* creature
Krücke ['krʏkə] **die; ~, ~n a)** *(Stock)* crutch; **an** *od.* **auf ~n** *(Dat.)* **gehen** walk on crutches; **b)** *(Griff)* crook; handle; **c)** *(ugs. abwertend) (Versager)* dead loss *(coll.)*; wash-out *(sl.)*; *(Gegenstand)* dead loss *(coll.)*
Krück·stock der walking-stick; *s. auch* **Blinde**

Krug [kru:k] **der; ~[e]s, Krüge** ['kry:gə] *(Gefäß für Flüssigkeiten)* jug; *(größer)* pitcher; *(Bier~)* mug; *(aus Ton)* mug; stein
Krume ['kru:mə] **die; ~, ~n** crumb
Krümel ['kry:ml̩] **der; ~s, ~** crumb
krümelig *Adj.* **a)** crumbly
krümeln *itr. V.* **a)** *(zerfallen)* crumble; be crumbly; **b)** *(Krümel machen)* make crumbs
krumm [krʊm] **1.** *Adj.* **a)** bent ⟨*nail etc.*⟩; crooked ⟨*stick, branch, etc.*⟩; bandy ⟨*legs*⟩; bent ⟨*back*⟩; **eine ~e Nase** a crooked nose; *(Hakennase)* a hooked nose; **~ sein/werden** ⟨*person*⟩ stoop/develop a stoop; **etw. ~ biegen** bend sth.; **jmdn. ~ und lahm schlagen** beat sb. black and blue; **sich ~ und schief lachen** *(ugs.)* fall about laughing; laugh one's head off *(coll.)*; **b)** *nicht präd. (ugs.: unrechtmäßig)* crooked; **ein ~es Ding drehen** get up to sth. crooked. **2.** *adv.* crookedly; **~ gewachsen** crooked ⟨*tree etc.*⟩; **~ dasitzen/gehen** slouch/walk with a stoop; **steh/sitz nicht so ~ da!** stand/sit up straight!; *s. auch* **Finger b**
krumm·beinig *Adj.* bandy[-legged]; bow-legged
krümmen ['krʏmən] **1.** *tr. V.* bend; **gekrümmt** curved ⟨*line, surface*⟩. **2.** *refl. V.* writhe; **sich vor Schmerzen/in Krämpfen ~:** double up with pain/cramp
krumm-: ~|lachen *refl. V. (ugs.) s.* **schieflachen; ~|legen** *refl. V. (ugs.)* scrimp and scrape; pinch and scrape; **~|nehmen** *unr. tr. V. (ugs.)* **etw. ~nehmen** take offence at sth.; take sth. the wrong way
Krümmung die; ~, ~en a) *(der Wirbelsäule)* curvature; *(der Nase usw.)* curve; **b)** *(Geom.)* curvature
Krüppel ['krʏpl̩] **der; ~s, ~:** cripple; **zum ~ werden** be crippled; **jmdn. zum ~ schlagen** beat sb. and leave him/her a cripple
Kruste ['krʊstə] **die; ~, ~n a)** crust; *(vom Braten)* crisp; **b)** *(Überzug)* coating
Kruzifix ['kru:tsifɪks] **das; ~es, ~e** crucifix
Kübel ['ky:bl̩] **der; ~s, ~ a)** pail; *(Pflanzen~)* tub; **es gießt wie aus ~n** *(ugs.)* it's bucketing down; **b)** *(Toiletteneimer)* [latrine] bucket
Kubik- [ku'bi:k-] cubic ⟨*metre, foot, etc.*⟩
kubisch ['ku:bɪʃ] *Adj.* **a)** cubical; cube-shaped; **b)** *(Math.)* cubic ⟨*equation etc.*⟩

Kubismus der; ~ *(Kunstw.)* cubism *no art.*
Küche ['kʏçə] **die; ~, ~n a)** kitchen; **b)** *(Einrichtung)* kitchen furniture *no indef. art.*; **c)** *(Kochk.)* cooking; cuisine; **die chinesische** *usw.* **~:** Chinese *etc.* cooking; **kalte/warme ~:** cold/hot meals *pl. or* food; *s. auch* **gutbürgerlich**
Kuchen ['ku:xn̩] **der; ~s, ~:** cake; *(Obst~)* flan; *(Torte)* gateau; cake
Küchen·abfälle *Pl.* kitchen scraps
Kuchen·blech das baking-sheet *or* -tray
Küchen·chef der chef
Kuchen-: ~form die cake-tin; **~gabel die** pastry-fork
Küchen-: ~gerät das kitchen utensil; *(als Kollektivum)* kitchen utensils *pl.*; **~messer das** kitchen knife; **~schabe die** cockroach; **~schrank der** kitchen cupboard
Kuchen-: ~teig der cake mixture; **~teller der** cake-plate
Küchen-: ~tisch der kitchen table; **~waage die** kitchen scales *pl.*; **~zettel der** menu
kuckuck ['kʊkʊk] *Interj.* cuckoo
Kuckuck der; ~s, ~e a) cuckoo; **[das] weiß der ~** heaven [only] knows; it's anybody's guess; **zum ~ [noch mal]!** *(salopp)* for crying out loud! *(coll.)*; **wo, zum ~, hast du nur die Zeitung hingelegt?** *(salopp)* where the hell did you put the newspaper? *(coll.)*; **b)** *(scherzh.: Siegel des Gerichtsvollziehers)* bailiff's seal *(placed on distrained goods)*
Kuckucks-: ~ei das *(fig. ugs.)* **sich als ~ei erweisen** turn out to be more of a liability than an asset; **jmdm./sich ein ~ei ins Nest legen** do sb./oneself a dubious service; **~uhr die** cuckoo clock
Kuddelmuddel ['kʊdl̩mʊdl̩] **der** *od.* **das; ~s** *(ugs.)* muddle; confusion
Kufe ['ku:fə] **die; ~, ~n** *(von Schlitten, Schlittschuh)* runner; *(von Flugzeug, Hubschrauber)* skid
Kugel ['ku:gl̩] **die; ~, ~n a)** ball; *(Geom.)* sphere; *(Kegeln)* bowl; *(beim Kugelstoßen)* shot; *(eines ~lagers)* ball[-bearing]; **eine ruhige ~ schieben** *(ugs.)* have a cushy number *(coll.)*; **b)** *(ugs.: Geschoß)* bullet; *(Kanonen~)* [cannon-]ball; *(Luftgewehr~)* pellet; **sich** *(Dat.)* **eine ~ durch den Kopf schießen** blow one's brains out
kugel-, Kugel-: ~blitz der

(Met.) ball lightning; ~**förmig** *Adj.* spherical; ~**gelenk das** *(Anat., Technik)* ball-and-socket joint; ~**hagel der** hail of bullets; ~**kopf der** golf ball; ~**kopfmaschine die** golf-ball typewriter; ~**lager das** *(Technik)* ball-bearing

kugeln 1. *tr. V.* roll. 2. *refl. V.* roll [about]; **sich [vor Lachen]** ~ *(ugs.)* double *or* roll up [laughing *or* with laughter]. 3. *itr. V.; mit sein* roll

kugel-, Kugel-: ~**rund** [--'-] *Adj.* **a)** round as a ball *postpos.;* **b)** *(scherzh.: dick)* rotund; plump; tubby; ~**schreiber der** ballpoint [pen]; ball-pen; Biro **(P)**; ~**sicher** *Adj.* bullet-proof; ~**stoßen das;** ~**s** shot[-put]; *(Disziplin)* shot-putting *no art.;* putting the shot *no art.;* ~**stoßer der,** ~**stoßerin die** shot-putter

Kuh [ku:] **die;** ~, **Kühe** ['ky:ə] **a)** cow; **heilige** ~ *(ugs.)* sacred cow; **b)** *(Elefanten~, Giraffen~, Flußpferd~)* cow; *(Hirsch~)* hind; **c)** *(salopp abwertend: Frau)* cow *(sl. derog.)*

Kuh-: ~**dorf das** *(salopp abwertend)* one-horse town *(sl.);* ~**fladen der** cow-pat; ~**glocke die** cow-bell; ~**handel der** *(ugs. abwertend)* shady horse-trading *no indef. art.;* ~**haut die** cowhide; **das geht auf keine** ~**haut** *(fig. salopp)* it's absolutely staggering *or* beyond belief; ~**herde die** herd of cows

kühl [ky:l] 1. *Adj.* **a)** cool; **mir ist/wird** ~: I feel/I'm getting chilly; **etw.** ~ **lagern/aufbewahren** store/keep sth. in a cool place; **b)** *(abweisend, nüchtern)* cool; **ein** ~**er Rechner** a cool, calculating person. 2. *adv. (abweisend, nüchtern)* coolly

Kühle ['ky:lə] **die;** ~ *(auch fig.)* coolness; **die** ~ **der Nacht** the cool of the night

kühlen 1. *tr. V.* cool; chill, cool ⟨*drink*⟩; refrigerate ⟨*food*⟩; **seinen Zorn/seine Rache [an jmdm.]** ~: vent one's rage/revenge oneself [on sb.]. 2. *itr. V.* ⟨*cold compress, ointment, breeze, etc.*⟩ have a cooling effect

Kühler der; ~**s,** ~ **a)** *(am Auto)* radiator; *(~haube)* bonnet *(Brit.);* hood *(Amer.);* **jmdn. auf den** ~ **nehmen** *(ugs.)* drive *or* run into *or* hit sb.; **b)** *(Sekt~)* ice-bucket

Kühler-: ~**figur die** radiator mascot; ~**haube die** bonnet *(Brit.);* hood *(Amer.)*

Kühl-: ~**haus das** cold store; ~**mittel das** *(Technik)* coolant; ~**raum der** cold store; cold-storage room; ~**schrank der** refrigerator; fridge *(Brit. coll.);* icebox *(Amer.);* ~**tasche die** cool bag; ~**truhe die** [chest] freezer; deep-freeze; *(im Lebensmittelgeschäft)* freezer [cabinet]; ~**turm der** *(Technik)* cooling tower

Kühlung die; ~, ~**en a)** cooling; **b)** *(Vorrichtung)* cooling system; *(für Lebensmittel)* refrigeration system; **c)** *o. Pl. (Frische)* coolness; **sich** *(Dat.)* ~ **verschaffen** cool down *or* off

Kühl-: ~**wagen der a)** *(Eisenb.)* refrigerated *or* refrigerator car *or* *(Brit.)* wagon; **b)** *(Lastwagen)* refrigerated *or* refrigerator truck *or* *(Brit.)* lorry; ~**wasser das** cooling water

Kuh-: ~**milch die** cow's milk; ~**mist der** cow dung

kühn [ky:n] 1. *Adj.* **a)** bold; *(gewagt)* daring; brave, fearless ⟨*warrior*⟩; **das übertraf meine** ~**sten Träume** that exceeded my wildest dreams; **b)** *(dreist)* audacious; impudent. 2. *adv.* **a)** boldly; *(gewagt)* daringly; **b)** *(dreist)* audaciously; impudently

Kühnheit die; ~ **a)** boldness; *(Gewagtheit)* daringness; **b)** *(Dreistigkeit)* audacity; impudence

Kuh·stall der cowshed

Küken ['ky:kn̩] **das;** ~**s,** ~: chick

kulant [ku'lant] *Adj.* obliging; accommodating; fair ⟨*terms*⟩

Kulanz [ku'lants] **die;** ~: readiness *or* willingness to oblige; **aus** ~: out of good will

Kuli ['ku:li] **der;** ~**s,** ~**s a)** coolie; **b)** *(ugs.: Kugelschreiber)* ballpoint; Biro **(P)**

kulinarisch [kuli'na:rɪʃ] *Adj.* culinary

Kulisse [ku'lɪsə] **die;** ~, ~**n** piece of scenery; flat; wing; *(Hintergrund)* backdrop; **die** ~**n** the scenery *sing.;* **die** ~ **für etw. bilden** *(fig.)* form the backdrop to sth.; **hinter den** ~**n** *(fig.)* behind the scenes

Kulissen·schieber der *(ugs. scherzh.)* scene-shifter

Kuller·augen *Pl. (ugs. scherzh.)* big, round eyes; **er machte** ~: his eyes nearly popped out of his head

kullern ['kʊlɐn] *(ugs.)* 1. *itr. V.* **a)** *mit sein* roll; **b) mit den Augen** ~: roll one's eyes. 2. *tr. V.* roll

Kult [kʊlt] **der;** ~**[e]s,** ~**e** *(auch fig.)* cult; **mit jmdm./etw. einen** ~ **treiben** make a cult out of sb./sth.

Kult-: ~**bild das** devotional image; ~**figur die** cult figure; ~**film der** cult film; ~**handlung die** ritual; ritualistic act

kultisch 1. *Adj.* cultic; ritual, cultic ⟨*object*⟩. 2. *adv.* ⟨*worship*⟩ cultically

kultivieren [kʊlti'vi:rən] *tr. V.* *(auch fig.)* cultivate

kultiviert 1. *Adj.* **a)** cultivated; cultured; **b)** *(fein)* refined. 3. *adv.* in a cultivated *or* cultured manner; *(fein)* in a refined manner; with refinement

Kultivierung die; ~, ~**en** *(auch fig.)* cultivation; improvement

Kult·stätte die centre of cult worship

Kultur [kʊl'tu:ɐ̯] **die;** ~, ~**en a)** *o. Pl. (geistiger Überbau)* culture; **b)** *(Zivilisation, Lebensform)* civilization; **c)** *o. Pl. (Kultiviertheit, geistiges Niveau)* **ein Mensch von** ~: a cultured person; **sie hat [keine]** ~: she is [un]cultured; **d)** *o. Pl. (kultivierte Lebensart)* refinement; ~ **haben** be refined; **e)** *(Landw., Gartenbau)* young crop; *(Forstw.)* young plantation; **f)** *(Biol., Med.)* culture

Kultur-: ~**attaché der** cultural attaché; ~**austausch der** cultural exchange; ~**banause der** *(abwertend, oft scherzh.)* philistine; ~**beutel der** sponge-bag *(Brit.);* toilet-bag

kulturell [kʊltu'rɛl] 1. *Adj.* cultural. 2. *adv.* culturally

kultur-, Kultur-: ~**epoche die** cultural epoch; ~**film der** documentary film; ~**geschichte die** *o. Pl.* history of civilization; *(einer bestimmten Kultur)* cultural history; ~**gut das** cultural possessions *pl.;* ~**hoheit die** autonomy *or* independence in cultural and educational matters; ~**kreis der** cultural area; ~**los** *Adj.* uncultured; lacking in culture *postpos.;* ~**magazin das** *(Ferns.)* arts magazine; ~**minister der** minister for the arts; ~**ministerium das** ministry for the arts; ~**politik die** cultural and educational policy; ~**revolution die** cultural revolution; ~**stufe die** level of civilization; ~**szene die;** *o. Pl.* *(ugs.)* cultural scene; ~**volk das** civilized people *sing.;* ~**zentrum das a)** cultural centre; centre of cultural life; **b)** *(Anlage)* arts centre

Kultus-: ~**minister der** minister for education and cultural affairs; ~**ministerium das** ministry of education and cultural affairs

Kümmel ['kʏml̩] **der; ~s, ~ a)** caraway [seed]; **b)** *(Branntwein)* kümmel

Kummer ['kʊmɐ] **der; ~s** sorrow; grief; *(Ärger, Sorgen)* trouble; **~ um** *od.* **über jmdn.** grief for *or* over sb.; **hast du ~?** is there a problem?; **jmdm. ~ machen** give sb. trouble *or* bother; **ich bin ~ gewohnt** *(ugs.)* it happens all the time; I'm used to it

kümmerlich 1. *Adj.* **a)** puny; stunted ⟨*vegetation, plants*⟩; **b)** *(ärmlich)* wretched; miserable; **c)** *(abwertend: gering)* miserable; meagre, scanty ⟨*knowledge, leftovers*⟩; very poor ⟨*effort*⟩. **2.** *adv.* **sich ~ ernähren** live on a poor *or* meagre diet; **sich ~ durchschlagen** eke out a bare/miserable existence

kümmern ['kʏmɐn] **1.** *refl. V.* **a) sich um jmdn./etw. ~:** take care of *or* look after sb./sth.; **sich darum ~, daß ...:** see to it that ...; **b)** *(sich befassen mit)* **sich nicht um das Geschwätz** *usw.* **~:** not worry *or* mind about the gossip *etc.*; **sich nicht um Politik ~:** not care about *or* be interested in politics; **kümmere dich um deine eigenen Angelegenheiten** mind your own business. **2.** *tr. V.* concern; **was kümmert dich das?** what concern *or* business is it of yours?; what's it to you?; **Was kümmert's mich?** What do I care?

Kummer·speck der *(ugs.)* **sie hat ~ angesetzt** all the worrying has made her eat too much, and she's [really] put on weight

kummer·voll 1. *Adj.* sorrowful; sad. **2.** *adv.* sorrowfully; sadly

Kumpan [kʊm'pa:n] **der; ~s, ~e, Kumpanin die; ~, ~nen** *(ugs.)* **a)** pal *(coll.)*; mate; buddy *(coll.)*; **b)** *(abwertend: Mittäter)* accomplice

Kumpel ['kʊmpl̩] **der; ~s, ~,** *ugs. auch:* **~s a)** *(Bergmannsspr.)* miner; collier; **b)** *(salopp: Kamerad)* pal; mate; buddy *(coll.)*

kumpelhaft 1. *Adj.* matey; chummy *(coll.)*. **2.** *adv.* matily; chummily *(coll.)*

kündbar *Adj.* terminable ⟨*contract*⟩; redeemable ⟨*loan, mortgage*⟩; **Beamte sind nicht ~:** established civil servants cannot be dismissed *or* given their notice

Kündbarkeit die; ~ *(von Verträgen)* terminability; *(von Anleihen, Hypotheken)* redeemability

[1]**Kunde** ['kʊndə] **der; ~n, ~n a)** customer; *(eines Architekten, einer Versicherung usw.)* client; **b)** *(ugs.: Kerl)* customer *(coll.)*

[2]**Kunde die; ~** *(geh.)* tidings *pl.* *(literary)*; **jmdm. von etw. ~ geben** *(veralt.)* bring sb. tidings of sth.

künden ['kʏndn̩] **1.** *tr. V.* *(geh.: ver~)* proclaim; **diese Zeichen ~ Unglück** these omens herald misfortune. **2.** *itr. V.* *(geh.)* **von etw. ~:** bear witness to *or* tell of sth.

Kunden-: ~dienst der a) *o. Pl.* service to customers; *(Wartung)* after-sales service; **b)** *(Abteilung)* service department; **~kreis der** customers *pl.*; *(eines Architekten-, Anwaltbüros, einer Versicherung usw.)* clientele; **~stamm der** regular clientele *or* trade

kund|geben *(geh.) unr. tr. V.* declare; announce; express, make known ⟨*opinion, feelings*⟩

Kundgebung die; ~, ~en rally

kundig 1. *Adj.* knowledgeable; well-informed; *(sachverständig)* expert; **einer Sache** *(Gen.)* **~/nicht ~ sein** *(geh.)* know about sth./have no knowledge of sth. **2.** *adv.* expertly

kündigen ['kʏndɪgn̩] **1.** *tr. V.* call in, cancel ⟨*loan*⟩; foreclose ⟨*mortgage*⟩; cancel, discontinue ⟨*magazine subscription, membership*⟩; terminate ⟨*contract, agreement*⟩; denounce ⟨*treaty*⟩; **seine Stellung ~:** give in *or* hand in one's notice (**bei** to); **ich bin gekündigt worden** *(ugs.)* I've been given my notice; **der Vermieter hat ihm die Wohnung gekündigt** the landlord gave him notice to quit the flat *(Brit.)* *or (Amer.)* apartment; **er hat seine Wohnung gekündigt** he's given notice that he's leaving his flat *(Brit.)* *or* *(Amer.)* apartment; **jmdm. die Freundschaft ~** *(fig.)* break off a friendship with sb. **2.** *unr. itr. V.* **a)** *(ein Mietverhältnis beenden)* ⟨*tenant*⟩ give notice [that one is leaving]; **jmdm. ~** ⟨*landlord*⟩ give sb. notice to quit; **zum 1. Juli ~:** give notice for 1 July; **b)** *(ein Arbeitsverhältnis beenden)* ⟨*employee*⟩ give in *or* hand in one's notice (**bei** to); **jmdm. ~** ⟨*employer*⟩ give sb. his/her notice

Kündigung die; ~, ~en a) *(eines Kredits)* calling-in; cancellation; *(einer Hypothek)* foreclosure; *(der Mitgliedschaft, eines Abonnements)* cancellation; discontinuation; *(eines Vertrags)* termination; **b)** *(eines Arbeitsverhältnisses)* **jmdm. die ~ aussprechen** give sb. his/her notice; dismiss sb.; **mit ~ drohen** ⟨*employee*⟩ threaten to give in *or* hand in one's notice *or* to quit; ⟨*employer*⟩ threaten dismissal; **fristlose ~:** dismissal without notice; **c)** *(eines Mietverhältnisses)* **sie mußte mit ~ rechnen** she had to reckon on being given notice to quit; **d)** *s.* **Kündigungsschreiben; e)** *(Kündigungsfrist)* [period *or* term of] notice

Kündigungs-: ~frist die period of notice; **~grund der** *(Arbeitsrecht)* grounds *pl.* for dismissal; grounds *pl.* for giving sb. his/her notice; *(Mietrecht)* grounds *pl.* for giving sb. notice to quit; **~schreiben das** written notice; notice in writing; **~schutz der** protection against wrongful dismissal

Kundin die; ~, ~nen customer/client; *s. auch* [1]**Kunde a**

Kundschaft die; ~, ~en *o. Pl.*; *s.* [1]**Kunde a:** customers *pl.*; clientele; **~!** service!

Kundschafter der; ~s, ~: scout

kund|tun *(geh.)* **1.** *unr. tr. V.* announce; make known. **2.** *unr. refl. V.* be revealed; show itself

künftig ['kʏnftɪç] **1.** *Adj.* future. **2.** *adv.* in future

Kunst [kʊnst] **die; ~, Künste** ['kʏnstə] **a)** art; **die Schwarze ~** *(Magie)* the black art; *(Buchdruck)* [the art of] printing; **die schönen Künste** [the] fine arts; fine art *sing.*; **was macht die ~?** *(ugs.)* how are things?; how's tricks? *(sl.)*; **b)** *(das Können)* skill; **die ärztliche ~:** medical skill; **die ~ des Reitens/der Selbstverteidigung** the art of riding/selfdefence; **das ist keine ~!** *(ugs.)* there's nothing 'to it; **mit seiner ~ am Ende sein** be at a complete loss; *s. auch* **Regel a**

kunst-, Kunst-: ~akademie die art college; college of art; **~ausstellung die** art exhibition; **~banause der** *(abwertend)* philistine; **~buch das** art book; **~denkmal das** artistic and cultural monument; **~druck der** [fine] art print; **~dünger der** chemical *or* artificial fertilizer; **~erzieher der, ~erzieherin die** art teacher; **~erziehung die** art education; *(Schulfach)* art; **~faser die** man-made *or* synthetic fibre; **~fehler der** professional error; **ein ärztlicher ~fehler** a professional error on the part of a doctor; **~fertig 1.** *Adj.* skilful; **2.** *adv.* skilfully; **~fertigkeit die** skill; skilfulness; **~flieger der** aerobatic pilot; stunt pilot *(coll.)*; **~flug der** aerobatics *sing.*; stuntflying *(coll.)*; **~gegenstand der** work of art; **~gerecht 1.** *Adj.* expert; skilful; **2.** *adv.* expertly; skilfully; **~geschichte die a)** *o.*

Pl. art history; history of art; **b)** *(Buch)* art history book; book on the history of art; **~geschichtlich 1.** *Adj.* art historical ‹*studies, evidence, expertise, point of view, etc.*›; **2.** *adv.* **~geschichtlich interessiert/versiert** interested/well versed in art history *or* the history of art; **~gewerbe das** arts and crafts *pl.*; **~gewerblich** *Adj.* craft *attrib.* ‹*objects, skills, etc.*›; **~gewerbliche Arbeiten** craftwork *sing.*; **~griff der** move; *(fig.)* trick; dodge; **~handel der** [fine-]art trade; **~händler der** [fine-]art dealer; **~handwerk das** craftwork; **~historiker der** art historian; **~historisch** *s.* **~geschichtlich; ~kenner der** art connoisseur *or* expert; **~kritik die** art criticism; **die ~:** the art critics *pl.*; **~kritiker der** art critic; **~leder das** artificial *or* imitation leather

Künstler ['kʏnstlɐ] **der; ~s, ~, Künstlerin die; ~, ~nen a)** artist; *(Zirkus~, Varieté~)* artiste; **ein bildender ~:** a visual artist; **b)** *(Könner)* genius (**in** + *Dat.* at); **ein ~ in seinem Fach** a genius in one's field/at one's trade

künstlerisch 1. *Adj.* artistic. **2.** *adv.* artistically; **ein ~ wertvoller Film** a film of great artistic worth

Künstler-: ~name der stage-name; **~pech das** *(ugs. scherzh.)* hard luck

künstlich ['kʏnstlıç] **1.** *Adj.* **a)** artificial; artificial, glass ‹*eye*›; false ‹*teeth, eyelashes, hair*›; synthetic, man-made ‹*fibre*›; imitation, synthetic ‹*diamond*›; **b)** *(gezwungen)* forced ‹*laugh, cheerfulness, etc.*›; enforced ‹*rest*›. **2.** *adv.* **a)** artificially; **b) sich ~ aufregen** *(ugs.)* get worked up *or* excited about nothing

kunst-, Kunst-: ~liebhaber der art lover; lover of the arts; **~los 1.** *Adj.* plain; **2.** *adv.* plainly; **~maler der** artist; painter; **~pause die** pause for effect; *(iron.: Stockung)* awkward pause; **~richtung die** trend in art; **~sammlung die** art collection; **~schatz der** art treasure; **~seide die** artificial silk; rayon; **~springen das** *(Sport)* springboard diving; **~stoff der** synthetic material; plastic; **~stück das** trick; **das ist kein ~stück** *(ugs.)* it's no great feat *or* achievement; **~stück!** *(ugs. iron.)* that's no great achievement; *(ist nicht verwunderlich)* it's hardly surprising; **~turnen das** gymnastics *sing.*; **~verstand der** artistic sense; feeling for art; **~voll 1.** *Adj.* ornate *or* elaborate and artistic; *(kompliziert)* elaborate; **2.** *adv.* **a)** ornately *or* elaborately and artistically; **b)** *(geschickt)* skilfully; **~werk das** work of art; **~wissenschaft die** aesthetics and art history

kunter·bunt ['kʊntɐ-] **1.** *Adj.* **a)** *(vielfarbig)* multi-coloured; **b)** *(abwechslungsreich)* varied; **c)** *(ungeordnet)* jumbled ‹*confusion, muddle, rows, etc.*›. **2.** *adv.* **a)** ‹*painted, printed*› in many colours; **b)** *(abwechslungsreich)* **ein ~ gestalteter Abend** an evening of varied entertainment; **c)** *(ungeordnet)* **~ durcheinander sein** be higgledy-piggledy *or* all jumbled up; **es ging ~ durcheinander** it was completely chaotic

Kunz [kʊnts] *s.* **Hinz**

Kupfer ['kʊpfɐ] **das; ~s a)** copper; **b)** *(~geschirr)* copperware; *(~geld)* coppers *pl.*

kupfer-, Kupfer-: ~geld das coppers *pl.*; **~münze die** copper coin; copper; **~rot** *Adj.* copper-red; copper-coloured; **~stich der a)** *o. Pl.* copperplate engraving *no art.*; **b)** *(Blatt)* copperplate print *or* engraving

Kupon *s.* **Coupon**

Kuppe ['kʊpə] **die; ~, ~n a)** [rounded] hilltop; **b)** *(Finger~)* tip; end

Kuppel ['kʊpl̩] **die; ~, ~n** dome; *(kleiner)* cupola

Kuppel-: ~bau der; *Pl.* **~bauten** domed building; **~dach das** domed *or* dome-shaped roof

Kuppelei die; ~, ~en a) *(veralt. abwertend)* match-making; **b)** *o. Pl. (Rechtsspr.)* procuring; procuration

kuppeln 1. *itr. V. (bei einem Kfz)* operate the clutch. **2.** *tr. V.* **a)** *(koppeln)* couple (**an** + *Akk.*, **zu** [on] to); **b)** *(Technik)* couple

Kuppler der; ~s, ~ *(abwertend)* procurer

Kupplerin die; ~, ~nen *(abwertend)* procuress

Kupplung die; ~, ~en a) *(Kfz-W.)* clutch; **b)** *(Technik: Vorrichtung zum Verbinden)* coupling

Kupplungs·pedal das clutch-pedal

Kur [kuːɐ̯] **die; ~, ~en** [health] cure; *(ohne Aufenthalt im Badeort)* course of treatment; **eine ~ machen** take a cure/a course of treatment; **in ~ gehen** go to a health resort *or* spa [to take a cure]

Kür [kyːɐ̯] **die; ~, ~en** *(Eiskunstlauf)* free programme; *(Turnen)* optional exercises *pl.*; **eine ~ laufen/tanzen** skate/dance one's free programme; **eine ~ turnen** perform one's optional exercises

Kuratorium [kura'toːri̯ʊm] **das; ~s, Kuratorien** board of trustees

Kur-: ~aufenthalt der stay at a health resort *or* spa; **~bad das** health resort; spa

Kurbel ['kʊrbl̩] **die; ~, ~n** crank [handle]; *(an Fenstern, Spieldosen, Grammophonen)* winder

kurbeln 1. *tr. V.* **etw. nach oben/unten ~:** wind sth. up/down. **2.** *itr. V.* turn *or* wind a/the handle

Kurbel·welle die *(Technik)* crankshaft

Kürbis ['kʏrbıs] **der; ~ses, ~se a)** pumpkin; **b)** *(salopp: Kopf)* nut *(sl.)*; bonce *(Brit. sl.)*

küren *regelm. (veralt. auch unr.) tr. V.* choose (**zu** as)

Kur-: ~fürst der *(hist.)* Elector; **~fürstentum das** *(hist.)* electorate; **~gast der** visitor to a/the health resort *or* spa; *(Patient)* patient at a/the health resort *or* spa; **~haus das** assembly rooms [at a health resort *or* spa]

Kurier [ku'riːɐ̯] **der; ~s, ~e** courier; messenger

Kurier·dienst der courier *or* messenger service

kurieren *tr. V. (auch fig.)* cure (**von** of)

kurios [ku'ri̯oːs] **1.** *Adj.* curious; strange; odd. **2.** *adv.* curiously; strangely; oddly

Kuriosität [kuri̯ozi'tɛːt] **die; ~, ~en a)** *o. Pl.* strangeness; oddity; peculiarity; **b)** *(Gegenstand)* curiosity; curio

Kuriositäten·kabinett das gallery of curios

Kur-: ~klinik die health clinic; **~konzert das** concert [at a health resort or spa]; spa concert

Kür·lauf der *(Eiskunstlauf)* free programme

Kur-: ~ort der health resort; spa; **~park der** gardens *pl.* [of a/the health resort *or* spa]; **~pfuscher der** *(ugs. abwertend)* quack; doctor; **~pfuscherei die** *(ugs. abwertend)* quackery

Kurs [kʊrs] **der; ~es, ~e a)** *(Richtung)* course; **auf [nördlichen] ~ gehen** set [a northerly] course; **ein harter/weicher ~** *(fig.)* a hard/soft line; **den ~ ändern/halten** *(auch fig.)* change *or* alter/hold *or* maintain course; **~ auf Hamburg** *(Akk.)* **nehmen** set course for *or* head for Hamburg; **b)** *(von Wertpapieren)* price; *(von Devisen)* rate of exchange; exchange rate; **zum ~ von ...:** at a rate of ...; **der ~**

des Dollars the dollar rate; **hoch im ~ stehen** ⟨*securities*⟩ be high; *(fig.)* be very popular (**bei** with); **c)** *(Lehrgang)* course; **ein ~ in Spanisch** *(Dat.)* a course in Spanish; a Spanish course; **d)** *(die Teilnehmer eines Kurses)* class; **e)** *(Sport: Rennstrecke)* course

Kurs-: ~änderung die *(auch fig.)* change of course; **~anstieg der** *(Börsenw.)* rise in prices/price; price rise; *(bei Devisen)* rise in exchange rates/the exchange rate; **~buch das** *(Eisenb.)* timetable

Kürschner ['kʏrʃnɐ] **der; ~s, ~, Kürschnerin die; ~, ~nen** furrier

kursieren *itr. V.; auch mit sein* circulate

kursiv [kʊr'zi:f] *(Druckw.)* **1.** *Adj.* italic. **2.** *adv.* **etw. ~ drucken** print sth. in italics; *(zur Hervorhebung)* italicize sth.

Kurs-: ~korrektur die *(auch fig.)* course correction; **~leiter der** course-leader; **~rück·gang der** *(Börsenw.)* fall in prices/price; price fall; *(bei Devisen)* fall in exchange rates/the exchange rate; **~schwankung die** *(Börsenw.)* fluctuation in prices/price; *(bei Devisen)* fluctuation in exchange rates/the exchange rate; **~teilnehmer der** course participant

Kursus ['kʊrzʊs] **der; ~, Kurse** *s.* **Kurs c, d**

Kurs-: ~wagen der *(Eisenb.)* through carriage *or* coach; **~wechsel der** *(auch fig.)* change of course

Kur·taxe die visitors' tax

Kurve ['kʊrvə] **die; ~, ~n a)** *(einer Straße)* bend; curve; **die Straße macht eine [scharfe] ~:** the road bends *or* curves [sharply]; **die ~ kratzen** *(ugs.)* quickly make oneself scarce *(coll.)*; **die ~ kriegen** *(ugs.)* manage to do it; *(etw. überwinden)* manage to do something decisive about it; **b)** *(Geom.)* curve; **c)** *(in der Statistik, Temperatur~ usw.)* graph; curve; **d)** *(Bogenlinie)* curve; **eine ~ fliegen** do a banking turn; **e)** *Pl. (ugs.: Körperformen)* curves

kurven *itr. V.; mit sein (ugs.: fahren)* **durch die Gegend ~:** drive/ride around; **durch ganz Europa ~:** drive/ride around the whole of Europe

kurven·reich *Adj.* **a)** winding; twisting; **„~reiche Strecke"** 'series of bends'; **b)** *(ugs. scherzh.)* curvaceous

Kur·verwaltung die administrative office/offices of a/the health resort *or* spa

kurz [kʊrts]; **kürzer** ['kʏrtsɐ], **kürzest...** ['kʏrtsəst...] **1.** *Adj.* **a)** *(räumlich)* short; **~e Hosen** short trousers; shorts; **etw. kürzer machen** make sth. shorter; shorten sth.; **jmdn. einen Kopf kürzer machen** *(ugs.)* chop sb.'s head off; **etw./alles ~ und klein schlagen** *od.* **hauen** *(ugs.)* smash sth./everything to bits *or* pieces; **den kürzeren ziehen** come off worst *or* second-best; get the worst of it; **nicht zu ~/zu ~ kommen** get one's/less than one's fair share; **b)** *(zeitlich)* short, brief ⟨*trip, journey, visit, reply*⟩; short ⟨*life, break, time*⟩; quick ⟨*look*⟩; **nach einer ~en Weile** after a short *or* little while; **es ~ machen** make *or* keep it short; be brief; **c)** *(knapp)* short, brief ⟨*outline, note, report, summary, introduction*⟩; **~ und bündig** *od.* **knapp** brief and succinct. **2.** *adv.* **a)** *(zeitlich)* briefly; for a short time *or* while; **die Freude währte nur ~** *(geh.)* his/her *etc.* joy was short-lived; **binnen ~em** shortly; soon; **über ~ oder lang** sooner or later; **vor ~em** a short time *or* while ago; recently; **sie lebt erst seit ~em in Bonn** she's only been living in Bonn [for] a short time *or* while; **b)** *(knapp)* **~ gesagt/~ und gut** in a word; **~ angebunden sein** be curt *or* brusque (**mit** with); **sich ~ fassen** be brief; **c)** *(rasch)* **ich muß mal ~ weg** I must leave you for a few minutes; **er schaute ~ herein** he looked *or* dropped in for a short while; **kann ich Sie ~ sprechen?** can I speak to you *or* have a word with you for a moment?; **~ und schmerzlos** *(ugs.)* quickly and smoothly *or* without any hitches; *s. auch* **entschlossen 3; d)** *(wenig)* just; **~ vor/hinter der Kreuzung** just before/past the crossroads; **~ vor/nach Pfingsten** just *or* shortly before/after Whitsun; **~ bevor .../nachdem ...:** just *or* shortly before .../after ...

kurz-, Kurz-: ~arbeit die short time; short-time working; **~|arbeiten** *itr. V.* work short time; **~arbeiter der** short-time worker; worker on short time; **~ärm[e]lig** [~|ɛrm(ə)lɪç] *Adj.* short-sleeved; **~atmig** [~a:tmɪç] *Adj. (auch fig.)* short-winded; **~atmig sein** be short of breath; be short-winded

Kurze der; *adj. Dekl. (ugs.)* **a)** *(Kurzschluß)* short *(coll.)*; **b)** *(Schnaps)* schnapps

Kürze ['kʏrtsə] **die; ~, ~n a)** *o. Pl.* shortness; **b)** *o. Pl. (geringe Dauer)* shortness; short duration; brevity; **in ~:** shortly; soon; **c)** *o. Pl. (Knappheit)* brevity; **in aller/gebotener ~:** very briefly/with due brevity; **in der ~ liegt die Würze** *(Spr.)* brevity is the soul of wit *(prov.)*

Kürzel ['kʏrtsl̩] **das; ~s, ~ a)** shorthand symbol; **b)** *(Abkürzung)* abbreviation

kürzen *tr. V.* **a)** shorten; shorten, take up ⟨*garment*⟩; **b)** *(verringern)* shorten ⟨*speech*⟩; shorten, abridge ⟨*article, book*⟩; reduce, cut ⟨*pension, budget, etc.*⟩; **c)** *(Math.)* cancel

kürzer *s.* **kurz**

kurzer·hand *Adv.* without more ado; **jmdn. ~ vor die Tür setzen** *(ugs.)* unceremoniously throw sb. out; **etw. ~ ablehnen** flatly reject sth.; reject sth. out of hand

kürzer|treten *unr. itr. V.* take things *or* it easier; *(sparsamer sein)* cut back; spend less

kürzest... *s.* **kurz**

kurz-, Kurz-: ~fassung die shortened *or* abridged version; **~film der** short; short film; **~fristig 1.** *Adj.* **a)** *(plötzlich)* ⟨*refusal, resignation, etc.*⟩ at short notice; **b)** *(für kurze Zeit)* short-term; **2.** *adv.* **a)** *(plötzlich)* at short notice; **sich ~fristig entschließen, etw. zu tun** make up one's mind within a short time to do sth.; **b)** *(für kurze Zeit)* for a short time *or* period; *(auf kurze Sicht)* in the short term; **c)** *(in kurzer Zeit)* without delay; **~geschichte die** *(Literaturw.)* short story; **~haarig** *Adj.* short-haired ⟨*dog, breed, etc.*⟩; ⟨*person*⟩ with short hair; **~|halten** *unr. tr. V.* **jmdn. ~halten** *(wenig Geld geben)* keep sb. short of money; *(wenig erlauben)* keep sb. on a tight rein; **~lebig** [~le:bɪç] *Adj. (auch fig.)* short-lived; *(wenig haltbar)* non-durable ⟨*goods, materials*⟩; with a short life *postpos.*

kürzlich *Adv.* recently; not long ago; **erst ~:** just *or* only recently; only a short time ago

kurz-, Kurz-: ~meldung die brief report; *(während einer anderen Sendung)* news flash; **~nachrichten** *Pl.* news *sing.* in brief; news summary *sing.*; **~|schließen** *unr. tr. V.* short-circuit; **~schluß der a)** *(Elektrot.)* short-circuit; **b)** *(fig. ugs.)* brainstorm; **c)** *(falscher Schluß)* fallacy; **~schluß·handlung die** sudden irrational act;

~schrift die shorthand; **~sichtig** *(auch fig.)* **1.** *Adj.* short-sighted; **2.** *adv.* short-sightedly; **~sichtigkeit die;** ~ *(auch fig.)* short-sightedness

Kurzstrecken-: ~flug der short-haul flight; **~läufer der** *(Sport)* sprinter

kurz-: ~|treten *unr. itr. V.* take things *or* it easy; *(sparsam sein)* retrench; cut back; **~um** [-'-] *Adv.* in short; in a word

Kürzung die; ~, ~en a) cut; reduction; **eine ~ des Gehaltes** a cut *or* reduction in salary; a salary cut; **b)** *(Streichung)* cut; *(das Streichen)* abridgement

kurz-, Kurz-: ~waren *Pl.* haberdashery *sing. (Brit.)*; notions *(Amer.)*; **~weilig** *Adj.* entertaining; **~welle die** *(Physik, Rundf.)* short wave; **~wellen·sender der** *(Funkt., Rundf.)* short-wave transmitter; **~zeitig 1.** *Adj.* brief; **2.** *adv.* briefly; for a short time

kuschelig *Adj.* cosy

kuscheln ['kʊʃl̩n] *refl. V.* **sich an jmdn. ~:** snuggle up *or* cuddle up to sb.; ‹*cat etc.*› snuggle up to sb.; **sich in etw.** *(Akk.)* **~:** snuggle up in sth.

kuschen ['kʊʃn̩] *itr. V.* **a)** ‹*dog*› lie down; **b)** *(fig.)* knuckle under **(vor +** *Dat.* to)

Kusine [ku'zi:nə] **die; ~, ~n** *s.* **Cousine**

Kuß [kʊs] **der; Kusses, Küsse** ['kʏsə] kiss

Küßchen ['kʏsçən] **das; ~s, ~:** little kiss

kuß·echt *Adj.* kissproof

küssen ['kʏsn̩] *tr., itr. V.* kiss; **jmdm. die Hand ~:** kiss sb.'s hand; **küss' die Hand** *(südd., österr.) (beim Kommen)* how do you do?; good day; *(beim Gehen)* goodbye; **sich** *od. (geh.)* **einander ~:** kiss [each other]

Kuß·hand die: jmdm. eine ~ zuwerfen blow sb. a kiss; **jmdn./etw. mit ~ nehmen** *(ugs.)* be only too glad *or* pleased to take sb./sth.

Küste ['kʏstə] **die; ~, ~n** coast

Küsten-: ~fischerei die inshore fishing; **~schiffahrt die** coastal shipping *no art.*; **~wache die** coastguard [service]

Küster ['kʏstɐ] **der; ~s, ~:** sexton

Kutsche ['kʊtʃə] **die; ~, ~n a)** coach; carriage; **b)** *(salopp: Auto)* jalopy *(coll.)*

Kutscher der; ~s, ~: coachman; coach-driver

kutschieren 1. *itr. V.; mit sein* **a)** drive, ride [in a coach *or* carriage]; **b)** *(ugs.)* **durch die Gegend/durch Europa ~:** drive around/drive around Europe. **2.** *tr. V.* **a) jmdn. ~:** drive sb. [in a coach *or* carriage]; **b)** *(ugs.)* **jmdn. nach Hause ~:** run sb. home

Kutte ['kʊtə] **die; ~, ~n** [monk's/nun's] habit

Kutteln ['kʊtl̩n] *Pl. (südd., österr., schweiz.)* tripe *sing.*

Kutter ['kʊtɐ] **der; ~s, ~:** cutter

Kuvert [ku've:ɐ̯] **das; ~s, ~s** *(landsch., veralt.)* envelope

Kuvertüre [kuvɛr'ty:rə] **die; ~, ~n** chocolate coating

kW [ka:'ve:] *(Physik) Abk.* **Kilowatt** kW

KW *Abk.* **Kurzwelle** SW

kWh *(Physik) Abk.* **Kilowattstunde** kWh

Kybernetik [kybɛr'ne:tɪk] **die; ~:** cybernetics *sing.*

kyrillisch [ky'rɪlɪʃ] *Adj.* Cyrillic

KZ [ka:'tsɛt] **das; ~[s], ~[s]** *Abk.* **Konzentrationslager**

L

l, L [ɛl] **das; ~, ~:** l/L; *s. auch* **a, A**

l *Abk.* **Liter** l.

labb[e]rig ['labərɪç] *Adj. (ugs. abwertend)* **a)** *(fade)* wishy-washy; **~ schmecken** taste of nothing; **b)** *(weich)* floppy, limp ‹*material*›; floppy ‹*trousers, dress, etc.*›

laben ['la:bn̩] *(geh.)* **1.** *tr. V.* **jmdn. ~:** give sb. refreshment; **ein ~der Trunk** a refreshing drink. **2.** *refl. V.* refresh oneself **(an +** *Dat.*, **mit** with)

labern ['la:bɐn] *(ugs. abwertend)* **1.** *tr. V.* talk; **was laberst du da?** what are you rabbiting *(Brit. coll.)* *or* babbling on about? **2.** *itr. V.* rabbit *(Brit. coll.)* *or* babble on

labil [la'bi:l] *Adj.* delicate, frail ‹*constitution, health*›; poor ‹*circulation*›; unstable ‹*person, character, situation, equilibrium, etc.*›

Labilität [labili'tɛ:t] **die; ~, ~en** *s.* **labil:** delicateness; frailness; poorness; instability

Labor [la'bo:ɐ̯] **das; ~s, ~s,** *auch:* **~e** laboratory

Laborant [labo'rant] **der; ~en, ~en, Laborantin die; ~, ~nen** laboratory *or (coll.)* lab assistant *or* technician

Laboratorium [labora'to:ri̯ʊm] **das; ~s, Laboratorien** laboratory

laborieren *itr. V. (ugs.)* **a)** *(leiden)* suffer **(an +** *Dat.* from); **er laboriert schon seit Wochen an einer Grippe** he's been trying to shake off the flu for weeks *(coll.)*; **b)** *(sich abmühen)* **an etw.** *(Dat.)* **~:** labour *or* toil away at sth.

Labyrinth [laby'rɪnt] **das; ~[e]s, ~e** maze; labyrinth

Lach·anfall der laughing-fit; fit of laughing

[1]**Lache** ['laxə] **die; ~, ~n** *(ugs.)* laugh

[2]**Lache** ['la(:)xə] **die; ~, ~n** puddle; *(von Blut, Öl)* pool

lächeln ['lɛçl̩n] *itr. V.* smile **(über +** *Akk.* at); **freundlich/verlegen ~:** give a friendly/an embarrassed smile

Lächeln das; ~s smile

lachen *itr. V.* laugh **(über +** *Akk.* at); **da kann man** *od.* **ich doch nur ~:** that's a laugh; **jmdn. zum Lachen bringen** make sb. laugh; **platzen/sterben vor Lachen** *(fig.)* split one's sides laughing/die laughing; **die Sonne** *od.* **der Himmel lacht** *(fig.)* the sun is shining brightly; **wer zuletzt lacht, lacht am besten** *(Spr.)* he who laughs last, laughs longest; **zum Lachen sein** *(ugs. abwertend)* be laughable *or* ridiculous; **daß ich nicht lache!** *(ugs.)* don't make me laugh *(coll.)*; **was gibt es denn zu ~?** what's so funny?; **es wäre ja** *od.* **doch gelacht, wenn ...** *(ugs.)* it would be ridiculous if ...; **nichts zu ~ haben** *(ugs.)* have a hard time of it; *s. auch* **dritt...;** [2]**Erbe**

Lachen das; ~s laughter; **ein lautes ~:** a loud laugh; **sie konnte sich das ~ kaum verbeißen** she could hardly stop herself laughing; **ihm wird das ~ noch vergehen** he'll be laughing on the other side of his face

Lacher der; ~s, ~ a) laugher; **die ~:** those who are/were laughing; **die ~ auf seiner Seite haben** score by making everybody laugh; **b)** *(ugs.: kurzes Lachen)* laugh

lächerlich ['lɛçɐlɪç] *(abwertend)* **1.** *Adj.* **a)** *(komisch)* ridiculous; **jmdn./sich [vor jmdm.] ~ machen** make a fool of sb./oneself *or* make sb./oneself look silly [in front of sb.]; **etw. ins Lächerliche ziehen** make a joke out of sth.; **b)** *(töricht)* ridiculous; ludicrous ‹*argument, statement*›; **c)** *(gering)* derisory, ridiculously *or* ludicrously small ⟨*sum, amount*⟩; ridiculously low ⟨*price, payment*⟩; **d)** *(geringfügig)* ridiculously trivial *or* trifling; **~e Kleinigkeiten** ridiculous trivialities. **2.** *adv.* ridiculously; **~ wenig** ridiculously *or* ludicrously little

Lächerlichkeit die; ~, ~en *(abwertend)* **a)** *o. Pl.* ridiculousness; *(von Argumenten, Behauptungen usw.)* ridiculousness; ludicrousness; **jmdn. der ~ preisgeben** make a laughing-stock of sb.; make sb. look ridiculous; **b)** *meist Pl.* ridiculous triviality

lach-, Lach-: ~gas das laughing-gas; **~haft** *(abwertend)* **1.** *Adj.* ridiculous; laughable; **2.** *adv.* ridiculously; **~krampf der** paroxysm of laughter; violent fit of laughter; **einen ~krampf bekommen** go [off] into fits of laughter

Lachs der; ~es, ~e salmon

Lachs·ersatz der rock salmon

lachs·farben *Adj.* salmon-pink; salmon-coloured

Lack [lak] **der; ~[e]s, ~e** varnish; *(für Metall, Lackarbeiten)* lacquer; *(Auto~)* paint; *(transparent)* lacquer; *(Nagel~)* varnish

Lack·affe der *(ugs. abwertend)* dandy

lackieren *tr. V.* varnish ⟨*wood*⟩; varnish, paint ⟨*finger-nails*⟩; spray ⟨*car*⟩; *(mit Emaillelack)* paint ⟨*metal*⟩; **einen Wagen neu ~:** respray a car

Lackierte der/die: der/die ~ sein *(ugs.)* have to carry the can *(sl.)*

Lackierung die; ~, ~en *(auf Holz)* varnish; *(auf Metall, Autos)* paintwork; *(auf Lackarbeiten)* lacquer

Lack·leder das patent leather

Lackmus·papier ['lakmʊs-] **das** *(Chemie)* litmus paper

Lack-: ~schaden der damage to the paintwork; **~schuh der** patent-leather shoe

Lade ['la:də] **die; ~, ~n** *(landsch.)* drawer

Lade·gerät das *(Elektrot.)* charger

¹laden 1. *unr. tr. V.* **a)** load; **etw. aus etw. ~:** unload sth. from sth. **die Schiffe ~ Getreide** the ships are taking on *or* are being loaded with grain; **der LKW hat Sand ge~:** the truck is loaded up with sand; **der Tanker hat Flüssiggas ge~:** the tanker has a cargo of *or* is carrying liquid gas; **b)** *(legen)* load; **sich** *(Dat.)* **einen Sack auf die Schultern ~:** load a sack on one's shoulders; **schwere Schuld auf sich ~** *(fig.)* incur a heavy burden of guilt; **c)** *(Munition einlegen)* load ⟨*gun, pistol, etc.*⟩; **d)** *(Physik)* charge; **er ist ge~** *(ugs.)* he's livid *(coll.)*; he's hopping mad *(coll.)*. **2.** *unr. itr. V.* load [up]; **der LKW hat schwer ge~:** the truck is heavily loaded

²laden *unr. tr. V.* **a)** *(Rechtsspr.)* summon; **b)** *(geh.: ein~)* invite

Laden der; ~s, Läden ['lɛ:dn̩] **a)** shop; store *(Amer.)*; **b)** *(ugs.: Unternehmung)* **der ~ läuft** business is good; **wie ich den ~ kenne** *(fig.)* if I know how things go in this outfit *(coll.)*; **den ~ dichtmachen** shut up shop; **den ~ schmeißen** manage *or* handle everything with no problem; **c)** *Pl. auch ~* *(Fenster~)* shutter

Laden-: ~dieb der shop-lifter; **~diebstahl der** shop-lifting; **~hüter der** *(abwertend)* non-seller; article/line which isn't/wasn't selling; **~preis der** shop price; **~schluß der** shop *or* *(Amer.)* store closing-time; **kurz vor/nach ~schluß** shortly before/after the shops *or* *(Amer.)* stores close/closed; **um 14 Uhr ist ~schluß** the shops *or* *(Amer.)* stores close at two o'clock; **~tisch der** [shop-]counter; **unterm ~tisch** *(ugs.)* under the counter

Lade-: ~rampe die loading ramp; **~raum der a)** *(beim Auto)* luggage-space; **b)** *(beim Flugzeug, Schiff)* hold; **c)** *(bei LKWs)* payload space

lädieren [lɛ'di:rən] *tr. V.* damage; *(fig.)* damage, harm ⟨*reputation etc.*⟩; undermine ⟨*confidence*⟩; **lädiert aussehen** *(ugs., scherzh.)* look battered

lädst [lɛ:tst] *2. Pers. Sg. Präsens v.* **laden**

lädt [lɛ:t] *3. Pers. Sg. Präsens v.* **laden**

Ladung die; ~, ~en a) *(Schiffs~, Flugzeug~)* cargo; *(LKW~)* load; **b)** *(beim Sprengen, Schießen)* charge; **c)** *(ugs.: Menge)* load *(coll.)*; **d)** *(Physik)* charge; **e)** *(Rechtsspr.: Vor~)* summons *sing.*

lag [la:k] *1. u. 3. Pers. Sg. Prät. v.* **liegen**

Lage ['la:gə] **die; ~, ~n a)** situation; location; **in ruhiger ~:** in a quiet location; **eine gute ~ haben** be peacefully/well situated; be in a good/peaceful location; **in höheren/tieferen ~n** *(Met.)* on high/low ground; **b)** *(Art des Liegens)* position; **jetzt habe ich eine bequeme ~:** now I'm lying comfortably; now I'm [lying] in a comfortable position; **c)** *(Situation)* situation; **er war nicht in der ~, das zu tun** he was not in a position to do that; **versetzen Sie sich in meine ~:** put yourself in my position *or* place; **nach ~ der Dinge** as matters stand/stood; **die ~ peilen** *od.* **spannen** *(ugs.)* see how the land lies; find out the lie of the land; *s. auch* **Herr; d)** *meist Pl. (Schwimmen)* **die 400 m ~n** the 400 m. individual medley; **die 4 × 100 m ~n** the 4 × 100 m. medley relay; **e)** *(Schicht)* layer; **f)** *(Stimm~)* register; **g)** *(ugs.: Runde)* round; **eine ~ ausgeben** *(ugs.)* *od.* **schmeißen** *(salopp)* get *or* stand a round

Lage-: ~besprechung die discussion of the situation; **~plan der** map of the area

Lager ['la:gɐ] **das; ~s, ~ a)** camp; **ein ~ aufschlagen** set up *or* pitch camp; **b)** *(Gruppe, politischer Block)* camp; **ins andere ~ überwechseln** change camps *or* sides; join the other side; **c)** *(Raum)* store-room; *(in Geschäften, Betrieben)* stock-room; **etw. auf** *od.* **am ~ haben** have sth. in stock; **etw. auf ~ haben** *(fig. ugs.)* be ready with sth.; **d)** *(Warenbestand)* stock; **e)** *(geh.)* bed; **an jmds. ~ treten** step up to sb.'s bedside

Lager-: ~bestand der *(Wirtsch.)* stock; **den ~bestand aufnehmen** do a stock-taking; **~feuer das** camp-fire; **~halle die, ~haus das** warehouse

Lagerist der; ~en, ~en storeman; storekeeper

Lageristin die; ~, ~nen storekeeper

Lager·koller der: einen ~ bekommen *od.* **kriegen** be driven to a frenzy by life in the camp

lagern 1. *tr. V.* **a)** store; **etw. kühl/trocken ~:** keep *or* store sth. in a cool/dry place; **b)** *(hinlegen)* lay down; **jmdn. flach/bequem ~:** lay sb. flat/in a comfortable position; **die Beine hoch ~:** rest one's legs in a raised position; **c)** *(Technik)* support; mount ⟨*machine-part, workpiece*⟩. **2.** *itr. V.* **a)** camp; be encamped; **b)** *(liegen)* lie; ⟨*foodstuffs, medicines, etc.*⟩ be stored *or* kept; *(sich ab~)* have settled; **c)** *(Geol.)* **hier ~ Ölvorräte** there are deposits of oil here; **d) ganz ähnlich/anders gelagert sein** ⟨*case*⟩ be quite similar/different [in nature]. **3.** *refl. V.* settle oneself/itself down

Lager-: ~raum a) store-room; *(im Geschäft, Betrieb)* stock-room; **b)** *o. Pl. (Kapazität)* storage space; *(in Lagerhallen)* warehouse space; **~stätte die** *(Geol.)* deposit

Lagerung die; ~, ~en a) storage; **bei ~ im Tiefkühlfach** if *or* when stored in a deep-freeze; **b) bei richtiger/falscher ~ des Verletzten** if the injured person is placed in the correct/wrong position

Lager·verwalter der storekeeper; stores supervisor

Lagune [la'gu:nə] **die; ~, ~n** lagoon

lahm [la:m] **1.** *Adj.* **a)** lame; crippled, useless ‹*wing*›; **ein ~es Bein haben** be lame in one leg; **b)** *(ugs.: unbeweglich)* stiff; **ihm wurde der Arm ~:** his arm became *or* got stiff; **c)** *(ugs. abwertend: schwach)* lame, feeble ‹*excuse, explanation, etc.*›; **d)** *(ugs. abwertend: matt)* dreary; dull; feeble ‹*protest*›; dull, dreary, lifeless ‹*discussion*›; **ein ~er Typ** a dull, lethargic [sort of] bloke *(Brit. coll.) or (sl.)* guy; *s. auch* **Ente a. 2.** *adv.* **a)** *(kraftlos)* feebly; **b)** *(ugs. abwertend)* lethargically

Lahme der/die; *adj. Dekl.* cripple

lahmen *itr. V.* be lame

lähmen ['lɛ:mən] *tr. V.* **a)** paralyse; **einseitig gelähmt sein** be paralysed down one side of one's body; **vor Angst wie gelähmt sein** be paralysed with fear; **b)** *(fig.)* cripple, paralyse ‹*economy, industry*›; bring ‹*traffic*› to a standstill; deaden ‹*enthusiasm*›; numb ‹*will*›

lahm|legen *tr. V.* bring ‹*traffic, production, industry*› to a standstill; paralyse ‹*industry*›

Lähmung die; ~, ~en a) paralysis; **b)** *(fig.) (der Wirtschaft, Industrie)* paralysis; *(der Begeisterung)* deadening; *(des Willens)* numbing

Laib [laip] **der; ~[e]s, ~e** loaf; **ein ~ Käse** a whole cheese

Laich [laiç] **der; ~[e]s, ~e** spawn

laichen *itr. V.* spawn

Laie ['laiə] **der; ~n, ~n a)** layman/laywoman; **da staunt der ~ [und der Fachmann wundert sich]** it's incredible; **b)** *(Kirche)* layman/laywoman; **die ~n** the laity *pl.*

laien-, Laien-: ~haft 1. *Adj.* amateurish; unprofessional; inexpert. **2.** *adv.* amateurishly; unprofessionally; inexpertly; **~prediger der** *(Rel.)* lay preacher; **~richter der** lay judge; **~schau·spieler der** amateur actor; **~spiel das;** *o. Pl.* amateur performance; **~theater das** amateur theatre group

Lakai [la'kai] **der; ~en, ~en a)** lackey; liveried footman; **b)** *(fig. abwertend)* lackey

Lake ['la:kə] **die; ~, ~n** brine

Laken ['la:kn̩] **das; ~s, ~** *(bes. nordd.)* sheet

lakonisch [la'ko:nɪʃ] **1.** *Adj.* laconic. **2.** *adv.* laconically

Lakritz [la'krɪts] **der** *od.* **das; ~es, ~e, Lakritze die; ~, ~n** liquorice

la la [la'la] *in* **so ~ ~** *(ugs.)* so-so

lallen ['lalən] *tr., itr. V.* ‹*baby*› babble; ‹*drunk/drowsy person*› mumble

Lama ['la:ma] **das; ~s, ~s** llama

Lamelle [la'mɛlə] **die; ~, ~n a)** *(einer Jalousie)* slat; **b)** *(eines Pilzes)* gill; lamella *(Bot.)*

lamentieren [lamɛn'ti:rən] *itr. V. (ugs.)* moan, complain **(über +** *Akk.* about)

Lametta [la'mɛta] **das; ~s a)** lametta; **b)** *(ugs. iron.: Orden)* gongs *pl. (coll.)*

Lamm [lam] **das; ~[e]s, Lämmer** ['lɛmɐ] lamb

lamm-, Lamm-: ~braten der roast lamb *no indef. art.; (Gericht)* roast of lamb; **~fell das** lambskin; **~fleisch das** lamb; **~fromm 1.** *Adj.* ‹*horse*› as gentle as a [little] lamb; ‹*person*› as meek as a [little] lamb; **2.** *adv.* **~fromm antworten** answer like a lamb; **~kotelett das** lamb chop

Lampe ['lampə] **die; ~, ~n a)** light; *(Tisch~, Öl~, Signal~)* lamp; *(Straßen~)* lamp; light; **b)** *(bes. fachspr.: Glüh~)* bulb; *s. auch* **Meister**

Lampen~: ~fieber das stage fright; **~schirm der** [lamp]shade

Lampion [lam'piɔŋ] **der; ~s, ~s** Chinese lantern

lancieren [lã'si:rən] *tr. V.* **a)** [deliberately] spread ‹*report, rumour, etc.*›; **b) jmdn. in eine Stellung ~:** get sb. into a position by pulling strings; **c)** *(bes. Wirtsch., Werbung)* launch

Land [lant] **das; ~es, Länder** ['lɛndɐ] *od. (veralt.)* **~e a)** *o. Pl.* land *no indef. art.;* **an ~:** ashore; **~ in Sicht!** *(Seemannsspr.)* land [ahead]!; **„~ unter!" melden** report that the land is flooded *or* under water; **[wieder] ~ sehen** *(fig.)* be able to see light at the end of the tunnel *(fig.);* **[sich** *(Dat.)***] eine Millionärin/antike Truhe/einen fetten Auftrag an ~ ziehen** *(ugs., oft scherzh.)* hook a millionairess/get one's hands on an antique chest/land a fat contract; **b)** *o. Pl. (Grund und Boden)* land; **ein Stück ~:** a plot *or* piece of land *or* ground; **das ~ bebauen/bestellen** farm/till the land; **c)** *Pl.* **~e** *(veralt.) (Gegend)* country; land; **Wochen/Jahre waren ins ~ gegangen** weeks/years had passed *or* gone by; **d)** *o. Pl. (dörfliche Gegend)* country *no indef. art.;* **auf dem ~ wohnen** live in the country; **aufs ~ ziehen** move into the country; **e)** *Pl.* **Länder** *(Staat)* country; **andere Länder, andere Sitten** *(Spr.)* every nation has its own ways of behaving; **~ und Leute kennenlernen** get to know the country and its people *or* inhabitants; **außer ~es gehen/sich außer ~es befinden** leave the country/be out of the country; **wieder im ~e sein** *(ugs.)* be back again; **f)** *(Bundesland)* Land; state; *(österr.)* province

land-, Land-: ~ab [-'-] *s.* **~auf; ~adel der** *(hist.)* landed aristocracy; **~arbeiter der** agricultural worker; farm worker; **~auf** [-'-] *Adv. in* **~auf, ~ab** *(geh.)* throughout the land; far and wide; **~besitz der** *s.* **Grundbesitz; ~bevölkerung die** rural population

Lande-: ~anflug der *(Flugw.)* [landing] approach; **~bahn die** *(Flugw.)* [landing] runway

land·einwärts [-'--] *Adv.* inland

Lande·klappe die *(Flugw.)* landing flap

landen 1. *itr. V.; mit sein* **a)** land; *(ankommen)* arrive; **weich ~:** make a soft landing; **bei jmdm. nicht ~ [können]** *(fig. ugs.)* not get anywhere *or* very far with sb.; **b)** *(ugs.: gelangen)* land *or* end up; **im Krankenhaus/Zuchthaus/Papierkorb ~:** land up in hospital/end up in prison/the waste-paper basket. **2.** *tr. V.* **a)** land ‹*aircraft, troops, passengers, fish, etc.*›; **b)** *(ugs.: zustande bringen)* pull off ‹*victory, coup*›; have ‹*smash hit*›; **c)** *(Boxen)* land ‹*punch*›

Lande·platz der *(Flugw.)* landing-strip; airstrip; *(Hubschrauber~)* landing-pad

Ländereien [lɛndə'raiən] *Pl.* estates

Länder-: ~kampf der *(Sport)* international match; **~spiel das** *(Sport)* international [match]

landes-, Landes-: ~grenze die national border *or* frontier; **~haupt·stadt die** capital; **~innere das** interior [of the country]; **~liste die** *(Politik)* regional list; **~regierung die** government of a/the Land/province; **~sprache die** language of the country; **~tracht die** national costume *or* dress; **~üblich** *Adj.*

usual *or* customary in a/the country; ~**verrat der** *(Rechtsw.)* treason; ~**verteidigung die;** *o. Pl.* national defence; ~**währung die** currency of a/the country

land-, Land-: ~**flucht die** migration from the land *or* countryside [to the towns]; ~**frau die** countrywoman; ~**friedensbruch der** *(Rechtsw.)* breach of the peace; ~**funk der** *(Sendefolge)* farming programmes *pl.* [on the radio]; ~**gericht das** regional court; Land court; ~**haus das** country house; ~**karte die** map; ~**kreis der** district; ~**krieg der** land warfare; ~**läufig** *Adj.* widely held *or* accepted; *(nicht fachlich)* popular; ~**leben das;** *o. Pl.* country life; ~**leute** *Pl.* country folk *or* people

ländlich ['lɛntlɪç] *Adj.* rural; country *attrib.* ⟨*life*⟩; **die ~e Ruhe** the quiet of the countryside

Land-: ~**luft die** country air; ~**maschine die** agricultural machine; farm machine; ~**plage die** plague [on the country]; *(fig.)* pest; nuisance; ~**ratte die** *(ugs., oft scherzh.)* landlubber; ~**regen der** steady rain

Landschaft die; ~, ~**en a)** landscape; *(ländliche Gegend)* countryside; **b)** *(Gemälde)* landscape; **c)** *(Gegend)* region

landschaftlich 1. *Adj.* **a) die ~e Schönheit** the scenic beauty; the beauty of the landscape; **b)** regional ⟨*accent, speech, expression, custom, usage, etc.*⟩. **2.** *adv.* ~ **herrlich gelegen sein** be in a glorious natural setting

Landschafts-: ~**bild das a)** *(Gemälde)* landscape [painting]; **b)** *(Aussehen)* landscape; ~**maler der** landscape painter; ~**pflege die;** *o. Pl.* landscape conservation *no art.;* ~**schutzgebiet das** conservation area

Land·sitz der country seat

Lands·mann der; *Pl.* ~**leute** fellow-countryman; compatriot

Land-: ~**straße die** country road; *(im Gegensatz zur Autobahn)* ordinary road; ~**streicher der** tramp; vagrant; ~**streicherei die;** *o. Pl.* vagrancy *no art.;* ~**streitkräfte** *Pl.* land forces; ~**strich der** area; ~**tag der** Landtag; state parliament; *(österr.)* provincial parliament

Landung die; ~, ~**en** landing; **zur ~ ansetzen** begin one's/its landing approach

Landungs-: ~**boot das** landing-craft; ~**brücke die** [floating] landing-stage; ~**steg der** landing-stage

Land-: ~**vermesser der** [land] surveyor; ~**weg der** overland route; **auf dem ~weg** overland; by the overland route; ~**wein der** ordinary local wine; vin du pays; ~**wirt der** farmer

Land·wirtschaft die a) *o. Pl.* agriculture *no art.;* farming *no art.;* **b)** *(Betrieb)* [small] farm

land·wirtschaftlich 1. *Adj.; nicht präd.* agricultural; agricultural, farm *attrib.* ⟨*machinery*⟩. **2.** *adv.* ~ **genutzt werden** be used for agricultural *or* farming purposes

Landwirtschaftsministerium das ministry of agriculture

Land·zunge die *(Geogr.)* tongue of land

[1]lang [laŋ]; **länger** ['lɛŋɐ], **längst...** ['lɛŋst...] **1.** *Adj.* **a)** *(räumlich)* long; **eine Bluse mit ~en Ärmeln** a long-sleeved blouse; **etw. länger machen** make sth. longer; lengthen sth.; **ein fünf Meter ~es Seil** a rope five metres long *or* in length; **b)** *(ugs.: groß)* tall; *s. auch* **Latte a; Lulatsch; c)** *(ausführlich)* long; **des ~en und breiten** *(geh.)* at great length; in great detail; **d)** *(zeitlich)* long; long, lengthy ⟨*speech, lecture, etc.*⟩; prolonged ⟨*thought*⟩; **seit ~er Zeit, seit ~em** for a long time. **2.** *adv.* **a)** *(zeitlich)* [for] a long time; **der ~ anhaltende Beifall** the lengthy *or* prolonged applause; **etw. nicht länger ertragen können** be unable to bear *or* stand sth. any longer; ~ **und breit** at great length; in great detail; **b) einen Augenblick/mehrere Stunden ~:** for a moment/several hours; **den ganzen Winter ~:** all through the winter; **sein Leben ~:** all one's life; *s. auch* **länger 2, 3**

[2]lang *(bes. nordd.)* **1.** *Präp. mit Akk.: s.* **entlang 1. 2.** *Adv. s.* **entlang 2; [nicht] wissen, wo es ~ geht** *(fig.)* [not] know what it's all about

lang-: ~**ärm[e]lig** [~ɛrm(ə)lɪç] *Adj.* long-sleeved; ~**atmig** [~a:tmɪç] **1.** *Adj.* long-winded; **2.** *adv.* long-windedly; **etw. ~atmig erzählen** relate sth. at great length

lange; länger, am längsten *Adv.* **a)** a long time; **er ist schon ~ fertig** he finished a long time ago; ~ **schlafen/arbeiten** sleep/work late; **bist du schon ~ hier?** have you been here long?; **es ist noch gar nicht ~ her, daß ich ihn gesehen habe** it's not long since I saw him; I saw him not long ago; **da kannst du ~ warten** you can wait for ever; **sie wird es nicht mehr ~ machen** *(ugs.)* she won't last much longer; *s. auch* **länger 3; b)** *(bei weitem)* **das ist [noch] ~ nicht alles** that's not all by any means; that's not all, not by a long chalk *or* shot *(coll.);* **ich bin noch ~ nicht fertig** I'm nowhere near finished; **er ist noch ~ nicht soweit** he's got a long time to go till then; **hier ist es ~ nicht so schön** it isn't nearly as nice here

Länge ['lɛŋə] **die;** ~, ~**n a)** *(auch zeitlich) (hoher Wuchs)* tallness; length; **eine ~ von zwei Metern haben** be two metres in length; **auf einer ~ von zwei Kilometern** for two kilometres; **sich zu seiner ganzen ~ aufrichten** draw oneself up to one's full height; **ein Film von einer Stunde ~:** a film one hour in length; an hour-long film; **etw./sich in die ~ ziehen** drag sth. out/drag on; go on and on; **b)** *(Sport)* length; **mit einer ~ [Vorsprung] siegen** win by a length; **c)** *Pl. (in einem Film, Theaterstück usw.)* long drawn-out *or* tedious scene; *(in einem Buch)* long drawn-out *or* tedious passage; **d)** *(Geogr.)* longitude

langen *(ugs.)* **1.** *itr. V.* **a)** *s.* **reichen 1a; b)** *(greifen)* reach **(in +** *Akk.* into; **auf +** *Akk.* on to; **nach** for); **c)** *s.* **reichen 1b. 2.** *tr. V.* **a)** *(landsch.) s.* **reichen 2a; b) jmdm. eine ~:** give sb. a clout [around the ear] *(coll.)*

Längen-: ~**grad der** *(Geogr.)* degree of longitude; ~**maß das** unit of length

länger ['lɛŋɐ] **1.** *s.* **[1]lang, lange. 2.** *Adj.* **eine ~e Abwesenheit/Behandlung** a fairly long *or* prolonged absence/period of treatment; **seit ~er Zeit** for quite some time. **3.** *adv.* for some time

Lange·weile die; ~ *od.* **Langenweile** boredom; ~ **haben** be bored

lang-, Lang-: ~**finger der** *(oft scherzh.) (Dieb)* thief; *(Taschendieb)* pickpocket; ~**fristig** [~frɪstɪç] **1.** *Adj.* long-term; long-dated ⟨*loan*⟩; **2.** *adv.* on a long-term basis; ~**fristig gesehen** in the long term; ~**haarig** *Adj.* long-haired; ~**jährig** *Adj.; nicht präd.* ⟨*customer, friend*⟩ of many years' standing; long-standing ⟨*friendship*⟩; ~**jährige Erfahrung** many years of experience; many years' experience; ~**lauf der** *(Skisport)* cross-country; ~**läufer der** cross-country skier; ~**lebig** [~le:bɪç] *Adj.* long-lived ⟨*animals,* *organisms*›; durable ‹*goods, materials*›; ~**lebige Gebrauchsgüter** consumer durables; ~**lebigkeit die;** ~ longevity; long-livedness; *(von Gebrauchsgütern)* durability; ~|**legen** *refl. V. (ugs.)* **a)** lie down; have a lie down; **b)** *(salopp: hinfallen)* fall flat [on one's face/back]

länglich ['lɛŋlɪç] *Adj.* oblong; long narrow ‹*opening*›; long [narrow] ‹*envelope*›

Lang·mut die; ~: forbearance

längs [lɛŋs] **1.** *Präp. + Gen. od. (selten) Dat.* along; ~ **des Flusses** *od.* **dem Fluß** along the river [bank]. **2.** *Adv.* **a)** lengthways; **stellt das Sofa hier ~ an die Wand** put the sofa along here against the wall; **b)** *(nordd.) s.* **entlang 2**

Längs·achse die longitudinal axis

langsam 1. *Adj.* **a)** slow; low ‹*speed*›; **b)** *(allmählich)* gradual. **2.** *adv.* **a)** slowly; **geh [etwas] ~er!** go [a bit] more slowly; slow down [a bit]!; ~, **aber sicher** *(ugs.)* slowly but surely; **b)** *(allmählich)* gradually; **es wird ~ Zeit, daß du gehst** it's about time you left *or* went

Langsamkeit die; ~: slowness

Lang·schläfer der late riser

Langspiel·platte die long-playing record; LP

längs-, Längs-: ~**richtung die** longitudinal direction; **in ~richtung** lengthways; ~**schnitt der** longitudinal section; ~**seite die** long side; ~**seits** *(Seemannsspr.)* **1.** *Präp. + Gen.* alongside; **2.** *Adv.* alongside

längst [lɛŋst] *Adv.* **a)** *(schon lange)* long since; **er ist [schon] ~ fertig** he finished a long time ago; **ich wußte das ~:** I've known that for a long time; I knew that a long time ago; **b)** *s.* **lange b**

längstens ['lɛŋstn̩s] *Adv. (ugs.)* at [the] most; ~ **eine Woche** a week at the most

Lang·strecke die a) long haul *or* distance; **b)** *(Sport)* long distance

Langstrecken-: ~**flug der** long-haul flight; ~**lauf der** *(Sport)* long-distance race; *(Disziplin)* long-distance running *no art.*; ~**läufer der** *(Sport)* long-distance runner

Languste [laŋ'gʊstə] **die;** ~, ~**n** spiny lobster; langouste

lang-, Lang-: ~**weilen 1.** *tr. V.* bore; **er sah gelangweilt aus dem Fenster** he gazed out of the window, feeling bored. **2.** *refl. V.* be bored; **sich tödlich** *od.* **zu Tode ~weilen** be bored to death; ~**weiler der** *(ugs. abwertend)* **a)** bore; **b)** *(schwerfälliger Mensch)* slowcoach; ~**weilig 1.** *Adj.* **a)** boring; dull ‹*place*›; **b)** *(ugs.: schleppend)* slow ‹*person*›; tedious ‹*business*›; **2.** *adv.* boringly; ~**welle die** *(Physik, Rundf.)* long wave; ~**wierig** [~vi:rɪç] *Adj.* lengthy; prolonged ‹*search*›; protracted, lengthy, long ‹*negotiations, treatment*›

Lanze ['lantsə] **die;** ~, ~**n** lance; *(zum Werfen)* spear; **für jmdn. eine ~ brechen** *(fig.)* take up the cudgels on sb.'s behalf

lapidar [lapi'da:ɐ̯] **1.** *Adj. (kurz, aber wirkungsvoll)* succinct; *(knapp)* terse. **2.** *adv.* succinctly/tersely

Lappalie [la'pa:li̯ə] **die;** ~, ~**n** trifle

Lappe ['lapə] **der;** ~**n,** ~**n** Lapp; Laplander

Lappen der; ~**s,** ~ **a)** cloth; *(Fetzen)* rag; *(Wasch~)* flannel; **b)** *(salopp: Geldschein)* [large] note; **c) jmdm. durch die ~ gehen** *(ugs.)* slip through sb.'s fingers

läppern ['lɛpɐn] *refl. V.* **es läppert sich** *(ugs.)* it's mounting up

läppisch ['lɛpɪʃ] *Adj.* silly

Lapp·land (das) Lapland

Lärche ['lɛrçə] **die;** ~, ~**n** larch

Larifari das; ~**s** *(ugs.)* nonsense; rubbish

Lärm [lɛrm] **der;** ~**[e]s** noise; *(fig.)* fuss (**um** about) to-do; ~ **schlagen** kick up *or* make a fuss

lärm-, Lärm-: ~**bekämpfung die;** *o. Pl.* noise abatement; ~**belästigung die** disturbance caused by noise; ~**empfindlich** *Adj.* sensitive to noise *postpos.*

lärmen *itr. V.* make a noise; ~**d** noisy

Lärm·schutz der a) protection against noise; **b)** *(Vorrichtung)* noise barrier; noise *or* sound insulation *no indef. art.*

Larve ['larfə] **die;** ~, ~**n** grub; larva

las [la:s] *1. u. 3. Pers. Sg. Prät. v.* **lesen**

lasch [laʃ] **1.** *Adj.* limp ‹*handshake*›; feeble ‹*action, measure*›; listless ‹*movement, gait*›; lax ‹*upbringing*›. **2.** *adv.: s. Adj.:* limply; feebly; listlessly; laxly; ~ **gewürzt sein** be insipid *or* tasteless

Lasche die; ~, ~**n** *(Gürtel~)* loop; *(eines Briefumschlags)* flap; *(Schuh~)* tongue

Laser ['leɪzə] **der;** ~**s,** ~ *(Physik)* laser

Laser·strahl der *(Physik)* laser beam

lasieren [la'zi:rən] *tr. V.* varnish

laß [las] *Imperativ Sg. v.* **lassen**

lassen ['lasn̩] **1.** *unr. tr. V.* **a)** *mit Inf. (2. Part. ~) (veranlassen)* **etw. tun ~:** have *or* get sth. done; **Wasser in die Wanne laufen ~:** run water into the bath; **das Licht über Nacht brennen ~:** keep the light on overnight; **jmdn. warten/erschießen ~:** keep sb. waiting/have sb. shot; **jmdn. grüßen ~:** send one's regards to sb.; **jmdn. kommen/rufen ~:** send for sb.; **jmdn. etw. wissen ~:** let sb. know sth.; **b)** *mit Inf. (2. Part. ~) (erlauben)* **jmdn. etw. tun ~:** let sb. do sth.; allow sb. to do sth.; **jmdn. ausreden ~:** let sb. finish speaking; allow sb. to finish speaking; **er läßt sich** *(Dat.)* **nichts sagen** you can't tell him anything; **c)** *(zugestehen, belassen)* **laß den Kindern den Spaß** let the children enjoy themselves; **jmdn. in Frieden ~:** leave sb. in peace; **laß ihn in seinem Glauben** don't disillusion him; **jmdn. unbeeindruckt ~:** leave sb. unimpressed; **das muß man ihm/ihr ~:** one must grant *or* give him/her that; **d)** *(hinein~/heraus~)* let *or* allow (**in** + *Akk.* into, **aus** out of); **jmdn. ins Zimmer ~:** let *or* allow sb. into the room; **e)** *(unterlassen)* stop; *(Begonnenes)* put aside; **laß das!** stop that *or* it!; **laß das Grübeln!** stop brooding!; **etw. nicht ~ können** be unable to stop sth.; **es nicht ~ können, etw. zu tun** be unable to stop doing sth.; **tu, was du nicht ~ kannst** go ahead and do what you want to do; **f)** *(zurück~; bleiben ~)* leave; **jmdn. allein ~:** leave sb. alone *or* on his/her own; **g)** *(überlassen)* **jmdm. etw. ~:** let sb. have sth.; **h) laß/laßt uns gehen/fahren!** let's go!; **i)** *(verlieren)* lose; *(ausgeben)* spend; **sein Leben für eine Idee ~:** lay down one's life for an idea; **j) laß sie nur erst einmal erwachsen sein** wait till she's grown up. **2.** *unr. refl. V. (2. Part. ~)* **a) die Tür läßt sich leicht öffnen** the door opens easily; **das läßt sich nicht beweisen** it can't be proved; **das läßt sich machen** that can be done; *s. auch* **hören 1 b, c; b)** *unpers.* **es läßt sich nicht leugnen/verschweigen, daß ...:** it cannot be denied *or* there's no denying that .../we/you *etc.* cannot hide the fact that ...; **hier läßt es sich leben/wohl sein** it's a good life here. **3.** *unr. itr. V.* **a)** *(ugs.)* **Laß mal. Ich mache das schon** Leave it. I'll do it; **Laß doch** *od.* **nur! Du kannst mir das Geld später zu-**

rückgeben That's all right. You can pay me back later; **b)** *(2. Part. ~) (veranlassen)* **ich lasse bitten** would you ask him/her/them to come in; **ich habe mir sagen ~, daß ...:** I've been told *or* informed that ...; **c)** *(veralt.: aufgeben)* **von jmdm./etw. ~:** part from sb./sth.

lässig ['lɛsɪç] **1.** *Adj.* casual. **2.** *adv.* **a)** *(ungezwungen)* casually; **b)** *(ugs.: leicht)* easily; effortlessly

Lässigkeit die; ~ a) casualness; **b)** *(ugs.: Leichtigkeit)* effortlessness

Lasso ['laso] **das** *od.* **der; ~s, ~s** lasso

läßt [lɛst] *3. Pers. Sg. Präsens v.* **lassen**

Last [last] **die; ~, ~en a)** load; *(Trag~)* load; burden; **b)** *(Gewicht)* weight; **c)** *(Bürde)* burden; **die ~ des Amtes/der Verantwortung** the burden of office/responsibility; **jmdm. zur ~ fallen/werden** be/become a burden on sb.; **jmdm. etw. zur ~ legen** charge sb. with sth.; accuse sb. of sth.; **d)** *Pl. (Abgaben)* charges; *(Kosten)* costs; **die steuerlichen ~en** the tax burden *sing.*; **die Verpackungskosten gehen zu ~en des Kunden** the cost of packaging will be charged to the customer

Last·auto das *(ugs.) s.* **~kraftwagen**

lasten *itr. V.* **a)** be a burden; **auf jmdm./etw. ~:** weigh heavily [up]on sb./sth.; **das Amt lastet auf seinen Schultern** *(fig.)* the burden of office rests on his shoulders; **b)** *(belastet sein mit)* **auf dem Haus ~ zwei Hypotheken** the house is encumbered with two mortgages

Lasten·auf·zug der goods lift *(Brit.)*; freight elevator *(Amer.)*

[1]**Laster der; ~s, ~** *(ugs.: Lkw)* truck; lorry *(Brit.)*

[2]**Laster das; ~s, ~:** vice

lasterhaft *Adj. (abwertend)* depraved

Laster·höhle die *(ugs. abwertend)* den of vice *or* iniquity

lästerlich 1. *Adj.* malicious ‹*remark*›; malevolent ‹*curse, oath*›. **2.** *adv.* ‹*curse*› malevolently; ‹*speak*› maliciously

Läster·maul das *(abwertend salopp)* **ein ~ sein/haben** have a malicious tongue; be constantly making malicious remarks

lästern ['lɛstɐn] *itr. V. (abwertend)* make malicious remarks **(über** + *Akk.* about)

Lästerung die; ~, ~en *(gegen Gott)* blasphemy

lästig ['lɛstɪç] *Adj.* tiresome ‹*person*›; tiresome, irksome ‹*task, duty, etc.*›; troublesome ‹*illness, cough, etc.*›; **jmdm. ~ sein** *od.* **fallen/werden** be/become a nuisance to sb.

Last-: ~kahn der [cargo] barge; **~kraftwagen der** heavy goods *(Brit.) or (Amer.)* freight vehicle; **~schrift die** debit; **~wagen der** truck; lorry *(Brit.)*; **~wagen·fahrer der** truck driver; lorry driver *(Brit.)*

Lasur [la'zu:ɐ̯] **die; ~, ~en** varnish; *(farbig)* glaze

Latein [la'tain] **das; ~s** Latin; **mit seinem ~ am Ende sein** be at one's wit's end; *s. auch* **Deutsch**

Latein·amerika (das) Latin America

latein·amerikanisch *Adj.* Latin-American

lateinisch *Adj.* Latin; *s. auch* **deutsch;** [2]**Deutsche**

latent [la'tɛnt] **1.** *Adj.* latent. **2.** *adv.* **~ vorhanden sein** be latent

Laterne [la'tɛrnə] **die; ~, ~n a)** *(Leuchte)* lamp; lantern *(Naut.)*; **b)** *(Straßen~)* street light; street lamp

Laternen·pfahl der lamppost

Latinum [la'ti:nʊm] **das; ~s: das kleine/große ~:** ≈ GCSE/'A' level Latin [examination]

Latrine [la'tri:nə] **die; ~, ~n** latrine

latschen ['la:tʃn̩] *itr. V.; mit sein (salopp)* trudge; *(schlurfend)* slouch

Latschen der; ~s, ~ *(ugs.)* old worn-out shoe; *(Hausschuh)* old worn-out slipper; **er ist bald aus den ~ gekippt, als er hörte, ...** *(salopp)* he was flabbergasted when he heard ...

Latte ['latə] **die; ~, ~n a)** lath; slat; *(Zaun~)* pale; **eine lange ~** *(ugs.)* a beanpole; **b)** *(Sport: des Tores)* [cross]bar; **c)** *(Leichtathletik)* bar; **d) eine [lange] ~ von Schulden/Vorstrafen** *(ugs.)* a [large] pile of debts/a [long] list *or* string of previous convictions

Latten-: ~kreuz das *(Fuß-, Handball)* angle of the [cross]bar and the post; **~rost der** *(auf dem Boden)* duckboards *pl.*; *(eines Bettes)* slatted frame; **~zaun der** paling fence

Latz [lats] **der; ~es, Lätze** ['lɛtsə] bib; **jmdm. eine[n] vor den ~ knallen** *od.* **ballern** *(salopp)* sock *(sl.) or* thump sb.

Lätzchen ['lɛtsçən] **das; ~s, ~:** bib

Latz·hose die bib and brace; *(für Kinder)* dungarees *pl.*

lau [lau] *Adj.* **a)** tepid, lukewarm ‹*water etc.*›; *(nicht mehr kalt)* warm ‹*beer etc.*›; **b)** *(mild)* mild ‹*wind, air, evening, etc.*›; mild and gentle ‹*rain*›; **c)** *(unentschlossen)* lukewarm; half-hearted

Laub [laup] **das; ~[e]s** leaves *pl.*; **dichtes/neues ~:** thick/new foliage

Laub·baum der broad-leaved tree

Laube die; ~, ~n summer-house; *(überdeckter Sitzplatz)* bower; arbour; *s. auch* **fertig**

Laub-: ~frosch der tree frog; **~säge die** fretsaw; **~wald der** deciduous wood/forest

Lauch [laux] **der; ~[e]s, ~e** *(Bot.)* allium; *(Porree)* leek

Lauer ['lauɐ] **die; ~** *(ugs.)* **auf der ~ liegen** *od.* **sein** *(ugs.)* lie in wait; **sich auf die ~ legen** settle down to lie in wait

lauern *itr. V.* **a)** *(auch fig.)* lurk; **auf jmdn./etw. ~:** lie in wait for sb./sth.; **ein ~der Blick** a sly look; **b)** *(ugs.: ungeduldig warten)* **auf jmdn./etw. ~:** wait [impatiently] for sb./sth.

Lauf [lauf] **der; ~[e]s, Läufe** ['lɔyfə] **a)** *o. Pl.* running; **b)** *(Sport: Wettrennen)* heat; **c)** *o. Pl. (Ver~, Entwicklung)* course; **im ~[e] der Zeit** in the course of time; **im ~[e] der Jahre** over the years; as the years go/went by; **im ~[e] des Tages** during the day; **einer Sache** *(Dat.)* **ihren** *od.* **freien ~ lassen** give free rein to sth.; **der ~ der Geschichte/Welt** the course of history/the way of the world; **seinen ~ nehmen** take its course; **d)** *(von Schußwaffen)* barrel; **etw. vor den ~ bekommen** get a shot at sth.; **e)** *o. Pl. (eines Flusses, einer Straße)* course; **der obere/untere ~ eines Flusses** the upper/lower reaches *pl.* of a river; **f)** *(Musik)* run; **g)** *(Jägerspr.)* leg

Lauf-: ~bahn die a) *(Werdegang)* career; **eine wissenschaftliche/künstlerische ~bahn einschlagen** take up a career in the sciences/as an artist; **b)** *(Leichtathletik)* running-track; **~bursche der** errand boy; messenger boy

laufen 1. *unr. itr. V.; mit sein* **a)** run; **ge~ kommen** come running up; **er lief, was er konnte** *(ugs.)* he ran as fast as he could; **b)** *(gehen)* go; *(zu Fuß gehen)* walk; **es sind noch/nur fünf Minuten zu ~:** it's another/only five minutes' walk; **in** *(Akk.)***/gegen etw. ~:** walk into sth.; **dauernd zum Arzt/in die Kirche ~** *(ugs.)* keep running to the doctor/be always going to church; **c)** *(in einem Wettkampf)* run; *(beim Eislauf)* skate; *(beim Ski~)* ski; **ein Pferd ~ lassen** run a horse; **d)** *(im Gang sein)* ‹*machine*› be running; ‹*radio, television, etc.*› be on; *(funktionieren)* ‹*machine*› run; ‹*radio, television, etc.*› work; **auf Hochtouren ~:** be running at full speed; **e)** *(sich bewegen, fließen; auch fig.)* **auf Schienen/ über Rollen ~:** run on rails/over pulleys; **von den Fließbändern ~:** come off the conveyor belts; **es lief mir eiskalt über den Rücken** a chill ran down my spine; **ihm lief der Schweiß über das Gesicht** the sweat ran down his face; **Wasser in die Wanne ~ lassen** run the bathwater; **deine Nase läuft** your nose is running; you've got a runny nose; **der Käse läuft** the cheese has gone runny *(coll.)*; **f)** *(gelten)* ‹*contract, agreement, engagement, etc.*› run; **g)** ‹*programme, play*› be on; ‹*film*› be on *or* showing; ‹*show*› be on *or* playing; **der Hauptfilm läuft schon** the main film has already started; **h)** *(fahren)* run; **auf Grund ~:** run aground; **i)** *(vonstatten gehen)* **parallel mit etw. ~:** run in parallel with sth.; **der Laden läuft/die Geschäfte ~ gut/schlecht** *(ugs.)* the shop is doing well/badly/ business is good/bad; **wie geplant/nach Wunsch ~:** go as planned *or* according to plan; **j)** ‹*negotiations, investigations*› be in progress *or* under way; **k)** *(registriert sein)* **auf jmds. Namen** *(Akk.)* **~:** be in sb.'s name; **l)** *(ugs.: gut verkäuflich sein)* go *or* sell well. **2.** *unr. tr. u. itr. V.* **a)** *mit sein* run; *(zu Fuß gehen)* walk; **b)** *mit sein* **einen Rekord ~:** set up a record; **über die 100 m 9,9 Sekunden ~:** run the 100m. in 9.9 seconds; **c)** *mit haben od. sein* **Ski/Schlittschuh/Rollschuh ~:** ski/skate/roller-skate; **d) sich** *(Dat.)* **die Füße wund ~:** get sore feet from running/walking; **sich** *(Dat.)* **ein Loch in die Schuhsohle ~:** wear a hole in one's shoe *or* sole. **3.** *unr. refl. V.* **a) sich warm ~:** warm up; **b)** *unpers.* **in diesen Schuhen läuft es sich sehr bequem** these shoes are very comfortable for running/walking in *or* to run/ walk in

laufend **1.** *Adj.; nicht präd.* **a)** *(ständig)* regular ‹*interest, income*›; recurring ‹*costs*›; **die ~en Arbeiten** the day-to-day *or* routine work *sing.*; **b)** *(gegenwärtig)* current ‹*issue, year, month, etc.*›; **c)** *(aufeinanderfolgend)* **zehn Mark der ~e Meter** ten marks a *or* per metre; **d) auf dem ~en sein/bleiben** be/keep *or* stay up-to-date *or* fully informed; **jmdn. auf dem ~en halten** keep sb. up-to-date *or* informed **2.** *adv.* constantly; continually; ‹*increase*› steadily

laufen|lassen *unr. tr. V. (ugs.)* **jmdn. ~:** let sb. go

Läufer ['lɔyfɐ] **der; ~s, ~ a)** *(Sport)* runner; *(Handball)* half-back; **b)** *(Fußball veralt.)* half-back; **c)** *(Teppich) (long narrow)* carpet; **d)** *(Schach)* bishop

Lauferei die; ~, ~en *(ugs.)* running around *no pl.*

Läuferin die; ~, ~nen runner

Lauf·feuer das brush fire; **wie ein ~:** like wildfire

läufig ['lɔyfɪç] *Adj.* on heat *postpos.*; in season *postpos.*

Lauf-: ~kundschaft die passing trade; **~masche die** ladder; **~paß der: jmdm. den ~paß geben** *(ugs.)* give sb. his/her marching orders *(coll.)*; **er hat seiner Freundin den ~paß gegeben** *(ugs.)* he finished with his girl-friend *(coll.)*; **~schritt der a) wir haben die ganze Strecke im ~schritt zurückgelegt** we ran all the way; **im ~schritt, marsch, marsch!** at the double, quick march!; **b)** *(Leichtathletik)* running step

läufst [lɔyfst] *2. Pers. Sg. Präsens v.* **laufen**

Lauf-: ~stall der playpen; **~steg der** catwalk

läuft [lɔyft] *3. Pers. Sg. Präsens v.* **laufen**

Lauf·zeit die term; **ein Kredit mit befristeter ~:** a limited-term loan

Lauge ['laugə] **die; ~, ~n a)** soapy water; soapsuds; **b)** *(Chemie)* alkaline solution

Laune ['launə] **die; ~, ~n a)** mood; **schlechte/gute ~ haben** be in a bad/good mood *or* temper; **jmdn. bei [guter] ~ halten** keep sb. in a good mood; keep sb. happy *(coll.)*; **bringt gute ~ mit!** come ready to enjoy yourselves; **b)** *meist Pl. (wechselnde Stimmung)* mood; **die ~n des Wetters** *(fig.)* the vagaries of the weather; **c)** *(spontane Idee)* whim; **aus einer ~ heraus** on a whim; on the spur of the moment

launenhaft, launisch *Adj.* temperamental; *(unberechenbar)* capricious

Laus [laus] **die; ~, Läuse** ['lɔyzə] louse; **ihm ist eine ~ über die Leber gelaufen** *(ugs.)* he has got out of bed on the wrong side; **jmdm./ sich eine ~ in den Pelz setzen** *(ugs.)* let sb./oneself in for something

Laus·bub der little rascal *or* devil; scamp

Lausbuben·streich der prank

lauschen ['lauʃn̩] *itr. V.* **a)** *(horchen)* listen *(so as to overhear sth.)*; **b)** *(zuhören)* listen [attentively] (*Dat.* to)

Lauscher der; ~s, ~ a) eavesdropper; **b)** *(Jägerspr.: Ohr)* ear

Lause-: ~bengel der, ~junge der *(salopp) s.* **Laus·bub**

lausen *tr. V.* delouse; **ich denk', mich laust der Affe!** *(salopp)* well, I'll be damned *or* blowed!

lausig *(ugs.)* **1.** *Adj. (abwertend)* lousy *(sl.)*; rotten *(coll.)*; **~e Zeiten** hard times; **b)** perishing *(Brit. sl.)*, freezing ‹*cold*›; terrible *(coll.)*, awful ‹*heat*›. **2.** *adv.* terribly *(coll.)*; awfully; **~ kalt** perishing cold *(Brit. sl.)*

[1]laut [laut] **1.** *Adj.* **a)** loud; *(fig.)* loud, garish ‹*colour*›; **der Motor ist zu ~:** the engine is too noisy; **spreche ich jetzt ~ genug?** can you hear me now?; **werden Sie bitte nicht ~!** there's no need to shout; **~ werden** *(fig.: bekannt werden)* be made known; **b)** *(geräuschvoll)* noisy. **2.** *adv.* **a)** loudly; **~er sprechen** speak louder; speak up; **~ lachen** laugh out loud; **etw. [nicht] ~ sagen [dürfen]** [not be allowed to] say sth. out loud; **~ denken** think aloud; **das kannst du aber ~ sagen** *(ugs.)* you can say 'that again; **b)** *(geräuschvoll)* noisily

[2]laut *Präp. + Gen. od. Dat. (Amtsspr.)* according to

Laut der; ~[e]s, ~e *(auch Phon.)* sound; **keinen ~ von sich geben** not make a sound; **fremde/heimatliche ~e** sounds of a foreign/ familiar tongue

Laute die; ~, ~n lute

lauten *itr. V.* ‹*answer, instruction, slogan*› be, run; ‹*letter, passage, etc.*› read, go; ‹*law*› state; **die Anklage lautet auf ...:** the charge is ...; **auf jmds. Namen** *(Akk.)* **~:** be in sb.'s name

läuten [lɔytn̩] **1.** *tr., itr. V.* ring; ‹*alarm clock*› go off; **Mittag ~:** strike midday; **ich habe davon ~ gehört** *od.* **hören, daß ...** *(fig. ugs.)* I have heard rumours that ... **2.** *itr. V. (bes. südd.: klingeln)* ring; **nach jmdm. ~:** ring for sb.; **es läutete** the bell rang *or* went

[1]lauter *Adj. (geh.)* honourable ‹*person, intentions, etc.*›; honest ‹*truth*›

[2]lauter *indekl. Adj.* nothing but; sheer, pure ‹*nonsense, joy, etc.*›;

das sind ~ Lügen that's nothing but lies; that's a pack of lies; **vor ~ Arbeit komme ich nicht ins Theater** I can't go to the theatre because of all the work I've got

läutern ['lɔytɐn] *tr. V. (geh.)* reform ⟨*character*⟩; purify ⟨*soul*⟩

laut·hals *Adv.* ⟨*sing, shout, etc.*⟩ at the top of one's voice; **~ lachen** roar with laughter

lautlich 1. *Adj.* phonetic. **2.** *adv.* phonetically

laut·los 1. *Adj.* silent; soundless; *(wortlos)* silent. **2.** *adv.* silently; soundlessly

Laut·schrift die *(Phon.)* phonetic alphabet

Laut·sprecher der loudspeaker; loud hailer *(esp. Naut.)*; *(einer Stereoanlage usw.)* speaker

Lautsprecher-: ~anlage die public address *or* PA system; loudspeaker system; **~box die** speaker cabinet

laut·stark 1. *Adj.* loud; vociferous, loud ⟨*protest*⟩. **2.** *adv.* loudly; ⟨*protest*⟩ vociferously, loudly

Laut·stärke die volume; **in/bei voller ~:** at full volume

lau·warm *Adj.* lukewarm ⟨*food*⟩; lukewarm, tepid ⟨*drink*⟩; *(nicht mehr kalt)* warm ⟨*beer etc.*⟩

Lava ['la:va] **die; ~, Laven** *(Geol.)* lava

Lavabo [la'va:bo] **das; ~[s], ~s** *(schweiz.) s.* **Waschbecken**

Lavendel [la'vɛndl̩] **der; ~s, ~:** lavender

lavieren [la'vi:rən] *tr. V., itr. V.* manœuvre

Lawine [la'vi:nə] **die; ~, ~n** *(auch fig.)* avalanche; **eine ~ von Protesten** *(fig.)* a storm of protest

Lawinen·gefahr die danger of avalanches

lax [laks] **1.** *Adj.* lax. **2.** *adv.* laxly

Lazarett [latsa'rɛt] **das; ~[e]s** military hospital

Leasing ['li:zɪŋ] **das; ~s, ~s** *(Wirtsch.)* leasing

leben ['le:bn̩] **1.** *itr. V.* live; *(lebendig sein)* be alive; **für jmdn./etw. ~** *od. (geh.)* **jmdm./einer Sache ~:** live for sb./sth.; **leb[e] wohl!** farewell!; **nicht mehr ~ wollen** not want to go on living; have lost the will to live; **von seiner Rente ~:** live on one's pension/salary; **von seiner Hände Arbeit ~:** live by the work of one's hands; **Wie geht es dir? – Man lebt!** *(ugs.)* How are you? – Oh, surviving *(coll.)*; **fleischlos ~:** not eat meat; **von Kartoffeln ~:** live on potatoes. **2.** *tr. V.* live; **ein glückliches Leben ~:** live a happy life

Leben das; ~s, ~ a) life; **das ~:** life; **jmdm. das ~ retten** save sb.'s life; **sich** *(Dat.)* **das ~ nehmen** take one's [own] life; **am ~ sein/bleiben** be/stay alive; **seines ~s nicht [mehr] sicher sein** not be safe [any more]; **um sein ~ rennen** run for one's life; **ums ~ kommen** lose one's life; **auf Tod und ~ kämpfen** be engaged in a life-and-death struggle; **etw. für sein ~ gern essen/tun** love sth./doing sth.; **etw. ins ~ rufen** bring sth. into being; **mit dem ~ davonkommen/das nackte ~ retten** escape/barely escape with one's life; **jmdm. nach dem ~ trachten** try to kill sb.; **ein/sein [ganzes] ~ lang** one's whole life long; **noch nie im ~/zum erstenmal im ~:** never in/for the first time in one's life; **mit beiden Beinen** *od.* **Füßen im ~ stehen** have one's feet firmly on the ground; **nie im ~, im ~ nicht!** *(ugs.)* not on your life! *(coll.)*; never in your life! *(coll.)*; **ein ~ in Wohlstand** a life of affluence; **wie das ~ so spielt** it's funny the way things turn out; **im öffentlichen ~ stehen** be in public life; **so ist das ~:** such is life; that's the way things go; **die Musik ist ihr ~:** music is her [whole] life; **b)** *(Betriebsamkeit)* **auf dem Markt herrschte ein reges ~:** the market was bustling with activity; **das ~ auf der Straße** the comings and goings in the street; **~ ins Haus bringen** bring some life into the house

lebend *Adj.* living; live ⟨*animal*⟩; **tot oder ~:** dead or alive

lebendig [le'bɛndɪç] **1.** *Adj.* **a)** *(auch fig.)* living; **jmdn. ~** *od.* **bei ~em Leibe verbrennen** burn sb. alive; **man fühlt sich hier wie ~ begraben** being stuck here is like being buried alive *(coll.)*; **die Erinnerung daran wurde in ihm wieder ~:** the memory of it came back to him vividly; **b)** *(lebhaft)* lively ⟨*account, imagination, child, etc.*⟩; gay, bright ⟨*colours*⟩. **2.** *adv.* *(lebhaft)* in a lively fashion *or* way

Lebendigkeit die; ~: liveliness

lebens-, Lebens-: ~abend der *(geh.)* evening *or* autumn of one's life *(literary)*; **~abschnitt der** stage of *or* chapter in one's life; **~art die** *o. Pl.* manners *pl.*; **~bedingungen** *Pl.* conditions of life; **~bejahend** *Adj.* ⟨*person*⟩ with a positive attitude *or* approach to life; **~bereich der** area of life; **~dauer die a)** life-span; **b)** *(von Maschinen)* [useful] life; **~echt 1.** *Adj.* true-to-life; **2.** *adv.* in a true-to-life way; **~ende das** end [of one's life]; **~erfahrung die** experience *no indef. art.* of life; **~erinnerungen** *Pl.* *(Memoiren)* memoirs; **~erwartung die** life expectancy; **~fähig** *Adj.* *(auch fig.)* viable; **~fremd** *Adj.* out of touch with *or* remote from everyday life *postpos.*; **~freude die;** *o. Pl.* zest for life; joie de vivre; **~froh** *Adj.* full of zest for life *or* joie de vivre *postpos.*; **~führung die** lifestyle; **~gefahr die** mortal danger; **„Achtung, ~gefahr!“** 'danger'; **sie schwebt in ~gefahr** she is in danger of dying; *(von einer Kranken)* her condition is critical; **außer ~gefahr sein** be out of danger; **~gefährlich 1.** *Adj.* highly *or* extremely dangerous; critical ⟨*injury*⟩; **2.** *adv.* critically ⟨*injured, ill*⟩; **~gefährte der, ~gefährtin die** *(geh.)* companion through life *(literary)*; **~gemeinschaft die a)** *(von Menschen)* long-term relationship; **b)** *(Biol.: von Tieren, Pflanzen)* biocoenosis; **~geschichte die** life-story; **~groß** *Adj.* life-size; **~größe die: eine Statue in ~größe** a life-size statue; **~haltungs·kosten** *Pl.* cost of living *sing.*; **~inhalt der** purpose in life; **ihre Familie ist ihr ~inhalt** her family is her whole life; **~jahr das** year of [one's] life; **in seinem 12. ~jahr** in his twelfth year; **mit dem vollendeten 18. ~jahr** on reaching the age of eighteen; **~kraft die** vitality; vital energy; **~künstler der: ein [echter/wahrer] ~künstler** a person who always knows how to make the best of things; **~lage die** situation [in life]; **~lang 1.** *Adj.* lifelong; **2.** *adv.* all one's life; **~länglich 1.** *Adj.* **~länglicher Freiheitsentzug** life imprisonment; **„~länglich“ bekommen** get life imprisonment *or* *(coll.)* life; **2.** *adv.* **jmdn. ~länglich gefangenhalten** keep sb. imprisoned for life; **~lauf der** curriculum vitae; c.v.; **~lustig** *Adj.* ⟨*person*⟩ full of the joys of life

Lebens·mittel das; *meist Pl.* food[stuff]; **~** *Pl.* food *sing.*; foods *(formal)*; foodstuffs *(formal)*; *(als Ware)* food *sing.*

Lebensmittel-: ~abteilung die food department; **~geschäft das** food shop

lebens-, Lebens-: ~müde *Adj.* weary of life *pred.*; **du bist wohl ~müde?** *(scherzh.)* you must be tired of living; **~mut der** courage to go on living; **~nah 1.** *Adj.* true-to-life ⟨*film, description, etc.*⟩; ⟨*teaching*⟩ closely related to life; **2.** *adv.* **etw. ~nah schildern** describe sth. in a true-to-life way; **~notwendig** *Adj.* essential ⟨*foodstuff*⟩; vital ⟨*organ*⟩; **~qualität die;** *o. Pl.* quality of life; **~raum der a)** *(Umkreis)* lebensraum; **b)** *(Biol.) s.* **Biotop; ~regel die** rule [of life]; maxim; **~retter der** rescuer; **sein ~retter** the person who saved his life; **~standard der** standard of living; **~stellung die** permanent position *or* job; job for life; **~stil der** lifestyle; **~tüchtig** *Adj.* able to cope with life *postpos.*; **~umstände** *Pl.* circumstances; **~unfähig** *Adj.* non-viable; **~unterhalt der: seinen ~unterhalt verdienen** earn one's living; **für jmds. ~unterhalt sorgen** support sb.; **~untüchtig** *Adj.* unable to cope with life *postpos.*; **~versicherung die** life insurance; life assurance; **eine ~versicherung abschließen** take out a life insurance *or* assurance policy; **~wandel der** way of life; **einen zweifelhaften/einwandfreien ~wandel führen** lead a dubious/an irreproachable life; **~weg der** [journey through] life; **~weise die** way of life; **~werk das** life's work; **~wert** *Adj.* **ein ~wertes Leben** a life worth living; **~wichtig** *s.* **~notwendig; ~wille der** will to live; **~zeichen das** sign of life; **kein ~zeichen [von sich] geben** show no sign of life; **~zeit die** life[-span]; **auf ~zeit** for life; **ein Beamter auf ~zeit** an established civil servant

Leber ['le:bɐ] **die; ~, ~n** liver; **es an der ~ haben** *(ugs.)* have [got] liver trouble; **frisch** *od.* **frei von der ~ weg sprechen** *od.* **reden** *(ugs.)* speak one's mind; *s. auch* **Laus**

leber-, Leber-: ~fleck der liver spot; **~käse der;** *o. Pl. meat loaf made with mincemeat, [minced liver,] eggs, and spices;* **~knödel der** *(südd., österr.) meat ball made from minced liver, onions, eggs, and flour;* **~krank** *Adj.* ⟨*patient etc.*⟩ suffering from a liver complaint *or* disorder; **~leiden das** *(Med.)* liver complaint *or* disorder; **~pastete die** liver pâté; **~tran der** fish-liver oil; *(des Kabeljaus)* cod-liver oil; **~wurst die** liver sausage; **die gekränkte** *od.* **beleidigte ~wurst spielen** *(ugs.)* get all huffy *(coll.)*

Lebe-: ~wesen das living being *or* thing *or* creature; **~wohl** [--'-] **das; [e]s, ~** *od.* **~e** *(geh.)* farewell; **jmdm. ~wohl sagen** bid sb. farewell

lebhaft 1. *Adj.* **a)** lively ⟨*person, gesture, imagination, bustle, etc.*⟩; lively, animated ⟨*conversation, discussion*⟩; lively, brisk ⟨*activity*⟩; busy ⟨*traffic*⟩; brisk ⟨*business*⟩; vivid ⟨*idea, picture, etc.*⟩; **b)** *(kräftig)* lively ⟨*interest*⟩; lively, gay ⟨*pattern*⟩; bright, gay ⟨*colour*⟩; vigorous ⟨*applause, opposition*⟩. **2.** *adv.* **a)** in a lively way *or* fashion; ⟨*remember sth.*⟩ vividly; **sich ~ unterhalten** have a lively *or* animated conversation; **sich** *(Dat.)* **etw. ~ vorstellen können** have a vivid picture of sth.; **b)** *(kräftig)* brightly, gaily ⟨*coloured*⟩; gaily ⟨*patterned*⟩

Leb·kuchen der ≈ gingerbread

leb-, Leb-: ~los *Adj.* lifeless ⟨*body, eyes*⟩; **[wie] ~los daliegen** lie there as if dead; **~tag der: [all] mein/dein** *usw.* **~tag** *(ugs.)* all my/your *etc.* life; **so was habe ich mein ~tag nicht erlebt** *(ugs.)* I've never seen anything like it in all my life *or* in all my born days; **~zeiten** *Pl.* **zu jmds. ~zeiten** while sb. is/was still alive; during sb.'s lifetime

lechzen ['lɛçtsn̩] *itr. V. (geh.)* **nach einem Trunk/nach Kühlung ~:** long for a drink/to be able to cool off; **nach Rache** *usw.* **~:** thirst for revenge *etc.*

leck [lɛk] *Adj.* leaky; **~ sein** leak

Leck das; ~[e]s, ~s leak

[1]**lecken 1.** *tr. V.* lick; **sich** *(Dat.)* **die Lippen** *usw.* **~:** lick one's lips *etc.*; **jmdm. die Hand** *usw.* **~:** lick sb.'s hand *etc.*; **leck mich [doch]!** *(derb)* [why don't you] piss off! *(coarse); s. auch* **Arsch a; Finger b. 2.** *itr. V.* **an etw.** *(Dat.)* **~:** lick sth.

[2]**lecken** *itr. V. (leck sein)* leak

lecker *Adj.* tasty ⟨*meal*⟩; delicious ⟨*cake etc.*⟩; good ⟨*smell, taste*⟩; *(fig.: ansprechend)* lovely ⟨*girl*⟩

Lecker·bissen der delicacy; **ein musikalischer ~** *(fig.)* a musical treat

Leckerei die; ~, ~en *(ugs.)* dainty; *(Süßigkeit)* sweet [meat]

leck|schlagen *unr. itr V.; mit sein (Seemannsspr.)* be holed

led. *Abk.* **ledig**

Leder ['le:dɐ] **das; ~s, ~ a)** leather; **in ~ [gebunden]** leather-bound; **zäh wie ~ sein** ⟨*person*⟩ be as hard as nails; **jmdm. ans ~ gehen/wollen** *(ugs.)* go for sb./be out to get sb.; **b)** *(Fenster~)* leather; chamois *or* chammy [leather]; **c)** *(Fußballjargon: Ball)* ball; leather *(dated sl.)*

Leder-: ~garnitur die leather-upholstered suite; **~hand·schuh der** leather glove; **~hose die** leather shorts *pl.*; lederhosen *pl.*; *(lang)* leather trousers; **~jacke die** leather jacket; **~mantel der** leather [over]coat

[1]**ledern** *tr. V.* leather

[2]**ledern** *Adj.* **a)** *nicht präd. (aus Leder)* leather; **b)** *(wie Leder)* leathery

Leder-: ~sessel der leather[-upholstered] armchair; **~sohle die** leather sole; **~waren** *Pl.* leather goods

ledig ['le:dɪç] *Adj.* **a)** single; unmarried ⟨*mother*⟩; **b) einer Sache** *(Gen.)* **~ sein** *(geh.)* be free of sth.

Ledige der/die; *adj. Dekl.* single person

lediglich *Adj.* only; merely; simply

Lee [le:] **die** *od.* **das; ~** *(Seemannsspr.)* **nach ~ drehen/ in ~ liegen** turn to/tie to leeward

leer [le:ɐ̯] *Adj.* **a)** empty; blank, clean ⟨*sheet of paper*⟩; **sein Glas ~ trinken** empty *or* drain one's glass; **seinen Teller ~ essen** clear one's plate; **die Schachtel ~ machen** *(ugs.)* finish the box; **~ ausgehen** come away empty-handed; **~ laufen** ⟨*machine*⟩ idle; *s. auch* **leerlaufen; b)** *(menschenleer)* empty; empty, deserted ⟨*streets*⟩; **die Wohnung steht ~:** the house is standing empty *or* is unoccupied; **c)** *(abwertend: oberflächlich)* empty ⟨*words, promise, talk, display*⟩; vacant ⟨*expression*⟩; **mit ~en Augen/~em Blick starren** stare vacantly

Leere die; ~ *(auch fig.)* emptiness; **eine gähnende ~:** a gaping void; **eine innere ~** *(fig.)* a feeling of emptiness inside

leeren 1. *tr. V.* **a)** empty; empty, clear ⟨*post-box*⟩; **b)** *(österr.: gießen)* pour ⟨*water, milk, etc.*⟩. **2.** *refl. V.* ⟨*hall, theatre, etc.*⟩ empty

leer-, Leer-: ~gefegt *Adj.* deserted ⟨*street, town*⟩; **wie ~gefegt** deserted; **~gewicht das** unladen weight; **~gut das** empties *pl.*; **~lauf der;** *o. Pl.* **a) im ~lauf den Berg hinunterfahren** ⟨*driver*⟩ coast down the hill in neutral; ⟨*cyclist*⟩ freewheel *or* coast down the hill; **b)** *(fig.)* **es gab [viel] ~lauf im Büro** there were [long] slack periods in the office; **~|laufen** *unr. itr. V.; mit sein* **die Badewan-**

ne läuft ~: the bathwater is running out; **das Faß ist ~gelaufen** the barrel has run dry; *s. auch* **leer a; ~stehend** *Adj.* empty, unoccupied ‹*house, flat*›; **~taste die** space-bar

Leerung die; ~, ~en emptying; **nächste ~ um 12 Uhr** *(auf Briefkästen)* next collection at 12.00

Lefze ['lɛftsə] **die; ~, ~n** lip

legal [le'ga:l] **1.** *Adj.* legal. **2.** *adv.* legally

legalisieren *tr. V.* legalize

Legalität [legali'tɛ:t] **die; ~:** legality; **außerhalb der ~:** outside the law

Legasthenie [legaste'ni:] **die; ~, ~n** *(Psych., Med.)* difficulty in learning to read and write

Legastheniker [legas'te:nikɐ] **der; ~s, ~** *(Psych., Med.)* one who has difficulty with reading and writing

Lege·henne die laying hen

legen ['le:gn̩] **1.** *tr. V.* **a)** lay [down]; **jmdn. auf den Rücken ~:** lay sb. on his/her back; **etw. auf den Tisch ~:** lay sth. on the table; **etw. aus der Hand/beiseite ~:** put sth. down/aside *or* down; **etw. in den Kühlschrank ~:** put sth. in the refrigerator; **die Füße auf den Tisch ~:** put one's feet on the table; **b)** *(verlegen)* lay ‹*pipe, cable, railway track, carpet, tiles, etc.*›; **c)** *(in eine bestimmte Form bringen)* **etw. in Falten ~:** fold sth.; **sich** *(Dat.)* **die Haare ~ lassen** have one's hair set; *s. auch* **Falte c; d)** *(schräg hinstellen)* lean; **etw. an etw.** *(Akk.)* **~:** lean sth. [up] against sth. **2.** *tr., itr. V.* ‹*hen*› lay. **3.** *refl. V.* **a)** lie down; **sich auf etw.** *(Akk.)* **~:** lie down on sth.; **das Schiff/Flugzeug legte sich auf die Seite** the ship keeled over/the aircraft banked steeply; **sich in die Kurve ~:** lean into the bend; *s. auch* **Bett a; Ohr b; b)** *(nachlassen)* ‹*wind, storm*› die down, abate, subside; ‹*noise*› die down, abate; ‹*enthusiasm*› wear off, subside, fade; ‹*anger*› abate, subside; ‹*excitement*› die down, subside; **c)** *(sich herabsenken)* **sich auf** *od.* **über etw.** *(Akk.)* **~:** ‹*mist, fog*› descend *or* settle on sth., [come down and] blanket sth.

legendär [legɛn'dɛ:ɐ̯] *Adj.* legendary

Legende [le'gɛndə] **die; ~, ~n a)** legend; **zur ~ werden** *(fig.)* ‹*event, incident, etc.*› become legendary; ‹*person*› become a legend; **b)** *(Zeichenerklärung)* legend; key

leger [le'ʒe:ɐ̯] **1.** *Adj.* **a)** casual; relaxed; **b)** *(bequem)* casual ‹*jacket etc.*›. **2.** *adv.* **a)** casually; in a casual *or* relaxed manner; **b)** *(bequem)* ‹*dress*› casually

Legierung die; ~, ~en alloy

Legion [le'gio:n] **die; ~, ~en a)** *(Milit.)* legion; *(Fremden~)* Legion; **b)** *(Menge)* horde **(von** of)

Legionär [legio'nɛ:ɐ̯] **der; ~s, ~e** legionary

legislativ [legɪsla'ti:f] *(Politik)* **1.** *Adj.* legislative. **2.** *adv.* by legislation

Legislative [legɪsla'ti:və] **die; ~, ~n** *(Politik)* legislature

Legislatur·periode [legɪsla'tu:ɐ̯-] **die** *(Politik)* parliamentary term; legislative period

legitim [legi'ti:m] **1.** *Adj.* legitimate. **2.** *adv.* legitimately

Legitimation [legitima'tsio:n] **die; ~, ~en a)** *(auch Rechtsw.: Ehelicherklärung)* legitimation; **b)** *(Ausweis)* proof of identity; *(Bevollmächtigung)* authorization

legitimieren 1. *tr. V.* **a)** *(rechtfertigen)* justify; **b)** *(bevollmächtigen)* authorize; **c)** *(für legitim erklären)* legitimize ‹*child, relationship*›. **2.** *refl. V.* show proof of one's identity

Legitimität [legitimi'tɛ:t] **die; ~:** legitimacy

Lehm [le:m] **der; ~s** loam; *(Ton)* clay

Lehm-: ~boden der loamy soil; *(Tonerde)* clay soil; **~hütte die** mud hut

lehmig *Adj.* loamy ‹*soil, earth*›; *(tonartig)* clayey ‹*soil, shoes, etc.*›

Lehm·ziegel der clay brick

Lehne ['le:nə] **die; ~, ~n** *(Rükken~)* back; *(Arm~)* arm

lehnen 1. *tr. V.* lean **(an +** *Akk.*, **gegen** against); **den Kopf an etw.** *(Akk.)* **~:** lean one's head on sth. **2.** *refl. V.* lean **(an +** *Akk.*, **gegen** against; **über +** *Akk.* over); **sich aus dem Fenster ~:** lean out of the window. **3.** *itr. V.* be leaning **(an +** *Dat.* against)

Lehn-: ~stuhl der armchair; **~wort das** loan-word

Lehr-: ~amt das *(Schulw.)* teaching post; *(Beruf)* **das ~amt** the teaching profession; **~auftrag der** lectureship; **~beauftragte der/die** lecturer; **~buch das** textbook

Lehre ['le:rə] **die; ~, ~n a)** *(Berufsausbildung)* apprenticeship; **eine ~ machen** serve an apprenticeship **(als** as); **bei jmdm. in die ~ gegangen sein** *(fig.)* have learnt a lot from sb.; **b)** *(Weltanschauung)* doctrine; **die christliche ~:** Christian doctrine; **die ~ Kants/Hegels/Buddhas** the teachings *pl.* of Kant/Hegel/Buddha; **c)** *(Theorie, Wissenschaft)* theory; **die ~ vom Schall** the science of sound *or* acoustics; **d)** *(Erfahrung)* lesson; **laß dir das eine ~ sein!** let that be a lesson to you; **jmdm. eine [heilsame] ~ erteilen** teach sb. a [salutary] lesson

lehren *tr., itr. V. (auch fig.)* teach; **jmdn. lesen** *usw.* **~:** teach sb. to read *etc.*; **ich werde dich ~, so bockig zu sein!** *(ugs.)* I'll teach you to be so contrary *(coll.)*

Lehrer der; ~s, ~ a) *(auch fig.)* teacher; **er ist ~ für Geschichte** he teaches history; he is a history teacher; **b)** *(Ausbilder)* instructor

Lehrer·ausbildung die teacher training *no art.*

Lehrerin die; ~, ~nen teacher

Lehrer-: ~kollegium das teaching staff; faculty *(Amer.)*; **~mangel der** shortage of teachers

Lehrerschaft die; ~, ~en teachers *pl.*; *(einer Schule)* teaching staff; faculty *(Amer.)*

Lehrer-: ~schwemme die *(ugs.)* glut of teachers; **~zimmer das** staff-room

Lehr-: ~fach das a) subject; **b)** *(Beruf des Lehrens)* teaching profession; **im ~fach tätig sein** be a teacher; **~gang der** course **(für, in +** *Dat.* in); **einen ~gang machen, an einem ~gang teilnehmen** take a course; **~geld das** *(fig.)* **du kannst dir dein ~geld zurückgeben lassen!** your education was wasted on you; **~geld geben** *od.* **[be]zahlen [müssen]** learn the hard way; **~herr der** *(geh. veralt.)* master *(of an apprentice)*; **~jahr das** year as an apprentice; **~körper der** *(Amtsspr.)* teaching staff; faculty *(Amer.)*; **~kraft die** teacher

Lehrling ['le:ɐ̯lɪŋ] **der; ~s, ~e** apprentice; *(in kaufmännischen Berufen)* trainee

Lehrlings·ausbildung die training of apprentices

lehr-, Lehr-: ~mädchen das [girl] apprentice; *(in kaufmännischen Berufen)* [girl] trainee; **~meister der** teacher; *(Vorbild)* mentor; **~methode die** teaching method; **~mittel das** *(Schulw.)* teaching aid; **~mittel** *Pl.* teaching materials; **~reich** *Adj.* instructive, informative ‹*book, film, etc.*›; **es war eine ~reiche Erfahrung für ihn** the experience taught him a lot; **~stelle die** apprenticeship; *(in kaufmännischen Berufen)* trainee post; **~stoff der**

(Schulw.) syllabus; ~**stuhl der** *(Hochschulw.)* chair **(für** of); ~**veranstaltung die** *(Hochschulw.)* class; *(Vorlesung)* lecture; ~**zeit die** [period of] apprenticeship

Leib [laip] **der;** ~**[e]s,** ~**er** *(geh.)* **a)** body; **am ganzen** ~ **zittern** shiver all over; **bleib mir vom** ~**[e]!** keep away from me!; **etw. am eigenen** ~ **erfahren** *od.* **erleben** experience sth. for oneself; **er hat sich mit** ~ **und Seele der Musik verschrieben** he dedicated himself heart and soul to music; **mit** ~ **und Seele dabei sein** put one's whole heart into it; **jmdm. auf den** ~ *od.* **zu** ~**e rücken** *(ugs.)* chivvy sb.; *(mit Kritik)* get at sb. *(coll.)*; **sich** *(Dat.)* **jmdn. vom** ~**e halten** *(ugs.)* keep sb. at arm's length; **jmdm. mit einer Sache vom** ~**e bleiben** *(ugs.)* not pester sb. with sth.; **einer Sache** *(Dat.)* **zu** ~**e rücken** tackle sth.; set about sth.; **jmdm. auf den** ~ **geschnitten sein** be tailor-made for sb.; suit sb. down to the ground; **b)** *(geh., fachspr.: Bauch)* belly; *(Magen)* stomach; **c) eine Gefahr für** ~ **und Leben** *(veralt.)* a danger to life and limb

Leib·eigenschaft die; *o. Pl.* *(hist.)* serfdom *no def. art.*

leiben *itr. V.* **wie er/sie** *usw.* **leibt und lebt** to a T

Leibes·erziehung die *(Schulw.)* physical education; PE

Leib·gericht das favourite dish

leibhaftig [laip'haftıç] **1.** *Adj.* **a)** *(persönlich)* in person *postpos.*; **da stand er** ~ **vor uns** there he was, as large as life; **b)** *(echt)* real; **ein** ~**er Herzog** a real live duke; **der Leibhaftige** *(scherzh.)* the devil incarnate. **2.** *adv. (ugs.)* actually; believe it or not

leiblich *Adj.* **a)** physical ‹*well-being*›; **b)** *(blutsverwandt)* real ‹*mother, parents, etc.*›

Leib-: ~**rente die** life annuity; ~**speise die** *s.* ~**gericht;** ~**wache die,** ~**wächter der** bodyguard; ~**wäsche die** underwear; underclothes *pl.*

Leiche ['laiçə] **die;** ~**,** ~**n** [dead] body; *(bes. eines Unbekannten)* corpse; **er sieht aus wie eine lebende** *od.* **wandelnde** ~ *(salopp)* he looks like death warmed up *(sl.)*; **nur über meine** ~! over my dead body!; **über** ~**n gehen** *(abwertend)* be utterly ruthless *or* unscrupulous

leichen-, Leichen-: ~**blaß** *Adj.* deathly pale; white as a sheet *postpos.*; ~**halle die** mortuary; ~**hemd das** burial garment; ~**schändung die** desecration of a corpse; *(sexuell)* necrophilia *no art.*; ~**schau·haus das** morgue; ~**starre die** rigor mortis; ~**tuch das** *(veralt.)* winding-sheet; shroud; ~**wagen der** hearse; ~**zug der** *(geh.)* cortège; funeral procession

Leichnam ['laiçna:m] **der;** ~**s,** ~**e** *(geh.)* body; **jmds.** ~: sb.'s body *or* mortal remains *pl.*

leicht [laiçt] **1.** *Adj.* **a)** light; lightweight ‹*suit, material*›; ~**e Waffen** small-calibre arms; ~**e Kleidung** thin clothes; *(luftig)* light *or* cool clothes; **gewogen und zu** ~ **befunden** tried and found wanting; **mit** ~**er Hand** with ease; **etw. auf die** ~**e Schulter nehmen** *(ugs.)* take sth. casually; make light of sth.; **b)** *(einfach)* easy ‹*task, question, job, etc.*›; *(nicht anstrengend)* light ‹*work, duties, etc.*›; **ein** ~**es Leben haben** have an easy life; **es** ~**/nicht** ~ **haben** have/not have it easy *or* an easy time of it; **nichts** ~**er als das** nothing could be simpler *or* easier; **mit jmdm. [kein]** ~**es Spiel haben** find sb. is [not] easy meat; **c)** *(schwach)* slight ‹*accent, illness, wound, doubt, etc.*›; light ‹*wind, rain, sleep, perfume*›; **ein** ~**er Stoß [in die Rippen]** a gentle nudge [in the ribs]; **d)** *(bekömmlich)* light ‹*food, wine*›; mild ‹*cigar, cigarette*›; **e)** *(heiter)* light-hearted; **ihr wurde es etwas/viel** ~**er** she felt somewhat/much easier *or* relieved; **f)** *(unterhaltend)* light ‹*music, reading, etc.*›; **g) ein** ~**es Mädchen** *(veralt. abwertend)* a loose-living girl. **2.** *adv.* **a)** lightly ‹*built*›; ~ **bekleidet sein** be lightly *or* thinly dressed; **b)** *(einfach, schnell, spielend)* easily; ~ **verständlich** *od.* **zu verstehen sein** be easy to understand; be easily understood; **sie hat** ~ **reden** it's easy *or* all very well for her to talk; **das ist** ~**er gesagt als getan** that's easier said than done; **sie wird** ~ **böse** she has a quick temper; **das ist** ~ **möglich** that is perfectly possible; **ihr wird** ~ **schlecht** the slightest thing makes her sick; **c)** *(geringfügig)* slightly; ~ **gewürzt** lightly seasoned; **es regnete** ~: there was a light rain falling; **es hat** ~ **gefroren** there was a slight frost

leicht-, Leicht-: ~**athlet der** [track/field] athlete; ~**athletik die** [track and field] athletics *sing.*; ~**bekleidet** *Adj.; präd. getrennt geschrieben* lightly dressed; ~**benzin das** benzine; ~**entzündlich** *Adj.; präd. getrennt geschrieben* highly inflammable; ~**|fallen** *unr. itr. V.; mit sein* be easy; **das fällt mir** ~: it is easy for me; I find it easy; ~**fertig 1.** *Adj.* **a)** careless ‹*behaviour, person*›; rash ‹*promise*›; ill-considered, slapdash ‹*plan*›; **b)** *(veralt.: moralisch bedenkenlos)* promiscuous; loose ‹*woman*›; **2.** *adv.* carelessly; ~**fertigkeit die;** *o. Pl.* carelessness; ~**gewicht das** *(Schwerathletik)* **a)** *o. Pl.* lightweight; *s. auch* **Fliegengewicht; b)** *(ugs. scherzh.)* *(Mädchen)* sylph; *(Mann)* featherweight; ~**gläubig** *Adj.* gullible; credulous; ~**gläubigkeit die** gullibility; credulity; ~**hin** *Adv.* without [really] thinking; *(lässig)* casually; **etw.** ~**hin sagen** say sth. casually *or* unthinkingly

Leichtigkeit ['laiçtıçkait] **die;** ~ **a)** lightness; **b)** *(Mühelosigkeit)* ease; **mit** ~: with ease; easily

leicht-, Leicht-: ~**lebig** *Adj.* happy-go-lucky; ~**|machen** *tr. V.* **jmdm./sich etw.** ~**machen** make sth. easy for sb./oneself; **es sich** *(Dat.)* ~ **machen** make it *or* things easy for oneself; take the easy way out; ~**matrose der** ordinary seaman; ~**metall das** light metal; *(Legierung)* [light] alloy; ~**|nehmen** *unr. tr. V.* **etw.** ~**nehmen** make light of sth.; **seine Aufgabe nicht** ~**nehmen** take one's task seriously; **nimm's** ~: don't worry about it; ~**sinn der;** *o. Pl.* carelessness *no indef. art.*; *(mit Gefahr verbunden)* recklessness *no indef. art.*; *(Fahrlässigkeit)* negligence *no indef. art.*; ~**sinnig 1.** *Adj.* careless; *(sich, andere gefährdend)* reckless; *(fahrlässig)* negligent; **2.** *adv.* carelessly; *(gefährlich)* recklessly; ‹*promise*› rashly; ~**sinnig mit seinem Geld umgehen** be careless with one's money; ~**|tun** *unr. refl. V. (ugs.)* **sich mit etw.** ~**tun/nicht** ~**tun** manage sth. easily/have a hard time with sth.; ~**verdaulich** *Adj.; präd. getrennt geschrieben* [easily] digestible; ~**verletzt** *Adj.; präd. getrennt geschrieben* slightly injured; ~**verletzte der/die** slightly injured man/woman/person

leid [lait] *Adj.; nicht attr.* **a) es tut mir** ~**[, daß ...]** I'm sorry [that ...]; **so** ~ **es mir tut, aber ...:** I'm very sorry, but ...; **er tut mir** ~**/ es tut mir** ~ **um ihn** I feel sorry for him; **es tut mir** ~ **darum** I feel sorry *or* *(coll.)* bad about it; **b)** *(überdrüssig)* **etw./jmdn.** ~ **sein/werden**

(ugs.) be/get fed up with *(coll.) or* tired of sth./sb.

Leid das; ~[e]s a) *(Schmerz)* suffering; *(Kummer)* grief; sorrow; **großes** *od.* **schweres ~ erfahren** suffer greatly; *(Kummer)* suffer great sorrow; **geteiltes ~ ist halbes ~** *(Spr.)* a sorrow shared is a sorrow halved; **jmdm. sein ~ klagen** tell sb. all one's woes; **b)** *(Unrecht)* wrong; *(Böses)* harm; **jmdm. ein ~ zufügen** wrong/harm sb.; do sb. wrong/harm

leiden 1. *unr. itr. V.* **a)** suffer **(an, unter +** *Dat.* from); **unter jmdm. ~:** suffer because of sb.; **b)** *(Schaden nehmen)* suffer **(durch, unter +** *Dat.* from). **2.** *unr. tr. V.* **a) jmdn. [gut] ~ können** *od.* **[gern] ~ mögen** like sb.; **ich kann sie/das nicht ~:** I can't stand her/it; **b)** *(geh.: ertragen müssen)* suffer ‹*hunger, thirst, want, torment, etc.*›; **c)** *(dulden)* tolerate; **sie ist überall wohl gelitten** *(geh.)* she is liked by everybody

Leiden das; ~s, ~ a) *(Krankheit)* illness; *(Gebrechen)* complaint; **b)** *(Qual)* suffering; **Freud[en] und ~[en]** joy[s] and sorrow[s]

leidend *Adj.* **a)** *(krank)* ailing; in poor health *postpos.*; **~ aussehen** look sickly *or* poorly; **b)** *(schmerzvoll)* strained ‹*voice*›; martyred ‹*expression*›; ‹*look*› full of suffering

Leidenschaft die; ~, ~en passion; **mit ~:** fervently; passionately; **seine ~ für etw. entdecken** realize one's great love for sth.; **er ist Sammler aus ~:** he is a dedicated collector

leidenschaftlich 1. *Adj.* passionate; ardent, passionate ‹*lover*›; passionate[ly keen] ‹*skier, collector, etc.*›; violent, passionate ‹*hatred, quarrel*›; vehement ‹*protest*›. **2.** *adv.* **a)** passionately; *(eifrig)* dedicatedly; **~ diskutiert werden** be discussed heatedly; **b) etw. ~ gern tun** adore doing sth.

Leidens-: ~gefährte der, ~gefährtin die; ~genosse der, ~genossin die fellow-sufferer; **~geschichte die;** *o. Pl. (christl. Rel.)* **die ~geschichte Christi** Christ's Passion; **seine ~geschichte** *(fig.)* his tale of woe; **~miene die** woeful *or* martyred expression

leider *Adv.* unfortunately; **ich habe ~ keine Zeit** unfortunately *or* I'm afraid I haven't any time; **~ ja/nein** I'm afraid so/afraid not; **~ Gottes ist es nun einmal so** *(ugs.)* that's how it is, I'm afraid *or* worse luck; *(in förmlichen Briefen)* **wir müssen Ihnen ~ mitteilen ...:** we regret to inform you ...

leid·geprüft *Adj.* sorely tried; long-suffering

leidig *Adj.* tiresome; wretched

leidlich 1. *Adj.* reasonable; passable. **2.** *adv.* reasonably; fairly; **es geht mir [ganz] ~** *(ugs.)* I'm quite well *or* not too bad; **sie kann ~ Klavier spielen** she can play the piano reasonably well

Leid-: ~tragende der/die; *adj. Dekl.* victim; **der** *od.* **die ~tragende/die ~tragenden [dabei] sein** be the one/ones to suffer [in this]; **~wesen das: zu jmds. ~wesen** to sb.'s regret

Leier ['laiɐ] **die; ~, ~n** lyre; **[es ist] immer die alte/dieselbe ~** *(ugs. abwertend)* [it's] always the same old story

Leier·kasten der *(ugs.)* barrel-organ; hurdy-gurdy *(coll.)*

leiern *(ugs.)* **1.** *tr. V.* **a)** *(kurbeln)* **[nach oben/unten] ~:** wind [up/down]; **b)** *(monoton aufsagen)* drone through; *(schnell)* reel *or* rattle off. **2.** *itr. V.* **a) an etw.** *(Dat.)* **~:** wind away at sth.; **b)** *(monoton sprechen)* drone [on]

Leih-: ~bibliothek die, ~bücherei die lending library

leihen ['laiən] *unr. tr. V.* **a) jmdm. etw. ~:** lend sb. sth.; lend sth. to sb.; **b)** *(entleihen)* borrow; **[sich** *(Dat.)***] [von** *od.* **bei jmdm.] etw. ~:** borrow sth. [from sb.]; **ein geliehener Wagen** a borrowed car; **c)** *(geh.: gewähren)* lend, give ‹*support*›; give ‹*attention*›

Leih-: ~gabe die loan (*Gen.* from); **~gebühr die** hire *or (Amer.)* rental charge; *(bei Büchern)* lending charge; borrowing fee; **~haus das** pawnbroker's; pawnshop; **~mutter die** surrogate mother; **~wagen der** hire *or (Amer.)* rental car; **[sich** *(Dat.)***] einen ~wagen nehmen** hire *or (Amer.)* rent a car; **~weise** *Adv.* on loan; **hier hast du das Buch, aber nur ~weise** I'll give you the book, but only to borrow; **jmdm. etw. ~ überlassen** lend sth. to sb.

Leim [laim] **der; ~[e]s** glue; **aus dem ~ gehen** *(ugs.)* come apart; **jmdm. auf den ~ gehen** *od.* **kriechen** *(ugs.)* be taken in by sb.; fall for sb.'s trick/tricks; **jmdn. auf den ~ führen** *(ugs.)* take sb. in

leimen *tr. V.* **a)** glue **(an +** *Akk.* to); *(zusammen~)* glue [together]; **b) jmdn. ~** *(ugs.)* take sb. in

Leine ['lainə] **die; ~, ~n a)** rope; *(Zelt~)* guy-rope; **~ ziehen** *(ugs.)* clear off; **b)** *(Wäsche~, Angel~)* line; **c)** *(Hunde~)* lead *(esp. Brit.)*; leash; **den Hund an die ~ nehmen** put the dog on the lead/leash; **jmdn. an der [kurzen] ~ haben** *od.* **halten** *(ugs.)* keep sb. on a tight rein; **jmdn. an die ~ legen** *(ugs.)* get sb. under one's thumb

leinen *Adj.* linen

Leinen das; ~s a) *(Gewebe)* linen; **b)** *(Buchw.)* cloth; **Ausgabe in ~** cloth edition

Leinen-: ~band der cloth-bound volume; **~einband der** cloth binding; **~kleid das** linen dress; **~tuch das** linen cloth

Lein-: ~öl das linseed oil; **~samen der** linseed; **~wand die a)** *o. Pl.* linen; *(grob)* canvas; **b)** *(des Malers)* canvas; **c)** *(für Filme und Dias)* screen; **einen Roman** *usw.* **auf die ~wand bringen** *(fig.)* film a novel *etc.*

leise ['laizə] **1.** *Adj.* **a)** quiet; soft ‹*steps, music, etc.*›; faint ‹*noise*›; **sei ~!** be quiet!; **könnt ihr nicht ~r sein?** can't you make less noise?; **die Musik ~[r] stellen** turn the music down; **b)** *nicht präd. (leicht; kaum merklich)* faint; slight; slight, gentle ‹*touch*›; light ‹*rain*›; **nicht die ~ste Ahnung haben, nicht im ~sten ahnen** not have the faintest *or* slightest idea. **2.** *adv.* **a)** quietly; **sprich doch etwas ~r** lower your voice; **~ weinen** cry softly; **b)** *(leicht; kaum merklich)* slightly; ‹*touch, rain*› gently

Leise·treter der *(abwertend)* pussyfooter

Leiste ['laistə] **die; ~, ~n a)** strip; *(Holz~)* batten; *(profiliert)* moulding; *(halbrund)* beading; *(am Auto)* trim; *(Tapeten~)* [picture-]rail; picture moulding *(Amer.)*; **eine ~:** a piece *or* strip of moulding/beading/trim; *(Holz~)* a batten; **b)** *(Knopf~)* facing; **c)** *(Anat.)* groin; **d)** *(Weberei)* selvage

leisten 1. *tr. V.* **a)** do ‹*work*›; *(schaffen)* achieve ‹*a lot, nothing*›; **gute** *od.* **ganze Arbeit ~:** do good work *or* a good job; *(gründlich arbeiten)* do a thorough job; **der Motor leistet 80 PS** the engine develops *or* produces 80 b.h.p.; **b)** *(verblaßt od. als Funktionsverb)* **jmdm. Hilfe ~:** help sb.; **einen Eid ~:** swear *or* take an oath **2.** *refl. V. (ugs.)* **a) sich** *(Dat.)* **etw. ~:** treat oneself to sth.; **b) sich** *(Dat.)* **etw. [nicht] ~ können** [not] be able to afford sth.; **er kann es sich** *(Dat.)* **~, das zu tun** he can afford to do it; *(etw. Ris-*

kantes) he can get away with doing it; **c)** *(wagen)* **sich** *(Dat.)* **etw. ~:** get up to sth.; **was der sich** *(Dat.)* **leistet!** the things he gets away with!; **wer hat sich** *(Dat.)* **diese Frechheit geleistet?** who was it who had the cheek to do/say *etc.* that?

Leisten der; ~s, ~: last; **alles/alle über einen ~ schlagen** *(ugs.)* lump everybody/everything together

Leisten·bruch der rupture

Leistung die; ~, ~en a) *o. Pl. (Qualität bzw. Quantität der Arbeit)* performance; **Bezahlung nach ~:** payment according to performance *or* results; *(in der Industrie)* payment according to productivity; **b)** *(Errungenschaft)* achievement; *(im Sport)* performance; **eine große sportliche/technische ~:** a great sporting/technical feat; **c)** *o. Pl. (Leistungsvermögen, Physik: Arbeits~)* power; *(Ausstoß)* output; **die ~ einer Fabrik** the output *or* [production] capacity of a factory; **d)** *(Zahlung, Zuwendung)* payment; *(Versicherungsw.)* benefit; **die sozialen ~en der Firma** the firm's fringe benefits; **e)** *(Dienst~)* service; **f)** *o. Pl. (das Leisten)* carrying out; *(Eides~)* swearing

leistungs-, Leistungs-: ~druck der *(bei Arbeitnehmern)* pressure to work harder; *(bei Sportlern, Schülern)* pressure to achieve *or* to do well; **~fähig** *Adj.* capable ‹*person*›; *(körperlich)* able-bodied ‹*person*›; *(gute Arbeit leistend)* efficient ‹*worker, factory, industry, etc.*›; powerful ‹*engine, computer, etc.*›; *(konkurrenzfähig)* competitive ‹*firm, industry*›; **~fähigkeit die** *(eines Menschen)* capability; *(bei guter Arbeitsleistung)* efficiency; *(eines Betriebs, der Industrie)* productivity; *(Wirtschaftlichkeit)* efficiency; *(eines Motors, eines Computers usw.)* power; performance; **~gerecht 1.** *Adj.* ‹*salary, income*› based on performance *or* results; *(in der Industrie)* based on productivity; **2.** *adv.* **~gerecht bezahlt werden** receive a performance-related salary; **~gesellschaft die** [highly] competitive society; performance-oriented society; **~kurs der** *(Schulw.)* extension course; **~orientiert** *Adj.* achievement-oriented; [highly] competitive ‹*society*›; **~prämie die** productivity bonus; **~prinzip das;** *o. Pl.* achievement principle; competitive principle; **~schau die** *(Wirtsch., Landw.)* [product] exhibition; **~schwach** *Adj.* not performing well *pred.*; low-achieving *attrib.* ‹*worker, pupil*›; *(minderbegabt)* less able, lower-ability *attrib.* ‹*pupil*›; weak ‹*team*›; low-powered ‹*engine*›; **~sport der** competitive sport *no art.*; **~sportler der, ~sportlerin die** competitive sportsman/sportswoman; **~stark** *Adj.* high-performing *attrib.* ‹*athlete*›; able ‹*pupil, athlete, etc.*›; high-performance *attrib.*, powerful ‹*engine, car* ›; highly efficient ‹*business, power-station*›; *(sehr konkurrenzfähig)* highly competitive ‹*business, athlete*›; **~träger der, ~trägerin die** *(Sport)* key player; **~vermögen das** *s.* **~fähigkeit; ~zentrum das** *(Sport)* intensive training centre; **~zwang der** *(Soziol.) (bei Arbeitnehmern)* compulsion to work hard; *(bei Sportlern/Schülern)* compulsion to achieve *or* to do well

Leit-: ~artikel der *(Zeitungsw.)* leading article; leader; **~bild das** model

leiten ['laitn̩] *tr. V.* **a)** *(anführen)* lead, head ‹*expedition, team, discussion, etc.*›; be head of ‹*school*›; *(verantwortlich sein für)* be in charge of ‹*project, expedition, etc.*›; manage ‹*factory, enterprise*›; *(den Vorsitz führen bei)* chair ‹*meeting, discussion, etc.*›; *(Musik: dirigieren)* conduct ‹*orchestra, choir*›; direct ‹*small orchestra etc.*›; *(Sport: als Schiedsrichter)* referee ‹*game, match*›; **~der Angestellter** executive; manager; **~de Angestellte** senior *or* managerial staff; **~der Beamter** senior civil servant; **b)** *(begleiten, führen)* lead; **jmdn. auf die richtige Spur ~:** put sb. on the right track; **sich von etw. ~ lassen** [let oneself] be guided by sth.; **c)** *(lenken)* direct; route ‹*traffic*›; *(um~)* divert ‹*traffic, stream*›; **d)** *auch itr. (Physik)* conduct ‹*heat, current, sound*›; **etw. leitet gut/schlecht** sth. is a good/bad conductor

[1]Leiter der; ~s, ~ a) *(einer Gruppe)* leader; head; *(einer Abteilung)* head; manager; *(eines Instituts)* director; *(einer Schule)* head teacher; headmaster *(Brit.)*; principal *(esp. Amer.)*; *(einer Diskussion)* leader; *(Vorsitzender)* chair[man]; *(eines Chors)* choirmaster; *(Dirigent)* conductor; **kaufmännischer ~:** marketing manager; *(Verkaufs~)* sales manager; **b)** *(Physik)* conductor

[2]Leiter die; ~, ~n ladder; *(Steh~)* step-ladder

Leiterin die; ~, ~nen *s.* **[1]Leiter;** *(einer Schule)* head teacher; headmistress *(Brit.)*; principal *(esp. Amer.)*; *(eines Chors)* choirmistress

Leit-: ~faden der a) [basic] textbook; **~faden der Physik** basic course in physics; introduction to physics; **b)** *(Leitgedanke)* main idea *or* theme; **~hammel der** *(abwertend: Führer)* leader [of the herd]; boss-figure; **~linie die a)** *(Richtlinie)* guideline; **b)** *(Verkehrsw.)* lane marking; **~motiv das** *(Musik, Literaturw., fig.)* **a)** leitmotiv; **b)** *(Leitgedanke)* dominant *or* central theme; **~planke die** crash barrier; guardrail *(Amer.)*; **~satz der** guiding principle; **~spruch der** motto

Leitung die; ~, ~en a) *o. Pl. s.* **leiten a:** leading; heading; *(Schulw.)* working as a/the head; being in charge; management; chairing; *(Musik)* conducting; directing; *(Sport)* refereeing; **b)** *o. Pl. (einer Expedition usw.)* leadership; *(Verantwortung)* responsibility (*Gen.* for); *(einer Firma)* management; *(einer Sitzung, Diskussion)* chairmanship; *(Schulw.)* headship; *(Musik)* conductorship; *(Sport)* [task of] refereeing; **unter der ~ eines Managers stehen** be headed by a manager; **die ~ der Sendung/Diskussion hat X** the programme is presented/the discussion is chaired by X; **c)** *(leitende Personen)* management; *(einer Schule)* head and senior staff; **d)** *(Rohr~)* pipe; *(Haupt~)* main; **Wasser aus der ~ trinken** drink tap-water; **e)** *(Draht, Kabel)* cable; *(für ein Gerät)* lead; *(einzelne od. ohne Isolierung)* wire; **die ~en [im Haus** *usw.*] the wiring *sing.* [of the house *etc.*]; **f)** *(Telefon~)* line; **es ist jemand in der ~** *(ugs.)* there's somebody on the line; **gehen Sie aus der ~!** get off the line!; **auf einer anderen ~ sprechen** be [talking] on another line; *(fig.)* **eine lange ~ haben** *(ugs.)* be slow on the uptake; **er steht** *od.* **sitzt auf der ~** *(salopp)* he's not really with it *(coll.)*

Leitungs-: ~mast der *(für Strom)* pylon; *(Telefonmast)* telegraph-pole; **~rohr das** [water/gas] pipe; *(Haupt~)* main; **~wasser das** tap-water

Leit-: ~währung die *(Wirtsch.)*

base *or* key currency; **~werk das** *(Flugw., Waffent.)* control surfaces *pl.; (am Heck)* tail unit
Lektion [lɛk'tsi̯o:n] **die; ~, ~en** lesson; **jmdm. eine ~ erteilen** *(fig.)* teach sb. a lesson
Lektor ['lɛktɔr] **der; ~s, ~en** [lɛk'to:rən] **a)** *(Hochschulw.) junior university teacher in charge of practical or supplementary classes etc.;* **b)** *(Verlags~)* [publisher's] editor
Lektorat [lɛkto'ra:t] **das; ~[e]s, ~e a)** *(Hochschulw.)* post of 'Lektor'; **b)** *(im Verlag)* editorial department
Lektorin die; ~, ~nen *s.* **Lektor**
Lektüre [lɛk'ty:rə] **die; ~, ~n a)** *o. Pl.* reading; **bei der ~ des Romans** when reading the novel; **b)** *(Lesestoff)* reading [matter]; **etw. als ~ empfehlen** recommend sth. as a good read
Lende ['lɛndə] **die; ~, ~n** loin
Lenden-: ~schurz der loincloth; **~stück das** *(Kochk.)* piece of loin
Leninismus [leni'nɪsmʊs] **der; ~:** Leninism *no art.*
lenkbar *Adj.* **a) leicht/schwer ~ sein** be easy/difficult to steer; **b)** *(von Menschen)* acquiescent; obedient; manageable, controllable ‹*child*›
lenken ['lɛŋkn̩] *tr. V.* **a)** *auch itr.* steer ‹*car, bicycle, etc.*›; be at the controls of ‹*aircraft*›; guide ‹*missile*›; *(fahren)* drive ‹*car etc.*›; **wenn du geschickt lenkst** if you do some crafty steering; **b)** direct, guide ‹*thoughts etc.*› **(auf +** *Akk.* to); turn ‹*attention*› **(auf +** *Akk.* to); steer ‹*conversation*›; **die Diskussion auf etw./jmdn. ~:** steer *or* bring the discussion round to sth./sb.; **den Verdacht auf jmdn. ~:** throw suspicion on sb.; **c)** control ‹*person, press, economy*›; rule, govern ‹*state*›; **eine gelenkte Wirtschaft** a planned economy
Lenker der; ~s, ~ a) *(Lenkstange)* handlebars *pl.;* **b)** *(Fahrer)* driver
Lenkerin die; ~, ~nen *s.* **Lenker b**
Lenk-: ~rad das steering-wheel; **~stange die** handlebars *pl.*
Lenkung die; ~, ~en a) *o. Pl.* *(Leitung)* control; *(eines Staates)* ruling *no indef. art.;* governing *no indef. art.;* **b)** *(Kfz-W.)* steering
Lenz [lɛnts] **der; ~es, ~e** *(dichter. veralt.)* spring; **einen sonnigen** *od.* **ruhigen** *od.* **faulen ~ haben** *od.* **schieben** *(salopp)* have an easy time of it; *(eine leichte Arbeit haben)* have a cushy job *(coll.);* **sich** *(Dat.)* **einen schönen ~ machen** *(salopp)* take it easy
Leopard [leo'part] **der; ~en, ~en** leopard
Lepra ['le:pra] **die; ~:** leprosy *no art.*
Lepra·kranke der/die leper
Lerche ['lɛrçə] **die; ~, ~n** lark
lern-, Lern-: ~bar *Adj.* learnable; **das ist [für jeden] ~:** that can be learnt [by anybody]; **~begierig** *Adj.* eager to learn *postpos.;* **~behindert** *Adj. (Päd.)* educationally subnormal; with learning difficulties *postpos., not pred.;* **~eifer der** eagerness to learn
lernen ['lɛrnən] **1.** *itr. V.* **a)** study; **gut/schlecht ~:** be a good/poor learner *or* pupil; *(fleißig/nicht fleißig sein)* work hard/not work hard [at school]; **leicht ~:** find it easy to learn; find school-work easy; **mit jmdm. ~** *(ugs.)* help sb. with his/her [school-]work; **b)** *(Lehrling sein)* train. **2.** *tr. V.* **a)** learn; **schwimmen/Klavier ~:** learn to swim/play the piano; **er/ mancher lernt es nie** *(ugs.)* he/ some people [will] never learn; **von ihm kann man noch was ~:** you can learn a thing or two from him; **das will gelernt sein** that is something one has to learn; **gelernt ist gelernt** once learnt, never forgotten; **das Fürchten ~:** find out what it is to be afraid; **b) einen Beruf ~:** learn a trade; **Bäcker** *usw.* **~:** train to be *or* as a baker *etc.*
Lern-: ~hilfe die aid to learning; **~mittel das** learning aid; *(Lehrmittel)* teaching aid; **~mittel** *Pl.* teaching materials; **~prozeß der** learning process
lesbar *Adj.* **a)** legible; **b)** *(klar)* lucid ‹*style*›; *(verständlich)* comprehensible; **gut ~:** easy to read; very readable
Lesbe ['lɛsbə] **die; ~, ~n** *(ugs.)* Lesbian; dike *(sl.)*
Lesbierin ['lɛsbi̯ərɪn] **die; ~, ~nen** Lesbian
lesbisch *Adj.* Lesbian; **~ sein** be a Lesbian/Lesbians
Lese ['le:zə] **die; ~, ~n** grape-harvest
Lese-: ~brille die reading-glasses *pl.;* **~buch das** reader; **~gerät das** *(DV)* reader; **~lampe die** reading-lamp
[1]**lesen 1.** *unr. tr., itr. V.* **a)** read; **sie las in einem Buch** she was reading a book; **er liest aus seinem neuesten Werk** he is reading from his latest work; **ein Gesetz [zum ersten Mal] ~** *(Parl.)* give a bill a [first] reading; **die/eine Messe ~:** say Mass/a Mass; **b)** *(fig.)* **Gedanken ~ können** be a mind-reader; **jmds. Gedanken ~:** read sb.'s mind *or* thoughts; **aus der Hand ~:** read palms; **c)** *(Hochschulw.)* lecture **(über +** *Akk.* on); **er liest neue Geschichte** he lectures on modern history. **2.** *unr. refl. V.* read
[2]**lesen** *unr. tr. V.* **a)** *(sammeln, pflücken)* pick ‹*grapes, berries, fruit*›; gather ‹*firewood*›; glean ‹*ears of corn*›; **b)** *(aussondern)* pick over
lesens·wert *Adj.* worth reading *postpos.*
Leser der; ~s, ~: reader
Lese·ratte die *(ugs. scherzh.)* bookworm; voracious reader
Leser·brief der reader's letter; **~e** readers' letters; **„~e"** *(Zeitungsrubrik)* 'Letters to the editor'
Leserin die; ~, ~nen *s.* **Leser**
leserlich 1. *Adj.* legible. **2.** *adv.* legibly
Leserschaft die; ~: readership
Leser·zuschrift die *s.* **Leserbrief**
Lese-: ~saal der reading-room; **~stoff der** reading matter; **~zeichen das** bookmark
Lesung die; ~, ~en a) *(auch Parl.)* reading; **b)** *(christl. Kirche)* lesson
Lethargie [letar'gi:] **die; ~:** lethargy
lethargisch 1. *Adj.* lethargic. **2.** *adv.* lethargically
Lette ['lɛtə] **der; ~n, ~n, Lettin die; ~, ~nen** Latvian; Lett
lettisch *Adj.* Latvian; Lettish ‹*language*›
Lett·land (das); ~s Latvia
Letzt ['lɛtst] **zu guter ~:** in the end; *(endlich)* at long last
letzt... *Adj.; nicht präd.* **a)** last; **die ~e Reihe** the back row; **auf dem ~en Platz sein** be [placed] last; *(während des Rennens)* be in last place; *(in einer Tabelle)* be in bottom place; **er war** *od.* **wurde ~er, er ging als ~er durchs Ziel** he came last; **der/die ~e sein** be the last; **als ~er aussteigen** be the last [one] to get off; **er ist der ~e, dem ich das sagen würde** he's the last person I would tell [about it]; **am Letzten [des Monats]** on the last day of the month; **mein ~es Geld** the last of my money; **mit ~er Kraft** gathering all his/her remaining strength; **~en Endes** in the end; when all is said and done; **jmds./die ~e Rettung sein** *(fig.)* be sb.'s/the last hope; *s. auch* **Ölung; Wille; b)** *(äußerst...)* ultimate; **jmdm. das Letzte an ...** *(Dat.)* **abverlangen** demand of sb.

the utmost *or* maximum ...; **das Letzte hergeben** give one's all; **bis aufs ~e** totally; *(finanziell)* down to the last penny; **bis ins ~e** down to the last detail; **bis zum ~en** to the utmost; **c)** *(gerade vergangen)* last; *(neuest...)* latest ⟨*news*⟩*;* **in den ~en Wochen/Jahren** in the last few weeks/in recent years; **in der ~en Zeit** recently; *s. auch* **Schrei; c)** *(ugs. abwertend) (schlechtest...)* worst; *(entsetzlichst...)* most dreadful; **er ist der ~e Mensch** he is the lowest of the low; **die Show war das Letzte** *(ugs.)* the show was the end *(coll.) or* the pits *(coll.);* **das ist doch das Letzte!** *(ugs.)* that really is the limit!

lẹtzte·mal: das ~: [the] last time

lẹtzt·ẹndlich *Adv.* in the end; *(schließlich doch)* ultimately

lẹtzten·mal: beim ~: last time; **zum ~:** for the last time

lẹtzter... *Adj.* **~er/der ~e/~es/das ~e** *usw.* the latter

lẹtzt·genannt *Adj.; nicht präd.* last-mentioned; last-named ⟨*person*⟩

lẹtztlich *Adv.* ultimately; in the end

lẹtzt·möglich *Adj.; nicht präd.* latest possible

Leuchte ['lɔyçtə] **die; ~, ~n a)** light; *(Tischlampe)* lamp; **b)** *(fig. ugs.)* **er ist eine ~ auf diesem Gebiet** he is a leading light in this field

leu̲chten *itr. V.* **a)** ⟨*moon, sun, star, etc.*⟩ be shining; ⟨*fire, face*⟩ glow; **grell ~:** give a glaring light; glare; **in der Sonne ~:** ⟨*hair, sea, snow*⟩ gleam in the sun; ⟨*mountains etc.*⟩ glow in the sun; **golden ~:** have a golden glow; **seine Augen leuchteten vor Freude** *(fig.)* his eyes were shining *or* sparkling with joy; **b)** shine a/the light; **jmdm. ~:** light the way for sb.; **mit etw. in etw.** *(Akk.)* **~:** shine sth. into sth.; **jmdm. mit etw. ins Gesicht ~:** shine sth. into sb.'s face

leu̲chtend *Adj.; nicht präd.* **a)** shining ⟨*eyes*⟩*;* brilliant, luminous ⟨*colours*⟩*;* bright ⟨*blue, red, etc.*⟩*;* **grell ~:** glaring; **etw. in den ~sten Farben schildern** *(fig.)* paint sth. in glowing colours; **b)** *(großartig)* shining ⟨*example*⟩

Leu̲chter der; ~s, ~ candelabrum; *(für eine Kerze)* candlestick; *(Kron~)* chandelier

Leu̲cht-: ~farbe die luminous paint; **~feuer das** *(Seew.)* beacon; light; *(Flugw.)* runway light; **~käfer der** firefly; *(Glühwürmchen)* glow-worm; **~kugel die** flare; **~rakete die** rocket flare; **~reklame die** neon [advertising] sign; **~röhre die** neon tube; **~schrift die** neon letters *pl.;* **~stoff·röhre die** fluorescent tube; *(für ~reklame)* neon tube; **~turm der** lighthouse; **~ziffer die** luminous numeral

leugnen ['lɔygnən] **1.** *tr. V.* deny; **er leugnete die Tat/das Verbrechen** he denied doing the deed/committing the crime; **es ist nicht zu ~:** it is undeniable. **2.** *itr. V.* deny it; *(alles ~)* deny everything

Leukämie [lɔykɛ'mi:] **die; ~, ~n** *(Med.)* leukaemia

Leukoplast ⓦⓩ [lɔyko'plast] **das** sticking-plaster

Leumund ['lɔymʊnt] **der; ~[e]s** *(geh.)* reputation; **jmdm. einen guten ~ bescheinigen** vouch for sb.'s good character

Leute ['lɔytə] *Pl.* **a)** people; **die reichen/alten ~:** the rich/the old; **wir sind hier bei feinen ~n** we are in a respectable household; **die kleinen ~:** the ordinary people; the man *sing.* in the street; **was werden die ~ sagen?** *(ugs.)* what will people say?; **wir sind geschiedene ~:** I will have no more to do with you/him *etc.;* **vor allen ~n** in front of everybody; **unter die ~ bringen** *(ugs.)* spread ⟨*rumour*⟩*;* tell everybody about ⟨*suspicions etc.*⟩*;* **b)** *(ugs.: als Anrede)* **los, ~!** come on, everybody! *(coll.);* c'mon, folks! *(Amer.);* **c)** *(ugs.: Arbeiter)* people; *(Milit.: Soldaten)* men; **die Hälfte der ~:** half the staff

Leutnant ['lɔytnant] **der; ~s, ~s** *od. selten:* **~e** second lieutenant *(Milit.)*

leu̲t·selig 1. *Adj.* affable. **2.** *adv.* affably

Leu̲t·seligkeit die; *o. Pl.* affability

Leviten [le'vi:tn̩] *Pl. in* **jmdm. die ~ lesen** *(ugs.)* read sb. the Riot Act *(coll.)*

Lexikon ['lɛksikɔn] **das; ~s, Lexika a)** encyclopaedia; **b)** *(veralt.: Wörterbuch)* dictionary

Liane ['li̲a:nə] **die; ~, ~n** *(Bot.)* liana

Libanese [liba'ne:zə] **der; ~n, ~n, Libane̲sin die; ~, ~nen** Lebanese

Libanon ['li:banɔn] **(das)** *od.* **der; ~s** Lebanon

Libelle [li'bɛlə] **die; ~, ~n** dragonfly

liberal [libə'ra:l] **1.** *Adj.* liberal. **2.** *adv.* liberally; **jmdn. ~ erziehen** give sb. a liberal education

Libera̲le der/die; *adj. Dekl.* liberal

liberalisi̲eren *tr. V.* liberalize

Liberalisi̲erung die; ~, ~en liberalization

Liberalismus [libəra'lɪsmʊs] **der; ~:** liberalism

Libero ['li:bəro] **der; ~s, ~s** *(Fußball)* sweeper

Libretto [li'brɛto] **das; ~s, ~s** *od.* **Librẹtti** libretto

Libyen ['li:by̆ən] **(das); ~s** Libya

libysch ['li:bʏʃ] *Adj.* Libyan

licht [lɪçt] *Adj.* **a)** *(geh.)* light; light, pale ⟨*colour*⟩*;* **es war ~er Tag** it was broad daylight; **einen ~en Moment** *od.* **Augenblick/~e Momente haben** *(fig.)* have a lucid moment/lucid moments; *(scherzh.)* have a bright moment/bright moments; **b)** *(dünn bewachsen)* sparse; thin; **~es Haar haben** be thin on top; **die Reihen der alten Kameraden/der Zuschauer werden ~er** *(fig.)* the ranks of old comrades are dwindling/the rows of spectators are emptying; **c)** *(bes. Technik)* **die ~e Höhe/Weite** the [overall] internal height/width

Lịcht das; ~[e]s, ~er a) *o. Pl.* light; **das ~ des Tages** the light of day; **etw. gegen das ~ halten** hold sth. up to the light; **bei ~ besehen** *(fig.)* seen in the light of day; **jmdm. im ~ stehen** stand in sb.'s light; **das ~ der Welt erblicken** *(geh.)* see the light of day; *(fig.)* **ein zweifelhaftes/ungünstiges ~ auf jmdn. werfen** throw a dubious/unfavourable light on sb.; **~ in etw.** *(Akk.)* **bringen** shed some light on sth.; **jmdn. hinters ~ führen** fool sb.; pull the wool over sb.'s eyes; **jmdn./etw./sich ins rechte ~ rücken** *od.* **setzen** *od.* **stellen** show sb./sth. in the correct light/appear in the correct light; **in einem guten** *od.* **günstigen/schlechten ~ erscheinen** appear in a good *or* a favourable/a bad *or* an unfavourable light; **in ein falsches ~ geraten** give the wrong impression; **das ~ scheuen** shun the light; **ans ~ kommen** come to light; be revealed; **b)** *(elektrisches ~)* light; **das ~ anmachen/ausmachen** switch *or* turn the light on/off; **c)** *Pl. auch* **~e** *(Kerze)* candle; *(fig.)* **kein** *od.* **nicht gerade ein großes ~ sein** *(ugs.)* be no genius; be not exactly brilliant; **mir ging ein ~ auf** *(ugs.)* it dawned on me; I realized what was going on; **sein**

~ [nicht] unter den Scheffel stellen [not] hide one's light under a bushel; **d)** *o. Pl. (ugs.: Strom)* electricity

licht-, Licht-: ~anlage die lighting installation; **~beständig** *Adj.* light-fast; **~bild das a)** [small] photograph *(for passport etc.)*; **b)** *(veralt.) (Diapositiv)* slide; *(Fotografie)* photograph; **~bilder·vortrag der** slide lecture; **~blick der** bright spot; **~durchlässig** *Adj.* translucent; **~empfindlich** *Adj.* sensitive to light; **~empfindlichkeit die** sensitivity to light

¹lichten 1. *tr. V.* thin out ⟨*trees etc.*⟩; *(fig.)* reduce ⟨*number*⟩. **2.** *refl. V.* ⟨*trees*⟩ thin out; ⟨*hair*⟩ grow thin; ⟨*fog, mist*⟩ clear, lift; **die Reihen ~ sich** *(fig.)* the numbers are dwindling; *(im Theater usw.)* the rows are emptying

²lichten *tr. V. (Seemannsspr.)* **den/die Anker ~:** weigh anchor

lichterloh ['lıçtɐ'lo:] **1.** *Adj.; nicht präd.* blazing ⟨*fire*⟩; fierce, leaping ⟨*flames*⟩. **2.** *adv.* **~ brennen** be blazing fiercely

Lichter·meer das sea of lights

licht-, Licht-: ~geschwindigkeit die speed of light; **~hupe die** headlight flasher; **~jahr das** *(Astron.)* light-year; **~kegel der** beam; **~maschine die** *(Kfz-W.) (für Gleichstrom)* dynamo; *(für Wechselstrom)* alternator; generator *(esp. Amer.)*; **~mast der** lamp-standard; **~pause die** photostat (*Brit.* **P**) *(of transparent original)*; **~quelle die** light-source; **~schacht der** light-shaft; **~schalter der** light-switch; **~schein der** gleam [of light]; *(~strahl)* beam of light; **~scheu** *Adj.* **a)** shadeloving ⟨*plant*⟩; ⟨*animal*⟩ that shuns the light; **b)** *(fig.)* shady ⟨*riff-raff*⟩; **~schranke die** photoelectric beam; **~seite die** bright *or* good side; **alles hat seine Licht- und Schattenseiten** everything has its good and bad sides; **~strahl der** beam [of light]; **~undurchlässig** *Adj.* light-proof

Lichtung die; ~, ~en clearing; **auf dieser ~:** in this clearing

Licht-: ~verhältnisse *Pl.* light conditions; **~zeichen das** light signal

Lid [li:t] **das; ~[e]s, ~er** eyelid

Lid·schatten der eye-shadow

lieb [li:p] **1.** *Adj.* **a)** *(liebevoll)* kind ⟨*words, gesture*⟩; **viele ~e Grüße [an ... (Akk.)]** much love [to ...] *(coll.)*; **das ist ~ von dir** it's sweet of you; **b)** *(liebenswert)* likeable; nice; *(stärker)* lovable, sweet ⟨*child, girl, pet*⟩; **seine Frau/ihr Mann ist sehr ~:** his wife/her husband is a dear; **c)** *(artig)* good, nice ⟨*child, dog*⟩; **sei schön ~!** be a good girl/boy!; **sich bei jmdm. ~ Kind machen** *(ugs. abwertend)* get on the right side of sb.; **d)** *(geschätzt)* dear; **sein liebstes Spielzeug** his favourite toy; **~e Karola, ~er Ernst!** *(am Briefanfang)* dear Karola and Ernst; **~er Gott** dear God; **der ~e Gott** the Good Lord; **wenn dir dein Leben ~ ist, ...:** if you value your life ...; **das ~e Geld** *(iron.)* the wretched money; **den ~en langen Tag** *(ugs.)* all the livelong day; **meine Lieben** *(Familie)* my people; my nearest and dearest *(joc.)*; *(als Anrede)* [you] good people; *(an Familie usw.)* my dears; **meine Liebe** my dear; *(herablassend)* my dear woman/girl; **mein Lieber** *(Mann an Mann)* my dear fellow; *(Frau/Mann an Jungen)* my dear boy; *(Frau an Mann)* my dear man; **~e Mitbürgerinnen und Mitbürger!** fellow citizens; **~e Kinder/Freunde!** children/friends; **~ Gemeinde, ~e Schwestern und Brüder!** *(christl. Kirche)* dearly beloved; **[ach] du ~e Güte** *od.* **~e Zeit** *od.* **~er Himmel** *od.* **~es bißchen!** *(ugs.) (erstaunt)* good grief!; good heavens!; [good] gracious!; *(entsetzt)* good grief!; heavens above!; **mit jmdm./etw. seine ~e Not haben** have no end of trouble with sb./sth.; **e)** *(angenehm)* welcome; **es wäre mir ~/~er, wenn ...:** I should be glad *or* should like it/should prefer it if ...; **am ~sten wäre mir, ich könnte heute noch abreisen** I should like it best if I could leave today; **wir hatten mehr Schnee, als mir ~ war** we had too much snow for my liking. **2.** *adv.* **a)** *(liebenswert)* kindly; **das hast du aber ~ gesagt** you 'did put that nicely; **b)** *(artig)* nicely

lieb·äugeln *itr. V.* **mit etw. ~:** have one's eye on sth.; fancy sth.

Liebchen ['li:pçən] **das; ~s, ~** *(abwertend)* lady-love

Liebe ['li:bə] **die; ~, ~n a)** *o. Pl.* love; **~ zu jmdm.** love for sb.; **~ zu etw.** love of sth.; **~ zu Gott** love of God; **aus ~ [zu jmdm.]** for love [of sb.]; **bei aller ~, aber das geht zu weit** much as I sympathize, that's going too far; *(Briefschluß)* **in ~ Dein Egon** [with] all my love, yours, Egon; **~ geht durch den Magen** *(scherzh.)* the way to a man's heart is through his stomach; **~ macht blind** *(Spr.)* love is blind *(prov.)*; **~ auf den ersten Blick** love at first sight; **seine ganze ~ gehört dem Meer** he adores the sea; **mit ~** *(liebevoll)* lovingly; with loving care; **b)** *(ugs.: geliebter Mensch)* love; **seine große ~:** his great love; the [great] love of his life; **c) tu mir die ~ und ...** do me a favour and ...

liebe·bedürftig *Adj.* in need of love *or* affection *postpos.*

Liebelei [li:bə'lai] **die; ~, ~en** *(abwertend)* flirtation

lieben 1. *tr. V.* **a) jmdn. ~:** love sb.; *(verliebt sein)* be in love with *or* love sb.; *(sexuell)* make love to sb.; **sich ~:** be in love; *(sexuell)* make love; **was sich liebt, das neckt sich** *(Spr.)* lovers always tease each other; **b) etw. ~:** be fond of sth.; like sth.; *(stärker)* love sth.; **es ~, etw. zu tun** like *or* enjoy doing sth.; *(stärker)* love doing sth.. **2.** *itr. V.* be in love; *(sexuell)* make love; **er ist unfähig zu ~:** he is incapable of love

liebend 1. *Adj.; nicht präd.* loving; **der/die Liebende** the lover. **2.** *adv.* **etw. ~ gerne tun** [simply] love doing sth.

lieben|lernen *tr. V.* learn to love

liebens-; Liebens-: ~wert *Adj.* likeable ⟨*person*⟩; *(stärker)* loveable ⟨*person*⟩; attractive, endearing ⟨*trait*⟩; **~würdig** *Adj.* kind; charming ⟨*smile*⟩; **~würdigerweise** *Adv.* kindly; **~würdigkeit die; ~, ~en** kindness; **würden Sie die ~würdigkeit haben, das Fenster zu schließen?** would you be so kind as to shut the window?

lieber 1. *Adj.: s.* **lieb. 2.** *Adv.: s.* **gern**

Liebes~: ~beziehung die [love] affair (**zu, mit** with); **~brief der** love-letter; **~dienst der** [act of] kindness; favour; **~erklärung die** declaration of love; **~film der** romantic film; **~geschichte die a)** love-story; **b)** *(~affäre)* [love] affair; **~heirat die** love-match; **~kummer der** lovesickness; **~kummer haben** be lovesick; be unhappily in love; **sich aus ~kummer umbringen** kill oneself for love; **~leben das;** *o. Pl.* love-life; **~lied das** love-song; **~müh[e] die: das ist vergebliche** *od.* **verlorene ~müh[e]** that is a waste of effort; **~paar das** courting couple; [pair of] lovers; **~roman der** romantic novel; **~spiel das** love-play

liebestoll *Adj.* love-crazed

liebe·voll 1. *Adj.* loving *attrib.* ‹*care*›; affectionate ‹*embrace, person*›. **2.** *adv.* **a)** lovingly; affectionately; **b)** *(mit Sorgfalt)* lovingly; with loving care
lieb|gewinnen *unr. tr. V.* grow fond of
lieb·geworden *Adj.; nicht präd.* ‹*habit*› of which one has grown very fond
lieb|haben *unr. tr. V.* love; *(gern haben)* be fond of
Liebhaber ['li:pha:bɐ] **der; ~s, ~ a)** lover; **b)** *(Interessierter, Anhänger)* enthusiast (*Gen.* for); *(Sammler)* collector; **ein ~ von schönen Teppichen/Oldtimern** a lover of beautiful carpets/a vintage-car enthusiast
Liebhaber·ausgabe die collector's edition; bibliophile edition
Liebhaberei die; ~, ~en hobby
Liebhaberin die; ~, ~nen *s.* **Liebhaber b**
Liebhaberstück das collector's item
lieb·kosen *tr. V. (geh.)* caress
Liebkosung die; ~, ~en *(geh.)* caress
lieblich 1. *Adj.* **a)** charming; appealing; gentle ‹*landscape*›; sweet ‹*scent, sound*›; fragrant ‹*flower*›; melodious ‹*sound*›; **b)** mellow ‹*red wine*›; [medium] sweet ‹*white wine*›. **2.** *adv.* charmingly; sweetly; *(angenehm)* pleasingly
Lieblichkeit die; ~ charm; sweetness; *(einer Landschaft)* gentleness
Liebling der; ~s, ~e a) *(geliebte Person; bes. als Anrede)* darling; **b)** *(bevorzugte Person)* favourite; *(des Publikums)* darling; **der ~ des Lehrers** teacher's pet
Lieblings- favourite
lieb·los 1. *Adj.* loveless; heartless, unfeeling ‹*treatment, behaviour*›. **2.** *adv.* **a)** without affection; **b)** *(ohne Sorgfalt)* carelessly; without proper care
Lieblosigkeit die; ~ unkindness; lack of feeling; *(Mangel an Sorgfalt)* lack of care
Liebschaft die; ~, ~en [casual] affair; *(Flirt)* flirtation
liebst... [li:pst...] **1.** *Adj.: s.* **lieb. 2.** *Adv.* **am ~en** *s.* **gern**
Liebste der/die; *adj. Dekl. (veralt.)* loved one; sweetheart; **meine ~:** my dearest
Lied [li:t] **das; ~[e]s, ~er** song; *(Kirchen~)* hymn; *(deutsches Kunst~)* lied; **es ist immer das alte** *od.* **das gleiche** *od.* **dasselbe ~** *(ugs.)* it's always the same old story; **davon kann ich ein ~ singen** I can tell you a thing or two about that
Lieder-: ~abend der [evening] song recital; *(mit deutschen Kunstliedern)* [evening] lieder recital; **~buch das** song-book
liederlich ['li:dɐlɪç] **1.** *Adj.* **a)** *(schlampig)* slovenly; messy ‹*hair-style, person*›; slipshod, slovenly ‹*work*›; **b)** *(verwerflich)* dissolute. **2.** *adv.* sloppily; messily; **~ angezogen sein** be slovenly dressed
Liederlichkeit die; ~ a) *(Schlampigkeit)* slovenliness; **b)** *(Verwerflichkeit)* dissoluteness
Lieder-: ~macher der; ~macherin die singer-song-writer *(writing satirical songs mainly on topical/political subjects)*; **~zyklus der** song-cycle
lief [li:f] *1. u. 3. Pers. Sg. Prät. v.* **laufen**
Lieferant [lifə'rant] **der; ~en, ~en** supplier
Lieferanten·eingang der goods entrance; *(bei Wohnhäusern)* tradesmen's entrance
lieferbar *Adj.* available; **sofort ~:** available for immediate delivery
Liefer-: ~bedingungen *Pl.* terms of delivery; **~frist die** delivery time; **bei Möbeln besteht eine ~frist von 6–8 Wochen** there is 6–8 weeks delivery on furniture
liefern ['li:fɐn] **1.** *tr. V.* **a)** deliver (**an** + *Akk.* to); *(zur Verfügung stellen)* supply; **jmdm. etw. ~:** supply sb. with sth.; deliver sth. to sb.; **b)** *(hervorbringen)* produce; *(geben)* provide ‹*eggs, honey, examples, raw material, etc.*›; **c) sich** *(Dat.)* **eine Schlacht ~:** fight a battle [with each other]; **d) geliefert sein** *(ugs.)* be sunk *(coll.)*; have had it *(coll.)*. **2.** *itr. V.* deliver; **wir ~ auch ins Ausland** we also supply our goods abroad *or* deliver to foreign destinations
Liefer-: ~schein der acknowledgement of delivery; delivery note; **~termin der** delivery date
Lieferung die; ~, ~en a) *o. Pl.* delivery; **b)** *(Ware)* consignment [of goods]; delivery; **c)** *(Buchw.)* instalment; *(eines Wörterbuchs usw.)* fascicle
Liefer-: ~wagen der [delivery] van; *(offen)* pick-up; **~zeit die** *s.* **~frist**
Liege ['li:gə] **die; ~, ~n** day-bed; *(zum Ausklappen)* bed-settee; sofa bed; *(als Gartenmöbel)* sun-lounger
liegen *unr. itr. V.* **a)** lie; ‹*person*› be lying down; **während der Krankheit mußte er ~:** while he was ill he had to lie down all the time; **auf den Knien ~:** be prostrate on one's knees; **im Krankenhaus/auf Station 6 ~:** be in hospital/in ward 6; **[krank] im Bett ~:** be [ill] in bed; **der Wagen liegt gut auf der Straße** the car holds the road well; **b)** *(vorhanden sein)* lie; **es liegt Schnee auf den Bergen** there is snow [lying] on the hills; **der Stoff liegt 80 cm breit** the material is 80 cm wide; **c)** *(sich befinden)* be; ‹*object*› be [lying]; ‹*town, house, etc.*› be [situated]; **die Preise ~ höher** prices are higher; **wie die Dinge ~:** as things are *or* stand [at the moment]; **die Stadt liegt an der Küste** the town is *or* lies on the coast; **das Dorf liegt sehr hoch** the village is very high up; **das liegt auf meinem Weg** it is on my way; **etw. rechts/links ~ lassen** leave sth. on one's right/left; **das Fenster liegt nach vorn/nach Süden/zum Garten** the window is at the front/faces south/faces the garden; **es liegt nicht in meiner Absicht, das zu tun** it is not my intention to do that; **das Essen lag mir schwer im Magen** the food/meal lay heavy on my stomach; **d)** *(zeitlich)* be; **das liegt noch vor mir/schon hinter mir** I still have that to come/that's all behind me now; **e) das liegt an ihm** *od.* **bei ihm** it is up to him; *(ist seine Schuld)* it is his fault; **die Verantwortung/Schuld liegt bei ihm** it is his responsibility/fault; **an mir soll es nicht ~:** I won't stand in your way; *(ich werde mich beteiligen)* I'm easy *(coll.)*; **ich weiß nicht, woran es liegt** I don't know what the reason is; **woran mag es nur ~, daß ...?** why ever is it that ...?; **f)** *(gemäß sein)* **es liegt mir nicht** it doesn't suit me; it isn't right for me; *(es spricht mich nicht an)* it doesn't appeal to me; *(ich mag es nicht)* I don't like it *or* care for it; **es liegt ihm nicht, das zu tun** he does not like doing that; *(so etwas tut er nicht)* it is not his way to do that; **g) daran liegt ihm viel/wenig/nichts** he sets great/little/no store by that; it means a lot/little/nothing to him; **an ihm liegt mir schon etwas** I do care about him [a bit]; **h)** *(bedeckt sein)* **der Tisch liegt voller Bücher** the desk is covered with books; **i)** *(bes. Milit.: verweilen)* be; ‹*troops*› be stationed; ‹*ship*› lie; *s. auch* **Straße a; liegend**

liegen|bleiben *unr. itr. V.; mit sein* **a)** stay [lying]; **[im Bett] ~:** stay in bed; **bewußtlos/bewegungslos ~:** lie unconscious/motionless; **b)** *(nicht tauen)* ‹*snow*› lie; **c)** *(bleiben)* ‹*things*› stay, be left; *(vergessen werden)* be left behind; *(nicht verkauft werden)* remain unsold; **d)** *(nicht erledigt werden)* be left undone; **diese Briefe können bis morgen ~:** these letters can wait until tomorrow; **e)** *(eine Panne haben)* break down

liegend reclining, recumbent ‹*figure, posture*›; prone ‹*position*›; horizontal ‹*position, engine*›

liegen|lassen *unr. tr. V.* leave; *(vergessen)* leave [behind]; **alles liegen- und stehenlassen** drop everything; **b)** *(unerledigt lassen)* leave ‹*work*› undone; leave ‹*letters*› unposted/unopened; *s. auch* **links 1 a**

Liegenschaft die; ~, ~en; *meist Pl. (bes. Rechtsspr.)* land holding; *(Gebäude)* property

Liege-: ~platz der mooring; **~sitz der** reclining seat; **~stuhl der** deck-chair; **~stütz der** press-up; **~wagen der** couchette car; **~wiese die** sunbathing lawn

lieh [li:] *1. u. 3. Pers. Sg. Prät. v.* **leihen**

lies [li:s] *Imperativ Sg. v.* **lesen**

Lieschen ['li:sçən] **a) ~ Müller** *(ugs.)* the average girl/woman *(coll.)*; **b) Fleißiges ~** *(Bot.)* busy Lizzie

ließ [li:s] *1. u. 3. Pers. Sg. Prät. v.* **lassen**

liest [li:st] *3. Pers. Sg. Präsens v.* **lesen**

Lift [lɪft] **der; ~[e]s, ~e** *od.* **~s a)** lift *(Brit.)*; elevator *(Amer.)*; **b)** *Pl.:* **~e** *(Ski~, Sessel~)* lift

liften *tr. V.* **sich ~ lassen** *(ugs.)* have a face-lift

Liga ['li:ga] **die; ~, Ligen** league; *(Sport)* division

liieren [li'i:rən] *refl. V.* start an affair; **mit jmdm. liiert sein** be having an affair with sb.

Likör [li'kø:ɐ̯] **der; ~s, ~e** liqueur

lila ['li:la] *indekl. Adj.* mauve; *(dunkel~)* purple

Lila das; ~s *od. (ugs.)* **~s** mauve; *(Dunkel~)* purple

Lilie ['li:li̯ə] **die; ~, ~n** lily

Liliput- ['li:lipʊt-] miniature

Liliputaner [lilipu'ta:nɐ] **der; ~s, ~:** dwarf; midget

Limit ['lɪmɪt] **das; ~s, ~s** limit

limitieren *tr. V.* limit; restrict

Limo ['lɪmo] **die,** *auch:* **das; ~, ~[s]** *(ugs.)*, **Limonade** [limo'na:də] **die; ~, ~n** fizzy drink; mineral; *(Zitronen~)* lemonade

Limone [li'mo:nə] **die; ~, ~n** lime

Limousine [limu'zi:nə] **die; ~, ~n** [large] saloon *(Brit.) or (Amer.)* sedan; *(mit Trennwand)* limousine

Linde ['lɪndə] **die; ~, ~n a)** *(Baum)* lime[-tree]; **b)** *o. Pl. (Holz)* limewood

Lindenblüten-: ~honig der lime-blossom honey; **~tee der** lime-blossom tea

lindern ['lɪndɐn] *tr. V.* alleviate, relieve ‹*suffering*›; ease, relieve ‹*pain*›; quench, slake ‹*thirst*›

Linderung die; ~ *(der Not)* relief; alleviation; *(des Schmerzes)* relief

lind·grün *Adj.* lime-green

Lineal [line'a:l] **das; ~s, ~e** ruler; **Striche mit einem ~ ziehen** rule lines

linear [line'a:ɐ̯] *Adj. (fachspr., geh.)* linear

Linguist [lɪŋ'gu̯ɪst] **der; ~en, ~en** linguist

Linguistik die; ~: linguistics *sing., no art.*

Linguistin die; ~, ~nen linguist

linguistisch 1. *Adj.* linguistic. **2.** *adv.* linguistically

Linie ['li:ni̯ə] **die; ~, ~n a)** line; **auf die [schlanke] ~ achten** *(ugs. scherzh.)* watch one's figure; **die feindliche[n] ~[n]** *(Milit.)* [the] enemy lines *pl.*; **in vorderster ~ stehen** *(fig.)* be in the front line; **b)** *(Verkehrsstrecke)* route; *(Eisenbahn~, Straßenbahn~)* line; route; **fahren Sie mit der ~ 4** take a *or* the number 4; **c)** *(allgemeine Richtung)* line; policy; **eine/keine klare ~ erkennen lassen** reveal a/no clear policy; **e)** *(Verwandtschaftszweig)* line; **in direkter ~ von jmdm. abstammen** be directly descended from *or* a direct descendant of sb.; **f) in erster ~ geht es darum, daß das Projekt beschleunigt wird** the first priority is to speed up the project; **auf der ganzen ~:** all along the line

linien-, Linien-: ~bus der regular bus; **~flug der** scheduled flight; **~richter der** *(Fußball usw.)* linesman; *(Tennis)* line judge; *(Rugby)* touch judge; **~treu** *(abwertend)* **1.** *Adj.* loyal to the party line *postpos.*; **2.** *adv.* ‹*act*› in accordance with the party line; **~verkehr der** regular services *pl.*; *(Flugw.)* scheduled *or* regular services *pl.*

liniert [li'ni:ɐ̯t] *Adj.* ruled; lined

link [lɪŋk] *(salopp)* **1.** *Adj.* underhand; shady, underhand ‹*deal*›. **2.** *adv.* in an underhand way

link... *Adj.* **a)** left; left[-hand] ‹*edge*›; **die ~e Spur** the left-hand lane; **~er Hand, auf der ~en Seite** on the left-hand side; **mit dem ~en Fuß** *od.* **Bein zuerst aufgestanden sein** *(fig. ugs.)* have got out of bed on the wrong side; **b)** *(innen, nicht sichtbar)* wrong, reverse ‹*side*›; **~e Maschen** *(Handarb.)* purl stitches; **c)** *(in der Politik)* left-wing; leftist *(derog.)*; **der ~e Flügel einer Partei** the left wing of a party

[1]**Linke der/die** *adj. Dekl.* left-winger; leftist *(derog.)*; **die ~n** the left *sing.*

[2]**Linke die; ~n, ~n a)** *(Hand)* left hand; **jmdm. zur ~n** on sb.'s left; to the left of sb.; **zur ~n** on the left; **b)** *(Politik)* left

linkisch 1. *Adj.* awkward. **2.** *adv.* awkwardly

links [lɪŋks] **1.** *Adv.* **a)** *(auf der linken Seite)* on the left; **~ von jmdm/etw.** on sb.'s left *or* to the left of sb./on *or* to the left of sth.; **von ~:** from the left; **nach ~:** to the left; **sich ~ halten** keep to the left; **er blickte weder nach ~ noch nach rechts, sondern rannte einfach über die Straße** he didn't look left or right, but just ran straight across the road; **sich ~ einordnen** move *or* get into the left-hand lane; **jmdn./etw. ~ liegenlassen** *(fig.)* ignore sb./sth.; **b)** *(Politik)* on the left wing; **~ stehen** *od.* **sein** be left-wing *or* on the left; **c)** *(Handarb.)* **zwei ~, zwei rechts** two purl, two plain; purl two, knit two; **ein ~ gestrickter Pullover** a purl[-knit] pullover. **2.** *Präp. mit Gen.* **~ des Rheins** on the left side *or* bank of the Rhine. **3. mit ~** *(fig. ugs.)* easily; with no trouble

links-, Links-: ~abbieger der *(Verkehrsw.)* motorist/cyclist/car *etc.* turning left; **~außen** *Adv.* **a)** *(Ballspiele)* ‹*run, break through*› down the left wing; **b)** *(Politik ugs.)* on the extreme left [wing]; **~außen der; ~, ~** *(Ballspiele)* left wing; outside left; **~extremismus der** *(Politik)* left-wing extremism; **~extremist der** *(Politik)* left-wing extremist; **~gerichtet** *Adj. (Politik)* left-wing orientated; **~gewinde das** *(Technik)* left-hand thread; **~händer** [-hɛndɐ] **der; ~s, ~, ~händerin die; ~, ~nen** left-hander; **~händer[in] sein** be left-handed; **~händig 1.** *Adj.* left-handed; **2.** *adv.* with one's left hand; **~herum** *Adv.* [round] to the left; **etw. ~herum drehen**

turn sth. anticlockwise *or* [round] to the left; **~intellektuelle der/die** left-wing intellectual; **~kurve die** left-hand bend; **~lastig** *Adj. (Politik ugs.)* leftist; **~liberal** *Adj.* left-wing liberal; **~radikal** *(Politik) Adj.* radical left-wing; **~radikalismus der** left-wing radicalism; **~ruck der** *(Politik ugs.)* shift to the left; **~rum** *Adv. (ugs.) s.* **~herum; ~seitig 1.** *Adj.* ⟨*paralysis*⟩ of the left side; **die ~seitige Uferbefestigung** the reinforcement of the left bank; **2.** *adv.* on the left [side]; **~verkehr der** driving *no art.* on the left; **in Irland ist ~verkehr** they drive on the left in Ireland

Linoleum [li'no:leʊm] **das; ~s** linoleum; lino

Linol·schnitt der linocut

Linse ['lɪnzə] **die; ~, ~n a)** lentil; **b)** *(Med., Optik)* lens

Linsen·suppe die lentil soup

Lippe ['lɪpə] **die; ~, ~n** lip; **sie brachte es nicht über die ~n** she couldn't bring herself to say it; **an jmds. Lippen** *(Dat.)* **hängen** hang on sb.'s every word; **eine [dicke** *od.* **große] ~ riskieren** *(salopp)* shoot one's mouth off *(sl.)*

Lippen-: ~bekenntnis das *(abwertend)* empty talk *no pl.;* **~bekenntnisse für etw. ablegen** pay lip-service to sth. **~stift der** lipstick

liquid [li'kvi:t] *Adj. (Wirtsch.)* liquid ⟨*funds, resources*⟩*;* solvent ⟨*business*⟩*;* **ich bin nicht ~:** I'm out of funds

Liquidation [likvida'tsio:n] **die; ~, ~en a)** *(verhüll.: Tötung)* liquidation; **b)** *(Wirtsch.)* liquidation *no indef. art.*

liquidieren [likvi'di:rən] *tr. V. (auch Wirtsch.)* liquidate

Liquidität [likvidi'tɛ:t] **die; ~, ~en** *(Wirtsch.)* liquidity; solvency

lispeln ['lɪspl̩n] *itr. V.* lisp; **er hat schon immer gelispelt** he's always had a lisp

List [lɪst] **die; ~, ~en a)** [cunning] trick *or* ruse; **b)** *(listige Art) o. Pl.* cunning; **mit ~ und Tücke** *(ugs.)* by cunning and trickery

Liste die; ~, ~n list; **schwarze ~:** blacklist

Listen-: ~platz der *(Politik)* place on the [party] list; **~preis der** list price; **~wahl die** *(Parl.)* list system

listig 1. *Adj.* cunning; crafty. **2.** *adv.* cunningly; craftily; **jmdn. ~ ansehen/angrinsen** look/grin at sb. slyly.

Litanei [lita'nai] **die; ~, ~en** *(Rel., auch fig. abwertend)* litany

Litauen ['li:tauən] **(das); ~s** Lithuania

Litauer der; ~s, ~, Litauerin die; ~, ~nen Lithuanian

litauisch *Adj.* Lithuanian

Liter ['li:tɐ] **der,** *auch:* **das; ~s, ~:** litre

literarisch [lɪtə'ra:rɪʃ] **1.** *Adj.* literary. **2.** *adv.* **~ interessiert/gebildet sein** be interested in literature/be well-read

Literat [lɪtə'ra:t] **der; ~en, ~en** writer; literary figure

Literatur [lɪtəra'tu:ɐ̯] **die; ~, ~en** literature; **belletristische ~:** belles-lettres *pl.*

Literatur-: ~an·gabe die [bibliographical] reference; **~gattung die** literary genre; **~geschichte die** literary history; history of literature; **~kritik die** literary criticism; **~kritiker der** literary critic; **~verzeichnis das** list of references; **~wissenschaft die** literary studies *pl., no art.;* study of literature

Liter·flasche die litre bottle

liter·weise *Adv.* by the litre; in litres

Litfaß·säule ['lɪtfas-] **die** advertising column *or* pillar

Lithographie [litogra'fi:] **die; ~, ~n a)** *o. Pl.* lithography *no art.;* **b)** *(Druck)* lithograph

litt [lɪt] *1. u. 3. Pers. Sg. Prät. v.* **leiden**

Liturgie [litʊr'gi:] **die; ~, ~n** *(christl. Kirche)* liturgy

liturgisch *Adj.* liturgical

Litze ['lɪtsə] **die; ~, ~n** braid

live [laif] *(Rundf., Ferns.)* **1.** *Adj.* live. **2.** *adv.* live

Live-Sendung die *(Rundf., Ferns.)* live programme; *(Übertragung)* live broadcast

Lizenz [li'tsɛnts] **die; ~, ~en** licence; **etw. in ~ herstellen** manufacture sth. under licence

Lizenz·ausgabe die *(Buchw.)* licensed edition

Lkw, LKW [ɛlka:'ve:] **der; ~[s], ~[s]** *Abk.* **Lastkraftwagen**

Lkw-Fahrer der truck *or (Brit.)* lorry driver; trucker *(Amer.)*

Lob [lo:p] **das; ~[e]s, ~e** praise *no indef. art.;* **ein ~ bekommen** receive praise; come in for praise; **ein ~ dem Küchenchef/der Hausfrau** my compliments to the chef/the hostess

Lobby ['lɔbi] **die; ~, ~s** *od.* **Lobbies** lobby

Lobbyist der; ~en, ~en lobbyist

loben *tr.V.* praise; **jmdn./etw. ~d erwähnen** commend sb./sth.

lobens·wert *Adj.* praiseworthy; laudable; commendable

Lobes·hymne die *(oft iron.)* hymn of praise

Lob-: ~gesang der song *or* hymn of praise; **~hudelei die** *(abwertend)* extravagant praise *no pl.* **(auf +** *Akk.* of)

löblich ['lø:plɪç] *Adj.* laudable; commendable

lob-, Lob-: ~lied das song of praise; **ein ~lied auf jmdn./etw. anstimmen** *(fig.)* sing sb.'s praises/the praises of sth.; **~rede die** eulogy; panegyric; **eine ~rede auf jmdn. halten** make a speech in praise of sb.; eulogize sb.

Loch [lɔx] **das; ~[e]s, Löcher** ['lœçɐ] **a)** hole; **ein ~ im Zahn/Kopf haben** have a hole *or* cavity in one's tooth/gash on one's *or* the head; **jmdm. ein ~** *od.* **Löcher in den Bauch fragen** *(salopp)* drive sb. up the wall with [all] one's questions *(coll.);* **Löcher in die Luft gucken** *od.* **starren** *(ugs.)* gaze into space; **auf dem letzten ~ pfeifen** be on one's/its last legs; **b)** *(salopp abwertend: Wohnraum)* hole

lochen *tr. V.* punch holes/a hole in; punch, clip ⟨*ticket*⟩*;* punch [holes in] ⟨*invoice, copy, bill*⟩ *(for filing); (perforieren)* perforate

Locher der; ~s, ~ punch

löcherig *Adj.* holey; full of holes *pred.*

löchern *tr. V. (ugs.)* **jmdn. ~:** pester sb. to death

Loch-: ~karte die *(Technik, DV)* punch[ed] card; **~streifen der** *(Technik, DV)* punch[ed] tape

Locke ['lɔkə] **die; ~, ~n** curl; **~n haben** have curly hair

locken *tr. V.* **a)** lure; *(fig.)* entice **(aus** out of, **in +** *Akk.* into); **b)** *(reizen)* tempt

Locken-: ~kopf der a) curly hair; **b)** *(Mensch)* curly head; **~pracht die** *(scherzh.)* magnificent head of curls; **~wickler der** [hair] curler *or* roller

locker 1. *Adj.* **a)** loose ⟨*tooth, nail, chair-leg, etc.*⟩*; s. auch* **Schraube; b)** *(durchlässig, leicht)* loose ⟨*soil, snow, fabric*⟩*;* light ⟨*mixture, cake*⟩*;* **c)** *(entspannt)* relaxed ⟨*position, muscles*⟩*;* slack ⟨*rope, rein*⟩*; (fig.: unverbindlich)* loose ⟨*relationship, connection, etc.*⟩*;* **das Seil/die Zügel ~ lassen** slacken the rope [off]/slacken the reins; **d)** *(leichtfertig)* loose ⟨*morals, life*⟩*;* frivolous ⟨*jokes, remarks*⟩*;* **sein ~es Mundwerk** *(salopp)* his big mouth *(coll.);* **ein ~er Vogel** *(ugs.)* a bit of a lad *(coll.).* **2.** *adv.* **a)** loosely; **b)** *(ent-*

spannt, ungezwungen) loosely; **dieses Gesetz wird ~ gehandhabt** this law is not strictly enforced

locker-: ~|lassen *unr. itr. V.* **nicht ~lassen** *(ugs.)* not give *or* let up; **~|machen** *tr. V. (ugs.)* fork up *or* out *(sl.); shell out (sl.)*

lockern 1. *tr. V.* **a)** loosen ⟨*screw, tie, collar, etc.*⟩; slacken [off] ⟨*rope, dog-leash, etc.*⟩; *(fig.)* relax ⟨*regulation, law, etc.*⟩; **b)** *(entspannen)* loosen up, relax ⟨*muscles, limbs*⟩; **c)** *(auf~)* loosen, break up ⟨*soil*⟩. **2.** *refl. V.* **a)** ⟨*brick, tooth, etc.*⟩ work itself loose; **b)** *(entspannen)* ⟨*person*⟩ loosen up; *(vor Spielbeginn)* loosen *or* limber up; *(fig.)* ⟨*tenseness, tension*⟩ ease

Lockerung die; ~, ~en a) loosening; *(fig.: von Bestimmung, Gesetz usw.)* relaxation; **b)** *(Entspannung)* loosening up; relaxation

lockig *Adj.* curly

Lock·mittel das enticement

Lockung die; ~, ~en *(geh.)* temptation; **jmds. ~en** *(Dat.)* **widerstehen** resist sb.'s enticements

Lock·vogel der decoy

Loden ['lo:dn̩] **der; ~s, ~:** loden

Loden·mantel der loden coat

lodern *itr. V. (geh.)* blaze

Löffel ['lœfl̩] **der; ~s, ~ a)** spoon; *(als Maßangabe)* spoonful; **b)** *(Jägerspr.)* ear; **jmdm. eins** *od.* **ein paar hinter die ~ geben** *(ugs.)* give sb. a clout round the ear

löffeln *tr. V.* spoon [up]

löffel·weise *Adv.* by the spoonful

log [lo:k] *1. u. 3. Pers. Sg. Prät. v.* **lügen**

Log [lɔk] **das; ~s, ~e** *(Seew.)* log

Logarithmen *s.* **Logarithmus**

Logarithmen·tafel die *(Math.)* log[arithmic] table

Logarithmus [loga'rɪtmʊs] **der; ~, Logarithmen** *(Math.)* logarithm; log

Log·buch das *(Seew.)* log [book]

Loge ['lo:ʒə] **die; ~, ~n a)** box; **b)** *(Freimaurer~)* lodge

Logen·platz der seat in a box

Loggia ['lɔdʒia] **die; ~, Loggien** balcony

logieren *(veralt.) itr. V.* stay

Logik ['lo:gɪk] **die; ~:** logic

Logis [lo'ʒi:] **das; ~** [lo'ʒi:(s)], **~** [lo'ʒi:s] lodgings *pl.*; room/rooms *pl.*; *s. auch* **Kost**

logisch ['lo:gɪʃ] **1.** *Adj.* logical; **[ist doch] ~** *(ugs.)* yes, of course. **2.** *adv.* logically

logischerweise *Adv.* logically; *(ugs.: selbstverständlich)* naturally

logo ['lo:go] *Adj. (salopp)* **[ist doch] ~!** you bet! *(coll.)*; of course!

Lohn [lo:n] **der; ~[e]s, Löhne** ['lø:nə] **a)** wage[s *pl.*]; pay *no indef. art., no pl.*; **b)** *o. Pl. (Belohnung, auch fig.)* reward

Lohn-: ~abhängige der/die wage-earner; **~aus·fall der** loss of earnings; **~ausgleich der** making-up of wages; **[eine] kürzere Arbeitszeit bei vollem ~ausgleich** shorter working hours with no loss of pay; **~buchhalter der** payroll clerk; **~buchhaltung die a)** *o. Pl.* payroll accounting; **b)** *(Abteilung)* payroll office; **~büro das** payroll office; **~empfänger der** wage-earner

lohnen 1. *refl., itr. V.* be worth it; be worthwhile; **die Anstrengung hat sich gelohnt** it was worth the effort; **das lohnt sich nicht für mich** it's not worth my while; **die Mühe hat [sich] gelohnt** it was worth the trouble *or* effort. **2.** *tr. V.* be worth; **die Ausstellung lohnt einen Besuch** the exhibition is worth a visit *or* is worth visiting; **das lohnt die Mühe nicht** it is not worth the trouble

löhnen *tr., itr. V. (salopp)* pay; fork out *or* up *(sl.)*

lohnend *Adj.* rewarding ⟨*task*⟩; worthwhile, rewarding ⟨*occupation*⟩; worthwhile ⟨*aim*⟩; *(einträglich)* financially rewarding; lucrative

Lohn-: ~erhöhung die wage *or* pay increase *or (Brit.)* rise; **~forderung die** wage demand *or* claim; **~fortzahlung die** continued payment of wages; **~kosten** *Pl.* wage costs; **~runde die** wage *or* pay round

Lohn·steuer die income tax

Lohnsteuer-: ~jahres·ausgleich der annual adjustment of income tax; **~karte die** income-tax card

Lohn·zettel der pay-slip

Loipe ['lɔypə] **die; ~, ~n** *(Skisport)* [cross-country] course

Lok [lɔk] **die; ~, ~s** engine; locomotive

lokal [lo'ka:l] *Adj.* local; **Lokales** *(Zeitungsw.)* local news *sing.*

Lokal das; ~s, ~e pub *(Brit. coll.)*; bar *(Amer.)*; *(Speise~)* restaurant

Lokal·blatt das *(Zeitungsw.)* local paper

lokalisieren *tr. V. (geh., fachspr.)* locate

Lokal-: ~kolorit das local colour; **~patriotismus der** local patriotism; **~redaktion die** *(Rundf., Ferns., Zeitungsw.)* local-news section; **~runde die** round for everyone [in the pub *(Brit. coll.) or (Amer.)* bar]; **~teil der** *(Zeitungsw.)* local section; **~termin der** *(Rechtsspr.)* visit to the scene [of the crime]; **~verbot das: [in einer Gaststätte] ~verbot haben/bekommen** be/get banned [from a pub *(Brit. coll.) or (Amer.)* bar]; **~zeitung die** local [news]paper

Lok·führer der *s.* **Lokomotivführer**

Lokomotive [lokomo'ti:və] **die; ~, ~n** locomotive; [railway] engine

Lokomotiv·führer der engine-driver *(Brit.)*; engineer *(Amer.)*

Lokus ['lo:kʊs] **der; ~** *od.* **~ses, ~** *od.* **~se** *(salopp)* loo *(Brit. coll.)*; john *(Amer. coll.)*

Lokus·papier das *(salopp)* loo paper *(Brit. coll.)*; toilet paper

London ['lɔndɔn] **(das); ~s** London

Londoner 1. *indekl. Adj.; nicht präd.* London. **2. der; ~s, ~:** Londoner; *s. auch* **Kölner**

Lorbeer ['lɔrbe:ɐ̯] **der; ~s, ~en a)** laurel; **b)** *(Gewürz)* bay-leaf; **c)** *(~kranz)* laurel wreath; **[sich] auf seinen ~en ausruhen** *(fig. ugs.)* rest on one's laurels

Lorbeer-: ~blatt das bay-leaf; **~kranz der** laurel wreath

Lore ['lo:rə] **die; ~, ~n** car; *(kleiner)* tub

los [lo:s] **1.** *Adj.; nicht attr.* **a) der Knopf ist ~:** the button has come off; **ich habe die Schraube/das Brett/das Rad ~:** I have got the screw out/the board/wheel off; **b) jmdn./etw. ~ sein** *(befreit sein von)* be rid *or (coll.)* shot of sb./sth.; *(verloren haben)* have lost sth.; **c) hier ist viel/wenig/immer etw. ~:** there is a lot/not much/always sth. going on here; **was ist hier ~?** *(was geschieht?)* what's going on here?; *(was ist nicht in Ordnung?)* what's the matter here?; what's up here? *(coll.)*; **mit jmdm./etw. ist nichts/nicht viel ~** *(ugs.)* sb./sth. isn't up to much *(coll.)*; **was ist denn mit dir ~?** what's up *or* wrong *or* the matter with you? **2.** *Adv.* **a)** *(als Aufforderung)* come on!; *(geh schon!)* go on!; **auf die Plätze** *od.* **Achtung, fertig, ~!** on your marks, get set, go; **nun aber ~!** [come on,] let's get moving *or* going!; **b) er ist mit dem Wagen ~** *(ugs.: losgefahren)* he's gone off in the car

Los das; ~es, ~e a) lot; **das ~ soll entscheiden** it shall be decided by drawing lots; **b)** *(Lotterie~)* ticket; **das Große ~:** [the] first prize; **mit jmdm./etw. das Große ~ ziehen** *(fig.)* hit the jackpot with sb./sth.; **c)** *(geh.: Schicksal)* lot

-los *Adj.* -less

lösbar *Adj.* soluble, solvable ⟨*problem, equation, etc.*⟩

los-: **~|bekommen** *unr. tr. V.* get ⟨*string, tape, ribbon, etc.*⟩ off; get ⟨*screw, nail, etc.*⟩ out; **~|binden** *unr. tr. V.* untie; **~|brechen 1.** *unr. itr. V.; mit sein* **a)** *(beginnen)* ⟨*storm*⟩ break; ⟨*cheering, laughter, etc.*⟩ break out; **b)** *(abbrechen)* break off. **2.** *unr. tr. V.* break off

Lösch·blatt das piece of blotting-paper

¹löschen ['lœʃn̩] *tr. V.* **a)** put out, extinguish ⟨*fire, candle, flames, etc.*⟩; **b)** close ⟨*bank account*⟩; delete, strike out ⟨*entry*⟩; erase, wipe out ⟨*recording, memory, etc.*⟩; **c)** quench ⟨*thirst*⟩

²löschen *tr. V. (Seemannsspr.)* unload

Lösch-: **~fahrzeug das** fire-engine; **~papier das** blotting paper; **~zug der** set of fire-fighting appliances

lose 1. *Adj.* **a)** *(nicht fest, auch fig.)* loose; **b)** *(nicht verpackt)* loose ⟨*sugar, cigarettes, sweets, sheets of paper, nails, etc.*⟩; unbottled ⟨*drink*⟩; **c)** *(ugs.: leichtfertig)* **er ist ein ~er Vogel** he is a bit of a lad; **d)** *(ugs.: vorlaut, frech)* cheeky; impudent; **einen ~n Mund haben** be a cheeky *or* impudent so-and-so *(coll.).* **2.** *adv. (auch fig.)* loosely; **~ herunterhängen** hang down loosely *or* loose

Löse·geld das ransom; **das ~ wurde in einer Telefonzelle hinterlegt** the ransom money was left in a telephone kiosk

los|eisen *tr. V.* **jmdn./sich ~** *(ugs.)* prise *or* get sb. away **(von** from)/get away **(von** from)

losen *itr. V.* draws lots **(um** for); **~, wer anfangen soll** draw lots to decide who will start

lösen ['lø:zn̩] **1.** *tr. V.* **a)** remove, take *or* get off ⟨*stamp, wallpaper*⟩; **etw. von etw. ~:** remove sth. from sth.; **das Fleisch von den Knochen ~:** take the meat off the bones; **b)** *(lockern)* take *or* let ⟨*handbrake*⟩ off; release ⟨*handbrake*⟩; undo ⟨*screw, belt, tie*⟩; remove, untie ⟨*string, rope, knot, bonds*⟩; loosen ⟨*phlegm*⟩; ease ⟨*cramp*⟩; *(fig.)* ease, relieve ⟨*[mental] pain, tension, etc.*⟩; remove ⟨*inhibitions*⟩; **jmds. Zunge ~** *(fig.)* loosen sb.'s tongue; **c)** *(klären)* solve ⟨*problem, puzzle, equation, etc.*⟩; resolve ⟨*contradiction, conflict*⟩; solve, resolve ⟨*difficulty*⟩; **d)** *(annullieren)* break off ⟨*engagement*⟩; cancel ⟨*contract*⟩; sever ⟨*connection, relationship*⟩; **e)** *(auflösen)* **etw. in etw.** *(Dat.)* **~:** dissolve sth. in sth.; **f)** *(kaufen)* buy; obtain ⟨*ticket*⟩. **2.** *refl. V.* **a)** come off; ⟨*avalanche*⟩ start; ⟨*wallpaper, plaster*⟩ come off *or* away; ⟨*packing, screw*⟩ come loose *or* undone; ⟨*paint, book-cover*⟩ come off; ⟨*phlegm, cough*⟩ get looser; ⟨*cramp*⟩ ease; **sich von etw. ~:** come off sth.; **sich von seinem Elternhaus ~** *(fig.)* break away from one's parental home; **b)** *(sich klären, entwirren)* ⟨*puzzle, problem*⟩ be solved; **sich von selbst ~** ⟨*problem*⟩ solve *or* resolve itself; **c)** *(zergehen)* **sich in etw.** *(Dat.)* **~:** dissolve in sth.; **d) aus seiner Pistole löste sich ein Schuß** *(geh.)* his pistol went off

Los·entscheid der: durch ~: by drawing lots; *(bei einem Preisausschreiben)* by [making *or* having] a draw

los-: **~|fahren** *unr. itr. V.; mit sein* **a)** *(starten)* set off; *(wegfahren)* move off; **b)** *(zufahren)* **auf jmdn./etw. ~fahren** drive/ride towards sb./sth.; **direkt auf jmdn./etw. ~fahren** drive/ride straight at sb./sth.; **~|gehen** *unr. itr. V.; mit sein* **a)** *(aufbrechen)* set off; **auf ein Ziel ~gehen** *(fig.)* go straight for a goal; **~ geht's!** let's be off; **b)** *(ugs.: beginnen)* start; **es geht ~:** it's starting; *(fangen wir an)* let's go; **~ geht's!** let's get started; **c)** *(ugs.: abgehen)* ⟨*button, handle, etc.*⟩ come off; **d)** *(angreifen)* **auf jmdn. ~gehen** go for sb.; **e)** *(abgefeuert werden)* ⟨*gun, mine, firework, etc.*⟩ go off; **~|kommen** *unr. itr. V.; mit sein (ugs.)* **a)** get away; **b)** *(freikommen)* get free; free oneself; *(freigelassen werden)* be freed; **von jmdm./etw. ~kommen** *(fig.)* get away from sb./get rid of sth.; **vom Alkohol ~kommen** *(fig.)* get off *or* give up alcohol; **~|kriegen** *tr. V. (ugs.)* **a)** get ⟨*screw, nail, etc.*⟩ out; get ⟨*lid*⟩ off; **b)** *(loswerden)* get rid *or (sl.)* shot of; **c)** *(verkaufen können)* get rid of; **~|lassen** *unr. tr. V.* **a)** let go of; **der Gedanke/das Bild ließ sie nicht mehr ~** *(fig.)* she could not get the thought/image out of her mind; **b)** *(freilassen)* let ⟨*person, animal*⟩ go; **c) jmdn. auf jmdn./etw. ~lassen** *(ugs. abwertend)* let sb. loose on sb./sth.; **d)** *(ugs.: äußern)* come out with ⟨*remark, joke, etc.*⟩; **e)** *(abschicken)* send off ⟨*letter, telegram, etc.*⟩; **~|laufen** *unr. itr. V.; mit sein* start running; **lauf schnell los und hol Brot** run out and get some bread; **~|legen** *itr. V. (ugs.)* get going *or* started

löslich *Adj.* soluble; **leicht/schwer ~:** readily/not readily *or* only slightly soluble

los-: **~|machen 1.** *tr. V. (ugs.)* let ⟨*animal*⟩ loose; untie, undo ⟨*string, line, rope*⟩; take out ⟨*plank*⟩; unhitch ⟨*trailer*⟩; **das Boot ~machen** cast off; **2.** *refl. V. (ugs., auch fig.)* free oneself **(von** from); **~|müssen** *unr. itr. V. (ugs.)* have to be off; have to go; **ich muß ~:** I must be off

Los·nummer die [lottery] ticket number

los|reißen 1. *unr. tr. V.* tear off; *(schneller, gewaltsamer)* rip off; pull ⟨*plank*⟩ off; ⟨*wind*⟩ rip ⟨*tile*⟩ off; **2.** *unr. refl. V.* break free *or* loose; **sich von etw.** *(Dat.)* **~reißen** break free *or* loose from sth.; *(fig.)* tear oneself away from sth.

Löß [lœs] **der; Lösses, Lösse** *(Geol.)* loess

los|sagen *refl. V.* **sich von jmdm./etw. ~sagen** renounce sb./sth.; break with sb./sth.

Löß·boden der loess soil

los-: **~|schicken** *tr. V. (ugs.)* send off ⟨*letter, telegram, etc.*⟩; **jmdn. ~schicken, um etw. zu holen** send sb. out to get sth.; **~|schießen** *unr. itr. V. (fig. ugs.)* fire away; **~|schlagen** *unr. itr. V. (bes. Milit.)* attack; launch one's attack

Los·trommel die [lottery] drum

Losung die; ~, ~en a) slogan; **b)** *(Milit.: Kennwort)* password

Lösung die; ~, ~en a) solution *(Gen.* to); *(eines Konflikts, Widerspruchs)* resolution; **des Rätsels ~:** the answer to the mystery; **b)** *(einer Verlobung)* breaking off; *(eines Vertrags)* cancellation; *(einer Verbindung, eines Verhältnisses)* severing; *(eines Arbeitsverhältnisses)* termination; **c)** *(Flüssigkeit)* solution

Lösungs-: **~mittel das** *(Physik, Chemie)* solvent; **~vorschlag der** proposed solution **(für** to)

Los·verkäufer der [lottery-] ticket seller

los-: **~|werden** *unr. tr. V.; mit sein* **a)** *(sich befreien können von)* get rid of; **ich werde den Gedanken/Verdacht nicht ~, daß ...:** I can't get the thought/suspicion/impression out of my mind that ...; **b)** *(ugs.: aussprechen, mit-*

teilen) tell; **er wollte etwas ~werden** he wanted to tell me/us *etc.* something; **c)** *(ugs.: verkaufen)* get rid of; flog *(Brit. sl.)*; **d)** *(ugs.: verlieren)* lose; **~|wollen** *unr. itr. V. (ugs.)* want to be off; **~|ziehen** *unr. itr. V.; mit sein (ugs.)* set off

Lot [lo:t] **das; ~[e]s, ~e a)** *(Senkblei)* plumb[-bob]; **b) im ~ stehen** be plumb; **nicht im ~ sein, außer ~ sein** be out of plumb; **nicht im ~ sein** *(fig.)* not be straightened *or* sorted out; **[wieder] ins ~ kommen** *(fig.)* be all right [again]; **c)** *(Geom.)* perpendicular

löten ['lø:tn̩] *tr. V.* solder

Lotion [lo'tsi̯o:n] **die; ~, ~en** *od.* **~s** lotion

Löt-: ~kolben der soldering-iron; **~lampe die** blowlamp

Lotos ['lo:tɔs] **der; ~, ~:** lotus

lot·recht 1. *Adj.* perpendicular; vertical. **2.** *adv.* perpendicularly; vertically

Lotse ['lo:tsə] **der; ~n, ~n** *(Seew.)* pilot; *(fig.)* guide

lotsen *tr. V.* **a)** *(Seew.)* pilot; *(Flugw.)* guide; **b)** *(leiten)* guide; **c)** *(ugs.: führen, leiten)* drag

Lotterie [lɔtə'ri:] **die; ~, ~n** lottery

Lotterie-: ~gewinn der win in the lottery; *(gewonnenes Geld)* lottery winnings *pl.*; **~los das** lottery-ticket

Lotter·leben das *o. Pl. (abwertend)* dissolute life

Lotto ['lɔto] **das; ~s, ~s a)** national lottery; **b)** *(Gesellschaftsspiel)* lotto

Lotto-: ~gewinn der win in the national lottery; *(gewonnenes Geld)* winnings *pl.* in the national lottery; **~schein der** national-lottery coupon; **~zahlen** *Pl.* winning national-lottery numbers

Löwe ['lø:və] **der; ~n, ~n a)** lion; **b)** *(Astrol.)* Leo; the Lion

Löwen-: ~anteil der lion's share; **~maul, ~mäulchen das** *o. Pl. (Bot.)* snapdragon; antirrhinum; **~zahn der;** *o. Pl. (Bot.)* dandelion

Löwin ['lø:vɪn] **die; ~, ~nen** lioness

loyal [loa'ja:l] **1.** *Adj.* loyal **(gegenüber** to). **2.** *adv.* loyally

Loyalität [loajali'tɛ:t] **die; ~:** loyalty **(gegenüber** to)

LP [ɛl'pe:] **die; ~, ~[s]** *Abk.* **Langspielplatte** LP

lt. *Abk.* [2]**laut**

Luchs [lʊks] **der; ~es, ~e** lynx; **wie ein ~ aufpassen** watch like a hawk

Lücke ['lʏkə] **die; ~, ~n a)** gap; *(Park~, auf einem Formular, in einem Text)* space; **b)** *(Mangel)* gap; *(in der Versorgung)* break; *(im Gesetz)* loophole

Lücken·büßer der *(ugs.)* stopgap

lückenhaft 1. *Adj.* ‹*teeth*› full of gaps; gappy ‹*teeth*›; sketchy ‹*knowledge*›; sketchy, vague ‹*memory*›; incomplete, sketchy ‹*report, account, etc.*›; incomplete ‹*statement*›; ‹*alibi*› full of holes; **sein Wissen/seine Erinnerung ist ~:** there are gaps in his knowledge/memory. **2.** *adv.* ‹*remember*› vaguely, sketchily

lücken·los 1. *Adj.* unbroken ‹*line, row, etc.*›; complete ‹*account, report, curriculum vitae*›; solid, cast-iron ‹*alibi*›; comprehensive, perfect ‹*knowledge*›; **sie hat ein strahlend weißes, ~es Gebiß** she has gleaming white teeth without any gaps. **2.** *adv.* without any gaps

lud [lu:t] *1. u. 3. Pers. Sg. Prät. v.* **laden**

Lude ['lu:də] **der; ~n, ~n** *(salopp)* pimp; ponce

Luder das; ~s, ~ *(salopp)* so-and-so *(coll.)*

Luft [lʊft] **die; ~, Lüfte** ['lʏftə] **a)** *o. Pl.* air; **an die frische ~ gehen/in der frischen ~ sein** get out in[to]/be out in the fresh air; **jmdn. an die [frische] ~ setzen** *od.* **befördern** *(ugs.: hinauswerfen)* show sb. the door; **die ~ anhalten** hold one's breath; **halt die ~ an!** *(ugs.) (hör auf zu reden!)* pipe down *(coll.)*; put a sock in it *(Brit. sl.)*; *(übertreib nicht so!)* come off it! *(coll.)*; **tief ~ holen** take a deep breath; **~ schnappen** *(ugs.)* get some fresh air; **er kriegte keine/kaum ~:** he couldn't breathe/could hardly breathe; **die ~ ist rein** *(fig.)* the coast is clear; **sich in ~ auflösen** *(ugs.)* vanish into thin air; ‹*plans*› go up in smoke *(fig.)*; **er ist ~ für mich** *(ugs.)* I ignore him completely; **da bleibt einem die ~ weg** *(ugs.)* it takes your breath away; **ihm/der Firma geht die ~ aus** *(fig. ugs.)* he's/the firm's going broke *(coll.)*; **b)** *(Himmelsraum)* air; **Aufnahmen aus der ~ machen** take pictures from the air; **etw. in die ~ sprengen** *od.* **jagen** *(ugs.)* blow sth. up; **in die ~ fliegen** *od.* **gehen** *(ugs.: explodieren)* go up; **in die ~ gehen** *(fig. ugs.)* blow one's top *(coll.)*; **aus der ~ gegriffen sein** *(fig.)* ‹*story, accusation*› be pure invention; **in der ~ liegen** *(fig.)* ‹*crisis, ideas, etc.*› be in the air; **in die ~ gehen** *(fig. ugs.)* blow one's top *(coll.)*; **etw. in der ~ zerreißen** *(fig. ugs.)* tear sth. to pieces; **c)** *o. Pl. (fig.: Spielraum)* space; room; **d) sich** *(Dat.) od.* **seinem Herzen ~ machen** get it off one's chest *(coll.)*; **seinem Zorn/Ärger** *usw.* **~ machen** *(ugs.)* give vent to one's anger

Luft-: ~abwehr die *(Milit.)* air defence; anti-aircraft defence; **~angriff der** *(Milit.)* air raid; **~ballon der** balloon; **~bild das** aerial photograph; **~blase die** air bubble; **~brücke die** airlift

Lüftchen ['lʏftçən] **das; ~s, ~:** breeze

luft-, Luft-: ~dicht *Adj.* airtight; **~druck der** air pressure; atmospheric pressure; **~durchlässig** *Adj.* pervious *or* permeable to air *postpos.*; well-ventilated ‹*shoes*›

lüften 1. *tr. V.* **a)** air; **b)** *(hochheben)* raise, lift ‹*hat, lid, veil, etc.*›; **c)** *(enthüllen)* reveal; disclose ‹*secret*›. **2.** *itr. V.* air the room/house *etc.*; **wir müssen hier mal ~:** we must let some [fresh] air in here

Lüfter der; ~s, ~: fan

Luft·fahrt die; *o. Pl.* aeronautics *sing., no art.*; *(mit Flugzeugen)* aviation *no art.*

Luftfahrt·gesellschaft die airline

luft-, Luft-: ~feuchtigkeit die [atmospheric] humidity; **~filter der** *od.* **das** *(Technik)* air filter; **~fracht die** air freight; **~gekühlt** *Adj.* air-cooled; **~gewehr das** air rifle; airgun; **~hoheit die** air sovereignty

luftig *Adj.* airy ‹*room, building, etc.*›; well ventilated ‹*cellar, store*›; light, cool ‹*clothes*›

Luftikus ['lʊftikʊs] **der; ~[ses], ~se** *(ugs. abwertend)* careless and unreliable sort *(coll.)*

Luftkissen-: ~boot das hovercraft; **~fahrzeug das** hovercraft

luft-, Luft-: ~krieg der air warfare *no art.*; aerial warfare *no art.*; **~kur·ort der** climatic health resort; **~leer** *Adj.* **ein ~leerer Raum** a vacuum; **im ~leeren Raum** *(fig.)* in a vacuum; **~linie die: 1000 km ~linie** 1,000 km. as the crow flies; **~loch das** air-hole; **~matratze die** air-bed; air mattress; Lilo **(P)**; **~pirat der** [aircraft] hijacker; **~post die** airmail; **etw. per** *od.* **mit ~ schicken** send sth. [by] airmail; **~post·brief der** airmail letter; **~pumpe die** air pump; *(für*

Fahrrad) [bicycle-]pump; **~raum der** airspace; **~röhre die** *(Anat.)* windpipe; trachea *(Anat.)*; **~sack der** *(Kfz-W.)* airbag; **~schacht der** ventilation shaft; *(einer Klimaanlage)* ventilation duct; **~schiff das** airship; **~schlacht die** air battle; aerial battle; **~schlange die** *meist Pl.* [paper] streamer; **~schloß das;** *meist Pl.* castle in the air

Luftschutz-: ~bunker, ~keller der air-raid shelter

Luft-: ~spiegelung die mirage; **~sprung der** jump in the air; **~streitkräfte** *Pl. (Milit.)* air force *sing.*; **~strom der** stream of air; **~strömung die** *(Met.)* airstream; air current; **~temperatur die** *(Met.)* air temperature

Lüftung die; ~, ~en a) ventilation; **b)** *(Anlage)* ventilation system

Lüftungs·klappe die ventilation flap

Luft-: ~veränderung die change of air; **~verpestung die** *(abwertend)*, **~verschmutzung die** air pollution; **~waffe die** air force; **~weg der a) auf dem ~:** by air; **b)** *Pl. (Anat.: Atemwege)* airways; air passages; **~widerstand der** *(Physik)* air resistance; **~zufuhr die** air supply; **~zug der** *o. Pl.* [gentle] breeze; *(in Zimmern, Gebäuden)* draught

Lug [lu:k] *in* **~ und Trug** lies *pl.* and deception

Lüge ['ly:gə] **die; ~, ~n** lie; **~n haben kurze Beine** *(Spr.)* [the] truth will out; **jmdn./etw. ~n strafen** prove sb. a liar/give the lie to sth.; *s. auch* **fromm 1c**

lügen 1. *itr. V.* lie; **~ wie gedruckt** lie like mad; be a terrible liar *(coll.)*. **2.** *tr. V.* **das ist gelogen!** that's a lie!

Lügen·detektor der lie-detector

Lügner der; ~s, Lügnerin die; ~, ~nen liar

[1]**Lukas** ['lu:kas] **(der)** Luke

[2]**Lukas der; ~, ~** try-your-strength machine; **hau den ~!** try your strength!

Luke ['lu:kə] **die; ~, ~n a)** *(Dach~)* skylight; **b)** *(bei Schiffen)* hatch; *(Keller~)* trap-door

lukrativ [lukra'ti:f] **1.** *Adj.* lucrative. **2.** *adv.* lucratively

Lulatsch ['lu:l(a:)tʃ] **der; ~[e]s, ~e** *(ugs.)* [long] lanky fellow; **ein langer ~:** a beanpole

Lümmel ['lyml] **der; ~s, ~ a)** *(abwertend: Flegel)* lout; **b)** *(ugs., fam.: Bengel)* rascal

lümmeln *refl. V. (ugs. abwertend) s.* **flegeln**

Lump [lʊmp] **der; ~en, ~en** scoundrel; rogue

lumpen *tr. V.* **sich nicht ~ lassen** *(ugs.)* splash out *(coll.)*

Lumpen der; ~s, ~: rag

Lumpen-: ~gesindel das *(abwertend)* rabble; riff-raff; **~sammler der** rag-and-bone man

lumpig *(abwertend) Adj. nicht präd.* paltry, miserable ‹*pay etc.*›

Lunge ['lʊŋə] **die; ~, ~n** lungs *pl.*; *(Lungenflügel)* lung; **er hat es auf der ~** *(ugs.)* he has got lung trouble *(coll.)*; **auf ~** *od.* **über die ~ rauchen** inhale; **sich** *(Dat.)* **die ~ aus dem Hals** *od.* **Leib schreien** *(ugs.)* yell one's head off *(coll.)*

lungen-, Lungen-: ~embolie die *(Med.)* pulmonary embolism; **~entzündung die** pneumonia *no indef. art.*; **~flügel der** lung; **~krank** *Adj.* suffering from a lung disease *postpos.*; **~kranke der/die** person with *or* suffering from a lung disease; **~krebs der** lung cancer; **~zug der** inhalation

Lunte ['lʊntə] **die; ~, ~n** fuse; match; **~ riechen** *(ugs.)* smell a rat

Lupe ['lu:pə] **die; ~, ~n** magnifying glass; **jmdn./etw. unter die ~ nehmen** *(ugs.)* examine sb./sth. closely; take a close look at sb./sth.

lupenrein *Adj.* **a)** flawless ‹*diamond, stone, etc.*›; ‹*diamond*› of the first water; **b)** *(musterhaft)* genuine ‹*amateur*›; unimpeachable ‹*record, reputation*›; perfect ‹*forgery, gentleman*›

lupfen ['lʊpfn̩] *(südd., schweiz., österr.)*, **lüpfen** ['lʏpfn̩] *tr. V.* raise; lift

Lurch [lʊrç] **der; ~[e]s, ~s** amphibian

Lusche ['lʊʃə] **die; ~, ~n** *(Kartenspiele ugs.)* low card

Lust [lʊst] **die; ~ a)** *(Bedürfnis)* **~ haben** *od.* **verspüren, etw. zu tun** feel like doing sth.; **große/keine ~ haben, etw. zu tun** really/not feel like doing sth.; **wir hatten nicht die geringste ~, das zu tun** we didn't feel in the least *or* slightest like doing it; **auf etw.** *(Akk.)* **~ haben** fancy sth.; **b)** *(Vergnügen)* pleasure; joy; **die ~ an etw.** *(Dat.)* **verlieren** lose interest in *or* stop enjoying sth.; **etw. mit ~ und Liebe tun** love doing sth.

Lüster ['lʏstɐ] **der; ~s, ~** chandelier

lüstern 1. *Adj.* lecherous; lascivious. **2.** *adv.* **a)** lecherously; lasciviously; **b)** *(begierig)* greedily

Lust-: ~gefühl das feeling of pleasure; **~gewinn der** *o. Pl.* pleasure

lustig 1. *Adj.* **a)** *(vergnügt)* merry; jolly; merry, jolly, jovial ‹*person*›; happy, enjoyable ‹*time*›; **das kann ja ~ werden!** *(ugs. iron.)* this/that is going to be fun! **sich über jmdn./etw. ~ machen** make fun of sb./sth.; **b)** *(komisch)* funny; amusing. **2.** *adv.* **a)** *(vergnügt)* ‹*laugh, play*› merrily, happily; **b)** *(komisch)* funnily; amusingly; **sie kann so ~ erzählen** she can tell such funny *or* amusing stories; **c)** *(unbekümmert)* gaily

-lustig ['lʊstɪç] *adj.* **[sehr] tanz~/sanges~/lese~ sein** be very fond of *or* keen on dancing/singing/reading

Lustigkeit die; ~ a) merriness; jolliness; *(Frohsinn auch)* joviality; **b)** *(Komik)* funniness

Lüstling ['lʏstlɪŋ] **der; ~s, ~e** *(veralt. abwertend, scherzh.)* lecher

lust-, Lust-: ~los 1. *Adj.* listless; *(ohne Begeisterung)* unenthusiastic [and uninterested]; **2.** *adv.* listlessly; *(ohne Begeisterung)* without enthusiasm [or interest]; **~spiel das** comedy; **~wandeln** *itr. V.; mit sein od. haben (geh. veralt.)* stroll; take a stroll

Lutheraner [lʊtə'ra:nɐ] **der; ~s, ~**Lutheran

Luther·bibel die Luther's Bible; Lutheran Bible

lutherisch *Adj.* Lutheran

lutschen ['lʊtʃn̩] **1.** *tr. V.* suck. **2.** *itr. V.* suck; **an etw.** *(Dat.)* **~:** suck sth.

Lutscher der; ~s, ~ a) lollipop; **b)** *(Schnuller)* dummy *(Brit.)*; pacifier *(Amer.)*

Luv [lu:f] **die** *od.* **das; ~** *(Seemannsspr.)* **in/nach ~:** to windward

Luxemburg ['lʊksm̩bʊrk] **(das); ~s** Luxembourg

luxuriös [lʊksu'ri̯ø:s] **1.** *Adj.* luxurious. **2.** *adv.* luxuriously

Luxus ['lʊksʊs] **der; ~** *(auch fig.)* luxury; **etw. ist reiner ~:** sth. is sheer extravagance

Luxus-: ~artikel der luxury article; **~ausführung die** de luxe version; **~hotel das** luxury hotel; **~jacht die** luxury yacht; **~klasse die** luxury class

Luzifer ['lu:tsifɛr] **(der)** Lucifer

LW *Abk.* **Langwelle** LW

Lymph·drüse die *(veralt.)*, **Lymph·knoten der** lymph node *or* gland

lynchen ['lʏnçn̩] *tr. V.* lynch; *(scherzh.)* lynch; kill

Lynch·justiz die lynch-law

Lyra ['lʏ:ra] **die;** ~, **Lyren** *(Mus.)* lyre

Lyrik ['lʏ:rɪk] **die;** ~: lyric poetry

Lyriker der; ~s, ~, **Lyrikerin die;** ~, ~nen lyric poet; lyricist

lyrisch 1. *Adj.* **a)** lyric ⟨*poem, poetry, epic, drama*⟩; lyrical ⟨*passage, style, description, etc.*⟩; **b)** *(gefühlvoll)* lyrical; **c)** *nicht präd. (Mus.)* lyric. **2.** *adv.* lyrically

Lyzeum [lʏ'tse:ʊm] **das;** ~s, **Lyzeen** [lʏ'tse:ən] girls' high school

M

m, M [ɛm] **das;** ~, ~: m/M; *s. auch* **a, A**

m *Abk.* **Meter** m

M *Abk.* [1]**Mark**

Maat [ma:t] **der;** ~[e]s, ~e[n] **a)** [ship's] mate; **b)** *(Dienstgrad)* petty officer

Mach·art die style; *(Schnitt)* cut

machbar *Adj.* feasible

Mache die; ~ *(ugs.)* **a)** *(abwertend)* sham; **b) etw. in der ~ haben** have sth. on the stocks; be working on sth.

machen ['maxn̩] **1.** *tr. V.* **a)** *(herstellen)* make; **aus diesen Äpfeln ~ wir Saft** we will make juice from these apples; **sich** *(Dat.)* **etw. ~ lassen** have sth. made; **Geld/ein Vermögen/einen Gewinn** ~: make money/a fortune/a profit; **dafür ist er einfach nicht gemacht** *(fig.)* he's just not cut out for it; **etw. aus jmdm.** ~: make sb. into sth.; *(verwandeln)* turn sb. into sth.; **jmdn. zum Präsidenten** *usw.* ~: make sb. president *etc.;* **b) jmdm. einen Kostenvoranschlag** ~: let sb. have *or* give sb. an estimate; **jmdm. einen guten Preis** ~ *(ugs.)* name a good price; **c)** *(zubereiten)* get, prepare ⟨*meal*⟩; **jmdm./sich [einen] Kaffee** ~: make [some] coffee for sb./oneself; **jmdm. einen Cocktail** ~: get *or* mix sb. a cocktail; **d)** *(verursachen)* **jmdm. Arbeit** ~: cause *or* make [extra] work for sb.; **jmdm. Sorgen** ~: cause sb. anxiety; worry sb.; **jmdm. Mut/Hoffnung** ~: give sb. courage/hope; **das macht Durst/Hunger** *od.* **Appetit** this makes one thirsty/hungry; this gives one a thirst/an appetite; **das macht das Wetter** that's [because of] the weather; **das macht das viele Rauchen** that comes from smoking a lot; **mach, daß du nach Hause kommst!** *(ugs.)* off home with you!; **ich muß ~, daß ich zum Bahnhof komme** *(ugs.)* I must see that I get to the station; **e)** *(ausführen)* do ⟨*job, repair, etc.*⟩; **seine Hausaufgaben** ~: do one's homework; **ein Foto** *od.* **eine Aufnahme** ~: take a photograph; **ein Examen** ~: take an exam; **einen Spaziergang** ~: go for *or* take a walk; **eine Reise** ~: go on a journey *or* trip; **einen Besuch [bei jmdm.]** ~: pay [sb.] a visit; **wie man's macht, macht man's falsch** *od.* **verkehrt** *(ugs.)* [however you do it,] there's always something wrong; **er macht es nicht unter 100 DM** he won't do it for under *or* less than 100 marks; **f) jmdn. glücklich/eifersüchtig** *usw.* ~: make sb. happy/jealous *etc.;* **etw. größer/länger/kürzer** ~: make sth. bigger/longer/shorter; **mach es dir gemütlich** *od.* **bequem!** make yourself comfortable *or* at home; **das Kleid macht sie älter** the dress makes her look older; **g)** *(tun)* do; **mußt du noch viel ~?** do you still have a lot to do?; **mach' ich, wird gemacht!** *(ugs.)* will do!; **was ~ Sie [beruflich]?** what do you do [for a living]?; **was soll ich nur ~?** what am I to do?; **so etwas macht man nicht** that [just] isn't done; **mit mir könnt ihr es ja ~** *(ugs.)* you can get away with it with me; **h) was macht ...?** *(wie ist um ... bestellt?)* how is ...?; **was macht die Arbeit?** how is the job [getting on]?; how are things at work?; **i)** *(ergeben) (beim Rechnen)* be; *(bei Geldbeträgen)* come to; **zwei mal zwei macht vier** two times two is four; **was** *od.* **wieviel macht das [alles zusammen]?** how much does that come to?; **das macht 12 DM** that is *or* costs 12 marks; *(Endsumme)* that comes to 12 marks; **j)** *(schaden)* **was macht das schon?** what does it matter?; **macht das was?** does it matter?; do you mind?; **macht nichts!** *(ugs.)* never mind!; it doesn't matter; **k)** *(teilnehmen an)* **einen Kursus** *od.* **Lehrgang** ~: take a course; **l)** *(ugs.: veranstalten)* organize, *(coll.)* do ⟨*trips, meals, bookings, etc.*⟩; **ein Fest** ~: give a party; **m) mach's gut!** *(ugs.)* look after yourself!; *(auf Wiedersehen)* so long!; **n)** *(ugs.: ordnen, saubermachen, renovieren);* do ⟨*room, stairs, washing, etc.*⟩; **das Bett** ~: make the bed; **sich** *(Dat.)* **die Haare/Fingernägel** ~: do one's hair/nails; **o)** *(ugs.: verhüll. seine Notdurft verrichten)* **sein Geschäft** ~: relieve oneself; **groß/klein** ~: do big jobs/small jobs *(child language).* **2.** *refl. V.* **a)** *mit Adj.* **sich hübsch** ~: smarten [oneself] up; **sich schmutzig** ~: get [oneself] dirty; **sich verständlich** ~: make oneself clear; **das macht sich bezahlt!** it's worth it!; **b)** *(beginnen)* **sich an etw.** *(Akk.)* ~: get down to sth.; **c)** *(ugs.: sich entwickeln)* do well; get on; **du hast dich aber gemacht!** you've made great strides!; **d)** *(passen)* **das macht sich gut hier** this fits in well; this looks good here; **e) mach dir nichts daraus!** *(ugs.)* don't let it bother you; **ich mache mir nichts daraus** it doesn't bother me; **sich** *(Dat.)* **nichts/wenig aus jmdm./etw.** ~ *(ugs.)* not care at all/much for sb./sth.; **f) wir wollen uns** *(Dat.)* **einen schönen Abend** ~: we want to have an enjoyable evening; **g) sich** *(Dat.)* **Feinde** ~: make enemies; **sich** *(Dat.)* **jmdn. zum Freund/Feind** ~: make a friend/an enemy of sb.; **h) wenn es sich [irgendwie] ~ läßt** if it can [somehow] be done; if it is [at all] possible. **3.** *itr. V.* **a)** *(ugs.: sich beeilen)* **mach schon!** get a move on! *(coll.);* look snappy! *(coll.);* **mach schneller!** hurry up!; **b) das macht müde** it makes you tired; it is tiring; **das macht hungrig/durstig** it makes you hungry/thirsty; **das Kleid macht dick** the dress makes one look fat; **c)** *(tun)* **laß mich nur** ~ *(ugs.)* leave it to me; **d)** *(ugs. verhüll.)* ⟨*child, pet*⟩ perform *(coll.);* **ins Bett/in die Hose** ~: wet one's bed/pants; **e)** *(ugs.)* **auf naiv** *usw.* ~: pretend to be naïve; **auf feine Dame** *usw.* ~: act the fine lady; **f)** *(landsch. ugs.: sich begeben)* go

Machenschaften *Pl. (abwertend)* machinations; wheeling and dealing *sing.*

Machete [ma'xe:tə] **die;** ~, ~n machete

Macht [maxt] **die;** ~, **Mächte** ['mɛxtə] **a)** *o. Pl.* power; *(Stärke)* strength; *(Befugnis)* authority; power; **mit aller** ~: with all one's might; **alles, was in seiner ~ steht, tun** do everything in one's power; **seine ~ ausspielen** use one's authority *or* power; **das liegt nicht in ihrer ~:** that is not within her power; that is outside her authority; **die ~ der Gewohnheit/der Verhältnisse** the force of habit/circumstances; **b)** *o. Pl. (Herrschaft)* power *no art.*; **die ~ ergreifen** *od.* **an sich reißen** seize power; **an die ~ kommen** come to power; **an der ~ sein** be in power; **c)** *(Staat)* power; **d) die Mächte der Finsternis** the powers of darkness; **böse Mächte** evil forces

Macht-: ~anspruch der claim *or* pretension to power; **~befugnis die** authority *no pl.*, *no art.*; power *no art.*; **~bereich der** sphere of influence; **~ergreifung die** *(Politik)* seizure of power (*Gen.* by); **~gier die** *(abwertend)* craving for power; **~haber der; ~s, ~:** ruler; **~hunger der** hunger for power

mächtig ['mɛçtɪç] **1.** *Adj.* **a)** powerful; **die Mächtigen dieser Welt** the high and mighty; the wielders of power; **b)** *(beeindrukkend groß)* mighty; powerful, mighty ⟨*voice, blow*⟩; tremendous, powerful ⟨*effect*⟩; *(ugs.)* terrific *(coll.)* ⟨*luck*⟩; terrible *(coll.)* ⟨*fright*⟩; **~en Hunger/~e Angst haben** be terribly hungry/afraid; **c)** *(landsch.: schwer)* heavy ⟨*food*⟩. **2.** *adv. (ugs.)* terribly *(coll.)*; extremely; **er ist ~ gewachsen** he has grown a lot; **ihr müßt euch ~ beeilen** you'll really have to step on it *(coll.)*

macht-, Macht-: ~kampf der *(bes. Politik)* power struggle; **~los** *Adj.* powerless; impotent; **gegen etw. ~los sein, einer Sache** *(Dat.)* **~los gegenüberstehen** be powerless in the face of sth.; **~losigkeit die** impotence **(gegen, gegenüber** in the face of); **~politik die** power politics *sing.*, *no art.*; **~probe die** trial of strength; **~streben das** ambition for power; **~voll 1.** *Adj.* powerful; *(imponierend)* impressive ⟨*demonstration, appearance*⟩; **2.** *adv.* powerfully; *(imponierend)* impressively; **~wechsel der** *(Politik)* change of government; **~wort das** word of command; decree; **ein ~wort sprechen** put one's foot down; lay down the law

Mach·werk das *(abwertend)* shoddy effort

Macke ['makə] **die; ~, ~n a)** *(salopp: Tick)* fad; **'ne ~ haben** have a fad; *(verrückt sein)* be off one's rocker *(sl.)*; **b)** *(ugs.: Defekt)* defect; *(optisch)* mark; blemish

Macker der; ~s, ~ a) *(Jugendspr.: Freund, Kerl)* guy *(sl.)*; bloke *(Brit. sl.)*; **b)** *(abwertend)* macho

Mädchen ['mɛːtçən] **das; ~s, ~ a)** girl; **b)** *(Haus~)* maid; **~ für alles** *(ugs.)* maid of all work; *(im Büro usw.)* girl Friday; *(Mann)* man Friday

mädchenhaft *Adj.* girlish

Mädchen-: ~handel der; *o. Pl.* white-slave traffic; **~händler der** white-slave trader; **~klasse die** girls' class; **~kleidung die** girls' clothes *pl.*; **~name der a)** *(Vorname)* girl's name; **b)** *(Geburtsname)* maiden name; **~schule die** girls' school; school for girls

Made ['maːdə] **die; ~, ~n** maggot; *(Larve)* larva; **leben wie die ~ im Speck** be living in the lap of luxury *or* off the fat of the land

madig *Adj.* maggoty; **jmdn./etw. ~ machen** *(ugs.)* run sb./sth. down

Madonna [ma'dɔna] **die; ~, Madonnen a)** *(christl. Rel.) o. Pl.* **die ~:** Our Lady; the Virgin Mary; **b)** *(Kunst)* madonna

Madrigal [madri'gaːl] **das; ~s, ~e** *(Literaturw., Musik)* madrigal

Maf[f]ia ['mafi̯a] **die; ~, ~s a)** *o. Pl.* Mafia; **b)** *(fig.)* mafia

mag [maːk] *1. u. 3. Pers. Sg. Präsens v.* **mögen**

Magazin [maga'tsiːn] **das; ~s, ~e a)** *(Lager)* store; *(für Waren)* stock-room; *(für Waffen u. Munition)* magazine; **b)** *(für Patronen, Dias, Film usw.)* magazine; **c)** *(Zeitschrift)* magazine; *(Rundf., Ferns.)* magazine programme

Magd [maːkt] **die; ~, Mägde** ['mɛːkdə] *(veralt.)* [female] farmhand; *(Vieh~)* milkmaid; *(Dienst~)* maidservant

Magen ['maːgn̩] **der; ~s, Mägen** ['mɛːgn̩] *od.* **~:** stomach; **mir knurrt der ~** *(ugs.)* my tummy is rumbling *(coll.)*; **sich** *(Dat.)* **den ~ verderben** get an upset stomach; **etw. auf nüchternen ~ essen/trinken** eat/ drink sth. on an empty stomach; **jmdm. auf den ~ schlagen** upset sb.'s stomach; **jmdm. schwer im ~ liegen** lie heavy on sb.'s stomach; **diese Sache liegt mir schwer auf dem ~** *(fig. ugs.)* this business is preying on my mind; **da dreht sich einem/mir der ~ um** *(ugs.)* it's enough to make *or* it makes one's/my stomach turn; *(fig.)* it makes you/me sick; *s. auch* **Liebe a**

magen-, Magen-: ~beschwerden *Pl.* stomach trouble *sing.*; **~bitter der; ~s, ~:** bitters *pl.*; **~gegend die** region of the stomach; **~geschwür das** stomach ulcer; **~grube die** pit of the stomach; **~knurren das** *(ugs.)* tummy rumbles *pl. (coll.)*; **~krampf der** stomach cramp; **~krank** *Adj.* **~ sein** have a stomach complaint; **~krebs der** cancer of the stomach; **~saft der** gastric juice; **~säure die** gastric acid; **~schmerzen** *Pl.* stomach-ache *sing.*

mager ['maːgɐ] **1.** *Adj.* **a)** *(dünn)* thin; **b)** *(fettarm)* low-fat; low in fat *pred.*; lean ⟨*meat*⟩; **c)** *(nicht ertragreich)* poor ⟨*soil, harvest*⟩; infertile ⟨*field*⟩; lean ⟨*years*⟩; *(fig.: dürftig)* meagre ⟨*profit, increase, success, report, etc.*⟩; thin ⟨*programme*⟩. **2.** *adv.* **~ essen** follow a low-fat diet; eat low-fat foods

Mager·käse der low-fat cheese

Magerkeit die; ~ a) thinness; **b)** *(fig.: Dürftigkeit)* meagreness

Mager-: ~milch die skim[med] milk; **~quark der** low-fat curd cheese

Magie [ma'giː] **die; ~:** magic; **Schwarze/Weiße ~:** black/white magic

Magier ['maːgi̯ɐ] **der; ~s, ~** *(auch fig.)* magician

magisch 1. *Adj.* magic ⟨*powers*⟩; *(geheimnisvoll)* magical ⟨*attraction, light, force, etc.*⟩; *(unwirklich)* eerie ⟨*light, half-light*⟩. **2.** *adv. (durch Zauber)* by magic; *(wie durch Zauber)* as if by magic; magically; *(unwirklich)* eerily

Magistrat [magɪs'traːt] **der; ~[e]s, ~e** City Council

Magma ['magma] **das; ~s, Magmen** *(Geol.)* magma

Magnat ['magnaːt] **der; ~en, ~en** magnate

Magnesium [ma'gneːzi̯ʊm] **das; ~s** *(Chemie)* magnesium

Magnet [ma'gneːt] **der; ~en** *od.* **~[e]s, ~e** *(auch fig.)* magnet

Magnet-: ~band das *Pl.:* **~bänder** magnetic tape; **~feld das** *(Physik)* magnetic field

magnetisch 1. *Adj. (auch fig.)* magnetic. **2.** *adv.* magnetically

magnetisieren *tr. V. (Physik)* magnetize

Magnetismus der; ~ *(Physik)* magnetism

Magnolie [ma'gnoːli̯ə] **die; ~, ~n** magnolia

mäh [mɛː] *Interj.* baa

Mahagoni [maha'goːni] **das; ~s** mahogany

Mäh·drescher der combine harvester

mähen ['mɛːən] **1.** *tr. V.* mow ⟨*grass, lawn, meadow*⟩; cut, reap

⟨*corn*⟩. **2.** *itr. V.* mow; *(Getreide ~)* reap
Mahl [ma:l] **das; ~[e]s, Mähler** [mɛ:lɐ] *(geh.)* meal; repast *(formal)*
mahlen *unr. tr., itr. V.* grind; **etw. fein/grob ~:** grind sth. fine/ coarsely
Mahl·zeit meal; ~! *(ugs.)* have a good lunch; bon appetit; **[na dann] prost ~!** *(ugs.)* what a delightful prospect! *(iron.)*
Mäh·maschine die [power] mower; *(für Getreide)* reaper
Mahn-: ~bescheid der writ for payment; **~brief der** *s.* **~schreiben**
Mähne ['mɛ:nə] **die; ~, ~n** mane; *(scherzh.: Haarschopf)* mane [of hair]
mahnen ['ma:nən] *tr., itr. V.* **a)** urge; **zur Eile/Vorsicht ~:** urge haste/caution; **jmdn. zur Eile/ Vorsicht~:** urge sb. to hurry/to be careful; **jmdn. eindringlich ~:** give sb. an urgent warning; **b)** *(erinnern)* remind (**an** + *Akk.* of); **einen Schuldner [schriftlich] ~:** send a debtor a [written] demand for payment *or* a reminder
Mahn-: ~gebühr die reminder fee; **~mal das** memorial *(erected as a warning to future generations);* **~schreiben das** reminder
Mahnung die; ~, ~en a) exhortation; *(Warnung)* admonition; **b)** *([Zahlungs]erinnerung)* reminder
Mähre ['mɛ:rə] **die; ~, ~n** *(veralt. abwertend)* jade *(dated)*
Mai [mai] **der; ~[e]s** *od.* **~, ~e** May; **der Erste ~:** the first of May; May Day
Mai-: ~baum der maypole; **~bowle die** *cup made of white wine and champagne with fresh woodruff;* **~glöckchen das** lily of the valley; **~käfer der** Maybug; **~kundgebung die** May Day rally
Mais [mais] **der; ~es** maize; corn *(esp. Amer.); (als Gericht)* sweet corn
Maische ['maiʃə] **die; ~, ~n** mash
Mais-: ~kolben der corn-cob; *(als Gericht)* corn on the cob; **~korn das** grain of maize *or (Amer.)* corn; **~mehl das** maize *or (Amer.)* corn flour
Majestät [majɛs'tɛ:t] **die; ~, ~en a)** *(Titel)* Majesty; **Seine/Ihre/ Eure** *od.* **Euer ~:** His/Her/Your Majesty; **b)** *o. Pl. (geh.)* majesty
majestätisch 1. *Adj.* majestic. **2.** *adv.* majestically
Majestäts·beleidigung die lèse-majesté
Major [ma'jo:ɐ] **der; ~s, ~e** *(Milit.)* major; *(Luftwaffe)* squadron leader *(Brit.);* major *(Amer.)*
Majoran ['ma:joran] **der; ~s, ~e** marjoram
majorisieren *tr. V. (geh.)* outvote
Majorität [majori'tɛ:t] **die; ~, ~en** majority
makaber [ma'ka:bɐ] *Adj.* macabre
Makel ['ma:kl̩] **der; ~s, ~** *(geh.)* **a)** *(Schmach)* stigma; taint; **an ihm haftet ein ~:** a stain *or* taint clings to him; **b)** *(Fehler)* blemish; flaw
makel·los 1. *Adj.* flawless, perfect ⟨*skin, teeth, figure, stone*⟩; spotless, immaculate ⟨*white, cleanness, clothes*⟩; *(fig.)* spotless, unblemished ⟨*reputation, character*⟩. **2.** *adv.* immaculately; spotlessly ⟨*clean*⟩; *(fehlerfrei)* flawlessly
Makellosigkeit die; ~ *s.* **makellos:** flawlessness; perfection; spotlessness; immaculateness
mäkeln ['mɛ:kl̩n] *itr. V. (abwertend)* carp
Make-up [me:k'/ap] **das; ~s, ~s** make-up
Makkaroni [maka'ro:ni] *Pl.* macaroni *sing.*
Makler ['ma:klɐ] **der; ~s, ~ a)** estate agent *(Brit.);* realtor *(Amer.);* **b)** *(Börsen~)* broker
Makler-: ~gebühr die, ~provision die agent's fee *or* commission; *(eines Börsenmaklers)* brokerage charges *pl.*
Makrele [ma'kre:lə] **die; ~, ~n** mackerel
makro-, Makro- [makro-]: **~biotisch 1.** *Adj.* macrobiotic; **2.** *adv.* on macrobiotic principles; **~kosmos der** macrocosm; **~molekül das** macromolecule
Makrone [ma'kro:nə] **die; ~, ~n** macaroon
Makulatur [makula'tu:ɐ] **die; ~, ~en a)** *(Druckw.)* spoilt sheets *pl.;* spoilage *no pl.;* **b)** *(Altpapier)* waste paper; **c) ~ reden** *(ugs.)* talk rubbish
mal [ma:l] **1.** *Adv.* times; *(bei Flächen)* by; **zwei ~ zwei** twice two; two times two; **der Raum ist 5 ~ 6 Meter groß** the room is five metres by six. **2.** *Partikel (ugs.)* **komm ~ her!** come here!; **hör ~ zu!** listen!; *s. auch* **einmal 2**
¹Mal das; ~[e]s, ~e time; **nur dies eine ~:** just this once; **kein einziges ~:** not once; not a single time; **beim ersten/letzten ~:** the first/last time; **zum zweiten/x-ten ~:** for the second/n-th time; **von ~ zu ~ heftiger werden/nachlassen** become more and more violent/decrease more and more [each time]; **mit einem ~[e]** *(plötzlich)* all at once; all of a sudden
²Mal das; ~[e]s, ~e *od.* **Mäler** ['mɛ:lɐ] mark; *(Muttermal)* birthmark; *(braun)* mole
Malachit [mala'xi:t] **der; ~s, ~e** malachite
Malaie [ma'laiə] **der; ~n, ~n** Malay
malaiisch *Adj.* Malayan
Malaria [ma'la:ria] **die; ~:** malaria
Mal·buch das colouring-book
malen *tr., itr. V.* paint ⟨*picture, portrait, person, etc.*⟩; *(mit Farbstiften)* draw with crayons; *(ausmalen)* colour; **sich ~ lassen** have one's portrait painted; **etw. in düsteren Farben ~** *(fig.)* paint *or* portray sth. in gloomy colours; **etw. allzu rosig/schwarz ~** *(fig.)* paint far too rosy/black a picture of sth.
Maler ['ma:lɐ] **der; ~s, ~** painter
Malerei die; ~, ~en a) *o. Pl.* painting *no art.;* **b)** *(Gemälde)* painting
Malerin die; ~, ~nen [woman] painter
malerisch 1. *Adj.* **a)** *(pittoresk)* picturesque; **b)** *(zur Malerei gehörend)* artistic ⟨*skill, talent*⟩; ⟨*skill, talent*⟩ as a painter. **2.** *adv. (pittoresk)* picturesquely ⟨*situated*⟩
Maler·meister der master painter [and decorator]
Malheur [ma'lø:ɐ] **das; ~s, ~e** *od.* **~s** mishap
Mal·kasten der paintbox
mal|nehmen *unr. tr., itr. V.* multiply (**mit** by)
Mal-: ~pinsel der paintbrush; **~technik die** painting technique
Malteser- [mal'te:zɐ-]: **~hilfsdienst der** ≈ St John Ambulance Brigade; **~kreuz das** *(auch Technik)* Maltese cross; **~orden der** *o. Pl.* Order of the Knights of St John
malträtieren [maltrɛ'ti:rən] *tr. V.* maltreat; ill-treat
Malve ['malvə] **die; ~, ~n** mallow
Malz [malts] **das; ~es** malt
Malz-: ~bier das malt beer; **~bonbon das** malted cough lozenge
Mal·zeichen das multiplication sign
Malz·kaffee der *coffee substitute made from germinated, dried, and roasted barley*
Mama ['mama, *geh. veralt.:* ma'ma:] **die; ~, ~s** *(fam.)* mamma

Mami ['mami] **die; ~, ~s** *(fam.)* mummy *(Brit. coll.)*; mommy *(Amer. coll.)*
Mammut ['mamʊt] **das; ~s, ~e** *od.* **~s** mammoth
Mammut- mammoth ‹*project, undertaking, etc.*›; *(lange dauernd)* marathon ‹*trial*›
mampfen ['mampfn̩] *tr., itr. V. (salopp)* munch; nosh *(sl.)*
man [man] *Indefinitpron. im Nom.* **a)** one; you; **~ kann nie wissen** one *or* you never can tell; **~ versteht sein eigenes Wort nicht** you can't hear yourself speak; **~ nehme 250 g Butter** take 250 grams of butter; **b)** *(irgend jemand)* somebody; *(die Behörden; die Leute dort)* they; **hat ~ dir das nicht mitgeteilt?** didn't anybody/they tell you that?; **~ vermutet/hat herausgefunden, daß ...:** it is thought/has been discovered that ...; **c)** *(die Menschen im allgemeinen)* people *pl.*; **das trägt ~ heute** that's what people wear *or* what is worn nowadays; **so etwas tut ~ nicht** that's not done
Management ['mænɪdʒmənt] **das; ~s, ~s** management
managen ['mɛnɪdʒn̩] *tr. V.* **a)** *(ugs.)* fix; organize; **ich manage das schon** I'll fix it; *(durch Tricks)* I'll fiddle it *(sl.)*; **b)** *(betreuen)* manage, act as manager for ‹*singer, artist, player*›; **von jmdm. gemanagt werden** have sb. as one's manager
Manager ['mɛnɪdʒɐ] **der; ~s, ~, Managerin die; ~, ~nen** manager; *(eines Fußballvereins)* club secretary
Manager·krankheit die stress disease *no def. art.*
manch [manç] *Indefinitpron.* **a)** *attr.* many a; **[so] ~er Beamte, ~ ein Beamter** many an official; **in ~er Beziehung** in many respects; **~ einer** many a person/man; **~ eine** many a woman; **b)** *(alleinstehend)* **~er** many a person/man; **~e** *(~ eine)* many a woman; *(~e Leute)* some; **[so] ~es** a number of things; *(allerhand Verschiedenes)* all kinds of things; **[so] ~es von dem, was wir lernten** much of what we learnt
mancherlei *indekl. unbest. Gattungsz.: attr.* various; a number of; *(alleinstehend)* various things; a number of things
manch·mal *Adv.* sometimes
Mandant [man'dant] **der; ~en, ~en, Mandantin die; ~, ~nen** *(Rechtsw.)* client
Mandarine [manda'ri:nə] **die; ~, ~n** mandarin [orange]; tangerine
Mandat [man'da:t] **das; ~[e]s, ~e a)** *(Parlamentssitz)* [parliamentary] seat; **b)** *(Auftrag) (eines Abgeordneten)* mandate; *(eines Anwalts)* brief
Mandats-: ~gebiet das mandated territory; mandate; **~träger der** *(Politik)* member of parliament; deputy
Mandel ['mandl̩] **die; ~, ~n a)** almond; **b)** *(Anat.)* tonsil
Mandel-: ~baum der almond[-tree]; **~entzündung die** tonsillitis *no indef. art.*; **~operation die** tonsillectomy
Mandoline [mando'li:nə] **die; ~, ~n** mandolin
Manege [ma'ne:ʒə] **die; ~, ~n** *(im Zirkus)* ring; *(in der Reitschule)* arena
Mangan [maŋ'ga:n] **das; ~s** *(Chemie)* manganese
[1]**Mangel** ['maŋl̩] **der; ~s, Mängel** ['mɛŋl̩] **a)** *o. Pl. (Fehlen)* lack **(an** + *Dat.* of); *(Knappheit)* shortage, lack **(an** + *Dat.* of); **~ an Vitaminen** vitamin deficiency; **wegen ~s an Beweisen** for lack of evidence; **aus ~ an Erfahrung** from *or* owing to lack of experience; **b)** *(Fehler)* defect; **geringfügige Mängel** minor flaws *or* imperfections
[2]**Mangel die; ~, ~n** *(Wäsche~)* [large] mangle; **jmdn. durch die ~ drehen** *od.* **in die ~ nehmen** *(fig. salopp)* put sb. through the hoop
Mangel-: ~beruf der understaffed profession; **~erscheinung die** *(Med.)* deficiency symptom
mangelhaft 1. *Adj. (fehlerhaft)* defective ‹*goods, memory*›; faulty ‹*goods, German, English, etc.*›; *(schlecht)* poor ‹*memory, lighting*›; *(unzulänglich)* inadequate ‹*knowledge, lighting*›; incomplete ‹*reports*›; *(Schulw.)* **die Note „~"** the mark 'unsatisfactory'; *(bei Prüfungen)* the fail mark. **2.** *adv. (fehlerhaft)* defectively; faultily; *(schlecht)* poorly; *(unzulänglich)* inadequately
Mangel·krankheit die *(Med.)* deficiency disease
[1]**mangeln** *itr. V.* **es mangelt an etw.** *(Dat.) (ist nicht vorhanden)* there is a lack of sth.; *(ist unzureichend vorhanden)* there is a shortage of sth.; sth. is in short supply; **jmdm./einer Sache mangelt es an etw.** *(Dat.)* sb./sth. lacks sth.; **seine ~de Menschenkenntnis** his inadequate understanding of people
[2]**mangeln 1.** *tr. V.* mangle. **2.** *itr. V.* do the mangling
mangels ['maŋls] *Präp. mit Gen.* in the absence of
Mangel·ware die: ~ sein be scarce *or* in short supply; ‹*article*› be a scarce commodity; **erfahrene Fachkräfte sind ~** *(fig. ugs.)* experienced skilled workers are thin on the ground *(coll.)*
Mango ['maŋgo] **die; ~, ~s** mango
Mangrove [maŋ'gro:və] **die; ~, ~n** mangrove forest
Manie [ma'ni:] **die; ~, ~n** mania; **bei jmdm. zur ~ werden** become an obsession with sb.
Manier [ma'ni:ɐ̯] **die; ~, ~en a)** manner; **in gewohnter ~:** in his/her usual way *or* manner; **b)** *Pl. (Umgangsformen)* manners
Manierismus der; ~ *(Kunstwiss., Literaturw.)* mannerism
manierlich 1. *Adj.* **a)** *(fam.)* well-mannered; well-behaved ‹*child*›; **b)** *(ugs.: einigermaßen gut)* reasonable; decent. **2.** *adv.* **a)** *(fam.)* properly; nicely; **b)** *(ugs.: einigermaßen gut)* **ganz/recht ~:** quite/really nicely *or* decently
Manifest das; ~[e]s, ~e manifesto
manifestieren *refl. V. (geh.)* be manifested; manifest itself
Maniküre [mani'ky:rə] **die; ~, ~n a)** *o. Pl.* manicure; **~ machen** manicure oneself; **b)** *(Person)* manicurist
maniküren *tr. V.* manicure
Manipulation [manipula'tsi̯o:n] **die; ~, ~en** manipulation
manipulierbar *Adj.* **leicht ~:** easy to manipulate
manipulieren [manipu'li:rən] *tr. V.* manipulate; rig ‹*election result, composition of a committee*›
manisch ['ma:nɪʃ] *(geh., Psych.)* **1.** *Adj.* manic. **2.** *adv.* maniacally
manisch-depressiv *Adj. (Psych., Med.)* manic-depressive
Manko ['maŋko] **das; ~s, ~s** *(Mangel)* shortcoming; deficiency; *(Nachteil)* handicap
Mann [man] **der; ~[e]s, Männer** ['mɛnɐ]; *s. auch* **Mannen a)** man; **ein ~, ein Wort** a man's word is his bond; **ein ~ der Tat** a man of action; **ein ~ aus dem Volk** a man of humble origins; **ein ~ des Volkes** a man of the people; **der geeignete** *od.* **richtige ~ sein** be the right man; **der böse** *od.* **schwarze ~:** the bogy man; **der ~ auf der Straße** the man in the street; **auf den ~ dressiert sein** ‹*dog*› be trained to attack people; **der ~ im Mond** the man in the moon; ; **[mein lieber] ~!** *(ugs.) (überrascht, bewundernd)* my goodness!; *(ver-*

ärgert) for goodness sake!; **seinen ~ stehen** do one's duty; **du hast wohl einen kleinen ~ im Ohr** *(salopp)* you must be out of your tiny mind *(sl.)*; **etw. an den ~ bringen** *(ugs.: verkaufen)* flog sth. *(Brit. sl.)*; push sth. *(Amer.)*; find a taker/takers for sth.; **Kämpfe** *od.* **der Kampf ~ gegen ~**: hand-to-hand fighting; **von ~ zu ~**: [from] man to man; **b)** *(Besatzungsmitglied)* man; **mit 1 000 ~ Besatzung** with a crew of 1,000 [men]; **alle ~ an Deck!** *(Seemannsspr.)* all hands on deck!; **~ über Bord!** *(Seemannsspr.)* man overboard!; **c)** *(Teilnehmer)* **uns fehlt der dritte/vierte ~ zum Skatspielen** we need a third/fourth person *or* player for a game of skat; **d)** *(Ehemann)* husband

Männchen ['mɛnçən] **das; ~s, ~ a)** little man; **~ malen** draw matchstick men; **b)** *(Tier~)* male; **~ machen** ⟨*animal*⟩ sit up and beg

Mannen *Pl.* *(scherzh.: Team, Mannschaft usw.)* troops

Mannequin ['manəkɛ̃] **das; ~s, ~s** mannequin; [fashion] model

Männer-: ~arbeit die a man's work; work for a man; **~bekanntschaft die** male *or* gentleman friend; **~beruf der** all-male profession; *(überwiegend von Männern ausgeübt)* male-dominated profession; **~chor der** male voice choir; **~sache die: das ist ~sache** that's men's business; **~stimme die** man's voice; male voice

Mannes-: ~alter das manhood *no art.*; **im besten ~alter sein** be in the prime of life *or* in one's prime; **~kraft die** *(geh.)* virility

Mannig·faltigkeit die; ~: [great] diversity

Männlein ['mɛnlain] **das; ~s, ~ a)** **[kleines] ~**: little man; **b)** *(ugs. scherzh.)* **~ und/oder Weiblein** men and/or women; *(bei jüngeren)* boys and/or girls

männlich 1. *Adj.* **a)** male ⟨*sex, line, descendant, flower, etc.*⟩; **~er Vorname** boy's *or* man's name; **b)** *(für den Mann typisch)* masculine ⟨*behaviour, characteristic, etc.*⟩; male ⟨*vanity*⟩; **c)** *(Sprachw.)* masculine. **2.** *adv.* in a masculine way

Männlichkeit die; ~ a) masculinity; manliness; **b)** *(Potenz)* virility

Manns·bild das *(ugs., bes. südd., österr.)* man

Mannschaft die; ~, ~en a) *(Sport, auch fig.)* team; **die erste/zweite ~** *(Fußball)* the first/second eleven; **b)** *(Schiffs-, Flugzeugbesatzung)* crew; **c)** *(Milit.: Einheit)* unit; **vor versammelter ~** *(fig.)* in front of everybody

Mannschafts-: ~aufstellung die *(Sport)* **a)** [composition of the] team; team line-up; **b)** *(das Aufstellen)* selection of the team; **~geist der;** *o. Pl.* *(Sport)* team spirit; **~kapitän der** *(Sport)* team captain; **~spiel das** *(Sport)* team game; **~wagen der** personnel carrier

manns-: ~hoch *Adj.* as tall as a man *postpos.*; six-foot-high; **~toll** *Adj.* *(ugs. abwertend)* man-mad *(coll.)*; nymphomaniac

Mann·weib das *(abwertend)* amazon

Manöver [ma'nø:vɐ] **das; ~s, ~ a)** *(Milit.)* exercise; **~** *Pl.* manœuvres; **ins ~ gehen** *od.* **ziehen** go on manœuvres; **b)** *(Bewegung; fig. abwertend: Trick)* manœuvre

Manöver·kritik die *(fig.)* post-mortem *(coll.)*

manövrieren *tr., itr. V.* manœuvre

manövrier-: ~fähig *Adj.* manœuvrable; **~unfähig** *Adj.* unmanœuvrable

Mansarde [man'zardə] **die; ~, ~n** attic; *(Zimmer)* attic room

Manschette [man'ʃɛtə] **die; ~, ~n a)** cuff; **[vor etw.** *(Dat.)***] ~n haben** *(fig. ugs.)* have got the willies *(sl.)* *or* have got the wind up *(Brit. sl.)* [about sth.]; **b)** *(Umhüllung)* paper frill

Manschetten·knopf der cuff-link

Mantel ['mantl̩] **der; ~s, Mäntel** ['mɛntl̩] **a)** coat; *(schwerer)* overcoat; **den ~ des Schweigens über etw.** *(Akk.)* **breiten** *(fig. geh.)* observe a strict silence about sth.; **b)** *(Technik)* *(Isolier~, Kühl~)* jacket; *(Rohr~)* sleeve; *(Kabel~)* sheath; *(Geschoß~)* [bullet-]casing; *(einer Granate)* [shell-]case; *(Reifen~)* [outer] cover; casing; **c)** *(Geom.: Zylinder~, Kegel~)* curved surface

Mäntelchen ['mɛntl̩çən] **das, ~s, ~:** little coat; *(für Kinder)* [child's] coat; *s. auch* **Wind**

Mantel-: ~tarif der *(Arbeitswelt)* terms of the *Manteltarifvertrag*; **~tarifvertrag der** *(Wirtsch.)* framework collective agreement [on working conditions]

Manual [ma'nu̯a:l] **das; ~s, ~e, Manuale das; ~[s], ~[n]** *(Musik)* keyboard; manual

manuell [ma'nu̯ɛl] **1.** *Adj.* manual. **2.** *adv.* manually; by hand

Manufaktur [manufak'tu:ɐ̯] **die; ~, ~en** [small] factory *(where goods are produced largely by hand)*

Manuskript [manu'skrɪpt] **das; ~[e]s, ~e a)** *(auch hist.)* manuscript; *(Typoskript)* typescript; *(zu einem Film/Fernsehspiel/Hörspiel)* script; **b)** *(Notizen eines Redners usw.)* notes *pl.*

Maoismus [mao'ɪsmʊs] **der; ~:** Maoism *no art.*

maoistisch 1. *Adj.* Maoist. **2.** *adv.* on Maoist lines

Mäppchen ['mɛpçən] **das; ~s, ~:** pencil-case

Mappe ['mapə] **die; ~, ~n a)** folder; *(größer, für Zeichnungen usw.)* portfolio; **b)** *(Aktentasche)* briefcase; *(Schul~)* school-bag

Marathon- ['ma(:)ratɔn]: **~lauf der** marathon; **~läufer der** marathon runner; **~sitzung die** marathon session

Märchen [~mɛ:ɐ̯çən] **das; ~s, ~ a)** fairy story; fairy-tale; **b)** *(ugs.: Lüge)* [tall] story *(coll.)*; **erzähl doch keine ~!** don't give me that story! *(coll.)*

Märchen-: ~buch das book of fairy stories; **~erzähler der** teller of fairy stories; **~figur die** fairy-tale figure; **~film der** film of a fairy story

märchenhaft 1. *Adj.* **a)** fairy-story *attrib.*; *(wie ein Märchen)* fairy-story-like; as in a fairy story *postpos.*; **b)** *(zauberhaft)* magical; *(feenhaft)* fairy-like; **~ sein** be sheer magic; be like a dream; **c)** *(ugs.)* *(großartig)* fabulous; *(sehr groß)* fantastic *(coll.)*, incredible *(coll.)* ⟨*speed, wealth*⟩. **2.** *adv. s. Adj.*: **a)** as in a fairy story; **b)** magically; **~ schön** bewitchingly beautiful; **c)** *(ugs.)* fantastically *(coll.)*; incredibly *(coll.)*

Märchen-: ~land das: das ~land the world of fairy-tale; fairyland; **~prinz der** fairy-tale prince; *(fig.)* Prince Charming; **~schloß das** fairy-tale castle

Marder ['mardɐ] **der; ~s, ~:** marten

Margarine [marga'ri:nə] **die; ~:** margarine

Marge ['marʒə] **die; ~, ~n** *(Wirtsch.)* margin

Margerite [margə'ri:tə] **die; ~, ~n** ox-eye daisy; *(als Zierpflanze)* marguerite

Maria [ma'ri:a] **(die); ~s** *od.* **Mariens** *od.* *(Rel.)* **Mariä** Mary

Marien·käfer der ladybird

Marihuana [mari'hu̯a:na] **das; ~s** marijuana

Marinade [mari'na:də] **die; ~, ~n a)** *(Beize)* marinade; **b)** *(Salatsauce)* [marinade] dressing

Marine [ma'ri:nə] **die; ~, ~n a)** *(Flotte)* fleet; **b)** *(Kriegs~)* navy
Marine-: ~soldat der marine; **~stütz·punkt der** naval base; **~uniform die** naval uniform
marinieren *tr. V.* marinade; **marinierte Heringe** soused herrings
Marionette [mario'nɛtə] **die; ~, ~n** puppet; marionette; *(fig. abwertend)* puppet
Marionetten-: ~regierung die *(abwertend)* puppet government; **~spieler der** puppet-master; puppeteer; **~theater das** puppet theatre
[1]**Mark** [mark] **die; ~, ~** mark; **Deutsche ~:** Deutschmark; German mark; **~ der DDR** GDR mark; **zwei ~ fünfzig** two marks fifty; **keine müde ~** *(ugs.)* not a penny; not a cent *(Amer.)*
[2]**Mark das; ~[e]s a)** *(Knochen~)* marrow; medulla *(Anat.)*; **das ging mir durch ~ und Bein** *(fig.)* it put my teeth on edge; it went right through me; **b)** *(Bot.)* *(Frucht~)* pulp
markant [mar'kant] *Adj.* striking; distinctive; prominent ⟨*figure, nose, chin*⟩
Marke die; ~, ~n a) *(Waren~)* brand; *(Fabrikat)* make; **b)** *(Brief~, Rabatt~, Beitrags~)* stamp; **c)** *(Garderoben~)* [cloakroom *or* *(Amer.)* checkroom] counter *or* tag; *(Zettel)* [cloakroom *or* *(Amer.)* checkroom] ticket; *(Essen~)* meal-ticket; **d)** *(Erkennungs~)* [identification] disc; *(Dienst~)* [police] identification badge; ≈ warrant card *(Brit.)* *or* *(Amer.)* ID card; **e)** *(Lebensmittel~)* coupon; **f)** *(Markierung)* mark; *(Sport: Rekord)* record [height/distance]; **g)** *(salopp)* **du bist mir vielleicht eine ~!** you are a fine one! *(iron.)*
Marken-: ~artikel der proprietary *or* *(Brit.)* branded article; **~artikel** *Pl.* proprietary *or* *(Brit.)* branded goods; **~erzeugnis das, ~fabrikat das** proprietary *or* *(Brit.)* branded product; **~name der** brand name; **~zeichen das** trade mark
Marketing ['markətɪŋ] **das; ~s** *(Wirtsch.)* marketing
markieren 1. *tr. V.* **a)** *(auch fig.)* mark; *(Sport)* mark out ⟨*course*⟩; **b)** *(ugs.)* sham ⟨*illness, breakdown, etc.*⟩; **c)** *(Sport)* mark ⟨*player*⟩. **2.** *itr. V.* *(ugs.)* sham; put it on *(coll.)*
Markierung die; ~, ~en a) *(Zeichen)* marking; **b)** *o. Pl.* *(das Markieren)* marking [out]
markig 1. *Adj.* *(kernig)* pithy ⟨*saying, style*⟩; *(kraftvoll)* vigorous, breezy ⟨*commands, manner*⟩; **~e Worte** strong words; *(iron.: große Reden)* big words. **2.** *adv.* pithily
Markise [mar'ki:zə] **die; ~, ~n** awning
Mark-: ~klößchen das *(Kochk.)* bone-marrow dumpling; **~knochen der** marrowbone; **~stück das** one-mark piece
Markt [markt] **der; ~[e]s, Märkte** ['mɛrktə] **a)** market; **heute/freitags ist ~:** today/Friday is market-day; **auf dem ~:** at the market; **b)** *s.* **~platz; c)** *(Super~)* supermarket; **d)** *(Warenverkehr, Absatzgebiet)* market; **eine Ware auf den ~ bringen** *od.* **werfen** market a product; **auf dem ~ sein** ⟨*article*⟩ be on the market
markt-, Markt-: ~anteil der share of the market; **~beherrschend** *Adj.* market-dominating *attrib.*; **~forschung die** market research *no def. art.*; **~frau die** market-woman; **~halle die** covered market; **~leiter der** supermarket manager; **~lücke die** gap in the market; **~platz der** market-place *or* -square; **~schreier der** barker; stall-holder who cries his wares; **~stand der** market stall; **~tag der** market-day; **~wert der** market value; **~wirtschaft die** market economy; **~wirtschaftlich 1.** *Adj.; nicht präd.* market-economy; free-market; **2.** *adv.* on market-economy lines
Markus ['markʊs] **(der); Markus'** Mark
Marmelade [marmə'la:də] **die; ~, ~n** jam; *(Orangen~)* marmalade
Marmelade[n]·glas das jam-jar
Marmor ['marmɔr] **der; ~s** marble
marmoriert *Adj.* marbled
Marmor·kuchen der marble cake
marmorn *Adj.* marble
marode *Adj.* *(ugs. abwertend)* clapped-out *(Brit. sl.)*
Marokko [ma'rɔko] **(das); ~s** Morocco
Marone [ma'ro:nə] **die; ~, ~n** [sweet] chestnut
Marotte [ma'rɔtə] **die; ~, ~n** fad
Mars [mars] **der; ~** *(Astron.)* Mars *no def. art.*
Mars·bewohner der Martian
marsch [marʃ] *Interj.* **a)** *(Milit.)* [forward] march; **b)** *(ugs.)* **~ ~!** off with you!; *(beeil dich!)* move it! *(coll.)*; look snappy! *(coll.)*; **~ ins Bett!** off to bed [with you]!
[1]**Marsch der; ~[e]s, Märsche** ['mɛrʃə] **a)** *(Milit.)* march; *(Wanderung)* [long] walk; hike; **jmdn. in ~ setzen** *(Milit.)* march sb. off; *(fig.)* mobilize sb.; **sich in ~ setzen** make a move; get moving; *(Milit.)* march off; **b)** *(Musikstück)* march
[2]**Marsch die; ~, ~en** fertile marshland
Marschall ['marʃal] **der; ~s, Marschälle** ['marʃɛlə] *(hist.)* marshal
Marsch-: ~flugkörper der cruise missile; **~gepäck das** *(Milit.)* marching pack
marschieren *itr. V.; mit sein* **a)** march; **b)** *(ugs.: mit großen Schritten gehen)* march; stalk; *(wandern)* walk; hike
Marsch-: ~musik die march music; **~route die** *(Milit.)* route; *(fig.)* line [of approach]; **~verpflegung die** *(Milit.)* marching rations *pl.*; *(fig. ugs.)* rations *pl.* [for the journey]
Mars-: ~mensch der Martian; **~sonde die** Mars probe
Marter ['martɐ] **die; ~, ~n** *(geh.)* *(Folter)* torture; *(fig.: seelisch)* torment
martern *tr. V.* *(geh.)* torture; *(fig.: seelisch)* torment
Marter·pfahl der stake
martialisch [mar'tsia:lɪʃ] *(geh.)* **1.** *Adj.* warlike ⟨*appearance, figure, etc.*⟩; martial ⟨*music*⟩. **2.** *adv.* in a warlike manner; *(drohend)* threateningly; aggressively
Martins·horn das *(volkst.)* siren *(of emergency vehicle)*
Märtyrer ['mɛrtyrɐ] **der; ~s, ~, Märtyrerin die; ~, ~nen** martyr
Martyrium [mar'ty:riʊm] **das; ~s, Martyrien** martyrdom; **ein ~** *(fig.)* sheer martyrdom
Marxismus [mar'ksɪsmʊs] **der; ~:** Marxism *no art.*
Marxist der; ~en, ~en, Marxistin die; ~, ~nen Marxist
marxistisch 1. *Adj.* Marxist. **2.** *adv.* ⟨*view, interpret*⟩ from a Marxist point of view; ⟨*think, act*⟩ in line with Marxism
März [mɛrts] **der; ~[es],** *dichter.:* **~en** March; *s. auch* **April**
Marzipan [martsi'pa:n, *österr.* '---] **das; ~s** marzipan
Marzipan·schwein das marzipan pig
Masche ['maʃə] **die; ~, ~n a)** stitch; *(Lauf~)* run; ladder *(Brit.)*; *(beim Netz)* mesh; **durch die ~n des Gesetzes schlüpfen** *(fig.)* slip through a loophole in the law; **b)** *(ugs.: Trick)* trick; **das ist die ~:** that's the way *or* trick;

c) *(ugs.: Mode, Gag)* **die neueste ~:** the latest fad *or* craze

Maschen·draht der wire netting

Maschine [ma'ʃi:nə] **die; ~, ~n a)** machine; **b)** *(ugs.: Automotor)* engine; **c)** *(Flugzeug)* [aero]plane; **d)** *(ugs.: Motorrad)* machine; **e)** *(Schreib~)* typewriter; **~ schreiben** type

maschine·geschrieben *Adj.* typed; typewritten

maschinell [maʃi'nɛl] **1.** *Adj.* machine *attrib.;* by machine *postpos.* **2.** *adv.* by machine; **~ hergestellt** machine-made

maschinen-, Maschinen-: ~bau der; *o. Pl.* **a)** machine construction *no art.;* mechanical engineering *no art.;* **b)** *(Lehrfach)* mechanical engineering *no art.;* **~bau·ingenieur der** mechanical engineer; **~geschrieben** *Adj. s.* **maschinegeschrieben; ~gewehr das** machine-gun; **~park der** plant; **~pistole die** sub-machine-gun; **~schlosser der** fitter; **~schreiben das** typing; **~schrift die** typing; *(Schriftart)* typeface; type

Maschinerie [maʃinə'ri:] **die; ~, ~n** machinery

maschine·schreiben *unr. itr. V.* type

Maschinist der; ~en, ~en a) machinist; **b)** *(Schiffs~)* engineer

Masern *Pl.* measles *sing. or pl.*

Maserung die; ~, ~en *(in Holz, Leder)* [wavy] grain; *(in Marmor)* vein; *(in Fell)* patterning

Maske ['maskə] **die; ~, ~n a)** *(auch fig.)* mask; **b)** *(Theater)* make-up

Masken·ball der masked ball; masquerade

Maskerade [maskə'ra:də] **die; ~, ~n** [fancy-dress] costume; **~ sein** *(fig.)* be a masquerade

maskieren 1. *tr. V.* **a)** mask; **b)** *(verkleiden)* dress up. **2.** *refl. V.* **a)** put on a mask/masks; **b)** *(sich verkleiden)* dress up

Maskierung die; ~, ~en a) *(das Verkleiden)* dressing up; **b)** *(Verkleidung)* disguise; **c)** *(Tarnung)* masking; disguising

Maskottchen [mas'kɔtçən] **das; ~s, ~:** [lucky] mascot

maskulin [masku'li:n, *auch* '---] **1.** *Adj.* *(auch Sprachw.)* masculine. **2.** *adv.* in a masculine way

Maskulinum ['maskuli:nʊm] **das; ~s, Maskulina** *(Sprachw.)* masculine noun

Masochismus [mazɔ'xɪsmʊs] **der; ~** *(Psych.)* masochism *no art.*

Masochist [mazɔ'xɪst] **der; ~en, ~en** *(Psych.)* masochist

masochistisch *(Psych.)* **1.** *Adj.* masochistic. **2.** *adv.* masochistically

maß *1. u. 3. Pers. Sg. Prät. v.* **messen**

[1]Maß [ma:s] **das; ~es, ~e a)** measure **(für** of); **~e und Gewichte** weights and measures; **b)** *(fig.)* **ein gerüttelt ~ [an** *(Dat.)* *od.* **von etw.]** *(geh.)* a good measure [of sth.]; **das ~ ist voll** enough is enough; **das ~ vollmachen** go too far; **mit zweierlei ~ messen** apply different [sets of] standards; **c)** *(Größe)* measurement; *(von Räumen, Möbeln)* dimension; measurement; **[bei] jmdm. ~ nehmen** take sb.'s measurements; measure sb. [up]; **d)** *(Grad)* measure, degree **(an +** *Dat.* of); **im höchsten ~[e]** extremely; exceedingly; **e) über die** *od.* **alle ~en** *(geh.)* beyond [all] measure

[2]Maß die; ~, ~[e] *(bayr., österr.)* litre [of beer]

Massage [ma'sa:ʒə] **die; ~, ~n** massage

Massage·gerät das massager

Massaker [ma'sa:kɐ] **das; ~s, ~:** massacre

massakrieren *tr. V.* massacre

Maß-: ~an·zug der made-to-measure suit; tailor-made suit; **~arbeit die a)** *(von Kleidungsstücken)* **[eine] ~arbeit sein** be made-to-measure; **b)** *(genaue Arbeit)* neat work

Masse ['masə] **die; ~, ~n a)** mass; *(Kochk.)* mixture; **b)** *(Menge)* mass; **die ~ macht's** *(ugs.)* it's quantity that's important; **sie kamen in ~n** they came in their masses *or* in droves; **das ist eine ganze ~** *(ugs.)* that's a lot *(coll.)* *or* a great deal; **c)** *(Menschen~)* **die breite ~:** the bulk *or* broad mass of the population; **d)** *(Physik)* mass

Maß·einheit die unit of measurement

Massen-: ~an·drang der crush; **~arbeitslosigkeit die** mass unemployment; **~aufgebot das** large body *or* contingent; **~bewegung die** mass movement; **~blatt das** mass-circulation paper; **~entlassungen** *Pl.* mass redundancies *pl.;* **~fabrikation die** mass production; **~grab das** mass grave

massenhaft 1. *Adj.; nicht präd.* in huge numbers *postpos.;* **das ~e Auftreten dieser Schädlinge** the appearance of huge numbers of these pests. **2.** *adv.* on a huge *or* massive scale; **~ Geld haben** *(ugs.)* have pots of money *(coll.)*

massen-, Massen-: ~hysterie die mass hysteria; **~karambolage die** multiple crash; [multiple] pile-up; **~kundgebung die** mass rally; **~medium das** mass medium; **~mord der** mass murder; **~mörder der** mass murderer; **~produktion die** mass production; **~schlägerei die** [grand] free-for-all; pitched battle *(fig.);* **~sport der** mass sport; **~tourismus der** mass tourism *no art.;* **~weise** *Adv.* in huge quantities; *(in großer Zahl)* in huge numbers

Masseur [ma'sø:ɐ] **der; ~s, ~e** masseur

Masseurin die; ~, ~nen masseuse

Masseuse [ma'sø:zə] **die; ~, ~n** *(auch verhüll.)* masseuse

Maß·gabe die: nach ~ (+ *Gen.*) *(geh.)* in accordance with

maß·gearbeitet *Adj.* custom-made; made-to-measure ‹*clothes*›

maß·gebend, maß·geblich 1. *Adj.* authoritative ‹*book, expert, opinion*›; definitive ‹*text*›; important, influential ‹*person, circles, etc.*›; decisive ‹*factor, influence, etc.*›; *(zuständig)* competent ‹*authority, person, etc.*›; **sein Urteil ist nicht ~:** his opinion carries no weight. **2.** *adv.* ‹*influence*› considerably, to a considerable extent; *(entscheidend)* decisively; **~ an etw.** *(Dat.)* **beteiligt sein** play a leading role in sth.

maß-: ~geschneidert *Adj.* made-to-measure; *(fig.)* tailor-made; **~|halten** *unr. itr. V.* exercise moderation

massieren *tr. V.* massage

massig 1. *Adj.* massive; bulky, massive ‹*figure*›. **2.** *adv.* *(ugs.)* **~ Geld verdienen** earn pots of money *(coll.)*

mäßig ['mɛ:sɪç] **1.** *Adj.* **a)** moderate; **b)** *(gering)* moderate, modest ‹*interest, income, talent, attendance*›; **c)** *(mittel~)* mediocre; indifferent; indifferent ‹*health*›. **2.** *adv.* **a)** in moderation; **~, aber regelmäßig** *(scherzh.)* in moderation but regularly; **b)** *(gering)* moderately ‹*gifted, talented*›; **c)** *(mittel~)* indifferently

mäßigen *(geh.)* *refl. V.* **a)** practise *or* exercise moderation **(bei** in); **b)** *(sich beherrschen)* control *or* restrain oneself

Mäßigung die; ~: moderation; restraint

massiv [ma'si:f] **1.** *Adj.* **a)** solid; **~ bauen** build solidly; **b)** *(heftig)* massive ‹*demand*›; crude ‹*accusation, threat*›; heavy, strong ‹*at-*

tack, criticism, pressure〉. **2.** *adv.* 〈*attack*〉 heavily, strongly; 〈*accuse, threaten*〉 crudely

Massiv das; ~s, ~e massif

Maß·krug der *(südd., österr.)* litre tankard *or* beer-mug; *(aus Steingut)* stein

maß·los 1. *Adj. (äußerst)* extreme; *(übermäßig)* inordinate; gross 〈*exaggeration, insult*〉; excessive 〈*demand, claim*〉; *(grenzenlos)* boundless 〈*ambition, greed, sorrow, joy*〉; extravagant 〈*spendthrift*〉. **2.** *adv. (äußerst)* extremely; *(übermäßig)* inordinately; 〈*exaggerate*〉 grossly

Maßlosigkeit die; ~ *s.* **maßlos:** extremeness; inordinateness; grossness; excessiveness; boundlessness

Maßnahme die; ~, ~n measure; **~n ergreifen** take measures

Maß·regel die regulation; *(Maßnahme)* measure

maßregeln *tr. V. (zurechtweisen)* reprimand; *(bestrafen)* discipline

Maß·reg[e]lung die *(Zurechtweisung)* reprimand; *(Bestrafung)* disciplinary measure

Maß·stab der a) standard; **einen hohen ~ anlegen/setzen** apply/set a high standard; **b)** *(Geogr.)* scale; **diese Karte hat einen großen/kleinen ~:** this is a large-/small-scale map; **im ~ 1:100** to a scale of 1:100

maßstab[s]gerecht, maßstab[s]·getreu 1. *Adj.* scale *attrib.* 〈*model, drawing, etc.*〉; [true] to scale *pred.;* **2.** *adv.* to scale

maß·voll 1. *Adj.* moderate; **2.** *adv.* in moderation

[1]**Mast** [mast] **der; ~[e]s, ~en,** *auch:* **~e** *(Schiffs~, Antennen~)* mast; *(Stange, Fahnen~)* pole; *(Hochspannungs~)* pylon

[2]**Mast die; ~, ~en** *(Landw.)* fattening

Mast·darm der *(Anat.)* rectum

mästen ['mɛstn̩] *tr. V.* fatten; *(fig. ugs.)* overfeed

Mast·schwein das fattening pig; *(gemästet)* fattened pig

Mästung die; ~: fattening

Masturbation [masturba'tsi̯o:n] **die; ~, ~en** masturbation

masturbieren [mastʊr'bi:rən] *itr., tr. V.* masturbate

Matador [mata'do:ɐ̯] **der; ~s, ~e a)** matador; **b)** *(fig.)* star

Match [mɛtʃ] **das** *od.* **der; ~[e]s, ~s** *od.* **~e** match

Match·ball der *([Tisch]tennis)* match point

Material [mate'ri̯a:l] **das; ~s, ~ien a)** material; *(Bau~)* materials *pl.;* **b)** *(Hilfsmittel, Utensilien)* materials *pl.; (für den Bau)* equipment; **c)** *(Beweis~)* evidence

Material·fehler der material defect

Materialismus der; ~ *(auch abwertend)* materialism

Materialist der; ~en, ~en, Materialistin die; ~, ~en *(auch abwertend)* materialist

materialistisch *(auch abwertend)* **1.** *Adj.* materialistic. **2.** *adv.* materialistically

Material-: ~kosten *Pl.* cost *sing.* of materials; **~sammlung die** collection *or* gathering of material; **~schlacht die** *(Milit.)* battle of matériel

Materie [ma'te:ri̯ə] **die; ~, ~n a)** matter; **b)** *(geh.: Thema, Gegenstand)* subject

materiell [mate'ri̯ɛl] **1.** *Adj.* **a)** *(stofflich)* material; physical; **b)** *(wirtschaftlich)* material 〈*value, damage*〉; *(finanziell)* financial. **2.** *adv. (wirtschaftlich)* materially; *(finanziell)* financially

Mathe ['matə] *o. Art. (Schülerspr.)* maths *sing. (Brit. coll.);* math *(Amer. coll.)*

Mathe·arbeit die *(Schülerspr.)* maths test *(coll.)*

Mathematik [matəma'ti:k] **die; ~:** mathematics *sing., no art.*

Mathematiker der; ~s, ~, Mathematikerin die; ~, ~nen mathematician

Mathematik·unterricht der mathematics teaching/lesson; *s. auch* **Englischunterricht**

mathematisch 1. *Adj.* mathematical. **2.** *adv.* mathematically

Matinee [mati'ne:] **die; ~, ~n** matinée

Matjes ['matjəs] **der; ~, ~:** matie [herring]

Matjes-: ~filet das filleted matie [herring]; **~hering der** salted matie [herring]

Matratze [ma'tratsə] **die; ~, ~n** mattress

Matriarchat [matriar'ça:t] **das; ~[e]s, ~e** matriarchy

Matrize [ma'tri:tsə] **die; ~, ~n** *(Druckw.)* **a)** matrix; **b)** *(Folie)* stencil

Matrone [ma'tro:nə] **die; ~, ~n** matron

Matrose [ma'tro:zə] **der; ~n, ~n a)** sailor; seaman; **b)** *(Dienstgrad)* ordinary seaman

Matrosen-: ~an·zug der sailor suit; **~mütze die** sailor's cap

Matsch der; ~[e]s *(ugs.)* **a)** *(aufgeweichter Boden)* mud; *(breiiger Schmutz)* sludge; *(Schnee~)* slush; **b)** *(Brei)* mush

matschig *Adj. (ugs.)* **a)** muddy; slushy 〈*snow*〉; **b)** *(weich)* mushy; squashy 〈*fruit*〉

matt [mat] **1.** *Adj.* **a)** weak; weary 〈*limbs, spirit, etc.*〉; weak, faint 〈*voice, smile, pulse*〉; feeble 〈*applause, reaction*〉; limp, feeble 〈*handshake*〉; faint 〈*echo*〉; **b)** *(glanzlos)* matt 〈*paper, polish, etc.*〉; dull 〈*metal, mirror, etc.*〉; dull, lustreless 〈*eyes, look*〉; **c)** *(undurchsichtig)* frosted 〈*glass*〉; pearl 〈*light-bulb*〉; **d)** *(gedämpft)* soft, subdued 〈*light*〉; soft, pale 〈*colour*〉; **e)** *(beim Schachspiel)* **[Schach und] ~!** checkmate!; **~ sein** be checkmated; **jmdn. ~ setzen** *(auch fig.)* checkmate sb. **2.** *adv.* **a)** *(kraftlos)* weakly; 〈*smile*〉 weakly, faintly; 〈*applaud, react*〉 feebly; **b)** *(gedämpft)* softly 〈*lit*〉; **c)** *(mäßig)* 〈*protest, contradict*〉 feebly, weakly

Matt das; ~s *(Schach)* [check]mate

matt·blau *Adj.* pale blue

Matte ['matə] **die; ~, ~n** mat

Matthäus [ma'tɛ:ʊs] **(der); Matthäus'** Matthew

Mattigkeit die; ~: weakness; *(Erschöpfung)* weariness

Matt·scheibe die *(ugs.)* telly *(Brit. coll.);* box *(coll.)*

Matura die; ~ *(österr., schweiz.) s.* **Abitur**

Matz [mats] **der; ~es, ~e** *od.* **Mätze** ['mɛtsə] *(fam.)* **kleiner ~:** little man

Mätzchen ['mɛtsçən] *Pl. (ugs.)* **laßt die ~:** stop fooling about *or* around; stop your antics; **~ machen** fool about *or* around

mau [mau̯] *(ugs.)* **1.** *Adj.; nicht attr. (flau)* queasy; *(unwohl)* poorly. **2.** *adv.* badly; **die Geschäfte gehen ~:** business is bad

Mauer ['mau̯ɐ] **die; ~, ~n** *(auch fig., Sport)* wall; **die [Berliner] ~** *(hist.)* the [Berlin] Wall; **die Chinesische ~:** the Great Wall of China

Mauer·blümchen das *(ugs.) (beim Tanz)* wallflower *(coll.); (unscheinbares Mädchen, auch fig.)* Cinderella

mauern 1. *tr. V.* build; **gemauert** *(aus Ziegeln)* brick 〈*chimney, wall, etc.*〉. **2.** *itr. V.* **a)** lay bricks; **b)** *(Ballspiele)* play defensively; **c)** *(Kartenspiele)* hold back one's good cards

Mauer-: ~segler der swift; **~vorsprung der** projecting section of a/the wall; **~werk das a)** *(aus Stein)* stonework; masonry; *(aus Ziegeln)* brickwork; **b)** *(Mauern)* walls *pl.*

Maul [maul] **das; ~[e]s, Mäuler** ['mɔylɐ] **a)** *(von Tieren)* mouth; **b)** *(derb: Mund)* gob *(sl.)*; **er hat fünf hungrige Mäuler zu stopfen** *(fig.)* he's got five hungry mouths to feed; **das** *od.* **sein ~ aufmachen** *(fig.)* say something; **ein großes ~ haben** *(fig.)* shoot one's mouth off *(fig. sl.)*; **halt's** *od.* **halt dein ~:** shut your trap *(sl.)*; shut up *(coll.)*; *s. auch* **stopfen 1 d; verbrennen 2 b**

Maul·beer·baum der mulberry tree

maulen *itr. V. (salopp)* grouse *(coll.)*; moan; grumble

maul-, Maul-: ~esel der mule; *(Zool.)* hinny; **~faul** *Adj. (ugs. abwertend)* uncommunicative; taciturn; **~held der** *(ugs. abwertend)* loudmouth; braggart; **~korb der** *(auch fig.)* muzzle; **einem Hund/***(fig.)* **jmdm. einen ~ anlegen** muzzle a dog/sb.; **~sperre die** *(salopp)* **die ~sperre kriegen** *(fig.)* gape in surprise; **~tasche die** *(Kochk.) filled pasta case [served in soup]*; **~tier das** mule; **~- und Klauen·seuche die** *(Tiermed.)* foot-and-mouth disease

Maul·wurf der mole

Maulwurfs-: ~haufen der, ~hügel der molehill

maunzen ['mauntsn̩] *itr. V. (ugs.)* ⟨*cat*⟩ miaow plaintively

Maurer ['maurɐ] **der; ~s, ~:** bricklayer

Maurer-: ~kelle die brick [-layer's] trowel; **~meister der** master bricklayer

Maus [maus] **die; ~, Mäuse** ['mɔyzə] **a)** mouse; **weiße Mäuse sehen** *(fig. ugs.)* see pink elephants; **eine graue ~** *(fig. ugs. abwertend)* a colourless nondescript sort of [a] person; *s. auch* **Katze; b)** *Pl. (salopp: Geld)* bread *sing. (sl.)*; dough *sing. (sl.)*

Mauschelei die; ~, ~en *(ugs. abwertend)* shady wheeling and dealing *no indef. art.*

mauscheln ['mauʃl̩n] *itr. V. (ugs. abwertend)* engage in shady wheeling and dealing

Mäuschen ['mɔysçən] **das; ~s, ~ a)** little mouse; **~ sein** *od.* **spielen** *(fig. ugs.)* be a fly on the wall *(coll.)*; **b)** *(fig. ugs.)* **mein ~:** my sweet

mäuschen·still *Adj., adv. s.* **mucksmäuschenstill**

Mause-: ~falle die mousetrap; *(fig.)* trap; **~loch das** mousehole

Mauser die; ~: moult; **in der ~ sein** be moulting

mausern *refl. V.* moult; **sich zur Dame ~** *(fig. ugs.)* blossom into a lady

mause·tot *Adj. (ugs.)* [as] dead as a doornail *pred.*; stone-dead

maus·grau *Adj.* mouse-grey

Mausoleum [mauzo'le:ʊm] **das; ~s, Mausoleen** mausoleum

Maut [maut] **die; ~, ~en** toll

Max [maks]: **strammer Max** fried egg on ham and bread

Maxi- ['maksi-] maxi-⟨*coat, skirt, etc., single*⟩

maximal [maksi'ma:l] **1.** *Adj.* maximum. **2.** *adv.* **bis zu ~ 85 °C/20 t** up to a maximum of 85 °C/20 t

Maximal·forderung die maximum demand

Maxime [ma'ksi:mə] **die; ~, ~n** maxim

maximieren *tr. V.* maximize

Maximierung die; ~, ~en maximization

Maximum ['maksimʊm] **das; ~s, Maxima** maximum **(an +** *Dat.* of)

Mayonnaise [majɔ'nɛ:zə] **die; ~, ~n** mayonnaise

Mäzen [mɛ'tse:n] **der; ~s, ~e** *(geh.)* patron

MdB, M.d.B. *Abk.* **Mitglied des Bundestages** Member of the Bundestag

MdL, M.d.L. *Abk.* **Mitglied des Landtages** Member of the Landtag

MdNR *Abk.* **Mitglied des Nationalrates** *(Österreich)* Member of the Nationalrat

m.E. *Abk.* **meines Erachtens** in my opinion *or* view

Mechanik [me'ça:nɪk] **die; ~ a)** *(Physik)* mechanics *sing., no art.*; **b)** *(Mechanismus)* mechanism; **c)** *(Funktion)* mechanics *sing. or pl.*

Mechaniker der; ~s, ~, Mechanikerin die; ~, ~nen mechanic

mechanisch 1. *Adj.* mechanical; power *attrib.* ⟨*loom, press*⟩. **2.** *adv.* mechanically

Mechanismus der; ~, Mechanismen *(auch fig.)* mechanism

Meckerei die; ~, ~en *(ugs. abwertend)* moaning; grousing *(sl.)*; grumbling

Meckerer der; ~s, ~ *(ugs. abwertend)* moaner; grouser *(sl.)*; grumbler

meckern ['mɛkɐn] *itr. V.* **a)** *(auch fig.)* bleat; **b)** *(ugs. abwertend: nörgeln)* grumble; moan; grouse *(sl.)*; **etw. zu ~ haben** have sth. to grumble *etc.* about

Medaille [me'daljə] **die; ~, ~n** medal; *s. auch* **Kehrseite a**

Medaillen·gewinner der medallist; medal winner

Medaillon [medal'jõ:] **das; ~s, ~s a)** locket; **b)** *(Kochk., bild. Kunst)* medallion

Medien *s.* **Medium**

Medien- media ⟨*concern, policy, syndicate, etc.*⟩

Medikament [medika'mɛnt] **das; ~[e]s, ~e** medicine; *(Droge)* drug; **ein ~ gegen Kopfschmerzen** a remedy for headaches

medikamentös [medikamɛn'tø:s] **1.** *Adj.* ⟨*treatment*⟩ with drugs. **2.** *adv.* ⟨*treat, cure*⟩ with drugs

Meditation [medita'tsio:n] **die; ~, ~en** meditation

meditieren [medi'ti:rən] *itr. V.* meditate **(über +** *Akk.* [up]on)

Medium ['me:diʊm] **das; ~s, Medien** medium

Medizin [medi'tsi:n] **die; ~, ~en a)** *o. Pl.* medicine *no art.*; **b)** *(Heilmittel)* medicine **(gegen** for)

Mediziner [medi'tsi:nɐ] **der; ~s, ~, Medizinerin die; ~, ~nen** doctor; *(Student)* medical student

medizinisch 1. *Adj.* **a)** medical ⟨*journal, problem, etc.*⟩; **~e Fakultät** faculty of medicine; **b)** *(heilend)* medicinal ⟨*bath etc.*⟩; medicated ⟨*toothpaste, soap, etc.*⟩. **2.** *adv.* medically

Medizin-: ~mann der; *Pl.* **~männer** medicine man; **~student der, ~studentin die** medical student

Meer [me:ɐ̯] **das; ~[e]s, ~e** *(auch fig.)* sea; *(Welt~)* ocean; **ans ~ fahren** go to the seaside; **am ~:** by the sea; **aufs ~ hinausfahren** go out to sea; **übers ~ fahren** cross the sea; **1 000 m über dem ~:** 1 000 m above sea-level

Meer-: ~busen der gulf; **~enge die** straits *pl.*; strait

Meeres-: ~biologie die marine biology *no art.*; **~boden der** sea bed *or* bottom *or* floor; **~bucht die** bay; **~fauna die** marine fauna; **~früchte** *Pl. (Kochk.)* seafood *sing.*; **~klima das** maritime climate; **~luft die** *(Met.)* maritime air; **~spiegel der** sea-level; **20 m über/unter dem ~spiegel** 20 m above/below sea-level; **~strömung die** current; *(im Weltmeer)* ocean current

meer-, Meer-: ~jungfrau die mermaid; **~katze die** guenon; **~rettich der** horse-radish; **~salz das** sea-salt; **~schaum der** meerschaum; **~schweinchen das** guinea-pig; **~wasser das** sea water

Mega- ['mɛga-] mega⟨*watt, -ton, -hertz, etc.*⟩
Megaphon das; ~s, ~e megaphone; loud hailer
Mehl [me:l] **das;** ~[e]s **a)** flour; *(gröber)* meal; **b)** *(Pulver)* powder; *(Knochen~, Fisch~)* meal
mehlig *Adj.* **a)** floury; **b)** *(wie Mehl)* powdery ⟨*sand etc.*⟩; **c)** mealy ⟨*potato, apple, etc.*⟩
Mehl-: ~schwitze die *(Kochk.)* roux; **~speise die** *(österr.)* sweet; dessert; **~tau der** mildew
mehr [me:ɐ̯] **1.** *Indefinitpron.* more; **ein Grund ~, es zu tun** one more *or* an additional reason for doing it; **das war ~ als unverschämt** that was impertinent, to say the very least; **das schmeckt nach ~** *(ugs.)* it's very moreish *(coll.)*; **~ nicht?** is that all?; **~ oder minder** *od.* **weniger** more or less. **2.** *adv.* **a)** *(in größerem Maße)* more; **b)** *(eher)* **~ schlecht als recht** after a fashion; **er ist ~ Künstler als Gelehrter** he is more of an artist than a scholar; **c) nicht ~:** not ... any more; no longer; **es war niemand ~ da** there was no one left; **es hat sich keiner ~ gemeldet** there was not another word from anyone; **ich erinnere mich nicht ~:** I no longer remember; **das wird nie ~ vorkommen** it will never happen again; **davon will ich nichts ~ hören** I don't want to hear any more about it; **da ist nichts ~ zu machen** there is nothing more to be done; **ich habe keine Lust/kein Interesse ~:** I have lost all desire/interest; **du bist doch kein Kind ~:** you're no longer a child; you're not a child any more; **sie hat ihren Großvater nicht ~ gekannt** she never hat the chance to know her grandfather; **d) ich habe nur ~ 5 Mark** *(südd., österr., schweiz.)* I've only 5 marks left
Mehr das; ~s: **ein ~ an Zeit** *(Dat.) usw.* more time *etc.*
mehr-, Mehr-: ~aufwand der additional expenditure *no pl.*; **~bändig** *Adj.* in several volumes *postpos.*; **~belastung die** extra *or* additional burden (*Gen.* on); **~deutig 1.** *Adj.* ambiguous; **2.** *adv.* ambiguously; **~deutigkeit die;** ~, ~en ambiguity; **~einnahme die** additional revenue
mehren *(geh.)* **1.** *tr. V.* increase. **2.** *refl. V.* increase
mehrer... *Indefinitpron.* **a)** *attr.* several; a number of; *(verschieden)* various; several; **~e hundert Bücher** several hundred[s of] books; **b)** *alleinstehend* **~e** several people; **sie kamen zu ~en** several of them came
Mehr·erlös der extra *or* additional proceeds *pl.*
mehr·fach 1. *Adj.; nicht präd.* multiple; *(wiederholt)* repeated; **ein Bericht in ~er Ausfertigung** several copies *pl.* of a report; **der ~e deutsche Meister** the player/sprinter *etc.* who has been German champion several times; **ein ~er Millionär** a multimillionaire. **2.** *adv.* several times; *(wiederholt)* repeatedly; **~ vorbestraft sein** have several previous convictions
mehr-, Mehr-: ~familienhaus das multiple dwelling *(formal)*; large house with several flats *(Brit.) or (Amer.)* apartments; **~farbig** *Adj.* multi-coloured; [multi-]colour *attrib.*; **~geschossig** *Adj. s.* **~stöckig**
Mehrheit die; ~, ~en majority; **in der ~ sein** be in the majority; **die ~ haben/erringen** have/win a majority; **die ~ verlieren** lose one's majority; **er wurde mit großer ~ gewählt** he was elected by a large majority; **die einfache/relative/absolute ~** *(Politik)* a simple/a relative/an absolute majority
mehrheitlich 1. *Adj.; nicht präd.* majority; of the majority *postpos.* **2.** *adv.* by a majority
Mehrheits-: ~entscheidung die majority decision; **~wahlrecht das** first-past-the-post electoral system
mehr-, Mehr-: ~jährig *Adj.; nicht präd.* lasting several years *postpos.*; **eine ~jährige Erfahrung** several years' experience; several years of experience; **~kampf der** *(Sport)* multi-discipline event; **~kosten** *Pl.* additional *or* extra costs; **~malig** *Adj.; nicht präd.* repeated; **~mals** *Adv.* several times; *(wiederholt)* repeatedly; **~parteien·system das** multi-party system; **~seitig** *Adj.* consisting of several pages *postpos., not pred.*; several pages long *postpos.*; **~silbig** *Adj.* polysyllabic; **~sprachig** *Adj.* multilingual; **~sprachigkeit die** multilingualism; **~stimmig** *(Musik)* **1.** *Adj.* for several voices *postpos.*; **ein ~stimmiges Lied** a part-song; **2.** *adv.* **~stimmig singen** sing in harmony; **~stöckig** *Adj.* several storeys high *postpos.*; *(vielstöckig)* multi-storey; **~stufig** *Adj.* consisting of several steps *postpos., not pred.*; multi-stage ⟨*rocket*⟩; **~stündig** *Adj.; nicht präd.* lasting several hours *postpos., not pred.*; ⟨*delay*⟩ of several hours; **~stündige Verhandlungen** several hours of negotiations; **~tägig** *Adj.; nicht präd.* lasting several days *postpos., not pred.*; **~teilig** *Adj.* in several parts *postpos.*; **~wert·steuer die** *(Wirtsch.)* value added tax *(Brit.)*; VAT *(Brit.)*; sales tax *(Amer.)*; **~wöchig** *Adj.; nicht präd.* lasting several weeks *postpos., not pred.*; ⟨*absence*⟩ of several weeks; **~zahl die;** *o. Pl.* **a)** *(Sprachw.)* plural; **b)** *(Mehrheit)* majority
Mehr·zweck-: multi-purpose
meiden ['maidn̩] *unr. tr. V. (geh.)* avoid
Meile ['mailə] **die;** ~, ~n mile; **das riecht man drei ~n gegen den Wind** *(abwertend)* you can smell it a mile off; *(fig.)* you can tell that a mile off; it stands out a mile
Meilen·stein der *(auch fig.)* milestone
meilen·weit 1. *Adj.* ⟨*distance*⟩ of many miles. **2.** *adv.* for miles; **~ entfernt** *(auch fig.)* miles away (**von** from)
Meiler der; ~s, ~ **a)** charcoal kiln; **b)** *(Atom~)* [atomic] pile
mein [main] *Possessivpron.* my; **~e Damen und Herren** ladies and gentlemen; **das Buch dort, ist das ~[e]s?** that book over there, is it mine?; **was ~ ist, ist auch dein** what's mine is yours; **das Meine** *(geh.: Eigentum)* my possessions *pl. or* property; **ich habe das Meine getan** *(was ich konnte)* I have done what I could; *(meinen Teil)* I have done my share; **sie kann ~ und dein nicht unterscheiden** *(scherzh.)* she doesn't understand that some things don't belong to her; **die Meinen** *(geh.)* my family
Mein·eid der perjury *no indef. art.*; **einen ~ schwören** perjure oneself; commit perjury
meinen 1. *itr. V.* think; **[ganz] wie Sie ~!** whatever you think; *(wie Sie möchten)* [just] as you wish; **~ Sie?** do you think so?; **ich meine ja nur [so]** *(ugs.)* it was just an idea *or* a thought. **2.** *tr. V.* **a)** *(denken, glauben)* think; **man sollte ~, ...:** one would think *or* would have thought ...; **das meine ich auch** I think so too; **b)** *(sagen wollen, im Sinn haben)* mean; **was meint er damit?** what does he mean by that?; **das habe ich nicht gemeint** that's not what I meant; **c)** *(beabsichtigen)* mean; intend; **er meint es gut/ehrlich** he means

well *or* his intentions are good/ his intentions are honest; **es gut mit jmdm. ~:** mean well by sb.; **er hat es nicht so gemeint** *(ugs.)* he didn't mean it like that; **d)** *(sagen)* say

meiner *Gen. von* **ich** *(geh.)* **gedenke ~:** remember me; **erbarme dich ~:** have mercy upon me

meinerseits *Adv. (von meiner Seite)* on my part; *(auf meiner Seite)* for my part; **ganz ~:** the pleasure is [all] mine

meinesgleichen *indekl. Pron.* people *pl.* like me *or* myself; *(abwertend)* the likes *pl.* of me; my sort *or* kind

meinetwegen *Adv.* **a)** because of me; on my account; *(für mich)* on my behalf; *(mir zuliebe)* for my sake; *(um mich)* about me; **b)** [*auch* --'--] *(ugs.)* as far as I'm concerned; **~!** if you like; **also gut, ~!** fair enough!; **c)** *(zum Beispiel)* for instance

meinetwillen *Adv. in* **um ~:** for my sake

Meinung die; ~, ~en opinion (**zu** on, **über** + *Akk.* about); **eine vorgefaßte/gegenteilige ~ haben** have preconceived ideas *pl.*/hold an opposite opinion; **anderer/geteilter ~ sein** be of a different opinion/differing opinions *pl.;* hold a different view/differing views *pl.;* **nach meiner ~, meiner ~ nach** in my opinion *or* view; **ganz meine ~:** I agree entirely; **einer ~ sein** be of *or* share the same opinion; **die öffentliche ~:** public opinion; **jmdm. [gehörig] die ~ sagen** give sb. a [good] piece of one's mind

meinungs-, Meinungs-: ~äußerung die [expression of] opinion; **das Recht auf freie ~äußerung** the right of free speech; **~austausch der** exchange of views; **~forscher der** opinion pollster *or* researcher; **~forschung die** opinion research; **~forschungs·institut das** opinion research institute; **~freiheit die** freedom to form and express one's own opinions; *(Redefreiheit)* freedom of speech; **~umfrage die** [public] opinion poll; **~umschwung der** swing of opinion; **~verschiedenheit die** *(auch verhüll.: Streit)* difference of opinion

Meise ['maizə] **die; ~, ~n** tit[mouse]; **eine ~ haben** *(salopp)* be nuts *(sl.);* be off one's head *(coll.)*

Meißel ['maisl̩] **der; ~s, ~:** chisel

meißeln 1. *tr. V.* chisel; carve ⟨*statue, sculpture*⟩ with a chisel. **2.** *itr. V.* chisel; work with a chisel; carve

meist [maist] *Adv.* mostly; usually; *(zum größten Teil)* mostly; for the most part; **er hat ~ keine Zeit** he doesn't usually have any time

meist... 1. *Indefinitpron. u. unbest. Zahlw.* most; **das ~e Geld haben** have [the] most money; **die ~en Leute haben ...:** most people have; **die ~en Leute, die da waren** most of the people who were there; **die ~e Zeit des Jahres** most of the year; **er hat das ~e vergessen** he has forgotten most of it. **2.** *Adv.* **am ~:** most; **die am ~en befahrene Straße** the most used road; **darüber habe ich mich am ~en gefreut** that pleased me [the] most

meist·bietend *adv.* **etw. ~bietend versteigern/verkaufen** *usw.* auction sth. off/sell sth. *etc.* to the highest bidder

meistens ['maistn̩s] *Adv. s.* **meist**

meistenteils *Adv.* for the most part

Meister ['maistɐ] **der; ~s, ~ a)** master craftsman; **seinen ~ machen** *(ugs.)* get one's master craftsman's diploma *or* certificate; **b)** *(Vorgesetzter) (in der Fabrik, auf der Baustelle)* foreman; *(in anderen Betrieben)* boss *(coll.);* **c)** *(geh.: Könner)* master; **es ist noch kein ~ vom Himmel gefallen** *(Spr.)* you can't always expect to get it right first time; **[in jmdm.] seinen ~ gefunden haben** have met one's match [in sb.]; **d)** *(Künstler, geh.: Lehrer)* master; **e)** *(Sport)* champion; *(Mannschaft)* champions *pl.;* **f)** *(salopp: Anrede)* chief *(coll.);* guv *(Brit. sl.);* **g) ~ Lampe** Master Hare; **~ Petz** Bruin the Bear

Meister·brief der master craftsman's diploma *or* certificate

meisterhaft 1. *Adj.* masterly. **2.** *adv.* in a masterly manner; **es ~ verstehen, etw. zu tun** be a [past]-master *or* an expert at doing sth.

Meister·hand die master-hand; **von ~:** by a master-hand

Meisterin die; ~, ~nen a) master craftswoman; **b)** *(geh.: Könnerin)* master; **c)** *(Sport)* [women's] champion

Meister·leistung die masterly performance; *(Meisterstück)* masterpiece; *(geniale Tat)* master-stroke

meistern *tr. V.* master; master, overcome ⟨*problem, difficulty*⟩; control ⟨*anger, excitement, etc.*⟩; **sein Schicksal/Leben ~:** cope with one's fate/with life

Meister·prüfung die examination for the/one's master craftsman's diploma *or* certificate

Meisterschaft die; ~, ~en a) *o. Pl.* mastery; **b)** *(Sport)* championship; *(Veranstaltung)* championships *pl.;* **die ~ erringen** take the championship

Meisterschafts·spiel das *(Sport)* championship match *or* game

Meister-: ~singer der Meistersinger; mastersinger; **~stück das a)** *piece of work executed to qualify as a master craftsman;* **b)** *(Meisterleistung)* masterpiece (**an** + *Dat.* of); *(geniale Tat)* masterstroke; **~titel der a)** *(Sport)* championship [title]; **b)** *(im Handwerksberuf)* title of master craftsman; **~werk das** masterpiece (**an** + *Dat.* of)

Mekka ['mɛka] **(das); ~s** Mecca

Melancholie [melaŋko:'li:] **die; ~, ~n** melancholy; *(Psych.)* melancholia

Melancholiker [melaŋ'ko:likɐ] **der; ~s, ~, Melancholikerin die; ~, ~nen** melancholic

melancholisch 1. *Adj.* melancholy; melancholy, melancholic ⟨*person, temperament*⟩. **2.** *adv.* melancholically

Melange [me'lã:ʒ(ə)] **die; ~, ~n** *(österr.) s.* **Milchkaffee**

melden ['mɛldn̩] **1.** *tr. V.* **a)** report; *(registrieren lassen)* register ⟨*birth, death, etc.*⟩ (*Dat.* with); **wie soeben gemeldet wird** *(Fernseh., Rundf.)* according to reports just coming in; **jmdn. als vermißt ~:** report sb. missing; **nichts/nicht viel zu ~ haben** *(ugs.)* have no/little say; **b)** *(ankündigen)* announce; **c)** *(Schülerspr.)* **jmdn. ~:** tell on sb. **2.** *refl. V.* **a)** report; **sich freiwillig ~:** volunteer (**zu** for); **sich auf eine Anzeige ~:** reply to *or* answer an advertisement; **sich zu einer Prüfung ~:** enter for an examination; **polizeilich gemeldet sein** be registered with the police; **b)** *(am Telefon)* answer; **es meldet sich niemand** there is no answer *or* reply; **c)** *(ums Wort bitten)* put one's hand up; **d)** *(von sich hören lassen)* get in touch (**bei** with); **wenn du etwas brauchst, melde dich** if you need anything let me/us know; **Otto 2, bitte ~!** Otto 2, come in please!

Melde·pflicht die *(Verwaltung)* obligation to register with the authorities; **polizeiliche ~** obligation to register with the police

mẹlde·pflichtig *Adj. (Gesundheitsw.)* notifiable ‹*disease*›

Mẹldung die; ~, ~en a) report; *(Nachricht)* piece of news; *(Ankündigung)* announcement; **~en vom Sport** sports news *sing.*; **b) ~ machen** *od.* **erstatten** *(Milit.)* report; make a report; **c)** *(Anmeldung) (bei einem Wettbewerb, Examen)* entry; *(bei einem Kurs)* enrolment; **wir bitten um freiwillige ~en** we are asking *or* calling for volunteers; **d)** *(Wort~)* request to speak; **gibt es noch weitere ~en?** does anyone else wish to speak?

meliert [mɛ'li:ɐt] *Adj.* mottled; **braun ~:** mottled brown; **[grau] ~es Haar** hair streaked with grey

Melisse [me'lɪsə] **die; ~, ~n** melissa; balm

melken ['mɛlkn̩] *regelm., unr. tr. V.* milk

Mẹlk·maschine die milking machine

Melodie [melo'di:] **die; ~, ~n** melody; *(Weise)* tune; melody; **nach einer ~:** to a melody/tune

Melodik [me'lo:dɪk] **die; ~** *(Musik)* melodic characteristics *pl.*; *(Lehre)* theory of melody

melọdisch 1. *Adj.* melodic; melodious. **2.** *adv.* melodically; melodiously; **~ sprechen** speak in a melodic *or* melodious voice

melodramạtisch *Adj.* melodramatic

Melone [me'lo:nə] **die; ~, ~n a)** melon; **b)** *(ugs. scherzh.)* bowler [hat]

Membran [mɛm'bra:n] **die; ~, ~en, Membrạne die; ~, ~n a)** *(Technik)* diaphragm; **b)** *(Biol., Chemie)* membrane

Memoiren [me'mo̯a:rən] *Pl.* memoirs

Memorandum [memo'randʊm] **das; ~s, Memorạnden** *od.* **Memorạnda** memorandum

Menge ['mɛŋə] **die; ~, ~n a)** *(Quantum)* quantity; amount; **die dreifache ~:** three times *or* triple the amount; **b)** *(große Anzahl)* large number; lot *(coll.)*; **eine ~ Leute** a lot *or* lots *pl.* of people *(coll.)*; **er weiß eine [ganze] ~** *(ugs.)* he knows [quite] a lot *(coll.) or* a great deal; **sie bildet sich eine ~ ein** *(ugs.)* she is very conceited; **jede ~ Arbeit/Alkohol** *usw. (ugs.)* masses *pl. or* loads *pl.* of work/alcohol *etc. (coll.)*; *s. auch* **rauh 1h; c)** *(Menschen~)* crowd; throng; **d)** *(Math.)* set

mẹngen *(veralt.) tr. V.* mix

mẹngen-, Mẹngen-: ~lehre die; *o. Pl.* set theory *no art.*; **~mäßig 1.** *Adj.* quantitative; **2.** *adv.* quantitatively; **~rabatt der** *(Wirtsch.)* bulk discount

Meningitis [menɪŋ'gi:tɪs] **die; ~, Meningitịden** *(Med.)* meningitis

Meniskus [me'nɪskʊs] **der; ~, Menịsken** *(Anat., Optik)* meniscus

Mennige ['mɛnɪgə] **die; ~:** red lead

Mensa ['mɛnza] **die; ~, ~s** *od.* **Mẹnsen** refectory, canteen *(of university, college)*

Mensch [mɛnʃ] **der; ~en, ~en a)** *(Gattung)* **der ~:** man; **die ~en** man *sing.*; human beings; mankind *sing. no art.*; **nur noch ein halber ~ sein** be just about all in; **wieder ein ~ sein** *(ugs.)* feel like a human being again; **b)** *(Person)* person; man/woman; **~en** people; **kein ~:** no one; **unter die ~en gehen** mix with people; **wie der erste ~/die ersten ~en:** extremely awkwardly; **von ~ zu ~:** man to man/woman to woman; **~, ärgere dich nicht** *(Gesellschaftsspiel)* ludo; **c)** *(salopp: Anrede) (bewundernd)* wow; *(erstaunt)* wow; good grief; *(vorwurfsvoll)* for heaven's sake; **~, war das ein Glück!** boy, that was a piece of luck!; **~ Meier!** good grief!

mẹnschen-, Mẹnschen-: ~affe der anthropoid [ape]; **~auflauf der** crowd [of people]; **~feindlich 1.** *Adj.* **a)** misanthropic; **b)** *(unmenschlich)* inhuman ‹*system, policy etc.*›; ‹*environment*› hostile to man; **2.** *adv.* **a)** misanthropically; **b)** *(unmenschlich)* inhumanly; **~fresser der** *(ugs.)* cannibal; *(Mythol.)* man-eater; **~freundlichkeit die;** *o. Pl.* philanthropy; **aus reiner ~freundlichkeit** out of the sheer goodness of one's heart; **~führung die** leadership; **~gedenken das: das wird seit ~gedenken so gemacht** it has been done that way for as long as anyone can remember; **der heißeste Sommer seit ~gedenken** the hottest summer in living memory; **~gestalt die** human form; **ein Engel/Teufel** *od.* **Satan in ~gestalt sein** be an angel in human form/the devil incarnate; **~hand die: von ~hand** *(geh.)* ‹*created*› by the hand of man, by human hand; **~handel der** trade *or* traffic in human beings; *(Sklavenhandel)* slave-trade; **~händler der** trafficker [in human beings]; *(Sklavenhändler)* slave-trader; **~kenntnis die;** *o. Pl.* ability to judge character *or* human nature; **~kette die** human chain; **~leben das** life; **der Unfall forderte vier ~leben** *(geh.)* the accident claimed four lives; **~leer** *Adj.* deserted; **~menge die** crowd [of people]; **~möglich** *Adj.*; *nicht attr.* humanly possible; **das ~mögliche tun** do all that is/was humanly possible; **~opfer das** human sacrifice; **~raub der** kidnapping; abduction; **~recht das** human right; **~rechts·konvention die** Human Rights Convention; **~schlag der** race *or* breed [of people]; **~seele die: keine ~seele** not a [living] soul

Mẹnschens·kind: ~! *(salopp) (erstaunt)* good heavens; good grief; *(vorwurfsvoll)* for heaven's sake

mẹnschen-, Mẹnschen-: ~unwürdig 1. *Adj.* ‹*accommodation*› unfit for human habitation; ‹*conditions*› unfit for human beings; ‹*behaviour*› unworthy of a human being; **2.** *adv.* ‹*treat*› in a degrading and inhumane way; ‹*live, be housed*› in conditions unfit for human beings; **~verachtung die** contempt for humanity *or* mankind; **~verstand der** human intelligence *or* intellect; *s. auch* **gesund; ~würde die** human dignity *no art.*; **~würdig 1.** *Adj.* humane ‹*treatment*›; ‹*accommodation*› fit for human habitation; ‹*conditions*› fit for human beings; **2.** *adv.* ‹*treat*› humanely; ‹*live, be housed*› in conditions fit for human beings

Mẹnschheit die; ~: mankind *no art.*; humanity *no art.*

Mẹnschheits-: ~entwicklung die evolution of man; **~traum der** dream of mankind

mẹnschlich 1. *Adj.* **a)** human; **~es Versagen** human error; *s. auch* **irren a; b)** *(annehmbar)* civilized; **c)** *(human)* humane ‹*person, treatment, etc.*›; human ‹*trait, emotion, etc.*›. **2.** *adv.* **a) er ist ~ sympathisch** I like him as a person; **sich ~ näherkommen** get on closer [personal] terms [with one another]; **b)** *(human)* humanely; in a humane manner

Mẹnschlichkeit die humanity *no art.*; **etw. aus reiner ~ tun** do sth. for purely humanitarian reasons

Mensen *s.* **Mensa**

Menstruation [mɛnstrua'tsio:n] **die; ~, ~en** menstruation; *(Periode)* [menstrual] period

Mentalität [mɛntali'tɛ:t] **die; ~, ~en** mentality

Menthol [mɛn'to:l] **das; ~s** menthol

Menü [me'ny:] **das; ~s, ~s** *(auch DV)* menu; *(im Restaurant)* set meal *or* menu

Meridian [meri'di̯a:n] **der; ~s, ~e** *(Geogr., Astron.)* meridian

Merino [me'ri:no] **der; ~s, ~s** *(Stoff)* merino

Merino-: ~schaf das merino [sheep]; **~wolle die** merino wool

merkbar 1. *Adj.* perceptible; noticeable; *(deutlich)* noticeable. **2.** *adv.* perceptibly; noticeably; *(deutlich)* noticeably

Merk·blatt leaflet; *(mit Anweisungen)* instruction leaflet

merken ['mɛrkn̩] **1.** *tr. V.* notice; **deutlich zu ~ sein** be plain to see; be obvious; **an seinem Benehmen merkt man, daß ...** you can tell by his behaviour that ...; **das merkt doch jeder/keiner** everybody/nobody will notice; **jmdn. etw. ~ lassen** let sb. see sth.; **du merkst aber auch alles!** *(ugs. iron.)* how very observant of you!; **merkst du was?** *(ugs.)* have you noticed something?. **2.** *refl., auch tr. V.* **sich** *(Dat.)* **etw. ~:** remember sth.; *(sich einprägen)* memorize; **hast du dir die Adresse gemerkt?** have you made a mental note of the address?; **diesen Mann muß man sich** *(Dat.)* **~:** this is a man to take note of; **ich werd' mir's** *od.* **werd's mir ~** *(ugs.)* I won't forget that; I'll remember that; **merk dir das** just remember that

merklich 1. *Adj.* perceptible; noticeable; *(deutlich)* noticeable. **2.** *adv.* perceptibly; noticeably; *(deutlich)* noticeably

Merkmal das; ~s, ~e feature; characteristic

Merkur [mɛr'ku:ɐ̯] **der; ~s** *(Astron.)* Mercury

merkwürdig 1. *Adj.* strange; odd; peculiar. **2.** *adv.* strangely; oddly; peculiarly

merkwürdiger·weise *Adv.* strangely *or* oddly *or* curiously enough

meschugge [me'ʃʊgə] *Adj.; nicht attrib. (salopp)* barmy *(Brit. sl.)*; nuts *pred. (sl.)*; off one's rocker *pred. (sl.)*

meßbar ['mɛsba:ɐ̯] *Adj.* measurable

Meß-: ~becher der measuring jug; **~diener der** *(kath. Kirche)* server

[1]**Messe** ['mɛsə] **die; ~, ~n** *(Gottesdienst, Musik)* mass; **die ~ halten** *od. (geh.)* **zelebrieren** say *or* celebrate mass; **für jmdn. eine ~ lesen** say a mass for sb.

[2]**Messe die; ~, ~n a)** *(Ausstellung)* [trade] fair; **auf der ~:** at the [trade] fair; **b)** *(landsch.: Jahrmarkt, Volksfest)* fair

[3]**Messe die; ~, ~n** *(Seew., Milit.)* mess; *(Raum)* mess-room

Messe-: ~gelände das site of a/the [trade] fair; *(mit festen ~hallen)* exhibition centre; **~halle die** exhibition hall

messen 1. *unr. tr. V.* **a)** measure; take ⟨*pulse, blood, pressure, temperature*⟩; **b)** *(beurteilen)* judge **(nach, an +** *Dat.* by); **jmdn. an jmdm. ~:** judge sb. by comparison with sb.; **ge~ an** (+ *Dat.*) having regard to. **2.** *unr. itr. V.* measure; **er mißt 1,85 m** he's 1.85 m [tall]; **genau ~:** make an exact measurement/exact measurements. **3.** *unr. refl. V. (geh.)* compete **(mit** with); **sich mit jmdm./etw. [in etw.** *(Dat.)***] [nicht] ~ können** [not] be as good as sb./sth. [in sth.]

Messer das; ~s, ~ a) knife; *(Hack~)* chopper; *(Rasier~)* [cut-throat] razor; **jmdm. das ~ an die Kehle setzen** *(fig. ugs.)* hold sb. at gunpoint; **auf des ~s Schneide stehen** *(fig.)* hang in the balance; be balanced on a knife-edge; **jmdn. ans ~ liefern** *(fig. ugs.)* inform on sb.; **bis aufs ~** *(fig. ugs.)* ⟨*fight etc.*⟩ to the bitter end; **jmdm. ins [offene] ~ laufen** *(fig. ugs.)* play right into sb.'s hands; **b)** *(ugs.: Skalpell)* **unters ~ müssen** have to go under the knife *(coll.)*

messer-, Messer-: ~scharf 1. *Adj.* razor-sharp; *(fig.)* trenchant ⟨*criticism*⟩; incisive ⟨*logic*⟩; razor-sharp ⟨*wit, intellect*⟩; **2.** *adv. (fig. ugs.)* ⟨*think*⟩ with penetrating insight; ⟨*argue*⟩ incisively; **~spitze die a)** point of a/the knife; **b)** *(Mengenangabe)* **eine ~spitze** just a trace; **eine ~spitze Salz** a large pinch of salt; **~stecherei** [-ʃtɛçə'rai̯] **die; ~, ~en** knife-fight; fight with knives; **~stich der** knife-thrust; *(Wunde)* knife-wound; stab wound

Meß·gerät das measuring device *or* instrument

Messias [mɛ'si:as] **der; ~:** Messiah

Messing ['mɛsɪŋ] **das; ~s** brass

Messing·waren *Pl.* brassware *sing.*

Meß-: ~instrument das measuring instrument; **~technik die** technology of measurement

Messung die; ~, ~en measurement

Meß·wert der measured value; *(Ableseergebnis)* reading

Mestize [mɛs'ti:tsə] **der; ~n, ~n** mestizo

Met [me:t] **der; ~[e]s** mead

Metall [me'tal] **das; ~s, ~e** metal

Metall·arbeiter der metalworker

metallen *Adj. nicht präd.* metal

Metaller der; ~s, ~, Metallerin die; ~, ~nen *(ugs.)* metalworker

metall·haltig *Adj.* metalliferous

metallic [me'talɪk] *indekl. Adj.* metallic [grey/blue/*etc.*]

Metall·industrie die metal-processing and metal-working industries *pl.*

metallisch *Adj.* metallic; metal *attrib.*, metallic ⟨*conductor*⟩

Metallurgie [metalʊr'gi:] **die; ~:** [extractive] metallurgy *no art.*

metall·verarbeitend *Adj.; nicht präd.* metalworking

Metamorphose [metamɔr'fo:zə] **die; ~, ~n** metamorphosis

Metapher [me'tafɐ] **die; ~, ~n** *(Stilk.)* metaphor

Metaphorik [meta'fo:rɪk] **die; ~** *(Stilk.)* imagery; metaphors *pl.*

metaphorisch *(Stilk.) Adj.* metaphorical

meta·physisch 1. *Adj.* metaphysical; **2.** *adv.* metaphysically

Meteor [mete'o:ɐ̯] **der; ~s, ~e** *(Astron.)* meteor

Meteorit [meteo'ri:t] **der; ~en** *od.* **~s, ~e[n]** *(Astron.)* meteorite

Meteorologe [meteoro'lo:gə] **der; ~n, ~n** meteorologist

Meteorologie die; ~: meteorology *no art.*

meteorologisch 1. *Adj.* meteorological. **2.** *adv.* meteorologically

Meter ['me:tɐ] **der** *od.* **das; ~s, ~:** metre; **drei ~ lang** three metres long; **in 100 ~ Höhe** at a height of 100 metres; **auf den letzten ~n** in the last few metres

meter-, Meter-: ~dick *Adj.* metres thick *postpos.*; **~hoch** *Adj.* metres high *postpos.*; ⟨*snow*⟩ metres deep; **der Schnee lag ~hoch** the snow was metres deep; **~lang** *Adj.* metres long *postpos.*; **~maß das** tape-measure; *(Stab)* [metre] rule; **~ware die** fabric/material *etc.* sold by the metre; **~weise** *Adv.* by the metre; **~weit** *Adj.* metres long *postpos.*

Methan [me'ta:n] **das; ~s** methane

Methanol [meta'no:l] **das; ~s** *(Chemie)* methanol

Methode [me'to:də] **die; ~, ~n** method

Methodik [me'to:dɪk] **die; ~, ~en** methodology

methodisch **1.** *Adj.* methodological; *(nach einer Methode vorgehend)* methodical. **2.** *adv.* methodologically; *(nach einer Methode)* methodically

Methodist [meto'dɪst] **der;** ~**en,** ~**en** Methodist

Metier [me'ti̯e:] **das;** ~**s,** ~**s** profession; **sein** ~ **beherrschen** know one's job

Metrik ['me:trɪk] **die;** ~, ~**en** metrics

metrisch **1.** *Adj.* **a)** *(Verslehre, Musik)* metrical; **b)** metric ⟨*ton, system, etc.*⟩. **2.** *adv.* metrically

Metronom [metro'no:m] **das;** ~**s,** ~**e** *(Musik)* metronome

Metropole [metro'po:lə] **die;** ~, ~**n** metropolis

Mett [mɛt] **das;** ~**[e]s** *(landsch.)* minced meat, mince *(pork)*

Mett·wurst **die** *soft smoked sausage made of minced pork and beef*

Metzelei [mɛtsə'lai̯] **die;** ~, ~**en** *(abwertend)* slaughter; butchery

Metzger ['mɛtsgɐ] **der;** ~**s,** ~ *(bes. westmd., südd., schweiz.)* butcher; *(im Schlachthof)* slaughterman

Metzger- *s.* **Fleischer-**

Metzgerei **die;** ~, ~**en** *(bes. westmd., südd., schweiz.)* butcher's [shop]

Meute ['mɔytə] **die;** ~, ~**n** **a)** *(Jägerspr.)* pack; **b)** *(ugs. abwertend: Menschengruppe)* mob

Meuterei [mɔytə'rai̯] **die;** ~, ~**en** mutiny; *(fig.)* revolt; mutiny

Meuterer ['mɔytərɐ] **der;** ~**s,** ~: mutineer; *(fig.)* rebel

meutern *itr. V.* **a)** mutiny; ⟨*prisoners*⟩ riot; **b)** *(fig. ugs.)* rebel; *(murren)* moan

Mexikaner [mɛksi'ka:nɐ] **der;** ~**s,** ~, **Mexikanerin** **die;** ~, ~**nen** Mexican

mexikanisch *Adj.* Mexican

Mexiko ['mɛksiko] **(das);** ~**s** Mexico

MEZ *Abk.* **mitteleuropäische Zeit** CET

mg *Abk.* **Milligramm** mg

MG [ɛm'ge:] **das;** ~**s,** ~**s** *Abk.* **Maschinengewehr**

Mi. *Abk.* **Mittwoch** Wed.

miau [mi'au̯] *Interj.* miaow

miauen *itr. V.* miaow

mich [mɪç] **1.** *Akk. des Personalpron.* **ich** me. **2.** *Akk. des Reflexivpron. der 1. Pers.* myself

mick[e]rig ['mɪk(ə)rɪç] *Adj. (ugs.)* miserable; measly *(sl.)*; puny ⟨*person*⟩; puny, stunted ⟨*plant, tree*⟩

Midi- ['mi:di-] midi⟨*-skirt, dress, coat*⟩

mied [mi:t] *1. u. 3. Pers. Sg. Prät. v.* **meiden**

Mieder ['mi:dɐ] **das;** ~**s,** ~ **a)** *(Korsage)* girdle; **b)** *(Leibchen)* bodice

Mieder·waren *Pl.* corsetry *sing.*

Mief [mi:f] **der;** ~**[e]s** *(salopp abwertend)* fug *(coll.)*

miefen *itr. V. (ugs. abwertend)* pong *(coll.)*; stink

Miene ['mi:nə] **die;** ~, ~**n** expression; face; **mit unbewegter** ~: with an impassive expression; impassively; **gute** ~ **zum bösen Spiel machen** grin and bear it

Mienen·spiel **das** facial expressions *pl.*

mies [mi:s] *(ugs.)* **1.** *Adj. (abwertend)* terrible *(coll.)*; lousy *(sl.)*; rotten *(sl.)*; lousy *(sl.)*, foul ⟨*mood*⟩. **2.** *adv.* **a)** *(abwertend: schlecht)* terribly badly *(coll.)*; lousily *(sl.)*; rottenly *(sl.)*; **b)** *(unwohl)* **ihm geht es** ~: he's in a terrible state *(coll.)*

Miese ['mi:zə] *Pl.; adj. Dekl. (salopp)* **2 000** ~ **auf dem Konto haben** be 2,000 marks in the red at the bank; **in den** ~**n sein** be in the red; *(beim Kartenspiel)* be down on points

mies-, Mies-: ~|**machen** *tr. V. (ugs. abwertend) (schlechtmachen)* **jmdn./etw.** ~**machen** run sb./sth. down; ~**macher** **der** *(ugs. abwertend)* carping critic; *(Spielverderber)* killjoy; ~**muschel** **die** [common] mussel

Miete ['mi:tə] **die;** ~, ~**n** **a)** rent; *(für ein Auto, Boot)* hire charge; *(für Fernsehgeräte usw.)* rental; **b)** *o. Pl. (das Mieten)* renting; **zur** ~ **wohnen** live in rented accommodation; rent a house/flat *(Brit.)* or *(Amer.)* apartment/room/rooms; **bei jmdm. zur** ~ **wohnen** lodge with sb.

Miet·einnahmen *Pl.* income *sing.* from rents

mieten *tr. V.* rent; *(für kürzere Zeit)* hire

Mieter ['mi:tɐ] **der;** ~**s,** ~: tenant

Miet·erhöhung **die** rent increase

Mieterin **die;** ~, ~**nen** tenant

miet-, Miet-: ~**frei** *Adj., adv.* rent-free; ~**kauf** **der** *(Wirtsch.)* ≈ hire purchase *(Brit.)* or *(Amer.)* installment plan *(with option to buy outright or terminate the agreement at a specified date)*; ~**partei** **die** tenant

Miets-: ~**haus** **das** block of rented flats *(Brit.)* or *(Amer.)* apartments; ~**kaserne** **die** *(abwertend)* tenement block

Miet-: ~**vertrag** **der** tenancy agreement; ~**wagen** **der** hire-car; ~**wohnung** **die** rented flat *(Brit.)* or *(Amer.)* apartment; ~**wucher** **der** charging of exorbitant rents; ~**zins** **der;** *Pl.* ~**e** *(südd., österr., schweiz., Amtsspr.)* rent

Mieze ['mi:tsə] **die;** ~, ~**n** **a)** *(fam.: Katze)* puss; pussy *(child lang.)*; **b)** *(salopp: Mädchen)* chick *(sl.)*; *(als Anrede)* sweetie

Migräne [mi'grɛ:nə] **die;** ~, ~**n** migraine

Mikado [mi'ka:do] **das;** ~**s** spillikins *sing.*; jack-straws *sing.*

Mikro ['mi:kro] **das;** ~**s,** ~**s** *(ugs.)* mike *(coll.)*

mikro-, Mikro- micro-

Mikrobe [mi'kro:bə] **die;** ~, ~**n** microbe

mikro-, Mikro-: ~**elektronik** **die** micro-electronics *sing., no art.*; ~**phon** [--'-] **das;** ~**s,** ~**e** microphone; ~**prozessor** [~pro'tsɛsɔr] **der;** ~**s,** ~**en** [...'so:rən] microprocessor; ~**skop** [~'sko:p] **das;** ~**s,** ~**e** microscope; ~**skopisch** **1.** *Adj.* microscopic; **2.** *adv.* microscopically; ~**wellen·herd** **der** microwave oven

Milbe ['mɪlbə] **die;** ~, ~**n** mite; *(Zecke)* tick

Milch [mɪlç] **die;** ~ milk; ~ **geben** give or yield milk

Milch-: ~**bar** **die** milk bar; ~**brötchen** **das** milk roll; ~**drüse** **die** mammary gland; ~**flasche** **die** **a)** milk-bottle; **b)** *(für Säuglinge)* feeding-bottle; baby's bottle; ~**gebiß** **das** milk-teeth *pl.*

milchig **1.** *Adj.* milky. **2.** *adv.* ~ **weiß** milky-white

Milch-: ~**kaffee** **der** coffee with plenty of milk; ~**kännchen** **das** milk-jug; ~**kanne** **die** milk-can; *(zum Transportieren von* ~*)* [milk-]churn; ~**kuh** **die** dairy *or* milk *or* milch cow; ~**mädchen·rechnung** **die** *(ugs.)* naïve miscalculation; ~**mix·getränk** **das** milk shake; ~**reis** **der** rice pudding; ~**schokolade** **die** milk chocolate; ~**straße** **die** Milky Way; Galaxy; ~**vieh** **das** dairy cattle; ~**wirtschaft** **die** dairying *no art.*; ~**zahn** **der** milk-tooth

mild [mɪlt], **milde** **1.** *Adj.* **a)** *(gütig)* lenient ⟨*judge, judgement*⟩; benevolent ⟨*ruler*⟩; mild, lenient, light ⟨*punishment*⟩; mild ⟨*words, accusation*⟩; mild, gentle ⟨*reproach*⟩; gentle ⟨*smile, voice*⟩; **jmdn.** ~ **stimmen** induce sb. to take a lenient attitude; **b)** *(nicht rauh)* mild ⟨*climate, air, winter,* *etc.*⟩; **c)** *(nicht scharf)* mild ⟨*spice, coffee, tobacco, cheese, etc.*⟩; ~ **schmecken** be mild; **d)** *(schonend)* mild ⟨*soap, shampoo, detergent*⟩; **e)** *nicht präd. (veralt.: mildtätig)* charitable; **eine ~e Gabe** alms *pl.* **2.** *adv.* **a)** *(gütig)* leniently; ⟨*smile, say*⟩ gently; **b)** *(gelinde)* mildly; ~ **ausgedrückt** to put it mildly; putting it mildly

Milde die; ~ **a)** *(Gnade, Güte)* leniency; **[jmdm. gegenüber]** ~ **walten lassen** be lenient [with sb.]; **b)** *(des Klimas usw.)* mildness; **c)** *(milder Geschmack)* mildness

mildern 1. *tr. V.* moderate ⟨*criticism, judgement*⟩; mitigate ⟨*punishment*⟩; soothe ⟨*anger*⟩; reduce ⟨*intensity, strength, effect*⟩; modify ⟨*impression*⟩; ease, soothe, relieve ⟨*pain*⟩; alleviate ⟨*poverty, need*⟩; **~de Umstände** *(Rechtsw.)* mitigating circumstances. **2.** *refl. V.* ⟨*anger, rage, agitation*⟩ abate

Milderung die; ~ *(eines Tadels, Urteils)* moderation; *(einer Strafe)* mitigation; *(von Schmerz)* easing; soothing; relief; *(von Armut, Not)* alleviation

mild·tätig *Adj.* charitable

Mild·tätigkeit die; *o. Pl.* charity

Milieu [mi'li̯ø:] **das;** ~s, ~s milieu; environment; *(fig.: Prostitution usw.)* world of pimps and prostitutes; **er stammt aus kleinbürgerlichem** ~: his background is petit bourgeois

militant [mili'tant] *Adj.* militant

[1]**Militär** [mili'tɛ:ɐ̯] **das;** ~s **a)** armed forces *pl.*; military; **beim** ~ **sein/vom** ~ **entlassen werden** be in/be discharged from the forces; **b)** *(Soldaten)* soldiers *pl.*; army

[2]**Militär der;** ~s, ~s [high-ranking military] officer

Militär-: ~arzt der medical officer; **~dienst der** military service; **seinen ~dienst ableisten** do one's military *or* national service; **~diktatur die** military dictatorship; **~fahrzeug das** military vehicle; **~flugzeug das** military aircraft; **~geistliche der** army chaplain; **~gericht das** military court; court martial; **vor ein ~gericht gestellt werden** be brought before *or* tried by a military court; be court-martialled

militärisch 1. *Adj.* military. **2.** *adv.* **jmdn.** ~ **grüßen** salute sb.

Militarisierung die; ~: militarization

Militarist der; ~en, ~en *(abwertend)* militarist

militaristisch *Adj. (abwertend)* militarist; militaristic

Militär-: ~junta die military junta; **~parade die** military parade; **~putsch der** military putsch

Military ['mɪlɪtərɪ] **die;** ~, ~s *(Reiten)* three-day event

Miliz [mi'li:ts] **die;** ~, ~en militia; *(Polizei)* police

milk [mɪlk], **milkst, milkt** *(veralt.) Imperativ Sg., 2. u. 3. Pers. Sg. Präsens v.* **melken**

Mill. *Abk.* **Million** m.

Mille ['mɪlə] **die;** ~, ~ *(salopp)* grand *(sl.)*; thousand marks/pounds *etc.*

Milliardär [mɪli̯ar'dɛ:ɐ̯] **der;** ~s, ~e, **Milliardärin die;** ~, ~nen multi-millionaire *(possessing at least a thousand million marks etc.)*; billionaire *(Amer.)*

Milliarde [mɪ'li̯ardə] **die;** ~, ~n thousand million; billion; *s. auch* **Million**

Milli-: ~bar das *(Met.)* millibar; **~gramm das** milligram

Milli·meter der *od.* **das** millimetre

Millimeter-: ~arbeit die; *o. Pl. (ugs.) (am Steuer)* delicate piece of manœuvring; *(bei Ballspielen)* [neat] piece of precision play; **~papier das** [graph] paper ruled in millimetre squares

Million [mɪ'li̯o:n] **die;** ~, ~en million; **eine/zwei [en** a/two million; **~en [von ...]** millions [of ...]

Millionär [mɪli̯o'nɛ:ɐ̯] **der;** ~s, ~e, **Millionärin die;** ~, ~nen millionaire

millionen-, Millionen-: ~auflage die *(Buchw.)* **dieses Buch erschien in ~auflage** [over] a million copies of this book were printed; **~fach 1.** *Adj.* millionfold ⟨*increase etc.*⟩; **2.** *adv.* a million times; **~gewinn der a)** *(Ertrag)* profit of a million/of millions; **b)** *(Lotteriegewinn)* prize of a million/of millions; **~schaden der** damage *no pl., no indef. art.* running into millions; **~schwer** *Adj. (ugs.)* worth millions *pred.*; **~stadt die** town with over a million inhabitants

millionst... *Ordinalz.* millionth; *s. auch* **hundertst...**

million[s]tel *Bruchz.* millionth; *s. auch* **hundertstel**

Million[s]tel das *od. (schweiz.)* **der;** ~s, ~: millionth

Milz [mɪlts] **die;** ~ *(Anat.)* spleen

Mime ['mi:mə] **der;** ~n, ~n *(geh.)* Thespian

mimen *tr. V.* put on a show of ⟨*admiration, efficiency*⟩; **den Kranken/Unschuldigen** ~: pretend to be ill/act the innocent

Mimik ['mi:mɪk] **die;** ~: gestures and facial expressions *pl.*

Mimikry ['mɪmikri] **die;** ~ *(Zool.)* mimicry; *(fig.)* camouflage

mimisch 1. *Adj.; nicht präd.* mimic. **2.** *adv.* ⟨*show*⟩ by means of gestures and facial expressions

Mimose [mi'mo:zə] **die;** ~, ~n **a)** mimosa; **b)** *(fig.)* over-sensitive person; **die reinste** ~ **sein** be extraordinarily sensitive

mimosenhaft *(fig.)* **1.** *Adj.* over-sensitive. **2.** *adv.* over-sensitively

Min. *Abk.* **Minute[n]** min.

Minarett [mina'rɛt] **das;** ~s, ~e *od.* ~s minaret

minder ['mɪndɐ] *Adv. (geh.)* less; **[nicht]** ~ **angenehm sein** be [no] less pleasant; *s. auch* **mehr 1**

minder... *Adj.; nicht präd.* inferior, lower ⟨*quality*⟩; **von ~er Bedeutung sein** be of less importance

minder-, Minder-: ~begabt *Adj.* less gifted *or* able; **~bemittelt** *Adj.* without much money *postpos., not pred.*; **~bemittelt sein** not have much money; **er ist doch geistig ~bemittelt** *(fig. salopp abwertend)* he isn't all that bright *(coll.)*; **~bemittelte** *Pl.* needy persons

Minderheit die; ~, ~en minority

Minderheits·regierung die minority government

minder-, Minder-: ~jährig *Adj. (Rechtsw.)* ⟨*child etc.*⟩ who is/was a minor *or* under age; **~jährig sein** be a minor *or* under age; **~jährige der/die;** *adj. Dekl. (Rechtsw.)* minor; person under age; **~jährigkeit die;** ~ *(Rechtsw.)* minority

mindern *(geh.) tr. V.* reduce ⟨*income, price, number of staff, tension, etc.*⟩; impair ⟨*performance, abilities*⟩; diminish, reduce ⟨*value, quality, dignity, pleasure, influence*⟩; detract from ⟨*reputation*⟩

Minderung die; ~, ~en *s.* **mindern:** reduction (*Gen.* in); impairment (*Gen.* of); diminution (*Gen.* of); detraction (*Gen.* from)

minder·wertig *Adj. (abwertend)* inferior, low-quality ⟨*goods, material*⟩; low-quality, low-grade ⟨*meat*⟩; *(fig.)* inferior

Minder·wertigkeit die; *o. Pl. s.* **minderwertig:** inferiority; low quality; low grade; *(fig.)* inferiority

Minderwertigkeits-: ~gefühl das *(Psych.)* feeling of inferiority; **~komplex der** *(Psych.)* inferiority complex

Minder·zahl die; *o. Pl.* minority

mindest... ['mɪndəst...] *Adj.; nicht präd.* slightest; least; **das ist das ~e, was du tun kannst** it is the least you can do; **nicht im ~en** not in the least

Mịndest-: ~alter das minimum age; **~anforderung die** minimum requirement

mindestens ['mɪndəstn̩s] *Adv.* at least

Mịndest-: ~gebot das reserve price; **~haltbarkeits·datum das** best-before date; **~lohn der** minimum wage; **~maß das** minimum (**an** + *Dat.*, **von** of)

Mine ['mi:nə] **die; ~, ~n a)** *(Erzbergwerk)* mine; **b)** *(Sprengkörper)* mine; **auf eine ~ laufen** strike a mine; **c)** *(Bleistift~)* lead; *(Kugelschreiber~, Filzschreiber~)* refill

Mịnen·such·boot das *(Milit.)* minesweeper

Mineral [mine'ra:l] **das; ~s, ~e** *od.* **Mineralien** mineral

Mineralogie die; ~: mineralogy *no art.*

Mineral·öl das mineral oil

Mineralöl-: ~gesellschaft die oil company; **~steuer die** tax on oil

Mineral-: ~quelle die mineral spring; **~wasser das** mineral water

Mini ['mɪni] **das; ~s, ~s** *(Mode)* mini *(coll.)*

Mịni- mini-

Miniatur [minia'tu:ɐ̯] **die; ~, ~en** miniature

Miniatur·aus·gabe die *(Buchw.)* abridged edition

Mịni·golf das minigolf; crazy golf

minimal [mini'ma:l] **1.** *Adj.* minimal; marginal ‹*advantage, lead*›; very slight ‹*benefit, profit*›. **2.** *adv.* minimally

Minimal·forderung die minimum demand

Minimum ['mi:nimʊm] **das; ~s, Minima** minimum (**an** + *Dat.* of)

Mịni·rock der miniskirt

Minister [mi'nɪstɐ] **der; ~s, ~:** minister (**für** for); *(eines britischen Hauptministeriums)* Secretary of State (**für** for); *(eines amerikanischen Hauptministeriums)* Secretary (**für** of)

ministeriell [minɪste'ri̯ɛl] **1.** *Adj.* ministerial. **2.** *adv.* by the minister

Minịsterin die; ~, ~nen *s.* **Minister**

Ministerium [minɪs'te:ri̯ʊm] **das; ~s, Ministerien** Ministry; Department *(Amer.)*

Minịster·präsident der a) *(eines deutschen Bundeslandes)* minister-president; prime minister *(Brit.)*; governor *(Amer.)*; **b)** *(Premierminister)* Prime Minister

Ministrant [minɪs'trant] **der; ~en, ~en** *(kath. Kirche)* server

Minna ['mɪna] **die; ~, ~s** *(ugs. veralt.)* maid; **jmdn. zur ~ machen** *(ugs.)* tear sb. off a strip *(Brit. sl.)*; bawl out *(coll.)*; **eine grüne ~** *(ugs.)* a Black Maria; a patrol wagon *(Amer.)*

Minorität [minori'tɛ:t] **die; ~, ~en** *s.* **Minderheit**

minus ['mi:nʊs] **1.** *Konj.* minus. **2.** *Adv.* **a)** minus; **~ fünf Grad, fünf Grad ~:** minus five degrees; five degrees below [zero]; **b)** *(Elektrot.)* negative. **3.** *Präp. mit Gen.* *(Kaufmannsspr.)* less; minus

Mịnus das; ~ a) *(Fehlbetrag)* deficit; *(auf einem Konto)* overdraft; **~ machen** make a loss; **im ~ sein** be in debit; be in the red; **b)** *(Nachteil)* minus; drawback; *(im Beruf)* disadvantage

Mịnus-: ~pol der negative pole; *(einer Batterie)* negative terminal; **~punkt der a)** minus *or* penalty point; **b)** *(Nachteil)* disadvantage; **~zeichen das** minus sign

Minute [mi'nu:tə] **die; ~, ~n** minute; **es ist neun Uhr [und] sieben ~n** it is seven minutes past nine *or* nine seven; **hast du ein paar ~n Zeit für mich?** can you spare me a few minutes *or* moments?; **in letzter ~:** at the last minute *or* moment; **auf die ~ pünktlich** punctual to the minute

minuten·lang 1. *Adj.; nicht präd.* ‹*applause, silence, etc.*› lasting [for] several minutes. **2.** *adv.* for several minutes

Minuten·zeiger der minute-hand

-minütig *Adj.* **ein fünf~er Heulton** a wail lasting five minutes; **eine fünfzehn~e Verspätung** a fifteen-minute delay

minuziös [minu'tsi̯ø:s] *(geh.)* **1.** *Adj.* minutely *or* meticulously precise *or* detailed ‹*account, description*›; minute ‹*detail*›; ‹*manœuvre*› requiring minute precision. **2.** *adv.* meticulously

Minze ['mɪntsə] **die; ~, ~n** mint

Mio. *Abk.* **Million[en]** m.

mir [mi:ɐ̯] **1.** *Dat. Sg. des Personalpron. ich* **a)** to me; *(nach Präpositionen)* me; **gib es ~:** give it to me; give me it; **Freunde von ~:** friends of mine; **gehen wir zu ~:** let's got to my place; **~ nichts, dir nichts** *(ugs.)* just like that; without so much as a 'by your leave'; **von ~ aus** as far as I'm concerned; **b) geht ~ nicht an meinen Schreibtisch!** keep away from my desk!; **und grüß ~ alle Verwandten!** and give my regards to all the relatives; **du bist ~ vielleicht einer!** *(ugs.)* a fine one you are!. **2.** *Dat. des Reflexivpron. der 1. Pers. Sg.* myself; **ich habe ~ gedacht, daß ...:** I thought that ...; **ich will ~ ein neues Kleid kaufen** I want to buy myself a new dress

Mirabelle [mira'bɛlə] **die; ~, ~n** mirabelle

Misanthrop [mizan'tro:p] **der; ~en, ~en** *(geh.)* misanthrope

Mịsch-: ~brot das bread made from wheat and rye flour; **~ehe die** mixed marriage

mischen ['mɪʃn̩] **1.** *tr. V.* mix; **etw. in etw.** *(Akk.)* **~:** put sth. into sth.; **Wasser und Wein ~:** mix water with wine; **die Karten ~:** shuffle the cards. **2.** *refl. V.* **a)** *(sich ver~)* mix (**mit** with); ‹*smell, scent*› blend (**mit** with); **in meine Freude mischte sich Angst** my joy was mingled with fear; **b)** *(sich ein~)* **sich in etw.** *(Akk.)* **~:** interfere *or* meddle in sth.; **c)** *(sich begeben)* **sich unters Publikum** *usw.* **~:** mingle with the audience *etc.* **3.** *itr. V. (Kartenspiel)* shuffle; *s. auch* **gemischt**

Mịscher der; ~s, ~ *(Bauw.)* [cement-]mixer

Mịsch-: ~farbe die non-primary colour; **~form die** mixture; **~gewebe das** mixture

Mischling ['mɪʃlɪŋ] **der; ~s, ~e** half-caste; half-breed

Mischmasch ['mɪʃmaʃ] **der; ~[e]s, ~e** *(ugs., meist abwertend)* hotchpotch; mishmash

Mịsch-: ~maschine die *(Bauw.)* cement-mixer; **~pult das** *(Film, Rundf., Ferns.)* mixing desk *or* console

Mịschung die; ~, ~en *(auch fig.)* mixture; *(Tee~, Kaffee~, Tabak~)* blend; *(Pralinen~)* assortment

Mịschungs·verhältnis das proportion in the mixture

Mịsch·wald der mixed [deciduous and coniferous] forest

miserabel [mizə'ra:bl̩] *(ugs.)* **1.** *Adj.* **a)** *(schlecht)* dreadful *(coll.)*, atrocious ‹*film, food*›; pathetic, miserable ‹*achievement*›; miserable, dreadful *(coll.)*, atrocious ‹*weather*›; **b)** *(elend)* miserable; wretched; **ich fühle mich ~:** I feel dreadful; **c)** *(niederträchtig)* abominable ‹*behaviour*›. **2.** *adv.* **a)** *(schlecht)* dreadfully *(coll.)*; atrociously; ‹*sleep*› dreadfully badly *(coll.)*; ~ **bezahlt werden** be very badly *or* poorly paid; **b)** *(elend)* **ihm geht es gesundheitlich** ~: he's in a bad way; **c)** *(niederträchtig)* abominably

Misere [mi'ze:rə] **die**; ~, ~**n** *(geh.)* wretched *or* dreadful state

Mispel ['mɪspl̩] **die**; ~, ~**n** medlar

miß [mɪs] *Imperativ Sg. v.* **messen**

mißachten *tr. V.* disregard; ignore

[1]**Miß·achtung die** disregard

[2]**Miß·achtung die** *(Geringschätzung)* disdain; contempt

Miß·behagen das [feeling of] unease; uncomfortable feeling

Miß·bildung die deformity

mißbilligen *tr. V.* disapprove of

Miß·billigung die disapproval

Miß·brauch der abuse; misuse; *(falsche Anwendung)* misuse; *(von Feuerlöscher, Notbremse)* improper use

mißbrauchen *tr. V.* abuse; misuse; abuse ⟨*trust*⟩; **jmdn. für** *od.* **zu etw.** ~: use sb. for sth.

mißbräuchlich [-brɔyçlɪç] **1.** *Adj.* ~**e Verwendung/Anwendung** misuse. **2.** *adv.* **etw.** ~ **verwenden/handhaben** misuse sth.

mißdeuten *tr. V.* misinterpret

missen *tr. V. (geh.)* do *or* go without; do without ⟨*person*⟩; **jmdn./etw. nicht** ~ **mögen** not want to be without sb./sth.

Miß·erfolg der failure

Miß·ernte die crop failure

Misse·tat ['mɪsə-] **die** *(geh. veralt.)* misdeed

Misse·täter der, Misse·täterin die *(geh. veralt.)* malefactor

mißfallen *unr. itr. V.* **etw. mißfällt jmdm.** sb. dislikes *or* does not like sth.

Mißfallen das; ~**s** displeasure (**über** + *Akk.* at); *(Mißbilligung)* disapproval (**über** + *Akk.* of); **jmds.** ~ **erregen** incur sb.'s displeasure/disapproval

Mißfallens·äußerung die expression of displeasure/disapproval

miß·gebildet *Adj.* deformed

Miß·geburt die *(Med.)* monster; monstrosity

Miß·geschick das mishap; *(Pech)* bad luck; *(Unglück)* misfortune; **jmdm. passiert ein** ~: sb. has a mishap/a piece *or* stroke of bad luck/a misfortune

miß·gestaltet *Adj.* misshapen; deformed ⟨*person, child*⟩

miß·glücken *itr. V.; mit sein s.* **mißlingen**

mißgönnen *tr. V.* **jmdm. etw.** ~: begrudge sb. sth.

Miß·griff der error of judgement

Miß·gunst die [envy and] resentment (**gegenüber** of)

miß·günstig 1. *Adj.* resentful. **2.** *adv.* resentfully

mißhandeln *tr. V.* maltreat; ill-treat

Miß·handlung die maltreatment; ill-treatment; ~**en** maltreatment *sing.*; ill-treatment *sing.*

Mission [mɪ'si̯o:n] **die**; ~, ~**en a)** *(geh.: Auftrag)* mission; **in geheimer** ~: on a secret mission; **b)** *o. Pl. (Rel.)* mission; **in der [äußeren/inneren]** ~ **tätig sein** do missionary work [abroad/in one's own country]; **c)** *(geh.: diplomatische Vertretung)* mission

Missionar [mɪsi̯o'na:ɐ̯] **der**; ~**s**, ~**e**, **Missionarin die**; ~, ~**nen** missionary

missionarisch *Adj.* missionary

missionieren 1. *itr. V.* do missionary work. **2.** *tr. V.* convert by missionary work; *(fig.)* convert to one's own ideas

Missionierung die; ~: **die** ~ **eines Landes/eines Volkes** missionary work in a country/among a people; *(Bekehrung)* the conversion of a country/people [by missionary work]

Miß·kredit der *in* **jmdn./etw. in** ~ **bringen** bring sb./sth. into discredit; bring discredit on sb./sth.

mißlang *1. u. 3. Pers. Sg. Prät. v.* **mißlingen**

mißlich *Adj. (geh.)* awkward, difficult ⟨*situation*⟩; difficult ⟨*conditions*⟩; unfortunate ⟨*incident*⟩

mißliebig ['mɪsli:bɪç] *Adj.* unpopular; ~**e Ausländer** unwanted foreigners

mißlingen [mɪs'lɪŋən] *unr. itr. V.; mit sein* fail; be unsuccessful; be a failure; **ein mißlungener Versuch** a failed *or* unsuccessful attempt

mißlungen [mɪs'lʊŋən] *2. Part. v.* **mißlingen**

miß·mutig 1. *Adj.* bad-tempered; sullen ⟨*face*⟩; **warum bist du heute so** ~**?** why are you in such a bad mood today? **2.** *adv.* bad-temperedly

mißraten *unr. itr. V.; mit sein* ⟨*cake, photo, etc.*⟩ turn out badly; **ein** ~**es Kind** a child who has turned out badly

Miß·stand der deplorable state of affairs *no pl.*; *(Übel)* evil; *(üble Praktiken)* abuse; **die Mißstände im Bildungswesen** the deplorable state of education

mißt *2. u. 3. Pers. Sg. Präsens v.* **messen**

Miß·ton der discordant note; *(fig.)* note of discord; discordant note

mißtrauen *itr. V.* **jmdm./einer Sache** ~: mistrust *or* distrust sb./sth.

Mißtrauen das; ~**s** mistrust, distrust (**gegen** of); **voll[er]** ~: extremely mistrustful *or* distrustful (**gegen** of)

Mißtrauens-: ~**antrag der** motion of no confidence; ~**votum das** vote of no confidence

mißtrauisch ['mɪstrau̯ɪʃ] **1.** *Adj.* mistrustful; distrustful; *(argwöhnisch)* suspicious. **2.** *adv.* mistrustfully; distrustfully; *(argwöhnisch)* suspiciously

Miß·verhältnis das disparity; *(an Größe)* disproportion

miß·verständlich 1. *Adj.* unclear; ⟨*formulation, concept, etc.*⟩ that could be misunderstood; ~ **sein** be liable to be misunderstood. **2.** *adv.* ⟨*express oneself, describe*⟩ in a way that could be misunderstood

Miß·verständnis das misunderstanding

miß|verstehen[1] *unr. tr. V.* misunderstand

Miß·wahl die *contest for the title of 'Miss Europe', 'Miss World' etc.*

Miß·wirtschaft die mismanagement

Mist [mɪst] **der**; ~**[e]s a)** dung; *(Dünger)* manure; *(mit Stroh usw. gemischt)* muck; *(~haufen)* dung/manure/muck heap; **das ist nicht auf ihrem** ~ **gewachsen** *(fig. ugs.)* that didn't come out of her own head; **b)** *(ugs. abwertend) (Schund)* rubbish, junk, trash *all no indef. art.*; *(Unsinn)* rubbish, nonsense, *(sl.)* rot *all no indef. art.*; *(lästige, dumme Angelegenheit)* nonsense; ~ **bauen** make a mess of things; mess things up; **mach bloß keinen** ~! just don't do anything stupid

Mistel ['mɪstl̩] **die**; ~, ~**n** mistletoe

Mistel·zweig der piece of mistletoe

Mist-: ~**forke die** *(nordd.)*, ~**gabel die** dung-fork; ~**haufen der** dung/manure/muck heap

mistig *(salopp)* **1.** *Adj.* rotten *(sl.)*. **2.** *adv.* in a rotten way *(sl.)*

Mist-: ~**käfer der** dung-beetle; ~**stück das** *(derb)* lousy good-for-nothing bastard *(sl.)*; *(Frau)* lousy good-for-nothing bitch *(sl.)*; ~**wetter das** *(salopp)* lousy weather *(sl.)*

[1] *ich mißverstehe, mißverstanden, mißzuverstehen*

mit [mɪt] **1.** *Präp. mit Dat.* **a)** *(Gemeinsamkeit, Beteiligung)* with; **b)** *(Zugehörigkeit)* with; **ein Haus ~ Garten** a house with a garden; **Herr Müller ~ Frau** Herr Müller and his wife; **c)** *(einschließlich)* with; including; **ein Zimmer ~ Frühstück** a room with breakfast included; **d)** *(Inhalt)* **ein Sack ~ Kartoffeln/Glas ~ Marmelade** a sack of potatoes/ pot of jam; **e)** *(Begleitumstände)* with; **etw. ~ Absicht tun/~ Nachdruck fordern** do sth. deliberately/demand sth. forcefully; **~ 50 [km/h] fahren** drive at 50 [k.p.h]; **f)** *(Hilfsmittel)* with; **~ der Bahn/dem Auto fahren** go by train/car; **~ der Fähre/„Hamburg"** on the ferry/the 'Hamburg'; **g)** *(allgemeiner Bezug)* with; **~ der Arbeit ging es recht langsam voran** the work went very slowly; **~ einer Tätigkeit beginnen/aufhören** take up/give up an occupation; **raus/fort ~ dir!** out/off you go!; **h)** *(zeitlich)* **~ Einbruch der Dunkelheit/Nacht** when darkness/night falls/fell; **~ 20 [Jahren]** at [the age of] twenty; **~ der Zeit/den Jahren** in time/as the years go/went by; **i)** *(gleichlaufende Bewegung)* with; **~ dem Strom/Wind** with the tide/wind. **2.** *Adv.* **a)** *(auch)* too; as well; **~ dabeisein** be there too; *s. auch* **Partie f; b)** *(neben anderen)* also; too; as well; **es lag ~ an ihm** it was partly his doing; **c)** *(ugs.)* **~ das wichtigste der Bücher** one of the most important of the books; **seine Arbeit war ~ am besten** his work was among the best; **d)** *(vorübergehende Beteiligung)* **ihr könntet ruhig einmal ~ anfassen** it wouldn't hurt you to lend a hand just for once; **e)** *s. auch* **damit 1c; womit b**

Mit·angeklagte der/die co-defendant; *(mit geringerer Strafandrohung)* defendant to a lesser charge

Mit·arbeit die; *o. Pl.* **a)** *(das Tätigsein)* collaboration (**bei, an** + *Dat.* on); **die ~ in der Praxis ihres Mannes** working in her husband's practice; **b)** *(Mithilfe)* assistance (**bei, in** + *Dat.* in); **seine zwanzigjährige ~ in der Organisation** his twenty years of service to the organization; **c)** *(Beteiligung)* participation (**in** + *Dat.* in)

mit|arbeiten *itr. V.* **a) bei einem Projekt/an einem Buch ~:** collaborate on a project/book; **im elterlichen Geschäft ~:** work in one's parents' shop; **b)** *(sich beteiligen)* participate (**in** + *Dat.* in); **im Unterricht besser ~:** take a more active part in lessons

Mit·arbeiter der a) *(Betriebsangehörige[r])* employee; **b)** *(bei einem Projekt, an einem Buch)* collaborator; **ein freier ~:** a freelance; a freelance worker

mit|bekommen *unr. tr. V.* **a) etw. ~:** be given *or* get sth. to take with one; *(fig.)* inherit sth.; **b)** *(wahrnehmen)* be aware of; *(durch Hören/Sehen)* hear/see; **es war so laut, daß ich nur die Hälfte mitbekam** it was so noisy that I only caught half of it; **c)** *(verstehen)* **ich war so müde, daß ich nicht viel ~ habe** I was so tired that I did not grasp very much

mit|benutzen *tr. V.* share; have the use of

Mit·besitzer der joint owner; co-owner

mit|bestimmen 1. *itr. V.* have a say (**in** + *Dat.* in). **2.** *tr. V.* have an influence on

Mit·bestimmung die; *o. Pl.* participation (**bei** in); *(der Arbeitnehmer)* co-determination

Mit·bestimmungs·recht das right of co-determination

Mit·bewerber der fellow applicant; *(Wirtsch.)* competitor; **ich hatte nur einen ~ um diese Stelle** there was only one other applicant for the job [besides me]

Mit·bewohner der fellow resident; *(in Wohnung)* flatmate

mit|bringen *unr. tr. V.* **a) etw. ~:** bring sth. with one; **etw. aus der Stadt/dem Urlaub/von dem Markt/der Reise ~:** bring sth. back from town/holiday/the market/one's trip; **jmdm./sich etw. ~:** bring sth. with one for sb./bring sth. back for oneself; **Gäste ~:** bring guests home; **b)** *(fig.: haben)* have, possess ‹*ability, gift, etc.*› (**für** for); **genügend Zeit ~:** come with enough time at one's disposal; *s. auch* **Laune a**

Mitbringsel [-brɪŋzl] **das; ~s, ~:** [small] present; *(Andenken)* [small] souvenir

Mit·bürger der fellow citizen; **ältere ~** *(Amtsspr.)* senior citizens

mit|denken *unr. itr. V.* follow [the argument/explanation/what is being said *etc.*]

mit|dürfen *unr. itr. V. (ugs.) (mitkommen dürfen)* be allowed to come along *or* too; *(mitgehen, mitfahren dürfen)* be allowed to go along *or* too

mit·einander *Adv.* **a)** with each other *or* one another; **~ sprechen/kämpfen** talk to each other *or* one another/fight with each other *or* one another; **b)** *(gemeinsam)* together; **ihr seid Gauner, alle ~!** you are all a pack of rogues!; you're all rogues, the lot of you!

Mit·einander das; ~[s] living and working together *no art.*

mit|erleben *tr. V.* **a)** witness ‹*events etc.*›; **b) er hat den Krieg noch miterlebt** he was still alive during the war

mit|essen *unr. tr. V.* eat ‹*skin etc.*› as well

Mit·esser der *(Pickel)* blackhead

mit|fahren *unr. itr. V.; mit sein* **bei jmdm. [im Auto] ~:** go with sb. [in his/her car]; *(auf einer Reise)* travel with sb. [in his/her car]; *(mitgenommen werden)* get *or* have a lift with sb. [in his/her car]

Mit·fahrer der fellow passenger; *(vom Fahrer aus gesehen)* passenger

Mitfahr·gelegenheit die lift

mit·fühlend 1. *Adj.* sympathetic. **2.** *adv.* sympathetically

mit|führen *tr. V.* **a)** *(Amtsspr.: bei sich tragen)* **etw. ~:** carry sth. [with one]; **b)** *(transportieren)* ‹*river, stream*› carry along

mit|geben *unr. tr. V.* **jmdm. etw. ~:** give sb. sth. to take with him/ her; **jmdm. eine gute Erziehung ~geben** *(fig.)* provide sb. with a good education

Mit·gefangene der/die fellow prisoner

Mit·gefühl das; *o. Pl.* sympathy

mit|gehen *unr. itr. V.; mit sein* **a)** go too; **mit jmdm. ~:** go with sb.; **etw. ~ lassen** *(fig. ugs.)* walk off with sth. *(coll.)*; pinch sth. *(sl.)*; **b)** *(sich mitreißen lassen)* **begeistert/ enthusiastisch ~:** respond enthusiastically (**bei, mit** to)

mit·genommen *Adj.* worn-out ‹*furniture, carpet*›; **~ sein/aussehen** ‹*book etc.*› be/look to be in a sorry state; *(fig.)* ‹*person*› be/look worn out

Mit·gift die; ~, ~en *(veralt.)* dowry

Mit·glied das member (*Gen.*, **in** + *Dat.* of); **~ im Ausschuß sein** be a member of *or* sit on the committee; **„Zutritt nur für ~er"** 'members only'

Mitglieder·versammlung die general meeting

Mitglieds-: ~ausweis der membership card; **~beitrag der** membership subscription; **~staat der** member state *or* country

mit|haben *unr. tr. V. (ugs.)* **etw. ~:** have got sth. with one

mit|halten *unr. itr. V.* keep up (**bei** in, **mit** with)

mit|helfen *unr. itr. V.* help (**bei, in** + *Dat.* with); **beim Bau der Garage ~:** help to build the garage

Mit·hilfe die; *o. Pl.* help; assistance

mit|hören 1. *tr. V.* listen to; *(zufällig)* overhear ‹*conversation, argument, etc.*›; *(abhören)* listen in on. **2.** *itr. V.* listen; *(zufällig)* overhear; *(jmdn. abhören)* listen in

Mit·inhaber der joint owner; co-owner; *(einer Firma, eines Restaurants auch)* joint proprietor

mit|kommen *unr. itr. V.; mit sein* **a)** come too; **kommst du mit?** are you coming [with me/us]?; **ich kann nicht ~:** I can't come; **bis zur Tür ~:** come with sb. to the door; **b)** *(Schritt halten)* keep up; **in der Schule/im Unterricht gut/schlecht ~:** get on well/badly at school/with one's lessons; **da komme ich nicht mehr mit!** *(fig. ugs.)* I can't understand it at all

mit|kriegen *tr. V. (ugs.) s.* **mitbekommen**

mit|laufen *unr. itr. V.; mit sein* **a) mit jmdm. ~:** run with sb.; **b)** *(Sport)* **beim 100-m-Lauf** *usw.* **~:** run in the 100 m. *etc.;* **c) ein Tonband ~ lassen** have a tape recorder running

Mit·läufer der *(abwertend)* [mere] supporter

Mit·laut der consonant

Mit·leid das pity, compassion (**mit** for); *(Mitgefühl)* sympathy (**mit** for); **mit jmdm. ~ haben** *od.* **empfinden** feel pity *or* compassion/have *or* feel sympathy for sb.

Mit·leidenschaft die: jmdn./etw. in ~ ziehen affect sb./sth.

mitleid·erregend *Adj.* pitiful

mit·leidig 1. *Adj.* compassionate; *(mitfühlend)* sympathetic. **2.** *adv.* compassionately; *(mitfühlend)* sympathetically; *(iron.)* pityingly

mit|machen 1. *tr. V.* **a)** go on ‹*trip*›; join in ‹*joke*›; follow ‹*fashion*›; fight in ‹*war*›; do ‹*course, seminar*›; **b)** *(ugs.: billigen)* **das mache ich nicht mit** I can't go along with it; **ich mache das nicht länger mit!** I'm not standing for it any longer; **c)** *(ugs.: zusätzlich erledigen)* **jmds. Arbeit ~:** do sb.'s work as well as one's own; **d)** *(ugs.: erleiden)* **zwei Weltkriege/viele Bombenangriffe mitgemacht haben** have been through two world wars/many bomb attacks. **2.** *itr. V.* **a)** *(sich beteiligen)* take part (**bei** in); **willst du ~?** do you want to join in?; **b)** *(ugs.)* **meine Beine machen nicht mehr mit** my legs are giving up on me *(coll.)*

Mit·mensch der fellow man; fellow human being

mit|mischen *itr. V. (ugs.)* be involved (**bei** in); **er will auch ~:** he wants to get involved, too

mit|nehmen *unr. tr. V.* **a) jmdn. ~:** take sb. with one; **etw. ~:** take sth. with one; *(verhüll.: stehlen)* walk off with sth. *(coll.)*; *(kaufen)* take sth.; **etw. wieder ~:** take sth. away [with one] again; **das Frachtschiff nimmt auch Passagiere mit** the cargo ship also carries passengers; **Essen/Getränke zum Mitnehmen** food/drinks to take away *or (Amer.)* to go; **jmdn. im Auto ~:** give sb. a lift [in one's car]; **b)** *(ugs.: streifen)* **der LKW hat die Hecke mitgenommen** the truck *or (Brit.)* lorry took the hedge with it; **c)** *(fig. ugs.: nicht verzichten auf)* do *(coll.)* ‹*sights etc.*›; **auch Soho ~:** take in Soho as well; **d)** *(in Mitleidenschaft ziehen)* **jmdn. ~:** take it out of sb.; **von etw. mitgenommen sein** be worn out by sth.; *(traurig gemacht)* be grieved by sth.

mit|rechnen 1. *itr. V.* work the sum out at the same time. **2.** *tr. V.* **etw. ~:** include sth. [in the calculation]

mit|reden *itr. V.* **a)** join in the conversation; **b)** *(mitbestimmen)* have a say

Mit·reisende der/die fellow passenger

mit|reißen *unr. tr. V.* **a)** ‹*avalanche, flood*› sweep away; **b)** *(begeistern)* **seine Rede hat alle Zuhörer mitgerissen** the audience was carried away by his speech; **die ~de Musik** the rousing music

mit·samt *Präp. mit Dat.* together with; **die ganze Familie ~ Hund und Katze** the whole family, complete with cat and dog

mit|schleppen *tr. V. (ugs.) (tragen)* **etw. ~:** lug *or (sl.)* cart sth. with one

mit|schneiden *unr. tr. V.* record [live]

mit|schreiben 1. *unr. tr. V.* **etw. ~:** take sth. down. **2.** *unr. itr. V.* write *or* take down what is/was said; *(in Vorlesungen usw.)* take notes

Mit·schuld die share of the blame *or* responsibility (**an** + *Dat.* for); *(an Verbrechen)* complicity (**an** + *Dat.* in)

mit·schuldig *Adj.* **an etw.** *(Dat.)* **~ sein/werden** be/become partly to blame *or* partly responsible for sth.; *(an Verbrechen)* be/become guilty of complicity in sth.; **sich ~ machen** put oneself in the position of being partly to blame *or* partly responsible for sth.; *(an Verbrechen)* become guilty of complicity as a result of one's own actions

Mit·schüler der, Mit·schülerin die schoolfellow

mit|sein *unr. itr. V.; mit sein; Zusammenschreibung nur im Inf. und Part. (ugs.)* **er ist beim letzten Ausflug nicht mitgewesen** he didn't come [with us] on our last trip; **waren eure Kinder im Urlaub mit?** did your children go on holiday with you?; **warst du auch mit im Konzert?** were you at the concert too?

mit|singen 1. *unr. tr. V.* join in ‹*song etc.*›. **2.** *unr. itr. V.* join in [the singing]; sing along

mit|spielen *itr. V.* **a)** join in the game; **wenn das Wetter mitspielt** *(fig.)* if the weather is kind; **b)** *(mitwirken)* **in einem Film/bei einem Theaterstück ~:** be *or* act in a film/play; **in einem Orchester ~:** play in an orchestra; **c)** *(sich auswirken)* play a part (**bei** in); **d)** *(zusetzen)* **jmdm. übel** *od.* **böse ~:** ‹*authorities*› treat sb. badly; ‹*opponent*› give sb. a rough time

Mit·spieler der, Mit·spielerin die player; *(in derselben Mannschaft)* team-mate

Mitsprache·recht das; *o. Pl.* **ein/kein ~ bei etw. haben** have a say/no say in sth.

mit|sprechen 1. *unr. tr. V.* join in [saying]. **2.** *unr. itr. V.: s.* **mitreden**

mittag ['mɪta:k] *Adv.* **heute/morgen/gestern ~:** at midday *or* lunch-time today/tomorrow/yesterday; **Montag ~:** at midday on Monday; Monday at midday; Monday lunch-time; **seit Montag ~:** since Monday midday; **was gibt es heute ~ zu essen?** what's for lunch today?

[1]**Mittag der; ~s, ~e a)** midday *no art.*; **gegen ~:** around midday *or* noon; **über ~:** at midday *or* lunch-time; **zu ~ essen** have lunch; **b) ~ machen** *(ugs.)* take one's lunch-hour *or* lunch-break

[2]**Mittag das; ~s** *(ugs.)* lunch; **~ essen** have lunch

Mittag·essen das lunch; midday meal; **beim ~ sitzen** be having [one's] lunch *or* one's midday meal

mittäglich ['mɪtɛ:klɪç] **1.** *Adj.; nicht präd.* midday; lunch-time ⟨*invitation*⟩. **2.** *adv.* at midday *or* lunch-time

mittags ['mɪta:ks] *Adv.* at midday *or* lunch-time; **12 Uhr ~:** 12 noon; 12 o'clock midday; **Dienstag ~** *od.* **dienstags ~:** Tuesday lunch-time

Mittags-: ~glut die, ~hitze die midday *or* noonday heat; heat of midday; **~pause die** lunch-hour; lunch-break; **~ruhe die** period of quiet after lunch; **~schlaf der** after-lunch sleep; **~sonne die** midday *or* noonday sun; **~tisch der: am ~tisch sitzen** be sitting at the table having lunch; **~zeit die** *o. Pl. (Zeit gegen 12 Uhr)* lunch-time *no art.*; midday *no art.*

Mit·täter der accomplice

Mitte ['mɪtə] **die; ~, ~n a)** middle; *(Punkt)* middle; centre; *(eines Kreises, einer Kugel, Stadt)* centre; **wir nahmen sie in die ~:** we had her between us; **die goldene ~** *(fig.)* the golden mean; **ab durch die ~!** *(fig. ugs.)* off you go; **b)** *(Zeitpunkt)* middle; **~ des Monats/Jahres** in the middle of the month/year; **~ Februar** in mid-February; in the middle of February; **er ist ~ [der] Dreißig** he's in his mid-thirties; **c)** *(Politik)* centre; **d) wir haben sie wieder in unserer ~ begrüßt** we welcomed her back into our midst *or* amongst us

mit|teilen *tr. V.* **jmdm. etw. ~:** tell sb. sth.; *(informieren)* inform sb. of *or* about sth.; communicate sth. to sb. *(formal)*; *(amtlich)* notify *or* inform sb. of sth.; **er teilte mit, daß ...** *(gab bekannt)* he announced that ...

mitteilsam *Adj.* communicative; *(gesprächig)* talkative

Mit·teilung die communication; *(Bekanntgabe)* announcement; **jmdm. eine vertrauliche ~ machen** give sb. confidential information; **ich muß dir eine traurige ~ machen** I have some sad news for you

Mitteilungs·bedürfnis das need to talk [to others]

Mittel ['mɪtl̩] **das; ~s, ~ a)** means *sing.*; *(Methode)* way; method; *(Werbe~, Propaganda~, zur Verkehrskontrolle)* device (+ *Gen.* for); **mit allen ~n versuchen, etw. zu tun** try by every means to do sth.; **[nur] ~ zum Zweck sein** be [just] a means to an end; **~ und Wege suchen/finden** look for/find ways and means; **b)** *(Arznei)* **ein ~ gegen Husten/Schuppen** *usw.* a remedy *or* cure for coughs *pl.*/dandruff *sing. etc.*; **c)** *(Substanz)* **ein ~ gegen Ungeziefer/Insekten** a pesticide/an insect repellent; **d)** *Pl. (Geld~)* funds; [financial] resources; *(Privat~)* means; resources; **mit öffentlichen ~n** from public funds

mittel-, Mittel-: ~alter das; *o. Pl.* Middle Ages *pl.*; **das sind Zustände wie im ~alter** *(ugs.)* it's positively medieval; **~alterlich** *Adj.* medieval; **~amerika (das)** Central America [and the West Indies]

mittelbar 1. *Adj.* indirect. **2.** *adv.* indirectly

mittel-, Mittel-: ~ding das; *o. Pl.* **ein ~ding zwischen Moped und Fahrrad** something between a moped and a bicycle; **~europa (das)** Central Europe; **~feld das a)** *(Fußball)* midfield; **b)** *(Sport: im Wettbewerb)* **im ~feld sein** be in the pack; *(in der Tabelle)* be in mid-table; **~feld·spieler der** *(Fußball)* midfield player; **~finger der** middle finger; **~fristig** [-frɪstɪç] **1.** *Adj.* medium-term ⟨*solution, financial planning*⟩; **2.** *adv.* **[etw.] ~fristig planen** plan [sth.] on a medium-term basis; **~gebirge das** low-mountain region; low mountains *pl.*; **~groß** *Adj.* medium-sized; ⟨*person*⟩ of medium height; **~klasse die** middle range; *(Größenklasse)* middle [size-]range; **~klasse·wagen der** car in the middle range; *(hinsichtlich der Größe)* medium-sized car; **~kreis der** *(Ballspiele)* centre circle; **~linie die** centre line; *(Fußball)* halfway line; **~los** *Adj.* without means *postpos.*; penniless; *(arm)* poor; *(verarmt)* impoverished; **~maß das: gutes ~maß sein** be a good average; **~mäßig** *(oft abwertend)* **1.** *Adj.* mediocre; indifferent; indifferent ⟨*weather*⟩; **2.** *adv.* indifferently; **~mäßigkeit die** *(oft abwertend)* mediocrity; **~meer das** Mediterranean [Sea]

Mittelmeer-: ~länder *Pl.* Mediterranean countries; **~raum der** Mediterranean [area]

mittel-, Mittel-: ~ohr·entzündung die *(Med.)* inflammation of the middle ear; **~prächtig** *(ugs. scherzh.) Adj.* **[nur] ~prächtig** not particularly marvellous; **~punkt der a)** centre; *(einer Strecke)* midpoint; **b)** *(Mensch/Sache im Zentrum)* centre *or* focus of attention; **ein kultureller ~punkt** a cultural centre; **etw. in den ~punkt stellen** focus on sth.

mittels *Präp. mit Gen. (Papierdt.)* by means of

Mittel·scheitel der centre parting

Mittels-: ~mann der *Pl.* **~männer** *od.* **~leute, ~person die** intermediary; go-between

mittel-, Mittel-: ~stand der; *o. Pl.* middle class; **~ständisch** *Adj.* middle-class; medium-sized ⟨*firm*⟩ *(in private ownership)*; **~streifen der** central reservation; median strip *(Amer.)*; **~stürmer der** *(Sport)* centre-forward; **~weg der** middle course; **der goldene ~weg** the happy medium; **~welle die** *(Physik, Rundf.)* medium wave; **~wert der** mean [value]

mitten *Adv.* **~ an/auf etw.** *(Akk./Dat.)* in the middle of sth.; **der Teller brach ~ durch** the plate broke in half; **~ in etw.** *(Akk./Dat.)* into/in the middle of sth.; **~ durch die Stadt** right through the town; **~ unter uns** *(Dat.)* in our midst; **der Schuß traf ihn ~ ins Herz** the shot hit him right in the heart; **~ im Pazifik** in mid-Pacific; **~ in der Aufregung** in the midst of the excitement

mitten-: ~drin *Adv.* [right] in the middle; **~durch** *Adv.* [right] through the middle

Mitter·nacht ['mɪtɐ-] **die;** *o. Pl.* midnight *no art.*

mitter·nächtlich *Adj.; nicht präd.* midnight; **zu ~er Stunde** at midnight

Mitt·fünfziger der man in his mid-fifties

mittler... ['mɪtlər...] *Adj.; nicht präd.* **a)** middle; **der/die/das ~e** the middle one; **die ~e Reife** *(Schulw.) standard of achievement for school-leaving certificate at a Realschule or for entry to the sixth form in a Gymnasium; s. auch* **Osten c**; **b)** *(einen Mittelwert darstellend)* average ⟨*temperature*⟩; moderate ⟨*speed*⟩; medium-sized ⟨*company, town*⟩; medium ⟨*quality, size*⟩; **ein Mann ~en Alters** a middle-aged man

Mittler ['mɪtlɐ] **der; ~s, ~:** mediator

Mittler·rolle die mediating role

mittler·weile ['mɪtlɐ'vaj̲lə] *Adv.* **a)** *(seitdem, allmählich)* since then; *(bis jetzt)* by now; **b)** *(unterdessen)* in the mean time

mit|tragen *unr. tr. V.* bear part of, share ⟨*responsibility, cost*⟩; take part of, share ⟨*blame*⟩

mit|trinken 1. *unr. tr. V.* **etw. ~:**

drink sth. with me/us *etc.*; **trinkst du einen mit?** are you going to have a drink with me/us *etc.?*

Mitt·sommernacht die midsummer's night; *(zur Sommersonnenwende)* Midsummer Night

Mittwoch ['mɪtvɔx] **der; ~[e]s, ~e** Wednesday; *s. auch* **Dienstag; Dienstag-**

mittwochs *Adv.* on Wednesday[s]; *s. auch* **dienstags**

mit·unter *Adv.* now and then; from time to time; sometimes

mit·verantwortlich *Adj.* partly responsible *pred.*; *(beide/alle zusammen)* jointly responsible *pred.*

Mit·verantwortung die share of the responsibility

mit|verdienen *itr. V.* go out to work as well

mit|versichern *tr. V.* include in one's insurance

mit·wirken *itr. V.* **an etw.** *(Dat.)***/bei etw. ~**: collaborate on/be involved in sth.

Mitwirkende der/die *adj. Dekl.* *(an einer Sendung)* participant; *(in einer Show)* performer; *(in einem Theaterstück)* actor

Mit·wirkung die; *o. Pl. s.* **mitwirken:** collaboration; involvement

Mit·wisser der; ~s ~, Mit·wisserin die; ~, ~nen: er hatte zu viele ~: there were too many people who knew about what he'd done

Mitwisserschaft die; ~: knowledge of the matter/crime

mit|wollen *unr. itr. V. (ugs.) (mitkommen wollen)* want to come with sb.; *(mitgehen, mitfahren wollen)* want to go with sb.

mit|zählen 1. *itr. V.* count; **die Sonntage zählen bei den Urlaubstagen nicht mit** Sundays don't count as holidays. **2.** *tr. V.* count in; include

mit|ziehen *unr. itr. V.; mit sein* **a)** *(mitgehen)* go with him/them *etc.*; **b)** *(ugs.: mitmachen)* go along with it; *(bei einer Klage, Initiative)* give it one's backing

Mix·becher ['mɪks-] **der** [cocktail-]shaker

mixen ['mɪksn̩] *tr. V.* *(auch Rundf., Ferns., Film)* mix

Mixer der; ~s, ~ a) *(Bar~)* barman; bartender *(Amer.)*; **b)** *(Gerät)* blender and liquidizer

Mix·getränk das mixed drink; cocktail

Mixtur [mɪks'tu:ɐ̯] **die; ~, ~en** *(Pharm., fig.)* mixture

mm *Abk.* **Millimeter** mm.

Mo. *Abk.* **Montag** Mon.

Mob [mɔp] **der; ~s** *(abwertend)* mob

Möbel ['mø:bl̩] **das; ~s, ~ a)** *Pl.* furniture *sing., no indef. art.*; **b)** piece of furniture

Möbel-: ~haus das furniture store; **~packer der** removal man; **~spedition die** furniture-removal firm; **~stück das** piece of furniture; **~wagen der** furniture van; removal van

mobil [mo'bi:l] *Adj. (auch Milit.)* mobile; *(einsatzbereit)* mobilized; **~ machen** mobilize

Mobile ['mo:bilə] **das; ~s, ~s** mobile

Mobiliar [mobi'li̯a:ɐ̯] **das; ~s** furnishings *pl.*

mobilisieren *tr. V. (Milit., fig.)* mobilize; **die Massen ~** *(fig.)* stir the masses into action

Mobilisierung die; ~, ~en *(Milit., fig.)* mobilization

Mobilität [mobili'tɛ:t] **die; ~** *(Soziol.)* mobility

Mobilmachung die; ~, ~en mobilization

möblieren *tr. V.* furnish

Mocca ['mɔka] *s.* **Mokka**

mochte ['mɔxtə] *1. u. 3. Pers. Sg. Prät. v.* **mögen**

möchte ['mœçtə] *1. u. 3. Pers. Sg. Konjunktiv II v.* **mögen**

Möchte·gern- would-be ⟨*poet, Casanova, etc.*⟩

modal [mo'da:l] *Adj. (Sprachw.)* modal

Modalität [modali'tɛ:t] **die; ~, ~en** *(geh.)* provision; condition

Modal·verb das *(Sprachw.)* modal verb

Mode ['mo:də] **die; ~, ~n a)** fashion; **jede ~ mitmachen** follow fashion's every whim; **mit der ~ gehen** follow the fashion; **nach der neuesten ~**: in the latest style; **in ~/aus der ~ kommen** come into/go out of fashion; **b)** *Pl.* *(~kleidung)* fashions

mode-, Mode-: ~bewußt *Adj.* fashion-conscious; **~farbe die** fashionable colour; **~journal das** fashion magazine

Modell [mo'dɛl] **das; ~s, ~e** *(auch fig.)* model; *(Technik: Entwurf)* [design] model; pattern; *(in Originalgröße)* mock-up; **jmdm. ~ sitzen** *od.* **stehen** model *or* sit for sb.

Modell-: ~eisen·bahn die model railway; **~flugzeug das** model aircraft

modellhaft *Adj.* exemplary; model *attrib.*; pilot ⟨*scheme*⟩

modellieren *tr. V.* model, mould ⟨*figures*⟩; mould ⟨*clay, wax*⟩

Modellier·masse die modelling material *(esp. clay or wax)*

Modell-: ~projekt das pilot scheme; **~versuch der** pilot scheme

Moden·schau die fashion show *or* parade

Mode-: ~püppchen das, ~puppe die fashion-crazy bird *(Brit. sl.) or (Amer. coll.)* dame

Moder ['mo:dɐ] **der; ~s** mould; *(~geruch)* mustiness; *(Verwesung, auch fig.)* decay

Moderation [modera'tsi̯o:n] **die; ~, ~en** *(Rundf., Ferns.)* presentation

Moderator [mode'ra:tɔr] **der; ~s, ~en** [-'to:rən], **Moderatorin die; ~, ~nen** *(Rundf., Ferns.)* presenter

moderieren [mode'ri:rən] **1.** *tr. V. (Rundf., Ferns.)* present ⟨*programme*⟩. **2.** *itr. V.* be the presenter

moderig *Adj.* musty

¹modern ['mo:dɐn] *itr. V.; auch mit sein* go mouldy; *(verwesen)* decay

²modern [mo'dɛrn] **1.** *Adj.* modern; *(modisch)* fashionable. **2.** *adv.* in a modern manner *or* style; *(modisch)* fashionably; *(aufgeschlossen)* progressively

Moderne die; ~ a) modern age; modern times *pl.*; **b)** *(Kunstrichtung)* modern arts *pl., no art.*

modernisieren 1. *tr. V.* modernize. **2.** *itr. V.* introduce modern methods

Modernisierung die; ~, ~en modernization

Modernität die; ~: modernity

Mode-: ~schau die *s.* **Modenschau; ~schmuck der** costume jewellery; **~schöpfer der** couturier; **~schöpferin die** couturière; **~tanz der** dance [briefly] in vogue; **~wort das;** *Pl.* **~wörter** vogue-word; 'in' expression *(coll.)*; **~zeit·schrift die** fashion magazine

Modi *s.* **Modus**

modifizieren [modifi'tsi:rən] *tr. V. (geh.)* modify

modisch ['mo:dɪʃ] **1.** *Adj.* fashionable. **2.** *adv.* fashionably

Modus ['mɔdʊs] **der; ~, Modi** *(Sprachw.)* mood

Mofa ['mo:fa] **das; ~s, ~s** [low-powered] moped

Mogelei die; ~, ~en *(ugs.)* cheating *no pl.*

mogeln *(ugs.)* **1.** *itr. V.* cheat. **2.** *tr. V.* **etw. in etw.** *(Akk.)* **~**: slip sth. into sth.

mögen ['mø:gn̩] **1.** *unr. Modalverb; 2. Part.* **~**: **a)** *(wollen)* want to; **das hätte ich sehen ~**: I would have liked to see that; **b)** *(geh.:*

sollen) **das mag genügen** that should be *or* ought to be enough; **c)** *(geh.: Wunschform)* **möge er bald kommen!** I do hope he'll come soon!; **d)** *(Vermutung, Möglichkeit)* **sie mag/mochte vierzig sein** she must be/must have been [about] forty; **Meier, Müller, Koch – und wie sie alle heißen ~:** Meier, Müller, Koch and [the rest,] whatever they're called; **wie viele Personen ~ das sein?** how many people would you say there are?; **was mag sie damit gemeint haben?** what can she have meant by that?; **[das] mag sein** maybe; **e)** *(geh.: Einräumung)* **es mag kommen, was will** come what may; **f)** *Konjunktiv II (den Wunsch haben)* **ich/sie möchte gern wissen ...:** I would *or* should/she would like to know ...; **ich möchte nicht stören, aber ...:** I don't want to interrupt, but ...; **ich möchte zu gerne wissen** I'd love to know ...; **man möchte meinen, er sei der Chef** one would [really] think he was the boss. **2.** *unr. tr. V.* **a) [gern] ~:** like; **sie mag keine Rosen** she does not like roses; **sie mag ihn sehr [gern]** she likes him very much; *(hat ihn sehr gern)* she is very fond of him; **ich mag lieber/am liebsten Bier** I like beer better/best [of all]; **b)** *Konjunktiv II (haben wollen)* **möchten Sie ein Glas Wein?** would you like a glass of wine?; **ich möchte lieber Tee** I would prefer tea *or* rather have tea. **3.** *unr. itr. V.* **a)** *(es wollen)* like to; **ich mag nicht** I don't want to; **magst du?** do you want to?; *(bei einem Angebot)* would you like one/some?; **b)** *Konjunktiv II (fahren, gehen usw. wollen)* **ich möchte nach Hause/in die Stadt/auf die Schaukel** I want *or* I'd like to go home/into town/on the swing; **er möchte zu Herrn A** he would like to see Mr A

möglich ['mø:klıç] *Adj.* possible; **es war ihm nicht ~ [zu kommen]** he was unable [to come]; it was not possible for him [to come]; **sobald/so gut es mir ~ ist** as soon/as well as I can; **das** *od.* **alles Mögliche tun, sein ~stes tun** do everything possible; do one's utmost; **dort kann man alles ~e kaufen** *(ugs.)* you can get all sorts of things there; **sie hatte alles ~e zu kritisieren** she criticized everything; **alle ~en Leute** *(ugs.)* all sorts of people; **das ist gut/leicht/durchaus ~:** that is very/wholly/entirely possible; **man sollte es nicht für ~ halten** one would not believe it possible; **[das ist doch] nicht ~!** impossible!; I don't believe it!

möglicherweise *Adv.* possibly

Möglichkeit die; ~, ~en possibility; *(Gelegenheit)* opportunity; chance; *(möglicher Weg)* way; **nach ~:** if possible; **es besteht die ~, daß ...:** there is a chance *or* possibility that ...; **ist es die** *od.* **ist [denn] das die ~!** *(ugs.)* well, I'll be damned! *(coll.)*; whatever next!; **die ~ haben, etw. zu tun** have an opportunity of doing sth. *or* to do sth.; **das übersteigt meine [finanziellen] ~en** that is beyond my [financial] means

möglichst *Adv.* **a)** *(so weit wie möglich)* as much *or* far as possible; *(wenn möglich)* if [at all] possible; **macht ~ keinen Lärm** don't make any noise if you can possibly help it; **b)** *(so ... wie möglich)* **~ groß/schnell/oft** as big/fast/often as possible; **mit ~ großer Sorgfalt** with the greatest possible care; *s. auch* **möglich**

Mohair [mo'hɛ:ɐ̯] **der; ~s** mohair

Mohammed ['mo:hamɛt] **(der)** Muhammad

Mohammedaner der; ~s, ~, Mohammedanerin, die; ~, ~nen Muslim; Muhammadan

mohammedanisch *Adj.* Muslim, Muhammadan

Mohikaner [mohi'ka:nɐ] **der; ~s, ~:** Mohican; **der letzte ~, der Letzte der ~** *(ugs. scherzh.)* the last one; the last survivor *(joc.)*

Mohn [mo:n] **der; ~s a)** poppy; **b)** *(Samen)* poppy seed; *(auf Brot, Kuchen)* poppy seeds *pl.*

Mohn-: ~blume die poppy; **~brötchen das** poppy-seed roll; **~kuchen der** poppy-seed cake

Mohr [mo:ɐ̯] **der; ~en, ~en** *(veralt.)* Moor

Möhre ['mø:rə] **die; ~, ~n** carrot

Mohren·kopf der chocolate marshmallow

Möhren·saft der carrot juice

Mohr·rübe die carrot

Mokassin [moka'si:n] **der; ~s, ~s** moccasin

mokieren [mo'ki:rən] *refl. V.* *(geh.)* **sich über etw.** *(Akk.)* **~:** mock *or* scoff at sth.; **sich über jmdn. ~:** mock sb.

Mokka ['mɔka] **der; ~s a)** mocha [coffee]; **b)** *(Getränk)* strong black coffee

Mokka-: ~löffel der [small] coffee-spoon; **~tasse die** small coffee-cup

Molch [mɔlç] **der; ~[e]s, ~e** newt

Mole ['mo:lə] **die; ~, ~n** [harbour] mole

Molekül [mole'ky:l] **das; ~s, ~e** *(Chemie)* molecule

Molekular- molecular

molk [mɔlk] *1. u. 3. Pers. Sg. Prät. v.* **melken**

Molkerei die; ~, ~en dairy

Molkerei·produkt das dairy product

Moll [mɔl] **das; ~** *(Musik)* minor [key]; **a-Moll** A minor

mollig ['mɔlıç] **1.** *Adj.* **a)** *(rundlich)* plump; **b)** *(warm)* cosy; snug. **2.** *adv.* cosily; snugly; **~ warm** warm and cosy

Moloch ['mo:lɔx] **der; ~s, ~e** *(geh.)* Moloch; voracious giant

Molotow·cocktail ['mɔlotɔf-] **der** Molotov cocktail

¹Moment [mo'mɛnt] **der; ~[e]s, ~e** moment; **einen ~ bitte!** just a moment, please!; **~ [mal]!** [hey!] just a moment!; wait a mo! *(coll.)*; **im nächsten/selben ~:** the next/at the same moment; **jeden ~** *(ugs.)* [at] any moment; **im ~:** at the moment

²Moment das; ~[e]s, ~e factor, element **(für** in); **das auslösende ~ für etw. sein** be the trigger for sth.

momentan [momɛn'ta:n] **1.** *Adj.* **a)** *nicht präd.* present; current; **b)** *(vorübergehend)* temporary; *(flüchtig)* momentary. **2.** *adv.* **a)** at the moment; at present; **b)** *(vorübergehend)* temporarily; *(flüchtig)* momentarily; for a moment

Monarch [mo'narç] **der; ~en, ~en** monarch

Monarchie die; ~, ~n monarchy

Monarchin die; ~, ~nen monarch

monarchisch 1. *Adj.* monarchical. **2.** *adv.* monarchically

Monarchist der; ~en, ~en monarchist

monarchistisch 1. *Adj.* monarchist ⟨*party, group*⟩; monarchistic ⟨*tendency, views*⟩. **2.** *adv.* monarchistically

Monat ['mo:nat] **der; ~s, ~e** month; **im ~ April** in the month of April; **Ihr Schreiben vom 22. dieses ~s** your letter of the 22nd [inst.]; **sie ist im vierten ~ [schwanger]** she is four months pregnant; **was verdienst du im ~?** how much do you earn per month?

monatelang 1. *Adj.; nicht präd.* lasting for months *postpos., not pred.*; **die ~en Verhandlungen** the negotiations, which lasted for several months; **nach ~er Krankheit** after months of illness. **2.** *adv.* for months [on end]

-monatig a) *(... Monate alt)* ...-month-old; **b)** *(... Monate dau-*

ernd) ... month's/months'; ...-month; **eine viermonatige Kur** a four-month course of treatment; **mit dreimonatiger Verspätung** three months late

monatlich 1. *Adj.* monthly. 2. *adv.* monthly; every month; *(je Monat)* per month

Monats-: ~anfang der, ~beginn der beginning of the month; **~binde die** sanitary towel *(Brit.)*; sanitary napkin *(Amer.)*; **~blutung die** [monthly] period; **~einkommen das** monthly income; **~ende das** end of the month; **~erste der** first [day] of the month; **~frist** *o. Art.; o. Pl.* **in** *od.* **innerhalb** *od.* **binnen ~frist** within [a period of] a *or* one month; **~gehalt das** month's salary; **vier ~gehälter** four months' salary *sing.*; **ein dreizehntes ~gehalt** an extra month's salary; **~karte die** monthly season-ticket; **~letzte der** last day of the month; **~lohn der** month's wages; **vier ~löhne** four months' wages *pl.*; **~miete die** month's rent; **zwei ~mieten** two months' rent; **eine ~miete von 1000 DM** a monthly rent of 1,000 marks; **~mitte die** middle of the month; **~rate die** monthly instalment

Mönch [mœnç] **der; ~[e]s, ~e** monk

Mönchs-: ~kloster das monastery; **~kutte die** monk's habit *or* cowl

Mönch[s]tum das; ~s monasticism

Mond [mo:nt] **der; ~[e]s, ~e** moon; **ich könnte ihn auf den ~ schießen** *(salopp)* I wish he'd get lost *(sl.)*; **hinter dem ~ leben** *(fig. ugs.)* be a bit behind the times *or* not quite with it *(coll.)*; **nach dem ~ gehen** *(ugs.)* ‹*clock, watch*› be hopelessly wrong

mondän [mɔn'dɛ:n] 1. *Adj.* [highly] fashionable; smart. 2. *adv.* fashionably; in a fashionable style

mond-, Mond-: ~auf·gang der moonrise; **~fähre die** *(Raumf.)* lunar module; **~finsternis die** *(Astron.)* lunar eclipse; eclipse of the moon; **~gesicht das** moonface; **~hell** *Adj. (geh.)* moonlit; **~kalb das** *(salopp)* dim-wit *(coll.)*; dope *(coll.)*; **~lande·fähre die** *(Raumf.)* lunar module; **~landschaft die** *(auch fig.)* lunar landscape; **~landung die** moon landing; **~licht das;** *o. Pl.* moonlight; **~los** *Adj.* moonless; **~oberfläche die** lunar surface; **~phase die** moon's phase; **~schein der;** *o. Pl.* moonlight; **der kann mir mal im ~schein begegnen** *(salopp)* he can get lost *(sl.)*; **~sichel die** crescent moon; **~süchtig** *Adj.* sleep-walking *attrib. (esp. by moonlight)*; **~umlaufbahn die** lunar orbit; **~untergang der** moonset

Monetarismus [moneta'rɪsmʊs] **der; ~** *(Wirtsch.)* monetarism *no art.*

Moneten [mo'ne:tn̩] *Pl. (ugs.)* cash *sing.*; dough *sing. (sl.)*

Mongole [mɔŋ'go:lə] **der; ~n, ~n a)** Mongol; **b)** *(Bewohner der Mongolei)* Mongolian

Mongolei [mɔŋgo'lai] **die; ~:** Mongolia *no art.*

mongolid [mɔŋgo'li:t] *Adj. (Anthrop.)* Mongoloid

mongolisch *Adj.* Mongolian

Mongolismus der; ~ *(Med.)* mongolism *no art.*

mongoloid *Adj. (Med.)* mongoloid

monieren [mo'ni:rən] *tr. V.* criticize; *(beanstanden)* find fault with

Monitor ['mo:nitɔr] **der; ~s, ~en** [-'to:rən] monitor

mono ['mo:no] *Adv. (ugs.)* ‹*hear, play, etc.*› in mono *(coll.)*

mono-, Mono- mono-

monogam [mono'ga:m] 1. *Adj.* monogamous. 2. *adv.* monogamously

Monogamie die; ~: monogamy

Mono·gramm das; ~s, ~e monogram

Monokel [mo'nɔkl̩] **das; ~s, ~:** monocle

Mono·kultur die *(Landw.)* monoculture

Monolog [mono'lo:k] **der; ~s, ~e** monologue

Monopol [mono'po:l] **das; ~s, ~e** monopoly (**auf** + *Akk.*, **für** in, of)

Monopol·stellung die [position of] monopoly

monotheistisch [monote'ɪstɪʃ] monotheistic

monoton [mono'to:n] 1. *Adj.* monotonous. 2. *adv.* monotonously

Monotonie die; ~, ~n monotony

Monster ['mɔnstɐ] **das; ~s, ~:** monster; *(häßlich)* [hideous] brute

Monster- *s.* **Mammut-**

Monstranz [mɔn'strants] **die; ~, ~en** *(kath. Kirche)* monstrance

Monstren *s.* **Monstrum**

monströs [mɔn'strø:s] *Adj. (geh., auch fig.)* monstrous; [huge and] hideous

Monstrum ['mɔnstrʊm] **das; ~s, Monstren a)** *(auch fig.: Mensch)* monster; **b)** *(Ungetüm)* hulking great thing *(coll.)*

Monsun [mɔn'zu:n] **der; ~s, ~e** *(Geogr.)* monsoon

Montag ['mo:nta:k] **der** Monday; *s. auch* **blau; Dienstag; Dienstag-**

Montage [mɔn'ta:ʒə] **die; ~, ~n a)** *(Bauw., Technik) (Zusammenbau)* assembly; *(Einbau)* installation; *(Aufstellen)* erection; *(Anbringen)* fitting (**an** + *Akk. od. Dat.* to); mounting (**auf** + *Akk. od. Dat.* on); **auf ~:** *(ugs.)* away on a job; **b)** *(Film, Fot., bild. Kunst, Literaturw.)* montage

Montage-: ~band das assembly line; **~halle die** assembly shop

montags *Adv.* on Monday[s]; *s. auch* **dienstags**

Montan·industrie die coal and steel industry

Monteur [mɔn'tø:ɐ̯] **der; ~s, ~e** mechanic; *(Installateur)* fitter; *(Elektro~)* electrician

montieren [mɔn'ti:rən] *tr. V.* **a)** *(zusammenbauen)* assemble (**aus** from); erect ‹*building*›; **b)** *(anbringen)* fit (**an** + *Akk. od. Dat.* to; **auf** + *Akk. od. Dat.* on); *(einbauen)* install (**in** + *Akk.* in); *(befestigen)* fix (**an** + *Akk. od. Dat.* to); **eine Lampe an die** *od.* **der Decke ~:** put up *or* fix a light on the ceiling

Montur [mɔn'tu:ɐ̯] **die; ~, ~en** *(ugs.)* outfit *(coll.)*; gear *no pl. (coll.)*

Monument [monu'mɛnt] **das; ~[e]s, ~e** *(auch fig.)* monument

monumental *Adj. (auch fig.)* monumental; *(massiv)* massive

Monumental- monumental

Moor [mo:ɐ̯] **das; ~[e]s, ~e** bog; *(Bruch)* marsh; *(Flach~)* fen; *(Hoch~)* high moor

Moor·bad das mud-bath

Moos [mo:s] **das; ~es, ~e a)** moss; **~ ansetzen** gather moss; **b)** *o. Pl. (salopp)* cash; dough *(sl.)*

moos-: ~bedeckt, ~bewachsen *Adj.* moss-covered; **~grün** *Adj.* moss-green

Mop der; ~s, ~s mop

Moped ['mo:pɛt] **das; ~s, ~s** moped

Moped·fahrer der moped-rider

Mops [mɔps] **der; ~es, Möpse** ['mœpsə] **a)** *(Hund)* pug [dog]; **b)** *(salopp: dicke Person)* podge *(coll.)*; fatty *(derog.)*

mopsen *tr. V. (fam.)* pinch *(sl.)*

Moral [mo'ra:l] **die; ~ a)** *(Norm)* morality; **gegen die ~ verstoßen** offend against morality *or* the code of conduct; **die herrschende**

~: [currently] accepted standards *pl.*; **doppelte** ~: double standards *pl.*; **b)** *(Sittlichkeit)* morals *pl.*; **keine ~ haben** have no sense of morals; **c)** *(Selbstvertrauen)* morale; **die ~ ist gut/schlecht** morale is high/low; **d)** *(Lehre)* moral; **e)** *(Philos.)* ethics *sing.*

Moral·apostel der *(abwertend)* upholder of moral standards

moralisch [mo'ra:lɪʃ] **1.** *Adj.* **a)** *nicht präd.* moral; **b)** *(sittlich einwandfrei)* moral; morally upright; *(tugendhaft)* virtuous. **2.** *adv.* morally; *(tugendhaft)* morally; virtuously

Moralist der; ~**en,** ~**en** moralist

Moral-: ~philosophie die moral philosophy; **~prediger der** *(abwertend)* moralizing prig; **~predigt die** *(abwertend)* [moralizing] lecture; homily; **[jmdm.] eine ~predigt halten** deliver a homily [to sb.]

Morast [mo'rast] **der;** ~**[e]s,** ~**e** *od.* **Moräste** [mo'rɛstə] **a)** bog; swamp; **b)** *o. Pl. (Schlamm)* mud; *(auch fig.)* mire

morastig *Adj.* muddy

Morchel ['mɔrçl̩] **die;** ~, ~**n** morel

Mord [mɔrt] **der;** ~**[e]s,** ~**e** murder (**an** + *Dat.* of); *(durch ein Attentat)* assassination; **einen ~ begehen** commit murder; **versuchter** ~: attempted murder; **~ aus Eifersucht** *(Schlagzeile)* jealousy killing; **dann gibt es ~ und Totschlag** *(fig. ugs.)* all hell is/will be let loose

Mord-: ~anklage die charge of murder; **~anschlag der** attempted murder (**auf** + *Akk.* of); *(Attentat)* assassination attempt (**auf** + *Akk.* on); **~drohung die** murder threat

morden *tr., itr. V.* murder; **das sinnlose Morden** the senseless killing

Mörder ['mœrdɐ] **der;** ~**s,** ~: murderer *(esp. Law)*; killer; *(politischer ~)* assassin

Mörderin die; ~, ~**nen** murderer; murderess; *(politische ~)* assassin

mörderisch 1. *Adj.* **a)** *(ugs.)* murderous; fiendish ⟨*cold*⟩; dreadful *(coll.)* ⟨*clamour, weather, storm*⟩; **b)** *(todbringend)* murderous. **2.** *adv.* *(ugs.)* dreadfully *(coll.)*; frightfully *(coll.)*

mord-, Mord-: ~fall der murder case; **~instrument das** *(fig. scherzh.)* murderous[-looking] weapon *or* device; **~kommission die** murder *or (Amer.)* homicide squad; **~prozeß der** murder trial

mords-, Mords- *(ugs.)* terrific *(coll.)*; tremendous *(coll.)*

mords-, Mords-: ~ding das *(ugs.)* whopper *(coll.)*; **~hunger der** *(ugs.)* terrific hunger *(coll.)*; **einen ~hunger haben** be ravenous *or* famished; **~krach der** *(ugs.)* **a)** terrible din *or* racket *(coll.)*; **b)** *(Streit)* terrific row *(coll.)*; **~mäßig** *(ugs.)* **1.** *Adj.*; *nicht präd.* terrific *(coll.)*; tremendous *(coll.)*; *(entsetzlich)* terrible *(coll.)*; infernal *(coll.)* ⟨*din, racket*⟩; **2.** *adv.* tremendously *(coll.)*; incredibly *(coll.)*; *(entsetzlich)* terribly *(coll.)*; **~stimmung die** *(ugs.)* terrific atmosphere *(coll.)*

Mord-: ~verdacht der suspicion of murder; **~versuch der** attempted murder; *(Attentat)* assassination attempt; **~waffe die** murder weapon

morgen ['mɔrgn̩] *Adv.* **a)** tomorrow; **~ früh/mittag/abend** tomorrow morning/lunchtime/evening; **~ in einer Woche/in vierzehn Tagen** tomorrow week/fortnight; a week/fortnight tomorrow; **~ um diese** *od.* **die gleiche Zeit** this time tomorrow; **bis ~!** until tomorrow!; see you tomorrow!; **~ ist auch [noch] ein Tag** tomorrow is another day; **die Mode/Technik von ~** *(fig.)* tomorrow's fashions *pl.*/technology; **b)** *(am Morgen)* **heute/gestern ~**: this/yesterday morning; **[am] Sonntag ~**: on Sunday morning

Morgen der; ~**s,** ~ **a)** morning; **am ~** in the morning; **am folgenden** *od.* **nächsten ~**: next *or* the following morning; **früh am ~, am frühen ~**: early in the morning; **eines [schönen] ~s** one [fine] morning; **gegen ~**: towards morning; **~ für ~**: every single morning; morning after morning; **den ganzen ~**: all morning; **guten ~!** good morning!; **~!** *(ugs.)* morning! *(coll.)*; **[jmdm.] guten ~ sagen** *od.* **wünschen** say good morning [to sb.]; wish [sb.] good morning; *(grüßen)* say hello [to sb.]; **b)** *(veralt.: Feldmaß)* ≈ acre; **fünf ~ Land** five acres of land

Morgen-: ~ausgabe die morning edition; **~dämmerung die** dawn; daybreak

morgendlich *Adj.*; *nicht präd.* morning; **die ~e Kühle/Stille** the cool/peace of [early] morning

Morgen-: ~grauen das daybreak; **im** *od.* **beim ~grauen** in the first light of day; **~gymnastik die** morning exercises *pl.*; daily dozen *(coll.)*; **~land das;** *o. Pl. (veralt.)* East; Orient; **~luft die** morning air; **~luft wittern** *(fig. scherzh.)* see one's chance; **~mantel der** dressing-gown; **~muffel der** *(ugs.)* **ein ~muffel sein** be grumpy in the mornings; **~rock der** dressing-gown; **~rot das** rosy/red dawn

morgens *Adv.* in the morning; *(jeden Morgen)* every morning; **~ um 7 Uhr, um 7 Uhr ~**: at 7 in the morning/every morning; **dienstags ~**: on Tuesday morning[s]; **von ~ bis abends** all day long; from morning to evening

Morgen-: ~sonne die morning sun; **~spaziergang der** *(esp. early)* morning walk; **~stern der** morning star; **~stunde die** hour of the morning; **~stunde hat Gold im Munde** *(Spr.)* the early bird catches the worm *(prov.)*; **~zeitung die** morning paper

morgig *Adj.*; *nicht präd.* tomorrow's; **der ~e Tag** tomorrow

Mormone [mɔr'mo:nə] **der;** ~**n,** ~**n, Mormonin die;** ~, ~**nen** Mormon

Morphium ['mɔrfi̯ʊm] **das;** ~**s** morphine

morphium·süchtig *Adj.* addicted to morphine *pred.*

morsch [mɔrʃ] *Adj.* *(auch fig.)* rotten; brittle ⟨*bones*⟩; crumbling ⟨*rock, masonry*⟩

Morse·alphabet ['mɔrzə-] **das** Morse code *or* alphabet

morsen ['mɔrzn̩] **1.** *itr. V.* send a message/messages in Morse. **2.** *tr. V.* send ⟨*signal, message*⟩ in Morse

Mörser ['mœrzɐ] **der;** ~**s,** ~ *(auch Milit.)* mortar

Morse·zeichen das Morse symbol

Mortadella [mɔrta'dɛla] **die;** ~, ~**s** mortadella

Mortalität [mɔrtali'tɛ:t] **die;** ~: mortality [rate]

Mörtel ['mœrtl̩] **der;** ~**s** mortar

Mosaik [moza'i:k] **das;** ~**s,** ~**en** *od.* ~**e** *(auch fig.)* mosaic

Mosambik [mozam'bi:k] **(das);** ~**s** Mozambique

Moschee [mɔ'ʃe:] **die;** ~, ~**n** mosque

Moschus ['mɔʃʊs] **der;** ~: musk

Mose ['mo:zə] **(der)** Moses; **das erste/zweite/dritte/vierte/fünfte Buch ~**: Genesis/Exodus/Leviticus/Numbers/Deuteronomy; *s. auch* **Buch a**

Mosel ['mo:zl̩] **die;** ~: Moselle

Mosel der; ~**s,** ~, **Mosel·wein der** Moselle [wine]

mosern ['mo:zɐn] *itr. V.* *(ugs.)* gripe *(coll.)* (**über** + *Akk.* about)

Moses (der) Moses
Moskau ['mɔskau] **(das); ~s** Moscow
Moskauer 1. *indekl. Adj.* Moscow *attrib.* **2. der; ~s, ~:** Muscovite; *s. auch* **Kölner**
Moskito [mɔs'ki:to] **der; ~s, ~s** mosquito
Moslem ['mɔslɛm] **der; ~s, ~s** Muslim
moslemisch *Adj.* Muslim
Most [mɔst] **der; ~[e]s, ~e a)** *(südd.: junger Wein)* new wine; **b)** *(südd.: Obstsaft)* [cloudy fermented] fruit-juice; **d)** *(südd., schweiz., österr.: Obstwein)* fruit-wine; *(Apfel~)* [rough] cider
Mostrich ['mɔstrɪç] **der; ~s** *(nordostd.)* mustard
Motel ['mo:tl] **das; ~s, ~s** motel
Motiv [mo'ti:f] **das; ~s, ~e a)** motive; **b)** *(Literaturw., Musik usw.: Thema)* motif; theme; **c)** *(bild. Kunst, Fot., Film: Gegenstand)* subject
Motivation [motiva'tsio:n] **die; ~, ~en** *(Psych., Päd.)* motivation
motivieren *tr. V. (geh.)* motivate
Motivierung die; ~, ~en *(geh.)* motivation
Motodrom [moto'dro:m] **das; ~s, ~e** autodrome; speedway *(Amer.)*
Motor ['mo:tɔr] **der; ~s, ~en** *(Verbrennungs~)* engine; *(Elektro~)* motor; *(fig.)* driving force (*Gen.* behind)
Motor·boot das motor boat; *(Rennboot)* power boat
Motoren-: ~geräusch das sound of the engine/engines; **~lärm der** engine noise
Motor·haube die *(Kfz-W.)* bonnet *(Brit.)*; hood *(Amer.)*
-motorig *adj.* -engined; **ein~/zwei~:** single-engined/twin-engined
Motorik [mo'to:rɪk] **die; ~** *(bes. Med.)* motor functions *pl.*
motorisch *Adj. (Psych.)* motor *attrib.*
motorisieren 1. *tr. V.* motorize
Motorisierung die; ~: motorization
Motor-: ~öl das *(Kfz-W.)* engine oil; **~rad das** motor cycle
Motorrad-: ~fahrer der motorcyclist; **~rennen das** motorcycle race; *(Sport)* motor-cycle racing; **~sport** motor-cycling
Motor-: ~roller der motor scooter; **~säge die** power saw; **~schaden der** engine trouble *no indef. art.*; *(Panne)* mechanical breakdown; **~sport der** motor sport *no art.*; **~wäsche die** engine wash-down
Motte ['mɔtə] **die; ~, ~n** moth; **von etw. angezogen werden wie die ~en vom Licht** be attracted by sth. as moths to the light; **[ach,] du kriegst die ~n!** *(ugs.)* my godfathers!
Motten-: ~kiste die a) *(fig.)* **Filme/Geschichten/Gags aus der ~kiste** ancient films/stories/gags; **~kugel die** mothball
Motto ['mɔto] **das; ~s, ~s** motto; *(Schlagwort)* slogan; **nach dem ~: ... leben** live according to the maxim: ...
motzen ['mɔtsn̩] *itr. V. (ugs.)* grouch *(coll.)*, bellyache *(sl.)* (**über** + *Akk.* about)
moussieren [mʊ'si:rən] *itr. V.* sparkle; *(als Eigenschaft)* be sparkling
Möwe ['mø:və] **die; ~, ~n** gull
MP ['ɛm'pi:] **die; ~, ~s** *Abk.* **a) Maschinenpistole** sub-machine-gun; **b) Militärpolizei** MPs *pl.*; military police *pl.*
Mrd. *Abk.* **Milliarde** bn.
Ms., MS *Abk.* **Manuskript** MS
MTA [ɛmte:'|a:] **die; ~, ~[s]** *Abk.* **medizinisch-technische Assistentin** medical-laboratory assistant
mtl. *Abk.* **monatlich** mthly.
Mücke ['mʏkə] **die; ~, ~n** midge; gnat; *(größer)* mosquito; **aus einer ~ einen Elefanten machen** *(ugs.)* make a mountain out of a molehill
mucken ['mʊkn̩] *itr. V. (ugs.)* grumble; mutter; **ohne zu ~:** without a murmur
Mucken *Pl. (ugs.)* whims; *(Eigenarten)* little ways *or* peculiarities; *(Launen)* moods; **[seine] ~ haben** ‹*person*› have one's little ways/one's moods; ‹*car, machine*› be a little unpredictable *or* temperamental
Mücken·stich der midge/mosquito bite
Mucks [mʊks] **der; ~es, ~e** *(ugs.)* murmur [of protest]; slight[est] sound; **keinen ~ sagen** not utter a [single] word *or* sound
mucksen *refl. V. (ugs.)* make a sound
Muckser der; ~s, ~ *(ugs.) s.* **Mucks**
mucks·mäuschen·still *(ugs.)* **1.** *Adj.; nicht attr.* utterly silent; ‹*person*› as quiet as a mouse *postpos.*. **2.** *adv.* in total silence; without making a sound
müde ['my:də] **1.** *Adj.* tired; *(ermattet)* weary; *(schläfrig)* sleepy; **Bier macht ~:** beer makes you feel sleepy; **ein ~s Lächeln** *(auch fig.)* a weary smile; **etw. ~ sein** *(geh.)* be tired of sth.; **einer Sache** *(Gen.)* **~ werden** *(geh.)* tire *or* grow tired of sth.; **nicht ~ werden, etw. zu tun** never tire of doing sth.; *s. auch* [1]**Mark. 2.** *adv.* wearily; *(schläfrig)* sleepily
-müde *adj.* tired of ...; **amts~:** tired of [holding] office
Müdigkeit die; ~: tiredness; **ich könnte vor ~ umfallen** I'm so tired I can hardly stand; **[nur] keine ~ vorschützen!** *(ugs.)* it's no use saying you're tired!
-müdigkeit die weariness of ...; **Zivilisations~:** weariness of civilized living; culture fatigue
[1]**Muff** [mʊf] **der; ~[e]s** *(nordd.)* musty smell; *(Gestank)* fug
[2]**Muff der; ~[e]s, ~e** muff
Muffel ['mʊfl] **der; ~s, ~** *(ugs.)* sourpuss *(coll.)*; grouch *(coll.)*
-muffel der; ~s, ~ *(ugs.)* person who is not into ... at all *(coll.)*; **er ist ein Fußball~:** he isn't into football at all *(ugs.)*
muffelig *(ugs.)* **1.** *Adj.* grumpy; surly **2.** *adv.* grumpily
[1]**muffig** *Adj. (modrig riechend)* musty; *(stickig; auch fig.)* stuffy
[2]**muffig** *(ugs.) s.* **muffelig**
Mufflon ['mʊflɔn] **der; ~s, ~s** *(Zool.)* moufflon
muh [mu:] *Interj.* moo
Müh [my:]: **mit ~ und Not** with great difficulty; only just
Mühe ['my:ə] **die; ~, ~n** trouble; **alle ~ haben, etw. zu tun** be hard put to do sth.; **mit jmdm./etw. seine ~ haben** have a lot of trouble *or* a hard time with sb./sth.; **keine ~ scheuen** spare no pains *or* effort; **sich** *(Dat.)* **viel ~ machen** go to *or* take a lot of trouble (**mit** over); **machen Sie sich [bitte] keine ~!** [please] don't put yourself out!; [please] don't bother!; **es hat viel ~ gekostet** it took much time and effort; **sich** *(Dat.)* **~ geben** make an effort *or* take pains; **sich** *(Dat.)* **mit jmdm./etw. ~ geben** take [great] pains *or* trouble over sb./sth.; **gib dir keine ~!** you needn't bother
mühelos 1. *Adj.* effortless. **2.** *adv.* effortlessly; without the slightest difficulty
Mühelosigkeit die; ~: effortlessness
muhen *itr. V.* moo
mühen *refl. V. (geh.)* strive; **sich mit etw. ~:** take pains over sth.
mühe·voll *Adj.* laborious; painstaking ‹*work*›; **ein ~er Weg** an arduous path
Mühle ['my:lə] **die; ~, ~n a)** mill; **in die ~ der Justiz geraten** *(fig.)* become enmeshed in the wheels *or* machinery of justice; **das ist**

Wasser auf seine ~ *(ugs.)* it's [all] grist to his mill; it just confirms what he has always thought; **b)** *(Kaffee~)* [coffee-] grinder; **c)** *(Spiel) o. Art., o. Pl.* nine men's morris; **d)** *(Konstellation beim ~spiel)* mill; **e)** *(ugs. abwertend) (Auto, Motorrad)* heap *(coll.)*; *(Auto, Flugzeug)* crate *(coll.)*; *(Fahrrad)* rattletrap; boneshaker

Mühle·spiel das nine men's morris

Mühl-: ~rad das mill-wheel; **~stein der** millstone

Mühsal ['my:za:l] **die; ~, ~e** *(geh.)* tribulation; *(Strapaze)* hardship; *(Arbeit)* toil *no pl.*

mühsam 1. *Adj.* laborious; **ein ~es Lächeln** a forced smile. **2.** *adv.* laboriously; *(schwierig)* with difficulty

müh·selig *(geh.)* **1.** *Adj.* laborious; arduous ‹*journey, life*›. **2.** *adv.* with [great] difficulty

Mulatte [mu'latə] **der; ~n, ~n, Mulattin die; ~, ~nen** mulatto

Mulde ['mʊldə] **die; ~, ~n** hollow

Muli ['mu:li] **das; ~s, ~s** mule

Mull [mʊl] **der; ~[e]s** *(Stoff)* mull; *(Verband~)* gauze

Müll [mʏl] **der; ~s** refuse; rubbish; garbage *(Amer.)*; trash *(Amer.)*; *(Industrie~)* [industrial] waste; **etw. in den ~ werfen** throw sth. in the dustbin *(Brit.)* or *(Amer.)* garbage can; **„~ abladen verboten"** 'no dumping'; 'no tipping' *(Brit.)*

Müll-: ~abfuhr die a) refuse *or (Amer.)* garbage collection; **b)** *(Unternehmen)* refuse *or (Amer.)* garbage collection [service]; **~beutel der** dustbin *(Brit.) or (Amer.)* garbage can liner

Mull·binde die gauze bandage

Müll-: ~deponie die *(Amtsspr.)* refuse disposal site; **~eimer der** rubbish *or* waste bin

Müller ['mʏlɐ] **der; ~s, ~:** miller

Müll-: ~halde die refuse dump; **~haufen der** heap of rubbish *or (Amer.)* garbage; **~kippe die** [refuse] dump *or (Brit.)* tip; **~mann der;** *Pl.* **~männer** *(ugs.)* dustman *(Brit.)*; garbage man *(Amer.)*; **~sack der** refuse bag; **~schlucker der** rubbish *or (Amer.)* garbage chute; **~tonne die** dustbin *(Brit.)*; garbage *or* trash can *(Amer.)*; **~tüte die** bin bag; **~verbrennung die** refuse *or (Amer.)* garbage incineration; **~wagen der** dust-cart *(Brit.)*; garbage truck *(Amer.)*

mulmig ['mʊlmɪç] *Adj. (ugs.: unbehaglich)* uneasy; **ein ~es Gefühl haben** feel uneasy

Multi ['mʊlti] **der; ~s, ~s** *(ugs.)* multinational

Multimedia-Show [-'me:di̯a-]: **die** multi-media presentation

multipel [mʊl'ti:pl̩] *Adj.; nicht präd. (fachspr.)* multiple

Multiple-choice-Verfahren ['mʌltɪpl'tʃɔɪs-] **das** *(fachspr.)* multiple-choice method

Multiplikation [mʊltiplika'tsi̯o:n] **die;** ~, **~en** multiplication

multiplizieren [mʊltipli'tsi:rən] *tr., itr. V. (auch fig.)* multiply **(mit** by)

Mumie ['mu:mi̯ə] **die; ~, ~n** mummy

mumienhaft 1. *Adj.* mummy-like. **2.** *adv.* like a mummy; as though mummified

mumifizieren [mumifi'tsi:rən] *tr. V.* mummify

Mumifizierung die; ~, ~en mummification

Mumm [mʊm] **der; ~s** *(ugs.) (Mut)* guts *pl. (coll.)*; spunk *(coll.)*; *(Tatkraft)* drive; zap *(sl.)*; *(Kraft)* muscle-power

mummeln *(nordd.)*, **mümmeln** *tr., itr. V. (fam.) (kauen)* chew; *(knabbern)* nibble

Mumpitz ['mʊmpɪts] **der; ~es** *(ugs. abwertend)* rubbish; tripe *(sl.)*

Mumps [mʊmps] **der** *od.* **die; ~:** mumps *sing.*

München ['mʏnçn̩] **(das); ~s** Munich

Münch[e]ner ['mʏnçnɐ] **1.** *indekl. Adj.* Munich *attrib.* **2. der; ~s, ~:** inhabitant/native of Munich; *s. auch* **Kölner**

Mund [mʊnt] **der; ~[e]s, Münder** ['mʏndɐ] mouth; **vor Staunen blieb ihm der ~ offenstehen** he gaped in astonishment; **er küßte ihren ~** *od.* **küßte sie auf den ~:** he kissed her on the lips; **von ~ zu ~ beatmet werden** be given mouth-to-mouth resuscitation *or* the kiss of life; **mit vollem ~ sprechen** speak with one's mouth full; **etw. aus jmds. ~ hören** hear *or* have sth. from sb.'s [own] lips; **sein ~ steht nicht** *od.* **nie still** *(ugs.)* he never stops talking; **den ~ nicht aufkriegen** *(fig. ugs.)* not open one's mouth; have nothing to say for oneself; **den ~ aufmachen/nicht aufmachen** *(fig. ugs.)* say something/not say anything; **den ~ voll nehmen** *(fig. ugs.)* talk big *(coll.)*; **nimm doch den ~ nicht so voll!** *(fig. ugs.)* don't be such a bighead!; **einen großen ~ haben** *(fig. ugs.)* talk big *(coll.)*; **den** *od.* **seinen ~ halten** *(ugs.) (schweigen)* shut up *(coll.)*; *(nichts sagen)* not say anything; *(nichts verraten)* keep quiet **(über** + *Akk.* about); **jmdm. den ~ verbieten** silence sb.; **jmdm. den ~ [ganz] wäßrig machen** *(fig. ugs.)* [really] make sb.'s mouth water; **er/sie ist nicht auf den ~ gefallen** *(fig. ugs.)* he's/she's never at a loss for words; **... ist in aller ~e** *(fig.)* everybody's talking about ...; **etw./ein Wort in den ~ nehmen** utter sth./use a word; **jmdm. nach dem ~ reden** *(fig.)* echo what sb. says; *(schmeichelnd)* butter sb. up; tell sb. what he/she wants to hear; **jmdm. über den ~ fahren** *(fig. ugs.)* cut sb. short

Mund·art die dialect

Mundart·dichter der dialect author; *(Lyriker)* dialect poet

mundartlich 1. *Adj.* dialectal ‹*forms, expressions, words*›; ‹*texts, poems, etc.*› in dialect. **2.** *adv.* ; **stark ~ gefärbt** strongly coloured by dialect

Mundart·sprecher der dialect speaker

Mund·dusche die water pick

Mündel ['mʏndl̩] **das; ~s, ~:** ward

munden *itr. V. (geh.)* taste good; **es mundete ihm nicht** he did not enjoy it; he did not like the taste of it; **das wird dir ~:** this will tickle your palate

münden ['mʏndn̩] *itr. V.; mit sein* ‹*river*› flow **(in** + *Akk.* into); ‹*corridor, street, road*› lead **(in** + *Akk. od. Dat.*, **auf** + *Akk. od. Dat.* into)

mund-, Mund-: ~faul *(ugs.)* **1.** *Adj.* uncommunicative; **2.** *adv.* uncommunicatively; **~gerecht** *Adj.* bite-sized; *(fig.)* easily digestible ‹*information*›; **~geruch der** bad breath *no indef. art.*; **~harmonika die** mouth-organ

mündig *Adj.* **a)** of age *pred.*; **~ werden** come of age; **b)** *(urteilsfähig)* responsible adult *attrib.*; **~ werden** become capable of mature judgement

mündlich 1. *Adj.* oral; **~e Vereinbarung** verbal agreement; **~e Verhandlung** *(Rechtsw.)* hearing. **2.** *adv.* orally; ‹*agree*› verbally; **alles weitere ~!** *(im Brief)* I'll tell you the rest when we meet

Mund-: ~raub der petty theft [of food/consumables]; **~schutz der a)** *(Med.)* face-mask; **b)** *(Boxen)* gum-shield

M-und-S-Reifen ['ɛm ʊnt 'ɛs-] **der** snow tyre

Mund·stück das a) *(bei Instrumenten, Pfeifen usw.)* mouthpiece; **b)** *(bei Zigaretten)* tip

mund·tot *Adj.* **jmdn. ~ machen** silence sb.
Mündung die; ~, ~en a) mouth; *(größere Trichter~)* estuary; **b)** *(bei Feuerwaffen)* muzzle
Mund-: ~werk das; *o. Pl. (ugs.)* **ein loses ~werk [haben]** [have] a loose tongue; **~winkel der** corner of one's mouth; **~-zu--Beatmung die** mouth-to-mouth resuscitation; kiss of life
Munition [muni'tsio:n] **die; ~** *(auch fig.)* ammunition
Munitions-: ~fabrik die munitions factory; **~lager das** ammunition dump
munkeln ['mʊŋkl̩n] *tr., itr. V. (ugs.)* **man munkelt, daß ...:** there is a rumour that ...
Münster ['mʏnstɐ] **das; ~s, ~:** minster; *(Dom)* cathedral
munter ['mʊntɐ] **1.** *Adj.* **a)** cheerful; merry; *(lebhaft)* lively ⟨*eyes, game* ⟩; **~ werden** cheer up; liven up; **[gesund und] ~ sein** be as fit as a fiddle; ⟨*elderly person*⟩ be hale and hearty; **b)** *(wach)* awake; **~ werden** wake up; come round *(joc.)*. **2.** *adv.* **a)** merrily; cheerfully; **b)** *(unbekümmert)* gaily; cheerfully
Munterkeit die; ~: cheerfulness; gaiety
Münz·automat der slot-machine; *(Telefon)* payphone; pay station *(Amer.)*
Münze ['mʏntsə] **die; ~, ~n a)** coin; **klingende** *od.* **bare ~** *(geh.)* cash; **etw. für bare ~ nehmen** *(fig.)* take sth. literally; **jmdm. [etw.] mit gleicher ~ heimzahlen** pay sb. back in the same coin [for sth.]; **b)** *(Münzanstalt)* mint
münzen *tr. V.* coin; **auf jmdn./etw. gemünzt sein** *(fig.)* ⟨*remark etc.*⟩ be aimed at sb./sth.
Münz-: ~fernsprecher der coin-box telephone; payphone; pay station *(Amer.)*; **~kunde die** numismatics *sing.*; **~sammlung die** coin collection; **~tankstelle die** coin-in-the-slot petrol *(Brit.)* or *(Amer.)* gas station
Muräne [mu'rɛ:nə] **die** moray eel
mürb [mʏrp] *(südd., österr.)*, **mürbe** ['mʏrbə] *Adj.* **a)** crumbly ⟨*biscuit, cake, etc.*⟩; tender ⟨*meat*⟩; soft ⟨*fruit*⟩; mealy ⟨*apple*⟩; **das Fleisch ~ machen** tenderize the meat; **b)** *(brüchig)* crumbling; *(morsch)* rotten; ⟨*leather*⟩ worn soft; **~ werden/sein** *(fig.: zermürbt)* get/be worn out; **jmdn. ~ machen** *(fig.)* wear sb. down
Mürbe·teig, *(südd., österr.)* **Mürb·teig der** short pastry
Murks [mʊrks] **der; ~es** *(salopp abwertend)* botch; mess; **~ machen** make a botch *or* mess [of it]
Murmel ['mʊrml̩] **die; ~, ~n** marble
murmeln *tr., itr. V.* mumble; mutter; *(sehr leise)* murmur; **etw. vor sich hin ~:** mutter *or* mumble/murmur sth. to oneself
Murmel·tier das marmot; *s. auch* **schlafen 1 a**
murren ['mʊrən] *itr. V.* grumble **(über +** *Akk.* about); **ohne zu ~:** without a murmur
mürrisch ['mʏrɪʃ] **1.** *Adj.* grumpy; surly, sullen ⟨*expression*⟩. **2.** *adv.* grumpily
Mus [mu:s] **das** *od.* **der; ~es, ~e** purée; **zu ~ kochen** cook to a pulp
Muschel ['mʊʃl̩] **die; ~, ~n a)** mussel; **b)** *(Schale)* [mussel-]shell; **c)** *(am Telefon) (Hör~)* ear-piece; *(Sprech~)* mouthpiece
Muse ['mu:zə] **die; ~, ~n** muse; **die leichte ~:** light [musical] entertainment; **von der ~ geküßt werden** *(scherzh.)* get some inspiration
Museen *s.* **Museum**
Musen·tempel der *(veralt., noch scherzh.)* temple of the Muses
Museum [mu'ze:ʊm] **das; ~s, Museen** museum
museums-, Museums-: ~führer der museum guide; **~reif** *Adj. (ugs. iron.)* fit for a museum *postpos.*; **~stück das** *(auch fig. ugs. iron.)* museum piece; **~wärter der** museum attendant
Musical ['mju:zi:kl̩] **das; ~s, ~s** musical
Musik [mu'zi:k] **die; ~, ~en a)** *o. Pl.* music; **~ im Blut haben** have music in one's blood; **~ in jmds. Ohren** *(Dat.)* **sein** *(fig. ugs.)* be music to sb.'s ears; **b)** *(Werk)* piece [of music]; *(Partitur)* score **(zu** for); **die ~ zu diesem Stück** the [incidental] music for this play; *s. auch* **Handkäse**
musikalisch [muzi'ka:lɪʃ] **1.** *Adj.* musical; **~e Leitung: ...:** conducted by ... **2.** *adv.* musically; **er ist ~ veranlagt** he is musical
Musikalität [muzikali'tɛ:t] **die; ~:** musicality
Musikant [muzi'kant] **der; ~en, ~en** musician
Musikanten·knochen der funny-bone
Musik·box die juke-box
Musiker der; ~s, ~, Musikerin die; ~, ~nen musician
Musik-: ~geschichte die history of music; **~hochschule die** academy *or* college of music; **~instrument das** musical instrument; **~lehrer der, ~lehrerin die** music-teacher; **~schule die** school of music; **~stück das** piece of music; **~stunde die** music-lesson; **~unterricht der a)** *o. Pl. (das Unterrichten)* music-teaching; **b)** *(Stunde)* music-lesson; *(Stunden)* music-lessons *pl.*; **c)** *(als Schulfach)* music
Musik-: ~wissenschaft die *o. Pl.* musicology; **~wissenschaftler der** musicologist
musisch 1. *Adj.* artistic ⟨*talent, person, family, etc.*⟩; ⟨*talent*⟩ for the arts; ⟨*education*⟩ in the arts. **2.** *adv.* artistically; **~ veranlagt sein** have an artistic disposition
musizieren [muzi'tsi:rən] *itr. V.* play music; *(bes. unter Laien)* make music
Muskat [mʊs'ka:t] **der; ~[e]s, ~e** nutmeg
Muskateller [mʊska'tɛlɐ] **der; ~s, ~:** muscatel [wine]
Muskat·nuß die nutmeg
Muskel ['mʊskl̩] **der; ~s, ~n** muscle
Muskel-: ~faser die muscle fibre; **~kater der** stiff muscles *pl.*; **~kraft die** muscle-power; **~krampf der** cramp; **~paket das** *(ugs.)* **a)** bulging muscles *pl.* **b)** *(ugs.) s.* **~protz; ~protz der** *(ugs.)* muscleman; Tarzan *(joc.)*; **~riß der** torn muscle; **~schwund der** *(Med.)* muscular atrophy; **~zerrung die** *(Med.)* pulled muscle
Muskulatur [mʊskula'tu:ɐ] **die; ~, ~en** musculature; muscular system
muskulös [mʊsku'lø:s] *Adj.* muscular
Müsli ['mʏs:li] **das; ~s, ~s** muesli
Muslim ['mʊslɪm] **der; ~s, ~e** Muslim
muslimisch *Adj.* Muslim
muß [mʊs] *1. u. 3. Pers. Sg. Präsens v.* **müssen**
Muß das; ~: necessity; must *(coll.)*
Muße ['mu:sə] **die; ~:** leisure; **etw. in** *od.* **mit ~ tun** do sth. at one's leisure; take one's time over sth.
Muß·ehe die *(ugs.)* shotgun marriage
müssen ['mʏsn̩] **1.** *unr. Modalverb; 2. Part.* **~ a)** *(gezwungen, verpflichtet sein)* have to; **er muß es tun** he must do it; he has to *or* *(coll.)* has got to do it; **er muß es nicht tun** he does not have to do it; he has not got to do it *(coll.)*; **er mußte es tun** *od.* **hat es tun ~:** he had to do it; **muß er es tun?** must

he do it?; does he have to *or (coll.)* has he got to do it?; **wir werden zurückkommen ~:** we shall have to come back; **muß das jetzt sein?** does it have to be now?; **muß das sein?** it is really necessary?; **es muß nicht sein** it is not essential; **so mußte es ja kommen** it was inevitable that it should come to this; **wir ~ Ihnen leider mitteilen, daß ...:** we regret to have to inform you that ...; **das muß 1968 gewesen sein** it must have been in 1968; **er muß gleich hier sein** he will be here *or* he is bound to be here at any moment; **b)** *Konjunktiv II* **es müßte doch möglich sein** it ought to be possible; **reich müßte man sein!** how nice it would be to be rich!; **man müßte nochmals zwanzig sein** oh to be twenty again!. **2.** *unr. itr. V.* **a)** *(gehen, fahren, gebracht werden usw. müssen)* have to go; **ich muß zur Arbeit/nach Hause** I have to *or* must go to work/go home; **b) ich muß mal** *(fam.)* I've got to *or* need to spend a penny *(Brit. coll.) or (Amer. coll.)* go to the john; **c)** *(gezwungen, verpflichtet sein)* **muß er?** does he have to?; has he got to? *(coll.)*; **er muß nicht** he doesn't have to *or (coll.)* hasn't got to

Muße·stunde die free hour; hour of leisure

müßig ['my:sɪç] **1.** *Adj.* **a)** idle ⟨*person*⟩; ⟨*hours, weeks, life*⟩ of leisure; **b)** *(zwecklos)* pointless. **2.** *adv.* idly

Müßig-: ~gang der *o. Pl.* leisure; *(Untätigkeit)* idleness; **~gänger der** idler

mußte ['mʊstə] *1. u. 3. Pers. Sg. Prät. v.* **müssen**

Mustang ['mʊstaŋ] **der; ~s, ~s** mustang

Muster ['mʊstɐ] **das; ~s, ~ a)** *(Vorlage)* pattern; **b)** *(Vorbild)* model (**an** + *Dat.* of); **er ist ein ~ an Fleiß** he is a model of industry; **er ist ein ~ von einem Ehemann** *(ugs.)* he is a model husband; **c)** *(Verzierung)* pattern; **d)** *(Probe)* specimen; *(Warenprobe)* sample

muster-, Muster-: ~beispiel das perfect example; *(Vorbild)* model; **~exemplar das a)** *(oft iron.: Vorbild)* perfect specimen; **b)** *(Probeexemplar)* specimen copy; **~gültig 1.** *Adj.* exemplary; perfect, impeccable ⟨*order*⟩; **2.** *adv.* in an exemplary fashion

musterhaft 1. *Adj.* exemplary; perfect, impeccable ⟨*order, condition*⟩; model ⟨*pupil*⟩. **2.** *adv.* in an exemplary fashion

Muster-: ~knabe der *(oft abwertend)* model child; **~koffer der** case of samples

mustern *tr. V.* **a)** eye; *(gründlich)* scrutinize; **b)** *(Milit.)* **jmdn. ~:** give sb. his medical

Muster-: ~schüler der, ~schülerin die model pupil

Musterung die; ~, ~en *(Milit.)* medical examination; medical

Musterungs·bescheid der summons to attend one's medical examination

Mut [mu:t] **der; ~[e]s a)** courage; **allen** *od.* **all seinen ~ zusammennehmen** take one's courage in both hands; screw up one's courage; **das gab** *od.* **machte ihr neuen ~:** that gave her new heart; **nur ~!** don't lose heart!; *(trau dich)* be brave!; **b)** *(veralt.) in* **guten** *od.* **frohen ~es sein** be in good spirits

Mutation [muta'tsi̯o:n] **die; ~, ~en** *(Biol.)* mutation

Mütchen ['my:tçən]: **sein ~ [an jmdm.] kühlen** *(ugs. [scherzh.])* vent one's wrath [on sb.]

mutig 1. *Adj.* brave; courageous, brave ⟨*words, decision, speech*⟩. **2.** *adv.* bravely; courageously

mut·los *Adj.* *(niedergeschlagen)* dejected; despondent; *(entmutigt)* disheartened; dispirited

Mut·losigkeit die; ~: dejection; despondency

mutmaßen ['mu:tma:sn̩] *tr., itr. V.* conjecture

mutmaßlich *Adj.; nicht präd.* supposed; presumed; suspected ⟨*terrorist etc.*⟩

Mutmaßung die; ~, ~en conjecture

Mut·probe die test of courage

[1]**Mutter** ['mʊtɐ] **die; ~, Mütter** ['mʏtɐ] mother; **sie wird ~** *(ist schwanger)* she is expecting a baby; **eine ~ von drei Kindern** a mother of three

[2]**Mutter die; ~, ~n** *(Schrauben~)* nut

Mutter-: ~boden der topsoil; **~brust die** mother's breast

Mütterchen ['mʏtɐçən] **das; ~s, ~: [altes] ~:** little old lady

Mutter·freuden *Pl.:* **~ entgegensehen** *(geh.)* be expecting a child

Mutter-: ~land das *Pl.* **~länder a)** *(Kolonialstaat)* mother country; **b)** *(Heimat)* original home; motherland; **~leib der** womb

mütterlich 1. *Adj.* **a)** *nicht präd.* the/his/her *etc.* mother's; maternal ⟨*line, love, instincts, etc.*⟩; **b)** *(fürsorglich)* motherly ⟨*woman, care*⟩. **2.** *adv.* in a motherly way

mütterlicher·seits *Adv.* on the/his/her *etc.* mother's side; **sein Großvater ~:** his maternal grandfather; his grandfather on his mother's side

Mutter-: ~liebe die motherly love *no art.*; **~mal das;** *Pl.* **~male** birthmark; **~milch die** mother's milk; **~mund der** neck of the womb; cervix; **~schaf das** mother ewe

Mutterschaft die; ~: motherhood

mutter-, Mutter-: ~schutz der *laws pl. protecting working pregnant women and mothers of newborn babies;* **~seelen·allein** *Adj.; nicht attr.* all alone; all on my *etc.* own; **~söhnchen das** *(abwertend)* mummy's *or (Amer.)* mama's boy; **~sprache die** native language; mother tongue; **~tag der;** *o. Pl.* Mother's Day *no def. art.*; **~tier das** mother [animal]; dam; **~witz der** *o. Pl.* **a)** *(Humor)* natural wit; **b)** *(Schläue)* native cunning

Mutti ['mʊti] **die; ~, ~s** mummy *(Brit. coll.)*; mum *(Brit. coll.)*; mommy *(Amer. coll.)*; mom *(Amer. coll.)*

mut-, Mut-: ~wille der; *o. Pl.* wilfulness; *(Übermut)* devilment; **aus [bloßem] ~n** from [sheer] devilment; **~willig 1.** *Adj.* wilful; wanton ⟨*destruction*⟩; *(übermütig)* high-spirited; **2.** *adv.* wilfully; wantonly; *(aus Übermut)* from devilment; **~willigkeit die; ~:** *s.* **Mutwille**

Mütze ['mʏtsə] **die; ~, ~n** cap; **etwas** *od.* **eins auf die ~ kriegen** *(fig. ugs.)* get told off; get a telling off

MW *Abk. (Rundf.)* **Mittelwelle** MW

Mw.-St., MwSt. *Abk.* **Mehrwertsteuer** VAT

Myrte ['mʏrtə] **die; ~, ~n** myrtle

Myrten·kranz der myrtle wreath

mysteriös [mʏste'ri̯ø:s] **1.** *Adj.* mysterious. **2.** *adv.* mysteriously

mystifizieren [mʏstifi'tsi:rən] *tr. V.* shroud in mystery

Mystik ['mʏstɪk] **die; ~:** mysticism

mystisch 1. *Adj.* mystical. **2.** *adv.* mystically

Mythologie [mytolo'gi:] **die; ~, ~n** mythology

mythologisch *Adj.* mythological

Mythos ['my:tɔs] **der; ~, Mythen a)** myth; **b)** *(glorifizierte Person od. Sache)* legend

N

n, N [ɛn] **das;** ~, ~: n/N; *s. auch* **a, A**

N *Abk.* **Nord[en]** N

na [na] *Interj. (ugs.)* **a)** *(als Frage, Anrede, Aufforderung)* well; **na, du?** oh, it's you?; **na los!** come on then!; **na, wird's bald?** come on, aren't you ready yet?; **na und?** *(wennschon)* so what?; **b)** *(beschwichtigend)* **na, na, na!** now, now, come along; **c)** *([zögernd] zustimmend)* **na schön!, na gut!** oh, OK *(coll.);* well, all right; **na, dann bis später** right, see you later then; **d)** *(bekräftigend)* **na und ob!** and how! *(coll.);* I'll say! *(coll.);* **na und wie!** and how! *(coll.);* **na eben!** exactly!; **na endlich!** at last!; **e)** *(triumphierend)* **Na also! Ich hatte doch recht!** There you are! I was right!; **f)** *(zweifelnd, besorgt)* **na, wenn das mal gutgeht** *od.* **klappt** well, let's hope it'll be OK *(coll.);* **na, wenn das dein Vater merkt!** oh dear, what if your father notices?; **na, ich weiß nicht** hmm, I'm not sure; **g)** *(staunend)* **na so [et]was!** well I never!; **h)** *(drohend)* **na warte!** just [you] wait!; *(auf einen nicht Anwesenden bezogen)* just let him wait!

Nabe ['na:bə] **die;** ~, ~**n** hub

Nabel ['na:bl̩] **der;** ~**s,** ~: navel

Nabel-: ~schau die *(salopp)* **~schau halten** bare one's soul; **~schnur die** *Pl.* **~schnüre** umbilical cord

nach [na:x] **1.** *Präp. mit Dat.* **a)** *(räumlich)* to; **ist das der Zug ~ Köln?** is that the train for Cologne *or* the Cologne train?; **~ Hause gehen** go home; **sich ~ vorn/hinten beugen** bend forwards/backwards; **komm ganz ~ vorn** come right to the front; **~ links/rechts** to the left/right; **~ allen Richtungen** in all directions; **~ Osten [zu]** eastwards; [towards the] east; **~ außen/innen** outwards/inwards; **ich bringe den Abfall ~ draußen** I am taking the rubbish outside; **b)** *(zeitlich)* after; **zehn [Minuten] ~ zwei** ten [minutes] past two; **c) ~ fünf Minuten** after five minutes; five minutes later; **d)** *(mit bestimmten Verben, bezeichnet das Ziel der Handlung)* for; **e)** *(bezeichnet [räumliche und zeitliche] Reihenfolge)* after; **f)** *(gemäß)* according to; **~ meiner Ansicht** *od.* **Meinung, meiner Ansicht** *od.* **Meinung ~:** in my view *or* opinion; **aller Wahrscheinlichkeit ~:** in all probability; **[frei] ~ Goethe** [freely] adapted from Goethe; **~ der neuesten Mode gekleidet** dressed in [accordance with] the latest fashion; **~ etw. schmecken/riechen** taste/smell of sth.; **sie kommt eher ~ dem Vater** *(ugs.)* she takes more after her father; **jmdn. nur dem Namen ~ kennen** know sb. by name only; **dem Gesetz ~:** in accordance with the law; by law. **2.** *Adv.* **a)** *(räumlich)* **[alle] mir ~!** [everybody] follow me!; **b)** *(zeitlich)* **~ und ~:** little by little; gradually; **~ wie vor** still; as always

nach|äffen [-ɛfn̩] *tr. V. (abwertend)* mimic

Nachäfferei die; ~ *(abwertend)* mimicry; mimicking

nach|ahmen [-a:mən] *tr. V.* imitate

nachahmens·wert *Adj.* worthy of imitation *postpos.;* exemplary

Nachahmung die; ~, ~**en** imitation

nach|arbeiten *tr. V.* **a)** *(nachholen)* **eine Stunde ~:** work an extra hour to make up; **sie muß die versäumten Stunden ~:** she has to make up for the hours she missed; **b)** *(überarbeiten)* go over, finish off ‹*workpiece*›

Nachbar ['naxba:ɐ̯] **der;** ~**n** *od. selten* ~**s,** ~**n** neighbour; **~s Hund** the neighbours'/neighbour's dog

Nachbar-: ~dorf das neighbouring village; **~haus das** house next door

Nachbarin die; ~, ~**nen** neighbour

Nachbar·land das; *Pl.* **...länder** neighbouring country

nachbarlich *Adj.* **a)** *nicht präd. (dem Nachbarn/den Nachbarn gehörend)* neighbour's/neighbours'; **b)** *(unter Nachbarn üblich)* neighbourly

Nachbarschaft die; ~ **a)** *(die Nachbarn)* **die [ganze] ~:** all the neighbours *pl.;* the whole neighbourhood; **b)** *(Beziehungen)* **gute ~:** good neighbourliness; **c)** *(Gegend)* neighbourhood; *(Nähe)* vicinity

nachbarschaftlich *Adj.* neighbourly

Nachbar·tisch der next *or* neighbouring table

nach|behandeln *tr. V.* **a)** treat again; **b)** *(nach ärztlicher Behandlung)* **jmdn./etw. ~:** give sb./sth. follow-up treatment

Nach·behandlung die follow-up treatment; after-care

nach|bestellen *tr. V.* **[noch] etw. ~:** order more of sth.; ‹*shop*› order further stock of sth., re-order sth.

nach|beten *tr. V. (ugs. abwertend)* repeat parrot-fashion; regurgitate

nach|bilden *tr. V.* reproduce, copy (+ *Dat.* from)

Nach·bildung die a) *o. Pl.* copying; **b)** *(Kopie)* copy; replica

nach|bringen *unr. tr. V.* bring along ‹*sth. left behind*›

nach·dem 1. *Konj.* after; **ich ging erst, ~ ich mich vergewissert hatte** I only left when I had made sure. **2.** *Adv.: s.* [1]**je 3b**

nach|denken *unr. itr. V.* think (**über** + *Akk.* about); *(lange u. erwägend)* reflect (**über** + *Akk.* on); **denk mal [scharf] nach** have a [good] think; think carefully; **ohne nachzudenken** without stopping to think

Nach·denken das thought; **Zeit zum ~:** time to think; **nach langem ~:** after thinking about it for a long time

nachdenklich 1. *Adj.* thoughtful; pensive. **2.** *adv.* thoughtfully; pensively

nach|drängen *itr. V.; mit sein* push from behind

Nach·druck der; *Pl.* ~**e a)** *o. Pl.* **mit ~:** emphatically; **auf etw.** *(Akk.)* **[besonderen] ~ legen** place [particular] emphasis on sth.; stress sth. [particularly]; **b)** *(Druckw.)* reprint

nach|drucken *tr. V.* reprint ‹*book*›; print more ‹*letterheads etc.*›

nachdrücklich 1. *Adj.; nicht präd.* emphatic ‹*warning, confirmation, advice*›; insistent ‹*demand*›; urgent ‹*request, appeal*›. **2.** *adv.* emphatically; **~ darauf hinweisen, daß ...:** emphasize that ...:

Nach·durst der morning-after thirst

nach|eifern *itr. V.* **jmdm. ~:** emulate sb.

nach·einander *Adv.* **a)** one after the other; **kurz/unmittelbar ~:**

one shortly/immediately after the other; **b) ~ sehen** keep an eye on each other; **sich ~ richten** co-ordinate with one another

nach|empfinden *unr. tr. V.* **a)** empathize with ‹*feeling*›; share ‹*delight, sorrow*›; **ich kann [dir] deinen Ärger gut ~:** I can well understand *or* appreciate your feeling of anger; **b)** *(nachmachen)* re-create ‹*expression, atmosphere, event*›; **einer Sache** *(Dat.)* **nachempfunden sein** take its inspiration from sth.; be modelled on sth.

nach|erzählen *tr. V.* retell

Nach·erzählung die retelling [of a story]; *(Schulw.)* reproduction

Nachfahr [-fa:ɐ̯] **der; ~en , ~en** *(geh.)* descendant

nach|fahren *unr. itr. V.; mit sein* follow [on]; **jmdm. ~:** follow sb.

nach|feiern *tr. V.* celebrate ‹*birthday, Christmas*› at a later date

Nach·folge die succession; **die ~ B.s regeln** settle who is to be B's successor; **jmds. ~ antreten** succeed sb.

nach·folgend *Adj.; nicht präd.* following; subsequent ‹*chapter, issue*›

Nachfolger der; ~s, ~, Nachfolgerin die; ~, ~nen successor

Nach·forderung die additional demand (*Gen.*, **von** for)

nach|forschen *itr. V.* make inquiries; investigate [the matter]

Nach·forschung die investigation; inquiry; **~en [nach etw.] anstellen** make inquiries [into sth.]

Nach·frage die a) *(Wirtsch.)* demand (**nach** for); **b) danke der [gütigen] ~** *(meist scherzh.)* how kind of you to inquire

nach|fragen *itr. V.* ask; inquire; **bei jmdm. ~:** ask sb.

nach|fühlen *tr. V.* empathize with; **das kann ich dir ~!** I know how you feel!

nach|füllen *tr. V.* refill ‹*glass, vessel, etc.*›; *(wenn nicht leer)* fill up; top up; **Salz/Wein ~:** put [some] more salt/wine in

nach|geben *unr. itr. V.* **a)** give way; *(aus Schwäche)* give in; **b)** *(sich dehnen)* stretch; **das Material gibt ein wenig nach** there is some give in the material; **c)** *(Bankw., Wirtsch.: sinken)* ‹*prices, currency*› weaken

Nach·gebühr die excess postage

Nach·geburt die afterbirth

nach|gehen *unr. itr. V.; mit sein* **a)** *(folgen)* **jmdm./einer Sache ~:** follow sb./sth.; **einer Sache/einer Frage/einem Problem** *usw.* **~** *(fig.)* look into a matter/question/problem *etc.*; **b)** *(nicht aus dem Kopf gehen)* **jmdm. ~:** remain on sb.'s mind; occupy sb.'s thoughts; **c) seinen Geschäften** *od.* **Beschäftigungen/seinem Tagewerk ~:** go about one's business/daily work; **einem Beruf ~:** practise a profession; **d)** ‹*clock, watch*› be slow; **[um] eine Stunde ~:** be an hour slow; **eine Stunde am Tag ~:** lose an hour a day

nach·gemacht *Adj.* imitation ‹*leather, gold*›

Nach·geschmack der aftertaste

nach·giebig [-gi:bɪç] *Adj.* indulgent; yielding; *(weich)* soft

Nachgiebigkeit die; ~: indulgence; *(Weichheit)* softness

nach|gießen 1. *unr. tr. V.* pour [in] some more; **jmdm. Wein ~:** top up sb.'s wine. **2.** *unr. itr. V.* **jmdm. ~:** pour sb. some more; top sb. up

nach|grübeln *itr. V.* ponder (**über** + *Akk.* over)

nach|gucken *tr., itr. V. (ugs.) s.* **nachsehen 1 a, b, c, 2 a, b**

Nach·hall der; ~[e]s, ~e reverberation; *(fig.)* reverberations *pl.*

nach|hallen *itr. V.* reverberate

nach·haltig 1. *Adj.* lasting. **2.** *adv. (auf längere Zeit)* for a long time; *(nachdrücklich)* persistently

Nach·hause·weg der way home

nach|helfen *unr. itr. V.* help

nach·her [*auch:* '--] *Adv.* **a)** afterwards; *(später)* later [on]; **bis ~!** see you later!; **b)** *(ugs.: womöglich)* then perhaps; *(sonst)* otherwise

Nach·hilfe die coaching

Nachhilfe-: ~lehrer der coach; **~stunde die** private lesson; **~unterricht der** coaching

nach·hinein: im ~ *(nachträglich)* afterwards; later; *(zurückblikkend)* with hindsight

Nach·hol·bedarf der need to catch up; **ein ~ an etw.** *(Dat.)* a need to make up for the shortage of sth.

nach|holen *tr. V.* **a)** catch up on ‹*work, sleep*›; make up for ‹*working hours missed*›; **den Schulabschluß ~:** take one's final school examination as a mature student; **b)** *(zu sich holen)* fetch

Nach·hut die; ~, ~en *(Milit.; auch fig.)* rearguard

nach|jagen *itr. V.; mit sein* **jmdm./einer Sache ~:** chase after sb./sth.; **dem Erfolg/Geld ~** *(fig.)* devote oneself to the pursuit of success/money

Nachkomme der; ~n, ~n descendant; *(eines Tieres)* offspring

nach|kommen *unr. itr. V.; mit sein* **a)** follow [later]; come [on] later; **seine Familie wird [später] ~:** his family will join him later; **b) seinen Verpflichtungen ~:** meet one's commitments; **einem Wunsch/Befehl/einer Bitte ~:** comply with a wish/an order/grant a request; **c)** *(Schritt halten können)* be able to keep up

Nachkommenschaft die; ~: descendants *pl.*; *(eines Tieres)* offspring

Nachkömmling [-kœmlɪŋ] **der; ~s, ~e** much younger child *(than the rest)*

Nach·kriegs-: ~generation die post-war generation; **~zeit die** post-war period

nach|laden *unr. tr., itr. V.* reload

Nach·laß der; Nachlasses, Nachlasse *od.* **Nachlässe a)** estate; *(hinterlassene Gegenstände)* personal effects *pl. (left by the deceased)*; **b)** *(Kaufmannsspr.: Rabatt)* discount; reduction

nach|lassen 1. *unr. itr. V.* let up; ‹*rain, wind*› ease, let up; ‹*storm, heat*› abate, die down; ‹*anger*› subside, die down; ‹*pain, stress, pressure*› ease, lessen; ‹*noise*› lessen; ‹*fever*› go down; ‹*effect*› wear off; ‹*interest, enthusiasm, strength, courage*› flag, wane; ‹*resistance*› weaken; ‹*health, hearing, eyesight, memory*› get worse, deteriorate; ‹*performance*› deteriorate, fall off; ‹*business*› drop off, fall off. **2.** *unr. tr. V. (Kaufmannsspr.)* give *or* allow a discount of

nach·lässig 1. *Adj.* careless; untidy ‹*dress*›. **2.** *adv.* carelessly

nachlässigerweise *Adv.* carelessly

Nach·lässigkeit die; ~, ~en *s.* **nachlässig 1:** carelessness; untidiness

nach|laufen *unr. itr. V.; mit sein* **a) jmdm./einer Sache ~:** run *or* chase after sb./sth.; **b)** *(fig.)* chase after, pursue ‹*illusion*›

nach|legen *tr., itr. V.* **[Holz/Kohlen] ~:** put some more wood/coal on

nach|lesen *unr. tr. V.* look up; *(überprüfen)* check; **in den Statistiken ist nachzulesen, daß ...:** the statistics show that ...

nach|liefern *tr. V. (später liefern)* supply later; *(zusätzlich liefern)* supply additionally; **der Rest wird nächste Woche nachgeliefert** the rest of the delivery will follow next week

nach|lösen 1. *tr. V.* **eine Fahrkarte ~:** buy a ticket [on the train/tram *(Brit.) or (Amer.)* streetcar]. **2.** *itr. V.* pay the excess [fare]

nach|machen *tr. V.* **a)** copy; *(imitieren)* imitate; do an impersonation of ⟨*politician etc.*⟩; forge ⟨*signature*⟩; forge, counterfeit ⟨*money*⟩; **jmdm. alles ~:** copy everything sb. does; **das soll mir einer ~!** follow that!; *s. auch* **nachgemacht; b)** *(ugs.: später machen)* do later

nach|messen 1. *unr. tr. V.* check the measurements of; check ⟨*distance, length, etc.*⟩. **2.** *itr. V.* check the measurements

nach·mittag *Adv.* **heute/morgen/gestern ~:** this/tomorrow/yesterday afternoon; **[am] Sonntag ~:** on Sunday afternoon

Nach·mittag der afternoon; **am ~:** in the afternoon; *(heute)* this afternoon; **am frühen/späten ~:** early/late in the afternoon; **am selben ~:** the same afternoon; **am ~ des 8. März** on the afternoon of 8 March

nach·mittags *Adv.* in the afternoon; *(heute)* this afternoon; **dienstags ~:** on Tuesday afternoons; **um vier Uhr ~:** at four in the afternoon; at 4 p.m.

Nach·mittags·vor·stellung die afternoon performance; [afternoon] matinée

Nachnahme die; ~, ~n per ~: cash on delivery; COD

Nach·name der surname; **wie heißt du mit ~n?** what is your surname?

nach|plappern *tr. V.* repeat parrot-fashion; **jmdm. alles ~:** repeat everything sb. says

Nach·porto das excess postage

nachprüfbar *Adj.* verifiable

nach|prüfen *tr. V.* check ⟨*document, statement, weight, alibi*⟩; verify ⟨*correctness*⟩

Nach·prüfung die checking

nach|rechnen 1. *tr. V.* check ⟨*figures*⟩. **2.** *itr. V. (zur Kontrolle)* check [the figures]

Nach·rede die: üble ~: malicious gossip; *(Rechtsw.)* defamation [of character]

nach|reichen *tr. V.* hand in subsequently

nach|rennen *unr. itr. V.: s.* **nachlaufen**

Nachricht ['na:xrɪçt] **die; ~, ~en a)** news *no pl.;* **das ist eine gute ~:** that is [a piece of] good news; **eine ~ hinterlassen** leave a message; **ich habe keine ~ von ihm** *(Brief usw.)* I haven't heard *or* had any word from him; **jmdm. ~ geben** inform sb.; **b)** *Pl. (Ferns., Rundf.)* news *sing.;* **~en hören** listen to the news; **Sie hören ~en** here is the news

Nachrichten-: ~agentur die news agency; **~dienst der** intelligence service; **~sendung die** news broadcast; **~sprecher der, ~sprecherin die** newsreader; **~technik die** telecommunications [technology] *no art.*

nach|rücken *itr. V.; mit sein* move up; **[auf den Posten] ~:** be promoted [to the post]; take over [the post]

Nach·ruf der; ~[e]s, ~e obituary **(auf +** *Akk.* of)

nach|rufen *unr. tr., itr. V.* **jmdm. [etw.] ~:** call [sth.] after sb.

nach|rüsten *itr. V. (Milit.)* counter-arm

Nach·rüstung die; ~ *(Milit.)* counter-arming

nach|sagen *tr. V.* **a)** *(wiederholen)* repeat; **b) jmdm. Schlechtes ~:** speak ill of sb.; **man sagt ihm nach, er verstehe etwas davon** he is said to know something about it; **du darfst dir nicht ~ lassen, daß ...:** you mustn't let it be said of you that ...

Nach·saison die late season

nach|salzen *unr., auch regelm. tr., itr. V.* **[etw.] ~:** put more salt in/on [sth.]

Nach·satz der postscript; *(gesprochen)* final remark

nach|schauen *(bes. südd., österr., schweiz.) s.* **nachsehen 1 a, b, c, 2 a, b**

nach|schenken 1. *tr. V.* **jmdm. Wein/Tee ~:** top up sb.'s glass with wine/cup with tea. **2.** *itr. V.* **jmdm. ~:** top up sb.'s glass/cup *etc.*

nach|schicken *tr. V.* **a)** *(durch die Post o. ä.)* forward; send on; **b)** *(folgen lassen)* **jmdm. jmdn. ~:** send sb. after sb.

Nach·schlag der *(ugs.: zusätzliche Portion)* second helping; seconds *pl.*

nach|schlagen 1. *unr. tr. V.* look up ⟨*word, reference, text*⟩. **2.** *unr. itr. V.* **im Lexikon/Wörterbuch ~:** consult the encyclopaedia/dictionary

Nachschlage·werk das work of reference

nach|schleichen *unr. itr. V.; mit sein* **jmdm. ~:** creep *or* steal after sb.

nach|schleifen *tr. V. (ugs.)* drag [along] behind one/it

Nach·schlüssel der duplicate key

nach|schmeißen *unr. tr. V. (ugs.)* **a)** *(billig o. ä. geben)* give away; **man bekommt sie nachgeschmissen** you get them for next to nothing; **b)** *s.* **nachwerfen**

Nach·schub der *(Milit.)* **a)** supply **(an +** *Dat.* of); *(fig.)* [provision of] further *or* fresh supplies *pl.* **(an +** *Dat.* of); **b)** *(~material)* supplies *pl.* **(an +** *Dat.* of); *(fig.)* further supplies *pl.*

nach|schütten *tr. V.* put on more ⟨*coal, coke, etc.*⟩; pour in more ⟨*water*⟩

nach|sehen 1. *unr. itr. V.* **a) jmdm./einer Sache ~:** look *or* gaze after sb./sth.; **b)** *(kontrollieren)* check *or* have a look [to see]; **c)** *(nachschlagen)* look it up; have a look. **2.** *unr. tr. V.* **a)** *(nachlesen)* look up ⟨*word, passage*⟩; **b)** *(überprüfen)* check [over]; look over; **c)** *(nicht verübeln)* overlook, let pass ⟨*remark*⟩

Nach·sehen das: das ~ haben not get a look-in; *(nichts abbekommen)* be left with nothing

nach|senden *unr. od. regelm. tr. V.: s.* **nachschicken a**

Nach·sicht die leniency; **mit jmdm. ~ haben** *od.* **üben** be lenient with sb.; make allowances for sb.

nachsichtig 1. *Adj.* lenient, forbearing **(gegen, mit** towards). **2.** *adv.* leniently

Nach·silbe die *(Sprachw.)* suffix

nach|sitzen *unr. itr. V. (Schulw.)* be in detention; **[eine Stunde] ~ müssen** have [an hour's] detention

Nach·speise die dessert; sweet

Nach·spiel das: die Sache wird noch ein ~ haben! this affair will have repercussions; **ein gerichtliches ~ haben** result in court proceedings

nach|spielen *itr. V. (Ballspiele, bes. Fußball)* **[einige Minuten] ~:** play [a few minutes of] time added on; **~ lassen** ⟨*referee*⟩ add on time

nach|spionieren *itr. V.* **jmdm. ~:** spy on sb.

nach|sprechen *unr. tr. V.* **[jmdm.] etw. ~:** repeat sth. [after sb.]

nächst... 1. *Sup. zu* **nahe. 2.** *Adj.* **a)** nearest *attrib.; (räumliche od. zeitliche Reihenfolge)* next *attrib.;* closest ⟨*relatives*⟩; **die ~e Straße links** the next street on the left; **am ~en Tag** the next day; **am ~en ersten** on the first of next month; **bei ~er Gelegenheit** at the next opportunity; **beim ~en Mal, das ~e Mal** the next time; **der ~e bitte!** next [one], please; **wer kommt als ~er**

dran? whose turn is it next?; **b) der ~e Weg zum Bahnhof** the shortest way to the station. **3.** *adv.* **am ~en** nearest; *s. auch* **best... b**

nächst·beste *Adj. s.* **erstbeste**

Nächst·beste der/die/das; *adj. Dekl.* the first one [to turn up]

Nächste der; ~n, ~n *(geh.)* neighbour

nach|stehen *unr. itr. V.* **jmdm./einer Sache in nichts ~:** be in no way inferior to sb./sth.

nach·stehend *Adj.* following

nach|steigen *unr. itr. V.; mit sein (ugs.)* **einem Mädchen ~:** try to get off with *(Brit. coll.) or (Amer. coll.)* make it with a girl

nach|stellen 1. *tr. V.* **a)** *(Sprachw.)* **A wird B** *(Dat.)* **nachgestellt** A is placed after B; **nachgestellte Präposition** postpositive preposition; **b)** *(zurückstellen)* put back ‹*clock, watch*›; **c)** *(neu/genauer einstellen)* [re]adjust; take up the adjustment on ‹*brakes, clutch*›. **2.** *itr. V. (geh.)* **einem Tier/einem Flüchtling ~:** hunt an animal/hunt *or* pursue a fugitive; **einem Mädchen ~** *(ugs.)* chase a girl

Nach·stellung die a) *(Sprachw.)* postposition; **b)** *Pl. (Verfolgung)* pursuit *sing.*

Nächsten·liebe die charity [to one's neighbour]; brotherly love

nächstens ['nɛ:çstn̩s] *Adv.* **a)** *(demnächst)* shortly; in the near future; **passen Sie ~ besser auf!** be more careful next time; **b)** *(ugs.: wenn es so weitergeht)* if it goes on like this

nächst-: ~gelegen *Adj.; nicht präd.* nearest; **~höher** *Adj.; nicht präd.* next higher; **die ~höhere Klasse** the next class [up]; **~liegend** *Adj.; nicht präd.* first, immediate ‹*problem*›; [most] obvious ‹*explanation etc.*›; **das Nächstliegende** the [most] obvious thing; **~möglich** *Adj.; nicht präd.* earliest possible

nach|suchen *itr. V.* **um etw. ~** *(geh.)* request sth.; *(bes. schriftlich)* apply for sth.

nacht [naxt] *Adv.* **gestern/morgen/Dienstag ~:** last night/tomorrow night/on Tuesday night; **heute ~:** tonight; *(letzte Nacht)* last night

Nacht die; ~, Nächte ['nɛçtə] night; **es wird/ist ~:** it is getting dark/it is dark; night is falling/has fallen; **bei ~, in der ~** at night[-time]; **eines ~s** one night; **letzte ~:** last night; **die halbe ~:** half the night; **die ganze ~ [hindurch]** all night long; **diese ~:** tonight; **mitten in der ~:** in the middle of the night; **bis tief in die ~ hinein, bis spät in der ~:** until late at night; *(bis in die Morgenstunden)* into the small hours; **in der ~ vom 12. auf den 13. Mai** on the night of 12 May; **in der ~ auf Montag** on Sunday night; **über ~ bleiben** stay overnight; **über ~ berühmt werden** *(fig.)* become famous overnight; **sich** *(Dat.)* **die ~ um die Ohren schlagen** *(ugs.)* stay up all night; **zu[r] ~ essen** *(südd., österr.)* have one's evening meal; **gute ~!** good night!; **[na,] dann gute ~!** *(iron.)* [well,] that's that; **bei ~ und Nebel** under cover of darkness; *(heimlich)* furtively; like a thief in the night; *s. auch* **heilig b; schwarz 1a**

nacht·aktiv *Adj. (Zool.)* nocturnal

nacht-, Nacht-: ~arbeit die; *o. Pl.* night work *no art.;* **~bar die** night-spot *(coll.);* **~blind** *Adj.* night-blind; **~dienst der** night duty; **~dienst haben** be on night duty; ‹*chemist's shop*› be open late

Nach·teil der disadvantage; **im ~ sein, sich im ~ befinden** be at a disadvantage; **sich zu seinem ~ verändern** change for the worse

nachteilig 1. *Adj.* detrimental; harmful; **über sie ist nichts Nachteiliges bekannt** nothing to her disadvantage is known about her. **2.** *adv.* detrimentally; harmfully; **sich ~ auswirken** have a detrimental *or* harmful effect

nächte·lang 1. *Adj.; nicht präd.* lasting several nights *postpos.; (ganze Nächte dauernd)* all-night. **2.** *adv.* night after night

Nacht-: ~essen das *(bes. südd., schweiz.) s.* **Abendessen; ~eule die** *(ugs. scherzh.)* night-owl *(coll.);* **~falter der** moth; **~frost der** night frost; **~gespenst das** [nocturnal] ghost; **~hemd das** nightshirt; **~himmel der;** *o. Pl.* night sky

Nachtigall ['naxtɪgal] **die; ~, ~en** nightingale

Nach·tisch der; *o. Pl.* dessert; sweet; **zum** *od.* **als ~:** as a *or* for dessert; **was gibt's zum ~?** what's for pudding *or (coll.)* afters?

Nacht·leben das night-life

nächtlich ['nɛçtlɪç] *Adj.; nicht präd.* nocturnal; night ‹*sky*›; ‹*darkness, stillness*› of the night

Nacht-: ~lokal das night-spot *(coll.);* **~mensch der** night-owl *(coll.);* **~portier der** night porter

nach|tragen *unr. tr. V.* **a)** *(schriftlich ergänzen)* insert, add; *(noch sagen)* add; **b) jmdm. etw. ~:** follow sb. carrying sth.; **c) jmdm. etw. ~** *(fig.)* hold sth. against sb.

nach·tragend *Adj.* unforgiving; *(rachsüchtig)* vindictive; **ich bin nicht ~:** I don't bear grudges

nachträglich [-trɛ:klɪç] **1.** *Adj.; nicht präd.* later; subsequent ‹*apology*›; *(verspätet)* belated ‹*greetings, apology*›; *(zusätzlich)* additional. **2.** *adv.* afterwards; subsequently; *(verspätet)* belatedly

nach|trauern *itr. V.* **jmdm./einer Sache ~:** bemoan *or* lament the passing of sb./sth.; *(sich sehnen nach)* pine for sb./sth.

Nacht·ruhe die night's sleep; **angenehme ~!** sleep well!

nachts *Adv.* at night; **Montag** *od.* **montags ~:** on Monday nights; **um 3 Uhr ~, ~ um 3 [Uhr]** at 3 o'clock in the morning

Nacht-: ~schicht die night shift; **~schicht haben** be on night shift; work nights; **~schwester die** night nurse; **~tisch der** bedside table; **~tisch·lampe die** bedside light; **~tresor der** night safe; **~wache die a)** *(Wachdienst)* night-watch; *(im Krankenhaus)* night-duty; *(eines Soldaten)* night guard-duty; **b)** *(Person)* night-guard; *(für Fabrik, Büro o.ä.)* night-watchman; **~wächter der** night-watchman; **~wanderung die** nocturnal ramble

Nach·untersuchung die follow-up examination; check-up

nachvollziehbar *Adj.* comprehensible; **leicht/schwer ~:** easy/difficult to comprehend

nach|vollziehen *unr. tr. V.* reconstruct ‹*train of thought*›; *(begreifen)* comprehend

nach|wachsen *unr. itr. V.; mit sein* **[wieder] ~:** grow again

Nach·wahl die by-election

Nach·wehen *Pl. (Med.)* afterpains; *(fig. geh.)* unpleasant after-effects

nach|weinen *itr. V.* **jmdm./einer Sache ~:** bemoan the loss of sb./sth.; *s. auch* **Träne**

Nachweis [-vais] **der; ~es, ~e** proof *no indef. art.* (*Gen.*, **über** + *Akk.* of); *(Zeugnis)* certificate (**über** + *Akk.* of)

nachweisbar 1. *Adj.* demonstrable ‹*fact, truth, error, defect, guilt*›; provable ‹*fact, guilt*›; detectable ‹*substance, chemical*›. **2.** *adv.* demonstrably

nach|weisen *unr. tr. V.* prove; **jmdm. einen Fehler/Diebstahl ~:** prove sb. made a mistake/committed a theft; **man konnte ihm nichts ~:** they could not prove anything against him
nachweislich 1. *Adj.; nicht präd.* demonstrable. **2.** *adv.* demonstrably; as can be proved
Nach·welt die; *o. Pl.* posterity *no art.;* future generations *pl., no art.*
nach|werfen *unr. tr. V.* **jmdm. etw. ~:** throw sth. after sb.; **eine Münze ~:** put in another coin
nach|winken *itr. V.* **jmdm./einer Sache ~:** wave after sb./sth.
nach|wirken *itr. V.* have a lasting effect (**bei** on); ‹*medicine*› continue to have an effect; ‹*literary work*› continue to have an influence
Nach·wirkung die after-effect; *(fig.: Einfluß)* influence
Nach·wort das; *Pl.* **~worte** afterword, postface (**zu** to)
Nach·wuchs der; *o. Pl* **a)** *(fam.: Kind[er])* offspring; **sie erwartet ~:** she's expecting [a baby]; **b)** *(junge Kräfte)* new blood; *(für eine Branche usw.)* new recruits *pl.*
nach|zahlen *tr., itr. V.* **a)** pay later; **1 000 DM Steuern ~:** pay 1,000 marks back tax; **b)** *(zusätzlich zahlen)* **25 DM ~:** pay another 25 marks
nach|zählen *tr., itr. V.* [re]count; check
Nach·zahlung die additional payment; *(spätere Zahlung)* deferred payment; *(Steuerzahlung)* back tax
nach|ziehen 1. *unr. itr. V.* **a)** *(ugs.: ebenso handeln)* do likewise; follow suit; **b)** *mit sein (hinterhergehen)* **jmdm./einer Sache ~:** follow sb./sth. **2.** *unr. tr. V.* **a)** *(hinter sich herziehen)* drag ‹*foot, leg*›; **b)** *(verstärkend)* retrace, go over ‹*line*›; pencil ‹*eyebrows*›; **c)** *(festziehen)* tighten [up] ‹*nut, bolt*›
Nachzügler [-tsy:klɐ] **der** straggler; *(spät Ankommender)* latecomer
Nackedei ['nakədai] **der; ~s, ~s a)** *(fam. scherzh.: Kind)* **[kleiner] ~:** naked little thing *or* monkey; little bare-bum *(Brit. coll.);* **b)** *(ugs. scherzh.: Person)* person in the buff
Nacken ['nakn̩] **der; ~s, ~:** back *or* nape of the neck; *(Hals)* neck; **den Kopf in den ~ werfen** throw one's head right back; **jmdm. im ~ sitzen** *(fig.)* be breathing down sb.'s neck; **die Furcht/Angst sitzt ihm im ~:** he is gripped by fear
nackend *(veralt., landsch.) s.* **nackt a**
Nacken·haar das hair on the back of one's neck; neck hair
nackert ['nakɐt] *(südd., österr.),* **nackig** *(bes. md.) s.* **nackt a**
nackt [nakt] *Adj.* **a)** *(unbekleidet)* naked; bare ‹*feet, legs, arms, skin, fists*›; **sich ~ ausziehen** strip naked; strip off completely; **~ baden** bathe in the nude; **b)** *(kahl)* bald ‹*head*›; hairless ‹*chin*›; featherless ‹*bird*›; bare ‹*rocks, island, tree, branch, walls, bulb*›; **auf dem ~en Boden schlafen** sleep on the bare floor; **c)** *(unverhüllt)* stark ‹*poverty, misery, horror*›; naked ‹*greed*›; plain ‹*fact, words*›; plain, unvarnished ‹*truth*›; **~e Angst** sheer *or* stark terror; **d)** bare ‹*existence*›; **das ~e Leben retten** barely manage to escape with one's life; save one's skin [and nothing more]
Nackt-: ~baden das; ~s nude bathing; **~bade·strand der** nudist beach
Nackte der/die; *adj. Dekl.* naked man/woman
Nackt·foto das nude photo
Nacktheit die; ~: nakedness; nudity; *(fig.: der Landschaft usw.)* bareness
Nadel ['na:dl̩] **die; ~, ~n** needle; *(Steck~, Hut~, Haar~)* pin; *(Häkel~)* hook; *(für Tonabnehmer)* stylus
Nadel-: ~baum der conifer; coniferous tree; **~holz das;** *Pl.* **~hölzer** softwood; pine-wood; **~kissen das** pincushion
nadeln *itr. V.* ‹*tree*› shed its needles
Nadel-: ~öhr das eye of a/the needle; **~stich der a)** needle-prick; *(einer Stecknadel usw.)* pinprick; *(fig.: Bosheit)* barbed *or (coll.)* snide remark; **b)** *(Nähstich)* stitch; **~streifen·anzug der** pin-stripe suit; **~wald der** coniferous forest
Nagel ['na:gl̩] **der; ~s, Nägel** ['nɛ:gl̩] **a)** nail; **b)** *(fig.)* **den ~ auf den Kopf treffen** *(ugs.)* hit the nail on the head *(coll.);* **Nägel mit Köpfen machen** *(ugs.)* do things properly; make a real job of it; **den Sport** *usw.* **/den Beruf an den ~ hängen** *(ugs.)* give up sport *etc./ (coll.)* chuck in one's job; **c)** *(Finger~, Zehen~)* nail; **das brennt mir auf** *od.* **unter den Nägeln** *(fig. ugs.)* it's so urgent I just have to get on with it *or* it just won't wait; **sich** *(Dat.)* **etw. unter den ~ reißen** *(fig. salopp)* make off with sth.
Nagel-: ~bürste die nailbrush; **~feile die** nail-file; **~lack der** nail varnish *(Brit.);* nail polish
nageln *tr. V.* nail (**an** + *Akk.* to, **auf** + *Akk.* on); *(Med.)* pin ‹*bone, leg, etc.*›
nagel-, Nagel-: ~neu *Adj. (ugs.)* brand-new; **~reiniger der** nail-cleaner; **~schere die** nail-scissors *pl.*
nagen ['na:gn̩] **1.** *itr. V.* gnaw; **an etw.** *(Dat.)* **~:** gnaw [at] sth. **2.** *tr. V.* gnaw off; **ein Loch ins Holz ~:** gnaw a hole in the wood
nagend *Adj.* gnawing ‹*pain, hunger, fear*›; nagging ‹*pain, doubts, uncertainty, etc.*›
Nage·tier das rodent
nah [na:] *s.* **nahe**
Näh·arbeit die [piece of] sewing; **~en** sewing jobs; sewing *sing.*
Nah·aufnahme die close-up [photograph]
nahe ['na:ə] **1.** *Adj.* **näher** ['nɛ:ɐ], **nächst...** [nɛ:çst...] **a)** *(räumlich)* near *pred.;* close *pred.;* nearby *attrib.;* **in der näheren Umgebung** in the neighbourhood; around here/there; *s. auch* **Osten c; b)** *(zeitlich)* imminent; near *pred.;* **in ~r Zukunft** in the near future; **c)** *(eng)* close ‹*relationship, relative, friend*›. **2.** *adv.* **näher, am nächsten a)** *(räumlich)* **~ an** (+ *Dat./ Akk.*), **~ bei** close to; **~ gelegen** nearby; **komm mir nicht zu ~!** don't come too close!; keep your distance!; **~ beieinander** close together; **von ~m** from close up; at close quarters; **aus** *od.* **von nah und fern** *(geh.)* from near and far; **jmdm. zu ~ treten** *(fig.)* offend sb.; **b)** *(zeitlich)* **~ daran sein, etw. zu tun** be on the point of doing sth.; **c)** *(eng)* closely; *s. auch* **näher. 3.** *Präp. mit Dat. (geh.)* near; close to; **den Tränen/dem Wahnsinn ~ sein** be on the brink of tears/on the verge of madness
Nähe ['nɛ:ə] **die; ~: a)** closeness; proximity; *(Nachbarschaft)* vicinity; **in der ~ der Stadt** near the town; **in meiner ~:** near me; **er wohnt in der ~/ganz in der ~:** he lives in the vicinity *or* nearby/ very near; **etw. aus der ~ betrachten** take a closer look at sth.; **aus der ~ betrachtet** *(auch fig.)* viewed more closely; *s. auch* **greifbar 1 a**
nahe-: ~bei *Adv.* nearby; close by; **~|bringen** *unr. tr. V.* **jmdm. die moderne Kunst** *usw.* **~bringen** make modern art *etc.* accessible to sb.; **~|gehen** *unr. itr. V.; mit sein* **jmdm. ~gehen** affect sb. deeply

nahe-: ~**|kommen** *unr. itr. V.; mit sein* **einer Sache** *(Dat.)* **~kommen** come close to sth.; ⟨*amount*⟩ approximate to sth.; ~**|legen** *tr. V.* suggest; give rise to ⟨*suspicion, supposition, thought*⟩; **jmdm. etw. ~legen** suggest sth. to sb.; ~**|liegen** *unr. itr. V.* ⟨*thought*⟩ suggest itself; ⟨*suspicion, question*⟩ arise; **~liegend** *Adj.* ⟨*question, idea*⟩ which [immediately] suggests itself; natural ⟨*suspicion*⟩; obvious ⟨*reason, solution*⟩

nahen *(geh.) itr. V.; mit sein* draw near; **sein/ihr Ende nahte** the end was near; **eine ~de Katastrophe** imminent disaster

nähen 1. *itr. V.* sew; *(Kleider machen)* make clothes. **2.** *tr. V.* **a)** sew ⟨*seam, hem*⟩; *(mit der Maschine)* machine ⟨*seam, hem*⟩; *(herstellen)* make ⟨*dress, coat, curtains, etc.*⟩; **b)** *(Med.)* stitch ⟨*wound etc.*⟩; *s. auch* **doppelt 2a**

näher 1. *Komp. zu* **nahe. 2.** *Adj.; nicht präd.* **a)** *(kürzer)* shorter ⟨*way, road*⟩; **b)** *(genauer)* further, more precise ⟨*information*⟩; closer ⟨*investigation, inspection*⟩; **die ~en Umstände** the precise circumstances; **bei ~em Hinsehen** on closer examination; **wissen Sie Näheres [darüber]?** do you know any more [about it]?; do you know any details?; **Näheres hierzu siehe unten** for further information on this see below. **3.** *adv.* **a) bitte treten Sie ~!** please come in/nearer/this way; **b)** *(genauer)* more closely; *(im einzelnen)* in [more] detail; **jmdn./etw. ~ kennenlernen** get to know sb./sth. better; **ich kenne ihn nicht ~:** I don't know him well

näher|bringen *unr. tr. V.* **jmdm. etw. ~:** make sth. more real *or* more accessible to sb.

Nah·erholungs·gebiet das nearby recreational area

Näherin die; ~, ~nen needlewoman

näher-: ~**|kommen** *unr. itr. V.; mit sein* **jmdm. [menschlich] ~kommen** get on closer terms with sb.; **sich** *(Dat.)/(geh.)* **einander ~kommen** become closer

nähern *refl. V.* approach; **die Tiere näherten sich bis auf wenige Meter** the animals came up to within a few metres; **sich jmdm./einer Sache** *(Dat.)* **~:** approach sb./sth.; draw nearer to sb./sth.; **sich dem Ziel der Reise ~:** near one's destination

nahe-: ~**|stehen** *unr. itr. V.* **jmdm. ~stehen** be on close *or* intimate terms with sb.; **einer Partei ~stehen** sympathize with a party; **~stehend** *Adj.* **eine der Witwe ~stehende Cousine** a cousin who is/was on close terms with the widow; **~zu** *Adv. (mit Adjektiven)* almost; nearly; wellnigh ⟨*impossible*⟩; all but ⟨*exhausted, impossible*⟩; *(mit Zahlenangabe)* close on

Näh·garn das [sewing] cotton

Nah·kampf der *(Milit.)* close combat

Näh-: ~**kästchen das a)** *s.* **~kasten; b) aus dem ~kästchen plaudern** *(ugs. scherzh.)* tell all; *(als Kenner, Fachmann)* tell the inside story; **~kasten der** sewing-box; work-box

nahm [na:m] *1. u. 3. Pers. Sg. Prät. v.* **nehmen**

Näh-: ~**maschine die** sewing-machine; **~nadel die** sewing-needle

Nähr·boden der culture medium; *(fig.)* breeding-ground

nähren ['nɛ:rən] *tr. V.* **a)** *(ernähren)* feed ⟨*animal, child*⟩ **(mit** on); **gut/schlecht genährt** well-fed/underfed; **b)** *(geh.: entstehen lassen)* nurture ⟨*hope, suspicion, hatred*⟩; cherish ⟨*desire, hope*⟩; foster ⟨*plan, hatred*⟩

nahrhaft *Adj.* nourishing; nutritious; **ein ~es Essen** *od. (geh.)* **Mahl** a square meal

Nahrung ['na:rʊŋ] **die; ~:** food; **dem Verdacht/den Gerüchten** *usw.* **~ geben** *od.* **bieten** *(fig.)* help to nurture *or* foster the suspicion/the rumours *etc.*

Nahrungs-: ~**aufnahme die** intake of food; **die ~aufnahme verweigern** refuse food; **~mittel das** food [item]; **~mittel** *Pl.* foodstuffs; **~suche die** search for food

Nähr·wert der nutritional value

Näh·seide die sewing silk

Naht [na:t] **die; ~, Nähte** ['nɛ:tə] **a)** seam; **aus den** *od.* **allen Nähten platzen** *(fig. ugs.)* ⟨*person, fig.: institution etc.*⟩ be bursting at the seams; **b)** *(Med., Anat.)* suture

naht·los 1. *Adj.* seamless; *(fig.)* perfectly smooth ⟨*transition*⟩. **2.** *adv.* **Studium und Beruf gehen nicht ~ ineinander über** there is not a perfectly smooth transition from study to work

Naht·stelle die a) *(Schweißnaht)* seam; **b)** *(Berührungsstelle)* point of contact, interface **(von** between); *(Grenzlinie)* borderline

Nah-: ~**verkehr der** local traffic; **~verkehrsmittel das** form of local transport; **~verkehrszug der** local train

Näh·zeug das sewing things *pl.*

Nah·ziel das short-term *or* immediate aim

naiv [na'i:f] **1.** *Adj.* naïve; ingenuous ⟨*look; child*⟩; unaffected ⟨*pleasure*⟩. **2.** *adv.* naïvely

Naivität [naivi'tɛ:t] **die; ~:** naïvety; *(eines Blickes, Kindes)* ingenuousness; *(von Vergnügen)* unaffectedness

Naivling der; ~s, ~e *(ugs. abwertend)* [naïve] simpleton

Name ['na:mə] **der; ~ns, ~n** name; **wie war gleich Ihr ~?** what was your name again?; **ich kenne ihn/es nur dem ~n nach** I know him/it only by name; **unter jmds. ~n** *(Dat.)* under sb.'s name; **das Konto/das Auto läuft auf meinen ~n** the account is in/the car is registered in my name; **ein Mann mit ~n Emil** a man by the name of Emil; **in jmds./einer Sache ~n, im ~n von jmdm./etw.** on behalf of sb./sth.; **in Gottes ~n!** *(ugs.)* for God's sake; *s. auch* **Hase a; Kind a**

namen-, Namen-: ~**gedächtnis das** memory for names; **~liste die** list of names; **~los** *Adj.* nameless; *(unbekannt)* unknown; anonymous ⟨*author, poet*⟩

namens 1. *Adv.* by the name of; called. **2.** *Präp. mit Gen. (Amtsspr.)* on behalf of

Namens-: ~**änderung die** change of name; **~schild das a)** *(an Türen usw.)* name-plate; **b)** *(zum Anstecken)* name-badge; **~tag der** name-day; **sie hat am ... ~tag** it is her name-day on the ...; **~vetter der** namesake

namentlich ['na:məntlɪç] **1.** *Adj.* by name *postpos.;* **eine ~e Abstimmung** a roll-call vote. **2.** *adv.* by name; **jmdn. ~ nennen** mention sb. by name; name sb.; **3.** *Adv. (besonders)* particularly; especially

namhaft *Adj.* **a)** *nicht präd. (berühmt)* noted; of note *postpos.;* **b)** *(ansehnlich)* noteworthy ⟨*sum, difference*⟩; notable ⟨*contribution, opportunity*⟩

nämlich ['nɛ:mlɪç] *Adv.* **a) er kann nicht kommen, er ist ~ krank** he cannot come, as he is ill; he can't come – he's ill[, you see] *(coll.);* **b)** *(und zwar)* namely; *(als Füllwort)* **das war ~ ganz anders** it was quite different in fact *or* actually

nannte ['nantə] *1. u. 3. Pers. Sg. Prät. v.* **nennen**

nanu [na'nu:] *Interj.* **~, was machst du denn hier?** hello, what are you doing here?; **~, wo ist**

denn der ganze Käse geblieben? that's funny, what's happened to all that cheese?; ~, **Sie gehen schon?** what, you're going already?

Napalm ['na:palm] **das;** ~s napalm

Napalm·bombe die napalm bomb

Napf [napf] **der;** ~[e]s, **Näpfe** ['nɛpfə] bowl

Napf·kuchen der gugelhupf; ring cake

Nappa ['napa] **das;** ~[s] ~s, **Nappa·leder das** nappa [leather]

Narbe ['narbə] **die;** ~, ~n **a)** scar; **b)** *(Bot.)* stigma

narbig *Adj.* scarred; *(von Pocken o. ä.)* pitted; pock-marked

Narkose [nar'ko:zə] **die;** ~, ~n *(Med.)* narcosis; **aus der ~ aufwachen** come round from the anaesthetic

Narkose·arzt der anaesthetist

Narkotikum [nar'ko:tikum] **das;** ~s, **Narkotika** *(Med.)* narcotic

narkotisieren *tr. V. (Med.)* anaesthetize ‹*patient*›; put ‹*patient*› under a general anaesthetic

Narr [nar] **der;** ~en, ~en fool; *(Hof~)* jester; fool; *(Fastnachts~)* carnival jester *or* reveller; **sich zum ~en machen** let oneself be fooled; **jmdn. zum ~en haben** *od.* **halten** play tricks on sb.; *(täuschen)* pull the wool over sb.'s eyes; **einen ~en an jmdm. gefressen haben** *(ugs.)* be dotty about sb. *(coll.)*

narren ['narən] *tr. V. (geh.)* **jmdn. ~:** make a fool of sb.; *(täuschen)* deceive sb.

Narren·freiheit die freedom to do as one pleases

narren·sicher *(ugs.)* **1.** *Adj.* foolproof; **2.** *adv.* in a foolproof way

Närrin ['nɛrɪn] **die;** ~, ~nen fool

närrisch ['nɛrɪʃ] **1.** *Adj.* **a)** *(verrückt)* crazy; *(wirr im Kopf)* scatter-brained; dotty *(coll.)*; **[ein] ~es Zeug reden** talk gibberish; **auf etw.** *(Akk.) od.* **nach etw. ganz ~ sein** be mad keen on sth. *(sl.)*; **b)** *nicht präd. (karnevalistisch)* carnival-crazy ‹*season*›; **das ~e Treiben [beim Karneval** *od.* **Fasching]** the mad *or* crazy carnival antics *pl.* **2.** *adv. (verrückt)* crazily terrifically *(coll.)*

Narzisse [nar'tsɪsə] **die;** ~, ~n narcissus; **gelbe ~:** daffodil

narzißtisch *Adj.* narcissistic

nasal [na'za:l] **1.** *Adj.* nasal. **2.** *adv.* nasally

Nasal der; ~s, ~e *(Sprachw.)* nasal

naschen ['naʃn̩] **1.** *itr. V.* **a)** eat sweet things; *(Bonbons essen)* eat sweets *(Brit.) or (Amer.)* candy; **gern ~:** have a sweet tooth; **b)** *(heimlich essen)* have a nibble. **2.** *tr. V.* eat ‹*sweets, chocolate, etc.*›; **er/sie hat Milch genascht** he/she has been at the milk

Nascherei die; ~, ~en **a)** *o. Pl.* [continually] eating sweet things; **b)** *(Süßigkeit)* ~en sweets

naschhaft *Adj.* fond of sweet things *postpos.*; sweet-toothed; **~ sein** have a sweet tooth

Nasch·katze die *(fam.)* compulsive nibbler; *(Süßigkeiten naschend)* compulsive sweet- *(Brit.) or (Amer.)* candy-eater

Nase ['na:zə] **die;** ~, ~n **a)** nose; **mir blutet die ~:** my nose is bleeding; I've got a nosebleed; **mir läuft die ~, meine ~ läuft** I've got a runny nose; **b)** *(fig.)* **der Bus ist mir vor der ~ weggefahren** *(ugs.)* I missed the bus by a whisker; **jmdm. die Tür vor der ~ zuschlagen** *(ugs.)* shut the door in sb.'s face; **die ~ voll haben** *(ugs.)* have had enough; **von jmdm./etw. die ~ [gestrichen] voll haben** *(ugs.)* be sick [to death] of sb./sth.; **seine ~ in etw./alles stecken** *(ugs.)* stick one's nose into sth./everything *(coll.)*; **jmdm. eine lange ~ machen** *od.* **eine ~ drehen** *(ugs.)* cock a snook at sb.; **immer der ~ nach** *(ugs.)* just follow your nose; **jmdn. an der ~ herumführen** *(ugs.)* pull the wool over sb.'s eyes; **auf die ~ fallen** *(ugs.)* come a cropper *(sl.)*; **jmdm. etw. auf die ~ binden** *(ugs.)* let sb. in on sth.; **jmdm. auf der ~ herumtanzen** *(ugs.)* play sb. up; **jmdm. eins** *od.* **was auf die ~ geben** *(ugs.)* put sb. in his/her place; **jmdm. etw. aus der ~ ziehen** *(ugs.)* worm sth. out of sb.; **das sticht mir schon lange in die ~** *(ugs.)* I've had my eye on that for a long time; **jmdn. mit der ~ auf etw.** *(Akk.)* **stoßen** *(ugs.)* spell sth. out to sb.; **pro ~** *(ugs.)* per head; **jmdm. unter die ~ reiben, daß ...** *(ugs.)* rub it in that ...; **c)** *(Geruchssinn, Gespür)* nose; **eine gute ~ für etw. haben** have a good nose for sth.; *(etw. intuitiv wissen)* have a sixth sense for sth.

näseln ['nɛ:zl̩n] *itr. V.* talk through one's nose

Nasen-: ~bein das nasal bone; **~bluten das;** ~s bleeding from the nose; **~bluten haben/bekommen** have/get a nosebleed; **~flügel der** side of the nose; *(einschl. ~loch)* nostril; **~länge die: mit einer ~länge** *(Pferdesport)*, **um eine ~länge** *(fig.)* by a head; **~loch das** nostril; **~rücken der** ridge of the/one's nose; **~spitze die** tip of the/one's nose; **jmdm. etw. an der ~spitze ansehen** *(fig. ugs.)* tell sth. by sb.'s face; **~stüber der** swat on the nose; **~tropfen** *Pl.* nose-drops

nase-, Nase-: ~rümpfend 1. *Adj.* disapproving; **2.** *adv.* disdainfully; **~weis 1.** *Adj.* precocious; pert ‹*remark, reply*›; **sei nicht so ~weis!** don't be such a little know-all!; **2.** *adv.* precociously; **~weis der;** ~es, ~e *(fam.)* [little] know-all; [little] clever Dick *(coll.)*

Nas·horn das rhinoceros

nas·lang *in* **alle ~:** constantly; all the time

naß [nas] **nasser** *od.* **nässer** ['nɛsɐ], **nassest...** *od.* **nässest...** ['nɛsəst...] **1.** *Adj.* wet; **~ machen** make wet; sprinkle ‹*washing*›; **sich/das Bett ~ machen** wet oneself/one's bed; **durch und durch** *od.* **bis auf die Haut ~:** wet through; soaked to the skin. **2.** *adv.* **sich ~ rasieren** have a wet shave; *(immer)* use a razor and shaving cream

Nässe ['nɛsə] **die;** ~: wetness; *(an Wänden usw.)* dampness; **bei ~:** in the wet; in wet weather

nässen 1. *itr. V.* ‹*wound, eczema*› suppurate. **2.** *tr. V. (geh.)* make wet; wet ‹*bed, feet, etc.*›

naß-, Naß-: ~forsch 1. *Adj.* brash. **2.** *adv.* brashly; **~kalt** *Adj.* cold and wet; raw; **~rasur die** wet shaving *no art.*

Nation [na'tsi̯o:n] **die;** ~, ~en nation

national [natsi̯o'na:l] **1.** *Adj.* **a)** national; **b)** *(patriotisch)* nationalist. **2.** *adv.* **a)** at a national level; nationally; **b)** *(patriotisch)* ‹*think, feel*› nationalistically

national-, National-: ~bewußt *Adj.* nationally conscious; **~bewußt sein** be conscious of one's nationality; have a sense of national identity; **~bewußtsein das** [sense of] national consciousness; sense of national identity; **~elf die** *(Fußball)* national team *or* side; **~feier·tag der** national holiday; **~flagge die** national flag; **~gericht das** national dish; **~hymne die** national anthem

Nationalismus der; ~: nationalism *usu. no art.*

nationalistisch 1. *Adj.* nationalist; nationalistic. **2.** *adv.* nationalistically

Nationalität [natsi̯onali'tɛ:t] **die;** ~, ~en nationality

National-: **~mannschaft die** national team; **~sozialismus der** National Socialism; **~sozialist der** National Socialist; **~spieler der** *(Sport)* national player; international; **~straße die** *(schweiz.)* national highway; **~versammlung die** National Assembly

NATO, Nato ['na:to] **die; ~:** NATO; Nato *no art.*

Natrium ['na:triʊm] **das; ~s** *(Chemie)* sodium

Natron ['na:trɔn] **das; ~s [doppeltkohlensaures] ~:** sodium bicarbonate; bicarbonate of soda; bicarb *(coll.)*; **[kohlensaures] ~:** sodium carbonate; soda

Natter ['natɐ] **die; ~, ~n** colubrid

Natur [na'tu:ɐ̯] **die; ~, ~en a)** *o. Pl.* nature *no art.*; **wider die ~:** unnatural; **die freie ~:** [the] open countryside; **Tiere in freier ~ sehen** see animals in the wild; **zurück zur ~:** back to nature; **b)** *(Art, Eigentümlichkeit)* nature; **eine gesunde/eiserne/labile ~ haben** *(ugs.)* have a healthy/cast-iron/delicate constitution; **das widerspricht ihrer ~:** it is not in her nature; **jmdm. zur zweiten ~ werden** become second nature to sb.; **in der ~ der Sache/der Dinge liegen** be in the nature of things; **c)** *(Mensch)* sort *or* type of person; sort *(coll.)*; type *(coll.)*; **d)** *o. Pl. (natürlicher Zustand)* **Möbel in Kiefer ~:** natural pine furniture; **sie ist von ~ aus blond/gutmütig** she is naturally fair/good-natured

Naturalien [natu'ra:li̯ən] *Pl.* natural produce *sing.* *(used as payment)*; **in ~** *(Dat.)* **bezahlen** pay in kind

Naturalismus der; ~: naturalism

naturalistisch 1. *Adj.* **a)** naturalistic; **b)** *(den Naturalismus betreffend)* naturalist. **2.** *adv.* **a)** naturalistically; **b)** *(den Naturalismus betreffend)* ‹*influenced*› by naturalism

natur-, Natur-: **~belassen** *Adj.* natural ‹*oils, foods, etc.*›; **~blond** *Adj.* naturally fair *or* blond; **~bursche der** child of nature; **~denkmal das** natural monument

Naturell das; ~s, ~e disposition; temperament; **das widerspricht seinem ~:** it's not in his nature

natur-, Natur-: **~ereignis das, ~erscheinung die** natural phenomenon; **~faser die** natural fibre; **~forscher der** naturalist; **~forschung die** natural-history research; **~freund der** nature-lover; **~gegeben** *Adj.* natural and inevitable ‹*state of affairs*›; **etw. als ~gegeben ansehen** regard sth. as part of the natural order [of things]; **~gemäß** *Adv.* naturally; **~geschichte die;** *o. Pl.* natural history; **~gesetz das** law of nature; **~getreu 1.** *Adj.*; lifelike ‹*portrait, imitation*›; faithful ‹*reproduction*›; **2.** *adv.* ‹*draw*› true to life; ‹*reproduce*› faithfully; **~gewalt die** force of nature; **~heilkunde die** naturopathy *no art.*; **~katastrophe die** natural disaster; **~kunde die;** *o. Pl. (veralt.)* nature study *no art.*

natürlich [na'ty:ɐ̯lɪç] **1.** *Adj.* natural; **eines ~en Todes sterben** die a natural death; die of natural causes; **ein Bild in ~er Größe** a life-size portrait; **das ist die ~ste Sache der Welt** it is the most natural thing in the world. **2.** *adv.* ‹*laugh, behave*› naturally. **3.** *Adv.* **a)** *(wie erwartet)* naturally; of course; **b)** *(zwar)* of course

natürlicher·weise *Adv.* naturally; of course

Natürlichkeit die; ~: naturalness

natur-, Natur-: **~produkt das** natural product; **~produkte** natural produce *sing.*; **~rein** *Adj.* pure ‹*honey, jam, fruit, juice, etc.*›; ‹*wine*› free of additives; **~schauspiel das** natural spectacle; **~schutz der** [nature] conservation; **unter ~schutz** *(Dat.)* **stehen** be protected by law; be a protected species/variety/area *etc.*; **~schutz·gebiet das** nature reserve; **~talent das** [great] natural talent *or* gift; *(begabter Mensch)* naturally talented *or* gifted person; **ein ~talent sein** have a [great] natural gift *or* talent; **~verbunden** *Adj.* ‹*person*› in tune with nature; **~wissenschaft die** natural science *no art.*; **~wissenschaftler der** [natural] scientist; **~wissenschaftlich 1.** *Adj.* scientific; **2.** *adv.* scientifically; **~wunder das** miracle *or* wonder of nature

nautisch ['nau̯tɪʃ] *Adj.* *(Seew.)* naval ‹*officer*›; navigational ‹*instrument, calculation*›

Navigation [naviga'tsi̯o:n] **die; ~** *(Seew., Flugw.)* navigation *no art.*

Navigations·fehler der *(Seew., Flugw.)* navigational error

Nazi ['na:tsi] **der; ~s, ~s** Nazi

Nazismus der; ~: Nazi[i]sm *no art.*

nazistisch *Adj.* Nazi

Nazi·zeit die Nazi period

n. Chr. *Abk.* **nach Christus** AD

Neandertaler [ne'andɐta:lɐ] **der; ~s, ~** *(Anthrop.)* Neanderthal man

Neapel [ne'a:pl̩] **(das); ~s** Naples

Nebel ['ne:bl̩] **der; ~s, ~** fog; *(weniger dicht)* mist; **bei ~:** in fog/mist; when it is foggy/misty; **ausfallen wegen ~[s]** *(ugs. scherzh.)* be cancelled; *s. auch* **Nacht**

Nebel-: **~bank die;** *Pl.:* **~bänke** *(über dem Meer)* fog-bank; *(über dem Land)* large patch of fog; **~feld das** mist/fog patch; patch of mist/fog

nebelhaft *Adj.* hazy ‹*idea, recollection, etc.*›

Nebel·horn das *Pl.:* **~hörner** fog-horn

nebelig *s.* **neblig**

Nebel-: **~scheinwerfer der** fog-lamp; **~schlußleuchte die** rear fog lamp; **~wand die** wall of fog

neben ['ne:bn̩] **1.** *Präp. mit Dat.* **a)** *(Lage)* next to; beside; **dicht ~ jmdm./etw. sitzen** sit close *or* right beside sb./sth.; **b)** *(außer)* apart from; aside from *(Amer.)*. **2.** *Präp. mit Akk.* **a)** *(Richtung)* next to; beside; **b)** *(verglichen mit)* beside; compared to *or* with

neben·an *Adv.* next door

Neben-: **~anschluß der** extension; **~bedeutung die** secondary meaning

neben·bei *Adv.* **a)** ‹*work*› on the side, as a sideline; *(zusätzlich)* as well; in addition; **für Geologie interessiert er sich nur ~:** his interest in geology is only secondary; **b)** *(beiläufig)* ‹*remark*› incidentally, by the way; ‹*ask*› by the way; ‹*inform*› by the by; ‹*mention*› in passing; **~ gesagt** *od.* **bemerkt** incidentally; by the way

neben-, Neben-: **~beruf der** second job; sideline; **er ist im ~beruf Fotograf** he has a second job *or* sideline as a photographer; **~beruflich 1.** *Adj.* **eine ~berufliche Tätigkeit** a second job; **2.** *adv.* on the side; **er arbeitet ~beruflich als Übersetzer** he translates as a sideline; **~beschäftigung die** second job; sideline; ; **~buhler der, ~buhlerin die** rival

neben·einander *Adv.* **a)** next to one another *or* each other; *(fig.: zusammen)* ‹*live, exist*› side by side; **~ wohnen** live next door to one another *or* each other; **sich zu zweit ~ aufstellen** line up two abreast; **b)** *(gleichzeitig)* together

nebeneinander·her *Adv.* alongside each other *or* one another; ‹*walk*› side by side

nebeneinander-: ~|legen *tr. V.* lay *or* place ‹*objects*› next to each other *or* side by side; **~|setzen** *tr. V.* put *or* place ‹*persons, objects*› next to each other *or* one another; **~|sitzen** *unr. itr. V.* sit next to each other *or* one another; **~|stellen** *tr. V.* put *or* place ‹*tables, chairs, etc.*› next to each other

Neben-: ~eingang der side entrance; **~einkünfte** *Pl.*, **~einnahme, die, ~einnahmen** *Pl.* additional *or* supplementary income *sing.*; **~erwerb der** second job; secondary occupation; **~fach das** subsidiary subject; minor *(Amer.)*; **etw. im ~fach studieren** study sth. as a subsidiary subject; minor in sth. *(Amer.)*; **~fluß der** tributary; **~frage die** side issue; secondary issue; **~gebäude das a)** annexe; outbuilding; **b)** *(Nachbargebäude)* adjacent *or* neighbouring building; **~geräusch das** background noise; **~geräusche** *(Funkw., Fernspr.)* interference *sing.*; noise *sing.*; *(bei Tonband, Plattenspieler)* [background] noise *sing.*; **~handlung die** subplot; **~haus das** house next door; neighbouring house

neben·her *Adv. s.* **nebenbei**

nebenher-: ~|fahren *unr. itr. V.; mit sein* drive alongside; *(mit dem Rad, Motorrad)* ride alongside; **~|laufen** *unr. itr. V.; mit sein* **a)** run alongside; **b)** *(zugleich ablaufen)* proceed at the same time

neben-, Neben-: ~höhle die *(Anat.)* paranasal sinus **~kosten** *Pl.* **a)** additional costs; **b)** *(bei Mieten)* heating, lighting, and services; **~mann der;** *Pl.:* **~männer** *od.* **~leute** neighbour; **sein ~mann** the person sitting/standing/walking next to him; his neighbour; **~produkt das** *(auch fig.)* by-product; **~raum der** next *or* adjoining room; room next door; *(kleiner, unwichtiger)* side-room; **~rolle die** supporting role; **eine ~rolle [in etw. *(Dat.)*] spielen** *(fig.)* play a secondary *or* minor role [in sth.]; **~sache die** minor *or* inessential matter; **das ist ~sache** *(ugs.)* that's beside the point; **~sächlich** *Adj.* of minor importance *postpos.*; unimportant; minor, trivial ‹*detail*›; **etw. als ~sächlich abtun** reject sth. as irrelevant *or* beside the point; **~sächlichkeit die; ~, ~en a)** *o. Pl.* unimportance; *(fehlender Bezug zur Sache)* irrelevance; **b)** *(Unwichtiges)* matter of minor importance; unimportant matter; *(nicht zur Sache Gehörendes)* irrelevancy; **~satz der** *(Sprachw.)* subordinate clause; **~stelle die a)** extension; **b)** *(Filiale)* branch; **~straße die** side street; *(außerhalb der Stadt)* minor road; **~tätigkeit die** second job; sideline; **~tisch der** next *or* neighbouring table; **~verdienst der** additional earnings *pl. or* income; **~wirkung die** side-effect; **~zimmer das** next room; **sie gingen in ein ~zimmer** they went into an adjoining room

neblig *Adj.* foggy; *(weniger dicht)* misty

Necessaire [nesɛˈsɛːɐ̯] **das; ~s, ~s** sponge-bag *(Brit.)*; toilet bag *(Amer.)*

necken [ˈnɛkn̩] *tr. V.* tease; **jmdn. mit jmdm./etw. ~:** tease sb. about sb./sth.; **sich ~:** tease each other *or* one another

Neckerei die; ~: teasing

neckisch 1. *Adj.* **a)** teasing; *(verspielt)* playful; *(schelmisch)* mischievous; **b)** *(keß)* jaunty, saucy ‹*cap*›; saucy, provocative ‹*dress, blouse, etc.*›. **2.** *adv.* ‹*smile, say*› saucily, cheekily

nee [neː] *(ugs.)* no; nope *(Amer. coll.)*

Neffe [ˈnɛfə] **der; ~n, ~n** nephew

Negation [negaˈtsi̯oːn] **die; ~, ~en** negation

negativ [ˈneːgatiːf] **1.** *Adj.* negative. **2.** *adv.* ‹*answer*› in the negative; **etw. ~ beeinflussen** have a negative influence on sth.; **etw. ~ bewerten** judge sth. unfavourably; **sich ~ äußern** comment negatively **(zu** on); **der Test/die Testbohrung verlief ~:** the test proved unsuccessful/the test well yielded nothing

Negativ das; ~s, ~e *(Fot.)* negative

Neger [ˈneːgɐ] **der; ~s, ~** Negro

Negerin die; ~, ~nen Negress

negieren *tr. V.* deny ‹*fact, assertion, guilt, etc.*›

Negligé [negliˈʒeː] **das; ~s, ~s** négligé, negligee

nehmen [ˈneːmən] *unr. tr. V.* **a)** take; **etw. in die Hand/unter den Arm ~:** take sth. in one's hand/take *or* put sth. under one's arm; **etw. an sich** *(Akk.)* **~:** pick sth. up; *(und aufbewahren)* take charge of sth.; **sich** *(Dat.)* **etw. ~:** take sth.; *(sich bedienen)* help oneself to sth.; **zu sich ~:** take in ‹*orphan*›; **sie nahm ihren Vater zu sich** she had her father come and live with her; **auf sich** *(Akk.)* **~:** take on ‹*responsibility, burden*›; take ‹*blame*›; **die Dinge ~, wie sie kommen** take things as they come; **b)** *(wegnehmen)* **jmdm./einer Sache etw. ~:** deprive sb./sth. of sth.; **jmdm. die Sicht/den Ausblick ~:** block sb.'s view; **die Angst von jmdm. ~:** relieve sb. of his/her fear; **es sich** *(Dat.)* **nicht ~ lassen, etw. zu tun** not let anything stop one from doing sth.; **c)** *(benutzen)* use ‹*ingredients, washing-powder, wool, brush, knitting-needles, etc.*›; **man nehme ...** *(in Rezepten)* take ...; **den Zug/ein Taxi** *usw.* **~:** take the train/a taxi *etc.*; **[sich** *(Dat.)***] einen Anwalt** *usw.* **~:** get a lawyer *etc.*; **d)** *(aussuchen)* take; **ich nehme die Pastete** I'll have the pâté; **e)** *(in Anspruch nehmen)* take ‹*lessons, holiday, etc.*›; **f)** *(verlangen)* charge; **g)** *(einnehmen, essen)* take ‹*medicines, tablets, etc.*›; **etwas [Richtiges] zu sich ~:** have something [decent] to eat; **sie nimmt die Pille** she's taking *or* she's on the pill *(coll.)*; **h)** *(auffassen)* take **(als** as); **etw./jmdn. ernst/etw. leicht ~:** take sth./sb. seriously/take sth. lightly; **jmdn. nicht für voll ~** *(ugs.)* not take sb. seriously; **i)** *(behandeln)* treat ‹*person*›; **j)** *(überwinden, militärisch einnehmen)* take ‹*obstacle, bend, incline, village, bridgehead, etc.*›; *(fig.)* take ‹*woman*›; **k)** *(Sport)* take ‹*ball, punch*›; **einen Spieler hart ~:** foul a player blatantly

Neid [nai̯t] **der; ~[e]s** envy; jealousy; **vor ~ platzen** *(ugs.)* die of envy *(coll.)*; **gelb** *od.* **grün vor ~ werden, vor ~ erblassen** turn *or* go green with envy; **das muß der ~ ihr lassen** *(ugs.)* you've got to give her that; you've got to say that much for her

neiden *tr. V. (geh.)* **jmdm. etw. ~:** envy sb. [for] sth.

Neider der; ~s, ~: envious person

Neid·hammel der *(salopp abwertend)* envious sod *(sl.)*

neidisch 1. *Adj.* envious; **auf jmdn./etw. ~ sein** be envious of sb./sth. **2.** *adv.* enviously

neid-: ~los 1. *Adj.* ungrudging ‹*admiration*›; ‹*joy*› without envy; **2.** *adv.* ‹*acknowledge, admire*› without envy; **~voll 1.** *Adj.* envious ‹*glance*›; ‹*person*› filled with *or* full of envy; ‹*admiration*› mixed with envy; **2.** *adv.* ‹*watch*› full of envy

Neige ['naigə] **die;** ~ *(geh.)* dregs *pl.;* lees *pl.;* **ein Glas bis zur ~ leeren** drain a glass to the dregs; **zur ~ gehen** *(aufgebraucht sein)* ‹*money, supplies, etc.*› run low; *(zu Ende gehen)* ‹*year, day, holiday*› draw to its close

neigen 1. *tr. V.* tip, tilt ‹*bottle, glass, barrel, etc.*›; incline ‹*head, upper part of body*›. **2.** *refl. V.* **a)** ‹*person*› lean, bend; ‹*ship*› heel over, list; ‹*scales*› tip; **b)** *(schräg abfallen)* ‹*meadows*› slope down; **c)** *(geh.: zu Ende gehen)* ‹*day, year, holiday*› draw to its close. **3.** *itr. V.* **a) zu Erkältungen ~:** be susceptible *or* prone to colds; **zur Korpulenz/Schwermut ~:** have a tendency to put on weight/tend to be melancholy; **b)** *(tendieren)* tend; **zu der Ansicht ~, daß ...:** tend towards the view that ...

Neigung die; ~, ~en a) *o. Pl. (des Kopfes)* nod; **b)** *o. Pl. (Geneigtsein)* inclination; *(eines Geländes)* slope; **c)** *(Vorliebe)* inclination; **seine politischen/künstlerischen ~en** his political/artistic leanings; **eine ~ für etw.** a penchant *or* fondness for sth.; **d)** *o. Pl. (Anfälligsein)* tendency; **e)** *o. Pl. (Lust)* inclination; **f)** *(Liebe)* affection; fondness; liking

Neigungs·winkel der angle of inclination

nein [nain] *Interj.* no; ~ **danke** no, thank you; **man muß auch ~ sagen können** one must be able to say no; **~, nicht!** no, don't!; **~ und abermals ~!** no, and that's final!

Nein das; ~[s], ~[s] no; **mit ~ stimmen** vote no

Nein·stimme die no-vote; vote against

Nektar ['nektar] **der; ~s, ~e** *(Bot.)* nectar

Nektarine [nekta'ri:nə] **die; ~, ~n** nectarine

Nelke ['nɛlkə] **die; ~, ~n a)** pink; *(Dianthus caryophyllus)* carnation; **b)** *(Gewürz)* clove

nennen ['nɛnən] **1.** *unr. tr. V.* **a)** call; **jmdn. nach jmdm. ~:** call *or* name sb. after sb.; **jmdn. beim Vornamen ~:** call sb. by his/her first *or* Christian name; **das nenne ich Mut/eine Überraschung** that's what I call courage/well, that 'is a surprise; **b)** *(mitteilen)* give ‹*name, date of birth, address, reason, price, etc.*›; **c)** *(anführen)* give, cite ‹*example*›; *(erwähnen)* mention ‹*person, name*›. **2.** *unr. refl. V.* ‹*person, thing*› be called; **er nennt sich Maler** *usw. (behauptet Maler usw. zu sein)* he calls himself a painter *etc.*; **und so was nennt sich nun ein Freund** *(ugs.)* and he/she has the nerve to call himself/herself a friend

nennens·wert *Adj.* considerable ‹*influence, changes, delays, damage*›; **nichts Nennenswertes** nothing worth mentioning *or* nothing of note

Nenner der; ~s, ~ *(Math.)* denominator; **etw. auf einen ~ bringen** *(fig.)* reduce sth. to a common denominator

Nennung die; ~, ~en *s.* **nennen 1 b, c:** giving; citing; mentioning

neo-, Neo-: neo-

Neon ['ne:ɔn] **das; ~s** *(Chemie)* neon

Neon-: ~licht das neon light; **~reklame die** neon sign; **~röhre die** neon tube; [neon] strip light

Nepp [nɛp] **der; ~s** *(ugs. abwertend)* daylight robbery *no art.;* rip-off *(sl.)*

neppen *tr. V. (ugs. abwertend)* rook; rip ‹*tourist, customer, etc.*› off *(sl.)*

Nepp·lokal das *(ugs. abwertend)* clip-joint *(sl.)*

[1]**Neptun** [nɛp'tu:n] **der; ~s** *(Astron.),* [2]**Neptun (der); ~s** *(Myth.)* Neptune

Nerv [nɛrf] **der; ~s, ~en a)** nerve; **jmdm. den ~ töten** *(fig. ugs.)* drive sb. up the wall *(coll.);* **b)** *Pl. (nervliche Konstitution)* nerves; **gute/schwache ~en haben** have strong/bad nerves; **die ~en [dazu] haben, etw. zu tun** have the nerve to do sth.; **die ~en bewahren** *od.* **behalten** keep calm; **die ~en verlieren** lose control [of oneself]; lose one's cool *(sl.);* **ich bin mit den ~en am Ende** my nerves cannot take any more; **du hast vielleicht ~en!** *(ugs.)* you've got a nerve!; **jmdm. auf die ~en gehen** *od.* **fallen** get on sb.'s nerves

nerven *(salopp)* **1.** *tr. V.* **jmdn. ~:** get on sb.'s nerves. **2.** *itr. V.* be wearing on the nerves

nerven-, Nerven-: ~anspannung die nervous strain; nervous tension *no indef. art.;* **~arzt der** neurologist; **~aufreibend** *Adj.* nerve-racking; **~belastung die** strain on the nerves; **~bündel das** *(ugs.)* bundle of nerves *(coll.);* **~gift das** neurotoxin; **~heilanstalt die** *(veralt.)* mental *or* psychiatric hospital; **~kitzel der** *(ugs.)* kick *(coll.);* **~krank** *Adj.* **a)** ‹*person*› suffering from a nervous disease *or* disorder; **b)** *(psychisch krank)* mentally ill; **~krankheit die a)** nervous disease *or* disorder; **b)** *(psychische Krankheit)* mental illness; **~krieg der** *(ugs.)* war of nerves; **~leiden das** nervous complaint *or* disorder; **~sache die** *in* **das ist reine ~sache** *(ugs.)* it's a matter *or* question of nerves; **~säge die** *(salopp)* pain in the neck *(coll.);* **~zusammenbruch der** nervous breakdown

nervlich 1. *Adj.* nervous ‹*strain*›. **2.** *adv.* **dieser ständigen Spannung war er ~ nicht gewachsen** his nerves were not up to this constant tension

nervös [nɛr'vø:s] **1.** *Adj.* **a)** nervy, jittery ‹*person*›; nervous ‹*haste, movement*›; **~ sein** be jittery *(coll.) or* on edge; **das macht mich ganz ~:** it really gets on my nerves; *(das beunruhigt mich)* it makes me really nervous; **b)** *(Med.)* nervous ‹*twitch, gastric disorder, etc.*›. **2.** *adv.* nervously

Nervosität [nɛrvozi'tɛ:t] **die; ~:** nervousness

nerv·tötend *Adj.* nerve-racking ‹*wait*›; nerve-shattering ‹*sound, noise*›; soul-destroying ‹*activity, work*›

Nerz [nɛrts] **der; ~es, ~e** mink

Nerz·mantel der mink coat

[1]**Nessel** ['nɛsl̩] **die; ~, ~n** nettle; **sich in die ~n setzen** *(fig. ugs.)* get [oneself] into hot water *(coll.)*

[2]**Nessel der; ~s, ~** *(Stoff)* coarse, untreated cotton cloth

Nessel·sucht die; ~: nettle-rash; hives

Nest [nɛst] **das; ~[e]s, ~er a)** nest; **das eigene ~ beschmutzen** *(fig.)* foul one's own nest; **er hat sich ins warme** *od.* **gemachte ~ gesetzt** *(fig. ugs.)* he had his future made for him; **b)** *(fam.: Bett)* bed; **c)** *(ugs. abwertend: kleiner Ort)* little place; **ein gottverlassenes ~:** a God-forsaken hole; **d)** *(Schlupfwinkel)* hide-out; den

Nest·beschmutzer der; ~s, ~ *(abwertend)* person who is/was guilty of fouling his/her own nest

nesteln ['nɛstl̩n] *itr. V.* fiddle, *(ungeschickt)* fumble **(an** + *Dat.* with)

Nest-: ~häkchen das *(fam.)* [spoilt] baby of the family; **~wärme die** warmth of a [happy] family upbringing *or* of [happy] family life

nett [nɛt] **1.** *Adj.* **a)** nice; *(freundlich)* nice; kind; **sei so ~ und hilf mir!** would you be so good *or* kind as to help me?; **~, daß du anrufst** it's nice *or* kind of you to ring; **etwas Nettes erleben/sagen** have a pleasant experience/say something nice; **b)** *(hübsch)* pretty ‹*girl, town, dress, etc.*›; nice, pleasant ‹*pub, house, town, etc.*›; **c)** *nicht präd. (ugs.: beträchtlich)* nice little *(coll.)* ‹*profit, extra earnings, income*›; **eine ~e Summe/eine ~e Stange Geld** a tidy sum *(coll.)*; **d)** *(ugs. iron.: unerfreulich)* nice *(coll.)* ‹*affair*›; nice *(coll.)*, fine ‹*state of affairs, mess*›; **das kann ja ~ werden!** that'll be fun *(coll.)*. **2.** *adv.* *(angenehm)* nicely; *(freundlich)* nicely; kindly; **sich ~ mit jmdm. unterhalten** have a pleasant conversation with sb.

netter·weise *Adv. (ugs.)* kindly

Nettigkeit die; ~, ~en a) *o. Pl.* kindness; goodness; **b)** *(Äußerung)* **jmdm. ein paar ~en sagen** say a few nice *or* kind things to sb.

netto ['nɛto] *Adv.* ‹*weigh, earn, etc.*› net

Netto- net ‹*income, salary, etc*›

Netz [nɛts] **das; ~es, ~e a)** *(auch Fischer~, Tennis~, Ballspiele)* net; *(Einkaufs~)* string bag; *(Gepäck~)* [luggage-]rack; **ein ~ von Lügen** *(fig.)* a web of lies; **jmdm. ins ~ gehen** *(fig.)* fall into sb.'s trap; **b)** *(Spinnen~)* web; **c)** *(Verteiler~, Verkehrs~, System von Einrichtungen)* network; *(für Strom, Wasser, Gas)* mains *pl.*

Netz-: ~anschluß der mains connection; **~ball der** *(Tennis, Volleyball)* net ball; **~gerät das** *(Elektrot.)* power pack; **~haut die** *(Anat.)* retina; **~hemd das** string vest; **~karte die** area season ticket; *(Eisenb.)* unlimited travel ticket; **~roller der** *(Tennis)* net-cord [stroke]; **~strumpf der** net stocking

neu [nɔy] **1.** *Adj.* **a)** new; **ein ganz ~es Fahrrad** a brand new bicycle; **die Neue Welt** the New World; **das Neue Testament** the New Testament; **die ~este Mode** the latest fashion; **die ~esten Nachrichten/Ereignisse** the latest news/most recent events; **viel Glück im ~en Jahr** best wishes for the New Year; Happy New Year; **das ist mir ~:** that is news to me; **das Neue daran ist ...:** what's new about it is ...; **das Neueste auf dem Markt** the latest thing on the market; **der/die Neue** the new man/woman/boy/girl; **was gibt es Neues?** what's new?; **weißt du schon das Neueste?** *(ugs.)* have you heard the latest?; **aufs ~e** anew; afresh; again; **auf ein ~es!** let's try again!; **von ~em** all over again; *(noch einmal)* [once] again; **seit ~estem werden dort keine Kreditkarten mehr akzeptiert** just recently they've started refusing to accept credit cards; **in ~erer/~ester Zeit** quite/just *or* very recently; **das ist ~eren Datums** that is of a more recent date; **die ~en** *od.* **~eren Sprachen** modern languages; **b)** *nicht präd. (kürzlich geerntet)* new ‹*wine, potatoes*›; **c)** *(sauber)* clean ‹*shirt, socks, underwear, etc.*›. **2.** *adv.* **a) ~ tapeziert/gespritzt/gestrichen/möbliert** repapered/resprayed/repainted/refurnished; **ein Geschäft ~ eröffnen** reopen a shop; **sich ~ einkleiden** provide oneself with a new set of clothes; **noch einmal ~ beginnen** start again from scratch; **b)** *(gerade erst)* **diese Ware ist ~ eingetroffen** this item has just come in *or* arrived; **~ erschienene Bücher** newly published books; books that have just come out *or* appeared; **3000 Wörter sind ~ hinzugekommen** 3,000 new words have been added

neu-, Neu-: ~ankömmling der new arrival; **~anschaffung die: die ~anschaffung von Produktionsanlagen** the acquisition of new production plant; **~anschaffungen machen** buy new items; **~artig** *Adj.* new; **ein ~er Staubsauger** a new type of vacuum cleaner; **~auflage die** new edition; **~ausgabe die** new edition

Neu·bau der; *Pl.* **Neubauten** new house/building

Neubau-: ~viertel das new district; **~wohnung die** flat *(Brit.) or (Amer.)* apartment in a new block/house; new flat *(Brit.) or (Amer.)* apartment

Neu-: ~bearbeitung die a) *(eines Buches, Textes)* revision; *(eines Theaterstücks)* adaptation; **b)** *(neue Fassung)* new version; **~beginn der** new beginning; **~druck der** reprint [with corrections]; **~einstellung die: eine ~einstellung vornehmen** take on a new employee; *(von Angestellten)* make a new appointment; **~entdeckung die a)** *(auch fig.)* new discovery; **b)** *(Wiederentdeckung)* rediscovery; **~entwicklung die a) die ~entwicklung von Heilmitteln** the development of new medicines; **b)** *(neu Entwickeltes)* new development

neuerdings ['nɔyɐ'dɪŋs] *Adv.* recently; **Fahrkarten gibt es ~ nur noch am Automaten** as of a short while ago one can only get tickets from a machine; **er trägt ~ eine Perücke** he has recently started wearing a wig

Neuerer der; ~s, ~ innovator

neu-, Neu-: ~eröffnet *Adj.; präd. getrennt geschrieben* **a)** newly-opened; **b)** *(wiedereröffnet)* reopened; **~eröffnung die a)** opening; **b)** *(Wiedereröffnung)* reopening; **~erscheinung die** new publication; *(Schallplatte)* new release

Neuerung die; ~, ~en innovation

neu-, Neu-: ~erwerbung die a) a) die ~ von Büchern *usw.* the acquisition of new books *etc.*; **b)** *(Gegenstand)* new acquisition; **~fassung die** revised version; *(eines Films)* remake; **~fundland (das); ~s** Newfoundland; **~fundländer der; ~s, ~** *(Hunderasse)* Newfoundland [dog]; **~geboren** *Adj.* newborn; **sich wie ~geboren fühlen** feel a new man/woman; **~geborene das;** *adj. Dekl.* newborn child

Neu·gier, Neugierde [-gi:ɐdə] **die; ~:** curiosity; *(Wißbegierde)* inquisitiveness; **aus [reiner] ~:** out of [sheer] curiosity

neu·gierig 1. *Adj.* curious; inquisitive; prying *(derog.)*, nosy *(coll. derog.)* ‹*person*›; **da bin ich aber ~!** *(iron.)* I'll believe it when I see it; I can hardly wait! *(iron.)*; **auf etw.** *(Akk.)* **~ sein** be curious about sth.; **viele Neugierige** many inquisitive people *or* spectators; **ich bin ~, was er dazu sagt** I'm curious to know what he'll say about it; **ich bin ~, ob er kommt** I wonder whether he'll come. **2.** *adv.* ‹*ask*› inquisitively; ‹*peer*› nosily *(coll. derog.)*; **jmdn. ~ mustern** eye sb. curiously

Neu·gründung die a) die ~ eines Vereins *usw.* the founding *or* establishment of a new club *etc.*; **b)** *(neu Gegründetes)* **eine ~ sein** have recently been founded *or* established; **c)** *(erneute Gründung)* refoundation; re-establishment

Neuheit die; ~, ~en a) *o. Pl.* novelty; **b)** *(Neues)* new product/gadget/article *etc.*

Neuigkeit die; ~, ~en piece of news; **~en** news *sing.*

Neu·jahr das New Year's Day; **~ feiern** celebrate New Year; *s. auch* **prosit**

Neujahrs-: ~nacht die New Year's night; **~tag der** New Year's Day

Neu·land das; *o. Pl.* **a)** newly reclaimed *or* new land; **b)** *(unerforschtes Land)* new *or* virgin territory; **wissenschaftliches ~ betreten** *(fig.)* break new ground in science

neulich *Adv.* recently; the other day

Neuling ['nɔylɪŋ] **der; ~s, ~e** newcomer (**in** + *Dat.* to); new man/woman/girl/boy; *(auf einem Gebiet)* novice

neu·modisch *(abwertend)* **1.** *Adj.* newfangled *(derog.)*. **2.** *adv.* ‹*dress*› in a newfangled way

Neu·mond der new moon; **heute ist/haben wir ~:** there's a new moon today

neun [nɔyn] *Kardinalz.* nine; **alle ~[e]** *(Kegeln)* a floorer; *s. auch* **acht**

Neun die; ~, ~en nine; **ach, du grüne ~e** *(ugs.)* oh, my goodness!; good grief!; *s. auch* [1]**Acht**

neun-, Neun- (*s. auch* **acht-, Acht-**): **~fach** *Vervielfältigungsz.* ninefold; *s. auch* **achtfach; ~fache das; ~n: das ~fache von 4 ist 36** nine fours are *or* nine times four makes thirty-six; *s. auch* **Achtfache; ~hundert** *Kardinalz.* nine hundred; **~jährig** *Adj. (9 Jahre alt)* nine-year-old *attrib.; (9 Jahre dauernd)* nine-year *attrib.;* **~mal** *Adv.* nine times; *s. auch* **achtmal; ~malklug 1.** *Adj.* smart-aleck *attrib. (coll.);* **2.** *adv.* in a smart-aleck way *(coll.)*

neunt [nɔynt] *in* **wir waren zu ~:** there were nine of us; *s. auch* [2]**acht**

neun·tausend *Kardinalz.* nine thousand

Neuntel das *(schweiz. meist* **der**); **~s, ~:** ninth

neuntens *Adv.* ninthly

neun·zehn *Kardinalz.* nineteen; *s. auch* **achtzehn**

neunzig ['nɔyntsɪç] *Kardinalz.* ninety; *s. auch* **achtzig**

neunziger *indekl. Adj.; nicht präd.* **die ~ Jahre** the nineties; *s. auch* **achtziger**

Neunziger der; ~s, ~: ninety-year-old

neu-, Neu-: ~ordnung die reorganization; **~reg[e]lung die a);** *o. Pl.* **die ~regelung der Arbeitszeit** *usw.* the revision of regulations governing working hours *etc.;* **b)** *(Bestimmung)* new regulation; **~reich** *(abwertend) Adj.* nouveau riche; **~reiche der/die** nouveau riche

Neurologe der; ~n, ~n neurologist

Neurologie die; ~ neurology

Neurologin die; ~, ~nen neurologist

Neurose [nɔy'ro:zə] **die; ~, ~n** *(Med., Psych.)* neurosis

Neurotiker [nɔy'ro:tikɐ] **der; ~s, ~, Neurotikerin die; ~, ~nen** *(Med., Psych., auch ugs.)* neurotic

neurotisch *Adj. (Med., Psych., auch ugs.)* neurotic

Neu·schnee der fresh snow

Neu·see·land (das); ~s New Zealand

Neuseeländer der; ~s, ~: New Zealander

neuseeländisch *Adj.* New Zealand

neu·sprachlich *Adj.* modern languages *attrib.* ‹*teaching*›*;* **ein ~sprachliches Gymnasium** a grammar school with emphasis on modern languages

neutral [nɔy'tra:l] *Adj. (auch Völkerr., Phys., Chem.)* neutral

neutralisieren *tr. V. (auch Völkerr., Chem., Elektrot.)* neutralize

Neutralität [nɔytrali'tɛ:t] **die; ~, ~en** *(auch Völkerr., Chem., Elektrot.)* neutrality

Neutron ['nɔytrɔn] **das; ~s, ~en** [-'tro:nən] *(Kernphysik)* neutron

Neutronen·bombe die neutron bomb

Neutrum ['nɔytrʊm] **das; ~s, Neutra** *(österr. nur so) od.* **Neutren** *(Sprachw.)* neuter

neu-, Neu-: ~vermählt *Adj. (geh.)* newly wed *or* married; **die Neuvermählten** the newly-weds; **~wahl** new election; **die ~wahl des Bundespräsidenten** the election of a new Federal President; **~wert der** value when new; original value; *(Versicherungsw.)* replacement value; **~wertig** *Adj.* as new; **~zeit die;** *o. Pl. (Zeit nach 1500)* modern era *or* age; *(Gegenwart)* modern times *pl.;* modern age; **~zeitlich 1.** *Adj.* modern; since the Middle Ages *postpos., not pred.; (modern)* modern ‹*device, equipment, methods, etc.*›*;* **2.** *adv. (modern)* ‹*equip, fit*› with all modern conveniences; **~zugang der** *(im Krankenhaus)* new admission; *(in der Bibliothek)* new accession; **~zulassung die: die ~zulassung von Kraftfahrzeugen** the registration of new vehicles

nicht [nɪçt] *Adv.* **a)** not; **sie raucht ~** *(im Moment)* she is not smoking; *(gewöhnlich)* she does not *or* doesn't smoke; **alle klatschten, nur sie ~:** they all applauded except for her; **Wer hat das getan? – Sie ~!** Who did that? – It wasn't her; **Gehst du hin? – Nein, ich gehe ~!** Are you going? – No, I'm not; **Ich mag ihn ~. – Ich auch ~:** I don't like him. – Neither do I; **ich kann das ~ mehr** *od.* **länger sehen** I can't stand the sight of it any more *or* longer; **~ einmal** *od. (ugs.)* **mal** not even; **~ mehr als** no more than; **b)** *(Bitte, Verbot o. ä. ausdrückend)* **~!** [no,] don't!; **„~ hinauslehnen!“** *(im Zug)* 'do not lean out of the window'; **bitte ~!** please don't!; **c)** *(Zustimmung erwartend)* **er ist dein Bruder, ~?** he's your brother, isn't he?; **du magst das, ~ [wahr]?** you like that, don't you?; **kommst du [etwa] ~?** aren't you coming[, then]?; **willst du ~ mitkommen?** won't you come too?; **d)** *(verwundert)* **was du ~ sagst!** you don't say!; **e)** *([bedingte] Anerkennung ausdrükkend)* **~ übel!** not bad!

nicht-, Nicht-: non-

Nicht-: ~achtung die a) *in* **jmdn. mit ~achtung strafen** punish sb. by ignoring him/her; send sb. to Coventry; **b)** *(Geringschätzung)* lack of regard *or* respect; **~angriffs·pakt der** non-aggression pact; **~beachtung die** non-observance; **~beachtung einer roten Ampel** failure to observe a red light; **~befolgung die: ~befolgung der Vorschriften** non-compliance *or* failure to comply with the regulations

Nichte die; ~, ~n niece

Nicht-: ~einmischung die *(Politik)* non-intervention; non-interference; **~gefallen das** *in* **bei ~gefallen** *(Kaufmannsspr.)* if not satisfied

nichtig *Adj.* **a)** *(geh.: wertlos, belanglos)* vain ‹*things, pleasures, etc.*›*;* trivial ‹*reason*›*;* petty ‹*quarrel*›*;* idle ‹*thoughts, chatter*›*;* empty ‹*pretext*›*;* **b)** *(Rechtsspr.: ungültig)* invalid, void ‹*contract, will, etc.*›

nicht-, Nicht-: ~mitglied das non-member; **~öffentlich** *Adj.* not open to the public *pred.;* closed, private ‹*meeting*›*;* **~raucher der** non-smoker; **ich bin ~raucher** I don't smoke; I'm a non-smoker; **~raucher·abteil das** non-smoking *or* no-smoking compartment; **~rostend** *Adj.* non-rusting ‹*blade*›*;* stainless ‹*steel*›

nichts [nɪçts] *Indefinitpron.* nothing; **er sieht ~:** he sees nothing; he doesn't see anything; **hast du ~ gegessen?** haven't you eaten anything?; **für ~ und wieder ~** *(ugs.)* for nothing at all; **~ zu ma-**

chen!/~ da! *(ugs.)* nothing doing *(coll.)*; **von mir bekommst du ~ mehr** you'll get nothing more from me; you won't get anything more from me; **~ anderes** nothing else; **jetzt interessiert er sich für ~ anderes mehr** he's now no longer interested in anything else; **~ als** nothing but; **~ wie ins Bett/weg!** quick into bed/let's go!; **~ wie hinterher!** put your skates on, after him/her/them! *(sl.)*; *s. auch* **danken 2 a**

Nichts das; ~, ~e a) *o. Pl. (Philos.: das Nicht-Sein)* nothingness *no art.*; **b)** *o. Pl. (leerer Raum)* void; **er war wie aus dem ~ aufgetaucht** he appeared from nowhere; **c)** *o. Pl. (wenig von etw.)* **etw. aus dem ~ aufbauen** built sth. up from nothing; **vor dem ~ stehen** be left with nothing; be faced with ruin; **d)** *(abwertend: Mensch)* nobody; nonentity

nichts·ahnend *Adj.* unsuspecting

Nicht-: ~schwimmer der nonswimmer; **er war ~schwimmer** he could not swim; **~schwimmer·becken das** non-swimmers' *or* learners' pool

nichts·desto·weniger *Adv.* nevertheless; none the less

Nicht·seßhafte der/die; *adj. Dekl. (Amtsspr.)* person of no fixed abode *(Admin. Lang.)*

nichts-, Nichts-: ~könner der *(abwertend)* incompetent; bungler; **~nutz der; ~es, ~e** *(veralt. abwertend)* good-for-nothing; **~nutzig** *Adj. (veralt. abwertend)* good-for-nothing *attrib.*; worthless ‹*existence*›; **~sagend 1.** *Adj.* meaningless, empty ‹*talk, phrases, etc.*›; *(fig.: ausdruckslos)* vacant ‹*smile*›; expressionless ‹*face*›; **2.** *adv.* meaninglessly ‹*formulated*›; ‹*smile*› vacantly; **~tuer** [~tu:ɐ] **der; ~s, ~** *(abwertend)* layabout; loafer; **~tun das a)** inactivity; doing nothing *no art.*; **b)** *(Müßiggang)* idleness *no art.*; lazing about *no art.*

Nicht·zutreffende das: ~s streichen delete as applicable

Nickel ['nɪkl̩] **das; ~s** nickel

Nickel·brille die metal-rimmed glasses *or* spectacles

nicken ['nɪkn̩] *itr. V.* **a)** nod; **zustimmend ~:** nod one's agreement; **mit dem Kopf ~:** nod one's head; **b)** *(fam.: schlafen)* doze; snooze *(coll.)*

Nickerchen das; ~s, ~ *(fam.)* nap; snooze *(coll.)*; **ein ~ halten** *od.* **machen** take *or* have forty winks *or* a nap

Nicki ['nɪki] **der; ~[s], ~s** velour pullover *or* sweater

nie [ni:] *Adv.* never; **mich besucht ~ jemand** nobody ever visits me; **~ mehr!** never again!; **~ und nimmer!** never!; **~ im Leben!** not on your life!; **das werde ich ~ im Leben vergessen** I shall never forget it as long as I live

nieder ['ni:dɐ] **1.** *Adj.; nicht präd.* **a)** lower ‹*class, intelligence*›; petty, minor ‹*official*›; lowly ‹*family, origins, birth*›; menial ‹*task*›; **b)** *(Biol.: nicht hoch entwickelt)* lower ‹*plant, animal, organism*›. **2.** *Adv. (hinunter)* down

nieder-, Nieder-: ~|brennen 1. *unr. itr. V.; mit sein (herunterbrennen)* ‹*fire*› burn low; ‹*building*› burn down; **2.** *unr. tr. V.* burn down ‹*building, village, etc.*›; **~|brüllen** *tr. V. (ugs.)* **jmdn. ~brüllen** shout sb. down; **~deutsch** *Adj.* Low German ‹*dialect*›; **~gang der** fall; decline; **~|gehen** *unr. itr. V.; mit sein* **a)** *(landen)* ‹*plane, spacecraft, balloonist*› come down; ‹*parachutist*› drop; ‹*birds*› land; **b)** *(fallen)* ‹*rain, satellite, avalanche*› come down; **c)** *(Boxen: zu Boden fallen)* go down; **~geschlagen** *Adj.* despondent; dejected; **~geschlagenheit die; ~:** despondency; dejection; **~|halten** *unr. tr. V.* oppress ‹*nation, people, class*›; keep ‹*nation, people, class*› in subjection; keep ‹*person*› down; **~|knien** *itr. V. (auch mit sein), refl. V.* kneel down; *(unterwürfig, demütig)* go down on one's knees; **~lage die** defeat; **jmdm. eine ~lage beibringen** inflict a defeat on sb.

Nieder·lande *Pl.*: **die ~:** the Netherlands

Niederländer der; ~s, ~: Dutchman; Netherlander

Niederländerin die; ~, ~nen Dutchwoman; Netherlander; *s. auch* **-in**

niederländisch *Adj.* Dutch; Netherlands *attrib.* ‹*government, embassy, etc.*›

nieder-, Nieder-: ~|lassen *unr. refl. V.* **a)** *(ein Geschäft, eine Praxis eröffnen)* set up *or* establish oneself in business; ‹*doctor, lawyer*› set up a practice *or* in practice; **b)** *(seinen Wohnsitz nehmen)* settle; **c)** *(geh.: sich setzen)* sit down; seat oneself; ‹*bird*› settle, alight; **~lassung die; ~, ~en a)** *o. Pl. s.* **~lassen a:** setting up in business; setting up of a practice *or* in practice; **b)** *(Ort)* settlement; **c)** *(Wirtsch.: Zweigstelle)* branch; **~|legen** *tr. V.* **a)** *(geh.: hinlegen)* lay *or* put *or* set down; lay ‹*wreath*›; lay down ‹*one's arms*›; **b)** *(nicht weitermachen)* lay down, resign [from] ‹*office*›; relinquish ‹*command*›; **c)** *(geh.: aufschreiben)* set down; **~legung die; ~, ~en a)** *(geh.: eines Kranzes)* laying; **b)** *(eines Amtes)* resignation (*Gen.* from); *(eines Kommandos)* relinquishing; **c)** *(geh.: Niederschrift)* setting down; **~|machen** *(ugs.)*, **~|metzeln** *tr. V.* butcher; **~|prasseln** *itr. V.; mit sein* ‹*rain, hail*› beat down; ‹*blows, rebukes, questions, etc.*› rain down; **~|reißen** *unr. tr. V.* **a)** *(abreißen)* pull down ‹*building, wall*›; **b)** *(zu Boden reißen)* **jmdn. ~reißen** knock sb. over

Nieder·sachsen (das) Lower Saxony

nieder-, Nieder-: ~|schießen *unr. tr. V.* gun down; **~schlag der a)** *(Met.)* precipitation; **b)** *(Boxen)* knock-down; **c)** *(Ausdruck)* **[seinen] ~schlag in etw.** *(Dat.)* **finden** find expression in sth.; **~|schlagen 1.** *unr. tr. V.* **a)** *(zu Boden schlagen)* **jmdn. ~schlagen** knock sb. down; **b)** *(umschlagen)* turn down ‹*hatbrim, collar*›; **c)** *(beenden)* suppress, put down ‹*revolt, uprising, etc.*›; put an end to ‹*strike*›; **d)** *(senken)* lower ‹*eyes, eyelids*›; *s. auch* **niedergeschlagen; 2.** *unr. refl. V.* **sich in etw.** *(Dat.)* **~schlagen** ‹*experience, emotion*› find expression in sth.; ‹*performance, hard work*› be reflected in sth.; **~schlags·frei** *Adj.* ‹*period*› without [any] precipitation *not pred.*; **~schlagung die** suppression; putting down; **~|strecken** *(geh.)* **1.** *tr. V.* **jmdn. ~strecken** knock sb. down; *(mit einem Schuß)* shoot sb. down; **einen Tiger/Hirsch ~strecken** bring down a tiger/stag; **2.** *refl. V. (sich hinlegen)* lie down; **auf das** *od.* **dem Sofa ~gestreckt** stretched out on the sofa; **~tracht die; ~** *(geh.)* malice; *(als Charaktereigenschaft)* vileness; despicableness; **~trächtig 1.** *Adj.* malicious ‹*person, slander, lie, etc.*›; *(verachtenswert)* vile, despicable ‹*person*›; base, vile ‹*misrepresentation, slander, lie*›; **2.** *adv.* ‹*betray, lie, treat*› in a vile *or* despicable way; ‹*smile*› maliciously; **~trächtigkeit die; ~, ~en a)** *o. Pl. s.* **~trächtig 1:** maliciousness; vileness; despicableness; baseness; **b)** *(gemeine Handlung)* vile or despicable act, ~|**treten** *unr. tr. V.* tread ‹*grass, flowers, carpet-pile, etc.*›; *(fig.)* trample ‹*person*› underfoot

Niederung die; ~, ~en low-lying area; *(an Flußläufen, Küsten)* flats *pl.*; *(Tal)* valley

nieder|werfen *unr. tr. V.* **a)** *(geh.: besiegen)* overcome, defeat ‹*enemy, rebels, etc.*›; **b)** *(geh.: beenden) s.* **niederschlagen 1c; c)** *(geh.: schwächen)* ‹*illness, fever*› lay ‹*person*› low

niedlich ['ni:tlɪç] **1.** *Adj.* sweet; cute *(Amer. coll.)*; sweet little *attrib.*; dear little *attrib.* **2.** *adv.* ‹*dance, nibble*› sweetly, prettily; ‹*babble, play*› sweetly, cutely *(Amer. coll.)*

niedrig 1. *Adj.* **a)** low; short ‹*grass*›; **b)** *(von geringem Rang)* lowly ‹*origins, birth*›; low ‹*rank, status, intellectual level*›; **c)** *(sittlich tiefstehend)* base ‹*instinct, desire, emotion, person*›; vile ‹*motive*›. **2.** *adv.* ‹*hang, fly*› low

niemals ['ni:ma:ls] *Adv.* never

niemand ['ni:mant] *Indefinitpron.* nobody; no one; ~ **war im Büro** there was nobody *or* no one in the office; there wasn't anybody *or* any one in the office; ~ **anders** *od.* **anderer** nobody *or* no one else; **es kann ~ anders** *od.* **anderer als du gewesen sein** it can't have been anybody *or* any one [else] but you; **das darfst du ~[em] sagen!** you mustn't tell anybody that!; **laß ~ Fremdes herein** don't let anybody *or* anyone in you don't know; don't let any strangers in

Niemands·land das; *o. Pl. (auch fig.)* no man's land

Niere ['ni:rə] **die; ~, ~n** kidney; **jmdm. an die ~n gehen** *(fig. ugs.)* get to sb. *(coll.)*

nieren, Nieren-: ~förmig *Adj.* kidney-shaped; **~leiden das** kidney disease; **~stein der** kidney stone; renal calculus *(Med.)*

nieseln ['ni:zl̩n] *unpers. V.* drizzle

Niesel·regen der drizzle

niesen ['ni:zn̩] *itr. V.* sneeze

Nies·pulver das sneezing-powder

¹Niete ['ni:tə] **die; ~, ~n a)** *(Los)* blank; **b)** *(ugs.: Mensch)* dead loss *(coll.)* **(in +** *Dat.* at)

²Niete die; ~, ~n rivet

nieten ['ni:tn̩] *tr. V.* rivet

Nieten·hose die [pair of] studded jeans

niet- und nagelfest *in* **[alles] was nicht ~ ~ ~ ist** *(ugs.)* [everything] that's not nailed *or* screwed down

Nihilismus [nihi'lɪsmʊs] **der; ~:** nihilism

Nihilist der; ~en, ~en nihilist

nihilistisch *Adj.* nihilistic

Nikolaus ['nɪkolaus] **der; ~, ~e** St Nicholas

Nikolaus·tag der St Nicholas' Day

Nikotin [niko'ti:n] **das; ~s** nicotine

nikotin-, Nikotin-: ~arm *Adj.* low-nicotine *attrib.*; low in nicotine *pred.*; **~gehalt der** nicotine content; **~vergiftung die** nicotine poisoning

Nil [ni:l] **der; ~[s]** Nile

Nil·pferd das hippopotamus; hippo *(coll.)*

nimmer-, Nimmer-: ~mehr *Adv.* **a)** *(veralt.: nie)* never; **b)** *(südd., österr.: nie wieder)* never again; **~satt** *Adj.; nicht präd. (fam.)* insatiable; **~wieder·sehen** *in* **auf ~wiedersehen verschwinden** *(ugs., oft scherzh.)* vanish never to be seen again

Nippel ['nɪpl̩] **der; ~s, ~ a)** *(Technik; ugs.: Brustwarze)* nipple; **b)** *(am Wasserball)* valve

nippen ['nɪpn̩] *itr. V. (trinken)* sip; take a sip/sips; *(essen)* nibble **(von** at); **am Glas ~:** sip from *or* take a sip/sips from the glass

Nippes ['nɪpəs], **Nipp·sachen** *Pl.* [porcelain] knick-knacks; small [porcelain] ornaments

nirgends ['nɪrgn̩ts], **nirgendwo** *Adv.* nowhere; **er war ~ zu finden** he was nowhere *or* wasn't anywhere to be found

nirgend-: ~woher *Adv.* from nowhere; **sie konnten die Medikamente ~woher bekommen** they couldn't get the medicines from anywhere; **~wohin** *Adv.* **wir gehen ~wohin** we're not going anywhere

Nische ['ni:ʃə] **die; ~, ~n a)** *(Einbuchtung)* niche; **b)** *(Erweiterung eines Raumes)* recess

nisten ['nɪstn̩] *itr. V.* nest

Nist·kasten der nest-box; nesting-box

Nitrat [ni'tra:t] **das; ~[e]s, ~e** *(Chemie)* nitrate

Nitro·glyzerin [ni:tro-] **das;** *o. Pl.* nitro-glycerine

Niveau [ni'vo:] **das; ~s, ~s a)** level; **eine Zeitung mit ~:** a quality newspaper; **er hat wenig ~:** he is not very cultured *or* knowledgeable; **b)** *(Qualitäts~)* standard

niveau·voll *Adj.* cultured and intelligent ‹*person*›; ‹*entertainment, programme*› of quality *postpos., not pred.*

nix [nɪks] *Indefinitpron. (ugs.) s.* **nichts**

Nixe die; ~, ~n *(germ. Myth.)* nixie; *(mit Fischschwanz)* mermaid

NO *Abk.* **Nordost[en]** NE

nobel ['no:bl̩] **1.** *Adj.* **a)** *(geh.: edel)* noble; noble[-minded] ‹*person*›; **b)** *(oft spött.: luxuriös)* elegant, *(coll.)* posh ‹*boutique, house, hotel*›; fine ‹*cigar*›; **c)** *(ugs.: freigebig)* lavish, generous ‹*tip, present*›; generous ‹*person*›. **2.** *adv.* **a)** *(geh.: edel)* nobly; **b)** *(oft spött.: luxuriös)* ‹*dress, live, eat*› in the grand style

Nobel-: ~herberge die *(salopp)* posh *or* swish hotel; **~kutsche die** *(salopp)* posh *or* swish car *(coll.)*

Nobel- [no'bɛl-]: **~preis der** Nobel prize; **~preisträger der** Nobel prize-winner

noch [nɔx] **1.** *Adv.* **a)** *([wie] bisher, derzeit)* still; ~ **nicht** not yet; **sie sind immer ~ nicht da** they're still not here; **ich sehe ihn kaum ~:** I hardly ever see him any more; ~ **nach Jahren** even years later; **b)** *(als Rest einer Menge)* **ich habe [nur] ~ zehn Mark** I've [only] ten marks left; **es dauert ~ fünf Minuten** it'll be another five minutes; **es fehlt [mir/dir** *usw.***] ~ eine Mark** I/you *etc.* need another mark; **c)** *(bevor etw. anderes geschieht)* just; **ich mache das [jetzt/dann] ~ fertig** I'll just get this finished; **d)** *(irgendwann einmal)* some time; one day; **du wirst ihn [schon] ~ kennenlernen** you'll get to know him yet; **er wird ~ kommen** he will still come; **e)** *(womöglich)* if you're/he's *etc.* not careful; **du kommst ~ zu spät!** you'll be late if you're not careful; **f)** *(drückt eine geringe zeitliche Distanz aus)* only; **sie war eben** *od.* **gerade ~ hier** she was here only a moment ago; **es ist ~ keine Woche her, daß ...:** it was less than a week ago that ...; **g)** *(nicht später als)* ~ **am selben Abend** the [very] same evening; **h)** *(drückt aus, daß etw. unwiederholbar ist)* **ich habe Großvater ~ gekannt** I'm old enough to have known grandfather; **i)** *(drückt aus, daß sich etw. im Rahmen hält)* **Er hat ~ Glück gehabt. Es hätte weit schlimmer kommen können** He was lucky. It could have been much worse; **das geht ~:** that's [still] all right *or (coll.)* OK; **das ist ~ lange kein Grund** that still isn't any sort of reason; **das ist ja ~ [ein]mal gutgegangen**

(ugs.) it was just about all right; **j)** *(außerdem, zusätzlich)* **wer war ~ da?** who else was there?; **er hat [auch/außerdem] ~ ein Fahrrad** he has a bicycle as well; **~ etwas Kaffee?** [would you like] some more coffee?; **~ ein/zwei Bier, bitte!** another beer/two more beers, please!; **ich habe das ~ einmal/~ einige Male gemacht** I did it again/several times more; **er ist frech und ~ dazu dumm** *od.* **dumm dazu** he's cheeky and stupid with it; **Geld/Kleider** *usw.* **~ und ~:** heaps and heaps of money/clothes *etc. (coll.)*; **k)** *(bei Vergleichen)* **er ist ~ größer [als Karl]** he is even taller [than Karl]; **er will ~ mehr haben** he wants even *or* still more; **das ist ~ besser** that's even better *or* better still; **und wenn er auch ~ so bittet** however much he pleads; **l)** *(nach etw. Vergessenem fragend)* **wie heißt/hieß sie [doch] ~?** [now] what's/what was her name again? **2.** *Partikel* **das ist ~ Qualität!** that's what I call quality; **du wirst es ~ bereuen!** you'll regret it!; **der wird sich ~ wundern** *(ugs.)* he's in for a surprise; **er kann ~ nicht einmal lesen** he can't even read. **3.** *Konj. (und auch nicht)* nor; **weder ... noch** neither ... nor

nochmalig ['nɔxma:lɪç] *Adj.; nicht präd.* further

noch·mals *Adv.* again

NOK [ɛn|o:'ka:] **das; ~[s]** *Abk.* **Nationales Olympisches Komitee** NOC

Nomade [no'ma:də] **der; ~n, ~n, Nomadin die, ~, ~nen** nomad

Nomen ['no:mən] **das; ~s, Nomina** noun; substantive

Nominativ ['no:minati:f] **der; ~s, ~e** *(Sprachw.)* nominative [case]

nominell [nomi'nɛl] *Adj.; nicht präd.* nominal ‹*member, leader*›; ‹*Christian*› in name only

nominieren [nomi'ni:rən] *tr. V.* **a)** nominate; **b)** *(Sport: aufstellen)* name ‹*player, team*›

Nonkonformismus [nɔnkɔnfɔr'mɪsmʊs] **der; ~:** nonconformism

nonkonformistisch **1.** *Adj.* nonconformist; unconventional ‹*dress*›. **2.** *adv.* ‹*think, behave, etc.*› in an unconventional way

Nonne ['nɔnə] **die; ~, ~n** nun

Nonnen·kloster **das** convent; nunnery

Nonsens ['nɔnzɛns] **der; ~[es]** nonsense

nonstop [nɔn'stɔp] *adv.* non-stop

Nonstop-: ~flug der non-stop flight; **~kino das** 24-hour cinema

Noppe ['nɔpə] **die; ~, ~n a)** *(in einem Faden, Gewebe)* knop; nub; **b)** *(auf einer Oberfläche)* bump; *(auf einem Tischtennisschläger)* pimple

Nord [nɔrt] *o. Art.; o. Pl.* **a)** *(bes. Seemannsspr., Met.: Richtung)* north; **nach ~:** northwards; **b)** *(Gebiet)* North; **c)** *(Politik)* North; **d)** *einem Subst. nachgestellt (nördlicher Teil, nördliche Lage)* North; **Autobahnkreuz Köln-~:** motorway intersection Cologne North

nord-, Nord-: ~amerika (das) North America; **~amerikaner der** North American; **~deutsch 1.** *Adj.* North German; **2.** *adv.* **etw. ~deutsch aussprechen** pronounce sth. with a North German accent; **~deutschland (das)** North Germany

Norden der; ~s a) *(Richtung)* north; **nach ~:** northwards; to the north; **im/aus dem** *od.* **von** *od.* **vom ~:** in/from the north; **die Grenze nach ~:** the northern border; **b)** *(Gegend)* northern part; **c)** *(Geogr., Politik)* North; **der hohe/höchste ~:** the far North

Nord-: ~england (das) the North of England; **~europa (das)** Northern Europe; **~irland (das)** Northern Ireland

nordisch *Adj. (auch Völkerk.)* Nordic; **~e Kombination** *(Skisport)* Nordic combined

Nord-: ~kap das North Cape; **~korea (das)** North Korea; **~küste die** north *or* northern coast

nördlich ['nœrtlɪç] **1.** *Adj.* **a)** *(im Norden)* northern; **15 Grad ~er Breite** 15 degrees north [latitude]; **b)** *(nach, aus dem Norden)* northerly; **c)** *(des Nordens)* Northern. **2.** *adv.* northwards; **~ von ...:** [to the] north of ...; **sehr [weit] ~ sein** be a long way north. **3.** *Präp. mit Gen.* [to the] north of

nord-, Nord-: ~licht ['--] **das** northern lights *pl.*; aurora borealis; **~osten der** *(Richtung, Gegend)* north-east; *s. auch* **Norden; ~östlich 1.** *Adj.* **a)** *(im ~osten gelegen)* north-eastern; **b)** *(nach ~osten gerichtet, aus ~osten kommend)* north-easterly; **2.** *adv.* north-eastwards; **~östlich von ...:** [to the] north-east of ...; *s. auch* **nördlich 2; 3.** *Präp. mit Gen.* [to the] north-east of; **~pol** ['--] **der** North Pole

Nord·rhein-Westfalen (das) North Rhine-Westphalia

Nord·see die; *o. Pl.* North Sea

nord-, Nord-: ~seite ['---] **die** northern side; **~südlich** *Adj.* **in ~südlicher Richtung** from north to south; **~wärts** ['--] *Adv. (nach Norden)* northwards; **~westen der** *(Richtung, Gegend)* north-west; *s. auch* **Norden a; ~westlich 1.** *Adj.* **a)** *(im ~westen gelegen)* north-western; **b)** *(nach ~westen gerichtet, aus ~westen kommend)* north-westerly; **2.** *adv. (nach ~westen)* northwestwards; **~westlich von ...:** [to the] north-west of ...; *s. auch* **nördlich 2; 3.** *Präp. mit Gen.* [to the] north-west of; **~wind** ['--] **der** north *or* northerly wind

Nörgelei die; ~, ~en *(abwertend)* **a)** *o. Pl. (das Nörgeln)* moaning; grumbling; *(das Kritteln)* carping; **b)** *(Äußerung)* moan; grumble

nörgeln ['nœrgl̩n] *itr. V. (abwertend)* moan, grumble **(an** + *Dat.* about); *(kritteln)* carp **(an** + *Dat.* about)

Nörgler der; ~s, ~ *(abwertend)* moaner; grumbler; *(Krittler)* carper; fault-finder

Norm [nɔrm] **die; ~, ~en a)** norm; **als ~ gelten** count as the norm; **b)** *(geforderte Arbeitsleistung)* quota; target; **die ~ erfüllen** fulfil one's/its quota; meet *or* achieve one's/its target; **c)** *(Sport)* qualifying standard; **d)** *(technische, industrielle ~)* standard; standard specifications *pl.*

normal [nɔr'ma:l] **1.** *Adj.* **a)** normal; **du bist doch nicht ~!** *(ugs.)* there must be something wrong with you!; **b)** *(ugs.: gewöhnlich)* ordinary. **2.** *adv.* **a)** normally; **b)** *(ugs.: gewöhnlich)* in the normal *or* ordinary way

Normal·benzin das ≈ two-star petrol *(Brit.)*; regular *(Amer.)*

normalerweise *Adv.* normally; usually

Normal-: ~fall der normal case; **im ~fall** normally, usually; **~gewicht das** normal weight

normalisieren 1. *tr. V.* normalize. **2.** *refl. V.* return to normal

Normalität [nɔrmali'tɛ:t] **die; ~:** normality *no def. art.*

Normal-: ~maß das a) *(normales Maß)* normal size; **b)** *(Meßwesen)* standard measure; **~verbraucher der a)** ordinary *or* average consumer; **b)** *(ugs.: Durchschnittsmensch)* **Otto ~verbraucher** *(scherzh.)* the average punter *(coll.)*; **~zustand der** normal state

Normandie [nɔrman'di:] **die; ~:** Normandy

normativ [nɔrma'ti:f] *Adj.* normative

normen, normieren *tr. V.* standardize

Normung die; ~, ~en standardization

Norwegen ['nɔrve:gn̩] **(das); ~s** Norway

Norweger der; ~s, ~, Norwegerin die; ~, ~nen Norwegian

norwegisch *Adj.* Norwegian

Nostalgie [nɔstal'gi:] **die; ~:** nostalgia

not [no:t] *in* **~ tun** *od.* **sein** *(geh., landsch.)* be necessary

Not die; ~, Nöte ['nø:tə] **a)** *(Bedrohung, Gefahr)* **Rettung in** *od.* **aus höchster ~:** rescue from extreme difficulties; **in ~ sein** be in desperate straits; **b)** *o. Pl. (Mangel, Armut)* need; poverty [and hardship]; **~ leiden** suffer poverty *or* want [and hardship]; **in ~ geraten** encounter hard times; **~ macht erfinderisch** necessity is the mother of invention *(prov.)*; **in der ~ frißt der Teufel Fliegen** beggars can't be choosers *(prov.)*; **c)** *o. Pl. (Verzweiflung)* anguish; distress; **d)** *(Sorge, Mühe)* trouble; **in Nöten sein** have many troubles; **seine [liebe] ~ mit jmdm./etw. haben** have a lot of trouble *or* a lot of problems with sb./sth.; **mit knapper ~:** by the skin of one's teeth; **e)** *o. Pl. (veralt.: Notwendigkeit)* necessity; **zur ~:** if need be; if necessary; **wenn ~ am Mann ist** when the need arises; **aus der ~ eine Tugend machen** make a virtue of necessity

Notar [no'ta:ɐ̯] **der; ~s, ~e** notary

Notariat [notar'i̯a:t] **das; ~[e]s, ~e a)** *(Amt)* notaryship; **b)** *(Kanzlei)* notary's office

notariell [nota'ri̯ɛl] **1.** *Adj.* notarial. **2.** *adv.* **~ beglaubigt** attested by a notary

Not-: ~arzt der doctor on [emergency] call; emergency doctor; **~arzt·wagen der** doctor's car for emergency calls; **~ausgang der** emergency exit; **~behelf der** makeshift; **~bremse die** emergency brake; **~dienst der** *s.* **Bereitschaftsdienst**

Notdurft [-dʊrft] **die; ~** *(geh.)* **seine [große/kleine] ~ verrichten** relieve oneself

not·dürftig 1. *Adj.* meagre ‹*payment, pension*›; rough and ready, makeshift ‹*shelter, repair*›; scanty ‹*cover, clothing*›. **2.** *adv.* scantily ‹*clothed*›; **etw. ~ reparieren** repair sth. in a rough and ready *or* makeshift way

Note ['no:tə] **die; ~, ~n a)** *(Zeichen)* note; **eine ganze/halbe ~:** a crotchet/quaver *(Brit.)*; a whole note/half note *(Amer.)*; **b)** *Pl. (Text)* music *sing.*; **nach/ohne ~n spielen** play from/without music; **c)** *(Schul~)* mark; **d)** *(Eislauf, Turnen)* score; **e)** *(Dipl.)* note; **f)** *o. Pl. (Flair)* touch

Noten-: ~blatt das sheet of music; **~heft das a)** *(Publikation)* book of music; **b)** *(Heft mit ~papier)* manuscript book; **~papier das** music-paper; **~schlüssel der** clef; **~schrift die** [musical] notation; **~ständer der** music-stand

not-, Not-: ~fall der a) *(Gefahr)* emergency; **für den ~fall** in case of emergency; **b)** *(Schwierigkeiten)* case of need; **im ~fall** if need be; **~falls** *Adv.* if need be; if necessary; **~gedrungen** *Adv.* of necessity; **ich habe ~gedrungen eine neue gekauft** I had no choice but to *or* I was forced to buy a new one; **~groschen der** nest-egg

notieren [no'ti:rən] **1.** *tr. V.* **a)** **[sich** *(Dat.)***] etw. ~:** note sth. down; make a note of sth.; **b)** *(Börsenw., Wirtsch.)* quote **(mit** at). **2.** *itr. V. (Börsenw., Wirtsch.)* be quoted **(mit** at); **die meisten Rohstoffe ~ unverändert** most commodity prices are unchanged

nötig ['nø:tɪç] **1.** *Adj.* necessary; **dafür** *od.* **dazu fehlt mir die ~e Geduld/das ~e Geld** I don't have the patience/money necessary *or* needed for that; **etw./jmdn. ~ haben** need sth./sb.; **es ~ haben, etw. zu tun** need to do sth.; **sich zu entschuldigen, hat er natürlich nicht ~** *(iron.)* of course he does not feel the need to apologize; **du hast/er hat** *usw.* **es gerade ~** *(ugs.)* you're/he's a fine one to talk *(coll.)*; **das wäre [doch] nicht ~ gewesen!** *(ugs.)* you shouldn't have!; **das Nötigste** the bare essentials *pl.* **2.** *adv.* **er braucht ~ Hilfe** he is in urgent need of *or* urgently needs help; **was er am ~sten braucht, ist ...:** what he most urgently needs is ...

nötigen *tr. V.* **a)** *(zwingen)* compel; force; *(Rechtsspr.)* intimidate; coerce; **jmdn. zur Unterschrift ~:** compel *or* force sb. to sign; **sich genötigt sehen, etw. zu tun** feel compelled to do sth.; **b)** *(geh.: auffordern)* press; urge

Nötigung die; ~, ~en *(bes. Rechtsspr.)* intimidation; coercion

Notiz [no'ti:ts] **die; ~, ~en a)** note; **sich** *(Dat.)* **eine ~ machen** make a note; **b)** *(Zeitungs~)* **eine [kurze] ~:** a brief report; **c)** *in* **von jmdm./etw. [keine] ~ nehmen** take [no] notice of sb./sth.

Notiz-: ~block der; *Pl.* **~blocks,** *schweiz.:* **~blöcke** notepad; **~buch das** notebook; **~zettel der** note

not-, Not-: ~lage die serious difficulties *pl.*; **~landen**[1] *itr. V.; mit sein* do an emergency landing; **~landung die** emergency landing; **~leidend** *Adj.* needy; impoverished; *(fig.)* ailing ‹*industry*›; **die Notleidenden** the [poor and] needy; **~lösung die** stopgap; **~lüge die** evasive lie; *(aus Rücksichtnahme)* white lie

notorisch [no'to:rɪʃ] **1.** *Adj.* notorious. **2.** *adv.* notoriously

not-, Not-: ~ruf der a) *(Hilferuf)* emergency call; *(eines Schiffes)* Mayday call; distress call; **b)** *(Nummer)* emergency number; **~ruf·säule die** emergency telephone *(mounted in a pillar)*; **~schlachten**[2] *tr., itr. V.* slaughter ‹*sick or injured animal*›; **~sitz der** extra seat; *(ausklappbar)* tip-up seat; fold-away seat; **~stand der a)** *(Krise, Übelstand)* crisis; **b)** *(Staatsrecht)* state of emergency; **~stands·gebiet das a)** *(auch fig.)* disaster area; **b)** *(Wirtsch.)* depressed area; **~stands·gesetz das** emergency law; **~unterkunft die** emergency accommodation *no pl., no indef. art.*; **~unterkünfte** emergency accommodation *sing.*; **~vorrat der** emergency supply; **~wehr die** *(Rechtsw.)* self-defence; **in** *od.* **aus ~wehr** in self-defence

not·wendig 1. *Adj.* necessary; *(unvermeidlich)* inevitable; **es ist ~, daß wir etwas tun** we must do something; **das Notwendigste** the bare essentials *pl.* **2.** *adv.* **a)** *s.* **nötig 2**; **b)** *(zwangsläufig, unbedingt)* necessarily

notwendiger·weise *Adv.* necessarily

Notwendigkeit die; ~, ~en necessity

Not-: ~zeit die time of emergency; *(Zeit des Mangels)* time of need; **~zucht die** *(Rechtsw. veralt.)* rape

Nougat ['nu:gat] **der;** *auch* **das; ~s** nougat

---

[1] *ich notlande, notgelandet, notzulanden*

[2] *ich notschlachte, notgeschlachtet, notzuschlachten*

Novelle [no'vɛlə] **die; ~, ~n a)** *(Literaturw.)* novella; **b)** *(Gesetzes~)* amendment

novellieren *tr. V. (Politik, Rechtsw.)* amend

November [no'vɛmbɐ] **der; ~[s], ~:** November; *s. auch* **April**

Novität [novi'tɛ:t] **die; ~, ~en** novelty; *(neue Erfindung)* innovation; *(neue Schallplatte)* new release; *(neues Buch)* new publication

Novize [no'vi:tsə] **der; ~n, ~n, Novizin die; ~, ~nen** novice

Nr. *Abk.* **Nummer** No

N.T. *Abk.* **Neues Testament** NT

Nu der *in* **im Nu, in einem Nu** in no time

Nuance ['nỹã:sə] **die; ~, ~n a)** *(Unterschied, Feinheit)* nuance; **b)** *(Grad)* shade; **eine ~ dunkler/schneller** a shade darker/faster

nüchtern ['nʏçtɐn] **1.** *Adj.* **a)** *(nicht betrunken)* sober; **wieder ~ werden** sober up; **b)** *(mit leerem Magen)* **der Patient muß ~ sein** the patient's stomach must be empty; **auf ~en Magen rauchen** smoke on an empty stomach; **c)** *(realistisch)* sober; sober, matter-of-fact ⟨*account, assessment, question, etc.*⟩; bare ⟨*figures*⟩; **d)** *(schmucklos, streng)* austere; bare ⟨*room*⟩; unadorned, bare ⟨*walls*⟩; *(ungeschminkt)* bare, plain ⟨*fact*⟩. **2.** *adv.* **a)** *(realistisch)* soberly; **b)** *(schmucklos, streng)* austerely

nuckeln *(ugs.) itr. V.* suck **(an +** *Dat.* at); **am Daumen/Schnuller ~:** suck one's thumb/a *or* one's dummy

Nudel ['nu:dl̩] **die; ~, ~n a)** piece of spaghetti/vermicelli/tortellini *etc.; (als Suppeneinlage)* noodle; **~n** *(Teigwaren)* pasta *sing.; (als Suppeneinlage, Reis ~n)* noodles; **b)** *(ugs.)* **eine komische ~:** a real character

Nudel-: ~salat der pasta salad; **~suppe die** soup with noodles

Nudist [nu'dɪst] **der; ~en, ~en** nudist; naturist

Nugat *s.* **Nougat**

nuklear [nukle'a:ɐ] **1.** *Adj.* nuclear. **2.** *adv.* **~ angetrieben** nuclear-powered

Nuklear·krieg der nuclear war

null [nʊl] **1.** *Kardinalz.* nought; **~ Komma sechs** [nought] point six; **sieben, ~, ~, sechs, ~, vier** *(Fernspr.)* seven double-O, six O four *(Brit.);* seven zero zero, six zero four *(Amer.);* **~ Grad Celsius** nought *or* zero degrees Celsius; **bei ~ Fehlern** if there are no mistakes; **fünf zu ~:** five–nil; **das Spiel endete ~ zu ~:** the game was a goalless draw; **fünfzehn ~** *(Tennis)* fifteen-love; **gegen ~ Uhr** around twelve midnight; **es ist ~ Uhr dreißig** it is twelve-thirty a.m.; **etw. für ~ und nichtig erklären** declare sth. null and void. **2.** *indekl. Adj. (ugs.)* **~ Ahnung** no idea at all

Null die; ~, ~en a) *(Ziffer)* nought; zero; **in ~ Komma nichts** *(ugs.)* in less than no time; **gleich ~ sein** *(fig.)* be practically zero; **b)** *o. Pl., o. Art. (Marke)* zero; **auf ~ stehen** ⟨*indicator, needle, etc.*⟩ be at zero; **fünf Grad unter/über ~:** five degrees below/above zero *or* freezing; **c)** *(abwertend) (Versager)* failure; dead loss *(coll.)*; *(unbedeutender Mensch)* nonentity

null·acht·fünfzehn, null·acht·fuffzehn *(ugs. abwertend)* **1.** *indekl. Adj.; nicht attr.* run-of-the-mill. **2.** *adv.* ⟨*dressed, furnished*⟩ in a run-of-the-mill way

Null-: ~punkt der zero; **die Temperatur ist auf den ~punkt abgesunken** the temperature has dropped to zero *or* to freezing-point; **~tarif der: zum ~tarif** free of charge

numerieren [numə'ri:rən] *tr. V.* number

numerisch [nu'me:rɪʃ] **1.** *Adj.* numerical. **2.** *adv.* numerically

Numerus ['nʊmərʊs] **der; ~, Numeri** *(Sprachw.)* number

Numerus clausus [~ 'klaʊzʊs] **der; ~ ~:** *fixed number of students admissible to a university to study a particular subject;* numerus clausus

Nummer ['nʊmɐ] **die; ~, ~n a)** number; **ein Wagen mit Münchner ~:** a car with a Munich registration; **ich bin unter der ~ 242679 zu erreichen** I can be reached on 242679; **bloß eine ~ sein** *(fig.)* be just a *or* nothing but a number; **[die] ~ eins** [the] number one; **auf ~ Sicher gehen** *(ugs.)* play safe; not take any chances; **b)** *(Ausgabe)* number; issue; **c)** *(Größe)* size; **d)** *(Darbietung)* turn; **e)** *(ugs.: Musikstück)* number; **f)** *(ugs.: Person)* character

Nummern·schild das number-plate; license plate *(Amer.)*

nun [nu:n] **1.** *Adv.* now; **von ~ an** from now on; **~, wo sie krank ist** now [that] she's ill; **~ erst** only now. **2.** *Partikel* now; **so wichtig ist es ~ auch wieder nicht** it's not all 'that important; **das hast du ~ davon!** it serves you right!; **~ gib schon her!** now hand it over!; **kommst du ~ mit oder nicht?** now are you coming or not?; **so ist das ~ [einmal/mal]** that's just the way it is *or* things are; **~ gut** *od.* **schön** [well,] all right; **~, ~!** now, come on; **ja ~!** oh, well!; **~?** well?; **~ ja ...:** well, yes ...; **~ denn!** *(also gut)* well, all right!; *(also los)* well then

nun·mehr *Adv. (geh.)* **a)** now; **b)** *(von ~ an)* from now on; henceforth

nur [nu:ɐ] **1.** *Adv.* **a)** *(nicht mehr als)* only; just; **ich habe ~ eine Stunde Zeit** I only have an hour; **er hat ~ einen einzigen Fehler gemacht** he made just a single mistake; **das ist ~ recht und billig** it is only right and proper; **b)** *(ausschließlich)* only; **alle durften mitfahren, ~ ich nicht** everyone was allowed to go, all except me; **er tut das mit Absicht, ~ um dich zu provozieren** he does it deliberately, just to provoke you; **nicht ~ ..., sondern auch ...:** not only ..., but also ...; **nicht ~, daß ...:** it's not just that ...; **ich male ~ so zum Spaß** I paint just for fun; **Warum fragst du? – Ach, ~ so** Why do you ask? – Oh, no particular reason; **~ daß ...:** except that ...; **das ist ~ zu wahr!** it's only too true! **2.** *Partikel* **a)** *(in Wünschen)* **wenn das ~ gut geht!** let's [just] hope it goes well; **wenn er ~ käme/hier wäre** if only he would come/he were here; **b)** *(ermunternd, tadelnd)* **~ keine Hemmungen!** don't be inhibited!; **~ zu!** go ahead; **c)** *(warnend)* **laß dich ~ nicht erwischen** just don't let me/him/her/them catch you; **~ Geduld/vorsichtig/langsam** just be patient/careful/take it easy; **~ nicht!** don't, for goodness' sake!; **d)** *(fragend)* just; **wie soll ich ihm das ~ erklären?** just how am I supposed to explain it to him?; **was sollen wir ~ tun?** what on earth are we going to do?; **was hat er ~?** whatever's the matter with him?; **e)** *(verallgemeinernd)* just; **er lief, so schnell er ~ konnte** he ran just as fast as he could; **f)** *(sogar)* only; just; **g) es wimmelte ~ so von Insekten** it was just teeming with insects; **er schlug auf den Tisch, daß es ~ so krachte** he crashed his fist [down] on the table. **3.** *Konj.* but

Nürnberg ['nʏrnbɛrk] **(das); ~s** Nuremberg

nuscheln ['nʊʃl̩n] *tr., itr. V. (ugs.)* mumble

Nuß [nʊs] **die; ~, Nüsse** ['nʏsə] **a)** nut; **eine harte ~ [für jmdn.]** *(fig.)*

a hard *or* tough nut [for sb. to crack]; **b)** *(salopp abwertend: Mensch)* so-and-so *(coll.)*

Nuß-: ~baum der a) walnut-tree; **b)** *o. Pl. (Holz)* walnut; **~knacker der** nutcrackers *pl.*; **~schale die** nutshell; *(fig.: Boot)* cockle-shell; **~schokolade die** nut chocolate

Nüster ['nʏstɐ] **die; ~, ~n** nostril

Nutte ['nʊtə] **die; ~, ~n** *(derb abwertend)* tart *(sl.)*; pro *(Brit. coll.)*; hooker *(Amer. sl.)*

nutz-: ~bar *Adj.* usable; exploitable, utilizable ‹*mineral resources, invention*›; cultivatable ‹*land, soil*›; **etw. praktisch ~bar machen** turn sth. to practical use; **die Sonnenenergie ~bar machen** harness solar energy **(für** for); **~bringend 1.** *Adj.* *(nützlich)* useful; *(gewinnbringend)* profitable. **2.** *adv.* profitably

nütze ['nʏtsə] *in* **zu etw. ~ sein** be good for sth.; **[jmdm.] zu nichts ~ sein** be no use *or* good [to sb.]

nutzen ['nʊtsn̩] **1.** *tr. V.* **a)** use; exploit, utilize ‹*natural resources*›; cultivate ‹*land, soil*›; use, harness ‹*energy source*›; exploit ‹*advantage*›; **eine Fläche landwirtschaftlich ~:** use an area for agriculture; **b)** *(be~, aus~)* use; make use of; **eine Gelegenheit ~, etw. zu tun** take [advantage of] an opportunity to do sth.; **seine Chance ~:** take one's chance. **2.** *itr. V. s.* **nützen 1**

Nutzen der; ~s a) benefit; **den ~ [von etw.] haben** benefit *or* gain [from sth.]; **~ aus etw. ziehen** benefit from sth.; exploit sth.; **[jmdm.] von ~ sein** be of use *or* useful [to sb.]; **b)** *(Profit)* profit

nützen ['nʏtsn̩] **1.** *itr. V.* be of use (*Dat.* to); **nichts ~:** be useless *or* no use; **jmdm. sehr ~:** be very useful *or* of great use to sb.; **was hat ihm das genützt?** what good did it do him?; **es würde nichts/wenig ~:** it wouldn't be any/much use *or* wouldn't do any/much good; **da nützt alles nichts** there's nothing to be done. **2.** *tr. V. s.* **nutzen 1**

Nutz-: ~fahrzeug das *(Lastwagen, Lieferwagen usw.)* commercial vehicle; goods vehicle; *(Bus, Straßenbahn usw.)* public-service vehicle; **~fläche die a)** *(von Gebäuden)* usable floor space; **b)** *(Landw.)* **landwirtschaftliche ~flächen** land *sing.* available for agriculture

nützlich ['nʏtslɪç] *Adj.* useful; **sich ~ machen** make oneself useful

Nützlichkeit die; ~: usefulness

nutzlos 1. *Adj.* useless; *(vergeblich)* futile; vain *attrib.*; in vain *pred.*; **es wäre ~, das zu tun** it would be useless *or* pointless *or* futile doing that. **2.** *adv.* uselessly; *(vergeblich)* futilely; in vain; **er hat das Geld ~ vergeudet** he squandered the money on useless items

Nutz·losigkeit die; ~: uselessness; *(Vergeblichkeit)* futility; vainness

Nutznießer der; ~s, ~, Nutznießerin die; ~, ~nen a) beneficiary; **b)** *(Rechtsw.)* usufructuary

Nutzung die; ~, ~en use; *(des Landes, des Bodens)* cultivation; *(von Bodenschätzen)* exploitation; utilization; *(einer Energiequelle)* use; harnessing; **die wirtschaftliche ~ einer Fläche** the use of an area for financial benefit; **jmdm. etw. zur ~ überlassen** give sb. the use of sth.

NVA *Abk.* **Nationale Volksarmee** *(ehem. DDR)* National People's Army

NW *Abk.* **Nordwest[en]** NW

Nylon Ⓦⓩ ['nailɔn] **das; ~s** nylon

Nylon·strumpf der nylon stocking

Nymphe ['nʏmfə] **die; ~, ~n** *(Myth., Zool.)* nymph

Nymphomanin die; ~, ~nen *(Psych.)* nymphomaniac

O

o, O [o:] **das; ~,** *(ugs.:)* **~s, ~,** *(ugs.:)* **~s** o/O; *s. auch* **a, A**

o [o:] *Interj.* **oh**

O *Abk.* **Ost[en]** E

ö, Ö [ø:] **das; ~,** *(ugs.:)* **~s, ~,** *(ugs.:)* **~s** o/O umlaut; *s. auch* **a, A**

Oase [o'a:zə] **die; ~, ~n** *(auch fig.)* oasis

ob [ɔp] *Konj.* **a)** whether; **ob wir es schaffen?** will we manage it?; **ob er will oder nicht** whether he wants to or not; **ob arm, ob reich** whether rich or poor; **b)** *in* **und ob!** of course!; you bet! *(coll.)*

OB *Abk.* **Oberbürgermeister**

Obacht ['o:baxt] **die; ~** *(bes. südd.)* caution; **~, da kommt ein Auto!** watch out! *or* look out! *or* careful!, there's a car coming; **~ auf jmdn./etw. geben** look after *or* take care of sb./sth.; *(aufmerksam sein)* pay attention to sb./sth.; **~ geben, daß ...** take care that ...

Obdach ['ɔpdax] **das; ~[e]s** *(geh.)* shelter

obdach·los *Adj.* homeless; **~ werden** be made homeless

Obdachlose der/die; *adj. Dekl.* homeless person/man/woman; **die ~n** the homeless

Obdachlosen·asyl das hostel for the homeless

Obduktion [ɔpduk'tsio:n] **die; ~, ~en** *(Med., Rechtsw.)* post-mortem [examination]; autopsy

O-Beine *Pl.* bandy legs; bow-legs

oben ['o:bn̩] *Adv.* **a)** *(an hoch/höher gelegenem Ort)* **hier/dort ~:** up here/there; **[hoch] ~ am Himmel** [high] up in the sky; **~ bleiben** stay up; **weiter ~:** further up; **nach ~:** upwards; **der Weg nach ~:** the way up; **warme Luft steigt nach ~:** warm air rises; **~ auf dem Dach** up on the roof; **von ~:** from above; **von ~ herab** *(fig.)* condescendingly; **b)** *(im Gebäude)* upstairs; **nach ~:** upstairs; **der Aufzug fährt nach ~:** the lift *(Brit.) or (Amer.)* elevator is going up; **c)** *(am oberen Ende, zum oberen Ende hin)* at the top; **~ im/auf dem Schrank** at the/up on top of the cupboard; **nach ~ [hin]** towards the top; **weiter ~ [im Tal]** further *or* higher up [the valley]; **von ~:** from the top; **~ links/rechts** at the top on the left/right; **~ [links/rechts]** *(in Bildunterschriften)* above [left/right]; **auf Seite 25 ~:** at the top of page 25; **die fünfte Zeile von ~:** the fifth line from the top; the fifth line down; **nach ~ kommen** *(an die Oberfläche)* come up; **„~"** 'this side up'; **wo** *od.* **was ist [bei dem Bild] ~:** which is the right way up [on the picture]?; which is the top [of the picture]?; **bis ~ hin voll sein** *(ugs.)* be full to the top; **von ~ bis unten** from top to bottom; **er musterte sie von ~ bis unten** he looked her up and down*(coll.)*; **~ ohne** topless; **~ an der Tafel** at the head of the table; **d)** *(an der Oberseite)* on top; **e)** *(in einer Hierarchie, Rangfolge)* at the top; **weit/ganz ~:** near the top/right at the top; **der Befehl kam von ~:** the order came from above; **die da ~** *(ugs.)* the high-ups *(coll.)*; **f)** *([weiter] vorn im Text)* above; **g)** *(im Norden)* up north; **hier/dort ~:** up here/there [in the north]

oben-: ~an *Adv.* at the top; **~drauf** *Adv. (ugs.)* on top; **~erwähnt, ~genannt** ['----] *Adj.; nicht präd.*, **~stehend** ['----] *Adj.* above-mentioned

ober... ['o:bɐ] *Adj.* **a)** upper *attrib.*; top *attrib.*; *(ganz oben liegend)* top *attrib.*; **die ~e rechte Ecke** the top right-hand corner; **am ~en Ende der Straße** at the top [end] of the street; **das Oberste zuunterst kehren** turn everything upside down; **das ~ste Stockwerk** the top[most] storey; **b)** *(der Quelle näher gelegen)* upper; **c) die ~en Klassen der Schule** the senior classes *or* forms of the school; **das ~ste Gericht des Landes** the highest court in the land; **der Oberste Sowjet** the Supreme Soviet

Ober der; ~s, ~ waiter; **Herr ~!** waiter!

Ober-: ~arm der upper arm; **~arzt der** *(Vertreter des Chefarztes)* assistant medical director; *(Leiter einer Spezialabteilung)* consultant; **~befehlshaber der, ~befehlshaberin die** *(Milit.)* supreme commander; commander-in-chief; **~begriff der** generic term; **~bekleidung die** outer clothing; **~bett das** duvet (Brit.); stuffed quilt *(Amer.)*; **~bürger·meister der** mayor; **~deck der** upper deck; *(eines Busses)* upper *or* top deck

Ober·fläche die surface; *(Flächeninhalt)* surface area; **die Diskussion blieb zu sehr an der ~** *(fig.)* the discussion remained far too superficial

oberflächlich 1. *Adj.* superficial; **eine erste, ~ Schätzung** a first, rough estimate. **2.** *adv.* superficially; **etw. nur ~ kennen** have only a superficial knowledge of sth.; **etw. ~ lesen** read sth. cursorily; **er arbeitet zu ~** he is too superficial in the way he works

Oberflächlichkeit die; ~: superficiality

Ober·geschoß das upper storey

ober·halb 1. *Adv.* above; **weiter ~:** further up; **~ von** above. **2.** *Präp. mit Gen.* above

Ober-: ~hand die *in* **die ~hand [über jmdn./etw.] haben/gewinnen** *od.* **bekommen** have/gain *or* get the upper hand [over sb./sth.]; **~haupt das** head; *(einer Verschwörung)* leader; **~haus das** *(Parl.)* upper house *or* chamber; *(in Großbritannien)* House of Lords; Upper House; **~hemd das** shirt; **~herrschaft die** *o. Pl.* sovereignty; supreme power; **~kellner der** head waiter; **~kiefer der** upper jaw; **~klasse die a)** *(Soziol.)* upper class; **b)** *(Schulw.)* senior class *or* form; **~kommandierende der/die;** *adj. Dekl. (Milit.) s.* **~befehlshaber; ~körper der** upper part of the body; **den ~körper frei machen** strip to the waist; **~leder das** upper; **~leitung die** *(elektrische Leitung)* overhead cable; **~leutnant der** *(beim Heer)* lieutenant *(Brit.)*; *(bei der Luftwaffe)* flying officer *(Brit.)*; first lieutenant *(Amer.)*; **~leutnant zur See** sub-lieutenant *(Brit.)*; lieutenant junior grade *(Amer.)*; **~licht das** high window; *(über einer Tür)* fanlight; **~lippe die** upper lip; **~schenkel der** thigh; **~schicht die** upper class; **~schule die** secondary school; **~seite die** top[side]; upper side; *(eines Stoffes)* right side

oberst... *s.* **ober...**

Ober·staats·anwalt der *senior public prosecutor (at a regional court)*

Oberst·leutnant der *(beim Heer)* lieutenant-colonel; *(bei der Luftwaffe)* wing commander; lieutenant-colonel *(Amer.)*

Ober-: ~stübchen das *s.* **richtig 1 b; ~studien·direktor der a)** headmaster *(Brit.)*; principal; **b)** *(ehem. DDR) highest honorary title for a teacher;* **~studien·rat der a)** senior teacher; **b)** *(ehem. DDR) honorary title for a teacher;* **~stufe die** *(Schulw.)* upper school; **~teil das** *od.* **der** top [part]; *(eines Bikinis, Anzugs, Kleids usw.)* top [half]; **~wasser das;** *o. Pl.* headwater; *(fig.)* **~wasser haben** feel in a strong position; **~wasser bekommen** have one's hand strengthened

ob·gleich *Konj. s.* **obwohl**

Ob·hut die; ~: *(geh.)* care; **jmdn./etw. jmds. ~** *(Dat.)* **anvertrauen** entrust sb./sth. to sb.'s care

obig ['o:bɪç] *Adj.; nicht präd.* above

Objekt [ɔp'jɛkt] **das; ~s, ~e a)** *(auch Sprachw., Kunstwiss.)* object; *(Fot., bei einem Experiment)* subject; **b)** *(Kaufmannsspr.: Immobilie)* property

objektiv [ɔpjɛk'ti:f] **1.** *Adj.* objective; real, actual ‹*cause, danger*›. **2.** *adv.* objectively

Objektiv das; ~s, ~e a) *(Optik)* objective; **b)** *(Fot.)* lens

Objektivität [ɔpjɛktivi'tɛ:t] **die; ~:** objectivity

Objekt·satz der *(Sprachw.)* object clause

obligatorisch [obliga'to:rɪʃ] **1.** *Adj.* **a)** obligatory; compulsory ‹*subject, lecture, etc.*›; necessary ‹*qualification*›; **b)** *(iron.: unvermeidlich)* obligatory. **2.** *adv.* obligatorily; compulsorily

Oboe [o'bo:ə] **die; ~, ~n** oboe

Obrigkeit ['o:brɪçkait] **die; ~, ~en** authorities *pl.*

Obrigkeits·staat der authoritarian state

ob·schon *Konj. (geh.) s.* **obwohl**

observieren *tr. V.* **a) jmdn./etw. ~:** keep sb./sth. under surveillance; **b)** *(wissenschaftlich)* observe

obskur [ɔps'ku:ɐ̯] *Adj. (geh.)* **a)** *(unbekannt, unklar)* obscure; **b)** *(dubios)* dubious

Obst [o:pst] **das, ~[e]s** fruit

Obst-: ~baum der fruit-tree; **~garten der** orchard; **~händler der** fruiterer; **~kuchen der** fruit flan

Obstler ['o:pstlɐ] **der; ~s, ~** *(bes. südd.)* fruit brandy

Obst-: ~saft der fruit juice; **~salat der** fruit salad; **~torte die** fruit flan

obszön [ɔps'tsø:n] **1.** *Adj.* obscene. **2.** *adv.* obscenely

Obszönität [ɔpstsøni'tɛ:t] **die; ~, ~en** obscenity

O·bus der trolley bus *(Brit.)*

ob·wohl *Konj.* although; though

Ochs [ɔks] **der; ~en, ~en** *(südd., österr., schweiz., ugs.)*, **Ochse** ['ɔksə] **der; ~n, ~n a)** ox; bullock; **~ am Spieß** roast ox; **b)** *(salopp)* numskull *(coll.)*; **ich ~!** what a numskull I am!

Ochsen-: ~brust die *(Kochk.)* brisket of beef; **~schwanz·suppe die** *(Kochk.)* oxtail soup; **~zunge die** *(Kochk.)* ox-tongue

öd [ø:t] *(geh.) s.* **öde**

od. *Abk.* **oder**

Ode ['o:də] **die; ~, ~n** ode (**an** + *Akk.* to, **auf** + *Akk.* on)

öde ['ø:də] *Adj.* **a)** *(verlassen)* deserted ‹*beach, village, house, street, etc.*›; *(unbewohnt)* desolate ‹*area, landscape*›; **b)** *(unfruchtbar)* barren; **c)** *(langweilig)* tedious; dreary

Öde die; ~, ~n a) *o. Pl. s.* **öde a–b:** desertedness; desolateness; barrenness; **b)** *(öde Gegend)* wasteland; waste; **c)** *(Langeweile)* tediousness; dreariness

oder ['o:dɐ] *Konj.* **a)** or; **~ aber** or

else; *s. auch* **entweder; b)** *(in Fragen)* **du kommst doch mit, ~?** you will come, won't you?; **er ist doch hier, ~?** he is here, isn't he? *(zweifelnd)* he is here – or isn't he?; **das ist doch erlaubt, ~ [etwa nicht]?** that is allowed, isn't it?

Oder-Neiße-Linie die Oder-Neisse Line

Ofen ['o:fn̩] **der; ~s, Öfen a)** heater; *(Kohle~)* stove; *(Öl~, Petroleum~)* stove; heater; *(elektrischer ~)* heater; fire; **wenn sie uns erwischen, ist der ~ aus** *(ugs.)* if they catch us, it's all over; **b)** *(Back~)* oven; **c)** *(Brenn~, Trokken~)* kiln; **d)** *(landsch.: Herd)* cooker

Ofen-: ~heizung die; *o. Pl.* heating *no art.* by stoves; **~rohr das** [stove] flue

offen ['ɔfn̩] **1.** *Adj.* **a)** open; unsealed ‹*envelope*›; ulcerated ‹*legs*›; **mit ~em Mund** with one's mouth open; **der Knopf/Schlitz ist ~:** the button is/one's flies are undone; **ein ~es Hemd** a shirt with the collar unfastened; **sie trägt ihr Haar ~:** she wears her hair loose; **~ haben** *od.* **sein** be open; **die Tür ist ~** *(nicht abgeschlossen)* the door is unlocked; **mit ~en Karten spielen** play with the cards face up on the table; *(fig.)* put one's cards on the table; **~es Licht/Feuer** a naked light/an open fire ; **das ~e Meer, die ~e See** the open sea; **~e Türen einrennen** *(fig.)* fight a battle that's/battles that are already won; **mit ~en Augen** *od.* **Sinnen durch die Welt** *od.* **durchs Leben gehen** go about/go through life with one's eyes open; **für neue Ideen** *od.* **gegenüber neuen Ideen ~ sein** be receptive *or* open to new ideas; **b)** *(lose)* loose ‹*sugar, flour, oats, etc.*›; **~er Wein** wine on tap *or* draught; **c)** *(frei)* vacant ‹*job, post*›; **~e Stellen** vacancies; *(als Rubrik)* 'Situations Vacant'; **d)** *(ungewiß, ungeklärt)* open, unsettled ‹*question*›; uncertain ‹*result*›; **der Ausgang des Spiels ist noch völlig ~:** the result of the match is still wide open; **e)** *(noch nicht bezahlt)* outstanding ‹*bill*›; **f)** *(freimütig, aufrichtig)* frank [and open] ‹*person*›; frank, candid ‹*look, opinion, reply*›; honest ‹*character, face*›; **~ zu jmdm. sein** be open *or* frank with sb.; **g)** *nicht präd. (unverhohlen)* open ‹*threat, mutiny, hostility, opponent, etc.*›; **h)** *(Sprachw.)* open ‹*vowel, syllable*›. **2.** *adv.* **a)** *(frei zugänglich, sichtbar, unverhohlen)* openly; **b)** *(freimütig, aufrichtig)* openly; frankly; **~ gesagt** frankly; to be frank *or* honest; **~ gestanden** to tell you the truth

offen·bar 1. *Adj.* obvious. **2.** *adv.* **a)** *(offensichtlich)* obviously; clearly; **b)** *(anscheinend)* evidently

offenbaren *(geh.)* **1.** *tr. V.* reveal. **2.** *refl. V.* **a)** *(sich erweisen)* **sich als etw. ~:** ‹*person*› show *or* reveal oneself to be sth.; **b)** *(sich mitteilen)* **sich jmdm. ~:** confide in sb.

Offenbarung die; ~, ~en revelation

Offenbarungs·eid der oath of disclosure

offen-: ~|bleiben *unr. itr. V.; mit sein* **a)** remain *or* stay open; **b)** *(ungeklärt bleiben)* remain open; ‹*decision*› be left open; **c)** *(unerfüllt bleiben)* ‹*wish*› remain unsatisfied; **~|halten** *unr. tr. V.* **etw. ~halten** keep sth. open

Offen·heit die; ~: *s.* **offen f:** frankness [and openness]; candidness; honesty

offen-, Offen-: ~herzig 1. *Adj.* frank, candid ‹*conversation, remark*›; frank and open ‹*person*›; **2.** *adv.* frankly; openly; **~herzigkeit die; ~:** frankness; candidness; candour; **~kundig 1.** *Adj.* obvious, evident **(für** to); obvious, patent, manifest ‹*lie, betrayal, misuse*›; **2.** *adv.* obviously; clearly; **~|lassen** *unr. tr. V.* **etw. ~lassen** leave sth. open; **~lassen, ob ...:** leave it open whether ...; **~sichtlich 1.** *Adj.* obvious; evident; **2.** *adv.* obviously; *(anscheinend)* evidently

offensiv [ɔfən'zi:f] **1.** *Adj.* **a)** offensive; **b)** *(Sport)* attacking. **2.** *adv.* **a)** offensively; **b)** *(Sport)* **~ spielen** play an attacking game

Offensive die; ~, ~n *(auch Sport)* offensive; **in der ~:** on the offensive; **die ~ ergreifen, in die ~ gehen** go on to the offensive

offen|stehen *unr. itr. V.* **a)** be open; **b)** *(zur Benutzung freigegeben sein)* be open (*Dat.* to); **c)** *s.* **freistehen a; d)** *(unbezahlt sein)* be outstanding

öffentlich ['œfn̩tlɪç] **1.** *Adj.* public; state *attrib.*, [state-] maintained ‹*school*›; **die ~e Meinung** public opinion; **Erregung ~en Ärgernisses** *(Rechtsw.)* creating a public nuisance; **der ~e Dienst** the civil service; **die Ausgaben der ~en Hand** public spending *sing.*; **eine Persönlichkeit des ~en Lebens** a public figure. **2.** *adv.* **a)** publicly; ‹*perform, appear*› in public; **~ tagen** meet in open session; **etw. ~ versteigern** sell sth. by public auction; **b)** *(vom Staat usw.)* publicly ‹*funded etc.*›

Öffentlichkeit die; ~: a) public; **unter Ausschluß der ~:** in private *or* secret; *(Rechtsw.)* in camera; **etw. an die ~ bringen** bring sth. to public attention; make sth. public; **vor die ~ treten** appear in public; **in aller ~:** [quite openly] in public; **b)** *(das Öffentlichsein)* **das Prinzip der ~ in der Rechtsprechung** the principle that justice be administered in open court

Öffentlichkeits·arbeit die; *o. Pl.* public relations work *no art.*

öffentlich-rechtlich *Adj.* under public law *postpos., not pred.*; **~es Fernsehen** state-owned television

Offerte [ɔ'fɛrtə] **die; ~, ~n** *(Kaufmannsspr.)* offer

offiziell [ɔfi'tsi̯ɛl] **1.** *Adj.* official. **2.** *adv.* officially

Offizier [ɔfi'tsi:ɐ̯] **der; ~s, ~e** officer

Offiziers·lauf·bahn die officer's career

öffnen ['œfnən] **1.** *tr. (auch itr.) V.* open; turn on ‹*tap*›; undo ‹*coat, blouse, button, zip*›; **die Bank ist** *od.* **hat über Mittag geöffnet** the bank is open at lunch-time; **„hier ~"** 'open here'; **jmdm. den Blick für etw. ~:** open sb.'s eyes to sth. **2.** *itr. V.* **a) [jmdm.] ~:** open the door [to sb.]; **wenn es klingelt, mußt du ~:** if there's a ring at the door, you must go and answer it; **b)** *(geöffnet werden)* ‹*shop, bank, etc.*› open; **c)** *(sich ~)* ‹*door*› open. **3.** *refl. V.* **a)** open; **die Erde öffnete sich** the ground opened up; **b)** *(sich erweitern)* ‹*valley, lane, forest, etc.*› open out **(auf +** *Akk.*, **zu** on to); ‹*view*› open up

Öffner der; ~s, ~: opener

Öffnung die; ~, ~en a) *(offene Stelle)* opening; *(Fot., Optik)* aperture; **b)** *o. Pl. (das Öffnen)* opening; **eine ~ der Leiche** a post-mortem on the body; **c)** *o. Pl. (das Aufgeschlossensein)* openness **(für** to); **eine ~ der Partei nach links anstreben** *(Pol.)* strive to open the party up to left-wing ideas

Öffnungs·zeiten *Pl. (eines Geschäfts, einer Bank)* opening times; hours of business; *(eines Museums, Zoos usw.)* opening times

Offsetdruck ['ɔfzɛt-] **der; -[e]s** offset printing

oft [ɔft] *Adv.* **öfter** ['œftɐ]; *(selten)* **öftest** ['œftəst] often; **wie oft soll**

ich dir noch sagen, daß ...? how many [more] times do I have to tell you that ...?
öfter ['œftɐ] *Adv.* now and then; [every] once in a while; **des ~en** *(geh.)* on many occasions
oftmals *Adv.* often; frequently
OG *Abk.* **Obergeschoß**
oh [o:] *Interj.* oh
OHG *Abk.* **Offene Handelsgesellschaft** general partnership
ohne ['o:nə] **1.** *Präp. mit Akk.* **a)** without; **~ mich!** [you can] count me out!; **der Versuch blieb ~ Erfolg** the attempt was unsuccessful; **ein Mann ~ jeglichen Humor** a man totally lacking in humour *or* without any sense of humour; **b)** *(mit Auslassung des Akkusativs)* **ich rauche nur ~:** I only smoke untipped *or* filterless cigarettes; **wir baden am liebsten ~:** we prefer to bathe in the nude; **er/sie ist [gar] nicht [so] ~** *(ugs.)* he's/she's quite something; *s. auch* **oben c; c) ~ weiteres** *(leicht, einfach)* easily; *(ohne Einwand)* readily; **das traue ich ihm ~ weiteres zu** I can quite *or* easily believe he's capable of that; **d)** excluding. **2.** *Konj.* **er nahm Platz, ~ daß er gefragt hätte** he sat down without asking; **~ zu zögern** without hesitating; without hesitation
ohne-: ~dies *Adv. (geh.)* in any case; **~einander** *Adv.* without each other; **~gleichen** *Adj.; attr. nachgestellt* unparalleled; **eine Frechheit ~gleichen** an unprecedented impertinence; **~hin** *Adv.* anyway; **er war ~hin schon überlastet** he was already overburdened as it was
Ohnmacht ['o:nmaxt] **die; ~, ~en a)** faint; swoon *(literary)*; **in ~ fallen** *od. (geh.)* **sinken** faint *or* pass out/swoon; **b)** *(Machtlosigkeit)* powerlessness; impotence
ohnmächtig 1. *Adj.* **a)** unconscious; **~ werden** faint; pass out; **~ sein** have fainted *or* passed out; be in a dead faint; **b)** *(machtlos)* powerless; impotent; impotent, helpless ‹*fury, rage*›; helpless ‹*bitterness, despair*›. **2.** *adv.* impotently; **~ zusehen** watch powerless *or* helplessly
Ohnmachts·anfall der fainting fit
oho [o'ho:] **1.** *Interj.* oho; *(protestierend)* oh no. **2.** *s.* **klein 1 a**
Ohr [o:ɐ̯] **das; ~[e]s, ~en a)** ear; **auf dem linken ~ taub sein** be deaf in one's left ear; **gute/schlechte ~en haben** have good/poor hearing *sing.*; **er hört nur auf einem ~:** he only has one good ear; **ich habe seine Worte/die Melodie noch im ~:** his words are still ringing in my ears/the tune is still going around my head; **b)** *(fig.)* **die ~en aufmachen** *od.* **aufsperren/spitzen** *(ugs.)* pin back/prick up one's ears; **die Wände haben ~en** the walls have ears; **ein offenes ~ für jmdn./etw. haben** be ready to listen to sb./be open to *or* ready to listen to sth.; **auf dem ~ hört er schlecht/nicht** *(ugs.)* he doesn't want to hear anything about that; **sich aufs ~ legen** *od. (ugs.)* **hauen** get one's head down *(coll.)*; **noch feucht/nicht [ganz] trocken hinter den ~en sein** *(ugs.)* be still wet behind the ears; **schreib dir das mal hinter die ~en!** *(ugs.)* just you remember that!; **eine** *od.* **eins/ein paar hinter die ~en kriegen** *(ugs.)* get a thick ear; **jmdm. [mit etw.] in den ~en liegen** *(ugs.)* pester sb. the whole time [with sth.]; **bis über beide ~en verliebt [in jmdn.]** *(ugs.)* head over heels in love [with sb.]; **bis über beide** *od.* **die ~en in etw. stekken** *(ugs.)* be up to one's ears in sth. *(coll.)*; **jmdn. übers ~ hauen** *(ugs.)* take sb. for a ride *(sl.)*; put one over on sb. *(coll.)*; **viel** *od.* **eine Menge um die ~en haben** *(ugs.)* have a lot on one's plate *(coll.)*; **zum einen ~ rein- und zum anderen wieder rausgehen** *(ugs.)* go in one ear and out the other *(coll.)*; *s. auch* **faustdick 2; Fell a; Floh a**
Öhr [ø:ɐ̯] **das, ~[e]s, ~e** eye
ohren-, Ohren-: ~arzt der otologist; ear specialist; **~betäubend 1.** *Adj.* ear-splitting; deafening; deafening ‹*applause*›; **2.** *adv.* deafeningly; **~sausen das** ringing in the *or* one's ears; **~schmalz das** ear-wax; **~schmaus der** *(ugs.) in* **ein ~schmaus sein** be a joy to hear; **~schmerz der** earache; **~schmerzen haben** have [an] earache *sing.*; **~schützer der** earmuff; **~sessel der** wing-chair
ohr-, Ohr-: ~feige die box on the ears; **jmdm. eine ~feige geben** *od. (ugs.)* **verpassen** box sb.'s ears; give sb. a box on the ears; **~feigen** [-faign̩] *tr. V.* **jmdn. ~feigen** box sb.'s ears; **ich könnte mich [selbst] ~feigen!** *(ugs.)* I could kick myself!; **~läppchen das** ear-lobe; **~muschel die** external ear; auricle; **~ring der** ear-ring; **~stecker der** ear-stud; **~wurm der a)** earwig; **b)** *(ugs.: Melodie)* catchy tune; **ein ~wurm sein** be really catchy
oje [o'je:] *Interj. (veralt.)*, **ojemine** [o'je:mine] *Interj. (veralt.)* oh dear; dear me
okay [o'ke] *(ugs.) Interj., Adj., adv.* OK *(coll.)*; okay *(coll.)*
okkupieren *tr. V.* occupy
öko-, Öko- [øko-]: eco-
Ökologe [øko'lo:gə] **der; ~n, ~n** ecologist
Ökologie die; ~: ecology
ökologisch 1. *Adj.* ecological. **2.** *adv.* ecologically
Ökonomie die; ~, ~n economics *sing.*; **politische ~:** political economy
ökonomisch 1. *Adj.* **a)** economic; **b)** *(sparsam)* economical. **2.** *adv.* economically
Öko·system das ecosystem
Oktave [ɔk'ta:və] **die; ~, ~n** octave
Oktober [ɔk'to:bɐ] **der; ~[s], ~:** October; *s. auch* **April**
Öl [ø:l] **das; ~[e]s, ~e** oil; **auf ~ stoßen** strike oil; **in ~ malen** paint in oils; **~ ins Feuer gießen** *(fig.)* add fuel to the flames
Öl-: ~baum der olive-tree; **~bild das** oil-painting
Oldtimer ['oʊldtaɪmə] **der; ~s, ~:** vintage car; *(vor 1905 gebaut)* veteran car
Oleander [ole'andɐ] **der; ~s, ~:** *(Bot.)* oleander
ölen *tr. V.* oil; oil, lubricate ‹*shaft, engine, etc.*›; **wie geölt** *(fig. ugs.)* like clockwork; *s. auch* **Blitz a**
Öl-: ~farbe die a) oil-based paint; **b)** *(zum Malen)* oil-paint; **mit ~farben malen** paint in oils; **~gemälde das** oil-painting; **~götze der** *in* **wie ein ~götze/wie die ~götzen** *(ugs.)* like a zombie/zombies; **~heizung die** oil-fired heating *no indef. art.*
ölig 1. *Adj.* oily. **2.** *adv.* **~ glänzen** have an oily sheen
Oligarchie [oligar'çi:] **die; ~, ~n** oligarchy
oliv [o'li:f] *Adj.* olive[-green]
Olive [o'li:və] **die; ~, ~n** olive
Oliven-: ~baum der olive-tree; **~öl das** olive oil
oliv·grün *Adj.* olive-green
Öl-: ~kanister der, ~kanne die oilcan; **~krise die** oil crisis
oll [ɔl] *Adj. (ugs., bes. nordd.)* old; **je ~er, je doller** *(ugs. scherzh.)* the older they get, the more they want to live it up
Öl-: ~lampe die oil-lamp; **~leitung die** oil-pipe; *(größer)* oil pipeline; **~malerei die** oil painting *no art.*; **~meß·stab der** *(bes. Kfz-W.)* dip-stick; **~ofen der** oil heater; **~pest die** oil pollution *no indef. art.*; **~sardine die** sardine in oil; **eine Dose ~n** a

tin of sardines; **~scheich der** *(ugs.)* oil sheikh; **~stand der** oil-level; **~tank der** oil-tank; **~tanker der** oil-tanker; **~teppich der** oil-slick

Ölung die; ~, ~en oiling; lubrication; **Letzte Ölung** *(kath. u. orthodoxe Kirche)* extreme unction

Öl-: ~wanne die *(bes. Kfz-W.)* sump; **~wechsel der** *(bes. Kfz-W.)* oil-change

Olymp [o'lʏmp] **der; ~s** Mount Olympus

Olympiade [olʏm'pi̯a:də] **die; ~, ~n a)** Olympic Games *pl.;* Olympics *pl.;* **b)** *(Wettbewerb)* Olympiad

Olympia-: ~mannschaft die Olympic team *or* squad; **~sieger der, ~s, ~, ~siegerin die; ~, ~nen** Olympic champion; **~stadion das** Olympic stadium

olympisch *Adj.* Olympic; **die Olympischen Spiele** the Olympic Games; the Olympics

Öl·zeug das oilskins *pl.*

Oma ['o:ma] **die; ~, ~s** *(fam.)* gran[ny] *(coll./child lang.);* grandma *(coll./child lang.)*

Omelett das; ~[e]s, ~e *od.* **~s,** *(Kochk.)* omelette

Omen ['o:mən] **das; ~s, ~** *od.* **Omina** ['o:mina] omen

ominös [omi'nø:s] **1.** *Adj.* **a)** ominous; **b)** *(bedenklich, zweifelhaft)* sinister. **2.** *adv.* ominously

Omnibus ['ɔmnibʊs] **der; ~ses, ~se** omnibus *(formal); (Privat- und Reisebus auch)* coach

Omnibus- *s.* **Bus-**

Onanie [ona'ni:] **die; ~:** onanism *no art.;* masturbation *no art.*

onanieren *itr. V.* masturbate

ondulieren [ɔndu'li:rən] *tr. V.* *(veralt.)* crimp; wave

Onkel ['ɔŋkl̩] **der; ~s, ~** *od. (ugs.)* **~s a)** uncle; **b)** *(Kinderspr.: Mann)* **sag dem ~ guten Tag!** say hello to the nice man; **der ~ Doktor** the nice doctor

OP [o:'pe:] **der; ~[s], ~[s]** *Abk.* **Operationssaal**

Opa ['o:pa] **der; ~s, ~s** *(fam.)* grandad *(coll./child lang.);* grandpa *(coll./child lang.)*

Opal [o'pa:l] **der; ~s, ~e** opal

Oper ['o:pɐ] **die; ~, ~n** opera; *(Institution, Ensemble)* Opera; **in die ~ gehen** go to the opera; **an die/zur ~ gehen** *(als Sänger)* become an opera-singer

Operation [opəra'tsi̯o:n] **die; ~, ~en** operation

Operations-: ~saal der operating-theatre *(Brit.) or* -room; **~schwester die** theatre sister *(Brit.);* operating-room nurse *(Amer.);* **~tisch der** operating-table

operativ [opəra'ti:f] **1.** *Adj.; nicht präd. (Med.)* operative. **2.** *adv.* *(Med.)* by operative surgery

Operette [opə'rɛtə] **die; ~, ~n** operetta

operieren 1. *tr. V.* operate on ‹*patient*›; **sich ~ lassen** have an operation. **2.** *itr. V.* operate; **vorsichtig ~** *(vorgehen)* proceed carefully

Opern-: ~arie die [operatic] aria; **~glas das** opera-glass[es *pl.*]; **~haus das** opera-house; **~sänger der, ~sängerin die** opera-singer

Opfer ['ɔpfɐ] **das; ~s, ~ a)** *(Verzicht)* sacrifice; **ein ~ [für etw.] bringen** make a sacrifice [for sth.]; **kein ~ scheuen** consider no sacrifice too great; **b)** *(Geschädigter)* victim; **jmdm./einer Sache zum ~ fallen** fall victim to sb./sth.; be the victim of sb./sth.; **c)** *(~gabe)* sacrifice; **jmdm./einer Sache etw. zum ~ bringen** sacrifice sth. to sb./sth.

opfer-, Opfer-: ~bereit *Adj.* ‹*person*› who is ready *or* willing to make sacrifices; **~bereitschaft die** readiness *or* willingness to make sacrifices; **~gabe die** [sacrificial] offering; **~lamm das** sacrificial lamb; **wie ein ~lamm** *(fig. ugs.)* like a lamb to the slaughter

opfern 1. *tr. V.* **a)** *(darbringen)* sacrifice; make a sacrifice of; offer up ‹*fruit, produce, etc.*›; **b)** *(fig.: hingeben)* sacrifice, give up ‹*time, holiday, money, life*›. **2.** *itr. V.* **[den Göttern] ~:** offer sacrifice [to the gods]. **3.** *refl. V.* **a) sich für jmdn./etw. ~:** sacrifice oneself for sb./sth.; **b)** *(ugs. scherzh.)* be the martyr; **wer opfert sich denn und ißt den Nachtisch auf?** who's going to volunteer to finish off the dessert?

Opfer·stock der; *Pl.* **...stöcke** offertory box

Opiat [o'pi̯a:t] **das; ~[e]s, ~e** opiate

Opium ['o:pi̯ʊm] **das; ~s** *(auch fig.)* opium

opponieren *itr. V.* take the opposite side; **gegen jmdn./etw. ~:** oppose sb./sth.

opportun [ɔpɔr'tu:n] *Adj.* *(geh.)* appropriate; *(günstig)* advantageous

Opportunismus der; ~: opportunism

Opportunist der; ~en, ~en opportunist

opportunistisch 1. *Adj.* opportunist; opportunistic. **2.** *adv.* opportunistically

Opposition [ɔpozi'tsi̯o:n] **die; ~, ~en** *(auch Politik, Sprachw., Astron., Schach, Fechten)* opposition; **etw. aus [reiner** *od.* **lauter] ~ tun** do sth. just to be contrary

oppositionell [ɔpozitsi̯o'nɛl] **1.** *Adj.* opposition *attrib.* ‹*group, movement, circle, etc.*›; ‹*newspaper, writer, artist, etc.*› opposed to the government. **2.** *adv.* **~ eingestellt sein** hold opposing views

Oppositions-: ~führer der opposition leader; *(in Großbritannien)* Leader of the Opposition; **~partei die** opposition party

Optik die; ~, ~en a) *o. Pl. (Wissenschaft)* optics *sing., no art.;* **b)** *(Fot. ugs.) (Linse)* lens; *(Linsen)* optics *pl.;* lens system; **das ist eine Frage der ~** *(fig.)* it depends on your point of view; **c)** *o. Pl. (Erscheinungsbild)* appearance; **der ~ wegen** for visual effect

Optiker der; ~s, ~, Optikerin die; ~, ~nen optician

optimal [ɔpti'ma:l] **1.** *Adj.* optimal; optimum *attrib.* **2.** *adv.* **jmdn. ~ beraten** give sb. the best possible advice

Optimismus der; ~: optimism

Optimist der; ~en, ~en, Optimistin die; ~, ~nen optimist

optimistisch 1. *Adj.* optimistic. **2.** *adv.* optimistically

optisch 1. *Adj.* optical; visual ‹*impression*›; **aus ~en Gründen** for [the sake of] optical *or* visual effect; *(fig.)* for [the sake of] effect. **2.** *adv.* optically; visually ‹*impressive, successful, effective*›; **~ wahrnehmbar sein** be perceivable with the eye

Orakel [o'ra:kl̩] **das; ~s, ~** oracle

orakeln 1. *tr. V.* **~, daß ...:** make mysterious prophecies that ... **2.** *itr. V.* make mysterious prophecies

oral [o'ra:l] **1.** *Adj.* oral. **2.** *adv.* orally

orange [o'rã:ʒ(ə)] *indekl. Adj.* orange

Orange die; ~, ~n orange

orange·farben, orange·farbig *Adj.* orange[-coloured]

orangen [o'rã:ʒn̩] *Adj.* orange

Orangen-: ~baum orange-tree; **~marmelade die** orange marmalade; **~saft der** orange-juice; **~schale die** orange-peel *no pl.*

Orang-Utan ['o:raŋ'|u:tan] **der; ~s, ~s** orang-utan

Oratorium [ora'to:ri̯ʊm] **das; ~s, Oratorien** oratorio

Orchester [ɔr'kɛstɐ] **das; ~s, ~:** orchestra

Orchidee [ɔrçi'de:(ə)] **die; ~, ~n** orchid

Orden ['ɔrdn̩] **der; ~s, ~ a)** *(Gemeinschaft)* order; **in einen ~ eintreten, einem ~ beitreten** join an order; become a member of an order; **b)** *(Ehrenzeichen, Milit.)* decoration; *(in runder Form)* medal; **jmdm. einen ~ [für etw.] verleihen** decorate sb. [for sth.]

ordentlich ['ɔrdn̩tlıç] **1.** *Adj.* **a)** *(ordnungsliebend)* [neat and] tidy; *(methodisch)* orderly; **b)** *(geordnet)* [neat and] tidy ⟨*room, house, desk, etc.*⟩; neat ⟨*handwriting, clothes*⟩; **c)** *(anständig)* respectable; proper ⟨*manners*⟩; **etwas Ordentliches lernen** learn a proper trade; **d)** *nicht präd. (planmäßig)* regular, ordinary ⟨*meeting*⟩; full ⟨*member*⟩; **~es Gericht** court exercising civil and criminal jurisdiction; *s. auch* **Professor a; e)** *(ugs.: richtig)* proper; real; **etwas Ordentliches essen** have some proper food; **eine ~e Tracht Prügel** a real good hiding *(coll.)*; **f)** *(ugs.: tüchtig)* **ein ~es Stück Kuchen** a nice big piece of cake; **ein ~es Stück Arbeit** a fair old bit of work *(coll.)*; **g)** *(ugs.: recht gut)* decent ⟨*wine, flat, marks, etc.*⟩. **2.** *adv.* **a)** *(geordnet)* tidily; neatly; ⟨*write*⟩ neatly; **~ aufgeräumt** neatly tidied; **b)** *(anständig)* properly; **c)** *(ugs.: gehörig)* **~ feiern** have a real good celebration *(coll.)*; **greift ~ zu!** tuck in!; **d)** *(ugs.: recht gut)* ⟨*ski, speak, etc.*⟩ really well

Ordentlichkeit die; ~: [neatness and] tidiness; *(der Schrift, Kleidung)* neatness; *(methodische Veranlagung)* orderliness

Order ['ɔrdɐ] **die; ~, ~s** *od.* **~n** order; **~ haben, etw. zu tun** have orders to do sth.

ordinär [ɔrdi'nɛ:ɐ̯] **1.** *Adj.* **a)** *(abwertend)* vulgar; common; vulgar ⟨*joke, song, expression, language*⟩; cheap and obtrusive ⟨*perfume*⟩; **b)** *nicht präd. (alltäglich)* ordinary. **2.** *adv.* vulgarly; in a vulgar manner

Ordinarius [ɔrdi'na:ri̯ʊs] **der; ~, Ordinarien** [full] professor **(für** of)

ordnen ['ɔrdnən] **1.** *tr. V.* **a)** arrange; **b)** *(regeln)* regulate ⟨*traffic*⟩; settle ⟨*one's affairs*⟩; **sein Leben/seine Finanzen ~:** straighten out one's life/put one's finances in order. **2.** *refl. V.* form up

Ordner der; ~s, ~ a) *(Hefter)* file; **b)** *(Aufsichtsperson)* steward

Ordnung die; ~, ~en a) *o. Pl. (ordentlicher Zustand)* order; tidiness; **~ halten** keep things tidy; **hier herrscht ~:** everything is neat and tidy here; **~ schaffen, für ~ sorgen** sort things out; **etw. in ~ bringen** sort sth. out; **ist dein Paß in ~?** is your passport in order?; **hier ist etw. nicht in ~:** there's something wrong here; **mit ihr ist etwas nicht in ~, sie ist nicht in ~** *(ugs.)* there's something wrong *or* the matter with her; **sie ist in ~** *(ugs.: ist nett, verläßlich o. ä.)* she's OK *(coll.)*; **alles [ist] in schönster** *od.* **bester ~:** everything's [just] fine; [things] couldn't be better; **[das] geht [schon] in ~** *(ugs.)* that'll be OK *(coll.) or* all right; **b)** *o. Pl. (geregelter Ablauf)* routine; **c)** *o. Pl. (System [von Normen])* order; *(Struktur)* structure; **d)** *o. Pl. (Disziplin)* order; **hier herrscht ~:** we have some discipline here; **~ halten** ⟨*teacher etc.*⟩ keep order; **e)** *(Formation)* formation; **f)** *(Biol.)* order; **g)** *o. Pl. (Rang)* **eine Straße zweiter ~:** a second-class road; **ein Reinfall erster ~** *(fig. ugs.)* a disaster of the first order *or* water; **h)** *o. Pl. (Math.)* order; **i)** *(Mengenlehre)* ordered set

ordnungs-, Ordnungs-: ~gemäß 1. *Adj.* ⟨*conduct etc.*⟩ in accordance with the regulations; **2.** *adv.* in accordance with the regulations; **~halber** *Adv.* as a matter of form; **~liebe die** liking for neatness and tidiness; **~liebend** *Adj.* ⟨*person*⟩ who likes to see things neat and tidy; **~ruf der** call to order; **~strafe die** *(Rechtsw.)* penalty for contempt of court; **~widrig** *(Rechtsw.)* **1.** *Adj.* ⟨*actions, behaviour, etc.*⟩ contravening the regulations; illegal ⟨*parking*⟩; **2.** *adv.* **~widrig handeln** act in contravention of the regulations; contravene *or* infringe the regulations; **~widrigkeit die** *(Rechtsw.)* infringement of the regulations

Organ [ɔr'ga:n] **das; ~s, ~e a)** *(Anat., Biol.)* organ; **b)** *(ugs.: Stimme)* voice; **c)** *(Zeitung)* organ *(formal)*; **d)** *(Institution)* organ; *(Mensch)* agent

Organisation [ɔrganıza'tsi̯o:n] **die; ~, ~en** organization

Organisations·talent das a) *(Fähigkeit)* talent for organization; **b)** *(Mensch)* person with a talent for organization

Organisator [ɔrgani'za:tɔr] **der; ~s, ~en** [-'to:rən] organizer

organisatorisch 1. *Adj.* organizational. **2.** *adv.* organizationally

organisch 1. *Adj.* **a)** *(auch Chemie)* organic; **b)** *(Med.)* organic; physical. **2.** *adv.* **a)** organically; **sich ~ in etw.** *(Akk.)* **einfügen** form an organic part of sth.; **b)** *(Med.)* organically; physically

organisieren 1. *tr. V.* **a)** *(vorbereiten, aufbauen)* organize; **b)** *(ugs.: beschaffen)* get [hold of]. **2.** *itr. V.* **gut ~ können** be a good organizer. **3.** *refl. V.* organize **(zu** into); **er will sich ~:** he wants to join the union *etc.*

Organismus der; Organismen organism

Organist der; ~en, ~en, Organistin die; ~, ~nen organist

Organ·verpflanzung die organ transplantation

Orgasmus [ɔr'gasmʊs] **der; ~, Orgasmen** orgasm

Orgel ['ɔrgl̩] **die; ~, ~n** organ

Orgel-: ~konzert das organ concerto; *(Solo)* organ recital; **~pfeife die** organ-pipe; **[dastehen] wie die ~pfeifen** *(scherzh.)* [stand in a row] from the tallest to the shortest

Orgie ['ɔrgi̯ə] **die; ~, ~n** *(auch fig.)* orgy; **eine ~ feiern** have an orgy

Orient ['o:ri̯ɛnt] **der; ~s a)** *(Vorder- u. Mittelasien)* Middle East and south-western Asia *(including Afghanistan and Nepal)*; **der Vordere ~:** the Middle East; **b)** *(veralt.: Osten)* Orient

Orientale [ori̯ɛn'ta:lə] **der; ~n, ~n, Orientalin die; ~, ~nen** *s.* **Orient a:** *man/woman from the Middle East [or south-western Asia]*

orientalisch *Adj.* oriental

orientieren 1. *refl. V.* **a)** *(sich zurechtfinden)* get one's bearings; **sich an etw.** *(Dat.)***/nach einer Karte ~:** get one's bearings by sth./using a map; **b)** *(sich unterrichten)* **sich über etw.** *(Akk.)* **~:** inform oneself about sth.; **c)** *(sich ausrichten)* **sich an etw.** *(Dat.)* **~:** be oriented towards sth.; ⟨*policy, advertising*⟩ be geared towards sth.; **politisch links/rechts orientiert sein** lean towards the left/right politically. **2.** *tr. V.* **a)** *(unterrichten)* inform **(über +** *Akk.* about); **b)** *(ausrichten)* **seine Ziele nach etw. ~:** base one's aims on sth. **3.** *itr. V.* **über etw.** *(Akk.)* **~:** report on sth.

Orientierung die; ~ a) *(Orientierungssinn, -möglichkeit)* **hier ist die ~ schwer** it's difficult to get your bearings here; **die ~ verlieren** lose one's bearings; **b)** *(Unterrichtung)* **zu Ihrer ~:** for your in-

formation; **c)** *(das Sichausrichten)* orientation (**auf** + *Akk.* towards, **an** + *Dat.* according to)
orientierungs-, Orientierungs-: ~hilfe die aid to orientation; **~los** *Adj. (auch fig.)* disoriented; **~sinn der** sense of direction
Orient·teppich der oriental carpet; *(Läufer)* oriental rug
original [origi'na:l] **1.** *Adj.* original; *(echt)* genuine; authentic. **2.** *adv.* **~ indische Seide** genuine Indian silk; **etw. ~ übertragen** broadcast sth. live
Original das; ~s, ~e a) *(Urschrift o. ä.)* original; **b)** *(eigenwilliger Mensch)* character
Original·fassung die original version
original·getreu 1. *Adj.* faithful *or* true [to the original] *postpos.* **2.** *adv.* in a manner faithful *or* true to the original
Originalität [originali'tɛ:t] **die; ~ a)** *(Echtheit)* genuineness; authenticity; **b)** *(Einmaligkeit)* originality
Original·ton der: Reportageausschnitte im ~ von 1936 excerpts from news reports with the original 1936 soundtrack; **„~ton UdSSR-Fernsehen"** 'USSR television commentary'
originell [origi'nɛl] **1.** *Adj. (ursprünglich)* original; *(neu)* novel; *(ugs.: witzig)* witty, funny, comical ‹*story*›; comical, funny ‹*costume*›. **2.** *adv.* *(ursprünglich)* ‹*write, argue*› with originality; *(ugs.: witzig)* ‹*write, argue*› wittily
Orkan [ɔr'ka:n] **der; ~[e]s, ~e** hurricane; *(fig.)* thunderous storm
orkan·artig *Adj.* ‹*winds, gusts*› of almost hurricane force
Ornament [ɔrna'mɛnt] **das; ~[e]s, ~e** *(Kunstw.)* ornament
Ornithologie die; ~: ornithology *no art.*
¹Ort [ɔrt] **der; ~[e]s, ~e a)** *(Platz)* place; **etw. an seinem ~ lassen** leave sth. where it is/was; **~ der Handlung: ...:** the scene of the action is ...; **an den ~ des Verbrechens zurückkehren** return to the scene of the crime; **an ~ und Stelle** there and then; **an ~ und Stelle sein/ankommen** *(an der gewünschten Stelle)* be/arrive there; **b)** *(~schaft) (Dorf)* village; *(Stadt)* town; **von ~ zu ~:** from place to place; **das beste Hotel am ~:** the best hotel in the place
²Ort *in* **vor ~:** on the spot; *(Bergmannsspr.)* at the [coal-]face
Örtchen ['œrtçən] **das; ~s, ~** *(ugs. verhüll.)* **das ~:** the smallest room *(coll. euphem.)*; **aufs ~ müssen** have to pay a visit *(coll. euphem.)*
orten *tr. V.* find the position of
orthodox [ɔrto'dɔks] *Adj.* **a)** *(Rel.)* orthodox; **b)** *(starr)* rigid; **c)** *(strenggläubig)* strict
Orthographie die; ~; ~n orthography
orthographisch *Adj.* orthographic; **~e Fehler** spelling mistakes
Orthopäde [ɔrto'pɛ:də] **der; ~n, ~n** orthopaedist; orthopaedic specialist
Orthopädie [ɔrtopɛ'di:] **die; ~, ~n a)** *o. Pl.* orthopaedics *sing., no art.*; **b)** *(ugs.: Abteilung)* orthopaedic department
Orthopädin die; ~, ~nen *s.* **Orthopäde**
orthopädisch 1. *Adj.* orthopaedic. **2.** *adv.* orthopaedically
örtlich ['œrtlıç] **1.** *Adj.* local. **2.** *adv.* locally; **~ betäubt werden** be given a local anaesthetic; **~ begrenzte Kampfhandlungen** [limited] local encounters
Örtlichkeit die; ~, ~en a) *(Gebiet)* locality; **b)** *(Stelle)* place; **c)** *s.* **Örtchen**
orts-, Orts-: ~angabe die indication of place; **~ansässig** *Adj.* local; **die ~ansässigen** the local residents; **~ausgang der** end of the village/town
Ortschaft die; ~, ~en *(Dorf)* village; *(Stadt)* town; **geschlossene ~:** built-up area
orts-, Orts-: ~eingang der entrance to the village/town; **~fremd** *Adj.* **a)** *(nicht ~ansässig)* **~fremde Personen** visitors to the village/town; **b)** *(nicht ~kundig)* **~fremd sein** be a stranger [to the village/town]; **~gespräch das** *(Fernspr.)* local call; **~kenntnis die** knowledge of the place; **[gute] ~kenntnisse haben** know the place [well]; **~kundig** *Adj.* **ein ~kundiger Führer/ein Ortskundiger** a guide/someone who knows the place well; **~name der** place-name; **~netz das** *(Fernspr.)* local exchange network; **~netz·kennzahl die** *(Fernspr.)* dialling code; area code *(Amer.)*; **~sinn der** sense of direction; **~verkehr der a)** *(Straßenverkehr)* local traffic; **b)** *(Telefon)* local telephone service; **~zeit die** local time
Öse ['ø:zə] **die; ~, ~n** eye; *(an Schuh, Stiefel)* eyelet
Ossi ['ɔsi] **der; ~s, ~s** *(salopp) s.* **Ostdeutsche**
Ost [ɔst] *o. Art.; o. Pl.* **a)** *(bes. Seemannsspr., Met.: Richtung)* east; *s. auch* **Osten a; b)** *(östliches Gebiet, Politik)* East; **c)** *einem Substantiv nachgestellt (östlicher Teil, östliche Lage)* East; **Autobahnausfahrt Köln-~:** motorway exit Cologne East
ost-, Ost-: ~asien (das) East *or* Eastern Asia; **~block der;** *o. Pl.* Eastern bloc; **~block·staat der** Eastern-bloc state; **~deutsch** *Adj.* Eastern German; *(hist.: auf die DDR bezogen)* East German; **~deutsche der/die** Eastern German; *(hist.: DDR-Bürger[in])* East German; **~deutschland (das)** Eastern Germany; *(hist.: DDR)* East Germany
Osten der; ~s a) *(Richtung)* east; **nach ~:** eastwards; **im/aus dem** *od.* **von ~:** in/from the east; **b)** *(Gegend)* eastern part; **c)** *(Geogr.)* **der Ferne ~:** the Far East; **der Mittlere ~:** south-western Asia *(including Afghanistan and Nepal)*; **der Nahe ~:** the Middle East; **d)** *(Politik)* **der ~** *(der Ostblock)* the East
ostentativ [ɔstɛnta'ti:f] *(geh.)* **1.** *Adj.* pointed ‹*absence, silence*›; overt ‹*hostility*›; exaggerated ‹*heartiness*›; ostentatious ‹*gesture*›. **2.** *adv.* pointedly; ‹*embrace*› ostentatiously
Oster-: ~ei das Easter egg; **~fest das** Easter [holiday]; **~glocke die** daffodil; **~hase der** Easter hare; **~lamm das** Paschal lamb
österlich ['ø:stɐlıç] **1.** *Adj.* Easter *attrib.* **2.** *adv.* **~ geschmückt** decorated for Easter
Oster-: ~marsch der Easter march *(against war and nuclear weapons)*; **~montag der** Easter Monday *no. def. art.; s. auch* **Dienstag**
Ostern ['o:stɐn] **das; ~, ~:** Easter; **Frohe** *od.* **Fröhliche ~!** Happy Easter!; **zu** *od.* *(bes. südd.)* **an ~:** at Easter; **wenn ~ und Pfingsten auf einen Tag fallen** *(ugs.)* not this side of doomsday *(coll.)*
Österreich ['ø:stəraiç] **(das); ~s** Austria
Österreicher der; ~s, ~, Österreicherin die; ~, ~nen Austrian
österreichisch *Adj.* Austrian; *s. auch* **deutsch; Deutsch**
Oster-: ~sonntag der Easter Sunday *no def. art.*; **~woche die** week before Easter
ost-, Ost-: ~europa (das) Eastern Europe; **~europäisch** *Adj.* East[ern] European
Ost·friesland (das); ~s East Friesland; Ostfriesland

Ost·küste die east[ern] coast

östlich ['œstlıç] **1.** *Adj.* **a)** *(im Osten)* eastern; **15 Grad ~er Länge** 15 degrees east [longitude]; **b)** *(nach, aus dem Osten)* easterly; **c)** *(des Ostens, auch Politik)* Eastern. **2.** *adv.* eastwards; **~ von ...:** [to the] east of...; **sehr [weit] ~ sein** be a long way east. **3.** *Präp. mit Gen.* [to the] east of

Ost·politik die Ostpolitik *(German policy towards Eastern Europe)*

Ost·see die; *o. Pl.* Baltic [Sea]

ost·wärts *Adv.* eastwards

Ost·wind der east[erly] wind

[1]**Otter** ['ɔtɐ] **der; ~s, ~** *(Fisch~)* otter

[2]**Otter die; ~, ~n** *(Viper)* adder; viper

Otto ['ɔto] **der; ~s, ~s** *(salopp)* whopper *(sl.); s. auch* **Normalverbraucher**

Otto·motor der Otto engine

Ouvertüre [uvɛr'tyːrə] **die; ~, ~n** *(auch fig.)* overture (*Gen.* to)

oval [o'vaːl] *Adj.* oval

Oval das; ~s, ~e oval

Ovation [ova'tsi̯oːn] **die; ~, ~en** ovation; **jmdm. ~en darbringen** give sb. an ovation

Overall ['oʊvərɔːl] **der; ~s, ~s** overalls *pl.*

Oxyd [ɔ'ksyːt] **das; ~[e]s, ~e** *(Chemie)* oxide

Oxydation [ɔksyda'tsi̯oːn] **die; ~, ~en** *(Chemie, Physik)* oxidation

oxydieren 1. *itr. V.; auch mit sein (Chemie, Physik)* oxidize. **2.** *tr. V. (Chemie)* oxidize

Ozean ['oːtseaːn] **der; ~s, ~e** *(auch fig.)* ocean

Ozean·dampfer der ocean-going steamer; *(für Passagiere)* ocean liner

Ozon [o'tsoːn] **der** *od.* **das; ~s** ozone

Ozon·loch das *(ugs.)* hole in the ozone layer

P

p, P [pe] **das; ~, ~:** p/P; *s. auch* **a, A**

paar [paːɐ̯] *indekl. Indefinitpron.* **ein ~ ...:** a few ...; *(zwei od. drei)* a couple of ...; a few ...; **ein ~ waren dagegen** a few [people]/a couple [of people] were against [it]; **deine ~ Mark** the few marks/couple of marks you've got; **alle ~ Minuten** every few minutes/every couple of minutes; **du kriegst gleich ein ~ [gelangt]** *(ugs.)* I'll stick one on you *(coll.)*

Paar das; ~[e]s, ~e pair; *(Mann und Frau, Tanz~)* couple; **ein ~ Würstchen** two sausages; a couple of sausages; **zwei ~ Socken** two pairs of socks; **ein ~ Hosen** *(ugs.)* a pair of trousers

paaren 1. *refl. V.* **a)** *(sich begatten)* ‹*animals*› mate; ‹*people*› couple, copulate; **b)** *(sich verbinden)* **sich mit etw. ~:** be combined with sth. **2.** *tr. V.* **a)** *(kreuzen)* mate; **b)** *(zusammenstellen)* pair; **c)** *(verbinden)* combine (**mit** with)

paar-, Paar-: ~hufer der; ~s, ~ *(Zool.)* even-toed ungulate *(fachspr.);* cloven-hoofed animal; **~lauf der, ~laufen das** pair-skating; pairs *pl.;* **~mal** *Adv.* **ein ~mal** a few times; *(zwei- oder dreimal)* a couple of *or* a few times

Paarung die; ~, ~en a) *(Zool.)* mating; **b)** *(das Zusammenstellen)* pairing; **c)** *(das Verbinden)* combination

paar·weise 1. *Adv.* in pairs. **2.** *adj.* ‹*arrangement etc.*› in pairs

Pacht [paxt] **die; ~, ~en a)** *(Nutzung)* **etw. in ~ nehmen/geben** lease sth.; take/let sth. on lease; **etw. in ~ haben** have sth. on lease; **b)** *(Vertrag)* lease; **c)** *(Miete)* rent

pachten *tr. V.* lease; take a lease on; **jmdn./etw. [für sich] gepachtet haben** *(fig. ugs.)* have got a monopoly on sb./sth. *(coll.)*

Pächter ['pɛçtɐ] **der; ~s, ~, Pächterin die; ~, ~nen** leaseholder; lessee; *(eines Hofes)* tenant

[1]**Pack** [pak] **der; ~[e]s, ~e** *od.* **Päkke** ['pɛkə] pile; *(zusammengeschnürt)* bundle; *(Packung)* pack

[2]**Pack das; ~[e]s** *(ugs. abwertend)* rabble; riff-raff

Päckchen ['pɛkçən] **das; ~s, ~ a)** *(kleines Paket)* package; small parcel; *(Postw.)* small parcel *(below a specified weight); (Bündel)* packet; bundle; **b)** *s.* **Packung a**

Pack·eis das pack-ice

packen 1. *tr. V.* **a)** pack; **etw. in einen Koffer/ein Paket ~:** pack *or* put sth. in[to] a suitcase/put sth. in[to] a parcel; **etw. aus etw. ~:** unpack sth. from sth.; **sich/jmdn. ins Bett ~** *(ugs.)* go to bed/put sb. to bed; **b)** *(fassen)* grab [hold of]; seize; **jmdn. am** *od.* **beim Kragen ~:** grab [hold of] *or* seize sb. by the collar; **c)** *(überkommen)* **Furcht/Angst** *usw.* **packte ihn** he was seized with fear *etc.;* **d)** *(fesseln)* enthral; ‹*thriller, crime story, etc.*› grip; **ein ~des Rennen** a thrilling race; **e)** *(ugs.: schaffen)* **ein Examen ~:** manage to get through an exam *(coll.);* **es ~:** make a go of it; **~ wir's noch?** are we going to make it?; **einen Gegner ~** *(Sportjargon: besiegen)* get the better of an opponent; **f)** *(ugs.: begreifen)* get *(coll.);* **g)** *(salopp: weggehen)* **~ wir's?** shall we push off? *(sl.).* **2.** *itr. V. (Koffer usw. ~)* pack. **3.** *refl. V. (ugs. veralt.)* beat it *(sl.); clear off (coll.)*

Packen der; ~s, ~: pile; *(zusammengeschnürt)* bundle; *(von Geldscheinen)* wad

Pack-: ~esel der *(ugs.)* pack-donkey; *(fig.)* pack-horse; **~papier das** [stout] wrapping-paper

Packung die; ~, ~en a) packet; pack *(esp. Amer.);* **eine ~ Zigaretten** a packet *or (Amer.)* pack of cigarettes; **b)** *(Med., Kosmetik)* pack

Pädagoge [pɛda'goːgə] **der; ~n, ~n a)** *(Erzieher, Lehrer)* teacher; **b)** *(Wissenschaftler)* educationalist; educational theorist

Pädagogik die; ~: [theory and methodology of] education

Pädagogin die; ~, ~nen *s.* **Pädagoge**

pädagogisch 1. *Adj.* educational; ‹*lecture, dissertation, etc.*› on education; ‹*training*› in education; **Pädagogische Hochschule** College of Education. **2.** *adv.* educationally ‹*sound, wrong*›

Paddel ['padl̩] **das; ~s, ~:** paddle

Paddel·boot das canoe

paddeln *itr. V.; mit sein; ohne Richtungsangabe auch mit haben* **a)** *(Paddelboot fahren)* paddle; canoe; *(als Sport)* canoe; **b)** *(ugs.: schlecht schwimmen)* dog-paddle

paffen 1. *tr. V.* puff at ‹*pipe etc.*›; puff out ‹*smoke*›. **2.** *itr. V.* puff away

Page ['paːʒə] **der; ~n, ~n** *(Hotel~)* page; bellboy

Pagen·kopf der page-boy cut *or* style

Paket [pa'keːt] **das; ~[e]s, ~e a)** pile; *(zusammengeschnürt)* bundle; *(Eingepacktes, Post~, Schachtel)* parcel; *(Packung)* packet; pack *(esp. Amer.);* **b)** *(fig.: Gesamtheit)* package

Paket-: ~annahme die a) *o. Pl.*

acceptance of parcels; **b)** *(Stelle)* parcels office; *(Schalter)* parcels counter; **~ausgabe die a)** *o. Pl.* issue of parcels; **b)** *s.* **~annahme b**; **~karte die** parcel dispatch form; **~sendung die** parcel

Pakistan ['pa:kɪsta:n] **(das); ~s** Pakistan

Pakistaner der; ~s, ~, Pakistani [pakɪs'ta:ni] **der; ~[s], ~[s]/die; ~, ~[s]** Pakistani

Pakt [pakt] **der; ~[e]s, ~e** pact; **einen ~ [ab]schließen** make *or* conclude a pact

Palast [pa'last] **der; ~[e]s, Paläste** [pa'lɛstə] palace

Palästina [palɛ'sti:na] **(das); ~s** Palestine

Palästinenser der; ~s, ~, Palästinenserin die; ~, ~nen Palestinian

Palaver [pa'la:vɐ] **das; ~s, ~** *(ugs. abwertend)* palaver

palavern *itr. V. (ugs. abwertend)* palaver

Palette [pa'lɛtə] **die; ~, ~n a)** *(Malerei)* palette; **b)** *(bes. Werbespr.: Vielfalt)* diverse range

Palme ['palmə] **die; ~, ~n** palm[-tree]; **jmdn. auf die ~ bringen** *(ugs.)* ‹*person*› rile sb. *(coll.)*; ‹*situation*› make sb. wild

Palmen·wedel der palm frond

Palm-: ~kätzchen das [willow] catkin **~sonntag** [auch: -'--] **der** *(christl. Kirche)* Palm Sunday; **~wedel der** palm frond

Pampe ['pampə] **die; ~** *(bes. nordd. u. md.)* **a)** *(Matsch)* mud; mire; **b)** *(Brei)* mush

Pampelmuse ['pampl̩mu:zə] **die; ~, ~n** grapefruit

Pamphlet [pam'fle:t] **das; ~[e]s, ~e** *(Streitschrift)* polemical pamphlet; *(Schmähschrift)* defamatory pamphlet

pampig 1. *Adj.* **a)** *(ugs. abwertend: frech)* insolent; **b)** *(bes. nordd., ostd.: breiig)* mushy. **2.** *adv. (ugs. abwertend: frech)* insolently

Panama ['panama] **(das)** Panama

Panama·kanal der; *o. Pl.* Panama Canal

panieren *tr. V. (Kochk.)* **etw. ~:** bread sth.; coat sth. with breadcrumbs

Panier·mehl das breadcrumbs *pl.*

Panik ['pa:nɪk] **die; ~, ~en** panic; **[eine] ~ brach aus** panic broke out; **jmdn. in ~** *(Akk.)* **versetzen** throw sb. into a state of panic; **nur keine ~!** don't panic!

Panik·mache die; *o. Pl. (abwertend)* panic-mongering

panisch *Adj.* panic *attrib.* ‹*fear, terror*›; panic-stricken ‹*voice, flight*›; **~e Angst vor etw.** *(Dat.)* **haben** have a panic fear of sth.

Panne ['panə] **die; ~, ~n a)** *(Auto~)* breakdown; *(Reifen~)* puncture; flat [tyre]; **ich hatte eine ~** my car broke down/my car *or* I had a puncture; **b)** *(Betriebsstörung)* breakdown; **c)** *(Mißgeschick)* slip-up; mishap; **bei der Organisation gab es viele ~n** there were many organizational hitches

Pannen·dienst der breakdown service

Panorama [pano'ra:ma] **das; ~s, Panoramen** panorama

panschen ['panʃn̩] **1.** *tr. V. (ugs. abwertend)* water down; adulterate. **2.** *itr. V.* **a)** *(ugs. abwertend: mischen)* water down *or* adulterate the wine/beer *etc.*; **b)** *(ugs.: planschen)* splash about

Panther der; ~s, ~: panther

Pantoffel [pan'tɔfl̩] **der; ~s, ~n a)** backless slipper; **b)** *(mit Absatz)* mule; **c)** *(fig.)* **unterm ~ stehen** *(ugs.)* be henpecked

Pantoffel-: ~held der *(ugs. abwertend)* henpecked husband; **~kino das** *(ugs.)* telly *(coll.)*; **~tierchen das** *(Biol.)* slipper animalcule

Pantomime [panto'mi:mə] **die; ~, ~n** mime

pantschen *usw.* ['pantʃn̩] *s.* **panschen** *usw.*

Panzer ['pantsɐ] **der; ~s, ~ a)** *(Milit.)* tank; **b)** *(Zool.)* amour *no indef. art.*; *(von Schildkröten, Krebsen)* shell; **c)** *(hist.: Rüstung)* armour *no indef. art.*; **ein ~:** a suit of armour; **d)** *(Panzerung)* armour-plating *or* -plate *no indef. art.*; *(eines Reaktors)* shielding

Panzer-: ~faust die *(Milit.)* anti-tank rocket launcher; bazooka; **~glas das** bullet-proof glass; **~kreuzer der** *(Marine hist.)* armoured cruiser

panzern *tr. V.* armour[-plate]

Panzer-: ~schrank der safe; **~wagen der** *(Milit.)* **a)** *s.* **Panzer a; b)** *(Waggon)* armoured wagon

Papa ['papa, *geh., veralt.* pa'pa:] **der; ~s, ~** *(ugs.)* daddy *(coll.)*

Papagei [papa'gai] **der; ~en** *od.* **~s, ~e[n]** parrot

Papi ['papi] **der; ~s, ~s** *(ugs.) s.* **Papa**

Papier [pa'pi:ɐ] **das; ~s, ~e a)** paper; **ein Blatt ~:** a sheet of paper; **[nur] auf dem ~** *(fig.)* [only] on paper; **etw. zu ~ bringen** get *or* put sth. down on paper; **b)** *Pl. (Ausweis[e])* [identity] papers; **c)** *(Finanzw.: Wert~)* security

Papier-: ~deutsch das *(abwertend)* officialese; **~fabrik die** paper-mill; **~geld das** paper money; **~handtuch das** paper towel; **~korb der** waste-paper basket; **~kram der** *(ugs. abwertend)* [tedious] paperwork; **~krieg der** *(ugs. abwertend)* tedious form-filling; *(Korrespondenz)* tiresome exchange of letters; **~schlange die** [paper] streamer; **~taschentuch das** paper handkerchief; **~waren** *Pl.* stationery *sing.*

papp [pap] *in* **ich kann nicht mehr ~ sagen** *(ugs.)* I'm full to bursting-point *(coll.)*

Papp-: ~becher der paper cup; **~deckel der** cardboard

Pappe ['papə] **die; ~, ~n a)** *(Karton)* cardboard; **b)** *(ugs.: Brei)* mush; **5000 Mark sind nicht von** *od.* **aus ~** *(ugs.)* 5,000 marks isn't chicken-feed *(coll.)*

Pappel ['papl̩] **die; ~, ~n** poplar

pappen *(ugs.)* **1.** *tr. V.* stick **(an, auf +** *Akk.* on). **2.** *itr. V. (haftenbleiben)* stick **(an +** *Dat.* to); *(klebrig sein)* be sticky

Pappen-: ~deckel der cardboard; **~stiel** *in* **das ist kein ~stiel** *(ugs.)* it's not chicken-feed *(coll.)*; **etw. für einen ~stiel kaufen/kriegen** *(ugs.)* buy/get sth. for a song *or* for next to nothing

papperlapapp [papɐla'pap] *Interj.* rubbish

pappig *Adj.* **a)** sticky; **b)** *(breiig)* mushy

Papp-: ~karton der cardboard box; **~teller der** paper *or* cardboard plate

Paprika ['paprika] **der; ~s, ~[s] a)** pepper; **b)** *o. Pl. (Gewürz)* paprika

Paprika·schnitzel das cutlet with paprika sauce

Paps [paps] **der; ~** *(ugs.)* dad *(coll.)*

Papst [pa:pst] **der; ~[e]s, Päpste** ['pɛ:pstə] pope; *(fig. iron.)* high priest

päpstlich ['pɛ:pstlɪç] *Adj.* papal; *(fig. abwertend)* pontifical

Parabel [pa'ra:bl̩] **die; ~, ~n a)** *(bes. Literaturw.)* parable; **b)** *(Math.)* parabola

Parade [pa'ra:də] **die; ~, ~n a)** parade; **b)** *(Ballspiele)* save

Parade-: ~beispiel das perfect example; **~marsch der** *(Milit.)* marching in parade-step; *(Stechschritt)* goose-stepping; **~stück das** show-piece

Paradies [para'di:s] **das; ~es, ~e** *(auch fig.)* paradise

paradiesisch 1. *Adj.* **a)** *(Rel.)*

paradisical; **b)** *(herrlich)* heavenly; magnificent ‹*view*›. **2.** *adv.* *(herrlich)* ~ **ruhig gelegen** in a wonderfully peaceful situation; **dort ist es ~ schön** it's beautiful there, a real paradise

paradox [para'dɔks] *Adj.* **a)** paradoxical; **b)** *(ugs.: merkwürdig)* odd; strange

Paradox das; ~es, ~e *(bes. Philos., Rhet.)* paradox

paradoxer·weise *Adv.* **a)** paradoxically; **b)** *(ugs.: merkwürdigerweise)* strangely *or* oddly enough

Paragraph [para'gra:f] **der; ~en, ~en** section; *(im Vertrag)* clause

Paragraphen·reiter der *(abwertend)* **a)** *(Jurist)* lawyer; **b)** *(Pedant)* stickler for the rules

parallel [para'le:l] **1.** *Adj.* *(auch fig.)* parallel. **2.** *adv.* ~ **verlaufen** *(auch fig.)* run parallel **(mit, zu** to); ~ **zu etw.** *(fig.)* in parallel with sth.

Parallele die; ~, ~n a) *(Math.)* parallel [line]; **eine ~ zu etw. ziehen** draw a line parallel to sth.; **b)** *(fig.)* parallel

Parallel·klasse die *(Schulw.)* parallel class

Parallelogramm [paralelo'gram] **das; ~s, ~e** *(Math.)* parallelogram

Parallel-: ~schaltung die *(Elektrot.)* parallel connection; **~schwung der** *(Skisport)* parallel swing; **~straße die** street running parallel **(von** to)

Paranoia [para'nɔya] **die; ~** *(Med.)* paranoia

paranoid [parano'i:t] *Adj.* *(Med.)* paranoid

Para·nuß die Brazil-nut

Parasit [para'zi:t] **der; ~en, ~en** *(Biol., fig. abwertend)* parasite

parasitär [parazi'tɛ:ɐ̯] *(Biol., fig. abwertend) Adj.* parasitic

parat [pa'ra:t] *Adj.* ready; **eine Ausrede/Antwort ~ haben** be ready with an excuse/answer; **ich habe kein passendes Beispiel ~:** I can't think of a suitable example

Pärchen ['pɛ:ɐ̯çən] **das; ~s, ~:** pair; *(Liebespaar)* couple

Parcours [par'ku:ɐ̯] **der; ~ [...ɐ̯(s)], ~ [...ɐ̯s]** course

Parfum [par'fœ̃:], **Parfüm** [par'fy:m] **das; ~s, ~s** perfume; scent

Parfümerie [parfʏmə'ri:] **die; ~, ~en** perfumery

parfümieren *tr. V.* perfume; scent; **sich [viel zu stark] ~:** put [too much] perfume *or* scent on

parieren [pa'ri:rən] *itr. V.* *(ugs.)* do what one is told; **aufs Wort ~:** jump to it *(coll.)*

Pariser [pa'ri:zɐ] **1.** *indekl. Adj.* Parisian; Paris *attrib.*; **die ~ Metro** the Paris Metro. **2. der; ~s, ~ a)** *(Einwohner)* Parisian; **b)** *(ugs.: Kondom)* French letter *(coll.)*

Pariserin die; ~, ~nen Parisian

Parität [pari'tɛ:t] **die; ~, ~en** parity; equality

paritätisch 1. *Adj.* equal; **~e Mitbestimmung** co-determination based on equal representation. **2.** *adv.* equally

Park [park] **der; ~s, ~s** park; *(Schloß~ usw.)* grounds *pl.*

Parka der; ~s, ~s parka

Park·anlage die park; *(bei Schlössern usw.)* grounds *pl.*

parken 1. *tr. V.* park. **2.** *itr. V.* **a)** park; **b)** *(stehen)* be parked

Parkett [par'kɛt] **das; ~[e]s, ~e a)** *(Bodenbelag)* parquet floor; **~ legen** lay parquet flooring; **sich auf jedem ~ bewegen können** *(fig.)* be able to move in any circles; **b)** *(Theater)* [front] stalls *pl.*; parquet *(Amer.)*; **c)** *in* **etw. aufs ~ legen** *(ugs.)* dance sth.; *s. auch* **Sohle a**

Parkett·platz der seat in the [front] stalls

Park-: ~gebühr die parking-fee; **~haus das** multi-storey car-park; **~kralle die** wheel-clamp; **~landschaft die** parkland; **~lücke die** parking-space; **~platz der a)** car-park; parking lot *(Amer.)*; **b)** *(für ein einzelnes Fahrzeug)* parking-space; place to park; **~scheibe die** parking-disc; **~schein der** car-park ticket; **~uhr die** parking-meter; **~verbot das** ban on parking; **hier ist ~verbot** you are not allowed to park here; **im ~verbot stehen** be parked illegally

Parlament [parla'mɛnt] **das; ~[e]s, ~e** parliament; *(ein bestimmtes)* Parliament *no def. art.*

Parlamentarier [parlamɛn'ta:ri̯ɐ] **der; ~s, ~, Parlamentarierin die; ~, ~nen** member of parliament; *(in Großbritannien)* Member of Parliament; MP; *(in den Vereinigten Staaten)* Congressman/Congresswoman

parlamentarisch *Adj.*; *nicht präd.* parliamentary

Parlaments-: ~ausschuß der parliamentary committee; **~gebäude das** parliament building[s *pl.*]; **~sitzung die** sitting [of parliament]; **~wahl die** parliamentary election

Parmesan [parme'za:n] **der; ~[s]** Parmesan

Parodie [paro'di:] **die; ~, ~n** parody; **eine ~ auf etw./jmdn.** a parody of sth./take-off of sb.

parodieren *tr. V.* parody ‹*literary work, manner*›; take off ‹*person*›; satirize ‹*event*›

Parole [pa'ro:lə] **die; ~, ~n a)** *(Wahlspruch)* motto; *(Schlagwort)* slogan; **b)** *(bes. Milit.: Kennwort)* password

Paroli [pa'ro:li] *in:* **jmdm./einer Sache ~ bieten** give sb. as good as one gets/pit oneself against sth.

Part [part] **der; ~s, ~s** *od.* **~e a)** *(Musik: Stimme, Partie)* part; **b)** *(Theater, Film: Rolle)* part; role; **den [entscheidenden] ~ bei etw.** *(Dat.)* **spielen** *(auch fig.)* play the [crucial] part *or* role in sth.

Partei [par'tai̯] **die; ~, ~en a)** *(Politik)* party; **in** *od.* **bei der ~ sein** be a party member; **die ~ wechseln** change parties; **b)** *(Rechtsw.)* party; **c)** *(Gruppe, Mannschaft)* side; **es mit beiden ~en halten** run with the hare and hunt with the hounds *(fig.)*; **jmds.** *od.* **für jmdn./für etw. ~ ergreifen** *od.* **nehmen** side with sb./take a stand for sth.; **d)** *(Miets~)* tenant; *(mehrere Personen)* tenants *pl.*

partei-, Partei-: ~buch das party membership book; **~chef der** party leader; **~führung die** party leadership; **~genosse der** *(hist.: Mitglied der NSDAP)* party member; *(einer Arbeiterpartei)* **~genosse X** Comrade X; **~intern 1.** *Adj.*; *nicht präd.* internal [party] ‹*conflict, matters, material, etc.*›; **2.** *adv.* within the party

parteiisch 1. *Adj.* biased. **2.** *adv.* in a biased manner

partei-, Partei-: ~linie die party line; **~los** *Adj.* *(Politik)* independent ‹*MP*›; **er ist ~los** he is not attached to *or* aligned with any party; **~lose der/die;** *adj. Dekl.* *(Politik)* independent; person not attached to a party; **~mitglied das** party member; **~nahme die; ~, ~n** partisanship; taking sides *no art.*; **~politik die** party politics *sing.*; **~politisch 1.** *Adj.* party political; **2.** *adv.* from a party political point of view; **~programm das** party manifesto *or* programme; **~tag der** party conference *or* *(Amer.)* convention; **~vorsitzende der/die** party leader; **~vorstand der** party executive

Parterre das; ~s, ~s a) *(Erdgeschoß)* ground floor; first floor *(Amer.)*; **im ~:** on the ground *or* *(Amer.)* first floor; **b)** *(Theater veralt.)* stalls *pl.* *(Brit.)*; parterre *(Amer.)*; parquet *(Amer.)*

Partie [par'ti:] **die; ~, ~n a)** *(Teil)* part; **b)** *(Spiel, Sport: Runde)*

game; *(Golf)* round; **eine ~ Schach spielen** play a game of chess; **c)** *(Musik)* part; **d)** *(Ehepartner)* **eine gute ~ [für jmdn.] sein** be a good match [for sb.]; **sie hat eine gute/glänzende ~ gemacht** she has married well/extremely well; **e) mit von der ~ sein** join in; *(bei einer Reise usw.)* go along too; **f)** *(Kaufmannsspr.)* batch

partiell [par'tsiɛl] **1.** *Adj.* partial. **2.** *adv.* partially

¹Partikel [par'ti:kl̩] **die; ~, ~n** *(Sprachw.)* particle

²Partikel das; ~s, ~ *od.* **die; ~, ~n** *(bes. Physik, Chemie, Technik)* particle

Partisan [parti'za:n] **der; ~s** *od.* **~en, ~en** guerrilla; *(gegen Besatzungstruppen im Krieg)* partisan

Partisanen·krieg der guerilla war; *(Kriegführung)* guerilla warfare

Partisanin die; ~, ~nen *s.* **Partisan**

Partitur [parti'tu:ɐ̯] **die; ~, ~en** *(Musik)* score

Partizip [parti'tsi:p] **das; ~s, ~ien** [-'tsi:pi̯ən] *(Sprachw.)* participle; **das 1. ~** *od.* **~ Präsens/das 2. ~** *od.* **~ Perfekt** the present/past participle

Partner ['partnɐ] **der; ~s, ~, Partnerin die; ~, ~nen** partner; *(Bündnis~)* ally; *(im Film/Theater)* co-star

Partnerschaft die; ~, ~en partnership

partnerschaftlich 1. *Adj.* ‹*cooperation etc.*› on a partnership basis. **2.** *adv.* in a spirit of partnership; *(als Partnerschaft)* as a partnership

Partner·stadt die twin town *(Brit.);* sister city *or* town *(Amer.)*

partout [par'tu:] *Adv. (ugs.)* at all costs

Party ['pa:ɐ̯ti] **die; ~, ~s** *od.* **Parties** party; **eine ~ [zu ihrem bestandenen Examen/zu seinem Geburtstag] geben** give a party [to celebrate her passing the exam/for his birthday]; **auf** *od.* **bei ~s** at parties

Pasch [paʃ] **der; ~[e]s, ~e** *u.* **Päsche a)** *(beim Würfelspiel)* **einen ~ werfen** *(bei zwei Würfeln)* throw doubles *pl.; (bei drei Würfeln)* throw triplets; **b)** *(beim Domino)* double

Pascha ['paʃa] **der; ~s, ~s a)** *(hist.)* pasha; **b)** *(fig. abwertend)* male chauvinist; **den ~ spielen** act the lord and master

Paß [pas] **der; Passes, Pässe** ['pɛsə] **a)** *(Reise~)* passport; **b)** *(Gebirgs~)* pass; **c)** *(Ballspiele)* pass

passabel [pa'sa:bl̩] **1.** *Adj.* reasonable; tolerable; fair ‹*report*›; presentable ‹*appearance*›. **2.** *adv.* reasonably *or* tolerably well

Passage [pa'sa:ʒə] **die; ~, ~n a)** *(Ladenstraße)* [shopping] arcade; **b)** *(Abschnitt) (im Text)* passage; *(im Film)* sequence; *(Musik)* [virtuoso] passage

Passagier [pasa'ʒi:ɐ̯] **der; ~s, ~e** passenger; **blinder ~:** stowaway

Passagier-: ~dampfer der passenger steamer; **~flugzeug das** passenger aircraft

Paß·amt das passport office

Passant [pa'sant] **der; ~en, ~en, Passantin die; ~, ~nen** passer-by

Paß·bild das passport photograph

passé [pa'se:] *Adj.; nicht attr. (ugs.)* passé; out of date

passen 1. *itr. V.* **a)** *(die richtige Größe/Form haben)* fit; **etw. paßt [jmdm.] gut/nicht** sth. fits [sb.] well/does not fit [sb.]; **der Schlüssel paßt nicht ins Schloß** the key does not fit the lock; **b)** *(geeignet sein)* be suitable, be appropriate **(auf +** *Akk.*, **zu** for); *(harmonieren)* ‹*colour etc.*› match; **dieses Bild paßt besser in die Diele** this picture goes better in the hall; **zu etw./jmdm. ~:** go well with sth./be well suited to sb.; **zueinander ~:** ‹*things*› go well together; ‹*two people*› be suited to each other; **dieses Benehmen paßt zu ihm/paßt nicht zu ihm** *(ugs.)* that's just like him *(coll.)*/that's not like him; **diese Beschreibung paßt [genau] auf sie** this description fits her [exactly]; **c)** *(genehm sein)* **jmdm. ~** ‹*time*› be convenient for sb., suit sb.; **jmdm. paßt etw. nicht** sth. is inconvenient for sb.; **das könnte dir so ~!** *(ugs.)* you'd just love that, wouldn't you?; **d)** *(Kartenspiel)* pass; **bei dieser Frage muß ich ~** *(fig.)* I'll have to pass on that question. **2.** *tr. (auch itr.) V. (Ballspiele)* pass ‹*ball*›

passend *Adj.* **a)** *(geeignet)* suitable ‹*dress, present, etc.*›; appropriate, right ‹*words, expression*›; right ‹*moment*›; **bei einer ~en Gelegenheit** at an opportune moment; **b)** *(harmonierend)* matching ‹*shoes etc.*›; **die zum Kleid ~en Schuhe** the shoes to go with *or* match the dress

Paß·foto das *s.* **~bild**

passierbar *Adj.* passable ‹*road*›; navigable ‹*river*›

passieren 1. *tr. V.* pass; **die Grenze ~:** cross the border; **die Zensur ~** *(fig.)* be passed by the censor; get past the censor. **2.** *itr. V.; mit sein* happen; **es ist ein Unglück/etwas Schreckliches passiert** there has been an accident/something dreadful has happened; **jmdm. ist etwas/nichts passiert** something/nothing happened to sb.; *(jmd. ist verletzt/nicht verletzt)* sb. was/was not hurt

Passier·schein der pass; permit

Passion [pa'si̯o:n] **die; ~, ~en a)** passion; **b)** *o. Pl. (christl. Rel., Kunst, Musik)* Passion

passioniert *Adj.; nicht präd.* ardent, passionate ‹*collector, cardplayer, huntsman*›

passiv ['pasi:f] **1.** *Adj.* passive; non-active ‹*member*›; **~e Handelsbilanz** balance of trade deficit; *s. auch* **Bestechung. 2.** *adv.* passively; **sich [bei** *od.* **in etw.** *(Dat.)*] **~ verhalten** take a passive stance [in sth.]; take no active part [in sth.]

Passiv das; ~s, ~e *(Sprachw.)* passive

Passiva [pa'si:va] *Pl. (Wirtsch.)* liabilities

Passivität [pasivi'tɛ:t] **die; ~:** passivity

Passiv-: ~posten der *(Kaufmannsspr.)* liability; **~saldo der** *(Kaufmannsspr.)* debit balance; **~seite die** *(Kaufmannsspr.)* liabilities side

Paß-: ~kontrolle die a) *(das Kontrollieren)* passport inspection *or* check; **b)** *(Stelle)* passport control; **~straße die** [mountain] pass road

Paste ['pastə] **die; ~, ~n** *(auch Pharm.)* paste

Pastell [pas'tɛl] **das; ~[e]s, ~e a)** *(Farbton)* pastel shade; **b)** *o. Pl. (Maltechnik)* pastel *no art.*

pastell-, Pastell-: ~farbe die pastel colour; **~farben** *Adj.* pastel-coloured; **~ton der** pastel shade

Pastete [pas'te:tə] **die; ~, ~n a)** *(gefüllte ~)* vol-au-vent; **b)** *(in einer Schüssel o. ä. gegart)* pâté; *(in einer Hülle aus Teig gebacken)* pie

pasteurisieren [pastøri'zi:rən] *tr. V.* pasteurize

Pastille [pas'tɪlə] **die; ~, ~n** pastille

Pastor ['pastɔr] **der; ~s, ~en** pastor

pastoral [pasto'ra:l] *Adj.* **a)** *(seelsorgerlich)* pastoral; **b)** *(salbungsvoll)* unctuous; **c)** *(idyllisch)* pastoral ‹*literature*›

Pastorin die; ~, ~nen pastor

Pate ['pa:tə] **der; ~n, ~n** *(Taufzeuge)* godparent; *(Patenonkel)* godfather; *(Patin)* godmother; **bei jmdm. ~ stehen** act as *or* be godfather/godmother to sb.; **bei etw. ~ stehen** *(fig.)* be [the influence/influences] behind sth.; *(Vorbild sein)* act as the model for sth.

Paten-: ~kind das godchild; **~onkel der** godfather

Patenschaft die; ~, ~en *(christl. Rel.)* godparenthood

Paten·stadt die s. Partnerstadt

patent [pa'tɛnt] *(ugs.)* **1.** *Adj.* **a)** *(tüchtig)* capable; **b)** *(zweckmäßig)* ingenious ⟨*device, method, idea*⟩; clever ⟨*slogan etc.*⟩. **2.** *adv.* ingeniously; cleverly

Patent das; ~[e]s, ~e a) *(Schutz)* patent; **ein ~ auf etw.** *(Akk.)* **haben/etw. zum** *od.* **als ~ anmelden** have/apply for a patent for sth.; **b)** *(Erfindung)* [patented] invention; **c)** *(Ernennungsurkunde)* certificate [of appointment]; *(eines Kapitäns)* master's certificate; *(eines Offiziers)* commission

Paten·tante die *s.* **Patin**

patentieren *tr. V.* patent; **jmdm. etw. ~:** grant sb. a patent for sth.; **sich** *(Dat.)* **eine Erfindung ~ lassen** have an invention patented

Patent-: ~inhaber der patentee; **~lösung die** patent remedy **(für, zu** for); **~schutz der** patent protection

Pater ['pa:tɐ] **der; ~s, ~** *od.* **Patres** ['pa:tre:s] *(kath. Kirche)* Father

Paternoster [pa:tɐ'nɔstɐ] **der; ~s, ~** *(Aufzug)* paternoster [lift]

pathetisch [pa'te:tɪʃ] **1.** *Adj.* emotional, impassioned ⟨*speech, manner*⟩; melodramatic ⟨*gesture*⟩; emotive ⟨*style*⟩; pompous ⟨*voice*⟩. **2.** *adv.* emotionally; with much emotion; *(dramatisch)* [melo]dramatically

pathologisch 1. *(Med.; auch fig.) Adj.* pathological. **2.** *adv.* pathologically

Pathos ['pa:tɔs] **das; ~** emotionalism; **ein unechtes/hohles ~:** false/empty pathos; **etw. mit ~ vortragen** recite sth. with much feeling

Patient [pa'tsiɛnt] **der; ~en, ~en, Patientin die; ~, ~nen** patient

Patin die; ~, ~nen godmother

Patriarch [patri'arç] **der; ~en, ~en** patriarch

patriarchalisch 1. *Adj.* patriarchal; *(fig.: autoritär)* authoritarian. **2.** *adv.* in a patriarchal *or* *(fig.)* authoritarian manner

Patriot [patri'o:t] **der; ~en, ~en, Patriotin die; ~, ~nen** patriot

patriotisch 1. *Adj.* patriotic. **2.** *adv.* patriotically

Patriotismus der; ~: patriotism *usu. no def. art.*

Patron [pa'tro:n] **der; ~s, ~e a)** *(Heiliger)* patron saint; **b)** *(Stifter einer Kirche)* patron; founder; **c)** *(ugs.: Kerl)* type *(coll.)*

Patrone [pa'tro:nə] **die; ~, ~n** cartridge

Patronen-: ~gurt der, ~gürtel der cartridge-belt; *(über der Schulter getragen)* bandoleer; **~hülse die** cartridge-case

Patronin die; ~, ~nen *(Schutzheilige)* patron saint

Patrouille [pa'trʊljə] **die; ~, ~n** patrol

patrouillieren [patrʊl'ji:rən] *itr. V.; auch mit sein* be on patrol; **durch die Straßen ~:** patrol the streets

Patsche ['patʃə] **die; ~, ~n a)** *(ugs.) s.* **Klemme b; b)** *(ugs.: Hand)* paw *(coll.)*

patschen *itr. V. (ugs.)* **a)** *(klatschen)* slap; **sich** *(Dat.)* **auf die Schenkel ~:** slap one's thighs; **b)** *mit sein (~d gehen/fallen)* splash

patsch-, Patsch-: ~hand die; ~händchen das *(fam.)* [little] hand; handy-pandy *(child lang.)*; **~naß** *Adj. (ugs.)* sopping wet; **~naß geschwitzt** soaked in sweat

Patt das; ~s, ~s *(Schach; auch fig.)* stalemate

patzen ['patsn̩] *itr. V. (ugs.)* slip up *(coll.)*; boob *(sl.)*

Patzer der; ~s, ~ *(ugs.)* slip *(coll.)*; boob *(sl.)*

patzig *(ugs. abwertend)* **1.** *Adj.* snotty *(coll.)*; *(frech)* cheeky. **2.** *adv.* snottily *(coll.)*; *(frech)* cheekily

Pauke ['paukə] **die; ~, ~n** kettledrum; **die ~ schlagen** beat the drum/drums; **auf die ~ hauen** *(ugs.) (feiern)* paint the town red *(sl.)*; *(großtun)* blow one's own trumpet; **mit ~n und Trompeten durchfallen** *(ugs.)* ⟨*candidate*⟩ fail resoundingly; ⟨*broadcast, film, etc.*⟩ be a resounding failure

pauken 1. *tr. V. (ugs.)* swot up *(Brit. sl.)*, bone up on *(Amer. coll.)* ⟨*facts, figures, etc.*⟩. **2.** *itr. V. (ugs.)* swot *(Brit. sl.)*; *(fürs Examen)* cram *(coll.)*

Pauken·schlag der a) drumbeat; **b)** *(fig. Eklat)* sensation; bombshell

Pauker der; ~s, ~ *(Schülerspr.)* teacher; teach *(school sl.)*

Paukerei die; ~: *(ugs. abwertend)* swotting *(Brit. sl.)*; boning up *(Amer. coll.)*; *(fürs Examen)* cramming

Paus·backen *Pl. (fam.)* chubby cheeks

pausbäckig ['pausbɛkɪç] *Adj.* chubby-cheeked; chubby-faced; chubby ⟨*face*⟩

pauschal [pau'ʃa:l] **1.** *Adj.* **a)** all-inclusive ⟨*price, settlement*⟩; **b)** *(verallgemeinernd)* sweeping ⟨*judgement, criticism, statement*⟩; indiscriminate ⟨*prejudice*⟩; wholesale ⟨*discrimination*⟩. **2.** *adv.* **a)** ⟨*cost*⟩ overall, all in all; ⟨*pay*⟩ in a lump sum; **b)** *(verallgemeinernd)* wholesale

Pauschale die; ~, ~n *od.* **das; ~s, Pauschalien** [-'ʃa:liən] flat-rate payment

Pauschal-: ~gebühr die flat-rate [charge]; **~preis der** *(Einheitspreis)* flat rate; *(Inklusivpreis)* inclusive *or* all-in price; **~reise die** package holiday; *(mit mehreren Reisezielen)* package tour; **~summe die** lump sum

[1]**Pause** ['pauzə] **die; ~, ~n a)** *(Unterbrechung)* break; *(Ruhe~)* rest; *(Theater)* interval *(Brit.)*; intermission *(Amer.)*; *(Kino)* intermission; *(Sport)* half-time interval; **kleine/große ~:** *(Schule)* short/[long] break; **[eine] ~ machen/eine ~ einlegen** take *or* have a break; *(zum Ausruhen)* have a rest; **b)** *(in der Unterhaltung o. ä.)* pause; *(verlegenes Schweigen)* silence; **c)** *(Musik)* rest

[2]**Pause die; ~, ~n** *(Kopie)* tracing; *(Licht~)* Photostat *(Brit.* **P)**

pausen *tr. V.* trace; *(eine Lichtpause machen)* Photostat *(Brit.* **P)**

Pausen·brot das sandwich *(eaten during break)*

pausen·los 1. *Adj.; nicht präd.;* incessant ⟨*noise, moaning, questioning*⟩; continous, uninterrupted ⟨*work, operation*⟩. **2.** *adv.* incessantly; ceaselessly; ⟨*work*⟩ non-stop

pausieren *itr. V.* **a)** *(innehalten)* pause; **b)** *(aussetzen)* have *or* take a rest

Paus·papier das tracing paper; *(Kohlepapier)* carbon paper

Pavian ['pa:via:n] **der; ~s, ~e** baboon

Pavillon ['pavɪljɔn] **der; ~s, ~s** *(Archit.)* pavilion

Pazifik [pa'tsi:fɪk] **der; ~s** Pacific

pazifisch *Adj.* Pacific ⟨*area*⟩; **der Pazifische Ozean** the Pacific Ocean

Pazifismus der; ~: pacifism *no art.*

Pazifist der; ~en, ~en, Pazifistin die; ~, ~nen pacifist

pazifistisch 1. *Adj.* pacifist. **2.** *adv.* in a pacifist way

PDS [pe:de:ˈ|ɛs] **die;** ~ *Abk.* **Partei des Demokratischen Sozialismus** Party of Democratic Socialism

Pech [pɛç] **das;** ~**[e]s,** ~**e a)** pitch; **zusammenhalten wie** ~ **und Schwefel** *(ugs.)* be inseparable; ⟨*friends*⟩ be as thick as thieves *(coll.)*; **b)** *o. Pl. (Mißgeschick)* bad luck; **großes/unerhörtes** ~**:** rotten *(sl.)*/ *(coll.)* terrible luck; **bei** *od.* **mit etw./ mit jmdm.** ~ **haben** have bad luck with sth./sb.; be unlucky with sth./sb.; **dein** ~**, wenn du nicht aufpaßt** *(ugs.)* that's just your hard luck *(coll.)* if you don't pay attention

pẹch-, Pẹch-: ~**schwạrz** *Adj. (ugs.) s.* **rabenschwarz;** ~**strähne die** run of bad luck; ~**vogel der** unlucky devil *(coll.)*; *(Opfer vieler Unfälle)* walking disaster area *(coll.)*

Pedal [peˈda:l] **das;** ~**s,** ~**e** pedal; **[kräftig] in die** ~**e treten** *(beim Fahrrad)* pedal [really] hard

Pedant [peˈdant] **der;** ~**en,** ~**en, Pedạntin, die;** ~**,** ~**nen** pedant

pedạntisch 1. *Adj.* pedantic. **2.** *adv.* pedantically

Pediküre [pediˈky:rə] **die;** ~**,** ~**n a)** *o. Pl.; s.* **Fußpflege; b)** *(Berufsbez.)* chiropodist

pediküren *tr. V.* pedicure ⟨*feet, nails*⟩

Pegel [ˈpe:gl̩] **der;** ~**s,** ~ **a)** *(Gerät)* water-level indicator; *(für die Gezeiten am Meer)* tide-gauge; **b)** *(Wasserstand)* water-level

Pegel·stand der water-level

peilen [ˈpailən] *tr. V.* **a)** take a bearing on ⟨*transmitter, fixed point*⟩; **b)** *(Wassertiefe messen)* sound ⟨*depth*⟩; take soundings in ⟨*bay etc.*⟩

Pein [pain] **die;** ~ *(geh.)* torment; **jmdm. [viel** *od.* **große]** ~ **bereiten** cause sb. [much] anguish

peinigen [ˈpainɪgn̩] *tr. V. (geh.)* torment; *(foltern)* torture; **von Durst/Kälte gepeinigt werden** suffer agonies from thirst/cold

peinlich 1. *Adj.* **a)** embarrassing; awkward ⟨*question, position, pause*⟩; **es ist mir sehr** ~**:** I feel very bad *(coll.) or* embarrassed about it; **b)** *nicht präd. (äußerst genau)* meticulous; scrupulous. **2.** *adv.* **a)** unpleasantly ⟨*surprised*⟩; **[von etw.]** ~ **berührt sein** be painfully embarrassed [by sth.]; **b)** *(überaus [genau])* scrupulously; meticulously

Peinlichkeit die; ~ **a)** embarrassment; **die** ~ **der Situation** the awkwardness of the situation; **b)** *(Genauigkeit)* scrupulousness; meticulousness

Peitsche [ˈpaitʃə] **die;** ~**,** ~**n** whip; **er knallte mit der** ~**:** he cracked the whip

peitschen 1. *tr. V.* whip; *(fig.)* ⟨*storm, waves, rain*⟩ lash. **2.** *itr. V.; mit sein* ⟨*rain*⟩ lash **(an, gegen** + *Akk.* against, **in** + *Akk.* into); ⟨*shot*⟩ ring out

Pekinese [pekiˈne:zə] **der;** ~**n,** ~**n** Pekinese

Pelikan [ˈpe:lika:n] **der;** ~**s,** ~**e** pelican

Pelle [ˈpɛlə] **die;** ~**,** ~**n** *(bes. nordd.)* skin; *(abgeschält)* peel; *s. auch* **rücken 2a; sitzen a**

pẹllen *(bes. nordd.)* **1.** *tr. V.* peel ⟨*potato, egg, etc.*⟩. **2.** *refl. V.* ⟨*person, skin*⟩ peel

Pẹll·kartoffel die potato boiled in its skin

Pelz [pɛlts] **der;** ~**es,** ~**e a)** fur; coat; *(des toten Tieres)* skin; pelt; **b)** *o. Pl. (gegerbt; als Material)* fur; **mit** ~ **gefüttert** fur-lined; **c)** fur; *(~mantel)* fur coat; **d)** *(ugs.: Haut)* **sich** *(Dat.)* **die Sonne auf den** ~ **brennen lassen** soak up the sun; *s. auch* **rücken 2a; sitzen a**

pẹlzig *Adj.* **a)** furry; downy ⟨*peach*⟩; **b)** *(bes. westd.: mehlig)* mealy ⟨*apple*⟩; *(holzig)* woody ⟨*radish*⟩; **c)** *(belegt)* furred, coated ⟨*tongue, mouth*⟩

Pẹlz-: ~**jacke die** fur jacket; ~**kragen der** fur collar; ~**mantel der** fur coat; ~**mütze die** fur hat

Pendel [ˈpɛndl̩] **das;** ~**s,** ~**:** pendulum

pẹndeln *itr. V.* **a)** *(hin u. her schwingen)* swing [to and fro] **(an** + *Dat.* by); *(mit weniger Bewegung)* dangle; **b)** *mit sein (hin- u. herfahren)* **zwischen X und Y** ~ ⟨*bus, ferry, etc.*⟩ operate a shuttle service between X and Y; ⟨*person*⟩ commute between X and Y

Pẹndel-: ~**tür die** swing-door; ~**uhr die** pendulum clock; ~**verkehr der a)** *(Berufsverkehr)* commuter traffic; **b)** *(mit Pendelzug o. ä.)* shuttle service

penetrant [peneˈtrant] *(abwertend)* **1.** *Adj.* **a)** *(durchdringend)* penetrating, pungent ⟨*smell, taste*⟩; overpowering ⟨*stink, perfume*⟩; **b)** *(aufdringlich)* pushing, *(coll.)* pushy ⟨*person*⟩; overbearing ⟨*tone, manner*⟩; aggressive, pointed ⟨*question*⟩. **2.** *adv.* **a)** *(durchdringend)* overpoweringly; **es riecht** ~ **nach ...:** there is an overpowering smell of ...; **b)** *(aufdringlich)* overbearingly; in an overbearing manner

peng [pɛŋ] *Interj.* bang

penibel [peˈni:bl̩] **1.** *Adj.* over-meticulous ⟨*person*⟩; *(pedantisch)* pedantic. **2.** *adv.* painstakingly; over-meticulously ⟨*dressed*⟩

Penis [ˈpe:nɪs] **der;** ~**,** ~**se** *od.* **Penes** [ˈpe:ne:s] penis

Penizillin [penitsɪˈli:n] **das;** ~**s,** ~**e** penicillin

Pennäler [pɛˈnɛ:lɐ] **der;** ~**s,** ~**s** *(ugs.)* [secondary] schoolboy

Pennälerin die; ~**,** ~**nen** *(ugs.)* [secondary] schoolgirl

Pẹnn·bruder der *(ugs. abwertend)* tramp *(Brit.)*; hobo *(Amer.)*

Penne [ˈpɛnə] **die;** ~**,** ~**n** [secondary] school; swot-shop *(Brit. sl.)*

pẹnnen *itr. V. (salopp)* **a)** *(schlafen)* kip *(sl.)*; **b)** *(fig.: nicht aufpassen)* be half asleep; **c)** *(koitieren)* **mit jmdm.** ~**:** sleep with sb.

Pẹnner der; ~**s,** ~**, Pẹnnerin die;** ~**,** ~**nen** *(salopp abwertend) (Stadtstreicher)* tramp *(Brit.)* hobo *(Amer.)*

Pension [pãˈzio:n] **die;** ~**,** ~**en a)** *o. Pl. (Ruhestand)* **[vorzeitig] in** ~ **gehen** retire [early]; **in** ~ **sein** be retired *or* in retirement; **b)** *(Ruhegehalt)* [retirement] pension; **c)** *(Haus für [Ferien]gäste)* guesthouse; *(auf dem Kontinent)* pension; **d)** *o. Pl. (Unterkunft u. Verpflegung)* board

Pensionär [pãzioˈnɛ:ɐ] **der;** ~**s,** ~**e, Pensionärin die;** ~**,** ~**nen** retired civil servant; *(ugs.: Rentner)* [old-age] pensioner

Pensionat [pãzioˈna:t] **das;** ~**[e]s,** ~**e** *(veralt.)* boarding-school *(esp. for girls)*

pensionieren *tr. V.* pension off; retire; **sich [vorzeitig]** ~ **lassen** retire [early]; take [early] retirement

Pensionierung die; ~**,** ~**en** retirement

pensions-, Pensions-: ~**alter das** retirement age; ~**anspruch der** pension entitlement; ~**berechtigt** *Adj.* entitled to a pension *postpos.*; ~**reif** *Adj. (ugs.)* ripe for retirement *pred.*

Pensum [ˈpɛnzʊm] **das;** ~**s, Pẹnsen a)** *(Arbeit)* amount of work; work quota; **b)** *(Päd. veralt.: Lehrstoff)* syllabus

Pep [pɛp] **der;** ~**[s]** *(ugs.)* pep *(sl.)*; zip; ~ **haben** be dynamic *or* full of zip

Peperoni [pepeˈro:ni] **die;** ~**,** ~**:** chilli

per [pɛr] *Präp. mit Akk.* **a)** *(mittels)* by; ~ **Adresse X** care of X; c/o X; **b)** *(Kaufmannsspr.: [bis] zum)* by; *(am)* on; ~ **sofort** immediately; as of now; **c)** *(Kaufmannsspr.: pro)* per; **etw.** ~ **Stück verkaufen** sell sth. by the piece *or* separately

perfekt [pɛr'fɛkt] **1.** *Adj.* **a)** *(hervorragend)* outstanding; first-rate; *(vollkommen)* perfect ‹*crime, host*›; faultless ‹*English, French, etc.*›; **b)** *nicht attr. (ugs.: abgeschlossen)* finalized; concluded; **~ sein/werden** ‹*contract, deal*› be concluded *or* finalized; ‹*scandal, defeat*› be complete. **2.** *adv.* **a)** *(hervorragend)* outstandingly well; *(vollkommen)* ‹*fit, work, etc.*› perfectly; **b)** *(ugs.: vollständig)* good and proper *(coll.)*

Perfekt ['pɛrfɛkt] **das; ~s, ~e** *(Sprachw.)* perfect [tense]

Perfektion [pɛrfɛk'tsi̯o:n] **die; ~:** perfection; **handwerkliche/technische ~:** mastery of a craft/technical mastery

Perfektionismus der; ~: perfectionism

perforieren *tr. V. (Technik, Med.)* perforate

Pergament [pɛrga'mɛnt] **das; ~[e]s, ~e** parchment

Pergament·papier das grease-proof paper

Periode [pe'ri̯o:də] **die; ~, ~n a)** *(auch Chemie, Physik, Technik, Astron., Met., Sprachw., Musik)* period; *(Geol.)* era; **b)** *(Math.)* repetend; period; **3,3 ~:** 3.3 recurring

periodisch 1. *Adj.* regular; ‹*meeting, statement of account*› at regular intervals; *(Chemie)* periodic ‹*system*›. **2.** *adv.* regularly; at regular intervals

peripher [peri'fe:ɐ̯] **1.** *Adj.* peripheral. **2.** *adv.* peripherally

Peripherie [perife'ri:] **die; ~, ~n** periphery; *(einer Stadt)* outskirts *pl.*; fringe; *(Geom.: Begrenzungslinie)* circumference

Perle ['pɛrlə] **die; ~, ~n a)** *(auch fig.)* pearl; **~n vor die Säue werfen** *(fig. ugs.)* cast pearls before swine; **b)** *(aus Holz, Glas o. ä.)* bead; *(Bläschen beim Sekt usw.)* bubble; **c)** *(ugs. scherzh.: Hausgehilfin)* [invaluable] home help

perlen *itr. V.* **a)** *auch mit sein* **auf etw.** *(Dat.)* **~:** form pearls on sth.; **b)** *mit sein* **von etw. ~:** ‹*dew, sweat*› trickle *or* drip from sth.; **c)** *(Bläschen bilden)* ‹*champagne etc.*› sparkle, bubble

Perlen-: ~fischer der pearl-fisher; **~kette die** string of pearls; pearl necklace; *(mit Holzperlen usw.)* string of beads; bead necklace

Perl-: ~huhn das guinea-fowl; **~mutt** [~mʊt] **das; ~s, ~mutter die; ~** *od.* **das; ~s** mother-of-pearl

Perlon ⓦ **das; ~s** ≈ nylon

Perl-: ~wein der sparkling wine; **~zwiebel die** pearl *or* cocktail onion

permanent [pɛrma'nɛnt] **1.** *Adj.* permanent ‹*institution, deficit, crisis*›; constant ‹*danger, threat, squabble*›. **2.** *adv.* constantly

perplex [pɛr'plɛks] *(ugs.) Adj.* baffled, puzzled (**über** + *Akk.* by); *(verwirrt)* bewildered

Perser ['pɛrzɐ] **der; ~s, ~ a)** Persian; **b)** *s.* **Perserteppich**

Perserin die; ~, ~nen Persian

Perser·teppich der Persian carpet; *(kleiner)* Persian rug

Persianer [pɛr'zi̯a:nɐ] **der; ~s, ~:** Persian lamb; *(~mantel)* Persian lamb coat

Persien ['pɛrzi̯ən] **(das); ~s** Persia

Persiflage [pɛrzi'fla:ʒə] **die; ~, ~n** [gentle] mocking *no indef. art.*; **eine ~ auf jmdn./etw.** a [gentle] satire of sb./sth.

persisch *Adj.* Persian; *s. auch* **deutsch; Deutsch**

Person [pɛr'zo:n] **die; ~, ~en a)** person; **eine männliche/weibliche ~:** a male/female; **~en** *(als Gruppe)* people; **die Familie besteht aus fünf ~en** it is a family of five; **ich für meine ~ ...:** I for my part ...; **der Minister in [eigener] ~:** the minister in person; **sie ist die Güte/Geduld in ~:** she is kindness/patience personified *or* itself; **Angaben zur ~ machen** give one's personal details; **b)** *(in der Dichtung, im Film)* character; **c)** *(emotional: Frau)* female *(derog./joc.)*; **d)** *o. Pl. (Sprachw.)* person

Personal [pɛrzo'na:l] **das; ~s a)** *(in einem Betrieb o. ä.)* staff; **b)** *(im Haushalt)* servants *pl.*; [domestic] staff *pl.*

Personal-: ~abbau der reduction in staff; *(in mehreren Abteilungen/Betrieben)* staff cuts *pl.*; **~abteilung die** personnel department; **~ausweis der** identity card; **~büro das** personnel office; **~chef der** personnel manager

Personalien [pɛrzo'na:li̯ən] *Pl.* personal details *or* particulars

Personal-: ~kosten *Pl. (Wirtsch., Verwaltung)* staff costs; **~mangel der** staff shortage; **~pronomen das** *(Sprachw.)* personal pronoun

personell [pɛrzo'nɛl] **1.** *Adj.; nicht präd.* staff ‹*changes, difficulties*›; ‹*savings*› in staff; ‹*questions, decisions*› regarding staff *or* personnel. **2.** *adv.* with regard to staff *or* personnel

Personen-: ~aufzug der passenger lift *(Brit.) or (Amer.)* elevator; **~beschreibung die** personal description; **~gedächtnis das** memory for faces; **~kraftwagen der** *(bes. Amtsspr.)* private car *or (Amer.)* automobile; **~kreis der** group [of people]; **~kult der** *(abwertend)* personality cult; **~schaden der** *(Versicherungsw.)* physical *or* personal injury; **Unfälle mit ~schaden** accidents in which injuries are/were sustained; **~wagen der a)** *(Auto)* [private] car; automobile *(Amer.)*; *(im Unterschied zum Lastwagen)* passenger car *or (Amer.)* automobile; **b)** *(bei Zügen)* passenger coach; **~zug der** slow *or* stopping train; *(im Unterschied zum Güterzug)* passenger train

persönlich [pɛr'zø:nlɪç] **1.** *Adj.* personal; **~ werden** get personal. **2.** *adv.* personally; *(auf Briefen)* 'private [and confidential]'; **nimm doch nicht gleich alles [so] ~!** don't take everything so personally!

Persönlichkeit die; ~, ~en personality; *(Mensch)* person of character; **eine ~ sein** have a strong personality; **~en des öffentlichen Lebens** public figures

Perspektive [pɛrspɛk'ti:və] **die; ~, ~n** *(Optik, bild. Kunst, auch fig.)* perspective; *(Blickwinkel)* angle; viewpoint; *(Zukunftsaussicht)* prospect; **aus soziologischer ~/aus der ~ des Soziologen** *(fig.)* from a sociological viewpoint/the viewpoint of a sociologist

perspektivisch 1. *Adj.* ‹*drawing etc.*› in perspective; ‹*effect, narrowing, etc.*› of perspective. **2.** *adv.* in perspective; **~ verkürzen** foreshorten

Peru [pe'ru:] **(das); ~s** Peru

Peruaner [pe'ru̯a:nɐ] **der; ~s, ~** Peruvian

Perücke [pɛ'rʏkə] **die; ~, ~n** wig

pervers [pɛr'vɛrs] *(abwertend)* **1.** *Adj.* perverted; *(fig.: gegen jede Vernunft)* perverse. **2.** *adv.* **~ veranlagt sein** be of a perverted disposition

Pessimismus [pɛsi'mɪsmʊs] **der; ~:** pessimism

Pessimist der; ~en, ~en, Pessimistin die; ~, ~nen pessimist

pessimistisch 1. *Adj.* pessimistic. **2.** *adv.* pessimistically; **etw. ~ sehen** *od.* **betrachten** take a pessimistic view of sth.

Pest [pɛst] **die; ~:** plague; *(fig.: Mensch, Ungeziefer)* pest; menace; **ich hasse ihn/es wie die ~** *(ugs.)* I hate his guts/can't stand it *(coll.)*; **wie die ~ stinken** *(salopp)* stink to high heaven *(coll.)*

Peter ['pe:tɐ] **der**; **~s**, **~**: *(ugs.)* fellow; **Schwarzer ~** *(Kartenspiel)* ≈ old maid *(with a black cat card instead of an old maid)*; **jmdm. den Schwarzen ~ zuschieben** *(fig.)* pass the buck to sb. *(coll.)*

Petersilie [petɐ'zi:li̯ə] **die**; **~**: parsley; **ihm ist die ~ verhagelt** *(ugs.)* he's down in the dumps

Petition [peti'tsi̯o:n] **die**; **~**, **~en** *(Amtsspr.)* petition

Petri Heil ['pe:tri-] good fishing!; make a good catch!

Petroleum [pe'tro:leʊm] **das**; **~s** paraffin *(Brit.)*; kerosene *(Amer.)*

Petrus ['pe:trʊs] **(der)**; **Petrus** *od.* **Petri a)** *(christl. Rel.: Apostel)* St Peter; **b)** *(Patron des Wetters)* the clerk of the weather

Petting ['pɛtɪŋ] **das**; **~[s]**, **~s** petting

Petze die; **~**, **~n** *(Schülerspr. abwertend)* tell-tale; sneak *(Brit. school sl.)*; tattle-tale *(Amer. school sl.)*

petzen *(Schülerspr.)* **1.** *itr. V.* tell tales; sneak *(Brit. school sl.)*. **2.** *tr. V.* **~, daß ...**: tell teacher/sb.'s parents that ...

Petzer der; **~s**, **~** *s.* **Petze**

Pf *Abk.* **Pfennig**

Pfad [pfa:t] **der**; **~[e]s**, **~e** path; **vom ~ der Tugend abweichen** *(fig. geh.)* stray from the path of virtue

Pfad-: **~finder der**, **~s**, **~** Scout; **er ist bei den ~n** he is in the Scouts; **~finderin die**; **~**, **~nen** Guide *(Brit.)*; girl scout *(Amer.)*; **sie ist bei den ~nen** she is in the Guides *(Brit.)* or *(Amer.)* girl scouts

Pfaffe ['pfafə] **der**; **~n**, **~n** *(abwertend)* cleric; Holy Joe *(derog.)*

Pfahl [pfa:l] **der**; **~[e]s**, **Pfähle** ['pfɛ:lə] post; stake; *(Bauw.: Stütze für Gebäude)* pile; **[jmdm.] ein ~ im Fleisch[e] sein** be a thorn in sb.'s flesh

Pfahl·bau der; **~[e]s**, **~ten** pile-dwelling

Pfand [pfant] **das**; **~[e]s**, **Pfänder** ['pfɛndɐ] **a)** security; pledge *(esp. fig.)*; **etw. als** *od.* **in ~ nehmen/etw. als** *od.* **zum** *od.* **in ~ geben** take/give sth. as [a] security; **b)** *(für leere Flaschen usw.)* deposit (**auf** + *Dat.* on); **c)** *(beim Pfänderspiel)* forfeit

pfänden ['pfɛndn̩] *tr. (auch itr.) V.* impound, seize [under distress] *(Law)* ⟨*goods, chattels*⟩; attach ⟨*wages etc.*⟩ *(Law)*; **er ist gepfändet worden** the bailiffs have been on to him; execution was levied against him *(Law.)*

Pfänder·spiel das [game of] forfeits

Pfand-: **~flasche die** returnable bottle *(on which a deposit is payable)*; **~leiher der** pawnbroker

Pfändung die; **~**, **~en** seizure; distraint *(Law)*; *(von Geldsummen, Vermögensrechten)* attachment *(Law)*

Pfanne ['pfanə] **die**; **~**, **~n** [frying-]pan; **sich** *(Dat.)* **ein paar Eier in die ~ schlagen** fry [up] some eggs; **jmdn. in die ~ hauen** *(ugs.)* *(kritisieren)* take sb. to pieces; *(vernichtend schlagen)* beat sb. hollow *(coll.)*

Pfann·kuchen der a) *(bes. südd.: Eierkuchen)* pancake; **b)** *(Berliner ~)* doughnut

Pfarr·amt das a) parish office; **b)** *(Stellung)* pastorate

Pfarrei [pfa'rai̯] **die**; **~**, **~en a)** *(Bezirk)* parish; **b)** *(Dienststelle)* parish office

Pfarrer ['pfarɐ] **der**; **~s**, **~** *(katholisch)* parish priest; *(evangelisch)* pastor; *(anglikanisch)* vicar; *(von Freikirchen)* minister; *(Militär~)* chaplain; padre

Pfarrerin die; **~**, **~nen** [woman] pastor; *(in Freikirchen)* [woman] minister

Pfarr·haus das vicarage; *(katholisch)* presbytery; *(in Schottland)* manse

Pfau [pfau̯] **der**; **~[e]s**, **~en** *(österr. auch:)* **~en**, **~e** peacock

Pfauen·auge das peacock butterfly

Pfeffer ['pfɛfɐ] **der**; **~s**, **~** pepper; **hingehen** *od.* **bleiben, wo der ~ wächst** *(ugs.)* go to hell *(coll.)*; get lost *(sl.)*; *s. auch* **Hase a**

Pfeffer-: **~korn das** peppercorn; **~kuchen der** ≈ gingerbread

Pfefferminz ['pfɛfɐmɪnts] *o. Art., indekl.* peppermint

Pfefferminz·bonbon der *od.* **das** peppermint [sweet]

Pfeffer·minze die; *o. Pl.* peppermint [plant]

Pfefferminz·tee der peppermint tea

Pfeffer·mühle die pepper-mill

pfeffern *tr. V.* **a)** *(würzen)* season with pepper; **b)** *(ugs.: werfen)* chuck *(coll.)*; *(mit Wucht)* fling; hurl; **jmdm. eine ~** *(salopp)* sock or biff sb. one *(sl.)*; *s. auch* **gepfeffert**

Pfeffer-: **~nuß die** [small round] gingerbread biscuit; **~steak das** steak au poivre; pepper steak; **~streuer der** pepper-pot

pfeffrig ['pfɛfrɪç] *Adj.* peppery

Pfeife ['pfai̯fə] **die**; **~**, **~n a)** *(Tabak~)* pipe; **~ rauchen** smoke a pipe; be a pipe-smoker; **b)** *(Musikinstrument)* pipe; *(der Militärkapelle)* fife; *(Triller~, an einer Maschine usw.)* whistle; *(Orgel~)* [organ-]pipe; **nach jmds. ~ tanzen** *(fig.)* dance to sb.'s tune; **c)** *(salopp abwertend: Versager)* wash-out *(sl.)*

pfeifen 1. *unr. itr. V.* **a)** whistle; ⟨*bird*⟩ sing; pipe; **dreimal kurz ~**: give three short whistles; **es pfeift in seiner Brust** he wheezes in his chest; *s. auch* **Loch a**; **b)** *mit sein* **die Kugeln pfiffen ihm um die Ohren** the bullets whistled around him; **c)** *(auf einer Trillerpfeife o. ä.)* ⟨*policeman, referee, etc.*⟩ blow one's whistle; *(Sport: als Schiedsrichter fungieren)* act as referee; **d)** *(salopp)* **auf jmdn./etw. ~**: not give a damn about sb./sth.; **ich pfeife auf dein Geld** you can keep your money *(coll.)*; **e)** *(salopp: geständig sein)* squeal *(sl.)*. **2.** *unr. tr. V.* **a)** whistle ⟨*tune etc.*⟩; ⟨*bird*⟩ pipe, sing ⟨*song*⟩; **b)** **sich** *(Dat.)* **eins ~** *(ugs.)* whistle [nonchalantly] to oneself; *(auf einer Pfeife)* pipe, play ⟨*tune etc.*⟩; *(auf einer Trillerpfeife o. ä.)* blow ⟨*signal etc.*⟩ on one's whistle; **einen Elfmeter ~** *(Sport)* blow [the whistle] for a penalty; **c)** *(salopp spött.)* **ich pfeif' dir was** go and get knotted *(sl.)*; **d)** *(Sport: als Schiedsrichter leiten)* referee ⟨*match*⟩; **e)** *(salopp: verraten)* let out ⟨*secret*⟩

Pfeifen-: **~reiniger der** pipe-cleaner; **~tabak der** pipe tobacco

Pfeifer der; **~s**, **~ a)** *(Musik)* *(bes. hist.)* piper; *(in einer Militärkapelle)* fife-player; **b)** *(jmd., der pfeift)* whistler

Pfeil [pfai̯l] **der**; **~[e]s**, **~e** arrow; **~ und Bogen** bow and arrow; **schnell wie ein ~**: as quick as lightning

Pfeiler der; **~s**, **~**: pillar; *(Brücken~)* pier

pfeil, Pfeil-: **~gerade 1.** *Adj.* [as] straight as an arrow *postpos.*; dead straight; **2.** *adv.* [as] straight as an arrow; **~gift das** arrow-poison; **~schnell 1.** *Adj.* lightning-swift; **2.** *adv.* like a shot; **~spitze die** arrowhead

Pfennig ['pfɛnɪç] **der**; **~s**, **~e** pfennig; **eine Briefmarke zu 60 ~**: a 60-pfennig stamp; **er hat keinen ~ [Geld]** he hasn't a penny *or* *(Amer.)* cent; **auf den ~ sehen** *(ugs.)* watch *or* count every penny *or* *(Amer.)* cent; **nicht für fünf ~ Verstand/Humor haben** *(ugs.)* have not an ounce of common sense/have no sense of humour whatsoever; **wer den ~ nicht ehrt, ist des Talers nicht wert** *(Spr.)* take care of the pennies and the pounds will look after themselves *(prov.)*; *s. auch* **Heller**

Pfennig·fuchser [-fʊksɐ] **der; ~s, ~** *(ugs.)* penny-pincher

Pferch [pfɛrç] **der; ~[e]s, ~e** pen

pferchen *tr. V.* cram; pack

Pferd [pfɛ:ɐt] **das; ~[e]s, ~e a)** horse; **aufs/vom ~ steigen** mount/dismount; **zu ~:** by horse; on horseback; **das hält ja kein ~ aus** *(ugs.)* that's more than flesh and blood can stand; **ich denk', mich tritt ein ~** *(salopp)* I'm absolutely flabbergasted; **man hat schon ~e kotzen sehen** *(salopp)* [you never know,] anything can happen; **wie ein ~ arbeiten** *(ugs.)* work like a Trojan; **ihm gehen die ~e durch** *(ugs.)* he flies off the handle *(coll.)*; **auf das falsche ~ setzen** *(fig.)* back the wrong horse; **die ~e scheu machen** *(ugs.)* put people off; **das ~ am** *od.* **beim Schwanze aufzäumen** *(ugs.)* put the cart before the horse; **mit ihr kann man ~e stehlen** *(ugs.)* she's game for anything; **b)** *(Turngerät)* horse; **c)** *(Schachfigur)* knight

Pferde-: ~äpfel *Pl. (ugs.)* horse-droppings; horse-dung; **~fuß der** *(fig.: Mangel, Nachteil)* snag; drawback; **~gebiß das** *(fig. ugs.)* **er hat ein ~gebiß** he has teeth *pl.* like a horse; **~gesicht das** *(ugs.)* horsy face; **~koppel die** paddock; **~pfleger der** groom; **~rasse die** breed of horse; **~rennbahn die** racecourse; **~rennen das** horse-race; *(Sportart)* horse-racing; **beim ~rennen sein** be at the races *pl.*; **~schwanz der** horse's tail; *(fig.: Frisur)* pony-tail; **~sport der** equestrian sport *no art.*; *(~rennen)* horse-racing *no art.*; **~stall der** stable; **~stärke die** horse-power; **~wagen der** *(für Güter)* cart; *(für Personen)* carriage; *(der amerikanischen Pioniere usw.)* wagon

pfiff [pfɪf] *1. u. 3. Pers. Sg. Prät. v.* **pfeifen**

Pfiff der; ~[e]s, ~e a) whistle; **b)** *(ugs.: besonderer Reiz)* style; **mit ~:** stylish; with style; *(adverbiell)* stylishly; ⟨*cook*⟩ with flair

Pfifferling ['pfɪfɐlɪŋ] **der; ~s, ~e** chanterelle; **keinen ~ wert sein** *(ugs.)* be not worth a bean *(sl.)*

pfiffig 1. *Adj.* smart; bright, clever ⟨*idea*⟩; artful, knowing ⟨*smile, expression*⟩. **2.** *adv.* artfully; cleverly; **jmdn. ~ ansehen** look knowingly *or* artfully at sb.

Pfingsten ['pfɪŋstn̩] **das; ~, ~:** Whitsun

Pfingst-: ~montag der Whit Monday *no def. art.*; **~ochse der** *in* **herausgeputzt wie ein ~ochse** *(ugs.)* dressed up like a dog's dinner *(coll.)*; **~rose die** peony; **~sonntag der** Whit Sunday *no def. art.*; **~woche die** week before Whitsun

Pfirsich ['pfɪrzɪç] **der; ~s, ~e** peach

Pflänzchen ['pflɛntsçən] **das; ~s, ~ a)** little plant; **b)** *(fig.: Mensch)* **ein [zartes] ~:** a delicate creature

Pflanze ['pflantsə] **die; ~, ~n** plant

pflanzen 1. *tr. V.* plant **(in +** *Akk.* in). **2.** *refl. V. (ugs.)* plant oneself

Pflanzen-: ~fresser der herbivore; **~kunde die** botany *no def. art.*; **~öl das** vegetable oil; **~reich das;** *o. Pl.* plant kingdom; **~schutz·mittel das** [crop] pesticide; *(für den Garten)* garden pesticide

pflanzlich *Adj.* plant *attrib.* ⟨*life, motif*⟩; vegetable ⟨*dye, fat*⟩; *(vegetarisch)* vegetarian

Pflaster ['pflastɐ] **das; ~s, ~ a)** *(Straßen~)* road surface; *(auf dem Gehsteig)* pavement; **b)** *(ugs.: Ort)* **ein teures/gefährliches** *od.* **heißes ~:** an expensive/dangerous place *or* spot to be; **c)** *(Wund~)* sticking-plaster

Pflaster·maler der pavement artist

pflastern *tr. (auch itr.) V.* surface ⟨*road, path*⟩; *(mit Kopfsteinpflaster, Steinplatten)* pave ⟨*street, path*⟩

Pflaster·stein der paving-stone; *(Kopfstein)* cobble-stone

Pflaume ['pflaumə] **die; ~, ~n a)** plum; **getrocknete ~n** [dried] prunes; **b)** *(ugs. abwertend: Versager)* dead loss *(coll.)*

Pflaumen-: ~baum der plum-tree; **~kuchen der** plum flan; **~mus das** plum purée

Pflege ['pfle:gə] **die; ~:** care; *(Maschinen~, Fahrzeug~)* maintenance; *(fig.: von Beziehungen, Kunst, Sprache)* cultivation; fostering; **jmdn./etw. in ~** *(Akk.)* **nehmen** look after sb./sth.; **jmdm. etw.** *od.* **etw. bei jmdm. in ~** *(Akk.)* **geben** give sb. sth. to look after; entrust sth. to sb.'s care; **ein Kind in ~** *(Akk.)* **nehmen** look after a child; *(als Pflegeeltern)* foster a child; **ein Kind bei jmdm. in ~** *(Akk.)* **geben** give sb. a child to look after; *(bei Pflegeeltern)* have a child fostered by sb.

pflege-, Pflege-: ~bedürftig *Adj.* needing care *or* attention *postpos.*; ⟨*person*⟩ in need of care; **~eltern** *Pl.* foster-parents; **~kind das** foster-child; **~leicht** *Adj.* easy-care *attrib.* ⟨*textiles, flooring*⟩; minimum-care *attrib.* ⟨*plant, pan*⟩

pflegen 1. *tr. V.* look after; care for, nurse ⟨*sick person*⟩; care for, take care of ⟨*skin, teeth, floor*⟩; look after, maintain ⟨*bicycle, car, machine*⟩; cultivate ⟨*relations, arts, interests*⟩; foster ⟨*contacts, co-operation*⟩; keep up, pursue ⟨*hobby*⟩; **jmdn./ein Tier gesund ~:** nurse sb./an animal back to health. **2.** *itr. V.; mit Inf. + zu* **etw. zu tun ~:** be in the habit of doing sth.; usually do sth.; **..., wie er zu sagen pflegt/pflegte ...,** as he is wont to say/as he used to say. **3.** *refl. V.* take care of oneself; *(gesundheitlich)* look after oneself; *s. auch* **gepflegt**

Pfleger der; ~s, ~ a) *(Kranken~)* [male] nurse; **b)** *(Tier~)* keeper

Pflegerin die; ~, ~nen a) *(Kranken~)* nurse; **b)** *(Tier~)* keeper

Pflege-: ~sohn der foster-son; **~tochter die** foster-daughter

pfleglich 1. *Adj.* careful. **2.** *adv.* carefully; with care

Pflicht [pflɪçt] **die; ~, ~en a)** duty; **~ sein** be obligatory; **es ist seine ~ und Schuldigkeit** it's his bounden duty; **b)** *(Sport)* compulsory exercises *pl.*

pflicht-, Pflicht-: ~bewußt 1. *Adj.* conscientious; **~bewußt sein** have a sense of duty; **2.** *adv.* conscientiously; with a sense of duty; **~bewußtsein das** sense of duty; **~eifrig 1.** *Adj.* zealous; **2.** *adv.* zealously; full of zeal; **~fach das** compulsory subject; **~gefühl das;** *o. Pl.* sense of duty; **~gemäß 1.** *Adj.* in accordance with one's duty *postpos.*; **2.** *adv.* in accordance with one's duty; **~lektüre die** required reading; *(Schulw.)* set books *pl.*; **~übung die a)** *(Sport)* compulsory exercise; **b)** *(fig.)* ritual exercise; *(Buch, Film usw.)* obligatory effort; **~versicherung die** compulsory insurance

Pflock [pflɔk] **der; ~[e]s, Pflöcke** ['pflœkə] peg; *(für Tiere)* stake

pflücken ['pflʏkn̩] *tr. V.* pick ⟨*flowers, fruit, hops*⟩

Pflücker der; ~s, ~, Pflückerin die; ~, ~nen picker

Pflug [pflu:k] **der; ~[e]s, Pflüge** ['pfly:gə] plough

pflügen ['pfly:gn̩] *tr., itr. V.* plough

Pflug·schar die ploughshare

Pforte ['pfɔrtə] **die; ~, ~n** *(Tor)* gate; *(Tür)* door; *(Eingang)* entrance

Pförtner ['pfœrtnɐ] **der; ~s, ~** porter; *(eines Wohnblocks, Büros)* door-keeper; *(am Tor)* gatekeeper

Pförtner·haus das gatehouse; porter's lodge

Pfosten ['pfɔstn̩] **der; ~s, ~ a)** post; *(Tür~)* jamb; **b)** *(Sport: Tor~)* [goal-]post

Pfötchen ['pfø:tçən] **das; ~s, ~:** [little] paw; **[gib] ~!** [give us a] paw!

Pfote ['pfo:tə] **die; ~, ~n a)** paw; **b)** *(ugs.: Hand)* paw *(coll.)*; mitt *(sl.)*; **sich** *(Dat.)* **die ~n verbrennen** *(fig.)* burn one's fingers *(fig.)*

Pfriem [pfri:m] **der; ~[e]s, ~e** awl

Pfropf [pfrɔpf] **der; ~[e]s, ~e** blockage; *(in der Vene)* clot

pfropfen *tr. V. (ugs.)* cram; stuff; **gepfropft voll** crammed [full]; packed

Pfropfen der *(für Flaschen)* stopper; *(Korken)* cork; *(für Fässer)* bung

Pfründe ['pfrʏndə] **die; ~, ~n a)** *(kath. Kirche)* living; benefice; **b)** *(fig.)* sinecure

pfui [pfʊi] *Interj.* **a)** *(Ekel ausdrükkend)* ugh; yuck *(sl.)*; *(zu Kindern, Hunden)* [ugh,] you mucky pup; **~ Teufel** *od.* **Deibel** *od.* **Spinne!** *(ugs.)* ugh *or (sl.)* yuck, how disgusting!; **b)** *(Mißbilligung, Empörung ausdrückend)* ugh; really; *(Ruf)* boo; **~, schäm dich!** shame on you!; **~ rufen** boo

Pfui·ruf der boo

Pfund [pfʊnt] **das; ~[e]s, ~e** *(bei Maßangaben ungebeugt)* **a)** *(Gewicht)* pound *(= 500 grams in German-speaking countries)*; **zwei ~ Kartoffeln** two pounds of potatoes; **b)** *(Währungseinheit)* pound; **100 ~:** £100; one hundred pounds

pfundig *(ugs.)* **1.** *Adj.* great *(coll.)*; fantastic *(coll.)*. **2.** *adv.* fantastically *(coll.)*

Pfunds·kerl der *(ugs.)* great bloke *(Brit. coll.)*; great guy *(sl.)*

pfund·weise *Adv.* by the pound

Pfusch [pfʊʃ] **der; ~[e]s** *o. Art. (ugs. abwertend)* a botch-up; **~ machen** botch it

pfuschen *itr. V. (ugs. abwertend)* botch it; do a botched-up job

Pfuscher der; ~s, ~, Pfuscherin die; ~, ~nen *(ugs. abwertend)* botcher; bungler

Pfütze ['pfʏtsə] **die; ~, ~n** puddle

PH [pe:'ha:] **die; ~, ~s** *Abk.* **Pädagogische Hochschule**

Phänomen [fɛno'me:n] **das; ~s, ~e** phenomenon

phänomenal [fɛnome'na:l] **1.** *Adj.* phenomenal. **2.** *adv.* phenomenally

Phantasie [fanta'zi:] **die; ~, ~n a)** *o. Pl.* imagination; **eine schmutzige ~ haben** have a dirty mind; **b)** *meist Pl. (Produkt der ~)* fantasy

phantasie·los 1. *Adj.* unimaginative. **2.** *adv.* unimaginatively

Phantasielosigkeit die; ~: lack of imagination; *(Eintönigkeit)* dullness

phantasieren 1. *itr. V.* **a)** indulge in fantasies, fantasize **(von** about); **b)** *(Med.: irrereden)* talk deliriously. **2.** *tr. V.* **was phantasierst du da?** what's all that nonsense?

phantasie·voll 1. *Adj.* imaginative. **2.** *adv.* imaginatively

phantastisch 1. *Adj.* **a)** fantastic; ‹*idea*› divorced from reality; **b)** *(ugs.: großartig)* fantastic *(coll.)*; terrific *(coll.)*. **2.** *adv.* *(ugs.)* fantastically *(coll.)*; **~ tanzen** *(ugs.)* dance fantastically *(coll.) or* incredibly well

Phantom [fan'to:m] **das; ~s, ~e** phantom; illusion; **einem ~ nachjagen** *(fig.)* chase [after] an illusion *or* a shadow

Pharisäer [fari'zɛ:ɐ] **der; ~s, ~** *(auch fig.)* Pharisee

Pharmakologie [farmakolo'gi:] **die; ~:** pharmacology *no art.*

pharmazeutisch 1. *Adj.* pharmaceutical. **2.** *adv.* pharmaceutically

Pharmazie [farma'tsi:] **die; ~:** pharmaceutics *sing., no art.*; pharmaceutical chemistry *no art.*

Phase ['fa:zə] **die; ~, ~n** phase

Philanthrop [filan'tro:p] **der; ~en, ~en** *(geh.)* philanthropist

Philharmonie [fɪlharmo'ni:] **die; ~, ~n a)** *(Orchester)* philharmonic [orchestra]; **b)** *(Gebäude, Saal)* philharmonic hall

Philharmoniker [fɪlhar'mo:nikɐ] **der; ~s, ~:** member of a/the philharmonic orchestra; **die Wiener ~:** the Vienna Philharmonic Orchestra

Philippinen [fɪlɪ'pi:nən] *Pl.* Philippines

Philologe [filo'lo:gə] **der; ~n, ~n** teacher/student of language and literature; philologist *(Amer.)*

Philologie die; ~, ~n study of language and literature; philology *no art. (Amer.)*

Philologin die; ~, ~nen *s.* **Philologe**

Philosoph [filo'zo:f] **der; ~en, ~en** philosopher

Philosophie die; ~, ~n philosophy

philosophieren *itr. (auch tr.) V.* philosophize

Philosophin die; ~, ~nen philosopher

philosophisch 1. *Adj.* philosophical; ‹*dictionary*› of philosophy. **2.** *adv.* philosophically

phlegmatisch [flɛg'matɪʃ] **1.** *Adj.* phlegmatic. **2.** *adv.* phlegmatically

Phobie [fo'bi:] **die; ~, ~n** *(Psych.)* phobia

Phonetik [fo'ne:tɪk] **die; ~:** phonetics *sing.*

phonetisch 1. *Adj.* phonetic. **2.** *adv.* phonetically

Phono- ['fo:no-] phono ‹*socket, input*›

Phono·typistin [-ty'pɪstɪn] **die; ~, ~nen** audio typist

Phosphat [fɔs'fa:t] **das; ~[e]s, ~e** *(Chemie)* phosphate

Phosphor ['fɔsfɔr] **der; ~s** phosphorus

Photo ['fo:to] **das; ~s, ~s** *s.* **Foto**

photo-, Photo- *s. auch* **foto-, Foto-**

Photo·zelle die photo[-electric] cell

Phrase ['fra:zə] **die; ~, ~n a)** *(abwertend)* [empty] phrase; cliché; **~n dreschen** *(ugs.)* spout clichés; dole out catch-phrases; **b)** *(Musik, Sprachw.)* phrase

phrasen-, Phrasen-: ~drescher der; ~s, ~ *(ugs. abwertend)* phrase-monger; cliché-monger; **~drescherei** [~drɛʃə'rai] **die; ~, ~en** *(ugs. abwertend)* phrase-mongering; cliché-mongering; **~haft** *(abwertend)* **1.** *Adj.* empty; trite; *(voller Klischees)* cliché-ridden; **2.** *adv.* in an empty *or* trite manner

pH-Wert [pe:'ha:-] **der** *(Chemie)* pH[-value]

Physik [fy'zi:k] **die; ~:** physics *sing., no art.*

physikalisch [fyzi'ka:lɪʃ] **1.** *Adj.* physics *attrib.* ‹*experiment, formula, research, institute*›; physical ‹*map, chemistry, therapy, process*›. **2.** *adv.* in terms of physics

Physiker der; ~s, ~, Physikerin die; ~, ~nen physicist

Physik·saal der *(Schulw.)* physics laboratory

Physiologie [fyziolo'gi:] **die; ~:** physiology

physiologisch 1. *Adj.* physiological. **2.** *adv.* physiologically

physisch 1. *Adj.* physical. **2.** *adv.* physically

Pianist [pi̯a'nɪst] **der; ~en, ~en, Pianistin die; ~, ~nen** pianist
Piano ['pi̯a:no] **das; ~s, ~s** piano
Pickel ['pɪkl̩] **der; ~s, ~ a)** *(auf der Haut)* pimple; **b)** *(Spitzhacke)* pickaxe; *(Eis~)* ice-axe
Pickel·haube die spiked helmet
pickelig *Adj.* pimply
picken ['pɪkn̩] **1.** *itr. V.* peck (**nach** at; **an** + *Akk.*, **gegen** on, against). **2.** *tr. V.* ‹*bird*› peck; *(ugs.)* ‹*person*› pick; *(aufheben)* pick up
picklig *s.* **pickelig**
Picknick ['pɪknɪk] **das; ~s, ~e** *od.* **~s** picnic; **~ machen** *od.* **halten** have a picnic
picknicken *itr. V.* picnic
pieken *s.* **piken**
piek·fein ['pi:k'fai̯n] *(ugs.)* **1.** *Adj.* posh *(coll.).* **2.** *adv.* poshly *(coll.)*; **~ angezogen** wearing posh clothes *(coll.)*; dressed to the nines
piep [pi:p] *Interj.* cheep
Piep der; ~s, ~e *(ugs.)* **a)** *(Ton)* peep; **keinen ~ [davon] sagen** not say a thing [about it]
piepe ['pi:pə], **piep·egal** *Adj.* in **[jmdm.] ~ sein** *(ugs.)* not matter at all [to sb.]; **es ist mir ~** *(ugs.)* I don't give a damn
piepen *itr. V. (ugs.)* squeak; ‹*small bird*› cheep; chirp; **bei dir piept's wohl** *(salopp)* you must be off your rocker *(sl.)*; **zum Piepen sein** be a hoot *or* a scream *(coll.)*
Piep·matz der *(Kinderspr.)* dicky-bird *(coll.)*
piepsen *(ugs.)* **1.** *itr. V. s.* **piepen. 2.** *itr., tr. V. (mit hoher Stimme sprechen)* pipe; *(aufgeregt)* squeal
piepsig *Adj. (ugs.)* squeaky
piesacken ['pi:zakn̩] *tr. V. (ugs.)* pester
Pietät [pi̯e'tɛ:t] **die; ~:** respect; *(Ehrfurcht)* reverence
pietät·los 1. *Adj.* irreverent; *(gefühllos)* unfeeling; *(respektlos)* disrespectful; lacking in respect *postpos.* **2.** *adv.* irreverently
Pigment [pɪ'gmɛnt] **das; ~[e]s, ~e** pigment
[1]**Pik** [pi:k] **der** *in* **einen ~ auf jmdn. haben** *(ugs.)* have it in for sb.
[2]**Pik das; ~[s], ~[s]** *(Kartenspiel)* **a)** *(Farbe)* spades *pl.*; **~ ziehen/ausspielen** draw/play spades; **b)** *(Karte)* spade
pikant [pi'kant] **1.** *Adj.* **a)** piquant; *(würzig)* spicy; well-seasoned; *(appetitanregend)* appetizing; **b)** *(fig.: witzig)* piquant; ironical; **c)** *(verhüll.: schlüpfrig)* racy ‹*joke, story*›. **2.** *adv.* **~ gewürzt** piquantly *or* appetizingly seasoned
Pik·as das ace of spades
Pike ['pi:kə] **die; ~, ~n** pike; **etw. von der ~ auf [er]lernen** learn sth. by working one's way up from the bottom
piken *tr., itr. V. (ugs.)* prick; **jmdm. mit einer Nadel in den Arm ~:** poke a needle into sb.'s arm
pikiert [pi'ki:ɐ̯t] **1.** *Adj.* piqued; nettled. **2.** *adv.* ‹*reply, say*› in an aggrieved tone *or* voice
Pikkolo·flöte die piccolo
piksen ['pi:ksn̩] *tr., itr. V. (ugs.) s.* **piken**
Pik·sieben die *(Kartenspiel)* seven of spades; **dastehen wie ~** *(ugs.)* stand there looking stupid
Pilger ['pɪlgɐ] **der; ~s, ~:** pilgrim
Pilger·fahrt die pilgrimage
pilgern *itr. V.* **a)** *(auch fig.)* go on *or* make a pilgrimage; **b)** *(ugs.: gehen)* traipse *(coll.)*
Pille ['pɪlə] **die; ~, ~n** pill; **sie nimmt die ~:** she's on the pill *(coll.)*; **eine bittere ~ [für jmdn.] sein** *(fig.)* be a bitter pill [for sb.] to swallow
Pilot [pi'lo:t] **der; ~en, ~en a)** pilot; **b)** *(Motorsport)* [racing] driver
Pils [pɪls] **das; ~, ~:** Pils; Pils[e]ner [beer]
Pilz [pɪlt̮s] **der; ~es, ~e a)** fungus; *(Speise~, auch fig.)* mushroom; **giftige ~e** poisonous fungi; **wie ~e aus dem Boden** *od.* **der Erde schießen** be springing up like mushrooms; **b)** *o. Pl. (ugs.: ~infektion)* fungus [infection]
Pilz-: ~krankheit die a) *(Mykose)* mycosis; **b)** *(bei Pflanzen)* fungus [disease]; **~vergiftung die** fungus poisoning *no art.*; *(durch verdorbene Pilze)* mushroom poisoning *no art.*
Pimmel ['pɪml̩] **der; ~s, ~** *(salopp)* willy *(sl.)*
pingelig ['pɪŋəlɪç] *(ugs.)* **1.** finicky; pernickety *(coll.)*; *(wählerisch)* fussy; choosy *(coll.).* **2.** *adv.* in a pernickety way *(coll.)*; *(pedantisch)* pedantically
Pingpong ['pɪŋpɔŋ] **das; ~s** *(ugs.)* ping-pong
Pinguin ['pɪŋgui:n] **der; ~s, ~e** penguin
Pinie ['pi:ni̯ə] **die; ~, ~n** [stone- *or* umbrella] pine
Pinke ['pɪŋkə] **die; ~** *(ugs. veralt.)* dough *(sl.)*; lolly *(Brit. sl.)*
Pinkel ['pɪŋkl̩] **der; ~s, ~** *(ugs. abwertend)* **ein [feiner] ~:** a stuck-up prig
pinkeln *itr. (auch tr.) V. (salopp)* pee *(coll.)*; ‹*esp. child*› wee *(sl.)*
Pinkel·pause die *(ugs.)* stop for a pee *(coll.)*; rest stop *(Amer.)*
Pinn·wand die pin-board
Pinscher ['pɪnʃɐ] **der; ~s, ~** pinscher
Pinsel ['pɪnzl̩] **der; ~s, ~ a)** brush; *(Mal~)* paintbrush; **b)** *(ugs. abwertend: Dummkopf)* nitwit *(coll.)*; idiot *(coll.)*
pinseln *tr. V.* **a)** *(ugs.: anstreichen)* paint ‹*room, house, etc.*›; **b)** *(malen)* paint ‹*landscape, picture*›; daub ‹*slogans*›; **c)** *(Med.: ein~)* paint ‹*wound, gums, throat, etc.*›
Pinte ['pɪntə] **die; ~, ~n** *(ugs.) s.* **Kneipe a**
Pinzette [pɪn'tsɛtə] **die; ~, ~n** tweezers *pl.*
Pionier [pi̯o'ni:ɐ̯] **der; ~s, ~e a)** *(Milit.)* sapper; engineer; **b)** *(fig.: Wegbereiter)* pioneer; **c)** *(bes. ehem. DDR)* **[Junger] ~:** [Young] Pioneer
Pipapo [pipa'po:] **das; ~s** *(salopp)* **mit allem ~:** with all the frills
Pipeline ['pai̯plai̯n] **die; ~, ~s** pipeline
Pipi [pi'pi:] **das; ~s** *(Kinderspr.)* **~ machen** do wee-wees *(sl.)*
Pirat [pi'ra:t] **der; ~en, ~en** pirate
Piraten·sender der pirate radio station
Piraterie [piratə'ri:] **die; ~, ~n** piracy *no art.*
Pirouette [pi'ru̯ɛtə] **die; ~, ~n** pirouette
Pirsch [pɪrʃ] **die; ~** *(Jägerspr.)* [deer-] stalking; **auf die ~ gehen** go [deer-]stalking
pirschen 1. *itr. V.* **a)** *(Jägerspr.)* stalk; go stalking; **b)** *(ugs.: schleichen)* creep [silently]; steal. **2.** *refl. V. (ugs.)* creep [silently]; steal
Pisse ['pɪsə] **die; ~** *(derb)* piss *(coarse)*
pissen *itr. (auch tr.) V. (derb)* piss *(coarse)*
Pistazie [pɪs'ta:tsi̯ə] **die; ~, ~n** pistachio
Piste ['pɪstə] **die; ~, ~n a)** *(Skisport)* piste; ski-run; *(Renn~)* course; **b)** *(Rennstrecke)* track; **c)** *(Flugw.)* runway
Pistole [pɪs'to:lə] **die; ~, ~n** pistol; **wie aus der ~ geschossen** like a shot *or* a flash; **jmdm. die ~ auf die Brust setzen** *(fig.)* hold a pistol to sb.'s head
pitsch·naß ['pɪtʃ'nas] *Adj. (ugs.)* dripping wet; wet through
Pizza ['pɪtsa] **die; ~, ~s** *od.* **Pizzen** pizza
Pizzeria [pɪtse'ri:a] **die; ~, ~s** *od.* **Pizzerien** pizzeria
Pkw, PKW ['pe:ka:ve:] **der; ~[s], ~[s]** [private] car; automobile *(Amer.)*

placken ['plakn̩] *refl. V. (ugs.)* slave away

plädieren [plɛ'di:rən] *itr. V. (Rechtsw.)* plead (**auf** + *Akk.*, **für** for); *(fig.)* argue (**für** for, in favour of)

Plädoyer [plɛdo̯a'je:] **das; ~s, ~s** *(Rechtsw.)* final speech, summing up *(for the defence/prosecution)*; *(fig.)* plea

Plage ['pla:gə] **die; ~, ~n a)** [cursed *or (coll.)* pestilential] nuisance; **b)** *(ugs.: Mühe)* bother; trouble; **seine ~ mit jmdm./etw. haben** find sb./sth. a real handful

plagen 1. *tr. V.* **a)** torment; plague; **b)** *(ugs.: bedrängen)* harass; *(mit Bitten, Fragen)* pester. **2.** *refl. V.* **a)** *(sich abmühen)* slave away; **b)** *(leiden)* **sich mit etw. ~:** be troubled *or* bothered by sth.

Plagiat [pla'gi̯a:t] **das; ~[e]s, ~e** plagiarism *no art.*

Plakat [pla'ka:t] **das; ~[e]s, ~e** poster; **„~e ankleben verboten"** 'post no bills'

Plakat·wand die [poster] hoarding; billboard

Plakette [pla'kɛtə] **die; ~, ~n** badge; *(Scheibe)* disc

[1]**Plan der; ~[e]s, Pläne** ['plɛ:nə] **a)** plan; **nach ~ verlaufen** go according to plan; **b)** *(Karte)* map; plan; *(Stadt~)* [street] plan

[2]**Plan der** *in* **auf den ~ treten** appear on the scene; **auf den ~ rufen** bring ‹*person*› on to the scene; bring ‹*opponent*› into the arena; arouse ‹*curiosity*›

Plane die; ~, ~n tarpaulin

planen *tr., itr. V.* plan

Planet [pla'ne:t] **der; ~en, ~en** planet

planieren *tr. V.* level; grade *(as tech. term)*

Planier·raupe die bulldozer

Planke ['plaŋkə] **die; ~, ~n** plank; board

Plänkelei die; ~, ~en *s.* **Geplänkel a**

Plankton ['plaŋktɔn] **das; ~s** *(Biol.)* plankton

plan-, Plan-: ~los 1. *Adj.* aimless; *(ohne System)* unsystematic; **2.** *adv. s.* **1:** aimlessly; unsystematically; **~mäßig 1.** *Adj.* **a)** regular, scheduled ‹*service, steamer*›; **~mäßige Ankunft/Abfahrt** scheduled time of arrival/departure; **b)** *(systematisch)* systematic. **2.** *adv.* **a)** *(wie geplant)* according to plan; as planned; *(pünktlich)* on schedule; **b)** *(systematisch)* systematically; **~quadrat das** grid square

Plansch·becken ['planʃ-] **das** paddling-pool

planschen *itr. V.* splash [about]

Plantage [plan'ta:ʒə] **die; ~, ~n** plantation

Planung die; ~, ~en a) *(das Planen)* planning; **bei der ~:** at the planning stage; **b)** *(Plan)* plan

Plan-: ~wagen der covered wagon; **~wirtschaft die** planned economy

plappern ['plapɐn] *(ugs.)* **1.** *itr. V.* chatter. **2.** *tr. V.* babble ‹*nonsense*›

plärren ['plɛrən] **1.** *tr. V.* bawl [out] ‹*song*›; ‹*radio etc.*› blare out. **2.** *itr. V.* **a)** bawl; yell; ‹*radio etc.*› blare; **b)** *(ugs.: weinen)* wail

Plasma ['plasma] **das; ~s, Plasmen** *(Med., Physik)* plasma; *(Proto~)* protoplasm

Plast [plast] **der; ~[e]s, ~e** *(regional)* plastic

[1]**Plastik** ['plastɪk] **die; ~, ~en** sculpture

[2]**Plastik das; ~s** *(ugs.)* plastic

Plastik-: ~beutel der plastic bag; **~tüte die** plastic bag

plastisch 1. *Adj.* **a)** *(knetbar)* plastic; workable; **b)** *nicht präd. (bildhauerisch)* sculptural; ‹*ability*› as a sculptor; **c)** *(dreidimensional)* three-dimensional ‹*effect, formation, vision*›; sculptural ‹*decoration*›; **d)** *(fig.: anschaulich)* vivid ‹*description, picture*›; **e)** *(Med.)* plastic ‹*surgery, surgeon*›. **2.** *adv.* **a)** *(bildhauerisch)* sculpturally; **b)** *(dreidimensional)* three-dimensionally; **c)** *(fig.: anschaulich)* vividly; **sich** *(Dat.)* **etw. ~ vorstellen können** have a clear picture of sth. [in one's mind]

Platane [pla'ta:nə] **die; ~, ~n** plane-tree

Platin ['pla:ti:n] **das; ~s** platinum

Platitüde [plati'ty:də] **die; ~, ~n** *(geh.)* platitude

Platon ['pla:tɔn] **(der)** Plato

platonisch [pla'to:nɪʃ] **1.** *Adj.* **a)** Platonic ‹*philosophy, state*›; **b)** *(nicht sinnlich)* platonic ‹*love, relationship*›. **2.** *adv.* platonically

platsch [platʃ] *Interj.* splash

platschen *itr. V.* **a)** splash; **b)** *mit sein (~d schlagen)* splash (**an** + *Akk.*, **gegen** against); **c)** *(planschen)* splash about

plätschern ['plɛtʃɐn] *itr. V.* **a)** splash; ‹*rain*› patter; ‹*stream*› burble; **b)** *(planschen)* splash about; **c)** *mit sein* ‹*stream*› burble along

platt [plat] *Adj.* **a)** *(flach)* flat; **ein Platter** *(ugs.)* a flat *(coll.)*; a flat tyre; *(mit Loch)* a puncture; **sie ist ~ wie ein [Bügel]brett** *(salopp)* she is flat-chested; **b)** *(geistlos)* dull, vapid ‹*conversation, book*›; vacuous, feeble ‹*poem, joke*›; shallow, empty ‹*materialism, argument, imitation*›; **c)** *nicht präd. (ausgesprochen)* downright ‹*lie, swindle, slander*›; sheer ‹*cynicism*›; **d)** *nicht attr. (ugs., bes. nordd.: erstaunt)* dumbfounded; flabbergasted

Platt das; ~[s] [local] Low German dialect; *(allgemein: Niederdeutsch)* Low German

platt·deutsch *Adj.: s.* **niederdeutsch**

Platte die; ~, ~n a) *(Stein~)* slab; *(Metall~)* plate; sheet; *(Mikroskopie usw.: Glas~)* slide; *(Paneel)* panel; *(Span~, Hartfaser~ usw.)* board; *(Styropor~ usw.)* sheet; *(Tisch~)* [table-] top; *(Grab~)* [memorial] slab; *(photographische ~)* [photographic] plate; *(Druck~)* [pressure] plate; *(Kachel, Fliese)* tile; *(zum Pflastern)* flagstone; paving-stone; **b)** *(Koch~)* hotplate; **c)** *(Schall~)* [gramophone] record; **etw. auf ~** *(Akk.)* **aufnehmen** make a record of sth.; **die ~ kenne ich [schon]** *(fig. ugs.)* I've heard that one before; **d)** *(Teller)* plate; *(zum Servieren, aus Metall)* dish; **e)** *(Speise)* dish; **kalte ~:** selection of cold meats [and cheese]

Platten-: ~cover das record sleeve; **~firma die** record company; **~hülle die** record sleeve; **~sammlung die** record collection; **~spieler der** *(als Baustein)* record deck; *(komplettes Gerät)* record-player; **~teller der** turntable

Platt-: ~form die a) platform; **b)** *(fig.: Basis)* basic programme; **~fuß der a)** flat foot; **b)** *(ugs.: Reifenpanne)* flat *(coll.)*; flat tyre

Plattheit die; ~, ~en a) *o. Pl.* flatness; **b)** *o. Pl. (fig.)* dullness; **c)** *(Platitüde)* platitude

Platz [plats] **der; ~es, Plätze** ['plɛtsə] **a)** *(freie Fläche)* space; area; *(Bau~, Ausstellungsgelände usw.)* site; *(umbaute Fläche)* square; **b)** *(Park~)* car park; [parking] lot *(Amer.)*; **c)** *(Sport~) (ganze Anlage)* ground; *(Spielfeld)* field; *(Tennis~, Volleyball~ usw.)* court; *(Golf~)* course; **einen Spieler vom ~ stellen/tragen** send/carry a player off [the field]; **d)** *(Stelle)* place; spot; *(Position)* location; position; *(wo jmd., etw. hingehört)* place; **auf die Plätze, fertig, los!** on your marks, get set, go!; **nicht** *od.* **fehl am ~[e] sein** *(fig.)* be out of place; be inappropriate; **am ~[e] sein** *(fig.)* be appropriate; be called for; **e)** *(Sitz~)* seat; *(am Tisch, Steh~ usw.; fig.: im Kurs, Krankenhaus, Kindergarten usw.)* place; ~ **nehmen** sit down; **nehmen Sie ~!** take a seat; ~ **behalten** *(geh.)* remain seated; **f)** *(bes. Sport: Plazierung)* place; **den dritten ~ belegen** come third; **g)** *(Ort)* place; locality; **am ~:** in the town/village; **das größte Hotel am ~:** the largest hotel in the place; **h)** *o. Pl. (Raum)* space; room; **er/es hat [noch] ~/keinen ~:** there is enough space *or* room [left] for him/it/no room for him/it; **der Saal bietet ~** *od.* **hat ~ für 3 000 Personen** the hall takes *or* holds 3,000 people; **im Viktoriasee hätte ganz Irland ~:** the whole of Ireland could fit into Lake Victoria; **[jmdm./einer Sache] ~ machen** make room [for sb./sth.]; ~ **da!** make way!; out of the way!

Platz-: ~angst die *(volkst.)* claustrophobia; agoraphobia; **~anweiser der; ~s, ~:** usher; **~anweiserin die; ~, ~nen** usherette

Plätzchen ['plɛtsçən] **das; ~s, ~ a)** little place *or* spot; *(kleiner Raum)* little space; **b)** *(Keks)* biscuit *(Brit.)*; cookie *(Amer.)*; *(Schokoladen~)* [chocolate] pastille

platzen *itr. V.; mit sein* **a)** burst; *(explodieren)* explode; **ihm war eine Augenbraue geplatzt** one of his eyebrows had split open; **vor Wut** *(Dat.)* ~ *(fig.)* be bursting with rage; **b)** *(ugs.: scheitern)* fall through; **geplatzt sein** ‹*concert, meeting, performance, holiday, engagement*› be off; **der Wechsel ist geplatzt** the bill has bounced *(sl.)*; **etw. ~ lassen** put the kibosh on sth. *(sl.)*; **c)** *(ugs.: hinein~)* **in eine Versammlung ~:** burst into a meeting

platz, Platz-: ~hirsch der a) *(Jägerspr.)* dominant stag; **b)** *(fig. ugs.: beherrschende Figur)* boss-type *(coll.)*; **er ist hier der ~hirsch** he's the big noise around here *(sl.)*; **~karte die** reserved-seat ticket; **~konzert das** open-air concert; **~mangel der** lack of space; **~patrone die** blank [cartridge]; **~regen der** downpour; cloudburst; **~sparend 1.** *Adj.* space-saving; **2.** *adv.* economically; in a space-saving manner; **~verweis der** *(Sport)* sending-off; **~vorteil der;** *o. Pl. (Sport)* home advantage; **~wart der** *(Sport)* groundsman; **~wunde die** lacerated wound

Plauderei die; ~, ~en chat

plaudern ['plaudɐn] *itr. V.* **a)** chat **(über +** *Akk.*, **von** about); **b)** *(etw. aus~)* let on *(sl.)*

plausibel [plau'zi:bl̩] **1.** *Adj.* plausible; **jmdm. etw. ~ machen** make sth. seem convincing to sb. **2.** *adv.* plausibly

Playback ['pleɪbæk] **das; ~s, ~s** pre-recorded version; recording; *(Begleitung)* [pre-recorded] backing; *(im Fernsehen)* miming to a recording

Play·boy ['pleɪbɔɪ] **der** playboy

plazieren [pla'tsi:rən] **1.** *tr. V.* **a)** place; position ‹*loudspeakers*›; **b)** *(Sport: gezielt werfen, schlagen usw.)* place ‹*shot, ball*›; *(Boxen, Fechten)* land ‹*blow, hit*›. **2.** *refl. V.* **a)** *(sich setzen)* place *or* seat oneself **(auf +** *Akk. od. Dat.* on); *(sich stellen)* take up position **(an +** *Akk. od. Dat.* at, by); **b)** *(Sport)* be placed; **er konnte sich nicht ~:** he was unplaced

Plazierung die; ~, ~en *(Sport)* placing; place

pleite ['plaitə] *(ugs.) in* ~ **sein** ‹*person*› be broke *(coll.)*; ‹*company*› have gone bust *(coll.)*; ~ **gehen** go bust *(coll.)*

Pleite die; ~, ~n *(ugs.)* **a)** bankruptcy *no def. art.*; **vor der ~ stehen** be faced with bankruptcy; ~ **machen** go bust *(coll.)*; **b)** *(Mißerfolg)* flop *(sl.)*; wash-out *(sl.)*

Plenum ['ple:nʊm] **das; ~s** *(Versammlung)* plenary meeting; *(Sitzung)* plenary session

Pleuel·stange ['plɔyəl-] **die** *(Technik)* connecting-rod

Plexi·glas Ⓦⓩ ['plɛksigla:s] **das;** *o. Pl.* ≈ Perspex **(P)**

Plissee [plɪ'se:] **das; ~s, ~s a)** *(Falten)* accordion pleats *pl.*; **b)** *(Stoff)* accordion-pleated material

Plissee·rock der accordion-pleated skirt

Plombe ['plɔmbə] **die; ~, ~n a)** *(Siegel)* [lead] seal; **b)** *(veralt.: Zahnfüllung)* filling

plombieren *tr. V.* **a)** *(versiegeln)* seal; **b)** *(veralt.)* fill ‹*tooth*›

plötzlich ['plœtslɪç] **1.** *Adj.* sudden. **2.** *adv.* suddenly; **..., aber etwas** *od.* **ein bißchen ~** *(salopp)* ..., and jump to it; ..., and make it snappy *(coll.)*

Pluder·hose ['plu:dɐ-] **die** pantaloons *pl.*; *(orientalischer Art)* Turkish trousers *pl.*

plump [plʊmp] **1.** *Adj.* **a)** *(dick)* plump; podgy; massive ‹*stone, lump*›; *(unförmig)* ungainly, clumsy ‹*shape*›; *(rundlich)* bulbous; **b)** *(schwerfällig)* awkward, clumsy ‹*movements, style*›; **c)** *(abwertend: dreist)* crude, blatant ‹*lie, deception, trick*›; *(leicht durchschaubar)* blatantly obvious; *(unbeholfen)* clumsy ‹*excuse, advances*›; crude ‹*joke, forgery*›. **2.** *adv.* **a)** *(schwerfällig)* clumsily; awkwardly; **b)** *(abwertend: dreist)* in a blatantly obvious manner

Plumpheit die; ~, ~en a) *o. Pl. (Dicke)* plumpness; podginess; *(Unförmigkeit)* ungainliness; clumsiness; *(Rundlichkeit)* bulbousness; **b)** *o. Pl. (Schwerfälligkeit)* clumsiness; awkwardness; *(eines dicken Menschen)* ponderousness; **c)** *o. Pl. (abwertend: Dreistigkeit)* blatant nature; *(primitive Art)* crudity; clumsiness

plumps *Interj.* bump; thud; *(ins Wasser)* splash

Plumps der; ~es, ~e *(ugs.)* bump; thud; *(ins Wasser)* splash

plumpsen *itr. V.* fall with a bump; thud; *(ins Wasser)* splash

Plumps·klo das *(ugs.)* earth-closet

plump-vertraulich 1. *Adj.* over-familiar. **2.** *adv.* with excessive familiarity

Plunder ['plʊndɐ] **der; ~s** *(ugs. abwertend)* junk; rubbish

plündern ['plʏndɐn] *itr., tr. V.* **a)** loot; plunder, pillage ‹*town*›; **b)** *(scherzh.: [fast] leeren)* raid ‹*larder, fridge, account*›; ‹*bird, animal*› strip ‹*tree, border*›

Plünderung die; ~, ~en looting; *(einer Stadt)* plundering; **~en** cases of looting/plundering

Plural ['plu:ra:l] **der; ~s, ~e a)** *o. Pl.* plural; **b)** *(Wort)* word in the plural; plural form; **im ~ stehen** be [in the] plural

plus [plʊs] **1.** *Konj. (Math.)* plus. **2.** *Adv.* **a)** *(bes. Math.)* plus; **b)** *(Elektrot.)* positive. **3.** *Präp. mit Dat. (Kaufmannsspr.)* plus

Plus das; ~ a) *(Überschuß)* surplus; *(auf einem Konto)* credit balance; *(Gewinn)* profit; **im ~ sein** be in credit; **b)** *(Vorteil)* advantage; [extra] asset; **das ist ein ~ für dich** it's a point in your favour

Plüsch [ply:ʃ] **der; ~[e]s, ~e** plush

Plüsch·tier das cuddly toy

Plus·pol der positive pole; *(einer Batterie)* positive terminal

Plusquam·perfekt ['plʊskvampɛrfɛkt] **das** pluperfect [tense]

Plus·zeichen das plus sign

pneumatisch *(Technik)* **1.** *Adj.* pneumatic. **2.** *adv.* pneumatically

Po [po:] **der; ~s, ~s** *(ugs.)* bottom

Pöbel ['pø:bl̩] **der; ~s** *(abwertend)* rabble

pöbelhaft *(abwertend)* **1.** *Adj.* loutish; uncouth. **2.** *adv.* in a loutish manner
pochen ['pɔxn̩] *itr. V.* **a)** *(meist geh.: klopfen)* knock **(gegen** at, on); *(kräftig)* rap; thump; **es pocht** somebody is knocking at *or* on the door; **b)** *(geh.: sich berufen)* **auf etw.** *(Akk.)* **~:** insist on sth.; **c)** *(geh.: pulsieren)* ⟨*heart*⟩ pound, thump; ⟨*blood*⟩ pound, throb
Pocken ['pɔkən] *Pl.* smallpox *sing.*
Pocken·schutz·impfung die smallpox vaccination
pockig *Adj.* pock-marked ⟨*face, surface*⟩; pimpled ⟨*leather*⟩
Podest [po'dɛst] **das** *od.* **der; ~[e]s, ~e** rostrum
Podium ['po:di̯ʊm] **das; ~s, Podien a)** *(Plattform)* platform; *(Bühne)* stage; **b)** *(trittartige Erhöhung)* rostrum; podium
Podiums·diskussion die panel discussion
Poesie [poe'zi:] **die; ~, ~n a)** *o. Pl.* poetry; **b)** *(Gedicht)* poem
Poesie·album das autograph album *(with verses or sayings contributed by friends)*
Poet [po'e:t] **der; ~en, ~en** *(veralt.)* poet; bard *(literary)*
poetisch 1. *Adj.* poetic[al]. **2.** *adv.* poetically
Pogrom [po'gro:m] **das** *od.* **der; ~s, ~e** pogrom
Pointe ['po̯ɛ̃:tə] **die; ~, ~n** *(eines Witzes)* punch line; *(einer Geschichte)* point; *(eines Sketches)* curtain line
pointiert [po̯ɛ̃'ti:ɐ̯t] **1.** *Adj.* pointed ⟨*remark*⟩. **2.** *adv.* pointedly
Pokal [po'ka:l] **der; ~s, ~e a)** *(Trinkgefäß)* goblet; **b)** *(Siegestrophäe, ~wettbewerb)* cup
Pokal-: ~sieger der *(Sport)* cup-winners *pl.*; **~spiel das** *(Sport)* cup-tie
Pökel·fleisch das salt meat
pökeln ['pø:kl̩n] *tr. V.* salt
Poker ['po:kɐ] **das** *od.* **der; ~s** poker; *(fig.)* manoeuvrings *pl.*
Poker·gesicht das poker-face
pokern *itr. V.* play poker; *(fig.)* **um etw. ~:** bid for sth.
Pol [po:l] **der; ~s, ~e** pole; **der ruhende ~** *(fig.)* the calming influence
polar [po'la:ɐ̯] *Adj.* polar
Polar-: ~kreis der polar circle; **nördlicher/südlicher ~kreis** Arctic/Antarctic Circle; **~licht das;** *Pl.* **~lichter** aurora; polar lights *pl.*; **~stern der** polar star; pole-star
Pole der; ~n, ~n Pole
Polemik [po'le:mɪk] **die; ~, ~en** polemic
polemisch 1. *Adj.* polemic[al]. **2.** *adv.* polemically
polemisieren *itr. V.* polemize
Polen (das); ~s Poland
Polente [po'lɛntə] **die; ~** *(salopp)* cops *pl.* *(coll.)*
Police [po'li:sə] **die; ~, ~n** *(Versicherungsw.)* policy
Polier [po'li:ɐ̯] **der; ~s, ~e** [site] foreman
polieren *tr. V.* polish; **jmdm. die Fresse ~** *(derb)* smash sb.'s face in
Poli·klinik die out-patients' department *or* clinic
Polin die; ~, ~nen Pole
Politesse [poli'tɛsə] **die; ~, ~n** [woman] traffic warden
Politik [poli'ti:k] **die; ~, ~en a)** *o. Pl.* politics *sing., no art.*; **b)** *(eine spezielle ~)* policy
Politiker [po'li:tikɐ] **der; ~s, ~, Politikerin die; ~, ~nen** politician
politisch 1. *Adj.* **a)** political; **b)** *(klug u. berechnend)* politic. **2.** *adv.* **a)** politically; **b)** *(klug u. berechnend)* politicly; judiciously
politisieren 1. *itr. V.* talk politics; politicize. **2.** *tr. V.* politicize; *(politisch aktivieren)* make politically active
Politisierung die; ~: politicization
Politologe [polito'lo:gə] **der; ~n, ~n, Politologin die; ~, ~nen** political scientist
Politur [poli'tu:ɐ̯] **die; ~, ~en** polish
Polizei [poli'tsai̯] **die; ~, ~en a)** police *pl.*; **er ist** *od.* **arbeitet bei der ~:** he is in the police force; **b)** *o. Pl.* *(Dienststelle)* police station
Polizei-: ~auto das police car; **~beamte der** police officer; **~chef der** chief of police; chief constable *(Brit.)*; **~einsatz der** police operation; **~funk der** police radio; **~hund der** police dog; **~kontrolle die** police check
polizeilich 1. *Adj., nicht präd.* police; **~e Meldepflicht** obligation to register with the police. **2.** *adv.* by the police
Polizei-: ~präsident der *s.* **~chef; ~präsidium das** police headquarters *sing. or pl.*; **~revier das** police station; **~schutz der** police protection; **~spitzel der** police informer; **~staat der** police state; **~streife die** police patrol; **~stunde die** closing time; **~wache die** police station
Polizist [poli'tsɪst] **der; ~en, ~en, Polizistin die; ~, ~nen** policeman/policewoman
Pollen ['pɔlən] **der; ~s, ~** *(Bot.)* pollen
polnisch ['pɔlnɪʃ] *Adj.* Polish; *s. auch* **deutsch; Deutsch**
Polo·hemd das short-sleeved shirt
Polster ['pɔlstɐ] **das; ~s, ~: a)** upholstery *no pl., no indef. art.*; **b)** *(Rücklage)* reserves *pl.*
Polster·möbel *Pl.* upholstered furniture *sing.*
polstern *tr. V.* upholster ⟨*furniture*⟩; pad ⟨*door*⟩; **sie ist gut gepolstert** *(fig. ugs. scherzh.)* she is well-upholstered *(joc.)*
Polster-: ~sessel der [upholstered] armchair; easy chair; **~stuhl der** upholstered chair
Polter·geist der poltergeist
poltern ['pɔltɐn] *itr. V.* **a)** *(lärmen)* crash *or* thump about; **es poltert** there is a bang *or* crash; **b)** *mit sein* **der Karren polterte über das Pflaster** the cart clattered over the cobble-stones; **c)** *(schimpfen)* rant [and rave]
Poly-: ~ester [~'lɛstɐ] **der; ~s, ~** *(Chemie)* polyester; **~gamie** [~ga'mi:] **die; ~:** polygamy
Polyp [po'ly:p] **der; ~en, ~en a)** *(Zool., Med.)* polyp; **b)** *(salopp: Polizist)* cop *(sl.)*; copper *(Brit. sl.)*
poly·technisch 1. *Adj.* polytechnic. **2.** *adv.* **er war ~ ausgebildet** he had a polytechnic training
Pomade [po'ma:də] **die; ~, ~n** pomade; hair-cream
Pommern (das); ~s Pomerania
Pommes frites [pɔm'frit] *Pl.* chips *pl.* *(Brit.)*; French fries *pl.* *(Amer.)*
Pomp [pɔmp] **der; ~[e]s** pomp
pompös [pɔm'pø:s] **1.** *Adj.* grandiose; ostentatious. **2.** *adv.* grandiosely; ostentatiously
Pontius ['pɔntsi̯ʊs] *in* **von ~ zu Pilatus laufen** *(ugs.)* rush from pillar to post
[1]**Pony** ['pɔni] **das; ~s, ~s** pony
[2]**Pony der; ~s, ~s** *(Frisur)* fringe
Pony·frisur die [hair-style with a] fringe
Pop [pɔp] **der; ~[s]** pop
Popanz der; ~es, ~e *(abwertend)* bogey; bugbear
Popcorn ['pɔpkɔrn] **das; ~s** popcorn
popelig *(ugs. abwertend)* **1.** *Adj.* crummy *(coll.)*; lousy *(sl.)*; *(durchschnittlich)* second-rate. **2.** *adv.* crummily *(sl.)*
Popeline·mantel der poplin coat

popeln *itr. V. (ugs.)* **[in der Nase]** **~:** pick one's nose

Pop-: ~festival das pop festival; **~gruppe die** pop group; **~musik die** pop music

Popper ['pɔpɐ] **der; ~s, ~:** *fashion-conscious, apolitical young person*

poppig ['pɔpɪç] **1.** *Adj.* trendy. **2.** *adv.* trendily

populär [popu'lɛ:ɐ̯] **1.** *Adj.* popular (**bei** with). **2.** *adv.* popularly

Popularität die; ~: popularity

Pore ['po:rə] **die; ~, ~n** pore

Porno ['pɔrno] **der; ~s, ~s** *(ugs.)* porn[o] *(coll.)*

Porno·graphie [-gra'fi:] **die; ~, ~n** pornography

porno·graphisch 1. *Adj.* pornographic. **2.** *adv.* pornographically

porös [po'rø:s] porous

Porree ['pɔre] **der; ~s, ~s** leek; **ich mag ~:** I like leeks

Portal [pɔr'ta:l] **das; ~s, ~e** portal

Portemonnaie [pɔrtmɔ'ne:] **das; ~s, ~s** purse

Porti *Pl. s.* **Porto**

Portier [pɔr'ti̯e:] **der; ~s, ~s** porter

Portion [pɔr'tsi̯o:n] **die; ~, ~en a)** portion; helping; **eine halbe ~** *(fig. ugs. spött.)* a feeble little titch *(coll.)*; **eine ~ Eis** one ice-cream; **b)** *(ugs.: Anteil)* amount

Porto ['pɔrto] **das; ~s, ~s** *od.* **Porti** postage (**für** on, for); **„~ zahlt Empfänger"** 'postage will be paid by licensee'

porträtieren [pɔrtrɛ'ti:rən] *tr. V.* paint a portrait of/take a portrait [photograph] of; *(fig.)* portray

Portugal ['pɔrtugal] **(das); ~s** Portugal

Portugiese [pɔrtu'gi:zə] **der; ~n, ~n, Portugiesin die; ~, ~nen** Portuguese

portugiesisch *Adj.* Portuguese

Portwein ['pɔrtvain] **der** port

Porzellan [pɔrtsɛ'la:n] **das; ~s** porcelain; china

Posaune [po'zaunə] **die; ~, ~n** trombone

posaunen 1. *itr. V. (musizieren)* play the trombone. **2.** *tr. V. (ugs.)* **etw. in die** *od.* **alle Welt ~:** tell the whole world about sth.

Posaunist der; ~en, ~en trombonist

Pose ['po:zə] **die; ~, ~n** pose

posieren *itr. V.* pose

Position [pozi'tsi̯o:n] **die; ~, ~en** position

positiv ['po:ziti:f] **1.** *Adj.* positive. **2.** *adv.* positively; **etw. ~ bewerten** judge sth. favourably; **der Test verlief ~:** the test proved successful

Positiv das; ~s, ~e *(Fot.)* positive

Positur [pozi'tu:ɐ̯] **die; ~, ~en** pose; posture; **sich in ~ setzen** *od.* **stellen** *od.* **werfen** *(ugs. leicht spött.)* strike a pose; take up a posture

Posse ['pɔsə] **die; ~, ~n** farce

Possen der; ~s, ~ *(veralt.) Pl.* pranks; tricks; **~ reißen** play tricks

possessiv ['pɔsɛsi:f] *Adj. (Sprachw.)* possessive

Possessiv·pronomen das *(Sprachw.)* possessive pronoun

possierlich [pɔ'si:ɐ̯lɪç] **1.** *Adj.* sweet; cute *(Amer.)*. **2.** *adv.* sweetly; cutely *(Amer.)*

Post [pɔst] **die; ~, ~en a)** post *(Brit.)*; mail; **er ist** *od.* **arbeitet bei der ~:** he works for the Post Office; **etw. mit der** *od.* **per ~ schikken** send sth. by post *or* mail; **mit gleicher/getrennter ~:** by the same post/under separate cover; **b)** *(~amt)* post office; **auf die** *od.* **zur ~ gehen** go to the post office

Post-: ~amt das post office; **~anweisung die a)** *(Geldsendung) remittance paid in at a post office and delivered to the addressee by a postman;* **b)** *(Formular)* postal remittance form; **~auto das** post-office *or* mail van; **~beamte der** post-office official; **~bote der** *(ugs.)* postman *(Brit.)*; mailman *(Amer.)*

Posten ['pɔstn̩] **der; ~s, ~ a)** *(bes. Milit.: Wach~)* post; **auf dem ~ sein** *(ugs.) (in guter körperlicher Verfassung sein)* be in good form; *(wachsam sein)* be on one's guard; **auf verlorenem ~ stehen** *od.* **kämpfen** be fighting a losing battle; **b)** *(bes. Milit.: Wachmann)* sentry; guard; **~ stehen** stand guard *or* sentry; **c)** *(Anstellung)* post; position; job; **d)** *(Funktion)* position; **e)** *(bes. Kaufmannsspr.: Rechnungs~)* item

Poster ['po:stɐ] **das** *od.* **der; ~s, ~[s]** poster

Post-: ~fach das a) *(im ~amt)* post-office *or* PO box; **b)** *(im Büro, Hotel o. ä.)* pigeon-hole; **~geheimnis das** secrecy of the post; **~giro·amt das** post-office giro office; ≈ national giro[bank] centre *(Brit.)*; **~girokonto das** post-office giro account; ≈ national giro[bank] account *(Brit.)*; **~horn das** post-horn

postieren *tr. V.* **a)** *(aufstellen)* post; station; **sich ~:** station *or* position oneself; **b)** *(stellen)* position

post-, Post-: ~karte die postcard; **~kutsche die** mail-coach; **~lagernd** *Adj., adv.* poste restante; general delivery *(Amer.)*; **~leit·zahl die** postcode; postal code; Zip code *(Amer.)*; **~minister der** Postmaster General; **~scheck der** post-office giro cheque; ≈ national giro[bank] cheque *(Brit.)*; **~spar·buch das** post-office savings book *(Brit.)*; **~spar·kasse die** post-office savings bank *(Brit.)*; **~stempel der a)** *(Gerät)* stamp [for cancelling mail]; **b)** *(Abdruck)* postmark

postum [pɔs'tu:m] **1.** *Adj.; nicht präd.* posthumous. **2.** *adv.* posthumously

post-, Post-: ~wendend *Adv.* by return [of post]; *(fig.)* immediately; **~wurf·sendung die** direct-mail item; **~zustellung die** postal delivery *no def. art.*

potemkinsch [po'tɛmki:nʃ] *Adj.; in* **Potemkinsche Dörfer** façade *sing.;* sham *sing.*

potent [po'tɛnt] *Adj.* **a)** potent; **b)** *(finanzstark)* [financially] strong

Potential [potɛn'tsi̯a:l] **das; ~s, ~e** potential

potentiell [potɛn'tsi̯ɛl] **1.** *Adj.* potential. **2.** *adv.* potentially

Potenz [po'tɛnts] **die; ~, ~en a)** *o. Pl.* potency; **b)** *(Stärke)* power; **c)** *(Math.)* power

potenzieren 1. *tr. V.* **a)** *(verstärken)* increase; **b)** *(Math.)* **mit 5 ~:** raise to the power [of] 5. **2.** *refl. V. (sich steigern)* increase

Potpourri ['pɔtpʊri] **das; ~s, ~s** pot-pourri, medley (**aus, von** of)

Präambel [prɛ'ambl̩] **die; ~, ~n** preamble

Pracht [praxt] **die; ~:** splendour; magnificence; **eine [wahre] ~ sein** *(ugs.)* be [really] marvellous *or (coll.)* great

Pracht·exemplar das *(ugs.)* magnificent specimen; beauty

prächtig ['prɛçtɪç] **1.** *Adj.* **a)** *(prunkvoll)* splendid; magnificent; **b)** *(großartig)* splendid; marvellous. **2.** *adv.* **a)** *(prunkvoll)* splendidly; magnificently; **b)** *(großartig)* splendidly; marvellously

Pracht-: ~kerl der *(ugs.)* great chap *or (Brit. coll.)* bloke *or (sl.)* guy; **~stück das** *s.* **~exemplar**

prädestinieren *tr. V.* predestine

Prädikat [prɛdi'ka:t] **das; ~[e]s, ~e a)** *(Auszeichnung)* rating; **b)** *(Sprachw.)* predicate

prädikativ [prɛdika'ti:f] *(Sprachw.)* **1.** *Adj.* predicative. **2.** *adv.* predicatively

Prag [pra:k] **(das); ~s** Prague

prägen ['prɛ:gn̩] *tr. V.* **a)** emboss

⟨*metal, paper, leather*⟩; mint, strike ⟨*coin*⟩; **b)** *(auf~) (vertieft)* impress; *(erhaben)* emboss; **c)** *(fig.: beeinflussen)* shape; mould; **d)** *(fig.: erfinden)* coin ⟨*word, expression, concept*⟩

pragmatisch 1. *Adj.* pragmatic. **2.** *adv.* pragmatically

prägnant [prɛ'gnant] **1.** *Adj.* concise; succinct. **2.** *adv.* concisely; succinctly

Prägnanz [prɛ'gnants] **die; ~:** conciseness; succinctness

Prägung die; ~, ~en a) *(von Papier, Leder, Metall)* embossing; *(von Münzen)* minting; striking; **b)** *(auf Metall, Papier) (vertieft)* impression; *(erhaben)* embossing; **c)** *(Eigenart)* character; **d)** *(eines sprachlichen Ausdrucks)* coining; *(geprägter Ausdruck)* coinage

prahlen ['pra:lən] *itr. V.* boast, brag (**mit** about)

Prahlerei die; ~, ~en *(abwertend)* boasting; bragging

Praktik ['praktɪk] **die; ~, ~en** practice

Praktika *s.* **Praktikum**

praktikabel [prakti'ka:bl̩] *Adj.* practicable; practical

Praktikant [prakti'kant] **der; ~en, ~en, Praktikantin die; ~, ~nen a)** *(in einem Betrieb)* student trainee; trainee student; **b)** *(an der Hochschule)* physics/chemistry student *(doing a period of practical training)*

Praktiker der; ~s, ~ practical person

Praktikum ['praktikʊm] **das; ~s, Praktika** period of practical instruction *or* training

praktisch ['praktɪʃ] **1.** *Adj.* **a)** *(auf die Praxis bezogen)* practical; **~er Arzt** general practitioner; **b)** *(wirklich)* practical ⟨*result, problem, matter, etc.*⟩; concrete ⟨*example*⟩; **c)** *(nützlich)* practical ⟨*furniture, clothes, etc.*⟩; useful ⟨*present*⟩; **d)** *(geschickt, realistisch)* practical. **2.** *adv.* **a)** *(auf die Praxis bezogen)* in practice; **~ arbeiten** do practical work; **b)** *(wirklich)* in practice; **c)** *(nützlich)* practically; **d)** *(geschickt, realistisch)* practically; **e)** *(ugs.: so gut wie)* practically; virtually

praktizieren [prakti'tsi:rən] *tr. V.* **a)** *(anwenden)* practise; **b)** *(ugs.: irgendwohin bringen)* conjure; **jmdm. etw. ins Essen ~:** slip sth. into sb.'s food

Praline [pra'li:nə] **die; ~, ~n** [filled] chocolate

prall [pral] **1.** *Adj.* **a)** *(fest und straff)* hard ⟨*ball*⟩; firm ⟨*tomato, grape*⟩; bulging ⟨*sack, wallet, bag*⟩; big strong *attrib.* ⟨*thighs, muscles, calves*⟩; full, well-rounded ⟨*breasts*⟩; full, chubby ⟨*cheeks*⟩; taut, full ⟨*sail*⟩; *(fig.)* intense ⟨*life*⟩; vivid ⟨*picture*⟩; fully inflated ⟨*balloon*⟩; **b)** *(intensiv)* blazing ⟨*sun*⟩; strong ⟨*light*⟩. **2.** *adv.* fully ⟨*inflated*⟩; **eine ~ gefüllte Brieftasche** a wallet bulging with banknotes

prallen *itr. V.* **a)** *mit sein (hart auftreffen)* crash (**gegen/auf/an** + *Akk.* into); collide (**gegen/auf/an** + *Akk.* with); **der Ball prallte an den Pfosten** the ball hit the post; **b)** *(scheinen)* blaze

prall·voll *Adj. (ugs.)* ⟨*suitcase, rucksack*⟩ full to bursting; packed ⟨*room*⟩

Prämie ['prɛ:mi̯ə] **die; ~, ~n a)** *(Leistungs~)* bonus; *(Belohnung)* reward; *(Spar~, Versicherungs~)* premium; **b)** *(einer Lotterie)* [extra] prize

prämieren [prɛ'mi:rən], **prämiieren** [prɛmi'i:rən] *tr. V.* award a prize to ⟨*person, film*⟩; give an award for ⟨*best essay etc.*⟩

Prämierung, Prämiierung die; ~, ~en a) *(Auszeichnung)* **er/der Film wurde zur ~ vorgeschlagen** it was proposed that he should be given a prize/that a prize should be given for the film; **b)** *(Preisverleihung)* **die ~ der besten Schüler/Filme** the presentation of prizes to the best pupils/for the best films

Prämisse [prɛ'mɪsə] **die; ~, ~n** premiss

prangen ['praŋən] *itr. V.* be prominently displayed

Pranger der; ~s, ~ *(hist.)* pillory; **jmdn./etw. an den ~ stellen** *(fig.)* pillory sb./sth.

Pranke ['praŋkə] **die; ~, ~n a)** *(Pfote)* paw; **b)** *(salopp: große Hand)* paw *(coll.)*

Präparat [prɛpa'ra:t] **das; ~[e]s, ~e** preparation

präparieren *tr. V.* **a)** *(Biol., Med.: konservieren)* preserve; **b)** *(Biol., Anat.: zerlegen)* dissect

Präposition [prɛpozi'tsi̯o:n] **die; ~, ~en** *(Sprachw.)* preposition

Prärie [prɛ'ri:] **die; ~, ~n** prairie

Präsens ['prɛ:zɛns] **das; ~, Präsentia** [prɛ'zɛntsi̯a] *od.* **Präsenzien** [prɛ'zɛntsi̯ən] *(Sprachw.)* present [tense]

präsent [prɛ'zɛnt] *Adj.* present

präsentieren *tr. V.* **a)** *(anbieten; überreichen)* offer; **b)** *(vorlegen)* present; **jmdm. die Rechnung [für etw.] ~:** present sb. with the bill [for sth.]

Präservativ [prɛzɛrva'ti:f] **das; ~s, ~e** condom

Präsident [prɛzi'dɛnt] **der; ~en, ~en** president

Präsidenten·wahl die presidential election

Präsidentin die; ~, ~nen president

Präsidentschaft die; ~, ~en presidency

Präsidien *Pl. s.* **Präsidium**

Präsidium [prɛ'zi:di̯ʊm] **das; ~s, Präsidien a)** *(Führungsgruppe)* committee; **b)** *(Vorsitz)* chairmanship; **c)** *(Polizei~)* police headquarters *sing. or pl.*

prasseln ['prasl̩n] *itr. V.* ⟨*rain, hail*⟩ pelt down; ⟨*shots*⟩ clatter; ⟨*fire*⟩ crackle

prassen ['prasn̩] *itr. V.* live extravagantly; *(schlemmen)* feast

Präteritum [prɛ'te:ritʊm] **das; ~s, Präterita** *(Sprachw.)* preterite [tense]

präventiv [prɛvɛn'ti:f] *Adj.* preventive

Praxis ['praksɪs] **die; ~, Praxen a)** *o. Pl. (im Unterschied zur Theorie)* practice *no art.*; **in der ~:** in practice; **etw. in die ~ umsetzen** put sth. into practice; **b)** *o. Pl. (Erfahrung)* [practical] experience; **c)** *(eines Arztes, Anwalts, Psychologen usw.)* practice; **d)** *(~räume) (eines Arztes)* surgery *(Brit.)*; office *(Amer.)*; *(eines Anwalts, Psychologen usw.)* office; **e)** *(Handhabung)* procedure

Präzedenz·fall [prɛtse'dɛnts-] **der** precedent

präzise [prɛ'tsi:zə] **1.** *Adj.* precise ⟨*definition, answer*⟩; specific ⟨*wishes, suspicion*⟩. **2.** *adv.* precisely

präzisieren *tr. V.* make more precise; state more precisely

Präzision [prɛtsi'zi̯o:n] **die; ~:** precision

Präzisions·arbeit die precision work; *(genau nach Zeitplan)* precise timing

predigen ['pre:dɪgn̩] **1.** *itr. V.* *(Predigt halten)* deliver *or* give a/the sermon. **2.** *tr. V.* **a)** *(verkündigen)* preach; **b)** *(ugs.: auffordern zu)* preach; **c)** *(ugs.: belehrend sagen)* **wie oft habe ich dir das schon gepredigt!** how often have I told you that!

Prediger der; ~s, ~, Predigerin die; ~, ~nen preacher

Predigt ['pre:dɪçt] **die; ~, ~en a)** sermon; **b)** *(ugs.: Ermahnung)* lecture; **jmdm. eine ~ halten** lecture sb.

Preis [prai̯s] **der; ~es, ~e a)** *(Kauf~)* price (**für** of); **etw. zum**

halben ~ erwerben buy sth. at half-price; **um jeden ~** *(fig.)* at all costs; **b)** *(Belohnung)* prize; **der Große ~ von Frankreich** *(Rennsport)* the French Grand Prix

preis-, Preis-: ~anstieg der rise *or* increase in prices; **~ausschreiben das** [prize] competition; **~bewußt 1.** *Adj.* price-conscious; **2.** *adv.* price-consciously; **~bindung die** *(Wirtsch.)* price-fixing

Preisel·beere ['praizlbe:rə] **die** cowberry; cranberry *(Gastr.)*

preisen *unr. tr. V. (geh.)* praise; **sich glücklich ~:** count *or* consider oneself lucky

Preis-: ~erhöhung die price increase *or* rise; **~frage die a)** *(bei einem ~ausschreiben)* [prize] question; **b)** *(Geldfrage)* question of price; **~gabe die** *(geh.)* **a)** *(Verzicht)* abandonment; **b)** *(von Geheimnissen)* revelation; giving away; **~|geben** *unr. tr. V. (geh.)* **a)** *(ausliefern)* **jmdn. einer Sache** *(Dat.)* **~geben** expose sb. to *or* leave sb. to be the victim of sth.; **b)** *(aufgeben)* relinquish ‹*ideal, independence*›; surrender ‹*territory*›; **c)** *(verraten)* betray; give away; **~gekrönt** *Adj.* prize- *or* awardwinning; **~gericht das** jury; panel of judges; **~günstig 1.** *Adj.* ‹*goods*› available at unusually low prices; ‹*purchases*› at favourable prices. **2.** *adv.* **etw. ~günstig verkaufen/bekommen** sell/get sth. at a low price; **hier kann man ~günstig einkaufen** their prices are very reasonable here; **~lage die** price range

preislich *Adj.; nicht präd.* price; in price *postpos.*

preis-, Preis-: ~nachlaß der price reduction; discount; **~rätsel das** [prize] competition; **~schild das** price-tag; **~schlager der** *(ugs.)* bargain [offer]; **~senkung die** price reduction *or* cut; **~steigerung die** rise *or* increase in prices; **~träger der, ~trägerin die** prizewinner; **~treiberei die** *(abwertend)* forcing up of prices; **~vergleich der** price comparison; **~vergleiche anstellen** compare prices; **~verleihung die** presentation [of prizes/awards]; award ceremony; **~wert 1.** *Adj.* good value *pred.* **2.** *adv.* **dort kann man ~wert einkaufen** you get good value for money there; **hier kann man ~wert essen** you can eat at a reasonable price here

prekär [pre'kɛ:ɐ̯] *Adj.* precarious

Prell·bock der *(Eisenb.)* buffer

prellen ['prɛlən] **1.** *tr. V.* **a)** *(betrügen)* cheat (**um** out of); **die Zeche ~:** avoid paying the bill; **b)** *(verletzen)* bash; bruise; **c)** *(Ballsport)* bounce. **2.** *refl. V. (sich verletzen)* bruise oneself

Prellung die; ~, ~en bruise

Premiere [prə'mi̯e:rə] **die; ~, ~n** opening night; first night; *(Uraufführung)* première; *(fig.)* first appearance

Premier- [prə'mi̯e:-]: **~minister der, ~ministerin die** prime minister

preschen ['prɛʃn̩] *itr. V.; mit sein* tear

Presse ['prɛsə] **die; ~, ~n a)** press; *(für Zitronen)* squeezer; **b)** *o. Pl. (Zeitungen, ~kritik)* press

Presse-: ~agentur die press agency; news agency; **~bericht der** press report; **~empfang der** press reception; **~erklärung die** press statement; **~freiheit die** freedom of the press; **~konferenz die** press conference; **~meldung die** press report

pressen *tr. V.* **a)** *(zusammendrükken)* press; **b)** *(auspressen)* press ‹*fruit, garlic*›; squeeze ‹*lemon*›; **c)** *(drücken)* press; **d)** *(herstellen)* press ‹*record*›; mould ‹*plastic object*›

Presse-: ~sprecher der spokesman; press officer; **~stimmen** *Pl.* press commentaries *or* reviews

pressieren *itr. V. (bes. südd.)* ‹*matter*› be urgent; **mir pressiert's sehr** I am in a great hurry

Preß·luft die; *o. Pl.* compressed air

Preß·luft-: ~bohrer der pneumatic drill; **~hammer der** pneumatic *or* air hammer

Prestige [prɛs'ti:ʒə] **das; ~s** prestige

preußisch ['prɔy̯sɪʃ] *Adj.* Prussian

prickeln ['prɪkl̩n] *itr. V.* **a)** *(kribbeln, kitzeln)* tingle; **b)** *(perlen)* sparkle; **c)** *(reizen)* **eine ~de Spannung** a tingling atmosphere

Priel [pri:l] **der; ~[e]s, ~e** narrow channel *(in mud-flats)*

Priem [pri:m] **der; ~[e]s, ~e** *(Kautabak)* chewing-tobacco

pries [pri:s] *1. u. 3. Pers. Sg. Prät. v.* **preisen**

Priester ['pri:stɐ] **der; ~s, ~:** priest

Priesterin die; ~, ~nen priestess

Priester·seminar das seminary

prima ['pri:ma] **1.** *indekl. Adj. (ugs.)* great *(coll.);* fantastic *(coll.).* **2.** *adv. (ugs.)* ‹*taste*› great *(coll.),* fantastic *(coll.);* ‹*sleep*› fantastically *(coll.) or* really well; **es geht mir ~:** I feel great *(coll.)*

Prima die; ~, Primen *(Schulw.) (veralt.)* eighth and ninth years *(of a Gymnasium)*

primär [pri'mɛ:ɐ̯] **1.** *Adj.* primary. **2.** *adv.* primarily

Primel ['pri:ml̩] **die; ~, ~n** primula; primrose; *(Schlüsselblume)* cowslip

primitiv [primi'ti:f] **1.** *Adj.* **a)** primitive; **b)** *(schlicht)* simple; crude *(derog.).* **2.** *adv.* **a)** primitively; **b)** *(schlicht)* in a simple manner; crudely *(derog.)*

Primus ['pri:mʊs] **der; ~, Primi** *od.* **~se** *(veralt.)* top of the class

Prim·zahl die prime [number]

Prinz [prɪnts] **der; ~en, ~en** prince

Prinzessin die; ~, ~nen princess

Prinzip [prɪn'tsi:p] **das; ~s, ~ien** [-'tsi:pi̯ən] principle; **aus/im ~:** on/in principle; **ein Mensch von ~ien** a man/woman of principle

prinzipiell [prɪntsi'pi̯ɛl] **1.** *Adj.; nicht präd.* in principle *postpos., not pred.;* ‹*rejection*› on principle; **eine ~e Frage/Frage von ~er Bedeutung** a question of principle/of fundamental importance. **2.** *adv.* **a)** *(im Prinzip)* in principle; **b)** *(aus Prinzip)* on principle; as a matter of principle

Priorität [priori'tɛ:t] **die; ~, ~en a)** *o. Pl. (Vorrang)* priority; precedence; **~ vor etw.** *(Dat.)* **haben** have *or* take precedence over sth.; **b)** *Pl. (Rangfolge)* priorities; **~en setzen** establish priorities

Prise ['pri:zə] **die; ~, ~n** pinch; **eine ~ Sarkasmus/Ironie** *(fig.)* a hint *or* touch of sarcasm/irony

Prisma ['prɪsma] **das; ~s, Prismen** *(Math., Optik)* prism

Pritsche ['prɪtʃə] **die; ~, ~n** plank bed

privat [pri:va:t] **1.** *Adj.* private; personal ‹*opinion, happiness, etc.*›; **an/von Privat** to/from private individuals *pl.* **2.** *adv.* privately; **jmdn. ~ sprechen** speak to sb. in private *or* privately

Privat-: ~adresse die private *or* home address; **~angelegenheit die** private affair *or* matter; **das ist seine ~angelegenheit** that's his own business *or* his own private affair; **~besitz der** private property; **sich im ~besitz befinden** be privately owned *or* in private ownership; **~eigentum das** private property; **~fernsehen das** privately operated television; ≈ commercial television; **~gespräch das** private conversation; *(Telefongespräch)* private call

privatisieren *tr. V.* privatize; transfer into private ownership

Privat-: **~klinik die** private clinic *or* hospital; **~leben das;** *o. Pl.* private life; **~patient der** private patient; **~person die** private individual; **~sache die** *s.* **~angelegenheit; ~schule die** private school; *(Eliteschule in Großbritannien)* public school; **~unterricht der** private tuition; private lessons *pl.*; **~wirtschaft die** private sector

Privileg [privi'le:k] **das; ~[e]s, ~ien** [-'le:gi̯ən] privilege

privilegieren *tr. V.* grant privileges to

privilegiert *Adj.* privileged

pro [pro:] *Präp. mit Akk.* per; **~ Jahr** per year *or* annum; **~ Kopf** per head; **~ Stück** each; apiece; **~ Nase** *(ugs.)* each; a head

Pro das *in* **[das] ~ und [das] Kontra** the pros and cons *pl.*

pro-: pro-; **~westlich** pro-western

Probe ['pro:bə] **die; ~, ~n a)** *(Prüfung)* test; **die ~ aufs Exempel machen** put it to the test; **auf ~ sein** be on probation; **jmdn./etw. auf die ~ stellen** put sb./sth. to the test; **b)** *(Muster, Teststück)* sample; **eine ~ seines Könnens zeigen** *od.* **geben** *(fig.)* show what one can do; **c)** *(Theater~, Orchester~)* rehearsal

Probe-: **~alarm der** practice alarm; *(Feueralarm)* fire-drill *or* -practice; **~exemplar das** specimen copy; **~fahrt die** trial run; *(vor dem Kauf, nach einer Reparatur)* test drive; **~lauf der** *(Technik)* test run

proben *tr., itr. V.* rehearse

probe·weise *Adv.* ‹*employ*› on a trial basis; **den Motor ~ laufen lassen** test[-run] the engine

Probe·zeit die probationary *or* trial period

probieren 1. *tr. V.* **a)** *(versuchen)* try; have a go *or* try at; **b)** *(kosten)* taste; try; sample; **c)** *(aus~)* try out; *(an~)* try on ‹*clothes, shoes*›. **2.** *itr. V.* **a)** *(versuchen)* try; have a go *or* try; **Probieren geht über Studieren** the proof of the pudding is in the eating *(prov.)*; **b)** *(kosten)* have a taste

Problem [pro'ble:m] **das; ~s, ~e** problem

Problematik [proble'ma:tɪk] **die; ~** *(Schwierigkeit)* problematic nature; *(Probleme)* problems *pl.*

problematisch *Adj.* problematic[al]

problem·los 1. *Adj.* problem-free. **2.** *adv.* without any problems

Produkt [pro'dʊkt] **das; ~[e]s, ~e** *(auch Math., fig.)* product

Produktion [prodʊk'tsi̯o:n] **die; ~, ~en** production

Produktions-: **~abteilung die** production department; **~anlage die;** *meist Pl.* production unit; **~anlagen** production plant *sing.*; **~leiter der** production manager; **~mittel** *Pl. (marx.)* means of production; **~prozeß der, ~verfahren das** production process

produktiv [prodʊk'ti:f] **1.** *Adj.* productive; prolific ‹*writer, artist, etc.*›. **2.** *adv.* ‹*co-operate, work*› productively

Produktivität [prodʊktivi'tɛ:t] **die; ~:** productivity

Produzent [produ'tsɛnt] **der; ~en, ~en, Produzentin die; ~, ~nen** producer

produzieren 1. *tr. V.* **a)** *auch itr. (herstellen)* produce; **b)** *(ugs.: hervorbringen)* make ‹*bow, noise*›; come up with ‹*excuse, report*›. **2.** *refl. V. (ugs.: großtun)* show off

Prof. *Abk.* **Professor** Prof.

profan [pro'fa:n] *Adj. nicht präd.* profane; secular

professionell [profɛsi̯o'nɛl] **1.** *Adj.* professional. **2.** *adv.* professionally

Professor [pro'fɛsɔr] **der; ~s, ~en** [-'so:rən] **a)** *(Hochschul~)* professor; **ordentlicher ~:** [full] professor *(holding a chair)*; **außerordentlicher ~:** extraordinary professor *(not holding a chair)*; **b)** *(österr., sonst veralt.: Gymnasial~)* [grammar school] teacher

Professorin [profɛ'so:rɪn] **die; ~, ~nen a)** *(Hochschul~)* professor; **b)** *(österr.: Studienrätin)* mistress

Professur [profɛ'su:ɐ̯] **die; ~, ~en** professorship, chair **(für** in)

Profi der; ~s, ~s *(ugs.)* pro *(coll.)*

Profil [pro'fi:l] **das; ~s, ~e a)** *(Seitenansicht)* profile; **im ~:** in profile; **b)** *(von Reifen, Schuhsohlen)* tread; **c)** *(ausgeprägte Eigenart)* image

profilieren *refl. V.* make one's name *or* mark; *(sich unterscheiden)* give oneself a clearer image

profiliert *Adj.* prominent

Profi-: **~spieler der** professional player; **~sport der** professional sport

Profit [pro'fi:t] **der; ~[e]s, ~e** profit; **aus etw. ~ ziehen** *od.* **herausschlagen** turn sth. to one's profit *or* advantage; **~ machen** make a profit; **mit/ohne ~ arbeiten** run/not run at a profit

Profit·gier die *(abwertend)* greed for profit

profitieren *itr. V.* profit **(von, bei** by); **ich kann dabei nur ~:** I can't lose

Profit·streben das *(abwertend)* profit-seeking

pro forma [pro: 'fɔrma] *Adv.* **a)** *(der Form halber)* as a matter of form; **b)** *(zum Schein)* for the sake of appearances

Prognose [pro'gno:zə] **die; ~, ~n** *(auch Med.)* prognosis; *(Wetter~, Wirtschafts~)* forecast

Programm [pro'gram] **das; ~s, ~e** programme; program *(Amer., Computing)*; *(Verlags~)* list; *(Ferns.: Sender)* channel; *(Sendefolge: Ferns., Rundfunk)* programmes *pl. or (Amer.)* programs *pl.*; *(Tagesordnung)* agenda; **etw. paßt jmdm. nicht ins** *od.* **in sein ~:** sth. doesn't fit in with sb.'s plans; **auf dem ~ stehen** *(fig.)* be on the programme *or* agenda; *(bei einer Sitzung)* be on the agenda

programm-, Programm-: **~gemäß** *adv.* according to programme *or* plan; **~heft das** programme; **~hinweis der** programme announcement

programmieren *tr. V.* **a)** *(DV)* program; **b)** *(auf etw. festlegen)* programme; condition

Programmierer der; ~s, ~, Programmiererin die; ~, ~nen *(DV)* programmer

Programm-: **~vorschau die** *(im Fernsehen)* preview [of the week's/evening's *etc.* viewing]; *(im Kino)* trailers *pl.*; **~zeitschrift die** radio and television magazine

progressiv [progrɛ'si:f] **1.** *Adj.* progressive. **2.** *adv.* progressively

Projekt [pro'jɛkt] **das; ~[e]s, ~e** project

Projektil [projɛk'ti:l] **das; ~s, ~e** projectile

Projektor [pro'jɛktɔr] **der; ~s, ~en** [-'to:rən] projector

projizieren [proji'tsi:rən] *tr. V. (Optik, Math.)* project

Proklamation [proklama'tsi̯o:n] **die; ~, ~en** proclamation

proklamieren *tr. V.* proclaim

Pro-Kopf- per head *or* capita *postpos.*

Prokura [pro'ku:ra] **die; ~, Prokuren** *(Kaufmannsspr.)* [full] power of attorney; procuration *(formal)*

Prolet [pro'le:t] **der; ~en, ~en** *(abwertend)* peasant; boor

Proletariat [proleta'ri̯a:t] **das; ~[e]s** proletariat

Proletarier [prole'ta:ri̯ɐ] **der; ~s, ~:** proletarian

proletarisch *Adj.* proletarian

proletenhaft *(abwertend)* **1.** *Adj.* boorish. **2.** *adv.* boorishly

Prolog [pro'lo:k] **der; ~[e]s, ~e** prologue

Promenade [promə'na:də] **die; ~, ~n** promenade

Promenaden-: ~deck das promenade deck; **~mischung die** *(scherzh.)* mongrel

Promille [pro'mɪlə] **das; ~s, ~ a)** *(Tausendstel)* **ein Blutalkoholgehalt von zwei ~:** a blood alcohol level of two parts per thousand; **bei 0,4/unter einem ~ liegen** be 0.4/less than one in a *or* per thousand; **b)** *(ugs.: Blutalkohol)* alcohol level

Promille·grenze die legal [alcohol] limit

prominent [promi'nɛnt] *Adj.* prominent

Prominente der/die; *adj. Dekl.* prominent figure

Prominenz [promi'nɛnts] **die; ~:** prominent figures *pl.; (das Prominentsein)* prominence

Promotion [promo'tsi̯o:n] **die; ~, ~en** gaining of a/one's doctorate; **er schloß sein Studium mit der ~ ab** he completed his studies by gaining *or* obtaining his doctorate

promovieren [promo'vi:rən] *itr. V.* **a)** *(die Doktorwürde erlangen)* gain *or* obtain a/one's doctorate; **b)** *(eine Dissertation schreiben)* do a doctorate (**über +** *Akk.* on)

prompt [prɔmpt] **1.** *Adj.* prompt. **2.** *adv.* **a)** *(umgehend)* promptly; **b)** *(ugs., meist iron.: wie erwartet)* [and] sure enough

Pronomen [pro'no:mən] **das; ~s, ~** *od.* **Pronomina** [pro'no:mina] *(Sprachw.)* pronoun

Propaganda [propa'ganda] **die; ~:** *(auch fig. ugs.)* propaganda

Propagandist der; ~en, ~en, Propagandistin die; ~, ~nen a) propagandist; **b)** *(Wirtsch.: Werbefachmann/-frau)* demonstrator

propagieren *tr. V. (geh.)* propagate ‹*idea, view, belief, etc.*›

Propan [pro'pa:n] **das, ~s, Propan·gas das;** *o. Pl.* propane

Propeller [pro'pɛlɐ] **der; ~s, ~:** propeller; airscrew; prop *(coll.)*

proper ['prɔpɐ] **1.** *Adj.* **a)** *(adrett)* smart; **b)** *(ordentlich und sauber)* neat and tidy; **c)** *(sorgfältig, genau)* meticulous. **2.** *adv.* **a)** *(ordentlich und sauber)* neatly and tidily; **b)** *(sorgfältig, genau)* meticulously

Prophet [pro'fe:t] **der; ~en, ~en** prophet

Prophetin die; ~, ~nen prophetess

prophetisch 1. *Adj.* prophetic. **2.** *adv.* prophetically

prophezeien [profe'tsai̯ən] *tr. V.* prophesy (*Dat.* for); predict ‹*result, weather*›

Prophezeiung die; ~, ~en *s.* **prophezeien:** prophecy; prediction

Proportion [propɔr'tsi̯o:n] **die; ~, ~en** proportion

proportional [propɔrtsi̯o'na:l] *(auch Math.)* **1.** *Adj.* proportional. **2.** *adv.* proportionally; in proportion

Proporz [pro'pɔrts] **der; ~es, ~e** *(Politik)* proportional representation *no art.*

proppen·voll *(ugs.) Adj.* jam-packed *(coll.)*

Prosa ['pro:za] **die; ~:** prose

prosaisch 1. *Adj.* prosaic. **2.** *adv.* prosaically

prosit ['pro:zɪt] *Interj.* your [very good] health; **~ Neujahr!** happy New Year!

Prosit das; ~s, ~s toast; **ein ~ dem Geburtstagskind!** here's to the birthday boy/girl!

Prospekt [pro'spɛkt] **der** *od. (bes. österr.)* **das ~[e]s, ~e** *(Werbeschrift)* brochure; *(Werbezettel)* leaflet; *(Verlags~)* illustrated catalogue; *(nur mit Neuerscheinungen)* seasonal list

prost [pro:st] *Interj. (ugs.)* cheers *(Brit. coll.)*; **na denn** *od.* **dann ~!** *(ugs. iron.)* that's brilliant! *(coll. iron.); s. auch* **Mahlzeit**

Prost das; ~[e]s, ~e *(ugs.) s.* **Prosit**

Prostituierte die/der; *adj. Dekl.* prostitute

Prostitution [prostitu'tsi̯o:n] **die; ~:** prostitution *no art.*

Protagonist [protago'nɪst] **der; ~en, ~en, Protagonistin die; ~, ~nen** *(geh.)* protagonist

protegieren [prote'ʒi:rən] *tr. V. (geh.)* sponsor; patronize ‹*artist, composer, etc.*›

Protektorat [protɛkto'ra:t] **das; ~[e]s, ~e a)** *(geh.: Schirmherrschaft)* patronage; **b)** *(Völkerr.: Schutzherrschaft, Schutzgebiet)* protectorate

Protest [pro'tɛst] **der; ~[e]s, ~e** protest; **[bei jmdm.] ~ gegen jmdn./etw. einlegen** make a protest [to sb.] against sb./sth.; **etw. aus ~ tun** do sth. as a *or* in protest

Protestant [protɛs'tant] **der; ~en, ~en, Protestantin die; ~, ~nen** Protestant

protestantisch *Adj.* Protestant

Protest-: ~bewegung die protest movement; **~demonstration die** protest demonstration

protestieren *itr. V.* protest, make a protest (**gegen** against, about)

Protest-: ~kundgebung die protest rally; **~marsch der** protest march; **~welle die** wave of protest

Prothese [pro'te:zə] **die; ~, ~n a)** artificial limb; prosthesis *(Med.)*; **b)** *(Zahn~)* set of dentures; dentures *pl.;* prosthesis *(Med.)*

Protokoll [proto'kɔl] **das; ~s, ~e a)** *(wörtlich)* transcript; *(Ergebnis~)* minutes *pl.; (bei Gericht)* record; records *pl.;* **[das] ~ führen** make a transcript [of the proceedings]; *(bei einer Sitzung Notizen machen)* take *or* keep the minutes; **etw. zu ~ geben/zu ~ geben, daß ...:** make a statement about sth./to the effect that ...; **zu ~ nehmen** take down ‹*statement etc.*›; *(bei Gericht)* enter ‹*objection, statement*› in the record; **b)** *(diplomatisches Zeremoniell)* protocol; **c)** *(Strafzettel)* ticket

Protokollant [protokɔ'lant] **der; ~en, ~en, Protokollantin die; ~, ~nen** transcript writer; *(eines Ergebnisprotokolls)* keeper of the minutes; *(bei Gericht)* court reporter

protokollieren 1. *tr. V.* take down; minute, take the minutes of ‹*meeting*›; minute, record in the minutes ‹*remark*›. **2.** *itr. V.* *(bei einer Sitzung)* take *or* keep the minutes; *(bei Gericht)* keep the record; *(bei polizeilicher Vernehmung)* keep a record

Proto·plasma [proto'plasma] **das;** *o. Pl. (Biol.)* protoplasm

Proto·typ ['pro:toty:p] **der a)** *(geh.: Inbegriff)* archetype; epitome; **b)** *(Urform, erste Ausführung; Motorsport)* prototype

protzen *itr. V. (ugs.)* swank *(coll.)*; show off; **mit etw. ~:** show sth. off

protzig *(ugs. abwertend)* **1.** *Adj.* swanky *(coll.)*; showy. **2.** *adv.* swankily *(coll.)*

Proviant [pro'vi̯ant] **der; ~s, ~e** provisions *pl.*

Provinz [pro'vɪnts] **die; ~, ~en a)** *(Verwaltungsbezirk)* province; **b)** *o. Pl. (abwertend: kulturell rückständige Gegend)* provinces *pl.*

provinziell [provɪn'tsi̯ɛl] *(meist abwertend)* **1.** *Adj.* provincial; parochial ‹*views*›. **2.** *adv.* provincially

Provinzler [pro'vɪntslɐ] **der; ~s, ~** *(ugs. abwertend)* [narrow-minded] provincial

Provinz·nest das *(ugs. abwertend)* [tiny] provincial backwater

Provision [provi'zi̯o:n] **die; ~, ~en** *(Kaufmannsspr.)* commission

provisorisch [provi'zo:rɪʃ] **1.** *Adj.* temporary ⟨*accommodation, filling, bridge, etc.*⟩; provisional ⟨*status, capital, etc.*⟩; provisional, caretaker *attrib.* ⟨*government*⟩; provisional, temporary ⟨*measure, regulation, etc.*⟩. **2.** *adv.* temporarily; **etw. ~ reparieren** do *or* effect a temporary repair on sth.

Provokation [provoka'tsi̯o:n] **die; ~, ~en** provocation

provokativ [provoka'ti:f] **1.** *Adj.* provocative. **2.** *adv.* provocatively

provozieren [provo'tsi:rən] *tr. V.* **a)** *(herausfordern)* provoke; **b)** *(auslösen)* provoke; cause ⟨*accident, fight*⟩

Prozedur [protse'du:ɐ̯] **die; ~, ~en** procedure

Prozent [pro'tsɛnt] **das; ~[e]s, ~e a)** *nach Zahlenangaben Pl. ungebeugt (Hundertstel)* per cent *sing.*; **ich bin mir zu 90 ~ sicher** I'm 90 per cent certain; **der Plan wurde zu 90 ~ erfüllt** 90 per cent of the plan was fulfilled; **etw. in ~en ausdrücken** express sth. as a percentage; **b)** *Pl. (ugs.: Gewinnanteil)* share *sing.* of the profits; *(Rabatt)* discount *sing.*; **auf etw.** *(Akk.)* **~e bekommen** get a discount on sth.

Prozent-: ~rechnung die percentage calculation; **~satz der** percentage

prozentual [protsɛn'tu̯a:l] **1.** *Adj.; nicht präd.* percentage. **2.** *adv.* **~ am Gewinn beteiligt sein** have a percentage share in the profits

Prozeß [pro'tsɛs] **der; Prozesses, Prozesse a)** trial; *(Fall)* [court] case; **jmdm. den ~ machen** take sb. to court; **einen ~ gewinnen/verlieren** win/lose a case *or* lawsuit; **b)** *(Entwicklung, Ablauf)* process; **c)** *(fig.)* **mit jmdm./etw. kurzen ~ machen** *(ugs.)* make short work of sb./sth.

prozessieren *itr. V.* go to court; **gegen jmdn. ~:** bring an action *or* a lawsuit against sb.; *(seit längerer Zeit)* be engaged in an action *or* a lawsuit against sb.

Prozession [protsɛ'si̯o:n] **die; ~, ~en** procession

Prozeß·kosten *Pl.* legal costs

prüde ['pry:də] *(abwertend)* **1.** *Adj.* prudish. **2.** prudishly

Prüderie [pry:də'ri:] **die; ~** *(abwertend)* prudery; prudishness

prüfen ['pry:fn̩] *tr. V.* **a)** *auch itr.* test ⟨*pupil*⟩ (**in** + *Dat.* in); *(beim Examen)* examine ⟨*pupil, student, etc.*⟩ (**in** + *Dat.* in); **mündlich/schriftlich geprüft werden** have an oral/a written test/examination; **b)** *(untersuchen)* examine (**auf** + *Akk.* for); check, examine ⟨*device, machine, calculation*⟩ (**auf** + *Akk.* for); investigate, look into ⟨*complaint*⟩; *(testen)* test (**auf** + *Akk.* for); **c)** *(kontrollieren)* check ⟨*papers, passport, application, calculation, information, correctness, etc.*⟩; audit, check, examine ⟨*accounts, books*⟩; **d)** *(vor einer Entscheidung)* check ⟨*price*⟩; examine ⟨*offer*⟩; consider ⟨*application*⟩; **drum prüfe, wer sich ewig bindet** *(Spr.)* marry in haste, repent at leisure *(prov.)*; **e)** *(geh.: großen Belastungen aussetzen)* try; **sie ist vom Leben schwer geprüft worden** her life has been a hard trial. **3.** *refl. V.* search one's heart

Prüfer der; ~s, ~, Prüferin die; ~, ~nen a) tester; inspector; *(Buch~)* auditor; **b)** *(im Examen)* examiner

Prüfling ['pry:flɪŋ] **der; ~s, ~e** examinee; [examination] candidate

Prüf-: ~stand der *(Technik)* test bed; test stand; **~stein der** touchstone (**für** for, of); measure (**für** of)

Prüfung die; ~, ~en a) *(Examen)* examination; exam *(coll.)*; **eine ~ machen** *od.* **ablegen** take *or* do an examination; **b)** *(das [Über]prüfen) s.* **prüfen b–d:** examination; investigation; *(Kontrolle)* check; *(das Kontrollieren)* checking *no indef. art.*; *(Test)* test; *(das Testen)* testing *no indef. art.*; **nach/bei ~ Ihrer Beschwerde** after/on examining *or* investigating your complaint; **c)** *(geh.: schicksalhafte Belastung)* trial

Prüfungs-: ~angst die examination phobia; *(im Einzelfall)* examination nerves *pl.*; **~ordnung die** examination regulations *pl.*

Prügel ['pry:gl̩] **der; ~s, ~ a)** stick; cudgel; **b)** *Pl. (Schläge)* beating *sing.*; *(als Strafe für Kinder)* hiding *(coll.)*; **~ beziehen** get a hiding *(coll.) or* beating

Prügelei die; ~, ~en *(ugs.)* punch-up *(coll.)*; fight

Prügel·knabe der whipping-boy

prügeln 1. *tr. (auch itr.) V.* beat **2.** *refl. V.* **sich ~:** fight; **sich mit jmdm. [um etw.] ~:** fight sb. [over *or* for sth.]

Prügel·strafe die corporal punishment *no art.*

Prunk [prʊŋk] **der; ~[e]s** splendour; magnificence

Prunk·stück das showpiece

prunk·voll 1. *Adj.* magnificent; splendid. **2.** *adv.* magnificently; splendidly; magnificently, splendidly, sumptuously ⟨*furnished, decorated*⟩

prusten ['pru:stn̩] *itr. V. (ugs.) (schnauben)* snort; **vor Lachen ~:** snort with laughter

PS [pe:'ɛs] **das; ~, ~:** *Abk.* **a) Pferdestärke** h.p.; **b) Postskript[um]** PS

Psalm [psalm] **der; ~s, ~en** psalm

pseudo-, Pseudo- [psɔydo-] *(abwertend)* pseudo-

Pseudonym [psɔydo'ny:m] **das; ~s, ~e** pseudonym; *(eines Schriftstellers)* pseudonym; nom de plume; pen-name

Psyche ['psy:çə] **die; ~, ~n** psyche

Psychiater [psy'çi̯a:tɐ] **der; ~s, ~:** psychiatrist

Psychiatrie [psyçi̯a'tri:] **die; ~** psychiatry *no art.*

psychiatrisch 1. *Adj.; nicht präd.* psychiatric. **2.** *adv.* **jmdn. ~ behandeln** give sb. a psychiatric treatment

psychisch 1. *Adj.* psychological; psychological, mental ⟨*strain, disturbance, process*⟩; mental ⟨*illness*⟩. **2.** *adv.* psychologically; **~ gesund/krank sein** be mentally fit/ill; **ein ~ bedingtes Leiden** an illness of psychological origin

psycho-, Psycho- [psy:ço-]: **~analyse die** psychoanalysis *no art.*; **~analytiker der, ~analytikerin die** psychoanalyst; **~analytisch 1.** *Adj.* psychoanalytical; **2.** *adv.* psychoanalytically; **~krimi** ['----] **der** *(ugs.)* psychological thriller; **~loge** [~'lo:gə] **der; ~n, ~n** psychologist; **~logie** [~lo'gi:] **die; ~:** psychology; **~login die** *s.* **~loge; ~logisch 1.** *Adj.* psychological; **2.** *adv.* psychologically; **~path** [~'pa:t] **der; ~en, ~en, ~pathin die; ~, ~nen** psychopath; **~terror** ['----] **der** psychological intimidation; **~therapeut der, ~therapeutin die** psychotherapist; **~therapeutisch 1.** *Adj.* psychotherapeutic; **2.** *adv.* **jmdn. ~therapeutisch behandeln** give sb. psychotherapeutic treatment; **~therapie die** psychotherapy *no art.*; **~thriller** ['----] **der** *s.* **~krimi**

PTA [pe:te:'|a:] **die; ~, ~[s]** *Abk.* **pharmazeutisch-technische Assistentin** pharmaceutical-laboratory assistent

pubertär [pubɛr'tɛ:ɐ̯] **1.** *Adj.*

pubertal. **2.** *adv.* ~ **bedingt** caused by puberty *postpos.*

Pubertät [pubɛr'tɛ:t] **die; ~:** puberty

Publicity [pʌ'blɪsɪtɪ] **die; ~:** publicity

publik [pu'bli:k] *Adj.* *in* ~ **sein/werden** be/become public knowledge; **etw. ~ machen** make sth. public

Publikation [publika'tsi̯o:n] **die; ~, ~en** publication

Publikum ['pu:blikʊm] **das; ~s a)** *(Zuschauer, Zuhörer)* audience; *(beim Sport)* crowd; **b)** *(Kreis von Interessierten)* public; **c)** *(Besucher)* clientele

publikums-, Publikums-: ~erfolg der success with the public; **~liebling der** idol of the public; **~wirksam** *Adj.* with public appeal *postpos., not pred.*; punchy ‹*headline*›; ‹*headline*› with a strong appeal; effective, compelling ‹*broadcast*›

publizieren [publi'tsi:rən] *tr. (auch itr.) V.* publish

Publizist der; ~en, ~en, Publizistin die; ~, ~nen commentator on politics and current affairs; publicist

Puck [pʊk] **der; ~s, ~s** *(Eishokkey)* puck

Pudding ['pʊdɪŋ] **der; ~s, ~e** *od.* **~s** ≈ blancmange

Pudding·pulver das ≈ blancmange powder

Pudel ['pu:dl̩] **der; ~s, ~** poodle; **das war also des ~s Kern** *(fig.)* so 'that's what was behind it; **wie ein begossener ~ dastehen** *(ugs.)* stand there sheepishly

Pudel·mütze die bobble *or* pom-pom hat

pudel·wohl *Adv.* *(ugs.)* **sich ~wohl fühlen** feel on top of the world

Puder ['pu:dɐ] **der; ~s, ~:** powder

Puder·dose die powder compact

pudern *tr. V.* powder; **sich** *(Dat.)* **die Nase ~:** powder one's nose

Puder-: ~quaste die powder-puff; **~zucker der** icing sugar *(Brit.)*; confectioners' sugar *(Amer.)*

[1]**Puff der; ~[e]s, Püffe** ['pʏfə] *(ugs.)* *(Stoß)* thump; *(leichter/kräftiger Stoß mit dem Ellenbogen)* nudge/dig

[2]**Puff der** *od.* **das; ~s, ~s** *(salopp: Bordell)* knocking-shop *(sl.)*; brothel

puffen *(ugs.)* **1.** *tr. V.* **a)** *(stoßen)* thump; *(mit dem Ellenbogen)* nudge; dig; **b)** *(irgendwohin befördern)* push; shove; *(mit dem Ellenbogen)* elbow. **2.** *itr. V.* ‹*locomotive*› puff

Puffer der; ~s, ~ *(Vorrichtung)* buffer

Puff·mutter die *(salopp)* madam

puh [pu:] *Interj.* ugh; *(erleichtert)* phew

Pulk [pʊlk] **der; ~[e]s, ~s** *od.* **~e a)** *(Milit.: Verband)* group; **b)** *(Menge)* crowd

Pulle ['pʊlə] **die; ~, ~n** *(salopp)* bottle; **volle ~** *(fig. salopp)* flat out

Pulli ['pʊli] **der; ~s, ~s** *(ugs.)*,

Pullover [pʊ'lo:vɐ] **der; ~s, ~:** pullover; sweater

Pullunder [pʊ'lʊndɐ] **der; ~s, ~:** slipover

Puls [pʊls] **der; ~es, ~e** pulse; **jmds. ~ fühlen/messen** feel/take sb.'s pulse

Puls·ader die artery; **sich** *(Dat.)* **die ~n aufschneiden** slash one's wrists

pulsieren *itr. V.* *(auch fig.)* pulsate; ‹*blood*› pulse

Puls-: ~schlag der *(auch fig.)* pulse; *(einzelner ~schlag)* beat; **~wärmer der** wristlet

Pult [pʊlt] **das; ~[e]s, ~e a)** desk; *(Lese~)* lectern; desk; **b)** *(Schalt~)* control desk; console

Pulver ['pʊlfɐ] **das; ~s, ~ a)** powder; **b)** *(Schieß~)* [gun]powder; **das ~ hat er [auch] nicht [gerade] erfunden** *(ugs.)* he'll never set the world *or (Brit.)* the Thames on fire; **sein ~ verschossen haben** *(fig. ugs.)* have shot one's bolt

Pulver·faß das barrel of gunpowder; *(kleiner)* powder-keg; **auf einem** *od.* **dem ~ sitzen** *(fig.)* be sitting on a powder-keg *or* on top of a volcano

pulverisieren *tr. V.* pulverize; powder

Pulver-: ~kaffee der instant coffee; **~schnee der** powder snow

Puma ['pu:ma] **der; ~s, ~s** puma

Pummel ['pʊml̩] **der; ~s, ~** *(ugs.)*,

Pummelchen das; ~s, ~ *(ugs.)* podge

pumm[e]lig ['pʊm(ə)lɪç] *Adj.* *(ugs.)* chubby

Pump [pʊmp] **der; ~s** *(salopp)* **auf ~:** on tick *(coll.)*

Pumpe ['pʊmpə] **die; ~, ~n** pump

pumpen *tr., itr. V.* **a)** *(auch fig.)* pump; **b)** *(ugs.: verleihen)* lend; **c)** *(ugs. entleihen)* borrow

Pumper·nickel der pumpernickel

Pumps [pœmps] **der; ~, ~:** court shoe

Punk [paŋk] **der; ~[s], ~s** punk

Punker ['paŋkɐ] **der; ~s, ~ a)** *(Musiker)* punk rocker; **b)** *(Anhänger)* punk

Punkt [pʊŋkt] **der; ~[e]s, ~e a)** *(Tupfen)* dot; *(größer)* spot; **das ist [nicht] der springende ~** *(fig.)* that's [not] the point; **ein dunkler ~ [in jmds. Vergangenheit]** a dark chapter [in sb.'s past]; **b)** *(Satzzeichen)* full stop; period; **nun mach [aber] mal einen ~!** *(fig. ugs.)* come off it! *(coll.)*; **ohne ~ und Komma reden** *(ugs.)* talk nineteen to the dozen *(Brit.)*; rabbit *(Brit. coll.) or* talk on and on; **c)** *(I-Punkt)* dot; **d)** *(Stelle)* point; **ein schwacher/wunder ~** *(fig.)* a weak/sore point; **die Verhandlungen waren an einem toten ~ angelangt** the talks had reached deadlock *or* an impasse; **e)** *(Gegenstand, Thema, Abschnitt)* point; *(einer Tagesordnung)* item; point; **f)** *(Bewertungs~)* point; *(bei einer Prüfung)* mark; **nach ~en siegen** win on points; **g)** *(Musik)* dot; **h)** *(Math.)* point; **i)** *(Zeit~)* point; **~ 12 Uhr** at 12 o'clock on the dot

punkt·gleich *Adj.* *(Sport)* level on points *pred.*

Punkt·gleichheit die *(Sport)* **bei ~:** if the same number of points have been scored

punktieren *tr. V.* **a)** dot ‹*line etc.*›; **b)** *(Med.)* puncture; **c)** *(Musik)* dot ‹*note*›

pünktlich ['pʏŋktlɪç] **1.** *Adj.* punctual; **der Zug ist ~/nicht ~:** the train is on time/is late. **2.** *adv.* punctually; on time; **~ um 20 Uhr** at 8 o'clock sharp

Pünktlichkeit die; ~: punctuality

Punkt-: ~niederlage die *(Sport)* defeat on points; **~richter der** *(Sport)* judge; **~sieg der** *(Sport)* win on points; points win

punktuell [pʊŋk'tu̯ɛl] **1.** *Adj.* isolated ‹*interventions, checks, approaches, initiatives, etc.*›. **2.** *adv.* **sich mit einem Thema nur ~ befassen** deal only with certain *or* particular points relating to a topic

Punkt·zahl die score; number of points

Punsch [pʊnʃ] **der; ~[e]s, ~e** *od.* **Pünsche** ['pʏnʃə] punch

Pupille [pu'pɪlə] **die; ~, ~n** pupil

Puppe ['pʊpə] **die; ~, ~n a)** doll[y]; **b)** *(Marionette)* puppet; marionette; *(fig.)* puppet; **die ~n tanzen lassen** *(fig. ugs.)* pull the strings; *(es hoch hergehen lassen)* paint the town red *(sl.)*; **c)** *(salopp: Mädchen)* bird *(sl.)*; **d)** *(Zool.)* pupa; **e) bis in die ~n** *(ugs.)* till all hours

Puppen-: ~haus das *s.* **~stube; ~spiel das** *(Stück)* puppet show; **~spieler der** puppeteer; **~stube die** doll's house; dollhouse *(Amer.)*; **~theater das** puppet theatre; **~wagen der** doll's pram

pur [pu:ɐ̯] *Adj.* **a)** *(rein)* pure; **b)** *(unvermischt)* neat; straight; **bitte einen Whisky ~!** a neat whisky, please; **c)** *(bloß)* sheer; pure

Püree [py're:] **das; ~s, ~s a)** purée; **b)** *s.* **Kartoffelbrei**

Puritaner [puri'ta:nɐ] **der; ~s, ~ a)** Puritan; **b)** *(fig.)* puritan

puritanisch *Adj.* **a)** Puritan; **b)** *(fig.)* puritanical

Purpur ['pʊrpʊr] **der; ~s** crimson

Purpur·mantel der crimson *or* purple robe

purpur·rot *Adj.* crimson

Purzel·baum der *(ugs.)* somersault; **einen ~ machen** *od.* **schlagen** do *or* turn a somersault

purzeln ['pʊrtsl̩n] *itr. V.; mit sein (fam.)* tumble

Puste ['pu:stə] **die; ~** *(salopp)* puff; breath; **ganz aus der** *od.* **außer ~ sein** be out of puff; be puffed [out]; *s. auch* **ausgehen 1 b**

Pustel ['pʊstl̩] **die; ~, ~n** pimple; spot; pustule *(Med.)*

pusten *(ugs.)* **1.** *itr. V.* **a)** ⟨*person, wind*⟩ blow; **b)** *(keuchen)* puff [and pant *or* blow]. **2.** *tr. V.* blow

Pute ['pu:tə] **die; ~, ~n a)** turkey hen; *(als Braten)* turkey; **b)** *(salopp abwertend: Mädchen, Frau)* **eine dumme ~:** a silly goose *or* creature; **eine eingebildete ~:** a stuck-up little madam

Puter der; ~s, ~: turkeycock; *(als Braten)* turkey

puter·rot *Adj.* scarlet; bright red

Putsch [pʊtʃ] **der; ~[e]s, ~e** putsch; coup [d'état]

putschen *itr. V.* organize a putsch *or* coup

Putschist der; ~en, ~en putschist; rebel

Putsch·versuch der attempted putsch *or* coup

Putte ['pʊtə] **die; ~, ~n** *(Kunstwiss.)* putto

Putz [pʊts] **der; ~es** plaster; *(Rauh~)* roughcast; *(für Außenmauern)* rendering; **auf den ~ hauen** *(fig. salopp) (angeben)* boast; brag; *(ausgelassen feiern)* have a rave-up *(Brit. sl.)*; **~ machen** *(fig. salopp)* cause aggro *(Brit. sl.)*

putzen *tr. V.* **a)** *(blank reiben)* polish; **b)** *(säubern)* clean; groom ⟨*horse*⟩; **[sich** *(Dat.)***] die Zähne/die Nase ~:** clean *or* brush one's teeth/blow one's nose; **sich ~** ⟨*cat*⟩ wash itself; ⟨*bird*⟩ preen itself; **c)** *auch itr. (bes. rhein., südd., schweiz.: saubermachen)* clean ⟨*room, shop, etc.*⟩; **~ gehen** work as a cleaner *or (Brit.)* char[woman]; **d)** *(zum Kochen vorbereiten)* wash and prepare ⟨*vegetables*⟩

Putz-: ~fimmel der; *o. Pl. (ugs. abwertend)* mania for cleaning; **~frau die** cleaner; char[lady] *(Brit.)*

putzig *(ugs.)* **1.** *Adj. (entzückend)* sweet; cute *(Amer.)*; *(possierlich)* funny; comical. **2.** *adv. (entzückend)* sweetly; cutely *(Amer.)*; *(possierlich)* comically

putz-, Putz-: ~lappen der [cleaning-]rag; cloth; **~munter** *(ugs.)* **1.** *Adj.* chirpy *(coll.)*; perky; **~munter sein** be as bright as a button; **2.** *adv.* chirpily *(coll.)*; perkily; **~tuch das** *s.* **~lappen**

Puzzle ['pazl̩] **das; ~s, ~s, Puzzle·spiel das** jigsaw [puzzle]

PVC [pe:fau'tse:] **das; ~[s]** *Abk.* **Polyvinylchlorid** PVC

Pygmäe [pʏ'gmɛ:ə] **der; ~n, ~n** pygmy

Pyjama [pʏ'dʒa:ma] **der** *(österr., schweiz. auch:* **das); ~s, ~s** pyjamas *pl.*

Pyramide [pʏra'mi:də] **die; ~, ~n** pyramid

Pyrenäen [pʏre'nɛ:ən] *Pl.* **die ~:** the Pyrenees

Pyromane [pʏro'ma:nə] **der; ~n, ~n** *(Psych.)* pyromaniac

Q

q, Q [ku:] **das; ~, ~:** q, Q; *s. auch* **a, A**

qm *Abk.* **Quadratmeter** sq. m.

Quacksalber ['kvakzalbɐ] **der; ~s, ~** *(abwertend)* quack [doctor]

Quaddel ['kvadl̩] **die; ~, ~n** [irritating] spot

Quader ['kva:dɐ] **der; ~s, ~** *od. (österr.:)* **~n a)** ashlar block; [rectangular] block of stone; **b)** *(Geom.)* cuboid

Quadrat [kva'dra:t] **das; ~[e]s, ~e a)** *(Geom., Math.)* square; **6 cm im ~:** 6 cm. square; **drei im** *od.* **zum ~:** three squared; **b)** *(bebaute Fläche)* block [of houses]

Quadrat- square ⟨*kilometre etc.*⟩

quadratisch *Adj.* **a)** square; **b)** *(Math.)* quadratic

Quadrat·meter der *od.* **das** square metre

Quadratur [kvadra'tu:ɐ̯] **die; ~, ~en** *(Math., Astron.)* quadrature; **die ~ des Kreises** *(geh.)* the achievement of the impossible

Quadrat-: ~wurzel die *(Math.)* square root **(aus** of); **~zahl die** square number

quadrieren *(Math.)* **1.** *tr. V.* square. **2.** *itr. V.* square numbers

quadro-, Quadro- ['kva:dro-] quadraphonic ⟨*system, effect*⟩

quaken ['kva:kn̩] *itr. V.* ⟨*duck*⟩ quack; ⟨*frog*⟩ croak

quäken ['kvɛ:kn̩] **1.** *tr. V.* squawk; bawl out ⟨*song*⟩. **2.** *itr. V.* **a)** ⟨*voice*⟩ squawk; *(kreischen)* screech; ⟨*radio*⟩ blare; **b)** *(klagen)* ⟨*child*⟩ whine, whinge

Qual [kva:l] **die; ~, ~en a)** *o. Pl.* torment; agony *no indef. art.*; **[für jmdn.] eine ~ sein** be agony *or* torment for sb.; **er macht uns** *(Dat.)* **das Leben zur ~:** he's making our lives *pl.* a misery; **er hat die ~ der Wahl** *(scherzh.)* he is spoilt for choice; **b)** *meist Pl. (Schmerzen)* agony; **~en** pain *sing.*; agony *sing.*; *(seelisch)* torment *sing.*; **jmdn. von seinen ~en erlösen** put sb. out of his/her agony

quälen ['kvɛ:lən] **1.** *tr. V.* **a)** *(körperlich, seelisch)* torment ⟨*person, animal*⟩; maltreat, be cruel to ⟨*animal*⟩; *(foltern)* torture; **~de Ungewißheit** agonizing uncertainty; **b)** *(plagen)* ⟨*cough etc.*⟩ plague; *(belästigen)* pester. **2.** *refl. V.* **a)** *(leiden)* suffer; **b)** *(sich abmühen)* struggle

Quälerei die; ~, ~en a) torment; *(Folter)* torture; *(Grausamkeit)* cruelty; **Tierversuche sind [eine] reine ~:** animal experiments are simply cruel; **b)** *(das Belästigen)* pestering; **c)** *o. Pl. (ugs.: große Anstrengung)* struggle

Quäl·geist der; *Pl.* **~er** *(fam.)* pest

Qualifikation [kvalifika'tsio:n] **die; ~, ~en a)** *(Ausbildung)* qualifications *pl.*; **b)** *(Befähigung)* capability; **c)** *(Sport)* qualification; *(Wettkampf)* qualifier; qualifying round

Qualifikations·spiel das *(Sport)* qualifier, qualifying match

qualifizieren [kvalifi'tsi:rən] *refl.*

V. **a)** gain qualifications; **sich zum Facharbeiter ~:** gain the qualifications needed to be a skilled worker; **b)** *(Sport)* qualify

qualifiziert 1. *Adj.* **a)** ‹*work, post*› requiring particular qualifications; **b)** *(sachkundig)* competent; skilled ‹*work*›. **2.** *adv. (sachkundig)* competently

Qualität [kvali'tɛ:t] **die; ~, ~en** quality

qualitativ [kvalita'ti:f] **1.** *Adj.* qualitative; ‹*difference, change*› in quality. **2.** *adv.* with regard to quality

Qualitäts-: ~arbeit die high-quality workmanship; **~erzeugnis das** quality product

Qualle ['kvalə] **die; ~, ~n** jellyfish; **~n** jellyfish

Qualm [kvalm] **der; ~[e]s** [thick] smoke

qualmen 1. *itr. V.* **a)** give off clouds of [thick] smoke; **aus dem Kamin qualmt es** clouds of [thick] smoke are coming from the fireplace; **b)** *(ugs.: rauchen)* puff away. **2.** *tr. V. (ugs.: rauchen)* puff away at ‹*cigarette etc.*›

qual·voll 1. *Adj.* agonizing. **2.** *adv.* agonizingly; **~ sterben** die in great pain

Quanten *Pl. (salopp)* dirty great feet *(sl.)*

Quantität [kvanti'tɛ:t] **die; ~, ~en** quantity; *(Zahl)* number

quantitativ [kvantita'ti:f] **1.** *Adj.* quantitative. **2.** *adv.* quantitatively

Quantum ['kvantʊm] **das; ~s, Quanten** quota (**an** + *Dat.* of); *(Dosis)* dose

Quarantäne [karan'tɛ:nə] **die; ~, ~n** quarantine

Quark [kvark] **der; ~s a)** quark; [sour skim milk] curd cheese; **b)** *(ugs. abwertend) s.* **Käse b**

Quart [kvart] **die; ~, ~en** *(Musik) s.* **Quarte**

Quarta ['kvarta] **die; ~, Quarten** *(Schulw.)* **a)** *(veralt.)* third year *(of a Gymnasium)*; **b)** *(österr.)* fourth year *(of a Gymnasium)*

Quartal [kvar'ta:l] **das; ~s, ~e** quarter [of the year]

Quarte die; ~, ~n *(Musik)* fourth

Quarten *s.* **Quarta**

Quartett [kvar'tɛt] **das; ~[e]s, ~e a)** *(Musik, fig.)* quartet; **b)** *o. Pl. (Spiel) card-game in which one tries to get sets of four;* ≈ Happy Families; **c)** *(Spielkarten)* pack *(Brit.) or (Amer.)* deck of cards for *Quartett; (Satz von vier Karten)* set of four *Quartett* cards

Quartier [kvar'ti:ɐ̯] **das; ~s, ~e** accommodation *no indef. art.;* accommodations *pl. (Amer.);* place to stay; *(Mil.)* quarters *pl.;* **bei jmdm. ~ beziehen** put up *or* move in with sb.

Quarz [kva:ɐ̯ts] **der; ~es, ~e** quartz

Quarz·uhr die quartz clock; *(Armbanduhr)* quartz watch

quasi ['kva:zi] *Adv.* **[so] ~:** more or less; *(so gut wie)* as good as

quasi-, Quasi- quasi-

quasseln ['kvasl̩n] *(ugs.)* **1.** *itr. V.* chatter; rabbit on *(Brit. sl.)* **(von** about). **2.** *tr. V.* spout, babble ‹*nonsense*›

Quassel·strippe die *(ugs. abwertend)* chatterbox

Quaste die; ~, ~n tassel

Quatsch [kvatʃ] **der; ~[e]s a)** *(ugs. abwertend) (Äußerung)* rubbish; *(Handlung)* nonsense; **so ein ~!** what rubbish!; **b)** *(ugs.: Unfug)* messing about; **laß den ~:** stop messing about; **mach keinen ~:** don't do anything stupid; **c)** *(ugs.: Jux)* lark *(coll.);* **~ machen** fool around; lark about *(coll.);* **aus ~:** for a laugh

quatschen 1. *itr. V.* **a)** *(ugs.) (dumm reden)* rabbit on *(Brit. coll.);* blather; *(viel reden)* chatter; natter *(Brit. coll.);* **b)** *(ugs.: klatschen)* gossip; **c)** *(ugs.: Geheimes ausplaudern)* blab; open one's mouth; **d)** *(ugs.: sich unterhalten)* [have a] chat *or (coll.)* natter (**mit** with). **2.** *tr. V.* spout ‹*nonsense, rubbish*›

Quatsch·kopf der *(salopp)* stupid chatterbox; *(Schwätzer, Schwafler)* windbag

Queck·silber ['kvɛk-] **das** mercury; *(fig.)* quicksilver

Quelle ['kvɛlə] **die; ~, ~n a)** spring; *(eines Baches, eines Flusses)* source; **b)** *(fig.)* source; **an der ~ sitzen** *(ugs.) (für Informationen)* have access to inside information; *(für günstigen Erwerb)* be at the source of supply

quellen *unr. itr. V.; mit sein* **a)** ‹*liquid*› gush, stream; *(aus der Erde)* well up; ‹*smoke*› billow; ‹*crowd*› stream, pour; *(fig.)* ‹*tears*› well up; **b)** *(sich ausdehnen)* swell [up]

Quell·wasser das; *Pl.* **~wasser** spring water

quengelig *(ugs.)* **1.** *Adj.* whining; fretful. **2.** *adv.* in a whining voice; fretfully

quengeln ['kvɛŋl̩n] *itr. V. (ugs.)* **a)** ‹*baby*› whimper, *(coll.)* grizzle; **b)** *(nörgeln)* carp

Quentchen ['kvɛntçən] **das; ~s, ~** *(veralt.)* scrap; **ein ~ Glück** *(fig. geh.)* a little bit of luck

quer [kve:ɐ̯] *Adv.* sideways; crosswise; *(schräg)* diagonally; at an angle; **~ zu etw.** at an angle to sth.; *(rechtwinklig)* at right angles to sth.; **der Wagen steht ~ auf der Fahrbahn** the car is standing sideways across the road; **~ durch/über** *(+ Akk.)* straight through/across

Quer-: ~achse die transverse axis; **~balken der a)** crossbeam; **b)** *(Musik)* stroke; **~denker der** lateral thinker

Quere die *in* **jmdm. in die ~ kommen** *od.* **geraten** *(fig.)* get in sb.'s way *(coll.)*

quer·feld·ein *Adv.* across country

quer-, Quer-: ~flöte die transverse flute; **~format das** landscape format; **~kopf der** *(ugs.)* awkward cuss *(coll.);* **~|legen** *refl. V. (ugs.)* make difficulties; **~paß der** *(Sport)* crossfield pass; cross; lateral pass *(Amer.);* **~schiff das** *(Archit.)* transept; **~schnitt der** *(auch fig.)* cross-section; **~schnitt[s]·gelähmt** *Adj.* *(Med.)* paraplegic; **~schnitt[s]·lähmung die;** *o. Pl. (Med.)* paraplegia *no indef. art.;* paraplegic condition; **~straße die** intersecting road; **~summe die** *(Math.)* sum of the digits (**von, aus** of); **~treiber der** *(ugs. abwertend)* trouble-maker; **~verbindung die** link; *(Straße)* link [road]

quetschen 1. *tr. V.* **a)** crush ‹*person, limb, thorax*›; **sich** *(Dat.)* **den Arm/die Hand ~:** get one's arm/hand caught; **sich** *(Dat.)* **den Finger/die Zehe ~:** pinch one's finger/toe; **b)** *(drücken, pressen)* squeeze, squash (**gegen, an** + *Akk.* against, **in** + *Akk.* into). **2.** *refl. V.* **sich in/durch etw.** *(Akk.)* **~** *(ugs.)* squeeze into/through sth.

Quetschung die; ~, ~en bruise; contusion *(Med.)*

quick·lebendig *Adj.* [very] lively; active; *(bes. im Alter)* sprightly; spry; frisky ‹*small animal*›

quiek[s]en ['kvi:k(s)n̩] squeak; ‹*piglet*› squeal

quietschen ['kvi:tʃn̩] *itr. V.* squeak; ‹*brakes, tyres, crane*› squeal, screech; *(ugs.)* ‹*person*› squeal; shriek

quietsch-: ~fidel *(ugs.) Adj.* **a)** [really] chirpy *(coll.) or (esp. Amer.)* chipper; **b)** *(gesund und munter)* bright-eyed and bushy-tailed *pred. (coll.);* **~vergnügt** *Adj. (ugs.) s.* **~fidel a**

Quint [kvɪnt] **die; ~, ~en** *(Musik) s.* **Quinte**

Quinta ['kvɪnta] **die; ~, Quịnten** *(Schulw.)* **a)** *(veralt.)* second year *(of a Gymnasium);* **b)** *(österr.)* fifth year *(of a Gymnasium)*

Quinte ['kvɪntə] **die; ~, ~n** *(Musik)* fifth

Quịnt·essenz die *(geh.)* essential point; essence

Quintett [kvɪn'tɛt] **das; ~[e]s, ~e** *(Musik, fig.)* quintet

Quirl [kvɪrl] **der; ~[e]s, ~e** *long-handled blender with a star-shaped head*

quịrlen *tr. V.* ≈ whisk

quịrlig *Adj.* lively

quitt [kvɪt] *Adj.; nicht attr. (ugs.)* quits; **damit sind wir ~:** that makes us quits

Quitte ['kvɪtə] **die; ~, ~n** quince

quittieren *tr. V.* **a)** acknowledge, confirm ‹*receipt, condition*›; receipt, give a receipt for ‹*sum, invoice*›; **b) etw. mit etw. ~** *(fig.)* react *or* respond to sth. with sth.; **etw. mit Pfiffen ~:** greet sth. with catcalls

Quịttung die; ~, ~en a) receipt **(für, über** + *Akk.* for); **b)** *(fig.)* come-uppance *(coll.);* deserts *pl.;* **nun hast du die ~:** you've got what you deserve

Quiz [kvɪs] **das; ~, ~:** quiz

quoll [kvɔl] *1. u. 3. Pers. Sg. Prät. v.* **quellen**

Quote ['kvo:tə] **die; ~, ~n** *(Anteil)* proportion; *(Zahl)* number

Quoten·regelung die *requirement that women should be adequately represented*

Quotient [kvo'tsi̯ɛnt] **der; ~en, ~en** *(Math.)* quotient **(aus** of)

R

r, R [ɛr] **das; ~, ~:** r, R; **er rollt das R** he rolls his r's; *s. auch* **a, A**

Rabatt [ra'bat] **der; ~[e]s, ~e** discount

Rabatte [ra'batə] **die; ~, ~n** border

Rabatz [ra'bats] **der; ~es** *(ugs.)* **a)** *(Lärm)* racket; din; **b)** *(Protest)* **~ machen** kick up a fuss, *(coll.)* raise a stink **(bei** with)

Rabauke [ra'baukə] **der; ~n, ~n** *(ugs.)* roughneck *(coll.); (Rowdy)* hooligan

Rabbi ['rabi] **der; ~[s], ~nen** [ra'bi:nən] *od.* **~s** rabbi; *(Titel)* Rabbi

Rabbiner [ra'bi:nɐ] **der; ~s, ~:** rabbi

Rabe ['ra:bə] **der; ~n, ~n** raven; **schwarz wie ein ~** *(ugs.) (schmutzig)* as black as soot; **stehlen wie ein ~** *(ugs.)* pinch everything one can lay one's hands on *(coll.)*

Raben·mutter die *(abwertend)* uncaring [brute of a] mother

raben·schwạrz *Adj.* jet-black; raven-black ‹*beard, hair*›; pitch-black ‹*night*›

rabiat [ra'bi̯a:t] **1.** *Adj.* violent; brutal; savage ‹*kick*›; ruthless ‹*methods*›. **2.** *adv.* violently; brutally

Rache ['raxə] **die; ~:** revenge; **[an jmdm.] ~ nehmen** take revenge [on sb.]; **aus ~:** in revenge

Rachen ['raxn̩] **der; ~s, ~ a)** *(Schlund)* pharynx *(Anat.);* **b)** *(Maul)* mouth; maw *(literary); (fig.)* jaws *pl.*

rächen ['rɛçn̩] **1.** *tr. V.* avenge ‹*person, crime*›; take revenge for ‹*insult, crime*›. **2.** *refl. V.* **a)** take one's revenge; **sich an jmdm. [für etw.] ~:** take one's revenge on sb. [for sth.]; get even with sb. [for sth.]; **b)** *(fig.)* ‹*mistake[s], bad behaviour*› take its/their toll

Rachitis [ra'xi:tɪs] **die; ~** *(Med.)* rickets *sing.*

Rạch·sucht die; *o. Pl. (geh.)* lust for revenge

rạch·süchtig *(geh.) Adj.* vengeful; **~ sein** be out for revenge

rạckern ['rakɐn] *itr. V. (ugs.)* drudge; toil

Rad [ra:t] **das; ~es, Räder** ['rɛ:dɐ] **a)** wheel; **fünftes** *od.* **das fünfte ~ am Wagen sein** *(fig. ugs.)* be superfluous; *(die Harmonie stören)* be in the way; **unter die Räder kommen** *(fig. ugs.)* fall into bad ways; **b) nur ein ~ im Getriebe sein** be just a small cog in the machine; **c)** *(Fahr~)* bicycle; bike *(coll.);* **mit dem ~ fahren** go by bicycle *or (coll.)* bike; *s. auch* **radfahren; d)** *(Turnen)* cartwheel; **ein ~ schlagen** do a cartwheel; *s. auch* **radschlagen**

Radar [ra'da:ɐ̯] **der** *od.* **das; ~s** *(Technik)* radar

Radạr: ~falle die *(ugs.)* [radar] speed trap; **~kontrolle die** [radar] speed check; **~schirm der** radar screen

Radau [ra'dau̯] **der; ~s** *(ugs.)* row *(coll.);* racket

Rädchen ['rɛ:tçən] **das; ~s, ~** [little] wheel; *(Zahnrad)* [small] cog

Rạd·dampfer der paddle-steamer

rạdebrechen 1. *tr. V.* **Französisch/Deutsch** *usw.* **~:** speak broken French/German *etc.* **2.** *itr. V.* speak pidgin

radeln ['ra:dl̩n] *itr. V.; mit sein (ugs., bes. südd.)* cycle

Rädels·führer ['rɛ:dls-] **der** *(abwertend)* ringleader

rädern ['rɛ:dɐn] *tr. V.* **jmdn. ~** *(hist.)* break sb. on the wheel; *s. auch* **gerädert**

rạd-, Rạd-: ~|fahren *unr. itr. V. (Zusammenschreibung nur im Inf. u. 2. Part.); mit sein* **a)** cycle; ride a bicycle *or (coll.)* bike; **b)** *(ugs. abwertend: unterwürfig sein)* suck up to people *(sl.);* **~fahrer der, ~fahrerin die a)** cyclist; **b)** *(ugs. abwertend: Schmeichler)* toady; crawler *(coll.)*

Radi ['ra:di] **der; ~s, ~** *(bayr., österr. ugs.)* [large white] radish

Radiator [ra'di̯a:tɔr] **der; ~s, ~en** [-'to:rən] radiator

Radien *s.* **Radius**

radieren [ra'di:rən] *tr. (auch itr.) V.* erase

Radier·gummi der rubber [eraser]

Radierung die; ~, ~en *(Graphik)* etching

Radieschen [ra'di:sçən] **das; ~s, ~:** radish

radikal [radi'ka:l] **1.** *Adj.* radical; drastic ‹*measure, method, cure*›. **2.** *adv.* radically; *(vollständig)* totally, completely

Radikale der/die; *adj. Dekl.* radical

radikalisieren 1. *tr. V.* make [more] radical. **2.** *refl. V.* become more radical

Radikalịsmus der; ~radicalism; *(Haltung)* radical attitude

Radikalität [radikali'tɛ:t] **die; ~:** radicalness; radical nature

Radio ['ra:di̯o] **das** *(südd., schweiz. auch:* **der**); **~s, ~s** radio; **im ~:** on the radio; **~ hören** listen to the radio

radio-, Radio-: ~aktiv 1. *Adj.* radioactive; **2.** *adv.* radioactively; **~aktiv verseucht** contaminated by radioactivity *postpos.;* **~aktivität die;** *o. Pl.* radioactivity; **~sender** ['----] **der** radio station; **~wecker** ['----] **der** radio alarm clock

Radius ['ra:di̯ʊs] **der; ~, Radien** ['ra:di̯ən] *(Math.)* radius

Rad-: ~**kappe die** hub-cap; ~**lager das** wheel bearing

Radler der; ~s, ~: cyclist

Radon ['ra:dɔn] **das; ~s** *(Chemie)* radon

Rad-: ~**renn·bahn die** cycle-racing track; *(Stadion)* velodrome; ~**rennen das** cycle race; *(Sport)* cycle-racing; ~**rennfahrer der** racing cyclist

-rädrig [-rɛ:drɪç] *Adj.* -wheeled

rad-, Rad-: ~**|schlagen** *unr. itr. V.* do a cart-wheel; *(mehrmals)* do cart-wheels; ~**sport der** cycling *no def. art.*; ~**tour die** cycling tour; ~**weg der** cycle-path *or* -track

raffen ['rafn̩] *tr. V.* **a)** snatch; grab; rake in *(coll.)* ‹*money*›; **etw. [an sich] ~** *(abwertend)* seize sth.; *(eilig)* snatch *or* grab sth.; **b)** *(zusammenhalten)* gather ‹*material, curtain*›; **c)** *(gekürzt wiedergeben)* condense ‹*text*›

Raffinerie [rafinə'ri:] **die; ~, ~n** refinery

Raffinesse [rafi'nɛsə] **die; ~, ~n a)** *o. Pl. (Schlauheit)* guile; ingenuity; **b)** *meist Pl. (Finesse)* refinement

raffinieren [rafi'ni:rən] *tr. V. (bes. Chemie, Geol.)* refine

raffiniert 1. *Adj.* **a)** ingenious ‹*plan, design*›; *(verfeinert)* refined, subtle ‹*colour, scheme, effect*›; sophisticated ‹*dish, cut (of clothes)*›; **b)** *(gerissen)* cunning, artful ‹*person, trick*›. **2.** *adv.* **a)** ingeniously; cleverly; *(verfeinert)* with great refinement/ sophistication; **eine ~ geschnittene Bluse** a blouse with a sophisticated cut; **b)** *(gerissen)* cunningly; artfully

Rage ['ra:ʒə] **die; ~** *(ugs.)* fury; rage; **in ~ sein** be livid *(Brit. coll.)* *or* furious; **jmdn. in ~ bringen** make sb. hopping mad *(coll.)* *or* absolutely furious; **in ~ kommen** fly into a rage

ragen ['ra:gn̩] *itr. V.* **a)** *(vertikal)* rise [up]; ‹*mountains*› tower up; **aus dem Wasser ~:** stick *or* jut right out of the water; **in den Himmel ~:** tower *or* soar into the sky; **b)** *(horizontal)* project, stick out (**in** + *Akk.* into; **über** + *Akk.* over)

Ragout [ra'gu:] **das; ~s, ~s** *(Kochk.)* ragout

Rahm [ra:m] **der; ~[e]s** *(bes. südd., österr., schweiz.)* cream; *s. auch* **abschöpfen**

rahmen *tr. V.* frame

Rahmen der; ~s, ~ a) frame; *(Kfz-W.: Fahrgestell)* chassis; **b)** *(fig.: Bereich)* framework; *(szenischer Hintergrund)* setting; *(Zusammenhang)* context; *(Grenzen)* bounds *pl.*; limits *pl.*; **aus dem ~ fallen** be out of place; stick out; ‹*behaviour*› be unsuited to the occasion; **im ~ einer Sache** *(Gen.)* *(in den Grenzen)* within the bounds of sth.; *(im Zusammenhang)* within the context of sth.; *(im Verlauf)* in the course of sth.; **den ~ sprengen** be out of proportion

Rain [rain] **der; ~[e]s, ~e** margin of a/the field

Rakete [ra'ke:tə] **die; ~, ~n** rocket; *(Lenkflugkörper)* missile

Rallye ['rali] **die; ~, ~s** *(Motorsport)* rally

rammen *tr. V.* ram

Rampe ['rampə] **die; ~, ~n a)** *(Lade~)* [loading] platform; **b)** *(schiefe Fläche)* ramp; *(Auffahrt)* [sloping] drive; **c)** *s.* **Startrampe; d)** *(Theater)* apron; forestage

Rampen·licht das: im ~ [der Öffentlichkeit] stehen be in the lime-light

ramponieren [rampo'ni:rən] *tr. V. (ugs.)* batter; **ramponiert** battered, knocked-about ‹*furniture, phone-box*›; run-down, down-at-heel ‹*dwelling, room*›; shabby ‹*suit*›; dented ‹*confidence*›

Ramsch [ramʃ] **der; ~[e]s, ~e** *(ugs. abwertend)* **a)** *(Ware)* trashy goods *pl.*; **b)** *(Kram)* junk

Ramsch·laden der *(ugs. abwertend)* shop selling trashy goods

ran [ran] *Adv. (ugs.)* **a)** *s.* **heran; b) ~!** *(fang an)* off you go; *(fangen wir an)* let's go; **los, ~ an die Arbeit!** come on, get down to work!; **c) ~!** *(greif[t] an)* go at him/them!

Ranch [rɛntʃ] **die; ~, ~[e]s** ranch

Rand [rant] **der; ~[e]s, Ränder** ['rɛndɐ] **a)** edge; *(Einfassung)* border; *(Hut~)* brim; *(Brillen~, Gefäß~, Krater~)* rim; *(eines Abgrunds)* brink; *(auf einem Schriftstück)* margin; *(Weg~)* verge; *(Stadt~)* edge; outskirts *pl.*; *(fig.)* **etw. am ~e erwähnen** mention sth. in passing; **außer ~ und Band geraten/sein** *(ugs.)* go/be wild (**vor** with); *(rasen)* go/be berserk (**vor** with); **mit etw. [nicht] zu ~e kommen** *(ugs.)* [not] be able to cope with sth.; *s. auch* **Grab; b)** *(Schmutz~)* mark; *(rund)* ring; *(in der Wanne)* tide-mark *(coll.)*; **dunkle Ränder unter den Augen haben** have dark lines under one's eyes; **c) den ~ halten** *(salopp)* shut one's gob *(sl.)* *or* trap *(sl.)*

Randale [ran'da:lə] **die** *(salopp)* riot; **~ machen** riot

randalieren *itr. V.* riot; rampage; *(Radau machen)* create an uproar; **~de Halbstarke** young hooligans on the rampage

Randalierer der; ~s, ~: hooligan

rand-, Rand-: ~**bemerkung die** marginal note *or* comment; ~**erscheinung die** peripheral phenomenon; ~**gebiet das** outlying district; *(fig.)* fringe area; **die ~gebiete einer Stadt** the outskirts of a town; ~**gruppe die** *(Soziol.)* fringe *or* marginal group; ~**streifen der** verge; ~**voll** *Adj.* ‹*glass etc.*› full to the brim (**mit** with); brim-full ‹*glass, cup, bowl*› (**mit** of)

rang [raŋ] *1. u. 3. Pers. Sg. Prät. v.* **ringen**

Rang der; ~[e]s, Ränge ['rɛŋə] **a)** rank; *(in der Gesellschaft)* status; *(in bezug auf Bedeutung, Qualität)* standing; **jmdm./einer Sache den ~ ablaufen** leave sb./sth. far behind; **alles, was ~ und Namen hat** everybody who is anybody; **ein Physiker von ~:** an eminent physicist; **ersten ~es** of the first order; **b)** *(im Theater)* circle; **erster ~:** dress circle; **zweiter ~:** upper circle; **dritter ~:** gallery; **c)** *(Sport) s.* **Platz f**

Range ['raŋə] **die; ~, ~n** *(bes. md.)* [young] tearaway

ran|gehen *unr. itr. V. (ugs.) s.* **herangehen**

Rangelei die; ~, ~en *(ugs.) s.* **Gerangel**

rangeln ['raŋl̩n] *itr. V. (ugs.)* wrestle; struggle; *(kämpfen)* ‹*children*› scrap; **um etw. ~:** scramble *or* tussle for sth.; *(fig.: argumentieren)* wrangle over sth.

Rang·folge die order of precedence

Rangier·bahn·hof der marshalling yard

rangieren [raŋ'ʒi:rən] **1.** *tr., itr. V.* shunt ‹*trucks, coaches*›; switch ‹*cars*› *(Amer.)*. **2.** *itr. V.* be placed; **an letzter Stelle/auf Platz zwei ~:** be placed last/second

Rangierer der; ~s, ~: shunter *(Brit.)*; switchman *(Amer.)*

Rang-: ~**liste die** *(Sport)* ranking list; ~**ordnung die** order of precedence; *(Verhaltensf.)* pecking order

ran|halten *unr. refl. V. (ugs.)* get a move on *(coll.)*; *(bei der Arbeit)* get stuck in *(coll.)*

Ranke ['raŋkə] **die; ~, ~n** *(Bot.)* tendril

Ränke ['rɛŋkə] *Pl. (geh. veralt.)* intrigues; **~ schmieden** *(geh.)* scheme; hatch plots

ranken *itr., refl. V.* climb; grow;

sich um etw. ~: entwine itself around sth.; *(fig. geh.)* ‹*legends, mysteries*› be woven around sth.

Ranken·gewächs das creeper

ran-: ~**|klotzen** *itr. V. (salopp)* get stuck in *(coll.)*; ~**|kommen** *unr. itr. V.; mit sein (ugs.) s.* **herankommen a, c**; ~**|lassen** *unr. tr. V. (ugs.)*; **jmdn. an etw.** *(Akk.)* **nicht ~lassen** not let sb. anywhere near sth.; **laß mich mal ~!** let me have a go!; ~**|machen** *refl. V. (ugs.) s.* **heranmachen a, b**

rann [ran] *1. u. 3. Pers. Sg. Prät. v.* **rinnen**

rannte ['rantə] *1. u. 3. Pers. Sg. Prät. v.* **rennen**

ran|schmeißen *unr. refl. V.* **sich an jmdn.** ~ *(ugs.)* throw oneself at sb.

Ranzen ['rantsn̩] **der; ~s, ~ a)** satchel; **b)** *(salopp: Bauch)* [fat] belly

ranzig *Adj.* rancid

rapid [ra'pi:t] *(südd., österr., schweiz.),* **rapide 1.** *Adj.* rapid. **2.** *adv.* rapidly

Rappe ['rapə] **der; ~n, ~n** black horse

Rappel ['rapl̩] **der; ~s, ~** *(ugs.)* crazy turn *(coll.)*; **du hast wohl einen ~?** are you crazy?

rappeln *itr. V. (ugs.)* rattle (**an** + *Dat.* at); ‹*alarm, telephone*› jangle

Rappen der; ~s, ~: [Swiss] centime

Rapport [ra'pɔrt] **der; ~s, ~e** *(veralt.)* report

Raps [raps] **der; ~es** rape

rar [ra:ɐ̯] *Adj. (knapp)* scarce; *(selten)* rare; **sich ~ machen** *(ugs.)* not be around much *(coll.)*

Rarität [rari'tɛ:t] **die; ~, ~en** rarity

rasant [ra'zant] **1.** *Adj.* **a)** *(ugs.) (schnell)* tremendously fast *(coll.)* ‹*car, horse, runner, etc.*›; tremendous *(coll.)*, lightning *attrib.* ‹*speed, acceleration, development, progress, growth*›; hairy *(coll.)* ‹*driving*›; *(schnittig)* racy ‹*car, styling*›; **b)** *(ugs.) (schwungvoll)* dynamic, lively ‹*show*›; action-packed, exciting ‹*film, story*›; *(rassig)* classy *(sl.)* ‹*woman*›; dashing ‹*style, dress*›. **2.** *adv.* **a)** *(ugs.) (schnell)* at terrific speed *(coll.)*; ‹*increase*› by leaps and bounds; **b)** *(ugs.) (schwungvoll)* dashingly; *(rassig)* stylishly

rasch [raʃ] **1.** *Adj.* quick; quick, rapid ‹*step, progress, decision, action*›; speedy, swift ‹*end, action, decision, progress*›; fast, quick ‹*service, work, pace, tempo, progress*›; **in ~er Folge** in rapid *or* swift succession. **2.** *adv.* quickly; ‹*drive, act*› quickly, fast; ‹*decide, end, proceed*› swiftly, rapidly

rascheln ['raʃl̩n] *itr. V.* rustle; **es raschelte im Stroh** there was a rustling in the straw

rasen ['ra:zn̩] *itr. V.* **a)** *mit sein (ugs.: eilen)* dash *or* rush [along]; *(fahren)* tear *or* race along; *(fig.)* ‹*pulse*› race; **gegen einen Baum ~:** crash [at full speed] into a tree; **b)** *(toben)* ‹*person*› rage; *(wie wahnsinnig)* rave; *(fig.)* ‹*storm, sea, war*› rage

Rasen der; ~s, ~: grass *no indef. art.*; *(gepflegte ~fläche)* lawn; *(eines Spielfeldes usw.)* turf

rasend 1. *Adj.* **a) in ~er Fahrt** at breakneck speed; **b)** *(tobend)* raging; *(wie wahnsinnig)* raving; *(verrückt)* mad; **[vor Wut** *usw.*] **~ werden** be beside oneself [with rage *etc.*]; **die Schmerzen machen mich ~:** the pain is driving me mad; **c)** *(heftig)* violent ‹*jealousy, rage, pain*›; tumultuous ‹*applause*›; **~e Kopfschmerzen haben** have a splitting headache. **2.** *adv. (ugs.)* incredibly *(coll.)* ‹*fast, funny, expensive*›; insanely ‹*jealous*›

Rasen-: ~**mäher** [~mɛ:ɐ] **der; ~s, ~:** lawn-mower; ~**platz** *(Fußball usw.)* pitch; *(Tennis)* grass court

Raserei die; ~, ~en a) *(ugs.: schnelles Fahren)* tearing along *no art.*; **b)** *o. Pl. (das Toben)* [insane] frenzy; *(Wut)* rage

Rasier·apparat der [safety] razor; *(elektrisch)* electric shaver *or* razor

rasieren [ra'zi:rən] *tr. V.* shave; **sich ~:** shave; **sich naß/trocken/elektrisch ~:** have a wet/dry shave/use an electric shaver; **sich** *(Dat.)* **die Beine** *usw.* ~: shave one's legs

Rasierer der; ~s, ~ *(ugs.)* [electric] shaver

Rasier-: ~**klinge die** razor-blade; ~**messer das** cutthroat razor; ~**wasser das** *(nach der Rasur)* aftershave; *(vor der Rasur)* pre-shave lotion; ~**zeug das** shaving things *pl.*

Räson [rɛ'zɔŋ] **die; ~** *in* **zur ~ kommen** come to one's senses; **jmdn. zur ~ bringen** make sb. see reason

Raspel ['raspl̩] **die; ~, ~n a)** rasp; **b)** *(Küchengerät)* grater

raspeln *tr. V.* **a)** *auch itr.* rasp; **an etw.** *(Dat.)* **~:** work away at sth. with a rasp; **b)** *(Kochk.)* grate

Rasse ['rasə] **die; ~, ~n a)** breed; *(Menschen~)* race; **b) ~ haben** *(ugs.)* be terrific *(coll.)*; *(Temperament haben)* have plenty of spirit *or* mettle

Rasse·hund der pedigree dog

Rassel ['rasl̩] **die; ~, ~n** rattle

rasseln *itr. V.* **a)** rattle; **mit seinem Schlüsselbund ~:** jangle one's bunch of keys; **der Wecker rasselt** the alarm goes off with a jangling sound; **b)** *mit sein (sich ~d fortbewegen)* clatter; **c)** *mit sein* **durch eine Prüfung ~** *(salopp)* come unstuck in *or (Amer.)* flunk an exam *(coll.)*

Rassen- racial ‹*discrimination, hatred, segregation, etc.*›

Rasse·pferd das thoroughbred [horse]

rassig *Adj.* spirited, mettlesome ‹*horse*›; spirited, vivacious ‹*woman*›; sporty ‹*car*›; tangy ‹*wine, perfume*›; *(markant)* striking ‹*face, features, beauty*›

Rassismus der; ~: racism; racialism

Rassist der; ~en, ~en, Rassistin die; ~, ~nen racist; racialist

rassistisch 1. *Adj.* racist; racialist. **2.** *adv.* racialistically

Rast [rast] **die; ~, ~en** rest; **~ machen** stop for a break

rasten *itr. V.* rest; take a rest *or* break

Raster der; ~s, ~ a) *(Druckw.)* screen; **b)** *(fig.)* [conceptual] framework

Raster·fahndung die *(Kriminologie) pinpointing of suspects by means of computer analysis of data on many people*

rast-, Rast-: ~**haus das** roadside café; *(an der Autobahn)* motorway restaurant; ~**hof der** [motorway] motel [and service area]; ~**los 1.** *Adj.* restless ‹*person, spirit, life*›; unremitting, ceaseless ‹*work, search*›; **2.** *adv. s. Adj.*: restlessly; unremittingly; ceaselessly; ~**platz der a)** place to rest; **b)** *(an Autobahnen)* parking place *(with benches and WCs)*; picnic area; ~**stätte die** service area

Rasur [ra'zu:ɐ̯] **die; ~, ~en** shave

Rat [ra:t] **der; ~[e]s, Räte** ['rɛ:tə] **a)** *o. Pl. (Empfehlung)* advice; **ein ~:** a piece *or* word of advice; **da ist guter ~ teuer** I/we *etc.* hardly know which way to turn; **ich gab ihm den ~ zu ...:** I advised him to ...; **bei jmdm. ~ suchen** seek sb.'s advice; **jmdm. mit ~ und Tat beistehen** stand by sb. with moral and practical support; **jmdn./etw. zu ~[e] ziehen** consult sb./sth.; **ich wußte [mir] keinen ~ mehr** I was at my wit's end *or* completely at a loss; **b)** *(Gremium)* council; *(So-*

wjet) soviet; **der ~ der Stadt** the town council; **c)** *(Ratsmitglied)* councillor; council member

rät [rɛ:t] *3. Pers. Sg. Präsens v.* **raten**

Rate ['ra:tə] **die, ~, ~n a)** *(Teilbetrag)* instalment; **etw. auf ~n kaufen** buy sth. by instalments *or (Brit.)* on hire purchase *or (Amer.)* on the installment plan; **b)** *(Statistik)* rate

raten 1. *unr. itr. V.* **a)** *(einen Rat, Ratschläge geben)* **jmdm. ~:** advise sb.; **wozu rätst du mir?** what do you advise me to do?; **b)** *(schätzen)* guess; **richtig/falsch ~:** guess right/wrong; **dreimal darfst du ~** *(ugs. iron.)* I'll give you three guesses. **2.** *tr. V.* **a)** *(an~)* **jmdm. ~, etw. zu tun** advise sb. to do sth.; **laß dir das gesagt sein!** you better had [do that]!; *(tu das nicht)* don't you dare do that!; **b)** *(er~)* guess

Raten·zahlung die payment by instalments

Rate·spiel das guessing-game

Rat-: ~geber der adviser; *(Buch)* guide; **~haus das** town hall

ratifizieren [ratifi'tsi:rən] *tr. V.* ratify

Rätin ['rɛ:tɪn] **die; ~, ~nen** councillor

Ration [ra'tsio:n] **die; ~, ~en** ration; *s. auch* **eisern 1d**

rational [ratsio'na:l] **1.** *Adj.* rational. **2.** *adv.* rationally

rationalisieren *tr., itr. V.* rationalize

rationell [ratsio'nɛl] **1.** *Adj.* efficient; *(wirtschaftlich)* economic. **2.** *adv.* efficiently; *(wirtschaftlich, kräftesparend)* economically

rationieren *tr. V.* ration

Rationierung die; ~, ~en rationing *no indef. art.*

rat·los 1. *Adj.* baffled; at a loss *pred.*; helpless ‹*look*›. **2.** *adv.* helplessly; in a baffled way

Ratlosigkeit die; ~: perplexity; helplessness

ratsam ['ra:tza:m] *Adj.; nicht attr.* advisable; *(weise)* prudent

ratsch [ratʃ] *Interj.* zip; *(beim Zerreißen)* rip

Rat·schlag der [piece of] advice; *(Hinweis)* tip; **~schläge** advice *sing.*/tips

Rätsel ['rɛ:tsl̩] **das; ~s, ~ a)** riddle; *(Bilder~, Kreuzwort~ usw.)* puzzle; **b)** *(Geheimnis)* mystery; enigma

rätselhaft 1. *Adj.* mysterious; *(unergründlich)* enigmatic ‹*smile, expression, person*›; baffling ‹*problem*›. **2.** *adv.* mysteriously; *(unergründlich)* enigmatically

rätseln *itr. V.* puzzle, rack one's brains (**über** + *Akk.* over)

Rätsel·raten das a) *s.* **Rätsel a:** solving puzzles/riddles *no art.*; **b)** *(das Rätseln)* puzzling; *(das Raten)* guessing; *(fig.)* guessing-game

Ratte ['ratə] **die; ~, ~n** *(auch fig.)* rat

Ratten·fänger der a) rat-catcher; **der ~ von Hameln** *(Lit.)* the Pied Piper of Hamelin; **b)** *(fig. abwertend)* pied piper

rattern ['ratɐn] *itr. V.* **a)** clatter; ‹*sewing-machine, machine-gun*› chatter; ‹*engine*› rattle; **b)** *mit sein (~d fahren)* clatter [along]

Raub [raup] **der; ~[e]s a)** robbery; *(Entführung)* kidnapping; **b)** *(Beute)* [robber's] loot; stolen goods *pl.*

rauben 1. *tr. V.* steal; kidnap ‹*person*›; **jmdm. etw. ~:** rob sb. of sth.; *(geh.: wegnehmen)* deprive sb. of sth.; **jmdm. den Atem/die Sprache ~:** take sb.'s breath away/render sb. speechless. **2.** *itr. V.* rob; *(plündern)* plunder

Räuber ['rɔybɐ] **der; ~s, ~ a)** robber; **b)** *(Zool.: Tier)* predator

Raub-: ~fisch der predatory fish; **~katze die** wild cat; **~mord der** *(Rechtsw.)* murder (**an** + *Dat.* of) in the course of a robbery *or* with robbery as motive; **~tier das** predator; beast of prey; **~über·fall der** robbery (**auf** + *Akk.* of); **~vogel der** bird of prey; **~zug der** plundering raid

Rauch [raux] **der; ~[e]s** smoke; **sich in ~ auflösen, in ~ aufgehen** *(fig.)* go up in smoke

Rauch·abzug der smoke outlet; *(Rohr, Schacht)* flue

rauchen 1. *itr. V.* smoke; **es rauchte in der Küche** there was smoke in the kitchen; **sonst raucht es!** *(ugs.)* or there'll be trouble. **2.** *tr. (auch itr.) V.* smoke ‹*cigarette, pipe, etc.*›; **eine ~** *(ugs.)* have a smoke; **stark** *od.* **viel ~:** be a heavy smoker; **„Rauchen verboten“** 'No smoking'

Raucher der; ~s, ~: smoker; **er ist [starker] ~:** he smokes [heavily]; he is a [heavy] smoker; **möchten Sie ~ oder Nichtraucher [fliegen]?** would you like smoking or no smoking?

Räucher·aal der smoked eel

Raucher·abteil das smoking-compartment; smoker

Räucher·hering der smoked herring

Raucher·husten der smoker's cough

Raucherin die; ~, ~nen smoker

räuchern ['rɔyçɐn] **1.** *tr. V.* smoke ‹*meat, fish*›. **2.** *itr. V.* burn incense/joss-sticks *etc.*

Räucher-: ~schinken der smoked ham; **~stäbchen das** joss-stick

rauchig *Adj.* smoky; husky ‹*voice*›

Rauch-: ~pilz der mushroom cloud; **~schwaden der** cloud of smoke; **~verbot das** ban on smoking; **es herrscht ~verbot** smoking is prohibited; **~vergiftung die** poisoning *no art.* by smoke inhalation; **~wolke die** cloud of smoke; **~zeichen das** smoke signal

räudig *Adj.* mangy; **du ~er Hund!** *(derb)* you dirty rat! *(sl.)*

rauf [rauf] *Adv. (ugs.)* up; **~ mit euch!** up you go!; *s. auch* **herauf; hinauf**

Raufbold [-bɔlt] **der; ~[e]s, ~e** *(veralt.)* ruffian

rauf|bringen *unr. tr. V. (ugs.)* *(her)* bring up; *(hin)* take up

raufen 1. *itr., refl. V.* fight; **[sich] wegen** *od.* **um etw. ~:** fight [each other] over sth. **2.** *tr. V.* **sich** *(Dat.)* **die Haare/den Bart ~:** tear one's hair/at one's beard

Rauferei die; ~, ~en fight

rauf-: ~|gehen *unr. itr. V.; mit sein (ugs.)* go up; **~|kommen** *unr. itr. V.; mit sein (ugs.)* come up

rauh [rau] **1.** *Adj.* **a)** *(nicht glatt)* rough; **in einer ~en Schale steckt oft ein weicher Kern** *(Spr.)* behind a rough exterior there often beats a heart of gold; **b)** *(nicht mild)* harsh, raw ‹*climate, winter*›; raw ‹*wind*›; **c)** *(unwirtlich)* bleak, inhospitable ‹*region, mountains, etc.*›; rough ‹*weather*›; **d)** *(kratzig)* husky, hoarse ‹*voice*›; **e)** *(entzündet)* sore ‹*throat*›; **f)** *(grob, nicht feinfühlig)* rough; harsh ‹*words, tone*›; **er ist ~, aber herzlich** he is a rough diamond; **g) in ~en Mengen** *(ugs.)* in huge *or* vast quantities. **2.** *adv.* **a)** *(kratzig)* ‹*speak etc.*› huskily, hoarsely; **b)** *(grob, nicht feinfühlig)* roughly

rauh-, Rauh-: ~beinig *Adj. (ugs.)* gruff; rough and ready; **~faser·tapete die** woodchip wallpaper; **~haar·dackel der** wire-haired dachshund; **~reif der** hoar-frost

Raum der; ~[e]s, Räume ['rɔymə] **a)** *(Wohn~, Nutz~)* room; **im ~ stehen** *(fig.)* be in the air; **b)** *(Gebiet)* area; region; **c)** *o. Pl. (Platz)* room; space; **d)** *(Math., Philos., Astron.)* space

räumen ['rɔymən] *tr. V.* **a)** clear [away]; clear ⟨*snow*⟩; **Minen ~:** clear mines; **etw. aus dem Weg ~:** clear sth. out [of] the way; **seine Sachen auf die Seite ~:** clear *or* move one's things to one side; **etw. in Schubfächer** *(Akk.)* **~:** put sth. away in drawers; **b)** *(frei machen)* clear ⟨*street, building, warehouse, stocks, etc.*⟩; **c)** *(verlassen)* vacate ⟨*hotel room, cinema, house, flat, military position, area*⟩

Raum~: ~fahrer der astronaut; *(Kosmonaut)* cosmonaut; **~fahrt die** *o. Pl.* space flight; space travel; **~fahrzeug das** spacecraft; **~flug der** space flight; **~gleiter der; ~s, ~:** space shuttle; **~inhalt der** *(Math.)* volume; **~kapsel die** space capsule

räumlich 1. *Adj.* **a)** *(den Raum betreffend)* spatial; **aus ~en Gründen** for reasons of space; **b)** *(dreidimensional)* three-dimensional; stereophonic ⟨*sound*⟩; stereoscopic ⟨*vision*⟩. **2.** *adv.* **a)** spatially; **b)** *(dreidimensional)* three-dimensionally

Räumlichkeit die; ~, ~en *meist Pl.* rooms

raum-, Raum-: ~pflegerin die cleaning lady; cleaner; **~schiff das** spaceship; **~sparend** *Adj. s.* **platzsparend; ~station die** space station

Räumung die; ~, ~en a) clearing; **b)** *(das Verlassen)* vacation; vacating; **c)** *(Evakuierung)* evacuation; **d)** *(eines Lagers)* clearance

raunen ['raunən] *tr., itr. V. (geh.)* whisper; **ein Raunen ging durch die Reihen** a murmur went through the ranks

Raupe ['raupə] **die; ~, ~n** caterpillar

raus [raus] *Adv. (ugs.)* out; **~ mit euch!** out you go!

Rausch [rauʃ] **der; ~[e]s, Räusche** ['rɔyʃə] **a)** *(durch Alkohol)* state of drunkenness; **sich** *(Dat.)* **einen ~ antrinken** get drunk; **einen ~ haben** be drunk; **etw. im ~ tun** do sth. while drunk; **b)** *(durch Drogen)* drugged state; **einen ~ haben** be drugged; be high *(coll.)* [on drugs]; **etw. im ~ tun** do sth. while drugged; **c)** *(starkes Gefühl)* transport; **ein wilder/blinder ~:** a wild/blind frenzy; **der ~ der Geschwindigkeit** the exhilaration *or* thrill of speed

rauschen *itr. V.* **a)** ⟨*water, wind, torrent*⟩ rush; ⟨*trees, leaves*⟩ rustle; ⟨*skirt, curtains, silk*⟩ swish; ⟨*waterfall, surf, sea, strong wind*⟩ roar; ⟨*rain*⟩ pour down; **~der Beifall** *(fig.)* resounding applause; **b)** *mit sein (sich bewegen)* ⟨*water, river, etc.*⟩ rush; **sie rauschte aus dem Zimmer** she swept out of the room

Rausch·gift das drug; narcotic; **~ nehmen** take drugs; be on drugs

rauschgift-, Rauschgift-: ~handel der drug-trafficking; **~händler der** drug-trafficker; **~süchtig** *Adj.* drug-addicted; addicted to drugs *postpos.*; **~süchtige der/die;** *adj. Dekl.* drug addict

rausch·haft *Adj.* ecstatic

Rausch·mittel das *s.* **Rauschgift**

raus-: ~|ekeln *tr. V. (ugs.) s.* **hinausekeln; ~|fahren** *unr. tr. V.; mit sein (ugs.) s.* **hinaus|-; ~|feuern** *tr. V. (ugs.)* chuck out *(coll.)*; **~|fliegen** *unr. itr. V.; mit sein (ugs.) s.* **heraus-, hinausfliegen; ~|gehen** *unr. itr. V.; mit sein (ugs.) s.* **heraus-, hinausgehen; ~|kommen** *unr. itr. V.; mit sein (ugs.) s.* **heraus-, hinauskommen; ~|kriegen** *tr. V. (ugs.)* get out **(aus** of); **ich habe das Rätsel/die Aufgabe nicht ~gekriegt** I couldn't do the puzzle/exercise; **~|nehmen** *unr. tr. V. (ugs.) s.* **herausnehmen; ~|pauken** *tr. V.* **jmdn. ~pauken** *(ugs.)* get sb. off the hook *(sl.)*

räuspern ['rɔyspɐn] *refl. V.* clear one's throat

raus-, Raus-: ~|schmeißen *unr. tr. V. (ugs.)* chuck *(coll.) or* sling *(coll.)* ⟨*objects*⟩ out *or* away; give ⟨*employee*⟩ the push *(coll.) or* sack *(coll.) or* boot *(sl.)*; chuck *(coll.) or* throw ⟨*customer, drunk, tenant*⟩ out **(aus** of); **das ist ~geschmissenes Geld** that's money down the drain *(coll.)*; **~schmeißer der; ~s, ~** *(ugs.)* chucker-out *(coll.)*; bouncer *(coll.)*; **~schmiß der** *(ugs.)* chucking out; throwing out; *(Entlassung)* sacking

Raute ['rautə] **die; ~, ~n** *(Geom.)* rhombus

rauten·förmig *Adj.* rhombic; diamond-shaped

Razzia ['ratsia] **die; ~, Razzien** raid

Re das; ~s, ~s *(Skat)* redouble

Reagenz·glas das test-tube

reagieren *itr. V.* **a)** react **(auf** + *Akk.* to); **b)** *(Chemie)* react

Reaktion [reak'tsio:n] **die; ~, ~en** reaction **(auf** + *Akk.* to)

reaktionär [reaktsio'nɛ:r] *Adj. (Politik abwertend)* reactionary

Reaktionär der; ~s, ~e, Reaktionärin die; ~, ~nen *(Politik abwertend)* reactionary

reaktions-, Reaktions-: ~fähigkeit die; *o. Pl.* ability to react; **jmds. ~fähigkeit überprüfen** test sb.'s reactions; **~schnell** ⟨*person*⟩ with quick reactions; **~schnell sein** have quick reactions; **~vermögen das** *s.* **~fähigkeit**

Reaktor [re'aktɔr] **der; ~s, ~en** [-'to:rən] reactor

real [re'a:l] *Adj.* real

realisierbar *Adj. s.* **realisieren a:** realizable; implementable

Realisierbarkeit die; ~: practicability; feasibility

realisieren [reali'zi:rən] *tr. V. (geh.)* realize ⟨*plan, idea, aim, proposals, project, wish*⟩; implement ⟨*plan, programme, decision*⟩

Realismus der; ~ realism

Realist der; ~en, ~en realist

realistisch 1. *Adj.* realistic. **2.** *adv.* realistically

Realität die; ~, ~en reality

Realitäts·sinn der; *o. Pl.* sense of reality

Real-: ~schule die ≈ secondary modern school *(Brit. Hist.)*; **~schüler der** ≈ secondary modern school pupil *(Brit. Hist.)*

Rebe ['re:bə] **die; ~, ~n a)** vine shoot; **b)** *(geh.: Weinstock)* [grape] vine

Rebell [re'bɛl] **der; ~en, ~en** rebel

rebellieren *itr. V.* rebel **(gegen** against)

Rebellin die; ~, ~nen *s.* **Rebell**

Rebellion [rebɛ'lio:n] **die; ~, ~en** rebellion

rebellisch 1. *Adj.* rebellious. **2.** *adv.* rebelliously

Reb-: ~huhn das partridge; **~stock der** vine

rechen ['rɛçn̩] *tr. V. (bes. südd.)* rake

Rechen der; ~s, ~ *(bes. südd.)* rake

Rechen-: ~art die type of arithmetical operation; **~aufgabe die** arithmetical problem; **~fehler der** arithmetical error; **~maschine die** calculator

Rechenschaft die; ~: account; **jmdm. über etw.** *(Akk.)* **~ geben** *od.* **ablegen** account to sb. for sth.; **jmdm. über etw.** *(Akk.)* **~ schuldig sein** have to account to sb. for sth.; **ich bin Ihnen keine ~ schuldig** I am not answerable to you; I owe you no explanation; **jmdn. für etw. zur ~ ziehen** call *or* bring sb. to account for sth.

Rechenschafts·bericht der report

Rechen-: ~schieber der, ~stab der slide-rule; **~stunde die** arithmetic lesson; **~unter-**

richt der teaching of arithmetic; *(Fach)* arithmetic *no art.*

recherchieren *itr., tr. V. (geh.)* investigate

rechnen ['rɛçnən] **1.** *tr. V.* **a) eine Aufgabe ~:** work out a problem; **b)** *(veranschlagen)* reckon; estimate; **wir müssen zwei Stunden ~:** we must reckon on two hours; **gut/rund gerechnet** at a generous/ rough estimate; **c)** *(berücksichtigen)* take into account; **d)** *(einbeziehen)* count; **jmdn. zu seinen Freunden ~:** count sb. among *or* as one of one's friends. **2.** *itr. V.* **a)** do *or* make a calculation/calculations; **gut/schlecht ~ können** be good/bad at figures *or* arithmetic; **b)** *(zählen)* reckon; **vom 1. April an gerechnet** reckoning from 1 April; **in Schillingen ~:** reckon in shillings; **c)** *(ugs.: berechnen)* calculate; estimate; **er ist ein klug ~der Kopf** he is a shrewdly calculating person; **d)** *(wirtschaften)* budget carefully; **mit jeder Mark** *od.* **jedem Pfennig ~ müssen** have to count *or* watch every penny; **e) auf jmdn./etw.** *od.* **mit jmdm./etw. ~:** reckon *or* count on sb./sth.; **f) mit etw. ~** *(etw. einkalkulieren)* reckon with sth.; *(etw. erwarten)* expect sth.; **mit dem Schlimmsten ~:** be prepared for the worst. **3.** *refl. V.* pay

Rechnen das; ~s arithmetic

Rechner der; ~s, ~ a) ein guter/ schlechter ~ sein be good/bad at figures *or* arithmetic; **ein nüchterner ~ sein** *(fig.)* be shrewdly calculating; **b)** *(Gerät)* calculator; *(Computer)* computer

rechnerisch 1. *Adj.* arithmetical; ‹*value*› in figures. **2.** *adv.* ‹*determine*› by calculation, mathematically

Rechnung die; ~, ~en a) calculation; **[jmdm.] eine ~ aufmachen** work it out [for sb.]; **nach meiner ~** *(auch fig.)* according to my calculations; **seine ~ geht [nicht] auf** *(fig.)* his plans [do not] work out; **b)** *(schriftliche Kosten~)* bill; invoice *(Commerc.)*; **eine hohe/ niedrige ~:** a large/small bill; **eine ~ über 500 Mark** a bill for 500 marks; **das geht auf meine ~:** I'm paying for that; **diese Runde geht auf meine ~:** this round's on me; **auf eigene ~:** on one's own account; *(auf eigenes Risiko)* at one's own risk; **[jmdm.] etw. in ~ stellen** charge [sb.] for sth.; **c) einer Sache** *(Dat.)* **~ tragen** take sth. into account; *s. auch* **begleichen**

recht [rɛçt] **1.** *Adj.* **a)** *(geeignet)* right; **b)** *(richtig)* right; **ganz ~!** quite right!; **das ist ~, so ist es ~,** *(ugs.)* **~ so** that's fine; **c)** *(gesetzmäßig, anständig)* right; proper; **alles, was ~ ist** *(das geht zu weit)* there is a limit; **d)** *(wunschgemäß)* **jmdm. ~ sein** be all right with sb.; **e)** *(wirklich, echt)* real; **keine ~e Lust haben, etw. zu tun** not particularly *or* really feel like doing sth. **2.** *adv.* **a)** *(geeignet)* **du kommst gerade ~, um zu ...:** you are just in time to ...; **du kommst mir gerade ~** *(auch iron.)* you're just the person I needed; **b)** *(richtig)* correctly; **wenn ich es mir ~ überlege, dann ...:** if I really stop and think about it; **verstehen Sie mich bitte ~:** please don't misunderstand me; **gehe ich ~ in der Annahme, daß ...?** am I right in assuming that ...?; **c)** *(gesetzmäßig, anständig)* **~ handeln/leben** act/live properly; **d)** *(wunschgemäß)* **man kann ihm nichts ~ machen** there's no pleasing him; **man kann es nicht allen ~ machen** you can't please everyone; **e)** *(wirklich, echt)* really; rightly; **f)** *(ziemlich)* quite; rather; *s. auch* **erst 2; Recht d**

recht... *Adj.* **a)** right; right[-hand] ‹*edge*›; **die ~e Spur** the right-hand lane; **~er Hand, auf der ~en Seite** on the right-hand side; **b)** *(außen, sichtbar)* right ‹*side*›; **~e Maschen** *(Handarb.)* knit stitches; **c)** *(in der Politik)* right-wing; rightist *(derog.)*; **der ~e Flügel einer Partei** the right wing of a party; **d)** *(Geom.)* **ein ~er Winkel** a right angle

Recht das; ~[e]s, ~e a) *(Rechtsordnung)* law; **das ~ brechen/beugen** break/bend the law; **~ sprechen** administer the law; administer justice; **von ~s wegen** by law; *(ugs.: eigentlich)* by rights; **b)** *(Rechtsanspruch)* right; **das ~ des Stärkeren** the law of the jungle; **das ist sein gutes ~:** that is his right; **alle ~e vorbehalten** all rights reserved; **sein ~ fordern** *od.* **verlangen** demand one's rights; **zu seinem ~ kommen** *(fig.)* be given due attention; **c)** *o. Pl.* *(Berechtigung)* right **(auf** + *Akk.* to); **gleiches ~ für alle!** equal rights for all!; **im ~ sein** be in the right; **zu ~:** rightly; with justification; **d) recht haben** be right; **recht behalten** be proved right; **jmdm. recht geben** concede *or* admit that sb. is right

¹Rechte der/die; *adj. Dekl.* right-winger; rightist *(derog.)*; **die ~n** the right *sing.*

²Rechte die; *adj. Dekl.* **a)** *(Hand)* right hand; **jmdm. zur ~n** to the right of sb.; **zur ~n** on the right; **b)** *(Politik)* right

Recht·eck das rectangle

recht·eckig *Adj.* rectangular

recht·fertigen 1. *tr. V.* justify **(vor** + *Dat.* to). **2.** *refl. V.* justify oneself **(vor** + *Dat.* to)

Recht·fertigung die justification

recht-, Recht-: ~haber der; ~s, ~ *(abwertend)* self-opinionated person; **~haberei die; ~** *(abwertend)* self-opinionatedness; **~haberisch** *(abwertend) Adj.* self-opinionated

rechtlich 1. *Adj.* legal. **2.** *adv.* legally

recht·los *Adj.* without rights *postpos.*

Rechtlosigkeit die; ~: lack of rights

rechtmäßig 1. *Adj.* lawful; rightful; legitimate ‹*claim*›. **2.** *adv.* lawfully; rightfully; **das steht ihm ~ zu** that is his by right *or* rightfully his

Rechtmäßigkeit die; ~: legality; lawfulness; *(eines Anspruchs)* legitimacy

rechts 1. *Adv.* **a)** *(auf der rechten Seite)* on the right; **~ von jmdm./ etw.** on sb.'s right *or* the right of sb./on *or* to the right of sth.; **von ~:** from the right; **nach ~:** to the right; **sich ~ halten** keep to the right; **sich ~ einordnen** move *or* get into the right-hand lane; *s. auch* **links 1 a; b)** *(Politik)* on the right wing; **~ stehen** *od.* **sein** be right-wing *or* on the right; **c)** *(Handarb.)* **ein glatt ~ gestrickter Pullover** a pullover in stocking stitch; *s. auch* **links 1 c. 2.** *Präp. mit Gen.* **~ des Rheins** on the right side *or* bank of the Rhine

Rechts-: ~abbieger der *(Verkehrsw.)* motorist/cyclist/car *etc.* turning right; **~an·spruch der** legal right *or* entitlement; **einen ~anspruch auf etw.** *(Akk.)* **haben** have a legal right to *or* be legally entitled to sth.; **~anwalt der, ~anwältin die** lawyer; solicitor *(Brit.)*; attorney *(Amer.)*; *(vor Gericht)* barrister *(Brit.)*; attorney[-at-law] *(Amer.)*; advocate *(Scot.)*

rechts-, Rechts-: ~außen *Adv.* **a)** *(Ballspiele)* ‹*run, break through*› down the right wing; **b)** *(Politik ugs.)* on the extreme right [wing]; **~außen der; ~, ~** *(Ballspiele)* right wing; outside right; **~beistand der** legal adviser; **~beratung die** legal advice

recht-, Recht-: ~schaffen 1.

Adj. honest; upright; honest, decent ‹*work*›; **2.** *adv.* **a)** honestly; uprightly; **b)** *(intensivierend)* really ‹*tired, full, etc.*›; **~schaffenheit die;** ~: honesty, uprightness; **~schreib[e]·buch das** spelling-book; speller; *(Wörterbuch)* spelling dictionary; **~schreibfehler der** spelling mistake; **~schreibung die** orthography; **er ist in ~schreibung schwach** he's poor at spelling

rechts-, Rechts-: **~empfinden das** sense of [what is] right and wrong; **~extremismus der** *(Politik)* right-wing extremism; **~extremist der** *(Politik)* right-wing extremist; **~frage die** *(Rechtsw.)* legal question *or* issue; **~gelehrte der/die** jurist; **~gerichtet** *Adj. (Politik)* right-wing orientated; **~grundsatz der** *(Rechtsw.)* legal principle; **~gültig** *(Rechtsw.) Adj.* legally valid; **~händer** [~hɛndɐ] **der;** **~s, ~, ~händerin die; ~, ~nen** right-hander; **~händer[in] sein** be right-handed; **~händig 1.** *Adj.* right-handed; **2.** *adv.* right-handed; with one's right hand; **~herum** *Adv.* [round] to the right; **etw. ~herum drehen** turn sth. clockwise *or* [round] to the right; **~kräftig** *(Rechtsw.)* **1.** *Adj.* final [and absolute] ‹*decision, verdict*›; **~kräftig sein/werden** ‹*contract, agreement*› be/come into force; **2.** *adv.* **jmdn. ~kräftig verurteilen** pass a final sentence on sb.; **~kurve die** right-hand bend; **~lage die** *(Rechtsw.)* legal situation; **~lastig** *Adj. (Politik ugs.)* rightist; **~liberal** *Adj.* right-wing liberal; **~mittel das** *(Rechtsw.)* appeal; **~mittel einlegen** lodge an appeal; appeal; **~ordnung die** legal system; **~pflege die** *(Rechtsw.)* administration of justice

Rechtsprechung die; ~, ~en administration of justice; *(eines Gerichts)* jurisdiction

rechts-, Rechts-: **~radikal** *(Politik) Adj.* radical right-wing; **~radikalismus der** right-wing radicalism; **~ruck der** *(Politik ugs.)* shift to the right; **~rum** *Adv. (ugs.) s.* **~herum; ~seitig 1.** *Adj.* ‹*paralysis*› of the right side; **die ~seitige Uferbefestigung** the reinforcement of the right bank; **2.** *adv.* on the right [side]; **~staat der** [constitutional] state founded on the rule of law; **~staatlich** *Adj.* founded on the rule of law *postpos.*; **~staatlichkeit die;** ~: rule of law; **~verkehr der** driving *no art.* on the right; **in Frankreich ist ~verkehr** they drive on the right in France; **~weg der** *(Rechtsw.)* recourse to legal action *or* the courts *or* the law; **~widrig 1.** *Adj.* unlawful; illegal; **2.** *adv.* unlawfully; illegally; **~widrigkeit die** *o. Pl.* unlawfulness; illegality; **~wissenschaft die** jurisprudence

recht-: **~wink[e]lig** *Adj.* right-angled; **~zeitig 1.** *Adj.* timely; *(pünktlich)* punctual. **2.** *adv.* in time; *(pünktlich)* on time; **~zeitig zu/zum/zur** in [good] time for

Reck [rɛk] **das; ~[e]s, ~e** *od.* **~s** horizontal bar; high bar

recken 1. *tr. V.* stretch; **den Hals/Kopf ~:** crane one's neck. **2.** *refl. V.* stretch oneself; **sich ~ und strecken** have a good stretch

Reckturnen das horizontal-bar exercises *pl.*

Recorder [re'kɔrdɐ] **der; ~s, ~:** recorder

Redakteur [redak'tø:ɐ̯] **der; ~s, ~e, Redakteurin die; ~, ~nen** editor; **~ für Politik/Wirtschaft** political/economics editor

Redaktion [redak'tsi̯o:n] **die; ~, ~en a)** *(Redakteure)* editorial staff; **b)** *(Büro)* editorial department *or* office/offices *pl.*; **c)** *o. Pl. (das Redigieren)* editing

redaktionell [redaktsi̯o'nɛl] **1.** *Adj.* editorial. **2.** *adv.* editorially

Redaktions·schluß der time of going to press

Rede ['re:də] **die; ~, ~n a)** speech; *(Ansprache)* address; speech; **eine ~ halten** give *or* make a speech; **b)** *o. Pl. (Vortrag)* rhetoric; **die Kunst der ~:** the art of rhetoric; **c)** *(Äußerung)* **der langen ~ kurzer Sinn ist, daß ...:** the long and the short of it is that ...; **es ist die ~ davon, daß ...:** it is being said *or* people are saying that ...; **davon kann keine ~ sein** it's out of the question; **nicht der ~ wert sein** be not worth mentioning; **jmdm. ~ und Antwort stehen** give a full explanation [of one's actions] to sb.; **jmdn. zur ~ stellen** make someone explain himself/herself; **d)** *o. Pl. (Sprachw.)* **direkte** *od.* **wörtliche/indirekte ~:** direct/indirect speech

rede-, Rede-: **~freiheit die;** *o. Pl.* freedom of speech; **~gewandt** *Adj.* eloquent; **~gewandtheit die** eloquence

reden 1. *tr. V.* talk; **Unsinn ~:** talk nonsense; **kein Wort ~:** not say *or* speak a word. **2.** *itr. V.* **a)** *(sprechen)* talk; speak; **viel/wenig ~:** talk a lot *(coll.)*/not talk much; **b)** *(sich äußern, eine Rede halten)* speak; **er läßt mich nicht zu Ende ~:** he doesn't let me finish what I'm saying; *s. auch* **gut 2b; c)** *(sich unterhalten)* talk; **mit jmdm./über jmdn. ~:** talk to/about sb.; **miteinander ~:** have a talk [with one another]; **sie ~ nicht mehr miteinander** they are no longer on speaking terms; **mit sich ~ lassen** *(bei Geschäften)* be open to offers; *(bei Meinungsverschiedenheiten)* be willing to discuss the matter. **3.** *refl. V.* **sich heiser/in Wut ~:** talk oneself hoarse/into a rage

Redens·art die a) expression; *(Sprichwort)* saying; **b)** *Pl. (Phrase)* empty *or* meaningless words; **allgemeine ~en** empty generalizations

Rede-: **~schwall der** *(abwertend)* torrent of words; **~wendung die** *(Sprachw.)* idiom; idiomatic expression

redlich 1. *Adj.* honest; honest, upright ‹*person*›. **2.** *adv.* **a)** honestly; **sich ~ durchs Leben schlagen** make an honest living; **b)** *(intensivierend)* really

Redlichkeit die; ~: honesty

Redner der; ~s, ~ a) speaker; **b)** *(Rhetoriker)* orator

Rednerin die; ~, ~nen *s.* **Redner**

Redner·pult das lectern

red·selig *Adj.* talkative

Red·seligkeit die talkativeness

Reduktion [reduk'tsi̯o:n] **die; ~, ~en** reduction

reduzieren [redu'tsi:rən] **1.** *tr. V.* reduce **(auf +** *Akk.* to). **2.** *refl. V.* decrease; diminish

Reeder der; ~s, ~: shipowner

Reederei die; ~, ~en shipping firm *or* company

reell [re'ɛl] **1.** *Adj.* honest, straight ‹*person, deal, etc.*›; sound, solid ‹*business, firm, etc.*›; straight ‹*offer*›; decent; realistic ‹*price*›. **2.** *adv.* honestly; **~ einschenken** pour [out] a decent measure

Reet [re:t] **das; ~s** *(nordd.)* reeds *pl.*

reet·gedeckt *adj.* thatched

Referat [refe'ra:t] **das; ~[e]s, ~e a)** paper; **ein ~ halten** give *or* present a paper; **b)** *(kurzer schriftlicher Bericht)* report **(***Gen.* on); **c)** *(Abteilung)* department

Referendar [referɛn'da:ɐ̯] **der; ~s, ~e, Referendarin die; ~, ~nen** *candidate for a higher civil-service post who has passed the first state examination and is undergoing in-service training*

Referent [refe'rɛnt] **der; ~en, ~en, Referentin die; ~, ~nen**

person presenting a/the paper; *(Redner)* speaker

Referenz [refe'rɛnts] **die; ~, ~en a)** *(Empfehlung)* reference; **b)** *(Person, Stelle)* referee

referieren [refe'ri:rən] **1.** *itr. V.* **über etw.** *(Akk.)* **~:** give *or* present a paper on sth.; *(zusammenfassend)* give a report on sth. **2.** *tr. V.* **etw. ~:** give *or* present a paper on sth.; *(zusammenfassend)* give a report on sth.

reflektieren [reflɛk'ti:rən] **1.** *tr. V.* **a)** *auch itr. (zurückstrahlen)* reflect; **b)** *(geh.: nachdenken über)* *reflect or* ponder [up]on. **2.** *itr. V. (geh.: nachdenken)* reflect, ponder (**über** + *Akk.* [up]on)

Reflex [re'flɛks] **der; ~es, ~e a)** *(Physiol.)* reflex; **bedingter ~:** conditioned reflex; **b)** *(Licht~)* reflection

Reflexion [reflɛ'ksio:n] **die; ~, ~en** reflection

reflexiv [reflɛ'ksi:f] *Adj. (Sprachw.)* reflexive

Reflexiv·pronomen das *(Sprachw.)* reflexive pronoun

Reform [re'fɔrm] **die; ~, ~en** reform

Reformation die; ~ *(hist.)* Reformation

Reformator [refɔr'ma:tɔr] **der; ~s, ~en** [-ma'to:rən] reformer

reform·bedürftig *Adj.* in need of reform *postpos.*

Reformer der; ~s, ~: reformer

Reform·haus das health-food shop

reformieren *tr. V.* reform

Refrain [rə'frɛ̃:] **der; ~s, ~s** chorus; refrain

Regal [re'ga:l] **das; ~s, ~e** [set *sing.* of] shelves *pl.*; **ein Buch aus dem ~ nehmen** take a book from the shelf

Regatta [re'gata] **die; ~, Regatten** *(Sport)* regatta

rege ['re:gə] **1.** *Adj.* **a)** *(betriebsam)* busy ‹*traffic*›; brisk ‹*demand, trade, business, etc.*›; good ‹*participation; attendance*›; lively ‹*correspondence*›; **b)** *(lebhaft)* lively; lively, animated ‹*discussion, conversation*›; keen ‹*interest*›; **geistig ~:** mentally alert *or* active. **2.** *adv.* **a)** *(betriebsam)* actively; **~ an etw.** *(Akk.)* **teilnehmen** take an active part in sth.; **b)** *(lebhaft)* actively

Regel ['re:gl̩] **die; ~, ~n a)** rule; **die ~n eines Spiels/des Anstands** the rules of a game/of decency; **nach allen ~n der Kunst** *(fig.)* well and truly; **b) die ~ sein** be the rule; **in der** *od.* **aller ~:** as a rule; **c)** *(Menstruation)* period

regelbar *Adj.* adjustable

regel-, Regel-: ~los 1. *Adj.* disorderly; **2.** *adv.* in a disorderly manner; **~mäßig 1.** *Adj.* regular; **2.** *adv.* regularly; **~mäßigkeit die** regularity

regeln 1. *tr. V.* **a)** settle ‹*matter, question, etc.*›; put ‹*finances, affairs, etc.*› in order; **etw. durch Gesetz ~:** regulate sth. by law; **wir haben die Sache so geregelt, daß ...:** we've arranged things so that ...; **b)** *(einstellen, regulieren)* regulate; *(steuern)* control; *s. auch* **Verkehr a. 2.** *refl. V.* take care of itself; **die Sache hat sich [von selbst] geregelt** the matter has sorted itself out *or* resolved itself

regel·recht *(ugs.)* **1.** *Adj.; nicht präd.* proper *(coll.)*; real; real ‹*shock*›; real, absolute ‹*scandal*›; complete, utter ‹*flop, disaster*›; real, downright ‹*impertinence, insult*›. **2.** *adv.* really

Regelung die; ~, ~en a) *o. Pl. s.* **regeln 1a, b:** settlement; putting in order; regulation; control; **b)** *(Vorschriften)* regulation

regel·widrig 1. *Adj.* that is against the rules *postpos.*; **~ sein** be against the rules**2.** *adv.* **sich ~ verhalten** break the rules; **den Stürmer ~ attackieren** *(Ballspiele)* foul the forward

regen ['re:gn̩] **1.** *tr. V. (geh.)* move. **2.** *refl. V.* **a)** *(sich bewegen)* move; **kein Lüftchen regte sich** not a breath of air stirred; **b)** *(geh.: sich bemerkbar machen)* ‹*hope, doubt, desire, conscience*› stir

Regen der; ~s, ~ a) rain; **bei strömendem ~:** in pouring rain; **es wird ~ geben** it will rain; it is going to rain; **ein warmer ~** *(fig.)* a windfall; **vom ~ in die Traufe kommen** *(fig.)* jump out of the frying-pan into the fire; **jmdn. im ~ stehen lassen** *(fig. ugs.)* leave sb. in the lurch; **b)** *(fig.)* shower

regen·arm *Adj.* ‹*period, region, etc.*› with little rain[fall], with low rainfall

Regen·bogen der rainbow

Regen·bogen-: ~haut die *(Anat.)* iris; **~presse die** *(abwertend)* gossip magazines *pl.*

Regen-: ~cape das rain cape; **~dach das** rain-canopy

Regeneration die regeneration

regenerieren [regene'ri:rən] *(fachspr.)* **1.** *refl. V.* regenerate; *(geh.: sich erholen)* recuperate. **2.** *tr. V.* regenerate

regen-, Regen-: ~guß der downpour; **~haut die** [light] plastic mackintosh *or (coll.)* mac; **~mantel der** raincoat; mackintosh; mac *(coll.)*; **~reich** *Adj.* ‹*period, region, etc.*› with high rainfall; **~rinne die** gutter; **~schauer der** shower [of rain]; rain-shower; **~schirm der** umbrella

Regent [re'gɛnt] **der; ~en, ~en a)** *(Herrscher)* ruler; *(Monarch)* monarch; **b)** *(Stellvertreter)* regent

Regen·tag der rainy day

Regentin die; ~, ~nen *s.* **Regent**

Regen·tonne die water-butt

Regentschaft die; ~, ~en regency

Regen-: ~wasser das; *o. Pl.* rainwater; **~wetter das;** *o. Pl.* rainy *or* wet weather; **~wolke die** rain cloud; **~wurm der** earthworm; **~zeit die** rainy season

Regie [re'ʒi:] **die; ~ a)** *(Theater, Film, Ferns., Rundf.)* direction; **bei etw. ~ führen** direct sth.; **unter der ~ von ...:** directed by ...; **b)** *(Leitung, Verwaltung)* management; **unter staatlicher ~:** under state control

regieren [re'gi:rən] **1.** *itr. V.* rule (**über** + *Akk.* over); ‹*monarch*› reign, rule (**über** + *Akk.* over); ‹*party, administration*› govern. **2.** *tr. V.* **a)** rule; govern; ‹*monarch*› reign over, rule; **b)** *(Sprachw.)* govern, take ‹*case*›

Regierung die; ~, ~en a) *o. Pl. (Herrschaft)* rule; *(eines Monarchen)* reign; **die ~ übernehmen** *od.* **antreten** take over; come to power; **b)** *(Kabinett)* government

regierungs-, Regierungs-: ~bildung die formation of a/the government; **~chef der** head of government; **~erklärung die** government statement; **~feindlich** *Adj.* anti-government; **~freundlich** *Adj.* pro-government; **~gewalt die** government power *no art.*; **~krise die** government crisis; **~rat der** senior civil servant; **~sitz der** seat of government; **~sprecher der** government spokesman; **~umbildung die** government reshuffle; **~wechsel der** change of government

Regime [re'ʒi:m] **das; ~s, ~** [re'ʒi:mə] *(abwertend)* regime

Regime-: ~gegner der opponent of a/the regime; **~kritiker der** critic of a/the regime

Regiment [regi'mɛnt] **das; ~[e]s, ~e** *od.* **~er a)** *Pl.* **~e** *(Herrschaft)* rule; **das ~ führen** *(fig.)* give the orders; **ein strenges ~ führen** *(fig.)* be strict; **b)** *Pl.* **~er** *(Milit.)* regiment

Region [re'gio:n] **die; ~, ~en** region

regional [regio'na:l] **1.** *Adj.* regional. **2.** *adv.* regionally; **~ verschieden sein** differ from region to region

Regisseur [reʒɪ'sø:ɐ] **der; ~s, ~e, Regisseurin die; ~, ~nen** *(Theater, Film)* director; *(Ferns., Rundf.)* director; producer

Register [re'gɪstɐ] **das; ~s, ~ a)** index; **b)** *(amtliche Liste)* register; **c)** *(Musik) (bei Instrumenten)* register; *(Orgel~)* stop; **alle ~ ziehen** *(fig.)* pull out all the stops

registrieren [regɪs'tri:rən] *tr. V.* **a)** register; **b)** *(bewußt wahrnehmen)* note; register

Registrierung die; ~, ~en registration

reglementieren *tr. V.* regulate; regiment ⟨*people, life*⟩

Reglementierung die; ~, ~en regulation; *(Bevormundung)* regimentation

Regler der; ~s, ~ *(Technik)* regulator; *(Kybernetik)* control

reg·los *Adj.* motionless

Reglosigkeit die; ~: motionlessness

regnen ['re:gnən] **1.** *itr., tr. V. (unpers.)* rain; **es regnet** it is raining; **es regnete Steine** *(fig.)* stones rained down. **2.** *itr. V.; mit sein (fig.)* rain down

regnerisch *Adj.* rainy

regulär [regu'lɛ:ɐ] *Adj.* **a)** proper; regular ⟨*troops*⟩; normal, regular ⟨*working hours, flight*⟩; **b)** *(ugs.: regelrecht)* proper *(coll.)*; regular *(coll.)*

regulierbar *Adj.* regulable; adjustable ⟨*backrest*⟩

regulieren [regu'li:rən] *tr. V.* regulate

Regulierung die; ~, ~en regulation

Regung die; ~, ~en *(geh.: Gefühl)* stirring; **seine erste ~ war Unmut** his first emotion was displeasure; **sie folgte einer ~ ihres Herzens** she followed the promptings of her heart

regungs·los *Adj.* motionless

Regungslosigkeit die; ~: motionlessness

Reh [re:] **das; ~[e]s, ~e** roe-deer

reh-, Reh-: ~bock der roebuck; **~braun** *Adj.* light reddish brown; **~kitz das** fawn *or* kid [of a/the roe-deer]

Reibach ['raibax] **der; ~s** *(ugs.)* profits *pl.*; **einen [kräftigen] ~ machen** make a killing *(coll.)*

Reibe ['raibə] **die; ~, ~n** grater

Reib·eisen das grater; **eine Stimme wie ein ~:** a voice like a rasp

Reibe·kuchen der *(landsch.) s.* **Kartoffelpuffer**

reiben 1. *unr. tr. V.* **a)** rub; **etw. blank ~:** rub sth. until it shines; **sich** *(Dat.)* **den Schlaf aus den Augen ~:** rub the sleep from one's eyes; **b)** *(zerkleinern)* grate. **2.** *unr. itr. V.* rub **(an +** *Dat.* on). **3.** *unr. refl. V.* rub oneself/itself **(an +** *Dat.* against); **sie ~ sich ständig aneinander** *(fig.)* there is constant friction between them

Reiberei die; ~, ~en friction *no pl.*; **es gab ständig ~en mit seinem Sohn** there was constant friction between him and his son

Reibung die; ~, ~en *(Physik, fig.)* friction

reibungs·los 1. *Adj.* smooth; **2.** *adv.* smoothly

reich [raiç] **1.** *Adj.* **a)** *(vermögend)* rich; **~ heiraten** marry [into] money; **die Reichen** the rich; **b)** *(prächtig)* costly ⟨*goods, gifts*⟩; rich ⟨*décor, ornamentation, finery, furnishings*⟩; **c)** *(üppig)* rich; rich, abundant ⟨*harvest*⟩; lavish, sumptuous ⟨*meal*⟩; abundant ⟨*mineral resources*⟩; **~ an etw.** *(Dat.)* **sein** be rich in sth.; **d)** *(vielfältig)* rich ⟨*collection, possibilities, field of activity*⟩; wide, large, extensive ⟨*selection, choice*⟩; wide ⟨*knowledge, experience*⟩. **2.** *adv.* richly

-reich rich in ...; **kontrast~:** rich in contrast; **wasser~ sein** have abundant water

Reich das; ~[e]s, ~e a) empire; *(König~)* kingdom; realm; **das [Deutsche] ~** *(hist.)* the German Reich *or* Empire; **das Dritte ~** *(hist.)* the Third Reich; **b)** *(fig.)* realm; **ins ~ der Fabel gehören** belong to the realm[s] of fantasy; **das ~ der Pflanzen/Tiere** the plant/animal kingdom; **Dein ~ komme** *(bibl.)* thy Kingdom come

reichen 1. *itr. V.* **a)** *(aus~)* be enough; **das Geld reicht nicht** I/we *etc.* haven't got enough money; **das Brot muß noch bis Montag ~:** the bread must last till Monday; **die Farbe hat gerade gereicht** there was just enough paint; **das Seil reicht nicht** the rope's not long enough; **jetzt reicht's mir aber!** now I've had enough!; **danke, das reicht** that's enough, thank you; **b)** *(sich erstrecken)* reach; ⟨*forest, fields, etc.*⟩ extend; **bis zu etw. ~:** extend as far as sth.; **sein Einfluß reicht sehr weit** his influence extends a long way; **jmdm. bis an die Schultern ~:** come up to sb.'s shoulder; **c)** *(ugs.) s.* **auskommen a. 2.** *tr. V. (geh.)* **a)** pass; hand; **jmdm. die Hand ~:** hold out one's hand to sb.; **sich** *(Dat.)* **die Hand ~:** shake hands; **b)** *(servieren)* serve ⟨*food, drink*⟩

reich·haltig *Adj.* extensive; varied ⟨*programme*⟩; substantial ⟨*meal*⟩

reichlich 1. *Adj.* large; substantial; ample ⟨*space, time, reward*⟩; good ⟨*hour, litre, etc.*⟩; generous ⟨*tip*⟩. **2.** *adv.* **a)** amply; **b)** *(in großer Menge)* **~ Trinkgeld geben** tip generously; **Fleisch ist noch ~ vorhanden** there is still plenty of meat left; **~ Zeit/Platz/Gelegenheit haben** have plenty of *or* ample time/room/opportunity; **c)** *(mehr als)* over; more than; **~ 5000 Mark** a good 5,000 marks. **3.** *Adv. (ugs.: ziemlich, sehr)* **~ frech** a bit too cheeky

Reichs·tag der *o. Pl. (hist.)* Reichstag; *(des Heiligen Römischen Reichs)* Imperial Diet

Reichtum der; ~s, Reichtümer ['raiçty:mɐ] **a)** *o. Pl. (auch fig.)* wealth **(an +** *Dat.* of**)**; **der ~ an Vögeln** the abundance of birds; **b)** *Pl. (auch fig.)* riches

Reich·weite die reach; *(eines Geschützes, Senders, Flugzeugs)* range ; **in ~ sein** be within reach/range; **Geschütze mit großer ~:** long-range guns

reif [raif] *Adj.* **a)** ripe ⟨*fruit, grain, cheese*⟩; mature ⟨*brandy, cheese*⟩; **~ für etw. sein** *(ugs.)* be ready for sth.; **die Zeit ist noch nicht ~:** the time is not yet ripe; **b)** *(erwachsen, erfahren)* mature; **die ~eren Jahrgänge** those of mature age; **c)** *(ausgewogen, durchdacht)* mature; **eine ~e Leistung** *(ugs.)* a solid achievement

-reif ready for ...; **test-/olympia~:** ready for testing/for the Olympics; **aufführungs~:** ready to be performed

[1]Reif der; ~[e]s hoar-frost

[2]Reif der; ~[e]s, ~e *(geh.)* ring; *(Arm~)* bracelet; *(Diadem)* circlet

Reife die; ~ a) ripeness; *(von Menschen, Gedanken, Produkten)* maturity; **Zeugnis der ~:** Abitur certificate; **mittlere ~** *(Schulw.) school-leaving certificate usually taken after the fifth year of secondary school*; **b)** *(Reifung)* ripening; **während der ~:** during ripening

reifen 1. *itr. V.; mit sein* **a)** ⟨*fruit, cereal, cheese*⟩ ripen; ⟨*ovum, embryo, cheese*⟩ mature; **b)** *(geh.: älter, reifer werden)* mature **(zu** into**)**; **ein gereifter Mann** *(geh.)* a mature man; **c)** ⟨*idea, plan, decision*⟩ mature. **2.** *tr. V.* ripen ⟨*fruit, cereal*⟩

Reifen der; ~s, ~ a) *(Metallband, Sportgerät)* hoop; **b)** *(Gummi~)* tyre; **c)** *s.* [2]**Reif**

Reifen-: ~panne die flat tyre; puncture; **~profil das** [tyre] tread; **~wechsel der** tyre change

Reife-: ~prüfung die *school-leaving examination for university entrance qualification;* **~zeugnis das** Abitur certificate

Reif·glätte die ice on the roads

reiflich 1. *Adj.* [very] careful; **bei/nach ~er Überlegung** on mature consideration/after [very] careful consideration. **2.** *adv.* [very] carefully

Reifung die; ~ *s.* **reifen 1:** ripening; maturing; maturation

Reigen ['raign̩] **der; ~s, ~** round dance; **den ~ eröffnen** *(fig.)* start off; **ein bunter ~ von Melodien** a medley of tunes

Reihe ['raiə] **die; ~, ~n a)** row; **in ~n** *(Dat.)* **antreten** line up; *(Milit.)* fall in; **sich in fünf ~n aufstellen** line up in five rows; form five lines; **in Reih und Glied** *(Milit.)* in rank and file; **aus der ~ tanzen** *(fig. ugs.)* be different; **etw. in die ~ bringen** *(fig. ugs.)* put sth. straight *or* in order; **b)** *o. Pl. (Reihenfolge)* series; **die ~ ist an ihm/ihr** *usw.*, **er/sie** *usw.* **ist an der ~:** it's his/her *etc.* turn; **der ~ nach, nach der ~:** in turn; one after the other; **c)** *(größere Anzahl)* number; **d)** *(Gruppe)* ranks *pl.*; **aus den eigenen ~n** from one's/its own ranks; **e)** *(Math., Musik)* series

reihen *(geh.)* **1.** *tr. V. (auf~)* string; thread; **Perlen auf eine Schnur ~:** string pearls [on a thread]. **2.** *refl. V.* **sich an etw.** *(Akk.)* **~:** follow sth.

reihen-, Reihen-: ~folge die order; **~haus das** terraced house; **~untersuchung die** *(Med.)* mass screening; **~weise** *Adv. (ugs.)* by the dozen

Reiher der; ~s, ~: heron

reihern *itr. V. (salopp)* puke *(coarse)*

reih·um *Adv.* **etw. ~ gehen lassen** pass sth. round

Reim [raim] **der; ~[e]s, ~e** rhyme; **sich** *(Dat.)* **keinen ~ auf etw.** *(Akk.)* **machen [können]** *(fig.)* not [be able] to see rhyme or reason in sth.

reimen 1. *itr. V.* make up rhymes. **2.** *tr. V.* rhyme; **ein Wort auf ein anderes ~:** rhyme one word with another. **3.** *refl. V.* rhyme **(auf +** *Akk.* with); **das reimt sich nicht** *(fig.)* that makes no sense

reim·los *Adj.* unrhymed; rhymeless

[1]**rein** [rain] *Adv. (ugs.)* **~ mit dir!** in you go/come!

[2]**rein 1.** *Adj.* **a)** *(unvermischt)* pure; **b)** *(nichts anderes als)* pure; sheer; **etw. aus ~em Trotz tun** do sth. out of sheer *or* pure contrariness; **~e Theorie** pure theory; **die ~e Wahrheit sagen** tell the plain *or* unvarnished truth; **es war eine ~e Männersache** it was exclusively a men's affair; **eine ~e Arbeitergegend** a purely *or* entirely working-class district; **der ~ste Quatsch** *(ugs.)* pure *or* sheer *or* absolute nonsense; **dein Zimmer ist der ~ste Saustall** *(derb)* your room is a real pigsty; **c)** *(meist geh.: frisch, sauber)* clean; fresh ⟨*clothes, sheet of paper, etc.*⟩; pure, clean ⟨*water, air*⟩; clear ⟨*complexion*⟩; *(fig.)* **ein ~es Gewissen haben** have a clear conscience; **etw. ins ~e schreiben** make a fair copy of sth.; **etw. ins ~e bringen** clear sth. up; put sth. straight; **mit jmdm./etw. ins ~e kommen** get things straightened out with sb./get sth. sorted *or* straightened out. **2.** *Adv.* purely; **~ zufällig** purely *or* quite by chance; **~ gar nichts** *(ugs.)* absolutely nothing

Reine·machen das; ~s *(bes. nordd.)* cleaning session

Rein·fall der *(ugs.)* let-down; **das Stück war ein absoluter** *od.* **totaler ~:** the play was a complete flop *(coll.)*

rein|fallen *unr. itr. V.; mit sein (ugs.) s.* **hereinfallen a**

rein|gehen *unr. itr. V.; mit sein (ugs.) s.* **hineingehen**

Rein-: ~gewinn der net profit; **~haltung die: die ~ der Seen/der Luft** keeping the lakes/air clean *or* pure

rein|hauen 1. *unr. tr. V.* **jmdm. eine ~** *(salopp)* thump sb. [one] *(coll.)*. **2.** *unr. itr. V. (essen)* tuck in *(coll.)*

Reinheit die; ~ a) purity; **b)** *(Sauberkeit)* cleanness; *(des Wassers, der Luft)* purity; *(der Haut)* clearness

reinigen ['rainɪgn̩] *tr. V.* clean; clean, cleanse ⟨*wound, skin*⟩; purify ⟨*effluents, air, water, etc.*⟩; **Kleider [chemisch] ~ lassen** have clothes [dry-]cleaned

Reinigung die; ~, ~en a) *s.* **reinigen:** cleaning; cleansing; purification; dry-cleaning; **b)** *(Betrieb)* [dry-]cleaner's

rein-: ~|knien *refl. V. (ugs.) s.* **hineinknien; ~|kommen** *unr. itr. V.; mit sein (ugs.) s.* **hereinkommen; ~kriechen** crawl into sth.; **~|kriegen** *tr. V. (ugs.) s.* **hereinbekommen; hinein|-; ~|legen** *tr. V. (ugs.) s.* **hereinlegen**

reinlich *Adj.* cleanly

Reinlichkeit die; ~: cleanliness

rein-, Rein-: ~rassig *Adj.* purebred, thoroughbred ⟨*animal*⟩; **~|reden** *itr. V.* **jmdm. ~** *(ugs.)* interfere in sb.'s affairs; **~|reißen** *unr. tr. V. (ugs.)* **jmdn. ~reißen** drag sb. in *(fig.)*; **~|reiten** *unr. tr. V. (ugs.)* **jmdn. ~reiten** drag sb. in *(fig.)*; **~|schlagen** *unr. tr. V.* **a)** *(ugs.)* knock in; **etw. in etw.** *(Akk.)* **~schlagen** knock sth. into sth.; **b) jmdm. eine ~schlagen** *(salopp)* thump sb. [one] *(coll.)*; **~schrift die** fair copy; **~|steigern** *refl. V. (ugs.)* work oneself up; become worked up; **~|treten** *(ugs.)* **1.** *unr. itr. V.; mit sein* **in etw.** *(Akk.)* **~treten** step in[to] sth.; **2.** *unr. tr., itr. V.* **jmdm.** *od.* **jmdn. hinten ~treten** kick sb. up the backside; **~|waschen** *(ugs.)* **1.** *unr. tr. V.* **jmdn.** *od.* **jmds. Namen ~waschen** clear sb.; clear sb.'s name; **2.** *unr. refl. V.* clear oneself *or* one's name **~|wollen** *unr. itr. V. (ugs.)* want to come/go in; **~|würgen** *tr. V. (ugs.)* **jmdm. eine** *od.* **eins ~würgen** come down on sb. like a ton of bricks *(coll.)*

Reis [rais] **der; ~es** rice

Reise ['raizə] **die; ~, ~n** journey; *(kürzere Fahrt, Geschäfts~)* trip; *(Ausflug)* outing; excursion; trip; *(Schiffs~)* voyage; **eine ~ mit dem Auto/der Eisenbahn** a journey by car/train; a car/train journey; **eine ~ zur See** a sea voyage; *(Kreuzfahrt)* a cruise; **eine ~ machen** make a journey/go on a trip/an outing; **auf ~n sein** travel; *(nicht zu Hause sein)* be away; **glückliche** *od.* **gute ~!** have a good journey

reise-, Reise-: ~andenken das souvenir; **~begleiter der** *(~gefährte)* travelling companion; *(~leiter)* courier; *(für Kinder)* chaperon; **~büro das** travel agent's; travel agency; **~bus der** coach; **~fieber das** *(ugs.)* nervous excitement about the journey; **~führer der** *(Buch)* guidebook; **~gepäck das** luggage *(Brit.)*; baggage *(Amer.)*; *(am Flughafen)* baggage; **~gesellschaft die, ~gruppe, die** party of tourists; **~kosten** *Pl.* travel expenses; **~leiter der, ~leiterin**

die courier; **~lektüre die** reading matter for the journey; **~lustig** *Adj.* **~lustig sein** be a keen traveller

reisen *itr. V.; mit sein* **a)** travel; **er reist für einige Tage nach Paris** he's going to Paris for a few days; **b)** *(ab~)* leave; set off

Reisende der/die; *adj. Dekl.* traveller; *(Fahrgast)* passenger

Reise-: ~paß der passport; **~ruf der** SOS message for travellers; **~scheck der** traveller's cheque; **~tasche die** hold-all; **~verkehr der** holiday traffic; **~wecker der** travel alarm; **~welle die** surge of holiday traffic; **~wetterbericht der** holiday weather forecast; **~ziel das** destination

Reis·feld das paddy-field

Reisig das; ~s brushwood

Reisig·besen der besom

Reis·korn das grain of rice

Reiß-: ~aus der: ~aus nehmen *(ugs.)* scram *(sl.)*; scarper *(Brit. sl.)*; **~brett das** drawing-board

reißen ['raisn̩] **1.** *unr. tr. V.* **a)** tear; **sich** *(Dat.)* **ein Loch in die Hose ~:** tear *or* rip a hole in one's trousers; **jmdm. etw. aus den Händen/Armen ~:** snatch *or* tear sth. from sb.'s hands/arms; **sich** *(Dat.)* **die Kleider vom Leibe ~:** tear one's clothes off; **jmdn. aus seinen Gedanken ~** *(fig.)* awaken sb. rudely from his/her thoughts; **b)** *(ziehen an)* pull; *(heftig)* yank *(coll.)*; **c)** *(werfen, ziehen)* **eine Welle riß ihn zu Boden** a wave knocked him to the ground; **jmdn. in die Tiefe ~:** drag sb. down into the depths; **[innerlich] hin und her gerissen sein** *od.* **werden** *(fig.)* be torn [two ways]; **d)** *(töten)* ‹*wolf, lion, etc.*› kill, take ‹*prey*›; **e) etw. an sich ~** *(fig.)* seize sth.; **f)** *(ugs.: machen)* crack ‹*joke*›; make ‹*remark*›; **g)** *(Leichtathletik)* **die Latte/eine Hürde ~:** knock the bar down/knock a hurdle over. **2.** *unr. itr. V.* **a)** *mit sein* ‹*paper, fabric*› tear, rip; ‹*rope, thread*› break, snap; ‹*film*› break; ‹*muscle*› tear; **wenn alle Stricke** *od.* **Stränge ~** *(fig.)* if all else fails; **b)** *(ziehen)* **an etw.** *(Dat.)* **~:** pull at sth.; **c)** *(Leichtathletik)* bring the bar down/knock the hurdle over. **3.** *unr. refl. V.* **a)** tear oneself/itself **(aus, von** from**)**; **b)** *(ugs.: sich bemühen um)* **ich reiße mich nicht um diese Arbeit** I'm not all that keen on this work *(coll.)*; **sie ~ sich um die Eintrittskarten** they are scrambling to *or* fighting each other to get tickets

reißend *Adj.* rapacious ‹*animal*›; stabbing ‹*pain*›; **~en Absatz finden** sell like hot cakes; **ein ~er Fluß** a raging torrent

Reißer der; ~s, ~ *(ugs., oft abwertend)* thriller

reißerisch *(abwertend)* **1.** *Adj.* sensational; lurid ‹*headline*›. **2.** *adv.* sensationally

Reiß-: ~leine die *(Flugw.)* ripcord; **~nagel der** *s.* **~zwecke**; **~verschluß der** zip [fastener]; **~wolf der** shredder; **~zwecke die** drawing-pin *(Brit.)*; thumbtack *(Amer.)*

Reit·bahn die riding arena

reiten ['raitn̩] **1.** *unr. itr. V.; meist mit sein* ride. **2.** *unr. tr. V.; auch mit sein* ride; **Schritt/Trab/Galopp ~:** ride at a walk/trot/gallop; **ein Turnier ~:** ride in a tournament

Reiten das; ~s riding *no art.*

Reiter der; ~s, ~, **Reiterin die;** ~, ~nen rider

Reit-: ~hose die riding breeches *pl.*; **~peitsche die** riding whip; **~pferd das** saddle-horse; **~sport der** [horse-] riding; **~stall der** riding stable; **~stiefel der** riding boot; **~turnier das** riding event

Reiz [raits] **der;** ~es, ~e **a)** *(Physiol.)* stimulus; **b)** *(Attraktion)* attraction; appeal *no pl.*; *(des Verbotenen, Fremdartigen, der Ferne usw.)* lure; **ich kann dem keinen ~ abgewinnen** this has no appeal for me; **c)** *(Zauber)* charm; **weibliche ~e** female charms

reizbar *Adj.* irritable; **leicht ~ sein** be very irritable

Reizbarkeit die; ~: irritability

reizen 1. *tr. V.* **a)** annoy; tease ‹*animal*›; *(herausfordern, provozieren)* provoke; *(zum Zorn treiben)* anger; *s. auch* **gereizt; b)** *(Physiol.)* irritate; **c)** *(Interesse erregen bei)* **jmdn. ~:** attract sb.; appeal to sb.; **es würde mich sehr ~, das zu tun** I'd love to do that; **das Angebot reizt mich** I find the offer tempting; **d)** *(Kartenspiele)* bid. **2.** *itr. V.* **a) das reizt zum Lachen** it makes people laugh; **b)** *(Kartenspiele)* bid; **hoch ~** *(fig.)* play for high stakes

reizend 1. *Adj.* charming; delightful, lovely ‹*child*›; **das ist ja ~!** *(iron.)* [that's] charming! *(iron.)*. **2.** *adv.* charmingly; **wir haben uns ~ unterhalten** we had a delightful chat

Reiz·husten der *(Med.)* dry cough

reizlos *Adj.* unattractive; ‹*landscape, scenery*› lacking in charm

Reizung die; ~, ~en *(Physiol., Med.)* irritation

reiz·voll *Adj.* **a)** *(hübsch)* charming; delightful; **b)** *(interessant)* attractive; **die Aussicht ist nicht gerade ~:** the prospect isn't exactly enticing

Reiz·wort das emotive word

rekapitulieren *tr. V.* recapitulate

rekeln ['re:kl̩n] *refl. V. (ugs.)* stretch; **sich in der Sonne ~:** stretch out in the sun

Reklamation [reklama'tsi̯o:n] **die;** ~, ~en complaint **(wegen** about**)**; **spätere ~[en] ausgeschlossen** money cannot be refunded after purchase

Reklame [re'kla:mə] **die;** ~, ~n **a)** *o. Pl. (Werbung)* advertising *no indef. art.*; *(Ergebnis)* publicity *no indef. art*; **~ für jmdn./etw. machen** promote sb./advertise *or* promote sth.; **b)** *(ugs.: Werbemittel)* advert *(Brit. coll.)*; ad *(coll.)*; advertisement; *(im Fernsehen, Radio auch)* commercial

Reklame·trommel die: für jmdn./etw. die ~ rühren *(ugs.)* promote sb./sth. in a big way

reklamieren 1. *itr. V.* complain; make a complaint. **2.** *tr. V.* **a)** *(beanstanden)* complain about, make a complaint about **(bei** to, **wegen** on account of**)**; **b)** *(beanspruchen)* claim

rekonstruieren *tr. V.* reconstruct

Rekonstruktion die reconstruction

Rekord [re'kɔrt] **der;** ~[e]s, ~e record; **einen ~ aufstellen/innehaben** set up/hold a record

Rekord- record ‹*harvest, temperature, fee*›

Rekord-: ~lauf der record-breaking run; **~leistung die** record; **~zeit die** record time

Rekrut [re'kru:t] **der;** ~en, ~en *(Milit.)* recruit

Rektor ['rɛktɔr] **der;** ~s, ~en [-'to:rən] **a)** *(einer Schule)* head[master]; **b)** *(Universitäts~)* Rector; ≈ Vice-Chancellor *(Brit.)*; *(einer Fachhochschule)* principal

Rektorin die; ~, ~nen **a)** *(einer Schule)* head[mistress]; **b)** *s.* **Rektor b**

Relais [rə'lɛ:] **das;** ~ [rə'lɛ:(s)], ~ [rə'lɛ:s] *(Elektrot.)* relay

Relation [rela'tsi̯o:n] **die;** ~, ~en relation; **in einer/keiner ~ zu etw. stehen** bear a/no relation to sth.

relativ [rela'ti:f] **1.** *Adj.* relative. **2.** *adv.* relatively; **~ zu** relative to

relativieren *tr. V.* relativize

Relativität [relativi'tɛ:t] **die; ~, ~en** relativity

Relativitäts·theorie die; *o. Pl. (Physik)* theory of relativity

Relativ-: ~pronomen das *(Sprachw.)* relative pronoun; **~satz der** *(Sprachw.)* relative clause

relaxed [ri'lɛkst] *(salopp) Adj.; nicht attr.* laid-back *(coll.)*

relevant [rele'vant] *Adj.* relevant **(für** to)

Relief [re'liɛf] **das; ~s, ~s** *od.* **~e** *(bild. Kunst)* relief

Religion [reli'gio:n] **die; ~, ~en a)** *(auch fig.)* religion; **b)** *o. Pl.; o. Art. (Unterrichtsfach)* religious instruction *or* education; RI; RE

Religions-: ~freiheit die; *o. Pl.* religious freedom; **~krieg der** religious war; **~unterricht der** *s.* **Religion b; ~zugehörigkeit die** religion; religious confession

religiös [reli'giø:s] **1.** *Adj.* religious. **2.** *adv.* in a religious manner; **~ erzogen werden** have *or* receive a religious upbringing

Religiosität [religiozi'tɛ:t] **die; ~:** religiousness

Relikt [re'lɪkt] **das; ~[e], ~e** relic

Reling ['re:lɪŋ] **die; ~, ~s** *od.* **~e** *(Seew.)* [deck-]rail

Reliquie [re'li:kviə] **die; ~, ~n** *(Rel., bes. kath. Kirche)* relic

remis [rə'mi:] *(bes. Schach)* **1.** *indekl. Adj.; nicht attr.* drawn; **~ enden/ausgehen** end in a draw. **2.** *adv.* **~ spielen** draw

Remis das; ~ [rə'mi:(s)], **~** [rə'mi:s] *od.* **~en** *(bes. Schach)* draw; **~ anbieten** offer a draw

Remmidemmi [rɛmi'dɛmi] **das; ~** *(ugs.)* row *(coll.)*; racket

Rempelei die; ~, ~en *(ugs.)* pushing and shoving; jostling; *(Sport)* pushing

rempeln ['rɛmpl̩n] *(ugs.)* push; shove; jostle; *(Sport)* push

Ren [rɛn] **das; ~s, ~s** *od.* **~e** reindeer

Renaissance [rənɛ'sã:s] **die; ~, ~n a)** *o. Pl.* Renaissance; **b)** *(fig.)* revival; **eine ~ erleben** enjoy a renaissance

Rendezvous [rãde'vu:] **das; ~** [...'vu:(s)], **~** ['rãde'vu:s] rendezvous

Reneklode [re:nə'klo:də] **die; ~, ~n** greengage

renitent [reni'tɛnt] **1.** *Adj.* refractory. **2.** *adv.* refractorily

Renitenz [reni'tɛnts] **die; ~** refractoriness

Renn-: ~auto das racing car; **~bahn die** *(Sport)* race-track; *(für Pferde)* racecourse; racetrack; **~boot das** *(Motorboot)* power-boat; *(Segelboot)* racing yacht

rennen ['rɛnən] **1.** *unr. itr. V.; mit sein* run; **um die Wette ~:** have a race; race each other; **in sein Verderben ~** *(fig.)* rush headlong to one's doom; **dauernd zur Polizei ~** *(ugs.)* be always running to the police; **an/gegen jmdn./etw. ~:** run *or* bang into sb./sth. **2.** *unr. tr. V.* **a) sich** *(Dat.)* **an etw.** *(Dat.)* **ein Loch in den Kopf ~:** run *or* bang into sth. and hurt one's head; **b)** *(ugs.: stoßen)* **jmdm. etw. in die Rippen ~:** run sth. into sb.'s ribs

Rennen das; ~s, ~ running; *(Pferde~, Auto~)* racing; *(einzelner Wettbewerb)* race; **zum ~ gehen** *(Pferde~)* go to the races; *(Auto~)* go to the racing; **das ~ machen** *(ugs.)* win

Renner der; ~s, ~ *(ugs.)* big seller

Rennerei die; ~, ~en *(ugs.)* running around; chasing around

Renn-: ~fahrer der racing driver / cyclist / motor-cyclist; **~pferd das** racehorse; **~platz der** *s.* **~bahn; ~rad das** racing cycle; **~sport der** racing *no art.*; **~strecke die** race-track; **~wagen der** racing car

Renommee [renɔ'me:] **das; ~s, ~s** *(geh.)* reputation

renommieren [renɔ'mi:rən] *itr. V.* show off; **mit etw. ~:** brag about sth.

renommiert *Adj.* renowned **(wegen** for)

renovieren [reno'vi:rən] *tr. V.* renovate; redecorate ‹*room, flat*›

Renovierung die; ~, ~en renovation; *(eines Zimmers, einer Wohnung)* redecoration

rentabel [rɛn'ta:bl̩] **1.** *Adj.* profitable. **2.** *adv.* profitably

Rente ['rɛntə] **die; ~, ~n a)** pension; **auf** *od.* **in ~ gehen** *(ugs.)* retire; **auf** *od.* **in ~ sein** *(ugs.)* be retired; **b)** *(Kapitalertrag)* annuity

Renten·alter das pensionable age *no art.*

Ren·tier das reindeer

rentieren [rɛn'ti:rən] *refl. V.* be profitable; ‹*machinery, equipment*› pay its way; ‹*effort, visit, etc.*› be worth while

Rentner ['rɛntnɐ] **der; ~s, ~, Rentnerin die; ~, ~nen** pensioner

reparabel [repa'ra:bl̩] *Adj.* repairable; **nicht mehr ~ sein** be beyond repair

Reparationen [repara'tsio:nən] *Pl. (Politik)* reparations; **~ leisten/zahlen** make/pay reparations

Reparatur [repara'tu:ɐ] **die; ~, ~en** repair **(an + *Dat.*** to); **in ~ sein** be being repaired

reparatur-, Reparatur-: ~anfällig *Adj.* prone to break down *postpos.*; **~arbeit die** repair work; **~en** repair work *sing.*; repairs; **~bedürftig** *Adj.* ‹*device, appliance, vehicle, etc.*› [which is] in need of repair; **~werkstatt die** repair [work]shop; *(für Autos)* garage

reparieren [repa'ri:rən] *tr. V.* repair; mend

Repertoire [repɛ'toa:ɐ] **das; ~s, ~s** *(auch fig.)* repertoire

Report [re'pɔrt] **der; ~[e]s, ~e** report

Reportage [repɔr'ta:ʒə] **die; ~, ~n** report

Reporter [re'pɔrtɐ] **der; ~s, ~, Reporterin die; ~, ~nen** reporter

Repräsentant [reprɛzɛn'tant] **der; ~en, ~en, Repräsentantin die; ~, ~nen** representative

repräsentativ [reprɛzɛnta'ti:f] **1.** *Adj.* **a)** *(auch Politik)* representative **(für** of); **b)** *(ansehnlich)* imposing; *(mit hohem Prestigewert)* prestigious

Repräsentativ·umfrage die *(Statistik)* representative survey

repräsentieren [reprɛzɛn'ti:rən] **1.** *tr. V.* represent. **2.** *itr. V.* attend official and social functions

Repressalie [reprɛ'sa:liə] **die; ~, ~n** repressive measure

Repression [reprɛ'sio:n] **die; ~, ~en** repression

repressiv [reprɛ'si:f] **1.** *Adj.* repressive. **2.** *adv.* repressively

Reproduktion die reproduction

reproduzieren *tr. V. (fachspr., geh.)* reproduce

Reptil [rɛp'ti:l] **das; ~s, ~ien** [rɛp'ti:liən] reptile

Republik [repu'bli:k] **die; ~, ~en** republic

Republikaner [republi'ka:nɐ] **der; ~s, ~ a)** republican; **b)** *(Parteimitglied)* Republican

republikanisch *Adj.* republican

Requiem ['re:kviɛm] **das; ~, ~s** requiem

Requisit [rekvi'zi:t] **das; ~[e]s, ~en a)** *(Theater)* prop *(coll.)*; property; **b)** *(fig.)* requisite

Reservat [rezɛr'va:t] **das; ~[e]s, ~e a)** reservation; **b)** *(Naturschutzgebiet)* reserve

Reserve [re'zɛrvə] **die; ~, ~n a)** reserve **(an + *Dat.*** of); **etw. in ~ haben** have sth. in reserve; *s. auch* **eisern 1 d; still 1 f; b)** *(Milit., Sport)* reserves *pl.*; **c)** *o. Pl. (Zurückhaltung)* reserve; **jmdn. aus**

der ~ locken *(ugs.)* bring sb. out of his/her shell

Reserve-: ~bank die *Pl.* **~bänke** *(Sport)* substitutes' bench; **~kanister der** spare [petrol *(Brit.)* or *(Amer.)* gasoline] can; **~offizier der** reserve officer; **~rad das** spare wheel; **~reifen der** spare tyre; **~spieler der** *(Sport)* substitute; reserve; **~tank der** reserve [fuel] tank

reservieren *tr. V.* reserve

reserviert 1. *Adj.* reserved. **2.** *adv.* in a reserved way

Reservierung die; ~, ~en reservation

Reservist der; ~en, ~en *(Milit.)* reservist

Reservoir [rezɛr'vo̯a:ɐ̯] **das; ~s, ~e** *(auch fig.)* reservoir **(an +** *Dat.* of)

Residenz [rezi'dɛnts] **die; ~, ~en a)** residence; **b)** *(Stadt)* [royal] capital

residieren [rezi'di:rən] *itr. V.* reside

Resignation [rezɪgna'tsi̯o:n] **die; ~, ~en** resignation

resignieren [rezɪ'gni:rən] *itr. V.* give up

resigniert 1. *Adj.* resigned. **2.** *adv.* resignedly

resistent [rezɪs'tɛnt] *Adj. (Biol., Med.)* resistant **(gegen** to)

resolut [rezo'lu:t] **1.** *Adj.* resolute. **2.** *adv.* resolutely

Resolution [rezolu'tsi̯o:n] **die; ~, ~en** resolution

Resonanz [rezo'nants] **die; ~, ~en a)** *(Physik, Musik)* resonance; **b)** *(Reaktion)* response **(auf +** *Akk.* to); **~/keine ~ finden** meet with a/no response

Resopal ⓦ [rezo'pa:l] **das; ~s** ≈ melamine

resozialisieren *tr. V.* reintegrate into society

Resozialisierung die; ~, ~en reintegration into society

Respekt [re'spɛkt] **der; ~[e]s a)** respect; **~ vor jmdm./etw. haben** have respect for sb./sth.; **jmdm. ~ abnötigen** command sb.'s respect; **bei allem ~:** with all due respect **(vor +** *Dat.* to); **allen ~!, ~, ~!** good for you!; well done!; **b)** *(Furcht)* **jmdm. ~ einflößen** intimidate sb.; **vor jmdm./etw. [größten] ~ haben** be [much] in awe of sb./sth.

respektabel [respɛk'ta:bl̩] **1.** *Adj.* respectable. **2.** *adv.* respectably

respektieren *tr. V.* respect

respekt·los 1. *Adj.* disrespectful. **2.** *adv.* disrespectfully

Respektlosigkeit die; ~, ~en a) *o. Pl.* disrespectfulness; lack of respect; **b)** *(Äußerung)* disrespectful remark; *(Handlung)* impertinence

respekt·voll 1. *Adj.* respectful. **2.** *adv.* respectfully

Ressentiment [rɛsãti'mã:] **das; ~s, ~s** antipathy **(gegen** towards)

Ressort [rɛ'so:ɐ̯] **das; ~s, ~s** area of responsibility; *(Abteilung)* department

Rest [rɛst] **der; ~[e]s, ~e a)** rest; **~e** *(historische ~e, Ruinen)* remains; *(einer Kultur)* relics; **jmdm./einer Sache den ~ geben** *(ugs.)* finish sb./sth. off; **ein ~ Wein ist noch da** there's still a little bit *or* a drop of wine left; **morgen gibt es ~e** tomorrow we're having left-overs; **das ist der ~ vom Schützenfest** *(ugs.)* that's all there is left; **b)** *(Endstück, Stoff~ usw.)* remnant; **c)** *(Math.)* remainder; **20 durch 6 ist 3, ~ 2** 20 divided by 6 is 3 with *or* and 2 left over

Restaurant [rɛsto'rã:] **das; ~s, ~s** restaurant

Restauration [restaura'tsi̯o:n] **die; ~, ~en** *(auch Politik)* restoration

restaurieren [restau'ri:rən] *tr. V.* restore

restlich *Adj.; nicht präd.* remaining; **die ~e Butter** the rest of the butter

rest·los 1. *Adj.; nicht präd.* complete; total. **2.** *adv.* completely; totally

Rest·posten der *(Kaufmannsspr.)* remaining stock *no indef. art.*

Resultat [rezʊl'ta:t] **das; ~[e]s, ~e** result; **zu dem ~ kommen, daß ...:** come to the conclusion that ...

resultieren [rezʊl'ti:rən] *itr. V.* result **(aus** from); **daraus resultiert, daß ...:** the result *or* upshot of this is that ...

Resümee [rezʏ'me:] **das; ~s, ~s** résumé

resümieren [rezʏ'mi:rən] **1.** *tr. V.* summarize; give a résumé of. **2.** *itr. V.* sum up

Retorte [re'tɔrtə] **die; ~, ~n** *(Chemie)* retort

retour [re'tu:ɐ̯] *Adv. (bes. südd., österr., schweiz.)* back

Retrospektive [retrospɛk'ti:və] **die; ~, ~n a)** *(geh.)* retrospective view; **in der ~:** in retrospect; **b)** *(Ausstellung)* retrospective

retten ['rɛtn̩] **1.** *tr. V.* save; *(vor Gefahr)* save; rescue; *(befreien)* rescue; **jmdm. das Leben ~:** save sb.'s life; **jmdn. vor jmdm./etw. ~:** save sb. from sb./sth.; **ist er noch zu ~?** *(ugs. fig.)* has he gone [completely] round the bend? *(coll.)*; **das alte Haus/der Patient ist nicht mehr zu ~:** the old house is past saving/the patient is beyond help. **2.** *refl. V. (fliehen)* escape **(aus** from); **sich vor etw.** *(Dat.)* **~:** escape [from] sth.; **sich vor jmdm./etw. nicht** *od.* **kaum [noch] ~ können** be besieged by sb./be swamped with sth.. **3.** *itr. V. (Ballspiele)* save

Retter der; ~s, ~, Retterin die; ~, ~nen rescuer; *(eines Landes, einer Bewegung o. ä.)* saviour; **Christ der ~:** Christ the Saviour

Rettich ['rɛtɪç] **der; ~s, ~e** radish

Rettung die a) rescue; *(Rel., eines Landes usw.)* salvation; *(vor Zerstörung)* saving; **auf ~ warten/hoffen** wait for rescue/hope to be rescued; **es war jmds. ~, daß ...:** sb. was saved by the fact that ...; **das war meine ~:** that was my salvation

rettungs-, Rettungs-: ~boot das lifeboat; **~dienst der** ambulance service; *(Bergwacht, Seerettungsdienst, bei Katastrophen)* rescue service; **~hubschrauber der** rescue helicopter; **~los 1.** *Adj.* hopeless; **2.** *adv.* hopelessly; **~ring der** lifebelt; **~schwimmer der, ~schwimmerin die** life-saver; *(am Strand, im Schwimmbad)* life-guard; **~wagen der** ambulance

retuschieren *tr. V. (Fot., Druckw.)* retouch; *(fig.)* gloss over

Reue ['rɔyə] **die; ~:** remorse **(über +** *Akk.* for); *(Rel.)* repentance

reuen *tr. V. (meist geh.)* **etw. reut jmdn.** sb. regrets sth.

reu·mütig 1. *Adj.; nicht präd.* remorseful; repentant, penitent ⟨*sinner*⟩. **2.** *adv.* remorsefully; **du wirst ~ zurückkehren** you'll be back, saying you're sorry

Reuse [:rɔyzə] **die; ~, ~n** fish-trap

Revanche [re'vã:ʃ(ə)] **die; ~, ~n** revenge; *(Sport: Rückkampf, ~spiel)* return match/fight/game

revanchieren *refl. V.* **a)** get one's revenge, *(coll.)* get one's own back **(bei** on); **b)** *(ugs.: sich erkenntlich zeigen)* **sich bei jmdm. für eine Einladung/seine Gastfreundschaft ~:** return sb.'s invitation/repay sb.'s hospitality

Revers [rə've:ɐ̯] **das** *od. (österr.)* **der; ~** [rə'vɛ:ɐ̯(s)], **~** [rə'vɛ:ɐ̯s] lapel

revidieren [revi'di:rən] *tr. V.* revise; amend ⟨*law, contract*⟩

Revier [re'vi:ɐ̯] **das; ~s, ~e a)** *(Aufgabenbereich)* province; **b)** *(Zool.)* territory; **c)** *(Polizei~)*

(Dienststelle) [police] station; *(Bereich)* district; *(des einzelnen Polizisten)* beat; **d)** *(Forst~)* district; **e)** *(Jagd~)* preserve; shoot; **f)** *(Bergbau)* coalfield; **das ~:** the Ruhr/Saar coalfields *pl.*

Revision [revi'zi̯o:n] **die; ~, ~en a)** revision; *(Änderung)* amendment; **b)** *(Rechtsw.)* appeal [on a point/points of law]; **~ einlegen, in die ~ gehen** lodge an appeal [on a point/points of law]

Revolte [re'vɔltə] **die; ~, ~n** revolt

revoltieren *itr. V.* revolt, rebel (**gegen** against); *(fig.)* ⟨*stomach*⟩ rebel

Revolution [revolu'tsi̯o:n] **die; ~, ~en** *(auch fig.)* revolution

revolutionär [revolutsi̯o'nɛ:ɐ̯] **1.** *Adj.* revolutionary. **2.** *adv.* in a revolutionary way

Revolutionär der; ~s, ~e, Revolutionärin die; ~, ~nen revolutionary

revolutionieren *tr. V.* revolutionize

Revoluzzer [revo'lʊtsɐ] **der; ~s, ~** *(abwertend)* phoney revolutionary

Revolver [re'vɔlvɐ] **der; ~s, ~** revolver

Revolver·held der *(abwertend)* gun-slinger

Rezensent [retsɛn'zɛnt] **der; ~en, ~en** reviewer

rezensieren *tr. V.* review

Rezension die; ~, ~en review

Rezept [re'tsɛpt] **das; ~[e]s, ~e a)** *(Med.)* prescription; *(fig.)* remedy (**gegen** for); **b)** *(Anleitung)* recipe; *(fig.)* formula

rezept·frei 1. *Adj.* ⟨*medicine, drug, etc.*⟩ obtainable without a prescription. **2.** *adv.* **etw. ~ verkaufen** sell sth. without a prescription *or* over the counter

Rezeption [retsɛp'tsi̯o:n] **die; ~, ~en** reception *no art.*

rezept·pflichtig *Adj.* ⟨*medicine, drug, etc.*⟩ obtainable only on prescription

rezessiv [retsɛ'si:f] *(Biol.)* **1.** *Adj.* recessive. **2.** *adv.* recessively

reziprok [retsi'pro:k] *(bes. Math., Sprachw.) Adj.* reciprocal

Rezitativ [retsita'ti:f] **das; ~s, ~e** *(Musik)* recitative

rezitieren [retsi'ti:rən] *tr., itr. V.* recite

R-Gespräch ['ɛr-] **das** *(Fernspr.)* reverse-charge call *(Brit.)*; collect call *(Amer.)*

Rhabarber [ra'barbɐ] **der; ~s** rhubarb

Rhapsodie [rapso'di:] **die; ~, ~n** *(Musik, Literaturw.)* rhapsody

Rhein [rain] **der; ~[e]s** Rhine

Rhein·fall der Rhine Falls

rheinisch *Adj.* Rhenish; **eine ~e Spezialität** a speciality of the Rhine region

Rhein·land das; ~[e]s Rhineland

Rheinland-Pfalz (das); Rheinland-Pfalz' the Rhineland-Palatinate

rheinland-pfälzisch *Adj.* ⟨*capital, citizen, etc.*⟩ of the Rhineland-Palatinate

Rhesus- ['re:zʊs]: **~affe der** rhesus monkey; **~faktor der;** *o. Pl.* *(Med.)* rhesus factor; Rh factor

Rhetorik [re'to:rɪk] **die; ~, ~en** rhetoric

Rhetoriker der; ~s, ~ rhetorician

rhetorisch 1. *Adj.* rhetorical. **2.** *adv.* rhetorically

Rheuma ['rɔy̯ma] **das; ~s** *(ugs.)* rheumatism; rheumatics *pl. (coll.)*

Rheumatiker [rɔy̯'ma:tikɐ] **der; ~s, ~** *(Med.)* rheumatic

rheumatisch *(Med.)* **1.** *Adj.* rheumatic. **2.** *adv.* rheumatically

Rheumatismus [rɔy̯ma'tɪsmʊs] **der; ~** *(Med.)* rheumatism

Rhinozeros [ri'no:tserɔs] **das; ~[ses], ~se** rhinoceros; rhino *(coll.)*

Rhododendron [rodo'dɛndrɔn] **der** *od.* **das; ~s, Rhododendren** rhododendron

Rhomben *s.* **Rhombus**

rhombisch *Adj.* *(bes. Math.)* rhombic

Rhombus ['rɔmbʊs] **der; ~, Rhomben** ['rɔmbn̩] rhombus

Rhythmen *s.* **Rhythmus**

rhythmisch 1. *Adj.* rhythmical; rhythmic. **2.** *adv.* rhythmically

Rhythmus ['rʏtmʊs] **der; ~, Rhythmen** ['rʏtmən] rhythm; **aus dem ~ kommen** lose the rhythm

Rhythmus-: rhythm ⟨*guitar, section, etc.*⟩

richten ['rɪçtn̩] **1.** *tr. V.* **a)** direct ⟨*gaze*⟩ (**auf** + *Akk.* at, towards); turn ⟨*eyes, gaze*⟩ (**auf** + *Akk.* towards); point ⟨*torch, telescope, gun*⟩ (**auf** + *Akk.* at); aim, train ⟨*gun, missile, telescope, searchlight*⟩ (**auf** + *Akk.* on); *(fig.)* direct ⟨*activity, attention*⟩ (**auf** + *Akk.* towards); address ⟨*letter, remarks, words*⟩ (**an** + *Akk.* to); direct, level ⟨*criticism*⟩ (**an** + *Akk.* at); send ⟨*letter of thanks, message of greeting*⟩ (**an** + *Akk.* to); **b)** *(gerade~)* straighten; set ⟨*fracture*⟩; **c)** *(einstellen)* aim ⟨*cannon, missile*⟩; direct ⟨*aerial*⟩; **d)** *(aburteilen)* judge; *(verurteilen)* condemn; *s. auch* **zugrunde a. 2.** *refl. V.* **a)** *(sich hinwenden)* **sich auf jmdn./etw. ~** *(auch fig.)* be directed towards sb./sth.; **b) sich an jmdn./etw. ~** ⟨*person*⟩ turn on sb./sth.; ⟨*appeal, explanation*⟩ be directed at sb./sth.; **sich gegen jmdn./etw. ~** ⟨*person*⟩ criticize sb./sth.; ⟨*criticism, accusations, etc.*⟩ be aimed *or* levelled *or* directed at sb./sth.; **c)** *(sich orientieren)* **sich nach jmdm./jmds. Wünschen ~:** fit in with sb./sb.'s wishes; **sich nach den Vorschriften ~:** keep to the rules; **d)** *(abhängen)* **sich nach jmdm./etw. ~:** depend on sb./sth. **3.** *itr. V.* *(urteilen)* judge; pass judgement; **über jmdn. ~:** judge sb.; pass judgement on sb.; *(zu Gericht sitzen)* sit in judgement over sb.

Richter der; ~s, ~ judge; **jmdn. vor den ~ bringen** take sb. to court

Richterin die; ~, ~nen judge

richterlich *Adj.; nicht präd.* judicial

Richt-: **~fest das** topping-out ceremony; **~geschwindigkeit die** *(Verkehrsw.)* recommended maximum speed

richtig 1. *Adj.* **a)** right; *(zutreffend)* right; correct; correct ⟨*realization*⟩; accurate ⟨*prophecy, premonition*⟩; **bin ich hier ~ bei Schulzes?** is this the Schulzes' home?; **das ist genau das ~e für mich** that's just right for me; **ja ~!** yes, that's right; **b)** *(ordentlich)* proper; **nicht ganz ~ [im Kopf** *od.* *(ugs.)* **im Oberstübchen] sein** be not quite right in the head *(coll.)* *or* not quite all there *(coll.)*; **c)** *(wirklich, echt)* real; **du bist ein ~er Esel** you're a right *or* proper idiot *(coll.)*. **2.** *adv.* **a)** right; correctly; **~ sitzen** *od.* **passen** ⟨*clothes*⟩ fit properly; **meine Uhr geht ~:** my watch is right; **b)** *(ordentlich)* properly; **~ ausschlafen** have a good sleep; **c)** *(richtiggehend)* really

¹Richtige der/die; *adj. Dekl.* right man/ woman/person; **sie sucht noch den ~n** she's still looking for Mr Right

²Richtige der; *adj. Dekl.* **drei/sechs ~ im Lotto** three/six right in the lottery

³Richtige das; *adj. Dekl.* right thing

richtig·gehend *Adj.; nicht präd., adv. s.* **regelrecht**

Richtigkeit die; ~: correctness; **etw. hat seine ~, mit etw. hat es seine ~:** sth. is right

richtig-, Richtig-: **~|liegen** *unr. itr. V. (ugs.)* be right; **~|stellen** *tr. V.* correct; **~stellung die** correction

Richt-: ~linie die guideline; **~platz der** place of execution; **~schnur die;** *Pl.* **~schnuren** *(fig.)* guiding principle

Richtung die; ~, ~en a) direction; **die ~ ändern** *od.* **wechseln** change direction; ‹*ship, aircraft*› change course; **nach/aus allen ~en** in/from all directions; **der Zug/die Autobahn ~ Ulm** the train to Ulm/the motorway in the direction of Ulm; **wir gehen in diese ~:** we're going this way; **b)** *(fig.: Tendenz)* movement; trend; *(die Vertreter einer ~) (in der Kunst, Literatur)* movement; *(in einer Partei)* faction; *(Denk~)* school of thought

richtung·weisend *Adj.* ‹*idea, resolution, paper, speech*› that points the way ahead; *(in der Mode)* trend-setting

rieb [ri:p] *1. u. 3. Pers. Sg. Prät. v.* **reiben**

riechen ['ri:çn̩] **1.** *unr. tr. V.* **a)** smell; **jmdn./etw. nicht ~ können** *(fig. salopp)* not be able to stand sb./sth.; **b)** *(wittern)* ‹*dog etc.*› scent, pick up the scent of ‹*animal*›; **ich konnte ja nicht ~, daß ...** *(fig.)* [I'm not psychic,] I couldn't know that ... **2.** *unr. itr. V.* **a)** smell; **Hunde können sehr gut ~:** dogs have a very good sense of smell; **an jmdm./etw. ~:** smell sb./sth.; **laß mich mal [daran] ~:** let me have a sniff; **b)** *(einen Geruch haben)* smell **(nach** of); **gut/schlecht ~:** smell good/bad; **er roch aus dem Mund** he had bad breath; his breath smelt

Riecher der; ~s, ~ *(salopp)* **a)** *(Nase)* conk *(sl.)*; **b)** *(fig.: Gespür)* nose; **einen guten ~ für etw. haben** have a sixth sense for sth.

Ried [ri:t] **das; ~[e]s, ~e a)** *o. Pl. (Schilf)* reeds *pl.*; **b)** *(Gebiet)* reedy marsh

rief [ri:f] *1. u. 3. Pers. Sg. Prät. v.* **rufen**

Riege ['ri:gə] **die; ~, ~n** *(Turnen)* squad

Riegel ['ri:gl̩] **der; ~s, ~: a)** bolt; **einer Sache** *(Dat.)* **einen ~ vorschieben** *(fig.)* put a stop to sth.; *(etw. verhindern)* not let sth. happen; **b) ein ~ Schokolade** a bar of chocolate

Riemchen das; ~s, ~: [small] strap *or* belt

Riemen ['ri:mən] **der; ~s, ~ a)** strap; *(Treib~, Gürtel)* belt; **sich am ~ reißen** *(ugs.)* pull oneself together; **den ~ enger schnallen** *(fig. ugs.)* tighten one's belt; **b)** *(Ruder)* [long] oar

Riese ['ri:zə] **der; ~n, ~n** giant

rieseln ['ri:zl̩n] *itr. V.; mit Richtungsangabe mit sein* trickle; ‹*sand, lime*› trickle [down]; ‹*snow*› fall gently *or* lightly

Riesen- giant ‹*building, tree, salamander, tortoise, etc.*›; enormous ‹*task, selection, profit, sum, portion*›; tremendous *(coll.)* ‹*effort, rejoicing, success, hit*›; *(abwertend: schrecklich)* terrific *(coll.)*, terrible *(coll.)* ‹*stupidity, mess, scandal, fuss*›

riesen-, Riesen-: ~groß *Adj.* enormous; huge; gigantic; terrific *(coll.)* ‹*surprise*›; **~rad das** big wheel; Ferris wheel; **~schritt der** giant stride; **~slalom der** *(Skisport)* giant slalom

riesig 1. *Adj.* **a)** enormous; huge; gigantic; vast ‹*country*›; tremendous ‹*joy, enthusiasm, effort, progress, strength*›; terrific *(coll.)*, terrible *(coll.)* ‹*hunger, thirst*›; **b)** *(ugs.: großartig)* fabulous *(coll.)*, tremendous *(coll.)* ‹*party, film, etc.*›. **2.** *adv. (ugs.)* tremendously *(coll.)*; terribly *(coll.)*

Riesling ['ri:slɪŋ] **der; ~s, ~e** Riesling

riet [ri:t] *1. u. 3. Pers. Sg. Prät. v.* **raten**

Riff [rɪf] **das; ~[e]s, ~e** reef

rigoros [rigo'ro:s] **1.** *Adj.* rigorous. **2.** *adv.* rigorously

Rille ['rɪlə] **die; ~, ~n** groove

Rind [rɪnt] **das; ~[e]s, ~er a)** *(Kuh)* cow; *(Bulle)* bull; **~er** cattle *pl.*; **20 ~er** twenty head of cattle; **Hackfleisch/ein Steak vom ~:** minced *or (Amer.)* ground beef/a beef steak; **b)** *(~fleisch)* beef; **c)** *(Zool.)* bovine

Rinde die; ~, ~n a) *(Baum~)* bark; **b)** *(Brot~)* crust; *(Käse~)* rind

Rinder-: ~braten der roast beef *no indef. art.*; *(roh)* roasting beef *no indef. art.*; **ein ~braten** a joint of roast beef; *(roh)* a joint of [roasting] beef; **~leber die** ox liver; **~zucht die** cattle-breeding *or* -rearing *no art.*

Rind·fleisch das beef

Rinds-: ~braten der *(bes. südd., österr.) s.* **Rinderbraten; ~leder das** cowhide; oxhide

Rind·vieh das; *Pl.* **Rindviecher a)** *o. Pl.* cattle *pl.*; **b)** *(ugs. abwertend)* ass; [stupid] fool

Ring [rɪŋ] **der; ~[e]s, ~e a)** ring; **b)** *(Box~)* ring; **~ frei zur zweiten Runde** seconds out for the second round

Ring·buch das ring binder

Ringel·blume die marigold

ringeln 1. *tr. V.* curl; coil ‹*tail*›. **2.** *refl. V.* curl

Ringel-: ~natter die ring-snake; **~reihen der** ring-a-ring-o'-roses; **~schwanz der** curly tail

ringen 1. *unr. tr. V. (Sport, fig.)* wrestle; *(fig.: kämpfen)* struggle, fight **(um** for); *(fig.)* **mit den Tränen ~:** fight back one's tears; **die Ärzte ~ um sein Leben** the doctors are struggling *or* fighting to save his life; **nach Atem ~:** struggle for breath. **2.** *unr. tr. V.* **a) den Gegner zu Boden ~** *(auch fig.)* bring one's opponent down; **b) jmdm. etw. aus den Händen ~:** wrest sth. from sb.'s hands; **c) die Hände ~:** wring one's hands

Ringen das; ~s *(Sport)* wrestling *no art.*

Ringer der; ~s, ~: wrestler

ring-, Ring-: ~finger der ring-finger; **~förmig 1.** *Adj.* in the shape of a ring *postpos.*; circular; **2.** *adv.* ‹*arrange*› in a ring *or* circle; **~kampf der a)** [stand-up] fight; **b)** *(Sport)* wrestling bout; **~kämpfer der** wrestler; **~richter der** *(Boxen)* referee

rings [rɪŋs] *Adv.* all around

Ring·schlüssel der ring spanner

rings·herum *Adv.* all around [it/them *etc.*]

Ring·straße die ring road

rings-: ~um, ~umher *Adv.* all around

Rinne ['rɪnə] **die; ~, ~n** channel; *(Dach~, Rinnstein)* gutter; *(Rille)* groove

rinnen *unr. itr. V.* **a)** *mit sein* run; **b)** *(südd.: undicht sein)* leak

Rinnsal ['rɪnza:l] **das; ~[e]s, ~e** *(geh.)* rivulet

Rinn·stein der gutter; **im ~ landen** *(fig.)* end up in the gutter

Rippchen das; ~s, ~ *(Kochk. südd.)* rib [of pork]

Rippe ['rɪpə] **die; ~, ~n** *(auch Bot., Technik, Textilw., fig.)* rib; **sie hat nichts auf den ~n** *(ugs.)* she is only skin and bone

Risiko ['ri:ziko] **das; ~s, Risiken** *od. österr.* **Risken** risk; **ein/kein ~ eingehen** take a risk/not take any risks; **auf dein ~:** at your own risk

risiko-: ~freudig 1. *Adj.* risky ‹*driving*›; ‹*player, speculator*› who likes taking risks; **2.** *adv.* **er fährt/spielt sehr ~freudig** he likes to take [a lot of] risks when he drives/plays; **~los 1.** *Adj.* safe; without risk *postpos.*; **2.** *adv.* safely; without taking risks

riskant [rɪs'kant] **1.** *Adj.* risky. **2.** *adv.* riskily

riskieren [rɪs'ki:rən] *tr. V.* risk; venture ‹*smile, remark*›; run the risk of ‹*accident, thrashing, etc.*›; put ‹*reputation, job*› at risk; **et-**

was/nichts ~: take a risk/not take any risks

riß [rɪs] *1. u. 3. Pers. Sg. Prät. v.* **reißen**

Riß [rɪs] **der; Risses, Risse a)** *(in Stoff, Papier usw.)* tear; **b)** *(Spalt, Sprung)* crack; *(fig.: Kluft)* rift; split

rissig *Adj.* cracked; chapped ‹*lips*›

Riten *s.* **Ritus**

ritsch *Interj.* rip; zip; **~, ratsch** rip, rip

ritt *1. u. 3. Pers. Sg. Prät. v.* **reiten**

Ritt der; ~[e]s, ~e ride

Ritter der; ~s, ~ knight; **jmdn. zum ~ schlagen** *(hist.)* knight sb.; dub sb. [a] knight

ritterlich *Adj.* chivalrous

Ritter-: ~schlag der *(hist.)* knightly accolade; **~sporn der** *(Bot.)* larkspur; *(Gartenrittersporn)* delphinium

rittlings ['rɪtlɪŋs] *Adv.* astride

Ritual [ri'tu̯a:l] **das; ~s, ~e** *od.* **Ritualien** [-li̯ən] *(Rel., fig.)* ritual

rituell [ri'tu̯ɛl] *(Rel., fig.)* **1.** *Adj.* ritual. **2.** *adv.* ritually

Ritus ['ri:tʊs] **der; ~, Riten** *(Rel., fig.)* rite

Ritz [rɪts] **der; ~es, ~e, Ritze die; ~, ~n** crack; [narrow] gap

ritzen *tr. V.* **a)** scratch; *(tiefer)* cut; **b)** *(einritzen)* carve ‹*name etc.*› **(in** + *Akk.* in)

Rivale [ri'va:lə] **der; ~n, ~n, Rivalin die; ~, ~nen** rival

rivalisieren *itr. V.* **mit jmdm. um etw. ~:** compete with sb. for sth.; **~de Gruppen** rival groups

Rivalität [rivali'tɛ:t] **die; ~, ~en** rivalry *no indef. art.*

Roastbeef ['ro:stbi:f] **das; ~s, ~s** roast [sirloin *(Brit.)* of] beef

Robbe ['rɔbə] **die; ~, ~n** seal

robben *itr. V.; mit sein* crawl

Robe ['ro:bə] **die; ~, ~n** robe; *(schwarz)* gown

Roboter ['rɔbɔtɐ] **der; ~s, ~:** robot

robust [ro'bʊst] *Adj.* robust

Robustheit die; ~: robustness; *(Gesundheit)* robust constitution

roch [rɔx] *1. u. 3. Pers. Sg. Prät. v.* **riechen**

Rochade [rɔ'xa:də] **die; ~, ~n** *(Schach)* castling; **kleine/große ~:** short/long castling; **eine ~ ausführen** castle

röcheln ['rœçl̩n] *itr. V.* breathe stertorously; ‹*dying person*› give the death-rattle

Rochen ['rɔxn̩] **der; ~s, ~** *(Zool.)* ray

[1]Rock [rɔk] **der; ~[e]s, Röcke** ['rœkə] **a)** skirt; **b)** *(landsch.: Jacke)* jacket

[2]Rock der; ~[s] rock [music]

Rocker der; ~s, ~: rocker

rockig *Adj.* rock-like ‹*jazz etc.*›

Rock·musik die rock music

Rodel ['ro:dl̩] **der; ~s, ~** *(südd.) s.* **Rodelschlitten**

Rodel·bahn die toboggan-run; *(Sport)* luge-run

rodeln ['ro:dl̩n] *itr. V.; mit sein* sledge; toboggan; *(Sport)* luge

Rodeln das; ~s sledging *no art.;* tobogganing *no art.; (Sport)* luge

Rodel·schlitten der sledge; toboggan; *(Sport)* luge

roden ['ro:dn̩] **1.** *tr. V.* **a)** clear ‹*wood, land*›; *(ausgraben)* grub up ‹*tree*›; **b)** *(landsch.)* lift ‹*potatoes etc.*›. **2.** *itr. V.* clear the land

Rogen ['ro:gn̩] **der; ~s, ~:** roe

Roggen ['rɔgn̩] **der; ~s** rye

Roggen-: ~brot das rye bread; **ein ~brot** a loaf of rye bread; **~brötchen das** rye-bread roll

roh [ro:] **1.** *Adj.* **a)** raw ‹*food*›; **jmdn./etw. wie ein ~es Ei behandeln** handle sb./sth. with kid gloves; **b)** *(nicht bearbeitet)* rough, unfinished ‹*wood*›; rough, uncut ‹*diamond*›; rough-hewn, undressed ‹*stone*›; crude ‹*ore, metal*›; untreated ‹*skin*›; raw ‹*silk, sugar*›; **c)** *(brutal)* brutish ‹*person, treatment, etc.*›; *(grausam)* callous ‹*person, treatment*›; *(grob)* coarse, uncouth ‹*manners, words, joke*›; brute *attrib.* ‹*force*›. **2.** *adv. (brutal)* brutishly; *(grausam)* callously; *(grob)* coarsely; in an uncouth manner

Roh·bau der shell [of a/the building]

Roheit ['ro:hait] **die; ~, ~en a)** *o. Pl. (Brutalität)* brutishness; *(Grausamkeit)* callousness; *(Grobheit)* corseness; uncouthness; **b)** *(Handlung)* brutish/callous deed

Roh·kost die raw fruit and vegetables *pl.*

Rohling ['ro:lɪŋ] **der; ~s, ~e** *(abwertend: Mensch)* brute

Roh-: ~material das raw material; **~öl das** crude oil

Rohr [ro:ɐ̯] **das; ~[e]s, ~e a)** *(Leitungs~)* pipe; *(als Bauteil)* tube; *(Geschütz~)* barrel; **b)** *o. Pl. (Röhricht)* reeds *pl.;* **c)** *o. Pl. (Schilf usw. als Werkstoff)* reed

Rohr·bruch der burst pipe

Röhrchen ['rø:ɐ̯çən] **das; ~s, ~:** small pipe; *(Behälter)* small tube; **ins ~ blasen** take the breathalyser test

Röhre ['rø:rə] **die; ~, ~n a)** *(auch Neon~, Bild~, Tabletten~)* tube; *(Elektronen~)* valve *(Brit.);* tube *(Amer.);* **vor der ~ sitzen** *(ugs.)* sit in front of the box *(coll.);* **b)** *(Leitungs~)* pipe; **c)** *(eines Ofens)* oven; **in die ~ gucken** *(fig. ugs.)* be left out [in the cold]

röhren *itr. V.* ‹*stag etc.*› bell; *(fig.)* roar

röhren·förmig *Adj.* tubular

Röhren·hose die drainpipe trousers

Rohr-: ~flöte die reed-pipe; **~post die** pneumatic dispatch; **etw. mit ~post befördern** convey sth. by pneumatic tube; **~spatz der** *in* **schimpfen wie ein ~spatz** *(ugs.)* really create *(coll.);* **~stock der** cane [walking-stick]; **~zucker der** cane-sugar

Roh·stoff der raw material

Rokoko ['rɔkoko] **das; ~[s]** rococo; *(Zeit)* rococo period

Rolladen ['rɔla:dn̩] **der; ~s, Rolläden** ['rɔlɛ:dn̩] [roller] shutter

Roll·bahn die *(Flugw.)* taxiway

Rolle ['rɔlə] **die; ~, ~n a)** *(Spule)* reel; spool; **b)** *(zylindrischer [Hohl]körper; Zusammengerolltes)* roll; *(Schrift~)* scroll; **eine ~ Bindfaden/Markstücke/Kekse** a reel of string/roll of one-mark pieces/[round] packet of biscuits; **c)** *(Walze)* roller; *(Teig~)* rolling-pin; **d)** *(Rad)* [small] wheel; *(an Möbeln usw.)* castor; *(für Gardine, Schiebetür usw.)* runner; **e)** *(Turnen, Kunstflug)* roll; **f)** *(Theater, Film usw., fig.)* role; part; *(Soziol.)* role; **[bei jmdm./einer Sache] eine entscheidende ~ spielen** be of crucial importance [to sb./for sth.]; **es spielt keine ~:** it is of no importance; *(es macht nichts aus)* it doesn't matter; **aus der ~ fallen** forget oneself

rollen 1. *tr. V.* roll; **das R ~:** roll one's r's; **sich** *(Dat.)* **eine Zigarette ~:** roll oneself a cigarette. **2.** *itr. V.* **a)** *mit sein* ‹*ball, wheel, etc.*› roll; ‹*vehicle*› move; ‹*aircraft*› taxi; **etw. ins Rollen bringen** set sth. in motion; get sth. going *(lit. or fig.); (unbeabsichtigt)* set sth. moving; **b)** *mit Richtungsangabe mit sein* ‹*thunder, guns, echo*› rumble. **3.** *refl. V.* **a)** roll; **b)** ‹*paper, carpet*› curl [up]

Rollen·spiel das *(Sozialpsych.)* role-playing *no pl., no art.;* role-play *no pl., no art.*

Roller der; ~s, ~ scooter

Roll-: ~feld das [operational] airfield; landing-field; **~kommando das** party of bully-boys; **~kragen der** polo-neck; **~laden der** *s.* **Rolladen; ~mops der** rollmops

Rollo ['rɔlo] **das; ~s, ~s** [roller] blind

Roll·schuh der roller-skate; **~laufen** roller-skate

Rollschuh·bahn die roller-skating rink

Roll-: ~splitt der loose chippings *pl.*; **~stuhl der** wheelchair; **~treppe die** escalator

Rom [ro:m] *(das)*; ~s Rome; **Zustände wie im alten ~** *(fig.)* everything in chaos

Roman [ro'ma:n] **der; ~s, ~e** novel

Roman·figur die character from *or* in a novel

Romanik [ro'ma:nɪk] **die; ~:** Romanesque; *(Zeit)* Romanesque period

romanisch *Adj.* **a)** Romance ⟨*language, literature*⟩; Latin ⟨*people, country, charm*⟩; **b)** *(der Romanik)* Romanesque

Romanistik [roma'nɪstɪk] **die; ~:** Romance studies *pl., no art.*; *(Sprache und Literatur)* Romance languages and literature *no art.*

Romantik [ro'mantɪk] **die; ~ a)** romanticism; romantic nature; **b)** *(Literaturw., Musik usw.)* Romanticism *no art.*; *(Epoche)* Romantic period

romantisch 1. *Adj.* **a)** romantic; **b)** *(Literaturw., Musik usw.)* Romantic. **2.** *adv.* romantically

Romanze [ro'mantsə] **die; ~, ~n** *(auch fig.)* romance

Römer ['rø:mɐ] **der; ~s, ~, Römerin die; ~, ~nen** Roman

römisch *Adj.* Roman

römisch-katholisch 1. *Adj.* Roman Catholic. **2.** *adv.* **~ getauft** baptized into the Roman Catholic church

röm.-kath. *Abk.* **römisch-katholisch** RC

Rommé ['rɔme] **das; ~s, ~s** *(Kartenspiele)* rummy *no art.*

Rondo ['rɔndo] **das; ~s, ~s** *(Musik)* rondo

röntgen ['rœntgn̩] *tr. V.* X-ray; **sich** *(Akk.)*/**sich** *(Dat.)* **den Magen ~ lassen** have an X-ray/have one's stomach X-rayed

Röntgen- X-ray ⟨*picture, screen, apparatus, etc.*⟩

Röntgen·strahlen *Pl.* X-rays

rosa ['ro:za] *indekl. Adj., adv.* pink

Rosa das; ~s, ~ *od.* **~s** pink

rosa-: ~farben, ~farbig *Adj.* pink; **~rot** *Adj.* [deep] pink

Röschen ['rø:sçən] **das; ~s, ~** [little] rose

Rose ['ro:zə] **die; ~, ~n** rose

rosé [ro'ze:] *indekl. Adj., adv.* pale pink

Rosé der; ~s, ~s rosé [wine]

rosen-, Rosen-: ~beet das rose-bed; **~duft der** scent of roses; **~garten der** rose-garden; **~kohl der;** *o. Pl.* [Brussels] sprouts *pl.*; **~kranz der** *(kath. Kirche)* rosary; **einen ~kranz beten** say a rosary; **~montag der** the day before Shrove Tuesday; **~montags·zug der** *carnival procession on the day before Shrove Tuesday*; **~stock der** rose-tree; standard rose

Rosette [ro'zɛtə] **die; ~, ~n a)** *(Archit.)* rose-window; **b)** *(Verzierung, Bot.)* rosette

rosig 1. *Adj.* **a)** rosy ⟨*face, complexion, etc.*⟩; pink ⟨*piglet etc.*⟩; **b)** *(fig.)* rosy; optimistic ⟨*mood*⟩. **2.** *adv.* **ihm geht es nicht gerade ~:** things aren't too good with him

Rosine [ro'zi:nə] **die; ~, ~n** raisin; *(Korinthe)* currant; **[große] ~n im Kopf haben** *(fig. ugs.)* have big ideas

Rosmarin ['ro:smari:n] **der; ~s** rosemary

Roß [rɔs] **das; Rosses, Rosse** *od.* **Rösser** ['rœsɐ] *(geh.; südd., österr., schweiz.)* horse; steed *(poet./joc.)*; **hoch zu ~:** on horseback; **auf dem hohen ~ sitzen** be on one's high horse; **von seinem** *od.* **vom hohen ~ herunterkommen** *od.* **-steigen** get down off one's high horse

Roß-: ~haar das horsehair; **~kastanie die** horse-chestnut; **~kur die** *(ugs.)* drastic cure *or* remedy; **b)** *(Bett~)* base; frame

[1]Rost [rɔst] **der; ~[e]s, ~e a)** *(Gitter)* grating; grid; *(eines Ofens, einer Feuerstelle)* grate; *(Brat~)* grill; **b)** *(Bett~)* base; frame

[2]Rost der; -[e]s rust

rost-, Rost-: ~beständig *Adj.* rust-resistant; *(absolut)* rust-proof; **~braten der** grilled steak; *(österr.: Entrecote)* entrecôte; rib steak; **~braun** *Adj.* reddish-brown; russet; auburn ⟨*hair*⟩

rosten *itr. V.; auch mit sein* rust; *(auch fig.)* get rusty

rösten ['rœstn̩, 'rø:stn̩] *tr. V.* **a)** roast; toast ⟨*bread*⟩; **sich [in der Sonne] ~ lassen** roast oneself in the sun; **b)** *(bes. südd., österr., schweiz.) s.* **braten 1**

rost-, Rost-: ~farben, ~farbig *Adj.* rust-coloured; russet; **~fleck der** rust stain; **~frei** *Adj.* **a)** *(nicht rostend)* stainless ⟨*steel*⟩; **b)** *(ohne Rost)* rust-free

Rösti ['rø:sti] **die; ~** *(schweiz. Kochk.)* thinly sliced fried potatoes *pl.*

rostig *Adj.* rusty

rost·rot *Adj.* rust-coloured; russet

Rost·stelle die patch of rust; *(kleiner)* rust spot

rot [ro:t] **1.** *Adj.* red; **ein Roter** *(ugs.) (Wein)* a red [wine]; *(Rothaariger)* a redhead; *(Sozialist)* a red *(coll.)*; a leftie *(coll.)*; **~ werden** turn red; ⟨*person*⟩ go red; blush; ⟨*traffic-light*⟩ change to red. **2.** *adv.* **etw. ~ anstreichen** mark sth. in red; **[im Gesicht] ~ anlaufen** go red in the face; blush

Rot das; ~s, ~ *od.* **~s** red; *(Schminke)* rouge; **die Ampel zeigt ~:** the traffic-lights are red; **bei ~ über die Kreuzung fahren** cross the junction on the red

Rotation [rota'tsio:n] **die; ~, ~en** rotation

rot-, Rot-: ~backig, ~bäckig *Adj.* rosy-cheeked ⟨*child, girl*⟩; ruddy-cheeked ⟨*old man, farmer, etc.*⟩; **~barsch der** rose-fish; **~blond** *Adj.* sandy ⟨*hair*⟩; sandy-haired ⟨*person*⟩; **~braun** *Adj.* reddish-brown; russet; **~buche die** [European] beech

Röte ['rø:tə] **die; ~:** red[ness]

Röteln ['rø:tl̩n] *Pl.* **[die] ~:** German measles *sing.*

röten *refl. V.* go *or* turn red

rot-, Rot-: ~fuchs der a) *(Tier, Pelz)* red fox; **b)** *(Pferd)* chestnut; *(heller)* sorrel; **~glühend** *Adj.* red-hot; **~haarig** *Adj.* red-haired; **~haut die** *(ugs. scherzh.)* redskin; **~hirsch der** red deer

rotieren [ro'ti:rən] *itr. V.* **a)** rotate; **b)** *(ugs.: hektisch sein)* flap *(coll.)*; get into a flap *(coll.)*

Rot-: ~käppchen (das) Little Red Riding Hood; **~kehlchen das; ~s, ~:** robin [redbreast]; **~kohl der,** *(bes. südd., österr.)* **~kraut das** red cabbage

rötlich ['rø:tlɪç] *Adj.* reddish

rot-, Rot-: ~licht das; *o. Pl.* red light; **bei ~licht** under a red light; **~schwanz der, ~schwänzchen das** redstart; **~|sehen** *unr. itr. V. (ugs.)* see red; **~stift der** red pencil; **dem ~stift zum Opfer fallen** *(aufgegeben werden)* be scrapped; *(gestrichen werden)* be deleted

Rotte ['rɔtə] **die; ~, ~n** gang; mob

Rötung die; ~, ~en reddening

Rot-: ~wein der red wine; **~wild das** *(Jägerspr.)* red deer

Rotz [rɔts] **der; ~es** *(salopp)* snot *(sl.)*; **frech wie ~** *(salopp)* cheeky as anything; **~ und Wasser heulen** *(salopp)* cry one's eyes out

rotzen *(derb) itr. V.* **a)** blow one's nose loudly; **b)** *(Schleim in den Mund ziehen)* sniff back one's snot *(sl.)*; **c)** *(ausspucken)* gob *(coarse)*

rotz·frech *(salopp)* **1.** *Adj.* insolent; snotty *(sl.).* **2.** *adv.* insolently; snottily *(sl.)*
rotzig *Adj. (derb)* snotty *(sl.)*
Rotz·nase die a) *(derb)* snotty nose *(sl.);* **b)** *(salopp abwertend: Bengel)* snotty little brat *(sl.)*
Rouge [ru:ʒ] **das; ~s, ~s** rouge
Roulade [ru:la:də] **die; ~, ~n** *(Kochk.)* [beef/veal/pork] olive
Route ['ru:tə] **die; ~, ~n** route
Routine [ru'ti:nə] **die; ~ a)** *(Erfahrung)* experience; *(Übung)* practice; *(Fertigkeit)* proficiency; expertise; **b)** *(gewohnheitsmäßiger Ablauf)* routine *no def. art.*
routine·mäßig 1. *Adj.* routine; **2.** *adv.* as a matter of routine
routiniert [ruti'ni:ɐt] **1.** *Adj. (gewandt)* expert; skilled; *(erfahren)* experienced. **2.** *adv.* expertly; skilfully
Rowdy ['raudi] **der; ~s, ~s** *(abwertend)* hooligan
rubbeln ['rʊbl̩n] *tr., itr. V. (bes. nordd.)* rub [vigorously]
Rübe ['ry:bə] **die; ~, ~n a)** turnip; **rote ~:** beetroot; **gelbe ~** *(südd.)* carrot; **b)** *(salopp: Kopf)* nut *(sl.);* **eins auf die ~ kriegen** get a bonk *or* bash on the nut *(sl.)*
Rubel ['ru:bl̩] **der; ~s, ~:** rouble; **der ~ rollt** *(fig. ugs.)* the money keeps rolling in
Rüben·zucker der beet sugar
rüber ['rʏbɐ] *Adv. (ugs.)* over
rüber-: ~|gehen *unr. itr. V.; mit sein* go over; *(über die Straße)* cross over; **~|kommen** *unr. itr. V.; mit sein* **a)** come over; **b)** *(~können)* manage to get over/across; **~|schicken** *tr. V.* send over; **~|wollen** *unr. itr. V.* want to get over *or* across
Rubin [ru'bi:n] **der; ~s, ~e** ruby
Rubrik [ru'bri:k] **die; ~, ~en** *(Spalte)* column; *(Zeitungs~)* column; section; *(fig.: Kategorie)* category; **unter der ~ ...:** under the heading [of] ...
Ruck [rʊk] **der; ~[e]s, ~e** jerk; **sich** *(Dat.)* **einen ~ geben** *(fig.)* pull oneself together
ruck·artig 1. *Adj.* jerky. **2.** *adv.* with a jerk
rück-, Rück-: ~bezüglich *(Sprachw.)* **1.** *Adj.* reflexive; **2.** *adv.* reflexively; **~blende die** flashback; **~blick der** look back **(auf +** *Akk.* at); retrospective view **(auf +** *Akk.* of); **im ~blick** in retrospect; **~blickend 1.** *Adj.* retrospective; **2.** *adv.* retrospectively; in retrospect
rücken ['rʏkn̩] **1.** *tr. V.* move; **den Tisch an die Wand ~:** move *or* push the table against the wall. **2.** *itr. V.; mit sein* move; **mit seinem Stuhl näher an den Tisch ~:** move one's chair closer to the table; **jmdm. auf den Pelz** *od.* **die Pelle ~** *(ugs.)* squeeze right up to sb; **kannst du ein bißchen ~?** could you move up/over a bit?; **hört auf, mit den Stühlen zu ~:** stop shifting your chairs
Rücken der; ~s, ~: a) back; **ein Stück vom ~** *(Rindfleisch)* a piece of chine; *(Hammel, Reh)* a piece of saddle; **auf dem ~ liegen** lie on one's back; **es lief mir [heiß und kalt] über den ~:** [hot and cold] shivers ran down my spine; **den Wind im ~ haben** have the wind behind one; **verlängerter ~** *(scherzh.)* backside; posterior *(joc.);* **jmdm./einer Sache den ~ kehren** *(fig.)* turn one's back on sb./sth.; **jmdm. den ~ stärken** *(fig.)* give sb. moral support; **jmdm. den ~ freihalten** *(fig.)* ensure sb. is not troubled with other problems; **hinter jmds. ~** *(Dat.)* *(fig.)* behind sb.'s back; **jmdm. in den ~ fallen** *(fig.)* stab sb. in the back; **mit dem ~ an der** *od.* **zur Wand** *(fig.)* with one's back to the wall; **b)** *(Rückseite)* back; *(Buch~)* spine; *(des Berges)* ridge
rücken-, Rücken-: ~deckung die a) *(bes. Milit.)* rear cover; **b)** *(fig.)* backing; **jmdm. ~deckung geben** give sb. one's backing; **~flosse die** dorsal fin; **~frei** *Adj.* backless ⟨*dress*⟩; **~lehne die** [chair/seat] back; **~mark das** *(Anat.)* spinal marrow *or* cord; **~schmerzen** *Pl.* backache *sing.;* **~schwimmen das** backstroke; **~stärkung die** [moral] support; **~wind der** tail *or* following wind; **~wind haben** have a tail *or* following wind
rück-, Rück-: ~|erstatten *tr. V.; nur im Inf. u. 2. Part.* repay; **~erstattung die** repayment; reimbursement; **~fahrkarte die; ~fahrschein der** return [ticket]; **~fahrt die** return journey; **auf der ~fahrt** on the return journey *or* way back; **~fall der** *(Med., auch fig.)* relapse; **~fällig** *Adj.* **a)** *(Med., auch fig.)* relapsed ⟨*patient, alcoholic, etc.*⟩; **~fällig werden** have a relapse; ⟨*alcoholic etc.*⟩ go back to one's old ways; **b)** *(Rechtsspr.);* **~fällig werden** commit a second offence; **~flug der** return flight; **~frage die** query; **~|fragen** *itr. V.; nur im Inf. u. 2. Part.* query it; **~gabe die a)** return; **b)** *(Ballspiele)* back pass; **~gang der** drop, fall (*Gen.* in); **~gängig** *Adj.* **~gängig machen** cancel ⟨*agreement, decision, etc.*⟩; break off ⟨*engagement*⟩; **einen Kauf ~gängig machen** return what one has bought; **~grat das** spine; *(bes. fig.)* backbone; **~grat haben/kein ~grat haben** have guts *(coll.)*/be spineless; **jmdm. das ~grat brechen** *(fig.)* break sb.'s resistance; **~halt der** support; backing; **~halt·los 1.** *Adj.* unreserved, unqualified ⟨*support*⟩; complete, absolute ⟨*frankness*⟩; **2.** *adv.* unreservedly; without reservation; ⟨*trust*⟩ completely, absolutely; ⟨*confess*⟩ with complete frankness; **~hand die** *(Sport)* backhand; **mit [der] ~hand** on one's backhand
Rückkehr ['rʏkke:ɐ] **die; ~:** return
rück-, Rück-: ~lage die savings *pl.;* **eine kleine ~lage haben** have a small sum saved up; have a small nest-egg; **~lauf der a)** *(~fluß)* return flow; **b)** *(beim Tonbandgerät)* rewind; **~läufig** *Adj.* decreasing ⟨*number*⟩; declining ⟨*economic growth etc.*⟩; falling ⟨*rate, production, etc.*⟩; **~licht das** rear- *or* tail-light
rücklings *Adv.* on one's back
Rück-: ~porto das return postage; **~reise die** return journey; **~ruf der a)** *(Fernspr.)* return call; **b)** *(das Zurückbeordern)* recall
Ruck·sack der rucksack; *(Touren~)* back-pack
rück-, Rück-: ~schau die review **(auf +** *Akk.* of); **~schau halten** look back; **~schlag der a)** set-back; **b)** *(Tennis, Tischtennis usw.)* return; **~schluß der** conclusion **(auf +** *Akk.* about); **aus etw. ~schlüsse auf etw.** *(Akk.)* **ziehen** draw conclusions from sth. about sth.; **~schritt der** backward step; **~seite die** back; *(eines Gebäudes usw.)* back; rear; *(einer Münze usw.)* reverse; *(des Mondes)* far side; **siehe ~seite** see over[leaf]; **~sicht die** consideration; **mit ~sicht auf etw.** *(Akk.)* taking sth. into consideration; in view of sth.; **~sicht auf jmdn. nehmen** show consideration for *or* towards sb.; *(Verständnis haben)* make allowances for sb.; **ohne ~sicht auf etw.** *(Akk.)* with no regard for *or* regardless of sth.; **ohne ~sicht auf Verluste** *(ugs.)* regardless; **~sicht·nahme die; ~:** consideration
rücksichts-, Rücksichts-: ~los 1. *Adj.* **a)** inconsiderate; thoughtless; **ein ~loser Autofahrer** an inconsiderate driver; *(verantwortungslos)* a reckless driver;

b) *(schonungslos)* ruthless; **2.** *adv.* **a)** inconsiderately; thoughtlessly; *(verantwortungslos)* recklessly; **b)** *(schonungslos)* ruthlessly; **~losigkeit die; ~, ~en a)** lack of consideration; thoughtlessness; *(Verantwortungslosigkeit)* recklessness; **so eine ~losigkeit!** how inconsiderate *or* thoughtless; **b)** *(Schonungslosigkeit)* ruthlessness; **~voll** *Adj.* considerate; thoughtful; **2.** *adv.* considerately; thoughtfully

rück-, Rück-: ~sitz der back seat; **~spiegel der** rear-view mirror; **~spiel das** *(Sport)* second *or* return leg; **~sprache die** consultation; **[mit jmdm.] ~sprache nehmen** *(Papierdt.)* consult [sb.]; **~stand der a)** *(Übriggebliebenes, Rest)* residue; **b)** *(offener Rechnungsbetrag)* **~stände/ein ~stand** arrears *pl.*; **~stände eintreiben** collect outstanding debts; **c)** *(Zurückbleiben hinter dem gesetzten Ziel, Soll usw.)* backlog; *(bes. Sport: hinter dem Gegner)* deficit; **[mit etw.] im ~stand sein/in ~stand** *(Akk.)* **geraten** be/get behind [with sth.]; **die Mannschaft lag mit 0:3 im ~stand** *(Sport)* the team was trailing by three to nil; **~ständig** *Adj.* **a)** backward; **~ständig sein** be behind the times; **b)** *(schon länger fällig)* outstanding ⟨*payment, amount*⟩; ⟨*wages*⟩ still owing; **~stau der** *(von Wasser)* backing up; backwater; *(von Fahrzeugen)* tailback; **~stellung die** postponement **(um** by); **eine ~stellung vom Wehrdienst** a temporary exemption from military service; **~strahler der** reflector; **~taste die** backspacer; backspace key; **~tritt der a)** resignation **(von** from); *(von einer Kandidatur, von einem Vertrag usw.)* withdrawal **(von** from); **b)** *(ugs.) s.* **~trittbremse; ~tritt·bremse die** back-pedal brake; **~vergütung die** refund; **~|versichern** *refl. V.; nur im Inf. u. 2. Part.* cover oneself [two ways]; hedge one's bets; **~versicherung die** *(fig.)* safeguard; protection

rückwärtig [-vɛrtɪç] *Adj.* back; rear; **die ~e Seite** the back *or* rear

rückwärts [-vɛrts] *Adv.* backwards; **ein Blick [nach] ~:** a look back; a backward look; **ein Salto/eine Rolle ~:** a back somersault/backward roll; **~ einparken** reverse into a parking-space

Rückwärts·gang der reverse [gear]; **den ~ einlegen** *(auch fig.)* go into reverse; **im ~:** in reverse

Rückweg der way back; **jmdm. den ~ abschneiden** cut off sb.'s line of retreat

ruck·weise 1. *Adv.* in [a series of] jerks. **2.** *adj.* jerky

rück-, Rück-: ~wirkend 1. *Adj.* retrospective; backdated ⟨*pay increase*⟩; **2.** retrospectively; **~wirkung die a)** *(zeitlich)* retrospective force; **mit ~wirkung vom ...:** [retrospectively] as from ...; **b)** *(Auswirkung)* repercussion **(auf +** *Akk.* on); **~zahlung die** repayment; **~zieher der; ~s, ~ a)** *(Fußball)* overhead kick; **b)** *(fig. ugs.)* backing out *no art.*; **einen ~zieher machen** back out; **~zug der** retreat; **auf dem ~zug sein** be retreating

rüde ['ry:də] **1.** *Adj.* uncouth; coarse ⟨*language*⟩. **2.** *adv.* in an uncouth manner

Rüde der; ~n, ~n [male] dog

Rudel ['ru:dl̩] **das; ~s, ~** *(von Hirschen, Gemsen)* herd; *(von Wölfen, Hunden)* pack

Ruder ['ru:dɐ] **das; ~s, ~ a)** *(Riemen)* oar; **b)** *(Steuer~)* rudder; *(Steuerrad)* helm; **am ~ sein** *(fig.)* be at the helm; **das ~ herumwerfen** *(fig.)* change course *or* tack; **ans ~ kommen** *(fig.)* ⟨*party, leader*⟩ come to power; **aus dem ~ laufen** *(fig.)* go off course

Ruder·boot das row-boat; rowing-boat *(Brit.)*

Ruderer der; ~s, ~: oarsman; rower

Ruderin die; ~, ~nen oarswoman; rower

rudern 1. *itr. V.; mit sein* row; **mit den Armen ~** *(fig.)* swing one's arms [about]. **2.** *tr. V.* row

Ruder·regatta die rowing regatta

rudimentär [rudimɛn'tɛ:ɐ̯] *(Biol., geh.)* **1.** *Adj.* rudimentary. **2.** *adv.* in a rudimentary form

Ruf [ru:f] **der; ~[e]s, ~e a)** call; *(Schrei)* shout; cry; *(Tierlaut)* call; **b)** *o. Pl. (fig.: Aufforderung, Forderung)* call **(nach** for); **c)** *(Leumund)* reputation; **ein Mann von gutem/schlechtem ~:** a man with a good/bad reputation; **jmdn./etw. in schlechten ~ bringen** give sb./sth. a bad name; **er/es ist besser als sein ~:** he/it is not as bad as he/it is made out to be

rufen 1. *unr. itr. V.* call **(nach** for); *(schreien)* shout **(nach** for); ⟨*animal*⟩ call; **die Pflicht/die Arbeit ruft** *(fig.)* duty calls **2.** *unr. tr. V.* **a)** call; *(schreien)* shout; **b)** *(herbei-)* call; **jmdn. zu Hilfe ~:** call to sb. to help; **jmdm./sich** *(Dat.)* **etw. ins Gedächtnis** *od.* **in Erinnerung ~:** remind sb. of sth./recall sth.; **[jmdm.] wie gerufen kommen** *(ugs.)* come at just the right moment; **c)** *(telefonisch)* call; **~ Sie 88 86 66** ring 888 666; **d)** *(nennen)* **jmdn. etw. ~:** call sb. sth.

Rüffel ['rʏfl̩] **der; ~s, ~** *(ugs.)* ticking-off *(coll.)*

Ruf-: ~mord der character assassination **~name der** first name *(by which one is generally known)*; **~nummer die** telephone number

Rugby ['rakbi] **das; ~[s]** rugby [football]

Rüge ['ry:gə] **die; ~, ~n** reprimand; **eine ~ erhalten** be reprimanded

rügen *tr. V.* reprimand ⟨*person*⟩ **(wegen** for); censure ⟨*carelessness etc.*⟩

Ruhe ['ru:ə] **die; ~ a)** *(Stille)* silence; **~ [bitte]!** quiet *or* silence [please]!; **jmdn. um ~ bitten** ask sb. to be quiet; **~ geben** be quiet; **b)** *(Ungestörtheit)* peace; **in ~ [und Frieden]** in peace [and quiet]; **die [öffentliche] ~ wiederherstellen** restore [law and] order; **jmdn. in ~ lassen** leave sb. in peace; **jmdn. mit etw. in ~ lassen** stop bothering sb. with sth.; **keine ~ geben** not stop pestering; *(nicht nachgeben)* not give up; *(weiter protestieren)* go on protesting; **c)** *(Unbewegtheit)* rest; **zur ~ kommen** come to rest; **d)** *(Erholung, das Sichausruhen)* rest *no def. art.*; **angenehme ~** *(geh.)* sleep well; **sich zur ~ begeben** *(geh.)* retire [to bed]; **sich zur ~ setzen** *(in den Ruhestand treten)* take one's retirement; retire **(in +** *Dat.* to); **e)** *(Gelassenheit)* calm[ness]; composure; **er ist die ~ selbst** *(ugs.)* he is calmness itself; **[die] ~ bewahren/die ~ verlieren** keep calm/lose one's composure; keep/lose one's cool *(coll.)*; **in [aller] ~:** [really] calmly; **die ~ weghaben** *(ugs.)* be completely unflappable *(coll.)*; **immer mit der ~!** *(nur keine Panik)* don't panic!; *(nichts überstürzen)* no need to rush

ruhe-, Ruhe-: ~bedürfnis das; *o. Pl.* need of rest; **~bedürftig** *Adj.* in need of rest *postpos.*; **~los 1.** *Adj.* restless; **2.** *adv.* restlessly; **~losigkeit die; ~:** restlessness

ruhen *itr. V.* **a)** *(aus~)* rest; **b)** *(geh.: schlafen)* sleep; **c) im Grabe ~:** lie in one's grave; **„Ruhe sanft** *od.* **in Frieden!"** 'Rest in Peace'; **„Hier ruht ..."** 'Here lies ...'; **d)** *(stillstehen)* ⟨*work, business*⟩ have stopped; ⟨*production, firm*⟩ be at a standstill; ⟨*employment, insurance*⟩ be suspended; **der Verkehr ruht fast völlig** there is hardly any traffic; **nicht ~, bis ...:** not rest until ...; **e)** *(liegen)* rest; **in sich** *(Dat.)* **[selbst] ~:** be a well-balanced [and harmonious] person

Ruhe-: **~pause die** break; **eine ~pause einlegen** take a break; **~stand der;** *o. Pl.* retirement; **in den ~stand gehen/versetzt werden** go into retirement/be retired; **Versetzung in den ~stand** retirement; **~störung die** disturbance; *(Rechtsw.)* disturbance of the peace; **~tag der** closing day; **„Dienstag ~tag"** 'closed on Tuesdays'

ruhig ['ru:ɪç] **1.** *Adj.* **a)** *(still, leise)* quiet; **seid doch mal ~!** do be quiet!; **b)** *(friedlich, ungestört)* peaceful ⟨*times, life, scene, valley, spot, etc.*⟩; quiet ⟨*talk, reflection, life, spot*⟩; **er hat keine ~e Minute** he doesn't have a moment's peace; **c)** *(unbewegt)* calm ⟨*sea, weather*⟩; still ⟨*air*⟩; *(fig.)* peaceful ⟨*melody*⟩; quiet ⟨*pattern*⟩; *(gleichmäßig)* steady ⟨*breathing, hand, flame, steps*⟩; smooth ⟨*flight, crossing*⟩; **d)** *(gelassen)* calm ⟨*voice etc.*⟩; quiet, calm ⟨*person*⟩; **~ bleiben** keep calm; keep one's cool *(coll.)*. **2.** *adv.* **a)** *(still, leise)* quietly; **wir wohnen sehr ~:** we live in a very quiet area; **sich ~ verhalten** keep quiet; **b)** *(friedlich, ohne Störungen)* ⟨*sleep*⟩ peacefully; ⟨*go off*⟩ smoothly, peacefully; *(ohne Zwischenfälle)* uneventfully; ⟨*work, think*⟩ in peace; **c)** *(unbewegt)* ⟨*sit, lie, stand*⟩ still; *(gleichmäßig)* ⟨*burn, breathe*⟩ steadily; ⟨*run, fly*⟩ smoothly; **d)** *(gelassen)* ⟨*speak, watch, sit*⟩ calmly. **3.** *Adv.* by all means; **du kannst ~ mitkommen** by all means come along; you're welcome to come along; **lach mich ~ aus** all right *or* go ahead, laugh at me[, I don't care]; **soll er ~ meckern** *(ugs.)* let him moan[, I don't care]

Ruhm [ru:m] **der; ~[e]s** fame; **sich mit ~ bedecken** *(geh.)* cover oneself with glory

rühmen ['ry:mən] **1.** *tr. V.* praise. **2.** *refl. V.* **sich einer Sache** *(Gen.)* **~:** boast about sth.

Ruhmes·blatt das: das war kein ~ für ihn it did not reflect any credit on him; it did him no credit

rühmlich *Adj.* laudable; praiseworthy; notable ⟨*exception*⟩

ruhm·reich 1. *Adj.* glorious ⟨*victory, history*⟩; celebrated ⟨*general, army, victory*⟩. **2.** *adv.* **~ siegen** win a glorious victory

Ruhr [ru:ɐ̯] **die; ~, ~en** dysentery *no art.*

Rühr·ei das scrambled egg[s *pl.*]

rühren ['ry:rən] **1.** *tr. V.* **a)** *(um~)* stir ⟨*sauce, dough, etc.*⟩; *(ein~)* stir ⟨*egg, powder, etc.*⟩ ⟨**an, in** + *Akk.* into⟩; **b)** *(bewegen)* move ⟨*limb, fingers, etc.*⟩; *s. auch* **Finger b; c)** *(fig.)* move; touch; **jmdn. zu Tränen ~:** move sb. to tears; **es rührte ihn überhaupt nicht, daß ...:** it didn't bother him at all that ... **2.** *itr. V.* **a)** *(um~)* stir; **in etw.** *(Dat.)* **~:** stir sth. **b)** *(geh.: her~)* **das rührt daher, daß ...:** that stems from the fact that ... **3.** *refl. V.* **a)** *(sich bewegen)* move; **niemand rührte sich** nobody moved *or* stirred; *(fig.: unternahm etwas)* nobody did anything; **b)** *(Milit.)* **rührt euch!** at ease!

rührend 1. *Adj.* touching; **das ist ~ von Ihnen** *(ugs.)* that is terribly sweet *or* kind of you *(coll.)* **2.** *adv.* touchingly; **sie sorgt ~ für ihn** it is touching the way she looks after him

Ruhr·gebiet das; *o. Pl.* Ruhr [district]

rührig 1. *Adj.* active; *(mit Unternehmungsgeist)* enterprising; go-ahead; *(emsig)* busy; industrious. **2.** *adv.* actively; *(mit Unternehmungsgeist)* enterprisingly; *(emsig)* busily; industriously

rühr·selig *Adj.* **a)** emotional ⟨*person*⟩; **b)** *(allzu gefühlvoll)* over-sentimental ⟨*manner, mood, etc.*⟩; maudlin, *(coll.)* tear-jerking ⟨*play, song, etc.*⟩

Rührung die; ~: emotion; **von tiefer ~ ergriffen** deeply moved

Ruin [ru'i:n] **der; ~s** ruin

Ruine die; ~, ~n ruin

ruinieren *tr. V.* ruin; **sich finanziell ~:** ruin oneself [financially]

rülpsen ['rʏlpsn̩] *itr. V.* *(ugs.)* belch

Rülpser der; ~s, ~ *(ugs.)* belch

rum [rʊm] *Adv.* *(ugs.)* *s.* **herum**

rum|- *(ugs.)* *s.* **herum|-**

Rum [rʊm] **der; ~s, ~s** rum

Rumäne [ru'mɛ:nə] **der; ~n, ~n** Romanian

Rumänien (das); ~s Romania

Rumänin die; ~, ~nen Romanian

rumänisch *Adj.* Romanian

rum-: **~|ballern** *itr. V.* *(ugs.)* blast away; **~|fliegen** *(ugs.: herumliegen)* lie about *or* around; **~|gammeln** *itr. V.* *(ugs.)* *s.* **gammeln b; ~|hampeln** *itr. V.* *(ugs.)* hop *or* jig about; **~|hängen** *unr. itr. V.* *(ugs. abwertend: nichts Sinnvolles tun)* hang about *or* around; **~|kriegen** *tr. V.* *(ugs.)* *s.* **herumkriegen; ~|labern** *itr. V.* *(salopp abwertend)* natter *(Brit. coll.)* *or* chatter away *(coll.)*; rabbit on *(Brit. sl.)*; **~|machen** *itr. V.* *(salopp)* **a)** *s.* **herummachen; b)** *s.* **herumfummeln a; c)** *(sich [sexuell] einlassen)* play around; *(schmusen)* neck *(sl.)*

Rummel ['rʊml̩] **der; ~s** *(ugs.)* **a)** *(laute Betriebsamkeit)* commotion; *(Aufhebens)* fuss, to-do (**um** about); **b)** *(bes. nordd.: Jahrmarkt)* fair

Rummel·platz der *(bes. nordd.)* fairground

rumoren [ru'mo:rən] *itr. V.* *(ugs.)* make a noise; *(poltern)* ⟨*person*⟩ bang about; **es rumorte in seinem Bauch** *(fig.)* his stomach rumbled

Rumpel·kammer die *(ugs.)* box-room *(Brit.)*; junk-room

rumpeln ['rʊmpl̩n] *itr. V.* *(ugs.)* **a)** *(poltern)* bump and bang about; **b)** *mit sein (sich rumpelnd fortbewegen)* rumble; bump and bang

Rumpf [rʊmpf] **der; ~[e]s, Rümpfe** ['rʏmpfə] **a)** trunk [of the body]; **den ~ drehen/beugen** turn one's body/bend from the hips; **b)** *(beim Schiff)* hull; **c)** *(beim Flugzeug)* fuselage

rümpfen ['rʏmpfn̩] *tr. V.* **die Nase [bei etw.] ~:** wrinkle one's nose at sth.; **über jmdn./etw. die Nase rümpfen** *(fig.)* look down one's nose at sb./turn up one's nose at sth.

Rump·steak ['rʊmp-] **das** rump steak

rums [rʊms] *Interj.* bump; *(lauter, heller)* bang; *(beim Zusammenstoß)* crash

rum|toben *itr. V.* *(ugs.)* **a)** *auch mit sein* ⟨*child*⟩ charge *or* romp [noisily] about; ⟨*students etc.*⟩ rag; **b)** *(wüten)* rant and rave

Run [rʌn] **der; ~s, ~s** [big] rush; **ein [starker] ~ auf etw.** *(Akk.)* a [big] run on sth.

rund [rʊnt] **1.** *Adj.* **a)** round; **ein Gespräch am ~en Tisch** *(fig.)* a round-table conference; **b)** *(dicklich)* plump ⟨*arms etc.*⟩; chubby ⟨*cheeks*⟩; fat ⟨*stomach*⟩; **c)** *(ugs.: ganz)* round ⟨*dozen, number, etc.*⟩; **~e drei Jahre** three years or as near as makes no difference. **2.** *Adv.* **a)** *(ugs.: etwa)* about; approximately; **b) ~ um jmdn./etw.** [all] around sb./sth.; *s. auch* **Uhr**

Rund-: **~blick der** panorama; view in all directions; **~brief der** circular [letter]

Runde ['rʊndə] **die;** ~, **~n a)** *(Sport: Strecke)* lap; **die schnellste ~ fahren** do the fastest lap; **seine ~n ziehen** *od.* **drehen** do one's laps; **b)** *(Sport: Durchgang, Partie; Boxen: Abschnitt)* round; **eine ~ Golf/Skat** a round of golf/skat; **über die ~n kommen** *(fig. ugs.)* get by; manage; **c)** *(Personenkreis)* circle; *(Gesellschaft)* company; **d)** *(Rundgang)* round; **die ~ machen** *(ugs.)* ‹*drink, rumour*› go the rounds *pl.;* circulate; **e)** *(Lage)* round; **eine ~ Bier schmeißen** *(ugs.)* buy *or* stand a round of beer

rund-, Rund-: ~erneuern *tr. V. (Kfz-W.)* remould; retread; **~erneuerte Reifen** remoulds; retreads; **~fahrt die** *(auch Sport)* tour **(durch** of); **~flug der** [short] circular flight; circuit

Rund·funk der a) radio; **im ~:** on the radio; **b)** *s.* **Rundfunkanstalt**

Rundfunk-: ~anstalt die broadcasting corporation; *(Sender)* radio station; **~gebühren** *Pl.* radio licence fees; **~gerät das** radio set; **~programm das a)** *(Sendefolge)* [schedule *sing.* of] radio programmes *pl.;* **b)** *(Programmheft)* radio programme guide; **~reporter der** radio reporter; **~sender der** radio station; *(technische Anlage)* radio transmitter; **~sendung die** radio programme; **~sprecher der** radio announcer; **~übertragung die** radio broadcast

rund-, Rund-: ~gang der *(des Wachmanns, Chefarztes usw.)* round **(durch** of); **~|gehen** *unr. itr. V.; mit sein* **a)** *unpers. (ugs.)* **es geht rund** *(es ist viel Betrieb)* it's all go *(coll.); (es geht flott zu)* things are going with a swing; **b)** *(herumgereicht werden)* be passed round; *(fig.)* ‹*story, rumours*› go *or* do the rounds; **~heraus** *Adv.* straight out; ‹*say, ask*› bluntly; ‹*refuse*› flatly; **~herum** *Adv.* **a)** *(ringsum)* all around; *(darum herum)* all round it; **b)** *(völlig)* completely; *(fig.)* entirely ‹*satisfied*›

rundlich *Adj.* **a)** roundish; **b)** *(mollig)* plump; chubby

rund-, Rund-: ~reise die [circular] tour **(durch** of); **eine ~reise durch den Schwarzwald machen** tour the Black Forest; **~schreiben das** *s.* **~brief; ~um** *Adv. s.* **~herum**

Rundung die; ~, **~en** curve; *(hervortretend)* bulge

rund·weg *Adv.* ‹*refuse, deny*› flatly, point-blank

Rund·weg der circular path *or* walk

Rune ['ru:nə] **die;** ~, **~n** rune

runter ['rʊntɐ] *Adv. (ugs.)* **~ [da, das ist mein Platz]!** get off [there, that's my seat]; **~ mit den Klamotten** off with your clothes; get those clothes off; **Kopf ~!** head/heads down; *s. auch* **herunter; hinunter**

runter- *s.* **herunter-, hinunter-**

runter-: ~|dürfen *unr. itr. V. (ugs.)* be allowed to come down; *(hinausgehen dürfen)* be allowed out; **~|fallen** *unr. itr. V.; mit sein (ugs.)* fall down; *(von der Leiter usw.)* fall off; **die Leiter/von der Leiter ~fallen** fall off the ladder; **die Kreide fiel ihm ~:** he dropped the chalk; **~|gehen** *unr. itr. V.; mit sein (ugs.)* **a)** *(nach unten gehen)* go down; **b)** *(niedriger werden)* ‹*price, temperature, pressure, etc.*› go down, drop; **c)** *(die Höhe senken)* go down **(auf +** *Akk.* to); **wir müssen mit den Preisen ~gehen** we must reduce our prices; **~|hauen** *unr. tr. V.* **jmdm. eine/ein paar ~hauen** *(salopp)* give sb. a clip/a couple of clips round the ear; **~|rutschen** *itr. V.; mit sein (ugs.) s.* **herunterrutschen;** *s. auch* **Buckel a**

Runzel ['rʊntsl̩] **die;** ~, **~n** wrinkle

runz[e]lig *Adj.* wrinkled

runzeln *tr. V.* **die Stirn/die Brauen ~:** wrinkle one's brow/knit one's brows; *(ärgerlich)* frown; **mit gerunzelter Stirn** with wrinkled brow; *(ärgerlich)* frowning

Rüpel ['ry:pl̩] **der;** **~s,** ~ *(abwertend)* lout

rüpelhaft *(abwertend)* **1.** *Adj.* loutish. **2.** *adv.* in a loutish manner

rupfen ['rʊpfn̩] *tr. V.* **a)** pluck ‹*goose, hen, etc.*›; *s. auch* **Hühnchen; b)** *(abreißen)* pull up ‹*weeds, grass*›; pull off ‹*leaves etc.*›; **c)** *(ugs.: übervorteilen)* fleece ‹*person*› [of his/her money]

ruppig ['rʊpɪç] **1.** *Adj. (abwertend)* gruff ‹*person, behaviour*›; sharp ‹*tone*›. **2.** *adv. (abwertend)* gruffly

Rüsche ['ry:ʃə] **die;** ~, **~n** ruche; frill

Ruß [ru:s] **der;** **~es** soot

Russe ['rʊsə] **der;** **~n, ~n** Russian

Rüssel ['rysl̩] **der;** **~s,** ~ **a)** *(des Elefanten)* trunk; *(des Schweins)* snout; *(bei Insekten u. ä.)* proboscis; **b)** *(salopp: Nase)* conk *(sl.)*

rußen *itr. V.* give off sooty smoke

rußig *Adj.* sooty

Russin die; ~, **~nen** Russian

russisch 1. *Adj.* Russian; *s. auch* **Ei a. 2.** *adv.* **a) ~ verwaltet/besetzt** administered/occupied by Russia; **b)** *(auf ~)* in Russian

Russisch das; **~[s]** Russian

Ruß·land (das); **~s** Russia

rüsten ['rystn̩] **1.** *itr. V. (sich bewaffnen)* arm; **zum Krieg ~:** arm for war. **2.** *itr., refl. V. (geh.: sich bereit machen, auch fig.)* get ready; prepare

rüstig 1. *Adj.* sprightly; active; **er ist noch ~:** he is still hale and hearty

rustikal [rʊsti'ka:l] **1.** *Adj.* country-style ‹*food, inn, clothes, etc.*›; farmhouse *attrib.* ‹*food*›; rustic ‹*pattern*›; rustic, farmhouse *attrib.* ‹*furniture*›; *(als Nachahmung)* rustic-style ‹*furniture etc.*›. **2.** *adv.* in [a] country style

Rüstung die; ~, **~en a)** *(Bewaffnung)* armament *no art.; (Waffen)* arms *pl.;* weapons *pl.;* **b)** *(hist.: Schutzbekleidung)* suit of armour; **in voller ~:** in full armour

Rüstungs-: ~betrieb der armaments factory; **~industrie die** armaments *or* arms industry; **~kontrolle die** arms control; **~stopp der** arms freeze; **~wettlauf der** arms race

Rüst·zeug das a) *(Wissen)* requisite know-how; **b)** *(Ausrüstung)* equipment [for the job *or* task]

Rute ['ru:tə] **die;** ~, **~n** *(Stock)* switch; *(Birken~, Angel~, Wünschel~)* rod; *(zum Züchtigen)* cane; *(Bündel)* birch

Rutsch [rʊtʃ] **der;** **~[e]s, ~e** slide; **in einem** *od.* **auf einen ~** *(fig. ugs.)* in one go; **guten ~ [ins neue Jahr]!** happy New Year!

Rutsch·bahn die slide

Rutsche die; ~, **~n** chute

rutschen *itr. V.; mit sein* slide; ‹*clutch, carpet*› slip; *(aus~)* ‹*person*› slip; ‹*car etc.*› skid; *(nach unten)* slip [down]; **rutsch mal zur Seite!** *(ugs.)* move up a bit *(coll.)*

rutschig *Adj.* slippery

rütteln ['rytl̩n] **1.** *tr. V.* shake; **jmdn. aus dem Schlaf** *od.* **wach ~:** shake sb. out of his/her sleep. **2.** *itr. V.* shake; **an der Tür ~:** shake the door; ‹*wind*› make the door rattle; **daran ist nicht** *od.* **gibt es nichts zu ~** *(fig.)* there's nothing you can do about that

S

s, S [ɛs] **das; ~, ~:** s, S; *s. auch* **a, A**
s *Abk.* **Sekunde** sec.; s.
S *Abk.* **a) Süd, Süden** S.; **b)** *(österr.)* **Schilling** Sch.
s. *Abk.* **siehe**
S. *Abk.* **Seite** p.
s. a. *Abk.* **siehe auch**
Sa. *Abk.* **Samstag** Sat.
Saal [za:l] **der; ~[e]s, Säle** ['zɛ:lə] **a)** hall; *(Ball~)* ballroom; **b)** *(Publikum)* audience
Saal·ordner der steward
Saar·land ['za:ɐ̯lant] **das; ~[e]s** Saarland; Saar *(esp. Hist.)*
Saar·länder [-lɛndɐ] **der; ~s, ~, Saarländerin die; ~, ~nen** Saarlander
Saat [za:t] **die; ~, ~en a)** *(das Gesäte)* [young] crops *pl.*; **b)** *o. Pl. (das Säen)* sowing; **mit der ~ beginnen** start sowing; **c)** *(Samenkörner)* seed[s *pl.*]
Saat-: ~gut das; *o. Pl.* seed[s *pl.*]; **~kartoffel die** seed-potato
sabbern *itr. V.* ⟨*dog, person*⟩ slaver, slobber; ⟨*baby*⟩ dribble
Säbel ['zɛ:bl̩] **der; ~s, ~:** sabre
Säbel·rasseln das; ~s *(abwertend)* sabre-rattling
Sabotage [zabo'ta:ʒə] **die; ~, ~n** sabotage *no art.*
Sabotage·akt der act of sabotage
sabotieren *tr. V.* sabotage
sach-, ~Sach-: ~bearbeiter der person responsible **(für** for); **~bezogen 1.** *Adj.* relevant; pertinent ⟨*remark*⟩; **2.** *adv.* to the point; **~buch das** [popular] non-fiction *or* informative book; **~dienlich** *Adj. (Papierdt.)* useful; helpful
Sache ['zaxə] **die; ~, ~n a)** *Pl.* things; **scharfe ~n trinken** drink the hard stuff *(coll.)*; **b)** *(Angelegenheit)* matter; business *(esp. derog.)*; **es ist beschlossene ~, daß ...:** it's [all] arranged *or* settled that ...; **es ist die einfachste ~ [von] der Welt** it's the simplest thing in the world; **das ist so eine ~:** it's a bit tricky; **[mit jmdm.] gemeinsame ~ machen** join forces [with sb.]; **[sich** *(Dat.)*] **seiner ~ sicher** *od.* **gewiß sein** be sure one is right; **bei der ~ sein** concentrate on it; **zur ~ kommen** come to the point; **das tut nichts zur ~:** that's irrelevant; that's got nothing to do with it; **c)** *(Rechts~)* case; **Fragen zur ~:** questions about the case; **d)** *o. Pl. (Anliegen)* cause; **e)** *Pl. (ugs.: Stundenkilometer)* kilometres per hour
sach-, Sach-: ~gebiet das subject [area]; field; **~gemäß, ~gerecht 1.** *Adj.* proper; correct; **2.** *adv.* properly; correctly; **~kenntnis die** expertise; knowledge of the subject; **~kunde die a)** *s.* **~kenntnis; b)** *(Schulw.)* ≈ general subjects *pl.*; **~kundig 1.** *Adj.* with a knowledge of the subject *postpos., not pred.*; **sich ~kundig machen** acquaint oneself with the subject; **2.** *adv.* expertly; **~lage die;** *o. Pl.* situation
sachlich 1. *Adj.* **a)** *(objektiv)* objective; *(nüchtern)* functional ⟨*building, style, etc.*⟩; matter-of-fact, down-to-earth ⟨*letter etc.*⟩; **b)** *nicht präd. (sachbezogen)* factual ⟨*error*⟩; material ⟨*consideration*⟩; **aus ~en Gründen** for practical reasons. **2.** *adv.* *(objektiv)* objectively; ⟨*state*⟩ as a matter of fact; *(nüchtern)* ⟨*furnished*⟩ in a functional style; ⟨*written*⟩ in a matter-of-fact way; **b)** *(sachbezogen)* factually ⟨*wrong*⟩
sächlich ['zɛçlɪç] *Adj. (Sprachw.)* neuter
Sachlichkeit die; ~: objectivity; *(Nüchternheit)* functionalism
Sach·schaden der damage [to property] *no indef. art.*
Sachse ['zaksə] **der; ~n, ~n** Saxon
Sachsen (das); ~s Saxony
Sachsen-Anhalt (das); ~s Saxony-Anhalt
sächsisch ['zɛksɪʃ] *Adj.* Saxon
sacht [zaxt] **1.** *Adj.* gentle. **2.** *adv.* gently
sachte *Adv. (ugs.)* **~[, ~]** take it easy; *(nicht so hastig)* not so fast
sach-, Sach-: ~verhalt der; ~[e]s, ~e facts *pl.* [of the matter]; **~verstand der** expertise; grasp of the subject; **~verständig 1.** *Adj.* expert ⟨*opinion etc.*⟩; knowledgeable ⟨*person*⟩; **2.** *adv.* expertly; knowledgeably; **~wissen das** specialist knowledge; **~zwang der** [factual *or* material] constraint
Sack [zak] **der; ~[e]s, Säcke** ['zɛkə] **a)** sack; *(aus Papier, Kunststoff)* bag; **drei ~ Zement/Kartoffeln** three bags of cement/sacks of potatoes; **jmdn. in den ~ stecken** *(ugs.)* put sb. in the shade; **mit ~ und Pack** with bag and baggage; **b)** *(Hautfalte)* **Säcke unter den Augen haben** have bags under one's eyes; **c)** *(derb: Hoden~)* balls *pl. (coarse)*; **d)** *(derb abwertend: Mensch)* sod *(Brit. sl.)*
sacken *itr. V.; mit sein* ⟨*person*⟩ slump; ⟨*ship etc.*⟩ sink; ⟨*plane*⟩ drop rapidly, plummet
Sack-: ~gasse die cul-de-sac; *(fig.)* impasse; **~hüpfen das; ~s** sack race
Sadismus [za'dɪsmʊs] **der; ~:** sadism *no art.*
Sadist der; ~en, ~en, Sadistin die; ~, ~nen sadist
sadistisch 1. *Adj.* sadistic. **2.** *adv.* sadistically; **~ veranlagt sein** have sadistic tendencies
säen ['zɛ:ən] *tr. (auch itr.) V. (auch fig.)* sow; **dünn gesät sein** *(fig.)* be thin on the ground
Safari [za'fa:ri] **die; ~, ~s** safari
Safe [seɪf] **der** *od.* **das; ~s, ~s a)** safe; **b)** *(Schließfach)* safe-deposit box
Safran ['zafran] **der; ~s, ~e** saffron
Saft [zaft] **der; ~[e]s, Säfte** ['zɛftə] **a)** juice; **b)** *(in Pflanzen)* sap; **ohne ~ und Kraft** *(abwertend)* weak and lifeless; **c)** *(salopp: Elektrizität)* juice *(sl.)*
saftig *Adj.* **a)** juicy; sappy ⟨*stem*⟩; lush ⟨*meadow, green*⟩; *(fig.: lebensvoll)* lusty; **b)** *(ugs.)* hefty ⟨*slap, blow*⟩; steep *(coll.)* ⟨*prices, bill*⟩; terrific, big ⟨*surprise, punch-up*⟩; crude, coarse ⟨*joke, song, etc.*⟩; strongly-worded ⟨*letter etc.*⟩; strong, juicy ⟨*curse*⟩
Saft·laden der *(salopp abwertend)* lousy outfit *(sl.)*
saft·los *Adj. (fig.)* feeble, anodyne ⟨*language*⟩; **saft- und kraftlos** feeble; wishy-washy
Sage ['za:gə] **die; ~, ~n** legend; *(bes. nordische)* saga; **es geht die ~, daß ...** *(fig.: es heißt, daß ...)* there's a rumour going round that ...
Säge ['zɛ:gə] **die; ~, ~n** saw
Säge-: ~blatt das saw-blade; **~mehl das** sawdust
sagen 1. *tr. V.* **a)** say; **das kann jeder ~:** anybody can claim that; it's easy to talk; **sag das nicht!** *(ugs.)* don't [just] assume that; not necessarily; **dann will ich nichts gesagt haben** in that case forget I said anything; **was ich**

noch ~ wollte [oh] by the way; before I forget; **unter uns gesagt** between you and me; **wie gesagt** as I've said *or* mentioned; **das kann man wohl ~:** you can say 'that again; **heute abend, ~ wir, um acht** tonight, say, eight o'clock; **sage und schreibe** *(ugs.)* believe it or not; would you believe; **b)** *(meinen)* say; **was ~ Sie dazu?** what do you think about that?; **c)** *(mitteilen)* **jmdm. etw. ~:** say sth. to sb.; *(zur Information)* tell sb. sth.; **[jmdm.] seinen Namen/seine Gründe ~:** give [sb.] one's name/reasons; **[jmdm.] die Wahrheit ~:** tell [sb.] the truth; **das sag' ich dir** *(ugs.)* I'm telling *or* warning you; **ich hab's [dir] ja gleich gesagt!** *(ugs.)* I told you so!; *(habe dich gewarnt)* I warned you!; **ich will dir mal was ~:** let me tell you something; **laß dir das gesagt sein** *(ugs.)* make a note of *or* remember what I'm saying; **wem ~ Sie das!** *(ugs.)* you don't need to tell me [that]!; **was Sie nicht ~!** *(ugs., oft iron.)* you don't say!; **er läßt sich** *(Dat.)* **nichts ~:** he won't be told; you can't tell him anything; **d)** *(nennen)* **zu jmdm./etw. X ~:** call sb./sth. X; **e)** *(formulieren, ausdrücken)* say; **so kann man es auch ~:** you could put it like that; **etw. in aller Deutlichkeit ~:** make sth. perfectly clear; **du sagst es!** very true!; **willst du damit ~, daß ...?** are you trying to say *or* do you mean [to say] that ...?; **f)** *(bedeuten)* mean; **hat das etwas zu ~?** does that mean anything?; **g)** *(anordnen, befehlen)* tell; **du hast mir gar nichts zu ~:** you've no right to order me about; **etwas/nichts zu ~ haben** ‹*person*› have a/no say. **2.** *refl. V.* **sich** *(Dat.)* **etw. ~:** say sth. to oneself. **3.** *itr. V.* **wie sagt man [da]?** what does one say?; what's the [right] word?; **sag bloß!** *(ugs.)* you don't say!

sägen ['zɛ:gn̩] **1.** *itr. V.* **a)** saw; **b)** *(ugs. scherzh.: schnarchen)* snore loudly. **2.** *tr. V.* saw; *(zersägen)* saw up ‹*tree etc.*›

sagenhaft *(ugs.)* **1.** *Adj.* incredible *(coll.)*; fabulous *(coll.)* ‹*party, wealth*›. **2.** *adv.* incredibly *(coll.)*

Säge-: ~späne *Pl.* wood shavings; **~werk das** sawmill

sah [za:] *1. u. 3. Pers. Sg. Prät. v.* **sehen**

Sahne ['za:nə] **die; ~:** cream

Saison [zɛ'zõ:] **die; ~, ~s** season; **während/außerhalb der ~:** during the season/out of season *or* in the off-season; **~ haben** have one's busy time *or* season

saison-, Saison-: ~arbeit die seasonal work; **~arbeiter der** seasonal worker; **~bedingt 1.** *Adj.* seasonal; **2.** *adv.* due to seasonal influences

Saite ['zaitə] **die; ~, ~n** string; **andere** *od.* **strengere ~n aufziehen** *(fig.)* take stronger measures; get tough *(coll.)*

Saiten·instrument das stringed instrument

Sakko ['zako] **der** *od.* **das; ~s, ~s** jacket

Sakrament [zakra'mɛnt] **das; ~[e]s, ~e a)** *(bes. kath. Kirche)* sacrament; **b) ~ [noch mal]!** *(salopp)* for Heaven's sake!

Sakristei [zakrɪs'tai] **die; ~, ~en** sacristy

Salamander [zala'mandɐ] **der; ~s, ~:** salamander

Salami [za'la:mi] **die; ~, ~[s]** salami

Salat [za'la:t] **der; ~[e]s, ~e a)** salad; **b)** *o. Pl.* **[grüner] ~:** lettuce; **ein Kopf ~:** a [head of] lettuce; **c)** *o. Pl. (ugs.: Wirrwarr)* muddle; mess; **jetzt haben wir den ~!** *(ugs. iron.)* now we're in a right mess

Salat-: ~besteck das salad-servers *pl.*; **~soße die** salad-dressing

Salbe ['zalbə] **die; ~, ~n** ointment

Salbei ['zalbai] **der** *od.* **die; ~:** sage

salben *tr. V.* **a)** put ointment on ‹*part of body*›; **b)** *(kath. Kirche)* anoint ‹*sick or dying person, (Hist.) king, emperor, etc.*›

Saldo ['zaldo] **der; ~s, ~s** *od.* **Saldi** *(Buchf., Finanzw.)* balance

Säle *s.* **Saal**

Saline [za'li:nə] **die; ~, ~n** salt-works *sing. or pl.*

Salm [zalm] **der; ~[e]s, ~e** *(bes. rhein.)* salmon

Salmiak [zal'miak] **der** *od.* **das; ~:** sal ammoniac

Salmiak-: ~geist der [liquid] ammonia; ammonia water; **~pastille die** sal ammoniac pastille

salomonisch *(geh.)* **1.** *Adj.* Solomon-like; **ein ~es Urteil** a judgment of Solomon. **2.** *adv.* with the wisdom of Solomon

Salon [za'lõ:] **der; ~s, ~s a)** *(Raum)* drawing-room; salon; **b)** *(Geschäft)* [hair- *etc.*] salon

salon·fähig *Adj.* socially acceptable

salopp [za'lɔp] **1.** *Adj.* casual ‹*clothes*›; free and easy, informal ‹*behaviour*›; very colloquial, slangy ‹*saying, expression, etc.*›. **2.** *adv.* ‹*dress*› casually; informally; **~ reden** use slangy *or* [very] colloquial language

Salpeter [zal'pe:tɐ] **der; ~s** saltpetre

Salto ['zalto] **der; ~s, ~s** *od.* **Salti** somersault; *(beim Turnen auch)* salto

Salut [za'lu:t] **der; ~[e]s, ~e** *(Milit.)* salute; **~ schießen** fire a salute

salutieren *itr. V. (bes. Milit.)* salute; **vor jmdm. ~:** salute sb.

Salve ['zalvə] **die; ~, ~n** *(Milit.)* salvo; *(aus Gewehren)* volley

Salz [zalts] **das; ~es, ~e** salt

salz·arm 1. *Adj.* low in salt *postpos.*; low-salt. **2.** *adv.* **~ essen** eat food containing little salt

Salz·brezel die [salted] pretzel

salzen *unr., auch regelm. tr. V.* salt; **die Suppe ist stark gesalzen** the soup has a lot of salt in it

salzig *Adj.* salty

salz-, Salz-: ~kartoffel die; *meist Pl.* boiled potato; **~los 1.** *Adj.* salt-free; *(nicht gesalzen)* unsalted; **2.** *adv.* ‹*cook*› without any salt; **~los essen** eat unsalted food; **~lösung die** saline solution; **~säure die;** *o. Pl. (Chemie)* hydrochloric acid; **~stange die** salt stick; **~streuer der; ~s, ~:** salt-sprinkler; salt-shaker *(Amer.)*; **~wasser das;** *Pl.* **~wässer a)** *o. Pl. (zum Kochen)* salted water; **b)** *(Meerwasser)* salt water

Samariter [zama'ri:tɐ] **der; ~s, ~ [barmherziger] ~:** good Samaritan

Sambia ['zambia] **(das); ~s** Zambia

Samen ['za:mən] **der; ~s, ~ a)** *(~korn)* seed; **b)** *o. Pl. (~körner)* seed[s. *pl.*]; **c)** *o. Pl. (Sperma)* sperm; semen

Sammel-: ~band der; *Pl.* **~bände** anthology; **~becken das** collecting basin; reservoir; *(fig.)* gathering-point *or* -place; **~bestellung die** joint order; **~büchse die** collecting-box; **~fahrschein der** group ticket

sammeln ['zamln̩] **1.** *tr. (auch itr.) V.* collect; gather ‹*honey, firewood, material, experiences, impressions, etc.*›; gather, pick ‹*berries, herbs, mushrooms, etc.*›; gather ‹*people*› [together]; assemble ‹*people*›; cause ‹*light rays*› to converge; **gesammelte Werke** collected works. **2.** *refl. V.* **a)** gather [together]; ‹*light rays*› converge; **sich um jmdn./etw. ~:** gather round sb./sth.; **b)** *(sich konzentrieren)* collect oneself; gather oneself together

Sammel·platz der collection *or* collecting point; *(für Menschen)* assembly point

Sammelsurium [zaml̩'zu:ri̯ʊm] **das; ~s, Sammelsurien** *(abwertend)* hotchpotch

Sammler ['zamlɐ] **der; ~s, ~, Sammlerin die; ~, ~nen** collector; *(von Pilzen, Kräutern, Beeren usw.)* gatherer; picker

Sammlung die; ~, ~en a) collection; **b) [innere] ~:** composure

Samstag ['zamsta:k] **der; ~[e]s, ~e** Saturday; **langer ~:** Saturday on which the shops stay open late; *s. auch* **Dienstag; Dienstag-**

samstags *Adv.* on Saturdays

samt [zamt] **1.** *Präp. mit Dat.* together with. **2.** *Adv.* **~ und sonders** without exception

Samt der; ~[e]s, ~e velvet

samten *Adj.; nicht präd.* **a)** velvet; **b)** *(wie Samt)* velvety

Samt·hand·schuh der velvet glove; **jmdn. mit ~en anfassen** *(fig.)* handle sb. with kid gloves

samtig *Adj.* velvety

sämtlich ['zɛmtlɪç] *Indefinitpron. u. unbest. Zahlwort* **a)** *attr.* all the; **~e Werke** complete works; **b)** *alleinstehend* all

samt·weich *Adj.* velvety[-soft]; soft as velvet *postpos.*

Sanatorium [zana'to:ri̯ʊm] **das; ~s, Sanatorien** sanatorium

Sand [zant] **der; ~[e]s** sand; **... gibt es wie ~ am Meer** *(ugs.)* there are countless ...; ... are pretty thick on the ground *(coll.)*; **da ist ~ im Getriebe** *(fig. ugs.)* there's something gumming up the works *(coll.)*; **jmdm. ~ in die Augen streuen** *(fig.)* pull the wool over sb.'s eyes; **im ~[e] verlaufen** *(fig. ugs.)* come to nothing; **etw. [total] in den ~ setzen** *(fig. ugs.)* make a [complete] mess of sth.

Sandale [zan'da:lə] **die; ~, ~n** sandal

Sandalette [zanda'lɛtə] **die; ~, ~n** [high-heeled] sandal

Sand-: ~bank die; *Pl.* **~bänke** sandbank; **~burg die** sandcastle

sandig *Adj.* sandy

Sand-: ~kasten der [child's] sand-pit; sand-box *(Amer.)*; **~korn das;** *Pl.* **~körner** grain of sand; **~kuchen der** Madeira cake; **~männchen das;** *o. Pl.* sandman; **~papier das** sandpaper; **~sack der a)** sandbag; **b)** *(Boxen)* punching-bag; **~stein der** sandstone; **~strand der** sandy beach

sandte ['zantə] *1. u. 3. Pers. Sg. Prät. v.* **senden**

Sand·uhr die sand-glass

Sandwich ['zɛntvɪtʃ] **der** *od.* **das; ~s, ~[e]s** sandwich

sanft [zanft] **1.** *Adj.* gentle; *(leise, nicht intensiv)* soft ‹*music, colour, light*›; *(friedlich)* peaceful; **es auf die ~e Tour versuchen** *(ugs.)* try the gentle approach. **2.** *adv.* gently; *(leise)* ‹*speak, play*› softly; *(friedlich)* peacefully; **ruhe ~** *(auf Grabsteinen)* rest in peace

Sänfte ['zɛnftə] **die; ~, ~n** litter; *(geschlossen)* sedan-chair

Sanftheit die; ~: gentleness; *(von Klängen, Licht, Farben)* softness

sang [zaŋ] *1. u. 3. Pers. Sg. Prät. v.* **singen**

Sänger ['zɛŋɐ] **der; ~s, ~:** singer

Sängerin die; ~, ~nen singer

sang·los *Adv. in* **sang- und klanglos** *(ugs.)* simply; without any ado *or* fuss

sanieren [za'ni:rən] **1.** *tr. V.* **a)** redevelop ‹*area*›; rehabilitate ‹*building*›; *(renovieren)* renovate [and improve] ‹*flat etc.*›; **b)** *(Wirtsch.)* restore ‹*firm*› to profitability; rehabilitate ‹*agriculture, coal mining, etc.*›. **2.** *refl. V.* ‹*company etc.*› restore itself to profitability, get back on its feet again; ‹*person*› get oneself out of the red

Sanierung die; ~, ~en *s.* **sanieren: a)** redevelopment; rehabilitation; renovation; **b)** restoration to profitability

sanitär [zani'tɛ:ɐ̯] *Adj.; nicht präd.* sanitary ‹*installations*›

Sanitäter [zani'tɛ:tɐ] **der; ~s, ~** first-aid man; *(im Krankenwagen)* ambulance man

Sanitäts·wagen der ambulance

sank [zaŋk] *1. u. 3. Pers. Sg. Prät. v.* **sinken**

Sanktion [zaŋk'tsi̯o:n] **die; ~, ~en** sanction

sanktionieren *tr. V.* sanction

sann [zan] *1. u. 3. Pers. Sg. Prät. v.* **sinnen**

Saphir ['za:fɪr] **der; ~s, ~e** sapphire

Sardelle [zar'dɛlə] **die; ~, ~n** anchovy

Sardine [zar'di:nə] **die; ~, ~n** sardine

Sarg [zark] **der; ~[e]s, Särge** ['zɛrgə] coffin

Sarkasmus [zar'kasmʊs] **der; ~:** sarcasm

sarkastisch 1. *Adj.* sarcastic. **2.** *adv.* sarcastically

saß [za:s] *1. u. 3. Pers. Sg. Prät. v.* **sitzen**

Satan ['za:tan] **der; ~s, ~e a)** *o. Pl. (bibl.)* Satan *no def. art.*; **b)** *(ugs. abwertend: Mensch)* fiend

Satellit [zatɛ'li:t] **der; ~en, ~en** *(auch fig.)* satellite

Satelliten·staat der *(abwertend)* satellite [state]

Satire [za'ti:rə] **die; ~, ~n** satire

Satiriker der; ~s, ~: satirist

satirisch 1. *Adj.* satirical. **2.** *adv.* satirically; with a satirical touch

satt [zat] **1.** *Adj.* **a)** full [up] *pred.*; well-fed; **~ sein** be full [up]; have had enough [to eat]; **sich ~ essen** eat as much as one wants; **etw. macht ~:** sth. is filling; **b)** *(selbstgefällig)* smug, self-satisfied ‹*person, smile, expression, etc.*›; **c) jmdn./etw. ~ haben/kriegen** *(ugs.)* be/get fed up with sb./sth. *(coll.)*; **d)** *(intensiv)* rich, deep ‹*colour*›; rich, pure ‹*sound*›. **2.** *adv.* **a)** *(selbstgefällig)* smugly; complacently; **b)** *(reichlich)* **nicht ~ zu essen haben** not have enough to eat; **Tennis ~** *(fig.)* as much tennis as one could possibly want

Sattel ['zatl̩] **der; ~s, Sättel** ['zɛtl̩] saddle

sattel·fest *Adj.* experienced; **in etw.** *(Dat.)* **~ sein** be au fait with sth.; be well up in sth.

satteln *tr. V.* saddle

Sattel·tasche die saddle-bag

sättigen ['zɛtɪgn̩] **1.** *itr. V.* be filling. **2.** *tr. V.* **a)** *(geh.)* fill; **b)** *(fig.)* saturate

sättigend *Adj.* filling

Sättigung die; ~, ~en *(fig.)* saturation

Sattler ['zatlɐ] **der; ~s, ~:** saddler

sattsam *Adv.* ad nauseam; **~ bekannt** only too well known; notorious

Saturn [za'tʊrn] **der; ~s** Saturn *no def. art.*

Satz [zats] **der; ~es, Sätze** ['zɛtsə] **a)** *(sprachliche Einheit)* sentence; *(Teil~)* clause; **in** *od.* **mit einem ~:** in one sentence; briefly; **b)** *(Musik)* movement; **c)** *(Tennis, Volleyball)* set; *(Tischtennis, Badminton)* game; **d)** *(Sprung)* leap; jump; **einen ~ über etw.** *(Akk.)* **machen** jump *or* leap across sth.; **e)** *(Amtsspr.: Tarif)* rate; **f)** *(Set)* set; **g)** *(Boden~)* sediment; *(Kaffee~)* grounds *pl.*; **h)** *o. Pl. (Druckw.) (das Setzen)* setting; *(Gesetztes)* type matter

Satz-: ~aussage die *(Sprachw.)* predicate; **~ergänzung die** *(Sprachw.)* complement; **~gefüge das** *(Sprachw.)* complex sentence; **~gegen·stand der** *(Sprachw.)* subject [of a/the sentence]; **~glied das** *(Sprachw.)* component part [of a/the sentence]; **~teil der** *(Sprachw.) s.* **~glied**

Satzung ['zatsʊŋ] **die; ~, ~en** articles of association *pl.*; statutes *pl.*

satzungs·gemäß *Adj., adv.* in accordance with the articles of association *or* the statutes

Satz·zeichen das *(Sprachw.)* punctuation mark

Sau [zau̯] **die; ~, Säue** ['zɔy̯ə] **a)** *(weibliches Schwein)* sow; **b)** *(bes. südd.: Schwein)* pig; **jmdn. zur ~ machen** *(derb)* tear a strip off sb. *(sl.)*; **wie eine gesengte Sau fahren** *(derb)* drive like a madman; **c)** *(derb abwertend: schmutziger Mensch) (Mann)* dirty pig; *(Frau)* dirty cow *(sl.)*; **d)** *(derb abwertend: gemeiner Mensch)* swine

Sau·bande die *(salopp)* wretched swine *(derog.) pl.*; *(mehr scherzh.)* bunch of good-for-nothings *(sl.)*

sauber ['zau̯bɐ] **1.** *Adj.* **a)** clean; **b)** *(sorgfältig)* neat ‹*handwriting, division, work, etc.*›; **c)** *(einwandfrei)* perfect, faultless ‹*accent, technique, etc.*›; **d)** *(anständig)* upstanding ‹*attitude, person*›; fair ‹*solution*›; unsullied ‹*character*›; **~ bleiben** *(ugs.)* keep one's hands clean *(coll.)*; **e)** *nicht präd. (iron.: unanständig)* nice, fine *(iron.)*. **2.** *adv.* **a)** *(sorgfältig)* neatly ‹*written, dressed, mended, etc.*›; **b)** *(fehlerlos)* **[sehr] ~:** [quite] perfectly *or* faultlessly; **c)** *(anständig)* conscientiously; *(gerecht)* ‹*judge etc.*› fairly; **d)** *(iron.)* nicely *(iron.)*; **das hast du ~ hingekriegt** a fine job you made of that

sauber|halten *unr. tr. V.* keep ‹*room, floor, etc.*› clean

Sauberkeit die; ~: cleanness; *(bes. der Person)* cleanliness

säuberlich ['zɔy̯bɐlɪç] **1.** *Adj.* neat. **2.** *adv.* **[fein] ~:** neatly

sauber|machen 1. *tr. V.* clean. **2.** *itr. V.* clean; do the cleaning

säubern ['zɔy̯bɐn] *tr. V.* **a)** clean; **die Schuhe vom Lehm ~:** clean the mud off the shoes; **b)** *(fig.)* clear, rid **(von** of**)**; purge ‹*party, government, etc.*› **(von** of**)**

Säuberung die; ~, ~en a) cleaning; **b)** *(fig.)* purging

sau·blöd[e] *(salopp abwertend)* **1.** *Adj.* bloody silly *or* stupid. **2.** *adv.* in an bloody silly *or* stupid manner *(sl.)*

Sauce *s.* **Soße a**

Saudi [zau̯di] **der; ~s, ~s, Saudi·araber der** Saudi

Saudi-Arabien (das); ~s Saudi Arabia

saudiarabisch, saudisch *Adj.* Saudi Arabian; Saudi

sau·dumm *s.* **saublöd**

sauer ['zau̯ɐ] **1.** *Adj.* **a)** sour; sour, tart ‹*fruit*›; pickled ‹*herring, gherkin, etc.*›; acid[ic] ‹*wine, vinegar*›; **saurer Regen** acid rain; *s. auch* **Apfel a**; **b)** *nicht attr. (ugs.) (verärgert)* cross, annoyed **(auf** + *Akk.* with**)**; *(verdrossen)* sour; **c)** *(mühselig)* hard; difficult; **gib ihm Saures!** *(ugs.)* let him have it! *(coll.)*. **2.** *adv.* **a)** *(in Essig)* in vinegar; **b)** crossly; **~ reagieren** *(ugs.)* get annoyed *or* cross **(auf** + *Akk.* with**)**

Sauer-: ~ampfer [~ampfɐ] **der** sorrel; **~braten der** *braised beef marinated in vinegar and herbs*; sauerbraten *(Amer.)*

Sauerei die; ~, ~en *(salopp abwertend)* **a)** *(Unflätigkeit)* obscenity; **b)** *(Gemeinheit)* bloody *(Brit. sl.) or (coll.)* damn scandal *(sl.)*

Sauer-: ~kirsche die sour cherry; **~kraut das** *o. Pl.* sauerkraut; pickled cabbage

säuerlich 1. *Adj.* **[leicht] ~:** slightly sour. **2.** *adv.* *(mißvergnügt)* somewhat sourly

Sauer-: ~milch die sour milk; **~stoff der** *o. Pl.* oxygen

Sauerstoff- oxygen ‹*cylinder, apparatus, mask, etc.*›

Sauer·teig der leaven

saufen ['zau̯fn̩] **1.** *unr. itr. V.* **a)** ‹*animal*› drink; **b)** *(salopp) (trinken)* drink; swig *(coll.)*; *(Alkohol trinken)* drink; booze *(coll.)*; **~ wie ein Loch** drink like a fish. **2.** *unr. tr. V.* **a)** ‹*animal*› drink; **b)** *(salopp: trinken)* drink; **einen ~ gehen** go for a drink

Säufer ['zɔy̯fɐ] **der; ~s, ~** *(salopp, oft abwertend)* boozer *(coll.)*; piss artist *(sl.)*

Sauferei die; ~, ~en *(salopp)* booze-up *(coll.)*

Säuferin die; ~, ~nen *(salopp, oft abwertend)* boozer *(coll.)*; drunkard

säufst [zɔy̯fst] *2. Pers. Sg. Präsens v.* **saufen**

säuft [zɔy̯ft] *3. Pers. Sg. Präsens v.* **saufen**

saugen ['zau̯gn̩] **1.** *tr. V.* **a)** *auch unr.* suck; *s. auch* **Finger b**; **b)** *auch itr.* *(staub~)* vacuum; hoover *(coll.)*. **2.** *regelm. (auch unr.) itr. V.* **an etw.** *(Dat.)* **~:** suck [at] sth.. **3.** *unr. (auch regelm.) refl. V.* **sich voll etw. ~:** become soaked with sth.

säugen ['zɔy̯gn̩] *tr. V.* suckle

Sauger der; ~s, ~ a) *(auf Flaschen)* teat; **b)** *(Saugheber)* siphon

Säuger ['zɔy̯gɐ] **der; ~s, ~, Säuge·tier das** *(Zool.)* mammal

saug·fähig *Adj.* absorbent

Säugling ['zɔy̯klɪŋ] **der; ~s, ~e** baby; infant

Säuglings-: ~alter das; *o. Pl.* infancy; babyhood; **~pflege die** baby care; **~schwester die** infant *or* baby nurse

Saug·napf der *(Zool.)* sucker

Sau·haufen der *(salopp abwertend)* bunch of layabouts *(sl.)*

säuisch ['zɔy̯ɪʃ] *(salopp)* **1.** *Adj.* **a)** *(abwertend: unanständig)* obscene ‹*phone call*›; **b)** *(stark, groß)* hellish *(coll.)*. **2.** *adv.* *(sehr)* hellishly *(coll.)*

sau·kalt *Adj.* *(salopp)* bloody cold *(Brit. sl.)*; damn cold *(coll.)*

Sau·laden der *(salopp abwertend)* dump *(coll.)*

Säule ['zɔy̯lə] **die; ~, ~n** column; *(nur als Stütze, auch fig.)* pillar

Säulen-: ~gang der colonnade; **~halle die** columned hall

Saum [zaum] **der; ~[e]s, Säume** ['zɔy̯mə] hem

Sau·magen der *(Kochk.)* stuffed pig's stomach

sau·mäßig *(salopp)* **1.** *Adj.* **a)** *(sehr groß)* **das ist eine ~e Arbeit/Hitze** that's a hell of a job/temperature *(coll.)*; **~es Glück haben** be damned lucky *(coll.)*; **b)** *(abwertend: schlecht)* lousy *(sl.)*. **2.** *adv.* **a)** *(sehr)* damned *(coll.)*; **b)** *(abwertend: schlecht)* lousily *(sl.)*

säumen ['zɔy̯mən] *tr. V.* hem; *(fig. geh.)* line

säumig *(geh.) Adj.* tardy; dilatory

Sauna ['zau̯na] **die; ~, ~s** *od.* **Saunen** sauna

Säure ['zɔy̯rə] **die; ~, ~n a)** *o. Pl.* *(von Früchten)* sourness; tartness; *(von Wein, Essig)* acidity; *(von Soßen)* sharpness; **b)** *(Chemie)* acid

säure-: ~arm *Adj.* low in acid *postpos.*; **~beständig** *Adj.* acid-resistant; **~frei** *Adj.* acid-free

Saure·gurken·zeit die *(ugs.)* silly season *(Brit.)*

säure·haltig *Adj.* acid[ic]

Saurier ['zau̯ri̯ɐ] **der; ~s, ~:** large prehistoric reptile

Saus [zau̯s] *in* **in ~ und Braus leben** live the high life

säuseln ['zɔy̯zl̩n] **1.** *itr. V.* ‹*leaves etc.*› rustle; ‹*wind*› murmur. **2.** *tr. V.* *(iron.: sagen)* whisper

sausen *itr. V.* **a)** ‹*wind*› whistle; ‹*storm*› roar; ‹*head, ears*› buzz; ‹*propeller, engine, etc.*› whirr; **b)** *mit sein (fahren, gehen)* ‹*person*› rush; ‹*vehicle*› roar; ‹*whip, bullet, etc.*› whistle

sausen|lassen *unr. tr. V.* *(salopp)* not bother to follow up ‹*plan*›; let ‹*business deal*› go; give ‹*concert, show, etc.*› a miss

Sau·wetter das *(salopp abwertend)* lousy weather *(sl.)*

sau·wohl *Adj.* **sich ~ fühlen** *(salopp)* feel bloody *(Brit. sl.) or (coll.)* damn good *or* great

Savanne [za'vanə] **die; ~, ~n** savannah

Saxophon [zakso'fo:n] **das; ~s, ~e** saxophone

S-Bahn ['ɛs-] **die** city and suburban railway; S-bahn

S-Bahn-: ~hof der, ~-Station die S-bahn station; **~-Zug der** city and suburban train; S-bahn train

SB- [ɛs'be:-]: **~-Laden der** self-service shop; **~-Tankstelle die** self-service petrol *(Brit.) or (Amer.)* gasoline station

sch [ʃʃ] *Interj.* **a)** *(ruhig)* sh[h]; hush; **b)** *(weg da)* shoo

Schabe ['ʃa:bə] **die; ~, ~n** cockroach

Schabe·fleisch das minced beef

schaben 1. *tr. V.* **a)** *(schälen)* scrape ⟨*carrots, potatoes, etc.*⟩; *(glätten)* shave ⟨*leather, hide, etc.*⟩; plane ⟨*wood, surface, etc.*⟩; **b)** *(scheuern)* rub; **c)** *(entfernen)* scrape. **2.** *itr. V.* scrape; **an/auf etw.** *(Dat.)* **~:** scrape against sth./scrape sth.

Schaber der; ~s, ~: scraper

Schabernack ['ʃa:bɐnak] **der; ~[e]s, ~e a)** *(Streich)* prank; **jmdm. einen ~ spielen, mit jmdm. seinen ~ treiben** play a prank on sb.; **b)** *o. Pl. (Scherz, Spaß)* **aus ~ etw. tun** do sth. for a joke

schäbig ['ʃɛ:bɪç] **1.** *Adj.* **a)** *(abgenutzt)* shabby; **b)** *(jämmerlich, gering)* pathetic; miserable; **~e Gehälter** paltry wages; **c)** *(gemein)* shabby; mean. **2.** *adv.* **a)** *(abgenutzt)* shabbily; **b)** *(jämmerlich)* miserably; **~ bezahlen** pay poorly; **c)** *(gemein)* meanly

Schäbigkeit die; ~ shabbiness; *(des Gehalts)* paltriness

Schablone [ʃa'blo:nə] **die; ~, ~n a)** pattern; **b) in ~n denken** *(fig. abwertend)* think in stereotypes

schablonen·haft 1. *Adj.* stereotyped ⟨*thinking*⟩. **2.** *adv.* ⟨*think, argue, etc.*⟩ in a stereotyped manner

Schach [ʃax] **das; ~s, ~s a)** *o. Pl. (Spiel)* chess; **b)** *(Stellung)* check; **~ bieten** give check; **dem Gegner ~ bieten** check the opponent; **jmdn./etw. in ~ halten** *(fig. ugs.)* keep sb./sth. in check; *s. auch* **matt 1 e**

Schach·brett das chessboard

schachern *itr. V.* haggle **(um** over)

schach-, Schach-: ~figur die chess piece; chessman; **~matt** *Adj.* **a)** *(Schachspiel)* **~matt!** checkmate; **~matt sein** be checkmated; **jmdn. ~matt setzen** checkmate sb.; *(fig.)* render sb. powerless; **b)** *(ugs.: erschöpft)* exhausted; **~spiel das a)** *o. Pl. (Spiel)* chess; *(das Spielen)* chess-playing; **b)** *(Brett und Figuren)* chess set

Schacht [ʃaxt] **der; ~[e]s, Schächte** ['ʃɛçtə] shaft

Schachtel ['ʃaxtl̩] **die; ~, ~n a)** box; **eine ~ Zigaretten** a packet *or (Amer.)* pack of cigarettes; **b) alte ~** *(salopp abwertend)* old bag *(sl.)*

Schachtel·halm der *(Bot.)* horsetail

Schach·zug der move [in chess]; *(fig.)* move

schade ['ʃa:də] *Adj.; nicht attr.* **[wie] ~!** [what a] pity *or* shame; **das ist [sehr] ~!** that's a [terrible] pity *or* shame; **[es ist] ~ um jmdn./etw.** it's a pity *or* shame about sb./sth.; **für jmdn./für** *od.* **zu etw. zu ~ sein** be too good for sb./sth.

Schädel ['ʃɛ:dl̩] **der; ~s, ~ a)** skull; *(Kopf)* head; **jmdm. eins auf** *od.* **über den ~ geben** *(ugs.)* hit *or* knock sb. over the head; **mir brummt der ~** *(ugs.)* my head is throbbing; **einen dicken** *od.* **harten ~ haben** *(fig.)* be stubborn *or* pigheaded; **b)** *(fig.: Verstand)* **streng deinen ~ mal an!** tax your brains a bit; **es geht** *od.* **will nicht in seinen ~ [hinein], daß ...** *(ugs.)* he can't get it into his head that ...

Schädel-: ~basis·bruch der *(Med.)* basal skull fracture; **~bruch der** *(Med.)* skull fracture

schaden *itr. V.* **jmdm./einer Sache ~:** damage *or* harm sb./sth.; **Rauchen schadet der Gesundheit/dir** smoking damages your health/is bad for you; **jmds. Ansehen [sehr] ~:** do [great] damage to sb.'s reputation; **das schadet nichts** *(ugs.)* that doesn't matter; *(ist ganz gut)* that won't do any harm

Schaden der; ~s, Schäden ['ʃɛ:dn̩] **a)** damage *no pl., no indef. art.;* **ein kleiner/großer ~:** little/major damage; **jmdm. [einen] ~ zufügen** harm sb.; **das Haus weist einige Schäden auf** the house has some defects; **zu ~ kommen** *(verletzt werden)* be hurt *or* injured; **b)** *(Nachteil)* disadvantage; **zu ~ kommen** suffer; be adversely affected

schaden-, Schaden-: ~ersatz der damages *pl.;* **~freude die** *o. Pl.* malicious pleasure; **..., sagte er voller ~freude** ... he said gloatingly; **~froh 1.** *Adj.* gloating; **~froh sein** gloat; **2.** *adv.* with malicious pleasure

schadhaft ['ʃa:thaft] *Adj.* defective

schädigen ['ʃɛ:dɪgn̩] *tr. V.* damage ⟨*health, reputation, interests*⟩; harm, hurt ⟨*person*⟩; cause losses to ⟨*firm, industry, etc.*⟩

Schädigung die; ~: damage *no pl., no indef. art.* (*Gen.* to)

schädlich ['ʃɛ:tlɪç] *Adj.* harmful; **~e Folgen** damaging *or* detrimental consequences

Schädlichkeit die; ~: harmfulness

Schädling ['ʃɛ:tlɪŋ] **der; ~s, ~e** pest

Schädlings-: ~bekämpfung die pest control; **~bekämpfungsmittel das** pesticide

schadlos *Adj. in* **sich an jmdm./etw. ~ halten** take advantage of sb./sth.

schadstoff·arm *Adj. (bes. Kfz-W.)* low in harmful substances *postpos.;* clean-exhaust ⟨*vehicle*⟩; *(mit Katalysator)* ⟨*vehicle*⟩ with exhaust emission control

Schaf [ʃa:f] **das; ~[e]s, ~e a)** sheep; *s. auch* **schwarz b; b)** *(ugs.: Dummkopf)* twit *(Brit. sl.);* idiot *(coll.)*

Schaf·bock der ram

Schäfchen ['ʃɛ:fçən] **das; ~s, ~ a)** [little] sheep; *(Lamm)* lamb; **sein[e] ~ ins trockene bringen** *(ugs.)* take care of number one *(coll.);* **b)** *Pl. (ugs.: Schutzbefohlene)* flock *sing. or pl.*

Schäfer der; ~s, ~: shepherd

Schäfer·hund der sheep-dog; **[deutscher] ~:** Alsatian *(Brit.);* German shepherd

Schäfer·stündchen das lovers' tryst

Schaf·fell das sheepskin

schaffen ['ʃafn̩] **1.** *unr. tr. V.* **a)** *(er~)* create; **für jmdn./etw.** *od.* **zu jmdm./etw. wie geschaffen sein** be made *or* perfect for sb./sth.; **b)** *auch regelm. (herstellen)* create ⟨*conditions, jobs, situation, etc.*⟩; make ⟨*room, space, fortune*⟩; **klare Verhältnisse ~:** clear things up; straighten things out. **2.** *tr. V.* **a)** *(bewältigen)* manage; **es ~, etw. zu tun** manage to do sth.; **wenn wir uns beeilen, ~ wir es vielleicht noch** we might still make it if we hurry; **er hat die Prüfung nicht geschafft** *(ugs.)* he didn't pass the exam; **b)** *(ugs.: erschöpfen)* wear out; **die Hitze/Arbeit hat mich geschafft** the heat/work took it out of me; **c) etw. aus etw./in etw.** *(Akk.)* ~: get sth. out of/into sth.. **3.** *itr. V.* **a)** *(südd.: arbeiten)* work; **b) sich** *(Dat.)* **zu ~ machen** busy oneself; **mit ihm will ich nichts zu ~ haben** I don't want to have anything to do with him; **jmdm. zu ~ machen** cause sb. trouble

Schaffen das; ~s *(geh.)* work; **im Zenit seines ~s** at the peak of his creative work

Schaffens·kraft die *o. Pl. (geh.)* energy for work; *(eines Künstlers)* creativity; creative power

Schaffner ['ʃafnɐ] **der;** ~s ~ *(im Bus)* conductor; *(im Zug)* guard *(Brit.)*; conductor *(Amer.)*

Schaffnerin die; ~, ~nen *(im Bus)* conductress *(Brit.)*; *(im Zug)* guard *(Brit.)*; conductress *(Amer.)*

Schaffung die; ~: creation

Schaf-: ~garbe die yarrow; **~herde die** flock of sheep; **~hirt der** shepherd

Schafott [ʃa'fɔt] **das;** ~[e]s, ~e scaffold

Schafs-: ~käse der sheep's milk cheese; **~kopf** der **a)** *o. Pl. (Kartenspiel)* sheep's head; **b)** *(ugs.: Trottel)* dope *(coll.)*; idiot *(coll.)*

Schaf·stall der sheep-fold

Schaft [ʃaft] **der;** ~[e]s, **Schäfte** ['ʃɛftə] **a)** shaft; *(eines Gewehrs usw.)* stock; **b)** *(am Stiefel)* leg

Schaft·stiefel der high boot

Schakal [ʃa'ka:l] **der;** ~s, ~e jackal

schäkern *itr. V. (veralt.)* **a)** *(spaßen)* fool about; **b)** *(flirten)* flirt

schal [ʃa:l] *Adj.* stale ‹*drink, taste, joke*›; empty ‹*words, feeling*›

Schal der; ~s, ~s *od.* ~e scarf

Schale ['ʃa:lə] **die;** ~, ~n **a)** *(Obst~)* skin; *(abgeschält)* peel *no pl.*; **b)** *(Nuß~, Eier~, Muschel~ usw.)* shell; **c)** *(Schüssel)* bowl; *(flacher)* dish; *(Waag~)* pan; scale; **d) sich in ~ werfen** *od.* **schmeißen** *(ugs.)* get dressed [up] to the nines

schälen ['ʃɛ:lən] **1.** *tr. V.* peel ‹*fruit, vegetable*›; shell ‹*egg, nut, pea*›; skin ‹*tomato, almond*›; **einen Baumstamm ~:** remove the bark from a tree-trunk; **etw. aus etw. ~:** get sth. out of sth. **2.** *refl. V.* ‹*person, skin, nose, etc.*› peel; **du schälst dich am Rücken** your back is peeling

Schalk [ʃalk] **der;** ~[e]s, ~e *od.* **Schälke** ['ʃɛlkə] rogue; prankster; **jmdm. sitzt der ~ im Nacken** *(fig.)* sb. is really roguish *or* mischievous

schalkhaft *(geh.)* **1.** *Adj.* roguish; mischievous. **2.** *adv.* roguishly; mischievously

Schall [ʃal] **der;** ~[e]s, ~e *od.* **Schälle** ['ʃɛlə] **a)** *(geh.)* sound; **mit lautem ~:** loudly; **Name ist ~ und Rauch** names mean nothing; **b) der ~** *(Physik)* sound

schall-, Schall-: ~dämmend *Adj.* sound-deadening; sound-absorbing; **~dämpfer der a)** silencer; **b)** *(Musik)* mute; **~dicht** *Adj.* sound-proof

schallen *regelm., auch unr. itr. V.* ring out; *(nachhallen)* resound; echo; **~des Gelächter** ringing laughter; **~d lachen** roar with laughter

Schall-: ~geschwindigkeit die speed *or* velocity of sound; **~mauer die** sound *or* sonic barrier; **~platte die** record

Schallplatten-: *s.* **Platten-**

Schall·welle die *(Physik)* sound-wave

Schalotte [ʃa'lɔtə] **die;** ~, ~n shallot

schalt [ʃalt] *1. u. 3. Pers. Sg. Prät. v.* **schelten**

schalten ['ʃaltn̩] **1.** *tr. V.* **a)** switch; **b)** *(Elektrot.: verbinden)* connect; **c)** *(Zeitungsw.)* place ‹*advertisement*›. **2.** *itr. V.* **a)** *(Schalter betätigen)* switch, turn **(auf +** *Akk.* to); **b)** ‹*machine*› switch **(auf +** *Akk.* to); ‹*traffic light*› change **(auf +** *Akk.* to); **c)** *(im Auto)* change [gear]; **in den 4. Gang ~:** change into fourth gear; **d) sie kann ~ und walten, wie sie will, sie kann frei ~ und walten** she can manage things as she pleases; **e)** *(ugs.: begreifen)* twig *(coll.)*; catch on *(coll.)*

Schalter ['ʃaltɐ] **der;** ~s, ~ **a)** *(Strom~)* switch; **b)** *(Post~, Bank~ usw.)* counter

Schalter-: ~beamte der counter clerk; *(im Bahnhof)* ticket clerk; **~halle die** hall; *(im Bahnhof)* booking-hall *(Brit.)*; ticket office

Schalt-: ~getriebe das *(Kfz-W.)* [manual] gearbox; **~jahr das** leap year; **alle ~jahre [ein]mal** *(ugs.)* once in a blue moon; **~knüppel der** [floor-mounted] gear-lever

Schaltung die; ~, ~en **a)** *(Rundfunk: Verbindung)* link-up; **b)** *(Gang~)* manual gear change; **c)** *(Elektrot.)* circuit; wiring system

Scham [ʃa:m] **die;** ~ shame; **nur keine falsche ~!** no need for any false modesty

schämen ['ʃɛmən] *refl. V.* be ashamed *(Gen.*, **für, wegen** of); **du solltest dich [was** *(ugs.)*] **~!** you [really] should be ashamed of yourself; **schäm dich** shame on you

Scham·gefühl das; *o. Pl.* sense of shame

schamhaft 1. *Adj.* bashful. **2.** *adv.* bashfully

scham·los 1. *Adj.* **a)** *(skrupellos, dreist)* shameless; barefaced, shameless ‹*lie, slander*›; **b)** *(unanständig)* indecent ‹*gesture, remark, etc.*›; shameless ‹*person*›. **2.** *adv.* **a)** *(skrupellos, dreist)* shamelessly; **b)** *(unanständig)* indecently

Scham·losigkeit die; ~, ~en: *s.* **schamlos 1a, b:** shamelessness; indecency

Schampon ['ʃampɔn] *s.* **Shampoo**

scham·rot *Adj.* red with shame *postpos.*

Scham·röte die: ihm stieg die ~ ins Gesicht he blushed with shame

Schande ['ʃandə] **die;** ~: disgrace; shame; **es ist eine [wahre] ~:** it is a[n absolute] disgrace; **jmdm./einer Sache [keine] ~ machen** [not] disgrace sb./sth.; bring [no] disgrace *or* shame on sb./sth.

schänden ['ʃɛndn̩] *tr. V.* defile ‹*memorial, work of art, etc.*›; desecrate, defile ‹*holy place, grave, relic*›; violate ‹*corpse*›

schändlich 1. *Adj.* shameful; disgraceful. **2.** *adv.* shamefully; disgracefully

Schändlichkeit die; ~, ~en **a)** *o. Pl.* shamefulness; disgracefulness; **b)** *(Tat)* shameful action

Schand·tat die disgraceful *or* abominable deed; **zu jeder ~** *od.* **allen ~en bereit sein** *(ugs. scherzh.)* be game for anything

Schändung die; ~, ~en *s.* **schänden:** desecration; defilement

Schank·wirtschaft die public house *(Brit.)*; bar *(Amer.)*

Schar [ʃa:ɐ̯] **die;** ~, ~en crowd; horde; *(von Vögeln)* flock; **in [hellen] ~en** in swarms *or* droves

Schäre ['ʃɛ:rə] **die;** ~, ~n skerry

scharen 1. *refl. V.* gather. **2.** *tr. V.* **die Kinder um sich ~:** gather the children around one[self]

scharen·weise *Adv.* in swarms *or* hordes

scharf [ʃarf]; **schärfer** ['ʃɛrfɐ], **schärfst...** ['ʃɛrfst...] **1.** *Adj.* **a)** sharp; **b)** *(stark gewürzt, brennend, stechend)* hot; strong ‹*drink, vinegar, etc.*›; caustic ‹*chemical*›; pungent, acrid ‹*smell*›; **c)** *(durchdringend)* shrill; *(hell)* harsh; biting ‹*wind, air, etc.*›; sharp ‹*frost*›; **d)** *(deutlich* *wahrnehmend)* keen; sharp; **e)** *(deutlich hervortretend)* sharp ⟨*contours, features, nose, photograph*⟩; **f)** *(schonungslos)* tough, fierce ⟨*resistance, competition, etc.*⟩; sharp ⟨*criticism, remark, words, etc.*⟩; strong, fierce ⟨*opponent, protest, etc.*⟩; severe, harsh ⟨*sentence, law, measure, etc.*⟩; fierce ⟨*dog*⟩; **eine ~e Zunge haben** have a sharp tongue; **g)** *(schnell)* fast; hard ⟨*ride, gallop, etc.*⟩; **h)** *(explosiv)* live; *(Ballspiele)* powerful ⟨*shot*⟩; **~e Schüsse abgeben** fire live bullets; **i) das ~e S** *(bes. österr.)* the German letter 'ß'; **j)** *(ugs.: geil)* sexy ⟨*girl, clothes, pictures, etc.*⟩; randy ⟨*fellow, thoughts, etc.*⟩; **k) ~ auf jmdn./etw. sein** *(ugs.)* really fancy sb. *(coll.)*/be really keen on sth. **2.** *adv.* **a) ~ würzen/abschmecken** season/flavour highly; **~ riechen** smell pungent *or* strong; **b)** *(durchdringend)* shrilly; *(hell)* harshly; *(kalt)* bitingly; **c)** *(deutlich wahrnehmend)* ⟨*listen, watch, etc.*⟩ closely, intently; ⟨*think, consider, etc.*⟩ hard; **~ aufpassen** pay close attention; **d)** *(deutlich hervortretend)* sharply; **e)** *(schonungslos)* ⟨*attack, criticize, etc.*⟩ sharply, strongly; ⟨*contradict, oppose, etc.*⟩ strongly, fiercely; ⟨*watch, observe, etc.*⟩ closely; **f)** *(schnell)* fast; **~ bremsen** brake hard *or* sharply; **g) ~ schießen** shoot with live ammunition

Schạrf·blick der; *o. Pl.* perspicacity

Schärfe ['ʃɛrfə] **die; ~, ~n a)** *o. Pl.* sharpness; **b)** *o. Pl. (von Geschmack)* hotness; *(von Chemikalien)* causticity; *(von Geruch)* pungency; **c)** *o. Pl. (Intensität)* shrillness; *(von Licht, Farbe usw.)* harshness; *(des Windes)* bitterness; *(des Frostes)* sharpness; **d)** *o. Pl. s.* **scharf 1d:** sharpness; keenness; **e)** *o. Pl. (Klarheit)* clarity; sharpness; **f)** *o. Pl. s.* **scharf 1f:** toughness; ferocity; sharpness; strength; **g)** *(Heftigkeit)* harshness

schärfen 1. *tr. V. (auch fig.)* sharpen. **2.** *refl. V.* become sharper *or* keener

schärfer *s.* **scharf**

schạrf-, Schạrf-: ~kantig *Adj.* sharp-edged; **~|machen** *tr. V. (ugs.)* stir up; **einen Hund ~machen** urge a dog on; **~schütze der** marksman; **~sichtig** [~zıçtıç] *Adj.* sharp-sighted; perspicacious; **~sinn der** *o. Pl.* astuteness; acumen; **~sinnig 1.** *Adj.* astute; **2.** *adv.* astutely

schạ̈rfst... *s.* **scharf**

schạrfzüngig [-tsʏŋıç] **1.** *Adj.* sharp-tongued. **2.** *adv.* sharply

Scharlach ['ʃarlax] **der; ~s** *(Med.)* scarlet fever

Scharnier [ʃar'niːɐ̯] **das; ~s, ~e** hinge

Schärpe ['ʃɛrpə] **die; ~, ~n** sash

scharren ['ʃarən] *itr. V.* **a)** scrape; **mit den Füßen ~:** scrape one's feet; **b)** *(wühlen)* scratch. **2.** *tr. V.* scrape, scratch out ⟨*hole, hollow, etc.*⟩

Scharte ['ʃartə] **die; ~, ~n** nick

schạrtig *Adj.* nicked; jagged

Schaschlik ['ʃaʃlık] **der** *od.* **das; ~s, ~s** *(Kochk.)* shashlik

Schatten ['ʃatn̩] **der; ~s, ~ a)** shadow; **man kann nicht über seinen [eigenen] ~ springen** a leopard cannot change its spots *(prov.)*; **b)** *o. Pl. (schattige Stelle)* shade; **in jmds. ~ stehen** *(fig.)* be in sb.'s shadow; **jmdn./etw. in den ~ stellen** *(fig.)* put sb./sth. in the shade; **c)** *(dunkle Stelle, fig.)* shadow

schạttenhaft *Adj.* shadowy

Schạtten-: ~morelle [~morɛlə] **die; ~, ~n** morello cherry; **~seite die** shady side; **die ~seiten des Lebens kennenlernen** *(fig.)* get to know the dark side of life

schạttig *Adj.* shady

Schatz [ʃats] **der; ~es, Schätze** ['ʃɛtsə] **a)** treasure *no indef. art.*; **b)** *(ugs.: Liebling)* love *(coll.)*; darling; **c)** *(ugs.: hilfsbereiter Mensch)* treasure *(coll.)*

schätzen ['ʃɛtsn̩] **1.** *tr. V.* **a)** estimate; **wie alt schätzt du ihn?** how old do you think he is?; **sich glücklich ~:** deem oneself lucky; **grob geschätzt** at a rough estimate; **ein Haus ~:** value a house; **b)** *(ugs.: annehmen)* reckon; think; **c)** *(würdigen, hochachten)* **jmdn. ~:** hold sb. in high regard *or* esteem; **etw. zu ~ wissen** appreciate sth.; **ich weiß es zu ~, daß ...:** I appreciate the fact that ... **2.** *itr. V.* guess; **schätz mal** guess; have a guess

schạ̈tzen|lernen *tr. V.* come to appreciate *or* value

Schätzung die; ~, ~en estimate; **nach grober/vorsichtiger ~:** at a rough/cautious estimate

Schau [ʃau̯] **die; ~, ~en a)** *(Ausstellung)* exhibition; **b)** *(Vorführung)* show; **eine ~ machen** *od.* **abziehen** *(ugs.) (sich in Szene setzen)* put on a show; *(sich aufspielen)* show off; *(sich lautstark ereifern)* make a scene *or* fuss; **jmdm. die ~ stehlen** steal the show from sb.; **c) jmdn./etw. zur ~ stellen** exhibit *or* display sb./sth.; *(fig.)* display sb./sth.; **etw. zur ~ tragen** make a show of sth.

Schau·bild das chart

Schauder ['ʃau̯dɐ] **der; ~s, ~** *(vor Kälte, Angst)* shiver; *(vor Angst)* shudder; **mir lief ein ~ den Rücken hinunter** a shiver/shudder ran down my spine

schauderhaft 1. *Adj.* terrible; dreadful; awful. **2.** *adv.* terribly; dreadfully

schaudern *itr. V. (vor Kälte)* shiver; *(vor Angst)* shudder; *unpers.* **es schauderte ihn** he shivered/shuddered

schauen *(bes. südd., österr., schweiz.)* **1.** *itr. V.* **a)** look; **auf jmdn./etw. ~:** look at sb./sth.; *(fig.)* look to sb./sth.; **um sich ~:** look around [one]; **schau, schau!** well, well; **schau [mal], ich finde, du solltest ...:** look, I think you should ...; **b)** *(sich kümmern um)* **nach jmdm./etw. ~:** take *or* have a look at sb./sth.; **c)** *(achten)* **auf etw.** *(Akk.)* **~:** set store by sth.; **er schaut darauf, daß alle pünktlich sind** he sets store by everybody being punctual; **d)** *(ugs.: sich bemühen)* **schau, daß du ...:** see *or* mind that you ...; **e)** *(nachsehen)* have a look. **2.** *tr. V.* **Fernsehen ~:** watch television

Schauer der; ~s, ~ shower

Schauer·geschichte die horror story

schauerlich 1. *Adj.* **a)** horrifying; ghastly; **b)** *(ugs.: fürchterlich)* terrible *(coll.)*; dreadful *(coll.)*. **2.** *adv. (ugs.: fürchterlich)* dreadfully *(coll.)*, terribly *(coll.)*

Schaufel ['ʃau̯fl̩] **die; ~, ~n** shovel; *(für Mehl usw.)* scoop; *(Kehr~)* dustpan; **zwei ~n Erde** two shovelfuls of soil

schaufeln [ʃau̯fl̩n] *tr. V.* shovel; *(graben)* dig

Schau·fenster das shop-window

Schaufenster~: ~aus·lage die window display; **~bummel der** window-shopping expedition; **einen ~bummel machen** go window-shopping; **~puppe die** mannequin

Schau-: ~kampf der *(Boxen)* exhibition fight; **~kasten der** display case; show-case

Schaukel ['ʃau̯kl̩] **die; ~, ~n a)** swing; **b)** *(Wippe)* see-saw

schaukeln 1. *itr. V.* **a)** swing; *(im Schaukelstuhl)* rock; **auf einem Stuhl ~:** rock one's chair backwards and forwards; **b)** *(sich hin und her bewegen)* sway [to and fro]; *(sich auf und ab bewegen)* ⟨*ship, boat*⟩ pitch and toss;

‹*vehicle*› bump [up and down]; *unpers.* **es hat ganz schön geschaukelt** *(auf dem Boot)* the boat pitched and tossed quite a bit. **2.** *tr. V.* **a)** rock; **ein Kind auf den Knien ~:** dandle a child on one's knee; **b)** *(ugs.: fahren)* take; **jmdn. durch die Gegend ~:** drive sb. round the area; **c)** *(ugs.: bewerkstelligen)* manage

Schaukel-: ~pferd das rocking-horse; **~stuhl der** rocking-chair

schau·lustig *Adj.* curious

Schau·lustige der/die; *adj. Dekl.* curious onlooker

Schaum [ʃaum] **der; ~s, Schäume** ['ʃɔymə] **a)** foam; *(von Seife usw.)* lather; *(von Getränken, Suppen usw.)* froth; **etw. zu ~ schlagen** *(Kochk.)* beat sth. until frothy; **b)** *(Geifer)* foam; froth; **~ vor dem Mund haben** *(auch fig.)* foam *or* froth at the mouth

Schaum·bad das bubble bath

schäumen ['ʃɔymən] *itr. V.* foam; froth; ‹*soap etc.*› lather; ‹*beer, fizzy drink, etc.*› froth [up]

Schaum·gummi der foam rubber

schaumig *Adj.* frothy ‹*drink, dessert, etc.*›; sudsy, lathery ‹*water*›

Schaum-: ~stoff der [plastic] foam; **~wein der** sparkling wine

Schau·platz der scene

schaurig ['ʃaurɪç] **1.** *Adj.* **a)** dreadful; frightful; *(unheimlich)* eerie; **b)** *(ugs.: gräßlich, geschmacklos)* hideous; dreadful *(coll.).* **2.** *adv.* **a)** *(fürchterlich)* dreadfully; *(unheimlich)* eerily; **b)** *(ugs.: gräßlich, geschmacklos)* hideously; horribly *(coll.);* **c)** *(ugs.: überaus)* dreadfully *(coll.)*

Schau-: ~spiel das a) *o. Pl. (Drama)* drama *no art.;* **b)** *(ernstes Stück)* play; **c)** *(geh.: Anblick)* spectacle; **~spieler der** *(auch fig.)* actor; **~spielerin die** *(auch fig.)* actress

Schau-: ~steller [-ʃtɛlɐ] **der; ~s, ~:** showman; **~tafel die** illustrated chart

Scheck [ʃɛk] **der; ~s, ~s** cheque

Scheckheft das cheque-book

scheckig *Adj. s.* **gescheckt**

Scheck·karte die cheque card

scheel [ʃe:l] *(ugs.)* **1.** *Adj.* disapproving; *(mißtrauisch)* suspicious; *(neidisch)* envious; jealous. **2.** *adv.* disapprovingly; *(mißtrauisch)* suspiciously; *(neidisch)* enviously; jealously

scheffeln *tr. V. (ugs.)* rake in *(coll.)* ‹*money etc.*›; pile up, accumulate ‹*medals, awards, etc.*›

Scheibe ['ʃaibə] **die; ~, ~n a)** disc; *(Sportjargon: Puck)* puck; *(Schieß~)* target; **b)** *(abgeschnittene ~)* slice; **etw. in ~n schneiden** slice sth. up; cut sth. [up] into slices; **sich** *(Dat.)* **von jmdm./etw. eine ~ abschneiden können** *(fig.)* be able to learn a thing or two from sb./sth.; **c)** *(Glas~)* pane; *(Fenster~)* [window-] pane; *(Windschutz~)* windscreen *(Brit.);* windshield *(Amer.)*

Scheiben-: ~bremse die *(Kfz-W.)* disc brake; **~kleister der** *(ugs. verhüll.)* **~kleister!** blast [it]! *(coll.);* damn it! *(coll.);* **~wischer der** windscreen-wiper

Scheich [ʃaiç] **der; ~[e]s, ~s** *od.* **~e** sheikh

Scheide ['ʃaidə] **die; ~, ~n a)** sheath; **b)** *(Anat.)* vagina

scheiden 1. *unr. tr. V.* **a)** dissolve ‹*marriage*›; divorce ‹*married couple*›; **sich ~ lassen** get divorced *or* get a divorce; **b)** *(geh.: trennen)* divide; separate; **c)** *(geh.: unterscheiden)* distinguish. **2.** *unr. itr. V.; mit sein (geh.)* **a)** *(auseinandergehen)* part; **b)** *(sich entfernen)* depart; leave; **von jmdm. ~:** part from sb.; **aus dem Dienst/Amt ~:** retire from service/one's post *or* office; **aus dem Leben ~:** depart this life

Scheidung die; ~, ~en divorce; **die ~ einreichen** file [a petition] for divorce; **in ~ leben** be in the process of getting a divorce

Schein [ʃain] **der; ~[e]s, ~e a)** *o. Pl. (Licht~)* light; **der ~ des brennenden Hauses/der sinkenden Sonne** the glow of the burning house/setting sun; **b)** *o. Pl. (An~)* appearances *pl., no art.; (Täuschung)* pretence; **den ~ wahren** keep up appearances; **der ~ trügt** appearances are deceptive; **etw. nur zum ~ tun** [only] pretend to do sth.; make a show of doing sth.; **c)** *(Bescheinigung)* certificate; *(Gepäck~)* ticket; *(Tipp~)* coupon; **d)** *(Geld~)* note

scheinbar 1. *Adj.* apparent; seeming. **2.** *adv.* seemingly

scheinen *unr. itr. V.* **1. a)** shine; **b)** *(den Eindruck erwecken)* seem; appear; **mir scheint, [daß] ...:** it seems *or* appears to me that ...; **wie es scheint ...:** apparently. **2.** *unr. mod. V.* seem; appear; **jmd. scheint etw. nicht tun zu können** sb. doesn't seem *or* appear to be able to do sth.; sb. can't seem to do sth. *(coll.)*

schein-, Schein-: ~heilig 1. *Adj. (heuchlerisch)* hypocritical; *(Nichtwissen vortäuschend)* innocent; **2.** *adv. (heuchlerisch)* hypocritically; *(Nichtwissen vortäuschend)* innocently; **~heiligkeit die** hypocrisy; **~werfer der** floodlight; *(am Auto)* headlight; *(im Theater, Museum usw.)* spotlight; **~werfer·licht das** floodlight; *(des Autos)* headlights *pl.; (im Theater, Museum usw.)* spotlight [beam]

scheiß-, Scheiß- *(derb)* bloody *(Brit. sl.)*

Scheiße ['ʃaisə] **die; ~** *(derb, auch fig.)* shit *(coarse);* crap *(coarse);* **[bis zum Hals] in der ~ sitzen** *od.* **stecken** *(fig.)* be in the shit *(coarse);* be up shit creek *(coarse)*

scheißen *unr. itr. V. (derb)* [have *or (Amer.)* take a] shit *(coarse);* **auf jmdn./etw. ~** *(fig.)* not give a shit *(coarse) or (sl.)* damn about sb./sth.

Scheit [ʃait] **der; ~[e]s, ~e** *od.* **~er** *s.* **Holzscheit**

Scheitel ['ʃaitl̩] **der; ~s, ~ a)** parting; **einen ~ ziehen** make a parting; **vom ~ bis zur Sohle** from head to toe; **b)** *(höchster Punkt, Math.)* vertex; *(eines Winkels)* apex; vertex

scheiteln *tr. V.* part ‹*hair*›

Scheiter·haufen der: auf dem ~ sterben/verbrannt werden die/be burned at the stake

scheitern ['ʃaitɐn] *itr. V.; mit sein* fail; ‹*talks, marriage*› break down; ‹*plan, project*› fail, fall through; **eine gescheiterte Existenz sein** be a failure

Schelle ['ʃɛlə] **die; ~, ~n** bell

schellen *itr. V. (westd.) s.* **klingeln a, b**

Schellen·baum der Turkish crescent; pavillon chinois

Schell·fisch der haddock

Schelm [ʃɛlm] **der; ~[e]s, ~e** rascal; rogue

schelmisch 1. *Adj.* roguish; mischievous. **2.** *adv.* roguishly; mischievously

Schelte ['ʃɛltə] **die; ~, ~n** *(geh.)* scolding; **~ bekommen** be given *or* get a scolding

schelten *(südd., geh.)* **1.** *unr. itr. V.* **auf** *od.* **über jmdn./etw. ~:** moan about sb./sth.; **[mit jmdm.] ~:** scold [sb.]. **2.** *unr. tr. V.* **a)** *(tadeln)* scold; **b)** *(geh.: nennen)* call

Schema ['ʃe:ma] **das; ~s, ~s** *od.* **~ta** *od.* **Schemen a)** *(Muster)* pattern; *s. auch* **F; b)** *(Skizze)* diagram

schematisch 1. *Adj.* **a)** diagrammatic; **b)** *(mechanisch)* mechanical. **2.** *adv.* **a)** in diagram form; **b)** *(mechanisch)* mechanically

Schemel ['ʃe:ml̩] **der; ~s, ~:** stool; *(südd.: Fußbank)* footstool

[1]**Schemen** *s.* Schema

[2]**Schemen** ['ʃe:mən] **der** *od.* **das; ~s, ~:** shadowy figure

schemenhaft 1. *Adj.* shadowy. **2.** *adv.* **etw. ~ sehen** see only the outline *or* silhouette of sth.

Schenkel ['ʃɛŋkl̩] **der; ~s, ~ a)** thigh; **b)** *(Math.)* side; **c)** *(von einer Zange, Schere)* shank; *(vom Zirkel)* leg

schenken 1. *tr. V.* **a)** give; **jmdm. etw. [zum Geburtstag] ~:** give sb. sth. *or* sth. to sb. [as a birthday present *or* for his/her birthday]; **das ist ja geschenkt!** *(ugs.)* it's a gift!; **jmdm./einer Sache Beachtung/Aufmerksamkeit ~:** give sb./sth. one's attention; **jmdm. das Leben ~:** spare sb.'s life; *s. auch* **Gaul; b)** *(ugs.: erlassen)* **jmdm. etw. ~:** spare sb. sth.; **ihr ist im Leben nichts geschenkt worden** she has never had it easy in life. **2.** *refl. V.* **sich** *(Dat.)* **etw. ~** *(ugs.)* give sth. a miss

scheppern ['ʃɛpɐn] *itr. V. (ugs.)* clank; ⟨*bell*⟩ clang; **es hat gescheppert** *(hat einen Autounfall gegeben)* there was a smash *or* crash

Scherbe ['ʃɛrbə] **die; ~, ~n** fragment; **die ~n zusammenkehren** sweep up the [broken] pieces; **in tausend ~n zerspringen** be smashed to smithereens; **~n bringen Glück** *(Spr.)* break a thing, mend your luck

Schere ['ʃe:rə] **die; ~, ~n a)** scissors *pl.*; **eine ~:** a pair of scissors; **b)** *(Zool.)* claw

[1]**scheren** *unr. tr. V. (kürzen)* crop; *(von Haar befreien)* shear, clip ⟨*sheep*⟩; clip ⟨*dog*⟩

[2]**scheren** *tr., refl. V.* **sich um jmdn./etw. nicht ~:** not care about sb./sth.

[3]**scheren** *refl. V.* **scher' dich in dein Zimmer** go *or* get [off] to your room

Scheren·schnitt der silhouette

Scherereien *Pl. (ugs.)* trouble *no pl.*

Scherz [ʃɛrts] **der; ~es, ~e** joke; **seine ~e mit jmdm. treiben** play jokes on sb.; **etw. aus** *od.* **zum ~ sagen** say sth. as a joke *or* in jest; **~ beiseite** joking aside *or* apart

scherzen *itr. V.* joke; **über etw.** *(Akk.)* **~:** joke about sth.; **mit jmdm./etw. ist nicht zu ~** *(fig.)* sb./sth. is not to be trifled with

Scherz·frage die riddle

scherzhaft 1. *Adj.* jocular; joking *attrib.* **2.** *adv.* jocularly; jokingly

scheu [ʃɔy] **1.** *Adj.* shy; timid ⟨*animal*⟩; *(ehrfürchtig)* awed; **~ machen** frighten ⟨*animal*⟩; *s. auch* **Pferd. 2.** *adv.* shyly; *(von Tieren)* timidly

Scheu die; ~ a) shyness; *(von Tieren)* timidity; *(Ehrfurcht)* awe; **ohne jede ~:** without any inhibitions

scheuchen *tr. V.* **a)** *(treiben)* shoo; drive; **b)** *(fig. ugs.)* force; **jmdn. zum Arzt/an die Arbeit ~:** make sb. go *or* urge sb. to go to the doctor/to work

scheuen 1. *tr. V.* shrink from; shun ⟨*people, light, company, etc.*⟩; **weder Kosten noch Mühe ~:** spare neither expense nor effort. **2.** *refl. V.* **sich ~, etw. zu tun** shrink from doing sth. **3.** *itr. V.* ⟨*horse*⟩ shy **(vor +** *Dat.* at)

scheuern 1. *tr., itr. V.* **a)** *(reinigen)* scour; scrub; **b)** *(reiben)* rub; chafe. **2.** *tr. V. (reiben an)* rub. **3.** *refl. V. (reiben)* **sich** *(Akk.)* **wund ~:** rub oneself raw; chafe oneself; **sich** *(Dat.)* **das Knie [wund] ~:** rub one's knee raw; chafe one's knee

Scheuer-: ~pulver das scouring powder; **~tuch das;** *Pl.* **~tücher** scouring cloth

Scheune ['ʃɔynə] **die; ~, ~n** barn

Scheusal ['ʃɔyzal] **das; ~s, ~e** *(abwertend)* monster

scheußlich ['ʃɔyslɪç] **1.** *Adj.* **a)** dreadful; **b)** *(ugs.: äußerst unangenehm)* terrible *(coll.)*; dreadful *(coll.)*; dreadful *(coll.)*, ghastly *(coll.)* ⟨*weather, taste, smell*⟩. **2.** *adv.* **a)** dreadfully; **b)** *(ugs.: sehr)* terribly *(coll.)*; dreadfully *(coll.)*

Scheußlichkeit die; ~, ~en a) *o. Pl.* dreadfulness; **b)** *meist Pl. (etw. Scheußliches)* dreadful thing; *(Grausamkeit)* atrocity

Schi [ʃi:] *usw. s.* **Ski** *usw.*

Schicht [ʃɪçt] **die; ~, ~en a)** layer; *(Geol.)* stratum; *(von Farbe)* coat; *(sehr dünn)* film; **b)** *(Gesellschafts~)* stratum; **breite ~en [der Bevölkerung]** broad sections of the population; **in allen ~en** at all levels of society; **c)** *(Abschnitt eines Arbeitstages, Arbeitsgruppe)* shift; **~ arbeiten** work shifts; be on shift work

Schicht·arbeiter der shift worker

schichten *tr. V.* stack

Schicht·wechsel der change of shifts; **~ ist um 6** we/they *etc.* change shifts at 6

schicht·weise *Adv.* **a)** in layers; layer by layer; **b)** *(in Gruppen)* in shifts

schick [ʃɪk] **1.** *Adj.* **a)** stylish; stylish, chic ⟨*clothes, fashions*⟩; smart ⟨*woman, girl, man*⟩; **b)** *(ugs.: großartig, toll)* great *(coll.)*; fantastic *(coll.)*. **2.** *adv.* stylishly; stylishly, smartly ⟨*furnished, decorated*⟩

Schick der; ~[e]s style

schicken 1. *tr. V.* send; **jmdm. etw. ~, etw. an jmdn. ~:** send sth. to sb.; send sb. sth.; **jmdn. nach Hause/in den Krieg ~:** send sb. home/to war; **jmdn. einkaufen ~:** send sb. to do the shopping. **2.** *itr. V.* **nach jmdm. ~:** send for sb. **3.** *refl. V.* **a)** *(veralt.: sich ziemen)* be proper *or* fitting; **b) sich in etw.** *(Akk.)* **~:** resign *or* reconcile oneself to sth.

Schicki[micki] ['ʃɪkɪ('mɪkɪ)] **der; ~s, ~s** *(ugs.)* trendy *(coll.)*

schicklich *(veralt.)* **1.** *Adj.* proper; fitting; *(dezent)* seemly. **2.** *adv.* fittingly; *(dezent)* in a seemly way

Schicksal ['ʃɪkza:l] **das; ~s, ~e** fate; destiny; *(schweres Los)* fate; **das ~:** fate; destiny; **[das ist] ~** *(ugs.)* it's just fate; **das ~ hat es mit ihm gut gemeint** fortune smiled on him; **~ spielen** play the role of fate *or* destiny

schicksalhaft 1. *Adj.* fateful. **2.** *adv.* **~ verbunden** linked by fate

Schicksals·schlag der stroke of fate

Schiebe·dach das sliding roof; sunroof

schieben ['ʃi:bn̩] **1.** *unr. tr. V.* **a)** push; push, wheel ⟨*bicycle, pram, shopping trolley*⟩; *(drängen)* push; shove; **b)** *(stecken)* put; *(gleiten lassen)* slip; **den Riegel vor die Tür ~:** slip the bolt across; **c) etw. auf jmdn./etw. ~:** blame sb./sth. for sth.; **die Schuld/die Verantwortung auf jmdn. ~:** put the blame on sb. *or* lay the blame at sb.'s door/lay the responsibility at sb.'s door; **d)** *(salopp: handeln mit)* traffic in; push ⟨*drugs*⟩. **2.** *unr. refl. V.* **a)** *(sich zwängen)* **sich durch die Menge ~:** push one's way through the crowd; **b)** *(sich bewegen)* move; **ihr Rock schob sich nach oben** her skirt slid up. **3.** *unr. itr. V.* **a)** push; *(heftig)* push; shove; **b)** *mit sein (salopp: gehen)* mooch *(sl.)*; **c)** *(ugs.: mit etw. handeln)* **mit etw. ~:** traffic in sth.; **d)** *(Skat)* shove

Schieber der; ~s, ~ *(ugs.: Schwarzhändler)* black marketeer

Schiebe·tür die sliding door

Schieb·lehre die *(Technik)* vernier [calliper] gauge

Schiebung die; ~, ~en *(ugs.)* **a)** *(betrügerisches Geschäft)* shady deal; **b)** *(o. Pl.: Begünstigung)* pulling strings; *(bei einer Wahl,* *einem Wettbewerb)* rigging; *(bei einem Wettlauf, -rennen)* fixing; „~!" '[it's a] fix!'

schied [ʃi:t] *1. u. 3. Pers. Sg. Prät. v.* **scheiden**

Schieds-: ~gericht das a) *(Rechtsw.)* arbitration tribunal; **b)** *(Sport)* panel of judges; **~richter der** *(Sport)* referee; *(Tennis, Tischtennis, Hockey, Kricket, Federball)* umpire; *(Eislauf, Ski, Schwimmen)* judge; **~richter·ball der** *(Fußball)* drop ball; *(Basketball)* jump ball

schief [ʃi:f] **1.** *Adj.* **a)** *(schräg)* leaning ‹*wall, fence, post*›; *(nicht parallel)* crooked; not straight *pred.;* crooked ‹*nose*›; sloping, inclined ‹*surface*›; worn[-down] ‹*heels*›; **er hält den Kopf ~:** he holds his head to one side; **der Schiefe Turm von Pisa** the Leaning Tower of Pisa; **eine ~e Ebene** *(Phys.)* an inclined plane; **b)** *(fig.: verzerrt)* distorted ‹*picture, presentation, view, impression*›; false ‹*comparison*›. **2.** *adv.* **a)** *(schräg)* **das Bild hängt/der Teppich liegt ~:** the picture/carpet is crooked; **der Tisch steht ~:** the table isn't level; **jmdn. ~ ansehen** *(ugs.)* look at sb. askance; *s. auch* **Haussegen; b)** *(fig.: verzerrt)* **etw. ~ darstellen** give a distorted account of sth.

Schiefer ['ʃi:fɐ] **der; ~s** *(Gestein)* slate

schief-: ~|gehen *unr. itr. V.; mit sein (ugs.)* go wrong; **es wird schon ~gehen** *(iron.)* it'll all turn out OK *(coll.)*; **~gewickelt** *Adj. in* **~gewickelt sein** *(ugs.)* be very much mistaken; **~|lachen** *refl. V. (ugs.)* kill oneself laughing *(coll.)*; laugh one's head off; **~|laufen** *unr. itr. V.; mit sein (ugs.: ~gehen)* go wrong; **~|liegen** *unr. itr. V. (ugs.)* be on the wrong track

schielen ['ʃi:lən] *itr. V.* **a)** squint; have a squint; **leicht/stark ~:** have a slight/pronounced squint; **auf dem rechten Auge ~:** have a squint in one's right eye; **b)** *(ugs.: blicken)* look out of the corner of one's eye; **nach etw. ~:** steal a glance at sth.; *(fig.)* have one's eye on sth.; **c)** *(ugs.: spähen)* peep

schien [ʃi:n] *1. u. 3. Pers. Sg. Prät. v.* **scheinen**

Schien·bein das shinbone; **sich am ~ stoßen** bang one's shin

Schiene ['ʃi:nə] **die; ~, ~n a)** rail; **b)** *(Gleit~)* runner; **c)** *(Med.: Stütze)* splint

schienen *tr. V.* **jmds. Arm ~:** put sb.'s arm in a splint

Schienen-: ~bus der railbus; **~fahrzeug das** track vehicle

schier [ʃi:ɐ̯] *Adv.* wellnigh; almost

Schieß-: ~bude die shooting-gallery; **~eisen das** *(ugs.)* shooting-iron *(sl.)*

schießen ['ʃi:sn̩] **1.** *unr. itr. V.* **a)** shoot; ‹*pistol, rifle*› shoot, fire; **auf jmdn./etw. ~:** shoot/fire at sb./sth.; **gut/schlecht ~** ‹*person*› be a good/bad shot; **b)** *(Fußball)* shoot; **c)** *mit sein (ugs.: schnellen)* shoot; **ein Gedanke schoß ihr durch den Kopf** *(fig.)* a thought flashed through her mind; **zum Schießen sein** *(ugs.)* be a scream *(sl.)*; **d)** *mit sein (fließen, heraus~)* gush; *(spritzen)* spurt; **ich spürte, wie mir das Blut in den Kopf schoß** I felt the blood rush to my head; **e)** *mit sein (schnell wachsen)* shoot up; **die Preise ~ in die Höhe** prices are shooting up *or* rocketing. **2.** *unr. tr. V.* **a)** shoot; fire ‹*bullet, missile, rocket*›; **jmdn. zum Krüppel ~:** shoot and maim sb.; **b)** *(Fußball)* score ‹*goal*›; **den Ball ins Netz ~:** put the ball in the net; **das 3:2 ~:** make it 3–2; **c)** *(ugs.: fotografieren)* **einige Aufnahmen ~:** take a few snaps

Schießerei die; ~, ~en a) shooting *no indef. art., no pl.;* **b)** *(Schußwechsel)* gun-battle; **die ~ am Ende des Films** the shoot-out at the end of the film

Schieß-: ~pulver das gunpowder; **er hat das ~pulver [auch] nicht erfunden** *(ugs.)* he's not exactly a genius; **~scharte die** crenel; **~scheibe die** target; **~sport der** shooting *no art.;* **~stand der** shooting range

Schiff [ʃɪf] **das; ~[e]s, ~e a)** ship; **mit dem ~:** by ship *or* sea; **b)** *(Archit.: Kirchen~) (Mittel~)* nave; *(Quer~)* transept; *(Seiten~)* aisle

Schiffahrt die; *o. Pl. (Schiffsverkehr)* shipping *no indef. art.; (Schiffahrtskunde)* navigation; **die ~ einstellen** suspend all shipping movements

Schiffahrts-: ~linie die shipping route; **~weg der** [navigable] waterway

schiff-, Schiff-: ~bau der; *o. Pl.* shipbuilding *no art.;* **~bruch der** *(veralt.)* shipwreck; **~bruch erleiden** ‹*ship*› be wrecked; ‹*person*› be shipwrecked; **[mit etw.] ~bruch erleiden** *(fig.)* fail [in sth.]; **~brüchig** *Adj.* shipwrecked; **ein Schiffbrüchiger** a shipwrecked man

Schiffchen das; ~s, ~ a) [little] boat; **b)** *(ugs.: Kopfbedeckung)* forage cap; **c)** *(Weberei, Handarbeit, Nähen)* shuttle

Schiffer der; ~s, ~: boatman; *(eines Lastkahns)* bargee; *(Kapitän)* skipper

Schiffer·klavier das accordion

Schiffs-: ~arzt der ship's doctor; **~brücke die** pontoon bridge

Schiff·schaukel die swing-boat

Schiffs-: ~fahrt die boat trip; *(länger)* cruise; **~junge der** ship's boy; **~modell das** model ship; **~reise die** voyage; *(Vergnügungsreise)* cruise; **~schraube die** ship's propeller *or* screw; **~verkehr der** shipping traffic

Schiit [ʃi'i:t] **der; ~en, ~en** Shiite

schiitisch *Adj.* Shiite

Schikane [ʃi'ka:nə] **die; ~, ~n a)** harassment *no indef. art.;* **das ist eine ~:** that amounts to *or* is harassment; **aus reiner ~:** purely in order to harass him/her *etc.;* **b) mit allen ~n** *(ugs.)* ‹*kitchen, house*› with all mod cons *(Brit. coll.)*; ‹*car, bicycle, stereo*› with all the extras

schikanieren *tr. V.* **jmdn. ~:** harass sb.; mess sb. about *(coll.)*; **Rekruten ~:** bully recruits

[1]**Schild** [ʃɪlt] **der; ~[e]s, ~e a)** shield; **etw./nichts im ~e führen** be up to something/not be up to anything; **etwas gegen jmdn./etw. im ~e führen** be plotting sth. against sb./sth.; **b)** *(Wappen~)* shield; escutcheon; **c)** *s.* **Schirm c**

[2]**Schild das; ~[e]s, ~er** *(Verkehrs~)* sign; *(Nummern~)* number-plate; *(Namens~)* nameplate; *(Plakat)* placard; *(an einer Mütze)* badge; *(auf Denkmälern, Gebäuden, Gräbern)* plaque; *(Etikett)* label

Schild·drüse die *(Med.)* thyroid [gland]

schildern ['ʃɪldɐn] *tr. V.* describe

Schilderung die; ~, ~en description; *(von Ereignissen)* account; description

Schild·kröte die tortoise; *(Seeschildkröte)* turtle

Schilf [ʃɪlf] **das; ~[e]s a)** reed; *(Röhricht)* reeds *pl.*

schillern ['ʃɪlɐn] *itr. V.* shimmer

Schilling ['ʃɪlɪŋ] **der; ~s, ~e** schilling

schilt [ʃɪlt] *3. Pers. Sg. Präsens v.* **schelten**

Schimmel ['ʃɪml̩] **der; ~s, ~ a)** *o. Pl.* mould; *(auf Leder, Papier)* mildew; **b)** *(Pferd)* white horse

schimmelig *Adj.* mouldy; mildewy ‹*paper, leather*›

schimmeln *itr. V.; auch mit sein* go mouldy; ‹*leather, paper*› get covered with mildew

Schimmel·pilz der mould

Schimmer ['ʃɪmɐ] **der; ~s** *(Schein)* gleam; *(von Perlmutt)* lustre; shimmer; *(von Seide)* shimmer; sheen; *(von Haar)* sheen; **keinen [blassen]** *od.* **nicht den leisesten ~ [von etw.] haben** *(ugs.)* not have the faintest *or* foggiest idea [about sth.] *(coll.)*

schimmern *itr. V.* **a)** gleam; ‹*water, sea*› glisten, shimmer; ‹*metal*› glint, gleam; ‹*mother-of-pearl, silk*› shimmer; **der Stoff/die Seide schimmert rötlich** the material has a reddish tinge/the silk has a reddish sheen; **b)** *(durch~)* show **(durch** through)

schimmlig *s.* **schimmelig**

Schimpanse [ʃɪm'panzə] **der; ~n, ~n** chimpanzee

schimpfen 1. *itr. V.* **a)** carry on *(coll.)* **(auf, über +** *Akk.* about); *(meckern)* grumble, moan **(auf, über +** *Akk.* at); **b) mit jmdm. ~:** tell sb. off; scold sb. **2.** *tr. V.* **jmdn. dumm/faul ~:** call sb. stupid/lazy

Schimpf·wort das *(Beleidigung)* insult; *(derbes Wort)* swear-word

Schindel ['ʃɪndl̩] **die; ~, ~n** shingle

schinden ['ʃɪndn̩] **1.** *unr. tr. V.* **a)** maltreat; ill-treat; *(ausbeuten)* slave-drive; **jmdn./ein Tier zu Tode ~:** work sb./an animal to death; **b)** *(ugs.: herausschlagen)* **[bei jmdm.] Eindruck ~:** make an impression [on sb.]; **Zeit ~:** play for time. **2.** *unr. refl. V. (ugs.: sich abplagen)* slave away

Schinderei die; ~, ~en a) ill-treatment *no pl.; (Ausbeutung)* slave-driving *no pl.;* **b)** *(Strapaze, Qual)* struggle; *(Arbeit)* toil

Schind·luder das: mit etw. ~ treiben *(ugs.) (ausbeuten)* take advantage of *or* abuse sth.; *(vergeuden)* squander sth.

Schinken ['ʃɪŋkn̩] **der; ~s, ~ a)** ham; **b)** *(ugs.) (Buch)* great tome; *(Gemälde)* enormous painting; *(Film, Theaterstück)* epic

Schinken·speck der bacon

Schippe ['ʃɪpə] **die; ~, ~n a)** *(nordd., md.: Schaufel)* shovel; **~ und Handfeger** dustpan and brush; **jmdn. auf die ~ nehmen** *(fam.)* kid sb. *(sl.);* pull sb.'s leg; **b)** *(Kartenspiel) s.* **[2]Pik**

schippen *tr. V. (nordd., md.)* shovel; *(graben)* dig

schippern *(ugs.)* **1.** *itr. V.; mit sein* cruise. **2.** *tr. V.* ship ‹*goods, materials*›; skipper ‹*ship*›

Schirm [ʃɪrm] **der; ~[e]s, ~e a)** umbrella; brolly *(Brit. coll.); (Sonnen~)* sunshade; parasol; **b)** *(Lampen~)* shade; **c)** *(Mützen~)* peak

Schirm-: ~herr der patron; **~herrin die** patroness; **~herrschaft die** patronage; **~mütze die** peaked cap; **~ständer der** umbrella stand

schiß [ʃɪs] *1. u. 3. Pers. Sg. Prät. v.* **scheißen**

Schiß der; Schisses *(salopp: Angst)* **[vor etw.] ~ haben** be shit-scared [of sth.] *(coarse);* **~ kriegen** get the shits *(coarse)*

schlabberig *Adj. (ugs.)* baggy ‹*clothes*›; loose, limp ‹*material*›

schlabbern ['ʃlabɐn] **1.** *tr. V. (ugs.) (schlürfen)* ‹*person*› slurp; ‹*animal*› lap up. **2.** *itr. V.* **a)** *(abwertend)* slobber; **b)** *(schlenkern)* ‹*dress*› flap; ‹*trousers*› be baggy

Schlacht [ʃlaxt] **die; ~, ~en** battle; **die ~ bei** *od.* **von/um X** the battle of/for X; **in die ~ ziehen** go into battle; **sich eine ~ liefern** do battle; **sich eine erbitterte ~ liefern** *(fig.)* fight fiercely

schlachten *tr. (auch itr.) V.* slaughter; kill ‹*rabbit, chicken, etc.*›

Schlachtenbummler [-bʊmlɐ] **der** *(Sportjargon)* away supporter

Schlachter der; ~s, ~, Schlächter der; ~s, ~ *(nordd.)* butcher

Schlachterei die; ~, ~en, Schlächterei die; ~, ~en *(nordd.: Fleischerei)* butcher's [shop]

Schlacht-: ~feld das battlefield; **~haus das** slaughterhouse; **~hof der** slaughterhouse; abattoir; **~plan der** *(fig.)* plan of action; **~platte die** *dish with assorted cooked meats, sausages, and sauerkraut;* **~schiff das** *(Milit.)* battleship; **~vieh das** animals *pl.* kept for meat; *(kurz vor der Schlachtung)* animals *pl.* for slaughter

Schlacke ['ʃlakə] **die; ~, ~n a)** cinders *pl.; (größere Stücke)* clinker; **b)** *(Hochofen~)* slag

schlackern ['ʃlakɐn] *itr. V. (nordd., westmd.)* **a)** ‹*dress*› flap; ‹*bag*› dangle; ‹*trousers*› be baggy; **b)** *(wackeln, zittern)* shake; tremble; **mit den Armen ~:** flap one's arms about

Schlaf [ʃla:f] **der; ~[e]s** sleep; **einen leichten/festen/gesunden ~ haben** be a light/heavy/good sleeper; **jmdn. um den** *od.* **seinen ~ bringen** ‹*worry etc.*› give sb. sleepless nights/a sleepless night; ‹*noise*› stop sb. from sleeping; **jmdn. in den ~ singen/wiegen** sing/rock sb. to sleep; **das kann** *od.* **mache ich im ~** *(fig.)* I can do that with my eyes closed *or* shut; **halb im ~:** half asleep

Schlaf·anzug der pyjamas *pl.;* **ein ~:** a pair of pyjamas

Schläfchen ['ʃlɛ:fçən] **das; ~s, ~:** nap; snooze *(coll.);* **ein ~ halten** have a nap *or (coll.)* snooze

Schlaf·couch die bed-settee; sofa-bed

Schläfe ['ʃlɛ:fə] **die; ~, ~n** temple; **er hat graue ~n** his hair has gone grey at the temples

schlafen *unr. itr. V.* **a)** *(auch fig.)* sleep; **tief** *od.* **fest ~** *(zur Zeit)* be sound asleep; *(gewöhnlich)* sleep soundly; be a sound sleeper; **lange ~:** sleep for a long time; *(am Morgen)* sleep in; **~ wie ein Murmeltier** *(ugs.)* sleep like a log *or* top; **~ gehen** go to bed; **im Hotel/bei Bekannten ~:** stay in a hotel/with friends; **darüber muß ich noch ~:** I'd like to sleep on it; **bei jmdm. ~:** sleep at sb.'s house/in sb.'s room *etc.;* **mit jmdm. ~** *(verhüll.)* sleep with sb. *(euphem.);* **b)** *(ugs.: nicht aufpassen)* be asleep

Schlafens·zeit die bedtime

Schläfer ['ʃlɛ:fɐ] **der; ~s, ~, Schläferin die; ~, ~nen** sleeper

schlaff [ʃlaf] **1.** *Adj.* **a)** *(nicht straff, nicht fest)* slack ‹*cable, rope, sail*›; flaccid, limp ‹*penis*›; loose, slack ‹*skin*›; sagging ‹*breasts*›; flabby ‹*stomach, muscles*›; **b)** *(schlapp, matt)* limp ‹*body, hand, handshake*›; shaky ‹*knees*›; feeble ‹*blow*›; **c)** *(abwertend: träge)* lethargic. **2.** *adv.* **a)** *(locker, nicht straff)* slackly; **das Segel hing ~:** the sail hung limply; **b)** *(schlapp, matt)* limply

schlaf-, Schlaf-: ~gast der overnight guest; **~gelegenheit die** place to sleep; **~lied das** lullaby; **~los** *Adj.* sleepless ‹*night*›; **~losigkeit die; ~:** sleeplessness; insomnia; **~mittel das** sleep-inducing drug; soporific [drug]; **~mütze die** *(ugs.)* sleepyhead; *(jmd., der unaufmerksam ist)* day-dreamer

schläfrig ['ʃlɛ:frɪç] **1.** *Adj.* sleepy; **~ sein/werden** ‹*person*› be/become sleepy *or* drowsy. **2.** *adv.* sleepily

Schlaf-: ~saal der dormitory; **~sack der** sleeping-bag

schläfst [ʃlɛ:fst] *2. Pers. Sg. Präsens v.* **schlafen**

schläft [ʃlɛ:ft] *3. Pers. Sg. Präsens v.* **schlafen**

schlaf-, Schlaf-: ~tablette die

sleeping-pill *or* -tablet; **~wagen der** sleeping-car; sleeper; **~wandeln** *itr. V.; auch mit sein* sleepwalk; **~wandler der; ~s, ~, ~wandlerin die; ~, ~nen** sleepwalker; **~zimmer das** bedroom; *(Einrichtung)* bedroom suite

Schlag [ʃla:k] **der; ~[e]s, Schläge** ['ʃlɛ:gə] **a)** blow; *(Faust~)* punch; blow; *(Klaps)* slap; *(leichter)* pat; *(als Strafe für ein Kind)* smack; *(Peitschenhieb)* lash; *(Tennis~, Golf~)* stroke; shot; **Schläge kriegen** *(ugs.)* get *or* be given a thrashing *or* beating; **alles ging ~ auf ~**: everything went quickly; **keinen ~ tun** *(ugs.)* not do a stroke [of work]; **jmdm. einen ~ versetzen** deal sb. a blow; *(fig.)* be a blow to sb.; **auf einen ~** *(ugs.)* at one go; all at once; **b)** *(Auf~, Aufprall)* bang; *(dumpf)* thud; *(Klopfen)* knock; **c)** *o. Pl. (des Herzens, Pulses, der Wellen)* beating; *(eines Pendels)* swinging; **d)** *(einzelne rhythmische Bewegung) (Herz~, Puls~, Takt~)* beat; *(eines Pendels)* swing; *(Ruder~, Kolben~)* stroke; **e)** *o. Pl. (Töne) (einer Uhr)* striking; *(einer Glokke)* ringing; *(einer Trommel)* beating; *(eines Gongs)* clanging; **f)** *(einzelner Ton) (Stunden~)* stroke; *(Glocken~)* ring; *(Trommel~)* beat; *(Gong~)* clang; **~** *od. (österr., schweiz.)* **schlag acht Uhr** on the dot *or* stroke of eight; **g)** *o. Pl. (Vogelgesang)* song; **h)** *(Blitz~)* flash [of lightning]; **i)** *(Stromstoß)* shock; **j)** *(ugs.: ~anfall)* stroke; **jmdn. trifft** *od.* **rührt der ~** *(ugs.)* sb. is flabbergasted; **wie vom ~ getroffen** *od.* **gerührt** *(ugs.)* as if thunderstruck; **k)** *(Schicksals~)* blow; **l)** *(Tauben~)* cote; **m)** *(ugs.: Portion)* helping; **n)** *o. Pl. (österr.: ~sahne)* whipped cream

schlag-, Schlag-: ~ader die artery; **~anfall der** stroke; **einen ~anfall bekommen [haben]** have [had] a stroke; **~artig 1.** *Adj.; nicht präd.* very sudden; *(innerhalb kürzester Zeit geschehend)* instantaneous; **2.** *adv.* quite suddenly; *(innerhalb kürzester Zeit)* instantly; **~baum der** barrier; **~bohrer der** percussion drill; hammer drill

schlagen 1. *unr. tr. V.* **a)** hit; beat; strike; *(mit der Faust)* punch; hit; *(mit der flachen Hand)* slap; *(mit der Peitsche)* lash; **ein Kind ~**: smack a child; *(aufs Hinterteil)* spank a child; **jmdn. bewußtlos/zu Boden ~**: beat sb. senseless/to the ground; *(mit einem Schlag)* knock sb. senseless/to the ground; **die Hände vors Gesicht ~**: cover one's face with one's hands; **ein Loch ins Eis ~**: break *or* smash a hole in the ice; *s. auch* **grün a; b)** *(mit Richtungsangabe)* hit ‹*ball*›; *(mit dem Fuß)* kick; **einen Nagel in etw.** *(Akk.)* **~**: knock a nail into sth.; **etw. durch ein Sieb ~**: press sth. through a sieve; **c)** *(rühren)* beat ‹*mixture*›; whip ‹*cream*›; *(mit einem Schneebesen)* whisk; **die Sahne steif ~**: beat the cream till stiff; **d)** *(läuten)* ‹*clock*› strike; ‹*bell*› ring; **die Uhr schlägt acht** the clock strikes eight; **eine geschlagene Stunde** *(ugs.)* a whole hour; *s. auch* **dreizehn; Stunde a; e)** *(legen)* throw; **die Decke zur Seite ~**: throw aside the blanket; **f)** *(einwickeln)* wrap (**in** + *Akk.* in); **g)** *(besiegen, übertreffen)* beat; **jmdn. in etw.** *(Dat.)* **~**: beat sb. at sth.; **eine Mannschaft [mit] 2:0 ~**: beat a team [by] 2–0; **h)** *auch itr. (bes. Schach)* take ‹*chessman*›; **i)** *(fällen)* fell ‹*tree*›; **j)** *(spielen)* beat ‹*drum*›; *(geh.)* play ‹*lute, zither, harp*›; **den Takt/Rhythmus ~**: beat time; **k) etw. in etw./auf etw.** *(Akk.)* **~**: add sth. to sth. **2.** *unr. itr. V.* **a)** *(hauen)* **er schlug mit der Faust auf den Tisch** he beat the table with his fist; **jmdm. auf die Hand/ins Gesicht ~**: slap sb.'s hand/hit sb. in the face; **um sich ~**: lash *or* hit out; **b) mit den Flügeln ~** ‹*bird*› beat or flap its wings; **c)** *mit sein (prallen)* bang; **mit dem Kopf auf etw.** *(Akk.)***/gegen etw. ~**: bang one's head on/against sth.; **auf den Boden ~**: land with a thud on the floor; **d)** *mit sein* **jmdm. auf den Magen ~**: affect sb.'s stomach; **e)** *(pulsieren)* ‹*heart, pulse*› beat; *(heftig)* ‹*heart*› pound; ‹*pulse*› throb; **f)** *(läuten)* ‹*clock*› strike; ‹*bell*› ring; ‹*funeral bell*› toll; **g)** *auch mit sein (auftreffen)* **gegen/an etw.** *(Akk.)* **~** ‹*rain, waves*› beat against sth.; **h)** *meist mit sein (einschlagen)* **in etw.** *(Akk.)* **~** ‹*lightning, bullet, etc.*› strike *or* hit sth.; **i)** *mit sein* **nach dem Onkel** *usw.* **~**: take after one's uncle *etc.*. **3.** *unr. refl. V.* **a)** *(sich prügeln)* fight; **sich mit jmdm. ~**: fight with sb.; **sich um etw. ~** *(auch fig.)* fight over sth.; **b)** *(ugs.: sich behaupten)* hold one's own; **sich tapfer ~**: hold one's own well; put up a good showing; **c)** *(sich schädlich auswirken)* **sich auf die Leber ~**: affect the liver

schlagend 1. *Adj.* cogent, compelling ‹*argument, reason*›; cogent ‹*comparison*›; conclusive ‹*proof, evidence*›; *s. auch* **Wetter c. 2.** *adv.* ‹*prove, disprove*› conclusively; ‹*formulate*› cogently

Schlager der; ~s, ~ a) *(Lied)* pop song; *(Hit)* hit; **b)** *(Erfolg) (Buch)* best seller; *(Ware)* best-selling line; *(Film, Stück)* hit

Schläger ['ʃlɛ:gɐ] **der; ~s, ~ a)** *(abwertend: Raufbold)* tough; thug; **b)** *(Tennis~, Federball~, Squash~)* racket; *(Tischtennis~, Kricket~)* bat; *([Eis]hockey~, Polo~)* stick; *(Golf~)* club

Schlägerei die; ~, ~en brawl; fight

Schlager-: ~musik die; *o. Pl.* popular music; pop music; **~sänger der** pop singer

schlag-, Schlag-: ~fertig 1. *Adj.* quick-witted ‹*reply*›; ‹*person*› who is quick at repartee; **er ist ~fertig** he is quick at repartee; **2.** *adv.* **~fertig antworten/parieren** give a quick-witted reply/riposte; **~fertigkeit die;** *o. Pl.* quickness at repartee; **~instrument das** percussion instrument; **~kräftig 1.** *Adj.* **a)** *(Milit.)* powerful; **b)** *(überzeugend)* compelling ‹*argument*›; convincing ‹*example*›; **c)** *(effektiv)* strong, effective ‹*support, back-up, team*›; **2.** *adv. (überzeugend)* ‹*argue*› compellingly; **~loch das** pothole; **~obers** [~o:bɐs] **das; ~** *(österr.)*, **~rahm der** *(bes. südd., österr., schweiz.)* *s.* **~sahne; ~sahne die** whipping cream; *(geschlagen)* whipped cream; **~seite die** list; **[starke** *od.* **schwere] ~seite haben/bekommen** be listing [heavily] *or* have a [heavy] list/develop a [heavy] list; **~seite haben** *(ugs. scherzh.)* be rolling drunk; **~stock der** cudgel; *(für Polizei)* truncheon; **~wort das;** *Pl. meist* **~worte a)** *(Parole)* slogan; catch-phrase; **b)** *(abwertend: Redensart)* cliché; **~zeile die** *(Zeitungsw.)* headline; **~zeug das** drums *pl.*; **~zeuger der; ~s, ~, ~zeugerin die** drummer

schlaksig ['ʃla:ksɪç] *(ugs.) Adj.* gangling; lanky

Schlamassel [ʃla'masl̩] **der** *od.* **das; ~s** *(ugs.)* mess; **da haben wir den ~!** a right *or* fine mess we're in now!

Schlamm [ʃlam] **der; ~[e]s, ~e** *od.* **Schlämme** ['ʃlɛmə] **a)** mud; **b)** *(Schlick)* sludge; silt

schlammig *Adj.* **a)** muddy; **b)** *(schlickig)* sludgy; muddy

Schlampe ['ʃlampə] **die; ~, ~n** *(ugs. abwertend)* slut

schlampen *itr. V. (ugs. abwertend)* be sloppy; **bei etw. ~:** do sth. sloppily

Schlamperei die; ~, ~en *(ugs. abwertend)* sloppiness

schlampig *(ugs. abwertend)* **1.** *Adj.* **a)** *(liederlich)* slovenly; **b)** *(nachlässig)* sloppy, slipshod ‹*work*›. **2.** *adv.* **a)** *(liederlich)* in a slovenly way; **b)** *(nachlässig)* sloppily; in a sloppy *or* slipshod way

schlang [ʃlaŋ] *1. u. 3. Pers. Sg. Prät. v.* **schlingen**

Schlange die; ~, ~n a) snake; **b)** *(Menschen~)* queue; line *(Amer.)*; **~ stehen** queue; stand in line *(Amer.)*; **c)** *(Auto~)* tailback *(Brit.)*; backup *(Amer.)*

schlänge ['ʃlɛŋə] *1. u. 3. Pers. Sg. Konjunktiv II v.* **schlingen**

schlängeln ['ʃlɛŋl̩n] *refl. V.* **a)** ‹*snake*› wind [its way]; ‹*road*› wind, snake [its way]; **eine geschlängelte Linie** a wavy line; **b)** *(sich irgendwo hindurch bewegen)* wind one's way

Schlangen·linie die wavy line; **~linien fahren** ‹*cyclist*› weave along

schlank [ʃlaŋk] *Adj.* slim ‹*person*›; slim, slender ‹*build, figure*›; slender ‹*column, tree, limbs*›; **~ werden** get slimmer; slim down; **dieser Rock macht [dich] ~:** this skirt makes you look slim; *s. auch* **Linie a**

Schlankheit die; ~ *s.* **schlank:** slimness; slenderness

Schlankheits·kur die slimming diet; **eine ~ machen/beginnen** be/go on a slimming diet

schlank·weg *Adv. (ugs.)*‹*refuse*› flatly, point-blank; ‹*accept*› straight away

schlapp [ʃlap] *Adj.* **a)** worn out; tired out; *(wegen Schwüle)* listless; *(wegen Krankheit)* run-down; listless; **b)** *(ugs.: ohne Schwung)* wet *(sl.)*; feeble; **c)** slack ‹*rope, cable*›; loose, slack ‹*skin*›; flabby ‹*stomach, muscles*›

Schlappe die; ~, ~n setback; **eine [schwere] ~ erleiden** suffer a [severe] setback

schlappen *itr. V.; mit sein (schlurfend gehen)* shuffle

schlapp|machen *itr. V. (ugs.)* flag; *(zusammenbrechen)* flake out *(coll.)*; *(aufgeben)* give up

Schlapp·schwanz der *(salopp abwertend)* weed; wet *(sl.)*

Schlaraffen·land [ʃla'rafn̩-] **das;** *o. Pl.* Cockaigne

schlau [ʃlau̯] **1.** *Adj.* **a)** shrewd; astute; *(gerissen)* wily; crafty; cunning; **b)** *(ugs.: gescheit)* clever; bright; smart; **aus etw. nicht ~ werden** *(ugs.)* not be able to make head or tail of sth.; **aus jmdm. nicht ~ werden** *(ugs.)* not be able to make sb. out; *s. auch* **Buch a. 2.** *adv.* shrewdly; astutely; *(gerissen)* craftily; cunningly

Schlauch [ʃlau̯x] **der; ~[e]s, Schläuche** ['ʃlɔy̯çə] **a)** hose; **das war ein [ganz schöner] ~!** *(fig. ugs.)* it was a [real] slog; **b)** *(Fahrrad~, Auto~)* tube

Schlauch·boot das rubber dinghy; inflatable [dinghy]

schlauchen *(ugs.) tr., auch itr. V.* **jmdn. ~:** take it out of sb.

schlauch·los *Adj.* tubeless ‹*tyre*›

Schläue ['ʃlɔy̯ə] **die; ~:** shrewdness; astuteness; *(Gerissenheit)* wiliness; craftiness; cunning

Schlaufe ['ʃlau̯fə] **die; ~, ~n** loop; *(zum Festhalten)* strap

schlecht [ʃlɛçt] **1.** *Adj.* **a)** bad; poor, bad ‹*food, quality, style, harvest, health, circulation*›; poor ‹*salary, eater, appetite*›; poor-quality ‹*goods*›; bad, weak ‹*eyes*›; **in Mathematik ~ sein** be bad at mathematics; **das wäre nicht ~** that wouldn't be a bad idea; **mit jmdm.** *od.* **um jmdn./mit etw. steht es ~:** sb./sth. is in a bad way; **b)** *(böse)* bad; wicked; **das Schlechte im Menschen** the evil in man; **sie ist nicht die Schlechteste** she's not too bad; **c)** *nicht attr. (ungenießbar)* off; **das Fleisch ist ~ geworden** the meat has gone off. **2.** *adv.* **a)** badly; **sie spricht ~ Englisch** she speaks poor English; **er sieht/hört ~:** his sight is poor/he has poor hearing; **die Geschäfte gehen im Moment ~:** business is bad at the moment; **über jmdn.** *od.* **von jmdm. ~ sprechen** speak ill of sb.; **b)** *(schwer)* **heute geht es ~:** today is difficult; **das kann ich ~ sagen** I can't really say; **das wird sich ~ vermeiden lassen** it can hardly be avoided; **c)** *in* **~ und recht, mehr ~ als recht** after a fashion; **sie hat sich ~ und recht durchs Leben geschlagen** she got by in life as best she could

schlecht-: ~bezahlt *Adj. (präd. getrennt geschrieben)* badly *or* poorly paid; **~|gehen** *unr. itr. V.; unpers.; mit sein* **es geht ihr ~:** she is doing badly; things are going badly for her; *(gesundheitlich)* she is ill *or* unwell *or* poorly; **wenn sie das herausfindet, geht's dir ~!** *(ugs.)* if she finds out, you'll be [in] for it; **~gelaunt** [~gəlau̯nt] *Adj. (präd. getrennt geschrieben)* ill-tempered; bad-tempered; **~hin** *Adv.* **a)** *einem Subst. nachgestellt* **er war der Romantiker ~hin** he was the quintessential Romantic *or* the epitome of the Romantic; **b)** *(ganz einfach)* quite simply

Schlechtigkeit die; ~ badness; wickedness

schlecht|machen *tr. V.* **jmdn. ~:** run sb. down; disparage sb.

schlecken ['ʃlɛkn̩] *(bes. südd., österr.)* **1.** *tr. V.* lap up. **2.** *itr. V.* **an etw.** *(Dat.)* **~:** lick sth.

Schlegel ['ʃle:gl̩] **der; ~s, ~ a)** *(Werkzeug)* mallet; **b)** *(für Schlaginstrumente)* stick; **c)** *(südd., österr.) s.* **Keule c**

Schlehe ['ʃle:ə] **die; ~, ~n** sloe

schleichen ['ʃlai̯çn̩] **1.** *unr. itr. V.; mit sein* creep; *(heimlich)* creep; steal; sneak; ‹*cat*› slink, creep; *(langsam fahren)* crawl along; **die Zeit schlich** time crept by. **2.** *unr. refl. V.* creep; steal; sneak; ‹*cat*› slink, creep; **schleich dich!** *(ugs., bes. österr.)* get lost! *(sl.)*; buzz off! *(sl.)*

schleichend *Adj.* insidious ‹*disease*›; slow[-acting], insidious ‹*poison*›; creeping ‹*inflation*›; gradual ‹*crisis*›

Schleich·weg der secret path

Schleie ['ʃlai̯ə] **die; ~, ~n** *(Zool.)* tench

Schleier ['ʃlai̯ɐ] **der; ~s, ~ a)** veil; **b)** *(von Dunst)* veil of mist

schleier·haft *Adj.* **jmdm. [völlig** od. **vollkommen] ~ sein/bleiben** be/remain a [total *or* complete] mystery to sb.

Schleife ['ʃlai̯fə] **die; ~, ~n a)** bow; *(Fliege)* bow-tie; **b)** *(starke Biegung)* loop; *(eines Flusses)* loop; horseshoe bend

[1]**schleifen** *unr. tr. V.* **a)** *(schärfen)* sharpen; grind, sharpen ‹*axe*›; **b)** *(glätten)* grind; cut ‹*diamond, glass*›; *(mit Sand-/Schmirgelpapier)* sand; **c)** *(bes. Soldatenspr.: drillen)* **jmdn. ~:** drill sb. hard

[2]**schleifen 1.** *tr. V.* **a)** *(auch fig.)* drag; **b)** *(niederreißen)* **etw. ~:** raze sth. [to the ground]. **2.** *itr. V.; auch mit sein* drag; **die Kette schleift am Schutzblech** the chain scrapes the guard; **die Kupplung ~ lassen** *(Kfz-W.)* slip the clutch; **etw. ~ lassen** *(fig.)* let sth. slide; *s. auch* **Zügel**

Schleif·stein der grindstone

Schleim [ʃlai̯m] **der; ~[e]s, ~e a)** mucus; *(im Hals)* phlegm; *(von Schnecken, Aalen)* slime; **b)** *(sämiger Brei)* gruel

Schleim·haut die mucous membrane

schleimig *Adj. (auch fig.)* slimy; *(Physiol., Zool.)* mucous
schlemmen ['ʃlɛmən] **1.** *itr. V. (prassen)* have a feast. **2.** *tr. V. (verzehren)* feast on
Schlemmer der; ~s, ~: gourmet
Schlemmer·lokal das gourmet restaurant
schlendern ['ʃlɛndɐn] *itr. V.; mit sein* stroll
Schlenker ['ʃlɛŋkɐ] **der; ~s, ~** *(ugs.)* swerve; **einen ~ machen** swerve
schlenkern 1. *itr. V.* swing; dangle; **mit den Armen/mit den Beinen ~:** swing *or* dangle one's arms/legs. **2.** *tr. V.* swing, dangle ⟨*arms, legs*⟩
schlenzen ['ʃlɛntsn̩] *tr. V. (Sport, bes. [Eis]hockey, Fußball)* flick
Schlepp [ʃlɛp] **der** *in* **ein Fahrzeug in ~ nehmen** take a vehicle in tow
Schlepp·bügel der *(Skisport)* T-bar
Schleppe die; ~, ~n train
schleppen 1. *tr. V.* **a)** *(ziehen)* tow ⟨*vehicle, ship*⟩; **b)** *(tragen)* carry; lug; **c)** *(ugs.: mitnehmen)* drag. **2.** *refl. V.* drag *or* haul oneself
schleppend 1. *Adj.* **a)** *(schwerfällig)* shuffling, dragging ⟨*walk, steps*⟩; **b)** *(gedehnt)* dragging ⟨*speech*⟩; slow ⟨*song, melody*⟩; **c)** *(nicht zügig)* slow ⟨*service*⟩. **2.** *adv.* **a)** *(schwerfällig)* **~ gehen** shuffle along; **b)** *(gedehnt)* ⟨*speak*⟩ in a dragging voice; ⟨*sing, play*⟩ slowly; **c)** *(nicht zügig)* **die Arbeiten gehen nur ~ voran** the work is progressing slowly
Schlepper der; ~s, ~ a) *(Schiff)* tug; **b)** *(Traktor)* tractor; **c)** *(ugs.: jmd., der Kunden zuführt)* tout
Schlepp-: ~lift der T-bar [lift]; **~tau das** tow-line; row-rope; *(aus Draht)* tow-line; tow-cable; **etw. ins ~tau nehmen** take sth. in tow; **in jmds. ~tau** *(fig.)* in sb.'s wake
Schlesien ['ʃleːzi̯ən] **(das); ~s** Silesia
Schlesier ['ʃleːzi̯ɐ] **der; ~s, ~, Schlesierin die; ~, ~nen** Silesian
Schleuder ['ʃlɔydɐ] **die; ~, ~n** sling; *(mit Gummiband)* catapult *(Brit.)*; slingshot *(Amer.)*
schleudern 1. *tr. V.* **a)** *(werfen)* hurl; fling; **der Wagen wurde aus der Kurve geschleudert** the car was sent skidding off the bend; **b)** *(rotieren lassen)* centrifuge; spin ⟨*washing*⟩. **2.** *itr. V. mit sein (rutschen)* skid; *(fig. ugs.)* run into trouble
Schleuder-: ~preis der *(ugs.)* knock-down price; **~sitz der** ejector seat
schleunigst *Adv.* **a)** *(auf der Stelle)* at once; immediately; straight away; **b)** *(eilends)* hastily; with all haste
Schleuse ['ʃlɔyzə] **die; ~, ~n a)** sluice[-gate]; **b)** *(Schiffs~)* lock
schleusen *tr. V.* **a) ein Schiff ~:** pass a ship through a/the lock; **b)** *(geleiten)* shepherd; **c)** *(schmuggeln)* smuggle ⟨*secrets*⟩; infiltrate ⟨*spy, agent, etc.*⟩ **(in +** *Akk.* into)
schlich [ʃlɪç] *1. u. 3. Pers. Sg. Prät. v.* **schleichen**
Schlich der; ~[e]s, ~e trick; **jmdm. auf die ~e** *od.* **hinter jmds. ~e kommen** get on to sb.
schlicht [ʃlɪçt] **1.** *Adj.* **a)** simple; plain, simple ⟨*pattern, furniture*⟩; **b)** *(unkompliziert)* simple, unsophisticated ⟨*person, view, etc.*⟩; **ein ~es Ja oder Nein** a simple yes or no. **2.** *adv.* simply; simply, plainly ⟨*dressed, furnished*⟩; **~ und einfach** *(ugs.)* quite *or* just simply
schlichten 1. *tr. V.* settle ⟨*argument, difference of opinion*⟩; settle ⟨*industrial dispute etc.*⟩ by mediation. **2.** *itr. V.* mediate **(in +** *Dat.* in, **zwischen** between)
Schlichtheit die; ~ *s.* **schlicht a, b:** simplicity; plainness; unsophisticatedness
Schlick [ʃlɪk] **der; ~[e]s, ~e** silt
schlief [ʃliːf] *1. u. 3. Pers. Sg. Prät. v.* **schlafen**
Schließe ['ʃliːsə] **die; ~, ~n** clasp; *(Schnalle)* buckle
schließen 1. *unr. tr. V.* **a)** *(zumachen)* close; shut; put the top on ⟨*bottle*⟩; turn off ⟨*tap*⟩; fasten ⟨*belt, bracelet*⟩; do up ⟨*button, zip*⟩; close ⟨*street, route, electrical circuit*⟩; close off ⟨*pipe*⟩; *(fig.)* close ⟨*border*⟩; fill, close ⟨*gap*⟩; **b)** *(unzugänglich machen)* close, shut ⟨*shop, factory*⟩; *(außer Betrieb setzen)* close [down] ⟨*shop, school*⟩; **c)** *(ein~)* **etw./jmdn./sich in etw.** *(Akk.)* **~:** lock sth./sb./oneself in sth.; **d)** *(beenden)* close ⟨*meeting, proceedings, debate*⟩; end, conclude ⟨*letter, speech, lecture*⟩; **e)** *(befestigen)* **etw. an etw.** *(Akk.)* **~:** connect sth. to sth.; *(mit Schloß)* lock sth. to sth.; **f)** *(eingehen, vereinbaren)* conclude ⟨*treaty, pact, cease-fire, agreement*⟩; reach ⟨*settlement, compromise*⟩; enter into ⟨*contract*⟩; **wann wurde Ihre Ehe geschlossen?** when did you get married?; **Freundschaft mit jmdm. ~:** make friends with sb.; **g)** *(umfassen)* **jmdn. in die Arme ~:** take sb. in one's arms; embrace sb.; **h)** *(folgern)* **etw. aus etw. ~:** infer *or* conclude sth. from sth. **2.** *unr. itr. V.* **a)** close; shut; **der Schlüssel/das Schloß schließt schlecht** the key won't turn properly/the lock doesn't work properly; **b)** ⟨*shop*⟩ close, shut; ⟨*stock exchange*⟩ close; *(den Betrieb einstellen)* close [down]; **c)** *(enden)* end; conclude; **d)** *(urteilen)* **[aus etw.] auf etw.** *(Akk.)* **~:** infer *or* conclude sth. [from sth.]; **die Symptome lassen auf Hepatitis ~:** the symptoms indicate hepatitis; **von sich auf andere ~:** judge others by one's own standards. **3.** *unr. refl. V.* ⟨*door, window*⟩ close, shut; ⟨*wound, circle*⟩ close; ⟨*flower*⟩ close [up]
Schließ·fach das locker; *(bei der Post)* post-office box; PO box; *(bei der Bank)* safe-deposit box
schließlich *Adv.* **a)** finally; in the end; *(bei Erwünschtem auch)* at last; **~ und endlich** *(ugs.)* in the end; finally; **b)** *(immerhin, doch)* after all; **er ist ~ mein Freund** he is my friend, after all
Schließ·muskel der *(Anat.)* sphincter
Schließung die; ~, ~en a) *(der Geschäfte, Büros usw.)* closing; shutting; *(Stillegung, Einstellung)* closure; closing; *(fig.: einer Grenze)* closing; **b)** *(Beendigung)* **vor/nach ~ der Versammlung** before/after the meeting was closed; before/after the conclusion of the meeting; **c)** *s.* **schließen 1f:** conclusion; reaching
schliff [ʃlɪf] *1. u. 3. Pers. Sg. Prät. v.* **schleifen**
Schliff der; ~[e]s, ~e a) *o. Pl. (das Schleifen)* cutting; *(von Messern, Sensen usw.)* sharpening; **b)** *(Art, wie etw. geschliffen wird)* cut; *(von Messern, Scheren, Schneiden)* edge; **c)** *o. Pl. (Lebensart)* refinement; polish; **d)** *o. Pl. (Vollkommenheit)* **einem Text** *usw.* **den letzten ~ geben** put the finishing touches *pl.* to a text *etc.*
schlimm [ʃlɪm] **1.** *Adj.* **a)** grave, serious ⟨*error, mistake, accusation, offence*⟩; bad, serious ⟨*error, mistake*⟩; **das ist ~ für ihn** that's serious for him; **b)** *(übel)* bad; nasty, bad ⟨*experience*⟩; **[das ist alles] halb so ~:** it's not as bad as all that; **es ist nichts Schlimmes** it's nothing serious; **ist nicht ~!** [it] doesn't matter; **es gibt Schlimmeres** there are worse things; **c)** *(schlecht, böse)* wicked; *(ungezo-*

gen) naughty ‹*child*›; **d)** *(fam.: schmerzend)* bad; sore; bad, nasty ‹*wound*›. **2.** *adv.* ~ **d[a]ran sein** *(körperlich, geistig)* be in a bad way; *(in einer ~en Situation)* be in dire straits; **es hätte ~er ausgehen können** things could have turned out worse

schlimmsten·falls *Adv.* if the worst comes to the worst; ~ **kriegt man eine Verwarnung** at worst you'll get a caution

Schlinge ['ʃlɪŋə] **die; ~, ~n a)** *(Schlaufe)* loop; *(für den gebrochenen Arm o. ä.)* sling; *(zum Aufhängen)* noose; **b)** *(Fanggerät)* snare; **sich in der eigenen ~ fangen** *(fig.)* be hoist with one's own petard

Schlingel ['ʃlɪŋl̩] **der; ~s, ~:** rascal; rogue

schlingen 1. *unr. tr. V.* **a)** *(winden)* **etw. um etw. ~:** loop sth. round sth.; *(und zusammenbinden)* tie sth. round sth.; **die Arme um jmdn./etw. ~:** wrap one's arms round sb./sth.; **b)** *(binden)* tie ‹*knot*›; **etw. zu einem Knoten ~:** tie sth. up in a knot. **2.** *unr. refl. V. (sich winden)* **sich um etw. ~** ‹*snake*› wind *or* coil itself round sth.; ‹*plant*› wind *or* twine itself round sth. **3.** *unr. itr. V.* bolt one's food; wolf one's food [down]

schlingern ['ʃlɪŋɐn] *itr. V.; mit sein* ‹*ship, boat*› roll; ‹*train, vehicle*› lurch from side to side

Schling·pflanze die creeper

Schlips [ʃlɪps] **der; ~es, ~e** tie; **jmdm. auf den ~ treten** *(fig. ugs.)* tread on sb.'s toes

Schlitten ['ʃlɪtn̩] **der; ~s, ~ a)** sledge; sled; *(Pferde~)* sleigh; *(Rodel~)* toboggan; **~ fahren** go tobogganing; **die Kinder fuhren mit dem ~ den Hang hinunter** the children tobogganed down the slope; **mit jmdm. ~ fahren** *(fig. ugs.)* bawl sb. out *(coll.)*; **b)** *(salopp: Auto)* car; motor *(Brit.)*

Schlitten-: ~fahrt die sleigh ride; **~hund der** sled dog

schlittern ['ʃlɪtɐn] *itr. V.* **a)** *auch mit sein (rutschen)* slide; **b)** *mit sein (ins Rutschen kommen)* slip; slide; ‹*vehicle*› skid; ‹*wheel*› slip; **c)** *mit sein (fig.)* **in die Pleite ~:** slide into bankruptcy

Schlitt-: ~schuh der [ice-]skate; **~schuh laufen** *od.* **fahren** [ice-]skate; **~schuh·laufen das** [ice-]skating *no art.*; **~schuh·läufer der** [ice-]skater

Schlitz [ʃlɪts] **der; ~es, ~e a)** slit; *(Briefkasten~, Automaten~)* slot; **b)** *(Hosen~)* flies *pl.*; fly

Schlitz-: ~auge das; *meist Pl.* slit eye; **~ohr das** *(ugs.)* wily *or* crafty devil

schloß [ʃlɔs] *1. u. 3. Pers. Sg. Prät. v.* **schließen**

Schloß das; Schlosses, Schlösser ['ʃlœsɐ] **a)** *(Tür~, Gewehr~)* lock; **b)** *(Vorhänge~)* padlock; **hinter ~ und Riegel** *(ugs.)* behind bars; **c)** *(Verschluß)* clasp; **d)** *(Wohngebäude)* castle; *(Palast)* palace; *(Herrschaftshaus)* mansion

Schlosser der; ~s, ~: metalworker; *(Maschinen~)* fitter; *(für Schlösser)* locksmith; *(Auto~)* mechanic

Schlosserei die; ~, ~en a) *(Werkstatt)* metalworking shop; *(für Schlösser)* locksmith's workshop; **b)** *o. Pl.; s.* **~handwerk**

Schlosser-: ~handwerk das; *o. Pl.; s.* **Schlosser:** metalworking; fitter's trade; locksmithery; mechanic's trade; **~werkstatt die** *s.* **Schlosserei a**

Schloß-: ~park der castle *etc.* grounds *pl.*; **~ruine die** ruined castle *etc.*

Schlot [ʃlo:t] **der; ~[e]s, ~e** *od.* **Schlöte** ['ʃlø:tə] *(bes. md.: Schornstein)* chimney[-stack]; *(eines Schiffes)* funnel; **rauchen** *od.* **qualmen wie ein ~** *(ugs.)* smoke like a chimney

schlottern ['ʃlɔtɐn] *itr. V.* **a)** shake; tremble; **jmdm. ~ die Knie** sb.'s knees are shaking *or* trembling; **b)** ‹*clothes*› hang loose

Schlucht [ʃlʊxt] **die; ~, ~en** ravine; gorge

schluchzen ['ʃlʊxtsn̩] *itr. V.* sob; **in heftiges Schluchzen ausbrechen** burst into heavy sobbing

Schluchzer der; ~s, ~: sob

Schluck [ʃlʊk] **der; ~[e]s, ~e** *od.* **Schlücke** ['ʃlʏkə] swallow; mouthful; *(großer ~)* gulp; *(kleiner ~)* sip; **einen tüchtigen ~ [Bier] trinken** take a good *or* long swig [of beer] *(coll.)*; **hast du einen ~ zu trinken für uns?** have you got a drop of something for us to drink?

Schluck·auf der; ~s hiccups *pl.*; hiccoughs *pl.*

Schlückchen ['ʃʏkçən] **das; ~s, ~:** sip

schlucken 1. *tr. V.* **a)** *(auch fig. ugs.)* swallow; **etw. hastig ~:** gulp sth. down; **b)** *(ugs.: einatmen)* swallow ‹*dust*›; breathe in ‹*gas*›. **2.** *itr. V. (auch fig.)* swallow

Schlucker der; ~s, ~: *in* **armer ~** *(ugs.)* poor devil *or (Brit. coll.)* blighter

Schluck·impfung die oral vaccination

schluderig *s.* **schludrig**

schludern ['ʃlu:dɐn] *itr. V. (ugs. abwertend)* work sloppily

schludrig *(ugs. abwertend)* **1.** *Adj.* **a)** *(nachlässig)* slipshod ‹*work, examination*›; botched ‹*job*›; slapdash ‹*person, work*›; **b)** *(schlampig [aussehend])* scruffy. **2.** *adv.* **a)** *(nachlässig)* in a slipshod *or* slapdash way; **b)** *(schlampig [aussehend])* scruffily

schlug [ʃlu:k] *1. u. 3. Pers. Sg. Prät. v.* **schlagen**

Schlummer ['ʃlʊmɐ] **der; ~s** *(geh.)* slumber *(poet./rhet.)*; *(Nickerchen)* doze

schlummern *itr. V. (geh.)* slumber *(poet./rhet.)*; *(dösen)* doze

Schlund [ʃlʊnt] **der; ~[e]s, Schlünde** ['ʃlʏndə] [back of the] throat; pharynx *(Anat.)*; *(eines Tieres)* maw

schlüpfen ['ʃlʏpfn̩] *itr. V.; mit sein* slip; **in ein/aus einem Kleid** *usw.* **~:** slip into *or* slip on/slip out of *or* slip off a dress *etc.*; **[aus dem Ei] ~:** ‹*chick*› hatch out

Schlüpfer der; ~s, ~ *(für Damen)* knickers *pl. (Brit.)*; panties *pl.*; *(für Herren)* [under]pants *pl. or* trunks *pl.*; **ein ~:** a pair of knickers/underpants

Schlupf·loch ['ʃlʊpf-] **das a)** *(Schlupfwinkel)* hiding-place; **b)** *(Durchschlupf)* hole; *(Lücke im Gesetz usw.)* loophole

schlüpfrig ['ʃlʏpfrɪç] *Adj.* **a)** *(feucht u. glatt)* slippery; **b)** *(abwertend: anstößig)* lewd

Schlupf·winkel der hiding-place; *(von Banditen, Flüchtlingen usw.)* hide-out

schlurfen ['ʃlʊrfn̩] *itr. V.; mit sein* shuffle

schlürfen ['ʃlʏrfn̩] *tr. V.* *(geräuschvoll)* slurp [up] *(coll.)*; drink noisily; *(genußvoll)* savour; *(in kleinen Schlucken)* sip

Schluß [ʃlʊs] **der; Schlusses, Schlüsse** ['ʃlʏsə] **a)** *o. Pl. (Endzeitpunkt)* end; *(eines Vortrags o. ä.)* conclusion; *(Dienst~)* knocking-off time; **nach/gegen ~ der Aufführung** after/towards the end of the performance; **mit etw. ist ~:** sth. is at an end *or* over; *(ugs.: etw. ist ruiniert)* sth. has had it *(coll.)*; **mit dem Rauchen ist jetzt ~:** there's to be no more smoking; you must stop smoking; *(auf sich bezogen)* I've given up smoking; **~ jetzt!, ~ damit!** stop it!; that'll do!; **~ für heute!** that's it *or* that'll do for today; **am** *od.* **zum ~:** at the end; *(schließlich)* in the end; finally; **~ machen** *(ugs.)* stop; *(Feierabend machen)* knock off; *(seine Stellung aufgeben)* pack in one's job *(sl.)*; *(eine Freundschaft usw. lösen)* break it off; *(sich das Leben nehmen)* end it all *(coll.)*; **ich mache ~ für heute** I'm calling it a day; **mit etw. ~ machen** stop sth.; **mit jmdm. ~ machen** finish with sb.; break it off with sb.; **b)** *(letzter Abschnitt)* end; *(eines Zuges)* back; *(eines Buchs, Schauspiels usw.)* ending; **c)** *(Folgerung)* conclusion **(auf +** *Akk.* regarding); *(Logik)* deduction; **Schlüsse aus etw. ziehen** draw conclusions from sth.

Schluß-: ~abstimmung die *(Parl.)* final vote; **~akkord der** *(Musik)* final chord; **~akte die** *(Dipl.)* final communiqué

Schlüssel ['ʃlʏsl̩] **der; ~s, ~ a)** key; **der ~ zur Wohnungstür** the front door key; **b)** *(Schrauben~)* spanner; **c)** *(Lösungsweg, Lösungsheft)* key; *(Kode)* code; cipher; **d)** *(Musik)* clef

Schlüssel-: ~bart der bit [of a/the key]; **~anhänger der** key-fob; **~bein das** collar-bone; **~blume die** cowslip; *(Primel)* primula; **~bund der** *od.* **das** bunch of keys; **~kind das** *(ugs.)* latchkey child; **~loch das** keyhole; **~ring der** key-ring; **~roman der** *(Literaturw.)* roman à clef

schluß-: ~endlich *Adv. (bes. schweiz.)* finally; **~folgern** *tr. V.* conclude **(aus** from)

schlüssig ['ʃlʏsɪç] **1.** *Adj.* **a)** conclusive ‹*proof, evidence*›; convincing, logical ‹*argument, conclusion, statement*›; **b) sich** *(Dat.)* **[darüber] ~ werden** make up one's mind. **2.** *adv.* conclusively

Schluß-: ~kapitel das *(auch fig.)* final *or* closing chapter; **~läufer der** *(Leichtathletik)* last runner, anchor man *(in a relay team)*; **~licht das;** *Pl.* **~lichter a)** *(an Fahrzeugen)* tail- *or* rear-light; **b)** *(ugs.: letzter einer Kolonne)* **das ~licht machen/sein** bring up the rear; **c)** *(ugs.: Letzter, Schlechtester)* **das ~licht der Klasse sein** be bottom of the class; **~mann der;** *Pl.* **~männer** *(Ballspiele)* goalie *(coll.)*; **~pfiff der** *(Ballspiele)* final whistle; **~punkt der a)** *(Satzzeichen)* full stop; **b)** *(Abschluß)* conclusion; *(einer Feier)* finale; **~runde die** *(Sport: eines Rennens)* final *or* last lap; *(Boxen, Ringen, fig.: des Wahlkampfes usw.)* final *or* last round; **~strich der** [bottom] line; **einen ~strich ziehen/unter etw.** *(Akk.)* **ziehen** *(fig.)* make a clean break/draw a line under sth.; **~verkauf der** [end-of-season] sale[s *pl.*]

Schmach [ʃma:x] **die; ~** *(geh.)* ignominy; shame; *(Demütigung)* humiliation; **[mit] ~ und Schande** [in] deep disgrace

schmachten ['ʃmaxtn̩] *itr. V.* *(geh.)* **a)** *(leiden)* languish; **b)** *(spött.: sich sehnen)* **nach jmdm./etw. ~:** pine *or* yearn for sb./sth.

schmachtend *(spött.) Adj.* soulful *(coll.)* ‹*look, song*›; schmaltzy *(coll.)* ‹*song, music*›

schmächtig *Adj.* slight; weedy *(coll. derog.)*

schmach·voll *(geh.) Adj.* ignominious; *(erniedrigend)* humiliating

schmackhaft ['ʃmakhaft] **1.** *Adj.* tasty; **jmdm. etw. ~ machen** *(fig. ugs.)* make sth. palatable to sb. **2.** *adv.* in a tasty way; **etw. ~ zubereiten** make sth. tasty

schmähen ['ʃmɛ:ən] *tr. V. (geh.)* revile

schmählich 1. *Adj.* shameful; *(verächtlich)* despicable. **2.** *adv.* shamefully; *(in verächtlicher Weise)* despicably

Schmäh·wort das; *Pl.* **~worte** term of abuse

schmal [ʃma:l]; **schmaler** *od.* **schmäler** ['ʃmɛ:lɐ]; **schmalst...** *od.* **schmälst...:** *Adj.* narrow; slim, slender ‹*hips, hands, figure, etc.*›; thin ‹*lips, face, nose, etc.*›

schmälern *tr. V.* diminish; reduce; restrict, curtail ‹*rights*›; *(herabsetzen)* belittle

Schmal-: ~film der 8 mm/16 mm cine film; **~film·kamera die** 8 mm/16 mm cine camera; **~seite die** short side; *(eines Korridors usw.)* end

Schmalspur- *(ugs.)* small-time *(coll.)* ‹*politician, academic*›; *(dilettantisch)* lightweight ‹*academic*›; amateur ‹*engineer*›

[1]**Schmalz** [ʃmalts] **das; ~es** *(Schweine~)* lard

[2]**Schmalz der; ~es** *(abwertend)* schmaltz *(coll.)*; **mit viel ~:** with plenty of slushy *or* soppy sentimentality *(coll.)*

Schmalz·brot das slice of bread and dripping

schmalzig *(abwertend)* **1.** *Adj.* schmaltzy *(coll.)*; slushy[-sentimental]. **2.** *adv.* with schmaltzy *(coll.)* *or* slushy sentimentality

Schmankerl ['ʃmaŋkɐl] **das; ~s, ~n** *(bayr., österr.)* delicacy; *(fig.)* treat

schmarotzen [ʃma'rɔtsn̩] *itr. V.* **a)** *(abwertend)* sponge; free-load *(sl.)*; **bei jmdm. ~:** sponge on sb.; **b)** *(Biol.)* live as a parasite **(in/auf +** *Dat.* in/on)

Schmarotzer der; ~s, ~ a) *(abwertend)* sponger; free-loader *(sl.)*; **b)** *(Biol.)* parasite

Schmarren ['ʃmarən] **der; ~s, ~ a)** *(österr., auch südd.) pancake broken up with a fork after frying;* **b)** *(ugs. abwertend: Unsinn)* trash; rubbish

schmatzen *itr. V.* smack one's lips; *(geräuschvoll essen)* eat noisily

Schmaus ['ʃmaus] **der; ~es, Schmäuse** ['ʃmɔyzə] *(veralt., noch scherzh.)* [good] spread *(coll.)*; *(reichhaltig)* feast

schmausen *(veralt.)* **1.** *itr. V.* eat with relish. **2.** *tr. V.* eat ‹*food*› with relish

schmecken ['ʃmɛkn̩] **1.** *itr. V.* taste **(nach** of); **[gut] ~:** taste good; **das hat geschmeckt** that was good; *(war köstlich)* that was delicious; **schmeckt es [dir]?** are you enjoying it *or* your meal?; [how] do you like it?; **laßt es euch ~!** enjoy your food!; tuck in! *(coll.)*. **2.** *tr. V.* taste; *(kosten)* sample

Schmeichelei die; ~, ~en flattering remark; blandishment

schmeichelhaft 1. *Adj.* flattering; complimentary ‹*words, speech*›; **wenig ~:** not very flattering. **2.** *adv.* flatteringly

schmeicheln ['ʃmaiçl̩n] *itr. V.* **jmdm. ~:** flatter sb.

Schmeichler ['ʃmaiçlɐ] **der; ~s, ~, Schmeichlerin die; ~, ~nen** flatterer

schmeichlerisch *Adj.* flattering; honeyed ‹*words, tone*›

schmeißen ['ʃmaisn̩] *(ugs.)* **1.** *unr. tr. V.* **a)** *(werfen)* chuck *(coll.)*; sling *(coll.)*; *(schleudern)* fling; hurl; **etw. nach jmdm. ~:** throw *or (coll.)* chuck sth. at sb.; **b)** *(abbrechen, aufgeben)* chuck in *(coll.)* ‹*job, studies, etc.*›; **c)** *(spendieren)* stand ‹*drink*›; **[für jmdn.] eine Party ~:** throw a party [for sb.] *(coll.)*; **d)** *(bewältigen)* handle; deal with; **wir werden den Laden schon ~:** we'll manage OK *(coll.)*. **2.** *unr. refl. V.* **a)** *(sich werfen)* throw oneself; *(mit Wucht)* hurl oneself; **b) sich in seinen Smoking** *usw.* **~:** get togged up *(sl.)* in one's dinner-jacket *etc.* **3.** *unr. itr. V.* **mit Steinen/Tomaten** *usw.* **[nach jmdm.] ~:** chuck stones/tomatoes *etc.* [at sb.] *(coll.)*; **mit Geld um sich ~** *(fig.)* throw one's money around; lash out *(coll.)*

Schmeiß·fliege die blowfly; *(blaue ~)* bluebottle

Schmelze die; ~, ~n [process of] melting

schmelzen ['ʃmɛltsn̩] **1.** *unr. itr. V.; mit sein* melt; *(fig.)* ⟨*doubts, apprehension, etc.*⟩ dissolve, fade away. **2.** *unr. tr. V.* melt; smelt ⟨*ore*⟩; render ⟨*fat*⟩

Schmelz-: ~käse der processed cheese; **~punkt der** melting-point; **~tiegel der** crucible; melting-pot *(esp. fig.)*

Schmerbauch ['ʃme:ɐ̯-] **der** *(ugs.)* pot-belly

Schmerle ['ʃmɛrlə] **die; ~, ~n** *(Zool.)* loach

Schmerz [ʃmɛrts] **der; ~es, ~en a)** *(physisch)* pain; *(dumpf u. anhaltend)* ache; **wo haben Sie ~en?** where does it hurt?; **~en im Arm** pain in one's arm; *(an verschiedenen Stellen)* pains in one's arm; **~en haben** be in pain; **vor ~[en] weinen/sich vor ~en winden** cry with/writhe in pain *or* agony; **b)** *(psychisch)* pain; *(Kummer)* grief; **ein seelischer ~:** mental anguish *or* suffering; **jmdm. ~en bereiten** cause sb. pain/grief

schmerz·empfindlich *Adj.* sensitive to pain *pred.*; **~ sein** have a low pain threshold

schmerzen 1. *tr. V.* **jmdn. ~:** hurt sb.; *(Kummer bereiten)* grieve sb.; cause sb. sorrow; **es schmerzt mich, daß ...:** it grieves *or* pains me that ... **2.** *itr. V.* hurt; **heftig ~:** be intensely painful

Schmerzens-: ~geld das compensation *(for pain and suffering caused)*; exemplary damages *(Law)*; **~schrei der** cry of pain; *(laut)* scream [of pain]

schmerz·frei *Adj.* free of pain *pred.*; painless ⟨*operation*⟩

Schmerz·grenze die *(fig.)* **jetzt/dann ist die ~ erreicht** this/that is the absolute limit

schmerzhaft *Adj.* painful; *(wund)* sore

schmerzlich 1. *Adj.* painful; distressing; **die ~e Gewißheit haben, daß ...:** be painfully aware that ... **2.** *adv.* painfully

schmerz-, Schmerz-: ~lindernd 1. *Adj.* pain-relieving; **2.** *adv.* **~lindernd wirken** relieve pain; **~los 1.** *Adj.* painless; **2.** *adv.* painlessly; *s. auch* **kurz 2c; ~stillend 1.** *Adj.* painkilling; analgesic *(Med.)*; **~stillendes Mittel** pain-killer; analgesic; **2.** *adv.* **~stillend wirken** have a painkilling *or* analgesic effect; **~tablette die** painkilling *or (Med.)* analgesic tablet; **~verzerrt** *Adj.* ⟨*face, smile*⟩ distorted *or* twisted with pain

Schmetter·ball der *(Tennis usw.)* smash

Schmetterling der; ~s, ~e butterfly; *(Nachtfalter)* moth

Schmetterlings·stil der; *o. Pl. (Schwimmen)* butterfly [stroke]

schmettern ['ʃmɛtɐn] **1.** *tr. V.* **a)** *(schleudern)* hurl (**an** + *Akk.* at, **gegen** against); **jmdn./etw. zu Boden ~:** send sb./sth. crashing to the ground; **b)** *(laut spielen, singen usw.)* blare out ⟨*march, music*⟩; ⟨*person*⟩ sing lustily ⟨*song*⟩; bellow ⟨*order*⟩; **einen Tusch ~:** unleash a loud flourish; **c)** *(Tennis usw.)* smash ⟨*ball*⟩. **2.** *itr. V.* **a)** *mit sein (aufprallen)* crash; smash; **b)** *(schallen)* ⟨*trumpet, music, etc.*⟩ blare out

Schmetter·schlag der *(bes. Faustball, Volleyball)* smash

Schmied [ʃmi:t] **der; ~[e]s, ~e** blacksmith; *s. auch* **Glück b**

Schmiede die; ~, ~en smithy; forge

schmiede·eisern *Adj.* wrought-iron

schmieden *tr. V. (auch fig.)* forge (**zu** into, **aus** from, out of); **Pläne/ein Komplott ~** *(fig.)* hatch plans/a plot

schmiegen ['ʃmi:gn̩] **1.** *refl. V.* snuggle, nestle (**in** + *Akk.* in); **sich an jmdn. ~:** snuggle [close] up to sb.; **sie schmiegte sich eng an seine Seite** she pressed *or* nestled close to his side. **2.** *tr. V.* press (**an** + *Akk.* against)

schmiegsam *Adj.* supple ⟨*leather, material*⟩

¹Schmiere die; ~, ~n a) *(Schmierfett)* grease; **b)** *(schwieriger Schmutz)* greasy *or* slimy mess; **c)** *(ugs. abwertend: Provinztheater)* flea-pit *(sl.)* of a provincial theatre

²Schmiere die; ~: *in* **[bei etw.] ~ stehen** *(ugs.)* act as look-out [while sth. takes place]

schmieren ['ʃmi:rən] **1.** *tr. V.* **a)** *(mit Schmiermitteln)* lubricate; *(mit Schmierfett)* grease; **[gehen** *od.* **laufen] wie geschmiert** *(ugs.)* [go] like clockwork *or* without a hitch; **b)** *(streichen, auftragen)* spread ⟨*butter, jam, etc.*⟩ (**auf** + *Akk.* on); **Salbe auf eine Wunde ~:** apply ointment to a wound; **sich** *(Dat.)* **Creme ins Gesicht ~:** rub cream into one's face; **c)** *(mit Aufstrich)* **Brote ~:** spread slices of bread; **d)** *(abwertend: unsauber schreiben)* scrawl ⟨*essay, school work*⟩; *(schnell und nachlässig schreiben)* scribble, dash off ⟨*article, play, etc.*⟩; **e) jmdm. eine ~** *(salopp)* give sb. a clout *(coll.)*. **2.** *itr. V.* **a)** ⟨*oil, grease*⟩ lubricate; **b)** *(ugs. unsauber schreiben)* ⟨*person*⟩ scrawl, scribble; ⟨*pen, ink*⟩ smudge, make smudges

Schmieren·komödiant der *(abwertend)* cheapjack play-actor

Schmiererei die; ~, ~en *(ugs. abwertend)* **a)** *o. Pl. (unsauberes Schreiben)* scrawling; scribbling; **b)** *(unsauber Geschriebenes)* scrawl; scribble

Schmier-: ~fett das grease; **~fink der** *(ugs. abwertend) (im Schreiben)* messy writer; *(jmd., der Wände beschmiert)* graffiti-writer; *(jmd., der Diffamierendes schreibt)* muck-raker; **~heft das** rough-book

schmierig *Adj.* **a)** greasy ⟨*surface, clothes, hands, step, etc.*⟩; slimy ⟨*earth, surface*⟩; **b)** *(abwertend: widerlich freundlich)* slimy, *(coll.)* smarmy ⟨*person*⟩

Schmier-: ~mittel das lubricant; **~papier das** scrap paper; **~seife die** soft soap

schmilzt [ʃmɪltst] *2. u. 3. Pers. Sg. Präsens v.* **schmelzen**

Schminke ['ʃmɪŋkə] **die; ~, ~n** make-up

schminken 1. *tr. V.* make up ⟨*face, eyes*⟩; **die Lippen ~:** put lipstick on. **2.** *refl. V.* make oneself up; put on make-up

schmirgeln ['ʃmɪrgl̩n] *tr. V.* rub down; *(bes. mit Sandpapier)* sand; remove ⟨*paint, rust*⟩ with emery-paper/sandpaper

Schmirgel·papier das emery-paper; *(Sandpapier)* sandpaper

schmiß [ʃmɪs] *1. u. 3. Pers. Sg. Prät. v.* **schmeißen**

Schmöker ['ʃmø:kɐ] **der; ~s, ~** *(ugs.)* lightweight adventure story/romance; **ein dicker ~:** a thick tome of light reading

schmökern *(ugs.)* **1.** *itr. V.* bury oneself in a book. **2.** *tr. V.* bury oneself in ⟨*book*⟩

schmollen ['ʃmɔlən] *itr. V.* sulk; **mit jmdm. ~:** be in a huff and refuse to speak to sb.

Schmoll-: ~mund der pouting mouth; **einen ~mund machen** *od.* **ziehen** pout

schmolz [ʃmɔlts] *1. u. 3. Pers. Sg. Prät. v.* **schmelzen**

Schmor·braten der pot roast; braised beef

schmoren ['ʃmo:rən] **1.** *tr. V.* braise; **jmdn. [im eigenen Saft] ~ lassen** *(ugs.)* leave sb. to stew in his/her own juice. **2.** *itr. V.* **a)** *(garen)* braise; **b)** *(ugs.: schwitzen)* swelter

Schmu [ʃmu:] **der;** ~s *(ugs.)* *in* ~ **machen** cheat; work a fiddle *(sl.)*
schmuck [ʃmʊk] *Adj.* attractive; pretty; *(schick)* smart ⟨*clothes, house, ship, etc.*⟩
Schmuck der; ~[e]s **a)** *(~stücke)* jewelry; jewellery *(esp. Brit.)*; **b)** *(~stück)* piece of jewelry/jewellery; **c)** *(Zierde)* decoration
schmücken ['ʃmʏkn̩] *tr. V.* decorate; embellish ⟨*writings, speech*⟩
schmuck-, Schmuck-: ~**käst·chen das,** ~**kasten der** jewelry *or (esp. Brit.)* jewellery box; ~**los** *Adj.* plain; bare ⟨*room*⟩
Schmucklosigkeit die; ~: plainness; *(eines Zimmers)* bareness
Schmuck·stück das piece of jewelry *or (esp. Brit.)* jewellery; **das** ~ **seiner Sammlung** *(fig.)* the jewel of his collection
schmuddelig ['ʃmʊdəlɪç] *Adj. (ugs. abwertend)* grubby; mucky *(coll.)*; *(schmutzig u. unordentlich)* messy; grotty *(Brit. sl.)*
Schmuggel ['ʃmʊgl̩] **der;** ~s smuggling *no art.*; ~ **treiben** smuggle
schmuggeln *tr., itr. V.* smuggle (**in** + *Akk.* into; **aus** out of)
Schmuggel·ware die smuggled goods *pl.*; contraband *no pl.*
Schmuggler der; ~s, ~, **Schmugglerin die;** ~, ~**nen** smuggler
schmunzeln ['ʃmʊntsl̩n] *itr. V.* smile to oneself
Schmus [ʃmu:s] **der;** ~es *(ugs.)* waffle; *(Schmeichelei)* soft soap
schmusen ['ʃmu:zn̩] *itr. V. (ugs.)* cuddle; ⟨*couple*⟩ kiss and cuddle; **mit jmdm.** ~: cuddle sb.; ⟨*lover*⟩ kiss and cuddle *or (sl.)* neck with sb.
Schmuser der; ~s, ~ *(ugs.)* affectionate type; cuddly sort *(coll.)*
Schmutz [ʃmʊts] **der;** ~es **a)** dirt; *(Schlamm)* mud; **etw. macht viel/keinen** ~: sth. makes a great deal of/leaves no mess; **jmdn./etw. durch den** ~ **ziehen** *(fig.)* drag sb./sth. through the mud *(fig.)*; **b)** *(abwertend: Literatur usw.)* filth; ~ **und Schund** trash and filth
schmutzen *itr. V.* get dirty
Schmutz-: ~**fänger der a)** *(etw., das Schmutz anzieht)* dirt-trap; **b)** *(bei Fahrzeugen)* mud-flap; ~**fink der;** ~en *od.* ~s, ~en *(ugs.)* **a)** *(unsauberer Mensch)* [dirty] pig *(coll.)*; *(Kind)* dirty brat; **b)** *(unmoralischer Mensch)* depraved type *(coll.)*; ~**fleck der** dirty mark (**in** + *Dat.* on); *(in der Landschaft usw.)* blot
schmutzig 1. *Adj.* **a)** *(unsauber)* dirty; *(ungepflegt)* dirty, slovenly ⟨*person, restaurant, etc.*⟩; **sich** ~ **machen** get [oneself] dirty; **b)** *(abwertend)* cocky ⟨*remarks*⟩; smutty ⟨*joke, song, story*⟩; dirty ⟨*thoughts, business, war*⟩; crooked, shady ⟨*practices, deal*⟩. **2.** *adv. (abwertend)* ~ **grinsen** smirk
Schmutzigkeit die; ~: dirtiness
Schmutz-: ~**titel der** *(Druckw.)* half-title; ~**wäsche die** dirty washing
Schnabel ['ʃna:bl̩] **der;** ~s, **Schnäbel** ['ʃnɛ:bl̩] **a)** beak; **b)** *(ugs.: Mund)* gob *(sl.)*; **reden, wie einem der** ~ **gewachsen ist** say just what one thinks
Schnabel·tier das duck-billed platypus
Schnack [ʃnak] **der;** ~[e]s, ~s *od.* **Schnäcke** ['ʃnɛkə] *(nordd.)* **a)** *(Unterhaltung)* chat; **b)** *(abwertend: Gerede)* [idle] chatter; gossip; **c)** *(witziger Spruch)* witty saying; bon mot
Schnake ['ʃna:kə] **die;** ~, ~**n a)** daddy-long-legs; crane-fly; **b)** *(Stechmücke)* mosquito
Schnaken·stich der *(bes. südd.)* mosquito bite
Schnalle ['ʃnalə] **die;** ~, ~**n a)** *(Gürtel~)* buckle; **b)** *(österr.: Türklinke)* door-handle
schnallen *tr. V.* **a)** *(mit einer Schnalle festziehen)* buckle ⟨*shoe, belt*⟩; fasten ⟨*strap*⟩; **den Gürtel enger/weiter** ~: tighten/loosen one's belt; **b)** *(mit Riemen/Gurten befestigen)* strap (**auf** + *Akk.* on to); **c)** *(los~)* **etw. von etw.** ~: unstrap sth. from sth.; **d)** *(salopp: begreifen)* twig *(coll.)*
schnalzen *itr. V.* [**mit der Zunge/den Fingern**] ~: click one's tongue/snap one's fingers; **mit der Peitsche** ~: crack the whip
schnappen ['ʃnapn̩] **1.** *itr. V.* **a)** **nach jmdm./etw.** ~ ⟨*animal*⟩ snap *or* take a snap at sb./sth.; **nach Luft** ~ *(fig.)* gasp for breath *or* air; **b)** *mit sein* **ins Schloß** ~ ⟨*door*⟩ click shut; ⟨*bolt*⟩ snap home. **2.** *tr. V.* **a)** ⟨*dog, bird, etc.*⟩ snatch; [**sich** *(Dat.)*] **jmdn./etw.** ~ *(ugs.)* ⟨*person*⟩ grab sb./sth.; *(mit raschem Zugriff)* snatch sb./sth.; **b)** *(ugs.: festnehmen)* catch, *(sl.)* nab ⟨*thief etc.*⟩
Schnapp-: ~**messer das a)** clasp-knife; **b)** *(Stichwaffe)* flick-knife; ~**schloß das** spring lock; ~**schuß der** snapshot
Schnaps [ʃnaps] **der;** ~es, **Schnäpse** ['ʃnɛpsə] **a)** spirit; *(Klarer)* schnapps; **zwei Schnäpse** two glasses of spirit/schnapps; **b)** *o. Pl. (Spirituosen)* spirits *pl.*
Schnaps·idee die *(ugs.)* hare-brained idea
schnarchen ['ʃnarçn̩] *itr. V.* snore
Schnarcher der; ~s, ~ *(ugs.)* snorer
schnarren ['ʃnarən] *itr. V.* ⟨*alarm clock, telephone, doorbell*⟩ buzz [shrilly]
schnattern ['ʃnatɐn] *itr. V.* **a)** ⟨*goose etc.*⟩ cackle, gaggle; **b)** *(ugs.: eifrig schwatzen)* jabber [away]; chatter
schnauben ['ʃnaubn̩] *regelm. (auch unr.) itr. V.* ⟨*person, horse*⟩ snort ⟨**vor** with⟩; *(fig.)* ⟨*steam locomotive*⟩ puff, chuff
schnaufen ['ʃnaufn̩] *itr. V.* puff, pant (**vor** with); *(fig.)* ⟨*steam locomotive*⟩ puff, chuff
Schnaufer der; ~s, ~ *(ugs.)* breath; **den letzten** ~ **tun** *(verhüll.)* breathe one's last
Schnauz·bart ['ʃnauts-] **der** large moustache; mustachio *(arch.)*; *(an den Seiten herabhängend)* walrus moustache
Schnauze die; ~, ~**n a)** *(von Tieren)* muzzle; *(der Maus usw.)* snout; *(Maul)* mouth; **eine kalte** ~: a cold nose; **b)** *(derb: Mund)* gob *(sl.)*; **jmdm. in die** ~ **hauen** smack sb. in the gob *(sl.)*; **die** ~ **voll haben** *(salopp)* be fed up to the back teeth *(coll.)*; **eine große** ~ **haben** shoot one's mouth off *(sl.)*; [**halt die**] ~! shut your trap! *(sl.)*; **frei** [**nach**] ~, **nach** ~ *(salopp)* as one thinks fit; as the mood takes one; **c)** *(ugs.)* *(eines Flugzeugs)* nose; *(eines Fahrzeugs)* front
schnauzen *tr., itr. V. (ugs.)* bark; *(ärgerlich)* snap; snarl
Schnauzer der; ~s, ~ **a)** *(Hund)* schnauzer; **b)** *(ugs.)* *s.* **Schnauzbart**
Schnecke ['ʃnɛkə] **die a)** *(Tier)* snail; *(Nackt~)* slug; **jmdn.** [**so**] **zur** ~ **machen** *(ugs.)* give sb. [such] a good carpeting *(coll.)*; **b)** *(ugs.: Gebäck)* Belgian bun
Schnecken-: ~**haus das** snail-shell; ~**tempo das** *(ugs.)* snail's pace
Schnee [ʃne:] **der;** ~s **a)** snow; **in tiefem** ~ **liegen** lie under deep snow; **b)** *(Eier~)* beaten egg-white; **das Eiweiß zu** ~ **schlagen** beat the egg-white until stiff; **c)** *(Jargon: Kokain)* snow *(sl.)*
Schnee·ball der a) snowball; **b)** *(Strauch)* snowball-tree; guelder rose
Schneeball·schlacht die snowball fight

schnee-, Schnee-: ~bedeckt *Adj.* snow-covered; **~besen der** whisk; **~blind** *Adj.* snow-blind; **~brille die** snow-goggles *pl.*; **~fall der** snowfall; fall of snow; **~flocke die** snowflake; **~frei** *Adj.* free of snow *postpos.*; **~gestöber das** snow flurry; **~glätte die** [slippery surface due to] packed snow; **~glöckchen das** snowdrop; **~grenze die** snow-line; *(beweglich)* snow limit; **~kette die** snow-chain; **~könig der** *in* **sich freuen wie ein ~könig** *(ugs.)* be as pleased as Punch; **~mann der** snowman; **~matsch der** slush; **~pflug der** *(auch Ski)* snow-plough; **~sturm der** snowstorm; **~treiben das** driving snow; **~verhältnisse** *Pl.* snow conditions; **~verwehungen** *Pl.* snow-drifts *pl.*

Schneewittchen [-'vɪtçən] **das;** **~s** Snow White

Schneid der; **~[e]s,** *südd., österr.:* **die;** **~** *(ugs.)* guts *pl.* *(coll.)*

Schneid·brenner der *(Technik)* cutting torch; oxy-acetylene cutter

Schneide ['ʃnaidə] **die;** **~, ~n** [cutting] edge; *(Klinge)* blade

schneiden 1. *unr. itr. V.* **a)** cut (**in** + *Akk.* into); **b)** *(Medizinerjargon)* operate; **c) ~d** biting ‹*wind, cold, voice, sarcasm*›. **2.** *unr. tr. V.* **a)** cut; cut, reap ‹*corn etc.*›; cut, mow ‹*grass*›; *(in Scheiben)* slice ‹*bread, sausage, etc.*›; *(klein ~)* cut up, chop ‹*wood, vegetables*›; *(zu~)* cut out ‹*dress*›; *(stutzen)* prune ‹*tree, bush*›; trim ‹*beard*›; **sich** *(Dat.)* **von jmdm. die Haare ~ lassen** have one's hair cut by sb.; **hier ist eine Luft zum Schneiden** *(fig.)* there's a terrible fug in here *(coll.)*; **ein eng/weit/gut geschnittenes Kleid** a tight-fitting/loose-fitting/well-cut dress; **b)** *(Medizinerjargon: auf~)* operate on ‹*patient*›; cut [open] ‹*tumour, ulcer, etc.*›; lance ‹*boil, abscess*›; **c)** *(Film, Rundf., Ferns.: cutten)* cut, edit ‹*film, tape*›; **d)** *(beim Fahren)* **eine Kurve ~:** cut a corner; **jmdn./einen anderen Wagen ~:** cut in on sb./another car; **e)** *(kreuzen)* ‹*line, railway, etc.*› intersect, cross; **die Linien/Straßen ~ sich** the lines/roads intersect; **f)** *(Tennis usw.)* slice, put spin on ‹*ball*›; *(Fußball)* curve ‹*ball, free kick*›; *(Billard)* put side on ‹*ball*›; **g) eine Grimasse ~:** grimace; **h)** *(ignorieren)* **jmdn. ~:** cut sb. dead; send sb. to Coventry *(Brit.)*. **3.** *refl. V.* **ich habe mir** *od.* **mich in den Finger geschnitten** I've cut my finger

Schneider der; **~s, ~ a)** tailor; *(Damen~)* dressmaker; **frieren wie ein ~** *(ugs.)* be frozen stiff; **b)** *(ugs.: Schneidegerät)* cutter; *(für Scheiben)* slicer; **c)** *(Skat)* schneider; **aus dem ~ sein** *(fig.)* be in the clear; be clear of trouble

Schneiderei die; **~, ~en a)** tailor's shop; *(Damen~)* dressmaker's shop; **b)** *o. Pl.* *(das Schneidern)* tailoring; *(von Damenkleidern)* dressmaking

Schneiderin die; **~, ~nen** tailor; *(Damen~)* dressmaker

schneidern 1. *tr. V.* make ‹*dress, clothes*›; make, tailor ‹*suit*›. **2.** *itr. V.* make clothes/dresses; *(beruflich)* work as a tailor; *(als Schneiderin)* work as a dressmaker

Schneider-: ~puppe die tailor's dummy; **~sitz der** cross-legged position; **im ~sitz** cross-legged

Schneide·zahn der incisor

schneidig 1. *Adj.* **a)** *(forsch, zackig)* dashing; *(waghalsig)* daring; bold; rousing, brisk ‹*music*›; **b)** *(flott, sportlich)* dashing ‹*appearance, fellow*›; trim ‹*figure*›. **2.** *adv.* briskly

schneien ['ʃnaiən] *itr. V.* **a)** *(unpers.)* **es schneit** it is snowing; **es schneit jeden Tag** it snows every day; **b)** *mit sein (fig.)* ‹*blossom, confetti, etc.*› rain down, fall like snow

Schneise ['ʃnaizə] **die;** **~, ~n a)** *(Wald~)* aisle; *(als Feuerschutz)* firebreak; **b)** *(Flug~)* [air] corridor

schnell [ʃnɛl] **1.** *Adj.* quick ‹*journey, decision, service, etc.*›; fast ‹*car, skis, road, track, etc.*›; quick, rapid, swift ‹*progress*›; quick, swift ‹*movement, blow, action*›; **ein ~es Tempo** a high speed; a fast pace; **auf die ~e** *(ugs.)* in a trice; *(übereilt)* in [too much of] a hurry; in a rush; *(kurzfristig)* at short notice; quickly. **2.** *adv.* quickly; ‹*drive, move, etc.*› fast, quickly; ‹*spread*› quickly, rapidly; *(bald)* soon ‹*sold, past, etc.*›; **nicht so ~!** not so fast!; **mach ~!** *(ugs.)* move it! *(coll.)*; **wie heißt er noch ~?** *(ugs.)* what's his name again?

Schnell·bahn die *(Verkehrsw.)* municipal railway

Schnelle die; **~, ~n a)** *o. Pl.* rapidity; **b)** *(Geog.: Strom~)* rapids *pl.*

schnellebig [-lebɪç] *Adj.* fast-moving ‹*age*›

schnellen 1. *itr. V.; mit sein* shoot (**aus** + *Dat.* out of; **in** + *Akk.* into); **in die Höhe ~:** ‹*person*› leap to one's feet *or* up; ‹*rocket, fig.: prices etc.*› shoot up. **2.** *tr. V.* send ‹*ball, stone, etc.*› flying; hurl ‹*ball, stone, etc.*›; whip ‹*fishing-line*›

Schnell·hefter der loose-leaf binder; quick-release file

Schnelligkeit die; **~, ~en** speed

Schnell-: ~imbiß der snack-bar; **~koch·topf der** pressure-cooker; **~kurs der** crash course; **~reinigung die** express cleaner's

schnellstens *Adv.* as quickly as possible; *(möglichst bald)* as soon as possible

Schnell-: ~straße die expressway *(on which slow-moving vehicles are prohibited)*; **~zug der** express [train]

Schnepfe ['ʃnɛpfə] **die;** **~, ~n** snipe; *(Wald~)* woodcock

schnetzeln ['ʃnɛtsl̩n] *(bes. südd.)* cut ‹*meat*› into thin strips

schneuzen ['ʃnɔytsn̩] **1.** *tr. V.* **einem Kind die Nase ~:** blow a child's nose; **sich** *(Dat.)* **die Nase ~:** blow one's nose. **2.** *refl. V.* *(geh.)* blow one's nose

Schnick·schnack der *(ugs.)* **a)** *(wertloses Zeug)* trinkets *pl.*; *(Zierat)* frills *pl.* *(fig.)*; **b)** *(Geschwätz)* waffle; *(Unsinn)* drivel

schniegeln ['ʃni:gl̩n] *refl. V.* spruce oneself up

schnipp *Interj.* snip; **~, schnapp!** snip, snip

Schnippchen das; **~s** trick; **jmdm. ein ~ schlagen** *(ugs.)* outsmart sb. *(coll.)*; put one over on sb. *(sl.)*

Schnippel der *od.* **das;** **~s, ~** *(ugs.)* *s.* **Schnipsel**

schnippeln *(ugs.)* **1.** *itr. V.* *(mit der Schere)* snip [away] (**an** + *Dat.* at); **an der Wurst ~** *(mit dem Messer)* cut little snippets of sausage. **2.** *tr. V.* **a)** *(ausschneiden)* snip [out]; **b)** *(zerkleinern)* shred ‹*vegetables*›; chop ‹*beans etc.*› [finely]

schnippen ['ʃnɪpn̩] **1.** *itr. V.* **a)** *(mit den Fingern)* snap one's fingers (**nach** at). **2.** *tr. V.* **a)** *(wegschleudern)* flick (**von** off, from); **die Asche von der Zigarette ~:** flick the ash off one's cigarette; **b)** *(herausschleudern)* tap ‹*cigarette, card, etc.*› (**aus** out of)

schnippisch 1. *Adj.* pert ‹*reply, tone, etc.*›; *(anmaßend)* cocky ‹*girl, tone, expression*›. **2.** *adv.* pertly; *(anmaßend)* cockily

Schnipsel ['ʃnɪpsl̩] **der** *od.* **das;** **~s, ~:** scrap; *(Papier~, Stoff~)* snippet; shred

schnipseln *s.* **schnippeln**
schnitt [ʃnɪt] *1. u. 3. Pers. Sg. Prät. v.* **schneiden**
Schnitt der; ~[e]s, ~e a) cut; **b)** *(das Mähen) (von Gras)* mowing; cut; *(von Getreide)* harvest; **einen** *od.* **seinen ~ [bei etw.] machen** *(fig. ugs.)* make a profit [from sth.]; **c)** *(Film, Ferns.)* editing; cutting; **d)** *(~muster)* [dressmaking] pattern; **e)** *(Längs~, Quer~, Schräg~)* section; **f)** *(ugs.: Durch~)* average; **er fährt einen ~ von 200 km/h** he is driving at *or* doing an average [speed] of 125 m.p.h.; **im ~:** on average; **g)** *(Math.) s.* **golden 1c; h)** *(Geom.: ~fläche)* intersection; **i)** *(Ballspiele: Drall)* spin
Schnitt-: ~blume die cut flower; **~bohne die** French bean; **~brot das** cut *or* sliced bread
Schnittchen das; ~s, ~: canapé; [small] open sandwich
Schnitte die; ~, ~n a) *(bes. nordd.: Scheibe)* slice; **eine ~ [Brot]** a slice of bread; **eine [belegte] ~:** an open sandwich; **b)** *meist Pl. (österr.: Waffel)* wafer
Schnitt·fläche die cut surface
schnittig 1. *Adj.* stylish, smart ‹*suit, appearance, etc.*›; *(sportlich)* racy ‹*car, yacht, etc.*›. **2.** *adv.* stylishly; *(sportlich)* racily
Schnitt-: ~käse der cheese suitable for slicing; hard cheese; *(in Scheiben)* cheese slices *pl.*; **~lauch der** chives *pl.*; **~muster das** [dressmaking] pattern; **~muster·bogen der** pattern chart; **~punkt der** intersection; *(Geom.)* point of intersection; **~stelle die** *(DV)* interface; **~wunde die** cut; *(lang u. tief)* gash
Schnitzel ['ʃnɪtsl̩] **das; ~s, ~ a)** *(Fleisch)* [veal/pork] escalope; **b)** *(Stückchen) (von Papier)* scrap; snippet; *(von Holz)* shaving
Schnitzel·jagd die paper-chase
schnitzeln *tr. V.* chop up ‹*vegetables etc.*› [into small pieces]; shred ‹*cabbage*›
schnitzen *tr., itr. V.* carve; **an etw.** *(Dat.)* **~:** carve away at sth.
Schnitzer der; ~s, ~ a) *(Handwerker)* carver; **b)** *(ugs.: Fehler)* boob *(Brit. sl.)*; goof *(sl.)*; **sich** *(Dat.)* **einen groben ~ leisten** make an awful boob *(Brit. sl.) or (sl.)* goof; *(mit einer Bemerkung)* drop an awful clanger *(sl.)*
Schnitzerei die; ~, ~en carving
Schnitz·messer das wood-carving knife
schnob [ʃno:p] *1. u. 3. Pers. Sg. Prät. v.* **schnauben**
schnodderig ['ʃnɔdərɪç] *(ugs.)* **1.** *Adj.* brash. **2.** *adv.* brashly
Schnodderigkeit die; ~, ~en *(ugs.)* **a)** *o. Pl. (Art, Wesen)* brashness; **b)** *(Äußerung/Handlung)* brash remark/action
schnöde *(geh. abwertend)* **1.** *Adj.* **a)** *(verachtenswert)* despicable; contemptible; base ‹*cowardice*›; **b)** *(gemein)* contemptuous, scornful ‹*glance, reply, etc.*›; harsh ‹*reprimand*›. **2.** *adv. (gemein)* contemptuously; ‹*reprimand*› harshly; ‹*exploit, misuse*› flagrantly, blatantly
Schnorchel ['ʃnɔrçl̩] **der; ~s, ~:** snorkel
Schnörkel ['ʃnœrkl̩] **der; ~s, ~:** scroll; curlicue; *(der Handschrift, in der Rede)* flourish
schnorren ['ʃnɔrən] *tr., itr. V. (ugs.)* scrounge *(coll.)*; **etw. bei** *od.* **von jmdm. ~:** scrounge *(coll.) or* cadge sth. off sb.
Schnorrer der; ~s, ~ *(ugs.)* scrounger *(coll.)*; sponger
Schnösel ['ʃnø:zl̩] **der; ~s, ~** *(ugs. abwertend)* young whippersnapper
schnüffeln ['ʃnʏfl̩n] **1.** *itr. V.* **a)** *(riechen)* sniff; **an etw.** *(Dat.)* **~:** sniff sth.; **b)** *(ugs. abwertend: heimlich suchen; spionieren)* snoop [about] *(coll.)*; **in etw.** *(Akk.)* **~:** pry into sth.; stick one's nose into sth. *(coll.)*; **in jmds. Papieren ~:** nose about in sb.'s papers; **c)** *(Drogenjargon: Dämpfe ~)* sniff [glue/paint *etc.*]; **d)** *(ugs.: die Nase hochziehen)* sniff. **2.** *tr. V. (Drogenjargon)* sniff ‹*glue etc.*›
Schnüffler der; ~s, ~ a) *(ugs. abwertend)* Nosey Parker; *(Spion)* snooper *(coll.)*; **b)** *(Drogenjargon)* [glue-, paint-, *etc.*]sniffer
Schnuller ['ʃnʊlɐ] **der; ~s, ~:** dummy *(Brit.)*; pacifier *(Amer.)*
Schnulze ['ʃnʊltsə] **die; ~, ~n** *(ugs. abwertend) (Lied/Melodie)* slushy song/tune; *(Theaterstück, Film, Fernsehspiel)* tear-jerker *(coll.)*; slushy play
schnupfen ['ʃnʊpfn̩] **1.** *itr. V.* **a)** *(Tabak ~)* take snuff; **b)** *(bei Tränen, Nasenschleim)* sniff. **2.** *tr. V.* take a sniff of ‹*cocaine etc.*›; *(gewohnheitsmäßig)* sniff ‹*cocaine etc.*›; **Tabak ~:** take snuff
Schnupfen der; ~s, ~: [head] cold; **[den** *od.* **einen] ~ haben** have a [head] cold
Schnupf·tabak der snuff
schnuppe ['ʃnʊpə] **das/er ist mir ~/mir völlig ~** *(ugs.)* I don't care/I couldn't care less about it/him *(coll.)*
schnuppern ['ʃnʊpɐn] **1.** *itr. V.* sniff; **an etw.** *(Dat.)* **~:** sniff sth.. **2.** *tr. V.* sniff
Schnur ['ʃnu:ɐ̯] **die; ~, Schnüre** ['ʃny:rə] *od.* **Schnuren a)** *(Bindfaden)* piece of string; *(Kordel)* piece of cord; *(Zelt~)* guy[-rope]; **b)** *(ugs.: Kabel)* flex *(Brit.)*; lead; cord *(Amer.)*
schnüren ['ʃny:rən] **1.** *tr. V.* **a)** tie ‹*bundle, string, sb.'s hands, etc.*›; tie [up] ‹*parcel, person*›; tie, lace up ‹*shoe, corset, etc.*›; **etw. zu Bündeln ~:** tie sth. up in bundles; **b) Angst schnürte ihm die Kehle** *(fig.)* fear constricted his throat. **2.** *refl. V.* **sich in das Fleisch** *usw.* **~:** cut into the flesh *etc.*
schnur·gerade *Adj., adv.* dead straight
Schnurr·bart der moustache
schnurr·bärtig *Adj.* with a moustache *postpos.*; **~ sein** have a moustache
schnurren ['ʃnʊrən] *itr. V.* ‹*cat*› purr; ‹*machine*› hum; ‹*camera, spinning-wheel, etc.*› whirr
Schnurr·haar das *(Zool.)* whiskers *pl.*
Schnür-: ~schuh der lace-up shoe; **~senkel** [~zɛŋkl̩] **der** *(bes. nordd.)* [shoe-]lace; *(für Stiefel)* bootlace; **sich** *(Dat.)* **die ~senkel binden** tie one's shoe-laces; **~stiefel der** lace-up boot
schnur·stracks *Adv. (ugs.)* straight
schnurz [ʃnʊrts] *Adj. (ugs.) s.* **schnuppe**
Schnute ['ʃnu:tə] **die; ~, ~n** *(fam., bes. nordd.: Mund)* mouth; gob *(sl.)*; **eine ~ ziehen** *od.* **machen** make *or* pull a [sulky] face
schob [ʃo:p] *1. u. 3. Pers. Prät. v.* **schieben**
Schober ['ʃo:bɐ] **der; ~s, ~ a)** open-sided barn; **b)** *(Heuhaufen)* [hay-]stack; [hay-]rick
Schock der; ~[e]s, ~s *(auch Med.)* shock; **jmdm. einen [schweren/leichten] ~ versetzen** *od.* **geben** give sb. a [nasty/slight] shock *or* a [nasty/bit of a] fright; **unter ~ stehen** be in [a state of] shock; be suffering from shock
schocken *tr. V. (ugs.)* shock
Schocker der; ~s, ~ *(ugs.) (Roman/Film)* sensational book/film; shocker *(coll.)*
schockieren *tr. V.* shock; **über etw.** *(Akk.)* **schockiert sein** be shocked at sth.
Schöffe ['ʃœfə] **der; ~n, ~n** lay judge *(acting together with another lay judge and a professional judge)*
Schöffen·gericht das *court*

presided over by a professional judge and two lay judges

Schöffin die; ~, ~nen *s.* **Schöffe**

Schokolade [ʃoko'la:də] **die; ~, ~n a)** *(Süßigkeit)* chocolate; **b)** *(Getränk)* [drinking] chocolate

schokolade[n]-, Schokolade[n]-: ~braun *Adj.* chocolate[-brown]; **~eis das** chocolate ice-cream; **~guß der** chocolate icing

Schokoladen-: ~pudding der chocolate blancmange; **~seite die** *(ugs.)* best side; **~torte die** chocolate cake *or* gateau

scholl [ʃɔl] *1. u. 3. Pers. Sg. Prät. v.* **schallen**

Scholle ['ʃɔlə] **die; ~, ~n a)** *(Erd~)* clod [of earth]; **die heimatliche ~** *(fig.)* one's native soil; **b)** *(Eis~)* [ice-]floe; **c)** *(Fisch)* plaice; **die ~n** the plaice

Scholli ['ʃɔli] *in* **mein lieber ~!** *(ugs.)* my goodness!; good heavens!

schon [ʃo:n] **1.** *Adv.* **a)** *(bereits) (oft nicht übersetzt)* already; *(in Fragen)* yet; **er kommt ~ heute/ist ~ gestern gekommen** he's coming today/he came yesterday; **er ist ~ da/[an]gekommen** he is already here/has already arrived; **er ist ~ gestern angekommen** he arrived as early as yesterday; **wie lange bist du ~ hier?** how long have you been here?; **~ damals/jetzt** even at that time *or* in those days/even now; **~ [im Jahre] 1926** as early as 1926; back in 1926; **b)** *(fast gleichzeitig)* there and then; **er schwang sich auf das Fahrrad, und ~ war er weg** he jumped on the bicycle and was away [in a flash]; **c)** *(jetzt)* **~ [mal]** now; *(inzwischen)* meanwhile; **d)** *(selbst, sogar)* even; *(nur)* only; **das bekommt man ~ für 150 Mark** you can get it for as little as 150 marks; **e)** *(ohne Ergänzung, ohne weiteren Zusatz)* on its own; **[allein] ~ der Gedanke daran ist schrecklich** the mere thought *or* just the thought of it is dreadful; **~ der Name ist bezeichnend** the very name is significant; **~ darum** *od.* **aus diesem Grund** for this reason alone. **2.** *Partikel* **a)** *(verstärkend)* really; *(gewiß)* certainly; **du wirst ~ sehen!** you'll see!; **b)** *(ugs. ungeduldig: endlich)* **nun komm ~!** come on!; hurry up!; **und wenn ~!** so what; what if he/she/it does/did/was *etc.*; **c)** *(beruhigend: wahrscheinlich)* all right; **er wird sich ~ wieder erholen** he'll recover all right; he's sure to recover; **d)** *(zustimmend, aber etwas einschränkend)* **~ gut** OK *(coll.)*; **Lust hätte ich ~, nur keine Zeit** I'd certainly like to, but I've no time; **das ist ~ möglich, nur ...:** that is quite possible, only ...; **e)** *(betont: andererseits)* **er ist nicht besonders intelligent, aber sein Bruder ~:** he's not particularly intelligent, but his brother is; **f)** *(einschränkend, abwertend)* **was weiß der ~!** what does 'he know [about it]!; **was soll das ~ heißen?** what's 'that supposed to mean?

schön [ʃø:n] **1.** *Adj.* **a)** *(anziehend, reizvoll)* beautiful; handsome ⟨*youth, man*⟩; **die ~en Künste** the fine arts; **~e Literatur** belles-lettres *pl.*; **ich finde das Buch ~:** the book appeals to me; **b)** *(angenehm, erfreulich)* pleasant, nice ⟨*day, holiday, dream, relaxation, etc.*⟩; fine ⟨*weather*⟩; *(nett)* nice; **das war eine ~e Zeit** those were wonderful days; **einen ~en Tod haben** die peacefully; **das ist ~ von dir** it's nice of you; **das Schöne daran/an ihm** the nice thing about it/him; **das ist zu ~, um wahr zu sein** that is too good to be true; **c)** *(gut)* good ⟨*wine, beer, piece of work, etc.*⟩; **d)** *(in Höflichkeitsformeln)* **~e Grüße** best wishes; **recht ~en Dank für ...:** thank you very much for ...; many thanks for ...; **e)** *(ugs.: einverstanden)* OK *(coll.)*; all right; **also ~:** right then; **~ und gut** *(ugs.)* all well and good; **f)** *(iron.: leer)* **~e Worte** fine[-sounding] words; *(schmeichlerisch)* honeyed words; **g)** *(ugs.: beträchtlich)* handsome, *(coll.)* tidy ⟨*sum, fortune, profit*⟩; considerable ⟨*quantity, distance*⟩; pretty good ⟨*pension*⟩; **das hat ein ganz ~es Gewicht** it's quite a weight; **h)** *(iron.: unerfreulich)* nice *(coll. iron.)*; **das sind ja ~e Aussichten!** this is a fine look-out *sing. (iron.)*; what a delightful prospect! *sing. (iron.)*; **eine ~e Bescherung** a nice *or* fine mess *(coll. iron.)*. **2.** *adv.* **a)** *(anziehend, reizvoll)* beautifully; **sich ~ zurechtmachen** make oneself look nice; **b)** *(angenehm, erfreulich)* nicely; **~ warm/weich/langsam** nice and warm/soft/slow; **wir haben es ~ hier** we're very well off here; **c)** *(gut, ausgezeichnet)* well; **d)** *(in Höflichkeitsformeln)* **bitte ~, können Sie mir sagen, ...:** excuse me, could you tell me ...; **grüß deine Mutter ~ von mir** give your mother my kind regards; **e)** *(iron.)* **wie es so ~ heißt, wie man so ~ sagt** as they say; **f)** *(ugs.: beträchtlich)* really; *(vor einem Adjektiv)* pretty; **ganz ~ arbeiten müssen** have to work jolly hard *(Brit. coll.)*; **[ganz] ~ dämlich** damned stupid. **3.** *Partikel (ugs. verstärkend)* **~ ruhig bleiben/~ langsam fahren** be nice and quiet/drive nice and slowly; **bleib ~ liegen!** lie there and be good; **sei ~ brav** be a good boy/girl

schonen 1. *tr. V.* treat ⟨*clothes, books, furniture, etc.*⟩ with care; *(schützen)* protect ⟨*hands, furniture*⟩; *(nicht strapazieren)* spare ⟨*voice, eyes, etc.*⟩; conserve ⟨*strength*⟩; *(nachsichtig behandeln)* go easy on, spare ⟨*person*⟩; **jmdm. eine Nachricht ~d beibringen** break news gently to sb.; **eine ~de Behandlung** gentle treatment. **2.** *refl. V.* take care of oneself; *(sich nicht überanstrengen)* take things easy; **sich mehr ~:** take things easier

Schoner der; ~s, ~ *(Seemannsspr.)* schooner

Schön·färberei die; ~, ~en embellishment; **ohne jede ~:** without any whitewashing

Schon-: ~frist die period of grace; *(nach einer Operation)* period of convalescence; **~gang der a)** *(Kfz-W.)* high gear; *(Overdrive)* overdrive; **b)** *(bei Waschmaschinen)* programme for delicate fabrics

schön·geistig *Adj.* aesthetic; **die ~e Literatur** belletristic literature

Schönheit die; ~, ~en beauty

Schönheits-: ~chirurgie die cosmetic surgery *no art.*; **~farm die** health farm; **~fehler der** blemish; *(fig.)* minor defect; *(Nachteil)* slight drawback; **~königin die** beauty queen; **~pflege die** beauty care *no art.*; **~reparatur die** cosmetic repair; *(in einem Haus/einer Wohnung)* redecorating *no pl.*; **~wettbewerb der** beauty contest

Schon·kost die light food

schön|machen *(ugs.)* **1.** *tr. V.* smarten ⟨*person, thing*⟩ up; make ⟨*person, thing*⟩ look nice. **2.** *refl. V.* smarten oneself up; make oneself look smart

schön-, Schön-: ~|reden *itr. V. (abwertend)* turn on the smooth talk; sweet-talk *(Amer.)*; **~redner der** *(abwertend)* smooth *or (Amer.)* sweet talker; **~schrift die a)** *(Zierschrift)* calligraphy; *(sorgfältige Schrift)* neat handwriting; **etw. in ~schrift abschreiben** copy sth. out neatly *or* in one's best handwriting; **b)** *(ugs.: Reinschrift)* neat *or* clean copy;

~**tun** *unr. itr. V. (ugs.)* **jmdm. ~tun** soft-soap sb.; butter sb. up
Schonung die; ~, ~en a) *o. Pl. (Nachsicht)* consideration; *(nachsichtige Behandlung)* considerate treatment; *(nach Krankheit/Operation)* [period of] rest; *(von Gegenständen)* careful treatment; **b)** *(Jungwald)* [young] plantation
schonungs·los 1. *Adj.* unsparing, ruthless ‹*criticism etc.*›; blunt ‹*frankness*›. **2.** *adv.* unsparingly; ‹*say*› without mincing one's words
Schön·wetter·periode die spell of fine weather; fine spell
Schon·zeit die *(Jagdw.)* close season
Schopf [ʃɔpf] **der; ~[e]s, Schöpfe** ['ʃœpfə] shock of hair; **die Gelegenheit beim ~[e] packen** *od.* **ergreifen** *(ugs.)* seize *or* grasp the opportunity with both hands
schöpfen ['ʃœpfn̩] *tr. V.* **a)** scoop [up] ‹*water, liquid*›; *(mit einer Kelle)* ladle ‹*soup*›; **Wasser aus einem Brunnen ~:** draw water from a well; **b)** *(geh.: einatmen)* draw, take ‹*breath*›; **frische Luft ~:** take a breath of fresh air; **c)** *(geh.: für sich gewinnen)* draw ‹*wisdom, strength, knowledge*› **(aus** from); **neuen Mut/neue Hoffnung ~:** take fresh heart/find fresh hope
Schöpfer der; ~s, ~: creator; *(Gott)* Creator
Schöpferin die; ~, ~nen creator
schöpferisch 1. *Adj.* creative; **eine ~e Pause** a pause for inspiration. **2.** *adv.* creatively; **~ tätig sein** be creative
Schöpferkraft die creative powers *pl.;* creativity
Schöpf-: ~kelle die, ~löffel der ladle
Schöpfung die; ~, ~en a) *o. Pl. (geh.: Erschaffung)* creation; *(Erfindung)* invention; **b)** *(geh.: ~ der Welt)* **die ~:** the Creation; *(von Gott Erschaffenes)* Creation; **c)** *(geh.: Kunstwerk, ~ der Mode usw.)* creation; *(Werk)* work
Schöpfungs·geschichte die Creation story
Schöppchen ['ʃœpçən] **das, ~s, ~:** small glass of wine/beer
Schoppen ['ʃɔpn̩] **der; ~s, ~ a)** [quarter-litre/half-litre] glass of wine/beer; **b)** *(veralt.: Hohlmaß)* **ein ~:** ≈ half a litre
schor [ʃoːɐ̯] *1. u. 3. Pers. Sg. Prät. v.* **[1]scheren**
Schorf [ʃɔrf] **der; ~[e]s, ~e** scab
Schorle ['ʃɔrlə] **die; ~, ~n** wine with mineral water; ≈ spritzer; *(mit Apfelsaft)* apple juice with mineral water

Schorn·stein ['ʃɔrn-] **der** chimney; *(Schiffs~, Lokomotiv~)* funnel; **der ~ raucht** *(fig.)* things are ticking over nicely; business is good; **Geld in den ~ schreiben** *(fig. ugs.)* write off money
Schornstein·feger der; ~s, ~: chimney-sweep
schoß [ʃɔs] *1. u. 3. Pers. Sg. Prät. v.* **schießen**
Schoß [ʃoːs] **der; ~es, Schöße** ['ʃøːsə] **a)** lap; **ein Kind auf den ~ nehmen** take *or* sit a child on one's lap; **die Hände in den ~ legen** *(fig.)* sit back and do nothing; **jmdm. in den ~ fallen** *(fig.)* just fall into sb.'s lap; **im ~ der Familie/der Kirche** *(fig.)* in the bosom of the family/of Mother Church; *s. auch* **Hand f; b)** *(geh.: Mutterleib)* womb
Schoß·hund der, Schoß·hündchen das lap-dog
Schößling ['ʃœslɪŋ] **der; ~s, ~e a)** *(Trieb)* shoot; **b)** *(Ableger zum Pflanzen)* cutting
Schote ['ʃoːtə] **die; ~, ~n** pod; siliqua *(as tech. term)*; **fünf ~n Paprika** five peppers
Schott [ʃɔt] **das; ~[e]s, ~en** *(Seemannsspr.)* bulkhead
Schotte ['ʃɔtə] **der; ~n, ~n** Scot; Scotsman; **er ist ~:** he's a Scot; he's Scottish; **die ~n** the Scots; the Scottish
Schotten·rock der tartan skirt; *(Kilt)* kilt
Schotter ['ʃɔtɐ] **der; ~s, ~ a)** *(für Straßen)* [road-]metal; gravel; *(für Schienen)* ballast; **b)** *(Geol.)* gravel
Schotter·straße die road with [loose] gravel surface
Schottin die; ~, ~nen Scot; Scotswoman
schottisch 1. *Adj.* Scottish; Scots, Scottish ‹*dialect, accent, voice, etc.*›; **~er Whisky** Scotch whisky. **2.** *adv.* ‹*speak*› with a Scots *or* Scottish accent
Schottland (das); ~s Scotland
schraffieren [ʃraˈfiːrən] *tr. V.* hatch; *(feiner)* shade ‹*drawing*›
schräg [ʃrɛːk] **1.** *Adj.* **a)** diagonal ‹*line, beam, cut, etc.*›; sloping ‹*surface, roof, wall, side, etc.*›; slanting, slanted ‹*writing, eyes, etc.*›; tilted ‹*position of the head etc., axis*›; **b)** *(ugs.: unseriös)* offbeat; **c)** *(ugs.: zweifelhaft)* shady, *(coll.)* dodgy ‹*type, firm, etc.*›. **2.** *adv.* at an angle; *(diagonal)* diagonally; **den Kopf ~ halten** hold one's head to one side; tilt one's head; **~ stehende Augen** slanting eyes; **~ gegenüber** diagonally opposite; **er saß ~ vor/hinter mir** he was sitting in front of/behind me and to one side; **~ gedruckt** [printed] in italics *postpos.*; **jmdn. ~ angucken** *(fig. ugs.)* look askance at sb.
Schräge die; ~, ~n a) *(schräge Fläche)* sloping surface; *(Hang)* slope; **das Zimmer hat eine ~:** the room has a sloping wall; **b)** *(Neigung)* slope
Schräg-: ~lage die angle; *(eines Schiffes)* list; *(eines Flugzeugs)* bank; **das Schiff hat ~lage** the ship is listing *or* is at an angle; **~streifen der** diagonal stripe; **~strich der** oblique stroke
schrak [ʃraːk] *1. u. 3. Pers. Sg. Prät. v.* **schrecken**
Schramme ['ʃramə] **die; ~, ~n** scratch
Schrammel·musik ['ʃraml̩-] **die** *Viennese popular music played on violins, guitar, and accordion;* Schrammeln ensemble music
schrammen *tr. V.* scratch **(an +** *Dat.* on)
Schrank [ʃraŋk] **der; ~[e]s, Schränke** ['ʃrɛŋkə] cupboard; closet *(Amer.)*; *(Glas~; kleiner Wand~)* cabinet; *(Kleider~)* wardrobe; *(Bücher~)* bookcase; *(im Schwimmbad, am Arbeitsplatz usw.)* locker
Schränkchen ['ʃrɛŋkçən] **das; ~s, ~:** cabinet
Schranke ['ʃraŋkə] **die; ~, ~n a)** *(auch fig.)* barrier; **jmdn. in die ~ fordern** *(geh.)* throw down the gauntlet to sb.; **b)** *(fig.: Grenze)* limit
schrankenlos 1. *Adj.* boundless, unbounded ‹*admiration, confidence, loyalty, etc.*›; unlimited, limitless ‹*power, freedom, etc.*›. **2.** *adv.* boundlessly
Schranken·wärter der level-crossing *(Brit.) or (Amer.)* grade-crossing attendant; crossing-keeper
Schrank-: ~koffer der wardrobe trunk; **~wand die** shelf *or* wall unit
Schraub·deckel *der* screw-top
Schraube ['ʃraubə] **die; ~, ~n a)** *(Schlitz~)* screw; *(Sechskant~/Vierkant~)* bolt; **bei ihm ist eine ~ locker** *od.* **los** *(fig. salopp)* he has [got] a screw loose *(coll.)*; **b)** *(Schiffs~)* propeller; screw; **c)** *(Turnen)* twist; *(Kunstspringen)* twist dive; **d)** *(Kunstflug)* vertical spin
schrauben 1. *tr. V.* **a)** *(befestigen)* screw **(an, auf +** *Akk.* on to); *(mit Sechskant-/Vierkantschrauben)* bolt **(an, auf +** *Akk.* [on] to); *(entfernen)* unscrew/

unbolt (**von** from); **b)** *(drehen)* screw ⟨*nut, hook, light-bulb, etc.*⟩ (**auf** + *Akk.* on to; **in** + *Akk.* into); *(lösen)* unscrew ⟨*cap etc.*⟩ (**von** from); **c) die Preise/Erwartungen in die Höhe ~:** push prices up *or* make prices spiral/raise expectations. **2.** *refl. V.* **sich [in die Höhe] ~:** spiral upwards

Schrauben-: ~schlüssel der spanner; **~zieher der; ~s, ~:** screwdriver

Schraub-: ~stock der vice; **~verschluß der** screw-top

Schreber- ['ʃre:bɐ-]: **~garten der** ≈ allotment *(cultivated primarily as a garden)*; **~gärtner der** ≈ allotment-holder

Schreck [ʃrɛk] **der; ~[e]s, ~e** fright; scare; *(Schock)* shock; **jmdm. einen ~ einjagen** give sb. a fright *or* scare/shock; **vor ~:** with fright; ⟨*run away*⟩ in one's fright; **ach du ~!** *(ugs.)* oh my God!; **[oh] ~, laß nach!** *(scherzh.)* God help us!; oh no, not that!

schrecken 1. *tr. V.* **a)** *(geh.)* frighten; scare; **b)** *(auf~)* startle (**aus** out of); make ⟨*person*⟩ jump. **2.** *regelm. (auch unr.) itr. V.* start [up]; **aus dem Schlaf ~:** awake with a start; start from one's sleep

Schrecken der; ~s, ~ a) *(Schreck)* fright; scare; *(Entsetzen)* horror; *(große Angst)* terror; **jmdn. in Angst und ~ versetzen** terrify sb.; **mit dem [bloßen] ~ davonkommen** escape with no more than a scare *or* fright; **b)** *(Schrecklichkeit, Schrecknis)* horror; **ein Bild des ~s** a terrible *or* terrifying picture; **c)** *(fig.: gefürchtete Sache, Person)* **der ~ des Volkes/***(scherzh.)* **der Schule** *usw.* the terror of the nation/*(joc.)* the school *etc.*

schrecken·erregend 1. *Adj.* terrifying. **2.** *adv.* terrifyingly

Schreckens·herrschaft die reign of terror

Schreck·gespenst das spectre; *(gegenwärtig)* nightmare

schreckhaft *Adj.* easily scared

Schreckhaftigkeit die; ~: easily scared nature; tendency to take fright

schrecklich 1. *Adj.* **a)** terrible; **b)** *(ugs.: unerträglich)* terrible *(coll.)*; **c)** *(ugs.: sehr groß)* **es hat ihm ~en Spaß gemacht** he found it terrific fun *(coll.)*. **2.** *adv.* **a)** terribly; horribly; **b)** *(ugs. abwertend: unerträglich)* terribly *(coll.)*; dreadfully *(coll.)*; **c)** *(ugs.: sehr, äußerst)* terribly *(coll.)*

Schreck-: ~schraube die *(ugs. abwertend)* battleaxe; **~schuß der** *(auch fig.)* warning shot; **~schuß·pistole die** blank [cartridge] gun *or* pistol; **~sekunde die** moment of terror/shock; *(Reaktionszeit)* reaction time

Schrei [ʃrai] **der; ~[e]s, ~e** cry; *(lauter Ruf)* shout; *(durchdringend)* yell; *(gellend)* scream; *(kreischend)* shriek; *(des Hahns)* crow; **der letzte ~** *(fig. ugs.)* the latest thing

Schreib-: ~automat der word processor; **~block der;** *Pl.* **~blocks** *od.* **~blöcke** writing-pad

schreiben ['ʃraibn̩] **1.** *unr. itr. V.* write; ⟨*typewriter*⟩ type; **orthographisch richtig ~:** spell correctly; **auf** *od.* **mit der Maschine ~:** type; **hast du mal was zum Schreiben?** have you got anything to write with?; **er hat großes Talent zum Schreiben** he has great talent as a writer; **an einem Roman** *usw.* **~:** be writing a novel *etc.*; **jmdm.** *od.* **an jmdn. ~:** write to sb. **2.** *unr. tr. V.* **a)** write; **etw. mit der Hand/Maschine ~:** write sth. by hand *or* in longhand/type sth.; **wie schreibt man dieses Wort?** how is this word spelt?; **eine Klausur/Klassenarbeit ~:** do an exam/a class test; **die Zeitungen ~ viel Unsinn** the newspapers print a lot of nonsense; **Karl hat geschrieben. – So, was schreibt er denn?** I've had a letter from Karl. – Oh, what does he say?; **b)** *(veralt.)* **wir ~ heute den 21. September** today is 21 September; **c)** *(erklären für)* **jmdn. gesund/krank ~:** certify sb. as fit/give sb. a doctor's certificate. **3.** *unr. refl. V.* be spelt

Schreiben das; ~s, ~ a) *o. Pl.* writing *no def. art.*; **b)** *(Brief)* letter

Schreiber der; ~s, ~ a) writer; *(Verfasser)* author; **b)** *(veralt.: Sekretär, Schriftführer)* secretary; clerk

Schreiberin die; ~, ~en writer; *(Verfasserin)* authoress

schreib-, Schreib-: ~faul *Adj.* lazy about [letter-]writing *postpos.*; **~fehler der** spelling mistake; *(Versehen)* slip [of the pen]; **~heft das** *(usu. lined)* exercise-book; **~maschine die** typewriter; **etw. mit [der]** *od.* **auf der ~maschine schreiben** type sth.; **mit [der] ~maschine geschrieben** typewritten; typed; **~maschinen·papier das** typing paper; **~papier das** writing-paper; **~schrift die** cursive writing; *(gedruckt)* [cursive] script; **~stube die** *(Milit.)* orderly room; **~tisch der** desk; **~tisch·täter der** mastermind behind the scenes; *(Beamter)* desk-bound director of operations

Schreibung die; ~, ~en spelling

Schreib-: ~verbot das writing ban; **~waren** *Pl.* stationery *sing.*; writing materials; **~waren·geschäft das** stationer's; stationery shop *or (Amer.)* store; **~weise die** spelling; **~zeug das** writing things *pl.*

schreien 1. *unr. itr. V.* ⟨*person*⟩ cry [out]; *(laut rufen/sprechen)* shout; *(durchdringend)* yell; *(gellend)* scream; ⟨*baby*⟩ yell, bawl; ⟨*animal*⟩ scream; ⟨*owl, gull, etc.*⟩ screech; ⟨*cock*⟩ crow; ⟨*donkey*⟩ bray; ⟨*crow*⟩ caw; ⟨*monkey*⟩ shriek; **zum Schreien sein** *(ugs.)* be a scream *(sl.)*; **nach etw. ~:** yell for sth.; *(fig.)* cry out for sth.; *(fordern)* demand sth. **2.** *unr. tr. V.* shout

schreiend *(fig.) Adj.* **a)** *(grell)* garish ⟨*colour, poster, etc.*⟩; loud ⟨*pattern*⟩; **b)** *(empörend)* glaring, flagrant ⟨*injustice, anomaly*⟩; blatant ⟨*wrong*⟩

Schrei·hals der *(ugs.) (Kind)* bawler

Schreiner der; ~s, ~ *(bes. südd.)* *s.* **Tischler**

schreinern *(bes. südd.)* **1.** *itr. V.* do joinery. **2.** *tr. V.* make ⟨*furniture etc.*⟩

schreiten ['ʃraitn̩] *unr. itr. V.; mit sein (geh.)* **a)** walk; *(mit großen Schritten)* stride; *(marschieren)* march; **auf und ab ~:** pace up and down; **b) zu etw. ~** *(fig.)* proceed to sth.; **zur Tat ~:** go into action/get down to work

schrickt [ʃrɪkt] *3. Pers. Sg. Präsens v.* **schrecken**

schrie [ʃri:] *1. u. 3. Pers. Sg. Prät. v.* **schreien**

schrieb [ʃri:p] *1. u. 3. Pers. Sg. Prät. v.* **schreiben**

Schrieb der; ~[e]s, ~e *(ugs.)* missive *(coll.)*

Schrift [ʃrɪft] **die; ~, ~en a)** *(System)* script; *(Alphabet)* alphabet; **b)** *(Hand~)* [hand]writing; **c)** *(Druckw.: ~art)* [type-]face; **d)** *(Text)* text; *(wissenschaftliche Abhandlung)* paper; *(Werk)* work; **die [Heilige] ~:** the Scriptures *pl.*

schrift·deutsch *Adj.* **a)** written German; **b)** *s.* **hochdeutsch**

Schrift·führer der secretary

schriftlich 1. *Adj.* written; **das Schriftliche** written work; *(ugs.: die ~e Prüfung)* the written exam; **ich habe [darüber] leider nichts**

Schriftliches I'm afraid I haven't got anything in writing. **2.** *adv.* in writing; **jmdn. ~ einladen** send sb. a written invitation; **das lasse ich mir ~ geben** I'll get that in writing
schrift-, Schrift-: ~satz der *(Druckw.)* type matter; **~setzer der** typesetter; **~sprache die** written language; **~sprachlich 1.** *Adj.* used in the written language *postpos.;* **2.** *adv.* in the written language; **~steller der** writer; **~stellerei die; ~:** writing *no def. art.;* **~stellerin die; ~, ~nen** writer; **~stellerisch** *Adj.* literary ⟨*work, activity*⟩; ⟨*talent*⟩ as a writer; **~stellerisch begabt sein** be talented as a writer; **~stück das** [official] document
Schrift-: ~verkehr der; *o. Pl.* correspondence; **~wechsel der** correspondence; **~zeichen das** character
schrill [ʃrɪl] **1.** *Adj.* shrill; *(fig.)* strident ⟨*propaganda, colours, etc.*⟩. **2.** *adv.* shrilly
schrillen *itr. V.* shrill; sound shrilly
Schrippe [ˈʃrɪpə] **die; ~, ~n** *(bes. berlin.)* long [bread] roll
schritt [ʃrɪt] *1. u. 3. Pers. Sg. Prät. v.* **schreiten**
Schritt der; ~[e]s, ~e a) step; **einen ~ zur Seite machen** *od.* **tun** take a step sideways; **~ für ~** *(auch fig.)* step by step; **den ersten ~ machen** *od.* **tun** *(fig.) (den Anfang machen)* take the first step; *(als erster handeln)* make the first move; **auf ~ und Tritt** wherever one goes; at every step; **b)** *Pl. (Geräusch)* footsteps; **c)** *(Entfernung)* pace; **nur ein paar ~e von uns entfernt** only a few yards away from us; **d)** *(Gleich~)* **aus dem ~ kommen** *od.* **geraten** get out of step; **im ~ gehen** walk in step; **e)** *o. Pl. (des Pferdes)* walk; **im ~:** at a walk; **f)** *o. Pl. (Gangart)* walk; **seinen ~ verlangsamen/beschleunigen** slow/quicken one's pace; **[mit jmdm./etw.] ~ halten** *(auch fig.)* keep up *or* keep pace [with sb./sth.]; **g)** *(~geschwindigkeit)* walking pace; **[im] ~ fahren** go at walking pace *or* a crawl; **„~ fahren"** 'dead slow'; **h)** *(fig.: Maßnahme)* step; measure; **i)** *(Teil der Hose, Genitalbereich)* crotch
Schrittempo das; ~s walking pace; **[im] ~ fahren** go at walking pace *or* a crawl
Schritt·macher der *(Sport, Med.; auch fig.)* pace-maker
schritt·weise 1. *Adv.* step by step; gradually; **2.** *adj.; nicht präd.* step by step; gradual
schroff [ʃrɔf] **1.** *Adj.* **a)** precipitous, sheer ⟨*rock etc.*⟩; **b)** *(plötzlich)* sudden, abrupt ⟨*transition, change*⟩; *(kraß)* stark ⟨*contrast*⟩; **c)** *(barsch)* abrupt, curt ⟨*refusal, manner*⟩; brusque ⟨*manner, behaviour, tone*⟩. **2.** *adv.* **a)** ⟨*rise, drop*⟩ sheer; ⟨*fall away*⟩ precipitously; **b)** *(plötzlich, unvermittelt)* suddenly; abruptly; **c)** *(barsch)* curtly; ⟨*interrupt*⟩ abruptly; ⟨*treat*⟩ brusquely
Schroffheit die; ~, ~en a) *o. Pl.* precipitousness; **b)** *o. Pl. (Plötzlichkeit)* suddenness; abruptness; *(Kraßheit)* starkness; **c)** *o. Pl. (Barschheit)* curtness; abruptness; brusqueness; **mit ~:** curtly
schröpfen [ˈʃrœpfn̩] *tr. V. (ugs.)* fleece
Schrot [ʃroːt] **der** *od.* **das; ~[e]s, ~e a)** coarse meal; *(aus Getreide)* whole meal *(Brit.);* whole grain; **b)** *(aus Blei)* shot; **c)** *in* **ein Mann von echtem/bestem ~ und Korn** a man of sterling qualities
schroten *tr. V.* grind ⟨*grain etc.*⟩ [coarsely]; crush ⟨*malt*⟩ [coarsely]
Schrot-: ~flinte die shotgun; **~kugel die** pellet; **~ladung die** round of shot; small-shot charge
Schrott [ʃrɔt] **der; ~[e]s, ~e a)** scrap [-metal]; **ein Auto zu ~ fahren** *(ugs.)* write a car off; **b)** *o. Pl. (salopp fig.)* rubbish; junk
schrott-, Schrott-: ~händler der scrap-dealer; scrap-merchant; **~platz der** scrap-yard; **~reif** *Adj.* ready for the scrap-heap *postpos.;* fit for scrap *postpos.;* **~wert der** scrap value
schrubben [ˈʃrʊbn̩] *tr. (auch itr.) V.* scrub
Schrubber der; ~s, ~: [long-handled] scrubbing-brush
Schrulle [ˈʃrʊlə] **die; ~, ~n** cranky idea; *(Marotte)* quirk
schrullig *Adj.* cranky ⟨*person, idea*⟩; zany *(coll.)* ⟨*story etc.*⟩
schrumpelig *Adj. (ugs.)* wrinkly; wrinkled
schrumpeln [ˈʃrʊmpl̩n] *itr. V.; mit sein (ugs.)* ⟨*skin*⟩ go wrinkled; ⟨*apple etc.*⟩ shrivel
schrumpfen [ˈʃrʊmpfn̩] *itr. V.; mit sein* shrink; ⟨*metal, rock*⟩ contract; ⟨*apple etc.*⟩ shrivel; ⟨*skin*⟩ go wrinkled; *(abnehmen)* decrease; ⟨*supplies, capital, hopes*⟩ dwindle
Schrumpf·kopf der *(Völkerk.)* shrunken head
Schub [ʃuːp] **der; ~[e]s, Schübe** [ˈʃyːbə] **a)** *(Physik: ~kraft)* thrust; **b)** *(Med.: Phase)* phase; stage; **c)** *(Gruppe, Anzahl)* batch; **d)** *(bes. ostmd.: ~lade)* drawer
Schuber [ˈʃuːbɐ] **der; ~s, ~** slip-case
Schub·fach das *s.* **Schublade**
Schub-: ~karre die, ~karren der wheelbarrow; **~kasten der** drawer; **~kraft die** thrust; **~lade die** drawer; *(fig.: Kategorie)* pigeon-hole
Schubs [ʃʊps] **der; ~es, ~e** *(ugs.)* shove; *(fig.: Ermunterung)* prod
schubsen *tr. (auch itr.) V. (ugs.)* push; shove
schüchtern [ˈʃʏçtɐn] **1.** *Adj.* **a)** shy ⟨*person, smile, etc.*⟩; shy, timid ⟨*voice, knock, etc.*⟩; **b)** *(fig.: zaghaft)* tentative, cautious ⟨*attempt, beginnings, etc.*⟩; cautious ⟨*hope*⟩. **2.** *adv.* **a)** shyly; ⟨*knock, ask, etc.*⟩ timidly; **b)** *(fig.: zaghaft)* tentatively; cautiously
Schüchternheit die; ~: shyness
Schuft [ʃʊft] **der; ~[e]s, ~e** *(abwertend)* scoundrel; swine
schuften *(ugs.) itr. V.* slave *or* slog away; **er schuftet für zwei** he does the work of two [people]
schuftig 1. *Adj.* mean; despicable. **2.** *adv.* meanly; despicably
Schuh [ʃuː] **der; ~[e]s, ~e** shoe; *(hoher ~, Stiefel)* boot; **umgekehrt wird ein ~ draus** *(fig. ugs.)* the reverse *or* opposite is true; **wissen, wo jmdn. der ~ drückt** *(fig ugs.)* know where sb.'s problems lie; **jmdm. etw. in die ~e schieben** *(fig. ugs.)* pin the blame for sth. on sb.
Schuh-: ~bürste die shoe-brush; **~creme die** shoe-polish; **~größe die** shoe size; **welche ~größe hast du?** what size shoe[s] do you take?; **~löffel der** shoehorn; **~macher der; ~s, ~:** shoemaker; *s. auch* **Bäcker; ~macherei die; ~, ~en a)** shoemaker's; **b)** *o. Pl. (Handwerk)* shoemaking *no art.;* **~plattler** [~platlɐ] **der; ~s, ~:** *folk dance in Tirol, Bavaria and Carinthia, involving the slapping of the thighs, knees, and shoe soles;* **~sohle die** sole [of a/one's shoe]
Schul-: ~abgänger der; ~s, ~: school-leaver; **~abschluß der** school-leaving qualification; **~alter das;** *o. Pl.* school age; **~amt das** education authority; **~anfang der** first day at school; **um 8 Uhr ist ~anfang** school starts at 8 o'clock; **~anfänger der** child [just] starting school; **~arbeit die a)** *s.* **~aufgabe; b)** *(österr.: Klassenarbeit)* [written] class test; **~aufgabe die** item of

homework; **~aufgaben** homework *sing.*; **~aufsatz der** school essay; **~ausflug der** school outing; **die ~bank drücken** *(ugs.)* go to *or* be at school; **~beginn der** *s.* **~anfang; ~beispiel das** textbook example **(für** of); **~besuch der** school attendance; **~bildung die;** *o. Pl.* [school] education; schooling; **~brot das** sandwich *(eaten during break)*; **~buch das** school-book; **~bus der** school bus

schuld *s.* **Schuld b**

Schuld [ʃʊlt] **die; ~, ~en a)** *o. Pl. (das Schuldigsein)* guilt; **er ist sich** *(Dat.)* **keiner ~ bewußt** he is not conscious of having done any wrong; **~ und Sühne** crime and punishment; **b)** *o. Pl. (Verantwortlichkeit)* blame; **es ist [nicht] seine ~:** it is [not] his fault; **jetzt hat er durch deine ~ seinen Zug verpaßt** now he has missed his train because of you; **[an etw.** *(Dat.)*] **schuld haben** *od.* **sein** be to blame [for sth.]; **sie ist an allem ~:** it's all her fault; **c)** *(Verpflichtung zur Rückzahlung)* debt; *(Hypothek)* mortgage; **in ~en geraten/sich in ~en stürzen** get into debt/into serious debt; **d)** *in* **[tief] in jmds. ~ stehen** *od.* **sein** *(geh.)* be [deeply] indebted to sb.

Schụld·bekenntnis das confession [of guilt]

schụld·bewußt *Adj.* guilty ⟨*look, face, etc.*⟩; **jmdn. ~ ansehen** give sb. a guilty look

schụlden *tr. V.* owe ⟨*money, respect, explanation*⟩; **was schulde ich Ihnen?** how much do I owe you?

schụlden·frei *Adj.* debt-free ⟨*person etc.*⟩; unmortgaged ⟨*house etc.*⟩; **ich bin/das Haus ist ~:** I am free of debt/the house is free of mortgage

Schụld·gefühl das feeling of guilt; guilty feeling; **~ haben/bekommen** feel/start to feel guilty

Schu̱l·dienst der; *o. Pl.* [school-]teaching *no art.*; **in den ~ gehen** go into teaching

schụldig *Adj.* **a)** guilty; **jmdn. ~ sprechen** *od.* **für ~ erklären** find sb. guilty; **er bekennt sich ~:** he admits his guilt; **auf ~ plädieren** ⟨*public prosecutor*⟩ ask for a verdict of guilty; **der [an dem Unfall] ~e Autofahrer** the driver to blame *or* responsible [for the accident]; **b) jmdm. etw. ~ sein/bleiben** owe sb. sth.; **was bin ich Ihnen ~?** what *or* how much do I owe you?; **d)** *nicht präd. (gebührend)* due; proper

Schụldige der/die; *adj. Dekl.* guilty person; *(im Strafprozeß)* guilty party; **einer muß ja der ~ sein** 'someone must have done it

Schụldigkeit die; ~, ~en duty; **meine [verdammte] Pflicht und ~:** my bounden duty

schụld·los *Adj.* innocent **(an +** *Dat.* of)

Schụldner der; ~s, ~debtor

Schụld·spruch der verdict of guilty

Schule [ˈʃuːlə] **die; ~, ~n a)** school; **zur** *od.* **in die ~ gehen, die ~ besuchen** go to school; **zur** *od.* **in die ~ kommen** come to school; *(als Schulanfänger)* start school; **auf** *od.* **in der ~:** at school; **aus der ~ plaudern** *(fig.)* reveal [confidential] information; spill the beans *(sl.)*; **~ machen** *(fig.)* become the accepted thing; form a precedent; **b)** *o. Pl. (Ausbildung)* training; **Hohe ~** *(Reiten)* haute école; **c)** *(Lehr-, Übungsbuch)* manual; handbook

schu̱len *tr. V.* train; **ein geschultes Auge** a practised *or* expert eye

Schu̱l·englisch das school English

Schüler [ˈʃyːlɐ] **der; ~s, ~ a)** pupil; *(Schuljunge)* schoolboy; **er ist noch ~:** he is still at school; **b)** *(fig.: eines Meisters)* pupil; *(Jünger)* disciple

Schü̱ler-: ~austausch der school exchange; **~ausweis der** schoolchild's pass

Schü̱lerin die; ~, ~nen pupil; schoolgirl

Schü̱ler-: ~karte die schoolchild's season ticket; **~lotse der** *pupil trained to help other schoolchildren to cross the road;* **~mitverwaltung die** pupil participation *no art.* in school administration; **~sprache die** school slang; **~zeitung die** school magazine

schu̱l-, Schu̱l-: ~fach das school subject; **~ferien** *Pl.* school holidays *or (Amer.)* vacation *sing.*; **~fest das** school open day; **~frei** *Adj.* ⟨*day*⟩ off school; **morgen ist/haben wir ~frei** there is/we have no school tomorrow; **~freund der** school-friend; **~funk der** schools broadcasting *no art.*; *(Sendungen)* [radio] programmes *pl.* for schools; **~gelände das** school grounds *pl. or* premises *pl.*; **~geld das** school fees *pl.*; **laß dir dein ~geld wiedergeben!** *(ugs.)* they can't have taught you a thing at school; **~heft das** exercise book; **~hof der** school yard

schu̱lisch *Adj.; nicht präd.* ⟨*conflicts, problems, etc.*⟩ at school; school ⟨*work etc.*⟩; **seine ~en Leistungen** [the standard of] his school work *sing.*

schu̱l-, Schu̱l-: ~jahr das a) school year; **b)** *(Klasse)* year; **ein zehntes ~jahr** a tenth-year class; **~jugend die** schoolchildren *pl.*; **~junge der** schoolboy; **~kind das** schoolchild; **~klasse die** [school] class; **~landheim das** [school's] country hostel *(visited by school classes)*; **~leiter der** headmaster; head teacher; **~lektüre die** school reading [material]; *(einzelner Text)* school text; **~mädchen das** schoolgirl; **~medizin die;** *o. Pl.* orthodox *or* traditional medicine *no art.*; **~meister der** *(veralt., scherzh.)* schoolmaster; **~ordnung die** school rules *pl.*; **~pflichtig** *Adj.* required to attend school *postpos.*; **im ~pflichtigen Alter** of school age; **~ranzen der** [school] satchel; **~rat der** schools inspector; **~reif** *Adj.* ready for school *postpos.*; **~schiff das** training ship; **~schluß der;** *o. Pl.* end of school; **nach ~schluß** after school; **~schwänzer der** *(ugs.)* truant; **~sprecher der** pupils' representative; ≈ head boy; **~stunde die** [school] period; lesson; **~tag der** school day; **der erste/letzte ~tag** the first/last day of school; **~tasche die** schoolbag; *(Ranzen)* [school] satchel

Schulter [ˈʃʊltɐ] **die; ~, ~n** shoulder; **~ an ~** *(auch fig.)* shoulder to shoulder; **jmdm. auf die ~ klopfen** pat sb. on the shoulder *or (fig.)* back; *s. auch* **kalt 1; leicht a**

schụlter-, Schụlter-: ~blatt das *(Anat.)* shoulder-blade; **~frei** *Adj.* off-the-shoulder ⟨*dress*⟩; **~klappe die** shoulder-strap; epaulette; **~lang** *Adj.* shoulder-length

schụltern *tr. V.* shoulder; **das Gewehr ~:** shoulder arms

Schu̱l·tüte die *cardboard cone of sweets given to a child on its first day at school*

Schu̱lung die; ~, ~en training; *(Veranstaltung)* training course; **politische ~:** political schooling

Schu̱l-: ~unterricht der school lessons *pl., no art.*; **~weg der** way to school; **~zeit die** schooldays *pl.*; **~zeugnis das** school report

schụmmeln *itr., tr., refl. V. (ugs.)* *s.* **mogeln**

schummerig [ˈʃʊmərɪç] **1.** *Adj.*

dim ⟨*light etc.*⟩; dimly lit ⟨*room etc.*⟩. **2.** *adv.* dimly

Schund der; ~[e]s *(abwertend)* trash

schunkeln ['ʃʊŋkl̩n] *itr. V.* rock to and fro together *(in time to music, with linked arms)*

Schuppe ['ʃʊpə] **die; ~, ~n a)** scale; **es fiel ihm wie ~n von den Augen** he had a sudden, blinding realization; the scales fell from his eyes; **b)** *Pl. (auf dem Kopf)* dandruff *sing.; (auf der Haut)* flaking skin *sing.*

schuppen 1. *tr. V.* scale ⟨*fish*⟩. **2.** *refl. V.* ⟨*skin*⟩ flake; ⟨*person*⟩ have flaking skin

Schuppen der; ~s, ~ a) shed; **b)** *(ugs.: Lokal)* joint *(sl.)*

Schur [ʃuːɐ̯] **die; ~, ~en a)** *(das Scheren)* shearing; **b)** *(Landw.: das Mähen, Schneiden)* cut

Schür·eisen das poker

schüren ['ʃyːrən] *tr. V.* **a)** poke ⟨*fire*⟩; *(gründlich)* rake ⟨*fire, stove, etc.*⟩; **b)** *(fig.)* stir up ⟨*hatred, envy, etc.*⟩; fan the flames of ⟨*passion*⟩; **jmds. Hoffnung ~:** raise sb.'s hopes

schürfen ['ʃʏrfn̩] **1.** *itr. V.* **a)** scrape; **b)** *(Bergbau)* dig [experimentally] **(nach** for); **nach Gold** *usw.* **~:** prospect for gold *etc.*. **2.** *tr. V.* **a) sich** *(Dat.)* **das Knie** *usw.* **[wund/blutig] ~:** graze one's knee *etc.* [and make it sore/bleed]; **b)** *(Bergbau)* mine ⟨*ore etc.*⟩ open-cast *or (Amer.)* open-cut. **3.** *refl. V.* graze oneself

Schürf·wunde die graze; abrasion

Schür·haken der poker *(with hooked end); (für den Ofen)* rake

Schurke ['ʃʊrkə] **der; ~n, ~n** *(abwertend)* rogue; villain

Schur·wolle die: [reine] ~: pure new wool

Schurz [ʃʊrts] **der; ~es, ~e a)** apron; **b)** *(Lenden~)* loincloth

Schürze ['ʃʏrtsə] **die; ~, ~n** apron; *(Frauen~, Latz~)* pinafore; **jmdm. an der ~ hängen** *(fig.)* be tied to sb.'s apron-strings

schürzen *tr. V.* **a)** gather up; **b)** *(aufwerfen)* purse ⟨*lips, mouth*⟩; **c)** *(geh.: binden)* tie ⟨*knot*⟩

Schürzen-: ~band *s.* **Schürzenzipfel; ~zipfel der** apron-string; **jmdm. am ~zipfel hängen** *(fig. ugs.)* be tied to sb.'s apron-strings

Schuß [ʃʊs] **der; Schusses, Schüsse** ['ʃʏsə] *(bei Maßangaben ungebeugt)* **a)** shot **(auf +** *Akk.* at); **21 ~ Salut** a 21-gun salute; **weit** *od.* **weitab vom ~** *(fig. ugs.)* well away from the action; at a safe distance; *(abseits)* far off the beaten track; **der ~ kann nach hinten losgehen** *(fig. ugs.)* it could backfire *or* turn out to be an own goal; **ein ~ in den Ofen** *(fig.)* a complete waste of effort; **b)** *(Menge Munition/Schießpulver)* round; **drei ~ Munition** three rounds of ammunition; **keinen ~ Pulver wert sein** *(fig. ugs.)* be worthless *or* not worth a thing; **c)** *(~wunde)* gunshot wound; **d)** *(mit einem Ball, Puck usw.)* shot **(auf +** *Akk.* at); **e)** *(kleine Menge)* dash; **Cola** *usw.* **mit ~:** Coke **(P)** *etc.* with something strong; brandy/rum *etc.* and Coke **(P)** *etc.*; **f)** *(Drogenjargon)* shot; fix *(sl.)*; **g)** *(Skisport)* **~ fahren** schuss; **h)** *(ugs.)* **in etw. in ~ bringen/halten** get sth. into/keep sth. in[good] shape

Schussel ['ʃʊsl̩] **der; ~s, ~** *(ugs.)* scatter-brain; wool-gatherer

Schüssel ['ʃʏsl̩] **die; ~, ~n** bowl; *(flacher)* dish

schusselig *(ugs.)* **1.** *Adj.* scatter-brained; *(fahrig)* dithery. **2.** *adv.* in a scatter-brained way

Schusseligkeit die; ~ *(ugs.)* wool-gathering; muddle-headedness; *(schusselige Art)* scatter-brained way

Schuß-: ~fahrt die *(Ski)* schuss; **~linie die** line of fire; **in die/jmds. ~linie geraten** *(auch fig.)* come under fire/come under fire from sb.; **~verletzung die** gunshot wound; **~waffe die** weapon *(firing a projectile); (Gewehr usw.)* firearm; **~wechsel der** exchange of shots

Schuster ['ʃuːstɐ] **der; ~s, ~** *(ugs.)* shoemaker; *(Flick~)* shoe-repairer; cobbler *(dated)*; **auf ~s Rappen** *(scherzh.)* on Shanks's pony

Schuster·handwerk das; *o. Pl.* shoemaking *no art.*

Schutt [ʃʊt] **der; ~[e]s a)** rubble; **„~ abladen verboten"** 'no tipping'; 'no dumping'; **in ~ und Asche liegen/sinken** *(geh.)* lie in ruins/be reduced to rubble; **b)** *(Geol.)* debris; detritus

Schutt·ablade·platz der rubbish dump *or (Brit.)* tip; garbage dump *(Amer.)*

Schütte ['ʃʏtə] **die; ~, ~n a)** *(Behälter)* [kitchen] drawer-container *(for flour etc.)*; **b)** *(Rutsche)* chute

Schüttel·frost der [violent] shivering fit; **~ haben** have violent shivers

schütteln ['ʃʏtl̩n] **1.** *tr. (auch itr.) V.* **a)** shake; **den Kopf [über etw.** *(Akk.)***]/die Faust [gegen jmdn.] ~:** shake one's head [over sth.]/one's fist [at sb.]; **jmdm. die Hand ~:** shake sb.'s hand; shake sb. by the hand; **das Fieber/die Angst/das Grauen schüttelte ihn** he was shivering *or* shaking with fever/fear/gripped with horror; **b)** *(unpers.)* **es schüttelte ihn [vor Kälte]** he was shaking [with *or* from cold]. **2.** *refl. V.* shake oneself/itself. **3.** *itr. V.* **mit dem Kopf ~:** shake one's head

schütten ['ʃʏtn̩] **1.** *tr. V.* pour ⟨*liquid, flour, grain, etc.*⟩; *(unabsichtlich)* spill ⟨*liquid, flour, etc.*⟩; tip ⟨*rubbish, coal, etc.*⟩; **jmdm./sich Wein über den Anzug ~:** spill wine on sb.'s/one's suit. **2.** *itr. V. (unpers.) (ugs.: regnen)* pour [down]

schütter ['ʃʏtɐ] *Adj.* sparse; thin

Schutt-: ~halde die pile *or* heap of rubble; **~haufen der** pile of rubble; *(Abfallhaufen)* rubbish heap; **~platz der** [rubbish] dump *or (Brit.)* tip; garbage dump *(Amer.)*

Schutz [ʃʊts] **der; ~es** protection **(vor +** *Dat.*, **gegen** against); *(Feuer~)* cover; *(Zuflucht)* refuge; **im ~ der Dunkelheit/Nacht** under cover of darkness/night; **unter einem Baum ~ [vor dem Regen** *usw.***] suchen/finden** seek/find shelter *or* take refuge [from the rain *etc.*] under a tree; **jmdn. [vor jmdm./gegen etw.] in ~ nehmen** defend sb. *or* take sb.'s side [against sb./sth.]

schutz-, Schutz-: ~bedürftig *Adj.* in need of protection *postpos.*; **~blech das** mudguard; **~brief der** *(Kfz-W.)* travel insurance; *(Dokument)* travel insurance certificate; **~brille die** [protective] goggles *pl.*

Schütze ['ʃʏtsə] **der; ~n, ~n a)** marksman; **b)** *(Fußball usw.: Tor~)* scorer; **c)** *(Milit.: einfacher Soldat)* private; **d)** *(Astrol., Astron.)* Sagittarius; **er/sie ist [ein] ~:** he/she is a Sagittarian

schützen 1. *tr. V.* protect **(vor +** *Dat.* from, **gegen** against); safeguard ⟨*interest, property, etc.*⟩ **(vor +** *Dat.* from); **sich ~d vor jmdn./etw. stellen** stand protectively in front of sb./sth.; **gesetzlich geschützt** registered [as a trademark]; **„vor Wärme/Kälte/Licht ~":** 'keep away from heat/cold/light'. **2.** *itr. V.* provide *or* give protection **(vor +** *Dat.* from, **gegen** against); *(vor Wind, Regen)* provide *or* give shelter **(vor +** *Dat.* from)

Schützen·fest das *shooting competition with fair*

Schụtz·engel der guardian angel
Schụtzen-: ~graben der *(Milit.)* trench; **~hilfe die** *(ugs.)* support; **~panzer der** armoured personnel carrier; **~verein der** shooting *or* rifle club
Schụtz-: ~film der protective film; **~gebühr die a)** token *or* nominal charge; **b)** *(verhüll.: erpreßte Zahlung)* protection money *no pl., no indef. art.*; **~heilige der/die** *(kath. Rel.)* patron saint; **~helm der** helmet; *(bei Renn-, Motorradfahrern)* crash-helmet; *(bei Bauarbeitern usw.)* safety helmet; **~hülle die** [protective] cover; *(für Dokumente usw.)* folder; **~hütte die a)** *(Unterstand)* shelter; **b)** *(Berghütte)* mountain hut; **~impfung die** vaccination; inoculation; **~kleidung die** protective clothing; **~leute** *s.* **~mann**
Schützling ['ʃʏtslɪŋ] **der; ~s, ~e** protégé; *(Anvertrauter)* charge
schụtz-, Schụtz-: ~los *Adj.* defenceless; unprotected; **~mann der;** *Pl.* **~männer** *od.* **~leute** *(ugs. veralt.)* [police] constable; copper *(Brit. coll.)*; **~patron der** patron saint; **~schicht die** protective layer **(aus** of); *(flüssig aufgetragen)* protective coating; **~suchend** *Adj.* seeking protection *postpos.*; **~umschlag der** dust-jacket; *(für Papiere)* cover
schwabbelig ['ʃvabəlɪç] *Adj.* flabby ⟨*stomach, person, etc.*⟩; wobbly ⟨*jelly etc.*⟩
schwạbbeln *itr. V. (ugs.)* wobble
Schwabe ['ʃva:bə] **der; ~n, ~n** Swabian
Schwaben (das); ~s Swabia
Schwaben·streich der *(scherzh.)* piece of folly
Schwäbin ['ʃvɛ:bɪn] **die; ~, ~nen** Swabian
schwäbisch 1. *Adj.* Swabian. **2.** *adv.* in Swabian dialect
schwach [ʃvax]; **schwächer** ['ʃvɛçɐ], **schwächst...** ['ʃvɛçst...] **1.** *Adj.* **a)** *(kraftlos)* weak; weak, delicate ⟨*child, woman*⟩; frail ⟨*invalid, old person*⟩; low-powered ⟨*engine, car, bulb, amplifier, etc.*⟩; weak, poor ⟨*eyesight, memory, etc.*⟩; poor ⟨*hearing*⟩; delicate ⟨*health, constitution*⟩; **~ werden** grow weak; *(fig.: schwanken)* weaken; waver; *(nachgeben)* give in; **mir wird [ganz] ~:** I feel [quite] faint; **in einer ~en Stunde** in a weak moment; **b)** *(nicht gut)* poor ⟨*pupil, player, runner, performance, result, effort, etc.*⟩; weak ⟨*candidate, argument, opponent, play, film, etc.*⟩; **er ist in Latein sehr ~:** he is very bad at Latin; **das ist aber ein ~es Bild!** *(fig. ugs.)* that's a poor show *(coll.)*; **c)** *(gering, niedrig, klein)* poor, low ⟨*attendance etc.*⟩; sparse ⟨*population*⟩; slight ⟨*effect, resistance, gradient, etc.*⟩; light ⟨*wind, rain, current*⟩; faint ⟨*groan, voice, pressure, hope, smile, smell*⟩; weak, faint ⟨*pulse*⟩; lukewarm ⟨*applause, praise*⟩; faint, dim ⟨*light*⟩; pale ⟨*colour*⟩; **das Licht wird schwächer** the light is fading; **d)** *(wenig konzentriert)* weak ⟨*solution, acid, tea, coffee, beer, poison, etc.*⟩; **e)** *(Sprachw.)* weak ⟨*conjugation, verb, noun, etc.*⟩. **2.** *adv.* **a)** *(kraftlos)* weakly; **b)** *(nicht gut)* poorly; **c)** *(in geringem Maße)* poorly ⟨*attended, developed*⟩; sparsely ⟨*populated*⟩; slightly ⟨*poisonous, acid, alcoholic, sweetened, salted, inclined, etc.*⟩; ⟨*rain*⟩ slightly; ⟨*remember, glow, smile, groan*⟩ faintly; lightly ⟨*accented*⟩; ⟨*beat*⟩ weakly; **d)** *(Sprachw.)* **~ gebeugt/konjugiert** weak
Schwäche ['ʃvɛçə] **die; ~, ~n** weakness; **eine ~ für jmdn./etw. haben** have a soft spot for sb./a weakness for sth.
Schwạ̈che·anfall der sudden feeling of faintness
schwạ̈chen *tr. V.* weaken
schwächer *s.* **schwach**
Schwạchheit die; ~: weakness; **die ~ des Greises/des Alters** the frailty of the old man/of old age
Schwạch·kopf der *(salopp)* bonehead *(sl.)*; dimwit *(coll.)*
schwạ̈chlich *Adj.* weakly, delicate ⟨*person*⟩; frail ⟨*old person, constitution*⟩; delicate ⟨*nerves, stomach, constitution*⟩
Schwächling ['ʃvɛçlɪŋ] **der; ~s, ~e** weakling
schwạch-, Schwạch-: ~punkt der weak point; **~sichtig** *Adj. (Med.)* weak-sighted; **~sinn der;** *o. Pl.* **a)** *(Med.)* mental deficiency; **b)** *(ugs. abwertend: Unsinn)* [idiotic *(coll.)*] rubbish *or* nonsense; **~sinnig 1.** *Adj.* **a)** *(Med.)* mentally deficient; **b)** *(ugs.: unsinnig)* idiotic *(coll.)*, nonsensical ⟨*measure, policy, etc.*⟩; rubbishy ⟨*film etc.*⟩; **2.** *adv. (ugs.)* idiotically *(coll.)*; stupidly
schwächst... *s.* **schwach**
Schwạch·strom der *(Elektrot.)* current of low amperage; *(mit niedriger Spannung)* low-voltage current
Schwạ̈chung die; ~, ~en weakening
Schwaden ['ʃva:dn̩] **der; ~s, ~:** [thick] cloud
Schwadron [ʃva'dro:n] **die; ~, ~en** *(Milit. hist.)* squadron
schwadronieren *itr. V. (abwertend)* bluster
Schwafelei die; ~, ~en *(ugs. abwertend)* **a)** *o. Pl.* rabbiting on *(Brit. sl.)*; **b)** *(Bemerkung)* rubbishy remark; **~en** blether *sing.*
schwafeln ['ʃva:fl̩n] *(ugs. abwertend)* **1.** *itr. V.* rabbit on *(Brit. sl.)*, waffle **(von** about). **2.** *tr. V.* blether ⟨*nonsense*⟩
Schwager ['ʃva:gɐ] **der; ~s, Schwäger** ['ʃvɛ:gɐ] brother-in-law
Schwägerin ['ʃvɛ:gərɪn] **die; ~, ~nen** sister-in-law
Schwalbe ['ʃvalbə] **die; ~, ~n a)** swallow; **eine ~ macht noch keinen Sommer** *(Spr.)* one swallow does not make a summer *(prov.)*; **b)** *(ugs.: Papierflieger)* paper aeroplane
Schwall [ʃval] **der; ~[e]s, ~e** torrent; flood
schwamm [ʃvam] *1. u. 3. Pers. Sg. Prät. v.* **schwimmen**
Schwạmm der; ~[e]s, Schwämme ['ʃvɛmə] **a)** sponge; **~ drüber!** *(ugs.)* [let's] forget it; **b)** *(südd., österr.: Pilz)* mushroom
Schwammerl ['ʃvamɐl] **das; ~s, ~[n]** *(bayr., österr.)* mushroom
schwạmmig 1. *Adj.* **a)** spongy; **b)** *(aufgedunsen)* flabby, bloated ⟨*face, body, etc.*⟩; **c)** *(unpräzise)* woolly ⟨*concept, manner of expression, etc.*⟩. **2.** *adv. (unpräzise)* vaguely
Schwạmmigkeit die; ~ a) sponginess; **b)** *(Aufgedunsenheit)* flabbiness; bloated appearence; **c)** *(Vagheit)* woolliness
Schwan [ʃva:n] **der; ~[e]s, Schwäne** ['ʃvɛ:nə] swan; **mein lieber ~!** *(ugs.)* my goodness!; good heavens!
schwand [ʃvant] *1. u. 3. Pers. Sg. Prät. v.* **schwinden**
schwanen *itr. V. (ugs.)* **jmdm. schwant etw.** sb. senses sth.
schwang [ʃvaŋ] *1. u. 3. Pers. Sg. Prät. v.* **schwingen**
schwanger ['ʃvaŋɐ] *Adj.* pregnant **(von** by); **sie ist im vierten Monat ~:** she is in her fourth month [of pregnancy]
Schwạngere die; *adj. Dekl.* expectant mother; pregnant woman
schwängern ['ʃvɛŋɐn] *tr. V.* make ⟨*woman*⟩ pregnant; **sich von jmdm. ~ lassen** get [oneself] pregnant by sb.
Schwạngerschaft die; ~, ~en pregnancy

Schwangerschafts·abbruch der termination of pregnancy; abortion

Schwank [ʃvaŋk] **der; ~[e]s, Schwänke** ['ʃvɛŋkə] **a)** *(Literaturw.: Erzählung)* comic tale; *(auf der Bühne)* farce; **b)** *(komische Episode)* comic event

schwanken *itr. V. (mit Richtungsangabe mit sein)* **a)** sway; ⟨*boat*⟩ rock; *(heftiger)* roll; ⟨*ground, floor*⟩ shake; **b)** *(fig.: unbeständig sein)* ⟨*prices, temperature, etc.*⟩ fluctuate; ⟨*number, usage, etc.*⟩ vary; **c)** *(fig.: unentschieden sein)* waver; *(zögern)* hesitate; **er schwankt noch, ob** he is still undecided [as to] whether

Schwankung die; ~, ~en variation; *(der Kurse usw.)* fluctuation

Schwanz [ʃvants] **der; ~es, Schwänze** ['ʃvɛntsə] **a)** tail; **kein ~** *(fig. salopp)* not a bloody *(Brit. sl.) or (coll.)* damn soul; **den ~ einklemmen** *(fig. salopp)* draw in one's horns; **b)** *(salopp: Penis)* prick *(coarse);* cock *(coarse)*

schwänzeln ['ʃvɛntsl̩n] *itr. V.* wag its tail/their tails

schwänzen ['ʃvɛntsn̩] *tr., itr. V. (ugs.)* skip, cut ⟨*lesson etc.*⟩; **[die Schule] ~:** play truant *or (Amer.)* hookey; **den Dienst ~:** skive [off] *(Brit. sl.)*

Schwanz·flosse die *(Zool., Flugw.)* tail-fin; *(des Wals)* tail flukes *pl.*

schwappen ['ʃvapn̩] *itr. V.* **a) [hin und her] ~:** slosh [around]; **an die Bordwand ~:** splash *or* slap against the side of the boat; **b)** *mit Richtungsangabe mit sein* splash, slosh (**über** + *Akk.* over, **aus** out of)

Schwäre ['ʃvɛːrə] **die; ~, ~n** *(geh.)* [festering] ulcer

Schwarm [ʃvarm] **der; ~[e]s, Schwärme** ['ʃvɛrmə] **a)** swarm; **ein ~ Krähen/Heringe** a flock of crows/shoal of herrings; **b)** *(fam.: Angebetete[r])* idol; heart-throb

schwärmen ['ʃvɛrmən] *itr. V.* **a)** *mit Richtungsangabe mit sein* swarm; **b)** *(begeistert sein)* **für jmdn./etw. ~:** be mad about *or* really keen on sb./sth.; **sie schwärmt für ihren Skilehrer** she has a crush on her skiing instructor *(sl.);* **von etw. ~:** go into raptures about sth.

Schwärmer der; ~s, ~ a) *(Phantast)* dreamer; *(Begeisterter)* [passionate] enthusiast; **b)** *(Zool.: Schmetterling)* hawk-moth

schwärmerisch 1. *Adj.* rapturous ⟨*enthusiasm, admiration, letter, etc.*⟩; effusive ⟨*person, language*⟩; *(begeistert)* wildly enthusiastic. **2.** *adv.* rapturously; ⟨*speak*⟩ effusively

Schwarte ['ʃvartə] **die; ~, ~n a)** *(Speck~)* rind; *(Haut~)* skin; **b)** *(ugs.: dickes Buch)* **[dicke] ~:** thick *or* weighty tome

Schwarten·magen der brawn

schwarz [ʃvarts]; **schwärzer** ['ʃvɛrtsɐ], **schwärzest...** ['ʃvɛrtsəst...] **1.** *Adj.* **a)** black; Black ⟨*person*⟩*;* filthy[-black] ⟨*hands, fingernails, etc.*⟩*;* **~ wie die Nacht/wie Ebenholz** as black as pitch/jet-black; **mir wurde ~ vor den Augen** everything went black; **b)** *(fig.)* **der ~e Erdteil** *od.* **Kontinent** the Dark Continent; **die ~e Rasse** the Blacks *pl.;* **das ~e Schaf sein** be the black sheep; **~e Liste** blacklist; **das habe ich ~ auf weiß** *(fig.)* I've got it in black and white *or* in writing; **er kann warten, bis er ~ wird** *(ugs.)* he can wait till the cows come home *(coll.);* **das Schwarze Meer** the Black Sea; *s. auch* **Mann a; c)** *(illegal)* illicit, shady ⟨*deal, exchange, etc.*⟩; **der ~e Markt** the black market; **d)** *(ugs.: katholisch)* Catholic; **e)** *(ugs.: christdemokratisch)* Christian Democrat. **2.** *adv.* **a)** ⟨*write, underline, etc.*⟩ in black; **~ gestreift** with black stripes; **b)** *(illegal)* illegally; illicitly; **etw. ~ kaufen** buy sth. illegally *or* on the black market

Schwarz das; ~[es], ~: black; **in ~ gehen, ~ tragen** wear black

schwarz-, Schwarz-: ~afrika (das) Black Africa; **~arbeit die;** *o. Pl.* work done on the side *(and not declared for tax); (abends)* moonlighting *(coll.);* **~|arbeiten** *itr. V.* do work on the side *(not declared for tax); (abends)* moonlight *(coll.);* **~arbeiter der** person who does work on the side; *(abends)* moonlighter *(coll.);* **~bär der** black bear; **~brot das** black bread

[1]Schwarze der; *adj. Dekl.* **a)** *(Neger)* Black; *(Dunkelhaariger)* dark-haired man/boy; **b)** *(österr.: Kaffee)* black coffee

[2]Schwarze die; *adj. Dekl. (Negerin)* Black [woman/girl]; *(Dunkelhaarige)* dark haired woman/girl

[3]Schwarze das; *adj. Dekl. (der Zielscheibe)* bull's eye; **ins ~ treffen** hit the bull's eye; *(fig.)* hit the nail on the head

schwärzen *tr. V.* blacken; black out ⟨*words*⟩

schwarz-, Schwarz-: ~|fahren *unr. itr. V.; mit sein* travel without a ticket *or* without paying; dodge paying the fare; **~fahrer der** fare-dodger; **~haarig** *Adj.* black-haired; **~handel der** black market (**mit** in); *(Tätigkeit)* black marketeering (**mit** in); **~händler der** black marketeer; **~hörer der** radio user without a licence; radio licence dodger; **~markt der** black market; **~|sehen** *unr. itr. V.* **a)** *(pessimistisch sein)* look on the black side; be pessimistic; **für jmdn./etw. ~sehen** be pessimistic about sb./sth.; **b)** *(Ferns.)* watch television without a licence; **~seher der a)** *(ugs.)* pessimist; **b)** *(Ferns.)* [television] licence dodger

Schwarz·wald der; ~[e]s Black Forest

Schwarzwälder [-'vɛldɐ] **die; ~, ~** *(Torte)* Black Forest gateau

schwarz·weiß *Adj.* black and white

schwarzweiß-, Schwarzweiß-: ~fernseh·gerät das black and white television [set]; **~foto das** black and white photo; **~|malen** *itr. V.* paint *or* put things in [crude] black-and-white terms

Schwarz·wurzel die black salsify

Schwatz [ʃvats] **der; ~es, ~e** *(fam.)* chat; natter *(coll.);* **einen ~ halten** have a chat *or (coll.)* natter

Schwätzchen ['ʃvɛtsçən] **das; ~s, ~** *(fam.)* [little] chat; [little] natter *(coll.)*

schwatzen, *(bes. südd.)* **schwätzen** ['ʃvɛtsn̩] **1.** *itr. V.* **a)** chat; **b)** *(über belanglose Dinge)* chatter; natter *(coll.);* **c)** *(etw. ausplaudern)* talk; blab; **d)** *(in der Schule)* talk. **2.** *tr. V.* say; talk ⟨*nonsense, rubbish*⟩

Schwätzer der; ~s, ~, Schwätzerin die; ~, ~nen *(abwertend)* chatterbox; *(klatschhafter Mensch)* gossip

schwatzhaft *Adj. (abwertend)* talkative; garrulous; *(klatschhaft)* gossipy

Schwebe ['ʃveːbə] **die** *in* **in der ~ sein/bleiben** *(fig.)* be/remain in the balance

Schwebe-: ~bahn die *(Seilbahn)* cableway; *(Hängebahn)* [overhead] monorail; *(Magnetschwebebahn)* levitation railway; **~balken der** *(Turnen)* [balance] beam

schweben *itr. V.* **a)** ⟨*bird, balloon, etc.*⟩ hover; ⟨*cloud, balloon, mist*⟩ hang; *(im Wasser)* float; **in Gefahr ~** *(fig.)* be in danger; **b)** *mit sein (durch die Luft)* float;

(herab~) float [down]; *(mit dem Fahrstuhl)* glide; *(wie schwerelos gehen)* ‹*dancer etc.*› glide; **c)** *(unentschieden sein)* be in the balance; **alle ~den Fragen** all outstanding questions

Schwede ['ʃve:də] **der; ~n, ~n** Swede

Schweden ['ʃve:dn̩] **(das); ~s** Sweden

Schwedin die; ~, ~nen Swede

schwedisch ['ʃve:dɪʃ] *Adj.* **a)** Swedish; *s. auch* **deutsch; Deutsch; b)** *in* **hinter ~en Gardinen** *(ugs.)* behind bars *(coll.)*

Schwefel ['ʃve:fl̩] **der; ~s** sulphur

schwefel·haltig *Adj.* containing sulphur *postpos., not pred.;* sulphurous ‹*Quelle, Boden*›; **schwach ~ sein** have a low sulphur content

schwefelig *s.* **schweflig**

schwefeln *tr. V.* sulphurize

Schwefel-: ~säure die *(Chemie)* sulphuric acid; **~wasserstoff der** *(Chemie)* hydrogen sulphide

schweflig *Adj.* sulphurous ‹*acid*›

Schweif [ʃvaif] **der; ~[e]s, ~e** *(auch fig.: eines Kometen)* tail; *(eines Fuchses)* brush

schweifen *itr. V.; mit sein (geh.; auch fig.)* wander; roam

Schweige-: ~geld das hush money; **~marsch der** silent [protest-]march; **~minute die** minute's silence

schweigen *unr. itr. V.* **a)** *(nicht sprechen)* remain *or* stay *or* keep silent; say nothing; **kannst du ~?** can you keep a secret?; **~ Sie!** be silent *or* quiet! hold your tongue!; **ganz zu ~ von ...:** not to mention ...; let alone ...; **die ~de Mehrheit** the silent majority; **b)** *(aufhören zu tönen usw.)* ‹*music, noise, etc.*› stop

Schweigen das; ~s silence; **sich in ~ hüllen** maintain one's silence; **jmdn. zum ~ bringen** *(auch verhüll.)* silence sb.

Schweige·pflicht die *(eines Priesters)* obligation of secrecy; *(eines Arztes, Anwalts)* duty to maintain confidentiality

schweigsam *Adj.* silent; quiet; *(verschwiegen)* discreet

Schweigsamkeit die; ~: silence; quietness; *(Verschwiegenheit)* discretion

Schwein [ʃvain] **das; ~[e]s, ~e a)** pig; **Hackfleisch vom ~:** pork mince; **b)** *o. Pl. (Fleisch)* pork; **c)** *(salopp abwertend) (gemeiner Mensch)* swine; *(Schmutzfink)* dirty *or* mucky devil *(coll.);* mucky pig *(coll.);* **d)** *(salopp: Mensch)* **ein armes ~:** a poor devil; **kein ~ war da** there wasn't a bloody *(Brit. sl.) or (coll.)* damn soul there; **e)** *(ugs.: Glück)* **[großes] ~ haben** have a [big] stroke of luck; *(davonkommen)* get away with it *(coll.)*

Schweine-: ~bauch der *(Kochk.)* belly pork; **~braten der** *(Kochk.)* roast pork *no indef. art.;* **ein ~braten** a joint of pork; **~fleisch das** pork; **~fraß der** *(derb abwertend)* pigswill *(coll.);* **~hund der** *(derb abwertend)* bastard *(sl.);* swine; **der innere ~hund** lack of will-power; **~lende die** *(Kochk.)* loin of pork

Schweinerei die; ~, ~en *(ugs. abwertend)* **a)** *(Schmutz)* mess; **b)** *(Gemeinheit)* mean *or* dirty trick; **es ist eine ~, daß das nicht erlaubt ist** it's disgusting that that's not allowed; **c)** *(Zote)* dirty *or* smutty joke; *(Handlung)* obscene act

Schweine-: ~schmalz das lard; *(zum Streichen)* dripping; **~schnitzel das** *(Kochk.)* escalope of pork; **~stall der** *(auch fig.)* pigsty; pigpen *(Amer.)*

schweinisch *(ugs. abwertend)* **1.** *Adj.* **a)** *(schmutzig)* filthy; **b)** *(unanständig)* dirty; smutty. **2.** *adv. (unanständig)* ‹*behave*› obscenely, disgustingly

schweins-, Schweins-: ~hachse die, *(bes. südd.)* **~haxe die** *(Kochk.)* knuckle of pork; **~leder das** pigskin; **~ledern** *Adj.* pigskin

Schweiß [ʃvais] **der; ~es** sweat; *(höflicher: Transpiration)* perspiration; **mir brach der ~ aus** I broke out in a sweat; **ihm brach der kalte ~ aus** he came out in a cold sweat

schweiß-, Schweiß-: ~ausbruch der sweat; **einen ~ausbruch bekommen** start to sweat; **~bedeckt** *Adj.* covered in *or* with sweat *postpos.;* **~brenner der** *(Technik)* welding torch

schweißen *tr., itr. V.* weld

Schweißer der; ~s, ~, Schweißerin die; ~, ~nen welder

schweiß-, Schweiß-: ~fuß der sweaty foot; **~gebadet** *Adj.* bathed in sweat *postpos.;* **~naht die** *(Technik)* weld; **~naß** *Adj.* sweaty; damp with sweat *pred.;* **~perle die** bead of sweat

Schweiz ['ʃvaits] **die; ~:** Switzerland *no art.;* **in die ~ reisen** travel to Switzerland; **aus der ~ stammen** come from Switzerland

Schweizer der; ~s, ~ Swiss

schweizer·deutsch *Adj.* Swiss German; *s. auch* **deutsch; Deutsch**

Schweizerin die; ~, ~nen Swiss

schweizerisch *Adj.* Swiss

Schweizer Käse der Swiss cheese

Schwel·brand der smouldering fire

schwelen ['ʃve:lən] *(auch fig.)* smoulder

schwelgen ['ʃvɛlgn̩] *itr. V.* **a)** *(essen u. trinken)* feast; **in etw.** *(Dat.)* **~:** feast on sth.; **b) in Erinnerungen** *usw.* **~:** wallow in memories *etc.;* **in Farben ~** *(geh.)* revel in colours

schwelgerisch *Adj.* sumptuous, opulent ‹*meal, grandeur*›; luxuriant ‹*blossom*›

Schwelle ['ʃvɛlə] **die; ~, ~n a)** *(auch Physiol., Psych., fig.)* threshold; **ich werde keinen Fuß mehr über seine ~ setzen** *(fig. geh.)* I shall not set foot in his house/flat *etc.* again; **b)** *(Eisenbahn~)* sleeper *(Brit.);* [cross-]tie *(Amer.);* **c)** *(Geogr.)* swell

schwellen *unr. itr. V.; mit sein* swell; ‹*limb, face, cheek, etc.*› swell [up], become swollen; ‹*river*› become swollen, rise

Schwellen-: ~angst die; *o. Pl.* fear of entering a place; **~land das** *country at the stage of economic take-off*

Schwellung die; ~, ~en *(Med.)* swelling

Schwemme ['ʃvɛmə] **die; ~, ~n** *(Wirtsch.)* glut **(an +** *Dat.* of); **b)** *(für Tiere)* watering-place

schwemmen *tr. V.* wash

Schwemm·land das; *o. Pl.* alluvial land

Schwengel ['ʃvɛnl̩] **der; ~s, ~ a)** *(Glocken~)* clapper; **b)** *(Pumpen~)* handle

Schwenk [ʃvɛŋk] **der; ~s, ~s a)** *(Drehung)* swing; **b)** *(Film, Ferns.)* pan; **die Kamera machte einen ~ auf den Helden** the camera panned to the hero

schwenken 1. *tr. V.* **a)** *(schwingen)* swing; wave ‹*flag, handkerchief*›; **b)** *(spülen)* rinse; **c)** *(drehen)* swing round; swivel; pan ‹*camera*›; swing, traverse ‹*gun*›. **2.** *itr. V.; mit sein* ‹*marching column*› swing, wheel; ‹*camera*› pan; ‹*road, car*› swing; **rechts schwenkt!** *(Milit.)* right wheel!

schwer [ʃve:ɐ̯] **1.** *Adj.* **a)** heavy; heavy[-weight] ‹*fabric*›; *(massiv)* solid ‹*gold*›; **2 Kilo ~ sein** weigh two kilos; **wie ~ bist du?** how much do you weigh?; **b)** *(anstren-*

gend, mühevoll) heavy ‹*work*›; hard, tough ‹*job*›; hard ‹*day*›; difficult ‹*birth*›; **es ~/nicht ~ haben** have it hard/easy; **Schweres durchmachen** go through hard times; **c)** *(schlimm)* severe ‹*shock, disappointment, strain, storm*›; serious, grave ‹*wrong, injustice, error, illness, blow, reservation*›; serious ‹*accident, injury*›; heavy ‹*punishment, strain, loss, blow*›; grave ‹*suspicion*›; **ein ~er Junge** *(ugs.)* a crook with a record *(coll.)*. **2.** *adv.* **a)** heavily ‹*built, laden, armed*›; **~ wiegen** be heavy; **~ tragen** be carrying sth. heavy [with difficulty]; **~ heben** lift heavy weights; **~ auf jmdm./etw. liegen** *od.* **lasten** *(auch fig.)* weigh heavily on sb./sth.; **b)** *(anstrengend, mühevoll)* ‹*work*› hard; ‹*breathe*› heavily; **~ erkämpft sein** be hard won; **~ erkauft** dearly bought; bought at great cost *postpos.*; **~ hören** be hard of hearing; **c)** *(sehr)* seriously ‹*injured*›; greatly, deeply ‹*disappointed*›; ‹*punish*› severely, heavily; **~ aufpassen** *(ugs.)* take great care; **~ verunglücken** have a serious accident; **~ im Irrtum sein** *(ugs.)* be very much mistaken; **das will ich ~ hoffen** *(ugs.)* I should jolly well think so *(Brit. coll.)*; **er ist ~ in Ordnung** *(ugs.)* he's a good bloke *(Brit. coll.) or (sl.)* guy

schwer-, Schwer-: ~arbeit die; *o. Pl.* heavy work; **~arbeiter der** worker engaged in heavy physical work; **~athletik die** weightlifting *no art.*/combat sports *no art.*/shot-putting *no art.*/discus-throwing *no art.*; **~behindert** *Adj.* severely handicapped; *(körperlich auch)* severely disabled; **~behinderte der/die** severely handicapped person; *(körperlich auch)* severely disabled person; **die ~behinderten** the severely handicapped/disabled; **~behinderten·ausweis der** disabled person's pass; **~beschädigt** *Adj.* **a)** *präd. getrennt geschrieben* badly damaged; **b)** *(veralt.: schwerbehindert)* severely disabled; **~bewaffnet** *Adj.; präd. getrennt geschrieben* heavily armed

Schwere die; ~: a) weight; **b)** *(Physik: Schwerkraft)* gravity; **c)** *s.* **schwer 1c**: severity; seriousness; gravity; heaviness; **d)** *(Schwierigkeitsgrad)* difficulty

schwere·los 1. *Adj.* weightless. **2.** *adv.* weightlessly

Schwerelosigkeit die; ~: weightlessness

Schwerenöter ['ʃve:rənø:tɐ] **der; ~s, ~** *(ugs. scherzh.)* lady-killer *(coll.)*

schwer-, Schwer-: ~erziehbar *Adj.; präd. getrennt geschrieben* difficult ‹*child*›; **~|fallen** *unr. itr. V.; mit sein* **jmdm. fällt etw. ~:** sb. finds sth. difficult; **auch wenn's schwerfällt** whether you like it or not; **~fällig 1.** *Adj.* ponderous, heavy ‹*movement, steps*›; *(fig.)* cumbersome ‹*bureaucracy, procedure*›; ponderous ‹*style, thinking*›; **2.** *adv.* ponderously; **~fälligkeit die;** *o. Pl. s.* **~fällig:** ponderousness; heaviness; *(fig.)* cumbersomeness; ponderousness; **~gewicht das a)** *o. Pl.* heavyweight; **die Meisterschaften im ~gewicht** the heavyweight championships; **b)** *(Sportler)* heavyweight; **c)** *o. Pl. (Schwerpunkt)* main focus; emphasis; **~gewichtig** *Adj.* heavyweight *attrib.*; **~gewichtler** ['gəvɪçtlɐ] **der; ~s, ~** *(Schwerathletik)* heavyweight; **~hörig** *Adj.* hard of hearing *pred.*; **~hörigkeit die; ~:** hardness of hearing; **~industrie die** heavy industry; **~kraft die;** *o. Pl. (Physik, Astron.)* gravity; **~krank** *Adj.; präd. getrennt geschrieben* seriously ill

schwerlich *Adv.* hardly

schwer-, Schwer-: ~|machen *tr. V.* **jmdm./sich etw. ~machen** make sth. difficult for sb./oneself; **~metall das** heavy metal; **~mut die** melancholy; **~mütig 1.** *Adj.* melancholic; **2.** *adv.* melancholically; **~|nehmen** *unr. tr. V.* **etw. ~nehmen** take sth. seriously; **~punkt der** *(Physik)* centre of gravity; *(fig.)* main focus; *(Hauptgewicht)* main stress

Schwert [ʃve:ɐ̯t] **das; ~[e]s, ~er** sword; **das ~ ziehen** *od.* **zücken** draw one's sword

Schwert-: ~fisch der swordfish; **~lilie die** iris

schwer|tun *unr. refl. V. (ugs.)* **sich** *(Akk. od. Dat.)* **mit** *od.* **bei etw. ~:** have trouble with sth.; **sich** *(Akk. od. Dat.)* **mit jmdm. ~:** not get along with sb.

schwer-, Schwer-: ~verbrecher der serious offender; **~verdaulich** *Adj.; präd. getrennt geschrieben (auch fig.)* hard to digest *pred.*; **~verletzt** *Adj.; präd. getrennt geschrieben* seriously injured; **~verletzte der/die** seriously injured person; serious casualty; **~verwundet** *Adj.; präd. getrennt geschrieben* seriously wounded; **~wiegend** *Adj.* serious, grave ‹*reservation, consequence, objection, accusation, etc.*›; momentous ‹*decision*›; serious ‹*case, problem*›

Schwester ['ʃvɛstɐ] **die; ~, ~n a)** sister; **b)** *(Nonne)* nun; *(als Anrede)* Sister; **~ Petra** Sister Petra; **c)** *(Kranken~)* nurse; *(als Anrede)* Nurse; *(zur Oberschwester)* Sister

schwesterlich 1. *Adj.* sisterly. **2.** *adv.* **~ handeln** act in a sisterly way

Schwestern-: ~helferin die nursing auxiliary; auxiliary nurse; **~schülerin** *die* probationer

schwieg [ʃvi:k] *1. u. 3. Pers. Prät. v.* **schweigen**

Schwieger-: ['ʃvi:gɐ-] **~eltern** *Pl.* parents-in-law; **~mutter die** mother-in-law; **~sohn der** son-in-law; **~tochter die** daughter-in-law; **~vater der** father-in-law

Schwiele ['ʃvi:lə] **die; ~, ~n** callus

schwielig *Adj.* callused; **~e Hände** horny hands

schwierig ['ʃvi:rɪç] *Adj.* difficult

Schwierigkeit die; ~, ~en difficulty; **in ~en** *(Akk.)* **geraten** get into difficulties; **~en bekommen** have problems *or* trouble; **jmdn./sich in ~en** *(Akk.)* **bringen** get sb./oneself into trouble; **ohne ~en** without difficulty

Schwierigkeits·grad der degree of difficulty; *(von Lehrmaterial usw.)* level of difficulty

Schwimm-: ~bad das swimming-baths *pl. (Brit.)*; swimming-pool; **~becken das** swimming-pool; **~dock das** floating dock

schwimmen 1. *unr. itr. V.* **a)** *meist mit sein* swim; **~ gehen** go swimming; **b)** *meist mit sein (treiben, nicht untergehen)* float; **c)** *(ugs.: unsicher sein)* be all at sea; **ins Schwimmen geraten** start to flounder; **d)** *(überschwemmt sein)* be awash; **f)** *mit sein (triefen von)* **in etw.** *(Dat.)* **~:** be swimming in sth.; **im Geld ~** *(fig.)* be rolling in money *or* in it *(coll.)*; **g)** *mit sein (ver~)* swim. **2.** *unr. tr. V.; auch mit sein* swim

Schwimmen das; ~: swimming *no art.*

Schwimmer der; ~s, ~: a) swimmer; **b)** *(der Angel, Technik)* float

Schwimmer·becken das swimmers' pool

Schwimm-: ~flosse die flipper; **~fuß der** webbed foot; **~gürtel der** swimming-belt;

~kran der floating crane; **~lehrer der** swimming instructor; **~vogel der** web-footed bird; **~weste die** life-jacket

Schwindel ['ʃvɪndl̩] **der; ~s a)** *(Gleichgewichtsstörung)* dizziness; giddiness; vertigo; **b)** *(Anfall)* dizzy *or* giddy spell; attack of dizziness *or* giddiness *or* vertigo; **c)** *(abwertend) (Betrug)* swindle; fraud; *(Lüge)* lie; **den ~ kenne ich** *(ugs.)* that's an old trick; I know that trick

Schwindelei die; ~, ~en *(ugs.)* **a)** *o. Pl.* fibbing; **b)** *(Lüge)* fib

schwindel-: ~erregend vertiginous ‹*height, speed, depths*›; *(fig.)* meteoric ‹*career, success*›; **~frei** *Adj.* **~frei sein** have a head for heights; not suffer from vertigo

schwindelig *s.* **schwindlig**

schwindeln *itr. V.* **a) mich** *od.* **mir schwindelt** I feel dizzy *or* giddy; **in ~der Höhe** at a dizzy height; **b)** *(lügen)* tell fibs

schwinden ['ʃvɪndn̩] *unr. itr. V.; mit sein* fade; ‹*supplies, money*› run out, dwindle; ‹*effect*› wear off; ‹*interest*› fade, wane, fall off; ‹*fear, mistrust*› lessen, diminish; ‹*powers, influence*› wane, decline; ‹*courage, strength*› fail; **ihm schwand der Mut** his courage failed him

Schwindler der; ~s, ~ *(Lügner)* liar; *(Betrüger)* swindler; *(Hochstapler)* confidence trickster; con man *(coll.)*

schwindlig *Adj.* dizzy; giddy; **jmdm. wird es ~:** sb. gets dizzy *or* giddy

Schwind·sucht die *(veralt.)* consumption; tuberculosis

schwind·süchtig *Adj.* *(veralt.)* consumptive; tubercular

Schwinge ['ʃvɪŋə] **die; ~, ~n** *(geh.; auch fig.)* wing

schwingen 1. *unr. itr. V.* **a)** *mit sein (sich hin- u. herbewegen)* swing; **b)** *(vibrieren)* vibrate; **c)** *(Physik)* ‹*wave*› oscillate. **2.** *unr. tr. V. (hin- u. herbewegen)* swing; wave ‹*flag, wand*›; *(fuchteln mit)* brandish ‹*sword, axe, etc.*›; **große Reden ~** *(ugs.)* talk big; *s. auch* **Tanzbein. 3.** *unr. refl. V. (sich schnell bewegen)* **sich aufs Pferd/Fahrrad ~:** swing oneself *or* leap on to one's horse/bicycle; **der Vogel schwang sich in die Luft** *(fig.)* the bird soared [up] into the air

Schwinger der; ~s, ~ *(Boxen)* swing

Schwing·tür die swing-door

Schwingung die; ~, ~en a) swinging; *(Vibration)* vibration; **etw. in ~ versetzen** set sth. swinging/vibrating; **b)** *(Physik)* oscillation

Schwips [ʃvɪps] **der; ~es, ~e** *(ugs.)* **einen [kleinen] ~ haben** be [a bit] tipsy *or (coll.)* merry

schwirren ['ʃvɪrən] *itr. V.* **a)** *(tönen)* ‹*insect*› buzz; ‹*bowstring*› twang; **b)** *mit sein* ‹*arrow, bullet, etc.*› whiz; ‹*bird*› whir; ‹*insect*› buzz; **allerlei schwirrte mir durch den Kopf** *(fig.)* all sorts of things buzzed through my head

schwitzen 1. *itr. V.* **a)** *(auch fig.)* sweat; **ins Schwitzen kommen** *(auch fig.)* start to sweat; **b)** *(beschlagen)* steam up. **2.** *refl. V.* **sich bei der Arbeit klatschnaß ~:** get soaked with sweat from working

schwitzig *Adj.* *(ugs.)* sweaty

schwor [ʃvoːɐ̯] *1. u. 3. Pers. Sg. Prät. v.* **schwören**

schwören ['ʃvøːrən] **1.** *unr. tr., itr. V.* swear ‹*fidelity, allegiance, friendship*›; swear, take ‹*oath*›; **ich schwöre es[, so wahr mir Gott helfe]** I swear it[, so help me God]; **jmdm./sich etw. ~:** swear sth. to sb./oneself. **2.** *unr. itr. V.* swear an/the oath; **auf die Bibel ~:** swear on the Bible

schwul [ʃvuːl] *Adj.* *(ugs.)* gay *(coll.)*

schwül [ʃvyːl] *Adj.* **a)** *(feuchtwarm)* sultry; close; **b)** *(beklemmend)* oppressive

Schwule der; *adj. Dekl.* *(ugs.)* gay *(coll.)*; *(abwertend)* queer *(sl.)*

Schwüle die; ~: sultriness

schwülstig ['ʃvʏlstɪç] **1.** *Adj.* bombastic, pompous; over-ornate ‹*art, architecture*›. **2.** *adv.* bombastically; pompously

schwumm[e]rig ['ʃvʊmərɪç] *Adj.* *(ugs.)* **a)** *(unwohl)* queasy; funny *(coll.)*; **b)** *(bang)* jittery *(coll.)*; nervous; apprehensive

Schwund [ʃvʊnt] **der; ~[e]s a)** decrease, drop *(Gen.* in); *(an Interesse)* waning; falling off

Schwung [ʃvʊŋ] **der; ~[e]s, Schwünge** ['ʃvʏŋə] **a)** *(Bewegung)* swing; **b)** *(Linie)* sweep; **der elegante ~ ihrer Brauen/ihrer Nase** the elegant arch of her eyebrows/curve of her nose; **c)** *o. Pl. (Geschwindigkeit)* momentum; **~ holen** build *or* get up momentum; *(auf einer Schaukel usw.)* work up a swing; **etw. in ~ bringen** *(fig. ugs.)* get sth. going; **in ~ sein** *(fig. ugs.) (in guter Stimmung)* have livened up; *(wütend)* be worked up; *(gut laufen)* ‹*business, practice*› do a lively trade; *(gut vorankommen)* be getting on well; be right in the swing [of it]; **in ~ kommen** *(fig. ugs.) (in gute Stimmung kommen)* get going; liven up; *(wütend werden)* get worked up; *(gut vorankommen)* get right in the swing [of it]; ‹*business*› pick up; **d)** *o. Pl. (Antrieb)* drive; energy; **e)** *o. Pl. (mitreißende Wirkung)* sparkle; vitality; **f)** *o. Pl. (ugs.: größere Menge)* stack *(coll.)*; *(von Menschen)* crowd; bunch *(sl.)*

schwung·haft *Adj.* thriving; brisk, flourishing ‹*trade, business*›

schwung-: ~los 1. *Adj.* **a)** *(antriebsschwach)* lacking in energy *or* drive *postpos.*; listless; **b)** *(langweilig)* lack-lustre ‹*speech, performance, etc.*›; **2.** *adv.* ‹*sing, dance, etc.*› in a lack-lustre way; **~voll a)** *(mitreißend)* lively; spirited; spirited ‹*words*›; lively, *(coll.)* snappy ‹*tune*›; **b)** *(kraftvoll)* vigorous; **ein ~voller Handel** a roaring trade; **c)** *(elegant)* sweeping ‹*movement, gesture*›; bold ‹*handwriting, line, stroke*›; **2.** *adv.* **a)** *(mitreißend)* spiritedly; with verve; ‹*speak*› spiritedly; **b)** *(kraftvoll)* with great vigour

Schwur [ʃvuːɐ̯] **der; ~[e]s, Schwüre** ['ʃvyːrə] **a)** *(Gelöbnis)* vow; **b)** *(Eid)* oath; **die Hand zum ~ erheben** raise one's hand to take the oath

Schwur·gericht das *court with a jury*

Science-fiction ['saɪənsˈfɪkʃən] **die; ~:** science fiction

sechs [zɛks] *Kardinalz.* six; *s. auch* **acht**

Sechs die; ~, ~en six; **eine ~ schreiben/bekommen** *(Schulw.)* get a 'fail' mark; *s. auch* **¹Acht a, b, d, e; Zwei b**

Sechs·eck das hexagon

sechs·eckig *Adj.* hexagonal

Sechser der; ~s, ~ *(ugs.) (Ziffer, beim Würfeln)* six; *(Bahn, Bus)* [number] six; *(im Lotto)* six winning numbers

sechs-, Sechs-: ~fach *Vervielfältigungsz.* sixfold; *s. auch* **achtfach; ~fache das;** *adj. Dekl.* **etw. um ein ~faches/um das ~fache erhöhen** increase sth. by a factor of six; *s. auch* **Achtfache; ~hundert** *Kardinalz.* six hundred; **~jährig** *Adj. (6 Jahre alt)* six-year-old *attrib.*; six years old *postpos.*; *(6 Jahre dauernd)* six-year *attrib.*; **~kant·mutter die** hexagon nut; **~köpfig** *Adj.* six-headed ‹*monster*›; ‹*family, committee*› of six

sechs·mal *Adv.* six times; *s. auch* **achtmal**

sechst [zɛkst] *in* **wir waren zu ~**: there were six of us; *s. auch* **²acht**

sẹchst... *Ordinalz.* sixth; *s. auch* **acht...**

sẹchs-, Sẹchs-: ~tage·rennen das *(Radsport)* six-day race; **~tägig** *Adj. (6 Tage alt)* six-day-old *attrib.; (6 Tage dauernd)* six-day[-long] *attrib.;* **~tausend** *Kardinalz.* six thousand

sẹchs·teilig *Adj.* six-piece ‹*tool-set etc.*›; six-part ‹*serial*›

sẹchstel *Bruchz.* sixth

Sẹchstel das, *schweiz. meist* **der; ~s, ~**: sixth; *s. auch* **Achtel**

sẹchstens *Adv.* sixthly

Sẹchs-: ~und·sẹchzig das; ~: sixty-six; **~zylịnder·motor der** six-cylinder engine

sechzehn ['zɛçtse:n] *Kardinalz.* sixteen; *s. auch* **achtzehn**

Sẹchzehn·meter·raum der *(Fußball)* penalty area

sechzig ['zɛçtsɪç] *Kardinalz.* sixty; *s. auch* **achtzig**

sẹchziger *indekl. Adj.; nicht präd.* **die ~ Jahre** the sixties; **zwei ~ Briefmarken/Zigarren** two sixty-pfennig stamps/cigars; **eine ~ Glühbirne** a 60-watt bulb

¹Sẹchziger der; ~s, ~: sixty-year-old

²Sẹchziger die; ~, ~ a) *(Briefmarke)* sixty-pfennig/schilling *etc.* stamp; **b)** *(Zigarre)* sixty-pfennig cigar; *(Glühbirne)* 60-watt bulb

sechzigst... ['zɛçtsɪçst] *Ordinalz.* sixtieth; *s. auch* **achtzigst...**

SED [ɛsle:'de:] **die; ~**: *Abk. (ehem. DDR)* **Sozialistische Einheitspartei Deutschlands** Socialist Unity Party of Germany

Sediment [zedi'mɛnt] **das; ~[e]s, ~e** *(Geol., Chemie)* sediment

¹See [ze:] **der; ~s, ~n** lake; **der Baikalsee** Lake Baikal

²See die; ~, ~n a) *o. Pl. (Meer)* sea; **an die ~ fahren** go to the seaside; **an der ~**: by the sea[side]; **auf ~**: at sea; **er ist auf ~**: he is away at sea; **auf hoher ~**: on the high seas; **in ~ gehen** *od.* **stechen** put to sea; **Leutnant/Kapitän zur ~** *(Marine)* sub-lieutenant/[naval] captain; **zur ~ fahren** be a seaman; **b)** *o. Pl. (Seemannsspr.: ~gang)* **ruhige/rauhe** *od.* **schwere ~**: calm/rough *or* heavy sea

see-, See-: ~adler der sea eagle; white-tailed [sea] eagle; **~bad das** seaside health resort; **~beben das** seaquake; **~fahrt die a)** *o. Pl.* seafaring *no art.;* sea travel *no art.; (~fahrtskunde)* navigation; **b)** *(~reise)* voyage; *(Kreuzfahrt)* cruise; **~fisch der** sea fish; salt-water fish; **~gang der;** *o. Pl.* **leichter/starker** *od.* **hoher** *od.* **schwerer ~gang** light/heavy *or* rough sea; **~gefecht das** naval engagement; sea battle; naval battle; **~hafen der a)** *(Hafenanlagen)* harbour; **b)** *(Stadt)* seaport; **~handel der** maritime trade; **~herrschaft die;** *o. Pl.* maritime supremacy; **~hund der a)** common seal; **b)** *o. Pl. (Pelz)* seal[skin]; **~igel der** sea-urchin; **~karte die** sea chart; **~klima das** *(Geogr.)* maritime climate; **~krank** *Adj.* seasick; **~krankheit die;** *o. Pl.* seasickness; **~krieg der** naval war; *(Kriegsführung)* naval warfare; **~lachs der** pollack

Seele ['ze:lə] **die; ~, ~n** *(auch Rel., fig.)* soul; *(Psyche)* mind; **sich** *(Dat.)* **die ~ aus dem Leib schreien** *(ugs.)* shout/scream one's head off *(coll.);* **jmdm. auf der ~ liegen** *(geh.)* weigh on sb.['s mind]; **jmdm. aus der ~ sprechen** *(ugs.)* take the words out of sb.'s mouth; **aus tiefster ~**: with all one's heart; ‹*thank*› from the bottom of one's heart; **sich** *(Dat.)* **etw. von der ~ reden** unburden oneself about sth.; **die ~ von etw. sein** be the heart of sth.; **eine ~ von Mensch sein** be a good [-hearted] soul

seelen-, Seelen-: ~friede[n] der peace of mind; **~heil das** *(christl. Rel.)* salvation of one's/sb.'s soul; **~leben das;** *o. Pl. (geh.)* inner life; **~ruhe die** calmness; **in aller ~ruhe** calmly; **~ruhig 1.** *Adj.* calm; unruffled; **2.** *adv.* calmly

seelisch 1. *Adj.* psychological ‹*cause, damage, tension*›; mental ‹*equilibrium, breakdown, illness, health*›. **2.** *adv.* **~ bedingt sein** have psychological causes; **~ krank** mentally ill

See·löwe der sea-lion

Seel·sorge die; *o. Pl.* pastoral care

Seelsorger der; ~s, ~: pastoral worker; *(Geistlicher)* pastor

See-: ~macht die maritime *or* naval power; sea power; **~mann der;** *Pl.* **~leute** seaman; sailor

seemännisch ['ze:mɛnɪʃ] *Adj.* nautical

Seemanns·garn das; *o. Pl.* seaman's yarn

See-: ~meile die nautical mile; **~not die;** *o. Pl.* distress [at sea]; **jmdn. aus ~ retten** rescue sb. in distress; **~pferd[chen] das** sea-horse; **~räuber der** pirate; **~recht das;** *o. Pl.* maritime law; **~reise die** voyage; *(Kreuzfahrt)* cruise; **~rose die** water-lily; **~sack der** kitbag; **~schiffahrt die** maritime shipping *no art.;* sea shipping *no art.;* **~schlacht die** sea battle; naval battle; **~stern der** starfish; **~streitkräfte** *Pl.* naval forces; **~tang der** seaweed; **~tüchtig** *Adj.* seaworthy; **~ufer das** lake shore; shore of a/the lake; **~vogel der** sea-bird; **~weg der** sea route; **auf dem ~weg** by sea; **~zunge die** sole

Segel ['ze:gl̩] **das; ~s, ~**: sail; **die ~ streichen** strike sail; *(fig.)* throw in the towel **(vor +** *Dat.* in the face of)

Segel-: ~boot das sailing-boat; **~fliegen das;** *o. Pl.* gliding *no art.;* **~flieger der** glider pilot; **~flugzeug das** glider; **~jacht die** sailing-yacht

segeln *itr. V. mit sein* sail; **~ gehen** go sailing; go for a sail

Segel-: ~regatta die sailing regatta; **~schiff das** sailing ship; **~tuch das** sailcloth

Segen ['ze:gn̩] **der; ~s, ~ a)** blessing; *(Gebet in der Messe)* benediction; **über jmdn./etw. den ~ sprechen** bless sb./sth.; **[jmdm.] seinen ~ [zu etw.] geben** *(ugs.)* give [sb.] one's blessing [on sth.]; **meinen ~ hat er!** *(ugs.)* I have no objection [to his doing that]; *(iron.)* the best of luck to him!; **b)** *o. Pl. (Wohltat)* blessing; **etw. zum ~ der Menschheit nutzen** exploit sth. to the benefit of mankind

Segens·wünsche *Pl.* good wishes

Segler der; ~s, ~ a) *(Schiff)* sailing-ship *or* -vessel; **b)** *(Sportler)* yachtsman

Seglerin die; ~, ~nen yachtswoman

Segment [zɛ'gmɛnt] **das; ~[e]s, ~e** segment

segnen ['ze:gnən] *tr. V.* **a)** bless; **b)** *(ausstatten mit)* **mit jmdm./etw. gesegnet sein** *(auch iron.)* be blessed with sb./sth.; **im gesegneten Alter von 88 Jahren** at the venerable age of 88 years

Segnung die; ~, ~en blessing; *(iron.)* dubious blessing

sehen ['ze:ən] **1.** *unr. itr. V.* **a)** see; **schlecht/gut ~**: have bad *or* poor/good eyesight; **hast du ge~?** did you see?; **mal ~, wir wollen** *od.* **werden ~** *(ugs.)* we'll see; **siehste!** *(ugs.),* **siehst du wohl!** there, you see!; **laß mal ~**: let me *or* let's see; let me *or* let's have a look; **siehe oben/unten/Seite 80** see above/below/page 80; **da kann man** *od. (ugs.)* **kannste mal**

~, ...: that just goes to show ...; **b)** *(hin~)* look; **auf etw.** *(Akk.)* ~: look at sth.; **nach der Uhr** ~: look at one's watch; **sieh mal** *od.* **doch!** look!; **siehe da!** lo and behold!; **alle Welt sieht auf Washington** *(fig.)* all eyes are turned on Washington; **c)** *(zeigen, liegen)* **nach Süden/Norden** ~: face south/north; **d)** *(nach~)* have a look; see; **e) nach jmdm.** ~ *(betreuen)* keep an eye on sb.; *(besuchen)* drop by to see sb.; *(nach~)* look in on sb.; **nach etw.** ~ *(betreuen)* keep an eye on sth.; *(nach~)* take a look at sth.. **2.** *unr. tr. V.* **a)** *(erblicken)* see; **jmdn./etw. [nicht] zu** ~ **bekommen** [not] get to see sb./sth.; **von ihm/davon ist nichts zu** ~: he/it is nowhere to be seen; **ich habe ihn kommen [ge]~**: I saw him coming; **das sieht man** you can see that; **sieht man das?** does it show?; **hat man so was schon ge~!** did you ever see anything like it!; **[überall] gern ge~ sein** be welcome [everywhere]; **jmdn. vom Sehen kennen** know sb. by sight; **etw. gern** ~: approve of sth.; **jmdn./etw. nicht mehr** ~ **können** *(fig. ugs.)* not be able to stand the sight of sb./sth. any more; **kein Blut** ~ **können** *(ugs.)* not be able to stand the sight of blood; **er kann sich in dieser Gegend nicht mehr** ~ **lassen** he can't show his face around here any more; **b)** *(an~, betrachten)* watch ‹*television, performance*›; look at ‹*photograph, object*›; **c)** *(treffen)* see; **wir** ~ **uns morgen!** see you tomorrow!; **d)** *(sich vorstellen)* see; **e)** *(feststellen, erkennen)* see; **ich möchte doch einmal ~, ob er es wagt** I'd just like to see whether he dares [to]; **das sieht man an der Farbe** you can tell by the colour; **wir sahen, daß wir nicht mehr helfen konnten** we saw that we could not help any more; **etw. in jmdm.** ~: see sth. in sb.; **das wollen wir [doch] erst mal** ~! we'll 'see about that; **man wird** ~ **[müssen]** we'll [just have to] see; **da sieht man es [mal] wieder** it's the same old story; **f)** *(beurteilen)* see; **das sehe ich anders** I see it differently; **so darf man das nicht** ~: you mustn't look at it that way *or* like that; **so ge~**: looked at that way *or* in that light; **rechtlich ge~**: seen from a legal point of view; **ich werde ~, was ich für Sie tun kann** I'll see what I can do for you. **3.** *unr. refl. V.* **a) er kann sich nicht satt** ~: he can't see enough **(an** + *Dat.* of); **b)** *(sich betrachten als)* **sich genötigt/veranlaßt** ~, **... zu ...**: feel compelled to ...; **sich in der Lage ~, ... zu ...**: feel able to ...; think one is able to ...

sehens·wert *Adj.* worth seeing *postpos.*

Sehens·würdigkeit die sight; **die ~en [der Stadt] besichtigen** go sightseeing [in the town]; see the sights [of the town]

Seher der; ~s, ~: seer; prophet

Seherin die; ~, ~**nen** seer; prophetess

Seh-: ~fehler der sight defect; defect of vision; **~kraft die;** *o. Pl.* sight

Sehne ['ze:nə] **die;** ~, ~**n a)** *(Anat.)* tendon; sinew; **b)** *(Bogen~)* string; **c)** *(Geom.)* chord

sehnen *refl. V.* **sich nach jmdm./etw.** ~: long *or* yearn for sb./sth.; **sich [danach]** ~, **etw. zu tun** long *or* yearn to do sth.; **er sehnt sich nach Hause** he longs to go home

Seh·nerv der *(Anat.)* optic nerve

sehnig *Adj.* **a)** stringy ‹*meat*›; **b)** *(kräftig)* sinewy ‹*figure, legs, arms, etc.*›

sehnlichst 1. *Adj.; nicht präd.* **das ist mein ~es Verlangen/mein ~er Wunsch** that's what I long for most/that's my dearest wish. **2.** *adv.* **sich** *(Dat.)* **etw.** ~ **wünschen** long *or* yearn for sth.

Sehn·sucht die longing; yearning; ~ **nach jmdm. haben** long *or* yearn to see sb.

sehn·süchtig 1. *Adj.* longing *attrib.*, yearning *attrib.* ‹*desire, look, gaze, etc.*›; *(wehmütig verlangend)* wistful ‹*gaze, sigh, etc.*›. **2.** *adv.* longingly; *(wehmütig verlangend)* wistfully; **jmdn./etw.** ~ **erwarten** look forward longingly to seeing sb./to sth.; long for sb. to come/for sth.

sehnsuchts·voll *(geh.) s.* **sehnsüchtig**

sehr [ze:ɐ̯] *Adv.* **a)** *mit Adj. u. Adv.* very; ~ **viel** a great deal; **ich bin** ~ **dafür/dagegen** I'm very much in favour/against [it]; **ich bin Ihnen** ~ **dankbar** I'm most grateful to you; **jmdn.** ~ **gern haben** like sb. a lot *(coll.) or* a great deal; **b)** *mit Verben* very much; greatly; **er hat** ~ **geweint** he cried a great deal *or (coll.)* a lot; **danke** ~! thank you *or* thanks [very much]; **bitte** ~, **Ihr Schnitzel!** here's your steak, sir/madam; **Danke** ~! – **Bitte** ~! Thank you – You're welcome; **ja,** ~! yes, very much!; **nein, nicht** ~! no, not much!; **zu** ~: too much

Seh-: ~schwäche die weak vision *or* sight *no indef. art.;* **~test der** eye test

sei [zai] *1. u. 3. Pers. Sg. Präsens Konjunktiv u. Imperativ Sg. v.* **sein**

seicht [zaiçt] **1.** *Adj.* shallow; *(fig.)* shallow; superficial. **2.** *adv. (fig.)* shallowly; superficially

seid [zait] *2. Pers. Pl. Präsens u. Imperativ Pl. v.* **sein**

Seide ['zaidə] **die;** ~, ~**n** silk

Seidel ['zaidl̩] **das;** ~s, ~ *(half-litre)* beer-mug

Seidel·bast der daphne

seiden 1. a) *nicht präd. (aus Seide)* silk; **b)** *(wie Seide)* silky. **2.** *adv.* silkily

Seiden-: ~papier das tissue paper; **~raupe die** silkworm; **~strumpf der** silk stocking

seidig 1. *Adj.* silky. **2.** *adv.* silkily

Seife ['zaifə] **die;** ~, ~**n** soap

Seifen-: ~blase die soap bubble; *(fig.)* bubble; **~blasen machen** blow bubbles; **~lauge die** [soap]suds *pl.;* **~pulver das** soap powder; **~schale die** soap-dish; **~schaum der;** *o. Pl.* lather

seifig *Adj.* soapy

seihen ['zaiən] *tr. V.* strain

Seil [zail] **das;** ~s, ~**e** rope; *(Draht~)* cable; **auf dem** ~ **tanzen** dance on the high wire

Seil·bahn die cableway

Seiler der; ~s, ~: rope maker

seil-, Seil-: ~springen *unr. itr. V.; nur im Inf. u. im 2. Part.; mit sein* skip; **~tänzer der, ~tänzerin die** tightrope-walker; **~winde die** cable winch

¹sein [zain] **1.** *unr. itr. V.* **a)** be; **wie ist das Wetter?** what is the weather like?; **wie wäre es mit einem Schnaps?** how about a schnaps?; **ist das kalt heute!** it's so cold today; **wie dem auch sei** be that as it may; **seien Sie bitte so freundlich und geben Sie mir ...**: [would you] be so kind as to give me ...; **das Buch ist meins** *od. (ugs.)* **mir** the book is mine; **das wär's** that's that; *(beim Einkaufen)* that's all; that's it *(coll.);* **er ist Schwede/Lehrer** he is Swedish *or* a Swede/a teacher; **was ist er [von Beruf]?** what does he do [for a living]?; **wo warst du so lange?** where have you been all this time?; **bist du es?** is that you?; **Karl war's** *(ist verantwortlich)* it was Karl [who did/said *etc.* it]; **b)** *(unpers.)* **mir ist kalt/besser** I am *or* feel cold/better; **mir ist schlecht** I feel sick; **jmdm. ist, als [ob] ...**: sb. feels as if ...; *(jmd. hat den Eindruck [als])* sb. has a feeling that ...; **jmdm. ist nach etw.** *(ugs.)* sb. feels like *or* fancies sth.; **c)** *(ergeben)* be; make; **drei**

und vier ist *od. (ugs.)* **sind sieben** three and four is *or* makes seven; **d)** *(unpers.) (bei Zeitangabe)* be; **es ist drei Uhr/Mai/Winter** it is three o'clock/May/winter; **e)** *(sich befinden)* be; **sind noch Tomaten da?** are there any tomatoes left?; **bist du schon mal bei Eva gewesen?** have you ever been to Eva's?; **f)** *(stammen)* be; come; **er ist aus Berlin** he is *or* comes from Berlin; **g)** *(stattfinden)* be; *(sich ereignen)* be; happen; **es war an einem Sonntag im April** it was on a Sunday in April; **muß das ~?** is that really necessary?; **was darf es ~?** *(im Geschäft)* what can I get you?; **das kann doch nicht ~!** that's just not possible!; **h)** *(existieren)* be; exist; **ist was?** *(ugs.)* is anything wrong *or* the matter?; **das war einmal** that's all past now; **es war einmal ein Prinz** once upon a time there was a prince; **wenn du nicht gewesen wärest** if it hadn't been for you. **2.** *Hilfsverb* **a)** *(... werden können)* **es ist niemand zu sehen** there's no one to be seen; **das war zu erwarten** that was to be expected; **die Schmerzen sind kaum zu ertragen** the pain is hardly bearable; **es ist zu verkaufen** it is for sale; **b)** *(... werden müssen)* **die Richtlinien sind strengstens zu beachten** the guidelines are to be strictly followed; **c)** *(zur Perfektumschreibung)* have; **er ist gestorben** he has died; **sie sind gerade mit dem Wagen in die Stadt** *(ugs.)* they've just driven off into town; **d)** *(zur Bildung des Zustandspassivs)* be; **wir waren gerettet** we were saved

[2]**sein** *Possessivpron.* **a)** *(vor Substantiven) (bei Männern)* his; *(bei Mädchen)* her; *(bei Dingen, Abstrakta)* its; *(bei Tieren)* its; *(bei Männchen auch)* his; *(bei Weibchen auch)* her; *(bei Ländern)* its; her; *(bei Städten)* its; *(bei Schiffen)* her; its; *(nach man)* one's; his *(Amer.)*; **b)** *o. Subst.* his; **er hat das Seine getan** *(was er konnte)* he has done what *or* all he could; *(sein Teil)* he has done his part *or (coll.)* bit

Sein das; ~s *(Philos.)* being; *(Dasein)* existence; **~ und Schein** appearance and reality

seiner *(geh.)* **1.** *Gen. von* **er: sich ~ erbarmen** have pity on him; **~ gedenken** remember him. **2.** *Gen. von* **es: das Tier lag dort, bis sich jemand ~ annahm** the animal lay there until somebody came and looked after it

seiner-: ~seits *Adv.* for his part; *(von ihm)* on his part; **~zeit** *Adv.* at that time; in those days

seines·gleichen *indekl. Pron.* **a)** *(nach er)* his own kind; people *pl.* like himself; **er verkehrt am liebsten mit ~:** he prefers to associate with his own kind; **der König hat mich wie ~ behandelt** the King treated me as an equal; **b)** *(nach man)* one's own kind; **c)** *(nach es)* **das Kind soll mit ~ spielen** the child should play with others its own age; **das sucht ~:** it is without equal *or* is unequalled

seinet-: ~wegen *Adv.* **a)** because of him; on his account; **b)** *(ihm zuliebe, für ihn)* for his sake; for him; **c)** *(von ihm aus)* **er sagte, ~wegen sollten wir ruhig gehen** he said as far as he was concerned we could go; **~willen** *Adv. in* **um ~willen** for his sake; for him

sein|lassen *unr. tr. V. (ugs.)* stop; **laß das sein!** stop it!

Seismograph [zaismo'gra:f] **~ der; ~en, ~en** seismograph

seit [zait] **1.** *Präp. mit Dat. (Zeitpunkt)* since; *(Zeitspanne)* for; **~ 1955/dem Unfall** since 1955/the accident; **~ Wochen/einiger Zeit** for weeks/some time [past]; **ich bin ~ zwei Wochen hier** I've been here [for] two weeks; **er geht ~ vier Wochen zur Schule** he has been going to school for four weeks; **~ damals, ~ der Zeit** since then; **~ wann hast du ihn nicht mehr gesehen?** when was the last time you saw him?. **2.** *Konj.* since; **~ du hier wohnst** since you have been living here

seit·dem 1. *Adv.* since then; **das Haus steht ~ leer** since then the house has stood empty. **2.** *Konj. s.* **seit 2**

Seite ['zaitə] **die; ~, ~n a)** side; **auf** *od.* **zu beiden ~n der Straße/des Tores** on both sides of the road/gate; **die hintere/vordere ~:** the back/front; **zur** *od.* **auf die ~ gehen** *od.* **treten** move aside *or* to one side; move out of the way; **zur ~!** make way!; **jmdn. zur ~ nehmen** take sb. aside; **etw. auf die ~ schaffen** *(ugs.)* help oneself to sth.; **etw. auf die ~ legen** *(ugs.: sparen)* put sth. away *or* aside; **alles** *od.* **jedes Ding hat seine zwei ~n** *(fig.)* there are two sides to everything; **~ an ~:** side by side; **jmdm. zur ~ stehen** stand by sb.; **b)** *(Richtung)* side; **von allen ~n** *(auch fig.)* from all sides; **nach allen ~n** in all directions; *(fig.)* on all sides; **c)** *(Buch~, Zeitungs~)* page; **d)** *(Eigenschaft, Aspekt)* side; **auf der einen ~, ... auf der anderen ~ ...:** on the one hand ... on the other hand ...; **etw. ist jmds. schwache ~** *(ugs.)* sth. is not exactly sb.'s forte; *(ist jmds. Schwäche)* sb. has a weakness for sth.; **jmds. starke ~ sein** *(ugs.)* be sb.'s forte *or* strong point; **sich von der besten ~ zeigen** show one's best side; **e)** *(Partei)* side; **sich auf jmds. ~** *(Akk.)* **schlagen** take sb.'s side; **auf jmds. ~ stehen** *od.* **sein** be on sb.'s side; **jmdn. auf seine ~ bringen** *od.* **ziehen** win sb. over; **auf/von seiten der Direktion** on/from the management side; **von anderer ~ verlautete, daß ...:** it was learned from other sources that ...; **f)** *(Familie)* side

seiten *in* **auf/von ~:** *s.* **Seite e**

Seiten-: ~ausgang der side exit; **~blick der** sidelong look; *(kurzer Blick)* sidelong glance; **~eingang der** side entrance; **~flügel der a)** *(eines Gebäudes)* wing; **b)** *(eines Flügelaltars)* side panel; **~gebäude das** annex; **~hieb der** *(fig.)* side-swipe **(auf** + *Akk.* at); **~linie die a)** *(Geneal.)* offset; offshoot; **b)** *(Fußball, Rugby)* touch-line; *(Tennis, Hockey, Federball)* sideline; **~ruder das** *(Flugw.)* rudder

seitens *Präp. mit Gen. (Papierdt.)* on the part of

seiten-, Seiten-: ~schiff das [side] aisle; **~sprung der** infidelity; **einen ~sprung machen** have an affair; **~stechen das;** *o. Pl.* **~stechen haben/bekommen** have/get a stitch; **~straße die** side-street; **eine ~straße der Schillerstraße** a side-street off the Schillerstrasse; **~streifen der** verge; *(einer Autobahn)* hard shoulder; **„~streifen nicht befahrbar"** 'Soft Verges'; **~verkehrt** *Adj.* reversed; **~wand die** side wall; **~wechsel der** *(Ballspiele)* change of ends; **~wind der;** *o. Pl.* side wind; cross-wind; **~zahl die a)** page number; **b)** *(Anzahl der Seiten)* number of pages

seit·her *Adv.* since then

seitlich 1. *Adj.* **der Eingang ist ~:** the entrance is at the side. **2.** *adv.* *(an der Seite)* at the side; *(von der Seite)* from the side; *(nach der Seite)* to the side; **~ von jmdm. stehen** stand to the side of sb.

seit·wärts *Adv.* sideways

sek., Sek. *Abk.* **Sekunde** sec.

Sekret [ze'kre:t] **das; ~[e]s, ~e** *(Med., Biol.)* secretion

Sekretär [zekre'tɛ:ɐ̯] **der; ~s, ~e a)** secretary; **b)** *(Beamter)* middle-ranking civil servant; **c)**

(Schreibschrank) secretaire; secretary; bureau *(Brit.)*

Sekretariat [zekreta'ria:t] **das; ~[e]s, ~e** [secretary's/secretaries'] office

Sekretärin die; ~, ~nen secretary

Sekt [zɛkt] **der; ~[e]s, ~e** high-quality sparkling wine; ≈ champagne

Sekte ['zɛktə] **die; ~, ~n** sect

Sektion [zɛk'tsio:n] **die; ~, ~en** *(Abteilung)* section; *(im Ministerium)* department

Sektor ['zɛktɔr] **der; ~s, ~en** [-'to:rən] **a)** *(Fachgebiet)* field; sphere; **industrieller/wirtschaftlicher ~:** industrial/economic sector; **b)** *(Geom.)* sector

Sekunda [ze'kʊnda] **die; ~, Sekụnden** *(Schulw.)* **a)** *(veralt.)* sixth and seventh years *(of a Gymnasium)*; **b)** *(österr.)* second year *(of a Gymnasium)*

Sekundant [zekʊn'dant] **der; ~en, ~en** second *(in a duel or match)*

sekundär [zekun'dɛ:ɐ̯] **1.** *Adj.* secondary. **2.** *adv.* secondarily

Sekundär·literatur die secondary literature

Sekunde [ze'kʊndə] **die; ~, ~n a)** *(auch Math., Musik)* second; **es ist auf die ~ 12 Uhr** it is twelve o'clock precisely; **b)** *(ugs.: Augenblick)* second; moment

sekụndenlang 1. *Adj.* momentary. **2.** *adv.* for a moment; momentarily

Sekụnden-: ~schnelle die; *o. Pl.* **in ~schnelle** in a matter of seconds **~zeiger der** second hand

selber ['zɛlbɐ] *indekl. Demonstrativpron. s.* **selbst 1**

Sẹlber·machen das; ~s *(ugs.)* do-it-yourself *no art.*

selbst [zɛlpst] **1.** *indekl. Demonstrativpron.* **ich/du/er ~:** I myself/you yourself/he himself; **wir/ihr ~:** we ourselves/you yourselves; **sie ~:** she herself; *(Pl.)* they themselves; **Sie ~:** you yourself; *(Pl.)* you yourselves; **das Haus ~:** the house itself; **du hast es ~ gesagt** you said so yourself; *(betonter)* you yourself said so; **Wie geht's dir? – Gut! Und ~?** *(ugs.)* How are you? – Fine! And how about you?; **von ~:** automatically; **es versteht sich von ~:** it goes without saying; **die Ruhe ~ sein** *(ugs.)* be calmness itself. **2.** *adv.* even

Sẹlbst·achtung die self-respect; self-esteem

sẹlb·ständig 1. *Adj.* independent; **ein ~er Unternehmer** a self-employed business man; **sich ~ machen** set up on one's own. **2.** *adv.* independently; **~ denken** think for oneself

Sẹlbständige der/die; *adj. Dekl.* self-employed [business] person

Sẹlbständigkeit die; ~: independence

sẹlbst-, Sẹlbst-: ~auslöser der *(Fot.)* delayed-action shutter release; **~bedienungs·laden der** self-service shop; **~befriedigung die** masturbation *no art.*; **~beherrschung die** self-control *no art.*; **~bestätigung die** *(Psych.)* selfaffirmation *no art.*; **~bestimmungs·recht das;** *o. Pl.* right of selfdetermination; **~beteiligung die** *(Versicherungsw.)* [personal] excess; **~betrug der** self-deception *no art.*; **~bewußt 1.** *Adj.* self-confident; self-possessed; **2.** *adv.* self-confidently; **~bewußt·sein das** self-confidence *no art.*; *(einer sozialen Schicht o. ä.)* self-assurance; **~bildnis das** self-portrait; **~disziplin die;** *o. Pl.* self-discipline *no art.*; **~erhaltungs·trieb der** instinct for self-preservation; survival instinct; **~erkenntnis die;** *o. Pl.* self-knowledge *no art.*; **~gebacken** *Adj.* home-made; home-baked; **~gedreht** *Adj.* **~gedrehte Zigaretten, Selbstgedrehte** [one's own] rolled cigarettes; **~gefällig** *(abwertend)* **1.** *Adj.* self-satisfied; smug; **2.** *adv.* smugly; in a self-satisfied way; **~gefälligkeit die;** *o. Pl.* self-satisfaction; smugness; **~gemacht** *Adj.* home-made ⟨*jam, liqueur, sausage, basket, etc.*⟩; self-made ⟨*dress, pullover, etc.*⟩; ⟨*dress, pullover, etc.*⟩ one has made oneself; **~gerecht** *(abwertend)* **1.** *Adj.* self-righteous; **2.** *adv.* self-righteously; **~gespräch das** conversation with oneself; **~gespräche führen** talk to oneself; **~gestrickt** *Adj.* home-made; hand-knitted; **~herrlich 1.** *Adj.* high-handed; **2.** *adv.* high-handedly; in a high-handed manner; **~hilfe die;** *o. Pl.* self-help *no art.*; **~justiz die** self-administered justice; **~justiz üben** take the law into one's own hands; **~klebe·folie die** self-adhesive plastic sheeting; **~klebend** *Adj.* self-adhesive; **~kosten·preis der** *(Wirtsch.)* cost price; **zum [reinen] ~kostenpreis** at [no more than] cost; **~kritik die;** *o. Pl.* self-criticism; **~kritisch 1.** *Adj.* self-critical; **2.** *adv.* self-critically; **~laut der** vowel; **~los 1.** *Adj.* selfless; **2.** *adv.* selflessly; unselfishly; **~mord der** suicide *no art.*; **~mord begehen** commit suicide; **~mörder der** suicide; **~mörderisch** *Adj.* suicidal

sẹlbstmord·gefährdet *Adj.* potentially suicidal

Sẹlbstmord·versuch der suicide attempt

sẹlbst-, Sẹlbst-: ~redend *Adv.* naturally; of course; **~sicher 1.** *Adj.* self-confident; **2.** *adv.* in a self-confident manner; full of self-confidence; **~sucht die;** *o. Pl.* selfishness; self-interest; **~süchtig 1.** *Adj.* selfish; **2.** *adv.* selfishly; **~tätig 1.** *Adj.* automatic; **2.** *adv.* automatically; **~verständlich 1.** *Adj.* natural; **etw. für ~verständlich halten** regard sth. as a matter of course; *(für gegeben hinnehmen)* take sth. for granted; **das ist doch ~verständlich** that goes without saying; **2.** *adv.* naturally; of course; **~verständlich nicht!** of course not!; **~verständlichkeit die** matter of course; **etw. mit der größten ~verständlichkeit tun** do sth. as if it were the most natural thing in the world; **~verteidigung die** self-defence *no art.*; **~vertrauen das** self-confidence; **~verwaltung die** self-government *no art.*; **~wähl·ferndienst der** *(Postw.)* direct dialling; STD; **~zerstörung die** self-destruction; **~zweck der;** *o. Pl.* end in itself; **~zweck sein/zum ~zweck werden** be/become an end in itself

Selen [ze'le:n] **das; ~s** *(Chemie)* selenium

selig ['ze:lɪç] **1.** *Adj.* **a)** *(Rel.)* blessed; **Gott hab' ihn ~:** God rest his soul; *s. auch* **geben a; glauben b; b)** *(tot)* late [lamented]; **c)** *(kath. Kirche: seliggesprochen)* **die ~e Dorothea** the blessed Dorothy; **d)** *(glücklich)* blissful ⟨*idleness, slumber, etc.*⟩; blissfully happy ⟨*person*⟩; **~ [über etw. (Akk.)] sein** be overjoyed *or (coll.)* over the moon [about sth.]. **2.** *adv.* blissfully

Seligkeit die; ~, ~en a) *o. Pl. (Rel.)* [state of] blessedness; beatitude; **die ewige ~:** eternal bliss; **b)** *(Glücksgefühl)* bliss *no pl.*; [blissful] happiness *no pl.*

Selig·sprechung die; ~, ~en *(kath. Kirche)* beatification

Sellerie ['zɛləri] **der; ~s, ~[s]** *od.* **die; ~, ~** *(Stauden~)* celeriac; *(Stangen~)* celery

selten ['zɛltn̩] **1.** *Adj.* rare; infrequent ‹*visit, visitor*›; **in den ~sten Fällen** very rarely. **2.** *adv.* **a)** rarely; **wir sehen uns nur noch ~:** we seldom *or* hardly ever see each other now; **b)** *(sehr)* exceptionally; uncommonly

Seltenheit die; ~, ~**en** rarity; **es ist eine ~, daß ...:** it is rare that ...

Seltenheits·wert der; ~[e]s rarity value

Selters·wasser ['zɛltɐs-] **das** seltzer [water]

seltsam 1. *Adj.* strange; peculiar; odd. **2.** *adv.* strangely; peculiarly

seltsamerweise *Adv.* strangely enough

Semantik [ze'mantɪk] **die;** ~ *(Sprachw.)* semantics *sing., no art.*

semantisch *(Sprachw.)* **1.** *Adj.* semantic. **2.** *adv.* semantically

Semester [ze'mɛstɐ] **das;** ~s, ~ semester; **er hat 14 ~ Jura studiert** he studied law for seven years

Semikolon [zemi'ko:lɔn] **das;** ~s, ~s *od.* **Semikola** semicolon

Seminar [zemi'na:ɐ̯] **das;** ~s, ~e **a)** *(Lehrveranstaltung)* seminar (**über** + *Akk.* on); **b)** *(Institut)* department; **das juristische ~/~ für Alte Geschichte** the Law Department/Department of Ancient History; **c)** *(Priester~)* seminary; **d)** *(für Referendare) course for student teachers prior to their second state examination*

Seminar-: ~arbeit die seminar paper; **~schein der** certificate of attendance [at a seminar]

Semit [ze'mi:t] **der;** ~en, ~en, **Semitin die;** ~, ~nen Semite

semitisch *Adj.* Semitic

Semmel ['zɛml̩] **die;** ~, ~n *(bes. österr., bayr., ostmd.)* [bread] roll; **weggehen wie warme ~n** *(ugs.)* sell like hot cakes

Semmel-: ~brösel der ; *meist Pl.* breadcrumb; **~knödel der** *(bayr., österr.)* bread dumpling

sen. *Abk.* **senior** sen.

Senat [ze'na:t] **der;** ~[e]s, ~e *(Hist., Politik, Hochschulw.)* senate; **der US-~:** the US Senate

Senator der; ~s, ~en, **Senatorin die;** ~, ~nen senator

Sende- ['zɛndə-]: **~bereich der, ~gebiet das** *(Rundf., Ferns.)* transmitting area

[1]**senden** *unr. (auch regelm.) tr. V. (geh.)* send; **jmdm. etw. ~:** send sb. sth.; **etw. an jmdn. ~:** send sth. to sb.

[2]**senden** *regelm. (schweiz. unr.) tr., itr. V.* broadcast ‹*programme, play, etc.*›; transmit ‹*concert, signals, Morse, etc.*›; **Hilferufe ~:** send out distress signals

Sender der; ~s, ~: [broadcasting] station; *(Anlage)* transmitter

Sende-: ~reihe die *(Rundf., Ferns.)* series [of programmes]; **~schluß der** *(Rundf., Ferns.)* close-down; end of broadcasting; **zum ~schluß noch ein Krimi** now as our last programme, a thriller; **~station die** *(Funk, Rundf., Ferns.)* broadcasting station; **~zeit die** *(Rundf., Ferns.)* broadcasting time; **die ~zeit um zehn Minuten überschreiten** overrun by ten minutes

Sendung die; ~, ~en **a)** consignment; **b)** *o. Pl. (geh.: Aufgabe)* mission; **c)** *(Rundf., Ferns.: Darbietung)* programme; broadcast; **d)** *(Rundfunkt., Ferns.: Ausstrahlung)* transmission; broadcast[ing]

Senf [zɛnf] **der;** ~[e]s, ~e mustard

Senf-: ~gas das mustard gas; **~gurke die** *gherkin pickled with mustard seeds*

sengen 1. *tr. V.* singe. **2.** *itr. V.* **a)** *(brennen)* singe; **b)** *(heiß sein)* be scorching; **eine ~de Hitze** a scorching heat

senil [ze'ni:l] **1.** *Adj.* senile. **2.** *adv.* in a senile manner

Senilität die; ~: senility

senior ['ze:ni̯ɔr] *indekl. Adj.; nach Personennamen* senior

Senior der; ~s, ~en [ze'ni̯o:rən] **a)** *(Kaufmannsspr.)* senior partner; **b)** *(Sport)* senior [player]; **c)** *(Rentner)* senior citizen; **d)** *(Älteste̲r)* oldest member

Senioren·heim das home for the elderly

Seniorin [ze'ni̯o:rɪn] **die;** ~, ~nen *s.* **Senior**

Senk·blei das *(Bauw.)* plumb[-bob]

Senke ['zɛŋkə] **die;** ~, ~n hollow

senken 1. *tr. V.* **a)** lower; *(Bergbau)* sink ‹*shaft*›; lower ‹*flag*›; drop ‹*starting flag*›; **den Kopf ~:** bow one's head; **die Augen** *od.* **den Blick/die Stimme ~:** lower one's eyes *or* glance/voice; **b)** *(herabsetzen)* reduce ‹*fever, pressure, prices, etc.*›. **2.** *refl. V.* ‹*curtain, barrier, etc.*› fall, come down; ‹*ground, building, road*› subside, sink; ‹*water-level*› fall, sink

senk-, Senk-: ~fuß der *(Anat.)* flat foot; **~grube die** *(Bauw.)* cesspit; **~recht 1.** *Adj.* vertical; **in ~rechter Stellung** in an upright position; **2.** *adv.* vertically; **~rechte die a)** *(Geom.)* perpendicular; **b)** vertical line; vertical; upright; **~recht·starter der;** ~s, ~ **a)** *(Flugzeug)* vertical take-off aircraft; **b)** *(ugs.) (Aufsteiger)* whizz-kid *(coll.)*; *(Sache)* instant success

Senkung die; ~, ~en **a)** *o. Pl.* lowering; **b)** *o. Pl. (Reduzierung)* reduction; lowering

Senner der; ~s, ~ *(bayr., österr.) Alpine herdsman and dairyman*

Sennerin die; ~, ~nen *(bayr., österr.) Alpine herdswoman and dairywoman*

Senn·hütte die *(bayr., österr.)* Alpine hut

Sensation [zɛnza'ts̯i̯o:n] **die;** ~, ~en sensation; *(Darbietung)* sensational performance

sensationell 1. *Adj.* sensational. **2.** *adv.* in a sensational manner; sensationally; **eine ~ aufgemachte Story** a sensationalized story

Sense ['zɛnzə] **die;** ~, ~n **a)** scythe; **b)** *(salopp)* **jetzt ist ~:** this really is [the end of] it *(coll.)*

sensibel [zɛn'zi:bl̩] **1.** *Adj.* sensitive. **2.** *adv.* sensitively

Sensibilität die; ~: sensitivity

Sensor ['zɛnzɔr] **der;** ~s, ~en [-'zo:rən] *(Technik)* sensor

Sensor·taste die *(Technik)* touch key

sentimental [zɛntimɛn'ta:l] **1.** *Adj.* sentimental. **2.** *adv.* sentimentally

Sentimentalität die; ~, ~en sentimentality; **das sind bloße ~en** that is mere sentimentality

separat [zepa'ra:t] **1.** *Adj.* separate; **~e Wohnung** self-contained flat *(Brit.) or (Amer.)* apartment. **2.** *adv.* separately; **er wohnt ~:** he has self-contained accommodation

Separatismus der; ~: separatism *no art.*

September [zɛp'tɛmbɐ] **der;** ~[s], ~: September; *s. auch* **April**

Septime [zɛp'ti:mə] **die;** ~, ~n *(Musik)* seventh

Serbe ['zɛrbə] **der;** ~n, ~n Serb; Serbian

Serbien ['zɛrbi̯ən] **(das);** ~s Serbia

Serbin die; ~, ~nen *s.* **Serbe**

serbisch *Adj.* Serbian; *s. auch* **deutsch; Deutsch**

serbo·kroatisch [zɛrbokro'a:tɪʃ] *Adj.* Serbo-Croat; *s. auch* **deutsch; Deutsch**

Serenade [zere'na:də] **die;** ~, ~n *(Musik)* serenade

Serie ['ze:ri̯ə] **die;** ~, ~n series; **eine ~ Briefmarken** a set of stamps

serien·mäßig 1. *Adj.* standard ‹*product, model, etc.*›; *(immer ein-*

gebaut) ⟨*feature, accessory*⟩ fitted as standard. **2.** *adv.* **a)** ~ **gefertigt** *od.* **gebaut** produced in series; **b)** *(nicht als Sonderausstattung)* ⟨*fitted, supplied, etc.*⟩ as standard; ~ **mit etw. ausgerüstet sein** have sth. as a standard fitting

Serien·produktion die series production

seriös [ze'ri̯øːs] **1.** *Adj.* **a)** *(solide)* respectable ⟨*person, hotel, etc.*⟩; *(vertrauenswürdig)* reliable, trustworthy ⟨*firm, partner, etc.*⟩; **b)** *(ernstgemeint)* serious ⟨*offer, applicant, artist, etc.*⟩. **2.** *adv.* *(solide)* respectably; *(vertrauenswürdig)* in a trustworthy manner

Seriosität [zeri̯ozi'tɛːt] **die;** ~ *(geh.)* respectability; *(Vertrauenswürdigkeit)* reliability; trustworthiness; *(eines Geschäftsmanns, einer Firma)* probity

Serpentine [zɛrpɛn'tiːnə] **die;** ~, ~**n a)** *(Weg)* zigzag mountain road *(with numerous hairpin bends)*; **b)** *(Kehre)* hairpin bend

Serum ['zeːrʊm] **das;** ~**s, Seren** *od.* **Sera** *(Med., Physiol.)* serum

[1]**Service** [zɛr'viːs] **das;** ~, ~: [dinner *etc.*] service

[2]**Service** ['zøːɐ̯vɪs] **der** *od.* **das;** ~, ~**s** ['zøːɐ̯vɪsɪs] **a)** *o. Pl.* service; *(Abteilung)* service department; **b)** *(Tennis: Aufschlag)* serve; service

servieren [zɛr'viːrən] **1.** *tr. V.* **a)** *(auftragen)* serve ⟨*food, drink*⟩; *(fig.)* serve up ⟨*information*⟩; deliver ⟨*line, punchline, etc.*⟩; **jmdm. etw.** ~: serve sb. sth.; **b)** *(Ballspiele)* **jmdm. den Ball** ~: feed/*(Tennis)* serve the ball to sb. **2.** *itr. V.* **a)** serve [at table]; **b)** *(Fußball)* pass; make a pass; *(Tennis)* serve

Serviererin die; ~, ~**nen** waitress

Servier·wagen der [serving-] trolley

Serviette [zɛr'vi̯ɛtə] **die;** ~, ~**n** napkin; serviette *(Brit.)*

Servo- ['zɛrvo-]: ~**bremse die** servo [-assisted] brake; ~**lenkung die** power [-assisted] steering *no indef. art.*

Servus ['zɛrvʊs] *Interj.* *(bes. südd., österr.)* *(beim Abschied)* goodbye; so long *(coll.)*; *(zur Begrüßung)* hello

Sesam ['zeːzam] **der;** ~**s,** ~**s a)** *(Pflanze)* sesame; *(Samen)* sesame seeds *pl.*; **b)** ~, **öffne dich!** open sesame!

Sessel ['zɛsl̩] **der;** ~**s,** ~ **a)** easy chair; *(mit Armlehne)* armchair; **b)** *(österr.: Stuhl)* chair

Sessel·lift der chair-lift

seßhaft ['zɛshaft] *Adj.* settled ⟨*tribe, way of life*⟩; ~ **werden** settle [down]

Set [zɛt] **das** *od.* **der;** ~**[s],** ~**s a)** *(Satz)* set, combination (**aus** of); **b)** *(Deckchen)* table- *or* place-mat

setzen ['zɛt͡sn̩] **1.** *refl. V.* **a)** *(hin~)* sit [down]; **setz dich/setzt euch/setzen Sie sich** sit down; take a seat; **sich aufs Sofa** *usw.* ~: sit on the sofa *etc.*; **sich zu jmdm.** ~: [go and] sit with sb.; join sb.; **b)** *(sinken)* ⟨*coffee, solution, froth, etc.*⟩ settle; ⟨*sediment*⟩ sink to the bottom; **c)** *(in präp. Verbindungen)* **sich mit jmdm. ins Einvernehmen** ~: come to an agreement with sb.; d) *(dringen)* **der Staub setzt sich in die Kleider** the dust gets into one's clothes. **2.** *tr. V.* **a)** *(plazieren)* put; **eine Figur/einen Stein** ~: move a piece/man; **b)** *(einpflanzen)* plant ⟨*tomatoes, potatoes, etc.*⟩; **c)** *(aufziehen)* hoist ⟨*flag etc.*⟩; set ⟨*sails, navigation lights*⟩; **d)** *(Druckw.)* set ⟨*manuscript etc.*⟩; **e)** *(schreiben)* put ⟨*name, address, comma, etc.*⟩; **seinen Namen unter etw.** *(Akk.)* ~: put one's signature to sth.; sign sth.; **f)** *(in präp. Verbindungen)* **in/außer Betrieb** ~: start up/stop ⟨*machine etc.*⟩; put ⟨*lift etc.*⟩ into operation/take ⟨*lift etc.*⟩ out of service; *(ein-/ausschalten)* switch on/off; **g)** *(aufstellen)* put up, build ⟨*stove*⟩; stack ⟨*logs, bricks*⟩; **h) sein Geld auf etw.** *(Akk.)* ~: put one's money on sth.; *s. auch* **Akzente; Ende; i)** *(ugs.)* **es setzt was** *od.* **Prügel** *od.* **Hiebe** he/she *etc.* gets a hiding *(coll.) or* thrashing. **3.** *itr. V.* **a)** *meist mit sein (im Sprung)* leap, jump (**über** + *Akk.* over); **über einen Fluß** ~ *(mit einer Fähre o. ä.)* cross a river; **b)** *(beim Wetten)* bet; **auf ein Pferd/auf Rot** ~: back a horse/put one's money on red

Setzer der; ~**s,** ~, **Setzerin die;** ~, ~**nen** *(Druckw.)* [type]setter

Setz·kasten der a) *(Gartenbau)* seedling box; **b)** *(Druckw.)* [type-]case

Setzling ['zɛt͡slɪŋ] **der;** ~**s,** ~**e** seedling

Setz·maschine die composing *or* typesetting machine

Seuche ['zɔy̯çə] **die;** ~, ~**n** epidemic

Seuchen-: ~**bekämpfung die** epidemic control *no art.*; ~**gefahr die;** *o. Pl.* danger of an epidemic

seufzen ['zɔy̯ft͡sn̩] *itr., tr. V.* sigh; **schwer/erleichtert** ~: give *or* heave a deep sigh/a sigh of relief

Seufzer der; ~**s,** ~: sigh

Sex [zɛks] **der;** ~**[es]** sex *no art.*

Sex-: ~**-Appeal** [~ ə'piːl] **der;** ~**s** sex appeal; ~**bombe die** *(salopp)* sex-bomb *(coll.)*; sexpot *(coll.)*; ~**film der** sex film

Sexismus der; ~: sexism *no art.*

sexistisch 1. *Adj.* sexist. **2.** *adv.* ⟨*behave, think, etc.*⟩ in a sexist manner

Sex·shop der; ~**s,** ~**s** sex shop

Sexta ['zɛksta] **die;** ~, **Sexten** *(Schulw.)* **a)** *(veralt.)* first year *(of a Gymnasium)*; **b)** *(österr.)* sixth year *(of a Gymnasium)*

Sextant [zɛks'tant] **der;** ~**en,** ~**en** sextant

Sexte ['zɛkstə] **die;** ~, ~**n** *(Musik)* sixth

Sextett [zɛks'tɛt] **das;** ~**[e]s,** ~**e** *(Musik)* sextet

Sexual- [zɛ'ksu̯aːl-]: ~**erziehung die** sex education; ~**hormon das** sex hormone

Sexualität [zɛksu̯ali'tɛːt] **die;** ~: sexuality *no art.*

Sexual-: ~**kunde die;** *o. Pl.* *(Schulw.)* sex education *no art.*; ~**leben das;** *o. Pl.* sex life; ~**verbrechen das** sex crime; ~**verbrecher der** sex offender

sexuell 1. *Adj.* sexual. **2.** *adv.* sexually

sexy ['zɛksi] *(ugs.)* **1.** *indekl. Adj.* sexy. **2.** *adv.* sexily

Sezessions·krieg der; *o. Pl.* [American] Civil War

sezieren [ze't͡siːrən] *tr. V.* dissect ⟨*corpse*⟩

sfr., *(schweiz. nur:)* **sFr.** *Abk.* **Schweizer Franken**

Shampoo [ʃam'puː], **Shampoon** [ʃam'poːn] **das;** ~**s,** ~**s** shampoo

Sheriff ['ʃɛrɪf] **der;** ~**s,** ~**s** sheriff

Sherry ['ʃɛri] **der;** ~**s,** ~**s** sherry

Shorts [ʃɔrt͡s] *Pl.* shorts

Show [ʃoʊ] **die;** ~, ~**s** show

Show·master ['-maːstə] **der;** ~**s,** ~: compère

Siam [ziːam] **(das);** ~**s** *(hist.)* Siam

Siamese [zia'meːzə] **der;** ~**n,** ~**n, Siamesin die;** ~, ~**nen** Siamese

siamesisch *Adj.* Siamese

Siam·katze die Siamese cat

Sibirien [zi'biːri̯ən] **(das);** ~**s** Siberia

sibirisch *Adj.* Siberian; ~**e Kälte** Arctic temperatures *pl.*

sich [zɪç] *Reflexivpron. der 3. Pers. Sg. und Pl. Akk. und Dat.* **a)** himself / herself / itself / themselves; *(auf* man *bezogen)* oneself; *(auf das Anredepron.* Sie *bezogen)* yourself/yourselves; *(mit reflexiven Verben)* ~ **freuen/wundern/schämen/täuschen** be pleased/surprised/ashamed/mistaken; ~

sorgen/verspäten/öffnen worry/ be late/open; *s. auch* **an 1 d; kommen l; b)** *reziprok* one another; each other

Sichel ['zɪçḷ] **die; ~, ~n** sickle

sicher ['zɪçɐ] **1.** *Adj.* **a)** *(ungefährdet)* safe ⟨*road, procedure, etc.*⟩; secure ⟨*job, investment, etc.*⟩; **vor jmdm./etw. ~ sein** be safe from sb./sth.; **~ ist ~**: it's better to be on the safe side; better safe than sorry; **b)** *(zuverlässig)* reliable ⟨*evidence, source*⟩; secure ⟨*income*⟩; certain, undeniable ⟨*proof*⟩; *(vertrauenswürdig)* reliable, sure ⟨*judgment, taste, etc.*⟩; **c)** *(selbstbewußt)* [self-]assured, [self-]confident ⟨*person, manner*⟩; **d)** *(gewiß)* certain; sure; **der ~e Sieg/Tod** certain victory/ death. **2.** *adv.* **a)** *(ungefährdet)* safely; **um ganz ~ zu gehen** to be quite sure; **b)** *(zuverlässig)* reliably; **~ [Auto] fahren** be a safe driver; **c)** *(selbstbewußt)* [self-]confidently; **~ auftreten** behave in a self-assured *or* self-confident manner. **3.** *Adv.* certainly; *(plädierend)* surely; **~ kommt er bald** he is sure to come soon

sicher|gehen *unr. itr. V.; mit sein* play safe; **um sicherzugehen** to be on the safe side

Sicherheit die; ~, ~en a) *o. Pl.* safety; *(der Öffentlichkeit)* security; **die ~ der Arbeitsplätze** job security; **in ~ sein** be safe; **jmdn./ etw. in ~ [vor etw.** *(Dat.)*] **bringen** save *or* rescue sb./sth. [from sth.]; **sich vor etw.** *(Dat.)* **in ~ bringen** escape from sth.; **b)** *o. Pl. (Gewißheit)* certainty; **mit an ~** *(Akk.)* **grenzender Wahrscheinlichkeit** with almost complete certainty; almost certainly; **c)** *(Wirtsch.: Bürgschaft)* security; **d)** *o. Pl. (Zuverlässigkeit, Vertrauenswürdigkeit)* reliability; soundness; **e)** *o. Pl. (Selbstbewußtsein)* [self-] confidence; [self-]assurance; **~ im Auftreten** [self-]confidence of manner

sicherheits-, Sicherheits-: ~abstand der *(Verkehrsw.)* safe distance between vehicles; **~bindung die** *(Ski)* safety binding; **~glas das** safety glass; **~gurt der a)** *(im Auto, Flugzeug)* seat-belt; **b)** *(für Bauarbeiter, Segler)* safety harness; **~halber** *Adv.* to be on the safe side; for safety's sake; **~kette die** safety *or* door chain; **~nadel die** safety-pin; **~rat der;** *o. Pl.* Security Council; **~schloß das** safety lock; **~vorkehrung die** [safety] precaution; safety measure; **~vorschrift die** safety regulation

sicherlich *Adv.* certainly

sichern *tr. V.* **a)** make ⟨*door etc.*⟩ secure; *(garantieren)* safeguard ⟨*rights, peace*⟩; *(schützen)* protect ⟨*rights etc.*⟩; **b)** *(verschaffen; polizeilich ermitteln)* secure ⟨*ticket, clue, etc.*⟩; **sich** *(Dat.)* **etw. ~**: secure sth.

sicher|stellen *tr. V.* **a)** *(beschlagnahmen)* impound ⟨*goods, vehicle*⟩; seize ⟨*stolen goods*⟩; confiscate ⟨*licence etc.*⟩; **b)** *(gewährleisten)* guarantee ⟨*supply, freedom, etc.*⟩

Sicherung die; ~, ~en a) *o. Pl. (das Sichern)* safeguarding **(vor** + *Dat.*, **gegen** from, against); *(das Schützen)* protection **(vor** + *Dat.*, **gegen** from, against); **b)** *(Elektrot.)* fuse; **c)** *(techn. Vorrichtung)* safety-catch

Sicherungs·kasten der fuse-box

Sicht [zɪçt] **die; ~, ~en a)** *o. Pl. (~weite)* visibility *no art.*; *(Ausblick)* view **(auf** + *Akk.*, **in** + *Akk.* of); **gute** *od.* **klare/schlechte ~**: good/poor visibility; **außer ~ sein** be out of sight; **Land in ~!** land ahoy!; **b)** *o. Pl. (Kaufmannsspr.)* **Wechsel auf ~**: bill payable on demand *or* at sight; **c) auf lange/kurze ~**: in the long/ short term; **auf lange** *od.* **weite ~ planen** plan on a long-term basis; **d)** *(Betrachtungsweise)* point of view; **aus meiner ~**: as I see it

sichtbar 1. *Adj.* visible; *(fig.)* apparent ⟨*reason*⟩. **2.** *adv.* visibly

sichten *tr. V.* **a)** *(erspähen)* sight; **b)** *(durchsehen)* sift [through]; *(prüfen)* examine

sichtlich 1. *Adj.* obvious; evident. **2.** *adv.* obviously; evidently; visibly ⟨*impressed*⟩

Sicht-: ~verhältnisse *Pl.* visibility *sing.*; **~vermerk der** visa; **~weite die** visibility *no art.*; **außer/in ~weite sein** be out of/in sight

sickern ['zɪkɐn] *itr. V.; mit sein* seep; *(spärlich fließen)* trickle; *(fig.)* ⟨*money*⟩ leak away

sie [zi:] **1.** *Personalpron.; 3. Pers. Sg. Nom. Fem. (bei weiblichen Personen und Tieren)* she; *(bei Dingen, Tieren)* it; *(bei Behörden)* they *pl.*; **Wer hat es gemacht? – Sie war es/Sie** Who did it? – It was her/She did; s. auch [1]**ihr; ihrer a. 2.** *Personalpron.; 3. Pers. Pl. Nom.* **a)** they; **Wer hat es gemacht? – Sie waren es/Sie** Who did it? – It was them/They did; *s. auch* **ihnen; ihrer b; b)** *(ugs.: man)* **hier wollen ~ das neue Rathaus bauen** here's where they are going to build the new town hall. **3.** *Akk. des Personalpron.* **sie 1** *(bei weiblichen Personen und Tieren)* her; *(bei Dingen und Tieren)* it; *(bei Behörden)* them *pl.* **4.** *Akk. des Personalpron.* **sie 2 a** them

Sie *Personalpron.* you; **jmdn. mit ~ anreden** address sb. as 'Sie'; use the polite form of address to sb.

Sieb [zi:p] **das; ~[e]s, ~e** sieve; *(Kaffee~, Tee~)* strainer; *(für Sand, Kies usw.)* riddle

[1]**sieben 1.** *tr. V.* **a)** *(durch~)* sieve ⟨*flour etc.*⟩; riddle ⟨*sand, gravel, etc.*⟩; **b)** *(auswählen)* screen ⟨*candidates, visitors, etc.*⟩. **2.** *itr. V.* **a)** use a sieve/strainer/riddle; **b)** *(auswählen)* pick and choose

[2]**sieben** *Kardinalz.* seven; *s. auch* **acht**

Sieben die; ~, ~en a) seven; **b)** *(ugs.: Bus-, Bahnlinie)* number seven

Siebener der; ~s, ~ *(ugs.) s.* **Sieben**

sieben-, Sieben-: ~fach *Vervielfältigungsz.* sevenfold; *s. auch* **achtfach; ~fache das;** *adj. Dekl.* **das ~fache** seven times as much; *s. auch* **Achtfache; ~hundert** *Kardinalz.* seven hundred; **~jährig** *Adj.* **a)** *(7 Jahre alt)* seven-year-old *attrib.*; seven years old *pred.*; **b)** *(7 Jahre dauernd)* seven-year *attrib.*; **der Siebenjährige Krieg** the Seven Years War; **~köpfig** *Adj.* seven-headed ⟨*monster*⟩; ⟨*family, committee*⟩ of seven; **~mal** *Adj.* seven times; *s. auch* **achtmal; ~meilenstiefel** *Pl. (scherzh.)* seven-league boots; **~meter der** *(Hockey)* penalty [shot]; *(Hallenhandball)* penalty [throw]; **~sachen** *Pl. (ugs.)* **meine/deine** *usw.* **~sachen** my/your *etc.* belongings *or (coll.)* bits and pieces; **~schläfer der** dormouse

siebent... *Ordinalz. s.* **siebt...**

sieben·tausend *Kardinalz.* seven thousand

sieben·teilig *Adj.* seven-piece ⟨*tool-set etc.*⟩; seven-part ⟨*serial*⟩

siebentel ['zi:bn̩tḷ] *s.* **siebtel**

Siebentel das; ~s, ~ *s.* **Siebtel**

siebentens *Adv. s.* **siebtens**

siebt [zi:pt] *in* **wir waren zu ~**: there were seven of us; *s. auch* [2]**acht**

siebt... [zi:pt...] *Ordinalz.* seventh; *s. auch* **acht...**

siebtel ['zi:ptḷ] *Bruchz.* seventh

Siebtel das (*schweiz. meist* **der); ~s, ~**: seventh

siebtens *Adv.* seventhly
sieb·zehn *Kardinalz.* seventeen
siebzig ['zi:ptsıç] *Kardinalz.* seventy; *s. auch* **achtzig**
[1]**Siebziger der; ~s, ~:** seventy-year-old
[2]**Siebziger die; ~, ~** *(ugs.)* **a)** *(Briefmarke)* seventy-pfennig/centimes *etc.* stamp; **b)** *(Zigarre)* seventy-pfennig cigar; **c)** *Pl.* *(ugs.: 70er Jahre)* seventies
siebzigst... *Ordinalz.* seventieth
siedeln ['zi:dl̩n] *itr. V.* settle
sieden ['zi:dn̩] *unr. od. regelm. itr. V.* boil
Siede·punkt der *(Physik; auch fig.)* boiling-point
Siedler der; ~s, ~, Siedlerin die; ~, ~nen settler
Siedlung die; ~, ~en a) *(Wohngebiet)* [housing] estate; **b)** *(Niederlassung)* settlement
Siedlungs·haus das house on an estate; estate house
Sieg [zi:k] **der; ~[e]s, ~e** victory, *(bes. Sport)* win (**über** + *Akk.* over); **den ~ davontragen** *od.* **erringen** *(geh.)* be victorious; *(Sport)* be the winner/winners; **ein ~ der Vernunft** *(fig.)* a victory for common sense
Siegel ['zi:gl̩] **das; ~s, ~:** seal; *(von Behörden)* stamp; **unter dem ~ der Verschwiegenheit** *(fig.)* under the seal of secrecy
siegeln *tr. V.* seal
Siegel·ring der signet-ring
siegen *itr. V.* win; **über jmdn. ~:** gain *or* win a victory over sb.; *(bes. Sport)* win against sb.; beat sb.; **mit 2:0 ~** *(Sport)* win 2–0 *or* by two goals to nil
Sieger der; ~s, ~: winner; *(Mannschaft)* winners *pl.*; *(einer Schlacht)* victor; **zweiter ~ sein** *(ugs.)* be runner-up/runners-up
Sieger·ehrung die presentation ceremony; awards ceremony
Siegerin die; ~, ~nen *s.* **Sieger**
sieges·sicher 1. *Adj.* certain *or* confident of victory *pred.*; *(erfolgssicher)* certain *or* confident of success *pred.*; **2.** *adv.* confident of victory; ⟨*say, smile*⟩ confidently
sieg·reich *Adj.* victorious; winning ⟨*team*⟩; successful ⟨*campaign*⟩; **nach einer ~en Schlacht** after winning a battle
sieh [zi:], **siehe** *Imperativ Sg. v.* **sehen**
siehst [zi:st] *2. Pers. Sg. Präsens v.* **sehen**
sieht [zi:t] *3. Pers. Sg. Präsens v.* **sehen**
Siel [zi:l] **der** *od.* **das; ~[e]s, ~e** *(nordd.)* dike sluice *or* floodgate
siezen ['zi:tsn̩] *tr. V.* call 'Sie' *(the polite form of address)*
Signal [zı'gna:l] **das; ~s, ~e** signal; **das ~ steht auf „Halt"** the signal is at 'stop'
Signal-: ~anlage die *(Verkehrsw.)* signals *pl.*; **~flagge die** *(Seew.)* signal flag
signalisieren *tr. V.* indicate ⟨*danger, change, etc.*⟩; *(fig.: übermitteln)* signal ⟨*message, warning, order*⟩ (+ *Dat.* to)
Signal-: ~lampe die indicator light; **~mast der a)** *(Seew.)* signalling mast; **b)** *(Eisenb.)* signal post *or* mast
Signatur [zıgna'tu:ɐ̯] **die; ~, ~en a)** initials *pl.*; *(Kürzel)* abbreviated signature; *(des Künstlers)* autograph; **b)** *(veralt.: Unterschrift)* signature; **c)** *(in einer Bibliothek)* shelf-mark; **d)** *(auf Landkarten)* [map] symbol
signieren *tr. V.* sign; autograph ⟨*one's own work*⟩
Silbe ['zılbə] **die; ~, ~n** syllable; **etw. mit keiner ~ erwähnen** not say a word about sth.
Silben-: ~rätsel das *puzzle in which syllables must be combined to form words;* **~trennung die** word-division *(by syllables)*
Silber ['zılbɐ] **das; ~s a)** *(Edelmetall, Farbe)* silver; **b)** *(silbernes Gerät)* silver[ware]
silber-, Silber-: ~besteck das silver cutlery; **~blond** *Adj.* silver-blond; **~fischchen das** silver-fish; **~geld das;** *o. Pl.* silver; **~geschirr das** silver plate; silverware; **~grau** *Adj.* silver-grey; **~haltig** *Adj.* silver-bearing; argentiferous; **~hochzeit die** silver wedding; **~medaille die** silver medal; **~mine die** silver mine; **~münze die** silver coin
silbern 1. *Adj.* **a)** silver; silvery ⟨*moonlight, shade, gleam, etc.*⟩. **2.** *adv.* ⟨*ornament, coat, etc.*⟩ with silver; ⟨*shine, shimmer, etc.*⟩ with a silvery lustre
Silber-: ~papier das silver paper; **~streif der, ~streifen der** *in* **ein ~streifen am Horizont** *(fig.)* a ray of hope on the horizon
Silhouette [zi'lu̯ɛtə] **die; ~, ~n a)** silhouette; **b)** *(Mode)* line
Silicium [zi'li:tsi̯ʊm] **das; ~s** silicon
Silikat [zili'ka:t] **das; ~[e]s, ~e** *(Chemie)* silicate
Silikon [zili'ko:n] **das; ~s, ~e** *(Chemie)* silicone
Silizium *s.* **Silicium**
Silo ['zi:lo] **der** *od.* **das; ~s, ~s** silo
Silvester [zıl'vɛstɐ] **der** *od.* **das; ~s, ~:** New Year's Eve; **~ feiern** see the New Year in
Silvester·nacht die night of New Year's Eve
Simbabwe [zım'ba:bvə] **(das); ~s** Zimbabwe
simpel ['zımpl̩] **1.** *Adj.* **a)** simple ⟨*question, task*⟩; **b)** *(abwertend: beschränkt)* simple-minded ⟨*person*⟩; simple ⟨*mind*⟩; **c)** *(oft abwertend: schlicht)* basic ⟨*toy, dress, etc.*⟩. **2.** *adv.* **a)** simply; **b)** *(abwertend: beschränkt)* in a simple-minded manner
Simpel der; ~s, ~ *(bes. südd. ugs.)* simpleton; fool
Sims [zıms] **der** *od.* **das; ~es, ~e** ledge; sill; *(Kamin~)* mantelpiece
Simulant [zimu'lant] **der; ~en, ~en, Simulantin die; ~, ~nen** malingerer
Simulator [zimu'la:tɔr] **der; ~s, ~en** [-'to:rən] *(Technik)* simulator
simulieren 1. *tr. V.* feign, sham ⟨*illness, emotion, etc.*⟩; simulate ⟨*situation, condition, etc.*⟩. **2.** *itr. V.* *(Krankheit vortäuschen)* feign illness; pretend to be ill; **er simuliert nur** he's just putting it on
simultan [zimʊl'ta:n] **1.** *Adj.* simultaneous. **2.** *adv.* simultaneously
Simultan·dolmetscher der simultaneous interpreter
Sinai·halb·insel ['zi:nai-] **die** Sinai Peninsula
sind [zınt] *1. u. 3. Pers. Pl. Präsens v.* **sein**
Sinfonie [zınfo'ni:] **die; ~, ~n** symphony
Sinfonie·orchester das symphony orchestra
sinfonisch *Adj.* symphonic
Singapur ['zıŋgapu:ɐ̯] **(das); ~s** Singapore
singen ['zıŋən] **1.** *unr. itr. V.* **a)** sing; **einen ~den Tonfall haben** have a lilting cadence; **b)** *(salopp: vor der Polizei aussagen)* squeal *(sl.)*. **2.** *unr. tr. V.* **a)** sing ⟨*song, aria, contralto, tenor, etc.*⟩; **b)** **jmdn. in den Schlaf ~:** sing sb. to sleep
[1]**Single** ['zıŋl̩] **die; ~, ~s** *(Schallplatte)* single
[2]**Single der; ~[s], ~s** single person; **~s** single people *no art.*
Sing·spiel das *(Musik)* Singspiel
Singular ['zıŋgula:ɐ̯] **der; ~s** singular
singulär *(geh.)* **1.** *Adj.* rare. **2.** *adv.* rarely
Sing·vogel der songbird
sinken ['zıŋkn̩] *unr. itr. V.; mit sein* **a)** ⟨*ship, sun*⟩ sink, go down; ⟨*plane, balloon*⟩ descend, go down; *(geh.)* ⟨*leaves, snowflakes*⟩

fall; **b)** *(nieder~)* fall; **den Kopf ~ lassen** let one's head drop; **auf** *od. (geh.)* **in die Knie ~:** sink *or* fall to one's knees; **c)** *(niedriger werden)* ‹*temperature, level*› fall, drop; **das Thermometer/Barometer sinkt** the temperature is falling/the barometer is going back; **d)** *(an Wert verlieren)* ‹*price, value*› fall, go down; **e)** *(nachlassen, abnehmen)* fall; go down; ‹*excitement, interest*› diminish, decline; **jmds. Mut sinkt** sb. loses courage

Sinn [zɪn] **der; ~[e]s, ~e a)** sense; **den** *od.* **einen sechsten ~ [für etw.] haben** have a sixth sense [for sth.]; **seine fünf ~e nicht beisammen haben** *(ugs.)* be not quite right in the head; **b)** *Pl. (geh.: Bewußtsein)* senses; mind *sing.*; **nicht bei ~en sein** be out of one's senses *or* mind; **wie von ~en** as if he/she had gone out of his/her mind; **c)** *o. Pl. (Gefühl, Verständnis)* feeling; **einen ~ für Gerechtigkeit/Humor** *usw.* **haben** have a sense of justice/humour *etc.*; **d)** *o. Pl. (geh.: Gedanken, Denken)* mind; **er hat ganz in meinem ~ gehandelt** he acted correctly to my mind *or* my way of thinking; **mir steht der ~ [nicht] danach/nach etw.** I [don't] feel like it/sth.; **sich** *(Dat.)* **etw. aus dem ~ schlagen** put [all thoughts of] sth. out of one's mind; **etw. im ~ haben** have sth. in mind; **jmdm. in den ~ kommen** come to sb.'s mind; **e)** *o. Pl. (~gehalt, Bedeutung)* meaning; **im strengen/wörtlichen ~:** in the strict/literal sense; **f)** *(Ziel u. Zweck)* point; **der ~ des Lebens** the meaning of life

Sinn·bild das symbol

sinnen *(geh.) unr. itr. V.* think; ponder; **auf Rache ~:** be out for revenge

sinn·entstellend 1. *Adj.* which distorts/distorted the meaning *postpos., not pred.* **2.** *adv.* ‹*translate, shorten*› so that the *or* its meaning is distorted

Sinnes-: ~eindruck der sense impression; sensation; **~organ das** sense-organ; sensory organ; **~täuschung die** trick of the senses; **~wandel der** change of mind *or* heart

Sinn·gedicht das epigram

sinn·gemäß 1. *Adj.* **eine ~e Übersetzung** a translation which conveys the general sense; **2.** *adv.* **etw. ~ übersetzen/wiedergeben** translate the general sense of sth./give the gist of sth.

sinnig *Adj. (meist spött. od. iron.)* clever; sensible *(iron.)*

sinnlich *Adj.* sensory ‹*impression, perception, stimulus*›; sensual ‹*love, mouth*›; sensuous ‹*pleasure, passion*›

Sinnlichkeit die; ~ sensuality

sinn·los 1. *Adj.* **a)** *(unsinnig)* senseless; **b)** *(zwecklos)* pointless; **c)** *(abwertend: übermäßig)* mad; wild. **2.** *adv.* **a)** *(unsinnig)* senselessly; **b)** *(zwecklos)* pointlessly; **c)** *(abwertend: übermäßig)* like mad; **~ betrunken** blind drunk

Sinnlosigkeit die a) senselessness; **b)** *(Zwecklosigkeit)* pointlessness

sinn-, Sinn-: ~spruch der saying; **~verwandt** *Adj. (Sprachw.)* synonymous ‹*words*›; **~voll 1.** *Adj.* **a)** *(vernünftig)* sensible; **b)** *(einen Sinn ergebend)* meaningful; **2.** *adv.* **a)** *(vernünftig)* sensibly; **b)** *(einen Sinn ergebend)* meaningfully; **~widrig** *Adj. (geh.)* nonsensical

Sint·flut ['zɪnt-] **die** Flood; Deluge; **nach mir/uns die ~:** I/we don't care what happens after I've/we've gone

sintflut·artig 1. *Adj.* torrential. **2.** *adv.* in torrents

Sinus ['zi:nʊs] **der; ~, ~** *od.* **~se** *(Math.)* sine

Siphon ['zi:fõ] **der; ~s, ~s a)** siphon; **b)** *(Geruchsverschluß)* [anti-siphon] trap

Sippe ['zɪpə] **die; ~, ~n a)** *(Völkerk.)* sib; **b)** *(meist scherzh. od. abwertend: Verwandtschaft)* clan

Sippschaft die; ~, ~en a) *(meist abwertend: Sippe)* clan; **b)** *(abwertend: Gesindel)* bunch *(coll.)*; crowd *(coll.)*

Sirene [zi're:nə] **die; ~, ~n** siren

Sirenen·geheul das wail of a/the siren/of sirens

sirren ['zɪrən] *itr. V.* buzz

Sirup ['zi:rʊp] **der; ~s, ~e** syrup; *(streichfähig auch)* treacle *(Brit.)*; molasses *sing. (Amer.)*

Sitte ['zɪtə] **die; ~, ~n a)** *(Brauch)* custom; tradition; **b)** *(moralische Norm)* common decency; **c)** *Pl. (Benehmen)* manners; **das sind ja feine ~n!** *(iron.)* that's a nice way to behave! *(iron.)*

sitten-, Sitten-: ~dezernat das vice squad; **~geschichte die** history of life and customs; **~lehre die** ethics *sing.*; moral philosophy; **~los 1.** *Adj.* immoral; **2.** *adv.* immorally; **~polizei die** *(volkst.)* vice squad; **~strolch der** *(Pressejargon)* [sexual] molester; **~verfall der** moral decline; decline in moral standards; **~widrig 1.** *Adj.* **a)** *(Rechtsw.)* illegal ‹*methods, advertising, etc.*›; **b)** *(unmoralisch)* immoral ‹*behaviour*›; **2.** *adv. s.* **1:** illegally; immorally

Sittich ['zɪtɪç] **der; ~s, ~e** parakeet

sittlich 1. *Adj.* moral. **2.** *adv.* morally

Sittlichkeit die; *o. Pl.* morality; morals *pl.*

Sittlichkeits-: ~verbrechen das sexual crime; **~verbrecher der** sex offender

Situation [zitu̯a'tsi̯o:n] **die; ~, ~en** situation

Sitz [zɪts] **der; ~es, ~e a)** seat; **b)** *(mit Stimmrecht)* seat; **~ und Stimme haben** have a seat and a vote; **c)** *(Regierungs~)* seat; *(Verwaltungs~)* headquarters *sing. or pl.*; **d)** *(sitzende Haltung)* sitting position; *(beim Reiten)* seat; **e)** *(von Kleidungsstücken)* fit; **f) auf einen ~** *(ugs.)* in *or* at one go

Sitz-: ~bad das sitz-bath; hip-bath; **~ecke die** sitting area

sitzen *unr. itr. V.; südd., österr., schweiz. mit sein* **a)** sit; **bleiben Sie bitte ~:** please don't get up; please remain seated; **er saß den ganzen Tag in der Kneipe** he spent the whole day in the pub *(Brit. coll.)*; **im Sattel ~:** be in the saddle; **jmdm. auf der Pelle** *od.* **dem Pelz ~** *(salopp)* keep bothering sb.; keep on at sb. *(coll.)*; **b)** *(sein)* be; **die Firma sitzt in Berlin** the firm is based in Berlin; **einen ~ haben** *(salopp)* have had one too many; **c)** *([gut]passen)* fit; **die Krawatte sitzt nicht** the tie isn't straight; **d)** *(ugs.: gut eingeübt sein)* **Lektionen so oft wiederholen, bis sie ~:** keep on repeating lessons till they stick *(coll.)*; **e)** *(ugs.: wirksam treffen)* hit home; **f)** *(Mitglied sein)* be, sit (**in** + *Dat.* on); **g)** *(ugs.: eingesperrt sein)* be in prison *or (sl.)* inside

sitzen-: ~|bleiben *unr. itr. V. (ugs.)* **a)** *(nicht versetzt werden)* stay down [a year]; have to repeat a year; **b)** *(abwertend: als Frau unverheiratet bleiben)* be left on the shelf; **c)** *(keinen Käufer finden)* **auf etw.** *(Dat.)* **~bleiben** be left *or (coll.)* stuck with sth.; **~|lassen** *unr. tr. V. (ugs.)* **a)** *(nicht heiraten)* jilt; **b)** *(im Stich lassen)* leave in the lurch; **er hat Frau und Kinder ~lassen** *od. (seltener:)* **~gelassen** he left his wife and children; **c)** *(hinnehmen)* **etw. nicht auf sich** *(Dat.)* **~lassen** not take sth.; not stand for sth.

Sitz-: ~fleisch das *(ugs. scherzh.)* **kein ~fleisch haben** not have the staying power; not be able to stick at it; *(nicht stillsitzen können)* not be able to sit still; **~gruppe die** group of seats; **~kissen das** *(im Sessel, Sofa)* [seat] cushion; *(auf dem Fußboden)* [floor] cushion; **~ordnung die** seating plan *or* arrangement; **~platz der** seat; **~streik der** sit-down strike

Sitzung die; ~, ~en meeting; *(Parlaments~)* sitting; session

Sitzungs-: ~periode die session; **~saal der** conference hall; *(eines Gerichts)* court-room

Sizilianer [zitsi'li̯a:nɐ] **der; ~s, ~, Sizilianerin die; ~, ~nen** Sicilian

Sizilien [zi'tsi:li̯ən] **(das); ~s** Sicily

Skala ['ska:la] **die; ~, Skalen a)** scale; **b)** *(Reihe)* range

Skalp [skalp] **der; ~s, ~e** scalp

Skalpell [skal'pɛl] **das; ~s, ~e** scalpel

skalpieren *tr. V.* scalp

Skandal [skan'da:l] **der; ~s, ~e** scandal

skandalös *Adj.* scandalous

skandieren [skan'di:rən] *tr. V.* **a)** chant; **b)** *(Verslehre)* scan

Skandinavien [skandi'na:vi̯ən] **(das); ~s** Scandinavia

Skandinavier [skandi'na:vi̯ɐ] **der; ~s, ~, Skandinavierin die; ~, ~nen** Scandinavian

skandinavisch *Adj.* Scandinavian

Skat [ska:t] **der; ~[e]s, ~e** *od.* **~s** skat; **~ dreschen** play skat

Skelett [ske'lɛt] **das; ~[e]s, ~e** skeleton

Skepsis ['skɛpsɪs] **die; ~:** scepticism

Skeptiker der; ~s, ~: sceptic

skeptisch 1. *Adj.* sceptical. **2.** *adv.* sceptically

Sketch [skɛtʃ] **der; ~[es], ~e[s]** *od.* **~s** sketch

Ski [ʃi:] **der; ~s, ~er** *od.* **~:** ski; **~ laufen** *od.* **fahren** ski

Ski-: ~bindung die ski binding; **~lauf der, ~laufen das** skiing *no art.*; **~läufer der** skier; **~lehrer der** ski-instructor; **~lift der** ski-lift; **~springen das; ~s, ~:** ski-jumping *no art.*; **~stiefel der** ski boot; **~stock der** ski stick; ski pole

Skizze ['skɪtsə] **die; ~, ~n a)** *(Zeichnung)* sketch; **b)** *(Konzept)* outline; **c)** *(kurze Aufzeichnung)* [brief] account

Skizzen·block der sketch-pad; sketch-block

skizzieren *tr. V.* **a)** *(zeichnen)* sketch; **b)** *(aufzeichnen)* outline; **c)** *(entwerfen)* draft

Sklave ['skla:və] **der; ~n, ~n** slave

Sklaven-: ~halter der slave-owner; **~händler der** *(auch fig. abwertend)* slave-trader

Sklaverei die; ~: slavery *no art.*

Sklavin die; ~, ~nen slave

sklavisch ['skla:vɪʃ] *(abwertend)* **1.** *Adj.* slavish. **2.** *adv.* slavishly

Sklerose [skle'ro:zə] **die; ~, ~n** *(Med.)* sclerosis *no art.*; **multiple ~:** multiple sclerosis

Skonto ['skɔnto] **der** *od.* **das; ~s, ~s** *(Kaufmannsspr.)* [cash] discount

Skorbut [skɔr'bu:t] **der; ~[e]s** *(Med.)* scurvy *no art.*

Skorpion [skɔr'pi̯o:n] **der; ~s, ~e a)** *(Tier)* scorpion; **b)** *(Astrol.)* Scorpio

Skrupel ['skru:pl̩] **der; ~s, ~:** scruple

skrupel·los *(abwertend)* **1.** *Adj.* unscrupulous. **2.** *adv.* unscrupulously

Skrupellosigkeit die; ~ *(abwertend)* unscrupulousness

Skulptur [skʊlp'tu:ɐ̯] **die; ~, ~en** sculpture

skurril [skʊ'ri:l] **1.** *Adj.* absurd; droll ⟨*person*⟩. **2.** *adv.* absurdly

Slalom ['sla:lɔm] **der; ~s, ~s** *(Ski-, Kanusport)* slalom; **im ~ fahren** *(fig.)* zigzag

Slawe ['sla:və] **der; ~n, ~n, Slawin die; ~, ~nen** Slav

slawisch *Adj.* Slav[ic]; Slavonic

Slip [slɪp] **der; ~s, ~s** briefs *pl.*

Slogan ['slo:gn̩] **der; ~s, ~s** slogan

Slowake [slo'va:kə] **der; ~n, ~n** Slovak

Slowakei die; ~: Slovakia *no art.*

Slowakin die; ~, ~nen Slovak

slowakisch *Adj.* Slovak; Slovakian

Slowene [slo've:nə] **der; ~n, ~n** Slovene; Slovenian

Slowenien (das); ~s Slovenia

Slowenin die; ~, ~nen Slovene; Slovenian

slowenisch *Adj.* Slovene; Slovenian

Slum [slam] **der; ~s, ~s** slum

Smaragd [sma'rakt] **der; ~[e]s, ~e** emerald

Smog [smɔk] **der; ~[s], ~s** smog

Smog·alarm der smog warning; **bei ~:** if there is a smog warning

Smoking ['smo:kɪŋ] **der; ~s, ~s** dinner-jacket *or (Amer.)* tuxedo and dark trousers

Snob [snɔp] **der; ~s, ~s** snob

Snobismus der; ~: snobbery; snobbishness

so [zo:] **1.** *Adv.* **a)** *meist betont (auf diese Weise; in, von dieser Art)* like this/that; this/that way; **so ist sie nun einmal** that's the way she is; **wenn dem so ist** if that's the case; **so ist es!** *(zustimmend)* that's correct *or* right!; **recht so!, gut so!** right!; that's fine!; **so oder so gerät der Minister unter Druck** either way the minister will come under pressure; **weiter so!** carry on in the same way!; **b)** *meist betont (dermaßen, überaus)* so; **er ist nicht so dumm, das zu tun** he is not so stupid as to do that; **c)** *(genauso)* as; **so gut ich konnte** as best I could; **er ist [nicht] so groß wie du** he is [not] as tall as you [are]; **d)** *meist betont (ugs.: solch[e])* such; **so ein Kind** such a child; a child like that; **so ein Pech/eine Frechheit!** what bad luck/a cheek!; **so ein Idiot!** what an idiot!; **ist sie nicht Kontoristin oder so was?** isn't she a clerk or something?; **[na** *od.* **nein** *od.* **also] so was!** *(überrascht/empört)* well, I never!; **so einer/eine/eins** one like that; one of those; **e)** *betont (eine Zäsur ausdrückend)* right; OK *(coll.)*; **f)** *(ugs.: schätzungsweise)* about; **g)** *unbetont (bei Zitaten od. Quellenangaben)* **die Religion, so Marx, ist ...:** religion, according to Marx is ...; **h)** *unbetont (ugs.: und/oder ähnliches)* **ich spiele ein bißchen Tischtennis, Billard und so** I play a bit of table tennis, billiards and that sort of thing; **i)** *betont (erstaunt, zweifelnd)* **so?** really?; **so, so** *(meist iron.)* oh, I see; **j)** *betont (ohne Hilfsmittel)* **ich brauche keine Leiter, da komme ich auch so ran** *(ugs.)* I don't need a ladder, I can reach it [without one]; **k)** *betont (ohne Zutaten)* just as it is; **l)** *betont (ugs.: umsonst)* for nothing. **2.** *Konj.* **a)** *(konsekutiv, in Verbindung mit „daß")* **so daß ...** *(damit)* so that ...; *(und deshalb)* and so ...; **b)** *(konzessiv)* however. **3.** *Partikel* **a)** just; **ich weiß nicht so recht, ob ich gehen soll** I'm not really sure if I should go; **Warum fragst du? – Ach, nur so** Why do you ask? – Oh, no particular reason; **b)** *(in Aufforderungssätzen verstärkend)* **so komm doch** come on now

So. *Abk.* **Sonntag** Sun.

s. o. *Abk.* **siehe oben**

SO *Abk.* **Südost[en]** SE

sobald *Konj.* as soon as

Socke ['zɔkə] **die; ~, ~n** sock; **sich auf die ~n machen** *(ugs.)* get going; **von den ~n sein** *(ugs.)* be flabbergasted

Sockel ['zɔkl̩] **der; ~s, ~ a)** *(einer Säule, Statue)* plinth; **b)** *(unterer Teil eines Hauses, Schrankes)* base; **c)** *(Elektrot.)* base

Sockel·betrag der *(Wirtsch.)* basic sum

Soda ['zo:da] **die; ~** *od.* **das; ~s** soda

so·dann *Adv.* **a)** *(danach)* then; thereupon; **b)** *(außerdem)* and furthermore

so daß *s.* **so 2 a**

Soda·wasser das; *Pl.* **Sodawässer** soda; soda-water

Sod·brennen das; ~s heartburn; pyrosis

so·eben *Adv.* just; **die Nachricht kam ~:** the news came just now

Sofa ['zo:fa] **das; ~s, ~s** sofa; settee

Sofa·kissen das [sofa] cushion; scatter cushion

so·fern *Konj.* provided [that]

soff [zɔf] *1. u. 3. Pers. Sg. Prät. v.* **saufen**

so·fort *Adv.* immediately; at once; **ich bin ~ fertig** I'll be ready in a moment; *(mit einer Arbeit)* I'll be finished in a moment

Sofort·bild·kamera die *(Fot.)* instant-picture camera

Sofort·hilfe die emergency relief *or* aid

sofortig *Adj.* *(unmittelbar)* immediate

Sofort·maßnahme die immediate measure

Soft-Eis ['zɔftl|ais] **das** soft ice-cream

Software ['zɔftvɛ:ɐ̯] **die; ~, ~s** *(DV)* software

sog [zo:k] *1. u. 3. Pers. Sg. Prät. v.* **saugen**

Sog der; ~[e]s, ~e suction; *(bei Schiffen)* wake; *(bei Fahr-, Flugzeugen)* slip-stream; *(von Wasser, auch fig.)* current

sog. *Abk.* **sogenannt**

so·gar *Adv.* even; **sie ist krank, ~ schwer krank** she is ill, in fact *or* indeed seriously ill

so·genannt *Adj.* so-called

so·gleich *Adv.* immediately; at once

Sohle ['zo:lə] **die; ~, ~n a)** *(Schuh~)* sole; **eine kesse** *od.* **heiße ~ aufs Parkett legen** *(ugs.)* put up a good show on the dance-floor; **auf leisen ~n** softly; noiselessly; **b)** *(Fuß~)* sole [of the foot]; **c)** *(Tal~)* bottom; *(eines Flusses)* bottom; bed; **d)** *(Einlege~)* insole

sohlen *tr. V.* sole

Sohn [zo:n] **der; ~es, Söhne** ['zø:nə] **a)** *(männlicher Nachkomme)* son; **der ~ Gottes** the Son of God; **der verlorene ~:** the prodigal son; **b)** *o. Pl. (fam.: Anrede an einen Jüngeren)* son; boy

Soja- ['zo:ja-]: **~bohne die** soy[a] bean; **~soße die** soy[a] sauce

Sokrates ['zo:kratɛs] **(der); Sokrates'** Socrates

so·lang[e] *Konj.* so *or* as long as; **~ du nicht alles aufgegessen hast** unless *or* until you have eaten everything up

solar [zo'la:ɐ̯] *Adj.* solar

Solarium [zo'la:ri̯ʊm] **das; ~s, Solarien** [...i̯ən] solarium

Solar-: ~technik die solar technology *no art.*; **~zelle die** *(Physik, Elektrot.)* solar cell

solch [zɔlç] *Demonstrativpron.* **a)** *attr.* such; **ich habe ~en Hunger** I am so hungry; **ich habe ~e Kopfschmerzen** I've got such a headache; **das macht ~en Spaß!** it's so much fun!; **b)** *selbständig* **~e wie die** people like that; **die Sache als ~e** the thing as such; **es gibt ~e und ~e** *(ugs.)* it takes all sorts *or* kinds [to make a world]; **c)** *ungebeugt (geh.: so [ein])* such; **bei ~ einem herrlichen Wetter** when the weather is so beautiful

Sold [zɔlt] **der; ~[e]s, ~e** [military] pay

Soldat [zɔl'da:t] **der; ~en, ~en** soldier; **~ auf Zeit** soldier serving for a fixed period

Soldaten·fried·hof der military *or* war cemetery

Soldatin die; ~, ~nen [female *or* woman] soldier; **sie ist ~:** she is a soldier

soldatisch 1. *Adj.* military ⟨*discipline, expression, etc.*⟩; soldierly ⟨*figure, virtue*⟩. **2.** *adv.* in a military *or* soldierly manner

Söldner ['zœldnɐ] **der; ~s, ~:** mercenary

Sole ['zo:lə] **die; ~, ~n** salt water; brine

Sol·ei das pickled egg

solid [zo'li:t] *s.* **solide**

solidarisch 1. *Adj.* **~es Verhalten zeigen** show one's solidarity **2.** *adv.* **~ handeln/sich ~ verhalten** act in/show solidarity

solidarisieren *refl. V.* show [one's] solidarity

Solidarität die; ~: solidarity

Solidaritäts·streik der solidarity strike

solide 1. *Adj.* **a)** solid ⟨*rock, wood, house*⟩; sturdy ⟨*shoes, shed, material, fabric*⟩; solid, sturdy ⟨*furniture*⟩; [good-]quality ⟨*goods*⟩; **b)** *(gut fundiert)* sound ⟨*work, workmanship, education, knowledge*⟩; solid ⟨*firm, business*⟩; **c)** *(anständig)* respectable ⟨*person, life, occupation, profession*⟩. **2.** *adv.* **a)** solidly ⟨*built*⟩; sturdily ⟨*made*⟩; **b)** *(gut fundiert)* soundly ⟨*educated, constructed*⟩; **c)** *(anständig)* ⟨*live*⟩ respectably, steadily

Solidität die; ~ *s.* **solide 1 a–c:** solidness; sturdiness; soundness; respectability

Solist [zo'lɪst] **der; ~en, ~en** soloist

soll *1. u. 3. Pers. Sg. Präsens v.* **sollen**

Soll [zɔl] **das; ~[s], ~[s] a)** *(Kaufmannsspr., Bankw.: Schulden)* debit; **~ und Haben** debit and credit; **im ~:** in debit; **b)** *(Kaufmannsspr.: linke Buchführungsseite)* debit side; **c)** *(Wirtsch.: Arbeits~)* quota; **sein ~ erfüllen** achieve *or* meet one's target; **d)** *(Wirtsch.: Plan~)* quota; target

sollen 1. *unr. Modalv.; 2. Part.* **~ a)** *(bei Aufforderung, Anweisung, Auftrag)* **solltest du nicht bei ihm anrufen?** were you not supposed to ring him?; **was soll ich als nächstes tun?** what should I do next?; what do you want me to do next?; **[sagen Sie ihm,] er soll hereinkommen** tell him to come in; **ich soll dir schöne Grüße von Herrn Meier bestellen** Herr Meier asked me to give you *or* sends his best wishes; **b)** *(bei Wunsch, Absicht, Vorhaben)* **du sollst alles haben, was du brauchst** you shall have everything you require; **das sollte ein Witz sein** that was meant to be a joke; **was soll denn das heißen?** what is that supposed to mean?; **c)** *(bei Ratlosigkeit)* **was soll ich nur machen?** what am I to do?; **was soll nur aus ihm werden?** what is to become of him?; **d)** *(Notwendigkeit ausdrückend)* **man soll so etwas nicht unterschätzen** it's not to be taken *or* it shouldn't be taken so lightly; **e)** *häufig im Konjunktiv II (Erwartung, Wünschenswertes ausdrückend)* **du solltest dich schämen** you ought to be ashamed of yourself; **das hättest du besser nicht tun ~:** it would have been better if you hadn't done that; **f)** *(jmdm. beschieden sein)* **er sollte seine Heimat nicht wiedersehen** he was never to see his homeland again; **es hat nicht sein ~** *od.* **nicht ~ sein** it was not to be; **g)** *im Konjunktiv II (eine Möglichkeit ausdrückend)* **wenn du ihn sehen solltest, sage ihm bitte ...:** if you should see him, please tell him ...; **h)** *im Präsens (sich für die Wahrheit nicht verbürgend)* **das Restaurant soll sehr teuer sein** the restaurant is supposed *or* said to be very expensive; **i)** *im Konjunktiv II (Zweifel ausdrückend)* **sollte das sein Ernst sein?** is he really being serious?; **j)** *(können)* **mir soll es gleich sein** it's all the same to me; it doesn't matter to me; **man sollte glauben, daß ...:** you would think that ... **2.** *tr., itr. V.* **was soll das?** what's the idea?; **was soll ich dort?** what would I do there?

solo ['zo:lo] *indekl. Adj.; nicht attr.* **a)** *(bes. Musik: als Solist)* solo; **b)** *(ugs., oft scherzh.: ohne Begleitung)* on one's own *postpos.*

Solo ['zo:lo] **das; ~s, ~s** *od.* **Soli** ['zo:li] *(bes. Musik)* solo

Sombrero [zɔm'bre:ro] **der; ~s, ~s** sombrero

so·mịt [auch: '--] *Adv.* consequently; therefore

Sommer ['zɔmɐ] **der; ~s, ~:** summer; *s. auch* **Frühling**

Sọmmer-: ~anfang der beginning of summer; **~ferien** *Pl.* summer holidays; **~frische die** *(veralt.)* **a)** summer holiday; **b)** *(Ort)* summer [holiday] resort

sọmmerlich 1. *Adj.* summer; summery ⟨*warmth, weather*⟩; summer's *attrib.* ⟨*day, evening*⟩. **2.** *adv.* **es war oft schon ~ warm** it was often as warm as summer

sọmmer-, Sọmmer-: ~reifen der standard tyre; **~saison die** summer season; **~schluß·verkauf der** summer sale/sales; **~semester das** summer semester; **~sprosse die** freckle; **~sprossig** *Adj.* freckled; **~zeit die a)** *o. Pl. (Jahreszeit)* summertime; **b)** *(Uhrzeit)* summer time

Sonate [zo'na:tə] **die; ~, ~n** *(Musik)* sonata

Sonde ['zɔndə] **die; ~, ~n a)** *(Med.) (zur Untersuchung)* probe; *(zur Ernährung)* tube; **b)** *(Raum~)* [space] probe

Sọnder·angebot das special offer; **etw. im ~angebot anbieten** have a special offer on sth.

sọnderbar 1. *Adj.* strange; odd. **2.** *adv.* strangely; oddly

Sọnderbarkeit die; ~: strangeness; oddness

Sọnder·fall der special case; exception

sọnder·gleichen *Adv., nachgestellt* **eine Frechheit/Unverschämtheit ~:** the height of cheek/impudence

sọnderlich 1. *Adj.* **a)** particular; [e]special; **b)** *(sonderbar)* strange; peculiar; odd. **2.** *adv.* **a)** particularly; especially; **b)** *(sonderbar)* strangely

Sọnderling der; ~s, ~e strange *or* odd person

Sọnder-: ~marke die special issue [stamp]; **~müll der** hazardous waste

[1]**sọndern** *tr. V. (geh.)* separate **(von** from)

[2]**sọndern** *Konj.* but; **nicht er hat es getan, ~ sie** 'he didn't do it, 'she did; **nicht nur ..., ~ [auch] ...:** not only ... but also ...

Sọnder·nummer die special edition *or* issue

sọnders *s.* **samt 2**

Sọnder-: ~schule die special school; **~urlaub der a)** *(Milit.)* special leave; **b)** *(zusätzlicher Urlaub)* special *or* extra holiday; **~zug der** special train

sondieren [zɔn'di:rən] *tr. V.* sound out; **das Terrain ~:** see *or* find out how the land lies

Sonett [zo'nɛt] **das; ~[e]s, ~e** *(Dichtk.)* sonnet

Sonn·abend ['zɔn|a:bn̩t] **der** *(bes. nordd.)* Saturday

sọnn·abends *Adv.* on Saturday[s]

Sọnne die; ~, ~n sun; *(Licht der ~)* sun[light]

sọnnen *refl. V.* sun oneself; **sich in etw.** *(Dat.)* **~** *(fig.)* bask in sth.

sọnnen-, Sọnnen-: ~aufgang der sunrise; **~bad das** sunbathing *no pl., no indef. art.;* **~baden** *itr. V.* sunbathe; **~blende die a)** *(Fot.)* lens-hood; **b)** *(im Auto)* sun visor; **~blume die** sunflower; **~brand der** sunburn *no indef. art.;* **~bräune die** sun-tan; **~brille die** sunglasses *pl.;* **~energie die** solar energy; **~finsternis die** solar eclipse; eclipse of the sun; **~fleck der** *(Astron.)* sunspot; **~gebräunt** *Adj.* sun-tanned; **~hut der** sun-hat; **~klar** *Adj. (ugs.)* crystal-clear; **~kollektor der** *(Technik)* solar collector; **~licht das** sunlight; **~öl das** sun-tan oil; sun-oil; **~schein der;** *o. Pl.* sunshine; **bei ~schein** in sunshine; **~schutz·creme die** sun-tan lotion; **~schirm der** sunshade; **~stich der** *(Med.)* sunstroke *no indef. art.;* **du hast wohl einen ~stich** *(fig. salopp)* you must be mad; **~strahl der** ray of sun[shine]; **~system das** *(Astron.)* solar system; **~uhr die** sundial; **~untergang der** sunset; **~zelle die** *(Physik, Elektrot.)* solar cell

sọnnig *Adj.* **a)** sunny; *(fig.)* happy ⟨*youth, childhood, time*⟩; cheerful ⟨*sense of humour, ways*⟩; **b)** *(iron.: naiv)* naive

Sọnn·tag der Sunday; *s. auch* **Dienstag; Dienstag-**

sọnn·täglich 1. *Adj.* Sunday *attrib.* **2.** *adv.* **~ gekleidet sein** be dressed in one's Sunday best

sọnntags *Adv.* on Sunday[s]

Sọnntags-: ~arbeit die; *o. Pl.* Sunday working *no art.;* **~dienst der** Sunday duty; **~fahrer der** *(abwertend)* Sunday driver; **~kind das: er ist ein ~kind** *(fig.)* he was born lucky *or* under a lucky star; **~predigt die** Sunday sermon; **~staat der;** *o. Pl. (scherzh.)* Sunday best; **im ~staat** in one's Sunday best

sonst [zɔnst] *Adv.* **a) der ~ so freundliche Mann ...:** the man, who is/was usually so friendly, ...; **er hat es besser als ~ gemacht** he did it better than usual; **alles war wie ~:** everything was [the same] as usual; **haben Sie ~ noch Fragen?** have you any other questions?; **hat er ~ nichts erzählt?** [apart from that,] he didn't say anything else?; **~ noch was?** *(ugs., auch iron.)* anything else?; **~ nichts, nichts ~:** nothing else; **wer/was/wie/wo [denn] ~?** who/what/how/where else?; **b)** *(andernfalls)* otherwise; or

sonstig... *Adj.; nicht präd.* other; further; **„Sonstiges"** 'miscellaneous'

sọnst-: ~was *Indefinitpron. (ugs.)* anything else; **er hat ~was unternommen** he has tried all sorts of things; **~wer** *Indefinitpron. (ugs.)* somebody else; *(fragend, verneinend)* anybody else; **er meint, er ist ~wer** he thinks he's really something *(coll.);* he thinks he's the bee's knees *(coll.);* **~wie** *Adv. (ugs.)* in some other way; *(fragend, verneinend)* in any other way; **~wo** *Adv. (ugs.)* somewhere else; *(fragend, verneinend)* anywhere else

so·ọft *Konj.* whenever

Sopran [zo'pra:n] **der; ~s, ~e** *(Musik)* **a)** *(Stimmlage)* soprano [voice]; **b)** *o. Pl. (im Chor)* sopranos *pl.;* **c)** *(Sängerin)* soprano

Sopranịst der; ~en, ~en sopranist

Sopranịstin die; ~, ~nen soprano

Sorge ['zɔrgə] **die; ~, ~n a)** *o. Pl. (Unruhe, Angst)* worry; **keine ~:** don't [you] worry; **in ~ um jmdn./etw. sein** be worried about sb./sth.; **b)** *(sorgenvoller Gedanke)* worry; **ich mache mir ~n um dich** I am worried about you; **lassen Sie das meine ~ sein** let 'me worry about that; **c)** *o. Pl. (Mühe, Für-*

sorge) care; **die ~ um das tägliche Brot** the worry of providing one's daily bread; **ich werde dafür ~ tragen, daß ...:** I will see to it *or* make sure that ...

sorgen 1. *refl. V.* worry, be worried (**um** about). **2.** *itr. V.* **a) für jmdn./etw. ~:** take care of *or* look after sb./sth.; **für die Zukunft der Kinder ist gesorgt** the children's future is provided for; **b)** *(bewirken)* **für etw. ~:** cause sth.

sorgen-, Sorgen-: ~frei 1. *Adj.* carefree ‹*person, future, existence, etc.*›; **2.** *adv.* **~frei leben** live in a carefree manner; **~kind das** *(auch fig.)* problem child; **~voll 1.** *Adj.* worried; anxious; **2.** *adv.* worriedly; anxiously

Sorge·recht das; *o. Pl.* *(Rechtsw.)* custody (**für** of)

Sorg·falt ['zɔrkfalt] **die; ~:** care; **große ~ auf etw.** *(Akk.)* **verwenden** *od.* **legen** take great *or* a great deal of care over sth.

sorg·fältig 1. *Adj.* careful. **2.** *adv.* carefully

Sorgfältigkeit die; ~: carefulness

sorg·los 1. *Adj.* **a)** *(ohne Sorgfalt)* careless; **b)** *(unbekümmert)* carefree. **2.** *adv.* **~ mit etw. umgehen** treat sth. carelessly

Sorglosigkeit die; ~ a) *(Mangel an Sorgfalt)* carelessness; **b)** *(Unbekümmertheit)* carefreeness

sorgsam 1. *Adj.* careful. **2.** *adv.* carefully

Sorte ['zɔrtə] **die; ~, ~n a)** sort; type; kind; *(Marke)* brand; **bitte ein Pfund von der besten ~:** a pound of the best quality, please; **b)** *Pl.* *(Devisen)* foreign currency *sing.*

sortieren *tr. V.* sort [out] ‹*pictures, letters, washing, etc.*›; grade ‹*goods etc.*›; *(fig.)* arrange ‹*thoughts*›

Sortiment [zɔrti'mɛnt] **das; ~[e]s, ~e** range (**an** + *Dat.* of)

Sortiments·buch·handel der retail book trade

so·sehr *Konj.* however much

Soße ['zo:sə] **die; ~, ~n** sauce; *(Braten~)* gravy; sauce; *(Salat~)* dressing

sott [zɔt] *1. u. 3. Pers. Sg. Prät. v.* **sieden**

Souffleur [zu'flø:ɐ̯] **der; ~s, ~e, Souffleuse** [zu'flø:zə] **die; ~, ~n** *(Theater)* prompter

soufflieren [zu'fli:rən] *tr. V.* prompt

Souterrain ['zu:tɛrɛ̃] **das; ~s, ~s** basement

Souvenir [zuvə'ni:ɐ̯] **das; ~s, ~s** souvenir

souverän [zuvə'rɛ:n] **1.** *Adj.* **a)** sovereign; **b)** *(überlegen)* superior. **2.** *adv.* **er siegte ganz ~:** he won in a very impressive way

Souverän der; ~s, ~e sovereign

Souveränität die; ~ sovereignty

so·viel 1. *Konj.* **a)** *(nach dem, was)* as *or* so far as; **~ mir bekannt ist** so far as I know; **b)** *(in wie großem Maße auch immer)* however much. **2.** *Indefinitpron.* **~ wie** *od.* **als** as much as; **halb/doppelt ~:** half/twice as much

so·weit 1. *Konj.* **a)** *(nach dem, was)* as *or* so far as; **~ mir bekannt ist** so far as I know; **b)** *(in dem Maße, wie)* [in] so far as. **2.** *Adv.* by and large; on the whole; *(bis jetzt)* up to now; **~ wie** *od.* **als möglich** as far as possible; **~ sein** *(ugs.)* be ready; **es ist ~:** the time has come

so·wenig 1. *Konj.* however little. **2.** *Indefinitpron.* **~ wie** *od.* **als möglich** as little as possible; **ich kann es ~ wie du** I can't do it any more than you can

so·wie *Konj.* **a)** *(und auch)* as well as; **b)** *(sobald)* as soon as

so·wie·so *Adv.* anyway; **das ~!** *(ugs.)* that goes without saying!; of course!

Sowjet [zɔ'vjɛt] **der; ~s, ~s** *(Behörde)* soviet; **der Oberste ~:** the Supreme Soviet

Sowjet·bürger der Soviet citizen

sowjetisch *Adj.* Soviet

Sowjet·union die Soviet Union

so·wohl *Konj.* **~ ... als** *od.* **wie [auch] ...:** both ... and ...; ... as well as ...

sozial [zo'tsi̯a:l] **1.** *Adj.* social; **~e Marktwirtschaft** social market economy. **2.** *adv.* socially; **~ handeln** act in a socially conscious *or* public-spirited way

sozial-, Sozial-: ~abgaben *Pl.* social welfare contributions; **~amt das** social welfare office; **~arbeit die;** *o. Pl.* social work; **~arbeiter der, ~arbeiterin die** social worker; **~demokrat der** Social Democrat; **~demokratie die** social democracy *no art.*; **~demokratisch** *Adj.* social democratic; **Sozialdemokratische Partei [Deutschlands]** [German] Social Democratic Party; **~hilfe die** social welfare

Sozialismus der; ~: socialism *no art.*

Sozialist der; ~en, ~en, Sozialistin die; ~, ~nen socialist

sozialistisch 1. *Adj.* socialist. **2.** *adv.* **~ regierte Länder** countries with socialist governments

sozial-, Sozial-: ~kunde die; *o. Pl.* social studies *sing., no art.*; **~leistungen** *Pl.* social welfare benefits; **~liberal** *Adj.* liberal socialist ‹*politician etc.*›; *(aus SPD und FDP)* liberal-social democrat ‹*coalition etc.*›; **~politik die** social policy; **~prestige das** social status; **~produkt das** *(Wirtsch.)* national product; **~staat der** welfare state; **~versicherung die** social security; **~wohnung die** ≈ council flat *(Brit.)*; municipal housing unit *(Amer.)*

Soziologe [zotsi̯o'lo:gə] **der; ~n, ~n** sociologist

Soziologie die; ~: sociology

Soziologin die; ~, ~nen sociologist

soziologisch 1. *Adj.* sociological. **2.** *adv.* sociologically

Sozius ['zo:tsi̯ʊs] **der; ~, ~se a)** *Pl. auch:* **Sozii** ['zo:tsii] *(Wirtsch.: Teilhaber)* partner; **b)** *(beim Motorrad)* pillion

so·zu·sagen *Adv.* so to speak; as it were

Spachtel ['ʃpaxtl̩] **der; ~s, ~** *od.* **die; ~, ~n a)** *(für Kitt)* putty-knife; *(zum Abkratzen von Farbe)* paint-scraper; *(zum Malen)* palette-knife; spatula; **b)** *(~masse)* filler

Spachtel·masse die filler

spachteln *tr. V.* **a)** stop, fill ‹*hole, crack, etc.*›; smooth over ‹*wall, panel, surface, etc.*›; **b)** *(ugs.: essen)* put away *(coll.)* ‹*food, meal*›

[1]**Spagat** [spa'ga:t] **der** *od.* **das; ~[e]s, ~e** splits *pl.*; **[einen] ~ machen** do the splits

[2]**Spagat der; ~[e]s, ~e** *(südd., österr.)* string

Spaghetti [ʃpa'gɛti] *Pl.* spaghetti *sing.*

spähen ['ʃpɛ:ən] *itr. V.* peer; *(durch eine Ritze usw.)* peep

Späher der; ~s, ~ *(Milit.)* scout; *(Posten)* look-out; *(Spitzel)* informer

Spalier [ʃpa'li:ɐ̯] **das; ~s, ~e a)** trellis; **b)** *(aus Menschen)* double line; *(Ehren~)* guard of honour; **~ stehen** line the route; ‹*soldiers*› form a guard of honour

Spalt [ʃpalt] **der; ~[e]s, ~e** opening; *(im Fels)* fissure; crevice; *(zwischen Vorhängen)* chink; gap; *(langer Riß)* crack; **die Tür einen ~ [weit] öffnen** open the door a crack *or* slightly

Spalte die; ~, ~n a) crack; *(Fels~)* crevice; cleft; **b)** *(Druckw.: Druck~)* column; **c)** *(österr.: Scheibe)* slice

spalten 1. *unr. (auch regelm.) tr.*

V. (auch Physik, fig.) split; **Holz ~:** chop wood. **2.** *unr. (auch regelm.) refl. V.* **a)** *(auch Physik, fig.)* split; **b)** *(Chemie)* split; break down

Span [ʃpa:n] **der; ~[e]s, Späne** ['ʃpɛ:nə] *(Hobel~)* shaving; *(Feil~)* filing *usu. in pl.; (beim Bohren)* boring *usu. in pl.; (beim Drehen)* turning *usu. in pl.;* **feine [Metall]späne** swarf *sing.;* **wo gehobelt wird, [da] fallen Späne** *(Spr.)* you cannot make an omelette without breaking eggs *(prov.)*

Span·ferkel das sucking pig

Spange ['ʃpaŋə] **die; ~, ~n** clasp; *(Haar~)* hair-slide *(Brit.);* barrette *(Amer.); (Arm~)* bracelet; bangle

Spaniel ['ʃpa:ni̯əl] **der; ~s, ~s** spaniel

Spanien ['ʃpa:ni̯ən] **(das); ~s** Spain

Spanier ['ʃpa:ni̯ɐ] **der; ~s, ~:** Spaniard; **die ~:** the Spanish *or* Spaniards

spanisch *Adj.* Spanish; **das kommt mir ~ vor** *(ugs.)* that strikes me as odd; *s. auch* **deutsch; Deutsch**

Span·korb der chip basket; chip

spann [ʃpan] *1. u. 3. P. Sing. Prät. v.* **spinnen**

Spann der; ~[e]s, ~e instep

Spann·beton der *(Bauw.)* prestressed concrete

Spanne die; ~, ~n a) *(Zeit~)* span of time; **b)** *(veralt.: Längenmaß)* span

spannen 1. *tr. V.* **a)** tighten, tauten ‹*violin string, violin bow, etc.*›; draw ‹*bow*›; tension ‹*spring, tennis net, drumhead, saw-blade*›; stretch ‹*fabric, shoe, etc.*›; draw *or* pull ‹*line*› tight *or* taut; tense, flex ‹*muscle*›; cock ‹*gun, camera shutter*›; **b)** *(befestigen)* put up ‹*washing-line*›; stretch ‹*net, wire, tarpaulin, etc.*› **(über +** *Akk.* over); **einen Bogen Papier in die Schreibmaschine ~:** insert *or* put a sheet of paper in the typewriter; **etw. in einen Schraubstock ~:** clamp sth. in a vice; **c)** *(schirren)* hitch up, harness **(vor, an +** *Akk.* to); **d)** *(bes. südd., österr.: merken)* notice. **2.** *refl. V.* **a)** become *or* go taut; ‹*muscles*› tense; **b)** *(geh.: sich wölben)* **sich über etw.** *(Akk.)* **~** ‹*bridge, rainbow*› span sth. **3.** *itr. V. (zu eng sein)* ‹*clothing*› be [too] tight; ‹*skin*› be taut

spannend 1. *Adj.* exciting; *(stärker)* thrilling; **mach's nicht so ~!** *(ugs.)* don't keep me/us in suspense. **2.** *adv.* excitingly; *(stärker)* thrillingly

Spanner der; ~s, ~ a) *(Schuh~)* shoe-tree; *(Stiefel~)* boot-tree; *(Hosen~)* [trouser-] hanger; **b)** *(Zool.)* geometer; **c)** *(ugs.: Voyeur)* peeping Tom

Spann·kraft die; *o. Pl.* vigour

Spannung die; ~, ~en a) *o. Pl.* excitement; *(Neugier)* suspense; tension; **jmdn. mit ~ erwarten** await sb. eagerly; **b)** *o. Pl. (eines Romans, Films usw.)* suspense; **c)** *(Zwistigkeit, Nervosität)* tension; **d)** *(das Straffsein)* tension; tautness; **e)** *(elektrische ~)* tension; *(Voltzahl)* voltage; **unter ~ stehen** be live; **f)** *(Mechanik)* stress

Spannungs·gebiet das *(Politik)* area of tension

spannungs·geladen *Adj.* **a)** *(gespannt)* ‹*atmosphere etc.*› charged with tension; **b)** *(spannend)* ‹*novel, film, etc.*› full of suspense

Spann·weite die *(Zool.: Flügel~)* [wing-]span; wing-spread; *(eines Flugzeugs)* [wing-]span

Span·platte die chipboard

Spar-: ~buch das *(Bankw.)* savings book; passbook; **~büchse die** money-box; **~einlage die** *(Bankw.)* savings deposit

sparen ['ʃpa:rən] **1.** *tr. V.* save; **deine Ratschläge kannst du dir ~:** you can keep your advice. **2.** *itr. V.* **a) für** *od.* **auf etw.** *(Akk.)* **~:** save up for sth.; **b)** *(sparsam wirtschaften)* economize **(mit** on); **er sparte nicht mit Lob** *(fig.)* he was unstinting *or* generous in his praise; **an etw.** *(Dat.)* **~:** be sparing with sth.; *(Billigeres nehmen)* economize on sth.

Sparer der; ~s, ~, Sparerin die; ~, ~nen saver

Spar·flamme die; *o. Pl.* low flame *or* heat; **auf ~:** on a low flame *or* heat

Spargel ['ʃpargl̩] **der; ~s, ~,** *schweiz. auch* **die; ~, ~n** asparagus *no pl., no indef. art.;* **ein ~:** an asparagus stalk

Spar-: ~groschen der *(ugs.)* nest-egg; savings *pl.;* **~guthaben das** credit balance *(in a savings account);* **~kasse die** savings bank; **~kassen·buch das** savings book; passbook; **~konto das** savings *or* deposit account

spärlich ['ʃpɛ:ɐlɪç] **1.** *Adj.* sparse ‹*vegetation, beard, growth*›; thin ‹*hair, applause*›; scanty ‹*leftovers, knowledge, news, evidence*›; scanty, skimpy ‹*clothing*›; slack ‹*demand*›; scattered ‹*remains, remnants*›; poor ‹*lighting, harvest, result, source*›; meagre ‹*income, salary*›. **2.** *adv.* sparsely, thinly ‹*populated, covered*›; poorly ‹*lit, attended*›; scantily, skimpily ‹*dressed*›

Spar·maßnahme die economy measure

Sparren ['ʃparən] **der; ~s, ~** *(Dach~)* rafter

sparsam 1. *Adj.* **a)** thrifty ‹*person*›; *(wirtschaftlich)* economical; **mit etw. ~ sein** be economical with sth.; **er ist mit Worten/Lob immer sehr ~** *(fig.)* he is a man of few words/he is very sparing in his praise; **b)** *(fig.: gering, wenig, klein)* sparse ‹*detail, decoration, interior, etc.*›; economical ‹*movement, manner of expression, etc.*›. **2.** *adv.* **a) ~ mit dem Papier umgehen** use paper sparingly; economize on paper; **~ leben** live frugally; **~ mit seinen Kräften umgehen** conserve one's energy; **b)** *(wirtschaftlich)* economically; **c)** *(fig.: in geringem Maße)* ‹*use*› sparingly; sparsely ‹*decorated, furnished*›

Sparsamkeit die; ~ a) thrift[iness]; **aus ~:** for the sake of economizing; **b)** *(Wirtschaftlichkeit)* economicalness; **c)** *(fig.: geringes Maß)* economy

Spar·schwein das piggy bank

Spartakiade [ʃparta'ki̯a:də] **die; ~, ~n** Spartakiad

spartanisch 1. *Adj. (auch fig.).* Spartan. **2.** *adv.* **~ leben** lead a Spartan life

Sparte ['ʃpartə] **die; ~, ~n a)** *(Teilbereich)* area; branch; *(eines Geschäfts)* line [of business]; *(des Wissens)* branch; field; speciality; **b)** *(Rubrik)* section

Spaß [ʃpa:s] **der; ~es, Späße** ['ʃpɛ:sə] **a)** *o. Pl. (Vergnügen)* fun; **wir hatten alle viel ~:** we all had a lot of fun *or* a really good time; we all really enjoyed ourselves; **~ an etw.** *(Dat.)* **haben** enjoy sth.; **[jmdm.] ~/keinen ~ machen** be fun/no fun [for sb.]; **ein teurer ~** *(ugs.)* an expensive business; **was kostet der ~?** *(ugs.)* how much will that little lot cost? *(coll.);* **viel ~!** have a good time!; **b)** *(Scherz)* joke; *(Streich)* prank; antic; **er macht nur ~:** he's only joking *or (sl.)* kidding; **~ beiseite!** joking aside *or* apart; **~ muß sein!** there's no harm in a joke; **da hört [für mich] der ~ auf** that's getting beyond a joke; **~/keinen ~ verstehen** be able/not be able to take a joke; have a/have no sense of humour; **in Gelddingen versteht er keinen ~:** he won't stand for any nonsense where money is concerned; **er ist immer zu Späßen aufgelegt** he's always ready for a laugh; **im** *od.* **zum** *od.* **aus ~**: as a joke; for fun; **sich** *(Dat.)* **einen ~ mit jmdm. erlauben** play a joke on sb.

spaßen ['ʃpa:sn̩] *itr. V.* joke; kid *(coll.)*; **er läßt nicht mit sich ~**: he won't stand for any nonsense; **mit ihm/damit ist nicht zu ~**: he/it is not to be trifled with

spaßes·halber *Adv.* for the fun of it; for fun

spaßig *Adj.* funny; comical; amusing

Spaß-: **~macher der** joker; **~vogel der** joker

spastisch *(Med.)* **1.** *Adj.* spastic. **2.** *adv.* **~ gelähmt sein** suffer from spastic paralysis

spät [ʃpɛ:t] **1.** *Adj.* late; belated ⟨*fame, repentance*⟩; **am ~en Abend** in the late evening; **bis in die ~e Nacht** until late into the night; **wie ~ ist es?** what time is it? **2.** *adv.* late; **~ am Abend** late in the evening; **du kommst aber ~!** you're very late; **wenn ich jetzt nicht losfahre, komme ich zu ~**: if I don't leave now I'll be late; **wir sind [schon ziemlich] ~ dran** *(ugs.)* we're late [enough already]

Spät·dienst der late duty; *(im Betrieb)* late shift

Spatel ['ʃpa:tl̩] **der; ~s, ~ a)** spatula; **b)** *s.* **Spachtel a**

Spaten ['ʃpa:tn̩] **der; ~s, ~**: spade

später **1.** *Adj.; nicht präd.* **a)** *(nachfolgend, kommend)* later ⟨*years, generations, etc.*⟩; **b)** *(zukünftig)* future ⟨*owner, wife, etc.*⟩. **2.** *Adv.* later; **was willst du denn ~ [einmal] werden?** what do you want to do when you grow up?; **[also dann] bis ~!** see you later!

spätestens *Adv.* at the latest; **~ [am] Freitag** [by] Friday at the latest

Spät-: **~lese die** late vintage; **~schicht die** late shift

Spatz [ʃpats] **der; ~en, ~en a)** sparrow; **er ißt wie ein ~**: he eats like a bird; **die ~en pfeifen es von den** *od.* **allen Dächern** it's common knowledge; **b)** *(fam.: Liebling)* pet; **c)** *(fam.: kleines Kind)* mite; tot *(coll.)*

Spätzle ['ʃpɛtslə] *Pl.* spaetzle; spätzle; *kind of noodles*

spazieren [ʃpa'tsi:rən] *itr. V.; mit sein* **a)** stroll; **b)** *(veralt.: spazierengehen)* go for a walk *or* a stroll

spazieren-: **~|fahren 1.** *unr. itr. V.; mit sein (im Auto)* go for a drive *or* ride *or* spin; *(im Bus usw., mit dem Fahrrad od. Motorrad)* go for a ride. **2.** *tr. V.* **jmdn. ~fahren** *(im Auto)* take sb. for a drive *or* ride *or* spin; **ein Kind [im Kinderwagen] ~fahren** take a baby for a walk [in a pram]; **~|gehen** *unr. itr. V.; mit sein* go for a walk *or* stroll; **ein Stück ~gehen** go for a little walk *or* stroll

Spazier-: **~fahrt die** *(mit dem Auto)* drive; ride; spin; *(mit dem Bus usw., mit dem Fahrrad od. Motorrad)* ride; **~gang der** walk; stroll; **~gänger der; ~s, ~, ~gängerin die; ~, ~nen** person out for a walk *or* stroll; **~weg der** footpath

SPD [ɛspe:'de:] **die; ~** *Abk.* **Sozialdemokratische Partei Deutschlands** SPD

Specht [ʃpɛçt] **der; ~[e]s, ~e** woodpecker

Speck [ʃpɛk] **der; ~[e]s, ~e a)** bacon fat; *(Schinken~)* bacon; **b)** *(von Walen, Robben)* blubber; **c)** *(ugs. scherzh.: Fettpolster)* fat; flab *(sl.)*; **er hat ganz schön ~ auf den Rippen** he's well padded

speckig *Adj.* greasy

Speck-: **~scheibe die** rasher *or* slice of bacon; **~schwarte die** bacon rind; **~seite die** side of bacon; **~stein der** *(Mineral)* lard stone; soapstone; steatite

Spediteur [ʃpedi'tø:ɐ̯] **der; ~s, ~e** carrier; haulier; haulage contractor; *(per Schiff)* carrier; *(Möbel~)* furniture-remover

Spedition [ʃpedi'tsio:n] **die; ~, ~en a)** *(Beförderung)* carriage; transport; **b)** *s.* **Speditionsfirma**

Speditions·firma die forwarding agency; *(per Schiff)* shipping agency; *(Transportunternehmen)* haulage firm; firm of hauliers; *(per Schiff)* firm of carriers; *(Möbelspedition)* removal firm

Speer [ʃpe:ɐ̯] **der; ~[e]s, ~e a)** spear; **b)** *(Sportgerät)* javelin

Speer-: **~spitze die** *(auch fig.)* spearhead; **~werfen das; ~s** *s.* **~wurf a**; **~werfer der** *(Sport)* javelin-thrower; **~wurf der a)** *o. Pl. (Disziplin)* javelin-throwing; **b)** *(Wurf)* javelin-throw

Speiche ['ʃpaiçə] **die; ~, ~n a)** spoke; **b)** *(Anat.)* radius

Speichel ['ʃpaiçl̩] **der; ~s** saliva; spittle

Speicher ['ʃpaiçɐ] **der; ~s, ~ a)** storehouse; *(Lagerhaus)* warehouse; *(~becken)* reservoir; *(fig.)* store; **b)** *(südd.: Dachboden)* loft; attic; **auf dem ~**: in the loft *or* attic; **c)** *(Elektronik)* memory; store

speichern *tr. V.* store

Speicherung die; ~, ~en storing; storage

speien ['ʃpaiən] *(geh.) unr. tr., itr. V.* **a)** spit; spew [forth] ⟨*lava, fire, etc.*⟩; belch ⟨*smoke*⟩; spout ⟨*water*⟩; **b)** *(erbrechen)* vomit

Speise ['ʃpaizə] **die; ~, ~n a)** *(Gericht)* dish; **b)** *o. Pl. (geh.: Nahrung)* food

Speise-: **~gaststätte die** restaurant; **~kammer die** larder; pantry; **~karte die** menu; **~lokal das** restaurant

speisen 1. *itr. V. (geh.)* eat; *(dinieren)* dine; **zu Mittag/Abend ~**: lunch *or* have lunch/dine *or* have dinner. **2.** *tr. V.* **a)** *(geh.: verzehren)* eat; *(dinieren)* dine on; **b)** *(geh.)* feed; **c)** *(Technik)* **etw. mit Strom/Wasser ~**: supply sth. with electricity/water

Speise-: **~öl das** edible oil; **~reste** *Pl.* left-overs; *(zwischen den Zähnen)* food particles; **~röhre die** *(Anat.)* gullet; oesophagus *(Anat.)*; **~saal der** dining-hall; *(im Hotel, in einer Villa usw.)* dining-room; **~schrank der** food-cupboard; **~wagen der** dining-car; restaurant car *(Brit.)*; **~zettel der** menu

spei·übel *Adj.; nicht attr.* **mir ist ~**: I think I'm going to be violently sick

Spektakel [ʃpɛk'ta:kl̩] **der; ~s, ~** *(ugs.)* **a)** *(Lärm)* row *(coll.)*; rumpus *(coll.)*; racket; **b)** *(laute Auseinandersetzung)* fuss; **einen ~ machen** kick up *or* make a fuss

spektakulär [ʃpɛktaku'lɛ:ɐ̯] **1.** *Adj.* spectacular. **2.** *adv.* spectacularly

Spektral-: **~analyse die** *(Technik)* spectral analysis; **~farbe die** colour of the spectrum

Spektrum ['ʃpɛktrʊm] **das; ~s, Spektren** *(auch fig.)* spectrum

Spekulation [ʃpekula'tsio:n] **die; ~, ~en a)** *(Mutmaßung, Erwartung; auch Philos.)* speculation; **b)** *(Wirtsch.)* speculation **(mit** in)

Spekulatius [ʃpeku'la:tsiʊs] **der; ~, ~**: *spiced biscuit in the shape of a human or other figure, eaten at Christmas*

spekulativ [ʃpekula'ti:f] **1.** *Adj.* speculative. **2.** *adv.* speculatively

spekulieren [ʃpeku'li:rən] *itr. V.* **a)** *(ugs.)* **darauf ~, etw. tun zu können** count on being able to do sth.; **er spekuliert auf den Laden** he's counting on getting the shop; **b)** *(mutmaßen)* speculate; **c)** *(Wirtsch.)* speculate **(mit** in)

Spelunke [ʃpe'lʊŋkə] **die; ~, ~n** *(ugs. abwertend)* dive *(coll.)*

Spelze ['ʃpɛltsə] **die; ~, ~n** *(des Getreidekorns)* husk

spendabel [ʃpɛn'da:bl̩] *Adj.* generous; open-handed

Spende ['ʃpɛndə] **die; ~, ~n** donation; contribution; **eine kleine ~ bitte!** would you like to make a small donation?

spenden *tr., itr. V.* **a)** donate; give; contribute; **[etw.] fürs Rote Kreuz ~:** contribute [sth.] to *or* for the Red Cross; **Blut/eine Niere ~:** give blood/donate a kidney; **b)** *(fig. geh.)* give ⟨*light, applause, comfort*⟩; afford, give ⟨*shade*⟩; give off ⟨*heat*⟩; provide ⟨*water*⟩; administer ⟨*communion, baptism*⟩; give, bestow ⟨*blessing*⟩; confer ⟨*holy orders*⟩

Spenden-: ~aktion die campaign for donations; **~aufruf der** appeal for donations

Spender der; ~s, ~, Spenderin die; ~, ~nen donor; donator; contributor

spendieren *tr. V. (ugs.)* get, buy ⟨*drink, meal, etc.*⟩; stand ⟨*round*⟩

Spendier·hosen *Pl. in* **die/seine ~ anhaben** be in a generous mood; be feeling generous

Spengler ['ʃpɛŋlɐ] **der; ~s, ~** *(südd., österr., schweiz.) s.* **Klempner**

Spenzer ['ʃpɛntsɐ] **der; ~s, ~ a)** *(Jacke)* spencer; **b)** *(Unterhemd)* tight-fitting short-sleeved vest

Sperber ['ʃpɛrbɐ] **der; ~s, ~:** sparrow-hawk

Sperling ['ʃpɛrlɪŋ] **der; ~s, ~e** sparrow

Sperma ['ʃpɛrma] **das; ~s, Spermen** *od.* **Spermata** sperm; semen

sperr·angel·weit *Adv. (ugs.)* **~ offen** *od.* **geöffnet** wide open

Sperr·bezirk der restricted *or* prohibited area

Sperre die; ~, ~n a) *(Barriere)* barrier; *(Straßen~)* road-block; **b)** *(Milit.)* obstacle; **c)** *(Eisenb.)* barrier; **d)** *(fig.: Verbot, auch Sport)* ban; *(Handels~)* embargo; *(Import~, Export~)* blockade; *(Nachrichten~)* [news] black-out; **e)** *(Psych.: Blockierung, Hemmung)* block; **f)** *(Technik)* locking device

sperren 1. *tr. V.* **a)** close ⟨*road, tunnel, bridge, entrance, border, etc.*⟩; close off ⟨*area*⟩; **etw. für jmdn./etw. ~:** close sth. to sb./sth.; **b)** *(blockieren)* block ⟨*access, entrance, etc.*⟩; **c)** *(Technik)* lock ⟨*mechanism etc.*⟩; **d)** cut off, disconnect ⟨*water, gas, electricity, etc.*⟩; **jmdm. den Strom/das Telefon ~:** cut off *or* disconnect sb.'s electricity/telephone; **e)** *(Bankw.)* stop ⟨*cheque, overdraft facility*⟩; freeze ⟨*bank account*⟩; **f)** *(ein~)* **ein Tier/jmdn. in etw.** *(Akk.)* **~:** shut *or* lock an animal/sb. in sth.; **jmdn. ins Gefängnis ~:** put sb. in prison; lock sb. up [in prison]; **g)** *(Sport: behindern)* obstruct; **h)** *(Sport: von der Teilnahme ausschließen)* ban; **i)** *(Druckw.: spationieren)* print ⟨*word, text*⟩ with the letters spaced. **2.** *refl. V.* **sich [gegen etw.] ~:** balk *or* jib [at sth.]. **3.** *itr. V. (Sport)* obstruct

Sperr·holz das plywood

sperrig *Adj.* unwieldy

Sperr-: ~konto das *(Bankw.)* blocked account; **~müll der** bulky refuse *(for which there is a separate collection service)*; **~sitz der** *(im Kino)* seat in the back stalls; *(im Zirkus)* front seat; *(im Theater)* seat in the front stalls; **~stunde die** closing time

Sperrung die; ~, ~en *s.* **sperren 1a–e, i:** closing; closing off; blocking; locking; cutting off; disconnection; stopping; freezing; banning

Sperr·vermerk der restriction note *(regarding sale of property, withdrawal of investment, disclosure of information, etc.)*

Spesen ['ʃpe:zn̩] *Pl.* expenses; **auf ~:** on expenses; **außer ~ nichts gewesen** *(scherzh.)* [it was] a waste of time and effort

Spezi ['ʃpe:tsi] **der; ~s, ~[s] a)** *(südd., österr., schweiz. ugs.)* [bosom] pal *(coll.)*; chum *(coll.)*; **b)** *(ugs.: Getränk)* lemonade and cola

Spezial-: ~gebiet das special *or* specialist field; **~geschäft das** specialist shop

spezialisieren *refl. V.* specialize **(auf +** *Akk.* in)

Spezialist der; ~en, ~en, Spezialistin die; ~, ~nen specialist

Spezialität [ʃpetsiali'tɛ:t] **die; ~, ~en** speciality; specialty

Spezial·slalom der *(Ski)* special slalom

speziell [ʃpe'tsiɛl] **1.** *Adj.* special; specific ⟨*question, problem, etc.*⟩; specialized ⟨*book, knowledge, etc.*⟩. **2.** *Adv. (besonders, gerade)* especially; *(eigens)* specially

spezifisch 1. *Adj.* specific; characteristic ⟨*smell, style*⟩; **~es Gewicht/~e Wärme** *(Phys.)* specific gravity/heat. **2.** *adv.* specifically

spezifizieren *tr. V.* specify; *(einzeln aufführen)* itemize ⟨*bill, expenses, etc.*⟩

Sphäre ['sfɛ:rə] **die; ~, ~n** *(auch fig.)* sphere; **in höheren ~n schweben** *(scherzh.)* have one's head in the clouds

sphärisch *Adj.* **a)** spherical; **b)** *(fig.: himmlisch)* heavenly

Sphinx [sfɪŋks] **die** *od.* **der; ~, ~e** *od.* **Sphingen** ['sfɪŋən] *(Ägyptologie, Kunstwiss.)* sphinx

Spick·aal der *(bes. nordd.)* smoked eel

spicken ['ʃpɪkn̩] *tr. V.* **a)** *(Kochk.)* lard; **b)** *(fig. ugs.: reichlich versehen)* **eine Rede mit Zitaten ~:** lard a speech with quotations

Spick·zettel der *(ugs.)* crib *(coll.)*

spie [ʃpi:] *1. u. 3. Pers. Sg. Prät. v.* **speien**

Spiegel ['ʃpi:gl̩] **der; ~s, ~ a)** mirror; **im ~ der Presse** *(fig.)* as mirrored *or* reflected in the press; **b)** *(Wasserstand, Blutzucker~, Alkohol~ usw.)* level; *(Wasseroberfläche)* surface

spiegel-, Spiegel-: ~bild das *(auch fig., Math.)* reflection; **~bildlich 1.** *Adj.* **eine ~bildliche Abbildung** a mirror image; **2.** *adv.* **~bildlich abgebildet** reproduced as a *or* in mirror image; **~blank** *Adj.* shining; **~ei das** fried egg; **~glas das** mirror glass; **~glatt** *Adj.* like glass *postpos.*; as smooth as glass *postpos.*

spiegeln 1. *itr. V.* **a)** *(glänzen)* shine; gleam; **b)** *(als Spiegel wirken)* reflect the light. **2.** *tr. V.* reflect; mirror. **3.** *refl. V. (auch fig.)* be mirrored *or* reflected

Spiegel-: ~reflex·kamera die reflex camera; **~schrift die** mirror writing

Spiegelung die; ~, ~en a) *(auch fig., Math.)* reflection; **b)** *(Med.)* speculum examination

spiegel·verkehrt *s.* **spiegelbildlich**

Spiel [ʃpi:l] **das; ~[e]s, ~e a)** *(das Spielen, Spielerei)* play; **für ihn ist alles nur ein ~:** everything's just a game to him; **ein ~ mit dem Feuer** *(fig.)* playing with fire; **b)** *(Glücks~; Gesellschafts~)* game; *(Wett~)* game; match; **gewonnenes ~ haben** be home and dry; **auf dem ~ stehen** be at stake; **etw. aufs ~ setzen** put sth. at stake; risk sth.; **jmdn./etw. aus dem ~ lassen** *(fig.)* leave sb./sth. out of it; **ins ~ kommen** *(fig.)* ⟨*factor*⟩ come into play; ⟨*person, authorities, etc.*⟩ become involved; ⟨*matter, subject, etc.*⟩ come into it; **im ~ sein** *(fig.)* be involved; **c)** *(Utensilien)* game; **d)** *o. Pl. (eines Schauspielers)* performance; **e)** *(eines Musikers)* performance; playing; **f)** *(Sport: ~weise)* game;

g) *(Schau~)* play; **h)** *(Technik: Bewegungsfreiheit)* [free] play

Spiel-: **~art die** variety; **~automat der** gaming-machine; *(Geschicklichkeitsspiel)* amusement machine; **~ball der a)** *(Sport) (Tennis)* game point; *(Volleyball)* match ball; **b)** *(Billard)* red [ball]; **c)** *(fig.)* plaything; **sie ist der ~ball ihrer Leidenschaften** she allows herself to be torn hither and thither by her passions; **~bank die** casino

spielen 1. *itr. V.* **a)** play; **um die Meisterschaft ~:** play for the championship; **sie haben 1:0 gespielt** the match ended 1–0; **auf der Gitarre ~:** play the guitar; **er kann vom Blatt/nach Noten ~:** he can sight-read/play from music; **b)** *(um Geld)* play; **er begann zu trinken und zu ~:** he began to drink and to gamble; **um Geld ~:** play for money; **c)** *(als Schauspieler)* act; perform; **d)** *(sich abspielen)* **der Film spielt in Berlin** the film is set in Berlin; **e)** *(fig.: sich bewegen)* ‹*wind, water, etc.*› play; **seine Muskeln ~ lassen** flex one's muscles; **seinen Charme/seine Beziehungen ~ lassen** *(fig.)* bring one's charm/connections to bear; **f)** *(fig.: übergehen)* **das Blau spielt ins Violette** the blue is tinged with purple. **2.** *tr. V.* **a)** play; **Räuber und Gendarm ~:** play cops and robbers; **Cowboy ~:** play at being a cowboy; **Geige** *usw.* **~:** play the violin *etc.*; **Trumpf/Pik/ein As ~:** play a trump/spades/an ace; **b)** *(aufführen, vorführen)* put on ‹*play*›; show ‹*film*›; perform ‹*piece of music*›; play ‹*record*›; **was wird hier gespielt?** *(fig. ugs.)* what's going on here?; **c)** *(schauspielerisch darstellen)* play ‹*role*›; **den Beleidigten/Unschuldigen ~** *(fig.)* act offended/play the innocent; **sein Interesse war [nur] gespielt** he [only] pretended to be interested; his interest was [merely] feigned; **d)** *(Sport: werfen, treten, schlagen)* play; **einen Ball mit Rückhand ~:** play a ball backhand. **3.** *refl. V.* **sich warm ~:** warm up

spielend 1. *Adj.; nicht präd.* **mit ~er Leichtigkeit** with consummate *or* effortless ease. **2.** *adv.* easily; **etw. ~ beherrschen** master sth. effortlessly

Spieler der; ~s, ~ player; *(Glücks~)* gambler

Spielerei die; ~, ~en a) *o. Pl.* playing *no art.*; *(im Glücksspiel)* gambling *no art.*; *(das Herumspielen)* playing *or* fiddling about or around **(an** + *Dat.* with); **b)** *(müßiges Tun, Spiel)* **eine ~ mit Zahlen** playing [around] with numbers; **c)** *(Kinderspiel, Leichtigkeit)* child's play *no art.*; **d)** *(Tand)* gadget

Spielerin die; ~, ~nen *s.* **Spieler**

spielerisch 1. *Adj.* **a)** playful; **mit ~er Leichtigkeit** with consummate *or* effortless ease; **b)** *nicht präd. (Sport)* **sein ~es Können** his skill as a player. **2.** *adv.* **a)** playfully; **b)** *(Sport)* in playing terms

Spiel-: **~feld das** *(Fußball, Hokkey, Rugby usw.)* field; pitch *(Brit.)*; *(Tennis, Squash, Federball, Volleyball usw.)* court; **~figur die** piece; **~film der** feature film; **~führer der** *(Sport)* [team] captain; **~geld das** play *or* toy money; **~hölle die** *(ugs. abwertend)* gambling-den; **~kamerad der** playmate; playfellow; **~karte die** playing-card; **~kasino das** casino; **~leiter der a)** *(im Fernsehen)* quiz-master; *(im Roulett)* tourneur; **b)** *s.* **Regisseur**; **~macher der** *(Sportjargon)* key player; **~marke die** chip; jetton; **~plan der** programme; **~platz der** playground; **~raum der a)** room to move *(fig.)*; scope; latitude; *(bei Ausgaben, Budget)* leeway; **b)** *(Technik)* clearance; **~regel die** *(auch fig.)* rule of the game; **gegen die ~regeln verstoßen** *(auch fig.)* break the rules; **~sachen** *Pl.* toys; **~stein der** piece; *(beim Damespiel, Schach)* piece; man; **~straße die** play street; **~uhr die a)** musical clock; **b)** musical box *(Brit.)*; music box *(Amer.)*; **~verderber der; ~s, ~, ~verderberin die; ~, ~nen** spoil-sport; **~waren** *Pl.* toys; **~waren·geschäft das** toyshop; **~zeit die a)** *(Theater: Saison)* season; **b)** *(Aufführungsdauer)* run; **c)** *(Sport)* playing time; **~zeug das a)** toy; *(fig.)* toy; plaything; **b)** *o. Pl.* *(~sachen, ~waren)* toys *pl.*; **~zug der** *(Sport, in einem Brettspiel)* move

Spieß [ʃpi:s] **der; ~es, ~e a)** *(Waffe)* spear; **den ~ umdrehen** *(ugs.)* turn the tables; **wie am ~ brüllen** *(ugs.)* scream one's head off; scream blue murder *(sl.)*; **b)** *(Brat~)* spit; *(Schaschlik~)* skewer; **ein am ~ gebratener Ochse** an ox roasted on the spit; a spit-roasted ox; **c)** *(Fleisch~)* kebab; **d)** *(Soldatenspr.)* [company] sergeant-major

Spieß·bürger der *(abwertend)* [petit] bourgeois

spieß·bürgerlich *s.* **spießig**

Spießchen das; ~s, ~ a) *(Cocktailspieß)* cocktail stick; **b)** *(Schaschlikspieß)* skewer; **c)** *(Fleischspieß)* kebab

spießen *tr. V.* **a) eine Olive auf einen Cocktailspieß ~:** spear an olive with a cocktail stick; **etw. in etw.** *(Akk.)* **~:** stick sth. in sth.

Spießer der; ~s, ~ *(abwertend)* [petit] bourgeois

Spieß·geselle der *(abwertend: Komplize)* accomplice

spießig *(abwertend)* **1.** *Adj.* [petit] bourgeois. **2.** *adv.* ‹*think, behave, etc.*› in a [petit] bourgeois way

Spießigkeit die; ~ *(abwertend)* [petit] bourgeois narrow-mindedness

Spieß·rute die *in* **~n laufen** *(auch fig.)* run the gauntlet

Spike [ʃpaik] **der; ~s, ~s a)** spike; **b)** *(eines Reifens)* stud

Spike[s]·reifen der studded tyre

Spinat [ʃpi'na:t] **der; ~[e]s, ~e** spinach

Spind [ʃpɪnt] **der** *od.* **das; ~[e]s, ~e** locker

Spindel ['ʃpɪndl̩] **die; ~, ~n** spindle

spindel·dürr *Adj.* skinny

Spinett [ʃpi'nɛt] **das; ~[e]s, ~e** spinet

Spinne ['ʃpɪnə] **die; ~, ~n** spider

spinnen ['ʃpɪnən] **1.** *unr. tr. V.* spin *(fig.)*; plot ‹*intrigue*›; think up ‹*idea*›; hatch ‹*plot*›. **2.** *unr. itr. V.* **a)** spin; **b)** *(ugs.: verrückt sein)* be crazy *or (sl.)* nuts *or (sl.)* crackers; **Ich soll bezahlen? Du spinnst wohl!** [What,] me pay? You must be joking *or (sl.)* kidding

Spinnen·netz das spider's web

Spinner der; ~s, ~ a) *(Beruf)* spinner; **b)** *(ugs. abwertend)* nutcase *(sl.)*; idiot

Spinnerei die; ~, ~en a) *o. Pl.* spinning *no art.*; **b)** *(Werkstatt)* spinning mill; **c)** *(ugs. abwertend)* crazy idea

Spinnerin die; ~, ~nen *(Beruf)* spinner

spinnert ['ʃpɪnɐt] *Adj. (ugs., bes. südd.)* slightly potty *(sl.)*

Spinn-: **~rad das** spinning-wheel; **~webe die; ~, ~n** cobweb

Spion [ʃpi'o:n] **der; ~s, ~e a)** spy; **b)** *(Guckloch)* spyhole; **c)** *(Spiegel am Fenster)* tell-tale mirror

Spionage [ʃpio'na:ʒə] **die; ~:** spying; espionage

Spionage·abwehr die a) counter-espionage; counter-intelligence; **b)** *(Dienst)* counter-

espionage *or* counter-intelligence service
spionieren *itr. V.* **a)** spy **(gegen** against); **b)** *(fig. abwertend)* spy; snoop [about] *(coll.)*
Spioniererei die; ~, ~**en** *(fig. abwertend)* snooping [about] *no pl. (coll.)*
Spionin die; ~, ~**nen** spy
Spiral·bohrer der twist drill *or* bit
Spirale [ʃpi'ra:lə] **die;** ~, ~**n a)** *(auch Geom., fig.)* spiral; **b)** *(zur Empfängnisverhütung)* coil
Spiral·feder die coil spring
spiral·förmig 1. *Adj.* spiral[-shaped]. **2.** *adv.* spirally
Spiritist der; ~**en,** ~**en** spiritualist; spiritist
spiritistisch *Adj.* spiritualist[ic]; spiritistic
Spiritual ['spɪrɪtjʊəl] **das** *od.* **der;** ~**s,** ~**s** [negro] spiritual
Spirituose [spiri'tu̯o:zə] **die;** ~, ~**n** spirit *usu. in pl.*
Spiritus ['ʃpi:ritʊs] **der;** ~, ~**se** spirit; ethyl alcohol; **mit** ~ **kochen** cook on a spirit stove
Spiritus·kocher der spirit stove
Spital [ʃpi'ta:l] **das;** ~**s, Spitäler** [ʃpi'tɛ:lɐ] *(bes. österr., schweiz.)* hospital
spitz [ʃpɪts] **1.** *Adj.* **a)** *(nicht stumpf)* pointed ⟨*tower, arch, shoes, nose, beard, etc.*⟩; sharp ⟨*pencil, needle, stone, etc.*⟩; fine ⟨*pen nib*⟩; *(Geom.)* acute ⟨*angle*⟩; **b)** *(schrill)* shrill ⟨*cry etc.*⟩; **c)** *(ugs.: abgezehrt)* haggard; **d)** *(boshaft)* cutting ⟨*remark, etc.*⟩. **2.** *adv.* **a)** ~ **zulaufen** taper to a point; ~ **zulaufend** pointed; **b)** *(boshaft)* cuttingly
Spitz der; ~**es,** ~**e** *(Hund)* spitz
spitz-, Spitz-: ~**bart der a)** goatee; pointed beard; **b)** *(Mann)* man with a/the goatee *or* pointed beard; ~**bube der** *(scherzh.: Schlingel)* rascal; scallywag; scamp; ~**bübisch 1.** *Adj.* roguish; mischievous; **2.** *adv.* roguishly; mischievously
spitze *indekl. Adj. (ugs.) s.* **klasse**
Spitze die; ~, ~**n a)** *(Nadel~, Bleistift~ usw.)* point; *(Pfeil~, Horn~ usw.)* tip; **b)** *(Turm~, Baum~, Mast~ usw.)* top; *(eines Dreiecks, Kegels, einer Pyramide)* top; apex; vertex *(Math.)*; *(eines Berges)* summit; top; **c)** *(Zigarren~, Haar~, Zweig~)* end; *(Schuh~)* toe; *(Finger~, Nasen~, Schwanz~, Flügel~, Spargel~)* tip; **d)** *(vorderes Ende)* front; **an der** ~ **liegen** *(Sport)* be in the lead *or* in front; **e)** *(führende Position)* top; **an der** ~ **[der Tabelle] stehen** *od.* **liegen** *(Sport)* be [at the] top [of the table]; **sich an die** ~ **[einer Bewegung] setzen** put oneself at the head [of a movement]; **f)** *(einer Firma, Organisation usw.)* head; *(einer Hierarchie)* top; *(leitende Gruppe)* management; **die** ~**n der Gesellschaft** the leading figures of society; **g)** *(Höchstwert)* maximum; peak; *(ugs.: Spitzenzeit)* peak period; **das Auto fährt 160 km** ~**:** the car has *or* does a top speed of 160 km. per hour; **h) [absolute/einsame]** ~ **sein** *(ugs.)* be [absolutely] great *(coll.)*; **i)** *(fig.: Angriff)* dig **(gegen** at); ~**n austeilen** make pointed remarks; **j)** *(Textilwesen)* lace
Spitzel der; ~**s,** ~ *(abwertend)* informer
spitzeln *itr. V. (abwertend)* act as an informer
spitzen *tr. V.* sharpen ⟨*pencil*⟩; purse ⟨*lips, mouth*⟩; **die Ohren** ~ ⟨*dog*⟩ prick up its ears; *(fig.)* ⟨*person*⟩ prick up one's ears
Spitzen-: ~**bluse die** lace blouse; ~**erzeugnis das** top-quality product; ~**geschwindigkeit die** top speed; ~**kandidat der** leading *or* top candidate; ~**klasse die a)** top class; **ein Hotel der** ~**klasse** a top-class hotel; **b)** ~**klasse sein** *(ugs.)* be really great *(coll.)*; ~**kraft die** top-class *or* top-flight professional; ~**leistung die** top-class performance; ~**politiker der** top *or* leading politician; ~**qualität die** top quality; ~**reiter der** top rider; *(fig.)* leader; *(Ware)* top *or* best seller; *(Mannschaft)* top team; ~**sportler der** top sportsman; ~**stellung die** top position; ~**technologie die** state-of-the-art technology; ~**wert der** peak; maximum [value]
Spitzer der; ~**s,** ~**:** [pencil-]sharpener
spitz-, Spitz-: ~**findig 1.** *Adj.*; hair-splitting, over-subtle; quibbling ⟨*distinction*⟩; pettifogging ⟨*quibble*⟩; **2.** *adv.* in an over-subtle way; ~**findigkeit die;** ~, ~**en a)** *o. Pl.* over-subtlety; *(Haarspalterei)* hair-splitting; **b)** *(etwas Spitzfindiges)* nicety; *(Äußerung)* hair-splitting remark; ~**hacke die** pick; pickaxe
spitz-, Spitz-: ~|**kriegen** *tr. V. (ugs.)* tumble to *(coll.)*; get wise to *(sl.)*; ~**maus die a)** shrew; ~**name der** nickname; ~**wegerich** [~ve:gərɪç] **der** *(Bot.)* ribwort; ~**winklig 1.** *Adj.* acute-angled ⟨*triangle*⟩; **2.** *adv.* at an acute angle; ~**züngig 1.** *Adj.* sharp-tongued; **2.** *adv.* ⟨*reply*⟩ sharply
Spleen [ʃpli:n] **der;** ~**s,** ~**e** *od.* ~**s** strange *or* peculiar habit; eccentricity; **du hast ja einen** ~**!** there must be something the matter with you!; you must be dotty *(coll.)*
spleenig *Adj.* eccentric; dotty *(coll.)*
Splitt [ʃplɪt] **der;** ~**[e]s,** ~**e** [stone] chippings *pl.*; *(zum Streuen)* grit
splitten *tr. V. (Wirtsch.)* **a)** split ⟨*shares*⟩; **b)** *(Politik)* **die Stimmen** ~**:** *give one's first vote to a particular candidate and one's second to a party other than that of the chosen candidate*
Splitter der; ~**s,** ~**:** splinter; *(Granat~, Bomben~)* splinter; fragment
Splitter·gruppe die splinter group
splittern *itr. V.* **a)** *(Splitter bilden)* splinter; **b)** *mit sein (in Splitter zerbrechen)* ⟨*glass, windscreen, etc.*⟩ shatter
splitter·nackt *Adj. (ugs.)* stark naked; starkers *pred. (Brit. sl.)*
Splitter·partei die splinter party
splittrig *s.* **splitterig**
SPÖ [ɛspe:'ø:] **die;** ~**:** *Abk.:* **Sozialistische Partei Österreichs** Austrian Socialist Party
Spoiler ['ʃpɔylɐ] **der;** ~**s,** ~ *(Kfz-W.)* spoiler
sponsern ['ʃpɔnzɐn] *tr. V.* sponsor
Sponsor ['ʃpɔnzɐ] **der;** ~**s,** ~**s** *od.* ~**en** [-'zo:rən] sponsor
spontan [ʃpɔn'ta:n] **1.** *Adj.* spontaneous. **2.** *adv.* spontaneously
sporadisch [ʃpo'ra:dɪʃ] **1.** *Adj.* sporadic. **2.** *adv.* sporadically
Spore ['ʃpo:rə] **die;** ~, ~**n** *(Biol.)* spore
Sporen *s.* **Spore, Sporn**
Sporen-: ~**pflanze die** *(Bot.)* cryptogam; ~**tierchen das** *(Zool.)* sporozoan
Sporn [ʃpɔrn] **der;** ~**[e]s,** ~**e** *od.* **Sporen** ['ʃpo:rən] *Pl.* **Sporen** spur; **einem Pferd die Sporen geben** spur a horse
spornen *tr. V.* spur ⟨*horse*⟩
Sport [ʃpɔrt] **der;** ~**[e]s a)** sport; *(als Unterrichtsfach)* sport; physical education; PE; ~ **treiben** do sport; **beim** ~**:** while doing sport; **b)** *(~art)* sport; **c)** *(Hobby, Zeitvertreib)* hobby; pastime
Sport-: ~**abzeichen das** sports badge; ~**anlage die** sports complex; ~**art die** [form of] sport; ~**artikel der** piece of sports equipment; ~**artikel** *Pl.* sports

equipment *sing.;* **~fest das** sports festival; *(einer Schule)* sports day; **~flieger der** sports pilot; **~flugzeug das** sports plane; **~freund der a)** sports fan; **b)** *(Kamerad)* sporting friend; **~funktionär der** sports official; **~geist der;** *o. Pl.* sportsmanship; sporting spirit; **~hochschule die** college of physical education; **~journalist der** sports journalist; **~kleidung die** sportswear; sports clothes *pl.;* **~lehrer der** sports instructor; *(in einer Schule)* PE *or* physical education teacher; games teacher

Sportler ['ʃpɔrtlɐ] **der; ~s, ~:** sportsman

Sportlerin die; ~, ~nen sportswoman

sportlich 1. *Adj.* **a)** sporting *attrib.* ‹*success, performance, interests, etc.*›; **~e Veranstaltungen** sports events; sporting events; **auf ~em Gebiet** in the field of sport; **b)** *(fair)* sportsmanlike; sporting; **c)** *(fig.: flott, rasant)* sporty ‹*car, driving, etc.*›; **d)** *(zu sportlicher Leistung fähig)* sporty, athletic ‹*person*›; **e)** *(jugendlich wirkend)* sporty, smart but casual ‹*clothes*›; smart but practical ‹*hair-style*›. **2.** *adv.* **a)** as far as sport is concerned; **~ aktiv sein** be an active sportsman/sportswoman; **b)** *(fair)* sportingly; **c)** *(fig.: flott, rasant)* in a sporty manner

sportlich-elegant 1. *Adj.* casually elegant. **2.** *adv.* casually but elegantly ‹*dressed*›

Sport-: ~platz der sports field; *(einer Schule)* playing field/fields *pl.;* **~schuh der a)** sports shoe; **b)** *(sportlicher Schuh)* casual shoe; **~sendung die** sports programme

Sports·freund der sports enthusiast; **Hallo, ~freund! Wie geht's?** *(ugs.)* hello, mate *(coll.)*

Sport-: ~stadion das [sports] stadium; **~student der** sports student; **~taucher der** skindiver; **~un·fall der** sporting *or* sports accident; **~verein der** sports club; **~verletzung die** sports injury; **~wagen der a)** *(Auto)* sports car; **b)** *(Kinderwagen)* push-chair *(Brit.);* stroller *(Amer.)*

Spott [ʃpɔt] **der; ~[e]s** mockery; *(höhnischer)* ridicule; derision; **~ und Hohn** scorn and derision

spott·billig *(ugs.)* **1.** *Adj.* dirt cheap; **2.** *adv.* **da kann man ~ einkaufen** you can get *or* buy things dirt cheap there

spötteln ['ʃpœtl̩n] *itr. V.* mock [gently]; poke *or* make [gentle] fun

spotten ['ʃpɔtn̩] *itr. V.* **a)** mock; poke *or* make fun; *(höhnischer)* ridicule; be derisive; **über jmdn./etw. ~:** mock sb./sth.; make fun of sb./sth.; *(höhnischer)* ridicule sb./sth.; be derisive about sb./sth.; **b)** *(fig.)* be contemptuous of; scorn; **er spottete der Gefahr** *(Gen.) (geh.)* he was contemptuous of *or* scorned the danger

Spötter ['ʃpœtɐ] **der; ~s, ~:** mocker

spöttisch ['ʃpœtɪʃ] **1.** *Adj.* mocking ‹*smile, remark, speech, etc.*›; *(höhnischer)* derisive, ridiculing ‹*remark, speech, etc.*›; **ein ~er Mensch** a person who likes poking fun. **2.** *adv.* mockingly; **~ lächeln** give a mocking smile

Spott-: ~lust die; *o. Pl.* love of *or* delight in mockery *or* poking fun; **~preis der** *(ugs.)* ridiculously low price; **etw. für einen** *od.* **zu einem ~preis bekommen** get sth. dirt cheap *or* for a song

sprach [ʃpra:x] *1. u. 3. Pers. Sg. Prät. v.* **sprechen**

Sprach·begabung die; *o. Pl.* talent *or* gift for languages

Sprache ['ʃpra:xə] **die; ~, ~n a)** language; **in englischer ~:** in English; **hast du die ~ verloren?** *(ugs.)* haven't you got a tongue in your head?; **b)** *(Sprechweise)* way of speaking; speech; *(Stil)* style; **c)** *(Rede)* **die ~ auf jmdn./etw. bringen** bring the conversation round to sb./sth.; **etw. zur ~ bringen** bring sth. up; raise sth.; **heraus mit der ~!** come on, out with it!

spräche ['ʃprɛ:çə] *1. u. 3. Pers. Sg. Konjunktiv II v.* **sprechen**

Sprachen-: ~schule die language school; **~studium das** language studies *pl., no art.*

sprach-, Sprach-: ~fehler der speech impediment *or* defect; **~führer der** phrase-book; **~gebrauch der** [linguistic] usage; **~kenntnisse** *Pl.* knowledge *sing.* of a language/languages; **seine französischen ~kenntnisse** his knowledge of French; **~kundig** *Adj.* proficient in *or* conversant with the language *postpos.;* **~kurs der** language course; **~labor das** language laboratory *or* *(coll.)* lab; **~lehre die a)** grammar; **b)** *(Buch)* grammar [book]

sprachlich 1. *Adj.* linguistic; **~e Feinheiten** subtleties of language. **2.** *adv.* linguistically

sprach-, Sprach-: ~los *Adj.* *(überrascht)* speechless; dumbfounded; **~philosophie die** philosophy of language; **~rohr das** *(Repräsentant)* spokesman; *(Propagandist)* mouthpiece; **~übung die** language exercise; linguistic exercise; **~unterricht der** language teaching *or* instruction; **~wissenschaft die** linguistics *sing., no art.;* **~wissenschaftlich 1.** *Adj.* linguistic; **eine ~wissenschaftliche Abhandlung** a linguistics dissertation; **2.** *adv.* linguistically

sprang [ʃpraŋ] *1. u. 3. Pers. Sg. Prät. v.* **springen**

spränge ['ʃprɛŋə] *1. u. 3. Pers. Sg. Konjunktiv II v.* **springen**

Spray [ʃpre:] **das** *od.* **der; ~s, ~s** spray

Spray·dose die aerosol [can]

sprayen *tr., itr. V.* spray

Sprech-: ~an·lage die intercom *(coll.);* **~blase die** balloon *(coll.);* **~chor der** chorus

sprechen ['ʃprɛçn̩] **1.** *unr. itr. V.* speak **(über +** *Akk.* about; **von** about, of); *(sich unterhalten, sich besprechen auch)* talk **(über +** *Akk.,* **von** about); ‹*parrot etc.*› talk; **deutsch/flüsternd ~:** speak German/in a whisper *or* whispers; **er spricht wenig** he doesn't say *or* talk much; **es spricht Pfarrer N.** the speaker is the Revd. N.; **für/gegen etw. ~:** speak in favour of/against sth.; **mit jmdm. ~:** speak *or* talk with *or* to sb.; **ich muß mit dir ~:** I must talk *or* speak with you; **er spricht mit sich selbst** he talks to himself; **mit wem spreche ich?** who is speaking please?; to whom am I speaking, please?; **~ Sie noch?** *(am Telefon)* are you still there?; **gut/schlecht von jmdm.** *od.* **über jmdn. ~:** speak well/ill of sb.; **für jmdn. ~:** speak for sb.; speak on *or (Amer.)* in behalf of sb.; **vor der Betriebsversammlung ~:** speak to *or* address a meeting of the workforce; **zu einem** *od.* **über ein Thema ~:** speak on *or* about a subject; **frei ~:** extemporize; speak without notes; **aus seinen Worten/seinem Blick sprach Angst** *usw.* his words/the look in his eyes expressed fear *etc.;* **auf jmdn./etw. zu ~ kommen** get to talking about sb./sth.; **für/gegen jmdn./etw. ~** *(in günstigem/ungünstigem Licht erscheinen lassen)* be a point in sb.'s/sth.'s favour/against sb./sth.; **was spricht denn dafür/dagegen?** what is there to be said for/against it? **2.** *unr. tr. V.* **a)** speak ‹*language,* *dialect*〉; say 〈*word, sentence*〉; ~ **Sie Französisch?** do you speak French?; **„Hier spricht man Deutsch"** 'German spoken'; 'we speak German'; **b)** *(rezitieren)* say, recite 〈*poem, text*〉; say 〈*prayer*〉; recite 〈*spell*〉; pronounce 〈*blessing, oath*〉; *s. auch* **Recht a; c) jmdn.** ~: speak to sb.; **Sie haben mich ~ wollen?** you wanted to see me *or* speak to me?; **ich bin heute für niemanden mehr zu ~:** I can't see anyone else today; **d)** *(aus~)* pronounce 〈*name, word, etc.*〉

Sprecher der; ~s, ~ a) spokesman; **b)** *(Ansager)* announcer; *(Nachrichten~)* newscaster; news-reader; **c)** *(Kommentator, Erzähler)* narrator; **d)** *(Sprachw.)* speaker

Sprecherin die; ~, ~nen a) spokeswoman; **b)** *s.* **Sprecher b, c, d**

sprech-, Sprech-: ~funk der radio-telephone system; **~funkgerät das** radio-telephone; *(Walkie-talkie)* walkie-talkie; **~stunde die** consultation hours *pl.; (eines Arztes)* surgery; consulting hours *pl.; (eines Rechtsanwalts usw.)* office hours *pl.;* **wann haben Sie ~stunde?** when are your consultation hours/when is your surgery *or* what are your surgery hours?; **~stunden·hilfe die** *(eines Arztes)* receptionist; *(eines Zahnarztes)* assistant; **~übung die** elocution *or* speech exercise; *(zu therapeutischen Zwecken)* speech exercise; **~weise die** manner of speaking; **~zeit die** visiting time; **~zimmer das** consulting-room

Spreiz·dübel der expanding anchor

spreizen 1. *tr. V.* spread 〈*fingers, toes, etc.*〉; **die Beine ~:** spread one's legs apart; open one's legs; **mit gespreizten Beinen stehen/sitzen** stand/sit with one's legs apart. **2.** *refl. V. (geh.) (sich zieren)* **sie spreizte sich erst dagegen, dann stimmte sie zu** she made a fuss at first, [but] then agreed

Spreiz·fuß der *(Med.)* spread foot

Sprengel ['ʃprɛŋl̩] **der; ~s, ~ a)** *(Kirchen~)* parish; *(Diözese)* diocese; **b)** *(österr.)* administrative district

sprengen ['ʃprɛŋn̩] *tr. V.* **a)** blow up; blast 〈*rock*〉; **etw. in die Luft ~:** blow sth. up; **b)** *(gewaltsam öffnen, aufbrechen)* force [open] 〈*door*〉; force 〈*lock*〉; break open 〈*burial chamber etc.*〉; burst, break 〈*bonds, chains*〉; *(fig.)* break up 〈*meeting, demonstration*〉; *s. auch* **Rahmen b; c)** *(be~)* water 〈*flower-bed, lawn*〉; sprinkle 〈*street, washing*〉 with water; *(verspritzen)* sprinkle; *(mit dem Schlauch)* spray

Spreng-: ~kraft die explosive power; **~ladung die** explosive charge; **~satz der** explosive charge; **~stoff der** explosive

Sprengung die; ~, ~en a) blowing-up; *(im Steinbruch)* blasting; **b)** *s.* **sprengen 1b:** forcing [open]; forcing; breaking open; bursting; breaking; *(fig.)* breaking up; **c)** *(das Besprengen)* sprinkling; *(mit dem Schlauch)* spraying

Sprenkel ['ʃprɛŋkl̩] **der; ~s, ~:** spot; dot; speckle

sprenkeln *tr. V.* sprinkle spots of 〈*colour*〉; sprinkle 〈*water*〉

Spreu [ʃprɔy] **die; ~:** chaff; **die ~ vom Weizen trennen** *(fig.)* separate the wheat from the chaff

sprich [ʃprɪç] *Imperativ Sg. v.* **sprechen**

sprichst [ʃprɪçst] *2. Pers. Sg. Präsens v.* **sprechen**

spricht [ʃprɪçt] *3. Pers. Sg. Präsens v.* **sprechen**

Sprich·wort das; *Pl.* **Sprichwörter** proverb

sprich·wörtlich *Adj.* proverbial. **2.** *adv.* proverbially

sprießen ['ʃpri:sn̩] *unr. itr. V.; mit sein* 〈*leaf, bud*〉 shoot, sprout; 〈*seedlings*〉 come *or* spring up; 〈*beard*〉 sprout; *(fig.)* 〈*club, organization, etc.*〉 spring up

Spriet [ʃpri:t] **das; ~[e]s, ~e** *(Seemannsspr.)* sprit

Spring·brunnen der fountain

springen ['ʃprɪŋən] **1.** *unr. itr. V.* **a)** *mit sein* jump; *(mit Schwung)* leap; spring; jump; 〈*frog, flea*〉 hop, jump; **vom Fünfmeterbrett ~:** dive from the five-metre board; **jmdm. an die Kehle ~:** leap at sb.'s throat; **auf die Beine** *od.* **Füße ~:** jump to one's feet; **b)** *meist mit sein (Sport)* jump; *(beim Stabhochsprung, beim Kasten, Pferd)* vault; *(beim Turm~, Kunst~)* dive; **c)** *mit sein (sich in Sprüngen fortbewegen)* bound; **d)** *(ugs.) in* **eine Runde Bier ~ lassen** stand a round of beer; **er könnte ruhig mal was ~ lassen** he could easily fork out something just once in a while *(sl.);* **e)** *mit sein (fig.: schnellen, hüpfen, fliegen)* 〈*pointer, milometer, etc.*〉 jump **(auf +** *Akk.* to); 〈*traffic-lights*〉 change **(auf +** *Akk.* to); 〈*spark*〉 leap; 〈*ball*〉 bounce; 〈*spring*〉 jump out; **[von etw.] ~** 〈*fan belt, bicycle-chain, button, tyre, etc.*〉 come off [sth.]; **f)** *mit sein* 〈*string, glass, porcelain, etc.*〉 break; *(Risse, Sprünge bekommen)* crack; **gesprungene Lippen** cracked *or* chapped lips. **2.** *unr. tr. V.; auch mit sein (Sport)* perform 〈*somersault, twist dive, etc.*〉; **5,20 m/einen neuen Rekord ~:** jump 5.20m/make a record jump

Springer der; ~s, ~ a) *(Weit~, Hoch~, Ski~)* jumper; *(Stabhoch~)* [pole-]vaulter; *(Kunst~, Turm~)* diver; *(Fallschirm~)* parachutist; **b)** *(Schachfigur)* knight

Springerin die; ~, ~nen *s.* **Springer a**

spring-, Spring-: ~flut die spring tide; **~kraut das** impatience; **~lebendig** *Adj.* extremely lively; full of beans *pred. (coll.);* **~pferd das** jumper; **~reiten das** show-jumping *no art.;* **~seil das** skipping-rope *(Brit.);* jump-rope *(Amer.);* **~turnier das** *(Reiten)* show-jumping competition

Sprint [ʃprɪnt] **der; ~s, ~s** *(auch Sport)* sprint

sprinten *itr. (auch tr.) V.; mit sein (Sport; ugs.: schnell laufen)* sprint

Sprinter der; ~s, ~, Sprinterin die; ~, ~nen *(Sport)* sprinter

Sprit [ʃprɪt] **der; ~[e]s, ~e a)** *(ugs.: Treibstoff)* gas *(Amer. coll.);* juice *(sl.);* petrol *(Brit.);* **b)** *(ugs.: Schnaps)* shorts *pl.*

Spritze ['ʃprɪtsə] **die; ~, ~n a)** *(zum Vernichten von Ungeziefer)* spray; *(Teig~, Torten~, Injektions~)* syringe; **b)** *(Injektion)* injection; jab *(coll.);* **eine ~ bekommen** have an injection *or (coll.)* jab; **c)** *(Feuer~)* hose; *(Löschfahrzeug)* fire engine

spritzen ['ʃprɪtsn̩] **1.** *tr. V.* **a)** *(versprühen)* spray; *(ver~)* splash 〈*water, ink, etc.*〉; spatter 〈*ink etc.*〉; *(in Form eines Strahls)* spray, squirt 〈*water, foam, etc.*〉; pipe 〈*cream etc.*〉; **b)** *(be~, besprühen)* water 〈*lawn, tennis-court*〉; water, spray 〈*street, yard*〉; spray 〈*plants, crops, etc.*〉; pump 〈*concrete*〉; *(mit Lack)* spray 〈*car etc.*〉; **jmdn. naß ~:** splash sb.; *(mit Wasserpistole, Schlauch)* spray sb.; **c)** *(injizieren)* inject 〈*drug etc.*〉; **d)** *(~d herstellen)* create 〈*ice-rink*〉 by spraying; pipe 〈*cake-decoration etc.*〉; produce 〈*plastic article*〉 by injection moulding; **e)** *(ugs.: einer Injektion unterziehen)* **jmdn./sich ~:** give sb. an injection/inject oneself; **jmdm. ein Schmerzmittel ~:** give sb. a pain-killing injection; **f)** *(verdünnen)* dilute ‹*wine etc.*› with soda-water/lemonade *etc.* **2.** *itr. V.* **a) die Kinder planschten und spritzten** the children splashed and threw water about; **b)** *mit Richtungsangabe mit sein* ‹*hot fat*› spit; ‹*mud etc.*› spatter, splash; ‹*blood, water*› spurt; **das Wasser spritzte ihm ins Gesicht** the water splashed up into his face; **c)** *mit sein (ugs.: rennen)* dash

Spritzer der; ~s, ~ *(kleiner Tropfen)* splash; *(von Farbe)* splash; spot; *(Schuß)* dash; splash

spritzig 1. *Adj.* **a)** sparkling ‹*wine*›; tangy ‹*fragrance, perfume*›; **b)** *(lebendig)* lively ‹*show, music, article*›; sparkling ‹*production, performance*›; racy ‹*style*›; **c)** *(temperamentvoll)* nippy *(coll.)*; zippy ‹*car, engine*›; **d)** *(flink)* agile, nimble ‹*person*›. **2.** *adv.* sparklingly ‹*produced, performed, etc.*›; racily ‹*written*›; **die Mannschaft spielte sehr ~:** the team played with great speed and agility

Spritz·tour die *(ugs.)* spin

spröd, spröde [ʃprø:t] *Adj.* **a)** brittle ‹*glass, plastic, etc.*›; dry ‹*hair, lips, etc.*›; *(rissig)* chapped ‹*lips, skin*›; *(rauh)* rough ‹*skin*›; **b)** *(fig.: rauh klingend)* harsh, rough ‹*voice*›; **c)** *(fig.: abweisend)* aloof ‹*person, manner, nature*›

Sprödheit, Sprödigkeit die; ~ a) *s.* **spröde a:** brittleness; dryness; roughness; **b)** *(fig.: rauher Klang)* harshness; roughness; **c)** *(fig.: abweisendes Wesen)* aloofness

sproß [ʃprɔs] *1. u. 3. Pers. Sg. Prät. v.* **sprießen**

Sproß der; Sprosses, Sprosse *od.* **Sprossen** *(Bot.)* shoot

Sprosse die; ~, ~n a) *(auch fig.)* rung; **b)** *(eines Fensters)* glazing bar; sash bar

Sprossen-: ~kohl der *(österr.) s.* **Rosenkohl; ~wand die** wall bars *pl.*

Sprößling ['ʃprœslɪŋ] **der; ~s, ~e** *(ugs. scherzh.)* offspring; **seine ~e** his offspring *pl.*

Sprotte ['ʃprɔtə] **die; ~, ~n** sprat; **Kieler ~n** smoked [Kiel] sprats

Spruch [ʃprʊx] **der; ~[e]s, Sprüche** ['ʃprʏçə] **a)** *(Wahl~)* motto; *(Sinn~)* maxim; adage; *(Aus~)* saying; aphorism; *(Zitat)* quotation; quote; *(Parole)* slogan; *(Bibel~)* quotation; saying; **b)** *Pl. (ugs. abwertend: Phrase)* **das sind doch alles nur Sprüche** that's just talk *or* empty words *pl.*; **Sprüche machen** *od.* **klopfen** talk big *(coll.)*

Spruch·band das; *Pl.* **~bänder** banner

Sprüche·klopfer der; ~s, ~ *(ugs. abwertend)* big mouth *(coll.)*

spruch·reif *Adj.* **das ist noch nicht ~:** that's not definite, so people mustn't start talking about it yet

Sprudel ['ʃpru:dl̩] **der; ~s, ~ a)** *(Selterwasser)* sparkling mineral water; **b)** *(österr.: Erfrischungsgetränk)* fizzy drink

sprudeln *itr. V.* **a)** *mit sein* ‹*spring, champagne, etc.*› bubble (**aus** out of); **b)** *(beim Kochen)* bubble; **c)** *(beim Entweichen von Gas)* ‹*lemonade, champagne, etc.*› fizz, effervesce

Sprudel·wasser das; *Pl.* **-wässer** sparkling mineral water

Sprüh·dose die aerosol [can]

sprühen ['ʃpry:ən] **1.** *tr. V.* spray; **Wasser auf die Blätter ~:** spray the leaves with water; **seine Augen sprühten Haß** *(fig.)* his eyes flashed hatred. **2.** *itr. V. mit Richtungsangabe mit sein* ‹*sparks, spray*› fly; ‹*flames*› spit; ‹*waterfall*› send out a fine spray; *(fig.)* ‹*eyes*› sparkle (**vor** + *Dat.* with); ‹*intellect, wit*› sparkle; **~der Witz** sparkling wit

Sprüh·regen der drizzle; fine rain

Sprung [ʃprʊŋ] **der; ~[e]s, Sprünge** ['ʃprʏŋə] **a)** *(auch Sport)* jump; *(schwungvoll)* leap; *(Satz)* bound; *(Sprung über das Pferd)* vault; *(Wassersport)* dive; *(fig.)* leap; **sein Herz machte vor Freude einen ~** *(fig.)* his heart leapt for joy; **ein [großer] ~ nach vorn** *(fig.)* a [great] leap forward; **keine großen Sprünge machen können** *(fig. ugs.)* not be able to afford many luxuries; **auf einen ~** *(fig. ugs.)* for a few minutes; **auf dem ~[e] sein** *(fig. ugs.)* be in a rush; **b)** *(ugs.: kurze Entfernung)* stone's throw; **c)** *(Riß)* crack; **einen ~ haben/bekommen** be cracked/crack; **d)** *in* **jmdm. auf die Sprünge helfen** *(ugs.)* help sb. on his/her way

Sprung-: ~brett das *(auch fig.)* springboard; **~feder die** [spiral] spring

sprunghaft 1. *Adj.* **a)** erratic ‹*person, character, manner*›; disjointed ‹*conversation, thoughts*›; **b)** *(unvermittelt)* sudden; abrupt; **c)** *(ruckartig)* rapid ‹*change*›; sharp ‹*increase*›. **2.** *adv.*; *s.* **1 b–c:** disjointedly; suddenly; abruptly; rapidly; sharply

Sprung-: ~lauf der *(Ski)* ski-jumping *no art.*; **~rahmen der** spring bed-frame; **~schanze die** *(Ski)* ski-jumping hill; **~seil das** skipping-rope *(Brit.)*; jump-rope *(Amer.)*; **~tuch das;** *Pl.* **~tücher** safety blanket; **~turm der** *(Sport)* diving-platform

Spucke die; ~: spit; **mir blieb die ~ weg** *(ugs.)* it took my breath away; I was speechless

spucken ['ʃpʊkn̩] **1.** *itr. V.* **a)** spit; **in die Hände ~** *(fig.: an die Arbeit gehen)* go to work with a will; **b)** *(ugs.: erbrechen)* throw up *(coll.)*; be sick *(Brit.)*. **2.** *tr. V.* spit; spit [up], cough up ‹*blood, phlegm*›; **Feuer ~:** breathe fire; ‹*volcano*› belch fire; *s. auch* [2]**Ton d**

Spuk [ʃpu:k] **der; ~[e]s, ~e a)** [ghostly *or* supernatural] manifestation; **b)** *(schreckliches Geschehen)* horrific episode

spuken *itr. V. auch unpers.* **hier/in dem Haus spukt es** this place/the house is haunted; **dieser Aberglaube spukt noch immer in den Köpfen vieler Menschen** *(fig.)* this superstition still lurks in many people's minds

Spül·becken das sink

Spule ['ʃpu:lə] **die; ~, ~n a)** spool; bobbin; *(für Tonband, Film)* spool; reel; **b)** *(Elektrot.)* coil

Spüle die; ~, ~n sink unit; *(Becken)* sink

spulen *tr., itr. V.* spool; *(am Tonbandgerät)* wind

spülen ['ʃpy:lən] **1.** *tr. V.* **a)** rinse; bathe ‹*wound*›; **b)** *(landsch.: abwaschen)* wash up ‹*dishes, glasses, etc.*›; **Geschirr ~:** wash up; **c)** *(schwemmen)* wash. **2.** *itr. V.* **a)** *(beim WC)* flush [the toilet]; **b)** *(den Mund ausspülen)* rinse out [one's mouth]; **c)** *(landsch.) s.* **abwaschen 2**

Spül-: ~maschine die dishwasher; **~mittel das** washing-up liquid

Spülung die, ~, ~en a) *(Med.)* irrigation; *(der Vagina)* douche; **b)** *(beim WC)* flush

Spund [ʃpʊnd] **der; ~[e]s, ~e/Spünde** ['ʃpʏndə] **a)** *Pl.* **Spünde** *(Zapfen)* bung; **b)** *Pl.* **~e** *(ugs.)* **junger ~:** young greenhorn *or* tiro

Spund·loch das bung-hole

Spur [ʃpu:ɐ] **die; ~, ~en a)** *(Abdruck im Boden)* track; *(Folge von Abdrücken)* tracks *pl.*; *(Blut~, Schleim~ usw.)* trail; **von dem Vermißten fehlt jede ~:** there is no trace of the missing person; **eine heiße ~** *(fig.)* a hot trail;

jmdm./einer Sache auf der ~ sein be on the track *or* trail of sb./sth.; **b)** *(Anzeichen)* trace; *(eines Verbrechens)* clue (*Gen.* to); **c)** *(sehr kleine Menge; auch fig.)* trace; **da fehlt noch eine ~ Paprika** it needs just a touch of paprika; **von Reue keine ~:** not a trace *or* sign of penitence; **keine** *od.* **nicht die ~** *(ugs.: als Antwort)* not in the slightest; **d)** *(Verkehrsw.: Fahr~)* lane; **die ~ wechseln** change lanes; **in** *od.* **auf der linken ~ fahren** drive in the left-hand lane; **e)** *(Fahrlinie)* **[die] ~ halten** stay on its line; **f)** *(Elektrot., DV)* track

spürbar 1. *Adj.* noticeable; perceptible; distinct, perceptible ⟨*improvement*⟩; evident ⟨*relief, embarrassment*⟩. **2.** *adv.* noticeably; perceptibly; *(sichtlich)* clearly ⟨*relieved, on edge*⟩

spuren *itr. V. (ugs.)* toe the line *(coll.)*; do as one's told

spüren [ˈʃpy:rən] *tr. V.* feel; *(instinktiv)* sense; *(merken)* notice; **die Peitsche zu ~ bekommen** get a taste of the whip

Spür·hund der tracker dog; *(fig.: Spitzel)* bloodhound; snooper *(coll.)*

spur·los 1. *Adj.; nicht präd.* total, complete ⟨*disappearance*⟩. **2.** *adv.* ⟨*disappear*⟩ completely *or* without trace; **es ist nicht ~ an ihm vorübergegangen** it has not failed to leave its mark on him

Spür·sinn der; *o. Pl. (feiner Instinkt)* intuition

Spurt [ʃpʊrt] **der; ~[e]s, ~s** *od.* **~e a)** spurt; **b)** *o. Pl. (Sport.: ~vermögen)* turn of speed

spurten *itr. V.* **a)** *mit Richtungsangabe mit sein* spurt; **b)** *mit sein (ugs.: schnell laufen)* sprint

Spur·wechsel der change of lane

sputen [ˈʃpu:tn̩] *refl. V. (veralt.)* make haste

Squaw [skwɔ:] **die; ~, ~s** squaw

St. *Abk.* **a) Sankt** St.; **b) Stück**

Staat [ˈʃta:t] **der; ~[e]s, ~en a)** state; **die ~en** *(die USA)* the States; **von ~s wegen** on the part of the [state] authorities; **b)** *o. Pl. (ugs.: Festkleidung, Pracht)* finery; **in vollem ~:** in all one's finery; **mit diesem Mantel ist kein ~ mehr zu machen** *(fig. ugs.)* this coat is past it *(coll.)*

staaten-, Staaten-: ~bund der confederation; **~los** *Adj.* stateless; **~lose der/die;** *adj. Dekl.* stateless person *or* subject

staatlich 1. *Adj.* state *attrib.* ⟨*sovereignty, institutions, authorities, control, etc.*⟩; ⟨*power, unity, etc.*⟩ of the state; state-owned ⟨*factory etc.*⟩; **~e Mittel** government *or* public money *sing.* **2.** *adv.* by the state; **~ anerkannt/geprüft/finanziert** state-approved/-certified/-financed; **~ subventioniert werden** receive a state subsidy

staats-, Staats-: ~akt der *(Festakt)* state ceremony; **~amt das** public office; **~angehörige der/die** national; **~angehörigkeit die** nationality; **~anwalt der** public prosecutor; **~anwaltschaft die** public prosecutor's office; **~bank die** national bank; **~beamte der** civil servant; **~besuch der** state visit; **~bürger der** citizen; **er ist deutscher ~bürger** he is a German citizen *or* national; **~bürgerkunde die** *(ehem. DDR) school subject involving ideological education of socialist citizens;* ≈ civics *sing. no art.*; **~bürgerlich** *Adj.; nicht präd.* civil ⟨*rights*⟩; civic ⟨*duties, loyalty*⟩; ⟨*education, attitude*⟩ as a citizen; **~bürgerschaft die** *s.* **~angehörigkeit; ~chef der** head of state; **~dienst der** civil service; **~examen das** *final university examination;* **~examen machen** ≈ take one's finals; **~form die** type of state; state system; **~gebiet das** territory [of a/the state]; **~geheimnis das** *(auch fig.)* state secret; **~gewalt die** *o. Pl.* authority of the state; *(Exekutive)* executive power; **~grenze die** state frontier *or* border; **~kanzlei die** *(BRD)* Minister-President's Office; *(Schweiz)* Cantonal Chancellory; **~kirche die** state *or* established church; **~mann der;** *Pl.* **-männer** statesman; **~männisch** [~mɛnɪʃ] **1.** *Adj.* statesmanlike ⟨*wisdom, farsightedness, etc.*⟩; ⟨*abilities, skill*⟩ of a statesman; **2.** *adv.* in a statesmanlike manner; **~minister der** minister of state; *(Minister ohne Ressort)* minister without portfolio; *(BRD: Staatssekretär)* secretary of state; **~oberhaupt das** head of state; **~präsident der** [state] president; **~raison die, ~räson die** reasons *pl.* of State *no def. art.*; **~sekretär der** permanent secretary; **~streich der** coup d'état; **~trauer die** national mourning *no indef. art.*

Stab [ʃta:p] **der; ~[e]s, Stäbe** [ˈʃtɛ:bə] **a)** rod; *(länger, für ~hochsprung o. ä.)* pole; *(eines Käfigs, Gitters, Geländers)* bar; *(Staffel~; geh.: Taktstock)* baton; *(Bischofs~)* crosier; *(Hirten~)* crook; **den ~ über jmdn./etw. brechen** *(geh.)* condemn sb./sth. out of hand; **b)** *(Milit.)* staff; **c)** *(Team)* team

Stäbchen [ˈʃtɛ:pçən] **das; ~s, ~ a)** *(kleiner Stab)* little rod; [small] stick; **b)** *(Eß~)* chopstick

Stab-: ~hoch·springer der pole-vaulter; **~hoch·sprung der a)** *o. Pl. (Disziplin)* pole-vaulting *no art.*; **im ~hochsprung** in the pole-vault; **b)** *(Sprung)* pole-vault

stabil [ʃtaˈbi:l] **1.** *Adj.* sturdy ⟨*chair, cupboard*⟩; robust, sound ⟨*health*⟩; stable ⟨*prices, government, economy, etc.*⟩. **2.** *adv.* **~ gebaut** solidly built

stabilisieren 1. *tr. V.* stabilize. **2.** *refl. V.* **a)** stabilize; become more stable; **b)** ⟨*health, circulation, etc.*⟩ become stronger

Stabilität [ʃtabiliˈtɛ:t] **die; ~** sturdiness; *(von Gesundheit, Konstitution usw.)* robustness; soundness; *(das Beständigsein)* stability

Stab·lampe die torch *(Brit.)*; flashlight *(Amer.)*

Stabs-: ~arzt der *(Milit.)* medical officer, MO *(with the rank of captain)*; **~feldwebel der** *(Milit.)* warrant-officer 2nd class; **~offizier der** *(Milit.)* staff officer

stach [ʃta:x] *1. u. 3. Pers. Sg. Prät. v.* **stechen**

Stachel [ˈʃtaçl̩] **der; ~s, ~n a)** spine; *(Dorn)* thorn; **b)** *(Gift~)* sting; **c)** *(spitzes Metallstück)* spike; *(von ~draht)* barb

Stachel-: ~beere die gooseberry; **~draht der** barbed wire

stachelig *Adj.* prickly

Stachel·schwein das porcupine

stachlig *s.* **stachelig**

Stadion [ˈʃta:diɔn] **das; ~s, Stadien** stadium

Stadium [ˈʃta:diʊm] **das; ~s, Stadien** stage

Stadt [ʃtat] **die; ~, Städte** [ˈʃtɛ(:)tə] **a)** town; *(Groß~)* city; **die ~ Basel** the city of Basel; **in die ~ gehen** go into town; go downtown *(Amer.)*; **b)** *(Verwaltung)* town council; *(in der Großstadt)* city council; city hall *no art. (Amer.)*; **bei der ~ [angestellt] sein/arbeiten** work for the council *or (Amer.)* for city hall

stadt-, Stadt-: ~auswärts *Adv.* out of town; **~auto·bahn die** urban motorway *(Brit.)* *or (Amer.)* freeway; **~bahn die** urban railway; **~bibliothek die** municipal library; **~bummel der** *(ugs.)* **einen ~bummel machen**

take a stroll through the town/city centre; **~einwärts** *Adv.* into town; downtown *(Amer.)*

Städte·partnerschaft die twinning *(Brit.) or (Amer.)* sister-city arrangement *(between towns/cities)*

Städter der; ~s, ~, Städterin die; ~, ~nen a) town-dweller; *(Großstädter, -städterin)* city-dweller; **b)** *(Stadtmensch)* townie *(coll.)*

Stadt-: ; ~führer der town/city guidebook; **~gespräch das a)** *(Telefongespräch)* local call; **b)** *in* **~gespräch sein** be the talk of the town; **~halle die** civic *or* municipal hall

städtisch 1. *Adj.* **a)** *(kommunal)* municipal; **b)** *(urban)* urban ‹*life, way of life, etc.*›; town ‹*clothes*›; ‹*manners, clothes*› of a town-dweller. **2.** *adv.* **a)** *(kommunal)* municipally; **~ verwaltet** run by the town/city council; **b)** *(urban)* **ausgesprochen ~ gekleidet** wearing clothes with a decidedly town style

Stadt-: ~kasse die a) *(Geldmittel)* municipal funds *pl., no art.*; **b)** *(Stelle)* town/city treasurer's office; **~kern der** *s.* **~mitte; ~mauer die** town/city wall; **~mensch der** townie *(coll.)*; **~mitte die** town centre; *(einer Großstadt)* city centre; downtown area *(Amer.)*; **~park der** municipal park; **~plan der** [town/city] street plan *or* map; **~rand der** outskirts *pl.* of the town/city; **am ~:** on the outskirts of the town/city; **~rat der a)** town/city council; **b)** *(Mitglied)* town/city councillor; **~rundfahrt die** sightseeing tour round a/the town/city; **~staat der** city-state; **~streicher der** town/city tramp; **~teil der** district; part [of a/the town]; **~theater das** municipal theatre; **~tor das** town/city gate; **~verkehr der** town/city traffic; **~verwaltung die** municipal authority; town/city council; **~viertel das** district; **~zentrum das** town/city centre; downtown area *(Amer.)*

Staffel [ˈʃtafl̩] **die; ~, ~n a)** *(Sport: Mannschaft)* team; *(für den ~lauf)* relay team; **b)** *(Sport: ~lauf)* relay race; **c)** *(Luftwaffe: Einheit)* flight; **d)** *(Formation von Schiffen, begleitenden Polizisten, usw.)* escort formation

Staffelei die; ~, ~en easel

Staffel-: ~lauf der *(Sport)* relay race; **~läufer der** *(Sport)* relay runner/skier

staffeln *tr. V.* **a)** *(aufstellen)* arrange in a stagger *or* in an echelon; **b)** *(abstufen)* grade ‹*salaries, fees, prices*›; stagger ‹*times, arrivals, starting-places*›

Staffelung die; ~, ~en a) *(Anordnung)* staggered arrangement; **b)** *(Abstufung) (von Gebühren, Gehältern, Preisen)* grad[u]ation; *(von Vorgängen)* staggering

Stagnation [ʃtagnaˈtsi̯oːn] **die; ~, ~en** stagnation

stagnieren *itr. V.* stagnate

stahl [ʃtaːl] *1. u. 3. Pers. Sg. Prät. v.* **stehlen**

Stahl der; ~[e]s, Stähle [ˈʃtɛːlə] *od.* **~e** steel; **Nerven wie** *od.* **aus ~ haben** have nerves of steel

Stahl-: ~bau der; *Pl.* **~bauten a)** *o. Pl. (Bautechnik)* steel construction *no art.*; **b)** *(Gebäude)* steel-frame building; **~beton der** *(Bauw.)* reinforced concrete; ferroconcrete; **~beton·bau der;** *o. Pl.* reinforced concrete construction; **~blech das** sheet steel

stählen *tr. V. (geh.)* toughen; harden

stählern *Adj.* **a)** *nicht präd. (aus Stahl)* steel; **b)** *(fig. geh.)* ‹*muscles, nerves*› of steel; ‹*will*› of iron

stahl-, Stahl-: ~grau *Adj.* steel-grey; **~hart** *Adj.* as hard as steel *postpos.*; **~helm der** *(Milit.)* steel helmet; **~roß das** *(ugs. scherzh.)* bike *(coll.)*; trusty steed *(coll. joc.)*; **~wolle die** steel wool

stak [ʃtaːk] *1. u. 3. Pers. Sg. Prät. v.* **stecken**

staksen [ˈʃtaːksn̩] *itr. V.; mit sein (ugs.)* stalk; *(taumelnd)* teeter

staksig *(ugs.)* **1.** *Adj.* spindly, shaky-legged ‹*foal etc.*›; teetering ‹*steps*›. **2.** *adv.* **~ gehen** walk as though on stilts; *(unsicher)* walk with teetering steps

Stalagmit [ʃtalakˈmiːt] **der; ~s** *od.* **~en, ~e[n]** *(Geol.)* stalagmite

Stalaktit [ʃtalakˈtiːt] **der; ~s** *od.* **~en, ~e[n]** *(Geol.)* stalactite

Stalinismus [stalinɪsmʊs] **der; ~:** Stalinism *no art.*

stalinistisch 1. *Adj.* Stalinist. **2.** *adv.* in a Stalinist way; along Stalinist lines

Stall [ʃtal] **der; ~[e]s, Ställe** [ˈʃtɛlə] *(Pferde~, Renn~)* stable; *(Kuh~)* cowshed; *(Hühner~)* [chicken-]coop; *(Schweine~)* [pig]sty; *(für Kaninchen, Kleintiere)* hutch; *(für Schafe)* pen

Stallaterne die stable lamp

Stall-: ~bursche der stable lad; **~dung der** *(von Kühen/Schweinen/Schafen)* cow/pig/sheep dung; *(von Pferden)* horse manure; **~hase der** *(ugs.)* domestic rabbit; **~knecht der** *(veralt.)* stable lad; *(für Kühe)* cowhand; **~laterne die** *s.* **Stallaterne; ~meister der** head groom; **~mist der** *s.* **~dung**

Stallung die; ~, ~en *(Pferdestall)* stable; *(Kuhstall)* cow-shed; *(Schweinestall)* [pig]sty

Stamm [ʃtam] **der; ~[e]s, Stämme a)** *(Baum~)* trunk; **eine Hütte aus rohen Stämmen** a hut of rough-hewn boles; **b)** *(Volks~, Geschlecht)* tribe; **der ~ Davids** the house of David; **c)** *o. Pl. (fester Bestand)* core; *(von Fachkräften, Personal)* permanent staff; **zum ~ gehören** be one of the regulars *(coll.)*; *(der Belegschaft einer Firma)* be a permanent member of staff; **d)** *(Sprachw.)* stem

Stamm-: ~baum der family tree; *(eines Tieres)* pedigree; *(Biol.)* phylogenetic tree; **~buch das** family album *(recording births, marriages, deaths, etc.)*

stammeln [ˈʃtamln̩] *tr., itr. V.* stammer

stammen *itr. V.* come **(aus, von** from); *(datieren)* date **(aus, von** from); **die Idee stammt nicht von ihm** the idea isn't his

Stammes-: ~geschichte die; *o. Pl. (Biol.)* phylogenesis *no art.*; **~häuptling der** tribal chief

Stamm-: ~essen das set meal; **~form die;** *meist Pl. (Sprachw.)* principal part; **~gast der** *(im Lokal/Hotel)* regular customer/visitor; regular *(coll.)*; **~gericht das** set dish; **~halter der** *(oft scherzh.)* son and heir *(esp. joc.)*; **~kneipe die** *(ugs.)* favourite *or* usual pub *(Brit. coll.) or (Amer.)* bar; **~kunde der** regular customer; **~lokal das** favourite *or* usual restaurant/pub *(Brit.) or* bar *(Amer.)*/café; **~platz der** *(auch fig.)* regular place; *(Sitz)* regular *or* usual seat; *(für Wohnwagen, Zelt usw.)* regular site; **~silbe die** *(Sprachw.)* stem syllable; **~tisch der a)** *(Tisch)* regulars' table *(coll.)*; **b)** *(~tischrunde)* group of regulars *(coll.)*; **c)** *(Treffen)* get-together with the regulars *(coll.)*

stampfen [ˈʃtampfn̩] **1.** *itr. V.* **a)** *(laut auftreten)* stamp; **mit den Füßen/den Hufen ~:** stamp one's feet/its hoofs; **b)** *mit sein (sich fortbewegen)* tramp; *(mit schweren Schritten)* trudge; **c)** *(mit wuchtigen Stößen sich bewegen)* ‹*machine, engine, etc.*› pound. **2.** *tr. V.* **a) mit den Füßen den Rhythmus ~:** tap the rhythm with one's feet; **b)** *(fest~)* compress; *(ram-*

men) drive ⟨*pile*⟩ **(in** + *Akk.* into); **c)** *(zerkleinern)* mash ⟨*potatoes*⟩; pulp ⟨*fruit*⟩; crush ⟨*sugar*⟩; pound ⟨*millet, flour*⟩

Stampfer der; ~s, ~ a) *(für Erde usw.)* tamper; *(Stößel)* pestle; **b)** *(Küchengerät)* masher

Stampf·kartoffeln *Pl. (nordd.)* mashed potatoes

stand [ʃtant] *1. u. 3. Pers. Sg. Prät. v.* **stehen**

Stand der; ~[e]s, Stände ['ʃtɛndə] **a)** *o. Pl. (das Stehen)* standing position; **keinen sicheren ~ haben** not have a secure footing; **ein Sprung/Start aus dem ~:** a standing jump/start; **[bei jmdm.** *od.* **gegen jmdn.] einen schweren ~ haben** *(fig.)* have a tough time [of it] [with sb.]; **etw. aus dem ~ [heraus] beantworten** *(ugs.)* answer sth. off the top of one's head *(coll.)*; **b)** *(~ort)* position; **c)** *(Verkaufs~; Box für ein Pferd)* stall; *(Messe~, Informations~)* stand; *(Zeitungs~)* [newspaper] kiosk; **d)** *o. Pl. (erreichte Stufe; Zustand)* state; **etw. auf den neu[e]sten ~ bringen** bring sth. up to date *or* update sth.; **e)** *(des Wassers, Flusses)* level; *(des Thermometers, Zählers, Barometers)* reading; *(der Kasse, Finanzen)* state; *(eines Himmelskörpers)* position; **den ~ des Thermometers ablesen** take the thermometer reading; **f)** *o. Pl. (Familien~)* status; **g)** *(Gesellschaftsschicht)* class; *(Berufs~)* trade; *(Ärzte, Rechtsanwälte)* [professional] group; **der geistliche ~:** the clergy

Standard ['ʃtandart] **der; ~s, ~s** standard

Standard-: standard ⟨*equipment, example, letter, form, solution, model, work, language*⟩

Standard·situation die *(Sport)* set piece

Standarte [ʃtan'dartə] **die; ~, ~n** standard

Stand·bild das statue

Ständchen ['ʃtɛntçən] **das; ~s, ~:** serenade; **jmdm. ein ~ bringen** serenade sb.

Stander ['ʃtandɐ] **der; ~s, ~:** pennant

Ständer ['ʃtɛndɐ] **der; ~s, ~ a)** *(Gestell, Vorrichtung)* stand; *(Kleider~)* coat-stand; *(Wäsche~)* clothes-horse; *(Kerzen~)* candle-holder; **b)** *(Elektrot.)* stator; **c)** *(salopp: erigierter Penis)* hard-on *(sl.)*

standes-, Standes-: ~amt das registry office; **~amtlich 1.** *Adj.; nicht präd.* registry office ⟨*wedding, document*⟩; **2.** *adv.* **~amtlich heiraten** get married in a registry office; **~beamte der** registrar; **~bewußt** *Adj.* conscious of one's social standing *or* rank *postpos.*; **~bewußtsein das** consciousness of one's social standing *or* rank; **~dünkel der** *(abwertend)* snobbery; **~gemäß 1.** *Adj.* befitting sb.'s station *or* social standing *postpos.*; **~gemäß sein** befit sb.'s station *or* social standing; **2.** *adv.* as befits one's station *or* social standing; **~unterschied der** difference of rank; class difference

stand-, Stand-: ~fest *Adj.* **a)** *(fest stehend)* steady; stable; strong ⟨*stalk, stem*⟩; **b)** *(standhaft)* steadfast; **~festigkeit die a)** stability; *(eines Gebäudes)* structural strength; **b)** *(Standhaftigkeit)* steadfastness; **~haft 1.** *Adj.* steadfast; **2.** *adv.* steadfastly; **~haftigkeit die; ~:** steadfastness; **~|halten** *unr. itr. V.* stand firm; **einer Sache** *(Dat.)* **~halten** withstand *or* stand up to sth.; **einer näheren Überprüfung nicht ~halten** not stand [up to] *or* bear closer scrutiny

ständig 1. *Adj.; nicht präd.* **a)** *(andauernd)* constant ⟨*noise, worry, pressure, etc.*⟩; **b)** *(fest)* permanent ⟨*residence, correspondent, staff, member, etc.*⟩; standing ⟨*committee*⟩; regular ⟨*income*⟩. **2.** *adv.* constantly; **mußt du sie ~ unterbrechen?** do you have to keep [on] interrupting her?; **sie kommt ~ zu spät** she's forever coming late

ständisch *Adj.* corporative

stand-, Stand-: ~licht das; *Pl.* **~lichter** *(Kfz-W.) (Beleuchtung)* sidelights *pl.*; *(Leuchte, Lampe)* sidelight; **~ort der;** *Pl.* **~orte a)** position; *(eines Betriebes o. ä.)* location; site; **b)** *(Milit.: Garnison)* garrison; base; **~punkt der** *(fig.)* point of view; viewpoint; **den ~punkt vertreten/auf dem ~punkt stehen, daß ...:** take the view that ...; **~quartier das** base; **~rechtlich 1.** *Adj.; nicht präd.* summary ⟨*execution, shooting*⟩; **2.** *adv.* **jmdn. ~rechtlich erschießen** shoot sb. summarily; **~spur die** *(Verkehrsw.)* hard shoulder; **~uhr die** grandfather clock

Stange ['ʃtaŋə] **die; ~, ~n a)** *(aus Holz)* pole; *(aus Metall)* bar; *(dünner)* rod; *(Kleider~)* rail; *(Vogel~)* perch; **Kleider/Anzüge von der ~** *(ugs.)* off-the-peg dresses/suits; **von der ~ kaufen** *(ugs.)* buy off-the-peg clothes; **bei der ~ bleiben** *(ugs.)* keep at it *(coll.)*; **eine ~ Zimt/Vanille/Lakritze** *usw.* a stick of cinnamon/vanilla/liquorice *etc.*; **eine ~ Zigaretten** *a carton containing ten packets of cigarettes*; **eine [schöne] ~ Geld** *(ugs.)* a small fortune *(coll.)*; **b)** *(bes. md.: zylindrisches Glas)* [straight] glass

Stangen-: ~brot das French bread; **~spargel der** asparagus spears *pl. or* stalks *pl.*

stank [ʃtaŋk] *1. u. 3. Pers. Sg. Prät. v.* **stinken**

Stänkerer der; ~s, ~ *(ugs. abwertend)* grouser *(coll.)*; stirrer

stänkern ['ʃtɛŋkɐn] *itr. V. (ugs. abwertend)* stir *(coll.)*; **gegen jmdn./etw. ~:** go on about sb./sth.

Stanniol [ʃta'ni̯o:l] **das; ~s, ~e** tin foil; *(Silberpapier)* silver paper

Stanniol·papier das silver paper

stanzen *tr. V.* press; *(prägen)* stamp; *(ausstanzen)* punch ⟨*numbers, holes, punch-cards, etc.*⟩

Stapel ['ʃta:pl̩] **der; ~s, ~ a)** pile; **ein ~ Holz** a pile *or* stack of wood; **b)** *(Schiffbau)* stocks *pl.*; **vom ~ laufen** be launched; **vom ~ lassen** launch ⟨*ship*⟩

Stapel·lauf der launch[ing]

stapeln 1. *tr. V. (schichten)* pile up; stack; *(fig.: ansammeln)* accumulate. **2.** *refl. V.* pile up; *(gestapelt sein)* be piled up

stapfen *itr. V.; mit sein* tramp

[1]Star [ʃta:ɐ̯] **der; ~[e]s, ~e** *od. (schweiz.)* **~en** *(Vogel)* starling

[2]Star der; ~s, ~s *(berühmte Persönlichkeit)* star

[3]Star der; ~[e]s, ~e: der graue ~: cataract; **der grüne ~:** glaucoma

Star-: star ⟨*conductor, guest singer, etc.*⟩; top ⟨*lawyer, model, agent*⟩

starb [ʃtarp] *1. u. 3. Pers. Sg. Prät. v.* **sterben**

stark [ʃtark]; **stärker** ['ʃtɛrkɐ], **stärkst...** ['ʃtɛrkst...] **1.** *Adj.* **a)** strong ⟨*man, current, structure, team, drink, verb, pressure, wind, etc.*⟩; potent ⟨*drink, medicine, etc.*⟩; powerful ⟨*engine, lens, voice, etc.*⟩; *(ausgezeichnet)* excellent ⟨*runner, player, performance*⟩; **sich für jmdn./etw. ~ machen** *(ugs.)* throw one's weight behind sb./sth.; *s. auch* **Seite d; Stück c; b)** *(dick)* thick; stout ⟨*rope, string*⟩; *(verhüll.: korpulent)* well-built *(euphem.)*; **c)** *(zahlenmäßig groß, umfangreich)* sizeable, large ⟨*army, police*⟩; big ⟨*demand*⟩; **eine 100 Mann ~e Truppe** a 100-strong unit; **d)** *(hef-*

tig, intensiv) heavy ⟨*rain, snow, traffic, smoke, heat, cold, drinker, smoker, demand, pressure*⟩; severe ⟨*frost, pain*⟩; strong ⟨*impression, influence, current, resistance, dislike*⟩; grave ⟨*doubt, reservations*⟩; great ⟨*exaggeration, interest*⟩; hearty ⟨*eater, appetite*⟩; loud ⟨*applause*⟩; **e)** *(Jugendspr.: großartig)* great *(coll.)*; fantastic *(coll.)*. **2.** *adv.* **a)** *(sehr, überaus, intensiv) (mit Adj.)* very; heavily ⟨*indebted, stressed*⟩; greatly ⟨*increased, reduced, enlarged*⟩; strongly ⟨*emphasized, characterized*⟩; badly ⟨*damaged, worn, affected*⟩; *(mit Verb)* ⟨*rain, snow, drink, smoke, bleed*⟩ heavily; ⟨*exaggerate, impress*⟩ greatly; ⟨*enlarge, reduce, increase*⟩ considerably; ⟨*support, oppose, suspect*⟩ strongly; ⟨*remind*⟩ very much; **~ wirkend** with a powerful effect *postpos.*; **~ riechen/duften** have a strong smell/scent; **~ gewürzt** strongly seasoned; **es ist ~/zu ~ gesalzen** it is very/too salty; **~ erkältet sein** have a heavy *or* bad cold; **er geht ~ auf die Sechzig zu** *(ugs.)* he's pushing sixty *(coll.)*; **b)** *(Jugendspr.: großartig)* fantastically *(coll.)*; **c)** *(Sprachw.)* **~ flektieren** *od.* **flektiert werden** be a strong noun/verb

Stark·bier das strong beer

Stärke ['ʃtɛrkə] **die; ~, ~n a)** *o. Pl.* strength; *(eines Motors)* power; *(einer Glühbirne)* wattage; **b)** *(Dicke)* thickness; *(Technik)* gauge; **c)** *o. Pl. (zahlenmäßige Größe)* strength; size; **d)** *(besondere Fähigkeit, Vorteil)* strength; **jmds. ~/nicht jmds. ~ sein** be sb.'s forte/not be sb.'s strong point; **e)** *(von Wind, Strömung, Einfluß, Empfindung, Widerstand usw.)* strength; *(von Hitze, Kälte, Druck, Regenfall, Sturm, Schmerzen, Abneigung)* intensity; *(von Frost)* severity; *(von Lärm, Verkehr)* volume; **f)** *(organischer Stoff)* starch

stärken 1. *tr. V.* **a)** *(kräftigen, festigen; auch fig.)* strengthen; boost ⟨*power, prestige*⟩; ⟨*drink, food, etc.*⟩ fortify ⟨*person*⟩; **jmds. Selbstbewußtsein ~** *(fig.)* give sb.'s self-confidence a boost; **b)** *(steif machen)* starch ⟨*washing etc.*⟩. **2.** *refl. V. (sich erfrischen)* fortify *or* refresh oneself. **3.** *itr. V.* **ein ~des Mittel** a tonic

stärker, stärkst... *s.* **stark**

Stark·strom der *(Elektrot.)* heavy current; *(mit hoher Spannung)* high-voltage current

Stärkung die; ~, ~en a) *o. Pl.* strengthening; **zur ~ trank er erst mal einen Whisky** he drank a whisky to fortify himself; **b)** *(Erfrischung)* refreshment

Stärkungs·mittel das tonic

starr [ʃtar] **1.** *Adj.* **a)** rigid; *(steif)* stiff **(vor +** *Dat.* with); fixed ⟨*expression, smile, stare*⟩; **~ vor Schreck** paralysed with terror; **b)** *(nicht abwandelbar)* inflexible, rigid ⟨*law, rule, principle*⟩; **c)** *(unnachgiebig)* inflexible, obdurate ⟨*person, attitude, etc.*⟩. **2.** *adv.* **a)** rigidly; *(steif)* stiffly; **jmdn. ~ ansehen** look at sb. with a fixed stare; **b)** *(unnachgiebig)* obdurately

starren *itr. V.* **a)** *(starr blicken)* stare **(in +** *Akk.* into, **auf, an, gegen +** *Akk.* at); **jmdm. ins Gesicht ~:** stare sb. in the face; **b)** *(ganz bedeckt sein mit)* **vor/von Schmutz** *od.* **Dreck ~:** be filthy; be covered in filth; **vor Waffen ~:** be bristling with weapons

Starrheit die; ~: *s.* **starr 1: a)** rigidity; stiffness; fixity; **b)** inflexibility; rigidity; **c)** inflexibility; obduracy

starr-, Starr-: ~köpfig *Adj. (abwertend)* pig-headed; **~sinn der;** *o. Pl.* pig-headedness; **~sinnig** *Adj. (abwertend)* pig-headed

Start [ʃtart] **der; ~[e]s, ~s a)** *(Sport; auch fig.)* start; **einen guten ~ haben** get off to *or* make a good start; **b)** *(Sport: ~platz)* start; **an den ~ gehen/am ~ sein** *(fig.: teilnehmen)* start; **c)** *(Sport: Teilnahme)* participation; **d)** *(eines Flugzeugs)* take-off; *(einer Rakete)* launch

start-, Start-: ~bahn die [take-off] runway; **~bereit** *Adj.* ready to start *postpos.*; ⟨*aircraft*⟩ ready for take-off; *(zum Aufbruch bereit)* ready to set off *postpos.*; **~block der;** *Pl.* **~blöcke** *(Sport)* starting-block

starten 1. *itr. V.; mit sein* **a)** start; ⟨*aircraft*⟩ take off; ⟨*rocket*⟩ blast off, be launched; **b)** *(an einem Wettkampf teilnehmen)* compete; *(bei einem Rennen)* start **(bei, in +** *Dat.* in); **c)** *(den Motor anlassen)* start the engine; **d)** *(aufbrechen)* set off; set out; **e)** *(beginnen)* start; begin. **2.** *tr. V.* start ⟨*race, campaign, tour, production, etc.*⟩; launch ⟨*missile, rocket, satellite, attack*⟩; start [up] ⟨*engine, machine, car*⟩

Starter der; ~s, ~ *(Sport, Kfz-W.)* starter

start-, Start-: ~erlaubnis die a) *(Sport)* authorization to compete; **b)** *(Flugw.)* clearance [for take-off]; **~hilfe die a)** *(Unterstützung)* financial help, backing *(to get a project off the ground)*; **b) ich brauche ~hilfe** I need help to get my car started; **~hilfe·kabel das** jump leads *pl.*; **~klar** *Adj.* ready to start *postpos.*; ⟨*aircraft*⟩ clear *or* ready for take-off; **~linie die** *(Sport)* starting-line; **~nummer die** *(Sport)* [start] number; **~rampe die** launching pad; **~schuß der** *(Sport)* **der den ~schuß zum 100-m-Lauf geben** fire the gun for the start of the 100 metres; **den ~schuß zu** *od.* **für etw. geben** *(fig.)* give sth. the go-ahead *or* the green light

Statik ['ʃta:tɪk] **die; ~ a)** *(Physik)* statics *sing., no art.*; **b)** *(Bauw.)* static equilibrium

Station [ʃta'tsio:n] **die; ~, ~en a)** *(Haltestelle)* stop; **b)** *(Bahnhof, Sender, Forschungs~, Raum~)* station; **c)** *(Zwischen~, Aufenthalt)* stopover; **~ machen** stop over *or* off; make a stopover; **d)** *(Kranken~)* ward; **e)** *(einer Entwicklung, Karriere usw.)* stage

stationär [ʃtatsio'nɛ:ɐ̯] **1.** *Adj. (Med.)* ⟨*admission, examination, treatment*⟩ in hospital, as an in-patient. **2.** *adv. (Med.)* in hospital; **jmdn. ~ behandeln/aufnehmen** treat/admit sb. as an in-patient

stationieren *tr. V.* station ⟨*troops*⟩; deploy ⟨*weapons, bombers, etc.*⟩

Stationierung die; ~, ~en stationing; *(von Waffen, Raketen usw.)* deployment

Stations-: ~arzt der ward doctor; **~schwester die** ward sister; **~taste die** *(Rundf.)* preset [tuning] button; preset; **~vorsteher der** *(Eisenb.)* station-master

statisch ['ʃta:tɪʃ] *Adj.* static; ⟨*laws*⟩ of statics; **~e Berechnungen** *(Bauw.)* calculations relating to static equilibrium

Statist [ʃta'tɪst] **der; ~en, ~en** *(Theater, Film)* extra; *(fig.)* bystander; supernumerary

Statistik [ʃta'tɪstɪk] **die; ~, ~en a)** *o. Pl. (Wissenschaft)* statistics *sing., no art.*; **b)** *(Zusammenstellung)* statistics *pl.*; **eine ~:** a set of statistics

statistisch 1. *Adj.* statistical. **2.** *adv.* statistically

Stativ [ʃta'ti:f] **das; ~s, ~e** tripod

statt [ʃtat] **1.** *Konj. s.* **anstatt 1. 2.** *Präp. mit Gen.* instead of; **~ dessen** instead [of this]

Statt die; ~ *(veralt., geh.)* abode *(arch.)*; **an jmds./einer Sache ~:**

in sb.'s place/in place of sth.; instead of sb./sth.; *s.* **auch Eid**

Stätte ['ʃtɛtə] **die; ~, ~n** *(geh.)* place; **eine heilige/historische ~:** a holy/historic site

statt-, Statt-: ~|finden *unr. itr. V.* take place; ‹*process, development*› occur; **~|geben** *unr. itr. V. (Amtsspr.)* **einer Sache** *(Dat.)* **~geben** accede to sth.; **einer Klage ~geben** uphold a complaint; **~haft** *Adj.; nicht attr.* permissible; **~halter der** *(hist.)* governor

stattlich 1. a) well-built; strapping ‹*lad*›; *(beeindruckend)* imposing ‹*figure, stature, building, etc.*›; impressive ‹*trousseau, collection*›; **b)** *(beträchtlich)* considerable; sizeable ‹*part*›; considerable, appreciable ‹*sum, number*›. **2.** *adv.* impressively; splendidly

Statue ['ʃta:tu̯ə] **die; ~, ~n** statue

statuieren *tr. V. (geh.)* establish ‹*principle, purpose*›; lay down ‹*right, principle*›; *s. auch* **Exempel**

Statur [ʃta'tu:ɐ̯] **die; ~, ~en** build; **kräftig von ~** *od.* **von kräftiger ~ sein** have a powerful build

Status ['ʃta:tʊs] **der; ~, ~** ['ʃta:tu:s] **a)** *(geh.: Stand)* state; **b)** *([rechtliche] Stellung)* status

Statut [ʃta'tu:t] **das; ~[e]s, ~en** statute

Stau der; ~[e]s, ~s *od.* **~e a)** *(von Wasser, Blut usw.)* build-up; **b)** *(von Fahrzeugen)* tailback *(Brit.)*; backup *(Amer.)*; **im ~ stehen** sit *or* be stuck in a jam

Staub [ʃtaup] **der; ~[e]s** dust; **[im ganzen Haus] ~ wischen** dust [the whole house]; **[im Wohnzimmer] ~ saugen** vacuum *or (Brit. coll.)* hoover [the sitting-room]; **[viel] ~ aufwirbeln** *(fig. ugs.)* stir things up [quite a bit] *(coll.)*; cause [a lot of] aggro *(Brit. sl.)*; **sich aus dem ~[e] machen** *(fig. ugs.)* make oneself scarce *(coll.)*

Stäubchen ['ʃtɔypçən] **das; ~s, ~:** speck of dust

stauben *itr. V.* cause dust; ‹*person*› cause *or* raise dust; **es staubt sehr** there is a lot of dust

stäuben ['ʃtɔybn̩] *tr. V.* **etw. auf/über etw.** *(Akk.)* **~:** sprinkle sth. on/over sth.

staubig *Adj.* dusty

staub-, Staub-: ~lappen der duster; **~saugen** *tr. V.* vacuum, *(Brit. coll.)* hoover ‹*room, carpet, etc.*›; **~sauger der** vacuum cleaner; Hoover *(Brit.* **P)**; **~tuch das;** *Pl.* **~tücher** duster; **~wedel der** feather duster; **~wolke die** cloud of dust

Stau·damm der dam

Staude ['ʃtaudə] **die; ~, ~n** *(Bot.)* herbaceous perennial

stauen ['ʃtauən] **1.** *tr. V.* dam [up] ‹*stream, river*›; staunch *or* stem flow of ‹*blood*›. **2.** *refl. V.* ‹*water, blood, etc.*› accumulate, build up; ‹*people*› form a crowd; ‹*traffic*› form a tailback/tailbacks *(Brit.) or (Amer.)* backup/backups; *(fig.)* ‹*anger*› build up

Stau·mauer die dam [wall]

staunen ['ʃtaunən] *itr. V.* be amazed *or* astonished **(über +** *Akk.* at); *(beeindruckt sein)* marvel **(über +** *Akk.* at); **er staunte nicht schlecht, als er das hörte** *(ugs.)* he was flabbergasted when he heard it; **da staunst du, was?** *(ugs.)* quite a shock, isn't it?; shattered, eh? *(coll.)*; **~d** with *or* in amazement; *s. auch* **Bauklotz**

Staunen das; ~s amazement, astonishment **(über +** *Akk.* at); *(staunende Bewunderung)* wonderment; **jmdn. in ~ [ver]setzen** astonish *or* amaze sb.; **er kam aus dem ~ nicht mehr heraus** he couldn't get over it

Stau·see der reservoir

Stauung die; ~, ~en a) *(eines Bachs, Flusses)* damming; *(des Blutes, Wassers)* stemming the flow; *(das Sichstauen)* build-up; **b)** *(Verkehrsstau)* tailback *(Brit.)*; backup *(Amer.)*; jam

Std. *Abk.* **Stunde** hr.

Steak [ste:k] **das; ~s, ~s** steak

stechen ['ʃtɛçn̩] **1.** *unr. itr. V.* **a)** ‹*thorn, thistle, spine, needle*› prick; ‹*wasp, bee*› sting; ‹*mosquito*› bite; ‹*fig.: sun*› be scorching; **sich** *(Dat.)* **in den Finger ~:** prick one's finger; **b)** *(hinein~)* **mit etw. in etw.** *(Akk.)* **~:** stick *or* jab sth. into sth.; **c)** *(die Stechuhr betätigen) (bei Arbeitsbeginn)* clock on; *(bei Arbeitsende)* clock off; **d)** *(Kartenspiel)* ‹*suit*› be trumps; **e)** *(Sport)* jump-off. **2.** *unr. tr. V.* **a)** *(mit dem Messer, Schwert)* stab; *(mit der Nadel, mit einem Dorn usw.)* prick; ‹*bee, wasp*› sting; ‹*mosquito*› bite; *(Fischereiw.: fangen)* spear ‹*eel, pike*›; *(ab~)* stick ‹*pig, calf*›; **sich an etw.** *(Dat.)* **~:** prick oneself on sth.; **b)** *(hervorbringen)* make ‹*hole, pattern*›; **c)** *(unpers.)* **es sticht mich in der Seite** I've got a stabbing pain in my side; **d)** *(herauslösen)* cut ‹*peat, turf, asparagus, etc.*›; pick ‹*lettuce, mushrooms*›; **e)** *(gravieren)* engrave ‹*design etc.*›; **f)** *(Kartenspiel)* take ‹*card*›

Stechen das; ~s, ~ *(Sport)* jump-off

stechend *Adj.* penetrating, pungent ‹*smell*›; penetrating ‹*glance, eyes*›

Stech-: ~karte die clocking-on card; **~mücke die** mosquito; gnat; **~uhr die** time clock

steck-, Steck-: ~brief der description [of a/the wanted person]; *(Plakat)* 'wanted' poster; **~brieflich** *Adv.* **der ~brieflich Gesuchte** the wanted man; **der Mörder wird ~brieflich gesucht** descriptions/'wanted' posters of the murderer have been circulated; **~dose die** socket; power point

stecken ['ʃtɛkn̩] **1.** *tr. V.* **a)** put; **etw. in die Tasche ~:** put *or (coll.)* stick sth. in one's pocket; **b)** *(mit Nadeln)* pin ‹*hem, lining, etc.*›; pin [on] ‹*badge*›; pin up ‹*hair*›; . **2.** *regelm. (geh. auch unr.) itr. V.* be; **der Schlüssel steckt [im Schloß]** the key is in the lock; **wo hast du denn so lange gesteckt?** *(ugs.)* where did you get to *or* have you been all this time?; **er steckt in Schwierigkeiten** *(ugs.)* he's having problems; **hinter etw.** *(Dat.)* **~** *(fig. ugs.)* be behind sth.

stecken-, Stecken-: ~|bleiben *unr. itr. V.; mit sein* get stuck; *(fig.)* ‹*negotiations etc.*› get bogged down; **es blieb in den Anfängen ~** *(fig.)* it never got beyond the early stages; **das Wort blieb ihm vor Angst im Halse** *od.* **in der Kehle ~:** he was speechless with fear; **~|lassen** *unr. tr. V.* **den Schlüssel [im Schloß] ~lassen** leave the key in the lock; **~pferd das a)** *(Spielzeug)* hobby-horse; **b)** *(Liebhaberei)* hobby

Stecker der; ~s, ~: plug

Steckling ['ʃtɛklɪŋ] **der; ~s, ~e** cutting

Steck-: ~nadel die pin; **jmdn./etw. suchen wie eine ~nadel** *(ugs.)* search high and low for sb./sth.; **~rübe die** *(bes. nordd.)* swede; **~schlüssel der** socket spanner

Steg [ʃte:k] **der; ~[e]s, ~e** *(schmale Brücke)* [narrow] bridge; *(Fußgänger~)* foot-bridge; *(Laufbrett)* gangplank; *(Boots~)* landing-stage

Steg·reif der: aus dem ~: impromptu; **er hielt aus dem ~ eine kleine Rede** he gave a short speech extempore *or* off the cuff

Stegreif·rede die impromptu *or* extempore speech

Steh·auf·männchen das tumbling figure; tumbler

stehen ['ʃte:ən] **1.** *unr. itr. V.; südd., österr., schweiz. mit sein* **a)** stand; **er arbeitet ~d** *od.* **im Ste-**

hen he works standing up; **mit jmdm./etw. ~ und fallen** *(fig.)* stand or fall with sb./sth.; **das Haus steht noch** the house is still standing; **b)** *(sich befinden)* be; ‹*upright object, building*› stand; **das Verb steht am Satzende** the verb comes at the end of the sentence; **wo steht dein Auto?** where is your car [parked]?; **Schweißperlen standen auf seiner Stirn** beads of sweat stood out on his brow; **ich tue alles, was in meinen Kräften** *od.* **meiner Macht steht** I'll do everything in my power; **vor dem Bankrott ~:** be faced with bankruptcy; **c)** *(einen bestimmten Stand haben)* **auf etw.** *(Dat.)* ~ ‹*needle, hand*› point to sth.; **das Barometer steht tief/auf Regen** the barometer is reading low/indicating rain; **die Ampel steht auf rot** the traffic lights are [on] red; **es steht mir bis zum Hals[e]** *od.* **bis oben** *od.* **bis hier[hin]** I'm fed up to the back teeth with it *(sl.);* I'm sick to death of it *(coll.);* **der Wind steht günstig/nach Norden** *(Seemannsspr.)* the wind stands fair/ is from the north; **wie steht es/das Spiel?** *(Sport)* what's the score?; **die Chancen ~ fifty-fifty** the chances are fifty-fifty; **die Sache steht gut** things are going well; **wie steht es mit deiner Gesundheit?** how is your health?; **der Weizen steht gut** the wheat is growing well; **d)** *(einen bestimmten Kurs, Wert haben)* ‹*currency*› stand (**bei** at); **wie steht das Pfund?** what is the rate for the pound?; how is the pound doing? *(coll.);* **die Aktie steht gut** the share price is high; **e)** *(nicht in Bewegung sein)* be stationary; ‹*machine etc.*› be at a standstill; **meine Uhr steht** my watch has stopped; **f)** *(geschrieben, gedruckt sein)* be; **was steht in dem Brief?** what does it say in the letter?; **in der Zeitung steht, daß ...:** it says in the paper that ...; **g)** *(Sprachw.: gebraucht werden)* ‹*subjunctive etc.*› occur; be found; **mit dem Dativ ~:** be followed by *or* take the dative; **h) zu jmdm./etw. ~:** stand by sb./sth.; **wie stehst du dazu?** what's your view on this?; **hinter jmdm./etw. ~** *(jmdn. unterstützen)* be [right] behind sb./sth.; support sb./sth.; **i) jmdm. [gut] ~** ‹*dress etc.*› suit sb. [well]; **Lächeln steht dir gut** *(fig.)* it suits you *or* you look nice when you smile; **j)** *(sich verstehen)* **mit jmdm. gut/ schlecht ~:** be on good/bad terms *or* get on well/badly with sb.; **k) auf etw.** *(Akk.)* **steht Gefängnis** sth. is punishable by imprisonment. 2. *unr. refl. V.; südd., österr., schweiz. mit sein (ugs.)* **a)** *(in bestimmten Verhältnissen leben)* **sich gut/schlecht ~:** be comfortably/badly off; **b)** *(sich verstehen)* **sich gut/schlecht mit jmdm. ~:** be on good/bad terms *or* get on well/badly with sb.

stehen: ~|bleiben *unr. itr. V.; mit sein* **a)** *(anhalten)* stop; ‹*traffic*› come to a standstill; *(fig.)* ‹*time*› stand still; **wo sind wir ~geblieben?** *(fig.)* where had we got to?; where were we?; **b)** *(unverändert gelassen werden)* stay; be left; *(zurückgelassen werden)* be left behind; *(der Zerstörung entgehen)* ‹*building*› be left standing; **~|lassen** *unr. tr. V.* **a)** *(belassen, nicht entfernen)* leave; **alles ~ und liegenlassen** drop everything; **sich** *(Dat.)* **einen Bart ~lassen** *(ugs.)* grow a beard; **b)** *(vergessen)* leave [behind]; **c)** *(sich abwenden von)* **jmdn. ~lassen** walk off and leave sb. standing there

Steh-: ~kneipe die stand-up bar; **~lampe die** standard lamp *(Brit.);* floor lamp *(Amer.);* **~leiter die** step-ladder

stehlen [ˈʃteːlən] **1.** *unr. tr., itr. V.* steal; **jmdm. etw. ~:** steal sth. from sb.; **jmdm. das Portemonnaie ~:** steal sb.'s purse. **2.** *unr. refl. V.* steal; creep

Steh-: ~platz der *(im Theater/ Stadion)* standing place; *(im Bus)* space to stand; **es gab nur noch ~plätze** there was standing-room only; **~vermögen das;** *o. Pl.* stamina; staying-power

steif [ʃtaif] **1.** *Adj.* **a)** stiff; *(ugs.: erigiert)* erect ‹*penis*›; **b)** *(förmlich)* stiff, formal ‹*person, greeting, style*›; formal ‹*reception*›; **c)** *(Seemannsspr.: stark)* stiff ‹*wind, breeze*›; **d)** *(ugs.: stark)* strong ‹*coffee*›; stiff, strong ‹*alcoholic drink*›. **2.** *adv.* **a)** stiffly; **b)** *(Seemannsspr.: stark)* **der Wind steht** *od.* **weht ~ aus Südost** there's a stiff wind blowing from the south-east; **c) ~ und fest behaupten/glauben, daß ...** *(ugs.)* swear blind/be completely convinced that ...

Steig·bügel der *(auch Anat.)* stirrup

steigen [ˈʃtaign̩] **1.** *unr. itr. V.; mit sein* **a)** ‹*person, animal, aircraft, etc.*› climb; ‹*mist, smoke, sun, object*› rise; ‹*balloon*› climb, rise; **Drachen ~ lassen** fly kites; **auf eine Leiter/die Leiter ~:** climb a ladder/get on to the ladder; **aus der Wanne/in die Wanne ~:** get out of/into the bath; **in den/aus dem Zug ~:** board *or* get on/get off *or* out of the train; **ins/aus dem Flugzeug ~:** board/leave the aircraft; **der Duft steigt mir in die Nase** the scent gets up my nose; *s. auch* **Kopf a; b)** *(ansteigen, zunehmen)* rise (**auf +** *Akk.* to, **um** by) ‹*price, cost, salary, output*› increase, rise; ‹*debts, tension*› increase, mount; ‹*chances*› improve; **in jmds. Achtung ~** *(fig.)* go up *or* rise in sb.'s estimation; **c)** *(ugs.: stattfinden)* be on; **morgen soll ein Fest ~:** there's to be a party tomorrow. **2.** *unr. tr. V.; mit sein* climb ‹*stairs, steps*›

Steiger der; ~s, ~ *(Bergbau)* overman

steigern 1. *tr. V.* **a)** increase ‹*speed, value, sales, consumption, etc.*› (**auf +** *Akk.* to); step up ‹*demands, production, pace, etc.*›; raise ‹*standards, requirements*›; *(verstärken)* intensify ‹*fear, tension*›; heighten, intensify ‹*effect*›; exacerbate ‹*anger*›; **b)** *(Sprachw.)* compare ‹*adjective*›. **2.** *refl. V.* **a)** ‹*confusion, speed, profit, etc.*› increase; ‹*pain, excitement, tension*› become more intense; ‹*excitement, tension*› mount; ‹*hate, anger*› grow, become more intense; ‹*costs*› escalate; ‹*effect*› be heightened *or* intensified; **sich** *od.* **seine Leistung[en] ~:** improve one's performance; **b)** *(hineinsteigern)* **sich [mehr und mehr] in einen Erregungszustand ~:** work oneself up into [more and more of] a state [of excitement]

Steigerung die; ~, ~en a) increase (*Gen.* in); *(Verstärkung)* intensification; *(einer Wirkung)* heightening; *(des Zorns)* exacerbation; *(Verbesserung)* improvement (*Gen.* in); *(bes. Sport: Leistungs~)* improvement [in performance]; **b)** *(Sprachw.)* comparison

Steigung die; ~, ~en gradient

steil [ʃtail] **1.** *Adj.* **a)** steep; upright, straight ‹*handwriting, flame*›; meteoric ‹*career*›; rapid ‹*rise*›; **b)** *nicht präd. (Jugendspr. veralt.: beeindruckend)* fabulous *(coll.);* super *(coll.).* **2.** *adv.* steeply

Steil-: ~hang der steep escarpment; **~küste die** *(Geogr.)* cliffs *pl.;* **~paß der** *(Fußball)* deep [forward] pass; **~wand die** rock wall

Stein [ʃtain] **der; ~[e]s, ~e a)** *o. Pl.* stone; *(Fels)* rock; **ihr Gesicht war**

zu ~ geworden *(fig.)* her face had hardened; **b)** *(losgelöstes Stück, Kern, Med., Edel~, Schmuck~)* stone; *(Kiesel~)* pebble; **eine Uhr mit 12 ~en** a 12-jewel watch; **der ~ der Weisen** *(geh.)* the philosophers' stone; **ein ~ des Anstoßes** *(geh.)* a bone of contention; **mir fällt ein ~ vom Herzen** that's a weight off my mind; **es friert ~ und Bein** *(ugs.)* it's freezing hard; **~ und Bein schwören** *(ugs.)* swear blind; **den ~ ins Rollen bringen** *(fig.)* set the ball rolling; **jmdm. [die** *od.* **alle] ~e aus dem Weg räumen** *(fig.)* smooth sb.'s path; make things easy for sb.; **jmdm. ~e in den Weg legen** *(fig.)* create obstacles *or* make things difficult for sb.; **c)** *(Bau~)* [stone] block; *(Ziegel~)* brick; **keinen ~ auf dem anderen lassen** not leave one stone upon another; **d)** *(Spiel~)* piece; *(rund, flach)* counter; **bei jmdm. einen ~ im Brett haben** *(fig.)* be in sb.'s good books

stein-, Stein-: ~alt *Adj.* aged; ancient; **~alt werden** live to a great age; **~bock der a)** *(Tier)* ibex; **b)** *(Astrol.)* Capricorn; the Goat; *s. auch* **Fisch c; ~bruch der** quarry

steinern *Adj.* **a)** *nicht präd.* stone ⟨*floor, bench, etc.*⟩; **b)** *(wie versteinert)* stony ⟨*face, features*⟩

stein-, Stein-: ~fuß·boden der stone floor; **~gut das** earthenware; **~hart** *Adj.* rock-hard

steinig *Adj.* stony

steinigen *tr. V.* stone ⟨*person*⟩

stein-, Stein-: ~kohle die [hard] coal; **~metz** [~mɛts] **der** stonemason; **~obst das** stone-fruit; **~pilz der** cep; **~reich** *Adj. (ugs.)* filthy rich; **~schlag der** *(Fachspr.)* rock fall; **„Achtung ~schlag"** 'beware falling rocks'; **~topf der** earthenware pot; **~wurf der: jmdn. mit ~würfen wegjagen** chase sb. away by throwing stones [at him/her]; **~zeit die** Stone Age; *(fig.)* stone age

Steiß [ʃtais] **der; ~es, ~e a)** *(Anat.: ~bein)* coccyx; **b)** *(ugs.: Gesäß)* backside; behind *(coll.)*

Steiß·bein das *s.* **Steiß a**

Stellage [ʃtɛ'la:ʒə] **die; ~, ~n** rack

Stell·dich·ein das; ~[s], ~[s] *(veralt.)* rendezvous; tryst *(arch./literary)*; **sich** *(Dat.)* **ein ~ geben** *(fig.)* gather; assemble

Stelle ['ʃtɛlə] **die; ~, ~n a)** place; **eine schöne ~ zum Campen** a nice spot for camping; **die Truhe ließ sich nicht von der ~ rücken** the chest could not be shifted *or* would not budge; **an jmds. ~ treten** take sb.'s place; **ich an deiner ~ würde das nicht machen** I wouldn't do it if I were you; **ich möchte nicht an deiner ~ sein** I shouldn't like to be in your place; **auf der ~:** immediately; **er war auf der ~ tot** he died instantly; **auf der ~ treten** *(ugs.)*, **nicht von der ~ kommen** *(fig.)* make no headway; not get anywhere; **zur ~ sein** be there *or* on the spot; **b)** *(begrenzter Bereich)* patch; *(am Körper)* spot; **eine kahle ~:** a bare patch; *(am Kopf)* a bald patch; **seine empfindliche ~** *(fig.)* his sensitive *or* sore spot; **c)** *(Passage)* passage; **an anderer ~:** elsewhere; in another passage; **d)** *(Punkt im Ablauf einer Rede usw.)* point; **an dieser/früherer ~:** at this point *or* here/earlier; **eine schwache ~ in der Argumentation** *(fig.)* a weak point in the argument; **e)** *(in einer Rangordnung, Reihenfolge)* place; **an achter ~ liegen** be in eighth place; **an erster ~ geht es hier um ...:** here it is primarily a question of ...; **f)** *(Math.)* figure; **die erste ~ hinter** *od.* **nach dem Komma** the first decimal place; **g)** *(Arbeits~)* job; *(formeller)* position; *(bes. als Beamter)* post; **ohne ~ sein** be unemployed; **eine freie ~:** a vacancy; **h)** *(Dienst~)* office; *(Behörde)* authority

stellen 1. *tr. V.* **a)** put; *(mit Sorgfalt, ordentlich)* place; *(aufrecht hin~)* stand; **jmdn. wieder auf die Füße ~** *(fig.)* put sb. back on his/her feet; **jmdn. vor eine Entscheidung ~** *(fig.)* confront sb. with a decision; **auf sich [selbst] gestellt sein** *(fig.)* be thrown back on one's own resources; **b)** *(ein~, regulieren)* set ⟨*points, clock, scales*⟩; set ⟨*clock*⟩ to the right time; **den Wecker auf 6 Uhr ~:** set the alarm for 6 o'clock; **das Radio lauter/leiser ~:** turn the radio up/down; **c)** *(bereit~)* provide; produce ⟨*witness*⟩; **d) jmdn. besser ~:** ⟨*firm*⟩ improve sb.'s pay; **gut/schlecht gestellt** comfortably/badly off; **e)** *(auf~)* set ⟨*trap*⟩; lay ⟨*net*⟩; **f) kalt ~:** put ⟨*food, drink*⟩ in a cold place; leave ⟨*champagne etc.*⟩ to chill; **warm ~:** put ⟨*plant*⟩ in a warm place; keep ⟨*food*⟩ warm *or* hot; **g)** *(fassen, festhalten)* catch ⟨*game*⟩; apprehend ⟨*criminal*⟩; **h)** *(aufrichten)* ⟨*dog, horse, etc.*⟩ prick up ⟨*ears*⟩; stick up ⟨*tail*⟩; **i)** *(erstellen)* prepare ⟨*horoscope, bill*⟩; make ⟨*diagnosis, prognosis*⟩; **j)** *(verblaßt)* put ⟨*question*⟩; set ⟨*task, essay, topic, condition*⟩; make ⟨*application, demand, request*⟩; **jmdm. eine Frage ~:** ask sb. a question. **2.** *refl. V.* **a)** place oneself; **stell dich neben mich/ans Ende der Schlange/in die Reihe** come and stand by me/go to the back of the queue *(Brit.) or (Amer.)* line/get into line; **sich auf die Zehenspitzen ~:** stand on tiptoe; **sich gegen jmdn./etw. ~** *(fig.)* oppose sb./sth.; **sich hinter jmdn./etw. ~** *(fig.)* give sb./sth. one's backing; **b) sich schlafend/taub** *usw.* **~:** feign sleep/deafness *etc.*; pretend to be asleep/deaf *etc.*; **c)** *(sich ausliefern)* **sich [der Polizei] ~:** give oneself up [to the police]; **d)** *(nicht ausweichen)* **sich einem Herausforderer/der Presse ~:** face a challenger/the press; **sich einer Disskusion ~:** consent to take part in a discussion; **e)** *(Stellung beziehen)* **sich positiv/negativ zu jmdm./etw. ~:** take a positive/negative view of sb./sth.; **sich mit jmdm. gut ~:** try to get on good terms with sb.

stellen-, Stellen-: ~angebot das offer of a job; *(Inserat)* job advertisement; **„~angebote"** 'situations vacant'; **~anzeige die** job advertisement; **~gesuch das** 'situation wanted' advertisement; **„~gesuche"** 'situations wanted'; **~suche die** job-hunting *no art.*; search for a job; **auf ~suche sein** be looking for a job; be job-hunting; **~weise** *Adv.* in places; **~wert der a)** *(Math.)* place value; **b)** *(fig.: Bedeutung)* standing; status

Stellung die; ~, ~en a) position; **in gebückter ~:** in a bent posture; **die ~ der Frau in der Gesellschaft** the position *or* standing of women in society; **in ~ gehen** *(Milit.)* take up [one's] position; **[zu/gegen etw.] ~ beziehen** *(fig.)* take a stand [on/against sth.]; **b)** *(Posten)* job; *(formeller)* position; *(bes. als Beamter)* post; **c)** *o. Pl. (Einstellung)* attitude **(zu** to, towards); **zu etw. ~ nehmen** express one's opinion *or* state one's view on sth.; **er hat zu dem Vorschlag offiziell ~ genommen** he made an official statement on the proposal

Stellungnahme die; ~, ~n opinion; *(kurze Äußerung)* statement; **eine ~ zu etw. abgeben** give one's opinion *or* views on sth.; *(sich kurz zu etw. äußern)* make a statement on sth.

stellungs·los *Adj.* unemployed; jobless

stell-, Stell-: ~vertretend 1. *Adj.; nicht präd.* acting; *(von Amts wegen)* deputy ‹*minister, director, etc.*›; **2.** *adv.* as a deputy; **~vertretend für jmdn.** deputizing for sb.; on sb.'s behalf; **~vertreter der** deputy; **der ~vertreter Christi** *(kath. Rel.)* the Vicar of Christ; **~werk das** *(Eisenb.)* signal-box *(Brit.)*; switchtower *(Amer.)*; *(Anlage)* control gear for signals and points *(Brit.)* or *(Amer.)* switches

Stelze die; ~, ~n *meist Pl.* stilt

stelzen *itr. V.; mit sein* strut; stalk

stemmen ['ʃtɛmən] **1.** *tr. V.* **a)** *(hoch~)* lift [above one's head]; *(Gewichtheben)* lift ‹*weight*›; **b)** *(drücken)* brace ‹*feet, knees*› **(gegen** against); **die Arme in die Hüften/Seiten ~:** place one's arms akimbo; put one's hands on one's hips; **c)** *(meißeln)* chisel ‹*hole etc.*›. **2.** *refl. V.* **sich in die Höhe ~:** haul oneself to one's feet; **sich gegen etw. ~:** brace oneself against sth.; *(fig.)* resist sth. **3.** *itr. V. (Skisport)* stem

Stempel ['ʃtɛmpl̩] **der; ~s, ~ a)** stamp; *(Post~)* postmark; **einer Sache** *(Dat.)* **seinen ~ aufdrücken** *(fig.)* leave one's mark on sth.; **b)** *(Punze)* hallmark; **c)** *(Bot.: Teil der Blüte)* pistil

stempeln *tr. V.* **a)** stamp ‹*passport, form*›; postmark ‹*letter*›; cancel ‹*postage stamp*›; **b)** hallmark ‹*gold, silver, ring, etc.*›

Stengel ['ʃtɛŋl̩] **der; ~s, ~:** stem; stalk

[1]**Steno** ['ʃte:no] **die; ~;** *meist o. Art. (ugs.)* shorthand

[2]**Steno das; ~s, ~s** *(ugs.) s.* **Stenogramm**

steno-, Steno-: ~block ['---] **der** shorthand pad; **~gramm das** shorthand text; **ein ~gramm aufnehmen** take a dictation in shorthand; **~graph der; ~en, ~en** stenographer; **~graphie die; ~, ~n** stenography *no art.*; shorthand *no art.*; **~graphieren** *itr. V.* do shorthand; **~typistin die** shorthand typist

Stepp·decke die quilt

Steppe ['ʃtɛpə] **die; ~, ~n** steppe

[1]**steppen** *tr. (auch itr.) V. (nähen)* backstitch

[2]**steppen** *itr. V. (tanzen)* tap-dance

Stepp·jacke die quilted jacket

Step-: ~tanz der tap-dance; **~tänzer der, ~tänzerin die** tap-dancer

Sterbe-: ~bett das death-bed; **~fall der** *s.* **Todesfall**

sterben ['ʃtɛrbn̩] **1.** *unr. itr. V.; mit sein* die; **im Sterben liegen** lie dying; **und wenn sie nicht gestorben sind, dann leben sie noch heute** and they lived happily ever after; **er ist für mich gestorben** *(fig.)* he's finished *or* he doesn't exist as far as I'm concerned; **vor Angst/Neugier ~** *(ugs.)* die of fright/be dying of curiosity. **2.** *unr. tr. V.; mit sein* **den Hungertod ~:** die of starvation; starve to death; **den Heldentod ~:** die a hero's death

sterbens-, Sterbens-: ~angst die terrible fear; **~elend** *Adj.* wretched; **~krank** *Adj.* **a)** *s.* **~elend; b)** *(sehr krank)* mortally ill; **~langweilig** *Adj.* deadly boring; **~wort, ~wörtchen das** *in* **kein** *od.* **nicht ein ~wort** *od.* **~wörtchen** not a [single] word

Sterbe-: ~sakramente *Pl. (kath. Kirche)* last rites; **~urkunde die** death certificate

sterblich *Adj.* mortal; *s. auch* **Überrest**

Sterbliche der/die; *adj. Dekl.* **a)** *(dichter.)* mortal; **b) ein gewöhnlicher ~r** an ordinary mortal *or* person

Sterblichkeit die; ~: mortality

stereo ['ʃte:reo] *Adv.* in stereo

Stereo das; ~s stereo

stereo-, Stereo-: ~anlage die stereo [system]; **~aufnahme die** stereo recording; **~phonie** [~fo'ni:] **die; ~** stereophony *no art.*; **~ton der** stereo sound; **~typ** [---'-] **1.** *Adj.* stereotyped ‹*discussion, pattern, etc.*›; stereotyped, stock ‹*question, reply, phrase, utterance*›; mechanical ‹*smile*›; **2.** *adv.* in a stereotyped way

steril [ʃte'ri:l] **1.** *Adj. (auch fig. abwertend)* sterile. **2.** *adv.* **a)** *(keimfrei)* **~ verpackt sein** be in a sterile pack/sterile packs; **b)** *(fig. abwertend: unschöpferisch, nüchtern)* sterilely

sterilisieren *tr. V.* sterilize

Sterling ['stɛ:lɪŋ] **der; ~s, ~e: 2 Pfund ~:** £2 sterling; **einen Betrag in Pfund ~ tauschen** change a sum into sterling

Stern [ʃtɛrn] **der; ~[e]s, ~e a)** star; **~e sehen** *(ugs.)* see stars; **in den ~en stehen** *(fig.)* be in the lap of the gods; **b)** *(Orden, Auszeichnung)* star; **ein Hotel mit fünf ~en** a five-star hotel

Stern·bild das constellation

Sternchen das; ~s, ~ a) [little] star; **b)** *(als Verweis)* asterisk

Sternen·banner das Star-spangled Banner, Stars and Stripes *pl.*

sternen·klar *Adj.* starlit, starry ‹*sky, night*›

stern-, Stern-: ~fahrt die rally; **~förmig** *Adj.* star-shaped; **~hagel·voll** *Adj. (salopp)* paralytic *(Brit. sl.)*; blotto *(sl.)*; **~himmel der** starry sky; **~klar** *Adj. s.* **sternenklar; ~kunde die;** *o. Pl.* astronomy *no art.*; **~marsch der** [protest] march; **~schnuppe die; ~, ~n** shooting star; **~stunde die** *(geh.)* great moment; **~warte die** observatory; **~zeichen das** *s.* **Tierkreiszeichen**

stet [ʃte:t] *Adj. (geh.)* **a)** constant ‹*goodwill, devotion, companion*›; steady ‹*rhythm*›; **b)** *(ständig)* constant; continous

Stethoskop [ʃteto'sko:p] **das; ~s, ~e** *(Med.)* stethoscope

stetig ['ʃte:tɪç] **1.** *Adj.* steady ‹*growth, increase, decline*›; constant, continuous ‹*movement, vibration*›. **2.** *adv.* ‹*grow, increase, drop*› steadily; ‹*move, vibrate*› constantly, continuously

stets [ʃte:ts] *Adv.* always

[1]**Steuer** ['ʃtɔyɐ] **das; ~s, ~** *(von Fahrzeugen)* [steering-]wheel; *(von Schiffen)* helm; **sich ans** *od.* **hinters ~ setzen** get behind the wheel; **das ~ übernehmen** take over the wheel *or* the driving; *(bei Schiffen, fig.)* take over the helm; **Trunkenheit am ~:** drunken driving; being drunk at the wheel

[2]**Steuer die; ~, ~n a)** tax; **~n zahlen** *(Lohn-/Einkommensteuer)* pay tax; **etw. von der ~ absetzen** set sth. off against tax; **b)** *o. Pl. (ugs.: Behörde)* tax authorities *pl.*

steuer-, Steuer-: ~berater der tax consultant *or* adviser; **~bord das** *od. österr.* **der;** *o. Pl. (Seew., Flugw.)* starboard; **~bord[s]** *Adv. (Seew., Flugw.)* to starboard; **~erhöhung die** tax increase; **~erklärung die** tax return; **~ermäßigung die** tax relief; **~frei** *Adj.* tax-free; free of tax *pred.*; **~freibetrag der** tax allowance; **~gelder** *Pl.* taxes; **~gerät das a)** *(Rundfunkt.)* receiver; **b)** *(Elektrot.)* control device *or* unit; **~gesetz das;** *meist Pl.* tax law; **~klasse die** tax category; **~knüppel der** control column; joystick *(coll.)*

steuerlich 1. *Adj.; nicht präd.* tax ‹*advantages, benefits, etc.*›. **2.** *adv.* **~ absetzbar** tax-deductible

steuer-, Steuer-: ~los *Adj.* out of control; **~mann der;** *Pl.* **~leute** *od.* **~männer a)** *(Seew. veralt.)*

helmsman; steersman; **b)** *(Rudersport)* cox; **Vierer mit/ohne ~mann** coxed/coxless fours; **~marke die** revenue stamp; *(für Hunde)* licence disc

steuern 1. *tr. V.* **a)** *(fahren)* steer; *(fliegen)* pilot, fly ‹*aircraft*›; fly ‹*course*›; **b)** *(Technik)* control; **c)** *(beeinflussen)* control, regulate ‹*process, activity, price, etc.*›; steer ‹*discussion etc.*›; influence ‹*opinion etc.*›. **2.** *itr. V.* **a)** *(im Fahrzeug)* be at the wheel; *(auf dem Schiff)* be at the helm; **b)** *mit sein (Kurs nehmen, ugs.: sich hinbewegen; auch fig.)* head

steuer-, Steuer-: ~pflicht die; *o. Pl. (Steuerw.)* liability to [pay] tax; **~pflichtig** *Adj. (Steuerw.)* ‹*person*› liable to [pay] tax; taxable ‹*goods, assets, income, profits, etc.*›; **~rad das a)** steering-wheel; **b)** *(Seew.)* [ship's] wheel; helm; **~recht das** tax law; **~schuld die** *(Steuerw.)* tax[es] owing *no indef. art.*; *(Verpflichtung)* tax liability; **~senkung die** *(Steuerw.)* tax cut

Steuerung die; ~, ~en a) *(System)* controls *pl.*; **b)** *o. Pl. s.* **steuern 1a, c, d:** steering; piloting; flying; control; regulation; steering; influencing

Steven ['ʃte:vn̩] **der; ~s, ~** *(Vorder~)* stem; *(Achter~)* stern-post

Steward ['stju:ɐt] **der; ~s, ~s** steward

Stewardeß ['stju:ɐdɛs] **die; ~, Stewardessen** stewardess

StGB *Abk.* **Strafgesetzbuch**

stibitzen [ʃti'bɪtsn̩] *tr. V. (fam.)* pinch *(sl.)*; swipe *(sl.)*

Stich der; ~[e]s, ~e a) *(mit einer Waffe)* stab; *(fig.: böse Bemerkung)* dig; gibe; **b)** *(Dornen~, Nadel~)* prick; *(von Wespe, Biene, Skorpion usw.)* sting; *(Mücken~ usw.)* bite; **c)** *(~wunde)* stab wound; **d)** *(beim Nähen)* stitch; **e)** *(Schmerz)* stabbing *or* shooting *or* sharp pain; **es gab mir einen ~ [ins Herz]** *(fig.)* I was cut to the quick; **f)** *(Kartenspiel)* trick; **g) jmdn./etw. im ~ lassen** leave sb. in the lurch/abandon sth.; **mein Gedächtnis hat mich im ~ gelassen** my memory has failed me; **h)** *(Fechten)* hit; **i)** *(bild. Kunst)* engraving; **j)** *o. Pl. (Farbschimmer)* tinge; **ein ~ ins Blaue** a tinge of blue; **k) einen [leichten] ~ haben** *(ugs.)* ‹*food, drink*› be off, have gone off; *(salopp)* ‹*person*› be nuts *(sl.)*; be round the bend *(coll.)*

Stichelei die; ~, ~en *(ugs. abwertend)* **a)** *(Bemerkung)* dig; gibe; **b)** *o. Pl.* **hör auf mit deiner ~:** stop getting at me/him *etc. (coll.)*

sticheln *itr. V.* make snide remarks *(coll.)* **(gegen** about)

stich-, Stich-: ~fest *s.* **hiebfest; ~flamme die** tongue *or* jet of flame; **~haltig,** *(österr.)* **~hältig 1.** *Adj.* sound, valid ‹*argument, reason*›; valid ‹*assertion, reply*›; conclusive ‹*evidence*›. **2.** *adv.* **etw. ~haltig begründen** back sth. with sound *or* valid reasons; **~haltigkeit,** *(österr.)* **~hältigkeit die; ~** *s.* **~haltig:** soundness; validity; conclusiveness.

Stichling ['ʃtɪçlɪŋ] **der; ~s, ~e** stickleback

Stich·probe die [random] sample; *(bei Kontrollen)* spot check

stichst [ʃtɪçst] *2. Pers. Sg. Präsens v.* **stechen**

sticht [ʃtɪçt] *3. Pers. Sg. Präsens v.* **stechen**

Stich-: ~tag der set date; *(letzter Termin)* deadline; **~wahl die** final *or* deciding ballot; run-off; **~wort das a)** *Pl.* **~wörter** headword; *(in Registern)* entry; **b)** *Pl.* **~worte** *(Theater)* cue; **~wunde die** stab wound

sticken ['ʃtɪkn̩] **1.** *itr. V.* do embroidery. **2.** *tr. V.* embroider

Stickerei die; ~, ~en *(Handarb.)* **a)** *(Verzierung)* embroidery *no pl.*; embroidered pattern; **b)** *(gestickte Arbeit)* piece of embroidery

stickig *Adj.* stuffy; stale ‹*air*›

Stick·stoff der nitrogen

stieben ['ʃti:bn̩] *unr. (auch regelm.) itr. V. (geh., veralt.)* **a)** *auch mit sein (auseinanderwirbeln)* ‹*dust, snow*› be thrown up in a cloud; ‹*sparks*› fly; ‹*water*› spray; **b)** *mit sein* **Schnee stiebt durch die Ritzen** snow blows through the cracks; **c)** *mit sein (davoneilen)* dash; **nach allen Seiten ~:** scatter in all directions

Stief·bruder ['ʃti:f-] **der** stepbrother; *(ugs.: Halbbruder)* half-brother

Stiefel ['ʃti:fl̩] **der; ~s, ~** boot

Stiefel·knecht der bootjack

stiefeln *itr. V.; mit sein (ugs.)* stride

stief-, Stief-: ~kind das stepchild; *(fig.)* poor relation *(fig.)*; **~mutter die** stepmother; **~mütterchen das** *(Bot.)* pansy; **~mütterlich 1.** *Adj.* poor, shabby ‹*treatment*›; **2.** *adv.* **~mütterlich behandeln** treat ‹*person*› poorly *or* shabbily; neglect ‹*pet, flowers, doll, problem*›; **~schwester die** stepsister; *(ugs.: Halbschwester)* half-sister; **~sohn der** stepson; **~tochter die** stepdaughter; **~vater der** stepfather

stieg [ʃti:k] *1. u. 3. Pers. Sg. Prät. v.* **steigen**

Stiege die; ~, ~n a) *(Holztreppe)* [wooden] staircase; [wooden] stairs *pl.*; **b)** *(südd., österr.: Treppe)* stairs *pl.*; steps *pl.*

Stieglitz ['ʃti:glɪts] **der; ~es, ~e** goldfinch

stiehlst [ʃti:lst], **stiehlt** *2. u. 3. Pers. Sg. Präsens v.* **stehlen**

Stiel [ʃti:l] **der; ~[e]s, ~e a)** *(Griff)* handle; *(Besen~)* [broom-]stick; *(für Süßigkeiten)* stick; **ein Eis am ~:** an ice-lolly *(Brit.)*; a Popsicle *(Amer.* **P)**; **b)** *(bei Gläsern)* stem; **c)** *(bei Blumen)* stem; stalk; *(an Obst, Obstblüten usw.)* stalk

Stiel·kamm der tail comb

stier 1. *Adj.* vacant. **2.** *adv.* vacantly

Stier [ʃti:ɐ̯] **der; ~[e]s, ~e a)** bull; **b)** *(Astrol.)* Taurus; the Bull

stieren *itr. V.* stare [vacantly] **(auf** + *Akk.* at); **vor sich hin ~:** stare [vacantly] into space

Stier-: ~kampf der bullfight; **~kämpfer der** bullfighter

stieß [ʃti:s] *1. u. 3. Pers. Sg. Prät. v.* **stoßen**

¹Stift [ʃtɪft] **der; ~[e]s, ~e a)** *(aus Metall)* pin; *(aus Holz)* peg; **b)** *(Blei~, Bunt~, Zeichen~)* pencil; *(Mal~)* crayon; *(Schreib~)* pen; **c)** *(ugs.: Lehrling)* apprentice

²Stift das; ~[e]s, ~e a) *(christl. Kirche: Institution)* foundation; **b)** *(österr.: Kloster)* monastery

stiften *tr. V.* **a)** found, establish ‹*monastery, hospital, prize, etc.*›; endow ‹*prize, professorship, scholarship*›; *(als Spende)* donate, give **(für** to); **b)** *(herbeiführen)* cause, create ‹*unrest, confusion, strife, etc.*›; bring about ‹*peace, order, etc.*›; arrange ‹*marriage*›

stiften|gehen *unr. itr. V.; mit sein (ugs.)* disappear; hop it *(sl.)*

Stifter der; ~s, ~: founder; *(Spender)* donor

Stiftung die; ~, ~en a) *(Rechtsspr.)* foundation; endowment; **b)** *(Anstalt)* foundation; **c)** *(Spende)* donation *(Gen.* by)

Stift·zahn der *(Zahnmed.)* post crown

Stil [ʃti:l] **der; ~[e]s, ~e** style; **in dem ~ ging es weiter** *(ugs.)* it went on in that vein

Stil·blüte die howler *(coll.)*

stilisieren *tr. V.* stylize

stilistisch 1. *Adj.* stylistic. **2.** *adv.* stylistically

still [ʃtɪl] **1.** *Adj.* **a)** *(ruhig, leise)* quiet; *(ganz ohne Geräusche)* silent; still; quiet, peaceful ‹*valley, area, etc.*›; **sei ~!** be quiet!; **im Saal wurde es ~:** the hall went quiet; **b)** *(reglos)* still; **~es [Mineral]wasser** still [mineral] water; **c)** *(ohne Aufregung, Hektik)* quiet ‹*day, life*›; quiet, calm ‹*manner*›; **d)** *(nicht gesprächig)* quiet; **e)** *(wortlos)* silent ‹*reproach, grief, etc.*›; **f)** *(heimlich)* secret; **~e Reserven** *(Wirtsch.)* secret *or* hidden reserves; *(ugs.)* [secret] savings; **g) der Stille Ozean** the Pacific [Ocean]. **2.** *adv.* **a)** *(ruhig, leise)* quietly; *(geräuschlos)* silently; **b)** *(zurückhaltend)* quietly; **c)** *(wortlos)* in silence

Stille die; ~ a) *(Ruhe)* quiet; *(Geräuschlosigkeit)* silence; stillness; **in der ~ der Nacht** in the still of the night; **b)** *(Regungslosigkeit) (des Meeres)* calm[ness]; *(der Luft)* stillness; **c) in aller ~ heiraten** have a quiet wedding; **die Beerdigung fand in aller ~ statt** it was a quiet funeral

Stilleben das still life

stillegen *tr. V.* close *or* shut down; close ‹*railway line*›; lay up ‹*ship, vehicle, fleet*›

Stillegung die; ~, ~en closure; shut-down; *(von Schiff, Fahrzeug, Flotte)* laying up; *(einer Eisenbahnstrecke)* closure

stillen 1. *tr. V.* **a) ein Kind ~:** breast-feed a baby; **ich muß das Baby jetzt ~:** I must feed the baby *or* give the baby a feed now; **b)** *(befriedigen)* satisfy ‹*hunger, desire, curiosity*›; quench ‹*thirst*›; still *(literary)* ‹*hunger, thirst, desire*›; **c)** *(eindämmen)* stop ‹*bleeding, tears, pain*›; stanch ‹*blood*›. **2.** *itr. V.* breast-feed

still-, Still-: ~|halten *unr. itr. V.* **a)** *(sich nicht bewegen)* keep *or* stay still; **b)** *(nicht reagieren)* keep quiet; **~schweigen das a)** *(Schweigen)* silence; **mit ~schweigen** in silence; **b)** *(Diskretion)* **~schweigen bewahren** maintain silence; keep silent; **~schweigend 1.** *Adj.; nicht präd.* **a)** *(wortlos)* silent; **b)** *(ohne Abmachung)* tacit ‹*assumption, agreement*›; **2.** *adv.* **a)** *(wortlos)* in silence; **b)** *(ohne Abmachung)* tacitly; **~|sitzen** *unr. itr. V.* sit still; **~stand der;** *o. Pl.* standstill; **die Entzündung/den Verkehr zum ~stand bringen** stop the inflammation/bring the traffic to a standstill; **die Blutung ist zum ~stand gekommen** the bleeding has stopped; **~|stehen** *unr. itr. V.* **a)** ‹*factory, machine*› be *or* stand idle; ‹*traffic*› be at a standstill; ‹*heart etc.*› stop; **b)** *(Milit.)* stand at *or* to attention; **~gestanden!** attention!

Still·zeit die lactation period

Stil·mittel das stylistic device

stil·voll 1. *Adj.* stylish. **2.** *adv.* stylishly

Stimm-: ~band das; *meist Pl.* vocal cord; **~bruch der: er ist im ~bruch** his voice is breaking

Stimme [ˈʃtɪmə] **die; ~, ~n a)** voice; **der ~ der Vernunft folgen** *(fig.)* listen to the voice of reason; **der ~ des Herzens/Gewissens folgen** *(fig. geh.)* follow [the dictates of] one's heart/conscience; **mit stockender ~:** in a faltering voice; **b)** *(Meinung)* voice; **die ~n in der Presse waren kritisch** press opinion was critical; **c)** *(bei Wahlen, auch Stimmrecht)* vote

stimmen 1. *itr. V.* **a)** *(zutreffen)* be right *or* correct; **stimmt es, daß ...?** is it true that ...?; **das kann unmöglich ~:** that can't possibly be right; **b)** *(in Ordnung sein)* ‹*bill, invoice, etc.*› be right *or* correct; **stimmt so** that's all right; keep the change; **hier stimmt etwas nicht** there's something wrong here; **bei ihm stimmt es** *od.* **etwas nicht** *(salopp)* there must be something wrong with him; **c)** *(seine Stimme geben)* vote; **mit Ja ~:** vote yes *or* in favour. **2.** *tr. V.* **a)** *(in eine Stimmung versetzen)* make; **das stimmt mich traurig** that makes me [feel] sad; **b)** *(Musik)* tune ‹*instrument*›; **eine Gitarre höher/tiefer ~:** raise/lower the pitch of a guitar

Stimmen·gewirr das babble of voices

Stimm·enthaltung die abstention

stimmhaft *(Sprachw.)* **1.** *Adj.* voiced. **2.** *adv.* **~ gesprochen werden** be voiced

stimm·los *(Sprachw.)* **1.** *Adj.* voiceless; unvoiced; **2.** *adv.* **~ ausgesprochen werden** not be voiced

Stimm·recht das right to vote

Stimmung die; ~, ~en a) mood; **in ~ sein** be in a good mood; **in ~ kommen** get in the mood; liven up; **jmdn. in ~ bringen** liven sb. up; **b)** *(Atmosphäre)* atmosphere; **c)** *(öffentliche Meinung)* opinion; **~ für/gegen jmdn./etw. machen** stir up [public] opinion in favour of/against sb./sth.

stimmungs-, Stimmungs-: ~kanone die *(ugs. scherzh.)* entertainer who is always the life and soul of the party; **~umschwung der** change of mood; **~voll 1.** *Adj.* atmospheric; **2.** *adv.* ‹*describe, light*› atmospherically; ‹*sing, recite*› with great feeling

Stimm·zettel der ballot-paper

stink-, Stink- *(salopp)* stinking *(sl.)* ‹*drunk, mood*›; terribly *(coll.)* ‹*bourgeois, posh*›

Stink·bombe die stink-bomb

stinken [ˈʃtɪŋkn̩] *unr. itr. V.* **a)** *(abwertend)* stink; pong *(coll.)*; **nach etw. ~:** stink *or* reek of sth.; **b)** *(ugs.: Schlechtes vermuten lassen)* **die Sache/es stinkt** it smells; it's fishy *(coll.)*; **c)** *(salopp: mißfallen)* **die Hausarbeit stinkt mir** I'm fed up to the back teeth with housework *(sl.)*; **mir stinkt's** I'm fed up to the back teeth *(sl.)*

stink-, Stink-: ~faul *Adj. (salopp abwertend)* bone idle *(coll.)*; **~langweilig** *(ugs.)* **1.** *Adj.* deadly boring; **2.** *adv.* in a deadly boring way; **~normal** *(salopp)* **1.** *Adj.* dead *(coll.) or* boringly ordinary; **2.** *adv.* in a dead ordinary way *(coll.)*; **~reich** *Adj. (salopp)* stinking rich *(sl.)*; **~tier das** skunk; **~wut die** *(salopp)* towering rage; **eine ~wut [auf jmdn.] haben** be livid *(Brit. coll.) or* furious [with sb.]

Stipendium [ʃtiˈpɛndi̯ʊm] **das; ~s, Stipendien** *(als Auszeichnung)* scholarship; *(als finanzielle Unterstützung)* grant

stirbst [ʃti:rbst], **stirbt** *2. u. 3. Pers. Sg. Präsens v.* **sterben**

Stirn [ˈʃtɪrn] **die; ~, ~en** forehead; brow; **jmdm./einer Sache die ~ bieten** *(fig.)* stand *or* face up to sb./sth.; **die ~ haben, etw. zu tun** *(fig.)* have the nerve *or* gall to do sth.

Stirn-: ~band das; *Pl.* **~bänder** headband; **~runzeln das; ~s** frown

stob [ʃto:p] *1. u. 3. Pers. Sg. Prät. v.* **stieben**

stöbern [ˈʃtø:bɐn] *itr. V. (ugs.)* rummage

stochern [ˈʃtɔxɐn] *itr. V.* poke; **mit dem Feuerhaken im Feuer ~:** poke the fire; **im Essen ~:** pick at one's food

[1]**Stock** [ʃtɔk] **der; ~[e]s, Stöcke** [ˈʃtœkə] **a)** *(Ast, Spazier~)* stick; *(Zeige~)* pointer; stick; *(Takt~)* baton; **steif wie ein ~:** as stiff as a poker; **am ~ gehen** walk with a stick; *(ugs.: erschöpft sein)* be whacked *(Brit. coll.) or* deadbeat; **b)** *(Ski~)* pole; stick; **c)** *(Pflanze) (Rosen~)* [rose-]bush;

(Reb~) vine; **d)** *(Eishockey, Hockey, Rollhockey)* stick

²Stọck der; ~[e]s, ~ *(Etage)* floor; storey; **das Haus hat vier ~:** the house is four storeys high; **im fünften ~:** on the fifth *(Brit.) or (Amer.)* sixth floor

stọck-: ~besọffen *Adj. (derb)* pissed as a newt/as newts *pred. (coarse);* blind drunk; **~blịnd** *Adj. (ugs.)* as blind as a bat *pred. (coll.);* totally blind; **~dụnkel** *Adj. (ugs.)* pitch-dark

Stọ̈ckel·schuh der high- *or* stiletto-heeled shoe; **~e** high heels; high- *or* stiletto-heeled shoes

stọcken *itr. V.* **a) ihm stockte das Herz/der Atem** his heart missed *or* skipped a beat/he caught his breath; **b)** *(unterbrochen sein)* ⟨*traffic*⟩ be held up, come to a halt; ⟨*conversation, production*⟩ stop; ⟨*talks negotiations, etc.*⟩ grind to a halt; ⟨*business*⟩ slacken *or* drop off; ⟨*journey*⟩ be interrupted; **die Antwort kam ~d** he/she gave a hesitant reply; **c)** *(innehalten)* falter

stọck-, Stọck-: ~fịnster *Adj. (ugs.) s.* **stockdunkel; ~fisch der a)** stockfish; **b)** *(ugs. abwertend: Mensch)* boring *or* dull old stick; **~nüchtern** *Adj. (ugs.)* stone-cold sober; **~sauer** *Adj.; nicht attr. (salopp)* pissed off *(Brit. sl.)* **(auf +** *Akk.* with); **~schirm der** walking-length umbrella; **~steif** *(ugs.)* **1.** *Adj.* extremely stiff ⟨*gait*⟩; **2.** *adv.* extremely stiffly; as stiff as a poker; **~taub** *Adj. (ugs.)* stone-deaf; as deaf as a post

Stọckung die; ~, ~en hold-up (*Gen.* in)

Stọckwerk das *s.* ²Stock

Stoff [ʃtɔf] **der; ~[e]s, ~e a)** *(für Textilien)* material; fabric; **b)** *(Materie)* substance; **c)** *o. Pl. (Philos.)* matter; **d)** *(Thema)* subject[-matter]; **~ für einen Roman sammeln** collect material for a novel; **e)** *(Gesprächsthema)* topic; **f)** *o. Pl. (salopp: Alkohol)* booze *(coll.);* **g)** *o. Pl. (salopp: Rauschgift)* stuff *(sl.);* dope *(sl.)*

Stoffel [ˈʃtɔfl̩] **der; ~s, ~** *(ugs. abwertend)* boor; churl

Stọff-: ~wechsel der metabolism; **~wechsel·krankheit die** metabolic disease

stöhnen [ˈʃtøːnən] *itr. V.* moan; *(vor Schmerz)* groan

stoisch 1. *Adj. (Philos.)* Stoic; *(fig.)* stoic. **2.** *adv.* stoically

Stola [ˈʃtoːla] **die; ~, Stolen** shawl; *(Pelz~)* stole

Stollen [ˈʃtɔlən] **der; ~s, ~ a)** *(Kuchen)* Stollen; **b)** *(unterirdischer Gang)* gallery; tunnel; **c)** *(Bergbau)* gallery; **d)** *(bei Sportschuhen)* stud

stolpern [ˈʃtɔlpɐn] *itr. V.; mit sein* **a)** stumble; trip; **ins Stolpern kommen** stumble; trip; *(fig.)* lose one's thread; **über jmdn. ~** *(fig. ugs.)* bump *or* run into sb.; **ich bin über dieses Wort gestolpert** *(fig.)* I was puzzled by that word; **b)** *(fig.: straucheln)* come to grief, *(coll.)* come unstuck **(über +** *Akk.* over)

stolz [ʃtɔlts] **1.** *Adj.* **a)** proud **(auf +** *Akk.* of); **b)** *(überheblich)* proud[-hearted]; **c)** *(imposant)* proud ⟨*building, castle, ship, etc.*⟩; **d)** *(ugs.: beträchtlich)* steep *(coll.),* hefty *(coll.)* ⟨*price*⟩; tidy *(coll.)* ⟨*sum*⟩; **~ wie ein Spanier** as proud as can be. **2.** *adv.* proudly

Stọlz der; ~es pride **(auf +** *Akk.* in); **die Rosen sind sein ganzer ~:** his roses are his pride and joy

stolzieren *itr. V.; mit sein* strut

stop [stɔp] *Interj.* stop; *(Verkehrsw.)* halt

stopfen [ˈʃtɔpfn̩] **1.** *tr. V.* **a)** darn ⟨*socks, coat, etc., hole*⟩; **b)** *(hineintun)* stuff; **jmdm./sich etwas in den Mund ~:** stuff sth. into sb.'s/one's mouth; **c)** *(füllen)* stuff ⟨*cushion, quilt, etc.*⟩; fill ⟨*pipe*⟩; **d)** *(ausfüllen, verschließen)* plug, stop [up] ⟨*hole, leak*⟩; **jmdm. das Maul ~** *(salopp)* shut sb. up. **2.** *itr. V.* **a)** *(den Stuhlgang hemmen)* cause constipation; **b)** *(ugs.: sehr sättigen)* be very filling

Stọpf-: ~garn das darning-cotton *or* -thread; **~nadel die** darning-needle

Stọpp der; ~s, ~s a) *(das Anhalten)* stop; **b)** *(Einstellung)* freeze (*Gen.* on)

Stọpp·ball der *(Badminton, [Tisch]tennis)* drop-shot

Stoppel [ˈʃtɔpl̩] **die; ~, ~n;** *meist Pl. (auch Bart~)* stubble *no pl.*

Stọppel-: ~bart der *(ugs.)* stubble; **~feld das** stubble-field

stọppelig *Adj.* stubbly

stọppen 1. *tr. V.* **a)** stop; **den Ball ~** *(Fußball)* trap *or* stop the ball; **b)** time ⟨*athlete, run*⟩. **2.** *itr. V.* stop; **der Angriff stoppte** *(fig.)* the attack got no further *or* fizzled out

Stọpper der; ~s, ~ *(Fußball)* centre-half; stopper

Stọpp·licht das; *Pl.* **~er** stop-light

stọpplig *Adj. s.* **stoppelig**

Stọpp-: ~schild das stop sign; **~uhr die** stop-watch

Stöpsel [ˈʃtœpsl̩] **der; ~s, ~ a)** plug; *(einer Karaffe usw.)* stopper; **b)** *(Elektrot.)* [jack-]plug

Stör [ʃtøːɐ̯] **der; ~s, ~e** sturgeon

Storch [ʃtɔrç] **der; ~[e]s, Störche** [ˈʃtœrçə] stork; **wie ein ~ im Salat gehen** walk clumsily and stiff-leggedly

Store [ʃtoːɐ̯] **der; ~s, ~s** net curtain

stören 1. *tr. V.* **a)** *(behindern)* disturb; disrupt ⟨*court proceedings, lecture, church service, etc.*⟩; **bitte lassen Sie sich nicht ~:** please don't let me disturb you; **b)** *(stark beeinträchtigen)* disturb ⟨*relation, security, law and order, peaceful atmosphere, etc.*⟩; interfere with ⟨*transmitter, reception*⟩; *(absichtlich)* jam ⟨*transmitter*⟩; **hier ist der Empfang oft gestört** there is often interference [with reception] here; **c)** *(mißfallen)* bother; **das stört mich nicht** I don't mind; that doesn't bother me; **das stört mich an ihr** that's what I don't like about her. **2.** *itr. V.* **a) darf ich reinkommen, oder störe ich?** may I come in, or am I disturbing you?; **entschuldigen Sie bitte, daß** *od.* **wenn ich störe** I'm sorry to bother you; **bitte nicht ~!** [please] do not disturb; **b)** *(als Mangel empfunden werden)* spoil the effect; **c)** *(Unruhe stiften)* make *or* cause trouble. **3.** *refl. V.* **sich an jmdm./etw. ~:** take exception to sb./sth.

Störenfried [ˈʃtøːrənfriːt] **der; ~[e]s, ~e Störer der; ~s, ~** *(abwertend)* trouble-maker

stornieren *tr. V.* **a)** *(Finanzw., Kaufmannsspr.)* reverse ⟨*wrong entry*⟩; **b)** *(Kaufmannsspr.)* cancel ⟨*order, contract*⟩

Storno [ˈʃtɔrno] **der** *od.* **das; ~s, Stọrni** *(Finanzw., Kaufmannsspr.)* reversal

störrisch [ˈʃtœrɪʃ] **1.** *Adj.* stubborn; obstinate; refractory ⟨*child, horse*⟩; unmanageable ⟨*hair*⟩. **2.** *adv.* stubbornly; obstinately

Stör·sender der jammer

Störung die; ~, ~en a) disturbance; *(einer Gerichtsverhandlung, Vorlesung, eines Gottesdienstes)* disruption; **b)** *(Beeinträchtigung)* disturbance; disruption; **eine technische ~:** a technical fault; **atmosphärische ~** *(Met.)* atmospheric disturbance; *(Rundf.)* atmospherics *pl.*

Story [ˈstɔrɪ] **die; ~, ~s** *od.* **Stọries** story

Stoß [ʃtoːs] **der; ~es, Stöße** [ˈʃtøːsə] **a)** *(mit der Faust)* punch; *(mit dem Fuß)* kick; *(mit dem*

Kopf, den Hörnern) butt; *(mit dem Ellbogen)* dig; **jmdm. einen kleinen ~ mit dem Ellenbogen geben** nudge sb.; give sb. a nudge; **b)** *(mit einer Waffe) (Stich)* thrust; *(Schlag)* blow; **c)** *(beim Schwimmen, Rudern)* stroke; **d)** *(Stapel)* pile; stack; **e)** *(beim Kugelstoßen)* put; throw; **f)** *(stoßartige Bewegung)* thrust; *(Atem~)* gasp; **g)** *(Erd~)* tremor

Stoß·dämpfer der *(Kfz-W.)* shock absorber

Stößel ['ʃtø:sl] **der; ~s, ~:** pestle

stoß·empfindlich *Adj.* sensitive to shock *postpos.*

stoßen 1. *unr. tr. V.* **a)** *auch itr. (mit der Faust)* punch; *(mit dem Fuß)* kick; *(mit dem Kopf, den Hörnern)* butt; *(mit dem Ellbogen)* dig; **jmdn.** *od.* **jmdm. in die Seite ~:** dig sb. in the ribs; *(leicht)* nudge sb. in the ribs; **b)** *(hineintreiben)* plunge, thrust ‹*dagger, knife*›*;* push ‹*stick, pole*›*;* **c)** *(stoßend hervorbringen)* knock, bang ‹*hole*›*;* **d)** *(schleudern)* push; **die Kugel ~:** *(beim Kugelstoßen)* put the shot; *(beim Billard)* strike the ball; **e)** *(zer~)* pound ‹*sugar, cinnamon, pepper*›. **2.** *unr. itr. V.* **a)** *mit sein (auftreffen)* bump (**gegen** into); **b)** *mit sein (begegnen)* **auf jmdn. ~:** bump *or* run into sb.; **c)** *mit sein (entdecken)* **auf etw.** *(Akk.)* **~:** come upon *or* across sth.; **auf Erdöl ~:** strike oil; **auf Ablehnung ~** *(fig.)* meet with disapproval; **d)** *mit sein* **zu jmdm. ~** *(jmdn. treffen)* meet up with sb.; *(sich jmdm. anschließen)* join sb.; **e)** *mit sein (zuführen)* **auf etw.** *(Akk.)* **~:** ‹*path, road*› lead [in]to sth.; **f)** *(grenzen)* **an etw.** *(Akk.)* **~** ‹*room, property, etc.*› be [right] next to sth. **3.** *unr. refl. V.* bump *or* knock oneself; **ich habe mich am Kopf gestoßen** I bumped *or* banged my head; **sich** *(Dat.)* **den Kopf blutig ~:** bang one's head and cut it; **sich an etw.** *(Dat.)* **~** *(fig.)* object to *or* take exception to sth.

Stoß-: ~gebet das quick prayer; **~seufzer der** heartfelt groan; **~stange die** bumper

stößt [ʃtø:st] *3. Pers. Sg. Präsens v.* **stoßen**

stoß-, Stoß-: ~verkehr der; *o. Pl.* rush-hour traffic; **~waffe die** thrust weapon; **~weise** *Adv.* **a)** *(ruckartig)* spasmodically; ‹*breathe*› spasmodically, jerkily; **b)** *(in Stapeln)* by the pile; in piles; **~zahn der** tusk

stottern ['ʃtɔtɐn] **1.** *itr. V.* stutter; stammer; **sie stottert stark** she has a strong *or* bad stutter *or* stammer; **ins Stottern kommen** *od.* **geraten** start stuttering *or* stammering. **2.** *tr. V.* stutter [out]; stammer [out]

Str. *Abk.* **Straße** St./Rd.

straf-, Straf-: ~anstalt die penal institution; prison; **~arbeit die** imposition *(Brit.);* **~bank die;** *Pl.* **~bänke** *(Eishockey, Handball)* penalty bench; **~bar** *Adj.* punishable; **das ist ~bar** that is a punishable offence; **sich ~bar machen** make oneself liable to prosecution

Strafe ['ʃtra:fə] **die; ~, ~n** punishment; *(Rechtsspr.)* penalty; *(Freiheits~)* sentence; *(Geld~)* fine; **sie empfand die Arbeit als ~:** she found the work a real drag *or (coll.)* bind; **etw. unter ~ stellen** make sth. punishable; **zur ~:** as a punishment

strafen *tr. V.* punish; **jmdn. ~d ansehen** give sb. a reproachful look; **jmdn. mit Verachtung ~:** treat sb. with contempt as a punishment; **mit ihm sind wir gestraft** he is a real pain; *s. auch* **Lüge**

Straf-: ~entlassene der/die; *adj. Dekl.* ex-convict; ex-prisoner; **~erlaß der** *(Rechtsw.)* remission [of a/the sentence]

straff [ʃtraf] **1.** *Adj.* **a)** *(fest, gespannt)* tight, taut ‹*rope, lines, etc.*›*;* firm ‹*breasts, skin*›*;* erect ‹*posture, figure*›*;* tight ‹*rein[s]*›*;* **b)** *(energisch)* tight ‹*organization, planning, etc.*›; strict ‹*discipline, leadership, etc.*›. **2.** *adv.* **a)** *(fest, gespannt)* **[zu] ~ sitzen** ‹*clothes*› be [too] tight; **~ zurückgekämmtes Haar** hair combed back tightly; **b)** *(energisch)* tightly, strictly ‹*organized, planned, etc.*›

straf·fällig *Adj.* **~ werden** commit a criminal offence; **die Zahl der Straffälligen** the number of offenders

straffen 1. *tr. V.* **a)** *(spannen)* tighten; **diese Creme strafft die Haut** this cream firms the skin; **b)** *(raffen)* tighten up ‹*text, procedure, organization, etc.*›. **2.** *refl. V.* ‹*person*› straighten oneself, draw oneself up; ‹*rope etc.*› tighten; ‹*body, back*› stiffen; ‹*posture, bearing*› straighten

straf-, Straf-: ~frei *Adj.* **~frei ausgehen** go unpunished; get off [scot-]free *(coll.);* **~freiheit die;** *o. Pl.* exemption from punishment; **~gefangene der/die** prisoner; **~gericht das** *(fig.)* judgement; **ein ~gericht des Himmels** divine judgement; **~gesetz das** criminal *or* penal law; **~gesetz·buch das** criminal *or* penal code; **~kolonie die** penal colony

sträflich ['ʃtrɛ:flɪç] **1.** *Adj.* criminal. **2.** *adv.* criminally

Sträfling ['ʃtrɛ:flɪŋ] **der; ~s, ~e** prisoner

straf-, Straf-: ~mandat das [parking, speeding, *etc.*] ticket; **~maß das** sentence; **~minute die a)** *(bes. Eishockey, Handball)* minute of penalty time; **b)** *(Rennsport, Springreiten, Biathlon, usw.)* penalty minute; **~porto das** surcharge; **~predigt die** *(ugs.)* lecture; **~punkt der** *(Sport)* penalty point; **~raum der** *(bes. Fußball)* penalty area; **~register das** criminal records *pl.;* **~richter der** *(Rechtsw.)* criminal judge; **~stoß der** *(Fußball) s.* **Elfmeter; ~tat die** criminal offence; **~täter der** offender; **~verfahren das** criminal proceedings *pl.;* **~versetzen** *tr. V. nur im Inf. u. Part. gebr.* transfer for disciplinary reasons; **~würdig** *Adj. (Rechtsw.)* punishable; **~zettel der** *(ugs.) s.* **~mandat**

Strahl [ʃtra:l] **der; ~[e]s, ~en a)** *(Licht, fig.)* ray; *(von Scheinwerfern, Taschenlampen)* beam; **b)** *(Flüssigkeit)* jet; **c)** *(Math., Phys.)* ray

strahlen *itr. V.* **a)** shine; **bei ~dem Wetter** in glorious sunny weather; **~d weiß** sparkling white; **b)** *(glänzen)* sparkle; **c)** *(lächeln)* beam (**vor +** *Dat.* with); **er strahlte über das ganze Gesicht** he was beaming all over his face; **d)** *(Physik)* radiate; emit rays

strahlen-, Strahlen-; ~belastung die radioactive contamination; **~förmig 1.** *Adj.* radial; **2.** *adv.* radially; **~unfall der** radiation accident

Strahl·triebwerk das jet engine

Strahlung die; ~, ~en radiation

Strähne ['ʃtrɛ:nə] **die; ~, ~n a)** *(Haare)* strand; **eine graue ~:** a grey streak; **b)** *(fig.: Zeitspanne)* streak

strähnig 1. *Adj.* straggly ‹*hair*›. **2.** *adv.* in strands

stramm [ʃtram] **1.** *Adj.* **a)** *(straff)* tight, taut ‹*rope, line, etc.*›; tight ‹*clothes*›; **b)** *(kräftig)* strapping ‹*girl, boy*›*;* sturdy ‹*legs, body*›*;* **c)** *(gerade)* upright, erect ‹*posture, etc.*›; **d)** *(energisch)* strict ‹*discipline*›*;* strict, staunch ‹*Marxist, Catholic, etc.*›*;* brisk ‹*step*›. **2.** *adv.* **a)** *(straff)* tightly; **die Hose saß ziemlich ~:** the trousers were rather tight; **b)** *(kräftig)* sturdily ⟨*built*⟩; **c)** *(energisch)* ⟨*bring up*⟩ strictly; strictly, staunchly ⟨*Marxist, Catholic, etc.*⟩; ⟨*hold out*⟩ resolutely; **d)** *(ugs.: zügig)* ⟨*work*⟩ hard; ⟨*walk, march*⟩ briskly; ⟨*drive*⟩ fast, hard

stramm|stehen *unr. itr. V.* stand to *or* at attention

Strampel·höschen **das**, **Strampel·hose** **die** rompers *pl.*; romper suit; playsuit

strampeln ['ʃtramp!n] *itr. V.* **a)** ⟨*baby*⟩ kick [his/her feet] [and wave his/her arms about]; **b)** *mit sein (ugs.: mit dem Rad)* pedal; **c)** *(ugs.: sich sehr anstrengen)* sweat; struggle

Strand [ʃtrant] **der; ~[e]s, Strände** ['ʃtrɛndə] beach; *(geh. veralt.: Seeufer)* shore; strand; **am ~:** on the beach

Strand·bad **das** bathing beach *(on river, lake)*

stranden *itr. V.; mit sein* **a)** *(festsitzen)* ⟨*ship*⟩ run aground; *(fig.)* be stranded; **b)** *(geh.: scheitern)* fail

Strand-: **~gut das;** *o. Pl.* flotsam and jetsam; **~hotel das** beach hotel; **~kleid das** beach dress; **~korb der** basket chair; **~promenade die** promenade

Strang [ʃtraŋ] **der; ~[e]s, Stränge** ['ʃtrɛŋə] **a)** *(Seil)* rope; **jmdn. zum Tod durch den ~ verurteilen** *(geh.)* sentence sb. to be hanged; **b)** *(von Wolle, Garn usw.)* hank; skein; **c)** *(Nerven~, Muskel~, Sehnen~)* cord; **d)** *(Leine)* trace; **über die Stränge schlagen** *(ugs.)* kick over the traces; *s. auch* **ziehen 2 a**

Strapaze [ʃtra'pa:tsə] **die; ~, ~n** strain *no pl.*

strapazieren **1.** *tr. V.* be a strain on ⟨*person, nerves*⟩; **die tägliche Rasur strapaziert die Haut** shaving daily is hard on the skin; **die Reise würde ihn zu sehr ~:** the journey would be too much [of a strain] for him; **jmds. Geduld ~** *(fig.)* tax sb.'s patience. **2.** *refl. V.* strain *or* tax oneself

strapazier·fähig *Adj.* hard-wearing ⟨*clothes, shoes*⟩; hard-wearing, durable ⟨*material*⟩

strapaziös [ʃtrapa'tsiø:s] *Adj.* wearing

Straße ['ʃtra:sə] **die; ~, ~n** **a)** *(in Ortschaften)* street; road; *(außerhalb)* road; **auf offener ~:** in [the middle of] the street; **Verkauf über die ~:** take-away sales *pl.*; *(von alkoholischen Getränken)* off-licence sales *pl.*; **mit Prostituierten kann man hier die ~n pflastern** *(ugs.)* the place is full of prostitutes *(coll.)*; **jmdn. auf die ~ setzen** *od.* **werfen** *(ugs.) (aus einer Stellung)* sack sb. *(coll.)*; give sb. the sack *(coll.)*; *(aus einer Wohnung)* turn sb. out on to the street; **auf der ~ liegen** *od.* **sitzen** *od.* **stehen** *(ugs.) (arbeitslos sein)* be out of work; *(ohne Wohnung sein)* be on the streets; **auf die ~ gehen** *(ugs.) (demonstrieren)* take to the streets; *(der Prostitution nachgehen)* go on *or* walk the streets; **b)** *(Meerenge)* strait[s *pl.*]

Straßen·bahn **die** tram *(Brit.)*; streetcar *(Amer.)*

Straßen·bahn-: **~halte·stelle die** tram stop *(Brit.)*; **~linie die** tram route *(Brit.)*; **~schaffner der** tram conductor *(Brit.)*

Straßen-: **~bau der;** *o. Pl.* road building *no art.*; road construction *no art.*; **~ecke die** street corner; **~glätte die** slippery road surface; **~graben der** ditch [at the side of the road]; **~händler der** street trader; **~kampf der a)** street fight; street battle; **b)** *o. Pl. (Taktik, Strategie)* street-fighting; **~karte die** road-map; **~kreuzung die** crossroads *sing.*; **~laterne die** street lamp; **~musikant der** street musician; busker; **~rennen das** *(Rennsport)* road race; **~sammlung die** street collection; **~schild das** street-name sign; **~schlacht die** street battle; **~schuh der** walking-shoe; **~seite die** side of the street/road; *(eines Gebäudes)* street side; **~sperre die** roadblock; **~überführung die** *(für Fußgänger)* footbridge; *(für Fahrzeuge)* road bridge; **~unterführung die** *(für Fußgänger)* subway; *(für Fahrzeuge)* underpass; **~verkäufer der** street vendor; **~verkehr der** traffic

Strategie [ʃtrate'gi:] **die; ~, ~n** strategy

strategisch **1.** *Adj.* strategic. **2.** *adv.* strategically

Strato·sphäre [ʃtrato-] **die** stratosphere

sträuben ['ʃtrɔybn̩] **1.** *tr. V.* ruffle [up] ⟨*feathers*⟩; bristle ⟨*fur, hair*⟩. **2.** *refl. V.* ⟨*hair, fur*⟩ bristle, stand on end; ⟨*feathers*⟩ become ruffled; **b)** *(sich widersetzen)* resist; **sich ~, etw. zu tun** resist doing sth.; **sie hat sich mit Händen und Füßen gegen die Versetzung gesträubt** she resisted the transfer with all her might

Strauch [ʃtraux] **der; ~[e]s, Sträucher** ['ʃtrɔyçɐ] shrub

straucheln ['ʃtraux!n] *itr. V.; mit sein (geh.)* **a)** *(stolpern)* stumble; **b)** *(scheitern)* fail; **c)** *(straffällig werden)* go astray

¹Strauß [ʃtraus] **der; ~es, Sträuße** ['ʃtrɔysə] bunch of flowers; *(bes. als Geschenk)* bouquet [of flowers]; *(von kleinen Blumen)* posy

²Strauß **der; ~es, ~e** *(Vogel)* ostrich

Strebe ['ʃtre:bə] **die; ~, ~n** brace; strut

streben *itr. V.* **a)** *mit sein (hinwollen)* make one's way briskly; **er strebte zur Tür** he made briskly for the door; **die Partei strebt an die Macht** the party is reaching out for power; **b)** *(trachten)* strive (**nach** for); **danach ~, etw. zu tun** strive to do sth.

Strebe·pfeiler **der** buttress

Streber **der; ~s, ~, Streberin** **die; ~, ~nen** *(abwertend)* over-ambitious *or* pushing *or (coll.)* pushy person; *(in der Schule)* swot *(Brit. sl.)*; grind *(Amer. sl.)*

strebsam *Adj.* ambitious and industrious

Strebsamkeit **die; ~:** ambition and industriousness

Strecke ['ʃtrɛkə] **die; ~, ~n** **a)** *(Weg~)* distance; **auf der ~ bleiben** *(ugs.)* fall by the wayside; **b)** *(Abschnitt, Route)* route; *(Eisenbahn~)* line; **der Zug hielt auf freier** *od.* **offener ~:** the train stopped between stations; **c)** *(Sport)* distance; **die Läufer gehen auf die ~:** the runners are setting off; **d)** *(Geom.)* line segment; **e)** *(Jägerspr.)* **ein Tier zur ~ bringen** bag *or* kill an animal; **jmdn. zur ~ bringen** *(fig.)* hunt sb. down

strecken **1.** *tr. V.* **a)** *(gerade machen)* stretch ⟨*arms, legs*⟩; **b)** *(dehnen)* stretch [out] ⟨*arms, legs, etc.*⟩; **c)** *(lehnen)* stick *(coll.)*; **den Kopf aus dem Fenster ~:** stick one's head out of the window *(coll.)*; **d)** *(größer, länger, breiter machen)* stretch; hammer/roll out ⟨*metal*⟩; **e)** *(verdünnen)* thin down; **f)** *(rationieren)* eke out ⟨*provisions, fuel, etc.*⟩. **2.** *refl. V.* stretch out

Strecken·netz **das** route network; *(Eisenbahnw.)* rail network

strecken·weise *Adv.* in places; *(fig.: zeitweise)* at times

Streich [ʃtraiç] **der; ~[e]s, ~e** **a)** *(geh.: Hieb)* blow; **auf einen ~** *(veralt.)* at one blow; *(fig.)* at one fell swoop; at one go; **b)** *(Schabernack)* trick; prank; **jmdm. einen ~ spielen** play a trick on sb.; **mein Gedächtnis hat mir wieder einen ~ gespielt** my memory has been playing tricks on me again

streicheln ['ʃtraiçln] *tr. (auch itr.) V.* stroke; *(liebkosen)* stroke; caress

streichen 1. *unr. tr. V.* **a)** stroke; **b)** *(an~)* paint; **„frisch gestrichen"** 'wet paint'; **c)** *(wegstreifen)* sweep ⟨*crumbs etc.*⟩; **sich** *(Dat.)* **das Haar aus der Stirn ~:** push *or* smooth the hair back from one's forehead; **d)** *(drücken)* **Kitt in die Fugen ~:** press putty into the joints; **Tomaten durch ein Sieb ~:** rub *or* press tomatoes through a sieve; **e)** *(auftragen)* spread ⟨*butter, jam, ointment, etc.*⟩; **f)** *(be~)* **ein Brötchen [mit Butter]/mit Honig ~:** butter a roll/spread honey on a roll; **g)** *(aus~, tilgen)* delete; cross out; cancel ⟨*train, flight*⟩; **jmdn. von der Liste ~:** cross sb. off the list; **Nichtzutreffendes bitte ~!** please delete as appropriate *or* applicable; **h)** *(Rudern)* **die Riemen ~:** back water. 2. *unr. itr. V.* **a)** stroke; **jmdm. durch die Haare/über den Kopf ~:** run one's fingers through sb.'s hair/ stroke sb.'s head; **b)** *(an~)* paint; **c)** *mit sein (umhergehen)* wander

Streicher der; ~s, ~ *(Musik)* string-player; **die ~:** the strings

Streich·holz das match; *(als Spielzeug)* matchstick

Streichholz·schachtel die matchbox

Streich-: ~instrument das string[ed] instrument; **~käse der** cheese spread; **~orchester das** string orchestra; **~quartett das** string quartet

Streichung die; ~, ~en a) *(Tilgung)* deletion; *(Kürzung)* cutting *no indef. art.*; **b)** *(gestrichene Stelle)* deletion; *(Kürzung)* cut

Streife die; ~, ~n patrol; **auf ~ gehen/sein** go/be on patrol

streifen 1. *tr. V.* **a)** touch; brush [against]; ⟨*shot*⟩ graze; **jmdn. am Arm/an der Schulter ~:** touch sb. on the arm *or* brush against sb.'s arm/touch sb. on the shoulder; **mit dem Auto eine Mauer ~:** scrape a wall with the car; **jmdn. mit einem Blick ~** *(fig.)* glance fleetingly at sb.; **b)** *(fig.)* touch [up]on ⟨*problem, subject, etc.*⟩; **c) den Ring auf den/vom Finger ~:** slip the ring on/off one's finger; **die Ärmel nach oben ~:** pull/push up one's sleeves; **die Butter vom Messer ~:** wipe the butter off the knife; **sich** *(Dat.)* **die Kapuze/den Pullover über den Kopf ~:** pull the hood/slip the pullover over one's head. **2.** *itr. V. mit sein* **durch die Wälder ~:** roam the forests

Streifen der; ~s, ~ a) *(Linie)* stripe; *(auf der Fahrbahn)* line; **ein heller ~ am Horizont** a streak of light on the horizon; **b)** *(Stück, Abschnitt)* strip; *(Speck~)* rasher; **c)** *(ugs.: Film)* film

Streifen-: ~dienst der patrol duty; **~wagen der** patrol car

streifig *Adj.* streaky

Streif-: ~licht das; *Pl.* **~er** streak of light; **ein ~licht auf etw.** *(Akk.)* **werfen** *(fig.)* highlight sth.; **~schuß der** grazing shot; *(Wunde)* graze; **~zug der** expedition; *(fig.)* expedition; journey; *(eines Tieres)* prowl

Streik [ʃtraik] **der; ~[e]s, ~s** strike; **in den ~ treten** come out *or* go on strike; **mit ~ drohen** threaten to strike; threaten strike action; *s. auch* **wild 1 b**

Streik·brecher der strikebreaker; blackleg *(derog.)*; scab *(derog.)*

streiken *itr. V.* **a)** strike; be on strike; *(in den Streik treten)* come out *or* go on strike; strike; **b)** *(ugs.: nicht mitmachen)* go on strike; **c)** *(ugs.: nicht funktionieren)* pack up *(coll.)*; **der Kühlschrank streikt** the fridge has packed up *(coll.)*

Streikende der/die; *adj. Dekl.* striker

Streik-: ~posten der picket; **~recht das** right to strike

Streit [ʃtrait] **der; ~[e]s ;** *(Zank)* squabble; quarrel; *(Auseinandersetzung)* dispute; argument; **~ anfangen** start a quarrel *or* an argument; **mit jmdm. ~ bekommen** get into an argument *or* a quarrel with sb.

Streit·axt die battleaxe

streiten *unr. itr., refl. V.* quarrel; argue; *(sich zanken)* squabble; quarrel; *(sich auseinandersetzen)* argue; have an argument; **die Erben stritten [sich] um den Nachlaß** the heirs argued *or* fought over *or* disputed the estate; **darüber läßt sich ~:** one can argue about that; that's a debatable point

Streit-: ~frage die disputed question *or* issue; **~gespräch das** debate; disputation

streitig *Adj.* disputed ⟨*question, issue*⟩; **jmdm. jmdn./etw. ~ machen** dispute sb.'s right to sb./sth.

Streitigkeit die; ~, ~en *meist Pl.* **a)** quarrel; argument; **b)** *(Streitfall)* dispute

streit-, Streit-: ~kräfte *Pl.* armed forces; **~macht die;** *o. Pl.* *(veralt.)* forces *pl.*; **~süchtig** *Adj.* quarrelsome

streng [ʃtrɛŋ] **1.** *Adj.* **a)** *(hart)* strict ⟨*teacher, parents, upbringing, principle*⟩; severe ⟨*punishment*⟩; stringent, strict ⟨*rule, regulation, etc.*⟩; stringent ⟨*measure*⟩; rigorous ⟨*examination, check, test, etc.*⟩; stern ⟨*reprimand, look*⟩; **b)** *nicht präd. (strikt)* strict ⟨*order, punctuality, diet, instruction, Catholic*⟩; absolute ⟨*discretion*⟩; complete ⟨*rest*⟩; **c)** *nicht präd. (schnörkellos)* austere, severe ⟨*cut, collar, style, etc.*⟩; severe ⟨*hairstyle*⟩; **d)** *(herb)* severe ⟨*face, features, etc.*⟩; **e)** *(durchdringend)* pungent, sharp ⟨*taste, smell*⟩; **f)** *(rauh)* severe ⟨*winter*⟩; sharp, severe ⟨*frost*⟩. **2.** *adv.* **a)** *(hart)* ⟨*mark, judge, etc.*⟩ strictly, severely; ⟨*punish*⟩ severely; ⟨*look, reprimand*⟩ sternly; **~ durchgreifen** take rigorous action; **b)** *(strikt)* strictly; **~ verboten** strictly prohibited; **c)** *(schnörkellos)* **ein ~ geschnittenes Kostüm** a severe suit; **d)** *(durchdringend)* ⟨*smell*⟩ strongly

Strenge die; ~ a) *s.* **streng a:** strictness; severity; stringency; rigour; sternness; **b)** *(Striktheit)* strictness; **c)** *(von [Gesichts]zügen)* severity; **d)** *(von Geruch, Geschmack)* pungency; sharpness; **e)** *s.* **streng f:** severity; sharpness; **f)** *(Schnörkellosigkeit)* austerity; severity

streng-: ~genommen *adv.* strictly speaking; **~gläubig** *Adj.* strict

Streß [ʃtrɛs] **der; Stresses, Stresse** stress; **im ~ sein** be under stress

stressen *(ugs.) tr. V.* **jmdn. ~:** put sb. under stress; **vollkommen gestreßt sein** be under an enormous amount of stress

stressig *Adj. (ugs.)* stressful

Streu [ʃtrɔy] **die; ~, ~en** straw

streuen *tr. V.* **a)** spread ⟨*manure, sand, grit*⟩; sprinkle ⟨*salt, herbs, etc.*⟩; strew, scatter ⟨*flowers*⟩; *(fig.)* spread ⟨*rumour*⟩; **weit gestreut** *(fig.)* scattered *or* spread over a wide area; **b)** *(auch itr.) (mit Streugut)* **die Straßen [mit Sand/Salz] ~:** grit/salt the roads; put grit/salt down on the roads

streunen *itr. V.; meist mit sein (oft abwertend)* wander *or* roam about *or* around; **~de Hunde** stray dogs; **durch die Straßen ~:** roam *or* wander the streets

Streusel der *od.* **das; ~s, ~** streusel

Streusel·kuchen der streusel cake

strich [ʃtrɪç] *1. u. 3. Pers. Sg. Prät. v.* **streichen**

Strich der; ~[e]s, ~e a) *(Linie)*

line; *(Gedanken~)* dash; *(Schräg~)* diagonal; slash; *(Binde~, Trennungs~)* hyphen; *(Markierung)* mark; **keinen ~ tun** *od.* **machen** *od.* **arbeiten** not do a stroke *or* a thing; **jmdm. einen ~ durch die Rechnung/durch etw.** *(Akk.)* **machen** *(ugs.)* mess up *or* wreck sb.'s plans/mess up sb.'s plans for sth.; **unter dem ~:** at the end of the day; all things considered; **unter dem ~ sein** *(ugs.)* not be up to scratch; be below par; **b)** *o. Pl.* **der ~** *(salopp) (Prostitution)* [street] prostitution; street-walking; *(Gegend)* the red-light district; **auf den ~ gehen** walk the streets; **c)** *(streichende Bewegung)* stroke; **d)** *o. Pl. (Pinselführung)* strokes *pl.;* **e)** *o. Pl. (Bogen~)* bowing *no indef. art.;* **f)** *o. Pl. (Haar~, Fell~)* lie; *(eines Teppichs)* pile; *(von Samt o. ä.)* nap; **gegen den/mit dem ~ bürsten** brush ‹*hair, fur*› the wrong/right way; **jmdm. gegen den ~ gehen** *(ugs.)* go against the grain [with sb.]; **nach ~ und Faden** *(ugs.)* good and proper *(coll.);* well and truly

stricheln ['ʃtrɪçl̩n] *tr. V.* **a)** *(zeichnen)* sketch in [with short lines]; **eine gestrichelte Linie** a broken line; **b)** *(schraffieren)* hatch

strich-, Strich-: ~junge der *(salopp)* [young] male prostitute; **~mädchen das** *(salopp)* street-walker; hooker *(Amer. sl.);* **~punkt der** semicolon; **~weise** *(bes. Met.)* **1.** *Adv.* ‹*rain etc.*› in places; **2.** *adj.; nicht präd.* in places *postpos.;* local

Strick [ʃtrɪk] **der; ~[e]s, ~e a)** cord; *(Seil)* rope; **jmdm. aus etw. einen ~ drehen** *(fig.)* use sth. against sb.; **da kann ich mir ja gleich einen ~ nehmen** *od.* **kaufen!** I might as well end it all now; *s. auch* **reißen 2a; ziehen 2a; b)** *(fam.: Schlingel)* rascal

stricken *tr., itr. V.* knit

Strickerei die; ~, ~en a) *o. Pl. (Tätigkeit)* knitting; **b)** *(Produkt)* piece of knitting

Strick-: ~jacke die cardigan; **~leiter die** rope-ladder; **~muster das** knitting-pattern; *(fig.)* formula; **~nadel die** knitting-needle; **~waren** *Pl.* knitwear *sing.;* **~zeug das** knitting

Striegel ['ʃtri:gl̩] **der; ~s, ~:** curry-comb

striegeln *tr. V.* groom ‹*horse*›; **gestriegelt und gebügelt** *(fig.)* all spruced up

Strieme ['ʃtri:mə] **die; ~, ~n, Striemen der; ~s, ~:** weal

strikt [ʃtrɪkt] **1.** *Adj.* strict; exact ‹*opposite*›. **2.** *adv.* strictly; **ich bin ~ dagegen** I am totally opposed to it

Strippe ['ʃtrɪpə] **die; ~, ~n** *(ugs.)* string; **an der ~ hängen** *(fig.)* be on the phone *(coll.); (dauernd)* hog the phone *(coll.);* **jmdn. an der ~ haben/an die ~ kriegen** *(fig.)* have sb./get sb. on the phone *(coll.) or* line

strippen *itr. V. (ugs.)* do strip-tease; strip

Stripperin die; ~, ~nen *(ugs.)* stripper

Striptease ['ʃtrɪpti:s] **der** *od.* **das; ~:** strip-tease

Striptease·tänzerin die strip-tease dancer

stritt [ʃtrɪt] *1. u. 3. Pers. Sg. Prät. v.* **streiten**

strittig *Adj.* contentious ‹*point, problem*›; disputed ‹*territory*›; ‹*question*› in dispute, at issue; **~ ist nur, ob ...:** the only point at issue is whether ...

Stroh [ʃtro:] **das; ~[e]s** straw; **mit ~ gedeckt** ‹*roof, cottage*› thatched with straw

stroh-, Stroh-: ~blume die a) *(Immortelle)* immortelle; everlasting [flower]; **b)** *(Korbblütler)* straw-flower; **~dumm** *Adj. (ugs.)* witless *(coll.);* thick-headed; **~feuer das: wie ein ~feuer aufflammen** flare up briefly; **das war nur ein ~feuer** *(fig.)* it was just a flash in the pan; **~halm der** straw; **sich [wie ein Ertrinkender] an einen ~halm klammern** *(fig.)* grasp at a straw [like a drowning man]; **~hut der** straw hat; **~kopf der** *(ugs. abwertend)* thickhead; **~sack der** palliasse; **[ach du] heiliger ~sack!** *(ugs.)* jeepers creepers! *(coll.);* goodness gracious [me]!; **~witwe die** *(ugs. scherzh.)* grass widow; **~witwer der** *(ugs. scherzh.)* grass widower

Strolch [ʃtrɔlç] **der; ~[e]s, ~e a)** *(veralt.)* ruffian; **b)** *(fam. scherzh.: Junge)* rascal

Strom [ʃtro:m] **der; ~[e]s, Ströme** ['ʃtrø:mə] **a)** river; *(von Blut, Schweiß, Wasser, fig.: Erinnerungen, Menschen, Autos usw.)* stream; **ein reißender ~:** a raging torrent; **in Strömen regnen** *od. (ugs.)* **gießen** pour with rain; **in Strömen fließen** *(fig.)* flow freely; **das Blut floß in Strömen** *(fig.)* there was heavy bloodshed; **b)** *(Strömung)* current; **mit dem/gegen den ~ schwimmen** *(fig.)* swim with/against the tide *(fig.);* **c)** *(Elektrizität)* current; *(~versorgung)* electricity; **das Kabel führt** *od.* **steht unter ~:** the cable is live; **der ~ ist ausgefallen** there has been a power failure

strom-: ~ab[wärts] *Adv.* downstream; **~auf[wärts]** *Adv.* upstream

strömen ['ʃtrø:mən] *itr. V.; mit sein* stream; *(intensiv)* pour; **~der Regen** pouring rain

strom-, Strom-: ~kreis der [electric] circuit; **~leitung die** power line *or* cable; **~linien·förmig** *Adj.* streamlined; **~schnelle die** rapids *pl.*

Strömung die; ~, ~en a) current; *(Met.)* airstream; **b)** *(fig.) (Bewegung)* movement; *(Tendenz)* trend

Strom·verbrauch der electricity consumption

Strophe ['ʃtro:fə] **die; ~, ~n** verse; *(einer Ode)* strophe

strotzen ['ʃtrɔtsn̩] *itr. V.* **von** *od.* **vor etw.** *(Dat.)* **~:** be full of sth.; **von** *od.* **vor Kraft/Gesundheit ~:** be bursting with strength/health

strubb[e]lig ['ʃtrʊb(ə)lɪç] *Adj.* tousled; **du bist ja so ~!** your hair is in such a mess

Strudel ['ʃtru:dl̩] **der; ~s, ~ a)** whirlpool; *(kleiner)* eddy; **b)** *(bes. südd., österr.: Gebäck)* strudel

strudeln *itr. V.* ‹*water*› eddy, swirl

Struktur [ʃtrʊk'tu:ɐ̯] **die; ~, ~en** structure

strukturell [ʃtrʊktu'rɛl] **1.** *Adj.* structural. **2.** *adv.* structurally

strukturieren *tr. V.* structure

Strumpf [ʃtrʊmpf] **der; ~[e]s, Strümpfe** ['ʃtrʏmpfə] stocking; *(Socke, Knie~)* sock; **auf Strümpfen** in stockinged feet/in one's socks

Strumpf·band das; *Pl.* **~bänder** garter; *(Straps)* suspender *(Brit.);* garter *(Amer.)*

Strumpf-: ~halter der suspender *(Brit.);* garter *(Amer.);* **~hose die** tights *pl. (Brit.);* pantyhose *(esp. Amer.);* **eine ~hose** a pair of tights *(Brit.)*

Strunk [ʃtrʊŋk] **der; ~[e]s, Strünke** ['ʃtrʏŋkə] stem; stalk; *(Baum~)* stump

struppig ['ʃtrʊpɪç] *Adj.* shaggy ‹*coat, dog, beard*›; tangled, tousled ‹*hair*›

Struwwel·peter ['ʃtrʊvl̩-] **der** tousle-head

Stube ['ʃtu:bə] **die; ~, ~n** *(veralt.: Wohnraum)* [living-]room; parlour *(dated);* **die gute ~:** the front room *or (dated)* parlour; **immer rein in die gute ~!** *(ugs.)* come on in!

stuben-, Stuben-: ~arrest der *(ugs.)* detention *(in one's room)*; **[zwei Tage] ~arrest bekommen** be kept in [for two days]; **~fliege die** [common] house-fly; **~hocker der** *(ugs. abwertend)* stay-at-home; **~rein** *Adj.* **a)** house-trained; **b)** *(scherzh.: nicht zotig)* clean ⟨*joke etc.*⟩

Stuck [ʃtʊk] **der; ~[e]s** stucco

Stück [ʃtʏk] **das; ~[e]s, Stücke a)** piece; *(kleines)* bit; *(Teil, Abschnitt)* part; **ein ~ Kuchen/Zucker/Seife** a piece *or* slice of cake/a piece *or* lump of sugar/ a piece *or* bar of soap; **ein [gutes] ~ weiterkommen** get a [good] bit further; **ein hartes ~ Arbeit** a really tough job; **alles in ~e schlagen** smash everything [to pieces]; **im** *od.* **am ~:** unsliced ⟨*sausage, cheese, etc.*⟩; **in einem ~** *(ugs.)* ⟨*talk, rain*⟩ non-stop; **b)** *(Einzel~)* item; article; *(Exemplar)* specimen; *(Möbel~)* piece [of furniture]; **zwanzig ~ Vieh** twenty head of cattle; **ich nehme 5 ~/5 ~ von den Rosen** I'll take five [of them]/five of the roses; **30 Pfennig das ~, das ~ 30 Pfennig** thirty pfennigs each; **~ für ~:** piece by piece; *(eins nach dem andern)* one by one; **das gute ~** *(oft iron.)* the precious thing; **c) das ist [ja] ein starkes** *od.* **tolles ~** *(ugs.)* that's a bit much *or* a bit thick *(coll.)*; **d)** *(salopp abwertend: Person)* **ein faules/freches ~:** a lazy/cheeky thing *or* devil; **e)** *(Bühnen~)* play; **f)** *(Musik~)* piece

stückeln 1. *tr. V.* put together ⟨*sleeve, curtain*⟩ with patches. **2.** *itr. V.* sew on patches

Stücke·schreiber der playwright

stück-, Stück-: ~lohn der *(Wirtsch.)* piece-work pay; *(Akkordsatz)* piece-rate; **~preis der** unit price; **~weise** *Adv.* piece by piece; *(einzeln)* ⟨*sell*⟩ separately; **~werk das: ~werk sein/bleiben** be/remain incomplete; ⟨*book, work of art*⟩ remain a torso

Student [ʃtu'dɛnt] **der; ~en, ~en** student

Studenten-: ~ausweis der student card; **~bude die** *(ugs.)* student's room; **~heim das** student hostel; students' [hall of] residence

Studentin die; ~, ~nen student

studentisch *Adj.; nicht präd.* student

Studie ['ʃtu:di̯ə] **die; ~, ~n** study

Studien-: ~aufenthalt der study visit **(in +** *Dat.* to); **~fach das** subject [of study]; **~gang der** course of study; **~gebühr die** tuition fee; **~kolleg das** *(Hochschulw.)* preparatory course *(esp. for foreign students)*; **~platz der** university/college place; **~rat der, ~rätin die** *established graduate secondary-school teacher (Brit.); graduate high-school teacher with tenure (Amer.)*; **~referendar der** probationary graduate teacher; **~reise die** study trip

studieren [ʃtu'di:rən] **1.** *itr. V.* study; **er studiert noch** he is still a student. **2.** *tr. V.* study

Studierende der/die; *adj. Dekl.* student

Studio ['ʃtu:di̯o] **das; ~s, ~s** studio

Studio·bühne die studio theatre

Studium ['ʃtu:di̯ʊm] **das; ~s, Studien a)** *o. Pl.* study; *(Studiengang)* course of study; **während seines ~s** *(als er Student war)* in his student days; **b)** *(Erforschung)* study; **Studien über etw.** *(Akk.)* **betreiben** carry out studies into sth.; **c)** *o. Pl. (genaues Lesen)* study; **beim ~ der Akten** while studying the files

Stufe ['ʃtu:fə] **die; ~, ~n a)** step; *(einer Treppe)* stair; **„Achtung** *od.* **Vorsicht, ~!"** 'mind the step'; **b)** *(Raketen~, Geol., fig.: Stadium)* stage; *(Niveau)* level; *(Steigerungs~, Grad)* degree; *(Rang)* grade; **auf der gleichen ~ stehen [wie ...]** be of the same standard [as ...]; have the same status [as ...]; *(gleichwertig sein)* be equivalent [to ...]; **sich mit jmdm./etw. auf eine** *od.* **auf die gleiche ~ stellen** put oneself on a level with sb./sth.

stufen-, Stufen-: ~barren der *(Turnen)* asymmetric bars *pl.*; **~heck das** *(Kfz-W.)* booted rear; **~los 1.** *Adj.* continuously variable; **2.** *adv.* **~los verstellbar** continuously adjustable; **~weise 1.** *Adv.* in stages *or* phases; **2.** *adj.; nicht präd.* phased

stufig 1. *Adj.* layered ⟨*hair [style]*⟩; terraced ⟨*terrain*⟩. **2.** *adv.* **~ geschnittenes Haar** layered hair; hair cut in layers

Stuhl [ʃtu:l] **der; ~[e]s, Stühle** ['ʃty:lə] **a)** chair; **b)** *(fig.)* **sein ~ wackelt** his position is threatened *or* no longer secure; **jmdm. den ~ vor die Tür setzen** kick sb. out; show sb. the door; **jmdn. vom ~ reißen** *od.* **jagen/hauen** *(ugs.)* get sb. excited/take sb.'s breath away; **das hat mich fast** *od.* **bald vom ~ gehauen** *(ugs.)* you could have knocked me down with a feather; *s. auch* **elektrisch 1; c)** *(kath. Kirche)* see; **der ~ Petri** the Holy See *or* See of Rome; **d)** *(Med.)* stool; **e)** *s.* **Stuhlgang a**

Stuhl-: ~bein das chair-leg; **~gang der;** *o. Pl.* **a)** bowel movement[s]; **b)** *(Kot)* stool; **~lehne die a)** *(Rückenlehne)* chair-back; **b)** *(Armlehne)* chair-arm

Stukkateur [ʃtʊka'tø:ɐ̯] **der; ~s, ~e** [stucco] plasterer

stülpen ['ʃtʏlpn̩] *tr. V.* **etw. auf** *od.* **über etw.** *(Akk.)* **~:** pull/put sth. on to *or* over sth.

stumm [ʃtʊm] *Adj.* dumb ⟨*person*⟩; *(schweigsam)* silent ⟨*person, reproach, greeting, prayer, etc.*⟩; *(wortlos)* wordless ⟨*greeting, complaint, prayer, gesture, dialogue*⟩; mute ⟨*glance, gesture*⟩; *(Theater)* non-speaking ⟨*part, character*⟩; **~ vor Schreck** speechless with fear; **sie sahen sich ~ an** they looked at one another without speaking *or* in silence

Stumme der/die; *adj. Dekl.* mute; **die ~n** the dumb

Stummel ['ʃtʊml̩] **der; ~s, ~:** stump; *(Bleistift~)* stub; *(Zigaretten~/Zigarren~)* [cigarette-/cigar-]butt

Stumm·film der silent film

Stümper ['ʃtʏmpɐ] **der; ~s, ~** *(abwertend)* botcher; bungler

Stümperei die; ~, ~en *(abwertend)* **a)** *o. Pl.* botching; incompetence; **b)** *(Ergebnis)* botched job; piece of incompetence

stümperhaft *(abwertend)* **1.** *Adj.* incompetent; botched ⟨*job*⟩; *(laienhaft)* amateurish ⟨*attempt, drawing*⟩. **2.** *adv.* incompetently; *(laienhaft)* amateurishly

stümpern *itr. V. (abwertend)* work incompetently; *(pfuschen)* bungle

stumpf [ʃtʊmpf] **1.** *Adj.* **a)** blunt ⟨*pin, needle, knife, etc.*⟩; snub ⟨*nose*⟩; flat-topped ⟨*tower*⟩; **b)** *(Math.)* truncated ⟨*cone, pyramid*⟩; obtuse ⟨*angle*⟩; **c)** *(glanzlos, matt)* dull ⟨*paint, hair, metal, colour, etc.*⟩; *(rauh)* rough ⟨*stone, wood*⟩; **d)** *(abgestumpft, teilnahmslos)* impassive, lifeless ⟨*person, glance*⟩; impassive, apathetic ⟨*indifference, resignation*⟩; dulled ⟨*senses*⟩; blank ⟨*look, despair*⟩. **2.** *adv. (abgestumpft)* ⟨*sit, stare*⟩ impassively

Stumpf der; ~[e]s, Stümpfe ['ʃtʏmpfə] stump; **etw. mit ~ und Stiel ausrotten** eradicate sth. root and branch

Stumpf·sinn der; *o. Pl.* **a)**

apathy; **b)** *(Monotonie)* monotony; tedium

stumpf·sinnig 1. *Adj.* **a)** apathetic; vacant ‹*look*›; **b)** *(monoton)* tedious; dreary; soul-destroying ‹*job, work*›. **2.** *adv.* **a)** apathetically; ‹*stare*› vacantly; **b)** *(monoton)* tediously

Stündchen ['ʃtʏntçən] **das; ~s, ~** *(fam.)* **[für** *od.* **auf] ein ~:** for an hour or so; **jmds. letztes ~ ist gekommen** *od.* **hat geschlagen** sb.'s last hour has come

Stunde ['ʃtʊndə] **die; ~, ~n a)** hour; **eine ~ Pause** an hour's break; a break of an hour; **drei ~n zu Fuß/mit dem Auto** three hours' walk/drive; **120 km in der ~ fahren** do 120 kilometres per hour; **jede ~:** once an hour; **~ um ~:** [for] hours; [for] hour after hour; **jmds. letzte ~ hat geschlagen** *od.* **ist gekommen** sb.'s last hour has come; **b)** *(geh.)* *(Zeitpunkt)* hour; *(Zeit)* time; *(Augenblick)* moment; **in ~n der Not** in times of need; **zu vorgerückter** *od.* **später ~:** at a late hour; **zur ~:** at the present time; c) *(Unterrichts~)* lesson; **in der dritten ~:** in the third period

stunden *tr. V.* **jmdm. einen Betrag** *usw.* **~:** give sb. [extra] time to pay *or* allow sb. to defer payment of a sum *etc.*

stunden-, Stunden-: ~geschwindigkeit die: bei/mit einer ~geschwindigkeit von 60 km at a speed of 60 k.p.h.; **~kilometer der** *(ugs.)* kilometre per hour; k.p.h.; **er fuhr 120 ~kilometer** he was driving at *or* doing 120 k.p.h.; **~lang 1.** *Adj.; nicht präd.* lasting hours *postpos.;* **2.** *adv.* for hours; **~lohn der** hourly wage; **sie bekommt 12 Mark ~lohn** she gets paid 12 marks an hour *or* per hour; **~plan der** timetable; **~weise 1.** *Adv.* for an hour or two [at a time]; **er wird ~weise bezahlt** he is paid by the hour; **2.** *adj.; nicht präd.* ‹*hiring, payment*› by the hour; **~zeiger der** hour-hand

stündlich 1. *Adj.* hourly. **2.** *adv.* **a)** hourly; once an hour; **b)** *(jeden Augenblick)* at any moment

stupid[e] [ʃtu'pi:də] *(abwertend)* **1.** *Adj.* **a)** moronic, empty-headed ‹*person*›; moronic, vacuous ‹*expression*›; **b)** *(monoton)* soul-destroying. **2.** *adv.* moronically

Stups [ʃtʊps] **der; ~es, ~e** *(ugs.)* push; shove; *(leicht)* nudge

stupsen *tr. V. (ugs.)* push; shove; *(leicht)* nudge

Stups·nase die snub nose

stur *(ugs.)* **1.** *Adj.* **a)** *(abwertend)* *(eigensinnig, unnachgiebig)* obstinate; pig-headed; obstinate, dogged ‹*insistence*›; *(phlegmatisch)* stolid; dour; **ein ~er Bock** a pig-headed so-and-so *(coll.);* **auf ~ schalten** dig one's heels in; **b)** *(unbeirrbar)* dogged; persistent; **c)** *(abwertend: stumpfsinnig)* tedious. **2.** *adv.* **a)** *(abwertend: eigensinnig, unnachgiebig)* obstinately; **b)** *(unbeirrbar)* doggedly; **sie las/redete ~ weiter** she carried on reading/kept on talking regardless; **c)** *(abwertend: stumpfsinnig)* tediously; ‹*learn, copy*› mechanically

Sturheit die; ~ *(ugs. abwertend)* **a)** *(Eigensinnigkeit, Unnachgiebigkeit)* obstinacy; pig-headedness; *(phlegmatisches Wesen)* stolidity; dourness; **b)** *(Stumpfsinnigkeit)* deadly monotony

Sturm [-] **der; ~[e]s, Stürme** ['ʃtʏrmə] **a)** storm; *(heftiger Wind)* gale; **bei** *od.* **in ~ und Regen** in the wind and rain; **ein ~ im Wasserglas** a storm in a teacup; b) *(Milit.: Angriff)* assault **(auf +** *Akk.* on); **etw. im ~ erobern** *od.* **nehmen** *(auch fig.)* take sth. by storm; **gegen etw. ~ laufen** *(fig.)* be up in arms against sth.; **~ klingeln** ring the [door]bell like mad; lean on the [door]bell; **c)** *(Sport: die Stürmer)* forward line

stürmen ['ʃtʏrmən] **1.** *itr. V.* **a)** *unpers.* **es stürmt [heftig]** it's blowing a gale; **b)** *mit sein (rennen)* rush; *(verärgert)* storm; **c)** *(Sport: als Stürmer spielen)* play up front *or* as a striker; **d)** *(Sport, Milit.: angreifen)* attack. **2.** *tr. V.* *(Milit.)* storm ‹*town, position, etc.*›; *(fig.)* besiege ‹*booking-office, shop, etc.*›

Stürmer ['ʃtʏrmɐ] **der: ~s, ~** *(Sport)* striker; forward

Sturm·flut die storm tide

stürmisch ['ʃtʏrmɪʃ] **1.** *Adj.* **a)** stormy; *(fig.)* tempestuous, turbulent ‹*days, life, times, years*›; **b)** *(ungestüm)* tempestuous ‹*nature, outburst, welcome*›; tumultuous ‹*applause, welcome, reception*›; wild ‹*enthusiasm*›; passionate ‹*lover, embrace, temperament*›; vehement ‹*protest*›; **nicht so ~!** calm down!; take it easy!; **c)** *(rasant)* meteoric ‹*development, growth*›; lightning, breakneck ‹*speed*›. **2.** *adv.* **a)** ‹*protest*› vehemently; ‹*embrace*› impetuously, passionately; ‹*demand*› clamorously; ‹*applaud*› wildly; **b)** *(rasant)* at a tremendous rate *or* speed; at lightning speed

Sturm·schritt der: im ~schritt at the double

Sturz [ʃtʊrts] **der; -es, Stürze** ['ʃtʏrtsə] **a)** fall **(aus, von** from); *(Unfall)* accident; **ein ~ in die Tiefe** a plunge into the depths; **bei einem ~ vom Pferd** falling off a horse; **b)** *(fig.: von Preis, Temperatur usw.)* [sharp] fall, drop (*Gen.* in); **c)** *(Verlust des Amtes, der Macht)* fall; *(Absetzung)* overthrow; *(Amtsenthebung)* removal from office

stürzen ['ʃtʏrtsn̩] **1.** *itr. V.; mit sein* **a)** fall **(aus, von** from); *(in die Tiefe)* plunge; plummet; **b)** *(fig.)* ‹*temperature, exchange rate, etc.*› drop [sharply]; ‹*prices*› tumble; ‹*government*› fall, collapse; **c)** *(laufen)* rush; dash; **er stürzte ins Zimmer** he burst into the room; **d)** *(fließen)* stream; pour. **2.** *refl. V.* **sich auf jmdn./etw. ~** *(auch fig.)* pounce on sb./sth.; **sich aus dem Fenster ~:** hurl oneself *or* leap out of the window; **sich in etw.** *(Akk.)* **~:** throw oneself *or* plunge into sth.; **sich ins Vergnügen ~:** abandon oneself to pleasure. **3.** *tr. V.* **a)** throw; *(mit Wucht)* hurl; **jmdn. ins Unglück ~:** plunge sb. into misfortune; **b)** *(umdrehen)* upturn, turn upside-down ‹*mould, pot, box, glass, cup*›; turn out ‹*pudding, cake, etc.*›; **c)** *(des Amtes entheben)* oust ‹*person*› [from office]; *(gewaltsam)* overthrow, topple ‹*leader, government*›

Sturz-: ~flug der *(Flugw.)* [nose-]dive; **~helm der** crash-helmet

Stuß [ʃtʊs] **der; Stusses** *(ugs. abwertend)* rubbish; twaddle *(coll.)*

Stute ['ʃtu:tə] **die; ~, ~n** mare

Stütze die; ~, ~n *(auch fig.)* support; *(für die Wäscheleine)* prop; **~n für Kopf, Arme und Füße** head-, arm-, and foot-rests

[1]stutzen ['ʃtʊtsn̩] *itr. V.* stop short

[2]stutzen *tr. V.* trim; dock ‹*tail*›; clip ‹*ear, hedge, wing*›; prune ‹*tree, bush*›

stützen ['ʃtʏtsn̩] **1.** *tr. V.* support; *(mit Pfosten o. ä.)* prop up; *(aufstützen)* rest ‹*head, hands, arms, etc.*› **(auf +** *Akk.* on); **die Hände in die Seiten/den Kopf in die Hände gestützt** hands on hips/head in hands; **wo sind die Beweise, auf die Sie Ihre Anschuldigungen ~?** where is the evidence to support your accusations *or* on which your accusations are based?. **2.** *refl. V.* **sich auf jmdn./etw. ~:**

lean *or* support oneself on sb./sth.; **sich auf Fakten** *(Akk.)* ~ *(fig.)* ⟨*theory, statement etc.*⟩ be based on facts

stutzig *Adj.* ~ **werden** begin to wonder; get suspicious; **jmdn.** ~ **machen** make sb. wonder *or* suspicious

Stütz·punkt der *(bes. Milit.)* base

s.u. *Abk.* **siehe unten** see below

Sub·dominante ['zʊp-] **die** *(Musik)* subdominant

Subjekt [zʊp'jɛkt] **das;** ~**[e]s,** ~**e a)** subject; **b)** *(abwertend: Mensch)* creature; type *(coll.)*

subjektiv [zʊpjɛk'ti:f] **1.** *Adj.* subjective. **2.** *adv.* subjectively

Subjektivität [zʊpjɛktivi'tɛ:t] **die;** ~: subjectivity

Subjekt·satz der *(Sprachw.)* subject clause

Sub-: ~**kontinent der** *(Geogr.)* subcontinent; ~**kultur die** *(Soziol.)* subculture

Subskription [zʊpskrɪp'tsi̯o:] **die;** ~, ~**en** *(Buchw.)* subscription

Substantiv ['zʊpstanti:f] **das;** ~**s,** ~**e** *(Sprachw.)* noun

Substanz [zʊp'stants] **die;** ~, ~**en a)** *(auch fig.)* substance; **b)** *(Grundbestand)* **die** ~: the reserves *pl.*; **etw. geht an die** ~ *(fig. ugs.)* *(seelisch, nervlich)* sth. gets you down; *(körperlich)* sth. takes it out of you

subtil [zʊp'ti:l] *(geh.)* **1.** *Adj.* subtle. **2.** *adv.* subtly

subtrahieren *tr., itr. V. (Math.)* subtract

Subtraktion [zʊptrak'tsi̯o:n] **die;** ~, ~**en** *(Math.)* subtraction

Sub·tropen *Pl. (Geogr.)* subtropics

sub·tropisch *Adj.* subtropical

Such·anzeige die a) missing-person report; **b)** *(in der Zeitung)* 'lost' advertisement

Suche ['zu:xə] **die;** ~, ~**n** search **(nach** for); **auf der** ~ **[nach jmdm./etw.] sein** be looking/*(intensiver)* searching [for sb./sth.]; **sich [nach jmdm./etw.] auf die** ~ **machen** start searching *or* start a search [for sb./sth.]

suchen 1. *tr. V.* **a)** look for; *(intensiver)* search for; **„Kellner gesucht"** 'waiter wanted'; **jemanden wie ihn kann man** ~ *(ugs.)* you don't come across someone like him every day; **seinesgleichen** ~: be without equal *or* unequalled; **b)** *(bedacht sein auf, sich wünschen)* seek ⟨*protection, advice, company, warmth, etc.*⟩; look for ⟨*adventure*⟩; **Kontakt** *od.* **Anschluß** ~: try to get to know people; **Streit** ~: seek a quarrel; **er hat hier nichts zu** ~ *(ugs.)* he has no business [to be] here; **c)** *(geh.: trachten)* ~, **etw. zu tun** seek *or* endeavour to do sth. **2.** *itr. V.* search; **nach jmdm./etw.** ~: look/search for sb./sth.; **sich** ~**d umsehen** look around

Sucher der; ~**s,** ~ *(Fot.)* viewfinder

Such-: ~**hund der** tracker dog; ~**meldung die** announcement about a missing or wanted person

Sucht [zʊxt] **die;** ~, **Süchte** [zʏçtə] *od.* ~**en a)** addiction **(nach** to); **[bei jmdm.] zur** ~ **werden** *(auch fig.)* become addictive [in sb.'s case]; **b)** *Pl.* **Süchte** *(übermäßiges Verlangen)* craving, obsessive desire **(nach** for)

süchtig ['zʏçtɪç] *Adj.* **a)** addicted; ~ **machen** *(auch fig.)* be addictive; ~ **[nach etw.] sein** be an addict *or* addicted [to sth.]; **b)** *(versessen, begierig)* obsessive; **nach etw.** ~ **sein** be obsessed with sth.

Süchtige der/die; *adj. Dekl.*, **Suchtkranke der/die** addict

Süd [zy:t] *o. Art.*; *o. Pl.* **a)** *(Seemannsspr., Met.: Richtung)* south; **b)** *(Gebiet)* South; **c)** *(Politik)* South; **d)** *einem Subst. nachgestellt (südlicher Teil, südliche Lage)* South

Süd-: ~**afrika (das)** South Africa; ~**amerika (das)** South America

Sudan [zu'da:n] **(das);** ~**s** *od.* **der;** ~**s** Sudan

süd·deutsch *Adj.* South German

Süd·deutschland (das) South Germany

Süden der; ~**s a)** *(Richtung)* south; *s. auch* **Norden a; b)** *(Gegend)* South; **c)** *(Geogr.)* South; **der tiefe/tiefste** ~: the far South

Süd-: ~**england (das)** Southern England; the South of England; ~**europa (das)** Southern Europe; ~**frucht die** tropical [or sub-tropical] fruit; ~**korea (das)** South Korea; ~**küste die** south coast

Südländer ['zy:tlɛndɐ] **der;** ~**s,** ~: Southern European; Mediterranean type

südländisch *Adj.* Southern [European]; Mediterranean; Latin ⟨*temperament*⟩; ~ **aussehen** have Latin looks; look like a Southern European

südlich 1. *Adj.* **a)** *(im Süden)* southern; *s. auch* **nördlich 1 a; Polarkreis; Wendekreis a; b)** *(nach, aus dem Süden)* southerly; **c)** *(des Südens)* Southern. **2.** *adv.* southwards; *s. auch* **nördlich 2. 3.** *Präp. mit Gen.* [to the] south of

süd-, Süd-: ~**osten der** south-east; *s. auch* **Norden;** ~**östlich 1.** *Adj.* south-eastern; south-easterly ⟨*direction, wind, course*⟩; **2.** *adv.* ~**östlich [von X] liegen** be to the south-east [of X]; **3.** *Präp. mit Gen.* [to the] south-east of; ~**pol** ['--] **der** South Pole; ~**see** ['--] **die;** ~: **die** ~: the South Seas *pl.*; ~**seite** ['---] **die** south side; ~**wärts** ['--] *Adv.* southwards; ~**westen der** south-west; *s. auch* **Norden a;** ~**westlich 1.** *Adj.* south-western; south-westerly ⟨*direction, wind, course*⟩; **2.** *adv.* ~ **[von X] liegen** be to the south-west [of X]; **3.** *Präp. mit Gen.* [to the] south-west of; ~**wind** ['--] **der** south *or* southerly wind

Sues·kanal ['zu:ɛs-] **der;** ~**s** Suez Canal

Suff [zʊf] **der;** ~**[e]s** *(salopp)* **a) im** ~: while under the influence *(coll.)*; **b)** *(Trunksucht)* boozing *(coll.)*; **sich dem** ~ **ergeben** become a victim of the demon drink; take to the bottle *(coll.)*

süffig *Adj. (ugs.)* [very] drinkable

süffisant [zʏfi'zant] *(geh. abwertend)* **1.** *Adj.* smug. **2.** *adv.* smugly

Suffix [zʊ'fɪks] **das;** ~**es,** ~**e** *(Sprachw.)* suffix

suggerieren [zʊge'ri:rən] *tr. V.* **a)** *(geh., Psych.)* suggest; **b)** *(geh.: den Eindruck erwecken)* suggest; give the *or* an impression of

Suggestion [zʊgɛs'ti̯o:n] **die;** ~, ~**en a)** *(geh., Psych.)* suggestion; **b)** *o. Pl. (geh.: suggestive Wirkung)* suggestive effect *or* power

suggestiv [zʊgɛs'ti:f] *(geh., Psych.)* **1.** *Adj.* suggestive. **2.** *adv.* suggestively

Suggestiv·frage die leading question

suhlen *refl. V.* wallow

Sühne ['zy:nə] **die;** ~, ~**n** *(geh.)* atonement; expiation; ~ **[für etw.] leisten** make atonement *or* atone [for sth.]

sühnen *tr., itr. V.* **[für] etw.** ~: atone for *or* pay the penalty for sth.

Sühne·termin der *(Rechtsw.)* conciliation hearing

Suite ['svi:t(ə)] **die;** ~, ~**n** suite

sukzessiv [zʊktsɛ'si:f] **1.** *Adj.* gradual. **2.** *adv.* gradually

Sulfat [zʊl'fa:t] **das;** ~**[e]s,** ~**e** *(Chemie)* sulphate

Sulfonamid [zʊlfona'mi:t] **das;** ~**[e]s,** ~**e** *(Pharm.)* sulphonamide

Sultan ['zʊlta:n] **der; ~s, ~e** sultan
Sultanine [zʊlta'ni:nə] **die; ~, ~n** sultana
Sülze ['zʏltsə] **die; ~, ~n a)** diced meat/fish in aspic; *(vom Schweinskopf)* brawn; **b)** *(Aspik)* aspic
summarisch [zʊ'ma:rɪʃ] *(geh.)* **1.** *Adj.* summary; brief ⟨*summary*⟩. **2.** *adv.* summarily; briefly
Sümmchen ['zʏmçən] **das; ~s, ~** *(ugs.)* **ein hübsches** *od.* **nettes ~:** a tidy little sum *(coll.)*
Summe ['zʊmə] **die; ~, ~n** sum
summen 1. *itr. V.* hum; *(lauter, heller)* buzz; **es summt** there's a hum/buzzing. **2.** *tr., auch itr. V.* hum ⟨*tune, song, etc.*⟩
Summer der; ~s, ~: buzzer
summieren *refl. V.* add up (**auf** + *Akk.* to)
Summ·ton der buzzing [tone]; *(leiser)* hum
Sumpf [zʊmpf] **der; ~[e]s, Sümpfe** ['zʏmpfə] marsh; *(bes. in den Tropen)* swamp; *(fig.)* morass; quagmire
sumpfig *Adj.* marshy
Sund [zʊnt] **der; ~[e]s, ~e** *(Geogr.)* sound
Sünde ['zʏndə] **die; ~, ~n** sin; *(fig.)* misdeed; transgression; **eine ~ wert sein** *(scherzh.)* be worth a little transgression; ⟨*food*⟩ be naughty but nice
Sünden-: ~bekenntnis das confession of one's sins; **~bock der** *(ugs.)* scapegoat
Sünder der; ~s, ~, Sünderin die; ~, ~nen sinner
sündhaft 1. *Adj.* **a)** sinful; **b)** *(ugs.)* **ein ~er Preis** an outrageous price. **2.** *adv.* **a)** sinfully; **b)** *(ugs.: sehr)* outrageously ⟨*expensive*⟩; stunningly ⟨*beautiful*⟩
sündig 1. *Adj.* sinful; *(lasterhaft)* wicked. **2.** *adv.* sinfully
sündigen *itr. V.* sin
super *(salopp)* **1.** *indekl. Adj.* super *(coll.)*; fantastic *(coll.)*; **~ aussehen/ sich ~ fühlen** look/feel great *(coll.)*. **2.** *adv.* fantastically *(coll.)*
Super ['zu:pɐ] **das; ~s, ~:** four star *(Brit.)*; premium *(Amer.)*
super- ultra-⟨*long, high, fast, modern, masculine, etc.*⟩
Super- super-⟨*hero, figure, car, group, etc.*⟩; terrific *(coll.)*, tremendous *(coll.)* ⟨*success, offer, chance, idea, etc.*⟩
Super-8-Film der super 8 film
super·klug *(iron.)* **1.** *Adj.* extra clever; smart-aleck *(coll. derog.)*. **2.** *adv.* in a smart-aleck way *(coll. derog.)*

Superlativ ['zu:pɐlati:f] **der; ~s, ~e** *(Sprachw.)* superlative
super-, Super-: ~markt der supermarket; **~schlau** *Adj. (iron.) s.* **~klug; ~schnell** *(ugs.)* **1.** *Adj.* ultra-fast; **2.** *adv.* at tremendous speed
Suppe ['zʊpə] **die; ~; ~n** soup; **jmdm. die ~ versalzen** *(ugs.)* put a spoke in sb.'s wheel; put a spanner in sb.'s works; **jmdm. in die ~ spucken** *(salopp)* mess things up for sb.; *s. auch* **auslöffeln a**
Suppen-: ~fleisch das beef for making soup; **~grün das** green vegetables for making soup; **~huhn das** boiling fowl; **~löffel der** soup-spoon; **~schüssel die** soup-tureen; **~tasse die** soup-bowl; **~teller der** soup-plate; **~terrine die** soup-tureen; **~würfel der** stock cube
Surf·brett ['sə:f-] **das** surf-board
surfen ['sə:fn̩] *itr. V.* surf
Surfer ['sə:fɐ] **der; ~s, ~:** surfer
Sur·realismus [zʊrea'lɪsmʊs] **der; ~:** surrealism *no art.*
surrealistisch 1. *Adj.* surrealist ⟨*movement, painting, literature*⟩; surrealistic ⟨*image, story, scene*⟩. **2.** *adv.* surrealistically; ⟨*paint*⟩ in a surrealistic style; ⟨*influenced*⟩ by surrealism
surren ['zʊrən] *itr. V.* **a)** *(summen)* hum; ⟨*camera, fan*⟩ whirr; **b)** *mit sein (schwirren)* whirr
suspekt [zʊs'pɛkt] **1.** *Adj.* suspicious; **jmdm. ~ sein** seem suspicious to sb.; arouse sb.'s suspicions. **2.** *adv.* suspiciously
suspendieren [zʊspɛn'di:rən] *tr. V.* suspend; *(entlassen)* dismiss; **jmdn. vom Dienst/von seinem Amt ~:** suspend/dismiss sb. from his/ her post
süß [zy:s] **1.** *Adj.* sweet; **er ißt gern Süßes** he likes sweet things; he has a sweet tooth; **na, mein Süßer/meine Süße?** well, sweetheart? **2.** *adv.* sweetly; **träum ~!** sweet dreams!
süßen 1. *tr. V.* sweeten. **2.** *itr. V.* sweeten things; **mit Saccharin** *usw.* **~:** use saccharine *etc.* as a sweetener
Süß·holz das; *o. Pl.* liquorice [plant]; **~ raspeln** *(fig. ugs.)* ooze charm
Süßigkeit die; ~, ~en *meist Pl.* sweet *(Brit.)*; candy *(Amer.)*; **~en** sweets *(Brit.)*; candy *sing. (Amer.)*; *(als Ware)* confectionery *sing.*
süßlich 1. *Adj.* **a)** [slightly] sweet; on the sweet side *pred.*; **ein widerlich ~er Geschmack** an unpleasantly sickly *or* cloying taste; **b)** *(abwertend) (sentimental)* sickly mawkish ⟨*film*⟩; *(heuchlerisch freundlich)* sugary ⟨*smile etc.*⟩; smarmy *(coll.)* ⟨*expression, manners*⟩; honeyed ⟨*words*⟩. **2.** *adv. (abwertend)* ⟨*write, paint*⟩ mawkishly *or* in a sickly-sentimental style; ⟨*smile*⟩ smarmily *(coll.)*

süß-, Süß-: ~most der unfermented fruit juice; **~rahm·butter die** sweet cream butter; **~-sauer 1.** *Adj.* sweet-and-sour; *(fig.)* wry ⟨*smile, face*⟩; **2.** *adv.* **etw. ~-sauer zubereiten** give sth. a sweet-and-sour flavour; *(fig.)* ⟨*smile*⟩ wryly; **~speise die** sweet; dessert; **~stoff der** sweetener; **~waren** *Pl.* confectionery *sing.*; candy *sing. (Amer.)*; **~wasser das;** *Pl.* **~wasser** fresh water
svw. *Abk.* **soviel wie**
SW *Abk.* **Südwest[en]** SW
Symbol [zʏm'bo:l] **das; ~s, ~e** symbol
symbolhaft 1. *Adj.* symbolic (**für** of). **2.** *adv.* symbolically
Symbolik [zʏm'bo:lɪk] **die; ~:** symbolism
symbolisch 1. *Adj.* symbolic. **2.** *adv.* symbolically
symbolisieren *tr. V.* symbolize
Symbolismus der; ~: symbolism; *(Kunstrichtung)* Symbolism *no art.*
Symmetrie [zʏme'tri:] **die; ~, ~n** symmetry
symmetrisch 1. *Adj.* symmetrical. **2.** *adv.* symmetrically
Sympathie [zʏmpa'ti:] **die; ~, ~n** sympathy; **~ für jmdn. haben** sympathize with *or* have sympathy with sb.; **sich** *(Dat.)* **jmds./ alle ~n verscherzen** forfeit sb.'s/ everybody's sympathy; **bei aller ~:** with the best will in the world
Sympathisant [zʏmpati'zant] **der; ~en, ~en, Sympathisantin die; ~, ~nen** sympathizer (*Gen.* with)
sympathisch 1. *Adj.* congenial, likeable ⟨*person, manner*⟩; appealing, agreeable ⟨*voice, appearance, material*⟩; **er war mir gleich ~:** I took to him at once; I took an immediate liking to him. **2.** *adv.* in a likeable *or* appealing way; *(angenehm)* agreeably
sympathisieren *itr. V.* sympathize (**mit** with); **mit einer Partei ~:** be sympathetic towards a party
Symphonie [zʏmfo'ni:] *usw. s.* **Sinfonie** *usw.*
Symptom [zʏmp'to:m] **das; ~s, ~e** *(Med., geh.)* symptom (*Gen.*, **für, von** of)
symptomatisch *(Med., geh.)* **1.**

Adj. symptomatic (**für** of). **2.** *adv.* symptomatically

Synagoge [zyna'go:gə] **die; ~, ~n** synagogue

synchronisieren *tr. V.* **a)** *(Film)* dub ⟨*film*⟩; **b)** *(Technik, fig.)* synchronize ⟨*watches, operations, etc.*⟩; **synchronisiertes Getriebe** synchromesh gearbox

Syndikus ['zʏndikʊs] **der; ~s, ~e** *od.* **Syndizi** ['zʏnditsi] *(Rechtsanwalt einer Firma)* company lawyer *or (Amer.)* attorney

Syndrom [zʏn'dro:m] **das; ~s, ~e** *(Med.)* syndrome

Synkope die; ~, ~n [zʏn'ko:pə]; *(Musik)* syncopation

synonym [zyno'ny:m] *(Sprachw.)* **1.** *Adj.* synonymous. **2.** *adv.* synonymously

Synonym [zyno'ny:m] **das; ~s, ~e** *(Sprachw.)* synonym

syntaktisch [zʏn'taktɪʃ] *(Sprachw.)* **1.** *Adj.* syntactic. **2.** *adv.* syntactically

Syntax ['zʏntaks] **die; ~, ~en** *(Sprachw.)* syntax

Synthese [zʏn'te:zə] **die; ~, ~n** synthesis (*Gen.*, **von, aus** of)

synthetisch 1. *Adj.* synthetic. **2.** *adv.* synthetically

Syphilis ['zy:filɪs] **die; ~** *(Med.)* syphilis

Syrer ['zy:rɐ] **der; ~s, ~, Syrerin die; ~, ~nen** Syrian

Syrien ['zy:riən] **(das); ~s** Syria

syrisch ['zy:rɪʃ] *Adj.* Syrian

System [zʏs'te:m] **das; ~, ~e** system; **~ in etw.** *(Akk.)* **bringen** introduce some system into sth.; **hinter etw.** *(Dat.)* **steckt ~:** there's method in sth.

Systematik [zʏste'ma:tɪk] **die; ~, ~en** systematics *sing.*

systematisch [zʏste'ma:tɪʃ] **1.** *Adj.* systematic. **2.** *adv.* systematically

System·zwang der pressure imposed by the system

Szene ['stse:nə] **die; ~, ~n a)** scene; **hinter der ~:** backstage; behind the scenes; **er erhielt Beifall auf offener ~:** he was applauded during the scene; **sich in ~ setzen** *(fig.)* put oneself in the limelight; **b)** *(Auseinandersetzung)* scene; **[jmdm.] eine ~ machen** make a scene [in front of sb.]; **c)** *(ugs.: bestimmtes Milieu)* scene *(coll.)*

Szenen·wechsel der *(Theater)* scene-change

Szenerie [stsenə'ri:] **die; ~, ~n a)** *(Bühnendekoration)* set (*Gen.* for); **b)** *(Schauplatz)* scene

Szepter ['stsɛptɐ] **das; ~s, ~** *s.* **Zepter**

T

t, T [te:] **das; ~, ~,** *(ugs.)* **~s, ~s** t/T; *s. auch* **a, A**

Tabak ['ta(:)bak] **der; ~s, ~e** tobacco

Tabaks-: ~beutel der tobacco-pouch; **~dose die** tobacco-tin; **~pfeife die** [tobacco-]pipe

tabellarisch [tabɛ'la:rɪʃ] **1.** *Adj.* tabular; **ein ~er Lebenslauf** a curriculum vitae in tabular form. **2.** *adv.* in tabular form

Tabelle [ta'bɛlə] **die; ~, ~n a)** *(Übersicht)* table; **b)** *(Sport)* [league/championship] table

Tabellen·führer der *(Sport)* top team/player in the [league/championship] table

Tablett [ta'blɛt] **das; ~[e]s, ~s** *od.* **~e** tray; **jmdm. etw. auf einem silbernen ~ servieren** *(fig.)* hand sth. to sb. on a silver platter

Tablette die; ~, ~n tablet

tabu [ta'bu:] *Adj.* taboo

Tabu das; ~s, ~s taboo

Tacho ['taxo] **der; ~s, ~s** *(ugs.)* speedo *(coll.)*

Tacho·meter der *od.* **das** speedometer

Tadel ['ta:dl̩] **der; ~s, ~ a)** censure; **b)** *(im Klassenbuch)* black mark; **c)** *(geh.: Mangel, Makel)* blemish; flaw; **ohne ~:** *s.* **tadellos 1 a**

tadel·los 1. *Adj.* **a)** *(makellos)* impeccable; immaculate ⟨*hair, clothing, suit, etc.*⟩; perfect ⟨*condition, teeth, pronunciation, German, etc.*⟩; **b)** *(ugs.: sehr gut)* excellent; **c)** ['--'-] *(ugs.: als Ausruf der Zustimmung)* splendid *(coll.).* **2.** *adv.* **a)** *(makellos)* ⟨*dress*⟩ impeccably, immaculately; ⟨*fit, speak, etc.*⟩ perfectly; ⟨*live, behave, etc.*⟩ irreproachably; **b)** *(ugs.: sehr gut)* **hier wird man ~ bedient** the service is excellent here

tadeln *tr. V.* **jmdn. [für sein Verhalten** *od.* **wegen seines Verhaltens] ~:** rebuke sb. [for his/her behaviour]; **jmds. Arbeit ~:** criticize sb.'s work; **~der Blick** reproachful look

tadelns·wert *Adj.* reprehensible

Tafel ['ta:fl̩] **die; ~, ~n a)** *(Schiefer~)* slate; *(Wand~)* blackboard; **b)** *(plattenförmiges Stück)* slab; **eine ~ Schokolade** a bar of chocolate; **c)** *(Gedenk~)* plaque; **d)** *(geh.: festlicher Tisch)* table; **die ~ aufheben** *(fig.)* rise from the table; **e)** *(Druckw.)* plate

tafel-, Tafel-: ~apfel der *(Kaufmannsspr.)* dessert apple; **~fertig** *Adj.* *(Kochk.)* ready to serve *postpos.*; **~lappen der** blackboard cloth

tafeln *itr. V.* *(geh.)* feast

täfeln ['tɛ:fl̩n] *tr. V.* panel

Tafel-: ~obst das [dessert] fruit; **~salz das** table salt

Täfelung die; ~, ~en a) *(das Täfeln)* panelling; **b)** *(Paneel)* [wooden] panelling

Tafel-: ~wasser das; *Pl.* **~wässer** [bottled] mineral water; **~wein der** table wine

Taft [taft] **der; ~[e]s, ~e** taffeta

Tag [ta:k] **der; ~[e]s, ~e a)** day; **es wird/ist ~:** it's getting/it is light; **der ~ bricht an** *od.* **erwacht/neigt sich** *(geh.)* the day breaks/draws to an end *or* a close; **am ~[e]** during the day[time]; **am hellichten ~:** in broad daylight; **er redet viel, wenn der ~ lang ist** *(ugs.)* you can't put any trust in what he says; **man soll den ~ nicht vor dem Abend loben** *(Spr.)* don't count your chickens before they're hatched *(prov.)*; **es ist noch nicht aller ~e Abend** we haven't yet seen the end of the matter; **guten ~!** hello; *(bei Vorstellung)* how do you do?; *(nachmittags auch)* good afternoon; **etw. an den ~ legen** display sth.; **etw. an den ~ bringen** *od.* *(geh.)* **ziehen** bring sth. to light; reveal sth.; **an den ~ kommen** come to light; **über/unter ~[e]** *(Bergmannsspr.)* above ground/underground; **b)** *(Zeitraum von 24 Stunden)* day; **welchen ~ haben wir heute?** *(Wochentag)* what day is it today? what's today?; *(Datum)* what date is it today?; **heute in/vor drei ~en** three days from today/three days ago today; **den ~ über** during the day; **an diesem ~:** on this day; **dreimal am ~:** three times a day; **am ~e vorher** on the previous day; the day before; **~ für ~:** every [single] day; **von ~ zu ~:** day by day; **in den nächsten ~en** in the next few days; **der ~ X** the great day; **am folgenden ~:** the next day; **er hatte heute einen schlechten ~:** today was one of his bad days; **sich** *(Dat.)* **einen schönen/faulen ~ machen** *(ugs.)* have a nice/lazy day; **den lieben langen ~:** all day long; **~ der offenen Tür** open day; **eines ~es** one day; some day; **eines schönen ~es** one of these days; **von einem ~ auf den anderen** from one day to the next; overnight; **c)** *(Ehren~, Gedenk~)* **~ der Deutschen Einheit** Day of German Unity; **d)** *Pl. ([Lebens]zeit)* days; **seine ~e sind gezählt** his days are numbered; **auf meine/deine** *usw.* **alten ~e** in my/your *etc.* old age; **e)** *Pl. (ugs.: verhüll.: Menstruation)* period *sing.*

tag·aus *Adv.* **~, tagein** day in, day out; day after day

tage-, Tage-: ~bau der; *Pl.* **~e** *(Bergbau)* **a)** *o. Pl. (Bergbau über Tage)* opencast mining *no art.;* **b)** *(Anlage)* opencast mine; **~buch das** diary; **[über etw.** *(Akk.)***] ~buch führen** keep a diary [about sth.]; **~dieb der** *(abwertend)* idler; lazy-bones *sing.;* **~lang 1.** *Adj.; nicht präd.* lasting for days *postpos.;* **das ~lange Warten** the days of waiting; **nach ~langem Regen** after days of rain; **2.** *adv.* for days [on end]; **~löhner** [~lø:nɐ] **der; ~s, ~:** day-labourer

tagen *itr. V.* **a)** *(konferieren)* meet; **das Gericht tagt** the court is in session; **b)** *(geh.: dämmern)* **es tagt** day is breaking *or* dawning

Tage·reise die day's journey; **nach Passau sind es zehn ~n** it's a ten-day journey to Passau

Tages-: ~ablauf der day; daily routine; **~anbruch der** daybreak; dawn; **~arbeit die** day's work; **~ausflug der** day's outing; **~bedarf der** daily requirement; **~creme die** *(Kosmetik)* day cream; **~fahrt die** day trip; day excursion; **~gespräch das** topic of the day; **~karte die a)** *(Gastron.)* menu of the day; **b)** *(Fahr-, Eintrittskarte)* day ticket; **~kasse die a)** box-office *(open during the day);* **b)** *(~einnahme)* day's takings *pl.;* **~licht das;** *o. Pl.* daylight; **etw. ans ~licht bringen** *od.* **ziehen** *(fig.)***/ans ~licht kommen** *(fig.)* bring sth./come to light; **~marsch der a)** *(Fußmarsch)* day's hike; **b)** *(Strecke eines ~marsches)* day's march; **~mutter die;** *Pl.* **~mütter** child-minder; **~ordnung die** agenda; **an der ~ordnung sein** *(fig.)* be the order of the day; **~ration die** daily ration; **~tour die** *s.* **~fahrt; ~zeit die** time of day; **um diese ~zeit** at this time; **zu jeder ~- und Nachtzeit** at any time of the day or night; **~zeitung die** daily newspaper; daily

tage·weise *Adv.* on some days

tag·hell 1. *Adj.; nicht attr.* **a)** *(durch Tageslicht)* [day]light; **b)** *(wie am Tag)* bright as daylight *postpos.* **2.** *adv.* **etw. ist ~ erleuchtet** sth. is very brightly lit [up]

täglich ['tɛ:klɪç] **1.** *Adj.; nicht präd.* daily. **2.** *adv.* every day; **zweimal ~:** twice a day; **~ zwei Stunden** for two hours a day

tags *Adv.* **a)** by day; in the daytime; **b) ~ zuvor/davor** the day before; **~ darauf** the next *or* following day; the day after

Tag·schicht die day shift; **~ haben** be on [the] day shift

tags·über *Adv.* during the day

tag·täglich 1. *Adj.* day-to-day; daily. **2.** *adv.* every single day

Tag-: ~träumer der day-dreamer; **~und·nacht·gleiche die; ~, ~n** equinox

Tagung die; ~, ~en conference

Tagungs·ort der; *Pl.* **~orte** venue [for a/the conference]

Taifun [tai'fu:n] **der; ~s, ~e** typhoon

Taiga die; ~: taiga

Taille ['taljə] **die; ~, ~n** waist; **in der ~:** at the waist

Taillen·weite die waist measurement

Takelage [takə'la:ʒə] **die; ~, ~n** *(Seew.)* masts and rigging

Takt [takt] **der; ~[e]s, ~e a)** *(Musik)* time; *(Einheit)* bar; measure *(Amer.);* **den ~ [ein]halten** keep in time; **aus dem ~ kommen/sich nicht aus dem ~ bringen lassen** Lose/not lose the beat; **mit ihm muß ich mal ein paar ~e reden** *(fig. ugs.)* I need to have a serious talk with him; **b)** *o. Pl. (rhythmischer Bewegungsablauf)* rhythm; **im/gegen den ~:** in/out of rhythm; **c)** *o. Pl. (Feingefühl)* tact; **d)** *(Verslehre)* foot

Takt·gefühl das; *o. Pl.* sense of tact

taktieren *itr. V.* proceed tactically; **vorsichtig/klug ~:** use caution/clever tactics

Taktik ['taktɪk] **die; ~, ~en: [eine] ~:** tactics *pl.*

Taktiker der; ~s, ~, Taktikerin die; ~, ~nen tactician

taktisch 1. *Adj.* tactical. **2.** *adv.* tactically; **~ klug vorgehen** use clever *or* good tactics

takt·los 1. *Adj.* tactless. **2.** *adv.* tactlessly

Taktlosigkeit die; ~, ~en a) *o. Pl.* tactlessness; **b)** *(taktlose Handlung)* piece of tactlessness

Takt·stock der baton

takt·voll 1. *Adj.* tactful. **2.** *adv.* tactfully

Tal [ta:l] **das; ~[e]s, Täler** ['tɛ:lɐ] valley

tal·abwärts *Adv.* down the valley

Talar [ta'la:ɐ̯] **der; ~s, ~e** robe

tal·aufwärts *Adv.* up the valley

Talent [ta'lɛnt] **das; ~[e]s, ~e a)** *(Befähigung)* talent **(zu, für** for); **b)** *(Mensch)* talented person; **junge ~e fördern** promote young talent

talentiert [talɛn'ti:ɐ̯t] *Adj.* talented

Talg [talk] **der; ~[e]s, ~e a)** *(Speisefett)* suet; *(zur Herstellung von Seife, Kerzen usw.)* tallow; **b)** *(Haut~)* sebum

Talisman ['ta:lɪsman] **der; ~s, ~e** talisman

Talk-Show ['tɔ:kʃoʊ] **die** *(Ferns.)* talk show; chat show

Talmi ['talmi] **das; ~s a)** *(wertloser Schmuck)* imitation *or* cheap jewellery; *(fig.)* tinsel; **b)** *(vergoldete Legierung)* pinchbeck

Tal-: ~sohle die valley floor *or* bottom; *(fig.)* depression; **~sperre die** dam *(with associated reservoir and power-station);* **~station die** valley station

Tambour ['tambu:ɐ̯] **der; ~s, ~e** *od. (schweiz.)* **~en** *(veralt.)* drummer

Tamburin [tambu'ri:n] **das; ~s, ~e** tambourine

Tampon ['tampɔn] **der; ~s, ~s a)** *(Med.: Wattebausch)* tampon; plug; **b)** *(Menstruations~)* tampon

Tamtam [tam'tam] **das; ~s** *(ugs. abwertend)* **[großes] ~:** [a big] fuss; **~ machen** make a fuss

Tand [tant] **der; ~[e]s** trumpery

tändeln ['tɛndl̩n] *itr. V.* dally

Tandem ['tandɛm] **das; ~s, ~s** tandem; *(fig.)* pair

Tang [taŋ] **der; ~[e]s, ~e** seaweed

Tanga ['taŋga] **der; ~s, ~s** tanga

Tangens ['taŋgɛns] **der; ~, ~** *(Math.)* tangent

Tangente [taŋ'gɛntə] **die; ~, ~n a)** *(Math.)* tangent; **b)** *(Straße)* ring road; bypass

tangieren [taŋ'gi:rən] *tr. V.* **a)** affect; **b)** *(Math.)* be tangent to

Tango ['taŋgo] **der; ~s, ~s** tango

Tank [taŋk] **der; ~s, ~s,** *(seltener)* **~e** tank

tanken *tr., itr. V.* fill up; **Öl ~:** fill up with oil; **er tankte dreißig Liter [Super]** he put in thirty litres [of four-star]

Tanker der; ~s, ~: tanker

Tank-: ~säule die petrol-pump *(Brit.);* gasoline pump *(Amer.);* **~stelle die** petrol station *(Brit.);* gas station *(Amer.);* **~wagen der** tanker; **~wart** [~vart] **der; ~s, ~e** petrol-pump attendant *(Brit.)*

Tanne ['tanə] **die; ~, ~n a)** fir[-tree]; **schlank wie eine ~:** slender as a reed; **b)** *(Holz)* fir

Tannen-: ~baum der a) *(ugs.: Tanne a)* fir-tree; **b)** *(Weihnachtsbaum)* Christmas tree; **~nadel die** fir-needle; **~wald der** fir forest; **~zapfen der** fir-cone

Tante ['tantə] **die; ~, ~n a)** aunt; **b)** *(Kinderspr.: Frau)* lady; **c)** *(ugs.: Frau)* woman

tantenhaft 1. *Adj.* old-maidish; *(belehrend)* nannyish. **2.** *adv.* like an old maid; *(belehrend)* nannyishly

Tantieme [tã'ti̯e:mə] **die; ~, ~n a)** *(Gewinnbeteiligung)* percentage of the profits; **b)** *(von Künstlern)* royalty

Tanz [tants] **der; ~es, Tänze** ['tɛntsə] **a)** dance; **jmdn. zum ~ auffordern** ask sb. to dance *or* for a dance; **heute Abend ist ~:** there is dancing this evening; **b)** *(Zank, Auftritt)* song and dance *(fig. coll.)*

Tanz-: ~abend der evening dance; **~bar die** night-spot *(coll.)* with dancing; **~bär der** dancing bear; **~bein das** *in* **das ~bein schwingen** *(ugs. scherzh.)* shake a leg *(coll.);* **~boden der** dance-floor; **~café das** coffee-house with dancing

tänzeln ['tɛntsl̩n] *itr. V.* **a)** prance; **b)** *mit sein* **sie tänzelte ins Zimmer** she skipped into the room

tanzen 1. *itr. V.* **a)** dance; **~ gehen** go dancing; **auf dem Seil ~:** walk the tightrope; **b)** *mit sein (sich ~d fortbewegen)* dance; skip. **2.** *tr. V.* **Walzer ~:** dance a waltz; waltz

Tänzer ['tɛntsɐ] **der; ~s, ~, Tänzerin die; ~, ~nen a)** dancer; **b)** *(Partner[in])* dancing-partner

tänzerisch 1. *Adj.* dance-like ‹*movement, rhythm, step*›; **~e Begabung** talent for dancing. **2.** *adv.* **~ begabt sein** have a talent for dancing

Tanz-: ~fläche die dance-floor; **~kapelle die** dance band; **~lehrer der** dancing-teacher; **~lokal das** café/restaurant with dancing; **~musik die** dance music; **~orchester das** *s.* **~kapelle; ~saal der** dance-hall; *(in hotel, castle, etc.)* ballroom; **~sport der** ballroom *or* competition dancing *no art.;* **~stunde die a)** *(~kurs)* dancing-class; **~stunde nehmen, in die ~stunde gehen** take dancing lessons; go to dancing-class; **b)** *(einzelne Stunde)* dancing lesson; **~tee der** tea dance; thé dansant

Tapet das *in* **aufs ~ kommen** *(ugs.)* be brought up; come up [for discussion]; **etw. aufs ~ bringen** *(ugs.)* bring sth. up; broach sth.

Tapete [ta'pe:tə] **die; ~, ~n** wallpaper

Tapeten-: ~rolle die roll of wallpaper; **~wechsel der** *(ugs.)* change of scene

tapezieren [tape'tsi:rən] *tr. V.* [wall]paper

Tapezierer der; ~s, ~: paper-hanger

tapfer ['tapfɐ] **1.** *Adj.* brave; courageous. **2.** *adv.* **a)** bravely; courageously; **sich ~ halten** be brave; **b)** *(kräftig)* ‹*eat, drink*› heartily

Tapferkeit die; ~: courage; bravery

tappen *itr. V.* **a)** *mit sein* patter; **in eine Falle ~** *(fig.)* stumble into a trap; **b)** *(tastend greifen)* grope **(nach** for)

tapsig ['tapsiç] *(ugs.)* **1.** *Adj.* awkward; clumsy. **2.** *adv.* awkwardly; clumsily

Tarif [ta'ri:f] **der; ~s, ~e a)** *(Preis, Gebühr)* charge; *(Post~, Wasser~)* rate; *(Verkehrs~)* fares *pl.; (Zoll~)* tariff; **b)** *(~verzeichnis)* list of charges/rates/fares; tariff; **c)** *(Lohn~)* [wage] rate; *(Gehalts~)* [salary] scale; **weit über/unter ~ verdienen** earn well above/far below the agreed rate

Tarif·gruppe die *(Lohngruppe)* wage group; *(Gehaltsgruppe)* salary group

tariflich 1. *Adj.; nicht präd.* wage ‹*demand, dispute, etc.*›. **2.** *adv.* **Löhne und Gehälter sind ~ festgelegt** there are fixed rates for wages and salaries

Tarif·lohn der wage under the collective agreement

tarnen ['tarnən] **1.** *tr., itr. V.* camouflage. **2.** *refl. V.* camouflage oneself

Tarn·farbe die camouflage [colour]

Tarnung die; ~, ~en *(auch fig.)* camouflage

Tartan·bahn ['tartan-] **die** Tartan track **(P)**

Tasche ['taʃə] **die; ~, ~n a)** bag; **b)** *(in Kleidung, Koffer, Rucksack usw.)* pocket; **c)** *(fig.)* **sich** *(Dat.)* **die eigenen ~n füllen** *(ugs.)* line one's own pockets *or* purse; **jmdm. auf der ~ liegen** *(ugs.)* live off sb.; **etw. aus eigener** *od.* **der eigenen ~ bezahlen** pay for sth. out of one's own pocket; **jmdm. etw. aus der ~ ziehen** *(ugs.)* wangle money out of sb. *(coll.);* **[für etw.] tief in die ~ greifen [müssen]** *(ugs.)* [have to] dig deep in *or* into one's pocket [for sth.]; **jmdn. in die ~ stecken** *(ugs.)* put sb. in the shade; **sich** *(Dat.)* **in die eigene ~ lügen** *(ugs.)* fool oneself

Taschen-: ~buch das paperback; **~dieb der** pickpocket; **~geld das** pocket-money; **~lampe die** [pocket] torch *(Brit.) or (Amer.)* flashlight; **~messer das** pocket-knife; penknife; **~rechner der** pocket calculator; **~tuch das;** *Pl.* **~tücher** handkerchief; **~uhr die** pocket-watch

Täßchen ['tɛsçən] **das; ~s, ~:** [small] cup

Tasse ['tasə] **die; ~, ~n a)** cup; **eine ~ Tee** a cup of tea; **trübe ~** *(ugs. abwertend)* drip *(coll.);* **b)** *(~ mit Untertasse)* cup and saucer; **nicht alle ~n im Schrank haben** *(ugs.)* not be right in the head *(coll.)*

Taste ['tastə] **die; ~, ~n a)** key; **b)** *(am Telefon, Radio, Fernsehgerät, Taschenrechner usw.)* button

tasten 1. *itr. V.* grope, feel **(nach** for); **~de Fragen** *(fig.)* tentative questions. **2.** *refl. V.* grope *or* feel one's way

Tasten-: ~instrument das keyboard instrument; **~telefon das** push-button telephone

tat [ta:t] *1. u. 3. Pers. Sg. Prät. v.* **tun**

Tat die; ~, ~en a) *(Handlung)* act; *(das Tun)* action; **zur ~ schreiten** proceed to action; **jmdn. auf frischer ~ ertappen** catch sb. red-handed *or* in the act; **etw. in die ~ umsetzen** put sth. into action *or* effect; **eine gute ~ vollbringen** do a good deed; **b) in der ~** *(verstärkend)* actually; *(zustimmend)* indeed

Tat·bestand der a) facts *pl.* [of the matter *or* case]; **b)** *(Rechtsw.)* elements *pl.* of an offence; **der ~bestand der vorsätzlichen Tötung** the offence of premedidated murder

Taten·drang der desire *or* thirst for action

taten·los 1. *Adj.* idle. **2.** *adv.* idly; **einer Sache** *(Dat.)* **~los zusehen** watch sth. without taking any action

Täter ['tɛ:tɐ] **der; ~s, ~:** culprit; **wer ist der ~?** who did it?; **der ~ hat sich der Polizei gestellt** the person who committed the crime gave himself/herself up to the police; **nach dem ~ fahnden** search *or* look for the person responsible [for the crime]

Täterin die; ~, ~nen *s.* **Täter**

tätig ['tɛ:tɪç] *Adj.* **a)** **~ sein** work; **~ werden** *(bes. Amtsspr.)* take action; **b)** *(rührig, aktiv)* active

tätigen ['tɛ:tɪgn̩] *tr. V. (Kaufmannsspr., Papierdt.)* transact ‹*business, deal, etc.*›; **Einkäufe ~:** effect purchases

Tätigkeit die; ~, ~en a) activity; *(Arbeit)* job; **eine ~ ausüben** do work; do a job; **b)** *o. Pl. (das In-Betrieb-Sein)* operation

Tat·kraft die energy; drive

tat·kräftig 1. *Adj.* energetic, active ‹*person*›; active ‹*help, support*›. **2.** *adv.* energetically; actively

tätlich ['tɛ:tlɪç] **1.** *Adj.* physical ‹*clash, attack, resistance, etc.*›; **gegen jmdn. ~ werden** become violent towards sb. **2.** *adv.* physically

Tat·ort der; *Pl.* **~e** scene of a/the crime

tätowieren [tɛto'vi:rən] *tr. V.* tattoo; **sich** *(Dat.)* **etw. ~ lassen** have oneself tattooed with sth.

Tätowierung die; ~, ~en a) tattoo; **b)** *(das Tätowieren)* tattooing

Tat·sache die fact; **~?** *(ugs.)* really?; is that true?; **nackte ~n** hard facts; *(scherzh.)* naked bodies; **vollendete ~n schaffen** create a fait accompli; *s. auch* **Vorspiegelung**

Tatsachen·bericht der factual report

tatsächlich ['ta:tzɛçlɪç] **1.** *Adj.* actual; real. **2.** *adv.* actually; really; **ist das ~ wahr?** is that really true?; **ich habe mich ~ geirrt** I was indeed mistaken

tätscheln ['tɛtʃl̩n] *tr. V.* pat

tatschen ['tatʃn̩] *itr. V. (ugs.)* **an/auf etw.** *(Akk.)* **~:** paw sth.

tatt[e]rig *Adj. (ugs.)* shaky ‹*hands, movements, etc.*›; doddery ‹*person*›

tat·verdächtig *Adj.* suspected

Tatze ['tatsə] **die; ~, ~n** paw

¹Tau [tau] **der; ~[e]s** dew

²Tau das; ~[e]s, ~e *(Seil)* rope

taub [taup] *Adj.* **a)** deaf; **b)** *(wie abgestorben)* numb; **c)** *(leer, unbefruchtet usw.)* empty ‹*nut*›; unfruitful ‹*ear of corn*›; dead ‹*rock*›

¹Taube die; ~, ~n pigeon; *(Turtel~; auch Politik fig.)* dove

²Taube der/die; *adj. Dekl.* deaf person; deaf man/woman; **die ~n** the deaf

Tauben·schlag der pigeon-loft; *(für Turteltauben)* dovecot; **hier geht es zu wie in einem** *od.* **im ~** *(ugs.)* it's like Piccadilly Circus here *(Brit. coll.)*; it's like being in the middle of Times Square *(Amer.)*

Taubheit die; ~: deafness

taub·stumm *Adj.* deaf and dumb

Taub·stumme der/die; *adj. Dekl.* deaf mute

tauchen 1. *itr. V.* **a)** *auch mit sein* dive (**nach** for); **er kann zwei Minuten [lang] ~:** he can stay under water for two minutes; **b)** *mit sein (ein~)* dive; *(auf~)* rise; emerge. **2.** *tr. V.* **a)** *(ein~)* dip; **b)** *(unter~)* duck

Taucher der; ~s, ~: diver; *(mit Flossen und Atemgerät)* skin-diver

Taucher-: ~anzug der diving-suit; **~brille die** diving-goggles *pl.*

Taucherin die; ~, ~nen *s.* **Taucher**

Tauch·sieder der; ~s, ~: portable immersion heater

tauen 1. *itr. V.* **a)** *(unpers.)* **es taut** it's thawing; **b)** *mit sein (schmelzen)* melt. **2.** *tr. V.* melt; thaw

Tauf·becken das font

Taufe die; ~, ~n a) *o. Pl. (christl. Rel.: Sakrament)* baptism; **b)** *(christl. Rel.: Zeremonie)* christening; baptism; **etw. aus der ~ heben** *(fig. ugs.)* launch sth.

taufen *tr. V.* **a)** *(die Taufe vollziehen an)* baptize; **katholisch getauft sein** be baptized a Catholic; **b)** *(einen Namen geben)* christen ‹*child, ship, animal, etc.*›; **ein Kind auf den Namen Peter ~:** christen a child Peter

Täufling ['tɔyflɪŋ] **der; ~s, ~e** child to be baptized; *(Erwachsener)* person to be baptized

Tauf-: ~pate der godparent; *(männlicher ~pate)* godfather; **~patin die** godmother; **~schein der** certificate of baptism; baptismal certificate

taugen ['taugn̩] *itr. V.* **nichts/wenig** *od.* **nicht viel/etwas ~:** be no/not much/some good *or* use; **zu** *od.* **für etw. ~** ‹*person*› be suited to sth.; ‹*thing*› be suitable for sth.; **nicht wissen, was etw. wirklich taugt** not know how useful sth. really is

Taugenichts der; ~[es], ~e *(veralt. abwertend)* good-for-nothing

tauglich *Adj.* **a)** **[nicht] ~:** [un]suitable; **b)** *(für Militärdienst)* fit [for service]

Taumel der; ~s a) *(Schwindel, Benommenheit)* [feeling of] dizziness *or* giddiness; **b)** *(Begeisterung, Rausch)* frenzy; fever; **ein ~ der Begeisterung** a fever of excitement

taumeln ['taumln̩] *itr. V.* **a)** *auch mit sein* reel, sway; **b)** *mit sein (sich ~d bewegen)* stagger

Tausch der; ~[e]s, ~e exchange; **ein guter/schlechter ~:** a good/bad deal; **im ~ gegen** *od.* **für etw.** in exchange for sth.

tauschen 1. *tr. V.* exchange (**gegen** for); **Briefmarken ~:** exchange *or* swap stamps; **sie tauschten die Partner/Plätze** they changed *or* swapped partners/places. **2.** *itr. V.* **mit jmdm. ~** *(fig.)* change *or* swap places with sb.

täuschen ['tɔyʃn̩] **1.** *tr. V.* deceive; **der Schein täuscht uns oft** appearances are often deceiving; **wenn mich nicht alles täuscht** unless I'm completely mistaken. **2.** *itr. V.* **a)** *(irreführen)* be deceptive; **b)** *(bes. Sport: ablenken)* make a feint. **3.** *refl. V.* be wrong *or* mistaken (**in +** *Dat.* about); **ich habe mich in ihm getäuscht** I was wrong about him; he disappointed me; **da täuschst du dich aber [gewaltig]** but that's where you're [very much] mistaken

täuschend 1. *Adj.* remarkable, striking ‹*similarity, imitation*›. **2.** *adv.* remarkably

Tausch-: ~geschäft das exchange [deal]; **~handel der a)** bartering; **b)** *(Wirtsch.)* trade by barter

Täuschung die; ~, ~en a) *(das Täuschen)* deception; **b)** *(Selbst~)* delusion; illusion; **optische ~:** optical illusion

tausend ['tauznt] *Kardinalz.* **a)** a *or* one thousand; **~ und aber ~ Ameisen** thousands and thousands of ants; **b)** *(ugs.: sehr viele)* thousands of; **~ Dank** a thousand thanks

Tausend das; ~s, ~e *od.* **~ a)** *nicht in Verbindung mit Kardinalzahlen; Pl.:~ (Einheit von tausend Stück)* thousand; **vom ~:** per thousand; **b)** *Pl. (eine unbestimmte große Zahl)* thousands; **~e Zuschauer** thousands of spectators; **die Tiere starben zu ~en** the animals died in [their] thousands

Tausender der; ~s, ~ a) *(ugs.) (Tausendmarkschein usw.)* thousand-mark/-dollar *etc.* note; *(Betrag)* thousand marks/dollars *etc.*; **b)** *(Math.)* thousand

tausenderlei *Gattungsz.; indekl. (ugs.)* **a)** *(von verschiedener Art)* a thousand and one different ‹*answers, kinds, etc.*›; **b)** *(viele)* a thousand and one

Tausend·jahr·feier die millenary; millennial
tausend·jährig *Adj.; nicht präd.* **a)** *(tausend Jahre alt)* [one-]thousand-year-old; **b)** *(tausend Jahre dauernd)* thousand-year[-long]
tausend·mal *Adv.* a thousand times
tausendst... *Ordinalz.* thousandth
tausendstel ['tauzn̩tstl̩] *Bruchz.* thousandth
Tau-: ~tropfen der dew-drop; **~wetter das** thaw
Taxi ['taksi] **das; ~s, ~s** taxi; **mit dem ~:** by taxi *or* in a taxi
taxieren *tr. V.* **a)** *(ugs.: schätzen)* estimate **(auf** + *Akk.* at); **etw. zu hoch/niedrig ~:** overestimate/underestimate sth.; **b)** *(den Wert ermitteln von)* value **(auf** + *Akk.* at); **etw. zu hoch/niedrig ~:** overvalue/undervalue sth.; **c)** *(ugs.: mustern, prüfen)* size up *(coll.)*; **d)** *(einschätzen)* assess
Taxi-: ~fahrer der taxi-driver; **~fahrt die** taxi ride; **~stand der** taxi-rank *(Brit.)*; taxi-stand
Tb, Tbc [te:'be:, te:be:'tse:] **die; ~** *Abk.* **Tuberkulose** TB
Tb-Kranke [te:'be:-] **der/die;** *adj. Dekl.* TB patient; patient with TB
Team [ti:m] **das; ~s, ~s** team
Teamwork ['ti:mwə:k] **das; ~s** team-work
Technik ['tɛçnɪk] **die; ~, ~en a)** *o. Pl.* technology; *(Studienfach)* engineering *no art.*; **auf dem neuesten Stand der ~:** incorporating the latest technical advances; **b)** *o. Pl. (Ausrüstung)* equipment; machinery; **c)** *(Arbeitsweise, Verfahren)* technique; **d)** *o. Pl. (eines Gerätes)* workings *pl.*
Techniker der; ~s, ~, Technikerin die; ~, ~nen a) technical expert; **b)** *(im Sport, in der Kunst)* technician
technisch ['tɛçnɪʃ] **1.** *Adj.* technical ⟨*fault*⟩; technological ⟨*progress, age*⟩. **2.** *adv.* technically; technologically ⟨*advanced*⟩; **~ begabt sein** have a technical flair
Technokrat [tɛçno'kra:t] **der; ~en, ~en, Technokratin die; ~, ~nen** technocrat
Technologie [tɛçnolo'gi:] **die; ~, ~n** technology
Technologie·park der science park
technologisch 1. *Adj.* technological. **2.** *adv.* technologically
Teddy·bär ['tɛdi-] **der** teddy bear
Tee [te:] **der; ~s, ~s** tea; **[einen] ~ machen** make some tea
Tee-: ~beutel der tea-bag; **~glas das;** *Pl.* **~gläser** tea-glass; **~kanne die** teapot; **~löffel der** teaspoon
Teenager ['ti:neɪdʒə] **der; ~s, ~:** teenager
Teer [te:ɐ̯] **der; ~[e]s,** *(Arten:)* **~e** tar
teeren *tr. V.* tar; **jmdn. ~ und federn** tar and feather sb.
Tee-: ~rose die tea-rose; **~service das** tea-service; tea-set; **~sieb das** tea-strainer; **~stube die** tea-room; **~tasse die** teacup
Teich [taiç] **der; ~[e]s, ~e** pond
Teig [taik] **der; ~[e]s, ~e** dough; *(Kuchen~, Biskuit~)* pastry; *(Pfannkuchen~, Waffel~)* batter; *(in Rezepten auch)* mixture
teigig *Adj.* **a)** *(wie Teig)* doughy; **b)** *(blaß u. schwammig)* pasty ⟨*face, skin, complexion*⟩
Teig·waren *Pl.* pasta *sing.*
Teil [tail] **a) der; ~[e]s, ~e** *(etw. von einem Ganzen)* part; **achter ~** *(Achtel)* eighth; **weite ~e des Landes** wide areas of the country; **ein [großer** *od.* **guter] ~ der Bevölkerung** a [large] section of the population; **zum ~:** partly; **den größten ~ des Weges hat er zu Fuß zurückgelegt** he walked most of the way; **b) der** *od.* **das; ~[e]s, ~e** *(Anteil)* share; **c) der** *od.* **das; ~[e]s, ~e** *(Beitrag)* share; **ich will gerne mein[en] ~ dazu beisteuern** I should like to do my share *or* bit; **d) der; ~[e]s, ~e** *(beteiligte Person[en]; Rechtsw.: Partei)* party; **e) das; ~[e]s, ~e** *(Einzel~)* part; **etw. in seine ~e zerlegen** take sth. apart *or* to pieces
teil·bar *Adj.* divisible **(durch** by)
Teilchen das; ~s, ~ a) *(kleines Stück)* [small] part; **b)** *(Partikel)* particle
teilen 1. *tr. V.* **a)** *(zerlegen, trennen)* divide [up]; **b)** *(dividieren)* divide **(durch** by); **c)** *(auf~)* share **(unter** + *Dat.* among); **d)** *(teilweise überlassen, gemeinsam nutzen, teilhaben an)* share; **e)** *(in zwei Teile ~)* divide. **2.** *refl. V.* **a) sich** *(Dat.)* **etw. [mit jmdm.] ~:** share sth. [with sb.]; **b)** *(auseinandergehen)* **der Weg teilt sich** the road forks; **geteilter Meinung sein** have different views *or* opinions. **3.** *itr. V.* share
teil|haben *unr. itr. V.* share **(an** + *Dat.* in)
Teilhaber der; ~s, ~, Teilhaberin die; ~, ~nen partner
Teil·kasko·versicherung die *insurance giving limited cover*
Teilnahme ['tailna:mə] **die; ~, ~n a)** *(das Mitmachen)* participation **(an** + *Dat.* in); **~ an einem Kurs** attendance at a course; **b)** *(Interesse)* interest **(an** + *Dat.* in); **c)** *(geh.: Mitgefühl)* sympathy
teilnahms-: ~los *Adj. (gleichgültig)* indifferent; *(apathisch)* apathetic; **~voll 1.** *Adj.* compassionate; **2.** *adv.* compassionately; **jmdn. ~ ansehen** look at sb. with compassion
teil|nehmen ['tailne:mən] *unr. itr. V.* **a)** *(dabei sein bei)* **[an etw.** *(Dat.)***] ~:** attend [sth.]; **b)** *(beteiligt sein)* **[an etw.** *(Dat.)***] ~:** take part [in sth.]; **am Krieg ~:** fight in the war; **c)** *(als Lernender)* **[an einem Lehrgang]/am Unterricht ~:** attend [a course]/lessons; **d)** *(Teilnahme zeigen)* **an jmds. Schmerz/Glück ~:** share sb.'s pain/happiness
Teilnehmer der; ~s, ~ a) participant *(Gen.,* **an** + *Dat.* in); *(bei Wettbewerb auch)* competitor, contestant **(an** + *Dat.* in); **b)** *(Fernspr.)* subscriber
teils [tails] *Adv.* partly; **Wie hat es dir gestern gefallen? – Teils, ~** *(ugs.)* How did you like it yesterday? – So so
Teil-: ~strecke die *(einer Straße)* stretch; *(einer Buslinie usw.)* stage; *(Rennsport)* stage; **~stück das** piece; part
Teilung die; ~, ~en division
teil·weise 1. *Adv.* partly. **2.** *adj.* partial
Teil-: ~zahlung die instalment; **~zeit·arbeit die** part-time work *no indef. art.*; *(Stelle)* part-time job
Teint [tɛ̃:] **der; ~s, ~s** complexion
tele-, Tele- ['te:le-] tele-
Telefon ['te:lefo:n, *auch* tele'fo:n] **das; ~s, ~e** telephone; phone *(coll.)*; **ans ~ gehen** answer the [tele]phone
Telefon-: ~anruf der [tele]phone call; **~apparat der** telephone
Telefonat [telefo'na:t] **das; ~[e]s, ~e** telephone call
Telefon-: ~buch das [tele]phone book *or* directory; **~gebühr die** telephone charge; **~gespräch das** telephone conversation
telefonieren [telefo'ni:rən] *itr. V.* make a [tele]phone call; **mit jmdm. ~:** talk to sb. [on the telephone]/be on the telephone to sb.; **nach Hause/England ~:** phone home/make a [tele]phone call to England; **er telefoniert gerade** he is on the phone at the moment
telefonisch 1. *Adj.* telephone; **die ~e Zeitansage** the speaking clock *(Brit. coll.)*; the telephone

time service. **2.** *adv.* by telephone; **jmdm. etw. ~ mitteilen** inform sb. of sth. over the *or* by telephone; **ich bin ~ zu erreichen** I can be contacted by telephone

Telefonist [telefo'nɪst] **der; ~en, ~en, Telefonistin die; ~, ~nen** telephonist; *(in einer Firma)* switchboard operator

Telefon-: ~nummer die [tele]phone number; **~rechnung die** [tele]phone bill; **~verbindung die** telephone line; **~verzeichnis das** telephone list; **~zelle die** [tele]phone-booth *or (Brit.)* -box; call-box *(Brit.)*

Telegraf [tele'gra:f] **der; ~en, ~en** telegraph

Telegrafie die; ~: telegraphy *no art.*

telegrafieren [telegra'fi:rən] *itr., tr. V.* telegraph; **jmdm. ~:** send a telegram to sb.

telegrafisch 1. *Adj.* telegraphic. **2.** *adv.* by telegraph *or* telegram; **~ überwiesenes Geld** money sent by telegram *or* cable

Telegramm das telegram

Tele-: ~kolleg das ≈ Open University *(Brit.)*; **~objektiv das** *(Fot.)* telephoto lens

Telepathie [telepa'ti:] **die; ~:** telepathy *no art.*

Tele·skop [tele'sko:p] **das; ~s, ~e** telescope

Telex ['te:lɛks] **das; ~, ~[e]** telex

Teller ['tɛlɐ] **der; ~s, ~** plate; **ein ~ Suppe** a plate of soup

Teller·wäscher der dishwasher; **vom ~ zum Millionär [werden]** [go] from rags to riches

Tempel ['tɛmpl̩] **der; ~s, ~:** temple

Temperament [tɛmpəra'mɛnt] **das; ~[e]s, ~e a)** *(Wesensart)* temperament; **b)** *o. Pl.* **eine Frau mit ~:** a lively *or* vivacious woman; a woman with spirit; **sein ~ reißt alle mit** his vivacity infects everyone; **das ~ geht oft mit mir durch** I often lose my temper

temperament-: ~los *Adj.* spiritless; lifeless; **~voll** *Adj.* spirited ‹*person, speech, dance, etc.*›; lively ‹*start etc.*›

Temperatur [tɛmpəra'tu:ɐ̯] **die; ~, ~en** temperature; **die richtige ~ haben** be [at] the right temperature; **[erhöhte] ~ haben** ‹*person*› have *or* be running a temperature; **jmds. ~ messen** take sb.'s temperature

Temperatur-: ~anstieg der rise in temperature; **~rückgang der** drop *or* fall in temperature; **~sturz der** [sudden] fall *or* drop in temperature

temperieren [tɛmpə'ri:rən] *tr. V.* bring to the right temperature; **das Wasser ist gut temperiert** the water is [at] the right temperature

Tempo ['tɛmpo] **das; ~s, ~s** *od.* **Tempi a)** *Pl.* **~s** speed; **das ~ erhöhen** speed up; accelerate; **in** *od.* **mit hohem ~:** at high speed; **hier gilt ~ 100** there is a 100 k.p.h. speed limit here; **~ [~]!** *(ugs.)*, **macht mal ein bißchen ~** *(ugs.)* get a move on; **b)** *(Musik)* tempo; time

Tempo·limit das *(Verkehrsw.)* speed limit

Tendenz [tɛn'dɛnts] **die; ~, ~en a)** trend; **es herrscht die ~/die ~ geht dahin, ... zu ...:** there is a tendency to ...; the trend is to ...; **b)** *(Hang, Neigung)* tendency; **c)** *(oft abwertend: Darstellungsweise)* slant; bias

tendenziös [...'tsiø:s] *Adj.* tendentious

tendieren [tɛn'di:rən] *itr. V.* tend (**zu** towards); **der nach links ~de Flügel dieser Partei** the branch of the party with left-wing leanings

Tennis ['tɛnɪs] **das; ~:** tennis *no art.*

Tennis-: ~ball der tennis-ball; **~platz der** tennis-court; **~schläger der** tennis-racket; **~spiel das a)** *o. Pl. (Tennis)* tennis *no art.*; **b)** *(Einzelspiel)* game of tennis; **~spieler der** tennis-player

Tenor [te'no:ɐ̯] **der; ~s, Tenöre** [te'nø:rə], *(österr. auch:)* **~e** *(Musik)* **a)** *(Stimmlage, Sänger)* tenor; **b)** *o. Pl. (im Chor)* tenors *pl.*; tenor voices *pl.*

Teppich ['tɛpɪç] **der; ~s, ~e** carpet; *(kleiner)* rug; **auf dem ~ bleiben** *(fig. ugs.)* keep one's feet on the ground; **etw. unter den ~ kehren** *(fig. ugs.)* sweep sth. under the carpet

Teppich-: ~boden der fitted carpet; **~kehrer der** carpet-sweeper; **~klopfer der** carpet-beater

Termin [tɛr'mi:n] **der; ~s, ~e a)** *(festgelegter Zeitpunkt)* date; *(Anmeldung)* appointment; *(Verabredung)* engagement; **sich** *(Dat.)* **einen ~ geben lassen** make an appointment; **b)** *(Rechtsw.)* hearing

Terminal ['tø:ɐ̯minəl] **das; ~s, ~s** terminal

termin·gemäß 1. *Adj.* on time *postpos.* **2.** *adv.* on time; on schedule

Termin·kalender der appointments book; *(für gesellschaftliche Termine)* engagements diary

Termite [tɛr'mi:tə] **die; ~, ~n** termite

Terpentin [tɛrpɛn'ti:n] **das,** *(österr. meist:)* **der; ~s a)** *(Harz)* turpentine; **b)** *(ugs.: Terpentinöl)* turps *sing. (coll.)*

Terrain [tɛ'rɛ̃:] **das; ~s, ~s a)** *(Gelände)* terrain; **es ist für ihn ein unbekanntes ~** *(fig.)* it is unknown territory to him; **das ~ sondieren** *(fig. geh.)* sound out the situation; **b)** *(Baugelände)* building land

Terrarium [tɛ'ra:ri̯ʊm] **das; ~s, Terrarien** terrarium

Terrasse [tɛ'rasə] **die; ~, ~n** terrace

terrassen·förmig 1. *Adj.* terraced. **2.** *adv.* in terraces

Terrier ['tɛri̯ɐ] **der; ~s, ~:** terrier

Terrine [tɛ'ri:nə] **die; ~, ~n** tureen

territorial [tɛrito'ri̯a:l] *Adj.* territorial

Territorium [tɛri'to:ri̯ʊm] **das; ~s, Territorien a)** *(Gebiet, Land)* land; territory; **b)** *(Hoheitsgebiet)* territory

Terror ['tɛrɔr] **der; ~s a)** terrorism *no art.*; **blutiger ~:** terror and bloodshed; **b)** *(ugs.: Zank u. Streit)* trouble; **c)** *(ugs.: großes Aufheben)* big row *(coll.) or* fuss; **~ machen** raise hell *(coll.)*

Terror·anschlag der terrorist attack

terrorisieren *tr. V.* **a)** *(durch Terror unterdrücken)* terrorize; **b)** *(ugs.: belästigen)* pester

Terrorismus der; ~: terrorism *no art.*

Terrorist der; ~en, ~en, Terroristin die; ~, ~nen terrorist

Tertia ['tɛrtsi̯a] **die; ~, Tertien** *(Schulw.) (veralt.)* fourth and fifth year *(of a Gymnasium)*

Terz [tɛrts] **die; ~, ~en** *(Musik)* third

Tesa·film Ⓦᶻ ['te:za-] **der; ~[e]s** Sellotape *(Brit.)* **(P)**; Scotch tape *(Amer.)* **(P)**

Test [tɛst] **der; ~[e]s, ~s** *od.* **~e** test

Testament [tɛsta'mɛnt] **das; ~[e]s, ~e a)** will; **das ~ eröffnen** read the will; **er kann sein ~ machen** *(fig. ugs.)* he is [in] for it *(coll.)*; **b)** *(christl. Rel.)* Testament; **das Alte/Neue ~:** the Old/New Testament

testamentarisch [tɛstamɛn'ta:rɪʃ] **1.** *Adj.; nicht präd.* testamentary. **2.** *adv.* **etw. ~ verfügen** write sth. in one's will

testen *tr. V.* test (**auf** + *Akk.* for)

Test-: ~fall der test case; **~frage die** test question

Tetanus ['te:tanʊs] **der; ~** *(Med.)* tetanus *no art.*

teuer ['tɔyɐ] **1.** *Adj.* **a)** expensive; dear *usu. pred.*; **wie ~ war das?** how much did that cost?; **Kaffee soll wieder teurer werden** coffee is supposed to be going up again; *s. auch* **Rat a; b)** *(veralt.: geschätzt)* dear; **teurer Freund!** [my] dear friend!; **[mein] Teuerster!** [my] dearest; *(von Mann zu Mann)* [my] dearest friend. 2. *adv.* expensively; dearly; **etw. ~ kaufen/verkaufen** pay a great deal for sth./sell sth. at a high price; **sie haben ihren Sieg ~ erkauft** they paid a high price for their victory

Teuerung die; ~, ~en rise in prices

Teufel ['tɔyfl̩] **der; ~s, ~:** devil; **der ~:** the Devil; **wie der ~ fahren** drive in daredevil fashion; **der ~ ist los** all hell's let loose *(coll.)*; **dich reitet wohl der ~!** what's got into you?; **hol' dich/ihn** *usw.* **der ~!/der ~ soll dich/ihn** *usw.* **holen!** *(salopp)* sod *(Brit. sl.) or (coll.)* damn you/him *etc.*; **das weiß der ~!** *(salopp)* God [only] knows; **hinter etw. hersein wie der ~ hinter der armen Seele** *(ugs.)* be greedy for sth.; **den ~ werde ich [tun]!** *(salopp)* like hell [I will]! *(coll.)*; **mal bloß nicht den ~ an die Wand!** *(ugs.)* don't invite trouble/*(stärker)* disaster by talking like that!; **des ~s sein** *(ugs.)* be mad; have taken leave of one's senses; **in ~s Küche kommen/jmdn. in ~s Küche bringen** *(ugs.)* get into/put sb. in a hell of a mess *(coll.)*; **warum mußt du den jetzt auf ~ komm raus überholen?** *(ugs.)* why are you so hell-bent on overtaking him now? *(coll.)*; **zum ~ gehen** *(ugs.: kaputtgehen)* be ruined; **er soll sich zum ~ scheren!** *(salopp)* he can go to hell *(coll.) or* blazes *(coll.)*; **wer/wo** *usw.* **zum ~ ...** *(salopp)* who/where *etc.* the hell ... *(coll.)*; **wenn man vom ~ spricht[, dann ist er nicht weit]** *(scherzh.)* speak *or* talk of the devil [and he will appear]

Teufels-: ~kerl der *(ugs.)* amazing fellow; **~werk das** devil's work *no indef. art.*

teuflisch 1. *Adj.* **a)** devilish, fiendish ‹*plan, trick, etc.*›; fiendish, diabolical ‹*laughter, pleasure, etc.*›; **b)** *(ugs.: groß, intensiv)* terrible *(coll.)*; dreadful *(coll.)*. **2.** *adv.* **a)** fiendishly; diabolically; **b)** *(ugs.)* terribly *(coll.)*; dreadfully *(coll.)*

Text [tɛkst] **der; ~[e]s, ~e a)** text; *(eines Gesetzes, auf einem Plakat)* wording; *(eines Theaterstücks)* script; *(einer Oper)* libretto; **weiter im ~!** *(ugs.)* [let's] carry on!; **b)** *(eines Liedes, Chansons usw.)* words *pl.*; *(eines Schlagers)* words *pl.*; lyrics *pl.*; **c)** *(zu einer Abbildung)* caption

Text-: ~aufgabe die *(Schule)* problem; **~buch das** libretto

texten *tr. V.* write ‹*song, advertisement, etc.*›

Texter der; ~s, ~: writer; *(in der Werbung)* copy-writer

Textil-: ~branche die, ~gewerbe das textile trade *or* industry

Textilien *Pl.* **a)** textiles; **b)** *(Fertigwaren)* textile goods

Textil-: ~industrie die textile industry **~strand der** *(ugs. scherzh.)* beach where there is no nude bathing; **~waren** *Pl.* textile goods

Text-: ~stelle die passage [in a/the text]; **~verarbeitung die** text processing; word processing

Thailand (das); ~s Thailand

Thailänder der; ~s, ~, Thailänderin die; ~, ~nen Thai

Theater [te'a:tɐ] **das; ~s, ~ a)** theatre; **ins ~ gehen** go to the theatre; **zum ~ gehen** *(ugs.)* go into the theatre; tread the boards; **beim** *od.* **am ~ sein** be *or* work in the theatre; **~ spielen** act; *(fig.)* play-act; pretend; put on an act; **b)** *o. Pl. (fig. ugs.)* fuss; **mach [mir] kein ~!** don't make a fuss; **das ist doch alles nur ~:** that's all just play-acting

Theater-: ~besucher der theatre-goer; **~karte die** theatre ticket; **~stück das** [stage] play

theatralisch [tea'tra:lɪʃ] *(auch fig.)* **1.** *Adj.* theatrical. **2.** *adv.* theatrically

Theke ['te:kə] **die; ~, ~n a)** *(Schanktisch)* bar; **b)** *(Ladentisch)* counter; **unter der ~** *(fig.)* under the counter

Thema ['te:ma] **das; ~s, Themen** *od.* **~ta** subject; topic; *(einer Abhandlung)* subject; theme; *(Leitgedanke)* theme; **das ~ wechseln** change the subject; **vom ~ abkommen** *od.* **abschweifen** wander off the subject *or* point

Thematik [te'ma:tɪk] **die; ~, ~en** theme; *(Themenkreis)* themes *pl.*; *(Themenkomplex)* complex of themes

thematisch 1. *Adj.* thematic; **etw. nach ~en Gesichtspunkten ordnen** arrange sth. according to subject. **2.** *adv.* thematically; *(was das Thema betrifft)* as regards subject matter

Theologe [teo'lo:gə] **der; ~n, ~n** theologian

Theologie [teolo'gi:] **die; ~, ~n** theology *no art.*

Theologin die; ~, ~nen theologian

theologisch 1. *Adj.* theological. **2.** *adv.* theologically

Theoretiker [teo're:tikɐ] **der; ~s, ~:** theoretician; theorist

theoretisch [teo're:tɪʃ] **1.** *Adj.* theoretical. **2.** *adv.* theoretically

Theorie [teo'ri:] **die; ~, ~n** theory

Therapeut [tera'pɔyt] **der; ~en, ~en, Therapeutin die; ~, ~nen** therapist; therapeutist

therapeutisch 1. *Adj.* therapeutic. **2.** *adv.* therapeutically

Therapie [tera'pi:] **die; ~, ~n** therapy **(gegen** for); **eine ~ machen** *(ugs.)* undergo *or* have therapy *or* treatment

Thermal·bad das a) *(Ort)* thermal spa; **b)** *(Bad)* thermal bath

Thermik ['tɛrmɪk] **die; ~** *(Met.)* thermal

Thermo·meter [tɛrmo-] **das** thermometer

Thermos·flasche Ⓦⓩ ['tɛrmɔs-] **die** Thermos flask **(P)**; vacuum flask

Thermostat [tɛrmo'sta:t] **der; ~[e]s** *od.* **~en, ~e** *od.* **~en** thermostat

These ['te:zə] **die; ~, ~n** thesis

Thomas ['to:mas] **der; ~, ~se** *in* **ungläubiger ~:** doubting Thomas

Thron [tro:n] **der; ~[e]s, ~e** throne; **sein ~ wackelt** *(fig.)* his position is becoming very shaky

thronen *itr. V.* sit enthroned; *(fig.: erhöht liegen)* tower

Thron·folger der; ~s, ~, Thron·folgerin die; ~, ~nen heir to the throne

Thun·fisch ['tu:n-] **der** tuna

Thüringen ['ty:rɪŋən] **(das); ~s** Thuringia

Thymian ['ty:mia:n] **der; ~s, ~e** thyme

Tibet ['ti:bɛt] **(das); ~s** Tibet

tibetisch *Adj.* Tibetan

Tick der; ~[e]s, ~s a) *(ugs.: Schrulle)* quirk; thing *(coll.)*; **du hast wohl einen kleinen ~:** you must be round the bend *(coll.)*; **b)** *(Med.)* tic

ticken *itr. V.* tick; **du tickst wohl nicht richtig** *(salopp)* you must be off your rocker *(sl.)*

Ticket ['tɪkət] **das; ~s, ~s** ticket

ticktack ['tɪk'tak] *Interj.* tick-tock

tief [ti:f] **1.** *Adj.* **a)** *(auch fig.)* deep; low ‹*neckline, bow*›; long ‹*fall*›; **b)** *(niedrig)* low ‹*table, chair, temperature, tide, level, cloud*›; **den Sattel etwas ~er stellen** lower the saddle a bit; **c)** *(intensiv, stark)* deep; intense ‹*pain,* *suffering*⟩; utter ⟨*misery*⟩; great ⟨*need, want*⟩; **d)** *(weit im Innern gelegen)* **im ~en/~sten Afrika** in the depths of/in darkest Africa; **es freut mich aus ~stem Herzen/~ster Seele** I really am delighted; **in ~er/~ster Nacht** in the *or* at dead of night; **im ~en/~sten Winter** in the depths of winter. **2.** *adv.* **a)** *(weit unten)* deep; **100 m ~ in/unter der Erde** 100 metres [down] under the earth; **er war ~ in Gedanken** he was deep in thought; **b)** *(weit nach unten)* ⟨*dig, drill*⟩ deep; ⟨*fall, sink*⟩ a long way; ⟨*stoop, bow*⟩ low; **~er graben** dig deeper *or* more deeply; **c)** *(in nur geringer Höhe)* ⟨*fly, hover, etc.*⟩ low; **~ liegen** be at a lower level; **d)** *(nach unten)* ⟨*hang etc.*⟩ low; **~er gehen** ⟨*pilot*⟩ go lower; **e)** *(weit innen)* deep; **~ im Dschungel** deep in the jungle; **f)** *(weit nach innen)* deep; ⟨*breathe, inhale*⟩ deeply; **er sah ihr ~ in die Augen** he looked deep into her eyes; **~er ins All vorstoßen** push deeper into space; **bis ~ in die Nacht/in den Winter** *(fig.)* until deep *or* late into the night/well into winter; **g) er sprach ganz ~:** he spoke in a deep voice; **zu ~ singen** sing flat; **h)** *(intensiv, stark)* ⟨*feel etc.*⟩ deeply; ⟨*sleep*⟩ deeply, soundly

Tief das; ~s, ~s *(Met.)* low; depression; *(fig.)* low

tief-, Tief-: ~bau der; *o. Pl.* civil engineering *no art. (at or below ground level)*; **~betrübt** *Adj. (präd. getrennt geschrieben)* deeply distressed *or* saddened; **~druck der; a)** *o. Pl. (Met.)* low pressure; **b) der;** *Pl.* **~drucke** intaglio *or* gravure [printing]; *(Erzeugnis)* intaglio *or* gravure [print]; **~druck·gebiet das** *(Met.)* area of low pressure; depression

Tiefe die; ~, ~n a) *(Ausdehnung, Entfernung nach unten)* depth; **b)** *(weit unten, im Innern gelegener Bereich; auch fig.)* depths *pl.*; **in die ~ stürzen** plunge into the depths; **in der ~ ihres Herzens** *(fig.)* deep down in her heart; *s. auch* **Höhe f; c)** *(Ausdehnung nach hinten)* depth; **d)** *o. Pl. s.* **tief 1c:** depth; intensity; greatness; **e)** *(von Tönen, Klängen, Stimmen)* deepness; **f)** *o. Pl. (fig.: Tiefgründigkeit)* depth; profundity

Tief·ebene die *(Georgr.)* lowland plain

Tiefen·psychologie die depth psychology *no art.*

tief-, Tief-: ~flug der low-altitude flight *no art.*; flying *no art.* at low altitude; **~gang der** *(Schiffbau)* draught; *(fig.)* depth; **~garage die** underground car park; **~greifend 1.** *Adj.* profound; far-reaching; profound, deep ⟨*crisis*⟩; far-reaching ⟨*improvement*⟩; **2.** *adv.* profoundly; **~gründig** [~grʏndɪç] **1.** *Adj.* profound; **2.** *adv.* ⟨*discuss, examine*⟩ in depth; **~kühlen** *tr. V.* [deep-]freeze

Tief·kühl-: ~fach das freezer [compartment]; **~kost die** frozen food; **~truhe die** [chest] freezer *or* deep-freeze

tief-, Tief-: ~land das; *Pl.* **~länder** *od.* **~lande** lowlands *pl.*; **~liegend** *Adj.*; *nicht präd.*; **tiefer liegend, am tiefsten liegend** *od.* **tiefstliegend** low-lying ⟨*area*⟩; deep-set ⟨*eyes*⟩; **~punkt der** low [point]; **~schlaf der** deep sleep; **~schlag der** *(Boxen)* low punch; punch below the belt *(lit. or fig.)*; **~schürfend 1.** *Adj.* profound; **2.** *adv.* profoundly; **~see die** *(Geogr.)* deep sea; **~sinn der;** *o. Pl.* profundity; **~sinnig 1.** *Adj.* profound; **2.** *adv.* profoundly; **~stand der** *(auch fig.) (tiefer Stand)* low level; *(tiefster Stand)* lowest level; **~|stapeln** [~ʃtaːpl̩n] *itr. V.* understate the case; *(aus Bescheidenheit)* be modest; **~verschneit** *Adj. (präd. getrennt geschrieben)* covered in deep snow *postpos.*; deep in snow *postpos.*

Tiegel [ˈtiːgl̩] **der; ~s, ~** *(zum Kochen)* pan; *(Schmelz~)* crucible; *(Behälter)* pot

Tier [tiːɐ̯] **das; ~[e]s, ~e** animal; *(in der Wohnung gehaltenes)* pet; **ein hohes** *od.* **großes ~** *(ugs.)* a big noise *(sl.) or* shot *(sl.)*

Tier-: ~art die animal species; species of animal; **~arzt der** veterinary surgeon; vet

Tierchen das; ~s, ~: [little] animal; **jedem ~ sein Pläsierchen** *(ugs.)* each to his own; if that's what he/she wants

Tier-: ~freund der animal-lover; **~garten der** zoo; zoological garden; **~handlung die** pet shop; ~heim **das** animal home

tierisch 1. *Adj.* **a)** animal *attrib.*; bestial, savage ⟨*cruelty, crime*⟩; **b)** *(ugs.: unerträglich groß)* terrible *(coll.)*; **~er Ernst** deadly seriousness. **2.** *adv.* **a)** ⟨*roar*⟩ like an animal; savagely ⟨*cruel*⟩; **b)** *(ugs.: unerträglich)* terribly *(coll.)*; deadly ⟨*serious*⟩; baking ⟨*hot*⟩; perishing *(coll.)* ⟨*cold*⟩

tier-, Tier-: ~kreis der; *o. Pl. (Astron., Astrol.)* zodiac; **~kreis·zeichen das** *(Astron., Astrol.)* sign of the zodiac; **~lieb** *Adj.* animal-loving *attrib.*; fond of animals *postpos.*; **~liebe die;** *o. Pl.* love of animals; **~liebend** *Adj. s.* **~lieb; ~medizin die;** *o. Pl.* veterinary medicine; **~park der** zoo; **~pfleger der** animal-keeper; **~quälerei** [---ˈ-] **die** cruelty to animals; **~reich das;** *o. Pl.* animal kingdom; **~welt die** fauna

Tiger [ˈtiːgɐ] **der; ~s, ~:** tiger

Tigerin die; ~, ~nen tigress

Tilde [ˈtɪldə] **die; ~, ~n** tilde

tilgen [ˈtɪlgn̩] *tr. V.* **a)** *(geh.)* delete ⟨*word, letter, error*⟩; erase ⟨*record, endorsement*⟩; *(fig.)* wipe out ⟨*shame, guilt, traces*⟩; **b)** *(Wirtsch., Bankw.)* repay; pay off

Tilgung die; ~, ~en a) *(geh.) s.* **tilgen a:** deletion; erasure; wiping out; **b)** *(Wirtsch., Bankw.)* repayment

Till [tɪl] **(der)** *in* **~ Eulenspiegel** Till Eulenspiegel; *(fig.)* practical joker

timen [ˈtaimən] *tr. V.* time

Timing [ˈtaimɪŋ] **das; ~s, ~s** timing

Tingeltangel [ˈtɪŋl̩taŋl̩] **das** *od.* **der; ~s, ~** *(veralt. abwertend) (Lokal)* cheap night-club/dance-hall; honky-tonk *(coll.)*

Tinnef [ˈtɪnɛf] **der; ~s** *(ugs. abwertend)* rubbish; junk

Tinte [ˈtɪntə] **die; ~, ~n** ink; **in der ~ sitzen** *(ugs.)* be in the soup *(coll.)*

Tinten-: ~faß das ink-pot; **~fisch der** cuttlefish; *(Kalmar)* squid; *(Krake)* octopus

Tip [tɪp] **der; ~s, ~s a)** *(ugs.: Fingerzeig)* tip; **b)** *(bei Toto, Lotto usw.)* [row of] numbers

tippen [ˈtɪpn̩] **1.** *itr. V.* **a) an/gegen etw.** *(Akk.)* **~:** tap sth.; **an seine Mütze ~:** touch one's cap; **sich** *(Dat.)* **an die Stirn ~:** tap one's forehead; **b)** *(ugs.: maschineschreiben)* type; **c)** *(ugs.: vermuten)* reckon; **auf jmds. Sieg ~:** tip sb. to win; **du hast gut/richtig getippt** you were right; **d)** *(wetten)* do the pools/lottery *etc.*; **im Lotto ~:** do the lottery. **2.** *tr. V.* **a)** tap; **jmdn. auf die Schulter ~:** tap sb. on the shoulder; **b)** *(ugs.: mit der Maschine schreiben)* type; **c)** *(bei der Registrierkasse)* ring up; **d)** *(setzen auf)* choose; **sechs Richtige ~:** make six correct selections

Tipp·schein der [pools/lottery *etc.*] coupon

tipp·topp *(ugs.)* **1.** *Adj.* *(tadellos)*

immaculate; *(erstklassig)* tip-top. **2.** *adv.* immaculately

Tipp·zettel der *(ugs.) s.* **Tippschein**

Tirol [ti'ro:l] **(das); ~s** [the] Tyrol

Tiroler der; ~s, ~, Tirolerin die; ~, ~nen Tyrolese; Tyrolean

Tisch [tɪʃ] **der; ~[e]s, ~e a)** table; *(Schreib~)* desk; **vor/nach ~:** before/after lunch/dinner/the meal *etc.;* **bei ~ sein** *od.* **sitzen** be at table; **zu ~ sein** be having one's lunch/dinner *etc.;* **vom ~ aufstehen** get up from the table; ⟨*child*⟩ get down [from the table]; **bitte zu ~:** please take your places for lunch/dinner; **es wird gegessen, was auf den ~ kommt!** [you'll] eat what's put on the table!; **b)** *(fig.)* **reinen ~ machen** *(ugs.)* clear things up; sort things out; **jmdn. über den ~ ziehen** *(ugs.)* outman œuvre sb.; **unter den ~ fallen** *(ugs.)* go by the board

Tisch-: ~bein das table-leg; leg of the table; **~dame die** dinner partner; **~decke die** table-cloth; **~gebet das** grace; **~herr der** dinner partner

Tischler der; ~s, ~: joiner; *(bes. Kunst~)* cabinet-maker

Tischlerei die; ~, ~en a) *(Werkstatt)* joiner's/cabinet-maker's [workshop]; **b)** *o. Pl. (Handwerk)* joinery/cabinet-making

tischlern 1. *itr. V.* do woodwork. **2.** *tr. V.* make ⟨*shelves, cupboard, etc.*⟩

Tisch-: ~manieren *Pl.* table manners; **~nachbar der** person next to one [at table]; **~platte die** table-top; **~rede die** after-dinner speech; **~tennis das** table tennis; **~tuch das;** *Pl.* **~tücher** table-cloth; **~wein der** table wine

Titel ['ti:tl̩] **der; ~s, ~ a)** title; **b)** *(ugs.: Musikstück, Song usw.)* number

Titel-: ~anwärter der *(Sport)* title contender; contender for the title; **~bild das** cover picture; **~blatt das** title-page; **~kampf der** *(Sport)* final; *(Boxen)* title fight; **~rolle die** title-role; **~seite die a)** *(einer Zeitung, Zeitschrift)* [front] cover; **b)** *(eines Buchs)* title-page; **~verteidiger der** *(Sport)* title-holder; *(Mannschaft)* title-holders *pl.*

titulieren [titu'li:rən] *tr. V.* **a)** *(bezeichnen)* call; **jmdn. als** *od.* **mit „Flasche" ~:** call sb. a dead loss *(coll.);* **b)** *(veralt.: mit dem Titel anreden)* address; **jmdn. [als** *od.* **mit] Herr Doktor ~:** address sb. as Doctor

tja [tja(:)] *Interj.* [yes] well; *(Resignation ausdrückend)* oh, well

Toast [to:st] **der; ~[e]s, ~e** *od.* **~s a)** toast; *(Scheibe ~)* piece of toast; **b)** *(Trinkspruch)* toast

toasten *tr. V.* toast

Toaster der; ~s, ~: toaster

toben ['to:bn̩] *itr. V.* **a)** go wild **(vor +** *Dat.* with); *(fig.)* ⟨*storm, sea, battle*⟩ rage; **b)** *(tollen)* romp *or* charge about; **c)** *mit sein (laufen)* charge

Tob·sucht die; *o. Pl.* frenzied *or* mad rage; [mad] frenzy

tob·süchtig *Adj.* frenzied; raving mad

Tochter ['tɔxtɐ] **die; ~, Töchter** ['tœçtɐ] daughter; **die ~ des Hauses** the daughter *or* young lady of the house; **höhere ~:** young lady

Tod [to:t] **der, ~[e]s, ~e** *(auch fig.)* death; **eines natürlichen/gewaltsamen ~es sterben** die a natural/violent death; **jmdn. zum ~ durch den Strang/zum ~ durch Erschießen verurteilen** sentence sb. to death by hanging/by firing-squad; **bis in den ~:** till death; **für jmdn./etw. in den ~ gehen** die for sb./sth.; **sich zu ~e stürzen/trinken** fall to one's death/drink oneself to death; **jmdn./etw. auf den ~ nicht leiden/ausstehen können** *(ugs.)* not be able to stand or abide sb./sth.; **sich zu ~e schämen/langweilen** be utterly ashamed/bored to death; **zu ~e betrübt** extremely distressed; **sich** *(Dat.)* **den ~ holen** *(ugs.)* catch one's death [of cold]

tod-: ~bringend *Adj.* fatal ⟨*illness, disease, etc.*⟩*;* deadly, lethal ⟨*poison etc.*⟩*;* **~elend** *Adj.* utterly miserable; **~ernst 1.** *Adj.* deadly serious; **2.** *adv.* deadly seriously

todes-, Todes-: ~angst die a) fear of death; **b)** *(große Angst)* extreme fear; **~ängste ausstehen** be scared to death; **~anzeige die a)** *(in einer Zeitung)* death notice; **b)** *(Karte) card announcing a person's death;* **~fall der** death; *(in der Familie)* bereavement; **~jahr das** year of death; **~mutig 1.** *Adj.* utterly fearless; **2.** *adv.* utterly fearlessly; **~opfer das** death; fatality; **der Unfall forderte drei ~opfer** the accident claimed three lives; **~spirale die** *(Eis-, Rollkunstlauf)* death spiral; **~stoß der** death-blow; **~strafe die** death penalty; **~stunde die** hour of death; **~tag der: sein ~tag** the date of his death; **Mozarts 200. ~tag** the 200th anniversary of Mozart's death; **~ursache die** cause of death; **~urteil das** death sentence; **~verachtung die** [utter] fearlessness in the face of death

Tod·feind der deadly enemy

tod·krank *Adj.* critically ill

tödlich ['tø:tlɪç] **1.** *Adj.* **a)** fatal ⟨*accident, illness, outcome, etc.*⟩*;* lethal, deadly ⟨*poison, bite, shot, trap, etc.*⟩*;* lethal ⟨*dose*⟩*;* deadly, mortal ⟨*danger*⟩*;* **b)** *(sehr groß, ausgeprägt)* deadly ⟨*hatred, seriousness, certainty, boredom*⟩. **2.** *adv.* **a)** fatally; **er ist ~ verunglückt/abgestürzt** he was killed in an accident/he fell to his death; **b)** *(sehr)* terribly *(coll.)*

tod-, Tod-: ~müde *Adj.* dead tired; **~schick** *(ugs.)* **1.** *Adj.* dead smart *(coll.);* **2.** *adv.* dead smartly *(coll.);* **~sicher** *(ugs.)* **1.** *Adj.* sure-fire *(coll.)* ⟨*system, method, tip, etc.*⟩*;* **eine ~sichere Sache** a dead certainty *or (coll.)* cert; **2.** *adv.* for certain *or* sure; **~sünde die** *(auch fig.)* deadly *or* mortal sin; **~unglücklich** *Adj. (ugs.)* extremely *or* desperately unhappy

Tohuwabohu ['to:huva'bo:hu] **das; ~s, ~s** chaos

Toilette [toa'lɛtə] **die; ~, ~n a)** toilet; lavatory; **auf die** *od.* **zur ~ gehen** go to the toilet *or* lavatory; **eine öffentliche ~:** a public lavatory *or* convenience; **b)** *o. Pl. (geh.: das Sichankleiden)* toilet

Toiletten-: ~artikel der toiletry; **~becken das** lavatory *or* toilet bowl *or* pan; **~frau die, ~mann der** lavatory attendant; **~papier das** toilet paper

toi, toi, toi ['tɔy 'tɔy 'tɔy] *Interj.* **a)** *(gutes Gelingen!)* good luck!; **b)** *(unberufen!)* touch wood!

tolerant [tole'rant] **1.** *Adj.* tolerant **(gegen** of**)**. **2.** *adv.* tolerantly

Toleranz die; ~, ~en tolerance

tolerieren [tole'ri:rən] *tr. V.* tolerate

toll [tɔl] **1.** *Adj.* **a)** *(ugs.) (großartig)* great *(coll.);* fantastic *(coll.);* *(erstaunlich)* amazing; *(heftig, groß)* enormous ⟨*respect*⟩*;* terrific *(coll.)*⟨*noise, storm*⟩*;* **b)** *(wild, ausgelassen, übermütig)* wild; wild, mad ⟨*tricks, antics*⟩*;* **c)** *(ugs.: schlimm, übel)* terrible *(coll.);* **d)** *(veralt.) s.* **verrückt 1a. 2.** *adv.* **a)** *(ugs.: großartig)* terrifically well *(coll.);* *(ugs.: heftig, sehr)* ⟨*rain, snow*⟩ like billy-o *(coll.);* **~ hast du das gemacht** you've made a great job of that *(coll.);* **b)** *(wild, übermütig)* **bei dem Fest ging es ~ zu** it was a wild party; **c)** *(ugs.: schlimm, übel)* **treibt es nicht zu ~:** don't go too mad

tollen *itr. V.* **a)** romp about; **b)** *mit sein* romp

toll-, Toll-: **~kühn** **1.** *Adj.* daredevil *attrib.;* daring; **2.** *adv.* daringly; **~wut** **die** rabies *sing.;* **~wütig** *Adj.* rabid

Tolpatsch [tɔlpatʃ] **der;** **~[e]s, ~e** *(ugs.)* clumsy *or* awkward creature

Tölpel ['tœlpl̩] **der;** **~s, ~** *(abwertend; einfältiger Mensch)* fool

tölpelhaft *(abwertend)* **1.** *Adj.* foolish. **2.** *adv.* foolishly

Tomahawk ['tɔmaha:k] **der;** **~s, ~s** tomahawk

Tomate [to'ma:tə] **die;** **~, ~n** tomato; **du hast wohl ~n auf den Augen!** *(salopp)* you must be blind!

Tomaten- tomato ⟨*juice, purée, salad, sauce, soup, etc.*⟩

tomaten·rot *Adj.* brilliant red

Tombola ['tɔmbola] **die;** **~, ~s** *od.* **Tombolen** raffle

[1]**Ton** [to:n] **der;** **~[e]s, ~e** clay

[2]**Ton** **der;** **~[e]s, Töne** ['tø:nə] **a)** *(auch Physik, Musik; beim Telefon)* tone; *(Klang)* note; **b)** *(Film, Ferns. usw., ~wiedergabe)* sound; **c)** *(Sprechweise, Umgangs~)* tone; **den richtigen ~ finden** strike the right note; **ich verbitte mir diesen ~!** I will not be spoken to like that!; **der gute ~:** good form; **d)** *(ugs.: Äußerung)* word; **er konnte keinen ~ herausbringen** he couldn't say a word; **hast du/hat der Mensch [da noch] Töne?** that's just unbelievable; **große Töne reden** *od.* **spucken** *(ugs.)* talk big; **e)** *(Farb~)* shade; tone; **~ in ~ gehalten** colour co-ordinated; **f)** *(Akzent)* stress

ton-, Ton-: **~abnehmer** **der;** **~s, ~:** pick-up; **~angebend** *Adj.* predominant; **~angebend sein** *(in der Mode, Kunst usw.)* set the tone; *(in einer Gruppe o. ä.)* have the most *or* greatest say; **~arm** **der** pick-up arm; **~art** **die** **a)** *(Musik)* key; **b)** *(fig.)* tone; **~band** **das;** *Pl.* **~bänder** **a)** tape; **b)** *(ugs.: Gerät)* tape recorder

Ton·band-: **~aufnahme** **die** tape recording; **~gerät** **das** tape recorder

Ton·blende **die** *(Rundf., Ferns.)* tone control

tönen ['tø:nən] **1.** *itr. V.* **a)** *(geh.)* sound; ⟨*bell*⟩ sound, ring; *(schallen, widerhallen)* resound; **b)** *(ugs. abwertend)* boast. **2.** *tr. V.* *(färben)* tint

Ton·erde **die** *s.* **essigsauer**

tönern ['tø:nɐn] *Adj.; nicht präd.* clay

Ton-: **~fall** **der** tone; *(Intonation)* intonation; **~folge** **die** sequence of notes; **~gefäß** **das** earthen[ware] vessel; **~höhe** **die** pitch

Tonika ['to:nika] **die;** **~, Toniken** *(Musik)* tonic

ton-, Ton-: **~ingenieur** **der** sound engineer; **~kopf** **der** head; **~leiter** **die** *(Musik)* scale; **~los** **1.** *Adj.* toneless; **2.** *adv.* tonelessly

Tonnage [tɔ'na:ʒə] **die;** **~, ~n** *(Seew.)* tonnage

Tonne ['tɔnə] **die;** **~, ~n** **a)** *(Behälter)* drum; *(Müll~)* bin; *(Regen~)* water-butt; **b)** *(Gewicht)* tonne; metric ton; **c)** *(ugs.: dicker Mensch)* fatty *(coll.)*

Ton-: **~spur** **die** sound-track; **~störung** **die** interference *no def. art.* on sound; **~system** **das** *(Musik)* tone *or* tonic system; **~tauben·schießen** **das** clay-pigeon shooting *no art.;* **~techniker** **der** sound technician

Tönung **die;** **~, ~en** **a)** tinting; **b)** *(Farbton)* tint; shade

Ton·ware **die** earthenware *no pl.*

Top [tɔp] **das;** **~s, ~s** *(Mode)* top

top- ultra ⟨*modern, topical*⟩

Top- top; outstanding ⟨*location, performance, time*⟩

Topas [to'pa:s] **der;** **~es, ~e** topaz

Topf [tɔpf] **der;** **~es, Töpfe** ['tœpfə] **a)** pot; *(Braten~, Schmor~)* casserole; *(Stielkasserolle)* saucepan; **alles in einen ~ werfen** *(fig. ugs.)* lump everything together; **b)** *(zur Aufbewahrung)* pot; jar; **c)** *(Krug)* jug; **d)** *(Nacht~)* chamber pot; po *(coll.)*; *(für Kinder)* potty *(Brit. coll.)*; **e)** *(Blumen~)* [flower]pot; **f)** *(salopp: Toilette)* loo *(Brit. coll.)*; john *(Amer. coll.)*

Topf·blume **die** [flowering] pot plant

Töpfer ['tœpfɐ] **der;** **~s, ~:** potter

Töpferei **die;** **~, ~en** **a)** *o. Pl.* pottery *no art.;* **b)** *(Werkstatt)* pottery; potter's workshop; **c)** *(Erzeugnis)* piece of pottery

Töpferin **die;** **~, ~nen** potter

töpfern **1.** *itr. V.* do pottery. **2.** *tr. V.* make ⟨*vase, jug, etc.*⟩; **getöpferte Teller** hand-made pottery plates

Töpfer-: **~scheibe** **die** potter's wheel; **~waren** *Pl.* pottery *sing.*

top·fit *Adj.* in *or* on top form *postpos.;* *(gesundheitlich)* in fine fettle; as fit as a fiddle

Topf-: **~lappen** **der** oven cloth; **~pflanze** **die** pot plant

Topographie [topogra'fi:] **die;** **~, ~n** *(Geogr.)* topography *no art.*

Topspin ['tɔpspɪn] **der;** **~s, ~s** *(bes. Golf, Tennis, Tischtennis)* top spin

Tor [to'ɐ̯] **das;** **~[e]s, ~e** **a)** gate; *(einer Garage, Scheune)* door; *(fig.)* gateway; **b)** *(Ballspiele)* goal; **c)** *(Ski)* gate

Tor·bogen **der** arch[way]

Torero [to're:ro] **der;** **~[s], ~s** torero

Tores·schluß **der** *in* **kurz vor ~** *(ugs.)* at the last minute *or* the eleventh hour

Torf [tɔrf] **der;** **~[e]s, ~e** peat

Torf-: **~ballen** **der** bale of peat; **~moor** **das** peat bog; **~stecher** **der** peat-cutter

Torheit **die;** **~, ~en** *(geh.)* **a)** *o. Pl.* foolishness; **b)** *(Handlung)* foolish act; **eine [große] ~ begehen** do something [extremely] foolish

Tor·hüter **der** *(Ballspiele)* goalkeeper

töricht ['tø:rɪçt] *(geh.)* **1.** *Adj.* foolish ⟨*behaviour, action, hope*⟩; stupid ⟨*person, question, smile, face*⟩. **2.** *adv.* ⟨*behave, act*⟩ foolishly; ⟨*smile, ask*⟩ stupidly

Tor·jäger **der** *(Ballspiele)* goalscorer

torkeln ['tɔrkl̩n] *itr. V.; mit sein* stagger; reel

Tor·mann **der;** *Pl.* **~männer** *od.* **~leute** *(Ballspiele)* goalkeeper

Tornado [tɔr'na:do] **der;** **~s, ~s** tornado

Tornister [tɔr'nɪstɐ] **der;** **~s, ~** **a)** knapsack; **b)** *(Schulranzen)* satchel

torpedieren [tɔrpe'di:rən] *tr. V.* *(Milit., fig.)* torpedo

Torpedo [tɔr'pe:do] **der;** **~s, ~s** torpedo

Tor-: **~pfosten** **der** *(Ballspiele)* [goal-] post; **~schluß** **der** *s.* **Toresschluß;** **~schluß·panik** **die** last-minute panic; *(Furcht, keinen Partner mehr zu finden)* fear of being left on the shelf; **~schütze** **der** *(Ballspiele)* [goal-] scorer

Törtchen ['tœrtçən] **das;** **~s, ~:** tartlet

Torte ['tɔrtə] **die;** **~, ~n** gateau; *(Obst~)* [fruit] flan

Torten-: **~boden** **der** flan case; *(ohne Rand)* flan base; **~guß** **der** glaze; **~heber** **der** cake-slice; **~platte** **die** cake-plate

Tortur [tɔr'tu:ɐ̯] **die;** **~, ~en** **a)** ordeal; **b)** *(veralt.: Folter)* torture

Tor-: **~verhältnis** **das** *(Ballspiele)* goal average **~wart** **der;** **~[e]s, ~e** *(Ballspiele)* goalkeeper

tosen ['to:zn̩] *itr. V.*⟨*sea, surf*⟩ roar, rage; ⟨*storm*⟩ rage; ⟨*torrent, waterfall*⟩ roar, thunder; ⟨*wind*⟩ roar; **~der Beifall** *(fig.)* thunderous applause

tot [to:t] *Adj.* **a)** dead; **das Kind wurde ~ geboren** the baby was stillborn; **er war auf der Stelle ~:** he died instantly; **~ zusammenbrechen** collapse and die; **~ umfallen** drop dead; **er ist politisch ein ~er Mann** *(fig.)* he is finished as a politician; **halb ~ vor Angst** *usw. (ugs.)* paralysed with fear *etc.*; **den ~en Mann machen** *(ugs.)* float on one's back; **b)** *(abgestorben)* dead ⟨*tree, branch, leaves, etc.*⟩; **c)** *(fig.)* dull ⟨*colour*⟩; bleak ⟨*region etc.*⟩; dead ⟨*town, telephone line, socket, language*⟩; disused ⟨*railway line*⟩; extinct ⟨*volcano*⟩; dead, quiet ⟨*time, period*⟩; useless ⟨*knowledge*⟩; *s. auch* **Punkt d; Winkel a**

total [to'ta:l] **1.** *Adj.* total. **2.** *adv.* totally

totalitär [totali'tɛ:ɐ̯] *(Politik)* **1.** *Adj.* totalitarian. **2.** *adv.* in a totalitarian way; ⟨*organized, run*⟩ along totalitarian lines

Total·schaden **der** *(Versicherungsw.)* **an beiden Fahrzeugen entstand ~:** both vehicles were a write-off

tot-: **~|arbeiten** *refl. V. (ugs.)* work oneself to death; **~|ärgern** *refl. V. (ugs.)* get livid *(coll.)*; **ich könnte mich ~ärgern** I'm livid *(coll.) or* really furious

Tote ['to:tə] **der/die;** *adj. Dekl.* dead person; dead man/woman; **die ~n** the dead; **es gab zwei ~:** two people died *or* were killed; there were two fatalities

töten ['tø:tn̩] *tr., itr. V.* kill; *s. auch* **Nerv a**

toten-, Toten-: **~amt das** *(kath. Kirche) s.* **~messe; ~blaß, ~bleich** *Adj.* deathly pale; pale as death *postpos.*; **~gräber der** grave-digger; **~hemd das** shroud; **~klage die** lamentation *or* bewailing of the dead; **~kopf der a)** skull; **b)** *(als Symbol)* death's head; *(mit gekreuzten Knochen)* skull and cross-bones; **~messe die** *(kath. Kirche)* requiem [mass]; **~schein der** death certificate; **~still** *Adj.* deathly quiet *or* silent; **~wache die** vigil by the body

tot-, Tot-: **~|fahren 1.** *unr. tr. V.* [run over and] kill; **2.** *unr. refl. V.* kill oneself; **~geboren** *Adj. (präd. getrennt geschrieben)* stillborn; **ein ~geborenes Kind sein** *(fig.)* ⟨*project*⟩ be stillborn, not get off the ground; **~geburt die a)** stillbirth; **b)** *(Kind)* stillbirth; stillborn baby; **~gesagte der/die;** *adj. Dekl.* person declared dead; **~|lachen** *refl. V. (ugs.)* kill oneself laughing; **zum Totlachen sein** be killing *(coll.)*; be killingly funny *(coll.)*; **~|laufen** *unr. refl. V. (ugs.)* ⟨*movement, trend, fashion*⟩ peter *or* die out; ⟨*talks, discussions*⟩ peter out

Toto ['to:to:] **das** *od.* **der; ~s, ~s a)** *(Pferde~)* tote *(sl.)*; **b)** *(Fußball~)* [football] pools *pl.*; **[im] ~ spielen** do the pools

Toto-: **~gewinn der** win on the pools/*(sl.)* tote; **~schein der** pools coupon/*(sl.)* tote ticket

tot-, Tot-: **~|sagen** *tr. V.* declare ⟨*person*⟩ dead **~|schießen** *unr. tr. V. (ugs.)* **jmdn. ~schießen** shoot sb. dead; **~schlag der** *(Rechtsw.)* manslaughter *no indef. art.*; **~|schlagen** *unr. tr. V.* beat to death; **die Zeit ~schlagen** kill time; **~schläger der a)** *(Mensch)* manslaughterer; **b)** *(Waffe)* cosh *(Brit. coll.)*; blackjack *(Amer.)*; **~|schweigen** *unr. tr. V.* hush up; **jmdn. ~schweigen** keep quiet about sb.; **~|stellen** *refl. V.* pretend to be dead; play dead; **~|treten** *unr. tr. V.* trample ⟨*person*⟩ to death; step on and kill ⟨*insect*⟩

Tötung die; ~, ~en killing; **fahrlässige ~** *(Rechtsspr.)* manslaughter by culpable negligence

Toupet [tu'pe:] **das; ~s, ~s** toupee

toupieren [tu'pi:rən] *tr. V.* backcomb

Tour [tu:ɐ̯] **die; ~, ~en a)** tour **(durch** of); *(Kletter~)* [climbing] trip; *(kürzere Fahrt, Ausflug)* trip; *(mit dem Auto)* drive; *(mit dem Fahrrad)* ride; **eine ~ machen** go on a tour/trip *or* outing; *(Zech~)* go on a pub-crawl *(Brit. coll.)*; bar-hop *(Amer.)*; **b)** *(feste Strekke)* route; **c)** *(Tournee)* tour; **auf ~ gehen** go on tour; **d)** *(ugs.: Methode)* ploy; **die ~ zieht bei mir nicht** that [one] won't work with me; **etw. auf die sanfte ~ erreichen** get sth. by soft-soaping; **e) jmdm. die ~ vermasseln** *(ugs.)* put paid to sb.'s [little] plans; **f)** *Pl. (Technik: Umdrehungen)* revolutions; revs *(coll.)*; **jmdn. auf ~en bringen** *(ugs.)* really get sb. going; **auf vollen/höchsten ~en laufen** *(ugs.)* ⟨*preparations, work, etc.*⟩ be in full swing; **g) in einer ~** *(ugs.)* the whole time

Touren·wagen der *(Motorsport)* touring car

Tourismus [tu'rɪsmʊs] **der; ~:** tourism *no art.*

Tourist der; ~en, ~en tourist

Touristik die; ~: tourism *no art.*; tourist industry *or* business

Touristin die; ~, ~nen tourist

Tournee [tʊr'ne:] **die; ~, ~s** *od.* **~n** [tʊr'ne:ən] *s.* **Tour c**

Trab [tra:p] **der; ~[e]s** trot; **im ~:** at a trot; **im ~ reiten** trot; **jmdn. auf ~ bringen** *(ugs.)* make sb. get a move on; **jmdn. in ~ halten** *(ugs.)* keep sb. on the go *(coll.)*

Trabant [tra'bant] **der; ~en, ~en** *(Astron.)* satellite

traben *itr. V.; mit sein (auch ugs.: laufen)* trot

Trab·rennen das trotting; *(einzelne Veranstaltung)* trotting race

Tracht [traxt] **die; ~, ~en a)** *(Volks~)* traditional *or* national costume; *(Berufs~)* uniform; **die ~ der Nonnen** the nuns' dress *or* habit; **b)** *in* **eine ~ Prügel** a beating *or* thrashing; *(als Strafe für ein Kind)* a hiding

trachten *itr. V. (geh.)* strive **(nach** for, after); **all sein Trachten** all his striving *or* endeavours

Trachten·anzug der *suit in the style of a traditional or national costume*

trächtig ['trɛçtɪç] pregnant

Tradition [tradi'tsio:n] **die; ~, ~en** tradition

traditionell [traditsio'nɛl] **1.** *Adj.* traditional. **2.** *adv.* traditionally

traf [tra:f] *1. u. 3. Pers. Sg. Prät. v.* **treffen**

Trafik [tra'fɪk] **die; ~, ~en** *(österr.)* tobacconist's [shop]

Trag·bahre die stretcher

tragbar *Adj.* **a)** portable; **b)** wearable ⟨*clothes*⟩; **c)** *(finanziell)* supportable ⟨*cost, debt, etc.*⟩; **d)** *(erträglich)* bearable; tolerable

Trage die; ~, ~n a) *(Bahre)* stretcher; **b)** *(Traggestell)* pannier

träge ['trɛ:gə] **1.** *Adj.* **a)** sluggish; *(geistig)* lethargic; **b)** *(Physik)* inert. **2.** *adv.* sluggishly; *(geistig)* lethargically

tragen [tra:gn̩] **1.** *unr. tr. V.* **a)** carry; **das Auto wurde aus der Kurve getragen** *(fig.)* the car went off the bend; **b)** *(bringen)* take; **vom Wind getragen** *(fig.)* carried by [the] wind; **c)** *(ertragen)* bear ⟨*fate, destiny*⟩; bear, endure ⟨*suffering*⟩; **d)** *(halten)* hold; **einen Arm in der Schlinge ~:** have one's arm in a sling; **e)** *(von unten stützen)* support; **zum Tragen kommen** ⟨*advantage, improvement, quality*⟩ become noticeable; *s. auch* **tragend a–c; f)** *(belastbar sein durch)* be able to carry *or* take ⟨*weight*⟩; **der Ast trägt dich nicht** the branch won't take your weight; **g)** *(übernehmen, aufkommen für)* bear, carry ⟨*costs etc.*⟩; take ⟨*blame, responsibility, consequences*⟩; *(unterhalten, finanzieren)* support; **h)** *(am Körper)* wear ⟨*clothes, wig, glasses, jewel-*

lery, etc.⟩; have ⟨*false teeth, beard, etc.*⟩; **getragene Kleider** second-hand clothes; **i)** *(fig.: haben)* have ⟨*label etc.*⟩; have, bear ⟨*title*⟩; bear, carry ⟨*signature, inscription, seal*⟩; **j)** *(hervorbringen)* ⟨*tree*⟩ bear ⟨*fruit*⟩; ⟨*field*⟩ produce ⟨*crops*⟩; *(fig.)* yield ⟨*interest*⟩; **gut/wenig ~** ⟨*tree*⟩ produce a good/poor crop; ⟨*field*⟩ produce a good/poor yield; **k)** *(geh.: schwanger sein mit)* be carrying. **2.** *unr. itr. V.* **a)** carry; **wir hatten schwer zu ~:** we were heavily laden; **schwer an etw.** *(Dat.)* **zu ~ haben** have difficulty carrying sth.; find sth. very heavy to carry; *(fig.)* find sth. hard to bear; **das Eis trägt noch nicht** the ice is not yet thick enough to skate/walk *etc.* on; **b)** *(am Körper)* **man trägt [wieder] kurz/lang** short/long skirts are in fashion [again]; **c) eine ~de Sau/Kuh** a pregnant sow/cow; *s. auch* **tragend d. 3.** *unr. refl. V.* **a) sich gut/schlecht** *usw.* **~** ⟨*load*⟩ be easy/difficult *or* hard *etc.* to carry; **b) der Mantel/Stoff trägt sich angenehm** the coat/material is pleasant to wear; **c)** *in* **sich mit etw. ~:** be contemplating sth.; **d)** *(sich kleiden)* dress

tragend *Adj.* **a)** *(Stabilität gebend)* load-bearing; supporting ⟨*wall, column, function, etc.*⟩; **b)** *(fig.: grundlegend)* basic, main ⟨*idea, motif*⟩; **c)** *(fig.: wichtig, zentral)* leading, major ⟨*role, figure*⟩; **d)** *(weithin hörbar)* ⟨*voice*⟩ that carries [a long way]

Träger ['trɛ:gɐ] **der; ~s, ~ a)** porter; *(Sänften~, Sarg~)* bearer; **b)** *(Zeitungs~)* paper boy/girl; delivery boy/girl; **c)** *(Bauw.)* girder; [supporting] beam; **d)** *(an Kleidung)* strap; (Hosen~) braces *pl.*; **e)** *(Inhaber) (eines Amts)* holder; *(eines Namens, Titels)* bearer; *(eines Preises)* winner; **f)** *(fig.: Urheber, treibende Kraft)* moving force; **g)** *(fig.: Unterhalter)* **~ der Arbeitslosenversicherung ist der Staat** unemployment insurance is financed *or* funded by the state; **h)** *(fig.: einer Substanz, eines Erregers usw.)* carrier; **i)** *(Flugzeug~)* carrier; **j)** *(jmd., der etw. als Kleidung, Schmuck usw. trägt)* wearer

Trägerin die; ~, ~nen *s.* **Träger a, b, e, f, g, h, j**

Träger-: ~kleid das pinafore dress; ; **~rakete die** carrier vehicle *or* rocket; **~rock der** skirt with straps

Trage-: ~tasche die carrier-bag

trag-, Trag-: ~fähig *Adj.* able to take a load *or* weight *postpos.*; **eine ~fähige Mehrheit** *(fig.)* a workable majority; **~fläche die** wing; *(eines Boots)* hydrofoil; **~flächen·boot das, ~flügel·boot das** hydrofoil

Trägheit die; ~, ~en a) *o. Pl. s.* **träge 1 a:** sluggishness; lethargy; **b)** *(Physik)* inertia

Tragik ['tra:gɪk] **die; ~** tragedy

Tragi·komödie [tra:gi-] **die** tragicomedy

tragisch ['tra:gɪʃ] **1.** *Adj.* tragic; **das ist nicht [so] ~** *(ugs.)* it's not the end of the world *(coll.)*; **etw. ~ nehmen** take sth. to heart *(coll.)*. **2.** *adv.* tragically; **der Film/die Tour endete ~:** the film had a tragic ending/the trip ended in tragedy

Tragödie [tra'gø:di̯ə] **die; ~, ~n** tragedy

Trag·weite die; *o. Pl.* consequences *pl.*; **ein Ereignis von weltpolitischer ~:** an event of moment in world politics

Trainer ['trɛ:nɐ] **der; ~s, ~, Trainerin die; ~, ~nen a)** coach; trainer; *(eines Schwimmers, Tennisspielers)* coach; *(einer Fußballmannschaft)* manager; **b)** *(Pferdesport)* trainer

trainieren 1. *tr. V.* **a)** train; coach; ⟨*swimmer, tennis-player*⟩; manage ⟨*football team*⟩; train ⟨*horse*⟩; exercise ⟨*muscles etc.*⟩; **jmdn./ein Tier darauf ~, etw. zu tun** train sb./an animal to do sth.; **ein trainierter Schwimmer/Radfahrer** *usw.* a swimmer/cyclist *etc.* [who is] in training; **b)** *(üben, einüben)* practise ⟨*exercise, jump, etc.*⟩; **c)** *(zu Trainingszwecken ausüben)* **Fußball/Tennis ~:** do football/tennis training. **2.** *itr. V.* train; *(Motorsport)* practise; **mit jmdm. ~:** ⟨*trainer*⟩ coach sb.; ⟨*player*⟩ train with sb.

Training ['trɛ:nɪŋ] **das; ~s, ~s** *(Fitneß~, auch fig.: Ausbildung)* training *no indef. art.*; *(Motorsport, fig.)* practice; **Radfahren ist ein gutes ~:** cycling is a good form of training *or* exercise

Trainings-: ~anzug der track suit; **~hose die** track-suit bottoms *pl.*; **~jacke die** track-suit top; **~schuh der** training-shoe; trainer

Trakt [trakt] **der; ~[e]s, ~e** section; *(Flügel)* wing

Traktat [trak'ta:t] **der** *od.* **das; ~[e]s, ~e a)** *(Abhandlung)* treatise; **b)** *(religiöse Flugschrift)* tract

traktieren *tr. V.* set about ⟨*person, thing*⟩; **jmdn. mit Ohrfeigen/Faustschlägen ~:** slap sb. round the face/punch sb.

Traktor ['traktɔr] **der; ~s, ~en** [-'to:rən] tractor

trällern ['trɛlɐn] *tr., itr. V.* warble

Tramp [trɛmp] **der; ~s, ~s** tramp; hobo *(Amer.)*

Trampel ['trampl̩] **der; ~s, ~** *(ugs. abwertend) s.* **Trampeltier b**

trampeln 1. *itr. V.* **a) [mit den Füßen] ~:** stamp one's feet; **b)** *mit sein (abwertend: treten)* trample. **2.** *tr. V.* trample

Trampel-: ~pfad der [beaten] path; **~tier das a)** *(Kamel)* Bactrian camel; **b)** *(salopp abwertend)* clumsy clot *(Brit. sl.) or* oaf

trampen ['trɛmpn̩] *itr. V.; mit sein* hitch-hike

Tramper der; ~s, ~, Tramperin die; ~, ~nen hitch-hiker

Trampolin ['trampoli:n] **das; ~s, ~e** trampoline

Tran [tra:n] **der; ~[e]s a)** train-oil; **b) im ~** *(ugs.)* befuddled; in a daze; *(im Rausch)* stoned *(sl.)*

Trance ['trã:s(ə)] **die; ~, ~n** trance; **in ~:** in a trance; **in ~ fallen** go into a trance

tranchieren *tr. V. (Kochk.)* carve

Tranchier·messer das carving-knife

Träne ['trɛ:nə] **die; ~, ~n** tear; **seine ~n trocknen** dry one's eyes; **~n lachen** laugh till one cries *or* till the tears run down one's cheeks; **in ~n aufgelöst sein** be in floods of tears; **jmdm./einer Sache keine ~ nachweinen** not shed any tears over sb./sth.

tränen *itr. V.* ⟨*eyes*⟩ water

Tränen-: ~drüse die *(Anat.)* tear-gland; **auf die ~drüsen drücken** *(fig.)* lay on the agony; **~gas das** tear-gas

Tran·funzel die *(ugs. abwertend)* **a)** *(trübe Lampe)* miserable lamp; **b)** *(langweiliger Mensch)* ponderous dim-wit *(coll.)*; *(langsamer Mensch)* slowcoach; slowpoke *(Amer.)*

tranig *Adj.* **a)** ⟨*meat, fish*⟩ full of train-oil; **~ schmecken** taste like *or* of train-oil; **b)** *(ugs.: langsam)* sluggish; slow

trank [traŋk] *1. u. 3. Pers. Sg. Prät. v.* **trinken**

Tränke ['trɛŋkə] **die; ~, ~n** watering-place

tränken *tr. V.* **a)** *(auch fig.)* water; **b)** *(sich vollsaugen lassen)* soak

Trans·aktion [trans-] **die** transaction

trans·atlantisch *Adj.* transatlantic; across the Atlantic *postpos.*

Transfer [trans'fe:ɐ] **der; ~s, ~s** *(bes. Wirtsch., Sport)* transfer

trans-, Trans-: ~formator [~fɔr'ma:tɔr] **der; ~s, ~en** [-'to:rən] transformer; **~formieren** *tr. V.* transform **(in +** *Akk.* into, **auf +** *Akk.* to); **~fusion die** *(Med.)* transfusion

Transistor [tran'zɪstɔr] **der; ~s, ~en** [-'to:rən] transistor

[1]**Transit** [*auch:* '--] **der; ~s, ~e** transit

[2]**Transit** [tran'zi:t, *auch:* 'tranzɪt] **das; ~s, ~s** transit visa

transitiv ['tranziti:f] *(Sprachw.)* **1.** *Adj.* transitive. **2.** *adv.* transitively

Transit·verkehr der transit traffic

transparent [transpa'rɛnt] *Adj.* **a)** transparent; *(Licht durchlassend)* translucent, diaphanous ‹*curtain, fabric, etc.*›; **b)** *(fig.: verständlich)* intelligible

Transparent das; ~[e]s, ~e a) *(Spruchband)* banner; **b)** *(Bild)* transparency

Transparenz die; ~ a) transparency; *(von Gewebe, Porzellan usw.)* translucence; **b)** *(fig.: Verständlichkeit)* intelligibility

transpirieren [transpiri:rən] *itr. V. (bes. Med.)* perspire

Transplantation [transplanta'tsio:n] **die; ~, ~en** *(Med.)* transplant; *(von Haut)* graft

Transport [trans'pɔrt] **der; ~[e]s, ~e a)** *(Beförderung)* transportation; **beim** *od.* **auf dem ~:** during carriage; **b)** *(beförderte Lebewesen od. Sachen) (mit dem Zug)* train-load; *(mit mehreren Fahrzeugen)* convoy; *(Fracht)* consignment; shipment

transportabel [transpɔr'ta:bl̩] *Adj.* transportable; *(tragbar)* portable

Transporter der; ~s, ~ *(Flugzeug)* transport aircraft; *(Schiff)* cargo ship

Transporteur [...'tø:ɐ] **der; ~s, ~e** carrier

transportfähig *Adj.* moveable

transportieren 1. *tr. V.* transport ‹*goods, people*›; move ‹*patient*›

Transport-: ~kosten *Pl.* carriage *sing.*; transport costs; **~unter·nehmen das** haulage firm *or* contractor

Tran·suse die *(ugs. abwertend) s.* **Tranfunzel b**

Transvestit [transvɛs'ti:t] **der; ~en, ~en** transvestite

transzendental [transtsɛndɛn'ta:l] *Adj. (Philos.)* transcendental

Transzendenz [transtsɛn'dɛnts] **die; ~ a)** transcendency; transcendent nature; **b)** *(Philos.)* transcendence

Trapez [tra'pe:ts] **das; ~es, ~e a)** *(Geom.)* trapezium *(Brit.)*; trapezoid *(Amer.)*; **b)** *(im Zirkus o. ä.)* trapeze

trappeln ['trapl̩n] *itr. V.; mit sein* patter [along]; ‹*feet*› patter; ‹*hoofs*› go clip-clop

trara [tra'ra:] *Interj.* tantara

Trara [tra'ra:] **das; ~s** *(ugs. abwertend)* razzmatazz *(coll.)*; **viel ~ um etw.** *(Akk.)* **machen** make a great song and dance about sth. *(coll.)*

Trasse ['trasə] **die; ~, ~n a)** *(Verkehrsweg)* [marked-out] route *or* line; **b)** *(Damm)* [railway/road] embankment

trat [tra:t] *1. u. 3. Pers. Sg. Prät. v.* **treten**

Tratsch [tra:tʃ] **der; ~[e]s** *(ugs. abwertend)* gossip; tittle-tattle

tratschen *itr. V. (ugs. abwertend)* gossip; *(schwatzen)* chatter

Trau·altar der: [mit jmdn.] vor den ~ treten *(geh.)* enter into matrimony [with sb.]

Traube ['traubə] **die; ~, ~n a)** *(Beeren)* bunch; *(von Johannisbeeren o. ä.)* cluster; **b)** *(Wein~)* grape; **c)** *(Menschenmenge)* bunch; cluster

Trauben-: ~lese die grape harvest; **~saft der** grape-juice; **~zucker der** glucose

trauen ['trauən] **1.** *itr. V.* **jmdm./einer Sache ~:** trust sb./sth.; *s. auch* **Auge a. 2.** *refl. V.* dare; **du traust dich ja nicht!** you haven't the courage *or* nerve; **sich irgendwohin ~:** dare [to] go somewhere. **3.** *tr. V. (verheiraten)* ‹*vicar, registrar, etc.*› marry

Trauer ['trauɐ] **die; ~ a)** grief **(über +** *Akk.* over); *(um einen Toten)* mourning **(um +** *Akk.* for); **~ haben, in ~ sein** be in mourning; **b)** *(~kleidung)* mourning

Trauer-: ~fall der bereavement; **~feier die** memorial ceremony; *(beim Begräbnis)* funeral ceremony; **~flor der** mourning-band; black [crape] ribbon; **~karte die** [pre-printed] card of condolence; **~kleidung die** mourning clothes *pl.*; mourning; **~kloß der** *(ugs. scherzh.)* wet blanket; **~marsch der** *(Musik)* funeral march

trauern *itr. V.* **a)** mourn; **um jmdn. ~:** mourn for sb.; **die ~den Hinterbliebenen** the bereaved; **b)** *(Trauer tragen)* be in mourning

Trauer-: ~rand der black border *or* edging; **~spiel das** tragedy; *(fig. ugs.)* deplorable business; **es ist doch ein ~ spiel, daß ...:** it's quite pathetic that ...; **~weide die** weeping willow; **~zug der** funeral procession

Traufe ['traufə] **die; ~, ~n** eaves *pl.*

träufeln ['trɔyfl̩n] *tr. V.* [let] trickle **(in +** *Akk.* into); drip ‹*ear-drops etc.*›

traulich ['traulɪç] **1.** *Adj.* cosy; **in ~er Runde** in a friendly *or* an intimate circle. **2.** *adv.* cosily; *(vertraut)* intimately

Traum [traum] **der; ~[e]s, Träume** ['trɔymə] dream; **nicht im ~ habe ich mit der Möglichkeit gerechnet, zu gewinnen** I didn't imagine in my wildest dreams that I could win

Trauma ['trauma] **das; ~s, Traumen** *od.* **~ta** *(Psych., Med.)* trauma

träumen ['trɔymən] **1.** *itr. V.* dream **(von** of, about); *(unaufmerksam sein)* [day-]dream. **2.** *tr. V.* dream; **etwas Schreckliches ~:** have a terrible dream; **ich hätte mir nie ~ lassen, daß ...:** I should never have imagined it possible that ...; I never imagined that ...

Träumer der; ~s, ~, Träumerin die; ~, ~nen dreamer

träumerisch 1. *Adj.* dreamy; *(sehnsüchtig)* wistful. **2.** *adv.* dreamily; *(sehnsüchtig)* wistfully

traumhaft 1. *Adj.* **a)** dreamlike; **b)** *(ugs.: schön)* marvellous; fabulous *(coll.)*. **2.** *adv.* **a)** as if in a dream; **b)** *(ugs.: schön)* fabulously *(coll.)*

Traum·tänzer der *(abwertend)* wooly-headed idealist; fantasizer

traurig ['traurɪç] **1.** *Adj.* **a)** sad; sad, sorrowful ‹*eyes, expression*›; unhappy ‹*childhood, youth*›; unhappy, painful ‹*duty*›; **b)** *(kümmerlich)* sorry, pathetic ‹*state etc.*›; miserable ‹*result*›; **eine ~e Rolle** an unfortunate role. **2.** *adv.* sadly

Traurigkeit die; ~: sadness; sorrow

Trau-: ~ring der wedding-ring; **~schein der** marriage certificate

Trauung die; ~, ~en wedding [ceremony]

Trau·zeuge der witness *(at wedding ceremony)*

Traveller·scheck ['trɛvəlɐʃɛk] **der** traveller's cheque

Travestie [travɛs'ti:] **die; ~, ~n** travesty

Treck [trɛk] **der; ~s, ~s** train, column *(of refugees etc.)*

Treff [trɛf] **der; ~s, ~s** *(ugs.)* **a)** *(Treffen)* rendezvous; *(bes. von*

mehreren Personen) get-together *(coll.)*; **b)** *(Ort)* meeting-place

treffen 1. *unr. tr. V.* **a)** *(erreichen [und verletzen/schädigen])* hit; ‹*punch, blow, object*› strike; **jmdn. am Kopf/ins Gesicht ~:** hit *or* strike sb. on the head/in the face; **vom Blitz getroffen** struck by lightning; **ihn trifft keine Schuld** *(fig.)* he is in no way to blame; **b)** *(erraten)* hit on; hit ‹*right tone*›; **auf dem Foto ist er gut getroffen** the photo is a good likeness of him; that's a good photo of him; **c)** *(erschüttern)* affect [deeply]; *(verletzen)* hurt; **es hat ihn in seinem Stolz getroffen** it hurt his pride; **d)** *(schaden)* hit; damage; **warum muß es immer mich ~?** why does it always have to be me [who is affected *or* gets it]?; **e)** *(begegnen)* meet; **f)** *(vorfinden)* come upon, find ‹*anomalies etc.*›; **es gut/schlecht ~:** be *or* strike lucky/be unlucky; **g)** *(als Funktionsverb)* make ‹*arrangements, choice, preparations, decision, etc.*›; **eine Vereinbarung** *od.* **Absprache ~:** conclude an agreement. **2.** *unr. itr. V.* **a)** ‹*person, shot, etc.*› hit the target; **nicht ~:** miss [the target]; **b)** *mit sein* **auf etw.** *(Akk.)* **~:** come upon sth.; **auf Widerstand/Ablehnung ~:** meet with *or* encounter resistance/rejection; **auf jmdn./eine Mannschaft ~** *(Sport)* come up against sb./a team. **3.** *unr. refl. V.* **a) sich mit jmdm. ~:** meet sb.; **b)** *(unpers.)* **es trifft sich gut/schlecht** it is convenient/inconvenient

Treffen das; ~s, ~ a) meeting; **b)** *(Sport)* encounter

treffend 1. *Adj.* apt. **2.** *adv.* aptly

Treffer der; ~s, ~ a) *(Milit., Boxen, Fechten usw.)* hit; *(Schlag)* blow; *(Ballspiele)* goal; **b)** *(Gewinn)* win; *(Los)* winner

trefflich *(geh.)* **1.** *Adj.* excellent; splendid ‹*person*›; first-rate ‹*scholar*›. **2.** *adv.* excellently; splendidly

Treff·punkt der a) *(Stelle, Ort)* meeting-place; rendezvous; **b)** *(Geom.)* point of incidence

treff·sicher 1. *Adj.* with a sure aim *postpos., not pred.*; accurate ‹*marksman*›; *(fig.)* accurate ‹*language, mode of expression*›; unerring ‹*judgement*›; **2.** *adv.* *(auch fig.)* accurately; with unerring accuracy

Treib·eis das drift-ice

treiben ['traibn̩] **1.** *unr. tr. V.* **a)** drive ‹*animals, people, leaves, etc.*›; **er ließ sich von der Strömung ~:** he let himself be carried along by the current; **die Preise in die Höhe ~:** push *or* force up prices; **jmdn. zur Raserei/zur Verzweiflung/in den Tod ~:** drive sb. mad/to despair/to his/her death; **b)** *(an~)* drive ‹*wheels etc.*›; **jmdn. zur Eile ~:** make sb. hurry up; **c)** *(einschlagen)* drive ‹*nail, wedge, stake, etc.*› **(in +** *Akk.* into); **d)** *(durch Bohrung schaffen)* drive, cut ‹*tunnel, gallery*› **(in +** *Akk.* into; **durch** through); sink ‹*shaft*› **(in +** *Akk.* into); **e)** *(durchpressen)* force; press; **f)** *(sich beschäftigen mit)* go in for ‹*farming, cattle-breeding, etc.*›; study ‹*French etc.*›; carry on, pursue ‹*studies, trade, craft*›; **viel Sport ~:** do a lot of sport; go in for sport in a big way; **Handel ~:** trade; **was treibt ihr denn hier?** *(ugs.)* what are you up to *or* doing here?; **g)** *(ugs. abwertend: in Verbindung mit „es“:)* **es wüst/übel/toll ~:** lead a dissolute/bad life/live it up; **es zu toll ~:** overdo it; take things too far; **er hat es zu weit getrieben** he overstepped the mark; he went too far; **es [mit jmdm.] ~** *(ugs. verhüll.: koitieren)* have it off [with sb.] *(sl.)*; **h)** *(formen)* beat ‹*metal, object*›; chase ‹*silver, gold*›; **i)** *(Gartenbau)* force ‹*plants*›. **2.** *unr. itr. V.* **a)** *meist, mit Richtungsangabe nur, mit sein* drift; **b)** *(ugs.) (harntreibend sein)* get the bladder going; *(schweißtreibend sein)* make you sweat; **c)** *(ausschlagen)* ‹*tree, plant*› sprout

Treiben das; ~s, ~ a) *o. Pl.* bustle; **in der Fußgängerzone herrscht ein lebhaftes ~:** the pedestrian precinct is full of bustling activity; **b)** *o. Pl. (Tun)* activities *pl.*; doings *pl.*; *(Machenschaften)* wheelings and dealings *pl.*

Treiber der; ~s, ~ *(Jägerspr.)* beater

Treib-: ~gas das a) *(für Motoren)* liquefied petroleum gas; LPG; **b)** *(in Spraydosen)* propellant; **~haus das** hothouse; **~haus·effekt der** greenhouse effect; **~jagd die** *(Jägerspr.)* battue; shoot *(in which game is sent up by beaters)*; *(fig.)* witch-hunt; **~mittel das** *(Kochk.)* raising agent; **~sand der** quicksand; **~stoff der** fuel

Trend [trɛnt] **der; ~s, ~s** trend **(zu +** *Dat.* towards); *(Mode)* vogue

trennen ['trɛnən] **1.** *tr. V.* **a)** separate **(von** from); *(abschneiden)* cut off; sever ‹*head, arm*›; **b)** *(auf~)* unpick ‹*dress, seam*›; **c)** *(teilen)* divide ‹*word, parts of a room etc., fig.: people*›; **uns ~ Welten** *(fig.)* we are worlds apart; **d)** *(beim Telefon)* **wir wurden getrennt** we were cut off; **e)** *(zerlegen)* separate ‹*mixture*›; **f)** *(auseinanderhalten)* differentiate *or* distinguish between; make a distinction between ‹*terms*›. **2.** *refl. V.* **a)** *(voneinander weggehen)* part [company]; *(fig.)* **die Mannschaften trennten sich 0:0** the game ended in a goalless draw; the two teams drew 0:0; **die Firma hat sich von ihm getrennt** the company has dispensed with his services; **b)** *(eine Partnerschaft auflösen)* ‹*couple, partners*› split up; **sich in Güte ~:** part on good terms; **c)** *(hergeben)* **sich von etw. ~:** part with sth.

Trennung die; ~, ~en a) *(von Menschen)* separation **(von** from); **in ~ leben** have separated; **b)** *(von Gegenständen)* parting **(von** with); **c)** *(von Wörtern)* division; **d)** *(von Begriffen)* distinction **(von** between)

Trennungs-: ~linie die *(auch fig.)* dividing line; **~strich der a)** hyphen; **b)** *(fig.)* **einen ~strich ziehen** *od.* **machen** make a [clear] distinction; draw a [clear] line

trepp- [trɛp'-]: **~ab** *Adv.* down the stairs; **~auf** *Adv.* up the stairs

Treppe ['trɛpə] **die; ~, ~n** staircase; [flight *sing.* of] stairs *pl.*; *(im Freien, auf der Bühne)* [flight *sing.* of] steps *pl.*; **~n steigen** climb stairs; **eine ~ höher/tiefer** one floor *or* flight up/down

Treppen-: ~absatz der half-landing; **~geländer das** banisters *pl.*; **~haus das** stair-well; **das Licht im ~haus** the light on the staircase; **~stufe die** stair; *(im Freien)* step

Tresen ['tre:zn̩] **der; ~s, ~** *(bes. nordd.)* **a)** *(Theke)* bar; **b)** *(Ladentisch)* counter

Tresor [tre'zo:ɐ̯] **der; ~s, ~e** safe

Tret·boot das pedalo

treten ['tre:tn̩] **1.** *unr. itr. V.* **a)** *mit sein (einen Schritt, Schritte machen)* step **(in +** *Akk.* into, **auf +** *Akk.* on to); **ins Zimmer ~:** enter the room; **ans Fenster ~:** go to the window; **von einem Fuß auf den anderen ~:** shift from one foot to the other; **der Schweiß ist ihm auf die Stirn getreten** *(fig.)* the sweat came to his brow; **der Fluß ist über die Ufer getreten** *(fig.)* the river has overflowed its banks; **b)** *(seinen Fuß setzen)* **auf etw.** *(Akk.)* **~** *(absichtlich)* tread on sth.; *(unabsichtlich; meist mit* *sein)* step *or* tread on sth.; **jmdm. auf den Fuß ~**: step/tread on sb.'s foot *or* toes; **auf das Gas[pedal] ~**: step on the accelerator; **c)** *mit sein* **in jmds. Dienste ~**: enter sb.'s service; **d)** *(ausschlagen)* kick; **jmdm. an** *od.* **gegen das Schienbein ~**: kick sb. on the shin; **gegen die Tür ~**: kick the door. **2.** *unr. tr. V.* **a)** kick ‹*person, ball, etc.*›; **b)** *(trampeln)* trample, tread ‹*path*›; **c)** *(mit dem Fuß niederdrücken)* step on ‹*brake, pedal*›; operate ‹*clutch*›

Tret-: **~mine die** anti-personnel mine; **~mühle die** *(fig. ugs. abwertend)* treadmill

treu [trɔy] **1.** *Adj.* **a)** faithful, loyal ‹*friend, dog, customer, servant, etc.*›; faithful ‹*husband, wife*›; loyal ‹*ally, subject*›; staunch, loyal ‹*supporter*›; **jmdm. ~ sein/bleiben** be/remain true to sb.; *(fig.)* **sich selbst** *(Dat.)* **~ bleiben** be true to oneself; **seinen Grundsätzen ~ bleiben** stick to one's principles; **das Glück/der Erfolg ist ihm ~ geblieben** his luck has held out/success keeps coming his way; **c)** *(ugs.: ~herzig)* ingenuous, trusting ‹*eyes, look*›. **2.** *adv.* **a)** faithfully; loyally; **b)** *(ugs.: ~herzig)* trustingly

Treue die; ~ a) loyalty; *(von [Ehe]partnern)* fidelity; **b)** *(Genauigkeit)* accuracy

Treue·gelöbnis das pledge of loyalty; *(von Ehepartnern)* pledge of fidelity

treu-, Treu-: **~herzig 1.** *Adj.* ingenuous; *(naiv)* naïve; *(unschuldig)* innocent; **2.** *adv.* ingenuously; *(naiv)* naïvely; *(unschuldig)* innocently; **~los 1.** *Adj.* disloyal, faithless ‹*friend, person*›; unfaithful ‹*husband, wife, lover*›; **2.** *adv.* faithlessly; **~losigkeit die; ~**: disloyalty; faithlessness; *(von [Ehe]partnern)* infidelity

Triangel ['tri:aŋl] **der;** *österr.* **das; ~s, ~** *(Mus.)* triangle

Tribunal [tribu'na:l] **das; ~s, ~e** tribunal

Tribüne [tri'by:nə] **die; ~, ~n** [grand]stand

Tribut [tri'bu:t] **der; ~[e]s, ~e a)** *(hist.)* tribute *no indef. art.*; **b)** *(fig.)* due

Trichter ['trɪçtɐ] **der; ~s, ~ a)** funnel; **b)** *(Bomben~, Geogr.)* crater

Trick [trɪk] **der; ~s, ~s** trick; *(fig.: List)* ploy; **technische ~s** cunning techniques

Trick·film der animated cartoon [film]

tricksen ['trɪksn̩] *(ugs., bes. Sportjargon)* **1.** *itr. V.* use tricks; work a fiddle *(sl.)*; ‹*footballer*› play trickily. **2.** *tr. V.* fiddle *(sl.)*

trieb [tri:p] *1. u. 3. Pers. Sg. Prät. v.* **treiben**

Trieb der; ~[e]s, ~e a) *(innerer Antrieb)* impulse; *(Drang)* urge; *(Verlangen)* [compulsive] desire; **b)** *(Sproß)* shoot

trieb-, Trieb-: **~feder die** mainspring; *(fig.)* driving *or* motivating force; **~haft 1.** *Adj.* compulsive ‹*need, behaviour, action, etc.*›; carnal ‹*sensuality*›; **2.** *adv.* compulsively; **~wagen der** *(Eisenb.)* railcar; **~werk das** engine

triefen ['tri:fn̩] *unr. od. regelm. itr. V.* **a)** *mit sein (fließen) (in Tropfen)* drip; *(in kleinen Rinnsalen)* trickle; **b)** *(naß sein)* be dripping wet; ‹*nose*› run; **~d naß** dripping wet; *(durchnäßt)* wet through; **von Fett ~**: be dripping with fat

triezen ['tri:tsn̩] *tr. V. (ugs.)* torment; *(plagen)* pester; plague

trifft *3. Pers. Sg. Präsens v.* **treffen**

triftig *Adj.* good ‹*reason, excuse*›; valid, convincing ‹*motive, argument*›

Trigonometrie [trigonome'tri:] **die; ~**: trigonometry *no art.*

¹Trikot [tri'ko] **der** *od.* **das; ~s, ~s** *(Stoff)* cotton jersey

²Trikot das; ~s, ~ *(ärmellos)* singlet; *(eines Tänzers)* leotard; *(eines Fußballspielers)* shirt; **das gelbe ~** *(Radsport)* the yellow jersey

Triller ['trɪlɐ] **der; ~s, ~**: trill

trillern 1. *itr. V. (Musik)* trill; *(mit Trillern singen)* ‹*bird, person*› warble. **2.** *tr. V.* warble ‹*song*›

Triller·pfeife die police/referee's whistle

Trillion [trɪ'lio:n] **die; ~, ~en** trillion *(Brit.)*; quadrillion *(Amer.)*

Trilogie [trilo'gi:] **die; ~, ~n** trilogy

Trimester [tri'mɛstɐ] **das; ~s, ~** *(Hochschulw.)* term

Trimm-dich-Pfad der keep-fit *or* trim trail

trimmen ['trɪmən] *tr. V.* **a)** *(durch Sport)* get ‹*person*› into shape; **trimm dich durch Sport** keep fit with sport; **b) etw. auf alt** *usw.* **~**: do sth. up to look old *etc.*; **c)** *(durch Scheren)* clip ‹*dog*›; *(durch Bürsten)* groom ‹*dog*›

trinken ['trɪŋkn̩] **1.** *unr. itr. V.* drink; **jmdm. etw. zu ~ geben** give sb. sth. to drink; **was ~ Sie?** what are you drinking?; *(was möchten Sie ~?)* what would you like to drink?; **auf jmdn./etw. ~**: drink to sb./sth. **2.** *unr. tr. V.* drink; **einen Kaffee** *usw.* **~**: have a coffee *etc.*; **einen Schluck Wasser ~**: have a drink of water; **einen ~**: have a drink; **einen ~ gehen** *(ugs.)* go for a drink. **3.** *refl. V.* **sich satt ~**: drink one's fill

Trinker der; ~s, ~, Trinkerin die; ~, ~nen alcoholic; **ein heimlicher/starker ~**: a secret/heavy drinker

trink-, Trink-: **~fest** *Adj.* **~fest sein** be able to hold one's drink; **~gelage das** *(oft scherzh.)* drinking spree; **~geld das** tip; **wieviel ~geld gibst du ihm?** how much do you tip him?; **~halle die a)** *(in einem Heilbad)* pump-room; **b)** *(Kiosk)* refreshment kiosk; *(größer)* refreshment stall; **~halm der** [drinking-]straw; **~milch die** low-fat pasteurized milk; **~wasser das;** *Pl.* **~wässer** drinking-water; **„kein ~wasser"** 'not for drinking'

Trio ['tri:o] **das; ~s, ~s** *(Musik, fig.)* trio

Triole [tri'o:lə] **die; ~, ~n** *(Musik)* triplet

Trip [trɪp] **der; ~s, ~s a)** *(ugs.: Ausflug)* trip; jaunt; **b)** *(Drogenjargon: Rausch)* trip *(coll.)*; **auf dem ~ sein** be tripping *(coll.)*

trippeln ['trɪpl̩n] *itr. V.; mit sein* trip; ‹*child*› patter; *(affektiert)* mince

Tripper ['trɪpɐ] **der; ~s, ~**: gonorrhoea

trist [trɪst] *Adj.* dreary; dismal

tritt [trɪt] *Imperativ Sg. u. 3. Pers. Sg. Präsens v.* **treten**

Tritt der; ~[e]s, ~e a) *(Aufsetzen des Fußes)* step; *(einmalig)* [foot]step; **b)** *(Gleichschritt)* **im ~ marschieren/aus dem ~ geraten** *od.* **kommen** march in/get out of step; **~ fassen** fall in step; *(fig.: sich fangen)* recover oneself; **c)** *(Fuß~)* kick; **jmdm. einen ~ versetzen** give sb. a kick; kick sb.; **d)** *(~brett)* step; **e)** *(Bergsteigen) (Halt für Füße)* foothold; *(im Eis)* step; **f)** *(Gestell)* small step-ladder

Tritt-: **~brett das** step; *(an älterem Auto)* running-board; **~leiter die** step-ladder

Triumph [tri'ʊmf] **der; ~[e]s, ~e** triumph; **einen großen ~ feiern** have a great triumph *or* success; be huge success

triumphieren *itr. V.* **a)** *(Genugtuung empfinden)* exult; **b)** *(siegen)* be triumphant *or* victorious; triumph *(lit. or fig.)* **(über + Akk.** over)

Triumph·zug der *(hist.)* triumph; **im ~** *(fig.)* in a triumphal procession

trivial [tri'vi̯a:l] **1.** *Adj.* **a)** *(platt)* banal; trite; *(unbedeutend)* trivial; **b)** *(alltäglich)* humdrum ‹*life, career*›. **2.** *adv.* *(platt)* banally; ‹*say etc.*› tritely; ‹*written*› in a banal style

Trivialität die; ~, ~**en a)** *o. Pl.* *(Plattheit, Alltäglichkeit)* banality; triteness; **b)** *(platte Äußerung)* banality; *(Gemeinplatz)* commonplace [remark]

trocken ['trɔkn̩] **1.** *Adj.* **a)** dry; **etw.** ~ **bügeln/reinigen** dry-iron/dry-clean sth.; **sich** ~ **rasieren** use an electric razor; **auf dem trock[e]nen sitzen** *od.* **sein** *(ugs.)* be completely stuck *(coll.)*; *(pleite sein)* be skint *(Brit. sl.)*; **b)** *(ohne Zutat)* ~**es** *od.* *(ugs.)* ~ **Brot essen** eat dry bread; **c)** *(sachlich-langweilig)* dry, factual ‹*account, report, treatise*›; bare ‹*words, figures*›; dull, dry ‹*person*›; **d)** *(unverblümt)* dry ‹*humour, remark, etc.*›; **e)** *(dem Klang nach)* dry ‹*laugh, cough, sound*›; sharp ‹*crack*›. **2.** *adv.* **a)** *(sachlich-langweilig)* ‹*speak, write*› drily, in a matter-of-fact way; **b)** *(unverblümt)* drily

Trọcken-: ~**blume die;** *meist Pl.* dried flower; ~**gebiet das** *(Geogr.)* arid region; ~**haube die** [hood-type] hair-drier

Trọckenheit die; ~, ~**en a)** *o. Pl.* *(auch fig.)* dryness; **b)** *(Dürreperiode)* drought

trọcken-, Trọcken-: ~**kurs der** dry-skiing course; ~|**legen** *tr. V.* **a) ein Baby** ~**legen** change a baby's nappies *(Brit.)* or *(Amer.)* diapers; **b)** *(entwässern)* drain ‹*marsh, pond, etc.*›; ~**milch die** dried milk; ~|**reiben** *unr. tr. V.* rub ‹*hair, child, etc.*› dry; wipe ‹*crockery, window, etc.*› dry; ~**schwimmen das** preparatory swimming exercises *pl.* [on land]; ~**zeit die** dry season

trọcknen 1. *itr. V.; meist mit sein* dry. **2.** *tr. V.* dry

Troddel ['trɔdl̩] **die;** ~, ~**n** tassel

Trödel ['trø:dl̩] **der;** ~**s** *(ugs., oft abwertend)* junk; *(für den Flohmarkt)* jumble

trödeln *itr. V.* **a)** *(ugs., oft abwertend)* dawdle (**mit** over); **b)** *mit sein (ugs.: schlendern)* saunter

Trödler der; ~**s,** ~, **Trödlerin die;** ~, ~**nen a)** *(ugs. abwertend)* dawdler; slowcoach; slowpoke *(Amer.)*; **b)** *(ugs.: Händler[in])* junk-dealer

troff [trɔf] *1. u. 3. Pers. Sg. Prät. v.* **triefen**

trog [tro:k] *1. u. 3. Pers. Sg. Prät. v.* **trügen**

Trog der; ~**[e]s, Tröge** ['trø:gə] trough

Troika ['trɔyka] **die;** ~, ~**s** troika; *(fig.: Führungsgruppe)* triumvirate

trollen ['trɔlən] *(ugs.) refl. V.* push off *(coll.)*; **der Junge trollte sich in sein Zimmer** the boy took himself off to his room

Trommel ['trɔml̩] **die;** ~, ~**n a)** *(Schlaginstrument)* drum; **b)** *(Behälter; Kabel~, Seil~)* drum

Trọmmel-: ~**bremse die** *(Kfz-W.)* drum brake; ~**fell das a)** *(bei Trommeln)* drumhead; **b)** *(im Ohr)* ear-drum; ~**feuer das** *(Milit.; auch fig.)* [constant] barrage

trọmmeln 1. *itr. V.* **a)** beat the drum; *(als Beruf, Hobby usw.)* play the drums; **b)** *([auf etw.] schlagen, auftreffen)* drum (**auf** + *Akk.* on, **an** + *Akk.* against); **sie trommelte mit den Fäusten gegen die Tür** she hammered the door with her fists. **2.** *tr. V.* beat [out] ‹*march, rhythm, etc.*›

Trọmmel-: ~**schlag der** drumbeat; ~**schlegel der,** ~**stock der** drumstick; ~**wirbel der** drum-roll

Trọmmler der; ~**s,** ~, **Trọmmlerin die;** ~, ~**nen** drummer

Trompete [trɔm'pe:tə] **die;** ~, ~**n** trumpet

trompeten 1. *itr. V.* play the trumpet; *(fig.)* ‹*elephant*› trumpet. **2.** *tr. V.* play ‹*piece*› on the trumpet

Trompeter der; ~**s,** ~, **Trompeterin die;** ~, ~**nen** trumpeter

Tropen *Pl.* tropics

Tropen- tropical

Tropen·helm der sun-helmet

[1]**Tropf** [trɔpf] **der;** ~**[e]s, Tröpfe** ['trœpfə] *(abwertend)* twit *(Brit. sl.)*; moron *(coll.)*

[2]**Trọpf der;** ~**[e]s,** ~**e** *(Med.)* drip; **am** ~ **hängen** be on a drip

tröpfeln ['trœpfl̩n] **1.** *itr. V.* **a)** *mit sein* drip (**auf** + *Akk.* on to, **aus, von** from); **b)** *(unpers.) (ugs.: leicht regnen)* **es tröpfelt** it's spitting [with rain]. **2.** *tr. V.* let ‹*sth.*› drip (**in** + *Akk.* into, **auf** + *Akk.* on to)

trọpfen 1. *itr. V.; mit Richtungsangabe mit sein* drip; ‹*tears*› fall; **seine Nase tropft** his nose is running; *(unpers.)* **es tropft [vom Dach** *usw.*] water is *or* it's dripping from the roof *etc.*. **2.** *tr. V.* let ‹*sth.*› drip (**in** + *Akk.* into, **auf** + *Akk.* on to); **jmdm. eine Tinktur auf die Wunde** ~: pour drops of a tincture into sb.'s wound

Trọpfen der; ~**s,** ~ **a)** drop; **es regnet dicke** ~: the rain is falling in large drops *or* spots; **er hat keinen** ~ **[Alkohol] getrunken** he hasn't touched a drop; **ein** ~ **auf den heißen Stein sein** *(fig. ugs.)* be a drop in the ocean; **b) ein guter/edler** ~: a good/fine vintage

trọpf·naß *Adj.* dripping *or* soaking wet

Trọpf·stein·höhle die limestone cave with stalactites and/or stalagmites

Trophäe [tro'fɛ:ə] **die;** ~, ~**n** *(hist., Jagd, Sport)* trophy

tropisch *Adj.* tropical

Tropo·sphäre [tropo'sfɛ:rə] **die;** ~ *(Meteor.)* troposphere

Trosse ['trɔsə] **die;** ~, ~**n** hawser *(Naut.)*

Trost [tro:st] **der;** ~**[e]s** consolation; *(bes. geistlich)* comfort; **jmdm.** ~ **zusprechen** *od.* **spenden** comfort *or* console sb.; **nicht [ganz** *od.* **recht] bei** ~ **sein** *(ugs.)* be out of one's mind; have taken leave of one's senses

trösten ['trø:stn̩] **1.** *tr. V.* comfort, console (**mit** with); ~**de Worte** words of comfort; comforting words; **etw. tröstet jmdn.** sth. is a comfort to sb.. **2.** *refl. V.* console oneself; **sich mit einer anderen Frau** ~: find consolation with another woman

tröstlich *Adj.* comforting

trost-, Trost-: ~**los** *Adj.* **a)** *(ohne Trost)* hopeless; without hope *postpos.*; *(verzweifelt)* in despair *postpos.*; **b)** *(deprimierend, öde)* miserable, dreary ‹*time, weather, area, food, etc.*›; hopeless ‹*situation*›; ~**pflaster das** *(scherzh.)* consolation; ~**preis der** consolation prize

Trott [trɔt] **der;** ~**[e]s,** ~**e a)** *(Gangart)* trot; **b)** *(leicht abwertend: Ablauf)* routine; **in den alten** ~ **verfallen** fall back into the same old rut

Trọttel der; ~**s,** ~ *(ugs. abwertend)* fool; wally *(sl.)*

trọtten *itr. V.; mit sein* trot [along]; *(freudlos)* trudge

Trottoir [trɔ'to̯a:ɐ̯] **das;** ~**s,** ~**e** *od.* ~**s** pavement

trotz [trɔts] *Präp. mit Gen., seltener mit Dat.* in spite of; despite; ~ **allem** in spite of everything

Trọtz der; ~**es** defiance; *(Oppositionsgeist)* cussedness *(coll.)*; contrariness; **jmdm./einer Sache zum** ~: in defiance of sb./sth.

trọtz·dem [*auch:* '-'-] *Adv.* nevertheless; **er tat es** ~: he did it all *or* just the same

trọtzen *itr. V.* **a)** *(geh.: widerstehen)* **jmdm./einer Sache** ~ *(auch fig.)* defy sb./sth.; **Gefahren/der**

Kälte ~: brave dangers/the cold; **b)** *(trotzig sein)* be contrary

trotzig 1. *Adj.* defiant; *(widerspenstig)* contrary; bolshie *(coll.)*; difficult ‹*child*›. **2.** *adv.* defiantly

trüb[e] ['try:b(ə)] **1.** *Adj.* **a)** *(nicht klar)* murky ‹*stream, water*›; cloudy ‹*liquid, wine, juice*›; *(schlammig)* muddy ‹*puddle*›; *(schmutzig)* dirty ‹*glass, windowpane*›; dull ‹*eyes*›; **im trüben fischen** *(ugs.)* fish in troubled waters; **b)** *(nicht hell)* dim ‹*light*›; dull, dismal ‹*day, weather*›; grey, overcast ‹*sky*›; dull, dingy ‹*red, yellow*›; **c)** *(gedrückt)* gloomy ‹*mood, voice, etc.*›; dreary ‹*time*›; *s. auch* **Tasse a; d)** *(unerfreulich)* unfortunate, bad ‹*experience etc.*›. **2.** *adv.* **a)** *(nicht hell)* ‹*shine, light*› dimly; **b)** *(gedrückt)* ‹*smile, look*› gloomily; **c)** *(unerfreulich)* **~ laufen** go badly

Trubel ['tru:bl̩] **der; ~s** [hustle and] bustle; **im ~ der Ereignisse** *(fig.)* in the excitement of the moment; in the rush of events

trüben 1. *tr. V.* **a)** make ‹*liquid*› cloudy; cloud ‹*liquid*›; **b)** *(beeinträchtigen)* dampen, cast a cloud over ‹*mood*›; mar ‹*relationship*›; cloud ‹*judgement*›; **jmds. Blick [für etw.] ~:** blind sb. [to sth.]. **2.** *refl. V.* **a)** ‹*liquid*› become cloudy; ‹*eyes*› become dull; ‹*sky*› darken; **b)** *(sich verschlechtern)* ‹*relationship*› deteriorate; ‹*awareness, memory, etc.*› become dulled *or* dim

Trübsal ['try:pza:l] **die; ~, ~e** *(geh.)* **a)** *(Leiden)* affliction; **b)** *o. Pl.* *(Kummer)* grief; **~ blasen** *(ugs.)* mope **(wegen** over, about)

trüb·selig 1. *Adj.* **a)** *(öde)* dreary, depressing ‹*place, area, colour*›; dismal ‹*house*›; **b)** *(traurig)* gloomy, melancholy ‹*thoughts, mood*›; gloomy, miserable ‹*face*›; **2.** *adv.* *(traurig)* gloomily

Trüb·sinn der; *o. Pl.* melancholy; gloom

trudeln ['tru:dl̩n] *itr. V. mit sein* *(rollen)* roll; **das Flugzeug geriet ins Trudeln** the plane went into a spin

Trüffel ['tryfl̩] **die; ~, ~n** *od.* *(ugs.)* **der; ~s, ~:** truffle

trug [tru:k] *1. u. 3. Pers. Prät. v.* **tragen**

trügen 1. *unr. tr. V.* deceive; **wenn mich nicht alles trügt** unless I am very much mistaken. **2.** *unr. itr. V.* be deceptive; ‹*feeling, deception*› be a delusion; *s. auch* **Schein b**

trügerisch 1. *Adj.* **a)** deceptive; false ‹*hope, sign, etc.*›; treacherous ‹*ice*›; **b)** *(veralt.: auf Betrug zielend)* deceitful; **in ~er Absicht** with intent to deceive. **2.** *adv.* **a)** deceptively; **b)** *(veralt.: auf Betrug zielend)* deceitfully

Trug·schluß der wrong conclusion; *(Irrtum)* fallacy

Truhe ['tru:ə] **die; ~, ~n** chest

Trümmer ['trymɐ] *Pl.* *(eines Gebäudes)* rubble *sing.*; *(Ruinen)* ruins; *(eines Flugzeugs usw.)* wreckage *sing.*; *(kleinere Teile)* debris *sing.*; **die Stadt lag in ~n** the town lay in ruins; **eine Stadt in ~ legen** reduce a town to rubble; flatten a town [completely]

Trümmer-: ~feld das expanse of rubble; **~haufen der** pile *or* heap of rubble

Trumpf [trʊmpf] **der; ~[e]s, Trümpfe** ['trympfə] *(auch fig.)* trump [card]; *(Farbe)* trumps *pl.*; **was ist ~?** what are trumps?; **alle Trümpfe in der Hand haben** *(fig.)* hold all the [trump] cards; **~ sein** *(fig.)* *(das Nötigste sein)* be what matters; be the order of the day; *(Mode sein)* be the in thing

trumpfen *itr. V.* play a trump

Trumpf·karte die *(auch fig.)* trump card

Trunk [trʊŋk] **der; ~[e]s, Trünke** ['tryŋkə] *(geh.)* **a)** *(Getränk)* drink; beverage *(formal)*; **b)** *(das Trinken)* **sich dem ~ ergeben** take to drink

Trunkenbold [-bɔlt] **der; ~[e]s, ~e** *(abwertend)* drunkard

Trunkenheit die; ~ a) drunkenness; **~ am Steuer** drunken driving; **b)** *(geh.: Begeisterung)* [state of] intoxication

Trunk·sucht die; *o. Pl.* alcoholism *no art.*

Trupp [trʊp] **der; ~s, ~s** troop; *(von Arbeitern, Gefangenen)* gang; *(von Soldaten, Polizisten)* detachment; squad

Truppe die; ~, ~n a) *(Einheit der Streitkräfte)* unit; **b)** *Pl.* *(Soldaten)* troops; **c)** *o. Pl.* *(Streitkräfte)* [armed] forces *pl.*; **d)** *(Gruppe von Schauspielern, Artisten)* troupe; company; *(von Sportlern)* squad; *(Mannschaft)* team

Truppen-: ~gattung die arm [of the service]; corps; **~parade die** military parade; **~übungsplatz der** military training area

Trut- ['tru:t-]: **~hahn der** turkey [cock]; *(als Braten)* turkey; **~henne die** turkey [hen]

Tscheche ['tʃɛçə] **der; ~n, ~n** Czech

tschechisch 1. *Adj.* Czech. **2.** *adv.* **~ sprechend** Czech-speaking; *s. auch* **deutsch; Deutsch; ²Deutsche**

Tschechoslowakei [tʃɛçɔslova'kai] **die; ~:** Czechoslovakia *no art.*

tschilpen ['tʃɪlpn̩] *itr. V.* chirp

tschüs [tʃy:s] *Interj.* *(ugs.)* bye *(coll.)*; so long *(coll.)*

T-Shirt ['ti:ʃə:t] **das; ~s, ~s** T-shirt

Tuba ['tu:ba] **die; ~, Tuben** tuba

Tube ['tu:bə] **die; ~, ~n** tube; **auf die ~ drücken** *(fig. ugs.)* step on it *(coll.)*; put one's foot down

Tuberkulose [tubɛrku'lo:zə] **die; ~, ~n** *(Med.)* tuberculosis *no art.*

Tuch [tu:x] **das; ~[e]s, Tücher** ['ty:çɐ] *od.* **~e a)** *Pl.* **Tücher** cloth; *(Bade~)* [bath-]towel; *(Kopf~, Hals~)* scarf; **b)** *Pl.* **~e** *(Gewebe)* cloth

Tuch·fühlung die *(scherzh.)* physical contact; *(fig.: Kontakte)* [close] contact; **auf** *od.* **mit ~:** close together

tüchtig ['tʏçtɪç] **1.** *Adj.* **a)** efficient ‹*secretary, assistant, worker, etc.*›; *(fähig)* capable, competent **(in +** *Dat.* at); **freie Bahn dem Tüchtigen!** let ability win through; **b)** *(von guter Qualität)* excellent ‹*performance, piece of work, etc.*›; **~, ~!** *(auch iron.)* well done!; **c)** *nicht präd.* *(ugs.: beträchtlich)* sizeable ‹*piece, portion*›; big ‹*gulp*›; hearty ‹*eater, appetite*›. **2.** *adv.* **a)** efficiently; *(fähig)* competently; **~ arbeiten** work hard; **b)** *(ugs.: sehr)* really ‹*cold, warm*›; ‹*snow, rain*› good and proper *(coll.)*; ‹*eat*› heartily

Tüchtigkeit die; ~ a) efficiency; *(Fähigkeit)* ability; competence; *(Fleiß)* industry; **b) körperliche ~:** physical fitness

Tücke ['tʏkə] **die; ~, ~n a)** *o. Pl.* *(Hinterhältigkeit)* deceit[fulness]; *(List)* guile; scheming *no indef. art.*; **die ~ des Objekts** the perversity *or* *(coll.)* cussedness of inanimate objects; *s. auch* **List b; b)** *meist Pl.* *(hinterhältige Handlung)* wile; ruse; **c)** *meist Pl.* *([verborgene] Gefahr/Schwierigkeit)* [hidden] danger/difficulty; *(unberechenbare Eigenschaft)* vagary

tückisch 1. *Adj.* **a)** *(hinterhältig)* wily; *(betrügerisch)* deceitful; **b)** *(gefährlich)* treacherous ‹*bend, slope, spot, etc.*›; *(Gefahr signalisierend)* menacing ‹*look, eyes*›. **2.** *adv.* **a)** *(hinterhältig)* craftily; **b)** *(Gefahr signalisierend)* menacingly

tüfteln ['tʏftl̩n] *itr. V.* *(ugs.)* fiddle **(an +** *Dat.* with); do finicky work **(an +** *Dat.* on); *(geistig)* rack one's brains, puzzle (**an** + *Dat.* over)

Tüftler der; ~s, ~ *(ugs.) person who likes finicky jobs/niggling problems; (jmd., der gern Rätselspiele macht)* puzzle freak *(coll.)*

Tugend ['tu:gn̩t] **die; ~, ~en** virtue

tugendhaft 1. *Adj.* virtuous. **2.** *adv.* virtuously; **~ leben** live a life of virtue

Tüll [tʏl] **der; ~s, ~e** tulle

Tulpe ['tʊlpə] **die; ~, ~n a)** *(Pflanze)* tulip; **b)** *(Glas)* tulip glas

Tulpen·zwiebel die tulip-bulb

tummeln ['tʊml̩n] *refl. V.* **a)** *(umhertollen)* romp [about]; *(im Wasser)* splash about; **b)** *(bes. westmd., österr., sich beeilen)* stir one's stumps *(coll.);* get a move on *(coll.)*

Tummel·platz der *(auch fig.)* playground

Tümmler ['tʏmlɐ] **der; ~s, ~** *(Delphin)* bottle-nosed dolphin

Tumor ['tu:mɔr] **der; ~s, ~en** [tu'mo:rən], *ugs. auch* **~e** [tu'mo:rə] *(Med.)* tumour

Tümpel ['tʏmpl̩] **der; ~s, ~:** pond

Tumult [tu'mʊlt] **der; ~[e]s, ~e** tumult; commotion; *(Protest)* uproar

tun [tu:n] **1.** *unr. tr. V.* **a)** *(machen)* do; **ich weiß nicht, was ich ~ soll** I don't know what to do; **so etwas tut man nicht** that is just not done; **er hat sein möglichstes getan** he did his [level] best; **was ~?** what is to be done?; **man tut, was man kann** one does what one can; one tries one's best; **b)** *(erledigen)* do ‹*work, duty, etc.*›; **ich muß noch etwas [für die Schule] ~:** I've still got some [school-]work to do; **nach getaner Arbeit** when the work is/was done; **mit Geld/einer Entschuldigung** *usw.* **ist es nicht getan** money/an apology *etc.* is not enough; **es ~** *(ugs. verhüll.: koitieren)* do it *(sl.);* **c) [etwas] zu ~ haben** have something to do ; **ich hatte dort zu ~/dort geschäftlich zu ~:** I had things/business to do there; **[es] mit jmdm. zu ~ bekommen** *od. (ugs.)* **kriegen** get into trouble with sb./sth.; **mit sich [selbst] zu ~ haben** have problems [of one's own]; **[etwas] mit etw./jmdm. zu ~ haben** be concerned with sth./have dealings with sb.; **er hat noch nie [etwas] mit der Polizei zu ~ gehabt** he has never been involved with the police; **mit etw. nichts zu ~ haben** have nothing to do with sth.; not be concerned with sth.; **mit jmdm./etw. nichts zu ~ haben wollen** not want [to have] anything to do with sb./sth.; **d)** *nimmt die Aussage eines vorher gebrauchten Verbs auf* **es sollte am nächsten Tag regnen, und das tat es dann auch** it was expected to rain the next day, and it did [so]; **e)** *als Funktionsverb* make ‹*remark, catch, etc.*›; take ‹*step, jump*›; do ‹*deed*›; *(unpers.)* **plötzlich tat es einen furchtbaren Knall** suddenly there was a dreadful bang; **f)** *(bewirken)* work, perform ‹*miracle*›; **seine Wirkung ~:** have its effect; **g)** *(an~)* **jmdm. etw. ~:** do sth. to sb.; **er tut dir nichts** he won't hurt *or* harm you; **h) es ~** *(ugs.: genügen)* be good enough; **i)** *(ugs.: irgendwohin bringen)* put. **2.** *unr. itr. V.* **a)** *(ugs.: funktionieren)* work; **b) freundlich/geheimnisvoll ~:** pretend to be *or (coll.)* act friendly/act mysteriously; **er tut [so], als ob** *od.* **als wenn** *od.* **wie wenn er nichts wüßte** he pretends not to know anything. **3.** *unr. refl. V. (unpers.) (geschehen)* **es hat sich einiges getan** quite a bit has happened; **es tut sich nichts** there's nothing happening. **4.** *Hilfsverb zur Umschreibung des Konjunktivs (ugs.)* **das täte mich interessieren/freuen** I'd be interested in/pleased about that

Tun das; ~s action; activity

Tünche ['tʏnçə] **die; ~, ~n a)** *(Farbe)* distemper; wash; **[weiße] ~:** whitewash; **b)** *o. Pl., (abwertend: Oberfläche)* veneer *(fig.)*

tünchen *tr. (auch itr.) V.* distemper; **weiß ~:** whitewash

Tunesien [tu'ne:zi̯ən] **(das); ~s** Tunisia

Tu·nicht·gut der; ~ *od.* **~[e]s, ~e** good-for-nothing; ne'er-do-well

Tunke ['tʊŋkə] **die; ~, ~n** *(bes. ostmd.)* sauce; *(Bratensoße)* gravy

tunken *tr. V. (bes. ostmd.)* dip; dip, dunk ‹*biscuit, piece of bread, etc.*›

tunlichst ['tu:nlıçst] *Adv. (geh.)* **a)** *(möglichst)* as far as possible; **b)** *(unbedingt)* at all costs

Tunnel ['tʊnl̩] **der; ~s, ~** *od.* **~s** tunnel

Tunte ['tʊntə] **die; ~, ~n a)** *(ugs. abwertend: Frau)* female; **b)** *(salopp, auch abwertend: Homosexueller)* queen *(sl.)*

Tüpfelchen ['tʏpflçən] **das; ~s, ~:** dot; **das ~ auf dem i** the final touch

tupfen ['tʊpfn̩] *tr. V.* **a)** dab; **sich** *(Dat.)* **den Schweiß von der Stirn ~:** dab the sweat from one's brow; **etw. auf etw.** *(Akk.)* **~:** dab sth. on to sth.; **b)** *(mit Tupfen versehen)* dot; **ein getupftes Kleid** a spotted dress

Tupfen der; ~s, ~: dot; *(größer)* spot

Tupfer der; ~s, ~ a) *(ugs.) s.* **Tupfen; b)** *(Med.)* swab

Tür [ty:ɐ̯] **die; ~, ~en** door; *(Garten~)* gate; **an die ~ gehen** *(öffnen)* [go and] answer the door; **in der ~ stehen** stand in the doorway; **den Kopf zur ~ hereinstecken** put one's head round the door; **jmdm. die ~ einlaufen** *od.* **einrennen** *(fig. ugs.)* keep badgering sb.; **jmdm. die ~ vor der Nase zuschlagen** *(fig.)* slam the door in sb.'s face; **einer Sache** *(Dat.)* **~ und Tor öffnen** *(fig.)* open the door *or* way to sth.; **mit der ~ ins Haus fallen** *(fig. ugs.)* blurt out what one is after; **vor verschlossener ~ stehen** be locked out; **zwischen ~ und Angel** *(fig. ugs.)* in passing; **jmdm. die ~ weisen** *(fig. geh.)* show sb. the door; **jmdn. vor die ~ setzen** *(fig. ugs.)* chuck *(coll.) or* throw sb. out; **vor seiner eigenen ~ kehren** *(fig. ugs.)* set one's own house in order

Tür·angel die door-hinge

Turban ['tʊrba:n] **der; ~s, ~e** turban

Turbine [tʊr'bi:nə] **die; ~, ~n** *(Technik)* turbine

Turbinen·flugzeug das turbo-jet aircraft

turbinen·getrieben *Adj.* turbine-propelled ‹*ship, aircraft*›; turbine-driven ‹*generator*›

Turbo- ['tʊrbo-]: *(Technik)* turbo-

turbulent [tʊrbu'lɛnt] **1.** *Adj. (auch Physik, Astron., Met.)* turbulent. **2.** *adv. (auch Physik, Astron., Met.)* turbulently

Turbulenz [tʊrbu'lɛnts] **die; ~, ~en** *(auch Physik, Astron., Met.)* turbulence *no pl.*

Tür-: ~drücker der; ~s, ~ a) doorknob; **b)** *(Türöffner)* [automatic] door-opener; **~griff der** door-handle

Türke ['tʏrkə] **der; ~n, ~n** Turk

Türkei die; ~: Turkey *no art.*

türkis [tʏr'ki:s] *indekl. Adj.* turquoise

Türkis der; ~es, ~e *(Mineral.)* turquoise

türkisch *Adj.* Turkish; *s. auch* **deutsch; Deutsch; ²Deutsche**

Tür-: ~klinke die door-handle; **~klopfer der** door-knocker

Turm [tʊrm] **der; ~[e]s, Türme** ['tʏrmə] **a)** tower; *(spitzer Kirch~)* spire; steeple; **b)** *(Schach)* rook; castle; **c)** *s.* **Sprung~**

[1]**türmen** **1.** *tr. V. (stapeln)* stack up; *(häufen)* pile up. **2.** *refl. V.* be piled up; ‹*clouds*› gather

[2]**türmen** *itr. V.; mit sein (salopp)* scarper *(Brit. sl.);* do a bunk *(Brit. sl.)*

Turm-: **~falke** **der** kestrel; **~springen** **das;** *o. Pl.* high diving *no art.;* **~uhr** **die** tower clock

turnen ['tʊrnən] **1.** *itr. V.* **a)** *(Sport)* do gymnastics; *(Schulw.)* do gym *or* PE; **sie turnt gut** she's a good gymnast; *(Schulw.)* she's good at gym *or* PE; **er turnte am Reck** he was doing *or* performing exercises *or* was working on the horizontal bar; **b)** *mit sein (ugs.: klettern)* clamber; **c)** *(ugs.: herumklettern)* clamber about. **2.** *tr. V. (Sport)* do, perform ‹*exercise, routine*›

Turnen **das;** ~s gymnastics *sing., no art.; (Schulw.)* gym *no art.;* PE *no art.*

Turner **der;** ~s, ~, **Turnerin** **die;** ~, ~nen gymnast

Turn-: **~halle** **die** gymnasium; **~hemd** **das** [gym] singlet; *(für Turnunterricht)* gym *or* PE vest; **~hose** **die** *(mit langem Bein)* gym trousers *pl.; (mit kurzem Bein)* gym shorts *pl.; (für Turnunterricht)* gym *or* PE shorts *pl.*

Turnier [tʊr'ni:ɐ̯] **das;** ~s, ~e *(auch hist.)* tournament; *(Reit~)* show; *(Tanz~)* competition

Turn~: **~lehrer** **der** gym *or* PE teacher; **~schuh** **der** gym shoe; *(Trainingsschuh)* training shoe; trainer *(coll.);* **~unterricht** **der** gym *no art.;* PE *no art.*

Turnus ['tʊrnʊs] **der;** ~, ~se regular cycle; **er führt das Amt im ~ mit seinen Kollegen** he and his colleagues hold the office in rotation

Turn-: **~verein** **der** gymnastics club; **~zeug** **das** gym *or* PE kit

Tür-: **~öffner** **der** door-opener; **~rahmen** **der** door-frame; **~schild** **das** sign on a/the door; *(Namensschild)* name-plate; door-plate; **~schloß** **das** door-lock; **~schwelle** **die** threshold

turteln *itr. V. (scherzh.: zärtlich sein)* bill and coo

Turtel·taube ['tʊrtl̩-] **die** turtle-dove; *(fig.)* love-bird

Tür·vorleger **der** doormat

Tusch [tʊʃ] **der;** ~[e]s, ~e fanfare

Tusche **die;** ~, ~n Indian *(Brit.) or (Amer.)* India ink

tuscheln ['tʊʃl̩n] *itr., tr. V.* whisper

tuschen *tr. V.* **sich** *(Dat.)* **die Wimpern ~:** put one's mascara on

Tusch·zeichnung **die** pen-and-ink drawing

Tussi [:tʊsi] **die;** ~, ~s *(salopp)* female *(derog.); (Mädchen)* bird *(sl.);* chick *(coll.)*

Tüte ['ty:tə] **die;** ~, ~n **a)** bag; **~n kleben** *od.* **drehen** *(fig. ugs.)* be doing time; **das kommt nicht in die ~!** *(fig. ugs.)* not on your life! *(coll.);* no way!; **b)** *(Eis~)* cone; cornet; **c)** *(ugs.: beim Alkoholtest)* bag; **in die ~ blasen müssen** be breathalysed

tuten ['tu:tn̩] *itr. V.* hoot; ‹*siren, [fog-]horn*› sound; **er tutete auf seiner Spielzeugtrompete** he tooted on his toy trumpet

Twen [tvɛn] **der;** ~[s], ~s twenty-to-thirty-year-old

Typ [ty:p] **der;** ~s, ~en **a)** type; **sie ist genau mein ~** *(ugs.)* she's just my type; **dein ~ wird verlangt** *(salopp)* you're wanted; **er ist ein dunkler/blonder ~:** he's dark/fair; **b)** *Gen. auch* ~en *(ugs.: Mann)* bloke *(Brit. sl.);* guy *(sl.);* **c)** *(Technik: Modell) (Auto)* model; *(Flugzeug)* type

Type ['ty:pə] **die;** ~, ~n **a)** *(Druck~, Schreibmaschinen~)* type; **b)** *(ugs.) (Person)* type; sort; character

Typhus ['ty:fʊs] **der;** ~ typhoid [fever]

typisch **1.** *Adj.* typical (**für** of). **2.** *adv.* typically; **das ist ~ Mann** that's just typical of a man; **~ Gisela!** that's Gisela all over!

Typographie [typogra'fi:] **die;** ~, ~n *(Druckw.)* typography

Tyrann [ty'ran] **der;** ~en, ~en *(auch fig.)* tyrant

Tyrannei **die;** ~, ~en *(auch fig.)* tyranny

tyrannisch **1.** *Adj.* tyrannical. **2.** *adv.* tyrannically

tyrannisieren *tr. V.* tyrannize

U

u, U [u:] **das;** ~, ~: u, U; *s. auch* **a, A; X**

ü, Ü [y:] **das;** ~, ~: u umlaut; *s. auch* **a, A**

u. *Abk.* **und**

u. a. *Abk.* **unter anderem**

U-Bahn **die** underground *(Brit.);* subway *(Amer.); (bes. in London)* tube

U-Bahnhof **der,** **U-Bahn-Station** **die** underground station *(Brit.);* subway station *(Amer.); (bes. in London)* tube station

übel ['y:bl̩] **1.** *Adj.* **a)** foul, nasty ‹*smell, weather*›; bad, nasty ‹*headache, cold, taste*›; nasty ‹*situation, consequences*›; sorry ‹*state, affair*›; foul, *(coll.)* filthy ‹*mood*›; **nicht ~** *(ugs.)* not bad at all; **b)** *(unwohl)* **jmdm. ist/wird ~:** sb. feels sick; **c)** *(verwerflich)* bad; wicked; nasty, dirty ‹*trick*›; **ein übler Bursche** a bad sort *(coll.) or* lot. **2.** *adv.* **a)** **~ gelaunt sein** be in a bad mood; **er spielt nicht ~:** he plays pretty well; **b)** *(nachteilig, schlimm)* badly; **er ist ~ dran** he's in a bad way; **etw. ~ vermerken** take sth. amiss; **jmdn. ~ zurichten** give sb. a working over *(coll.)*

Übel **das;** ~s, ~ **a)** *(Mißstand, Ärgernis)* evil; **zu allem ~:** on top of everything else; to make matters [even] worse; **das kleinere ~:** the lesser evil; **b)** *(meist geh.: Krankheit)* illness; malady; **c)** *(geh., veralt.: das Böse)* evil *no art.;* **von** *od.* **vom ~ sein** be an evil

übel·gelaunt **1.** *Adj. (präd. getrennt geschrieben)* ill-humoured; ill-tempered. **2.** *adv.* ill-humouredly; ill-temperedly

Übelkeit **die;** ~, ~en nausea

übel-, Übel-: **~|nehmen** *unr. tr. V.* **jmdm. etw. ~nehmen** hold sth. against sb.; take sth. amiss; **etw. ~nehmen** take offence at sth.; take sth. amiss; **~riechend** *Adj.; nicht präd.* foul-smelling; evil-smelling; **~täter** **der** wrongdoer; *(Verbrecher)* criminal; *(Verantwortlicher)* culprit; **~|wollen** *unr. itr. V.* **jmdm. ~wollen** wish sb. ill

üben ['y:bn̩] **1.** *tr. V.* **a)** *(auch itr.)* practise; rehearse ‹*scene, play*›; practise on ‹*musical instrument*›; **b)** *(trainieren, schulen)* exercise ‹*fingers*›; train ‹*memory*›; **mit geübten Händen** with practised hands; **c)** *(geh.: bekunden, tun)* exercise ‹*patience, restraint, etc.*›; commit ‹*treason*›; take ‹*revenge, retaliation*›; **Kritik an etw.** *(Dat.)* **~:** criticize sth. **2.** *refl. V.* **sich in etw.** *(Dat.)* **~:** practise sth.

über ['y:bɐ] **1.** *Präp. mit Dat.* **a)** *(Lage, Standort)* over; above; *(in einer Rangfolge)* above; **~ jmdm. wohnen** live above sb.; **zehn Grad ~ Null** ten degrees above zero; ~

jmdm. stehen *(fig.)* be above sb.; **b)** *(während)* during; ~ **dem Lesen einschlafen** fall asleep over one's book/magazine *etc.;* **c)** *(infolge)* because of; as a result of; ~ **der Aufregung vergaß ich, daß ...:** in all the excitement I forgot that ... **2.** *Präp. mit Akk.* **a)** *(Richtung)* over; *(quer hinüber)* across; ~ **die Straße gehen** go across the road; cross the road; ~ **Karlsruhe nach Stuttgart** via Karlsruhe to Stuttgart; **Tränen liefen ihr ~ die Wangen** tears ran down her cheeks; **sich** *(Dat.)* **die Mütze ~ die Ohren** he pulled the cap down over his ears; **bis ~ die Knöchel im Schlamm versinken** sink up past one's ankles in mud; **seine Tochter geht ihm ~ alles** his daughter means more to him than anything; **b)** *(während)* over; ~ **Mittag** over lunchtime; ~ **Wochen/Monate** for weeks/months; ~ **Weihnachten** over Christmas; **die ganze Zeit ~:** the whole time; **die Woche/den Sommer ~:** during the week/summer; **den ganzen Winter/Tag ~:** all winter/day long; **c)** *(betreffend)* about; ~ **etw. reden/schreiben** talk/write about sth.; **ein Buch ~ die byzantinische Kunst** a book about *or* on Byzantine art; **d)** *(in Höhe von)* **ein Scheck/eine Rechnung ~ 1000 Mark** a cheque/bill for 1,000 marks; **e)** *(von mehr als)* **Kinder ~ 10 Jahre** children over ten [years of age]; **f) Gewalt ~ jmdn. haben** have power over sb.; **Wellingtons Sieg ~ Napoleon** Wellington's victory over Napoleon; **g) das geht ~ meine Kraft** that's too much for me; **h) sie macht Fehler ~ Fehler** she makes mistake after mistake; **i)** *(mittels, durch)* through ⟨*person*⟩; by ⟨*post, telex, etc.*⟩; over ⟨*radio, loudspeaker*⟩; **ich bin ~ die Autobahn gekommen** I came along the motorway; **etw. ~ alle Sender bringen/ausstrahlen** broadcast sth. on all stations **3.** *Adv.* **a)** *(mehr als)* over; **b) ~ und ~:** all over. **4.** *Adj.; nicht attr. (ugs.)* **jmdm. ~ sein** have the edge on sb. *(coll.)*

über·all [*od.* --'-] *Adv.* **a)** everywhere; **sie weiß ~ Bescheid** *(fig.)* she knows about everything; **b)** *(bei jeder Gelegenheit)* always

überall-: ~her *Adv.* from all over the place; **~hin** *Adv.* everywhere

Über·angebot das surplus **(an** + *Dat.* of); *(Schwemme)* glut **(an** + *Dat.* of)

über·ängstlich 1. *Adj.* over-anxious. **2.** *adv.* over-anxiously

über·anstrengen *tr. V.* overtax ⟨*person, energy*⟩; strain ⟨*eyes, nerves, heart*⟩; **sich ~:** overstrain *or* over-exert oneself

Über·anstrengung die over-exertion; ~ **der Augen/des Herzens** strain on the eyes/heart

über·arbeiten 1. *tr. V.* rework; revise ⟨*text, edition*⟩. **2.** *refl. V.* overwork

Über·arbeitung die; ~, ~en reworking; *(von Text, Manuskript, Ausgabe usw.)* revision; *(überarbeitete Fassung)* revised version

über·aus *Adv. (geh.)* extremely

über·backen *unr. tr. V.* **etw. mit Käse** *usw.* **~:** top sth. with cheese *etc.* and brown it lightly [under the grill/in a hot oven]; **ein mit Käse ~er Auflauf** a soufflé au gratin

überbeanspruchen[1] *tr. V.* put too great a strain on ⟨*heart, circulation, etc.*⟩; strain ⟨*nerves*⟩; overstrain, overstress ⟨*material*⟩; overburden, overstretch ⟨*facilities, services*⟩; overload ⟨*machine*⟩; make excessive use of ⟨*right, privilege*⟩; overtax ⟨*person, body, strength*⟩

Über·bau der; *Pl.* **~e** *(marx.)* superstructure

überbelichten[2] *tr. V. (Fot.)* overexpose

Über·beschäftigung die; *o. Pl. (Wirtsch.)* over-employment

überbetonen[3] *tr. V.* overstress

überbewerten[4] *tr. V.* overvalue; *(überschätzen)* overvalue; overrate; mark ⟨*pupil, piece of work, gymnast, skater, etc.*⟩ too high

über·bieten *unr. tr. V.* **a)** outbid **(um** by); **b)** *(übertreffen)* surpass; outdo ⟨*rival*⟩; break ⟨*record*⟩ **(um** by); exceed ⟨*target*⟩ **(um** by); **das ist kaum noch zu ~:** that takes some beating

Überbleibsel [-blaipsl̩] **das; ~s, ~:** remnant; *(einer Kultur)* relic

Über·blick der a) view; **einen guten ~ über etw.** *(Akk.)* **haben** have a good view over sth.; **b)** *(Abriß)* survey; **c)** *o. Pl. (Einblick)* overall view *or* perspective; **den ~ über etw.** *(Akk.)* **verlieren** lose track of sth.

über·blicken *tr. V. s.* **übersehen a, b**

[1] *ich überbeanspruche, überbeansprucht, überzubeanspruchen*

[2] *ich überbelichte, überbelichtet, überzubelichten*

[3] *ich überbetone, überbetont, überzubetonen*

[4] *ich überbewerte, überbewertet, überzubewerten*

über·bringen *unr. tr. V.* deliver; convey ⟨*greetings, congratulations*⟩

Über·bringer der; ~s, ~: bearer

über·brücken *tr. V.* bridge ⟨*gap, gulf*⟩; reconcile ⟨*difference*⟩

Überbrückungs·kredit der *(Finanzw.)* bridging loan

über·dachen *tr. V.* roof over; **überdacht** covered ⟨*terrace, station platform, etc.*⟩

über·dauern *tr. V.* survive ⟨*war, separation, hardship*⟩

über|decken *tr. V.* **a)** *(bedecken)* cover; **b)** *(verdecken)* cover up

über·denken *unr. tr. V.* **etw. ~:** think sth. over

über·dies *Adv.* moreover; what is more

über·dimensional 1. *Adj.* inordinately large ⟨*spectacles, table, statue, etc.*⟩; inordinate ⟨*love, influence*⟩. **2.** *adv.* enormously ⟨*enlarged*⟩

Über·dosis die overdose

über·drehen *tr. V.* **a)** overwind ⟨*watch*⟩; over-tighten ⟨*screw, nut*⟩; **b)** *(Technik)* over-rev *(coll.)* ⟨*engine*⟩

überdreht *Adj. (ugs.)* wound up; *(verrückt)* crazy

Über·druck der; *Pl.* **~drücke** excess pressure

Überdruß [-drʊs] **der; Überdrusses** surfeit **(an** + *Dat.* of); **etw. bis zum ~ tun** do sth. until one has wearied of it

überdrüssig [-drysɪç] *Adj.* **jmds./einer Sache ~ sein/werden** be/grow tired of sb./sth.

über·durchschnittlich 1. *Adj.* above average. **2.** *adv.* **sie ist ~ begabt** she is more than averagely gifted *or* talented; **er verdient ~ gut** he earns more than the average

über·eifrig 1. *Adj.* over-eager; *(zu emsig)* over-zealous. **2.** *adv.* over-eagerly; *(zu emsig)* over-zealously

über·eignen *tr. V.* **jmdm. etw. ~:** transfer sth. *or* make sth. over to sb.

über·eilen *tr. V.* rush; **übereilt** over-hasty

über·einander *Adv.* **a)** *(räumlich)* one on top of the other; **sie wohnen ~:** they live one above the other; **b)** *(fig.: voneinander)* about each other; about one another

übereinander-: ~|legen *tr. V.* **Holzscheite** *usw.* **~legen** lay pieces of wood *etc.* one on top of the other; **~|liegen** *unr. itr. V.; südd., österr. schweiz. mit sein* lie one on top of the other; **~|schla-**

gen *unr. tr. V.* **die Enden des Tuches ~schlagen** fold the edges of the cloth over; **die Arme/Beine ~schlagen** fold one's arms/cross one's legs

übereinǀkommen *unr. itr. V.; mit sein* agree; come to an agreement

Überein·kommen das; ~s, ~, **Übereinkunft** [-'ainkʊnft] **die;** ~, **Übereinkünfte** agreement; **ein Übereinkommen** *od.* **eine Übereinkunft treffen** enter into *or* make an agreement

übereinǀstimmen *itr. V.* **a)** *(einer Meinung sein)* agree; **mit jmdm. in etw.** *(Dat.)* ~: agree with sb. on sth.; **b)** *(sich gleichen)* ‹*colours, styles*› match; ‹*figures, statements, reports, results*› tally, agree; ‹*views, opinions*› coincide

übereinstimmend 1. *Adj.; nicht präd.* concurrent ‹*views, opinions, statements, reports*›. **2.** *adv.* **sie stellten ~ fest, daß ...:** they agreed in stating that ...; **wir sind ~ der Meinung, daß ...:** we share the view that ...

Überein·stimmung die a) *(von Meinungen)* agreement **(in** + *Dat.* on); **b)** *(Einklang, Gleichheit)* agreement (*Gen.* between); **etw. mit etw. in ~ bringen** reconcile sth. with sth.

über·empfindlich *Adj.* oversensitive **(gegen** to); *(Med.)* hypersensitive **(gegen** to)

¹überǀfahren 1. *unr. tr. V.* **jmdn. ~:** ferry *or* take sb. over. **2.** *unr. itr. V.; mit sein* cross over

²über·fahren *unr. tr. V.* **a)** run over; **b)** *(übersehen u. weiterfahren)* go through ‹*red light, stopsignal, etc.*›; **c)** *(hinwegfahren über)* cross; go over ‹*crossroads*›; **d)** *(ugs.: überrumpeln)* **jmdn. ~:** catch *or* take sb. unawares

Über·fahrt die crossing **(über** + *Akk.* of)

Über·fall der attack **(auf** + *Akk.* on); *(aus dem Hinterhalt)* ambush **(auf** + *Akk.* on); *(mit vorgehaltener Waffe)* hold-up; *(auf eine Bank o. ä.)* raid **(auf** + *Akk.* on); *(fig. ugs.)* surprise visit

über·fallen *unr. tr. V.* **a)** attack; raid ‹*bank, enemy position, village, etc.*›; *(hinterrücks)* ambush; *(mit vorgehaltener Waffe)* hold up; *(fig.: besuchen)* descend on; **jmdn. mit Fragen ~** *(fig.)* bombard sb. with questions; **b)** *(überkommen)* ‹*tiredness, homesickness, fear*› come over

über·fällig *Adj.* overdue

Überfall·kommando das flying squad

über·fliegen *unr. tr. V.* **a)** *(hinwegfliegen über)* fly over; overfly *(formal)*; **b)** *(flüchtig lesen)* skim [through]

überǀfließen *unr. itr. V.; mit sein s.* **¹überlaufen a, b**

über·flügeln *tr. V.* outshine; outstrip

Über·fluß der; *o. Pl.* abundance **(an** + *Dat.* of); *(Wohlstand)* affluence; **etw. im ~ haben** have sth. in abundance; **im ~ vorhanden sein** be in abundant *or* plentiful supply; **zu allem ~:** to cap *or* crown it all

über·flüssig *Adj.* superfluous; unnecessary ‹*purchase, words, work*›; *(zwecklos)* pointless

über·fluten *tr. V. (auch fig.)* flood

über·fordern *tr. V.* **jmdn. [mit etw.] ~:** overtax sb. [with sth.]; ask *or* demand too much of sb. [with sth.]

über·fragen *tr. V.* **da bin ich überfragt** I don't know the answer to that

über·fremden *tr. V.* **überfremdet werden/sein** ‹*language, culture, etc.*› be swamped [by foreign influences]; ‹*economy*› be dominated [by foreign firms/capital]; ‹*country*› be dominated [by foreign influences]

über·frieren *unr. itr. V.; mit sein* freeze over; **~de Nässe** black ice

¹überǀführen *tr. V.* transfer; **der Tote wurde in seine Heimat übergeführt** the body of the dead man was brought back to his home town/country

²über·führen *tr. V.* **a)** *s.* **¹überführen; b) jmdn. [eines Verbrechens] ~:** find sb. guilty [of a crime]; convict sb. [of a crime]

Über·führung die a) transfer; **b)** *(eines Verdächtigen)* conviction; **c)** *(Brücke)* bridge; *(Hochstraße)* overpass; *(Fußgänger~)* [foot-] bridge

über·füllt *Adj.* crammed full, chock-full **(von** with); *(mit Menschen)* overcrowded, packed **(von** with); over-subscribed ‹*course*›

Über·gabe die a) handing over **(an** + *Akk.* to); *(einer Straße, eines Gebäudes)* opening; *(von Macht)* handing over; transfer; **b)** *(Auslieferung an den Gegner)* surrender **(an** + *Akk.* to)

Über·gang der a) crossing; *(Bahn~)* level crossing *(Brit.)*; grade crossing *(Amer.)*; *(Fußgängerbrücke)* foot-bridge; *(an der Grenze, eines Flusses)* crossing-point; **b)** *(Wechsel, Überleitung)* transition **(zu, auf** + *Akk.* to)

übergangs-, Übergangs-: ~erscheinung die transitional phenomenon; **~los 1.** *Adj.; nicht präd.* without any transition *postpos.;* **2.** *adv.* without any transition; **~lösung die** interim *or* temporary solution; **~zeit die a)** transitional period; **b)** *(Frühling)* spring; *(Herbst)* autumn; *(Frühling und Herbst)* spring and autumn

über·geben 1. *unr. tr. V.* **a)** hand over; pass ‹*baton*›; **b)** *(übereignen)* transfer, make over (*Dat.* to); **c)** *(ausliefern)* surrender (*Dat.*, **an** + *Akk.* to); **d) eine Straße dem Verkehr ~:** open a road to traffic; **e)** *(abgeben, überlassen)* **er hat sein Amt ~:** he has handed over his position; **jmdm. etw. ~:** entrust sb. with sth.. **2.** *unr. refl. V. (sich erbrechen)* vomit

¹überǀgehen *unr. itr. V.; mit sein* **a)** pass; **an jmdn./in jmds. Besitz ~:** become sb.'s property; **b) zu etw. ~/ dazu ~, etw. zu tun** go over to sth./to doing sth.; **c) in etw.** *(Akk.)* ~ *(zu etw. werden)* turn into sth.; **in Gärung/Verwesung ~:** begin to ferment/decompose; **ineinander ~** *(sich vermischen)* merge; **d) uns gingen die Augen über** we were overwhelmed by the sight

²über·gehen *unr. tr. V.* **a)** *(nicht beachten)* ignore; *(nicht eingehen auf)* **etw. [mit Stillschweigen] ~:** pass sth. over in silence; **b)** *(auslassen, überspringen)* skip [over]; **c)** *(nicht berücksichtigen)* pass over; **jmdn. bei der Beförderung ~:** pass sb. over for promotion

über·genau 1. *Adj.* overmeticulous. **2.** *adv.* overmeticulously

über·geordnet *Adj.* higher ‹*authority, position, court*›; greater ‹*significance*›; superordinate ‹*concept*›

Über·gepäck das *(Flugw.)* excess baggage

Über·gewicht das a) excess weight; *(von Person)* overweight; **[5 kg] ~ haben** ‹*person*› be [5 kilos] overweight; **b)** *(fig.)* predominance; **das ~ [über jmdn./etw.] haben/gewinnen** be/become predominant [over sb./sth.]; **c) das ~ bekommen** *od.* **kriegen** *(ugs.)* ‹*person*› overbalance

über·gießen *unr. tr. V.* **etw. mit Wasser/Soße ~:** pour water/sauce over sth.

über·glücklich *Adj.* blissfully happy; *(hoch erfreut)* overjoyed

überǀgreifen *unr. itr. V.* **a)** *(bes. beim Klavierspiel, Turnen)* cross one's hands over; **b)** *(sich ausdehnen)* **auf etw.** *(Akk.)* **~:** spread to sth.

übergreifend *Adj.* predominant; *(allumfassend)* all-embracing

Über·griff der *(unrechtmäßiger Eingriff)* encroachment **(auf** + *Akk.* on); infringement **(auf** + *Akk.* of); *(Angriff)* attack **(auf** + *Akk.* on)

Über·größe die outsize

über|haben *unr. tr. V. (ugs.)* be fed up with *(coll.)*

überhand|nehmen *unr. itr. V.* get out of hand; ‹*attacks, muggings, etc.*› increase alarmingly; ‹*weeds*› run riot

[1]**über|hängen** *unr. itr. V.; südd., österr., schweiz. mit sein* ‹*part of building*› overhang; ‹*branch*› hang over; ‹*rock face*› form an overhang

[2]**über|hängen** *tr. V.* **sich** *(Dat.)* **eine Jacke/eine Tasche ~:** put a jacket round one's shoulders/ hang *or* sling a bag over one's shoulder

über·häufen *tr. V.* **jmdn. mit etw. ~:** heap *or* shower sth. on sb.; **jmdn. mit Ratschlägen/Vorwürfen ~:** bombard sb. with advice/pour reproaches on sb.

überhaupt *Adv.* **a)** *(insgesamt, im allgemeinen)* in general; **soweit es ~ Zweck hat** as far as there's any point in it at all; **b)** *(meist bei Verneinungen: gar)* **~ nicht** not at all; **das ist ~ nicht wahr** that's not true at all; **~ keine Zeit haben** have no time at all; not have any time at all; **das kommt ~ nicht in Frage** it's quite *or* completely out of the question; **~ nichts** nothing at all; nothing what[so]ever; **wenn ~:** if at all; **c)** *(überdies, außerdem)* besides. **2.** *Partikel* anyway; **was willst du hier ~?** what ar you doing here anyway?; **wie konnte das ~ passieren?** how could it happen in the first place?; **wissen Sie ~, mit wem Sie reden?** do you realize who you're talking to?

überheblich [-'he:plɪç] **1.** *Adj.* arrogant; supercilious ‹*grin*›. **2.** *adv.* arrogantly; ‹*grin*› superciliously

Überheblichkeit die; ~: arrogance

über·hitzen *tr. V. (auch fig.)* overheat

überhöht [y:bɐ'hø:t] *Adj. (zu hoch)* excessive

über·holen 1. *tr. V.* **a)** overtake *(esp. Brit.)*; pass *(esp. Amer.)*; **b)** *(übertreffen)* outstrip; **c)** *(wieder instand setzen)* overhaul. **2.** *itr. V.* overtake *(esp. Brit.)*; pass *(esp. Amer.)*

Überhol: ~manöver das overtaking *(esp. Brit.) or (esp. Amer.)* passing manœuvre; **~spur die** overtaking lane *(esp. Brit.)*; pass lane *(esp. Amer.)*; **~verbot das** prohibition of overtaking

über·hören *tr. V.* not hear; **das möchte ich überhört haben** I'll pretend I didn't hear that

über·irdisch 1. *Adj.* celestial; heavenly; *(übernatürlich)* supernatural; ethereal ‹*beauty*›. **2.** *adv.* celestially; *(übernatürlich)* supernaturally; ethereally ‹*beautiful*›

über·kandidelt [-kandi:dl̩t] *Adj. (ugs.)* affected

über·kleben *tr. V.* **die alten Plakate mit neuen ~:** stick new posters over the old ones; **wir überklebten die Anschrift** we stuck something over the address; we covered the address by sticking something over it

über|kochen *itr. V.; mit sein (auch fig. ugs.)* boil over

über·kommen *unr. tr. V.* **Ekel/ Furcht überkam mich** I was overcome by revulsion/fear

über|kriegen *tr. V. (ugs.)* **jmdn./ etw. ~:** get fed up with sb./sth. *(coll.)*

[1]**über·laden** *unr. tr. V. (auch fig.)* overload

[2]**überladen** *Adj.* over-ornate ‹*façade, style, etc.*›; overcrowded ‹*shop window*›

über·lagern *tr. V.* **a)** overlie; *(fig.)* combine with; **sich ~:** combine; **b)** *(Physik)* ‹*wave*› interfere with; ‹*force, field*› be superimposed on; **sich ~** ‹*waves*› interfere; ‹*forces, fields*› be superimposed

über·lappen *tr. V. (auch fig.)* overlap

über·lassen 1. *unr. tr. V.* **a)** *(geben)* **jmdm. etw. ~:** let sb. have sth.; **b) jmdn. jmds. Fürsorge ~:** leave sb. in sb.'s care; **sich** *(Dat.)* **selbst ~ sein** be left to one's own devices; **c) etw. jmdm. ~** *(etw. jmdn. entscheiden/tun lassen)* leave sth. to sb.; **das bleibt [ganz] dir ~:** that's [entirely] up to you; **überlaß das bitte mir** let that be my concern; let me worry about that; **etw. dem Zufall ~:** leave sth. to chance. **2.** *unr. refl. V.* **sich der Leidenschaft/den Träumen** *usw.* **~:** abandon oneself to one's passions/dreams *etc.*

über·lasten *tr. V.* overload; overburden, overstretch ‹*facilities, authorities*›; put too great a strain on ‹*heart, circulation, etc.*›; overstress ‹*structure, material*›; overtax ‹*person*›; *(mit Arbeit)* overwork ‹*person*›; *(psychisch)* put too great a strain *or* much on ‹*person*›

[1]**über|laufen** *unr. itr. V.; mit sein* **a)** ‹*liquid, container*› overflow; **b)** *(auf die gegnerische Seite überwechseln)* defect; ‹*partisan*› go over to the other side

[2]**über·laufen** *unr. tr. V.* seize; **ein Frösteln/Schauer überlief mich, es überlief mich [eis]kalt** a cold shiver ran down my spine; **es überlief sie heiß** a hot flush came over her

[3]**überlaufen** *Adj.* overcrowded; over-subscribed ‹*course, subject*›

Über·läufer der *(auch fig.)* defector

über·leben 1. *tr., auch itr. V.* survive; **das überleb' ich nicht!** I'll never get over it!; **jmdn. ~:** survive *or* outlive sb. **(um** by). **2.** *refl. V.* become outdated *or* outmoded; **sich überlebt haben** have become outdated; have had its day

Über·lebende der/die; *adj. Dekl.* survivor

über·lebens·groß *Adj.* larger than life-size

[1]**über|legen** *tr. V.* **jmdm. etw. ~:** put sth. over sb.

[2]**über·legen 1.** *tr. V.* consider; think over *or* about; **etw. noch einmal ~:** reconsider sth.; **es sich anders ~:** change one's mind; **wenn ich es mir recht überlege, ...:** now I come to think of it, ... **2.** *itr. V.* think; reflect; **ohne zu ~** *(unbedacht)* without thinking; *(spontan)* without a moment's thought; **ohne lange zu ~:** without much reflection; **laß mich mal ~:** let me think

[3]**überlegen 1.** *Adj.* **a)** superior; clear, convincing ‹*win, victory*›; **jmdm. ~ sein** be superior to sb. **(an** + *Dat.* in); **b)** *(herablassend)* supercilious; superior. **2.** *adv.* **a)** in a superior manner; ‹*play*› much the better: ‹*win, argue*› convincingly; **b)** *(herablassend)* superciliously; superiorly

Überlegenheit die; ~: superiority

überlegt [y:bɐ'le:kt] **1.** *Adj.* carefully considered. **2.** *adv.* in a carefully considered way

Überlegung die; ~, ~en a) *o. Pl.* thought; reflection; **nach reiflicher ~:** on careful consideration; **b)** *(Gedanke)* idea; **~en** *(Gedankengang)* thoughts; reflections

über|leiten *itr. V.* **zum nächsten/ zu einem neuen Thema ~**

⟨*speaker*⟩ move on to the next topic; **in etw.** *(Akk.)* **~:** lead into sth.

Über·leitung die transition

über·lesen *unr. tr. V.* overlook; miss

über·liefern *tr. V.* hand down

Über·lieferung die a) *(etw. Überliefertes)* tradition; **schriftliche ~nen** written records; **b)** *(Brauch)* tradition; custom

überlisten *tr. V.* outwit

überm *Präp. + Art.* = **über dem**

Über·macht die; *o. Pl.* superior strength; *(zahlenmäßig)* superior numbers *pl.*; **in der ~ sein** be superior in strength/numbers

über·mächtig *Adj.* **a)** superior; **b)** *(nicht mehr bezähmbar)* overpowering ⟨*desire, hatred, urge, etc.*⟩

über·malen *tr. V.* **etw. ~:** paint sth. over

Über·maß das; *o. Pl.* excessive amount, excess (**an** + *Dat.* of); **ein ~ an Arbeit** *od.* **Arbeit im ~ haben** have an excessive amount of work *or* more than enough work

über·mäßig 1. *Adj.* excessive. **2.** *adv.* excessively; **~ viel essen** eat to excess *or* excessively; **nicht ~ attraktiv** not especially attractive

Über·mensch der superman

über·menschlich *Adj.* superhuman

über·mitteln *tr. V.* send; *(als Mittler weitergeben)* pass on, convey ⟨*greetings, regards, etc.*⟩

über·morgen *Adv.* the day after tomorrow

über·müden *tr. V.* overtire

Über·mut der high spirits *pl.*; **etw. aus [lauter]** *od.* **im ~ tun** do sth. out of [pure] high spirits

übermütig ['y:bɐmy:tɪç] **1.** *Adj.* high-spirited; in high spirits *pred.* **2.** *adv.* high-spiritedly

über·nächst... *Adj.; nicht präd.* **im ~en Jahr, ~es Jahr** the year after next; **am ~en Tag** two days later; the next day but one; **~en Montag** a week on Monday; Monday week; **er wohnt im ~en Haus** he lives in the next house but one *or* lives two doors away

über·nachten *itr. V.* stay overnight; **bei jmdm. ~:** stay *or* spend the night at sb.'s house/flat *(Brit.) or (Amer.)* apartment *etc.*; **im Hotel ~:** stay the night at the hotel; **im Freien ~:** sleep in the open air

übernächtigt [y:bɐ'nɛçtɪçt] *Adj.* ⟨*person*⟩ tired *or* worn out [through lack of sleep]; tired ⟨*face, look, etc.*⟩

Übernachtung die; ~, ~en overnight stay; **~ und Frühstück** bed and breakfast

Übernahme ['y:bɐna:mə] **die; ~, ~n a)** *o. Pl. (von Waren, einer Sendung)* taking delivery *no art.*; *(einer Idee, eines Themas, von Methoden)* adoption, taking over *no indef. art.*; *(einer Praxis, eines Geschäfts, der Macht)* takeover; *(von Wörtern, Ausdrücken)* borrowing (**von** from); **b)** *(etw. Übernommenes)* borrowing

über·natürlich *Adj.* supernatural

über·nehmen 1. *unr. tr. V.* **a)** take delivery of ⟨*goods, consignment*⟩; receive ⟨*relay baton*⟩; take over ⟨*power, practice, business, building, school class*⟩; take on ⟨*job, position, task, role, case, leadership*⟩; undertake to pay ⟨*costs*⟩; **das laß mich ~:** let me do that; **b)** *(bei sich einstellen)* take on ⟨*staff*⟩; **c)** *(sich zu eigen machen)* adopt, take over ⟨*ideas, methods, subject, etc.*⟩ (**von** from); borrow ⟨*word, phrase*⟩ (**von** from). **2.** *unr. refl. V.* overdo things *or* it; **sich mit etw. ~:** take on too much with sth.; **übernimm dich nur nicht** *(iron.)* don't strain yourself!

über|ordnen *tr. V.* **a) etw. einer Sache** *(Dat.)* **~:** give sth. precedence over sth.; **b) jmdn. jmdm. ~:** place sb. above sb.; *s. auch* **übergeordnet**

über·prüfen *tr. V.* **a)** check (**auf** + *Akk.* for); check [over], inspect, examine ⟨*machine, device*⟩; **b)** *(kontrollieren)* check, inspect, examine ⟨*papers, luggage*⟩; review ⟨*issue, situation, results*⟩; *(Finanzw.)* examine, inspect ⟨*accounts, books*⟩

Über·prüfung die a) *o. Pl. s.* **überprüfen a:** checking *no indef. art.* (**auf** + *Akk.* for); checking [over] *no indef. art.*; inspection; examination; **b)** *(Kontrolle)* check; *(des Ausweises, der Geschäftsbücher)* examination; inspection; *(einer Lage, Frage, der Ergebnisse)* review

über|quellen *unr. itr. V.; mit sein* **a)** spill over; **b)** *(zu voll sein)* be brimming

über·queren *tr. V.* cross

über·ragen *tr. V.* **a)** *(hinausragen über)* **jmdn./etw. ~:** tower above sb./sth.; **der Berg überragt die Ebene** the mountain towers over the plain; **b)** *(übertreffen)* **jmdn. an etw.** *(Dat.)* **~:** be head and shoulders above sb. in sth.

überragend 1. *Adj.* outstanding. **2.** *adv.* outstandingly

überraschen *tr. V.* surprise; ⟨*storm, earthquake*⟩ take by surprise; *(durch einen Angriff)* take by surprise; catch unawares; **jmdn. beim Rauchen/Stehlen ~:** catch sb. smoking/stealing; **jmdn. überrascht ansehen** look at sb. in surprise

überraschend 1. *Adj.* surprising; surprise *attrib.* ⟨*attack, visit*⟩; *(unerwartet)* unexpected. **2.** *adv.* surprisingly; *(unerwartet)* unexpectedly; **die Nachricht kam ~:** the news came as a surprise

Überraschung die; ~, ~en surprise; **zu meiner [großen] ~:** to my [great] surprise

über·reden *tr. V.* persuade; **jmdn. ~, etw. zu tun** persuade sb. to do sth.; talk sb. into doing sth.

Überredung die; ~: persuasion

über·regional 1. *Adj.* national ⟨*newspaper, radio station*⟩; **~e Veranstaltungen** events involving several regions. **2.** *adv.* nationally; **~ bekannt werden** become known outside one's/its own region

über·reich 1. *Adj.* lavish ⟨*meal, decoration*⟩; abundant, very rich ⟨*harvest*⟩. **2.** *adv.* **jmdn. ~ beschenken/belohnen** lavish gifts on sb./reward sb. lavishly

über·reichen *tr. V.* **[jmdm.] etw. ~:** present sth. [to sb.]

über·reichlich 1. *Adj.* over-ample **2.** *adv.* over-amply

über·reif *Adj.* over-ripe

über·reizen *tr. V.* overtax ⟨*person*⟩; overstrain ⟨*eyes, nerves, etc.*⟩

über·rennen *unr. tr. V. (Milit.)* overrun

Über·rest der; *meist Pl.* remnant; **~e** *(eines Gebäudes)* remains; ruins; *(einer Mahlzeit)* left-overs; **die sterblichen ~e** *(geh. verhüll.)* the mortal remains

über·rollen *tr. V.* **a)** *(Milit.)* overrun; *(fig.)* overwhelm ⟨*person*⟩; ⟨*fashion, craze*⟩ sweep through ⟨*country*⟩; **b)** *(hinwegrollen über)* run down

über·rumpeln *tr. V.* **jmdn. ~:** take sb. by surprise; *(bei einem Angriff)* catch sb. unawares; take sb. by surprise; **jmdn. mit etw. ~:** take sb. by surprise with sth.

über·runden *tr. V.* **a)** *(Sport)* lap; **b)** *(übertreffen)* outstrip

übers *Präp. + Art.* **a)** = **über das**; **b) ~ Jahr** one year later

übersät [y:bɐ'zɛ:t] *Adj.* **mit** *od.* **von etw. ~ sein** be covered with sth.

über·sättigen *tr. V.* supersaturate ⟨*solution*⟩; glut ⟨*market*⟩; satiate ⟨*public*⟩

Überschall-: supersonic
über·schätzen *tr. V.* overestimate; overrate ⟨*writer, performer, book, performance, talent, ability*⟩
überschaubar *Adj.* **eine ~e Menge/Zahl** a manageable quantity/number; **ein ~er Zeitraum/~es Gebiet** a reasonably short period/small area
über·schauen *tr. V. s.* **übersehen a, b**
über|schäumen *itr. V.; mit sein* froth over; **~de Begeisterung** bubbling enthusiasm
über·schlafen *unr. tr. V.* sleep on ⟨*matter, problem, etc.*⟩
Über·schlag der a) rough calculation *or* estimate; **b)** *(Turnen)* handspring
¹über|schlagen 1. *unr. tr. V.* **die Beine ~:** cross one's legs. **2.** *unr. itr. V.; mit sein* ⟨*wave*⟩ break; ⟨*spark*⟩ jump
²über·schlagen 1. *unr. tr. V.* **a)** *(auslassen)* skip ⟨*chapter, page, etc.*⟩; **b)** *(ungefähr berechnen)* calculate *or* estimate roughly; make a rough calculation *or* estimate of. **2.** *unr. refl. V.* **a)** go head over heels; ⟨*car*⟩ turn over; **b)** ⟨*voice*⟩ crack
³überschlagen *Adj. (bes. md.)* lukewarm ⟨*liquid*⟩; moderately warm ⟨*room*⟩
über|schnappen *itr. V.; mit sein* **a)** *(ugs.: den Verstand verlieren)* go crazy; go round the bend *(coll.)*; **b)** *(ugs.: sich überschlagen)* ⟨*voice*⟩ crack
über·schneiden *unr. refl. V.* cross; intersect; *(fig.)* ⟨*problems, events, etc.*⟩ overlap
über·schreiben *unr. tr. V.* **a)** entitle; head ⟨*chapter, section*⟩; **b)** *(übertragen)* **etw. jmdm.** *od.* **auf jmdn. ~:** transfer sth. to sb.; make sth. over to sb.
über·schreiten *unr. itr. V.* **a)** cross; *(fig.)* pass; **b)** *(hinausgehen über)* exceed ⟨*authority, powers, budget, speed, limit, deadline, etc.*⟩
Über·schrift die heading; *(in einer Zeitung)* headline; *(Titel)* title
Über·schuß der surplus **(an** + *Dat.* of)
überschüssig ['y:bɐʃysɪç] *Adj.* surplus
über·schütten *tr. V.* cover; **jmdn./etw. mit Wasser ~:** throw water over sb./sth.; **jmdn. mit Vorwürfen/Lob ~** *(fig.)* heap reproach/praise on sb.
Überschwang der; ~[e]s exuberance
über|schwappen *itr. V.; mit sein* ⟨*liquid, container*⟩ slop over
über·schwemmen *tr. V. (auch fig.)* flood; **den Markt mit Waren ~** *(fig.)* flood *or* swamp the market with goods
Überschwemmung die; ~, ~en flood; *(das Überschwemmen)* flooding *no pl.*
Überschwemmungs·katastrophe die disastrous floods *pl.*
über·schwenglich [-ʃvɛŋlɪç] **1.** *Adj.* effusive ⟨*words, manner, etc.*⟩; wild ⟨*joy, enthusiasm*⟩. **2.** *adv.* effusively
Über·see *o. Art.* **aus** *od.* **von ~:** from overseas; **nach ~ auswandern** emigrate overseas; **Exporte nach ~:** overseas exports
Übersee-: ~dampfer der ocean-going steamer; **~hafen der** international port; **~handel der** overseas trade
überseeisch ['y:bɐze:ɪʃ] *Adj.; nicht präd.* overseas
übersehbar *Adj. (abschätzbar)* assessable; **der Schaden ist noch nicht ~:** the damage cannot yet be assessed
über·sehen *unr. tr. V.* **a)** look out over; *(fig.)* survey ⟨*subject*⟩; **b)** *(abschätzen)* assess ⟨*damage, situation, consequences, etc.*⟩; **c)** *(nicht sehen)* overlook; miss ⟨*turning, signpost*⟩; **d)** *(ignorieren)* ignore
über·senden *unr. (auch regelm.) tr. V.* send; remit, send ⟨*money*⟩
¹über|setzen 1. *tr. V.* ferry over. **2.** *itr. V.; auch mit sein* cross [over]
²über·setzen *tr., itr. V. (auch fig.)* translate; **etw. ins Deutsche/aus dem Deutschen ~:** translate sth. into/from German
Über·setzer der, Übersetzerin die; ~, ~nen translator
Übersetzung die; ~, ~en a) translation; **b)** *(Technik)* transmission ratio
Übersetzungs·büro das translation agency
Über·sicht die a) *o. Pl.* overall view, overview **(über** + *Akk.* of); **die ~ [über etw.** *(Akk.)*] **verlieren** lose track [of sth.]; **b)** *(Darstellung)* survey; *(Tabelle)* summary
über·sichtlich 1. *Adj.* clear; ⟨*crossroads*⟩ which allows a clear view. **2.** *adv.* clearly
¹über|siedeln, ²über·siedeln *itr. V.; mit sein* move **(nach** to)
über·sinnlich *Adj.* supersensory; *(übernatürlich)* supernatural
über·spannen *tr. V.* **a)** *(bespannen)* cover; **b)** *(zu stark spannen)* over-tension, over-tighten ⟨*string, cable*⟩; overdraw ⟨*bow*⟩; over-tension ⟨*spring*⟩; *s. auch* **Bogen c**
überspannt *Adj.* exaggerated ⟨*ideas, behaviour, gestures*⟩; extreme ⟨*views*⟩; inflated ⟨*demands, expectations*⟩
über·spielen *tr. V.* **a)** *(hinweggehen über)* cover up; cover up, gloss over ⟨*mistake*⟩; smooth over ⟨*difficult situation*⟩; **b)** **[auf ein Tonband] ~:** transfer ⟨*record*⟩ to tape; put ⟨*record*⟩ on tape; **[auf ein anderes Tonband] ~:** transfer to another tape; **c)** *(Funkw., Ferns.)* transfer
über·spitzen *tr. V.* **etw. ~:** push *or* carry sth. too far; **überspitzt ausgedrückt, könnte man sagen, daß ...:** to exaggerate, one might say that ...
¹über|springen *unr. itr. V.; mit sein* **a)** ⟨*spark, fire*⟩ jump across; **seine Begeisterung sprang auf uns alle über** *(fig.)* his enthusiasm communicated itself to all of us; **b)** *(unvermittelt übergehen zu)* **auf etw.** *(Akk.)* **~:** switch abruptly to sth.
²über·springen *unr. tr. V.* **a)** jump ⟨*obstacle*⟩; **b)** *(auslassen)* miss out; skip; **eine Klasse ~:** jump a class
über·stehen *unr. tr. V.* come through ⟨*danger, war, operation*⟩; get over ⟨*illness*⟩; withstand ⟨*heat, strain*⟩; ⟨*boat*⟩ weather, ride out ⟨*storm*⟩; *(überleben)* survive
über·steigen *unr. tr. V.* **a)** climb over; **b)** *(fig.: hinausgehen über)* exceed; **jmds. Fähigkeiten/Kräfte ~:** be beyond sb.'s abilities/strength
über·steuern 1. *tr. V. (Elektrot.)* overdrive. **2.** *itr. V. (Kfz-W.)* ⟨*vehicle*⟩ oversteer
über·stimmen *tr. V.* outvote
über·streichen *unr. tr. V.* paint over
über|streifen *tr. V.* **[sich** *(Dat.)*] **etw. ~:** slip sth. on
¹über|strömen *itr. V.; mit sein* overflow
²über·strömen *tr. V.* flood; **von Blut überströmt [sein]** [be] streaming with blood; **eine Welle des Glücks überströmte ihn** *(fig.)* a wave of happiness flooded over him
über|stülpen *tr. V.* pull on ⟨*hat etc.*⟩
Über·stunde die: er hat eine ~/drei ~n gearbeitet he did one hour's/three hours' overtime; **~n machen** *od.* **leisten** *od.* *(salopp)* **schieben** do overtime
über·stürzen 1. *tr. V.* rush; **nur nichts ~:** don't rush things; take it easy. **2.** *refl. V.* rush; *(rasch auf-*

einanderfolgen) ‹*events, news, etc.*› come thick and fast

überstürzt 1. *Adj.* hurried ‹*escape, departure*›; over-hasty ‹*decision*›. **2.** *adv.* ‹*decide, act*› over-hastily; ‹*depart*› hurriedly

über·tariflich 1. *Adj.* **~e Bezahlung/Zulagen** payment/bonuses above agreed rates. **2.** *adv.* **jmdn. ~ bezahlen** pay sb. above agreed rates

übertölpeln *tr. V.* dupe; con *(coll.)*

über·tönen *tr. V.* drown out

Übertrag ['y:bɐtra:k] **der; ~[e]s, Überträge** [-trɛ:gə] *(bes. Buchf.)* carry-over

über·tragbar *Adj.* transferable **(auf** + *Akk.* to); *(auf etw. anderes anwendbar)* applicable **(auf** + *Akk.* to); *(übersetzbar)* translatable; *(ansteckend)* communicable, infectious ‹*disease*›

über·tragen 1. *unr. tr. V.* **a)** transfer **(auf** + *Akk.* to); transmit ‹*power, torque, etc.*› **(auf** + *Akk.* to); communicate ‹*disease, illness*› **(auf** + *Akk.* to); carry over ‹*subtotal*›; *(auf etw. anderes anwenden)* apply **(auf** + *Akk.* to); *(übersetzen)* translate; render; **etw. ins reine** *od.* **in die Reinschrift ~:** make a fair copy of sth.; **in ~er Bedeutung, im ~en Sinne** in a transferred sense; **b)** *(senden)* broadcast ‹*concert, event, match, etc.*›; *(im Fernsehen)* televise; **c)** *(geben)* **jmdm. Aufgaben/Pflichten** *usw.* **~:** hand over tasks/duties *etc.* to sb.; *(anvertrauen)* entrust sb. with tasks/duties *etc.*; **jmdm. ein Recht ~:** confer a right on sb. **2.** *refl. V.* **sich auf jmdn. ~** ‹*disease, illness*› be communicated *or* be passed on to sb.; *(fig.)* ‹*enthusiasm, nervousness, etc.*› communicate itself to sb.

Übertragung die; ~, ~en a) *s.* **übertragen 1 a:** transference; transmission; communication; carrying over; application; translation; rendering; **b)** *(das Senden)* broadcasting; *(Programm, Sendung)* broadcast; *(im Fernsehen)* televising/television broadcast; **c)** *(von Aufgaben, Pflichten usw.)* entrusting; *(von Rechten)* conferral

Übertragungs·wagen der outside broadcast vehicle; OB vehicle

über·treffen *unr. tr. V.* **a)** surpass, outdo **(an** + *Dat.* in); break ‹*record*›; **jmdn. an Ausdauer ~:** be superior to sb. in stamina; **jmdn. an Fleiß/Intelligenz ~:** be more diligent/intelligent than sb.; **jmdn. in einem Fach ~:** be better than sb. at a subject; **sich selbst ~:** excel oneself; **b)** *(übersteigen)* exceed

über·treiben *unr. tr. V.* **a)** *auch itr.* exaggerate; **b)** *(zu weit treiben)* overdo; take *or* carry too far; **man kann es auch ~:** you can take things *or* go too far

Übertreibung die; ~, ~en exaggeration

[1]über|treten *unr. itr. V.; mit sein* **a)** *auch mit haben (Sport)* step over the line/step out of the circle; **b)** *(überwechseln)* change sides; **zu einer anderen Partei ~:** join another party; switch parties; **zum Katholizismus ~:** convert to Catholicism

[2]über·treten *unr. tr. V.* break, contravene ‹*law*›; infringe, violate ‹*regulation, prohibition*›

Übertretung die; ~, ~en a) *s.* **[2]übertreten:** breaking; contravention; infringement; violation; **b)** *(Vergehen)* misdemeanour

übertrieben [-tri:bn] **1.** *Adj.* exaggerated; *(übermäßig)* excessive ‹*care, thrift, etc.*›. **2.** *adv.* excessively

Über·tritt der change of allegiance, switch **(zu** to); *(Rel.)* conversion **(zu** to)

über·trumpfen *tr. V.* outdo

über·tünchen *tr. V.* cover with whitewash; *(fig.)* cover up

Übervölkerung die; ~: overpopulation

über·voll *Adj.* overfull; overcrowded, packed ‹*room, train, tram, etc.*›; packed ‹*theatre, cinema*›

über·vorteilen *tr. V.* cheat

über·wachen *tr. V.* watch, keep under surveillance ‹*suspect, agent, area, etc.*›; supervise ‹*factory, workers, process*›; control ‹*traffic*›; monitor ‹*progress, production process, experiment, patient*›

überwältigen [-'vɛltɪgn̩] *tr. V.* **a)** overpower; **b)** *(fig.)* ‹*sleep, emotion, fear, etc.*› overcome; ‹*sight, impressions, beauty, etc.*› overwhelm

überwältigend 1. *Adj.* overwhelming ‹*sight, impression, victory, majority, etc.*›; overpowering ‹*smell*›; stunning ‹*beauty*›. **2.** *adv.* stunningly ‹*beautiful*›

über|wechseln *itr. V.; mit sein* **a)** cross over **(auf** + *Akk.* to); **auf eine andere Spur ~:** change lanes; move to another lane; **b)** *(übertreten)* change sides; **ins feindliche Lager~:** go over to the enemy; **c)** *(mit etw. anderem beginnen)* **zu etw. ~:** change over to sth.; **zu einem anderen Thema ~:** turn to another topic

über·weisen *unr. tr. V.* **a)** transfer ‹*money*› **(an, auf** + *Akk.* to); **b)** *(zu einem anderen Arzt schikken)* refer **(an** + *Akk.* to); **c)** *(zuleiten)* refer ‹*proposal*› **(an** + *Akk.* to); pass on ‹*file, application*› **(an** + *Akk.* to)

Über·weisung die a) *o. Pl.* transfer **(an, auf** + *Akk.* to); **b)** *(Summe)* remittance; **c)** *(eines Patienten)* referral **(an** + *Akk.* to); **d)** *s.* **Überweisungsschein**

Überweisungs-: ~formular das *(Bankw.)* [credit] transfer form; **~schein der** *(Med.)* certificate of referral

[1]über|werfen *unr. tr. V.* throw on ‹*clothes*›

[2]über·werfen *unr. refl. V.* **sich mit jmdm. ~:** fall out with sb.

über·wiegen 1. *unr. itr. V.* predominate. **2.** *unr. tr. V.* ‹*advantages, disadvantages, etc.*› outweigh; ‹*emotion, argument*› prevail over

überwiegend 1. [*auch* --'--] *Adj.* overwhelming; **der ~e Teil der Bevölkerung** the majority of the population. **2.** *adv.* mainly

über·winden 1. *unr. tr. V.* **a)** overcome ‹*resistance*›; overcome, surmount ‹*difficulty, obstacle, gradient*›; conquer ‹*capitalism, apartheid, etc.*›; overcome, get over ‹*fear, inhibitions, disappointment, grief*›; get past ‹*stage*›; **b)** *(aufgeben)* overcome ‹*doubt, misgivings, reservations*›; give up ‹*way of thinking, point of view*›; **c)** *(geh.: besiegen)* overcome; vanquish *(literary)*. **2.** *unr. refl. V.* overcome one's reluctance; **sich [dazu] ~, etw. zu tun** bring oneself to do sth.

Über·windung die a) *s.* **überwinden 1 a:** overcoming; surmounting; conquest; getting over/past; **b)** *(Besiegung)* overcoming; vanquishing *(literary)*; **c)** *(das Sichüberwinden)* **es war eine große ~ für ihn** it cost him a great effort; **das hat mich viel ~ gekostet** that was a real effort of will for me

über·wintern 1. *itr. V.* [over]-winter; spend the winter. **2.** *tr. V.* overwinter ‹*plant*›

über·wuchern *tr. V.* overgrow

Über·zahl die; *o. Pl.* majority; **in der ~ sein** be in the majority; ‹*army, enemy*› be superior in numbers

überzählig [-tsɛlɪç] *Adj.* surplus; spare

überzeugen 1. *tr. V.* convince; **jmdn. von etw. ~:** convince/persuade sb. of sth. **2.** *itr. V.* be convincing. **3.** *refl. V.* convince *or* satisfy oneself; **sich persönlich** *od.* **mit eigenen Augen [von etw.] ~:** see [sth.] for oneself
überzeugend 1. *Adj.* convincing; convincing, persuasive ‹*arguments, proof, words, speech*›. **2.** *adv.* convincingly; ‹*argue, speak*› convincingly, persuasively
überzeugt *Adj.* **a)** *nicht präd.* convinced; **b) von etw. ~ sein** *(etw. hoch einschätzen)* be convinced by sth.; **er ist sehr von sich [selbst] ~:** he's very sure of himself
Über·zeugung die a) *o. Pl.* convincing; *(das Umstimmen)* persuasion; **b)** *(feste Meinung)* conviction; **zu der ~ kommen** *od.* **gelangen, daß ...:** become convinced that ...; **meiner ~ nach ...:** I am convinced that ...
Überzeugungs·kraft die; *o. Pl.* power[s] of persuasion; persuasiveness
[1]über|ziehen *unr. tr. V.* **a)** pull on ‹*clothes*›; **b) jmdm. eins** *od.* **ein paar ~** *(ugs.)* give sb. a clout
[2]über·ziehen 1. *unr. tr. V.* **a) etw. mit etw. ~:** cover sth. with sth.; **die Betten frisch ~:** put clean sheets on the beds; change the sheets on the beds; **b)** overdraw ‹*account*› **(um** by); **sie hat ihr Konto [um 300 Mark] überzogen** she is [300 marks] overdrawn; **die vorgesehene Sendezeit ~:** overrun the programme time. **2.** *unr. itr. V.* **a)** overdraw one's account; go overdrawn; **b)** *(bei einer Sendung, einem Vortrag)* overrun. **3.** *unr. refl. V.* ‹*sky*› cloud over, become overcast
Überzieher der; ~s, ~ a) *(veralt.: Herrenmantel)* [light] overcoat; **b)** *(salopp: Kondom)* johnny *(Brit. sl.)*; rubber *(sl.)*
Überziehungs·kredit der *(Finanzw.)* overdraft facility
überzüchtet [y:bɐˈtsʏçtət] *Adj.* overbred; over-sophisticated ‹*engines, systems*›
über·zuckern *tr. V.* sugar
Überzug der a) *(Beschichtung)* coating; **b)** *(Bezug)* cover
üblich [ˈy:plɪç] *Adj.* usual; *(normal)* normal; *(gebräuchlich)* customary; **das ist hier so ~:** that's the accepted *or (coll.)* done thing here; **wie ~:** as usual
U-Boot das submarine; sub *(coll.)*
übrig [ˈy:brɪç] *Adj.* remaining *attrib.*; *(ander...)* other; **das/alles ~e erzähle ich dir später** I'll tell you the rest/all the rest later; **die/alle ~en** the/all the rest *or* others; **im ~en** besides; **ich habe noch Geld ~:** I [still] have some money left; *(ich habe mehr Geld, als ich brauche)* I [still] have some money to spare; **für jmdn./etw. wenig/nichts ~ haben** have little/no time for sb./sth. *(fig.)*
übrig|bleiben *unr. itr. V.; mit sein* be left; remain; ‹*food, drink*› be left over; **ihm bleibt nichts [anderes** *od.* **weiter] ~, als zu ...:** he has no [other] choice but to ...; there is nothing he can do but to ...
übrigens [ˈy:brɪgn̩s] *Adv.* by the way; incidentally
übrig|lassen *unr. tr. V.* leave; leave ‹*food, drink*› over; **sehr** *od.* **viel/nichts zu wünschen ~:** leave much *or (coll.)* a lot/nothing to be desired
Übung [ˈy:bʊŋ] **die; ~, ~en a)** exercise; **b)** *o. Pl. (das Üben, Geübtsein)* practice; **aus der ~ kommen/außer ~ sein** get/be out of practice; **~ macht den Meister** *(Spr.)* practice makes perfect *(prov.)*; **c)** *(Lehrveranstaltung)* class; seminar
Übungs·buch das book of exercises; *(Lehrbuch)* textbook with exercises
UdSSR [u:de:|ɛs|ɛs|ˈɛr] *Abk.* **die; ~: Union der Sozialistischen Sowjetrepubliken** USSR
UEFA [u:ˈe:fa:] *Abk.* **die; ~** *(Fußball)* UEFA
Ufer [ˈu:fɐ] **das; ~s, ~:** bank; *(des Meers)* shore; **ans ~ gespült werden** be washed ashore; **der Fluß trat über die ~:** the river burst its banks
ufer-, Ufer-: ~befestigung die bank reinforcement; **~böschung die** [river/canal] embankment; **~los** *Adj.* limitless; boundless ‹*love, indulgence, etc.*›; endless ‹*discussions, talks, quarrel, subject*›; **ins ~lose gehen** ‹*plans, ambitions, etc.*› know no bounds; **~promenade die** riverside walk; *(am Meer)* promenade
UFO, Ufo [ˈu:fo] **das; ~[s], ~s** UFO
Uganda [uˈganda] **(das); ~s** Uganda
Ugander [uˈgandɐ] **der; ~s, ~, Uganderin die; ~, ~nen** Ugandan
Uhr [u:ɐ] **die; ~, ~en a)** clock; *(Armband~, Taschen~)* watch; *(Wasser~, Gas~)* meter; *(an Meßinstrumenten)* dial; gauge; **auf die** *od.* **nach der ~ sehen** look at the time; **nach meiner ~:** by *or* according to my clock/watch; **jmds. ~ ist abgelaufen** *(fig.)* the sands of time have run out for sb.; **wissen, was die ~ geschlagen hat** *(fig.)* know what's what; know how things stand; **rund um die ~** *(ugs.)* round the clock; **b)** *(bei Uhrzeitangaben)* **acht ~:** eight o'clock; **acht ~ dreißig** half past eight; 8.30 [eɪtˈθɛ:tɪ]; **wieviel ~ ist es?** what's the time?; what time is it?; **um wieviel ~ treffen wir uns?** [at] what time shall we meet?; when shall we meet?
Uhr-: ~armband das watchstrap; **~kette die** watch-chain; **~macher der** watchmaker/clockmaker; **~werk das** clock/watch mechanism; **~zeiger der** clock-/watch-hand; **~zeigersinn der: im ~zeigersinn** clockwise; **entgegen dem ~zeigersinn** anticlockwise; **~zeit die** time
Uhu [ˈu:hu] **der; ~s, ~s** eagle owl
Ukraine [ukraˈi:nə] **die; ~:** Ukraine
Ukrainer der; ~s, ~, Ukrainerin die; ~, ~nen Ukrainian
UKW-Sender [u:ka:ˈve:-] **der** VHF station; ≈ FM station
Ulk [ʊlk] **der; ~s, ~e** lark *(coll.)*; *(Streich)* trick; [practical] joke
ulkig *(ugs.)* **1.** *Adj.* funny. **2.** *adv.* in a funny way
Ulme [ˈʊlmə] **die; ~, ~n** elm
ultimativ [ʊltimaˈti:f] **1.** *Adj.* ‹*demand*› made as an ultimatum. **2.** *adv.* **etw. ~ fordern** demand sth. in [the form of] an ultimatum
Ultimatum [ʊltiˈma:tʊm] **das; ~s, Ultimaten** ultimatum; **[jmdm.] ein ~ stellen** give *or* set [sb.] an ultimatum
Ultra·kurz·welle [ʊltraˈkʊrtsvɛlə] **die a)** *(Phys., Funkw., Rundf.)* ultra-short wave; **b)** *(Rundf.: Wellenbereich)* very high frequency; VHF
Ultra·schall [ˈ---] **der** *(Physik, Med.)* ultrasound
ultra·violett *Adj. (Physik)* ultraviolet
um [ʊm] **1.** *Präp. mit Akk.* **a)** *(räumlich)* [a]round; **um etw. herum** [a]round sth.; **das Rad dreht sich um seine Achse** the wheel turns on its axle; **um die Ecke** round the corner; **um sich schlagen** lash *or* hit out; **b)** *(zeitlich) (genau)* at; *(etwa)* around [about]; **um acht [Uhr]** at eight [o'clock]; **um den 20. August [herum]** around [about] 20 August; **c) Tag um Tag/Stunde um Stunde** day after day/hour after hour; **Meter um Meter/Schritt um Schritt** metre by metre/step by

step; **d)** *(bei Maß- u. Mengenangaben)* by; **die Temperatur stieg um 5 Grad** the temperature rose [by] five degrees; **um nichts/einiges/vieles besser sein** be no/somewhat/a lot better. **2.** *Adv.* around; about; **um [die] 50 Personen [herum]** around *or* about *or* round about 50 people. **3.** *Konj.* **a)** *(final)* **um ... zu** [in order] to; **b)** *(konsekutiv)* **er ist groß genug, um ... zu ...:** he is big enough to ...; **c)** *(desto)* **je länger ..., um so besser ...:** the longer ..., the better ...; **um so besser/schlimmer!** all the better/worse!; **um so mehr, als ...** *(zumal, da ...)* all the more so, as *or* since ...

umländern *tr. V.* change; alter; revise ⟨*text, novel*⟩; alter ⟨*garment*⟩

umlarbeiten *tr. V.* alter ⟨*garment*⟩; revise, rework ⟨*text, novel, music*⟩

umarmen *tr. V.* embrace; put one's arms around; *(an sich drükken)* hug; **sie umarmten sich** they embraced/hugged

Um·bau der; ~[e]s, ~ten rebuilding; reconstruction; *(kleinere Änderung)* alteration; *(zu etw. anderem)* conversion; *(fig.: eines Systems, einer Verwaltung)* reorganization

¹umlbauen *tr., auch itr. V.* rebuild; reconstruct; *(leicht ändern)* alter; *(zu etw. anderem)* convert **(zu** into); *(fig.)* reorganize ⟨*system, administration, etc.*⟩; **das Bühnenbild ~:** change the set

²um·bauen *tr. V.* surround; **umbauter Raum** interior space

umlbenennen *unr. tr. V.* change the name of, rename ⟨*street, square, etc.*⟩; **etw. in etw.** *(Akk.)* **~:** change the name of sth. to sth.; rename sth. sth.

umlbilden *tr. V.* reorganize, reconstruct ⟨*department etc.*⟩; reshuffle ⟨*government, cabinet*⟩

umlbinden *unr. tr. V.* put on ⟨*tie, apron, scarf, etc.*⟩

umlblättern 1. *tr. V.* turn [over] ⟨*page*⟩. **2.** *itr. V.* turn the page/pages

umlblicken *refl. V.* **a)** look around; **b)** *(zurückblicken)* [turn to] look back **(nach** at)

umlbringen *unr. tr. V.* kill; **diese Packerei bringt mich fast um** *(fig. ugs.)* all this packing's nearly killing me *(coll.)*

Um·bruch der a) radical change; *(Umwälzung)* upheaval; **b)** *o. Pl. (Druckw.)* make-up; *(Ergebnis)* page proofs *pl.*

umlbuchen 1. *tr. V.* **a)** change ⟨*flight, journey route*⟩ **(auf +** *Akk.* to); **b)** *(Finanzw.)* transfer **(auf +** *Akk.* to). **2.** *itr. V.* change one's booking **(auf +** *Akk.* to)

umldenken *unr. itr. V.* revise one's thinking; rethink

umldisponieren *itr. V.* change one's arrangements; make new arrangements

umldrehen 1. *tr. V.* turn round; turn over ⟨*coin, hand, etc.*⟩; turn ⟨*key*⟩; **jeden Pfennig [dreimal] ~** *(ugs.)* watch every penny. **2.** *refl. V.* turn round; *(den Kopf wenden)* turn one's head; **sich nach jmdm. ~:** turn/turn one's head to look at sb.. **3.** *itr. V.; auch mit sein (ugs.: umkehren)* turn back; *(ugs.: wenden)* turn round

Um·drehung die turn; *(eines Motors usw.)* revolution; rev *(coll.)*; *(eines Planeten)* rotation

um·einander *Adv.* **sich ~ kümmern/sorgen** take care of/worry about each other *or* one another; **sich ~ drehen** revolve around each other

¹umlfahren *unr. tr. V.* knock over *or* down

²um·fahren *unr. tr. V.* go round; make a detour round ⟨*obstruction, busy area*⟩; *(im Auto)* drive *or* go round; *(im Schiff)* sail *or* go round; *(auf einer Umgehungsstraße)* bypass ⟨*town, village, etc.*⟩

umlfallen *unr. itr. V.; mit sein* **a)** *(umstürzen)* fall over; **b)** *(zusammenbrechen)* collapse; **tot ~:** fall down dead; **vor Hunger fast ~:** be faint with hunger; **c)** *(ugs. abwertend: seine Meinung ändern)* do an about-face; do a U-turn

Um·fang der a) circumference; *(eines Quadrats usw.)* perimeter; *(eines Baums, Menschen usw.)* girth; circumference; **b)** *(Größe)* size; **der Band hat einen ~ von 250 Seiten** the volume contains 250 pages *or* is 250 pages thick; **c)** *(Ausmaß)* extent; *(von Wissen)* range; extent; *(einer Stimme)* range; *(einer Arbeit, Untersuchung)* scope; **in vollem ~:** fully; completely; **in großem ~:** on a large scale

umfang·reich *Adj.* extensive; substantial ⟨*book*⟩

um·fassen *tr. V.* **a)** grasp; *(umarmen)* embrace; **b)** *(enthalten)* contain; *(einschließen)* include; take in; span, cover ⟨*period*⟩; **c)** *(umgeben)* enclose; surround; **d)** *(Milit.: umzingeln)* surround; encircle

umfassend 1. *Adj.* full ⟨*reply, information, survey, confession*⟩; extensive, wide, comprehensive ⟨*knowledge, powers*⟩; broad ⟨*education*⟩; extensive ⟨*preparations, measures*⟩. **2.** *adv.* ⟨*inform*⟩ fully

Um·feld das a) *(Psych., Soziol.)* milieu; **b)** *s.* **Umgebung a**

umlformen *tr. V.* **a)** reshape; remodel; recast, revise ⟨*poem, novel*⟩; transform ⟨*person*⟩; **b)** *(Elektrot.)* convert

Um·frage die survey; *(Politik)* opinion poll; **eine ~ machen** *od.* **veranstalten** carry out a survey/conduct an opinion poll

umlfüllen *tr. V.* **etw. in etw.** *(Akk.)* **~:** transfer sth. into sth.

umlfunktionieren *tr. V.* change the function of; **etw. zu etw. ~:** turn sth. into sth.

Um·gang der a) *o. Pl. (gesellschaftlicher Verkehr)* contact; dealings *pl.*; **jmd. hat guten/schlechten ~:** sb. keeps good/bad company; **mit jmdm. ~ haben/pflegen** associate with sb.; **mit jmdm. keinen ~ haben** have nothing to do with sb.; **er ist kein ~ für dich!** he is not suitable *or* fit company for you; **b)** *o. Pl. (das Umgehen)* **den ~ mit Pferden lernen** learn how to handle horses; **im ~ mit Kindern erfahren sein** be experienced in dealing with children

umgänglich ['ʊmgɛŋlɪç] *Adj.* *(verträglich)* affable; friendly; *(gesellig)* sociable

umgangs-, Umgangs-: ~form die; *meist Pl.* **gute/schlechte/keine ~formen haben** have good/bad/no manners; **~sprache die** colloquial language; **~sprachlich 1.** *Adj.* colloquial; **2.** *adv.* colloquially

um·geben *unr. tr. V.* **a)** surround; ⟨*hedge, fence, wall, etc.*⟩ enclose; ⟨*darkness, mist, etc.*⟩ envelop; **b) etw. mit etw. ~:** surround sth. with sth.; *(einfrieden)* enclose sth. with sth.; **sich mit jmdm./etw. ~:** surround oneself with sb./sth.

Umgebung die; ~, ~en a) surroundings *pl.*; *(Nachbarschaft)* neighbourhood; *(eines Ortes)* surrounding area; **die nähere/weitere ~ Mannheims** the immediate/broader environs *pl* of Mannheim; **b)** *(fig.)* milieu; **jmds. nähere ~:** those *pl.* close to sb.

¹umlgehen *unr. itr. V.; mit sein* **a)** *(im Umlauf sein)* ⟨*list, rumour, etc.*⟩ go round, circulate; ⟨*illness, infection*⟩ go round; **Angst geht in der Bevölkerung um** fear is spreading in the population; **b)** *(spuken)* **im Schloß geht ein Ge-**

spenst um a ghost haunts this castle; the castle is haunted; **c)** *(behandeln)* **mit jmdm. freundlich** *usw.* **~:** treat sb. kindly *etc.;* **mit etw. sorgfältig** *usw.* **~:** treat sth. carefully *etc.;* **er kann mit Geld nicht ~:** he can't handle money

²um·gehen *unr. tr. V.* **a)** *(herumgehen, -fahren um)* go round; make a detour round ‹*obstruction, busy area*›; *(auf einer Umgehungsstraße)* bypass ‹*town, village, etc.*›; **b)** *(vermeiden)* avoid; avoid, get round ‹*problem, difficulty*›; evade ‹*question, issue*›; **c)** *(nicht befolgen)* get round, circumvent ‹*law, restriction, etc.*›; evade ‹*obligation, duty*›

umgehend 1. *Adj.; nicht präd.* immediate. **2.** *adv.* immediately

Umgehungs·straße die bypass

umgekehrt 1. *Adj.* inverse ‹*ratio, proportion*›; reverse ‹*order*›; opposite ‹*sign*›; **es verhält sich** *od.* **ist genau ~:** the very opposite *or* reverse is true *or* the case. **2.** *adv.* inversely ‹*proportional*›; **vom Englischen ins Deutsche und ~ übersetzen** translate from English into German and vice versa

um|gestalten *tr. V.* reshape; remodel; redesign ‹*square, park, room, etc.*›; rework ‹*text, music, etc.*›; *(reorganisieren)* reorganize; *(verändern)* change

um|graben *unr. tr. V.* dig over

um|gruppieren *tr. V.* rearrange

Um·hang der cape

um|hängen *tr. V.* **a) etw. ~:** hang sth. somewhere else; **b) jmdm./sich einen Mantel ~:** drape a coat round sb.'s/one's shoulders; **jmdm. eine Medaille ~:** hang a medal round sb.'s neck

Umhänge·tasche die shoulder-bag

um|hauen *unr. tr. V.* **a)** *(fällen)* fell; **b)** *(ugs.: niederwerfen)* knock down; floor; **es hat mich fast umgehauen, als ich davon hörte** *(salopp)* I was flabbergasted when I heard

um·her *Adv.* around; **weit ~:** all around

umher-: *s.* **herum-**

umhin|können *unr. itr. V.* **sie konnte nicht/kaum umhin, das zu tun** she had no/scarcely had any choice but to do it; *(einem inneren Zwang folgend)* she couldn't help/could scarcely help but do it

um|hören *refl. V.* keep one's ears open; *(direkt fragen)* ask around

um·hüllen *tr. V.* wrap; *(fig.)* ‹*mist, fog, etc.*› shroud; **jmdn./etw. mit etw. ~:** wrap sb./sth. in sth.

Umkehr ['ʊmke:ɐ̯] **die; ~** *(auch fig.)* turning back; **zur ~ gezwungen werden** be forced to turn back

um|kehren 1. *itr. V.; mit sein* turn back; *(fig. geh.: sich wandeln)* change one's ways; **auf halbem Wege ~** *(fig.)* stop half-way. **2.** *tr. V.* **a)** turn upside down; turn over ‹*sheet of paper*›; *(nach links drehen)* turn ‹*garment etc.*› inside out; *(nach rechts drehen)* turn ‹*garment etc.*› right side out; **das ganze Haus [nach etw.] ~** *(fig.)* turn the whole house upside down [looking for sth.]; **b)** *(ins Gegenteil verkehren)* reverse; invert ‹*ratio, proportion*›

um|kippen 1. *itr. V.; mit sein* **a)** fall over; ‹*boat*› capsize, turn over; ‹*vehicle*› overturn; **b)** *(ugs.: ohnmächtig werden)* keel over; **c)** *(ugs. abwertend) s.* **umfallen c; d)** *(ugs.: umschlagen)* ‹*wine*› go off; **e)** *(Ökologie)* ‹*river, lake*› reach the stage of biological collapse. **2.** *tr. V.* tip over; knock over ‹*lamp, vase, glass, cup*›; capsize ‹*boat*›; turn ‹*boat*› over; overturn ‹*vehicle*›

um·klammern *tr. V.* clutch; clasp; **etw./jmdn. fest umklammert halten** keep a firm grip on sth./clutch sb. tightly

um|klappen *tr. V.* fold down

Umkleide·kabine die changing-cubicle

um|kleiden *(geh.)* **1.** *refl. V.* change; change one's clothes. **2.** *tr. V.* **jmdn. ~:** change sb.; change sb.'s clothes

Umkleide·raum der changing-room *(Brit.)*; *(im Theater)* green-room

um|knicken 1. *itr. V.; mit sein* **a) [mit dem Fuß] ~:** go over on one's ankle; **b)** ‹*tree, stalk, blade of grass, etc.*› bend; ‹*branch*› bend and snap. **2.** *tr. V.* **a)** *(falten)* fold ‹*page, sheet of paper*› over; **b)** *(abknicken)* bend over; break ‹*flower, stalk*›

um|kommen *unr. itr. V.; mit sein* die; *(bei einem Unglück, durch Gewalt)* get killed; die; **ich komme um vor Hitze/Hunger** *(fig. ugs.)* I'm dying in this heat/of hunger *(coll.)*; **vor Langeweile ~** *(fig. ugs.)* be bored to death *(coll.)*; die of boredom *(coll.)*

Um·kreis der *o. Pl.* surrounding area; **im ~ von 5 km** within a radius of 5 km.; **im [näheren] ~ der Stadt** in the [immediate] vicinity of the town

um·kreisen *tr. V.* circle; ‹*spacecraft, satellite*› orbit; ‹*planet*› revolve [a]round; **seine Gedanken umkreisten das Thema** *(fig.)* he kept turning the matter over in his mind

um|krempeln *tr. V.* **a)** *(aufkrempeln)* turn up ‹*cuff*›; roll up ‹*sleeve, trouser-leg*›; **b)** *(ugs.: von Grund auf ändern)* **etw. ~:** give sth. a shake-up; **jmdn. ~:** [completely] change sb.

um|laden *unr. tr. V.* transfer ‹*goods etc.*›

Um·lage die: ~[n] share of the cost[s]; *(bei einer Wohnung)* share of the bill[s]; **die ~ beträgt 30 Mark pro Person** the cost is 30 marks per person

um·lagern *tr. V.* besiege

Um·land das; *o. Pl.* surrounding area; **das ~ von Köln** the area around Cologne

Um·lauf der a) rotation; **ein ~ [der Erde um die Sonne] dauert ein Jahr** one revolution [of the earth around the sun] takes a year; **b)** *o. Pl. (Zirkulation)* circulation; **in** *od.* **im ~ sein** ‹*magazine, report, etc.*› be circulating; ‹*coin, banknote*› be in circulation; **in ~ bringen** *od.* **setzen** circulate ‹*report, magazine, etc.*›; circulate, put about, start ‹*rumour*›; bring ‹*coin, banknote*› into circulation; **c)** *(Rundschreiben)* circular

Umlauf·bahn die *(Astron., Raumf.)* orbit

Um·laut der *(Sprachw.)* umlaut

um|legen *tr. V.* **a)** *(um einen Körperteil)* put on; **b)** *(auf den Boden, die Seite legen)* lay down; flatten ‹*corn, stalks, etc.*›; *(fällen)* fell; **c)** *(umklappen)* fold down; turn down ‹*collar*›; throw ‹*lever*›; turn over ‹*calendar-page*›; **d)** *(ugs.: zu Boden werfen)* floor, knock down ‹*person*›; **e)** *(salopp: ermorden)* **jmdn. ~:** do sb. in *(sl.)*; bump sb. off *(sl.)*; **f)** *(verlegen)* transfer ‹*patient, telephone call*›; **den Termin ~:** change the date **(auf +** *Akk.* to); **g)** *(anteilmäßig verteilen)* split, share ‹*costs*› **(auf +** *Akk.* between)

um|leiten divert; re-route; divert ‹*river, stream*›

Um·leitung die diversion; re-routing

um|lernen *itr. V.* **a)** *(beruflich)* retrain; **b)** *(seine Anschauungen ändern)* learn to think differently

umliegend *Adj.* surrounding ‹*area, district*›; *(nahe)* nearby ‹*building*›

Umnachtung die; ~, ~en *(geh.)* derangement

um|pflügen *tr. V.* plough up

um|quartieren *tr. V.* re-accommodate ‹*person*› **(in +**

Akk. in); re-quarter, re-billet ‹*troops*› (**in** + *Akk.* in); move ‹*patient*›

um|rahmen *tr. V.* frame ‹*face etc.*›; **eine Feier mit Musik** *od.* **musikalisch** ~ *(fig.)* begin and end a ceremony with music; give a ceremony a musical framework

umrändert [ʊm'rɛndɐt] *Adj.* **schwarz** ~: with a black border; **rot ~e Augen** red-rimmed eyes

um|räumen 1. *tr. V.* rearrange. **2.** *itr. V.* rearrange things

um|rechnen *tr. V.* convert (**in** + *Akk.* into)

Um·rechnung die conversion (**in** + *Akk.* into)

Umrechnungs·kurs der exchange rate

¹um|reißen *unr. tr. V.* pull ‹*mast, tree*› down; knock ‹*person*› down; ‹*wind*› tear ‹*tent etc.*› down

²um·reißen *unr. tr. V.* outline; summarize ‹*subject, problem, situation*›; **fest** *od.* **klar** *od.* **scharf umrissen** clearly defined ‹*programme*›; clear-cut ‹*ideas, views*›

um|rennen *unr. tr. V.* [run into and] knock down

um·ringen *tr. V.* surround; *(in großer Zahl)* crowd round

Um·riß der *(auch fig.)* outline

um|rühren *tr. (auch itr.) V.* stir

um|rüsten *tr. V.* **a)** *(Technik)* convert (**auf** + *Akk.* to, **zu** into); **b)** *(Milit.)* **eine Armee [auf Atomwaffen]** ~: re-equip an army [with nuclear weapons]

ums [ʊms] *Präp.* + *Art.* **a)** = **um das**; **b)** ~ **Leben kommen** lose one's life

um|satteln *itr. V. (ugs.)* change jobs; ‹*student*› change courses

Um·satz der turnover; *(Verkauf)* sales *pl.* (**an** + *Dat.* of); **1000 Mark ~ machen** turn over 1,000 marks

Umsatz·steuer die turnover *or (Amer.)* sales tax

¹um|säumen *tr. V.* hem

²um·säumen *tr. V. (fig.)* surround

um|schalten 1. *tr. V. (auch fig.)* switch [over] (**auf** + *Akk.* to); move ‹*lever*›. **2.** *itr. V.* **a)** *(auch fig.)* switch *or* change over (**auf** + *Akk.* to); **in den zweiten Gang** ~: change into second gear; **wir schalten jetzt ins Stadion um** now we're going over to the stadium; **b)** *(umgeschaltet werden)* **die Ampel schaltet [auf Grün] um** the traffic lights are changing [to green]

Um·schlag der a) cover; **b)** *(Brief~)* envelope; **c)** *(Schutz~)* jacket; *(einer Broschüre, eines Heftes)* cover; **d)** *(Med.: Wickel)* compress; *(warm)* poultice; **e)** *(Hosen~)* turn-up; *(Ärmel~)* cuff; **f)** *(Veränderung)* [sudden] change (*Gen.* in); **g)** *(Wirtsch.: Güter~)* transfer; trans-shipment

um|schlagen 1. *unr. tr. V.* **a)** *(umklappen)* turn up ‹*sleeve, collar, trousers*›; turn over ‹*page*›; **b)** *(umladen, verladen)* turn round, trans-ship ‹*goods*›. **2.** *unr. itr. V.; mit sein* ‹*weather, mood*› change (**in** + *Akk.* into); ‹*wind*› veer [round]; ‹*voice*› break; **ins Gegenteil** ~: change completely; become the opposite

Umschlag-: ~hafen der port of trans-shipment; **~platz der** trans-shipment centre

um·schließen *unr. tr. V.* **a)** ‹*river, wall*› surround; ‹*shell, husk, etc.*› enclose; ‹*hand, fingers, tentacles*› clasp, hold; **b)** *(einschließen, umzingeln)* surround, encircle ‹*position, enemy*›

um·schlingen *unr. tr. V.* **jmdn./etw. [mit den Armen]** ~: put one's arms around sb./sth.; embrace sb./sth.

um|schmeißen *unr. tr. V. (ugs.) s.* **umwerfen a, b**

¹um|schreiben *unr. tr. V.* **a)** rewrite; **b)** *(übertragen)* transfer ‹*money, property*› (**auf** + *Akk.* to); **c)** *(transkribieren)* transcribe

²um·schreiben *unr. tr. V.* **a)** *(in Worte fassen)* describe; *(definieren)* define ‹*meaning, sb.'s task, etc.*›; *(paraphrasieren)* paraphrase ‹*word, expression*›; **b)** *(mit einer Linie umgeben)* outline; *(andeuten)* indicate

Um·schreibung die description; *(Definition)* definition; *(Verhüllung)* circumlocution (*Gen.* for)

um|schulen 1. *tr. V.* **a) ein Kind [auf eine andere Schule]** ~: transfer a child [to another school]; **b)** *(beruflich)* retrain. **2.** *itr. V.* retrain (**auf** + *Akk.* as)

Umschulung die *(beruflich)* retraining *no pl.* (**auf** + *Akk.* as)

um|schütten *tr. V.* **a)** pour [into another container]; decant ‹*liquid*›; **b)** *(verschütten)* spill

Um·schweif der circumlocution; **ohne ~e** without beating about the bush

Um·schwung der complete change; *(in der Politik usw.)* U-turn; volte-face

um|sehen *unr. refl. V.* **a)** look (**nach** for); **sich im Zimmer** ~: look [a]round the room; **sehen Sie sich ruhig um** *(im Geschäft usw.)* by all means have a look round; **du wirst dich noch** ~! *(ugs.)* you're in for a [nasty] shock; **b)** *(zurücksehen)* look round *or* back

umseitig 1. *Adj.* ‹*text, illustration, etc.*› overleaf. **2.** *adv.* overleaf

um|setzen 1. *tr. V.* **a)** move; *(auf anderen Sitzplatz)* move to another seat/other seats; *(auf anderen Posten, Arbeitsplatz usw.)* move, transfer (**in** + *Akk.* to); *(umpflanzen)* transplant ‹*bush etc.*›; *(in anderen Topf)* repot ‹*plant*›; **b)** *(verwirklichen)* implement ‹*plan*›; translate ‹*plan, intention, etc.*› into action *or* reality; realize ‹*ideas*›; **c)** *(Wirtsch.)* turn over, have a turnover of ‹*x marks etc.*›; sell ‹*goods*›. **2.** *refl. V. (den Sitzplatz wechseln)* move to another seat/other seats; change seats; *(den Tisch wechseln)* move to another table; change tables

Um·sicht die; *o. Pl.* circumspection; prudence

um·sichtig 1. *Adj.* circumspect; prudent. **2.** *adv.* circumspectly; prudently

um|siedeln 1. *tr. V.* resettle; **nach X umgesiedelt werden** be moved to X. **2.** *itr. V.; mit sein* move (**in** + *Akk.*, **nach** to)

Um·siedler der resettled person; *(freiwillig)* resettler

Um·siedlung die resettlement

um·sonst *Adv.* **a)** *(unentgeltlich)* free; for nothing; **für** ~ *(ugs.)* free, gratis, and for nothing *(joc.)*; **b)** *(vergebens)* in vain; **c) nicht ~ hat er davor gewarnt** not for nothing did he warn of that

um·spannen *tr. V.* clasp ‹*hand, wrist, ankle, etc.*›; put one's hands round ‹*neck etc.*›

um|springen *unr. itr. V.; mit sein* **a)** ‹*wind*› veer round (**auf** + *Akk.* to); ‹*traffic-light*› change; **b)** *(ugs. abwertend)* **mit jmdm. grob** *usw.* ~: treat sb. roughly *etc.*

um|spulen *tr. V.* rewind ‹*tape, film*›

Um·stand der a) *(Gegebenheit)* circumstance; *(Tatsache)* fact; **die näheren Umstände** the particular circumstances; *(Einzelheiten)* the details; **ein glücklicher** ~: a lucky *or* happy chance; **unter allen Umständen** whatever happens; **unter Umständen** possibly; **in anderen Umständen sein** *(ugs.)* be expecting; be in the family way *(coll.)*; **b)** *(Aufwand)* business; hassle *(coll.)*; **macht keine [großen] Umstände** please don't go to any bother *or* trouble

umständlich ['ʊmʃtɛntlɪç] **1.** *Adj.*

involved, elaborate ‹*procedure, method, description, explanation, etc.*›; awkward, difficult ‹*journey, job*›; *(kompliziert)* involved; complicated; *(weitschweifig)* long-winded; *(Umstände machend)* awkward, *(coll.)* pernickety ‹*person*›. **2.** *adv.* in an involved *or* roundabout way; *(weitschweifig)* ‹*explain etc.*› at great length *or* in a long-winded way; **warum einfach, wenn's auch ~ geht?** *(iron.)* why do things the easy way if you can make them difficult? *(iron.)*

Umstands-: ~kleid das maternity dress; **~wort das** *(Sprachw.)* adverb

umstehend *Adj.; nicht präd.* standing round *postpos.*; **die Umstehenden** the bystanders

um|steigen *unr. itr. V.* **a)** change (**in** + *Akk.* [on] to); **nach Frankfurt ~** *(ugs.)* change for Frankfurt; **b)** *(fig. ugs.)* change over, switch (**auf** + *Akk.* to)

[1]um|stellen 1. *tr. V.* **a)** *(anders stellen)* rearrange, change round ‹*furniture, books, etc.*›; reshuffle ‹*team*›; **b)** *(anders einstellen)* reset ‹*lever, switch, points, clock*›; **c)** *(ändern)* change *or* switch over (**auf** + *Akk.* to). **2.** *refl. V.* **a)** adjust; **b)** *s.* **umsteigen b**

[2]um·stellen *tr. V.* surround

um|stimmen *tr. V.* win ‹*person*› round; **er ließ sich nicht ~:** he was not to be persuaded; he refused to change his mind

um|stoßen *unr. tr. V.* **a)** knock over; **b)** *(fig.)* reverse ‹*judgement, decision*›; change ‹*plan, decision*›

umstritten *Adj.* disputed; controversial ‹*bill, book, author, etc.*›

um|strukturieren *tr. V.* restructure

um|stülpen *tr. V.* turn inside out

Um·sturz der coup

um|stürzen 1. *tr. V.* overturn; knock over; *(fig.)* topple, overthrow ‹*political system, government*›. **2.** *itr. V.* overturn; ‹*wall, building, chimney*› fall down

Umstürzler ['ʊmʃtʏrtslɐ] **der; ~s, ~** *(abwertend)* subversive agent

umstürzlerisch *(abwertend)* **1.** *Adj.* subversive. **2.** *adv.* **sich ~ betätigen** engage in subversive activities

Umsturz·versuch der attempted coup

Um·tausch der exchange; **reduzierte Ware ist vom ~ ausgeschlossen** sale goods cannot be exchanged

um|tauschen *tr. V.* exchange ‹*goods, article*› (**gegen** for); change ‹*dollars, pounds, etc.*› (**in** + *Akk.* into)

Um|triebe *Pl. (abwertend)* [subversive] intrigues; subversion *sing.*

Um·trunk der communal drink

um|tun *unr. refl. V. (ugs.)* look [a]round; **sich nach etw. ~:** be on the look-out *or* looking for sth.

U-Musik die; *o. Pl.* light music

um|wälzen *tr. V.* **a)** roll over; **~d** *(fig.)* revolutionary ‹*ideas, effect*›; epoch-making ‹*events*›; **b)** circulate ‹*water, air*›

Umwälzung die; ~, ~en *(fig.)* revolution

um|wandeln *tr. V.* convert ‹*substance, building, etc.*› (**in** + *Akk.* into); commute ‹*sentence*› (**in** + *Akk.* to); *(ändern)* change; alter; **er ist wie umgewandelt** he is a changed man

Um·wandlung die conversion (**in** + *Akk.* into); *(einer Strafe)* commutation (**in** + *Akk.* to); *(der Gesellschaft usw.)* transformation

um|wechseln *tr. V.* change ‹*money*› (**in** + *Akk.* into)

Um·weg der detour; **auf ~en** by a circuitous *or* roundabout route; *(fig.)* in a roundabout way; **auf dem ~ über** (+ *Akk.*) *(fig.)* [indirectly] via

Um·welt die a) environment; **b)** *(Menschen)* people *pl.* around sb.

umwelt-, Umwelt-: ~bedingt *Adj.* caused by the *or* one's environment *postpos.*; **~belastung die** environmental pollution *no indef. art.*; **~feindlich 1.** *Adj.* inimical to the environment *postpos.*; ecologically undesirable; **2.** *adv.* in an ecologically undesirable way; **~forschung die;** *o. Pl.* ecology; **~freundlich 1.** *Adj.* environment-friendly; ecologically desirable; **2.** *adv.* in an ecologically desirable way; **~politik die** ecological policy; **~schäden** *Pl.* environmental damage *sing.*; damage *sing.* to the environment; **~schädlich** *Adj.* harmful to the environment *postpos.*; ecologically harmful; **~schutz der** environmental protection *no art.*; conservation of the environment; **~schützer der** environmentalist; conservationist; **~sünder der** *(ugs.)* deliberate polluter of the environment; **~verschmutzung die** pollution [of the environment]

um|werfen *unr. tr. V.* **a)** knock over; *(fig. ugs.: aus der Fassung bringen)* bowl ‹*person*› over; stun ‹*person*›; **b)** *(fig. ugs.; umstoßen)* knock ‹*plan*› on the head *(coll.)*

umwerfend *(ugs.)* **1.** *Adj.* fantastic *(coll.)*; stunning *(coll.)*. **2.** *adv.* fantastically [well] *(coll.)*; brilliantly; **~ komisch** hilariously funny

um·wickeln *tr. V.* wrap; bind; *(mit einem Verband)* bandage; **etw. mit Draht ~:** wind wire round sth.

Umzäunung die; ~, ~en fence, fencing (*Gen.* round)

um|ziehen 1. *unr. itr. V.; mit sein* move (**an** + *Akk.*, **in** + *Akk.*, **nach** to). **2.** *unr. tr. V.* **jmdn. ~:** change sb.; get sb. changed; **sich ~:** change; get changed

um·zingeln [ʊm'tsɪŋl̩n] *tr. V.* surround; encircle

Um·zug der a) move; *(von Möbeln)* removal; **b)** *(Festzug)* procession

UN [u:'ɛn] *Pl.* UN *sing.*

unabänderlich [ʊnlap'lɛndɐlɪç] **1.** *Adj.* unalterable; irrevocable ‹*decision*›. **2.** *adv.* irrevocably; **das steht ~ fest** that is absolutely certain

unabdingbar [ʊnlap'dɪŋba:ɐ̯] *Adj. (geh.)* indispensable

unabhängig 1. *Adj.* independent (**von** of). **2.** *adv.* independently (**von** of); **~ davon, ob ...** *usw.* irrespective *or* regardless of whether ... *etc.*

Unabhängigkeit die independence

unabkömmlich [ʊnlap'kœmlɪç] *Adj.* indispensable; **sie ist im Moment ~:** she is otherwise engaged

unablässig ['ʊnlaplɛsɪç] **1.** *Adj.* incessant; constant ‹*repetition*›; unremitting ‹*effort*›. **2.** *adv.* incessantly; constantly

unabsehbar 1. *Adj.* **a)** *(fig.)* incalculable, immeasurable ‹*extent, damage, etc.*›; **b)** *(noch nicht vorauszusehen)* unforeseeable ‹*consequences*›. **2.** *adv.* **a)** incalculably; immeasurably; **b)** *(in einem noch nicht erkennbaren Ausmaß)* to an unforeseeable extent

unabsichtlich 1. *Adj.* unintentional. **2.** *adv.* unintentionally

unabwendbar *Adj.* inevitable

unachtsam 1. *Adj.* **a)** inattentive; **einen Augenblick ~ sein** let one's attention wander for a moment; **b)** *(nicht sorgfältig)* careless. **2.** *adv. (ohne Sorgfalt)* carelessly

Unachtsamkeit die; ~ a) inattentiveness; **b)** *(mangelnde Sorgfalt)* carelessness

unangebracht *Adj.* inappropriate; misplaced

unangefochten *Adj.* unchallenged ‹*victor, leadership, etc.*›;

(unbestritten) undisputed, unchallenged ‹*assertion, thesis*›
ụnangemeldet *Adj.* unexpected ‹*visit, guest*›; unauthorized ‹*demonstration*›
ụnangemessen 1. *Adj.* unsuitable; inappropriate; unreasonable, disproportionate ‹*demand, claim, sentence, etc.*›. **2.** *adv.* unsuitably; inappropriately; disproportionately ‹*high, low*›
ụnangenehm 1. *Adj.* unpleasant (*Dat.* for); *(peinlich)* embarrassing, awkward ‹*question, situation*›; **es ist mir sehr ~, daß ich mich verspätet habe** I am most upset about being late; **~ werden** ‹*person*› get *or* turn nasty. **2.** *adv.* unpleasantly; **~ auffallen** make a bad impression
ụnangetastet *Adj.* untouched
unangreifbar *Adj. (auch fig.)* unassailable; impregnable ‹*fortress*›; *(unanfechtbar)* irrefutable ‹*argument, thesis*›; incontestable ‹*judgement etc.*›
unannehmbar *Adj.* unacceptable
Ụnannehmlichkeit die trouble; **jmdm. ~en bereiten** cause sb. [a lot of *(coll.)*] problems *or* difficulties
ụnansehnlich *Adj.* unprepossessing; plain ‹*girl*›
ụnanständig 1. *Adj.* **a)** improper; *(anstößig)* indecent ‹*behaviour, remark*›; dirty ‹*joke*›; rude ‹*word, song*›; **b)** *(verwerflich)* immoral. **2.** *adv.* **a)** improperly; indecently; **b)** *(verwerflich)* immorally
Ụnanständigkeit die *o. Pl.* impropriety; indecency; *(Obszönität)* obscenity
unantạstbar *Adj.* inviolable
ụnappetitlich 1. *Adj.* unappetizing; *(fig.)* unsavoury ‹*joke*›; unsavoury-looking ‹*person*›; disgusting ‹*wash-basin, nails, etc.*›. **2.** *adv.* unappetizingly
Ụnart die bad habit
ụnartig *Adj.* naughty
ụnästhetisch *Adj.* unpleasant, unsavoury ‹*sight etc.*›; ugly ‹*building etc.*›; *(abstoßend)* disgusting
ụnauffällig 1. *Adj.* inconspicuous; unobtrusive ‹*scar, defect, skill, behaviour, surveillance, etc.*›; discreet ‹*signal, elegance*›. **2.** *adv.* inconspicuously; unobtrusively; ‹*behave, follow, observe, disappear, leave*› unobtrusively, discreetly
unauffịndbar *Adj.* untraceable; **~ sein** be nowhere to be found
ụnaufgefordert *Adv.* without being asked; **~ eingesandte Manuskripte** unsolicited manuscripts
unaufhạltsam 1. *Adj.* inexorable. **2.** *adv.* inexorably
unaufhörlich 1. *Adj.; nicht präd.* constant; incessant; continuous ‹*rain*›. **2.** *adv.* constantly; ‹*rain, snow*› continuously
ụnaufmerksam *Adj.* inattentive
Ụnaufmerksamkeit die inattentiveness
ụnaufrichtig *Adj.* insincere; **jmdm. gegenüber ~ sein** not be honest with sb.
Ụnaufrichtigkeit die *o. Pl.* insincerity
unausbleiblich *Adj.* inevitable; unavoidable
ụnausgeglichen *Adj.* **a)** [emotionally] unstable ‹*person, behaviour*›; **b)** *(Wirtsch.)* ‹*balance of payments*› not in balance; unsettled ‹*account, debt*›
unausstehlich 1. *Adj.* unbearable ‹*person, noise, smell, etc.*›; insufferable ‹*person*›; intolerable ‹*noise, smell*›. **2.** *adv.* unbearably; intolerably
unbändig ['ʊnbɛndɪç] **1.** *Adj.* **a)** boisterous ‹*person, horse, temperament*›; **b)** unbridled, unrestrained ‹*desire, longing, joy, merriment*›; unbridled, uncontrollable ‹*fury, hate, anger*›. **2.** *adv.* **a)** wildly; **b)** *(sehr)* unrestrainedly; tremendously *(coll.)*; **sich ~ freuen** jump for joy
ụnbarmherzig 1. *Adj. (auch fig.)* merciless; remorseless, unsparing ‹*severity*›; *(fig.)* very severe ‹*winter, cold*›. **2.** *adv.* mercilessly; without mercy
ụnbeabsichtigt 1. *Adj.* unintentional. **2.** *adv.* unintentionally
ụnbeachtet *Adj.* unnoticed; obscure ‹*existence*›; **jmdn./etw. ~ lassen** not take any notice of sb./sth.
ụnbeanstandet 1. *Adj.* **etw. ~ lassen** let sth. pass. **2.** *adv.* without objection
ụnbebaut *Adj.* **a)** undeveloped ‹*site, land*›; **b)** *(unbestellt)* uncultivated ‹*land, area*›
ụnbedacht 1. *Adj.* rash; thoughtless. **2.** *adv.* rashly; thoughtlessly
ụnbedarft *Adj. (ugs.)* **a)** inexpert; lay; **b)** *(naiv)* naïve; *(dümmlich)* gormless *(coll.)*
ụnbedenklich *Adj.* harmless, safe ‹*substance, drug*›; unobjectionable ‹*joke, reading matter*›
ụnbedeutend 1. *Adj.* insignificant; minor ‹*artist, poet*›; slight, minor ‹*improvement, change, error*›. **2.** *adv.* slightly
ụnbedingt 1. *Adj.* absolute ‹*trust, faith, reliability, secrecy, etc.*›; complete ‹*rest*›. **2.** *Adv.* absolutely; *(auf jeden Fall)* whatever happens; **nicht ~:** not necessarily
ụnbeeinflußt *Adj.* uninfluenced
ụnbefangen 1. *Adj.* **a)** *(ungehemmt)* uninhibited; natural, uninhibited ‹*behaviour*›; **b)** *(unvoreingenommen)* impartial. **2.** *adv.* **a)** freely; without inhibition; ‹*behave*› naturally; **b)** *(unvoreingenommen)* **jmdm./einer Sache ~ gegenübertreten** approach sb./sth. with an open mind
Ụnbefangenheit die *s.* **unbefangen 1 a, b:** uninhibitedness; naturalness; impartiality
ụnbefriedigend *Adj.* unsatisfactory
ụnbefriedigt *Adj.* dissatisfied (**von** with); unsatisfied ‹*need, curiosity, desire, etc.*›; *(unausgefüllt)* unfulfilled (**von** by); *(sexuell)* [sexually] frustrated
ụnbefristet 1. *Adj.* for an indefinite *or* unlimited period *postpos.*; indefinite ‹*strike*›; unlimited ‹*visa*›. **2.** *adv.* for an indefinite *or* unlimited period
ụnbefugt 1. *Adj.* unauthorized. **2.** *adv.* without authorization
ụnbegabt *Adj.* ungifted; untalented
unbegreiflich 1. *Adj.* incomprehensible (*Dat.*, **für** to); incredible ‹*love, goodness, stupidity, carelessness, etc.*›
ụnbegreiflicherweise *Adv.* inexplicably
ụnbegrenzt 1. *Adj.* unlimited. **2.** *adv.* ‹*stay, keep, etc.*› indefinitely; **ich habe nicht ~ Zeit** I don't have unlimited time
ụnbegründet *Adj.* unfounded; groundless
Ụnbehagen das uneasiness, disquiet; **ein körperliches ~:** a physical discomfort; **das bereitet mir ~:** it makes me feel uneasy
ụnbehaglich 1. *Adj.* uneasy, uncomfortable ‹*feeling, atmosphere*›; uncomfortable ‹*thought, room*›; **mir war ~ zumute** I was *or* felt uneasy. **2.** *adv.* uneasily; uncomfortably
ụnbehelligt *Adj.* unmolested
ụnbeherrscht 1. *Adj.* uncontrolled; intemperate, wild ‹*reaction, behaviour, remark*›. **2.** *adv.* without any self-control
Ụnbeherrschtheit die; ~: lack of self-control
ụnbeholfen 1. *Adj.* clumsy; awkward. **2.** *adv.* clumsily; awkwardly

unbekannt 1. *Adj.* **a)** unknown; unidentified ⟨*caller, donor, flying object*⟩; *(nicht vertraut)* unfamiliar; **sie ist hier ~:** she is not known here; **~e Täter** unknown *or* unidentified culprits; **[Straf]anzeige gegen Unbekannt** *(Rechtsw.)* charge against person *or* persons unknown; **b)** *(nicht vielen bekannt)* little known; obscure ⟨*poet, painter, etc.*⟩; **c) jmd./etw. ist jmdm. ~:** sb. does not know sb./sth.; **„Empfänger ~"** 'not known at this address'. **2.** *adv.* **„Empfänger ~ verzogen"** 'moved'; 'address unknown'

[1]**Unbekannte der/die;** *adj. Dekl.* unknown *or* unidentified man/woman; *(Fremde[r])* stranger; **der große ~** *(scherzh.)* the mystery man *or* person

[2]**Unbekannte die;** *adj. Dekl. (Math.; auch fig.)* unknown

unbekleidet *Adj.* without any clothes on *postpos.*; bare ⟨*torso etc.*⟩; naked ⟨*corpse*⟩

unbekümmert 1. *Adj.* carefree; *(ohne Bedenken, lässig)* casual; **sie ist [ziemlich] ~:** she doesn't worry [much]; she is [pretty] unconcerned. **2.** *adv.* **a)** in a carefree way; without a care in the world; **b)** *(ohne Bedenken)* without caring *or* worrying

Unbekümmertheit die; ~ a) carefree manner *or* attitude; carefreeness; **b)** *(Bedenkenlosigkeit)* lack of concern

unbelastet *Adj.* **a)** not under load *postpos.*; **b)** *(sorgenfrei)* free from care *or* worries *postpos.*; **c)** *(ohne Schuld)* **~ sein** have a clean record; **d)** *(schuldenfrei)* unmortgaged ⟨*property, land*⟩

unbeleuchtet *Adj.* unlit; ⟨*vehicle*⟩ without [any] lights

unbeliebt *Adj.* unpopular (**bei** with)

Unbeliebtheit die unpopularity

unbemannt *Adj.* unmanned

unbemerkt *Adj.* unnoticed

unbenommen *Adj.* **es ist/bleibt jmdm. ~, zu ...:** sb. is/remains free *or* at liberty to ...

unbenutzt *Adj.* unused

unbeobachtet *Adj.* unobserved; **in einem ~en Augenblick** when no one is/was watching

unbequem 1. *Adj.* **a)** uncomfortable; **b)** *(lästig)* awkward, embarrassing ⟨*question, opinion*⟩; awkward, troublesome ⟨*politician etc.*⟩; unpleasant ⟨*criticism, truth, etc.*⟩. **2.** *adv.* uncomfortably

Unbequemlichkeit die; ~ a) lack of comfort; **b)** *(Lästigkeit)* awkwardness

unberechenbar 1. *Adj.* unpredictable. **2.** *adv.* unpredictably

unberechtigt 1. *Adj.* **a)** *(ungerechtfertigt)* unjustified; **b)** *(unbefugt)* unauthorized. **2.** *adv. (unbefugt)* without authorization

unberücksichtigt *Adj.* unconsidered; **etw. ~ lassen** leave sth. out of consideration; ignore sth.

unberührt *Adj.* **a)** untouched; virgin ⟨*snow, forest, wilderness*⟩; **ein Stück ~er Natur** a stretch of unspoilt countryside; **b)** *(geh.: jungfräulich)* in the virgin state *postpos.*; **sie ist noch ~:** she is still a virgin; **c)** *(unbeeindruckt)* unmoved (**von** by)

Unberührtheit die; ~ a) unspoiled state; **b)** *(geh.: Jungfräulichkeit)* virginity; **c)** *(das Unbeeindrucktsein)* lack of emotion; impassivity

unbeschadet *Präp. mit Gen.* regardless of; notwithstanding

unbeschädigt *Adj.* undamaged

unbescheiden *Adj.* presumptuous

unbescholten *Adj.* respectable; **~ sein** *(Rechtsspr.)* have no [previous] convictions

unbeschrankt *Adj.* ⟨*crossing*⟩ without gates, with no gates

unbeschränkt 1. *Adj.* unlimited; limitless ⟨*possibilities, power*⟩. **2.** *adv.* **für etw. ~ haften** have unlimited liability for sth.

unbeschreiblich 1. *Adj.* indescribable; unimaginable ⟨*fear, beauty*⟩; ⟨*fear, beauty*⟩ beyond description. **2.** *adv.* indescribably ⟨*beautiful*⟩; unbelievably ⟨*busy*⟩

unbeschrieben *Adj.* blank, empty ⟨*piece of paper, page*⟩; *s. auch* **Blatt b**

unbeschwert 1. *Adj.* carefree. **2.** *adv.* free from care; ⟨*dance, play*⟩ with a light heart

unbesehen 1. *Adj.* unquestioning ⟨*acceptance*⟩. **2.** *adv.* without hesitation

unbesiegbar *Adj.* invincible

unbesiegt *Adj.* undefeated ⟨*army*⟩; unbeaten ⟨*team, player*⟩

unbesonnen 1. *Adj.* impulsive ⟨*person, nature*⟩; unthinking ⟨*remark*⟩; *(übereilt)* ill-considered, rash ⟨*decision, action*⟩. **2.** *adv.* ⟨*act*⟩ without thinking; *(übereilt)* rashly

unbesorgt *Adj.* unconcerned; **seien Sie ~!** don't [you] worry; you can set your mind at rest

unbespielt *Adj.* blank ⟨*tape, cassette*⟩

unbeständig *Adj.* changeable, unsettled ⟨*weather*⟩; erratic, inconsistent ⟨*performance, person*⟩

unbestätigt *Adj.* unconfirmed

unbestechlich *Adj.* **a)** incorruptible; **b)** *(fig.)* uncompromising ⟨*critic*⟩; incorruptible ⟨*character*⟩; unerring ⟨*judgement*⟩

unbestimmt 1. *Adj.* **a)** indefinite; indeterminate ⟨*age, number*⟩; *(ungewiß)* uncertain; **b)** *(ungenau)* vague; **c)** *(Sprachw.)* indefinite ⟨*article, pronoun*⟩. **2.** *adv. (ungenau)* vaguely

unbestritten 1. *Adj.* undisputed. **2.** *adv.* indisputably

unbeteiligt 1. *Adj.* **a)** uninvolved; **ein Unbeteiligter** someone who is/was not involved; an outsider; *(ein Unschuldiger)* an innocent party; **b)** *(gleichgültig)* indifferent; detached ⟨*manner, expression*⟩. **2.** *adv.* with a detached *or* indifferent air

unbetont *Adj.* unstressed

unbeugsam *Adj.* uncompromising; tenacious; indomitable, unshakeable ⟨*will, pride*⟩; unwavering, resolute ⟨*character*⟩

unbewacht *Adj.* unsupervised ⟨*prisoners etc.*⟩; unattended ⟨*carpark*⟩

unbewaffnet *Adj.* unarmed

unbewältigt *Adj.* unmastered, uncompleted ⟨*task*⟩; unresolved ⟨*conflict, problem*⟩

unbeweglich *Adj.* **a)** *(bewegungslos)* motionless; still ⟨*air, water*⟩; fixed ⟨*gaze, expression*⟩; **b)** *(starr)* immovable, fixed ⟨*part, joint, etc.*⟩; **c)** *(nicht mobil)* immobile; **d)** *(schwerfällig) (geistig)* ponderous; *(körperlich)* slow-moving; slow on one's feet *pred.*

unbewiesen *Adj.* unproved

unbewohnt *Adj.* uninhabited ⟨*area*⟩; unoccupied ⟨*house, flat*⟩

unbewußt 1. *Adj.* unconscious. **2.** *adv.* unconsciously

unbezahlbar *Adj.* prohibitively expensive; priceless ⟨*painting, china*⟩; **meine Sekretärin ist einfach ~** *(ugs.)* my secretary is worth her weight in gold

unbezahlt *Adj.* unpaid; ⟨*goods etc.*⟩ not [yet] paid for

unblutig 1. *Adj.* bloodless. **2.** *adv.* without bloodshed

unbrauchbar *Adj.* unusable; *(untauglich)* useless ⟨*method, person*⟩

unbürokratisch 1. *Adj.* unbureaucratic; **auf möglichst ~e Weise** with as little red tape as possible. **2.** *adv.* unbureaucratically; without a great deal of red tape

und [ʊnt] *Konj.* **a)** *(nebenordnend)* and; *(folglich)* [and] so; **das deutsche ~ das französische Volk** the German and French peoples;

zwei ~ drei ist fünf two and *or* plus three makes five; **es wollte ~ wollte nicht gelingen** it simply *or* just wouldn't work; **hoch ~ höher** higher and higher; **~ ich?** [and] what about me?; **der ~ der** so-and-so; **zu der ~ der Zeit** at such-and-such a time; **so ~ so ist es gewesen** it was like this; *s. auch* **na a; ob b; wie 1 c; b)** *(unterordnend) (konsekutiv)* **sei so gut ~ mach das Fenster zu** be so good as to shut the window; **es fehlte nicht viel, ~ der Deich wäre gebrochen** it wouldn't have taken much to breach the dike; *(konzessiv)* **du mußt es tun, ~ fällt es dir noch so schwer** you must do it however difficult you may find it

Ụndank der ingratitude; **~ ist der Welt Lohn** *(Spr.)* that's all the thanks you get

ụndankbar 1. *Adj.* **a)** ungrateful ‹*person, behaviour*›; **b)** thankless ‹*task*›; unrewarding ‹*role, job, etc.*›. **2.** *adv.* ungratefully

undefinierbar *Adj.* **a)** indefinable; **b)** *(nicht bestimmbar)* unidentifiable; indeterminable ‹*feeling*›; indeterminate ‹*colour*›

undẹnkbar *Adj.* unthinkable; inconceivable

undẹnklich *Adj. in* **vor ~er Zeit** *od.* **~en Zeiten** an eternity ago; **seit ~er Zeit** *od.* **~en Zeiten** since time immemorial

ụndeutlich 1. *Adj.* unclear; indistinct; *(ungenau)* vague ‹*idea, memory, etc.*›. **2.** *adv.* indistinctly; *(ungenau)* vaguely

ụndicht *Adj.* leaky; leaking; **~ werden** start to leak; develop a leak; **eine ~e Stelle** *(auch fig.)* a leak; **~e Fenster** windows which do not fit tightly

Ụnding das *in* **ein ~ sein** be preposterous *or* ridiculous

ụndiszipliniert 1. *Adj.* undisciplined; ‹*pupils, class*› lacking in discipline. **2.** *adv.* in an undisciplined way

undurchdrịnglich *Adj.* impenetrable; pitch-dark ‹*night*›

ụndurchlässig *Adj.* impermeable; *(wasserdicht)* watertight; waterproof; *(luftdicht)* airtight

ụndurchsichtig *Adj.* **a)** opaque ‹*glass*›; non-transparent ‹*fabric etc.*›; **b)** *(fig.)* unfathomable, inscrutable ‹*plan, intention, role*›; shady ‹*character, business*›

ụneben *Adj.* uneven; *(holprig)* bumpy ‹*road, track*›

ụnecht *Adj.* artificial ‹*fur, hair*›; false ‹*teeth*›; imitation ‹*jewellery, marble, etc.*›; *(gefälscht)* counterfeit ‹*notes*›; bogus, fake ‹*painting*›; false, insincere ‹*friendliness, sympathy, smile, etc.*›

ụnehelich *Adj.* illegitimate ‹*child*›; **~ geboren sein** be born out of wedlock

ụnehrenhaft 1. *Adj.* dishonourable. **2.** *adv.* dishonourably

ụnehrlich 1. *Adj.* dishonest. **2.** *adv.* dishonestly; by dishonest means

Ụnehrlichkeit die dishonesty

ụneigennützig 1. *Adj.* unselfish; selfless. **2.** *adv.* unselfishly; selflessly

ụneinig *Adj.* ‹*party*› divided by disagreement; **[sich** *(Dat.)***] ~ sein** disagree; be in disagreement

Ụneinigkeit die disagreement **(in +** *Dat.* on)

ụneins *Adj.; nicht attr.* **~ sein** be divided **(in +** *Dat.* on); ‹*persons, bodies*› be at variance *or* at cross purposes **(in +** *Dat.* over); **mit jmdm. ~ sein[, wie ...]** be unable to agree with sb. [how ...]

ụnempfänglich *Adj.* unreceptive **(für** to)

ụnempfindlich *Adj.* **a)** insensitive **(gegen** to); **b)** *(nicht anfällig, immun)* immune **(gegen** to, against); **c)** *(strapazierfähig)* hard-wearing; *(pflegeleicht)* easy-care *attrib.*

unẹndlich 1. *Adj.* infinite, boundless ‹*space, sea, expanse, fig.: love, care, patience, etc.*›; *(zeitlich)* endless; never-ending; **das Unendliche** the infinite *(Philos.)*; infinity *(Math.)*; **auf ~ stellen** *(Fot.)* focus ‹*lens*› on infinity. **2.** *adv.* infinitely ‹*lovable, sad*›; immeasurably ‹*happy*›; **sich ~ freuen** be tremendously pleased

Unẹndlichkeit die; ~ a) infinity *no def. art.*; **b)** *(geh.: Ewigkeit)* eternity *no def. art.*

unentbẹhrlich *Adj.* indispensable (*Dat.*, **für** to)

unentgẹltlich [*od.* '----] **1.** *Adj.* free. **2.** *adv.* free of charge; ‹*work*› for nothing, without pay

ụnentschieden 1. *Adj.* **a)** unsettled ‹*case, matter*›; undecided ‹*question*›; **b)** *(Sport, Schach)* drawn ‹*game, match*›; **c)** *(unentschlossen)* indecisive ‹*person*›. **2.** *adv.* *(Sport, Schach)* **~ spielen** draw; **~ enden** end in a draw; **das Spiel steht 0 : 0 ~**: the game is a goalless draw [so far]

Ụnentschieden das; ~s, ~ *(Sport, Schach)* draw

ụnentschlossen *Adj.* **a)** undecided; **b)** *(entschlußunfähig)* indecisive

unentschụldbar 1. *Adj.* inexcusable. **2.** *adv.* inexcusably

ụnentschuldigt 1. *Adj.* without giving any reason *postpos., not pred.*; **~es Fernbleiben vom Unterricht** absence from school. **2.** *adv.* without giving any reason

ụnentwegt [*od.* --'-] **1.** *Adj.* **a)** persistent ‹*fighter, champion, efforts*›; **ein paar Unentwegte** a few stalwarts; **b)** *(unaufhörlich)* constant; incessant. **2.** *adv.* **a)** persistently; **b)** *(unaufhörlich)* constantly; incessantly

unerbịttlich *(auch fig.)* **1.** *Adj.* inexorable; unsparing, unrelenting ‹*critic*›; relentless ‹*battle, struggle*›; implacable ‹*hate, enemy*›; **gegen jmdn. ~ sein** be completely unyielding towards sb. **2.** *adv.* inexorably; relentlessly; **~ durchgreifen** take uncompromising action

ụnerfahren *Adj.* inexperienced

unerfịndlich *Adj.* *(geh.)* unfathomable; inexplicable

ụnerfreulich 1. *Adj.* unpleasant; bad ‹*news*›. **2.** *adv.* unpleasantly

ụnergiebig *Adj.* *(auch fig.)* unproductive; *(fig.: nicht lohnend)* unrewarding ‹*work, subject*›

unergrụndlich *Adj.* unfathomable, inscrutable ‹*motive, mystery, etc.*›; inscrutable ‹*expression, smile*›

ụnerheblich 1. *Adj.* insignificant. **2.** *adv.* insignificantly; [very] slightly

unerhọ̈rt 1. *Adj.* **a)** enormous, tremendous ‹*sum, quantity, etc.*›; incredible *(coll.)*, phenomenal ‹*speed, effort, performance, increase*›; incredible *(coll.)*, fantastic *(coll.)* ‹*splendour, luck*›; **b)** *(empörend)* outrageous; scandalous. **2.** *adv.* **a)** *(überaus)* incredibly *(coll.)*; **b)** *(empörend)* outrageously

ụnerkannt *Adj.* unrecognized; *(nicht identifiziert)* unidentified

unerklä̱rlich *Adj.* inexplicable

unerläßlich [ʊnlɛɐ̯'lɛslɪç] *Adj.* indispensable; essential

ụnerlaubt 1. *Adj.* ‹*entry, parking, absenteeism*› without permission; unauthorized ‹*parking, entry*›; *(illegal)* illegal. **2.** *adv.* without authorization *or* permission; *(illegal)* illegally

unermẹßlich 1. *Adj.* *(geh.)* immeasurable ‹*expanse, distance*›; boundless ‹*spaces*›; immeasurable, immense ‹*wealth, fortune*›; untold ‹*suffering, misery, damage*›; inestimable ‹*value, importance*›. **2.** *adv.* immeasurably; ‹*rich*› beyond measure

unermü̱dlich 1. *Adj.* tireless; untiring. **2.** *adv.* tirelessly

unersạttlich *Adj.* insatiable

unerschọ̈pflich *Adj.* inexhaustible

ụnerschrocken 1. *Adj.* intrepid; fearless. **2.** *adv.* intrepidly; fearlessly

unerschụ̈tterlich 1. *Adj.* unshakeable; imperturbable ⟨*calm, equanimity*⟩. **2.** *adv.* unshakeably

unerschwịnglich *Adj.* prohibitively expensive

unersẹtzlich *Adj.* irreplaceable; irretrievable, irrecoverable ⟨*loss*⟩; irreparable ⟨*harm, damage, loss of person*⟩

unerträglich [*od.* '----] **1.** *Adj.* unbearable ⟨*pain, heat, person, etc.*⟩; intolerable ⟨*situation, conditions, moods, etc.*⟩. **2.** *adv.* unbearably

ụnerwartet 1. *Adj.* unexpected; **es kam für alle ~:** it came as a surprise to everybody. **2.** *adv.* unexpectedly

ụnerwünscht *Adj.* unwanted; unwelcome ⟨*interruption, visit, visitor*⟩; undesirable ⟨*side-effects*⟩; **Sie sind hier ~:** you are not wanted *or* welcome here

ụnfähig *Adj.* **a) ~ sein, etw. zu tun** *(ständig)* be incapable of doing sth.; *(momentan)* be unable to do sth.; **b)** *(abwertend)* incompetent

Ụnfähigkeit die a) inability; **b)** *(abwertend)* incompetence

ụnfair 1. *Adj.* unfair **(gegen** to). **2.** *adv.* unfairly

Ụnfall der accident; **bei einem ~:** in an accident

ụnfall-, Ụnfall-: ~arzt der casualty doctor; **~flucht die** *(Rechtsspr.)* **~flucht begehen** fail to stop after [being involved in] an accident; **~folge die** consequence *or* effect of an/the accident; **er starb an den ~folgen** he died as a result of the accident; **~frei 1.** *Adj.* accident-free; free from accidents *postpos.*; **2.** *adv.* without an accident; **~station die** accident *or* casualty department; **~stelle die** scene of an/the accident; **~versicherung die** accident insurance; **~wagen der a)** *(Krankenwagen)* ambulance; **b)** *(beschädigter Wagen)* car [that has been] damaged in an/the accident

unfạßbar 1. *Adj.* incomprehensible; *(unglaublich)* incredible, unimaginable ⟨*poverty, cruelty, etc.*⟩. **2.** *adv.* incomprehensibly; *(unglaublich)* incredibly; unimaginably

unfẹhlbar *Adj.* infallible

Unfẹhlbarkeit die; ~: infallibility

ụnfein 1. *Adj.* ill-mannered, unrefined ⟨*behaviour etc.*⟩; unrefined, coarse ⟨*manner, word*⟩; bad ⟨*manners*⟩. **2.** *adv.* ⟨*behave*⟩ badly, in an ill-mannered way

unflätig ['ʊnflɛːtɪç] *(geh. abwertend)* **1.** *Adj.* coarse ⟨*behaviour, manners, speech, etc.*⟩; obscene ⟨*expression, word, curse*⟩; dirty ⟨*song*⟩. **2.** *adv.* coarsely; obscenely

ụnförmig *Adj.* shapeless ⟨*lump, shadow, etc.*⟩; huge ⟨*legs, hands, body*⟩; bulky, ungainly ⟨*shape, shoes, etc.*⟩

ụnfrei *Adj.* not free *pred.*; subject, dependent ⟨*people*⟩

Ụnfreiheit die; *o. Pl.* slavery *no art.*; bondage *(esp. Hist./literary) no art.*; **ein Leben in ~:** a life of bondage *or* without freedom

ụnfreiwillig 1. *Adj.* involuntary; *(erzwungen)* enforced ⟨*stay*⟩; *(nicht beabsichtigt)* unintended ⟨*joke, humour*⟩. **2.** *adv.* involuntarily; without wanting to; *(unbeabsichtigt)* unintentionally

ụnfreundlich 1. *Adj.* **a)** unfriendly **(zu, gegen** to); unkind ⟨*words, remark*⟩; **b)** *(fig.)* unpleasant; cheerless ⟨*room*⟩. **2.** *adv.* in an unfriendly way

Ụnfreundlichkeit die a) *o. Pl.* unfriendliness; **b)** *(Handlung)* unfriendly act; *(Äußerung)* unkind remark

Ụnfriede[n] der discord; **in ~ leben/auseinandergehen** live in a state of strife/part in hostility

ụnfrisiert *Adj.* ungroomed ⟨*hair*⟩; **sie war ~:** she had not done her hair

ụnfruchtbar *Adj.* **a)** infertile ⟨*soil, field, land*⟩; **b)** *(Biol.)* infertile; sterile

Ụnfruchtbarkeit die a) infertility; **b)** *(Biol.)* infertility; sterility

Unfug ['ʊnfuːk] **der; ~[e]s a)** [piece of] mischief; **allerlei ~ anstellen** get up to all kinds of mischief *or (coll.)* monkey business; **grober ~:** public nuisance; **b)** *(Unsinn)* nonsense

Ungar ['ʊŋgar] **der; ~n, ~n, Ụngarin die; ~, ~nen** Hungarian

ụngarisch 1. *Adj.* Hungarian. **2.** *adv.* in Hungarian

Ụngarisch das; ~[s] Hungarian

Ụngarn (das); ~s Hungary

ụngastlich 1. *Adj.* inhospitable. **2.** *adv.* inhospitably

ụngeachtet *Präp. mit Gen. (geh.)* notwithstanding; despite

ụngeahnt 1. *Adj.* unsuspected; *(stärker)* undreamt-of *attrib.* **2.** *adv.* unexpectedly

ụngebeten *Adj.* uninvited

ụngebildet *Adj.* uneducated

ụngeboren *Adj.* unborn

ụngebräuchlich *Adj.* uncommon; rare; rarely used ⟨*method*⟩

ụngebührlich *(geh.)* **1.** *Adj.* improper, unseemly ⟨*behaviour*⟩; unreasonable ⟨*demand*⟩. **2.** *adv.* ⟨*behave*⟩ improperly; unreasonably ⟨*high, long, etc.*⟩

ụngedeckt *Adj.* **a)** uncovered ⟨*cheque, bill of exchange, etc.*⟩; **b)** unlaid ⟨*table*⟩; **c)** *(ungeschützt)* unprotected; **d)** *(Ballspiele)* unmarked ⟨*player*⟩

Ụngeduld die impatience

ụngeduldig 1. *Adj.* impatient. **2.** *adv.* impatiently

ụngeeignet *Adj.* unsuitable; *(für eine Aufgabe, Stellung)* unsuited

ungefähr ['ʊŋgəfɛːɐ̯] **1.** *Adj.; nicht präd.* approximate; rough ⟨*idea, outline*⟩. **2.** *Adv.* **a)** approximately; roughly; **~ 100** about *or* roughly 100; **so ~** *(ugs.)* more or less; **wo ~ ...?** whereabouts ...?; **b)** **[wie] von ~:** [as if] by chance; **es kommt nicht von ~[, daß ...]** it's no accident [that ...]

ụngefährlich 1. *Adj.* safe; harmless ⟨*animal, person, illness, etc.*⟩; **nicht ~ sein** be not without danger. **2.** *adv.* safely

ụngehalten *(geh.)* **1.** *Adj.* annoyed **(über +** *Akk.*, **wegen** about); *(entrüstet)* indignant. **2.** *adv.* indignantly; ⟨*reply, say*⟩ in an aggrieved tone

ụngeheizt *Adj.* unheated

ụngeheuer 1. *Adj.* enormous; immense; tremendous ⟨*strength, energy, effort, enthusiasm, fear, success, pressure, etc.*⟩; vast, immense ⟨*fortune, knowledge*⟩; terrible *(coll.)*, terrific *(coll.)* ⟨*pain, rage*⟩. **2.** *adv.* tremendously; terribly *(coll.)* ⟨*difficult, clever*⟩

Ụngeheuer das; ~s, ~ *(auch fig.)* monster

ungeheuerlich 1. *Adj.* monstrous; outrageous. **2.** *adv. (ugs.)* terribly *(coll.)*

ụngehindert *Adj.* unimpeded

ụngehobelt *Adj. (fig.)* uncouth

ụngehörig 1. *Adj.* improper ⟨*behaviour*⟩; *(frech)* impertinent ⟨*tone, answer*⟩. **2.** *adv.* improperly; *(frech)* impertinently

ụngehorsam *Adj.* disobedient

Ụngehorsam der disobedience

ụngeklärt *Adj.* unsolved ⟨*question, problem*⟩; unknown ⟨*cause*⟩

ụngekündigt *Adj.* **in ~er Stellung** not under notice *postpos.*

ụngekünstelt 1. *Adj.* natural; unaffected. **2.** *adv.* naturally; unaffectedly

ungekürzt *Adj.* unabridged ⟨*edition, book*⟩; uncut ⟨*film, speech*⟩

ungeladen *Adj.* **a)** unloaded ⟨*gun, camera*⟩; **b)** uninvited ⟨*guest*⟩

ungelegen *Adj.* inconvenient, awkward ⟨*time*⟩; **das kommt mir sehr ~/nicht ~:** that is very inconvenient *or* awkward/quite convenient for me

ungelernt *Adj.* unskilled

ungelogen *Adv. (ugs.)* honestly

ungemein 1. *Adj.; nicht präd.* exceptional ⟨*progress, popularity*⟩; tremendous ⟨*advantage, pleasure*⟩. **2.** *adv.* exceptionally; **das freut mich ~:** that pleases me no end *(coll.)*

ungemütlich 1. *Adj.* **a)** uninviting, cheerless ⟨*room, flat*⟩; uncomfortable, unfriendly ⟨*atmosphere*⟩; **b)** *(unangenehm)* unpleasant ⟨*situation*⟩; **es wird jetzt ~:** things are getting nasty. **2.** *adv.* uncomfortably ⟨*furnished*⟩

ungenau 1. *Adj.* inaccurate ⟨*measurement, estimate, thermometer, translation, etc.*⟩; imprecise, inexact ⟨*definition, formulation, etc.*⟩; *(undeutlich)* vague ⟨*memory, idea, impression*⟩. **2.** *adv.* **a)** inaccurately; ⟨*define*⟩ imprecisely, inexactly; ⟨*remember*⟩ vaguely; **b)** *(nicht sorgfältig)* ⟨*work*⟩ carelessly

Ungenauigkeit die inaccuracy; *(einer Definition)* imprecision; inexactness

ungeniert ['ʊnʒeni:ɐt] **1.** *Adj.* free and easy; uninhibited; **er war ganz ~:** he was not at all embarrassed *or* concerned. **2.** *adv.* openly; ⟨*yawn*⟩ unconcernedly; *(ohne Scham)* ⟨*undress etc.*⟩ without any embarrassment

ungenießbar *Adj.* **a)** *(nicht eßbar)* inedible; *(nicht trinkbar)* undrinkable; **b)** *(fig. ugs.)* unbearable ⟨*person*⟩; **er ist heute ~:** he's in a foul mood today *(sl.)*

ungenügend 1. *Adj.* inadequate; **die Note „~"** *(Schulw.)* the 'unsatisfactory' [mark]. **2.** *adv.* inadequately

ungepflegt *Adj.* neglected ⟨*garden, park, car, etc.*⟩; unkempt ⟨*person, appearance, hair*⟩; uncared-for ⟨*hands*⟩

ungerade *Adj.* odd ⟨*number*⟩

ungerecht 1. *Adj.* unjust, unfair. **2.** *adv.* unjustly; unfairly

ungerechtfertigt *Adj.* unjustified; unwarranted

Ungerechtigkeit die; ~, ~en injustice

ungeregelt *Adj.* irregular; disorganized

ungereimt *Adj. (fig.)* inconsistent; illogical; *(ugs. abwertend: verworren)* muddled

ungern *Adv.* reluctantly; **etw. ~ tun** not like *or* dislike doing sth.

ungeschält *Adj.* unpeeled ⟨*fruit*⟩; **~er Reis** paddy rice

ungeschehen *Adj.* **in etw. ~ machen** undo sth.

Ungeschicklichkeit die clumsiness; ineptitude

ungeschickt 1. *Adj.* clumsy; awkward ⟨*movement, formulation, etc.*⟩. **2.** *adv.* clumsily; ⟨*bow, express oneself, etc.*⟩ awkwardly; **sich ~ anstellen** show a lack of skill; show oneself to be inept

ungeschminkt *Adj.* **a)** not made-up *pred.;* without make-up *postpos.;* **b)** *(fig.)* unvarnished ⟨*truth*⟩; uncoloured ⟨*account*⟩

ungeschoren *Adj.* **a)** unshorn; **b) ~ bleiben** *(fig).* be left in peace; be spared

ungeschützt *Adj.* unprotected; *(Wind und Wetter ausgesetzt)* exposed

ungesehen *Adj.; nicht attr.* unseen

ungesetzlich 1. *Adj.* unlawful; illegal. **2.** *adv.* unlawfully; illegally

ungestört *Adj.* undisturbed; uninterrupted ⟨*development*⟩; **~ arbeiten** work in peace *or* without interruption

ungestraft 1. *Adj.; nicht attr.* unpunished. **2.** *adv.* with impunity

ungestüm ['ʊngəʃty:m] *(geh.)* **1.** *Adj.* impetuous, tempestuous ⟨*person, embrace, nature, etc.*⟩. **2.** *adv.* impetuously

ungesund 1. *Adj. (auch fig.)* unhealthy; *(fig.: übermäßig)* excessive ⟨*ambition, activity*⟩; **Rauchen ist ~:** smoking is bad for you *or* for your health. **2.** *adv.* unhealthily; **~ leben** lead an unhealthy life

ungetrübt *Adj.* unclouded, perfect ⟨*happiness*⟩; unalloyed ⟨*pleasure*⟩; unspoilt, perfect ⟨*days, relationship*⟩

Ungetüm ['ʊngəty:m] **das; ~s, ~e** monster

ungeübt *Adj.* unpractised ⟨*hand*⟩; **in etw. ~ sein** lack practice in sth.

ungewiß *Adj.* uncertain; **jmdn. [über etw.** *(Akk.)*] **im ungewissen lassen** leave sb. in the dark *or* keep sb. guessing [about sth.]

Ungewißheit die uncertainty

ungewöhnlich 1. *Adj.* **a)** unusual; **b)** *(enorm)* exceptional ⟨*strength, beauty, ability, etc.*⟩; outstanding ⟨*achievement, success*⟩. **2.** *adv.* **a)** *(unüblich)* ⟨*behave*⟩ abnormally, strangely; **b)** *(enorm)* exceptionally

ungewohnt 1. *Adj.* unaccustomed; unfamiliar ⟨*method, work, surroundings, etc.*⟩. **2.** *adv.* unusually

ungewollt 1. *Adj.* unwanted; *(unbeabsichtigt)* unintentional; inadvertent. **2.** *adv.* unintentionally; inadvertently

Ungeziefer ['ʊngətsi:fɐ] **das; ~s** vermin *pl.*

ungezogen 1. *Adj.* naughty; badly behaved; bad ⟨*behaviour*⟩; *(frech)* cheeky. **2.** *adv.* naughtily; ⟨*behave*⟩ badly

ungezwungen 1. *Adj.* natural, unaffected ⟨*person, behaviour, cheerfulness*⟩; informal, free and easy ⟨*tone, conversation, etc.*⟩. **2.** *adv.* ⟨*behave*⟩ naturally, unaffectedly; ⟨*talk*⟩ freely

Ungezwungenheit die; ~ *s.* **ungezwungen 1:** naturalness; unaffectedness; informality

ungläubig 1. *Adj.* **a)** disbelieving; incredulous; **b)** *(Rel.)* unbelieving. **2.** *adv.* incredulously; in disbelief

Ungläubige der/die *(Rel.)* unbeliever

unglaublich 1. *Adj.* **a)** incredible; **b)** *(ugs.: sehr groß)* incredible *(coll.)*, fantastic *(coll.)* ⟨*speed, amount, luck, etc.*⟩. **2.** *adv.* *(ugs.: äußerst)* incredibly *(coll.)*

unglaubwürdig *Adj.* implausible; untrustworthy, unreliable ⟨*witness etc.*⟩

ungleich 1. *Adj.* unequal; odd, unmatching ⟨*socks, gloves, etc.*⟩; odd ⟨*couple*⟩. **2.** *adv.* **a)** unequally; **b)** *(ungleichmäßig)* unevenly. **3.** *Adv.* far ⟨*larger, more difficult, etc.*⟩

ungleichmäßig 1. *Adj.* uneven. **2.** *adv.* unevenly

Unglück das; ~[e]s, ~e a) *(Unfall)* accident; *(Flugzeug~, Zug~)* crash; accident; *(Mißgeschick)* mishap; **das ist [doch] kein ~!** that's not a disaster; it doesn't really matter; **b)** *o. Pl. (Not)* misfortune; *(Leid)* suffering; distress; **sich ins ~ stürzen, in sein ~ rennen** rush headlong into disaster *or* to one's ruin; **c)** *(Pech)* bad luck; misfortune; **das bringt ~:** that's unlucky; **zu allem ~:** to make matters worse; **ein ~ kommt selten allein** *(ugs.)* it never rains but it pours

unglücklich 1. *Adj.* **a)** unhappy; **mach dich nicht ~!** don't do it!; **b)** *(nicht vom Glück begünstigt)* unfortunate ⟨*person*⟩; *(bedauerns-*

wert, arm) hapless ⟨*person, animal*⟩; **c)** *(ungünstig, ungeschickt)* unfortunate ⟨*moment, combination, meeting, etc.*⟩; unhappy ⟨*end, choice, solution*⟩; unfortunate, unhappy ⟨*coincidence, formulation*⟩. **2.** *adv.* **a)** unhappily; ~ **verliebt sein** be unhappy in love; **b)** *(ungünstig)* unfortunately; *(ungeschickt)* unhappily, clumsily ⟨*translated, expressed*⟩
ụnglücklicherweise *Adv.* unfortunately
Ụnglücks-: ~fall der accident; **~zahl die** unlucky number
Ụngnade die *in* **[bei jmdm.] in** ~ *(Akk.)* **fallen/in** ~ *(Dat.)* **sein** fall/be out of favour [with sb.]
ụngnädig 1. *Adj.* bad-tempered; grumpy. **2.** *adv.* in a bad-tempered way; grumpily
ụngültig *Adj.* invalid; void *(esp. Law)*; spoilt ⟨*vote, ballot-paper*⟩
ụngünstig 1. *Adj.* **a)** unfavourable; unfavourable, poor ⟨*climate, weather*⟩; *(unglücklich)* unfortunate ⟨*consequence*⟩; unfortunate, bad ⟨*shape, layout*⟩; *(unvorteilhaft)* unfavourable, unflattering ⟨*light, perspective, impression*⟩; unflattering ⟨*cut of dress*⟩; inconvenient ⟨*position*⟩; *(schädlich)* harmful ⟨*effect*⟩; **b)** *(unpassend)* inconvenient ⟨*time*⟩; *(ungeeignet)* inappropriate, inconvenient ⟨*time, place*⟩; unsuitable ⟨*colour etc.*⟩. **2.** *adv.* **a)** unfavourably; badly ⟨*designed, laid out*⟩; *(unvorteilhaft)* unflatteringly ⟨*cut*⟩; **sich** ~ **auswirken** have a harmful effect; **b)** *(unpassend, ungeeignet)* inconveniently
ụngut *Adj.* **a)** uneasy ⟨*feeling, premonition*⟩; unpleasant ⟨*aftertaste, recollection, memories*⟩; **b) nichts für** ~! no offence [meant]! *(coll.)*
unhạltbar *Adj.* **a)** untenable ⟨*thesis, statement, etc.*⟩; **b)** *(unerträglich)* unbearable, intolerable ⟨*conditions, situation*⟩; **c)** *(Ballspiele)* unstoppable
ụnhandlich *Adj.* unwieldy
Ụnheil das disaster; ~ **anrichten** *od.* **stiften** wreak havoc
unheilbar 1. *Adj.* incurable. **2.** *adv.* incurably
ụnheil·voll *Adj.* disastrous; *(verhängnisvoll)* fateful; ominous ⟨*development*⟩
ụnheimlich 1. *Adj.* **a)** eerie ⟨*story, figure, place, sound*⟩; eerie, uncanny ⟨*feeling*⟩; **jmdm.** ~ **sein** give sb. an eerie feeling *or (coll.)* the creeps; **mir ist/wird [es]** ~: I have an eerie *or* uncanny feeling; **b)** *(ugs.)* *(schrecklich)* terrible *(coll.)* ⟨*coward, idiot, hunger, headache, etc.*⟩; terrific *(coll.)* ⟨*fun, sum, etc.*⟩. **2.** *adv.* **a)** eerily; uncannily; **b)** *(ugs.)* terribly *(coll.)* ⟨*fat, nice, etc.*⟩; terrifically *(coll.)* ⟨*important, large*⟩; incredibly *(coll.)* ⟨*quick, long*⟩
ụnhöflich 1. *Adj.* impolite. **2.** *adv.* impolitely
Ụnhöflichkeit die impoliteness
Ụnhold der; ~**[e]s,** ~**e a)** fiend; demon; **b)** *(abwertend: böser Mensch)* monster
ụnhygienisch 1. *Adj.* unhygienic. **2.** *adv.* unhygienically
uni ['ʏni] *indekl. Adj.* plain, single-colour ⟨*material etc.*⟩; plain ⟨*tie*⟩
Uni ['ʊni] **die;** ~, ~**s** *(ugs.)* university
Uniform ['ʊnifɔrm] **die;** ~, ~**en** uniform
Unikum ['u:nikʊm] **das;** ~**s, Unika** *od.* ~**s** *(ugs.: Original)* [real] character
ụninteressant *Adj.* **a)** uninteresting; *(nicht von Belang)* of no interest *postpos.*; unimportant; **nicht** ~: quite interesting; **b)** *(nicht lohnend, nicht attraktiv)* untempting, unattractive ⟨*offer*⟩
ụninteressiert *Adj.* uninterested; not interested **(an** + *Dat.* in)
Union [u'ni̯o:n] **die;** ~, ~**en** union
universell [univɛr'zɛl] **1.** *Adj.* universal. **2.** *adv.* universally
Universität [univɛrzi'tɛ:t] **die;** ~, ~**en** university
Universitäts-: ~klinik die university hospital; **~stadt die** university town; **~studium das** study *no art.* at university
Universum [uni'vɛrzʊm] **das;** ~**s** universe
unken ['ʊŋkn̩] *itr. V. (ugs.)* prophesy doom [and destruction] *(joc.)*
ụnkenntlich *Adj.* unrecognizable ⟨*person, face*⟩; indecipherable ⟨*writing, stamp*⟩
Ụnkenntnis die; *o. Pl.* ignorance
ụnklar 1. *Adj.* **a)** *(undeutlich)* unclear; indistinct; *(fig.: unbestimmt)* vague ⟨*feeling, recollection, idea*⟩; **b)** *(nicht klar verständlich)* unclear; **c)** *(nicht durchschaubar)* unclear ⟨*origin, situation, etc.*⟩; *(ungewiß)* uncertain ⟨*outcome*⟩; **sich** *(Dat.)* **über etw.** *(Akk.)* **im** ~**en sein** be unclear *or* unsure about sth.; **jmdn. über etw.** *(Akk.)* **im** ~**en lassen** keep sb. guessing about sth.
Ụnklarheit die a) *o. Pl.* *(Undeutlichkeit)* lack of clarity; indistinctness; *(Unverständlichkeit)* lack of clarity (*Gen.* in); *(Ungewißheit)* uncertainty; **b)** *(unklarer Punkt)* unclear point
ụnkollegial *Adj.* inconsiderate *or* unhelpful [to one's colleagues]
ụnkompliziert 1. *Adj.* uncomplicated, straightforward ⟨*person, mechanism, etc.*⟩. **2.** *adv.* ⟨*express*⟩ straightforwardly, simply
ụnkontrollierbar *Adj.* impossible to check *or* supervise *postpos.*; *(nicht zu beherrschen)* uncontrollable
ụnkontrolliert *Adj.* **a)** unsupervised; **b)** *(unbeherrscht)* uncontrolled ⟨*emotions, outburst*⟩
ụnkonzentriert *Adj.* lacking in concentration *postpos.*
Ụnkosten *Pl.* **a)** [extra] expense *sing.*; expenses; **sich in** ~ **stürzen** dig deep into one's pocket; **b)** *(Kosten)* costs; expenditure *sing.*
Ụnkraut das *o. Pl.* weeds *pl.*; ~ **vergeht nicht** *(ugs. scherzh.)* it would take a great deal to finish off his/her/our sort *(coll.)*
unkündbar *Adj.* permanent; **er ist** ~: he cannot be given notice
ụnkundig *Adj. (geh.)* ignorant; **des Lesens/Schreibens/Deutschen** ~: unable to read/to write/to speak German
ụnlängst *Adv. (geh.)* not long ago; recently
ụnlauter *Adj. (geh.)* dishonest; **~er Wettbewerb** *(Rechtsspr.)* unfair competition
ụnleserlich 1. *Adj.* illegible. **2.** *adv.* illegibly
unleugbar 1. *Adj.* undeniable; indisputable. **2.** *adv.* undeniably; indisputably
ụnliebsam 1. *Adj.* unpleasant. **2.** *adv.* **er ist** ~ **aufgefallen** he made a bad impression
Ụnlust die *(Widerwille)* reluctance; *(Lustlosigkeit)* lack of enthusiasm
ụnmaßgeblich *Adj.* of no consequence *postpos.*; inconsequential
ụnmäßig 1. *Adj.* **a)** immoderate; excessive; **b)** *(enorm)* tremendous ⟨*desire, thirst, fear, etc.*⟩. **2.** *adv.* **a)** excessively; ⟨*eat, drink*⟩ to excess; **b)** *(überaus, sehr)* tremendously *(coll.)*
Ụnmenge die mass; enormous number/amount
Ụnmensch der brute; **man ist ja kein** ~ *(ugs.)* I'm not inhuman
ụnmenschlich 1. *Adj.* **a)** inhuman; brutal; subhuman, appalling ⟨*conditions*⟩; **b)** *(entsetzlich)* terrible *(coll.)*, appalling ⟨*pain, heat, suffering, etc.*⟩. **2.** *adv.* **a)** in an inhuman way; **b)** *(entsetzlich)* appallingly *(coll.)*

unmerklich 1. *Adj.* imperceptible. **2.** *adv.* imperceptibly
unmißverständlich 1. *Adj.* **a)** *(eindeutig)* unambiguous; **b)** *(offen, direkt)* blunt ‹*answer, refusal*›; unequivocal ‹*language*›. **2.** *adv.* **a)** *(eindeutig)* unambiguously; **b)** *(offen, direkt)* bluntly; unequivocally
unmittelbar 1. *Adj.* **a)** *nicht präd.* immediate ‹*vicinity, past, future*›; **aus ~er Nähe schießen** shoot at close quarters *or* from point-blank range; **b)** *(direkt)* direct ‹*contact, connection, influence, etc.*›; immediate ‹*cause, consequence, predecessor, successor*›. **2.** *adv.* **a)** immediately; right ‹*behind, next to*›; **~ bevorstehen** be imminent; be almost upon us *etc.*; **b)** *(direkt)* directly
unmodern 1. *Adj.* old-fashioned; *(nicht modisch)* unfashionable. **2.** *adv.* in an old-fashioned way; *(nicht modisch)* unfashionably
unmöglich 1. *Adj.* **a)** impossible; **ich verlange ja nichts Unmögliches [von dir]** I'm not asking [you] for the impossible; **b)** *(ugs.: nicht akzeptabel, unangebracht)* impossible ‹*person, behaviour, colour, ideas, place, etc.*›; **sich ~ machen** make a fool of oneself; make oneself look ridiculous; **c)** *(ugs.: erstaunlich, seltsam)* incredible. **2.** *adv.* *(ugs.)* ‹*behave*› impossibly; ‹*dress*› ridiculously. **3.** *Adv.* *(ugs.: unter keinen Umständen)* **ich/es** *usw.* **kann ~ ...:** I/it *etc.* can't possibly ...
unmoralisch 1. *Adj.* immoral. **2.** *adv.* immorally
unmotiviert 1. *Adj.* unmotivated. **2.** *adv.* without reason; for no reason
unmündig *Adj.* **a)** under-age; **~ sein** be under age *or* a minor; **b)** *(fig.)* dependent
unmusikalisch *Adj.* unmusical
Unmut der *(geh.)* displeasure; annoyance; **seinen ~ an jmdm. auslassen** take it out on sb.
unnachahmlich 1. *Adj.* inimitable. **2.** *adv.* inimitably
unnachgiebig *Adj.* intransigent
unnachsichtig 1. *Adj.* merciless; unmerciful; unrelenting ‹*severity*›. **2.** *adv.* mercilessly; ‹*punish*› unmercifully
unnatürlich 1. *Adj.* unnatural; forced ‹*laugh*›. **2.** *adv.* unnaturally; ‹*laugh*› in a forced way; ‹*speak*› affectedly
unnormal 1. *Adj.* abnormal. **2.** *adv.* abnormally
unnötig 1. *Adj.* unnecessary; needless, pointless ‹*heroism*›. **2.** *adv.* unnecessarily
unnütz 1. *Adj.* useless ‹*stuff, person, etc.*›; pointless ‹*talk*›; wasted ‹*words*›; pointless, wasted ‹*expense, effort*›; vain ‹*attempt*›. **2.** *adv.* needlessly
UNO ['u:no] **die: die ~:** the UN
unordentlich 1. *Adj.* untidy. **2.** *adv.* untidily; ‹*tie, treat, etc.*› carelessly
Unordnung die disorder; mess
unparteiisch 1. *Adj.* impartial. **2.** *adv.* impartially
Unparteiische der/die; *adj. Dekl. (Sport) s.* **Schiedsrichter a**
unpassend 1. *Adj.* inappropriate; unsuitable ‹*dress etc.*›. **2.** *adv.* inappropriately; unsuitably ‹*dressed etc.*›
unpersönlich 1. *Adj.* impersonal; distant, aloof ‹*person*›. **2.** *adv.* impersonally; ‹*answer, write*› in impersonal terms
unpraktisch 1. *Adj.* unpractical. **2.** *adv.* in an unpractical way
unpünktlich 1. *Adj.* **a)** unpunctual ‹*person*›; **b)** *(verspätet)* late, unpunctual ‹*payment*›. **2.** *adv.* late
Unpünktlichkeit die lack of punctuality
unrasiert *Adj.* unshaven
Unrecht das; *o. Pl.* **a) unrecht haben** be wrong; **jmdm. unrecht tun** do sb. an injustice; do wrong by sb.; **b)** wrong; **im ~ sein** be [in the] wrong
unrechtmäßig 1. *Adj.* unlawful; illegal. **2.** *adv.* unlawfully; illegally
unredlich *(geh.)* **1.** *Adj.* dishonest. **2.** *adv.* dishonestly
unregelmäßig 1. *Adj.* irregular. **2.** *adv.* irregularly
Unregelmäßigkeit die irregularity
unreif *Adj.* **a)** unripe; **b)** *(nicht erwachsen)* immature
unrein *Adj.* **a)** *(auch fig.)* impure; bad ‹*breath, skin*›; *(nicht sauber)* dirty, polluted ‹*water, air*›; **b) etw. ins ~e schreiben** make a rough copy of sth.; write sth. [out] in rough
unrentabel 1. *Adj.* unprofitable. **2.** *adv.* unprofitably
unrichtig 1. *Adj.* incorrect; inaccurate. **2.** *adv.* incorrectly
Unruhe die a) *(auch fig.)* unrest; *(Lärm)* noise; commotion; *(Unrast)* restlessness; agitation; *(Besorgnis)* anxiety; disquiet; **b)** *(Unfrieden)* unrest; **~ stiften** stir up trouble; **c)** *Pl. (Tumulte)* disturbances; unrest *sing.*
unruhig 1. *Adj.* **a)** restless; *(besorgt)* anxious; *(nervös)* agitated; jittery; *(fig.)* choppy ‹*sea*›; busy ‹*pattern*›; busy, eventful ‹*life*›; unsettled, troubled ‹*time*›; **b)** *(ungleichmäßig)* uneven ‹*breathing, pulse, running, etc.*›; fitful ‹*sleep, motion*›; disturbed ‹*night*›; unsettled ‹*life*›. **2.** *adv.* **a)** restlessly; *(besorgt)* anxiously; **b)** *(ungleichmäßig)* ‹*breathe, run*› unevenly; ‹*sleep*› fitfully
uns [ʊns] **1.** *Dat. u. Akk. von* **wir** us; gib es **~:** give it to us; **gib ~ das Geld** give us the money; **Freunde von ~:** friends of ours; **bei ~:** at our home *or (coll.)* place; *(in der Heimat)* where I/we live *or* come from; **bei ~ gegenüber** opposite us *or* our house. **2.** *Dat. u. Akk. des Reflexivpron. der 1. Pers. Pl.* **a)** *refl.* ourselves; **wir schämen ~:** we are ashamed [of ourselves]; **von ~ aus** *(aus eigenem Antrieb)* on our own initiative; **b)** *reziprok* one another; **wir haben ~ gestritten** we had an argument *or* quarrel
unsachgemäß 1. *Adj.* improper. **2.** *adv.* improperly
unsachlich 1. *Adj.* unobjective; **~ werden** lose one's objectivity. **2.** *adv.* without objectivity
unsanft 1. *Adj.* rough; hard ‹*push, impact*›. **2.** *adv.* roughly; **~ geweckt werden** be rudely awoken
unsauber 1. *Adj.* **a)** *(schmutzig)* dirty; **b)** *(nachlässig)* untidy, sloppy ‹*work, writing, etc.*›; **c)** *(unlauter)* shady ‹*practice, deal, character, etc.*›; underhand, dishonest ‹*method, means, intention*›; *(Sport: unfair)* unsporting, unfair ‹*play*›. **2.** *adv.* **a)** *(nachlässig)* untidily; carelessly; **b)** *(unklar)* ‹*sing, play*› inaccurately; **c)** *(Sport: unfair)* unsportingly; unfairly
unschädlich *Adj.* harmless; **~ machen** render harmless, neutralize ‹*toxic substance, germ, etc.*›; put ‹*weapon, person*› out of action; render ‹*bomb etc.*› safe; *(verhüll.: töten)* eliminate ‹*person*›
unscharf *Adj.* blurred, fuzzy ‹*photo, picture*›
unschätzbar *Adj.* inestimable ‹*value etc.*›; invaluable ‹*service*›; priceless ‹*riches etc.*›
unscheinbar *Adj.* inconspicuous; nondescript; unspectacular ‹*plumage, blossom*›
unschlagbar *Adj.* unbeatable ‹*opponent, prices, etc.*›
unschlüssig *Adj.* undecided *pred.*; undecisive ‹*gesture, attitude*›

Unschuld die; *o. Pl.* **a)** innocence; **seine Hände in ~ waschen** *(fig.)* wash one's hands in innocence; **b)** *(Jungfräulichkeit)* virginity

unschuldig 1. *Adj.* **a)** innocent; **an etw.** *(Dat.)* **~ sein** be not guilty of sth.; **b) er/sie ist noch ~:** he/she is still a virgin. **2.** *adv.* innocently

Unschulds-: ~lamm das *(spött.)* little innocent; **~miene die** innocent expression

unselbständig *Adj.* dependent [on other people]; **sei doch nicht immer so ~!** try to be a bit more independent!

[1]**unser** ['ʊnzɐ] *Possessivpron. der 1. Pers. Pl.* our; **das ist ~s** *od. (geh.)* **~es** *od. (geh.)* **das ~e** that is ours; **sein Wagen stand neben ~[e]m** *od.* **unsrem** his car was next to ours; **die Unseren** our family; **wir müssen das Uns[e]re dazu tun** we must do our share

[2]**unser** *Gen. von* **wir** *(geh.)* of us; **in ~ aller Interesse** in the interest of all of us

unser·einer, unser·eins *Indefinitpron. (ugs.)* the likes of us *pl.;* our sort *(coll.)*

unserer·seits *Adv.* for our part

unseres·gleichen *indekl. Indefinitpron.* people *pl.* like us

unseret-: ~halben, ~wegen *Adv. s.* **unsertwegen; ~willen** *Adv. s.* **unsertwillen**

unsert-: ~wegen *Adv.* **a)** because of us; on our account; *(für uns)* on our behalf; *(uns zuliebe)* for our sake; **b)** *(was uns angeht)* as far as we are concerned; **~willen** *Adv.* **um ~willen** for our sake[s]

unsicher 1. *Adj.* **a)** *(gefährlich)* unsafe; dangerous; *(gefährdet)* at risk *pred.;* insecure ⟨*job*⟩; **einen Ort ~ machen** *(scherzh.)* honour a place with one's presence *(joc.); (sich vergnügen)* have a good time in a place; *(sein Unwesen treiben)* get up to one's tricks in a place; **b)** *(unzuverlässig)* uncertain, unreliable ⟨*method*⟩; unreliable ⟨*source, person*⟩; **c)** *(zögernd)* uncertain, hesitant ⟨*step*⟩; *(zitternd)* unsteady, shaky ⟨*hand*⟩; *(nicht selbstsicher)* insecure; diffident; unsure of oneself *pred.;* **jmdn. ~ machen** put sb. off his/her stroke; **d)** *(keine Gewißheit habend)* unsure; uncertain; **e)** *(ungewiß)* uncertain. **2.** *adv.* **a)** ⟨*walk, stand, etc.*⟩ unsteadily; ⟨*drive*⟩ without [much] confidence; **b)** *(nicht selbstsicher)* ⟨*smile, look*⟩ diffidently

Unsicherheit die a) *o. Pl. (Gefährlichkeit)* dangerousness; *(Gefahren)* dangers *pl.;* **b)** *o. Pl. (Unzuverlässigkeit)* uncertainty; unreliability; **c)** *o. Pl. (Zaghaftigkeit)* unsureness; *(der Schritte o.ä.)* unsteadiness; **d)** *o. Pl. (fehlende Selbstsicherheit)* insecurity; lack of [self-]confidence; **e)** *o. Pl. (Ungewißheit)* uncertainty; **f)** *o. Pl. (der Arbeitsplätze)* insecurity; *(des Friedens)* instability

Unsicherheits·faktor der element of uncertainty

unsichtbar *Adj.* invisible (**für** to)

Unsinn der; *o. Pl.* **a)** nonsense; **b)** *(Unfug)* tomfoolery; fooling about *no art.;* **mach [ja] keinen ~:** don't do anything silly; no messing about

unsinnig *Adj.* nonsensical ⟨*statement, talk, etc.*⟩; absurd, ridiculous ⟨*demand etc.*⟩

Unsitte die bad habit; *(allgemein verbreitet)* bad practice

unsittlich 1. *Adj.* indecent. **2.** *adv.* indecently

unsozial 1. *Adj.* unsocial ⟨*policy, measure, rent, etc.*⟩; antisocial ⟨*behaviour*⟩. **2.** *adv.* unsocially; ⟨*behave*⟩ antisocially

unsportlich 1. *Adj.* **a)** unathletic ⟨*person*⟩; **b)** *(unfair)* unsporting; unsportsmanlike. **2.** *adv. (unfair)* in an unsporting way

unsr... *s.* [1]**unser**

unsrer·seits *s.* **unsererseits**

unsres·gleichen *s.* **unseresgleichen**

unstatthaft *Adj.* inadmissible

unsterblich 1. *Adj.* immortal; *(fig.)* undying ⟨*love*⟩. **2.** *adv. (ugs.: außerordentlich)* incredibly *(coll.)*

unstet *(geh.)* **1.** *Adj.* **a)** *(ruhelos)* restless ⟨*person, glance, thoughts, etc.*⟩; unsettled ⟨*life*⟩; **b)** *(unbeständig)* vacillating ⟨*person, nature*⟩; *(labil)* unstable ⟨*person, character*⟩. **2.** *adv. (ruhelos)* restlessly

Unstimmigkeit die; ~, ~en a) *o. Pl.* inconsistency; **b)** *(etw. Unstimmiges)* discrepancy; **c)** *(Meinungsverschiedenheit)* difference [of opinion]

Unsumme die vast *or* huge sum

unsympathisch *Adj.* uncongenial, disagreeable ⟨*person*⟩; unpleasant ⟨*characteristic, nature, voice*⟩; **er ist mir ~/nicht ~:** I find him disagreeable/quite likeable

Untat die misdeed; evil deed

untätig *Adj.* idle

untauglich *Adj.* **a)** unsuitable ⟨*applicant*⟩; **b)** *(für Militärdienst)* unfit [for service] *postpos.*

unteilbar *Adj.* indivisible

unten ['ʊntn̩] *Adv.* **a)** down; **hier/da ~:** down here/there; **nach ~** *(auch fig.)* downward; **mit dem Gesicht nach ~:** face downwards; **von ~:** from below; **~ liegen** be down below; *(darunter)* lie underneath; **b)** *(in Gebäuden)* downstairs; **nach ~:** downstairs; **der Aufzug fährt nach ~/kommt von ~:** the lift *(Brit.) or (Amer.)* elevator is going down/coming up; **c)** *(am unteren Ende, zum unteren Ende hin)* at the bottom; **nach ~ [hin]** towards the bottom; *(als Bildunterschrift)* **„~ [rechts]"** 'below [right]'; *(auf einem Karton o.ä.)* **„~"** 'other side up'; **d)** *(an der Unterseite)* underneath; **e)** *(in einer Hierarchie, Rangfolge)* at the bottom; **ziemlich weit ~ auf der Liste** rather a long way down/right at the bottom of the list; **f)** *([weiter] hinten im Text)* below; **weiter ~:** further on; below; **g)** *(im Süden)* down south; **hier/dort ~:** down here/there [in the south]

unten·drunter *Adv. (ugs.)* underneath [it/them]

unten-: ~durch *Adv.* through underneath; **~erwähnt, ~genannt** *Adj.; nicht präd.* undermentioned *(Brit.);* mentioned below *postpos.;* **~herum, ~rum** *Adv. (ugs.)* down below; **~stehend** *Adj.; nicht präd.* following; given below *postpos.*

unter ['ʊntɐ] **1.** *Präp. mit Dat.* **a)** *(Lage, Standort, Abhängigkeit, Unterordnung)* under; **~ jmdm. wohnen** live below sb.; **b)** *(weniger, niedriger usw. als)* **Mengen ~ 100 Stück** quantities of less than 100; **c)** during; *(modal)* **~ Angst/Tränen** in *or* out of fear/in tears; **~ dem Beifall der Menge** applauded by the crowd; **d)** *(aus einer Gruppe)* among[st]; **~ anderem** among[st] other things; **e)** *(zwischen)* among[st]; **~ sich** by themselves; **~ uns gesagt** between ourselves *or* you and me; **f)** *(Zustand)* under; **~ Strom stehen** be live; *s. auch* **Tag a, Woche. 2.** *Präp. mit Akk.* **a)** *(Richtung, Ziel, Abhängigkeit, Unterordnung)* under; **b)** *(niedriger als)* **~ Null sinken** drop below zero; **c)** *(zwischen)* among[st]; **~ Strom/Dampf setzen** switch on/put under steam. **3.** *Adv.* less than

unter... *Adj.* **a)** lower; bottom; *(ganz unten)* bottom; **das ~e/~ste Stockwerk** the lower/bottom storey; **das Unterste zuoberst kehren** *(ugs.)* turn everything upside down; **b)** lower ⟨*Rhine, Nile,* *etc.*⟩; **c)** *(in der Rangfolge o. ä.)* lower; lesser ⟨*authority*⟩; **die ~en Klassen der Schule** the junior classes *or* forms of the school

unter-, Unter-: ~abteilung die department; **~arm der** forearm; **~belichten** *tr. V. (Fot.)* underexpose; **~bewerten** *tr. V.* undervalue; underrate; mark ⟨*gymnast, skater*⟩ too low

unter-, Unter-: ~bieten *unr. tr. V.* **a)** *(weniger fordern)* undercut **(um** by); **b)** *(bes. Sport)* beat ⟨*record*⟩; **~binden** *unr. tr. V.* stop; **~bleiben** *unr. itr. V.; mit sein* **etw. ~bleibt** sth. does not occur *or* happen; **das hat zu ~bleiben!** this must stop; **~brechen** *unr. tr. V.* interrupt; break ⟨*journey, silence*⟩; terminate ⟨*pregnancy*⟩; **wir sind unterbrochen worden** *(im Telefongespräch)* we've been cut off; **~brechung die** *s.* **~brechen:** interruption; break *(Gen.* in); termination; **~breiten** *tr. V. (geh.)* present

unter-, Unter-: ~|bringen *unr. tr. V.* **a)** put; **b)** *(beherbergen)* put up; **c) jmdn. bei einer Firma ~bringen** *(ugs.)* get sb. a job in a company; **~bringung die; ~, ~en** accommodation *no indef. art.*; **~|buttern** *tr. V. (ugs.: unterdrücken)* push aside *(fig.)*; **~deck das** lower deck

unter·der·hand *Adv.* on the quiet ; **etw. ~ erfahren** hear sth. on the grape-vine

unter-, Unter-: ~dessen *s.* **inzwischen; ~drücken** *tr. V.* **a)** suppress; hold back ⟨*comment, question, answer, criticism, etc.*⟩; **b)** *(niederhalten)* suppress ⟨*revolution etc.*⟩; oppress ⟨*minority etc.*⟩; **~drücker der** *(abwertend)* oppressor; **~drückung die; ~, ~en a)** suppression; **b)** *(das Unterdrücktwerden, -sein)* oppression

unter·durchschnittlich 1. *Adj.* below average. **2.** *adv.* below the average

unter·einander *Adv.* **a)** *(räumlich)* one below the other; **b)** *(miteinander)* among[st] ourselves/themselves *etc.*

untereinander-: ~|liegen *unr. itr. V.* lie *or* be one below *or* underneath the other; **~|stehen** *unr. itr. V.* be one below the other

unter-, Unter-: ~entwickelt *Adj.* underdeveloped; **~ernährt** *Adj.* undernourished; suffering from malnutrition *postpos.*; **~ernährung die** malnutrition

Unter·fangen das; ~s *(geh.)* venture; undertaking

unter|fassen *tr. V. (ugs.)* **a) jmdn. ~:** take sb.'s arm; **b)** *(stützen)* support

Unter·führung die underpass; *(für Fußgänger)* subway *(Brit.)*; [pedestrian] underpass *(Amer.)*

Unter·gang der a) *(Sonnen~, Mond~ usw.)* setting; **b)** *(von Schiffen)* sinking; **c)** *(das Zugrundegehen)* decline; *(plötzlich)* destruction; *(von Personen)* downfall; *(der Welt)* end

unter·geben *Adj.* subordinate

Untergebene der/die; *adj. Dekl.* subordinate

unter-, Unter-: ~|gehen *unr. itr. V.; mit sein* **a)** ⟨*sun, star, etc.*⟩ set; ⟨*ship*⟩ sink, go down; ⟨*person*⟩ drown, go under; **seine Worte gingen in dem Lärm ~:** his words were drowned by *or* lost in the noise; **b)** *(zugrunde gehen)* come to an end; **~geordnet** *Adj.* **a)** secondary ⟨*role, importance, etc.*⟩; subordinate ⟨*position, post, etc.*⟩; **b)** *(Sprachw.)* subordinate; **~geschoß das** basement; **~gewicht das;** *o. Pl.* underweight; **~gewichtig** *Adj.* underweight

unter-: ~gliedern *tr. V.* subdivide; **[1]~graben** *unr. tr. V.* undermine *(fig.)*

unter-, Unter-: [2]~graben *unr. tr. V.* dig in; **~grenze die** lower limit; **~grund der a)** *(bes. Landw.)* subsoil; **b)** *(Bauw.: Baugrund)* foundation; **c)** *(Farbschicht)* background; **d)** *o. Pl. (bes. Politik)* underground

Untergrund-: ~bahn die underground [railway] *(Brit.)*; subway *(Amer.)*

unter-, Unter-: ~|haken *tr. V. (ugs.)* **jmdn. ~haken** take sb.'s arm; **~gehakt gehen** walk arm in arm; **~halb 1.** *Adv.* below; **weiter ~halb** further down; **~halb von** below; **2.** *Präp. mit Gen.* below; **~halt der;** *o. Pl.* **a)** living; **b)** *(~haltszahlung)* maintenance; **c)** *(Instandhaltung[skosten])* upkeep

unter·halten 1. *unr. tr. V.* **a)** *(versorgen)* support; **b)** *(instand halten)* maintain ⟨*building*⟩; **c)** *(betreiben)* run, keep ⟨*car, hotel*⟩; **d)** *(pflegen)* maintain, keep up ⟨*contact, correspondence*⟩; **e)** entertain ⟨*guest, audience*⟩. **2.** *unr. refl. V.* **a)** talk; converse; **b)** *(sich vergnügen)* enjoy oneself; **habt ihr euch gut ~?** did you have a good time?

unterhaltsam *Adj.* entertaining

Unterhalts-: ~anspruch der maintenance claim; claim for maintenance; **~kosten** *Pl.* maintenance *sing.*; **~pflicht die** obligation to pay maintenance

Unterhaltung die a) *o. Pl. (Versorgung)* support; **b)** *o. Pl. (Instandhaltung)* maintenance; upkeep; **c)** *o. Pl. (Aufrechterhaltung)* maintenance; **d)** *(Gespräch)* conversation; **e)** *(Zeitvertreib)* entertainment; **ich wünsche gute ~:** enjoy yourself/yourselves

Unterhaltungs-: ~lektüre die light reading *no art.*; **~musik die** light music

Unter-: ~händler der *(bes. Politik)* negotiator; **~haus das** *(Parl.)* lower house *or* chamber; *(in Großbritannien)* House of Commons; Lower House; **~hemd das** vest *(Brit.)*; undershirt *(Amer.)*

unter·höhlen *tr. V.* hollow out; erode

unter-, Unter-: ~holz das; *o. Pl.* underwood; undergrowth; **~hose die** *(für Männer)* [under]pants *pl.*; *(für Frauen)* panties; knickers *(Brit.)*; briefs *pl.*; **~irdisch 1.** *Adj.* underground; **2.** *adv.* underground

unter·jochen *tr. V.* subjugate

unter-, Unter-: ~|jubeln *tr. V.* **jmdm. etw. ~jubeln** *(ugs.)* palm sth. off on sb.; **~kiefer der** lower jaw; **~|kommen** *unr. itr. V.; mit sein* **a)** *(Unterkunft finden)* find accommodation; **b)** *(ugs.: eine Stelle finden)* find *or* get a job; **c)** *(bes. südd., österr.: begegnen)* **so etwas ist mir noch nicht ~gekommen** I've never come across 'anything like it; **~körper der** lower part of the body; **~|kriegen** *tr. V.* **sich nicht ~kriegen lassen** *(ugs.)* not let things get one down

unterkühlt *Adj.* **a) ~ sein** be suffering from hypothermia *or* exposure; **b)** *(fig.)* dry, factual ⟨*style*⟩; cool ⟨*person*⟩; icy ⟨*tone*⟩

Unter·kühlung die reduction of body temperature

Unterkunft ['ʊntɐkʊnft] **die; ~, Unterkünfte** [...kʏnftə] accommodation *no indef. art.*; lodging *no indef. art.*; **~ und Frühstück** bed and breakfast; **~ und Verpflegung** board and lodging; **die Unterkünfte der Soldaten** the soldiers' quarters

Unter·lage die a) *(Schreib~, Matte o. ä.)* pad; *(für eine Schreibmaschine usw.)* mat; *(unter einer Matratze, einem Teppich)* underlay; *(zum Schlafen usw.)* base; **b)** *Pl. (Akten, Papiere)* documents; papers

Unter·laß der: ohne ~: incessantly

unter·lassen *unr. tr. V.* refrain from [doing]; *(versäumen)* omit, fail to [do]

Unterlassung die; ~, ~en omission; failure

unter·laufen 1. *unr. tr. V.; mit sein* occur; **jmdm. ist ein Fehler/ Irrtum ~:** sb. made a mistake. 2. *unr. tr. V.* evade; get round

[1]**unter|legen** *tr. V.* **a)** put under[neath]; **b) einem Text einen anderen Sinn ~:** read another meaning into a text

[2]**unter·legen** *tr. V.* **a)** underlay (**mit** with); **b) einem Film Musik ~:** put music to a film

[3]**unter·legen** *Adj.* inferior; **jmdm. zahlenmäßig ~ sein** be outnumbered by sb.

Unterlegene der/die; *adj. Dekl.* loser

Unter·leib der lower abdomen

unter·liegen *unr. itr. V.* **a)** *mit sein* lose; be beaten *or* defeated; **in einem Kampf ~:** lose a fight; **b) einer Sache** *(Dat.)* **~:** be subject to sth.

Unter·lippe die lower lip

unterm *Präp. + Art.* = **unter dem**

Unter·malung die; ~, ~en accompaniment (*Gen.* to)

unter·mauern *tr. V. (fig.)* back up; support

unter-, Unter-: ~|mengen *tr. V.* mix in; **~miete die;** *o. Pl.* subtenancy; sublease; **bei jmdm. zur ~miete wohnen** be sb.'s subtenant; lodge with sb.; **~mieter der** subtenant; lodger

unterminieren [ʊntɐmi'ni:rən] *tr. V.* undermine

untern *(ugs.) Präp. + Art.* = **unter den**

unter·nehmen *unr. tr. V.* **a)** undertake; make; make ⟨*attempt*⟩; take ⟨*steps*⟩; **b) etwas ~:** do something

Unter·nehmen das; ~s, ~ a) enterprise; venture; undertaking; **b)** *(Firma)* enterprise; concern

unternehmend *Adj.* enterprising; active

Unternehmer der; ~s, ~: employer; *(in der Industrie)* industrialist

unternehmerisch 1. *Adj.* entrepreneurial. 2. *adv.* ⟨*think*⟩ in an entrepreneurial *or* business-like way

Unternehmung die; ~, ~en *s.* **Unternehmen**

Unternehmungs·geist der; *o. Pl.* spirit of enterprise; **er war voller ~:** he was full of initiative

unternehmungs·lustig *Adj.* active

unter-, Unter-: ~offizier der a) non-commissioned officer; **b)** *(Dienstgrad)* corporal; **~|ordnen** 1. *tr. V.* subordinate; 2. *refl. V.* accept a subordinate role; **~ordnung die** subordination; **~prima die** *(Schulw. veralt.)* eighth year *(of a Gymnasium)*; **~privilegiert** *Adj. (geh.)* underprivileged; **~privilegierte der/die;** *adj. Dekl. (geh.)* underprivileged person; **~punkt der** subsidiary point

Unterredung die; ~, ~en discussion; **er bat ihn um eine ~:** he asked to see him to discuss something [with him]

Unterricht ['ʊntɐrɪçt] **der; ~[e]s, ~e** instruction; *(Schul~)* teaching; *(Schulstunden)* classes *pl.*; lessons *pl.*; **jmdm. ~ [in Musik** *usw.*] **geben** give sb. [music *etc.*] lessons; teach sb. [music]

unterrichten 1. *tr. V.* **a)** *(lehren)* teach; **b)** *(informieren)* inform (**über** + *Akk.* of, about). 2. *itr. V. (Unterricht geben)* teach. 3. *refl. V. (sich informieren)* inform oneself (**über** + *Akk.* about)

Unterrichts-: ~fach das subject; **~stoff der** subject-matter; **~stunde die** lesson; period

Unterrichtung die; ~, ~en instruction; *(Information)* information

Unter·rock der [half] slip

unter|rühren *tr. V.* stir in

unters *Präp. + Art.* = **unter das**

unter·sagen *tr. V.* forbid; prohibit

Unter·satz der *s.* **Untersetzer**

unter-, Unter-: ~schätzen *tr. V.* underestimate ⟨*amount, effect, meaning, distance, etc.*⟩; underrate ⟨*writer, performer, book, performance, talent, ability*⟩; **~scheiden** 1. *unr. tr. V.* distinguish; **die Zwillinge sind kaum zu ~scheiden** you can hardly tell the twins apart; 2. *unr. itr. V.* distinguish; differentiate; **zwischen Richtigem und Falschem ~scheiden** tell the difference between right and wrong; 3. *unr. refl. V.* differ (**durch** in, **von** from); **~scheidung die** *(Vorgang)* differentiation; *(Resultat)* distinction

unter-, Unter-: ~schenkel der shank; lower leg; **~schicht die** *(Soziol.)* lower class; [1]**~|schieben** *unr. tr. V.* push under[neath]

[2]**unter·schieben** *unr. tr. V.* **jmdm. etw. ~** *(fig.)* attribute sth. falsely to sb.

Unter·schied der; ~[e]s, ~e difference; **es ist [schon] ein [großer] ~, ob ...:** it makes a [big] difference whether ...; **ohne ~ der Rasse/des Geschlechts** without regard to *or* discrimination against race/sex; **im ~ zu ihm/ zum ~ von ihm** in contrast to him

unter·schieden *Adj.* different

unterschiedlich 1. *Adj.* different; *(uneinheitlich)* variable; varying. 2. *adv.* in different ways

Unterschiedlichkeit die; ~, ~en difference (*Gen.* between); *(Uneinheitlichkeit)* variability

unterschieds·los 1. *Adj.* uniform; equal ⟨*treatment*⟩. 2. *adv.* ⟨*treat*⟩ equally; *(ohne Benachteiligung)* without discrimination

unter·schlagen 1. *unr. tr. V.* embezzle, misappropriate ⟨*money, funds, etc.*⟩; *(unterdrükken)* intercept ⟨*letter*⟩; withhold, suppress ⟨*fact, news, information, etc.*⟩. 2. *unr. itr. V.* **er hat ~:** he embezzled money

Unterschlagung die; ~, ~en *s.* **unterschlagen:** embezzlement; misappropriation; withholding; suppression

Unter·schlupf der; ~[e]s, ~e shelter; *(Versteck)* hiding-place; hide-out

unter|schlüpfen *itr. V.; mit sein (ugs.)* hide out; *(Obdach finden)* take shelter (**vor** + *Dat.* from)

unter·schreiben 1. *unr. itr. V.* sign. 2. *unr. tr. V.* sign; *(fig.)* approve; subscribe to

unter·schreiten *unr. tr. V.* fall below

Unter·schrift die a) signature; **seine ~ unter etw.** *(Akk.)* **setzen** put one's signature to sth.; sign sth.; **b)** *(Bild~)* caption

unterschwellig [-ʃvɛlɪç] 1. *Adj.* subliminal. 2. *adv.* subliminally

unter-, Unter-: ~see·boot das submarine; **~seite die** underside; *(eines Stoffes)* wrong side; **~sekunda die** *(Schulw. veralt.)* sixth year *(of a Gymnasium)*; **~|setzen** *tr. V.* put underneath; **~setzer der** mat; *(für Gläser)* coaster

untersetzt *Adj.* stocky

unter·spülen *tr. V.* undermine and wash away

unterst ... *s.* **unter ...**

Unter·stand der a) *(Bunker)* dug-out; **b)** *(Unterschlupf)* shelter

unter·stehen 1. *unr. itr. V.* **jmdm. ~:** be subordinate *or* answerable to sb.; **jmdm. untersteht eine Abteilung** sb. is responsible for a department. 2. *unr. refl. V.* dare; **untersteh dich!** [don't] you dare!

[1]**unter|stellen** 1. *tr. V.* **a)** *(zur*

Aufbewahrung) keep; store ‹*furniture*›; **b)** *(unter etw.)* put underneath. **2.** *refl. V.* take shelter

²unter·stellen *tr. V.* **a)** *(jmdm. unterordnen, übertragen)* **jmdm. eine Abteilung ~:** put sb. in charge of a department; **die Behörde ist dem Ministerium unterstellt** the office is under the ministry; **b)** *(annehmen)* assume; **c)** *(unterschieben)* **jmdm. böse Absichten ~:** insinuate *or* imply that sb.'s intentions are bad

Unter·stellung die a) subordination **(unter** + *Akk.* to); **b)** *(falsche Behauptung)* insinuation

unter·streichen *unr. tr. V.* **a)** underline; **b)** *(fig.: hervorheben)* emphasize

Unter·streichung die; ~, ~en a) underlining; **b)** *(fig.)* emphasizing

Unter·stufe die *(Schulw.)* lower school

unter·stützen *tr. V.* support

Unter·stützung die a) support; **b)** *(finanzielle Hilfe)* allowance; *(für Arbeitslose)* [unemployment] benefit *no art.*; **staatliche ~:** state aid

unter·suchen *tr. V.* **a)** examine; **etw. auf etw.** *(Akk.)* **~:** test sth. for sth.; **sich ärztlich ~ lassen** have a medical examination *or* check-up; **b)** *(aufzuklären suchen)* investigate

Untersuchung die; ~, ~en a) *s.* **untersuchen:** examination; test; investigation; **b)** *(wissenschaftliche Arbeit)* study

Untersuchungs-: ~gefängnis das prison *(for people awaiting trial)*; **~haft die** imprisonment *or* detention while awaiting trial; **jmdn. in ~haft nehmen** commit sb. for trial

untertan [-ta:n] *Adj.* **sich** *(Dat.)* **jmdn./etw. ~ machen** *(geh.)* subjugate sb./dominate sth.

Untertan der; ~s *od.* **~en, ~en** subject

untertänig [-tɛ:nɪç] **1.** *Adj.* subservient; **Ihr ~ster Diener** *(veralt.)* your most obedient *or* humble servant. **2.** *adv.* subserviently

unter-, Unter-: ~tasse die saucer; **fliegende ~tasse** *(fig.)* flying saucer; **~tauchen 1.** *itr. V.; mit sein* **a)** dive [under]; **b)** *(fig.)* go underground; **2.** *tr. V.* duck; **~teil das** *od.* **der** bottom part

unter·teilen *tr. V.* subdivide; *(aufteilen)* divide

Unter·teilung die; ~, ~en [sub]division

Unter-: ~tertia die *(Schulw. veralt.)* fourth year *(of a Gymnasium)*; **~titel der** subtitle

unter·treiben *unr. itr. V.* play things down

Untertreibung die; ~, ~en understatement

unter-, Unter-: ~vermieten *tr., itr. V.* sublet; **~vermietung die** subletting; **~versorgung die** under-supply **(mit** of)

unter·wandern *tr. V.* infiltrate

Unter·wanderung die infiltration *no indef. art.*

Unter·wäsche die; *o. Pl.* underwear

unterwegs *Adv.* on the way; **~ sein** be on the *or* one's/its way **(nach** to); *(nicht zu Hause sein)* be out [and about]; **sie waren vier Wochen ~:** they travelled for four weeks

unter·weisen *unr. tr. V. (geh.)* instruct **(in** + *Dat.* in)

Unter·weisung die instruction

Unter·welt die; *o. Pl.* underworld

unter·werfen 1. *unr. tr. V.* **a)** subjugate ‹*people, country*›; **b)** *(unterziehen)* subject *(Dat.* to); **c) einer Sache** *(Dat.)* **unterworfen sein** be subject to sth. **2.** *unr. refl. V.* **sich** [**jmdm./einer Sache**] **~:** submit [to sb./sth.]

Unterwerfung die; ~, ~en a) *(das Unterwerfen)* subjugation **(unter** + *Akk.* to); **b)** *(das Sichunterwerfen)* submission **(unter** + *Akk.* to)

unterwürfig [-vʏrfɪç] *(abwertend)* **1.** *Adj.* obsequious. **2.** *adv.* obsequiously

Unterwürfigkeit die; ~ *(abwertend)* obsequiousness

unter·zeichnen *tr. V.* sign

Unterzeichnung die signing

¹unter|ziehen *unr. tr. V.* [**sich** *(Dat.)*] **etw. ~:** put sth. on underneath

²unter·ziehen 1. *unr. refl. V.* **sich einer Sache** *(Dat.)* **~:** undertake sth.; **sich einer Operation** *(Dat.)* **~:** undergo *or* have an operation. **2.** *unr. tr. V.* **etw. einer Untersuchung/Überprüfung** *(Dat.)* **~:** examine/check sth.

Untiefe die; ~, ~en shallow

Untier das monster

untragbar *Adj.* unbearable; intolerable

untrennbar *Adj.* inseparable

untreu *Adj.* **a)** disloyal; **jmdm. ~ werden** be disloyal to sb.; **seinen Grundsätzen ~ werden** abandon one's principles; **b)** *(in der Ehe, Liebe)* unfaithful

Untreue die a) disloyalty; **b)** *(in der Ehe, Liebe)* unfaithfulness

untröstlich *Adj.* inconsolable; **ich bin ~, daß ...:** I am extremely sorry that ...

untrüglich *Adj.* unmistakable

Untugend die bad habit

unüberlegt 1. *Adj.* rash. **2.** *adv.* rashly

unübersehbar *Adj.* **a)** *(offenkundig)* conspicuous; obvious; **b)** *(sehr groß)* enormous; immense

unübersichtlich 1. *Adj.* unclear; confusing ‹*arrangement*›; blind ‹*bend*›. **2.** *adv.* unclearly; confusingly ‹*arranged*›

unübertrefflich 1. *Adj.* superb. **2.** *adv.* superbly

unübertroffen *Adj.* unsurpassed

unüblich *Adj.* not usual *or* customary *pred.*; unusual

unumgänglich *Adj.* [absolutely] necessary

unumschränkt ['ʊn|ʊmʃrɛŋkt] *Adj.* absolute

unumwunden ['ʊn|ʊmvʊndn̩] **1.** *Adj.* frank. **2.** *adv.* frankly; openly

ununterbrochen 1. *Adj.* incessant. **2.** *adv.* incessantly

unveränderlich *Adj.* unchangeable, unchanging ‹*law, principle*›; constant ‹*quantity etc.*›; permanent ‹*mark*›

Unveränderlichkeit die; ~: *s.* **unveränderlich:** unchangeableness; unchangingness; constancy; permanence

unverändert *Adj.* unchanged ‹*appearance, weather, condition*›; unaltered, unrevised ‹*edition etc.*›

unverantwortlich. 1. *Adj.* irresponsible. **2.** *adv.* irresponsibly

unveräußerlich *Adj. (geh.)* inalienable ‹*rights, principles*›

unverbesserlich *Adj.* incorrigible

unverbindlich 1. *Adj.* not binding *pred.*; without obligation *postpos*; non-committal ‹*answer, words*›; detached, impersonal ‹*attitude, person*›. **2.** *adv.* ‹*send, reserve*› without obligation

Unverbindlichkeit die; ~, ~en a) *o. Pl.* freedom from obligation; *(einer Person)* detached *or* impersonal manner; **b)** *(Äußerung)* non-committal remark

unverblümt [ʊnfɛɐ̯'bly:mt] **1.** *Adj.* blunt. **2.** *adv.* bluntly

unverdächtig 1. *Adj.* free from suspicion *postpos.* **2.** *adv.* in a way that does/did not arouse suspicion

unverdaulich *Adj.* indigestible

unverdient 1. *Adj.* undeserved. **2.** *adv.* undeservedly

unverdorben *Adj.* unspoilt
unverdrossen *Adj.* undeterred; *(unverzagt)* undaunted
unverdünnt *Adj.* undiluted
unvereinbar *Adj.* incompatible
unverfälscht 1. *Adj.* genuine; unadulterated ‹*wine etc.*›; pure ‹*dialect*›; unaltered ‹*custom, text*›. **2.** *adv.* in pure/unaltered form
unverfänglich *Adj.* harmless
unverfroren 1. *Adj.* insolent; impudent. **2.** *adv.* insolently; impudently
Unverfrorenheit die; ~, ~en a) *o. Pl.* insolence; impudence; **b)** *(Äußerung)* insolent remark; impertinence
unvergänglich *Adj.* immortal ‹*fame*›; unchanging ‹*beauty*›; abiding ‹*recollection*›
Unvergänglichkeit die *s.* **unvergänglich:** immortality; unchangingness; abidingness
unvergeßlich *Adj.* unforgettable; **dieses Erlebnis wird mir ~ bleiben** *od.* **sein** I shall never forget this experience
unvergleichlich 1. *Adj.* incomparable. **2.** *adv.* incomparably
unverhältnismäßig *Adv.* unusually
unverheiratet *Adj.* unmarried
unverhofft ['ʊnfɛɐ̯hɔft] **1.** *Adj.* unexpected. **2.** *adv.* unexpectedly
unverhohlen 1. *Adj.* unconcealed. **2.** *adv.* openly
unverkäuflich *Adj.* **a)** *(nicht zum Verkauf bestimmt)* **diese Vase ist ~:** this vase is not for sale; **~es Muster** free sample; **b)** *(nicht absetzbar)* unsaleable
unverkennbar 1. *Adj.* unmistakable. **2.** *adv.* unmistakably
unvermeidlich *Adj.* **a)** *(nicht vermeidbar)* unavoidable; **b)** *(sich als Folge ergebend)* inevitable
unvermindert *Adj., adv.* undiminished
unvermittelt 1. *Adj.* sudden; abrupt. **2.** *adv.* suddenly; abruptly
Unvermögen das lack of ability; inability
unvermutet 1. *Adj.* unexpected. **2.** *adv.* unexpectedly
Unvernunft die stupidity
unvernünftig 1. *Adj.* stupid; foolish. **2.** *adv.* **er raucht/trinkt ~ viel** he smokes/drinks more than is good for him
unverrichtet *Adj.* **~er Dinge** without having achieved anything
unverschämt 1. *Adj.* **a)** impertinent, impudent ‹*person, manner, words, etc.*›; barefaced, blatant ‹*lie*›; **b)** *(ugs.: sehr groß)* outrageous ‹*price, luck, etc.*›. **2.** *adv.* **a)** impertinently; impudently; ‹*lie*› barefacedly; blatantly; **b)** *(ugs.: sehr)* outrageously ‹*expensive*›
Unverschämtheit die; ~, ~en a) *o. Pl.* impertinence; impudence; *(einer Lüge)* barefacedness; blatancy; **b)** *(Äußerung o. ä.)* [piece of] impertinence; **das ist eine ~!** that's outrageous!
unversehens *Adv.* suddenly
unversehrt *Adj.* unscathed; unhurt; *(unbeschädigt)* undamaged
Unversehrtheit die; ~: intactness
unversöhnlich *Adj.* irreconcilable
unverständlich *Adj.* incomprehensible; *(undeutlich)* unclear ‹*pronunciation, presentation, etc.*›; **es ist [mir] ~, warum ...:** I cannot *or* do not understand why ...
Unverständnis das lack of understanding
unversucht *Adj.* **nichts ~ lassen** try everything; leave no stone unturned
unverträglich *Adj.* **a)** *(unbekömmlich)* indigestible; unsuitable ‹*medicine*›; **b)** *(streitsüchtig)* quarrelsome; **c)** *(nicht harmonierend)* incompatible ‹*blood groups, medicines, transplant tissue*›
Unverträglichkeit die *s.* **unverträglich:** indigestibility; unsuitability; quarrelsomeness; incompatibility
unvertretbar *Adj.* unjustifiable
unverwechselbar *Adj.* unmistakable; distinctive
unverwundbar *Adj.* invulnerable
unverwüstlich *Adj.* indestructible; *(fig.)* irrepressible ‹*nature, humour*›; robust ‹*health*›
unverzeihlich *Adj.* unforgivable; inexcusable
unverzüglich 1. *Adj.* prompt; immediate. **2.** *adv.* promptly; immediately
unvollkommen 1. *Adj.* **a)** imperfect; **b)** *(unvollständig)* incomplete ‹*collection, account, etc.*›. **2.** *adv.* **a)** imperfectly; **b)** *(unvollständig)* incompletely
unvollständig *Adj.* incomplete
unvorbereitet *Adj.* unprepared
unvorhergesehen 1. *Adj.* unforeseen ‹*difficulty, event, expenditure*›; unexpected ‹*visit*›. **2.** *adv.* unexpectedly
unvorhersehbar *Adj.* unforeseeable
unvorsichtig 1. *Adj.* careless; *(unüberlegt)* rash. **2.** *adv.* carelessly; *(unüberlegt)* rashly
Unvorsichtigkeit die *s.* **unvorsichtig 1: a)** *o. Pl. (Art)* carelessness; rashness; **b)** *(Handlung usw.)* **eine ~ begehen** do sth. careless/rash
unvorstellbar 1. *Adj.* inconceivable; unimaginable. **2.** *adv.* unimaginably; **~ leiden** suffer terribly
unvorteilhaft *Adj.* **a)** *(nicht attraktiv)* unattractive ‹*figure, appearance*›; **b)** *(ohne Vorteil)* unfavourable, poor ‹*purchase, exchange*›; unprofitable ‹*business*›
unwahr *Adj.* untrue
Unwahrheit die a) *o. Pl.* untruthfulness; **b)** *(Äußerung)* untruth
unwahrscheinlich 1. *Adj.* **a)** improbable; unlikely; **b)** *(ugs.: sehr viel)* incredible *(coll.)*. **2.** *adv. (ugs.: sehr)* incredibly *(coll.)*
Unwahrscheinlichkeit die improbability
unweigerlich [ʊn'vaigɐlɪç] **1.** *Adj.* inevitable. **2.** *adv.* inevitably
unweit 1. *Präp. mit Gen.* not far from. **2.** *Adv.* not far (**von** from)
Unwesen das; *o. Pl.* dreadful state of affairs; **sein ~ treiben** *(abwertend)* be up to one's mischief *or* one's tricks
unwesentlich 1. *Adj.* unimportant; insignificant. **2.** *adv.* slightly; marginally
Unwetter das [thunder]storm
unwichtig *Adj.* unimportant
unwiderruflich 1. *Adj.* irrevocable. **2.** *adv.* irrevocably
unwidersprochen *Adj.* unchallenged
unwiderstehlich *Adj.* irresistible
Unwiderstehlichkeit die; ~: irresistibility
unwiederbringlich *(geh.)* **1.** *Adj.* irretrievable. **2.** *adv.* irretrievably
Unwille[n] der; *o. Pl.* displeasure; indignation
unwillig 1. *Adj.* indignant; angry; *(widerwillig)* unwilling; reluctant. **2.** *adv.* indignantly; angrily; *(widerwillig)* unwillingly; reluctantly
unwillkommen *Adj.* unwelcome
unwillkürlich 1. *Adj.* **a)** spontaneous ‹*cry, sigh*›; instinctive ‹*reaction, movement, etc.*›; **b)** *(Physiol.)* involuntary ‹*movement etc.*›. **2.** *adv.* **a)** ‹*shout etc.*› spontaneously; ‹*react, move, etc.*› instinctively; **b)** *(Physiol.)* ‹*move etc.*› involuntarily
unwirklich *(geh.) Adj.* unreal

unwirksam *Adj.* ineffective
Unwirksamkeit die ineffectiveness
unwirsch 1. *Adj.* surly; ill-natured. 2. *adv.* ill-naturedly
Unwissen das ignorance
Unwissenheit die; ~ ignorance; ~ **schützt nicht vor Strafe** ignorance is no defence
unwissenschaftlich *Adj.* unscientific
unwissentlich 1. *Adj.* unconscious. 2. *adv.* unknowingly; unwittingly
unwohl *Adv.* **a)** unwell; **mir ist ~:** I don't feel well; **b)** *(unbehaglich)* uneasy
Unwohlsein das; ~: indisposition; **ein heftiges ~ überkam ihn** he suddenly felt very unwell
unwürdig *Adj.* **a)** undignified ‹*person, behaviour*›; degrading ‹*treatment*›; **b)** *(unangemessen)* unworthy
Unzahl die; *o. Pl.* huge *or* enormous number
unzählbar *Adj.* uncountable
unzählig *Adj.* innumerable; countless
unzähmbar *Adj.* untameable
Unze ['ʊntsə] **die;** ~, ~n ounce
Unzeit die: zur ~ *(geh.)* at an inopportune moment
unzeitgemäß *Adj.* anachronistic
unzensiert *Adj.* **a)** uncensored; **b)** *(unbenotet)* unmarked; ungraded *(Amer.)*
unzerbrechlich *Adj.* unbreakable
unzerstörbar *Adj.* indestructible
unzertrennlich *Adj.* inseparable
unzivilisiert *Adj. (abwertend)* uncivilized
Unzucht die; *o. Pl. (veralt.)* ~ **mit Abhängigen** illicit sexual relations *pl.* with dependants; **widernatürliche ~:** unnatural sexual act[s]; **gewerbsmäßige ~:** prostitution
unzüchtig 1. *Adj.* obscene ‹*letter, gesture*›. 2. *adv.* ‹*touch, approach, etc.*› indecently; ‹*speak*› obscenely
unzufrieden *Adj.* dissatisfied; *(stärker)* unhappy
Unzufriedenheit die dissatisfaction; *(stärker)* unhappiness
unzugänglich *Adj.* inaccessible; *(fig.)* unapproachable ‹*character, person, etc.*›
unzulänglich *(geh.)* 1. *Adj.* insufficient; inadequate. 2. *adv.* insufficiently; inadequately
Unzulänglichkeit die; ~, ~en **a)** *o. Pl.* insufficiency; inadequacy; **b)** *(etw. Unzulängliches)* inadequacy; shortcoming
unzulässig *Adj.* inadmissible; undue ‹*influence, interference, delay*›; improper ‹*method, use, etc.*›
unzumutbar *Adj.* unreasonable
Unzumutbarkeit die; ~ unreasonableness
unzurechnungsfähig *Adj.* not responsible for one's actions *pred.; (geistesgestört)* of unsound mind *postpos;* **für ~ erklärt werden** be certified insane
unzureichend *Adj.* insufficient; inadequate
unzustellbar *Adj. (Postw.)* undeliverable ‹*mail*›; **„falls ~, bitte zurück an Absender"** 'if undelivered, please return to sender'
unzutreffend *Adj.* inappropriate; inapplicable; *(falsch)* incorrect; **„Unzutreffendes bitte streichen"** 'please delete as appropriate'
unzuverlässig *Adj.* unreliable
Unzuverlässigkeit die unreliability
unzweckmäßig 1. *Adj.* unsuitable; *(unpraktisch)* impractical. 2. *adv.* unsuitably; *(unpraktisch)* impractically
unzweifelhaft 1. *Adj.* unquestionable; undoubted. 2. *adv.* unquestionably; undoubtedly
üppig 1. *Adj.* lush, luxuriant ‹*vegetation*›; thick ‹*hair, beard*›; sumptuous, opulent ‹*meal*›; full ‹*bosom*›; voluptuous ‹*figure, woman*›. 2. *adv.* luxuriantly; ‹*dine, eat*› sumptuously
Üppigkeit die; ~ *s.* **üppig 1:** lushness; luxuriance; thickness; sumptuousness; opulence; fullness; voluptuousness
Ur·abstimmung die [*esp.* strike] ballot
Ur·ahn[e] der oldest known ancestor
ur·alt *Adj.* very old; ancient
Uran [u'ra:n] **das;** ~s *(Chemie)* uranium
ur·auf·führen *tr. V.* première, give the first performance of ‹*play, concerto, etc.*›; première ‹*film*›
Ur·auf·führung die première; first night *or* performance; *(eines Films)* première; first showing
urbar *Adj.* ~ **machen** cultivate ‹*land*›; reclaim ‹*swamp,desert*›
Urbarmachung die; ~ *s.* **urbar:** cultivation; reclamation
Ur·bevölkerung die native population; native inhabitants *pl.*
Ur·bild das a) *(Vorbild)* archetype; prototype; **b)** *(Inbegriff, Ideal)* perfect example; epitome
ur·eigen *Adj.; nicht präd.* very own; **seine ~en Interessen** his own best interests
Ur·einwohner der native inhabitant; **die australischen ~:** the Australian Aborigines
Ur·enkel der great-grandson
Ur·enkelin die great-granddaughter
ur·gemütlich *(ugs.)* 1. *Adj.* extremely cosy; *(bequem)* extremely comfortable. 2. *adv.* extremely cosily/comfortably
Urgroß-: ~eltern *Pl.* great-grandparents; **~mutter die** great-grandmother; **~vater der** great-grandfather
Ur·heber der; ~s, ~ **a)** originator; initiator; **b)** *(bes. Rechtsspr.: Verfasser, Autor)* author
urig ['u:rɪç] *Adj.* natural ‹*person*›; real ‹*beer*›; cosy ‹*pub*›
Urin [u'ri:n] **der;** ~s urine
urinieren *itr. V.* urinate
ur·komisch *Adj.* extremely funny; hilarious
Ur·kunde die; ~, ~n document; *(Bescheinigung, Sieger~, Diplom~ usw.)* certificate
Urlaub der; ~[e]s, ~e holiday[s] *(Brit.);* vacation; *(bes. Milit.)* leave; ~ **haben** have a holiday/have leave; **auf** *od.* **in** *od.* **im ~ sein** be on holiday/leave; **in ~ gehen/fahren** go on holiday
Urlauber der; ~s, ~, **Urlauberin die;** ~, ~nen holiday-maker
Urlaubs-: ~ort der; *Pl.* **~orte** holiday resort; **~reise die** holiday [trip]; **eine ~reise ans Meer/ins Gebirge machen** go on holiday to the seaside/go for a holiday in the mountains; **~sperre die a)** *(Milit.)* ban on leave; **b)** *(österr.)* holiday closure; **~tag der** day of holiday; **~zeit die** holiday period *or* season
Urne ['ʊrnə] **die;** ~, ~n **a)** urn; **b)** *(Wahl~)* [ballot-]box; **c)** *(Verlosungs~)* box; *(Lostrommel)* drum
Urologe [uro'lo:gə] **der;** ~n, ~n, **Urologin die;** ~, ~nen urologist
Ur·oma die *(fam.)* great-granny *(coll./child lang.)*
Ur·opa der *(fam.)* great-grandpa *(coll./child lang.)*
Ur·sache die cause **(für** of); **keine ~!** don't mention it; you're welcome
ur·sächlich *Adj.* causal; **in ~em Zusammenhang stehen** be causally related **(mit** to)
Ur·schrei der *(Psych.)* primal scream
Ur·sprung der origin; **seinen ~ in etw.** *(Dat.)* **haben** originate from sth.
ur·sprünglich 1. *Adj.* **a)** original

⟨*plan, price, form, material, etc.*⟩; initial ⟨*reaction, trust, mistrust, etc.*⟩; **b)** *(natürlich)* natural. **2.** *adv.* **a)** originally; initially; **b)** *(natürlich)* naturally

Ursprünglichkeit die; ~: naturalness

Urteil das; ~**s,** ~**e** judgement; *(Meinung)* opinion; *(Strafe)* sentence; *(Gerichts~)* verdict; **das** ~ **lautete auf 10 Jahre Freiheitsstrafe** the sentence was ten years' imprisonment

urteilen *itr. V.* form an opinion; judge; **über etw./jmdn.** ~: judge sth./sb.; give one's opinion on sth./sb.; **fachmännisch** ~: give an expert opinion

urteils-, Urteils-: ~**fähig** *Adj.* competent *or* able to judge *postpos.*; ~**fähigkeit die;** *o. Pl.* competence *or* ability to judge; ~**kraft die;** *o. Pl.* [power of] judgement; ~**vermögen das;** *o. Pl.* competence *or* ability to judge

Ur·text der original

urtümlich ['u:ɐ̯ty:mlɪç] *Adj.* natural ⟨*landscape etc.*⟩; primitive ⟨*culture etc.*⟩

Ur·ur-: ~**enkel der** great-great-grandson; ~**enkelin** *die* great-great-granddaughter; ~**groß·mutter die** great-great-grandmother; ~**groß·vater der** great-great-grandfather

Ur·wald der primeval forest; **tropischer** ~: tropical forest; jungle

ur·wüchsig ['u:ɐ̯vy:ksɪç] *Adj.* natural ⟨*landscape, power*⟩; earthy ⟨*language, humour*⟩

Urwüchsigkeit die; ~ *s.* **urwüchsig:** naturalness; earthiness

Ur·zeit die primeval times *pl.*; **vor** ~**en** in ages past; **seit** ~**en** since primeval times; *(ugs.: seit längerer Zeit)* since the year dot *(coll.)*

USA [u:|ɛs'|a:] *Pl.* USA

Usurpator [uzʊr'pa:tɔr] **der;** ~**s,** ~**en** [...pa'to:rən] usurper

usw. *Abk.* **und so weiter** etc.

Utensil [utɛn'zi:l] **das;** ~**s,** ~**ien** [... i̯ən] piece of equipment; ~**ien** equipment *sing.*

Utopie [uto'pi:] **die;** ~, ~**n** Utopia; *(Idealvorstellung)* Utopian dream

utopisch *Adj.* Utopian

UV *Abk.* **Ultraviolett** UV

UV-Strahlen *Pl.* UV rays; ultraviolet rays

Ü-Wagen der *(Rundf., Ferns.)* OB van or vehicle

u. Z. *Abk. (bes. ehem. DDR)* **unserer Zeitrechnung** AD

V

v, V [fau̯] **das;** ~, ~: v, V

v. *Abk.* **von** *(in Familiennamen)* von

V *Abk.* **Volt** V

Vagabund [vaga'bʊnt] **der;** ~**en,** ~**en** *(veralt.)* vagabond

vagabundieren *itr. V.* **a)** live as a vagabond/as vagabonds; **b)** *mit sein (umherziehen)* wander *or* travel around

vage 1. *Adj.* vague. **2.** *adv.* vaguely

Vagina [va'gi:na] **die;** ~, **Vaginen** *(Anat.)* vagina

vakant [va'kant] *Adj.* vacant

Vakanz [va'kants] **die;** ~, ~**en** *(geh.)* vacancy

Vakuum ['va:kuʊm] **das;** ~**s, Vakuen** [... kuən] *(auch fig.)* vacuum

vakuum·verpackt *Adj.* vacuum-packed

Valentins·tag ['va:lɛnti:ns-] **der** [St] Valentine's Day

Valuta [va'lu:ta] **die;** ~, **Valuten** *(Finanzw.)* foreign currency

Vamp [vɛmp] **der;** ~**s,** ~**s** vamp

Vampir ['vampi:ɐ̯] **der;** ~**s,** ~**e** vampire

Vanille [va'nɪljə] **die;** ~: vanilla

Vanille-: ~**eis das** vanilla ice-cream; ~**pudding der** vanilla pudding; ~**zucker der** vanilla sugar

variabel [va'ri̯a:bl̩] **1.** *Adj.* variable. **2.** *adv.* variably

Variable [va'ri̯a:blə] **die;** *adj. Dekl. (Math., Physik)* variable

Variante [va'ri̯antə] **die;** ~, ~**n** *(geh.)* variant; variation

Variation [varia'tsi̯o:n] **die;** ~, ~**en** *(auch Musik)* variation (*Gen.*, **über, zu** on)

Varieté [vari̯e'te:] **das;** ~**s,** ~**s** variety theatre; *(Aufführung)* variety show

variieren *tr., itr. V.* vary

Vase ['va:zə] **die;** ~, ~**n** vase

Vater ['fa:tɐ] **der;** ~**s, Väter** ['fɛ:tɐ] **a)** father; **er ist** ~ **von drei Kindern** he is the father of three children; **er ist der [geistige]** ~ **dieser Idee** *(fig.)* he thought up this idea; this idea is his; ~ **Staat** *(scherzh.)* the State; **Heiliger** ~ *(kath. Kirche)* Holy Father; **b)** *(Tier)* sire; **c)** *o. Pl. (Rel.)* Father; **Gott** ~: God the Father

Vater-: ~**freuden** *Pl.* ~**freuden entgegensehen** *(meist scherzh.)* be expecting a happy event; be going to be a father; ~**haus das** *(geh.)* parental home; ~**land das;** *Pl.* ~**länder** fatherland

väterlich ['fɛ:tɐlɪç] **1.** *Adj.* **a)** *nicht präd.* the/one's father's; paternal ⟨*line, love, instincts, etc.*⟩; **b)** *(fürsorglich)* fatherly. **2.** *adv.* in a fatherly way

väterlicherseits *Adv.* on the/one's father's side; **meine Großeltern** ~: my paternal grandparents; my grandparents on my father's side

vater-, Vater-: ~**schaft die;** ~, ~**en** fatherhood; *(bes. Rechtsw.)* paternity; ~**schafts·klage die** paternity suit; ~**stadt die** *(geh.)* home town; ~**stelle die: bei** *od.* **an jmdm.** ~**stelle vertreten** take the place of a father to sb.; ~**tag der** Father's Day *no def. art.*; ~**unser das;** ~**s,** ~: Lord's Prayer

Vati ['fa:ti] **der;** ~**s,** ~**s** *(fam.)* dad[dy] *(coll.)*

Vatikan [vati'ka:n] **der;** ~**s** Vatican

vatikanisch *Adj.; nicht präd.* Vatican

Vatikan·stadt die; *o. Pl.* Vatican City

V-Ausschnitt ['fau̯-] **der** V-neck

v. Chr. *Abk.* **vor Christus** BC

VEB [fau̯e:'be:] *Abk.* **Volkseigener Betrieb**

Vegetarier [vege'ta:ri̯ɐ] **der;** ~**s,** ~: vegetarian

vegetarisch 1. *Adj.* vegetarian. **2.** *adv.* **er lebt** *od.* **ernährt sich** ~: he is a vegetarian; he lives on a vegetarian diet

Vegetation [...'tsi̯o:n] **die;** ~, ~**en** vegetation *no indef. art.*

vegetieren *itr. V.* vegetate

vehement [vehe'mɛnt] *(geh.)* **1.** *Adj.* vehement. **2.** *adv.* vehemently

Vehemenz [...'mɛnts] **die;** ~ *(geh.)* vehemence

Vehikel [ve'hi:kl̩] **das;** ~**s,** ~ *(oft abwertend, auch fig. geh.)* vehicle; **ein altes/klappriges** ~: an old crock *(sl.)*

Veilchen ['failçən] **das;** ~**s,** ~: violet

veilchen·blau *Adj.* violet

Vektor ['vɛktɔr] **der;** ~**s,** ~**en** [-'to:rən] *(Math., Physik)* vector

Velo ['ve:lo] **das; ~s, ~s** *(schweiz.)* bicycle; bike *(coll.)*
Vene ['ve:nə] **die; ~, ~n** *(Anat.)* vein
Venedig [ve'ne:dıç] **(das); ~s** Venice
Venezianer [vene'tsi̯a:nɐ] **der; ~s, ~:** Venetian
venezi̱anisch *Adj.* Venetian
Ventil [vɛn'ti:l] **das; ~s, ~e** valve; *(fig.)* outlet; **b)** *(der Orgel)* pallet
Ventilation [vɛntila'tsi̯o:n] **die; ~, ~en** ventilation
Ventilator [vɛnti'la:tɔr] **der; ~s, ~en** [...la'to:rən] ventilator
Venus ['ve:nʊs] **die; ~:** Venus *no def. art.*
verạbreden 1. *tr. V.* arrange; **am verabredeten Ort** at the agreed place. **2.** *refl. V.* **sich im Park/zum Tennis ~:** arrange to meet in the park/for tennis; **sich mit jmdm. ~:** arrange to meet sb.; **ich bin mit ihm verabredet** I am meeting him
Verạbredung die; ~, ~en a) *(Absprache)* arrangement; **eine ~ treffen** arrange to meet *or* a meeting; **b)** *(verabredete Zusammenkunft)* appointment; **eine ~ absagen** call off a meeting *or* an engagement; **ich habe eine ~:** I am meeting somebody; *(formell)* I have an appointment; *(mit meinem Freund/meiner Freundin)* I have a date *(coll.)*
verạbreichen *tr. V.* administer ⟨*medicine*⟩; give ⟨*injection, thrashing*⟩
Verạbreichung die; ~, ~en administration; administering
verạbscheuen *tr. V.* detest; loathe
verạbscheuenswürdig *Adj.* *(geh.)* detestable; loathsome
verạbschieden 1. *tr. V.* **a)** say goodbye to; **b)** *(aus dem Dienst)* retire; **c)** *(annehmen)* adopt ⟨*plan, budget*⟩; pass ⟨*law*⟩. **2.** *refl. V.* **sich [von jmdm.] ~:** say goodbye [to sb.]; *(formell)* take one's leave [of sb.]
Verạbschiedung die; ~, ~en a) leave-taking; **b)** *(aus dem Dienst)* retirement; **c)** *(eines Plans, Etats)* adoption; *(eines Gesetzes)* passing
verạchten *tr. V.* despise
Verächter [fɛɐ̯'|ɛçtɐ] **der; ~s, ~:** opponent; critic
verächtlich [fɛɐ̯'|ɛçtlıç] **1.** *Adj.* **a)** *(abschätzig)* contemptuous; **b)** *(verachtenswürdig)* contemptible; despicable; **jmdn./etw. ~ machen** disparage sb./sth.; run sb./sth. down. **2.** *adv.* contemptuously
Verạ̈chtlichkeit die; ~: contempt; contemptuousness
Verạchtung die; ~: contempt; **jmdn. mit ~ strafen** treat sb. with contempt
verạlbern *tr. V.* make fun of
verallgeme̱inern *tr., itr. V.* generalize
Verallgeme̱inerung die; ~, ~en generalization
verạlten *itr. V.; mit sein* become obsolete; **veraltete Methoden** obsolete *or* antiquated methods
Veranda [vɛ'randa] **die; ~, Verạnden** veranda; porch
verạ̈nderlich *Adj.* **a)** changeable ⟨*weather*⟩; variable ⟨*character, star*⟩; **b)** *(veränderbar)* variable
Verạ̈nderlichkeit die; ~, ~en *s.* **veränderlich:** changeability; variability
verạ̈ndern 1. *tr. V.* change; **der Bart verändert ihn stark** the beard makes him look very different. **2.** *refl. V.* **a)** change; **sich zu seinem Vorteil/Nachteil ~:** change for the better/worse; **b)** *(die Stellung wechseln)* **sich [beruflich] ~:** change one's job
Verạ̈nderung die change *(Gen.* in); **an etw.** *(Dat.)* **eine ~ vornehmen** change sth.
verạ̈ngstigen *tr. V.* frighten; scare
verạnkern *tr. V.* fix ⟨*tent, mast, pole, etc.*⟩; *(mit einem Anker)* anchor; *(fig.)* embody ⟨*right etc.*⟩
Verạnkerung die; ~, ~en a) fixing; *(mit einem Anker)* anchoring; *(fig.)* embodiment; **b)** *(Halterung)* anchorage; fixture
verạnlagen *tr.V.* *(Steuerw.)* assess **(mit** at)
verạnlagt *Adj.* **künstlerisch/praktisch/romantisch ~ sein** have an artistic bent/be practically minded/have a romantic disposition; **ein homosexuell ~er Mann** a man with homosexual tendencies
Verạnlagung die; ~, ~en a) [pre]disposition; **seine homosexuelle/künstlerische/praktische/romantische ~:** his homosexual tendencies *pl.*/artistic bent/practical nature/romantic disposition; **b)** *(Steuerw.)* assessment
verạnlassen *tr. V.* **a)** cause; induce; **b)** *(erledigen lassen)* **etw. ~:** see to it that sth. is done *or* carried out; **ich werde alles Weitere/das Nötige ~:** I will take care of *or* see to everything else/I will see [to it] that the necessary steps are taken
Verạnlassung die; ~, ~en a) reason, cause **(zu** for); **b) auf jmds. ~ [hin]** on sb.'s orders
verạnschaulichen *tr. V.* illustrate
Verạnschaulichung die; ~, ~en illustration
verạnschlagen *tr. V.* estimate **(mit** at); **etw. zu hoch/niedrig ~:** overestimate/underestimate sth.
Verạnschlagung die; ~, ~en estimate
verạnstalten *tr. V.* **a)** organize; hold, give ⟨*party*⟩; **b)** *(ugs.)* make ⟨*noise, fuss*⟩
Verạnstalter der; ~s, ~, Verạnstalterin die; ~, ~nen organizer
Verạnstaltung die; ~, ~en a) *(das Veranstalten)* organizing; organization; **b)** *(Ereignis)* event
verạntworten 1. *tr. V.* take responsibility for; **ich kann das vor Gott/mir selbst/meinem Gewissen nicht ~:** I cannot be responsible for that before God/I cannot justify it to myself/I cannot square that with my conscience. **2.** *refl. V.* **sich für etw. ~:** answer for sth.; **sich vor jmdm. ~:** answer to sb.; **sich vor Gericht ~:** answer to the courts
verạntwortlich *Adj.* responsible; **jmdn. für etw. ~ machen** hold sb. responsible for sth.
Verạntwortlichkeit die; ~, ~en responsibility
Verạntwortung die; ~, ~en responsibility **(für** for); **die ~ für etw. übernehmen** take *or* accept [the] responsibility for sth.; **ich tue es auf deine ~:** you must take responsibility; on your own head be it; **jmdn. [für etw.] zur ~ ziehen** call sb. to account [for sth.]
verạntwortungs-, Verạntwortungs-: ~bewußt *Adj.* responsible; **~bewußtsein das;** *o. Pl.*, **~gefühl das;** *o. Pl.* sense of responsibility; **~los** *Adj.* irresponsible; **~losigkeit die; ~:** irresponsibility; **~voll** *Adj.* responsible
veräppeln [fɛɐ̯'ɛpl̩n] *tr. V.* **jmdn ~** *(ugs.)* have *(Brit. coll.) or (Amer. coll.)* put sb. on
verạrbeiten *tr. V.* **a)** use; **etw. zu etw. ~:** make sth. into sth.; use sth. to make sth.; **b)** *(verdauen)* digest ⟨*food*⟩; **c)** *(geistig bewältigen)* digest, assimilate ⟨*film, experience, impressions*⟩; come to terms with ⟨*disappointment*⟩
verạrbeitet *Adj.* **gut/schlecht** *usw.* **~:** well/badly *etc.* finished ⟨*suit, dress, car, etc.*⟩
Verạrbeitung die; ~, ~en a) use; **b)** *(Art der Fertigung)* finish
verạ̈rgern *tr. V.* annoy
Verạ̈rgerung die; ~, ~en annoyance
verạrmen *itr. V.; mit sein* become poor *or* impoverished

verarschen *tr. V.* **jmdn.** ~ *(derb)* take the piss *(coarse) or (Brit. sl.)* mickey out of sb.
verarzten *tr. V. (ugs.)* patch up *(coll.)* ‹*person*›; fix *(coll.)* ‹*wound, injury, etc.*›
verästeln [fɛɐ̯'ɛstl̩n] *refl. V.* branch out
Veräst[e]lung die; ~, ~**en** ramification
verätzen *tr. V.* corrode ‹*metal etc.*›; burn ‹*skin, face, etc.*›
Verätzung die corrosion; *(der Haut)* burn
verausgaben 1. *tr. V. (Papierdt.)* spend. **2.** *refl. V.* wear oneself out
veräußern *tr. V.* dispose of, sell ‹*property*›
Verb [vɛrp] **das;** ~**s,** ~**en** verb
verbal [vɛr'ba:l] *Adj.* **1.** *(auch Sprachw.)* verbal. **2.** *adv.* verbally
Verband der a) *(Binde)* bandage; dressing; **b)** *(Vereinigung)* association
Verband-: ~**kasten der** first-aid box; ~**material das** dressing materials *pl.*; ~**päckchen das** packet of dressings; ~**zeug das** first-aid things *pl.*
verbannen *tr. V. (auch fig.)* banish
Verbannung die; ~, ~**en** banishment; exile
verbarrikadieren 1. *tr. V.* barricade. **2.** *refl. V.* barricade oneself
verbauen *tr. V.* **a)** *(versperren)* obstruct; block; **sich die Zukunft** ~ *(fig.)* spoil one's prospects for the future; **b)** *(zum Bauen verwenden)* use
verbeißen *unr. tr. V.* suppress; hold back ‹*tears etc.*›
verbergen *unr. tr. V.* **a)** *(auch fig.)* hide; conceal; **sich** ~: hide; **b)** *(verheimlichen)* hide; **jmdm. etw.** ~, **etw. vor jmdm.** ~: keep sth. from sb.
verbessern 1. *tr. V.* **a)** improve ‹*machine, method, quality*›; improve [up]on, better ‹*achievement*›; beat ‹*record*›; reform ‹*schooling, world*›; **b)** *(korrigieren)* correct. **2.** *refl. V.* **a)** improve; **b)** *([beruflich] aufsteigen)* better oneself
Verbesserung die a) improvement; **eine** ~ **der Lage** an improvement in the situation; **b)** *(Korrektur)* correction
verbeugen *refl. V.* bow (**vor** + *Dat.* to)
Verbeugung die; ~, ~**en** bow; **eine** ~ **vor jmdm. machen** bow to sb.
verbeulen *tr. V.* dent
verbiegen 1. *unr. tr. V.* bend. **2.** *unr. refl. V.* bend; buckle
verbieten 1. *unr. tr. V.* **a)** forbid; **jmdm. etw.**~: forbid sb. sth.; **sie hat ihm das Haus verboten** she forbade him to enter the house; **„Betreten des Rasens/Rauchen verboten"!** 'keep off the grass'/'no smoking'; **b)** *(für unzulässig erklären)* ban. **2.** *unr. refl. V.* **sich [von selbst]** ~: be out of the question
verbilligen 1. *tr. V.* bring down *or* reduce the cost of; bring down *or* reduce the price of, reduce ‹*goods*›. **2.** *refl. V.* become *or* get cheaper; ‹*goods*› come down in price, become *or* get cheaper
verbinden 1. *unr. tr. V.* **a)** *(bandagieren)* bandage; dress; **jmdm./sich den Fuß** ~: bandage *or* dress sb.'s/one's foot; **jmdn./sich** ~: dress sb.'s/one's wounds; **b)** *(zubinden)* bind; **jmdm. die Augen** ~: blindfold sb.; **mit verbundenen Augen** blindfold[ed]; **c)** *(zusammenfügen)* join ‹*wires, lengths of wood, etc.*›; join up ‹*dots*›; **d)** *(zusammenhalten)* hold ‹*parts*› together; **e)** *(in Beziehung bringen)* connect (**durch** by); link ‹*towns, lakes, etc.*› (**durch** by); **f)** *(verknüpfen)* combine ‹*abilities, qualities, etc.*›; **die damit verbundenen Anstrengungen/Kosten** *usw.* the effort/cost *etc.* involved; **g)** *auch itr. (telefonisch)* **jmdn. [mit jmdm.]** ~: put sb. through [to sb.]; **Moment, ich verbinde** one moment, I'll put you through; **falsch verbunden sein** have got the wrong number; **h)** *auch itr.* **er war ihr freundschaftlich verbunden** he was bound to her by ties of friendship; **uns verbindet nichts mehr** nothing holds us together any longer; **i)** *(assoziieren)* associate (**mit** with). **2.** *unr. refl. V.* **a)** *(auch Chemie)* combine (**mit** with); **b)** *(sich zusammentun)* join [together]; join forces; **c)** *(in Gedanken)* be associated (**mit** with)
verbindlich 1. *Adj.* **a)** *(freundlich)* friendly; *(entgegenkommend)* forthcoming; ~**sten Dank!** *(geh.)* a thousand thanks; **b)** *(bindend)* obligatory; compulsory; binding ‹*agreement, decision, etc.*›. **2.** *adv.* **a)** *(freundlich)* in a friendly manner; *(entgegenkommend)* in a forthcoming manner; **b)** ~ **zusagen** definitely agree; **jmdm. etw.** ~ **anbieten** make sb. a firm offer of sth.
Verbindlichkeit die; ~, ~**en a)** *o. Pl. s.* **verbindlich 1 a:** friendliness; forthcomingness; **b)** *o. Pl.: s.* **verbindlich 1 b:** obligatory *or* compulsory nature; **c)** *Pl. (Kaufmannsspr.: Schulden)* liabilities (**gegen** to)
Verbindung die a) *(das Verknüpfen)* linking; **b)** *(Zusammenhalt)* join; connection; **c)** *(verknüpfende Strecke)* link; **d)** *(durch Telefon, Funk)* connection; **keine** ~ **mit jmdm. bekommen** not be able to get through to sb.; **e)** *(Verkehrs~)* connection (**nach** to); **die** ~ **zur Außenwelt** connections *pl.* with the outside world; **f)** *(Kombination)* combination; **in** ~ **mit etw.** in conjunction with sth.; **g)** *(Bündnis)* association; **eheliche** ~ *(geh.)* marriage; **h)** *(Kontakt)* contact; **sich mit jmdm. in** ~ **setzen,** ~ **mit jmdm. aufnehmen** get in touch *or* contact with sb.; contact sb.; **in** ~ **bleiben** keep in touch; **seine** ~**en spielen lassen** pull a few strings *(coll.)*; **i)** *(Zusammenhang)* connection; **jmdn. mit etw. in** ~ **bringen** connect sb. with sth.; **j)** *(Studenten~)* society; **k)** *(Chemie)* compound
Verbindungs-: ~**mann der;** *Pl.* ~**männer** *od.* ~**leute** intermediary; *(Agent)* contact [man] (**zu** with); ~**tür die** connecting door
verbissen 1. *Adj.* **a)** *(hartnäckig)* dogged; doggedly determined; **b)** *(verkrampft)* grim. **2.** *adv.* **a)** *(hartnäckig)* doggedly; with dogged determination; **b)** *(verkrampft)* grimly
Verbissenheit die; ~: doggedness; dogged determination
verbitten *unr. refl. V.* **sich** *(Dat.)* **etw.** ~: refuse to tolerate sth.; **ich verbitte mir diesen Ton** I will not be spoken to in that tone of voice
verbittern *tr. V.* embitter; make bitter; **verbittert** embittered; bitter
Verbitterung die; ~, ~**en** bitterness; embitterment
verblassen *itr. V.; mit sein (auch fig. geh.)* fade
Verbleib der; ~**[e]s** *(geh.)* **a)** *(Ort)* whereabouts *pl.*; **b)** *(das Verbleiben)* staying; **ein weiterer** ~: a longer stay
verbleiben *unr. itr. V.; mit sein* **a)** *(sich einigen)* **wie seid ihr denn nun verblieben?** what did you arrange?; **wir sind so verblieben, daß er sich bei mir meldet** we left it that he would contact me; **b)** *(geh.: bleiben)* remain; stay; **c)** *(im Briefschluß)* remain; **ich verbleibe mit freundlichen Grüßen Ihr ...:** I remain, Yours sincerely, ...; **d)** *(übrigbleiben)* remain; **etw. verbleibt jmdm.** sb. has sth. left

verbleichen *unr. od. regelm. itr. V.; mit sein (auch fig.)* fade

verblendet *Adj.* blind

Verblendung die; ~, ~**en** blindness

verblöden *itr. V.; mit sein (ugs.)* become a zombie *(coll.)*

verblüffen [fɛɐ̯'blʏfn̩] *tr., auch itr. V.* astonish; amaze; astound; *(verwirren)* baffle

verblüffend 1. *Adj.* astonishing; amazing; astounding. **2.** *adv.* astonishingly; amazingly; astoundingly

Verblüffung die; ~, ~**en** astonishment; amazement

verblühen *itr. V.; mit sein (auch fig.)* fade

verbluten *itr., auch refl. V.; mit sein* bleed to death

verbohren *refl. V. (ugs.)* become obsessed (**in** + *Akk.* with)

verbohrt *Adj. (abwertend)* pig-headed; stubborn; obstinate; *(unbeugsam)* inflexible

verborgen *Adj.* hidden; **es wird ihm nicht ~ bleiben** he shall hear of it; *(nicht entgehen)* it will not escape his notice; **im Verborgenen** out of the public eye

Verbot das; ~[e]s, ~e ban (*Gen.*, **von** on); **trotz ärztlichen ~s** against doctor's orders

Verbots·schild das; *Pl.* ~**schilder** *(Verkehrsw.)* prohibitive sign

Verbrauch der; ~[e]s consumption (**von, an** + *Dat.* of); **zum alsbaldigen ~ bestimmt** for immediate consumption

verbrauchen *tr. V.* **a)** use; consume ⟨*food, drink*⟩; use up ⟨*provisions*⟩; spend ⟨*money*⟩; consume, use ⟨*fuel*⟩; *(fig.)* use up ⟨*strength, energy*⟩; **das Auto verbraucht 10 Liter [auf 100 Kilometer]** the car does 10 kilometres to the litre; **b)** *(verschleißen)* wear out ⟨*clothing, shoes, etc.*⟩; **die Luft in den Räumen ist verbraucht** the air in the rooms is stale

Verbraucher der; ~s, ~, **Verbraucherin die;** ~, ~**nen** consumer

verbrechen *unr. tr. V. (scherzh.)* **ich habe nichts verbrochen!** I haven't been up to *or* haven't done anything!; **wer hat denn dieses Gedicht verbrochen?** who's responsible for *or* who's the perpetrator of this poem?

Verbrechen das; ~s, ~ crime (**an** + *Dat.*, **gegen** against)

Verbrecher der; ~s, ~: criminal

Verbrecher·bande die gang *or* band of criminals

Verbrecherin die; ~, ~**nen** criminal

verbrecherisch *Adj.* criminal

verbreiten 1. *tr. V.* **a)** *(bekannt machen)* spread ⟨*rumour, lies, etc.*⟩; **etw. über den Rundfunk ~:** broadcast sth. on the radio; **b)** *(weitertragen)* spread ⟨*disease, illness, etc.*⟩; **c)** *(erwecken)* radiate ⟨*optimism, happiness, calm, etc.*⟩; spread ⟨*fear*⟩. **2.** *refl. V.* **a)** *(bekannt werden)* ⟨*rumour*⟩ spread; **b)** *(sich ausbreiten)* ⟨*smell, illness, religion, etc.*⟩ spread

verbreitern 1. *tr. V.* widen; *(fig.)* broaden ⟨*basis*⟩. **2.** *refl. V.* widen out; get wider

Verbreitung die; ~, ~**en a)** *s.* **verbreiten 1 a, b, c:** spreading; broadcasting; radiation; **b)** *(Ausbreitung)* spread

verbrennen 1. *unr. itr. V.; mit sein* **a)** burn; ⟨*person*⟩ burn to death; **die Dokumente sind verbrannt** the documents were destroyed by fire; **es riecht verbrannt** *(ugs.)* there's a smell of burning; **der Kuchen ist verbrannt** the cake got burnt. **2.** *tr. V.* **a)** burn; burn, incinerate ⟨*rubbish*⟩; cremate ⟨*dead person*⟩; **b)** *(verletzen)* burn; **sich** *(Dat.)* **an der heißen Suppe die Zunge ~:** burn *or* scald one's tongue on the hot soup; **sich** *(Dat.)* **den Mund** *od.* *(derb)* **das Maul ~** *(fig.)* say too much; *s. auch* **Finger b**

Verbrennung die; ~, ~**en a)** *s.* **verbrennen 2 a:** burning; incineration; cremation; **b)** *(Kfz-W.)* combustion; **c)** *(Wunde)* burn

verbringen *unr. tr. V.* spend ⟨*time, holiday, etc.*⟩

verbrüdern *refl. V.* avow friendship and brotherhood

Verbrüderung die; ~, ~**en** avowal of friendship and brotherhood

verbrühen *tr. V.* scald; **sich** *(Dat.)* **den Arm ~:** scald one's arm

verbuchen *tr. V.* enter; *(fig.)* notch up ⟨*success, score, etc.*⟩; **etw. auf einem Konto ~:** credit sth. to an account

verbuddeln *tr. V. (ugs.)* bury

verbummeln *tr. V. (ugs., oft abwertend)* **a)** *(verbringen)* waste, fritter away ⟨*time, day, afternoon, etc.*⟩; **b)** *(vergessen)* forget [all] about; clean forget; **c)** *(verlieren)* lose; *(verlegen)* mislay

verbünden [fɛɐ̯'bʏndn̩] *refl. V.* form an alliance

verbündet *Adj.* **[miteinander] ~:** in alliance *postpos.*

Verbündete der/die; *adj. Dekl.* ally

verbüßen *tr. V.* serve ⟨*sentence*⟩

Verbüßung die; ~: serving

Verdacht [fɛɐ̯'daxt] **der;** ~[e]s, ~e *od.* **Verdächte** [fɛɐ̯'dɛçtə] suspicion; ~ **schöpfen** become suspicious; **wen hast du in ~?** who do you suspect?; **ich geriet in [den] ~, das Geld gestohlen zu haben** I was suspected of having stolen the money

verdächtig [fɛɐ̯'dɛçtɪç] **1.** *Adj.* suspicious; **sich ~ machen** arouse suspicion. **2.** *adv.* suspiciously

Verdächtige der/die; *adj. Dekl.* suspect

verdächtigen *tr. V.* suspect (*Gen.* of)

Verdächtigung die; ~, ~**en** suspicion

verdammen *tr. V.* condemn; *(Rel.)* damn ⟨*sinner*⟩; **b) dazu verdammt sein, etw. zu tun** *(fig.)* be condemned to do sth.

verdammt 1. *Adj.; nicht präd.* **a)** *(salopp abwertend)* bloody *(Brit. sl.)*; damned *(coll.)*; ~ **[noch mal** *od.* **noch eins]!** damn [it all] *(coll.)*; bloody hell *(Brit. sl.)*; ~ **und zugenäht!** damn and blast [it]! *(coll.)*; **b)** *(ugs.: sehr groß)* **[ein] ~es Glück haben** be damn[ed] lucky *(coll.)*. **2.** *adv.* *(ugs.: sehr)* damn[ed] *(coll.)* ⟨*cold, heavy, beautiful, etc.*⟩; **ich mußte mich ~ beherrschen** I had to keep a bloody good grip on myself *(Brit. coll.)*

Verdammung die; ~, ~**en** condemnation; damnation

verdampfen 1. *itr. V.; mit sein* evaporate; vaporize. **2.** *tr. V.* evaporate; vaporize

verdanken *tr. V.* **jmdm./einer Sache etw. ~:** owe sth. to sb./sth.

verdarb [fɛɐ̯'darp] *1. u. 3. Pers. Sg. Prät. v.* **verderben**

verdauen [fɛɐ̯'dauən] *tr., itr. V.* *(auch fig.)* digest; *(fig.)* get over ⟨*bad experience, shock, etc.*⟩

verdaulich [fɛɐ̯'daulɪç] *Adj.* digestible

Verdauung die; ~: digestion

Verdauungs-: ~**beschwerden** *Pl.* digestive trouble *sing.*; ~**störung die** poor digestion *no pl.*

Verdeck das; ~[e]s, ~e top; hood *(Brit.)*; *(bei Kinderwagen)* hood

verdecken *tr. V.* **a)** *(nicht sichtbar sein lassen)* hide; cover; **jmdm. die Sicht ~:** block sb.'s view; **b)** *(verbergen)* cover; conceal; *(fig.)* conceal ⟨*intentions etc.*⟩

verdenken *unr. tr. V.* **jmdm. etw. nicht ~ [können]** not [be able to] hold sth. against sb.

Verderb *s.* **Gedeih**

verderben 1. *unr. itr. V.; mit sein* ⟨*food, harvest*⟩ go bad *or* off, spoil. **2.** *unr. tr. V.* **a)** spoil; *(stärker)* ruin; *(fig.)* ruin ‹*evening*›; spoil ‹*appetite, enjoyment, fun, good mood, etc.*›; **b)** *(geh.: negativ beeinflussen)* corrupt; deprave; **er will es mit niemandem ~:** he tries to please everybody. **3.** *unr. refl. V.* **sich** *(Dat.)* **den Magen/die Augen ~:** give oneself an upset stomach/ruin one's eyesight

Verderben das; ~s undoing; ruin; **in sein** *od.* **ins ~ rennen** rush headlong towards ruin

verderblich *Adj.* **a)** perishable ‹*food*›; **leicht ~:** highly perishable; **b)** *(unheilvoll)* pernicious; *(moralisch schädlich)* corrupting; pernicious ‹*influence, effect, etc.*›

Verderblichkeit die; ~ *s.* **verderblich:** perishableness; perniciousness; corrupting effect

verdeutlichen *tr. V.* **etw. ~:** make sth. clear; *(erklären)* explain sth.

verdichten 1. *refl. V.* ‹*fog, smoke*› thicken, become thicker; *(fig.)* ‹*suspicion, rumour*› grow; ‹*feeling*› intensify. **2.** *tr. V. (fig.)* condense ‹*events etc.*› **(zu** into)

Verdickung die; ~, ~en thickened section; *(Schwellung)* swelling

verdienen 1. *tr. V.* **a)** earn; **b)** *(wert sein)* deserve; **er verdient kein Vertrauen** he doesn't deserve to be trusted; **womit habe ich das verdient?** what have I done to deserve that? **2.** *itr. V.* **beide Eheleute ~:** husband and wife are both wage-earners *or* are both earning; **gut ~:** have a good income

Verdiener der; ~s, ~: wage-earner

[1]**Verdienst der** income; earnings *pl.*

[2]**Verdienst das; ~[e]s, ~e** merit

verdienst·voll 1. *Adj.* commendable. **2.** *adv.* commendably

verdient 1. *Adj.* **a)** ‹*person*› of outstanding merit; **sich um etw. ~ machen** render outstanding services to sth.; **b)** *(gerecht, zustehend)* well-deserved. **2.** *adv.* deservedly

verdientermaßen *Adv.* deservedly

verdonnern *tr. V. (salopp)* sentence; **jmdn. dazu ~, etw. zu tun** order *or* make sb. do sth. [as a punishment]

verdoppeln 1. *tr. V.* double; *(fig.)* double, redouble ‹*efforts etc.*›. **2.** *refl. V.* double

Verdoppelung die; ~, ~en doubling

verdorben [fɛɐ̯'dɔrbn̩] *2. Part. v.* **verderben**

Verdorbenheit die; ~: depravity

verdorren [fɛɐ̯'dɔrən] *itr. V.; mit sein* wither [and die]; ‹*meadow*› scorch

verdrängen *tr. V.* **a)** drive out ‹*inhabitants*›; *(fig.: ersetzen)* displace; **jmdn. aus seiner Stellung ~:** oust sb. from his/her job; **b)** *(Psych.)* repress/*(bewußt)* suppress ‹*experience, desire, etc.*›

Verdrängung die; ~, ~en a) *s.* **verdrängen a:** driving out; displacement; ousting; **b)** *(Psych.)* repression; *(bewußt)* suppression

verdrecken *(ugs. abwertend)* **1.** *tr. V.* make filthy dirty. **2.** *itr. V.; mit sein* get *or* become filthy dirty

verdrehen *tr. V.* **a)** twist ‹*joint*›; roll ‹*eyes*›; **den Hals ~:** twist one's head round; **sich** *(Dat.)* **den Hals ~:** crick one's neck; **b)** *(ugs. abwertend: entstellen)* twist ‹*words, facts, etc.*›; distort ‹*sense*›

verdreifachen *refl., tr. V.* treble; triple

verdreschen *unr. tr. V. (ugs.)* thrash

verdrießen [fɛɐ̯'dri:sn̩] *unr. tr. V. (geh.)* irritate; annoy

verdrießlich 1. *Adj.* morose. **2.** *adv.* morosely

Verdrießlichkeit die; ~: moroseness

verdroß [fɛɐ̯'drɔs] *1. u. 3. Pers. Sg. Prät. v.* **verdrießen**

verdrossen 1. *2. Part. v.* **verdrießen. 2.** *Adj.* morose; *(mißmutig und lustlos)* sullen. **3.** *adv.* morosely; *(mißmutig und lustlos)* sullenly

Verdrossenheit die; ~ *s.* **verdrossen 2:** moroseness; sullenness

verdrücken *(ugs.)* **1.** *tr. V.* **a)** *(essen)* polish off *(coll.)*; **b)** *(verknautschen)* crumple ‹*clothes*›. **2.** *refl. V.* slip away

Verdruß [fɛɐ̯'drʊs] **der; Verdrusses, Verdrusse** annoyance; *(Unzufriedenheit)* dissatisfaction; discontentment; **jmdm. ~ bereiten** annoy sb.

verduften *itr. V.; mit sein (salopp: sich entfernen)* hop it *(Brit. sl.)*; clear off *(coll.)*

Verdummung die; ~ a) die ~ der Massen zum Ziel haben be aimed at dulling the mind of the masses; **b)** *(das Dummwerden)* stultification

verdunkeln 1. *tr. V.* **a)** darken; *(vollständig)* black out ‹*room, house, etc.*›; **b)** *(verdecken)* darken; *(fig.)* cast a shadow on ‹*happiness etc.*›. **2.** *refl. V.* darken; grow darker; *(fig.)* ‹*expression etc.*› darken

Verdunk[e]lung die; ~, ~en darkening; *(vollständig)* blackout

verdünnen *tr. V.* **a)** dilute; *(mit Wasser)* water down; dilute; thin [down] ‹*paint etc.*›

Verdünnung die; ~, ~en dilution

verdunsten 1. *itr. V.; mit sein* evaporate. **2.** *tr. V.* evaporate; ‹*plant*› transpire ‹*water*›

Verdunstung die; ~: evaporation

verdursten *itr. V.; mit sein* die of thirst

verdüstern 1. *tr. V.* darken; *(fig. geh.)* cast a shadow across. **2.** *refl. V.* darken; grow dark; *(fig.)* darken

verdutzt [fɛɐ̯'dʊtst] *Adj.* taken aback *pred.*; nonplussed; *(verwirrt)* baffled

veredeln *tr. V.* **a)** *(geh.)* ennoble; improve ‹*taste*›; **b)** *(Technik)* refine

Vered[e]lung die; ~, ~en a) *(geh.)* ennoblement; **b)** *(Technik)* refinement

verehren *tr. V.* **a)** *(vergöttern)* venerate; revere; **verehrte Frau Müller!** Dear Frau Müller; **b)** *(geh.: bewundern)* admire; *(ehrerbietig lieben)* worship; adore; **c)** *(scherzh.: schenken)* give

Verehrer der; ~s, ~, Verehrerin die; ~, ~en admirer

Verehrung die; *o. Pl.* **a)** veneration; reverence; **b)** *(Bewunderung)* admiration

vereidigen [fɛɐ̯'|aidɪgn̩] *tr. V.* swear in; **jmdn. auf etw.** *(Akk.)* **~:** make sb. swear to sth.; **ein vereidigter Sachverständiger** a sworn expert

Vereidigung die; ~, ~en swearing in

Verein organization; *(zur Förderung der Denkmalspflege usw.)* society; *(der Kunstfreunde usw.)* association; society; *(Sport~)* club

vereinbar *Adj.; nicht attr.* compatible **(mit** with)

vereinbaren *tr. V.* **a)** *(festlegen)* agree; arrange ‹*meeting etc.*›; **b)** *(in Einklang bringen)* reconcile; **[nicht] zu ~ sein** be [in]compatible *or* [ir]reconcilable

Vereinbarung die; ~, ~en agreement; **eine ~ treffen** come to an agreement

vereinfachen *tr. V.* simplify

Vereinfachung die; ~, ~en simplification

vereinheitlichen *tr. V.* standardize

vereinigen 1. *tr. V.* unite; merge ⟨*businesses*⟩; *(zusammenfassen)* bring together; **die Mehrheit der Stimmen auf sich ~:** receive the majority of the votes. **2.** *refl. V.* unite; ⟨*organizations, firms*⟩ merge; *(fig.)* be combined

vereinigt *Adj.* united; **Vereinigtes Königreich [Großbritannien und Nordirland]** United Kingdom [of Great Britain and Northern Ireland]; **Vereinigte Staaten [von Amerika]** United States *sing.* [of America]

Vereinigung die a) *(Rechtsw.)* organization; **b)** *(Zusammenschluß)* uniting; *(von Unternehmen)* merging

vereinsamen *itr. V.; mit sein* become [increasingly] lonely *or* isolated

Vereinsamung die; ~: loneliness; isolation

vereinzelt 1. *Adj.; nicht präd.* occasional; isolated, occasional ⟨*shower, outbreak of rain, etc.*⟩. **2.** *adv. (zeitlich)* occasionally; now and then; *(örtlich)* here and there

vereisen *itr. V.; mit sein* freeze *or* ice over; ⟨*wing*⟩ ice up; ⟨*lock*⟩ freeze up; **eine vereiste Fahrbahn** an icy carriageway

vereiteln *tr. V.* thwart; prevent; thwart, foil ⟨*attempt, plan, etc.*⟩; thwart, frustrate ⟨*efforts, intentions, etc.*⟩

Vereitelung die; ~ *s.* **vereiteln:** thwarting; prevention; foiling; frustrating

vereitern *itr. V.; mit sein* go septic; **vereitert sein** be septic

verenden *itr. V.; mit sein* perish; die

verengen 1. *refl. V.* narrow; become narrow; ⟨*pupils*⟩ contract; ⟨*blood-vessel*⟩ constrict, become constricted. **2.** *tr. V.* make narrower; narrow; restrict, narrow ⟨*field of vision etc.*⟩; make ⟨*circle, loop*⟩ smaller

vererben 1. *tr. V.* **a)** leave, bequeath ⟨*property*⟩ (*Dat.*, **an** + *Akk.* to); **b)** *(Biol.)* transmit, pass on ⟨*characteristic, disease*⟩; pass on ⟨*talent*⟩ (*Dat.*, **auf** + *Akk.* to). **2.** *refl. V. (Biol.)* ⟨*disease, tendency*⟩ be passed on *or* transmitted (**auf** + *Akk.* to)

Vererbung die; ~, ~en *(Biol.)* heredity *no art.;* **das ist ~:** it runs in the family

verewigen *refl. V. (ugs.)* leave one's mark

[1]**verfahren 1.** *unr. refl. V.* lose one's way. **2.** *unr. tr. V.* use up ⟨*petrol*⟩. **3.** *unr. itr. V.; mit sein* proceed; **mit jmdm./etw. ~:** deal with sb./sth.

[2]**verfahren** *Adj.* dead-end ⟨*situation*⟩

Verfahren das; ~s, ~ a) procedure; *(Technik)* process; *(Methode)* method; **b)** *(Rechtsw.)* proceedings *pl.*

Verfall der; *o. Pl.* **a)** decay; dilapidation; **b)** *(Auflösung)* decline; **c)** *(das Ungültigwerden)* expiry

verfallen *unr. itr. V.; mit sein* **a)** *(baufällig werden)* fall into disrepair; become dilapidated; **b)** *(körperlich)* ⟨*strength*⟩ decline; **c)** *(untergehen)* ⟨*empire*⟩ decline; ⟨*morals, morale*⟩ deteriorate; **d)** *(ungültig werden)* expire; **e) jmdm. ~:** become a slave; **dem Alkohol ~:** become addicted to alcohol; **f)** *(übergehen)* **in seinen Dialekt ~:** lapse into one's dialect; **das Pferd verfiel in [einen] Trab** the horse broke into a trot; **g) auf jmdn./etw. ~:** think of sb./sth.; **auf einen sonderbaren Gedanken ~:** hit upon a strange idea

verfälschen *tr. V.* distort, misrepresent ⟨*statement, message*⟩; falsify, misrepresent ⟨*facts, history, truth*⟩; falsify ⟨*painting, banknote*⟩; adulterate ⟨*wine, milk, etc.*⟩

verfänglich [fɛɐ̯ˈfɛŋlɪç] *Adj.* awkward, embarrassing ⟨*situation, question, etc.*⟩; incriminating ⟨*evidence, letter, etc.*⟩

verfärben 1. *refl. V.* change colour; ⟨*washing*⟩ become discoloured; ⟨*leaves*⟩ turn. **2.** *tr. V.* discolour

Verfärbung die a) change of colour; **b)** *(verfärbte Stelle)* discoloration; discoloured patch

verfassen *tr. V.* write, compose ⟨*poetry*⟩; write, draw up ⟨*document, law*⟩

Verfasser der; ~s, ~, Verfasserin die; ~, ~nen author; writer

Verfassung die a) *(Politik)* constitution; **b)** *o. Pl. (Zustand)* state [of health/mind]; **in guter/schlechter ~ sein** be in good/poor shape; **in bester ~:** on top form

verfaulen *itr. V.; mit sein* rot; *(fig.)* ⟨*system, social order*⟩ decay; *(fig.: moralisch)* degenerate

verfechten *unr. tr. V. (eintreten für)* advocate, champion ⟨*theory, hypothesis, etc.*⟩; uphold ⟨*view*⟩; *(verteidigen)* defend

Verfechter der, Verfechterin die; ~, ~nen advocate; champion

verfehlen *tr. V.* **a)** miss ⟨*train, person, etc.*⟩; **b)** miss ⟨*target etc.*⟩

Verfehlung die; ~, ~en misdemeanour; *(Rel.)* transgression

verfeinden *refl. V.* **sich ~ mit** make an enemy of; **verfeindet sein** be enemies

verfeinern [fɛɐ̯ˈfainɐn] **1.** *tr. V.* improve; refine ⟨*method, procedure, sense*⟩. **2.** *refl. V.* improve; ⟨*method, procedure, sense*⟩ be refined

Verfeinerung die; ~, ~en *s.* **verfeinern 1, 2:** improvement; refinement

verfestigen 1. *tr. V.* harden. **2.** *refl. V.* harden

verfeuern *tr. V.* **a)** burn; **alles Holz verfeuert haben** have used up all the wood; **b)** *(verschießen)* fire

verfilmen *tr. V.* film; make a film of; **der Roman wird jetzt verfilmt** the novel is now being made into a film

Verfilmung die; ~, ~en a) filming; **b)** *(Film)* film [version]

verfilzen *itr. V.; mit sein* ⟨*fabric, garment*⟩ felt; become felted; ⟨*hair*⟩ become matted

verfinstern 1. *tr. V.* obscure ⟨*sun etc.*⟩. **2.** *refl. V. (auch fig.)* darken

verflachen 1. *itr. V.; mit sein* ⟨*ground*⟩ flatten *or* level out, become flatter; ⟨*water*⟩ become shallow; *(fig.)* ⟨*discussion*⟩ become superficial *or* trivial. **2.** *refl. V.* ⟨*ground*⟩ flatten *or* level out. **3.** *tr. V.* flatten; level

verflechten *unr. tr. V.* interweave; intertwine; interlace; *(verwickeln)* involve; **miteinander verflochten sein** *(fig.)* be interlinked

Verflechtung die; ~, ~en interconnection; *(Verwicklung)* involvement

verfliegen 1. *unr. refl. V.* ⟨*pilot*⟩ lose one's way; ⟨*aircraft*⟩ get off course. **2.** *unr. itr. V.; mit sein* ⟨*smoke*⟩ disperse, vanish; ⟨*scent, smell*⟩ fade, disappear; ⟨*mood, tiredness, alcohol etc.*⟩ evaporate; **b)** ⟨*time*⟩ fly by; ⟨*anger*⟩ pass

verflixt [fɛɐ̯ˈflɪkst] *(ugs.)* **1.** *Adj.* **a)** *(ärgerlich)* awkward, unpleasant ⟨*situation, business, etc.*⟩; **b)** *(abwertend: verdammt)* blasted *(Brit.);* blessed; confounded; **~ [noch mal]!, ~ noch eins!, ~ und zugenäht!** [damn and] blast *(Brit. coll.);* **c)** *s.* **verdammt 1b. 2.** *adv. (sehr)* damned *(coll.)*

verflossen *Adj. (ugs.)* former; **seine ~e Freundin** his ex-girlfriend

verfluchen *tr. V.* curse

verflucht 1. *Adj. (salopp)* **a)** damned *(coll.);* bloody *(Brit. sl.);*

~ [noch mal]!, ~ **und zugenäht!** *(derb)* damn [it] *(coll.)*; **b)** *nicht präd. (sehr groß)* **wir hatten ~es Glück** we were damned lucky *(coll.)*. **2.** *adv. (sehr)* damned *(coll.)*

verflüchtigen *refl. V.* ‹*alcohol etc.*› evaporate; ‹*smoke*› disperse; ‹*smell*› disappear; *(fig.)* ‹*fear, astonishment*› subside; ‹*cheerfulness, mockery*› vanish; ‹*time of youth*› be dissipated

verfolgen *tr. V.* **a)** pursue; hunt, track ‹*animal*›; **jmdn. auf Schritt und Tritt ~:** follow sb. wherever he/she goes; **der Gedanke daran verfolgte ihn** *(fig.)* the thought of it haunted him; **vom Pech verfolgt sein** *(fig.)* be dogged by bad luck; **b)** *(bedrängen)* plague; **c)** *(bedrohen)* persecute; **d)** *(zu verwirklichen suchen)* pursue ‹*policy, plan, career, idea, purpose, etc.*›; **e)** *(beobachten)* follow ‹*conversation, events, trial, developments, etc.*›; **f) etw. [strafrechtlich] ~:** prosecute sth.

Verfolger der; ~s, ~, **Verfolgerin die;** ~, ~nen pursuer; *(Häscher)* persecutor

Verfolgte der/die; *adj. Dekl.* victim of persecution

Verfolgung die; ~, ~en **a)** pursuit; **die ~ aufnehmen** take up the chase; **b)** *(Bedrohung)* persecution; **c)** *s.* **verfolgen d:** pursuance; **d) [strafrechtliche] ~:** prosecution

Verfolgungs-: ~**jagd die** pursuit; chase; ~**wahn der** persecution mania

verfressen *Adj. (salopp abwertend)* piggish *(coll.)*; greedy

verfroren *Adj.* sensitive to the cold

verfügen 1. *tr. V. (anordnen)* order; *(dekretieren)* decree. **2.** *itr. V.* **a)** *(bestimmen)* **über etw.** *(Akk.)* **[frei] ~ können** be free to decide what to do with sth.; **über jmdn. ~:** tell sb. what to do; **b)** *(haben)* **über etw.** *(Akk.)* ~: have sth. at one's disposal; have ‹*good connections, great experience*›

Verfügung die; ~, ~en **a)** order; *(Dekret)* decree; **b)** *o. Pl.* **etw. zur ~ haben** have sth. at one's disposal; **jmdm. etw. zur ~ stellen** put sth. at sb.'s disposal; **sein Amt zur ~ stellen** offer to give up one's post *or* office; **jmdm. zur ~ stehen** be at sb.'s disposal

verführen 1. *tr. V.* **a)** tempt; **jmdn. zum Trinken ~:** encourage sb. to take up drinking; **b)** *(sexuell)* seduce. **2.** *itr. V.* **zu etw. ~:** be a temptation to sth.

Verführer der seducer

Verführerin die seductress

verführerisch 1. *Adj.* **a)** tempting; **b)** *(aufreizend)* seductive. **2.** *adv.* **a)** temptingly; **b)** *(aufreizend)* seductively

Verführung die temptation; *(sexuell)* seduction

verfüttern *tr. V.* **a)** *(zu fressen geben)* feed (*Dat.* to); **b)** *(verbrauchen)* use [up] as animal/bird food

vergällen [fɛɐ̯'gɛlən] *tr. V.* spoil ‹*enjoyment etc.*›; sour ‹*life*›

vergammeln *(ugs.)* **1.** *itr. V.; mit sein* ‹*food*› go bad. **2.** *tr. V.* waste ‹*time*›

vergammelt *Adj. (ugs. abwertend)* scruffy *(coll.)*; tatty *(coll.)*; tatty *(coll.)*, decrepit ‹*vehicle*›

vergangen [fɛɐ̯'gaŋən] *Adj.; nicht präd.* **a)** *(vorüber, vorbei)* bygone, former ‹*times, years, etc.*›; **b)** *(letzt...)* last ‹*year, Sunday, week, etc.*›

Vergangenheit die; ~, ~en **a)** past; **die jüngste ~:** the recent past; **etw. gehört der ~ an** sth. is a thing of the past; **b)** *(Grammatik)* past tense

vergänglich [fɛɐ̯'gɛŋlɪç] *Adj.* transient; transitory; ephemeral

Vergänglichkeit die; ~: transience; transitoriness

Vergaser der; ~s, ~ *(Kfz-W.)* carburettor

vergaß [fɛɐ̯'ga:s] *1. u. 3. Pers. Sg. Prät. v.* **vergessen**

vergattern *tr. V.* **jmdn. zu etw. ~:** swear sb. to sth.; **jmdn. dazu ~, etw. zu tun** enjoin sb. to do sth.

vergeben *unr. tr. V.* **a)** *auch itr. (geh.: verzeihen)* forgive; **jmdm. etw. ~:** forgive sb. [for] sth.; **b)** throw away ‹*chance, goal, etc.*›; **einen Elfmeter ~:** waste a penalty; **c)** *(geben)* place ‹*order*› (**an** + *Akk.* with); award ‹*grant, prize*› (**an** + *Akk.* to); **d) sich** *(Dat.)* **etwas/nichts ~:** lose/not lose face

vergebens 1. *Adv.* in vain; vainly. **2.** *adj.* **es war ~:** it was of *or* to no avail

vergeblich 1. *Adj.* futile; vain, futile ‹*attempt, efforts*›; **~ sein** be of *or* to no avail. **2.** *adv.* in vain; vainly

Vergeblichkeit die; ~: futility

Vergebung die; ~, ~en *(geh.)* forgiveness

vergegenwärtigen [od. ---'---] *refl. V.* **sich** *(Dat.)* **etw. ~:** imagine sth.; *(erinnern)* recall sth.

vergehen 1. *unr. itr. V.; mit sein* **a)** ‹*time*› pass [by], go by; **es vergeht kein Tag, an dem er nicht anruft** not a day passes by without him ringing up *(Brit.) or (coll.)* phoning; **b)** ‹*pain*› wear off, pass; ‹*pleasure*› fade; **ihr verging der Appetit** she lost her appetite; **c)** ‹*cloud, scent*› disappear; ‹*fog*› lift. **2.** *unr. refl. V.* **a) sich gegen das Gesetz ~:** violate the law; **b) sich an jmdm. ~:** commit indecent assault on sb.; indecently assault sb.

Vergehen das; ~s, ~: crime; *(Rechtsspr.)* offence

vergelten *unr. tr. V.* repay (**durch** with); **jmdm. etw. ~:** repay sb. for sth.

Vergeltung die a) repayment; **b)** *(Rache)* revenge; **~ an jmdm. üben** take revenge on sb.

vergessen [fɛɐ̯'gɛsn̩] **1.** *unr. tr. (auch itr.) V.* forget; *(liegenlassen)* forget; leave behind; **das kannst du ~!** *(ugs.)* forget it!; you can forget about that! **2.** *refl. V.* forget oneself

Vergessenheit die; ~: oblivion; **in ~ geraten** fall into oblivion

vergeßlich [fɛɐ̯'gɛslɪç] *Adj.* forgetful

Vergeßlichkeit die; ~: forgetfulness

vergeuden [fɛɐ̯'gɔydn̩] *tr. V.* waste; squander, waste ‹*money*›

Vergeudung die; ~, ~en waste; squandering

vergewaltigen *tr. V.* **a)** rape; **b)** *(fig.)* oppress ‹*nation, people*›; violate ‹*truth, conscience, law, language, etc.*›

Vergewaltigung die; ~, ~en **a)** rape; **b)** *s.* **vergewaltigen b:** oppression; violation

vergewissern [fɛɐ̯gə'vɪsɐn] *refl. V.* make sure (*Gen* of)

vergießen *unr. tr. V.* **a)** spill; **b) Tränen ~:** shed tears; **viel Schweiß ~:** sweat blood *(fig.)*; *s. auch* **Blut**

vergiften *tr. V. (auch fig.)* poison

Vergiftung die; ~, ~en poisoning

vergiß [fɛɐ̯'gɪs] *Imper. Sg. v.* **vergessen**

Vergiß·mein·nicht das; ~[e]s, ~[e] forget-me-not

vergißt *2. u. 3. Pers. Sg. Präs. v.* **vergessen**

vergittert *Adj.* barred

Vergleich der; ~[e]s, ~e **a)** comparison; **dieser ~ hinkt** this is a poor comparison; **das ist doch kein ~!** there is no comparison; **im ~ zu** *od.* **mit etw.** in comparison with sth.; compared with *or* to sth.; **b)** *(Sprachw.)* simile; **c)** *(Rechtsw.)* settlement

vergleichbar *Adj.* comparable

vergleichen 1. *tr. V.* compare

(**mit** with, to); **die Uhrzeit ~:** check that one has the correct time; **das ist [doch gar] nicht zu ~:** that [really] doesn't stand comparison *or* compare. **2.** *refl. V.* **a) sich mit jmdm. ~:** compete with sb.; **b)** *(Rechtsw.)* reach a settlement; settle

vergleichs-, Vergleichs-: ~form die *(Sprachw.)* comparative/superlative form; **~möglichkeit die** opportunity for comparison; **~weise** *Adv.* comparatively

verglühen *itr. V.; mit sein* ‹*log, wick, fire, etc.*› smoulder and go out; ‹*glow of sunset*› fade; ‹*satellite, rocket, wire, etc.*› burn out

vergnügen [fɛɐ̯'gny:gn̩] *refl. V.* enjoy oneself; have a good time

Vergnügen das; ~s, ~: pleasure; *(Spaß)* fun; **ein teueres ~** *(ugs.)* an expensive bit of fun *(coll.)*; **es ist mir ein ~:** it's a pleasure; **mit wem habe ich das ~?** with whom do I have the pleasure of speaking?; **etw. macht jmdm. [großes] ~:** sth. gives sb. [great] pleasure; sb. enjoys sth. [very much]; **viel ~!** *(auch iron.)* have fun!; **mit [dem größten] ~:** with [the greatest of] pleasure

vergnüglich 1. *Adj.* amusing, entertaining ‹*play, programme*›. **2.** *adv.* amusingly; entertainingly

vergnügt 1. *Adj.* **a)** cheerful; happy ‹*smile*›; merry ‹*group of people*›; **b)** *(unterhaltsam)* enjoyable. **2.** *adv.* cheerfully; ‹*smile*› happily

Vergnügung die; ~, ~en pleasure

Vergnügungs-: ~lokal das bar providing entertainment; *(Nachtlokal)* night-club; **~reise die** pleasure-trip; **~viertel das** pleasure district

vergolden *tr. V.* gold-plate ‹*jewellery etc.*›; *(mit Blattgold)* gild ‹*statue, dome, etc.*›; *(mit Gold bemalen)* paint ‹*statue, dome, etc.*› gold; *(fig.)* ‹*evening sun*› bathe ‹*roof-tops etc.*› in gold

vergönnen *tr. V.* grant

vergöttern [fɛɐ̯'gœtɐn] *tr. V.* idolize

vergraben 1. *unr. tr. V. (auch fig.)* bury. **2.** *unr. refl. V.* ‹*animal*› bury itself (**in** + *Akk. od. Dat.* in); *(fig.)* withdraw from the world; hide oneself away

vergraulen *tr. V. (ugs.)* put off

vergreifen *unr. refl. V.* **a) sich im Ton/Ausdruck ~:** adopt the wrong tone/use the wrong expression; **b) sich an etw.** *(Dat.)* **~** *(an fremdem Eigentum)* misappropriate sth.; **c) sich an jmdm. ~:** assault sb.

vergriffen *Adj.* out of print *pred.*

vergrößern [fɛɐ̯'grø:sɐn] **1.** *tr. V.* **a)** extend ‹*room, area, building, etc.*›; increase ‹*distance*›; **sein Repertoire ~:** extend *or* increase *or* enlarge one's repertoire; **b)** *(vermehren)* increase; **c)** *(größer reproduzieren)* enlarge ‹*photograph etc.*›. **2.** *refl. V.* **a)** ‹*firm, business, etc.*› expand; **eine krankhaft vergrößerte Leber** a pathologically enlarged liver; **b)** *(zunehmen)* increase. **3.** *itr. V.* ‹*lens etc.*› magnify

Vergrößerung die; ~, ~en a) *s.* **vergrößern 1, 2:** extension; increase; enlargement; expansion; **b)** *(Foto)* enlargement; **in 100facher ~:** enlarged 100fold

Vergrößerungs·glas das magnifying glass

vergucken *refl. V. (ugs.)* **a) sich in jmdn./etw. ~:** fall for sb./sth. *(coll.)*; **b)** *(falsch sehen)* be mistaken [about what one saw]

Vergünstigung die; ~, ~en privilege

vergüten *tr. V.* **a) jmdm. etw. ~:** reimburse sb. for sth.; **b)** *(bes. Papierdt.: bezahlen)* remunerate, pay for ‹*work, services*›

Vergütung die; ~, ~en a) *(Rückerstattung)* reimbursement; **b)** *(Geldsumme)* remuneration

verhaften *tr. V.* arrest

Verhaftete der/die; *adj. Dekl.* person under arrest; man/woman under arrest; arrested man/woman

Verhaftung die; ~, ~en arrest

verhaken *refl. V.* ‹*person*› get hooked *or* caught up; ‹*zip*› get caught

verhallen *itr. V.; mit sein* ‹*sound*› die away; **[ungehört] ~** *(fig.)* ‹*call, words, etc.*› go unheard *or* unheeded

¹verhalten *unr. refl. V.* **a)** behave; *(reagieren)* react; **sich still** *od.* **ruhig ~:** keep quiet; **ich verhielt mich abwartend** I decided to wait and see; **b)** *(beschaffen sein)* be; **die Sache verhält sich nämlich so** this is how things stand *or* the matter stands; **c)** *(im Verhältnis stehen)* **a verhält sich zu b wie x zu y** a is to b as x is to y

²verhalten 1. *Adj.* **a)** *(unterdrückt)* restrained; **mit ~em Tempo** at a measured pace; **b)** *(dezent)* restrained, subdued, muted ‹*colours*›; muted, soft ‹*notes, voice, etc.*›; **c)** *(zurückhaltend)* reserved; **eine ~e Fahrweise** a cautious way of driving. **2.** *adv.* **a)** *(unterdrückt)* in a restrained manner; **b)** *(zurückhaltend)* in a reserved manner; ‹*drive*› cautiously; **c)** *(dezent)* ‹*speak, play, etc.*› softly

Verhalten das; ~s behaviour; *(Vorgehen)* conduct

Verhaltens-: ~maß·regel die; *meist Pl.* rule of conduct; **~weise die** behaviour; **~weisen** behaviour patterns; patterns of behaviour

Verhältnis [fɛɐ̯'hɛltnɪs] **das; ~ses, ~se a) ein ~ von drei zu eins** a ratio of three to one; **im ~ zu früher** in comparison with *or* compared to earlier times; **b)** *(persönliche Beziehung)* relationship (**zu** with); **ein gutes ~ zu jmdm. haben** get on well with sb.; **c)** *(ugs.: intime Beziehung)* affair; relationship; **d)** *(ugs.) (Geliebte)* lady-friend; *(Geliebter)* man; **e)** *Pl. (Umstände)* conditions; **in bescheidenen** *od.* **einfachen/gesicherten ~sen leben** live in modest circumstances/be financially secure; **aus einfachen ~sen kommen** come from a humble background; **über seine ~se leben** live beyond one's means

verhältnis-, Verhältnis-: ~gleichung die *(Math.)* proportion; **~mäßig** *Adv.* relatively; comparatively; **~wort das;** *Pl.* **~wörter** *(Sprachw.)* preposition

verhandeln 1. *itr. V.* **a)** negotiate (**über** + *Akk.* about); **b)** *(strafrechtlich)* try a case; *(zivilrechtlich)* hear a case. **2.** *tr. V.* **a) etw. ~:** negotiate over sth.; **b)** *(strafrechtlich)* try ‹*case*›; *(zivilrechtlich)* hear ‹*case*›

Verhandlung die a) ~en negotiations; **mit jmdm. in ~ stehen** be negotiating with sb.; be [involved *or* engaged] in negotiations *pl.* with sb.; **zu ~en bereit sein** be open to negotiation *sing.*; **b)** *(strafrechtlich)* trial (**gegen** of); *(zivilrechtlich)* hearing

verhandlungs-, Verhandlungs-: ~bereit *Adj.* ready *or* willing to negotiate *pred.*; **~grundlage die** basis for negotiation[s]; **~tisch der** negotiating table

verhangen *Adj.* overcast

verhängen *tr. V.* **a)** cover (**mit** with); **b)** impose ‹*fine, punishment*› (**über** + *Akk.* on); declare ‹*state of emergency, state of siege*›; *(Sport)* award, give ‹*penalty etc.*›

Verhängnis [fɛɐ̯'hɛŋnɪs] **das; ~ses, ~se** undoing; **jmdm. zum ~ werden** be sb.'s undoing

verhängnis·voll *Adj.* disastrous; fatal, disastrous ‹*mistake, weakness, hesitation, etc.*›
Verhängung die; ~ *s.* **verhängen b:** imposition; declaration
verharmlosen *tr. V.* play down
verhärmt [fɛɐ̯'hɛrmt] *Adj.* careworn
verharren *itr. V. (geh.)* remain; *(plötzlich, kurz)* pause; **in Resignation ~:** remain resigned
verhärten [fɛɐ̯'hɛrtn̩] **1.** *tr. V.* **a)** harden ‹*material etc.*›; **b)** *(unbarmherzig machen)* harden; make ‹*person*› hard. **2.** *refl. V.* **a)** *(hart werden)* ‹*tissue*› become hardened; ‹*tumour*› become scirrhous; **b)** *(gefühllos werden)* harden one's heart **(gegen** against); **die Fronten haben sich verhärtet** the positions of the opposing parties have become entrenched
verhaspeln *refl. V. (ugs.)* stumble over one's words
verhaßt *Adj.* hated; detested; **es war ihm ~:** he hated *or* detested it; **nichts ist mir so ~ wie ...:** there is nothing I detest so much as ...
verhätscheln *tr. V. (ugs.)* pamper
verhauen *(ugs.)* **1.** *unr. tr. V.* **a)** beat up; *(als Strafe)* beat; **b)** *(falsch machen)* make a mess of; muck up *(Brit. sl.)*. **2.** *unr. refl. V.* make a mistake *or* slip
verheben *unr. refl. V.* do oneself an injury [while lifting sth.]
verheddern [fɛɐ̯'hɛdɐn] *refl. V.* **sich in etw.** *(Dat.)* **~:** get tangled up in sth.
verheerend *Adj.* **a)** devastating; disastrous; **b)** *(ugs.: scheußlich)* ghastly *(coll.)*; dreadful *(coll.)*
verhehlen *tr. V. (geh.)* conceal, hide (*Dat.* from); **ich kann/will [es] nicht ~, daß ...:** there is no denying/I have no wish to deny that ...
verheilen *itr. V.; mit sein* ‹*wound*› heal [up]
verheimlichen *tr. V.* **[jmdm.] etw. ~:** keep sth. secret [from sb.]; conceal *or* hide sth. [from sb.]
verheiraten 1. *refl. V.* get married **(mit** to). **2.** *tr. V. (veralt.)* marry **(mit, an** + *Akk.* to)
Verheiratete der/die; *adj. Dekl.* married person; married man/woman; ~ *Pl.* married people; married men/women
verheißen *unr. tr. V. (geh.; auch fig.)* promise; **nichts Gutes ~:** not bode *or* augur well
verheißungs·voll 1. *Adj.* promising. **2.** *adv.* full of promise
verheizen *tr. V.* burn; use as fuel
verhelfen *unr. itr. V.* **jmdm. zu etw. ~:** help sb. to get/achieve sth.; **jmdm. zur Flucht/zum Sieg ~:** help sb. to escape/win
verherrlichen *tr. V.* glorify ‹*war, violence, deed, etc.*›; extol ‹*virtues, leader, etc.*›; celebrate ‹*nature, freedom, peace, etc.*›
verhetzen *tr. V.* incite; stir up
verheult [fɛɐ̯'hɔ̯ylt] *Adj. (ugs.)* ‹*eyes*› red from crying; ‹*face*› puffy *or* swollen from crying
verhexen *tr. V. s.* **verzaubern a**
verhindern *tr. V.* prevent; prevent, avert ‹*war, disaster, etc.*›; **er ist verhindert** he is prevented from coming; **ein verhinderter Künstler** *(ugs.)* a would-be artist
Verhinderung die; ~, ~**en** *s.* **verhindern:** prevention; averting
verhohlen [fɛɐ̯'ho:lən] *Adj.* concealed; **kaum ~e Neugier** ill-concealed curiosity
verhöhnen [vɛr'hœnen] *tr. V.* mock; deride; ridicule
verhökern [fɛɐ̯'hø:kɐn] *tr. V. (salopp)* flog *(Brit. sl.)*
Verhör [fɛɐ̯'hø:ɐ̯] **das;** ~**[e]s,** ~**e** interrogation; questioning; *(bei Gericht)* examination; **jmdn. ins ~ nehmen** interrogate *or* question sb.; *(fig.)* grill *or* quiz sb.
verhören 1. *tr. V.* interrogate; question; *(bei Gericht)* examine. **2.** *refl. V.* mishear; hear wrongly
verhüllen *tr. V.* cover; *(fig.)* disguise; mask; **eine verhüllte Drohung** *(fig.)* a veiled threat
verhüllend *Adj. (Literaturw.)* euphemistic
verhungern *itr. V.; mit sein* die of starvation; starve [to death]; **ich bin am Verhungern** *(ugs.)* I'm starving *(fig. coll.)*
verhüten *tr. V.* prevent; prevent, avert ‹*disaster*›; **der Himmel verhüte, daß ...:** heaven forbid that ...
Verhütung die; ~, ~**en** prevention; *(Empfängnis~)* contraception
Verhütungs·mittel das contraceptive
verhutzelt [fɛɐ̯'hʊts̩lt] *Adj. (ugs.)* wizened ‹*person, face*›; shrivelled ‹*fruit, plant*›
verirren *refl. V.* **a)** get lost; lose one's way; ‹*animal*› stray; **b)** *(gelangen)* stray **(in, an** + *Akk.* into)
Verirrung die; ~, ~**en** aberration
verjagen *tr. V.* chase away
verjähren [fɛɐ̯'jɛ:rən] *itr. V.; mit sein* come under the statute of limitations
Verjährung die; ~, ~**en** limitation
verjubeln *tr. V. (ugs.)* blow *(sl.)* ‹*money*›
verjüngen [fɛɐ̯'jʏŋən] **1.** *tr. V.* rejuvenate ‹*person, skin, etc.*›; *(jünger aussehen lassen)* make ‹*person*› look younger; recruit younger blood into ‹*team, company, etc.*›. **2.** *refl. V. (schmaler werden)* taper; become narrower; narrow
verkabeln *tr. V.* connect up [by cable]
verkalken *itr. V.; mit sein* **a)** ‹*tissue*› calcify, become calcified; ‹*arteries*› become hardened; ‹*bone*› thicken; ‹*pipe, kettle, coffee-machine, etc.*› fur up; **b)** *(ugs.: senil werden)* become senile; **er ist schon ziemlich verkalkt** he is already pretty gaga *(sl.)*
verkalkulieren *refl. V.* miscalculate
Verkalkung die; ~, ~**en a)** *s.* **verkalken a:** calcification; hardening; thickening; furring-up; **b)** *(ugs.: Senilität)* senility
verkappt *Adj.* disguised
verkatert [fɛɐ̯'ka:tɐt] *Adj. (ugs.)* hung-over *(coll.)*
Verkauf der a) sale; *(das Verkaufen)* sale; selling; **zum ~ stehen** be [up] for sale; **b)** *o. Pl. (Abteilung)* sales *sing. or pl., no art.*
verkaufen 1. *tr. V. (auch fig.)* sell (*Dat.*, **an** + *Akk.* to); **„zu verkaufen"** 'for sale'. **2.** *refl. V.* **a)** ‹*goods*› sell; **b)** *(ugs.: falsch kaufen)* make a bad buy
Verkäufer der, Verkäuferin die a) seller; vendor *(formal)*; **b)** *(Berufsbez.)* sales *or* shop assistant; salesperson; *(im Außendienst)* salesman/saleswoman; salesperson
verkäuflich *Adj.* saleable; marketable; **schwer/leicht ~ sein** be hard/easy to sell
verkaufs-, Verkaufs-: ~offen *Adj.* **der ~offene Samstag** *od.* **Sonnabend** Saturday on which *or* when the shops are open all day; **~personal das** sales staff; **~preis der** retail price
Verkehr der; ~**s a)** traffic; **den ~ regeln** regulate *or* control the [flow of] traffic; **aus dem ~ ziehen** take ‹*coin, banknote*› out of circulation; take ‹*product*› off the market; **jmdn. aus dem ~ ziehen** *(ugs. scherzh.)* put sb. out of circulation *(joc.)*; **b)** *(Umgang)* contact; communication; **c)** *(Sexual~)* intercourse
verkehren 1. *itr. V.* **a)** *auch mit sein (fahren)* run; ‹*aircraft*› fly; **der Dampfer verkehrt zwischen Hamburg und Helgoland** the steamer plies *or* operates *or* goes between Hamburg and Heligo-

land; **b) mit jmdm. ~:** associate with sb.; **bei jmdm. ~:** visit sb. regularly; **in einem Lokal ~:** frequent a pub *(Brit. coll.)*; **in den besten Kreisen ~:** move in the best circles. **2.** *tr. V.* turn (**in** + *Akk.* into); **den Sinn einer Aussage ins Gegenteil ~:** twist the meaning of a statement right round. **3.** *refl. V.* turn (**in** + *Akk.* into); **sich ins Gegenteil ~:** change to the opposite

verkehrs-, Verkehrs-: **~ampel die** traffic lights *pl.*; **~amt das** tourist information office; **~aufkommen das** volume of traffic; **~büro das** tourist office; **~gefährdung die** constituting *no art.* a hazard to other traffic; **eine ~gefährdung darstellen** be *or* constitute a hazard to other traffic; **~hindernis das** obstruction to traffic; **~knotenpunkt der** [traffic] junction; **~kontrolle die** traffic check; **~ministerium das** ministry of transport; **~mittel das** means of transport; **die öffentlichen ~mittel** public transport *sing.*; **~regel die;** *meist Pl.* traffic regulation; **~schild das;** *Pl.* **~schilder** traffic sign; road sign; **~sicherheit die;** *o. Pl.* road safety; *(eines Fahrzeugs)* roadworthiness; **~teilnehmer der** road-user; **~unfall der** road accident; **~widrig 1.** *Adj.* contrary to road traffic regulations *postpos.*; **2.** *adv.* contrary to road traffic regulations; **~zeichen das** traffic sign; road sign

verkehrt 1. *Adj.* wrong; **das ist gar nicht so ~:** that's not such a bad idea; **an den Verkehrten geraten** *(ugs.)* come to the wrong person. **2.** *adv.* wrongly; **alles ~ machen** do everything wrong; *s. auch* **herum a**

verkeilen 1. *tr. V.* wedge. **2.** *refl. V.* become wedged (**in** + *Akk.* in); **sich ineinander ~:** become wedged together

verkennen *unr. tr. V.* fail to recognize; misjudge ⟨*situation*⟩; fail to appreciate ⟨*efforts, achievement, etc.*⟩; **es ist nicht zu ~, daß ...:** it cannot be denied *or* is undeniable that ...; **ein verkanntes Genie** an unrecognized genius

Verkettung die; ~, ~en *(von Zufällen usw.)* chain

verklagen *tr. V.* sue (**auf** + *Akk.* for); take proceedings against; take to court

verklären 1. *tr. V. (auch Rel.)* transfigure. **2.** *refl. V. (auch fig.)* be transfigured; ⟨*eyes*⟩ shine blissfully

verklausulieren [fɛɐ̯klauzuˈliːrən] *tr. V.* **a)** *(mit Klauseln versehen)* hedge ⟨*contract etc.*⟩ with qualifying clauses; **b)** *(verbergen)* hedge ⟨*admission of guilt etc.*⟩ round with qualifications

verkleben 1. *itr. V.; mit sein* stick together. **2.** *tr. V.* **a)** *(zusammenkleben)* stick together; **verklebte Hände** sticky hands; **b)** *(zukleben)* seal up ⟨*hole*⟩; **c)** *(festkleben)* stick [down] ⟨*floor-covering etc.*⟩

verkleiden *tr. V.* **a)** disguise; *(kostümieren)* dress up; **sich ~:** disguise oneself/dress [oneself] up; **b)** *(verdecken)* cover; *(auskleiden)* line; face ⟨*façade*⟩

Verkleidung die a) *o. Pl.* disguising; *(das Kostümieren)* dressing up; **b)** *(Kostüm) (als Tarnung)* disguise; *(bei einer Party usw.)* fancy dress; **c)** *s.* **verkleiden b:** covering; lining; facing; **d)** *(Umhüllung)* cover

verkleinern [fɛɐ̯ˈklainɐn] **1.** *tr. V.* **a)** make smaller; reduce the size of; reduce ⟨*size, number, etc.*⟩; **b)** *(kleiner reproduzieren)* reduce ⟨*photograph etc.*⟩. **2.** *refl. V.* become smaller; ⟨*number*⟩ decrease, grow smaller. **3.** *itr. V.* ⟨*lens etc.*⟩ make things look *or* appear smaller

Verkleinerung die; ~, ~en reduction in size; making smaller; *(des Formats, der Anzahl, durch eine Linse)* reduction

Verkleinerungs·form die *(Sprachw.)* diminutive form

verklemmt *(fig. ugs.)* **1.** *Adj.* inhibited. **2.** *adv.* in an inhibited manner

verklingen *unr. itr. V.; mit sein* ⟨*sound, voice, song, etc.*⟩ fade away; *(fig.)* ⟨*mood*⟩ wear off

verknappen 1. *tr. V.* cut back [on] ⟨*imports*⟩. **2.** *refl. V.* run short

verkneifen *unr. refl. V. (ugs.)* **a) sich** *(Dat.)* **eine Frage/Bemerkung ~:** bite back a question/remark; **ich konnte mir das Lachen kaum ~:** I could hardly keep a straight face; I could hardly stop myself laughing; **b)** *(verzichten)* manage *or* do without; **es sich** *(Dat.)* **~, etw. zu tun** stop oneself doing sth.

verkniffen *Adj.* strained ⟨*expression*⟩; pinched ⟨*mouth, lips*⟩

verknittern *tr. V.* crumple

verknoten 1. *tr. V.* **a)** *(verknüpfen)* tie; knot; **miteinander ~:** tie together; **b)** *(festbinden)* tie (**an** + *Akk.* to). **2.** *refl. V.* become knotted

verknüpfen *tr. V.* **a)** *(knoten)* tie; knot; **die beiden Fäden miteinander ~:** tie *or* knot the two threads together; **b)** *(zugleich tun)* combine; **c)** *(in Beziehung setzen)* link; *(unwillkürlich)* associate

Verknüpfung die; ~, ~en a) *s.* **verknüpfen 1:** tying; knotting; combination; linking; association; **b)** *(Knoten)* knots *pl.*

verkochen *itr. V.; mit sein* **a)** *(verdampfen)* boil away; **b)** *(breiig werden)* boil down to a pulp

verkohlen *itr. V.* char; become charred

[1]**verkommen** *unr. itr. V.; mit sein* **a)** *(verwahrlosen)* go to the dogs; *(moralisch, sittlich)* go to the bad; ⟨*child*⟩ go wild; **b)** *(verfallen)* ⟨*building etc.*⟩ go to rack and ruin, fall into disrepair, become dilapidated; ⟨*garden*⟩ run wild; ⟨*area*⟩ become run down; **c)** *(herabsinken)* degenerate (**zu** into); **d)** *(verderben)* ⟨*food*⟩ go bad

[2]**verkommen** *Adj.* depraved; **ein ~es Subjekt** a dissolute character

verkomplizieren *tr. V.* complicate

verkonsumieren *tr. V. (ugs.)* get through; consume

verkorken *tr. V.* cork [up]

verkörpern *tr. V.* **a)** *(als Schauspieler)* play [the part of]; **b)** *(bilden)* ⟨*person*⟩ embody, personify

Verkörperung die; ~, ~en embodiment; personification

verköstigen [fɛɐ̯ˈkœstɪgn̩] *tr. V.* feed; provide with meals

Verköstigung die; ~, ~en a) *o. Pl.* feeding; **b)** *(Kost)* foods; meals *pl.*

verkraften *tr. V.* cope with

verkrampfen *refl. V.* ⟨*muscle*⟩ become cramped; ⟨*person*⟩ go tense, tense up; **verkrampft lächeln** smile tensely

verkriechen *unr. refl. V.* ⟨*animal*⟩ creep [away]; ⟨*person*⟩ hide [oneself away]; **am liebsten hätte ich mich [in den hintersten Winkel] verkrochen** I'd have liked to crawl away and hide in a corner; I wished the ground would open and swallow me up

verkrümeln *refl. V. (ugs.: sich entfernen)* slip off *or* away

verkrümmt *Adj.* bent ⟨*person*⟩; crooked ⟨*finger*⟩; curved ⟨*spine*⟩

Verkrümmung die crookedness; **~ der Wirbelsäule** curvature of the spine

verkrüppeln 1. *itr. V.; mit sein* ⟨*tree*⟩ become stunted; **verkrüppelt** stunted. **2.** *tr. V.* cripple ⟨*person*⟩; **verkrüppelte Arme/Füße** deformed arms/crippled feet

Verkrüppelung die; ~, ~en deformity

verkrusten *itr. V.; mit sein* form a crust; ⟨*wound*⟩ form a scab

verkümmern *itr. V.; mit sein* **a)** ‹*person, animal*› go into a decline; ‹*plant etc.*› become stunted; ‹*muscle, limb*› waste away, atrophy; **b)** ‹*talent, emotional life, etc.*› wither away; ‹*strength*› decline, fade; ‹*relationship*› become less close; ‹*trade, initiative*› dwindle

Verkümmerung die; ~, ~en *s.* **verkümmern b:** withering away; declining; fading; becoming less close; dwindling

verkünden *tr. V.* announce; pronounce ‹*judgement*›; promulgate ‹*law, decree*›; ‹*omen*› presage

verkündigen *tr. V. (geh.)* **a)** *(predigen)* preach; **b)** *(bekanntmachen)* announce; proclaim

Verkündigung die a) *(das Predigen)* preaching; **b)** *(Bekanntmachung)* announcement; proclamation

Verkündung die; ~, ~en announcement; *(von Urteilen)* pronouncement; *(von Gesetzen, Verordnungen)* promulgation

verkuppeln *tr. V.* pair off

verkürzen 1. *tr. V.* **a)** *(verringern)* reduce; *(abkürzen)* shorten; **b)** *(abbrechen)* cut short ‹*stay, life*›; put an end to, end ‹*suffering*›; **verkürzte Arbeitszeit** reduced *or* shorter working hours *pl.*; **c) sich** *(Dat.)* **die Zeit ~:** while away the time; make the time pass more quickly. **2.** *refl. V. (kürzer werden)* become shorter; shorten; ‹*perspective*› become foreshortened. **3.** *itr. V. (Ballspiele)* close the gap (**auf** + *Akk.* to)

Verkürzung die *s.* **verkürzen 1 a, b:** reduction; shortening; cutting short; ending

verlachen *tr. V.* laugh at

verladen *unr. tr. V.* **a)** *(laden)* load; **b)** *(ugs.: betrügen)* **jmdn. ~:** take sb. for a ride *(sl.)*; con sb. *(sl.)*; *(Ballspiele)* out-trick sb.

Verlade·rampe die loading platform

Verladung die loading

Verlag [fɛɐ̯'la:k] **der; ~[e]s, ~e** publishing house *or* firm; publisher's

verlagern 1. *tr. V.* shift ‹*weight, centre of gravity*›; *(an einen anderen Ort)* move; *(fig.)* transfer; shift ‹*emphasis*›. **2.** *refl. V. (auch fig.)* shift; ‹*area of high/low pressure etc.*› move

Verlagerung die moving; **eine ~ des Schwergewichts** *(fig.)* a shift in emphasis

Verlags-: ~haus das publishing house *or* firm; **~programm das** [publisher's] list

verlangen 1. *tr. V.* **a)** *(fordern)* demand; *(wollen)* want; **das ist zuviel verlangt** that's asking too much; that's too much to expect; **von jedem wird Pünktlichkeit verlangt** everyone is required *or* expected to be punctual; **b)** *(nötig haben)* ‹*task etc.*› require, call for ‹*patience, knowledge, experience, skill, etc.*›; **c)** *(berechnen)* charge; **sie verlangte 200 Mark von ihm** she charged him 200 marks; **d)** *(sehen wollen)* ask for, ask to see ‹*passport, driving-licence, etc.*›; **e)** *(am Telefon)* ask for; ask to speak to; **du wirst am Telefon verlangt** you're wanted on the phone *(coll.)*. **2.** *itr. V. (geh.)* **a)** *(bitten)* **nach einem Arzt/Priester** *usw.* **~:** ask for a doctor/priest *etc.*; **nach einem Glas Wasser ~:** ask for a glass of water; **b)** *(sich sehnen)* **nach jmdm./etw. ~:** long for sb./sth.

Verlangen das; ~s, ~ a) *(Bedürfnis)* desire (**nach** for); **b)** *(Forderung)* demand; **auf ~:** on request; **auf jmds. ~:** at sb.'s request

verlängern [fɛɐ̯'lɛŋɐn] **1.** *tr. V.* **a)** lengthen, make longer ‹*skirt, sleeve, etc.*›; extend ‹*flex, cable, road, etc.*›; **b)** *(länger gültig sein lassen)* renew ‹*passport, driving-licence, etc.*›; extend, renew ‹*contract*›; **c)** *(länger dauern lassen)* extend, prolong ‹*stay, life, suffering, etc.*› (**um** by); **ein verlängertes Wochenende** a long weekend; **d)** *(verdünnen)* add water *etc.* to ‹*sauce, gravy, etc.*› *(to make it go further)*. **2.** *refl. V. (länger werden)* become longer; ‹*stay, life, suffering, etc.*› be prolonged (**um** by); *(länger gültig bleiben)* ‹*contract etc.*› be extended

Verlängerung die; ~, ~en a) *s.* **verlängern 1 a–c:** lengthening; renewal; extension; prolongation; **b)** *(Ballspiele)* extra time *no indef. art.*; *(nachgespielte Zeit)* injury time *no indef. art.*; **c)** *(Teilstück)* extension

verlangsamen 1. *tr. V.* **das Tempo/seine Schritte ~:** reduce speed/slacken one's pace; slow down. **2.** *refl. V.* slow down; ‹*pace*› slacken

Verlaß der *in* **auf jmdn./etw. ist [kein] ~:** sb./sth. can[not] be relied *or* depended [up]on

[1]**verlassen 1.** *unr. refl. V. (vertrauen)* rely, depend (**auf** + *Akk.* on); **er verläßt sich darauf, daß du kommst** he's relying on you to come; **worauf du dich ~ kannst** you can depend on *or* be sure of that. **2.** *unr. tr. V.* leave; **Großvater hat uns für immer ~** *(verhüll.)* grandfather has been taken from us *(euphem.)*

[2]**verlassen** *Adj.* deserted ‹*street, square, village, etc.*›; empty ‹*house*›; *(öd)* desolate ‹*region etc.*›; **einsam und ~:** all alone

verläßlich [fɛɐ̯'lɛslıç] **1.** *Adj.* reliable, dependable ‹*person*›. **2.** *adv.* reliably

Verläßlichkeit die; ~: reliability

Verlaub [fɛɐ̯'laup] **der: mit ~** *(geh.)* with your permission

Verlauf der; ~[e]s, Verläufe course; **der glückliche ~ der Revolution** the fortunate outcome of the revolution

verlaufen 1. *unr. itr. V.; mit sein* **a)** *(sich erstrecken)* run; **b)** *(ablaufen)* ‹*test, rehearsal, etc.*› go; ‹*party etc.*› go off; **c)** ‹*butter, chocolate, etc.*› melt; ‹*make-up, ink*› run. **2.** *unr. itr. (auch refl.) V.; mit sein (sich verlieren)* ‹*track, path*› disappear (**in** + *Dat.* in). **3.** *unr. refl. V.* **a)** *(sich verirren)* get lost; lose one's way; **b)** *(auseinandergehen)* ‹*crowd etc.*› disperse; **c)** *(abfließen)* ‹*floods*› subside

Verlaufs·form die *(Sprachw.)* progressive *or* continuous form

verlautbaren 1. *tr. V.* announce [officially]. **2.** *itr. V.; mit sein (geh.)* become known

Verlautbarung die; ~, ~en announcement; *(inoffizielle Meldung)* [unofficial] report

verlauten 1. *tr. V.* announce. **2.** *itr. V.; mit sein* be reported; **aus amtlicher Quelle verlautet, daß ...:** official reports say that ...

verleben *tr. V.* **a)** *(verbringen)* spend; **b)** *(ugs.: verbrauchen)* spend ‹*money*› on everyday needs

verlebt *Adj.* dissipated

[1]**verlegen 1.** *tr. V.* **a)** *(nicht wiederfinden)* mislay; **b)** *(verschieben)* postpone (**auf** + *Akk.* until); *(vor~)* bring forward (**auf** + *Akk.* to); **c)** *(umlegen)* move; transfer ‹*patient*›; **d)** *(legen)* lay ‹*cable, pipe, carpet, etc.*›; **e)** *(veröffentlichen)* publish. **2.** *refl. V. (sich ausrichten)* take up ‹*subject, activity, occupation, etc.*›; resort to ‹*guesswork, flattery, silence, lying, etc.*›

[2]**verlegen 1.** *Adj.* **a)** embarrassed; **b) um etw. ~ sein** *(etw. nicht zur Verfügung haben)* be short of sth.; *(etw. benötigen)* be in need of sth. **2.** *adv.* in embarrassment

Verlegenheit die; ~, ~en a) *o. Pl. (Befangenheit)* embarrassment; **in ~ geraten** get *or* become embarrassed; **jmdn. in ~ bringen** embarrass sb.; **b)** *(Unannehmlichkeit)* embarrassing situation

Verleger der; ~s, ~, **Verlegerin die;** ~, ~nen publisher

Verlegung die; ~, ~en **a)** *(Verschiebung)* postponement; *(Vorverlegung)* bringing forward *no art.*; **um eine ~ des Termins bitten** ask to change the appointment; **b)** *s.* [1]**verlegen 1 c:** moving; transfer; **c)** *(von Kabeln, Rohren, Teppichen usw.)* laying

verleiden *tr. V.* **jmdm. etw. ~:** spoil sth. for sb.

Verleih der; ~[e]s, ~e **a)** *o. Pl. (das Verleihen)* hiring out; *(von Autos)* renting *or* hiring out; **b)** *(Unternehmen)* hire firm *or* company

verleihen *unr. tr. V.* **a)** hire out; rent *or* hire out ‹*car*›; *(umsonst)* lend [out]; **b)** *(überreichen)* award; bestow, confer ‹*award, honour*›; **jmdm. einen Orden ~:** decorate sb.; **c)** *(verschaffen)* give; lend

Verleiher der; ~s, ~: hirer; *(Film~)* distributor

Verleihung die; ~, ~en **a)** *s.* **verleihen a:** hiring out; renting out; lending [out]; **b)** *s.* **verleihen b:** awarding; bestowing; conferring; *(Zeremonie)* award; conferment; bestowal

verleiten *tr. V.* **jmdn. dazu ~, etw. zu tun** lead *or* induce sb. to do sth.; *(verlocken)* tempt *or* entice sb. to do sth.; **jmdn. zum Stehlen ~:** lead sb. into stealing

verlernen *tr. V.* forget; **das Kochen ~:** forget how to cook

[1]**verlesen 1.** *unr. tr. V.* read out. **2.** *unr. refl. V. (falsch lesen)* make a mistake/mistakes in reading; **er hat sich wohl ~:** he must have read it wrongly

verletzen [fɛɐ̯'lɛtsn̩] *tr. V.* **a)** *(beschädigen)* injure; *(durch Schuß, Stich)* wound; **b)** *(kränken)* hurt, wound ‹*person, feelings*›; **c)** *(verstoßen gegen)* violate; infringe ‹*regulation*›; break ‹*agreement, law*›; **d)** *(eindringen in)* violate ‹*frontier, airspace, etc.*›

verletzlich *Adj.* vulnerable

Verletzlichkeit die; ~: vulnerability

Verletzte der/die; *adj. Dekl.* injured person; casualty; *(durch Schuß, Stich)* wounded person

Verletzung die; ~, ~en **a)** *(Wunde)* injury; **eine ~ am Knie haben** have an injury to one's knee *or* an injured knee; **b)** *(Kränkung)* hurting; wounding; **c)** *s.* **verletzen c:** violation; infringement; breaking; **d)** *(Grenz~, Luftraum~ usw.)* violation

verleugnen *tr. V.* deny; disown ‹*friend, relation*›; **er kann seine Herkunft nicht ~:** it is obvious where he comes from; **sich selbst ~:** go against *or* betray one's principles

Verleugnung die denial; *(eines Freundes, Verwandten)* disownment

verleumden [fɛɐ̯'lɔymdn̩] *tr. V.* slander; *(schriftlich)* libel

Verleumder der; ~, ~: slanderer; *(schriftlich)* libeller

verleumderisch *Adj.* slanderous; *(in Schriftform)* libellous

Verleumdung die; ~, ~en **a)** *o. Pl.* slander; *(in Schriftform)* libelling

verlieben *refl. V.* fall in love **(in + Akk.** with); **ein verliebtes Pärchen** a pair of lovers

Verliebte der/die; *adj. Dekl.* lover

Verliebtheit die; ~: being *no art.* in love

verlieren [fɛɐ̯'li:rən] **1.** *unr. tr. V.* lose; ‹*plant, tree*› lose, shed ‹*leaves*›; **die Katze verliert Haare** the cat is moulting. **2.** *unr. itr. V.* lose; **an etw.** *(Dat.)* **~:** lose sth. **3.** *unr. refl. V.* **a)** *(weniger werden)* ‹*enthusiasm*› subside; ‹*reserve etc.*› disappear; **b)** *(entschwinden)* vanish; ‹*sound*› die away

Verlierer der; ~s, ~: loser

Verlies [fɛɐ̯'li:s] **das;** ~es, ~e dungeon

verlischt [fɛɐ̯'lɪʃt] *3. Pers. Sg. Präsens v.* **verlöschen**

verloben *refl. V.* become *or* get engaged, *(arch.)* become betrothed **(mit** to)

Verlobte der; *adj. Dekl.* fiancé

Verlobte die; *adj. Dekl.* fiancée

Verlobung die; ~, ~en engagement; betrothal *(arch.)*; *(Feier)* engagement party

verlocken *tr. V. (geh.)* tempt; entice

verlockend *Adj.* tempting; enticing

Verlockung die; ~, ~en temptation; enticement

verlogen [fɛɐ̯'lo:gn̩] **1.** *Adj.* lying, mendacious ‹*person*›; false ‹*morality, phrases, romanticism, etc.*›; insincere ‹*compliment*›. **2.** *adv.* mendaciously; falsely

Verlogenheit die; ~, ~en *(eines Menschen)* mendacity; *(einer Moral, von Phrasen usw.)* falseness; *(von Komplimenten)* insincerity

verlor [fɛɐ̯'lo:ɐ̯] *1. u. 3. Pers. Sg. Prät. v.* **verlieren**

verloren 1. *2. Part. v.* **verlieren. 2.** *Adj.* lost; **[eine] ~e Mühe** a wasted effort; **die Sache ist ~:** it's hopeless; **er ist ~:** that's the end of him now

verloren|gehen *unr. itr. V.; mit sein* **a)** *(abhanden kommen)* get lost; **durch diesen Umweg ging uns/ging viel Zeit verloren** we lost a lot of time/a lot of time was lost by this detour; **b)** *(nicht gewonnen werden)* ‹*war, battle, etc.*› be lost

verlosch [fɛɐ̯'lɔʃ] *1. u. 3. Pers. Sg. Prät. v.* **verlöschen**

verloschen *2.Part. v.* **verlöschen**

verlöschen *unr. itr. V.; mit sein* ‹*light, fire, etc.*› go out

verlosen *tr. V.* raffle

Verlosung die; ~, ~en raffle; draw; *(Ziehung)* draw; *(Vorgang)* raffling

verlottern [fɛɐ̯'lɔtɐn] *itr. V.; mit sein (abwertend)* ‹*building, town, area, etc.*› become run-down; ‹*person*› go to seed; ‹*firm, business*› go downhill, go to the dogs

Verlust der; ~[e]s, ~e loss **(an + Dat.** of); **etw. mit ~ verkaufen** sell sth. at a loss

Verlust·meldung die casualty report

verlust·reich *Adj.* **a)** *(mit vielen Toten)* ‹*battle etc.*› involving heavy losses; **b)** *(finanziell)* heavily loss-making ‹*product, project, etc.*›

vermachen *tr. V.* **jmdm. etw. ~:** leave *or* bequeath sth. to sb.; *(fig.: schenken, überlassen)* give sth. to sb.; let sb. have sth.

vermählen [fɛɐ̯'mɛ:lən] *(geh.) refl. V.* **sich [jmdm.** *od.* **mit jmdm.] ~:** marry *or* wed [sb.]

Vermählung die; ~, ~en *(geh.)* **a)** marriage; wedding; **b)** *(Fest)* wedding ceremony

vermasseln [fɛɐ̯'masl̩n] *tr. V. (salopp)* **a)** *(verderben)* muck up *(Brit. sl.)*; mess up; ruin; **b)** *(verhauen)* make a cock-up *(Brit. sl.)* or mess of ‹*exam etc.*›

vermehren 1. *tr. V.* increase **(um** by). **2.** *refl. V.* **a)** increase; **b)** *(sich fortpflanzen)* reproduce

Vermehrung die; ~, ~en **a)** increase *(Gen.* in); **b)** *(Fortpflanzung)* reproduction

vermeidbar *Adj.* avoidable; **die Niederlage wäre ~ gewesen** the defeat could have been avoided

vermeiden *unr. tr. V.* avoid; **es läßt sich nicht ~:** it is unavoidable; **es ~, etw. zu tun** avoid doing sth.

Vermeidung die; ~, ~en avoidance

vermeintlich [fɛɐ̯'maintlɪç] *adv.* supposedly

vermengen 1. *tr. V. (mischen)* mix (**miteinander** together). 2. *refl. V. (sich mischen)* mingle

Vermerk [fɛɐ̯'mɛrk] **der; ~[e]s, ~e** note; *(amtlich)* remark; *(Stempel)* stamp; *(im Kalender)* entry

vermerken *tr. V.* **a)** *(notieren)* make a note of; note [down]; *(in Akten, Wachbuch usw.)* record; **das sei aber nur am Rande vermerkt** but that is only by the way; **b)** *(feststellen)* note

[1]**vermessen** *unr. tr. V.* measure; survey ⟨*land, site*⟩

[2]**vermessen** *Adj. (geh.)* presumptuous

Vermessenheit die; ~, ~en *(geh.)* presumption; presumptuousness

Vermessung die measurement; *(Land~)* surveying

vermiesen *tr. V. (ugs.)* **jmdm. etw. ~:** spoil sth. for sb.

vermieten *tr. (auch itr.) V.* rent [out], let [out] ⟨*flat, room, etc.*⟩ (**an** + *Akk.* to); hire [out] ⟨*boat, car, etc.*⟩; **„Zimmer zu ~"** 'room to let'

Vermieter der landlord

Vermieterin die landlady

Vermietung die; ~, ~en *s.* **vermieten:** renting [out]; letting [out]; hiring [out]

vermindern 1. *tr. V.* reduce; decrease; reduce, lessen ⟨*danger, stress*⟩; lessen ⟨*admiration, ability*⟩; reduce ⟨*debt*⟩. 2. *refl. V.* decrease; ⟨*influence, danger*⟩ decrease, diminish

vermindert *Adj.* **~e Zurechnungsfähigkeit** *(Rechtsw.)* diminished responsibility

Verminderung die *s.* **vermindern 1:** reduction; decreasing; lessening

vermischen 1. *tr. V.* mix (**miteinander** together); blend ⟨*teas, tobaccos, etc.*⟩; **Wahres und Erdachtes miteinander ~:** mingle truth and fiction. 2. *refl. V.* mix; *(fig.)* mingle

vermissen *tr. V.* **a)** *(sich sehnen nach)* miss; **b)** *(nicht haben)* **ich vermisse meinen Ausweis** my identity card is missing; **er gilt als** *od.* **ist vermißt** *(fig.)* he is listed as a missing person

Vermißte der/die; *adj. Dekl.* missing person

vermitteln 1. *itr. V.* mediate, act as [a] mediator (**in** + *Dat.* in). 2. *tr. V.* **a)** *(herbeiführen)* arrange; negotiate ⟨*transaction, cease-fire, compromise*⟩; **b)** *(besorgen)* **jmdm. eine Stelle ~:** find sb. a job; find a job for sb.; **c)** *(weitergeben)* impart ⟨*knowledge, insight, values, etc.*⟩; communicate, pass on ⟨*message, information, etc.*⟩; convey, give ⟨*feeling*⟩; pass on ⟨*experience*⟩

Vermittler der; ~s, ~ a) *s.* **vermitteln 1:** mediator; **b)** *s.* **vermitteln 2c:** imparter; communicator; conveyer; **c)** *(von Berufs wegen)* agent

Vermittler·rolle die role of mediator

Vermittlung die; ~, ~en a) *s.* **vermitteln 1:** mediation; **b)** *s.* **vermitteln 2a:** arrangement; negotiation; **durch die ~ eines Beamten** through the good offices of an official; **c)** *(das Besorgen)* **die ~ einer Stelle** finding a job for sb.; **d)** *s.* **vermitteln 2c:** imparting; communicating; passing on; conveying; **e)** *(Telefonzentrale)* exchange; *(in einer Firma)* switchboard; *(Telefonist)* operator

vermöbeln *tr. V. (ugs.)* beat up; *(als Strafe)* thrash

vermodern *itr. V.; mit sein* decay; rot

vermögen *(geh.) unr. tr. V.* **etw. zu tun ~:** be able to do sth.; be capable of doing sth.

Vermögen das; ~s, ~ a) *o. Pl. (geh.: Fähigkeit)* ability; **b)** *(Besitz)* fortune; **er hat ~:** he has money; he is a man of means; **sein ganzes ~:** all his money

vermögend *Adj.* wealthy; well-off

Vermögens-: ~steuer die wealth tax; **~verhältnisse** *Pl.* financial circumstances

vermummen [fɛɐ̯'mʊmən] *tr. V.* **a)** *(einhüllen)* wrap up [warmly]; **b)** *(verbergen)* disguise

Vermummungs·verbot, das ban on wearing masks [during demonstrations]

vermurksen [fɛɐ̯'mʊrksn̩] *tr. V. (ugs.)* mess up; muck up *(Brit. sl.)*

vermuten *tr. V.* suspect; **das ist zu ~:** that is what one would suppose *or* expect; we may assume that; **ich vermutete ihn in der Bibliothek** I supposed *or* presumed he was in the library

vermutlich 1. *Adj.* probable, likely ⟨*result*⟩. 2. *Adv.* presumably; *(wahrscheinlich)* probably

Vermutung die; ~, ~en supposition ; *(Verdacht)* suspicion

vernachlässigen *tr. V.* neglect; *(unberücksichtigt lassen)* ignore; disregard

Vernachlässigung die; ~, ~en neglect; *(das Nichtberücksichtigen)* disregard

vernageln *tr. V.* nail up, cover ⟨*hole etc.*⟩; **mit Brettern vernagelt** boarded up

vernarben *itr. V.; mit sein* [form a] scar; heal *(lit. or fig.)*

vernarren *refl. V.* **in jmdn./etw. vernarrt sein** be infatuated with *or (coll.)* crazy about sb./be crazy *(coll.)* abouth sth.

vernaschen *tr. V.* **a)** spend on sweets *(Brit.)* *or (Amer.)* candy; **b)** *(salopp)* lay ⟨*girl*⟩ *(sl.)*

vernebeln *tr. V.* shroud ⟨*area*⟩ in fog; *(mit Rauch)* cover ⟨*area*⟩ with a smoke-screen

vernehmbar *Adj. (geh.)* audible

vernehmen *unr. tr. V.* **a)** *(geh.: hören, erfahren)* hear; **b)** *(verhören)* question; *(vor Gericht)* examine

Vernehmen das: dem/allem ~ nach from what/all that one hears

vernehmlich 1. *Adj.* [clearly] audible. 2. *adv.* audibly; **laut und ~:** loud and clear

Vernehmung die; ~, ~en questioning; *(vor Gericht)* examination

vernehmungsfähig *Adj.* in a condition *or* fit to be questioned/examined *postpos.*

verneigen *refl. V. (geh.)* bow (**vor** + *Dat.* to, *(literary)* before)

Verneigung die; ~, ~en *(geh.)* bow

verneinen *tr. (auch itr.) V.* **a)** say 'no' to ⟨*question*⟩; answer ⟨*question*⟩ in the negative; **er schüttelte ~d den Kopf** he shook his head to say 'no'; **b)** *(Sprachw.)* negate

Verneinung die; ~, ~en a) ~ einer Frage negative answer to a question; **b)** *(Sprachw.)* negation

vernetzen *tr. V. (Chemie, Technik)* interlink

vernichten *tr. V.* destroy; exterminate ⟨*pests, vermin*⟩

vernichtend 1. *Adj.* crushing ⟨*defeat*⟩; shattering ⟨*blow*⟩; *(fig.)* devastating ⟨*criticism*⟩. 2. *adv.* **den Feind ~ schlagen** inflict a crushing defeat on the enemy

Vernichtung die; ~, ~en destruction; *(von Schädlingen)* extermination

verniedlichen *tr. V.* trivialize ⟨*matter, situation, etc.*⟩; play down ⟨*guilt, error*⟩

Verniedlichung die; ~, ~en trivialization

Vernissage [vɛrnɪ'sa:ʒə] **die; ~, ~n** *(geh.)* private view *(of contemporary artist's exhibition)*

Vernunft [fɛɐ̯'nʊnft] **die; ~:** reason; **~ annehmen** see reason; come to one's senses; **jmdn. zur ~ bringen** make sb. see reason

vernunft·begabt *Adj.* rational

vernünftig [fɛɐ̯'nynftɪç] **1.** *Adj.* **a)** sensible; **b)** *(ugs.: ordentlich, richtig)* decent. **2.** *adv.* **a)** sensibly; **b)** *(ugs.: ordentlich, richtig)* ‹*talk, eat*› properly; ‹*dress*› sensibly
veröden 1. *itr. V.; mit sein* become deserted. **2.** *tr. V. (Med.)* treat ‹*varicose veins*› by injection
veröffentlichen *tr. V.* publish
Veröffentlichung die; ~, ~en publication
verordnen *tr. V.* **[jmdm. etw.] ~:** prescribe [sth. for sb.]
Verordnung die prescribing; prescription
verpachten *tr. V.* lease
Verpachtung die; ~, ~en leasing
verpacken *tr. V.* pack; wrap up ‹*present, parcel*›; **etw. als Geschenk ~:** gift-wrap sth.
Verpackung die a) *o.Pl.* packing; **b)** *(Umhüllung)* packaging *no pl.;* wrapping
verpassen *tr. V.* **a)** miss ‹*train, person, entry (Mus.), chance, etc.*›; **b)** *(ugs.)* **jmdm. eins ~:** clout sb. one *(coll.)*
verpatzen *tr. V. (ugs.)* make a mess of; muck up *(Brit. sl.);* botch ‹*job*›
verpennen *(salopp)* **1.** *itr. V.* oversleep. **2.** *tr. V.* **a)** *(vergessen)* forget; **b)** *(verschlafen)* sleep through ‹*morning etc.*›
verpesten *tr. V. (abwertend)* pollute
verpetzen *tr. V.* **jmdn. [beim Lehrer** *usw.*] **~:** tell *or (sl.)* split on sb. [to the teacher *etc.*]
verpfänden *tr. V.* pawn ‹*article*›; mortgage ‹*house*›; *(fig.)* pledge ‹*word, honour*›
Verpfändung die pawning; *(von Hausbesitz)* mortgaging; mortgage
verpfeifen *unr. tr. V. (ugs. abwertend)* grass *or* split on ‹*person*› *(sl.)* **(bei** to)
verpflanzen *tr. V.* **a)** transplant ‹*tree, bush*›; **b)** *(Med.)* transplant ‹*heart etc.*›; graft ‹*skin*›
Verpflanzung die; ~, ~en a) transplanting; **b)** *(Med.)* transplant[ing]; *(von Haut)* graft
verpflegen *tr. V.* cater for; feed
Verpflegung die; ~, ~en a) *o.Pl.* catering *no indef. art.* (*Gen.* for); **b)** *(Nahrung)* food; **Unterkunft und ~:** board and lodging
Verpflegungs·kosten *Pl.* cost *sing.* of food *or* meals
verpflichten 1. *tr. V.* **a)** oblige; commit; *(festlegen, binden)* bind; **zur Verschwiegenheit verpflichtet** sworn to secrecy; **jmdm. verpflichtet sein** be indebted to sb.; **b)** *(einstellen, engagieren)* engage ‹*actor, manager, etc.*›; *(Sport)* sign ‹*player*›. **2.** *refl. V.* undertake; promise; **sich zu einer Zahlung ~:** commit oneself to making a payment; **sich vertraglich ~:** sign a contract; bind oneself by contract
Verpflichtung die; ~, ~en a) obligation; commitment; **[finanzielle] ~en** [financial] commitments; liabilities; **b)** *(Engagement)* engaging; engagement; *(Sport: eines Spielers)* signing
verpfuschen *tr. V. (ugs.)* make a mess of; muck up *(Brit. sl.)*
verplanen *tr. V.* **a)** get the plans wrong for; **b)** *(festlegen, einteilen)* book ‹*person, time*› up; commit ‹*money, reprint*›
verplappern *refl. V. (ugs.)* blab *(coll.);* let the cat out of the bag
verplempern [fɛɐ̯'plɛmpɐn] *(ugs.) tr. V.* fritter away
verpönt *Adj.:* scorned; *(tabu)* taboo
verprassen *tr. V.* squander, *(sl.)* blow ‹*money, fortune*›
verprügeln *tr. V.* beat up; *(zur Strafe)* thrash
verpuffen *itr. V.; mit sein* go phut; *(fig.)* fizzle out
verpulvern *tr. V. (ugs.)* blow *(sl.)* ‹*money*›; *(allmählich)* fritter away ‹*money*›
verpuppen *refl. V. (Zool.)* pupate
Verputz der plaster; *(auf Außenwänden)* rendering
verputzen *tr. V.* **a)** *(mit Putz versehen)* plaster; render ‹*outside wall*›; **b)** *(ugs.: aufessen)* polish off *(coll.)* ‹*food*›
Verputzer der; ~s, ~: plasterer
verquer 1. *Adj.* **a)** *(schief)* angled, crooked ‹*position*›; **b)** *(absonderlich)* weird, outlandish ‹*idea*›. **2.** *adv.* **a)** *(schief)* at an angle; crookedly; **b)** *(absonderlich)* ‹*behave*› weirdly
verquicken [fɛɐ̯'kvɪkn̩] *tr. V.* combine
Verquickung die; ~, ~en combination
verquirlen *tr. V.* mix [with a whisk]; whisk
verquollen [fɛɐ̯'kvɔlən] *Adj.* swollen
verrammeln *tr. V.* barricade
verramschen [fɛɐ̯'ramʃn̩] *s.* **verschleudern a**
verrannt [fɛɐ̯'rant] *Adj.* obsessed
Verrat der; ~[e]s betrayal **(an +** *Dat.* of); **~ begehen** *(Politik)* commit [an act of] treason
verraten 1. *unr. tr. V.* **a)** betray ‹*person, cause*›; betray, give away ‹*secret, plan, etc.*› **(an +** *Akk.* to); **~ und verkauft sein** be well and truly in the soup *or* sunk *(coll.);* **b)** *(ugs.: mitteilen)* **jmdm. den Grund** *usw.* **~:** tell sb. the reason *etc.;* **c)** *(erkennen lassen)* show, betray ‹*feelings, surprise, fear, etc.*›; show ‹*influence, talent*›; **d)** *(zu erkennen geben)* give ‹*person*› away. **2.** *unr. refl. V.* **a)** ‹*person*› give oneself away; **b)** *(sich zeigen)* show itself; be revealed
Verräter [fɛɐ̯'rɛːtɐ] **der; ~s, ~:** traitor (*Gen.*, **an** + *Dat.* to)
Verräterin die; ~, ~en traitress
verräterisch *Adj.* **a)** treacherous ‹*plan, purpose, act, etc.*›; **b)** *(erkennen lassend)* tell-tale, give-away ‹*look, gesture*›
verräuchern *tr. V.* fill with smoke
verraucht *Adj.* smoke-filled; smoky
verrechnen 1. *tr. V.* include, take into account ‹*amount etc.*›; *(gutschreiben)* credit ‹*cheque etc.*› to another account. **2.** *refl. V. (auch fig.)* miscalculate; make a mistake/mistakes
Verrechnung die settlement **(mit** by means of); **„nur zur ~“** *(Bankw.)* 'not negotiable'; 'a/c payee [only]'
Verrechnungs·scheck der *(Wirtsch.)* crossed cheque
verrecken *itr. V.; mit sein (salopp)* die [a miserable death]
verregnen *itr. V.; mit sein* be spoilt *or* ruined by rain; **verregnet** rainy, wet ‹*spring, summer, holiday, etc.*›
verreiben *unr. tr. V.* rub in
verreisen *itr. V.; mit sein* go away; **verreist sein** be away
verreißen *unr. tr. V.(ugs.)* tear ‹*book, play, etc.*› to pieces
verrenken [fɛɐ̯'rɛŋkn̩] *tr. V.* **a)** *(verletzen)* dislocate; **sich** *(Dat.)* **den Fuß ~:** twist one's ankle; **b)** *(biegen)* **sich** *od.* **seine Glieder ~:** go into *or* perform contortions
Verrenkung die; ~, ~en a) *(Verletzung)* dislocation; **b)** *(Biegung des Körpers)* contortion
verrennen *unr. refl. V.* get on the wrong track *or* off course; **sich in etw.** *(Akk.)* **~:** become obsessed with sth.
verrichten *tr. V.* perform ‹*work, duty, etc.*›
Verrichtung die carrying out; performance
verriegeln *tr. V.* bolt
Verriegelung die; ~, ~en a) *(das Verriegeln)* bolting; **b)** *(Vorrichtung)* bolt mechanism
verringern [fɛɐ̯'rɪŋɐn] **1.** *tr. V.* reduce. **2.** *refl. V.* decrease

Verringerung die; ~: reduction; decrease (*Gen.*, **von** in)

Verriß der *(ugs.)* damning review *or* criticism (**über** + *Akk.* of)

verrosten *itr. V.; mit sein* rust

verrotten *itr. V.; mit sein* rot; ⟨*building etc.*⟩ decay

verrücken *tr. V.* move; shift

verrückt *(ugs.)* **1.** *Adj.* **a)** mad; **~ werden** go mad *or* insane; **jmdn. ~ machen** drive sb. mad; **du bist wohl ~!** you must be mad *or* crazy!; **wie ~:** like mad *or (coll.)* crazy; **ich werde ~!** I'll be blowed *(sl.) or (coll.)* damned; **~ spielen** *(salopp)* ⟨*person*⟩ act crazy *(coll.)*; ⟨*car, machine, etc.*⟩ play up *(coll)*; ⟨*watch, weather*⟩ go crazy; **b)** *(überspannt, ausgefallen)* crazy ⟨*idea, fashion, prank, day, etc.*⟩; **so was Verrücktes!** what a crazy idea!; **c)** *(begierig)* crazy; **auf jmdn.** *od.* **nach jmdm/auf etw.** *(Akk.)* **~ sein** be crazy *(coll.) or* mad about sb./sth. **2.** *adv.* crazily; ⟨*behave*⟩ crazily *or* like a madman; ⟨*paint, dress, etc.*⟩ in a mad *or* crazy way

Verrückte der/die; *adj. Dekl. (ugs.)* madman/madwoman; lunatic

Verrücktheit die; ~, ~en a) *o.Pl.* madness; insanity; *(Überspanntheit)* craziness; **b)** *(überspannte Idee)* crazy idea

Verruf der *in* **in ~ kommen** *od.* **geraten** fall into disrepute

verrufen *Adj.* disreputable

verrühren *tr. V.* stir together; mix

verrutschen *itr. V.* slip

Vers [fɛrs] **der; ~es, ~e** verse; *(Zeile)* line; **~e schreiben** *od.* *(ugs.)* **schmieden** write verse *or* poetry; **sich** *(Dat.)* **einen ~ auf etw.** *(Akk.)***/darauf machen** *(fig.)* make sense of sth./put two and two together

versagen 1. *itr. V.* fail; ⟨*machine, engine*⟩ stop [working], break down; **menschliches Versagen** human error; **ihre Stimme versagte** her voice failed. **2.** *tr. V. (geh.) (nicht gewähren)* **jmdm. etw. ~:** deny *or* refuse sb. sth.; **ein Kind blieb ihr versagt** a child was denied her; **ich konnte es mir nicht ~, darauf zu antworten** I could not refrain from answering. 3. *refl. V.* **sich jmdm. ~:** refuse to give oneself *or* surrender to sb.

Versager der; ~s, ~: failure

versalzen *unr. tr. V.* **a)** put too much salt in/on; **die Suppe ist versalzen** there is too much salt in the soup; the soup is too salty; **b)** *(fig. ugs.)* spoil (*Dat.* for)

versammeln 1. *tr. V.* assemble; gather [together]; **seine Leute um sich ~:** gather one's people around one. **2.** *refl. V.* assemble; *(weniger formell)* gather

Versammlung die a) meeting; *(Partei~)* assembly; *(unter freiem Himmel, bes. politisch)* rally; **auf einer ~ sprechen** speak at a meeting/rally; **b)** *(Gremium)* assembly

Versand der; ~[e]s a) dispatch; **b)** *(Abteilung)* dispatch department

versanden *itr. V.; mit sein* fill with sand; ⟨*harbour etc.*⟩ silt up; *(mit Sand bedeckt werden)* be covered with sand

Versand·haus das mail-order firm

versauern *itr. V.; mit sein (ugs.)* waste away; stagnate

versaufen 1. *unr. tr. V. (salopp)* drink one's way through

versäumen *tr. V.* **a)** *(verpassen)* miss; lose ⟨*time, sleep*⟩; **b)** *(vernachlässigen, unterlassen)* neglect ⟨*duty, task*⟩; **das Versäumte/Versäumtes nachholen** make up for *or* catch up on what one has neglected *or* failed to do

Versäumnis das; ~ses, ~se omission

verschaffen *tr. V.* **jmdm. Arbeit/Geld/Unterkunft** *usw.* **~:** provide sb. with work/money/accommodation *etc.*; get sb. work/money/accommodation *etc.*; **sich** *(Dat.)* **etw. ~:** get hold of sth.; obtain sth.; **sich** *(Dat.)* **Respekt ~:** gain respect; **was verschafft mir die Ehre?** *(iron.)* to what do I owe this honour?

verschämt [fɛɐ̯ˈʃɛːmt] **1.** *Adj.* bashful. **2.** *adv.* bashfully

verschandeln *tr. V. (ugs.)* spoil; ruin

Verschandelung die; ~, ~en *(ugs.)* ruination *no indef. art.*

verschanzen *refl. V. (Milit.)* take up a [fortified] position; *(in einem Graben)* entrench oneself; dig [oneself] in; **sich hinter einer Zeitung ~** *(fig.)* take cover *or* hide behind a newspaper

verschärfen 1. *tr. V.* **a)** intensify ⟨*conflict, difference, desire, etc.*⟩; increase, step up ⟨*pace, pressure*⟩; tighten ⟨*law, control, restriction, etc*⟩; make ⟨*penalty*⟩ more severe; **b)** make ⟨*unemployment etc.*⟩ worse; aggravate ⟨*situation, crisis, etc.*⟩. **2.** *refl. V.* **a)** ⟨*pace, pressure, etc.*⟩ increase; ⟨*pain, tension, conflict, difference*⟩ intensify; **b)** *(sich verschlimmern)* get worse

verschärft 1. *Adj.; nicht präd.* **a)** increased ⟨*pressure*⟩; intensified ⟨*conflict*⟩; more intense ⟨*training*⟩; tighter, stricter ⟨*control, check, restriction*⟩; more severe ⟨*reprimand, punishment*⟩; **b)** *(schlimmer geworden)* aggravated. **2.** *adv. (strenger)* more strictly

verscharren *tr. V.* bury *(just below the surface)*

verschätzen *refl. V.* **sich in etw.** *(Dat.)* **~:** misjudge sth.

verscheiden *unr.itr. V.; mit sein (geh.)* pass away

verschenken *tr. V.* **a)** give away; **etw. an jmdn. ~:** give sth. to sb.; **b)** *(ungewollt vergeben)* waste ⟨*space*⟩; give away ⟨*points*⟩

verscherzen *refl. V.* **sich** *(Dat.)* **etw. ~:** lose *or* forfeit sth. [through one's own folly]

verscheuchen *tr. V.* chase away *(lit. or fig.)*; *(durch Erschrecken)* frighten *or* scare away

verschicken *tr. V. s.* **versenden**

verschieben 1. *unr. tr. V.* **a)** shift; move; **b)** *(aufschieben)* put off, postpone (**auf** + *Akk.* till); **etw. auf unbestimmte Zeit ~:** postpone sth. indefinitely; **c)** *(ugs.: illegal verkaufen)* traffic in ⟨*goods*⟩. **2.** *unr. refl. V.* **a)** get out of place; *(rutschen)* slip; **b)** *(erst später stattfinden)* be postponed (**um** for)

Verschiebung die a) movement; *(fig.: Änderung)* alteration, shift (*Gen.* in); **b)** *(zeitlich)* postponement

verschieden 1. *Adj.* **a)** *(nicht gleich)* different (**von** from); **er hat zwei ~e Socken an** he is wearing two odd socks *or* two socks that don't match; **das ist von Fall zu Fall ~:** that varies from one case to another; **b)** *nicht präd. (vielfältig)* various; **auf ~e Weise** in various ways; **die ~sten ...:** all sorts of ...; **in den ~sten Farben** in the most varied colours; in a whole variety of colours; **~es** various things *pl.*; **„Verschiedenes"** 'miscellaneous'; *(Tagesordnungspunkt)* 'any other business'. **2.** *adv.* differently; **~ groß** of different sizes *postpos.*; different-sized; ⟨*people*⟩ of different heights

verschieden·artig 1. *Adj.* different in kind *pred.*; *(mehr als zwei)* diverse. **2.** *adv.* diversely; *(auf verschiedene Weise)* in various different ways

Verschiedenartigkeit die; ~: difference in nature; *(unter mehreren)* diversity

Verschiedenheit die; ~, ~en difference; dissimilarity; *(unter mehreren)* diversity

verschiedentlich *Adv.* on various occasions

verschießen *unr. tr. V.* **a)** *(als Geschoß verwenden)* fire ⟨*shell, cartridge, etc.*⟩; **b)** *(verbrauchen)* use up ⟨*ammunition*⟩; *s. auch* **Pulver b; c) einen Strafstoß ~** *(Fußball)* miss with a penalty

verschimmeln *itr. V.; mit sein* go mouldy; **verschimmelt** mouldy

[1]**verschlafen 1.** *unr.itr. (auch refl.)V.* oversleep. **2.** *unr. tr. V.* **a)** *(schlafend verbringen)* sleep through ⟨*morning, journey, concert, etc.*⟩; **b)** *(versäumen)* not wake up in time to catch ⟨*train, bus*⟩; **c)** *(ugs.: vergessen)* forget about ⟨*appointment etc.*⟩

[2]**verschlafen** *Adj.* **a)** half-asleep; **b)** *(fig.: ruhig, langweilig)* sleepy ⟨*town, village*⟩

Verschlag der shed; *(für Kaninchen)* hutch

[1]**verschlagen** *unr. tr. V.* **a) [jmdm.] die Seite ~:** lose sb.'s place *or* page; **die Seite ~** *(im eigenen Buch)* lose one's place *or* page; **b) jmdm. die Sprache** *od.* **Rede/den Atem ~:** leave sb. speechless/take sb.'s breath away; **c)** *(Ballspiele)* mishit ⟨*ball*⟩; **d) das Leben hat ihn nach X ~:** the vagaries of life caused him to end up in X

[2]**verschlagen 1.** *Adj. (abwertend: gerissen)* sly; shifty. **2.** *adv. (abwertend: gerissen)* slyly; shiftily

Verschlagenheit die; ~ *(abwertend)* slyness; shiftiness

verschlampen *(ugs., bes. südd.) tr. V.* succeed in losing *(iron.)*

verschlechtern 1. *tr. V.* make worse. **2.** *refl. V.* get worse; deteriorate; **sich [finanziell/wirtschaftlich** *usw.*] **~:** be worse off [financially/economically *etc.*]

Verschlechterung die; ~, ~en worsening, deterioration (*Gen.* in)

verschleiern *tr. V.* **a)** veil; **b)** *(fig.: verbergen)* draw a veil over, cover up ⟨*deception, facts, scandal, etc.*⟩; hide ⟨*intentions*⟩

verschleiert 1. *Adj.* veiled; misty ⟨*vision etc.*⟩; fogged ⟨*photograph*⟩. **2.** *adv.* **ohne Brille sieht er alles nur ~:** without [his] glasses he sees everything as in a mist

Verschleierung die; ~, ~en a) *o. Pl.* veiling; **b)** *(fig.: von Sachverhalten, Motiven)* covering up

verschleimt *Adj.* congested with phlegm *postpos.*

Verschleimung die; ~, ~en mucous congestion

Verschleiß [fɛɐ̯ˈʃlais] **der; ~es, ~e a)** *(Abnutzung)* wear *no indef. art.;* wear and tear *sing., no indef. art;* **einen höheren ~ haben** wear more rapidly; have a higher rate of wear; **b)** *(Verbrauch)* consumption **(an +** *Dat.* of)

verschleißen 1. *unr. itr. V.; mit sein* wear out. **2.** *unr. tr. V.* wear out; *(fig.)* run down, ruin ⟨*one's nerves, one's health*⟩; use up ⟨*energy, ability, etc.*⟩; **verschlissen** worn ⟨*material, suit, etc.*⟩; worn out ⟨*machine parts etc.*⟩

verschleppen *tr. V.* **a)** carry off ⟨*valuables, animals*⟩; take away ⟨*person*⟩; *(bes. nach Übersee)* transport ⟨*convicts, slaves, etc.*⟩; **b)** *(weiterverbreiten)* carry, spread ⟨*disease, bacteria, mud, etc.*⟩; **c)** *(verzögern)* delay; *(in die Länge ziehen)* draw out; **d)** *(unbehandelt lassen)* let ⟨*illness*⟩ drag on [and get worse]; **verschleppte Krankheit** illness aggravated by neglect

Verschleppung die; ~, ~en *s.* **verschleppen: a)** carrying off; transportation; **b)** carrying; spreading; **c)** delaying; drawing out; **d)** aggravation by neglect

verschleudern *tr. V.* **a)** *(billig verkaufen)* sell dirt cheap; *(mit Verlust)* sell at a loss; **b)** *(abwertend: verschwenden)* squander

verschließbar *Adj.* **a)** closable; **[luftdicht] ~:** sealable ⟨*container etc.*⟩; **b)** *(abschließbar)* lockable

verschließen 1. *unr. tr. V.* **a)** close ⟨*package, tin, pores, mouth, etc*⟩; close up ⟨*blood-vessel, aperture, etc.*⟩; stop ⟨*bottle*⟩; *(mit einem Korken)* cork ⟨*bottle*⟩ ; **etw. luftdicht ~:** make sth. airtight; put an airtight seal on sth.; **die Augen/Ohren [vor etw.** *(Dat.)*] **~** *(fig.)* close one's eyes *or* be blind/turn a deaf ear *or* be deaf [to sth.]; **b)** *(abschließen)* lock ⟨*door, cupboard, drawer, etc.*⟩; lock up ⟨*house etc.*⟩; *s. auch* **Tür; c)** *(wegschließen)* lock away **(in +** *Dat. od. Akk.* in); **d)** *(versperren)* bar ⟨*way etc.*⟩. **2.** *unr. refl. V.* **a) sich jmdm. ~:** be closed to sb.; ⟨*person*⟩ shut oneself off from sb.; **b)** *in* **sich einer Sache** *(Dat.)* **~:** close one's mind to sth.; *(ignorieren)* ignore sth.

verschlimmbessern *tr. V. (ugs. scherzh.)* make worse with so-called corrections

verschlimmern 1. *tr. V.* make worse; aggravate ⟨*state of health*⟩. **2.** *refl. V.* get worse; worsen; ⟨*position, conditions*⟩ deteriorate, worsen

Verschlimmerung die; ~, ~en worsening

verschlingen 1. *unr. tr. V.* **a)** [inter]twine ⟨*threads, string, etc.*⟩ **(zu** into); **b)** *(essen, fressen)* devour ⟨*food*⟩; *(fig.)* devour, consume ⟨*novel, money, etc.*⟩. **2.** *unr. refl. V.* **sich ineinander ~:** become entwined *or* intertwined

verschlissen *2. Part. v.* **verschleißen 2**

verschlossen *Adj. (wortkarg)* taciturn, tight-lipped; *(zurückhaltend)* reserved

Verschlossenheit die; ~ *(Wortkargheit)* taciturnity; *(Zurückhaltung)* reserve

verschlucken 1. *tr. V.* swallow ⟨*food, bone, word, etc.*⟩; *(fig.)* absorb, deaden ⟨*sound*⟩; absorb, eliminate ⟨*rays*⟩. **2.** *refl. V.* choke **(an +** *Dat.* over)

verschlungen *Adj.* entwined ⟨*ornamentation*⟩; winding ⟨*path etc.*⟩; **er saß mit ~en Armen da** he sat there with arms folded

Verschluß der a) *(am BH, an Schmuck usw.)* fastener; fastening; *(an Taschen, Schmuck)* clasp; *(an Schuhen, Gürteln)* buckle; *(am Schrank, Fenster, Koffer usw.)* catch; *(an Flaschen)* top; *(Stöpsel)* stopper; *(Schraub~)* [screw-]top; [screw-]cap; *(Tank~)* cap; **b) unter ~:** under lock and key

Verschluß·sache die [item of] confidential information

verschmachten *itr. V.; mit sein (geh.)* fade away **(vor +** *Dat.* from); *(vor Sehnsucht)* pine away

verschmähen *tr. V. (geh.)* spurn; **verschmähte Liebe** unrequited love

verschmerzen *tr. V.* get over ⟨*defeat, disappointment*⟩

verschmieren *tr. V.* **a)** smear ⟨*window etc.*⟩; *(beim Schreiben)* mess up ⟨*paper*⟩; scrawl all over ⟨*page*⟩; **b)** *(verteilen)* spread ⟨*butter etc.*⟩; smudge ⟨*ink*⟩

verschmitzt [fɛɐ̯ˈʃmɪtst] **1.** *Adj.* mischievous; roguish. **2.** *adv.* mischievously; roguishly

verschmoren *itr. V.; mit sein (ugs.)* burn

verschmust [fɛɐ̯ˈʃmu:st] *Adj. (ugs.)* ⟨*child, cat, etc.*⟩ that always wants to be cuddled

verschmutzen 1. *itr. V.; mit sein* ⟨*material*⟩ get dirty; ⟨*river etc.*⟩ become polluted. **2.** *tr. V.* dirty, soil ⟨*carpet, clothes*⟩; pollute ⟨*air, water, etc.*⟩

Verschmutzung die; ~, ~en a) *(der Umwelt)* pollution; **b)** *(von Stoffen, Teppichen usw.)* soiling; **c)** *(Schmutz)* dirt *no. pl.;* **~en** [cases *pl.* of] soiling *sing.*

verschnaufen *itr. (auch refl.) V.* have *or* take a breather
Verschnauf·pause die rest; breather
verschneit *Adj.* snow-covered *attrib.;* covered with snow *postpos.*
verschnörkelt *Adj.* ornate
verschnupft [fɛɐ̯ˈʃnʊpft] *Adj.* suffering from a cold *postpos.*
verschnüren *tr. V.* tie up (**zu** into)
verschollen [fɛɐ̯ˈʃɔlən] *Adj.* missing; **er ist ~:** he has disappeared; *(wird vermißt)* he is missing; **er galt seit langem als ~:** for a long time it had been thought he had disappeared
verschonen *tr. V.* spare; **von etw. verschont bleiben** be spared by sth.; escape sth.; **jmdn. mit etw. ~:** spare sb. sth.
verschönern [fɛɐ̯ˈʃø:nɐn] *tr. V.* brighten up
verschränken [fɛɐ̯ˈʃrɛŋkn̩] *tr. V.* fold ‹*arms*›; cross ‹*legs*›; clasp ‹*hands*›
verschrauben *tr. V.* screw on; **[miteinander] ~:** screw together
verschrecken *tr. V.* frighten *or* scare [off *or* away]
verschreiben 1. *unr. tr. V.* **a)** *(verbrauchen)* use up ‹*paper, ink, pencils, etc.*›; **b)** *(Med.: verordnen)* prescribe ‹*medicine, treatment, etc.*›; **jmdm. ein Medikament ~:** prescribe a medication for sb.; **c)** *(falsch schreiben)* write incorrectly *or* wrongly. **2.** *unr. refl. V.* **a)** *(einen Fehler machen)* make a slip of the pen; **b)** *(sich widmen)* **sich einer Sache** *(Dat.)* **~:** devote oneself to sth.
Verschreibung die; ~, ~en prescription
verschreibungs·pflichtig *Adj.* available only on prescription *postpos.*
verschrie[e]n [fɛɐ̯ˈʃri:[ə]n *Adj.* notorious (**wegen** for)
verschroben [fɛɐ̯ˈʃro:bn̩] **1.** *Adj.* eccentric, cranky ‹*person*›; cranky, weird ‹*ideas*›. **2.** *adv.* eccentrically; weirdly
Verschrobenheit die; ~, ~en eccentricity
verschrotten *tr. V.* scrap
verschrumpeln *itr. V.; mit sein (ugs.)* go shrivelled; **verschrumpelt** shrivelled
verschüchtern *tr. V.* intimidate; **verschüchtert** timid; *(adverbial)* timidly
verschulden 1. *tr. V.* be to blame for ‹*accident, death, etc.*›; *(Fußball usw.)* give away ‹*goal, corner*›. **2.** *refl. V.* get into debt; **er hat sich dafür hoch ~ müssen** he had to borrow heavily to do that
Verschulden das; ~s guilt; **durch eigenes/fremdes ~:** through one's own/someone else's fault
verschuldet *Adj.* **a)** in debt *postpos.* (**bei** to); **hoch ~:** deeply in debt; **b)** *(belastet)* mortgaged; **hoch ~:** heavily mortgaged
Verschuldung die; ~, ~en indebtedness *no. pl.*
verschütten *tr. V.* **a)** spill; **b)** *(begraben)* bury ‹*person*› [alive]; submerge, bury ‹*road etc.*›; *(fig.)* submerge; **ein Verschütteter** one of those buried/trapped
verschwägert [fɛɐ̯ˈʃvɛ:gɐt] *Adj.* related by marriage *postpos.*
verschweigen *unr. tr. V.* conceal ‹*truth etc.*›; *(verheimlichen)* keep quiet about; **jmdm. etw. ~:** hide *or* conceal sth. from sb.; **du verschweigst mir doch etwas** you're keeping something from me; *s. auch* **verschwiegen 2**
verschwenden *tr. V.* waste (**an** + *Akk.* on)
Verschwender der; ~s, ~ *(von Geld)* spendthrift; *(von Dingen)* wasteful person
verschwenderisch 1. *Adj.* **a)** wasteful, extravagant ‹*person*›; ‹*life*› of extravagance; **b)** *(üppig)* lavish; sumptuous. **2.** *adv.* **a)** wastefully, extravagantly; **b)** *(üppig)* lavishly; sumptuously
Verschwendung die; ~, ~en wastefulness; extravagance; **so eine ~!** what a waste!
verschwiegen 1. *2. Part. v.* **verschweigen. 2.** *Adj.* **a)** *(diskret)* discreet; **b)** *(still, einsam)* secluded ‹*place, bay*›; quiet ‹*restaurant etc.*›
Verschwiegenheit die; ~: secrecy; *(Diskretion)* discretion
verschwimmen *unr. itr. V.; mit sein* blur; become blurred; **die Zeilen/Buchstaben verschwammen mir vor den Augen** the lines/letters swam in front of my eyes
verschwinden *unr. itr. V.; mit sein* **a)** disappear; vanish; ‹*pain, spot, etc.*› disappear, go [away]; **es ist besser, wir ~/laß uns hier ~:** we'd better/let's make ourselves scarce *(coll.);* **verschwinde [hier]!** off with you!; go away!; hop it! *(sl.);* **ich muß mal ~** *(ugs. verhüll.)* I have to pay a visit *(coll.) or (Brit. coll.)* spend a penny; **jmdn. ~ lassen** take sb. away; *(ermorden)* eliminate sb.; do away with sb.; **etw. ~ lassen** *(wegzaubern)* ‹*conjurer*› make sth. disappear *or* vanish; *(stehlen)* help oneself to sth. *(coll.); (unterschlagen, beiseite schaffen)* dispose of sth.; **b) neben jmdm./etw. ~** *(sehr klein wirken)* be dwarfed by sb./sth.; *(unbedeutend wirken)* pale into insignificance beside sb./sth.
verschwindend 1. *Adj.* tiny. **2.** *adv.* **~ klein** tiny; minute
verschwistert [fɛɐ̯ˈʃvɪstɐt] *Adj.* **[miteinander] ~ sein** *(Bruder u. Schwester sein)* be brother and sister; *(Brüder u. Schwestern sein)* be brothers and sisters; *(Brüder/Schwestern sein)* be brothers/sisters
verschwitzen *tr. V.* **a)** make ‹*shirt, dress, etc.*› sweaty; **verschwitzt** sweaty; **b)** *(ugs.: vergessen)* forget
verschwollen [fɛɐ̯ˈʃvɔlən] *Adj.* swollen
verschwommen 1. *2. Part. v.* **verschwimmen. 2.** *Adj.* blurred ‹*photograph, vision*›; blurred, hazy ‹*outline*›; vague, woolly ‹*idea, concept, formulation, etc.*›; vague ‹*hope*›. **3.** *adv.* ‹*express, formulate, refer*› vaguely; ‹*remember*› hazily; **ich sehe alles ganz ~:** everything looks blurred to me
verschwören *unr. refl. V.* conspire, plot (**gegen** against)
Verschwörer der; ~s, ~, Verschwörerin die; ~, ~nen conspirator
Verschwörung die; ~, ~en conspiracy; plot
versehen 1. *unr. tr. V.* **a)** *(ausstatten)* provide; equip ‹*car, factory, machine, etc.*›; **b)** *(ausüben, besorgen)* perform ‹*duty etc.*›; **bei jmdm. den Haushalt ~:** keep house for sb.; **c)** *(innehaben)* hold ‹*post, job*›. **2.** *unr. refl. V.* **a)** *(einen Fehler machen)* make a slip; slip up; **b)** *in* **ehe man sich's versieht** before you know where you are
Versehen das; ~s, ~: oversight; slip; **aus ~:** by mistake; inadvertently
versehentlich 1. *Adv.* by mistake; inadvertently. **2.** *adj.; nicht präd.* inadvertent
versehrt [fɛɐ̯ˈze:rt] *Adj.* disabled
Versehrte der/die; *adj. Dekl.* disabled person
verselbständigen *refl. V.* become independent
versenden *unr. (auch regelm.) tr. V.* send ‹*letter, parcel*›; send out ‹*invitations*›; dispatch ‹*goods*›
Versendung die *s.* **versenden:** sending; sending out; dispatch
versengen *tr. V.* scorch; singe ‹*hair*›

versenken *tr. V.* **a)** sink ‹*ship*›; lower ‹*body, coffin*›; **b)** *(verschwinden lassen)* lower, retract ‹*aerial, rostrum, etc.*›

Versenkung die a) *s.* **versenken a, b:** sinking; lowering; retraction; **b) in der ~ verschwinden** *(fig. ugs.)* vanish from the scene; sink into oblivion

versessen [fɛɐ̯'zɛsn̩] *Adj.* **auf jmdn./etw. ~ sein** be dead keen on *or* crazy about sb./sth. *(coll.)*; **darauf ~ sein, etw. zu tun** be dying to do sth.

versetzen 1. *tr. V.* **a)** move; transfer, move ‹*employee*›; *(in die nächsthöhere Klasse)* move ‹*pupil*› up, *(Amer.)* promote ‹*pupil*› **(in +** *Akk.* to); *(fig.)* transport **(in +** *Akk.* to); **b)** *(nicht geradlinig anordnen)* stagger; **c)** *(verpfänden)* pawn; **d)** *(verkaufen)* sell; **e)** *(ugs.: vergeblich warten lassen)* stand ‹*person*› up *(coll.)*; **f)** *(vermischen)* mix; **g)** *(erwidern)* retort; **h) etw. in Bewegung ~:** set sth. in motion; **jmdn. in Erstaunen/Unruhe/Angst/Begeisterung ~:** astonish sb./make sb. uneasy/frighten sb./fill sb. with enthusiasm; **jmdn. in die Lage ~, etw. zu tun** put sb. in a position to do sth.; **jmdm. einen Stoß/Fußtritt/Schlag** *usw.* **~:** give sb. a push/kick/deal sb. a blow *etc.* **2.** *refl. V.* **sich an jmds. Stelle** *(Akk.)* *od.* **in jmds. Lage** *(Akk.)* **~:** put oneself in sb.'s position *or* place

Versetzung die; ~, ~en a) moving; *(einer Pflanze)* transplanting; *(eines Schülers)* moving up, *(Amer.)* promotion **(in +** *Akk.* to); *(eines Angestellten)* transfer; move; **b)** *(Verpfändung)* pawning; **c)** *(Verkauf)* selling; sale; **d)** *(das Mischen)* mixing; *s. auch* **Ruhestand**

Versetzungs·zeugnis das *(Schulw.)* end-of-year report *(confirming pupil's move to a higher class)*

verseuchen *tr. V. (auch fig.)* contaminate

Versicherer der; ~s, ~: insurer

versichern 1. *tr. V.* **a)** *(als wahr hinstellen)* assert, affirm ‹*sth.*›; **etw. hoch und heilig/eidesstattlich ~:** swear blind to sth./attest sth. in a statutory declaration; **b)** *(vertraglich schützen)* insure **(bei** with); **sein Leben ist hoch versichert** his life is assured *or* insured for a large sum. **2.** *refl. V. (geh.)* **sich einer Sache** *(Gen.)* **~:** make sure *or* certain of sth.

Versicherte der/die; *adj. Dekl.* insured [person]

Versicherung die a) *(Beteuerung)* assurance; **eine eidesstattliche ~:** a statutory declaration; **b)** *(Schutz durch Vertrag)* insurance; *(Vertrag)* insurance [policy] **(über +** *Akk.* for); **eine ~ abschließen** take out an insurance [policy]; **c)** *(Gesellschaft)* insurance [company]

versicherungs-, Versicherungs-: ~beitrag der insurance premium; **~fall der** event giving rise to a claim; **~gesellschaft die** insurance company; **~karte die a)** *(Sozialversicherung)* insurance *or* contribution card; **b)** *(Kfz-Versicherung)* **die grüne ~karte** the green card; **~nehmer der** policy-holder; **~pflichtig** *Adj.* **a)** subject to compulsory insurance *postpos.*; **b)** *(Sozialversicherung)* ‹*person*› liable for [insurance] contributions; ‹*earnings*› subject to [insurance] contributions; **~police die** *s.* **~schein; ~prämie die** *s.* **~beitrag; ~schein der** insurance policy

versickern *itr. V.; mit sein* ‹*river etc.*› drain *or* seep away

versiegeln *tr. V.* seal

Versiegelung die; ~, ~en a) seal; **b)** *o. Pl. (das Versiegeln)* sealing

versiegen *itr. V.; mit sein (geh.)* dry up; run dry; ‹*tears*› cease [to flow]; *(fig.)* peter out; ‹*energy*› run out

versiert [vɛr'ziːɐ̯t] *Adj.* experienced [and knowledgeable]; **in etw.** *(Dat.)* **~ sein** be well versed in sth.

versilbern *tr. V.* **a)** silver-plate; **b)** *(ugs.: verkaufen)* turn into cash; flog *(Brit. sl.)*

versinken *unr. itr. V.; mit sein* **a)** sink; **im Schlamm/Schnee ~:** sink into the mud/snow; **im Moor ~:** be sucked into the bog; **ich wäre am liebsten im Erdboden versunken** I wished the ground would [open and] swallow me up; **b)** *(fig.)* **~ in (+** *Akk.)* become immersed in *or* wrapped up in ‹*memories, thoughts*›; subside, lapse into ‹*melancholy, silence, etc.*›

versinnbildlichen *tr. V.* symbolize

Version [vɛr'zi̯oːn] **die; ~, ~en** version

versklaven *tr. V.* enslave

versnoben *itr. V.; mit sein (abwertend)* become snobbish; turn into a snob

versohlen *tr. V. (ugs.)* belt ‹*person, backside, etc.*›

versöhnen [fɛɐ̯'zøːnən] **1.** *refl. V.* **sich [miteinander] ~:** become reconciled; make it up. **2.** *tr. V.* reconcile; **jmdn. mit seinem Schicksal ~:** reconcile sb. to his/her fate

versöhnlich 1. *Adj.* **a)** conciliatory; **b)** *(erfreulich)* positive; optimistic. **2.** *adv.* **a)** in a conciliatory way; ‹*say*› in a conciliatory tone; **b)** *(erfreulich)* ‹*end*› positively, optimistically

Versöhnung die; ~, ~en reconciliation

versonnen 1. *Adj.* dreamy; *(in Gedanken versunken)* lost in thought *postpos.* **2.** *adv.* dreamily; *(in Gedanken)* lost in thought

versorgen *tr. V.* **a)** supply; **hast du den Hund/die Blumen schon versorgt?** have you fed the dog/watered the flowers?; **b)** *(unterhalten, ernähren)* provide for ‹*children, family*›; **c)** *(sorgen für)* look after; attend to, see to ‹*heating, garden, etc.*›; **jmdn. ärztlich ~:** give sb. medical care; *(kurzzeitig)* give sb. medical attention

Versorger der; ~s, ~, Versorgerin die; ~, ~nen breadwinner; provider

Versorgung die; ~, ~en a) *o. Pl.* supply[ing]; **die ~ einer Stadt mit etw.** the supply of sth. to a town; **b)** *(Unterhaltung, Ernährung)* support[ing]; **c)** *(Bedienung, Pflege)* care; **ärztliche ~:** medical care *or* treatment; *(kurzzeitig)* medical attention; **d)** *(Bezüge)* maintenance

verspannen *refl. V.* ‹*muscle*› tense up; **verspannt** taut ‹*muscle*›; *(verkrampft)* seized-up ‹*back*›

Verspannung die *(Med.: der Muskulatur)* tension

verspäten *refl. V.* be late

verspätet *Adj.* late ‹*arrival, rose, butterfly*›; belated ‹*greetings, thanks*›

Verspätung die; ~, ~en lateness; *(verspätetes Eintreffen)* late arrival; **[fünf Minuten] ~ haben** be [five minutes] late; **eine fünfminütige ~:** a five-minute delay; **seine** *od.* **die ~ aufholen** make up the lost time; **mit dreimonatiger ~:** three months late

verspeisen *tr. V. (geh.)* consume; partake of

versperren *tr. V.* block ‹*road, entrance*›; obstruct ‹*view*›; **jmdm. den Weg/die Sicht ~:** block sb.'s path/block *or* obstruct sb.'s view

verspielen 1. *tr. V.* gamble away; *(fig.: verwirken)* squander, throw away ‹*opportunity, chance*›; forfeit ‹*right, credibility, sb.'s trust, etc*›. **2.** *itr. V.* **in [bei jmdm.] verspielt haben** *(ugs.)* have had it [so far as sb. is concerned] *(coll.)*. **3.** *refl. V.* play a wrong note/wrong notes

verspielt 1. *Adj. (auch fig.)* playful; fanciful, fantastic ‹*form, design, etc.*›. **2.** *adv.* playfully *(lit. or fig.)*; ‹*dress, designed*› fancifully, fantastically

verspinnen *unr. tr. V.* spin ‹*wool*› *(***zu** into)

versponnen 1. *2. Part. v.* **verspinnen. 2.** *Adj.* eccentric, odd ‹*person*›; odd, weird ‹*idea*›

verspotten *tr. V.* mock; ridicule

Verspottung die; ~, ~en mocking; ridiculing

versprechen 1. *unr. tr. V.* **a)** promise; **was er verspricht, hält er auch** he keeps his promises; **sein Blick versprach nichts Gutes** his glance was ominous; **b) sich** *(Dat.)* **etw. von etw./jmdm. ~:** hope for sth. *or* to get sth. from sth./sb. **2.** *unr. refl. V.* make a slip/slips of the tongue

Versprechen das; ~s, ~: promise

Versprecher der; ~s, ~: slip of the tongue

Versprechung die; ~, ~en promise

versprengen *tr. V.* **a)** *(bes. Milit.)* disperse; scatter; **b)** *(verspritzen)* sprinkle ‹*water*›

verspritzen *tr. V.* **a)** spray; **b)** *(bespritzen)* spatter ‹*windscreen, coat, etc.*›

versprühen *tr. V.* spray; **Geist** *od.* **Witz ~** *(fig.)* show sparkling wit; scintillate

verspüren *tr. V.* feel

verstaatlichen *tr. V.* nationalize

Verstaatlichung die; ~, ~en nationalization

verstädtern [fɛɐ̯ˈʃtɛːtɐn] *itr. V.; mit sein* become urbanized

Verstand der; ~[e]s *(Fähigkeit zu denken)* reason *no art.*; *(Fähigkeit, Begriffe zu bilden)* mind; *(Vernunft)* [common] sense *no art.*; **Tiere haben keinen ~:** animals do not have the power *or* faculty of reason; **der menschliche ~:** the human mind; **wenn du deinen ~ gebraucht hättest** if you had used your brain *or* had been thinking; **ich hätte ihm mehr ~ zugetraut** I thought he would have had more sense; **manchmal zweifle ich an seinem ~:** I sometimes doubt his sanity; **hast du denn den ~ verloren** *(ugs.)* have you taken leave of your senses?; are you out of your mind?; **das geht über meinen ~:** that's beyond me

verstandes·mäßig 1. *Adj.* rational; intellectual ‹*inferiority, superiority*›. **2.** *adv.* rationally; intellectually ‹*inferior, superior*›

verständig [fɛɐ̯ˈʃtɛndɪç] **1.** *Adj.* sensible; intelligent. **2.** *adv.* sensibly; intelligently

verständigen [fɛɐ̯ˈʃtɛndɪgn̩] **1.** *tr. V.* notify, inform (**von, über** + *Akk.* of). **2.** *refl. V.* **a)** make oneself understood; **sich mit jmdm. ~:** communicate with sb.; **b)** *(sich einigen)* **sich [mit jmdm.] über/auf etw.** *(Akk.)* **~:** come to an understanding *or* reach agreement [with sb.] about *or.* on sth.

Verständigkeit die; ~: understanding; intelligence

Verständigung die; ~, ~en a) notification; **b)** *(das Sichverständlichmachen)* communication *no art.*; **c)** *(Einigung)* understanding

Verständigungs·schwierigkeit die difficulty of communication

verständlich 1. *Adj.* **a)** comprehensible; *(deutlich)* clear ‹*pronunciation, presentation, etc.*›; **[leicht] ~:** easily understood; **schwer ~:** difficult to understand; **sich ~ machen** make oneself understood; **jmdm. etw. ~ machen** make sth. clear to sb.; **b)** *(begreiflich, verzeihlich)* understandable. **2.** *adv.* comprehensibly; in a comprehensible way; *(deutlich)* ‹*speak, express oneself, present*› clearly

verständlicher·weise *Adv.* understandably

Verständnis das; ~ses, ~se understanding; **ein ~ für Kunst/ Musik** an appreciation of *or* feeling for art/music; **ich habe volles ~ dafür, daß ...:** I fully understand that ...; **für so etwas habe ich kein ~:** I have no time for that kind of thing; **für die Unannehmlichkeiten bitten wir um [Ihr] ~:** we ask for your forbearance *or* we apologize for the inconvenience caused

verständnis-, Verständnis-: ~los 1. *Adj.* uncomprehending; **2.** *adv.* uncomprehendingly; **~losigkeit die** incomprehension; **voller ~losigkeit** uncomprehendingly; with a complete lack of understanding; **~voll 1.** *Adj.* understanding; **2.** *adv.* understandingly

verstärken 1. *tr. V.* **a)** strengthen; **b)** *(zahlenmäßig)* reinforce ‹*troops, garrison, etc.*› (**um** by); enlarge, augment ‹*orchestra, choir*› (**um** by); **c)** *(intensiver machen)* intensify, increase ‹*effort, contrast, impression, suspicion*›; *(lauter machen)* amplify ‹*signal, sound, guitar, etc.*›. **2.** *refl. V.* increase

Verstärker der; ~s, ~: amplifier

verstärkt 1. *Adj.; nicht präd.* **a)** increased; *(größer)* greater ‹*efforts, vigilance, etc.*›; **in ~em Maße** to a greater *or* increased extent; **b)** *(zahlenmäßig)* enlarged, augmented ‹*orchestra, choir, etc.*›; reinforced *(Mil.)* ‹*unit*›. **2.** *adv.* to an increased extent

Verstärkung die; ~, ~en a) strengthening; **b)** *(zahlenmäßig)* reinforcement *(esp. Mil.)*; *(eines Orchesters usw.)* enlargement; **c)** *(Intensivierung, Zunahme)* increase (*Gen.* in); *(der Lautstärke)* amplification; **d)** *(zusätzliche Person[en])* reinforcements *pl.*

verstauben *itr. V.; mit sein* get dusty; gather dust *(lit. or fig.)*

verstaubt *Adj.* dusty; covered in dust *postpos.*; *(fig. abwertend)* old-fashioned; outmoded

verstauchen *tr. V.* sprain; **sich** *(Dat.)* **den Fuß/die Hand ~:** sprain one's ankle/wrist

Verstauchung die; ~, ~en sprain

verstauen *tr. V.* pack (**in** + *Dat. od. Akk.* in[to]); *(bes. im Boot/ Auto)* stow (**in** + *Dat. od. Akk.* in); **etw. in einem Schrank ~:** put *or (coll.)* stash sth. away in a cupboard

Versteck das; ~[e]s, ~e hiding-place; *(eines Flüchtlings, Räubers usw.)* hide-out: **~ spielen** play hide-and-seek; **[mit jmdm./miteinander] ~ spielen** *(fig.)* hide *or* keep things [from sb./one another]

verstecken 1. *tr. V.* hide (**vor** + *Dat.* from). **2.** *refl. V.* **sich [vor jmdm./etw.] ~:** hide [from sb./ sth.]; **sich versteckt halten** be [in] hiding; *(versteckt bleiben)* remain in hiding; **sich vor** *od.* **neben jmdm. nicht zu ~ brauchen** *(fig.)* not need to fear comparison with sb.; **sich hinter seinen Vorschriften ~** *(fig.)* use one's rules and regulations to hide behind

versteckt *Adj.* hidden; concealed ‹*polemics*›; veiled ‹*threat*›; *(heimlich)* secret ‹*malice, activity, etc.*›; disguised ‹*foul*›; *(verstohlen)* furtive ‹*glance, smile*›

verstehen 1. *unr. tr. V.* **a)** *(wahrnehmen)* understand; make out; **er war am Telefon gut/schlecht/ kaum zu ~:** it was easy/difficult/ barely possible to understand *or* make out what he was saying on the telephone; **b)** *auch itr. (begrei-*

fen, interpretieren) understand; **ich verstehe** I understand; I see; **wir ~ uns schon** we understand each other; we see eye to eye; **du bleibst hier, verstehst du!** you stay here, understand!; **jmdm. etw. zu ~ geben** give sb. to understand sth.; **das ist in dem Sinne** *od.* **so zu ~, daß ...:** it is supposed to mean that...; **wie soll ich das ~?** how am I to interpret that?; what am I supposed to make of that?; **jmdn./etw. falsch ~:** misunderstand sb./sth.; **versteh mich bitte richtig** *od.* **nicht falsch** please don't misunderstand me *or* get me wrong; **etw. unter etw.** *(Dat.)* **~:** understand sth. by sth.; **jmdn./sich als etw. ~:** see sb./oneself as sth.; consider sb./oneself to be sth.; *s. auch* **Spaß b; c)** *(beherrschen, wissen)* **es ~, etw. zu tun** know how to do sth.; **er versteht eine Menge von Autos** he knows a lot about cars. **2.** *unr. refl. V.* **a) sich mit jmdm. ~:** get on with sb.; **sie ~ sich** they get on well together; **b)** *(selbstverständlich sein)* **das versteht sich [von selbst]** that goes without saying; **c)** *(Kaufmannsspr.: gemeint sein)* **der Preis versteht sich einschließlich Mehrwertsteuer** the price is inclusive of VAT; **d) sich auf Pferde/Autos** *usw. (Akk.)* **~:** know what one is doing with horses/cars; know all about horses/cars

versteifen 1. *tr. V.* stiffen ⟨*collar, part of body, etc.*⟩. **2.** *itr. V.; mit sein* stiffen [up]; become stiff. **3.** *refl. V.* **a)** stiffen [up]; become stiff; **b)** *in* **sich auf etw.** *(Akk.)* **~:** insist on sth.

versteigen *unr. refl. V.* **a)** *(sich verirren)* get lost [while climbing]; *(nicht mehr herunterkönnen)* get stuck; get into difficulties; **b) sich zu einer Behauptung gegen jmdn.** *usw.* **~:** have the presumption to make an assertion on sb. *etc.*

versteigern *tr. V.* auction; **etw. ~ lassen** put sth. up for auction

Versteigerung die auction *no indef. art.;* **zur ~ kommen** *od.* **gelangen** *(Amtsspr.)* be auctioned

versteinern 1. *itr. V.; mit sein* ⟨*plant, animal*⟩ fossilize, become fossilized; ⟨*wood etc.*⟩ petrify, become petrified; *(fig. geh.)* ⟨*person*⟩ go rigid; ⟨*expression, face*⟩ harden, become stony. **2.** *refl. V. (geh.)* ⟨*face, features*⟩ harden

verstellbar *Adj.* adjustable

verstellen 1. *tr. V.* **a)** *(falsch plazieren)* misplace; put [back] in the wrong place; **b)** *(anders einstellen)* adjust ⟨*seat etc.*⟩; alter [the adjustment of] ⟨*mirror etc.*⟩; reset ⟨*alarm clock, points, etc.*⟩; **der Sitz läßt sich in der Höhe ~:** the seat can be adjusted for height; **c)** *(versperren)* block, obstruct ⟨*entrance, exit, view, etc.*⟩; **d)** *(zur Täuschung verändern)* disguise, alter ⟨*voice, handwriting*⟩. **2.** *refl. V.* **a)** *(seine Einstellung, Position verändern)* alter; *(so daß es falsch eingestellt ist)* get out of adjustment; **b)** *(sich anders geben als man ist)* pretend; play-act; **sich vor jmdm. ~:** pretend to sb.

Verstellung die play-acting; pretence; *(der Stimme, Schrift)* disguising; alteration

versterben *unr. itr. V.; mit sein (geh.)* die; pass away; **mein verstorbener Mann** my late husband

verstiegen *Adj.* whimsical ⟨*person*⟩; extravagant, fantastic ⟨*idea, expectation, etc.*⟩

verstimmen *tr. V.* **a)** *(Musik)* put ⟨*instrument*⟩ out of tune; **b)** *(schlechtgelaunt machen)* put ⟨*person*⟩ in a bad mood; *(verärgern)* annoy

verstimmt *Adj.* **a)** *(Musik)* out of tune *pred.;* **b)** *(verärgert)* put out, peeved, disgruntled **(über +** *Akk.* by, about); **ein ~er Magen** an upset stomach

Verstimmung die disgruntled *or* bad mood; *(Verärgerung)* annoyance

verstockt 1. *Adj.* obdurate; stubborn. **2.** *adv.* obdurately; stubbornly

Verstocktheit die; ~: obduracy; stubbornness

verstohlen [fɛɐ̯'ʃto:lən] **1.** *Adj.* furtive; surreptitious. **2.** *adv.* furtively; surreptitiously

verstopfen 1. *tr. V.* block. **2.** *itr. V.; mit sein* become blocked

Verstopfung die; ~, ~en *(Med.: Stuhl~)* constipation

Verstorbene [fɛɐ̯'ʃtɔrbənə] **der/die;** *adj. Dekl. (geh.)* deceased

verstört *Adj.* distraught

Verstörtheit die; ~: distressed *or* distraught state; distress

Verstoß der violation, infringement **(gegen** of)

verstoßen 1. *unr. tr. V.* disown; **aus dem Elternhaus ~ werden** be turned out of one's parents' house; **ein Verstoßener** an outcast. **2.** *unr. itr. V.* **gegen etw. ~:** infringe *or* contravene sth.; **gegen die Etikette ~:** commit a breach of etiquette

verstrahlen *tr. V.* **a)** radiate; **b)** *(radioaktiv verseuchen)* contaminate with radiation

verstreichen 1. *unr. tr. V.* **a)** *(verteilen)* apply, put on ⟨*paint*⟩; spread ⟨*butter etc.*⟩; **b)** *(verbrauchen)* use [up] ⟨*paint*⟩. **2.** *unr. itr. V.; mit sein (geh.)* ⟨*time*⟩ pass [by]

verstreuen *tr. V.* **a)** *(verteilen)* scatter; put down ⟨*bird food, salt*⟩; *(unordentlich)* strew; **b)** *(versehentlich)* spill

verstricken 1. *tr. V.* **jmdn. in etw.** *(Akk.)* **~:** involve sb. in sth.; draw sb. into sth. **2.** *refl. V.* **sich in etw.** *(Akk.)* **~:** become entangled *or* caught up in sth.

Verstrickung die; ~, ~en involvement **(in +** *Akk.* in)

verströmen *tr. V.* exude

verstümmeln *tr. V.* mutilate; *(fig.)* garble ⟨*report*⟩; chop, mutilate ⟨*text*⟩; mutilate, do violence to ⟨*name*⟩; **sich selbst ~:** maim oneself

Verstümmelung die; ~, ~en mutilation; *(fig.: einer Meldung usw.)* garbling

verstummen *itr. V.; mit sein (geh.)* fall silent; ⟨*music, noise, conversation*⟩ cease; *(allmählich)* die *or* fade away; *(fig.)* ⟨*rumour, question*⟩ go away

Versuch der; ~[e]s, ~e a) attempt; **beim ~, etw. zu tun** in attempting to do sth.; **das käme auf einen ~ an** we'll have to try it and see; **b)** *(Experiment)* experiment **(an +** *Dat.* on); *(Probe)* test

versuchen 1. *tr. V.* **a)** try; attempt; **versuch's doch!** *(drohend)* just you try!; *(ermunternd)* just try it!; **es mit jmdm./etw. ~:** give sb./sth. a try; **es bei jmdm. ~:** try sb.; **versuchter Mord** *(Rechtsspr.)* attempted murder; *s. auch* **Glück a; b)** *(auch bibl.: in Versuchung führen)* tempt. **2.** *tr., itr. V. (probieren)* **den Kuchen/von dem Kuchen ~:** try the cake/some of the cake. **3.** *refl. V.* **sich in/an etw.** *(Dat.)* **~:** try one's hand at sth.

versuchs-, Versuchs-: ~anordnung die set-up for an/the experiment/for experiments; **~kaninchen das** *(fig.)* guinea-pig; **~person die** *(bes. Med., Psych.)* test *or* experimental subject; **~weise 1.** *Adv.* on a trial basis; as an experiment; **2.** *adj.; nicht präd.* experimental

Versuchung die; ~, ~en temptation; **in ~** *(Akk.)* **kommen** *od.* **geraten[, etw. zu tun]** be *or* feel tempted [to do sth.]

versündigen *refl. V.* **sich an jmdm./etw. ~:** sin against sb./sth.

Versunkenheit die; ~ *(geh.)* [state of] contemplation; deep meditation

versüßen *tr. V.* **jmdm./sich etw. ~** *(fig.)* make sth. more pleasant for sb./oneself

vertagen 1. *tr. V.* adjourn ‹*meeting, debate, etc.*› (**auf** + *Akk.* until); postpone ‹*decision, verdict*› (**auf** + *Akk.* until). **2.** *refl. V.* ‹*court*› adjourn; ‹*meeting*› be adjourned

vertauschen *tr. V.* **a)** exchange; switch; reverse, switch ‹*roles*›; reverse, transpose ‹*poles*›; **etw. mit etw. ~:** exchange sth. for sth.; **b)** *(verwechseln)* mix up

Vertauschung die; ~, ~en a) exchange; *(von Buchstaben, Polen usw.)* transposition; *(von Rollen)* reversal; switching; **b)** *(Verwechslung)* mixing up

verteidigen [fɛɐ̯'taidɪgn̩] **1.** *tr. V.* defend. **2.** *itr. V. (Ballspiele)* defend

Verteidiger der; ~s, ~, Verteidigerin, die; ~, ~nen a) *(auch Sport)* defender; **b)** *(Rechtsw.)* defence counsel

Verteidigung die, ~, ~en *(auch Sport, Rechtsw.)* defence

Verteidigungs-: ~minister der minister of defence; **~ministerium das** ministry of defence

verteilen 1. *tr. V.* **a)** *(austeilen)* distribute, hand out ‹*exercise books, leaflets, prizes, etc.*› (**an** + *Akk.* to, **unter** + *Akk.* among); share [out], distribute ‹*money, food*› (**an** + *Akk.* to, **unter** + *Akk.* among); allocate ‹*work*›; **b)** *(an verschiedene Plätze bringen)* distribute ‹*weight etc.*› (**auf** + *Akk.* over); spread ‹*cost*› (**auf** + *Akk.* among); **c)** *(verstreichen, verstreuen, verrühren usw.)* distribute, spread ‹*butter, seed, dirt, etc.*›. **2.** *refl. V.* **a)** spread out; **b)** *(sich ausbreiten, verteilt sein)* be distributed (**auf** + *Akk.* over)

Verteilung die distribution; *(der Rollen, der Arbeit)* allocation

vertelefonieren *tr. V. (ugs.)* spend ‹*time*› telephoning *or* on the phone; spend ‹*money*› on telephoning

verteuern 1. *tr. V.* make ‹*goods*› more expensive. **2.** *refl. V.* become more expensive

Verteuerung die increase *or* rise in price

verteufeln *tr. V.* condemn; denigrate

vertiefen 1. *tr. V.* **a)** deepen (**um** by); make deeper; **b)** *(intensivieren)* deepen ‹*knowledge, understanding, love*›; deepen, strengthen ‹*dislike, friendship, collaboration, etc.*›. **2.** *refl. V.* **a)** deepen; become deeper; **b)** *(sich konzentrieren)* **sich ~ in** (+ *Akk.*) bury oneself in ‹*book, work, etc.*›; become deeply involved in ‹*conversation*›; **in Gedanken vertieft** deep in thought; **c)** *(intensiver werden)* ‹*friendship*› deepen; ‹*relations*› become closer

Vertiefung die; ~, ~en a) deepening; *(von Zusammenarbeit, Beziehungen)* strengthening; *(von Wissen)* consolidation; reinforcement; **b)** *(in Gedanken)* absorption (**in** + *Akk.* in); **c)** *(Mulde)* depression; hollow

vertikal [vɛrti'ka:l] **1.** *Adj.* vertical. **2.** *adv.* vertically

Vertikale die; ~; ~n a) *(Linie)* vertical line; **b)** *o. Pl. (Lage)* **die ~:** the vertical *or* perpendicular

vertilgen *tr. V.* **a)** *(vernichten)* exterminate ‹*vermin*›; kill off ‹*weeds*›; **b)** *(ugs.: verzehren)* devour, *(joc.)* demolish ‹*food*›

vertippen 1. *refl. V.* **a)** make a typing mistake/typing mistakes; *(auf der Rechenmaschine, dem Tastentelefon usw.)* press the wrong number; **b)** *(im Lotto, Toto, bei Vorhersagen)* get it wrong. **2.** *tr. V.* mistype ‹*word*›; type ‹*word, letter*› wrongly

vertonen *tr. V.* set ‹*text, poem*› to music; set, write the music to ‹*libretto*›

Vertonung die; ~, ~en setting [to music]; **die ~ eines Librettos** writing the music to a libretto

vertrackt [fɛɐ̯'trakt] *(ugs.) Adj.* complicated, involved ‹*situation, business, etc.*›; tricky, intricate ‹*job*›

Vertrag der; ~[e]s, Verträge [...trɛ:gə] contract; *(zwischen Staaten)* treaty; **mündlicher ~:** verbal agreement; **[bei jmdm.] unter ~ stehen** be under contract [to sb.]

vertragen 1. *unr. tr. V.* **a)** endure; tolerate *(esp. Med.)*; *(aushalten, leiden können)* stand; bear; take ‹*joke, criticism, climate, etc.*›; **sie verträgt dieses Medikament nicht** this medicine does not agree with her at all; **ich könnte jetzt einen Whisky ~** *(ugs.)* I could do with *or* wouldn't say no to a whisky. **2.** *unr. refl. V.* **a)** **sich mit jmdm. ~:** get on *or* along with sb.; **sich gut [miteinander] ~:** get on well together; **sie ~ sich wieder** they are friends again; they have made it up; **b)** *(passen)* **sich mit etw. ~:** go with sth.

vertraglich 1. *Adj.* contractual. **2.** *adv.* contractually; by contract

verträglich [fɛɐ̯'trɛ:klɪç] *Adj.* **a)** digestible ‹*food*›; **leicht/schwer ~:** easily digestible/indigestible; **ein gut ~es Medikament** a drug which has no side-effects; **b)** *(umgänglich)* good-natured; easy to get on with *pred.*

Verträglichkeit die; ~, ~en a) digestibility; **b)** *(Umgänglichkeit)* good nature

Vertrags·entwurf der draft contract/treaty

vertrauen *itr. V.* **jmdm./einer Sache ~:** trust sb./sth.; **auf etw.** *(Akk.)* **~:** [put one's] trust in sth.; **auf sein Glück ~:** trust to luck

Vertrauen das; ~s trust; confidence; **~ zu jmdm./etw. haben/fassen** have/come to have confidence in sb./sth.; trust/come to trust sb./sth.; **jmdm. [sein] ~ schenken** put one's trust in sb.; **sein ~ in jmdn./etw. setzen** put *or* place one's trust in sb./sth.; **im ~ [gesagt]** [strictly] in confidence; between you and me; **im ~ auf etw.** *(Akk.)* trusting to *or* in sth.; **jmdn. ins ~ ziehen** take sb. into one's confidence

vertrauen·erweckend *Adj.* inspiring *or* that inspires confidence *postpos.*

vertrauens-, Vertrauens-: ~arzt der independent examining doctor *(working for health service, health insurance, etc.)*; **~bruch der** breach of trust; *(wenn man Vertrauliches weitersagt)* breach of confidence; **~frage die** *(Parl.)* question of confidence; **die ~frage stellen** ask for a vote of confidence; **~frau die a)** spokeswoman (*Gen.* for); representative; **b)** *s.* **~mann; ~mann der a)** *Pl.* **~männer** *od.* **~leute** spokesman (*Gen.* for); representative; **b)** *Pl.* **~leute** *(in der Gewerkschaft)* [union] representative; *(in einer Fabrik o. ä.)* shop steward; **~person die** person in a position of trust; **~sache die** matter *or* question of trust; **~selig** *Adj.* all too trustful *or* trusting; **~stellung die** position of trust; **~verhältnis das** relationship based on trust; **~voll 1.** *Adj.* trusting ‹*relationship*›; ‹*collaboration, co-operation*› based on trust; *(zuversichtlich)* confident; **2.** *adv.* trustingly; *(zuversichtlich)* confidently; **sich ~voll an jmdn. wenden** turn to sb. with complete confidence; **~würdig** *Adj.* trustworthy; **~würdigkeit die** trustworthiness

vertraulich 1. *Adj.* **a)** confidential; **b)** *(freundschaftlich, intim)* familiar ‹*manner, tone, etc.*›; intimate ⟨*mood, conversation, whisper*⟩. **2.** *adv.* **a)** confidentially; in confidence; **b)** *(freundschaftlich, intim)* in a familiar way; familiarly

Vertraulichkeit die; ~, ~en a) *o. Pl.* confidentiality; **b)** *(vertrauliche Information)* confidence; **c)** *o. Pl. (distanzloses Verhalten)* familiarity; *(Intimität)* intimacy

verträumen *tr. V.* [day-]dream away ⟨*time*⟩

verträumt 1. *Adj.* dreamy. **2.** *adv.* dreamily

vertraut [fɛɐ̯'traut] *Adj.* **a)** close ⟨*friend etc.*⟩; intimate ⟨*circle, conversation, etc.*⟩; **mit jmdm. ~ werden** become very friendly *or* close friends with sb.; **b)** *(bekannt)* familiar; **mit etw. gut/wenig ~ sein** be well acquainted with sth./have little knowledge of sth.; **sich mit etw. ~ machen** familiarize oneself with sth.

Vertraute der/die; *adj. Dekl.* close friend; **enger ~r** intimate friend

Vertrautheit die; ~ *s.* **vertraut:** closeness; intimacy; familiarity

vertreiben *unr. tr. V.* **a)** drive out **(aus** of); *(wegjagen)* drive away ⟨*animal, smoke, clouds, etc.*⟩ **(aus** from); **die vertriebenen Juden** the exiled *or* expelled Jews; **die Müdigkeit/Sorgen ~** *(fig.)* fight off tiredness/drive troubles away; **b)** *(verkaufen)* sell

Vertreibung die; ~, ~en driving out; *(das Wegjagen)* driving away; *(aus der Heimat)* expulsion

vertretbar *Adj.* defensible ⟨*risk etc.*⟩; tenable, defensible ⟨*standpoint*⟩; justifiable ⟨*costs*⟩

vertreten 1. *unr. tr. V.* **a)** stand in *or* deputize for ⟨*colleague etc.*⟩; ⟨*teacher*⟩ cover for ⟨*colleague*⟩; **b)** *(repräsentieren)* represent ⟨*person, firm, interests, constituency, country, etc.*⟩; *(Rechtsw.)* act for ⟨*person, prosecution, etc.*⟩; **schwach/stark ~:** poorly/well represented; **c)** *(einstehen für, verfechten)* support ⟨*point of view, principle*⟩; hold ⟨*opinion*⟩; advocate ⟨*thesis etc.*⟩; pursue ⟨*policy*⟩; **etw. zu ~ haben** be responsible for sth. **2.** *unr. refl. V.* **sich** *(Dat.)* **die Füße** *od.* **Beine ~** *(ugs.: sich Bewegung verschaffen)* stretch one's legs

Vertreter der; ~s, ~, Vertreterin die; ~, ~nen a) *(Stell~)* deputy; stand-in; *(eines Arztes)* locum *(coll.)*; **b)** *(Interessen~, Repräsentant)* representative; *(Handels~)* sales representative; commercial traveller; **ein ~ für Staubsauger** a traveller in vacuum cleaners; **c)** *(Verfechter, Anhänger)* supporter; advocate

Vertretung die; ~, ~en a) deputizing; **jmds. ~ übernehmen** stand in *or* deputize for sb.; ⟨*doctor*⟩ act as locum for sb. *(coll.)*; **in ~ von Herrn N.** in place of *or* standing in for Mr. N.; **b)** *(Vertreter[in])* deputy; stand-in; *(eines Arztes)* locum *(coll.)*; **c)** *(Delegierte[r])* representative; *(Delegation)* delegation; **eine diplomatische ~:** a diplomatic mission; **d)** *(Handels~)* [sales] agency; *(Niederlassung)* agency; branch; **e)** *(Interessen~)* representation; **f)** *(Verfechtung)* advocacy

Vertretungs·stunde die *(Schulw.)* cover lesson

vertretungs·weise *Adv.* as a [temporary] replacement *or* stand-in

Vertriebene der/die; *adj. Dekl.* expellee [from his/her homeland]

vertrinken *unr. tr. V.* spend ⟨*money*⟩ on drink

vertrocknen *itr. V.; mit sein* dry up

vertrödeln *tr. V. (ugs. abwertend)* dawdle away, waste ⟨*time*⟩

vertrösten *tr. V.* put ⟨*person*⟩ off **(auf** + *Akk.* until)

vertun 1. *unr. tr. V.* waste; **die Mühe war vertan** it was a waste of effort. **2.** *unr. refl. V. (ugs.)* make a slip

vertuschen *tr. V.* hush up ⟨*scandal etc.*⟩; keep ⟨*truth etc.*⟩ secret

verübeln *tr. V.* **jmdm. eine Äußerung** *usw.* **~:** take sb.'s remark *etc.* amiss; **das kann man ihm kaum ~:** one can hardly blame him for that

verüben *tr. V.* commit ⟨*crime etc.*⟩

verulken *tr. V. (ugs.)* make fun of; take the mickey out of *(Brit. coll.)*

verunglimpfen [fɛɐ̯'ʊnglimpfn̩] *tr. V. (geh.)* denigrate ⟨*person, etc.*⟩; sully ⟨*name, memory*⟩

verunglücken *itr. V.; mit sein* **a)** have an accident; ⟨*car etc.*⟩ be involved in an accident; **mit dem Auto/Flugzeug ~:** be in a car/an air accident *or* crash; **b)** *(scherzh.: mißlingen)* go wrong; ⟨*attempt*⟩ fail; ⟨*cake, sauce, etc.*⟩ be a disaster

Verunglückte der/die; *adj. Dekl.* accident victim; casualty

verunreinigen *tr. V.* **a)** pollute; contaminate ⟨*water, milk, flour, oil*⟩; **~de Stoffe** pollutants/contaminants; **b)** *(geh.: beschmutzen)* dirty, soil ⟨*clothes, floor, etc.*⟩; *(durch Fäkalien)* foul ⟨*pavement etc.*⟩

Verunreinigung die a) *o. Pl.* pollution; *(von Wasser, Milch, Mehl, Öl)* contamination; **b)** *o. Pl. (von Kleidern, Fußböden usw.)* soiling; *(von Straßen usw.)* fouling

verunsichern *tr. V.* **jmdn. ~:** make sb. feel unsure *or* uncertain; *(so daß er sich gefährdet fühlt)* undermine sb.'s sense of security; **verunsichert** insecure; *(nicht selbstsicher)* unsure of oneself

Verunsicherung die *(Unsicherheit)* [feeling of] insecurity

verunstalten [fɛɐ̯'|ʊnʃtaltn̩] *tr. V.* disfigure

verursachen *tr. V.* cause

Verursacher der; ~s, ~: cause; person responsible; **der ~ des Unfalls** the person responsible for the accident

verurteilen *tr. V.* **a)** pass sentence on; sentence; **jmdn. zu Gefängnis** *od.* **einer Haftstrafe ~:** sentence sb. to imprisonment; **jmdn. zu einer Geldstrafe ~:** impose a fine on sb.; **jmdn. zum Tode ~:** sentence *or* condemn sb. to death; **der zum Tode Verurteilte** the condemned man; **zum Scheitern verurteilt sein** *(fig.)* be condemned to failure *or* bound to fail; **b)** *(fig.: negativ bewerten)* condemn ⟨*behaviour, action*⟩

Verurteilte der/die; *adj. Dekl.* convicted man/woman

Verurteilung die; ~, ~en a) sentencing; **b)** *(fig.)* condemnation

vervielfachen 1. *tr. V.* greatly increase; *(multiplizieren)* multiply ⟨*number*⟩. **2.** *refl. V.* multiply [several times]

vervielfältigen *tr. V.* duplicate, make copies of ⟨*document etc.*⟩

Vervielfältigung die; ~, ~en a) duplicating; copying; **b)** *(Kopie)* copy

vervollkommnen [fɛɐ̯'fɔlkɔmnən] **1.** *tr. V.* perfect. **2.** *refl. V.* become perfected

Vervollkommnung die; ~, ~en perfecting; *(Zustand)* perfection

vervollständigen *tr. V.* complete; *(vollständiger machen)* make ⟨*library etc.*⟩ more complete

Vervollständigung die; ~, ~en completion/making more complete

verwachsen *Adj.* deformed

verwackeln *(ugs.)* **1.** *tr. V.* make ⟨*picture*⟩ blurred; **verwackelt** blurred; shaky. **2.** *itr. V.; mit sein* turn out blurred

verwählen *refl. V.* misdial; dial the wrong number

verwahren 1. *tr. V.* keep [safe]; *(verstauen)* put away [safely]. 2. *refl. V.* protest

verwahrlosen *itr. V.; mit sein* a) get in a bad state; ⟨*house, building*⟩ fall into disrepair, become dilapidated; ⟨*garden, hedge*⟩ grow wild, become overgrown; ⟨*person*⟩ let oneself go, *(coll.)* go to pot; **etw. ~ lassen** neglect sth.; allow sth. to get in a bad state; **b)** *(sittlich ~)* fall into bad ways; **[sittlich] verwahrlost** depraved

Verwahrlosung die; ~ *(eines Gebäudes)* dilapidation; *(einer Person)* advancing decrepitude; *(sittliche ~)* decline into depravity

Verwahrung die keeping [in a safe place]; **etw. in ~ nehmen/haben** take sth. into safe keeping/hold sth. in safe keeping

verwaist *Adj.* orphaned ⟨*child*⟩; *(fig.)* lonely, deserted ⟨*person, place*⟩; unoccupied ⟨*house*⟩

verwalten *tr. V.* **a)** *(betreuen)* administer, manage ⟨*estate, property, etc.*⟩; run, look after ⟨*house*⟩; hold ⟨*money*⟩ in trust; **b)** *(leiten)* run, manage ⟨*hostel, kindergarten, etc.*⟩; *(regieren)* administer ⟨*area, colony, etc.*⟩; govern ⟨*country*⟩

Verwalter der; ~s, ~, **Verwalterin die;** ~, ~nen administrator; *(eines Amts usw.)* manager; *(eines Nachlasses)* trustee

Verwaltung die; ~, ~en **a)** *(Betreuung, Leitung)* administration; management; **die öffentliche/staatliche ~:** the public/state authority; **b)** *(eines Gebiets)* administration; *(eines Landes)* government

Verwaltungs-: ~beamte der administrative official; administrator; **~bezirk der** administrative district; **~gebühr die** administrative charge *or* fee

verwandelbar *Adj.* convertible

verwandeln 1. *tr. V.* **a)** convert (**in** + *Akk.*, **zu** into); *(völlig verändern)* transform (**in** + *Akk.*, **zu** into); **ich fühlte mich wie verwandelt** I felt a different person *or* transformed; **b)** *(Ballspiele)* score from ⟨*corner, free kick*⟩; convert ⟨*penalty*⟩. 2. *refl. V.* **sich in etw.** *(Akk.)* *od.* **zu etw. ~:** turn *or* change into sth.; *(bei chemischen Vorgängen usw.)* be converted into sth. 3. *itr. V.* *(Ballspiele)* **er verwandelte [zum 2:0]** he scored [to make it 2–0]

Verwandlung die; ~, ~en *(das Verwandeln)* conversion (**in** + *Akk.*, **zu** into); *(völlige Veränderung)* transformation (**in** + *Akk.*, **zu** into)

[1]**verwandt** [fɛɐ̯'vant] *2. Part. v.* **verwenden**

[2]**verwandt** *Adj.* **a)** related (**mit** to); **b)** *(fig.: ähnlich)* similar ⟨*views, ideas, forms*⟩

Verwandte der/die; *adj. Dekl.* relative; relation

Verwandtschaft die; ~, ~en **a)** relationship (**mit** to); *(fig.: Ähnlichkeit)* affinity; **zwischen ihnen besteht keine ~:** they are not related [to one another]; **b)** *o. Pl.* *(Verwandte)* relatives *pl.*; relations *pl.*; **die ganze ~:** all one's relatives

verwandtschaftlich 1. *Adj.* family ⟨*ties, relationships, etc.*⟩. 2. *adv.* **~ miteinander verbunden sein** be related [to each other]

Verwandtschafts·verhältnis das family relationship

verwanzen 1. *itr. V.; mit sein* **verwanzt** bug-ridden. 2. *tr. V.* *(fig.)* bug

verwarnen *tr. V.* warn, caution (**wegen** for)

Verwarnung die; ~, ~en warning; caution; **eine gebührenpflichtige ~:** a fine and a caution

verwaschen *Adj.* **a)** washed out, faded ⟨*jeans, material, inscription, etc.*⟩; **b)** *(blaß)* washy, watery ⟨*colour*⟩; blurred ⟨*lines, contours*⟩

verweben *tr. V.* **a)** weave with; use [for weaving]; **b)** *auch unr.* **[miteinander] ~:** interweave ⟨*threads*⟩; **etw. in etw.** *(Akk.)* ~ *(auch fig.)* weave sth. into sth.

verwechselbar *Adj.* mistakable (**mit** for)

verwechseln *tr. V.* **a) [miteinander] ~:** confuse ⟨*two things/people*⟩; **er verwechselt immer rechts und links** he always gets mixed up between *or* mixes up right and left; **etw. mit etw./jmdn. mit jmdm. ~:** mistake sth. for sth./sb. for sb.; confuse sth. with sth./sb. with sb.; **Entschuldigung, ich habe Sie [mit jemandem] verwechselt/ich habe die Tür[en] verwechselt** sorry, I thought you were *or* I mistook you for somebody else/I've got the wrong door; **jmdm. zum Verwechseln ähnlich sehen** be the spitting image of sb.; **b)** *(vertauschen)* mix up; **jemand hat meinen Regenschirm verwechselt** somebody has taken my umbrella by mistake

Verwechslung die; ~, ~en **a)** [case of] confusion; mistake; **b)** *(Vertauschung)* mixing up

verwegen 1. *Adj.* daring; *(auch fig.)* audacious. 2. *adv.* *(auch fig.)* audaciously

Verwegenheit die; ~: daring; *(auch fig.)* audacity

verwehen *tr. V.* **a)** *(zudecken)* cover [over] ⟨*track, path*⟩; **b)** *(wegwehen)* blow away; scatter; **vom Winde verweht** *(fig.)* gone with the wind

verwehren *tr. V.* **jmdm. etw. ~:** refuse *or* deny sb. sth.

Verwehung die; ~, ~en [snow]drift

verweichlichen 1. *itr. V.; mit sein* grow soft. 2. *tr. V.* make soft; **ein verweichlichter Junge** a mollycoddled boy

Verweichlichung die; ~, ~en **a)** *(Vorgang)* **die ~ der Jugendlichen verhindern** prevent young people from becoming soft; **b)** *(Zustand)* softness

verweigern 1. *tr. V.* refuse; **die Aussage/einen Befehl/die Nahrungsaufnahme ~:** refuse to make a statement/to obey an order/to take food; **den Kriegsdienst ~:** refuse to do military service; be a conscientious objector. 2. *refl. V.* object; refuse to co-operate; **sich jmdm./einer Sache ~:** refuse to accept sb./sth. 3. *itr. V.* **a)** *(ugs.: den Kriegsdienst ~)* refuse [to do military service]; be a conscientious objector; **b)** *(Pferdesport)* refuse

Verweigerung die; ~, ~en refusal; *(Protest)* protest

verweilen *itr. V.* *(geh.)* stay; *(länger als nötig)* linger

verweint [fɛɐ̯'vaint] *Adj.* tear-stained ⟨*face*⟩; ⟨*eyes*⟩ red with tears *or* from crying; ⟨*person*⟩ with a tear-stained face

Verweis der; ~es, ~e **a)** reference (**auf** + *Akk.* to); *(Quer~)* cross-reference; **b)** *(Tadel)* reprimand; rebuke; **jmdm. einen ~ erteilen** reprimand *or* rebuke sb.

verweisen *unr. tr. V.* **a) jmdn./einen Fall** *usw.* **an jmdn./etw. ~** *(auch Rechtsspr.)* refer sb./a case *etc.* to sb./sth.; **b)** *(wegschicken)* **jmdn. von der Schule/aus dem Saal ~:** expel sb. from the school/send sb. out of the room; **jmdn. des Landes ~:** exile *or* *(Hist.)* banish sb.; **c) jmdn. auf den zweiten Platz ~** *(Sport)* relegate sb. to *or* push sb. into second place; **d)** *auch itr.* *(hinweisen)* **[jmdn.] auf etw.** *(Akk.)* **~:** refer [sb.] to sth.; *(durch Querverweis)* cross-refer [sb.] to sth.

verwelken *itr. V.; mit sein* wilt; *(fig.)* ⟨*fame*⟩ fade

verweltlichen 1. *tr. V.* secularize. 2. *itr. V.; mit sein (geh.)* become worldly *or* secularized
Verweltlichung die; ~, ~en secularization
verwendbar *Adj.* usable (**zu, für** for)
Verwendbarkeit die; ~: usability
verwenden 1. *unr. od. regelm. tr. V.* **a)** use (**zu, für** for); **b)** *(aufwenden)* spend ‹*time*› (**auf** + *Akk.* on); **viel Energie/Mühe auf etw.** *(Akk.)* **~:** put a lot of energy/effort into sth. 2. *unr. od. regelm. refl. V. (geh.)* **sich [bei jmdm.] für jmdn./etw. ~:** intercede [with sb.] for sb./use one's influence [with sb.] on behalf of sth.
Verwendung die; ~, ~en use; **~ finden** be used; **unter ~ einer Sache** *(Gen.) od.* **von etw.** using sth.
Verwendungs-: ~möglichkeit die [possible] application *or* use; **~zweck der** application; purpose; **„~zweck"** *(auf Zahlkarten usw.)* 'as payment for'
verwerfen *unr. tr. V.* **a)** reject; dismiss ‹*thought*›; **etw. als unsittlich ~:** condemn sth. as [being] immoral; **b)** *(Rechtsw.)* dismiss ‹*appeal, action*›; overturn, quash ‹*judgement*›
verwerflich *(geh.)* 1. *Adj.* reprehensible. 2. *adv.* reprehensibly
Verwerflichkeit die; ~ *(geh.)* reprehensibility; reprehensible *or* despicable nature
verwertbar *Adj.* utilizable; usable
Verwertbarkeit die; ~: usability
verwerten *tr. V.* utilize, use (**zu** for); make use of, exploit ‹*suggestion, experience, knowledge*›
Verwertung die utilization; use; *(bes. kommerziell)* exploitation
verwesen *itr. V.; mit sein* decompose
Verwesung die; ~: decomposition; **in ~ übergehen** start to decompose
verwetten *tr. V.* spend ‹*money*› on betting
verwickeln 1. *refl. V.* **a)** get tangled up *or* entangled; **b)** *(sich verfangen)* **sich in etw.** *(Akk. od. Dat.)* **~:** get caught [up] in sth.; **sich in Widersprüche ~** *(fig.)* tie oneself up in contradictions. 2. *tr. V.* involve; **in etw.** *(Akk.)* **verwickelt werden/sein** get/be mixed up *or* involved in sth.
verwickelt *Adj.* involved; complicated
Verwicklung die; ~, ~en complication
verwildern *itr. V.* ‹*garden*› become overgrown, go wild; ‹*domestic animal*› go wild, return to the wild
verwildert *Adj.* overgrown ‹*garden*›; ‹*animal*› which has gone wild
Verwilderung die; ~, ~en a) return to the wild [state]; **b)** *(geh.: von Menschen)* reversion to a primitive state
verwirken *tr. V. (geh.)* forfeit
verwirklichen 1. *tr. V.* realize ‹*dream*›; realize, put into practice ‹*plan, proposal, idea, etc.*›; carry out ‹*project, intention*›. 2. *refl. V.* **a)** ‹*hope, dream*› be realized *or* fulfilled; **b)** *(sich voll entfalten)* **sich [selbst] ~:** realize one's [full] potential; fulfil oneself
Verwirklichung die; ~, ~en realization; *(eines Wunsches, einer Hoffnung)* fulfilment
verwirren 1. *tr. V.* entangle, tangle up ‹*thread etc.*›; tousle, ruffle ‹*hair*›. 2. *tr. (auch itr.) V.* confuse; bewilder; **~d viele Möglichkeiten** a bewildering number of possibilities. 3. *refl. V.* ‹*thread etc.*› become entangled; ‹*hair*› become tousled *or* ruffled; ‹*person, mind*› become confused
Verwirrung die; ~, ~en confusion; **jmdn. in ~ bringen** make sb. confused *or* bewildered; **in ~ geraten** become confused *or* bewildered; **im Zustand geistiger ~:** in a disturbed *or* confused mental state
verwirtschaften *tr. V.* squander ‹*money*› by mismanagement
verwischen 1. *tr. V.* smudge ‹*signature, writing, etc.*›; smear ‹*paint*›; **alle Spuren ~** *(fig.)* cover up all [one's] tracks. 2. *refl. V.* become blurred
verwittern *itr. V.; mit sein* weather
verwitwet [fɛɐ̯'vɪtvət] *Adj.* widowed
verwöhnen [fɛɐ̯'vø:nən] *tr. V.* spoil; **das Schicksal hat ihn nicht gerade verwöhnt** *(fig.)* fate has not exactly smiled upon him
verwöhnt *Adj.* spoilt; *(anspruchsvoll)* discriminating; ‹*taste, palate*› of a gourmet
verworren [fɛɐ̯'vɔrən] *Adj.* confused, muddled ‹*ideas, situation, etc.*›
verwundbar *Adj.* open to injury *pred.*; *(fig.)* vulnerable
Verwundbarkeit die vulnerability
verwunden *tr. V.* wound; injure
verwunderlich *Adj.* surprising
verwundern 1. *tr. V.* surprise; *(erstaunen)* astonish; **verwundert** surprised/astonished (**über** + *Akk.* at). 2. *refl. V.* be surprised (**über** + *Akk.* at); *(erstaunt sein)* be astonished (**über** + *Akk.* at)
Verwunderung die; ~: surprise; *(Staunen)* astonishment
Verwundete der/die; *adj. Dekl.* wounded person; casualty
Verwundung die; ~, ~en a) wounding; **b)** *(Wunde, Verletzung)* wound
verwunschen *Adj.* enchanted; bewitched
verwünschen *tr. V.* curse
Verwünschung die; ~, ~en a) *(das Verfluchen)* cursing; **b)** *(Fluch)* curse; oath
verwurschteln [fɛɐ̯'vʊrʃtl̩n], **verwursteln** *(ugs.)* 1. *tr. V.* get ‹*thing*› in a muddle *or* a tangle. 2. *refl. V.* get in a muddle *or* a tangle
verwüsten *tr. V.* devastate
Verwüstung die; ~, ~en devastation
verzählen *refl. V.* miscount; **ich verzähle mich dauernd** I keep losing count
verzärteln [fɛɐ̯'tsɛ:ɐ̯tl̩n] *tr. V.* mollycoddle
verzaubern *tr. V.* **a)** cast a spell on; bewitch; **jmdn. in etw.** *(Akk.)* **~:** transform sb. into sth.; **b)** *(fig.)* enchant
Verzauberung die; ~, ~en a) casting of a/the spell (*Gen.* on); **b)** *(fig.)* enchantment
Verzehr [fɛɐ̯'tse:ɐ̯] **der; ~[e]s** consumption; **zum alsbaldigen ~ bestimmt** for immediate consumption
verzehren . *tr. V.* consume
verzeichnen *tr. V.* **a)** *(falsch zeichnen)* draw wrongly; **b)** *(aufführen)* list; *(eintragen)* enter; *(registrieren)* record; **der Ort ist auf der Karte nicht verzeichnet** the place is not [marked] on the map; **große Erfolge/Verluste zu ~ haben** have scored great successes/suffered great losses
Verzeichnis [fɛɐ̯'tsaiçnɪs] **das; ~ses, ~se** list; *(Register)* index
verzeihen *unr. tr., itr. V.* forgive; *(entschuldigen)* excuse ‹*behaviour, remark, etc.*›; **jmdm. [etw.] ~:** forgive sb. [sth. *or* for sth.]; **~ Sie [bitte] die Störung** pardon the intrusion; [please] excuse me for disturbing you
verzeihlich *Adj.* forgivable; excusable; **kaum ~:** almost unforgivable
Verzeihung die; ~: forgiveness; **~, können Sie mir sagen, ...?** ex-

cuse me, could you tell me ...?; ~! sorry!; **jmdn. um ~ bitten** apologize to sb.; **ich bitte vielmals um ~**: I do apologize *or* [do] beg your pardon

verzerren 1. *tr. V.* **a)** contort ⟨*face etc.*⟩ (**zu** into); **b)** *(akustisch, optisch)* distort ⟨*sound, image*⟩. **2.** *itr. V.* ⟨*loudspeaker, mirror, etc.*⟩ distort. **3.** *refl. V.* ⟨*face, features*⟩ become contorted (**zu** into)

Verzerrung die; ~, ~en a) *(des Gesichts usw.)* contortion; **b)** *(des Klangs, eines Bildes, der Realität usw.)* distortion

verzetteln *refl. V.* dissipate one's energies; try to do too many things at once

Verzicht [fɛɐ̯'tsɪçt] **der; ~[e]s, ~e a)** renunciation (**auf** + *Akk.* of); **auf etw.** *(Akk.)* **~ leisten** *(geh.)* renounce sth.; **b)** *(auf Reichtum, ein Amt usw.)* relinquishment (**auf** + *Akk.* of)

verzichten *itr. V.* do without; **~ auf** (+ *Akk.*) *(sich enthalten)* refrain *or* abstain from; *(aufgeben)* give up ⟨*share, smoking, job, etc.*⟩; renounce ⟨*inheritance*⟩; renounce, relinquish ⟨*right, privilege*⟩; *(opfern)* sacrifice ⟨*holiday, salary*⟩; **ich verzichte auf deine Hilfe/Ratschläge** I can do without *or* you can keep your help/advice; **darauf kann ich ~** *(iron.)* I can do without that; **auf eine Strafanzeige ~**: not bring a charge

[1]**verziehen** *2. Part. v.* **verzeihen**

[2]**verziehen 1.** *unr. tr. V.* **a)** screw up ⟨*face, mouth, etc.*⟩; **b)** *(schlecht erziehen)* spoil. **2.** *unr. refl. V.* **a)** twist; be contorted; **b)** *(aus der Form geraten)* go out of shape; ⟨*wood*⟩ warp; **ein verzogener Rahmen** a distorted frame; **c)** *(wegziehen)* ⟨*clouds, storm*⟩ move away, pass over: ⟨*fog, mist*⟩ disperse; **d)** *(ugs.: weggehen)* take oneself off; **verzieh dich!** *(salopp)* clear *(coll.)* *or (sl.)* push off. **3.** *unr. itr. V.; mit sein* move [away]; **„Empfänger [unbekannt] verzogen"** 'no longer at this address'

verzieren *tr. V.* decorate

Verzierung die; ~, ~en decoration

verzögern 1. *tr. V.* **a)** delay (**um** by); **b)** *(verlangsamen)* slow down. **2.** *refl. V.* be delayed (**um** by)

Verzögerung die; ~, ~en a) delaying; delay (*Gen.* in); **b)** *(Verlangsamung)* slowing down; *(Technik)* deceleration; **c)** *(Verspätung)* delay; hold-up

verzollen *tr. V.* pay duty on

Verzug der; ~[e]s a) delay; **[mit etw.] im ~ sein/in ~ kommen** *od.* **geraten** be/fall behind [with sth.]; **jmdn./etw. in ~ bringen** delay sb./sth.; hold sb. up/put sth. back; **b) es ist Gefahr im ~** *(ugs.)* danger is imminent

verzweifeln *itr. V.; meist mit sein* despair; **über etw./jmdn. ~**: despair at sth./of sb.; **am Leben/an den Menschen ~**: despair of life/humanity; **es ist zum Verzweifeln!** it's enough to drive you to despair

verzweifelt 1. *Adj.* **a)** despairing ⟨*person, animal*⟩; **~ sein** be in despair *or* full of despair; **b)** desperate ⟨*situation, attempt, effort, struggle, etc*⟩. **2.** *adv.* **a)** *(entmutigt)* despairingly; **b)** *(sehr angestrengt)* desperately

Verzweiflung die; ~: despair; **etw. aus ~ tun** do sth. out of despair; **jmdn. zur ~ treiben/bringen** drive sb. to despair

verzweigen *refl. V.* branch [out]; **das Unternehmen ist stark verzweigt** *(fig.)* the firm is very diversified

verzwickt [fɛɐ̯'tsvɪkt] *(ugs.) Adj.* tricky; complicated

Veteran [vete'ra:n] **der; ~en, ~en** *(auch fig.)* veteran

Veterinär [veteri'nɛ:ɐ̯] **der; ~s, ~e** veterinary surgeon

Veto ['ve:to] **das; ~s, ~s** veto; **ein ~ gegen etw. einlegen** veto sth.

Veto·recht das right of veto

Vetter ['fɛtɐ] **der; ~s, ~n** cousin

Vettern·wirtschaft die; *o. Pl. (abwertend)* nepotism

vgl. *Abk.* **vergleiche** cf.

v. H. *Abk.* **vom Hundert** per cent

via ['vi:a] *Präp.* via

Viadukt [via'dʊkt] **das** *od.* **der; ~[e]s, ~e** viaduct

Vibration [vibra'tsio:n] **die; ~, ~en** vibration

vibrieren [vi'bri:rən] *itr. V.* vibrate; ⟨*voice*⟩ quiver, tremble

Video das; ~s, ~s *(ugs.)* video

Video-: ~auf·zeichnung die video recording; **~clip** [~klɪp] **der; ~s, ~s** video; **~film der** video [film]; **~kamera die** video camera; **~kassette die** video cassette; **~recorder der** video recorder

Vieh [fi:] **das; ~[e]s a)** *(Nutztiere)* livestock *sing. or pl.*; **jmdn. wie ein Stück ~ behandeln** treat sb. like an animal; **b)** *(Rind~)* cattle *pl.*; **c)** *(derb abwertend: Mensch)* bastard

Vieh-: ~bestand der stocks *pl.* of animals/cattle; **~futter das** animal/cattle feed *or* fodder; **~händler der** livestock/cattle dealer

viehisch 1. *Adj.* terrible *(coll.)* ⟨*fear, stupidity, pain*⟩. **2.** *adv.(ugs.)* ⟨*hurt*⟩ like hell *(coll.)*

Vieh-: ~stall der cowshed; **~zucht die;** *o. Pl.* [live]stock/cattle breeding *no art.*; **~züchter der** [live]stock/cattle breeder

viel [fi:l] **1.** *Indefinitpron. u. unbest. Zahlw.* **a)** *Sg.* a great deal of; a lot of *(coll.)*; **so/wie/nicht/zu ~**: that/how/not/too much; **~[es]** *(viele Dinge, vielerlei)* much; **ich kann mich an ~es nicht mehr erinnern** there's much I can't remember; **der ~e Regen** all the rain; **gleich ~ Geld** the same amount of money; **um ~es jünger** a great deal younger; **~ Erfreuliches** a great many pleasant things; **er hat in ~em recht** he is right on many points; **er ist nicht ~ über fünfzig** he is not much more than *or* much over fifty; **b)** *Pl.* many; **gleich ~[e]** the same number of; **~e hundert** many hundreds of; **die ~en Menschen** all the people; **seine ~en Kinder** all his children. **2.** *Adv.* **a)** *(oft, lange)* a great deal; a lot *(coll.)*; **b)** *(wesentlich)* much; a great deal; a lot *(coll.)*; **~ mehr/weniger** much more/less; **~ zu klein** much too small

viel-: ~beschäftigt *Adj.* very busy; **~deutig** [~dɔytɪç] **1.** *Adj.* ambiguous; **2.** *adv.* ambiguously

vielerlei *indekl. unbest. Gattungsz.* **a)** *attr.* many different; all kinds *or* sorts of; **b)** *subst.* all kinds *or* sorts of things

viel-, Viel-: ~fach 1. *Adj.* **a)** multiple; **die ~fache Menge** many times the amount; **auf ~fachen Wunsch unserer Zuschauer** at the request of many of our viewers; **b)** *(vielfältig)* manifold; many kinds of; **2.** *adv.* many times; **~fache das; ~n;** *adj. Dekl.* **a) ein ~faches** many times the amount/number; **um ein ~faches** many times over; **um ein ~faches schneller/teurer** many times faster/more expensive; **~falt die; ~**: diversity; wide variety; **~fältig** [~fɛltɪç] **1.** *Adj.* many and diverse; **2.** *adv.* in many different ways; **~fraß der a)** *(ugs.: Mensch)* glutton; [greedy-]guts *sing. (sl.)*; **b)** *(Tier)* wolverine

vielleicht [fi'laiçt] **1.** *Adv.* **a)** perhaps; maybe; **hast du den Schirm ~ im Büro liegenlassen?** could it be that you left your umbrella in the office?; **b)** *(ungefähr)* perhaps; about. **2.** *Partikel* **a) kannst**

du mir ~ sagen, ...? could you possibly tell me ...?; **hast du ~ meinen Bruder gesehen?** have you seen my brother by any chance? **b)** *(wirklich)* really; **ich war ~ aufgeregt** I was terribly excited *or* as excited as anything *(coll.)*; **du bist ~ ein Blödmann!** what a stupid idiot you are! *(coll.)*

viel-, Viel-: ~mals *Adv.* **ich bitte ~mals um Entschuldigung** I'm very sorry; I do apologize; **sie läßt ~mals grüßen** she sends her best regards *or* wishes; **danke ~mals** thank you very much; many thanks; **~mehr** [*od.* -'-] *Konj. u. Adv.* rather; *(im Gegenteil)* on the contrary; **~sagend 1.** *Adj.* meaningful; **2.** *adv.* meaningfully; **~seitig 1.** *Adj.* versatile ⟨*person*⟩; varied ⟨*work, programme, etc.*⟩; **auf ~seitigen Wunsch** by popular request; **diese Küchenmaschine ist sehr ~seitig** this food processor has many uses; **2.** *adv.* **~seitig begabt sein** be versatile; **~versprechend 1.** *Adj.* [very] promising; **2.** *adv.* [very] promisingly; **~zahl die;** *o. Pl.* large number; multitude

vier [fi:ɐ̯] *Kardinalz.* four; **alle ~e von sich strecken** *(ugs.)* put one's feet up; **auf allen ~en** *(ugs.)* on all fours; *s. auch* **acht**

Vier die; ~, **~en** four; **eine ~ schreiben/bekommen** *(Schulw.)* get a D; *s. auch* [1]**Acht; Zwei**

vier-, Vier-: *(s. auch* **acht-, Acht-); ~beiner der;** ~s, ~ *(ugs.)* four-legged friend; **~eck das** quadrilateral; *(Rechteck)* rectangle; *(Quadrat)* square; **~eckig** *Adj.* quadrilateral; *(rechteckig)* rectangular; *(quadratisch)* square

Vierer der; ~s ~ **a)** *(Rudern)* four; **b)** *(ugs.: im Lotto)* four winning numbers *pl.*; **c)** *(ugs.: Ziffer, beim Würfeln)* four; **d)** *(landsch.: Schulnote)* D; **e)** *(ugs.: Autobus)* [number] four

vier·fach *Vervielfältigungsz.* fourfold; quadruple

Vier·fache das; ~n; *adj. Dekl.* **um das ~:** fourfold; by four times the amount; **die Preise sind um das ~ gestiegen** the prices have quadrupled *or* increased four times

vier-: ~hundert *Kardinalz.* four hundred; **~jährig** *Adj.* *(4 Jahre alt)* four-year-old *attrib.*; four years old *pred.*; *(4 Jahre dauernd)* four-year *attrib.*; **~köpfig** *Adj.* four-headed ⟨*monster*⟩; ⟨*family, staff*⟩ of four

Vierling der; ~s, ~e quadruplet

vier-, Vier-: ~mal *Adv.* four times; *s. auch* **achtmal; ~radantrieb der** *(Kfz-W.)* four-wheel drive; **~räd[e]rig** [~rɛ:d[ə]rɪç] *Adj.* four-wheeled; **~spurig 1.** *Adj.* four-lane ⟨*road, motorway*⟩; **~spurig sein** have four lanes; **2.** *adv.* **~spurig befahrbar sein** have all four lanes open; **eine Straße ~spurig ausbauen** widen a road into four lanes; **~stellig** *Adj.* four-figure *attrib.*; *s. auch* **achtstellig; ~sterne·hotel** [-'----] **das** four-star hotel; **~stöckig** *Adj.* four-storey; *s. auch* **achtstöckig**

viert [fi:ɐ̯t] *in* **wir waren zu ~:** there were four of us; *s. auch* [2]**acht**

viert... *Ordinalz.* fourth; *s. auch* **acht...**

vier-, Vier-: ~tägig *Adj.* four-day *attrib.*; *s. auch* **achttägig; ~takter der;** ~s, ~ *(Auto)* car with a four-stroke engine; *(Motor)* four-stroke engine; **~tausend** *Kardinalz.* four thousand; **~teilen** *tr. V.* quarter

viertel ['fɪrtl̩] *Bruchz.* quarter; **ein ~ Pfund/eine ~ Million** a quarter of a pound/million

Viertel ['fɪrtl̩] **das** (*schweiz. meist* **der**); ~s, ~ **a)** quarter; **ein ~ Wein** *(ugs.)* a quarter-litre of wine; **~ vor/nach eins** [a] quarter to/past one; **drei ~:** three-quarters; **b)** *(Stadtteil)* quarter; district

viertel-, Viertel-: ~finale das *(Sport)* quarter-final; **sich für das ~finale qualifizieren** qualify for the quarter-finals; **~jahr das** three months *pl.*; **~jährlich 1.** *Adj.* quarterly; **2.** *adv.* quarterly; every three months; **~liter der** quarter of a litre; **~note die** *(Musik)* crotchet *(Brit.)*; quarter note *(Amer.)*; **~pfund das** quarter [of a] pound; **~stunde die** quarter of an hour; **~stündlich 1.** *Adj.; nicht präd.* quarter-hourly; every quarter of an hour *postpos.*; **2.** *adv.* every quarter of an hour

viertens ['fi:ɐ̯tn̩s] *Adv.* fourthly; *s. auch* **zweitens**

viertürig [-ty:rɪç] *Adj.* four-door *attrib.*; **~ sein** have four doors

vier-: ~wöchig *Adj.* four-week [-long]; **~zehn** ['fɪr-] *Kardinalz.* fourteen; *s. auch* **achtzehn; ~zehnjährig** *Adj.* *(14 Jahre alt)* fourteen-year-old *attrib.*; fourteen years old *pred.*; *(14 Jahre dauernd)* fourteen-year *attrib.*

vier-['fɪr-]: **~zehn·tägig** *Adj.; nicht präd.* two-week; **~zehn·täglich 1.** *Adj.; nicht präd.* fortnightly; **2.** *adv.* fortnightly; every two weeks

vierzig ['fɪrtsɪç] *Kardinalz.* forty; *s. auch* **achtzig**

vierziger ['fɪrtsɪgɐ] *indekl. Adj.; nicht präd.* **die ~ Jahre** the forties; *s. auch* **achtziger**

Vierziger ['fɪrtsɪgə] **der;** ~s, **~:** forty-year-old

vierzig·jährig ['fɪrtsɪç-] *Adj.* *(40 Jahre alt)* forty-year-old *attrib.*; forty years old *pred.*; *(40 Jahre dauernd)* forty-year *attrib.*

vierzigst... ['fɪrtsɪçst...] *Ordinalz.* fortieth; *s. auch* **acht...**

Vierzig·stunden·woche die forty-hour week

Vier-: ~zimmer·wohnung die four-room flat *(Brit.) or (Amer.)* apartment; **~zylinder der** *(ugs.)* four-cylinder

Vietnam [viɛt'nam] **(das);** ~s Vietnam

Vietnamese [viɛtna'me:zə] **der;** ~n, **~n, Vietnamesin, die;** ~, **~nen** Vietnamese

vietnamesisch 1. *Adj.* Vietnamese. **2.** *adv.* **wir waren ~ essen** we went to a Vietnamese restaurant *or* for a Vietnamese meal

Vietnam·krieg der; *o. Pl.* Vietnam War

Vikar [vi'ka:ɐ̯] **der;** ~s, **~e a)** *(kath. Kirche)* locum tenens; **b)** *(ev. Kirche)* ≈ [trainee] curate

Viktoria [vɪk'to:ria] **(die)** Victoria

viktorianisch *Adj.* Victorian

Villa ['vɪla] **die;** ~, **Villen** villa

Villen·viertel das exclusive residential district

violett [vio'lɛt] purple; violet

Violett das; ~s, **~e** *od. ugs.* **~s** purple; violet; *(im Spektrum)* violet

Violine [vio'li:nə] **die;** ~, **~n** *(Musik)* violin

Violin-: ~konzert das violin concerto; **~schlüssel der** treble clef

Violon·cello [violɔn'tʃɛlo] **das** violoncello

Viper ['vi:pɐ] **die;** ~, **~n** viper; adder

Viren *s.* **Virus**

virtuos [vɪr'tuo:s] **1.** *Adj.* virtuoso ⟨*performance etc.*⟩. **2.** *adv.* in a virtuoso manner

Virtuose [vɪr'tuo:zə] **der;** ~n, **~n** virtuoso

Virtuosität die; **~:** virtuosity

Virus ['vi:rʊs] **das;** ~, **Viren** ['vi:rən] virus

Virus·infektion die virus infection

Visa *s.* **Visum**

Visen *s.* **Visum**

Visier [vi'zi:ɐ̯] **das;** ~s, **~e a)** *(am*

Helm) visor; **b)** *(an der Waffe)* backsight

Vision [vi'zi̯o:n] **die; ~, ~en** vision

Visite [vi'zi:tə] **die; ~, ~n** round; **um 10 Uhr war ~:** at 10 o'clock, the doctor did his round

Visiten·karte die visiting-card

Viskose [vɪs'ko:zə] **die; ~** *(Chemie)* viscose

visuell [vi'zu̯ɛl] *(geh.)* **1.** *Adj.* visual. **2.** *adv.* visually

Visum ['vi:zʊm] **das; ~s, Visa** ['vi:za] *od.* **Visen** ['vi:zn̩] visa

vital [vi'ta:l] **1.** *Adj.* **a)** *(voller Energie)* vital; energetic; vigorous; **~ sein** be full of life *or* vigour; **b)** *(wichtig)* vital. **2.** *adv.* *(voller Energie)* energetically

Vitalität die; ~: vitality

Vitamin [vita'mi:n] **das; ~s, ~e** vitamin

vitamin-, Vitamin-: ~arm *Adj.* ⟨*food, diet, etc.*⟩ low in vitamins; **~mangel der;** *o. Pl.* vitamin deficiency; **~reich** *Adj.* rich in vitamins *postpos.*; vitamin-rich

Vitrine [vi'tri:nə] **die; ~, ~n** display case; show-case; *(Möbel)* display cabinet

Vize- ['vi:tsə]: **~kanzler der** vice-chancellor; **~könig der** viceroy; **~präsident der** vice-president

Vogel ['fo:gl̩] **der; ~s, Vögel** ['fø:gl̩] **a)** bird; **[mit etw.] den ~ abschießen** *(ugs.)* take the biscuit [with sth.] *(coll.)*; **einen ~ haben** *(salopp)* be off one's rocker *or* head *(sl.)*; **jmdm. den ~ zeigen** tap one's forehead at sb. *(as a sign that one thinks he/she is stupid)*; **b)** *(salopp, oft scherzh.: Mensch)* character; **ein komischer ~:** an odd bird *or* character

vogel-, Vogel-: ~beere die rowan-berry; **~ei das** bird's egg; **~frei** *Adj.* *(hist.)* outlawed; **jmdn./etw. für ~frei erklären** outlaw sb./sth.; **~futter das** bird food; **~käfig der** birdcage; **~kunde die;** *o. Pl.* ornithology *no art.*

vögeln ['fø:gl̩n] *tr., itr. V.* *(derb)* screw *(coarse)*

Vogel-: ~nest das bird's nest; **~perspektive die** bird's eye view; **~scheuche** [~ʃɔyçə] **die; ~, ~n** scarecrow

Vogesen [vo'ge:zn̩] *Pl.* Vosges [Mountains]

Vokabel [vo'ka:bl̩] **die; ~, ~n** *od. österr. auch* **das; ~s, ~:** word; vocabulary item; **~n** vocabulary *sing.*; vocab *sing.* *(Sch. coll.)*

Vokabel·heft das vocabulary *or* *(coll.)* vocab book

Vokabular das; ~s, ~e vocabulary

Vokal [vo'ka:l] **der; ~s, ~e** *(Sprachw.)* vowel

Volk [fɔlk] **das; ~[e]s, Völker** ['fœlkɐ] **a)** people; **das ~ der Kurden** the Kurdish people; **das irische und das deutsche ~:** the Irish and German peoples; **b)** *o. Pl.* *(Bevölkerung)* people *pl.*; *(Nation)* people *pl.*; nation; **im ~e** among the people; **das arbeitende/unwissende ~:** the working people/the ignorant masses *pl.*; **c)** *o. Pl.* *(einfache Leute)* people *pl.*; **ein Mann aus dem ~:** a man of the people; **d)** *o. Pl.* *(ugs.: Leute)* people *pl.*; **viel junges ~:** many young people

völker-, Völker-: ~bund der; *o. Pl.* League of Nations; **~kunde die;** *o. Pl.* ethnology *no art.*; **~mord der** genocide; **~recht das;** *o. Pl.* international law *no art.*; **~rechtlich 1.** *Adj.*; *nicht präd.* ⟨*issue, problem, etc.*⟩ of international law; **~rechtliche Verträge** agreements in *or* under international law; **2.** *adv.* ⟨*settle*⟩ in accordance with international law; ⟨*control, regulate*⟩ by international law; ⟨*recognize*⟩ under international law

Völker-: ~verständigung die international understanding; understanding between nations; **~wanderung die a)** *(hist.)* migration of peoples; **b)** *(ugs.)* mass migration; *(Zug)* mass progression

volks-, Volks-: ~ab·stimmung die plebiscite; **~armee die** People's Army; **~befragung die** *(Politik)* referendum; **~begehren das** *(Politik)* petition for a referendum; **~eigen** *Adj.* *(ehem. DDR)* publicly *or* nationally owned; **~er Betrieb** publicly *or* nationally owned company; **~eigentum das** *(ehem. DDR)* national[ly owned] property; **~entscheid der** *(Politik)* referendum; **~fest das** public festival; *(Jahrmarkt)* fair; **~held der** folk hero; **~hoch·schule die** adult education centre; **ein Kurs an der ~hochschule** an adult education class; **~kammer die** *(ehem. DDR)* **die ~kammer** the Volkskammer; the People's Chamber; **~kunde die** folklore; **~lied das** folk-song; **~märchen das** folk-tale; **~mund der;** *o. Pl.* **im ~mund wird das ... genannt** in the vernacular it is called ...; **~musik die** folk-music; **~polizei die;** *o. Pl.* *(ehem. DDR)* People's Police; **~polizist der** *(ehem. DDR)* People's Policeman; member of the People's Police; **~republik die** People's Republic; **die ~republik China** the People's Republic of China; **~schule die a)** *(Bundesrepublik Deutschland und Schweiz veralt.)* school providing basic primary and secondary education; **b)** *(österr.)* primary school; **~stamm der** tribe; **~stück das** *(Theater)* folk play; **~tanz der** folk-dance; **~tracht die** traditional costume; *(eines Landes)* national costume; **~trauer·tag der** *(Bundesrepublik Deutschland)* national remembrance day

volkstümlich ['fɔlksty:mlɪç] **1.** *Adj.* popular; **ein ~er Politiker** a politician of the people *or* with the common touch; **der ~e Name einer Pflanze** the vernacular name of a plant; **~e Preise** popular prices. **2.** *adv.* **~ schreiben** write in terms readily comprehensible to the layman

volks-, Volks-: ~verdummung die *(ugs. abwertend)* deliberate deception of the public; **~vertreter der** representative of the people; **~vertretung die** representative body of the people; ; **~wirt der** economist; **~wirtschaft die** national economy; *(Fach)* economics *sing., no art.*; **~wirtschaftler der; ~s, ~:** economist; **~wirtschaftlich 1.** *Adj.* economic; **2.** *adv.* economically; **~zählung die** [national] census; **~zorn der** public anger

voll [fɔl] **1.** *Adj.* **a)** full; **der Saal ist ~ Menschen** the room is full of people; **~ von** *od.* **mit etw. sein** be full of sth.; **das Glas ist halb ~:** the glass is half full; **jeder bekam einen Korb ~:** everybody received a basketful; **mit ~en Backen kauen** eat with bulging cheeks; **aus dem ~en schöpfen** draw on abundant *or* plentiful resources; **~e Pulle** *od.* **~[es] Rohr** *(salopp)* ⟨*drive*⟩ flat out; *s. auch* **Mund; b)** *(salopp: betrunken)* plastered *(sl.)*; canned *(Brit. sl.)*; **c)** *(üppig)* full ⟨*figure, face, lip*⟩; thick ⟨*hair*⟩; ample ⟨*bosom*⟩; **d)** *(ganz, vollständig)* full; complete ⟨*seriousness, success*⟩; **etw. mit ~em Recht tun** be quite right to do sth.; **in ~er Fahrt** at full speed; **in ~em Gange sein** be in full swing; **die ~ Wahrheit** the full *or* whole truth; **mit dem ~en Namen unterschreiben** sign one's full name *or* one's name in full; **das Dutzend ist ~:** it's a round dozen; **jmdn. nicht für ~ nehmen**

not take sb. seriously; s. **auch Hals b; e)** *(kräftig)* full, rich ⟨*taste, aroma*⟩; rich ⟨*voice*⟩. **2.** *adv.* fully; ~ **und ganz** completely; **etw.** ~ **auslasten** make full use of sth.; ~ **verantwortlich für etw. sein** be wholly responsible *or* bear full responsibility for sth.

voll·auf [*od.* '--] *Adv.* completely; fully; ~ **genügen/reichen** be quite enough

vollaufen *unr. itr. V., trennbar* fill up; **etw.** ~ **lassen** fill sth. [up]; **sich** ~ **lassen** *(salopp)* get completely paralytic *or* canned *(Brit. sl.)*

voll-, Voll-: ~automatisch 1. *Adj.* fully automatic; **2.** *adv.* fully automatically; **~bad das** bath; **~bart der** full beard; **~beschäftigung die;** *o. Pl. (Wirtsch.)* full employment *no art.;* **~blut das** thoroughbred; **~bremsung die: eine ~bremsung machen** put the brakes full on; **~bringen** [-'--] *unr. tr. V. (geh.)* accomplish; achieve; **~dampf der;** *o. Pl. (Seemannsspr.)* full steam; **mit ~dampf** at full steam *or* speed; *(fig. ugs.)* flat out

Völle·gefühl ['fœlə-] **das;** *o. Pl.* feeling of fullness

voll·enden *tr. V.* complete; finish; **mit vollendetem** *od.* **dem vollendeten 16. Lebensjahr** on reaching the age of 16 *or* completing one's sixteenth year

vollendet 1. *Adj.* accomplished ⟨*performance*⟩; perfect ⟨*gentleman, host, manners, reproduction*⟩. **2.** *adv.* ⟨*play*⟩ in an accomplished manner; perfectly

vollends ['fɔlɛnts] *Adv.* completely

Voll·endung die completion; **kurz vor der ~ stehen** be nearing completion; **mit/nach ~ des 65. Lebensjahres** on reaching the age of 65 *or* completing one's sixty-fifth year

voller *indekl. Adj.* full of; filled with; **sein Anzug war ~ Flecken** his suit was covered with stains

Volley·ball der volleyball

voll-, Voll-: ~führen [-'--] *tr. V.* perform, execute ⟨*somersault, movement*⟩; perform ⟨*dance, deed*⟩; **~|füllen** *tr. V.* fill up; **~gas das;** *o. Pl.* **~gas geben** put one's foot down; **~gas fahren** drive flat out; **~|gießen** *unr. tr. V. (füllen)* fill [up]

völlig ['fœlɪç] **1.** *Adj.; nicht präd.* complete; total. **2.** *adv.* completely; totally; **du hast ~ recht** you are absolutely right; **das ist ~ unmöglich** that is absolutely impossible; **mit etw. ~ einverstanden sein** be in complete agreement with sth.

voll-, Voll-: ~jährig *Adj.* of age *pred.;* **~jährig werden** come of age; attain one's majority; **sie hat zwei ~jährige Kinder** she has two children who are of age; **~jährigkeit die;** ~: majority *no art.;* **~kasko·versicherung die** fully comprehensive insurance; **~klimatisiert** *Adj.* fully air-conditioned

voll·kommen 1. *Adj.* **a)** [-'-- *od.* '---] *(vollendet)* perfect; **b)** ['---] *(vollständig)* complete; total. **2.** ['---] *adv.* completely; totally

Vollkommenheit die; ~: perfection

voll-, Voll-: ~korn·brot das wholemeal *(Brit.) or (Amer.)* wholewheat bread; **~|machen 1.** *tr. V.* **a)** *(ugs.: füllen)* fill up; **um das Maß ~zumachen** *(fig.)* to crown *or* cap it all; **b)** *(ugs.: beschmutzen)* **etw. ~machen** get or make sth. dirty; **[sich** *(Dat.)***] die Hosen/Windeln ~machen** mess one's pants/nappy; **c)** *(vollständig machen)* complete; **~macht die;** ~, **~en a)** authority; **jmdm. [die] ~macht geben/erteilen** give/grant sb. power of attorney; **in ~macht** per procurationem; **b)** *(Urkunde)* power of attorney; **~milch die** full-cream milk; **~mond der;** *o. Pl.* full moon; **~|packen** *tr. V.* pack full; **~pension die;** *meist o. Art.; o. Pl.* full board *no art.;* **~|pumpen** *tr. V.* pump up ⟨*tyre*⟩; fill up ⟨*reservoir*⟩; **~|saugen** *regelm. (auch unr.) refl. V.* ⟨*leech*⟩ suck itself full **(mit** of); ⟨*sponge*⟩ become saturated **(mit** with); **~schlank** *Adj.* with a fuller figure *postpos., not pred.;* **~schlank sein** have a fuller figure; **~|schmieren** *(ugs.) tr. V.* **a)** *(beschmutzen)* smear; **sich** *(Dat.)* **das ganze Gesicht mit etw. ~schmieren** get *or* smear sth. all over one's face; **b)** *(abwertend: ~schreiben, ~malen)* scrawl/draw all over ⟨*wall etc.*⟩; fill ⟨*exercise book etc.*⟩ with scrawl; **~|schreiben** *unr. tr. V.* fill [with writing]; **~sperrung die** *(Verkehrsw.)* complete closure; **~|spritzen** *tr. V.* **jmdn./etw. ~spritzen** splash water *etc.* all over sb./sth.; *(mit Schlauch usw.)* spray water *etc.* all over sb./sth.; **~ständig 1.** *Adj.* complete; full ⟨*text, address, etc.*⟩; **2.** *adv.* completely; ⟨*list*⟩ in full; **~ständigkeit die;** ~: completeness; **~|stopfen** *tr. V. (ugs.)* stuff *or* cram full; **~strecken** [-'--] *tr. V.* enforce ⟨*penalty, fine, law*⟩; carry out ⟨*sentence*⟩ **(an** + *Dat.* on); **ein Testament ~strecken** execute a will; **~strecker** [-'--] **der;** ~**s,** ~ *(des Gesetzes)* enforcer; *(eines Testaments)* executor; **~streckung** [-'--] **die;** ~, **~en** *s.* **vollstrecken:** enforcement; carrying out; execution

Vollstreckungs·befehl der *(Rechtsw.)* enforcement order; writ of execution

voll-, Voll-: ~|tanken *tr. (auch itr.) V.* fill up; **bitte ~tanken** fill it up, please; **~treffer der** direct hit; **ein ~treffer sein** *(fig.)* hit the bull's eye; **~trunken** *Adj.* completely *or* blind drunk; **in ~trunkenem Zustand** in a state of total inebriation; **~trunkenheit die** total inebriation *or* intoxication; **~verb das** *(Sprachw.)* full verb; **~versammlung die** general meeting; **~waise die** orphan; **~wertig** *Adj.* full ⟨*job, member*⟩; [fully] adequate ⟨*replacement, substitute, nourishment, diet*⟩; **~zählig** [~tsɛːlɪç] *Adj.* complete; **als wir ~zählig [versammelt] waren** when everyone was present

voll·ziehen 1. *unr. tr. V.* carry out; perform ⟨*sacrifice, ceremony, sexual intercourse*⟩; **die Ehe ~:** consummate the marriage; **die ~de Gewalt** the executive [power]. **2.** *unr. refl. V.* take place

Voll·zug der *s.* **vollziehen 1:** carrying out; performance

Vollzugs-: ~anstalt die penal institution; **~beamte der** [prison] warder

Volontär [volɔn'tɛːɐ̯] **der;** ~**s,** ~**e** trainee *(receiving a low salary in return for training)*

Volontariat [volɔnta'ri̯aːt] **das;** ~**[e]s,** ~**e a)** *(Zeit)* period of training; **b)** *(Stelle)* traineeship

Volontärin die; ~, ~**nen** *s.* **Volontär**

volontieren *itr. V.* work as a trainee **(bei** with)

Volt [vɔlt] **das;** ~ *od.* ~**[e]s,** ~ *(Physik, Elektrot.)* volt

Volt·meter das; ~**s,** ~ *(Elektrot.)* voltmeter

Volumen [vo'luːmən] **das;** ~**s,** ~**:** volume

vom [fɔm] *Präp.* + *Art.* **a)** = **von dem; b)** *(räumlich, zeitlich)* from the; **links/rechts ~ Eingang** to the left/right of the entrance; **~ Stuhl aufspringen** jump up out of one's chair; ~ **ersten Januar an** [as] from the first of January; **c)** *(zur*

Angabe der Ursache) **das kommt ~ Rauchen/Alkohol** that comes from smoking/drinking alcohol; **jmdn. ~ Sehen kennen** know sb. by sight

von [fɔn] *Präp. mit Dat.* **a)** *(räumlich)* from; **nördlich/südlich ~ Mannheim** to the north/south of Mannheim; **rechts/links ~ mir** on my right/left; **~ hier an** *od. (ugs.)* **ab** from here on[ward]; **~ Mannheim aus** from Mannheim; **etw. ~ etw. [ab]wischen/[ab]brechen/[ab]reißen** wipe/break/tear sth. off sth.; *s. auch* **aus 2 c; her a; ¹vorn; b)** *(zeitlich)* from; **~jetzt an** *od. (ugs.)* **ab** from now on; **~ heute/morgen an** [as] from today/tomorrow; starting today/tomorrow; **~ Kindheit an** from *or* since childhood; **in der Nacht ~ Freitag auf** *od.* **zu Samstag** during Friday night *or* the night of Friday to Saturday; **das Brot ist ~ gestern** it's yesterday's bread; *s. auch* **her b; c)** *(anstelle eines Genitivs)* of; **ein Stück ~ dem Kuchen** a slice of the cake; **acht ~ zehn** eight out of ten; **d)** *(zur Angabe des Urhebers, der Ursache, beim Passiv)* by; **müde ~ der Arbeit sein** be tired from work[ing]; **etw. ~ seinem Taschengeld kaufen** buy sth. with one's pocket-money; **sie hat ein Kind ~ ihm** she has a child by him; *s. auch* **wegen 2; e)** *(zur Angabe von Eigenschaften)* of; **eine Fahrt ~ drei Stunden** a three-hour drive; **Kinder [im Alter] ~ vier Jahren** children aged four; **~ bester Qualität** of the best quality; **f)** *(bestehend aus)* of; **g)** *(als Adelsprädikat)* von; **h)** *(in bezug auf)* **er ist ~ Beruf Lehrer** he is a teacher by profession; **i)** *(über)* about; **~ diesen Dingen spricht man besser nicht** it's better not to speak of such things

von·einạnder *Adv.* from each other *or* one another; ⟨*disappointed*⟩ in each other *or* one another

vonnöten [fɔn'nø:tn̩] *Adj.* **~ sein** be necessary

vonstatten [fɔn'ʃtatn̩] *Adv.* **~ gehen** proceed

Vopo ['fo:po] **der; ~s, ~s** *(ugs.) s.* **Volkspolizist**

vor [fo:ɐ̯] **1.** *Präp. mit Dat.* **a)** *(räumlich)* in front of; *(weiter vorn als)* ahead of; in front of; *(nicht ganz so weit wie)* before; *(außerhalb)* outside; **~ einem Hintergrund** against a background; **200 m ~ der Abzweigung** 200 m. before the turn-off; **~ der Stadt** outside the town; **etw. ~ sich haben** *(fig.)* have sth. before one; **b)** *(zeitlich)* before; **es ist fünf [Minuten] ~ sieben** it is five [minutes] to seven; **c)** *(bei Reihenfolge, Rangordnung)* before; **knapp ~ jmdm. siegen** win just ahead *or* in front of sb.; **d)** *(in Gegenwart von)* before; in front of; **~ Zeugen** before *or* in the presence of witnesses; **e)** *(auf Grund von)* with; **~ Kälte zittern** shiver with cold; **~ Hunger/Durst umkommen** die of hunger/thirst; **~ Arbeit/Schulden nicht mehr aus und ein wissen** not know which way to turn for work/debts; **f) ~ fünf Minuten/Jahren** five minutes/years ago; **heute/gestern/morgen ~ einer Woche** a week ago today/yesterday/tomorrow. **2.** *Präp. mit Akk.* in front of; **keinen Schritt ~ die Tür setzen** not set foot outside the door; **er fuhr bis ~ die Haustür** he drove right up to the front door; **~ sich hin** to oneself; **still ~ sich hin arbeiten** work away quietly. **3.** *Adv.* forward; **Freiwillige ~!** volunteers to the front!; **~ und zurück** backwards and forwards

vor·ạb *Adv.* beforehand

Vo̱r·abend der evening before; *(fig.)* eve

Vo̱r·ahnung die premonition; presentiment; **dunkle/schlimme ~en** dark forebodings

vor·an [fo'ran] *Adv.* forward[s] ahead; first

vorạn-: ~|gehen *unr. itr. V.; mit sein* **a)** go first *or* ahead; **jmdm. ~gehen** go ahead of sb.; **[jmdm.] mit gutem Beispiel ~gehen** *(fig.)* set [sb.] a good example; **b)** *(Fortschritte machen)* make progress; **rasch/nur schleppend ~gehen** make rapid/only slow progress; **es geht mit der Arbeit nicht [so recht] ~:** the work is not making [much] progress; **c)** *s.* **vorausgehen b; ~|kommen** *unr. itr. V.; mit sein* **a)** make headway; **gut ~kommen** make good headway *or* progress; **b)** *(fig.)* make progress; **die Arbeit kommt gut/nicht ~:** the work is making good progress *or* coming along well/not making any progress; **beruflich ~kommen** get on in one's job

Vo̱r·ankündigung die advance announcement; **ohne ~:** without any advance *or* prior notice

Vo̱r·arbeit die preliminary work *no pl.*

vo̱r|arbeiten *refl. V.* work one's way forward

Vo̱r·arbeiter der foreman

vor·aus 1. [-'-] *Präp. mit Dat., nachgestellt* in front; **jmdm./seiner Zeit ~ sein** *(fig.)* be ahead of sb./one's time. **2.** ['--] *Adv.* **im ~** in advance

vorau̱s-, Vorau̱s-: ~|berechnen *tr. V. (auch fig.)* calculate in advance; **~|gehen** *unr. itr. V.; mit sein* **a)** go [on] ahead; **ihm geht der Ruf ~, sehr streng zu sein** *(fig.)* he has the reputation of being very strict; **b)** *(zeitlich)* **einem Ereignis ~gehen** precede an event; **dem Entschluß gingen lange Überlegungen ~:** the decision was preceded by *or* followed lengthy deliberations; **~|haben** *unr. tr. V.* **jmdm./einer Sache etw. ~haben** have the advantage of sth. over sb./sth.; **er hat ihm viel/nichts ~:** he has a great/no advantage over him; **~sage die** *s.* **Vorhersage; ~|sagen** *tr. V.* predict; **jmdm. die Zukunft ~sagen** foretell *or* predict sb.'s future; **~|schauen** *itr. V.* look ahead; **~schauende Planung/Politik** foresighted planning/policy; **~|schicken** *tr. V.* **a)** send [on] ahead; **b)** *(einleitend sagen)* say first; **ich muß folgendes ~schicken** I must start *or* begin by saying the following; **~sehbar** *Adj.* foreseeable; **~|sehen** *unr. tr. V.* foresee; **das war/war nicht ~zusehen** that was foreseeable/unforeseeable; **~|setzen** *tr. V.* **a)** assume; **etw. als bekannt ~setzen** assume sth. is known; **er setzte stillschweigend ~, daß ...:** he took it for granted that ...; **Ihr Einverständnis ~ gesetzt** provided that you agree; **~gesetzt, [daß] ...:** provided [that] ...; **b)** *(erfordern)* require ⟨*skill, experience, etc.*⟩; presuppose ⟨*good organization, planning, etc.*⟩; **~setzung die; ~, ~en a)** *(Annahme)* assumption; *(Prämisse)* premiss; **b)** *(Vorbedingung)* prerequisite; **unter der ~setzung, daß ...:** on condition *or* on the pre-condition that ...; **er hat die besten ~setzungen für den Job** he has the best qualifications for the job; **~sicht die** foresight; **aller ~sicht nach** in all probability; **in weiser ~sicht** *(scherzh.)* with great foresight; **~sichtlich 1.** *Adj.; nicht präd.* anticipated; expected; **2.** *adv.* probably; **der Abflug wird sich ~sichtlich verzögern** the departure is expected to be delayed

Vo̱r·bau der; *Pl.* **~ten** porch

vo̱r|bauen *itr. V.* make provision

Vo̱r·bedacht der: mit ~: intentionally; deliberately

Vo̱rbehalt ['fo:ɐ̯bəhalt] **der; ~[e]s,**

~e reservation; **unter ~:** with reservations; **unter dem ~, daß ...:** with the reservation that ...; **ohne ~:** unreservedly; without reservation

vor|behalten *unr. tr. V.* **sich** *(Dat.)* **etw. ~:** reserve oneself sth.; reserve sth. [for oneself]; **„Änderungen ~"** 'subject to alterations'; **alle Rechte ~** *(Druckw.)* all rights reserved; **jmdm. ~ sein/bleiben** be left to sb.; ‹*decision*› be left [up] to sb.

vorbehaltlich *Präp. mit Gen. (Papierdt.)* subject to

vorbehalt·los 1. *Adj.* unreserved. **2.** *adv.* unreservedly; without reservation[s]

vor·bei *Adv.* **a)** *(räumlich)* past; by; **der Wagen war schon [an uns] ~:** the car was already past [us] *or* had already gone past *or* by [us]; **an etw.** *(Dat.)* **~:** past sth.; **[wieder] ~!** missed [again]!; **b)** *(zeitlich)* past; over; *(beendet)* finished; over; **es ist acht Uhr ~** *(ugs.)* it is past *or* gone eight o'clock; *s. auch* **aus 2a**

vorbei-: ~|fahren 1. *unr. itr. V.; mit sein* **a)** drive/ride past; pass; **an jmdm. ~fahren** drive/ride past sb.; pass sb.; **b) [bei jmdm./der Post] ~fahren** *(ugs.)* drop in *(coll.)* [at sb.'s/at the post office]; **2.** *tr. V. (ugs.)* **kannst du mich schnell beim Bahnhof ~fahren?** can you just run me to the station?; **~|gehen** *unr. itr. V.; mit sein* **a)** pass; go past; **an jmdm./etw. ~gehen** pass sb./sth. go past sb./sth.; **im Vorbeigehen** in passing; **b) [bei jmdm./der Post] ~gehen** *(ugs.)* drop in *(coll.)* [at sb.'s/at the post office]; **c)** *(nicht treffen)* miss; **am Ziel ~gehen** miss its mark *or* target; **d)** *(vergehen)* pass; **~|kommen** *unr. itr. V.; mit sein* **a)** pass; **an etw.** *(Dat.)* **~kommen** pass sth.; **b) [bei jmdm.] ~kommen** *(ugs.)* drop in *(coll.)* [at sb.'s]; **c)** *(vorbeigehen, -fahren können)* get past *or* by; **daran kommt man nicht ~** *(fig.)* there's no getting around *or* away from that; **~|lassen** *unr. tr. V. (ugs.)* let past *or* by; **~|reden** *itr. V.* **an etw.** *(Dat.)* **~reden** miss sth.; **aneinander ~reden** talk at cross purposes; **~|schießen** *unr. itr. V.* **a)** miss; **am Ziel ~schießen** miss the target; **b) am Tor ~schießen** shoot wide of the goal

vor·belastet *Adj.* handicapped (**durch** by); **erblich ~ sein** have an inherited defect

Vor·bemerkung die preliminary remark

vor|bereiten *tr. V.* **a)** prepare (**auf** + *Akk.*, **für** for); **b)** prepare for ‹*trip, party, etc.*›

Vor·bereitung die; ~, ~en preparation; **~en für etw. treffen** make preparations for sth.

Vor·besitzer der previous owner

vor|bestellen *tr. V.* order in advance

Vor·bestellung die advance order

vor·bestraft *Adj. (Amtsspr.)* with a previous conviction/previous convictions *postpos., not pred.;* **[mehrfach] ~ sein** have [several] previous convictions

vor|beugen 1. *tr. V.* bend ‹*head, upper body*› forward. **2.** *refl. V.* lean *or* bend forward. **3.** *itr. V.* **einer Sache** *(Dat.) od.* **gegen etw. ~:** prevent sth.; **~de Maßnahmen** preventive measures; **Vorbeugen ist besser als Heilen** *(Spr.)* prevention is better than cure *(prov.)*

Vor·beugung die prevention (**gegen** of); **zur ~:** as a preventive

Vor·bild das model; **jmdm. ein gutes ~ sein** be a good example to sb.; set sb. a good example; **sich** *(Dat.)* **jmdn./etw. zum ~ nehmen** take sb. as a model *or* model oneself on sb./take sth. as a model

vor·bildlich 1. *Adj.* exemplary. **2.** *adv.* in an exemplary way *or* manner

Vor·bote der harbinger

vor|bringen *unr. tr. V.* say; **eine Frage/Forderung/ein Anliegen ~:** ask a question/make a demand/express a desire; **Argumente ~:** present *or* state arguments; **Beweise ~:** produce evidence

Vor·dach das canopy

Vor·denker der *(Politikjargon)* guiding intellectual force

vorder... ['fɔrdɐ...] *Adj.; nicht präd.* front; **die ~sten Reihen** the rows at the very front; **der Vordere Orient** the Middle East

vorder-, Vorder-: ~achse die front axle; **~ansicht die** front view; **~bein das** foreleg; **~gebäude das** front building; **~grund der** foreground; **im ~grund stehen** *(fig.)* be prominent *or* to the fore; **etw. in den ~grund stellen** *od.* **rücken** *(fig.)* give priority to sth.; place special emphasis on sth.; **in den ~grund treten** *od.* **rücken** *(fig.)* come to the fore; **sich in den ~grund drängen** *(fig.)* push oneself forward; **~gründig** [~grʏndɪç] **1.** *Adj.* superficial; **2.** *adv.* superficially

Vorder-: ~mann der; *Pl.* **~männer** person in front; **ihr/sein ~mann** the person in front of her/him; **jmdn. auf ~mann bringen** *(ugs.)* lick sb. into shape; **den Garten auf ~mann bringen** *(ugs.)* get the garden ship-shape; **~pfote die** front paw; **~rad das** front wheel; **~rad·antrieb der** front-wheel drive; **~seite die** front; *(einer Münze, Medaille)* obverse; **~sitz der** front seat

vorderst... *s.* **vorder...**

vor|drängen *refl. V.* push [one's way] forward *or* to the front; *(fig.)* push oneself forward

vor|dringen *unr. itr. V.; mit sein* push forward; advance; **bis zu jmdm. ~** *(fig.)* reach sb.; get as far as sb.

vor·dringlich 1. *Adj.* **a)** priority *attrib.* ‹*treatment*›; **~ sein** be a matter of priority; **b)** *(dringlich)* urgent; **unser ~stes Anliegen** our main *or* overriding concern. **2.** *adv.* **a)** as a matter of priority; **b)** as a matter of urgency

Vor·druck der; *Pl.* **Vordrucke** form

vor·ehelich *Adj.* pre-marital

vor·eilig 1. *Adj.* rash. **2.** *adv.* rashly; **~ schließen, daß ...:** jump to the conclusion that ...

vor·einander *Adv.* **a)** one in front of the other; **b) Geheimnisse ~ haben** have secrets from each other; **Hochachtung/Furcht ~ haben** have great respect for each other/be afraid of each other

vor·eingenommen *Adj.* biased; prejudiced (**für** in favour of, **gegen** against, **gegenüber** towards)

Voreingenommenheit die; ~: prejudice; bias

vorenthalten[1] *unr. tr. V.* **jmdm. etw. ~:** withhold sth. from sb.

Vor·entscheidung die preliminary decision

vor·erst [*od.* -'-] *Adv.* for the present; for the time being

Vorfahr [-fa:ɐ̯] **der; ~en** *od. selten* **~s, ~en** forefather; ancestor

vor|fahren *unr. itr. V.; mit sein* **a) [vor dem Hotel/Haus] ~:** drive/ride up [outside the hotel/house]; **b)** *(nach vorn fahren)* ‹*person*› drive *or* move forward; ‹*car*› move forward; **c)** *(vorausfahren)* drive *or* go on ahead

Vor·fahrt die; *o. Pl.* right of way; **„~ beachten/gewähren!"** 'give way'; **die ~ nicht beachten** fail to give way; **jmdm. die ~ nehmen** fail to give way to sb.

[1] *ich enthalte vor (od. seltener: vorenthalte), vorenthalten, vorzuenthalten*

Vorfahrt[s]-: ~schild das right-of-way sign; **~straße die** main road

Vor·fall der incident; occurrence

vor|fallen *unr. itr. V.; mit sein* **a)** happen; occur; **ist etwas [Besonderes] vorgefallen?** has anything [special] happened?; **b)** *(nach vorn fallen)* fall forward

vor|finden *unr. tr. V.* find

Vor·freude die anticipation

Vor·frühling der early spring

vor|fühlen *itr. V.* **bei jmdm. ~:** sound sb. out

vor|führen *tr. V.* **a)** bring forward; **jmdn. dem Richter ~:** bring sb. before the judge; **b)** *(zeigen)* show; **wann führst du uns deinen Freund vor?** when are you going to introduce your boy-friend to us?; **c)** *(demonstrieren)* demonstrate; **jmdm. etw. ~:** demonstrate sth. to sb.; **d)** *(darbieten)* show ‹*film, slides, etc.*›; present ‹*circus act, programme*›; perform ‹*play, trick, routine*›

Vor·führung die a) bringing forward; **der Richter ordnete ihre ~ an** the judge ordered her to be brought forward; **b)** *(das Zeigen)* showing; exhibiting; **c)** *(das Demonstrieren)* demonstration; **d)** *(das Darbieten) s.* **vorführen d:** showing; presentation; performance; **e)** *(Veranstaltung) s.* **vorführen d:** show; presentation; performance

Vorführ·wagen der demonstration car *or* model

Vor·gabe die *(Sport)* handicap

Vor·gang der a) occurrence; event; *(Prozeß)* process; **b)** *(Amtsspr.)* file; **der ~ XY** the file on XY

Vorgänger [-gɛŋɐ] **der; ~s, ~, Vorgängerin die; ~, ~nen** *(auch fig.)* predecessor

Vor·garten der front garden

vor|gaukeln *tr. V.* **jmdm. ~, daß ...:** lead sb. to believe that ...; **jmdm. eine heile Welt ~:** lead sb. to believe in a perfect world

vor|geben *unr. tr. V.* **a)** *(vortäuschen)* pretend; **b)** *(Sport)* **jmdm. eine Runde/50 m/15 Punkte ~:** give sb. a lap [start]/[a start of] 50 m/[a lead of] 15 points; **c)** *(im voraus festlegen)* set in advance

Vor·gebirge das promontory

vor·gedruckt *Adj.* pre-printed

vor·gefaßt *Adj.; nicht präd.* preconceived

vorgefertigt *Adj.* pre-fabricated

Vor·gefühl das presentiment

vor|gehen *unr. itr. V.; mit sein* **a)** *(ugs.: nach vorn gehen)* go forward; **zum Altar ~:** go up to the altar; **b)** *(vorausgehen)* go on ahead; **jmdn. ~ lassen** let sb. go first; **c)** ‹*clock*› be fast; **d)** *(einschreiten)* **gegen jmdn./etw. ~:** take action against sb./sth.; **e)** *(verfahren)* proceed; **f)** *(sich abspielen)* happen; go on; **in jmdm. ~:** go on inside sb.; **mit ihm war eine Veränderung vorgegangen** there had been a change in him; a change had taken place in him; **g)** *(Vorrang haben)* have priority; come first

Vor·geschichte die a) *o. Pl.* prehistory *no art.*; **b)** *(eines Vorgangs)* history

vor·geschichtlich *Adj.* prehistoric

Vor·geschmack der; *o. Pl.* foretaste

Vor·gesetzte der/die; *adj. Dekl.* superior

vor·gestern *Adv.* the day before yesterday; **~ mittag/~ abend/~ morgen** *od.* **früh** the day before yesterday at midday/the evening before last/the morning of the day before yesterday; **er ist von ~** *(ugs.)* he is old-fashioned *or* behind the times; **Ansichten von ~** *(ugs.)* old-fashioned *or* outdated views

vor·gestrig *Adj.; nicht präd.* of the day before yesterday *postpos.*

vor|greifen *unr. itr. V.* **jmdm. ~:** anticipate sb.; jump in ahead of sb.; **einer Sache** *(Dat.)* **~:** anticipate sth.

Vor·griff der anticipation **(auf +** *Akk.* of); *(bei einer Erzählung)* jump *or* leap ahead **(auf +** *Akk.* to)

vor|haben *unr. tr. V.* intend; plan; **er hat eine Reise vor** *od.* **er hat vor, eine Reise zu machen** he intends going on a journey/plans to go on a journey; **hast du heute abend etwas vor?** have you anything planned *or* any plans for this evening?; are you doing anything this evening?; **er hat Großes mit seinem Sohn vor** he has great plans for his son

Vor·haben das; ~s, ~ *(Plan)* plan; *(Projekt)* project

Vor·halle die entrance hall

vor|halten 1. *unr. tr. V.* **a)** hold up; **sich** *(Dat.)* **etw. ~:** hold sth. [up] in front of oneself; **mit vorgehaltener Waffe** at gunpoint; *s. auch* **Hand c; b) jmdm. etw. ~** *(fig.)* reproach sb. for sth. **2.** *unr. itr. V. (ugs., auch fig.)* last

Vor·haltungen *Pl.* **jmdm. ~ machen** reproach sb. **(wegen** for)

Vor·hand die a) *(Sport, bes. Tennis)* forehand; **mit [der] ~:** on one's forehand; **b)** *(beim Pferd)* forehand

vorhanden [-'handn̩] *Adj.* existing; *(verfügbar)* available; **~ sein** exist *or* be in existence/be available

Vorhanden·sein das; ~s existence

Vor·hang der curtain

Vorhänge·schloß das padlock

Vor·haut die foreskin; prepuce

vor·her [*od.* -'-] beforehand; *(davor)* before; **eine Woche ~:** a week earlier *or* before

vorher|gehen *unr. itr. V.; mit sein* **in den ~den Wochen** in the preceding weeks; in the weeks before; **wie im ~den erläutert** as explained above

vorherig [*od.* '---] *Adj.; nicht präd.* prior ‹*notice, announcement, warning*›; previous ‹*discussion, agreement*›: **~e Bezahlung** payment in advance

Vor·herrschaft die; *o. Pl.* supremacy; dominance

vor|herrschen *itr. V.* predominate; **~d** predominant

vorher-, Vorher-: ~sage die prediction; *(des Wetters)* forecast; **~|sagen** *tr. V.* predict; forecast ‹*weather*›; **~sehbar** *Adj.* foreseeable; **~|sehen** *unr. tr. V.* foresee

vor|heucheln *tr. V.* feign (*Dat.* to); **er heuchelte ihr Liebe vor** he pretended to love her

vor·hin [*od.* -'-] *Adv.* a short time *or* while ago; **der Junge von ~** *(ugs.)* the boy who we saw a short time ago *or* just now

Vor·hut die; ~, ~en advance guard; *(fig.)* vanguard

vorig... *Adj.; nicht präd.* last

Vor·jahr das previous year

Vor·kämpfer der pioneer; **ein ~ der Freiheit** a pioneering champion of freedom

Vorkehrungen *Pl.* precautions; **~ treffen** take precautions

Vor·kenntnis die background knowledge

vor|knöpfen *tr. V.* **sich** *(Dat.)* **jmdn. [ordentlich] ~** *(ugs.)* give sb. a [proper] talking-to *(coll.)*

vor|kommen *unr. itr. V.; mit sein* **a)** happen; **daß mir so etwas nicht wieder vorkommt!** I hope I never experience anything like that again; **b)** *(vorhanden sein)* occur; **die Pflanze kommt nur im Gebirge vor** the plant is found only in the mountains; **in einer Erzählung ~** ‹*character, figure*› appear in a story; **c)** *(erscheinen)* seem; **das Lied kommt mir bekannt vor** I seem to know the song; **es kam**

mir [so] vor, als ob ...: I felt *or* it seemed as if ...; **du kommst dir wohl schlau vor** I suppose you think you're clever; **ich komme mir überflüssig vor** I feel [as if I am] superfluous; **d)** *(ugs.: nach vorne kommen)* come forward; **e)** *(hervorkommen)* come out; **hinter/unter etw.** *(Dat.)* **~:** come out from behind/under sth.

Vorkommen das; ~s, ~ a) *o. Pl.* occurrence; **b)** *(Geol.)* deposit

Vorkommnis [-kɔmnɪs] **das; ~ses, ~se** incident; occurence

Vor·kriegszeit die pre-war period

vor|laden *unr. tr. V.* summon

Vor·ladung die summons

Vor·lage die a) *o. Pl.; s.* **vorlegen a:** presentation; showing; production; submission; tabling; introduction; **gegen ~ einer Sache** *(Gen.)* on production *or* presentation of sth.; **b)** *(Gesetzentwurf)* bill; **c)** *(Muster)* pattern; *(Modell)* model; **nach einer/ohne ~ zeichnen** draw from/without a model; **d)** *(Ballspiele, bes. Fußball)* forward pass

vor|lassen *unr. tr. V.* **a) jmdn. ~** *(ugs.)* let sb. go first *or* in front; **b)** *(empfangen)* admit; let in

Vor·läufer der precursor; forerunner

vor·läufig 1. *Adj.* temporary; provisional ‹*diagnosis, settlement, result, successor*›; interim ‹*order, agreement*›. **2.** *adv.* for the time being; for the present

vor·laut 1. *Adj.* forward. **2.** *adv.* forwardly

Vor·leben das; *o. Pl.* past life; past

Vorlege- serving ‹*cutlery, fork, spoon*›

vor|legen *tr. V.* **a)** present; show, produce ‹*certificate, identity card, etc.*›; show ‹*sample*›; submit ‹*evidence*›; table, introduce ‹*parliamentary bill*›; **b)** *(anbringen vor)* **eine Kette/einen Riegel ~:** put a chain on *or* across/a bolt across; **c)** *(geh.: aufgeben)* serve ‹*food*›; **jmdm. etw. ~:** serve sb. with sth.; serve sth. to sb.

Vorleger der; ~s, ~ mat; *(Bett~)* rug

vor|lesen *unr. tr., itr. V.* read aloud *or* out; read ‹*story, poem, etc.*› aloud; **jmdm. [etw.] ~:** read [sth.] to sb.; **lies schon vor!** read it out!; read out what it says!

Vor·lesung die lecture; *(~sreihe)* series *or* course of lectures

Vorlesungs·verzeichnis das lecture timetable

vor·letzt... *Adj.; nicht präd.* last but one; next to last; penultimate ‹*page, episode, etc.*›; **mein ~es Exemplar** my last copy but one; my next to last copy; **~e Woche** the week before last

Vor·liebe die preference; [special] fondness *or* liking; **eine ~ für etw. haben** be fond of *or* partial to sth.; **etw. mit ~ tun** particularly like doing sth.

vorlieb|nehmen *unr. itr. V.* **mit jmdm./etw. ~:** make do with sb./sth.

vor|liegen *unr. itr. V.* **a) jmdm. ~** ‹*application, complaint, plans, etc.*› be with sb.; **das Beweismaterial liegt dem Gericht vor** the evidence is before *or* has been submitted to the court; **die Ergebnisse liegen uns noch nicht vor** we do not have the results yet; **die mir ~de Ausgabe/~den Ergebnisse** the edition/results in front of me; **im ~den Fall** in the present case; **b)** *(bestehen)* be [present]; exist; ‹*symptom*› be present; ‹*book*› be available; **gegen ihn liegt nichts vor** there is nothing against him; **hier liegt ein Irrtum vor** there is a mistake here; **ein Verschulden des Fahrers liegt nicht vor** the driver is/was not to blame

vorm [fo:ɐ̯m] *Präp. + Art.* **a) = vor dem; b)** *(räumlich)* in front of the; **c)** *(zeitlich, bei Reihenfolge, Rangordnung)* before the

vor|machen *tr. V. (ugs.)* **a) jmdm. etw. ~:** show sb. sth.; **ihm macht niemand was vor** there is no one better than him; no one can teach him anything; **b)** *(vortäuschen)* **jmdm./sich etwas ~:** kid *(coll.) or* fool sb./oneself; **der läßt sich nichts ~:** he's nobody's fool

Vormacht·stellung die; *o. Pl.* [position of] supremacy *no art.*

Vor·marsch der *(auch fig.)* advance; **auf dem** *od.* **im ~ sein** be advancing *or* on the advance; *(fig.)* be gaining ground

vor|merken *tr. V.* make a note of; **ich habe Sie für den Kurs vorgemerkt** I've put you down for the course

vor·mittag *Adv.* **heute/morgen/Freitag ~:** this/tomorrow/Friday morning

Vor·mittag der morning

vor·mittags *Adv.* in the morning

Vor·monat der previous *or* preceding month

Vor·mund der; *Pl.* **Vormunde** *od.* **Vormünder** guardian; **ich brauche keinen ~** *(fig.)* I don't need anyone telling me what to do

Vormundschaft die; ~, ~en guardianship

vorn ['fɔrn] *Adv.* at the front; **das Zimmer liegt nach ~ [raus]** *(ugs.)* the room faces the front; **~ am Haus** at the front of the house; **~ im Bild** in the foreground of the picture; **nach ~ sehen** look in front *or* to the front; **nach ~ gehen** go to the front; **da ~:** over there; **der Wind kam von ~:** the wind came from the front; **noch einmal von ~ anfangen** start afresh; start from the beginning again; **von ~ bis hinten** *(ugs.)* from beginning to end

Vor·name der first *or* Christian name

vorne *Adv. s.* **vorn**

vornehm ['fo:ɐ̯ne:m] **1.** *Adj.* **a)** noble ‹*character, behaviour, gesture, etc.*›; **~e Gesinnung** noblemindedness; **b)** *(der Oberschicht angehörend, kultiviert)* distinguished; **die ~e Welt/die ~en Kreise** high society; **c)** *(adlig)* noble; **d)** *(elegant)* exclusive, *(coll.)* posh ‹*district, hotel, resort*›; elegant, *(coll.)* posh ‹*villa, clothes*›. **2.** *adv.* **a)** nobly; **b)** *(elegant)* elegantly

vor|nehmen 1. *unr. refl. V.* **sich** *(Dat.)* **etw. ~:** plan sth.; **sich** *(Dat.)* **~, etw. zu tun** plan to do sth.; **sich** *(Dat.)* **~, mit dem Rauchen aufzuhören** resolve to give up smoking. **2.** *unr. tr. V.* **a)** carry out, make ‹*examination, search, test*›; perform ‹*action, ceremony*›; make ‹*correction, change, division, choice, selection*›; take ‹*measurements*›; **b) [sich** *(Dat.)***] ein Buch/eine Arbeit ~** *(ugs.)* get down to reading a book/to a piece of work; **c) sich** *(Dat.)* **jmdn. ~** *(ugs.)* give sb. a talking-to *(coll.)*

Vornehmheit die; ~: *s.* **vornehm 1a–d:** nobility; exclusivity; elegance; **seine ~ beeindruckte sie** she was impressed by his distinguished manner

vornehmlich *Adv. (geh.)* above all; primarily

vor|neigen *tr., refl. V.* lean forward

vorn-: ~herein: von ~herein from the start *or* outset *or* beginning; **~über** *Adv.* forwards

Vor·ort der suburb

Vor·platz der forecourt

Vor·posten der *(Milit., fig.)* outpost

vor|preschen *itr. V.; mit sein (fig.)* rush ahead

Vor·programm das supporting programme

vor|programmieren *tr. V. (auch fig.)* pre-programme

vorlragen *itr. V.* project; jut out

Vor·rang der; *o. Pl.* **a) [den] ~ [vor jmdm./etw.] haben** have priority *or* take precedence [over sb./sth.]; **jmdm./einer Sache den ~ geben** give sb./sth. priority; **b)** *(bes. österr.: Vorfahrt)* right of way

vorrangig [-raŋɪç] **1.** *Adj.* priority *attrib.* ⟨*treatment*⟩; **~ sein** be a matter of priority *or* of prime importance; **unser ~es Anliegen** our primary concern. **2.** *adv.* as a matter of priority; **jmdn. ~ behandeln** give sb. priority treatment

Vor·rat der supply, stock **(an +** *Dat.* of); **solange der ~ reicht** while stocks last

vorrätig [-rɛ:tɪç] *Adj.* in stock *postpos.*; **etw. nicht mehr ~ haben** be out of [stock of] sth.

Vorrats-: ~kammer die pantry; larder; **~keller der** cellar store-room; **~raum der** store-room

Vor·raum der anteroom

vorlrechnen *tr. V.* **jmdm. etw. ~:** work sth. out *or* calculate sth. for sb.; **jmdm. seine Fehler ~** *(fig.)* enumerate sb.'s mistakes

Vor·recht das privilege

Vor·rede die a) introductory remarks *pl.*; **sich nicht lange mit der ~ aufhalten** not take long over the introductions; **b)** *(Vorwort)* preface; foreword

Vor·richtung die device

vorlrücken 1. *tr. V.* move forward; advance ⟨*chess piece*⟩. **2.** *itr. V.*; *mit sein* move forward; *(Milit.)* advance; **mit dem Turm ~** *(Schach)* advance the rook; **auf den 5. Platz ~:** move up to fifth place; **zu vorgerückter Stunde** *(geh.)* at a late hour

Vor·ruhestand der early retirement

Vor·runde die *(Sport)* preliminary *or* qualifying round

vors *Präp. + Art.* **a) = vor das; b)** in front of the; **jmdm. ~ Auto laufen** run in front of sb.'s car

vorlsagen *tr. V.* **a)** *auch itr.* **jmdm. [die Antwort] ~:** tell sb. the answer; *(flüsternd)* whisper the answer to sb.; **b)** *(aufsagen)* recite (*Dat.* to)

Vor·saison die start of the season; early [part of the] season

Vor·satz der intention; **den ~ fassen, etw. zu tun** resolve to do sth.; make a resolution to do sth.; **den ~ haben, etw. zu tun** intend to do sth.; have the intention of doing sth.; **mit ~:** with intent

vorsätzlich [-zɛtslɪç] **1.** *Adj.* intentional; deliberate; wilful ⟨*murder, arson, etc.*⟩. **2.** *adv.* intentionally; deliberately

Vor·schau die preview

Vor·schein der: etw. zum ~ bringen reveal sth.; bring sth. to light; **zum ~ kommen** appear; *(entdeckt werden)* come to light; **wieder zum ~ kommen** reappear

vorlschieben 1. *unr. tr. V.* **a)** push ⟨*bolt*⟩ across; *s. auch* **Riegel a; b)** *(nach vorn schieben)* push forward. **2.** *unr. refl. V.* push forward

vorlschießen *unr. tr. V.* **jmdm. Geld ~:** advance sb. money

Vorschlag der suggestion; proposal

vorlschlagen *unr. tr. V.* suggest, propose (*Dat.* to)

Vorschlag·hammer der sledgehammer

vor·schnell *Adj., adv.*: *s.* **voreilig**

vor·schreiben *unr. tr. V.* stipulate, lay down, set ⟨*conditions*⟩; lay down ⟨*rules*⟩; prescribe ⟨*dose*⟩; **er wollte uns ~, was wir zu tun hätten** he wanted to tell us *or* dictate to us what to do; **die vorgeschriebene Geschwindigkeit/Dosis** the prescribed speed/dose

Vor·schrift die instruction; order; *(gesetzliche od. amtliche Bestimmung)* regulation; **ich lasse mir von dir keine ~en machen** I won't be told what to do by you; I won't be dictated to by you; **das ist ~:** that's/those are the regulations; **das ist gegen die ~:** it's against the rules *or* regulations; **die Medizin nach ~ einnehmen** take the medicine as directed

vorschrifts·mäßig 1. *Adj.* correct; proper. **2.** *adv.* correctly; properly

Vor·schub der: jmdm./einer Sache ~ leisten encourage sb./encourage *or* promote *or* foster sth.

Vorschul- pre-school ⟨*age, education*⟩

Vor·schule die nursery school

Vor·schuß der advance; **50 Mark ~:** an advance of 50 marks

vorlschützen *tr. V.* plead as an excuse; **wichtige Geschäfte/Krankheit ~:** pretend one has important business/feign illness; *s. auch* **Müdigkeit**

vorlschweben *itr. V.* **jmdm. schwebt etw. vor** sb. has sth. in mind

vorlsehen 1. *unr. tr. V.* **a)** plan; **etw. für/als etw. ~:** intend sth. for/as sth.; **jmdn. für/als etw. ~:** designate sb. for/as sth.; **b)** ⟨*law, plan, contract, etc.*⟩ provide for. **2.** *unr. refl. V.* **sich [vor jmdm./etw.] ~:** be careful [of sb./sth.]; **sieh dich vor dem Hund vor** be careful of *or* mind the dog; **sieh dich vor, daß du nicht krank wirst** be careful *or* take care you don't become ill

Vorsehung die; ~: Providence *no art.*

vorlsetzen *tr. V.* **a)** move forward; **den rechten/linken Fuß ~:** put one's right/left foot forward; **b) jmdm. etw. ~:** serve sb. sth.

Vor·sicht die; *o. Pl.* care; *(bei Risiko, Gefahr)* caution; care; *(Umsicht)* circumspection; caution; **zur ~:** as a precaution; to be on the safe side; **~!** be careful!; watch *or* look out!; **„~, Glas"** 'glass – handle with care'; **„~, bissiger Hund"** 'beware of the dog'; **~ an der Bahnsteigkante** stand back from the edge of the platform; **„~, Stufe!"** 'mind the step!'; **„~, Steinschlag"** 'danger, falling rocks'; **„~, frisch gestrichen"** 'wet paint'

vorsichtig [-zɪçtɪç] **1.** *Adj.* careful; *(bei Risiko, Gefahr)* cautious; careful; *(umsichtig)* circumspect; cautious; guarded ⟨*remark, hint, question, optimism*⟩; cautious, conservative ⟨*estimate*⟩; **sei ~!** be careful!; take care! **2.** *adv.* carefully; with care; **~ optimistisch** guardedly *or* cautiously optimistic; **etw. ~ andeuten** hint at sth. cautiously; **~ geschätzt** at a conservative estimate

vorsichts·halber *Adv.* as a precaution; to be on the safe side

Vorsichts·maßnahme die precautionary measure; precaution

Vor·silbe die *(Präfix)* prefix

vorlsingen 1. *unr. tr. V.* sing (*Dat.* to). **2.** *unr. itr. V.* **a)** sing (*Dat.* to); **wenn er ~ soll** when he has to sing in public *or* in front of people; **b)** *(zur Prüfung)* have *or* take a singing test; **bei der Oper ~:** audition for *or* have an audition with the opera company

vor·sintflutlich *Adj. (ugs.)* antiquated

Vor·sitz der chairmanship; **den ~ haben** be the chairman; be in the chair; *(im Gericht)* preside over the trial

Vorsitzende der/die; *adj. Dekl.* chair[person]; *(bes. Mann)* chairman; *(Frau auch)* chairwoman

Vor·sorge die; *o. Pl.* precautions *pl.*; *(für den Todesfall, Krankheit, Alter)* provisions *pl.*; *(Vorbeugung)* prevention; **~ treffen** take precautions **(gegen** against)/make provisions **(für** for)

vorlsorgen *itr. V.* make provisions

Vorsorge·untersuchung die *(Med.)* medical check-up

vorsorglich 1. *Adj.* precautionary ⟨*measure, check-up, etc.*⟩. 2. *adv.* as a precaution

Vor·spann der *(Film, Ferns.)* opening credits *pl.*

Vor·speise die starter; hors d'œuvre

Vor·spiegelung die: unter ~ falscher Tatsachen under false pretences

Vor·spiel das a) *(Theater)* prologue; *(Musik)* prelude; **b)** *(vorm Geschlechtsakt)* foreplay

vor|spielen 1. *tr. V.* **a)** play ⟨*piece of music*⟩ (*Dat.* to, for); act out, perform ⟨*scene*⟩ (Dat. for, in front of); **b)** *(vorspiegeln)* feign (*Dat.* to); **spiel uns doch nichts vor!** don't try and fool us!; **jmdm. Theater/eine Komödie ~:** put on an act for sb. **2.** *itr. V.* **a)** play (*Dat.* to, for); **b)** *(bei einer Bewerbung)* audition, have an audition (**bei** for)

Vor·sprache die *(österr.)* visit (**bei** to)

vor|sprechen 1. *unr. tr. V.* **a) jmdm. etw. ~:** pronounce *or* say sth. first for sb.; **b)** *(zur Prüfung)* recite. **2.** *unr. itr. V.* **a)** *(zur Prüfung)* recite one's examination piece; *(bei Bewerbungen)* audition; **am Theater ~:** audition for the Theatre; **jmdn. ~ lassen** audition sb.; **b)** *(einen Besuch machen)* **bei jmdm. [in einer Angelegenheit] ~:** call on sb. about a matter; **bei** *od.* **auf einer Behörde ~:** call at an office

vor·springen *unr. itr. V.; mit sein* **a)** *(ugs.)* jump out (**hinter** + *Dat.* from behind); **b)** *(vorstehen)* jut out; project; **ein ~des Kinn** a prominent chin

Vor·sprung der a) lead; **einen ~ [vor jmdm.] haben** have a lead [over sb.]; be ahead [of sb.]; **jmdm. zehn Schritte/Minuten ~ geben** give sb. ten paces'/minutes' start; **b)** *(vorspringender Teil)* projection; *(Fels~)* ledge

Vor·stadt die suburb; **in der ~ wohnen** live in the suburbs

vor·städtisch *Adj.* suburban

Vor·stand der a) *(einer Firma)* board [of directors]; *(eines Vereins)* executive committee; *(einer Partei)* executive; **im ~ sein** be on the board/executive committee/executive; **b)** *s.* **Vorstandsmitglied**

Vorstands·mitglied das *s.* **Vorstand a:** member of the board; board member; member of the executive committee; member of the executive

vor|stehen *unr. itr. V.* **a)** ⟨*house, roof, etc.*⟩ project, jut out; ⟨*teeth, chin*⟩ stick out; ⟨*cheek-bones*⟩ be prominent; **~de Zähne** buck-teeth; projecting teeth; **b)** *(geh.)* **einer Institution/dem Haushalt ~:** be the head of an institution/the household; **einer Abteilung ~:** be in charge of *or* run a department

vorstehend *Adj.* above *attrib;* **im ~en** above; **das Vorstehende** the above

Vorsteher der; ~s, ~: head; *(einer Gemeinde)* chairman; *(eines Klosters)* abbot

Vorsteher·drüse die prostate [gland]

Vorsteherin die; ~, ~nen head; *(eines Klosters)* abbess

vorstell·bar *Adj.* conceivable; imaginable

vor|stellen 1. *tr. V.* **a)** put ⟨*leg, foot, etc.*⟩ out *or* forward; **die Uhr ~:** put the clock forward; **b)** *(bekannt machen; auch fig.)* introduce; **jmdn./sich jmdm. ~:** introduce sb./oneself to sb.; **c) sich ~** *(bei Bewerbung)* come/go for [an] interview (**bei** with); **d)** *(darstellen)* represent; **er stellt etwas vor** *(ugs.) (sieht gut aus)* he looks good; *(gilt als Persönlichkeit)* he is somebody. **2.** *refl. V.* **a) sich** *(Dat.)* **etw. ~:** imagine sth.; **ich habe mir das Wochenende ganz anders vorgestellt** the weekend was not at all what I had imagined; **ich kann ihn mir gut als Lehrer ~:** I can easily imagine *or* see him as a teacher; **man stelle sich** *(Dat.)* **bitte einmal vor, daß ...:** just imagine that ...; **b) sich** *(Dat.)* **unter etw.** *(Dat.)* **etw. ~:** understand sth. by sth.; **darunter kann ich mir nichts ~:** it doesn't mean anything to me

Vor·stellung die a) *(Begriff)* idea; **er macht sich** *(Dat.)* **keine ~ [davon], welche Mühe das kostet** he has no idea how much effort that costs; **das entspricht ganz/nicht meinen ~en** that is exactly/not what I had in mind; **b)** *o. Pl.* *(Phantasie)* imagination; **das geht über alle ~ hinaus** it is unimaginable; **c)** *(Aufführung)* performance; *(im Kino)* showing; **eine schwache ~ geben** *(fig.)* perform badly; **d)** *(das Bekanntmachen)* introduction; **e)** *(Präsentation)* presentation; **f)** *(bei Bewerbung)* interview

Vorstellungs-: ~gespräch das interview; **~kraft die;** *o. Pl.,* **~vermögen das;** *o. Pl.* [powers *pl.* of] imagination

Vor·stopper der *(Fußball)* central defender

Vor·stoß der advance; **einen ~ unternehmen** push forward; advance

vor|stoßen *unr. itr. V.; mit sein* advance; push forward

Vor·strafe die *(Rechtsw.)* previous conviction

Vorstrafen·register das criminal records *pl.*

vor|strecken *tr. V.* **a)** stretch ⟨*arm, hand*⟩ out; stick out ⟨*stomach*⟩; **den Kopf/Hals ~:** crane one's neck forward; **b)** *(auslegen)* advance ⟨*money, sum*⟩

Vor·stufe die preliminary stage

Vor·tag der day before; previous day; **am ~ der Prüfung** the day before *or* on the eve of the examination

vor|tanzen 1. *tr. V.* **er tanzte ihnen den Foxtrott vor** he showed them *or* demonstrated how to dance the foxtrot. **2.** *itr. V.* demonstrate one's dancing ability

vor|tasten *refl. V. (auch fig.)* feel one's way forward

vor|täuschen *tr. V.* feign ⟨*interest, illness, etc.*⟩; simulate ⟨*reality etc.*⟩; fake ⟨*crime*⟩

Vor·täuschung die *s.* **vortäuschen:** feigning; simulation; faking

Vor·teil [*od.* 'fɔrtail] **der** advantage; **jmdm. gegenüber im ~ sein** have an advantage over sb.; **[für jmdn.] von ~ sein** be advantageous [to sb.]; **sich zu seinem ~ verändern** change for the better

vorteilhaft 1. *Adj.* advantageous. **2.** *adv.* advantageously; **sich ~ auswirken** have a favourable *or* beneficial effect; **sich ~ kleiden** wear clothes that suit one

Vortrag [-tra:k] **der; ~[e]s, Vorträge** [-trɛ:gə] **a)** *(Rede)* talk; *(wissenschaftlich)* lecture; **einen ~ halten** give a talk/lecture; **b)** *(Darbietung)* presentation; performance; *(eines Gedichts)* recitation; rendering

vor|tragen *unr. tr. V.* **a)** perform ⟨*gymnastic routine etc.*⟩; sing ⟨*song*⟩; perform, play ⟨*piece of music*⟩; recite ⟨*poem*⟩; **b)** present ⟨*case, matter, request, demands*⟩; lodge, make ⟨*complaint*⟩; express ⟨*wish, desire*⟩

Vortragende der/die; *adj. Dekl.: s.* **Vortrag a:** speaker; lecturer

Vortrags-: ~reihe die series of lectures/talks; **~reise die** lecture tour

vor·trefflich 1. *Adj.* excellent; splendid; superb ⟨*singer, swimmer, etc.*⟩. **2.** *adv.* excellently; splendidly; ⟨*sing, swim, etc.*⟩ superbly

vo̱r|treiben *unr. tr. V.* drive ⟨*tunnel, shaft*⟩
vo̱r|treten *unr. itr. V.; mit sein* step forward
Vo̱r·tritt der: jmdm. den ~ lassen *(auch fig.)* let sb. go first
vorü̱ber *Adv.* **a)** *(zeitlich)* over; **~ sein** be over; ⟨*pain*⟩ be gone; ⟨*danger*⟩ be past; **das ist aus und ~** *(ugs.)* that is [all] over and done with; **b)** *(räumlich)* past; **an etw.** *(Dat.)* **~:** past sth.
vorü̱ber|gehen *unr. itr. V.; mit sein* **a)** go *or* walk past; pass by; **an jmdm./etw. ~:** go past sb./sth.; pass sb./sth.; *(achtlos)* pass sb./sth. by; *(fig.)* ignore sb./sth.; **im Vorübergehen** in passing; *(fig.: nebenbei)* in a trice; **b)** *(vergehen)* pass; ⟨*pain*⟩ go; ⟨*storm*⟩ pass, blow over; **das geht vorüber** *(ugs.) (tröstend)* it'll pass; *(scherzh. iron.)* that won't last long
vorü̱bergehend 1. *Adj.* temporary; passing ⟨*interest etc.*⟩; brief ⟨*illness*⟩. **2.** *adv.* temporarily; *(kurz)* for a short time; briefly
Vo̱r·urteil das bias; *(voreilige Schlußfolgerung)* prejudice (**gegen** against, towards); **gegen etw. ~e haben** be biased/prejudiced against *or* towards sth.
Vo̱r·väter *Pl.* forefathers
Vo̱r·vergangenheit die *(Sprachw.)* pluperfect
Vo̱r·verkauf der advance sale of tickets; advance booking
vo̱r|verlegen *tr. V. (zeitlich)* bring forward (**um** by)
Vo̱rwahl die a) *(Politik)* preliminary election; *(in den USA)* primary; **b)** *(Fernspr.)* dialling code
Vo̱rwand der; ~[e]s, Vo̱rwände pretext; *(Ausrede)* excuse; **etw. zum ~ nehmen** use sth. as a pretext/an excuse
vo̱r·wärts *Adv.* forwards; *(weiter)* onwards; *(mit der Vorderseite voran)* facing forwards; **ein Schritt ~** *(auch fig.)* a step forwards; **ein Salto ~:** a forward somersault
vo̱rwärts|kommen *unr. itr. V.; mit sein* make progress; *(im Beruf, Leben)* get on; get ahead
Vo̱r·wäsche die prewash
vor·wẹg *Adv.* **a)** *(vorher)* beforehand; **b)** *(voraus)* in front; ahead
vorwẹg|nehmen *unr. tr. V.* anticipate; **um das Ergebnis vorwegzunehmen, ...:** to come straight to the result ...
vo̱r|weisen *unr. tr. V.* produce; **etw. ~ können, etw. vorzuweisen haben** *(fig.)* possess sth.
vo̱r|werfen *unr. tr. V.* **a) jmdm. etw. ~:** reproach sb. with sth.; *(beschuldigen)* accuse sb. of sth.; **jmdm. ~, etw. getan zu haben** reproach sb. with *or* accuse sb. of doing *or* having done sth.; **jmdm. Parteilichkeit ~:** accuse sb. of being biased; **ich habe mir nichts vorzuwerfen** I've nothing to reproach myself for; **b) etw. den Tieren ~** *(hinwerfen)* throw sth. to the animals
vo̱r·wiegend *Adv.* mainly
vo̱r·witzig *Adj.* bumptious; pert ⟨*child*⟩; *(neugierig)* curious
Vo̱r·wort das; *Pl.* **~e** foreword; preface
Vo̱r·wurf der reproach; *(Beschuldigung)* accusation; **jmdm. etw. zum ~ machen** reproach sb. with sth.; **jmdm. [wegen etw.] einen ~/Vorwürfe machen** reproach sb. [for sth.]; **sich** *(Dat.)* **[wegen etw.] Vorwürfe machen** reproach *or* blame oneself [for sth.]
vo̱rwurfs·voll 1. reproachful. **2.** *adv.* reproachfully
Vo̱r·zeichen das a) *(Omen)* omen; **b)** *(Math.)* [algebraic] sign; **c)** *(Musik)* sharp/flat [sign]; *(für Tonart)* key signature
vo̱r|zeigen *tr. V.* show; produce, show ⟨*passport, ticket, etc.*⟩
Vo̱r·zeit die prehistory; **in grauer ~:** in the dim and distant past
vo̱rzeitig 1. *Adj.* premature; early ⟨*retirement*⟩. **2.** *adv.* prematurely; ⟨*be retired*⟩ early
vo̱r|ziehen *unr. tr. V.* **a)** *(lieber mögen)* prefer (*Dat.* to); *(bevorzugen)* favour, give preference to ⟨*person*⟩; **b)** *(zuziehen)* draw ⟨*curtain*⟩; **c)** *(vorverlegen)* bring forward ⟨*date*⟩ (**um** by); **d)** *(nach vorn ziehen)* pull forward
Vo̱r·zimmer das outer office; ante-room
Vo̱r·zug der a) *o. Pl.* preference (**gegenüber** over); **jmdm./einer Sache den ~ geben** prefer sb./sth.; **b)** *(gute Eigenschaft)* good quality; merit; *(Vorteil)* advantage; **c)** *(österr. Schulw.)* distinction
vorzüglich [fo:ɐ̯'tsy:klɪç] **1.** *Adj.* excellent; first-rate. **2.** *adv.* excellently; **~ speisen** have an excellent meal
vo̱rzugs·weise *Adv.* preferably
Voten *s.* **Votum**
Votum ['vo:tʊm] **das; ~s, Vo̱ten** vote
vulgär [vʊl'gɛ:ɐ̯] **1.** *Adj.* vulgar. **2.** *adv.* in a vulgar way; **sich ~ ausdrücken** use vulgar language
Vulkan [vʊl'ka:n] **der; ~s, ~e** volcano
vulka̱nisch *Adj.* volcanic
vulkanisi̱eren *tr. V.* vulcanize
v. u. Z. *Abk.* **vor unserer Zeit[rechnung]** BC

w, W [ve:] **das; ~s, ~:** w, W
W *Abk.* **a) West, Westen** W.; **b) Watt** W.
WAA [ve:a:'a:] **die; ~, ~s** *Abk.* **Wiederaufbereitungsanlage**
Waage ['va:gə] **die; ~, ~n a)** [pair *sing.* of] scales *pl.*; *(Gold~, Apotheker~ usw.)* balance; **er bringt 80 kg auf die ~** *(ugs.)* he tips the scales at 80 kilos; **sich** *(Dat.)* **die ~ halten** balance out; balance one another; **b)** *(Astrol., Astron.)* Libra
wa̱age·recht 1. *Adj.* horizontal. **2.** *adv.* horizontally
Wa̱age·rechte die; *adj. Dekl.: s.* **Horizontale**
Wa̱ag·schale die scale pan; **etw. in die ~ werfen** *(fig.)* bring sth. to bear
wabb[e]lig ['vab(ə)lɪç] *Adj. (ugs.)* wobbly; flabby ⟨*muscles*⟩
Wabe ['va:bə] **die; ~, ~n** honeycomb
Wa̱ben·honig der comb honey
wach [vax] **1.** *Adj.* **a)** awake; **in ~em Zustand** in a state of wakefulness; **jmdn. ~ machen** wake sb. up; **~werden** wake up; **b)** *(aufmerksam, rege)* alert ⟨*mind, eyes, etc.*⟩; attentive ⟨*audience*⟩; lively, keen ⟨*interest*⟩. **2.** *adv.* alertly; attentively
Wạch·ablösung die changing of the guard/watch
Wache ['vaxə] **die; ~, ~n a)** *(Wachdienst) (Milit.)* guard *or* sentry duty; *(Seew.)* watch; **~ haben** *od.* **halten** *(Milit.)* be on guard *or* sentry duty; *(Seew.)* be on watch; have the watch; **~ stehen** *(Milit.)* stand on guard; **b)** *(Wächter)* guard; *(Milit.: Posten)* sentry; **c)** *(Mannschaft) (Milit.)* guard; *(Seew.)* watch; **d)** *(Polizei~)* police station
wạchen *itr. V.* **a)** *(geh.)* be awake; **b) bei jmdm. ~:** stay up at

sb.'s bedside; sit up with sb.; **c) über etw.** *(Akk.)* ~: watch over *or* keep an eye on sth.; **er wachte darüber, daß ...:** he watched carefully to ensure that ...

wạch-, Wạch-: ~|**halten** *unr. tr. V.* keep ⟨*interest, memory, etc.*⟩ alive; ~**hund der** guard-dog; watch-dog; ~**lokal das** *(Milit.)* guardroom; ~**mann der;** *Pl.* ~**männer** *od.* ~**leute a)** watchman; **b)** *(österr.: Polizist)* policeman; ~**mannschaft die** *(Milit.)* guard detachment

Wacholder [va'xɔldɐ] **der; ~s, ~ a)** juniper; **b)** *(Schnaps) spirit from juniper berries;* ≈ gin

wạch-, Wạch-: ~**posten der** *(Milit.)* guard; sentry; ~|**rufen** *unr. tr. V.* awaken, rouse ⟨*enthusiasm, ambition, etc.*⟩; evoke, bring back ⟨*memory, past*⟩; ~|**rütteln** *tr. V.* rouse *or* shake ⟨*sb.*⟩ out of his/her apathy; stir ⟨*conscience*⟩

Wachs [vaks] **das; ~es, ~e** wax

wachsam ['vaxza:m] **1.** *Adj.* watchful; vigilant; **sei ~!** be on your guard!. **2.** *adv.* vigilantly

Wạchsamkeit die; ~: vigilance

[1]**wachsen** ['vaksn̩] *unr. itr. V.; mit sein* **a)** *(auch fig.)* grow; ⟨*building*⟩ rise; **sich** *(Dat.)* **einen Bart ~ lassen** grow a beard; **sich** *(Dat.)* **die Haare ~ lassen** let one's hair grow long; **b)** *(fig.: allmählich entstehen)* evolve [naturally]; **eine gewachsene Ordnung** an organic order

[2]**wạchsen** *tr. V.* wax

wächsern ['vɛksɐn] *Adj. (geh.: bleich)* waxen

Wạchs-: ~**figur die** waxwork; wax figure; ~**figuren·kabinett das** waxworks *sing. or pl.;* waxworks museum

wächst [vɛkst] *2. u. 3. Pers. Sg. Präsens v.* **wachsen**

Wạch·stube die *(Milit.)* guardroom; *(Polizeiwache)* duty room

Wachstum ['vakstu:m] **das; ~s** growth

wạchstums-, Wạchstums-: ~**fördernd** *Adj.* promoting growth *postpos.;* ~**hemmend** *Adj.* inhibiting growth *postpos.;* ~**hormon das** growth hormone

Wachtel ['vaxtl̩] **die; ~, ~n** quail

Wächter ['vɛçtɐ] **der; ~s, ~:** guard; *(Leib~)* bodyguard; *(Nacht~, Turm~)* watchman; *(Park~)* [park-]keeper

Wạcht-: ~**meister der** constable *(Brit.);* patrolman *(Amer.);* **Herr ~meister** *(Anrede)* officer; ~**posten der** *s.* **Wachposten**

Wạch·traum der day-dream; waking dream

Wạch[t]·turm der watch-tower

wạckelig 1. *Adj.* **a)** *(nicht stabil)* wobbly ⟨*chair, table, etc.*⟩; loose ⟨*tooth*⟩; shaky, rickety ⟨*structure*⟩; rickety ⟨*car, furniture*⟩; **b)** *(ugs.: kraftlos, schwach)* frail ⟨*person*⟩; frail, doddery ⟨*old person*⟩; **~ auf den Beinen sein** be a bit shaky on one's feet; **c)** *(fig. ugs.: gefährdet, bedroht)* dodgy *(Brit. coll.)* ⟨*business*⟩; insecure, shaky ⟨*job*⟩; **er steht in der Schule/in Latein ziemlich ~:** things are dodgy for him at school *(Brit. coll.)*/his Latin is somewhat shaky. **2.** *adv.* **~ stehen** be wobbly

Wạckel·kontakt der *(Elektrot.)* loose connection

wackeln ['vakl̩n] *itr. V.* **a)** wobble; ⟨*post etc.*⟩ move about; ⟨*tooth etc.*⟩ be loose; ⟨*house, window, etc.*⟩ shake; **mit dem Kopf/den Hüften ~:** waggle *or* wag one's head/wiggle one's hips; **mit dem Schwanz ~:** wag its tail; **b)** *mit sein (ugs.: gehen)* ⟨*person*⟩ totter; **c)** *(ugs.: gefährdet, bedroht sein)* ⟨*job, government*⟩ be insecure; ⟨*firm*⟩ be in a dodgy *(Brit. coll.) or* shaky state

wacker ['vakɐ] *(veralt.)* **1.** *Adj.* **a)** *(rechtschaffen)* upright; decent; *(iron.)* trusty; worthy; **b)** *(tapfer)* valiant; **c)** *(tüchtig)* hearty ⟨*drinker, eater*⟩. **2.** *adv.* **a)** *(tapfer)* valiantly; **sich ~ schlagen** put up a good show; **b)** *(tüchtig)* ⟨*eat, drink, etc.*⟩ heartily

wacklig ['vaklɪç] *s.* **wackelig**

Wade ['va:də] **die; ~, ~n** calf

Waffe ['vafə] **die; ~, ~n** *(auch fig.)* weapon; *(Feuer~)* firearm; **~n tragen** bear arms; **Kriegsdienst mit der ~:** service under arms; **unter ~n** under arms; **die ~n strecken** lay down one's arms; *(fig.)* give up the struggle

Waffel ['vafl̩] **die; ~, ~n a)** waffle; *(dünne ~, Eis~)* wafer; **b)** *(Eistüte)* cone

Wạffel·eisen das waffle-iron

Wạffen-: ~**gewalt die: mit ~gewalt** by force of arms; ~**handel der** arms trade; arms trading; ~**lager das** arsenal; ~**ruhe die** cease-fire; ~**schein der** firearms licence; ~**still·stand der** armistice; [permanent] cease-fire

Wa̱ge·mut der daring; audacity

wa̱ge·mutig *Adj.* daring; audacious

wagen ['va:gn̩] **1.** *tr. V.* risk; **[es] ~, etw. zu tun** dare to do sth.; **einen Versuch ~:** dare to make an attempt; risk an attempt; *s. auch* **gewagt 2. 2.** *refl. V.* **sich irgendwohin/nicht irgendwohin ~:** venture somewhere/not dare to go somewhere

Wa̱gen der; ~s, ~: a) *(PKW)* car; *(Omnibus)* bus; *(LKW)* truck; lorry *(Brit.); (Liefer~)* van; **b)** *(Pferde~)* cart; *(Kutsche)* coach; carriage; *(Plan~)* wagon; *(Zirkus~, Wohn~)* caravan *(Brit.);* trailer *(esp. Amer.);* **der Große ~** *(Astron.)* the Plough; the Big Dipper *(Amer.);* **jmdm. an den ~ fahren** *(fig. ugs.)* give sb. what for *(sl.);* pitch into sb. *(coll.);* **c)** *(Eisenbahn~) (Personen~)* coach; carriage; *(Güter~)* truck; wagon; car *(Amer.); (Straßenbahn~)* car; **d)** *(Kinder~, Puppen~)* pram *(Brit.);* baby carriage *(Amer.); (Sport~)* push-chair *(Brit.);* stroller *(Amer.);* **e)** *(Hand~)* handcart; **f)** *(Einkaufs~)* [shopping] trolley

Wa̱gen-: ~**heber der** jack; ~**park der** vehicle pool; ~**rad das** cartwheel

Waggon [va'gɔŋ, *südd., österr.:* va'go:n] **der; ~s, ~s,** *südd., österr.:* **~s, ~e** wagon; truck *(Brit.);* car *(Amer.)*

wa̱ghalsig 1. *Adj.* daring; risky ⟨*speculation*⟩; *(leichtsinnig)* reckless ⟨*driver, rider*⟩. **2.** *adv.* daringly; ⟨*speculate*⟩ riskily; *(leichtsinnig)* recklessly

Wagnis ['va:knɪs] **das; ~ses, ~se** daring exploit *or* feat; *(Risiko)* risk

Wahl [va:l] **die; ~, ~en a)** *o. Pl.* choice; **eine/seine ~ treffen** make a/one's choice; **jmdm. die ~ lassen** let sb. choose; **mir bleibt** *od.* **ich habe keine [andere] ~:** I have no choice *or* alternative; **es stehen drei Menüs zur Wahl** there are three set meals to choose from; **die ~ fiel auf ihn** the choice fell on him; **in die engere ~ kommen** be short-listed *or* put on the short list *(Brit.);* **b)** *(in ein Gremium, Amt)* election; **in Hessen ist ~** *od.* **sind ~en** there are elections in Hessen; **sich zur ~ stellen** stand *or (Amer.)* run for election; **geheime ~:** secret ballot; **c)** *(Güteklasse)* quality; **die Socken sind zweite ~:** the socks are seconds

wä̱hlbar *Adj.*eligible for election *postpos.*

wa̱hl-, Wa̱hl-: ~**berechtigt** *Adj.* eligible *or* entitled to vote *postpos.;* ~**berechtigte der/die;** *adj. Dekl.* person entitled to vote; ~**beteiligung die** turn-out; ~**bezirk der** ward

wählen ['vɛ:lən] **1.** *tr. V.* **a)** choose; select ⟨*station, programme, etc.*⟩; **seine Worte [sorg-**

fältig] ~: choose one's words [carefully]; **b)** *(Fernspr.)* dial ⟨*number*⟩; **c)** *(durch Stimmabgabe)* elect **(in** + *Akk.* to); **jmdn. zum Vorsitzenden** ~: elect sb. as chairman; **d)** *(stimmen für)* vote for ⟨*party, candidate*⟩. **2.** *itr. V.* **a)** choose **(zwischen** + *Dat.* between); **haben Sie schon gewählt?** *(im Lokal)* are you ready to order?; **b)** *(Fernspr.)* dial; **c)** *(stimmen)* vote; **wann wird gewählt?** when are the elections?

Wähler der; ~**s,** ~: voter

Wahl·ergebnis das election result

Wählerin die; ~, ~**nen** voter

wählerisch *Adj.* choosy; particular **(in** + *Dat.* about)

Wählerschaft die; ~, ~**en** electorate

wahl-, Wahl-: ~**fach das** *(Schulw.)* optional subject; ~**gang der** ballot; ~**geheimnis das** secrecy *or* confidentiality of the ballot; ~**geschenk das** pre-election bonus; ~**heimat die** adopted country/place of residence; ~**kabine die** polling-booth; ~**kampf der** election campaign; ~**kreis der** constituency; ~**lokal das** polling-station; ~**los 1.** *Adj.* indiscriminate; random; **2.** *adv.* indiscriminately; at random; ~**programm das** election manifesto; ~**propaganda die** election propaganda; ~**recht das a)** *o.Pl.* **[aktives]** ~**recht** right to vote; *(einer Gruppe)* franchise; **passives** ~**recht** right to stand [as a candidate] for election; **b)** *(Rechtsvorschriften)* electoral law; ~**rede die** election speech

Wähl·scheibe die *(Fernspr.)* dial

wahl-, Wahl-: ~**schein der** voting permit *(esp. for postal voter)*; ~**sieg der** election victory; ~**spruch der** motto; ~**urne die** ballot-box; ~**weise** *Adv.* as desired; to choice; ~**weise ... oder ...:** either ... or ... [as desired]

Wahn [va:n] **der;** ~**[e]s a)** mania; **b)** *(Täuschung)* delusion; **er lebt in dem** ~**, daß ...:** he is labouring under the delusion that ...

wähnen ['vɛ:nən] *tr. V. (geh.)* think [mistakenly]; imagine; **jmdn. in Sicherheit** *od.* **sicher** ~: imagine *or* think sb. is safe

wahn-, Wahn-: ~**sinn der;** *o. Pl.* **a)** insanity; madness; **b)** *(ugs.: Unvernunft)* madness; lunacy; **das ist ja** ~**sinn!** that's just crazy!; **c)** ~**sinn!** *(salopp)* incredible! *(coll.)*; ~**sinnig 1.** *Adj.* **a)** insane; mad; ~**sinnig werden** go insane; **wie** ~**sinnig** *(ugs.)* like mad *or* *(coll.)* crazy; **ich werde** ~**sinnig!** *(ugs.)* fantastic! *(coll.)*; **b)** *(ugs.: ganz unvernünftig)* mad; crazy; **c)** *(ugs.: groß, heftig, intensiv)* terrific *(coll.)* ⟨*effort, speed, etc.*⟩; terrible *(coll.)* ⟨*fright, job, pain*⟩. **2.** *adv. (ugs.)* incredibly *(coll.)*; terribly *(coll.)*

Wahnsinnige der/die; *adj. Dekl.* maniac; madman/madwoman

wahr [va:ɐ̯] *Adj.* **a)** true; **[das] ist ja gar nicht** ~**!** that's just not true!; **du hast Hunger, nicht** ~**?** you're hungry, aren't you?; **nicht** ~**, er weiß es doch?** he does know, doesn't he?; **das darf [doch] nicht** ~ **sein!** I don't believe it!; **etw.** ~ **machen** carry sth. out; **b)** *nicht präd. (wirklich)* real ⟨*reason, motive, feelings, joy, etc.*⟩; actual ⟨*culprit*⟩; true, real ⟨*friend, friendship, love, art*⟩; veritable ⟨*miracle*⟩; **im** ~**sten Sinne des Wortes** in the truest sense of the word; **das ist das einzig Wahre** *(ugs.)* that's just what the doctor ordered *(coll.)*

wahren *tr. V. (geh.)* preserve, maintain ⟨*balance, equality, neutrality, etc.*⟩; maintain, assert ⟨*authority, right*⟩; keep ⟨*secret*⟩; defend, safeguard ⟨*interests, rights, reputation*⟩; *s. auch* **Distanz b; Form e**

währen ['vɛ:rən] *itr. V. (geh.)* last; **ein lange** ~**der Prozeß** a process of long duration; **was lange währt, wird endlich gut** *(Spr.)* it will be/was worth it in the end

während ['vɛ:rənt] **1.** *Konj.* **a)** *(zeitlich)* while; **b)** *(adversativ)* whereas. **2.** *Präp. mit Gen.* during; *(über einen Zeitraum von)* for; ~ **des ganzen Tages/Abends** all day [long]/throughout the [entire] evening

während·dessen *Adv.* in the mean time; meanwhile

wahr|haben *unr. tr. V.* **etw. nicht** ~ **wollen** not want to admit sth.

wahrhaftig 1. *Adj. (geh.)* truthful ⟨*person*⟩; **der** ~**e Gott** the true God. **2.** *adv.* really; genuinely

Wahrheit die; ~, ~**en** truth

wahrheits-, Wahrheits-: ~**gemäß 1.** *Adj.* truthful; accurate ⟨*information*⟩; **2.** *adv.* truthfully; ~**getreu 1.** *Adj.* truthful; **2.** *adv.* truthfully; ~**liebe die** love of truth

wahrnehmbar *Adj.* perceptible

wahr|nehmen *unr. tr. V.* **a)** perceive; discern; *(spüren)* feel; detect ⟨*sound, smell*⟩; *(bemerken)* notice; be aware of; *(erkennen, ausmachen)* make out; discern; detect, discern ⟨*atmosphere, undertone*⟩; **b)** *(nutzen)* take advantage of ⟨*opportunity*⟩; exploit ⟨*advantage*⟩; exercise ⟨*right*⟩; **c)** *(vertreten)* look after ⟨*sb.'s interests, affairs*⟩; **d)** *(erfüllen, ausführen)* carry out, perform ⟨*function, task, duty*⟩; fulfil ⟨*responsibility*⟩

Wahrnehmung die; ~, ~**en a)** perception; *(eines Sachverhalts)* awareness; *(eines Geruchs, eines Tons)* detection; **b)** *(Nutzung)* *(eines Rechts)* exercise; *(einer Gelegenheit, eines Vorteils)* exploitation; **c)** *(Vertretung)* representation; **d)** *(einer Funktion, Aufgabe, Pflicht)* performance; execution; *(einer Verantwortung)* fulfilment

wahr·sagen, wahr|sagen 1. *itr. V.* tell fortunes; **aus den Karten/den Handlinien** ~: read the cards/palms. **2.** *tr. V.* predict, foretell ⟨*future*⟩; **sie hat ihm gewahrsagt, daß er ...:** she predicted that he ...

Wahrsager der; ~**s,** ~, **Wahrsagerin die;** ~, ~**nen** fortune-teller

wahrscheinlich 1. *Adj.* probable; likely; ~ **klingen** sound plausible; **wenig** ~: not very likely. **2.** *adv.* probably

Wahrscheinlichkeit die; ~, ~**en** probability *(also Math.)*; likelihood; **mit einiger/hoher** *od.* **großer** ~: quite/very probably; **aller** ~ **nach** in all probability

Wahrung die; ~: preservation; maintenance; *(eines Geheimnisses)* keeping; *(von Interessen, Rechten, Ruf)* defence; safeguarding

Währung die; ~, ~**en** currency

Währungs-: ~**einheit die** currency unit; monetary unit; ~**reform die** currency reform

Wahr·zeichen das symbol; *(einer Stadt)* [most famous] landmark

Waise ['vaizə] **die;** ~, ~**n** orphan; **er/sie ist** ~: he/she is an orphan

Waisen-: ~**haus das** orphanage; ~**kind das** orphan; ~**rente die** orphan's [social] benefit

Wal [va:l] **der;** ~**[e]s,** ~**e** whale

Wald [valt] **der;** ~**[e]s, Wälder** ['vɛldɐ] wood; *(größer)* forest; **viel** ~: a great deal of woodland; **den** ~ **vor [lauter] Bäumen nicht sehen** *(fig.)* not see the wood for the trees

Wald-: ~**arbeiter der** forestry worker; ~**brand der** forest fire

Wäldchen ['vɛltçən] **das** copse; spinney

waldig *Adj.* wooded

wald-, Wald-: ~**lauf der: [einen]**

~**lauf machen** go jogging through the woods; ~**meister der**; *o. pl. (Bot.)* woodruff; ~**reich** *Adj.* densely wooded; ~**sterben das** death of the forest [as a result of pollution]; ~**weg der** forest path; *(für Fahrzeuge)* forest track

Wal-: ~**fang der**; *o. Pl.* whaling *no def. art.*; ~**fänger der** whaler; ~**fisch der** *(ugs.)* whale

Waliser [va'li:zɐ] **der**; ~**s**, ~: Welshman

Waliserin die; ~, ~**nen** Welshwoman

walisisch [va'li:zɪʃ] *Adj.* Welsh; **das Walisische** Welsh

Wall [val] **der**; ~**[e]s**, **Wälle** ['vɛlə] earthwork; embankment; rampart *(esp. Mil.)*

Wallach ['valax] **der**; ~**[e]s**, ~**e** gelding

wallen *itr. V.* **a)** boil; **b)** ~**des Haar**/~**de Gewänder** *(geh.)* flowing hair/robes

wall-, Wall-: ~**fahren** *itr. V.; mit sein* make a pilgrimage; ~**fahrer der** pilgrim; ~**fahrt die** pilgrimage

Wall·fahrts-: ~**kirche die** pilgrimage church; ~**ort der** place of pilgrimage

Wal·nuß ['valnʊs] **die** walnut

Wal·roß ['valrɔs] **das**; *Pl.* ~**rosse** walrus

walten ['valtn̩] *itr. V. (geh.)* ⟨*good sense, good spirit*⟩ prevail; ⟨*peace, silence, harmony, etc.*⟩ reign; **Vorsicht/Gnade** *usw.* ~ **lassen** exercise caution/mercy *etc.*; **Vernunft** ~ **lassen** be reasonable; *s. auch* **Amt b, schalten 2d**

Walze ['valtsə] **die**; ~, ~**n a)** roller; *(Schreib~)* platen; **b)** *(eines mechanischen Musikinstruments)* barrel

walzen *tr. V.* roll ⟨*field, road, steel, etc.*⟩

wälzen ['vɛltsn̩] **1.** *tr. V.* **a)** roll ⟨*round object*⟩; heave ⟨*heavy object*⟩; *(drehen)* roll ⟨*person etc.*⟩ over; **etw. in Mehl** ~ *(Kochk.)* toss sth. in flour; **b)** *(fig. ugs.)* **Bücher** ~: pore over books; **Probleme** ~: mull over problems. **2.** *refl. V.* roll; *(auf der Stelle)* roll about *or* around; *(im Krampf, vor Schmerzen)* writhe around; **sich schlaflos im Bett** ~: toss and turn in bed, unable to sleep

Walzer der; ~**s**, ~: waltz; **kannst du** ~ **tanzen?** can you waltz?

Wälzer der; ~**s**, ~ *(ugs.)* hefty tome

Wampe ['vampə] **die**; ~, ~**n** *(ugs. abwertend)* pot belly

wand *1. u. 3. Pers. Sg. Prät. v.* **winden**

Wand [vant] **die**; ~, **Wände** ['vɛndə] **a)** wall; *(Trenn~)* partition; **die eigenen vier Wände** one's own four walls; **jmdn. an die** ~ **stellen** *(verhüll. ugs.)* put sb. up against a wall *(euphem.)*; ~ **an** ~ **wohnen** live next door to one another; be neighbours; **b) spanische** ~: folding screen; **c)** *(eines Behälters, Schiffs)* side; *(eines Zeltes)* wall; side; **d)** *(Fels~)* face; wall

Wandalismus [vanda'lɪsmʊs] **der**; ~: vandalism

Wand·behang der wall hanging

Wandel ['vandl̩] **der**; ~**s** change; **im** ~ **der Zeiten** through the ages

wandeln 1. *refl. V.* change (**in** + *Akk.* into). **2.** *tr. V.* change. **3.** *itr. V.; mit sein (geh.)* stroll

Wander-: ~**ausstellung die** touring exhibition; ~**bühne die** touring company; ~**düne die** wandering dune

Wanderer der; ~**s**, ~: walker; *(der weite Wege zurücklegt)* rambler; hiker

Wander-: ~**falke der** peregrine falcon; ~**gewerbe das** itinerant trade; ~**heuschrecke die** migratory locust; ~**karte die** rambler's [path] map

wandern ['vandɐn] *itr. V.; mit sein* **a)** hike; ramble; *(ohne Angabe des Ziels)* go hiking *or* rambling; **b)** *(ugs.: gehen)* wander *(lit. or fig.)*; *(fig.)* ⟨*glance, eyes, thoughts*⟩ roam, wander; **c)** *(ziehen, reisen)* travel; *(ziellos)* roam; ⟨*exhibition, circus, theatre*⟩ tour, travel; ⟨*animal, people, tribe*⟩ migrate; *(fig.)* ⟨*cloud, star*⟩ drift; ~**de Stämme** nomadic tribes; **d)** ⟨*glacier, dune, island*⟩ move, shift; ⟨*kidney etc.*⟩ be displaced; **e)** *(ugs.: befördert werden)* land; **in den Papierkorb** ~: land *or* be thrown in the waste-paper basket

Wander-: ~**pokal der** challenge cup; ~**ratte die** brown rat

Wanderschaft die; ~: travels *pl.*

Wanderung die; ~, ~**en a)** hike; walking tour; *(sehr lang)* trek; **eine** ~ **machen** go on a hike/tour/trek; **b)** *(Zool., Soziol.)* migration

Wander·weg der footpath *(for ramblers)*

Wandlung die; ~, ~**en** change; *(Ver~)* transformation

Wand·malerei die mural painting; wall-painting; *(Bild)* mural; wall-painting

Wand-: ~**schrank der** *s.* **Einbauschrank**; ~**tafel die** [wall] blackboard

wandte ['vantə] *1. u. 3. Pers. Prät. v.* **wenden**

Wand-: ~**teppich der** wall-hanging; tapestry; ~**zeitung die** wall newspaper

Wange ['vaŋə] **die**; ~, ~**n** *(geh.)* cheek; ~ **an** ~: cheek to cheek

Wankel·motor der Wankel engine

Wankel·mut der *(geh.)* vacillation

wankelmütig [-my:tɪç] *Adj.* *(geh.)* vacillating

wanken ['vaŋkn̩] *itr. V.* **a)** sway; ⟨*person*⟩ totter; *(unter einer Last)* stagger; **b)** *mit sein (unsicher gehen)* stagger; totter; **c)** *(geh.: bedroht sein)* ⟨*government, empire, etc.*⟩ totter; **ins Wanken geraten** begin to totter; ⟨*theory, faith, etc.*⟩ become shaky; **ins Wanken bringen** make ⟨*monarchy, government, etc.*⟩ totter; shake ⟨*resolve, faith*⟩

wann [van] *Adv.* when; ~ **kommst du morgen?** when *or* [at] what time are you coming tomorrow?; ~ **ist dieses Jahr Ostern?** when *or* on what date does Easter fall this year?; **seit** ~ **wohnst du dort?** how long have you been living there?; **bis** ~ **kann ich noch anrufen?** until when *or* how late can I still phone?; **von** ~ **an?** from when?; **von** ~ **bis** ~ **gilt es?** for what period is it valid?; **bis** ~ **ist das Essen fertig?** [by] when will the food be ready?; **ich weiß nicht**, ~: I don't know when; **du kannst kommen**, ~ **du willst** you can come when[ever] you like; ~ **[auch] immer** *(geh.)* whenever

Wanne die; ~, ~**n** bath[tub]

Wanst [vanst] **der**; ~**[e]s**, **Wänste** ['vɛnstə] *(ugs. abwertend)* belly; **sich den** ~ **vollschlagen** stuff oneself *(coll.)*

Wanze ['vantsə] **die**; ~, ~**n** bug; *(ugs.: Abhör~)* bug *(coll.)*

Wappen ['vapn̩] **das**; ~**s**, ~: coat of arms

Wappen-: ~**kunde die**; *o. Pl.* heraldry *no art.*; ~**spruch der** motto; ~**tier das** heraldic beast

wappnen *refl. V. (geh.)* forearm oneself

war [va:ɐ̯] *1. u. 3. Pers. Sg. Prät. v.* **sein**

warb [varp] *1. u. 3. Pers. Sg. Prät. v.* **werben**

ward [vart] *(geh.) 1. u. 3. Pers. Sg. Prät. v.* **werden**

Ware ['va:rə] **die**; ~, ~**n a)** ~**[n]** goods *pl.*; wares *pl.*; *(einzelne ~)* article; commodity *(Econ., fig.)*; *(Erzeugnis)* product; **heiße** ~ *(ugs.)* hot goods; **b)** *(Kaufmannsspr.: Stoff)* material

Waren-: ~**angebot das** range of goods; ~**annahme die**: „~an-

nahme" 'goods in'; **~haus das** department store; **~lager das** *(einer Fabrik o.ä.)* stores *pl.; (eines Geschäftes)* stock-room; *(größer)* warehouse; *(Bestand)* stocks *pl.;* **~muster das** sample; **~zeichen das** trade mark

warf [varf] *1. u. 3. Pers. Sg. Prät. v.* **werfen**

warm [varm]; **wärmer** ['vɛrmɐ], **wärmst ...** ['vɛrmst ...] **1.** *Adj.* **a)** warm; hot ‹*meal, food, bath, spring*›; hot, warm ‹*climate, country, season, etc.*›; **~e Küche** hot food; **das Essen ~ machen/stellen** heat up the food/keep the food warm *or* hot; **im Warmen sitzen** sit in the warm; **~ halten** ‹*coat, blanket, etc.*› keep one warm; **etw. ~ halten** keep sth. warm; **mir ist/wird ~:** I feel warm/I'm getting warm; *(zu ~)* I feel hot/I'm getting hot; **sich ~ laufen** warm up; **b)** *(herzlich)* warm ‹*sympathy, appreciation, words, etc.*›; **[mit jmdm./etw.] ~ werden** *(ugs.)* warm [to sb./sth.]. **2.** *adv.* warmly; **~ essen/duschen** have a hot meal/ shower

Wärme ['vɛrmə] **die; ~:** warmth; *(Hitze; Physik)* heat

Wärme-: ~isolation die thermal insulation; **~kraftwerk das** thermal power station; **~lehre die** *(Physik)* theory of heat; *(Thermodynamik)* thermodynamics *sing., no art.*

wärmen 1. *tr. V.* warm; *(aufwärmen)* warm up ‹*food, drink*›; **jmdn./sich ~:** warm sb./oneself up. **2.** *itr. V.* be warm; *(warm halten)* keep one warm; **die Sonne wärmt kaum** the sun has hardly any warmth

Wärme-: ~pumpe die *(Technik)* heat pump; **~strahlung die** thermal radiation

Wärm·flasche die hot-water bottle

warm-, Warm-: ~front die *(Met.)* warm front; **~|halten** *unr. refl. V. (ugs.)* **sich** *(Dat.)* **jmdn. ~halten** keep on the right side of sb.; **~herzig 1.** *Adj.* warm-hearted; **2.** *adv.* warm-heartedly; **~luft die;** *o. Pl.* warm air; **~miete die** *(ugs.)* rent inclusive of heating

wärmstens ['vɛrmstn̩s] *Adv.* warmly ‹*recommend sth.*›

Warm·wasser das; *o. Pl.* hot water

Warm·wasser-: ~heizung die hot-water heating; **~versorgung die** hot-water supply

Warn-: ~blink·anlage die, *(ugs.)* **~blinker der** *(Kfz-W.)* hazard warning lights *pl.;* **~dreieck das** *(Kfz-W.)* hazard warning triangle

warnen ['varnən] *tr. (auch itr.) V.* warn **(vor +** *Dat.* of, about); **jmdn. [davor] ~, etw. zu tun** warn sb. against doing sth.; **die Polizei warnt vor Nebel/vor Taschendieben** the police have issued a fog warning/a warning against pickpockets; **ein ~des Beispiel** a cautionary example

Warn-: ~schild das warning sign; **~schuß der** warning shot; **~signal das** warning signal; **~streik der** token strike

Warnung die; ~, ~en warning **(vor +** *Dat.* of, about); **das ist meine letzte ~:** that's the last warning I shall give you; I shan't warn you again

Warn·zeichen das warning sign

Warschau ['varʃau] **(das); ~s** Warsaw

Warschauer 1. der; ~s, ~: citizen of Warsaw. **2.** *indekl. Adj.* Warsaw

Warte ['vartə] **die; ~, ~n** *(geh.)* vantage-point; **von jmds. ~ aus [gesehen]** *(fig.)* [seen] from sb.'s standpoint

Warte-: ~halle die waiting room; *(Flugw.)* departure lounge; **~liste die** waiting list; **auf ~liste** *(Flugw.)* on stand-by

warten ['vartn̩] **1.** *itr. V.* wait **(auf +** *Akk.* for); **warte mal!** wait a moment!; just a moment!; **na warte!** *(ugs.)* just you wait!; **„bitte warten!"** 'wait'; *(am Telefon)* 'hold the line please'; **da kannst du lange ~!** *(iron.)* you'll have a long wait; you'll be lucky *(iron.);* **nicht lange auf sich ~ lassen** not be long in coming; **sie wollen mit dem Heiraten noch [etwas] ~:** they want to wait a little before getting married; **darauf habe ich schon lange gewartet** *(iron.)* I've seen that coming [for a long time]. **2.** *tr. V.* service ‹*car, machine, etc.*›

Wärter ['vɛrtɐ] **der; ~s, ~:** attendant; *(Tier~, Zoo~, Leuchtturm~)* keeper; *(Kranken~)* orderly; *(Gefängnis~)* warder

Warte·raum der waiting-room

Wärterin die; ~, ~nen *s.* **Wärter**

Warte-: ~saal der waiting-room; **~zeit die a)** wait; **nach einer ~zeit von einer Stunde** after waiting for an hour; **b)** *(festgesetzte Frist)* waiting period; **~zimmer das** waiting-room

-wärts [-vɛrts] *adv.* ‹*north-, south-, up-, down-, etc.*›wards; **seit~:** sideways

Wartung die; ~, ~en service; *(das Warten)* servicing; *(Instandhaltung)* maintenance

warum [va'rʊm] *Adv.* why; **~ nicht gleich so?** why not do that in the first place?

Warze ['vartsə] **die; ~, ~n a)** wart; **b)** *(Brust~)* nipple

was [vas] **1.** *Interrogativpron. Nom. u. Akk. u. (nach Präp.) Dat. Neutr.; s. auch (Gen.)* **wessen 1b** what; **~ kostet das?** what *or* how much does that cost?; **~ ist er [von Beruf]?** what's his job?; **[das ist] gut, ~?** *(ugs.: nicht?)* not bad, eh?; **~ ist?, ~ denn?** *(was ist denn los?)* what is it?; what's up?; **ach ~!** *(ugs.)* oh, come on!; of course not!; **für** *od.* **zu ~ brauchst du es?** *(ugs.)* what do you need it for?; **~ der alles weiß!** what a lot he knows!; **~ es [nicht] alles gibt!** *(Ding)* what will they think of next?; *(Ereignis)* the things people will do!; **und ~ nicht alles** *(ugs.)* and so on ad infinitum; **~ [auch] immer** whatever; **~ für [ein] ...:** what sort *or* kind of ... **2.** *Relativpron. Nom. u. Akk. u. (nach Präp.) Dat. Neutr.; s. auch (Gen.)* **wessen 2b: a) [das] ~:** what; **alles, was ...:** everything *or* all that ...; **alles, ~ ich weiß** all [that] I know; **das Beste, was du tun kannst** the best thing that you can do; **vieles/manches/nichts/dasselbe/etwas, ~ ...:** much/many things/nothing/the same one/ something that ...; **~ mich betrifft/das anbelangt, [so] ...:** as far as I'm/that's concerned, ...; **b)** *weiterführend* which; **er hat zugesagt, ~ mich gefreut hat** he agreed, which pleased me; **es hat geregnet, ~ uns aber nicht gestört hat** it rained, but that didn't bother us. **3.** *Indefinitpron. Nom. u. Akk. u. (nach Präp.) Dat. Neutr. (ugs.)* **a)** *(etwas)* something; *(in Fragen, Verneinungen)* anything; **er hat kaum ~ gesagt** he hardly said anything *or* a thing; **ist ~?** is anything wrong?; **so ~:** such a thing; something like that; **nein, so ~!** you don't say!; **so ~ könnte mir nicht passieren** nothing like that could happen to me; **so ~ Dummes!** how stupid!; **gibt es ~ Neues?** Is there any news?; **aus ihm wird mal/wird nie ~:** he'll make something of himself/he'll never come to anything; **b)** *(ein Teil)* some. **4.** *Adv. (ugs.)* **a)** *(warum, wozu)* why; what ... for; **~ stehst du hier herum?** what are you standing around here for?; **b)** *(wie)* how; **~ hast du dich verändert!** how you've changed!

wasch-, Wasch-: ~anlage die; *(Autowaschanlage)* car-wash; **~anleitung die** washing instructions *pl.;* **~automat der** washing-machine; **~bar** *Adj.* washable; **~bär der** racoon; **~becken das** wash-basin

Wäsche ['vɛʃə] **die; ~, ~n a)** *o. Pl. (zu waschende Textilien)* washing; *(für die Wäscherei)* laundry; **schmutzige ~ waschen** *(fig.)* wash [one's] dirty linen in public; **b)** *o. Pl. (Unter~)* underwear; **dumm/verdutzt aus der ~ gucken** *(ugs.)* look stupid/flabbergasted; **c)** *(das Waschen)* washing *no pl.;* **bei/nach der ersten ~:** when washed for the first time/after the first wash; **in der ~ sein** be in the wash; **große ~ haben** be doing a big wash

wasch·echt *Adj.* **a)** colour-fast ⟨*textile, clothes*⟩; fast ⟨*colour*⟩; **b)** *(fig.: echt)* genuine; pukka *(coll.)*

Wäsche-: ~klammer die clothes-peg *(Brit.);* clothes-pin *(Amer.);* **~korb der** laundry-basket; *(für nasse Wäsche)* clothes-basket; **~leine die** clothes-line

waschen 1. *unr. tr. V.* wash; **sich ~:** wash [oneself]; have a wash; **jmdm./sich die Hände/das Gesicht** *usw.* **~:** wash sb.'s/one's hands/face *etc.;* **Wäsche ~:** do the/some washing; **sich ge~ haben** *(fig. ugs.)* be quite something. **2.** *unr. itr. V.* do the washing

Wäscherei die; ~, ~en laundry

Wäsche-: ~schleuder die spin-drier; **~ständer der** clothes-airer; **~trockner der a)** *(Maschine)* tumble-drier; **b)** *(Gestell)* clothes-airer

Wasch-: ~gelegenheit die washing facilities *pl.;* **~küche die a)** laundry-room; **b)** *(ugs.: Nebel)* pea-souper; **~lappen der a)** [face] flannel; washcloth *(Amer.);* **b)** *(ugs. abwertend) (Weichling)* softie *(coll.); (Feigling)* sissy; **~maschine die** washing-machine; **~mittel das** detergent; **~muschel die** *(österr.)* wash-basin; **~pulver das** washing-powder; **~raum der** washing-room; **~salon der** launderette; laundromat *(Amer.);* **~straße die** [automatic] car-wash

wäscht [vɛʃt] *3. Pers. Sg. Präsens v.* **waschen**

Wasch-: ~wasser das; *o. Pl.* washing water; **~zeug das;** *o. Pl.* washing things *pl.*

Wasser ['vasɐ] **das; ~s, ~/Wässer** ['vɛsɐ] **a)** *o. Pl.* water; **ins ~ gehen** *(zum Schwimmen)* go for a swim; *(verhüll.: sich ertränken)* drown oneself; **direkt am ~:** right by the water; *(am Meer)* right by the sea; **ein Boot zu ~ lassen** put out *or* launch a boat; **unter ~ stehen** be under water; be flooded; **etw. unter ~ setzen** flood sth.; **zu ~:** by sea; **b)** *Pl.* **~** *(fig.)* **sich über ~** *(Dat.)* **halten** keep one's head above water; **ins ~ fallen** fall through; **bis dahin fließt noch viel ~ den Fluß** *od.* **Rhein** *usw.* **hinunter** a lot of water will have flowed under the bridge by then; **mit allen ~n gewaschen sein** know all the tricks; **jmdm. das ~ abgraben** pull the carpet from under sb.'s feet; leave sb. high and dry; **jmdm. nicht das ~ reichen können** not be able to hold a candle to sb.; not be a patch on sb. *(coll.);* **c)** *Pl.* **Wässer** *(Mineral~, Tafel~)* mineral water; *(Heil~)* water; **d)** *o. Pl. (Gewässer)* **ein fließendes/stehendes ~:** a moving/stagnant stretch of water; **e)** *o. Pl. (Schweiß)* sweat; *(Urin)* water; urine; *(Speichel)* saliva; *(Gewebsflüssigkeit)* fluid; **~ lassen** pass water; **ihm lief das ~ im Munde zusammen** his mouth watered; **~ in den Beinen haben** have fluid in one's legs; *s. auch* **Blut; Rotz a; f)** *Pl.* **Wässer** *(Lösung, Lotion usw.)* lotion; *(Duft~)* scent

wasser-, Wasser-: ~arm *Adj.* ⟨*area*⟩ suffering from a water shortage; **~bad das** *(Kochk.)* bain-marie; **~ball der a)** beach-ball; **b)** *o. Pl. (Spiel)* water polo; **~becken das** pool; *(~tank)* water-tank

Wässerchen ['vɛsɐçən] **das; ~s, ~ a) er sieht aus, als könnte er kein ~ trüben** *(fig.)* he looks as though butter wouldn't melt in his mouth; **b)** *s.* **Wasser f**

wasser-, Wasser-: ~dampf der steam; **~dicht** *Adj.* **a)** waterproof ⟨*clothing, watch, etc.*⟩; watertight ⟨*container, seal, etc.*⟩; **b)** *(fig. ugs.)* watertight ⟨*alibi, contract*⟩; **~fahrzeug das** vessel; water-craft; **~fall der** waterfall; **reden wie ein ~** *(ugs.)* talk non-stop; **~farbe die** water-colour; **~floh der** water-flea; **~flugzeug das** seaplane; **~gekühlt** *Adj.* water-cooled; **~glas das** *(Gefäß)* glass; tumbler; *s. auch* **Sturm a; ~graben der a)** ditch; *(um eine Burg)* moat; **b)** *(Sport)* water-jump; **~hahn der** water-tap; faucet *(Amer.)*

wässerig ['vɛsərɪç] *s.* **wäßrig**

Wasser-: ~kessel der kettle; **~kraft die** water-power; **~kraftwerk das** hydroelectric power-station; **~lache die** puddle [of water]; **~leiche die** *(ugs.)* body of a drowned person; **~leitung die a)** water-pipe; *(Hauptleitung)* water-main; **b)** *(Aquädukt)* aqueduct; **~mann der;** *Pl.* **~männer** *(Astron., Astrol.)* Aquarius; *(Astrol.: Mensch)* Aquarian; **~melone die** water-melon

wassern *itr. V.; mit sein* land [on the water]

wässern ['vɛsɐn] *tr. V.* **a)** *(einweichen)* soak; *(Phot.)* wash ⟨*negative, print*⟩; **b)** *(bewässern)* water

wasser-, Wasser-: ~ober·fläche die surface of the water; **~pfeife die** hookah; water-pipe; **~pflanze die** aquatic plant; **~pistole die** water-pistol; **~rad das** water-wheel; **~ratte die a)** water-rat; **b)** *(ugs. scherzh.)* keen swimmer; *(Kind)* water-baby; **~rohr das** water-pipe; **~scheide die** *(Geogr.)* watershed; **~scheu** *Adj.* scared of water; **~schlauch der** [water-]hose; **~schutz·polizei die** river/lake police; **[1]~ski der** water-ski; **~ski fahren** water-ski; **[2]~ski das; ~s** water-skiing *no art.;* **~spiegel der** water-level; **~sport der** water-sport *no art.;* **~sportler der** water-sports enthusiast; **~spülung die** flush; flushing system; **~stand der** water-level; **~stelle die** watering-place; **~stoff der;** *o. Pl.* hydrogen; **~stoff·bombe die** hydrogen bomb; **~strahl der** jet of water; **~straße die** waterway; **~tank der** water-tank; **~tropfen der** drop of water; **~turm der** water-tower; **~uhr die** *(volkst.) s.* **~zähler; ~verschmutzung die** water-pollution; **~versorgung die** water-supply; **~vogel der** water-bird; aquatic bird; **~vorrat der** water-reserves *pl.;* water-supply; **~waage die** water-level; **~weg der** water-route; **auf dem ~weg** by water; **~werfer der** water-cannon; **~werk das** waterworks *sing.;* **~zähler der** water meter; **~zeichen das** watermark

wäßrig ['vɛs(ə)rɪç] *Adj.* **a)** watery; **b)** *(Chemie)* aqueous ⟨*solution*⟩

waten ['va:tn̩] *itr. V.; mit sein* wade

watscheln ['vatʃl̩n] *itr. V.; mit sein* waddle

[1]Watt [vat] **das; ~[e]s, ~en** mud-flats *pl.*

[2]Watt das; ~s, ~ *(Technik, Physik)* watt

Watte ['vatə] **die; ~, ~n** cotton wool

Watte·bausch der wad of cotton wool

Watten·meer das tidal shallows *pl. (covering mud-flats)*

wattieren *tr. V.* wad; *(gesteppt)* quilt ‹*garment*›; pad ‹*shoulder etc.*›

WC [ve:'tse:] **das; ~[s], ~[s]** toilet; WC

weben ['ve:bn̩] *regelm., (geh., fig.) auch unr. tr., itr. V.* weave

Weber der; ~s, ~: weaver

Weberei die; ~, ~en a) *o. Pl.* weaving *no art.;* **b)** *(Betrieb)* weaving-mill

Web·stuhl der loom

Wechsel ['vɛksl̩] **der; ~s, ~ a)** change; *(Geld~)* exchange; *(Spieler~)* substitution; **b)** *(das Sichabwechseln)* alternation; **der ~ der Jahreszeiten** the rotation *or* succession of the seasons; **im ~:** alternately; *(bei mehr als zwei)* in rotation; **c)** *(das Überwechseln)* move; *(Sport)* transfer; **d)** *(Bankw.)* bill of exchange **(über** + *Akk.* for)

wechsel-, Wechsel-: ~beziehung die interrelation; **in [einer] ~beziehung zueinander stehen** be interrelated; **~fälle** *Pl.* vicissitudes; ups and downs *(coll.);* **~geld das;** *o. Pl.* change; **~haft** *Adj.* changeable; **~jahre** *Pl.* change of life *sing.;* menopause *sing.;* **in die ~jahre kommen** reach the menopause; **~kurs der** exchange rate

wechseln 1. *tr. V.* **a)** change; **die Wohnung ~:** move home; **ein Hemd zum Wechseln** a spare shirt; *s. auch* **Besitzer a; b)** *([aus]tauschen)* exchange ‹*letters, words, glances, etc.*›; **c)** *(um~)* change ‹*money, note, etc.*› **(in** + *Akk.* into); **kannst du mir 100 Mark ~?** can you change 100 marks for me? **2.** *itr. V.* **a)** *(sich ändern)* change; **mit ~dem Erfolg** with varying success; **~de Bewölkung, ~d wolkig** *(bei Wettervorhersagen)* variable cloud; **b)** *mit sein (über~)* move

wechsel-, Wechsel-: ~objektiv das *(Fot.)* interchangeable lens; **~seitig 1.** *Adj.* mutual; **~seitige Abhängigkeit** interdependence; **2.** *adv.* mutually; **~spiel das** interplay; **~strom der** *(Elektrot.)* alternating current; **~stube die** bureau de change; **~weise** *Adv.* alternately; **~wirkung die** interaction

wecken ['vɛkn̩] *tr. V.* **a) jmdn. [aus dem Schlaf] ~:** wake sb. [up]; **b)** *(fig.)* arouse, awaken ‹*interest, curiosity*›; arouse ‹*anger*›; awaken ‹*desire, misgiving*›

Wecker der; ~s, ~ alarm clock; **jmdm. auf den ~ gehen** *od.* **fallen** *(ugs.)* get on sb.'s nerves

Wedel der; ~s, ~ a) *(Staub~)* feather-duster; **b)** *(Palm~, Farn~)* [palm/fern] frond

wedeln *itr. V.* **a)** ‹*tail*› wag; **[mit dem Schwanz] ~** ‹*dog*› wag its tail; *(winken)* **mit der Hand/einem Tuch ~:** wave one's hand/a handkerchief; **b)** *mit Richtungsangabe mit sein (Ski)* wedel

weder ['ve:dɐ] *Konj.:* **~ ... noch ...:** neither ... nor ...

weg [vɛk] *Adv.* **a)** away; *(verschwunden, ~gegangen)* gone; **er ist schon seit einer Stunde ~:** he left an hour ago; **sie ist schon ~:** she has already gone or left; **~ sein** *(fig. ugs.) (eingeschlafen sein)* have dropped off; *(bewußtlos sein)* be out [cold]; **[immer] ~ damit!** [let's] chuck it away! *(coll.);* **~ mit dir!** away *or* off with you!; **~ da!** get away from there!; **Hände ~ [von meiner Kamera]!** hands off [my camera]!; **Kopf ~!** move your head!; **[nur] ~ von hier!, nichts wie ~!** let's hop it *(sl.);* let's make ourselves scarce *(coll.);* **weit ~:** far away; a long way away; **b) von ... ~** *(ugs.: unmittelbar von)* straight off *or* from; **von der Schule ~ eingezogen werden** be conscripted straight from school; **c) über einen Schock/Schrecken** *usw.* **~ sein** *(ugs.)* have got over a shock/fright *etc.*

Weg [ve:k] **der; ~[e]s,~e a)** *(Fuß~)* path; *(Feld~)* track; **„kein öffentlicher ~"** 'no public right of way'; **am ~[e]** by the wayside; **b)** *(Zugang)* way; *(Passage, Durchgang)* passage; **sich** *(Dat.)* **einen ~ durch etw. bahnen** clear a path *or* way through sth.; **geh [mir] aus dem ~[e]:** get out of the *or* my way; **jmdm. den ~ abschneiden** head sb. off; **jmdm. im ~[e] stehen** *od. (auch fig.)* **sein** be in sb.'s way; *(fig.)* **einer Sache** *(Dat.)* **im ~[e] stehen** stand in the way of sth.; **jmdm. aus dem ~[e] gehen** keep out of sb's way; avoid sb.; **einer Diskussion aus dem ~[e] gehen** avoid a discussion; **jmdn./etw. aus dem ~[e] räumen** get rid of sb./sth.; **c)** *(Route, Verbindung)* way; route; **[jmdn.] nach dem ~ fragen** ask [sb.] the way; **wir haben denselben ~:** we're going the same way; **das liegt auf dem/meinem ~:** that's on the/my way; *(fig.)* **er ist mir über den ~ gelaufen** *(ugs.)* I ran *or* bumped into him; **jmdm. nicht über den ~ trauen** not trust sb. an inch; **den ~ des geringsten Widerstands gehen** take the line of least resistance; **seinen ~ machen** make one's way [in the world]; **d)** *(Strecke, Entfernung)* distance; *(Gang)* walk; *(Reise)* journey; **es sind 2 km/10 Minuten ~:** it is a distance of two kilometres/it is ten minutes' walk; **er hat noch einen weiten ~ vor sich** *(Dat.)* he still has a long way to go; **auf dem kürzesten ~:** by the shortest route; **auf halbem ~[e]** *(auch fig.)* half-way; **sich auf den ~ machen** set off; *(fig.)* **jmdm. einen guten Ratschlag mit auf den ~ geben** give sb. some good advice for his/her future life; **etw. in die ~e leiten** get sth. under way; **auf dem besten ~ sein, etw. zu tun** *(meist iron.)* be well on the way towards doing sth.; **er ist auf dem ~[e] der Besserung** he's on the road to recovery; **e)** *(ugs.: Besorgung)* errand; **einen ~ machen** do *or* run an errand; **jmdm. einen ~ abnehmen** run an errand for sb.; **f)** *(Methode)* way; *(Mittel)* means; **ich sehe keinen anderen ~:** I can't see any alternative; **auf schnellstem ~[e]** as speedily as possible; **auf schriftlichem ~[e]** by letter

Weg·bereiter der; ~s, ~: forerunner

weg-: ~|blasen *unr. tr. V.* blow away; **wie weggeblasen sein** have vanished; **~|bleiben** *unr. itr. V.; mit sein* **a)** stay away; *(von zu Hause)* stay out; **b)** *(ugs.: aussetzen)* ‹*engine*› stop; ‹*electricity*› go off; **mir blieb die Luft ~:** I was left gasping; **c)** *(ugs.: weggelassen werden)* be left out; **~|bringen** *unr. tr. V.* **a)** take away; *(zur Reparatur, Wartung usw.)* take in; **b)** *(ugs., bes. südd.) s.* **~kriegen; ~|denken** *unr. tr. V.* **sich** *(Dat.)* **etw. ~denken** imagine sth. is not there; **er ist aus unserem Team nicht ~zudenken** I/we can't imagine our team without him

Wegelagerer der; ~s, ~: highwayman

wegen 1. *Präp. mit Gen., in bestimmten Fällen auch mit Dat./mit endungslosem Nomen* **a)** *(zur Angabe einer Ursache, eines Grundes)* because of; owing to; **~ des schlechten Wetters** because of the bad weather; **[nur] ~ Peter/***(ugs.)* **euch** all because of Peter/you; **~ Hochwasser[s]** owing to flooding; **von Berufs ~:** for professional reasons; ~ **mir** *(ugs., bes. südd.)* because of me; *(was mich betrifft)* as far as I'm concerned; ~ **Umbau[s] geschlossen** closed for alterations; ~ **Mangel[s] an Beweisen** owing to lack of evidence; **b)** *(zur Angabe eines Zwecks, Ziels)* for [the sake of]; **er ist ~ dringender Geschäfte verreist** he's away on urgent business; **c)** *(um ... willen)* for the sake of; ~ **der Kinder/** *(ugs.)* **dir** for the children's/your sake; **d)** *(bezüglich)* about; regarding. **2.** *(ugs.)* **von ~!** you must be joking!; **von ~ billig!** cheap? not on your life!

Wegerich ['ve:gərıç] **der; ~s, ~e** *(Bot.)* plantain

weg-: ~|fahren 1. *unr. itr. V.; mit sein* **a)** leave; *(im Auto)* drive off; **wann seid ihr in Kiel ~gefahren?** when did you leave Kiel?; **b)** *(irgendwohin fahren)* go away. **2.** *unr. tr. V.* take away; *(mit dem Auto)* drive away; **~|fallen** *unr. itr. V.; mit sein* be discontinued; *(nicht mehr zutreffen)* ⟨*reason*⟩ no longer apply; *(entfallen)* be omitted; **~|fegen** *tr. V. (auch fig.)* sweep away; **~|fliegen** *unr. itr. V.; mit sein* fly away; *(~geschleudert werden)* fly off; **~|führen** *tr., itr. V.* lead away; **das führt vom Thema weg** this takes us away from the subject

Weg·gang der departure

weg|geben *unr. tr. V.* **a) ich gebe meine Wäsche weg** I send my washing to the laundry; **b)** *(verschenken)* give away; **~|gehen** *unr. itr. V.* **a)** leave; *(ugs.: ausgehen)* go out; *(ugs.: ~ziehen)* move away; **von jmdm. ~gehen** leave sb.; **geh ~!** go away!; **geh mir [bloß] ~ damit!** *(ugs.)* you can keep that!; **b)** *(verschwinden)* ⟨*spot, fog, etc.*⟩ go away; **c)** *(sich entfernen lassen)* ⟨*stain*⟩ come out; **d)** *(ugs.: verkauft werden)* sell; **~|gießen** *unr. tr. V.* pour away; **~|haben** *unr. tr. V. (ugs.)* have got rid of ⟨*dirt, stain, etc.*⟩; **etw. ~haben wollen** want to get rid of sth.; **~|holen** *tr. V. (ugs.)* take away; **~|jagen** *tr. V.* chase away; **~|kommen** *unr. itr. V.; mit sein* **a)** get away; *(~gehen können)* manage to go out; **mach, daß du [hier] ~kommst!** *(ugs.)* come on, hop it! *(sl.)*; make yourself scarce! *(coll.)*; **b)** *(abhanden kommen)* go missing; **c) gut/schlecht** *usw.* **[bei etw.] ~kommen** *(ugs.)* come off well/badly *etc.* [in sth.]; **~|können** *unr. itr. V.* **a)** be able to leave *or* get away; *(ausgehen können)* be able to go out; **b) die Zeitung kann weg** the paper can be thrown away; **~|kriegen** *tr. V.* get rid of ⟨*cold, pain, etc.*⟩; get out, get rid of ⟨*stain*⟩; shift, move ⟨*stone, tree-trunk*⟩; **~|lassen** *unr. tr. V.* **a) jmdn. ~lassen** let sb. go; **b)** *(auslassen)* leave out; omit; **~|laufen** *unr. itr. V.; mit sein* run away **(von, vor +** *Dat.* from); **von zu Hause ~laufen** run away from home; **seine Frau ist ihm ~gelaufen** *(ugs.)* his wife has gone *or* run off and left him *(coll.)*; **die Arbeit läuft [dir] nicht ~** *(ugs.)* the work will keep; **~|legen** *tr. V.* put away; *(aus der Hand legen)* put down; **~|machen** *tr. V. (ugs.)* remove; **~|müssen** *unr. itr.V.* **a)** have to leave; *(loskommen müssen)* have to get away; **b)** *(entfernt werden müssen)* have to be removed; *(~gebracht werden müssen)* ⟨*letter etc.*⟩ have to go; **du mußt da ~:** you'll have to move; **der Diktator muß ~:** the dictator must go; **~|nehmen** *unr. tr. V.* **a)** take away; remove; move ⟨*head, arm*⟩; **nimm die Finger da ~!** [keep your] fingers off!; **b) jmdm. etw. ~nehmen** take sth. away from sb.; **jmdm. die Freundin ~nehmen** pinch sb.'s girl-friend *(coll.)*; **jmdm. die Dame** *usw.* **~nehmen** *(Schach)* take sb.'s queen *etc.*; **~|räumen** *tr. V.* clear away; *(an seinen Platz tun)* tidy *or* put away; **~|schaffen** *tr. V.* take away; **~|scheren** *refl. V.* clear off *(coll.)*; **~|schicken** *tr. V.* **a)** send off ⟨*letter, parcel*⟩; **b)** send ⟨*person*⟩ away; **~|schieben** *unr. tr. V.* push away; **~|schleichen** *unr. itr., refl. V.; itr. mit sein* creep away; **~|schleppen** *tr. V.* **a)** carry *or* lug off *or* away; **b)** tow ⟨*car, rig, etc.*⟩ away; **~|schmeißen** *unr. tr. V. (ugs.)* chuck away *(coll.)*; **~|schnappen** *tr. V. (ugs.)* snatch away (*Dat.* from); **jmdm. die Freundin ~schnappen** pinch sb.'s girl-friend *(coll.)*; **~|schütten** *tr. V.* pour away; **~|sehen** *unr. itr. V.* **a)** look away; **b)** *(ugs.) s.* **hinwegsehen; ~|spülen** *tr. V.* wash away; **~|stecken** *tr. V.* **a)** put away; **b)** *(fig. ugs.: hinnehmen)* take, accept ⟨*blow*⟩; swallow ⟨*insult*⟩; **~|stellen** *tr. V.* put away; **~|stoßen** *unr. tr. V.* push *or* shove away

Weg·strecke die stretch [of road]; *(Entfernung)* distance

weg-: ~|tragen *unr. tr. V.* carry away; **~|treten 1.** *unr. tr. V.* kick away; **2.** *unr. itr. V.; mit sein* step away; *(Milit.)* dismiss; **[etwas] ~getreten sein** *(fig. ugs.)* be [somewhat] distracted; **~|tun** *unr. tr. V. (ugs.)* **a)** put away; **b)** *(wegwerfen)* throw away

Wegweiser der; ~s, ~ signpost

weg-: ~|werfen *unr. tr. V. (auch fig.)* throw away; **~werfend** *Adj.* dismissive ⟨*gesture, remark*⟩

Weg·werf- disposable ⟨*nappy, towel, cup, lighter, etc.*⟩

Wegwerf·gesellschaft die *(abwertend)* throw-away society

weg-: ~|wischen *tr. V.* wipe away; *(fig.)* erase ⟨*memory*⟩; dispel ⟨*fear, doubt*⟩; dismiss ⟨*objection*⟩; **~|wollen** *unr. itr. V.* want to go *or* leave; *(loskommen wollen)* want to get away; *(ausgehen wollen)* want to go out; **~|zerren** *tr. V.* drag away; **~|ziehen 1.** *unr.tr. V.* pull away; pull off ⟨*blanket*⟩; **2.** *unr. itr. V.; mit sein (umziehen)* move away; **aus X ~ziehen** leave X; move from X

weh [ve:] **1.** *Adj.* **a)** *nicht präd. (ugs.: schmerzend)* sore; **einen ~en Finger haben** have a sore *or* bad finger; **b)** *(geh.: schmerzlich)* painful; **ein ~es Lächeln/Gefühl** a sad smile/an aching feeling; **ihr ist ~ ums Herz** her heart aches; she is sore at heart. **2.** *adv. (ugs.)* **~ tun** hurt; **mir tut der Kopf ~:** my head is aching *or* hurts; **jmdm./sich ~ tun** hurt sb./oneself

Weh das; ~[e]s *(geh.)* sorrow; grief

wehe *Interj.* woe betide you/him *etc.*; **~ [dir], wenn du ...:** woe betide you if you ...

[1]**Wehe** ['ve:ə] **die; ~, ~n: die ~n setzten ein** the contractions started; she went into labour; **~n haben** have contractions; **in den ~n liegen** be in labour

[2]**Wehe die; ~, ~n** drift

wehen 1. *itr. V.* **a)** blow; **b)** *(flattern)* flutter; **ihre Haare wehten im Wind** her hair was blowing about in the wind; **c)** *mit sein* ⟨*leaves, snowflakes, scent*⟩ waft. **2.** *tr. V.* blow

weh-, Weh-: ~klage die *(geh.)* lamentation; **~klagen** *itr. V. (geh.)* lament; **über etw.** *(Akk.)* **~klagen** lament *or* bewail sth.; **~leidig** *(abwertend)* **1.** *Adj.* soft; *(weinerlich)* whining *attrib.*; **sei nicht so ~leidig!** don't be so soft *or* such a sissy; **2.** *adv.* self-pityingly; *(weinerlich)* whiningly; **~mut die; ~** *(geh.)* melancholy *or* wistful nostalgia; **~mütig 1.** *Adj.* melancholically *or* wistfully nostalgic; **2.** *adv.* with melancholy *or* wistful nostalgia

[1]**Wehr die; ~, ~en a) sich [gegen jmdn./etw.] zur ~ setzen** make a stand [against sb./sth.]; resist [sb./sth.]; **b)** *s.* **Feuerwehr**

[2]**Wehr das; ~[e]s, ~e** weir

Wehr·dienst *der; o. Pl.* military service *no art.*; **zum ~ einberufen werden** be called up; **seinen ~ ableisten** do one's military service

Wehr·dienst-: ~verweigerer der; ~s, ~: conscientious objector; **~verweigerung die** conscientious objection

wehren 1. *refl. V.* **a)** defend oneself; put up a fight; **sich tapfer ~:** defend oneself *or* resist bravely; **sich gegen etw. ~:** fight against sth.; **b) sich [dagegen] ~, etw. zu tun** resist having to do sth. **2.** *itr. V. (geh.)* **jmdm./einer Sache ~:** fight sb./fight [against] sth.

wehr-, Wehr-: ~haft *Adj.* **a)** able to defend oneself *postpos.*; **b)** *(befestigt)* fortified; **~los** *Adj.* defenceless; **jmdm./einer Sache ~los ausgeliefert sein** be defenceless against sb./sth.; **~losigkeit die; ~:** defencelessness; **~paß der** service record [book]; **~pflicht die;** *o. Pl.* military service; conscription; **die allgemeine ~pflicht** compulsory military service; **~pflichtig** *Adj.* liable for military service *postpos;* **~pflichtige der;** *adj. Dekl.* person liable for military service; **~sold der** military pay; **~übung die** reserve duty [re]training exercise

Wehwehchen das; ~s, ~: little complaint

Weib [vaịp] **das; ~[e]s, ~er a)** *(veralt., ugs.: Frau)* woman; female *(derog.)*; **sie ist ein tolles ~:** she's a bit of all right *(coll.)*; **b)** *(veralt., scherzh.: Ehefrau)* wife

Weibchen das; ~s, ~: female

Weiber·held der *(ugs.)* ladykiller

weibisch *(abwertend)* **1.** *Adj.* womanish; effeminate. **2.** *adv.* womanishly; effeminately

weiblich 1. *Adj.* **a)** female; **b)** *(für die Frau typisch)* feminine; **c)** *(Sprachw.)* feminine. **2.** *adv.* femininely

Weiblichkeit die; ~: femininity

Weibs-: ~bild das *(ugs.)* woman; *(abwertend)* female; **~stück das** *(abwertend)* bitch *(sl.)*

weich [vaịç] **1.** *Adj. (auch fig.)* soft; soft, mellow ‹*sound, voice*›; soft, gentle ‹*features*›; gentle ‹*mouth, face*›; **ein Ei ~ kochen** soft-boil an egg; **ein ~es Herz haben** be soft-hearted; **~ werden** *(ugs.)* soften; weaken. **2.** *adv.* softly; ‹*brake*› gently; **~ landen** *od.* **aufsetzen** make a soft landing

[1]**Weiche die; ~, ~n** *(Flanke)* flank

[2]**Weiche die; ~, ~n** points *pl. (Brit.)*; switch *(Amer.)*; **die ~n [für etw.] stellen** *(fig.)* set the course [for sth.]

[1]**weichen** *itr. V.; mit sein* soak

[2]**weichen** *unr. itr. V.; mit sein* move; **nicht von jmds. Seite ~:** not move from *or* leave sb.'s side; **dem Feind ~:** retreat from the enemy; **vor jmdm./etw. zur Seite ~:** step *or* move out of the way of sb./sth.; **die Angst wich von ihm** *(fig. geh.)* the fear left him

weich·gekocht *Adj. (präd. getrennt geschrieben)* soft-boiled

weich-, Weich-: ~herzig 1. *Adj.* soft-hearted; **2.** *adv.* soft-heartedly; **~herzigkeit die; ~:** soft-heartedness; **~käse der** soft cheese

weichlich 1. *Adj.* soft. **2.** *adv.* softly

Weichling der; ~s, ~e *(abwertend)* weakling

weich|machen *tr. V. (ugs.)* soften up

Weich·teile *Pl.* **a)** *(Anat.)* soft parts; **b)** *(ugs.: Genitalien)* privates

[1]**Weide** ['vaidə] **die; ~, ~n** *(Wiese)* pasture; **die Kühe auf die ~ treiben** drive the cows to pasture

[2]**Weide die; ~, ~n** *(Baum)* willow

Weide·land das pasture [land]; grazing land

weiden 1. *itr., tr. V.* graze. **2.** *refl. V.* **a)** *(geh.)* **er** *od.* **sein Auge weidete sich an dem Anblick** he feasted his eyes on the sight; **b) sich an jmds. Schmerz** *(Dat.)* **~:** gloat over sb.'s pain

Weiden-: ~gerte die willow rod; *(zum Korbflechten)* osier; *(kleiner)* wicker; **~kätzchen das** willow catkin

weidlich *Adv.* **etw. ~ ausnutzen** make full use of sth.

Weid·mann der; *Pl.* **~männer** *(geh.)* huntsman; hunter

weid·männisch 1. *Adj.* hunting, huntsman's *attrib.* **2.** *adv.* in the manner of a huntsman; like a huntsman

weigern ['vaigṇ] *refl. V.* refuse

Weigerung die; ~, ~en refusal

[1]**Weihe die; ~, ~n a)** *(Rel.: Einweihung)* consecration; dedication; **b)** *(kath. Kirche: Priester~, Bischofs~)* ordination; **die niederen/höheren ~n** *(hist.)* the minor/major orders

[2]**Weihe die; ~, ~n** *(Zool.)* harrier

weihen *tr. V.* **a)** *(Rel.)* consecrate; **b)** *(kath. Kirche: ordinieren)* ordain; **jmdn. zum Priester/Bischof ~:** ordain sb. priest/consecrate sb. bishop; **c)** *(Rel.: durch Weihe zueignen)* dedicate (*Dat.* to); **d) dem Tod[e]/dem Untergang geweiht sein** *(geh.)* be doomed to die/to fall

Weiher der; ~s, ~ *(bes. südd.)* [small] pond

Weihnachten das; ~, ~: Christmas; **frohe** *od.* **fröhliche ~!** Merry *or* Happy Christmas!; **grüne ~:** Christmas without snow; **zu** *od. (bes. südd.)* **an/über ~:** at *or* for/over Christmas

weihnachtlich 1. *Adj.* Christmassy. **2.** *adv.* **~ geschmückt** decorated for Christmas

Weihnachts-: ~abend der Christmas Eve; **~baum der** Christmas tree; **~feiertag der: der erste/zweite ~feiertag** Christmas Day/Boxing Day; **~fest das** Christmas; **~geld das** Christmas bonus; **~geschenk das** Christmas present *or* gift; **~lied das** Christmas carol; **~mann der;** *Pl.* **~männer a)** Father Christmas; Santa Claus; **b)** *(ugs.: Dummkopf)* silly idiot *(coll.)*; **~stern der a)** Christmas star; **b)** *(Pflanze)* poinsettia **~tag der** *s.* **~feiertag; ~zeit die** Christmas time; **in der ~zeit** at Christmas time

Weih-: ~rauch der incense; **~wasser das** *(kath. Kirche)* holy water; **~wasser·becken das** *(kath. Kirche)* stoup

weil [vaịl] *Konj.* because

Weile die; ~: while; **eine ganze ~:** a good while

weilen *itr. V. (geh.)* stay; *(sein)* be

Weiler der; ~s, ~: hamlet

Wein [vaịn] **der; ~[e]s, ~e a)** wine; **jmdm. reinen ~ einschenken** *(fig.)* tell sb. the truth; **b)** *o. Pl. (Reben)* vines *pl.*; *(Trauben)* grapes *pl.*; **c) wilder ~:** Virginia creeper

Wein-: ~[an]bau der; *o. Pl.* wine-growing *no art.*; **~bauer der** wine-grower; **~beere die** grape; **~berg der** vineyard; **~berg·schnecke die** [edible] snail; **~brand der** brandy

weinen 1. *itr. V.* cry; **um jmdn. ~:** cry *or* weep for sb.; **über jmdn./etw. ~:** cry over *or* about sb./sth.; **vor Freude ~:** cry *or* weep for *or* with joy; **es ist zum Weinen** it's enough to make you weep; **leise ~d abziehen** *(fig. ugs.)* leave with one's tail between one's legs. **2.** *tr. V.* shed ‹*tears*›. **3.** *refl. V.* **sich in den Schlaf ~:** cry oneself to sleep

weinerlich 1. *Adj.* tearful; weepy. 2. *adv.* tearfully
wein-, Wein-: ~**essig der** wine vinegar; ~**flasche** winebottle; ~**garten der** vineyard; ~**geist der;** *o. Pl.* ethyl alcohol; ethanol; ~**glas das** wineglass; ~**gut das** vineyard; ~**handlung die** wine-merchant's; ~**karte die** wine-list; ~**keller der** wine-cellar; ~**krampf der** crying fit; fit of crying; ~**lese die** grape harvest; ~**lokal das** wine bar; ~**probe die** wine-tasting [session]; ~**rebe die a)** grapevine; **b)** *(Ranke)* [grapevine] shoot; ~**rot** *Adj.* wine-red; wine-coloured; ~**schaum·creme die** *(Kochk.)* zabaglione; ~**stock der;** *Pl.* ~**stöcke** [grape]vine; ~**stube die** wine bar; ~**traube die** grape
weise ['vaizə] 1. *Adj.* wise; **ein Weiser** a wise man. 2. *adv.* wisely
Weise die; ~, ~**n a)** way; **auf diese/andere** ~: this way/ [in] another way; **auf meine** ~: in my own way; **auf geheimnisvolle** ~: in a mysterious manner; mysteriously; **in gewisser** ~: in certain respects; **b)** *(Melodie)* tune; melody
weisen 1. *unr. tr. V.* **a)** *(geh.)* show; **jmdm. etw.** ~: show sb. sth.; *s. auch* **Tür; b) jmdn. aus dem Zimmer** ~: send sb. out of the room; **etw. von sich** ~ *(fig.)* reject sth.. 2. *unr. itr. V.* point
Weisheit die; ~, ~**en a)** *o. Pl.* wisdom; **er hat die** ~ **[auch] nicht mit Löffeln gefressen** *(ugs.)* he is not all that bright; **mit seiner** ~ **am Ende sein** be at one's wit's end; **b)** *(Erkenntnis)* wise insight; *(Spruch)* wise saying
Weisheits·zahn der wisdom tooth
weis|machen *tr. V. (ugs.)* **das kannst du mir nicht** ~! you can't expect me to swallow that!; **das kannst du anderen** ~! tell that to the marines *(coll.)*
[1]**weiß** [vais] *1. u. 3. Pers. Sg. Präsens v.* **wissen**
[2]**weiß** *Adj.* white
Weiß das; ~**[e]s,** ~: white
weis-, Weis-: ~**sagen** *tr. V., auch itr. V.* prophesy; foretell; ~**sager der** prophet; ~**sagerin die** prophetess; ~**sagung die;** ~, ~**en** prophecy
weiß-, Weiß-: ~**bier das** *light, highly effervescent, top-fermented beer made from wheat and barley;* weiss beer; ~**blond** *Adj.* ash-blond/-blonde; ~**brot das** white bread; **ein** ~**brot** a white loaf; ~**dorn der** hawthorn
[1]**Weiße die;** ~, ~**n** *s.* [1]**Berliner 1**
[2]**Weiße der/die;** *adj. Dekl.* white; white man/woman
weißeln *(südd., österr., schweiz.),* **weißen** *tr. V.* paint white; *(tünchen)* whitewash
weiß-, Weiß-: ~**glut die** white heat; **jmdn. zur** ~**glut bringen** *(ugs.)* make sb. livid *(Brit. coll.);* ~**haarig** *Adj.* white-haired; ~**herbst der** rosé wine; ~**kohl der,** *(bes. südd., österr.)* ~**kraut das** white cabbage
weißlich *Adj.* whitish
weißt *2. Pers. Sg. Präsens v.* **wissen**
weiß-, Weiß-: ~**|waschen** *unr. tr. V.* **jmdn./sich** ~**waschen** *(ugs.)* clear sb.'s/one's name; ~**wein der** white wine; ~**wurst die** veal sausage
Weisung die; ~, ~**en** *(geh., Amtsspr.)* instruction; *(Direktive)* directive; **auf [jmds.]** ~ *(Akk.)* on [sb.'s] instructions; ~ **haben, etw. zu tun** have instructions to do sth.
weit [vait] 1. *Adj.* **a)** wide; long ⟨*way, journey, etc.*⟩; *(fig.)* broad ⟨*concept*⟩; **die** ~**e Welt** the big wide world; **im** ~**eren Sinn** *(fig.)* in the broader sense; **das Weite suchen** *(fig.)* take to one's heels; **b)** *(locker sitzend)* wide; **jmdm. zu** ~ **sein** ⟨*clothes*⟩ be too loose on sb.; **einen Rock** ~**er machen** let out a skirt; *s. auch* **weiter... 2.** *adv.* **a)** ~ **geöffnet** wide open; ~ **verbreitet** widespread; ~ **herumgekommen sein** have got around a good deal; have travelled widely; ~ **und breit war niemand zu sehen** there was no one to be seen anywhere; **b)** *(eine große Strecke)* far; ~ **[entfernt** *od.* **weg] wohnen** live a long way away *or* off; live far away; **15 km** ~: 15 km. away; **von** ~**em** from a distance; **von** ~ **her** from far away; *(fig.)* **es würde zu** ~ **führen, das alles jetzt zu analysieren** it would be too much to analyse it all now; **das geht zu** ~: that is going too far; **etw. zu** ~ **treiben, es mit etw. zu** ~ **treiben** overdo sth.; carry sth. too far; **so** ~, **so gut** so far, so good; *s. auch* **entfernt a; hersein c; c)** *(lange)* ~ **nach Mitternacht** well past midnight; ~ **zurückliegen** be a long way back *or* a long time ago; **d)** *(in der Entwicklung)* far; **sehr** ~ **mit etw. sein** have got a long way with sth.; **wie weit seid ihr?** how far have you got?; **wir wollen es gar nicht erst so** ~ **kommen lassen** we do not want to let it come to that; **e)** *(weitaus)* far; **jmdn.** ~ **übertreffen** surpass sb. by far *or* by a long way; **bei** ~**em** by far; by a long way; **bei** ~**em nicht so gut wie ...:** nowhere near as good as ...; *s. auch* **gefehlt 2; weiter**
weit-, Weit-: ~**ab** *Adv.* far away; ~**aus** *Adv.* far ⟨*better, worse, etc.*⟩; **der** ~**aus beste Reiter** by far *or* far and away the best rider; ~**blick der;** *o. Pl.* far-sightedness; **politischen** ~**blick haben** be politically far-sighted; ~**blickend** *Adj.* far-sighted
Weite die; ~, ~**n a)** expanse; **b)** *(bes. Sport: Entfernung)* distance; **c)** *(eines Kleidungsstückes, einer Öffnung usw.)* width
weiten 1. *tr. V.* widen. 2. *refl. V.* widen; ⟨*pupil*⟩ dilate
weiter *Adv.* **a)** farther; farther; **zwei Häuser** ~ **wohnen** live two houses further *or* farther on; **halt, nicht** ~! stop, don't go any further; ~! go on!; **nur immer** ~ **so!** keep it up!; **und so** ~: and so on; **und so** ~ **und so fort** and so on and so forth; **was geschah** ~? what happened then *or* next?; **b)** *(außerdem, sonst)* ~ **nichts, nichts** ~: nothing more *or* else; **ich brauche** ~ **nichts** I don't need anything else; there's nothing else I need; **das ist nicht** ~ **schlimm** it isn't that important; it doesn't really matter
weiter... *Adj.; nicht präd.* further; **ohne** ~**e Umstände** without any fuss; **bis auf** ~**es** for the time being; **des** ~**en** *(geh.)* furthermore; *s. auch* **ohne 1c**
weiter-, Weiter-: ~**|arbeiten** *itr. V.* continue *or* carry on working; ~**|bestehen** *unr. itr. V.* continue to exist; ~**|bilden** *tr. V. s.* **fortbilden;** ~**bildung die;** *o. Pl. s.* **Fortbildung;** ~**|bringen** *unr. tr. V.* **das bringt uns nicht** ~: that does not get us any further *or* anywhere; ~**|entwickeln** *tr., refl. V.* develop [further]; ~**|erzählen** *tr. V.* **a)** continue telling; *itr.* **erzähl doch** ~: do carry *or* go on; **b)** *(~sagen)* pass on; **erzähl das nicht** ~: don't tell anyone; ~**|fahren** *unr. itr. V.; mit sein* continue [on one's way]; *(~ reisen)* travel on; ~**|führen** *tr., itr. V.* continue; ~**führende Schulen** secondary schools; ~**|geben** *unr. tr. V.* **a)** pass on; **b)** *(Sport)* pass; ~**|gehen** *unr. itr. V.; mit sein* **a)** go on; **bitte** ~**gehen!** please move along *or* keep moving!; **b)** *(sich fortsetzen)* continue; go on; **das Leben geht** ~: life goes on; **so kann es mit uns nicht** ~**gehen** we cannot go on like this; **wie soll es denn nun** ~**gehen?** what is going to happen now?; ~**|helfen**

unr. itr. V. **jmdm. [mit etw.] ~helfen** help sb. [with sth.]; **~hin** *Adv.* **a)** *(immer noch)* still; **b)** *(künftig)* in future; **etw. auch ~hin tun** continue to do sth. [in future]; **~|kommen** *unr. itr. V.; mit sein* **a)** get further; **mach, daß du ~kommst** *(ugs.)* clear off *(coll.)*; **b)** *(Fortschritte machen)* make progress *or* headway; *(Erfolg haben)* get on; **~|laufen** *unr. itr. V.; mit sein* **a)** *(in Betrieb bleiben, auch fig.)* keep going; **b)** *(fortgeführt werden)* continue; **~|leben** *itr. V.* **a)** continue *or* carry on one's life; **b)** *(am Leben bleiben)* go on living; **c)** *(fig.)* live on; **~|leiten** *tr. V.* pass on ⟨*news, information, etc.*⟩; forward ⟨*letter, parcel, etc.*⟩; **~|machen** *(ugs.)* **1.** *itr. V.* carry on; go on; **2.** *tr. V.* carry on with; **~|reden** *itr. V.* go on *or* carry on talking; **~|reichen** *tr. V.* pass on; **~|sagen** *tr. V.* pass on; **sag es nicht weiter** don't tell anyone; **~|schicken** *tr. V.* forward; send on; send ⟨*person*⟩ on; **~|sehen** *unr. itr. V.* see; **morgen werden wir ~sehen** we'll see what we can do tomorrow; **~|spielen** *tr., itr. V.* go on *or* carry on playing; **der Schiedsrichter ließ ~spielen** the referee allowed play to continue; **~|sprechen** *unr. itr. V.* go on *or* carry on speaking *or* talking; **~|verarbeiten** *tr. V.* process; **~|verfolgen** *tr. V.* follow up ⟨*clue, case, etc.*⟩; continue to follow ⟨*developments, events, etc.*⟩; pursue further ⟨*idea, line of thought, etc.*⟩; **~|wissen** *unr. itr. V.* **nicht [mehr] ~wissen** be at one's wit's end; **~|ziehen** *unr. itr. V., mit sein* move on

weit-, Weit-: ~gehend; weiter gehend *od.* **weitgehender, weitestgehend** *od.* **weitgehendst 1.** *Adj.* extensive, wide, sweeping ⟨*powers*⟩; far-reaching ⟨*support, concessions, etc.*⟩; wide ⟨*support, agreement, etc.*⟩; general ⟨*renunciation*⟩; **2.** *adv.* to a large *or* great extent; **~gereist** *Adj.* *(präd. getrennt geschrieben)* widely travelled; **~her** *Adv.* *(geh.)* from afar; **~herzig** *Adj.* generous; liberal ⟨*interpretation*⟩; **~hin** *Adv.* for miles around; **~läufig 1.** *Adj.* **a)** *(ausgedehnt)* extensive; *(geräumig)* spacious; **b)** *(entfernt)* distant; **c)** *(ausführlich)* lengthy; long-winded; **2.** *adv.* **a)** *(ausgedehnt)* spaciously; **b)** *(entfernt)* distantly; **c)** *(ausführlich)* at length; long-windedly; **~maschig** *Adj.* wide-meshed; **~räumig 1.** *Adj.* spacious ⟨*room, area, etc.*⟩; wide ⟨*gap, space*⟩; **2.** *adv.* spaciously; **etw. ~räumig umfahren** give sth. a wide berth; **~reichend 1.** *Adj.* **a)** long-range; **b)** *(fig.)* far-reaching ⟨*importance, consequences*⟩; sweeping ⟨*changes, powers*⟩; extensive ⟨*relations, influence*⟩; **2.** *adv.* extensively; to a large extent; **~schweifig 1.** *Adj.* long-winded; **2.** *adv.* long-windedly; **~sichtig 1.** *Adj.* long-sighted; *(fig.)* far sighted; **2.** *adv.* *(fig.)* far-sightedly; **~sichtigkeit die; ~** long-sightedness; *(fig.)* far-sightedness; **~sprung der** *(Sport)* long jump *(Brit.)*; broad jump *(Amer.)*; **~verbreitet** *Adj.* *(präd. getrennt geschrieben)* widespread; common; common ⟨*plant, animal*⟩; **~verzweigt** *Adj.* *(präd. getrennt geschrieben)* extensive ⟨*network*⟩; ⟨*firm*⟩ with many [different] branches; **~winkel·objektiv das** wide-angle lens

Weizen ['vaitsn̩] **der; ~s** wheat

Weizen-: ~bier das *s.* **Weißbier; ~mehl das** wheat flour

welch [vɛlç] **1.** *Interrogativpron.* **a)** *(bei Wahl aus einer unbegrenzten Menge)* what; **aus ~em Grund?** for what reason?; **um ~e Zeit?** [at] what time?; **b)** *(bei Wahl aus einer begrenzten Menge) attr.* which; *alleinstehend* which one; **an ~em Tag/in ~em Jahr?** on which day/in which year?; **~er/~e/~es auch immer** whichever one; **~er/~e/~es von [den] beiden** which of the two; **c)** *(geh.: was für ein)* what a; *oft unflektiert* **~ reizendes Geschöpf!** what a charming creature!; **~ ein Zufall/Glück!** what a coincidence/how fortunate! **2.** *Relativpron.* *(bei Menschen)* who; *(bei Sachen)* which. **3.** *Indefinitpron.* some; *(in Fragen)* any

welk [vɛlk] *Adj.* withered ⟨*skin, hands, etc.*⟩; wilted ⟨*leaves, flower*⟩; limp ⟨*lettuce*⟩

welken *itr. V.; mit sein* wilt

Well·blech das corrugated iron

Welle ['vɛlə] **die; ~, ~n a)** *(auch Haar~, Physik, fig.)* wave; **grüne ~** *(Verkehrsw.)* linked *or* synchronised traffic lights; **die weiche ~** *(fig. ugs.)* the soft approach *or* line; **b)** *(Rundf.: ~nlänge)* wavelength; **c)** *(Technik)* shaft; **d)** *(Boden~)* undulation

wellen 1. *tr. V.* wave ⟨*hair*⟩; corrugate ⟨*iron, metal*⟩. **2.** *refl. V.* ⟨*hair*⟩ be wavy; ⟨*ground, carpet*⟩ undulate

wellen-, Wellen-: ~bad das artificial wave pool; **~bereich der** *(Rundf.)* waveband; **~brecher der** breakwater; **~förmig 1.** *Adj.* wavy ⟨*line, outline, seam, etc.*⟩; wavelike ⟨*motion, movement, etc.*⟩; **2.** *adv.* ⟨*be propagated*⟩ in the form of waves *or* as waves; **~gang der;** *o. Pl.* swell; **bei starkem ~gang** in heavy seas; **~länge die** *(Physik)* wavelength; **[mit jmdm.] auf der gleichen ~länge liegen** *(fig. ugs.)* be on the same wavelength [as sb.]; **~linie die** wavy line; **~reiten das** surfing *no art.*; **~sittich der** budgerigar

Well·fleisch das boiled belly pork

wellig *Adj.* wavy ⟨*hair*⟩; undulating ⟨*scenery, hills, etc.*⟩; uneven ⟨*surface, track, etc.*⟩

Well·pappe die corrugated cardboard

Welpe ['vɛlpə] **der; ~n, ~n** *(Hund)* whelp; pup; *(Wolf, Fuchs)* whelp; cub

Wels [vɛls] **der; ~es, ~e** catfish

Welt [vɛlt] **die; ~, ~en a)** *o. Pl.* world; **auf der ~:** in the world; **in der ganzen ~ bekannt sein** be known world-wide *or* all over the world; **die schönste Frau der Welt** the most beautiful woman in the world; **nicht die ~ kosten** *(ugs.)* not cost the earth *(coll.)*; **davon geht die ~ nicht unter** *(ugs.)* it's not the end of the world; **auf die** *od.* **zur ~ kommen** be born; **auf der ~ sein** have been born; **aus aller ~:** from all over the world; **in aller ~:** throughout the world; all over the world; **in alle ~:** all over the world; **um nichts in der ~, nicht um alles in der ~:** not for anything in the world *or* on earth; **um alles in der ~** *(ugs.)* for heaven's sake; **die ganze ~** *(fig.)* the whole world; **alle ~** *(fig. ugs.)* the whole world; everybody; **eine verkehrte ~:** a topsy-turvy world; **Kinder in die ~ setzen** *(ugs.)* have children; **zur ~ bringen** bring into the world; give birth to; **eine Dame/ein Mann von ~:** a woman/man of the world; **b)** *(~all)* universe; **uns trennen ~en** *(fig.)* we are worlds apart

welt-, Welt-: ~all das universe; cosmos; **~anschauung die** world-view; Weltanschauung; **~atlas der** atlas of the world; **~aus·stellung die** world fair; **~bekannt** *Adj.* known all over the world *pred.*; world-famous ⟨*artist, author, etc.*⟩; **~berühmt** *Adj.* world-famous; **~bewegend** *Adj.* world-shaking; **nicht**

~**bewegend sein** *(ugs. spött.)* be nothing to write home about *(coll.)*; ~**bild das** world view; conception of the world

Welten·bummler der; ~s, ~: globe-trotter

Welt·erfolg der world-wide success

Welter·gewicht das welterweight

welt-, Welt-: ~**fremd 1.** *Adj.* unworldly; **2.** *adv.* unrealistically; ~**frieden der** world peace; ~**geltung die** international standing; ~**geschichte die;** *o. Pl.* world history *no art.*; ~**karte die** map of the world; ~**klasse die** world class; ~**krieg der** world war; **der Zweite ~krieg** the Second World War; World War II

weltlich *Adj.* worldly; *(nicht geistlich)* secular

welt-, Welt-: ~**literatur die** world literature *no art.*; ~**macht die** world power; ~**männisch** [~mɛnɪʃ] **1.** *Adj.* sophisticated; **2.** *adv.* in a sophisticated manner; ~**markt der** *(Wirtsch.)* world market; ~**meer das** ocean; ~**meister der** world champion; ~**meisterschaft die** world championship; ~**politik die** world politics *pl.*; ~**politisch** *Adj.* **das ~politische Klima** the climate in world politics; ~**raum der** space *no art.*; ~**reich das** empire; ~**reise die** world tour; ~**rekord der** world record; ~**religion die** world religion; ~**sprache die** world language; ~**stadt die** cosmopolitan city; ~**untergang der** end of the world; ~**weit 1.** *Adj.; nicht präd.* world-wide; **2.** *adv.* throughout *or* all over the world; ~**wirtschaft die** world economy; ~**wunder das: die Sieben ~wunder** the Seven Wonders of the World; ~**zeit·uhr die** clock showing times around the world

wem [ve:m] *Dat. von* **wer 1.** *Interrogativpron.* to whom; who ... to; *(nach Präp.)* whom; who ...; **wem hast du das Buch geliehen?** to whom did you lend the book?; who did you lend the book to? **2.** *Relativpron. (derjenige, dem/diejenige, der)* the person to whom ...; the person who ... to; *(jeder, dem)* anyone to whom. **3.** *Indefinitpron. (ugs.)* to somebody *or* someone; *(nach Präp.)* somebody; someone; *(fragend, verneint)* to anybody *or* anyone; *(nach Präp.)* anybody; anyone

Wem·fall der dative [case]

wen [ve:n] *Akk. von* **wer 1.** *Interrogativpron.* whom; who *(coll.)*; **an/für ~:** to/for whom ...; who ... to/for; **an ~ schreibst du?** to whom are you writing? who are you writing to?; **~ von ihnen kennst du?** which [one] of these do you know? **2.** *Relativpron. (derjenige, den/diejenige, die)* the person whom; *(jeder, den)* anyone whom. **3.** *Indefinitpron. (ugs.)* somebody; someone; *(fragend, verneint)* anybody; anyone

Wende die; ~, ~n **a)** change; **eine ~ zum Besseren/Schlechteren** a change for the better/worse; **b) um die ~ des Jahrhunderts** at the turn of the century; **c)** *(Turnen)* front vault

Wende·kreis der a) *(Geogr.)* tropic; **der nördliche ~, der ~ des Krebses** the Tropic of Cancer; **der südliche ~, der ~ des Steinbocks** the Tropic of Capricorn; **b)** *(Kfz-W.)* turning circle

Wendel·treppe die spiral staircase

[1]**wenden 1.** *tr., auch itr. V. (auf die andere Seite)* turn [over]; toss ⟨*pancake, cutlet, etc.*⟩; *(in die entgegengesetzte Richtung)* turn [round]; **bitte ~!** please turn over. **2.** *itr. V.* turn [round]. **3.** *refl. V.* **sich zum Besseren/Schlechteren ~:** take a turn for the better/worse

[2]**wenden 1.** *unr., auch regelm. tr. V.* turn; **den Kopf ~:** turn one's head; **keinen Blick von jmdm. ~:** not take one's eyes off sb. **2.** *unr., auch regelm. refl. V.* **a)** ⟨*person*⟩ turn; **b)** *(sich richten)* **sich an jmdn. ~:** turn to sb.; **sich mit einer Bitte an jmdn. ~:** ask a favour of sb.; **an wen soll ich mich ~?** whom should I approach?; **das Buch wendet sich an junge Leser** *(fig.)* the book is addressed to *or* intended for young readers

Wende·punkt der turning-point

wendig 1. *Adj.* **a)** agile; nimble; manœuvrable ⟨*vehicle, boat, etc.*⟩; **b)** *(gewandt)* astute. **2.** *adv.* **a)** agilely; nimbly; **b)** *(gewandt)* astutely

Wendigkeit die; ~: *s.* **wendig 1:** agility; nimbleness; manœuvrability; astuteness

Wendung die; ~, ~en **a)** turn; **eine ~ um 180°** a 180° turn; **b)** *(Veränderung)* change; **eine ~ zum Besseren/Schlechteren** a turn for the better/worse; **c)** *(Rede~)* expression

Wen·fall der accusative [case]

wenig ['ve:nɪç] **1.** *Indefinitpron. u. unbest. Zahlw.* **a)** *Sing.* little; **das ~e Geld reicht nicht aus** this small amount of money is not enough; **~ Zeit/Geld haben** not have much *or* have little time/money; **das ist ~:** that isn't much; **dazu kann ich ~ sagen** I can't say much about that; **zu ~ Zeit/Geld haben** not have enough time/money; **ein Exemplar/50 Mark zu ~:** one copy too few/50 marks too little; **nur ~es** only a little; **b)** *Pl.* a few; **nur ~ Leute waren unterwegs** only a few people were about; **sie hatte ~ Bücher/Freunde** she had few books/friends; **die ~en, die davon wußten** the few who knew about it; **nur ~e haben teilgenommen** only a few took part. **2.** *Adv.* little; **nur ~ besser** only a little better; **wir waren nicht ~ erstaunt** we were more than a little astonished; **~ mehr** not much more; **ein ~:** a little; *(eine Weile)* for a little while

weniger 1. *Komp. von* **wenig 1: a)** *Sing.* less; **b)** *Pl.* fewer. **2.** *Komp. von* **wenig 2:** less; **es kommt ~ auf Quantität als auf Qualität an** quantity is less important than quality; **das ist ~ angenehm** that is not very pleasant; *s. auch* **mehr 1. 3.** *Konj.* less; **fünf ~ drei** five, take away three

wenigst... 1. *Sup. von* **wenig 1:** least; **am ~en** least; **in den ~en Fällen/für die ~en Menschen** in very few cases/for very few people; **nur die ~en** only very few; **das ~e, was wir tun können** the least we can do. **2.** *Sup. von* **wenig 2: am ~en** the least; **das hätte ich am ~en erwartet** that's the last thing I should have expected

wenigstens *Adv.* at least

wenn [vɛn] *Konj.* **a)** *(konditional)* if; **außer ~:** unless; **und [selbst] ~:** even if; **~ es sein muß, komme ich mit** If I have to, I'll come along; **~ es nicht anders geht** if there's no other way; **~ du schon rauchen mußt** if you 'must smoke; **b)** *(temporal)* when; **jedesmal,** *od.* **immer, ~:** whenever; **c)** *(konzessiv)* **wenn ... auch** even though; **und ~ es [auch] noch so spät ist ...:** no matter how late it is ...; however late it is ...; **[und] ~ auch!** *(ugs.)* even so; all the same; **d)** *(in Wunschsätzen)* if only; **~ ich doch** *od.* **nur** *od.* **bloß wüßte, ob ...:** if only I knew whether ...

Wenn das; ~s, ~ *od. (ugs.)* ~s: **das ~ und Aber, die ~[s] und Aber[s]** the ifs and buts

wenn·gleich *Konj. (geh.)* even though; although

wenn·schon *Adv.* ~ **[nicht] ..., dann ...:** even if [not] ..., then ...; **[na** *od.* **und]** ~! *(ugs.)* so what?; ~, **dennschon** *(ugs.)* if you're going to do something, you may as well do it properly; no half measures!

wer [veːɐ̯] *Nom.:* **1.** *Interrogativpron.* who; ~ **alles ist dabeigewesen?** which people were there?; who was there?; ~ **von ...:** which of; **was glaubt sie eigentlich,** ~ **sie ist?** who does she think she is? **2.** *Relativpron. (derjenige, der/diejenige, die)* the person who; *(jeder, der)* anyone *or* anybody who; ~ **es auch [immer]** *od.* ~ **immer es getan hat** *(geh.)* whoever did it. **3.** *Indefinitpron. (ugs.)* someone; somebody; *(fragend, verneint)* anyone; anybody; *s. auch* **wem, wen, wessen**

Werbe-: ~**abteilung die** advertising *or* publicity department; ~**agentur die** advertising agency; ~**fernsehen das** television commercials *pl.;* ~**film der** advertising *or* promotional *or* publicity film; ~**funk der** radio commercials *pl.;* ~**geschenk das** [promotional] free gift

werben ['vɛrbn̩] **1.** *unr. itr. V.* **a)** advertise; **für etw.** ~: advertise sth.; **für eine Partei** ~: canvass for a party; **b)** *(geh.: sich bemühen)* **um Wählerstimmen** ~: seek to attract votes; **um jmds. Gunst/Freundschaft** ~: court sb.'s favour/friendship. **2.** *unr. tr. V.* attract ‹*customers etc.*›; recruit ‹*soldiers, members, staff, etc.*›

Werbe-: ~**slogan der** advertising slogan; ~**spot der** commercial; advertisement; ad *(coll.);* ~**trommel die: [für jmdn./etw.] die** ~ **rühren** *od.* **schlagen** beat *or* thump the drum [for sb./sth.]

Werbung die; ~, ~**en a)** *o. Pl.* advertising; **für etw.** ~ **machen** advertise sth.; **b)** *o. Pl.: s.* **Werbeabteilung**

Werde·gang der a) development; **b)** *(Laufbahn)* career

werden ['veːɐ̯dn̩] **1.** *unr. itr. V.; mit sein; 2. Part.* **geworden a)** become; get; **älter** ~: get *or* grow old[er]; **du bist aber groß/schlank geworden!** you've grown so tall/slim; **wahnsinnig** *od.* **verrückt** ~: go mad; **gut** ~: turn out well; **das muß anders** ~: things have to change; **wach** ~: wake up; **rot** ~: go *or* turn red; **er ist 70 [Jahre alt] geworden** he has had his 70th birthday *or* has turned 70; **heute soll es/wird es heiß** ~: it's supposed to get/it's going to be hot today; **mir wird übel/heiß/schwindelig** I feel sick/I'm getting hot/dizzy; **Arzt/Vater** ~: become a doctor/a father; **was willst du einmal** ~**?** what do you want to be when you grow up?; **erster/letzter** ~: be *or* come first/last; **was soll das** ~**?** what is that going to be?; **eine** ~**de Mutter** a mother-to-be; an expectant mother; **b)** *(sich entwickeln)* **zu etw.** ~: become sth.; **das Wasser wurde zu Eis** the water turned into ice; **was soll aus dir** ~**?** what is to become of you?; **aus ihm ist nichts/etwas geworden** he hasn't got anywhere/has got somewhere in life; **daraus wird nichts** ~ nothing will come of it/that!; **c)** *unpers.* **es wird [höchste] Zeit** it is [high] time; **es wird 10 Uhr** it is nearly 10 o'clock; **es wird Tag/Nacht/Herbst** day is dawning/night is falling/autumn is coming; **d)** *(entstehen)* come into existence; **es werde Licht** *(bibl.)* let there be light; **e)** *(ugs.)* **sind die Fotos [was] geworden?** have the photos turned out [well]?; **wird's bald?** *(ugs.)* hurry up!; **was soll nur** ~**?** what's going to happen now? . **2.** *Hilfsverb; 2. Part.* **worden a)** *(zur Bildung des Futurs)* **wir** ~ **uns um ihn kümmern** we will take care of him; **dir werd' ich helfen!** *(ugs.)* I'll give you what for *(coll.);* **es wird gleich regnen** it is going to rain any minute; **wir** ~ **nächste Woche in Urlaub fahren** we are going on holiday next week; *(als Ausdruck der Vermutung)* **es wird um die 80 Mark kosten** it will cost around 80 marks; **sie** ~ **[wohl] im Garten sein** they are probably in the garden; **er wird doch nicht [etwa] krank sein?** he wouldn't be ill, would he?; **sie wird schon wissen, was sie tut** she must know what she's doing; **b)** *(zur Bildung des Passivs)* be; **du wirst gerufen** you are being called; **er wurde gebeten/ist gebeten worden** he was asked; **ihm wurde gesagt** he was told; **es wurde gelacht/gesungen/getanzt** there was laughter/singing/dancing; **unser Haus wird renoviert** our house is being renovated; **c)** *(zur Umschreibung des Konjunktivs)* **was würdest du tun?** what would you do?; **würden Sie bitte ...?** would you please ...?

Wer·fall der nominative [case]

werfen ['vɛrfn̩] **1.** *unr. tr. V.* **a)** throw; drop ‹*bombs*›; **die Tür ins Schloß** ~: slam the door shut; **jmdn. aus dem Saal** ~ *(fig. ugs.)* throw sb. out of the hall; **b)** *(ruckartig bewegen)* throw; **den Kopf in den Nacken** ~: throw *or* toss one's head back; **die Arme in die Höhe** ~: throw one's arms up; **c)** *(erzielen)* throw; **eine Sechs** ~: throw a six; **ein Tor** ~: shoot *or* throw a goal; **d)** *(bilden)* **Falten** ~: wrinkle; crease; **[einen] Schatten** ~: cast [a] shadow; **e)** *(gebären)* give birth to. **2.** *unr. itr. V.* **a)** throw; **mit etw. [nach jmdm.]** ~: throw sth. [at sb.]; **mit Geld/Fremdwörtern um sich** ~ *(fig.)* throw [one's] money around/bandy foreign words about; **b)** *(Junge kriegen)* give birth; ‹*dog, cat*› litter. **3.** *unr. refl. V.* **a)** *(auch fig.)* throw oneself; **sich auf eine neue Aufgabe** ~ *(fig.)* throw oneself into a new task; **sich in die Kleider** ~ *(fig.)* throw on one's clothes; **b)** *(sich verziehen)* buckle; ‹*wood*› warp

Werfer der; ~**s,** ~: thrower; *(Baseball)* pitcher; *(Cricket)* bowler

Werft [vɛrft] **die;** ~, ~**en** shipyard; dockyard

Werk [vɛrk] **das;** ~**[e]s,** ~**e a)** *o. Pl. (Arbeit)* work; **am** ~**[e] sein** be at work; **sich ans** ~ **machen, ans** ~ **gehen** set to *or* go to work; **b)** *(Tat)* work; **das ist dein** ~! that is your doing *or* handiwork; **c)** *(geistiges, künstlerisches Erzeugnis)* work; **d)** *(Betrieb, Fabrik)* factory; plant; works *sing. or pl.;* **ab** ~: ex works; **e)** *(Mechanismus)* mechanism; **das** ~ **einer Uhr/Orgel** the works *pl.* of a clock/organ

werk-, Werk- *(betriebs-, Betriebs-) s.* **werk[s]-, Werk[s]-**

Werk·bank die; *Pl.* ~**bänke** work-bench

werkeln *itr. V.* potter around *or* about

Werken das; ~**s** *(Schulw.)* handicraft

Werk[s]-: ~**angehörige der/die** factory *or* works employee; ~**arzt der** factory *or* works doctor

Werk·schutz der a) factory *or* works security; **b)** *(Personen)* factory *or* works security service

Werk[s]-: ~**gelände das** factory *or* works premises *pl.;* ~**halle die** workshop; ~**spionage die** industrial espionage

werk-, Werk-: ~**statt die;** ~**statt,** ~**stätten,** ~**stätte die** workshop; *(Kfz-W.)* garage; ~**stoff der** material; ~**stück das** workpiece; ~**tag der** working day; workday; ~**tags** *Adv.* on weekdays; ~**tätig** *Adj.* working; ~**tätige der/die;** *adj. Dekl.* worker; ~**unterricht der**

[handi]craft instruction *no art.*; *(Unterrichtsstunde)* [handi]craft lesson; **~zeug das a)** *(auch fig.)* tool; **b)** *o. Pl. (Gesamtheit)* tools *pl.*

Werkzeug·kasten der tool-box

Wermut ['ve:ɐmu:t] **der; ~[e]s, ~s a)** *(Pflanze)* wormwood; **b)** *(Wein)* vermouth

wert [ve:ɐt] *Adj.* **a)** *(geh.)* esteemed; *(als Anrede)* **~e Genossen!** my dear comrades; **wie ist Ihr ~er Name, bitte?** *(geh.)* may I have your name, please?; **b) etw. ~ sein** be worth sth.; **das ist nichts ~:** this is worth nothing *or* worthless; **der Teppich ist sein Geld nicht ~:** the carpet is not worth the money; **das ist nicht der Erwähnung** *(Gen.)* **~:** this is not worth mentioning; *s. auch* **Rede c**

Wert der; ~[e]s, ~e a) value; **im ~ steigen/fallen** increase/decrease in value; **an ~ gewinnen/verlieren** gain/lose in value; **im ~[e] von ...:** worth ...; **etw. unter [seinem] ~ verkaufen** sell sth. for less than its value; **einer Sache** *(Dat.)* **großen ~ beimessen** attach great value to sth.; **sich** *(Dat.)* **seines [eigenen] ~es bewußt sein** be conscious of one's own importance; **das hat [doch] keinen ~!** *(ugs.: ist sinnlos)* there's no point; **~ auf etw.** *(Akk.)* **legen** set great store by *or* on sth.; **b)** *Pl.* objects of value; **c)** *(Briefmarke)* denomination

Wert·arbeit die high-quality workmanship

wert·beständig *Adj.* of lasting value *postpos.*; stable ‹*currency, investment, etc.*›; **~beständig sein** retain its value

werten *tr., itr. V.* **a)** judge; assess; **etw. als besondere Leistung ~:** rate sth. as a special achievement; **etw. als Erfolg ~:** regard sth. as *or* consider sth. a success; **b)** *(Sport)* **etw. hoch/niedrig ~:** award high/low points to sth.

wert·frei 1. *Adj.* detached; impartial; neutral ‹*term*›. **2.** *adv.* with detachment; impartially

Wert·gegenstand der valuable object; object of value; **Wertgegenstände** valuables

-wertig *(Chemie, Sprachw.)* -valent

Wertigkeit die; ~, ~en *(Chemie, Sprachw.)* valency *(Brit.)*; valence *(Amer.)*

wert-, Wert-: ~los *Adj.* worthless; valueless; **~maßstab der** standard [of value]; **~paket das** *(Postw.)* registered parcel; **~papier das** *(Wirtsch.)* security; **~sache die;** *meist Pl.* valuable item *or* object; **~sachen** valuables; **~schätzung die** *(geh.)* esteem; high regard; **~steigerung die** appreciation; increase in value

Wertung die; ~, ~en judgement

Wert·urteil das value judgement

wert·voll *Adj.* valuable; *(schätzenswert)* estimable

Wesen ['ve:zn̩] **das; ~s, ~ a)** *o. Pl. (Natur)* nature; *(Art, Charakter)* character; nature; **ein freundliches/kindliches ~ haben** have a friendly/childlike nature *or* manner; **b)** *(Mensch)* creature; soul; **ein weibliches/männliches ~:** a woman *or* female/a man *or* male; **c)** *(Lebe~)* being; creature

Wesens-: ~art die nature; character; **~zug der** trait; characteristic

wesentlich 1. *Adj.* fundamental **(für** to); **von ~er Bedeutung** of considerable importance; **im ~en** essentially. **2.** *adv. (erheblich)* considerably; much; **es wäre mir ~ lieber, wenn wir ...:** I would much rather we ...

Wes·fall der genitive [case]

wes·halb *Adv. s.* **warum**

Wespe ['vɛspə] **die; ~, ~n** wasp

Wespen-: ~nest das wasp's nest; **in ein ~nest stechen** *(fig. ugs.)* stir up a hornets' nest; **~stich der** wasp sting

wessen *Interrogativpron.* **a)** *(Gen. von* **wer***)* whose; **b)** *(Gen. von* **was***)* **~ wird er beschuldigt?** what is he accused of?

Wessi ['vɛsi] **der; ~s, ~s** *(salopp) s.* **Westdeutsche**

West [vɛst] *o. Art.; o. Pl.* **a)** *(bes. Seemannsspr., Met.)* west; **b)** *(Gebiet)* West; **c)** *(Politik)* West; **d)** *einem Subst. nachgestellt (westlicher Teil, westliche Lage)* West; *s. auch* **¹Ost, ¹Nord**

West-Berlin (das) *(hist.)* West Berlin

west-, West-: ~deutsch *Adj.* Western German; *(hist.: auf die alte BRD bezogen)* West German; **~deutsche der/die** Western German; *(hist.: Bürger der alten BRD)* West German; **~deutschland (das)** Western Germany; *(hist.: alte BRD)* West Germany

Weste ['vɛstə] **die; ~, ~n** waistcoat *(Brit.)*; vest *(Amer.)*; **eine weiße** *od.* **saubere ~ haben** *(fig. ugs.)* have a clean record; **jmdm. etw. unter die ~ jubeln** *(fig. ugs.)* shift *or* push sth. on to sb.

Westen der; ~s a) *(Richtung)* west; **nach ~:** westwards; to the west; **im/aus** *od.* **von** *od.* **vom ~:** in/from the west; **b)** *(Gegend)* West; **im ~:** in the West; **der Wilde ~:** the Wild West; **c)** *(Geogr., Politik)* **der ~** the West; *s. auch* **Osten, Norden**

Westen·tasche die waistcoat *(Brit.) or (Amer.)* vest pocket; **etw. wie seine ~ kennen** *(ugs.)* know sth. like the back of one's hand

Western der; ~[s], ~: western

West·europa (das) Western Europe

west·europäisch *Adj.* West *or* Western European

West·indien (das) the West Indies *pl*

west·indisch *Adj.* West Indian

West·küste die west[ern] coast

westlich 1. *Adj.* **a)** *(im Westen)* western; **15 Grad ~er Länge** 15 degrees west [longitude]; **das ~e Frankreich** western France; **b)** *(nach, aus dem Westen)* westerly; **c)** *(des Westens, auch Politik)* Western. **2.** *Adv.* westwards; **~ von ...:** [to the] west of ... **3.** *Präp. mit Gen.* [to the] west of; *s. auch* **östlich**

west-, West-: ~seite die western side; **~wärts** *Adv.* [to the] west; **~wind der** west[erly] wind

wes·wegen *Adv. s.* **warum**

Wett-: ~bewerb der; ~[e]s, ~e a) competition; **gut im ~bewerb liegen** have a good chance of winning the competition; **b)** *o. Pl. (Wirtsch.: Konkurrenz)* competition *no indef. art.*; **~bewerber der** competitor

Wette ['vɛtə] **die; ~, ~n** bet; **was gilt die ~?** how much do you want to bet?; what do you bet?; **eine ~ [mit jmdm.] abschließen** make a bet [with sb.]; **[ich gehe] jede ~ [ein], daß ...:** I bet you anything [you like] that ...; **mit jmdm. um die ~ laufen** race sb.; **sie schwammen um die ~:** they raced each other at swimming

Wett·eifer der competitiveness

wett·eifern *itr. V.* compete **(mit** with, **um** for)

wetten *itr. V.* bet; **mit jmdm. ~:** have a bet with sb.; **mit jmdm. um etw. ~:** bet sb. sth.; **auf etw.** *(Akk.)* **~:** bet on sth.; put one's money on sth.; **[wollen wir] ~?** [do you] want to bet?; **ich wette hundert zu eins, daß ...** *(ugs.)* I'll bet [you] a hundred to one that ...; **so haben wir nicht gewettet** *(ugs.)* that was not the deal *or* not what we agreed; **auf Platz/Sieg ~:** make a place bet/bet on a win. **2.** *tr. V.* **10 Mark ~:** bet 10 marks

Wetter das; ~, ~ a) *o. Pl.*

weather; **bei jedem ~:** in all weathers; **es ist schönes ~:** the weather is good *or* fine; **bei jmdm. gut ~ machen** *(fig. ugs.)* get on the right side of sb.; butter sb. up; **b)** *(Un~)* storm; **c)** *Pl. (Bergbau)* **schlagende ~:** firedamp

wetter-, Wetter-: ~amt das meteorological office; **~aussichten** *Pl.* weather outlook *sing.;* **~bericht der** weather report; *(Voraussage)* weather forecast; **~besserung die** improvement in the weather; **~beständig** *Adj.* weatherproof; **~dienst der** weather *or* meteorological service; **~fahne die** weather-vane; **~fest** *Adj.* weather-resistant; **~fühlig** *Adj.* sensitive to [changes in] the weather *postpos.;* **~fühligkeit die; ~:** sensitivity to [changes in] the weather; **~hahn der** weathercock; **~karte die** weather-chart; weather-map; **~lage die** weather situation; *(fig.)* situation; climate; **~leuchten** *itr. V.; unpers.* **es ~leuchtet** there is summer lightning; **~leuchten das; ~s** sheet *(esp. summer)* lightning *no indef. art.*

wettern *itr. V. (ugs.: schimpfen)* curse; **gegen** *od.* **über etw./jmdn. ~:** loudly denounce sth./sb.

wetter-, Wetter-: ~satellit der weather satellite; **~seite die** windward side; side exposed to the weather; **~station die** weather-station; **~vorhersage die** weather forecast

wett-, Wett-: ~fahrt die race; **~kampf der** competition; **~kämpfer der** competitor; **~lauf der** race; **einen ~lauf machen** run a race; **ein ~lauf mit der Zeit/dem Tod** *(fig.)* a race against time/with death; **~|machen** *tr. V. (ugs.)* make up for; **etw. durch etw. ~machen** make up for sth. with sth.; *(wiedergutmachen)* make good ‹*loss, mistake, etc.*›; **~rennen das** *(auch fig.)* race; **ein ~rennen machen** have *or* run a race; **~rüsten das; ~s** arms race; **~streit der** contest

wetzen ['vɛtsn̩] **1.** *tr. V.* sharpen; whet. **2.** *itr. V.; mit sein (ugs.)* dash

Wetz·stein der whetstone

WEZ *Abk.* **Westeuropäische Zeit** GMT

Whiskey ['vɪski] **der; ~s ~s** whiskey; [American/Irish] whisky

Whisky ['vɪski] **der; ~s, ~s** whisky

wich [vɪç] *1. u. 3. Pers. Sg. Prät. v.* **weichen**

Wichse ['vɪksə] **die; ~, ~n** *(ugs.)* [shoe-]polish

wichsen 1. *tr. V. (ugs.)* polish. **2.** *itr. V. (derb)* wank *(Brit. coarse);* jerk off *(coarse)*

Wicht [vɪçt] **der; ~[e]s, ~e a)** *(fam.: Kind)* little rascal *or* imp *(joc.);* **b)** *(abwertend: Mensch)* [insignificant] creature

Wichtel der; ~s, ~, Wichtel·männchen das gnome; *(Kobold)* goblin

wichtig ['vɪçtɪç] *Adj.* important; **es ist mir ~ zu wissen, ob ...:** it is important to me to know if ...; **nichts Wichtigeres zu tun haben[, als ...]** *(auch iron.)* have nothing better to do [than ...]; **sich ~ machen** *od.* **tun** *(ugs. abwertend)* be full of one's own importance

Wichtigkeit die; ~ importance

Wichtigtuer [-tu:ɐ] **der; ~s, ~** *(ugs. abwertend)* pompous ass

wichtigtuerisch 1. *Adj.* self-important; pompous. **2.** *adv.* in a self-important manner; ‹*behave, act*› pompously

Wicke ['vɪkə] **die; ~, ~n** vetch; *(im Garten)* sweet pea

Wickel der; ~s, ~: compress; **jmdn. am** *od.* **beim ~ haben/nehmen** *(fig. ugs.)* have/grab sb. by the scruff of his/her neck

Wickel-: ~kind das baby; infant; **~kommode die** baby's changing-table

wickeln *tr. V.* **a)** *(auf~)* wind; *(ab~)* unwind; **Wolle zu einem Knäuel ~:** wind wool into a ball; **etw. auf/um etw.** *(Akk.)* **~:** wind sth. on to sth./round sth.; **b)** *(eindrehen)* **sich/jmdm. die Haare ~:** put one's/sb.'s hair in curlers *or* rollers; **c)** *(ein~)* wrap; *(aus~)* unwrap; **etw./jmdn./sich in etw.** *(Akk.)* **~:** wrap sth./sb./oneself in sth.; **er hat sich [fest] in seinen Mantel gewickelt** he wrapped his coat tightly [a]round himself; **d)** *(windeln)* **ein Kind ~:** change a baby's nappy; **er ist frisch gewickelt** he has had his nappy changed; **e)** *(bandagieren)* bandage; **f) schief gewickelt sein** *(ugs.)* be very much mistaken

Wickel-: ~rock der wrapover skirt; **~tisch der** baby's changing-table

Widder ['vɪdɐ] **der; ~s, ~ a)** ram; **b)** *(Astron., Astrol.)* Aries

wider ['vi:dɐ] *Präp. mit Akk.* **a)** *(veralt.)* against; **b)** *(geh.: entgegen)* contrary to; **~ besseres Wissen/alle Vernunft** against one's better knowledge/all reason; **~ Willen** against one's will

wider-, Wider-: ~borstig 1. *Adj.* unruly, unmanageable ‹*hair*›; *(fig.)* rebellious ‹*person*›; unruly, rebellious ‹*child*›; **2.** *adv.* rebelliously; **~fahren** [--'--] *unr. itr. V.; mit sein (geh.)* **etw. ~fährt jmdm.** sth. happens to sb.; **ihm ist [ein] Unrecht ~fahren** he has been done an injustice; **~haken der** barb; **~hall der** echo; *(fig.)* **[bei jmdm.] ~hall finden** meet with a [positive] response [from sb.]; **~|hallen** *itr. V.* echo; resound **(von** with); **~legen** [--'--] *tr. V.* **etw./jmdn. ~legen** refute *or* disprove sth./prove sb. wrong; **~legung** [--'--] **die; ~, ~en** refutation

widerlich *(abwertend)* **1.** *Adj.* **a)** revolting; repulsive; **~ schmekken/riechen** taste/smell revolting; **b)** repugnant, repulsive ‹*person, behaviour, etc.*›; awful ‹*headache etc.*›. **2.** *adv.* revoltingly; ‹*behave, act*› in a repugnant *or* repulsive manner; awfully ‹*cold, hot, sweet, etc.*›

wider-, Wider-: ~natürlich *Adj.* unnatural; **~rechtlich 1.** *Adj.* illegal; unlawful; **~rechtliches Betreten eines Geländes/Gebäudes** trespass[ing] on a property/unlawful *or* illegal entry to a building; **2.** *adv.* illegally; unlawfully; **~rede die** argument; contradiction; **keine ~rede!** don't argue!; no arguing!; **~ruf der** *(einer Aussage)* retraction; *(eines Befehls, einer Anordnung, Erlaubnis usw.)* revocation; withdrawal; **[bis] auf ~ruf** until revoked *or* cancelled; **~rufen** [--'--] *unr. tr., auch itr. V.* retract, withdraw ‹*statement, claim, confession, etc.*›; revoke, cancel ‹*order, permission, etc.*›; repeal ‹*law*›; **~sacher der; ~s, ~, ~sacherin die; ~, ~nen** *(geh.)* adversary; opponent; **~schein der** *(geh.)* reflection; **~setzen** [--'--] *refl. V.* **sich jmdm./einer Sache ~setzen** oppose sb./sth.; **~setzlich** [*od.* --'--] **1.** *Adj.* rebellious; **2.** *adv.* rebelliously; **~sinnig** *Adj.* absurd; **~spenstig** [~ʃpɛnstɪç] **1.** *Adj.* unruly; rebellious; wilful; unruly, unmanageable ‹*hair*›; stubborn ‹*horse, mule, etc.*›; **2.** *adv.* wilfully; rebelliously; **~spenstigkeit die; ~:** unruliness; rebelliousness; wilfulness; *(von Haaren)* unruliness; unmanageableness; *(von Pferden usw.)* stubbornness; **~|spiegeln, ~spiegeln** [--'--] **1.** *tr. V.* reflect; *(als Spiegelbild)* mirror; *(fig.)* reflect; **2.** *refl. V.* be reflected; *(als Spiegelbild)* be mirrored; *(fig.)* be reflected; **~sprechen** [--'--] *unr.*

itr. V. **a)** contradict; **jmdm./einer Sache/sich [selbst] ~sprechen** contradict sb./sth./oneself; **b)** *(im Gegensatz stehen zu)* contradict, be inconsistent with ⟨*facts, truth, etc.*⟩; **sich** *(Dat.)* **~sprechende Aussagen/Nachrichten** conflicting statements/news reports; **~spruch der a)** *o. Pl.* opposition; protest; **es erhob sich allgemeiner ~spruch** there was a general protest; **auf ~spruch stoßen** meet with opposition *or* protests; **b)** *(etw. Unvereinbares)* contradiction; **sich** *(Akk.)* **in ~sprüche verwickeln** get entangled *or* caught up in contradictions; **in ~spruch zu** *od.* **mit etw. stehen** contradict sth.; be contradictory to sth.; **~sprüchlich** *Adj.* contradictory ⟨*news, statements, etc.*⟩; inconsistent ⟨*behaviour, attitude, etc.*⟩; **~spruchs·los 1.** *Adj.; nicht präd.* unprotesting; uncontradicting; **2.** *adv.* without opposition *or* protest

Wider·stand der a) resistance **(gegen** to); **jmdm./einer Sache ~ leisten** resist sb./sth.; put up resistance to sb./sth.; **bei jmdm. auf ~ stoßen** meet with *or* encounter resistance from sb.; **allen Widerständen zum Trotz** despite all opposition; **der ~** *(Politik)* the Resistance; *s. auch* **Weg c; b)** *(Elektrot.: Schaltungselement)* resistor

widerstands, Widerstands-: ~bewegung die resistance movement; **~fähig** *Adj.* robust; resistant ⟨*material etc.*⟩; hardy ⟨*animal, plant*⟩; **~fähig gegen etw. sein** be resistant to sth.; **~fähigkeit die;** *o. Pl.* robustness; *(von Material usw.)* resistance; *(von Tier, Pflanze)* hardiness; **~kämpfer der** resistance fighter; **~kraft die;** *o. Pl.* resistance; **~los 1.** *Adj.* without resistance *postpos.;* **2.** *adv.* without resistance

wider-, Wider-: ~stehen [--'--] *unr. itr. V.* **a)** *(nicht nachgeben)* **[jmdm./einer Sache] ~stehen** resist [sb./sth.]; **b)** *(standhalten)* **jmdm./einer Sache ~stehen** withstand sb./sth.; **~streben** [--'--] *itr. V.* **etw. ~strebt jmdm.** sb. dislikes *or* detests sth.; **es ~strebt jmdm., etw. zu tun** sb. dislikes doing sth. *or* is reluctant to do sth.; **~d nachgeben/einwilligen** give in/agree reluctantly; **~streben** [--'--] **das** reluctance; **~streit der;** *o. Pl.* conflict; **~streitend** [--'--] *Adj.; nicht präd.* conflicting; **~wärtig** [~vɛrtɪç] *(abwertend)* **1.** *Adj.* **a)** *(unangenehm)* disagreeable, unpleasant ⟨*conditions, situation, etc.*⟩; **b)** *(ekelhaft, abscheulich)* revolting, repugnant ⟨*smell, taste, etc.*⟩; objectionable, offensive ⟨*person, behaviour, attitude, etc.*⟩; **2.** *adv.* ⟨*behave, act, etc.*⟩ in an objectionable *or* offensive manner; **~wärtigkeit die; ~, ~en a)** *o. Pl.* offensiveness; objectionableness; repulsiveness; **b)** *(Umstand)* disagreeable *or* unpleasant circumstance; **~wille der** aversion **(gegen** to); **~willig 1.** *Adj.; nicht präd.* reluctant; unwilling; **2.** *adv.* reluctantly; unwillingly; **etw. nur ~willig tun** do sth. only with reluctance; **~wort das: keine ~worte dulden** not tolerate any argument; **keine ~worte!** no arguments!

widmen ['vɪtmən] **1.** *tr. V.* **a) jmdm. ein Buch/Gedicht** *usw.* **~:** dedicate a book/poem *etc.* to sb.; **b)** *(verwenden für/auf)* **etw. jmdm./einer Sache ~:** devote sth. to sb./sth. **2.** *refl. V.* **sich jmdm./einer Sache ~:** attend to sb./sth.; *(ausschließlich)* devote oneself to sb./sth.

Widmung die; ~, ~en dedication **(an +** *Akk.* to)

widrig ['vi:drɪç] *Adj.* unfavourable; adverse

Widrigkeit die; ~, ~en adversity

wie 1. *Interrogativadv.* **a)** *(auf welche Art u. Weise)* how; **~ heißt er/das?** what is his/its name?; what is he/that called?; **~ [bitte]?** [I beg your] pardon?; *(entrüstet)* I beg your pardon!; **~ war das?** *(ugs.)* what was that?; what did you say?; **~ kommt es, daß ...?** how is it that ...?; **~ das?** *(ugs.)* how did that come about?; **b)** *(durch welche Merkmale gekennzeichnet)* **~ war das Wetter?** what was the weather like?; how was the weather?; **~ ist dein neuer Chef?** what is your new boss like? *(coll.);* how is your new boss? *(coll.);* **~ geht es ihm?** how is he?; **~ war es in Spanien?** what was Spain like?; what was it like in Spain?; **~ findest du das Bild?** what do you think of the picture?; **~ gefällt er dir?** how do you like him?; **~ wär's mit ...:** how about ...; **c)** *(in welchem Grade)* how; **~ lange/groß/hoch/oft?** how long/big/high/often?; **~ spät ist es?** what time is it?; **wie alt bist du?** how old are you?; **und ~!** and how! *(coll.);* **d)** *(ugs.: nicht wahr)* **das hat dir Spaß gemacht, ~?** you enjoyed that, didn't you? **2.** *Relativadv.* **[die Art,] ~ er es tut** the way *or* manner in which he does it; **~ schon der Name sagt** as the name already implies. **3.** *Konj.* **a)** *Vergleichspartikel* as; **[so] ... ~ ...:** as ... as ...; **er macht es [genauso] ~ du** he does it [just] like you [do]; **ich fühlte mich ~ ...:** I felt as if I were ...; **„N" ~ „Nordpol"** N for November; **~ [zum Beispiel]** like; such as; **~ wenn** as if *or* though; **b)** *(und, sowie)* as well as; both; **Männer ~ Frauen** men as well as women; both men and women; **c)** *(temporal: als)* **~ ich an seinem Fenster vorbeigehe, höre ich ihn singen** as I pass by his window I hear him singing; **d)** *(ugs.: außer)* **nichts ~ Ärger** nothing but trouble

wieder ['vi:dɐ] *Adv.* **a)** *(erneut)* again; **je/nie ~:** ever/never again; **immer ~,** *(geh.)* **~ und ~:** again and again; time and [time] again; **nie ~ Krieg!** no more war!; **was ist denn jetzt schon ~ los?** what's happened 'now?; **b)** *(unterscheidend: noch)* **einige ..., andere ... und ~ andere ...:** some ..., others ..., and yet others ...; **das ist ~ etwas anderes** that is something else again; **c)** *(drückt Rückkehr in früheren Zustand aus)* again; **alles ist ~ beim alten** everything is back as it was before; **etw. ~ an seinen Platz zurückstellen** put sth. back in its place; **ich bin gleich ~ da** I'll be right back *(coll.);* I'll be back in a minute; **d)** *(andererseits, anders betrachtet)* **das ist auch ~ wahr** that's true enough; **da hast du auch ~ recht** you're right there; **e)** *s.* **wiederum c; f)** *(zur Vergeltung/zum Dank)* likewise; also; **g)** *(ugs.: noch)* **wie heißt er ~?** what's his name again?; **wo/wann war das [gleich] ~?** where/when was that again?

wieder-, Wieder-: ~aufbau der; *o. Pl.* reconstruction; rebuilding; **~auf|bauen**[1] *tr. V.* reconstruct; rebuild; **~aufbereitungs·anlage die** recycling plant; *(Kerntechnik)* reprocessing plant; **~aufnahme die** *s.* **~ aufnehmen:** resumption; re-establishment; readmittance; **~auf|nehmen**[1] *unr. tr. V.* **a)** *(erneut beginnen mit)* resume; take up ⟨*subject, idea*⟩ again; re-establish ⟨*relations, contact*⟩; **ein Verfahren ~aufnehmen** *(Rechtsspr.)* reopen a case; **b)** *(als Mitglied)* readmit; **~auf|richten**[1] *tr. V.* give fresh heart to ⟨*person*⟩; **~auf|tauchen**[1] *itr. V.; mit sein* turn up again

wieder-, Wieder-: ~beginn

der recommencement; resumption; ~|**bekommen** *unr. tr. V.* get back; ~|**beleben** *tr. V.* revive, resuscitate ⟨*person*⟩; *(fig.)* revive, resurrect ⟨*friendship, custom, etc.*⟩; ~**belebungs·versuch der** attempt at resuscitation; **bei jmdm. ~belebungsversuche machen** attempt to revive *or* resuscitate sb.; ~|**bringen** *unr. tr. V.* bring back; ~|**entdecken** *tr. V.* rediscover; ~|**erkennen** *unr. tr. V.* recognize; **er war kaum ~zuerkennen** he was almost unrecognizable; ~|**finden 1.** *unr. tr. V.* find again; *(fig.)* regain ⟨*composure, dignity, courage, etc.*⟩; **2.** *unr. refl. V.* be found; ~**gabe die a)** *(Bericht)* report; account; **b)** *(Übersetzung)* rendering; **c)** *(Reproduktion)* reproduction; ~|**geben** *unr. tr. V.* **a)** *(zurückgeben)* give back; return; **b)** *(berichten)* report; give an account of; *(wiederholen)* repeat; *(ausdrücken)* express; *(zitieren)* quote; **c)** *(übersetzen)* render; **d)** *(darstellen)* portray; depict; **e)** *(hörbar, sichtbar machen)* reproduce; ~|**gewinnen** *unr. tr. V.* recover ⟨*lost item, money, etc.*⟩; regain ⟨*composure, equilibrium*⟩

wieder·gut|machen[2] *tr. V.* make good; put right; **den Schaden ~** *(bezahlen)* pay for the damage; **ein nicht wiedergutzumachendes Unrecht** an irreparable injustice

wieder|haben *untr. tr. V. (auch fig.)* have back

wieder-, Wieder-: ~her|stellen[3] *tr. V.* **a)** re-establish ⟨*contact, peace*⟩; **b)** *(reparieren)* restore ⟨*building*⟩; **c) jmdn. ~herstellen** restore sb. to health; get sb. on his/her feet again; ~**herstellung die a)** re-establishment; **b)** *(Wiederinstandsetzung)* restoration; **c)** *(Genesung)* recovery; ~**holbar** *Adj.* repeatable; ~**holen 1.** *tr. V.* **a)** repeat; replay ⟨*football match*⟩; retake ⟨*penalty kick*⟩; resit, retake ⟨*exam*⟩; hold ⟨*election*⟩ again; **b)** *(nochmals sagen)* repeat, reiterate ⟨*question, demand, offer, etc.*⟩; **c)** *(repetieren)* revise ⟨*lesson, vocabulary, etc.*⟩; **2.** *refl. V.* **a)** *(wieder dasselbe sagen)* repeat oneself; **b)** *(erneut geschehen)* happen again; **c)** *(wiederkehren)* be repeated; recur

wieder|holen *tr. V.* fetch *or* get back

wiederholt 1. *Adj.; nicht präd.* repeated; **zum ~en Male** yet again. **2.** *adv.* repeatedly

Wiederholung die; ~, ~en a) repetition; *(eines Fußballspiels usw.)* replay; *(eines Freistoßes, Elfmeters usw.)* retaking; *(einer Sendung)* repeat; *(einer Aufführung)* repeat performance; **b)** *(des Schuljahrs, einer Prüfung usw.)* repeating; **c)** *(von Lernstoff)* revision

Wiederholungs-: ~zahlwort das multiplicative; ~**zeichen das** *(Musik)* repeat sign

Wieder·hören: [auf] ~! goodbye! *(at end of telephone call)*

wieder-, Wieder-: ~|käuen [-kɔyən] **1.** *itr. V.* ruminate; chew the cud; **2.** *tr. V.* **a)** chew again; **b)** *(fig. abwertend)* rehash; ~**käuer der; ~s, ~:** ruminant; ~**kehr die; ~** *(geh.)* **a)** *(Rückkehr)* return; **b)** *(Wiederholung)* recurrence; ~|**kehren** *itr. V.; mit sein (geh.)* **a)** *(zurückkehren)* return; **b)** *(sich noch einmal ereignen)* come again; **c)** *(sich wiederholen)* be repeated; recur; ~|**kommen** *unr. itr. V.; mit sein* **a)** *(zurückkommen)* return; come back; **b)** *(noch einmal kommen)* come back *or* again; **c)** *(sich noch einmal ereignen)* ⟨*opportunity, past*⟩ come again; ~|**kriegen** *tr. V. (ugs.)* get back; ~**schauen das: [auf] ~schauen!** *(südd., österr.)* goodbye!; ~|**sehen** *unr. tr. V. (auch fig.)* see again; **sich überraschend ~sehen** see each other *or* meet again unexpectedly; ~**sehen das; ~s, ~:** reunion; **[auf] ~sehen!** goodbye!; **jmdm. auf ~sehen sagen** say goodbye to sb.

Wiedersehens·freude die; *o. Pl.* pleasure of seeing sb./each other again

wieder-, Wieder-: ~|tun *unr. tr. V.* do again; ~**um** *Adv.* **a)** *(erneut)* again; **b)** *(andererseits)* on the other hand; **so weit würde ich ~um nicht gehen** I wouldn't, however, go that far; **c)** *(meiner-, deinerseits usw.)* in turn; ~|**vereinigen** *tr. V.* reunify ⟨*country*⟩; ~**vereinigung die** reunification; ~|**verwenden** *unr. od. regelm. tr. V.* reuse; ~**verwendung die** reuse; ~**wahl die** re-election; **sich zur ~wahl stellen** stand *or* run for re-election; ~|**wählen** *tr. V.* re-elect

Wiege ['vi:gə] **die; ~, ~n** *(auch fig.)* cradle; **von der ~ bis zur Bahre** *(scherzh.)* from the cradle to the grave

[1]**wiegen** *unr. tr., itr. V.* weigh; **was** *od.* **wieviel wiegst du?** how much do you weigh?; what weight *or* how heavy are you?; **schwer ~** *(fig.)* carry weight

[2]**wiegen 1.** *tr. V.* rock; shake ⟨*head*⟩ *(in doubt)*; **die Hüften ~:** sway one's hips; **einen ~den Gang haben** have a rolling gait. **2.** *refl. V.* ⟨*boat, cradle, etc.*⟩ rock; ⟨*person, branch, etc.*⟩ sway

Wiegen-: ~fest das *(geh.)* birthday; ~**lied das** lullaby; cradlesong

wiehern ['vi:ɐn] *itr. V.* **a)** whinny; *(lauter)* neigh; **b) [vor Lachen] ~** *(fig. ugs.)* roar with laughter

Wien ['vi:n] **(das); ~s** Vienna

[1]**Wiener 1. der; ~s, ~:** Viennese. **2.** *indekl. Adj.* Viennese; **~ Würstchen** wiener; frankfurter; **~ Schnitzel** Wiener schnitzel

[2]**Wiener die; ~, ~:** wiener [sausage]

Wienerin die; ~, ~nen Viennese

wienerisch *Adj.* Viennese

wies ['vi:s] *1. u. 3. Pers. Sg. Prät. v.* **weisen**

Wiese ['vi:zə] **die; ~, ~n** meadow; *(Rasen)* lawn

Wiesel ['vi:zl̩] **das; ~s, ~:** weasel; **wie ein ~ laufen** run like a hare; *s. auch* **flink 1**

wie·so *Adv.* why

wie·viel [*od.* '--] *Interrogativpron. Sing.* how much; *Pl.* how many; **~ Uhr ist es?** what time is it?

wie·viel·mal [*od.* -'--] *Adv.* how many times

wievielt... [*od.* '--] *Interrogativadj.* ; **als ~er Läufer ist er durchs Ziel gekommen?** in what position did he finish?; **der ~e Band?** which number volume?; **der Wievielte ist heute?** what is the date today?; **am Wievielten?** [on] what date?

wie·weit *Adv.* to what extent; how far

Wikinger ['vi:kɪŋɐ] **der; ~s, ~:** Viking

wild [vɪlt] **1.** *Adj.* **a)** wild; rugged, wild ⟨*countryside, area, etc.*⟩; wild, unruly ⟨*hair, beard, etc.*⟩; **~e Triebe** rank shoots; **b)** *(unerlaubt)* unauthorized; illegal; **~es Parken** illegal parking; **in ~er Ehe leben** *(veralt.)* live in sin; **~er Streik** wildcat strike; **c)** *(heftig, gewaltig)* wild ⟨*panic, flight, passion, desire, etc.*⟩; fierce ⟨*battle, anger, determination, look*⟩; **~ auf etw.** *(Akk.)* **sein** *(ugs.)* be mad *or* crazy about sth. *(coll.)*; **~ auf jmdn.** *od.* **nach jmdm. sein** *(ugs.)* be mad *or* crazy *or* wild about sb. *(coll.)*; **d)** *(wütend)* furious ⟨*curs-*

---
[1] *ich baue/bereite usw. wieder auf*
[2] *ich mache wieder gut*
[3] *ich stelle wieder her*

ing, shouting, etc.⟩; ~ **werden** get furious; **den ~en Mann spielen** *(ugs.)* get heavy *(coll.)*; **e)** *(unbändig, ungestüm)* wild, unruly ⟨*child*⟩; **f)** *(maßlos, wüst)* wild ⟨*speculation, claim, rumour, accusation*⟩; vile ⟨*oaths, curses*⟩; **halb so ~ sein** *(ugs.)* not be as bad as all that *(coll.)*; **g)** *nicht präd. (primitiv)* savage; wild. **2.** *adv.* **a)** wildly; ~ **entschlossen sein** *(ugs.)* be absolutely determined; ~ **um sich schlagen** hit out *or* lash out wildly; **wie ~** *(ugs.)* like mad *(coll.)*; **b)** *(unerlaubt)* illegally; ~ **zelten** camp in an unauthorized place

Wild das; ~[e]s game; *(einzelnes Tier)* [wild] animal

Wild-: ~bahn die: in freier ~bahn in the wild; **~bret** [~brɛt] **das;~s** *(geh.)* game; **~dieb der** poacher

Wilde der/die; *adj. Dekl.* savage; **wie ein ~r/die ~n** *(ugs.)* like a mad thing/like mad things *(coll.)*

Wild·ente die wild duck

Wilderei die; ~, ~en poaching *no pl., no art.*

Wilderer der; ~s, ~: poacher

wildern 1. *itr. V.* **a)** poach; go poaching; **b)** ⟨*cat, dog*⟩ kill game. **2.** *tr. V.* poach

wild·fremd *Adj.* completely strange; **ein ~er Mensch** a complete stranger

Wild·gans die wild goose

Wildheit die; ~: wildness; *(eines Volkes usw.)* savageness

wild-, Wild-: ~katze die wild cat; **~lebend** *Adj.; nicht präd.* wild; living in the wild *postpos.*; **~leder das** suede

Wildnis die; ~, ~se wilderness

wild-, Wild-: ~pflanze die wild plant; **~sau die** wild sow; **~schwein das** wild boar; **~wachsend** *Adj.; nicht präd.* wild; **~wechsel der a)** game path; **b)** *o. Pl. (Vorgang)* game crossing; **~westfilm** [-'--] **der** western; Wild West film

Wilhelm ['vɪlhɛlm] **(der)** William

will [vɪl] *1. u. 3. Pers. Sg. Präsens v.* **wollen**

Wille der; ~ns will; *(Wunsch)* wish; *(Absicht)* intention; **guter/böser ~:** goodwill/ill will; **aus freiem ~n** of one's own free will; **seinen ~n durchsetzen** get one's own way; **laß ihm seinen ~n** let him have his way; **beim besten ~n nicht** not with the best will in the world; **Letzter ~:** will; last will and testament *(formal)*; **ich mußte wider ~n lachen** I couldn't help laughing

willen *Präp. mit Gen.* **um jmds./einer Sache ~:** for sb.'s/sth.'s sake

Willen der; ~s *s.* **Wille**

willen·los 1. *Adj.* will-less; **völlig ~ sein** have no will of one's own. **2.** *adv.* will-lessly

willens *Adj.* ~ **sein, etw. zu tun** *(geh.)* be willing to do sth.

willens-, Willens-: ~kraft die; *o. Pl.* will-power; strength of will; **~schwach** *Adj.* weak-willed; **~stark** *Adj.* strong-willed

willentlich ['vɪləntlɪç] **1.** *Adj.; nicht präd.* deliberate. **2.** *adv.* deliberately; on purpose

willig 1. *Adj.* willing. **2.** *adv.* willingly

will·kommen *Adj.* welcome; **jmdm. ~ sein** be welcome to sb.; **jmdn. ~ heißen** welcome sb.

Will·kommen das *od. (selten)* **der; ~s, ~:** welcome

Will·kür die; ~: arbitrary use of power; **jmds. ~** *(Dat.)* **ausgeliefert sein** be at sb.'s mercy

Willkür-: ~akt der arbitrary act; **~herrschaft die** tyranny

willkürlich 1. *Adj.* arbitrary. **2.** *adv.* arbitrarily

wimmeln ['vɪmḷn] *itr. V.* **von Menschen ~:** be teeming *or* swarming with people; **von Fischen/Ungeziefer ~:** be teeming with fish/swarming with vermin; *unpers.* **in dem Artikel wimmelt es von Fehlern** the article is teeming with mistakes

wimmern ['vɪmɐn] *itr. V.* whimper

Wimpel ['vɪmpḷ] **der; ~s, ~:** pennant

Wimper ['vɪmpɐ] **die; ~, ~n** [eye]lash; **ohne mit der ~ zu zucken** without batting an eyelid

Wimpern·tusche die mascara

Wind [vɪnt] **der; ~[e]s, ~e a)** wind; **bei ~ und Wetter** in all weathers; **[schnell] wie der ~:** like the wind; **b)** *(fig.)* **wissen/merken, woher der ~ weht** *(ugs.)* know/notice which way the wind's blowing; ~ **machen** *(ugs.)* brag; **viel ~ um etw. machen** *(ugs.)* make a great fuss about sth.; ~ **von etw. bekommen** *(ugs.)* get wind of sth.; **jmdm. den ~ aus den Segeln nehmen** *(ugs.)* take the wind out of sb.'s sails; **etw. in den ~ schlagen** turn a deaf ear *or* pay no heed to sth.; **in alle [vier] ~e** in all directions; **sein Mäntelchen nach dem ~e hängen** be a trimmer

Wind·bö[e] die gust of wind

Winde die; ~, ~n a) winch; **b)** *(Bot.)* bindweed; convulvulus

Windel ['vɪndḷ] **die; ~, ~n** nappy *(Brit.)*; diaper *(Amer.)*

Windel·höschen das nappy pants *pl.*

windel·weich *Adj. (ugs.)* **jmdn. ~ schlagen** *od.* **hauen** beat the living daylights out of sb. *(coll.)*

[1]**winden 1.** *unr. tr. V. (geh.)* make ⟨*wreath, garland*⟩; **etw. um etw. ~:** wind sth. around sth.; **jmdm. etw. aus der Hand ~:** wrest sth. from sb.'s hand. **2.** *unr. refl. V.* **a)** ⟨*plant, tendrils*⟩ wind **(um** around); ⟨*snake*⟩ coil [itself], wind itself **(um** around); **b)** *(sich krümmen)* writhe; **sich ~ wie ein Aal** *(fig.)* try to wriggle out of it; **c)** *(sich schlängeln)* ⟨*path, river*⟩ wind [its way]

[2]**winden** *itr. V.; unpers.* **es windet** it's windy

Windes·eile die: in *od.* **mit ~:** in next to no time

wind-, Wind-: ~fang der porch; **~geschützt** *Adj.* sheltered from the wind *postpos.*; sheltered; **~hose die** *(Met.)* whirlwind; **~hund der a)** greyhound; **Afghanischer ~hund** Afghan hound; **b)** *(ugs. abwertend)* careless and unreliable sort *(coll.)*

windig *Adj.* **a)** windy; **b)** *(ugs. abwertend)* shady; dubious ⟨*excuse*⟩

wind-, Wind-: ~jacke die wind-cheater *(Brit.)*; windbreaker *(Amer.)*; **~jammer der; ~s, ~** *(Seemannsspr.)* windjammer; **~mühle die** windmill; **~pocken** *Pl.* chicken-pox *sing.*; **~rädchen das** windmill; **~richtung die** wind direction; **~rose die** compass card; **~schief** *Adj.* crooked; **~schutz·scheibe die** windscreen *(Brit.)*; windshield *(Amer.)*; **~spiel das** [small] greyhound; **~stärke die** wind force; **~still** *Adj.* windless; still; **es war völlig ~still** there was no wind at all; **~stille die** calm; **es herrschte völlige ~stille** there was no wind at all; **~stoß der** gust of wind; **~surfer der** windsurfer

Windung die; ~, ~en a) bend; *(eines Flusses)* meander; *(des Darms, Gehirns)* convolution; **b)** *(spiralförmiger Verlauf)* spiral; *(einer Spule o. ä.)* winding

Wink [vɪŋk] **der; ~[e]s, ~e a)** sign; **b)** *(Hinweis)* hint; *(Ratschlag)* tip; hint; **ein ~ mit dem Zaunpfahl** *(scherzh.)* a strong hint

Winkel ['vɪŋkḷ] **der; ~s, ~ a)** angle; **toter ~:** blind spot; **b)** *(Ecke; auch fig.)* corner; **c)** *(Ort)* corner; spot; **d)** *(Werkzeug)* [carpenter's] square; *(T-förmig)* T-square

Winkel-: ~messer der protractor; **~zug der;** *meist Pl.* shady trick *or* move

winken 1. *itr. V.* **a)** wave; **mit etw. ~:** wave sth.; **jmdm. ~:** wave to sb.; *(jmdn. heranwinken)* beckon sb. over; **b)** *(fig.)* **etw. winkt jmdm.** sth. is in prospect for sb.; **dem Sieger winkt eine Flasche Sekt** the winner will receive a bottle of champagne. **2.** *tr. V.* **a)** *(heran~)* beckon; **jmdn. zu sich ~:** beckon sb. over [to one]; **der Polizist winkte den Wagen zur Seite** the policeman waved the car over [to the side]; **b)** *(signalisieren)* semaphore ⟨*message*⟩

winseln ['vɪnzl̩n] *itr. V.* **a)** ⟨*dog*⟩ whimper; **b)** *(abwertend)* whine

Winter ['vɪntɐ] **der; ~s, ~:** winter; *s. auch* **Frühling**

winter-, Winter-: ~anfang der beginning of winter; **~fest** *Adj.* **a)** winter *attrib.* ⟨*clothing*⟩; **b)** *s.* **~hart; ~garten der** conservatory; **~hart** *Adj. (Bot.)* hardy; **~kleidung die** winter clothes *pl. or* clothing

winterlich 1. *Adj.* wintry; winter *attrib.* ⟨*clothing, break*⟩. **2.** *adv.* **~ kalt/öde** cold/bare and wintry

Winter-: ~mantel der winter coat; **~reifen der** winter tyre; **~sachen** *Pl.* winter things; **~saison die** winter season; **~schlaf der** *(Zool.)* hibernation; **~schlaf halten** hibernate; **~schluß·verkauf der** winter sale[s *pl.*]; **~semester das** winter semester; **~sport der** winter sports *pl.*; **in den ~sport fahren** go on a winter sports holiday; **~sportler der** winter sportsman; **~tag der** winter['s] day; **~urlaub der** winter holiday; **~zeit die;** *o. Pl.* winter-time

Winzer ['vɪntsɐ] **der; ~s, ~, Winzerin die; ~, ~nen** winegrower

winzig ['vɪntsɪç] **1.** *Adj.* tiny. **2.** *adv.* **~ klein** tiny; minute

Winzigkeit die; ~tininess; minuteness

Wipfel ['vɪpfl̩] **der; ~s, ~:** tree-top

Wippe ['vɪpə] **die; ~, ~n** see-saw

wippen *itr. V.* bob up and down; *(hin und her)* bob about; *(auf einer Wippe)* see-saw; **er wippte in den Knien** he bobbed up and down, bending at the knees

wir [vi:ɐ] *Personalpron.; 1. Pers. Pl. Nom.* we; **~ beide** *od.* **beiden** we two; the two of us; **Wer kommt mit? – Wir!** Who's coming? – We are!; **Wer ist es? – Wir sind's!** Who is it? – It's us!; *s. auch* **unser; uns**

wirb [vɪrp] *Imperativ Sg. v.* **werben**

Wirbel ['vɪrbl̩] **der; ~s, ~ a)** *(im Wasser)* whirlpool; vortex; *(in der Luft)* whirlwind; *(kleiner)* eddy; *(von Rauch, beim Tanz)* whirl; **b)** *(Trubel)* hurly-burly; **c)** *(Aufsehen)* fuss; **um jmdn./etw. ~ machen** make a fuss about sb./sth.; **d)** *(Anat.)* vertebra; **e)** *(Haar~)* crown; **f)** *(Trommel~)* [drum] roll

Wirbellose *Pl.; adj. Dekl. (Zool.)* invertebrates

wirbeln 1. *itr. V.; mit Richtungsangabe mit sein* whirl; ⟨*water, snowflakes*⟩ swirl. **2.** *tr. V.* swirl ⟨*leaves, dust*⟩; whirl ⟨*dancer*⟩

Wirbel-: ~säule die *(Anat.)* vertebral column; spinal column; **~sturm der** cyclone; **~tier das** *(Zool.)* vertebrate; **~wind der** whirlwind

wirbt [vɪrpt] *3. Pers. Sg. Präsens v.* **werben**

wird [vɪrt] *3. Pers. Sg. Präsens v.* **werden**

wirf [vɪrf] *Imperativ Sg. v.* **werfen**

wirft *3. Pers. Sg. Präsens v.* **werfen**

wirken ['vɪrkn̩] **1.** *itr. V.* **a)** have an effect; **es wirkte erst nach einer Stunde** it only took effect after an hour; **schmerzstillend ~:** have a pain-killing effect; **gegen etw. ~:** be effective against sth.; **ihre Heiterkeit wirkte ansteckend** her cheerfulness was infectious; **b)** *(erscheinen)* seem; appear; **sie wirkt sehr nett** she seems very nice; **er wirkt auf mich sehr sympathisch** I find him very congenial; **c)** *(beeindrucken)* ⟨*person*⟩ make an impression **(auf +** *Akk.* on); ⟨*picture, design, etc.*⟩ be effective; **d)** *(tätig sein)* work. **2.** *tr. V. (geh.)* bring about; *s. auch* **Wunder a**

wirklich 1. *Adj.* real; actual, real ⟨*event, incident, state of affairs*⟩; real, true ⟨*friend*⟩. **2.** *Adv.* really; *(in der Tat)* actually; really

Wirklichkeit die; ~, ~en reality; **~ werden** become a reality; ⟨*dream*⟩ come true; **in ~:** in reality

wirksam ['vɪrkza:m] **1.** *Adj.* effective; **mit dem 1. Juli ~ werden** *(Amtsspr.)* take effect from 1 July. **2.** *adv.* effectively

Wirksamkeit die; ~: effectiveness

Wirk·stoff der active agent

Wirkung die; ~, ~en effect **(auf +** *Akk.* on); **ohne ~ bleiben** have no effect; **seine ~ verfehlen** fail to have the desired effect; **mit ~ vom 1. Juli** *(Amtsspr.)* with effect from 1 July

wirkungs-, Wirkungs-: ~los 1. *Adj.* ineffective; **2.** *adv.* ineffectively; **~losigkeit die; ~:** ineffectiveness; **~voll 1.** *Adj.* effective; **2.** *adv.* effectively

wirr [vɪr] **1. a)** tousled ⟨*hair, beard*⟩; tangled ⟨*ropes, roots*⟩; **ein ~es Durcheinander** a chaotic muddle; **b)** *(verworren)* confused. **2.** *adv.* **a) das Haar hing ihr ~ ins Gesicht** her tousled hair hung over her face; **alles lag ~ durcheinander** everything lay in a chaotic muddle; **b)** *(verworren)* **sie träumte ~:** she had confused dreams

Wirren *Pl.* turmoil *sing.*

Wirrwarr [-var] **der; ~s** chaos; *(von Stimmen)* clamour; *(von Meinungen)* welter; *(von Haaren, Wurzeln, Vorschriften)* tangle

Wirsing ['vɪrzɪŋ] **der; ~s, Wirsing·kohl der** savoy [cabbage]

Wirt [vɪrt] **der; ~[e]s, ~e a)** landlord; **b)** *(Biol.)* host

Wirtin die; ~, ~nen landlady

Wirtschaft die; ~, ~en a) economy; *(Geschäftsleben)* commerce and industry; **in die ~ gehen** become a business man/woman; **b)** *(Gast~)* public house; pub *(Brit. coll.)*; bar *(Amer.)*; **c)** *o. Pl. (ugs. abwertend: Unordnung)* mess; shambles *sing.*

wirtschaften 1. *itr. V.* **a) mit dem Geld gut ~:** manage one's money well; **mit Verlust/Gewinn ~:** run at a loss/profit; **b)** *(sich zu schaffen machen)* busy oneself. **2.** *tr. V.* **eine Firma in den Ruin ~:** ruin a company

wirtschaftlich 1. *Adj.* **a)** *nicht präd.* economic; **b)** *nicht präd. (finanziell)* financial; **c)** *(sparsam, rentabel)* economical. **2.** *adv.; s. Adj.:* economically; financially

Wirtschaftlichkeit die; ~: economic viability

wirtschafts-, Wirtschafts-: ~beziehungen *Pl.* economic relations; **~gebäude** *Pl.* domestic offices; **~geld das;** *o. Pl. s.* **Haushaltsgeld; ~gemeinschaft die** economic community; **~hilfe die** economic aid *no indef. art.*; **~krise die** economic crisis; **~lage die** economic situation; **~minister der** minister for economic affairs; **~ministerium das** ministry of economic affairs; **~politik die** economic policy; **~system das** economic system; **~union die** economic union; **~wissenschaft die;** *meist Pl.* economics *sing., no art.*; economic science *no art.*; **~wissenschaftler der**

economist; **~wunder das** *(ugs.)* economic miracle; **~zeitung die** financial newspaper; **~zweig der** economic sector

Wirts: ~haus das pub *(Brit. coll.); (mit Unterkunft)* inn; pub *(Brit. coll.);* **~leute** *Pl.* landlord and landlady; **~stube die** bar

Wisch ['vɪʃ] **der; ~[e]s, ~e** *(salopp abwertend)* piece *or* bit of paper

wischen *itr., tr. V.* wipe; **etw. von etw. ~:** wipe sth. off *or* from sth.; **Staub ~:** do the dusting; dust

Wischer der; ~s, ~: wiper

Wischiwaschi [vɪʃi'vaʃi] **das; ~s** *(salopp abwertend)* wish-wash

Wisch-: ~lappen der, ~tuch das; *Pl.* **~tücher** cloth

Wisent ['vi:zɛnt] **der; ~s, ~e** wisent; aurochs

Wismut ['vɪsmu:t] **das; ~[e]s** bismuth

wispern ['vɪspɐn] *itr., tr. V.* whisper

wiß-, Wiß-: ~begier, ~begierde die; *o. Pl.* thirst for knowledge; **~begierig** *Adj.* eager for knowledge; ‹*child*› eager to learn

wissen ['vɪsn̩] **1.** *unr. tr., auch itr. V.* know; **ich weiß [es]** I know; **ich weiß [es] nicht** I don't know; **etw. genau ~:** know sth. for certain; **soviel ich weiß** as far as I know; **ich weiß ein gutes Lokal** I know [of] a good pub *(Brit. coll.);* **er weiß immer alles besser** he always knows better; **nicht, daß ich wüßte** not so far as I know; not that I know of; **woher soll ich das ~?** how should I know?; **weißt du was, wir fahren einfach hin** I'll tell you what, let's just go there; **jmdn. etw. ~ lassen** let sb. know sth.; **was weiß ich** *(ugs.)* I don't know; **man kann nie ~** *(ugs.)* you never know; **gewußt, wie!** *(ugs.)* it's easy when you know how; **von jmdm./etw. nichts [mehr] ~ wollen** want to have nothing [more] to do with sb./sth.; **er tut, als sei es wer weiß wie wichtig** *(ugs.)* he behaves as if it were incredibly important *(coll.);* **ich weiß ihren Namen nicht mehr** I can't remember her name; **weißt du noch, wie arm wir damals waren?** do you remember how poor we were then?; *s. auch* **Gott a, b. 2.** *unr. itr. V.* **von etw./um etw. ~:** know about sth.; **ich weiß von nichts** I don't know anything about it. **3.** *unr. mod. V.* **etw. zu tun ~:** know how to do sth.; **er wußte zu berichten, daß ...:** he was able to report that ...

Wissen das; ~s knowledge; **meines ~s:** to my knowledge; **wider** *od.* **gegen besseres ~:** against one's better judgement; **nach bestem ~ und Gewissen** to the best of one's knowledge and belief

Wissenschaft die; ~, ~en science; **die ~:** science; **etw. ist eine ~ für sich** *(ugs.)* there's a real art to sth.

Wissenschaftler der; ~s ~, Wissenschaftlerin die; ~, ~nen academic; *(Natur~)* scientist

wissenschaftlich 1. *Adj.* scholarly; *(natur~)* scientific; **~er Assistent** ≈ assistant lecturer. **2.** *adv.* in a scholarly manner; *(natur~)* scientifically

wissens-, Wissens-: ~durst der; *o. Pl.* thirst for knowledge; **~gebiet das** area *or* field of knowledge; **~wert** *Adj.* **~wert sein** be worth knowing; **viel Wissenswertes** a great deal of valuable and interesting information

wissentlich ['vɪsn̩tlɪç] **1.** *Adj.; nicht präd.* deliberate. **2.** *adv.* knowingly; deliberately

witschen ['vɪtʃn̩] *itr. V.; mit sein (ugs.)* slip

wittern ['vɪtɐn] *tr. V.* get wind of; scent; *(fig.: ahnen)* sense

Witterung die; ~, ~en a) weather *no indef. art;* **b)** *(Jägerspr.) (Geruchssinn)* sense of smell; *(Geruch)* scent

Witwe ['vɪtvə] **die; ~, ~n** widow; **~ werden** be widowed

Witwer der; ~s, ~: widower; **~ werden** be widowed

Witz [vɪts] **der; ~es, ~e a)** joke; **ich mache keine ~e** I'm not joking; **das soll wohl ein ~ sein** you/he *etc.* must be joking; **b)** *o. Pl. (Geist)* wit; **mit ~:** wittily

Witz·bold [~bɔlt] **der; ~es, ~e** joker; *(der jmdm. einen Streich spielt)* practical joker; prankster

Witzelei die; ~, ~en a) *o. Pl.* teasing; **b)** *(Bemerkung)* joke

witzeln ['vɪtsl̩n] *itr. V.* joke

Witz·figur die *(ugs. abwertend)* figure of fun

witzig 1. *Adj.* **a)** *(spaßig)* funny; amusing; **b)** *(ugs.: seltsam)* funny; odd; **c)** *(einfallsreich)* imaginative. **2.** *adv.; s. Adj.:* amusingly; oddly; imaginatively

witz·los 1. *Adj.* **a)** dull; **b)** *(ugs.: sinnlos)* pointless. **2.** *adv. (ohne Witz)* unimaginatively

wo [vo:] **1.** *Adv.* **a)** *(interrogativ)* where; **wo gibt's denn so was!** *(ugs.)* who ever heard of such a thing!; **b)** *(relativisch)* where; *(temporal)* when; **überall, wo** wherever; **wo immer er auch sein mag** wherever he may be. **2.** *Konj.* **a)** *(da, weil)* seeing that; **b)** *(obwohl)* although; when; **c)** *(falls)* **wo möglich** if possible

wo·anders *Adv.* somewhere else; elsewhere

wob [vo:p] *1. u. 3. Pers. Sg. Prät. v.* **weben**

wo·bei *Adv.* **a)** *(interrogativ)* **~ hast du sie ertappt?** what did you catch her doing?; **~ ist es kaputtgegangen?** how did it get broken?; **b)** *(relativisch)* **er gab sechs Schüsse ab, ~ einer der Täter getötet wurde** he fired six shots – one of the criminals was killed

Woche ['vɔxə] **die; ~, ~n** week; **in dieser/der nächsten/der letzten ~:** this/next/last week; **heute in/vor einer ~:** a week today/a week ago today; **zweimal die** *od.* **in der ~:** twice a week; **unter der ~** *(landsch.)* during the week

wochen-, Wochen-: ~bett das: im ~bett liegen be lying in; **~blatt das** weekly; **~ende das** weekend; **schönes ~ende!** have a nice weekend!; **~end·haus das** weekend house; **~karte die** weekly season ticket; **~lang 1.** *Adj.; nicht präd.* lasting weeks *postpos;* **2.** *adv.* for weeks [on end]; **~lohn der** weekly wages *pl.;* **~markt der** weekly market; **~tag der: welcher ~tag ist heute?** what day of the week is it?; **~tags** *Adv.* on weekdays [and Saturdays]

wöchentlich ['vœçn̩tlɪç] **1.** *Adj.* weekly. **2.** *adv.* weekly; **~ einmal** once a week

-wöchentlich 1. *Adj.* -weekly. **2.** *adv.* every ... weeks; *s. auch* **achtwöchentlich**

Wochen·zeitung die weekly newspaper

-wöchig [-vœçɪç] **a)** *(... Wochen alt)* ... -week-old; **ein achtwöchiges Kind** an eight-week-old baby; **b)** *(... Wochen dauernd)* ... -week; ... week's/weeks'; **eine vierwöchige Kur** a four-week course of treatment; **mit dreiwöchiger Verspätung** three weeks late

Wöchnerin ['vœçnərɪn] **die; ~, ~nen** woman who has just given birth

Wodka ['vɔtka] **der; ~s, ~s** vodka

wo·durch *Adv.* **a)** *(interrogativ)* how; **~ unterscheidet sie sich von den anderen?** in what way is she different from the others?; **b)** *(relativisch)* as a result of which; **alles, ~ er sich verletzt fühlen könnte** anything that might offend him

wo·für *Adv.* **a)** *(interrogativ)* for what; ~ **brauchst du es?** what do you need it for?; ~ **hältst du mich?** what do you take me for?; **b)** *(relativisch)* for which

wog [vo:k] *1. u. 3. Pers. Sg. Prät. v.* **wiegen**

Woge ['vo:gə] **die;** ~, ~**n** *(auch fig.)* wave

wo·gegen 1. *Adv.* **a)** *(interrogativ)* against what; what ... against; ~ **ist sie allergisch?** what is she allergic to?; **b)** *(relativisch)* against which; which ... against. **2.** *Konj.* whereas

wogen ['vo:gn̩] *itr. V. (geh.)* ⟨*sea*⟩ surge; *(fig.)* ⟨*corn*⟩ wave; ⟨*crowd*⟩ surge

wo·her *Adv.* **a)** *(interrogativ)* where ... from; ~ **weißt du das?** how do you know that?; ~ **kennst du ihn?** where do you know him from?; **[ach]** ~ **denn!, ach** ~! *(ugs.)* good heavens, no!; not at all!; **b)** *(relativisch)* where ... from

wohin *Adv.* **a)** *(interrogativ)* where [... to]; ~ **damit?** *(ugs.)* where shall I put it/them?; **b)** *(relativisch)* where; **c)** *(indefinit)* **ich muß mal** ~ *(ugs. verhüll.)* I've got to pay a visit *or* a call *(euphem.)*

wo·hinein *Adv. s.* **worein**

wo·hingegen *Konj.* whereas

wo·hinter *Adv.* **a)** *(interrogativ)* behind what; what ... behind; **b)** *(relativisch)* behind which

wohl [vo:l] **1.** *Adv.* **a)** *(gesund)* well; **jmdm. ist nicht** ~, **jmd. fühlt sich nicht** ~: sb. does not feel well; **b)** *(behaglich)* at ease; happy; **mir ist nicht recht** ~ **bei der Sache** the whole thing makes me a bit uneasy; **leb** ~**!/leben Sie** ~! farewell!; ~ **oder übel** whether I/you *etc.* want to or not; **c)** *(durchaus)* well; **ich bin mir dessen** ~ **bewußt** I'm quite *or* perfectly conscious of that; ~**wissend, daß ...** *(geh.)* knowing full well that ...; **d)** *(ungefähr)* about; **e) sehr** ~**[, der** *od.* **mein Herr]** certainly[, sir]; very good [, sir]; **f)** *(jedoch)* **...** , ~ **aber ...:** but ...; however...; **g)** ~ **dem, der ...!** *(geh. veralt.)* happy the man who ...; **h)** *s.* **zwar a. 2.** *Partikel* **a)** *(vermutlich)* probably; ~ **kaum** hardly; **du bist** ~ **nicht recht bei Verstand?** have you taken leave of your senses?; **na** *od.* **ja, was/warum/wie** ~? need you ask what/why/how?; **das mag** ~ **sein** that may well be; **das wird** ~ **so sein** that's probably the case; **ich habe** ~ **nicht recht gehört** I don't think I could have heard right; **b)** *(verstärkend)* **wirst du** ~ **herkommen!** will you come here!; **siehst du** ~! there, you see!; **man wird doch** ~ **fragen dürfen** there's nothing wrong in asking, is there?

Wohl das; ~**[e]s** welfare; well-being; **das allgemeine/öffentliche** ~: the public good; **auf jmds.** ~ **trinken** drink sb.'s health; **[auf] dein** ~! your health!; **zum** ~! cheers!

wohl-, Wohl-: ~**auf** [-'-] *Adj., nicht attr. (geh.)* ~**auf sein** be well *or* in good health; ~**bedacht** *(geh.)* **1.** *Adj.* well-considered; [carefully] considered ⟨*reply, judgement*⟩; **2.** *adv.* in a carefully considered way; with careful consideration; ~**befinden das** well-being; ~**begründet** *(geh.) Adj.* well-founded; *(berechtigt)* well-justified; ~**behagen das** sense of well-being; **etw. mit** ~**behagen essen** eat sth. with relish; ~**behalten** *Adj.* safe and well ⟨*person*⟩; undamaged ⟨*thing*⟩; ~**bekannt** *Adj.* well-known; ~**durchdacht** *Adj.* carefully thought-out; ~**ergehen das** *s.* ~**befinden;** ~**erzogen** *Adj.* well brought-up; ~**fahrt die;** *o. Pl.* **a)** *(geh.)* welfare; **b)** *(öffentliche Fürsorge)* welfare services *pl.*

Wohlfahrts·staat der welfare state

wohl-, Wohl-: ~**gefallen das** pleasure; ~**gefällig 1.** *Adj.* ⟨*smile, look*⟩ of pleasure; **2.** *adv.* with pleasure; ~**geformt** *Adj.* well-formed; ~**gelitten** *Adj. (geh.)* well-liked; ~**gemerkt** *Adv.* please note; mark you; ~**genährt** *Adj. (meist spött.)* well-fed; ~**geraten** *Adj. (geh.)* fine *attrib.* ⟨*child*⟩; successful ⟨*piece of work, translation, etc.*⟩; ~**gesinnt** *Adj.* well-disposed; **jmdm./einer Sache** ~**gesinnt sein** be well-disposed towards sb./sth.; ~**habend** *Adj.* prosperous

wohlig 1. *Adj.; nicht präd.* pleasant; agreeable; *(gemütlich)* cosy. **2.** *adv.* ⟨*sigh, purr, etc.*⟩ with pleasure; ⟨*stretch oneself*⟩ luxuriously

wohl, Wohl-: ~**meinend** *Adj. (geh.)* well-meaning; ~**proportioniert** *Adj. (geh.)* well-proportioned; ~**riechend** *Adj. (geh.)* fragrant; ~**schmeckend** *Adj. (geh.)* delicious; ~**sein das:** **[zum]** ~**sein!** your health!; ~**stand der;** *o. Pl.* prosperity

Wohlstands·gesellschaft die; *o. Pl. (abwertend)* affluent society

wohl, Wohl-: ~**tat die a)** favour; **jmdm. eine** ~**tat erweisen** do sb. a good turn; **b)** *(etw., was Erleichterung bringt)* blissful relief; ~**täter der** benefactor; ~**tätig** *Adj.* charitable; ~**tätigkeit die;** *o. Pl.* charity; charitableness

Wohltätigkeits- charity ⟨*event, concert, etc.*⟩

wohl-, Wohl-: ~**tuend 1.** *Adj.* agreeable; **2.** *adv.* agreeably; ~|**tun** *unr. itr. V.* **etw. tut jmdm.** ~: sth. does sb. good; ~**überlegt 1.** *Adj.* well-considered; **2.** *adv.* in a carefully considered way; with careful consideration; ~**verdient** *Adj.* well-earned, well-deserved ⟨*reward, honour, success, etc.*⟩; well-deserved ⟨*punishment, fate*⟩; ~**verhalten das** good behaviour *no indef. art.;* ~**weislich** [~vaislɪç] *Adv.* deliberately; ~|**wollen** *unr. itr. V.* **jmdm.** ~**wollen** wish sb. well; ~**wollen das;** ~**s** goodwill; **jmdn. mit** ~**wollen betrachten** regard sb. benevolently; ~**wollend 1.** *Adj.* benevolent; favourable ⟨*judgement, opinion*⟩; **2.** *adv.* benevolently; ⟨*judge, consider*⟩ favourably

Wohn-: ~**anhänger der** caravan; trailer *(Amer.);* ~**block der;** *Pl.* ~**s,** *(schweiz.)* ~**blöcke** residential block

wohnen *itr. V.* live; *(kurzfristig)* stay; **sie** ~ **sehr hübsch** they have a lovely home; *(der Lage nach)* they live in a lovely spot; **wo** ~ **Sie?** where do you live/where are you staying?

wohn-, Wohn-: ~**gebiet das,** ~**gegend die** residential area; ~**gemeinschaft die** group sharing a flat *(Brit.) or (Amer.)* apartment/house; **in einer** ~**gemeinschaft leben** live in a shared flat *(Brit.) or (Amer.)* apartment/house; share a flat *(Brit.) or (Amer.)* apartment/house; ~**haft** *Adj.* resident; ~**haus das** [dwelling-]house; ~**heim das** home; *(für Obdachlose, Lehrlinge)* hostel; *(für Studenten)* hall of residence; ~**lage die:** **unsere** ~**lage ist optimal** our house/flat *(Brit.) or (Amer.)* apartment is ideally situated; **in ruhiger/guter** ~**lage** in a quiet/good area

wohnlich 1. *Adj.* homely. **2.** *adv.* in a homely way

Wohnlichkeit die; ~: homeliness

Wohn-: ~**mobil das;** ~**s,** ~**e** motor home; motor caravan; ~**ort der;** *Pl.* ~**e** place of residence; ~**raum der a)** living-room; **b)** *o. Pl.* *(~fläche)* living space; ~**siedlung die** residential estate; *(mit gleichartigen Häusern)* housing estate; **~sitz der** place of residence; domicile *(formal);* **ohne festen ~sitz** of no fixed abode

Wohnung die; ~, ~en a) flat *(Brit.);* apartment *(Amer.);* **b)** *o. Pl. (Unterkunft)* lodging

Wohnungs-: ~schlüssel der key to the flat *(Brit.) or (Amer.)* apartment; **~suche die** search for a flat *(Brit.) or (Amer.)* apartment; **auf ~suche sein** be flat-hunting; **~tür die** door of the flat *(Brit.) or (Amer.)* apartment

Wohn-: ~verhältnisse *Pl.* living conditions; **~viertel das** residential district; **~wagen der** caravan; trailer *(Amer.);* **~zimmer das** living-room

wölben ['vœlbn̩] **1.** *tr. V.* curve; arch ⟨*brows, shoulders*⟩; cup ⟨*hand*⟩; bend ⟨*metal*⟩; vault, arch ⟨*roof, ceiling*⟩. **2.** *refl. V.* curve; ⟨*sky, bridge, ceiling*⟩ arch; ⟨*chest*⟩ swell; ⟨*stomach, muscles*⟩ bulge; ⟨*metal*⟩ bend, buckle

Wölbung die; ~, ~en curve; *(einer Decke, des Himmels)* arch; vault; *(von Augenbrauen)* arch; *(eines Bauches, Muskels)* bulge

Wolf [vɔlf] **der; ~[e]s, Wölfe** ['vœlfə] **a)** wolf; **ein ~ im Schafspelz sein** *(fig.)* be a wolf in sheep's clothing; **mit den Wölfen heulen** *(fig. ugs.)* run with the pack; **b)** *(ugs.: Fleisch~)* mincer

Wölfin ['vœlfɪn] **die; ~, ~nen** [wolf] bitch

Wolfram ['vɔlfram] **das; ~s** *(Chemie)* tungsten

Wölkchen ['vœlkçən] **das; ~s, ~:** small cloud

Wolke ['vɔlkə] **die; ~, ~n** cloud; **aus allen ~n fallen** *(fig. ugs.)* be completely stunned

wolken-, Wolken-: ~bruch der; *Pl.* **~brüche** cloudburst; **~bruch·artig** *Adj.* torrential; **~decke die** [unbroken] cloud *no indef. art.;* **~kratzer der** skyscraper; **~los** *Adj.* cloudless

wolkig *Adj.* cloudy

Woll·decke die [woollen] blanket

Wolle ['vɔlə] **die; ~, ~n** wool; *(fig.: Haar)* hair; **sich in die ~ kriegen** *(fig. ugs.)* quarrel **(wegen** over); **sich in der ~ haben** *(fig. ugs.)* be at loggerheads

¹wollen *Adj.; nicht präd.* woollen

²wollen 1. *unr. Modalverb; 2. Part.* **~ a) etw. tun ~** *(den Wunsch haben, etw. zu tun)* want to do sth.; *(die Absicht haben, etw. zu tun)* be going to do sth.; **wir wollten gerade gehen** we were just about to go; **was will man da machen?** *(ugs.)* what can you do?; **ohne es zu ~:** without intending to; **dann will ich nichts gesagt haben** *(ugs.)* I take it all back; **das will ich meinen!** *(ugs.)* I absolutely agree; **wir ~ sehen** we'll see; **b)** *(in Aufforderungen)* **wollt ihr wohl Ruhe geben/damit aufhören!** *(ugs.)* will you be quiet/stop that!; **~ Sie bitte so freundlich sein und das heute noch erledigen** would you be so kind as to do it today; **c) er will ein Dichter sein** he claims to be a poet; **sie will es [nicht] gesehen haben** she claims [not] to have seen it; **d) die Wunde will [einfach] nicht heilen** the wound [just] won't heal; **der Motor wollte nicht anspringen** the engine wouldn't start; **es will [einfach] nicht gelingen** it just won't work; **e) etw. will getan sein** sth. needs *or (coll.)* has got to be done; **das will gelernt sein** it has to be learned; **f) das Buch will unterhalten** the book is intended *or* meant to entertain; **g) das will nichts heißen/nicht viel sagen** that doesn't mean anything/much. **2.** *unr. itr. V.* **a) du mußt nur ~, dann geht es auch** you only have to want to enough *or* have the will, then it's possible; **ob du willst oder nicht** whether you want to or not; **[ganz] wie du willst** just as you like; **wenn du willst, könnten wir ...:** if you want [to], we could ...; **das ist, wenn man so will, ...:** that is, if you like, ...; **[na] dann ~ wir mal!** *(ugs.)* [right,] let's get started!; **b)** *(ugs.: irgendwohin zu gehen wünschen)* **ich will nach Hause/ans Meer** I want to go home/to go to the seaside; **ich will hier raus** I want to get out of here; **zu wem ~ Sie?** whom do you want to see?; **er wollte zum Theater** he wanted to become an actor; **c)** *verneint (ugs.: funktionieren)* **der Motor will nicht** the engine won't go; **seine Beine/Gelenke/Augen ~ nicht mehr** his legs/joints/eyes just aren't up to it any more. **3.** *unr. tr. V.* **a)** want; **das wollte ich nicht** I didn't mean to do that; **das habe ich nicht gewollt** I never meant that to happen; **~, daß jmd. etw. tut** want sb. to do sth.; **da ist nichts [mehr] zu ~** *(ugs.)* there's nothing we/you *etc.* can do about it; **ich wollte, er wäre hier/es wäre vorbei** I wish he were here/it were over; **b) jmdm. nichts ~ können** *(ugs.)* be unable to harm sb.

Woll·gras das cotton grass

wollig *Adj.* woolly

Woll-: ~jacke die woollen cardigan; **~kleid das** woollen dress; **~knäuel das** ball of wool; **~sachen** *Pl.* woollen things; woollies *(coll.);* **~socke die** woollen sock; **~stoff der** woollen cloth; **~strumpf der** woollen stocking; *(Kniestrumpf)* woollen sock

Wollust ['vɔlʊst] **die; ~, Wollüste** ['vɔlʏstə] *(geh.)* lust; *(Sinnlichkeit)* sensuality; **etw. mit wahrer ~ tun** take great delight in doing sth.

wollüstig ['vɔlʏstɪç] *(geh.)* **1.** *Adj.* lustful; *(sinnlich)* sensual. **2.** *Adj.* lustfully; *(sinnlich)* sensually

wo·mit *Adv.* **a)** *(interrogativ)* **~ schreibst du?** what do you write with?; *(more formal)* with what do you write?; **b)** *(relativisch)* **~ du schreibst** which *or* that you write with; *(more formal)* with which you write; **~ ich nicht sagen will, daß ...:** by which I don't mean to say that ...

wo·möglich *Adv.* possibly

wo·nach *Adv.* **a)** *(interrogativ)* after what; what ... after; **~ suchst du?** what are you looking for?; **~ riecht es?** what does it smell of?; **~ richtet ihr euch?** what do you go by?; **b)** *(relativisch)* after which; which ... after

Wonne ['vɔnə] **die; ~, ~n** *(geh.)* bliss *no pl.;* ecstasy; *(etw., was Freude macht)* joy; delight; **es war eine ~, ihr zuzuhören** she was a joy *or* delight to listen to

wonnig *Adj.* sweet

woran [vo'ran] *Adv.* **a)** *(interrogativ)* **~ hast du dich verletzt?** what did you hurt yourself on?; **man weiß nicht, ~ man ist** you don't know where you are; **~ ist sie gestorben?** what did she die of?; **~ denkst du?** what are you thinking of?; **b)** *(relativisch)* **nichts, ~ man sich verletzen/anlehnen könnte** nothing one could hurt oneself on/one could lean against

worauf [vo'rau̯f] **a)** *(interrogativ)* **~ sitzt er?** what is he sitting on?; **~ wartest du?** what are you waiting for?; **~ will er hinaus?** what is he getting at?; **b)** *(relativisch)* **es gab nichts, ~ er sich hätte setzen können** there was nothing for him to sit on; **das einzige, ~ es jetzt ankommt** the only thing that matters now; **c)** *(relativisch: woraufhin)* whereupon

worauf·hin *Adv.* **a)** *(interrogativ)* **~ hat er das getan?** what made him do it?; what was the cause of his doing it?; **b)** *(relativisch)* whereupon

woraus [vo'raus] *Adv.* **a)** *(interrogativ)* ~ **trinken wir den Wein?** what shall we drink the wine from?; ~ **ist das Gewebe?** what is the fabric made of?; ~ **schließt du das?** what do you infer that from?; **b)** *(relativisch)* **es gab nichts, ~ wir den Wein hätten trinken können** there was nothing for us to drink the wine out of; **es gab nichts, ~ sie Werkzeuge machen konnten** there was nothing for them to make tools from

worden ['vɔrdn̩] *2. Part. v.* **werden 2c**

worein [vo'rain] *Adv.* **a)** *(interrogativ)* in what; what ... in; **b)** *(relativisch)* in which; which ... in

worin [vo'rɪn] *Adv.* **a)** *(interrogativ)* in what; what ... in; **ich weiß nicht, ~ der Unterschied liegt** I don't know what the difference is; **b)** *(relativisch)* in which; which ... in

Work·shop ['wə:kʃɔp] **der; ~s, ~s** workshop

Wort [vɔrt] **das; ~[e]s, Wörter** ['vœrtɐ] *od.* **~e a)** *Pl.* **Wörter,** *(auch:)* **~e** word; ~ **für ~:** word for word; **b)** *Pl.* *~e (Äußerung)* word; **mir fehlen die ~e** I'm lost for words; **davon ist kein ~ wahr** not a word of it is true; **nicht viele ~e machen** not beat about the bush; **ich verstehe kein ~:** I don't understand a word [of it]; **auf jmds. ~e hören** listen to what sb. says; **mit einem ~:** in a word; **mit anderen ~en** in other words; **ich glaube dir aufs ~:** I can well believe it; **jmdn. [nicht] zu ~ kommen lassen** [not] let sb. get a word in; **etw. mit keinem ~ erwähnen** not say a word about sth.; not mention sth. at all; **man verstand sein eigenes ~ nicht** you could not hear yourself speak; **jmdm. aufs ~ gehorchen** obey sb.'s every word; **ein ~ gab das andere** one thing led to another; **hast du [da noch] ~e?** what do you say to that?; **das ist das letzte/mein letztes ~:** that's the/my last word on the matter; **[immer] das letzte ~ haben wollen/müssen** want to have/have to have the last word; **Dr. Meyer hat das ~:** it's Dr Meyer's turn to speak; **das ~ ergreifen** *od.* **nehmen** start to speak; **jmdm. das ~ geben** *od.* **erteilen/entziehen** call upon sb. to speak/to finish speaking; **für jmdn. ein [gutes] ~ einlegen** put in a [good] word for sb.; **jmdm. das ~ aus dem Munde nehmen** take the words out of sb.'s mouth; **jmdm. das ~ im Munde herumdrehen** twist sb.'s words; **kein weiteres ~ über etw.** *(Akk.)* **verlieren** not say another word about sth.; **jmdm. ins ~ fallen** interrupt sb.; **ums ~ bitten** ask to speak; **sich zu ~ melden** indicate one's wish to speak; **c)** *Pl.* **~e** *(Spruch)* saying; *(Zitat)* quotation; **geflügelte ~e** well-known sayings and quotations; **d)** *Pl.* **~e** *(geh.: Text)* words *pl.*; **in ~ und Bild** in words and pictures; **e)** *Pl.* **~e** *(Versprechen)* word; **[sein] ~ halten** keep one's word; **sein ~ brechen** break one's word; **jmdm. sein ~ [auf etw.** *(Akk.)***] geben** give sb. one's word [on sth.]; **auf mein ~!** I give you my word; **jmdn. beim ~ nehmen** take sb. at his/her word

Wort·art die *(Sprachw.)* part of speech

wort·brüchig *Adj.* ~ **werden** break one's word

Wörter·buch das dictionary

wort-, Wort-: ~führer der/~führerin die spokesman/spokeswoman; spokesperson; **~getreu 1.** *Adj.* word-for-word; **2.** *adv.* word for word; **~karg 1.** *Adj.* taciturn ⟨*person*⟩; laconic ⟨*reply, greeting, etc.*⟩; **ein ~karger Mann** a man of few words; **2.** *adv.* taciturnly; ⟨*reply, greet, etc.*⟩ laconically; **~klauberei** [~klaubə'rai] **die; ~, ~en** quibbling; **~laut der;** *o. Pl.* wording; **im [vollen] ~laut** verbatim

wörtlich ['vœrtlɪç] **1.** *Adj.* **a)** word-for-word; *s. auch* **Rede e; b)** *(der eigentlichen Bedeutung entsprechend)* literal. **2.** *adv.* **a)** word for word; **das hat sie ~ gesagt** those were her very words; **b)** *(der eigentlichen Bedeutung entsprechend)* literally

wort-, Wort-: ~los 1. *Adj.* silent; wordless; unspoken ⟨*agreement, understanding*⟩; **2.** *adv.* without saying a word; **~meldung die: es liegen keine weiteren ~meldungen vor** no one else wishes to speak; **~schatz der** vocabulary; **~spiel das** play on words; *(mit gleich od. ähnlich lautenden Wörtern)* pun; play on words; **~stellung die** *(Sprachw.)* word order; **~wahl die;** *o. Pl.* choice of word; **~wechsel der** exchange of words; **~wörtlich 1.** *Adj.* word-for-word; **2.** *adv.: s.* **wörtlich 2**

worüber *Adv.* **a)** *(interrogativ)* over what ...; what ... over; ~ **bist du gestolpert?** what did you trip over?; ~ **lachst du?** what are you laughing about?; **b)** *(relativisch)* over which; which ... over

worum *Adv.* **a)** *(interrogativ)* around what; what ... around; ~ **geht es denn?** what is it about then?; **b)** *(relativisch)* around which; which ... around

worunter *Adv.* **a)** *(interrogativ)* under what; what ... under; ~ **leidet er?** what is he suffering from?; **b)** *(relativisch)* under which; which ... under

wo·von *Adv.* **a)** *(interrogativ)* from where; where ... from; ~ **soll er leben?** what is he supposed to live on?; ~ **redest du?** what are you talking about?; ~ **ist er müde/krank?** what has made him tired/ill?; **b)** *(relativisch)* from which; which ... from

wo·vor *Adv.* **a)** *(interrogativ)* in front of what; what ... in front of; ~ **hast du Angst?** what are you afraid of?; **b)** *(relativisch)* in front of which; which ... in front of; **das einzige, ~ ich Angst habe** the only thing I am afraid of

wo·zu *Adv.* **a)** *(interrogativ)* to what; what ... to; *(wofür)* what ... for; ~ **brauchst du das Geld?** what do you need the money for?; ~ **hast du dich entschlossen?** what have you decided [on]?; ~ **diese Umstände?** why all this fuss?; ~ **denn?** what for?; *(als Ausdruck der Ablehnung)* why should I/you *etc.?*; **b)** *(relativisch)* **dann habe ich gebügelt, ~ ich keine Lust hatte** then I did some ironing, which I had no inclination to do; ~ **du dich auch entschließt** whatever you decide on

Wrack [vrak] **das; ~[e]s, ~s** *od.* **~e** *(auch fig.)* wreck

wrang [vraŋ] *1. und 3. Pers. Sg. Prät. v.* **wringen**

wringen ['vrɪŋən] *unr. tr. V. (bes. nordd.)* wring

Wucher ['vu:xɐ] **der; ~s** profiteering; *(mit Zinsen)* usury; **[mit etw.] ~ treiben** profiteer [on sth.]; *(mit Zinsen)* charge an extortionate rate/extortionate rates of interest [on sth.]

Wucherer der; ~s, ~: profiteer; *(beim Verleihen von Geld)* usurer

wuchern *itr. V.* **a)** *auch mit sein* *(stark wachsen)* ⟨*plants, weeds, etc.*⟩ proliferate, run wild; *(fig.)* be rampant; **krebsartig** *~(fig.)* grow like a cancer; **b)** *(Wucher treiben)* **[mit etw.] ~:** profiteer [on sth.]; *(mit Zinsen)* lend [sth.] at extortionate interest rates

Wucher·preis der extortionate price

Wucherung die; ~, ~en growth

wuchs [vu:ks] *1. u. 3. Pers. Sg. Prät. v.* **wachsen**

Wuchs der; ~es *(Gestalt)* stature; **klein/groß von ~ sein** ⟨*person*⟩ be small/tall in stature
Wucht [vʊxt] **die; ~ a)** force; *(von Schlägen)* power; weight; **mit voller ~:** with full force; ⟨*hit*⟩ with all one's might; **b) eine ~ sein** *(salopp)* be absolutely fantastic *(coll.)*
wuchten *tr. V.* heave
wuchtig *Adj.* massive
wühlen ['vy:lən] **1.** *itr. V.* **a)** dig; *(mit der Schnauze)* root **(nach** for); ⟨*mole*⟩ tunnel, burrow; **b)** *(ugs.: suchen)* rummage [around] **(nach** for). **2.** *tr. V.* burrow; tunnel out ⟨*burrow*⟩. **3.** *refl. V.* **sich in etw.** *(Akk.)***/durch etw. ~:** burrow into/through sth.
Wulst [vʊlst] **der; ~[e]s, Wülste** ['vʏlstə] *od.* **~e** bulge; *(Fett~)* roll of fat; *(an Flasche, Reifen)* bead
wulstig *Adj.* bulging; thick ⟨*lips*⟩
wund [vʊnt] *Adj.* sore; **sich ~ laufen** walk until one's feet are sore; *s. auch* **Fuß b; Punkt d**
Wunde die; ~, ~n wound; **der Krieg hat dem Land tiefe ~n geschlagen** *(fig.)* the war has left deep scars on the country
wunder *Adv. (ugs.)* **er denkt, er sei ~ wer** he thinks he's really something; **er glaubt, ~ was geleistet zu haben** he thinks he's achieved something fantastic *(coll.)*; **er bildet sich ~ was darauf ein** he's terribly pleased with himself about it *(coll.)*
Wunder das; ~s, ~ a) miracle; **~ wirken** *(fig. ugs.)* work wonders; **ein/kein ~ sein** *(ugs.)* be a/no wonder; **was ~, wenn ...?** small *or* no wonder that ...; **er wird sein blaues ~ erleben** *(ugs.)* he's in for a nasty shock; **b)** *(etw. Erstaunliches)* wonder; **ein ~ an ...** *(Dat.)* a miracle of ...; **ein technisches ~:** a technological marvel
wunderbar 1. *Adj.* **a)** *(übernatürlich erscheinend)* miraculous; **auf ~e Weise** miraculously; **b)** *(herrlich)* wonderful; marvellous. **2.** *adv. (herrlich)* wonderfully; marvellously; **b)** *(ugs.: sehr schön)* wonderfully ⟨*cosy, warm, etc.*⟩
wunder-, Wunder-: ~glaube der belief in miracles; **~heiler der** faith-healer; **~hübsch 1.** *Adj.* wonderfully pretty; **2.** *adv.* quite beautifully; **~kerze die** sparkler; **~kind das** child prodigy
wunderlich 1. *Adj.* strange; odd. **2.** *adv.* strangely; oddly
Wunder·mittel das miracle cure
wundern 1. *tr. V.* surprise; **mich wundert, daß ...:** I'm surprised that ...; **es würde** *od.* **sollte mich [nicht] ~, wenn ...:** I should [not] be surprised *or* it would [not] surprise me if ... **2.** *refl. V.* be surprised **(über + *Akk.*** at); **du wirst dich [noch mal] ~** *(ugs.)* you're in for a shock; you've got a surprise in store
wunder-, Wunder- : ~schön 1. *Adj.* simply beautiful; *(herrlich)* simply wonderful; **2.** *adv.* quite beautifully; *(einwandfrei)* perfectly; **~tätig** *Adj.* miraculous; **~tüte die** surprise packet; **~voll 1.** *Adj.* wonderful; marvellous; **2.** *adv.* wonderfully; marvellously; **~werk das** marvel
wund-, Wund-: ~|liegen *unr. refl. V.* get bedsores **(an + *Dat.*** on); **sich** *(Dat.)* **den Rücken ~liegen** get bedsores on one's back; **~sein das** soreness; **~starrkrampf der** *(Med.)* tetanus
Wunsch [vʊnʃ] **der; ~[e]s, Wünsche** ['vʏnʃə] **a)** wish **(nach** to have); *(Sehnen)* desire **(nach** for); **sich** *(Dat.)* **einen ~ erfüllen/versagen** grant/deny oneself something one wants; **haben Sie noch einen ~?** will there be anything else?; **auf jmds. ~:** at sb.'s wish; **alles geht nach ~:** everything's going as we want/he wants *etc.*; **b)** *Pl.* **mit den besten/herzlichsten Wünschen** with best/warmest wishes
wünschbar *Adj. (bes. schweiz.)* desirable
Wunsch·denken das wishful thinking
Wünschel·rute ['vʏnʃl̩-] **die** divining rod
wünschen ['vʏnʃn̩] *tr. V.* **a) sich** *(Dat.)* **etw. ~:** want sth.; *(im stillen)* wish for sth.; **jmdm. Erfolg/nichts Gutes ~:** wish sb. success/no good; **jmdm. den Tod ~:** wish sb. dead; **was wünschst du dir?** what would you like?; **ich wünschte, du wärest hier** I wish you were here; **jmdn. weit fort ~:** wish sb. far away; **b) jmdm. alles Gute/frohe Ostern ~:** wish sb. all the best/a happy Easter; **sie wünschte ihm gute Besserung** she said she hoped he would soon get better; **c)** *auch itr. (begehren)* want; **was ~ Sie?, Sie ~?** *(von einem Bediensteten gesagt)* yes, madam/sir?; *(von einem Kellner gesagt)* what would you like?; *(von einem Verkäufer gesagt)* can I help you?; **ganz, wie Sie ~:** just as you like; **die gewünschte Auskunft** the information asked for; **etw. läßt [viel]/läßt nichts zu ~ übrig** sth. leaves a great deal/nothing to be desired; **es verlief alles wie gewünscht** everything went as we/he *etc.* had wanted
wünschens·wert *Adj.* desirable
wunsch-, Wunsch-: ~gemäß *Adv.* as desired; *(einer Bitte gemäß)* as requested; **~kind das** wanted child; **~konzert das** request concert; *(im Rundfunk)* request programme; **~los 1.** *Adj.* [perfectly] contented; perfect ⟨*happiness*⟩; **2.** *adv.* **~los glücklich sein** be perfectly contented; **~satz der** *(Sprachw.)* optative sentence; **~traum der** wishful dream; *(unrealistisch)* pipe-dream; **~zettel der** list of presents one would like
wurde ['vʊrdə] *1. u. 3. Pers. Sg. Prät. v.* **werden**
würde ['vʏrdə] *1. u. 3. Pers. Sg. Konjunktiv II v.* **werden**
Würde die; ~, ~n a) *o. Pl.* dignity; **sich in seiner ~ verletzt fühlen** feel that one's dignity has been affronted; **unter jmds. ~ sein** be beneath sb.'s dignity; **unter aller ~ sein** be beneath contempt; **b)** *(Rang)* rank; *(Amt)* office; *(Titel)* title; *(Auszeichnung)* honour; **zu höchsten ~n gelangen** attain high office
würde·los 1. *Adj.* undignified. **2.** *adv.* in an undignified way
Würdelosigkeit die; ~: lack of dignity
Würden·träger der dignitary
würde·voll 1. *Adj.* dignified. **2.** *adv.* with dignity
würdig 1. *Adj.* **a)** dignified; **b)** *(wert)* worthy; suitable ⟨*occasion*⟩; **jmds./einer Sache [nicht] ~ sein** [not] be worthy of sb./sth. **2.** *adv.* **a)** with dignity; ⟨*dressed*⟩ in a dignified manner; **b)** *(angemessen)* worthily; ⟨*celebrate*⟩ in a/the appropriate manner
würdigen *tr. V.* **a)** *(anerkennen, beachten)* recognize; *(schätzen)* appreciate; *(lobend hervorheben)* acknowledge; **etw. zu ~ wissen** appreciate sth.; **b) jmdn. keines Blickes/keiner Antwort ~:** not deign to look at/answer sb.
Würdigung die; ~, ~en *s.* **würdigen a:** recognition; appreciation; acknowledgement
Wurf [vʊrf] **der; ~[e]s, Würfe** ['vʏrfə] **a)** throw; *(beim Kegeln)* bowl; *(gezielt aufs Tor)* shot; **b)** *o. Pl. (das Werfen)* throwing/pitching/bowling; **beim ~:** when throwing/pitching/bowling; **c)** *(Zool.)* litter
Würfel ['vʏrfl̩] **der; ~s, ~ a)** cube; **Gemüse/Fleisch in ~ schneiden**

dice vegetables/meat; **b)** *(Spiel~)* dice; die *(formal)*; **die ~ sind gefallen** *(fig.)* the die is cast

Würfel·becher der dice-cup

würfeln 1. *itr. V.* throw the dice; *(mit Würfeln spielen)* play dice; **hast du schon gewürfelt?** have you already thrown *or* had your throw?; **um etw. ~:** play dice for sth. **2.** *tr. V.* **a)** throw; **b)** *(in Würfel schneiden)* dice ‹*vegetables, meat*›

Würfel-: ~spiel das a) *(Glücksspiel)* dice; *(einzelne Partie)* game of dice; **b)** *(Brettspiel)* dice game; **~zucker der;** *o. Pl.* cube sugar; lump sugar

Wurf-: ~geschoß das missile; **~scheibe die** discus; **~sendung die** *s.* **Postwurfsendung**

würgen ['vʏrgn̩] **1.** *tr. V.* strangle; throttle; *(fig.)* ‹*tie, collar*› strangle **2.** *itr. V.* **a)** *(Brechreiz haben)* retch; **b)** *(mühsam schlucken)* **an etw.** *(Dat.)* **~:** have to force sth. down; *s. auch* [2]**hängen 1 d**

[1]**Wurm** [vʊrm] **der; ~[e]s, Würmer** ['vʏrmɐ] worm; *(Made)* maggot; **da ist der ~ drin** *(fig. ugs.)* there's something wrong there; **jmdm. die Würmer aus der Nase ziehen** *(fig. ugs.)* get sb. to spill the beans *(fig. sl.)*

[2]**Wurm das; ~[e]s, Würmer** *(fam.)* little mite

wurmen *tr., auch itr. V. (ugs.)* **jmdn. ~:** rankle with sb.

Wurm·fort·satz der *(Anat.)* appendix

wurmig *Adj.*, **wurm·stichig** *Adj.* worm-eaten; *(madig)* maggoty

Wurscht [vʊrʃt] *(ugs.)* **jmd./etw. ist jmdm. ~:** sb. doesn't care about sb./sth.; **das ist mir völlig ~:** I couldn't care less about that

wurscht·egal *Adj. (ugs.) s.* **wurstegal**

Wurst [vʊrst] **die; ~, Würste** ['vʏrstə] **a)** sausage; *(Streich~)* ≈ meat spread; **es geht um die ~** *(fig. ugs.)* the crunch has come; **b)** *(wurstähnliches Gebilde)* roll; **eine ~ machen** *(fam.)* do a big one *(child lang.)*; **c)** *(ugs.) s.* **Wurscht**

Wurst·brot das *s.* **Wurst a:** open sausage/meat-spread sandwich; *(zusammengeklappt)* sausage/meat-spread sandwich

Würstchen ['vʏrstçən] **das; ~s, ~ a)** [small] sausage; **Frankfurter/Wiener ~:** frankfurter/wienerwurst; **heiße ~:** hot sausages; **b)** *(ugs., oft abwertend)* nobody; **c) ein armes ~** *(ugs.)* a poor soul

wurst·egal *Adj. (ugs.)*: **~ sein** not matter in the slightest

wursteln *itr. V. (ugs.)* **1.** potter; **an etw.** *(Dat.)* **~:** potter about with sth. **2.** *refl. V.* **sich durchs Leben ~:** muddle [along] through life

wurstig *(ugs.)* **1.** *Adj.* couldn't-care-less *attrib.* ‹*attitude, behaviour, reply*›. **2.** *adv.* in a couldn't-care-less way

Wurstigkeit die; ~ *(ugs.)* couldn't-care-less attitude

Wurst-: ~salat der *piquant salad with pieces of sausage, onion rings, boiled eggs and/or cheese;* **~waren** *Pl.* sausages; **~zipfel der** end of a/the sausage

Würze ['vʏrtsə] **die; ~, ~n a)** *(Gewürz)* spice; seasoning; **b)** *(Aroma)* aroma; *(fig.)* spice; *s. auch* **Kürze c**

Wurzel ['vʊrtsl̩] **die; ~, ~n a)** *(auch fig.)* root; **~n schlagen** take root; *(fig.)* put down roots; **das Übel an der ~ fassen** *od.* **packen** *(fig.)* strike at the root of the problem; **b)** *(Math.)* root; **~n ziehen** calculate roots

wurzeln *itr. V. (fig.)* **das Mißtrauen wurzelt tief in ihm** his mistrust is deep-rooted; **in etw.** *(Dat.)* **~** *(seinen Ursprung haben in)* be rooted in sth.; *(verursacht sein durch)* have its roots in sth.

Wurzel·werk das; *o. Pl.* roots *pl.*

würzen ['vʏrtsn̩] *tr. V.* season; *(fig.)* spice

würzig *Adj.* tasty; full-flavoured ‹*beer, wine*›; aromatic ‹*fragrance, smell, tobacco*›; tangy ‹*air*›; *(scharf)* spicy

Würzigkeit die; ~ *s.* **würzig:** tastiness; full flavour; aromatic fragrance; tanginess; spiciness

wusch [vu:ʃ] *1. u. 3. Pers. Sg. Prät. v.* **waschen**

wuschelig *Adj. (ugs.)* frizzy; fuzzy

Wuschel·kopf der *(ugs.)* **a)** shock *or* mop of frizzy *or* fuzzy hair; **b)** *(Mensch)* frizzy-haired *or* fuzzy-haired man/girl *etc*

wuselig *Adj. (bes. südd., md.)* busy; bustling

wuseln ['vu:zl̩n] *itr. V.; mit sein (bes. südd., md.)* scurry

wußte ['vʊstə] *1. und 3. Pers. Sg. Prät. v.* **wissen**

wüßte ['vʏstə] *1. und 3. Pers. Sg. Konjunktiv II v.* **wissen**

Wust [vu:st] **der; ~[e]s** *(abwertend)* jumble; *(fig.)* welter

wüst [vy:st] **1.** *Adj.* **a)** *(öde)* desolate; **b)** *(unordentlich)* chaotic; tangled, tousled ‹*hair, beard, etc.*›; wild ‹*appearance*›; **c)** *(abwertend: wild, ungezügelt)* wild; furious ‹*fight, shoot-out*›; **d)** *(abwertend: unanständig)* rude; coarse ‹*oath, abuse*›; **e)** *(abwertend: furchtbar, abscheulich)* terrible; foul *(sl.)*, terrible *(coll.)* ‹*weather*›. **2.** *adv.* **a)** *(unordentlich)* chaotically; **das Haar hing ihr ~ ins Gesicht** her hair straggled down over her face; **b)** *(abwertend: wild, ungezügelt)* wildly; **d)** *(abwertend: furchtbar, abscheulich)* terribly

Wüste die; ~, ~n desert; *(Eis~)* waste; *(fig.)* wasteland; **jmdn. in die ~ schicken** *(fig. ugs.)* give sb. the push *(coll.)*

Wüstenei die; ~, ~en waste-land

Wüsten·sand der desert sand[s *pl.*]

Wüstling der; ~s, ~e *(abwertend)* lecher; debauchee

Wut [vu:t] **die; ~:** rage; fury; **eine ~ auf jmdn. haben** be furious with sb.; **in ~ geraten** get furious; **jmdn. in ~ bringen** infuriate sb.

Wut-: ~anfall der fit of rage; **~ausbruch der** outburst of rage *or* fury

wüten ['vy:tn̩] *itr. V. (auch fig.)* rage; *(zerstören)* wreak havoc

wütend 1. *Adj.* furious; angry ‹*voice, mob*›; **auf jmdn. ~ sein** be furious with sb.; **über etw.** *(Akk.)* **~ sein** be furious about sth. **2.** *adv.* furiously; in a fury

wut·entbrannt *Adj.* infuriated; furious; in a fury *pred.*

Wüterich ['vy:tərɪç] **der; ~s, ~e** *(abwertend)* hot-tempered person; *(Gewaltmensch)* brute

wut·schnaubend *Adj.* snorting with rage *pred.*

Wz *Abk.* **Warenzeichen** TM; ®

X

[1]**x, X** [ɪks] **das; ~, ~:** x, X; **jmdm. ein X für ein U vormachen** *(fig.)* dupe sb.; **er läßt sich** *(Dat.)* **kein X für ein U vormachen** you can't fool him; he's not easily fooled; *s. auch* **a, A**

[2]**x** *unbest. Zahlwort (ugs.)* umpteen *(coll.)*

X-Beine *Pl.* knock-knees; **~ ha-**

ben have knock-knees; be knock-kneed

x-beliebig *Adj. (ugs.)* **irgendein ~er/irgendeine ~e/irgendein ~es** any old *(coll. attrib.)*; **jeder ~e Ort** any old place *(coll.)*

x-fach *Vervielfältigungsz.* **die ~e Menge** x times the amount; *(ugs.)* umpteen times the amount *(coll.)*

X-fache das; ~n: das ~ einer Zahl X times a number; **das ~ seines Einkommens** *(ugs.)* umpteen times his income *(coll.)*

x-mal *Adv. (ugs.)* umpteen times *(coll.)*; any number of times

x-t... *Ordinalz. (ugs.)* umpteenth *(coll.)*

x-te·mal: das ~ *(ugs.) (beim x-tenmal)* the umpteenth time *(coll.)*; *(zum x-tenmal)* for the umpteenth time *(coll.)*

x-ten·mal: zum ~ *(ugs.)* for the umpteenth time *(coll.)*; **beim ~** *(ugs.)* the umpteenth time *(coll.)*

Xylophon [ksylo'fo:n] **das; ~s, ~e** xylophone

Y

y, Y ['ʏpsilɔn] **das; ~, ~:** y, Y; *s. auch* **a, A**

Yacht *s.* **Jacht**

Yankee ['jɛnki] **der; ~s, ~s** *(oft abwertend)* Yankee *(Brit. coll.)*; Yank *(Brit. coll.)*

Yoga *s.* **Joga**

Yogi[n] *s.* **Jogi[n]**

Ypsilon ['ʏpsilɔn] **das; ~[s], ~s** y, Y; *(im griechischen Alphabet)* upsilon

Z

z, Z [tsɛt] **das; ~, ~:** z, Z; *s. auch* **a, A**

zack [tsak] *Interj. (salopp)* **~! ~!** get a move on! *(coll.)*; make it snappy! *(coll.)*; **bei ihm muß alles ~, ~ gehen** he likes things done at the double

Zacke die; ~, ~n point; peak; *(einer Säge, eines Kamms)* tooth; *(einer Gabel, Harke)* prong

zacken *tr. V.* serrate

Zacken der; ~s, ~ a) *s.* **Zacke; b) sich** *(Dat.)* **keinen ~ aus der Krone brechen** *(fig. ugs.)* not lose face

zackig 1. *Adj.* **a)** jagged; **b)** *(schneidig)* dashing; smart; rousing ⟨*music*⟩; brisk ⟨*orders*⟩. **2.** *adv.* **a)** *(gezackt)* jaggedly; **b)** *(schneidig)* smartly; ⟨*play music*⟩ rousingly

zaghaft 1. *Adj.* timid; *(zögernd)* hesitant; tentative. **2.** *adv.* timidly; *(zögernd)* hesitantly; tentatively

Zaghaftigkeit die; ~: timidity; *(Zögern)* hesitancy

zäh [tsɛ:] **1.** *Adj.* **a)** *(fest)* tough; heavy ⟨*dough, soil*⟩; *(dickflüssig)* glutinous; viscous ⟨*oil*⟩; **b)** *(schleppend)* sluggish, dragging ⟨*conversation*⟩; **c)** *(widerstandsfähig)* tough ⟨*person*⟩; **d)** *(beharrlich)* tenacious; tough ⟨*negotiations*⟩; dogged ⟨*resistance*⟩. **2.** *adv.* **a)** *(schleppend)* sluggishly; **b)** *(beharrlich)* tenaciously; ⟨*resist*⟩ doggedly

Zäheit die; ~ a) *(Festigkeit)* toughness; *(des Teigs, Bodens)* heaviness; *(Dickflüssigkeit)* glutinousness; *(von Öl)* viscosity; **b)** *(schleppendes Tempo)* sluggishness; **c)** *(Widerstandsfähigkeit)* toughness; **d)** *(Beharrlichkeit)* tenacity; *(des Widerstands)* doggedness

zäh·flüssig *Adj.* glutinous; viscous ⟨*oil*⟩; heavy ⟨*dough*⟩; *(fig.: langsam)* slow-moving ⟨*traffic*⟩

Zäh·flüssigkeit die; *o. Pl.* glutinousness; *(von Öl)* viscosity

Zähigkeit die; ~ a) *(Widerstandsfähigkeit)* toughness; **b)** *(Beharrlichkeit)* tenacity

Zahl ['tsa:l] **die; ~, ~en** number; *(Ziffer)* numeral; *(Zahlenangabe)* figure; **in den roten/schwarzen ~en** in the red/black; **acht** *usw.* **an der ~:** eight *etc.* in number; **in großer ~:** in great numbers

Zahl·adjektiv das numeral adjective

zahlbar *Adj. (Kaufmannsspr.)* payable

zählbar *Adj.* countable

zählebig *Adj.* hardy ⟨*plant, animal*⟩; **ein ~es Vorurteil** a prejudice which dies hard

zahlen 1. *tr. V.* **a)** pay ⟨*price, amount, rent, tax, fine, etc.*⟩ **(an** + *Akk.* to); **einen hohen Preis ~** *(auch fig.)* pay a high price; **b)** *(ugs.: bezahlen)* pay for ⟨*taxi, repair, etc.*⟩; **jmdm. etw. ~:** give sb. the money for sth.; *(spendieren)* pay for sth. for sb. **2.** *itr. V.* pay; **er will nicht ~:** he won't pay [up]; **~ bitte!** *(im Lokal)* [can I/we have] the bill, please!; **die Firma zahlt gut** the firm pays well

zählen ['tsɛ:lən] **1.** *itr. V.* **a)** count; **ich zähle bis drei** I'll count up to three; **b)** *(gehören)* belong **(zu** to); **diese Tage zählten zu den schönsten seines Lebens** these days were among *or* were some of the most wonderful in his life; **c)** *(gültig/wichtig sein)* count; **d) auf jmdn./etw. ~:** count on sb./sth. **2.** *tr. V.* **a)** count; **Geld auf den Tisch ~:** count money out on to the table; **seine Tage sind gezählt** *(fig.)* his days are numbered; **b)** *(geh.)* **er zählt 90 Jahre** he is 90 years of age; **die Stadt zählt 500 000 Einwohner** the town has 500,000 inhabitants; **c) jmdn. zu seinen Freunden ~:** count sb. among one's friends; **d)** *(wert sein)* be worth

zahlen-, Zahlen-: ~lotterie die, ~lotto das *lottery in which entrants guess which set of figures will be drawn at random from a fixed sequence of numbers;* **~mäßig 1.** *Adj.; nicht präd.* numerical; **2.** *adv.* numerically

Zähler der; ~s, ~ a) *(Meßgerät)* meter; **b)** *(Math.)* numerator

zahl-, Zahl-: ~karte die *(Postw.)* paying-in slip; **~los** *Adj.* countless; innumerable; **~meister der** *(auch fig.)* paymaster; *(auf Schiffen)* purser; **~reich 1.** *Adj.* numerous. **2.** *adv.* in large numbers

Zahlung die; ~, ~en payment; **etw. in ~ nehmen/geben** *(Kauf-*

mannsspr.) take/give sth. in part-exchange; take sth. as a trade-in/trade sth. in

Zählung die; ~, ~en count

zahlungs-, Zahlungs-: ~bedingungen *Pl. (Wirtsch.)* terms of payment; **~empfänger der** payee; **~fähig** *Adj.* solvent; **~fähigkeit die;** *o. Pl.* solvency; **~kräftig** *Adj. (ugs.)* affluent; **~mittel das** means of payment; **~unfähig** *Adj.* insolvent; **~unfähigkeit die** insolvency; **~weise die** method of payment

Zahl-: ~wort das; *Pl.* **~wörter** *(Sprachw.)* numeral; **~zeichen das** numeral

zahm [tsa:m] *(auch fig.)* **1.** *Adj.* tame. **2.** *adv.* tamely

zähmen ['tsɛ:mən] *tr. V.* **a)** *(auch fig.)* tame; subdue ‹*forces of nature*›; **b)** *(geh.)* restrain ‹*curiosity, impatience, etc.*›

Zahn [tsa:n] **der; ~[e]s, Zähne** ['tsɛ:nə] **a)** tooth; *(an einer Briefmarke usw.)* serration; **sich** *(Dat.)* **einen ~ ziehen lassen** have a tooth out; **b)** *(fig.)* **der ~ der Zeit** *(ugs.)* the ravages *pl.* of time; **[jmdm.] die Zähne zeigen** *(ugs.)* show [sb.] one's teeth; **die Zähne zusammenbeißen** *(ugs.)* grit one's teeth; **sich** *(Dat.)* **an jmdm./etw. die Zähne ausbeißen** *(ugs.)* get nowhere with sb./sth.; **jmdm. auf den ~ fühlen** *(ugs.)* sound sb. out; **bis an die Zähne bewaffnet** armed to the teeth; **c)** *(ugs.: Tempo)* **einen ganz schönen ~ draufhaben** be going like the clappers *(sl.)*; **einen ~ zulegen** get a move on *(coll.)*

zahn-, Zahn-: ~arzt der dentist; **~ärztlich 1.** *Adj.; nicht präd.* dental ‹*treatment etc.*›; **2.** *adv.* **jmdn. ~ärztlich behandeln** give sb. dental treatment; **~bürste die** toothbrush; **~creme die** *s.* **~pasta**

zähne-: ~fletschend *Adj.* baring its/their teeth *postpos.*; **~klappernd** *Adj.* with chattering teeth *postpos.*; **~knirschend** *Adv.* gnashing one's teeth; cursing silently

zahnen *itr. V.* ‹*baby*› be teething

zahn-, Zahn-: ~fleisch das gum; *(als Ganzes)* gums *pl.* **~los** *Adj.* toothless; **~lücke die** gap in one's teeth; **~medizin die** dentistry *no art.*; **~pasta** [~pasta] **die; ~, ~pasten** toothpaste; **~prothese die** dentures *pl.*; [set *sing.* of] false teeth *pl.*; **~pulver das** tooth-powder; **~putz·becher der** tooth-mug; **~rad das** gearwheel; *(kleines)* cog; *(einer Uhr)* [toothed] wheel; *(für Ketten)* sprocket; **~rad·bahn die** rack-railway; **~schmerzen** *Pl.* toothache *sing.*; **~stein der;** *o. Pl.* tartar; **~stocher der; ~s, ~:** toothpick; **~weh das;** *o. Pl. (ugs.)* toothache

Zaire [za'i:r] **(das); ~s** Zaire

Zairer [za'i:rɐ] **der; ~s, ~:** Zairese

Zander ['tsandɐ] **der; ~s, ~:** zander

Zange ['tsaŋə] **die; ~, ~n a)** *(Werkzeug)* pliers *pl.*; *(Eiswürfel~, Wäsche~, Zucker~)* tongs *pl.*; *(Geburts~)* forceps *pl.*; *(Kneif~)* pincers *pl.*; *(Loch~)* punch; **eine ~:** a pair of pliers/tongs/forceps/pincers/a punch; **jmdn. in die ~ nehmen** *(fig. ugs.)* put the screws on sb.; *(Fußballjargon)* crowd sb. out; **b)** *(bei Tieren)* pincer

Zank [tsaŋk] **der; ~[e]s** squabble; row

Zank·apfel der bone of contention

zanken *refl., itr. V.* squabble, bicker (**um** *od.* **über** + *Akk.* over)

zänkisch *Adj.* quarrelsome

Zäpfchen ['tsɛpfçən] **das; ~s, ~** *(Pharm.)* suppository

zapfen ['tsapfn̩] *tr. V.* tap, draw ‹*beer, wine*›

Zapfen der; ~s, ~ a) *(Bot.)* cone; **b)** *(Stöpsel)* bung; **c)** *(Eis~)* icicle

Zapfen·streich der *(Milit.)* **a)** *(Signal)* last post *(Brit.)*; taps *pl. (Amer.)*; **der Große ~:** the tattoo; **b)** *o. Pl. (Ende der Ausgehzeit)* time for return to barracks

Zapf-: ~hahn der tap; **~säule die** petrol-pump *(Brit.)*; gasoline pump *(Amer.)*

zappelig *Adj. (ugs.)* **a)** wriggly; fidgety ‹*child*›; **b)** *(nervös)* jittery *(coll.)*

zappeln ['tsapl̩n] *itr. V.* wriggle; ‹*child*› fidget; **mit den Beinen/Armen ~:** wave one's legs/arms about; **jmdn. ~ lassen** *(fig. ugs.)* keep sb. on tenterhooks

Zar [tsa:ɐ̯] **der; ~en, ~en** *(hist.)* Tsar

Zaren·reich das *(hist.)* tsardom

Zarin die; ~, ~nen *(hist.)* Tsarina

zaristisch *Adj. (hist.)* tsarist

zart [tsa:ɐ̯t] **1.** *Adj.* **a)** delicate; soft ‹*skin*›; tender ‹*bud, shoot*›; fragile, delicate ‹*china*›; delicate, frail ‹*health, constitution, child*›; **b)** *(weich)* tender ‹*meat, vegetables*›; soft ‹*filling*›; fine ‹*biscuits*›; **c)** *(leicht)* gentle ‹*kiss, touch*›; delicate ‹*colour, complexion, fragrance, etc.*›; soft, gentle ‹*voice, sound, tune*›. **2.** *adv.* delicately ‹*coloured, fragrant*›; ‹*kiss, touch*› gently

zart-, Zart-: ~besaitet *Adj.* highly sensitive; **~bitter** *Adj.* bittersweet ‹*chocolate*›; **~blau** *Adj.* pale blue; **~fühlend 1.** *Adj.* tactful; **2.** *adv.* tactfully; **~gefühl das** tact; delicacy of feeling

Zartheit die; ~ a) delicacy; *(der Haut)* softness; *(von Porzellan)* fragility; *(von Spitzen, Seide)* fineness; *(der Gesundheit, Konstitution)* delicateness; *(Sensibilität)* sensitivity; **b)** *(von Fleisch, Gemüse)* tenderness; **c)** *(eines Kusses, einer Berührung)* gentleness; *(einer Farbe, eines Dufts)* delicacy; *(einer Stimme, eines Tons)* softness; gentleness

zärtlich ['tsɛ:ɐ̯tlɪç] **1.** *Adj.* tender; loving; **~ werden** *(verhüll.)* start petting. **2.** *adv.* tenderly; lovingly

Zärtlichkeit die; ~, ~en a) *o. Pl.* tenderness; affection; **b)** *meist Pl. (Liebkosung)* caress; **es kam zu ~en [zwischen ihnen]** *(verhüll.)* they became intimate

zart·rosa *Adj.* pale pink

Zäsur [tsɛ'zu:ɐ̯] **die; ~, ~en a)** *(Verslehre, Musik)* caesura; **b)** *(geh.: Einschnitt)* break

Zauber ['tsaubɐ] **der; ~s, ~ a)** *(auch fig.)* magic; *(magische Handlung)* magic trick; *(Bann)* [magic] spell; **einen großen ~ auf jmdn. ausüben** *(fig.)* have a great fascination for sb.; **b)** *o. Pl. (ugs. abwertend: Aufheben)* fuss; **ich halte nichts von dem ganzen ~:** the whole palaver means nothing to me *(coll.)*

Zauberei die; ~: magic

Zauberer der; ~s, ~: a) magician; **b)** *(Zauberkünstler)* conjurer

Zauber·formel die magic spell; *(fig.)* magic formula; panacea

zauber·haft 1. *Adj.* enchanting; delightful. **2.** *adv.* enchantingly; delightfully

Zauberin die; ~, ~nen a) sorceress; **b)** *s.* **Zauberer b**

zauberisch *(geh.)* **1.** *Adj.* magical. **2.** *adv.* magically

Zauber-: ~kunst die a) *o. Pl.* magic *no art.*; *(eines Bühnenkünstlers)* magic *no art.*; conjuring *no art.*; **b)** *meist Pl. (magische Fähigkeit)* magic; **~künstler der** conjurer; magician

zaubern 1. *itr. V.* **a)** do magic; **b)** *(Zaubertricks ausführen)* do conjuring tricks. **2.** *tr. V. (auch fig.)* conjure; conjure up ‹*palace, horse*›; **eine Taube aus dem Hut ~:** produce a dove out of a hat

Zauber-: ~stab der magic wand; **~trick der** conjuring trick

Zauderer der; ~s, ~: waverer; ditherer

zaudern ['tsaudɐn] *itr. V. (geh.)* delay; ~, **etw. zu tun** delay in doing sth.

Zaum [tsaum] **der; ~[e]s, Zäume** ['tsɔymə] bridle; **sich/seine Zunge/seine Leidenschaften im ~ halten** *(fig. geh.)* restrain *or* control oneself/control one's tongue/control one's passions

zäumen ['tsɔymən] *tr. V.* bridle

Zaum·zeug das bridle

Zaun [tsaun] **der; ~[e]s, Zäune** ['tsɔynə] fence; **einen Streit/Krieg vom ~ brechen** *(fig.)* suddenly start a quarrel/war

Zaun-: ~gast der onlooker; **~könig der** wren; **~pfahl der** fence-post; *s. auch* **Wink b**

zausen ['tsauzn̩] *tr. V. (auch fig.)* ruffle; ruffle, tousle ‹*hair*›

z. B. *Abk.* **zum Beispiel** e.g.

ZDF [tsɛtde:'ɛf] **das; ~** *Abk.* **Zweites Deutsches Fernsehen** Second German Television Channel

Zebra ['tse:bra] **das; ~s, ~s** zebra

Zebra·streifen der zebra crossing *(Brit.)*; pedestrian crossing

Zebu ['tse:bu] **der** *od.* **das; ~s, ~s** zebu

Zẹche die; ~, ~n a) bill *(Brit.)*; check *(Amer.)*; **die ~ prellen** *(ugs.)* leave without paying [the bill]; **die ~ bezahlen müssen** *(fig.)* have to foot the bill *or* pay the price; **b)** *(Bergwerk)* pit; mine

zechen ['tsɛçn̩] *itr. V. (veralt., scherzh.)* tipple

Zẹcher der; ~s, ~ *(veralt., scherzh.)* tippler

Zẹch-: ~preller der; ~s, ~ *person who leaves without paying the bill;* bill-dodger; **~prellerei die; ~, ~en** *leaving without paying the bill;* bill-dodging

Zecke ['tsɛkə] **die; ~, ~n** tick

Zeder ['tse:dɐ] **die; ~, ~n** cedar

Zeh [tse:] **der; ~s, ~en, Zehe die; ~, ~n a)** toe; **jmdm. auf die Zehen treten** *(auch fig.)* tread on sb.'s toes; **b)** *(Knoblauch~)* clove

Zehen-: ~nagel der toe-nail; **~spitze die: auf/auf die ~spitzen** *(Dat./Akk.)* on tiptoe

zehn [tse:n] *Kardinalz.* ten; *s. auch* [1]**acht**

Zehner der; ~s, ~ a) *(ugs.: Geldschein, Münze)* ten; **b)** *(ugs.: Autobus)* number ten; **c)** *(Math.)* ten; **d)** *(Sprungturm)* ten-metre platform

zehn·fach *Vervielfältigungsz.* tenfold; **die ~e Menge** ten times the quantity; *s. auch* **achtfach**

Zehnfache das; *adj. Dekl.* **das ~:** ten times as much; *s. auch* **Achtfache**

zehn-, Zehn-: ~jährig *Adj. (10 Jahre alt)* ten-year-old *attrib.;* ten years old *postpos.; (10 Jahre dauernd)* ten-year *attrib.; s. auch* **achtjährig; ~kampf der** *(Sport)* decathlon; **~kämpfer der** decathlete; **~mal** *Adv.* ten times; *s. auch* **achtmal; ~markschein der** ten-mark note; **~pfẹnnig·[brief]marke die** ten-pfennig stamp; **~pfẹnnig·stück das** ten-pfennig piece; **~stöckig** *Adj.* ten-storey ‹*building*›; *s. auch* **achtstöckig**

zehnt [tse:nt] *in* **wir waren zu ~:** there were ten of us; *s. auch* [2]**acht**

zehnt... *Ordinalz.* tenth; *s. auch* **acht...**

zehn-: ~tägig *Adj. (10 Tage alt)* ten-day-old *attrib.; (10 Tage dauernd)* ten-day *attrib.;* **~tausend** *Kardinalz.* ten thousand; **die oberen Zehntausend** *(fig.: die vornehmen Leute)* the élite of society

zehn·teilig *Adj.* ten-piece ‹*tool-set etc.*›; ten-part ‹*serial*›; *s. auch* **achtteilig**

zehntel ['tse:ntl̩] *Bruchz.* tenth

Zehntel das (*schweiz. meist* **der**); **~s, ~:** tenth

Zehntel-: ~liter der tenth of a litre; **~sekunde die** tenth of a second

zehntens *Adv.* tenthly

zehren ['tse:rən] *itr. V.* **a) von etw. ~:** live on *or* off sth.; **von Erinnerungen** *usw.* **~** *(fig.)* sustain oneself on memories *etc.;* **b) an jmdm./jmds. Kräften ~:** wear sb. down/sap sb.'s strength

Zeichen ['tsaiçn̩] **das; ~s, ~ a)** sign; *(Laut, Wink)* signal; **das ~ zum Angriff** the signal to attack; **jmdm. ein ~ geben** signal to sb.; **zum ~, daß ...:** to show that ...; as a sign that ...; **b)** *(Markierung)* mark; *(Waren~)* [trade] mark; *(am Briefkopf)* reference; **[ein] ~ setzen** *(fig.)* set an example; point the way; **c)** *(Symbol)* sign; *(Chemie, Math., auf Landkarten usw.)* symbol; *(Satz~)* punctuation mark; *(Musik)* accidental; **d)** *(An~)* sign; indication; *(einer Krankheit)* sign; symptom; **ein ~ dafür, daß ...:** a [sure] sign that ...; **die ~ der Zeit erkennen** see which way the wind's blowing *(fig.)*; **d)** *(Tierkreis~)* sign [of the zodiac]; **ich bin im ~ des Krebses geboren** I was born under the sign of Cancer

Zeichen-: ~block der; *Pl.* **~s** *od.* **~blöcke** sketch-pad; **~brett das** drawing-board; **~erklärung die** legend; **~feder die** drawing-pen; **~setzung die** punctuation; **~sprache die** sign language; **~stift der** drawing-pencil; **~trickfilm der** animated cartoon; **~unterricht der** drawing lessons *pl.*; *(Schulfach)* art *no art.*

zeichnen 1. *tr. V.* **a)** draw; *(fig.)* portray ‹*character*›; **b) das Fell ist schön/auffallend gezeichnet** the fur has beautiful/striking markings; **er war von der Krankheit gezeichnet** *(fig.)* sickness had left its mark on him; **c)** *(bes. Kaufmannsspr.)* sign ‹*cheque*›; subscribe for ‹*share, loan*›. **2.** *itr. V.* **a)** draw; **b)** *(bes. Kaufmannsspr.: unterschreiben)* sign; **für etw. [verantwortlich] ~** *(fig.)* be responsible for sth.

Zeichner der; ~s, ~, Zeichnerin die; ~, ~nen graphic artist; *(Technik)* draughtsman/-woman

zeichnerisch 1. *Adj.* ‹*talent*› as a draughtsman/-woman *or* for drawing. **2.** *adv.* **~ begabt sein** have a talent for drawing; **etw. ~ darstellen** make a drawing of sth.

Zeichnung die; ~, ~en a) drawing; *(fig.)* portrayal; **b)** *(bei Tieren und Pflanzen)* markings *pl.*

Zeige·finger der index finger; forefinger; **der erhobene ~** *(fig.)* the wagging *or* monitory finger

zeigen ['tsaign̩] **1.** *tr. V.* point; **[mit dem Finger/einem Stock] auf jmdn./etw. ~:** point [one's finger/a stick] at sb./sth.. **2.** *tr. V.* show; **jmdm. etw. ~:** show sb. sth.; show sth. to sb.; *(jmdn. zu etw. hinführen)* show sb. to sth.; **dem werd' ich's ~!** *(ugs.)* I'll show him!; **zeig mal, was du kannst** show [us] what you can do. **3.** *refl. V.* **a)** *(sich sehen lassen)* appear; **mit ihr kann man sich überall ~:** you can take her anywhere; **b)** *(sich erweisen)* **sich als etw. ~:** prove to be sth.; **es wird sich ~, wer schuld war** time will tell who was responsible; **es hat sich gezeigt, daß ...:** it turned out that ...

Zeiger der; ~s, ~: pointer; *(Uhr~)* hand

Zeige·stock der; *Pl.* **~stöcke** pointer

zeihen ['tsaiən] *unr. tr. V. (geh.)* **jmdn. einer Sache** *(Gen.)* **~:** indict sb. of sth.

Zeile ['tsailə] **die; ~, ~n a)** line; **jmdm. ein paar ~n schreiben** drop sb. a line; **zwischen den ~n** *(fig.)* between the lines; **b)** *(Reihe)* row

Zeisig ['tsaizɪç] **der; ~s, ~e** siskin

zeit [tsait] *Präp. mit Gen.* **~ meines** *usw.***/unseres** *usw.* **Lebens** all my *etc.* life/our *etc.* lives

Zeit die; ~, ~en a) *o. Pl.* time *no art.;* **im Laufe der ~:** in the course of time; **mit der ~:** with time; in time; *(allmählich)* gradually; **die ~ arbeitet für/gegen jmdn.** time is on sb.'s side/is against sb.; **keine ~ verlieren dürfen** have no time to lose; **die ~ drängt** time is pressing; there is [precious] little time; **sich** *(Dat.)* **die ~ [mit etw.] vertreiben** pass the time [with/doing sth.]; **jmdm. ~/drei Tage** *usw.* **~ lassen** give sb. time/three days *etc.*; **sich** *(Dat.)* **~ lassen** take one's time; **sich** *(Dat.)* **für jmdn./etw. ~ nehmen** make time for sb./sth.; **b)** *(~punkt)* time; **seit der** *od.* **dieser ~:** since that time; **um diese ~:** at this time; **vor der ~:** prematurely; early; **von ~ zu ~:** from time to time; **zur ~:** at the moment; at present; **c)** *(~abschnitt, Lebensabschnitt)* time; period; *(Geschichtsabschnitt)* age; period; **die erste ~:** at first; **auf ~:** temporarily; **ein Vertrag auf ~:** a fixed-term contract; **zu meiner ~:** in my day; d) *(Sport)* time; **die ~ bei etw. stoppen** time sth.; **über die ~ kommen** *(Boxen)* go the distance; **e)** *(Sprachw.)* tense

zeit-, Zeit-: ~abschnitt der period; **~alter das** age; era; **~ansage die** *(im Radio)* time check; *(am Telefon)* speaking clock; **~auf·wand der: viel ~aufwand erfordern** take up a great deal of time; **~bombe die** *(auch fig.)* time bomb; **~gefühl das;** *o. Pl.* sense of time; **~geist der;** *o. Pl.* spirit of the age; **~gemäß** *Adj.* contemporary ‹*views*›; *(modern)* up-to-date; **~genosse der, ~genossin die a)** contemporary; **b)** *(ugs.: Mensch)* individual *(coll.)*; **~genössisch** [~gənœsɪʃ] *Adj.* contemporary; **~geschehen das: das [aktuelle] ~geschehen** current events *pl.*; **~geschichte die;** *o. Pl.* contemporary history *no art.*; **~geschichtlich** *Adj.; nicht präd.* contemporary ‹*source etc.*›; contemporary-history ‹*teaching etc.*›; **~gleich 1.** *Adj.* **a)** simultaneous; **b)** *(Sport)* ‹*runners etc.*› with the same time; **2.** *adv.* simultaneously

zeitig *Adj., adv.* early

zeitigen *tr. V. (geh.)* produce, yield ‹*result, success, etc.*›; provoke, precipitate ‹*uproar*›

zeit-, Zeit-: ~karte die season ticket; **~kritik die;** *o. Pl.* appraisal *or* analysis of contemporary issues; **~kritisch** *Adj.* ‹*essay etc.*› analysing contemporary issues; **~lang die: eine ~lang** for a while *or* a time; **~läuf[t]e** [~lɔyf(t)ə] *Pl. (geh.)* times; **über alle ~läuf[t]e hinweg** for all time; **~lebens** *Adv.* all my *etc.* life/our *etc.* lives

zeitlich 1. *Adj.* ‹*length, interval*› in time; chronological ‹*order, sequence*›. **2.** *adv.* **a)** with regard to time; **ich kann es ~ nicht einrichten** I can't fit it in time-wise *(coll.)*

zeit-, Zeit-: ~los 1. *Adj.* timeless; classic ‹*fashion, shape*›; **2.** *adv.* timelessly; **~los eingerichtet** furnished in a classic *or* timeless style; **~lupe die;** *o. Pl. (Film)* slow motion; **~lupen·tempo das: im ~lupentempo** at a crawl; at a snail's pace; **~nah 1.** *Adj.* topical ‹*play etc.*›; ‹*teaching, syllabus*› relevant to the present day; **2.** *adv.* topically; **~not die;** *o. Pl.* **in ~not geraten** *od.* **kommen** become pressed for time; **~plan der** schedule; **~punkt der** moment; **zum jetzigen ~punkt** at the present moment; at this point in time; **~raffer der** *(Film)* time-lapse; **~raubend** *Adj.* time-consuming; **~raum der** period; **~rechnung die** calendar; **vor unserer ~rechnung** BC; before Christ; **unserer/christlicher ~rechnung** AD; Anno Domini; **~schrift die** magazine; *(bes. wissenschaftlich)* journal; periodical; **~soldat der** *soldier serving for a fixed period;* **~spanne die** period

Zeitung die; ~, ~en [news]paper; **[die] ~ lesen** read the paper

Zeitungs-: ~annonce die, ~anzeige die newspaper advertisement; **~artikel der** newspaper article; **~inserat das** newspaper advertisement; **~leser der** newspaper-reader; **~notiz die** newspaper item; **~papier das a)** *(alte Zeitung[en])* newspaper; **b)** *(unbedruckt)* newsprint

zeit-, ~Zeit-: ~verschwendung die waste of time; **~verschwendung sein** be a waste of time; **~vertreib der; ~[e]s, ~e** pastime; **zum ~vertreib** to pass the time; **~weilig 1.** *Adj.; nicht präd.* temporary; **2.** *adv.* temporarily; for a time; **~weise** *Adv.* **a)** *(gelegentlich)* occasionally; at times; *(von Zeit zu Zeit)* from time to time; **~weise Regen** occasional rain; **b)** *(vorübergehend)* for a time; for a while; **~wort das;** *Pl.* **~wörter** *(Sprachw.)* verb

zelebrieren [tsele'bri:rən] *tr. V.* celebrate

Zelle ['tsɛlə] **die; ~, ~n** cell; *(Telefon~)* [tele]phone booth *or (Brit.)* box

Zell-: ~stoff der cellulose; **~teilung die** *(Biol.)* cell division

Zelluloid [tsɛlu'lɔyt] **das; ~[e]s** celluloid

Zellulose [tsɛlu'lo:zə] **die; ~, ~n** cellulose

Zell·wolle die rayon

Zelt [tsɛlt] **das; ~[e]s, ~e** tent; *(Fest~)* marquee; *(Zirkus~)* big top

zelten *itr. V.* camp; **wir waren ~:** we went camping

Zelt-: ~lager das camp; **~plane die** tarpaulin; **~platz der** camping site; campsite

Zement [tse'mɛnt] **der; ~[e]s, ~e** cement

Zement·boden der concrete floor

zementieren *tr. V.* **a)** cement; **b)** *(fig.)* make ‹*situation etc.*› permanent

Zenit [tse'ni:t] **der; ~[e]s** zenith; **im ~ stehen** be at its zenith

zensieren [tsɛn'zi:rən] **1.** *tr. V.* **a)** *(Schulw.)* mark, *(Amer.)* grade ‹*essay etc.*›; **b)** *(der Zensur unterziehen)* censor ‹*article, film, etc.*›. **2.** *itr. V. (Schulw.)* **streng/milde ~:** mark *or (Amer.)* grade severely/leniently

Zensur [tsɛn'zu:ɐ̯] **die; ~, ~en a)** *(Schulw.)* mark; grade *(Amer.)*; **b)** *(Kontrolle)* censorship; **c)** *(Behörde)* censors *pl.*

Zenti·meter ['tsɛnti-, auch: --'--] **der,** *auch:* **das** centimetre

Zentner ['tsɛntnɐ] **der; ~s, ~ a)** centner; metric hundredweight; **b)** *(österr., schweiz.) s.* **Doppelzentner**

Zentner·last die hundredweight load

zentner·schwer 1. *Adj. (äußerst schwer)* massively heavy. **2.** *adv.* **~ auf jmdm. lasten** *(fig.)* weigh heavily on sb.

zentral [tsɛn'tra:l] **1.** *Adj.* central. **2.** *adv.* centrally

Zentrale die; ~, ~n a) head *or* central office; *(der Polizei, einer Partei)* headquarters *sing. or pl.*; *(Funk~)* control centre; **b)** *(Telefon~)* [telephone] exchange; *(eines Hotels, einer Firma o. ä.)* switchboard

Zentral·heizung die central heating

zentralisieren *tr. V.* centralize

Zentral-: ~komitee das Central Committee; **~nerven·system das** central nervous system

Zentren *s.* **Zentrum**

Zentrifugal·kraft [tsɛntrifu'ga:l-] **die** *(Physik)* centrifugal force

Zentrifuge die; ~, ~n centrifuge

Zentripetal·kraft [tsɛntripe'ta:l-] **die** *(Physik)* centripetal force

Zentrum ['tsɛntrʊm] **das; ~s, Zentren** centre; **im ~**: at the centre; *(im Stadt~)* in the town/city centre

Zeppelin ['tsɛpəli:n] **der; ~s, ~e** Zeppelin

Zepter ['tsɛptɐ] **das,** *auch:* **der; ~s, ~**: sceptre

zerbeißen *unr. tr. V.* bite in two; ⟨*flea, mosquito, etc.*⟩ bite ⟨*person etc.*⟩ all over

zerbersten *unr. itr. V.; mit sein* burst apart

zerbomben *tr. V.* bomb to pieces; destroy by bombing; **zerbombt** bombed ⟨*streets, houses*⟩

zerbrechen 1. *unr. itr. V.; mit sein* break [into pieces]; smash [to pieces]; ⟨*glass*⟩ shatter; *(fig.)* ⟨*marriage, relationship*⟩ break up. **2.** *unr. tr. V.* break; smash, shatter ⟨*dishes, glass*⟩

zerbrechlich *Adj.* **a)** fragile; **„Vorsicht, ~!"** 'fragile; handle with care'; **b)** *(zart, schwach)* frail

Zerbrechlichkeit die; ~ *s.* **zerbrechlich:** fragility; frailty

zerbröckeln 1. *itr. V.; mit sein (auch fig.)* crumble away. **2.** *tr. V.* break into small pieces

zerdeppern [tsɛɐ̯'dɛpɐn] *tr. V. (ugs.)* smash

zerdrücken *tr. V.* mash ⟨*potatoes, banana*⟩; squash ⟨*fly etc.*⟩; crease ⟨*clothes*⟩

Zeremonie [tseremo'ni:] **die; ~, ~n** ceremony; *(fig.)* ritual

Zeremoniell [tseremo'ni̯ɛl] **das; ~s, ~e** ceremonial

Zerfall der; ~[e]s disintegration; *(fig.: der Moral)* breakdown; *(einer Leiche)* decomposition; *(eines Gebäudes)* decay

zerfallen *unr. itr. V.; mit sein* **a)** *(auch fig.)* disintegrate (**in** + *Akk.*, **zu** into); ⟨*building*⟩ fall into ruin, decay; ⟨*corpse*⟩ decompose, decay; **~de Mauern** crumbling walls; **b)** *(unterteilt sein)* be divided (**in** + *Akk.* into)

zerfetzen *tr. V.* rip *or* tear to pieces; rip *or* tear up ⟨*letter etc.*⟩ (**in** + *Akk.* into); tear apart ⟨*body, limb*⟩

zerfleddern *tr. V.* wear out

zerfleischen *tr. V.* tear ⟨*person, animal*⟩ limb from limb

zerfließen *unr. itr. V.; mit sein* **a)** *(schmelzen)* melt [away]; **in** *od.* **vor Mitleid ~** *(fig.)* dissolve with pity; **b)** *(auseinanderfließen)* ⟨*paint, ink*⟩ run; ⟨*shapes*⟩ dissolve

zerfransen 1. *itr. V.; mit sein* fray. **2.** *tr. V.* fray

zerfressen *unr. tr. V.* **a)** eat away; ⟨*moth etc.*⟩ eat holes in; **von Motten ~**: moth-eaten; **b)** *(zersetzen)* corrode ⟨*metal*⟩

zerfurchen *tr. V.* **a)** rut ⟨*track etc.*⟩; **b)** furrow ⟨*brow, face*⟩

zergehen *unr. itr. V.; mit sein* melt; *(sich auflösen)* dissolve; **auf der Zunge ~**: melt in the mouth

zerhacken *tr. V.* chop up

zerhauen *unr. tr. V.* chop up

zerkauen *tr. V.* chew [up]

zerkleinern [tsɛɐ̯'klainɐn] *tr. V.* *(zerhacken)* chop up; *(zerkauen)* chew up ⟨*food*⟩; *(zermahlen)* crush ⟨*rock etc.*⟩

Zerkleinerung die; ~, ~en *s.* **zerkleinern:** chopping up; chewing; crushing

zerklüftet *Adj.* fissured ⟨*landscape*⟩; craggy ⟨*mountains*⟩; deeply indented ⟨*coastline*⟩

zerknallen *itr. V.; mit sein (ugs.)* burst [with a bang]

zerknautschen *tr. V. (ugs.)* crumple

zerknirscht 1. *Adj.* remorseful. **2.** *adv.* remorsefully

zerknittern *tr. V.* crease; crumple

zerknüllen *tr. V.* crumple up [into a ball]

zerkochen 1. *itr. V.; mit sein* get overcooked. **2.** *tr. V.* overcook

zerkratzen *tr. V.* scratch

zerkrümeln 1. *tr. V.* crumble up. **2.** *itr. V.; mit sein* break into crumbs; crumble

zerlassen *unr. tr. V. (Kochk.)* melt

zerlegen *tr. V.* **a)** dismantle; take to pieces; strip, dismantle ⟨*engine*⟩; **etw. in seine Bestandteile ~**: reduce sth. to its component parts; **b)** cut up ⟨*animal, meat*⟩

zerlumpt *Adj.* ragged ⟨*clothes, person*⟩; **~ sein** ⟨*clothes*⟩ be in tatters, be torn; **~ herumlaufen** go about in rags

zermahlen *unr. tr. V.* grind

zermalmen [tsɛɐ̯'malmən] *tr. V.* crush

zermürben *tr. V.* wear ⟨*person*⟩ down; **~d** wearing; trying

zernagen *tr. V.* gnaw away

zerpflücken *tr. V.* **a)** pick ⟨*flower, lettuce, etc.*⟩ apart; **b)** *(fig.)* pull ⟨*book etc.*⟩ to pieces; destroy ⟨*alibi*⟩

zerplatzen *itr. V.; mit sein* burst

zerquetschen *tr. V.* crush; mash ⟨*potatoes*⟩; **20 Mark und ein paar Zerquetschte** *(ugs.)* 20 marks and a bit

Zerr·bild das distorted image

zerreiben *unr. tr. V.* crush ⟨*spices, paint colours, etc.*⟩

zerreißen 1. *unr. tr. V.* **a)** tear up; *(in kleine Stücke)* tear to pieces; ⟨*animal*⟩ tear ⟨*prey*⟩ limb from limb; dismember ⟨*prey*⟩; break ⟨*thread*⟩; **ich kann mich nicht ~**: I can't be in two places at once; **b)** *(beschädigen)* tear ⟨*stocking, trousers, etc.*⟩ (**an** + *Dat.* on). **2.** *unr. itr. V.; mit sein* ⟨*thread, string, rope*⟩ break; ⟨*paper, cloth, etc.*⟩ tear; **ihre Nerven waren zum Zerreißen gespannt** her nerves were stretched to breaking-point

zerren ['tsɛrən] **1.** *tr. V.* **a)** drag; **b) sich** *(Dat.)* **einen Muskel/eine Sehne ~**: pull a muscle/tendon. **2.** *itr. V.* **an etw.** *(Dat.)* **~**: tug *or* pull at sth.

zerrinnen *unr. itr. V.* melt; *(fig.)* ⟨*time*⟩ pass; **jmdm. unter den Händen ~**: slip through sb.'s fingers

zerrissen [tsɛɐ̯'rɪsn̩] *Adj.* **[innerlich] ~**: at odds with oneself

Zerr·spiegel der distorting mirror

Zerrung die; ~, ~en *(Muskel~)* pulled muscle; *(Sehnen~)* pulled tendon

zerrupfen *tr. V.* tear to bits

zerrütten [tsɛɐ̯'rʏtn̩] *tr. V.* ruin ⟨*health*⟩; shatter ⟨*nerves*⟩; ruin, wreck ⟨*marriage*⟩

Zerrüttung die; ~, ~en *(der Gesundheit)* ruining; *(der Nerven)* shattering; *(einer Ehe)* [irretrievable] breakdown

zersägen *tr. V.* saw up

zerschellen *itr. V.; mit sein* be dashed *or* smashed to pieces

[1]**zerschlagen 1.** *unr. tr. V.* smash ⟨*plate, windscreen, etc.*⟩; smash up ⟨*furniture*⟩; *(fig.)* smash ⟨*spy ring etc.*⟩; crush ⟨*enemy, attack*⟩; break up ⟨*cartel*⟩. **2.** *unr. refl. V.* ⟨*plan, deal*⟩ fall through

[2]**zerschlagen** *Adj.* *(erschöpft)* worn out; whacked *(Brit. coll.)*; tuckered [out] *(Amer. coll.)*; shattered *(Brit. coll.)*

Zerschlagung die; ~, ~en smashing; destruction; *(eines Gegners)* crushing

zerschmelzen *unr. itr. V.; mit sein (auch fig.)* melt

zerschmettern *tr. V.* smash; shatter ⟨*glass, leg, bone*⟩; *(fig.)* crush ⟨*army, enemy*⟩

zerschneiden *unr. tr. V.* **a)** cut; *(in Stücke)* cut up; *(in zwei Teile)* cut in two; carve ⟨*joint*⟩; **b)** *(verletzen)* cut [into] ⟨*skin etc.*⟩

zerschnippeln *tr. V. (ugs.)* cut up *or* snip into small pieces

zerschunden [tsɛɐ̯'ʃʊndn̩] *Adj.* covered in scratches *postpos.*

zersetzen 1. *tr. V.* **a)** corrode

⟨*metal*⟩; decompose ⟨*organism*⟩; **b)** *(fig.)* subvert ⟨*ideals*⟩; undermine ⟨*morale*⟩; **~de Schriften** subversive writings. **2.** *refl. V.* decompose; ⟨*wood, compost*⟩ rot

Zersetzung die; ~, ~en a) *s.* **zersetzen 2:** decomposition; rotting; **b)** *s.* **zersetzen 1 b:** subversion; undermining

zerspalten *unr. (auch regelm.) tr. V. (auch fig.)* split [up]

zersplittern *itr. V.; mit sein* ⟨*wood, bone*⟩ splinter; ⟨*glass*⟩ shatter; **das Land war in viele Kleinstaaten zersplittert** the country was fragmented into many small states

zersprengen *tr. V.* blow up; *(in Stücke)* blow to pieces

zerspringen *unr. itr. V.; mit sein* shatter; *(Sprünge bekommen)* crack

zerstampfen *tr. V.* pound, crush ⟨*spices etc.*⟩; mash ⟨*potatoes*⟩

zerstäuben *tr. V.* spray

Zerstäuber der; ~s, ~: atomizer

zerstechen *unr. tr. V.* **a)** sting all over; ⟨*mosquitoes*⟩ bite all over; **b)** *(beschädigen)* jab holes in ⟨*cushion etc.*⟩; puncture, slit ⟨*tyre*⟩

zerstieben *unr. (auch regelm.) itr. V.; mit sein (geh.)* scatter; ⟨*crowd*⟩ disperse

zerstören *tr. V.* destroy; ruin ⟨*landscape, health, life*⟩; dash, destroy ⟨*hopes, dreams*⟩; wreck, destroy ⟨*marriage*⟩

Zerstörer der; ~s, ~ destroyer

zerstörerisch 1. *Adj.* destructive. **2.** *adv.* **~ wirken** have a destructive effect

Zerstörung die destruction; *(der Gesundheit, Existenz)* ruin[ation]; *(einer Ehe)* wrecking; destruction; *(von Hoffnungen)* dashing; destruction

zerstoßen *unr. tr. V.* crush ⟨*berries etc.*⟩; *(im Mörser)* pound, crush ⟨*peppercorns etc.*⟩

zerstreiten *unr. refl. V.* **sich mit jmdm. ~:** fall out with sb.

zerstreuen 1. *tr. V.* **a)** scatter; disperse ⟨*crowd*⟩; **b)** *(unterhalten)* **jmdn./sich ~:** entertain sb./oneself; *(ablenken)* take sb.'s/one's mind off things; **c)** *(beseitigen)* allay ⟨*fear, doubt, suspicion*⟩; dispel ⟨*worry, concern*⟩. **2.** *refl. V.* disperse; *(schneller)* scatter

zerstreut 1. *Adj.* absent-minded. **2.** *adv.* absent-mindedly

Zerstreutheit die; ~: absent-mindedness

Zerstreuung die; ~, ~en diversion; *(Unterhaltung)* entertainment; **~ suchen** look for a distraction [to take one's mind off things]

zerstückeln *tr. V.* break ⟨*sth.*⟩ up into small pieces; *(zerschneiden)* cut *or* chop sth. up into small pieces; dismember ⟨*corpse*⟩

Zerstückelung die; ~, ~en *s.* **zerstückeln:** breaking up; cutting *or* chopping up; dismembering

zerteilen 1. *tr. V.* divide into pieces; *(zerschneiden)* cut into pieces; cut up. **2.** *refl. V.* part

Zertifikat [tsɛrtifi'ka:t] **das; ~[e]s, ~e** certificate

zertrampeln *tr. V.* trample all over ⟨*flower-bed etc.*⟩; trample ⟨*child etc.*⟩ underfoot

zertrennen *tr. V.* take apart

zertreten *unr. tr. V.* stamp on; crush ⟨*insect*⟩ underfoot

zertrümmern *tr. V.* smash; smash, shatter ⟨*glass*⟩; smash up ⟨*furniture*⟩; wreck ⟨*car, boat*⟩; reduce ⟨*building, city*⟩ to ruins

Zervelat·wurst [tsɛrvə'la:t-] **die** cervelat [sausage]

zerwühlen *tr. V.* churn up ⟨*bedclothes, soil*⟩; make a mess of, tousle ⟨*hair*⟩

Zerwürfnis [tsɛɐ̯'vʏrfnɪs] **das; ~ses, ~se** *(geh.)* quarrel; dispute; *(Bruch)* rift

zerzausen *tr. V.* ruffle; ruffle, tousle ⟨*hair*⟩; **zerzaust aussehen** look dishevelled

zetern ['tse:tɐn] *itr. V. (abwertend)* scold [shrilly]

Zettel ['tsɛtl̩] **der; ~s, ~:** slip *or* piece of paper; *(mit einigen Zeilen)* note; *(Bekanntmachung)* notice; *(Formular)* form; *(Kassen~)* receipt; *(Stimm~)* [ballot-]paper

Zeug [tsɔʏk] **das; ~[e]s, ~e a)** *o. Pl. (ugs., oft abwertend: Sachen)* stuff; **sie hat das ~ zu etw.** *(fig.)* she has what it takes to be sth. *or* has the makings of sth.; **was das ~ hält** *(fig. ugs.)* for all one's worth; ⟨*drive*⟩ hell for leather; **sich [mächtig] ins ~ legen** *(fig.)* do one's utmost; **b)** *(ugs.)* **dummes/albernes ~** *(Gerede)* nonsense; rubbish; **dummes ~ machen** mess about; **c)** *(Kleidung)* things *pl.*

Zeuge ['tsɔʏgə] **der; ~n, ~n** witness; **~ einer Sache** *(Gen.)* **sein/werden** be a witness to sth.; witness sth.; **die ~n Jehovas** the Jehovah's Witnesses

[1]**zeugen** *itr. V.* **von etw. ~** *(fig.)* testify to sth.; *(etw. zeigen)* display sth.

[2]**zeugen** *tr. V.* father ⟨*child*⟩

Zeugen·aussage die testimony; witness's statement

Zeugen-: ~stand der *o. Pl.* witness-box *(Brit.)*; witness stand *(Amer.)*; **~vernehmung die** examination of the witness/witnesses

Zeugin die; ~, ~nen witness

Zeugnis das; ~ses, ~se a) *(Schulw.)* report; **b)** *(Arbeits~)* reference; testimonial; **c)** *(Gutachten)* certificate; **d)** *(geh.: Beweis)* evidence; **~se einer früheren Kulturstufe** evidence *or* testimony of an earlier stage of civilization

Zeugung die; ~, ~en fathering

z. Hd. *Abk.* **zu Händen** attn.

zickig *(ugs. abwertend)* **1.** *Adj.* prim; *(prüde)* prudish. **2.** *adv.* primly; *(prüde)* prudishly

Zicklein das; ~s, ~: kid

Zickzack der; ~[e]s, ~e zigzag; **im ~:** in a zigzag

Ziege ['tsi:gə] **die; ~, ~n a)** goat; **b)** *(Schimpfwort: Frau)* cow *(sl. derog.)*

Ziegel ['tsi:gl̩] **der; ~s, ~ a)** brick; **b)** *(Dach~)* tile

Ziegel·dach das tiled roof

Ziegelei die; ~, ~en brickworks *sing.*

Ziegel·stein der brick

Ziegen-: ~bart der goat's beard; *(ugs.: Spitzbart)* goatee beard; **~bock der** he- *or* billy-goat; **~käse der** goat's cheese; **~milch die** goat's milk; **~peter der; ~s, ~** *(ugs.)* mumps *sing.*

zieh [tsi:] *1. u. 3. Pers. Sg. Prät. v.* **zeihen**

ziehen ['tsi:ən] **1.** *unr. tr. V.* **a)** pull; *(sanfter)* draw; *(zerren)* tug; *(schleppen)* drag; **jmdn. an sich ~:** draw sb. to one; **jmdn. am Ärmel ~:** pull sb. by the sleeve; **Perlen auf eine Schnur ~:** thread pearls/beads on to a string; **den Hut ins Gesicht ~:** pull one's hat down over one's face; **b)** *(fig.)* **es zog ihn zu ihr/zu dem Ort** he felt drawn to her/to the place; **alle Blicke auf sich ~:** attract *or* capture all the attention; **jmds. Zorn/Unwillen** *usw.* **auf sich ~:** incur sb.'s anger/displeasure *etc.*; **etw. nach sich ~:** result in sth.; entail sth.; **c)** *(heraus~)* pull out ⟨*nail, cork, organ-stop, etc.*⟩; extract ⟨*tooth*⟩; take out, remove ⟨*stitches, splinter*⟩; draw ⟨*cord, sword, pistol*⟩; **den Hut ~:** raise *or* doff one's hat; **etw. aus der Tasche ~:** take sth. out of one's pocket; **Zigaretten/Süßigkeiten** *usw.* **~** *(ugs.: aus Automaten)* get cigarettes/sweets *etc.* from a slot-machine; **die [Quadrat]wurzel ~** *(Math.)* extract the square root; **d)** *(dehnen)* stretch ⟨*elastic etc.*⟩; stretch out ⟨*sheets etc.*⟩; **e)** *(Gesichtspartien* *bewegen)* make ‹*face, grimace*›; **die Augenbrauen nach oben ~:** raise one's eyebrows; **die Stirn in Falten ~:** wrinkle *or* knit one's brow; *(mißmutig)* frown; **f)** *(bei Brettspielen)* move ‹*chess-man etc.*›; **g) er zog den Rauch in die Lungen** he inhaled the smoke [into his lungs]; **h)** *(zeichnen)* draw ‹*line, circle, arc, etc.*›; **i)** *(anlegen)* dig ‹*trench*›; build ‹*wall*›; erect ‹*fence*›; put up ‹*washing-line*›; run, lay ‹*cable, wires*›; draw ‹*frontier*›; trace ‹*loop*›; follow ‹*course*›; **sich** *(Dat.)* **einen Scheitel ~:** make a parting [in one's hair]; **j)** *(auf~)* grow ‹*plants, flowers*›; breed ‹*animals*›; **k)** *(verblaßt; auch als Funktionsverb)* draw ‹*lesson, conclusion, comparison*›; *s. auch* **Konsequenz a; Rechenschaft; Verantwortung a. 2.** *unr. itr. V.* **a)** *(reißen)* pull; **an etw.** *(Dat.)* **~:** pull on sth.; **der Hund zieht an der Leine** the dog is straining at the leash; **an einem** *od.* **am selben Strang ~** *(fig.)* be pulling in the same direction; **b)** *(funktionieren)* ‹*stove, pipe, chimney*› draw; ‹*car, engine*› pull; **c)** *mit sein (um~)* move **(nach, in +** *Akk.* to); **zu jmdm. ~:** move in with sb.; **d)** *mit sein (gehen)* go; *(marschieren)* march; *(umherstreifen)* roam; rove; *(fortgehen)* go away; leave; ‹*fog, clouds*› drift; **durch etw. ~:** pass through sth.; **in den Krieg ~:** go *or* march off to war; **die Schwalben ~ nach Süden** the swallows are flying southwards; **e)** *(saugen)* draw; **an einer Zigarette/Pfeife ~:** draw on a cigarette/pipe; **an einem Strohhalm ~:** suck at a straw; **f)** ‹*tea, coffee*› draw; **g)** *(Kochk.)* simmer; **h)** *unpers.* **es zieht [vom Fenster her]** there's a draught [from the window]; **i)** *(ugs.: zum Erfolg führen)* ‹*trick*› work; **das zieht bei mir nicht** that won't wash *or* won't cut any ice with me *(coll.)*; **j)** *(schmerzen)* **es zieht [mir] im Rücken** I've got backache; **ein leichtes/starkes Ziehen im Bauch** a slight/intense stomach-ache. **3.** *unr. refl. V.* **a)** ‹*road*› run, stretch; ‹*frontier*› run; **b) der Weg** *o.ä.* **zieht sich** *(ugs.)* the journey *etc.* goes on and on

Zieh·harmonika die piano accordion

Ziehung die; ~, ~en draw; **die ~ des Hauptgewinns** the draw for the main prize

Ziel [tsi:l] **das; ~[e]s, ~e a)** destination; **am ~ der Reise anlangen** reach the end of one's journey; reach one's destination; **b)** *(Sport)* finish; **c)** *(~scheibe; auch Milit.)* target; **über das ~ hinausschießen** *(fig.)* go too far; **d)** *(Zweck)* aim; goal; **sein ~ erreichen** achieve one's objective *or* aim; **[das] ~ unserer Bemühungen ist es, ... zu ...:** the object of our efforts is to ...; **sich** *(Dat.)* **ein ~ setzen** *od.* **stecken** set oneself a goal; **sich** *(Dat.)* **etw. zum ~ setzen** set oneself *or* take sth. as one's aim; **etw. zum ~ haben** have sth. as its goal

Ziel·band das *Pl.* **~bänder** *(Sport)* finishing-tape

ziel·bewußt 1. *Adj.* purposeful; determined. **2.** *adv.* purposefully; determinedly

zielen *itr. V.* **a)** aim **(auf +** *Akk.* at); **b) auf jmdn./etw. ~** *(fig.)* ‹*reproach, plan, efforts, etc.*› be aimed at sb./sth.

ziel-, Ziel-: ~gerade die *(Sport)* finishing-straight; **~los 1.** *Adj.* aimless; **2.** *adv.* aimlessly; **~losigkeit die; ~:** aimlessness; **~scheibe die** *(auch fig.)* target *(Gen.* for); **~setzung die; ~, ~en** aims *pl.*; objectives *pl.*; **~sicher 1.** *Adj.* decisive, purposeful ‹*steps*›; **2.** *adv.* decisively; **~strebig 1.** *Adj.* **a)** purposeful; **b)** *(energisch)* single-minded ‹*person*›; **2.** *adv.* **a)** purposefully; **b)** *(energisch)* single-mindedly; **~strebigkeit die; ~** *s.* **~strebig: a)** purposefulness; **b)** single-mindedness

ziemlich 1. *Adj. (ugs.)* fair, sizeable ‹*quantity, number*›; **mit ~er Lautstärke** quite loudly. **2.** *adv.* **a)** quite; fairly; *(etwas intensiver)* pretty; **du kommst ~ spät** you're rather late; **~ viele Leute** quite a few people; **b)** *(ugs.: fast)* pretty well; more or less

Zier [tsi:ɐ] **die; ~** *(veralt.) s.* **Zierde**

Zierat ['tsi:ra:t] **der; ~[e]s, ~e** *(geh.)* ornament[ation]; **bloßer ~ sein** be purely ornamental

Zierde die; ~, ~n *(auch fig.)* ornament; embellishment; **zur ~:** as decoration; **jmdm. zur ~ gereichen** *(fig.)* be a credit to sb.

zieren 1. *tr. V. (geh.)* adorn; decorate ‹*room*›. **2.** *refl. V.* be coy; *(sich bitten lassen)* need some coaxing *or* pressing

zierlich 1. *Adj.* dainty; delicate; petite, dainty ‹*woman, figure*›. **2.** *adv.* daintily; delicately

Zierlichkeit die; ~: daintiness; delicateness; *(einer Frau, Gestalt)* petiteness; daintiness

Zier-: ~pflanze die ornamental plant; **~stich der** *(Handarb.)* ornamental stitch; **~strauch der** ornamental shrub

Ziffer ['tsɪfɐ] **die; ~, ~n** numeral; *(in einer mehrstelligen Zahl)* digit; figure

Ziffer·blatt das dial; face

zig [tsɪç] *unbest. Zahlwort (ugs.)* umpteen *(coll.)*

Zigarette [tsiga'rɛtə] **die; ~, ~n** cigarette

Zigaretten-: ~länge die: auf *od.* **für eine ~länge** *(ugs.)* just for a smoke; **~papier das** cigarette-paper; **~pause die** *(ugs.)* break for a smoke; **~raucher der** cigarette-smoker

Zigarillo [tsiga'rɪlo] **der** *od.* **das; ~, ~s** cigarillo; small cigar

Zigarre [tsi'garə] **die; ~, ~n** cigar

Zigarren·raucher der cigar-smoker

Zigeuner [tsi'gɔynɐ] **der; ~s, ~, Zigeunerin die; ~, ~nen** gypsy

zig·mal *Adv. (ugs.)* umpteen times *(coll.)*

zigst ... *Ordinalz. (ugs.)* umpteenth *(coll.)*

zig·tausend *unbest. Zahlwort (ugs.)* umpteen thousand *(coll.)*

Zikade [tsi'ka:də] **die; ~, ~n** cicada

Zimmer ['tsɪmɐ] **das; ~s, ~:** room

Zimmer-: ~flucht die; *Pl.* **~en** suite [of rooms]; **~lautstärke die** domestic listening level; **das Radio auf ~lautstärke stellen** turn the radio down to a reasonable volume [so as not to disturb the neighbours]; **~mädchen das** chambermaid; **~mann der;** *Pl.* **~leute** carpenter

zimmern 1. *tr. V.* make ‹*shelves, coffin, etc.*›. **2.** *itr. V.* do carpentry; **an einem Regal ~:** be making a bookshelf

Zimmer-: ~suche die room-hunt; **auf ~suche sein** be looking for a room; **~temperatur die** room temperature; **~vermittlung die** accommodation office

zimperlich ['tsɪmpɐlɪç] *(abwertend)* **1.** *Adj.* timid; *(leicht angeekelt)* squeamish; *(prüde)* prissy; *(übertrieben rücksichtsvoll)* over-scrupulous. **2.** *s. Adj.:* timidly; squeamishly; prissily; over-scrupulously

Zimperlichkeit die; ~, ~en *(abwertend) s.* **zimperlich 1:** timidity; squeamishness; prissiness; over-scrupulousness

Zimt [tsɪmt] **der; ~[e]s, ~e** cinnamon

Zink [tsɪŋk] **das; ~[e]s** zinc

Zinke die; ~, ~n prong; *(eines Kamms)* tooth

Zinn [tsɪn] **das; ~[e]s a)** tin; **b)** *(Legierung)* pewter; **c)** *(Gegenstände)* pewter[ware]

Zinne die; ~, ~n merlon; **~n** battlements

Zinnie ['tsɪni̯ə] **die; ~, ~n** *(Bot.)* zinnia

Zinn·soldat der tin soldier

Zins [tsɪns] **der; ~es, ~en** interest; *(~satz)* interest rate; **~en tragen** *od.* **bringen** earn interest; **jmdm. etw. mit ~en** *od.* **mit ~ und Zinseszins zurückzahlen** *(fig.)* make sb. pay dearly for sth.

Zinses·zins der compound interest

zins-, Zins-: ~los 1. *Adj.* interest-free; **2.** *adv.* free of interest; **~rechnung die** calculation of interest; **~satz der** interest rate

Zipfel ['tsɪpfl] **der; ~s, ~** *(einer Decke usw.)* corner; *(Wurst~)* [tail-]end; *(einer ~mütze)* point; *(Spitze eines Sees usw.)* tip

Zipfel·mütze die [long-]pointed cap

Zipp ⓦⓩ [tsɪp] **der; ~s, ~s,** **Zipp·verschluß der** *(österr.) s.* **Reißverschluß**

zirka ['tsɪrka] *Adv.* about; approximately

Zirkel ['tsɪrkl̩] **der; ~s, ~ a)** *(Gerät)* [pair *sing.* of] compasses *pl.;* **b)** *(Kreis, Gruppe)* circle

zirkeln *tr. V.* measure out precisely

Zirkulation [tsɪrkula'tsi̯o:n] **die, ~, ~en** circulation

zirkulieren *itr. V.; auch mit sein* circulate

Zirkus ['tsɪrkʊs] **der; ~, ~se a)** circus; **b) mach nicht so einen ~!** *(ugs.)* don't make such a fuss!

Zirkus·zelt das big top

zirpen ['tsɪrpn̩] *itr. V.* chirp

zischen ['tsɪʃn̩] **1.** *itr. V.* **a)** hiss; ⟨*hot fat*⟩ sizzle; **b)** *mit sein* hiss; *(ugs.: flitzen)* whizz. **2.** *tr. V.* **a)** *(zischend sprechen)* hiss; **b) ein Bier/einen ~** *(ugs.)* knock back a beer *(coll.)*/knock one back *(coll.)*

Zisterne [tsɪs'tɛrnə] **die; ~, ~n** [underground] tank *or* cistern

Zitadelle [tsita'dɛlə] **die; ~, ~n** citadel

Zitat [tsi'ta:t] **das; ~[e]s, ~e** quotation **(aus** from)

Zither ['tsɪtɐ] **die; ~, ~n** zither

zitieren [tsi'ti:rən] *tr. V.* **a)** *auch itr.* quote **(aus, nach** from); *(Rechtsspr.: anführen)* cite; **..., ich zitiere: „..."** ... and I quote: '...'; **falsch ~:** misquote; **b)** *(vorladen, rufen)* summon **(vor** before, **zu** to)

Zitronat [tsitro'na:t] **das; ~[e]s** candied lemon-peel

Zitrone [tsi'tro:nə] **die; ~, ~n** lemon; **jmdn. auspressen** *od.* **ausquetschen wie eine ~** *(ugs.: ausfragen)* pump sb.

zitronen-, Zitronen-: ~falter der brimstone butterfly; **~gelb** *Adj.* lemon-yellow; **~presse die** lemon-squeezer; **~saft der** lemon-juice; **~säure die** citric acid; **~schale die** lemon-peel

Zitrus·frucht ['tsitrʊs-] **die** citrus fruit

zittern ['tsɪtɐn] *itr. V.* **a)** tremble **(vor +** *Dat.* with); *(vor Kälte)* shiver; ⟨*needle, arrow, leaf, etc.*⟩ quiver; *(beben)* ⟨*walls, windows*⟩ shake; **mit ~der Stimme** in a trembling *or* quavering voice; **b)** *(fig.)* tremble; quake; **vor jmdm./etw. ~:** be terrified of sb./sth.

zittrig 1. *Adj.* shaky; doddery ⟨*old man*⟩. **2.** *adv.* shakily

Zitze ['tsɪtsə] **die; ~, ~n** teat

zivil [tsi'vi:l] **1.** *Adj.* **a)** civilian ⟨*life, population*⟩; non-military ⟨*purposes*⟩; civil ⟨*aviation, marriage, law, defence*⟩; **b)** *(annehmbar)* decent; reasonable. **2.** *adv.* decently; reasonably

Zivil das; ~s civilian clothes *pl.;* **Polizist in ~:** plain-clothes policeman

Zivil-: ~bevölkerung die civilian population; **~courage die** courage of one's convictions; **~dienst der;** *o. Pl. s.* **Ersatzdienst**

Zivilisation [tsiviliza'tsi̯o:n] **die; ~, ~en** civilization

Zivilisations·krankheit die disease of modern civilization *or* society

zivilisieren *tr. V.* civilize

zivilisiert 1. *Adj.* civilized. **2.** *adv.* in a civilized way

Zivilist der; ~en, ~en civilian

Zivil-: ~kleidung die civilian clothes *pl;* **~luft·fahrt die** civil aviation; **~person die** civilian

ZK [tsɛt'ka:] **das; ~s, ~s** *Abk.* **Zentralkomitee**

Zobel ['tso:bl̩] **der; ~s, ~:** sable

Zofe ['tso:fə] **die; ~, ~n** *(hist.)* lady's maid

Zoff [tsɔf] **der; ~s** *(ugs.)* rowing *(coll.);* squabbling; **~ machen** cause trouble

zog *1. u. 3. Pers. Sg. Prät. v.* **ziehen**

zögerlich ['tsø:gɐlɪç] **1.** *Adj.* hesitant; tentative. **2.** *adv.* hesitantly; tentatively

zögern *itr. V.* hesitate; **ohne zu ~:** without hesitation; **nach einigem Zögern** after a moment's hesitation; **~d vorangehen** proceed hesitantly

Zögling ['tsø:klɪŋ] **der; ~s, ~e** *(veralt.)* boarding pupil; boarder

Zölibat [tsøli'ba:t] **das** *od.* **der; ~[e]s,~e** celibacy *no art.*

¹Zoll [tsɔl] **der; ~[e]s, Zölle** ['tsœlə] **a)** *(Abgabe)* [customs] duty; **b)** *o. Pl. (Behörde)* customs *pl.*

²Zoll der; ~[e]s, ~: inch

Zoll-: ~amt das customs house *or* office; **~beamte der, ~beamtin die** customs officer

zollen *tr. V. (geh.)* **jmdm. etw. ~:** accord sb. sth.; **jmdm. Respekt/Bewunderung ~:** show sb. respect/admiration; **jmdm./einer Sache Tribut ~:** pay tribute to sb./sth.

zoll-, Zoll-: ~erklärung die customs declaration; **~frei 1.** *Adj.* duty-free; free of duty *pred.;* **2.** *adv.* free of duty; **~kontrolle die** customs examination *or* check

Zöllner ['tsœlnɐ] **der; ~s, ~ a)** *(ugs. veralt.)* customs officer; **b)** *(hist.: Steuereintreiber)* tax-collector

Zoll·stock der folding rule

Zombie ['tsɔmbi] **der; ~[s], ~s** zombie

Zone ['tso:nə] **die; ~, ~n** zone

Zoo [tso:] **der; ~s, ~s** zoo; **im/in den ~:** at/to the zoo

Zoo·handlung die pet shop

Zoologe [tsoo'lo:gə] **der; ~n, ~n** zoologist

Zoologie die; ~: zoology *no art.*

Zoologin die; ~, ~nen zoologist

zoologisch 1. *Adj.* zoological; **~er Garten** zoological gardens *pl.* **2.** *adv.* zoologically

Zoom [zu:m] **das; ~s, ~s** *(Film, Fot.)* zoom [lens]

Zoo-: ~tier das zoo animal; **~wärter der** zookeeper

Zopf [tsɔpf] **der; ~[e]s, Zöpfe** ['tsœpfə] plait; *(am Hinterkopf)* pigtail; **einen alten ~ abschneiden** *(fig.)* put an end to an antiquated custom *or* practice

Zopf·spange die hair-slide *(Brit.),* barnette *(Amer.) (for a pigtail)*

Zorn [tsɔrn] **der; ~[e]s** anger; *(stärker)* wrath; fury; **ihn packte der ~:** he flew into a rage; **einen ~ auf jmdn. haben** *(ugs.)* be furious with sb.; **im ~:** in a rage; in anger

Zorn·ausbruch der angry outburst; fit of rage

zornig 1. *Adj.* furious **(über +** *Akk.* about, **auf +** *Akk.* with)

Zote ['tso:tə] **die; ~, ~n** dirty joke

zotig 1. *Adj.* smutty; dirty ⟨*joke*⟩. **2.** *adv.* smuttily

Zottel ['tsɔtl̩] **die; ~, ~n** *(ugs. abwertend: Haare)* shaggy locks

zottelig *Adj.* shaggy

zotteln *itr. V.; mit sein (ugs.)* saunter; amble

zottig *Adj.* shaggy

Ztr. *Abk.* **Zentner** cwt.

zu [tsu:] **1.** *Präp. mit Dat.* **a)** *(Richtung)* to; **zu ... hin** towards ...; **er kommt zu mir** *(besucht mich)* he is coming to my place; **b)** *(zusammen mit)* with; **zu dem Käse gab es Wein** there was wine with the cheese; **das paßt nicht zu Bier/zu dem Kleid** that doesn't go with beer/with that dress; **c)** *(Lage)* at; **zu beiden Seiten** on both sides; **zu seiner Linken** *(geh.)* on his left; **er kam zu dieser Tür herein** he came in by this door; **der Dom zu Speyer** *(veralt.)* Speyer Cathedral; **das Gasthaus zu den drei Eichen** the Three Oaks Inn; **d)** *(zeitlich)* at; **zu Weihnachten** at Christmas; **was schenkst du ihnen zu Weihnachten** what will you give them for Christmas?; **er will zu Ostern verreisen** he wants to go away for Easter; **zu dieser Stunde** at this time; **e)** *(Art u. Weise)* **zu meiner Zufriedenheit/Überraschung** to my satisfaction/surprise; **zu seinem Vorteil/Nachteil** to his advantage/disadvantage; *(bei Mengenangaben o. ä)* **zu Dutzenden/zweien** by the dozen/in twos; **sie sind zu einem Drittel/zu 50% arbeitslos** a third/50% of them are jobless; **zu einem großen Teil** largely; to a large extent; **f)** *(ein Zahlenverhältnis ausdrükkend)* **ein Verhältnis von 3 zu 1** a ratio of 3 to 1; **das Ergebnis war 2 zu 1** the result was 2–1 *or* 2 to 1; **g)** *(einen Preis zuordnend)* at; for; **fünf Briefmarken zu fünfzig [Pfennig]** five 50-pfennig stamps; **h)** *(eine Zahlenangabe zuordnend)* **ein Faß zu zehn Litern** a ten-litre barrel; **Portionen zu je einem Pfund** portions weighing a pound each; **i)** *(Zweck)* for; **sie sagte das zu seiner Beruhigung** she said it to allay his fears; **j)** *(Ziel, Ergebnis)* into; **zu etw. werden** turn into sth.; **die Kartoffeln zu einem Brei zerstampfen** mash the potatoes into a puree; **k)** *(über)* about; on; **sich zu etw. äußern** comment on sth.; **zu welchem Thema spricht er?** what is he going to speak about?; **was sagst du zu meinem Vorschlag?** what do you say to my proposal?; **l)** *(gegenüber)* **freundlich/häßlich zu jmdm. sein** be friendly/nasty to sb.; *s. auch* **zum; zur. 2.** *Adv.* **a)** *(allzu)* too; **zu sehr** too much; **er ist zu alt, um diese Reise zu unternehmen** he is too old to undertake this journey; **das ist ja zu schön/komisch!** that's really wonderful/hilarious!; that's too wonderful/hilarious for words!; **b)** *nachgestellt (Richtung)* towards; **der Grenze zu** towards the border; **c)** *(ugs.) elliptisch* **Augen/Tür zu!** shut your eyes/the door!; **d)** *(ugs.: Aufforderung)* **nur zu!** *(fang/fangt an!)* get going!; get down to it!; *(mach/macht weiter!)* get on with it! **3.** *Konj.* **a)** *(mit Infinitiv)* to; **du hast zu gehorchen** you must obey; **was gibt's da zu lachen?** what is there to laugh about?; **das ist nicht zu glauben** it is unbelievable; **Haus zu verkaufen/vermieten** house for sale/to let; **b)** *(mit 1. Part.)* **die zu gewinnenden Preise** the prizes to be won; **die zu erledigende Post** the letters *pl.* to be dealt with

zu·aller·erst *Adv.* first of all; *(hauptsächlich)* above all else

zu·aller·letzt *Adv.* last of all

zu|bauen *tr. V. (ugs.)* block ‹*entrance, door*›; obstruct ‹*view*›

Zubehör ['tsu:bəhø:ɐ̯] **das; ~[e]s, ~e** *od. schweiz.* **~den** accessories *pl.; (eines Staubsaugers, Mixers o. ä.)* attachments *pl.; (Ausstattung)* equipment

zu|beißen *unr. itr. V.* bite

zu|bekommen *unr. tr. V.* get ‹*suitcase, door, etc.*› shut; get ‹*clothes, buttons*› done up; manage to repair ‹*leak*›*;* manage to mend ‹*hole*›

zu|bereiten *tr. V.* prepare ‹*meal, food, cocktail, etc.*›*; (kochen)* cook ‹*fish, meat, etc.*›

Zu·bereitung die; ~, ~en preparation; *(Kochen)* cooking

Zu·bett·gehen das; ~s: vorm/beim ~: before/on going to bed

zu|bewegen 1. *tr. V.* **etw. auf jmdn./etw. ~:** move sth. towards sb./sth. **2.** *refl. V.* **sich auf etw. ~:** move towards sth.

zu|billigen *tr. V.* **jmdm. etw. ~:** grant *or* allow sb. sth.; **jmdm. ~, daß er in gutem Glauben gehandelt hat** accept that sb. acted in good faith; **jmdm. mildernde Umstände ~:** allow sb.'s plea of extenuating circumstances

zu|binden *unr. tr. V.* tie [up]

zu|blinzeln *itr. V.* **jmdm. ~:** wink at sb.

zu|bringen *unr. tr. V.* **a)** *(verbringen)* spend; **b)** *(landsch.) s.* **zubekommen**

Zu·bringer der; ~s, ~ a) *(Straße)* access *or* feeder road; **b)** *(Verkehrsmittel)* shuttle; *(Flughafenbus o. ä.)* courtesy bus

Zu·brot das; *o. Pl.* bit extra *or* on the side

zu|buttern *tr., itr. V. (ugs.)* chip in *(coll.)*

Zucht [tsʊxt] **die; ~, ~en a)** *o. Pl. (von Tieren)* breeding; *(von Pflanzen)* cultivation; breeding; *(von Bakterien, Perlen)* culture; **ein Pferd aus deutscher ~:** a German-bred horse; **b)** *(Einrichtung)* breeding establishment; *(für Pferde)* stud; *(für Pflanzen)* plant-breeding establishment; **c)** *o. Pl. (geh.: Disziplin)* discipline; **für ~ und Ordnung sorgen** keep order

züchten ['tsʏçtn̩] *tr. V. (auch fig.)* breed; cultivate ‹*plants*›*;* culture ‹*bacteria, pearls*›

Züchter der; ~s, ~, Züchterin, die; ~, ~nen breeder; *(von Pflanzen)* grower [of new varieties]; plant-breeder

Zucht·haus das a) [long-stay] prison; penitentiary *(Amer.);* **b)** *o. Pl. (Strafe)* [severest form of] imprisonment; imprisonment in a penitentiary *(Amer.)*

züchtigen ['tsʏçtɪgn̩] *tr. V. (geh.)* beat; thrash

zucht·los *(veralt.)* **1.** *Adj.* undisciplined. **2.** *adv.* without discipline; in an undisciplined way

Zucht·perle die cultured pearl

Züchtung die; ~, ~en a) breeding; *(von Pflanzen)* cultivation; **b)** *(Zuchtergebnis)* strain

zuck [tsʊk] *s.* **ruck, zuck**

zuckeln ['tsʊkl̩n] *itr. V.; mit sein* saunter; amble; *(schleppend)* trail; ‹*cart etc.*› trundle

zucken *itr. V.; mit Richtungsangabe mit sein* twitch; ‹*body, arm, leg, etc.*› jerk; *(vor Schreck)* start; ‹*flames*› flicker, flare up; ‹*light, lightning*› flicker, flash; **er zuckte zur Seite** he jumped to one side; **mit den Achseln** *od.* **Schultern ~:** shrug one's shoulders

zücken ['tsʏkn̩] *tr. V.* draw ‹*sword, dagger, knife*›*; (scherzh.)* take out, produce ‹*wallet, notebook, camera, etc.*›

Zucker der; ~s, ~ a) sugar; **b)** *o. Pl. (ugs.: ~krankheit)* diabetes; **~ haben** be a diabetic

Zucker·brot das: mit ~ und Peitsche with a carrot and a stick

zucker-, Zucker-: ~dose die sugar bowl; **~hut der** sugar loaf; **~krank** *Adj.* diabetic; **~krankheit die** diabetes

Zuckerl das; ~s, ~[n] *(südd., österr.)* sweet *(Brit.);* candy *(Amer.); (fig.)* sweetener; enticement

zuckern *tr. V.* sugar

zucker-, Zucker-: ~rohr das

sugar cane; **~rübe die** sugar beet; **~stange die** stick of rock; **~streuer der** sugar-caster; **~süß 1.** *Adj.* as sweet as sugar *postpos.;* beautifully sweet; *(fig. abwertend)* saccharine, sugary ⟨*picture, smile, etc.*⟩; **2.** *adv.* **~süß lächeln** *(fig. abwertend)* give a saccharine *or* sugary smile; **~watte die** candy-floss

zuckrig *Adj.* sugary

Zuckung die; ~, ~en twitch

zu|decken *tr. V.* cover up; cover [over] ⟨*well, ditch*⟩; **sich ~:** tuck oneself up; **gut/warm zugedeckt** well/warmly tucked up

zu·dem *Adv. (geh.)* moreover; furthermore

zu|drehen 1. *tr. V.* **a)** *(abdrehen)* turn off ⟨*tap, heating, water, gas*⟩; *(schließen)* screw ⟨*valve, container*⟩ shut; **b)** *s.* **zuwenden 2a. 2.** *refl. V.* **sich jmdm./einer Sache ~:** turn to *or* towards sb./sth.

zu·dringlich *Adj.* pushy *(coll.),* pushing ⟨*person, manner*⟩; *(sexuell)* importunate ⟨*person, manner*⟩; prying ⟨*glance*⟩

Zu·dringlichkeit die; ~, ~en a) *o. Pl.* pushiness *(coll.); (sexuell)* importunate manner; **b)** *Pl.* insistent advances *or* attentions

zu|drücken *tr. V.* press shut; push ⟨*door*⟩ shut; **jmdm. die Kehle ~:** choke *or* throttle sb.; *s. auch* **Auge a**

zu|eignen *tr. V. (geh.)* **jmdm. etw. ~:** dedicate sth. to sb.

Zu·eignung die; ~, ~en dedication

zu|eilen *itr. V.; mit sein* **auf jmdn./etw. ~:** hurry *or* rush towards sb./sth.

zu·ein·ander *Adv.* to one another; **Liebe ~ empfinden** have feelings of love towards one another; **gut/schlecht ~ passen** ⟨*things*⟩ go well together/not match; ⟨*people*⟩ be well-/ill-suited

zueinander-: ~|finden *unr. itr. V.* come together; **~|halten** *unr. itr. V.* stick together; **~|stehen** *unr. itr. V.* stand by one another; stick together

zu|erkennen *unr. tr. V.* **jmdm. eine Entschädigung/einen Preis ~:** award sb. compensation/a prize; **jmdm. einen Titel ~:** confer a title on sb.

Zuerkennung die; ~, ~en *s.* **zuerkennen:** award; conferring

zu·erst *Adv.* **a)** first; **er war ~ da** he was here first; he was the first to come; **b)** *(anfangs)* at first; to start with; **c)** *(erstmals)* first; for the first time

zu|fahren *unr. itr. V.; mit sein* **a) auf jmdn./etw. ~:** head towards sb./sth.; **auf jmdn./etw. zugefahren kommen** come towards sb./sth.; **b)** *(ugs.: los-, weiterfahren)* get a move on *(coll.);* **fahr zu!** step on it! *(coll.)*

Zu·fahrt die a) *o. Pl.* access [for vehicles]; **b)** *(Straße, Weg)* access road; *(zum Haus)* driveway

Zufahrts·straße die access road

Zu·fall der chance; *(zufälliges Zusammentreffen von Ereignissen)* coincidence; **es war [ein] reiner ~:** it was pure chance *or* coincidence; **es ist kein ~, daß ...:** it is no accident that ...; **durch ~:** by chance *or* accident; **daß wir uns dort begegneten, war ~:** our meeting there was a coincidence

zu|fallen *unr. itr. V.; mit sein* **a)** ⟨*door etc.*⟩ slam shut; ⟨*eyes*⟩ close; **ihm fielen [vor Müdigkeit] die Augen zu** his eyelids were drooping [with tiredness]; **b)** *(zukommen)* **jmdm. ~** ⟨*task*⟩ fall to sb.; ⟨*prize, inheritance*⟩ go to sb.; **ihm fällt alles nur so zu** everything just drops into his lap

zu·fällig 1. *Adj.* accidental; chance *attrib.* ⟨*meeting, acquaintance*⟩; random ⟨*selection*⟩. **2.** *adv.* by chance; **ich bin ~ hier vorbeigekommen** I just happened to be passing; **wissen Sie ~, wie spät es ist?** *(ugs.)* do you by any chance know the time?

zufälliger·weise *Adv. s.* **zufällig 2**

Zufalls-: ~auswahl die random selection; **~bekanntschaft die** chance acquaintance; **~treffer der** fluke

zu|fassen *itr. V.* make a snatch *or* grab

zu|fliegen *unr. itr. V.; mit sein* **a) auf jmdn./etw. ~:** fly towards sb./sth.; **es kam auf mich zugeflogen** it came flying towards me; **b) jmdm. ~** ⟨*bird*⟩ fly into sb.'s house; **ihm fliegen die Herzen zu** *(fig.)* all hearts surrender to his charms; **c)** *(ugs.: zufallen)* ⟨*door etc.*⟩ slam shut

zu|fließen *unr. itr. V.; mit sein* **einer Sache** *(Dat.)* **~:** flow towards sth.; **jmdm./einer Sache ~** *(fig.)* ⟨*money etc.*⟩ go to sb./sth.

Zu·flucht die refuge (**vor** + *Dat.* from); *(vor Unwetter o. ä.)* shelter (**vor** + *Dat.* from); **[seine] ~ zu etw. nehmen** *(fig.)* resort to sth.

Zufluchts·ort der place of refuge; sanctuary

Zu·fluß der feeder stream/river

zu|flüstern *tr. V.* **jmdm. etw. ~:** whisper sth. to sb.

zu·folge *Präp. mit Dat.; nachgestellt* according to

zu·frieden 1. *Adj.* contented; *(befriedigt)* satisfied; **mit etw. ~ sein** be satisfied with sth.; **bist du jetzt ~?** are you satisfied [now]?; **ein ~es Gesicht machen** look contented *or* satisfied; **wir können ~ sein** we can't complain. **2.** *adv.* contentedly

zufrieden|geben *unr. refl. V.* be satisfied

Zufriedenheit die; ~: contentment; *(Befriedigung)* satisfaction; **zu meiner vollen ~:** to my complete satisfaction

zufrieden|stellen *tr. V.* satisfy

zufriedenstellend 1. *Adj.* satisfactory. **2.** *adv.* satisfactorily

zu|frieren *unr. itr. V.; mit sein* freeze over

zu|fügen *tr. V.* **jmdm. etw. ~:** inflict sth. on sb.; **jmdm. Schaden/[ein] Unrecht ~:** do sb. harm/an injustice

Zufuhr ['tsu:fu:ɐ̯] **die; ~:** supply; *(Material)* supplies *pl.;* **die ~ milder Meeresluft** the stream of mild sea air

zu|führen 1. *itr. V.* **auf etw.** *(Akk.)* **~:** lead towards sth. **2.** *tr. V.* **einer Sache** *(Dat.)* **etw. ~:** supply sth. to sth.; supply sth. with sth.; **einer Firma Kunden/einer Partei Mitglieder ~:** bring new customers to a firm/new members to a party; **jmdn. der gerechten Strafe ~:** ensure that sb. gets condign punishment

Zug [tsu:k] **der; ~[e]s, Züge** ['tsy:gə] **a)** *(Bahn)* train; **ich nehme lieber den ~** *od.* **fahre lieber mit dem ~:** I prefer to go by train *or* rail; **jmdn. vom ~ abholen/zum ~ bringen** meet sb. off/take sb. to the train; **b)** *(Kolonne)* column; *(Umzug)* procession; *(Demonstrations~)* march; **c)** *(das Ziehen)* pull; traction *(Phys.);* **das ist der ~ der Zeit** *(fig.)* this is the modern trend *or* the way things are going; **d)** *(Wanderung)* migration; *(Streif~, Beute~, Diebes~)* expedition; **e)** *(beim Brettspiel)* move; **du bist am ~:** it's your move; **zum ~e kommen** *(fig.)* get a chance; **f)** *(Schluck)* swig *(coll.);* mouthful; *(großer Schluck)* gulp; **das Glas auf einen** *od.* **in einem ~ leeren** empty the glass at one go; **einen Roman in einem ~ durchlesen** *(fig.)* read a novel at one sitting; **er hat einen guten ~** *(ugs.)* he can really knock it back *(coll.);* **etw. in vollen Zügen genießen** *(fig.)* enjoy sth. to the full; **g)** *(beim Rauchen)* pull; puff; drag *(coll.)*; **h)** *(Atem~)* breath; **in tiefen** *od.* **vollen Zügen** in deep breaths; **in den letzten Zügen liegen** *(ugs.)* be at death's door; *(fig. scherzh.)* ‹*car, engine, machine*› be at its last gasp; ‹*project etc.*› be on the last lap; **i)** *o. Pl. (Zugluft; beim Ofen)* draught; **im ~ sitzen** sit in a draught; **j)** *(Gesichts~)* feature; trait; *(Wesens~)* characteristic; trait; **seine Züge** his features; **die Stadt trägt noch dörfliche Züge** the town still has something of the village about it; **das war kein schöner ~ von ihr** that did her no credit; **k)** *(landsch.: Schublade)* drawer; **l)** *(Bewegung eines Schwimmers od. Ruderers)* stroke; **m)** *(Milit.: Einheit)* platoon; **n)** *(Schulw.: Zweig)* side; **o)** *(Höhen~)* range; chain

Zu·gabe die a) *(Geschenk)* [free] gift; **b)** *(im Konzert, Theater)* encore; **c)** *o. Pl. (das Zugeben)* addition

Zug·abteil das [train] compartment *(Brit.)*

Zu·gang der a) *(Weg)* access; *(Eingang)* entrance; **b)** *(das Betreten, Hineingehen)* access; **c)** *(fig.)* access; **~ zu jmdm./etw. finden** be able to relate to sb./sth.; **d)** *s.* **Neuzugang**

zu·gange: ~ sein *(ugs.)* be busy *or* occupied

zugänglich ['tsu:gɛŋlɪç] *Adj.* **a)** accessible; *(geöffnet)* open; **schwer ~:** difficult to reach *pred.*; **die Zimmer sind von der Terrasse her ~:** the rooms can be reached from the terrace; **b)** *(zur Verfügung stehend)* available *(Dat.,* **für** to); **c)** *(aufgeschlossen)* approachable ‹*person*›; **für neue Ideen** *usw.* **~ sein** be amenable *or* receptive to new ideas *etc.*

Zugänglichkeit die; ~ a) accessibility; **b)** *(Aufgeschlossenheit)* receptiveness **(gegenüber** to)

Zug-: ~an·schluß der [train] connection; **~brücke die** drawbridge

zu·geben *unr. tr. V.* **a)** *(hinzufügen)* add *(Dat.* to); **b)** *(gestehen, zugestehen)* admit; admit, confess ‹*guilt, complicity*›; admit to, confess to ‹*deed, crime*›; **sie gab zu, es gestohlen zu haben** she admitted stealing it *or* having stolen it; **es war, zugegeben, viel Glück dabei** true, there was a lot of luck involved

zu·gegebener·maßen *Adv.* admittedly

zu·gegen *Adj.*: **~ sein** *(geh.)* be present

zu|gehen *unr. itr. V.; mit sein* **a)** **auf jmdn./etw. ~:** approach sb./sth.; **aufeinander ~** *(fig.)* try to come together; **dem Ende ~:** be coming to an end; **b)** *(ugs.: vorangehen)* get a move on *(coll.)*; step on it *(coll.)*; **c)** *(Papierdt.)* **jmdm. ~:** be sent to sb.; **jmdm. etw. ~ lassen** send sth. to sb.; **d)** *unpers. (verlaufen)* **auf dem Fest ging es fröhlich zu** it was very jolly at the party; **es müßte seltsam ~, wenn das nicht gelänge** something remarkable would have to happen for that not to succeed; **es geht nicht mit rechten Dingen zu** there is something fishy going on *(coll.)*; **e)** *(ugs.: sich schließen)* close; shut; **f)** *(ugs.: sich schließen lassen)* **die Tür/der Knopf geht nicht zu** the door will not shut/the button will not fasten; **der Reißverschluß geht schwer zu** the zip is difficult to do up

zu|gehören *itr. V. (geh.)* **jmdm./einer Sache ~:** belong to sb./sth.

zu·gehörig *Adj.* belonging to it/them *postpos., not pred.*; **einer Sache** *(Dat.)* **~:** belonging to sth.; **sich jmdm./einer Sache** *(Dat.)* **~ fühlen** have a feeling of belonging [to sb./sth.]

Zugehörigkeit die; ~: belonging **(zu** to)

zu·geknöpft *Adj. (fig. ugs.)* tight-lipped; *(nicht zugänglich)* unapproachable

Zügel [tsy:gl̩] **der; ~s, ~ a)** rein; **ein Pferd am ~ führen** lead a horse by the reins; **einem Pferd in die ~ fallen** stop a horse by seizing the reins; **b)** *(fig.)* **die ~ [fest] in der Hand haben** be [firmly] in control; have things [firmly] under control; **die ~ straffer anziehen** tighten up on things; **die ~ schießen lassen** let things take their course; **die ~ schleifen lassen** *od.* **lockern** slacken the reins

zügel·los *(fig.)* **1.** *Adj.* unrestrained; unbridled ‹*rage, passion*›; **ein ~es Leben führen** live a life of licentious indulgence. **2.** *adv.* without restraint; **~ leben** live a life of licentious indulgence

Zügellosigkeit die; ~, ~en lack of restraint

zügeln *tr. V.* **a)** rein [in] ‹*horse*›; **b)** *(fig.)* curb, restrain ‹*feeling, desire, curiosity, etc.*›; **sich ~:** restrain oneself

Zügelung die; ~, ~en curbing; restraining

Zu·gereiste der/die; *adj. Dekl.* newcomer

zu|gesellen *refl. V.* **sich jmdm./einer Sache ~:** join sb./sth.

Zu·geständnis das concession **(an +** *Akk.* to)

zu|gestehen *unr. tr. V.* **a)** grant ‹*right, claim, share, etc.*›; allow ‹*discount, commission, time*›; **b)** *(zugeben)* admit; concede

zu·getan *Adj.*: **jmdm. [herzlich] ~ sein** *(geh.)* be [very] attached to sb.; **den schönen Künsten ~ sein** have a penchant for the fine arts

Zu·gewinn der gain **(an +** *Dat.* in)

Zu·gezogene der/die; *adj. Dekl.* newcomer

Zug·führer der a) *(Eisenb.)* guard; **b)** *(Milit.)* platoon sergeant

zu|gießen *unr. tr.* add

zugig *Adj.* draughty, *(im Freien)* windy ‹*corner etc.*›

zügig ['tsy:gɪç] **1.** speedy; rapid. **2.** *adv.* speedily; rapidly

Zügigkeit die; ~: speediness; rapidity

Zug·kraft die *(fig.)* attraction

zug·kräftig *Adj.* effective ‹*publicity*›; powerful ‹*argument*›; influential ‹*name*›; catchy ‹*title, slogan*›

zu·gleich *Adv.* at the same time; **er ist Maler und Dichter ~:** he is both a painter and a poet

Zug-: ~luft die; *o. Pl.* draught; **~luft [ab]bekommen** be in a draught; **~pferd das a)** draught-horse; **b)** *(fig.: Attraktion)* big draw; crowd-puller

zu|greifen *unr. itr. V.* **a)** take hold; **b)** *(sich bedienen)* help oneself; **c)** *(fleißig arbeiten)* **[hart** *od.* **kräftig] ~:** [really] knuckle down to it

Zu·griff der a) grasp; **sich dem ~ der Polizei entziehen** escape the clutches of the police; **b)** *(fig.: Zugang)* access **(auf +** *Akk.* to)

zu·grunde *Adv.* **a) ~ gehen** *(sterben)* die **(an +** *Dat.* of); *(zerstört werden)* be destroyed **(an +** *Dat.* by); ‹*marriage*› founder **(an +** *Dat.* owing to); ‹*person*› go under; *(finanziell)* be ruined; ‹*company*› go to the wall; **~ richten** destroy; *(finanziell)* ruin ‹*company, person*›; **b) etw. ~ legen** use sth. as a basis; **etw. einer Sache** *(Dat.)* **~ legen** base sth. on sth.; **A liegt B** *(Dat.)* **~:** B is based on A

Zugs- *(österr.) s.* **Zug-**

zu|gucken *itr. V. (ugs.) s.* **zusehen**

Zug·unglück das train crash

zu·gunsten 1. *Präp. mit Gen.* in favour of. **2.** *Adv.* **~ von** in favour of

zu·gut: etw. ~ haben *(schweiz., südd.)* be owed sth.; **du hast [bei**

mir] 10 Mark ~: you've got ten marks to come [from me]

zu·gute *Adv.*: **jmdm. seine Jugend/Unerfahrenheit** *usw.* **~ halten** *(geh.)* take sb.'s youth/inexperience *etc.* into consideration; make allowances for sb.'s youth/inexperience *etc.*; **jmdm./einer Sache ~ kommen** stand sb./sth. in good stead; **jmdm. etw. ~ kommen lassen** let sb. have the benefit of *or* let sb. benefit from sth.

Zug-: **~verbindung die** rail *or (Amer.)* railroad service; **~verkehr der** rail *or (Amer.)* railroad traffic; **~vogel der** migratory bird

zu|haben *unr. itr. V. (ugs.)* **a)** ⟨*shop, office*⟩ be shut *or* closed; **wir haben montags zu** we are closed on Mondays; **b) endlich hat sie den Koffer/Reißverschluß zu** at last she's managed to shut the suitcase/do up the zip

zu|haken *tr. V.* hook up; do up the hooks on

zu|halten 1. *unr. tr. V.* hold closed; *(nicht öffnen)* keep closed; **jmdm./sich die Augen/den Mund** *usw.* **~**: put one's hand[s] over sb.'s/one's eyes/mouth *etc.*; **sich** *(Dat.)* **die Nase ~**: hold one's nose. **2.** *itr. V.* **auf etw.** *(Akk.)* **~**: head for sth.

zu|hängen *tr. V.* cover ⟨*window, cage*⟩

zu|hauen 1. *unr. itr. V.* **a)** *(ugs.)* bang *or* slam ⟨*door, window*⟩ shut; **b)** *(behauen)* hew into shape. **2.** *unr. itr. V. (ugs.)* hit *or* strike out

Zu·hause das; ~s home

zu|heilen *itr. V.; mit sein* heal [over]

zu·hinterst *Adv.* right at the back

zu|hören *itr. V.* listen (*Dat.* to); **nun hör mal zu** now listen; *(drohend)* now [you] listen here; **er kann gut ~**: he's a good listener

Zu·hörer der, Zu·hörerin die listener

zu|jubeln *itr. V.* **jmdm. ~**: cheer sb. [on]

zu|kehren *tr. V. s.* **zuwenden 2a**

zu|klappen 1. *tr. V.* close; fold ⟨*penknife*⟩ shut. **2.** *itr. V.; mit sein* ⟨*window, lid, etc.*⟩ click to *or* shut

zu|kleben *tr. V.* **a)** seal ⟨*letter, envelope*⟩; **b)** *(vollkleben)* cover

zu|knallen *(ugs.)* **1.** *tr. V.* slam. **2.** *itr. V.; mit sein* slam

zu|kneifen *unr. tr. V.* squeeze ⟨*eye[s]*⟩ shut; shut ⟨*eye[s]*⟩ tight; shut ⟨*mouth*⟩ tightly

zu|knöpfen *tr. V.* button up

zu|knoten *tr. V.* knot; tie up

zu|kommen *itr. V.; mit sein* **a) auf jmdn. ~**: approach sb.; *(zu jmdm. kommen)* come up to sb.; **er/es kam direkt auf mich zu** he/it came straight towards me; **er ahnte nicht, was noch auf ihn ~ sollte** *(fig.)* he had no idea what he was in for; **die Dinge auf sich ~ lassen** *(fig.)* take things as they come; **b)** *(geh.)* **jmdm. etw. ~ lassen** *(schicken)* send sb. sth.; *(schenken)* give sb. sth.; **jmdm./einer Sache Pflege ~ lassen** devote care to sth.; **c)** *s.* **zustehen**; **d)** *(beizumessen sein)* **dieser Entdeckung kommt große Bedeutung zu** great significance must be attached to this discovery

zu|korken *tr. V.* cork

zu|kriegen *tr. V. (ugs.) s.* **zubekommen**

Zukunft ['tsu:kʊnft] **die**; ~, **Zukünfte** ['tsu:kynftə] **a)** future; **für alle ~**: for all time; **~/keine ~ haben** have a/no future; **in naher/ferner ~**: in the near *or* immediate/distant future; **in ~**: in future; **ich wünsche Ihnen alles Gute für Ihre weitere ~**: I wish you all the best for the future; **b)** *(Sprachw.)* future [tense]

zu·künftig 1. *Adj.* future. **2.** *Adv.* in future

Zukünftige der/die; *adj. Dekl. (ugs.)* **mein ~r/meine ~**: my husband/wife-to-be; my intended *(joc.)*

Zukunfts-: **~forschung die** futurology *no art.*; **~roman der** novel set in the future

zu|lächeln *itr. V.* **jmdm. ~**: smile at sb.

Zulage die *(vom Arbeitgeber)* extra pay *no indef. art.*; additional allowance *no indef. art.*; *(vom Staat)* benefit

zu|langen *itr. V. (ugs.) s.* **zugreifen b, c**

zu|länglich *(geh.)* **1.** *Adj.* adequate. **2.** *adv.* adequately

zu|lassen *unr. tr. V.* **a)** *(erlauben, dulden)* allow; permit; **ich lasse keine Ausnahme zu** I do not allow *or* permit any exceptions; **b)** *(teilnehmen lassen)* admit; **c)** *(mit einer Erlaubnis, Lizenz usw. versehen)* **jmdn. als Arzt ~**: register sb. as a doctor; **der Anwalt ist beim Amtsgericht Mannheim zugelassen** the lawyer is registered to practise at Mannheim district court; **jmdn. zu einer Prüfung ~**: allow *or* permit sb. to take an examination; **d)** *(zur Benutzung, zum Verkauf usw. freigeben)* allow; permit; **ein Medikament ~**: approve a medicine [for sale]; **für den öffentlichen Verkehr zugelassen sein** be authorized for use on public highways; **e)** *(Kfz-W.)* register ⟨*vehicle*⟩; **f)** *(geschlossen lassen)* leave closed *or* shut; leave ⟨*letter*⟩ unopened; leave ⟨*collar, coat*⟩ fastened [up]

zu·lässig *Adj.* permissible

Zulässigkeit die; ~ permissibility

Zulassung die; ~, ~en **a)** *(Erlaubnis, Lizenz)* ~ **als Arzt** registration as a doctor; ~ **zur Teilnahme/zur Prüfung beantragen** apply for permission to attend/to take *or (Brit.)* sit an examination; **ihm ist die ~ zum Studium/zum Medizinstudium erteilt worden** he has been accepted at university/to study medicine; **b)** *(Freigabe)* approval; authorization; **c)** *(Kfz-W.)* registration

Zu·lauf der *o. Pl.* **~ haben** ⟨*shop, restaurant, etc.*⟩ enjoy a large clientele, be very popular; ⟨*doctor, lawyer*⟩ have a large practice, be very much in demand

zu|laufen *unr. itr. V.; mit sein* **a) auf jmdn./etw. ~** *(auch fig.)* run towards sb./sth.; **auf jmdn./etw. zugelaufen kommen** come running towards sb./sth.; **b) jmdm. ~** ⟨*cat, dog, etc.*⟩ adopt sb. as a new owner; **c)** *(hinzulaufen)* ⟨*water etc.*⟩ run in; **d)** *(sich verjüngen)* taper; **spitz ~**: taper to a point; **e)** *(ugs.: schnell laufen)* get one's skates on *(Brit. sl.)*; get a move on *(coll.)*

zu|legen 1. *refl. V.* **sich** *(Dat.)* **etw. ~**: get oneself sth.; **er hat sich einen Bart zugelegt** *(ugs.)* he has grown a beard. **2.** *itr. V. (ugs.)* **a)** *(sein Tempo steigern)* step on it *(coll.)*; **b)** *(wachsen)* ⟨*sales, output, turnover, etc.*⟩ increase; **der Dollar hat [um vier Pfennige] zugelegt** the dollar has risen [four pfennigs]. **3.** *tr. V. (ugs.)* add; **einen Schritt/***(ugs.:)* **Zahn ~**: get a move on *(coll.)*

zu·leid[e]: **jmdm. etwas/nichts ~ tun** hurt *or* harm sb./not [do anything to] hurt *or* harm sb.; *s. auch* **Fliege a**

zu|leiten *tr. V.* **a)** feed; supply; supply ⟨*nourishment*⟩; **b)** *(schicken)* send; forward

Zu·leitung die a) *o. Pl. s.* **zuleiten**: supply; sending; forwarding; **b)** *(Rohr, Kabel usw.)* feed line

zu·letzt *Adv.* **a)** last [of all]; **an sich selbst denkt sie immer ~**: she always thinks of herself last; **er kommt immer ~**: he always comes last; he is always [the] last;

das ~ geborene Kind the child born last; **b)** *(fig.: am wenigsten)* least of all; **nicht ~:** not least; **c)** *(das letzte Mal)* last; **ich habe ihn ~ gestern abend gesehen** I last saw him yesterday evening; **d)** *(schließlich)* in the end; **bis ~:** [right up] to *or* until the end

zu·liebe *Adv.* **jmdm./einer Sache ~:** for sb.'s sake/for the sake of sth.

zu|löten *tr. V.* solder; solder up ⟨*hole*⟩

zum [tsʊm] *Präp.* + *Art.* **a)** = **zu dem; b)** *(räumlich: Richtung)* to the; **ein Fenster ~ Hof** a window on to *or* facing the yard; **wo geht es ~ Stadion?** which is the way to the stadium?; **c)** *(räumlich: Lage)* **etw. ~ Fenster hinauswerfen** throw sth. out of the window; **d)** *(Zusammengehörigkeit, Hinzufügung)* **Milch ~ Tee/Sahne ~ Kuchen nehmen** take milk with [one's] tea/have cream with one's cake; **e)** *(zeitlich)* at the; **spätestens ~ 15. April** by 15 April at the latest; **~ Schluß/richtigen Zeitpunkt** at the end/the right moment; **f)** *(Zweck)* **ein Gerät ~ Schneiden** an instrument for cutting [with]; **hol dir was ~ Schreiben** get something to write with; **~ Spaß/Vergnügen** for fun/pleasure; **~ Lesen braucht er eine Brille** he needs glasses for reading; **~ Schutz** as *or* for protection; **etw. ~ Essen/Lesen** *(österr.)* sth. to eat/read; **g)** *(Folge)* **~ Ärger/Leidwesen seines Vaters** to the annoyance/sorrow of his father; **h) jmdn. ~ Direktor ernennen/~ Kanzler wählen** appoint sb. director/elect sb. chancellor; **~ Dieb werden** become a thief; **i) ~ ersten, ~ zweiten, ~ dritten!** *(bei Versteigerung)* going, going, gone!

zu|machen 1. *tr. V.* close; shut; fasten, do up ⟨*dress*⟩; seal ⟨*envelope, letter*⟩; turn off ⟨*tap*⟩; put the top on ⟨*bottle*⟩; *(stillegen)* close *or* shut down ⟨*factory, mine, etc.*⟩; **ich habe kein Auge zugemacht** I didn't sleep a wink. **2.** *itr. V.* **a)** close; shut; **b)** *(ugs., bes. nordd.: sich beeilen)* get a move on *(coll.)*

zu·mal 1. *Adv.* especially; particularly; **~ da ...:** especially *or* particularly since ... **2.** *Konj.* especially *or* particularly since

zu|marschieren *itr. V.; mit sein* **auf jmdn./etw. ~:** march towards sb./sth.

zu|mauern *tr. V.* wall *or* brick up

zu·meist *Adj.* in the main; for the most part

zumindest *Adv.* at least

zumutbar *Adj.* reasonable; **das ist ihm kaum/durchaus/nicht ~:** one can scarcely/quite well/not expect that of him

Zumutbarkeit die; ~: reasonableness

zu·mute *Adj.* **jmdm. ist unbehaglich/elend** *usw.* **~:** sb. feels uncomfortable/wretched *etc.*; **mir war nicht danach ~:** I didn't feel like it *or* in the mood; **mir war zum Weinen ~:** I felt like crying

zu|muten *tr. V.* **a)** *(abverlangen)* **jmdm. etw. ~:** expect *or* ask sth. of sb.; **diese Arbeit möchte ich ihm nicht ~:** I would not like to ask him to do this work *or* impose this work on him; **das ist ihm durchaus/nicht zuzumuten** it is perfectly reasonable to/one cannot expect *or* ask that of him; **sich zuviel ~:** take on too much; overdo it; **b)** *(antun)* **jmdm. etw. ~:** expect sb. to put up with sth.; **diesen Anblick wollte ich ihm nicht ~:** I wanted to spare him this sight

Zumutung die; ~, ~en a) *(Ansinnen)* unreasonable demand; imposition; **eine ~ sein** be unreasonable; **etw. als [eine] ~ empfinden** consider sth. unreasonable; **b)** *(Belästigung)* imposition; **eine ~ für jmdn. sein** be an imposition on sb.; **das Essen war eine ~:** the meal was an affront

zu·nächst 1. *Adv.* **a)** *(als erstes)* first; *(anfangs)* at first; **~ einmal** first; **b)** *(vorläufig)* for the moment; for the time being

zu|nageln *tr. V.* nail up; **etw. mit Brettern ~:** board sth. up

zu|nähen *tr. V.* sew up; *s. auch* **verdammt 1 a; verflixt 1 b**

Zunahme ['tsu'na:mə] **die; ~, ~n** increase (*Gen.*, **an** + *Dat.* in)

Zu·name der surname; last name

zündeln ['tsʏndl̩n] *itr. V.* *(auch fig.)* play with fire

zünden ['tsʏndn̩] **1.** *tr. V.* ignite ⟨*gas, fuel, etc.*⟩; detonate ⟨*bomb, explosive device, etc.*⟩; let off ⟨*fireworks*⟩; fire ⟨*rocket*⟩. **2.** *itr. V.* ⟨*rocket, engine*⟩ fire; ⟨*candle, lighter, match*⟩ light; ⟨*gas, fuel, explosive*⟩ ignite; *(fig.)* arouse enthusiasm

zündend *Adj.* *(fig.)* stirring, rousing ⟨*speech, song, tune, etc.*⟩

Zunder ['tsʊndɐ] **der; ~s a)** tinder; **trocken wie ~:** dry as tinder; tinder-dry; **b)** *(fig. ugs.)* **jmdm. ~ geben** lay into sb. *(coll.)*; **~ kriegen** get it in the neck *(coll.)*

Zünder der; ~s, ~ igniter; *(für Bombe, Mine)* detonator

Zünd-: ~holz das *(bes. südd., österr.)* match; **~kerze die** spark[ing]-plug; **~schloß das** *(Kfz-W.)* ignition [lock]; **~schlüssel der** *(Kfz-W.)* ignition key; **~schnur die** fuse; **~stoff der** *(fig.)* fuel for conflict

Zündung die; ~, ~en a) *s.* **zünden 1:** ignition; detonation; letting off; firing; **b)** *(Kfz-W.: Anlage)* ignition

zu|nehmen 1. *unr. itr. V.* **a)** increase (**an** + *Dat.* in); ⟨*moon*⟩ wax; **in ~dem Maße** to an increasing extent *or* degree; increasingly; **mit ~dem Alter** with advancing age; **b)** *(schwerer werden)* put on *or* gain weight; **er hat [um] ein Kilo zugenommen** he has put on *or* gained a kilo. **2.** *unr. tr., auch itr. V.* *(Handarb.)* increase

zunehmend *Adv.* increasingly

Zu·neigung die; ~, ~en affection (**zu** for, towards); **~ zu jmdm. fassen** become fond of sb.

zünftig ['tsʏnftɪç] *(ugs.)* **1.** *Adj.* proper. **2.** *adv.* properly

Zunge ['tsʊŋə] **die; ~, ~n a)** tongue; **[jmdm.] die ~ herausstrecken** put one's tongue out [at sb.]; **auf der ~ zergehen** melt in one's mouth; **b)** *(fig.)* **eine spitze** *od.* **scharfe/lose ~ haben** have a sharp/loose tongue; **böse ~n behaupten, daß ...:** malicious gossip has it that ...; malicious tongues are saying that ...; **seine ~ hüten** *od.* **zügeln** *od.* **im Zaum halten** guard *or* mind one's tongue; **ich mußte mir auf die ~ beißen** I had to bite my tongue; **lieber beiße ich mir die ~ ab** *(ugs.)* I would bite my tongue off first; **der Name liegt mir auf der ~:** the name is on the tip of my tongue; **sich** *(Dat.)* **die ~ abbrechen** tie one's tongue in knots; **mit [heraus]hängender ~:** with [one's/its] tongue hanging out; **c)** *(eines Blasinstruments)* reed; *(einer Orgel)* tongue; *(eines Schuhs)* tongue

züngeln ['tsʏŋl̩n] *itr. V.* **a)** ⟨*snake etc.*⟩ dart its tongue in and out; **b)** *mit Richtungsangabe mit sein* ⟨*flame*⟩ flicker; dart

Zungen-: ~spitze die tip of the tongue; **~wurst die** tongue sausage

Zünglein ['tsʏnlain] **das; ~s, ~ a)** [little] tongue; **b)** *(einer Waage)* [small] needle *or* pointer; **das ~ an der Waage sein** *(fig.)* tip the scales

zu·nichte *Adj.* **etw. ~ machen** ruin sth.; **jmds. Hoffnungen ~ machen** shatter *or* dash sb.'s hopes

zu|nicken *itr. V.* **jmdm./sich ~:** nod to sb./one another

zu·nutze *Adj.* **sich** *(Dat.)* **etw. ~ machen** make use of sth.; *(ausnutzen)* take advantage of sth.

zu·oberst *Adv.* [right] on [the] top; *s. auch* **unter... a**

zu|ordnen *tr. V.* **a)** relate (*Dat.* to); **b)** *(zurechnen)* **jmdn./etw. einer Sache** *(Dat.)* **~:** classify sb./sth. as belonging to sth.; **c)** *(zuweisen)* assign (*Dat.* to)

zu|packen *itr. V.* **a)** grab it/them; **fest ~ können** be able to grab things and grip them tightly; **b)** *(fig.)* knuckle down to it; **er hat eine sehr ~de Art** he has a very vigorous, purposeful manner

zupfen ['tsʊpfn̩] **1.** *itr. V.* **an etw.** *(Dat.)* **~:** pluck *or* pull at sth.; **sich** *(Dat.)* **am Ohrläppchen ~:** pull [at] one's ear lobe. **2.** *tr. V.* **a) etw. aus/von** *usw.* **etw. ~:** pull sth. out of/from *etc.* sth.; **b)** *(auszupfen)* pull out; pluck ‹*eyebrows*›; pull up ‹*weeds*›; **c)** pluck ‹*string, guitar, tune*›; **d) jmdn. am Ärmel/Bart ~:** pull *or* tug [at] sb.'s sleeve/beard

zu|pflastern *tr.V.* pave over

zu|pressen *tr. V.* press shut

zu|prosten *itr. V.* **jmdm. ~:** drink sb.'s health; raise one's glass to sb.

zur [tsu:ɐ̯] *Präp. + Art.* **a) = zu der; b)** *(räumlich: Richtung)* to the; **ein Fenster ~ Straße** a window on to the street; **wo geht es ~ Post?** which is the way to the post office?; **c)** *(räumlich: Lage)* **~ Tür hereinkommen** come [in] through the door; **d)** *(Zusammengehörigkeit, Hinzufügung)* **~ Hasenkeule empfehle ich einen Rotwein** I recommend a red wine with the haunch of hare; **e)** *(zeitlich)* at the; **~ Stunde/Zeit** at the moment; at present; **~ Adventszeit** at Advent time; **~ Jahreswende** at New Year; **rechtzeitig ~ Buchmesse** in [good] time for the book fair; **f)** *(Zweck)* **~ Entschuldigung** by way of [an] excuse; **~ Inspektion in die Werkstatt müssen** have to go in for a check-up; **g)** *(Folge)* **~ vollen Zufriedenheit ihres Chefs** to the complete satisfaction of her boss; **~ allgemeinen Erheiterung** to everybody's amusement; **h) sie wurde ~ Direktorin ernannt/~ Präsidentin gewählt** she was appointed director/elected president; **~ Diebin werden** become a thief; **die Wahlen ~ Knesseth** elections to the Knesset

zu|raten *unr. itr. V.* **ich würde dir ~:** I would advise you to do so; **auf jmds. Zuraten [hin]** on sb.'s advice *or* recommendation; **ich möchte [dir] weder zu- noch abraten** I should not like to advise you one way or the other

Zürcher ['tsʏrçɐ] **1.** *indekl. Adj.; nicht präd.* Zurich *attrib.* **2. der; ~s, ~:** inhabitant/native of Zurich; **er ist ~:** he is from Zurich

zu|rechnen *tr. V.* **jmdn./etw. einer Sache** *(Dat.)* **~:** class sb./sth. as belonging to sth.

zurechnungs·fähig *Adj.* **a)** sound of mind *pred.;* **b)** *(Rechtsw.: schuldfähig)* responsible [for one's actions]

Zurechnungs·fähigkeit die; *o. Pl.* **a)** soundness of mind; **b)** *(Rechtsw.: Schuldfähigkeit)* responsibility [for one's actions]

zurecht-, Zurecht-: ~|biegen *unr. tr. V.* bend into shape; **er wird die Sache schon wieder ~biegen** *(fig.)* he will get things straightened out *or* sorted out again; **~|finden** *unr. refl. V.* find one's way [around]; **er findet sich im Leben/in der Welt nicht [mehr] ~:** he is not able to cope with life/the world [any longer]; **~|kommen** *unr. itr. V.; mit sein* get on **(mit** with); **mit etw. ~kommen** cope with sth.; **~|legen** *tr. V.* **a)** lay out [ready] (*Dat.* for); **b)** *(fig.)* get ready; prepare; **sich** *(Dat.)* **eine Erwiderung ~gelegt haben** have a reply ready; **~|machen** *tr. V. (ugs.)* **a)** *(vorbereiten)* get ready; **b)** *(herrichten)* do up; **c) jmdn./sich ~:** get sb. ready/get [oneself] ready; *(schminken)* make sb. up/put on one's make-up; **~|rücken** *tr. V.* put *or* set ‹*chair, crockery, etc.*› in place; straighten ‹*tie*›; adjust ‹*spectacles, hat, etc.*›; *(fig.: richtigstellen, korrigieren)* put straight; *s. auch* **Kopf a; ~|schneiden** *unr. tr. V.* cut to size/shape; trim ‹*fringe, beard, hedge*›; **~|stutzen** *tr. V.* trim ‹*hedge, beard, hair etc.*›; **jmdn./etw. ~stutzen** *(fig.)* sort *or* straighten sb. out/get *or* knock sth. into shape; **~|weisen** *unr. tr. V.* rebuke; reprimand ‹*pupil, subordinate, etc.*›; **~weisung die** *s.* **~weisen:** rebuke; reprimand

zu|reden *itr. V.* **jmdm. ~:** persuade sb.; *(ermutigen)* encourage sb.; **jmdm. gut ~:** encourage sb.

zu·reichend *(geh.) s.* **zulänglich**

Zürich ['tsy:rɪç] **(das); ~s** Zurich

zu|richten *tr. V.* **a)** *(verletzen)* injure; **sie haben ihn übel zugerichtet** they [really] knocked him about; **b)** *(beschädigen)* make a mess of

zu|riegeln *tr. V.* bolt

zu·rück *Adv.* **a)** back; **ich bin gleich [wieder] ~:** I'll be right back *(coll.);* **ein Schritt ~:** a step backwards; **~!** get *or* go back!; **„~ an Absender"** 'return to sender'; **... und 10 Pfennig ~:** ... and 10 pfennigs change; *s. auch* **Dank a; hin d; Natur a; b)** *(weiter hinten; auch fig.)* behind

Zurück das *in* **es gibt kein ~ [mehr]** there is no going back

zurück-, Zurück-: ~|begleiten *tr. V.* **jmdn. ~begleiten** accompany sb. back; **~|behalten** *unr. tr. V.* **a)** keep [back]; retain; **b)** *(nicht mehr loswerden)* be left with ‹*scar, heart defect, etc.*›; **~|bekommen** *unr. tr. V.* get back; **Sie bekommen noch 10 Mark ~:** you get 10 marks change; **~|beordern** *tr. V.* order back; **~|beugen 1.** *tr. V.* bend back; **2.** *refl. V.* lean *or* bend back; **~|bleiben** *unr. itr. V.; mit sein* **a)** remain *or* stay behind; **b)** *(nicht mithalten)* lag behind; *(fig.)* fall behind; **hinter den Erwartungen ~bleiben** fall short of expectations; **in seiner Entwicklung ~bleiben** ‹*child*› be retarded *or* backward in its development; **c)** *(bleiben)* remain; **von der Krankheit ist [bei ihm] nichts ~geblieben** the illness has left no lasting effects [on him]; **d)** *(wegbleiben)* stay *or* keep back; *s. auch* **zurückgeblieben; ~|blicken** *itr. V. (auch fig.)* look back **(auf +** *Akk.* at, *fig.:* on); *(sich umblikken)* look back *or* round; **~|bringen** *unr. tr. V.* bring/take back; return; **~|drängen** *tr. V.* force back; drive back ‹*enemy*›; **~|drehen** *tr. V.* **a)** turn back; turn down ‹*heating, volume, etc.*›; **b)** *(rückwärts drehen)* turn backwards; **~|eilen** *itr. V.; mit sein* hurry back; **~|erhalten** *unr. tr. V. (geh.)* be given back; get back; **anliegend erhalten Sie ihre Bewerbungsunterlagen ~:** please find enclosed your application, which we are returning to you; **~|erinnern** *refl. V.* **sich an etw.** *(Akk.)* **~erinnern** remember *or* recall sth.; **~|erobern** *tr. V.* win back ‹*votes, majority, etc.*›; regain ‹*power, position, etc.*›; recapture ‹*territory, town, etc.*›; **~|erstatten** *tr. V.* refund; **jmdm. etw. ~erstatten** refund sth. to sb.; **~|erwarten** *tr. V.* **jmdn. ~erwarten** expect sb. back; **~|fahren 1.** *unr. itr. V.; mit sein* **a)** go/drive/

ride back; return; **b)** *(nach hinten fahren)* go back[wards]; **2.** *unr. tr. V.* **jmdn./etw. ~fahren** drive sb./sth. back; **~|fallen** *unr. itr. V.; mit sein* **a)** fall back; **b)** *(nach hinten fallen)* fall back[wards]; **c)** *(fig.: in Rückstand geraten)* fall behind; **d)** *(fig.: auf einen niedrigeren Rang)* drop **(auf** + *Akk.* to); **e)** *(fig.: in einen früheren Zustand)* **in etw.** *(Akk.)* **~fallen** fall back into sth.; **f)** *(fig.)* **an jmdn. ~fallen** ⟨*property*⟩ revert to sb.; **g)** *(fig.)* **auf jmdn. ~fallen** ⟨*actions, behaviour, etc.*⟩ reflect [up]on sb.; **~|finden** *unr. itr. V.* find one's way back; **~|fliegen 1.** *unr. itr. V.; mit sein* fly back; **2.** *unr. tr. V.* **jmdn./etw. ~fliegen** fly sb./sth. back; **~|fordern** *tr. V.* **etw. ~fordern** ask for sth. back; *(nachdrücklicher)* demand sth. back; **~|fragen** *itr. V.* answer with a question; **„...?" fragte er ~:** '...?', he asked in return; **~|führen 1.** *tr. V.* **a) jmdn. ~führen** take sb. back; **b) etw. ~führen** move sth. back; return sth.; **c) etw. auf etw.** *(Akk.)* **~führen** *(auf Ursprung)* trace sth. back to sth.; *(auf Ursache)* attribute sth. to sth.; put sth. down to sth.; *(auf einfachere Form)* reduce sth. to sth.; **2.** *itr. V.* lead back; **es führt kein anderer Weg ~:** there is no other way back; **~|geben** *unr. tr. V.* **a)** give back; return; hand in ⟨*driver's licence, membership card*⟩; return ⟨*goods, unused ticket, etc.*⟩; relinquish ⟨*mandate, office, etc.*⟩; give back ⟨*freedom*⟩; **jmdm. etw. ~geben** give sth. back to sb.; return sth. to sb.; **b)** *(erwidern)* reply; **c)** *auch itr. (Ballspiele)* return ⟨*ball, puck, service, pass, throw*⟩; *(nach hinten geben)* pass ⟨*ball*⟩ back; **[den Ball] an jmdn. ~geben** return the ball to sb.; **~geblieben** *Adj.* retarded; **~|gehen** *unr. itr. V.; mit sein* **a)** go back; return; *(sich zurückbewegen)* ⟨*pick-up arm, indicator, needle, etc.*⟩ return; **b)** *(nach hinten gehen)* go back; ⟨*enemy*⟩ retreat; **c)** *(verschwinden)* ⟨*bruise, ulcer*⟩ disappear; ⟨*swelling, inflammation*⟩ go down; ⟨*pain*⟩ subside; **d)** *(sich verringern)* decrease; go down; ⟨*fever*⟩ abate; ⟨*flood*⟩ subside; ⟨*business*⟩ fall off; **e)** *(zurückgeschickt werden)* be returned *or* sent back; **ein Essen ~gehen lassen** send a meal back; **f) auf jmdn. ~gehen** *(jmds. Werk sein)* go back to sb.; *(von jmdm. abstammen)* originate from *or* be descended from sb.; **der Name geht auf ein lateinisches Wort ~:** the name comes from a Latin word; **g)** *(sich zurückbewegen lassen)* ⟨*lever etc.*⟩ go back; **~|gewinnen** *unr. tr. V.* **a)** win back; regain ⟨*confidence, title, strength, freedom, etc.*⟩; **b)** *(Wirtsch.)* reclaim, recover ⟨*raw materials etc.*⟩; **~gezogen 1.** *Adj.* secluded; **2.** *adv.* **~gezogen leben** live a secluded life; **~gezogenheit die; ~:** seclusion; **~|greifen** *unr. itr. V.* **auf jmdn./etw. ~greifen** fall back on sb./sth.; **~|haben** *unr. tr. V.* have back; **hast du es inzwischen ~?** have you got it back yet?; **~|halten 1.** *unr. tr. V.* **a) jmdn. ~halten** hold sb. back; **er war durch nichts ~zuhalten** there was no stopping him; nothing would stop him; **b)** *(am Vordringen hindern)* keep back ⟨*crowd, mob, etc.*⟩; **c)** *(behalten)* withhold ⟨*news, letter, parcel, etc.*⟩; **d)** *(nicht austreten lassen)* hold back ⟨*tears etc.*⟩; **e)** *(von etw. abhalten)* **jmdn. ~halten** stop sb.; **jmdn. von etw. ~halten** keep sb. from sth.; **2.** *unr. refl. V.* **a)** *(sich zügeln, sich beherrschen)* restrain *or* control oneself; **b)** *(nicht aktiv werden)* **sich in einer Diskussion ~halten** keep in the background in a discussion; **~haltend 1.** *Adj.* **a)** reserved; subdued, muted ⟨*colour*⟩; **b)** *(kühl, reserviert)* cool, restrained ⟨*reception, response*⟩; **c)** *(sparsam)* **mit etw. ~haltend** ⟨*person*⟩ who is/was sparing with sth.; sparing with sth. *pred.*; **2.** *adv.* **a)** ⟨*behave*⟩ with reserve *or* restraint; **b)** *(kühl, reserviert)* coolly; **~haltung die;** *o. Pl.* reserve; *(Kühle)* coolness; **~haltung üben** *(geh.)* exercise restraint; **ein Buch mit ~haltung aufnehmen** give a book a cool reception; **~|holen** *tr. V.* **a)** fetch back; get back ⟨*money*⟩; bring back ⟨*satellite, missile*⟩; **jmdn. ~holen** bring sb. back; **b)** *(~rufen)* call back; **~|kämmen** *tr. V.* comb back; back-comb; **~gekämmte Haare** back-combed hair; **~|kehren** *itr. V.; mit sein* return; come back; **~|kommen** *unr. itr. V.; mit sein* **a)** come back; return; ⟨*letter*⟩ come back, be returned; **b)** *(zurückgelangen)* get back; **c) ~kommen auf** (+ *Akk.*) come back to ⟨*subject, question, point, etc.*⟩; come back on ⟨*offer*⟩; **~|können** *unr. itr. V.* be able to go back *or* return; **jetzt können wir nicht mehr ~** *(fig.)* there's no going *or* turning back now; **~|kriegen** *tr. V. s.* **~bekommen; ~|lassen** *unr. tr. V.* **a)** leave; **b)** *(zurückkehren lassen)* **jmdn. ~lassen** allow sb. to return; let sb. return; **~|lassung die; ~: unter ~lassung einer Sache/jmds.** leaving sth./sb. behind; **~|laufen** *unr. itr. V.; mit sein* **a)** run back; **b)** *(ugs.: zurückgehen)* come/go back; **c)** *(sich zurückbewegen)* run back; **das Tonband ~laufen lassen** run the tape back; **~|legen 1.** *tr. V.* **a)** put back; **b)** *(nach hinten beugen)* lean *or* lay ⟨*head*⟩ back; **c)** *(reservieren)* put aside, keep (*Dat.*, **für** for); **d)** *(sparen)* put away; put by; **e)** *(hinter sich bringen)* cover ⟨*distance*⟩; **2.** *refl. V.* lie back; *(sich ~lehnen, ~neigen)* lean back; **~|lehnen** *refl. V.* lean back; **~|liegen** *unr. itr. V.* **a) das Ereignis/das liegt einige Jahre ~:** the event took place/that was several years ago; **b)** *(bes. Sport)* be behind; **~|melden** *refl. V.* report back **(bei** to); **~|müssen** *unr. itr. V.* **a)** have to go back *or* return; **b)** *(zurückbefördert werden müssen)* have to go back; **~|nehmen** *unr. tr. V.* **a)** take back; **b)** *(widerrufen)* take back; **c)** *(rückgängig machen)* revoke, rescind ⟨*decision, ban, etc.*⟩; withdraw ⟨*complaint*⟩; **~|pfeifen** *unr. tr. V.* **a)** whistle ⟨*dog*⟩ back; **b)** *(fig. salopp)* **jmdn. ~pfeifen** call sb. off; **~|prallen** *itr. V.; mit sein* **a)** bounce back **(von** off); ⟨*bullet*⟩ ricochet **(von** from); **b)** *(fig.)* start back; *(entsetzt)* recoil; **~|reichen 1.** *tr. V.* hand back; **2.** *itr. V.* go back; **~|rufen** *unr. tr. V.* **a)** call back; recall ⟨*ambassador*⟩; **b)** *auch itr. (anrufen)* call *or (Brit.)* ring back; **c) jmdm./sich etw. ins Gedächtnis** *od.* **in die Erinnerung ~rufen** remind sb. of sth./call sth. to mind; **d)** *(als Antwort, nach hinten rufen)* call *or* shout back; **e)** *(Wirtsch.)* recall ⟨*defective goods, car, etc.*⟩; **~|schalten** *itr. V. (in kleineren Gang schalten)* change down; **~|schaudern** *itr. V.; mit sein* shrink back **(vor** + *Dat.* from); **~|schauen** *itr. V. (bes. südd., österr., schweiz.) s.* **~blicken; ~|scheuen** *itr. V.; mit sein s.* **[2]~schrecken; ~|schicken** *tr. V.* send back; **~|schieben** *unr. tr. V.* **a)** push back; draw back ⟨*bolt, curtains*⟩; **b)** *(nach hinten schieben)* push back[wards]; **~|schlagen 1.** *unr. tr. V.* **a)** *(nach hinten schlagen)* fold back ⟨*cover, hood, etc.*⟩; turn down ⟨*collar*⟩; *(zur Seite schlagen)* pull *or* draw back ⟨*curtains*⟩; **b)** *(durch einen Schlag zurückbefördern)* hit back; *(mit dem Fuß)* kick back; **c)** *(zum Rückzug zwingen, abwehren)* beat off, repulse ⟨*enemy, attack*⟩; **2.** *unr. itr. V.* **a)** hit back; ⟨*enemy*⟩ strike back, retaliate; **b)** *mit sein* ⟨*pendulum*⟩ swing back; ⟨*starting-handle*⟩ kick back; **c) auf etw.** *(Akk.)* **~schlagen** *(fig.)* have repercussions on sth.; [1]**~|schrecken** *tr. V.* **jmdn. ~schrecken** deter sb.; [2]**~|schrekken** *regelm., veralt. unr. itr. V.; mit sein* **a)** shrink back; recoil; **b) vor etw.** *(Dat.)* **~schrecken** *(fig.)* shrink from sth.; **er schreckt vor nichts ~:** he will stop at nothing; **~|sehnen** *refl. V.* ; **sich zu jmdm./nach Italien ~sehnen** long to be back with sb./in Italy; **~|senden** *unr. od. regelm. tr. V. (geh.) s.* **~schicken; ~|setzen 1.** *tr. V.* **a)** put back; **b)** *(nach hinten setzen)* move back; **c)** *(zurückfahren)* move back; reverse; back; **d)** *(fig.)* **jmdn. ~setzen** neglect sb.; **sich ~gesetzt fühlen** feel neglected; **2.** *refl. V.* **a)** sit down again **(an +** *Akk.* at); **b)** *(sich nach hinten setzen)* move back; **3.** *itr. V. (zurückfahren)* move back[wards]; reverse; back; **~setzung die; ~, ~en** neglect; *(Kränkung)* insult; slight; **~|stecken 1.** *tr. V.* **a)** put back; **b)** *(nach hinten stecken)* move back; **2.** *itr. V. (ugs.)* lower one's sights; **~|stehen** *unr. itr. V.* **a)** stand back; be set back; **b)** *(fig.: übertroffen werden)* be left behind; **hinter jmdm. ~stehen** take second place to sb.; **c)** *(fig.: verzichten)* miss out; **~|stellen 1.** *tr. V.* **a)** put back; **b)** *(nach hinten stellen)* move back; **c)** *(niedriger einstellen)* turn down ⟨*heating*⟩; put back ⟨*clock*⟩; **d)** *(reservieren)* put aside, keep (*Dat.*, **für** for); **e) jmdn. vom Wehrdienst ~stellen** defer sb.'s military service; defer sb. *(Amer.)*; **f)** *(aufschieben)* postpone; defer; **g)** *(hintanstellen)* put aside ⟨*reservations, doubts, etc.*⟩; **~|stoßen 1.** *unr. tr. V.* **a)** push back; **b)** *(von sich stoßen)* push away; **2.** *unr. itr. V.; mit sein s.* **~setzen 3; ~|treten** *unr. itr. V.* **a)** *mit sein* step back; **bitte von der Bahnsteigkante ~treten** please stand back from the edge of the platform; **b)** *mit sein (von einem Amt)* resign; step down; ⟨*government*⟩ resign; **als Vorsitzender/von einem Amt ~treten** step down as chairman/resign from an office; **c)** *mit sein (von einem Vertrag usw.)* withdraw **(von** from); back out **(von** of); **d)** *mit sein (fig.: in den Hintergrund treten)* become less important; fade in importance; **hinter/gegenüber etw.** *(Dat.)* **~treten** take second place to sth.; **~|verlangen** *tr. V.* demand back; **~|versetzen 1.** *tr. V.* **a)** move *or* transfer back; **b)** *(fig.)* take *or* transport back; **2.** *refl. V.* think oneself back **(in +** *Akk.* to); **~|weichen** *unr. itr. V.; mit sein* draw back **(vor +** *Dat.* from); back away; *(~schrecken)* shrink back, recoil **(vor +** *Dat.* from); **er wich keinen Schritt/Zentimeter ~:** he stood his ground; **~|weisen** *unr. tr. V.* **a)** send back; **b)** *(abweisen, nicht akzeptieren)* reject ⟨*proposal, question, demand, application, etc.*⟩; turn down, refuse ⟨*offer, request, invitation, help, etc.*⟩; turn away ⟨*petitioner, unwelcome guest*⟩; **c)** *(sich verwahren gegen)* repudiate ⟨*accusation, claim, etc.*⟩; **~weisung die** *s.* **~weisen:** sending back; rejection; turning down; refusal; turning away; repudiation; **~|werfen 1.** *unr. tr. V.* **a)** throw back; **den Kopf/sein Haar ~werfen** throw *or* toss one's head back/toss one's hair back; **b)** *(reflektieren)* reflect ⟨*light, sound*⟩; **c)** *(Milit.)* repulse ⟨*enemy*⟩; **d)** *(fig.: in einer Entwicklung)* set back; **2.** *unr. refl. V.* throw oneself back; **~|wollen 1.** *unr. itr. V.* want to go back; **2.** *unr. tr. V. (ugs.)* **etw. ~wollen** want sth. back; **~|zahlen** *tr. V.* pay back; **~|ziehen 1.** *unr. tr. V.* **a)** pull back; draw back ⟨*bolt, curtains, one's hand, etc.*⟩; **es zieht ihn in die Heimat/zu ihr ~** *(fig.)* he is drawn back to his homeland/to her; **b)** *(abziehen, zurückbeordern)* withdraw, pull back ⟨*troops*⟩; withdraw, recall ⟨*ambassador*⟩; **c)** *(rückgängig machen)* withdraw; cancel ⟨*order, instruction*⟩; **d)** *(wieder aus dem Verkehr ziehen)* withdraw ⟨*coin, stamp, etc.*⟩; **2.** *unr. refl. V.* withdraw **(aus, von** from); ⟨*troops*⟩ withdraw, pull back; **sich aufs Land/in sein Zimmer ~:** retreat to the country/retire to one's room; *s. auch* **zurückgezogen; 3.** *unr. itr. V.; mit sein* go back; return; **~|zucken** *itr. V.; mit sein* flinch; *(erschrocken)* start back; **mit der Hand ~zucken** jerk one's hand away

Zu·ruf der shout

zu|rufen *unr. tr. V.* **jmdm. etw. ~:** shout sth. to sb.

Zu·sage die a) *(auf eine Einladung hin)* acceptance; *(auf eine Stellenbewerbung hin)* offer; **b)** *(Versprechen)* promise; undertaking; **jmdm. die** *od.* **seine ~ geben, etw. zu tun** promise sb. that one will do sth.

zu|sagen 1. *itr. V.* **a)** *(auf eine Einladung hin)* **[jmdm.] ~/fest ~:** accept/give sb. a firm acceptance; **b)** *(auf ein Angebot hin)* accept; **c)** *(gefallen)* **jmdm. ~:** appeal to sb. **2.** *tr. V.* **a)** promise; **jmdm. etw. ~:** promise sb. sth.; **b)** *s. auch* **Kopf a**

zusammen [tsu'zamən] *Adv.* together; **wir bestellten uns ~ eine Flasche Wein** we ordered a bottle of wine between us; **ihr seid alle ~ Feiglinge!** *(ugs.)* you're cowards, the whole lot of you *(coll.)*; **er verdient mehr als alle anderen ~:** he earns more than the rest of them put together

zusammen-, Zusammen-: ~arbeit die; *o. Pl.* co-operation *no indef. art.*; **~|arbeiten** *itr. V.* co-operate; work together; *(kollaborieren)* collaborate; **~|ballen 1.** *tr. V.* **[zu einem Klumpen] ~ballen** make into a ball; **2.** *refl. V.* mass together; **~|beißen** *unr. tr. V.* **die Zähne ~beißen** clench one's teeth together; *s. auch* **Zahn b; ~|bekommen** *unr. tr. V.* **a)** get together, raise ⟨*money, rent, etc.*⟩; manage to collect ⟨*signatures*⟩; **b)** *(zusammengesetzt/~gebaut usw. bekommen)* get together; **c)** *(fig. ugs.)* remember; **~|binden** *unr. tr. V.* tie together; **~|bleiben** *unr. itr. V.; mit sein* stay together; **~|brauen 1.** *tr. V. (ugs.)* concoct ⟨*drink*⟩; **2.** *refl. V. (fig.)* ⟨*storm, bad weather, trouble, etc.*⟩ be brewing; ⟨*disaster*⟩ loom; **da braut sich was ~:** there's something brewing there; **~|brechen** *unr. itr. V.; mit sein* **a)** *(einstürzen)* collapse; **b)** *(zu Boden sinken)* ⟨*person, animal*⟩ collapse; *(fig.)* ⟨*person*⟩ break down; **c)** *(fig.)* collapse; ⟨*order, communications, system, telephone network*⟩ break down; ⟨*traffic*⟩ come to a standstill, be paralysed; ⟨*attack, front, resistance*⟩ crumble; **~|bringen** *unr. tr. V.* bring together; **jmdn. mit jmdm. ~bringen** bring sb. together with sb.; **~bruch der a)** *(eines Menschen)* collapse; *(psychisch, nervlich)* breakdown; **dem ~bruch nahe sein** be near to collapse/breakdown; **b)** *(fig.) s.* **~brechen c:** collapse; breakdown; crumbling; **~|drängen 1.** *tr. V.* push together; herd ⟨*crowd*⟩ together; **2.** *refl. V.* crowd together; *(fig.)* be concentrated **(auf +** *Akk.* into); **~|drücken**

tr. V. **a)** press together; *(komprimieren)* compress ‹*gas*›; **b)** *(zerdrücken)* crush; **~|fahren** *unr. itr. V.; mit sein* **a)** collide **(mit** with); **b)** *(~zucken)* start; jump; **~fall der** coincidence; **~|fallen** *unr. itr. V.; mit sein* **a)** collapse; **das ganze Lügengebäude fiel in sich ~** *(fig.)* the whole tissue of lies fell apart; **b)** *(~sinken, schrumpfen)* **[in sich] ~fallen** ‹*cake*› sink [in the middle]; ‹*froth, foam, balloon, etc.*› collapse; **c)** ‹*person*› become emaciated; **d)** *(zeitlich)* **[zeitlich] ~fallen** coincide; fall at the same time; **e)** *(räumlich)* coincide; **~|falten** *tr. V.* fold up; **~|fassen** *tr. V.* **a)** put together; **b)** *(in eine kurze Form bringen)* summarize; **etw. in einem Satz ~fassen** sum sth. up *or* summarize sth. in one sentence; **~fassend kann man sagen ...:** to sum up *or* in summary, one can say ...; **~fassung die** *s.* **~fassen:** putting together; summary; **~|fegen** *tr. V. (bes. nordd.)* sweep together; **~|finden** *unr. refl. V.* **a)** get together; **b)** *(zusammentreffen)* meet up; **~|flicken** *tr. V.* patch up; **~|fließen** *unr. itr. V.; mit sein* ‹*rivers, streams*› flow into each other, join up; *(fig.)* ‹*colours*› run together; ‹*sounds*› blend together; **~fluß der** confluence; **~|fügen 1.** *tr. V.* fit together; **2.** *refl. V.* fit together; **~|führen** *tr. V.* bring together; **getrennte Familien wieder ~führen** reunite divided families; **~|gehen** *unr. itr. V.; mit sein* **a)** *(sich verbünden, sich zusammentun)* join forces **(mit** with); *(fusionieren)* ‹*firms*› merge; **b)** *(zusammenpassen)* go together; **c)** *(ugs.: zusammenlaufen, ~fließen usw.)* join up; meet; **d)** *(ugs.: sich zusammenfügen, verbinden usw. lassen)* fit together; meet; **~|gehören** *itr. V.* belong together; **~gehörig** *Adj.* [closely] related *or* connected ‹*subjects, problems, etc.*›; matching *attrib.* ‹*pieces of tea service, cutlery, etc.*›; **die ~gehörigen Teile** the parts which belong together; **~gehörigkeits·gefühl** *das; o. Pl.* sense *or* feeling of belonging together; **~genommen** *Adj.; nicht attr.* **alle diese Dinge ~genommen** all these things together; **~gewürfelt** *Adj.* oddly assorted; **ein bunt ~gewürfelter Haufen** a motley collection of people; **~halt der;** *o. Pl.* cohesion; **~|halten 1.** *unr. tr. V.* **a)** hold together; **b)** *(beisammenhalten)* keep together; **sein Geld ~halten** be careful with one's money; **2.** *unr. itr. V.* **a)** hold together; **b)** *(fig.)* ‹*friends, family, etc.*› stick together; **~hang der** connection; *(einer Geschichte, Rede)* coherence; *(Kontext)* context; **in [keinem] ~hang mit etw. stehen** be [in no way] connected with sth.; **etw. mit etw. in ~hang bringen** connect sth. with sth.; make a connection between sth. and sth.; **im ~hang mit ...:** in connection with ...; **etw. aus dem ~hang lösen/reißen** take sth. out of [its] context; **~|hängen** *unr. itr. V.* **a)** be joined [together]; **in ~hängenden Sätzen** in coherent sentences; **b)** *(fig.)* **mit etw. ~hängen** *(zu etw. eine Beziehung haben)* be related to sth.; *(durch etw. [mit] verursacht sein)* be the result of sth.; **das hängt damit ~, daß ...:** that is connected with *or* has to do with the fact that ...; **~hang·los 1.** *Adj.* incoherent, disjointed ‹*speech, story, etc.*›; **2.** *adv.* ‹*speak*› incoherently; **~|kehren** *tr. V. (bes. südd.) s.* **~fegen; ~klappbar** *Adj.* folding; **~klappbar sein** fold up; **~|klappen 1.** *tr. V.* fold up; **2.** *itr. V.; mit sein (ugs.)* collapse; **~|kleben** *tr., itr. V.* stick together; **~|kneifen** *unr. tr. V.* press ‹*lips*› together; screw ‹*eyes*› up; **~|knüllen** *tr. V.* crumple up; *(fest)* screw up; **~|kommen** *unr. itr. V.; mit sein* **a)** meet; **mit jmdm. ~kommen** meet sb.; **b)** *(zueinanderkommen; auch fig.)* get together; **c)** *(gleichzeitig auftreten)* occur *or* happen together; **d)** *(sich summieren)* accumulate; **da werden schon so an die 50 Leute ~kommen** there are sure to be getting on for 50 people there altogether; **~|koppeln** *tr. V.* couple together; dock ‹*spacecraft*›; **~|kriegen** *tr. V. (ugs.) s.* **~bekommen; ~|krümmen** *refl. V.* double up; writhe; **~kunft** [-kʊnft] **die; ~, ~künfte** [-kʏnftə] meeting; **~|laufen** *unr. itr. V.; mit sein* **a)** ‹*people, crowd*› gather, congregate; **b)** ‹*rivers, streams*› flow into each other, join up; **c)** ‹*water, oil, etc.*› collect; **d)** ‹*colours*› run together; **~|leben** *itr. V.* live together; **~|leben das;** *o. Pl.* living together *no art.*; **~|legen 1.** *tr. V.* **a)** put *or* gather together; **b)** *(zusammenfalten)* fold [up]; **c)** amalgamate, merge ‹*classes, departments, etc.*›; combine ‹*events*›; **d)** put ‹*patients, guests, etc.*› together [in the same room]; **2.** *itr. V.* club together; pool our/your/their money; **~|nähen** *tr. V.* **a)** sew together; **etw. mit etw. ~nähen** sew sth. to sth.; **b)** *(reparieren)* sew up; **~|nehmen 1.** *unr. tr. V.* **a)** summon *or* muster up ‹*courage, strength, understanding*›; collect ‹*wits*›; *s. auch* **~genommen; 2.** *unr. refl. V.* get *or* take a grip on oneself; **nimm dich ~!** pull yourself together!; **~|packen 1.** *tr. V.* pack up; *(zusammen verpacken)* pack up together; **2.** *itr. V.* pack up; **~|passen** *itr. V.* go together; ‹*persons*› be suited to each other; **mit etw. ~passen** go with sth.; **~prall der; ~prall[e]s, ~pralle** collision; *(fig.)* clash; **~|prallen** *itr. V.; mit sein* collide **(mit** with); *(fig.)* clash; **~|pressen** *tr. V.* **a)** squeeze; **b)** *(aneinanderpressen)* press ‹*lips, hands*› together; **~|raffen** *tr. V.* gather up ‹*possessions, papers, etc.*›; bundle up ‹*clothes*›; **~|rechnen** *tr. V.* add up; **etw. mit etw. ~rechnen** add sth. to sth.; **~|reißen** *unr. refl. V. (ugs.) s.* **~nehmen 2; ~|rollen 1.** *tr. V.* roll up; **2.** *refl. V.* ‹*cat, dog, etc.*› curl up; ‹*hedgehog*› roll [itself] up [into a ball]; **~|rotten** *refl. V. (abwertend)* ‹*crowds, groups, etc.*› band together; ‹*youths*› gang together *or* up; *(in Aufruhr)* form a mob; **~|rücken 1.** *tr. V.* move ‹*chairs, tables, etc.*› together; **2.** *itr. V.; mit sein (auch fig.)* move closer together; **~|rufen** *unr. tr. V.* call together; **~|sacken** *itr. V.; mit sein (ugs.)* collapse; **~|schlagen 1.** *unr. tr. V.* **a)** strike *or* bang together; clap ‹*hands*› [together]; **die Hacken ~schlagen** click one's heels; **b)** *(verprügeln)* beat up; **c)** *(zertrümmern)* smash up *or* to pieces; **d)** *(zusammenfalten)* fold up; **2.** *unr. itr. V.; mit sein* **über jmdm./etw. ~schlagen** *(fig.)* engulf sb./sth.; **~schluß der** joining together; union; *(von Firmen)* merger; amalgamation; **~|schnüren** *tr. V.* **a)** tie up **(zu** in); **b)** *(einschnüren)* lace in ‹*waist*›; **~|schreiben** *unr. tr. V.* **a)** write together; **b)** *(abwertend: verfassen)* dash off ‹*report, letter, etc.*›; **~|schrumpfen** *itr. V.; mit sein* shrivel [up]; *(fig.)* dwindle; **~|sein** *unr. itr. V.; mit sein; Zusschr. nur im Inf. u. Part.* **a)** be together; **b)** *(zusammenleben)* be *or* live together; **~sein das a)** being together *no art.*; **b)** *(Treffen)* get-together; **~|setzen 1.** *tr. V.* **a)** put together; **b)** *(herstellen)* make; **ein ~gesetztes Wort/Verb** a compound word/verb; **c)** *(zusammenbauen, -montieren)* assemble; put together; **d)** *(beieinander sitzen lassen)* seat *or* put together; **jmdn. mit jmdm. ~setzen** seat *or* put sb. next to sb.; **2.** *refl. V.* **a)** sit together; *(zu einem Gespräch)* get together; **b) sich aus etw. ~setzen** be made up *or* composed of sth.; **~setzung die; ~, ~en a)** *(Aufbau)* composition; **„~setzung: ...“** *(als Aufschrift auf Medikamentenpackung)* 'ingredients: ...'; **b)** *(Sprachw.)* compound; **~|sitzen** *unr. itr.V.* sit together; **~|spielen** *itr. V.* play together; ‹*actors*› act together; **mit jmdm. ~spielen** play/act with sb.; **~|stehen** *unr. itr. V.* **a)** stand together; **mit jmdm. ~stehen** stand with sb.; **b)** *(fig.: zusammenhalten)* stand by one another; **~|stellen 1.** *tr. V.* **a)** put together; **b)** *(aus Teilen gestalten)* put together ‹*programme, film, book, menu, exhibition, team, delegation*›; draw up ‹*list, timetable*›; compile ‹*report, broadcast*›; work out ‹*route, tour*›; make up ‹*flower arrangement*›; **c)** *(in einer Übersicht, Liste usw.)* draw together; compile ‹*facts, data*›; **d)** *(kombinieren)* combine; **2.** *refl. V.* stand together; **~stellung die a)** *s.* **~stellen 1b:** putting together; drawing up; compilation; working out; making up; **b)** *(Übersicht)* survey; *(von Tatsachen, Daten)* compilation; **c)** *(Kombination)* combination; **~stoß der** collision; *(fig.)* clash (**mit** with); **bei dem ~stoß [der beiden Züge]** in the collision [between the two trains]; **~|stoßen** *unr. itr. V.; mit sein* collide (**mit** with); **wir stießen mit den Köpfen ~:** we banged *or* bumped our heads; **~|strömen** *itr. V.; mit sein (fig.)* ‹*people*› congregate; **~|stürzen** *itr. V.; mit sein* collapse; **~|suchen** *tr. V.* collect bit by bit; hunt out ‹*information*› bit by bit; **~|tragen** *unr. tr. V.* collect; **~|treffen** *unr. itr. V.; mit sein* **a)** meet; **mit jmdm. ~treffen** meet sb.; **b)** *(zeitlich)* coincide; **~|tun** *(ugs.)* **1.** *unr. tr. V.* put together; **2.** *unr. refl. V.* get together; **~|zählen** *tr. V.* add up; **~|ziehen 1.** *unr. tr. V.* **a)** draw *or* pull together; draw *or* pull ‹*noose, net*› tight; **b)** *(konzentrieren)* mass ‹*troops, police*›; **2.** *unr. refl. V.* contract; **3.** *unr. itr. V.; mit sein* move in together; **mit jmdm. ~ziehen** move in with sb.; **~|zucken** *itr. V.; mit sein* start; jump

Zu̱·satz der a) addition; **ohne ~ von ...:** without the addition of ...; without adding ...; **b)** *(Additiv)* additive

Zusatz·bremsleuchte die *(Kfz-W.)* high-level brake light

zusätzlich ['tsu:zɛtslıç] **1.** *Adj.* additional; **2.** *adv.* in addition

zu·schạnden *Adv.* **etw. ~ machen** wreck *or* ruin sth.; **~ werden** be wrecked *or* ruined

zu̱|schanzen *tr. V. (ugs.)* **jmdm./sich etw. ~:** wangle sth. for sb./oneself *(coll.)*

zu̱|schauen *itr. V. (südd., österr., schweiz.) s.* **zusehen**

Zu̱·schauer der, Zu̱·schauerin die; ~, ~nen spectator; *(im Theater, Kino)* member of the audience; *(an einer Unfallstelle)* onlooker; *(Fernseh~)* viewer; **die ~:** the spectators; the crowd *sing.*/the audience *sing.*/the onlookers/the audience *sing.;* the viewers

zu̱|schaufeln *tr. V.* fill in [with a shovel/shovels]

zu̱|schicken *tr. V.* **jmdm. etw. ~:** send sth. to sb.; send sb. sth.

zu̱|schieben *unr. tr. V.* **a)** push ‹*drawer, door*› shut; **den Riegel ~:** put the bolt across; **b) jmdm. die Schuld/Verantwortung ~** *(fig.)* lay the blame/responsibility on sb.

zu̱|schießen 1. *unr. tr. V. (als Zuschuß geben)* contribute (**zu** towards). **2.** *unr. itr. V.; mit sein* **auf jmdn./etw. zugeschossen kommen** come shooting towards sb.

Zu̱·schlag der a) additional *or* extra charge; *(auf Entgelt)* additional *or* extra payment; *(auf Fahrpreis)* supplement; **b)** *(Eisenb.: Fahrschein)* supplement ticket; **c)** *(bei einer Versteigerung)* acceptance of a/the bid; **d)** *(bei Ausschreibung eines Auftrags)* acceptance of a/the tender; **den ~ bekommen** get the contract

zu̱|schlagen 1. *unr. tr. V.* bang or slam ‹*door, window, etc.*› shut; close ‹*book*›; *(heftig)* slam ‹*book*› shut. **2.** *unr. itr. V.* **a)** *mit sein* ‹*door, trap*› slam *or* bang shut; **b)** *(einen Schlag führen)* throw a blow/blows; *(losschlagen)* hit *or* strike out; *(fig.)* ‹*army, police, murderer*› strike; **schlag doch zu!** [go on,] hit it/me/him *etc.*!

zuschlag·pflichtig [-'pflıçtıç] *Adj. (Eisenb.)* ‹*train*› on which a supplement is payable

zu̱|schließen 1. *unr. tr. V.* lock. **2.** *unr. itr. V.* lock up

zu̱|schnappen *itr. V.* **a)** *mit sein* snap shut; **b)** *(zubeißen)* snap

zu̱|schneiden *unr. tr. V.* cut out ‹*material, dress, jacket, etc.*›; saw ‹*plank, slat*› to size; **auf jmdn./etw. zugeschnitten sein** *(fig.)* be tailor-made for sb./sth.

Zu̱·schnitt der a) cut; **b)** *o. Pl. (das Zuschneiden)* cutting [out]; **c)** *(fig.: Format)* calibre

zu̱|schnüren *tr. V.* tie up; tie *or* do ‹*shoes*› up

zu̱|schrauben *tr. V.* screw the lid *or* top on ‹*jar, flask*›; screw ‹*lid, top*› on

zu̱|schreiben *unr. tr. V.* **a) jmdm./einem Umstand etw. ~:** attribute sth. to sb./a circumstance; **jmdm. das Verdienst/die Schuld an etw.** *(Dat.)* **~:** credit sb. with/blame sb. for sth.; **das hast du dir selbst zuzuschreiben** you only have yourself to blame [for this]

Zu̱·schrift die letter; *(auf eine Anzeige)* reply

zu·schụlden *Adv.* **sich** *(Dat.)* **etwas ~ kommen lassen** do wrong

Zu̱·schuß der contribution (**zu** towards); *(regelmäßiger ~)* allowance; **[staatlicher] ~:** state subsidy (**für, zu** towards)

zu̱|sehen *unr. itr. V.* **a)** watch; **jmdm. beim Arbeiten** *usw.* **~:** watch sb. working *etc.*; **vom [bloßen] Zusehen** [simply] by watching; **b)** *(dafür sorgen)* make sure; see to it; **sieh zu, daß ...:** see that ...; make sure that ...; **er soll ~, wie er das hinkriegt** he'll just have to manage somehow; **sieh zu, wo du bleibst!** you're on your own!

zusehends ['tsu:ze:ənts] *Adv.* visibly

zu̱·sein *unr. itr. V. mit sein; Zusschr. nur im Inf. u. Partizip* ‹*door, window*› be shut; ‹*shop*› have shut

zu̱|senden *unr. od. regelm. tr. V. s.* **zuschicken**

Zu̱·sendung die sending

zu̱|setzen 1. *tr. V.* **a) einem Stoff etw. ~:** add sth. to a substance; **b)** *(zuzahlen)* pay out. **2.** *itr. V. (ugs.)* **jmdm. ~:** *(jmdn. angreifen)* go for sb.; *(jmdn. bedrängen)* pester *or* badger sb.; ‹*mosquitoes etc.*› plague sb.; ‹*illness, heat*› take a lot out of sb.; **einer Sache** *(Dat.)* **~** *(etw. beschädigen)* damage sth.

zu̱|sichern *tr. V.* **jmdm. etw. ~:** promise sb. sth.; assure sb. of sth.

Zu̱·sicherung die promise; assurance

Zu̱·spiel das; *o. Pl. (Ballspiele)* passing; *(einzelner Spielzug)* pass

zu̱|spielen *tr. V.* **a) jmdm. den**

Ball ~: pass the ball to sb.; **b) der Presse Informationen ~** *(fig.)* leak information to the press

zu|spitzen 1. *tr. V.* **a)** sharpen to a point; **b)** *(fig.)* aggravate ⟨*position, crisis*⟩; intensify ⟨*competition, conflict, etc.*⟩; **c)** *(fig.)* make ⟨*question, answer*⟩ pointed. **2.** *refl. V.* become aggravated

Zu·spitzung die; ~, ~en *(fig.) s.* **zuspitzen b:** aggravation; intensification

zu|sprechen *unr. tr. V.* **a) er sprach ihr Trost/Mut zu** his words gave her comfort/courage; **b) jmdm. ein Erbe** *usw.* **~:** award sb. an inheritance *etc.*

Zu·spruch der; *o. Pl.* **[bei jmdm.] ~ finden** *(geh.)* be popular [with sb.]

Zu·stand der a) condition; *(bes. abwertend)* state; **in flüssigem ~:** in liquid form; **in betrunkenem ~:** while under the influence of alcohol; **geistiger/gesundheitlicher ~:** state of mind/health; **der ~ des Patienten** the patient's condition; **Zustände kriegen** *(ugs.)* have a fit *(coll.)*; **b)** *(Stand der Dinge)* state of affairs; situation; **das sind ja [schöne] Zustände!** that's a fine state of affairs!; these are fine goings-on!; **das ist doch kein ~!** that just won't do *(coll.)*; *s. auch* **Rom**

zu·stande *Adv.* **etw. ~ bringen** [manage to] bring about sth.; **~ kommen** come into being; *(geschehen)* take place

zu·ständig *Adj.* appropriate, proper, relevant ⟨*authority, office, etc.*⟩; **von ~er Seite** by the proper authority; **[für etw.] ~ sein** *(verantwortlich)* be responsible [for sth.]; *(kompetent)* be competent [to deal with sth.]; ⟨*court*⟩ have jurisdiction [in sth.]

Zuständigkeit die; ~, ~en *(Verantwortlichkeit)* responsibility; *(Kompetenz)* competence; *(eines Gerichts)* jurisdiction

zu·statten *Adv.* **jmdm./einer Sache ~ kommen** be a help *or* be useful to sb./for sth.; *(von Vorteil sein)* be of advantage to sb./sth.

zu|stecken *tr. V.* **jmdm. etw. ~:** slip sb. sth.

zu|stehen *unr. itr. V.* **etw. steht jmdm. zu** sb. is entitled to sth.; **ein Urteil über ihn steht mir nicht zu** it is not for me to judge him

zu|steigen *unr. itr. V.; mit sein* get on; **ist noch jemand zugestiegen?** *(im Bus)* any more fares, please?; *(im Zug)* ≈ tickets, please!

zu|stellen *tr. V.* **a)** block ⟨*entrance, passage, etc.*⟩; **b)** *(bringen)* deliver ⟨*letter, parcel, etc.*⟩; **jmdm. etw. ~:** deliver sth. to sb.

Zu·stellung die delivery

zu|steuern 1. *itr. V.; mit sein* **auf jmdn./etw. ~:** head for sb./sth. **2.** *tr. V.* **etw. auf jmdn./etw. ~:** steer *or* drive sth. towards sb./sth.

zu|stimmen *itr. V.* agree; **jmdm. [in einem Punkt] ~:** agree with sb. [on a point]; **dem kann ich nur ~:** I quite agree

Zu·stimmung die *(Billigung)* approval **(zu** of); *(Einverständnis)* agreement **(zu** to, with); **~ finden** meet with approval; **jmdm. seine ~ zu etw. geben** give sb. one's consent to *or* for sth.

zu|stopfen *tr. V.* **a)** plug, stop up; plug ⟨*ears*⟩; **b)** *(mit Nadel und Faden)* darn, mend ⟨*hole*⟩

zu|stöpseln *tr. V.* **a)** put a stopper in ⟨*bottle*⟩; *(mit Korken)* put a cork in, cork ⟨*bottle*⟩; **b)** put a plug in ⟨*basin*⟩; plug ⟨*drain etc.*⟩

zu|stoßen 1. *unr. tr. V.* push ⟨*door etc.*⟩ shut. **2.** *unr. itr. V.* **a)** strike out; ⟨*snake etc.*⟩ strike; *(mit einem Messer usw.)* make a stab; stab; **b)** *mit sein* **jmdm. ~:** happen to sb.

zu|streben *itr. V.; mit sein* **einer Sache** *(Dat.) od.* **auf etw.** *(Akk.)* **~:** make for sth.; *(fig.)* strive for *or* aim at sth.

Zu·strom der a) *(auch fig.)* flow; **b)** *(von Menschen)* influx; stream

zu|tage *Adv.* **~ kommen** *od.* **treten** become visible *(lit. or fig.)*; ⟨*stream*⟩ come to the surface; *(fig.)* become evident; ⟨*story*⟩ come out, be made public; ⟨*differences etc.*⟩ come into the open; **etw. ~ bringen** *od.* **fördern** *(aus der Tasche usw.)* produce sth.; *(fig.)* bring sth. to light; reveal sth.; **offen ~ liegen** be perfectly clear *or* evident

Zu·tat die ingredient

zu·teil *Adv.* *(geh.)* **jmdm./einer Sache ~ werden** be granted *or* accorded to sb./sth.; **jmdm. etw. ~ werden lassen** accord sb. sth.; bestow sth. on sb.

zu|teilen *tr. V.* allot, assign *(Dat.* to); **jmdm. seine Portion ~:** mete out his/her share to sb.; **die zugeteilte Menge** the allocated amount

Zu·teilung die a) allotting, assigning; *(einer Ration)* sharing out, allocation; *(eines Mandats, Quartiers)* allocation, assignment; **b)** *(Ration)* allocation, ration

zu·tiefst *Adv.* profoundly; **~ verletzt** deeply hurt *or* offended

zu|tragen *unr. refl. V. (geh.)* take place; occur

zuträglich ['tsu'trɛ:klıç] *Adj.* healthy ⟨*climate*⟩; **jmdm./einer Sache ~ sein** be good for sb./sth.; be beneficial to sb./sth.

zu|trauen *tr. V.* **jmdm. etw. ~:** believe sb. [is] capable of [doing] sth.; **den Mut hätte ich ihm gar nicht zugetraut** I should never have thought he had the courage; **ich hätte ihm mehr Taktgefühl zugetraut** I should have thought he had more tact; **das ist ihm [durchaus] zuzutrauen** I could [well] believe it of him; **sich** *(Dat.)* **etw. ~:** think one can do *or* is capable of doing sth.; **er traut sich** *(Dat.)* **zuwenig zu** he has too little self-confidence

Zutrauen das; ~s confidence, trust **(zu** in)

zutraulich 1. *Adj.* trusting; trustful. **2.** *adv.* trustingly; trustfully

Zutraulichkeit die; ~: trust[fulness]

zu|treffen *unr. itr. V.* **a)** be correct; **b) auf etw. ~:** apply to sth.

zutreffend 1. *Adj.* **a)** correct; *(treffend)* accurate; **es ist ~, daß ...:** it is correct *or* the case that ...; **b)** *(geltend)* applicable; relevant; **Zutreffendes bitte ankreuzen** please mark with a cross where applicable. **2.** *adv.* correctly; *(treffend)* accurately

zu|trinken *unr. itr. V.* **jmdm. ~:** raise one's glass and drink to sb.

Zu·tritt der entry; admittance; **„kein ~“, „~ verboten“** 'no entry'; 'no admittance'; **~ [zu etw.] haben** have access [to sth.]

zu|tun *unr. tr. V.* **kein Auge ~:** not sleep a wink

Zu·tun das; ~s: ohne jmds. ~: without sb.'s being involved; **es geschah ohne mein ~:** I had nothing to do with it

zu·ungunsten *Präp. mit Gen.* to the disadvantage of

zu·unterst *Adv.* right at the bottom; *s. auch* **ober... a**

zuverlässig ['tsu:fɛɐ̯lɛsıç] **1.** *Adj.* reliable; *(verläßlich)* dependable ⟨*person*⟩. **2.** *adv.* **a)** reliably; **er arbeitet sehr ~:** he is a very reliable worker; **b)** *(mit Gewißheit)* ⟨*confirm*⟩ with certainty; ⟨*know*⟩ for sure, for certain

Zuverlässigkeit die; ~: reliability; *(Verläßlichkeit)* dependability

Zuversicht [tsu:'fɛɐ̯zıçt] **die; ~:** confidence

zuversichtlich 1. *Adj.* confident; **sich ~ geben** express one's confidence. **2.** *adv.* confidently

Zuversichtlichkeit die; ~: confidence

zuviel 1. *indekl. Indefinitpron.* **a)** too much; **viel ~:** far *or* much too much; ~ **kriegen** *(ugs.)* blow one's top *(coll.)*; *(bei jmds. Worten)* see red; **das ist ~ gesagt** that's going too far; that's an exaggeration; **b)** *(ugs.: zu viele)* too many. **2.** *adv.* too much

zu·vor *Adv.* before; **tags/im Jahr ~:** the day/year before

zuvor|kommen *unr. itr. V.; mit sein* **a) jmdm.** ~: beat sb. to it; get there first; **b) einer Sache** *(Dat.)* ~: anticipate *or* forestall sth.

zuvorkommend 1. *Adj.* obliging; *(höflich)* courteous. **2.** *adv.* obligingly; *(höflich)* courteously

Zuvorkommenheit die; ~: courteousness; courtesy

Zu·wachs der; ~es, **Zuwächse** [-vɛksə] increase

zu|wachsen *unr. itr. V.; mit sein* become overgrown

zu|warten *itr. V.* wait

zu·wege *Adv.* **etw. ~ bringen** [manage to] achieve sth.

zu·weilen *Adv. (geh.)* now and again; at times

zu|weisen *unr. tr. V.* **jmdm. etw.** ~: allocate *or* allot sb. sth.

zu|wenden 1. *unr. od. regelm. refl. V.* **sich jmdm./einer Sache ~** *(auch fig.)* turn to sb./sth.; *(sich widmen)* devote oneself to sb./sth.. **2.** *unr. od. regelm. tr. V.* **a) jmdm./einer Sache etw. ~:** turn sth. to[wards] sb./sth.; **jmdm. den Rücken ~:** turn one's back on sb.; **b) jmdm. Geld ~** *(geh.)* give *or* donate money to sb.

Zu·wendung die a) *o. Pl. (Aufmerksamkeit)* [loving] attention *or* care; **b)** *(Geldgeschenk)* gift of money; *(Unterstützung)* [financial] contribution; *(Geldspende)* donation

zu·wenig 1. *indekl. Indefinitpron.* **a)** too little; **viel ~:** far too little; **b)** *(ugs.: zu wenige)* too few; not enough. **2.** *adv.* too little

zuwider *Adj.* **jmdm. ~ sein** be repugnant to sb.; **Spinat ist mir äußerst ~:** I absolutely detest spinach

zuwider-: ~|handeln *itr. V.* **dem Gesetz/einer Vorschrift** *usw.* **~handeln** contravene *or* infringe the law/a regulation *etc.*; **einer Anordnung/einem Verbot ~handeln** defy an instruction/a ban; **~|laufen** *unr. itr. V.; mit sein* **einer Sache** *(Dat.)* **~laufen** go against *or* run counter to sth.

zu|winken *itr. V.* **jmdm. ~:** wave to sb.

zu|zahlen *tr. V.* pay ⟨*five marks etc.*⟩ extra; **einen Betrag ~:** pay an additional sum

zu|ziehen 1. *unr. tr. V.* **a)** pull ⟨*door*⟩ shut; draw ⟨*curtain*⟩; pull *or* draw ⟨*knot, net*⟩ tight; do up ⟨*zip*⟩; **b)** call in ⟨*expert, specialist*⟩. **2.** *unr. refl. V.* **a) sich** *(Dat.)* **eine Krankheit/Infektion ~:** catch an illness/contract an infection; **sich** *(Dat.)* **einen Schädelbruch ~:** sustain a fracture of the skull; **sich** *(Dat.)* **jmds. Zorn ~:** incur sb.'s anger; **b)** *(sich schließen)* ⟨*knot, noose*⟩ tighten, get tight. **3.** *unr. itr. V.; mit sein* move here *or* into the area

Zu·zug der influx

zuzüglich ['tsu:tsy:klɪç] *Präp. mit Gen.* plus

zu|zwinkern *itr. V.* **jmdm. ~:** wink at sb.

zwacken ['tsvakn̩] *tr., auch itr. V. (ugs.) s.* **zwicken**

zwang [tsvaŋ] *1. u. 3. Pers. Sg. Prät. v.* **zwingen**

Zwang der; ~[e]s, **Zwänge** ['tsvɛŋə] **a)** compulsion; **auf jmdn. ~ ausüben** exert pressure on sb.; **der ~ der Verhältnisse** the force of circumstance[s]; **soziale Zwänge** social constraints; the constraints of society; **b)** *(innerer Drang)* irresistible urge; **aus einem ~ [heraus] handeln** act under a compulsion *or* on an irresistible impulse; **c)** *o. Pl. (Verpflichtung)* obligation; **es besteht kein ~ zur Teilnahme/zum Kauf** there is no obligation to take part/to buy anything

zwängen ['tsvɛŋən] **1.** *tr. V.* squeeze. **2.** *refl. V.* squeeze [oneself]

zwanglos 1. *Adj.* **a)** informal; casual, free and easy ⟨*behaviour*⟩; **b)** *(unregelmäßig)* haphazard ⟨*arrangement*⟩. **2.** *adv.* **a)** informally; freely; **es ging dort ziemlich ~ zu** things were pretty free and easy there; **b)** *(unregelmäßig)* haphazardly ⟨*arranged*⟩

Zwanglosigkeit die; ~ **a)** informality; **b)** *(Unregelmäßigkeit)* haphazard *or* casual manner

zwangs-, Zwangs-: ~lage die predicament; **~läufig** [~lɔyfɪç] **1.** *Adj.* inevitable; **2.** *adv.* inevitably; **~maßnahme die** coercive measure; sanction; **~versteigern** *tr. V., nur im Inf. u. Part. (Rechtsw.)* put up for compulsory auction; **~versteigerung die** *(Rechtsw.)* [compulsory] auction

zwanzig ['tsvantsɪç] *Kardinalz.* twenty; *s. auch* **achtzig**

zwanziger *indekl. Adj.; nicht präd.* **die ~ Jahre** the twenties; *s. auch* **achtziger**

[1]Zwanziger der; ~s, ~ **a)** twenty-year-old; **b)** *(Geldschein)* twenty-mark/franc/schilling *etc.* note

[2]Zwanziger die; ~, ~ *(ugs.)* twenty-pfennig/schilling *etc.* stamp

zwanzig·jährig *Adj. (20 Jahre alt)* twenty-year-old *attrib.*; *(20 Jahre dauernd)* twenty-year *attrib.*

Zwanzig·mark·schein der twenty-mark note

zwanzigst ... *Ordinalz.* twentieth; *s. auch* **acht...**; **achtzigst...**

zwar [tsva:ɐ̯] *Adv.* **a)** admittedly; **ich weiß es ~ nicht genau, aber ...:** I'm not absolutely sure [I admit,] but ...; **b) und ~:** to be precise; **er ist Zahnarzt, und ~ ein guter** he is a dentist, and a good one at that

Zweck [tsvɛk] **der;** ~[e]s, ~e **a)** purpose; **zu diesem ~:** for this purpose; **was ist der ~ Ihrer Reise?** what is the purpose of your journey?; **seinen ~ erfüllen** serve its purpose; **Geld für einen guten/wohltätigen ~:** money for a good cause/for a charity; **der ~ der Übung** *(ugs.)* the object *or* point of the exercise; **b)** *(Sinn)* point; **es hat keinen/wenig ~ [, das zu tun]** it's pointless *or* there is no point/there is little *or* not much point [in doing that]; **ohne [jeden] Sinn und ~:** completely pointless

zweck-, Zweck-: ~bau der; *Pl.* **~bauten** functional building; **~dienlich** *Adj.* appropriate; helpful, relevant ⟨*information etc.*⟩; **~entfremden** *tr. V.* use for another purpose; *(für den falschen Zweck)* misuse; **~los** *Adj.* pointless; **~mäßig 1.** *Adj.* appropriate; expedient ⟨*behaviour, action*⟩; functional ⟨*building, fittings, furniture*⟩; **2.** *adv.* appropriately ⟨*arranged, clothed*⟩; ⟨*act*⟩ expediently; ⟨*equip, furnish*⟩ functionally; **~mäßigkeit die** appropriateness; *(einer Handlung)* expediency; *(eines Gebäudes)* functionalism

zwecks *Präp. mit Gen. (Papierdt.)* for the purpose of

zwei [tsvai] *Kardinalz.* two; **wir ~:** we two; the two of us; **sie waren/kamen zu ~en** there were two of them/two of them came; **für ~ essen/arbeiten** eat enough for two/do the work of two people; **dazu gehören immer noch ~!** *(ugs.)* it takes two [to do that]!; *s. auch* **[1]acht**

Zwei die; ~, ~en **a)** *(Zahl)* two; **b)** *(Schulnote)* B; **eine ~ schreiben/**

bekommen get a B; **er hat die Prüfung mit ~ bestanden** he got a B in the examination; *s. auch* [1]**Acht a, d, e, g**

zwei-, Zwei-: ~bändig *Adj.* two-volume; **~bettzimmer das** twin-bedded room; **~deutig** [~dɔytɪç] **1.** *Adj.* **a)** ambiguous; equivocal ⟨*smile*⟩; **b)** *(fig.: schlüpfrig)* suggestive ⟨*remark, joke*⟩; **2.** *adv.* **a)** ambiguously; ⟨*smile*⟩ equivocally; **b)** *(fig.: schlüpfrig)* suggestively; **~deutigkeit die; ~, ~en** *s.* **zweideutig 1: a)** *o. Pl.* ambiguity; suggestiveness; **b)** *(Äußerung)* ambiguity; double entendre; **~dimensional** [~dimɛnzi̯ona:l] **1.** *Adj.* two-dimensional; **2.** *adv.* two-dimensionally; in two dimensions; **~ein·halb** *Bruchz.* two and a half

Zweier der; ~s, ~ a) *(ugs.) s.* **Zwei b; b)** *(ugs.: Münze)* two-pfennig piece; **c)** *(Ruderboot)* pair

zweierlei *Gattungsz.; indekl.* **a)** *attr.* two sorts *or* kinds of; two different ⟨*sizes, kinds, etc.*⟩; **mit ~ Maß messen** use double standards; **b)** *alleinstehend* two [different] things; **es ist ~, ob man es sagt oder [ob man es] auch tut** it is one thing to say it and another [thing] to do it

zwei-, Zwei-: ~fach *Vervielfältigungsz.* double; *(~mal)* twice; **die ~fache Menge/Länge** double *or* twice the amount/length; **etw. ~fach vergrößern/verkleinern** enlarge sth. to twice its size/reduce sth. to half-size; *s. auch* **achtfach; ~fache das;** *adj. Dekl.* **das ~fache** twice as much; *s. auch* **Achtfache; ~familien·haus das** two-family house; duplex *(esp. Amer.)*; **~farbig 1.** *Adj.* two-coloured; two-tone ⟨*scarf, paintwork, etc.*⟩; **2.** *adv.* in two colours

Zweifel ['tsvaifl̩] **der; ~s, ~:** doubt (**an** + *Dat.* about); **~ bekommen** become doubtful; **ich habe da so meine ~:** I have my doubts about that; **ich bin mir noch im ~, ob ...:** I am still uncertain whether ...; **etw. in ~ ziehen** question sth.; **[für jmdn.] außer ~ stehen** be beyond doubt [as far as sb. is concerned]; **über jeden** *od.* **allen ~ erhaben sein** be beyond any shadow of a doubt; **kein ~, ...:** there is/was no doubt about it, ...; **ohne ~:** without [any] doubt; **im ~:** in case of doubt; if in doubt

zweifelhaft *Adj.* **a)** doubtful; **b)** *(fragwürdig)* dubious

zweifel·los *Adv.* undoubtedly; without [any] doubt

zweifeln *itr. V.* doubt; **wenn man zweifelt** if one is in doubt *or* has any doubts; **an jmdm./etw. ~:** doubt sb./sth.; have doubts about sb./sth.; **~ daran, daß ..., ~, ob ...:** doubt whether ...; **daran ist nicht zu ~:** there can be no doubt about it

zweifels-, Zweifels-: ~fall der case of doubt; doubtful *or* problematic case; **im ~fall[e]** in case of doubt; if in doubt; **~frei 1.** *Adj.* definite; **~frei sein** be beyond doubt; **2.** *adv.* beyond [any] doubt; **~ohne** *Adv.* undoubtedly; without doubt

Zweifler der; ~s, ~doubter

Zweig [tsvaik] **der; ~[e]s, ~e** [small] branch; *(meist ohne Blätter)* twig; **auf keinen grünen ~ kommen** *(ugs.)* not get anywhere; **b)** *(fig.)* branch

zwei-: ~geschossig *Adj., adv. s.* **~stöckig; ~geteilt** *Adj.* divided; divided in two *postpos.*; **~gleisig** [-glaizɪç] **1.** *Adj.* two-track; double-track; *(fig.)* two-way; **2.** *adv.* **a)** ⟨*run*⟩ on two tracks; **b) ~ fahren** *(fig.)* follow a dual-track policy

Zweig·stelle die branch [office]

zwei-, Zwei-: ~hundert *Kardinalz.* two hundred; **~jährig** *Adj.* *(zwei Jahre alt)* two-year-old *attrib.*; *(zwei Jahre dauernd)* two-year *attrib.*; **~kampf der a)** single combat; *(Duell)* duel; **b)** *(Sport)* man-to-man tussle; duel; **~köpfig** *Adj.* **a)** two-headed; **b)** *(aus zwei Personen bestehend)* two-person *attrib.*; of two [people] *postpos.*; **~mal** *Adv.* twice; **das wird er sich** *(Dat.)* **~mal überlegen** he'll think twice about that; *s. auch* **achtmal; ~mark·stück das** two-mark piece; **~motorig** *Adj.* twin-engined; **~pfennig·stück das** two-pfennig piece; **~polig** *Adj.* double-pole; two-core ⟨*cable*⟩; two-pin ⟨*plug, socket*⟩; **~reiher der** double-breasted suit/coat/ jacket; **~schneidig** *Adj.* double-edged; **ein ~schneidiges Schwert** *(fig.)* a double-edged sword; **~sprachig 1.** *Adj.* bilingual; ⟨*sign*⟩ in two languages; **2.** *adv.* bilingually; ⟨*labelled, written, printed, etc.*⟩ in two languages; **~spurig 1.** *Adj.* **a)** two-lane ⟨*road*⟩; **b)** two-track ⟨*vehicle*⟩; **c)** two- *or* twin-track ⟨*recording*⟩; **2.** *adv.* **a)** in two lanes; **b)** ⟨*record*⟩ on two tracks; **~stellig** *Adj.* two-figure *attrib.* ⟨*number, sum*⟩; **~stöckig 1.** *Adj.* two-storey *attrib.*; **~stöckig sein** have two storeys *or* floors; **2.** *adv.* ⟨*build*⟩ two storeys high; **~strahlig** *Adj.* twin-engined ⟨*jet aircraft*⟩; **~stündig** *Adj.* two-hour *attrib.*; *(Schulw.)* double-period *attrib.* ⟨*test, examination*⟩; **nach ~stündiger Wartezeit** after waiting for two hours

zweit ['tsvait] **wir waren zu ~:** there were two of us; **sie sind zu ~ verreist** the two of them went away together; *s. auch* [2]**acht**

zweit... *Ordinalz.* second; **jeder ~e Einwohner** every other *or* second inhabitant; **jeder ~e** every other one; **~er Klasse fahren/liegen** travel second-class/be in a second-class hospital bed; **ich habe noch einen ~en** I have a second one; *(als Ersatz)* I have a spare; **wie kein ~er** as no one else can; like nobody else; *s. auch* **erst...**

zwei-, Zwei-: ~tägig *Adj.* *(2 Tage alt)* two-day-old *attrib.*; *(2 Tage dauernd)* two-day *attrib.*; *s. auch* **achttägig; ~takter der; ~s, ~** *(Motor)* two-stroke engine; *(Fahrzeug)* two-stroke; **~takt·motor der** two-stroke engine

zweit·ältest ... *Adj.* second oldest; **der/die Zweitälteste** the second oldest

zwei·tausend *Kardinalz.* two thousand

Zwei·tausender der *mountain more than two thousand metres high*

zweit·best... *Adj.* second best

zwei·teilig *Adj.* two-piece ⟨*suit, bathing-suit, suite, etc.*⟩; two-part ⟨*film, programme*⟩

zweite·mal: das ~: for the second time

zweiten·mal: zum ~: for the second time; **beim ~:** the second time [round]

zweitens *Adv.* secondly; in the second place

Zweite[r]-Klasse-Abteil das second-class compartment

zweit-, Zweit-: ~frisur die wig; **~kläßler der; ~s, ~** *(südd., schweiz.)* pupil in second class of primary school; second-year pupil; **~rangig** [~raŋɪç] *Adj.* of secondary importance *postpos.*; secondary ⟨*importance*⟩; **~stimme die** second vote

zwei·türig *Adj.* two-door ⟨*car*⟩

Zweit-: ~wagen der second car; **~wohnung die** second home

zwei-, Zwei-: ~wertig *Adj.* *(fachspr.)* bivalent **~zeiler der; ~s, ~:** couplet; **~zimmerwoh-**

nung [-'----] **die** two-room flat *(Brit.) or (Amer.)* apartment
Zwerch·fell ['tsvɛrç-] **das** *(Anat.)* diaphragm
Zwerg [tsvɛrk] **der; ~[e]s, ~e a)** dwarf; *(Garten~)* gnome; **b)** *(abwertend: unbedeutender Mensch)* [little] squirt *(coll.)*; wretch
zwergenhaft *Adj.* dwarfish
Zwergin die; ~, ~nen dwarf
Zwerg·wuchs der dwarfism *no art.*; stunted growth *no art.*
Zwetsche ['tsvɛtʃə] **die; ~, ~n** damson plum
Zwetschen-: ~kuchen der plum-flan; **~wasser das;** *Pl.* **~wässer** plum brandy
Zwetschken·knödel ['tsvɛtʃkn̩-] **der** *(Kochk.)* plum dumpling
zwicken ['tsvɪkn̩] *tr., auch itr. V.* **a)** pinch; **jmdm.** *od.* **jmdn. in den Arm ~:** pinch sb.'s arm; **b)** *(plagen)* **es zwickte und zwackte ihn überall** he had twinges *or* little aches and pains all over
Zwick·mühle die a) double mill; **b)** *(fig.: Dilemma)* dilemma
Zwie·back ['tsvi:bak] **der; ~[e]s, ~e** *od.* **Zwiebäcke** ['tsvi:bɛkə] rusk; **~ essen** eat rusks
Zwiebel ['tsvi:bl̩] **die; ~, ~n a)** onion; **b)** *(Blumen~)* bulb
Zwiebel-: ~suppe die onion soup; **~turm der** onion tower
zwie-, Zwie-: ~gespräch das *(geh.)* dialogue; **~licht das;** *o. Pl.* **a)** twilight; **b)** *(Mischung von Dämmer- und Kunstlicht)* half-light *(that is unpleasant for the eye)*; **c) ins ~licht geraten** *(fig.)* become suspect; ⟨*person*⟩ come under suspicion; **~lichtig** *Adj.* shady; dubious; **~spalt der; ~[e]s, ~e** *od.* **~spälte** [~ʃpɛltə] [inner] conflict; **in einen ~spalt geraten** get into a state of conflict; **~spältig** [~ʃpɛltɪç] *Adj.* conflicting ⟨*mood, feelings*⟩; discordant ⟨*impression*⟩; **~tracht die** *(geh.)* discord; **~tracht säen** sow the seeds of discord
Zwilling ['tsvɪlɪŋ] **der; ~s, ~e a)** twin; **b)** *Pl. (Astron., Astrol.)* Gemini; the Twins; **c)** *(Astrol.: Mensch)* Gemini
Zwillings-: ~bruder der twin brother; **~paar das** pair of twins; **~schwester die** twin sister
zwingen ['tsvɪŋən] **1.** *unr. tr. V.* force; **jmdn. [dazu] ~, etw. zu tun** force *or* compel sb. to do sth.; make sb. do sth.; **jmdn. zu einem Geständnis ~:** force sb. into a confession *or* to make a confession; **sich gezwungen sehen, etw. zu tun** find oneself forced *or* compelled to do sth.; **man kann ihn nicht dazu ~:** he can't be forced *or* made to do it **2.** *unr. refl. V.* force oneself
zwingend *Adj.* compelling ⟨*reason, logic*⟩; imperative, absolute ⟨*necessity*⟩
Zwinger der; ~s, ~ a) *(Hunde~)* kennel; *(ganze Anlage, auch Zucht)* kennels *pl.*; **b)** *(Gehege)* compound; enclosure; *(für Bären)* bear-pit
zwinkern ['tsvɪŋkɐn] *itr. V.* **[mit den Augen] ~:** blink; *(als Zeichen)* wink
zwirbeln ['tsvɪrbl̩n] *tr. V.* twirl; twist
Zwirn [tsvɪrn] **der; ~[e]s, ~e** [strong] thread *or* yarn
Zwirns·faden der [strong] thread
zwischen ['tsvɪʃn̩] *Präp. mit Dat./Akk.* **a)** between; **b)** *(unter, inmitten)* among[st]
zwischen-, Zwischen-: ~aufenthalt der stopover; **~bemerkung die** interjection; **~durch** [-'-] *Adv.* **a)** *(zeitlich)* between times; *(von Zeit zu Zeit)* from time to time; **b)** *(räumlich)* here and there; **~fall der** incident; **~frage die** question; **~händler der** *(Wirtsch.)* middleman; *(fig.)* go-between; **~hirn das** *(Anat.)* diencephalon; **~hoch das** *(Met.)* ridge of high pressure; **~|landen** *itr. V.; mit sein* **in X ~landen** land in X on the way; **~landung die** stopover; **~lösung die** interim solution; **~mahlzeit die** snack [between meals]; **~menschlich 1.** *Adj.* interpersonal ⟨*relations*⟩; ⟨*contacts*⟩ between people; **2.** *adv.* on a personal level; **~prüfung die** intermediate examination; **~raum der** space; gap; *(Lücke)* gap; **~ruf der** interruption; **~rufer der** heckler; **~runde die** *(Sport)* intermediate round; **~spurt der** *(Sport)* spurt; burst [of speed]; **~stadium das** intermediate stage; **~stufe die** intermediate stage; **~ton der** shade; nuance; *(fig.)* nuance; **~tür die** connecting door; **~zeit die** interim; *(länger)* intervening period; **in der ~zeit** in the meantime
Zwist [tsvɪst] **der; ~[e]s, ~e** *(geh.)* strife *no indef. art.*; *(Fehde)* feud; dispute; **in** *od.* **im ~ leben** live in a state of strife
Zwistigkeit die; ~, ~en *(geh.)* dispute
zwitschern ['tsvɪtʃɐn] **1.** *itr., auch tr. V.* chirp. **2.** *tr. V.* **einen ~** *(salopp)* have a drink
Zwitter ['tsvɪtɐ] **der; ~s, ~** hermaphrodite; *(fig.)* cross **(aus** between)
zwittrig *Adj.* hermaphroditic
zwo [tsvo:] *Kardinalz. (ugs.; bes. zur Verdeutlichung) s.* **zwei**
zwölf [tsvœlf] *Kardinalz.* twelve; **~ Uhr mittags/nachts** [twelve o'clock] midday/midnight; **es ist fünf [Minuten] vor ~** *(fig.)* we are on the brink; *s. auch* [1]**acht**
zwölf-, Zwölf- twelve-; *s. auch* **acht-, Acht-**
Zwölfer der; ~s, ~: twelve; *s. auch* **Achter c, d**
zwölf-, Zwölf-: ~fach *Vervielfältigungsz.* twelvefold; **die ~fache Menge** twelve times the quantity; *s. auch* **achtfach; ~fache das;** *adj. Dekl.*; **das ~fache** twelve times as much; *s. auch* **Achtfache; ~fingerdarm der** *(Anat.)* duodenum; **~jährig** *Adj. (12 Jahre alt)* twelve-year-old *attrib.*; twelve years old *pred.*; *(12 Jahre dauernd)* twelve-year *attrib.*; *s. auch* **achtjährig; ~mal** *Adv.* twelve times; *s. auch* **achtmal**
zwölft [tsvœlft] **wir waren zu ~:** there were twelve of us; *s. auch* [2]**acht**
zwölft... *Ordinalz.* twelfth; *s. auch* **acht...**
zwölftel *Bruchz.* twelfth; *s. auch* **achtel**
Zwölftel das (*schweiz. meist* **der**) **~s, ~:** twelfth
zwot... [tsvo:t...] *(ugs.; bes. bei Datumsangaben) s.* **zweit...**
Zyklen *s.* **Zyklus**
zyklisch ['tsy:klɪʃ] **1.** *Adj.* cyclic[al]. **2.** *adv.* cyclically; as a cycle
Zyklon [tsy'klo:n] **der; ~s, ~e** *(Met.)* cyclone
Zyklus ['tsy:klʊs] **der; ~, Zyklen** cycle
Zylinder [tsi'lɪndɐ] **der; ~s, ~ a)** cylinder; chimney; **b)** *(Hut)* top hat
zylindrisch 1. *Adj.* cylindrical. **2.** *adv.* cylindrically
Zyniker der; ~s, ~, Zynikerin die; ~, ~nen cynic
zynisch ['tsy:nɪʃ] **1.** *Adj.* cynical. **2.** *adv.* cynically
Zynismus der; ~: cynicism
Zypern ['tsy:pɐn] **(das); ~s** Cyprus
Zypresse [tsy'prɛsə] **die; ~, ~n** cypress
Zypriot [tsypri'o:t] **der; ~en, ~en, Zypriotin die; ~, ~nen** Cypriot
zypriotisch, zyprisch *Adj.* Cypriot
Zyste ['tsʏtə] **die; ~, ~n** cyst
z. Z., z. Zt. *Abk.* **zur Zeit**

The Revision of German Orthography
die neue Regelung der Rechtschreibung

In July 1996, after much debate, wide-ranging changes to the spelling of German were agreed and ratified by the governments of Germany, Austria, and Switzerland. The following list, whilst not all-encompassing, details those changes which may be of interest to the user of this dictionary. It is worth noting that although these reforms are valid immediately, they will not be expected to be reflected in all written texts until 2005. Until that date, both old and new spellings will be acceptable.

The following list contains words which are not included in the A–Z text of this dictionary. Nonetheless, it is the editors' view that the learner of German will gain a better overview of the systematic changes involved by studying a more comprehensive list of words affected by the reforms.

The following summary lists the most important changes:

1. The ß character

The ß character, which is generally replaced in Switzerland by a double s, will be retained in Germany and Austria, but will only be written after a long vowel (as in Fuß, Füße) and after a diphthong (as in Strauß, Sträuße).

Fluß, Baß, keß, läßt, Nußknacker become in future: *Fluss, Bass, kess, lässt, Nussknacker*

2. Nominalized adjectives

Nominalized adjectives will be written with a capital even in set phrases.

sein Schäfchen ins trockene bringen, im trüben fischen, im allgemeinen become in future: *sein Schäfchen ins Trockene bringen, im Trüben fischen, im Allgemeinen*

3. Words from the same word family

In certain cases the spelling of words belonging to the same family will be made uniform.

numerieren, überschwenglich become in future: *nummerieren* (like Nummer), *überschwänglich* (being related to Überschwang)

4. The same consonant repeated three times

When the same consonant repeated three times occurs in compounds, all three will be written even when a vowel follows.

Brennessel, Schiffahrt become in future: *Brennnessel, Schifffahrt* (exceptions are dennoch, Drittel, Mittag)

5. Verb, adjective and participle compounds

Verb, adjective and participle compounds will be written more frequently than previously in two words.

spazierengehen, radfahren, ernstgemeint, erdölexportierend become in future: *spazieren gehen, Rad fahren, ernst gemeint, Erdöl exportierend*

6. Compounds containing numbers in figures

Compounds containing numbers in figures will in future be written with a hyphen.

24karätig, 8pfünder become in future: *24-karätig, 8-Pfünder*

7. The division of words containing *st*

The *st* will be treated like a normal combination of consonants and no longer be indivisible.

Ha-stig, Ki-ste become in future: *has-tig, Kis-te*

8. The division of words containing *ck*

The combination *ck* will not be divided and will go on to the next line.

Bäk-ker, schik-ken become in future: *Bä-cker, schi-cken*

9. The division of foreign words

Compound foreign words which are hardly recognized as such today may be divided by syllables, without regard to their original components.

He-li-ko-pter (from the Greek helix and pteron) may also become in future: *He-li-kop-ter*

10. The comma before *und*

Where two complete clauses are connected by *und* a comma will not be obligatory.

Karl war in Schwierigkeiten, und niemand konnte ihm helfen. may also become in future: *Karl war in Schwierigkeiten und niemand konnte ihm helfen.*

11. The comma with infinitives and participles

Even longer clauses containing an infinitive or participle will not have to be divided off with a comma.

Er begann sofort, das neue Buch zu lesen. Ungläubig den Kopf schüttelnd, verließ er das Zimmer. may also become in future: *Er begann sofort das neue Buch zu lesen. Ungläubig den Kopf schüttelnd verließ er das Zimmer.*

An asterisk * denotes new spelling variants which are to be regarded as preferred forms

A

OLD	NEW
[gestern, heute, morgen] abend	[gestern, heute, morgen] Abend
aberhundert	*also:* Aberhundert
Aberhunderte	*also:* aberhunderte
abertausend	*also:* Abertausend
Abertausende	*also:* abertausende
Abfluß	Abfluss
abgeblaßt	abgeblasst
Abguß	Abguss
Ablaß	Ablass
Abriß	Abriss
Abschluß	Abschluss
Abschuß	Abschuss
absein	ab sein
Abszeß	Abszess
abwärtsgehen	abwärts gehen
in acht nehmen	in Acht nehmen
außer acht lassen	außer Acht lassen
8achser	8-Achser
der/die achte, den/die ich sehe	der/die Achte, den/die ich sehe
jeder/jede achte kommt mit	jeder/jede Achte kommt mit
achtgeben	Acht geben
achthaben	Acht haben
8jährig	8-jährig
der/die 8jährige	der/die 8-Jährige
8mal	8-mal
achtmillionenmal	acht Millionen Mal
8tonner	8-Tonner
achtunggebietend	Achtung gebietend
über Achtzig	über achtzig
Mitte [der] Achtzig	Mitte [der] achtzig
in die Achtzig kommen	in die achtzig kommen
die achtziger Jahre	*also:* die Achtzigerjahre*
die Achtzigerjahre	*also:* die achtziger Jahre
ackerbautreibende Völker	Ackerbau treibende Völker
Action-painting	Actionpainting *also:* Action-Painting
ade sagen	*also:* Ade sagen*
Aderlaß	Aderlass
Adhäsionsverschluß	Adhäsionsverschluss
Adreßbuch	Adressbuch
afro-amerikanisch	afroamerikanisch
afro-asiatisch	afroasiatisch
Afro-Look	Afrolook
After-shave	Aftershave
ich habe ähnliches erlebt	ich habe Ähnliches erlebt
und/oder ähnliches (u. ä./o. ä.)	und/oder Ähnliches (u. Ä./o. Ä.)
Alkoholmißbrauch	Alkoholmissbrauch
alleinerziehend	allein erziehend
alleinseligmachend	allein selig machend
alleinstehend	allein stehend
es ist das allerbeste, daß ...	es ist das Allerbeste, dass ...
im allgemeinen	im Allgemeinen
allgemeingültig	allgemein gültig
allgemeinverständlich	allgemein verständlich
allzubald	allzu bald
allzufrüh	allzu früh
allzugern	allzu gern
allzulange	allzu lange
allzuoft	allzu oft
allzusehr	allzu sehr
allzuviel	allzu viel
allzuweit	allzu weit
Alma mater	Alma Mater
Alpdruck	*also:* Albdruck
Alptraum	*also:* Albtraum
als daß	als dass
aus alt mach neu	aus Alt mach Neu
für alt und jung	für Alt und Jung
er ist immer der alte geblieben	er ist immer der Alte geblieben
alles beim alten lassen	alles beim Alten lassen
Alter ego	Alter Ego
altwienerisch	alt-wienerisch
Amboß	Amboss
Anbiß	Anbiss
andersdenkend	anders denkend
andersgeartet	anders geartet
anderslautend	anders lautend
aneinanderfügen	aneinander fügen
aneinandergeraten	aneinander geraten
aneinandergrenzen	aneinander grenzen
aneinanderlegen	aneinander legen
aneinanderreihen	aneinander reihen
angepaßt	angepasst
Angepaßtheit	Angepasstheit
Anglo-Amerikaner	Angloamerikaner
jmdm. angst machen	jmdm. Angst machen
anheimfallen	anheim fallen
anheimstellen	anheim stellen
Anlaß	Anlass
anläßlich	anlässlich
Anriß	Anriss
Anschiß	Anschiss
Anschluß	Anschluss
ansein	an sein
der Archimedische Punkt	der archimedische Punkt
im argen liegen	im Argen liegen
bei arm und reich	bei Arm und Reich
Armee-Einheit	*also:* Armeeeinheit
Aschantinuß	Aschantinuss
As	Ass
aufeinanderbeißen	aufeinander beißen
aufeinanderfolgen	aufeinander folgen
aufeinandertreffen	aufeinander treffen
aufgepaßt!	aufgepasst!
aufgerauht	aufgeraut
Aufguß	Aufguss
Auflösungsprozeß	Auflösungsprozess
aufrauhen	aufrauen
Aufriß	Aufriss
Aufschluß	Aufschluss
aufschlußreich	aufschlussreich
ein aufsehenerregendes Ereignis	ein Aufsehen erregendes Ereignis
aufsein	auf sein
auf seiten	aufseiten *also:* auf Seiten
der aufsichtführende Lehrer	der Aufsicht führende Lehrer
aufwärtsgehen	aufwärts gehen
aufwendig	*also:* aufwändig
auseinanderbiegen	auseinander biegen
auseinanderfallen	auseinander fallen
auseinandergehen	auseinander gehen
auseinanderhalten	auseinander halten
auseinanderleben	auseinander leben
auseinanderreißen	auseinander reißen
auseinandersetzen	auseinander setzen
Ausfluß	Ausfluss
Ausguß	Ausguss
Ausschluß	Ausschluss
Ausschuß	Ausschuss
aussein	aus sein
aufs äußerste gespannt	*also:* aufs Äußerste gespannt
außerstande	*also:* außer Stande

B

OLD	NEW
Bajonettverschluß	Bajonettverschluss
Ballettänzerin	Balletttänzerin
	also: Ballett-Tänzerin
Ballokal	Balllokal
	also: Ball-Lokal
Bänderriß	Bänderriss
jmdm. [angst und] bange machen	jmdm. [Angst und] Bange machen
bankrott gehen	Bankrott gehen
Baroneß	Baroness
baselstädtisch	basel-städtisch
baß erstaunt	bass erstaunt
Baß	Bass
Baßgeige	Bassgeige
Baßsänger	Basssänger
	also: Bass-Sänger
Baukostenzuschuß	Baukostenzuschuss
beeinflußbar	beeinflussbar
Beeinflußbarkeit	Beeinflussbarkeit
beeinflußt	beeinflusst
befaßt	befasst
Begrüßungskuß	Begrüßungskuss
behende	behände
Behendigkeit	Behändigkeit
beieinanderhaben	beieinander haben
beieinandersein	beieinander sein
beieinandersitzen	beieinander sitzen
beieinanderstehen	beieinander stehen
beifallheischend	Beifall heischend
beisammensein	beisammen sein
Beischluß	Beischluss
belemmert	belämmert
jeder beliebige	jeder Beliebige
Beschiß	Beschiss
Beschluß	Beschluss
beschlußfähig	beschlussfähig
Beschlußfassung	Beschlussfassung
Beschuß	Beschuss
ich will im besonderen erwähnen ...	ich will im Besonderen erwähnen ...
bessergehen	besser gehen
es ist das beste, wenn ...	es ist das Beste, wenn ...
aufs beste geregelt sein	*also:* aufs Beste geregelt sein
zum besten geben	zum Besten geben
zum besten haben/halten	zum Besten haben/halten
das erste beste	das erste Beste
bestehenbleiben	bestehen bleiben
Bestelliste	Bestellliste
	also: Bestell-Liste
bestgehaßt	bestgehasst
bestußt	bestusst
Betelnuß	Betelnuss
um ein beträchtliches höher	um ein Beträchtliches höher
in betreff	in Betreff
betreßt	betresst
Bettuch [*zu: Bett*]	Betttuch
	also: Bett-Tuch
bevorschußt	bevorschusst
bewußt	bewusst
bewußtlos	bewusstlos
Bewußtlosigkeit	Bewusstlosigkeit
Bewußtsein	Bewusstsein
in bezug auf	in Bezug auf
bezuschußt	bezuschusst
Bibliographie	*also:* Bibliografie
Bierfaß	Bierfass
die Bismarckschen Sozialgesetze	die bismarckschen Sozialgesetze
	also: die Bismarck'schen Sozialgesetze
Biß	Biss
bißchen	bisschen
du sollst bitte sagen	*also:* du sollst Bitte sagen*
es ist bitter kalt	es ist bitterkalt
Bittag	Bitttag
	also: Bitt-Tag
Blackout	*also:* Black-out*
blankpoliert	blank poliert
blaß	blass
Bläßhuhn/Bleßhuhn	Blässhuhn/Blesshuhn
bläßlich	blässlich
blaßrosa	blassrosa
Blattschuß	Blattschuss
der blaue Planet [*die Erde*]	der Blaue Planet
blaugestreift	blau gestreift
bläulichgrün	bläulich grün
bleibenlassen	bleiben lassen
blendendweiß	blendend weiß
blondgefärbt	blond gefärbt
Bluterguß	Bluterguss
Bonbonniere	*also:* Bonboniere
Börsentip	Börsentipp
im bösen wie im guten	im Bösen wie im Guten
Boß	Boss
Bouclé	*also:* Buklee
braungebrannt	braun gebrannt
bräunlichgelb	bräunlich gelb
des langen und breiten	des Langen und Breiten
breitgefächert	breit gefächert
Brennessel	Brennnessel
	also: Brenn-Nessel
Bruderkuß	Bruderkuss
Brummbaß	Brummbass
brütendheiß	brütend heiß
buntgefiedert	bunt gefiedert
buntschillernd	bunt schillernd
Büroschluß	Büroschluss
Butterfaß	Butterfass

C

OLD	NEW
Cashewnuß	Cashewnuss
Centre Court	Centrecourt
	also: Centre-Court
Chansonnier	*also:* Chansonier
Choreographie	*also:* Choreografie
Cleverneß	Cleverness
Comeback	*also:* Come-back*
Common sense	Commonsense
	also: Common Sense
Corned beef	Cornedbeef
	also: Corned Beef
Corpus delicti	Corpus Delicti
Countdown	*also:* Count-down*

D

OLD	NEW
dabeisein	dabei sein
Dachgeschoß	Dachgeschoss [*in Austria still written with ß*]
dahinterklemmen	dahinter klemmen
dahinterkommen	dahinter kommen
Dampfschiffahrt	Dampfschifffahrt
Danaidenfaß	Danaidenfass
darauffolgend	darauf folgend
Darmverschluß	Darmverschluss
darüberstehen	darüber stehen
dasein	da sein
daß	dass
daß-Satz	dass-Satz
	also: Dasssatz
datenverarbeitend	Daten verarbeitend
Dein [*in letters*]	dein

OLD	NEW
mein und dein verwechseln	Mein und Dein verwechseln
die Deinen	*also:* die deinen
die Deinigen	*also:* die deinigen
Dekolleté	*also:* Dekolletee
Delikateßgurke	Delikatessgurke
Delikateßsenf	Delikatesssenf *also:* Delikatess-Senf
Delphin	*also:* Delfin
Denkprozeß	Denkprozess
wir haben derartiges nicht bemerkt	wir haben Derartiges nicht bemerkt
dessenungeachtet	dessen ungeachtet
des weiteren	des Weiteren
auf deutsch	auf Deutsch
deutschsprechend	Deutsch sprechend
das d'Hondtsche System	das d'hondtsche System *also:* das d'Hondt'sche System
diät leben	Diät leben
Dich [*in letters*]	dich
dichtbehaart	dicht behaart
dichtgedrängt	dicht gedrängt
Differential	*also:* Differenzial*
Diktaphon	*also:* Diktafon
Dir [*in letters*]	dir
Doppelpaß	Doppelpass
dortbleiben	dort bleiben
dortzulande	*also:* dort zu Lande
draufsein	drauf sein
Dreß	Dress
etwas aufs dringendste fordern	*also:* etwas aufs Dringendste fordern
drinsein	drin sein
jeder dritte, der mitwollte	jeder Dritte, der mitwollte
zum dritten	zum Dritten
die dritte Welt	die Dritte Welt
drückendheiß	drückend heiß
Du [*in letters*]	du
auf du und du stehen	auf Du und Du stehen
im dunkeln tappen	im Dunkeln tappen
im dunkeln bleiben	im Dunkeln bleiben
dünnbesiedelt	dünn besiedelt
Dünnschiß	Dünnschiss
durcheinanderbringen	durcheinander bringen
durcheinandergeraten	durcheinander geraten
durcheinanderlaufen	durcheinander laufen
Durchfluß	Durchfluss
Durchlaß	Durchlass
durchnumerieren	durchnummerieren
Durchschuß	Durchschuss
durchsein	durch sein
dußlig	dusslig
Dußligkeit	Dussligkeit
Dutzende Reklamationen	*also:* dutzende Reklamationen
Dutzende von Reklamationen	*also:* dutzende von Reklamationen

E

OLD	NEW
ebensogut	ebenso gut
ebensosehr	ebenso sehr
ebensoviel	ebenso viel
ebensowenig	ebenso wenig
an Eides Statt	an Eides statt
sein eigen nennen	sein Eigen nennen
sich zu eigen machen	sich zu Eigen machen
einbleuen	einbläuen
aufs eindringlichste warnen	*also:* aufs Eindringlichste warnen
das einfachste ist, wenn ...	das Einfachste ist, wenn ...
Einfluß	Einfluss
einflußreich	einflussreich
aufs eingehendste untersuchen	*also:* aufs Eingehendste untersuchen

OLD	NEW
einiggehen	einig gehen
Einlaß	Einlass
einläßlich	einlässlich
Einriß	Einriss
Einschluß	Einschluss
Einschuß	Einschuss
Einschußstelle	Einschussstelle *also:* Einschuss-Stelle
Einsendeschluß	Einsendeschluss
einwärtsgebogen	einwärts gebogen
der/die/das einzelne kann ...	der/die/das Einzelne kann ...
jeder einzelne von uns	jeder Einzelne von uns
bis ins einzelne geregelt	bis ins Einzelne geregelt
ins einzelne gehend	ins Einzelne gehend
einzelnstehend	einzeln stehend
der/die/das einzige wäre ...	der/die/das Einzige wäre ...
kein einziger war gekommen	kein Einziger war gekommen
er als einziger/sie als einzige hatte ...	er als Einziger/sie als Einzige hatte ...
das einzigartige ist, daß ...	das Einzigartige ist, dass ...
Eisenguß	Eisenguss
die eisenverarbeitende Industrie	die Eisen verarbeitende Industrie
eisigkalt	eisig kalt
eislaufen	Eis laufen
Eisschnellauf	Eisschnelllauf
Eisschnelläufer	Eisschnellläufer
energiebewußt	energiebewusst
aufs engste verflochten	*also:* aufs Engste verflochten
engbefreundet	eng befreundet
engbedruckt	eng bedruckt
Engpaß	Engpass
nicht im entferntesten beabsichtigen	*also:* nicht im Entferntesten beabsichtigen
auf das entschiedenste zurückweisen	*also:* auf das Entschiedenste zurückweisen
Entschluß	Entschluss
ein Entweder-Oder gibt es hier nicht	ein Entweder-oder gibt es hier nicht
Entwicklungsprozeß	Entwicklungsprozess
erblaßt	erblasst
Erdgeschoß	Erdgeschoss [*in Austria still written with* ß]
Erdnuß	Erdnuss
die erdölexportierenden Länder	die Erdöl exportierenden Länder
erfaßbar	erfassbar
erfaßt	erfasst
Erguß	Erguss
erholungsuchende Großstädter	Erholung suchende Großstädter
Erlaß	Erlass
ermeßbar	ermessbar
ernstgemeint	ernst gemeint
ernstzunehmend	ernst zu nehmend
erpreßbar	erpressbar
nicht den erstbesten nehmen	nicht den Erstbesten nehmen
der erste, der gekommen ist	der Erste, der gekommen ist
das reicht fürs erste	das reicht fürs Erste
zum ersten, zum zweiten, zum dritten	zum Ersten, zum Zweiten, zum Dritten
die Erste Hilfe	die erste Hilfe
das erstemal	das erste Mal
zum erstenmal	zum ersten Mal
Erstkläßler	Erstklässler
die Erstplazierten	die Erstplatzierten
eßbar	essbar
Eßbesteck	Essbesteck
Eßecke	Essecke
essentiell	*also:* essenziell*
Eßlöffel	Esslöffel
eßlöffelweise	esslöffelweise
Eßtisch	Esstisch
etlichemal	etliche Mal
Euch [*in letters*]	euch

OLD	NEW
Euer [*in letters*]	euer
die Euren	*also:* die euren
die Eurigen	*also:* die eurigen
Existentialismus	*also:* Existenzialismus*
existentialistisch	*also:* existenzialistisch*
existentiell	*also:* existenziell*
Exportüberschuß	Exportüberschuss
Exposé	*also:* Exposee
expreß	express
Expreßreinigung	Expressreinigung
Expreßzug	Expresszug
Exzeß	Exzess

F

OLD	NEW
Fabrikationsprozeß	Fabrikationsprozess
fahrenlassen	fahren lassen
Fairneß	Fairness
Fair play	Fairplay
	also: Fair Play
fallenlassen	fallen lassen
Fallinie	Falllinie
	also: Fall-Linie
Fallout	*also:* Fall-out*
Familienanschluß	Familienanschluss
Fangschuß	Fangschuss
Faß	Fass
faßbar	fassbar
Faßbier	Fassbier
Fäßchen	Fässchen
faßlich	fasslich
du faßt	du fasst
Fast food	Fastfood
	also: Fast Food
Faxanschluß	Faxanschluss
Feedback	*also:* Feed-back*
Fehlpaß	Fehlpass
Fehlschuß	Fehlschuss
jmdm. feind sein	jmdm. Feind sein
feingemahlen	fein gemahlen
fernliegen	fern liegen
fertigbringen	fertig bringen
fertigstellen	fertig stellen
Fertigungsprozeß	Fertigungsprozess
festangestellt	fest angestellt
festumrissen	fest umrissen
festverwurzelt	fest verwurzelt
fettgedruckt	fett gedruckt
feuerspeiende Drachen	Feuer speiende Drachen
die fischverarbeitende Industrie	die Fisch verarbeitende Industrie
Fitneß	Fitness
Flachschuß	Flachschuss
fleischfressende Pflanzen	Fleisch fressende Pflanzen
Flohbiß	Flohbiss
das Bier floß in Strömen	das Bier floss in Strömen
flötengehen	flöten gehen
Fluß	Fluss
flußabwärts	flussabwärts
flußaufwärts	flussaufwärts
Flußbett	Flussbett
Flüßchen	Flüsschen
Flußdiagramm	Flussdiagramm
flüssigmachen	flüssig machen
Flußsand	Flusssand
	also: Fluss-Sand
Flußschiffahrt	Flussschifffahrt
	also: Fluss-Schifffahrt
Flußspat	Flussspat
	also: Fluss-Spat
die Haare fönen	die Haare föhnen
folgendes ist zu beachten	Folgendes ist zu beachten
wie im folgenden erläutert	wie im Folgenden erläutert
Fraktionsausschuß	Fraktionsausschuss

OLD	NEW
Fraktionsbeschluß	Fraktionsbeschluss
Free climbing	Freeclimbing
	also: Free Climbing
Free Jazz	*also:* Freejazz
Freßgier	Fressgier
Freßpaket	Fresspaket
Freßsack	Fresssack
	also: Fress-Sack
Friedensschluß	Friedensschluss
frischgebacken	frisch gebacken
fritieren	frittieren
frohgelaunt	froh gelaunt
frühverstorben	früh verstorben
Full-time-Job	Fulltimejob
	also: Full-Time-Job
Fünfpaß	Fünfpass
funkensprühend	Funken sprühend
Funkmeßtechnik	Funkmesstechnik
fürbaß	fürbass
fürliebnehmen	fürlieb nehmen
Fußballänderspiel	Fußballländerspiel
	also: Fußball-Länderspiel

G

OLD	NEW
Gangsterboß	Gangsterboss
im ganzen gesehen	im Ganzen gesehen
im großen und ganzen	im Großen und Ganzen
Gärungsprozeß	Gärungsprozess
Gäßchen	Gässchen
gefangenhalten	gefangen halten
gefangennehmen	gefangen nehmen
gefaßt	gefasst
gefirnißt	gefirnisst
es ist das gegebene, schnell zu handeln	es ist das Gegebene, schnell zu handeln
gegeneinanderprallen	gegeneinander prallen
gegeneinanderstoßen	gegeneinander stoßen
von allen gehaßt	von allen gehasst
geheimhalten	geheim halten
gehenlassen	gehen lassen
Gelaß	Gelass
gutgelaunt	gut gelaunt
gelblichgrün	gelblich grün
Gemse	Gämse
wir haben gemußt	wir haben gemusst
die Wunde hat genäßt	die Wunde hat genässt
aufs genaueste festgelegt	*also:* aufs Genaueste festgelegt
genaugenommen	genau genommen
genausogut	genauso gut
genausowenig	genauso wenig
Generalbaß	Generalbass
sie genoß den Sonnenschein	sie genoss den Sonnenschein
Genuß	Genuss
genüßlich	genüsslich
Genußmittel	Genussmittel
genußsüchtig	genusssüchtig
Geographie	*also:* Geografie
es hat gut gepaßt	es hat gut gepasst
wir haben gepraßt	wir haben geprasst
frisch gepreßter Saft	frisch gepresster Saft
geradehalten	gerade halten
geradesitzen	gerade sitzen
geradestellen	gerade stellen
Gerichtsbeschluß	Gerichtsbeschluss
um ein geringes weniger	um ein Geringes weniger
es geht ihn nicht das geringste an	es geht ihn nicht das Geringste an
nicht im geringsten stören	nicht im Geringsten stören
geringachten	gering achten
geringschätzen	gering schätzen
Geruchsverschluß	Geruchsverschluss
Geschäftsschluß	Geschäftsschluss
er wurde geschaßt	er wurde geschasst

OLD	NEW
Geschichtsbewußtsein	Geschichtsbewusstsein
Geschirreiniger	Geschirrreiniger *also:* Geschirr-Reiniger
Geschoß	Geschoss [*in Austria still written with* ß]
gestern abend/morgen/nacht	gestern Abend/Morgen/Nacht
alle waren gestreßt	alle waren gestresst
getrenntlebend	getrennt lebend
Gewinnummer	Gewinnnummer *also:* Gewinn-Nummer
gewiß	gewiss
Gewissensbiß	Gewissensbiss
Gewißheit	Gewissheit
gewißlich	gewisslich
ich habe es gewußt	ich habe es gewusst
Ginkgo	*also:* Ginko
Glacéhandschuh	*also:* Glaceehandschuh
glänzendschwarz	glänzend schwarz
glattgehen	glatt gehen
glatthobeln	glatt hobeln
glattschleifen	glatt schleifen
glattstreichen	glatt streichen
das gleiche tun	das Gleiche tun
aufs gleiche hinauskommen	aufs Gleiche hinauskommen
gleich und gleich gesellt sich gern	Geich und Gleich gesellt sich gern
gleichlautend	gleich lautend
Gleisanschluß	Gleisanschluss
Glimmstengel	Glimmstängel
glühendheiß	glühend heiß
Gnadenerlaß	Gnadenerlass
die Goetheschen Dramen	die goetheschen Dramen *also:* die Goethe'schen Dramen
Graphit	*also:* Grafit
Graphologie	*also:* Grafologie
gräßlich	grässlich
graugestreift	grau gestreift
grellbeleuchtet	grell beleuchtet
Grenzfluß	Grenzfluss
Greuel	Gräuel
greulich	gräulich
griffest	grifffest
jmdn. aufs gröbste beleidigen	*also:* jmdn. aufs Gröbste beleidigen
grobgemahlen	grob gemahlen
ein Programm für groß und klein	ein Programm für Groß und Klein
im großen und ganzen	im Großen und Ganzen
das größte wäre, wenn ...	das Größte wäre, wenn ...
Großschiffahrtsweg	Großschifffahrtsweg
groß schreiben [*mit großem Anfangsbuchstaben*]	großschreiben
grünlichgelb	grünlich gelb
Guß	Guss
Gußeisen	Gusseisen
gußeisern	gusseisern
guten Tag sagen	*also:* Guten Tag sagen*
es im guten versuchen	es im Guten versuchen
gutaussehend	gut aussehend
gutbezahlt	gut bezahlt
gutgehen	gut gehen
gutgehend	gut gehend
gutgelaunt	gut gelaunt
gutgemeint	gut gemeint
guttun	gut tun
gutunterrichtet	gut unterrichtet
H	
haftenbleiben	haften bleiben
haltmachen	Halt machen
Hämorrhoide	*also:* Hämorride
händchenhaltend	Händchen haltend
handeltreibend	Handel treibend

OLD	NEW
Handkuß	Handkuss
Handout	*also:* Hand-out*
hängenbleiben	hängen bleiben
hängenlassen	hängen lassen
Happy-End	Happyend *also:* Happy End
Haraß	Harass
Hard cover	Hardcover *also:* Hard Cover
Hard-cover-Einband	Hardcovereinband *also:*Hard-Cover-Einband
hartgekocht	hart gekocht
Haselnuß	Haselnuss
Haselnußstrauch	Haselnussstrauch *also:* Haselnuss-Strauch
Haß	Hass
haßerfüllt	hasserfüllt
häßlich	hässlich
Häßlichkeit	Hässlichkeit
Haßliebe	Hassliebe
du haßt	du hasst
Hauptschulabschluß	Hauptschulabschluss
nach Hause	*in Austria and Switzerland also:* nachhause
zu Hause	*in Austria and Switzerland also:* zuhause
haushalten	*also:* Haus halten
Haushaltsausschuß	Haushaltsausschuss
Hawaii-Insel	*also:* Hawaiiinsel
heiligsprechen	heilig sprechen
Heilungsprozeß	Heilungsprozess
heimlichtun	heimlich tun
heißbegehrt	heiß begehrt
heißgeliebt	heiß geliebt
heißumkämpft	heiß umkämpft
helleuchtend	hell leuchtend
hellicht	helllicht
hellila	helllila
hellodernd	hell lodernd
heransein	heran sein
heraussein	heraus sein
herbstlichgelb	herbstlich gelb
Heringsfaß	Heringsfass
hersein	her sein
herumsein	herum sein
heruntersein	herunter sein
Herzas	Herzass
jmdn. auf das herzlichste begrüßen	*also:* jmdn. auf das Herzlichste begrüßen
heute abend/mittag/nacht	heute Abend/Mittag/Nacht
Hexenschuß	Hexenschuss
hierbleiben	hier bleiben
hierlassen	hierlassen
hiersein	hier sein
hierzulande	*also:* hier zu Lande
High-Fidelity	Highfidelity *also:* High Fidelity
High-Society	Highsociety *also:* High Society
hilfesuchend	Hilfe suchend
hinaussein	hinaus sein
es wurde etwas hineingeheimnißt	es wurde etwas hineingeheimnisst
hinsein	hin sein
hintereinanderfahren	hintereinander fahren
hintereinandergehen	hintereinander gehen
hintereinanderschalten	hintereinander schalten
hinterhersein	hinterher sein
hinübersein	hinüber sein
er hißt die Flagge	er hisst die Flagge
Hochgenuß	Hochgenuss
Hochschulabschluß	Hochschulabschluss
aufs höchste erfreut sein	*also:* aufs Höchste erfreut sein
hofhalten	Hof halten

OLD	NEW
die Hohe Schule	die hohe Schule
hohnlachen	*also:* Hohn lachen
das holzverarbeitende Gewerbe	das Holz verarbeitende Gewerbe
Hosteß	Hostess
Hot dog	Hotdog
	also: Hot Dog
ein paar hundert	*also:* ein paar Hundert
viele Hunderte	*also:* viele hunderte
Hunderte von Zuschauern	*also:* hunderte von Zuschauern
Hungers sterben	hungers sterben
hurra schreien	*also:* Hurra schreien*

I

OLD	NEW
auch Ihr seid herzlich eingeladen [*in letters*]	auch ihr seid herzlich eingeladen
im allgemeinen	im Allgemeinen
im besonderen	im Besonderen
Imbiß	Imbiss
Imbißstand	Imbissstand
	also: Imbiss-Stand
im einzelnen	im Einzelnen
im nachhinein	im Nachhinein
Impfpaß	Impfpass
imstande	*also:* im Stande
im übrigen	im Übrigen
im voraus	im Voraus
im vorhinein	im Vorhinein
in betreff	in Betreff
in bezug auf	in Bezug auf
Indizes	*also:* Indices
Indizienprozeß	Indizienprozess
ineinanderfließen	ineinander fließen
ineinandergreifen	ineinander greifen
inessentiell	*also:* inessenziell*
Informationsfluß	Informationsfluss
in Frage stellen	*also:* infrage stellen
in Frage kommen	*also:* infrage kommen
innesein	inne sein
insektenfressende Pflanzen	Insekten fressende Pflanzen
instand halten	*also:* in Stand halten
instand setzen	*also:* in Stand setzen
I-Punkt	i-Punkt
irgend etwas	irgendetwas
irgend jemand	irgendjemand
I-Tüpfelchen	i-Tüpfelchen

J

OLD	NEW
ja sagen	*also:* Ja sagen*
Jagdschloß	Jagdschloss
Jäheit	Jähheit
Jahresabschluß	Jahresabschluss
2jährig, 3jährig, 4jährig ...	2-jährig, 3-jährig, 4-jährig ...
ein 2jähriger, 3jähriger, 4jähriger kann das noch nicht verstehen	ein 2-Jähriger, 3-Jähriger, 4-Jähriger kann das noch nicht verstehen
Jaß	Jass
du jaßt	du jasst
Jauchefaß	Jauchefass
jedesmal	jedes Mal
Job-sharing	Jobsharing
Joghurt	*also:* Jogurt
Joint-venture	Jointventure
	also: Joint Venture
Judaskuß	Judaskuss
Julierpaß	Julierpass
Jumbo-Jet	Jumbojet
für jung und alt	für Jung und Alt

K

OLD	NEW
Kabelanschluß	Kabelanschluss
Kabinettsbeschluß	Kabinettsbeschluss
Kaffee-Ernte	*also:* Kaffeeernte
Kaffee-Ersatz	*also:* Kaffeeersatz
Kalligraphie	*also:* Kalligrafie
kalorienbewußt	kalorienbewusst
kaltlächelnd	kalt lächelnd
Kameraverschluß	Kameraverschluss
Kammacher	Kammmacher
	also: Kamm-Macher
Kämmaschine	Kämmmaschine
	also: Kämm-Maschine
Kammuschel	Kammmuschel
	also: Kamm-Muschel
Känguruh	Känguru
Kanonenschuß	Kanonenschuss
Kapselriß	Kapselriss
Karamel	Karamell
karamelisieren	karamellisieren
2karäter, 3karäter, 4karäter ...	2-Karäter, 3-Karäter, 4-Karäter...
2karätig, 3karätig, 4karätig ...	2-karätig, 3-karätig, 4-karätig ...
Karoas	Karoass
Kartographie	*also:* Kartografie
Kaßler	Kassler
Katarrh	*also:* Katarr
kegelschieben	Kegel schieben
Kellergeschoß	Kellergeschoss [*in Austria still written with* ß]
kennenlernen	kennen lernen
Kennummer	Kennnummer
	also: Kenn-Nummer
die Keplerschen Gesetze	die keplerschen Gesetze
	also: die Kepler'schen Gesetze
keß	kess
Keßheit	Kessheit
Ketchup	*also:* Ketschup*
Kickdown	*also:* Kick-down*
Kick-off	*also:* Kickoff
an Kindes Statt	an Kindes statt
Kindesmißhandlung	Kindesmisshandlung
Kißchen	Kisschen
sich über etwas im klaren sein	sich über etwas im Klaren sein
klardenkend	klar denkend
klarsehen	klar sehen
klarwerden	klar werden
Klassenbewußtsein	Klassenbewusstsein
Klassenhaß	Klassenhass
klatschnaß	klatschnass
Klausenpaß	Klausenpass
klebenbleiben	kleben bleiben
Klee-Einsaat	*also:* Kleeeinsaat
Klee-Ernte	*also:* Kleeernte
bis ins kleinste geregelt	bis ins Kleinste geregelt
ein Staat im kleinen	ein Staat im Kleinen
ein Programm für groß und klein	ein Programm für Groß und Klein
kleingedruckt	klein gedruckt
kleinschneiden	klein schneiden
klein schreiben [*mit kleinem Anfangsbuchstaben*]	kleinschreiben
Klemmappe	Klemmmappe
	also: Klemm-Mappe
Klettverschluß	Klettverschluss
klitschnaß	klitschnass
es wäre das klügste, wenn ...	es wäre das Klügste, wenn ...
knapphalten	knapp halten
Knockout	*also:* Knock-out*
kochendheiß	kochend heiß
kohleführende Flöze	Kohle führende Flöze
Kolanuß	Kolanuss
Kollektivbewußtsein	Kollektivbewusstsein
Kolophonium	*also:* Kolofonium
Koloß	Koloss

OLD	NEW
Kombinationsschloß	Kombinationsschloss
Kommiß	Kommiss
Kommißbrot	Kommissbrot
Kommißstiefel	Kommissstiefel *also:* Kommiss-Stiefel
Kommuniqué	*also:* Kommunikee
Kompaß	Kompass
kompreß	kompress
Kompromiß	Kompromiss
kompromißbereit	kompromissbereit
kompromißlos	kompromisslos
Kompromißlösung	Kompromisslösung
Komteß	Komtess
Konferenzbeschluß	Konferenzbeschluss
Kongreß	Kongress
Kongreßhalle	Kongresshalle
Kongreßsaal	Kongresssaal *also:* Kongress-Saal
Kongreßstadt	Kongressstadt *also:* Kongress-Stadt
Königsschloß	Königsschloss
Kontrabaß	Kontrabass
Kontrollampe	Kontrolllampe *also:* Kontroll-Lampe
Kontrolliste	Kontrollliste *also:* Kontroll-Liste
Kopfnuß	Kopfnuss
Kopfschuß	Kopfschuss
kopfstehen	Kopf stehen
Koppelschloß	Koppelschloss
krank schreiben	krankschreiben
kraß	krass
Kraßheit	Krassheit
krebserregende Substanzen	Krebs erregende Substanzen
Kreiselkompaß	Kreiselkompass
Kreppapier	Krepppapier *also:* Krepp-Papier
Kreuzas	Kreuzass
die kriegführenden Parteien	die Krieg führenden Parteien
Kriminalprozeß	Kriminalprozess
Kristallüster	Kristalllüster *also:* Kristall-Lüster
kroß	kross
krummnehmen	krumm nehmen
KSZE-Schlußakte	KSZE-Schlussakte
Kunststoffolie	Kunststofffolie *also:* Kunststoff-Folie
Küraß	Kürass
den kürzeren ziehen	den Kürzeren ziehen
kürzertreten	kürzer treten
kurzgebraten	kurz gebraten
kurzhalten	kurz halten
Kurzpaß	Kurzpass
Kurzschluß	Kurzschluss
kurztreten	kurz treten
Kuß	Kuss
Küßchen	Küsschen
kußecht	kussecht
Kußhand	Kusshand
du/er/sie küßt	du/er/sie/küsst
Küstenschiffahrt	Küstenschifffahrt
Kwaß	Kwass

L

OLD	NEW
Ladenschluß	Ladenschluss
die La-Fontaineschen Fabeln	die la-fontaineschen Fabeln *also:* die la-Fontaine'schen Fabeln
Lamé	*also:* Lamee
Lamellenverschluß	Lamellenverschluss
etwas des langen und breiten erklären	etwas des Langen und Breiten erklären
langgestreckt	lang gestreckt
länglichrund	länglich rund
langstengelig	langstängelig
langziehen	lang ziehen
Lapsus linguae	Lapsus Linguae
läßlich	lässlich
du läßt	du lässt
zu Lasten	*also:* zulasten
Lattenschuß	Lattenschuss
laubtragende Bäume	Laub tragende Bäume
auf dem laufenden sein	auf dem Laufenden sein
laufenlassen	laufen lassen
Laufpaß	Laufpass
Layout	*also:* Lay-out*
Lebensgenuß	Lebensgenuss
Leberabszeß	Leberabszess
die lederverarbeitende Industrie	die Leder verarbeitende Industrie
leerstehend	leer stehend
leichenblaß	leichenblass
es ist mir ein leichtes, das zu tun	es ist mir ein Leichtes, das zu tun
leichtentzündlich	leicht entzündlich
leichtfallen	leicht fallen
leichtmachen	leicht machen
leichtnehmen	leicht nehmen
leichtverderblich	leicht verderblich
leichtverständlich	leicht verständlich
jmdm. leid tun	jmdm. Leid tun
Lenkradschloß	Lenkradschloss
Lernprozeß	Lernprozess
der letzte, der gekommen ist	der Letzte, der gekommen ist
als letzter fertig sein	als Letzter fertig sein
das letzte, was sie tun würde	das Letzte, was sie tun würde
bis ins letzte geklärt	bis ins Letzte geklärt
letzteres trifft zu	Letzteres trifft zu
zum letztenmal	zum letzten Mal
leuchtendblau	leuchtend blau
Lichtmeß	Lichtmess
es wäre uns das liebste, wenn ...	es wäre uns das Liebste, wenn ...
liebenlernen	lieben lernen
liebgewinnen	lieb gewinnen
liebhaben	lieb haben
liegenbleiben	liegen bleiben
liegenlassen	liegen lassen
Live-Mitschnitt	*also:* Livemitschnitt
Lizentiat	*also:* Lizenziat*
Lorbaß	Lorbass
Löß	*also:* Löss [*when pronounced with a short* ö]
Lößboden	*also:* Lössboden [*when pronounced with a short* ö]
Lößschicht	*also:* Lössschicht oder Löss-Schicht [*when pronounced with a short* ö]
Lötschenpaß	Lötschenpass
Love-Story	*also:* Lovestory
Luftschiffahrt	Luftschifffahrt
Luftschloß	Luftschloss

M

OLD	NEW
Magistratsbeschluß	Magistratsbeschluss
2mal, 3mal, 4mal ...	2-mal, 3-mal, 4-mal ...
Malaise	*also:* Maläse
Marschkompaß	Marschkompass
maschineschreiben	Maschine schreiben
maßhalten	Maß halten
Matrizes	*also:* Matrices
Maulkorberlaß	Maulkorberlass
Megaphon	*also:* Megafon
Mehrheitsbeschluß	Mehrheitsbeschluss
Meldeschluß	Meldeschluss
Meniskusriß	Meniskusriss

OLD	NEW
wir haben das menschenmögliche getan	wir haben das Menschenmögliche getan
Mesner	*also:* Messner
Meßband	Messband
meßbar	messbar
Meßbecher	Messbecher
Meßbuch	Messbuch
Meßdaten	Messdaten
Meßdiener	Messdiener
Meßfühler	Messfühler
Meßgewand	Messgewand
Meßinstrument	Messinstrument
Meßopfer	Messopfer
Meßstab	Messstab *also:* Mess-Stab
Meßtischblatt	Messtischblatt
Metallguß	Metallguss
Metallegierung	Metalllegierung *also:* Metall-Legierung
die metallverarbeitende Industrie	die Metall verarbeitende Industrie
Midlife-crisis	Midlifecrisis *also:* Midlife-Crisis
Milchgebiß	Milchgebiss
millionenmal	Millionen Mal
Milzriß	Milzriss
nicht im mindesten	*also:* nicht im Mindesten
mißachten	missachten
Mißbildung	Missbildung
mißbilligen	missbilligen
Mißbrauch	Missbrauch
Mißerfolg	Misserfolg
Mißernte	Missernte
mißfallen	missfallen
Mißfallenskundgebung	Missfallenskundgebung
Mißgeburt	Missgeburt
Mißgeschick	Missgeschick
mißglücken	missglücken
Mißgunst	Missgunst
mißgünstig	missgünstig
Mißklang	Missklang
Mißkredit	Misskredit
mißlich	misslich
mißlingen	misslingen
mißmutig	missmutig
mißraten	missraten
Mißstand	Missstand
Mißtrauen	Misstrauen
mißtrauisch	misstrauisch
Mißverständnis	Missverständnis
Mißwirtschaft	Misswirtschaft
mit Hilfe	*also:* mithilfe
[gestern, heute, morgen] mittag	[gestern, heute, morgen] Mittag
Mixed Pickles	*also:* Mixedpickles*
modebewußt	modebewusst
wir sprachen über alles mögliche	wir sprachen über alles Mögliche
sein möglichstes tun	sein Möglichstes tun
3monatig, 4monatig, 5monatig ...	3-monatig, 4-monatig, 5-monatig ...
3monatlich, 4monatlich, 5monatlich ...	3-monatlich, 4-monatlich, 5-monatlich ...
Monographie	*also:* Monografie
Mop	Mopp
Mordprozeß	Mordprozess
morgen abend, mittag, nacht	morgen Abend, Mittag, Nacht
[gestern, heute] morgen	[gestern, heute] Morgen
Moto-Cross	*also:* Motocross
Mückenschiß	Mückenschiss
Mulläppchen	Mullläppchen *also:* Mull-Läppchen
Multiple-choice-Verfahren	Multiplechoiceverfahren *also:* Multiple-Choice-Verfahren

OLD	NEW
Muskatnuß	Muskatnuss
Muskelriß	Muskelriss
ich muß	ich muss
du mußt	du musst
ich müßte	ich müsste
du müßtest	du müsstest
Mußheirat	Mussheirat
müßiggehen	müßig gehen
Musterprozeß	Musterprozess
Myrrhe	*also:* Myrre

N

OLD	NEW
nachfolgendes gilt auch ...	Nachfolgendes gilt auch ...
nach Hause	*in Austria and Switzerland* *also:* nachhause
im nachhinein	im Nachhinein
Nachlaß	Nachlass
Nachlaßverwalter	Nachlassverwalter
[gestern, heute, morgen] nachmittag	[gestern, heute, morgen] Nachmittag
Nachschuß	Nachschuss
der nächste, bitte!	der Nächste, bitte!
als nächstes wollen wir ...	als Nächstes wollen wir ...
im nachstehenden heißt es ...	im Nachstehenden heißt es ...
[gestern, heute, morgen] nacht	[gestern, heute, morgen] Nacht
nahebringen	nahe bringen
nahelegen	nahe legen
naheliegen	nahe liegen
naheliegend	nahe liegend
etwas des näheren erläutern	etwas des Näheren erläutern
näherliegen	näher liegen
nahestehen	nahe stehen
nahestehend	nahe stehend
Narziß	Narziss
Narzißmus	Narzissmus
narzißtisch	narzisstisch
naß	nass
naßforsch	nassforsch
naßgeschwitzt	nass geschwitzt
naßkalt	nasskalt
Naßrasur	Nassrasur
Naßschnee	Nassschnee *also:* Nass-Schnee
nationalbewußt	nationalbewusst
Nationaldreß	Nationaldress
Nebelschlußleuchte	Nebelschlussleuchte
Nebenanschluß	Nebenanschluss
nebeneinandersitzen	nebeneinander sitzen
nebeneinanderstehen	nebeneinander stehen
nebeneinanderstellen	nebeneinander stellen
Nebenfluß	Nebenfluss
im nebenstehenden wird gezeigt ...	im Nebenstehenden wird gezeigt ...
Necessaire	*also:* Nessessär
Negligé	*also:* Negligee
nein sagen	*also:* Nein sagen*
Netzanschluß	Netzanschluss
es aufs neue versuchen	es aufs Neue versuchen
auf ein neues!	auf ein Neues!
neueröffnet	neu eröffnet
New Yorker	*also:* New-Yorker
nichtrostend	*also:* nicht rostend
Nichtseßhafte	Nichtsesshafte
nichtssagend	nichts sagend
No-future-Generation	No-Future-Generation
die notleidende Bevölkerung	die Not leidende Bevölkerung
in Null Komma nichts	in null Komma nichts
das Thermometer steht auf Null	das Thermometer steht auf null
Nullage	Nulllage *also:* Null-Lage
Nulleiter	Nullleiter *also:* Null-Leiter

OLD	NEW
Nullösung	Nulllösung
	also: Null-Lösung
numerieren	nummerieren
Numerierung	Nummerierung
Nuß	Nuss
Nüßchen	Nüsschen
Nußknacker	Nussknacker
Nußschale	Nussschale
	also: Nuss-Schale
Nußschinken	Nussschinken
	also: Nuss-Schinken
Nußschokolade	Nussschokolade
	also: Nuss-Schokolade
Nußstrudel	Nussstrudel
	also: Nuss-Strudel
Nußtorte	Nusstorte

O

OLD	NEW
O-beinig	*also:* o-beinig
obenerwähnt	oben erwähnt
obenstehend	oben stehend
Obergeschoß	Obergeschoss [*in Austria still written with* ß]
offenbleiben	offen bleiben
offenlassen	offen lassen
offenstehen	offen stehen
O-förmig	*also:* o-förmig
des öfteren	des öfteren
ölmeßstab	ölmessstab
Ordonnanz	*also:* Ordonanz
Orthographie	*also:* Orthografie

P

OLD	NEW
Panther	*also:* Panter
die papierverarbeitende Industrie	die Papier verarbeitende Industrie
Pappmaché	*also:* Pappmaschee
parallellaufend	parallel laufend
parallelschalten	parallel schalten
Paranuß	Paranuss
Parlamentsbeschluß	Parlamentsbeschluss
Parnaß	Parnass
Parteikongreß	Parteikongress
Parteitagsbeschluß	Parteitagsbeschluss
Paß	Pass
Paßbild	Passbild
passé	*also:* passee
Paßform	Passform
Paßgang	Passgang
paßgerecht	passgerecht
Paßkontrolle	Passkontrolle
Paßstelle	Passstelle
	also: Pass-Stelle
Paßstraße	Passstraße
	also: Pass-Straße
Paßwort	Passwort
Patentverschluß	Patentverschluss
patschnaß	patschnass
Perkussionsschloß	Perkussionsschloss
Personenschiffahrt	Personenschifffahrt
Petitionsausschuß	Petitionsausschuss
Pfeffernuß	Pfeffernuss
Pferdegebiß	Pferdegebiss
pflichtbewußt	pflichtbewusst
Pflichtbewußtsein	Pflichtbewusstsein
Pfostenschuß	Pfostenschuss
Pikas	Pikass
Pimpernuß	Pimpernuss
er pißt	er pisst
Pistolenschuß	Pistolenschuss
pitschnaß	pitschnass
Platitüde	Plattitüde
	also: Platitude
Playback	*also:* Play-back*
plazieren	platzieren
pleite gehen	Pleite gehen
polyphon	*also:* polyfon
Pornographie	*also:* Pornografie
Portemonnaie	*also:* Portmonee
Potemkinsche Dörfer	potemkinsche Dörfer
	also: Potemkin'sche Dörfer
potentiell	*also:* potenziell*
potthäßlich	potthässlich
Poussierstengel	Poussierstängel
präferentiell	*also:* präferenziell*
er praßt	er prasst
preisbewußt	preisbewusst
Preisnachlaß	Preisnachlass
Preßform	Pressform
Preßluftbohrer	Pressluftbohrer
Preßsack	Presssack
	also: Press-Sack
Preßschlag	Pressschlag
	also: Press-Schlag
Preßspan	Pressspan
	also: Press-Span
du preßt	du presst
Preßwehe	Presswehe
Prinzeßbohne	Prinzessbohne
privatversichert	privat versichert
probefahren	Probe fahren
Problembewußtsein	Problembewusstsein
Produktionsprozeß	Produktionsprozess
Profeß	Profess
Programmusik	Programmmusik
	also: Programm-Musik
Progreß	Progress
Prozeß	Prozess
Prozeßkosten	Prozesskosten
Prozeßbevollmächtigte	Prozessbevollmächtigte
prozeßführend	prozessführend
Prozeßkosten	Prozesskosten
Prozeßrechner	Prozessrechner
pudelnaß	pudelnass
Pulverfaß	Pulverfass
pußlig	pusslig

Q

OLD	NEW
Quadrophonie	*also:* Quadrofonie
qualitätsbewußt	qualitätsbewusst
Quartalsabschluß	Quartalsabschluss
Quellfluß	Quellfluss
Quentchen	Quäntchen
Querpaß	Querpass
Quickstep	Quickstepp

R

OLD	NEW
radfahren	Rad fahren
Radikalenerlaß	Radikalenerlass
radschlagen	Rad schlagen
Rammaschine	Rammmaschine
	also: Ramm-Maschine
zu Rande kommen	*also:* zurande kommen
Rassenhaß	Rassenhass
ich raßle mit den Ketten	ich rassle mit den Ketten
zu Rate ziehen	*also:* zurate ziehen
Räterußland	Räterussland
Ratsbeschluß	Ratsbeschluss
Ratschluß	Ratschluss
Rauchfaß	Rauchfass
rauh	rau
rauhbeinig	raubeinig
Rauhfasertapete	Raufasertapete

OLD	NEW
Rauhfrost	Raufrost
Rauhhaardackel	Rauhaardackel
Rauhnächte	Raunächte
Rauhputz	Rauputz
Rauhreif	Raureif
Rausschmiß	Rausschmiss
recht haben	Recht haben
recht behalten	Recht behalten
recht bekommen	Recht bekommen
jmdm. recht geben	jmdm. Recht geben
Rechtens sein	rechtens sein
Rechtsbewußtsein	Rechtsbewusstsein
Redaktionsschluß	Redaktionsschluss
Regenguß	Regenguss
regennaß	regennass
Regreß	Regress
Regreßanspruch	Regressanspruch
Regreßpflicht	Regresspflicht
regreßpflichtig	regresspflichtig
reichgeschmückt	reich geschmückt
reichverziert	reich verziert
Reifungsprozeß	Reifungsprozess
Reisepaß	Reisepass
Reißverschluß	Reißverschluss
Reißverschlußsystem	Reißverschlusssystem *also:* Reißverschluss-System
Reschenpaß	Reschenpass
Rettungsschuß	Rettungsschuss
Rezeß	Rezess
Rhein-Main-Donau-Großschiffahrtsweg	Rhein-Main-Donau-Großschifffahrtsweg
das ist genau das richtige für mich	das ist genau das Richtige für mich
mit etwas richtigliegen	mit etwas richtig liegen
richtigstellen	richtig stellen
Riß	Riss
rißfest	rissfest
Roheit	Rohheit
Rolladen	Rollladen *also:* Roll-Laden
Rommé	*also:* Rommee
rosigweiß	rosig weiß
Roß	Ross
Roßbreiten	Rossbreiten
Roßhaarmatratze	Rosshaarmatratze
Roßkastanie	Rosskastanie
Roßkur	Rosskur
Rößl	Rössl
Roßtäuscherei	Rosstäuscherei
de rote Planet [*Mars*]	der Rote Planet
rotgestreift	rot gestreift
rotglühend	rot glühend
rötlichbraun	rötlich braun
die Rubensschen Gemälde	die rubensschen Gemälde *also:* die Rubens'schen Gemälde
Rückfluß	Rückfluss
Rückpaß	Rückpass
Rückschluß	Rückschluss
rückwärtsgewandt	rückwärts gewandt
Ruhegenuß	Ruhegenuss
ruhenlassen	ruhen lassen
ruhigstellen	ruhig stellen
Runderlaß	Runderlass
Rußland	Russland

S

OLD	NEW
Säbelraßler	Säbelrassler
Saisonnier	*also:* Saisonier
Saisonschluß	Saisonschluss
Salutschuß	Salutschuss
Salzfaß	Salzfass
Samenerguß	Samenerguss
Sammelanschluß	Sammelanschluss
Sankt Gallener	*also:* Sankt-Gallener
sanktgallisch	sankt-gallisch
Sanmarinese	San-Marinese
sanmarinesisch	san-marinesisch
sauberhalten	sauber halten
saubermachen	sauber machen
sausenlassen	sausen lassen
Saxophon	*also:* Saxofon
sein Schäfchen ins trockene bringen	sein Schäfchen ins Trockene bringen
Schalenguß	Schalenguss
Schallehre	Schalllehre *also:* Schall-Lehre
Schalloch	Schallloch *also:* Schall-Loch
Schalterschluß	Schalterschluss
etwas auf das schärfste verurteilen	*also:* etwas auf das Schärfste verurteilen
er schaßte ihn	er schasste ihn
ein schattenspendender Baum	ein Schatten spendender Baum
schätzenlernen	schätzen lernen
Schauprozeß	Schauprozess
Scheidungsprozeß	Scheidungsprozess
schießenlassen	schießen lassen
Schiffahrt	Schifffahrt *also:* Schiff-Fahrt
Schippenas	Schippenass
Schiß	Schiss
Schlachtroß	Schlachtross
Schlagfluß	Schlagfluss
Schlammasse	Schlammmasse *also:* Schlamm-Masse
schlechtgehen	schlecht gehen
schlechtgelaunt	schlecht gelaunt
das schlimmste ist, daß ...	das Schlimmste ist, dass ...
sie haben ihn auf das schlimmste getäuscht	*also:* sie haben ihn auf das Schlimmste getäuscht
er schliß Federn	er schliss Federn
Schlitzverschluß	Schlitzverschluss
Schloß	Schloss
Schlößchen	Schlösschen
Schloßherr	Schlossherr
Schloßpark	Schlosspark
Schluß	Schluss
Schlußbemerkung	Schlussbemerkung
schlußendlich	schlussendlich
schlußfolgern	schlussfolgern
Schlußfolgerung	Schlussfolgerung
Schlußlicht	Schlusslicht
Schlußpfiff	Schlusspfiff
Schlußpunkt	Schlusspunkt
Schlußsatz	Schlusssatz *also:* Schluss-Satz
Schlußspurt	Schlussspurt *also:* Schluss-Spurt
Schlußstrich	Schlussstrich *also:* Schluss-Strich
Schlußverkauf	Schlussverkauf
Schlußwort	Schlusswort
Schmerfluß	Schmerfluss
sie schmiß mit Steinen	sie schmiss mit Steinen
Schmiß	Schmiss
Schmuckblattelegramm	Schmuckblatttelegramm *also:* Schmuckblatt-Telegramm
schmutziggrau	schmutzig grau
Schnappschloß	Schnappschloss
Schnappschuß	Schnappschuss
Schnee-Eifel	*also:* Schneeeifel
Schnee-Eule	*also:* Schneeeule
Schneewächte	Schneewechte
Schnellimbiß	Schnellimbiss
Schnelläufer	Schnellläufer *also:* Schnell-Läufer

OLD	NEW
schnellebig	schnelllebig
Schnellebigkeit	Schnelllebigkeit
Schnellschuß	Schnellschuss
Schnepper	*also:* Schnäpper
schneppern	*also:* schnäppern
schneuzen	schnäuzen
Schokoladenguß	Schokoladenguss
aufs schönste übereinstimmen	*also:* aufs Schönste übereinstimmen
er schoß	er schoss
Schoß [einer Pflanze]	Schoss
Schräglaufend	schräg laufend
Schraubverschluß	Schraubverschluss
schreckensblaß	schreckensblass
Schreckschußpistole	Schreckschusspistole
Schrittempo	Schritttempo
	also: Schritt-Tempo
Schrotschuß	Schrotschuss
Schulabschluß	Schulabschluss
an etwas schuld haben	an etwas Schuld haben
sich etwas zuschulden kommen lassen	*also:* sich etwas zu Schulden kommen lassen
schuldbewußt	schuldbewusst
Schuldenerlaß	Schuldenerlass
Schulschluß	Schulschluss
Schulstreß	Schulstress
Schulterschluß	Schulterschluss
Schuß	Schuss
schußbereit	schussbereit
schußfest	schussfest
schußlig	schusslig
Schußlinie	Schusslinie
Schußschwäche	Schussschwäche
	also: Schuss-Schwäche
Schußwaffe	Schusswaffe
Schußwechsel	Schusswechsel
schwachbetont	schwach betont
schwachbevölkert	schwach bevölkert
aus schwarz weiß machen	aus Schwarz Weiß machen
Schwarze Magie	schwarze Magie
schwarzgefärbt	schwarz gefärbt
schwarzrotgolden	*also:* schwarz-rot-golden
schwerfallen	schwer fallen
schwernehmen	schwer nehmen
schwertun	schwer tun
schwerverständlich	schwer verständlich
Schwimmeister	Schwimmmeister
	also: Schwimm-Meister
Science-fiction	Sciencefiction
	also: Science-Fiction
Sechspaß	Sechspass
See-Elefant	*also:* Seeelefant
jedem das Seine	*also:* jedem das seine
das Seine beitragen	*also:* das seine beitragen
die Seinen	*also:* die seinen
die Seinigen	*also:* die seinigen
seinlassen	sein lassen
Seismograph	*also:* Seismograf
auf seiten	aufseiten
	also: auf Seiten
von seiten	vonseiten
	also: von Seiten
selbständig	*also:* selbstständig
Selbständigkeit	*also:* Selbstständigkeit
selbstbewußt	selbstbewusst
Selbstbewußtsein	Selbstbewusstsein
selbsternannt	selbst ernannt
selbstgebacken	selbst gebacken
selbstgemacht	selbst gemacht
selbstgestrickt	selbst gestrickt
Selbstschuß	Selbstschuss
selbstverdient	selbst verdient
seligpreisen	selig preisen
seligsprechen	selig sprechen

OLD	NEW
Senatsbeschluß	Senatsbeschluss
Sendeschluß	Sendeschluss
Sendungsbewußtsein	Sendungsbewusstsein
Sensationsprozeß	Sensationsprozess
Séparée	*also:* Separee
sequentiell	*also:* sequenziell*
seßhaft	sesshaft
Seßhaftigkeit	Sesshaftigkeit
S-förming	*also:* s-förmig
die Shakespeareschen Sonette	die shakespeareschen Sonette
	also: die Shakespeare'schen Sonette
Short story	Shortstory
	also: Short Story
Showbusineß	Showbusiness
Showdown	*also:* Show-down*
Shrimp	*also:* Schrimp
auf Nummer Sicher gehen	*also:* auf Nummer sicher gehen
das sicherste ist, wenn …	das Sicherste ist, wenn …
Sicherheitsschloß	Sicherheitsschloss
Sicherheitsverschluß	Sicherheitsverschluss
siedendheiß	siedend heiß
siegesbewußt	siegesbewusst
siegesgewiß	siegesgewiss
Simplonpaß	Simplonpass
die Singende Säge	die singende Säge
Siphonverschluß	Siphonverschluss
sitzenbleiben	sitzen bleiben
sitzenlassen	sitzen lassen
Skipaß	Skipass
Small talk	Smalltalk
	also: Small Talk
so daß	sodass
	also: so dass
Sommerschlußverkauf	Sommerschlussverkauf
alles sonstige besprechen wir morgen	alles Sonstige besprechen wir morgen
Soufflé	*also:* Soufflee
soviel du willst	so viel du willst
soviel wie	so viel wie
noch einmal soviel	noch einmal so viel
es ist soweit	es ist so weit
soweit wie möglich	so weit wie möglich
ich kann das sowenig wie du	ich kann das so wenig wie du
Sowjetrußland	Sowjetrussland
hier gilt kein Sowohl-Als-auch	hier gilt kein Sowohl-als-auch
Spaghetti	*also:* Spagetti
Spantenriß	Spantenriss
spazierenfahren	spazieren fahren
spazierengehen	spazieren gehen
Speichelfluß	Speichelfluss
Sperrad	Sperrrad
	also: Sperr-Rad
Sperriegel	Sperrriegel
	also: Sperr-Riegel
Spliß	Spliss
du splißt	du splisst
eine sporenbildende Pflanze	eine Sporen bildende Pflanze
Sportdreß	Sportdress
Sprenggeschoß	Sprenggeschoss [*in Austria still written with* ß]
Spritzguß	Spritzguss
es sproß neues Grün	es spross neues Grün
Sproß	Spross
Sproßachse	Sprossachse
Sprößchen	Sprösschen
Sprößling	Sprössling
staatenbildende Insekten	Staaten bildende Insekten
Stahlroß	Stahlross
Stallaterne	Stalllaterne
	also: Stall-Laterne
Stammutter	Stammmutter
	also: Stamm-Mutter
standesbewußt	standesbewusst

OLD	NEW
Standesbewußtsein	Standesbewusstsein
Startschuß	Startschuss
steckenbleiben	stecken bleiben
steckenlassen	stecken lassen
Steckschloß	Steckschloss
Steckschuß	Steckschuss
stehenbleiben	stehen bleiben
stehenlassen	stehen lassen
Stehimbiß	Stehimbiss
Steilpaß	Steilpass
Stemmeißel	Stemmmeißel
	also: Stemm-Meißel
Stendelwurz	Ständelwurz
Stengel	Stängel
Step	Stepp
Steptanz	Stepptanz
Stereophonie	*also:* Stereofonie
Steuererlaß	Steuererlass
Steuermeßbetrag	Steuermessbetrag
Stewardeß	Stewardess
stiftengehen	stiften gehen
etwas im stillen vorbereiten	etwas im Stillen vorbereiten
Stilleben	Stillleben
	also: Still-Leben
stillegen	stilllegen
Stillegung	Stilllegung
Stoffarbe	Stofffarbe
	also: Stoff-Farbe
Stoffetzen	Stofffetzen
	also: Stoff-Fetzen
Stoffülle	Stofffülle
	also: Stoff-Fülle
Stop	Stopp
Straferlaß	Straferlass
Strafprozeß	Strafprozess
Strafprozeßordnung	Strafprozessordnung
Straß	Strass
Streifschuß	Streifschuss
Streitroß	Streitross
strenggenommen	streng genommen
strengnehmen	streng nehmen
aufs strengste unterschieden	*also:* aufs Strengste unterschieden
Streß	Stress
der Lärm streßt	der Lärm stresst
Streßsituation	Stresssituation
	also: Stress-Situation
2stündig, 3stündig, 4stündig ...	2-stündig, 3-stündig, 4-stündig ...
2stündlich, 3stündlich, 4stündlich ...	2-stündlich, 3-stündlich, 4-stündlich ...
Stuß	Stuss
substantiell	*also:* substanziell*
Sustenpaß	Sustenpass

T

OLD	NEW
Tablettenmißbrauch	Tablettenmissbrauch
tabula rasa machen	Tabula rasa machen
zutage treten	*also:* zu Tage treten
2tägig, 3tägig, 4tägig ...	2-tägig, 3-tägig, 4-tägig ...
Tankschloß	Tankschloss
Tarifabschluß	Tarifabschluss
Täßchen	Tässchen
ein paar tausend	*also:* ein paar Tausend
Tausende von Zuschauern	*also:* tausende von Zuschauern
T-bone-Steak	T-Bone-Steak
Tee-Ei	*also:* Teeei
Tee-Ernte	*also:* Teeernte
Teerfaß	Teerfass
Telephon	Telefon
Telephonanschluß	Telefonanschluss
Thunfisch	*also:* Tunfisch
Tie-Break	*also:* Tiebreak
aufs tiefste gekränkt	*also:* aufs Tiefste gekränkt
tiefbewegt	tief bewegt
tiefempfunden	tief empfunden
tiefverschneit	tief verschneit
Tintenfaß	Tintenfass
Tip	Tipp
todblaß	todblass
Todesschuß	Todesschuss
Tolpatsch	Tollpatsch
tolpatschig	tollpatschig
Tomatenketchup	*also:* Tomatenketschup
Topographie	*also:* Topografie
Torschlußpanik	Torschlusspanik
Torschuß	Torschuss
totenblaß	totenblass
totgeboren	tot geboren
traditionsbewußt	traditionsbewusst
Tränenfluß	Tränenfluss
tränennaß	tränennass
Traß	Trass
Trekking	*also:* Trecking
treuergeben	treu ergeben
triefnaß	triefnass
auf dem trockenen sitzen	auf dem Trockenen sitzen
sein Schäfchen ins trockene bringen	sein Schäfchen ins Trockene bringen
tropfnaß	tropfnass
Troß	Tross
im trüben fischen	im Trüben fischen
Truchseß	Truchsess
Trugschluß	Trugschluss
Trumpfas	Trumpfass
Tuffelsen	Tufffelsen
	also: Tuff-Felsen
Türschloß	Türschloss

U

OLD	NEW
übelgelaunt	übel gelaunt
übelnehmen	übel nehmen
übelriechend	übel riechend
Überbiß	Überbiss
Überdruß	Überdruss
übereinanderlegen	übereinander legen
übereinanderliegen	übereinander liegen
übereinanderwerfen	übereinander werfen
Überfluß	Überfluss
Überflußgesellschaft	Überflussgesellschaft
Überguß	Überguss
überhandnehmen	überhand nehmen
übermorgen abend, nachmittag	übermorgen Abend, Nachmittag
Überschuß	Überschuss
überschwenglich	überschwänglich
überwächtet	überwechtet
ein übriges tun	ein Übriges tun
im übrigen wissen wir doch alle ...	im Übrigen wissen wir doch alle ...
alles übrige später	alles Übrige später
die übrigen kommen nach	die Übrigen kommen nach
übrigbehalten	übrig behalten
übrigbleiben	übrig bleiben
übriglassen	übrig lassen
U-förmig	*also:* u-förmig
Ultima ratio	Ultima Ratio
Umdenkprozeß	Umdenkprozess
die Liste umfaßt alles Wichtige	die Liste umfasst alles Wichtige
Umriß	Umriss
Umrißzeichnung	Umrisszeichnung
Umschichtungsprozeß	Umschichtungsprozess
Umschluß	Umschluss
umsein	um sein
um so [mehr, größer, weniger ...]	umso [mehr, größer, weniger ...]
Umstellungsprozeß	Umstellungsprozess

OLD	NEW
Umwandlungsprozeß	Umwandlungsprozess
Umwelteinfluß	Umwelteinfluss
sich ins unabsehbare ausweiten	sich ins Unabsehbare ausweiten
unangepaßt	unangepasst
Unangepaßtheit	Unangepasstheit
unbeeinflußbar	unbeeinflussbar
unbeeinflußt	unbeeinflusst
Anzeige gegen Unbekannt	Anzeige gegen unbekannt
unbewußt	unbewusst
und ähnliches (u. ä.)	und Ähnliches (u. Ä.)
unendlichemal	unendliche Mal
unerläßlich	unerlässlich
unermeßlich	unermesslich
Unfairneß	Unfairness
unfaßbar	unfassbar
unfaßlich	unfasslich
ungewiß	ungewiss
Ungewißheit	Ungewissheit
unigefärbt	uni gefärbt
im unklaren bleiben	im Unklaren bleiben
im unklaren lassen	im Unklaren lassen
unmißverständlich	unmissverständlich
unpäßlich	unpässlich
Unpäßlichkeit	Unpässlichkeit
unplaziert	unplatziert
unrecht haben	Unrecht haben
unrecht behalten	Unrecht behalten
unrecht bekommen	Unrecht bekommen
Unrechtsbewußtsein	Unrechtsbewusstsein
unselbständig	*also:* unselbstständig
Unselbständigkeit	*also:* Unselbstständigkeit
die Unseren	*also:* die unseren
die Unsrigen	*also:* die unsrigen
untenerwähnt	unten erwähnt
untenstehend	unten stehend
unterbewußt	unterbewusst
Unterbewußtsein	Unterbewusstsein
unterderhand	unter der Hand
untereinanderstehen	untereinander stehen
Untergeschoß	Untergeschoss [*in Austria still written with* ß]
ohne Unterlaß	ohne Unterlass
Untersuchungsausschuß	Untersuchungsausschuss
unvergeßlich	unvergesslich
unerläßlich	unerlässlich
unzähligemal	unzählige Mal

V

OLD	NEW
va banque spielen	*also:* Vabanque spielen
Varieté	*also:* Varietee
veranlaßt	veranlasst
verantwortungsbewußt	verantwortungsbewusst
Verantwortungsbewußtsein	Verantwortungsbewusstsein
Verbiß	Verbiss
verblaßt	verblasst
verbleuen	verbläuen
im verborgenen blühen	im Verborgenen blühen
das verdroß uns	das verdross uns
Verdruß	Verdruss
du verfaßt	du verfasst
vergeßlich	vergesslich
Vergeßlichkeit	Vergesslichkeit
Vergißmeinnicht	Vergissmeinnicht
du vergißt	du vergisst
verhaßt	verhasst
auf jmdn. ist Verlaß	auf jmdn. ist Verlass
verläßlich	verlässlich
Verläßlichkeit	Verlässlichkeit
verlorengehen	verloren gehen
vermißt	vermisst
Vermißtenanzeige	Vermisstenanzeige
er hat den Zug verpaßt	er hat den Zug verpasst
das Geld wurde verpraßt	das Geld wurde verprasst

OLD	NEW
Verriß	Verriss
verschiedenes war noch unklar	Verschiedenes war noch unklar
verschiedenemal	verschiedene Mal
Verschiß	Verschiss
Verschluß	Verschluss
Verschlußkappe	Verschlusskappe
Verschlußsache	Verschlusssache *also:* Verschluss-Sache
verselbständigen	*also:* verselbstständigen
Versorgungsengpaß	Versorgungsengpass
Vertragsabschluß	Vertragsabschluss
Vertragsschluß	Vertragsschluss
V-förmig	*also:* v-förmig
Vibraphon	*also:* Vibrafon
viel zuviel	viel zu viel
viel zuwenig	viel zu wenig
vielbefahren	viel befahren
vielgelesen	viel gelesen
Vierpaß	Vierpass
aus dem vollen schöpfen	aus dem Vollen schöpfen
voneinandergehen	voneinander gehen
von seiten	vonseiten *also:* von Seiten
vorangehendes gilt auch …	Vorangehendes gilt auch …
im vorangehenden heißt es …	im Vorangehenden heißt es …
im voraus	im Voraus
vorgefaßt	vorgefasst
vorgestern abend, mittag, morgen	vorgestern Abend, Mittag, Morgen
Vorhängeschloß	Vorhängeschloss
vorhergehendes gilt auch …	Vorhergehendes gilt auch …
im vorhergehenden heißt es …	im Vorhergehenden heißt es …
im vorhinein	im Vorhinein
das vorige gilt auch …	das Vorige gilt auch …
im vorigen heißt es …	im Vorigen heißt es …
Vorlegeschloß	Vorlegeschloss
vorliebnehmen	vorlieb nehmen
[gestern, heute, morgen] vormittag	[gestern, heute, morgen] Vormittag
Vorschlußrunde	Vorschlussrunde
Vorschuß	Vorschuss
Vorschußlorbeeren	Vorschusslorbeeren
vorstehendes gilt auch …	Vorstehendes gilt auch …
im vorstehenden heißt es …	im Vorstehenden heißt es …
vorwärtsgehen	vorwärts gehen
vorwärtskommen	vorwärts kommen

W

OLD	NEW
ein wachestehender Soldat	ein Wache stehender Soldat
Wachsabguß	Wachsabguss
Wächte	Wechte
Waggon	*also:* Wagon
Wahlausschuß	Wahlausschuss
Walkie-talkie	Walkie-Talkie
Walnuß	Walnuss
Walroß	Walross
Wandlungsprozeß	Wandlungsprozess
Warnschuß	Warnschuss
Wasserschloß	Wasserschloss
wäßrig	wässrig
Wehrpaß	Wehrpass
weichgekocht	weich gekocht
Weinfaß	Weinfass
aus schwarz weiß machen	aus Schwarz Weiß machen
weißgekleidet	weiß gekleidet
Weißrußland	Weißrussland
des weiteren wurde gesagt …	des Weiteren wurde gesagt …
weitgereist	weit gereist
weitreichend	weit reichend
weitverbreitet	weit verbreitet
Werkstattage	Werkstatttage *also:* Werkstatt-Tage
Werkstofforschung	Werkstoffforschung *also:* Werkstoff-Forschung

OLD	NEW
es besteht im wesentlichen aus …	es besteht im Wesentlichen aus …
Wetteufel	Wettteufel *also:* Wett-Teufel
Wetturnen	Wettturnen *also:* Wett-Turnen
widereinanderstoßen	widereinander stoßen
wieviel	wie viel
Winterschlußverkauf	Winterschlussverkauf
Wißbegierde	Wissbegierde
wißbegierig	wissbegierig
ihr wißt	ihr wisst
du wußtest	du wusstest
wir wüßten gern …	wir wüssten gern …
Witterungseinfluß	Witterungseinfluss
Wollappen	Wolllappen *also:* Woll-Lappen
Wollaus	Wolllaus *also:* Woll-Laus
als ob er wunder was getan hätte	als ob er Wunder was getan hätte
sich wundliegen	sich wund liegen
Wurfgeschoß	Wurfgeschoss [*in Austria still written with* ß]

X, Y

OLD	NEW
X-beinig	*also:* x-beinig
X-förmig	*also:* x-förmig
zum x-tenmal	zum x-ten Mal

Z

OLD	NEW
Zäheit	Zähheit
Zahlenschloß	Zahlenschloss
Zäpfchen-R	*also:* Zäpfchen-r
Zaubernuß	Zaubernuss
Zechenstillegung	Zechenstilllegung
Zeilengußmaschine	Zeilengussmaschine
2zeilig, 3zeilig, 4zeilig …	2-zeilig, 3-zeilig, 4-zeilig …
eine Zeitlang	eine Zeit lang
zur Zeit [*derzeit*]	zurzeit
Zellehre	Zelllehre *also:* Zell-Lehre
Zellstoffabrik	Zellstofffabrik *also:* Zellstoff-Fabrik
Zersetzungsprozeß	Zersetzungsprozess
zielbewußt	zielbewusst
Zierat	Zierrat
zigtausend	*also:* Zigtausend
zigtausende	*also:* zigtausende
Zippverschluß	Zippverschluss
Zirkelschluß	Zirkelschluss
Zivilprozeß	Zivilprozess
Zivilprozeßordnung	Zivilprozessordnung
Zoo-Orchester	*also:* Zooorchester
sich zu eigen machen	sich zu Eigen machen
zueinanderfinden	zueinander finden
Zufluß	Zufluss
sich zufriedengeben	sich zufrieden geben
zufriedenlassen	zufrieden lassen
zufriedenstellen	zufrieden stellen
zugrunde gehen	*also:* zu Grunde gehen
zugrunde legen	*also:* zu Grunde legen
zugrunde liegen	*also:* zu Grunde liegen
zugrundeliegend	zugrunde liegend *also:* zu Grunde liegend
zugrunde richten	*also:* zu Grunde richten
zugunsten	*also:* zu Gunsten
Zu Hause	*in Austria and Switzerland also:* zuhause
bei uns zulande	bei uns zu Lande
zulasten	*also:* zu Lasten
jmdm. etwas zuleide tun	*also:* jmdm. etwas zu Leide tun
zumute sein	*also:* zu Mute sein
Zündschloß	Zündschloss
Zungenkuß	Zungenkuss
Zungen-R	*also:* Zungen-r

OLD	NEW
sich etwas zunutze machen	*also:* sich etwas zu Nutze machen
jmdm. zupaß kommen	jmdm. zupass kommen
zugepreßt	zugepresst
zu Rande kommen	*also:* zurande kommen
jmdn. zu Rate ziehen	*also:* jmdn. zurate ziehen
sie hat zurückgemußt	sie hat zurückgemusst
zur Zeit [*derzeit*]	zurzeit
Zusammenfluß	Zusammenfluss
zusammengefaßt	zusammengefasst
zusammengepaßt	zusammengepasst
zusammengepreßt	zusammengepresst
Zusammenschluß	Zusammenschluss
zusammensein	zusammen sein
zuschanden werden	*also:* zu Schanden werden
sich etwas zuschulden kommen lassen	*also:* sich etwas zu schulden kommen lassen
Zuschuß	Zuschuss
Zuschußbetrieb	Zuschussbetrieb
zusein	zu sein
zustande bringen	*also:* zu Stande bringen
zustande kommen	*also:* zu Stande kommen
zutage födern	*also:* zu Tage fördern
zutage treten	*also:* zu Tage treten
zuungunsten	*also:* zu Ungunsten
zuviel	zu viel
Zuwege bringen	*also:* zu Wege bringen
zuwenig	zu wenig
die zwanziger Jahre	*also:* die Zwanzigerjahre*
die Zwanzigerjahre	*also:* die zwanziger Jahre
das Zweite Gesicht	das zweite Gesicht
er hat wie kein zweiter gearbeitet	er hat wie kein Zweiter gearbeitet
jeder zweite war krank	jeder Zweite war krank
Zweitkläßler	Zweitklässler
Zwischengeschoß	Zwischengeschoss [*in Austria still written with* ß]

Phonetic symbols used in transcriptions of English words / Die für das Englische verwendeten Zeichen der Lautschrift

ɑ:	bah	bɑ:
ɑ̃	ensemble	ɑ̃'sɑ̃mbl
æ	fat	fæt
æ̃	lingerie	'læ̃ʒərɪ
aɪ	fine	faɪn
aʊ	now	naʊ
b	bat	bæt
d	dog	dɒg
dʒ	jam	dʒæm
e	met	met
eɪ	fate	feɪt
eə	fairy	'feərɪ
əʊ	goat	gəʊt
ə	ago	ə'gəʊ
ɜ:	fur	fɜ:(r)
f	fat	fæt
g	good	gʊd
h	hat	hæt
ɪ	bit, lately	bɪt, 'leɪtlɪ
ɪə	nearly	'nɪəlɪ
i:	meet	mi:t
j	yet	jet
k	kit	kɪt
l	lot	lɒt
m	mat	mæt
n	not	nɒt
ŋ	sing	sɪŋ
ɒ	got	gɒt
ɔ:	paw	pɔ:
ɔ̃	fait accompli	feɪt æ'kɔ̃pli:
ɔɪ	boil	bɔɪl
p	pet	pet
r	rat	ræt
s	sip	sɪp
ʃ	ship	ʃɪp
t	tip	tɪp
tʃ	chin	tʃɪn
θ	thin	θɪn
ð	the	ðə
u:	boot	bu:t
ʊ	book	bʊk
ʊə	tourist	'tʊərɪst
ʌ	dug	dʌg
v	van	væn
w	win	wɪn
x	loch	lɒx
z	zip	zɪp
ʒ	vision	'vɪʒn

: Length sign, indicating that the preceding vowel is long, e.g. boot [bu:t]. / Längezeichen, bezeichnet Länge des unmittelbar davor stehenden Vokals, z. B. boot [bu:t].

' Stress mark, immediately preceding a stressed syllable, e.g. ago [ə'gəʊ]. / Betonung, steht unmittelbar vor einer betonten Silbe, z. B. ago [ə'gəʊ].

(r) An 'r' in parentheses is pronounced only when immediately followed by a vowel sound, e.g. pare [peə(r)]; pare away [peər ə'weɪ]. / Ein „r" in runden Klammern wird nur gesprochen, wenn im Textzusammenhang ein Vokal unmittelbar folgt, z. B. pare [peə(r)]; pare away [peər ə'weɪ].

Phonetic symbols used in transcriptions of German words / Die für das Deutsche verwendeten Zeichen der Lautschrift

a	hat	hat
a:	Bahn	ba:n
ɐ	Ober	'o:bɐ
ɐ̯	Uhr	u:ɐ̯
ã	Grand Prix	grã'pri:
ã:	Abonnement	abɔnə'mã:
ai̯	weit	vai̯t
au̯	Haut	hau̯t
b	Ball	bal
ç	ich	ıç
d	dann	dan
d͜ʒ	Gin	d͜ʒın
e	Methan	me'ta:n
e:	Beet	be:t
ɛ	mästen	'mɛstn̩
ɛ:	wählen	'vɛ:lən
ɛ̃	Ragoût fin	ragu'fɛ̃
ɛ̃:	Timbre	'tɛ̃:br(ə)
ə	Nase	'na:zə
f	Faß	fas
g	Gast	gast
h	hat	hat
i	vital	vi'ta:l
i:	viel	fi:l
i̯	Studie	'ʃtu:di̯ə
ı	Birke	'bırkə
j	ja	ja:
k	kalt	kalt
l	Last	last
l̩	Nabel	'na:bl̩
m	Mast	mast
n	Naht	na:t
n̩	baden	'ba:dn̩
ŋ	lang	laŋ
o	Moral	mo'ra:l
o:	Boot	bo:t
o̯	loyal	lo̯a'ja:l
õ	Fondue	fõ'dy:
õ:	Fond	fõ:
ɔ	Post	pɔst
ø	Ökonom	øko'no:m
ø:	Öl	ø:l
œ	göttlich	'gœtlıç
œ̃:	Parfum	par'fœ̃:
ɔy̯	Heu	hɔy̯
p	Pakt	pakt
p͜f	Pfahl	p͜fa:l
r	Rast	rast
s	Hast	hast
ʃ	schal	ʃa:l
t	Tal	ta:l
t͜s	Zahl	t͜sa:l
t͜ʃ	Matsch	mat͜ʃ
u	kulant	ku'lant
u:	Hut	hu:t
u̯	aktuell	ak'tu̯ɛl
ʊ	Pult	pʊlt
v	was	vas
x	Bach	bax
y	Physik	fy'zi:k
y:	Rübe	'ry:bə
y̆	Nuance	'ny̆ã:sə
ʏ	Fülle	'fʏlə
z	Hase	'ha:zə
ʒ	Genie	ʒe'ni:

| Glottal stop, e.g. Aa [a'|a]. / Stimmritzenverschlußlaut („Knacklaut"), z. B. Aa [a'|a].

: Length sign, indicating that the preceding vowel is long, e.g. Chrom [kro:m]. / Längezeichen, bezeichnet Länge des unmittelbar davor stehenden Vokals, z. B. Chrom [kro:m].

˜ Indicates a nasal vowel, e.g. Fond [fõ:]. / Zeichen für nasale Vokale, z. B. Fond [fõ:].

' Stress mark, immediately preceding a stressed syllable, e.g. Ballon [ba'lɔŋ]. / Betonung, steht unmittelbar vor einer betonten Silbe, z. B. Ballon [ba'lɔŋ].

Sign placed below a syllabic consonant, e.g. Büschel ['bʏʃl̩]. / Zeichen für silbischen Konsonanten, steht unmittelbar unter dem Konsonanten, z. B. Büschel ['bʏʃl̩].

Placed above or below a symbol indicates a non-syllabic vowel, e.g. Milieu [mi'li̯ø:]./ Halbkreis, untergesetzt oder übergesetzt, bezeichnet unsilbischen Vokal, z. B. Milieu [mi'li̯ø:].

Englische unregelmäßige Verben

Die im englisch-deutschen Wörterverzeichnis mit einer hochgestellten Ziffer versehenen unregelmäßigen Verben haben diese Ziffer auch in dieser Liste. Ein Sternchen (*) weist darauf hin, daß die korrekte Form von der jeweiligen Bedeutung abhängt.

Infinitive *Infinitiv*	Past Tense *Präteritum*	Past Participle *2. Partizip*
abide	abided, abode	abided, abode
arise	arose	arisen
awake	awoke	awoken
be	was *sing.*, were *pl.*	been
bear	bore	borne
beat	beat	beaten
begin	began	begun
behold	beheld	beheld
bend	bent	bent
beseech	besought, beseeched	besought, beseeched
bet	bet, betted	bet, betted
bid	*bade, bid	*bidden, bid
bind	bound	bound
bite	bit	bitten
bleed	bled	bled
bless	blessed, blest	blessed, blest
blow	*blew, blowed	*blown, blowed
break	broke	broken
breed	bred	bred
bring	brought	brought
broadcast	broadcast	broadcast
build	built	built
burn	burnt, burned	burnt, burned
burst	burst	burst
bust	bust, busted	bust, busted
buy	bought	bought
cast	cast	cast
catch	caught	caught
chide	chided, chid	chided, chid, chidden
choose	chose	chosen
cleave	cleaved, clove, cleft	cleaved, cloven, cleft
cling	clung	clung
come	came	come
cost	*cost, costed	*cost, costed
countersink	countersunk	countersunk
creep	crept	crept
cut	cut	cut
deal	dealt	dealt
dig	dug	dug
dive	dived, *(Amer.)* dove	dived
[1]do	did	done
draw	drew	drawn
dream	dreamt, dreamed	dreamt, dreamed
drink	drank	drunk
drive	drove	driven
dwell	dwelt	dwelt
eat	ate	eaten
fall	fell	fallen
feed	fed	fed
feel	felt	felt
fight	fought	fought
find	found	found
flee	fled	fled
fling	flung	flung
floodlight	floodlit	floodlit
fly	flew	flown
forbear	forbore	forborne
forbid	forbade, forbad	forbidden
forecast	forecast, forecasted	forecast, forecasted
foretell	foretold	foretold
forget	forgot	forgotten
forgive	forgave	forgiven
forsake	forsook	forsaken
freeze	froze	frozen

Infinitive *Infinitiv*	Past Tense *Präteritum*	Past Participle *2. Partizip*
get	got	*got, *(Amer.)* gotten
give	gave	given
go	went	gone
grind	ground	ground
grow	grew	grown
hamstring	hamstrung, hamstringed	hamstrung, hamstringed
hang	*hung, hanged	*hung, hanged
have	had	had
hear	heard	heard
heave	*heaved, hove	*heaved, hove
hew	hewed	hewn, hewed
hide	hid	hidden
hit	hit	hit
hold	held	held
hurt	hurt	hurt
input	input, inputted	input, inputted
keep	kept	kept
kneel	knelt, *(esp. Amer.)* kneeled	knelt, *(esp. Amer.)* kneeled
knit	*knitted, knit	*knitted, knit
know	knew	known
lay	laid	laid
lead	led	led
lean	leaned, *(Brit.)* leant	leaned, *(Brit.)* leant
leap	leapt, leaped	leapt, leaped
learn	learnt, learned	learnt, learned
leave	left	left
lend	lent	lent
let	let	let
^{2}lie	lay	lain
light	lit, lighted	lit, lighted
lose	lost	lost
make	made	made
mean	meant	meant
meet	met	met
mow	mowed	mown, mowed
output	output, outputted	output, outputted
outshine	outshone	outshone
overhang	overhung	overhung
pay	paid	paid
plead	pleaded, *(esp. Amer., Scot., dial.)* pled	pleaded, *(esp. Amer., Scot., dial.)* pled
prove	proved	*proved, *(esp. Amer., Scot., dial.)* proven
put	put	put
quit	quitted, *(Amer.)* quit	quitted, *(Amer.)* quit
read [ri:d]	read [red]	read [red]
rid	rid	rid

Infinitive *Infinitiv*	Past Tense *Präteritum*	Past Participle *2. Partizip*
ride	rode	ridden
^{2}ring	rang	rung
rise	rose	risen
run	ran	run
saw	sawed	sawn, sawed
say	said	said
see	saw	seen
seek	sought	sought
sell	sold	sold
send	sent	sent
set	set	set
sew	sewed	sewn, sewed
shake	shook	shaken, *(arch./coll.)* shook
shear	sheared	shorn, sheared
shed	shed	shed
shine	*shone, shined	*shone, shined
shit	shitted, shit, shat	shitted, shit, shat
shoe	shod	shod
shoot	shot	shot
show	showed	shown, showed
shrink	shrank	shrunk
shut	shut	shut
sing	sang	sung
sink	sank, sunk	sunk
sit	sat	sat
slay	*slew, slayed	*slain, slayed
sleep	slept	slept
slide	slid	slid
sling	slung	slung
slink	slunk	slunk
slit	slit	slit
smell	smelt, smelled	smelt, smelled
smite	smote	smitten
sow	sowed	sown, sowed
speak	spoke	spoken
speed	*sped, speeded	*sped, speeded
spell	spelled, *(Brit.)* spelt	spelled, *(Brit.)* spelt
spend	spent	spent
spill	spilt, spilled	spilt, spilled
spin	spun	spun
spit	spat, spit	spat, spit
split	split	split
spoil	spoilt, spoiled	spoilt, spoiled
spread	spread	spread
spring	sprang, *(Amer.)* sprung	sprung
stand	stood	stood
stave	*staved, stove	*staved, stove
steal	stole	stolen
stick	stuck	stuck
sting	stung	stung

Infinitive *Infinitiv*	Past Tense *Präteritum*	Past Participle *2. Partizip*
stink	stank, stunk	stunk
strew	strewed	strewed, strewn
stride	strode	stridden
strike	struck	struck, *(arch.)* stricken
string	strung	strung
strive	strove	striven
sublet	sublet	sublet
swear	swore	sworn
sweep	swept	swept
swell	swelled	swollen, swelled
swim	swam	swum
swing	swung	swung
take	took	taken
teach	taught	taught
tear	tore	torn
tell	told	told
think	thought	thought
thrive	thrived, throve	thrived, thriven
throw	throw	thrown
thrust	thrust	thrust
tread	trod	trodden, trod
understand	understood	understood
undo	undid	undone
wake	woke, *(arch.)* waked	woken, *(arch.)* waked
wear	wore	worn
[1]weave	wove	woven
weep	wept	wept
wet	wet, wetted	wet, wetted
win	won	won
[2]wind [waɪnd]	wound [waʊnd]	wound [waʊnd]
work	worked, *(arch., literary)* wrought	worked, *(arch., literary)* wrought
wring	wrung	wrung
write	wrote	written

German irregular verbs

Irregular and partly irregular verbs are listed alphabetically by infinitive. 1st, 2nd, and 3rd person present and imperative forms are given after the infinitive, and preterite subjunctive forms after the preterite indicative, where they take an umlaut, change *e* to *i,* etc.
Verbs with a raised number in the German-English section of the Dictionary have the same number in this list.
Compound verbs (including verbs with prefixes) are only given if a) they do not take the same forms as the corresponding simple verb, e.g. *befehlen,* or b) there is no corresponding simple verb, e.g. *bewegen.*

An asterisk (*) indicates a verb which is also conjugated regularly.

Infinitive *Infinitiv*	Preterite *Präteritum*	Past Participle *2. Partizip*
[1]backen (du bäckst, er bäckt; *auch:* du backst, er backt)	backte, *älter:* buk (büke)	gebacken
befehlen (du befiehlst, er befiehlt; befiehl!)	befahl (beföhle, befähle)	befohlen
beginnen	begann (begänne, *seltener:* begönne)	begonnen
beißen	biß	gebissen
bergen (du birgst, er birgt; birg!)	barg (bärge)	geborgen
bersten (du birst, er birst; birst!)	barst (bärste)	geborsten
[2]bewegen	bewog (bewöge)	bewogen
biegen	bog (böge)	gebogen
bieten	bot (böte)	geboten

Infinitive *Infinitiv*	Preterite *Präteritum*	Past Participle *2. Partizip*
binden	band (bände)	gebunden
bitten	bat (bäte)	gebeten
blasen (du bläst, er bläst)	blies	geblasen
bleiben	blieb	geblieben
bleichen*	blich	geblichen
braten (du brätst, er brät)	briet	gebraten
brechen (du brichst, er bricht; brich!)	brach (bräche)	gebrochen
brennen	brannte (brennte)	gebrannt
bringen	brachte (brächte)	gebracht
denken	dachte (dächte)	gedacht
dingen*	dang (dänge)	gedungen
dreschen (du drischst, er drischt; drisch!)	drosch (drösche)	gedroschen
dringen	drang (dränge)	gedrungen
dünken* (es dünkt *auch:* deucht)	deuchte	gedeucht
dürfen (ich darf, du darfst, er darf)	durfte (dürfte)	gedurft
empfehlen (du empfiehlst, er empfiehlt, empfiehl!)	empfahl (empföhle, *seltener:* empfähle)	empfohlen
erlöschen (du erlischst, er erlischt, erlisch!)	erlosch (erlösche)	erloschen
erschallen	erscholl (erschölle)	erschollen
[1, 3]erschrecken (du erschrickst, er erschrickt, erschrick!)	erschrak (erschräke)	erschrocken
essen (du ißt, er ißt, iß!)	aß (äße)	gegessen
fahren (du fährst, er fährt)	fuhr (führe)	gefahren
fallen (du fällst, er fällt)	fiel	gefallen
fangen (du fängst, er fängt)	fing	gefangen
fechten (du fichtst, er ficht; ficht!)	focht (föchte)	gefochten
finden	fand (fände)	gefunden
flechten (du flichtst, er flicht; flicht!)	flocht (flöchte)	geflochten
fliegen	flog (flöge)	geflogen
fliehen	floh (flöhe)	geflohen
fließen	floß (flösse)	geflossen
fressen (du frißt, er frißt, friß!)	fraß (fräße)	gefressen
frieren	fror (fröre)	gefroren
gären*	gor (göre)	gegoren
gebären (*geh.:* du gebierst, sie gebiert; gebier!)	gebar (gebäre)	geboren
geben (du gibst, er gibt; gib!)	gab (gäbe)	gegeben
gedeihen	gedieh	gediehen
gehen	ging	gegangen
gelingen	gelang (gelänge)	gelungen
gelten (du giltst, er gilt; gilt!)	galt (gölte, gälte)	gegolten
genesen	genas (genäse)	genesen
genießen	genoß (genösse)	genossen
geschehen (es geschieht)	geschah (geschähe)	geschehen
gewinnen	gewann (gewönne, gewänne)	gewonnen
gießen	goß (gösse)	gegossen
gleichen	glich	geglichen
gleiten	glitt	geglitten

Infinitive *Infinitiv*	Preterite *Präteritum*	Past Participle *2. Partizip*
glimmen	glomm (glömme)	geglommen
graben (du gräbst, er gräbt)	grub (grübe)	gegraben
greifen	griff	gegriffen
haben (du hast, er hat)	hatte (hätte)	gehabt
halten (du hältst, er hält)	hielt	gehalten
[1]hängen	hing	gehangen
hauen*	hieb	gehauen
heben	hob (höbe)	gehoben
heißen	hieß	geheißen
helfen (du hilfst, er hilft; hilf!)	half (hülfe, *selten:* hälfe)	geholfen
kennen	kannte (kennte)	gekannt
kiesen*	kor (köre)	gekoren
klimmen*	klomm (klömme)	geklommen
klingen	klang (klänge)	geklungen
kneifen	kniff	gekniffen
kommen	kam (käme)	gekommen
können (ich kann, du kannst, er kann)	konnte (könnte)	gekonnt
kreischen*	krisch	gekrischen
kriechen	kroch (kröche)	gekrochen
küren*	kor (köre)	gekoren
[1]laden (du lädst, er lädt)	lud (lüde)	geladen
[2]laden (du lädst, er lädt; *veralt., landsch.:* du ladest, er ladet)	lud (lüde)	geladen
lassen (du läßt, er läßt)	ließ	gelassen
laufen (du läufst, er läuft)	lief	gelaufen
leiden	litt	gelitten
leihen	lieh	geliehen
[1, 2]lesen (du liest, er liest; lies!)	las (läse)	gelesen
liegen	lag (läge)	gelegen
lügen	log (löge)	gelogen
mahlen	mahlte	gemahlen
meiden	mied	gemieden
melken* (du milkst, er milkt; milk!; du melkst, er melkt; melke!)	molk (mölke)	gemolken
messen (du mißt, er mißt; miß!)	maß (mäße)	gemessen
mißlingen	mißlang (mißlänge)	mißlungen
mögen (ich mag, du magst, er mag)	mochte (möchte)	gemocht
müssen (ich muß, du mußt, er muß)	mußte (müßte)	gemußt
nehmen (du nimmst, er nimmt; nimm!)	nahm (nähme)	genommen
nennen	nannte (nennte)	genannt
pfeifen	pfiff	gepfiffen
pflegen*	pflog (pflöge)	gepflogen
preisen	pries	gepriesen
quellen (du quillst, er quillt; quill!)	quoll (quölle)	gequollen
raten (du rätst, er rät)	riet	geraten
reiben	rieb	gerieben
reißen	riß	gerissen

Infinitive *Infinitiv*	Preterite *Präteritum*	Past Participle *2. Partizip*
reiten	ritt	geritten
rennen	rannte (rennte)	gerannt
riechen	roch (röche)	gerochen
ringen	rang (ränge)	gerungen
rinnen	rann (ränne, *seltener:* rönne)	geronnen
rufen	rief	gerufen
salzen*	salzte	gesalzen
saufen (du säufst, er säuft)	soff (söffe)	gesoffen
saugen*	sog (söge)	gesogen
schaffen*	schuf (schüfe)	geschaffen
schallen*	scholl (schölle)	geschallt
scheiden	schied	geschieden
scheinen	schien	geschienen
scheißen	schiß	geschissen
schelten (du schiltst, er schilt; schilt!)	schalt (schölte)	gescholten
[1]scheren	schor (schöre)	geschoren
schieben	schob (schöbe)	geschoben
schießen	schoß (schösse)	geschossen
schinden	schindete	geschunden
schlafen (du schläfst, er schläft)	schlief	geschlafen
schlagen (du schlägst, er schlägt)	schlug (schlüge)	geschlagen
schleichen	schlich	geschlichen
[1]schleifen	schliff	geschliffen
schließen	schloß (schlösse)	geschlossen
schlingen	schlang (schlänge)	geschlungen
schmeißen	schmiß	geschmissen
schmelzen (du schmilzt, er schmilzt; schmilz!)	schmolz	geschmolzen
schnauben*	schnob (schnöbe)	geschnoben
schneiden	schnitt	geschnitten
schrecken* (du schrickst, er schrickt; schrick!)	schrak (schräke)	geschreckt
schreiben	schrieb	geschrieben
schreien	schrie	geschrie[e]n
schreiten	schritt	geschritten
schweigen	schwieg	geschwiegen
schwellen (du schwillst, er schwillt; schwill!)	schwoll (schwölle)	geschwollen
schwimmen	schwamm (schwömme, *seltener:* schwämme)	geschwommen
schwinden	schwand (schwände)	geschwunden
schwingen	schwang (schwänge)	geschwungen
schwören	schwor (schwüre)	geschworen
sehen (du siehst, er sieht; sieh[e]!)	sah (sähe)	gesehen
sein (ich bin, du bist, er ist, wir sind, ihr seid, sie sind; sei!)	war (wäre)	gewesen
senden*	sandte (sendete)	gesandt
sieden*	sott (sötte)	gesotten
singen	sang (sänge)	gesungen
sinken	sank (sänke)	gesunken

Infinitive *Infinitiv*	Preterite *Präteritum*	Past Participle *2. Partizip*
sinnen	sann (sänne, *veralt.:* sönne)	gesonnen
sitzen	saß (säße)	gesessen
sollen (ich soll, du sollst, er soll)	sollte	gesollt
spalten*	spaltete	gespalten
speien	spie	gespie[e]n
spinnen	spann (spönne, spänne)	gesponnen
spleißen*	spliß	gesplissen
sprechen (du sprichst, er spricht; sprich!)	sprach (spräche)	gesprochen
sprießen	sproß (sprösse)	gesprossen
springen	sprang	gesprungen
stechen (du stichst, er sticht; stich!)	stach (stäche)	gestochen
stecken*	stak (stäke)	gesteckt
stehen	stand (stünde, *auch:* stände)	gestanden
stehlen (du stiehlst, er stiehlt; stiehl!)	stahl (stähle, *seltener:* stöhle)	gestohlen
steigen	stieg	gestiegen
sterben (du stirbst, er stirbt; stirb!)	starb (stürbe)	gestorben
stieben	stob (stöbe)	gestoben
stinken	stank (stänke)	gestunken
stoßen (du stößt, er stößt)	stieß	gestoßen
streichen	strich	gestrichen
streiten	stritt	gestritten
tragen (du trägst, er trägt)	trug (trüge)	getragen
treffen (du triffst; er trifft; triff!)	traf (träfe)	getroffen
treiben	trieb	getrieben
treten (du trittst, er tritt; tritt!)	trat (träte)	getreten
triefen*	troff (tröffe)	getroffen
trinken	trank (tränke)	getrunken
trügen	trog (tröge)	getrogen
tun	tat (täte)	getan
verderben (du verdirbst, er verdirbt; verdirb!)	verdarb (verdürbe)	verdorben
verdrießen	verdroß (verdrösse)	verdrossen
vergessen (du vergißt, er vergißt, vergiß!)	vergaß (vergäße)	vergessen
verlieren	verlor (verlöre)	verloren
verlöschen (du verlischst, er verlischt; verlisch!)	verlosch (verlösche)	verloschen
verschleißen*	verschliß	verschlissen
[1]wachsen (du wächst, er wächst)	wuchs (wüchse)	gewachsen
wägen	wog (wöge)	gewogen
waschen (du wäschst, er wäscht)	wusch (wüsche)	gewaschen
weben*	wob (wöbe)	gewoben
weichen	wich	gewichen
weisen	wies	gewiesen
wenden*	wandte (wendete)	gewandt
werben (du wirbst, er wirbt; wirb!)	warb (würbe)	geworben

Infinitive *Infinitiv*	Preterite *Präteritum*	Past Participle *2. Partizip*
werden (du wirst, er wird; werde!)	wurde, *dichter.:* ward (würde)	geworden; *als Hilfsv.:* worden
werfen (du wirfst, er wirft; wirf!)	warf (würfe)	geworfen
[1]wiegen	wog (wöge)	gewogen
[1]winden	wand (wände)	gewunden
wissen (ich weiß, du weißt, er weiß)	wußte (wüßte)	gewußt
wollen (ich will, du willst, er will)	wollte	gewollt
wringen	wrang (wränge)	gewrungen
zeihen	zieh	geziehen
ziehen	zog (zöge)	gezogen
zwingen	zwang (zwänge)	gezwungen

Weight / Gewichte

1,000 milligrams (mg.) *1 000 Milligramm (mg)*	= 1 gram (g.) = *1 Gramm (g)*	= 15.43 grains
1,000 grams *1 000 Gramm*	= 1 kilogram (kg.) = *1 Kilogramm (kg)*	= 2.205 pounds
1,000 kilograms *1 000 Kilogramm*	= 1 tonne (t.) = *1 Tonne (t)*	= 19.684 hundredweight
	1 grain (gr.)	= 0.065 g
437½ grains	= 1 ounce (oz.)	= 28.35 g
16 ounces	= 1 pound (lb.)	= 0.454 kg
14 pounds	= 1 stone (st.)	= 6.35 kg
112 pounds	= 1 hundredweight (cwt.)	= 50.8 kg
20 hundredweight	= 1 ton (t.)	= 1,016.05 kg

Length / Längenmaße

10 millimetres (mm.) *10 Millimeter (mm)*	= 1 centimetre (cm.) = *1 Zentimeter (cm)*	= 0.394 inch
100 centimetres *100 Zentimeter*	= 1 metre (m.) = *1 Meter (m)*	= 39.4 inches / 1.094 yards
1,000 metres *1 000 Meter*	= 1 kilometre (km.) = *1 Kilometer (km)*	= 0.6214 mile ≈ ⅝ mile
	1 inch (in.)	= 25.4 mm
12 inches	= 1 foot (ft.)	= 30.48 cm
3 feet	= 1 yard (yd.)	= 0.914 m
220 yards	= 1 furlong	= 201.17 m
8 furlongs	= 1 mile (m., mi.)	= 1.609 km
1,760 yards	= 1 mile	= 1.609 km

Vulgar fractions and mixed number / Brüche (gemeine Brüche) und gemischte Zahlen

$\frac{1}{2}$	a/one half	*ein halb*
$\frac{1}{3}$	a/one third	*ein drittel*
$\frac{1}{4}$	a/one quarter	*ein viertel*
$\frac{1}{10}$	a/one tenth	*ein zehntel*
$\frac{2}{3}$	two-thirds	*zwei drittel*
$\frac{5}{8}$	five-eighths	*fünf achtel*
$\frac{1}{100}$	a/one hundredth	*ein hundertstel*
$1\frac{1}{2}$	one and a half	*ein[und]einhalb*
$2\frac{1}{4}$	two and a quarter	*zwei[und]einviertel*
$5\frac{3}{10}$	five and three-tenths	*fünf[und]dreizehntel*

Decimal numbers / Dezimalzahlen

0.1	*0,1*	nought point one	*null Komma eins*
0.015	*0,015*	nought point nought one five	*null Komma null eins fünf*
1.43	*1,43*	one point four three	*eins Komma vier drei*
11.70	*11,70*	eleven point seven o [əʊ]	*elf Komma sieben null*

Erläuterungen zum Text

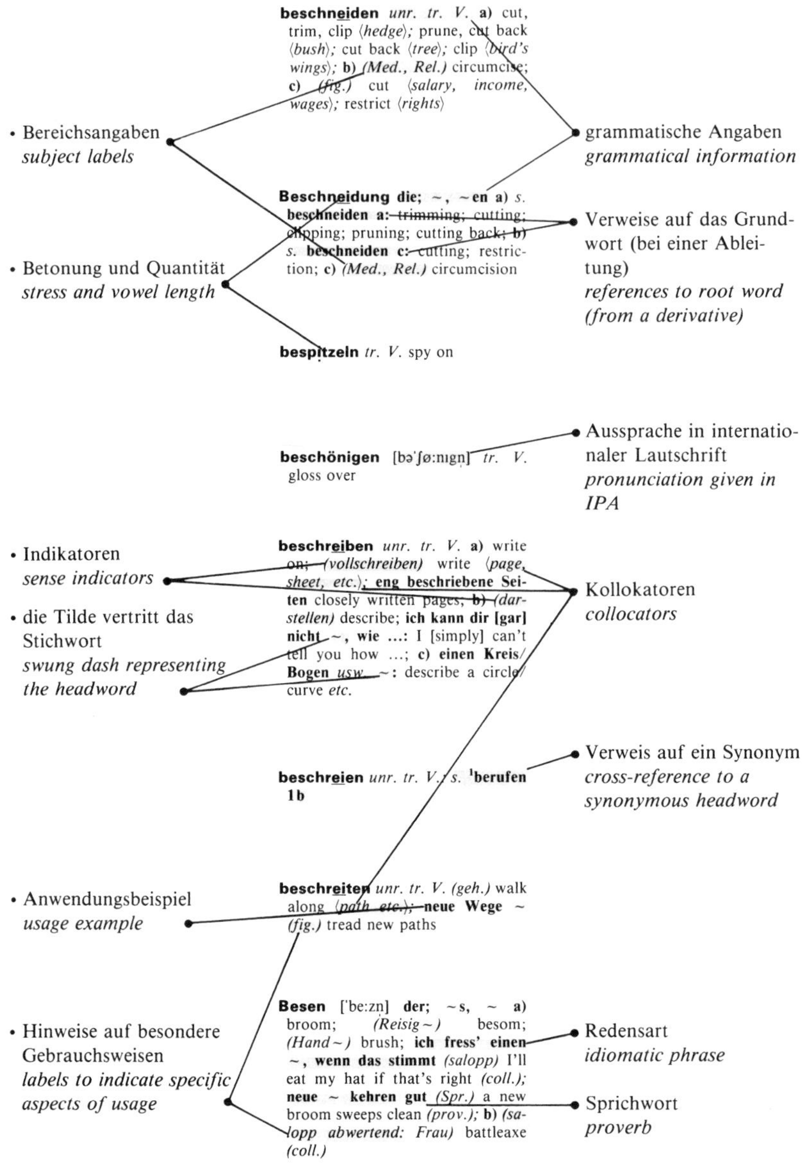

beschneiden unr. tr. V. a) cut, trim, clip ⟨hedge⟩; prune, cut back ⟨bush⟩; cut back ⟨tree⟩; clip ⟨bird's wings⟩; b) (Med., Rel.) circumcise; c) (fig.) cut ⟨salary, income, wages⟩; restrict ⟨rights⟩
• Bereichsangaben
subject labels
grammatische Angaben
grammatical information
Beschneidung die; ~, ~en a) s. beschneiden a: trimming; cutting; clipping; pruning; cutting back; b) s. beschneiden c: cutting; restriction; c) (Med., Rel.) circumcision
Verweise auf das Grundwort (bei einer Ableitung)
references to root word (from a derivative)
• Betonung und Quantität
stress and vowel length
bespitzeln tr. V. spy on
Aussprache in internationaler Lautschrift
pronunciation given in IPA
beschönigen [bəˈʃø:nɪgn̩] tr. V. gloss over
• Indikatoren
sense indicators
beschreiben unr. tr. V. a) write on; (vollschreiben) write ⟨page, sheet, etc.⟩; eng beschriebene Seiten closely written pages; b) (darstellen) describe; ich kann dir [gar] nicht ~, wie ...: I [simply] can't tell you how ...; c) einen Kreis/Bogen usw. ~: describe a circle/curve etc.
Kollokatoren
collocators
• die Tilde vertritt das Stichwort
swung dash representing the headword
beschreien unr. tr. V.: s. [1]berufen 1b
Verweis auf ein Synonym
cross-reference to a synonymous headword
• Anwendungsbeispiel
usage example
beschreiten unr. tr. V. (geh.) walk along ⟨path etc.⟩; neue Wege ~ (fig.) tread new paths
• Hinweise auf besondere Gebrauchsweisen
labels to indicate specific aspects of usage
Besen [ˈbe:zn̩] der; ~s, ~ a) broom; (Reisig~) besom; (Hand~) brush; ich fress' einen ~, wenn das stimmt (salopp) I'll eat my hat if that's right (coll.); neue ~ kehren gut (Spr.) a new broom sweeps clean (prov.); b) (salopp abwertend: Frau) battleaxe (coll.)
Redensart
idiomatic phrase
Sprichwort
proverb